49 50

ELECTRONICS ENGINEERS' HANDBOOK

OTHER McGRAW-HILL HANDBOOKS OF INTEREST

ELECTRONICS ENGINEERS' HANDBOOK

DONALD G. FINK, Editor-in-Chief

Executive Director and General Manager, Institute of Electrical and Electronics Engineers (Ret.);
Formerly Vice President—Research, Philco Corporation,
President of the Institute of Radio Engineers,
Editor of the Proceedings of the IRE;
Member, National Academy of Engineering;
Fellow of the Institute of Electrical and Electronics Engineers;
Fellow of the Institution of Electrical Engineers (London);
Eminent Member, Eta Kappa Nu

ALEXANDER A. McKENZIE, Assistant Editor

Contributing Editor, IEEE Spectrum,
Formerly Senior Editor, Electronics,
Editorial Manager, McGraw-Hill Book Company,
Member of the Technical Staff, M.I.T. Radiation Laboratory;
Senior Member, Institute of Electrical and Electronics Engineers

First Edition

McGRAW-HILL BOOK COMPANY

New York St. Louis San Francisco Auckland Düsseldorf
Johannesburg Kuala Lumpur London Mexico Montreal
New Delhi Panama Paris São Paulo
Singapore Sydney Tokyo Toronto

Library of Congress Cataloging in Publication Data

Fink, Donald G
  Electronics engineers' handbook.

    1. Electronics—Handbooks, manuals, etc.   I. Title.
TK7825.F56   621.381'02'02   74-32456
ISBN 0-07-020980-4

    567890   QBQB   784321098

The editors for this book at McGraw-Hill were Harold B. Crawford, Robert E. Curtis,
and Robert Braine, and the production supervisor was Stephen J. Boldish.
It was set in Times Roman by Rocappi/Lehigh, Inc.

It was printed and bound by Quinn & Boden Company, Inc.

CONTENTS

Elementary particles, quanta and photons, energy levels, states of matter, chemical phenomena; emission, transport, control and collection of charged particles; steady-state and time-varying phenomena; dielectric, magnetic, and electromagnetic phenomena; radiant energy, acoustic and optical phenomena; human hearing and vision; definitions, units, and symbols

Differential and integral calculus; series and expansions, transforms, probability, matrixes; Boolean algebra and symbolic logic

Circuit concepts and functions; lumped-constant and distributed circuits; network interconnections and switching; magnetic and dielectric circuits; glossary of criteria, laws, and theorems

Concepts, sources and measures of information; codes and coding; the communications channel, noise and interference

Definitions and concepts; exterior and interior system design; human factors; techniques

Conductive and resistive, dielectric and insulating, magnetic materials; semiconductors; electron-emitting, radiation-emitting materials; optical and photosensitive materials

Resistors, capacitors, inductors, transformers, electron tubes, cathode-ray tubes, semiconductor devices, transistors; batteries; ferromagnetic, ferroelectric, piezoelectric devices; fluidic devices; modular assemblies; relays, switches, insulators

CONTENTS

CONTRIBUTORS

Byron S. Anderson, Communications Consultant, formerly Chief, Information Acquisition Technical Area, United States Army Electronics Command (Sec. 22)

Howard P. Apple, Assistant Professor of Biomedical Engineering, Case Western Reserve University (Sec. 26)

Richard Babbitt, United States Army Electronics Command (Sec. 7)

Henry Ball, Research and Engineering, RCA Corporation (Sec. 20)

David K. Barton, Consulting Scientist, Missile Systems Division, Raytheon Co. (Sec. 25)

Ilan A. Blech, Associate Professor, Department of Materials Engineering, Technion-Israel Institute of Technology, Haifa (Sec. 6)

W. Calder, Chairman, Corporate Products Safety Committee, The Foxboro Company (Sec. 24)

Robert A. Castrignano, EVR Technology Department, CBS Laboratories (Sec. 20)

Joseph L. Chovan, Senior Engineer, Electronics Laboratory, General Electric Company (Sec. 14)

Wils L. Cooley, Assistant Professor of Electrical Engineering, West Virginia University (Sec. 3)

Munsey E. Crost, United States Army Electronics Command (Sec. 7, 11)

William F. Croswell, Head, Antenna Research Section, Langley Research Center, National Aeronautics & Space Administration (Sec. 18)

N. A. Diakides, United States Army Electronics Command (Sec. 11)

Milton Dishal, Senior Scientist, International Telephone and Telegraph Corporation Laboratories (Sec. 12)

Sam Di Vita, United States Army Electronics Command (Sec. 7)

Sven H. Dodington, Assistant Technical Director, International Telephone and Telegraph Corporation (Sec. 25)

C. Nelson Dorny, Associate Professor, Moore School of Electrical Engineering, University of Pennsylvania (Sec. 5)

Beverly Dudley, Consultant, formerly Staff Member, Institute for Defense Analyses (Sec. 1)

P. D. Dunn, Professor of Engineering Science, Department of Engineering and Cybernetics, The University of Reading, England (Sec. 27)

Myron D. Egtvedt, Senior Engineer, Electronics Laboratory, General Electric Company (Sec. 14)

Stanley L. Ehrlich, Consulting Engineer, Submarine Signal Division, Raytheon Company (Sec. 25)

J. C. Engel, Westinghouse Research Laboratories (Sec. 15)

C. H. Evans, Research Laboratories, Eastman Kodak Company (Sec. 20)

David G. Falconer, Research Physicist, Stanford Research Institute (Sec. 20)

George K. Farney, Varian Associates (Sec. 9)

Kenneth A. Fegley, Professor, Moore School of Electrical Engineering, University of Pennsylvania (Sec. 5)

Joseph Feinstein. Varian Associates (Sec. 9)

C. S. Fox. United States Army Electronics Command (Sec. 11)

R. E. Franseen. United States Army Electronics Command (Sec. 11)

Donald A. Fredenberg. Principal Engineer, Submarine Signal Division, Raytheon Company (Sec. 25)

Richard W. French. ELEMEK, Incorporated (Sec. 13)

E. Gaddy. Goddard Space Flight Center (Sec. 11)

Rodger L. Gamblin. Area Manager, Chemical Products Development, Office Products Division, IBM Corporation (Sec. 23)

Glenn B. Gawler. Senior Engineer, Barker Manufacturing Company (Sec. 14)

E. A. Gerber. Consultant, United States Army Electronics Command (Sec. 7)

Robert A. Gerhold. United States Army Electronics Command (Sec. 7)

Joseph M. Giannotto. United States Army Electronics Command (Sec. 7)

Charles H. Gibbs. Assistant Clinical Professor and Senior Research Associate, School of Dentistry and Engineering Design Center, Case Western Reserve University (Sec. 26)

S. B. Gibson. United States Army Electronics Command (Sec. 11)

Emanuel Gikow. Late United States Army Electronics Command (Sec. 7)

L. F. Gillespie. United States Army Electronics Command (Sec. 11)

B. Goldberg. Formerly United States Army Electronics Command (Sec. 11)

Thomas S. Gore, Jr., United States Army Electronics Command (Sec. 7)

R. D. Graft. United States Army Electronics Command (Sec. 11)

Alan B. Grebene. Vice-President Engineering, Exar Integrated Systems, Incorporated (Sec. 8)

W. A. Gutierrez. United States Army Electronics Command (Sec. 11)

Gunter K. Guttwein. United States Army Electronics Command (Sec. 7)

L. Gyugyi. Westinghouse Research Laboratories (Sec. 15)

Edward B. Hakim. United States Army Electronics Command (Sec. 7)

P. D. Hansen. Senior Technical Consultant, The Foxboro Company (Sec. 24)

G. Burton Harrold. Senior Engineer, Electronics Laboratory, General Electric Company (Sec. 13)

Jack H. Heimann. Principal Engineer, Submarine Signal Division, Raytheon Company (Sec. 25)

T. M. Heinrich. Westinghouse Research Laboratories (Sec. 15)

William P. Heising. Systems Development Division, IBM Corporation (Sec. 23)

Joseph P. Hesler. Consulting Engineer, Electronics Laboratory, General Electric Company (Sec. 13, 14)

Gerhard E. Hoernes. Systems Products Division, IBM Corporation (Sec. 23)

D. J. Horowitz. United States Army Electronics Command (Sec. 11)

William L. Hughes. Professor and Head of the School of Electrical Engineering, Oklahoma State University (Sec. 20)

G. M. Janney. Hughes Aircraft Company (Sec. 11)

Paul G. A. Jespers. Professor of Electrical Engineering, Institut d'Electricité, Catholic University of Louvain, Belgium (Sec. 16)

C. A. Johnson. United States Army Electronics Command (Sec. 11)

Earle D. Jones. Manager Electronics and Optics Group, Stanford Research Institute (Sec. 20)

Peter G. Katona. Associate Professor of Biomedical and Electrical Engineering, Case Western Reserve University (Sec. 26)

Chang S. Kim. Consulting Engineer, Electronics Laboratory, General Electric Company (Sec. 13)

Raymond S. Kiraly. Director of Engineering, Department of Artificial Organs, Research Division, Cleveland Clinic Foundation (Sec. 26)

Richard C. Kirby, Director, International Radio Consultative Committee, Geneva, Switzerland (Sec. 18)

Joseph H. Kirshner, United States Army Electronics Command (Sec. 7)

Myron W. Klein, Associate Director for Research and Development, Night Vision Laboratory, United States Army Electronics Command (Sec. 11)

W. Klein, United States Army Electronics Command (Sec. 11)

Wen H. Ko, Director, Engineering Design Center, Professor of Electrical Engineering, Professor of Biomedical Engineering, Case Western Reserve University (Sec. 26)

Granino A. Korn, Professor of Electrical Engineering, The University of Arizona (Sec. 2)

Theresa M. Korn, Tucson, Arizona (Sec. 2)

Samuel M. Korzekwa, Senior Engineer, Electronics Laboratory, General Electric Company (Sec. 13)

Henry N. Kozanowski, Commercial Electronic Systems Division, RCA Corporation (Sec. 20)

Ezra S. Krendel, Professor, Wharton School, University of Pennsylvania (Sec. 5)

Joseph A. Kuzneski, Senior Engineer, Submarine Signal Division, Raytheon Company (Sec. 25)

J. L. Lee, Manager, Program Management Services, The Foxboro Company (Sec. 24)

David Linden, United States Army Electronics Command (Sec. 7)

Harold W. Lord, Consulting Engineer (Sec. 13)

John W. Lunden, Senior Engineer, Electronics Laboratory, General Electric Company (Sec. 13)

Gregory J. Malinkowski, United States Army Electronics Command (Sec. 7)

P. R. Manzo, United States Army Electronics Command (Sec. 11)

Daniel W. Martin, Research Director, Engineering Research Department, D. H. Baldwin Company (Sec. 19)

Robert J. McFadyen, Consulting Engineer, Electronics Laboratory, General Electric Company (Sec. 13)

C. A. McKay, Corporate Director of Research Development and Engineering, The Foxboro Company (Sec. 24)

Floro Miraldi, Professor of Engineering, Case Western Reserve University (Sec. 26)

J. Thomas Mortimer, Associate Professor of Biomedical Engineering, Case Western Reserve University (Sec. 26)

J. W. Motto, Westinghouse Research Laboratories (Sec. 15)

Conrad E. Nelson, Senior Consulting Engineer, Heavy Military Electronic Systems, General Electric Company (Sec. 13)

Richard B. Nelson, Varian Associates (Sec. 9)

Michael R. Neuman, Associate Professor of Biomedical Engineering, Case Western Reserve University (Sec. 26)

W. E. Newell, Westinghouse Research Laboratories (Sec. 15)

Harry N. Norton, Member, Technical Staff, Jet Propulsion Laboratory, California Institute of Technology (Sec. 10)

Yukihiko Nosé, Department Head, Research Division, Department of Artificial Organs, Cleveland Clinic Foundation (Sec. 26)

R. M. Oates, Westinghouse Research Laboratories (Sec. 15)

Norman W. Parker, Staff Scientific Adviser, Motorola Incorporated (Sec. 20, 21)

B. R. Pelley, Westinghouse Research Laboratories (Sec. 15)

George F. Pfeifer, Senior Engineer, Electronics Laboratory, General Electric Company (Sec. 14)

P. F. Pittman, Westinghouse Research Laboratories (Sec. 15)

Robert Plonsey, Professor Biomedical Engineering, Case Western Reserve University (Sec. 26)

H. K. Pollehn, United States Army Electronics Command (Sec. 11)

Noble R. Powell, Consulting Engineer, Electronics Laboratory, General Electric Company (Sec. 14)

J. R. Predham, United States Army Electronics Command (Sec. 11)

Donald A. Priest, Varian Associates (Sec. 9)

E. G. Ramberg, RCA Laboratories, RCA Corporation (Sec. 20)

I. Reingold, United States Army Electronics Command (Sec. 7, 11)

Frederick M. Remley, Jr., Television Center, University of Michigan (Sec. 20)

Keith Y. Reynolds, International Video Corporation (Sec. 20)

Charles W. Rhodes, Manager, Television Products Engineering, Tektronix, Incorporated (Sec. 20)

D. A. Richardson, Development Project Engineer, The Foxboro Company (Sec. 24)

S. P. Rodak, United States Army Electronics Command (Sec. 11)

G. A. Rosica, Manager of Electronic Development and Engineering, The Foxboro Company (Sec. 24)

L. E. Scharmann, United States Army Electronics Command (Sec. 7)

E. J. Sharp, United States Army Electronics Command (Sec. 11)

R. R. Shurtz II, United States Army Electronics Command (Sec. 11)

Paul Skitzki, Technical Consultant, Submarine Signal Division, Raytheon Company (Sec. 25)

Jack Spergel, Late United States Army Electronics Command (Sec. 7)

Hans H. Stellrecht, Signetics Corporation (Sec. 8)

Joseph L. Stern, Vice President and Director of Engineering, Goldmark Communications Corporation (Sec. 21)

George W. Taylor, United States Army Electronics Command (Sec. 7)

Stephen W. Tehon, Consulting Engineer, Electronics Laboratory, General Electric Company (Sec. 13)

John B. Thomas, Professor of Electrical Engineering, Princeton University (Sec. 4)

Francis T. Thompson, Director, Electronics and Electromagnetics Research, Research Laboratories, Westinghouse Electric Corporation (Sec. 17)

Howard W. Town, Video Systems Division, Ampex Corporation (Sec. 20)

Harold R. Ward, Principal Engineer, Missile Systems Division, Raytheon Company (Sec. 25)

Gunter K. Wessel, Professor, Physics Department, Syracuse University (Sec. 13)

Everard M. Williams, Late George Westinghouse Professor of Engineering, Carnegie-Mellon University (Sec. 3)

R. A. Williamson, Jr., Senior Research Engineer, The Foxboro Company (Sec. 24)

R. S. Wiseman, Director, Research, Development and Engineering, and Director of Laboratories, United States Army Electronics Command (Sec. 11)

PREFACE TO THE FIRST EDITION

This new Handbook is the first to be devoted to the field of electronics engineering at large. Earlier important handbooks, notably those of Terman and of Henney, treated the field primarily from the point of view of the first important application in the field—radio engineering. Segments of the electronics field are treated exhaustively in excellent current handbooks, the majority of which appear in the bibliographies of this volume. But—surprisingly in view of the pervasive influence of electronics in so many areas of human endeavor—no one has heretofore attempted to bring together in one volume the essential principles, data, and design information on the components, circuits, equipment, and systems of electronics engineering as a whole. The present work is intended to fill the evident need for such a comprehensive single volume. Assembling its contents has proved to be a major undertaking, involving material contributed by 128 experts in their individual fields.

The present Handbook is a companion volume to the *Standard Handbook for Electrical Engineers,* the tenth edition of which was planned and edited by the undersigned. This "Electrical Handbook" is devoted primarily to the techniques of electrical power engineering, that is, the generation, transmission, distribution and utilization of electricity in macroscopic ("heavy current") forms. The many applications of electronics to the electrical power field are covered extensively in that volume. But, as I wrote in the Preface to the Tenth Edition, "to provide comprehensive design data on electronics circuits, systems and equipment would require another volume of equal size." This new "Electronics Handbook" is that volume and it is indeed of equal size: 2150 pages, one million words of text, 340 tables, 2000 illustrations, and 2500 bibliographic entries.

Aside from the different focus of subject matter, the aim of the Electronics Handbook is the same as that of its sister Electrical Handbook: to contain in a single volume all pertinent data within its scope, to be accurate and comprehensive in technical treatment, to be used in engineering practice (as well as in study in preparation for practice), and to be oriented toward application rather than theory. Sections on basic principles are included, but the predominant thrust is the practical use of those principles in engineering practice.

The Handbook is organized in four major divisions: *Principles Employed in Electronics Engineering,* Sections 1-5 inclusive; *Components, Devices and Assemblies,* Sections 6-11; *Circuits and Functions,* Sections 12-18; and *Systems and Applications,* Sections 19-27. The reader's attention is directed particularly to Section 6, *Properties of Materials,* prepared by Professor Blech of the Israel Institute of Technology. Contained in this 138-page section is, to the editor's knowledge, the most comprehensive compilation of data on materials used in electronics ever to appear in print.

While great care has been exercised in proofreading by the contributors and editors to check and recheck the data presented, it is inevitable that, in a first edition of a work of this size, some errors remain. The editor will appreciate these being brought to his attention.

The substantial effort made by all the contributors, not only to cover their special fields comprehensively, but to present their work in the most compact fashion consistent with informed and ready use, is gratefully acknowledged. I wish to thank particularly Assistant Editor Alexander A. McKenzie for the care with which he has guided the contributors in the final stages of editing and for his aid in the production of the book.

Donald G. Fink
Editor-in-Chief

SECTION 1

BASIC PHENOMENA OF ELECTRONICS

BY

B. DUDLEY Consultant, formerly Staff Member, Institute for Defense Analyses, Editor Technology Review, Massachusetts Institute of Technology; Senior Member, IEEE

CONTENTS

Numbers refer to paragraphs

BASIC PHENOMENA OF ELECTRONICS

SECTION 1

BASIC PHENOMENA OF ELECTRONICS

ELECTRONIC PROPERTIES AND STRUCTURE OF MATTER

1. Elementary Particles. The charged elementary particles of principal interest in electronics are the electron and the proton, designated e^- and p^+, respectively. The hydrogen atom, for example, consists of one electron and one proton. The mass, charge, and charge-to-mass ratios of these particles are as follows (where C stands for coulomb and kg for kilogram):

Mass at rest, of electron . . . . 9.1096×10^{-31} kg
　Of proton . . . . 1.6726×10^{-27} kg
Charge of electron . . . . -1.6022×10^{-19} C
　Of proton . . . . $+1.6022 \times 10^{-19}$ C
Charge-to-mass ratio, for electron . . . . 1.7588×10^{11} C/kg
　For proton . . . . 9.5791×10^7　C/kg

The elementary particles whose existence has been experimentally verified or postulated on theoretical grounds are listed in Table 1-1.

2. Atomic Structure. The atoms of each element consist of a dense nucleus around which electrons travel in well-defined orbits, or shells. The total mass of the nucleons (protons and neutrons) is taken to be equal to the mass of the atom. The number of nuclear protons is equal to the *atomic number Z* of the element. The number of nucleons is equal to the *mass number A* of the atom, and $A - Z$ is the number of neutrons in the nucleus. Heavy atoms have more neutrons than protons; excess of neutrons over protons is important in determining the stability of atoms, i.e., their radioactive properties. Atoms having the same atomic number but different mass numbers have the same chemical properties but different atomic weights. They are called *isotopes* of the chemical element.

The diameter of the atomic nucleus is between 10^{-15} and 10^{-16}m, whereas the diameter of the outer orbiting electrons (taken to be the diameter of the atom) is of the order of 10^{-10} m.

The nucleus carries a positive charge equal to the atomic number Z of the element times $+1.6 \times 10^{-19}$C, the charge of a proton. In the normal (un-ionized) atom there are Z orbiting electrons, each with negative charge $e^- = -1.6 \times 10^{-19}$C. At distances large compared with the atomic radius, the atom shows no net electric charge.

The extranuclear (electronic) structure of the atom is characteristic of the element. The orbiting electrons are arranged in successive *shells*. In order of increasing distance from the nucleus these shells are designated K, L, M, N, O, P, and Q. The number of electrons each shell can contain is limited. The electrons of the inner shells of complex atoms are tightly bound to the nucleus, and their paths can be altered only by high-energy particles, such as gamma rays. In the more complex atoms, electrons of the outer shells are relatively loosely bound to the nucleus. The outer shells account for the chemical and electrical properties of the elements.

3. Electron Orbits, Shells, and Energy States. Each orbiting electron in an atom has energy which is uniquely characterized by four *quantum numbers*. According to Pauli's exclusion principle, the wave functions describing the electrons must differ by at least one quantum number in the complete set required for their description.

An electron within an atom may be specified in terms of (1) a *principal quantum number n*, (2) an *azimuthal quantum number l*, (3) a *spatial quantum number m_l*, and (4) a *spin quantum number m_s or s*. The principal quantum number n specifies the shell in which an electron is located and hence principally specifies the energy state of the electron. Electrons lodged in the K, L, M, N, O, P, and Q shells have principal quantum numbers $n = 1, 2, 3, 4, 5, 6$, or 7, respectively.

Table 1-1. Elementary Particles

Family name	Name of particle	Symbol	Mass ($e = 1.0$)	Mass, MeV	Lifetime, s	Spin	Charge ($e^- = -1.0$)	Anti-particle
	Photon	γ	0	0	∞	1	0	γ
Electron	Electron	e^-	1	0.51098	∞	$\frac{1}{2}$	-1	e^+
	Electron neutrino	ν_e	0	0	∞	$\frac{1}{2}$	0	$\bar{\nu}_e$
Muon	Muon	μ^-	206.768	105.654	2.212×10^{-6}	$\frac{1}{2}$	-1	μ^+
	Muon neutrino	ν_μ	....	0	∞	$\frac{1}{2}$	0	$\bar{\nu}_\mu$
Meson	Pion, positive	π^+	273.18	139.59	2.55×10^{-8}	0	-1	π^-
	Neutral	π^0	264.20	135.0	1.9×10^{-16}	0	0	π^0
	Kaon, positive	K^+	966.6	493.9	1.22×10^{-8}	0	-1	K^-
	Neutral	K^0	974.2	497.8	1.0×10^{-10} 6.1×10^{-8}	0	0	$\bar{K}^0$
Baryons	Nucleon, proton	p^+	1,836.12	938.213	∞	$\frac{1}{2}$	-1	$\bar{p}^-$
	Nucleon, neutron	n^0	1,838.65	939.507	1.013×10^3	$\frac{1}{2}$	0	$\bar{n}^0$
	Lambda	Λ^0	2,182.8	1,115.36	2.51×10^{-10}	$\frac{1}{2}$	-1	$\bar{\Lambda}^0$
	Sigma, positive	Σ^+	2,327.7	1,189.40	8.1×10^{-11}	$\frac{1}{2}$	0	$\bar{\Sigma}^+$
	Neutral	Σ^0	2,332	1,191.5	$\sim 10^{-20}$	$\frac{1}{2}$	-1	$\bar{\Sigma}^0$
	Negative	Σ^-	2,340.5	1,195.6	1.6×10^{-10}	$\frac{1}{2}$	0	$\bar{\Sigma}^-$
	Xi, neutral	Ξ^0	2,566	1,311	1.5×10^{-10}	$\frac{1}{2}$	-1	$\bar{\Xi}^0$
	Negative	Ξ^-	2,580	1,318	1.28×10^{-10}	$\frac{1}{2}$	-1	$\bar{\Xi}^-$

The azimuthal quantum number l specifies the angular orbital momentum of an electron in each orbital state in various subshells. Together with n, the value of l designates the eccentricity of an electron orbit; the smaller the value of l the greater the eccentricity of the orbits for any given shell. The magnitudes of l may be any integer from 0 to $n - 1$. Electrons whose values of l are 0, 1, 2, 3, 4, and 5, respectively, are referred to as the s, p, d, g, and f electrons. The number of electrons in a subshell is determined by restrictions on m_l and m_s imposed by Pauli's exclusion principle.

The spatial quantum number m_l specifies differently oriented orbits having the same general shape; it specifies the orientation of the magnetic field of the electron orbit. This quantity is the projection of l on the magnetic axis; it may have $\pm(2l - 1)$ integral values from $-l$ to $+l$ including 0.

The spin quantum number, m_s or s, specifies the direction of spin of an electron on its own axis. Corresponding to spin in opposite directions, the two spin quantum numbers are $+h/2$ and $-h/2$, where $h/2\pi$ is Planck's constant ($= 6.626 \times 10^{-34}$ joule $\cdot$ s).

In a normal atom, orbiting electrons are arranged in the set of allowed states having the lowest total energy. As the complexity of atoms increases from hydrogen to uranium (the latter having 92 protons, and 146 neutrons), the electrons fill the shells and subshells by taking those states having the lowest total energy. Sometimes the energy state of an inner shell is less than that of a state in the outermost shell, and this accounts for the fact that some shells may begin to be filled before inner shells are totally filled.

4. Chemical Valence. The chemical properties of the elements are determined by the electrons in the outermost shell (valence electrons). Atoms with completely filled outer shells (the rare gases: helium, argon, krypton, xenon, and radon) are chemically inert. They contain eight electrons in their outer shells.

Atoms with a single electron in the outer shell (lithium, sodium, potassium, rubidium, cesium, franconium, and hydrogen) can easily lose their outer electron. They then become positive ions with completely filled shells.

Atoms with seven outer-shell electrons (the halogens: fluorine, chlorine, bromine, iodine, and astatine) readily pick up an electron from other atoms and become negative ions; they form molecules by sharing electrons and are said to have ionic bonding. Atoms with other numbers of outer-shell electrons tend to unite with other atoms in such ways that each atom has eight outer-shell electrons. Partially filled inner shells have an important bearing on the magnetic properties of the elements.

5. Conduction Electrons. When electrons are in close proximity in crystalline solids, the presence of nearby atoms affects their behavior, and their energies are no longer uniquely determined. The single energy level of an electron in a free or isolated atom is thereby spread into a band, or range, of energy levels. Whether or not the band of allowed energies is completely filled with electrons determines its properties as an electric conductor or insulator.

The *conduction band* is a range of states in the free-energy spectrum of a solid in which electrons can move freely; i.e., the electrons must be capable of effecting transitions among energy states. The valence electrons in metals, for example, are not firmly attached to individual atoms but are free to travel within the crystal lattice. Such electrons are called *conduction electrons*. There is one such conduction electron per atom in silver, copper, gold, and the alkali metals, all of which are good conductors.

An insulator or dielectric is a material in which every energy level, or quantum state, is filled and the electrons are unable to effect the transitions among states required for electric conduction.

6. Chemical Bonds and Compound Formation. Chemical bonds occur when the total energy of an aggregate is less with atoms near each other than separated. The charges of the atom play an important role in bonding, especially electrons in the outer shells.

Electrostatic or *ionic bonds* result from attractive forces between positive and negative ions or between pairs of oppositely charged ions. *Covalent bonds* occur when atoms share two or more electrons; i.e., shared electrons are attracted simultaneously to two atoms, and the resulting energy stability produces the bond. *Metallic bonds* are those in which the attractive forces result from the exchange interaction of the electron gas with the ionic lattice. *Van der Waals bonds* occur when molecules are formed, giving each atom an outer shell of eight atoms, as in an inert gas.

7. Energy Conservation. In a system in which all types of energy can be determined quantitatively, the algebraic sum of all forms of energy entering and leaving the system (or stored within it) is zero, and energy is conserved. The law of conservation of energy is often called the *first law of thermodynamics*.

8. Energy Conversion. Energy in a system can be neither created nor destroyed (except in nuclear processes when energy is converted to its equivalent form, mass). However, one form of energy can be converted into other forms. Thus, the potential energy of a system can be converted into kinetic energy and vice versa, and the energy of particles of one kind can be converted into energy of particles of quite another kind. Heat, produced by the random motion of elementary particles and their aggregates, is the form of energy into which other forms are ultimately converted.

9. Electromagnetic Effects. The dynamic behavior of elementary particles possessing both mass and charge produces electromagnetic phenomena. In free space the effects depend only on the nature of the charges and their motions, but such effects are greatly modified by atomic and molecular structure when charges move in material substances. Thus, the properties of materials modify the effects observed in free space.

The three major classes of electromagnetic materials are *conductors* or *semiconductors*, through which charges can flow more or less readily; *dielectrics* or *insulators*, through which charges are prevented from flowing; and *magnetic materials*, in which the motion of charges produces enhanced transverse forces.

Electromagnetic phenomena may be described by relationships between a stimulus and the response (the nature and magnitude of the effect produced). In conductors and semiconductors the stimulus is an electric field, and the response is the current, i.e., the directed motion of electric charges. In magnetic materials, the stimulus is current, and the response is the transverse force acting on, or produced by, the moving charges. The cause-effect relationships may be linear or nonlinear, constant or time-varying, single-valued or multiple-valued, and may be functions of temperature and other physical conditions.

10. Conduction Effects (see Par. 1-58). Electric conduction is the effect produced in a substance or system having mobile charges by the application of an electric force such that the charged particles flow through the conductor in the direction of the applied force. The phenomenon is attributed to electrons in the outer shells of atoms that are so loosely bound that they can be released by small electric forces. In good conductors free electrons can also be released by chemical, thermal, or other kinds of forces.

The release of charges in a conductor or semiconductor can arise from several mechanisms: (1) loss of loosely held conduction electrons in the outer shells of metals, (2) migration of holes and electrons in semiconductors, (3) release of electrons by thermal energy, (4) ejection of electrons from metals by radiant energy, (5) spontaneous ionization in electrolytes, (6) release of electrons from a conducting surface having high electric field at its surface, (7) impact of electrons on a surface to release secondary electrons, and (8) surface contact between substances of dissimilar atomic structure.

11. Dielectric Phenomena. Displacement of electric charges occurs in a substance or system having bound charges in which the application of an electric force produces a directed motion of charged particles in the direction of the applied force but of such limited extent that the charges do not separate from their parent atoms and hence do not move through the dielectric (see Par. 1-60).

Dielectric phenomena are attributed to electrons in the outer shells of atoms which are so tightly bound that they cannot become mobile except through application of electric forces so strong as to destroy the dielectric properties of the material.

The displacement of charges in a dielectric substance can arise from (1) electronic polarization, resulting from a shift of electron distributions within an atom with respect to its nucleus, (2) ionic polarization, in which the atoms having one or more outer-shell electrons combine with atoms that have an incomplete outer shell of electrons, (3) permanent molecular polarization in liquids and gases and arising from alignment of electric dipole moments of molecules, and (4) ferroelectric polarization, in which many electric dipole moments in crystals remain oriented in the direction in which they were aligned when the electric field was initially applied.

12. Magnetic Phenomena (see Pars. 1-73 to 1-98). Magnetic phenomena include a number of effects of substances or systems in which the motions of charged particles set up forces transverse to the motion of charges.

The phenomena are attributed to the electric fields carried by moving charges. Any directed motion of charges produces magnetic effects. In magnetic materials the phenomena of magnetism arise from the orientation of orbiting electrons and, to a smaller extent, from electron spins in atoms, molecules, and crystal domains.

The interaction of matter with a magnetic field produced by moving charges arises from several mechanisms: (1) ferromagnetism, produced by exchange forces between atomic moments, (2) diamagnetism, produced by electron spins in antiparallel pairs in closed electronic shells, (3) paramagnetism, produced by the orbital or spin moments of electrons, or both, as well as by moments of free electrons, (4) antiferrimagnetism, produced by exchange forces between atomic moments, and (5) ferrimagnetism, produced by the moment resulting from two antiferromagnetic lattices.

Ferromagnetic materials may have values of permeability many thousands of times greater than that of free space. Such materials are extensively used to enhance the magnetic force produced by a given magnetizing force. At high values of magnetizing force the permeability becomes essentially that of free space as the magnetic material becomes saturated. Ferromagnetic materials are characterized by marked hysteresis effects (see Par. **1-106**).

13. Thermoelectric Phenomena. Three important relations between electricity and heat are the Seebeck effect, the Thomson effect, and the Peltier effect.

The *Seebeck effect* is the potential difference generated by a temperature difference between the junctions in a circuit composed of two homogeneous electric conductors of dissimilar composition or, in nonhomogeneous conductors, by a temperature gradient in a nonhomogeneous region. The magnitude of the effect depends upon the metals and the distribution of temperature in them. In a circuit of several junctions connected in series in the same order throughout, the resultant thermoelectric power is the algebraic sum of the voltages due to the several conjoined pairs.

The *Thomson effect* is defined as the absorption or evolution of thermal energy produced by the interaction of electric current and a temperature gradient in a homogeneous electric conductor. The magnitude and direction of the electromotance depend on the conducting material and on the temperature gradient.

The *Peltier effect* describes the absorption or evolution of heat, in addition to the Joule heat, at a junction through which an electric current flows. In a nonhomogeneous, isothermal conductor it is the absorption or evolution of heat, in addition to the Joule heat produced by an electric current.

Table 1-2.   Electrochemical Series of Metals

Electrode material	Electrode potential, V	Electrode material	Electrode potential, V
Lithium	2.96	Cobalt	0.29
Rubidium	2.924	Nickel	0.231
Potassium	2.923	Tin	0.136
Barium	2.85	Lead	0.122
Sodium	2.713	Hydrogen (reference)	0.000
Strontium	2.7	Bismuth	−0.20
Calcium	2.5	Arsenic	−0.30
Magnesium	1.55	Copper	−0.345
Manganese	1.0	Titanium	−0.37
Zinc	0.758	Mercury	−0.799
Chromium	0.557	Silver	−0.800
Iron	0.441	Gold	−1.30
Cadmium	0.398		

14. Electrochemical Series. The electrochemical series, also called the electromotive force series, is a list of the standard potentials of specified electrochemical reactions in order of the voltages produced relative to a reference value. The electrode potential is the potential difference between the electrode material and a normal solution (1 ion/l) of the ion in equilibrium. The electrochemical series of metals used in electronic applications is given in Table 1-2.

15. Work Function. Work function is a term applied to the amount of energy required to transfer electrons, ions, molecules, etc., from the interior of one substance across an interface boundary into an adjacent substance or space. It is commonly expressed by the transfer of an electron across the boundary in units of electron volts. Work functions are common in thermionic emission and photoelectric emission (see Pars. **1-111** and **1-112**).

The *contact work function* is the energy required to transport an electron between two dissimilar conductors placed in contact (see Table 1-3).

Table 1-3.   Contact Work Function

Element	Work function, V	Element	Work function, V
Aluminum	3.38	Mercury	4.50
Barium	1.73	Nickel	4.96
Bismuth	4.17	Platinum	5.36
Cadmium	4.0	Potassium	1.60
Cobalt	4.21	Rubidium	4.52
Copper	4.46	Silicon	4.2
Germanium	4.5	Silver	4.44
Gold	4.46	Tantalum	4.1
Iron	4.40	Tin	4.09
Lead	3.94	Tungsten	4.38
Magnesium	3.63	Zinc	3.78
Manganese	4.14		

ELECTROSTATICS

16. Coulomb's Law.   The law of force between charges at rest may be stated as follows: The force of repulsion between like charges concentrated at points in an isotropic medium is proportional to the product of their charges, inversely proportional to the square of the distance between them, and inversely proportional to the dielectric coefficient of the medium in which the charges reside.   The force acts in a straight line between the centers of the charges.   If the charges are of like sign or polarity, the electric force is one of repulsion; if the charges are of opposite sign, the electric force is one of attraction.

In the rationalized system of units the Coulomb force is

$$F = \frac{Q_1 Q_2}{4\pi r_{12}^2 \epsilon}$$

where Q_1 and Q_2 are the magnitudes of the two concentrated charges, r_{12} is the distance between centers of the two charges, and ϵ is the dielectric coefficient (permittivity) of the medium in which the charges are situated.

In the mks system of units F is measured in newtons (N), distance in meters, charge in coulombs (1 C being the charge equal to 6.24×10^{18} elementary charges), and permittivity in coulombs^2 per newton-meter^2 (see Par. 1-21).

17. Principle of Superposition.   In any linear system, (i.e., one in which the effect produced is proportional to the cause), each cause acting separately produces its separate effect.   When several causes act simultaneously, the resultant effect is the sum of the effects produced by each cause acting separately.   Coulomb's law implies that the principle of superposition applies for electric charges.   If the principle of superposition holds for any system, that system is linear.   The total resultant effect at any point is then the vector sum of the individual component effects at that point.

18. Electric Field.   The region in which an electric charge exerts a measurable force on another (above and beyond the gravitational force between the two) is called an *electric field.* Since the force has magnitude and direction, it is a vector field.   The direction of the field is represented by lines designating the direction of the force which the field exerts on a small, isolated charge acting as a test body.

The magnitude of the field may be indicated graphically by the closeness or concentration of the field lines.   Quantitatively, the effect of the field is measured by the *electric field strength*.   The properties of the field may be expressed in terms of the force exerted on a charge in the field or in terms of the energy required to move a charge between two points in the field.   The first of these specifies a vector field of force, the second a scalar field of potential energy.

19. Test Body for Electric Field.   An electric field can be detected and its direction and magnitude can be determined by measuring the force acting on an isolated charge used as a test body.   Since a charged test body has a field of its own and modifies the field to be measured, the electric field can be uniquely ascertained only if the charge on the test body is as small as possible and the polarity of the test charge is stated.

20. Lines and Tubes of Force. In an electric field, a *line of electric force* is a curve drawn so that at every point it has the direction of the force acting on a charge used as a test body; i.e., it has the same direction as that of the field.

A *tube of force* is obtained by drawing a number of lines of force through the boundary of any small closed curve. The lines then form a tubular surface which is not cut by any other lines of force. A tube of force behaves as though it had inertia, and so work must be done to move it. The extremities of tubes of force enclose equal and opposite charges. Lines and tubes of force are introduced to aid in visualizing electrical phenomena, but no physical existence is to be attributed to them.

21. Permittivity, Dielectric Coefficient, Electric Susceptibility. For a given distribution of charges, the forces between them depend on the environment in which they are located. The greatest force occurs when charges are in free space (vacuum); the presence of material bodies reduces the Coulomb force.

The *permittivity* of any dielectric may be expressed as the product of two terms. One accounts for the dielectric properties of material bodies. The other may be regarded as accounting for the dielectric properties of free space but is actually a factor that provides a consistent set of units for Coulomb's law when units of force, charge, and distance are set up arbitrarily.

If ϵ is the permittivity of any substance, and ϵ_0 is the permittivity of free space (vacuum),

$$\epsilon = \epsilon_0 \epsilon_r$$

where ϵ_r is the *relative permittivity* of a material substance referred to that of vacuum. In general, ϵ_r is a complex number. Its quadrature (imaginary) term may usually be neglected, except at very high frequencies. The real component of ϵ_r is often called the *dielectric constant*. Since its value depends upon frequency and other factors, the term *dielectric coefficient* is preferable.

The permittivity may also be written as

$$\epsilon = \epsilon_0(1 + \chi_e)$$

where χ_e is the *electric susceptibility* of a dielectric. The electric susceptibility is a numeric measure of the polarization or displacement of electrons in atoms or molecules of a dielectric. The dielectric properties of a material substance may be expressed in terms of relative permittivity ϵ_r or electric susceptibility χ_e in simple numerics. Such properties may also be expressed in terms of the absolute permittivity of the substance ϵ in units of coulombs^2 per newton-meter^2 or farads per meter (F/m).

In the rationalized mks system of units the permittivity of free space is $\epsilon_0 = 8.854 \times 10^{-12} \text{C}^2/\text{Nm}^2$ (or 8.854×10^{-12} F/m).

22. Surface and Volume Charge Density. A region of space containing a large number of closely spaced charges is equivalent to a continuous distribution of charges when viewed at a distance large compared to the separation between charges. The effects of such charges may be described in terms of charge density.

The *surface charge density* is defined as the limiting value of the ratio of the charge increment to the increment of surface as the latter approaches zero. In mks units the surface charge density is measured in coulombs per square meter.

The *volume charge density* is defined as the limiting value of the ratio of the charge increment to the increment of volume as the latter approaches zero. In mks units the volume charge density is measured in coulombs per cubic meter.

If the charge density is known, the total charge on a surface or within a volume can be obtained by integrating over the surface or volume. The charge is in coulombs if surface area is in square meters or the volume is in cubic meters.

23. Electric Field Strength. The electric field strength (electric field intensity) at a given point is a vector quantity defined as the quotient of the force (that a small stationary test-body charge at that point will experience), to the charge as the charge approaches zero.

If an increment of force ΔF is produced on an increment of charge ΔQ at a point in an electric field, the electric field strength at the point is

$$E = \lim_{\Delta Q \to 0} \frac{\Delta F}{\Delta Q}$$

Since F is a vector quantity whereas Q is a scalar, the field strength E is a vector quantity having the same direction as that of the force on the test body. In mks units, the electric field strength is measured in units of newtons per coulomb. The common unit is volts per meter.

24. Dielectric Strength. The dielectric strength is the maximum field strength that a dielectric can sustain without breakdown. It is determined experimentally by tests on materials and in mks units is expressed in units of volts per meter of thickness of the sample tested. The results obtained depend upon the method and conditions of the test as well as upon the thickness of sample being tested, being greater for thin rather than thick samples. Dielectric strength is usually expressed in units of volts per centimeter, volts per millimeter, kilovolts per centimeter, or volts per mil (1 mil = 0.001 in.).

25. Field Strength Produced by Point Charges. Coulomb forces obey the principle of linear superposition in free space, as well as in many dielectrics. Therefore the electric field intensity at a point p produced by charges $Q_1 Q_2, \ldots, Q_n$ whose distances from p are $r_1, r_2, \ldots, r_n$ is the vector sum of the field strengths produced at the point by each charge individually. In a medium whose permittivity is ϵ, if the component field strengths are $E_1, E_2, \ldots, E_n$, the resultant field intensity of point charges is

$$F = E_1 + E_2 + \cdots + E_n$$

$$= \frac{Q_1}{4\pi\epsilon r_1^2} + \frac{Q_2}{4\pi\epsilon r_2^2} + \cdots + \frac{Q_n}{4\pi\epsilon r_n^2}$$

In the mks system of units, the field strength is in volts per meter when charge is specified in coulombs, distance in meters, and permittivity in the units given in Par. 1-21.

26. Field Strength Produced by Distributed Charges. If charges are distributed continuously through a volume (instead of being located individually at discrete points), the total charge in volume v produced by charge density ρ is $Q = \int \rho \, dv$ and the field strength at a distance R (large compared to the dimensions of the volume in which charges are distributed) is given by the volume integral

$$E = \iiint_{\text{vol}} \frac{\rho \, dv}{4\pi\epsilon R^2}$$

In general ρ and R may be functions of the triple integration over the volume.

27. Electrostatic Potential and Potential Difference. In general, energy must be expended to move a charge through an electric field. The amount of energy required per unit of charge is called *electric potential*. While it is not possible to specify uniquely the *absolute potential* of any point in an electric field, the *difference in potential* between two points in the field can be determined when the manner in which the electric field strength varies between the points is known.

The electrostatic potential difference between two points in an electric field resulting from a static distribution of electric charge is the scalar-product line integral of the electric field strength. It may be integrated along any path from one point to the other.

The potential difference between two points a and b in an electric field of strength E is, then

$$V_{ab} = -\int_b^a E \cdot dl = -\int_b^a E \cos\theta \, dl$$

where θ is the angle between the direction of the electric field and the direction of the line element dl. The potential difference V_{ab} is taken to be positive if work is done in carrying a positive charge from b to a. In the mks system of measurements, potential difference is measured in volts (equivalent to joules per coulomb).

28. Field Strength and Potential Difference. Potential difference is the change in energy per unit charge as a charge is moved between two points in an electric field in the direction of the field. The energy is negative if work is done in carrying a positive charge between points a and b, and the field strength is force per unit charge.

The differential potential difference may be expressed as

$$dV = -\frac{F \cdot dl}{Q} = -E \cdot dl$$

from which

$$E = -\frac{dV}{dl}$$

Hence, electric field strength is the negative rate of change of potential difference with respect to the distance a charge moves in an electric field; i.e., it is the gradient of the potential field. In the mks system, field strength is measured in units of volts per meter.

29. Voltage Rises and Voltage Drops. The amount of energy per charge between two points, whether due to separation of charges (as in electrostatics) or to changes of energy in dynamic processes, is commonly called voltage and is measured in volts.

A voltage drop exists when the energy of charges is diminished as charges do work. A voltage rise exists when the energy of charges is increased by some source of energy that does work on them. The total voltage rise is equal to the total voltage drop when voltages are measured in the same direction around a closed loop.

30. Potential Field of Point Charges. At any point the potential due to a number of charges establishing an electric field is the scalar sum of the potentials produced at that point by each charge acting alone, since potential fields obey the principle of linear superposition. The result is simple to compute since scalar addition of potential fields replaces vector addition of electric field strengths.

If $Q_1, Q_2, \ldots, Q_n$ are the charges whose distances from a point at which the potential of the field is to be determined are $r_1, r_2, \ldots, r_n$, then the potential at that point is

$$V = V_1 + V_2 + \cdots + V_n = \frac{1}{4\pi\epsilon}\left(\frac{Q_1}{r_1} + \frac{Q_2}{r_2} + \cdots + \frac{Q_n}{r_n}\right)$$

where ϵ is the permittivity of the medium in which charges reside.

31. Potential Field of Distributed Charges. If charges are distributed continuously through an element of volume dv in which the charge density is ρ, the potential at a point whose distance from dv is r is

$$V = \frac{1}{4\pi\epsilon}\iiint_{\text{vol}}\frac{\rho\,dv}{r}$$

where ϵ is the permittivity of the medium. Zero potential is taken to be at an infinitely remote surface (or point in the case of point charges).

32. Electric Dipole and Dipole Moment. A combination of two point charges of equal magnitude and opposite polarity separated by a distance small compared to that at which the field of the dipole is to be determined is called an *electric dipole* (or, a *doublet*). The dipole moment is the product of the magnitude of either charge of a dipole and the distance separating the two charges. In mks units dipole moment is measured in coulomb-meters; its direction is the direction of the line joining the two charges.

33. Field of a Dipole. Let a dipole consist of charges $+Q$ and $-Q$ separated by a distance d in a medium of permittivity ϵ. The center of the dipole is at the origin, and the field is to be determined at a point P (so remote from the origin that we may take $R_1 = r = R_2$ with negligible error) and at a direction measured by the angle θ. It is desired to determine the dipole field at point P, relative to that at the origin 0 (see Fig. 1-1).

The potential produced at point P is given by

$$V_{OP} = \frac{Qd}{4\pi\epsilon r^2}$$

whereas the electric field strength of the dipole at point P is

$$E_{OP} = \frac{Qd}{4\pi\epsilon r^3}(2\cos\theta + \sin\theta)$$

In mks units, V is in volts, E in volts per meter, Q in coulombs, r in meters, and permittivity ϵ in coulombs^2 per newton-meter^2.

34. Electric Polarization. At any point in a dielectric the electric polarization is the electric dipole moment per unit volume. It may also be expressed as the amount of bound charge per unit area perpendicular to the direction in which the charges of the dipole are displaced. If Q is the amount of bound charge displaced by distance d, the dipole moment is $p = Qd$. If $v = sd$ is the volume of a dielectric, d the distance between dipole charges, and s the area perpendicular to the direction in which bound charges move, the polarization is given by

$$P = \frac{p}{v} = \frac{Qd}{v} = \frac{Q}{s}$$

Electric polarization, in mks units, is measured in coulomb-meters per cubic meter (coulombs per square meter).

35. Electric Flux (Electric Displacement). The electric flux or displacement is a quantity associated with the amount of bound charge Q displaced in a dielectric which is subject to an electric field. The magnitude of the flux is proportional to the amount of bound charge displaced. In the mks system of units, the magnitude of the electric flux is equal to the amount of bound charge and is measured in coulombs. It is a scalar quantity. It may be expressed in terms of charge, flux density, or electric field strength and the dielectric properties of the dielectric, as well as the surface through which the charges are displaced.

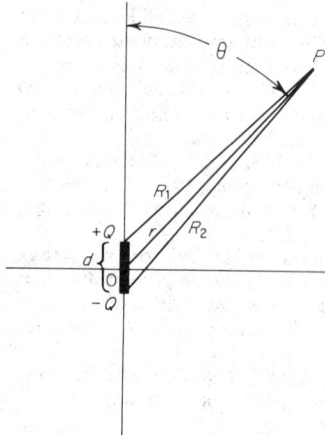

Fig. 1-1. Geometry for determining the field at point P of an electric dipole consisting of two equal charges of opposite sign, $+Q$ and $-Q$, separated by a small distance d.

36. Electric Flux Density. The electric flux density (also called *displacement density*) at a point in a dielectric is a vector quantity whose magnitude is proportional to the amount of bound charge per unit area of surface normal to the direction of charge displacement. It is expressed as the ratio of a small change in displacement to a small change in surface area through which the flux passes as the surface becomes vanishingly small. The electric flux density may also be regarded as the ratio of the electric force F per unit charge Q and can be found from Coulomb's law

$$D = \frac{F}{Q} = \frac{Q}{4\pi r^2 \epsilon} = \frac{Q}{s\epsilon} = \frac{E}{\epsilon}$$

where $s = 4\pi r^2$ is the closed area surrounding a test charge and $E = F/Q$ is the electric field strength.

37. Gauss' Theorem for Electrostatics. The electrostatic theorem of Gauss is a necessary consequence of Coulomb's law of electrostatics. It states that the amount of electric flux passing through any closed surface is proportional to (and in the rationalized systems of units is equal to) the amount of charge contained within the surface.

If D is the amount of electric flux density crossing any element of surface ds, θ the angle between the direction of D and the normal to the surface element, and ρ the electric charge density in volume element dv, the electric flux in rationalized units is

$$\Psi = \int_{\substack{\text{closed} \\ \text{surface}}} D \cos \theta \, ds = \int_{\text{vol}} \rho \, dv = Q$$

The surface integral is taken over a surface enclosing the volume expressed by the volume integral.

38. Divergence of Electric Flux Density. The divergence of the vector electric flux density is a scalar quantity equal to the ratio of the flux passing through a closed surface to the volume contained therein, as the volume becomes vanishingly small, or

$$\text{div } D = \lim_{\Delta v \to 0} \frac{\int_{\substack{\text{closed} \\ \text{surface}}} D \cos \theta \, ds}{\Delta v}$$

If D is expressed in terms of rectangular components D_x, D_y, and D_z, the divergence is expressed as the sum of the partial derivatives of each component in the direction of its axis, or

$$\text{div } D = \frac{\partial D_x}{\partial x} + \frac{\partial D_y}{\partial y} + \frac{\partial D_z}{\partial z}$$

39. Electric Field Vectors. The properties of electric fields in dielectrics may be expressed in terms of three field vectors: (1) electric field strength E, related to all charges, (2) electric flux density D, related only to free charges on the conducting surfaces at a boundary between dielectrics and conductors, and (3) polarization, or electric dipole moment per unit volume of dielectric P, related to the bound charges within a material dielectric. All three field vectors may be expressed in terms of a surface charge density Q/s. In addition, electric field strength is related to the electric force per charge exerted on a charge by the electric field, $E = F/Q$.

Electric field strength may be defined as the charge Q per unit surface s modified by the permittivity ϵ of the medium. The simplest situation arises for free space, and the properties of material substances need not be considered. In free space, $E = Q/s\epsilon_0 = D/\epsilon_0$, where D is the electric flux density produced by the free charge Q collected on a conducting surface s at the boundary of a dielectric and ϵ_0 is the permittivity of free space.

The situation is more complicated for material dielectrics having bound charges. For this case, $E = Q/s\epsilon$, but $\epsilon = \epsilon_0\epsilon_r = \epsilon_0(1 + \chi_e)$ (see Par. **1-21**).

The three electric field vectors are connected by the general relation

$$D = \epsilon_0 E + P$$

For free space $P = 0$, and $D = \epsilon_0 E$. For dielectric materials

$$D = \epsilon E = \epsilon_0 \epsilon_r E$$
$$P = (\epsilon_r - 1)\epsilon_0 E$$

The last equations are generally true for homogeneous, isotropic dielectric media, and the relative permittivity is usually independent of electric field strength E. For certain dielectrics, called *ferroelectrics*, ϵ_r is a function of field strength E and the ferroelectrics show properties of electric hysteresis.

40. Field Vectors at Dielectric Boundaries. The electrostatic field in a single dielectric is continuous, and the electric field strength changes an infinitesimal amount in an infinitesimal distance. The electric field may change abruptly, both in magnitude and direction, at the boundary of a dielectric or at the interface between two different media. The field at the boundary may then be analyzed in terms of two field components, one tangent to the boundary and the other normal to the boundary.

The tangential components of electric field strength are continuous across the boundary, (i.e., are the same on both sides of the boundary). At the interface of a conductor and a dielectric the tangential component of the electric field is zero.

The normal or perpendicular component of electric flux density changes at a boundary between two charged dielectrics by an amount equal to the surface charge density. The normal component of flux density is continuous across a charge-free boundary between the two dielectrics.

41. Potential Gradient. In an electric field, the maximum value of change of potential V with respect to distance l is equal to the magnitude of the electric field strength E and is obtained when the direction of the electric field strength is opposite that in which the potential increases most rapidly with distance.

Thus, the potential gradient at a point in an electric field is the limiting value of the ratio of a change in potential to a change in distance as the latter becomes vanishingly small:

$$\nabla V = \lim_{\Delta l \to 0} \frac{\Delta V}{\Delta l} = \frac{dV}{dl}$$

If dl is at an angle θ with respect to the direction in which the electric field strength changes most rapidly with distance, then $dV/dl = -E \cos \theta$. This is a maximum when $\theta = 0$. Consequently the potential gradient is then

$$\nabla V = \frac{dV}{dl} = -E$$

A negative sign is used because a rise in potential requires the positive test charge to be moved in a direction opposite to that taken to be the direction of the electric field.

42. Force on a Charge in an Electric Field. The electric field strength is defined as the force per charge acting on a charge in an electric field, or $E = F/Q$. Therefore, the force acting on a charge Q in an electric field of intensity E is

$$F = QE$$

In mks units, force is measured in newtons, charge in coulombs, and field strength in newtons per coulomb (volts per meter).

43. Work Done on Moving Charges. When a charge Q in an electric field of intensity E is moved a distance dl at an angle θ with respect to the direction of the field, the amount of work done is

$$dW = QE \cos \theta \, dl$$

and the total work done in moving the charge between two points a and b in the field is

$$W = - \int_b^a QE \cos \theta \, dl$$

In the mks system, work is expressed in joules, charge in coulombs, and field strength in volts per meter.

44. Conservative Properties of Electrostatic Fields. No work is done in carrying a charge around a closed path in an electrostatic field, for

$$W = - \oint QE \cdot dl = -\left(\int_b^a QE \cdot dl - \int_a^b QE \cdot dl \right) = 0$$

This result is true for electrostatic fields but must be modified for time-varying fields. In a conservative field the work done in moving a particle between two points is independent of the path taken. A field satisfying the equation given above is said to be a *conservative field*.

45. Storage of Electric Charges. If charges $+Q$ and $-Q$ are collected on conductors separated by free space or by a dielectric material, the mutual attraction between the two charges tends to hold the charges in place indefinitely. The charges are then said to be stored. They produce a stress in the intervening dielectric, measured by the electric field strength, and a voltage difference exists between the two charge-bearing conductors.

46. Capacitance and Elastance. Capacitance is the property of a system of conductors and dielectrics that permits the storage of electrically separated charges when a potential difference exists between the conductors. Elastance is the reciprocal of capacitance and is defined to emphasize the attributes of charge storage.

Elastance may be defined as the amount of energy U per charge Q required to transfer unit charge between two conductors separated by a dielectric. Quantitatively elastance is given by

$$S = \frac{U/Q}{Q} = \frac{V}{Q}$$

where V is the voltage difference required to transfer the charge Q between the two conductors.

Capacitance is given by the inverse relation

$$C = \frac{1}{S} = \frac{Q}{V}$$

Capacitors. A capacitor is a device consisting of electrodes separated by a dielectric (which may be air or vacuum) for introducing capacitance into an electric circuit or system or for providing for the storage of electric charge.

47. Capacitance between Two Conductors. For systems of simple geometry, the capacitance between two conductors separated by a single, homogeneous, isotropic dielectric may be calculated in terms of the physical dimensions of the conducting electrodes and the permittivity of the dielectric substance (Table 1-4).

Table 1-4. Capacitance Relationships

Geometry	Capacitance C
Parallel plates, closely spaced	$\epsilon s/l$
Coaxial cylinders of length l	$\dfrac{2\pi\epsilon l}{ln(r_2/r_1)}$
Concentric spheres	$\dfrac{4\pi\epsilon}{1/r_1 - 1/r_2}$
Isolated sphere	$4\pi\epsilon r$

In Table 1-4 $\epsilon = \epsilon_0\epsilon_r$ is the permittivity of the dielectric in coulombs^2 per newton-meter^2, and capacitance is in farads when the dimensions are expressed in meters. The distance between parallel plates is l, and the plates have an area of s, whereas r_1 and r_2 are the radii of inner and outer electrodes. The effects of fringing on the lines of force are neglected.

48. Energy in a Charged Capacitor. The total energy of the electric field in a capacitor whose capacitance is C when charged to potential difference V is

$$U = VQ$$

If the plates were initially at the same potential ($V = 0$), the average energy needed to charge the capacitor to voltage V is

$$U_{av} = \tfrac{1}{2}VQ$$

The voltage V may be expressed in terms of capacitance C between conductors across which charge Q is transferred, since $V = Q/C$. Therefore the average electric energy stored in the electric field of a charged capacitor is

$$U_{av} = \tfrac{1}{2}VQ = \tfrac{1}{2}V^2C = \frac{1}{2}\frac{Q^2}{C}$$

The energy is in joules if the charge is in coulombs and the potential difference in volts.

49. Energy Density of an Electric Field. In a dielectric of permittivity ϵ, in the form of a rectangular solid with surface area s and length l, a voltage difference V is needed to transfer a charge Q between the surfaces. The average energy stored in the dielectric in this charge transfer is

$$U_{av} = \tfrac{1}{2}VQ = \tfrac{1}{2}(El)(Ds) = \tfrac{1}{2}EDls = \tfrac{1}{2}EDv$$

where D is the flux density on the charged surfaces and $v = sl$ is the volume of the dielectric. Since $E = D/\epsilon$,

$$U_{av} = \tfrac{1}{2}\epsilon E^2 v = \frac{1}{2}\frac{D^2}{\epsilon}v$$

The average energy density at a point in the field is

$$u_{av} = \frac{U_{av}}{v} = \tfrac{1}{2}\epsilon E^2 = \frac{1}{2}\frac{D^2}{\epsilon}$$

ELECTROKINETICS

50. Force Components of Moving Charges. Any isolated charge in free space has a radially symmetrical field of force resolvable into components along each of three mutually perpendicular axes. If the charge is mobile, an electric force will cause the charge to move in the direction of the field of force, say the x direction. The field has no effect on the field of the charge in the y and z directions. Accordingly a charge moving in the x direction has field components capable of producing a transverse force, in the y and z directions, on other moving charges. Such a transverse force is the source of magnetic forces between two beams of charges or between two current-carrying conductors (see also Par. **1-118**).

Suppose charges Q_1 and Q_2 move in free space along parallel paths in a plane. The force between them can be resolved into a longitudinal component $F_\parallel$ in the direction of motion and a transverse component $F_\perp$ perpendicular to the direction of motion. The total Coulomb force may then be expressed as

$$F = F_\parallel + F_\perp = \frac{Q_1 Q_2}{4\pi r^2 \epsilon}$$

If θ is the angle between the line connecting the two charges and the direction of motion of charges moving in parallel paths, then the force between the charges in space is

$$F = F_\parallel + F_\perp = F \cos \theta + F \sin \theta$$

In material substances the forces acting on charges moving in parallel paths will be modified by the atomic structure of the material in which the charges move.

51. Moving Charges. Under the influence of an electric field of force, free or mobile charge carriers will move in the direction of the force exerted upon them. If motion is restricted, charges will move along the restricted path rather than in the direction of the maximum value of the field. Charge carriers of opposite polarity move in opposite directions under the influence of the same electric field.

In a material having no net or overall charge density, such as a good conductor, the motion of charges can be derived from the forces acting upon mobile charges. Such particles, in general, may have charges Q^+ and Q^- as well as masses m^+ and m^-, respectively. Assuming no relativistic change in the mass of the charge carriers, the electric force acting on them in a field of intensity E is

$$F = QE = (Q^+ - Q^-)E = Q^+ E - Q^- E$$

If the charges do not move too rapidly, the force is $F = ma = m(dv/dt)$, where $a = dv/dt$ is the acceleration imparted to them; the drift velocity of the charges is

$$v = \int a\, dt = \int \frac{F}{m}\, dt = \int \frac{Q^+}{m^+} E\, dt - \int \frac{Q^-}{m^-} E\, dt$$

In the mks system of units, speed is in meters per second, force F in newtons, charge Q in coulombs, mass m in kilograms, field strength E in volts per meter, and time t in seconds. In these expressions, Q^+ and m^+ are the charge and mass, respectively, of positively charged carriers, and Q^- and m^- are corresponding quantities for negatively charged carriers.

52. Speed and Mobility of Charge Carriers. The relations given above apply when charge carriers move with speed v less than about one-tenth the speed of electromagnetic propagation in free space ($c = 300,000$ km/s). For greater speeds the mass m of the moving particle becomes appreciably greater than its rest mass m_0, in accordance with relativity principles. As the intensity of the accelerating field increases, the speed of the particle increases and approaches c as a limit as indicated in Table 1-5.

Table 1-5. Mass and Velocity Ratios

Energy, MeV	Electron mass ratio m/m_0	Electron velocity ratio v/c
0.1	1.20	0.548
1.0	2.96	0.94
10	20.6	0.9988
10^2	196.6	0.9998
10^3	1,960	0.999999
10^4	19,580	0.9999999

The *mobility* of moving charges μ is defined to be the ratio of the drift velocity of charge carriers v to the intensity of the electric field E accelerating them, or $\mu = v/E$. If charge carriers of both polarities are simultaneously available, the net mobility of both kinds of mobile charge carriers is $\mu = \mu^+ - \mu^-$.

53. Electric Current. In a medium having mobile charged particles of volume density

ρ in which the particles move across a surface of area s with average speed v_{av}, the *current* is defined to be the product $\rho s v_{av}$. If the moving particles are electric charges $Q = ne$, the volume charge density is $\rho = Q/V$ and the electric current is

$$I = \rho s v_{av} = \frac{Q}{V} s v_{av} = \frac{ne}{V} s v_{av}$$

The electric current may also be expressed in terms of electric flux, since charge and flux are related. In either case, an electric current is the directed, i.e., nonrandom, motion of charge carriers in a charge-bearing medium.

In mks units, current is in amperes, charge in coulombs, volume in cubic meters, surface area in square meters, speed in meters per second, volume charge density in coulombs per cubic meter, and time in seconds.

The current may also be expressed in coulombs per second, as shown by the relation

$$I = \frac{Q}{V} s v_{av} = \frac{Q}{sl} s \frac{l}{t} = \frac{Q}{t}$$

where $V = sl$ and $v = l/t$. The charges producing the electric current may be of either polarity or of opposite polarities. In the latter case, the charges of unlike sign move simultaneously in opposite directions.

If the charges do not move at constant speed but a charge increment dQ passes across surface s in time dt the instantaneous value of the time-varying current is

$$i = \frac{dQ}{dt}$$

54. Current and Charge. Transfer of charge Q in time t at a steady rate produces a current $I = Q/t$, and $Q = It$. When an infinitesimal amount of charge dQ is transported in time interval dt, the current is $i = dQ/dt$ and the amount of charge transferred in the interval from t_1 to t_2 is

$$Q = \int_{t_1}^{t_2} i \, dt$$

55. Current Density. Current density J is defined as the quotient of the amount of current ΔI transferred normal to a surface area Δs_n as the latter becomes vanishingly small. In the limit

$$J = \frac{dI}{ds_n}$$

If θ is the angle between the motion of charges and the normal to the surface across which charges pass, the effective surface element is $ds_n = \cos \theta \, ds$ and the current density is

$$J = \frac{dI}{\cos \theta \, ds}$$

J is a vector quantity measured in amperes per square meter, in mks units.

56. Current Element. A quantity of charge remains constant whether stationary or in motion. Charges in motion pass through a conductor of length dl in time interval dt and have velocity $v = dl/dt$. Hence a quantity of moving charge may be expressed in terms of current as

$$Q = \int I \, dt = \int \frac{I}{v} dl$$

On the basis of this relation it is convenient to define a current element as

$$Qv = \int I \, dl$$

In the mks system of units, the current element is measured in ampere-meters, equivalent to coulomb-meters per second.

57. Sense of Electric Current. Mobile charges move in the direction of the electric force acting on them. Thus it is possible to have a current I^- due to the motion of mobile negative charges or a current I^+ due to motion of mobile positive charges. If charges of both polarity are mobile, the total current is the sum $I = I^- + I^+$.

When it is important to specify the direction of an electric current, the polarity of charge carriers constituting the current must be stated.

58. Conduction Current. Conduction current is measured by the amount of mobile charge transferred per unit time through a conductor. The transferred charges may be electrons (in metals and semiconductors), holes (in semiconductors), or ions (in gases and electrolytes).

Current is expressed as $I = (Q/V)sv_{av}$ (see Par. 1-53). For conduction current, Q/V is the number of charges per unit volume capable of moving and contributing to conduction; it is not the ordinary volume charge density by which charges of one polarity exceed those of the opposite polarity per unit volume.

The conduction current may be expressed as

$$I = \sigma \frac{s}{l} V = GV$$

where σ is a property of a unit volume of the conducting material called its *conductivity*, s is the area of the surface through which charges pass, and l is the distance along the conductor across which the difference of potential is V. The conductivity σ depends on properties of the conducting material (see Par. 1-64). The quantity G is called the *conductance* of the conductor.

59. Convection Current. The convection current is usually regarded as the motion of charge density, as in the motion of electrons in formed beams in electron tubes. It is expressed as

$$I_{conv} = \rho vs$$

where ρ is the volume charge density of charge carriers moving with velocity v across a surface area s. In mks units, current is in amperes, volume charge density is in coulombs per cubic meter, velocity is in meters per second, and area is in square meters.

60. Displacement Current. Electric fields of elementary charges extend considerably beyond the dimensions of charge carriers and may be expressed in terms of electric flux. The expression for current, $I = \rho sv$, may be applied to show that a time-changing electric flux in a dielectric produces a *displacement current*.

The transfer dQ of free charge to conducting surfaces on opposite sides of a dielectric sets up flux $d\psi$ in the dielectric, and the charge density is $\rho = dQ/dV = d\psi/dV$. Since the volume V is the product of the surface area s and length l, and since $dV = s\,dl$, the displacement current is

$$I_{dis} = \rho\,sv = \frac{d\psi}{dV} = \frac{d\psi}{s\,dl}s\frac{dl}{dt} = \frac{d\psi}{dt}$$

Since $\psi = Ds$ and $D = \epsilon E$, the displacement current in the dielectric may be written in terms of electric field strength as

$$I_{dis} = \frac{d\psi}{dt} = \frac{d(Ds)}{dt} = s\frac{dD}{dt} = s\frac{d(\epsilon E)}{dt} = s\epsilon\frac{dE}{dt}$$

The displacement current is zero if the electric flux remains constant with respect to time.

61. Total Current and Continuity of Current. Total current is the sum of the conduction, convection, and displacement components in a circuit or system. The conduction and convection components are called *true currents* since they represent transfer of charges. Displacement current is a *virtual* current, arising from the time rate of change of electric flux of charges that do not pass through a dielectric.

Continuity of Current. Charges may be neither created nor destroyed. Thus, for steady flow of charges, the outward motion of positive charge through any closed surface must be equal to the decrease of positive charges within the surface. If Q denotes the charge inside a closed surface and J_n is the current density normal to the surface s across which charges move, the equation of continuity of current is

$$I = \int J_n\,ds = -\frac{dQ}{dt}$$

If ρ is the volume charge density, the divergence of steady current through a closed surface is the time rate of decrease of charge density, which is zero over a closed surface. For time-varying currents i, the displacement current $\partial D/\partial t$ must also be taken into account, and the divergence becomes

$$\text{div}\left(i + \frac{\partial D}{\partial t}\right) = 0$$

62. Conduction in Crystals. Because of their wave nature, electrons are believed to be able to pass through a perfect crystal without encountering any resistance. Opposition to motion of electrons arises from deviations in periodicity of the potential in which electrons move through a crystal. Such deviations may be due to (1) lattice vibrations, (2) vacancies, holes, or other defects in the lattice, (3) foreign impurity atoms, and (4) grain boundaries.

In a pure metal at finite temperature, the ions are not at rest but vibrate thermally about their positions of equilibrium. Such vibrations cause scattering of electrons, which accounts for resistance to motion of charges. The ion vibrations and scattering increase with rise in temperature, and electric resistance also increases with rise in temperature of the conductor.

The crystal structure of highly conducting metals is such that the outer-shell electrons are shared by neighboring atoms and the electrons are free to wander throughout the substance. The number of free electrons per cubic centimeter is of the order of 10^{23}. Electric conduction takes place as a result of free electrons, unattached to any particular atoms, under the action of an electric field producing a force on the electrons.

63. Current and Voltage Relations in Conductors. The electric current through a conductor may be regarded as the effect produced by the energy expended to keep charges moving through the conductor. For a given applied voltage V the current I depends upon the nature of the material of which the conductor is made and its dimensions. In most metals and other good conductors, the current may be expressed as

$$I = \left(\sigma \frac{s}{l}\right)V = GV = \frac{V}{R}$$

where $G = \sigma s/l$ is the conductance and $R = V/I$ is the resistance of the conductor or system. In this equation, σ is the electric conductivity of the conducting material, l is the length of conductor across which voltage difference V exists when the current I passes through it, and s is the cross-sectional area of the conductor, assumed to be constant.

In the mks system of units, current is measured in amperes, potential difference is in volts, and conductance in amperes per volt, to which the name mho (abbreviated Ω^{-1}) is given. The resistance is measured in volts per ampere, or ohms (abbreviated Ω).

Many cases occur in which the current through a conductor is not directly proportional to voltage across it. Such devices are said to be nonlinear, and the voltage-current relations are best shown by graphs relating V and I.

Frequently current depends upon some parameter other than voltage, such as temperature, pressure, incident radiation, etc.

64. Resistivity and Conductivity. *Conductivity* is a factor expressing the ease with which an electric current is able to pass through a conducting material. It is given by

$$\sigma = \frac{Il}{Vs} = \frac{J}{E}$$

where J is the current density in the conductor and E is the electric field strength along the conductor, s is the area of the conductor across which current passes, and l is the length of the conductor. In mks units it is measured in amperes per square meter per volt per meter, which is equivalent to mhos per meter.

Resistivity is a factor expressing the opposition to the motion of charges through a conductor of unit length and unit cross sectional area in the system of units used. It is the reciprocal of conductivity and is given by

$$\rho = \frac{1}{\sigma} = \frac{Vs}{Il} = \frac{E}{J}$$

where the symbols have the meaning given above. Resistivity is measured in volts per meter per current per square meter in mks units, or in ohm-meters.

65. Temperature Variation of Resistivity. The resistivity of highly conducting solid metals increases with temperature. Over a fairly wide range, the increase in resistivity is proportional to the change in absolute temperature.

If ρ_{T_0} is the resistivity of a conducting material at a reference temperature T_0, and if ρ_{T_1} is the resistivity of the same material at another temperature T_1, then for linear change of resistivity with temperature,

$$\rho_{T_1} = \rho_{T_0}\left[1 + \frac{k}{\rho_{T_0}}(T_1 - T_0)\right]$$

where $k = (\rho_{T_1} - \rho_{T_0})/(T_1 - T_0)$ is the change in resistivity from ρ_{T_0} to ρ_{T_1} as the temperature changes from T_0 to T_1. The coefficient $\alpha = k/T_0$ is called the *temperature coefficient of resistivity* at the temperature T_0. It is experimentally determined, and values are listed in Sec. 6. At room temperatures, the temperature coefficient of resistivity of most metallic conductors is in the range from 0.002 to 0.005.

66. Resistance. The resistance of a linear conductor, in which current is proportional to applied voltage, is regarded as the opposition offered to the transport of charges through it. It is expressed as the ratio of the constant voltage difference between the ends of a conductor, through which current passes, to the resulting current. Since the voltage V across a conductor is the amount of energy U expended per unit charge Q and the current through the conductor I is *the amount of charge Q* transferred per unit time t, the resistance is given by

$$R = \frac{V}{I} = \frac{U/Q}{Q/t}$$

Resistance is measured in units of ohms, or volts per ampere.

The resistance of a nonlinear conductor, in which current is not directly proportional to applied voltage, is called the *variational* or *differential resistance*. It is measured as the limiting value of the quotient of a small change ΔV in voltage to the resulting small change ΔI in current as the latter becomes vanishingly small.

From another point of view, resistance is the factor by which the mean-square conduction current must be multiplied to give the corresponding power lost by dissipation as heat or some other form of radiated electromagnetic energy.

67. Conductance. The conductance of a linear conductor, in which the voltage drop is proportional to the current, is regarded as the ease with which charges are transferred through it. It is expressed as the ratio of the current through the conductor to the voltage difference across it. By analogy to resistance, of which conductance is the reciprocal, conductance is given by

$$G = \frac{I}{V} = \frac{Q/t}{U/Q}$$

Conductance is measured in amperes per volt, or mhos.

The conductance of a nonlinear conductor, in which voltage is not directly proportional to current, is called the *variational* or *differential conductance*. It is measured as the limiting value of the quotient of a small change ΔI in current to the small change ΔV in voltage producing it as the latter becomes vanishingly small.

Conductance is also the factor by which the mean-square voltage must be multiplied to give the corresponding power lost by dissipation as heat or some other form of radiated electromagnetic energy.

68. Resistance of a Conductor. The resistance of a conductor of length l, uniform cross sectional area s, and material resistivity ρ_0 at temperature T_0 is

$$R_0 = \frac{V}{I} = \frac{El}{Js} = \rho_0\frac{l}{s}$$

The resistance of the conductor is in ohms when l is in meters, cross-sectional area s is in square meters, resistivity ρ is in ohm-meters, and temperature is expressed in kelvins (K).

Resistor. A resistor is a device the primary purpose of which is to introduce resistance into an electric circuit or system. The ideal resistor dissipates electric energy without storing electric or magnetic energy.

69. Work Done by Electric Current. Work is done on charge carriers when they are accelerated by an electric field. Conversely, the charges do work or dissipate energy when they are decelerated. Their energy is dissipated as heat (or other lost energy) or in work done by the system in which they are decelerated.

If a charge Q is decelerated and produces a steady voltage drop V when a steady current I passes through a conductor, the work done in time t is

$$W = U = VQ = VIt$$

If charges do not move at a uniform rate, the voltage and current vary with respect to time and may be represented by v and i, respectively. In this case, the amount of work done in the time interval from $t = t_1$ to $t = t_2$ is

$$W = U = \int_{t_1}^{t_2} vi\, dt$$

The work done is in joules when the potential difference is in volts, current in amperes, and time in seconds.

Energy Dissipated in a Resistor. Since voltage and current are related by resistance R, work done by an electric current may be expressed in terms of resistance. Thus, the energy dissipated is

$$U = VIt = \frac{V^2}{R}t = I^2Rt$$

for charges transported at steady rate, or

$$U = \int_{t_1}^{t_2} vi\, dt = \int_{t_1}^{t_2} \frac{v^2}{R}\, dt = \int_{t_1}^{t_2} i^2R\, dt$$

for charges transferred at a nonuniform rate, where symbols have the meanings assigned to them above.

70. Electric Power. If a steady current I and a steady voltage V produce work W in time interval t, the time rate of energy conversion (electric power) is

$$P = \frac{W}{t} = \frac{VIt}{t} = VI = \frac{V^2}{R} = I^2R$$

If the current and voltage vary with time and are expressed as i and v, the instantaneous power p in a circuit of resistance R is

$$p = vi = \frac{v^2}{R} = i^2R$$

In systems in which voltage and current vary with time according to simple harmonic motion, the average value of power is

$$P_{\mathrm{av}} = \tfrac{1}{2}V_m I_m \cos\theta = VI \cos\theta$$

where the subscripts m designate maximum or amplitude values of voltage and current whose effective values are V and I, $\cos\theta$ is the power factor, and θ is the phase difference between the current and voltage.

71. Joule's Law of Heating Effect. Joule's law states that the rate at which heat is produced in an electric circuit of resistance R is equal to the product of the resistance and the square of the current. If I is the steady current through a resistor, the steady power dissipated in it is

$$P = I^2R$$

1-21

If i is the instantaneous value of time-varying current, the instantaneous power developed in resistor R is

$$p = i^2 R$$

Power is in watts when current is in amperes and resistance in ohms.

72. Faraday's Law of Electrolysis. In the process of electrolytic change, the mass of a substance deposited at one electrode or liberated at the other varies (1) directly as the quantity of electricity, (2) directly as the atomic weight of the substance deposited, and (3) inversely as its chemical valence. The Faraday electrolytic constant is the amount of charge required to liberate one gram atom of any univalent element. This quantity of electricity is 96,487 coulombs per gram equivalent.

Faraday's law of electrolysis may be stated quantitatively as

$$m = \frac{a}{e} \frac{WQ}{v} = \frac{1}{F} \frac{WQ}{v}$$

where m is the mass, in grams, of substance deposited, a is the mass, in grams, of atom having unit atomic weight, Q is the quantity of electricity passed through the cell, W is the atomic weight, v is the valence of atoms deposited or liberated, e is the charge of an electron in coulombs, and $a/e = 1/F$, where F is the Faraday constant = 96,487C.

MAGNETOSTATICS

73. Transverse Forces between Moving Charges. Mobile charges may be set into motion when acted upon by an electric field, and in motion, the charges carry their own electric fields with them. The three-dimensional field of charges may be resolved into a longitudinal component in the direction of motion and transverse components perpendicular to the direction of motion. The perpendicular components produce transverse, magnetic forces between moving charges. The forces also act between the conductors in which the charges travel. A *magnetostatic field* is produced when the charges move with uniform motion, a *magnetokinetic field* when the charges move nonuniformly.

The law of magnetic force between current elements is attributed to Biot and Savart and also to Ampere and Laplace. This law is as basic to magnetostatics and magnetokinetics as Coulomb's law is to electrostatics and electrokinetics.

74. Biot-Savart/Ampere-Laplace Law. As a result of experiment and theoretical reasoning it is concluded that the increment of force between two parallel current elements[1] is expressible by an equation of the form

$$d(dF) = \frac{\mu I_1 \, dl_1 \, I_2 \, dl_2}{4\pi r_{12}^2} \sin \theta$$

where I_1 and I_2 are values of steady currents in conductors whose elements of length are dl_1 and dl_2, respectively, r_{12} is the straight-line distance between dl_1 and dl_2, μ is a property of the medium in which the current-carrying conductors reside, and θ is the angle between the direction of r_{12} and the perpendicular from a current element and the direction of the parallel conductors. This equation is similar in form to Coulomb's equation for the force between two stationary point charges and indeed is closely related to Coulomb's law.

Experimentally it is not feasible to check the validity of the Biot-Savart law since differential current elements cannot be isolated. Currents flow in closed paths of finite length. The force between two closed current-carrying conductors can be measured experimentally. The force produced by currents in closed paths can be expressed mathematically by integrating the expression for $d(dF)$ over the total path lengths l_1 and l_2 by the double line integral

$$F = \oint \left(\oint \frac{\mu I_1}{4\pi r_{12}^2} \, dl_1 \right) I_2 \sin \theta \, dl_2$$

---

[1] For discussion of the transverse force of charges moving in nonparallel paths, see M. Mason and W. Weaver, "The Electromagnetic Field," Dover, New York, 1929, or A. O'Rahilly, "Electromagnetic Theory: A Critical Examination of Fundamentals," 2 Vols., Dover, New York, 1965.

While this equation has theoretical importance, the integrations required are too difficult (except for conductors of simple geometry) for the equation to have value in practical situations.

If the first and second current elements contain N_1 and N_2 turns with currents I_1 and I_2, respectively, the force must be multiplied by $N_1 N_2$. The force equation then takes the form

$$F = \oint \left(\oint \frac{\mu N_1 I_1}{4\pi r_{12}^2} dl_1 \right) N_2 I_2 \sin \theta \, dl_2 = \oint B_1 N_2 I_2 \sin \theta \, dl_2$$

The last expression on the right defines a field vector,

$$B_1 = \oint \frac{\mu N_1 I_1}{4\pi r_{12}^2} dl_1 = \frac{\mu N_1 I_1 l_1}{s}$$

called the *magnetic flux density.* The quantity $s = 4\pi r^2$ is a spherical surface of radius r along each element of length of conductor l_1, and so B_1 represents a surface density of magnetic field produced by current I_1 flowing through N_1 turns of conductor, each of length l_1, and related to the force F.

75. Principle of Superposition. Magnetic forces are produced by the interaction of a pair of current-carrying conductors or their equivalent. One of these establishes a magnetic field and the other exhibits a force reaction to the field. The magnitude of the magnetic field is proportional to the total current setting up the field, provided the permeability is constant. For a given field, the force on a current-carrying conductor is proportional to the current in the interacting conductor.

For magnetic fields, the principle of linear superposition applies when the permeability of the medium is constant, as it is for free space and most nonmagnetic substances. The permeability of iron and other ferromagnetic materials is not constant. In general, therefore, the principle of linear superposition does not apply for current-carrying conductors in the vicinity of ferromagnetic materials.

76. Magnetic Field. A magnetic field may be defined as a vector property of space or material bodies capable of exerting a force on a charge moving with directed motion in a conductor to accelerate the charge in proportion to its magnitude and its speed relative to that of the field. A magnetic field may be produced by a permanent magnet or by charges moving in a conductor with directed motion. In general the magnetic field may vary with time and is a vector function of the position at which a force acts in a transverse direction on the moving charge; it also depends upon the properties of the medium in which the charges move.

77. Permeability and Magnetic Susceptibility. The force between two current-carrying conductors depends in part upon the properties of the medium in which they are located. For most substances, the surrounding materials produce little magnetic effect, but substantial effects are produced by ferromagnetic materials composed of iron with other materials such as cobalt, nickel, aluminum, manganese, etc. Usually the presence of ferromagnetic materials substantially increases the magnetic force produced by currents flowing in conductors.

The magnetic effect of material substances may be taken into account by the *magnetic permeability* μ, which may be written as the product of two terms. One of the terms may be taken to express the magnetic properties of free space but is better regarded as a constant relating units in mechanical and electromagnetic systems. The other term accounts for the magnetic properties of material substances.

If μ is the permeability of any substance and μ_0 is the permeability of free space (or vacuum), we may write

$$\mu = \mu_0 \mu_r$$

where μ_r is the relative permeability of a material substance referred to that of vacuum. In rationalized mks units, $\mu_0 = 4\pi \times 10^{-7}$ henry per meter (H/m).

The permeability may also be written as the sum

$$\mu = \mu_0(1 + \chi_m)$$

where χ_m is a dimensionless quantity called the *magnetic susceptibility* of a substance and is a measure of the alignment of atoms or molecules in a magnetic material.

78. Magnetic Flux Density. Magnetic flux density may be looked upon from two different points of view. It may be defined as the force which a current-carrying conductor exerts on a test body so oriented that the maximum force is exerted upon it by the field. The Biot-Savart law shows that the maximum force occurs when the current element constituting the test body is perpendicular to the line connecting it with the current element establishing the magnetic field. The magnetic flux density may also be defined in terms of the properties of the current-carrying conductor originating the magnetic field.

In terms of the force which the first current element $N_1 I_1 dl_1$ exerts on a second current element $N_2 I_2 dl_2$, the physical significance of magnetic flux density may be expressed as

$$B_1 = \left(\frac{dF}{N_2 I_2 \sin\theta\, dl_2}\right)_{max} = \frac{dF_{max}}{N_2 I_2 dl_2}$$

where F_{max} is the maximum force exerted on the second current element for which, in the Biot-Savart law, $\sin\theta = \sin 90° = 1$.

In terms of the magnetic field produced by a current element $N_1 I_1 dl_1$, the magnetic flux density may be interpreted as

$$B_1 = \oint \frac{\mu N_1 I_1 dl_1}{4\pi r^2} = \frac{\mu N_1 I_1 l_1}{s}$$

where $s = 4\pi r^2$ is the surface area of a sphere of radius r positioned at each point along the conductor of length l_1. In mks units, magnetic flux density is measured in webers per square meter (Wb/m^2), which is equivalent to teslas (T).

79. Magnetic Flux. Magnetic flux through an area is defined as the surface integral of the normal component of magnetic flux density over the area. Thus the magnetic flux over area s is

$$\phi = \int_{surface} B\cos\theta\, ds = \int_{surface} B \cdot ds$$

where B is the magnetic flux density at the surface element ds and θ is the angle between the surface element ds and the direction of B at the element of area.

It is sometimes helpful to visualize the magnetic flux as the total quantity of lines of magnetic force set up around a current-carrying conductor in a medium whose permeability is μ. The magnetic flux produced by a current I in each of N closely spaced conductors in a medium whose permeability is μ is

$$\phi = \mu N I l = Bs$$

where l is the length of the conductor.

In the mks system of units, magnetic flux is measured in webers, current in amperes, length in meters, and the permeability in henrys per meter.

80. Gauss' Theorem for Magnetic Flux. The theorem of Gauss for magnetic flux is a necessary consequence of the Biot-Savart law and states that the amount of magnetic flux passing through any closed surface is equal to zero. This is interpreted to mean that as many lines of magnetic flux enter any closed surface as leave it. This result follows since magnetic fields arise from the transverse fields of moving charges and do not represent a point source or sink of magnetic effect (whereas charges are point sources or sinks of electrical phenomena).

If B is the amount of flux density crossing any element of surface ds, and if θ is the angle between the direction of B and the normal to the surface element, the magnetic flux is

$$\phi = \int_{\substack{closed\\surface}} B\cos\theta\, ds = \int_{\substack{closed\\surface}} B \cdot ds = 0$$

The theorem of Gauss and the statement of the divergence for magnetic flux (Par. **1-81**) given here as related to magnetostatics are equally valid for time-varying magnetic fields.

81. Divergence of Magnetic Flux. The lines of magnetic flux around a current-carrying conductor are closed loops concentric about the conductor. For any closed surface surrounding a segment of such a conductor, as many lines of magnetic flux will cross the surface in an outward direction as recross it elsewhere in an inward direction.

The divergence of magnetic flux density is zero. The divergence is equal to the ratio of the flux passing through a closed surface to the volume contained therein as the volume becomes vanishingly small, or

$$\text{div } B = \lim_{\Delta v \to 0} \frac{\int_{\substack{\text{closed} \\ \text{surface}}} B \cos \theta \, ds}{\Delta v} = 0$$

If B is expressed in terms of rectangular components, B_x, B_y, and B_z, each of which is a function of coordinates x, y, and z, the divergence is expressed as the sum of the partial derivatives of each component in the direction of its axis, or

$$\text{div } B = \frac{\partial B_x}{\partial x} + \frac{\partial B_y}{\partial y} + \frac{\partial B_z}{\partial z} = 0$$

This result applies for time-varying as well as for time-invariant magnetic fields.

82. Magnetic Field Strength. The magnetic effects of a current may be defined in terms of a quantity called the *magnetic field strength H*. The magnetic field strength is a field vector specifying the amount of force F produced per unit of magnetic flux ϕ set up by a conductor of N closely spaced turns each carrying the same current I through a conductor of length l in a medium whose permeability is μ. From this point of view, $H = F/\phi$, but in terms of fluxes produced by current-carrying conductors, the force is

$$F = \frac{\phi_1 \phi_2}{4\pi r_{12}^2 \mu} = \frac{\phi_1 \phi_2}{s\mu}$$

where s is the area of a closed sphere about each element of the flux produced. Hence:

$$H = \frac{\phi_1 \phi_2}{\mu s \phi_1} = \frac{\phi_2}{\mu s} = \frac{B}{\mu} = \frac{NI}{l}$$

The magnetizing force H is taken around a closed path of length l. This length is the circumference of a circular loop of magnetic lines centered on an element of the current-carrying conductor, when we deal with current-carrying conductors.

The magnetic field strength is a vector point function whose curl is the current density. It is proportional to magnetic flux density in regions free of magnetized substances.

83. Magnetization. Magnetization (the intensity of magnetization at a point in a medium) is defined as the intrinsic magnetic flux density at the point divided by the permeability of free space in the system of units employed. It is expressed as

$$M = (B - \mu_0 H)/\mu_0$$

The magnetization can be interpreted as the volume density of magnetic moment. It is a scalar quantity.

84. Ampere's Law for Magnetic Intensity. Ampere's circuital law (also referred to as Ampere's work law) can be derived from the Biot-Savart law of force. The circuital law states that the line integral of magnetic field intensity H about any closed path is equal to the current crossing the surface enclosed by the path. If H is the magnetic field intensity and I is the current crossing a surface bounded by a closed loop, then

$$\oint H \cdot dl = \int_{\substack{\text{bounded} \\ \text{surface}}} J \cdot ds = I$$

In this result, positive current is defined as that in which charges flow in the direction of advance of a right-hand screw when it is turned in the direction in which the closed loop is traced. The application of Ampere's circuital law requires determination of the total current enclosed by a closed loop. The line integral is defined as the *magnetomotance* of the closed loop.

85. Magnetic Vectors at Boundaries. The magnetic field may be defined in terms of (1) the magnetic induction or flux density B, (2) the magnetic field strength H or (3) the magnetization M of a magnetic substance.

The magnetic flux density, $B = F/NIl$, is the force per magnetic test body of a magnetic field acting on currents, whether the currents are *true currents*, i.e., due to charges moving with directed motion in a circuit, or *orbital currents*, due to the motions of orbiting electrons in magnetic substances.

The magnetic field strength, $H = NI/l$, is produced by the directed motion of free or mobile charges and applies only to "true" currents.

The magnetization M is the magnetic dipole moment per unit of volume. It applies to material substances under magnetic stress and is zero for free space.

For the three magnetic vectors, the following general relation holds:

$$B = \mu_0 H + M$$

For homogeneous, isotropic magnetic media of relative permeability μ_r the magnetic field vectors are related by the equations

$$B = \mu_0 \mu_r H = \mu H$$

$$M = (\mu_r - 1) H$$

86. Vectors at Magnetic Boundaries. The tangential components of magnetic field strength are continuous across the boundary of magnetic substance and are the same on either side of the boundary. The normal, or perpendicular, component of magnetic flux density is continuous between two media.

87. Magnetomotance. Magnetomotance V_m may be defined as the energy W per unit magnetic flux or the work done in moving a charge around a closed loop surrounding a current-carrying conductor.

In terms of the magnetizing force H produced by a current I flowing in N closely bunched conductors (each having the same current) and the distance l of a closed loop surrounding the current-carrying conductor, the magnetomotance is

$$V_m = \frac{W}{\phi} = \frac{F \cdot l}{\phi} = \frac{F}{\phi} l = Hl = \frac{NI}{l} l = NI$$

The magnetomotance is measured in ampere-turns, the magnetic field strength H in ampere-turns per meter, and distance l in meters. The notion of magnetomotance is especially useful when dealing with magnetic circuits. In this case N is the number of turns of a winding, I is the current through the winding, and l is the mean length of path of magnetic flux in the magnetic circuit.

88. Magnetic Potential. The magnetic (scalar) potential, the integral between two points of the dot product of the magnetic field strength H and the length of path l, is defined as the difference in magnetic potential between the two points. Thus, if the path avoids all currents

$$\int_1^2 H \cdot dl = U_2 - U_1$$

For a closed path which does not enclose current

$$\oint H \cdot dl = 0$$

For a closed path which encloses N turns carrying current I

$$\oint H \cdot dl = NI = V_m$$

89. Flux Linkages. When a closed line of magnetic flux links a closed turn of a conductor (or vice versa), the result is called a *flux link*. For an amount of magnetic flux ϕ linking N turns (closely spaced) of a conductor, the total number of flux linkages is

$$\Lambda = N\phi$$

In the mks system of units, flux linkages are measured in weber-turns and flux in webers.

90. Self-Inductance. Inductance may be defined as the property of an electric circuit by virtue of which a varying current induces an electromotance in that circuit or a neighboring circuit. It may also be defined as the property of a conductor or circuit that establishes magnetic flux linkages.

Self-inductance is the property of an electric circuit whereby an electric potential difference is induced in that circuit by a change of current in the circuit. It is the property of a conductor or circuit for establishing magnetic flux linkages with its own winding.

The coefficient of self-inductance L of a winding is given by

$$L = \frac{\partial \Lambda}{\partial I}$$

where Λ is the number of flux linkages and I is the current in the winding.

The definition of self-inductance is restricted to circuits that are small compared to a wavelength of an oscillation, so that the charges flowing are essentially functions of time but not of spatial distribution.

91. Mutual Inductance. Mutual inductance is the property of two electric circuits whereby an electromotive force is induced in one circuit by a change of current in the other. It may also be defined in terms of the number of flux linkages set up in one circuit per unit of current in the other. Mutual inductance may be either positive or negative.

The coefficient of mutual inductance M between two windings a and b is given by

$$M = \frac{\partial \Lambda_b}{\partial I_a}$$

where Λ_b is the total flux linkages in one winding and I_a is the current in the other winding.

The mutual inductance between two linear circuits or two linear current-carrying conductor arrangements is independent of the direction in which the flux linkages are taken. Hence, for linear systems,

$$M = M_{ab} = M_{ba} = \frac{\partial \Lambda_b}{\partial I_a} = \frac{\partial \Lambda_a}{\partial I_b}$$

92. Force on a Moving Charge in a Magnetic Field. A charged particle moving in a magnetic field is subject to a force in a direction perpendicular to its direction of travel and also perpendicular to the direction of the field. If Q is the magnitude of the charge, v its velocity, B the flux density of the magnetic field, and ϕ the angle between the direction of B and the direction of travel, the force on the particle is

$$F = QvB \sin \phi = Qv \times B$$

The magnetic field tends to accelerate the charge in a direction at right angles to its motion. The acceleration changes only the direction of the particle velocity, not its magnitude. The force F is in newtons when Q is in coulombs, v in meters per second, and B in webers per square meter.

93. Lorenz Force on a Moving Charge in a Combined Electric and Magnetic Field. If a charge Q is accelerated by an electric field of intensity E and moves with velocity v in a magnetic field of flux density B, the total force acting on it is the vector sum of the electric and magnetic forces, F_e and F_m, respectively, and is given by the relation

$$F = F_e + F_m = QE + QvB \sin \phi = Q(E + v \times B)$$

where ϕ is the angle between the direction of motion imparted by the electric field and the direction of B. The force is in newtons when Q is in coulombs, E is in volts per meter, v is meters per second, and B is in teslas or webers per square meter.

94. Magnetic Force on Current-Carrying Conductor. A current I flowing in a conductor element of length dl is equivalent to a mobile charge Q in the element moving with speed v. By combining this with the expression for the magnetic force on a moving charge, it follows that the force acting on a conductor of length dl carrying current I situated in a magnetic field of flux density B is

$$dF = I(dl \times B)$$

where dF is a vector element of force, in newtons, I is the current in amperes, dl is an element of conductor length in meters, and B is the flux density of the field in which the conductor is located, in webers per square meter.

95. Torque on Current-Carrying Loop. If a small current-carrying loop is immersed in a magnetic field, equal and opposite forces act on opposite sides of the loop, tending to cause the loop to rotate as a result of the torque thus acting on it. If the loop is a rectangle of length l and width w and is located in a field of flux density B, and if ϕ is the angle between the plane of the loop and the direction of the magnetic field, the torque is

$$T = 2F\frac{w}{2} = IBlw \cos \phi$$

or since lw is the area of the loop A, the torque is

$$T = IAB \cos \phi$$

The torque is in newton-meters when the current is in amperes, flux density in webers per square meter, the loop dimensions l and w in meters, and A in square meters.

96. Magnetic Moment. The product of the current I and the area A of a current-carrying loop is called the *magnetic moment* of the loop and is designated by $m = IA$.

The torque acting on the loop is expressed as

$$T = mB \cos \phi = mB \sin \theta$$

In the last term, θ is the angle between the direction of the flux density B and the normal to the plane of the coil.

The torque is in newton-meters when B is in webers per square meter and m is in ampere-square meters.

Magnetic Moment of Magnetized Body. The magnetic moment of a magnetized body is the volume integral of the magnetization

$$m = \int M \, dv$$

The magnetization M is the magnetic dipole moment per unit volume. It is a measure of the magnetic effect of material substances, independent of the "true" currents. It is zero in a vacuum.

97. Energy Stored in a Magnetic Field. When ϕ is the magnetic flux, the average energy of a magnetic field is $U_{av} = \frac{1}{2}\phi NI$ and self-inductance is defined as $L = N\phi/I$. Hence, the average energy stored in the magnetic field of an inductor is

$$U_{av} = \frac{1}{2}\phi NI = \frac{1}{2}LI^2$$

The energy stored is in joules when L is in henrys and I in amperes.

98. Magnetic Energy and Energy Density. Since the flux is $\phi = Bs$, $NI = Hl$ and l is the length of the volume $v = sl$, the average energy of a magnetic field is

$$U_{av} = \frac{1}{2}BHv = \frac{1}{2}\mu H^2 v = \frac{1}{2}\frac{B^2}{\mu}v$$

The energy density is the energy per unit volume, or

$$u_{av} = \frac{U_{av}}{v} = \frac{1}{2}\mu H^2 = \frac{1}{2}\frac{B^2}{\mu}$$

Energy density is in joules per cubic meter when H is in amperes per meter and B in webers per square meter.

Magnetic Energy. The energy U in any system in which a force F acts for a distance l_c is $U = F \cdot l_c$. The magnetic force is given by the relation $F = \phi NI/s$, and the magnetic energy is then

$$U = F \cdot l_c = \frac{\phi}{s}(NI)l_c = \phi(NI)\frac{ll_c}{s} = \phi NI$$

since $s = ll_c$. The average energy is

$$U_{av} = \frac{1}{2}U = \frac{1}{2}\phi NI$$

MAGNETOKINETICS

99. Electromagnetic Energy Conversions. When charges move with accelerated or nonuniform motion rather than with uniform motion, additional phenomena are introduced which do not become evident in the magnetostatic case. These phenomena are most easily interpreted in terms of the behavior of an ideal loss-free system capable of storing both electric and magnetic energy.

In a loss-free, energy-storing system in which charges move with directed nonuniform motion, the sum of the instantaneous values of electric and magnetic energy must remain constant as required by the law of energy conservation. Accordingly, any change in the electric energy of the system, $du_c = d(eq)$, must be accompanied by an equal and opposite change in the magnetic energy of the system, $-du_m = -d(\Lambda i)$. Since charges carry their fields with them when they move, the transport of an elemental amount of charge dq in the electric portion of the system requires a change in flux linkages by an amount $-d\Lambda$ in the magnetic portion of the system.

For the system as a whole, the net change in the two forms of energy is given by

$$du_e = e\,dq = -du_m = -i\,d\Lambda = -\frac{dq\,d\Lambda}{dt}$$

100. Induced Electromotance. By applying the principle of energy conservation to charges in nonuniform motion, it can be shown that the conversion of magnetic to electric energy establishes a voltage

$$e = -\frac{d\Lambda}{dt}$$

where $d\Lambda/dt$ is the time rate of change of flux linkages. Flux linkages may be expressed as

$$\Lambda = N\phi = NBs = NBlw$$

where N is the number of turns, ϕ is the magnetic flux, B is the flux density, and s is the area across which flux linkages occur; s may be given in terms of a length l and width w. It is possible, therefore, to establish an *induced electromotance* by varying any one of these factors with respect to time.

Only two methods of producing time-varying flux linkages are in common use: (1) the flux-changing method, used in transformers, and (2) the flux-cutting method, used in machinery.

In the flux-changing method, the magnetic flux in a fixed mechanical system changes with time, and the induced voltage is

$$e = -N\frac{d\phi}{dt}$$

In the flux-cutting method, usually the magnetic flux remains constant and a conductor cuts flux lines of the magnetic field perpendicular to lines of induction, with velocity v, to change flux linkages and generate a voltage

$$e = -NBlv$$

The induced voltage is in volts if the flux is in webers, flux density in webers per square meter, the length of the flux cutting loop in meters, time in seconds, and the speed with which the conductor cuts lines of flux in meters per second.

101. Faraday's Law for Induced Electromotance. Faraday's law for electromagnetic induction in a nonmoving circuit states that the electromotance (voltage) induced is proportional to the time rate of change of magnetic flux linked with the circuit.

For stationary circuits this law may be stated as

$$\oint E \cdot dl = -\frac{d}{dt}\int_{surface} B \cdot ds = -\int \frac{\partial B}{\partial t} \cdot ds$$

where E is the electric field intensity along a path length dl and B is the magnetic flux density normal to the surface ds.

If the voltage induced in a closed circuit is due to motion of the circuit with respect to the magnetic field, then

$$\oint E \cdot dl = \oint (v \times B) \cdot dl$$

If the voltage induced in a circuit is due to time rate of change of magnetic flux density, then

$$\oint E \cdot dl = - \int_{\text{surface}} \frac{\partial B}{\partial t} \cdot ds$$

If the electromotance is due to motion of the circuit with respect to the magnetic field as well as to time variations of flux density, then

$$\oint E \cdot dl = \oint (v \times B) \cdot dl - \int_{\text{surface}} \frac{\partial B}{\partial t} \cdot ds$$

The latter is the general case.

102. Voltage Induced in Inductors. The general equation for induced voltage may be used to derive the voltage induced in a self- or mutual inductance. The coefficient of self-inductance is defined as $L = N\phi/I$, so that the induced voltage is

$$e = -\frac{d(N\phi)}{dt} = -\frac{d(LI)}{dt} = -\left(L\frac{dI}{dt} + I\frac{dL}{dt} \right)$$

If the coefficient of inductance is time-invariant, as is often the case, the second term becomes zero and for a self-inductor

$$e = -L\frac{di}{dt}$$

The voltage induced is measured in volts when L is in henrys and the current change is in amperes per second.

The coefficient of mutual inductance M is defined to be $M = \Lambda_1/I_2$. Hence, the voltage e_2 induced in one winding by a current i_1 in another winding of a transformer or mutual inductor whose flux Λ_{12} links both circuits is

$$e_2 = -\frac{d\Lambda}{dt} = -\left(M\frac{di_1}{dt} + i_1\frac{dM}{dt} \right)$$

If the mutual inductance is constant with respect to time,

$$e_2 = -M\frac{di_1}{dt}$$

The induced voltage is in volts when M is in henrys and the current change is in amperes per second.

103. Current Induced in Conductors and Semiconductors. Time-varying flux linkages occurring in conducting materials induce a voltage in the conductor, and current will flow as a result in the conductor. Such a current is called an *induced current*. If a well-defined path is provided, the induced current flows in a definite path and can be made to perform useful functions. But if the conducting path extends over substantial volume, many paths are provided for the transfer of charges. Many current paths are then possible, and the currents may generate a considerable amount of heat in addition to any useful function.

104. Eddy (Foucault) Currents. Eddy, or Foucault, currents exist as a result of voltages induced in the body of a conducting mass by variations of the magnetic flux. The variation of magnetic flux is the result of a varying magnetic field or of the relative motion of the mass with respect to a magnetic field.

Eddy currents result in Joule heating of the conducting material. Losses in ferromagnetic materials (in which voltages are often induced in electric machinery) may be minimized by dividing the magnetically conducting substance into small segments (such as wires or laminations) to reduce the length of path in which a current exists. At high frequencies, finely divided particles of magnetic materials may be embedded in a nonconducting binder. The particles account for the magnetic properties of the material, but the nonconducting binder prevents eddy currents of appreciable magnitude from being generated.

105. Lenz's Law. The law of Lenz relates to the direction of the induced current and the direction of the magnetic flux producing the current. It may be stated as follows: the current in a conductor as a result of an induced voltage is in such a direction that the change in magnetic flux due to it is opposite to the change in flux that caused the induced voltage. This is in accordance with the principle of energy conservation.

106. Magnetic Hysteresis. If the magnetic field applied to a ferromagnetic material is gradually increased from zero and the flux density B produced is plotted for various values of the magnetizing force H, a nonlinear relation between B and H is usually observed. The curve of initial BH values when a material is first magnetized is shown between O and A in Fig. 1-2, the maximum value of B being called the *saturation flux density*. The curve OA is called the *initial* or *normal magnetization curve* of the material.

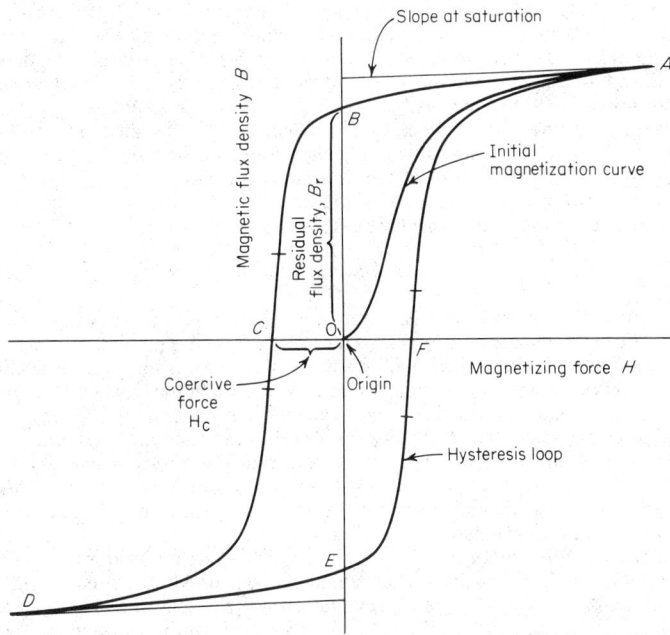

Fig. 1-2. Initial magnetization curve OA and cyclic curve $ABCDEFA$ of a ferromagnetic substance. The cyclic curve is known as a hysteresis loop.

If the magnetic field is then slowly decreased, the flux density decreases, but not in proportion to the decrease in magnetizing field strength. When H is reduced to zero, the specimen retains a certain magnetic flux B_r called the *residual flux density*. To reduce the magnetic flux to zero, the field strength H must be increased in the reverse direction from the initial magnetization. The magnetic field strength H_c at which B becomes zero is called the *coercive force*.

As H is further increased in a negative direction, the flux density also increases with opposite polarity and again reaches a saturation value at high values of H. When the field strength decreases from its maximum negative value, the flux density again decreases but not proportionally. It has residual flux density when $H = 0$, and further increase in H in a positive direction is needed to reduce B to zero. For further increases in H the value of B increases to its saturation value, completing the loop.

The phenomenon causing values of B to lag behind values of H so that the increasing and decreasing fields differ in magnitude is called *hysteresis*. The loop traced out by a complete cycle of BH values is called the *hysteresis loop*.

107. Magnetic Effect of Displacement Current in Dielectrics. The displacement current, treated in Par. **1-60**, occurs only in dielectrics and insulators, in which case conduction electrons are tightly bound to their atomic nuclei and are not able to move freely through the material. The displacement current I_d is found to produce a magnetic field intensity H which is proportional to the time rate of change of the electric field intensity E producing the displacement current. This situation is different from the case of conduction current, in which H and E are in time phase. Because H is proportional to dE/dt when a magnetic field is produced by displacement current and E is proportional to dH/dt, it is possible to generate electromagnetic waves in nonconducting materials.

Dielectric Hysteresis. Under ordinary conditions, the electric flux density in a dielectric is proportional to the electric field intensity E acting across a dielectric. But at very high frequencies, hysteresis effects occur in dielectric materials. The phenomenon is similar to that produced in magnetic materials at much lower frequencies. Hysteresis produces heat losses in dielectrics just as heat losses occur in magnetic materials displaying hysteresis phenomena.

108. Retardation. When an electromagnetic field varies with time, it is possible to determine field relations at different instants of time and at different locations. Such observations permit determination of the velocity with which electromagnetic effects are propagated and travel through media of different properties. These observations also permit determination of the retardation, or time delay, for a disturbance initiated at one point to reach another point where it is detected. If the medium through which electromagnetic disturbances are observed is homogeneous and isotropic, the speed with which the disturbance travels through the medium is constant.

If the disturbance is initiated at time t_0 and travels with velocity v, it will be detected at a point l distant from the source at a time t

$$t = t_0 + \frac{l}{v} = t_0 + t_d$$

The delay time t_d (the *retardation time*) is the interval required for the disturbance to travel the distance l from the source of the disturbance to the receiving point.

109. Induction and Radiation Field Components. Electromagnetic induction is the production of a voltage in a conducting circuit by a change in the magnetic flux linking that circuit. Hence, the *induction field* is the electromagnetic field set up by charges moving with accelerated movement, in which the energy produced is returned to the circuit. The voltage and current in a circuit subject to electromagnetic induction are connected by a time-rate-of-change relation (a derivative or integral with respect to time). If the circuit is sufficiently small to permit assuming that all phenomena occur "instantaneously," the finite velocity of electromagnetic propagation may be neglected.

The situation is different for the *radiation field.* The radiation field is that portion of the total electromagnetic field, set up by accelerated charges, in which the energy interchanges set up electromagnetic waves. The energy of such waves leaves the system in which they are produced, is propagated into space or through material substances (particularly dielectrics), and does not return to the source of the electromagnetic energy (see Pars. **1-135** to **1-149**).

EMISSION, TRANSPORT, CONTROL, AND COLLECTION OF CHARGE CARRIERS

110. Charges Released from Materials. The mobile charges required in the flow of electric current are produced by applying force to, and expending energy on, material substances. Mobile charges may be released by heat, electromagnetic radiation, impact of particles, intense electric fields, and the tunneling effect, in which an electron crosses a potential barrier.

111. Thermionic Emission. Thermionic emission occurs when the electrons in a substance have sufficiently large thermal energy to overcome the restraining forces at the surface of the substance and so escape into the surrounding space. Emitted electrons tend to cluster at the emitting surface, which carries a positive charge equal to the negative charge of the emitted electrons. The resulting *space charge* limits release of additional electrons until those already released are removed from the vicinity of the emitter.

If the electrons are removed from the emitting surface as rapidly as they are released, the emission is temperature-limited and the density of current of emitted electrons obeys the Richardson-Dushman equation

$$J = (1 - R)A_0 T^2 \varepsilon^{-\phi/kT}$$

where J is the emission current density in amperes per square centimeter, R is the coefficient of electrons reflected at the emitting surface (usually neglected since $R < 0.1$), A_0 is a constant whose value depends upon the emitting material, T is the temperature of the emitter on the absolute scale, $\varepsilon = 2.718 \cdots$ is the base of natural system of logarithms, k is Boltzmann's constant, and ϕ is the work function of emitting metal (see Par. 1-15).

The coefficient A_0 has the theoretical value

$$A_0 = \frac{4\pi mek^2}{h^3}$$

where m is the mass of the electron, e is the charge of the electron, k is Boltzmann's constant, and h is Planck's constant. Substitution of values for these constants yields the theoretical value $A_0 = 120 \text{A/cm}^2 \cdot \text{K}^2$.

For devices in which a space charge surrounds the emitter, the useful current is less than the temperature-limited value indicated above. Under space-charge limitation, the current depends upon the geometry of the device and upon the 3/2 power of the voltage difference between emitter (cathode) and collector (anode).

The Richardson-Dushman equation is based on the assumption that the electric field is sufficiently large for all electrons to be drawn to the collector. Thus, the current should not increase at any voltage greater than that required to produce saturation. This is not strictly true. The increase in current with voltage greater than saturation value is called the *Schottky effect* (Par. 1-115).

112. Photoelectric Emission. Photoelectric emission occurs when electrons are liberated from the surface of metals and compounds as a result of absorption of electromagnetic radiation in the ultraviolet, visible, or infrared portions of the spectrum. The emission current is proportional to the amount of radiant energy incident on the surface provided that photoelectric emission occurs for the spectral distribution of the incident radiation.

Photoemission cannot occur unless the energy of incident radiation, usually expressed in electron volts per electron, is equal to or greater than the minimum energy needed to release electrons (the work function). The threshold wavelength, in meters, at which photoemission can occur is

$$\lambda_0 = hc/e\phi$$

where $h = 6.6 \times 10^{-34} \text{J} \cdot \text{s}$ is Planck's constant, $c = 3 \times 10^8 \text{m/s}$ is the velocity of light, $e = 1.6 \times 10^{-19} \text{C}$ is the charge of the electron, and ϕ is the work function, in volts, of the emitting metal. For incident radiation of wavelength $\lambda < \lambda_0$, the maximum energy of the emitted electrons in eV is

$$\frac{h}{c\lambda} - \phi$$

113. Secondary Emission. Secondary emission occurs when primary electrons impinge on a surface with velocities greater than a critical value for each material and, by their impact, release other electrons, called *secondary electrons*. The energy of primary electrons is transferred to electrons in the bombarded surface, some of which acquire sufficient energy to overcome the potential barrier restraining them to the surface. Electrons emitted in this process include primary electrons elastically reflected from the surface, primary electrons inelastically reflected with some loss of energy, and true secondary electrons, whose mean energy is of the order of 10 eV and is substantially independent of the energy of primary electrons.

The effectiveness of a surface as a source of secondary electrons is expressed by the *secondary emission yield*, the ratio of the average number of secondary electrons released for each impinging primary electron. The maximum yield depends upon the energy of the primary electron and usually reaches a maximum value for a particular value of primary-electron energy. The maximum yield does not usually exceed about 2 for elements, but yields as great as 20 or more are obtained from some combination surfaces such as the alkali halides. The oxides and sulfides of metals usually provide yields of the order of 1 to 3. Secondary emission may occur from insulators as well as from conductors.

114. High-Field Emission. Field emission occurs when a high positive electric field exists at the surface of a cold metal-cathode surface. The electric field reduces the potential barrier at the surface of the metal, thereby lowering the energy needed to release electrons. The current density of electrons emitted from the cold cathode is of the form

$$J = AE^2\varepsilon^{-b/E}$$

where A is a coefficient depending on the material, E is the electric field intensity at the cathode surface, and b is a coefficient, approximately constant, whose value is determined by the work function of the metal.

115. Schottky Effect. The increase in current as the anode voltage is increased beyond the value at which the normal saturation value of emitted current occurs (the Schottky effect) results from reduction by the applied field of the work function. When this field effect is taken into account, the saturation current density is expressed as

$$J = A_0 T^2 \exp\left[-\frac{b}{T}(\phi - d\sqrt{V}\,)\right]$$

where A_0, T, and ϕ have values as given for the Richardson-Dushman equation (Par. 1-111), V is the collector voltage measured with respect to cathode, and d is a constant.

116. Ionization and Deionization. Ionization is the process by which a neutral atom or molecule acquires either a positive or negative charge and thereby becomes an ion. Deionization is the process by which an ion loses its charge and becomes a neutral atom or molecule.

The minimum amount of energy required to free an electron occupying the highest energy state in an atom is known as the *ionization energy*. It is usually expressed in electron volts. More than one electron can be removed from an atom, but the energy required to produce multiple ionization increases rapidly above that needed to remove the first electron. An important effect of ionization in gases is *space-charge neutralization*.

Ionizing agents include cosmic rays; natural radioactivity; x-rays; intense electric fields, giving rise to ionization by collision; high temperature, giving rise to thermal ionization; diffusion of charged particles; flames; chemical reactions and effects; photoelectric emission; thermionic emission; high field emission; and secondary emission.

Deionizing agents include electric field in a gas, electric field at electrodes, diffusion by neutral molecules, recombination of ions in gases, and recombination of ions at surfaces. In a gas, volume recombination usually occurs when electrons become attached to neutral atoms to form negative ions which then recombine with positive ions to form electrically neutral particles.

117. Unbound Electrons, Holes, and Electron-Hole Pairs. The highly conducting properties of common metals arise because electrons are not attached to any particular atom but form a mobile electron gas. In contrast, electrons in semiconductors are held together by covalent or valence bonds, along which electrons can move. Such bonds between atoms dominate the electrical behavior of semiconductor crystals.

At very low temperatures, valence electrons in a perfect crystal remain closely bound and the substance acts as a dielectric. As the temperature is increased, some of the valence bonds break, and bond disruption influences the motions of other bound electrons.

When a covalent bond in a semiconductor is broken, the vacancy left by the released electron behaves as though it were a newly formed mobile particle having positive charge whose magnitude is that of an electron and whose mass is slightly greater than that of an electron. The charge cavity thus formed is called a *hole*. Holes appear to move through the crystal structure much as a positive charge might be expected to move.

Actually, the hole itself does not move but becomes filled by an electron which leaves a hole elsewhere in the crystal to be subsequently filled by another electron leaving a hole behind it. The apparent motion of a positive hole in one direction is really produced by the release of electrons at other bonds and the motions of the newly released electrons in a direction opposite that in which the holes appear to travel. An electron filling a hole as an *electron-hole pair* neutralizes the local charge of both of these particles within the crystal.

118. Motion of Charges in Electromagnetic Fields. A particle of charge, e, in coulombs, located in a vacuum in which the electric field intensity, E, is in volts per meter, is acted upon by an accelerating force, in newtons,

$$F_{\parallel} = eE$$

If the particle moves in a magnetic field of flux density, B, in teslas, with velocity, v, in meters per second, it is acted upon by a transverse force, in newtons, of

$$F_{\perp} = (v \times B)e$$

If the particle is in an electromagnetic field, the total force acting upon it is the sum

$$F = F_{\parallel} + F_{\perp} = e(E + v \times B)$$

In a vacuum, a particle of charge $-e$, in coulombs, moving in an electric field, E, whose intensity is in volts per meter, from point x_1 to x_2, in meters, undergoes a change in energy, in joules, of

$$U = -e \int_{x_1}^{x_2} E \cdot dx = -e(V_1 - V_2)$$

The change of kinetic energy, in joules, of a particle of rest mass m_0 moving with velocity v m/s in vacuum is

$$U = m_0 c^2 \left[\frac{1}{1 - (v/c)^2} - 1 \right]$$

if m_0 is in kilograms and $c = 3 \times 10^8$ km/s is the speed of light propagation.

A charged particle moving with velocity v m/s in a magnetic field whose flux density is B T travels in a circular orbit whose radius, in meters, is

$$r = mv/eB$$

where m is mass of the particle in kilograms and e is its charge in coulombs.

119. Electric Conduction. Electric conduction is the transmission of charge carriers through, or by means of, electric conductors. The conductors may be in the solid, liquid, or gaseous state, each of which exhibits different mechanisms of charge transfer. The atomic structure of materials greatly influences the facility with which charges can be transmitted. The environment in which the conductor is located also modifies the conductivity, sometimes to a remarkable extent.

The conduction current passing through a surface s with drift velocity v is $I_{cond} = (Q/V)sv = \rho_f sv$, where $\rho_f = Q/V$ is the volume density of mobile charge carriers. The current density is $J = I/s = \rho_f v$, the drift velocity is $v = \mu E$, and $\mu = v/E$ is the mobility of charge carriers moving with drift speed v in an accelerating electric field of intensity E. Hence the current density may be expressed as

$$J = \rho_f v = \rho_f \mu E$$

In general, charge carriers may be of two kinds, one of charge Q^+ and mass m^+ and the other of charge Q^- and mass m^-. Both contribute to current, but positive and negative charges move in opposite directions, so that the conductivity is

$$\sigma = J/E = \rho_f^+ \mu^+ - \rho_f^- \mu^-$$

In mks units, conductivity is measured in ampere-meters per volt or coulomb-meters per second. This unit is the same as the mho per meter.

120. Superconductivity. At ordinary temperatures the electric conductivity of substances extends over a tremendous range, from about $1.6 \times 10^{-8} \ \Omega \cdot$ cm for silver to at least $10^{16} \ \Omega \cdot$ cm for such dielectrics as quartz. The range is even greater at very low temperatures.

Superconductivity is a property of materials characterized by zero electric resistivity (infinite conductivity) and, ideally, zero permeability. More than a score of elements, including semiconductors and many alloys, become superconducting at temperatures of $20°$K or less.

Superconductivity is found to occur in elements having two to five valence electrons outside the closed shell. It occurs at a *transition temperature* below which the material is superconducting and above which it is not. Superconductivity also depends upon the temperature and the magnetic field surrounding the conductor. It is explained on the hypothesis that the vibrating ions in a crystal lattice do not impede the flow of electrons at very low temperatures. At sufficiently low temperatures ions cease to vibrate, and a perfect, stationary crystal lattice freely allows the electrons to pass.

Currents once established in closed superconducting circuits have been found to continue undiminished for weeks but rapidly drop to zero when the temperature rises above the superconducting value.

121. Conduction in Metals. Metals, while free from net volume density of charge (of either polarity), contain approximately 10^{23} atoms per cubic centimeter. In ordinary metals (but not pure semiconductors) each atom is capable of releasing a free electron and leaving an immobile positive ion. Thus, highly conducting metals have a large volume density of unbound charges (electrons).

A voltage V applied between two points a and b of a metallic conductor of length l_{ab} produces a current I in the conductor perpendicular to the surface through which charges pass and along which an electric field E is established. In terms of the electric field E, the voltage between a and b is

$$V_{ab} = - \int_a^b E \cdot dl = -El_{ba}$$

Since the current density is $J = \rho_f v$ and the drift velocity is $v = \mu \epsilon$ (Par. **1-119**), the current through the conductor is

$$I = \int_{\text{surface}} J \cdot ds = J \cdot s = \rho_f v s = \rho_f \mu E s = (\rho_f \mu) s E = \left(\sigma \frac{s}{l} \right) V$$

In metals the conductivity σ is a constant for each metal and is independent of the current through it or the electric field intensity to which it is subjected.

122. Conduction in Semiconductors. Semiconductors are electronic conducting materials in which the conductivity depends, in large measure, upon impurities, lattice defects, and the temperature of the material. Two kinds of mobile charges are present simultaneously in a semiconductor, electrons and holes. They are oppositely charged and move in opposite directions in a given electric field. Holes and electrons have different masses and mobilities and are scattered in different ways.

The current density of charge carriers moving in a semiconductor is

$$J = e(\mu_e n + \mu_h p) E = (\sigma_e + \sigma_n) E$$

where e is the electron charge, μ is mobility, subscripts e and h refer to electrons and holes, respectively, n is the number of electrons per unit volume, p is the number of holes per unit volume, σ_e is the electron conductivity, and σ_h is the hole conductivity.

The conductivity of semiconductors is highly temperature dependent. The acceleration imparted to a mobile particle of mass m and charge Q is $a = F/m = QE/m$, where $F = EQ$ is the force acting upon the charge Q by the electric field intensity E. The velocity acquired by the charge is $v = QEt/m = \mu E$, where t is the square root of mean-square time during which the particles are accelerated. Thus the mobility is

$$\mu = \frac{Q}{m} t = \frac{Q}{m} \frac{l}{v}$$

where v is the thermal velocity and l is the mean length of the free path. Both v and l vary with temperature in a complicated way such that the conductivity increases with temperature.

123. Conduction in Liquids. According to their electric conductivity, liquids may be divided into three classes: those having conductivities of about 10^{-11} Ω/m (such as paraffin oils, xylol, etc.), which do not conduct and so serve as dielectrics or insulators, those having conductivities of about 10^{-4} Ω/m (alcohol and water), which are neither good conductors nor insulators, and those having conductivity of about 10 Ω/m (aqueous solutions of acids, bases, and salts), which are often regarded as good conductors.

The electric conductivity depends upon the number of ions per unit volume and the rate at which they move through the liquid. If there are n univalent molecules per unit volume of solution each with charge of an electron e and if k is the fraction of molecules dissociated, then the number of charged particles is kn. The mobilities μ^+ and μ^- are the velocities per unit field strength, and the current density is

$$J = ken(\mu^+ + \mu^-)E$$

The conductivity is

$$\sigma = J/E = ken(\mu^+ + \mu^-)$$

Electrolytic conduction approximately follows Ohm's law. The current is due to the movement of positive and negative ions (as well as electrons) that are abundant in ionized liquids. As ions move through the electrolyte, their transfer is evidenced by the deposit of neutralized ions as atoms at one electrode and the liberation of ions at the other electrode, which is used up (corroded) in time.

124. Conduction in Gases. Ionization is a dominant feature of electric conduction in gases; deionization is another. The different types of gas discharge depend upon the relative importance of ionization and deionization processes, which may be divided into self-maintaining and non-self-maintaining discharges. The first do not require an external source of ionization; the latter do.

The condition to be fulfilled for self-maintaining discharge is

$$\epsilon^{ad} = (1 + r)/r$$

where a is the number of ion pairs formed by an electron, per meter drift toward the anode, d is the distance between emitter and collector electrodes, and r is the average number of secondary electrons emitted from the cathode for each new positive ion formed in the gas. The establishment of a self-maintaining discharge requires that, on the average, each electron leaving the cathode must initiate sufficient ionization by collision for the positive ions so formed to produce at least one electron.

125. Conduction in Plasmas. A plasma is a highly ionized gas that does not easily fit the category of a solid, liquid, or gas. The ionization of a plasma is maintained primarily by electron collisions and by photoionization. The concentrations of positive and negative ions are relatively high and are roughly equal. Since a plasma is highly conducting, a relatively low voltage gradient exists across it. In general, the negative charge carriers are electrons. The neutral gas atoms, ions, and electrons are not necessarily in thermal equilibrium, and their velocities vary over a wide range.

Electrons in a plasma may vibrate longitudinally about their mean positions while the positive ions remain relatively fixed. This gives rise to electron plasma oscillations of frequency

$$f = \left(\frac{ne^2}{\pi m_e} \right)^{1/2}$$

where n is the number of electrons producing charge density, e is the charge of an electron, and m_e is the mass of an electron. If λ is the wavelength of oscillation, the velocity of propagation is

$$v = \lambda \left(\frac{ne^2}{\pi m_e} \right)^{1/2}$$

The frequency of electron plasma oscillations may be in the vicinity of 10^9 Hz.

Positive ions may also oscillate longitudinally, but at a lower frequency than electrons, and the frequency is temperature-dependent.

126. Current in Dielectric and Insulating Materials. In dielectric and insulating materials, electrons are so closely bound to their parent atoms that they cannot become free without physically damaging the material. Consequently, so long as the dielectric properties are maintained, no conduction current can occur in dielectrics. But a varying voltage applied across the faces of a dielectric can produce a varying displacement of charges within the dielectric, and the directed displacement behaves much as a time-varying conduction current does (see Par. 1-60).

If a time-varying electric field is applied across a dielectric, the displacement current in the dielectric is given as

$$I_{dis} = \frac{d\psi}{dt} = \frac{d(Ds)}{dt} = s\frac{dD}{dt} = s\frac{d(\epsilon E)}{dt} = s\epsilon\frac{dE}{dt}$$

where ψ is the electric flux displaced in the dielectric, D is the electric flux density, s is the surface area of dielectric subject to displacement, ϵ is the permittivity of the dielectric, E is the intensity of the applied electric field, and t is the time.

127. Mechanisms of Current Control.   Current can be controlled by changing the forces acting on, and the energies expended upon, the moving charges constituting the current. The motions of charge carriers can be controlled by an electric field, a magnetic field, or some combination of these components in an electromagnetic field.

In practice, the environment in which charges move, including the properties of surrounding materials, has great influence on mechanisms for current control.   The properties of the materials used in electron devices are themselves important in effecting certain controls, including unilateral conductivity, nonlinear voltage-current relations between electrodes, amplification, energy-conversion processes at input or output of the device, hysteresis of magnetic materials, and hysteresis of dielectric materials.

128. Control in Vacuum Tubes.   The current in vacuum tubes can be controlled by electric fields established by electrode potentials, magnetic fields through which charges pass, and by both electric and magnetic fields.

Electric Control.   When the current through a tube is limited by space charge, the current can be controlled by voltage applied to its electrodes, particularly one or more grids between emitter and collector.   For such conditions, the current to the collector (plate) is of the form

$$I_b = K \left(\frac{C_{gk}}{C_{pk}} E_c + E_p \right)^{3/2} = K(\mu E_c + E_p)^{3/2}$$

where K is a constant depending upon tube geometry, C_{gk} is capacitance between control electrode (grid) and emitter (cathode), C_{pk} is capacitance between collector (plate) and emitter (cathode), $\mu = C_{gk}/C_{pk}$ is the voltage amplification factor of the tube and depends on its geometry, E_c is the voltage applied to the control electrode relative to the cathode potential taken as reference value, and E_p is the voltage applied to the collector electrode measured with respect to the emitter (cathode) voltage.   The equation given above is for a tube having a single control electrode or grid. Similar equations can be developed for tubes having auxiliary grid electrodes.   The current is usually controlled by a time-varying voltage applied to one control electrode; sometimes control is effected by varying the potentials of more than one control electrode.

Magnetic Control.   The simplest electron tube having conduction characteristic controlled by a magnetic field is the early form of magnetron invented by A. W. Hull.   It consists of a filamentary electron-emitting cathode located on the axis of a cylindrical collector electrode.   When the filament is heated to incandescence and a positive potential V is applied to the collector relative to the emitter, electrons flow from filament to plate. If the tube is surrounded by a magnetic field of intensity H, the electrons do not travel from filament to plate in straight lines but move in a circle of radius r.   The electrons fail to strike the collector unless the magnetic field causes the electrons to move in a radius r equal to or greater than the distance between filament and cylindrical plate.   The magnetic field that just causes electrons to graze the cylindrical collector is

$$H = \left(\frac{8mV}{er} \right)^{1/2}$$

where m is the mass of the electron, e is the electron charge, r is the radius of the cylindrical collector, and V is the positive voltage of collector relative to that of the emitter.   For values of H greater than that indicated, for a given accelerating voltage V, electrons do not strike the collector, and current in the tube abruptly drops to zero.   Many other more sophisticated forms of magnetic control have been developed (see Sec. 9).

129. Control in Gaseous Devices.   Two methods of controlling current through gas tubes have been devised: those which control starting of a discharge by delaying the formation of the discharge until a desired time and those which control the starting of conduction by enhancing formation of the discharge at the desired time.

Devices of the first class can be controlled by a magnetic field, an external grid wound around the tube, or an electrode partially surrounding the collector electrode in a tube with cold cathode, as in the grid-glow tube.

Devices of the second class can be controlled by an igniter electrode in a tube having a mercury-pool cathode, application of voltage to an electrode insulated from the mercury-pool cathode, or one or more auxiliary electrodes to initiate discharge between cold electrodes in a gas.

In all cases, gas-discharge tubes operate by control of the *average current* through them rather than control of the instantaneous current, as in vacuum tubes. Since control periods are usually of the order of milliseconds, gas tubes are not generally useful for high-frequency applications.

Ionization of the gas neutralizes space charge within the device, so that the voltage drop across the gas tube is relatively small. Current density in vacuum tubes is proportional to the ³⁄₂ power of the emitter-collector voltage; in high-pressure gas tubes, current density is proportional to the square of the cathode-anode voltage.

130. Control in Semiconductors. In semiconductor devices, such as transistors, control of current in the output electrode circuit is possible because the terminal currents depend upon voltages applied across *semiconductor junctions.* Emitter and collector currents are of nearly equal magnitude and increase exponentially with voltage between emitter and base. The current in the base is smaller than the emitter and collector currents, and it also depends to a marked extent upon the emitter-base voltage.

When a small, slowly varying voltage is applied across emitter and base, the concentration of holes in the base, at the edge of the emitter-junction and space-charge layer, increases. The distribution of holes in the base changes in such a way that the collector current increases and the base current decreases. The ratio of change in collector current to change in base current is called the *current gain* of the device. Complete discussions are presented in Secs. **7** and **8.**

131. Auxiliary Methods of Current Control. While current control is achieved basically by changing the electric or magnetic field, or both, auxiliary methods of control exist in which the field itself is controlled by some energy-converting device responding to pressure, temperature, radiation, or other nonelectromagnetic quantity. For example, the current through a phototube may be controlled by mechanically varying the amount of radiation falling upon its electron-emitting surface.

132. Collection of Charges: Impact Phenomena. After being released, transported, and controlled, the mobile charges collected at an electrode convert their kinetic energy into other forms of energy upon impact. The kinetic energy of collected charges may be converted into useful current in an external circuit having the electrode as a terminal, heat energy that raises the temperature of the collector electrode, or radiant energy as the impacting electrons become absorbed by atoms of the collector electrode material, e.g., produce x-rays.

133. Electrode Heating. The collector electrode receiving current from a thermionically heated cathode must radiate power of amount

$$P = E_b I_b + A E_c I_c$$

where E_b is the voltage at which the collector is operated measured relative to the emitter potential, I_b is the current flowing to the collector, A is a coefficient whose value (less than 1.0) depends upon the extent to which the collector encloses the emitter, E_c is the voltage drop across the emitter, and I_c is the current taken by the emitter electrode.

If the collector is cooled only by radiation, the rise in temperature at the surface of the anode is found, from the Stefan-Boltzmann law, to be of the form

$$\Delta T = \left(\frac{P}{sek} \right)^{1/4} = \left(\frac{E_b I_b + A E_c I_c}{sek} \right)^{1/4}$$

where P is the amount of power to be dissipated, s is the total area of radiating surface, k is the Stefan-Boltzmann constant, e is the total radiation emissivity, and ΔT is the rise in temperature on the absolute scale. The other symbols have the meanings given in the equation for P, above. For metals, e has typical values of 0.1 at 1000°K and 0.2 at 2000°K. For radiation cooling alone, the permissible power dissipation is approximately 5 W/cm².

134. X-Ray Production. The sudden deceleration of high-speed electrons by a solid, as at the collector electrode, may under certain conditions produce x-rays. In this case a bombarding electron ionizes the atom it strikes by removing an electron in the innermost K shell. The vacancy left by this process is filled by an electron in a more remote shell (L, M, N, O, or P) which jumps to the K shell. The consequent release of energy appears as an x-ray quantum. The emitted x-rays extend toward longer wavelengths from a minimum wavelength corresponding to the maximum frequency

$$f_{max} = \tfrac{1}{2} m v^2 / h$$

where m is the mass of the electron, v is the maximum speed of the incident electron, and h is Planck's constant. Appreciable x-rays can be produced in tubes having accelerating voltages as low as 20,000 V. Tubes used for the intentional production of x-rays operate at much higher voltages, usually in the range of from 200,000 to 500,000 V.

ELECTROMAGNETIC FIELDS

135. Equations for Electromagnetic Field Vectors. The basis of electromagnetic field phenomena on a macroscopic (nonquantum) state is given by the four equations of Maxwell. The electric and magnetic intensities E and H are related to the electric and magnetic flux densities D and B and to the current density J. Maxwell's equations specify how D, B, and J depend upon the space and time variations of E and H, for fields in free space as well as in material substances.

The electromagnetic properties of a medium are specified in terms of auxiliary equations for the field vectors previously defined in this section, as follows:

$$D = \epsilon E = \epsilon_0 E + P = \epsilon_0 (1 + \chi_e) E$$

$$B = \mu H = \mu_0 H + \mu_0 M = \mu_0 (1 + \chi_m) H$$

$$J = \sigma E = \rho v$$

where ϵ is the permittivity of the medium, ϵ_0 is the permittivity of free space, P is the dielectric polarization, χ_e is the electric susceptibility, μ is the permeability of the medium, μ_0 is the permeability of free space, M is the magnetization, χ_m is the magnetic susceptibility, σ is the electric conductivity of the medium, ρ is the volume charge density, and v is the volume in which charge density ρ occurs.

These equations for the electromagnetic properties of a medium, together with the Lorentz force equation (see Par. 1-93)

$$F = F_e + F_m = QE + Q(v \times B)$$

are used in applying the Maxwell field equations relating the field vectors E, D, B, H, and J.

136. Maxwell's Equations. The four fundamental Maxwell relations between the electromagnetic field vectors are stated here in words as well as in integral and differential form, in the latter case in terms of the notation of vector analysis.

1. The total electric displacement passing through a closed surface is equal to the total charge within the volume enclosed (in free space, volume charge density is $\rho = 0$).

$$\int_{surface} D \cdot ds = \int_{vol} \rho \, dv \qquad\qquad \nabla \cdot D = \rho$$

2. The net magnetic flux passing through any closed surface is zero:

$$\int_{surface} B \cdot ds = 0 \qquad\qquad \nabla \cdot B = 0$$

3. The magnetomotance around any closed path is equal to the sum of the conduction and convection current and to the time rate of change of electric displacement passing through the surface bounded by the closed path.

$$\oint H \cdot dl = \int_{surface} \left(J + \frac{\partial D}{\partial t} \right) \cdot ds \qquad \nabla \times H = \left(J + \frac{\partial D}{\partial t} \right)$$

4. The electromotance taken around any closed path is equal to the time rate of change of the magnetic displacement passing through the surface bounded by the closed path.

$$\oint E \cdot dl = - \int_{surface} \frac{\partial B}{\partial t} \cdot ds \qquad \nabla \times E = -\frac{\partial B}{\partial t}$$

The equations at the left are the integral forms of Maxwell's relations over finite lengths and surface areas. The equations at the right are the differential forms of Maxwell's equations, expressing the field relationships at a point.

137. Restricted Applications of Maxwell's Equations. In the general forms given above, Maxwell's equations are difficult to apply except to extremely simple physical and geometrical conditions. The Maxwell equations can be reduced to less formidable equivalents under certain conditions for which some of the terms become zero or take simple forms. Modifications of the general form of the Maxwell equations are given for various special conditions in Table 1-6.

Table 1-6. Special Conditions for Maxwell's Equations

Condition	Modifications to Maxwell's Equations
1. Electrostatics	$J = 0$ $\quad$ $\partial D/\partial t = 0$ $\quad$ $\partial B/\partial t = 0$ Only the electric-divergence equation is applicable
2. Stationary electromagnetic field; magnetostatics	$J = 0$ $\quad$ $\partial D/\partial t = 0$ $\quad$ $\partial B/\partial t = 0$ Only electric and magnetic-divergence equations applicable
3. Quasi-stationary state for closed circuits	All Maxwell equations are applicable, but displacement current can usually be neglected, except in capacitors
4. Quasi-stationary state for open circuits	All Maxwell equations are applicable, but displacement current can sometimes be neglected
5. Electromagnetic field in vacuum or perfect dielectric ...	$J_{cond} = 0$ $\quad$ $\rho = Q/v = 0$ $\quad$ $\partial D/\partial t = J_{dis}$
6. Harmonic time variations of electromagnetic field of angular frequency ω	$\nabla \cdot D = \rho$ $\quad$ $\nabla \times H = (\sigma + j\omega\epsilon)E$ $\nabla \cdot B = 0$ $\quad$ $\nabla \times E = -j\omega\mu H$

138. Electromagnetic Waves. The wave equation is a quantitative expression for the dynamic behavior of some physical quantity that varies with time at a particular point in space and varies with space coordinates at any instant of time. The Maxwell equations can be used to derive the wave equation for time-varying electromagnetic quantities for an infinite, homogeneous, linear, isotropic medium. For such a situation, the properties of the medium, as specified by ϵ, μ, σ, and ρ, are constant and have the same values in all directions over the region considered. For this case, $D = \epsilon E$, $B = \mu H$, $J_{cond} = \sigma E$, and $\rho = dQ/dv$.

The divergence equations of Maxwell play no role in specifying wave properties since they are time-invariant. But the curl equations are used in their general form. The differential forms of the Maxwell equations are of interest since it is helpful to know field variations at a point in space or in a material medium. By a process of elimination, the variables of the time-dependent curl equations can be separated, to yield the relations

$$\nabla^2 E = \mu\epsilon\frac{\partial^2 E}{\partial t^2} + \mu\sigma\frac{\partial E}{\partial t} + \frac{1}{\epsilon}\nabla\rho$$

$$\nabla^2 H = \mu\epsilon\frac{\partial^2 H}{\partial t^2} + \mu\sigma\frac{\partial H}{\partial t}$$

Subject to the boundary conditions in any particular situation, these are the equations that must be satisfied by E and H in order for electromagnetic waves to be produced.

Important special cases, having considerable practical application, occur when $\rho = 0$ (as in free space) and when the time variations of the field vectors are expressible by the exponential $\epsilon^{j\omega t}$, where ω is a constant angular frequency in radians per second and t is the time. For this special case the wave equations are

$$\nabla^2 E = (\mu\epsilon\omega^2 - j\sigma\mu\omega)E = \gamma^2 E$$

$$\nabla^2 H = (\mu\epsilon\omega^2 - j\sigma\mu\omega)H = \gamma^2 H$$

Solutions of these equations take the form

$$E = [f_1(l - v_0 t)\epsilon^{-\gamma l} + f_2(l + v_0 t)\epsilon^{+\gamma l}]\epsilon^{j\omega t}$$
$$H = [f_1(l - v_0 t)\epsilon^{-\gamma l} + f_2(l + v_0 t)\epsilon^{+\gamma l}]\epsilon^{j\omega t}$$

where γ is a propagation constant expressing the change in amplitude and phase as the disturbance travels in a medium with speed v_0. The expression f_1 represents a wave of E and H flowing in the $+l$ direction with speed v_0, whereas f_2 represents a wave of E or H flowing in the opposite $-l$ direction and having the same time variation as f_1. As might be expected, E and H are related by the properties of the medium. This relation is usually expressed in terms of the intrinsic impedance of the medium (Par. 1-141) for effective rather than instantaneous values of E and H.

139. Propagation Coefficient. When a homogeneous, isotropic medium whose properties are specified by $D = \epsilon E$, $B = \mu H$, $J_c = \sigma E$, and $\rho = 0$ is subject to harmonic time variations, the E and H field components are found to be proportional to $\gamma^2 = -(\mu\epsilon\omega^2 - j\sigma\mu\omega) = j\omega\mu(\sigma - j\omega\epsilon)$. The quantity γ^2 is given in terms ϵ, μ, σ, and ρ specifying the physical properties of the medium and the factor ω expressing the time rate of change of the field vectors. The quantity γ relates the distance an electromagnetic wave travels in unit time, and is called the *propagation factor*.

The propagation factor may be expressed as a complex quantity

$$\gamma = \pm[j\omega\mu(\sigma - j\omega\epsilon)]^{1/2}$$

$$= \pm j\omega\sqrt{\mu\epsilon}\left(1 - j\frac{\sigma}{\omega\epsilon}\right)^{1/2} = \alpha + j\beta$$

If this equation is expanded by the binomial theorem,

$$\gamma = \alpha + j\beta = j\omega\sqrt{\mu\epsilon}\left[1 - \frac{j}{2}\frac{\sigma}{\omega\epsilon} + \frac{1}{8}\left(\frac{\sigma}{\omega\epsilon}\right)^2 + \cdots\right]$$

so that

$$\alpha \approx j\omega\sqrt{\mu\epsilon}\left(-\frac{j\sigma}{2\omega\epsilon}\right) = +\frac{\sigma}{2}\sqrt{\frac{\mu}{\epsilon}}$$

and

$$j\beta \approx j\omega\sqrt{\mu\epsilon}\left[1 + \frac{1}{8}\left(\frac{\sigma}{\omega\epsilon}\right)^2\right]$$

The propagation factor is seen to comprise an *attenuation factor* α and a *phase factor* β.

The attenuation factor α is a measure of the change in amplitude of E or H per unit length of the propagation path, in units of nepers per meter (or in most engineering work, in decibels per meter).

The phase factor β is a measure of the change in phase per unit length as the field vectors E and H are propagated through the medium. The phase factor is measured in radians per meter.

140. Velocity of Wave Propagation. If the E and H vectors of an electromagnetic wave vary with time according to the function $\epsilon^{j\omega t}$, where $\omega = 2\pi f$ is the angular velocity in radians per second, the disturbance from equilibrium conditions at the source will be 0 when $t_0 = 0$. At some later time t, the disturbance at the source will have produced a relative phase change of $\theta = \omega t$. If the electromagnetic wave travels with speed $v = l/t$, the phase shift at the source can be expressed in terms of the distance l the wave has traveled. Since the phase shift relates time and distance, it follows that $\theta = \omega t = \omega l/v = (\omega/v)l = \beta l$. Here $\beta = \theta/l$ is the phase change per unit length of wave propagation whereas $\omega = \theta/t$ is the phase shift per unit time. When ω and β are known, the velocity of propagation is

$$v = \frac{l}{t} = \frac{\omega}{\beta}$$

The speed of propagation is related to the electromagnetic constants of a substance by the relation $v = 1/(\mu\epsilon)^{1/2}$. Since $\mu = \mu_0\mu_r$ and $\epsilon = \epsilon_0\epsilon_r$, the velocity can be expressed as

$$v = \sqrt{\frac{1}{\mu\epsilon}} = \sqrt{\frac{1}{(\mu_0\epsilon_0)(\mu_r\epsilon_r)}} = \frac{c}{\sqrt{\mu_r\epsilon_r}}$$

where $c = \sqrt{1/\mu_0\epsilon_0}$ is the speed of propagation in free space. It has the approximate value 300,000 km/s, or 186,300 mi/s.

The velocity with which waves travel in material substances is always less than their speed in free space or a vacuum since μ_r and ϵ_r always have values greater than 1 (except for paramagnetic materials). Propagation is rapidly attenuated in paramagnetic and other conducting materials.

141. Intrinsic Impedance of a Medium. The field vectors of an electromagnetic wave are interrelated by the *intrinsic impedance* of the medium, $Z_0 = E/H$, where E and H are the effective, or root-mean-square (rms), values of the electric and magnetic field intensities. By making use of the relation, $\nabla \times E = -j\omega\mu H$ for harmonic time variations of the field, it can be shown that

$$H = \frac{E}{Z_0}\epsilon^{-\gamma l}$$

where

$$Z_0 = \sqrt{\frac{j\omega\mu}{\sigma + j\omega\epsilon}} = \sqrt{\frac{\mu}{\epsilon}}\frac{1}{1 - j(\sigma/\omega\epsilon)}$$

This expression can be expanded by the binomial theorem to express Z_0 as a complex quantity

$$Z_0 = R_0 + jX_0 \approx \sqrt{\frac{\mu}{\epsilon}}\left[1 + \frac{j}{2}\frac{\sigma}{\omega\epsilon} - \frac{3}{8}\left(\frac{\sigma}{\omega\epsilon}\right)^2 + \cdots\right]$$

The reference, or in-phase, component is the resistance

$$R_0 \approx \sqrt{\frac{\mu}{\epsilon}}\left[1 - \frac{3}{8}\left(\frac{\sigma}{\omega\epsilon}\right)^2 + \cdots\right]$$

and the quadrature component is the reactance

$$X_0 \approx \sqrt{\frac{\mu}{\epsilon}}\frac{\sigma}{2\omega\epsilon}$$

both of which are measured in ohms. For free space the properties of the medium are $\sigma = 0$, $\mu = \mu_0$, and $\epsilon = \epsilon_0$, and so the intrinsic impedance is

$$Z_0 = R_0 = \sqrt{\frac{\mu_0}{\epsilon_0}} = 379.7 \ \Omega$$

142. Radiation from a Simple Dipole. The basic features of electromagnetic radiation can be illustrated by considering radiation from a current element $I\,dl$ located at the center of a system of spherical coordinates, in which the instantaneous current is $i = I \cos \omega t$.

Any disturbance created by the dipole at time t_0 will be observed at a point whose distance from the origin is r at time $t = t + r/v$. That is, the effect observed at r at time t must have originated at the source at an earlier time $t_0 - r/v$, where v is the velocity of wave propagation.

At a remote point r at a distance very large compared to the dipole length l and at a frequency ω whose wavelength λ is also very large compared to the dipole length, the components of the radiated electromagnetic field are given, in spherical coordinates, as

$$E_\phi = 0 \qquad H_r = 0 \qquad H_\theta = 0$$

$$E_\theta = \frac{I\,dl \sin \theta}{4\pi\epsilon}\left(\underbrace{\frac{\sin \omega t}{\omega r^3}}_{\text{static}} + \underbrace{\frac{\cos \omega t}{r^2 v}}_{\substack{\text{induction} \\ \text{(quasi-steady)}}} - \underbrace{\frac{\omega \sin \omega t}{r v^2}}_{\text{radiation}}\right)$$

$$E_r = \frac{I\,dl\,\cos\theta}{2\pi\epsilon}\left(\underbrace{\frac{\sin\omega t}{\omega r^3}}_{\substack{\text{induction}\\\text{(quasi-steady)}}} + \underbrace{\frac{\cos\omega t}{r^2 v}}_{\text{radiation}}\right)$$

$$H_\phi = \frac{I\,dl\,\sin\theta}{4\pi}\left(\underbrace{\frac{\cos\omega t}{r^2}}_{\substack{\text{induction}\\\text{(quasi-steady)}}} - \underbrace{\frac{\omega\sin\omega t}{rv}}_{\text{radiation}}\right)$$

In these equations, $v = (1/\mu\epsilon)^{1/2}$ is the velocity of wave propagation.

With respect to distance from the source, the field may be divided into three kinds. One of these components, indicated as *static*, is the field to be expected if the charges changed very slowly. Another component, marked *induction*, represents time variations of the electric and field components that are returned to the system; the energy associated with such components does not permanently leave the system. The third kind of component, marked *radiation*, represents time variations of the electric and magnetic field components whose energy leaves the radiator and is propagated through the medium.

143. Poynting's Vector. For electromagnetic energy flowing into or out of any closed region, at any instant the rate of flow is proportional to the surface integral of the vector product of the electric field strength and the magnetizing force. This product, known as *Poynting's vector*, is

$$S = E \times H$$

The energy density S is in watts per square meter when the electric field intensity E is in volts per meter and the magnetic field intensity H in amperes per meter.

144. Transmission of Electromagnetic Waves. Absorption may be regarded as the transformation of energy from its original form to some other form as waves pass through a material substance. When a beam of radiant energy passes through a transparent medium whose size is large compared to the wavelength of the radiation, some of the radiant flux is continuously absorbed in the medium and the intensity of the beam steadily decreases. The decrease in intensity of the radiant energy in passing through the medium, called *absorption*, is given quantitatively by Lambert's law.

If I_0 is the intensity of monochromatic energy entering an absorbing medium, the intensity of the emerging energy in passing through a transmission path of length x is

$$I = I_0\epsilon^{-kx}$$

where k is a characteristic of the medium called the *absorbing coefficient*. The value of k is determined experimentally and, in general, is dependent upon the wavelength of the radation. The intensity of the beam of radiant energy emerging from the medium of thickness x is a measure of the transmission properties of the medium and is given by

$$I_{em} = I_0 - I = I_0(1 - \epsilon^{-kx})$$

145. Reflection. Reflection is the change in direction of an electromagnetic wavefront, with a sudden change in its phase. The reflection coefficient is the ratio of the radiant energy reflected from a surface to the total amount of energy incident on the surface.

The laws of reflection depend upon the shape of the reflecting surface. If the reflecting surface is a smooth, flat plane, the laws of reflection of a plane wave, (i.e., one having infinite radius of curvature) are: (1) the reflected beam lies in the plane of the incident ray and the normal of the reflecting surface at the point of incidence, and (2) the angle of reflection, i.e., the angle between the reflected ray and the surface normal, is equal to the angle of incidence.

146. Refraction. The index of refraction of a substance is the ratio of the velocity of radiant energy in a vacuum to its velocity in the substance. The index of refraction varies with the wavelength of the radiation.

Snell's law states that if i is the angle of incidence, r the angle of refraction, v_1 the velocity of propagation in the first medium, and v_2 the velocity of propagation in the second medium, the index of refraction is

$$n = \frac{v_1}{v_2} = \frac{\sin i}{\sin r}$$

Refraction is used to designate the change in the direction of a ray as it passes from one medium in which its velocity is v_1 to another in which its velocity is v_2.

147. Diffraction. Wave diffraction refers to the bending of rays around an obstacle such as a sharp edge or point over which radiant energy passes. It results from the spherical spreading of electromagnetic energy from a point source and indicates that the straight-line or rectilinear propagation of electromagnetic waves is an approximate concept.

If a source of radiant energy is small enough to be regarded as a point, the shadow of an object has its maximum sharpness. A certain amount of radiant energy from the source is always found, however, within the geometrical shadow as a result of diffraction at the edge of the shadow-casting object.

148. Wave Interference. The principle of linear superposition applies to radiation in a linear propagation medium. Hence, the intensity of two rays of monochromatic radiation in a linear medium is the sum or difference of their individual effects, taking account of their amplitude, frequency, and phase. If two beams have the same frequency, the additive and subtractive effects are said to produce interference when the beams lie in the same plane and travel substantially in the same path, large compared to the wavelength of the radiation. The phenomenon of cancellation and reinforcement of waves of the same frequency is called *wave interference.*

If the intensity of one beam of frequency $\omega = 2\pi f$ is

$$i_1 = I_1 \sin \omega t$$

and the properties of the second beam are given by

$$i_2 = I_2 \sin(\omega t + \phi)$$

where ϕ is the relative phase displacement between the two, then the net intensity of the two beams has the same frequency as that of the component rays, and its amplitude is

$$I = (I_1^2 + I_2^2 + 2 I_1 I_2 \cos \phi)^{1/2}$$

149. Dispersion and Scattering. Dispersion is the process by which radiation is separated in accordance with some characteristic (such as frequency, wavelength, or energy) into components which have different directions. The difference between the indices of refraction of any substance for two different wavelengths is a measure of dispersion for these two wavelengths and is called the *coefficient of dispersion,* or the dispersive power.

If n_1 and n_2 are the refractive indices of radiations whose wavelengths are λ_1 and λ_2 respectively, within the spectral range considered, and if n_m is the refractive index for some specified wavelength λ_m between λ_1 and λ_2, the dispersive power for wavelength λ_m is the ratio

$$d = \frac{n_1 - n_2}{n_m - 1}$$

The wavelength λ_m is commonly taken to be the mean of λ_1 and λ_2.

Scattering is the process of diffusing, in various directions, electromagnetic radiation by reflection from molecules, atoms, electrons, or other particles.

THE ELECTROMAGNETIC SPECTRUM

150. Regions of the Spectrum. It is useful to divide the electromagnetic spectrum into regions exhibiting common properties useful to science and technology. Near the lower end of the frequency spectrum, radiations are designated in terms of frequency bands which for the most part are useful in radio communication. The limits of frequency of bands used for radio communication are given in Sec. 22, Table 22-1.

Beyond the radio spectrum lie other frequency bands for which frequency assignments are not necessary. These bands include the infrared, visible, ultraviolet, x-ray, gamma-ray, and cosmic-ray portions of the spectrum. The relations between the various portions of the spectrum are shown in Fig. 1-3.

151. Sources and Detectors of Electromagnetic Energy (See Sec. 11.) The greater portion of the spectrum is known to man only through the effects produced by energy sources and observed by detectors of radiant energy.

Sources and detectors are of two major classes: frequency-selective devices responding to a band of frequencies that is small compared to the mean frequency of response, and broadband devices, capable of producing and responding to a range of frequencies that is large compared to the frequency of the band in which energy is generated or detected.

A wide variety of natural and artificial sources and detectors of radiant energy are available for use in different portions of the spectrum. Typical sources and detectors, together with the spectral region for which they are normally useful, are shown in Fig. 1-4.

152. Spectrum Utilization. The use to which the electromagnetic spectrum is put depends primarily upon the frequency (or wavelength) of the radiation and the propagation properties of the medium in which the waves travel. Because the electromagnetic spectrum covers a range of more than 22 decades, it is desirable to divide the spectrum into regions having similar properties.

Lowest-Frequency Bands: 1 to 10^3 Hz. In this lowest-frequency portion of the spectrum, restricted bandwidths prevent future use for communications, although proposals have been made to use this band for worldwide communications with submarines. Electric power is generated at frequencies of 50 or 60 Hz (occasionally at 25 Hz or other low frequency). For aviation use, electric power is used at 400 Hz, to reduce weight of iron-core apparatus.

This band also includes a portion of the audible (voice) frequencies, but electromagnetic energy is not radiated in this application.

Low Radio Frequencies: 10^3 to 2×10^5 Hz. This band is particularly useful for long-distance communication where reliability of transmission is important and sufficient radiation power can be made available. This band is usually used for radio telegraphy. As frequency is decreased in this band, reliability and signal strength improve, there are fewer interruptions from diurnal, seasonal, and solar causes, but static and other radio noises tend to increase.

Medium Frequencies: 2×10^5 to 2×10^6 Hz. The lower-frequency portion of this band is useful for services requiring reasonably stable transmission, day and night, over moderately long distances. Since appreciable atmospheric noise occurs in this band, substantial field strength is required for reliable communication. The upper-frequency portion of the band is characterized by relatively weak ground wave, but the sky wave prevailing at night is relatively large. This band is commonly used for sound broadcasting. The band is also used for fixed services, maritime mobile service, maritime radio navigation including loran, aeronautical radio navigation, and amateur communication.

High Frequencies: 2×10^6 to 3×10^7 Hz. Useful but somewhat erratic long-range propagation is possible with low power in this frequency range. When the transmission path is entirely in darkness and the ionosphere is undisturbed, frequencies below a maximum usable frequency are propagated over long distances. Transmission is dependent upon vagaries of the ionosphere. Fading and multiple-path effects often limit speed of communication. The large interference range limits the number of emissions that can be simultaneously radiated at a given frequency. The band is used for fixed services, mobile services, amateur transmissions, broadcasting, maritime mobile service, and telemetering.

Very High and Ultra High Frequencies: 3×10^7 to 3×10^9 Hz. Lumped circuits, useful at lower frequencies, give way to transmission lines and other distributed circuits in this band. The band is suitable for relatively short-distance communication for services transmitting a large amount of detail, including radar and television. Directive antenna systems of small size are economical and effective. If powerful transmitters and high-gain antennas are used, reliable long-distance propagation is possible by using waves scattered by turbulence in the troposphere. The band is used for fixed services, mobile services, space research, radio astronomy, telemetering and tracking, amateur transmissions, satellite communication, meteorological aids, radio location, radar, and television. It is also used for radio-frequency spectroscopy and in studies of magnetic resonance and nuclear quadruple resonance.

Microwaves: 3×10^9 to 3×10^{11} Hz. Transition from circuit to optical techniques

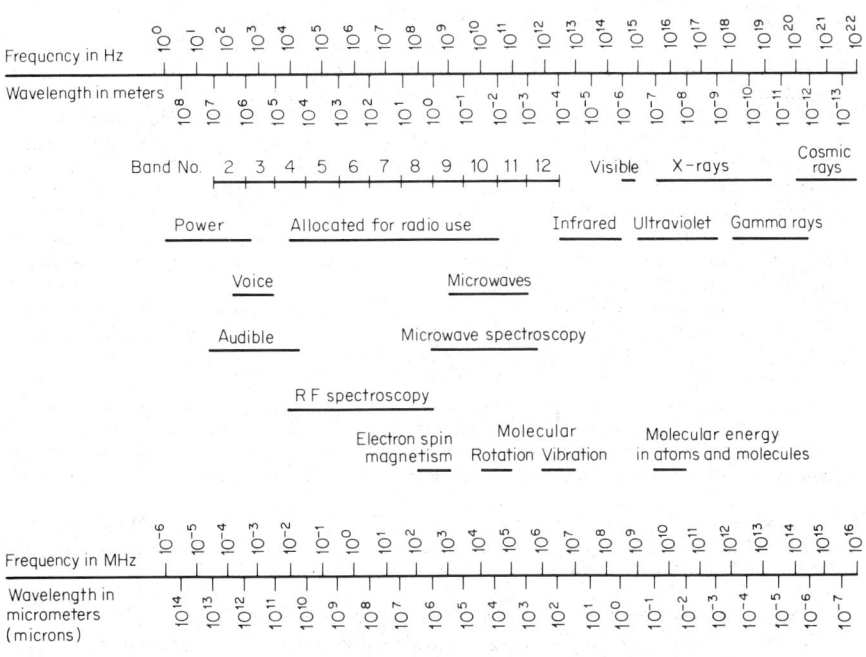

Fig. 1-3. The electromagnetic spectrum. The horizontal lines indicate the approximate spectral ranges of various physical phenomena and practical applications.

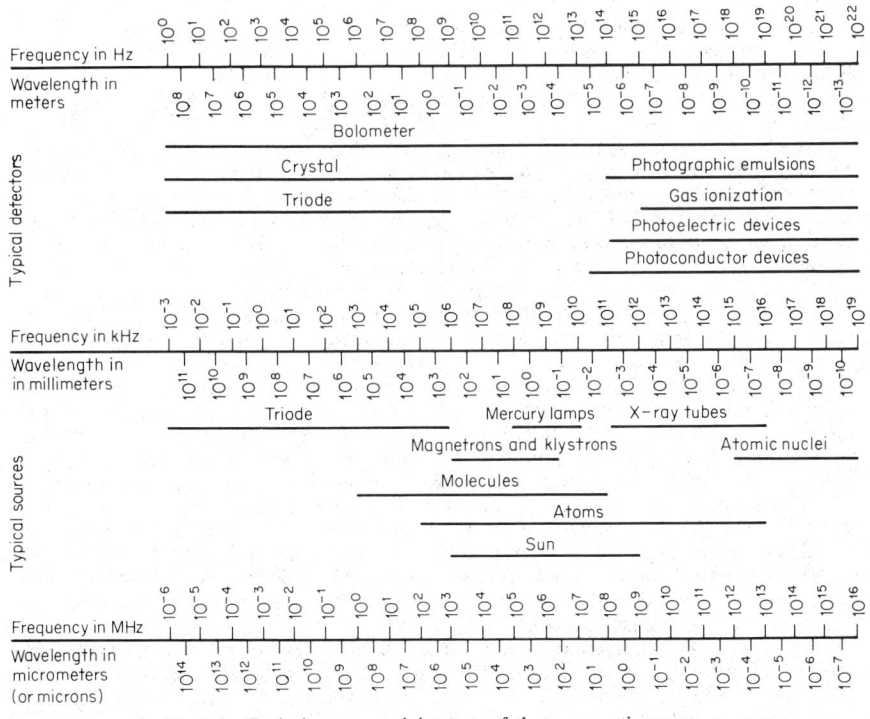

Fig. 1-4. Typical sources and detectors of electromagnetic energy.

characterizes microwave bands. Electromagnetic waves in this band are short enough to be propagated by highly directive antennas and waveguides. Long-distance communication can be carried out by a series of automatic relay stations mounted at high elevations within line of sight of each other. This portion of the spectrum is used for high-definition radar, television, and similar services requiring extensive bandwidths to convey a considerable amount of information. Microwave spectroscopy, dealing with electron-spin resonance and molecular rotation, is carried out in this band.

Infrared Region: 3×10^{11} *to* 4×10^{14} *Hz*. The infrared, visible, and ultraviolet portions of the spectrum have long been used for identifying molecules by their spectral emissions and as a subsequent means of qualitative analysis, for determining the geometry of simple molecules, and for quantitative analysis where spectral-line intensity is related to the concentration of substance in question. The low-frequency limit is not well defined and often extends into the microwave region of the spectrum.

Visible Spectrum: 3.95×10^{14} *to* 7.90×10^{14} *Hz*. This is a relatively narrow part of the spectrum, covering only one octave, but it is extremely important. This band contains all the radiant energy visible to the human eye; it is commonly divided into six or seven regions identified with different spectral hues. Spectroscopy conducted in the visible spectrum is usually identified with molecular vibrations or with electronic transitions of outer electrons of atoms.

Ultraviolet Spectrum: 7×10^{14} *to* 3×10^{16} *Hz*. Ultraviolet radiations have found bactericidal applications and, along with infrared radiations, are used in some specialized communications. In this portion of the spectrum electronic transitions of outer-shell electrons occur, and the spectrum measured is often referred to as the electronic spectrum.

X-Ray Spectrum: 3×10^{16} *to* 10^{19} *Hz*. Extensive medical, biological, and industrial applications have been made of x-rays. X-rays have made important contributions to our understanding of the structure of matter. X-ray spectroscopy is a valuable tool in studying electronic transitions of inner-shell electrons occurring when high energy is absorbed.

Gamma-Ray Spectrum: 10^{19} *to* 10^{22} *Hz*. This portion of the spectrum is useful in studying nuclear transitions. The upper part of this spectrum, from about 10^{20} Hz and beyond, is sometimes called the cosmic-ray spectrum.

153. Frequency Tolerances. For successful communication to be carried out by electromagnetic waves, stations must operate on an assigned carrier frequency, or within a band of frequencies, which must be maintained within specified limits of frequency tolerances.

Frequency tolerance is defined to be the maximum permissible departure of the center frequency of the band from an assigned frequency. Frequency tolerance is usually expressed in parts per million so that, for a specified frequency tolerance, the absolute deviation, in hertz, tends to be greater for stations operating on a higher frequency than those operating on lower frequencies. Sometimes, however, the maximum permissible tolerance is specified in absolute values of frequency, in hertz.

For stations of different classes of service, the permissible frequency tolerances also depend upon the peak power radiated from the antenna. In general, stations with low values of radiated power may have larger frequency tolerances than high-power stations for the same class of service.

154. Spurious Emissions. Spurious emissions are those occurring outside the frequency band assigned for a particular type of emission and quality of service. They do not contribute to the amount of information conveyed but often produce interference with other services. For these reasons, the amplitude of spurious emissions should be kept as low as possible. The absolute level of spurious-emission intensity depends upon the amount of power radiated. Accordingly, it may be specified in terms of absolute power levels or in terms of power levels relative to that of the radiated power. Spurious emissions may include harmonics, intermodulation products, cross modulation, and parasitic emission.

155. Standard-Frequency Transmissions. A number of stations throughout the world radiate waves whose carrier and modulation frequencies are very precisely established and so can be used for frequency standardization. The standard-frequency transmissions are usually radiated by government stations whose precision of frequency is established by national laboratories. In the United States authority for establishing national standards resides in the National Bureau of Standards, and standard-frequency transmissions are made from Bureau of Standards stations in Fort Collins (near Boulder), Colo., and in Puunene

Maui, Hawaii. Transmissions suitable as frequency standards for use in North America and the Pacific are listed in Table 1-7.

Transmissions from the National Bureau of Standards also include time signals and telegraphic transmissions regarding the ionospheric conditions to be expected in the near future. Details of such transmissions are subject to change, from time to time, but can be determined from the Frequency-Time Broadcast Services, National Bureau of Standards, Boulder, Colo., 80302.

Table 1-7. Standard-Frequency Transmissions

Frequency	Error, parts in 10^{10}	Comment
440 Hz*	0.5	WWV, Fort Collins, Colo., WWVH, Hawaii
600 Hz*	0.5	WWV, Fort Collins, Colo., WWVH, Hawaii
1,000 Hz*	0.5	WWV, Fort Collins, Colo.
17.8 kHz	0.5	NAA, Cutler, Maine
18.6 kHz	0.5	NPG/NLK, Jim Creek, Wash.
20.0 kHz	0.5	WWVL, Fort Collins, Colo.
21.4 kHz	0.5	NSS, Annapolis, Md.
24.0 kHz	0.5	NBA, Balboa, Panama, C.Z.
26.1 kHz	0.5	NPM, Hawaii
60.0 kHz	0.5	WWVB, Fort Collins, Colo.
100.0 kHz	0.5	Loran C, Carolina Beach, N.C.
2.5 MHz	0.5	WWV, Fort Collins, Colo., WWVH, Hawaii
3.33 MHz	50	CHU, Ottawa, Canada
5.0 MHz	0.5	WWV, Fort Collins, Colo., WWVH, Hawaii
7.335 MHz	50	CHU, Ottawa, Canada
10.0 MHz	0.5	WWV, Fort Collins, Colo., WWVH, Hawaii
14.67 MHz	50	CHU, Ottawa, Canada
15.0 MHz	0.5	WWV, Fort Collins, Colo., WWVH, Hawaii
20.0 MHz	0.5	WWV, Fort Collins, Colo.
25.0 MHz	0.5	WWV, Fort Collins, Colo.

* Modulation frequencies; all others are carrier frequencies.

SPEECH, HEARING, AND VISION

156. Sensory Perceptions in Electronics Engineering. In designing electronic devices it is often necessary to employ data on the relations between physical stimuli and the corresponding sensations evoked in man. To do so in terms of concepts and parameters capable of objective and mathematical formulation is difficult, since human sensations depend upon the situation as a whole and a given situation cannot be split into quantified fragments without seriously modifying the human response. Nor is it possible to measure sensations or impressions objectively, for these depend upon individual experiences, which are unique.

Fortunately it is possible to establish procedures for carrying out operations in which the magnitudes of the measurements of physical quantities (stimuli) are related to the sensitivity, or acuity, of human senses. When properly carried out and interpreted such psychophysical measurements can be very useful in the design of electronic systems and devices.

157. Cognition. The objective relations existing between entities of the physical world and man's subjective evaluation (cognition) of such relations are quite different and distinct.

The relationships between events described by the physical sciences and those described by the behavioral sciences must be examined with caution. There are no direct, immediate correlations between these fields in the sense that the relations between events of physical objects are of the same class as those represented by our subjective evaluations of them, or even of our sensory perceptions of them. Hence, psychophysical data can be easily misinterpreted and misused. It is essential, therefore, that the engineer understand in some detail the relations between stimuli and the psychophysical data he uses in the design of systems and devices.

158. Evaluation of Physical Stimuli. In assessing man's response to, and subjective evaluation of, various stimuli, there are three classes of evaluative concepts: *physical concepts*, which are quantitative and independent of the particular observer making measurements or observations; *psychological concepts*, which are subjective and uniquely related to individual experience; they have no quantitative significance because they cannot be expressed in operational terms; and *psychophysical concepts*, which can be made quantitative to the extent that they depend upon measurable aspects of the sense perceptions of the observer.

While different observers usually obtain different results when making psychophysical measurements of a given kind, it is often found that observations made with a large number of persons under identical measurement conditions tend to cluster around a set of values. Persons whose measured characteristics fall within small deviations of the mean of measurements on a large number of observers are regarded as having *normal sense responses*.

Particular sets of data may be adopted for general use by competent groups who examine, evaluate, and agree to accept the best data available as a standard. In this way, visibility curves, audibility curves, and similar data have been adopted and used in the design of systems or equipment.

159. Stimulus-Response Relations.   The change in a given stimulus that produces a just detectable change in sensory response is called the *difference threshold*.   Experiments show that the difference threshold is neither a fixed difference nor a fixed ratio of a change in stimulus but depends in a complicated way on the magnitude of the stimulus and technique of measurement.

The psychophysical response to stimuli activating different human senses can be expressed, roughly, by the law of psychology established by Weber and Fechner.   The Weber-Fechner law states that the least noticeable change of a stimulus is proportionally related to the magnitude of the stimulus.   Thus, if Δp is the minimum recognizable change in the perception of a stimulus as its magnitude is changed from s_1 to s_2 over a range $\Delta s = s_2 - s_1$, then, by the Weber-Fechner law,

$$\Delta p = k \frac{s_2 - s_1}{s} = k \frac{\Delta s}{s}$$

where k is a constant of proportionality.   By letting the increments become vanishingly small and integrating, the response is

$$p = k \log s + C$$

where C is a constant of integration.   This equation states that the response in perception to a stimulus is proportional to the logarithm of the magnitude of the stimulus.   Note that p represents a measure of the *perception* of a stimulus and is not a measure of subjective *sensation*.

The value of k varies considerably from sense to sense; it even varies for different magnitudes of a stimulus for the same sense.   Some experiments tend to show that the Weber-Fechner law might more accurately describe situations if written

$$\Delta p = k \frac{\Delta s}{s + A}$$

where A is a small value of the stimulus related to the threshold value but not identical to it. At small values of s, the constant A is a significant factor, but it becomes increasingly less important as values of s increase.

While the Weber-Fechner law is not beyond criticism, it is a pragmatically useful tool in dealing with perception and recognition processes.   It provides a rationale for a variety of logarithmic response measurements commonly used in communications and electronics engineering.

160. Logarithmic Response Units.   A number of logarithmic units have found extensive use in science and engineering.   Most of them have been used to designate power ratios. J. W. Horton has suggested a definition of logarithmic units to relate ratios of any two physical quantities, in units of *logits*.

A relative change in a quantity may be written

$$R = r^m$$

where R is a ratio of two quantities that expresses the change in their relative magnitude, r is a standard ratio in relative magnitude, and m is the exponent indicating the power to which r must be raised to yield the given ratio R.   From this relation,

$$m = \log_r R = \frac{\log R}{\log r}$$

In many physical situations, a barely perceptible change occurs when the quantities have a ratio of about 1.25, and so it is convenient to take $r = 10^{0.1} = 1.2589 \cdots$. If this value is adopted

$$m = 10 \log R \qquad \text{(logits)}$$

and m is a numeric, expressed in units of logits. This definition may be used and applied to any physical unit.

The *bel* (B), named in honor of Alexander Graham Bell, is defined as the common logarithm of the ratio of two powers, P_1 and P_2. Thus the number of bels N_B is

$$N_B = \log(P_2/P_1)$$

If P_2 is greater than P_1, N_B is positive, representing a gain in power; if $P_2 = P_1$, N_B is zero and the power level of the system remains unchanged; if P_2 is less than P_1, N_B is negative and the power level is diminished.

A smaller and more convenient unit for engineering purposes is the decibel (dB), whose magnitude is $1/10$ B. Thus

$$N_{dB} = 10 N_B = 10 \log(P_2/P_1)$$

In terms of electrical quantities, power may be expressed as $P = I^2 R$ or $P = E^2/R$. Accordingly, the change in power level, in terms of changes in current and voltage, is

$$N_{dB} = 10 \log \frac{P_2}{P_1} = 10 \log \left(\frac{I_2}{I_1}\right)^2 \frac{R_2}{R_1}$$

$$= 20 \log \frac{I_2}{I_1} + 10 \log \frac{R_2}{R_1}$$

$$= 20 \log \frac{E_2^2}{E_1} + 10 \log \frac{R_1}{R_2}$$

The power-level change, expressed in decibels, is correctly given in terms of voltage and current ratios alone only for the special case for which $R_1 = R_2$.

The *neper* (N_p) is defined to be the natural logarithm of the ratio of two currents, I_1 and I_2, or two voltages, E_1 and E_2. Thus, the number of nepers expressing a given change in voltage or current level is

$$N_{N_p} = \ln \frac{I_2}{I_1} = \ln \frac{E_2}{E_1}$$

The neper may also be defined as one-half the natural logarithm of two power ratios measured under such circumstances that the resistance of the circuit remains the same as the level is changed.

The *volume unit* (vu) is defined to be 10 times the common logarithm of the power ratio, P_2/P_1, where $P_1 = 1$ mW (0.001 W). If P_2 is expressed in watts,

$$N_{vu} = 10 \log (P_2/0.001) = 10 \log P_2 + 30$$

Note that the expression for volume unit specifies a reference level (0.001 W) so that the number of volume units represents the actual power level in a circuit.

Logarithmic expressions for power ratios cannot be used to indicate the amount of power in a circuit or system unless and until some reference level or zero power level is stated.

The *phon* is a unit of loudness level. The level of a sound, in phons, is numerically equal to the intensity level (in decibels) of a pure 1-kHz tone which is judged by the listener to be of equivalent loudness.

161. Adaptive Processes. Adaptive processes include a wide range of adjustments to changes in environmental conditions. Specifically, the term is used to refer to the adjustment of sense organs, such as the eye or receptors of the skin, to the intensity or quality of stimulation such as light, temperature or pressure prevailing at the moment, by changes in the sensitivity of the sense organs.

In visual processes, *dark adaptation* is the gradual adjustment and increased visual sensitivity to light of reduced intensity.

Light adaptation requires adjustment to highly intense visual excitation from previous excitation at low light intensities. *Visual adaptation* is a function of a number of variables, including the wavelength and intensity of the stimulus before and after adaptation and the length of time the eye has been stimulated. Light adaptation usually occurs within a few minutes but complete adaptation to the dark may require half an hour or more.

Similar effects occur in hearing. The ear appears to adapt rather quickly to new conditions of stimuli.

162. Components and Frequencies of Speech. On the average, the syllables of speech have a duration of about $\frac{1}{8}$ s, and the interval between syllables is about $\frac{1}{10}$ s. The frequency spectrum of speech sounds depends upon the resonant cavities formed by the throat, mouth, lips, and teeth as well as upon the syllables and words spoken. In general, the frequency spectrum for speech of women tends to be about an octave higher than that of men.

Frequencies below 200 Hz and above 7,000 Hz make little contribution to speech intelligibility. Consonants are more essential to speech intelligibility than vowels, but vowels account for the greater portion of the power generated in speech.

The dynamic power range of normally spoken speech is about 30 dB. Peaks of speech power average about 12 dB above the average level of speech, whereas the weakest speech sounds lie about 18 dB below average speech levels. For men, maximum power in speech is produced at frequencies near 400 Hz; the power decreases gradually with frequency up to about 4,000 Hz and then rapidly drops for high frequencies.

In large measure, the frequency spectrum to which an audio system should respond for the reproduction of speech depends upon the degree of naturalness required. Since it is also influenced by personal preferences and the noise and other interfering effects at the receiving point, it is likely to be a highly subjective choice. See also Sec. 19.

163. Power Levels of Speech. Over several minutes, the average power of connected speech is about 10 μW. For as loud shouting as possible, average speech sounds may reach levels of 1 mW (30 dB above normal speech levels). In faint whispers, speech power may drop to levels of 0.001 μW, 40 dB below the normal speech level. Hence, speech may vary over a dynamic range of 70 dB.

Measurements of the speech power levels of conversational speech for men's voices show that the average long-time speech power occurs in the octave between 500 and 1,000 Hz. Relative to the reference-level value of 10^{-13} W, the normal male voice produces sound levels of about 66 dB in the octave between 30 and 75 Hz. The level rises to about 80 dB in the octave between 500 and 1,000 Hz and falls to about 52 dB in the octave between 5 and 10 kHz.

164. Peak Factor Statistics of Speech. The natural reproduction of speech requires that peak values of speech power throughout the spectrum be taken into account as well as average levels. The data of Fig. 1-5 show results of some measurements of peak power pressures for men and women. The speech power at $\frac{1}{8}$-s intervals was measured in a number of frequency bands, and the value for the band plotted at the midfrequency. The percentages on the three sets of curves give the fraction of intervals having peak values of greater intensity than that indicated. For example, only about 1% of the peak sounds of speech lie above the 1% curves.

Measurements of the effective or rms values of speech sound pressures show variations similar to those given in Fig. 1-5 for peak values, except that the range or spread in power levels is reduced. For example, the dynamic range for rms values is about 18 dB for women and 25 dB for men, whereas the dynamic range for peak values is about 35 to 40 dB for both men and women.

165. Speech Intelligibility. The most common tests for speech intelligibility are articulation tests, in which a participant is asked to report, in writing, sounds heard during a test. The articulation index is the ratio of the number of correctly reported sounds or syllables to the total number presented to the observers. Such an index takes account of the speech clarity of the speaker as well as the speech perception of the listener.

Articulation tests are influenced by many factors, including the intensity of speech, background noise, the listener's familiarity with words or syllables spoken, the nature of the spoken sounds, the frequency spectrum of the transmission system, removal or absence of

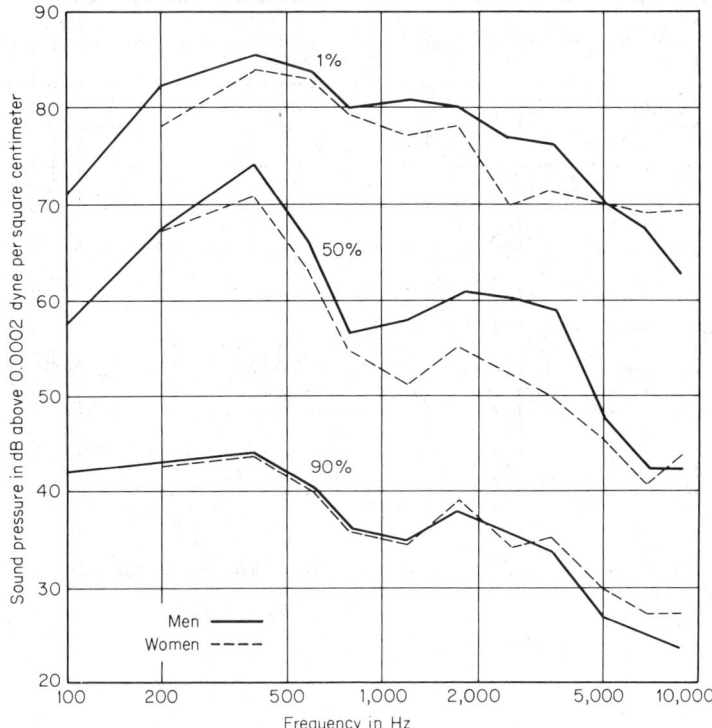

Fig. 1-5. Peak sound pressures in normal speech vs. frequency. The percentages indicate the fraction of the range lying above the respective curves.

certain frequency bands in the transmitted spectrum, and removal of peak amplitudes of speech power by clipping or amplitude limiting.

If lower frequencies of speech are removed, the articulation index does not change markedly until frequencies above 500 Hz are removed. If low frequencies between 500 and 2,500 Hz are removed, the articulation index drops sharply. On the other hand, if high frequencies beyond 2,500 Hz are removed, 80% articulation remains, but removal of frequencies above 1,000 Hz leads to impractical communication systems since only 40% of the words spoken are correctly identified.

166. Audibility. To a significant extent the threshold of audibility depends upon how audibility is determined as a psychophysical response. The curves of Fig. 1-6 show determinations of the threshold of audibility for persons having good hearing, as measured by three different techniques. In general, the absolute threshold values are different for the three observational methods used, but all curves follow the same general variation with regard to frequency. Moreover, determinations made by different observers with different subjects are in reasonably close agreement with the data of Fig. 1-6.

The threshold of audibility for a specified sound is the minimum effective sound pressure capable of providing an auditory sensation, in the absence of noise, in a specified fraction of all observers, the specified fraction usually being taken to be the 1% of subjects having the most acute hearing. The audibility threshold is usually taken to be 0.0002 microbar, or 0.0002 dyn/cm^2.

167. Loudness. The human ear is responsive to frequencies from about 20 to 20,000 Hz, covering a range of 10 octaves, but the sensitivity varies greatly with frequency. The human ear is most sensitive to frequencies from about 2 to 5 kHz and least sensitive to sounds at the extreme frequencies of the audible range. The subjective response to sounds of increasing sound-pressure intensity is described by the loudness of sounds. The loudness

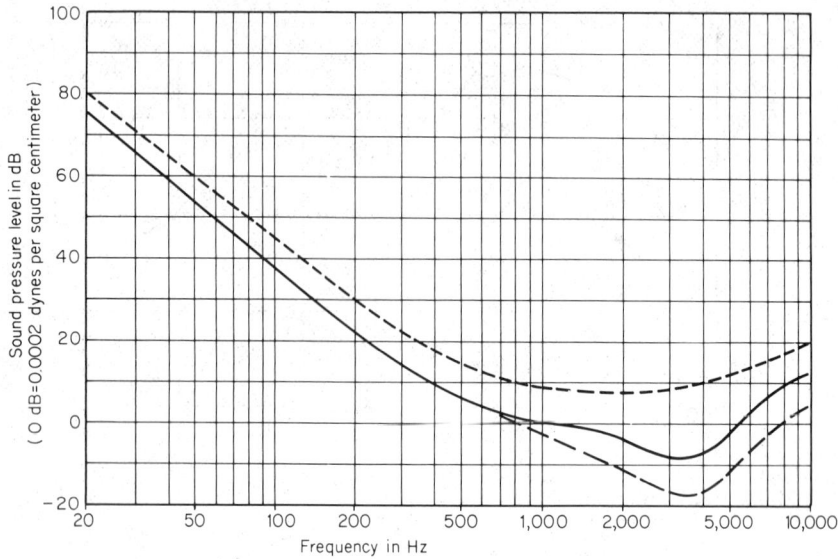

Fig. 1-6. Threshold of hearing of normal observer. The different curves arise from three different measurement methods.

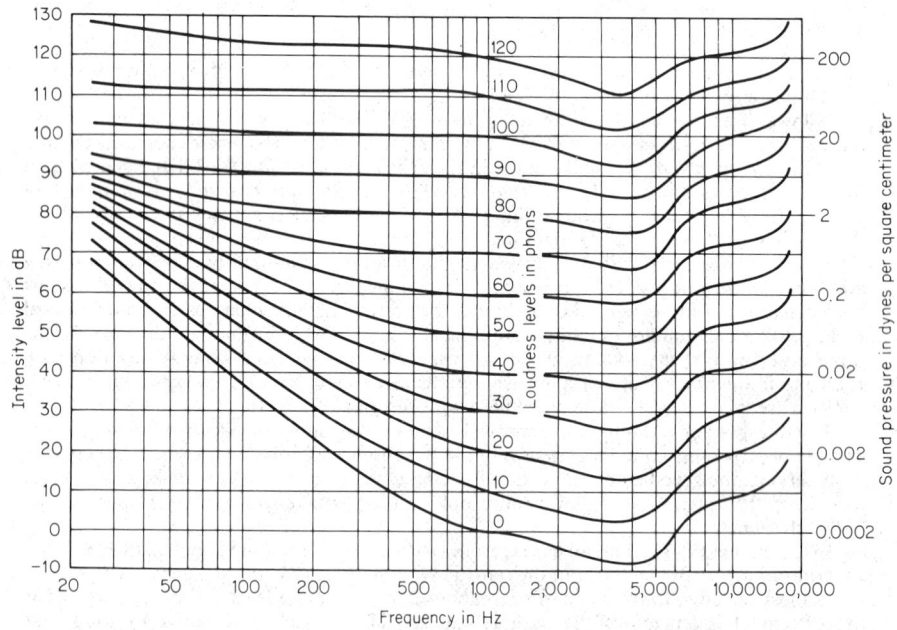

Fig. 1-7. Contours of equal loudness for the normal human ear. The curves are normalized at 1,000 Hz, and the corresponding loudness levels in phons are indicated at this frequency.

of sounds depends upon their intensity and their frequency and is specified quantitatively in terms of loudness levels, expressed in phons and measured at 1 kHz, which is arbitrarily chosen as the reference frequency (see Par. **1-160**). The loudness of a pure tone of a frequency other than 1 kHz is defined in terms of the sound-pressure level of a sound that appears to be as loud as the 1-kHz pure tone.

As shown in Fig. 1-7, the loudness levels of pure tones are a group of equal-loudness contours. In general the intensity level and the loudness level are described by the same number only at a frequency of 1 kHz. The loudness is a subjective evaluation; the intensity level is a physical measure.

168. Minimum Perceptible Changes in Intensity and Pitch. A change in sound-pressure level of about 1 dB can be detected for pure tones in the frequency range between 50 and 10,000 Hz by persons with normal hearing provided the sound level is 50 dB or more above the threshold value. If the sound-pressure level is less than 40 dB above the threshold value, changes of up to 3 dB are required to be perceptible. Under favorable conditions, changes in sound level as small as 0.3 dB can be detected by persons with normal hearing in the mid-audio-frequency range.

When the sound level is 40 dB or more above the threshold value, persons with normal hearing are able to detect changes in frequency of about 0.3 of 1% at frequencies of 1,000 Hz or higher. A frequency change of about 3 Hz is detectable at frequencies less than 1,000 Hz.

169. Space Perception in Hearing (Stereophony). The ability to localize the direction of a sound source depends, in part, upon facility in recognizing differences in loudness or sound intensity, phase in pure tones, and quality of sounds in the case of complex sounds for each of the two ears. In addition to these purely physical factors, localization of sound by binaural listening depends upon the ability to associate these differences with the direction of a sound source relative to the observer. Thus, experience is a factor in sound localization, quite independent of the hearing process itself.

Localization of sounds can be accomplished in different frequency ranges by different hearing mechanisms. At frequencies of about 300 to 400 Hz or less, the ability to evaluate the direction of a sound source depends upon detecting phase differences and usually is not well developed. Between 400 and 1,000 Hz, sound sources can be localized by detecting differences in phase of sound reaching the two ears. At frequencies above 1,000 Hz differences in the quality of sound reaching the two ears are used, and for frequencies greater than about 3,000 Hz, sound can be localized by differences in loudness at the two ears of an observer.

170. Vision. The structure of the human eye is shown in Fig. 1-8. Light is admitted through the cornea and aqueous humor, where it is bent to enter the eye through a variable-area aperture called the *iris*. Light travels through the lens whose shape can be controlled by muscles to focus the admitted light on the retina. The retina consists of several layers having two types of nerve-end cells. One set, the *cones*, numbers about 6.5×10^6 and is most heavily concentrated in the fovea. The other set, the *rods*, may number as many as 125×10^6 and is most numerous about 20° from the fovea. The rods and cones contain photosensitive chemicals which release electric charges when activated by light.

The cones are primarily sensitive to color or wavelength of the visible spectrum. The rods are more or less unresponsive to color and play a major role in perceiving shape under conditions of low ambient-light intensity.

Cone (photopic) vision is used in the discrimination of fine detail and color. The cones are of small diameter and are closely packed; they transmit images of appreciable detail but make small contribution to visual sensations at low values of field brightness.

Rod (scotopic) vision comes into play when field brightness is less than 0.01 foot-lambert. Sharp images are not transmitted by the rods; neither is color to any appreciable extent.

The rods and cones of the retina are able to distinguish between different amounts of light intensity within a range of about 380 to 760 nm and between hues primarily through functioning of the cones in the foveal region; they transmit stimuli of hues and intensity to the optic nerve. The response to activation is faster than the response to deactivation, so that *persistence of vision* occurs in the presence of interrupted light.

In addition to the above-mentioned functions of the rods and cones, the retina is also able to perceive geometrical arrangements of images focused on it through the simultaneous excitation of many rods and cones, to distinguish detail in the image as each rod and cone (or small group of them) responds to energy of an element of the image, to focus on images

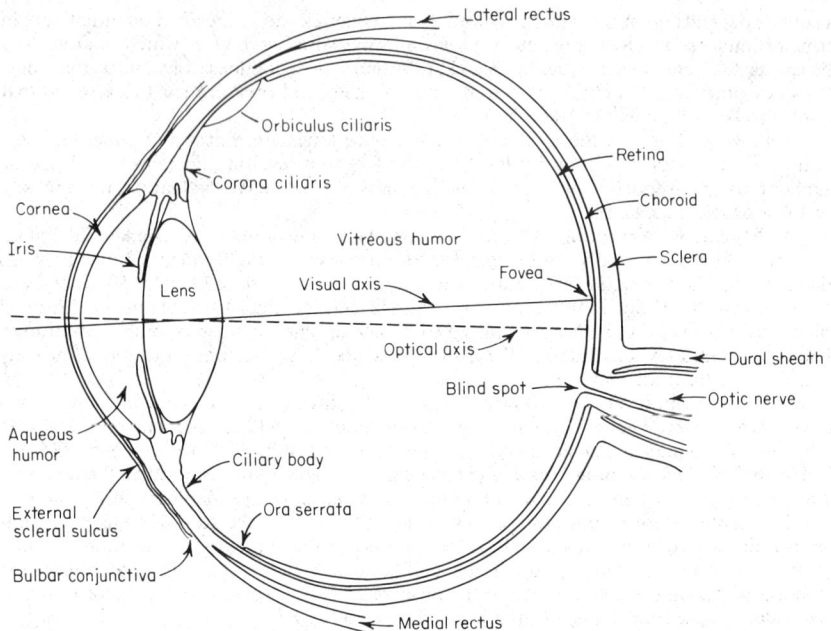

Fig. 1-8. Structure of the human eye.

at different distances by changing the shape of the lens, to produce sharp images on the retina, to distinguish motion as different rods and cones are activated in sequence by a changing pattern of energy in the image, and to perceive depth perspective or stereoscopic effects by combining in the brain the slightly different images seen by the two eyes.

171. Luminous Energy.   Radiant energy evaluated by the human eye is called *luminous energy*.   It is measured in physical quantities similar to those used for radiant energy but is weighted, according to frequency or wavelength, by the properties of the normal human eye.

Quantity of light Q (or Q_v) is the quantity of radiant energy as evaluated by the normal human eye.   In the mks system of units it is specified in units of lumen-seconds (lm.s).

Luminous flux ϕ (or ϕ_v) is the time rate of flow of light.   In the mks system of units it is measured in units of lumens (lm) or in units of candela-steradians.   The lumen is the unit of luminous flux and is equal to the flux through a unit solid angle (1 steradian) from a uniform point source of one candela (candle).   Alternatively, it is the flux on a unit surface all points of which are at unit distance from a uniform point source of one candela luminous intensity.

Luminous emittance or *exitance M* (or M_v) is the luminous flux density at a surface. In the mks system it is measured in units of lumens per square meter.

Illuminance E (or E_v) is the density of luminous flux incident on a surface and is the ratio of the luminous flux to the area of the surface when the latter is uniformly illuminated. The SI unit is the lux.

Luminous intensity I (or I_v) of a source of light in a given direction is the luminous flux per solid angle in the direction in question.   It is the luminous flux on a small surface normal to that direction divided by the solid angle in steradians that the surface subtends at the source.   In SI units (Par. **180**), it is measured in candelas, approximately equal to the older unit, the *candle*.

The *candela* is the luminous intensity, in the perpendicular direction, of a surface of $1/600,000$ m^2 of a blackbody at a temperature of freezing platinum under a pressure of 101,325 N/m^2.

172. Spectral Sensitivity of the Normal Observer.   (See Sec. **20**.)   The relative sensitivity of human eyes varies with frequency of monochromatic light and also with intensity of the radiation.   It also varies from person to person.   For the so-called normal observer, the relative sensitivity of light is the function of wavelength shown in Fig. **20-1**. The curve shows the relative spectral response for daytime (photopic) vision in which the cones play the dominant visual role.

For photopic vision, maximum response occurs at 554 nm. For nighttime vision, the maximum response occurs at 507 nm. The shift of 47 nm is known as the *Purkinje shift* and occurs at light levels between 0.001 and 0.01 lm/ft^2, when the transition takes place between rod and cone vision.

173. Brightness and Brightness Sensitivity. Brightness is the visual sensation produced by the emission or reflection of light. It is a measure of the light emitted from a surface in a given direction per unit area per unit solid angle and can be measured in units such as lumens per square meter.

The sensitivity of the human eye to brightness is measured by the minimum change in illumination needed to produce a just perceptible difference. Brightness sensitivity of the human eye for changes in illumination is nearly constant for moderate light levels. The least perceptible change of brightness, or the brightness threshold, is about 1.6 to 1.8% for all colors at normal light levels.

At low light levels, the brightness sensitivity decreases, first in the red end of the spectrum and finally in the blue, which is the last hue to be perceived. At the lowest levels at which objects can be seen, all sensation of hue or color disappears.

174. Contrast and Contrast Sensitivity. *Visual contrast* is the ratio of the difference in brightness between an object and its background to the brightness of the background. Such a definition is not entirely adequate since our normal notions of contrast vary depending upon the type and size of the object to be detected. Thus, it is common to divide contrast sensitivity into two classes: those in which small objects are to be contrasted and detected against their backgrounds and those in which the contrast between large contiguous surfaces is to be determined. The first of these involves variations of contrast with size of the object as well as contrast with illumination. Size is not a factor in determining contrast between large contiguous areas.

The *contrast sensitivity* is a measure of the ability to discriminate slight differences in contrast. It is the reciprocal of *visual contrast*, as defined above.

The contrast sensitivity for an object against its background when both are of the same color and are illuminated by white light is illustrated in Fig. 1-9. The reduced contrast sensitivity for objects having a dark surround may result from glare which makes the object more difficult to detect.

175. Tonal Discrimination. As used in photography, television, and the visual arts, tone discrimination is the ability to discern a change in the brightness of an image reproduced from an original scene. If the original scene and its reproduction have the same values of brightness in corresponding elements, they have the same brightness range. In this case the tone discrimination is the same as the brightness discrimination.

The full range of brightness values cannot usually be reconstructed in a reproduction; i.e., the brightness range of the reproduced scene is less than that of the original. If the brightness of all elements in the reproduction is proportional to the brightness of the corresponding elements of the original, the brightness range of the reproduction is reduced. If no other property than brightness is modified, there is no loss of detail in the bright or dark portions of the reproduced image, although the detail may be more difficult to detect because of the reduced brightness range or contrast.

If the brightness of the image is not proportional to that of the original, then in addition to brightness compression, details may be lost in the reproduction, either in the shadows, the highlights, or both. Tone discrimination can be improved by increasing the contrast of the reproduction, but this is accompanied by brightness or amplitude distortion in the dark or light areas of the reproduced image.

176. Resolving Power. Resolving power is the ability to detect small objects and distinguish fine detail in the presence of large contrast between an object and its background. It varies widely with the type of test object, the spectral distribution of luminous energy, background luminosity, contrast between the object and its background, and (most important) the criteria used in making determinations of the quantities or operations being performed or measured.

The resolving power of the eye is sometimes taken to be a measure of the distinctness with which the images of two point sources of light can be detected separately. It is the minimum angular separation of two objects which appear distinct and separate.

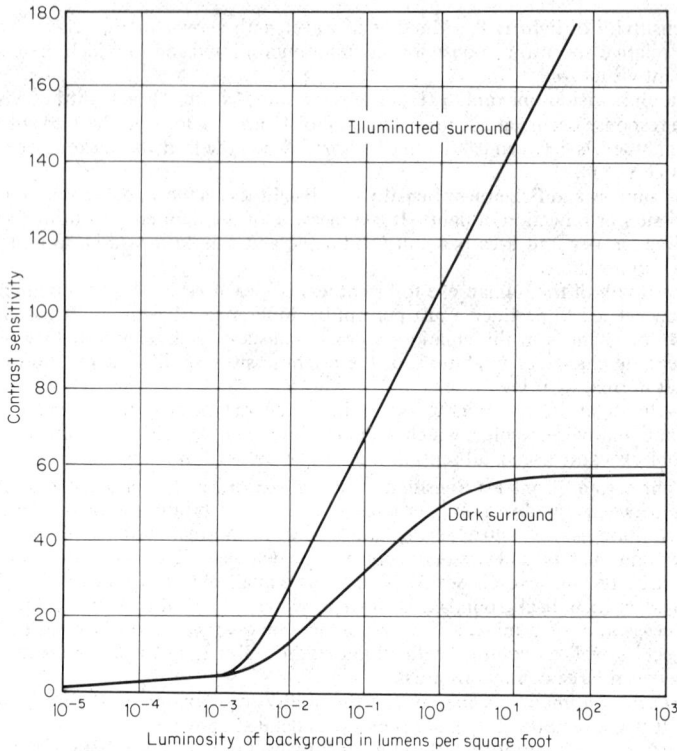

Fig. 1-9. Contrast-sensitivity curves for illuminated- and dark-background conditions.

177. Visual Acuity. Visual acuity is the ability to distinguish fine detail. It is usually expressed either as the ratio of the distance at which a line of letters on a Snellen test chart can be read by the observer being tested to the distance at which an observer with normal vision could read the line or as a visual efficiency rating whose percentage figures are related to the size of characters that can be read in each line of a chart such as that of the American Medical Association. Visual acuity may also be expressed as the reciprocal of the angle (measured in minutes of arc) subtended at the eye by a specified test object, such as two black dots on a white background.

Visual acuity increases with the luminosity to which the eye is adapted (Fig. 1-10). For values of luminosity less than about 10^{-2} lm/ft^2, the increase in acuity is small. From about 10^{-2} to about 10 lm/ft^2, the increase in acuity is proportional to the increase in luminosity of the background. Above 10 lm/ft^2, there is a tendency toward saturation, and visual acuity then increases gradually less rapidly.

The measurement of visual acuity depends upon the luminosity of the entire field, background as well as foreground. Visual acuity is greatest with white or nearly white light.

178. Flicker and Persistence of Vision. The eye does not respond instantly to a visual stimulus, nor does the sensation of vision disappear instantly when the stimulus is removed. Experiments show that the time an image is retained after the stimulus is removed is somewhat greater than the time needed to produce a visual sensation when the stimulus is suddenly applied.

This situation makes possible the viewing, in rapid succession, of related but slightly different scenes in such a way that a sensation of apparently continuous motion is conveyed. If the rate at which different images is presented is too small, the effect is accompanied by *flicker*. At sufficiently high projection rates, flicker disappears. The frequency at which flicker disappears depends upon the brightness of the field being viewed, as indicated in Fig. 1-11.

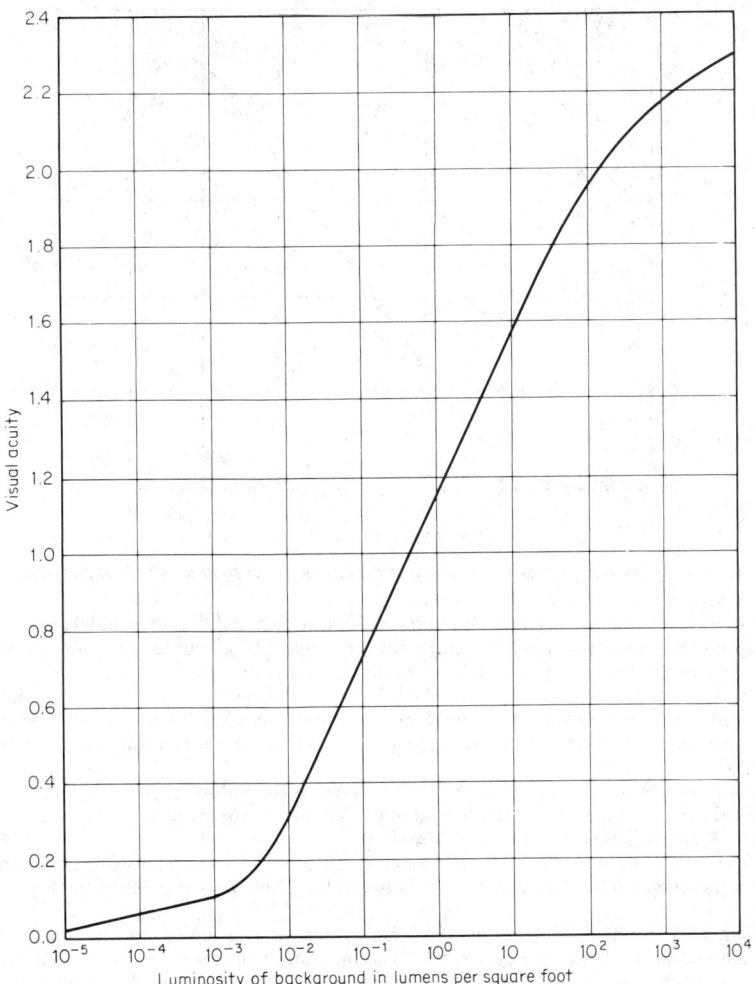

Fig. 1-10. Visual acuity as a function of background illumination.

179. Depth Perception and Stereoscopy. The images of an object, especially one nearby, are seen in a slightly different manner because the pupils of the two eyes are not coincident. This dissimilarity of two simultaneous images enables the brain to fuse the two so that a mental picture of a scene in three-dimensional space results. The perception of depth gained in this way is called *stereoscopic vision* and has its audio counterpart in stereophony.

ELECTRONIC QUANTITIES

180. Systems of Units. In 1960 the Eleventh General Congress on Weights and Measures promulgated the International System of Units (abbreviated SI). This comprises a universal coherent system of units in which the following six quantities are considered to be basic: (1) meter of length, (2) kilogram of mass, (3) second of time, (4) ampere of current, (5) Kelvin degree of temperature, and (6) candela of luminous intensity. In 1967 the General Conference on Weights and Measures gave the name kelvin to the International Standard of temperature (which had previously been called the degree Kelvin) and assigned to this temperature unit the symbol K without the associated symbol, °, for the degree. The SI units are of convenient size for most branches of science and engineering, especially

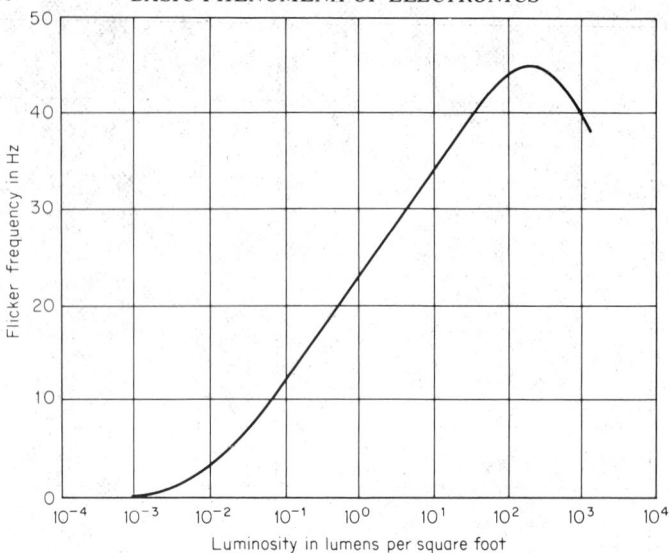

Fig. 1-11.   Flicker frequency as a function of the luminosity of successively projected images.

when used with suitable metric multipliers.   It should be noted that the candela is a unit of radiant energy evaluated in accordance with the visual energy curve for a normal human observer and, to this extent, depends upon subjective factors.

The SI proposals are described in "The International System of Units" by E. A. Matchly, a publication of the Office of Technology Utilization, National Aeronautics and Space Administration (NASA SP-7012), available from the Government Printing Office, Washington, D.C.

Definitions of the most important SI units given in the following paragraphs have been extracted from the records of the International Committee of Weights and Measures and the General Conferences of Weights and Measures.

Meter (m).   The *meter* is the length equal to 1 650 763.73 wavelengths in vacuum of the radiation corresponding to the transition between the levels $2p_{10}$ and $5d_5$ of the krypton-86 atom.

Kilogram (kg).   The *kilogram* is the unit of mass; it is equal to the mass of the international prototype of the kilogram.   (The international prototype of the kilogram is a particular cylinder of platinum-iridium alloy which is preserved in a vault at Sèvres, France, by the International Bureau of Weights and Measures.)

Second (s).   The *second* is the duration of 9 192 631 770 periods of the radiation corresponding to the transition between the two hyperfine levels of the ground state of the cesium-133 atom.

Ampere (A).   The *ampere* is that constant current which, if maintained in two straight parallel conductors of infinite length, of negligible circular cross section, and placed 1 meter apart in vacuum, would produce between these conductors a force equal to 2×10^{-7} newton per meter of length.

Kelvin (K).   The *kelvin*, unit of thermodynamic temperature, is the fraction 1/273.16 of the thermodynamic temperature of the triple point of water.

Candela (cd).   The *candela* is the luminous intensity, in the perpendicular direction, of a surface of 1/600 000 square meter of a blackbody at the temperature of freezing platinum under a pressure of 101 325 newtons per square meter.

Newton (N).   The *newton* is that force which gives to a mass of 1 kilogram an acceleration of 1 meter per second per second.

Joule (J).   The *joule* is the work done when the point of application of 1 newton is displaced a distance of 1 meter in the direction of the force.

Watt (W).   The *watt* is the power which gives rise to the production of energy at the rate of 1 joule per second.

Volt (V).   The *volt* is the difference of electric potential between two points of a conducting wire carrying a constant current of 1 ampere, when the power dissipated between these points is equal to 1 watt.

Ohm(Ω). The *ohm* is the electric resistance between two points of a conductor when a constant difference of potential of 1 volt, applied between these two points, produces in this conductor a current of 1 ampere, this conductor not being the source of any electromotive force.

Coulomb(C). The *coulomb* is the quantity of electric charge transported in 1 second by a current of 1 ampere.

Farad(F). The *farad* is the capacitance of a capacitor between the plates of which there appears a difference of potential of 1 volt when it is charged by a quantity of electricity equal to 1 coulomb.

Henry(H). The *henry* is the inductance of a closed circuit in which an electromotive force of 1 volt is produced when the electric current in the circuit varies uniformly at a rate of 1 ampere per second.

Weber(Wb). The *weber* is the magnetic flux which, linking a circuit of one turn, produces in it an electromotive force of 1 volt as it is reduced to zero at a uniform rate in 1 second.

Lumen(lm). The *lumen* is the luminous flux emitted in a solid angle of 1 steradian by a uniform point source having an intensity of 1 candela.

181. Values of Physical Constants. Table 1-8 lists physical constants from the work of B. N. Taylor, W. H. Parker, and D. N. Langenberg (*Rev. Mod. Phy.*, July 1969).

182. Symbols. Table 1-9 lists symbols, dimensions, and units for physical quantities.

Table 1-8. Physical Constants

Quantity	Symbol	Value	Error, ppm	Prefix	Unit
Speed of light in vacuum	c	2. 997 925 0	0.33	$\times 10^8$	m/s
Gravitational constant	G	6. 673 2	460	10^{-11}	N·m²/kg²
Avogadro constant	N_A	6. 022 169	6.6	10^{26}	kmol^{-1}
Boltzmann constant	k	1. 380 622	43	10^{-23}	J/K
Gas constant	R	8. 314 34	42	10^3	J/kmol·K
Volume of ideal gas, standard conditions .	V_0	2. 241 36	...	10^1	m³/kmol
Faraday constant	F	9. 648 670	5.5	10^7	C/kmol
Unified atomic mass unit	u	1. 660 531	6.6	10^{-27}	kg
Planck constant	h	6. 626 196	7.6	10^{-34}	J·s
	$h/2\pi$	1. 054 591 9	7.6	10^{-34}	J·s
Electron charge	e	1. 602 191 7	4.4	10^{-19}	C
Electron rest mass	m_e	9. 109 558	6.0	10^{-31}	kg
		5. 485 930	6.2	10^{-4}	u
Proton rest mass	m_p	1. 672 614	6.6	10^{-27}	kg
		1. 007 276 61	0.08	...	u
Neutron rest mass	m_n	1. 674 920	6.6	10^{-27}	kg
		1. 008 665 20	0.10	...	u
Electron charge to mass ratio	e/m_e	1. 758 802 8	3.1	10^{11}	C/kg
Stefan-Boltzmann constant	σ	5. 669 61	170	10^{-8}	W/m²·K⁴
First radiation constant	$8\pi hc$	4. 992 579	7.6	10^{-24}	J/m
Second radiation constant	hc/k	1. 438 833	43	10^{-2}	m/K
Rydberg constant	R_∞	1. 097 373 12	0.10	10^7	m^{-1}
Fine-structure constant	α	7. 297 351	1.5	10^{-3}	...
	α^{-1}	1. 370 360 2	1.5	10^{+2}	...
Bohr radius	a_0	5. 291 771 5	1.5	10^{-11}	m
Classical electron radius	r_e	2. 817 939	4.6	10^{-15}	m
Compton wavelength of electron	λ_C	2. 426 309 6	3.1	10^{-12}	m
	$\lambda_C/2\pi$	3. 861 592	3.1	10^{-13}	m
Proton	$\lambda_{C,p}$	1. 321 440 9	6.8	10^{-15}	m
	$\lambda_{C,p}/2\pi$	2. 103 139	6.8	10^{-16}	m
Neutron	$\lambda_{C,n}$	1. 319 621 7	6.8	10^{-15}	m
	$\lambda_{C,n}/2\pi$	2. 100 243	6.8	10^{-16}	m
Electron magnetic moment	μ_e	9. 284 851	7.0	10^{-24}	J/T
Proton magnetic moment	μ_p	1. 410 620 3	7.0	10^{-26}	J/T
Bohr magneton	μ_B	9. 274 096	7.0	10^{-24}	J/T
Nuclear magneton	μ_n	5. 050 951	10	10^{-27}	J/T
Gyromagnetic ratio of protons in H₂0 ...	γ'_p	2. 675 127 0	3.1	10^8	rad/s·T
	$\gamma'_p/2\pi$	4. 257 597	3.1	10^7	Hz/T
	γ_p	2. 675 196 5	3.1	10^8	rad/s·T
Corrected for diamagnetism of H₂0	$\gamma_p/2\pi$	4. 257 707	3.1	10^7	Hz/T
Magnetic flux quantum	Φ_0	2. 067 853 8	3.3	10^{-15}	Wb
Quantum of circulation	$h/2m_e$	3. 636 947	3.1	10^{-4}	J·s/kg
	h/m_e	7. 273 894	3.1	10^{-4}	J·s/kg

Table 1-9. Symbols for Physical Quantities
(From IEEE Standards 280, by permission)

Name of quantity	Symbol	Dimensions	Mks Unit	Abbreviation for mks Unit
Basic entities				
Mass	m	M	Kilogram	kg
Length	l	L	Meter	m
Time	t	T	Second	s
Electric current	I, i	QT^{-1}	Ampere	A
Temperature	T, θ	θ	Kelvin	K
Luminous intensity	I, I_v	I_v	Candela	cd
Space and time				
Plane angle	$\alpha, \beta, \gamma, \theta, \Theta, \phi$	...	Radian	rad
Solid angle	ω, Ω	...	Steradian	sr
Area	A, S	L^2	Square meter	m²
Volume	V, v	L^3	Cubic meter	m³
Period	T	T	Second	s
Frequency	f, ν	T^{-1}	Hertz	Hz
Angular frequency	ω	T^{-1}	Radian/second	rad/s
Velocity	v	LT^{-1}	Meter/second	m/s
Angular velocity	ω	T^{-1}	Radian/second	rad/s
Linear acceleration	a	LT^{-2}	Meter/second²	m/s²
Angular acceleration	α	T^{-2}	Radian/second²	rad/s²
Displacement	r, d	L	Meter	m
Angular displacement	θ	...	Radian	rad
Distance	l	L	Meter	m
Wavelength	λ	L	Meter	m
Radius	r	L	Meter	m
Diameter	d	L	Meter	m
Wave number	$\sigma, \tilde{\nu}$	L^{-1}	Meter⁻¹	m⁻¹
Time constant	τ, T	T^{-1}	Hertz	Hz
Rotational frequency	n	T^{-1}	Revolution/second	r/s
Mechanics				
Volume density	ρ	ML^{-3}	Kilogram/meter³	kg/m³
Momentum	p	MLT^{-1}	Kilogram-meter/second	kg·m/s
Moment of inertia	I, J	ML^2	Kilogram-meter²	kg·m²

Quantity	Symbol	Dimensions	Unit	Abbreviation
Force	F	MLT^{-2}	Newton	N (kg·m/s²)
Weight	W	MLT^{-2}	Newton	N
Weight density	γ	$ML^{-2}T^{-2}$	Newton/cubic meter	N/m³
Pressure	p	$ML^{-1}T^{-2}$	Newton/meter²	N/m²
Moment of force	M, T	ML^2T^{-2}	Newton-meter	N·m
Torque	T, M	ML^2T^{-2}	Newton-meter	N·m
Normal stress	σ	$ML^{-1}T^{-2}$	Newton/meter²	N/m²
Shear stress	τ	$ML^{-1}T^{-2}$	Newton/meter²	N/m²
Young's modulus	E	$ML^{-1}T^{-2}$	Newton/meter²	N/m²
Shear modulus	G	$ML^{-1}T^{-2}$	Newton/meter²	N/m²
Bulk modulus	K	$ML^{-1}T^{-2}$	Newton/meter²	N/m²
Energy	E, W	ML^2T^{-2}	Joule	J
Energy (volume) density	w	$ML^{-1}T^{-2}$	Joule/cubic meter	J/m³
Potential energy	U	ML^2T^{-2}	Joule	J
Kinetic energy	K	ML^2T^{-2}	Joule	J
Work	W	ML^2T^{-2}	Joule	J (N·m)
Power	P	ML^2T^{-3}	Watt	W (J/s)
Efficiency	η	$\ldots$	(Numeric)	
Angular momentum	L	ML^2T^{-1}	Kilogram-meter²/second	kg·m²/s
Gravitational field strength	g	LT^{-2}	Newton/kilogram	N/kg
Gravitational potential	V	L^2T^{-2}	Joule/kg	J/kg

Thermodynamics and heat

Quantity	Symbol	Dimensions	Unit	Abbreviation
Quantity of heat	Q	ML^2T^{-2}	Joule	J
Work	W	ML^2T^{-2}	Joule	J
Temperature (absolute)	T	θ	Kelvin	K
Temperature	t, θ		Degree Celsius	°C
Entropy	S	ML^2T^{-2}	Joule/kelvin	J/K
Internal energy	U	ML^2T^{-2}	Joule	J
Free energy	F	ML^2T^{-2}	Joule	J
Enthalpy	H	ML^2T^{-2}	Joule	J
Coefficient of linear expansion	α_l	LT^{-1}	Meter/kelvin	m/K
Coefficient of cubic expansion	α_v	L^3T^{-1}	Meter³/kelvin	m³/K
Thermal conductivity	λ, k	$L^2T^{-3}\theta^{-1}$	Watt/meter-kelvin	W/m·K
Specific heat capacity	c	$L^2T^{-2}\theta^{-1}$	Joule/kilogram-kelvin	J/kg·K
Joule-Thomson coefficient	$\ldots$	$ML^{-1}T^3\theta$	Kelvin-meter²/newton	K·m²/N

Electricity and magnetism

Quantity	Symbol	Dimensions	Unit	Abbreviation
Quantity of electricity (charge)	Q, q	Q	Coulomb	C (A·S)
Linear charge density	λ	$L^{-1}Q$	Coulomb/meter	C/m
Surface charge density	σ	$L^{-2}Q$	Coulomb/meter²	C/m²

Table 1-9. **Symbols for Physical Quantities** (*Continued*)
(From IEEE Standards 280, by permission)

Name of quantity	Symbol	Dimensions	Mks Unit	Abbreviation for mks Unit
Volume charge density	ρ	$L^{-3}Q$	Coulomb/meter³	C/m³
Electric dipole moment	p	LQ	Coulomb-meter	C·m
Polarization	P	$L^{-2}Q$	Coulomb/meter²	C/m²
Electric field intensity	E, K	$MLT^{-2}Q^{-1}$	Newton/coulomb or Volt/meter	N/C V/m
Permittivity of free space	ϵ_0, ϵ_v	$M^{-1}L^{-3}T^2Q^2$	Farad/meter	F/m
Relative permittivity	ϵ_r	...	(Numeric)	...
Electric susceptibility	χ_e, ϵ	...	(Numeric)	...
Electric flux	ψ, ϕ_e	Q	Coulomb	C
Electric flux density (electric displacement)	D	$L^{-2}Q$	Coulomb/meter²	C/m²
Electric potential difference	V	$ML^2T^{-2}Q^{-1}$	Volt	V (W/A·s)
Electromotance	E, V	$ML^2T^{-2}Q^{-1}$	Volt	V
Voltage	V, E	$ML^2T^{-2}Q^{-1}$	Volt	V
Elastance	S	$ML^2T^{-2}Q^{-2}$	Farad⁻¹	F⁻¹
Capacitance	C	$M^{-1}L^{-2}T^2Q^2$	Farad	F
Electric current (conduction)	I, i	QT^{-1}	Ampere	A
Convection current	I_c, i_c	QT^{-1}	Ampere	A
Displacement current	I_d, i_d	QT^{-1}	Ampere	A
Current density	J, s	$L^{-2}T^{-1}Q$	Ampere/meter²	A/m²
Resistance	R, r	$ML^2T^{-1}Q^{-2}$	Ohm	Ω (V/A)
Resistivity	ρ	$ML^3T^{-1}Q^{-2}$	Ohm-meter	Ω-m
Conductance	G, g	$M^{-1}L^{-2}TQ^2$	Mho (siemens)	mho or S
Conductivity	γ, σ	$M^{-1}L^{-3}TQ^2$	Mho/meter (siemens/meter)	mho/m or S/m
Mobility	μ_e, μ_n	$M^{-1}TQ$	Meter²/volt-second	m²/V·s
Electric energy	U_e	ML^2T^{-2}	Joule	J
Electric energy density	u_e	$ML^{-1}T^{-2}$	Joule/cubic meter	J/m³
Magnetic field intensity	H	$L^{-1}T^{-1}Q$	Ampere/meter	A/m
Magnetic scalar potential	U_m	$T^{-1}Q$	Ampere	A
Magnetomotance	$V_m, F, \mathcal{F}$	$ML^2T^{-1}Q^{-1}$	Ampere or ampere-turn	A
Magnetic flux	Φ	...	Weber or volt-second	A·t Wb (V·s)
Magnetic flux density	B	$MT^{-1}Q^{-1}$	Tesla or weber/meter²	T (Wb/m²)
Magnetic flux linkages	λ	$T^{-1}Q$	Ampere-turn	A·t
Magnetic vector potential	A	$MLT^{-1}Q^{-1}$	Weber/meter	Wb/m
Permeability of free space	μ_0	MLQ^{-2}	Henry/meter	H/m
Relative permeability	μ_r	...	(Numeric)	...
Magnetic permeability	μ	MLQ^{-2}	Henry/meter	H/m
Magnetic susceptibility	χ_m	...	(Numeric)	...
Electromagnetic dipole moment	m, μ	$L^2T^{-1}Q$	Ampere-meter²	A·m²
Self-inductance	L	ML^2Q^{-2}	Henry⁻¹	H (V·s/A)
Reciprocal inductance	Γ	$M^{-1}L^{-2}Q^2$	Henry⁻¹	H⁻¹
Mutual inductance	L_{ij}, M_{ij}	ML^2Q^{-2}	Henry	H

1-64

Quantity	Symbol	Dimensions	Unit	Unit symbol
Reluctivity	ν	$M^{-1}L^{-1}Q^2$	Meter/henry	m/H
Reluctance	$R, R_m, \mathscr{R}$	$M^{-1}L^{-3}Q^2$	Henry^{-1} or ampere-turn/weber	H^{-1} (A·t/Wb)
Permeance	$P, P_m, \mathscr{P}$	ML^3Q^{-2}	Henry	H
Magnetic polarization	J, B_i	$MT^{-1}Q^{-1}$	Tesla	T
Current element	X	$MT^{-1}Q$	Ampere-meter	A·m
Reactance	Z	$ML^2T^{-1}Q^{-2}$	Ohm	Ω
Impedance	B	$ML^2T^{-1}Q^{-2}$	Ohm	Ω
Susceptance	Y	$M^{-1}L^{-2}TQ^2$	Mho (siemens)	mho or S
Admittance	S	$M^{-1}L^{-2}TQ^2$	Mho (siemens)	mho or S
Poynting vector	P	MT^{-3}	Watt/meter^2	W/m^2
Electric power	W, U	ML^2T^{-3}	Watt	W
Electric energy		ML^2T^{-2}	Joule	J
Radian frequency	ω	T^{-1}	Radian/second	rad/s
Cyclic frequency	f	T^{-1}	Hertz	Hz
Wavelength	λ	L	Meter	m
Speed of propagation	c	LT^{-1}	Meter/second	m/s
Characteristic impedance	Z_0	$ML^2T^{-1}Q^{-2}$	Ohm	Ω
Intrinsic impedance	η	$ML^2T^{-1}Q^{-2}$	Ohm	Ω
Propagation coefficient	γ	L^{-1}	Complex neper/meter	m^{-1}
Attenuation coefficient	α	L^{-1}	Neper/meter	Np/m
Phase coefficient	β	L^{-1}	Radian/meter	rad/m
Resonant frequency	f_r	T^{-1}	Hertz	Hz
Skin depth	δ	L	Meter	m
Reflection coefficient	Γ		(Numeric)	
Voltage standing-wave ratio	S		(Numeric)	
Quality factor	Q		(Numeric)	
Loss angle	δ		Radian	rad
Number of turns	N, n		(Numeric)	
Time constant	τ, T	T^{-1}	Second	s
Damping coefficient	δ	T^{-1}	Neper/second	Np/s
Logarithmic decrement	Λ		(Numeric)	
Retarded scalar potential	V_r	$ML^2T^{-2}Q^{-1}$	Volt	V
Retarded vector potential	A_r	$MLT^{-1}Q^{-1}$	Weber/meter	Wb/m
Hysteresis coefficient	k_H		(Numeric)	
Eddy-current coefficient	k_e		(Numeric)	
Coupling coefficient	k, K		(Numeric)	
Leakage coefficient	σ		(Numeric)	

Sound and acoustics

Quantity	Symbol	Dimensions	Unit	Unit symbol
Velocity of sound	c	LT^{-1}	Meter/second	m/s
Sound energy flux	P	ML^2T^{-3}	Watt	W
Sound intensity	I	$M^{-1}LT^{-2}$	Watt/meter	W/m
Reverberation time	L_N	T	Second	s

Table 1-9. Symbols for Physical Quantities *(Concluded)*
(From IEEE Standards 280, by permission)

Name of quantity	Symbol	Dimensions	Mks Unit	Abbreviation for mks Unit
Specific acoustic impedance	Z_s	$M^{-2}LT^{-1}$	Newton-second/meter³	N·s/m³
Acoustic impedance	Z_a	$M^{-4}LT^{-1}$	Newton-second/meter⁵	N·s/m⁵
Mechanical impedance	Z_m	LT^{-1}	Newton-second/meter	N·s/m
Reflection factor	ρ	. . . .	(Numeric)	
Acoustic absorption factor	α	. . . .	(Numeric)	
Transmission factor	τ	. . . .	(Numeric)	
Dissipation factor	δ	. . . .	(Numeric)	
Loudness level	T	. . . .	Phon	

Light and radiation

Name of quantity	Symbol	Dimensions	Mks Unit	Abbreviation for mks Unit
Quantity of radiant energy	W, Q, Q_e	ML^2T^{-2}	Joule	J
Radiant flux	P, Φ, Φ_e	ML^2T^{-3}	Watt	W
Radiant exitance	M, M_e	$M^{-1}L^2T^{-3}$	Watt/meter²	W/m²
Irradiance	E, E_e	$M^{-1}L^2T^{-3}$	Watt/meter²	W/m²
Radiant intensity	I, I_e	ML^2T^{-3}	Watt/steradian	W/sr
Radiance	L, L_e	$M^{-1}L^2T^{-3}$	Watt/steradian-meter²	W/sr·m²
Quantity of light	Q, Q_v	$I_v T$	Lumen-second	lm·s
Luminous flux	Φ, Φ_v	I_v	Lumen	lm (cd·sr)
Luminous exitance	M, M_v	$I_v L^{-2}$	Lumen/meter²	lm/m²
Illuminance	E, E_v	$I_v L^{-1}$	Lux	lx
Luminous intensity	I, I_v	I_v	Candela	cd
Luminance	L, L_v	$I_v L^{-2}$	Candela/meter²	cd/m² (nit)
Luminous efficacy	$K(\lambda)$	$I_v M^{-1}L^{-2}T^3$	Lumen/watt	lm/W
Total efficacy	K, K_t	$I_v M^{-1}L^{-2}T^3$	Lumen/watt	lm/W
Absorptance	$\alpha(\lambda)$	. . .	(Numeric)	
Reflectance	$\rho(\lambda)$		(Numeric)	
Transmittance	$\tau(\lambda)$		(Numeric)	
Linear attenuation coefficient	. . .	L^{-1}	meter⁻¹	m⁻¹
Linear absorption coefficient	. . .	L^{-1}	Meter⁻¹	m⁻¹
Refractive index	n	. . .	(Numeric)	
Luminous efficacy	. . .	$I_v M^{-1}L^{-2}T^3$	Lumen/watt	lm/W

183. Bibliography
1. BESANCON, R. M. "Encyclopedia of Physics," Reinhold, New York, 1966.
2. CONDON, E. U., and H. ODISHAW "Handbook of Physics," McGraw-Hill, New York, 1958.
3. DARMOIS, G. "Matter, Electricity, and Energy," Walker, New York, 1964.
4. HAM, J. M., and G. R. SLEMON "Scientific Basis of Electrical Engineering," Wiley, New York, 1961.
5. HEMENWAY, C. L., R. W. HENRY, and M. CAULTON "Physical Electronics," 2d ed., Wiley, New York, 1967.
6. RICE, F. O., and E. TELLER "The Structure of Matter," Wiley, New York, 1949.
7. STANLEY, J. "Electric and Magnetic Properties of Metals" American Society of Metals, Metals Park, Ohio, 1963.
8. CORSON, D., and P. LORRAIN "Introduction to Electromagnetic Fields and Waves," Freeman, San Francisco, 1963.
9. HAYT, W. H. "Engineering Electromagnetics," McGraw-Hill, New York, 1958.
10. JACKSON, J. D. "Classical Electrodynamics," Wiley, New York, 1962.
11. JORDAN, E. C. "Electromagnetic Waves and Radiating Systems," Prentice-Hall, Englewood Cliffs, N.J., 1958.
12. ROTERS, E. C. "Electromagnetic Devices," Wiley, New York, 1941.
13. SCHELKUNOFF, S. A. "Electromagnetic Waves," Van Nostrand, Princeton, N.J., 1943.
14. ADLER, R. B., A. C. SMITH, and R. K. LONGINI "Introduction to Semiconductor Physics," Wiley, New York, 1964.
15. COBINE, J. D. "Gaseous Conductors," McGraw-Hill, New York, 1941; Dover, New York, 1958.
16. DeWITT, D., and A. L. ROSSOFF "Transistor Electronics," McGraw-Hill, New York, 1957.
17. SPANGENBERG, K. "Vacuum Tubes," McGraw-Hill, New York, 1948.
18. ADLER, R. B., L. J. CHU, and R. M. FANO "Electromagnetic Energy Transmission and Radiation," Wiley, New York, 1960.
19. ANDREWS, C. L. "Optics of the Electromagnetic Spectrum," Prentice-Hall, Englewood Cliffs, N.J., 1960.
20. BRONWELL, A. B., and R. E. BEAM "Theory and Application of Microwaves," McGraw-Hill, New York, 1947.
21. FANO, R. M., L. J. CHU, and R. B. ADLER "Electromagnetic Fields, Energy and Forces," M.I.T. Press, Cambridge, Mass., 1960.
22. BILLMEYER, F. W., and M. SALTZMAN "Principles of Color Technology," Wiley, New York, 1966.
23. BRAMSON, M. A. "Infrared Radiation: A Handbook for Application," Plenum, New York, 1968.
24. KOLLER, L. R. "Ultraviolet Radiation," Wiley, New York, 1965.
25. CANDLAND, D. K. "Psychology: The Experimental Approach," McGraw-Hill, New York, 1968.
26. SWETS, J. A. (ed.) "Signal Detection and Recognition by Human Observers," Wiley, New York, 1964.
27. OLSON, H. F. "Acoustical Engineering," Van Nostrand, Princeton, N.J., 1957.
28. ZEMLIN, W. R. "Speech and Hearing Science," Prentice-Hall, Englewood Cliffs, N.J., 1968.
29. HARRIS, C. M. "Handbook of Noise Control," McGraw-Hill, New York, 1957.
30. HUNT, F. V. "Electroacoustics," Harvard University Press, Cambridge, Mass., 1954.
31. RICHARDSON, E. G. "Technical Aspects of Sound," 2 vols., Elsevier, Amsterdam, 1953.
32. DAVSON, H. "The Eye," 4 vols., Academic, New York, 1962.
33. LeGRAND, Y. "Light, Color, and Vision," 2d ed., Wiley, New York, 1968.
34. RUBIN, M. L., and G. L. WALLS "Fundamentals of Visual Science," Charles C Thomas, Springfield, Ill., 1969.
35. IEEE Standard and American National Graphic Symbols for Electrical and Electronics Diagrams, *IEEE No. 315*, New York, 1971.

36. "IEEE Standard Dictionary of Electrical and Electronic Terms," IEEE and Wiley, New York, 1972.

37. Letter Symbols for Quantities Used for Electrical Science and Electrical Engineering, *IEEE No.* 280, ASME, 1968.

38. Letter Symbols for Units Used in Science and Technology, *IEEE No.* 260, ASME, 1967.

39. MECHTLY, E. A.   The International System of Units, *NASA SP* 7012, 1969.

40. Recommended Practice for Units in Published Scientific and Technical Work, *IEEE No.* 268, 1966.

41. HORTON, J. W.,   "Fundamentals of Sonar," U. S. Naval Institute, Annapolis, Md., 1957.

SECTION 2

MATHEMATICS: FORMULAS, DEFINITIONS, AND THEOREMS USED IN ELECTRONICS ENGINEERING

BY

GRANINO A. KORN Professor of Electrical Engineering, The University of Arizona
THERESA M. KORN

CONTENTS

Numbers refer to paragraphs

SECTION 2

MATHEMATICS: FORMULAS, DEFINITIONS, AND THEOREMS USED IN ELECTRONICS ENGINEERING

1. Introduction. This section contains a selection of reference material believed to be most useful for practicing electronics engineers. Topics generally understood by such workers, e.g., college algebra and plane and solid geometry, have been omitted; these are treated in the references listed in Par. **2-42**, notably in the "Mathematical Handbook for Scientists and Engineers," by G. A. Korn and T. M. Korn (McGraw-Hill, New York, 1968). The major portion of this section has been taken, by permission of the publisher, from that handbook and from the "Manual of Mathematics," by G. A. Korn and T. M. Korn (McGraw-Hill, New York, 1967).

DIFFERENTIAL CALCULUS

2. Derivatives and Differentiation. Let $y = f(x)$ be a real, single-valued function of the real variable x throughout a neighborhood of the point x. The *(first, first-order) derivative* or *(first-order) differential coefficient of $f(x)$ with respect to x at the point x is the limit*

$$\lim_{\Delta x \to 0} \frac{f(x + \Delta x) - f(x)}{\Delta x} \equiv \lim_{\Delta x \to 0} \frac{\Delta y}{\Delta x} \equiv \frac{dy}{dx} \equiv \frac{d}{dx} f(x) \equiv f'(x) \equiv y' \qquad (2\text{-}1)$$

The function $dy/dx \equiv f'(x)$ is a measure of the *rate of change of y with respect to x* at each point x where the limit (2-1) exists. On a graph of $y = f(x)$, $f'(x)$ corresponds to the *slope of the tangent*. Table 2-1 lists derivatives of frequently used functions.

3. Partial Derivatives. Let $y = f(x_1, x_2, \dots, x_n)$ be a real single-valued function of the real variables $x_1, x_2, \dots, x_n$ in a neighborhood of the point $(x_1, x_2, \dots, x_n)$. The *(first-order) partial derivative of $f(x_1, x_2, \dots, x_n)$ with respect to x_1 at the point $(x_1, x_2, \dots, x_n)$ is the* limit

$$\lim_{\Delta x_1 \to 0} \frac{f(x_1 + \Delta x_1, x_2, x_3, \dots, x_n) - f(x_1, x_2, \dots, x_n)}{\Delta x_1} \equiv \frac{\partial}{\partial x_1} f \equiv \frac{\partial y}{\partial x_1} \equiv f_{x_1}(x_1, x_2, \dots, x_n)$$

$$(2\text{-}2)$$

The function $\partial y / \partial x_1 \equiv (\partial y / \partial x_1)_{x_1, x_2, \dots, x_n} \equiv f_{x_1}(x_1, x_2, \dots, x_n)$ is a measure of the *rate of change of y with respect to x_1 for fixed values of the remaining independent variables* at each point $(x_1, x_2, \dots, x_n)$ where the limit (2-2) exists. The partial derivatives $\partial y / \partial x_2$, $\partial y / \partial x_3$, $\dots$, $\partial y / \partial x_n$ are defined in an analogous manner. *Each partial derivative $\partial y / \partial x_k$ can be found by differentiation of $f(x_1, x_2, \dots, x_n)$ with respect to x_k while the remaining $n - 1$ independent variables are regarded as constant parameters [partial differentiation of $f(x_1, x_2, \dots, x_n)$ with respect to x_k].*

4. Differentiation Rules. Table 2-2 summarizes the most important differentiation rules. The formulas of Table 2-2a and b apply to *partial differentiation* if $\partial / \partial x_k$ is substituted for d/dx in each case. Thus, if $u_i = u_i(x_1, x_2, \dots, x_n)(i = 1, 2, \dots, m)$,

$$\frac{\partial}{\partial x_k} f(u_1, u_2, \dots, u_m) = \sum_{i=1}^{m} \frac{\partial f}{\partial u_i} \frac{\partial u_i}{\partial x_k} \qquad k = 1, 2, \dots, n \qquad (2\text{-}3)$$

2-2

Table 2-1 Derivatives of Frequently Used Functions

$f(x)$	$f'(x)$	$f^{(r)}(x) = \dfrac{d^r}{dx^r}[f(x)]$
x^a	ax^{a-1}	$a(a-1)(a-2)\cdots(a-r+1)x^{a-r}$
e^x	e^x	e^x
a^x	$a^x \log_e a$	$a^x(\log_e a)^r$
$\log_e x$	$\dfrac{1}{x}$	$(-1)^{r-1}(r-1)!\dfrac{1}{x^r}$
$\log_a x$	$\dfrac{1}{x}\log_a e$	$(-1)^{r-1}(r-1)!\dfrac{1}{x^r}\log_a e$
$\sin x$	$\cos x$	$\sin\left(x + \dfrac{\pi r}{2}\right)$
$\cos x$	$-\sin x$	$\cos\left(x + \dfrac{\pi r}{2}\right)$

$f(x)$	$f'(x)$	$f(x)$	$f'(x)$
$\tan x$	$\dfrac{1}{\cos^2 x}$	$\arcsin x$	$\dfrac{1}{\sqrt{1-x^2}}$
$\cot x$	$-\dfrac{1}{\sin^2 x}$	$\arccos x$	$-\dfrac{1}{\sqrt{1-x^2}}$
$\sec x$	$\dfrac{\sin x}{\cos^2 x}$	$\arctan x$	$\dfrac{1}{1+x^2}$
$\operatorname{cosec} x$	$-\dfrac{\cos x}{\sin^2 x}$	$\operatorname{arccot} x$	$-\dfrac{1}{1+x^2}$
$\sinh x$	$\cosh x$	$\sinh^{-1} x$	$\dfrac{1}{\sqrt{x^2+1}}$
$\cosh x$	$\sinh x$	$\cosh^{-1} x$	$\dfrac{1}{\sqrt{x^2-1}}$
$\tanh x$	$\dfrac{1}{\cosh^2 x}$	$\tanh^{-1} x$	$\dfrac{1}{1-x^2}$
$\coth x$	$-\dfrac{1}{\sinh^2 x}$	$\coth^{-1} x$	$\dfrac{1}{1-x^2}$
$\operatorname{vers} x$	$\sin x$	x^x	$x^x(1 + \log_e x)$

Table 2-2   Differentiation rules*

a. Basic rules

$$\frac{d}{dx}f[u_1(x), u_2(x), \ldots, u_m(x)] = \frac{\partial f}{\partial u_1}\frac{du_1}{dx} + \frac{\partial f}{\partial u_2}\frac{du_2}{dx} + \cdots + \frac{\partial f}{\partial u_m}\frac{du_m}{dx}$$

$$\frac{d}{dx}f[u(x)] = \frac{df}{du}\frac{du}{dx} \qquad \frac{d^2}{dx^2}f[u(x)] = \frac{d^2f}{du^2}\left(\frac{du}{dx}\right)^2 + \frac{df}{du}\frac{d^2u}{dx^2}$$

b. Sums, products, and quotients; logarithmic differentiation

$$\frac{d}{dx}[u(x) + v(x)] = \frac{du}{dx} + \frac{dv}{dx} \qquad \frac{d}{dx}[\alpha u(x)] = \alpha\frac{du}{dx}$$

$$\frac{d}{dx}[u(x)v(x)] = v\frac{du}{dx} + u\frac{dv}{dx} \qquad \frac{d}{dx}\left[\frac{u(x)}{v(x)}\right] = \frac{1}{v^2}\left(v\frac{du}{dx} - u\frac{dv}{dx}\right)[v(x) \neq 0]$$

$$\frac{d}{dx}\log_e y(x) = \frac{y'(x)}{y(x)} \qquad \left[\text{logarithmic derivative of } y(x)\right]$$

NOTE: To differentiate functions of the form $y = \dfrac{u_1(x)u_2(x)\cdots}{v_1(x)v_2(x)\cdots}$, it may be convenient to find the logarithmic derivative first.

$$\frac{d^r}{dx^r}(\alpha u + \beta v) = \alpha\frac{d^r u}{dx^r} + \beta\frac{d^r v}{dx^r} \qquad \frac{d^r}{dx^r}(uv) = \sum_{k=0}^{r}\binom{r}{k}\frac{d^{r-k}u}{dx^{r-k}}\frac{d^k v}{dx^k}$$

c. Inverse function given. If $y = y(x)$ has the unique inverse function $x = x(y)$, and $dx/dy \neq 0$,

$$\frac{dy}{dx} = \left(\frac{dx}{dy}\right)^{-1} \qquad \frac{d^2 y}{dx^2} = -\frac{d^2 x}{dy^2}\bigg/\left(\frac{dx}{dy}\right)^3$$

d. Implicit functions. If $y = y(x)$ is given implicitly in terms of a suitably differentiable relation $F(x,y) = 0$, where $F_y \neq 0$,

$$\frac{dy}{dx} = -\frac{F_x}{F_y} \qquad \frac{d^2 y}{dx^2} = -\frac{1}{F_y^3}(F_{xx}F_y^2 - 2F_{xy}F_x F_y + F_{yy}F_x^2)$$

e. Function given in terms of a parameter t. Given $x = x(t)$,

$$y = y(t) \text{ and } \dot{x}(t) \equiv \frac{dx}{dt} \neq 0, \quad \dot{y}(t) \equiv \frac{dy}{dt}, \quad \ddot{x}(t) \equiv \frac{d^2 x}{dt^2}, \quad \ddot{y}(t) \equiv \frac{d^2 y}{dt^2},$$

$$\frac{dy}{dx} = \frac{\dot{y}(t)}{\dot{x}(t)} \qquad \frac{d^2 y}{dx^2} = \frac{\dot{x}(t)\ddot{y}(t) - \ddot{x}(t)\dot{y}(t)}{[\dot{x}(t)]^3}$$

* Existence of continuous derivatives is assumed in each case

INTEGRALS AND INTEGRATION

5. Definite Integrals (Riemann Integrals). A real function $f(x)$ bounded on the bounded closed interval $[a,b]$ is *integrable over* (a,b) *in the sense of Riemann* if and only if the sum $\sum_{i=1}^{m} f(\xi_i)(x_i - x_{i-1})$ tends to a unique finite limit I for every sequence of partitions $a = x_0 < \xi_1 < x_1 < \xi_2 < x_2 \cdots < \xi_m < x_m = b$ as max $|x_i - x_{i-1}| \to 0$.   In this case

$$I = \lim_{\max|x_i - x_{i-1}| \to 0} \sum_{i=1}^{m} f(\xi_i)(x_i - x_{i-1}) = \int_a^b f(x)\, dx \qquad (2\text{-}4)$$

is the *definite integral of $f(x)$ over (a,b) in the sense of Riemann (Riemann integral)*. $f(x)$ is called the *integrand*; a and b are the *limits of integration*. Table 2-3 summarizes important properties of definite integrals.

$\int_a^b f(x)\, dx$ represents the *area* bounded by the curve $y = f(x)$ and the x axis between the lines $x = a$ and $x = b$; areas below the x axis are represented by negative numbers.

6. Indefinite Integrals. A given single-valued function $f(x)$ has an *indefinite integral* $F(x)$ in $[a,b]$ if and only if there exists a function $F(x)$ such that $F'(x) = f(x)$ in $[a,b]$. In this case $F(x)$ is uniquely defined in $[a,b]$ except for an arbitrary additive constant C (*constant of integration*); one writes

$$F(x) = \int f(x)\, dx + C \qquad a \leq x \leq b \qquad (2\text{-}5)$$

Note that $F(x) - F(a) \equiv F(x)]_a^x$ is uniquely defined for $a \leq x \leq b$.

7. Fundamental Theorem of the Integral Calculus.   *If $f(x)$ is single-valued, bounded, and integrable on $[a,b]$ and there exists a function $F(x)$ such that $F'(x) = f(x)$ for $a \leq x \leq b$, then*

$$\int_a^x f(\xi)\, d\xi = F(x)]_a^x = F(x) - F(a) \qquad a \leq x \leq b \tag{2-6}$$

In particular, *if $f(x)$ is continuous in $[a, b]$,*

$$\frac{d}{dx}\int_a^x f(\xi)\, d\xi = f(x) \qquad a \leq x \leq b$$

and Eq. (2-6) applies.

NOTE: The fundamental theorem of the integral calculus enables one (1) to evaluate definite integrals by reversing the process of differentiation and (2) to solve differential equations by numerical evaluation of definite integrals.

8. Integration of Polynomials

$$\int (a_n + a_{n-1}x + a_{n-2}x^2 + \cdots + a_0 x^n)\, dx$$

$$\equiv a_n x + \tfrac{1}{2}a_{n-1}x^2 + \tfrac{1}{3}a_{n-2}x^3 + \cdots + \frac{1}{n+1}a_0 x^{n+1} + C$$

Table 2-3   Properties of Integrals

a. Elementary properties.   *If the integrals exist,*

$$\int_a^b f(x)\, dx = -\int_b^a f(x)\, dx \qquad \int_a^b f(x)\, dx = \int_a^c f(x)\, dx + \int_c^b f(x)\, dx$$

$$\int_a^b [u(x) + v(x)]\, dx = \int_a^b u(x)\, dx + \int_a^b v(x)\, dx \qquad \int_a^b \alpha u(x)\, dx = \alpha \int_a^b u(x)\, dx$$

b. Integration by parts.   *If $u(x)$ and $v(x)$ are differentiable for $a \leq x \leq b$, and if the integrals exist,*

$$\int_a^b u(x)v'(x)\, dx = u(x)v(x)\Big]_a^b - \int_a^b v(x)u'(x)\, dx$$

or

$$\int_a^b u\, dv = uv\Big]_a^b - \int_a^b v\, du$$

c. Change of Variable (integration by substitution).   *If $u = u(x)$ and its inverse function $x = x(u)$ are single-valued and continuously differentiable for $a \leq x \leq b$, and if the integral exists,*

$$\int_a^b f(x)\, dx = \int_{u(a)}^{u(b)} f[(x(u)]\frac{dx}{du}\, du = \int_{u(a)}^{u(b)} f[x(u)]\left(\frac{du}{dx}\right)^{-1} du$$

d. Differentiation with respect to a parameter.   *If $f(x,\lambda)$, $u(\lambda)$, and $v(\lambda)$ are continuously differentiable with respect to λ,*

$$\frac{\partial}{\partial\lambda}\int_a^b f(x,\lambda)\, dx = \int_a^b \frac{\partial}{\partial\lambda}f(x,\lambda)\, dx$$

$$\frac{\partial}{\partial\lambda}\int_{u(\lambda)}^{v(\lambda)} f(x,\lambda)\, dx = \int_{u(\lambda)}^{v(\lambda)} \frac{\partial}{\partial\lambda}f(x,\lambda)\, dx + f(v,\lambda)\frac{\partial v}{\partial\lambda} - f(u,\lambda)\frac{\partial u}{\partial\lambda} \qquad \text{(Leibnitz' rule)}$$

provided that the integrals exist and, in the case of improper integrals, converge uniformly in a neighborhood of the point λ.

The second case can often be reduced to the first by a suitable change of variables. Note also

$$\frac{\partial}{\partial\lambda}\int_a^\lambda f(x,\lambda)\, dx = \frac{1}{\lambda - a}\int_a^\lambda \left[f(x,\lambda) + (\lambda - a)\frac{\partial f}{\partial\lambda} + (x - a)\frac{\partial f}{\partial x}\right] dx$$

e. Inequalities.   *If the integrals exist,*

$$f(x) \leq g(x) \text{ in } (a, b) \text{ implies } \int_a^b f(x)\, dx \leq \int_a^b g(x)\, dx$$

If $|f(x)| \leq M$ on the bounded interval (a,b), the existence of $\int_a^b f(x)\, dx$ implies the existence of $\int_a^b |f(x)|\, dx$, and

$$\left|\int_a^b f(x)\, dx\right| \leq \int_a^b |f(x)|\, dx \leq M(b - a)$$

9. Integration of Rational Fractions.   Every rational integrand reduces to a sum of a polynomial and a set of partial fractions.   Partial-fraction terms are integrated with the aid of the following formulas:

$$\int \frac{dx}{(x - x_1)^m} \equiv \begin{cases} -\dfrac{1}{(m - 1)(x - x_1)^{m-1}} + C & m \neq 1 \\ \log_e(x - x_1) + C & m = 1 \end{cases}$$

$$\int \frac{dx}{[(x - a)^2 + \omega^2]} \equiv \frac{1}{\omega} \arctan \frac{x - a}{\omega} + C$$

$$\int \frac{dx}{[(x - a)^2 + \omega^2]^{m+1}} \equiv \frac{x - a}{2\,m\omega^2[(x - a)^2 + \omega^2]^m} + \frac{2m - 1}{2\,m\omega^2} \int \frac{dx}{[(x - a)^2 + \omega^2]^m}$$

$$\int \frac{x\,dx}{[(x - a)^2 + \omega^2]^{m+1}} \equiv \frac{a(x - a) - \omega^2}{2\,m\omega^2[(x - a)^2 + \omega^2]^m} + \frac{(2m - 1)a}{2\,m\omega^2} \int \frac{dx}{[(x - a)^2 + \omega^2]^m}$$

10. Integrands Reducible to Rational Functions by a Change of Variables   (Table 2-3c).
1.   *If the integrand $f(x)$ is a rational function of* sin x *and* cos x, introduce $u = \tan(x/2)$, so that

$$\sin x = \frac{2u}{1 + u^2} \qquad \cos x = \frac{1 - u^2}{1 + u^2} \qquad dx = \frac{2\,du}{1 + u^2}$$

2.   *If the integrand $f(x)$ is a rational function of* sinh x *and* cosh x, introduce $u = \tanh(x/2)$, so that

$$\sinh x = \frac{2u}{1 - u^2} \qquad \cosh x = \frac{1 + u^2}{1 - u^2} \qquad dx = \frac{2\,du}{1 - u^2}$$

NOTE: If $f(x)$ is a rational function of $\sin^2 x$, $\cos^2 x$, $\sin x \cos x$, and $\tan x$ (or of the corresponding hyperbolic functions), one simplifies the calculation by first introducing $v = x/2$, so that $u = \tan v$(or $u = \tanh v$).
3.   *If the integrand $f(x)$ is a rational function of x and either $\sqrt{1 - x^2}$ or $\sqrt{x^2 - 1}$*, reduce the problem to case 1 or 2 by the respective substitutions $x = \cos v$ or $x = \cosh v$.
4.   *If the integrand $f(x)$ is a rational function of x and $\sqrt{x^2 + 1}$*, introduce $u = x + \sqrt{x^2 + 1}$, so that

$$x = \tfrac{1}{2}\left(u - \frac{1}{u}\right) \qquad \sqrt{x^2 + 1} = \tfrac{1}{2}\left(u + \frac{1}{u}\right) \qquad dx = \tfrac{1}{2}\left(1 + \frac{1}{u^2}\right)du$$

5.   *If the integrand $f(x)$ is a rational function of x and $\sqrt{ax^2 + bx + c}$*, reduce the problem to case 3 ($b^2 - 4ac < 0$) or to case 4 ($b^2 - 4ac > 0$) through the substitution

$$v = \frac{2ax + b}{\sqrt{|4ac - b^2|}} \qquad x = \frac{v\sqrt{|4ac - b^2|} - b}{2a}$$

6.   *If the integrand $f(x)$ is a rational function of x and $u = \sqrt{(ax + b)/(cx + d)}$*, introduce u as a new variable.
7.   *If the integrand $f(x)$ is a rational function of x, $\sqrt{ax + b}$, and $\sqrt{cx + d}$*, introduce $u = \sqrt{ax + b}$ as a new variable.
Many other substitution methods apply in special cases.   Note that the integrals may not be real for all values of x.
Integrands of the form $x^n e^{ax}$, $x^n \log_e x$, $x^n \sin x$, $x^n \cos x(n \neq -1)$; $\sin^m x \cos^n x(n + m \neq 0)$; $e^{ax} \sin^n x$, $e^{ax} \cos^n x$ yield to repeated *integration by parts* (Table 2-3b).

11. Some Frequently Used Limits (Values of Indeterminate Forms)

$$\lim_{n\to\infty}\left(1+\frac{1}{n}\right)^n = e \approx 2.71828 \qquad n = 1, 2, \ldots \qquad \lim_{x\to 0}(1+x)^{1/x} = e$$

$$\lim_{x\to 0}\frac{c^x - 1}{x} = \log_e c \qquad \lim_{x\to 0} x^x = 1$$

$$\lim_{x\to 0}\frac{\sin x}{x} = \lim_{x\to 0}\frac{\tan x}{x} = \lim_{x\to 0}\frac{\sinh x}{x} = \lim_{x\to 0}\frac{\tanh x}{x} = 1$$

$$\lim_{x\to 0}\frac{\sin \omega x}{x} = \omega \qquad -\infty < \omega < \infty$$

$$\lim_{x\to 0} x^a \log_e x = \lim_{x\to 0} x^{-a}\log_e x = \lim_{x\to\infty} x^a e^{-x} = 0 \qquad a > 0$$

FOURIER SERIES AND FOURIER INTEGRALS

12. Fourier Series.

$$\tfrac{1}{2}a_0 + \sum_{k=1}^{\infty}(a_k \cos k\omega_0 t + b_k \sin k\omega_0 t) \equiv \sum_{k=-\infty}^{\infty} c_k e^{ik\omega_0 t} \qquad \omega_0 = \frac{2\pi}{T}$$

with

$$a_k = \frac{2}{T}\int_{-T/2}^{T/2} f(\tau)\cos k\omega_0\tau\, d\tau \qquad b_k = \frac{2}{T}\int_{-T/2}^{T/2} f(\tau)\sin k\omega_0\tau\, d\tau$$

$$c_k = c_{-k}^* = \tfrac{1}{2}(a_k - ib_k) = \frac{1}{T}\int_{-T/2}^{T/2} f(\tau)e^{-ik\omega_0 t}\, d\tau \qquad \omega_0 = \frac{2\pi}{T}$$

$$k = 0, 1, 2, \ldots$$

13. Properties of Fourier Transforms. Let

$$\mathcal{F}[f(t)] \equiv \int_{-\infty}^{\infty} f(t)e^{-2\pi i\nu t}\, dt \equiv c(\nu) \equiv F_F(i\omega) \equiv \sqrt{2\pi}\,C(\omega) \qquad \omega = 2\pi\nu$$

$$f(t) \equiv \int_{-\infty}^{\infty} c(\nu)e^{2\pi i\nu t}\, d\nu \equiv \int_{-\infty}^{\infty} F_F(i\omega)e^{i\omega t}\frac{d\omega}{2\pi} \equiv \frac{1}{\sqrt{2\pi}}\int_{-\infty}^{\infty} C(\omega)e^{i\omega t}\, d\omega$$

and assume that the Fourier transforms in question exist.

a. $\mathcal{F}[\alpha f_1(t) + \beta f_2(t)] \equiv \alpha\mathcal{F}[f_1(t)] + \beta\mathcal{F}[f_2(t)]$ linearity

$$\mathcal{F}[f^*(t)] \equiv c^*(-\nu) \equiv F_F^*(-i\omega)$$

$\mathcal{F}[f(\alpha t)] \equiv \dfrac{1}{\alpha}c\left(\dfrac{\nu}{\alpha}\right) \equiv \dfrac{1}{\alpha}F_F\left(\dfrac{i\omega}{\alpha}\right)$ change of scale, similarity theorem

$\mathcal{F}[f(t+\tau)] \equiv e^{2\pi i\nu\tau}c(\nu) \equiv e^{i\omega\tau}F_F(i\omega)$ shift theorem

b. Continuity Theorem. $\mathcal{F}[f(t,\alpha)] \to \mathcal{F}[f(t)]$ as $\alpha \to a$ implies $f(t,\alpha) \to f(t)$ wherever $f(t)$ is continuous. Analogous theorems apply to Fourier cosine and sine transforms.

c. Borel's Convolution Theorem. $\mathcal{F}[f_1(t)]\mathcal{F}[f_2(t)] \equiv \mathcal{F}[f_1(t) * f_2(t)]$, where

$$f_1(t) * f_2(t) \equiv \int_{-\infty}^{\infty} f_1(\tau)f_2(t-\tau)\, d\tau \equiv \int_{-\infty}^{\infty} f_1(t-\tau)f_2(\tau)\, d\tau$$

$$\mathcal{F}[f_1(t)f_2(t)] \equiv \int_{-\infty}^{\infty} c_1(\lambda)c_2(\nu-\lambda)\, d\lambda \equiv \int_{-\infty}^{\infty} c_1(\nu-\lambda)c_2(\lambda)\, d\lambda$$

$$\equiv \int_{-\infty}^{\infty} F_{F1}(i\lambda)F_{F2}[i(\omega-\lambda)]\frac{d\lambda}{2\pi}$$

$$\equiv \int_{-8}^{\infty} F_{F1}[i(\omega-\lambda)]F_{F2}(i\lambda)\frac{d\lambda}{2\pi}$$

Table 2-4 Fourier Coefficients and Mean-Square Values of Periodic Functions $\left(\operatorname{sinc} x \equiv \dfrac{\sin \pi x}{\pi x}\right)$

Periodic function, $f(t) = f(t+T)$	Fourier coefficients (for phasing as shown in diagram)	Average value $\langle f \rangle = \dfrac{a_0}{2}$	Mean-square value $\langle f^2 \rangle$
1. Rectangular pulses	$a_n = 2A\dfrac{T_0}{T}\operatorname{sinc}\dfrac{nT_0}{T}$ $b_n = 0$	$A\dfrac{T_0}{T}$	$A^2\dfrac{T_0}{T}$
2. Symmetrical triangular pulses	$a_n = A\dfrac{T_0}{T}\operatorname{sinc}^2\dfrac{nT_0}{2T}$ $b_n = 0$	$A\dfrac{T_0}{2T}$	$A^2\dfrac{T_0}{3T}$
3. Symmetrical trapezoidal pulses	$a_n = 2A\dfrac{T_0+T_1}{T}\operatorname{sinc}\dfrac{nT_1}{T}\operatorname{sinc}\dfrac{n(T_0+T_1)}{T}$ $b_n = 0$	$A\dfrac{T_0+T_1}{T}$	$A^2\dfrac{3T_0+2T_1}{3T}$
4. Half-sine pulses*†	$a_n = A\dfrac{T_0}{T}\left\{\operatorname{sinc}\left[\dfrac{1}{2}\left(\dfrac{2nT_0}{T}-1\right)\right] + \operatorname{sinc}\left[\dfrac{1}{2}\left(\dfrac{2nT_0}{T}+1\right)\right]\right\}$ $b_n = 0$	$\dfrac{2}{\pi}A\dfrac{T_0}{T}$	$A^2\dfrac{T_0}{2T}$
5. Clipped sinusoid	$a_n = \dfrac{A_0T_0}{T}\left\{\operatorname{sinc}\left[(n-1)\dfrac{T_0}{T}\right]\right.$ $\left.+\operatorname{sinc}\left[(n+1)\dfrac{T_0}{T}\right] - 2\cos\dfrac{\pi T_0}{T}\operatorname{sinc}\dfrac{nT_0}{T}\right\}$	$\dfrac{1}{\pi}A_0\left(\sin\dfrac{\pi T_0}{T}\right.$ $\left.-\dfrac{\pi T_0}{T}\cos\dfrac{\pi T_0}{T}\right)$	$\dfrac{1}{2\pi}A_0^2\left(\dfrac{\pi T_0}{T}-\dfrac{3}{2}\sin\dfrac{2\pi T_0}{T}\right.$ $\left.+\dfrac{2\pi T_0}{T}\cos^2\dfrac{\pi T_0}{T}\right)$
6. Triangular waveform	$a_n = 0$ $b_n = -\dfrac{A}{n\pi}\qquad n = 1, 2, \ldots$	$\dfrac{A}{2}$	$\dfrac{A^2}{3}$

*For $T_0 = \dfrac{T}{2} = \dfrac{\pi}{\omega}$, $f(t) = \dfrac{2}{\pi}A\left(\frac{1}{2} + \dfrac{\pi}{4}\cos\omega t + \frac{1}{3}\cos 2\omega t - \frac{1}{15}\cos 4\omega t + \frac{1}{35}\cos 6\omega t - \cdots\right)$ half-wave-rectified sinusoid.

†For $T_0 = T = \dfrac{2\pi}{\omega}$, $f(t) = -\dfrac{4}{\pi}A\left(\frac{1}{2} + \frac{1}{3}\cos 2\omega t - \frac{1}{15}\cos 4\omega t + \frac{1}{35}\cos 6\omega t - \cdots\right)$ full-wave-rectified sinusoid.

2-8

Table 2-5 Fourier Transform Pairs*

	$f(t) = \int_{-\infty}^{\infty} F(j\omega) e^{j\omega t} \frac{d\omega}{2\pi}$	$F(j\omega) = \int_{-\infty}^{\infty} f(t) e^{-j\omega t}\, dt$	
	$\text{rect}\,\dfrac{t}{T} = \begin{cases} 1 & (\|t\| < T/2) \\ 0 & (\|t\| > T/2) \end{cases}$	$T\,\text{sinc}\,\dfrac{\omega T}{2\pi} \equiv T\dfrac{\sin\frac{\omega T}{2}}{\frac{\omega T}{2}}$	
	$\text{sinc}\,\dfrac{t}{T} \equiv \dfrac{\sin\frac{\pi t}{T}}{\frac{\pi t}{T}}$	$T\,\text{rect}\,\dfrac{\omega T}{2\pi} = \begin{cases} 0 & \left(\|\omega\| < \frac{\pi}{T}\right) \\ T & \left(\|\omega\| > \frac{\pi}{T}\right) \end{cases}$	
	$\begin{cases} 1 - \dfrac{\|t\|}{T} & (\|t\| < T) \\ 0 & (\|t\| \ge T) \end{cases}$	$T\,\text{sinc}^2\,\dfrac{\omega T}{2\pi} \equiv T\left(\dfrac{\sin\frac{\omega T}{2}}{\frac{\omega T}{2}}\right)^2$	
	$e^{-\frac{\|t\|}{T}}$	$\dfrac{2T}{(\omega T)^2 + 1}$	
	$e^{-\frac{1}{2}\left(\frac{t}{T}\right)^2}$	$\sqrt{2\pi}\;T e^{-\frac{1}{2}(\omega T)^2}$	
	$\delta(t - T)$	$e^{-j\omega T}$	(Complex)
	$\cos \omega_0 t$	$\pi\left[\delta(\omega - \omega_0) + \delta(\omega + \omega_0)\right]$	
	$\sin \omega_0 t$	$\dfrac{\pi}{j}\left[\delta(\omega - \omega_0) - \delta(\omega + \omega_0)\right]$	(Imaginary)
	$\displaystyle\sum_{k=-\infty}^{\infty} \delta(t - kT)$ $\equiv \dfrac{1}{T}\displaystyle\sum_{j=-\infty}^{\infty} e^{2\pi i j \frac{t}{T}}$	$\dfrac{2\pi}{T}\displaystyle\sum_{i=-\infty}^{\infty} \delta\left(\omega - \dfrac{2\pi i}{T}\right)$ $\equiv \displaystyle\sum_{k=-\infty}^{\infty} e^{jk\omega T}$	

* Reprinted from G. A. Korn, "Basic Tables in Electrical Engineering," McGraw-Hill, New York, 1965, by permission.

d. Parseval's Theorem. If $\int_{-\infty}^{\infty} |f_1(t)|^2 \, dt$ and $\int_{-\infty}^{\infty} |f_2(t)|^2 \, dt$ exist, then

$$\int_{-\infty}^{\infty} \mathscr{F}^*[f_1(t)]\mathscr{F}[f_2(t)] \, d\nu = \int_{-\infty}^{\infty} f_1^*(t) f_2(t) \, dt$$

e. Modulation Theorem

$$\mathscr{F}[f(t)e^{i\omega_0 t}] \equiv F_F[i(\omega - \omega_0)] = c(\nu - \nu_0)$$

$$\mathscr{F}[f(t)\cos \omega_0 t] \equiv \tfrac{1}{2}\{F_F[i(\omega - \omega_0)] + F_F[i(\omega + \omega_0)]\} \equiv \tfrac{1}{2}[c(\nu - \nu_0) + c(\nu + \nu_0)]$$

$$\mathscr{F}[f(t)\sin \omega_0 t] \equiv \frac{1}{2i}\{F_F[i(\omega - \omega_0)] - F_F[i(\omega + \omega_0)]\} \equiv \frac{1}{2i}[c(\nu - \nu_0) - c(\nu + \nu_0)]$$

f. Differentiation Theorem

$$\mathscr{F}[f^{(r)}(t)] = (2\pi i \nu)^r \mathscr{F}[f(t)] \qquad r = 0, 1, 2, \ldots$$

provided that $f^{(r)}(t)$ exists for all t, and that all derivatives of lesser order vanish as $|t| \to \infty$.

VECTOR ALGEBRA

14. Vector Addition and Multiplication of Vectors by (Real) Scalars.   Vectors, for example, **a**, **b**, are shown in boldface type.

$$\mathbf{a} + \mathbf{b} = \mathbf{b} + \mathbf{a}$$

$$\mathbf{a} + (\mathbf{b} + \mathbf{c}) = (\mathbf{a} + \mathbf{b}) + \mathbf{c} = \mathbf{a} + \mathbf{b} + \mathbf{c}$$

$$\alpha(\beta\mathbf{a}) = (\alpha\beta)\mathbf{a} \qquad (\alpha + \beta)\mathbf{a} = \alpha\mathbf{a} + \beta\mathbf{a}$$

$$\alpha(\mathbf{a} + \mathbf{b}) = \alpha\mathbf{a} + \alpha\mathbf{b}$$

$$(1)\mathbf{a} = \mathbf{a} \qquad (-1)\mathbf{a} = -\mathbf{a} \qquad (0)\mathbf{a} = \mathbf{0}$$

$$\mathbf{a} - \mathbf{a} = \mathbf{0} \qquad \mathbf{a} + \mathbf{0} = \mathbf{a}$$

15. Scalar Product (Dot Product, Inner Product).   The *scalar product (dot product, inner product)* **a · b** [alternative notation (ab)] of two euclidean vectors **a** and **b** is the scalar **a · b** $= |\mathbf{a}||\mathbf{b}|\cos \gamma$ where γ is the angle $\sphericalangle$ **a**, **b**.

$$\mathbf{a} \cdot \mathbf{b} = \mathbf{b} \cdot \mathbf{a} \qquad \mathbf{a} \cdot (\mathbf{b} + \mathbf{c}) = \mathbf{a} \cdot \mathbf{b} + \mathbf{a} \cdot \mathbf{c} \qquad (\alpha\mathbf{a}) \cdot \mathbf{b} = \alpha(\mathbf{a} \cdot \mathbf{b})$$

$$\mathbf{a} \cdot \mathbf{a} = a^2 = |\mathbf{a}^2| \geq 0 \qquad |\mathbf{a} \cdot \mathbf{b}| \leq |\mathbf{a}||\mathbf{b}| \qquad \cos \gamma = \frac{\mathbf{a} \cdot \mathbf{b}}{\sqrt{\mathbf{a}^2\mathbf{b}^2}}$$

$$\mathbf{i} \cdot \mathbf{i} = \mathbf{j} \cdot \mathbf{j} = \mathbf{k} \cdot \mathbf{k} = 1 \qquad \mathbf{i} \cdot \mathbf{j} = \mathbf{j} \cdot \mathbf{k} = \mathbf{k} \cdot \mathbf{i} = 0$$

$$\mathbf{a} \cdot \mathbf{b} = (a_x\mathbf{i} + a_y\mathbf{j} + a_z\mathbf{k}) \cdot (b_x\mathbf{i} + b_y\mathbf{j} + b_z\mathbf{k}) = a_x b_x + a_y b_y + a_z b_z$$

$$a_x = \mathbf{a} \cdot \mathbf{i} \qquad a_y = \mathbf{a} \cdot \mathbf{j} \qquad a_z = \mathbf{a} \cdot \mathbf{k}$$

16. Vector (Cross) Product.   The *vector (cross) product* **a** × **b** (alternative notation [ab]) of two vectors **a** and **b** is the vector of magnitude

$$|\mathbf{a} \times \mathbf{b}| = |\mathbf{a}||\mathbf{b}|\sin \gamma$$

where γ is the angle $\sphericalangle$ **a**, **b**.   The direction of the vector is perpendicular to both **a** and **b** and such that the axial motion of a right-handed screw turning **a** into **b** is in the direction of **a** × **b**.

17. Scalar Triple Product (Box Product)

$$\mathbf{a} \cdot (\mathbf{b} \times \mathbf{c}) \equiv [\mathbf{abc}] = [\mathbf{bca}] = [\mathbf{cab}] = -[\mathbf{bac}] = -[\mathbf{cba}] = -[\mathbf{acb}]$$

$$[\mathbf{abc}]^2 = [(\mathbf{a} \times \mathbf{b})(\mathbf{b} \times \mathbf{c})(\mathbf{c} \times \mathbf{a})] = a^2 b^2 c^2 - a^2(\mathbf{b} \cdot \mathbf{c})^2$$

$$- b^2(\mathbf{a} \cdot \mathbf{c})^2 - c^2(\mathbf{a} \cdot \mathbf{b})^2 + 2(\mathbf{a} \cdot \mathbf{b})(\mathbf{b} \cdot \mathbf{c})(\mathbf{a} \cdot \mathbf{c})$$

$$= \begin{vmatrix} a \cdot a & a \cdot b & a \cdot c \\ b \cdot a & b \cdot b & b \cdot c \\ c \cdot a & c \cdot b & c \cdot c \end{vmatrix} \qquad \text{Gram's determinant}$$

$$[abc][def] = \begin{vmatrix} a \cdot d & a \cdot e & a \cdot f \\ b \cdot d & b \cdot e & b \cdot f \\ c \cdot d & c \cdot e & c \cdot f \end{vmatrix}$$

$$[abc] = \begin{vmatrix} a_x & b_x & c_x \\ a_y & b_y & c_y \\ a_z & b_z & c_z \end{vmatrix} \qquad > 0 \text{ if } \mathbf{a}, \mathbf{b}, \mathbf{c} \text{ are directed like righthanded cartesian axes}$$

Table 2-6   Relations Involving Vector (Cross) Products

a. Basic Relations

$$\mathbf{a} \times \mathbf{b} = -(\mathbf{b} \times \mathbf{a})$$

$$\mathbf{a} \times \mathbf{a} = 0 \qquad \mathbf{a} \cdot (\mathbf{a} \times \mathbf{b}) = \mathbf{b} \cdot (\mathbf{a} \times \mathbf{b}) = 0$$

$$(\alpha \mathbf{a}) \times \mathbf{b} = \alpha(\mathbf{a} \times \mathbf{b}) \qquad \mathbf{a} \times (\mathbf{b} + \mathbf{c}) = \mathbf{a} \times \mathbf{b} + \mathbf{a} \times \mathbf{c}$$

$$[(\alpha + \beta)\mathbf{a}] \times \mathbf{b} = (\alpha + \beta)(\mathbf{a} \times \mathbf{b}) = \alpha(\mathbf{a} \times \mathbf{b}) + \beta(\mathbf{a} \times \mathbf{b})$$

b. In terms of any basis $\mathbf{e}_1, \mathbf{e}_2, \mathbf{e}_3$

$$\mathbf{a} = \alpha_1 \mathbf{e}_1 + \alpha_2 \mathbf{e}_2 + \alpha_3 \mathbf{e}_3 \qquad \mathbf{b} = \beta_1 \mathbf{e}_1 + \beta_2 \mathbf{e}_2 + \beta_3 \mathbf{e}_3$$

$$\mathbf{a} \times \mathbf{b} = \begin{vmatrix} \mathbf{e}_2 \times \mathbf{e}_3 & \alpha_1 & \beta_1 \\ \mathbf{e}_3 \times \mathbf{e}_1 & \alpha_2 & \beta_2 \\ \mathbf{e}_1 \times \mathbf{e}_2 & \alpha_3 & \beta_3 \end{vmatrix}$$

c. In terms of right-handed rectangular cartesian components

$$\mathbf{i} \times \mathbf{i} = \mathbf{j} \times \mathbf{j} = \mathbf{k} \times \mathbf{k} = 0 \qquad \mathbf{i} \times \mathbf{j} = \mathbf{k} \qquad \mathbf{j} \times \mathbf{k} = \mathbf{i} \qquad \mathbf{k} \times \mathbf{i} = \mathbf{j}$$

$$\mathbf{a} \times \mathbf{b} = \begin{vmatrix} \mathbf{i} & a_x & b_x \\ \mathbf{j} & a_y & b_y \\ \mathbf{k} & a_z & b_z \end{vmatrix} = \mathbf{i} \begin{vmatrix} a_y & a_z \\ b_y & b_z \end{vmatrix} + \mathbf{j} \begin{vmatrix} a_z & a_x \\ b_z & b_x \end{vmatrix} + \mathbf{k} \begin{vmatrix} a_x & a_y \\ b_x & b_y \end{vmatrix}$$

$$= \mathbf{i}(a_y b_z - a_z b_y) + \mathbf{j}(a_z b_x - a_x b_z) + \mathbf{k}(a_x b_y - a_y b_x)$$

18. Other Products of More than Two Vectors.

$$\mathbf{a} \times (\mathbf{b} \times \mathbf{c}) = (\mathbf{a} \cdot \mathbf{c})\mathbf{b} - (\mathbf{a} \cdot \mathbf{b})\mathbf{c} = \begin{vmatrix} \mathbf{b} & \mathbf{c} \\ \mathbf{a} \cdot \mathbf{b} & \mathbf{a} \cdot \mathbf{c} \end{vmatrix} \qquad \text{vector triple product}$$

$$(\mathbf{a} \times \mathbf{b}) \cdot (\mathbf{c} \times \mathbf{d}) = (\mathbf{a} \cdot \mathbf{c})(\mathbf{b} \cdot \mathbf{d}) - (\mathbf{a} \cdot \mathbf{d})(\mathbf{b} \cdot \mathbf{c}) = \begin{vmatrix} \mathbf{a} \cdot \mathbf{c} & \mathbf{b} \cdot \mathbf{c} \\ \mathbf{a} \cdot \mathbf{d} & \mathbf{b} \cdot \mathbf{d} \end{vmatrix}$$

$$(\mathbf{a} \times \mathbf{b})^2 = \mathbf{a}^2 \mathbf{b}^2 - (\mathbf{a} \cdot \mathbf{b})^2$$

$$(\mathbf{a} \times \mathbf{b}) \times (\mathbf{c} \times \mathbf{d}) = [acd]\mathbf{b} - [bcd]\mathbf{a} = [abd]\mathbf{c} - [abc]\mathbf{d}$$

VECTOR ANALYSIS: DIFFERENTIAL OPERATORS

19. The Operator ∇.   In terms of rectangular cartesian coordinates, the linear operator ∇ (*del or nabla*) is defined by

$$\nabla \equiv \mathbf{i}\frac{\partial}{\partial x} + \mathbf{j}\frac{\partial}{\partial y} + \mathbf{k}\frac{\partial}{\partial z}$$

Its application to a scalar point function $\Phi(\mathbf{r})$ or a vector point function $\mathbf{F}(\mathbf{r})$ corresponds formally to a noncommutative multiplication operation with a vector having the rectangular cartesian components $\partial/\partial x$, $\partial/\partial y$, $\partial/\partial z$; thus, in terms of right-handed rectangular cartesian coordinates x, y, z,

$$\nabla\Phi(x,y,z) \equiv \text{grad } \Phi(x,y,z) \equiv \mathbf{i}\frac{\partial\Phi}{\partial x} + \mathbf{j}\frac{\partial\Phi}{\partial y} + \mathbf{k}\frac{\partial\Phi}{\partial z}$$

$$\nabla \cdot \mathbf{F}(x,y,z) \equiv \text{div } \mathbf{F}(x,y,z) \equiv \frac{\partial F_x}{\partial x} + \frac{\partial F_y}{\partial y} + \frac{\partial F_z}{\partial z}$$

$$\nabla \times \mathbf{F}(x,y,z) \equiv \text{curl } \mathbf{F}(x,y,z)$$

$$\equiv \mathbf{i}\left(\frac{\partial F_z}{\partial y} - \frac{\partial F_y}{\partial z}\right) + \mathbf{j}\left(\frac{\partial F_x}{\partial z} - \frac{\partial F_z}{\partial x}\right) + \mathbf{k}\left(\frac{\partial F_y}{\partial x} - \frac{\partial F_x}{\partial y}\right) \equiv \begin{vmatrix} \mathbf{i} & \dfrac{\partial}{\partial x} & F_x \\ \mathbf{j} & \dfrac{\partial}{\partial y} & F_y \\ \mathbf{k} & \dfrac{\partial}{\partial z} & F_z \end{vmatrix}$$

$$(\mathbf{G} \cdot \nabla)\mathbf{F} \equiv G_x\frac{\partial\mathbf{F}}{\partial x} + G_y\frac{\partial\mathbf{F}}{\partial y} + G_z\frac{\partial\mathbf{F}}{\partial z}$$

$$\equiv \mathbf{i}(\mathbf{G} \cdot \nabla F_x) + \mathbf{j}(\mathbf{G} \cdot \nabla F_y) + \mathbf{k}(\mathbf{G} \cdot \nabla F_z)$$

Table 2-7   Rules* for Operations Involving the Operator ∇

a. Linearity

$$\nabla(\Phi + \Psi) = \nabla\Phi + \nabla\Psi \qquad\qquad \nabla(\alpha\Phi) = \alpha\nabla\Phi$$

$$\nabla \cdot (\mathbf{F} + \mathbf{G}) = \nabla \cdot \mathbf{F} + \nabla \cdot \mathbf{G} \qquad \nabla \cdot (\alpha\mathbf{F}) = \alpha\nabla \cdot \mathbf{F}$$

$$\nabla \times (\mathbf{F} + \mathbf{G}) = \nabla \times \mathbf{F} + \nabla \times \mathbf{G} \qquad \nabla \times (\alpha\mathbf{F}) = \alpha\nabla \times \mathbf{F}$$

b. Operations on products

$$\nabla(\Phi\Psi) = \Psi\nabla\Phi + \Phi\nabla\Psi$$

$$\nabla(\mathbf{F} \cdot \mathbf{G}) = (\mathbf{F} \cdot \nabla)\mathbf{G} + (\mathbf{G} \cdot \nabla)\mathbf{F} + \mathbf{F} \times (\nabla \times \mathbf{G}) + \mathbf{G} \times (\nabla \times \mathbf{F})$$

$$\nabla \cdot (\Phi\mathbf{F}) = \Phi\nabla \cdot \mathbf{F} + (\nabla\Phi) \cdot \mathbf{F}$$

$$\nabla \cdot (\mathbf{F} \times \mathbf{G}) = \mathbf{G} \cdot \nabla \times \mathbf{F} - \mathbf{F} \cdot \nabla \times \mathbf{G}$$

$$(\mathbf{G} \cdot \nabla)\Phi\mathbf{F} = \mathbf{F}(\mathbf{G} \cdot \nabla\Phi) + \Phi(\mathbf{G} \cdot \nabla)\mathbf{F}$$

$$\nabla \times (\Phi\mathbf{F}) = \Phi\nabla \times \mathbf{F} + (\nabla\Phi) \times \mathbf{F}$$

$$\nabla \times (\mathbf{F} \times \mathbf{G}) = (\mathbf{G} \cdot \nabla)\mathbf{F} - (\mathbf{F} \cdot \nabla)\mathbf{G} + \mathbf{F}(\nabla \cdot \mathbf{G}) - \mathbf{G}(\nabla \cdot \mathbf{F})$$

$$(\mathbf{G} \cdot \nabla)\mathbf{F} = \tfrac{1}{2}[\nabla \times (\mathbf{F} \times \mathbf{G}) + \nabla(\mathbf{F} \cdot \mathbf{G}) - \mathbf{F}(\nabla \cdot \mathbf{G}) + \mathbf{G}(\nabla \cdot \mathbf{F})$$

$$- \mathbf{F} \times (\nabla \times \mathbf{G}) - \mathbf{G} \times (\nabla \times \mathbf{F})]$$

* Note that vector equations involving $\nabla\Phi$, $\nabla \cdot \mathbf{F}$, and/or $\nabla \times \mathbf{F}$ have a meaning independent of the coordinate system used.

20. The Laplacian Operator. The *Laplacian operator* $\nabla^2 \equiv (\nabla \cdot \nabla)$ (sometimes denoted by Δ), expressed in terms of rectangular cartesian coordinates by

$$\nabla^2 \equiv (\nabla \cdot \nabla) \equiv \frac{\partial^2}{\partial x^2} + \frac{\partial^2}{\partial y^2} + \frac{\partial^2}{\partial z^2}$$

may be applied to both scalar and vector point functions by noncommutative scalar "multiplication," so that

$$\nabla^2\Phi \equiv \left(\frac{\partial^2}{\partial x^2} + \frac{\partial^2}{\partial y^2} + \frac{\partial^2}{\partial z^2}\right)\Phi$$

$$\nabla^2\mathbf{F} \equiv \mathbf{i}\nabla^2 F_x + \mathbf{j}\nabla^2 F_y + \mathbf{k}\nabla^2 F_z$$

Note

$$\nabla^2(\alpha\Phi + \beta\Psi) = \alpha\,\nabla^2\Phi + \beta\,\nabla^2\Psi \qquad \text{linearity}$$

$$\text{and } \nabla^2(\Phi\Psi) = \Psi\,\nabla^2\Phi + 2(\nabla\Phi) \cdot (\nabla\Psi) + \Phi\,\nabla^2\Psi$$

2-12

21. Repeated Operations. Note the following rules for repeated operations with the operator ∇:

$$\text{div grad } \Phi = \nabla \cdot (\nabla\Phi) = \nabla^2\Phi$$

$$\text{grad div } \mathbf{F} = \nabla(\nabla \cdot \mathbf{F}) = \nabla^2\mathbf{F} + \nabla \times (\nabla \times \mathbf{F})$$

$$\text{curl curl } \mathbf{F} = \nabla \times (\nabla \times \mathbf{F}) = \nabla(\nabla \cdot \mathbf{F}) - \nabla^2\mathbf{F}$$

$$\text{curl grad } \Phi = \nabla \times (\nabla\Phi) = 0$$

$$\text{div curl } \mathbf{F} = \nabla \cdot (\nabla \times \mathbf{F}) = 0$$

THE LAPLACE TRANSFORMATION

22. Introduction. The Laplace transformation associates a unique function $F(s)$ of a complex variable s with each suitable function $f(t)$ of a real variable t. This correspondence is essentially reciprocal one to one for most practical purposes (Par. **2-26**); corresponding pairs of functions $f(t)$ and $F(s)$ can often be found by reference to tables. The Laplace transformation is defined so that many relations between, and operations on, the functions $f(t)$ correspond to simpler relations between, and operations on, the functions $F(s)$ (Table **2-9**). This applies particularly to the solution of differential and integral equations. It is thus often useful to transform a given problem involving functions $f(t)$ into an equivalent problem expressed in terms of the associated Laplace transforms $F(s)$ (*operational calculus* based on Laplace transformations or *transformation calculus*, Par. **2-29**).

23. Definition. The *one-sided Laplace transformation*

$$F(s) \equiv \mathfrak{L}[f(t)] \equiv \int_0^\infty f(t)e^{-st}\,dt$$

$$\equiv \lim_{\substack{a \to 0 \\ b \to \infty}} \int_a^b f(t)e^{-st}\,dt \qquad 0 < a < b$$

associates a unique *result or image function* $F(s)$ of the complex variable $s = \sigma + i\omega$ with every single-valued *object or original function* $f(t)$ (t real) such that the improper integral exists. $F(s)$ is called the (*one-sided*) *Laplace transform* of $f(t)$. The more explicit notation $\mathfrak{L}[f(t);s]$ is also used.

The Laplace transform exists for $\sigma \geq \sigma_0$, and the improper integral converges absolutely and uniformly to a function $F(s)$ analytic for $\sigma > \sigma_0$ if

$$\int_0^\infty |f(t)|e^{-\sigma t}\,dt = \lim_{\substack{a \to 0 \\ b \to \infty}} \int_a^b |f(t)|e^{-\sigma t}\,dt \qquad 0 < a < b$$

exists for $\sigma = \sigma_0$. The greatest lower bound σ_a of the real number σ_0 for which this is true is called the *abscissa of absolute convergence* of the Laplace transform $\mathfrak{L}[f(t)]$.

Although certain theorems relating to Laplace transforms require only the existence (simple convergence) of the transforms, *the existence of an abscissa of absolute convergence will be implicitly assumed throughout the following sections.* Wherever necessary, it is customary to specify the region of absolute convergence associated with a relation involving Laplace transforms by writing $\sigma > \sigma_a$ to the right of the relation in question.

The region of definition of the analytic function

$$F(s) = \mathfrak{L}[f(t)] \qquad \sigma > \sigma_a$$

can usually be extended so as to include the entire s plane with the exception of singular points situated to the left of the abscissa of absolute convergence. Such an extension of the region of definition is implied wherever necessary.

24. Inverse Laplace Transformation. The *inverse Laplace transform* $\mathfrak{L}^{-1}[F(s)]$ of a (suitable) function $F(s)$ of the complex variable $s = \sigma + i\omega$ is a function $f(t)$ whose Laplace transform is $F(s)$. *Not every function $F(s)$ has an inverse Laplace transform.*

25. The Inversion Theorem.   *Given $F(s) = \mathcal{L}[f(t)]$, $\sigma > \sigma_a$, then throughout every open interval where $f(t)$ is bounded and has a finite number of maxima, minima, and discontinuities,*

$$f_I(t) = \frac{1}{2\pi i} \lim_{R \to \infty} \int_{\sigma_1 - iR}^{\sigma_1 + iR} F(s)e^{st} ds$$

$$= \begin{cases} \frac{1}{2}[f(t-0) + f(t+0)] & \text{for } t > 0 \\ \frac{1}{2}f(0+0) & \text{for } t = 0 \quad \sigma_1 > \sigma_a \\ 0 & \text{for } t < 0 \end{cases}$$

In particular, for every $t > 0$ where $f(t)$ is continuous

$$f_I(t) = \frac{1}{2\pi i} \lim_{R \to \infty} \int_{\sigma_1 - iR}^{\sigma_1 + iR} F(s)e^{st} ds = f(t) \qquad \sigma_1 > \sigma_a \qquad (2\text{-}7)$$

The path of integration in Eq. (2-7) lies to the right of all singularities of $F(s)$. The inversion integral $f_I(t)$ reduces to $1/2\pi i \int_{\sigma_1 - i\infty}^{\sigma_1 + i\infty} F(s)e^{st} ds$ if the integral exists.

26. Uniqueness of the Laplace Transform and Its Inverse.   *The Laplace transform (1) is unique for each function $f(t)$ having such a transform.   Conversely, two functions $f_1(t)$ and $f_2(t)$ possessing identical Laplace transforms are identical for all $t > 0$ where both functions are continuous (Lerch's theorem).   A given function $F(s)$ cannot have more than one inverse Laplace transform continuous for all $t > 0$.*

Different discontinuous functions may have the same Laplace transform.   In particular, the generalized unit step function defined by $f(t) = 0$ for $t < 0$, $f(t) = 1$ for $t > 0$ has the Laplace transform $1/s$ regardless of the value assigned to $f(t)$ for $t = 0$.

LINEAR DIFFERENTIAL EQUATIONS

27. Homogeneous Linear Equations with Constant Coefficients.   The *first-order differential equation*

$$a_0 \frac{dy}{dt} + a_1 y = 0 \qquad a_0 \neq 0 \qquad (2\text{-}8)$$

has the solution

$$y = Ce^{-(a_1/a_0)t} \qquad C = y(0)$$

For $a_0/a_1 > 0$

$$y\left(\frac{a_0}{a_1}\right) = \frac{1}{e} y(0) \approx 0.37 y(0) \qquad y\left(\frac{4a_0}{a_1}\right) \approx 0.02 y(0)$$

a_0/a_1 is often referred to as the *time constant.*

The *second-order equation*

$$a_0 \frac{d^2 y}{dt^2} + a_1 \frac{dy}{dt} + a_2 y = 0 \qquad a_0 \neq 0 \qquad (2\text{-}9)$$

has the solution

$$y = C_1 e^{s_1 t} + C_2 e^{s_2 t} \qquad (2\text{-}10a)$$

$$s_{1,2} = \frac{-a_1 \pm \sqrt{a_1^2 - 4a_0 a_2}}{2a_0} \qquad a_1^2 - 4a_0 a_2 \neq 0 \quad (2\text{-}10b)$$

$$y = (C_1 + C_2 t)e^{-(a_1/2a_0)t} \qquad a_1^2 - 4a_0 a_2 = 0 \quad (2\text{-}10c)$$

If a_0, a_1, and a_2 are real, s_1 and s_2 become complex for $a_1^2 - 4a_0 a_2 < 0$; in this case, Eq. (2-10a) can be written as

$$y = e^{\sigma_1 t}(A \cos \omega_N t + B \sin \omega_N t) = Re^{\sigma_1 t}\sin(\omega_N t + \alpha)$$

Table 2-8 Theorems Relating Volume Integrals and Surface Integrals

	Theorem	Vector formula	Sufficient conditions	
			a. Throughout V	*b.* On S
1	Divergence theorem (Gauss' integral theorem)	$\displaystyle\int_V \nabla \cdot \mathbf{F}(\mathbf{r})\,dV = \int_S d\mathbf{A} \cdot \mathbf{F}(\mathbf{r})$	$\mathbf{F}(\mathbf{r})$, $\Phi(\mathbf{r})$ differentiable with continuous partial derivatives	Existence of integrals is sufficient
2	Theorem of the rotational	$\displaystyle\int_V \nabla \times \mathbf{F}(\mathbf{r})\,dV = \int_S d\mathbf{A} \times \mathbf{F}(\mathbf{r})$		
3	Theorem of the gradient	$\displaystyle\int_V \nabla\Phi(\mathbf{r})\,dV = \int_S d\mathbf{A}\,\Phi(\mathbf{r})$		
4	Green's theorems	$\displaystyle\int_V \nabla\Phi \cdot \nabla\Psi\,dV + \int_V \Psi\,\nabla^2\Phi\,dV$ $\displaystyle = \int_S d\mathbf{A} \cdot (\Psi\,\nabla\Phi)$ $\displaystyle = \int_S \Psi \frac{\partial\Phi}{\partial n}\,dA$	$\Phi(\mathbf{r})$, $\psi(\mathbf{r})$ differentiable with continuous partial derivatives; $\Phi(\mathbf{r})$ twice differentiable with continuous second partial derivatives	$\psi(\mathbf{r})$ continuous; $\Phi(\mathbf{r})$ differentiable with continuous partial derivatives
5		$\displaystyle\int_V (\Psi\,\nabla^2\Phi - \Phi\,\nabla^2\Psi)\,dV$ $\displaystyle = \int_S d\mathbf{A} \cdot (\Psi\,\nabla\Phi - \Phi\,\nabla\Psi)$ $\displaystyle = \int_S \left(\Psi\frac{\partial\Phi}{\partial n} - \Phi\frac{\partial\Psi}{\partial n}\right) dA$	$\Phi(\mathbf{r})$, $\Psi(\mathbf{r})$ twice differentiable with continuous second partial derivatives	$\Phi(\mathbf{r})$, $\Psi(\mathbf{r})$ differentiable with continuous partial derivatives
6	Special cases	$\displaystyle\int_V \nabla^2\Phi\,dV = \int_S d\mathbf{A} \cdot \nabla\Phi$ $\displaystyle = \int_S \frac{\partial\Phi}{\partial n}\,dA \quad \text{(Gauss' theorem)}$		
7		$\displaystyle\int_V |\nabla\Phi|^2\,dV + \int_V \Phi\,\nabla^2\Phi\,dv$ $\displaystyle = \int_S d\mathbf{A} \cdot (\Phi\,\nabla\Phi) = \int_S \Phi\frac{\partial\Phi}{\partial n}\,dA$		

Table 2-9 Theorems Relating Corresponding Operations on Object and Result Functions*†

Theorem number	Operation	Object function	Result function
1	Linearity $(\alpha, \beta$ constant)	$\alpha f_1(t) + \beta f_2(t)$	$\alpha F_1(s) + \beta F_2(s)$
2a 2b	Differentiation of object function‡ ... if $f'(t)$ exists for all $t > 0$ ... if $f^{(r)}(t)$ exists for all $t > 0$	$f'(t)$ $f^{(r)}(t)$ $r = 1, 2, \ldots$	$sF(s) - f(0+0)$ $s^r F(s) - s^{r-1}f(0+0)$ $\quad - s^{r-2}f'(0+0)$ $\quad - \cdots - f^{(r-1)}(0+0)$
2c	... if $f(t)$ is bounded for $t > 0$, and $f'(t)$ exists for $t > 0$ except for $t = t_1, t_2, \ldots$, where $f(t)$ has unilateral limits	$f'(t)$	$sF(s) - f(0+0)$ $\quad - \sum_k e^{-t_k s}[f(t_k+0) - f(t_k-0)]$
3	Integration of object function ... if $f'(t)$ exists for $t > 0$	$\int_0^t f(\tau)\,d\tau + C$	$\dfrac{F(s)}{s} + \dfrac{C}{s}$
4	Change of scale	$f(at)$ $a > 0$	$\dfrac{1}{a}F\left(\dfrac{s}{a}\right)$
5	Translation (shift) of object function ... if $f(t) = 0$ for $t \leq 0$	$f(t - b)$ $b \geq 0$	$e^{-bs}F(s)$
6	Convolution of object function¶	$f_1 * f_2 = \int_0^\infty f_1(\tau)f_2(t-\tau)\,d\tau$ $\equiv f_2 * f_1$	$F_1(s)F_2(s)$

7	Corresponding limits of object and result function (continuity theorem; α is independent of t and s)	$\lim_{\alpha \to a} f(t,a)$	$\lim_{\alpha \to a} F(s,a)$
8a	Differentiation and integration with respect to a parameter α independent of t and s	$\dfrac{\partial}{\partial \alpha} f(t,\alpha)$	$\dfrac{\partial}{\partial \alpha} F(s,\alpha)$
8b		$\displaystyle\int_{a_1}^{a_2} f(t,\alpha)\,d\alpha$	$\displaystyle\int_{a_1}^{a_2} F(s,\alpha)\,d\alpha$
9a	Differentiation of result function	$-t f(t)$	$F'(s)$
9b		$(-1)^r t^r f(t)$	$F^{(r)}(s)$
10	Integration of result function (path of integration situated to the right of the abscissa of absolute convergence)	$\dfrac{1}{t} f(t)$	$\displaystyle\int_s^\infty F(s)\,ds$
11	Translation of result function	$e^{at} f(t)$	$F(s-a)$

*From G. A. Korn and T. M. Korn, "Mathematical Handbook for Scientists and Engineers," McGraw-Hill, New York, 1961.

†The following theorems are valid whenever the Laplace transforms $F(s) = \mathfrak{L}[f(t)]$ in question exist in the sense of absolute convergence.)

‡The abscissa of absolute convergence for $\mathfrak{L}[f^{(r)}(t)]$ is 0 or σ_a, whichever is greater.

¶The existence of $f_1 * f_2$ is assumed; absolute convergence of $\mathfrak{L}[f_1(t)]$ and $\mathfrak{L}[f_2(t)]$ is a sufficient condition for the absolute convergence of $\mathfrak{L}[f_1 * f_2]$.

where the quantities

$$\sigma_1 = -\frac{a_1}{2a_0} \qquad \omega_N = \frac{\sqrt{4a_0 a_2 - a_1^2}}{2a_0}$$

are respectively known as the *damping constant* and the *natural (characteristic) circular frequency.* The constants C_1, C_2, A, B, R, and α are chosen so as to match given initial or boundary conditions (see also Fig. 2-1).

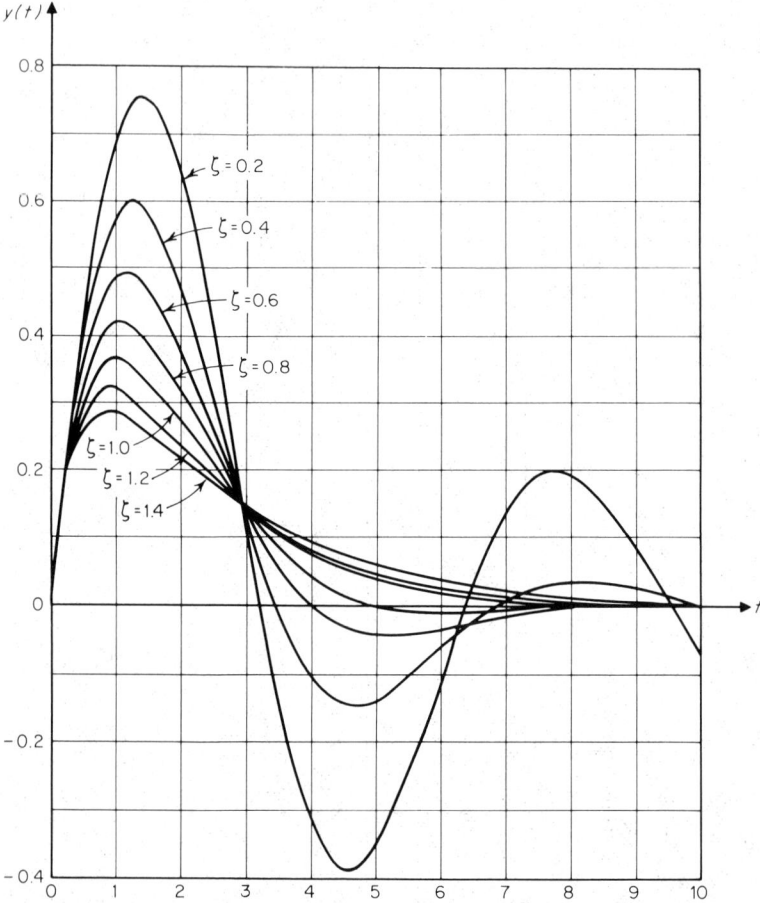

Fig. 2-1. Solution of the second-order differential equation

$$\frac{d^2 y}{dt^2} + 2\zeta \frac{dy}{dt} + v = 0$$

for $y(0) = 0$, $dy/dt_0 = 1$. Response is overdamped for $\zeta > 1$, critically damped for $\zeta = 1$, and underdamped for $\zeta < 1$.

If $a_0 a_2 > 0$, the quantity $\zeta = a_1/2\sqrt{a_0 a_2}$ is called the *damping ratio*; for $\zeta > 1$, $\zeta = 1$, $0 < \zeta < 1$ one obtains, respectively, an *overdamped* solution (2-10a), a *critically damped* solution (2-10b), or an *underdamped (oscillatory)* solution (2-10c). In the latter case, the *logarithmic decrement* $2\pi\sigma_1/\omega_N$ is the natural logarithm of the ratio of successive maxima of $y(t)$.

Equation (2-9) is often written in the *nondimensional form*

$$\frac{1}{\omega_1^2}\frac{d^2y}{dt^2} + 2\frac{\zeta}{\omega_1}\frac{dy}{dt} + y = 0 \qquad \text{with } s_{1,2} = -\omega_1\zeta \pm \omega_1\sqrt{\zeta^2 - 1}$$

$\omega_1 = \sqrt{a_2/a_0}$ is called the *undamped natural circular frequency*; for weak damping ($\zeta^2 \ll 1$), $\omega_1 \approx \omega_N$.

To solve the *rth-order differential equation*

$$\mathbf{L}y \equiv a_0\frac{d^r y}{dt^r} + a_1\frac{d^{r-1}y}{dt^{r-1}} + \cdots + a_r y = 0 \qquad a_0 \neq 0 \tag{2-11}$$

find the roots of the *r*th-degree algebraic equation

$$a_0 s^r + a_1 s^{r-1} + \cdots + a_0 = 0 \qquad \text{characteristic equation} \tag{2-12}$$

obtained, for example, on substitution of a trial solution $y = e^{st}$. If the *r* roots $s_1, s_2, \ldots$ of the characteristic equation (2-12) are distinct, the given differential equation (2-11) has the general solution

$$y = C_1 e^{s_1 t} + C_2 e^{s_2 t} + \cdots + C_r e^{s_r t} \tag{2-13a}$$

If a root s_k is of multiplicity m_k, replace the corresponding term in Eq. (2-13a) by

$$(C_k + C_{k1}t + C_{k2}t^2 + \cdots + C_{km_k-1}t^{m_k-1})e^{s_k t} \tag{2-13b}$$

The various terms of the solution (2-13b) are known as *normal modes* of the given differential equation. The *r* constants C_k and C_{kj} must be chosen so as to match given initial or boundary conditions.

If the given differential equation (2-11) is real, complex roots of the characteristic equation appear as pairs of complex conjugates $\sigma \pm i\omega$. The corresponding pairs of solution terms will also be complex conjugates and may be combined to form real terms:

$$t^m C e^{(\sigma+i\omega)t} + t^m C^* e^{(\sigma-i\omega)t} = t^m e^{\sigma t}(A\cos\omega t + B\sin\omega t)$$
$$= Rt^m e^{\sigma t}\sin(\omega t + \alpha) \tag{2-13c}$$

where *A* and *B*, or *R* and α, are new real constants of integration.

Given a *system of n homogeneous linear differential equation with constant coefficients*

$$\varphi_{j1}\left(\frac{d}{dt}\right)y_1 + \varphi_{j2}\left(\frac{d}{dt}\right)y_2 + \cdots + \varphi_{jn}\left(\frac{d}{dt}\right)y_n = 0 \qquad j = 1, 2, \ldots, n \tag{2-14}$$

where the $\varphi_{jk}(d/dt)$ are polynomials in d/dt, each of the *n* solution functions $y_k = y_k(t)(k = 1, 2, \ldots, n)$ has the form (2-13a); the s_k are now the roots of the algebraic equation

$$D(s) \equiv \det[\varphi_{jk}(s)] = 0 \qquad \text{characteristic equation of system (2-14)} \tag{2-15}$$

The constants of integration must again be matched to the given initial or boundary conditions.

28. Nonhomogeneous Equations. The general solution of the nonhomogeneous differential equation

$$\mathbf{L}y \equiv a_0\frac{d^r y}{dt^r} + a_1\frac{d^{r-1}y}{dt^{r-1}} + \cdots + a_r y = f(t) \tag{2-16}$$

can be expressed as the sum of the general solution (2-13a) of the reduced equation (2-11) and any particular integral of Eq. (2-16).

Table 2-10 Laplace Transform Pairs Involving Rational Algebraic Functions $F(s) = D_1(s)/D(s)$

Each formula holds for complex as well as for real polynomials $D_1(s)$ and $D(s)$, but the latter case is of greater practical interest. In this case the roots of $D(s) = 0$ are either real or they occur as pairs of complex conjugates, and the functions $f(t)$ are real. Note $(s-a)^2 + \omega_1^2 = [s - (a + i\omega_1)][s - (a - i\omega_1)]$ and $K_1 \sin \omega t + K_2 \cos \omega t = \sqrt{K_1^2 + K_2^2} \sin(\omega t + \alpha)$, with $\alpha = \arctan K_2/K_1$

No.	$F(s)$	$f(t)\,(t > 0)$	
1.1	$\dfrac{1}{s}$	1	
1.2	$\dfrac{1}{s-a}$	e^{at}	
1.3	$\dfrac{1}{s(s-a)}$	$Ae^{at} + K$	$A = \dfrac{1}{a} \quad K = -\dfrac{1}{a}$
1.4	$\dfrac{s+d}{s(s-a)}$	$Ae^{at} + K$	$A = \left(1 + \dfrac{d}{a}\right) \quad K = -\dfrac{d}{a}$
1.5	$\dfrac{1}{(s-a)(s-b)}$	$Ae^{at} + Be^{bt}$	$A = \dfrac{1}{a-b} \quad B = \dfrac{1}{b-a}$
1.6	$\dfrac{s+d}{(s-a)(s-b)}$	$Ae^{at} + Be^{bt}$	$A = \dfrac{a+d}{a-b} \quad B = \dfrac{b+d}{b-a}$
1.7	$\dfrac{1}{s(s-a)(s-b)}$	$Ae^{at} + Be^{bt} + K$	$A = \dfrac{1}{a(a-b)} \quad B = \dfrac{1}{b(b-a)} \quad K = \dfrac{1}{ab}$
1.8	$\dfrac{s+d}{s(s-a)(s-b)}$	$Ae^{at} + Be^{bt} + K$	$A = \dfrac{a+d}{a(a-b)} \quad B = \dfrac{b+d}{b(b-a)} \quad K = \dfrac{d}{ab}$
1.9	$\dfrac{s^2 + gs + d}{s(s-a)(s-b)}$		$A = \dfrac{a^2 + ga + d}{a(a-b)} \quad B = \dfrac{b^2 + gb + d}{b(b-a)} \quad K = \dfrac{d}{ab}$

No.	$F(s)$	$f(t)$	Coefficients
1.10	$\dfrac{1}{(s-a)(s-b)(s-c)}$	$Ae^{at}+Be^{bt}+Ce^{ct}$	$A=\dfrac{1}{(a-b)(a-c)}\quad B=\dfrac{1}{(b-a)(b-c)}\quad C=\dfrac{1}{(c-a)(c-b)}$
1.11	$\dfrac{s+d}{(s-a)(s-b)(s-c)}$		$A=\dfrac{a+d}{(a-b)(a-c)}\quad B=\dfrac{b+d}{(b-a)(b-c)}\quad C=\dfrac{c+d}{(c-a)(c-b)}$
1.12	$\dfrac{s^2+gs+d}{(s-a)(s-b)(s-c)}$		$A=\dfrac{a^2+ag+d}{(a-b)(a-c)}\quad B=\dfrac{b^2+bg+d}{(b-a)(b-c)}\quad C=\dfrac{c^2+cg+d}{(c-a)(c-b)}$
2.1	$\dfrac{1}{(s-a)^2+\omega_1^2}$	$Ae^{at}\sin(\omega_1 t+\alpha)$	$A=\dfrac{1}{\omega_1}\quad \alpha=0$
2.2	$\dfrac{s+d}{(s-a)^2+\omega_1^2}$		$A=\dfrac{1}{\omega_1}[(a+d)^2+\omega_1^2]^{1/2}\quad \alpha=\arctan\dfrac{\omega_1}{a+d}$
2.3	$\dfrac{1}{s[(s-a)^2+\omega_1^2]}$		$A=\dfrac{1}{\omega_1}\dfrac{1}{(a^2+\omega_1^2)^{1/2}}\quad \alpha=-\arctan\dfrac{\omega_1}{a}\quad K=\dfrac{1}{a^2+\omega_1^2}$
2.4	$\dfrac{s+d}{s[(s-a)^2+\omega_1^2]}$	$Ae^{at}\sin(\omega_1 t+\alpha)+K$	$A=\dfrac{1}{\omega_1}\left[\dfrac{(a+d)^2+\omega_1^2}{a^2+\omega_1^2}\right]^{1/2}\quad \alpha=\arctan\dfrac{\omega_1}{a+d}-\arctan\dfrac{\omega_1}{a}\quad K=\dfrac{d}{a^2+\omega_1^2}$
2.5	$\dfrac{s^2+gs+d}{s[(s-a)^2+\omega_1^2]}$		$A=\dfrac{1}{\omega_1}\left[\dfrac{(a^2-\omega_1^2+ag+d)^2+\omega_1^2(2a+g)^2}{a^2+\omega_1^2}\right]^{1/2}\quad \alpha=\arctan\dfrac{\omega_1(2a+g)}{a^2-\omega_1^2+ag+d}-\arctan\dfrac{\omega_1}{a}\quad K=\dfrac{d}{a^2+\omega_1^2}$

Table 2-10 Laplace Transform Pairs Involving Rational Algebraic Functions $F(s) = D_1(s)/D(s)$—Continued

Each formula holds for complex as well as for real polynomials $D_1(s)$ and $D(s)$, but the latter case is of greater practical interest. In this case the roots of $D(s) = 0$ are either real or they occur as pairs of complex conjugates, and the functions $f(t)$ are real. Note $(s - a)^2 + \omega_1^2 = [s - (a + i\omega_1)][s - (a - i\omega_1)]$ and $K_1 \sin \omega t + K_2 \cos \omega t = \sqrt{K_1^2 + K_2^2}\, \sin(\omega t + \alpha)$, with $\alpha = \arctan K_2/K_1$

No.	$F(s)$	$f(t)\,(t > 0)$	
2.6	$\dfrac{1}{(s - b)[(s - a)^2 + \omega_1^2]}$		$A = \dfrac{1}{\omega_1}\dfrac{1}{[(a - b)^2 + \omega_1^2]^{1/2}}$ $\qquad B = \dfrac{1}{(a - b)^2 + \omega_1^2}$ $\\[2mm] \alpha = -\arctan \dfrac{\omega_1}{a - b}$
2.7	$\dfrac{s + d}{(s - b)[(s - a)^2 + \omega_1^2]}$	$Ae^{at}\sin(\omega_1 t + \alpha) + Be^{bt}$	$A = \dfrac{1}{\omega_1}\left[\dfrac{(a + d)^2 + \omega_1^2}{(a - b)^2 + \omega_1^2}\right]^{1/2}$ $\qquad B = \dfrac{b + d}{(a - b)^2 + \omega_1^2}$ $\\[2mm] \alpha = \arctan \dfrac{\omega_1}{a + d} - \arctan \dfrac{\omega_1}{a - b}$
2.8	$\dfrac{s^2 + gs + d}{(s - b)[(s - a)^2 + \omega_1^2]}$		$A = \dfrac{1}{\omega_1}\left[\dfrac{(a^2 - \omega_1^2 + ag + d)^2 + \omega_1^2(2a + g)^2}{(a - b)^2 + \omega_1^2}\right]^{1/2}$ $\qquad B = \dfrac{b^2 + bg + d}{(a - b)^2 + \omega_1^2}$ $\\[2mm] \alpha = \arctan \dfrac{\omega_1(2a + g)}{a^2 - \omega_1^2 + ag + d} - \arctan \dfrac{\omega_1}{a - b}$

		$Ae^{at}\sin(\omega_1 t + \alpha) + Be^{bt} + K$	
2.9	$\dfrac{1}{s(s-b)[(s-a)^2+\omega_1^2]}$		$A = \dfrac{1}{\omega_1(a^2+\omega_1^2)^{1/2}[(a-b)^2+\omega_1^2]^{1/2}}$ $B = \dfrac{1}{b[(b-a)^2+\omega_1^2]} \qquad K = -\dfrac{1}{b(a^2+\omega_1^2)}$ $\alpha = -\arctan\dfrac{\omega_1}{a-b} - \arctan\dfrac{\omega_1}{a}$
2.10	$\dfrac{s+d}{s(s-b)[(s-a)^2+\omega_1^2]}$		$A = \dfrac{1}{\omega_1(a^2+\omega_1^2)^{1/2}}\left[\dfrac{(d+a)^2+\omega_1^2}{(a-b)^2+\omega_1^2}\right]^{1/2}$ $B = \dfrac{b+d}{b[(b-a)^2+\omega_1^2]} \qquad K = -\dfrac{d}{b(a^2+\omega_1^2)}$ $\alpha = \arctan\dfrac{\omega_1}{a+d} - \arctan\dfrac{\omega_1}{a-b} - \arctan\dfrac{\omega_1}{a}$
2.11	$\dfrac{s^2+gs+d}{s(s-b)[(s-a)^2+\omega_1^2]}$		$A = \dfrac{1}{\omega_1}\left\{\dfrac{(a^2-\omega_1^2+ag+d)^2+\omega_1^2(2a+g)^2}{(a^2+\omega_1^2)[(a-b)^2+\omega_1^2]}\right\}^{1/2}$ $B = \dfrac{b^2+bg+d}{b[(b-a)^2+\omega_1^2]} \qquad K = -\dfrac{d}{b(a^2+\omega_1^2)}$ $\alpha = \arctan\dfrac{\omega_1(2a+g)}{a^2-\omega_1^2+ag+d} - \arctan\dfrac{\omega_1}{a-b} - \arctan\dfrac{\omega_1}{a}$

Table 2-10 Laplace Transform Pairs Involving Rational Algebraic Functions $F(s) = D_1(s)/D(s)$ —Concluded

Each formula holds for complex as well as for real polynomials $D_1(s)$ and $D(s)$, but the latter case is of greater practical interest. In this case the roots of $D(s) = 0$ are either real or they occur as pairs of complex conjugates, and the functions $f(t)$ are real. Note $(s - a)^2 + \omega_1^2 = [s - (a + i\omega_1)][s - (a - i\omega_1)]$ and $K_1 \sin \omega t + K_2 \cos \omega t = \sqrt{K_1^2 + K_2^2} \sin(\omega t + \alpha)$, with $\alpha = \arctan K_2/K_1$

No.	$F(s)$	$f(t)(t > 0)$	
2.12	$\dfrac{1}{[(s-a)^2 + \omega_1^2](s^2 + \omega_2^2)}$		$A = \dfrac{1}{\omega_1}\dfrac{1}{[(a^2 + \omega_1^2 - \omega_2^2)^2 + 4a^2\omega_2^2]^{1/2}}$ $\qquad \alpha = -\arctan\dfrac{2a\omega_1}{a^2 - \omega_1^2 + \omega_2^2}$ $B = \dfrac{1}{\omega_2}\dfrac{1}{[(a^2 + \omega_1^2 - \omega_2^2)^2 + 4a^2\omega_2^2]^{1/2}}$ $\qquad \beta = \arctan\dfrac{2a\omega_2}{a^2 + \omega_1^2 - \omega_2^2}$
2.13	$\dfrac{s+d}{[(s-a)^2 + \omega_1^2](s^2 + \omega_2^2)}$	$Ae^{at}\sin(\omega_1 t + \alpha) + B\sin(\omega_2 t + \beta)$	$A = \dfrac{1}{\omega_1}\left[\dfrac{(a+d)^2 + \omega_1^2}{(a^2 + \omega_1^2 - \omega_2^2)^2 + 4a^2\omega_2^2}\right]^{1/2}$ $\alpha = \arctan\dfrac{\omega_1}{a+d}$ $\qquad\qquad - \arctan\dfrac{2a\omega_1}{a^2 - \omega_1^2 + \omega_2^2}$ $B = \dfrac{1}{\omega_2}\left[\dfrac{d^2 + \omega_2^2}{(a^2 + \omega_1^2 - \omega_2^2)^2 + 4a^2\omega_2^2}\right]^{1/2}$ $\beta = \arctan\dfrac{\omega_2}{d}$ $\qquad\qquad + \arctan\dfrac{2a\omega_2}{a^2 + \omega_1^2 - \omega_2^2}$
2.14	$\dfrac{s^2 + gs + d}{[(s-a)^2 + \omega_1^2](s^2 + \omega_2^2)}$		$A = \dfrac{1}{\omega_1}\left[\dfrac{(a^2 - \omega_1^2 + ag + d)^2 + \omega_1^2(2a+g)^2}{(a^2 + \omega_1^2 - \omega_2^2)^2 + 4a^2\omega_2^2}\right]^{1/2}$ $\alpha = \arctan\dfrac{\omega_1(2a+g)}{a^2 - \omega_1^2 + ag + d}$ $\qquad\qquad - \arctan\dfrac{2a\omega_1}{a^2 - \omega_1^2 + \omega_2^2}$ $B = \dfrac{1}{\omega_2}\left[\dfrac{(d - \omega_2^2)^2 + g^2\omega_2^2}{(a^2 + \omega_1^2 - \omega_2^2)^2 + 4a^2\omega_2^2}\right]^{1/2}$ $\beta = \arctan\dfrac{g\omega_2}{d - \omega_2^2}$ $\qquad\qquad + \arctan\dfrac{2a\omega_2}{a^2 + \omega_1^2 - \omega_2^2}$
3.1	$\dfrac{1}{s^2}$	t	

3.2	$\dfrac{1}{(s-a)^2}$	$(A+A_1 t)e^{at}$	$A=0 \quad A_1=1$
3.3	$\dfrac{s+d}{(s-a)^2}$		$A=1 \quad A_1=a+d$
3.4	$\dfrac{1}{s^2(s-a)}$	$Ae^{at}+K+K_1 t$	$A=\dfrac{1}{a^2} \quad K=-A \quad K_1=-\dfrac{1}{a}$
3.5	$\dfrac{s+d}{s^2(s-a)}$		$A=\dfrac{a+d}{a^2} \quad K=-A \quad K_1=-\dfrac{d}{a}$
3.6	$\dfrac{s^2+gs+d}{s^2(s-a)}$		$A=\dfrac{a^2+ag+d}{a^2} \quad K=1-A \quad K_1=-\dfrac{d}{a}$
3.7	$\dfrac{1}{s(s-a)^2}$	$(A+A_1 t)e^{at}+K$	$A=-\dfrac{1}{a^2} \quad A_1=\dfrac{1}{a} \quad K=-A$
3.8	$\dfrac{s+d}{s(s-a)^2}$		$A=-\dfrac{d}{a^2} \quad A=\dfrac{a+d}{a} \quad K=-A$
3.9	$\dfrac{s^2+gs+d}{s(s-a)^2}$		$A=\dfrac{a^2-d}{a^2} \quad A_1=\dfrac{a^2+ag+d}{a} \quad K=1-A$
3.10	$\dfrac{1}{(s-a)^2(s-b)}$	$(A+A_1 t)e^{at}+Be^{bt}$	$A=-\dfrac{1}{(a-b)^2} \quad A_1=\dfrac{1}{a-b} \quad B=-A$
3.11	$\dfrac{s+d}{(s-a)^2(s-b)}$		$A=-\dfrac{b+d}{(a-b)^2} \quad A_1=\dfrac{a+d}{a-b} \quad B=-A$

29. Laplace Transform Method of Solution. (See also Pars. 2-22 to 2-26.) To solve a linear differential equation (2-16) with given initial values $y(0 + 0)$, $y'(0 + 0)$, $y''(0 + 0)$, $\ldots y^{(r-1)}(0 + 0)$, apply the Laplace transformation to both sides, and let $\mathcal{L}[y(t)] \equiv Y(s)$, $\mathcal{L}[f(t)] \equiv F(s)$. The resulting linear *algebraic* equation (*subsidiary equation*)

$$(a_0 s^r + a_1 s^{r-1} + \cdots + a_r)Y(s) = F(s) + G(s)$$

$$G(s) \equiv y(0 + 0)(a_0 s^{r-1} + a_1 s^{r-2} + \cdots + a_{r-1})$$
$$+ y'(0 + 0)(a_0 s^{r-2} + a_1 s^{r-3} + \cdots + a_{r-2}) \qquad (2\text{-}17)$$
$$+ \cdots\cdots\cdots\cdots\cdots$$
$$+ y^{(r-2)}(0 + 0)(a_0 s + a_1) + a_0 y^{(r-1)}(0 + 0)$$

is easily solved to yield the Laplace transform of the desired solution $y(t)$ in the form

$$Y(s) = \frac{F(s)}{a_0 s^r + a_1 s^{r-1} + \cdots + a_r} + \frac{G(s)}{a_0 s^r + a_1 s^{r-1} + \cdots + a_r} \qquad (2\text{-}18)$$

The second term represents the effects of nonzero initial values of $y(t)$ and its derivatives.

In the same manner, one applies the Laplace transformation to a system of linear differential equations

$$\varphi_{j1}\left(\frac{d}{dt}\right)y_1 + \varphi_{j2}\left(\frac{d}{dt}\right)y_2 + \cdots + \varphi_{jn}\left(\frac{d}{dt}\right)y_n = f_j(t) \qquad j = 1, 2, \ldots, n \qquad (2\text{-}19)$$

to obtain

$$\varphi_{j1}(s)Y_1(s) + \varphi_{j2}(s)Y_2(s) + \cdots + \varphi_{jn}(s)Y_n(s) = F_j(s) + G_j(s) \qquad j = 1, 2, \ldots, n \qquad (2\text{-}20)$$

where the functions $G_j(s)$ depend on the given initial conditions. The linear *algebraic* equations (2-20) are solved by Cramer's rule to yield the unknown solution transforms

$$Y_k(s) = \sum_{j=1}^{n} \frac{A_{jk}(s)}{D(s)} F_j(s) + \sum_{j=1}^{n} \frac{A_{jk}(s)}{D(s)} G_j(s) \qquad k = 1, 2, \ldots, n \qquad (2\text{-}21)$$

where $A_{jk}(s)$ is the cofactor of $\varphi_{jk}(s)$ in the *system determinant* $D(s) \equiv \det[\varphi_{jk}(s)]$.

30. Sinusoidal Forcing Functions and Solutions; The Phasor Method. *Every system of linear differential equations (2-19) with sinusoidal forcing functions of equal frequency,*

$$f_j(t) \equiv B_j \sin(\omega t + \beta_j) \qquad j = 1, 2, \ldots, n \qquad (2\text{-}22a)$$

admits a unique particular solution of the form

$$y_k(t) \equiv A_k \sin(\omega t + \alpha_k) \qquad k = 1, 2, \ldots, n \qquad (2\text{-}22b)$$

In particular, *if all roots of the characteristic equation* $(D(s) = 0)$ *have negative real parts* (*stable systems*), *the sinusoidal solution (2-22b) is the unique steady-state solution obtained after all transients have died out.*

Given a system of linear differential equations (2-19) relating sinusoidal forcing functions and solutions (2-22), *one introduces a reciprocal one-to-one representation of these sinusoids by corresponding complex numbers* (*vectors, phasors*)

$$F_j = \frac{B_j}{\sqrt{2}} e^{i\beta_j} = \frac{B_j}{\sqrt{2}} \underline{/\beta_j} \qquad j = 1, 2, \ldots, n$$

$$\vec{Y}_k = \frac{A_k}{\sqrt{2}} e^{i\alpha_k} = \frac{A_k}{\sqrt{2}} \underline{/\alpha_k} \qquad k = 1, 2, \ldots, n \qquad (2\text{-}23)$$

The absolute value of each phasor equals the root-mean-square value of the corresponding sinusoid, while the phasor argument defines the phase of the sinusoid. *The phasors (2-23) are related by the* (*complex*) *linear algebraic equations* (*phasor equations*)

$$\varphi_{j1}(i\omega)\vec{Y}_1 + \varphi_{j2}(i\omega)\vec{Y}_2 + \cdots + \varphi_{jn}(i\omega)\vec{Y}_n = \vec{F}_j \qquad j = 1, 2, \ldots, n \qquad (2\text{-}24)$$

which correspond to Eq. (2-19) and may be solved for the unknown phasors

$$\vec{Y}_k = \sum_{j=1}^{n} \frac{A_{jk}(i\omega)}{D(i\omega)} \vec{F}_j \qquad k = 1, 2, \ldots, n \tag{2-25}$$

MATRICES AND LOGIC

31. Rectangular Matrices. An array

$$A \equiv \begin{bmatrix} a_{11} & a_{12} & \cdots & a_{1n} \\ a_{21} & a_{22} & \cdots & a_{2n} \\ \vdots & \vdots & \vdots & \vdots \\ a_{m1} & a_{m2} & \cdots & a_{mn} \end{bmatrix} \equiv [a_{ik}]$$

of real or complex numbers (scalars) a_{ik} will be called a real or complex *rectangular* $m \times n$ *matrix* whenever one of the matrix operations defined in Par. 2-32 is to be used. The elements a_{ik} are called *matrix elements*; the matrix element a_{ik} is situated in the ith *row* and in the kth *column* of the matrix; m is the number of rows, and n is the number of columns.

32. Basic Operations. Operations on matrices are defined in terms of operations on the matrix elements.

1. Two $m \times n$ matrices $A \equiv [a_{ik}]$ and $B \equiv [b_{ik}]$ are *equal* $(A = B)$ if and only if $a_{ik} = b_{ik}$ for all i, k.

2. The *sum of two $m \times n$ matrices* $A \equiv [a_{ik}]$ and $B \equiv [b_{ik}]$ is the $m \times n$ matrix

$$A + B \equiv [a_{ik}] + [b_{ik}] \equiv [a_{ik} + b_{ik}]$$

3. The *product of the $m \times n$ matrix* $A \equiv [a_{ik}]$ *by the scalar* α is the $m \times n$ matrix

$$\alpha A \equiv \alpha[a_{ik}] \equiv [\alpha a_{ik}]$$

4. The *product of the $m \times n$ matrix* $A \equiv [a_{ij}]$ *and the $n \times r$ matrix* $B \equiv [b_{jk}]$ is the $m \times r$ matrix

$$AB \equiv [a_{ij}][b_{jk}] \equiv \left[\sum_{j=1}^{n} a_{ij} b_{jk} \right]$$

In every matrix product AB the number n of columns of A must match the number of rows of B (A and B must be *conformable*). The existence of AB implies that of BA if and only if A and B are square matrices; in general $BA \neq AB$. Note

$$A + B = B + A \qquad A + (B + C) = (A + B) + C$$
$$\alpha(\beta A) = (\alpha\beta)A \qquad \alpha(AB) = (\alpha A)B = A(\alpha B)$$
$$A(BC) = (AB)C$$
$$\alpha(A + B) = \alpha A + \alpha B \qquad (\alpha + \beta)A = \alpha A + \beta A$$
$$A(B + C) = AB + AC \qquad (B + C)A = BA + CA$$

33. Identities and Inverses. Note the following definitions:

1. The $m \times n$ *null matrix* (*additive identity*) $[0]$ is the $m \times n$ matrix all of whose elements are equal to zero. Then

$$A + [0] = A \qquad [0]A = [0]$$
$$[0]B = C[0] = [0]$$

where A is any $m \times n$ matrix, B is any matrix having n rows, and C is any matrix having m columns.

2. The *additive inverse* (*negative*) $-A$ of the $m \times n$ matrix $A \equiv [a_{ik}]$ is the $m \times n$ matrix

$$-A \equiv (-1)A \equiv [-a_{ik}]$$

with $A + (-A) = A - A = [0]$.

3. The *identity matrix (unit matrix, multiplicative identity) I of order n* is the $n \times n$ diagonal matrix with unit diagonal elements:

$$I \equiv [\delta_k^i] \text{ where}^1 \qquad \delta_k^i = \begin{cases} 0 & \text{if } i \neq k \\ 1 & \text{if } i = k \end{cases}$$

Then

$$IB = B \qquad CI = C$$

where B is any matrix having n rows, and C is any matrix having n columns. For any $n \times n$ matrix A

$$IA = AI = A$$

4. A (necessarily square) matrix *A* is *nonsingular (regular)* if and only if it has a (necessarily unique) *multiplicative inverse* or *reciprocal A^{-1}* defined by

$$AA^{-1} = A^{-1}A = I$$

Otherwise *A* is a *singular* matrix.

A finite $n \times n$ matrix $A \equiv [a_{ik}]$ is nonsingular if and only if $\det(A) \equiv \det[a_{ik}] \neq 0$; in this case A^{-1} is the $n \times n$ matrix

$$A^{-1} \equiv [a_{ik}]^{-1} \equiv \left[\frac{A_{ki}}{\det[a]_{ik}} \right]$$

where A_{ik} is the cofactor of the element a_{ik} in the determinant $\det[a_{ik}]$.

---
[1] The symbol δ_k^i (or δ_i^k) is known as the *Kronecker delta.*

34. Matrix Notation for Simultaneous Linear Equations.   A set of simultaneous linear equations

$$\sum_{k=1}^{n} a_{ik} x_k = b_i \qquad i = 1, 2, \ldots, m$$

is equivalent to the matrix equation

$$Ax = b \qquad \text{or} \qquad \begin{bmatrix} a_{11} & a_{12} & \cdots & a_{1n} \\ a_{21} & a_{22} & \cdots & a_{2n} \\ \vdots & \vdots & \vdots & \vdots \\ a_{m1} & a_{m2} & \cdots & a_{mn} \end{bmatrix} \begin{bmatrix} x_1 \\ x_2 \\ \vdots \\ x_n \end{bmatrix} = \begin{bmatrix} b_1 \\ b_2 \\ \vdots \\ b_m \end{bmatrix} \qquad (2\text{-}26)$$

The unknowns x_k may be regarded as components of an unknown vector such that the transformation (2-26) yields the vector represented by the b_i. If, in particular, the matrix $[a_{ik}]$ is nonsingular (Par. 2-33), then the matrix equation (2-26) can be solved to yield the unique result

$$x = A^{-1}b$$

which is equivalent to Cramer's rule.

35. Boolean Algebras.   A *Boolean algebra* is a class S of objects $A, B, C \ldots$ admitting two binary operations, denoted as (*logical*) *addition and multiplication*, with the following properties.

For all A, B, C in S

1. S contains $A + B$ and AB (closure)
2. $A + B = B + A$
    $AB = BA$ (commutative laws)
3. $A + (B + C) = (A + B) + C$
    $A(BC) = (AB)C$ (associative laws)
4. $A(B + C) = AB + AC$
    $A + BC = (A + B)(A + C)$ (distributive laws)
5. $A + A = AA = A$ (idempotency)

6. $A + B = B$ if and only if $AB = A$ (consistency)
In addition,

7. $\mathcal{S}$ contains elements I and 0 such that, for every A in $\mathcal{S}$,

$$A + 0 = A \qquad AI = A$$
$$A0 = 0 \qquad A + I = I$$

8. For every element A, $\mathcal{S}$ contains an element $\tilde{A}$ (*complement* of A, also written $\bar{A}$ or $I - A$) such that

$$A + \tilde{A} = I \qquad A\tilde{A} = 0$$

In every Boolean algebra

$$A(A + B) \equiv A + AB \equiv A \qquad \text{(laws of absorption)}$$

$$\left. \begin{array}{c} \widetilde{(A + B)} \equiv \tilde{A}\tilde{B} \\ \\ \widetilde{(AB)} \equiv \tilde{A} + \tilde{B} \end{array} \right\} \quad \text{(dualization, or De Morgan's laws)}$$

$$\tilde{\tilde{A}} \equiv A \qquad \tilde{I} = 0 \qquad \tilde{0} = I$$

$$A + \tilde{A}B \equiv A + B \qquad AB + AC + B\tilde{C} \equiv AC + B\tilde{C}$$

If $A + B = B$, one may write AB as $B - A$ (*complement of A with respect to B*). Two or more objects A, B, C, ... of a Boolean algebra are *disjoint* if and only if every product involving distinct elements of the set equals 0.

The symbols $\cup$ (cup) and $\cap$ (cap) are frequently employed to denote logical addition and multiplication in any Boolean algebra, so that $A \cup B$ stands for $A + B$, and $A \cap B$ stands for AB.

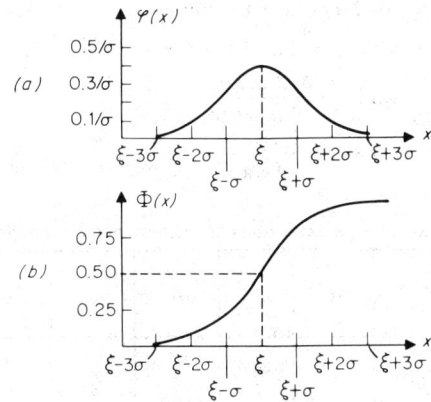

Fig. 2-2. (*a*) The normal frequency function; (*b*) the normal distribution function.

FORMULAS FOR TRIGONOMETRIC AND HYPERBOLIC FUNCTIONS

36. Relations between the Trigonometric Functions.   The basic relations

$$\sin^2 z + \cos^2 z = 1 \quad \text{and} \quad \frac{\sin z}{\cos z} = \tan z = \frac{1}{\cot z}$$

yield

$$\sin z = \pm\sqrt{1 - \cos^2 z}$$

$$= \frac{\tan z}{\pm\sqrt{1 + \tan^2 z}} = \frac{1}{\pm\sqrt{1 + \cot^2 z}}$$

2-29

$$\cos z = \pm\sqrt{1 - \sin^2 z}$$

$$= \frac{1}{\pm\sqrt{1 + \tan^2 z}} = \frac{\cot z}{\pm\sqrt{1 + \cot^2 z}}$$

$$\tan z = \frac{\sin z}{\pm\sqrt{1 - \sin^2 z}} = \frac{\pm\sqrt{1 - \cos^2 z}}{\cos z} = \frac{1}{\cot z}$$

$$\cot z = \frac{\pm\sqrt{1 - \sin^2 z}}{\sin z} = \frac{\cos z}{\pm\sqrt{1 - \cos^2 z}} = \frac{1}{\tan z}$$

37. Addition and Multiple-Angle Formulas.   The basic relation

$$\sin(A + B) = \sin A \cos B + \sin B \cos A$$

yields

$$\sin(A \pm B) = \sin A \cos B \pm \cos A \sin B$$

$$\cos(A \pm B) = \cos A \cos B \mp \sin A \sin B$$

$$\tan(A \pm B) = \frac{\tan A \pm \tan B}{1 \mp \tan A \tan B}$$

$$\cot(A \pm B) = \frac{\cot A \cot B \mp 1}{\cot A \pm \cot B}$$

$$\sin 2A = 2 \sin A \cos A$$

$$\cos 2A = \cos^2 A - \sin^2 A$$

$$= 2 \cos^2 A - 1 = 1 - 2 \sin^2 A$$

$$\tan 2A = \frac{2 \tan A}{1 - \tan^2 A}$$

$$\cot 2A = \frac{\cot^2 A - 1}{2 \cot A} = \tfrac{1}{2}(\cot A - \tan A)$$

Table 2-11   Special Values of Trigonometric Functions

A (degrees)	$0°$ $360°$	$30°$	$45°$	$60°$	$90°$	$180°$	$270°$
A (radians)	0	$\dfrac{\pi}{6}$	$\dfrac{\pi}{4}$	$\dfrac{\pi}{3}$	$\dfrac{\pi}{2}$	π	$\dfrac{3\pi}{2}$
$\sin A$	0	$\tfrac{1}{2}$	$\dfrac{1}{\sqrt{2}}$	$\tfrac{1}{2}\sqrt{3}$	1	0	-1
$\cos A$	1	$\tfrac{1}{2}\sqrt{3}$	$\dfrac{1}{\sqrt{2}}$	$\tfrac{1}{2}$	0	-1	0
$\tan A$	0	$\dfrac{1}{\sqrt{3}}$	1	$\sqrt{3}$	$\pm\infty$	0	$\pm\infty$
$\cot A$	$\pm\infty$	$\sqrt{3}$	1	$\dfrac{1}{\sqrt{3}}$	0	$\pm\infty$	0

Table 2-12 Relations between Trigonometric Functions of Different Arguments

	$-A$	$90° \pm A$	$180° \pm A$	$270° \pm A$	$n360° \pm A$
sin	$-\sin A$	$\cos A$	$\mp \sin A$	$-\cos A$	$\pm \sin A$
cos	$\cos A$	$\mp \sin A$	$-\cos A$	$\pm \sin A$	$\cos A$
tan	$-\tan A$	$\mp \cot A$	$\pm \tan A$	$\mp \cot A$	$\pm \tan A$
cot	$-\cot A$	$\mp \tan A$	$\pm \cot A$	$\mp \tan A$	$\pm \cot A$

$$\sin \frac{A}{2} = \pm \sqrt{\frac{1 - \cos A}{2}}$$

$$\cos \frac{A}{2} = \pm \sqrt{\frac{1 + \cos A}{2}}$$

$$\tan \frac{A}{2} = \frac{\sin A}{1 + \cos A} = \frac{1 - \cos A}{\sin A}$$

$$\cot \frac{A}{2} = \frac{\sin A}{1 - \cos A} = \frac{1 + \cos A}{\sin A}$$

$$a \sin A + b \cos A = r \sin(A + B) = r \cos(90° - A - B)$$

$$r = +\sqrt{a^2 + b^2} \qquad \tan B = \frac{b}{a}$$

$$\sin A \pm \sin B = 2 \sin \frac{A \pm B}{2} \cos \frac{A \mp B}{2}$$

$$\cos A + \cos B = 2 \cos \frac{A + B}{2} \cos \frac{A - B}{2}$$

$$\cos A - \cos B = -2 \sin \frac{A + B}{2} \sin \frac{A - B}{2}$$

$$\tan A \pm \tan B = \frac{\sin(A \pm B)}{\cos A \cos B}$$

$$\cot A \pm \cot B = \frac{\sin(B \pm A)}{\sin A \sin B}$$

$$2 \cos A \cos B = \cos(A - B) + \cos(A + B)$$

$$2 \sin A \sin B = \cos(A - B) - \cos(A + B)$$

$$2 \sin A \cos B = \sin(A - B) + \sin(A + B)$$

$$2 \cos^2 A = 1 + \cos 2A$$

$$2 \sin^2 A = 1 - \cos 2A$$

38. Hyperbolic Functions

$$\sinh z = \frac{e^z - e^{-z}}{2} \qquad \cosh z = \frac{e^z + e^{-z}}{2}$$

$$\tanh z = \frac{\sinh z}{\cosh z} \qquad \coth z = \frac{\cosh z}{\sinh z}$$

$$\text{sech } z = \frac{1}{\cosh z} \qquad \text{cosech } z = \frac{1}{\sinh z}$$

$$\cosh^2 z - \sinh^2 z = 1$$

$$\frac{\sinh z}{\cosh z} = \tanh z = \frac{1}{\coth z}$$

yield

$$\sinh z = \pm\sqrt{\cosh^2 z - 1} = \frac{\tanh z}{\pm\sqrt{1 - \tanh^2 z}} = \frac{1}{\pm\sqrt{\coth^2 z - 1}}$$

$$\cosh z = \pm\sqrt{1 + \sinh^2 z} = \frac{1}{\pm\sqrt{1 - \tanh^2 z}} = \frac{\coth z}{\pm\sqrt{\coth^2 z - 1}}$$

$$\tanh z = \frac{\sinh z}{\pm\sqrt{1 + \sinh^2 z}} = \frac{\pm\sqrt{\cosh^2 z - 1}}{\cosh z} = \frac{1}{\coth z}$$

$$\coth z = \frac{\pm\sqrt{1 + \sinh^2 z}}{\sinh z} = \frac{\cosh z}{\pm\sqrt{\cosh^2 z - 1}} = \frac{1}{\tanh z}$$

39. Formulas Relating Hyperbolic Functions of Compound Arguments

$$\sinh(A \pm B) = \sinh A \cosh B \pm \cosh A \sinh B$$

$$\cosh(A \pm B) = \cosh A \cosh B \pm \sinh A \sinh B$$

$$\tanh(A \pm B) = \frac{\tanh A \pm \tanh B}{1 \pm \tanh A \tanh B}$$

$$\coth(A \pm B) = \frac{\coth A \coth B \pm 1}{\coth B \pm \coth A}$$

$$\sinh 2A = 2 \cosh A \sinh A \qquad \cosh 2A = \cosh^2 A + \sinh^2 A$$

$$\tanh 2A = \frac{2 \tanh A}{1 + \tanh^2 A} \qquad \coth 2A = \frac{\coth^2 A + 1}{2 \coth A}$$

$$\sinh \frac{A}{2} = \pm\sqrt{\frac{\cosh A - 1}{2}} \qquad \cosh \frac{A}{2} = \pm\sqrt{\frac{\cosh A + 1}{2}}$$

$$\tanh \frac{A}{2} = \frac{\sinh A}{\cosh A + 1} = \frac{\cosh A - 1}{\sinh A}$$

$$\coth \frac{A}{2} = \frac{\sinh A}{\cosh A - 1} = \frac{\cosh A + 1}{\sinh A}$$

$$\sinh A \pm \sinh B = 2 \sinh \frac{A \pm B}{2} \cosh \frac{A \mp B}{2}$$

$$\cosh A + \cosh B = 2 \cosh \frac{A + B}{2} \cosh \frac{A - B}{2}$$

$$\cosh A - \cosh B = 2 \sinh \frac{A + B}{2} \sinh \frac{A - B}{2}$$

$$\tanh A \pm \tanh B = \frac{\sinh(A \pm B)}{\cosh A \cosh B} \qquad \coth A \pm \coth B = \frac{\sinh(B \pm A)}{\sinh A \sinh B}$$

$$2 \cosh A \cosh B = \cosh(A + B) + \cosh(A - B)$$

$$2 \sinh A \sinh B = \cosh(A + B) - \cosh(A - B)$$

$$2 \sinh A \cosh B = \sinh(A + B) + \sinh(A - B)$$

$$2 \cosh^2 A = 1 + \cosh 2A \qquad 2 \sinh^2 A = \cosh 2A - 1$$

40. Relations between Exponential, Trigonometric, and Hyperbolic Functions

$$e^{iz} = \cos z + i \sin z \qquad \cos z = \frac{e^{iz} + e^{-iz}}{2} \qquad \sin z = \frac{e^{iz} - e^{-iz}}{2i}$$

$$e^{-iz} = \cos z - i \sin z \qquad e^z = \cosh z + \sinh z \qquad e^{-z} = \cosh z - \sinh z$$

$$\cosh z = \frac{e^z + e^{-z}}{2} \qquad \sinh z = \frac{e^z - e^{-z}}{2}$$

$$\cos z = \cosh iz \qquad \cosh z = \cos iz$$

$$\sin z = -i \sinh iz \qquad \sinh z = -i \sin iz$$

$$\tan z = -i \tanh iz \qquad \tanh z = -i \tan iz$$

$$\cot z = i \coth iz \qquad \coth z = i \cot iz$$

$$\log_e z = \log_e |z| + i \arg (z)$$

$$\log_e (ix) = \log_e x + (2n + 1/2)\pi i$$
$$\log_e (-x) = \log_e x + (2n + 1)\pi i$$
$$n = 0, \pm 1, \pm 2, \ldots$$

$$\arccos z = i \cosh^{-1} z \qquad \cosh^{-1} z = i \arccos z$$

$$\arcsin z = -i \sinh^{-1} iz \qquad \sinh^{-1} z = -i \arcsin iz$$

$$\arctan z = -i \tanh^{-1} iz \qquad \tanh^{-1} z = -i \arctan iz$$

$$\text{arccot } z = i \coth^{-1} iz \qquad \coth^{-1} z = i \text{ arccot } iz$$

$$\arccos z = i \log_e (z + i\sqrt{1 - z^2}) \qquad \cosh^{-1} z = \log_e (z + \sqrt{z^2 - 1})$$

$$\arcsin z = -i \log_e (iz + \sqrt{1 - z^2}) \qquad \sinh^{-1} z = \log_e (z + \sqrt{z^2 + 1})$$

$$\arctan z = -\frac{i}{2} \log_e \frac{1 + iz}{1 - iz} \qquad \tanh^{-1} z = \tfrac{1}{2} \log_e \frac{1 + z}{1 - z}$$

$$\text{arccot } z = -\frac{i}{2} \log_e \frac{iz - 1}{iz + 1} \qquad \coth^{-1} z = \tfrac{1}{2} \log_e \frac{z + 1}{z - 1}$$

41. Power-Series Expansions

$$\frac{1}{1 - z} = 1 + z + z^2 + \cdots \qquad |z| < 1 \qquad \text{geometric series}$$

$$(1 + z)^p = 1 + \binom{p}{1} z + \binom{p}{2} z^2 + \cdots \qquad |z| < 1 \qquad \text{binomial series}$$

$$e^z = 1 + z + \frac{z^2}{2!} + \frac{z^3}{3!} + \cdots \qquad z \neq \infty$$

$$\sin z = z - \frac{z^3}{3!} + \frac{z^5}{5!} \mp \cdots \qquad \cos z = 1 - \frac{z^2}{2!} + \frac{z^4}{4!} \mp \cdots \qquad z \neq \infty$$

$$\sinh z = z + \frac{z^3}{3!} + \frac{z^5}{5!} + \cdots \qquad \cosh z = 1 + \frac{z^2}{2!} + \frac{z^4}{4!} + \cdots \qquad z \neq \infty$$

$$\log_e (1 + z) = z - \frac{z^2}{2} + \frac{z^3}{3} - \frac{z^4}{4} \pm \cdots \qquad |z| < 1$$

$$\arcsin z = z + \frac{1}{2} \cdot \frac{z^3}{3} + \frac{1}{2} \cdot \frac{3}{4} \cdot \frac{z^5}{5} + \frac{1}{2} \cdot \frac{3}{4} \cdot \frac{5}{6} \cdot \frac{z^7}{7} + \cdots$$

$$\sinh^{-1} z = z - \frac{1}{2} \cdot \frac{z^3}{3} + \frac{1}{2} \cdot \frac{3}{4} \cdot \frac{z^5}{5} - \frac{1}{2} \cdot \frac{3}{4} \cdot \frac{5}{6} \cdot \frac{z^7}{7} \pm \cdots$$
$$|z| < 1$$

$$\arctan z = z - \frac{z^3}{3} + \frac{z^5}{5} \mp \cdots$$
$$|z| < 1$$

$$\tanh^{-1} z = \tfrac{1}{2} \log_e \frac{1 + z}{1 - z} = z + \frac{z^3}{3} + \frac{z^5}{5} + \cdots$$

BIBLIOGRAPHY

42. Mathematical Reference Books

ABRAMOWITZ, M., and I. A. STEGUN   "Handbook of Mathematical Functions," National Bureau of Standards, Washington, D.C., 1964.

BIERENS DE HAAN, D.   "Nouvelles Tables d'intégrales Définies," Stechert, New York, 1939.

Byrd, P. F., and M. D. Friedman   "Handbook of Elliptic Integrals for Engineers and Physicists," Springer, Berlin, 1954.

Gröbner and Hofreiter   "Integral Tafel," 2d ed., Springer, Vienna, 1958.

Korn, G. A., and T. M. Korn   "Mathematical Handbook for Scientists and Engineers," 2d ed., McGraw-Hill, New York, 1968.

Lindman, C. F.   "Examen des novelles tables d'intégrales définies de M. Bierens de Haan," 1944.

Luke, Y. I.   "Integrals of Bessel Functions," McGraw-Hill, New York, 1962.

Petit Bois, G.   "Tables of Indefinite Integrals," Dover, New York, 1961.

Ryshik, I. M., and I. S. Gradstein   "Tables of Series, Products, and Integrals," Academic, New York, 1964.

Short tables of transcendental functions

Abramowitz, M., and I. A. Stegun   "Handbook of Mathematical Functions," National Bureau of Standards, Washington, D.C., 1964.

Dwight, H. B.   "Mathematical Tables," McGraw-Hill, New York, 1941.

Flügge, W.   "Four-Place Tables of Transcendental Functions," McGraw-Hill, New York, 1954.

Jahnke, E., and F. Emde   "Tables of Functions with Formulae and Curves," Dover, New York, 1945.

Statistical tables

Beyer, W. H.   "CRC Handbook of Tables for Probability and Statistics," Chemical Rubber Co., Cleveland, Ohio, 1966.

Burington, R. S., and D. C. May   "Handbook of Probability and Statistics," 2d ed., McGraw-Hill, New York, 1967.

Hald, A.   "Statistical Tables and Formulas," Wiley, New York, 1952.

Meredith, W.   "Mathematical and Statistical Tables," McGraw-Hill, New York, 1967.

Owen, D. B.   "Handbook of Statistical Tables," Addison-Wesley, Reading, Mass., 1962.

Pearson, E. S., and H. O. Hartley   "Biometrika Tables for Statisticians," Cambridge University Press, New York, 1956.

Indices to numerical tables

Etherington, Harold (ed.)   "Nuclear Engineering Handbook," McGraw-Hill, New York, 1958.

Fletcher, A.   Guide to Tables of Elliptic Functions, "Mathematical Tables and Other Aids to Computation," Vol. 3, No. 24, 1948.

———, J. C. P. Miller, and L. Rosenhead   "Index of Mathematical Tables," Addison-Wesley, Reading, Mass., 1962.

Greenwood, J. A., and H. O. Hartley   "Guide to Tables in Mathematical Statistics," Princeton University Press, Princeton, N.J. 1962.

SECTION 3

CIRCUIT PRINCIPLES

BY

EVERARD M. WILLIAMS Late George Westinghouse Professor of Engineering, Carnegie-Mellon University; Fellow, IEEE

WILS L. COOLEY Assistant Professor of Electrical Engineering, West Virginia University; Member, IEEE

CONTENTS

Numbers refer to paragraphs

SECTION 3

CIRCUIT PRINCIPLES

ELECTRIC-CIRCUIT CONCEPTS

1. Electric-circuit theory deals with the behavior of electric apparatus (components, devices, equipment, systems, etc.) as measured by the values and fluctuations of the *circuit variables*, the voltages between various terminals and between various conductors and the currents at and in various terminals and conductors.

Electric-circuit theory has two principal divisions: *equivalent circuits* for such circuit components and devices as can be treated by circuit theory and the *mathematical analysis* of the equivalent circuits.

For digital systems, the final equivalent circuit constitutes a logic diagram. In either case, the equivalent circuit represents a concise summary of the elements of a mathematical model for the behavior of the apparatus.

2. Equivalent-circuit theory is also widely known as *network theory*; the term *network*, in fact, is often applied synonymously with *circuit*. However, simple configurations of apparatus are more commonly known as circuits, while more complicated configurations (many connections and many devices or elements) are more likely to be referred to as networks. In network *topology*, the field concerned with the geometric configurations of networks, the term *circuit* often designates a *loop*.

Equivalent circuits are classed as *linear* or *nonlinear*, *reciprocal* or *nonreciprocal*, and *unilateral* or *bilateral*. There are various (essentially synonymous) descriptions for the property of linearity, one of which is as follows: if the response of a circuit, circuit element, etc., to an input $f_1(t)$ is $g_1(t)$ and its response to an input $f_2(t)$ is $g_2(t)$, where $f_1(t)$ and $f_2(t)$ are arbitrary functions (including arbitrary amplitude), the circuit is *linear* if its response to an input $f_1(t) + f_2(t)$ is equal to $g_1(t) + g_2(t)$. This description is an illustration of the *superposition principle*.

A *bilateral* circuit is one in which the circuit behavior is not changed if the terminal connections of the elements are interchanged. A rectifier conducts current easily in one direction and with difficulty in the other and is therefore *unilateral*. Other typical unilateral devices are vacuum tubes and transistors. Linear circuits are necessarily bilateral.

Unless specifically stated otherwise the treatment of analog equivalent circuits at every point in this section is limited to circuits which are linear. The question of linearity does not arise in the logic diagrams, which are mathematical models of digital circuits.

Reciprocal circuits are circuits which satisfy the reciprocity theorem (see Par. **3-13**). All linear bilateral equivalent circuits are reciprocal unless they contain *gyrators*.

3. Circuit Elements and Configurations. Equivalent circuits are composed of interconnected basic building blocks, e.g., resistances, capacitances, inductances, and voltage and current sources. The following, with examples given in Figs. 3-1 and 3-2, are common terms for describing circuits. *Elements*, or *circuit elements*, are the smallest units into which circuits can be subdivided; in Fig. 3-1, examples of *two-terminal* circuit elements are resistances R_1, $R_2, \ldots, R_6$ and capacitance C_3; magnetically coupled inductors L_a and L_b are an example of a *four-terminal* circuit element.

Any point on an element, component, device, etc., to which a conductor is attached is called a *terminal*. Any portion of a circuit between two terminals, like that between terminals a and b, b and e, or f and b, is a *branch*. The portion between a and f is also a branch because the induced voltage in L_b caused by current in L_a acts in series with the circuit.

A point at which conductors from two or more elements join is sometimes called a *node*; this term, however, is more often restricted to points at which conductors from three or more branches join.

Any specified sequence of branches, e.g., in Fig. 3-1 from *a* to *b* to *c* or from *a* to *b* to *d* to *e*, is a *path*. Any path, such as *a* to *b* to *f* to *a*, or *a* to *b* to *d* to *e* to *f* to *a*, which ends at the point at which it starts is a *closed path*, also known in circuit theory (in contrast with network topology) as a *loop*.

An equivalent circuit which can be laid out on a plane without any crossing of conductors is a *planar circuit*. Any loop in a circuit which does not enclose another loop, such as loops *abfga* and *bdefb* in Fig. 3-1, is also a *mesh*, also known as a *window* (the term mesh is sometimes limited to planar circuits).

Any element which receives energy from some source other than a signal generator is an *active element*. The term is also sometimes applied to any element which supplies energy to the rest of the circuit. Any element to which energy is supplied by other parts of the circuit is a *passive element*, also, in principle, a *load*, although the latter term is often applied only to a passive element to which energy is intentionally supplied.

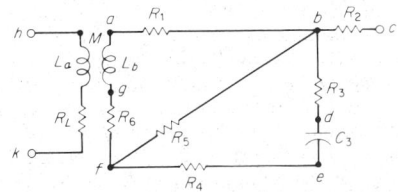

Fig. 3-1. Example of an equivalent circuit. Inductances L_a and L_b are magnetically coupled. Each is associated with one of the two separate parts in this circuit, i.e., with parts not physically connected with one another.

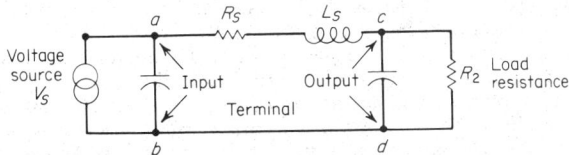

Fig. 3-2. Example of a complete equivalent circuit with source and load. Terminal pairs *a, b* and *c, d* are input and output terminals, or ports, respectively.

Many circuits transmit, shape, amplify, or modify the energy (or signals) supplied by a source before delivering the energy (or signals) onward to a load. Each pair of terminals of such a circuit to which a source or load is connected is termed a *port*, an appropriate terminology because such terminal pairs acts as points of entry or exit from the circuit. In a circuit with but one port, the circuit itself is the load; circuits with two ports commonly have an input and output port. A circuit may have *n* ports; when only one port of a *n*-port circuit is an input, this port is also known as the *driving point* (see Fig. 3-2).

4. Circuit Calculations. Once an equivalent circuit is formulated, circuit theory provides a method for calculating the behavior of the currents and voltages. For a circuit with *n* elements, there are, to start with, 2*n* unknowns, since there is a voltage drop and current associated with each element. When the known voltage-drop vs. current characteristics of each element are taken into account, there remain *n* unknowns.

The *n* equations required to determine these unknowns are obtained by applying Kirchhoff's laws. *Kirchhoff's current law* (KCL) states that the algebraic sum of the instantaneous currents entering a node (in this case a node should be the common terminal of three or more branches) is zero. A necessary and sufficient condition for independence of the resulting current equations is that each equation contain at least one current which does not appear in another current-law equation. *Kirchhoff's voltage law* (KVL) states that the algebraic sum of the instantaneous voltages around any loop is zero. The number of independent voltage-law equations is equal to the number of meshes in the circuit.

In applying Kirchhoff's laws, two strategies are used in writing final circuit equations to minimize the number of simultaneous equations to be solved. In *loop (or mesh) analysis*, the currents in the various branches are treated as unknowns, and the voltage drops are written in terms of these currents, with only the minimum number of unknown currents necessary (on the basis of KCL) specified at each node. In *nodal analysis*, the branch voltages are treated as the unknowns, with only the minimum number of unknown voltages necessary (on the basis of KVL) specified in each loop.

The result of either strategy for writing the individual equations is a system of simultaneous equations which can be solved to obtain a single equation in a single current or voltage as a function of the parameters of the circuit elements and the source or sources. If the original equations have been written with the instantaneous voltages $v_i(t)$ or currents $i_i(t)$ as variables, this single equation is a differential equation *of order equal to the number of independent energy-storing circuit elements in the circuit.* If the original equations have been written with voltages or currents in *sinor* form (for which the properties of the elements necessarily appear in *phasor* form), the single equation is an algebraic equation.

Solution of the differential equation referred to above, with appropriate initial conditions (which depend upon the initial energy states of the energy-storing circuit elements), yields an equation in the dependent variable or variables (voltages or currents) as a function of time; it contains a *transient* component, which generally decays with time (unless the circuit is unstable), and a *steady-state* component, which continues indefinitely. Solution of the algebraic equation with sinor and phasor terms yields an equation for only the steady-state behavior of the dependent variable or variables. Except first- or second-order equations, the differential-equation solutions for most circuits, however, are formidably difficult and rarely feasible without numerical values of the circuit elements, which makes it difficult to test various values of design parameters in order to obtain a desired performance. For high-order differential equations the most practicable solution is with an analog computer or an analog simulation on a digital computer. The sinor and phasor steady-state solution, on the other hand, is straightforward, albeit possibly algebraically complicated, regardless of the circuit complexity.

5. Circuit Excitations and Responses. The conventional analysis of those equivalent circuits of interest to electronic engineers deals with three classes of systems: *power-supply systems*, in which the circuits deliver voltage and current at appropriate values required to operate electronic equipment; *signal systems*, apparatus and networks associated with the origination, transmission, distribution, and reproduction of electric signals, e.g., telephone apparatus and the circuit portions of radio, television, and radar apparatus; and *control systems*, apparatus in which information is sensed (or fed in the form of commands) and used to control the application of energy, e.g., automatic controls for machine tools, automatic manufacturing processes, and automatic positioning systems.

In treating the equivalent circuits of *signal* and *control systems*, the engineer is basically concerned in determining the *output performance* of the circuit, i.e., the *output* which results from some *input*. Some examples of output characteristics of interest are shown in Table 3-1.

6. Frequency-Domain Analysis. The most common periodic excitation is of the form $g_1(t) = A_1 \sin 2\pi f_n t$, a sinusoid of amplitude A_1 (voltage or current, as the case may be) and frequency f_n in hertz. In this case, analysis of any circuit starts with determining the corresponding steady-state output $g_2(t) = A_2 \sin(2\pi f_n t - \theta_n)$ produced by the input $g_1(t)$. This output, in general, is a sinusoid of a different amplitude, A_2, and lags the input by a phase angle θ_n, where both A_2 and θ_n are functions of frequency f_n. In this analysis, the

Table 3-1. Output Characteristics of Some Typical Circuit Types

Type of system	Typical input	Typical outputs or responses of interest
Analog signal	Analog voltage or current	Signal-to-noise ratios, fidelity, information rates, channel capacities, channel requirements
Digital signal	Digital voltage or current pulses	Logic behavior, error rates, information rates, channel capacities, spectrum requirements
Regulator	Step change in regulated output as a function of an input or load fluctuation	Response time, shape of response, limits of regulation, noise, stability
Control	Control instruction	Shape of response, response time, accuracy, stability, noise

frequency f_n is varied over a range which is determined by the *frequency spectrum* of the signals with which the circuit will be used (this frequency spectrum may include a dc component at zero frequency, known as a *continuous excitation*). The significant signal-handling properties of a circuit are usually described by the *amplitude characteristic*, the variation in the ratio of the steady-state peak output A_2 to the input A_1 as a function of frequency, and the *phase characteristic*, the variation of the phase angle θ_n as a function of frequency. The phase characteristic is sometimes expressed in terms of a time-delay characteristic; the *phase (time) delay* is the apparent time delay between the instant at which the input $g_1(t)$ reaches a particular phase condition, e.g., a maximum, minimum, or axis crossing, and the instant at which the output function $g_2(t)$ reaches the same phase condition. This phase delay τ_ϕ at frequency f_n is computed as $\tau_\phi = \theta_n/2\pi f_n$ (in seconds). The *group delay* is the time corresponding to the apparent delay of a sinusoidal amplitude-envelope variation modulated on the input function. The group delay τ_g at frequency f_n is computed as $\tau_g = (1/2\pi)\,d\theta_n/df_n$ (in seconds).

7. Fourier Transforms. The time-function expression for any time-varying signal $h(t)$ is termed its *time-domain* representation. The mathematical equivalent of the same signal in the *frequency domain* is the sum of the infinite number of infinitesimal sinusoidal signals with the relative amplitude and phase at each frequency f_n given by the *Fourier transform* $F(\omega_n)$ of the signal $h(t)$, or

$$F(\omega_n) = \frac{1}{2\pi}\int_{-\infty}^{\infty} h(t)e^{-j\omega_n t}\,dt \qquad (3\text{-}1)$$

where ω_n, the angular frequency, is equal to $2\pi f_n$.

In the above form of the Fourier integral, the relative amplitude of each component $F(\omega_n)\,d\omega$ is actually the amplitude of an exponential excitation, $F(\omega_n)e^{j\omega_n t}\,d\omega$, rather than the amplitude of a sinusoidal excitation, and the steady-state components are understood to exist in the spectrum space from $\omega_n = -\infty$ to $\omega_n = +\infty$. With sinusoidal excitation, no significance is attached to a negative frequency. In terms of the above $F(\omega)$, however, the relative component, in the conventional sinusoidal sense, at each angular frequency ω_n comprises the sum $F(+\omega_n)e^{+j\omega_n t} + F(-\omega_n)e^{-j\omega_n t}$. The exponential form of excitation, with frequencies from $\omega_n = -\infty$ to $\omega_n = +\infty$, is quite convenient analytically. It has the further advantage that in the form $e^{(+\alpha+j\omega)t}$ it can represent damped (for α negative) trains of sine waves. In this representation, the coefficient $+\alpha + j\omega$ is known as the *complex frequency*. The plot of $F(\omega_n)$ as a function of frequency is known as the *frequency spectrum* of the signal $h(t)$.

As an example of an idealized signal spectrum, consider the pulse-excitation function $h(t)$ shown in Fig. 3-3. The Fourier transform $F(\omega)$ of this time function is $F(\omega) = 2/\omega \sin(\omega\tau_1/2)$, with the frequency spectrum plotted in Fig. 3-4. The plot shows the relative amplitude of the infinitesimal amplitude component at all angular frequencies ω_n. Any practical circuit to the input of which the pulse function of Fig. 3-3 is applied will

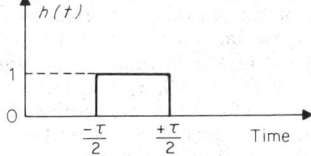

Fig. 3-3. Ideal pulse signal of duration τ_1 and unit amplitude.

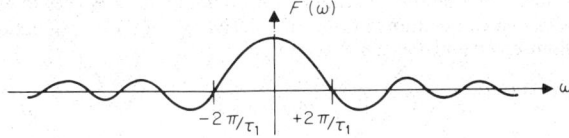

Fig. 3-4. Plot of $F(\omega) = 2/\omega \sin(\omega\tau_1/2)$, the Fourier transform of the pulse of Fig. 3-3.

CIRCUIT PRINCIPLES

necessarily cut off at some frequency less than infinity, and the resulting output pulse will have finite rise and decay times, also, probably an overshoot.

If the frequency and phase characteristics of a signal circuit are expressed as a *transfer function*, such as

$$T(\omega_n) = A_2/A_1\, e^{\,j\theta_n}$$

where A_2 is the amplitude of the output function

$$A_2\, e^{\,j(\omega_n t + \theta_n)}$$

corresponding to an input $A_1\, e^{\,j\omega_n t}$, the actual time-varying output $H(t)$ corresponding to an input $h(t)$ [for which the Fourier transform is $F(\omega_n)$] is given by the *inverse Fourier transform*

$$H(t) = \int_{-\infty}^{\infty} F(\omega_n) e^{\,j\omega_n t}\, T(\omega_n)\, d\omega \tag{3-2}$$

The procedure for determining the response of a linear circuit to a time-varying input function by the resolution of the input function into so-called *frequency components*, each of which is a steady-state sinusoid, and the calculation of the response by superposing the circuit response to each frequency taken separately is valid because the circuit is linear, in accordance with the definition of linearity given in Par. 3-2. It is thus an application of the so-called principle of *superposition* which applies to all linear systems.

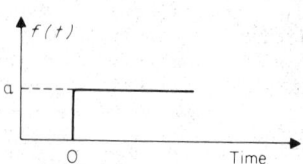

Fig. 3-5. The step function. For $t < 0$, the amplitude is zero. For $t > 0$, the amplitude is a constant, a. If $a = 1$, it is a unit step function, frequently designated as $u(t)$.

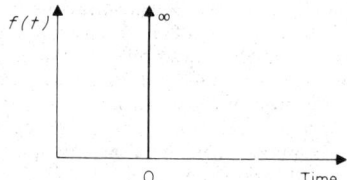

Fig. 3-6. The impulse function, of zero duration and infinite amplitude.

8. **Time-Domain Analysis.** The direct solution for a time-varying output of an electric circuit in response to a time-varying input is referred to as *time-domain* analysis. In the time domain, for example, a signal voltage v is specified by its actual variation, for example, $v = h(t)$, rather than by its equivalent frequency spectrum of steady-state sinusoids, as in the frequency-domain analysis discussed above. The direct solutions of the circuit differential equations for the (transient) response of a circuit to the nonperiodic time-varying function which corresponds to a realistic information carrying signal, however, are seldom undertaken for any but the simplest circuits and signals. While the solutions are not impossible (since the advent of the internally programmed digital computer, any calculation of circuit behavior for numerically specified excitation and circuit elements is inherently feasible), considerable information concerning the response of a circuit to realistic signals can be inferred more simply from its response to one or more of the three basic *time-domain* excitations, the *step* (Fig. 3-5), the *impulse* (Fig. 3-6), and the *ramp* (Fig. 3-7) functions. Any other time-varying excitation can be approximated by a superposition of step functions of appropriate amplitudes and starting instants. In accordance with the superposition principle, if the circuit is linear, its response to the series of steps which approximate the signal $h(t)$ can be determined by summing its responses to each step.

The *impulse function* can be thought of as the derivative of the step function.

When the impulse is defined as a $\delta(t)$ such that

$$\int_{t=0^-}^{t=0^+} \delta(t)\, dt = 1$$

it is a *unit impulse*, also known as a *Dirac function* or *delta function*. An impulse excitation of an electric circuit excites the circuit into its so-called *natural response*. Also, since the Fourier transform of the impulse yields a continuous frequency spectrum of constant amplitude from zero frequency to infinite frequency, the response of a circuit to an impulse can also be regarded as representation of its response to simultaneous excitations of a continuum of frequencies.

The *ramp function* can be regarded as the integral of a step function.

9. Two-Port Parameters and Transfer Functions. The term *transfer function*, used in the frequency-domain discussion above, represents the modification in the output imposed by a circuit on its input. If the input is a voltage or current, the output is a voltage or current, respectively, altered in amplitude and shifted in phase. In this context, the voltage-transfer function and current-transfer function are measures of the complete *two-port system performance*, starting with the *input signal source*, with its internal impedance, and ending with the output signal delivered to particular *load*, the effect of the impedance of which is included.

In another approach to circuit behavior, the properties of two-ports are described in such a way as to be independent of the impedance of the source and load connected to the input and output, respectively; in this latter context, which follows, transfer functions have a different meaning.

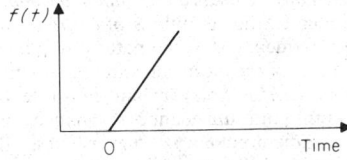

Fig. 3-7. The ramp function. For $t > 0$, $f(t) = kt$. When $k = 1$, that is, when the slope is unity, this is known as a unit ramp.

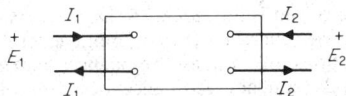

Fig. 3-8. Conventional representation of the two-port circuit. Since the input is understood to be connected to a source independent of the output, the instantaneous current I_1 entering the upper input terminal is equal and opposite to the instantaneous current I_1 leaving the lower input terminal. A similar situation occurs at the output port.

The relations between the various quantities in the circuit of Fig. 3-8, regardless of what may be connected at the input and output ports, can be shown to be given by any one pair of the following pairs of equations:

$$I_1 = y_{11}E_1 + y_{12}E_2 \tag{3-3}$$

$$I_2 = y_{21}E_1 + y_{22}E_2 \tag{3-4}$$

$$E_1 = Z_{11}I_1 + Z_{12}I_2 \tag{3-5}$$

$$E_2 = Z_{21}I_1 + Z_{22}I_2 \tag{3-6}$$

$$E_1 = h_{11}I_1 + h_{12}E_2 \tag{3-7}$$

$$I_2 = h_{21}I_1 + h_{22}E_2 \tag{3-8}$$

$$I_1 = g_{11}E_1 + g_{12}I_2 \tag{3-9}$$

$$E_2 = g_{21}E_1 + g_{22}I_2 \tag{3-10}$$

$$E_1 = AE_2 - BI_2 \tag{3-11}$$

$$I_1 = CE_2 - DI_2 \tag{3-12}$$

$$E_2 = DE_1 - BI_1 \tag{3-13}$$

$$I_2 = CE_1 - AI_1 \tag{3-14}$$

The coefficients in Eqs. (3-3) and (3-4) and (3-5) and (3-6) are known as the *short-circuit admittance* and *open-circuit impedance* parameters, respectively. The coefficients y_{21} (Z_{21}) and y_{12} (Z_{12}) are often termed the *forward transfer admittance* (*impedance*) and *reverse transfer admittance* (*impedance*), respectively. The coefficients y_{11}, y_{22} and Z_{11}, Z_{22} are often referred to as *driving-point admittances* and *driving-point impedances*, respectively.

The coefficients A, B, C, and D are known simply as the *ABCD* parameters or the *transfer* or *chain* parameters. The expressions $1/A$ and $-1/D$ are called the *open-circuit forward-voltage gain* and *short-circuit forward-current gain*, respectively. The expressions $-1/B$ and $1/C$ are called the *short-circuit forward-transfer admittance* and *open-circuit forward-transfer impedance* respectively. It can be shown by the reciprocity theorem that $AD - BC = 1$.

10. Resonant Frequencies and Resonance. Many single-port (two-terminal) circuits have singular properties at certain frequencies; examples are frequencies at which the impedance between the terminals of the circuit is at a maximum or minimum, frequencies at which this impedance is a pure resistance, and the natural frequency, or frequency of self-oscillation when the circuit is excited with a step function of current or voltage. Each of these singular frequencies is known as a *resonant frequency*.

The condition in which the terminal impedance is resistive at the applied frequency is known as *in-phase resonance*.   The conditions in which the terminal impedance is a maximum or minimum are known as *maximum impedance resonance* and *minimum impedance resonance*, respectively; these were formerly known as the conditions of *antiresonance* and *resonance*.   The condition in which the applied frequency is at the natural frequency of the circuit is the *natural-frequency resonance*.   In many of the electronic circuits in which resonance characteristics or, more generally, *tuned characteristics* play a significant role, the conditions of in-phase resonance, maximum or minimum impedance resonance, and natural-frequency resonance occur at or near the same frequency. Nevertheless, the differences in these frequencies are often significant for the behavior of the circuits.

Some examples of specific resonant frequencies are discussed in Pars. 3-20 to 3-24.

11. Filter Responses. As shown above, signals (which are fundamentally nonperiodic time-varying functions) are represented in the frequency domain by an equivalent or approximately equivalent frequency spectrum. To separate signals from one another and/ or to limit system noise and/or to limit the width of frequency spectrums, etc., *filters* are used. The basic filter-response characteristics (typically, a response in some form of output/input function) are the *low-pass*, the *high-pass*, the *bandpass*, and the *band-stop responses*, sketched in Fig. 3-9.

A *low-pass filter* passes steady-state sinusoids of all frequencies from zero (continuous, or dc excitation) up to some cutoff frequency f_1.   Ideally, such a filter would pass no sinusoids of frequency higher than f_1; in practice, the attenuation characteristic for unwanted signals falls off above the cutoff frequency with a finite slope.   It is physically impossible (indeed

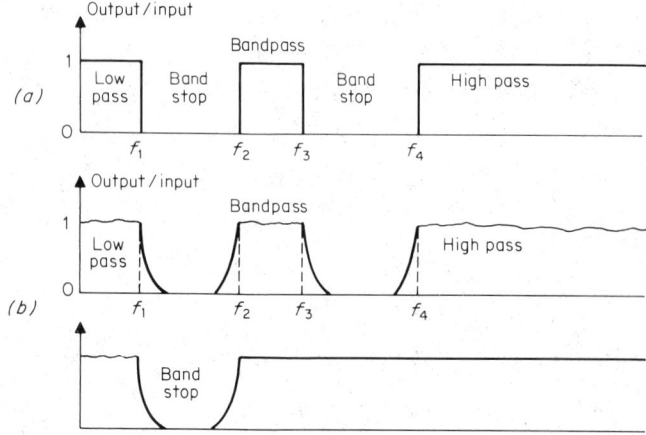

Fig. 3-9.   Filter responses: (*a*) ideal filter responses and (*b*) examples of actual responses.

undesirable) to obtain an infinitely steep cutoff above f_1 because such a cutoff entails an infinite phase shift in accordance with the minimum phase rule (see Par. **3-13**). A *bandpass filter* passes all frequencies from some f_2 to f_3, and a *high-pass filter* passes all frequencies above some f_4. A *band-stop filter* passes all frequencies except those between frequency f_1 and f_2. (Figure 3-9 shows a second stopband between f_3 and f_4.) All the attenuation characteristics outside the desired band have a finite slope for the same reason as that of the low-pass filter. The synthesis of filters to obtain desired characteristics is discussed in Sec. 12.

12. Isolation of Dc and Ac Components. The functioning of most active circuits usually requires a design for isolation of dc from ac components at certain points and the superposition of such components at others.

The dc component of a periodic time-varying function $h(t)$ is, by definition, equal to its average value $\overline{h(t)}$, given by

$$\overline{h(t)} = \frac{1}{T}\int_0^T h(t)\,dt \tag{3-15}$$

where T is the duration of one period, or cycle, of the time-varying function. Everything that remains in a periodic time-varying function after the dc component is removed, i.e.,

$$h'(t) = h(t) - \overline{h(t)} \tag{3-16}$$

is the sum of all ac components. The individual ac components can be determined, in amplitude, frequency, and phase, by expanding $h'(t)$ in a Fourier series (see Par. **3-13**).

Various elementary circuits for isolating, removing, or combining ac and dc components are illustrated in Fig. 3-10. In Fig. 3-10a, a dc source and ac source in series result in superposed ac and dc components. A capacitor, represented by capacitance C_0, passes the ac component and stops the dc component. A capacitor is the most common means for isolation of an ac component.

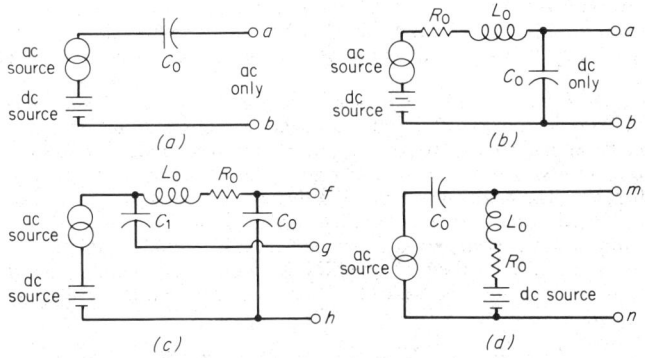

Fig. 3-10. Some elementary equivalent circuits for isolating and/or superposing dc and ac components.

In Fig. 3-10b, an inductor, represented by its equivalent inductance L_0 and resistance R_0, passes the dc component relatively easily but produces, in conjunction with the capacitance C_0 of a parallel capacitor, a large voltage drop in the ac component. The isolating properties of the circuits represented by Figs. 3-10a and b are combined in the circuit of Fig. 3-10c, in which the dc component is extracted and fed to terminals f and h and the ac component to terminals g and h. In the circuit represented by Fig. 3-10d, the dc and ac components are superposed at terminals m and n without either source interfering with the other, by the use of an inductor and a capacitor. This method of superposing the components is preferable to the series-source arrangements of Fig. 3-10a to c because the current from the ac source does not need to flow through the dc source and vice versa.

13. Circuit Theorems, Laws, and Principles. Circuit theorems, laws, and principles of primary utility in circuit analysis are listed below. Certain other theorems appear at appropriate points in other portions of this section (see also the Glossary, Par. **3-69**).

a. The Thevenin theorem states that in so far as the behavior of the circuit at its terminals is concerned, such a circuit can be replaced by a single sinor voltage source E in series with an impedance Z. The Thevenin equivalent source is illustrated in Fig. 3-11a; the value of the E is the open-circuit voltage developed between the terminals a and b, and Z can be found from the impedance observed between terminals a and b when all voltage sources are short-circuited and all current sources are open-circuited. In a dc case, E is a constant dc voltage and Z becomes a resistance R.

b. The Norton theorem states that insofar as the circuit behavior at its terminals is concerned, such a circuit can be replaced by a sinor current source I in parallel with an impedance Z. The Norton equivalent source is shown in Fig. 3-11b; the value of I is equal to the short-circuit current between terminals a and b, and the impedance Z is equal to the Thevenin impedance Z and is found in the same way. In the dc case, the Norton source becomes a constant dc current, and Z becomes a resistance R.

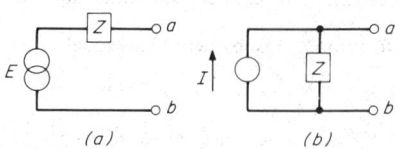

Fig. 3-11. (*a*) The Thevenin equivalent source and (*b*) the Norton equivalent source.

In certain simple cases, in which Z reduces to a non-frequency-dependent parameter, the Thevenin and Norton theorems are applicable in transient as well as steady-state calculations.

c. Superposition Principle. The superposition principle, in a form useful in simplifying circuit calculations, is as follows: in a linear circuit, the current at any point (or the voltage drop between any two points) caused by action of any number of current or voltage sources at various points in the circuit is the sum of the currents (or the sum of the voltage drops) that would be caused by each source separately; when the effect of each single source is being computed, all other sources are short-circuited if voltage sources (or open-circuited if current sources).

d. The compensation theorem, also known as the *substitution theorem,* states that any branch in a circuit can be replaced by a substitute branch, so long as the branch voltage and current remain the same, without affecting the voltages and currents in other branches of the circuit.

e. Fourier's theorem is used in harmonic analysis; by means of such analysis, any nonsinusoidal periodic function (that satisfies certain conditions which are satisfied for all practical voltage and current waveforms) can be resolved into a dc component plus sinusoidal components, which consist of a component of fundamental frequency f_0 and harmonics thereof. (The components found in this way are physically real in every sense, in that they will excite circuits tuned to the harmonics, etc.)

f. The Fourier series for a periodic function $h(t)$, of period T, in seconds, is most conveniently determined by scaling (in time) the function to a function $f(x)$ with a period equal to 2π units of time. For the scaled function, the series in trigonometric form is

$$f(x) = a_0/2 + a_1 \cos x + a_2 \cos 2x + \cdots + a_n \cos nx$$
$$+ b_1 \sin x + b_2 \sin 2x + \cdots + b_n \sin n_x \tag{3-17}$$

where

$$a_n = \frac{1}{\pi} \int_0^{2\pi} f(x) \cos nx \, dx \qquad b_n = \frac{1}{\pi} \int_0^{2\pi} f(x) \sin nx \, dx \tag{3-18}$$

In exponential form, the series is

$$f(x) = A_{-n} e^{-jnx} + \cdots + A_{-2} e^{-j2x} + A_{-1} e^{-jx} + A_0 + A_1 e^{jx}$$
$$+ A_2 e^{j2x} + \cdots + A_n e^{jnx} \tag{3-19}$$

where

$$A_n = \frac{1}{2\pi} \int_0^{2\pi} f(x) e^{-jnx} \, dx \tag{3-20}$$

The nth harmonic component of the function $f(x)$ is made up of the two terms $a_n \cos nx$ and $b_n \sin nx$, which can be combined into a single component by the identity

$$a_n \cos nx + b_n \sin nx = \sqrt{a_n^2 + b_n^2}(\sin nx + \theta_n) \tag{3-21}$$

where $\theta_n = \tan^{-1} a_n/b_n$. However, the series for many simple waveforms can be reduced to a series in sine terms only (or cosine terms only) by a judicious choice of the zero of the independent variable (x in the scaled series, t in the actual function). For example, if the zero can be chosen so that the function $f(t)$, as in Fig. 3-12a and b, is an odd or even function, the series contains only the sine or only the cosine terms, respectively.

g. Parseval's Theorem. The rms or *effective value* of a nonsinusoidal periodic function $h(t)$ is determined from the coefficients of the Fourier series for $h(t)$ as

$$[h(t)]_{rms} =$$

$$\sqrt{\left(\frac{a_0}{2}\right)^2 + \left(\frac{a_1}{2}\right)^2 + \left(\frac{a_2}{2}\right)^2 + \cdots + \left(\frac{a_n}{2}\right)^2 + \left(\frac{b_1}{2}\right)^2 + \left(\frac{b_2}{2}\right)^2 + \cdots + \left(\frac{b_n}{2}\right)^2}$$

$$\tag{3-22}$$

h. The reciprocity theorem applies to any linear bilateral circuit; it states that if a potential source is applied in one branch and a current observed in a second branch, then the same current will be observed in the first branch if the same potential is applied in the second.

i. The maximum-power-transfer-condition, sometimes termed the maximum-power-transfer theorem, describes the condition under which maximum possible power will be delivered to a load by a linear source. There are certain relations between Z_L and Z_i (Fig. 3-13) for which maximum power will be transferred from the source E to the load Z_L.

If the source is dc and Z_i and Z_L are resistances R_i and R_L, respectively, maximum power transfer will occur for $R_i = R_L$. If the source is a steady-state sinusoidal voltage source and the load impedance Z_L can be varied in any way whatever, maximum power transfer occurs for a phasor value of Z_L given by $Z_L = Z_i^*$, where the phasor Z_i^* is the *complex conjugate impedance* of the phasor Z_i; that is, if $Z_i = R_i + jX_i$, than $Z_L = R_i - jX_i$. *This yields the maximum possible power transfer under ac conditions.*

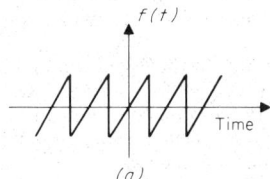

(a)

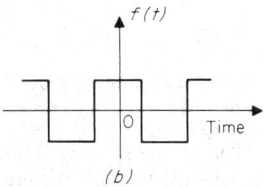

(b)

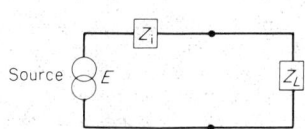

Fig. 3-12. (*a*) A sawtooth wave positioned so that it is an odd function; that is $f(t) = -f(-t)$; (*b*) a square wave positioned so that it is an even function, that is, $f(t) = f(-t)$.

Fig. 3-13. Equivalent circuit of a source (represented by its Thevenin equivalent sinor voltage E and phasor impedance Z_i) connected to a load (represented by its impedance phasor Z_L).

If the phasor Z_L can be varied in magnitude only but its phase angle is fixed, then maximum power transfer takes place when the magnitudes of the impedances are equal, i.e., when $|Z_L| = |Z_i|$. This condition results in less power transferred, in general, than $Z_L = Z_i^*$.

Finally, if the magnitude of Z_L is fixed but its phase angle θ_L can be varied, the maximum power transfer will take place in terms of the phase angle θ_i of the source impedance when

$$\sin \theta_L = -\frac{2|Z_L||Z_i|}{|Z_L|^2 + |Z_i|^2} \sin \theta_i \qquad (3\text{-}23)$$

This latter condition also results, in general, in less power transfer than $Z_L = Z_i^*$.

j. Foster's reactance theorem describes the possible reactance properties of any two-terminal circuit which contains purely reactive elements, no matter how these elements may be connected. The basic properties of interest are the resonances, represented as the poles (reactance infinite) and zeros (reactance zero) of the impedance between the terminals. For a pole at the origin ($\omega = 0$), the impedance is always represented by

$$Z = \pm j \frac{H(\omega^2 - \omega_1^2)(\omega^2 - \omega_3^2) \cdots (\omega^2 - \omega_p^2)}{\omega(\omega^2 - \omega_2^2)(\omega^2 - \omega_4^2) \cdots (\omega^2 - \omega_q^2)} \qquad (3\text{-}24)$$

For a zero at the origin,

$$Z = \pm j\omega \frac{H(\omega^2 - \omega_1^2)(\omega^2 - \omega_3^2) \cdots (\omega^2 - \omega_p^2)}{(\omega^2 - \omega_2^2)(\omega^2 - \omega_4^2) \cdots (\omega^2 - \omega_q^2)} \qquad (3\text{-}25)$$

The angular frequencies $\omega_1, \omega_3, \ldots, \omega_p$ are the frequencies at which Z is zero and $\omega_2, \omega_4, \ldots, \omega_p$ those at which there is a pole. The properties of a Foster circuit are often described by a plot like Fig. 3-14.

In addition to the frequencies of the poles and zeros, there is one other choice that must be made to specify the network completely. This amounts to specifying the constant H, which determines the numerical value of the reactance at points between the poles and zeros, and can be chosen to give the reactance a particular desired value at some one frequency. The reactances at all other frequencies are then determined. Figure 3-15 shows Foster networks with a minimum number of elements.

The constraints of the Foster function expressions can be evaded by the use in reactance networks of *negative capacitances* and/or *negative inductances* in conjunction with the positive capacitances and/or positive inductances. Such negative elements are synthesized by active circuits.[1]

k. The Minimum-Phase Rule and Phase-Area Theorem. In the discussion of frequency-domain characteristics in Par. **3-6**, *transfer function* was defined for a two-port circuit with an amplitude component equal to output/input and a phase component equal to the phase shift from the input to the output. Two important relations relate the minimum phase shift which must be developed by a two-port to the amplitude attenuation.

The first is the *phase-area theorem*, which states that the relation between attenuation and phase shift in any two-port circuit connected between a source with a resistive internal impedance and a resistive load is

$$\int_{-\infty}^{\infty} B \, d\mu = \frac{\pi}{17.37}(A_\infty - A_0) \qquad (3\text{-}26)$$

where B is the phase shift in radians, $\mu = \log_e(f/f_0)$, where f is the actual frequency and f_0 is any convenient reference frequency, A_∞ is the attenuation in decibels, at infinite frequency, and A_0 is the attenuation in decibels at zero frequency.

The above constrains the *minimum phase shift* that can occur; this is the phase shift which will result from the use of any *minimum-phase circuit*. Any actual two-port circuit will be a minimum-phase circuit unless it contains an *all-pass circuit* section, i.e., a circuit that passes all frequencies.

---

[1] See, for example, C. I. Jones and E. M. Williams, Hybrid Positive and Negative Parameter Delay-Line Synthesis, *Radio Electron. Eng.*, 1965, Vol. 29, No. 4, pp. 255–259.

An expression for the minimum phase shift that can be realized with a given attenuation characteristic is

$$B_n = \frac{\pi}{12}\left(\frac{dA}{d\mu}\right)_n + \frac{1}{6\pi}\int_{-\infty}^{\infty}\left[\frac{dA}{d\mu} - \left(\frac{dA}{d\mu}\right)_n\right]\log_e \coth\left|\frac{\mu}{2}\right| d\mu \qquad (3\text{-}27)$$

where B_n is the phase shift, in radians, at frequency f_n, at which B_n is desired, $dA/d\mu$ is the slope of the attenuation curve, in decibels per octave, $\mu = \log_e(f/f_n)$, where f is frequency, and $\log \coth|\mu/2|$ denotes the real part of the hyperbolic cotangent of $\mu/2$, which is complex when μ is negative.

A wide variety of detailed examples of phase or amplitude functions is available.[1]

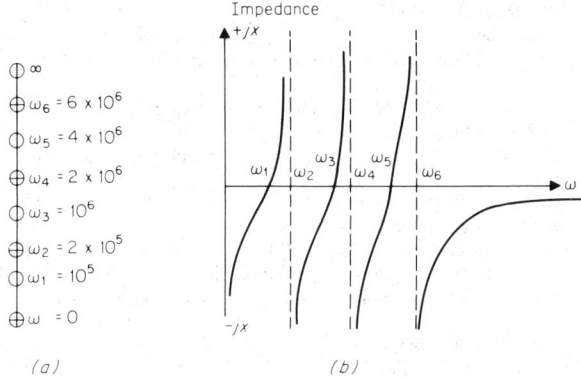

Fig. 3-14. (a) Example of the specification of the poles and zeros of a Foster circuit. (b) The actual reactance function.

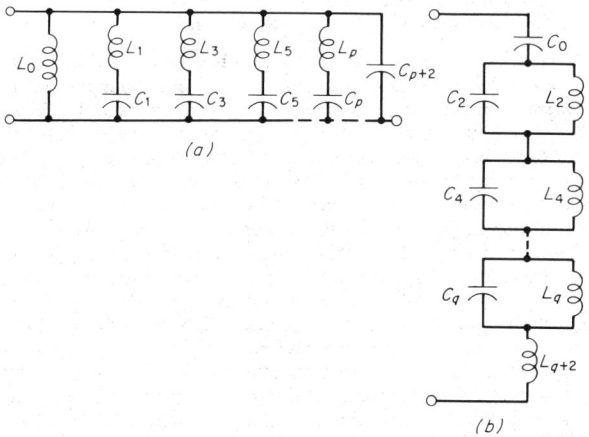

Fig. 3-15. The two Foster reactance configurations by which the reactance functions can be synthesized with a minimum number of elements. In (a) the zero-reactance frequencies are each determined by one of the combinations L_1C_1, L_3C_3, . . . or at $f = 0$ by L_0 or at $f = \infty$ by C_{p+2}. In (b) the pole-reactance frequencies are each determined by one of the combinations L_2C_2, L_4C_4, . . . or at $f = 0$ by C_0 or at $f = \infty$ by L_{q+2}.

---

[1] For example, see D. E. Thomas, Tables of Phase Associated with a Semi-Infinite Slope of Attenuation, *Bell Syst. Tech. J.*, 1947, Vol. 26, pp. 870–899, or H. W. Bode, Network Analysis and Feedback Amplifier Design, Van Nostrand, New York, 1945.

14. Circuit Functions. The individual circuits which constitute the building blocks of complete electronic systems are listed in the following outline, with cross-references to the sections of this handbook where they are treated in detail.

1. Analog circuit functions
   a. Amplification (Sec. **13**), e.g., dc, audio, video, rf, if, voltage, power amplifiers
   b. Restoration of dc component (Sec. **20**)
   c. Oscillation (Sec. **13**)
   d. Modulation (Secs. **4** and **14**), e.g., amplitude (AM), vestigial-carrier, suppressed-carrier, single-sideband, angle modulation, frequency modulation (FM), phase modulation (PM), preemphasis
   e. Demodulation (Secs. **4** and **14**), e.g., detection, conversion, frequency conversion, frequency multiplication, frequency division, parametric frequency multiplication and division, regenerative frequency division
   f. Synchronization (Secs. **13**, **20** and **23**), e.g., phase locking
   g. Multiplexing (Sec. **22**), e.g., time division, frequency division, diplexing, demultiplexing
   h. Rectification (Sec. **15**), e.g., rectifier and inversion circuits
2. Pulsed circuit functions (Analog Type)
   a. Position, duration, interval, width, and repetition-frequency modulation (Secs. **4** and **14**.)
   b. Pulse-code modulation (Secs. **4** and **14**)
3. Digital circuit functions (pulsed circuits)
   a. Binary pulsed circuits (Secs. **16** and **23**), e.g., binary switch circuits, bistable and monostable switches, sequential switches, counters, destructive, nondestructive read-out memory circuits, random-access memory circuits, read-only memory circuits
4. Time-delay functions
   a. Delay lines (Secs. **3** and **16**)
   b. Time-delay circuits (Secs. **3** and **16**)
5. Dc circuit functions
   a. Production of heat (Sec. **1**)
   b. Production of light (Sec. **11**)
   c. Magnetic holding force (Sec. **1**)
   d. Power supplies (Secs. **15** and **27**), regulated, unregulated

LUMPED-CONSTANT PARAMETERS AND CIRCUITS

15. Passive Lumped Parameters. The defining equations and the most useful time-domain and frequency-domain expressions for the lumped-constant circuit parameters are tabulated in Table 3-2. In the absence of any other circuit parameters in the same element, the defining equations given define a *constant* property. For example, the definition of resistance implies that the potential drop across a resistance is proportional to the current therein (with the polarities of voltage and current associated with the symbol), which is valid if the element in question is *ohmic*, or *linear*, i.e., conforms to Ohm's law.

Negative Parameters. Circuit elements, circuits, and devices which can be represented in the equivalent circuit by *negative resistance, negative inductance,* and/or *negative capacitance* are quite useful. Negative *incremental* resistance occurs in passive devices such as tunnel diodes and thermistors. Negative inductances and capacitances, however, can be realized only with circuits containing active devices.[1]

Quality Factor. Except for the resistance parameter in steady-state dc circuits, no circuit parameter is present as the sole parameter of a circuit element. The equivalent circuit of an inductor, for example, must include with its inductance some resistance to correspond to the power losses in the windings (and also in the core if a core is used). To describe the *quality* of coils and capacitors, a *quality factor* is used, defined as

$$Q_L = 2\pi f L / R(f) \qquad Q_c = 1/2\pi f C R(f)$$

in which L, C, and $R(f)$ are the equivalent series inductance, series capacitance, and series resistance measured at the frequency f for which the corresponding quality factor Q is specified.

---

[1] J. L. Merrill, Theory of the Negative-Impedance Converter, *Bell Syst. Tech. J.,* January 1951, Vol. 30, No. 1, pp. 88–109.

Table 3-2 Lumped-Constant Circuit Parameters

Parameter	Defining equation	Time-domain behavior	Frequency-domain behavior
Resistance, Ω	$R = \dfrac{v_{ab}}{I_{ab}}$	$V_{ab} = RI_{ab}$	$v_{ab} = RI_{ab}$ that is, $Z = R + j(0)$
Conductance, mhos $\left(\begin{array}{c}\text{reciprocal of}\\\text{resistance}\end{array}\right)$	$G = \dfrac{I_{ab}}{v_{ab}}$	$i_{ab} = Gv_{ab}$	$I_{ab} = Gv_{ab}$ that is, $Y = G + j(0)$
Incremental resistance, Ω	$R_{\mathrm{inc}} = \left(\dfrac{dv_{ab}}{di_{ab}}\right)_{V_0, I_0}$	$i_{ab} \approx I_0 + \dfrac{v_{ab} - V_0}{(R_{\mathrm{inc}})_{i=I_0}}$	$I_{ab} = \dfrac{(V_{ab})^{\mathrm{ac}}_{\mathrm{comp}}}{(R_{\mathrm{inc}})_{i=I_0}}$
Capacitance, F	$C = \dfrac{Q_a}{V_{ab}}$	$i_{ab} = \dfrac{C\,dv_{ab}}{dt}$	$I_{ab} = 2\pi f C E_{ab}$ that is, $jY = 2\pi f C$
Incremental capacitance	$(C_{\mathrm{inc}})_{V_0} = \left(\dfrac{dq_{ab}}{dv_{ab}}\right)_{V=V_0}$	$i_{ab} \approx (C_{\mathrm{inc}})_{V_0}\dfrac{dv_{ab}}{dt}$	$I_{ab} \approx j2\pi f C_{\mathrm{inc}}\,(V_{ab})^{\mathrm{ac}}_{\mathrm{comp}}$
Self-inductance, H	$L = \dfrac{\psi_{ab}}{I_{ab}}$	$v_{ab} = \dfrac{L\,di_{ab}}{dt}$	$V_{ab} = j2\pi f L I_{ab}$ that is, $Z = j2\pi f L$
Incremental inductance, H	$(L_{\mathrm{inc}})_{I_0} = \left(\dfrac{d\psi_{ab}}{di_{ab}}\right)_{i=I_0}$	$v_{ab} = (L_{\mathrm{inc}})_{I_0}\dfrac{di_{ab}}{dt}$	$V_{ab} = j2\pi f (L_{\mathrm{inc}})_{I_0}(I_{ab})^{\mathrm{ac}}_{\mathrm{comp}}$
Mutual inductance, H	$M_{12} = M_{21} = \dfrac{\psi_{ab}}{I_{cd}}$	$v_{ab} = L_1\dfrac{di_{ab}}{dt} + M_{12}\dfrac{di_{cd}}{dt}$ $v_{cd} = L_2\dfrac{di_{cd}}{dt} + M_{21}\dfrac{di_{ab}}{dt}$	$V_{ab} = j2\pi f(L_1 I_{ab} + M I_{cd})$ $V_{cd} = j2\pi f(L_2 I_{cd} + M I_{ab})$

16. Active Lumped Parameters. The equivalent circuits of active networks are constructed from the passive-circuit building blocks, defined in Table 3-2, the so-called voltage and current sources, various incremental parameters peculiar to each type of device, and/ or from their graphical characteristics.

The active devices, such as transistors, electron tubes, converter diodes, and parametric diodes, are inherently nonlinear; their terminal characteristics are described by graphical data. For small-signal operation, they can be modeled by linear equivalent circuits, using incremental parameters. For large signals, their operation is represented by *load lines* on the graphical characteristics, with supplementary linear parameters, e.g., capacitances. The load-line treatment of active devices appears in Par. 7-59.

17. Small-Signal Representation of Active Devices. Every transistor or electron tube operated under small-signal conditions has two equivalent circuits. The first is a dc equivalent circuit, which deals with the various resistors, voltage supplies, and/or current supplies required to provide bias and to power the device so that it can function at the desired operating point. One step in the design of a small-signal amplifier (after the selection of the device and circuit configurations to be used) is a dc circuit calculation for the selection of the various resistors and power source or sources, using this dc equivalent circuit.

The second is an ac equivalent circuit, which includes the incremental ac parameters of the device, capacitances and inductances (where appropriate), and the ac circuit equivalents of the passive elements used for coupling, loads, isolation, etc.

A wide variety of small-signal ac equivalent circuits is found in practice, particularly for transistors. The following discussion summarizes the most commonly used, together with definitions of the incremental parameters.

18. Small-Signal Transistor Equivalent Circuits. The symbols commonly used are e emitter; b base; c collector; v_{ce} (V_{ce}) variable (fixed) voltage, collector to emitter; v_{be} (V_{be}) variable (fixed) voltage, base to emitter; v_{cb} (V_{cb}) variable (fixed) voltage, collector to base; i_b (I_b) variable (fixed) current, base; i_e (I_e) variable (fixed) current, emitter; and i_c (I_c) variable (fixed) current, collector.

The h or *hybrid parameters* can be defined for the common-emitter, common-base, or common-collector circuits. For example, for the common-emitter transistor circuit, they are

$$h_{11} = h_{ie}(\text{input resistance}) = (\partial v_{be}/\partial i_b)_{V_{ce}=\text{const}}$$

$$h_{22} = h_{oe}(\text{output admittance}) = (\partial i_c/\partial v_{ce})_{I_b=\text{const}}$$

$$h_{12} = h_{re}(\text{feedback factor}) = (\partial v_{be}/\partial v_{ce})_{I_b=\text{const}}$$

$$h_{21} = h_{fe}(\text{current-amplification factor}) = (\partial i_c/\partial i_b)_{V_{ce}=\text{const}}$$

The corresponding h parameters for the common-base and common-collector amplifier circuits are designated as $h_{11} = h_{ib}$ and $h_{11} = h_{ic}$, respectively, and defined consistently with the circuits:

$$h_{11} = h_{ib} = \text{input resistance (common-base circuit)}$$

$$= (\partial v_{eb}/\partial i_e)_{V_{cb}=\text{const}}$$

Use of the h parameters leads to the equivalent circuit of Fig. 3-16, useful for low-frequency calculations.

Other Transistor Parameters. α_0 is the short-circuit common-base current amplification at low frequencies. (The symbol α, however, is also sometimes used for the *transport factor*, the fraction of minority carriers in the base which manage to arrive, through diffusion, in the collector region.)

$$\alpha_0 = \left(\frac{-\partial i_c}{\partial i_e}\right)_{V_{cb}=\text{const}} \tag{3-28}$$

α_f is the short-circuit common-base amplification at a frequency f

$$\alpha_f = \frac{\alpha_0}{1 + jf/f_\alpha} \tag{3-29}$$

in which f_α is the frequency at which the magnitude $|\alpha_f|$ of α_f has fallen to 0.707 of its low-frequency value. f_α is known as the alpha cutoff frequency or common-base alpha cutoff frequency. The symbol f_{ab} is also used for this frequency.

f_T (also sometimes f_1) is the current-gain bandwidth product, also known as the common-emitter frequency f_{ae}. This is the frequency at which the common-emitter current gain of the device has dropped to unity.

β (beta) is a symbol used sometimes for h_{fe}, the current-amplification factor (common emitter).

These are the transistor parameters commonly given in the data sheets for commercial transistors. The h parameters are not convenient for the formulation of equivalent circuits for calculating high-frequency performance, however. For this purpose, the T and pi parameters are useful. These follow, defined for the common-base amplifier, using the circuits of Fig. 3-17a and b.

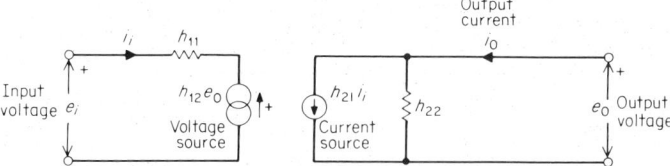

Fig. 3-16. Two-source equivalent circuit, using the hybrid (h) incremental parameters, a circuit valid for low-frequency small-signal transistor amplifiers. The feedback voltage source $h_{12}e_o$ is often omitted. For a common-emitter amplifier, for example, $i_i = i_b$, $h_{11} = h_{ie}$, $h_{22} = h_{oe}$, and $h_{21} = h_{fe}$.

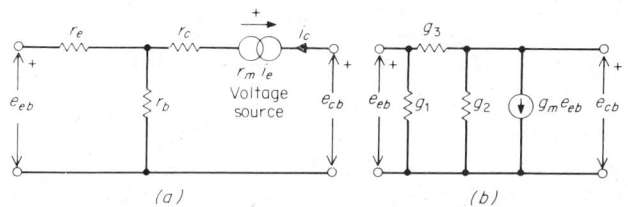

(a) (b)

Fig. 3-17. (a) T equivalent circuit for the common-base transistor small-signal amplifier. The resistors r_e and r_c are the forward and backward resistance of the emitter-base and collector-base diodes, respectively. Resistor r_b represents a collector-emitter interaction effect, and $r_m i_e$ provides a voltage source to correspond to the amplification effect. This voltage source in series with r_c is often replaced with a current source in parallel with r_c, of magnitude ai_e, where $a = r_m/r_c$. The constant a is approximately equal to α, defined in the text. (b) Pi equivalent circuit for the common-base amplifier. The conductances g_1 and g_2 correspond to the input and output diode conductances, and g_3 represents the interaction. The current source $g_m e_{eb}$ provides for the amplification effect.

Corresponding to a, as defined in the caption for Fig. 3-17a, an approximate current-amplification factor b for common-emitter operation is defined as $b = a/(1 - a)$. This in turn is approximately equal to h_{fe}, the hybrid-parameter current-amplification factor.

Examples of some further equivalent circuits for high-frequency operation are shown in Fig. 3-18. The capacitance C_c is the collector-to-base capacitance, which, incidentally, varies with the collector voltage.

Since the device parameters normally given are the h parameters, the following equivalence relations are needed to determine the parameters in these equivalent circuits:

$$r_c = \frac{1}{h_{oe}} \qquad r_e = \frac{h_{re}}{h_{oe}} \qquad r_b = h_{ie} = (1 + h_{fe})\frac{h_{re}}{h_{oe}} \tag{3-30}$$

At the highest frequencies, it is necessary to take into account the parameters of the leads in the transistors, resulting in even more complicated equivalent circuits.

The y parameters. For high-frequency transistors, the y equivalent-circuit parameters are sometimes specified. The complete y equivalent for a bipolar-junction transistor is shown in Fig. 3-19.

19. Small-Signal Electron-Tube Circuits. As with transistors, there are two equivalent circuits to be considered in each electron-tube application. The first circuit is a dc steady-state circuit, in which only resistances and the electron tube are considered. The electron tube is represented by its actual terminal dc characteristics, usually expressed graphically; capacitors are open circuits, and inductors and/or transformer windings are replaced by their dc resistances. This dc equivalent circuit is designed to provide proper-operating quiescent dc electrode potentials (grid bias, screen and anode voltages, etc.).

The second, or ac, equivalent circuit determines the transfer function of the electron tube and its associated components under signal conditions. Large-signal calculations are

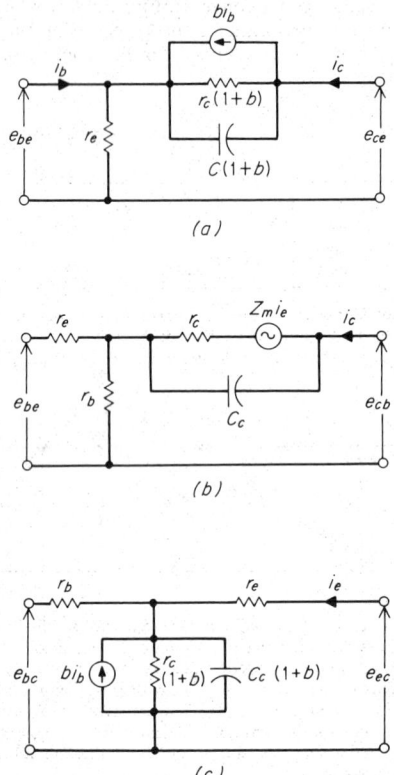

Fig. 3-18. Examples of high-frequency equivalent circuits for (a) common-emitter, (b) common-base, and (c) common-collector small-signal transistor amplifiers. The impedance Z_m in (b) is equal to αr_c.

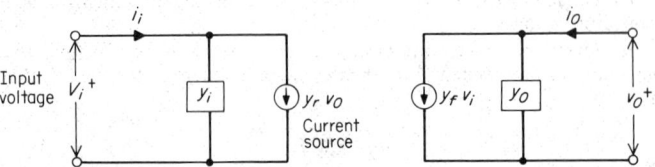

Fig. 3-19. The y-parameter equivalent circuit. For a common-base or common-emitter circuit, for example, the notation y_{ib} or y_{ie} respectively indicates the corresponding forward transfer admittance. The complex components of these admittances are often given in terms of capacitances because the capacitance values are frequency-independent.

developed by graphical plots on the charts of electron-tube characteristics and are not treated herein. The following is a brief treatment of *small-signal linear equivalent circuits*.

The following symbols are used: e_{ak}, E_{ak} instantaneous anode-to-cathode voltage, dc anode-to-cathode supply voltage; e_{gk}, E_{gk} instantaneous grid-to-cathode voltage, dc grid-bias (grid-to-cathode) voltage; i_a, I_a instantaneous anode current, average anode current; C_{gk} capacitance grid to cathode; C_{ga} capacitance grid to anode; and C_{ak} capacitance anode to cathode. If the electron tube has four or five elements (*cathode, control grid, screen grid, suppressor grid, anode*), the capacitances are defined as follows: C_{gk} ac measured capacitance grid to cathode with screen and suppressor (if present) at cathode potential; C_{ak} ac measured capacitance anode to cathode with screen and suppressor (if present) at ac cathode potential; and C_{ga} capacitance grid to anode with screen, suppressor (if used), and cathode tied together and grounded.

The small-signal electron-tube parameters are defined as follows:

$$\text{Amplification factor } \mu = (\partial e_{ak}/\partial e_{gk})_{i_a=\text{const}}$$

$$\text{Anode (plate) resistance } r_a = (\partial e_{ak}/\partial i_a)_{e_{gk}=\text{const}}$$

$$\text{Grid-anode transconductance (mutual conductance) } g_m \text{ or } g_{ag} = (\partial i_a/\partial e_{gk})_{e_{ak}=\text{const}}$$

In general, each of the quantities μ, r_a, and g_m is a function of the actual values of i_a, e_{gk}, and e_{ak}, respectively. Figure 3-20 gives the commonly used small-signal equivalent circuits for common-cathode three-element (triode) electron tubes.

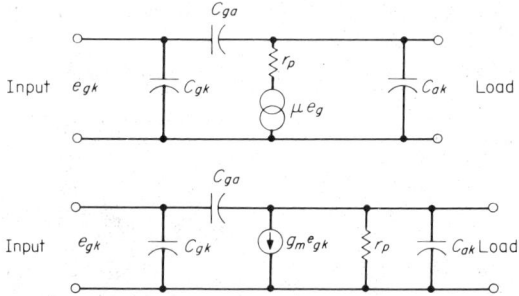

Fig. 3-20. Two alternative ac equivalent circuits for the common-cathode triode electron-tube amplifier. The load may, for example, be a resistor, transformer primary, tuned circuit, or the input to a subsequent stage. The incidental (stray) capacitances associated with the input and output must be added to C_{gk} and C_{ak}.

The performance of the triode electron-tube amplifier in the common-cathode mode is dominated at all but the lowest frequencies by the effect of the large difference between input and output voltages which is developed across the grid-anode capacitance C_{ga}. This results in an effectively greatly reduced input impedance (known as the *Miller effect*), so that small-signal operation of triodes in the common-cathode mode is restricted to very low frequencies.

20. Frequency- and Time-Domain Properties of Single-Port Circuits. The frequency- and time-domain properties of examples of simple equivalent circuits of resistors, capacitors, and/or inductors in one-port (two-terminal) configurations are shown in Table 3-3. The time-domain response given is the response current to a step function of voltage E_0, for cases *a, b,* and *c,* and the response voltage to a step function of current I_0 for case *d.* The frequency-domain description comprises the terminal (phasor) impedance Z_{ab} and its phase angle θ_{ab}.

Time Constant. The characteristic exponential time-domain response of the *RL* and *RC* circuits has led to the extremely important concept of the *time constant.* The time constants L/R (for the *RL* circuit) and RC (for the *RC* circuit) represent the amount of time, in seconds, before the current reaches within a certain fraction of its steady-state value; for the *RL* circuit, this is the time required for a rise to about 63% of the final value and for the *RC* circuit a drop to about 37% of the initial value. The time-constant concept, i.e., the time which must elapse before a physical phenomenon comes close to its final value, is widely employed not only in circuits but also in dealing with mechanical and thermal effects.

The *RC* circuit is often used to produce a controlled time delay; e.g., in analog timers.

Resonance. The *RLC* circuit displays *resonances* in the frequency domain. In the series *RLC* circuit, the minimum impedance and in-phase resonances occur when $2\pi fL = 1/2\pi fC$

Table 3-3.   Frequency- and Time-Domain Properties

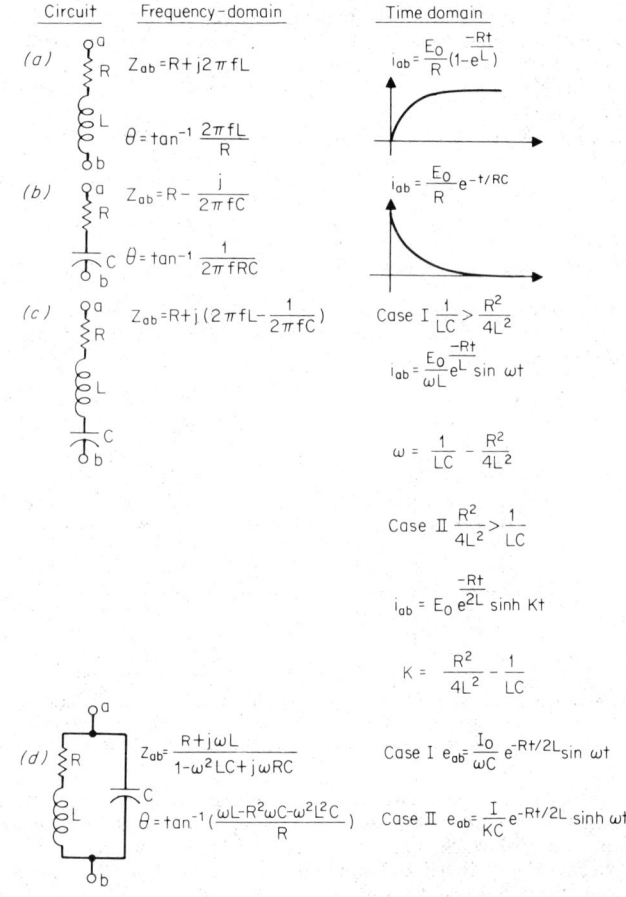

or for a frequency $f = 1/(2\pi\sqrt{LC}\,)$; the natural frequency of oscillation f_n in the time domain $f_n = \sqrt{1/LC - R^2/4L^2}/2\pi$ differs from the other resonant frequencies only slightly for small values of R. The series *RLC* circuit can, in principle, be used as a tuned circuit to select a signal of a particular frequency (or a signal with a narrow frequency spectrum); such use is limited because few practical electronic circuits are equivalent to true voltage sources. It has been shown that the selective properties of the series resonant circuit can be represented by the normalized universal[1] resonance curve shown in Fig. 3-21, based on the approximation that

$$\frac{\text{Current at frequency } f}{\text{Current at resonance } f_r} = \frac{1}{1 + \dfrac{a}{Q} + ja\dfrac{2 + a/Q}{1 + a/Q}} \approx \frac{1}{1 + j2a} \tag{3-31}$$

---

[1] Originally derived by F. E. Terman.

where

$$a = Q\frac{f - f_r}{f_r} = Q\frac{\text{cycles off resonance}}{\text{resonant frequency}}$$

$a/Q \ll 1$ for quite practical values of Q, and Q is a constant, independent of frequency.

The parallel RLC circuit of Table 3-3, entry d, is of considerably more importance than the series RLC circuit; it is widely used as a *frequency-discriminating circuit*, and its characteristics for the case in which it is excited with a current source often correspond to realistic situations. Typical resonance characteristics are shown in Fig. 3-22. The maximum-impedance resonant frequency and "in-phase" resonant frequency differ from one another, and neither coincides with the *nominal resonant frequency* $f_r = 1/(2\pi\sqrt{LC})$ or the natural frequency $f_n = \sqrt{1/LC - R^2/4L^2}/2\pi$. For physically realizable high values of

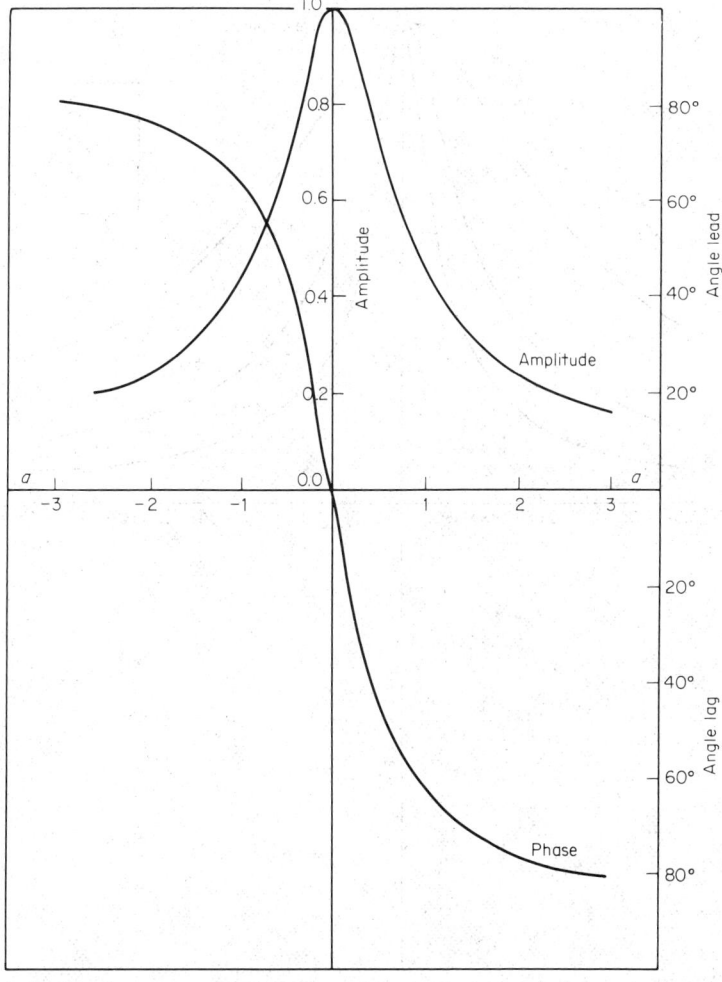

Fig. 3-21. Universal resonance curve for constant Q. Ratio of current and phase angle at frequency f to current at f_r as a function of normalized variable $a = Q(f - f_r)/f_r$, $f_r = 1/(2\pi\sqrt{LC})$.

Q, the differences between the various resonant frequencies become insignificant. The normalized universal resonance curve of Fig. 3-21 can be applied with reasonable accuracy to the parallel resonant circuit excited with a current source by connecting the ordinate of the curve to the ratio of voltage developed at frequency f to that developed at f_r.

21. Frequency- and Time-Domain Properties of Two-Port Circuits. The RL circuit and, particularly, the RC circuit can appear in equivalent circuits as two-port networks; the simple two-port configurations are shown in Fig. 3-23.

The amplitudes and phase angles of the transfer functions of the RC networks appear in Fig. 3-24 c and d (the independent variable is normalized as $K = 2\pi fRC$ so that the curves are of universal application). The time-domain responses for a typical pulse signal are given in Fig. 3-25.

Some limiting conditions are of special interest. For $K \gg 1$, the frequency-domain transfer function of the RC network of Fig. 3-23c becomes very nearly $E_{out}/E_{in} = 1/K$; this corresponds to an output inversely proportional to the frequency of the input. This is a

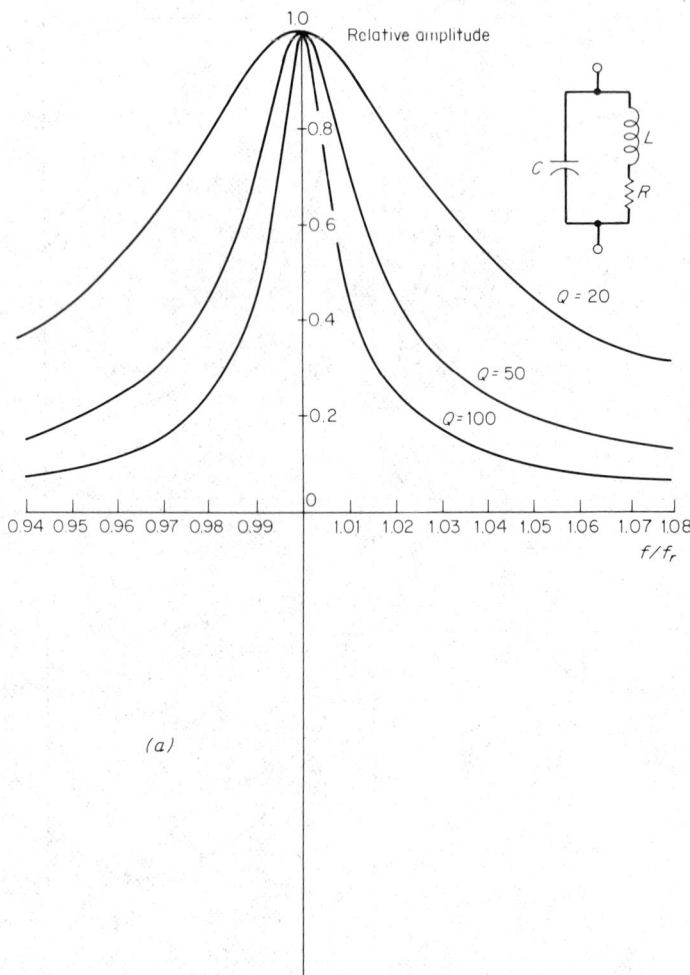

(a)

Fig. 3-22. (*a*) Parallel-circuit resonance curve (for constant Q). The magnitude of the impedance relative to the maximum impedance is plotted vs. the normalized frequency parameter f/f_r for various values of

deemphasis characteristic, which can be used, for instance, to process a signal fed to a phase-modulated transmitter in order to produce a carrier which is, in effect, actually frequency-modulated.

Analogously for $K \ll 1$, the transfer function of the circuit of Fig. 3-23d is $E_{out}/E_{in} \approx K$, resulting in a *preemphasis* of the signal frequency components proportional to frequency, which can be used for the inverse operation in an FM transmitter. Both network transfer functions have phase-angle variations which are inconsequential for speech, music, etc., program material. Their amplitude and phase characteristics can be disastrous in the distortion of pulse signals, as shown by Fig. 3-25.

The circuits of Fig. 3-23c and d are also commonly used as *integrating* and *differentiating* circuits, respectively. For example, if the circuit of Fig. 3-23c has been designed so that it operates with $v_{cd} \ll v_{ab}$ ($K \gg 1$), v_{cd} is proportional to the *time integral* of v_{ab}. Similarly, in the circuit of Fig. 3-23d, if $v_{cd} \gg V_{ab}$ ($K \ll 1$), v_{cd} is proportional to the *time derivative* of v_{ab}. In either case, if an output voltage is desired which is exactly equal to the time integral or

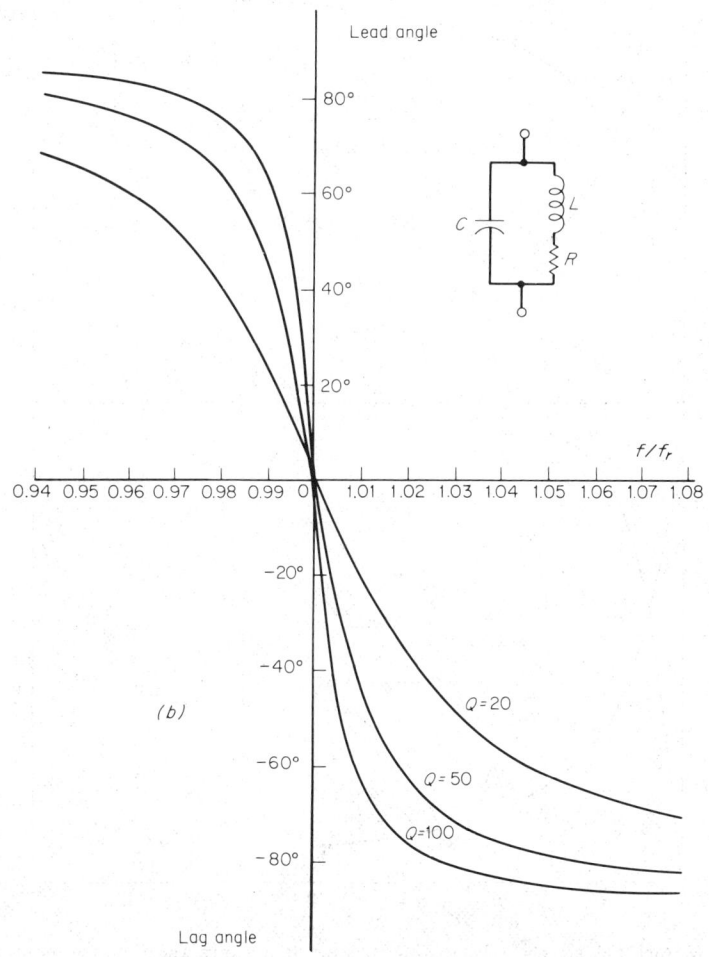

Q. (*b*) The phase angle of the impedance relative to the phase at resonance is plotted vs. the normalized frequency parameter f/f_r for various values of Q.

3-23

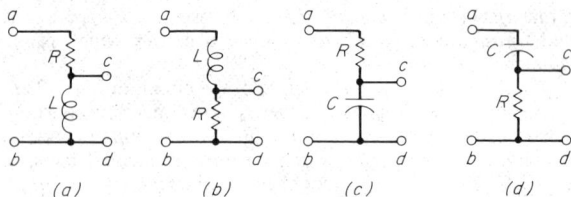

Fig. 3-23. Arrangements of RL and RC circuits as two-port networks. Cases (a) and (b) occur infrequently in equivalent circuits because an inductor can seldom be realistically treated as pure, i.e., without an equivalent resistance. On the other hand, (c) and (d) represent realistic equivalent circuits for a number of important cases.

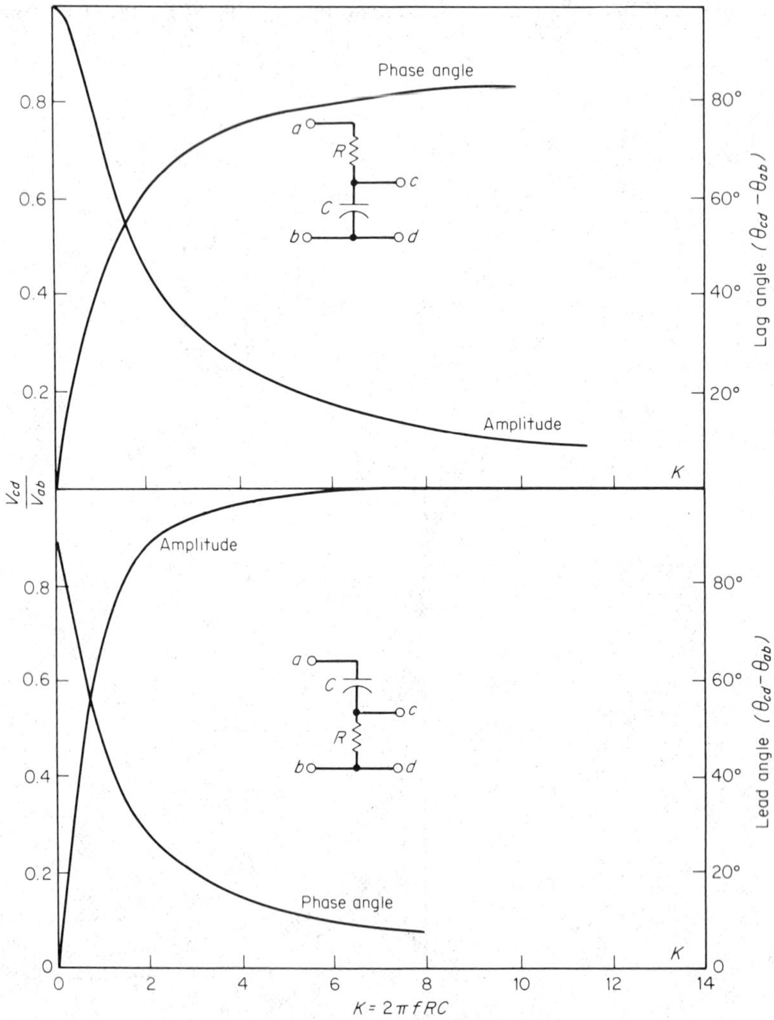

Fig. 3-24. Amplitude and phase characteristics of series RC networks. The plots show the magnitude of the ratio of output voltage V_{cd} to input voltage V_{ab} as well as the phase shift between output and input $\theta_{cd} - \theta_{ab}$.

time derivative of the input, the RC network must be followed by a distortionless amplifier of appropriate gain to increase the proportionality factor to unity.

The equivalent circuits of substantially all RC interstage coupling networks in electronic amplifiers reduce to the $K \ll 1$ configuration and condition of Fig. 3-23a at their low-frequency cutoff and the $K \gg 1$ configuration and condition at their high-frequency cutoff; hence, the time-domain signal distortions of such amplifiers can be interpreted in the light of such differentiating and integrating effects.

22. Coupling and Coupling Networks. Coupling networks are used to interconnect devices, to interconnect apparatus and devices, etc. Their functions include (1) isolation of dc components, a necessary feature for cascading electron-tube or transistor amplifiers (except for transistors in a complementary configuration), (2) shaping the (frequency-domain) amplitude or phase-angle transfer function, (3) impedance matching, and (4) waveform preservation or correction, i.e., direct control (as an alternative to the indirect frequency-domain approach) of time-domain characteristics. In each case, the circuit function depends not only upon the equivalent-circuit parameters of the coupling network itself but also upon the equivalent-circuit parameters of the devices or apparatus being coupled (since it is rare in electronics engineering to encounter devices or apparatus which are perfect current or voltage sources).

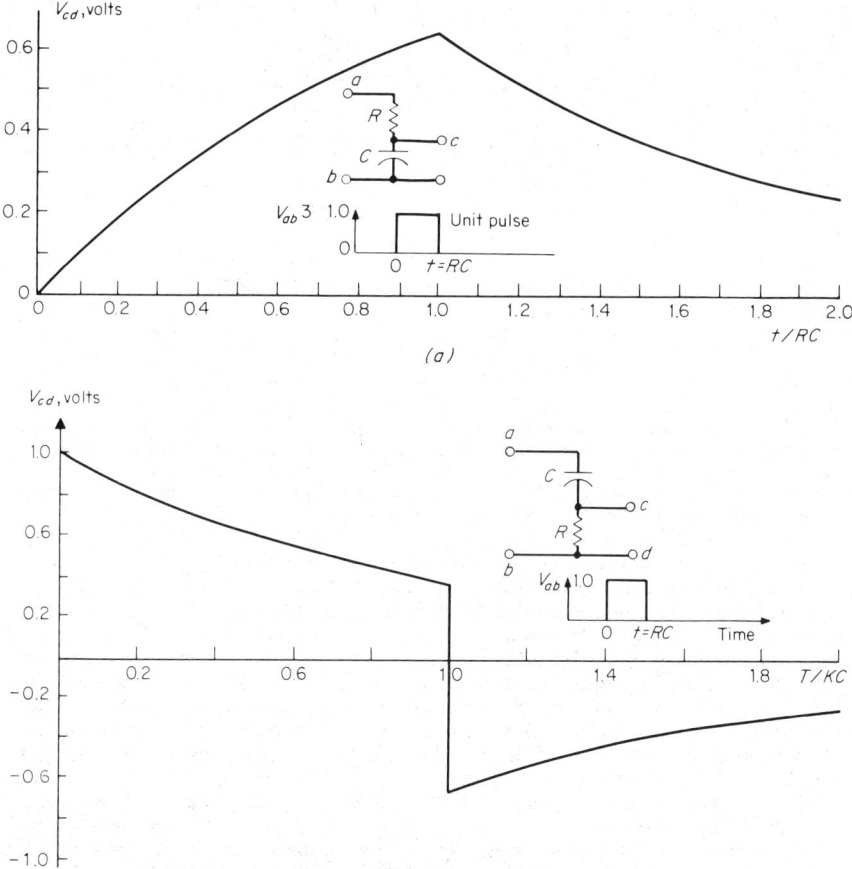

Fig. 3-25. Time-domain response of RC series-circuit output for input pulse of unit amplitude and duration RC s.

    Coupling Networks for Audio and Video Amplifiers. Audio and video amplifiers are essentially low-pass amplifiers, the response of which excludes a narrow range of frequencies from zero frequency (dc) up to some low-frequency cutoff. Equivalent circuits for some examples of such amplifiers appear in Fig. 3-26.

    Figure 3-26c is a simple example of a compensated video-amplifier coupling circuit in which an inductor L_L is inserted to compensate for the effect of the shunt capacitances C_{pk} and C_{gk} and increase the cutoff frequency.

    Coupling Networks for Bandpass Amplifiers (If and Rf Amplifiers). Since most electronic amplifiers are used to process signals rather than a single-frequency sinusoidal ac voltage or current, their transfer functions must be essentially bandpass in order to provide for the reproduction of both the signal carrier and its side frequencies. At the same time the passband must be restricted in its frequency width to reject the components of *adjacent* (in frequency) unwanted signals.

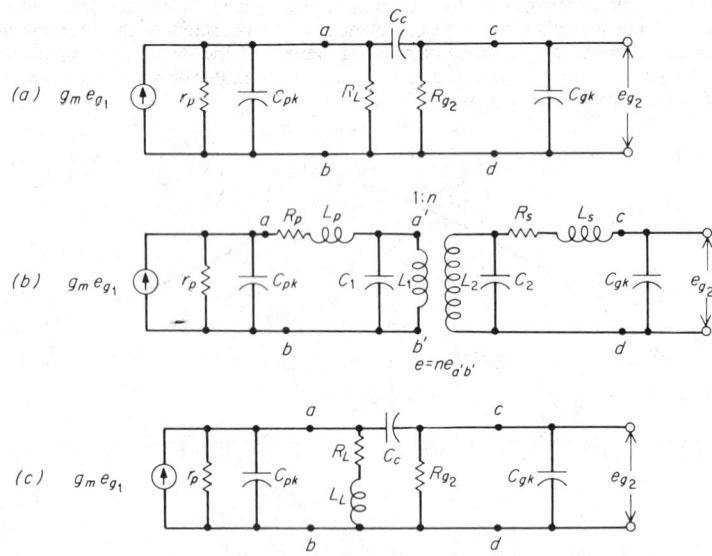

Fig. 3-26. Equivalent circuits for the interstage coupling in three examples of electron-tube low-pass (excluding dc) amplifier. In each case the portion of the circuit to the left of the terminals *a* and *b* is the equivalent circuit of the input electron tube, where e_{g1} is the signal at its grid. The two-port network appearing between terminals *a* and *b* (input) and *c* and *d* (output) is the coupling network, and the circuit to the right of terminals *c* and *d* is the equivalent circuit of the input to the next stage.

    The coupling networks in such amplifiers are universally of the tuned type, comprising capacitors and inductors in various configurations. Fig. 3-27 shows some examples of simple bandpass coupling networks. The coupling arrangement of Fig. 3-27a has an optimum response to a single frequency and can be expected to attenuate side frequencies. The overall amplifier response can be converted to a bandpass response by staggering the resonant frequencies of successive stages. The coupling network of Fig. 3-27b lends itself better to a bandpass response. The characteristic is highly dependent upon the coefficient of coupling between the primary (L_1) and secondary (L_2) inductors. Examples of the effect of coupling coefficient upon the frequency-domain amplitude and phase components of the transfer functions of the double-tuned circuit are shown in Fig. 3-28. By using different coefficients of coupling in successive stages, a relatively flat amplitude transfer function is quite readily obtainable. [1]

    23. Microelectronic Configurations. The circuit parameters that can be constructed in microelectronic configurations are limited to resistances and capacitances, excluding, for

---

[1] For a detailed treatment of tuned amplifiers see R. F. Shea (ed.), "Amplifier Handbook," McGraw-Hill, New York, 1966, Chap. 24, pp. 24-1 to 24-21.

practical purposes, inductors. Tuned circuits, however, may be achieved by the use of piezoelectric (electromechanical) components. A single-tuned circuit, for example, can be realized with a piezoelectric resonator with an electric-to-mechanical transducer at the input and a mechanical-to-electric transducer at the output. A double-tuned electric coupling network can be realized with two loosely coupled mechanical elements.

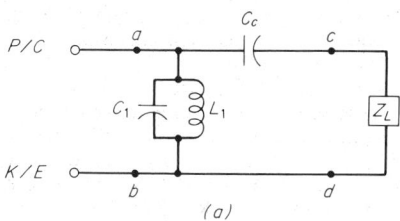

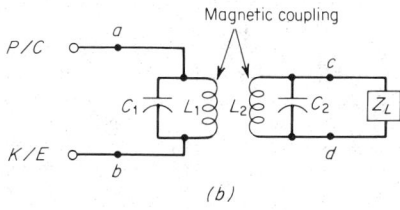

Fig. 3-27. Simplified schematic diagrams of (*a*) single-tuned and (*b*) double-tuned bandpass coupling networks. The impedance Z_L represents the impedance presented by the load or by the input to a following stage. The corresponding equivalent circuits would include the parameters of the input and output electron-tube or transistor input and the resistances of the inductors L_1 and L_2. *P/C* refers to the plate of an electron tube or collector of a transistor. *K/E* refers to the cathode of an electron tube or emitter of a transistor.

In microelectronic devices, neither very low nor very high values of resistance can be fabricated. The equivalent circuits of resistors include significant and unavoidable shunt capacitances, which at high frequencies must be treated as distributed elements.

24. Impedance Matching. As pointed out in Par. **3-13**, *maximum transfer of power from a source to a load takes place when the source impedance and load impedance are complex conjugates.* In certain cases in which a source and its load do not have complex-conjugate impedances (because their design properties are dictated by other factors, for example) and an impedance match is desired, an *impedance-matching device* or *impedance-matching network* may be used.

The essential problem is illustrated in Fig. 3-29. A generator of internal impedance $Z_g = R_g + jX_g$ is to be coupled to a load of impedance $Z_L = R_L + jX_L$. A device or network is to be connected between the source and load so that the impedance seen by the source is $Z'_L = R_g - jX_g$; that is, R_L is to be transformed to a resistance R_g and X_L to a reactance $-X_g$.

For an audio or video source with a relatively constant, largely resistive internal impedance when the load is similarly largely resistive and constant, a transformer is used with a *transformation ratio n* given by $n = \sqrt{R_L/R_G}$. Such transformers are of the untuned conventional ferromagnetic core, closely coupled winding types and can be designed to operate reasonably well from a signal-handling point of view over a frequency range such that the ratio of high-frequency cutoff to low-frequency cutoff can approach 1,000:1.

For a high-frequency bandpass application in which the bandwidth is a small fraction of the center frequency, a *tuned impedance-matching network* is feasible. Examples of such networks are shown in Fig. 3-30.

The components of the selected network (one of those shown in Fig. 3-30 or one of a wide variety of alternatives) are selected so that the impedance Z_{ab}, with the actual load ($Z_L = R_L + jX_L$) connected between terminals *c* and *d*, has the desired resistive and reactive components. Analytically, the design values are determined by solving two simultaneous equations, such as (in the case of Fig. 3-30*c*, for example)

$$R_{ab} = \frac{X_1^2 R_L + X_1 X_L R_L + X_1 X_2 R_L}{(X_1 - X_L - X_2)^2 + R_L^2} \qquad X_{ab} = \frac{X_1^2(X_1 + X_2) - X_1 R_L^2}{(X_1 - X_L - X_2)^2 + R_L^2} \qquad (3\text{-}32)$$

with $R_{ab} = R_g$ and $X_{ab} = -X_g$, the specified internal impedance components of the source. (In this particular network, physically realizable values of X_1 and X_2 restrict the application to cases in which $R_L < R_g$; for $R_L > R_g$ the network of Fig. 3-30*d* can be used.) The two-element impedance-matching networks have uniquely determined values of components since there are two variables and two conditions to be met. With networks having three or more components, the components can be chosen from a range for convenience, on account of some particular component availability or to obtain some desired phase shift. For cases in which the load impedance is a pure resistance, charts of values of the components[1] are available.

CIRCUITS WITH DISTRIBUTED PARAMETERS[2]

25. Distributed-Parameter Concepts.   At low frequencies, where the dimensions of the circuit components are small compared to the wavelengths of the signals considered, the *lumped-parameter* concept provides an adequate description for analysis.   The *displacement-current* term in Maxwell's equations (see Par. **3-69**) is negligible compared to the *conduction-current* term, and the variations of the magnetic and electric fields in space can be neglected. At very high frequencies, however, this is not the case, since the dimensions of the device are comparable to the wavelengths of the propagating signals.   We associate the energy stored in the magnetic field with the *distributed inductance* of the structure, the energy stored in the electric field with the *distributed capacitance*, and the power loss with the *distributed resistance*. The concept of distributed parameters is useful together with the concepts of equivalent voltage and current because it enables us to apply many of the techniques and properties of low-frequency analysis to high-frequency structures.

A *transmission line* is a structure consisting of two or more parallel conductors which guides *transverse electromagnetic* (TEM) *waves*.   TEM waves have their electric and magnetic fields perpendicular to the direction of propagation.   A *waveguide* is a hollow structure with conducting walls which supports transverse electric (TE) and transverse magnetic (TM) waves.   A waveguide is a more desirable structure to use than a transmission line for applications where the wavelength is less than 10 cm.   Transmission lines and waveguides are treated in detail in Sec. **9**.

26. Modes and Boundary Conditions.   The term *mode* is used to describe the electric and magnetic *field pattern* in a waveguide.   Specific modes are identified by specifying the

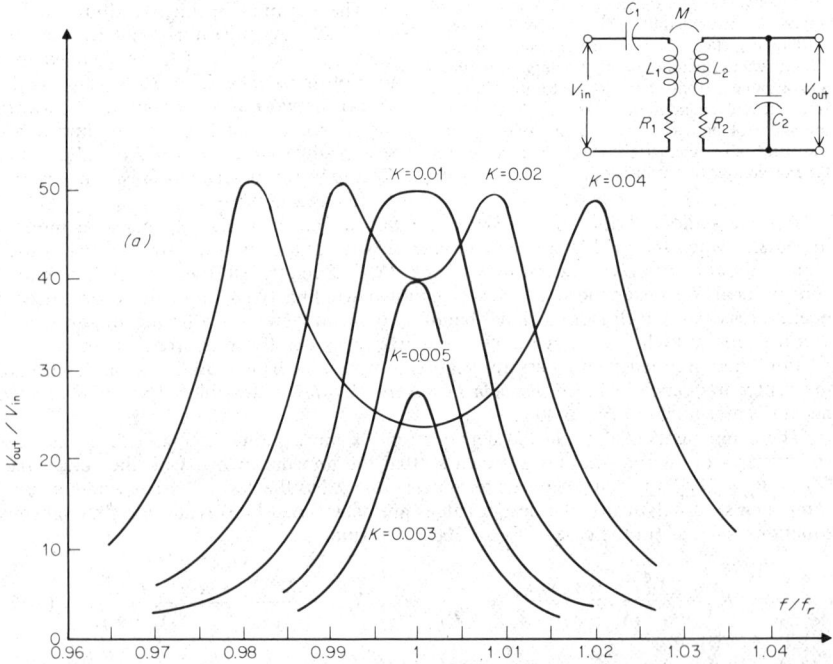

Fig. 3-28.   Frequency-response characteristics of a double-tuned circuit. The magnitude ratio (*a*) and the phase difference (*b*) between the output voltage and the input voltage are plotted vs. a normalized frequency parameter f/f_r for several values of coupling coefficient $k = M/\sqrt{L_1 L_2}$.

---

[1] See for example F. E. Terman, "Radio Engineers' Handbook," McGraw-Hill, New York, 1943, pp. 206-215.

[2] The authors received valuable assistance in preparing this material from Andrew T. Perlik.

type of wave, for example, TM, and by specifying the number of relative maxima occurring in the field configuration of the cross section with subscripts. For a *rectangular waveguide* (rectangular cross section) TE_{mn} denotes that the electric field is transverse to the direction of propagation and that the electric field has *m* relative maxima occurring along the width of the cross section and *n* relative maxima occurring along the height of the cross section.

For *circular waveguides* the subscript *m* denotes the number of relative maxima occurring in the radial field component in the angular direction, and the subscript *n* denotes the total number of relative maxima and minima of the angular field component in the radial direction.

The *boundary conditions* on the electric and magnetic fields at the boundary between two different media are derived by applying Maxwell's equations to elemental volumes containing the boundary or closed contours cutting the boundary. For perfect conductors, i.e., resistanceless conductors, the *normal component of the magnetic field intensity* **H** is zero and the *tangential component of the electric field* **E** is zero at the surface of the conductor.

27. Calculation of Distributed Parameters. A two-wire transmission line is shown together with its distributed-parameter representation in Fig. 3-31. A given transmission line is characterized by the values of its *distributed parameters r, l, g,* and *c.* These quantities are given per unit length of the line. Formulas are available for substantially all simple

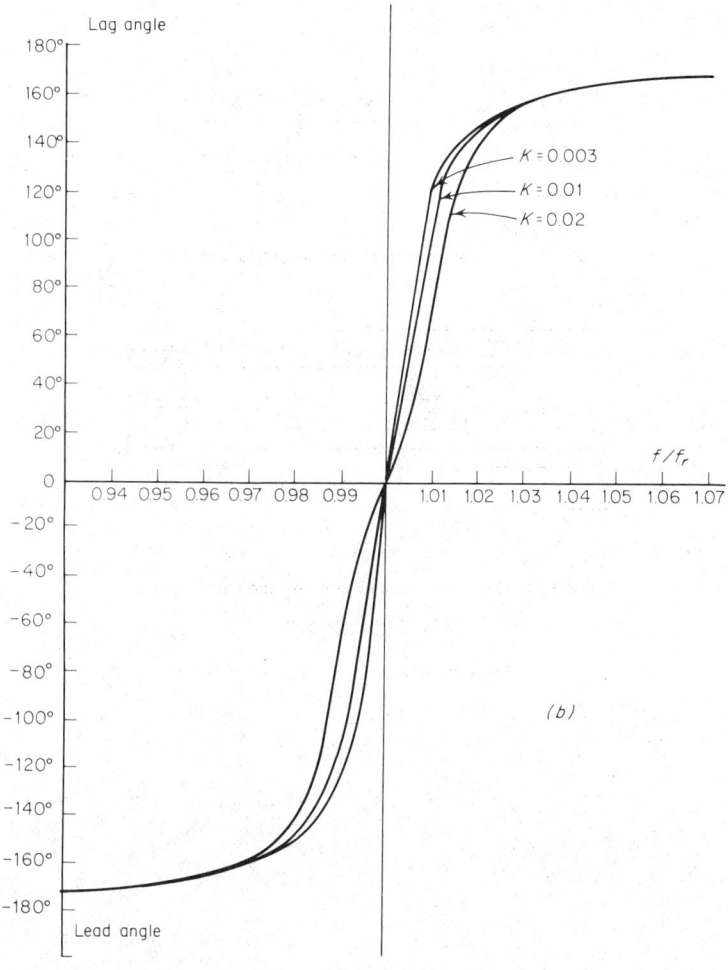

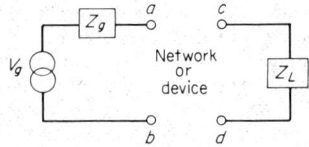

Fig. 3-29. General form of the impedance-matching configuration. The network or device ideally transforms Z_L so that the driving-point impedance $Z_{ab} = R_g - jX_g$. $Z_g = R_g + jX_g$, $Z_L = R_L + jX_L$.

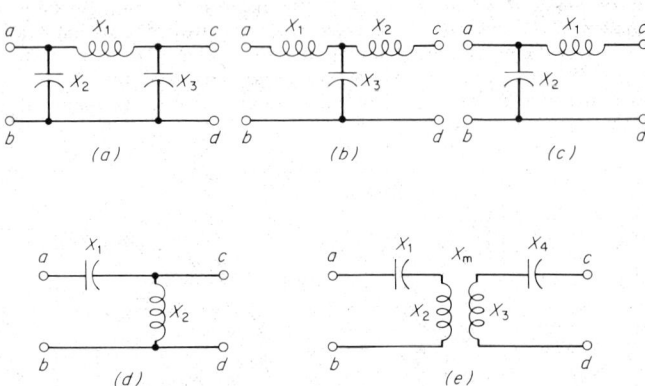

Fig. 3-30. Examples of impedance-matching networks.

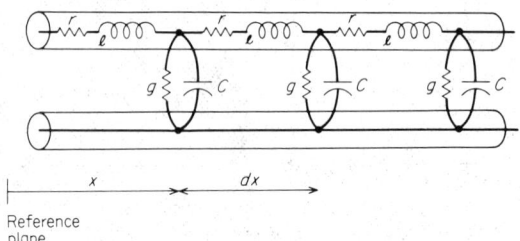

Fig. 3-31. Two-wire transmission line and parameters.

two-wire transmission-line configurations;[1] however, for any particular configuration these parameters are, in general, frequency-dependent because of the skin effect.

The *skin effect* is the phenomenon of increasing current-density concentration in the surface layers of the conductors as frequency increases; it arises because the reactance associated with the possible current paths is smallest near the conductor surface. The *skin depth* is given by $\sqrt{\sigma\pi\mu f}$, where f is the frequency of the current, μ the permeability of the medium, and σ its conductivity; it is the depth at which the current density has decreased to $1/\epsilon$ (37%) of its value at the surface of the conductor. This effect is not restricted to the conductors of transmission lines only but occurs in all conductors, including the conductors of circuit elements and those which interconnect circuit elements.

In Fig. 3-31, the voltage $e(x)$ is attenuated because of the series impedance $Z_s = r\,dx + j\omega l\,dx$, and the current $i(x)$ is attenuated by the shunt admittance $g\,dx + j\omega c\,dx$. Applying Kirchhoff's laws yields

$$e(x + dx) - e(x) = -\left[r\,dx\,i(x) + l\,dx\frac{\partial i(x)}{\partial t} \right]$$

$$i(x + dx) - i(x) = -\left[g\,dx\,e(x) + c\,dx\frac{\partial e(x)}{\partial t} \right]$$

(3-33)

Dividing by dx and taking the limit as $dx \rightarrow 0$ gives

$$\frac{\partial e(x)}{\partial x} = -ri(x) - l\frac{\partial i(x)}{\partial t} \qquad \frac{\partial i(x)}{\partial x} = -ge(x) - c\frac{\partial e(x)}{\partial t}$$

(3-34)

which are the equations in terms of the voltage and current as a function of the distance from some reference position.

This equivalent circuit gives excellent results in comparison to experimental data for low-frequency power transmission, communication, and some high-frequency applications for lines of uniform spacing and conductor size in which end effects are neglected. The *distributed series resistance r* in ohms per unit length is the resistance of a unit length of the two conductors. The *distributed shunt conductance g* in mhos per unit length is the equivalent conductance of a unit length of the lines caused by dielectric losses in the medium between the conductors. The *distributed series inductance l* in henrys per unit length is calculated by assuming that a unit current travels along an infinite extent of transmission line, going in one conductor and returning in the other, and calculating the resulting flux linkages in the section of unit length. The *distributed shunt capacitance c* in farads per unit length is calculated by assuming a constant voltage V between the two conductors, extended to infinity in either direction from the unit section, and determining the corresponding charge q in coulombs on the unit section.

Voltage and current definitions are introduced on an equivalent basis and are of value because many of the circuit-analysis techniques valid at low frequencies are also applicable at microwave frequencies.

Propagating *waveguide modes* have the following properties:[2] (1) power transmitted is given by an integral involving the transverse electric and transverse magnetic fields only; (2) in a loss-free guide supporting several *modes* of propagation, the power transmitted is the sum of that contributed by each mode individually; (3) the transverse fields vary with distance along the guide according to a propagation factor $e^{\pm j\beta z}$ only; (4) the transverse magnetic field is related to the transverse electric field by a simple constant, the *wave impedance* of the mode. The equivalent voltage and current for a waveguide supporting N modes can be written using superposition. The impedance-matrix description for a two-port waveguide for each propagating mode is

$$\begin{bmatrix} V_1 \\ V_2 \end{bmatrix} = \begin{bmatrix} Z_{11} & Z_{12} \\ Z_{21} & Z_{22} \end{bmatrix} \begin{bmatrix} I_1 \\ I_2 \end{bmatrix}$$

(3-35)

28. Delay Lines. Transmission lines used to obtain pulse time delays are one class of

---

[1] R. W. P. King, "Transmission Line Theory," Dover, New York, 1965, Chap. 1.

[2] R. E. Collin, "Fundamentals for Microwave Engineering," McGraw Hill, New York, 1966, Chap. 4, p. 145.

structure known as *delay lines*. The drawback is that the line must be rather long even for small time delays since the electromagnetic waves propagate at a speed close to the speed of light. Special compact low-velocity lines have been developed to avoid this inconvenience. The most common type is a coaxial line, in which the inner conductor is a helix. The vast majority of the so-called "electric" delay lines are artificial transmission lines consisting of lumped capacitors and inductors. The limitations of physically realizable amplitude- and phase-transfer functions are such that the practical delays obtained do not exceed the order of a few pulse periods. Longer time delays are achieved with *acoustic delay lines*, employing acoustic wave propagation and electromechanical transducers at the input and output.

29. Pulse-Forming Lines. A particularly useful application of the transient phenomena in transmission lines is the *pulse-forming lines*, an example of which appears in Fig. 3-32. In this example, the transmission line is slowly charged, by means of a voltage source in series with a relatively high resistance $R_s \gg R_0$, to an initial steady-state voltage V_0. When the load is connected by closing the switch S, the line develops a load voltage equal to $V_0/2$ and a wave of voltage with an amplitude $-V_0/2$ travels back toward the source. Since the source appears as an open

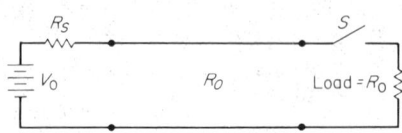

Fig. 3-32. Schematic diagram of an application of a transmission line as a pulse-forming line. In practical applications, a charging inductor is usually used in place of the charging resistor R_s.

circuit to the traveling wave, it is reflected as a voltage wave $-V_0/2$, which brings the resultant voltage to zero. The net effect is that the line delivers a rectangular voltage pulse to the load of amplitude $V_0/2$ and of duration equal to the round-trip time delay of the line. The charging source can alternatively be connected at the load end; in this case, the pulse-forming line is a one-port device. One-port "artificial" pulse-forming lines synthesized with lumped capacitors and inductors are called *Guillemin lines*.

30. Scattering-Matrix Description. One difficulty that arises in the impedance-matrix formulation is that of measuring the impedance parameters. The quantities that are directly measurable are the amplitudes and phase angles of the reflected waves. Since the field equations are linear, the reflected-wave amplitudes are linearly related to the incident-wave amplitudes. In matrix form we write $V^- = AV^+$. The equivalent voltages are chosen so that the *characteristic impedance* is 1. Thus the *scattering matrix* S is *symmetric* for *reciprocal structures*, and the power transmitted to each port is $\frac{1}{2}V_n^+ V_n^{+*}$. For a given structure the impedance matrix has definite values which are, of course, a function of frequency. If the terminal plane of the nth port is changed by l_n, the resulting scattering matrix is easily calculated:

$$[S'] = \begin{bmatrix} e^{-j\theta_1} & \cdots & \cdots \\ \cdots & e^{-j\theta_n} & \cdots \\ \cdots & \cdots & e^{-j\theta_N} \end{bmatrix} [S] \begin{bmatrix} e^{-j\theta_1} & & \cdots \\ \cdots & \cdots & \cdots \\ \cdots & \cdots & e^{-j\theta_N} \end{bmatrix} \tag{3-36}$$

where $\theta_i = \beta_i l_i$ is the electric phase shift and S is the scattering matrix of the original configuration (see Par. **3-44**).

Several properties of the scattering matrix can be easily derived. For a lossless structure $\sum_{n=1}^{N} S_{ns} S_{nr}^* = \delta_{sr}$, where δ_{sr} is the *Kronecker delta* function. If the most common microwave structure, the two-port, in addition to being lossless is also reciprocal, it is easily shown that $|S_{11}| = |S_{22}|$ and $|S_{12}| = \sqrt{1 - |S_{11}|^2}$. With $S_{11} = |S_{11}|e^{j\theta_1}$, $S_{22} = |S_{22}|e^{j\theta_2}$, and $S_{12} = (1 - |S_{11}|^2)^{1/2}e^{j\phi}$ we get

$$\phi = (\theta_1 + \theta_2 + \pi)/2 \pm n\pi$$

If the waveguide is not lossless, the power dissipation is given by $P_{\text{DISS}} = \frac{1}{2}V^+(I - SS^*)V^{+*}$.

31. Transmission-Matrix Description. When two-port microwave circuits are *cascaded*, it is convenient to represent each circuit by a *transmission matrix*, the transmission matrix of the entire structure being the product of the individual transmission matrices (see Par. **3-49**). When the independent variables are chosen to be the *port voltage* V_1 and the *port* *current* I_1, the transmission matrix is called the *voltage-current transmission matrix*. This approach is invaluable in the study of periodic structures.

$$\begin{bmatrix} V_2 \\ I_2 \end{bmatrix} = [T] \begin{bmatrix} V_1 \\ I_1 \end{bmatrix} \tag{3-37}$$

32. Resonators. At high frequencies short-circuit or open-circuit transmission lines can be used as *resonant circuits*. It is usually assumed that such lines are air-filled since dielectric-filled lines introduce additional losses resulting in a lower Q (see Par. **3-15**). For a short-circuited line of length L, with series resistance r Ω per unit length, l H per unit length, and c F per unit length, the input impedance is given by $Z_{in}(\omega) = rL + j\omega lL$, which is analogous to the result for the lumped-parameter series RLC circuit. Typical values for Q are from 100 to 10,000. *Microwave cavities can also be used as resonators and are of practical importance at frequencies above 1,000 MHz.*

The *resonant cavity* is a completely enclosed structure with a resonant frequency given by $f_{mnl} = c\sqrt{(l/2d)^2 + (n/2b)^2 + (m/2a)^2}$, where c is the velocity of light, a, b, and d the dimensions of the structure, and l,m,n are integers. Each set of integers l, m, n corresponds to a different resonant frequency.

33. Tuning. When a load is connected to a source through a transmission line or waveguide, the impedance of both the source and the load is matched to the impedance of the line. This reduces the sensitivity to frequency variations. Matching the source to the loaded waveguide or transmission line for maximum power transfer is not practical since a small change in the frequency of the source changes the electric length of the propagating structure, which greatly modifies the effective load impedance. In addition, if the load is not matched to the propagating structure, *standing waves* are present and additional losses are incurred. When the propagating structure is a transmission line, impedance matching is accomplished with tuning stubs. A *tuning stub* is a short-circuited transmission line of variable length. Tuning is accomplished by adjusting the length of the stub. When the losses are small, the effect of the tuning stubs can be studied on the Smith chart (see **Par. 9-2**).

34. Antennas and Radiators. An *antenna* is an impedance-matching device used to absorb or radiate electromagnetic waves. Its purpose is to *match* the impedance of the *propagating medium* (usually air or free space) to the source. The field configuration produced by a radiating antenna is separated into two parts, the *induction field*, which attenuates rapidly in the vicinity of the antenna, and the *radiating field*, which attenuates as $1/r^2$.

To transmit energy to the propagating medium most effectively the antenna must appear as a resistive load to the source at the frequency to be propagated. This is the *radiation resistance R_r*; it and the ohmic resistance R_0 are the only resistances that are usually considered, the loss due to eddy currents, leakage, etc., being neglected.

Antennas capable of functioning over a wide range of frequencies are called *aperiodic*, and those designed for a particular frequency, *tuned*. The geometry of the antenna determines the radiation pattern of the propagating signal in space.

Three-dimensional plots of the radiation pattern in space are difficult to construct. The normal procedure is to plot the radiation pattern for vertical planes, the *elevation patterns*, and horizontal planes, the *azimuth patterns*.

The two most important properties in antenna design are the *gain* and the *directivity*. Antenna *gain* is defined as $G = 4\pi P_r/P_T$, where P_T is the total power delivered to the antenna and P_r is the power radiated per unit solid angle in a given direction. The *directivity* is defined as $D = P_r/P_A$, where P_A is the average power radiated per unit solid angle.

When antennas are separated at distances comparable to the wavelength of the signals propagated, the mutual coupling becomes significant. The composite radiation pattern has properties that are not achievable with single-element radiators.

At microwave frequencies a variety of antenna types and arrays can be designed. The *horn radiator*, for example, is easier to couple to a waveguide than a *dipole* antenna, has a large power capacity, and provides more control over the radiation pattern. The design criteria involve a trade-off between the physical size of the radiator and the phase change it introduces. For details of antenna design, see Sec. **18**.

35. Distributed Parameters in Integrated Circuits. The technology of integrated circuits arose from the need to construct small, reliable, low-cost components for modern electronic systems (see Sec. **8**). The components are mounted on a common insulating or semiconducting substrate. The latter is more desirable because active elements can be synthesized more easily, but the use of a semiconducting substrate reduces the isolation between the circuit components. As a result, even at low frequencies, the distributed resistance and capacitance are important and affect the circuit operation. These distributed effects are not always detrimental; in particular, the *RC ladder* is a useful structure. A greater phase shift can be achieved over conventional lumped-ladder networks.

The equivalent circuit for a general type of *RC* ladder is shown in Fig. 3-33, where *r* and *c* are the resistance and capacitance per unit length. The factor α accounts for the fact that the substrates may have different resistivities. If the voltages and currents are assumed to be sinusoidal, we derive, using Kirchhoff's laws,

$$\frac{d^2v}{dx^2} = j\omega(1 + \alpha)rcv \qquad \frac{di}{dx} = j\omega cv \qquad (3\text{-}38)$$

One application for the distributed *RC* ladder network is the *notch filter*.

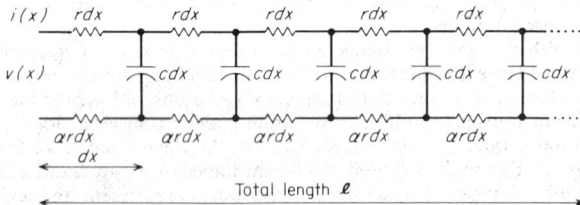

Fig. 3-33. Equivalent circuit for a general *RC* ladder.

NETWORK CONCEPTS

36. Network Topology. Network topology is concerned with the interconnection of the elements of a network. We study how the branches are connected without really considering the elements themselves. Here we define a *branch* as a directed line segment, i.e., a line with an assigned direction. Each end of the line is called a *node*. Branches may be connected to each other only at their nodes. A group of branches is called a *network* or a *graph*. It is properly called a *directed graph* because all the lines have assigned directions. Although shown as a directed line segment only, a branch can represent a *circuit element*, such as a resistor, generator, capacitor, etc. It can also simply represent a wire. Branches are usually identified by numbering them (see Fig. 3-34).

A *path* is defined as the record of a journey through a network in which one follows branches from node to node, never passing through any particular node more than once. If a path exists between every pair of nodes in a network, the network is *connected*.

37. Analysis of Connected Networks. A *subnetwork* can be identified as a part of a larger graph. Branches (with their nodes) which are not in the subnetwork are in the *complement* of the subnetwork. In a connected graph, it follows that a subnetwork will have at least one node in common with its complement (it will have no branches in common, however).

A *mesh* or *loop* is a path which ends upon the node from which it began. The branches of a mesh necessarily form a connected network (see Fig. 3-35). A *tree* is a subnetwork which contains all the nodes of the original connected network. It must be connected, and it must contain *no* meshes. A tree is not necessarily a path and in any event is not a closed

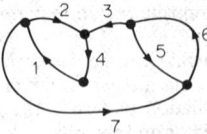

Fig. 3-34. Directed-graph representation of a simple network.

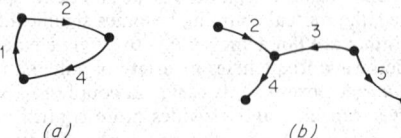

Fig. 3-35. A network loop (*a*) and tree (*b*).

path. Every connected network has at least one tree, which may be formed from the original network by removing one branch from each closed path subject to the constraint that the network remain connected. Those branches which have been removed to form the tree are called the *chords* or *links* of that tree.

It can be shown by a construction technique[1] that for a connected network

$$T = N - 1 \quad \text{and} \quad C = B - T$$

where T is the number of tree branches, N the number of nodes, B the number of branches, and C the number of chords. It can also be demonstrated that every tree has at least two nodes to which *only one* tree branch is connected.

We can use Kirchoff's laws (see Par. **3-4**) to state some useful conditions in determining how much information is necessary to specify a circuit uniquely. We begin by realizing that Kirchhoff's current law (KCL) is applicable at every node, and that Kirchhoff's voltage law (KVL) can be written around every mesh in the connected circuit. From this and the above definitions it can be stated that all branch voltages can be uniquely determined from the tree-branch voltages. Since the tree contains all the nodes of a network, it is possible to determine the voltage between any two nodes by applying KVL along a path connecting the nodes, thereby establishing any unknown branch voltage in terms of tree-branch voltages. Similarly, all the network currents can be uniquely determined from the chord currents. Remembering that there are at least two nodes of the tree which are connected to only one branch, it is clear that at least one node of the network contains only one current in addition to chord currents. This one current can then be determined by KCL. Let us now consider the tree subnetwork corresponding to the tree, with the branch we have just solved for removed from it. This new tree (and its network) has at least one node with only one unknown current (by previous arguments). Using KCL, we solve for this and remove its branch. It can be shown that all currents can be systematically found in this manner. These two developments tell us that there are no more than T independent branch voltages in a connected network, nor more than C independent currents.

38. Mesh analysis is a systematic network-analysis procedure which chooses chord currents as unknowns, further defining them in such a way that KCL is automatically satisfied at every node. Analysis rests on the selection of a number of *basic-mesh* or *fundamental-loop* units. A basic mesh is a loop comprising one chord and the path in the tree between the nodes of the chord (see Fig. 3-36). The number of basic mesh units is C. Fictitious loop currents are drawn through the C basic-mesh units choosing the direction so that the mesh current flows through the chord in the same direction as the original chord current. These currents satisfy KCL at every node, and it is unnecessary to write any KCL equations.

An example of a circuit topologically equivalent to Fig. 3-36 is shown in Fig. 3-37. Three simultaneous equations for this circuit can be written using KVL around each loop, assuming all initial conditions $= 0$:

$$-V_4 = i_1 R_1 + 1/C_1 \int i_1 \, dt - 1/C_1 \int i_2 \, dt$$

$$0 = i_2 R_3 + 1/C_1 \int i_2 \, dt + L_7 \, di_2/dt + L_5 \, di_2/dt - 1/C_1 \int i_1 \, dt \qquad (3\text{-}39)$$

$$\quad - L_5 \, di_3/dt$$

$$V_6 = L_5 \, di_3/dt - L_5 \, di_2/dt$$

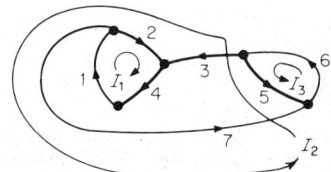

Fig. 3-36. Defining mesh currents. The heavy lines indicate a tree.

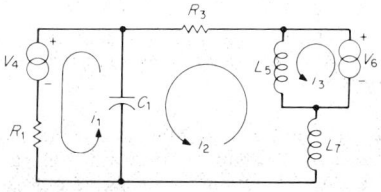

Fig. 3-37. An electric network topologically equivalent to Fig. 3-36.

[1] Paul M. Chirlian, "Basic Network Theory," McGraw-Hill, New York, 1969.

These are the necessary and sufficient equations to solve for the three unknown currents (and hence all circuit unknowns).

For a given loop, all the voltage drops caused by the flow of its circulating current are called *self-voltage drops*, whereas drops caused by currents from other loops are called *mutual* or *coupled voltage drops*. Thus in Fig. 3-37 for loop 2, $i_2 R_3$, $1/C_1 \int i_1\, dt$, $L_7 di_2/dt$, and $L_5 di_2/dt$ are all self-voltage drops, and $-1/C_1 \int i_1\, dt$ and $L_5 di_3/dt$ are coupled voltage drops.

39. Nodal analysis is a systematic network-analysis procedure which chooses nodal voltages as unknowns. One begins by arbitrarily picking a node which is labeled 0 and called the *reference node* or *datum node* or *ground*. All other circuit nodes are numbered systematically. Branch voltages are then determined in terms of the unknown node voltages by subtracting the voltage of the node at the tail of the branch arrow from the voltage of the node at the head of the branch arrow. The voltage of the reference node is defined as being zero.

Figure 3-38 shows such a network and circuit. Each of the four following equations is the result of applying KCL to each of the four nodes:

$$I_2 + I_7 = C_1 \frac{d}{dt}(V_{N_2} - V_{N_1}) \tag{3-40}$$

$$0 = -\frac{V_{N_2} - V_{N_3}}{R_4} - C_1 \frac{d}{dt}(V_{N_2} - V_{N_1}) \tag{3-41}$$

$$I_2 = -\frac{V_{N_2} - V_{N_3}}{R_4} + \frac{V_{N_3} - V_{N_4}}{R_3} \tag{3-42}$$

$$0 = -\frac{V_{N_4} - V_{N_3}}{R_3} - \frac{1}{L_5}\int V_{N_4}\, dt - \frac{V_{N_4}}{R_6} \tag{3-43}$$

If a branch is connected between the node for which KCL is being written and the reference node, the current through this branch is due only to the voltage at the node of interest and is therefore called a *self-current*. The two currents represented by the second and third terms of Eq. (3-43) are self-currents. If the branch connects the node of interest with a node other than a reference node, the current in that branch is called a *coupled* or *mutual current*. All other terms in the above equations are of this type.

These four equations can be solved for the four unknown node voltages (and hence all circuit unknowns).

40. Choice of Analysis Type. The chief virtue of the mesh and nodal analysis procedures is that they are systematic and hence guarantee a solution (although not necessarily with the application of the least possible work). Because they are systematic approaches, they are mutually exclusive. The decision to choose one approach above the other is based on:

1. *Number of unknowns.* Mesh analysis will produce C simultaneous equations, whereas nodal analysis will produce $N - 1$. That method which produces the fewer equations is preferred.

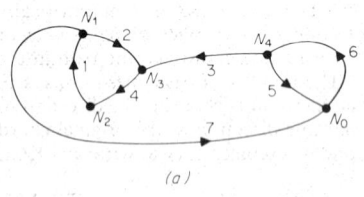

(a)

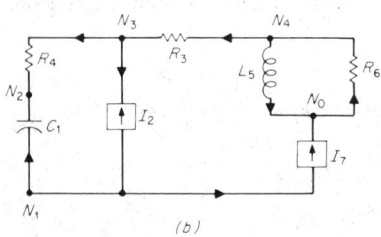

(b)

Fig. 3-38. Preliminary procedures for nodal analysis of (*a*) a directed graph and (*b*) its topological equivalent.

2. *Type of sources.* Mesh analysis cannot be done with current sources in the network. Nodal analysis cannot be done with voltage sources. Thevenin or Norton equivalents can be found for sources of the undesired type, but this represents more calculations.

3. *Mutual inductance.* Nodal analysis can be used with coupled coils only by finding noncoupled equivalent circuits for the coils.

41. Cut Sets and Cut-Set Analysis. Nodal analysis is a special case of *cut-set analysis*. A *fundamental cut set* or *basic cut set* consists of one (and only one) branch of the network tree together with any links which must be cut to divide the network into two parts. An isolated node, if it is formed, is a *circuit part*. Removal of the cut set should divide the circuit into only two parts. A set of fundamental cut sets includes those cut sets formed by applying the cut-set division for each of the branches of the network tree. The number of cut sets is therefore equal to the number of tree branches of the network. Once the cut sets have been determined, KCL is applied to the interior of each. By convention, a positive current is one which crosses the cut-set boundary in the same direction as the tree branch. The circuit unknowns are the cut-set voltages, the voltage assigned to the node which is within the cut set.

Figure 3-39 shows the cut-set analysis of a circuit together with the resulting equations. A branch voltage is computed as the algebraic sum of all the *cut-set voltages* of the cut sets through which it passes. If the positive branch direction passes out of the cut set, the cut-set voltage is positive. If the branch passes into the cut set (in the same direction as the arrows on the boundary corners), it is taken negative. Application of KVL to Fig. 3-39 yields the following results: $V_{B_1} = V_2 - V_1$, $V_{B_2} = V_1$, $V_{B_3} = V_3$, $V_{B_4} = -V_2$, $V_{B_5} = V_3 - V_4$, $V_{B_6} = V_4 - V_3$, and $V_{B_7} = V_1 - V_3 - V_4$. Application of KCL yields $i_2 - i_1 + i_7 = 0$, $i_4 - i_1 = 0$, $i_3 - i_7 = 0$, and $i_5 + i_7 - i_6 = 0$. The circuit unknowns are V_1, V_2, V_3, and V_4. Also needed are branch equations for the unknown currents: $i_1 = C_1 \, dV_{B_1}/dt$, $i_2 = I_2$, $i_3 = 1/R_3 \, V_{B_3}$, etc. If these branch equations are substituted into the cut-set KCL, then by substituting V_1, V_2, V_3, V_4 for all branch voltages, one has four equations in four unknowns.

42. Treatment of Dependent Generators. A *dependent generator* is a current or voltage source whose output is a function of (depends on) some other circuit variable. The value of this variable is usually unknown, and hence the value of the generator output is unknown. These generators therefore require special treatment. Dependent generators are a necessary part of all transistor and vacuum-tube equivalent circuits. Analysis begins with the standard techniques described earlier, where it is the custom to write all the source voltages

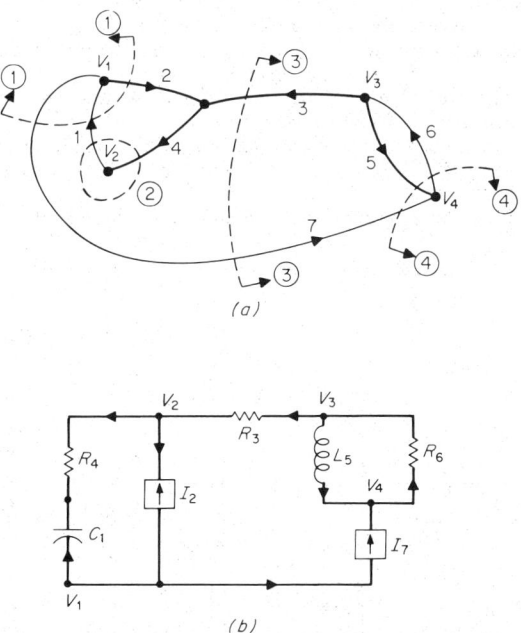

(a)

(b)

Fig. 3-39. Example of preliminary procedures for cut-set analysis. In (*a*) the heavy solid lines represent a tree, while the dashed lines are the cut-set boundaries. Each cut-set boundary is given a number and a direction. The topological equivalent for which example equations are given is shown in (*b*).

or currents on the left side of the equation, since they are known values. In spite of their being unknown, dependent sources appear on the left side. Then an additional series of equations is written (one for each dependent source) which describes the source dependency in terms of the circuit unknowns. These are substituted in, and the equations are regrouped with all unknowns on the right.

Figure 3-40 shows a simple two-mesh circuit with dependent voltage generators. The mesh equations based on Fig. 3-40 are

$$V_1 - r_m i_{ab} = i_1 R_2 + L_2 di_1/dt + 1/C \int i_1 dt - 1/C \int i_2 dt$$

$$\mu V_{ad} = L_1 di_2/dt + R_1 i_2 + 1/C \int i_2 dt - 1/C \int i_1 dt$$

$$i_{ab} = -i_2$$

$$V_{ad} = -V_1 + i_1 R_2 \tag{3-44}$$

$$V_1 = i_1 R_2 + L_2 di_1/dt + 1/C \int i_1 dt - 1/C \int i_2 dt - r_m i_2$$

$$-\mu V_1 = L_1 di_2/dt + R_1 i_2 + 1/C \int i_2 dt - 1/C \int i_1 dt - i_1 \mu R_2$$

43. Treatment of Nonlinear Elements. Analysis of networks containing nonlinear elements may be handled in a number of ways. Probably the most common approach is *linearization with small-signal analysis.* This technique approximates complex relationships as linear functions in the neighborhood of some operating point. Once the linearization has been accomplished, analysis can proceed according to the standard methods. If the circuit cannot be linearized, the nonlinear elements must be dealt with directly by developing an equation for voltage and current. Analysis can then be carried out almost as before, but subject to a number of constraints:

1. It is not valid to use superposition, i.e., to say that if current i_1 produces voltage drop V_1 and i_2 produces drop V_2, then $i_1 + i_2$ will produce drop $V_1 + V_2$.

2. The circuit may contain a nonlinear element which is multivalued in its relationship. If the branch voltage cannot be determined uniquely from the branch current, the circuit cannot be analyzed using nodal or cut-set analysis.

3. If the branch current cannot be determined uniquely from the branch voltage, the circuit cannot be analyzed using loop or mesh analysis.

Figure 3-41 shows an example of circuit analysis with a nonlinear capacitance. The node equations are $I = V_A/R_1 + (V_A - V_B)/R_2$ and $0 = (V_A - V_B)/R_2 + f_c(V_B)|_{V_B} dV_B/dt$. The capacitor equations are $i = C \, dV_c/dt$, $C = f_c(V_c)|_{V_c}$, $i = f_c(V_c)|_{V_c} dV_c/dt$, and $V_c = V_B$.

If f_c is a fairly simple relationship, the equations can be solved analytically by a number

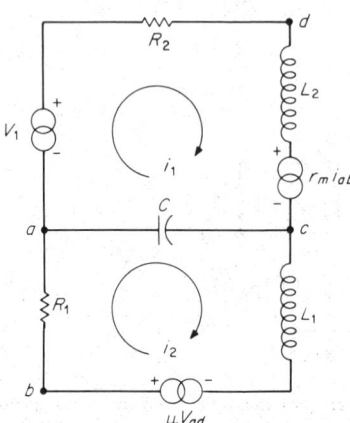

Fig. 3-40. A circuit with dependent sources.

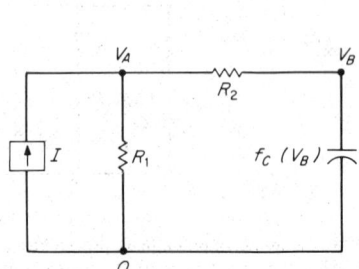

Fig. 3-41. Simple circuit containing a non-linear capacitance.

of techniques. If the relationship is complex, solution must be accomplished by graphical techniques, state-variable techniques (see Sec. 3) or numerical analysis.

44. Matrix Formulation. Circuit-analysis techniques are systematic. Indeed, if the equations are set up according to the system definitions, algorithms exist to carry out the solution completely. The systematic form of the equations is called a *matrix formulation*. A *matrix* is an array of numbers, symbolically denoted by square brackets, with m rows and n columns:

$$\begin{bmatrix} z_{11} & z_{12} & \cdots & z_{1n} \\ z_{21} & z_{22} & \cdots & z_{2n} \\ \cdots & \cdots & \cdots & \cdots \\ z_{m1} & z_{m2} & \cdots & z_{mn} \end{bmatrix}$$

Formulation of the *loop-analysis matrix* representation of Fig. 3-38 is carried out by writing the source voltages in a single column matrix on the left. The rest of the equations are arranged to the left. All terms in Eq. (3-40) which contain i_1 are grouped as the z_{11} term. All terms of Eq. (3-41) which contain i_1 are grouped as the z_{21} term. All terms of Eq. (3-40) which contain current i_2 are grouped as the z_{12} term, and so on. The currents themselves are placed in a column matrix to the right of this matrix.

$$\begin{bmatrix} -V_4 \\ 0 \\ V_6 \end{bmatrix} = \begin{bmatrix} R_1 + 1/C_1 \int & -1/C_1 \int & 0 \\ -1/C_1 \int & R_3 + 1/C_1 \int + L_5\,d/dt + L_7\,d/dt & -L_5\,d/dt \\ 0 & -L_5\,d/dt & L_5\,d/dt \end{bmatrix} \begin{bmatrix} i_1 \\ i_2 \\ i_3 \end{bmatrix} \quad (3\text{-}45)$$

This representation is equivalent to the system of equations and can be manipulated in representative form $[V] = [Z][I]$.

Notice that the matrix is symmetric about its main diagonal. The main-diagonal elements are the self-voltage drops; all other elements are coupled-voltage drops. Notice also the unusual forms of terms like $R_1 + 1C_1 \int$. These forms are called *operators* (*impedance operators*). When this term multiplies i_1, for instance, it implies an operation

$$(R_1 + 1/C_1 \int)i_1 \Rightarrow R_1 i_1 + 1/C_1 \int i_1 \, dt$$

This notation is usually made more compact by the definition of the *differential operator* p, such that $pf(t) \equiv d/dt f(t)$; in addition, the *integral operator* p^{-1} is defined as $p^{-1}f(t) \equiv f(t)/p = \int f(t)\,dt$. Thus $(R_1 + 1/C_1 \int)i_1 = (R_1 + 1/pC_1)i_1$.

By analogy, nodal (cut-set) analysis yields a matrix formulation for the circuit of Fig. 3-38:

$$\begin{bmatrix} I_2 + I_7 \\ 0 \\ I_2 \\ 0 \end{bmatrix} = \begin{bmatrix} -pC_1 & pC_1 & 0 & 0 \\ pC_1 - 1/R_4 & -pC_1 & pC_1 - 1/R_4 & 0 \\ 0 & 1/R_4 & 1/R_3 - 1/R_4 & 1/R_3 \\ 0 & 0 & 1/R_3 & -1/R_3 - 1/pL_5 - 1/R_6 \end{bmatrix} \begin{bmatrix} V_{N_1} \\ V_{N_2} \\ V_{N_3} \\ V_{N_4} \end{bmatrix} \quad (3\text{-}46)$$

$[I] = [Y][V]$, where the terms of the $[Y]$ matrix are *admittance* operators.

45. Laplace Transforms. The Laplace transform is an *integral transformation*, since it transforms a function of one variable into a function of another variable through the process of integration. It is exceedingly important to linear-circuit analysis because it transforms the solution of simultaneous differential circuit equations with constant coefficients into an *algebra* problem. Although the transformation process is not always possible, all functions ordinarily encountered in practical networks can be handled using Laplace transforms. The Laplace transform converts a function of *time* into a function of a variable s, which is a *complex number* traditionally expressed as $s = \sigma + j\omega$ ($j = \sqrt{-1}$), and called the *complex frequency*.

$$F(s) = \mathcal{L}[f(t)] = \int_0^\infty f(t)e^{-st}\,dt \quad (3\text{-}47)$$

$F(s)$ is in general a complex function. Note that the lower limit of integration is zero. It is assumed that $f(t) = 0$ for $t < 0$. If this is not the case, it will be necessary to redefine the time-axis origin or to use the *two-sided Laplace transform*.

It is also necessary to define the *inverse Laplace transform* to form the *Laplace transform pair*. This transformation function is a complex integration which results in a time function

$$f(t) = \mathcal{L}^{-1}[F(s)] = \frac{1}{2\pi j} \int_{s=C-j\infty}^{C+j\infty} F(s)e^{st}\,ds \tag{3-48}$$

This integration is usually exceedingly difficult to carry out for any worthwhile function. It is therefore not done, but tables are generated which express various common functions of time in terms of their Laplace transforms. These are called *Laplace transform tables*. Table 3-4 shows some of the more common transform pairs. The Laplace transformation is *unique*; i.e., for any $F(s)$, there exists one and only one corresponding $f(t)$, and vice versa.

The Laplace transform exhibits a number of useful properties:
1. It is linear. If α is a constant and $f_1(t)$ is some time function

$$\mathcal{L}[\alpha f_1(t)] = \alpha\mathcal{L}[f_1(t)] = \alpha F_1(s)$$

It is also true that if there exists another time function $f_2(t)$, then

$$\mathcal{L}[f_1(t) + f_2(t)] = \mathcal{L}[f_1(t)] + \mathcal{L}[f_2(t)] = F_1(s) + F_2(s)$$

2. It can be integrated

$$\int_0^t f_1(t)\,dt = \frac{1}{s}[f_1(t)] = \frac{1}{s}F_1(s)$$

3. It can be differentiated.

$$\mathcal{L}\left[\frac{df_1(t)}{dt}\right] = s\mathcal{L}[f_1(t)] - f_1(0) = sF_1(s) - f_1(0_+)$$

Note the extra term $f_1(0_+)$. This is an *initial condition* and is the numerical value of $f_1(t)$ at $t = 0_+$. Laplace transforms automatically handle initial conditions which occur in circuit analysis.

4. It can be differentiated two or more times by a repetitive process.

$$\frac{d^2 f_1(t)}{dt^2} = s\left(\left\{\frac{d}{dt}[f_1(t) - f_1(0_+)]\right\}\right)$$

$$d^2 f_1(t) = s^2 F_1(s) - s\frac{d}{dt}f_1(0_+) - f_1(0_+)$$

5. It can be *shifted in time*.

$$\mathcal{L}[f(t) \cdot u(t - y)] = e^{-sy}F(s)$$

6. *Multiplication* of Laplace transforms is equivalent to *convolution* of their time functions.

$$\mathcal{L}[f_1(t) * f_2(t)] = F_1(s) \cdot F_2(s)$$

46. Example of Laplace Transformation. Referring to the network and circuit of Figs. 3-36 and 3-37, the mesh equations 3-39 become

$$\begin{bmatrix} -V_4(s) \\ 0 \\ V_6(s) \end{bmatrix} = \begin{bmatrix} R_1 + 1/sC_1 & -1/sC_1 & 0 \\ -1/sC_1 & R_3 + sL_5 + sL_7 + 1/sC_1 & -sL_5 \\ 0 & -sL_5 & sL_5 \end{bmatrix}\begin{bmatrix} I_1(s) \\ I_2(s) \\ I_3(s) \end{bmatrix} \tag{3-49}$$

If one wishes to solve for $I_1(s)$ for instance, *Cramer's rule* can be applied to give

$$I_1(s) = \frac{s[-V_4(s)L_5L_7] + [-V_4(s)L_5R_3]}{s^2(R_1 L_5 L_7) + s\left(R_1 R_3 L_5 + \dfrac{L_5 L_7}{C_1}\right) + \left(\dfrac{R_1 L_5}{C_1} + \dfrac{R_3 L_5}{C_1}\right)} \tag{3-50}$$

Table 3-4   Laplace Transform Pairs

$f(t)$ $t \geq 0$	$F(s)$
$af(t)$	$aF(s)$
$f_1(t) \pm f_2(t)$	$F_1(s) \pm F_2(s)$
$\dfrac{df(t)}{dt} \triangleq f'(t)$	$sF(s) - f(0+)$
$\int f(t)\,dt \triangleq f^{(-1)}(t)$	$\dfrac{F(s)}{s} + \dfrac{f^{(-1)}(0+)}{s}$
1 or unit step at $t = 0$	$\dfrac{1}{s}$
t	$\dfrac{1}{s^2}$
$\dfrac{1}{(n-1)!} t^{n-1}$ n a positive integer	$\dfrac{1}{s^n}$
$e^{-\alpha t}$	$\dfrac{1}{s + \alpha}$
$\sin \beta t$	$\dfrac{\beta}{s^2 + \beta^2}$
$\cos \beta t$	$\dfrac{s}{s^2 + \beta^2}$
$e^{-\alpha t} \cos \beta t$	$\dfrac{s + \alpha}{(s + \alpha)^2 + \beta^2}$
$\dfrac{1}{a}(1 - e^{-at})$	$\dfrac{1}{s(s + a)}$
$\dfrac{1}{s_1 - s_2}(e^{s_1 t} - e^{s_2 t})$	$\dfrac{1}{(s - s_1)(s - s_2)}$
te^{-at}	$\dfrac{1}{(s + a)^2}$
t^n	$\dfrac{n!}{s^{n+1}}$
$e^{-at} \sin \omega_1 t$	$\dfrac{\omega_1}{(s + a)^2 + \omega_1^2}$
$e^{-at} \cos \omega_1 t$	$\dfrac{s + a}{(s + a)^2 + \omega_1^2}$
δ(delta function)	1

$V_4(s)$ and $V_6(s)$ can (in almost every instance) be expressed as the ratio of two polynomials in s. The final expression for $I_1(s)$ is then, in general, the ratio of two polynomials in s.

$$I_1(s) = \frac{\alpha_n s^n + \alpha_{n-1} s^{n-1} + \cdots + \alpha_1 s + \alpha_0}{\beta_m s^m + \beta_{m-1} s^{m-1} + \cdots + \beta_1 s + \beta_0} \qquad (3\text{-}51)$$

To determine $i_1(t)$, it is necessary to find the inverse transform of $I_1(s)$.

Partial-Fraction Expansion. Laplace transform tables do not cover such complicated expressions, and so $I_1(s)$ must be broken down by *partial-fraction expansion.* Partial-fraction expansion requires that $I_1(s)$ be a *proper fraction*, that is, $m > n$. If it is not, first use long division to reduce the power of the numerator.

$$I_1(s) = \gamma_{m-n} s^{m-n} + \gamma_{m-n-1} s^{m-n-1} + \cdots + \gamma_0 + \frac{\xi_{n-1} s^{n-1} + \xi_{n-2} s^{n-2} + \cdots + \xi_0}{\beta_n s^n + \beta_{n-1} s^{n-1} + \cdots + \beta_0} \qquad (3\text{-}52)$$

Then the fraction denominator must be factored:

$$I_1(s) = \gamma_{m-n} s^{m-n} + \cdots + \gamma_0 \frac{\xi_{n-1} s^{n-1} + \cdots + \xi_0}{\beta_n (s - s_1)(s - s_2) \cdots (s - s_n)} \qquad (3\text{-}53)$$

where $s_1, s_2, \ldots, s_n$ are the roots of the denominator. $I_1(s)$ can be converted from this form to the *partial-fraction expansion*

$$I_1(s) = \gamma_{m-n} s^{m-n} + \cdots + \gamma_0 + \frac{K_1}{s - s_1} + \frac{K_2}{s - s_2} + \cdots + \frac{K_n}{s - s_n} \qquad (3\text{-}54)$$

if all the s_i are different. If some roots are repeated ($s_2 = s_3$, for instance), the form is modified. The new form is best explained by example:

$$F(s) = \frac{\xi_5 s^5 + \xi_4 s^4 + \xi_3 s^3 + \xi_2 s^2 + \xi_1 s + \xi_0}{(s - s_1)(s - s_2)^3 (s - s_3)^2} \qquad (3\text{-}55)$$

$$F(s) = \frac{K_1}{s - s_1} + \frac{K_{21}}{(s - s_2)^3} + \frac{K_{22}}{(s - s_2)^2} + \frac{K_{23}}{(s - s_2)} + \frac{K_{31}}{(s - s_3)^2} + \frac{K_{32}}{(s - s_3)} \qquad (3\text{-}56)$$

The unknown constants K_i can be evaluated by a simple process. To find K_1, for instance, simply multiply both sides of $F(s)$ by $s - s_1$ and evaluate the function for $s = s_1$

$$K_1 = \left. \frac{\xi_{n-1} s^{n-1} + \cdots + \xi_0}{\beta_n (s - s_2) \cdots (s - s_n)} \right|_{s=s_1} \qquad (3\text{-}57)$$

To find the K's for a multiple root, one multiplies both sides by the highest power and then differentiates to find the K's associated with lower powers. For instance, in Eq. (3-56)

$$K_{21} = (s - s_3)^3 F(s)|_{s=s_1}$$

$$K_{22} = \frac{d}{ds}[(s - s_2)^3 F(s)]|_{s=s_1} \qquad (3\text{-}58)$$

$$K_{23} = \frac{1}{2} \frac{d^2}{ds^2}[(s - s_2)^3 F(s)]|_{s=s_1}$$

The final result is that $I_1(s)$ becomes a series of simple terms such as $r_i s^i$, γ_0, and $K_{rp}/(s - s_r)^p$.

All these terms are documented in a complete table of Laplace transforms so that $I_1(s)$ becomes the sum of a number of smaller, easier to find functions: $I_1(s) = I_{1a}(s) + I_{1b}(s) + \cdots + I_{1l}(s)$. Since each $I_{1k}(s)$ has a corresponding time function $i_{1k}(t)$, $i_1(t) = i_{1a}(t) + i_{1b}(t) + \cdots + i_{1l}(t)$.

47. Poles and Zeros. Functions of s are often defined in terms of their *poles* and *zeros*. $I_1(s)$ for instance [Eq. (3-51)] has n roots in its numerator polynomial and m roots in its denominator. For the n values which correspond to the roots of the numerator, $I_1(s) \to 0$. These are called *zeros* of the function $I_1(s)$. For the m values which correspond to the roots of the denominator, $I_1(s) \to \infty$. These are called *poles* of $I_1(s)$. These can be located on a graph of the s plane $(s = \alpha + j\omega)$.

This graph (Fig. 3-42) of the poles (x's) and zeros (0s) is equivalent to the polynomial representation. It is particularly useful in systems engineering (Sec. 5) and filter design (Sec. 12).

48. n-Port Networks. All networks have sites at which they can be connected to the outside world, called *terminals*. A simple resistor has two terminals, a transistor has three, and a stereo amplifier often has more than ten. The most powerful general statement which can be made about an *m-terminal network* results from considering all the currents which flow into the network. Potentials may also be defined for each terminal (see Fig. 3-43).

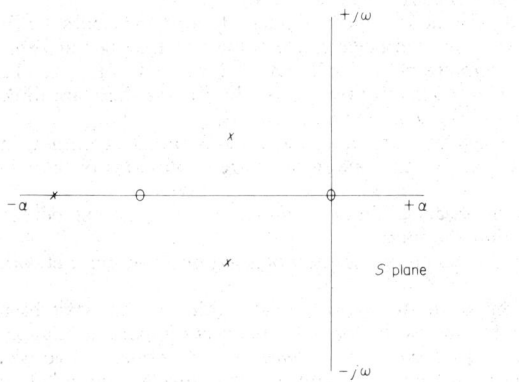

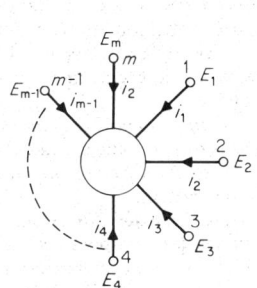

Fig. 3-42. Pole-zero representation of the ratio of two polynomials in s.   Fig. 3-43. An m-terminal network.

A simple application of the principle of conservation of charge to the network yields the result

$$\sum_{k=1}^{m} i_k = 0$$

The network can therefore be treated as a giant node obeying KCL.

If it is possible to exhaustively separate the m terminals into *pairs*, such that for each pair (say i_3 and i_7) one can state $i_3 = -i_7$, these pairs are called *ports*. This can nearly always be done, thereby producing an n-port network. A transistor is a two-port network, although it has three terminals. Connecting an extra wire to one of the terminals provides the extra terminal without violating any network laws.

All useful networks can therefore be expressed as n-port networks with paired currents (see Fig. 3-44); n-port voltages can also be derived from the m-terminal potentials. If port 1 was formed by pairing terminal 3 with terminal 7, then $V_1 = E_3 - E_7$. Convention assigns the positive direction of a port voltage such that port current flows into the port at the positive terminal and out of the negative terminal.

It can be shown that every n-port network can be uniquely characterized by *driving-point functions*. These driving-point functions relate voltage and currents at each of the ports and may be any of a number of types, e.g., *impedances, admittances,* or *hybrid* parameters.

A typical set of impedance parameters for an n-port network is

$$[Z] = \begin{bmatrix} z_{11} & z_{12} & \cdots & z_{1n} \\ z_{21} & & \cdots & \\ z_{31} & & \cdots & \\ \cdots & \cdots & \cdots & \cdots \\ z_{n1} & & \cdots & z_{nn} \end{bmatrix}$$

3-43

where z_{ii} is the voltage of port i divided by the current flowing in port i with all other ports open-circuited. This is the *self-impedance* of port i. The other impedances z_{ij}, $i \neq j$, are called *transfer impedances*. Z_{ij} is obtained by shorting port j, causing its port voltage to be zero. The ratio is then taken relating the voltage at port i to the current in port j with all other ports open.

By far the most frequently considered n-port network is the two-port. Further discussion will concern two-port networks, although the results apply to more complex n-ports.

49. Two-Port Networks. Two-port networks are commonly described in a number of ways. Table 3-5 shows these representations. They are also called *coupling networks, electric transducers, four-pole networks,* or *two-terminal pair networks.*

The scattering-matrix representation seems unduly complicated but is of great use in distributed-parameter systems such as transmission lines (see Sec. **9**).

A two-port network may have a very large number of internal loops and nodes, but it may be completely defined by knowing four parameters, one of the sets described in Table 3-5. Thus, problems involving very complex networks are handled with ease if they can be represented as a combination of a small number of two-ports. In general, there are three types of manipulations which commonly arise in two-port networks.

Transfer Calculations. It is necessary to find one of the voltages (input or output) in terms of both currents, or it is necessary to find one of the currents in terms of the two voltages.

Transmission Calculations. It is necessary to find the voltage or current at one port in terms of the voltage and current at the other port.

Insertion Calculation. It is desired to ascertain the effect of inserting a two-port network into a system.

50. Transfer Function. In addition to the various relationships which have been defined for two-port networks, others can also be formed. Two very important relationships are *voltage transfer function* and *current transfer function.* Consider a two-port connected as in Fig. 3-45 with one port defined as input and the other as output. Two dimensionless quantities may be defined: voltage transfer function V_2/V_1, and current transfer function i_2/i_1. Other possible ratios between one input and one output variable are also correctly called transfer functions. These have already been dealt with. They are z_{12} and z_{21}, transfer impedances, and y_{12} and y_{21}, transfer admittances.

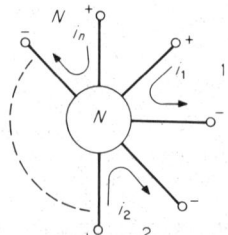
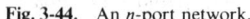
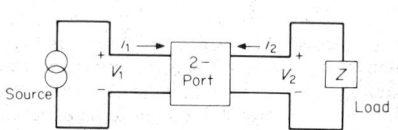

Fig. 3-44.   An n-port network.   Fig. 3-45.   Standard connections for a two-port.

51. Image Impedance. *Image impedance* or *iterative impedance* is the impedance which when connected to the output of a two-port causes the *driving-point impedance* (the input impedance) to be *identical* to the *terminating impedance* (that which has been connected to the output). The use of image impedances is particularly important in line-amplifier and filter design, in which it is desirable to insert a two-port in the middle of a network string. If the two-port is designed so that its image impedance is identical to the input of that which follows it, it can be inserted without changing *any* of the characteristics of the former network. This means that each two-port can be evaluated independently, and a network of many two-ports in series will have a transfer function which is the simple product of each individual transfer function.

Image-impedance terminations produce both maximum power transfer (subject to the constraints stated in Par. **3-13i**) and minumum circuit calculations.

52. Filters. A filter is a two-port network (in general) which has been designed to

Table 3-5 Matrix Representations of Two-Port Networks

Representation		Deriving equations	
Impedance	$\begin{bmatrix} z_{11} & z_{12} \\ z_{21} & z_{22} \end{bmatrix}$	$z_{11} = V_1/I_1$ $z_{12} = V_1/I_2$ $z_{21} = V_2/I_1$ $z_{22} = V_2/I_2$	$I_2 = 0$ $V_2 = 0$ $V_1 = 0$ $I_1 = 0$
Admittance	$\begin{bmatrix} y_{11} & y_{12} \\ y_{21} & y_{22} \end{bmatrix}$	$y_{11} = I_1/V_1$ $y_{12} = I_1/V_2$ $y_{21} = I_2/V_1$ $y_{22} = I_2/V_2$	$V_2 = 0$ $I_2 = 0$ $I_1 = 0$ $V_1 = 0$
Transmission or chain	$\begin{bmatrix} A & B \\ C & D \end{bmatrix}$	$A = V_1/V_2$ $B = -V_1/I_2$ $C = I_1/V_2$ $D = -I_1/I_2$	$I_2 = 0$ $V_2 = 0$ $I_2 = 0$ $V_2 = 0$
Hybrid	$\begin{bmatrix} h_{11} & h_{12} \\ h_{21} & h_{22} \end{bmatrix}$	$h_{11} = V_1/I_1$ $h_{12} = V_1/V_2$ $h_{21} = I_2/I_1$ $h_{22} = I_2/V_2$	$V_2 = 0$ $I_1 = 0$ $V_2 = 0$ $I_1 = 0$
Inverse hybrid	$\begin{bmatrix} g_{11} & g_{12} \\ g_{21} & g_{22} \end{bmatrix}$	$g_{11} = I_1/V_1$ $g_{12} = I_1/I_2$ $g_{21} = V_2/V_1$ $g_{22} = V_2/I_2$	$I_2 = 0$ $V_1 = 0$ $I_2 = 0$ $V_1 = 0$
Scattering	$\begin{bmatrix} s_{11} & s_{12} \\ s_{21} & s_{22} \end{bmatrix}$	$s_{11} = \dfrac{\dfrac{V_1}{\sqrt{R_{01}}} - I_1\sqrt{R_{01}}}{\dfrac{V_1}{\sqrt{R_{01}}} + I_1\sqrt{R_{01}}}$	$\dfrac{V_2}{\sqrt{R_{02}}} + I_2\sqrt{R_{02}} = 0$
		$s_{12} = \dfrac{\dfrac{V_1}{\sqrt{R_{01}}} - I_1\sqrt{R_{01}}}{\dfrac{V_2}{\sqrt{R_{02}}} + I_2\sqrt{R_{02}}}$	$\dfrac{V_1}{\sqrt{R_{01}}} + I_1\sqrt{R_{01}} = 0$
		$s_{21} = \dfrac{\dfrac{V_2}{\sqrt{R_{02}}} - I_2\sqrt{R_{02}}}{\dfrac{V_1}{\sqrt{R_{01}}} + I_1\sqrt{R_{01}}}$	$\dfrac{V_1}{\sqrt{R_{02}}} + I_2\sqrt{R_{02}} = 0$
		$s_{22} = \dfrac{\dfrac{V_2}{\sqrt{R_{02}}} - I_2\sqrt{R_{02}}}{\dfrac{V_2}{\sqrt{R_{02}}} + I_2\sqrt{R_{02}}}$	$\dfrac{V_1}{\sqrt{R_{01}}} + \dfrac{V_2}{\sqrt{R_{02}}} = 0$

transmit freely sinusoidal signals within one or more frequency bands and to attenuate substantially sinusoids of other frequencies.

Filters are usually characterized by their *voltage transfer function*. They are classed according to their voltage transfer function as low-pass, high-pass, bandpass, and band-stop. Filters with more complicated transfer characteristics are designed as a combination of these basic units.

Filters may be designed subject to a wide variety of constraints, depending on the intended use. See Sec. **12** for design procedures.

53. Dual Networks. The property of *duality* is extremely important in network analysis. One network is said to be the dual of another if they are described by the same equations except that all voltages have been replaced by currents, all currents by voltages, all resistances by conductances, etc. Figure 3-46 shows two networks which possess the property of duality.

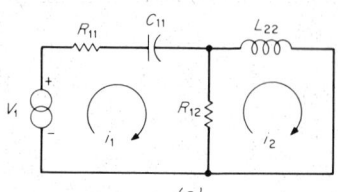

(a)

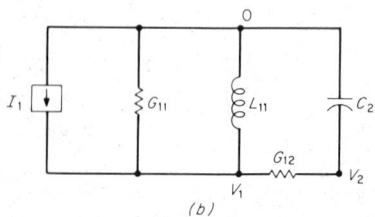

(b)

Fig. 3-46. Dual networks. (*a*) Loop analysis yields equations $V_1 = (R_{11} + R_{12} + 1/sC_{11})i_1 - R_{12} i_2$ and $0 = -R_{12}i_1 + (sL_{22} + R_{12})i_2$. (*b*) Node analysis yields equations $I_1 = (G_{11} + G_{12} + 1/sL_{11})V_1 - G_{12}V_2$ and $0 = -G_{12}V_1 + (sC_{22} + G_{12})V_2$.

For an arbitrary realizable network N and its dual network $\overline{N}$, if S is any true statement about N, then $\overline{S}$ is a true statement about $\overline{N}$. $\overline{S}$ is derived from S by replacing every electrical quantity or network description by its dual. Dual words and phrases are

Voltage	Current
Resistance	Conductance
Capacitance	Inductance
Impedance	Admittance
Reactance	Susceptance
Node	Loop
KVL	KCL
Short circuit	Open circuit
Charge	Flux
Tree branch	Chord
Node voltage	Loop current
Series addition	Parallel addition

Dual-network relationships are difficult to work with for circuits containing coupled inductors, transformers, and dependent sources. The circuit elements may be nonlinear and time-varying, however.

NETWORK INTERCONNECTIONS AND SWITCHING

54. Switching Networks. A switching network has, in general, n pairs of terminals, some of which are designated as inputs and some as outputs. In contrast to the networks considered above, however, it is assumed that each of the terminals of a switching circuit can exist only in a *finite number of states*; i.e., each terminal can take on only a finite number of values, which are discrete and identifiable. Since there are a finite number of terminals, it is theoretically possible to list all combinations of terminal values in a table, called a *truth table* or *state table*. Since the relationships between input states and output states can be identified in the truth table, it serves the same function as the transfer functions identified for analog n-port networks.

A switching network is composed only of *switching elements*. A switching element is a *resistive branch element* which exists only in states $R = 0$ and $R = \infty$.[1] Each individual switch element may be considered either by its *transmission properties* or *hindrance properties*. If the switch is *closed*, it will transmit signals freely and hence has a transmission of 1 and hindrance of 0. Conversely an *open* switch has 0 transmission and 1 hindrance. These two properties are *never mixed* in identifying switches in the same network. Switching networks

---

[1] Paul E. Wood, "Switching Theory," McGraw-Hill, New York, 1968.

are usually designed using transmission properties. Thus, a switch is defined as closed by a 1 and open by a 0.

Switches may be connected either in series or parallel. If two switches are in series, they will not transmit unless both are closed. If they are in parallel, transmission will be complete if *either* is closed. Figure 3-47 shows the results of series and parallel connections.

55. Uses of Switching Networks. Switching networks were first used to connect things together. An example of a network of this type is the telephone switching system, which by closing the proper switches in a large network can connect a caller to any other telephone in the system. Here one is concerned with finding a pathway through a maze, and a good deal of theory has been built up on this subject (see Sec. 22).

Switching networks are also used to do *logical manipulations*, using *Boolean algebra* (see Sec. 23). This is one of the primary functions of digital computers, especially those which are used to control processes of manufacture.

Switching networks are also used to do arithmetic. Switching networks add, subtract, multiply, and divide, usually in the binary system (see Sec. 23). This is also a primary function of many digital computers. Because the arithmetic is done with 0 and 1, like the logical manipulations, sometimes there is little distinction made between the two, and arithmetic becomes a type of logical manipulation.

Finally, switching networks are used to store information. Many switches are designed to remain in either the 0 or 1 state for extended periods of time. One can therefore set a particular pattern of 1s and 0s in a switch network, and it will be remembered for future reference. This is the primary function of computers which keep government and bank records, etc.

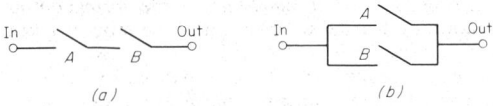

Fig. 3-47. Series and parallel switch connection. In (*a*) series connection transmits only when *A* and *B* are both closed. In (*b*) transmission is complete for either *A* or *B* closed.

The *state* of a mechanical switch is determined by the position of its contacts, and the state exists whether any voltage or current is applied to the switch or not. In order to use the switch state as a computational variable, it is necessary to ascertain its position. To determine its resistance or otherwise assess position, a voltage or current must be applied. If a current source or voltage source is permanently associated with a switch element, then the current through or the voltage across the switch element will reflect its state and may be considered as its *state value*. Since most electronic elements require permanent sources of voltage and current for their proper operations, it becomes convenient to consider a switch in terms of its output voltage or current.

56. Gate Elements. A gate element, or decision element, is a network with one or more input ports and a single output port. The state value of every port of a gate is either the voltage or the current associated with the port. The present output state is uniquely determined by the present input state. The use of voltage (current) as a state value at every port of a gate is called *voltage* (current) *logic*.

Gate elements are the basic building units in the construction of switching networks. They usually perform a simple and generally useful functional transformation.

A very small number of different gate operations are necessary to carry out any logic functional relationship. The set of operations AND, OR, and NOT are a *sufficient set*, for instance. Other sets are also possible. Table 3-6 shows three common sets of logic gate types together with their usual electrical symbols. It also shows the exclusive-or (XOR) function and an inverting input.

Set 1, 2, or 3 is necessary and sufficient to completely implement operations. They are therefore *minimal* and *complete*. More efficiency may result from mixing sets or using the XOR in addition to the usual set gates. In general, however, sets are not mixed, and design is carried out in one set.

It can be shown that there are 18 different minimal complete gate (or operation) sets which have two members. There are 6 minimal complete sets with three members, 4 with

Table 3-6. Gate Sets and Their Symbols

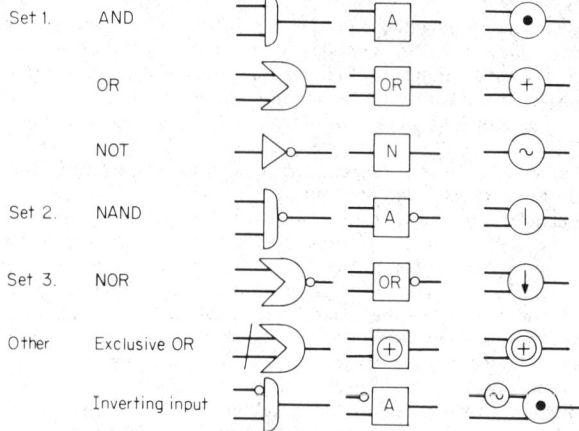

four members, and none with more than four.[1] As Table 3-6 implies, the NOR and NAND operations are each minimal sets in themselves. They are called *universal operations* because all logic functions can be performed by one type of gate. This makes NOR and NAND logic easy to manufacture and maintain since only a single type of gate circuit is used.

A small circle associated with a port is the usual symbol for *inversion*. If it is associated with the output, it provides the *logical complement* of the normal output. If it is associated with an input, this indicates that the complement of the input is taken before it enters the main-gate circuitry.

If the state variable x associated with a port is defined so that $x = 1$ for a high voltage or current value and $x = 0$ for a near-zero voltage or current value, the gate is called a *positive-logic gate*. Otherwise, it is a *negative-logic gate*.

As mentioned earlier, a switching network is described by its truth table. These tables actually serve to define the gate elements. Table 3-7 shows the truth tables for various gate elements.

Switching networks are further classified as being either *combinational* or *sequential*. A combinational network is a switching network whose present output port values (states) depend only on the present input port values.

Gate networks which contain no loops (a gate is never encountered twice along any network path) is called a *feedforward network*. A feedforward gate network is always a combinational network.

A switching network whose present output port values (states) depend on past input port states is called a *sequential network*.

Logic and arithmetic operations are generally considered combinational. The process of switching lines together or remembering information requires sequential networks. Sequential networks are built up from combinational networks by the introduction of *time-delay elements* or *memory elements* and by network interconnections which involve loops.

57. Combinational Circuit Interconnections. The major characteristic of combinational networks is that they contain *no loops*. Gates are interconnected for the purpose of carrying out logical or arithmetical manipulations of a more complex nature than a single gate can provide.

The central problem of a switching network is the *design problem*, in which some desired function is to be implemented in terms of a number of interconnected gates. Superficially, this process is quite easy, because it can be shown that any Boolean expression can be implemented directly in terms of circuit hardware; i.e., if the Boolean expression is written in a *standard or canonical form*, simple algorithms exist for translating the expression into gate hardware (see Sec. 23).

There are two canonical forms, the *canonic sum*, or *disjunctive canonical form*, and the *canonic product*, or *conjunctive canonical form*.

---

[1] See ibid., pp. 65-66.

Table 3-7. Truth Tables for Two-Input Gates

Gate — AND

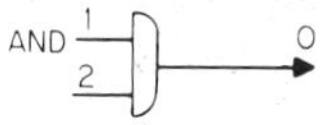

Input 1	Input 2	Output O
1	0	0
1	1	1
0	1	0
0	0	0

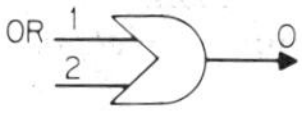

Gate — OR

1	2	0
1	0	1
1	1	1
0	1	1
0	0	0

Gate — NOT

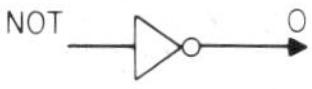

Input	Output
0	1
1	0

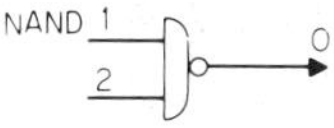

Gate — NAND

1	2	0
1	0	1
1	1	0
0	1	1
0	0	1

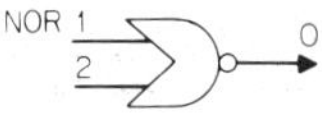

Gate — NOR

1	2	0
1	0	0
1	1	0
0	1	0
0	0	1

Gate — XOR

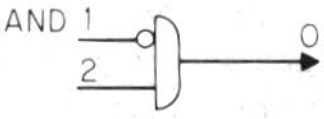

1	2	0
1	0	1
1	1	0
0	1	1
0	0	0

Gate — AND

1	2	0
1	0	0
1	1	0
0	1	1
0	0	0

The canonic-sum form is such that the function is expanded into a number of product terms (AND) in which each term is expressed as a product of all variables or their complements (where no product term has the same variable appearing twice) and these product terms are connected by the OR operation.[1] This is also known as a *sum of minterms,* where a minterm is a term which is the product of all variables or their complements.

The canonic-product form is such that the function is expanded into a number of sum terms (OR) in which each term is expressed as a sum of all variables or their complements (where no sum term has the same variable appearing twice) and these sum terms are connected by the AND operation. This is also known as a *product of maxterms,* where a

---

[1] See R. E. Miller, "Switching Theory," Vol. 1, Wiley, New York, 1965.

maxterm is a term which is the sum of all variables or their components. Designating a circuit from the canonic forms is a straightforward procedure, as Fig. 3-48 shows.

Problems of design of switching circuits are not concerned with whether or not a particular function can be realized but with how it can be realized with minimal cost or other additional constraints.

58. Minimum-Complexity Combinational Networks. The important design aim of reducing network complexity usually leads to lower cost and greater ease of construction. Minimum complexity may have several meanings, some of which are in opposition. A *minimally complex network* may be defined as:

1. That which contains the minimum number of gate elements.
2. That which utilizes a certain set of gate elements (see Table 3-7) but which contains the minimum number of elements from that set.

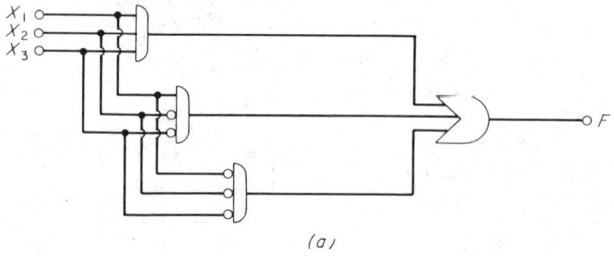

(a)

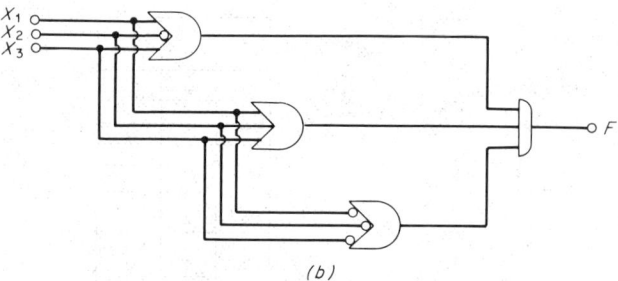

(b)

Fig. 3-48. Realization of canonical forms. (a) Realization of $F = (x_1 \cdot x_2 \cdot x_3) + (x_1 \cdot \bar{x}_2 \cdot \bar{x}_3) + (\bar{x}_1 \cdot \bar{x}_2 \cdot \bar{x}_3)$. ($b$) Realization of $F = (x_1 + \bar{x}_2 + x_3) \cdot (x_1 + x_2 + x_3) \cdot (\bar{x}_1 + \bar{x}_2 + \bar{x}_3)$.

3. That which has the fewest number of interconnections.
4. That which may be wired with the fewest crossovers on a circuit board.
5. That which minimizes the total cost of components used in its construction.
6. That which is easiest to maintain and repair.
7. That which operates at the highest speed.
8. That which has the highest reliability.

These and many other criteria may be of major importance in designing a switching network. It is known how to optimize the design in terms of some of these criteria but not others. Several minimization techniques will be discussed.

59. Map Techniques. The prime goal of *map minimization methods* is to produce a circuit realization which uses a minimum number of gate elements. The *map*, or *Karnaugh map* (Fig. 3-49), is an alternate way of presenting the data contained in the truth table. A map of function $f(x_1, x_2, \ldots, x_n)$ of n variables contains 2^n squares, one for every possible value of the n-tuple $(x_1, x_2, \ldots, x_n)$. If it is desired that some particular value of the n-tuple produce an output from the circuit, then a 1 is placed in the corresponding square. If no output is desired, a 0 goes in the square, and if one does not care what the output is, a d is placed in the square. A d can be counted as either a 1 or 0, depending on which condition produces greater minimization.

Minimization by the map method uses two basic laws of Boolean algebra:

$$x + \bar{x}y = x + y \quad \text{and} \quad xy + \bar{x}y = y \tag{3-59}$$

It is therefore necessary to arrange the map to make it easy to apply these laws by inspection. This is done by arranging the squares so that *any two adjacent squares* differ by *only one* of the variables in the *n*-tuple. Figure 3-49 shows an example, where special ordering is required in the 4 × 4 map to achieve properly adjacent squares.

Once the map is drawn, the output function is entered in all the squares. If one proceeds to minimize using the 1 and *d* entries, the result will be a minimal sum-of-products form, similar to the canonical sum. This is also called the *minterm canonical form* or *standard sum*. If one proceeds using the 0 and *d* entries, the result will be a minimal product-of-sums form, similar to the canonical product. This is also called the *maxterm canonical form* or *standard product*.

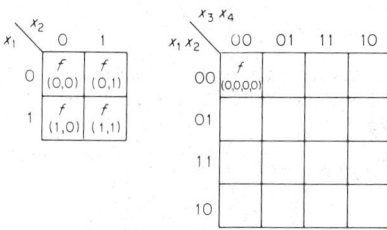

Fig. 3-49. 2 × 2 and 4 × 4 Karnaugh maps.

Minimization is accomplished by combining the map entries into groups. A *group* is a large square or rectangle (twice as long as wide) which contains squares of all the same value. For instance, to use the sum-of-products form, one groups squares containing 1s and *d*'s. The highest order of simplification is achieved when the 1s are combined into *as few groups as possible*, each of which contains as many adjacent cells as possible. Because of the form of the map, squares along its two sides are adjacent, as well as those on the top and bottom.

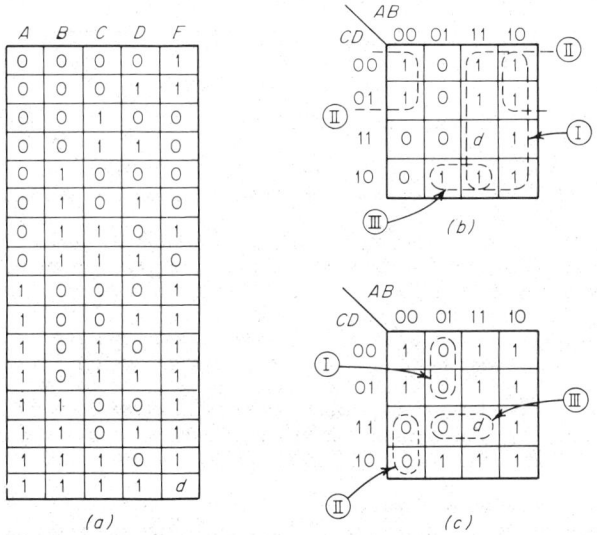

Fig. 3-50. Map minimization. (*a*) A truth-table representation of a function *F* of four variables. (*b*) A sum-of-minterms minimization. (*c*) A product-of-maxterms minimization.

Figure 3-50 shows such a grouping carried out in a 4-tuple. In Fig. 3-50*b*, it can be seen by looking at group I that $F = 1$ (or *d*) for all 4-tuples where $A = 1$. Therefore, group I $= A$. Group II is such that $F = 1$ for all occasions where *B* and *C* both $= 0$, and so on:

Group I: A

Group II: $\bar{B}\bar{C}$

Group III: $BC\bar{D}$

It is also possible to group the zeros, as in Fig. 3-50c. (In this case the d is considered a zero because it aids grouping: it need not be included if greater minimization is achieved by leaving it out.) Grouping of zeros makes use of two other Boolean algebra rules:

$$(x + y)(x + \bar{y}) = x \quad \text{and} \quad (x + y)(x + z + y) = (x + y)(x + z) \quad (3\text{-}60)$$

The zero groups represent maxterms. Group I in Fig. 3-50c is such that $F = 0$ only if A, $\bar{B}$, and C are each zero. The value of D does not matter. Similar conditions hold for groups II and III.

Group I: $A + \bar{B} + C$

Group II: $A + C + \bar{C}$

Group III: $\bar{B} + \bar{C} + \bar{D}$

Based on these two approaches, there are two minimum networks, shown in Fig. 3-51a and b. Clearly, if AND and OR gates are about the same price, Fig. 3-51a is preferred.

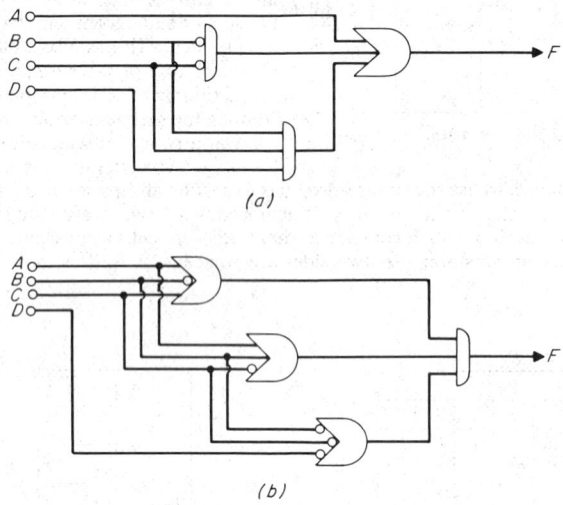

(a)

(b)

Fig. 3-51. (a) Realization of the reduction of Fig. 3-50b. (b) Realization of the reduction of Fig. 3-50c.

60. Prime Implicants. When a Karnaugh map is used to find a minimal representation, one tries to combine adjacent 1-squares into larger groups. Each group that can be made which is not properly contained in a larger group is a graphical example of a *prime implicant*. Thus in Fig. 3-50b, for instance, there are three prime implicants: A, $\bar{BC}$, and $BC\bar{D}$.

For large networks the map would become quite unwieldy, and so prime implicants are searched for directly from the truth table. The search process consists of the following steps:

"1. Form the canonical sum-of-product representation of the function to be minimized.

"2. Examine all product terms and apply the reduction $xy + x\bar{y} = x$ as many times as possible. All the new product terms so formed will have one less variable than the original terms

"3. Take the new set of product terms and repeat step 2 on this new set of terms. When no further reductions are possible, all the product terms that were generated by steps 1 and 2 and which cannot be further reduced are the prime implicants associated with the function to be minimized.

"4. The set of prime implicants is then inspected to choose a minimal set that can be ORed together to represent the function."[1]

A similar process can be carried out with the canonical product of sums.

61. Cubical Representation. This approach is another geometric representation similar to the Karnaugh map, in which the Boolean function of n variables is represented by an n-

---

[1] From Taylor L. Booth, "Digital Networks and Computer Systems." Copyright © 1971 by John Wiley and Sons, Inc. Reprinted by permission.

dimensional *unit cube.* Minimization is accomplished by combining cubes into higher-order cubes. A 0-*cube* is a single expression, equivalent to a 1-square on the Karnaugh map. A 1-*cube* is a pair of expressions (0-cubes) differing by only one variable. Together the two 0-cubes form a 1-cube. Two 1-cubes can form a 2-cube if they have one component which differs.

In Fig. 3-52, the X represents the variable which differs, called the *free component.* The 1s and 0s which remain are called *bound components.* The two 0-cubes which combine are called the *faces* of the resultant 1-cube, and so on. See Miller[1] for an explanation of minimization using cubical complexes.

62. Design Using Other Gate Sets. The techniques mentioned above all produce realizations utilizing AND, OR, and NOT operations. These are not necessarily the most desirable functions (indeed, NAND or NOR logic is usually much preferred because of the inherent stability and ease of construction of the gate circuitry). It therefore becomes necessary to translate the results of the minimization technique into other sets, most commonly *inverting* gates, NAND, NOR, NOT. The NAND operation can be represented in two ways, as shown in Fig. 3-53a. The NOR operation can also be represented in two ways, (see Fig. 3-53b). Each gate produces a *complementary,* or inverting, signal. If a signal passes through two inverting gates in series, it is twice complemented and regains its original character. An algorithm for generating NAND or NOR logic can be stated.

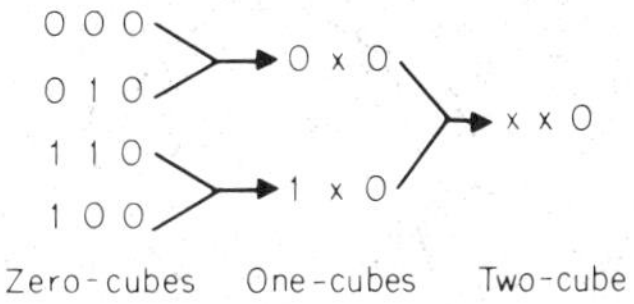

(a)

Zero-cubes   One-cubes   Two-cube

(b)

Fig. 3-52. Cubical minimization.

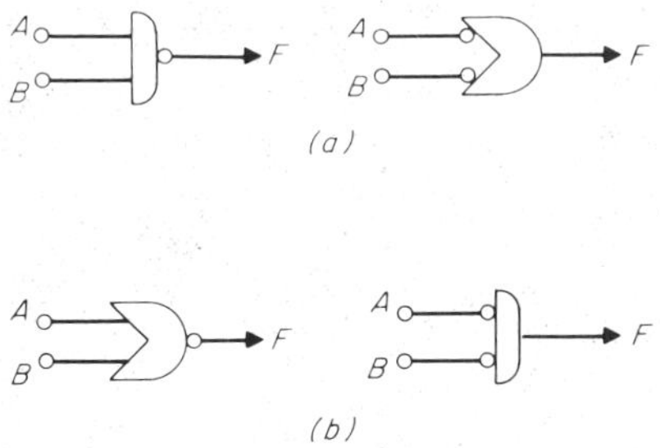

Fig. 3-53. Symbols for gates performing (a) the NAND operation and (b) the NOR operation.

A NAND (NOR) gate performs an OR (AND) operation on its complemented input variables if it is at an *odd* number of inversion levels; whereas an AND (OR) operation will be performed on its uncomplemented input variables if it is at an *even* number of inversion levels.

The foregoing statements lead to the following set of rules for obtaining the logic expression of the output signal of an interconnected array of inverting gates:

"1. Consider the gate from which the output signal will be obtained as the first (odd) level of inversion, the preceding gates as the second (even) level, etc.

"2. Consider all NAND gates in odd levels to perform the OR operation.

"3. Consider all NAND gates in even levels to perform the AND operation.

"4. Consider all NOR gates in odd levels to perform the AND operation.

"5. Consider all NOR gates in even levels to perform the OR operation.

"6. All input variables entering gates in odd levels should appear complemented in the logic expression of the output signal.

"7. All input variables entering gates in even levels should appear uncomplemented in the logic expression of the output signal."[2]

63. Sequential-Circuit Interconnections. Sequential circuits possess the property that their output states depend not only upon present input states but upon past values of inputs. This characteristic is achieved by including *delay elements* or *memory elements* into the circuit. Boolean algebra and the design techniques mentioned previously cannot express delay or memory, and so a different representation must be used. It is necessary to specify

---

[1] "Switching Theory," pp. 145–156.

[2] From R. L. Morris and J. R. Miller (eds.), "Designing with TTL Integrated Circuits," McGraw-Hill, New York, 1971, p. 115.

input states, output states, and internal states to describe a *sequential machine*, where the internal state describes the condition of all the memory or delay elements in the circuit. The input and internal states together are known as the *total state*. The present total state determines both the present output state and the next internal state. If the number of possible internal states is finite, the machine (the behavior representation of the network) is called a *finite-state sequential machine*.

A finite-state sequential machine can be uniquely and completely specified by its *state table* or a *state diagram*. The state table indicates the next state and present output for every possible combination of present inputs and internal states, as shown in Fig. 3-54a. The state diagram for the same machine is shown in Fig. 3-54b. It shows transitions from the present to next state as *directed line segments*. The segments are labeled to show the inputs which caused the transition and the present output.

Present inputs

	00	01	11	10
A	A,0	A,0	B,1	A,0
B	C,0	B,1	B,1	C,1
C	A,0	B,1	C,0	A,1

(Present state)

(a)

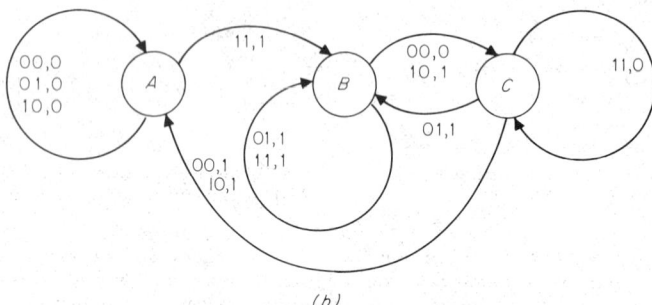

(b)

Fig. 3-54. (a) State table and (b) state diagram for sequential machine M.

64. Synchronous Sequential Machines. The most elementary form of sequential network is composed of AND, OR, NOT, and delay elements. A delay element is a one-input, one-output network such that its output state at time τ is equal to its input state at time $\tau - d$, where d is called the *delay time*. With these elements, one can form a synchronous sequential machine such that:

"1. Input changes occur only at instants in time iD, where i is a positive integer and D is a standard time interval.

"2. All delay elements have delays equal precisely to integer multiples of D.

"3. Gate elements and interconnections are delay-free.

"4. All input-, state-, and output-variable changes occur instantaneously."[1]

Any *finite-memory-span sequential machine* can be realized by a feedforward network with delay. A finite-memory-span machine is one in which the present state (and output) can be determined from knowledge of present and a finite number of past inputs alone, without knowledge of the original internal state of the machine. Such a machine can be realized with a network which can be reduced in complexity by techniques analogous to those used for combinational-circuit reduction. The following stages are necessary in this process:

"*Stage 1: Description of Desired Network Operation.* A complete set of specifications must be prepared describing the operation of the network. All inputs and outputs must be

[1] From P. Wood, "Switching Theory," McGraw-Hill, New York, 1968, p. 279.

identified, and the relationship between the quantities must be defined in a consistent manner.

"*Stage 2: Determination of State Table.* Using the specification established at stage 1, an initial state table or state diagram is defined for the network. The state table is checked to make sure that it satisfies all design criteria.

"*Stage 3: State-Table Minimization.* In the process of developing a state table to satisfy a given set of operational requirements an unnecessarily large number of states may be introduced. Since the number of information storage elements in a circuit increases as the number of states increases, it is often desirable to remove redundant states from the state table.

"*Stage 4: State Assignments.* The information contained in the state table must be encoded into binary form. This is not a unique process, and the encoding used can considerably influence the complexity of the resulting circuit. The result of this stage is to transform the state table into a *transition table.*

"*Stage 5: Network Realization.* Once a transition table has been constructed and a decision made concerning the type of storage elements to be used, the logic expressions relating the input and present state to the output can be obtained."[1]

Stages 1, 2, and 4 cannot usually be carried out by a straightforward algorithm, and therefore no more details of the procedure will be mentioned here.

A *memory element* may be distinguished from a *delay element.* A memory element produces delays of any desired length. Whereas the delay element produces an output equal to its input value d s earlier, the memory element is designed to hold as its output the most recently received input, regardless of the length of time involved.

65. Asynchronous Sequential Machines. Any sequential machine that violates one of the four synchronous criteria is considered *asynchronous.* Since the criteria are rather ideal, most real machines are asynchronous. Such machines are difficult to represent and construct. Usually, therefore, pseudo-synchronous behavior is achieved by incorporating a *clock. Clocked sequential systems* form the bulk of useful large-scale switching networks. The clock is a sequential machine that produces a series of pulses, usually occurring at regular intervals. The other network inputs and outputs can change at any time except during a clock pulse, at which time a new total state is defined. Thus, network activity exhibits the characteristics of a synchronous sequential network, and the clocked system can be designed in the same manner. In order to operate the clocked sequential system, it is necessary to distribute the clock pulse to virtually every element to assure synchronism.

66. Traffic and Loading. The previous concepts dealt with techniques used to simplify the physical design of a switching network. The network is designed to carry out a certain operation or to combine its inputs in a certain way to produce an output. The operation takes a certain time to perform, and hence there is a limit to the number of operations the circuit can perform in a given time. It certainly cannot meaningfully operate on two different sets of inputs at exactly the same time, for instance. Such situations regularly occur in telephone switching systems and computer systems. The system is called upon to perform an operation while it is busy working on some other operation and hence cannot respond. It then becomes necessary for those demanding services to (1) wait until facilities are available, (2) try again later, (3) share the system in some manner, or (4) use a different identical system.

The first three alternatives generally involve deterioration of service; i.e., the operation is not carried out as quickly as if the individual were the only user of the system. Providing multiple systems can reduce the deterioration of service in the multiple-demand situation but at the cost of maintaining the extra equipment necessary to provide faster service. *Traffic and loading theory* is concerned with the necessary trade-off between *quality* (speed) of service to the users and *cost* of equipment. It attempts to express quality of service and cost of equipment in comparable units (dollars, for instance) and *optimize* the system to provide the best service at the lowest cost. The optimization depends largely upon the traffic patterns the switching system is required to handle. Variations in traffic patterns are of several kinds.

a. Variation in Demand with Time. The system may be required to provide service under constant demand, or demands may come in rush-hour patterns. Nonconstant demand may vary predictably or randomly.

---

[1] From Taylor L. Booth, "Digital Networks and Computer Systems." Copyright © 1971 by John Wiley & Sons, Inc. Reprinted by permission.

b. Variation in Holding Time. Holding time refers to the length of time necessary to perform the service demanded. This can be fixed, vary in well-defined ways, or be randomly distributed.

c. Priority Considerations. Some systems allow the users to have priority assignments, in which a high-priority user can usurp the place of a lower-priority user.

Given the above variations in traffic, a theory has been developed which estimates (for a given situation) the expected quality of service to the user. The mathematics involved is that of probability. The calculations involved are not difficult; it is the determination of criteria for service which poses the greatest problem.

67. Queuing theory is a subset of traffic theory which has to do with *waiting for service.* If a user demands service and the facilities are busy, two things can happen: he is lost, i.e., he must try again later, or he is put on a waiting list and remains within the system awaiting service. Queuing theory deals with the characteristics of the waiting list (queue). It may be formed in a number of ways.

a. First in, First out. When the facility becomes available, the user who has been waiting longest is the next to receive service.

b. First in, Last out. Here the last person to request service gets it. This technique may seem "unfair," but it has wide usage in the storage of data, which do not mind the wait. Sometimes it is called a *push-down stack.*

c. Random Order of Service. As the facility becomes available, the next user is picked at random from a pool of waiting customers.

d. Nearest-Neighbor Service. This is the type of service usually on elevators, in which service is rendered to the waiting customer nearest to the location of the system at the time it becomes free.

e. Priority Queue. Here service is given on the basis of importance of the user or the loudness of his demands.

Unless there is a priority consideration, it would appear that a type *a* queue would provide the best service to its customers. Unfortunately, formation of this queue requires the system to *remember* the order in which the customers have demanded service, which in itself can require a sophisticated logic network. It is much cheaper for the system to form a type *c* or *d* queue. If this is done, calculation of the quality of service (waiting time) yields a probability function. Often the parameter of interest is average waiting time for service. This is a common criterion of service, but it can result in an unusually long wait for a particular customer. Therefore, an estimation of the maximum waiting time for service is also made. It is these waiting times which are compared to the dollar cost of additional service equipment to determine the required system size.

Scheduling Theory. Another important aspect of designing optimum switching systems concerns *routing,* or *scheduling,* of service. In completing a long-distance telephone call or a complex computer calculation, it becomes necessary to connect subsystems together temporarily to achieve the final result. Routing and scheduling theory concerns itself with the determination of the optimum patterns of interaction of these subsystems such that they are utilized as fully as possible without causing bottlenecks in the overall system. One tries to avoid as much as possible the situation where one part of the overall system is consistently limiting the quality of service.

68. Stored-Program Methods. In all the previous discussion, it has been implicit that the networks mentioned are designed to do a specific job by connecting various circuit elements together with wires, and indeed this is very often the case. Classical telephone switching circuits and many special-purpose computers are examples of large systems which are *hard-wired.* General-purpose digital computers and the newest telephone systems function in a very different manner, however. In these, the switching function is a *stored program,* and the switching network can be changed without moving a single physical wire of the system, i.e., by changing the program.

A *stored-program system* is divided into two main units, a *control unit* and a *processing unit.* The processing unit carries out the switching, logic, or arithmetic computations. It is constructed so that it can carry out a number of basic operations, such as AND, OR, NOT. The control unit tells the processing unit which operations to perform, in what order, what variables to consider as inputs, and what to do with the result. In order to do this, of course, the processing unit must have access to all inputs and all output lines, as well as the ability to perform any operation with any combination of variables. It must also be totally controllable by the control unit.

The control unit is basically a memory unit. The programmer enters a set of instructions into the memory in a binary coded form. During the operation, the control unit proceeds from one memory location to the next in an organized manner. At every location, the instruction contained within that location is sent to the processing unit, controlling its activity for a short time. In this way an extensive, easily altered switching system can be built up (see Sec. 23).

69. Glossary of Network Laws and Theorems

Ampere's Law

$$\oint \mathbf{H} \cdot d\mathbf{l} = \sum Ni$$

The integral of the magnetic field intensity **H** around a closed path is equal to the sum of all the currents whose paths are linked.

Compensation Theorem. If linear-circuit currents or voltages are altered by small changes in the value of one of the elements, call it R_x, the effects throughout the circuit can be calculated by placing a voltage source of value $v_x = i_x \delta R_x + \delta I_x \delta R_x$ in series with R_x and then, using superposition, calculating circuit currents and voltages produced by this source. These are added to the original quantities to produce the compensated values.

Faraday's Law

$$v = -d\phi/dt$$

The electromotive force v induced in a closed linear path surrounding a magnetic flux ϕ is equal to the negative of the rate of change of the flux with respect to time.

Fourier's Theorem

$$f(x) = \tfrac{1}{2}a_0 + \sum_{m=1}^{\infty} (A_m \cos mx + b_m \sin mx)$$

Any function $f(x)$ which is well behaved in the interval $a < x < b$, that is, it has only a finite number of finite discontinuities, can be expressed in the interval as an infinite sum of sine and cosine functions.

Joule's Law

$$P = i^2 R = v^2/R$$

If a circuit contains only an ohmic resistance, i.e., $R = V/I$ for all values of V, the instantaneous power dissipated in the resistance is equal to the square of the instantaneous current flow through the resistor times its resistance or, alternately, equal to the square of the instantaneous voltage applied divided by its resistance.

Kirchhoff's Current Law

$$\sum I = 0$$

The sum of all currents entering a circuit node is equal to zero.

Kirchhoff's Voltage Law

$$\sum V = 0$$

The sum of all voltage drops encountered in succession on a closed path within a circuit is equal to zero.

Laplace's Theorem

$$F(s) = \int_0^\infty f(t)e^{-st}\, dt$$

$$f(t) = \frac{1}{2\pi j}\int_{s=c-j\infty}^{c+j\infty} F(s)e^{st}\, ds$$

If $f(t)$ is any time function such that $f(t) = 0$ for $t < 0$, and if s can be made sufficiently large for $f(t)e^{-st} \to 0$ as $t \to \infty$, then $f(t)$ can be transformed from the real domain into a function $F(s)$ in the complex domain. This transformation enables one to transform linear integrodifferential equations with constant coefficients into algebraic equations, greatly simplifying their solution. The inverse transform also exists, whereby the time function can be recovered from $F(s)$.

Lenz's Law. If a constant current flows in the primary circuit A, and if, by the motion of A or of the secondary circuit B, a current is induced in B, the direction of this induced current will be such that by its electromagnetic action on A it tends to oppose the relative motion of the circuits.

Maximum-Power-Transfer Theorem

$$Z_s = Z_L^*$$

The maximum power may be transferred from a source to a load if the load impedance is the complex conjugate of the source impedance. See Par. **3-13** for more restrictive expressions.

Maxwell's Relations

$$\mathbf{\nabla} \cdot \mathbf{D} = \rho \quad \text{or} \quad \int_s \mathbf{D} \cdot d\mathbf{s} = \int_\tau \rho \, d\tau$$

The net flux of $\mathbf{D}$ out of any volume is equal to the algebraic sum of the real charges within that volume.

$$\mathbf{\nabla} \cdot \mathbf{B} = 0 \quad \text{or} \quad \int_s \mathbf{B} \cdot d\mathbf{s} = 0$$

The net flux of $\mathbf{B}$ out of any volume is zero, since $\mathbf{B}$ field lines are continuous and close upon themselves.

$$\mathbf{\nabla} \times \mathbf{E} = -\dot{\mathbf{B}} \quad \text{or} \quad \oint \mathbf{E} \cdot d\mathbf{l} = -\int_s \dot{\mathbf{B}} \cdot d\mathbf{s}$$

The line integral of $\mathbf{E}$ about any closed curve in the field is equal to the negative of the time rate of change of the magnetic flux through any surface spanning that curve.

$$\mathbf{\nabla} \times \mathbf{H} = \mathbf{J} + \dot{\mathbf{D}} \quad \text{or} \quad \oint \mathbf{H} \cdot d\mathbf{l} = \int_s (\mathbf{J} + \dot{\mathbf{D}}) \cdot d\mathbf{s}$$

The line integral of $\mathbf{H}$ about any closed curve in the field is equal to the vector sum of the amperian currents and the change in $\mathbf{D}$ through any surface spanning that curve.

Miller's Theorem. If an admittance Y is connected between the input and output terminals of a two-part network as shown in Fig. 3-55a, then it may be replaced by

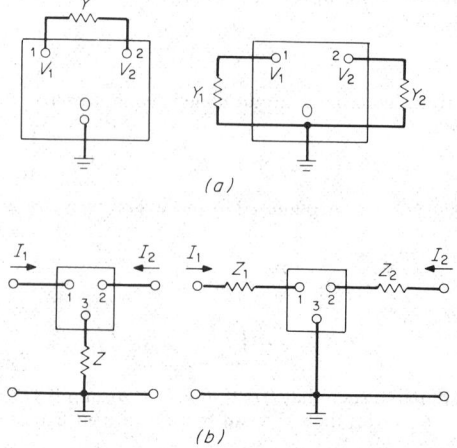

Fig. 3-55. Equivalent circuit transformation of Miller's Theorem.

3-58

$Y_1 = Y/(1 - K)$ and $Y_2 = [(K-1)/K]$, referred to ground, without altering the properties of the circuit. K is defined as V_2/V_1. If an impedance Z is connected to ground as in Fig. 3-55b, it may be replaced by $Z_1 = Z(1 - A)$ and $Z_2 = Z[(A - 1)/A]$ without altering the properties of the circuit. $A = -I_2/I_1$.

Millman's Theorem. (See Fig. 3-56.) Two or more nonideal voltage sources in parallel or nonideal current sources in series can be combined into a single equivalent source, where

$$E = \frac{E_1/Z_1 + E_2/Z_2}{1/Z_1 + 1/Z_2} \qquad Z = \frac{1}{1/Z_1 + 1/Z_2}$$

$$I = \frac{I_1/Y_1 + I_2/Y_2}{1/Y_1 + 1/Y_2} \qquad Y = \frac{1}{1/Y_1 + 1/Y_2}$$

Neumann's Law

$$v_1 = M_{12}\, di_2/dt \qquad v_2 = M_{21}\, di_1/dt \qquad M_{12} = M_{21}$$

When the magnetic medium is of uniform permeability, the mutual inductance of two circuits is the same whether current flows in the second coil and voltage is observed in the first or current flows in the first coil and voltage is observed in the second.

Norton's Theorem. (See Fig. 3-57.) As far as external conditions are concerned, i.e., observable characteristics at terminals A and B, any linear time-invariant one-port network (regardless of internal complexity) may be represented as an ideal current source in parallel with a single complex impedance.

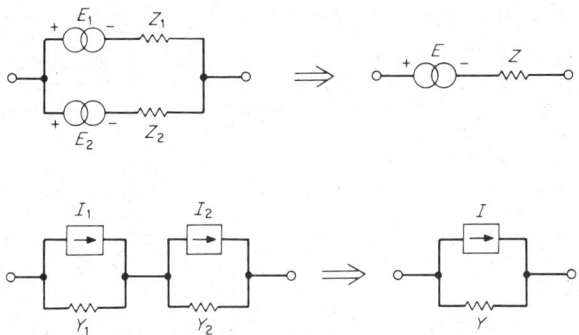

Fig. 3-56. Millman's reduction.

Ohm's Law

$$V = IR$$

When the current in a conductor is steady and there are no sources of emf within the conductor, then the voltage V between terminals of the conductor is proportional to the current I; the proportionality constant R is called resistance.

Poynting's Law

$$\mathbf{S} = \mathbf{E} \times \mathbf{H} \qquad \text{or} \qquad P = VI \cos \theta$$

The transfer of energy can be expressed as the vector cross product of the electric and magnetic fields. The transfer of energy takes place in a direction perpendicular to both fields. When the relation is expressed for sinusoidal voltage and current waveforms in a circuit, the power is equal to the value of the voltage waveform times the value of the current waveform times the cosine of the phase angle between the two waveforms

Reciprocity Theorem

$$I_j/V_K = I_K/V_j$$

In any linear, time-independent network with independent sources, the ratio of the current in a short circuit at one part of the network to the output of a voltage source at another part of the network does not change if the positions of the source and the short circuit are interchanged. Reciprocity also applies to a current source and the voltage developed across an open circuit.

Substitution Theorem. If a branch of a network is carrying current I_{AB} and the voltage across the branch is V_{AB}, a different branch may be substituted for it while all other parts of the network (including currents, voltages, and sources in other branches) remain unchanged provided the substituted branch also has voltage V_{AB} when current I_{AB} is flowing.

Superposition Theorem. In a linear network containing more than one independent source, the voltage across or the current through any element can be computed as follows.

1. Replace all the sources except one by their internal impedances (ideal current sources are replaced by open circuits, and ideal voltage sources by short circuits).

2. Compute the voltage across or the current through the element due to the single source.

3. Repeat steps 1 and 2 for each independent source in turn.

4. The sum of all the individual calculated voltages or currents provides the actual value of voltage or current for the complete circuit.

Tellegen's Theorem

$$\sum_{k=1}^{n} i_k v_k = 0$$

In any lumped network which contains any type of elements, the sum of the products of all the instantaneous branch voltages times their respective instantaneous branch currents is instantaneous power, and therefore the result is that the sum of the instantaneous power in all parts of a circuit is zero, which is consistent with the principle of conservation of energy.

Thevenin's Theorem. (See Fig. 3-58.) As far as external conditions are concerned, i.e., observable characteristics at terminals A and B, any linear time-invariant one-port network (regardless of internal complexity) can be represented as an ideal voltage source in series with a single complex impedance.

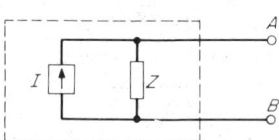

Fig. 3-57.   Norton equivalent.

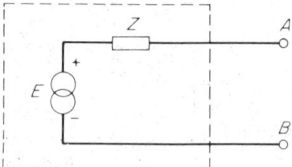

Fig. 3-58.   Thevenin equivalent.

70. Bibliography.

BECKMANN, P. "Introduction to Elementary Queuing Theory and Telephone Traffic," Golem Press, Boulder, Colo., 1968.

BELEVITCH, V. "Classical Network Theory," Holden-Day, San Francisco, 1968.

BODE, H. W. "Network Analysis and Feedback Amplifier Design," Van Nostrand, New York, 1945.

BOHN, E. V. "Introduction to Electromagnetic Fields and Waves," Addison-Wesley, Reading, Mass, 1968.

BOOTH, T. L. "Digital Networks and Computer Systems," Wiley, New York, 1971.

BRENNER, E., and M. JAVID "Analysis of Electric Circuits," McGraw-Hill, New York, 1959.

CHIRLIAN, P. M. "Basic Network Theory," McGraw-Hill, New York, 1968.

COLLIN, R. E. "Foundations for Microwave Engineering," McGraw-Hill, New York, 1966.

COWAN, J. D., and H. S. KIRSCHBAUM "Introduction to Circuit Analysis," Merrill, Columbus, Ohio, 1961.

CRUZ, J. B., and M. E. VAN VALKENBURG "Introductory Signals and Circuits," Blaisdell, Waltham, Mass., 1967.

DESOER, C. A., and E. S. KUH   "Basic Circuit Theory," McGraw-Hill, New York, 1969.

HUNT, W. T., JR., and R. STEIN   "Static Electromagnetic Devices," Allyn and Bacon, Boston, Mass., 1963.

KARNI, S.   "Network Theory: Analysis and Synthesis," Allyn and Bacon, Boston, Mass., 1966.

KING, R. W. P.   "Transmission Line Theory," McGraw-Hill, New York, 1955.

KRAUS, J. D.   "Antennas," McGraw-Hill, New York, 1950.

LIN, H. C.   "Integrated Electronics," Holden-Day, San Francisco, 1967.

MANNING, L. A.   "Electrical Circuits," McGraw-Hill, New York, 1966.

MATICK, R. E.   "Transmission Lines for Digital and Communication Networks," McGraw-Hill, New York, 1969.

McCLUSKEY, E. J.   "Introduction to the Theory of Switching Circuits," McGraw-Hill, New York, 1965.

MIDDENDORF, W. H.   "Analysis of Electric Circuits," Wiley, New York, 1956.

MILLER, R. E.   "Switching Theory," vol. 1, "Combinational Circuits," Wiley, New York, 1965.

MITTLEMAN, J.   "Circuit Theory Analysis," Hayden, New York, 1964.

MORRIS, R. L., and J. L. MILLER (eds.)   "Designing with TTL Integrated Circuits," McGraw-Hill, New York, 1971.

NEWCOMB, R. W.   "Active Integrated Circuit Synthesis," Prentice-Hall, Englewood Cliffs, N. J., 1968.

POPOVIĆ, B. D.   "Introductory Engineering Electromagnetics," Addison-Wesley, Reading, Mass., 1971.

RAMO, S., J. R. WHINNERY, and T. VANDUZER   "Fields and Waves in Communication Electronics," Wiley, New York, 1967.

RUBIN, M., and C. E. HALLER   "Communication Switching Systems," Reinhold, New York, 1966.

RUSTON, H., and J. BORDOGNA   "Electric Networks: Functions, Filters, Analysis," McGraw-Hill, New York, 1966.

SEELY, C.   "Electronic Circuits," Holt, Rinehart and Winston, New York, 1968.

SHEA, R. F. (ed.)   "Amplifier Handbook," McGraw-Hill, New York, 1966.

SKILLING, H. H.   "Electrical Engineering Circuits," 2d ed., Wiley, New York, 1965.

STEWART, J. L.   "Circuit Theory and Design," Wiley, New York, 1956.

TERMAN, F. E.   "Radio Engineer's Handbook," McGraw-Hill, New York, 1943.

THOUREL, L.   "The Antenna," Wiley, New York, 1960.

VAN VALKENBURG, M. E.   "Introduction to Modern Network Synthesis," Wiley, New York, 1960.

WEINBERG, L.   "Network Analysis and Synthesis," McGraw-Hill, New York, 1962.

WILLIAMS, E. M., and J. B. WOODFORD   "Transmission Circuits," Macmillan, New York, 1957.

WOOD, P. E., JR.   "Switching Theory," McGraw-Hill, New York, 1968.

SECTION 4

INFORMATION, COMMUNICATION, NOISE, AND INTERFERENCE

BY

JOHN B. THOMAS Professor of Electrical Engineering, Princeton University; Fellow, IEEE

CONTENTS

Numbers refer to paragraphs

SECTION 4

INFORMATION, COMMUNICATION, NOISE, AND INTERFERENCE

1. Introduction. Attempts began in the 1920s to develop a quantitative theory of information measure and to apply this measure to communication systems. One of the pioneering efforts was that of Hartley in 1928, who defined the information rate of a communication system as the logarithm of the number of possible messages that could be sent through the system, assuming that all messages were equally likely.

During World War II Nobert Wiener was largely responsible for the development of a general philosophy of communication and control called *cybernetics*, formalizing the concept that both desirable signals and undesirable signals (noises) could be defined in probabilistic terms as random processes. His work was well known to initiates by the end of World War II but did not become readily available until 1948.

Drawing on Wiener's concepts and taking into account the effect of noise and message probabilities, C. E. Shannon produced two classic papers in 1948. He introduced the concepts of entropy and channel capacity in communication systems and related them through the coding theorems. Wiener and Shannon might be considered the creators of modern communication and information theory.

The principal problem in most communication systems is the transmission of information in the form of messages or data from some originating *information source S* to some *destination* or *receiver D*. The method of transmission is frequently by means of electric signals more or less under the control of the sender. These signals are transmitted via a channel C, as shown in Fig. 4-1. The set of messages sent by the source will be denoted by $\{U\}$. If the channel were such that each member of U were received exactly, there would be no communication problem. However, due to channel limitations and noise, a corrupted version $\{U^*\}$ of $\{U\}$ is received at the information destination. It is generally desired that the distorting effects of channel imperfections and noise be minimized and that the number of messages sent over the channel in a given time be maximized.

These two requirements are interacting, since, in general, increasing the rate of message transmission increases the distortion or error. However, some forms of message are better suited for transmission over a given channel than others, in that they can be transmitted faster or with less error. Thus it may be desirable to modify the message set $\{U\}$ by a suitable *encoder E* to produce a new message set $\{A\}$ more suitable for a given channel. Then a decoder E^{-1} will be required at the destination to recover $\{U^*\}$ from the distorted set $\{A^*\}$. A typical block diagram of the resulting system is shown in Fig. 4-2.

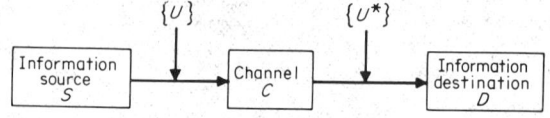

Fig. 4-1. Basic communication system.

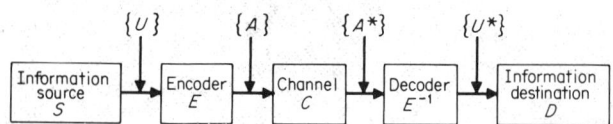

Fig. 4-2. Communication system with encoding and decoding.

4-2

2. Self-Information and Entropy. Information theory is concerned with the relative frequency of occurrence of messages and not with their meaning. In the model communication system given in Fig. 4-1, we presume that each member of the message set $\{U\}$ is expressible by means of some combination of a finite set of symbols called an *alphabet*. Let this source alphabet be denoted by the set $\{X\}$ with elements $x_1, x_2, \ldots, x_M$, where M is the size of the alphabet. The notation $p(x_i)$, $i = 1, 2, \ldots, M$, will be used for the probability of occurrence of the ith symbol x_i. In general the set of numbers $\{p(x_i)\}$ may be assigned arbitrarily so long, of course, as

$$p(x_i) \geq 0 \qquad i = 1, 2, \ldots, M \tag{4-1}$$

and
$$\sum_{i=1}^{M} p(x_i) = 1 \tag{4-2}$$

A measure of the amount of information contained in the ith symbol x_i will be defined based solely on the probability $p(x_i)$. Thus, for reasons that will become clearer later, the *self-information* $I(x_i)$ of the ith symbol x_i is defined as

$$I(x_i) = \log \frac{1}{p(x_i)} = -\log p(x_i) \tag{4-3}$$

This quantity is a decreasing function of $p(x_i)$ with the end-point values of infinity for the impossible event and zero for the certain event.

It follows directly from Eq. (4-3) that $I(x_i)$ is a discrete random variable, i.e., a real-valued function defined on the elements x_i of a probability space. Of the various statistical properties of this random variable $I(x_i)$, the most important is the expected value, or mean, given by

$$E\{I(x_i)\} = H(X) = \sum_{i=1}^{M} p(x_i) I(x_i) = -\sum_{i=1}^{M} p(x_i) \log p(x_i) \tag{4-4}$$

This quantity $H(X)$ will be called the *entropy* of the distribution $p(x_i)$. If $p(x_i)$ is interpreted as the probability of the ith state of a system in phase space, then this expression is identical to the entropy of statistical mechanics and thermodynamics. Furthermore, the relationship is more than a mathematical similarity. In statistical mechanics, entropy is a measure of the disorder of a system; in information theory, it is a measure of the uncertainty associated with a message source.

In the definitions of self-information and entropy, the choice of the base for the logarithm is arbitrary, but of course each choice results in a different system of units for the information measures. The most common bases used are base 2, base e (the natural logarithm), and base 10. When base 2 is used, the unit of $I(\cdot)$ is called the *binary digit* or *bit*; this base is most convenient when dealing with the binary case, i.e., the case where $M = 2$ and the alphabet consists of only two symbols, for example 0 and 1. When base e is used, the unit is the *nat*; this base is often used because of its convenient analytical properties in integration, differentiation, etc. The base 10 is encountered only rarely; the unit is the *Hartley*.

3. Entropy of Discrete Random Variables. The more elementary properties of the entropy of a discrete random variable can be inferred from a simple example. Consider the binary case, where $M = 2$, so that the alphabet consists of the symbols 0 and 1 with probabilities p and $1 - p$ respectively. It follows from Eq. (4-4) that

$$H_1(X) = -[p \log_2 p + (1 - p) \log_2 (1 - p)] \qquad \text{(bits)} \tag{4-5}$$

This equation can be plotted as a function of p, as shown in Fig. 4-3, and has the following interesting properties:

(1) $H_1(X) \geq 0$.
(2) $H_1(X)$ is zero only for $p = 0$ and $p = 1$.
(3) $H_1(X)$ is a maximum at $p = 1 - p = \frac{1}{2}$.

From the graph of Fig. 4-3 and from Eq. (4-4) it can be inferred or proved that the entropy $H(X)$ has the following properties for the general case of an alphabet of size M:

$$1 \quad H(X) \geq 0. \tag{4-6}$$

2 $H(X) = 0$ if and only if all of the probabilities are zero except for one, which must be unity. (4-7)

3 $H(X) \leq \log_b M$. (4-8)

4 $H(X) = \log_b M$ if and only if all the probabilities are equal so that $p(x_i) = 1/M$ for all i. (4-9)

4. Mutual Information and Joint Entropy. The usual communication problem concerns the transfer of information from a source S through a channel C to a destination D, as shown in Fig. 4-1. The source has available for forming messages an alphabet X of size M. A particular symbol x_i is selected from the M possible symbols and is sent over the channel C. It is the limitations of the channel that produce the need for a study of information theory.

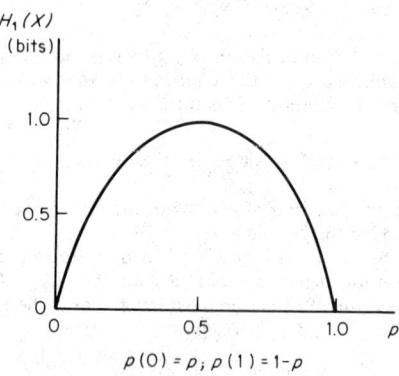

$$p(0) = p; \, p(1) = 1 - p$$

Fig. 4-3. Entropy in the binary case.

The information destination has available an alphabet Y of size N. For each symbol x_i sent from the source, a symbol y_j is selected at the destination. Two probabilities serve to describe the state of knowledge at the destination. Prior to the reception of a communication, the state of knowledge of the destination about the symbol x_j is the *a priori* probability $p(x_i)$ that x_i would be selected for transmission. After reception and selection of the symbol y_j, the state of knowledge concerning x_i is the conditional probability $p(x_i/y_j)$, which will be called the *a posteriori* probability of x_i. It is the probability that x_i was sent given that y_j was received. Ideally this a posteriori probability for each given y_j should be unity for one x_i and zero for all other x_i. In this case an observer at the destination is able to determine exactly which symbol x_i has been sent after the reception of each symbol y_j. Thus the uncertainty which existed previously and which was expressed by the a priori probability distribution of x_i has been removed completely by reception. In the general case it is not possible to remove all the uncertainty, and the best that can be hoped for is that it has been decreased. Thus the a posteriori probability $p(x_i/y_j)$ is distributed over a number of x_i but should be different. If the two probabilities are the same, then no uncertainty has been removed by transmission or no information has been transferred.

Based on this discussion and on other considerations that will become clearer later, the quantity $I(x_i; y_j)$ is defined as the information gained about x_i by the reception of y_j, where

$$I(x_i; y_j) = \log_b \frac{p(x_i/y_j)}{p(x_i)}$$ (4-10)

This measure has a number of reasonable and desirable properties.

Property 1. The information measure $I(x_i; y_j)$ is symmetric in x_i and y_j; that is,

$$I(x_i; y_j) = I(y_j; x_i)$$ (4-11)

Property 2. The mutual information $I(x_i; y_j)$ is a maximum when $p(x_i/y_j) = 1$, that is, when the reception of y_j completely removes the uncertainty concerning x_i:

$$I(x_i; y_j) \leq -\log p(x_i) = I(x_i)$$ (4-12)

Property 3. If two communications y_j and z_k concerning the same message x_i are received successively, and if the observer at the destination takes the a posteriori probability of the first as the a priori probability of the second, then the total information gained about x_i is the sum of the gains from both communications:

$$I(x_i;y_j,z_k) = I(x_i;y_j) + I(x_i;z_k/y_j) \qquad (4\text{-}13)$$

Property 4. If two communications y_j and y_k concerning two *independent* messages x_i and x_m are received, the total information gain is the sum of the two information gains considered separately:

$$I(x_i,x_m;y_j,y_k) = I(x_i;y_j) + I(x_m;y_k) \qquad (4\text{-}14)$$

These four properties of mutual information are intuitively satisfying and desirable. Moreover, if one begins by requiring these properties, it is easily shown that the logarithmic definition of Eq. (4-10) is the simplest form that can be obtained.

The definition of mutual information given by Eq. (4-10) suffers from one major disadvantage. When errors are present, an observer will not be able to calculate the information gain even after the reception of all the symbols relating to a given source symbol, since the same series of received symbols may represent several different source symbols. Thus the observer is unable to say which source symbol has been sent, and the best he can do is to compute the information gain with respect to each possible source symbol. In many cases it would be more desirable to have a quantity which is independent of the particular symbols. A number of quantities of this nature will be obtained in the remainder of this section.

The mutual information $I(x_i;y_j)$ is a random variable just as the self-information $I(x_i)$ was. However, two probability spaces X and Y are involved now, and several ensemble averages are possible. The *average mutual information* $I(X;Y)$ is defined as a statistical average of $I(x_i;y_j)$ with respect to the joint probability $p(x_i;y_j)$; that is,

$$I(X;Y) = E_{XY}\{I(x_i;y_j)\} = \sum_i \sum_j p(x_i,y_j)\log\frac{p(x_i/y_j)}{p(x_i)} \qquad (4\text{-}15)$$

This new function $I(X;Y)$ is the first information measure defined which does not depend on the individual symbols x_i or y_j. Thus it is a property of the whole communication system and will turn out to be only the first in a series of similar quantities used as a basis for the characterization of communication systems. This quantity $I(X;Y)$ has a number of expected properties. It is nonnegative; it is zero if and only if the ensembles X and Y are *statistically independent*; and it is symmetric in X and Y so that $I(X;Y) = I(Y;X)$.

A source entropy $H(X)$ was given by Eq. (4-4). It is obvious that a similar quantity, the destination entropy $H(Y)$, can be defined analogously by

$$H(Y) = - \sum_{j=1}^{N} p(y_j)\log p(y_j) \qquad (4\text{-}16)$$

This quantity will, of course, have all the properties developed for $H(X)$. In the same way the *joint or system entropy* $H(X, Y)$ may be defined by

$$H(X,Y) = - \sum_{i=1}^{M} \sum_{j=1}^{N} p(x_i,y_j)\log p(x_i,y_j) \qquad (4\text{-}17)$$

If X and Y are *statistically independent* so that $p(x_i, y_j) = p(x_i)p(y_j)$ for all i and j, then Eq. (4-17) can be written

$$H(X,Y) = H(X) + H(Y) \qquad (4\text{-}18)$$

On the other hand, if X and Y are not independent, then Eq. (4-17) becomes

$$H(X,Y) = H(X) + H(Y/X) = H(Y) + H(X/Y) \qquad (4\text{-}19)$$

where $H(Y/X)$ and $H(X/Y)$ are *conditional entropies* given by

$$H(Y/X) = -\sum_{i=1}^{M}\sum_{j=1}^{N} p(x_i, y_j)\log p(y_j/x_i) \qquad (4\text{-}20)$$

and

$$H(X/Y) = -\sum_{i=1}^{M}\sum_{j=1}^{N} p(x_i, y_j)\log p(x_i/y_j) \qquad (4\text{-}21)$$

These conditional entropies each satisfy an important inequality

$$0 \leq H(Y/X) \leq H(Y) \qquad (4\text{-}22)$$

and

$$0 \leq H(X/Y) \leq H(X) \qquad (4\text{-}23)$$

It follows from these last two expressions that Eq. (4-15) may be expanded to yield

$$I(X;Y) = -H(X, Y) + H(X) + H(Y) \geq 0 \qquad (4\text{-}24)$$

This equation can be rewritten in the two equivalent forms

$$I(X;Y) = H(Y) - H(Y/X) \geq 0 \qquad (4\text{-}25)$$

or

$$I(X/Y) = H(X) - H(X/Y) \geq 0 \qquad (4\text{-}26)$$

It is also clear, say from Eq. (4-24), that $H(X, Y)$ satisfies the inequality

$$H(X, Y) \leq H(X) + H(Y) \qquad (4\text{-}27)$$

Thus the joint entropy of two ensembles X and Y is a maximum when the ensembles are independent.

At this point it may be appropriate to comment on the meaning of the two conditional entropies $H(Y/X)$ and $H(X/Y)$. Let us refer first to Eq. (4-26). This equation expresses the fact that the average information gained about a message, when a communication is completed, is equal to the average source information less the average uncertainty that still remains about the message. From another point of view, the quantity $H(X/Y)$ is the average additional information needed at the destination after reception to completely specify the message sent. Thus $H(X/Y)$ represents the information lost in the channel. It is frequently called the *equivocation*. Let us now consider Eq. (4-25). This equation indicates that the information transmitted consists of the difference between the destination entropy and that part of the destination entropy that is not information about the source; thus the term $H(Y/X)$ can be considered a *noise entropy* added in the channel.

INFORMATION SOURCES

5. Message Sources. As shown in Fig. 4-1, an information source can be considered as emitting a given message u_i from the set $\{U\}$ of possible messages. In general, each message u_i will be represented by a sequence of symbols x_j from the source alphabet $\{X\}$, since the number of possible messages will usually exceed the size M of the source alphabet. Thus sequences of symbols replace the original messages u_i, which need not be considered further. When the source alphabet $\{X\}$ is of finite size M, the source will be called a *finite discrete source*. The problems of concern now are the interrelationships existing between symbols in the generated sequences and the classification of sources according to these interrelationships.

A random or stochastic process x_t, $t \in T$, can be defined as an indexed set of random variables where T is the *parameter set* of the process. If the set T is a sequence, then x_t is a stochastic process with discrete parameter (also called a *random sequence* or *series*). One way to look at the output of a finite discrete source is that it is a discrete-parameter stochastic process with each possible given sequence one of the ensemble members or realizations of the process. Thus the study of information sources can be reduced to a study of random processes.

The simplest case to consider is the *zero-memory source*, where the successive symbols obey the same fixed probability law so that the one distribution $p(x_i)$ determines the appearance of each indexed symbol. Such a source is called *stationary*. Let us consider sequences of length n, each member of the sequence being a realization of the random variable x_i with fixed probability distribution $p(x_i)$. Since there are M possible realizations of the random variable and n terms in the sequence, there must be M^n distinct sequences possible of length n. Let the random variable x_i in the jth position be denoted by x_{i_j} so that the sequence set (the message set) can be represented by

$$\{U\} = X^n = \{x_{i_1}, x_{i_2}, \cdots, x_{i_n}\} \qquad i = 1, 2, \ldots, M \tag{4-28}$$

The symbol X^n is sometimes used to represent this sequence set and is called the *nth extension of the zero-memory source X*. The probability of occurrence of a given message u_i is just the product of the probabilities of occurrence of the individual terms in the sequence so that

$$p\{u_i\} = p(x_{i_1}) p(x_{i_2}) \cdots p\{x_{i_n}\} \tag{4-29}$$

Now the entropy for the extended source X^n is

$$H(X^n) = -\sum_{X^n} p\{u_i\} \log p\{u_i\} = nH(X) \tag{4-30}$$

as expected. Note that, if base 2 logarithms are used, then $H(X)$ has units of bits per symbol, n is symbols per sequence, and $H(X^n)$ is in units of bits per sequence. For a zero-memory source, all sequence averages of information measure are obtained by multiplying the corresponding symbol by the number of symbols in the sequence.

6. Markov Information Source. The zero-memory source is not a general enough model in most cases. A constructive way to generalize this model is to assume that the occurrence of a given symbol depends on some number m of immediately preceding symbols. Thus the information source can be considered to produce an mth-order Markov chain and is called an *mth-order Markov source*.

For an mth-order Markov source, the m symbols preceding a given symbol position are called the *state* s_j of the source at that symbol position. If there are M possible symbols x_i, then the mth-order Markov source will have $M^m = q$ possible states s_j making up the *state set*

$$S = \{s_1, s_2, \cdots, s_q\} \qquad q = M^m \tag{4-31}$$

At a given time corresponding to one symbol position the source will be in a given state s_j. There will exist a probability $p(s_k/s_j) = p_{jk}$ that the source will move into another state s_k with the emission of the next symbol. The set of all such conditional probabilities is expressed by the *transition matrix T*, where

$$T = [p_{jk}] = \begin{bmatrix} p_{11} & p_{12} & \cdots & p_{1q} \\ p_{21} & p_{22} & \cdots & p_{2q} \\ \cdots & \cdots & \cdots & \cdots \\ p_{q1} & p_{q2} & \cdots & p_{qq} \end{bmatrix} \tag{4-32}$$

A *Markov matrix* or *stochastic matrix* is any square matrix with nonnegative elements such that the row sums are unity. It is clear that T is such a matrix since

$$\sum_{j=1}^{q} p_{ij} = \sum_{j=1}^{q} p(s_j/s_i) = 1 \qquad i = 1, 2, \ldots, q \tag{4-33}$$

Conversely, any stochastic matrix is a possible transition matrix for a Markov source of order m, where $q = M^m$ is equal to the number of rows or columns of the matrix.

A Markov chain is completely specified by its transition matrix T and by an *initial distribution vector* π giving the probability distribution for the first state occurring. For the zero-memory source, the transition matrix reduces to a stochastic matrix where all the rows are identical and are each equal to the initial distribution vector π, which is in turn equal to the vector giving the source alphabet a priori probabilities. Thus, in this case, we have

$$p_{jk} = p(s_k/s_j) = p(s_k) = p(x_k) \qquad k = 1, 2, \ldots, M \tag{4-34}$$

For each state s_i of the source an entropy $H(s_i)$ can be defined by

$$
\begin{aligned}
H(s_i) &= - \sum_{j=1}^{q} p(s_j/s_i) \log p(s_j/s_i) \\
       &= - \sum_{k=1}^{M} p(x_k/s_i) \log p(x_k/s_i)
\end{aligned}
\tag{4-35}
$$

The source entropy $H(S)$ in information units per symbol is the expected value of $H(s_i)$; that is,

$$
\begin{aligned}
H(S) &= - \sum_{i=1}^{q} \sum_{j=1}^{q} p(s_i) p(s_j/s_i) \log p(s_j/s_i) \\
     &= - \sum_{i=1}^{q} \sum_{k=1}^{M} p(s_i) p(x_k/s_i) \log p(x_k/s_i)
\end{aligned}
\tag{4-36}
$$

where $p(s_i) = p_i$ is the *stationary state probability* and is the ith element of the vector $\mathbf{P}$ defined by

$$
\mathbf{P} = [p_1 \quad p_2 \quad \cdots \quad p_q]
\tag{4-37}
$$

It is easy to show, as in Eq. (4-8), that the source entropy cannot exceed $\log M$, where M is the size of the source alphabet $\{X\}$. For a given source, the ratio of the actual entropy $H(S)$ to the maximum value it can have with the same alphabet is called the *relative entropy* of the source. The *redundancy* η of the source is defined as the positive difference between unity and this relative entropy:

$$
\eta = 1 - \frac{H(S)}{\log M}
\tag{4-38}
$$

The quantity $\log M$ is sometimes called the *capacity* of the alphabet.

CODES AND CODING

7. Noiseless Coding. The preceding discussion has emphasized the information source and its properties. We now consider the properties of the communication channel of Fig. 4-1. In general, an arbitrary channel will not accept and transmit the sequence of x_i's emitted from an arbitrary source. Instead the channel will accept a sequence of some other elements a_i chosen from a *code alphabet A* of size D, where

$$
A = \{a_1, a_2, \cdots, a_D\}
\tag{4-39}
$$

with D generally smaller than M. The elements of a_i of the code alphabet are frequently called *code elements* or *code characters*, while a given sequence of a_i's may be called a *code word*.

The situation is now describable in terms of Fig. 4-2, where an encoder E has been added between the source and channel. The process of *coding*, or *encoding*, the source consists of associating with each source symbol x_i a given code word, which is just a given sequence of a_i's. Thus the source emits a sequence of x_i's chosen from the source alphabet X, and the encoder emits a sequence of a_i's chosen from the code alphabet A. It will be assumed in all subsequent discussions that the code words are distinct, i.e., that each code word corresponds to only one source symbol.

Even though each code word is required to be distinct, sequences of code words may not have this property. An example is code A of Table 4-1, where a source of size 4 has been encoded in binary code with characters 0 and 1. In code A the code words are distinct, but sequences of code words are not. It is clear that such a code is not *uniquely* decipherable. On the other hand, a given sequence of code words taken from code B will correspond to a distinct sequence of source symbols. An examination of code B shows that in no case is a code word formed by adding characters to another word. In other words, no code word is a *prefix* of another. It is clear that this is a *sufficient* (but not necessary) condition for a code to be uniquely decipherable. That it is not necessary can be seen from an examination of codes C and D of Table 4-1. These codes are uniquely decipherable even though many of the code words are prefixes of other words. In these cases any sequence of code words can be decoded by subdividing the sequence of 0s and 1s to the left of every 0 for code C and to

Table 4-1. Four Binary Coding Schemes

Source symbol	Code A	Code B	Code C	Code D
x_1	0	0	0	0
x_2	1	10	01	10
x_3	00	110	011	110
x_4	11	111	0111	1110

NOTE: Code A is not uniquely decipherable; codes B, C, and D are uniquely decipherable; codes B and D are instantaneous codes; and codes C and D are comma codes.

the right of every 0 for code D. The character 0 is the first (or last) character of every code word and acts as a comma; therefore this type of code is called a *comma code.*

In general the channel will require a finite amount of time to transmit each code character. The code words should be as short as possible in order to maximize information transfer per unit time. The average length L of a code is given by

$$L = \sum_{i=1}^{M} n_i p(x_i) \tag{4-40}$$

where n_i is the length (number of code characters) of the code word for the source symbol x_i and $p(x_i)$ is the probability of occurrence of x_i. Although the average code length cannot be computed unless the set $\{p(x_i)\}$ is given, it is obvious that codes C and D of Table 4-1 will have a greater average length than code B unless $p(x_4) = 0$. Comma codes are not optimal with respect to minimum average length.

Let us encode the sequence $x_3 x_1 x_3 x_2$ into codes B, C, and D of Table 4-1 as shown below:

Code B: 110011010

Code C: 011001101

Code D: 110011010

Codes B and D are fundamentally different from code C in that codes B and D can be decoded word by word *without examining subsequent code characters* while code C cannot be so treated. Codes B and D are called *instantaneous codes* while code C is noninstantaneous. The instantaneous codes have the property (previously mentioned) that no code word is a prefix of another code word.

The aim of noiseless coding is to produce codes with the two properties of *unique decipherability* and *minimum average length L* for a given source S with alphabet X and probability set $\{p(x_i)\}$. Codes which have both these properties will be called *optimal.* It can be shown that if, for a given source S, a code is optimal among instantaneous codes, then it is optimal among all uniquely decipherable codes. Thus it is sufficient to consider instantaneous codes. A *necessary* property of optimal codes is that source symbols with higher probabilities must have shorter code words; i.e.,

$$p(x_i) > p(x_j) \Rightarrow n_i \leq n_j \tag{4-41}$$

The encoding procedure consists of the assignment of a code word to each of the M source symbols. The code word for the source symbol x_i will be of length n_i; that is, it will consist of n_i code elements chosen from the code alphabet of size D. It can be shown that a necessary and sufficient condition for the construction of a uniquely decipherable code is the *Kraft inequality*

$$\sum_{i=1}^{M} D^{-n_i} \leq 1 \tag{4-42}$$

8. Noiseless-Coding Theorem. It follows from Eq. (4-42) that the average code length L, given by Eq. (4-40), satisfies the inequality

$$L \geq \frac{H(X)}{\log D} \tag{4-43}$$

Equality (and minimum code length) occurs if and only if the source-symbol probabilities obey

$$p(x_i) = D^{-n_i} \qquad i = 1, 2, \ldots, M \tag{4-44}$$

A code where this equality applies is called *absolutely optimal.* Since an integer number of code elements must be used for each code word, the equality in Eq. (4-43) does not usually hold. However, by using one more code element, the average code length L can be bounded from above to give

$$\frac{H(X)}{\log D} \le L \le \frac{H(X)}{\log D} + 1 \qquad (4\text{-}45)$$

This last relationship is frequently called the *noiseless-coding theorem.*

9. Construction of Noiseless Codes. The easiest case to consider occurs when an absolutely optimal code exists; i.e., when the source-symbol probabilities satisfy Eq. (4-44). Note that code B of Table 4-1 is absolutely optimal if $p(x_1) = \frac{1}{2}$, $p(x_2) = \frac{1}{4}$, and $p(x_3) = p(x_4) = \frac{1}{8}$. In such cases, a procedure for realizing the code for arbitrary code-alphabet size ($D \ge 2$) is easily constructed as follows:

1. Arrange the M source symbols in the set x_i in order of decreasing probability.
2. Arrange the D code elements in an arbitrary but fixed order, that is, $a_1, a_2, \ldots, a_D$.
3. Divide the set of symbols x_i into D groups with equal probabilities of $1/D$ each. This division is always possible if Eq. (4-44) is satisfied.
4. Assign the element a_1 as the first digit for symbols in the first group, a_2 for the second, and a_i for the ith group.
5. After the first division each of the resulting groups contains a number of symbols equal to D raised to some integral power if Eq. (4-44) is satisfied. Thus a typical group, say group i, contains D^{k_i} symbols, where k_i is an integer (which may be zero). This group of symbols can be further subdivided k_i times into D parts of equal probabilities. Each division decides one additional code digit in the sequence. A typical symbol x_i is isolated after q divisions. If it belongs to the i_1 group after the first division, the i_2 group after the second division, etc., then the code word for x_i will be $a_{i_1} a_{i_2} \cdots a_{i_q}$.

An illustration of the construction of an absolutely optimal code for the case where $D = 3$ is given in Table 4-2. This procedure ensures that source symbols with high probabilities will have short code words and vice versa, since a symbol with probability D^{-n_i} will be isolated after n_i divisions and thus will have n_i elements in its code word, as required by Eq. (4-44).

Table 4-2. **Construction of an Optimal Code; $D=3$**

Source symbols x_i	A priori probabilities $p(x_i)$	Step			Final code		
		1	2	3			
x_1	$\frac{1}{3}$	1			1		
x_2	$\frac{1}{9}$	0	1		0	1	
x_3	$\frac{1}{9}$	0	0		0	0	
x_4	$\frac{1}{9}$	0	-1		0	-1	
x_5	$\frac{1}{27}$	-1	1	1	-1	1	1
x_6	$\frac{1}{27}$	-1	1	0	-1	1	0
x_7	$\frac{1}{27}$	-1	1	-1	-1	1	-1
x_8	$\frac{1}{27}$	-1	0	1	-1	0	1
x_9	$\frac{1}{27}$	-1	0	0	-1	0	0
x_{10}	$\frac{1}{27}$	-1	0	-1	-1	0	-1
x_{11}	$\frac{1}{27}$	-1	-1	1	-1	-1	1
x_{12}	$\frac{1}{27}$	-1	-1	0	-1	-1	0
x_{13}	$\frac{1}{27}$	-1	-1	-1	-1	-1	-1

NOTE: Average code length $L=2$ code elements per symbol; source entropy $H(X) = 2 \log_2 3$ bits per symbol

$$L = \frac{H(X)}{\log_2 3}$$

The code resulting from the process just discussed is sometimes called the *Shannon-Fano* code. It is apparent that the same encoding procedure can be followed whether or not the source probabilities satisfy Eq. (4-44). The set of symbols x_i are simply divided into D groups with probabilities as nearly equal as possible. The procedure is sometimes ambiguous, however, and more than one Shannon-Fano code may be possible. The ambiguity arises, of course, in the choice of approximately equiprobable subgroups.

For the general case where Eq. (4-44) is not satisfied, a procedure due to Huffman guarantees an optimal code, i.e., one with minimum average length. This procedure for code alphabet of arbitrary size D is as follows:

1. As before, arrange the M source symbols in the set x_i in order of decreasing probability.

2. As before, arrange the code elements in an arbitrary but fixed order, that is, $a_1, a_2, \ldots, a_D$.

3. Combine (sum) the probabilities of the D least likely symbols and reorder the resulting $M - (D - 1)$ probabilities; this step will be called *reduction* 1. Repeat as often as necessary until there are D ordered probabilities remaining. *Note*: For the binary case ($D = 2$), it will always be possible to accomplish this reduction in $M - 2$ steps. When the size of the code alphabet is arbitrary, the last reduction will result in exactly D ordered probabilities if and only if

$$M = D + n(D - 1)$$

where n is an integer. If this relationship is not satisfied, add *dummy* source symbols with zero probability. The entire encoding procedure is followed as before, and at the end the dummy symbols are thrown away.

4. Start the encoding with the last reduction which consists of exactly D ordered probabilities; assign the element a_1 as the first digit in the code words for all the source symbols associated with the first probability; assign a_2 to the second probability; and a_i to the ith probability.

5. Proceed to the next to the last reduction; this reduction consists of $D + (D - 1)$ ordered probabilities for a net gain of $D - 1$ probabilities. For the D new probabilities, the first code digit has already been assigned and is the same for all of these D probabilities; assign a_1 as the second digit for all source symbols associated with the first of these D new probabilities; assign a_2 as the second digit for the second of these D new probabilities, etc.

6. The encoding procedure terminates after $1 + n(D - 1)$ steps, which is one more than the number of reductions.

As an illustration of the Huffman coding procedure, a binary code is constructed in Table 4-3.

COMMUNICATION CHANNELS

10. Channel Capacity. The average mutual information $I(X;Y)$ between an information source and a destination was given by Eqs. (4-25) and (4-26) as

$$I(X;Y) = H(Y) - H(Y/X) = H(X) - H(X/Y) \geq 0 \qquad (4-46)$$

The average mutual information depends not only on the statistical characteristics of the channel but also on the distribution $p(x_i)$ of the input alphabet X. If the input distribution is varied until Eq. (4-46) is a maximum for a given channel, the resulting value of $I(X;Y)$ is called the *channel capacity* C of that channel; i.e.,

$$C = \max_{p(x_i)} I(X;Y) \qquad (4-47)$$

In general, $H(X)$, $H(Y)$, $H(X/Y)$, and $H(Y/X)$ all depend on the input distribution $p(x_i)$.

Hence, *in the general case*, it is not a simple matter to maximize Eq. (4-46) with respect to $p(x_i)$.

All the measures of information that have been considered in this treatment have involved only probability distributions on X and Y. Thus, for the model of Fig. 4-1, the joint distribution $p(x_i, y_j)$ is sufficient. Suppose the source [and hence the input distribution $p(x_i)$] is known; then it follows from the usual conditional-probability relationship

$$p(x_i, y_j) = p(x_i) p(y_j/x_i) \qquad (4-48)$$

Table 4-3. Construction of a Huffman Code; $D = 2$

Source symbols x_i	A priori probabilities $p(x_i)$	Final code	Reduction 1 / Step 5		Reduction 2 / Step 4		Reduction 3 / Step 3		Reduction 4 / Step 2		Reduction 5 / Step 1	
x_1	0.40	0	0.40	0	0.40	0	0.40	0	0.40	0	0.60	1
x_2	0.20	111	0.20	111	0.20	111	0.24	10	0.36	11	0.40	0
x_3	0.12	101	0.12	101	0.16	110	0.20	111	0.24	10		
x_4	0.08	1101	0.12	100	0.12	101	0.16	110				
x_5	0.08	1100	0.08	1101	0.12	100						
x_6	0.08	1001	0.08	1100								
x_7	0.04	1000										

Average code length $L = 1(0.40) + 3(0.20) + 3(0.12) + 4(0.08) + 4(0.08) + 4(0.08) + 4(0.04)$
$= 2.48$ code elements/symbol

that only the distribution $p(y_j/x_i)$ is needed for $p(x_i/y_j)$ to be determined. This conditional probability $p(y_j/x_i)$ can then be taken as a description of the information channel connecting the source X and the destination Y. Thus, a *discrete memoryless channel* can be defined as the probability distribution

$$p(y_j/x_i) \quad x_i \in X \quad \text{and} \quad y_j \in Y \tag{4-49}$$

or, equivalently, by the *channel matrix D*, where

$$D = [p(y_j/x_i)] = \begin{bmatrix} p(y_1/x_1) & p(y_2/x_1) & \cdots & p(y_N/x_1) \\ p(y_1/x_2) & p(y_2/x_2) & \cdots & p(y_N/x_2) \\ \cdots & \cdots & \cdots & \cdots \\ p(y_1/x_M) & & \cdots & p(y_N/x_M) \end{bmatrix} \tag{4-50}$$

A number of special types of channels are readily distinguished. Some of the simplest and/or most interesting are listed.

a. Lossless Channel. Here $H(X/Y) = 0$ for all input distribution $p(x_i)$, and Eq. (4-47) becomes

$$C = \max_{p(x_i)} H(X) = \log M \tag{4-51}$$

This maximum is obtained when the x_i are equally likely, so that $p(x_i) = 1/M$ for all i. The channel capacity is equal to the source entropy, and no source information is lost in transmission.

b. Deterministic Channel. Here $H(Y/X) = 0$ for all input distributions $p(x_i)$, and Eq. (4-47) becomes

$$C = \max_{p(x_i)} H(Y) = \log N \tag{4-52}$$

This maximum is obtained when the y_j are equally likely, so that $p(y_j) = 1/N$ for all j. Each member of the X set is uniquely associated with one, and only one, member of the destination alphabet Y.

c. Symmetric Channel. Here the rows of the channel matrix D are identical except for permutations, *and* the columns are identical except for permutations. If D is square, rows and columns are identical except for permutations. In the symmetric channel, the conditional entropy $H(Y/X)$ is independent of the input distribution $p(x_i)$ and depends only on the channel matrix D. As a consequence, the determination of channel capacity is greatly simplified and can be written

$$C = \log N + \sum_{j=1}^{N} p(y_j/x_i)\log p(y_j/x_i) \tag{4-53}$$

This capacity is obtained when the y_j are equally likely, so that $p(y_j) = 1/N$ for all j.

d. Binary Symmetric Channel (BSC). This is the special case of a symmetric channel where $M = N = 2$. Here the channel matrix can be written

$$D = \begin{bmatrix} p & 1-p \\ 1-p & p \end{bmatrix} \tag{4-54}$$

and the channel capacity is

$$C = \log 2 - G(p) \tag{4-55}$$

where the function $G(p)$ is defined as

$$G(p) = -[p \log p + (1-p)\log(1-p)] \tag{4-56}$$

This expression is mathematically identical to the entropy of a binary source as given in Eq. (4-5) and plotted in Fig. 4-3 using base 2 logarithms. For the same base, Eq. (4-55) is shown as a function of p in Fig. 4-4. As expected, the channel capacity is large if p, the probability of correct transmission, is either close to unity or to zero. If $p = 1/2$, there is no statistical evidence which symbol was sent, and the channel capacity is zero.

DECODING

11. Decision Schemes. A decision scheme or decoding scheme $\mathcal{B}$ is a partitioning of the Y set into M disjoint and exhaustive sets $B_1, B_2, \ldots, B_M$ such that when a destination symbol y_k falls into set B_i, it is decided that symbol x_i was sent. Implicit in this definition is a *decision rule* $d(y_j)$, which is a function specifying uniquely a source symbol for each destination symbol. Let $p(e/y_j)$ be the probability of error when it is decided that y_j has been received. Then the *total error probability* $p(e)$ is

$$p(e) = \sum_{j=1}^{N} p(y_j)\, p(e/y_j) \tag{4-57}$$

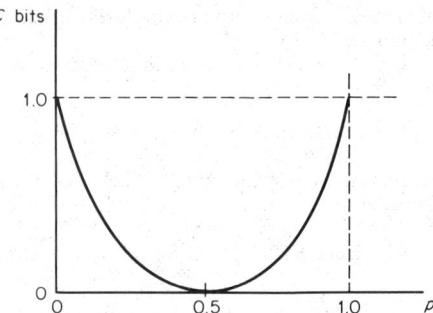

Fig. 4-4. Capacity of the binary symmetric channel.

For a given decision scheme $\mathcal{B}$, the conditional error probability $p(e/y_j)$ can be written

$$p(e/y_j) = 1 - p[d(y_j)/y_j] \tag{4-58}$$

where $p[d(y_j)/y_j]$ is the conditional probability $p(x_i/y_j)$ with x_i assigned by the decision rule; i.e., for a given decision scheme $d(y_j) = x_i$. The probability $p(y_j)$ is determined only by the source a priori probability $p(x_i)$ and by the channel matrix $D = [p(y_j/x_i)]$. Hence, only the term $p(e/y_j)$ in Eq. (4-57) is a function of the decision scheme. Since Eq. (4-57) is a sum of nonnegative terms, the error probability is a minimum when each summand is a minimum. Thus, the term $p(e/y_j)$ should be a minimum for each y_j. It follows from Eq. (4-58) that the minimum-error scheme is that scheme which assigns a decision rule

$$d(y_j) = x^* \qquad j = 1, 2, \ldots, N \tag{4-59}$$

where x^* is defined by

$$p(x^*/y_j) \geq p(x_i/y_j) \qquad i = 1, 2, \ldots, M \tag{4-60}$$

In other words, each y_j is decoded as the *a posteriori most likely* x_i. This scheme, which minimizes the probability of error $p(e)$, is usually called the *ideal observer*.

The ideal observer is not always a completely satisfactory decision scheme. It suffers from two major disadvantages: (1) For a given channel D, the scheme is defined only for a given input distribution $p(x_i)$. It would be preferable to have a scheme which was insensitive to input distributions. (2) The scheme minimizes average error but does not bound certain errors. For example, some symbols may always be received incorrectly. Despite these disadvantages, the ideal observer is a straightforward scheme which does minimize average error. It is also widely used as a standard with which other decision schemes may be compared.

Consider the special case where the input distribution is $p(x_i) = 1/M$ for all i, so that all x_i are equally likely. Now the conditional likelihood $p(x_i/y_j)$ is

$$p(x_i/y_j) = \frac{p(x_i)\, p(y_j/x_i)}{p(y_j)} = \frac{p(y_j/x_i)}{M p(y_j)} \tag{4-61}$$

For a given y_j, that input x_i is chosen which makes $p(y_j/x_i)$ a maximum, and the decision rule is

$$d(y_j) = x^\dagger \qquad j = 1, 2, \ldots, N \qquad (4\text{-}62)$$

where $x^\dagger$ is defined by

$$p(y_j/x^\dagger) \geq p(y_j/x_i) \qquad i = 1, 2, \ldots, M \qquad (4\text{-}63)$$

The probability of error becomes

$$p(e) = \sum_{j=1}^{N} p(y_j)\left[1 - \frac{p(y_j/x^\dagger)}{Mp(y_j)}\right] \qquad (4\text{-}64)$$

This decoder is sometimes called the *maximum-likelihood* decoder or decision scheme.

It would appear that a relationship should exist between the error probability $p(e)$ and the channel capacity C. One such relationship is the *Fano bound*, given by

$$H(X/Y) \leq G[p(e)] + p(e)\log(M - 1) \qquad (4\text{-}65)$$

and relating error probability to channel capacity through Eq. (4-47). Here $G(\cdot)$ is the function already defined by Eq. (4-56). The three terms in Eq. (4-65) can be interpreted as follows:

$H(X/Y)$ is the equivocation. It is the average additional information needed at the destination after reception to completely determine the symbol that was sent.

$G[p(e)]$ is the entropy of the binary system with probabilities $p(e)$ and $1 - p(e)$. In other words, it is the average amount of information needed to determine if the decision rule resulted in an error.

$\log(M - 1)$ is the maximum amount of information needed to determine which among the remaining $M - 1$ symbols was sent if the decision rule was incorrect; this information is needed with probability $p(e)$.

EFFECTS OF NOISE

12. The Noisy-Coding Theorem. The concept of channel capacity was discussed in Par. 4-10. Capacity is a fundamental property of an information channel in the sense that it is possible to transmit information through the channel at any rate less than the channel capacity with arbitrarily small probability of error. This result, which will be stated more precisely shortly, is called the *noisy-coding theorem* or *Shannon's fundamental theorem for a noisy channel*.

A proof of this theorem for an arbitrary channel is difficult and quite beyond the level of this treatment. However, a heuristic, i.e., nonrigorous, discussion is relatively straight-forward and affords considerable insight into the general problem. We begin by stating the assumptions made in order that confusion over what has been proved will be at a minimum.

1. The *source* will consist of a set $U = \{u_1, u_2, \cdots, u_m\}$ of m messages u_k. It will be assumed that these are *independent* and *equally likely*. This last assumption is a strong one, and it may be necessary to form this message set by encoding an original source alphabet $X = \{x_1, x_2, \cdots, x_M\}$ into long sequences; i.e., each sent message u_k may consist of a sequence of x_i's of length p, where p may be large. Now the source entropy $H(U)$ is given by

$$H(U) = \log m \qquad (4\text{-}66)$$

2. Each message u_k will be *encoded* into a code word consisting of a sequence of n binary digits for transmission over the binary channel. The number of possible code words of length n is 2^n; the number of messages u_k is m; thus the inequality

$$2^n \geq m \qquad \text{or} \qquad n \geq \log_2 m \qquad (4\text{-}67)$$

must be satisfied. The code words will be assigned at *random*; i.e., the probability is $m/2^n$ that a given n-place code word (of the 2^n possible code words) will be chosen to represent one of the set of m messages.

3. The *channel* will be taken to be the zero-memory binary symmetric channel (BSC) with probability p of correct transmission of a binary digit and probability of error $q = 1 - p$.

4. The *decision scheme* or *decoder* will operate as follows. At the destination, the received messages (there are 2^n possible) will be compared with all m of the messages that could have been sent. The message u_k will be considered sent which differs from the received message in the least number of binary digits. This decoder can be shown to be the ideal observer (actually the maximum-likelihood decoder since, in this particular case, the source probabilities are equally likely).

Consider a given code word of length n. The probability that exactly r digits will be altered in transmission (while $n - r$ digits are not altered) is just the probability of $n - r$ successes and r failures in n Bernoulli trials. This is the binomial distribution and is given by

$$p\{r \text{ errors}\} = \binom{n}{n - r} p^{n-r} q^r \qquad r = 0, 1, 2, \ldots, n \tag{4-68}$$

The mean number of errors is

$$E\{r\} = nq = n(1 - p) \tag{4-69}$$

Thus, the average number of binary digits altered by noise in the BSC will be nq.

There are $\binom{n}{r}$ code words (out of the total of 2^n) that differ *in exactly r digits* from a given code word. Therefore, the number K of code words that differ *in nq digits or less* from a given code word is given by the sum

$$K = \sum_{r=0}^{nq} \binom{n}{r} = \binom{n}{0} + \binom{n}{1} + \cdots + \binom{n}{nq - 1} + \binom{n}{nq} \tag{4-70}$$

For $q < \frac{1}{2}$, each term in this series for K is larger than the preceding term, so that K is bounded by

$$K \le (nq + 1)\binom{n}{nq} = (nq + 1)\frac{n!}{(nq)!(n - nq)!} \tag{4-71}$$

For large n, the factorial is given approximately by Stirling's formula

$$n! \approx \sqrt{2\pi}\, n^{n+1/2} e^{-n}$$

With this substitution, Eq. (4-71) can be written

$$K \le \sqrt{\frac{(nq + 1)^2}{2\pi npq}}\, q^{-nq} p^{-np} \tag{4-72}$$

Thus, on the average, a given sent code word can be received as any one of K code words differing in nq digits or less from the sent word. An error will result if any message u_k, other than the one originally transmitted, is represented by one of this group of K code words.

Recall that the code words were assigned at random from the set of 2^n possible code words. Therefore, the expected number of messages that could be changed into the original message by errors in the BSC and hence be confused with this original message is

$$L = K\frac{m}{2^n} \tag{4-73}$$

It follows from Eq. (4-72) that L is bounded by

$$L \le m\sqrt{\frac{(nq + 1)^2}{2\pi npq}}\, 2^{-nC_{s2}} \tag{4-74}$$

where C_{s2} is the capacity of the BSC and is given by

$$C_{s2} = 1 + p \log_2 p + q \log_2 q \tag{4-75}$$

For finite m and C_{s2}, the limit of L for large n is zero. Thus, for sufficiently long code words, the probability that a message will be decoded incorrectly approaches zero. Now let the number of messages m be fixed by the relation

$$m = \frac{2^{nC_{s2}}}{n} \tag{4-76}$$

Then Eq. (4-74) becomes

$$L \leq \sqrt{\frac{(nq + 1)^2}{2\pi n^3 pq}} \tag{4-77}$$

and, as before,

$$\lim_{n \to \infty} L = 0 \tag{4-78}$$

The channel input entropy per message is given by Eq. (4-66). This entropy per code digit is

$$H'(U) = \frac{\log_2 m}{n} \quad \text{(bits/digit)} \tag{4-79}$$

or, from Eq. (4-76),

$$H'(U) = C_{s2} - \frac{\log n}{n} \tag{4-80}$$

$$\text{and} \quad \lim_{n \to \infty} H'(U) = C_{s2} \quad \text{(bits/digit)} \tag{4-81}$$

Thus, for sufficiently long message sequences, information can be conveyed in the BSC at a rate approaching channel capacity and with arbitrarily small probability of error.

The noisy-coding theorem can be stated more precisely as follows: "Consider a discrete memoryless channel with nonzero capacity C; fix two numbers H and ϵ such that

$$0 < H < C \tag{4-82}$$

$$\text{and} \quad \epsilon > 0 \tag{4-83}$$

Let us transmit m messages $u_1, u_2, \cdots, u_m$ by code words each of length n binary digits. The positive integer n can be chosen so that

$$m \geq 2^{nH} \tag{4-84}$$

In addition, at the destination the m sent messages can be associated with a set $V = \{v_1, v_2, \cdots, v_m\}$ of received messages and with a decision rule $d(v_j) = u_j$ such that

$$p[d(v_j)/v_j] \geq 1 - \epsilon \tag{4-85}$$

i.e., decoding can be accomplished with a probability of error that does not exceed ϵ." The foregoing discussion of this theorem for the BSC is not a formal proof, of course, but merely a plausibility argument. However, a rigorous proof is quite involved and somewhat difficult. There is a converse to the noisy-coding theorem which states that it is not possible to produce an encoding procedure which allows transmission at a rate greater than channel capacity with arbitrarily small error.

ERROR CORRECTION

13. Error-Correcting Codes. The codes considered in Par. 4-7 were designed for minimum length in the noiseless-transmission case. For noisy channels, the coding theorem of Par. 4-12 guarantees the existence of a code which will allow transmission at any rate less than channel capacity and with arbitrarily small probability of error. However, the theorem does not provide a constructive procedure to devise such codes. Indeed, it implies that very long sequences of source symbols may have to be considered if reliable transmission at rates near channel capacity are to be obtained. In this section, we consider some of the elementary properties of simple *error-correcting codes*; i.e., codes which can be used to increase reliability in the transmission of information through noisy channels by correcting at least some of the errors that occur so that overall probability of error is reduced.

The discussion will be restricted to the BSC, and the notation of Par. **4-12** will be used. Thus, a source alphabet $X = \{x_1, x_2, \cdots, x_M\}$ of M symbols will be used to form a message set U of m messages u_k, where $U = \{u_1, u_2, \cdots, u_m\}$. Each u_k will consist of a sequence of the x_i's. Each message u_k will be encoded into a sequence of n binary digits for transmission over the BSC. At the destination, there exists a set $V = \{v_1, v_2, \cdots, v_{2^n}\}$ of all possible binary sequences of length n. The inequality $m \leq 2^n$ must hold. The problem is to associate with each sent message u_k a received message v_j so that $p(e)$, the overall probability of error, is reduced.

In the discussion of the noisy-coding theorem, a decoding scheme was used that examined the received message v_j and identified it with the sent message u_k, which differed from it in the least number of binary digits. In all the discussions of Par. **4-13** it will be assumed that this decoder is used. Let us define the *Hamming distance* $d(v_j, v_k)$ between two binary sequences v_j and v_k of length n as that integer which is the number of digits in which v_j and v_k disagree. Thus, if the distance between two sequences is zero, the two sequences are identical. It is easily seen that this distance measure has the following elementary properties:

$$(1) \qquad d(v_j, v_k) \geq 0 \text{ with equality if and only if } v_j = v_k. \qquad (4\text{-}86)$$

$$(2) \qquad d(v_j, v_k) = d(v_k, v_j). \qquad (4\text{-}87)$$

$$(3) \qquad d(v_j, v_l) \leq d(v_j, v_k) + d(v_k, v_l). \qquad (4\text{-}88)$$

$$(4) \qquad d(v_j, v_k) \leq n. \qquad (4\text{-}89)$$

The decoder we use is a *minimum-distance* decoder. As mentioned in Par. **4-12** the ideal-observer decoding scheme is a minimum-distance scheme for the BSC.

It is intuitively apparent that the sent messages should be represented by code words that all have the greatest possible distances between them. Let us investigate this matter in more detail by considering all binary sequences of length $n = 3$; there are $2^n = 2^3 = 8$ such sequences, viz.,

$$
\begin{array}{cccc}
000 & 001 & 011 & 111 \\
& 010 & 110 & \\
& 100 & 101 &
\end{array}
$$

It is convenient to represent these as the eight corners of a unit cube, as shown in Fig. 4-5a, where the x axis corresponds to the first digit, the y axis to the second, and the z axis to the third. Although direct pictorial representation is not possible, it is clear that binary sequences of length n greater than 3 can be considered as the corners of the corresponding n-cube.

Suppose that all eight binary sequences are used as code words to encode a source. If any binary digit is changed in transmission, an error will result at the destination since the sent message will be interpreted incorrectly as one of the three possible messages that differ in one code digit from the sent message. This situation is illustrated in Fig. 4-5b for the code words 000 and 111. A change of one digit in each of these code words produces one of three possible other code words.

Fig. 4-5b suggests that only two code words, say 000 and 111, should be used. The distance between these two words, or any other two words on opposite corners of the cube, is 3. If only one digit is changed in the transmission of each of these two code words, they can be correctly distinguished at the destination by a minimum-distance decoder. If two digits are changed in each word in transmission, it will not be possible to make this distinction.

It is clear that this reasoning can be extended to sequences containing more than three binary digits. For any $n \geq 3$, single errors in each code word can be corrected. If double errors are to be corrected without fail, it is necessary that there exist at least two code words with a minimum distance between them of 5; thus, for this case, binary code words of length 5 or greater must be used.

In the light of the previous discussion, it can be seen that the error-correcting properties of a code depend on the distance $d(v_j,v_k)$ between code words. Specifically, single errors can be corrected if all code words employed are at least a distance of 3 apart, double errors if the words are at a distance of 5 or more from each other, and, in general, q-fold errors can be corrected if

$$d(v_j,v_k) \geq 2q + 1 \qquad j \neq k \tag{4-90}$$

Of course errors involving less than q digits per code word can also be corrected if Eq. (4-90) is satisfied. If the distance between two code words is $2q$, there will always be a group of binary sequences which are in the middle, i.e., a distance q from *each* of the two words. Thus, by the proper choice of code words, q-fold errors can be *detected* but not corrected if

$$d(v_j,v_k) = 2q \qquad j \neq k \tag{4-91}$$

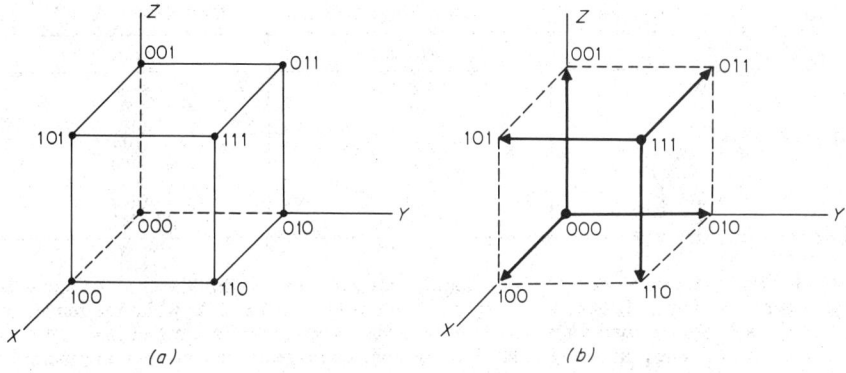

Fig. 4-5. Representation of binary sequences as the corners of an n-cube, $n = 3$: (*a*) binary sequences; (*b*) shift in sequence caused by a single error.

Let us now consider the maximum number of code words r that can be selected from the set of 2^n possible binary sequences of length n to form a code that will correct all single, double, $\ldots$, q-fold errors. In the example of Fig. 4-5, the number of code words selected was 2. In fact, it can be shown that there is no single-error-correcting code for $n = 3, 4$ containing more than two words. Suppose we consider a given code consisting of the words $\cdots, u_k, u_j, \ldots$. All binary sequences of distance q or less from u_k must "belong" to u_k, and to u_k only, if q-fold errors are to be corrected. Thus, associated with u_k are all binary sequences of distance $0, 1, 2, \ldots, q$ from u_k. The number of such sequences has been given previously by Eq. (4-70) and is

$$\binom{n}{0} + \binom{n}{1} + \binom{n}{2} + \cdots + \binom{n}{q} = \sum_{i=0}^{q} \binom{n}{i} \tag{4-92}$$

Since there are r of the code words, the total number of sequences associated with all the code words is

$$r \sum_{i=0}^{q} \binom{n}{i}.$$

This number can be no larger than 2^n, the total number of distinct binary sequences of length n. Therefore the following inequality must hold:

$$r \sum_{i=0}^{q} \binom{n}{i} \leq 2^n \qquad \text{or} \qquad r \leq \frac{2^n}{\sum_{i=0}^{q} \binom{n}{i}} \tag{4-93}$$

This is a *necessary* upper bound on the number of code words that can be used to correct all errors up to and including q-fold errors. It can be shown that it is not sufficient.

Let us consider the eight possible distinct binary sequences of length 3. Suppose we add one binary digit to each sequence in such a way that the total number of 1s in the sequence is *even* (or odd, if you wish). The result is shown in Table 4-4. Note that all the word sequences of length 4 differ from each other by a distance of at least 2. In accordance with Eq. (4-91), it should be possible now to *detect single errors* in all eight sequences. The detection method is straightforward. At the receiver, count the number of 1s in the sequence; if the number is odd, a single error (or, more precisely, an odd number of errors) has occurred; if the number is even, no error (or an even number of errors) has occurred. This particular scheme is a good one if only single errors are likely to occur and if only detection (rather than correction) is desired. The added digit is called a *parity-check digit*, and the scheme is a very simple example of a *parity-check code*.

Table 4-4. Parity-Check Code for Single-Error Detection

Message digits	Check digit	Word
000	0	0000
100	1	1001
010	1	0101
001	1	0011
110	0	1100
101	0	1010
011	0	0110
111	1	1111

14. Parity-Check Codes. More generally, in parity-check codes, the encoded sequence consists of n binary digits of which only $k < n$ are *information digits* while the remaining $l = n - k$ digits are used for error detection and correction and are called *check digits* or *parity checks*. The example of Table 4-4 is a single-error-detecting code, but, in general, q-fold errors can be detected and/or corrected. As the number of errors to be detected and/or corrected increases, the number l of check digits must increase. Thus, for fixed word length n, the number of information digits $k = n - l$ will decrease as more and more errors are to be detected and/or corrected. Also the total number of words in the code cannot exceed the right side of Eq. (4-93) or the number 2^k.

Parity-check codes are relatively easy to implement. The simple example given of a single-error-detecting code requires only that the number of 1s in each code word be counted. In this light, it is of considerable importance to note that these codes satisfy the noisy-coding theorem. In other words, it is possible to encode a source by parity-check coding for transmission over a BSC at a rate approaching channel capacity and with arbitrarily small probability of error. Then, from Eq. (4-84), we have

$$2^{nH} = 2^k \tag{4-94}$$

or H, the rate of transmission, is given by

$$H = k/n \tag{4-95}$$

As $n \to \infty$, the probability of error $p(e)$ approaches zero. Thus, in a certain sense, it is sufficient to limit a study of error-correcting codes to the parity-check codes.

As an example of a parity-check code, consider the simplest nondegenerate case where l, the number of check digits, is 2 and k, the number of information digits, is 1. This system is capable of single-error detection and correction, as we have already decided from geometric considerations. Since $l + k = 3$, each encoded word will be three digits long. Let us denote this word by $a_1 a_2 a_3$, where each a_i is either 0 or 1. Let a_1 represent the information digit and a_2 and a_3 represent the check digits.

Checking for errors is done by forming two independent equations from the three a_i, each equation being of the form of a modulo-2 sum, i.e., of the form

$$a_i \oplus a_j = \begin{cases} 0 & a_i = a_j \\ 1 & a_i \neq a_j \end{cases}$$

Take the two independent equations to be

$$a_2 \oplus a_3 = 0$$

and $\quad a_1 \oplus a_3 = 0$

for an *even-parity* check. For an odd-parity check, let the right sides of both of these equations be unity. If these two equations are to be satisfied, the only possible code words that can be sent are 000 and 111. The other six words of length 3 violate one or both of the equations.

Now suppose that 000 is sent and 100 is received. A solution of the two independent equations gives, for the received word,

$$\left.\begin{aligned} a_2 \oplus a_3 &= 0 \oplus 0 = 0 \\ a_1 \oplus a_3 &= 1 \oplus 0 = 1 \end{aligned}\right\} \Rightarrow 01$$

The check yields the binary check number 1, indicating that the error is in the first digit a_1, as indeed it is. If 111 is sent and 101 received, then

$$\left.\begin{aligned} a_2 \oplus a_3 &= 0 \oplus 1 = 1 \\ a_1 \oplus a_3 &= 1 \oplus 1 = 0 \end{aligned}\right\} \Rightarrow 10$$

and the binary check number is 10, or 2, indicating that the error is in a_2.

In the general case, a set of l independent linear equations is set up to derive a binary checking number whose value indicates the position of the error in the binary word. If more than one error is to be detected and corrected, the number l of check digits must increase, as discussed previously.

In the example just treated, the l check digits were used only to check the k information digits immediately preceding them. Such a code is called a *block code,* since all the information digits and all the check digits are contained in the block (code word) of length $n = k + l$. In some encoding procedures, the l check digits may also be used to check information digits appearing in preceding words. Such codes are called *convolutional* or *recurrent codes.* A parity-check code (either block or convolutional) where the word length is n and the number of information digits is k is usually called an (n,k) *code.*

15. Other Error-Detecting and Error-Correcting Codes. Unfortunately, a general treatment of error-detecting and error-correcting codes requires that the code structure be cast in a relatively sophisticated mathematical form. The commonest procedure is to identify the code letters with the elements of a finite (algebraic) field. The code words are then taken to form a vector subspace of n-tuples over the field. Such codes are called *linear codes* or, sometimes, *group codes.* Both the block codes and the convolutional codes mentioned in the previous paragraph fall into this category.

An additional constraint often imposed on linear codes is that they be *cyclic.* Let a code word $\bar{a}$ be represented by

$$\bar{a} = (a_0,a_1,a_2,\cdots,a_{n-1})$$

Then the ith *cyclic permutation* a^{-i} is given by $\bar{a}^i = (a_i,a_{i+1},\cdots,a_{n-1},a_0,a_1,\cdots,a_{i-1})$. A linear code is cyclic if, and only if, for every word $\bar{a}$ in the code, there is also a word $\bar{a}^i$ in the code. The permutations need not be distinct and, in fact, generally will not be. The eight code words

0000	0110	1001	1010
0011	1100	0101	1111

constitute a cyclic set. Included in the cyclic codes are some of those most commonly encountered such as the Bose and Ray-Chaudhuri (BCH) codes and shortened Reed-Muller codes. A more detailed discussion is beyond the scope of this treatment, but will be found in the references, Par. 4-41.

CONTINUOUS CHANNELS

16. Continuous-Amplitude Channels. The preceding discussion has concerned discrete message distributions and channels. Further, it has been assumed, either implicitly or explicitly, that the time parameter is discrete, i.e., that a certain number of messages, symbols, code digits, etc., are transmitted per unit time. Thus, we have been concerned with *continuous-amplitude, discrete-time* channels and with messages which can be modeled as *discrete random processes* with *discrete parameter*. There are three other possibilities, depending on whether the process amplitude and the time parameter have discrete or continuous distributions.

In Pars. 4-16 to 4-18 we consider the *continuous-amplitude, discrete-time* channel, where the input messages can be modeled as *continuous random processes* with *discrete parameter*. It will be shown later that continuous-time cases of engineering interest can be treated by techniques which amount to the replacement of the continuous parameter by a discrete parameter. The most straightforward method involves the application of the sampling theorem to band-limited processes. In this case the process is sampled at equispaced intervals of length $1/2W$, where W is the highest frequency of the process. Thus the continuous parameter t is replaced by the discrete parameter $t_k = k/2W, k = \cdots, -1, 0, 1, \cdots$.

Let us restrict our attention for the moment to continuous-amplitude, discrete-time situations. The discrete density $p(x_i)$, $i = 1, 2, \ldots, M$, of the source-message set is replaced by the continuous density $W_x(x)$, where, in general, $-\infty < x < \infty$, although the range of x may be restricted in particular cases. In the same way, other discrete densities are replaced by continuous densities. For example, the destination distribution $p(y_j)$, $j = 1, 2, \ldots, N$, becomes $W_y(y)$, and the joint distribution $p(x_i, y_j)$ will be $W_2(x, y)$.

A reasonable place to begin the study of continuous distribution is with the definition of source entropy as given by Eq. (4-4) for the discrete case:

$$H(X) = -\sum_i p(x_i) \log p(x_i) \tag{4-96}$$

The continuous density $W_x(x)$ may be approximated by

$$p(x_i) \approx W_x(x_i) \Delta x_i \tag{4-97}$$

where $p(x_i)$ is approximately the probability that the continuous random variable x with density $W_x(x)$ lies in an interval Δx_i which includes x_i. In the limiting case as $\Delta x_i \to 0$, this relationship will become exact, and Eq. (4-97) becomes

$$H(X) = \lim_{\Delta x_i \to 0} \left\{ -\sum_i W_x(x_i) \Delta x_i \log[W_x(x_i) \Delta x_i] \right\} \tag{4-98}$$

The logarithm can be expanded to yield

$$\begin{aligned} H(X) = &-\lim_{\Delta x_i \to 0} \sum_i W_x(x_i) \log[W_x(x_i)] \Delta x_i \\ &- \lim_{\Delta x_i \to 0} \sum_i W_x(x_i) \Delta x_i \log \Delta x_i \end{aligned} \tag{4-99}$$

The first term of this expression may be considered as a limiting form of the integral $\int_{-\infty}^{\infty} W_x(x) \log W_x(x) \, dx$, while the second term tends to infinity, e.g., if the intervals Δx_i are equal,

$$\begin{aligned} -\lim_{\Delta x \to 0} \sum_i W_x(x_i) \Delta x \log \Delta x &= -\lim_{\Delta x \to 0} \log \Delta x \int_{-\infty}^{\infty} W_x(x) \, dx \\ &= -\lim_{\Delta x \to 0} \log \Delta x = \infty \end{aligned} \tag{4-100}$$

This approach suggests that the entropy of a continuous distribution $W_x(x)$ might be defined in analogy with the discrete case by the first term of Eq. (4-99)

$$H(X) = -\int_{-\infty}^{\infty} W_x(x) \log W_x(x) \, dx \tag{4-101}$$

This definition is not completely satisfactory for a number of reasons having to do with the properties of this new $H(X)$.

   a. $H(X)$ May Be Negative, Positive, or Zero. In the discrete case, it was shown that

$H(X)$ was nonnegative. This is no longer necessarily true. For example, let $W_x(x)$ be uniformly distributed in the interval $(0, 1/a)$. Then we have

$$H(X) = -\int_0^{1/a} a \log a \, dx = -\log a = \begin{cases} > 0 & a < 1 \\ 0 & a = 1 \\ < 0 & a > 1 \end{cases} \tag{4-102}$$

b. H(X) Depends on the Coordinate System. In the interest of generality, consider the set of random variables $x_i, x_2, \ldots, x_n$ with joint distribution

$$W_{\bar{x}}(x_1, x_2, \ldots, x_n) = W_{\bar{x}}(\bar{x}) \tag{4-103}$$

Consider the entropy

$$H(X) = -\int_{-\infty}^{\infty} \cdots \int_{-\infty}^{\infty} W_{\bar{x}}(\bar{x}) \log W_{\bar{x}}(\bar{x}) \, d\bar{x} \tag{4-104}$$

where $d\bar{x} = dx_1 \, dx_2 \cdots dx_n$. Let us transform to a new coordinate system (set of random variables) $y_1, y_2, \ldots, y_n$ with joint distribution

$$W_{\bar{y}}(y_1, y_2, \ldots, y_n) = W_{\bar{y}}(\bar{y})$$

and entropy $H(\bar{Y})$, where

$$H(\bar{Y}) = -\int_{-\infty}^{\infty} \cdots \int_{-\infty}^{\infty} W_{\bar{y}}(\bar{y}) \log W_{\bar{y}}(\bar{y}) \, d\bar{y} \tag{4-105}$$

The densities $W_{\bar{x}}(\bar{x})$ and $W_{\bar{y}}(\bar{y})$ are related by

$$W_{\bar{y}}(\bar{y}) = W_{\bar{x}}(\bar{x}) J(\bar{x}/\bar{y})$$

where $J(\bar{x}/\bar{y})$ is the Jacobian of the transformation from $\bar{x}$ to $\bar{y}$. This last expression together with

$$d\bar{y} = J(\bar{y}/\bar{x}) \, d\bar{x}$$

can be substituted into Eq. (4-105) to yield

$$H(\bar{Y}) = -\int_{-\infty}^{\infty} \cdots \int_{-\infty}^{\infty} W_{\bar{x}}(\bar{x}) J(\bar{x}/\bar{y}) \log [W_{\bar{x}}(\bar{x}) J(\bar{x}/\bar{y})] J(\bar{y}/\bar{x}) \, d\bar{x}$$

On using the relationship

$$J(\bar{x}/\bar{y}) J(\bar{y}/\bar{x}) = 1$$

and expanding the logarithm, we obtain

$$H(\bar{Y}) = -\int_{-\infty}^{\infty} \cdots \int_{-\infty}^{\infty} W_{\bar{x}}(\bar{x}) \log W_{\bar{x}}(\bar{x}) \, d\bar{x}$$
$$- \int_{-\infty}^{\infty} \cdots \int_{-\infty}^{\infty} W_{\bar{x}}(\bar{x}) \log J(\bar{x}/\bar{y}) \, d\bar{x}$$

or $$H(\bar{Y}) = H(\bar{X}) - E_{\bar{x}}\{\log J(\bar{x}/\bar{y})\} \tag{4-106}$$

where $E_{\bar{x}}\{\cdot\}$ indicates the expectation operation with respect to the $\bar{x}$ distribution. Thus the entropy of a continuous distribution changes with the coordinate system. As an example, consider the linear transformation

$$y_i = \sum_{j=1}^{n} a_{ij} x_j \qquad i = 1, 2, \ldots, n$$

In this case the Jacobian is

$$J(\bar{y}/\bar{x}) = \begin{vmatrix} a_{11} & a_{12} & \cdots & a_{1n} \\ a_{21} & a_{22} & \cdots & a_{2n} \\ \cdots & \cdots & \cdots & \cdots \\ a_{n1} & a_{n2} & \cdots & a_{nn} \end{vmatrix} = \|a_{ij}\|$$

where $\|a_{ij}\|$ is the magnitude of the determinant $|a_{ij}|$.   Now Eq. (4-106) becomes

$$H(\bar{Y}) = H(\bar{X}) + \log\|a_{ij}\|$$

In one dimension, where $y = ax$, this becomes

$$H(Y) = H(X) + \log|a|$$

c. $H(X)$ Is Invariant to Translation.   Consider only the simple translation

$$y = x + b$$

The entropy $H(Y)$ is

$$H(Y) = -\int_{-\infty}^{\infty} W_y(y)\log W_y(y)\,dy = -\int_{-\infty}^{\infty} W_x(y-b)\log W_x(y-b)\,dy$$

After the change of variable $x = y - b$, we have

$$H(Y) = -\int_{-\infty}^{\infty} W_x(x)\log W_x(x)\,dx = H(X)$$

as expected.   Actually this result follows immediately from Eq. (4-106) since the Jacobian of a translation is unity.

It is clear that joint and conditional entropies can be defined in exact analogy to the discrete case discussed in Par. 4-4.   If the joint density $W_2(x,y)$ exists, and if

$$W_x(x) = \int_{-\infty}^{\infty} W_2(x,y)\,dy \qquad \text{and} \qquad W_y(y) = \int_{-\infty}^{\infty} W_2(x,y)\,dx$$

then the joint entropy $H(X,Y)$ is given by

$$H(X,Y) = -\int_{-\infty}^{\infty}\int W_2(x,y)\log W_2(x,y)\,dx\,dy \tag{4-107}$$

and the conditional entropies $H(X/Y)$ and $H(Y/X)$ are

$$H(X/Y) = -\int_{-\infty}^{\infty}\int W_2(x,y)\log \frac{W_2(x,y)}{W_y(y)}\,dx\,dy \tag{4-108}$$

and

$$H(Y/X) = -\int_{-\infty}^{\infty}\int W_2(x,y)\log \frac{W_2(x,y)}{W_x(x)}\,dx\,dy \tag{4-109}$$

The average mutual information follows from Eq. (4-15) and is

$$I(X;Y) = -\int_{-\infty}^{\infty}\int W_2(x,y)\log \frac{W_x(x)W_y(y)}{W_2(x,y)}\,dx\,dy \tag{4-110}$$

Although the entropy of a continuous distribution can be negative, positive, or zero, the average mutual information $I(X;Y)$ is nonnegative as in the discrete case.   Consider two continuous densities $p(x) \geq 0$ and $q(x) \geq 0$, where

$$\int_{-\infty}^{\infty} p(x)\,dx = \int_{-\infty}^{\infty} q(x)\,dx = 1$$

A well-known inequality is

$$\log_a z \geq \frac{1}{\ln a}\left(1 - \frac{1}{z}\right) \qquad z \geq 0$$

It follows that

$$\int_{-\infty}^{\infty} p(x)\log_a \frac{p(x)}{q(x)}\,dx \geq \frac{1}{\ln a}\int_{-\infty}^{\infty} p(x)\left[1 - \frac{q(x)}{p(x)}\right]dx = 0$$

with equality if and only if $p(x) = q(x)$.   It can be seen immediately from Eq. (4-110) that $I(X;Y) \geq 0$ with equality when x and y are statistically independent, i.e., when $W_2(x,y) = W_x(x)W_y(y)$.

17. Maximization of Entropy of Continuous Distributions. The entropy of a discrete distribution is a maximum when the distribution is uniform, i.e., when all outcomes are equally likely. In the continuous case, the entropy depends on the coordinate system, and it is possible to maximize this entropy subject to various constraints on the associated density function. The maximization itself is the so-called isoperimetric problem of the calculus of variations. For our purposes, the procedure is as follows. It is desired to find $y = y(x)$ so that the integral

$$I = \int_a^b F(x,y)\,dx \qquad (4\text{-}111)$$

is an extremum subject to the constraints that

$$\int_a^b F_1(x,y)\,dx = c_1$$

$$\cdots\cdots\cdots \qquad (4\text{-}112)$$

$$\int_a^b F_n(x,y)\,dx = c_n$$

where the c_i are preassigned constants and the F_i are functions determined by the problem. The function y is found by solving the equation

$$\frac{\partial F}{\partial y} + \lambda_1 \frac{\partial F_1}{\partial y} + \cdots + \lambda_n \frac{\partial F_n}{\partial y} = 0 \qquad (4\text{-}113)$$

The λ_i are constants (called *undetermined multipliers*) whose values are found by substituting the y found from Eq. (4-113) into the set of equations given by Eq. (4-112). We consider now several of the commonest and most useful entropy maximizations.

a. The Maximization of $H(X)$ for a Fixed Variance of x. We wish to maximize

$$H(X) = -\int_{-\infty}^{\infty} W_x(x) \log W_x(x)\,dx$$

subject to the constraints that

$$\int_{-\infty}^{\infty} W_x(x)\,dx = 1 \qquad (4\text{-}114)$$

and $$\int_{-\infty}^{\infty} x^2 W_x(x)\,dx = \sigma^2 \qquad (4\text{-}115)$$

Let us form the expression

$$F_0 = F + \lambda_1 F_1 + \lambda_2 F_2 = W_x \log W_x + \lambda_1 W_x + \lambda_2 x^2 W_x$$

where W_x has been used as a simplified notation for $W_x(x)$. Equation (4-113) now becomes

$$\frac{\partial F_0}{\partial y} = 0 = 1 + \log W_x + \lambda_1 + \lambda_2 x^2$$

or $$W_x(x) = e^{-(\lambda_1+1)} e^{-\lambda_2 x^2} \qquad (4\text{-}116)$$

The constants λ_1 and λ_2 can be found by substituting Eq. (4-116) into Eqs. (4-114) and (4-115). The result is

$$W_x(x) = \frac{1}{\sqrt{2\pi}\,\sigma} e^{-x^2/2\sigma^2} \qquad -\infty < x < \infty \qquad (4\text{-}117)$$

Thus, for fixed variance, the normal distribution has the largest entropy. It is clear that

$$\ln W_x(x) = -\tfrac{1}{2} \ln 2\pi\sigma^2 - \frac{x^2}{2\sigma^2}$$

and that the entropy in this case is

$$H(X) = \tfrac{1}{2} \ln 2\pi\sigma^2 + \tfrac{1}{2} \ln e = \tfrac{1}{2} \ln 2\pi e\sigma^2 \qquad (4\text{-}118)$$

This last result will be of considerable use later. Note that the normal distribution need not have a zero mean since entropy is invariant to a translation. For convenience, the natural logarithm has been used, and the units of H are nats.

   b. The Maximization of $H(X)$ for a Limited Peak Value of x. In this case, the single constraint is

$$\int_{-M}^{M} W_x(x)\,dx = 1 \tag{4-119}$$

and $W_x(x)$ is found from the equation

$$\frac{\partial}{\partial W_x}(-W_x \ln W_x + \lambda_1 W_x) = 0$$

$$\text{or} \quad W_x(x) = e^{\lambda_1 - 1} = \text{const}$$

This result may be used in Eq. (4-119) to obtain the uniform distribution

$$W_x(x) = \begin{cases} \dfrac{1}{2M} & |x| < M \\ 0 & |x| > M \end{cases}$$

The associated entropy is

$$H(X) = -\int_{-M}^{M} \frac{1}{2M} \log \frac{1}{2M}\,dx = \log 2M \tag{4-120}$$

   c. The Maximization of $H(X)$ for x Limited to Nonnegative Values and a Given Average Value. The constraints are

$$\int_{0}^{\infty} W_x(x)\,dx = 1 \tag{4-121}$$

$$\text{and} \quad \int_{0}^{\infty} x W_x(x)\,dx = \mu \tag{4-122}$$

and $W_x(x)$ is found from

$$\frac{\partial}{\partial W_x}(-W_x \ln W_x + \lambda_1 W_x + \lambda_2 x W_x) = 0$$

$$\text{or} \; W_x(x) = e^{\lambda_1 - 1} e^{\lambda_2 x}$$

As before, λ_1 and λ_2 can be eliminated through Eqs. (4-121) and (4-122) to yield

$$W_x(x) = \begin{cases} 0 & x < 0 \\ \dfrac{1}{\mu} e^{-(x/\mu)} & x \ge 0 \end{cases}$$

The entropy associated with this distribution is

$$H(X) = \ln \mu + 1 = \ln \mu e \tag{4-123}$$

   18. Gaussian Signals and Channels. Let us assume that the source symbol x and the destination symbol y are jointly gaussian, so that the joint density $W_2(x,y)$ is

$$W_2(x,y) = \frac{1}{2\pi\sigma_x\sigma_y\sqrt{1-\rho^2}} \exp\left\{-\frac{1}{2(1-\rho^2)}\left[\left(\frac{x}{\sigma_x}\right)^2 - 2\rho\frac{xy}{\sigma_x\sigma_y} + \left(\frac{y}{\sigma_y}\right)^2\right]\right\} \tag{4-124}$$

where σ_x^2 and σ_y^2 are the variances of x and y, respectively, and ρ is the correlation coefficient given by

$$\rho = \frac{E\{xy\}}{\sigma_x\sigma_y} \tag{4-125}$$

The univariate densities of x and y are given, of course, by

$$W_x(x) = \frac{1}{\sqrt{2\pi}\,\sigma_x}\exp\left[-\frac{1}{2}\left(\frac{x}{\sigma_x}\right)^2\right] \qquad -\infty < x < \infty \tag{4-126}$$

$$\text{and } W_y(y) = \frac{1}{\sqrt{2\pi}\,\sigma_y}\exp\left[-\frac{1}{2}\left(\frac{y}{\sigma_y}\right)^2\right] \qquad -\infty < y < \infty \tag{4-127}$$

Let us make use of Eq. (4-110) to calculate the average mutual information $I(X;Y)$. We have

$$\frac{W_2(x,y)}{W_x(x)W_y(y)} = \frac{1}{\sqrt{1-\rho^2}}\exp\left\{-\frac{\rho^2}{2(1-\rho^2)}\left[\left(\frac{x}{\sigma_x}\right)^2 - \frac{2xy}{\rho\sigma_x\sigma_y} + \left(\frac{y}{\sigma_y}\right)^2\right]\right\}$$

and

$$I(X;Y) = -\tfrac{1}{2}\ln(1-\rho^2)\int_{-\infty}^{\infty}\int_{-\infty}^{\infty} W_2(x,y)\,dx\,dy$$

$$-\frac{\rho^2}{2(1-\rho^2)}\int_{-\infty}^{\infty}\int_{-\infty}^{\infty} W_2(x,y)\left[\left(\frac{x}{\sigma_x}\right)^2 - \frac{2xy}{\rho\sigma_x\sigma_y} + \left(\frac{y}{\sigma_y}\right)^2\right]dx\,dy \tag{4-128}$$

This expression may be rewritten

$$I(X;Y) = -\tfrac{1}{2}\ln(1-\rho^2) - \frac{\rho^2}{2(1-\rho^2)}(1-2+1)$$

$$\text{or} \qquad I(X;Y) = -\tfrac{1}{2}\ln(1-\rho^2) \tag{4-129}$$

Thus the average mutual information in two jointly gaussian random variables is a function only of the correlation coefficient ρ and varies from zero to infinity.

The noise entropy $H(Y/X)$ can be written

$$H(Y/X) = H(Y) - I(X;Y) = \tfrac{1}{2}\ln 2\pi e\sigma_y^2(1-\rho^2) \tag{4-130}$$

Suppose that x and y are jointly gaussian as a result of independent zero-mean gaussian noise n being added in the channel to the gaussian input x, so that

$$y = x + n \tag{4-131}$$

In this case the correlation coefficient ρ becomes

$$\rho = \frac{E\{x^2 + nx\}}{\sigma_x\sigma_y} = \frac{\sigma_x^2}{\sigma_x\sigma_y} = \frac{\sigma_x}{\sigma_y} \tag{4-132}$$

and the noise entropy is

$$H(Y/X) = \tfrac{1}{2}\ln 2\pi e\sigma_n^2 \tag{4-133}$$

where σ_n^2 is the noise variance given by

$$\sigma_n^2 = E\{n^2\} = \sigma_y^2 - \sigma_x^2 \tag{4-134}$$

In this situation, Eq. (4-129) can be rewritten as

$$I(X;Y) = \tfrac{1}{2}\ln\left(1 + \frac{\sigma_x^2}{\sigma_n^2}\right) \tag{4-135}$$

It is conventional to define the signal power as $S = \sigma_x^2$ and the noise power as $N = \sigma_n^2$ and to rewrite this last expression as

$$I(X;Y) = \tfrac{1}{2}\ln\left(1 + \frac{S}{N}\right) \tag{4-136}$$

where S/N is the signal-to-noise power ratio.

In analogy with Par **4-10**, channel capacity C for the continuous-amplitude, discrete-time channel is

$$C = \max_{W_x(x)} I(X;Y) = \max_{W_x(x)} [H(Y) - H(Y/X)] \qquad (4\text{-}137)$$

Suppose the channel consists of an additive noise which is a sequence of independent gaussian random variables n each with zero mean and variance σ_n^2. In this case the conditional probability $W(y/x)$ at each time instant is normal with variance σ_n^2 and mean equal to the particular realization of x. The noise entropy $H(Y/X)$ is given by Eq. (4-133), and Eq. (4-137) becomes

$$C = \max_{W_x(x)} [H(Y)] - \tfrac{1}{2} \ln 2\pi e \sigma_n^2 \qquad (4\text{-}138)$$

If the input power is fixed at σ_x^2, then the output power is fixed at $\sigma_y^2 = \sigma_x^2 + \sigma_n^2$ and $H(Y)$ is a maximum if $y = x + n$ is a sequence of independent gaussian random variables. The value of $H(Y)$ is

$$H(Y) = \tfrac{1}{2} \ln 2\pi e(\sigma_x^2 + \sigma_n^2)$$

and the channel capacity becomes

$$C = \frac{1}{2} \ln\left(1 + \frac{\sigma_x^2}{\sigma_n^2}\right) = \frac{1}{2} \ln\left(1 + \frac{S}{N}\right) \qquad (4\text{-}139)$$

where S/N is the signal-to-noise power ratio. Note that the input x is a sequence of independent gaussian random variables and this last equation is identical to Eq. (4-136). Thus, for additive independent gaussian noise and an input power limitation, the discrete-time continuous-amplitude channel has a capacity given by Eq. (4-139). This capacity is realized when the input is an independent sequence of independent, identically distributed gaussian random variables.

BAND-LIMITED CHANNELS

19. Band-Limited Transmission. In this section, messages will be considered which can be modeled as continuous random processes $x(t)$ with continuous parameter t. The channels which transmit these messages will be called *amplitude-continuous, time-continuous* channels. Specifically attention will be restricted to signals (random processes) $x(t)$ which are *strictly band-limited*.

Suppose a given arbitrary (deterministic) signal $f(t)$ is available for all time. Is it necessary to know the amplitude of the signal for every value of time in order to characterize it uniquely? In other words, can $f(t)$ be represented (and reconstructed) from some set of *sample values* or *samples* $\ldots, f(t_1), f(t_0), f(t_1), \ldots$? Surprisingly enough, it turns out that, under certain fairly reasonable conditions, a signal can be represented exactly by samples spaced relatively far apart. The reasonable conditions are that the signal be strictly band-limited.

A (real) signal $f(t)$ will be called *strictly band-limited* $(-2\pi W, 2\pi W)$ if its Fourier transform $F(\omega)$ has the property

$$F(\omega) = 0 \qquad |\omega| > 2\pi W \qquad (4\text{-}140)$$

It is clear that this spectrum could be extended into a periodic frequency function with period $4\pi W$. In other words, a new function $F_e(\omega)$ can be defined by

$$F_e(\omega) = \sum_{n=-\infty}^{\infty} F(\omega + n4\pi W) \qquad (4\text{-}141)$$

This function is periodic with period $4\pi W$.

For reasonably well-behaved $F(\omega)$, the periodic function $F_e(\omega)$ can be expanded in a

Fourier series with period $4\pi W$, and, in the interval $-2\pi W \leq \omega \leq 2\pi W$, this Fourier series will converge to $F(\omega)$; that is,

$$F(\omega) = \sum_{k=-\infty}^{\infty} F_k e^{-jk\omega/2W} \qquad |\omega| \leq 2\pi W \qquad (4\text{-}142)$$

where $j = \sqrt{-1}$ and F_k is the Fourier coefficient given by

$$F_k = \frac{1}{4\pi W} \int_{-2\pi W}^{2\pi W} F(\omega) e^{jk\omega/2W} d\omega \qquad (4\text{-}143)$$

Since $F(\omega)$ is band-limited as described by Eq. (4-140), its inverse Fourier transform is

$$f(t) = \frac{1}{2\pi} \int_{-2\pi W}^{2\pi W} F(\omega) e^{j\omega t} d\omega \qquad (4\text{-}144)$$

If the *Nyquist instants* are defined as the set of times

$$\{t_n\} = \left\{ t_n / t_n = \frac{n}{2W} \qquad n = \cdots, -1, 0, 1, \ldots \right\} \qquad (4\text{-}145)$$

then it is clear that $f(t_n)$ is given from Eq. (4-144) as

$$f(t_n) = f\left(\frac{n}{2W}\right) = \frac{1}{2\pi} \int_{-2\pi W}^{2\pi W} F(\omega) e^{jn\omega/2W} d\omega \qquad (4\text{-}146)$$

A comparison of this last equation with Eq. (4-143) shows that the Fourier coefficient F_k is related to the sample value $f(k/2W)$ by

$$F_k = \frac{1}{2W} f\left(\frac{k}{2W}\right) \qquad (4\text{-}147)$$

If the sample values $f(n/2W)$ are given for all time, then the Fourier series

$$F(\omega) = \frac{1}{2W} \sum_{k=-\infty}^{\infty} f\left(\frac{k}{2W}\right) e^{-jk\omega/2W} \qquad |\omega| \leq 2\pi W \qquad (4\text{-}148)$$

determines $F(\omega)$ exactly and hence $f(t)$ through the inverse Fourier transform given by Eq. (4-144). This completes the proof of the existence of a sampling theorem: a function $f(t)$, strictly band-limited $(-2\pi W, 2\pi W)$ rad/s, is uniquely and exactly determined by its sample values spaced $1/2W$ apart throughout the time domain. Of course, there are an infinite number of such samples.

The reconstruction of $f(t)$ from its sample values is obtained on substituting Eq. (4-148) into Eq. (4-144):

$$f(t) = \frac{1}{2\pi} \int_{-2\pi W}^{2\pi W} \frac{1}{2W} \sum_{k=-\infty}^{\infty} f\left(\frac{k}{2W}\right) e^{-jk\omega/2W} e^{j\omega t} d\omega \qquad (4\text{-}149)$$

20. Sampling Theorem. We may interchange the order of summation and integration and evaluate the integral to obtain the *sampling representation*

$$f(t) = \sum_{k=-\infty}^{\infty} f\left(\frac{k}{2W}\right) \frac{\sin(2\pi Wt - k\pi)}{2\pi Wt - k\pi} \qquad (4\text{-}150)$$

This expression is sometimes called the *Cardinal series* or *Shannon's sampling theorem.* This expression, together with Eq. (4-146), relates the discrete time domain $\{k/2W\}$ with sample values $f(k/2W)$ to the continuous time domain $\{t\}$ of the function $f(t)$.

The interpolation function

$$k(t) = \frac{\sin 2\pi Wt}{2\pi Wt} \qquad (4\text{-}151)$$

has a Fourier transform $K(\omega)$ given by

$$K(\omega) = \begin{cases} \dfrac{1}{4\pi W} & |\omega| < 2\pi W \\ 0 & |\omega| > 2\pi W \end{cases} \tag{4-152}$$

Also the shifted function $k(t - k/2W)$ has the transform

$$\mathcal{F}\left\{ k\left(t - \frac{k}{2W}\right) \right\} = K(\omega) e^{j\omega k/2W} \tag{4-153}$$

Therefore, each term on the right side of Eq. (4-150) is a time function which is strictly bandlimited $(-2\pi W, 2\pi W)$, Note also that

$$k\left(t - \frac{k}{2W}\right) = \frac{\sin(2Wt - k\pi)}{2\pi Wt - k\pi} = \begin{cases} 1 & t = t_k = \dfrac{k\pi}{2W} \\ 0 & t = t_n, \quad n \neq k \end{cases} \tag{4-154}$$

Thus this sampling function $k(t - k/2W)$ is zero at all Nyquist instants except t_k, where it equals unity.

Suppose that a function $h(t)$ is not strictly bandlimited to at least $(-2\pi W, 2\pi W)$ rad/s and an attempt is made to reconstruct the function using Eq. (4-150) with sample values spaced $1/2W$ s apart. It is apparent that the reconstructed signal [which is strictly bandlimited $(-2\pi W, 2\pi W)$, as already mentioned] will differ from the original. Moreover a given set of sample values $\{f(k/2W)\}$ could have been obtained from a whole class of different signals. Thus it should be emphasized that the reconstruction of Eq. (4-150) is unambiguous only for signals strictly bandlimited to at least $(-2\pi W, 2\pi W)$ rad/s. The set of different possible signals with the same set of sample values $\{f(k/2W)\}$ are called the *aliases* of the bandlimited signal $f(t)$.

Let us now consider a signal (random process) $x(t)$ with *autocorrelation function* given by

$$R_x(\tau) = E\{x(t)x(t + \tau)\} \tag{4-155}$$

and *power spectral density*

$$\varphi_x(\omega) = \int_{-\infty}^{\infty} R_x(\tau) e^{-j\omega\tau} \, d\tau \tag{4-156}$$

which is just the Fourier transform of $R_x(\tau)$. The process will be assumed to have zero mean and to be strictly *bandlimited* $(-2\pi W, 2\pi W)$ in the sense that the power spectral density $\varphi_x(\omega)$ vanishes outside this interval; i.e.,

$$\varphi_x(\omega) = 0 \qquad |\omega| > 2\pi W \tag{4-157}$$

It has just been shown that a deterministic signal $f(t)$ bandlimited $(-2\pi W, 2\pi W)$ admits the *sampling representation* of Eq. (4-150). It can also be shown that the random process $x(t)$ admits the same expansion; i.e.,

$$x(t) = \sum_{k=-\infty}^{\infty} x\left(\frac{k}{2W}\right) \frac{\sin(2\pi Wt - k\pi)}{2\pi Wt - k\pi} \tag{4-158}$$

The right side of this expression is a random variable for each value of t and converges mean square to the process $x(t)$ for each t. The proof is straightforward. Since $\varphi_x(\omega)$ vanishes outside $(-2\pi W, 2\pi W)$, its Fourier transform $R_x(\tau)$ has a sampling representation

$$R_x(\tau) = \sum_{k=-\infty}^{\infty} R_x\left(\frac{k}{2W}\right) \frac{\sin(2\pi Wt - k\pi)}{2\pi Wt - k\pi} \tag{4-159}$$

We define a partial sum $x_N(t)$ by

$$x_N(t) = \sum_{k=-N}^{N} x\left(\frac{k}{2W}\right) \frac{\sin(2\pi Wt - k\pi)}{2\pi Wt - k\pi}$$

Then, after some manipulation, we find that

$$\lim_{N\to\infty} E\{|x(t) - x_N(t)|^2\} = R_x(0) - 2R_x(0) + R_x(0) = 0$$

and $x_N(t)$ converges mean square to $x(t)$. Thus the process $x(t)$ with continuous time parameter t may be represented by the process $x(k/2W)$, $k = \cdots, -2, -1, 0, 1, 2, \ldots$, with discrete time parameter $t_k = k/2W$. For bandlimited signals or channels it is sufficient, therefore, to consider the discrete-time case and to relate the results to continuous time through Eq. (4-158).

Suppose the continuous-time process $x(t)$ has a spectrum $\varphi_x(\omega)$ which is *flat and band-limited* so that

$$\varphi_x(\omega) = \begin{cases} N_0 & |\omega| \le \omega_b \\ 0 & |\omega| > \omega_b \end{cases} \tag{4-160}$$

Then the autocorrelation function can be found as the inverse Fourier transform of $\varphi_x(\omega)$, viz.,

$$R_x(\tau) = \frac{1}{2\pi} \int_{-\infty}^{\infty} \varphi_x(\omega) e^{j\omega\tau} d\omega \tag{4-161}$$

In this special case, Eq. (4-161) becomes

$$R_x(\tau) = 2N_0 W \frac{\sin 2\pi W\tau}{2\pi W\tau} \tag{4-162}$$

This function passes through zero at intervals of $1/2W$ so that

$$R_x\left(\frac{k}{2W}\right) = 0 \quad k = \cdots, -2, -1, 1, 2, \ldots \tag{4-163}$$

Thus, samples spaced $k/2W$ apart are *uncorrelated if the power spectral density is flat and band-limited* $(-2\pi W, 2\pi W)$. *If the process is gaussian, the samples are independent.* This implies that continuous-time band-limited $(-2\pi W, 2\pi W)$ gaussian channels, where the noise has a flat spectrum, have a capacity C given by Eq. (4-139) as

$$C = \tfrac{1}{2} \ln\left(1 + \frac{S}{N}\right) \quad \text{(nats/sample)} \tag{4-164}$$

Here N is the variance of the additive, flat, bandlimited gaussian noise and S is $R_x(0)$, the fixed variance of the input signal. The units of Eq. (4-164) are on a per sample basis. Since there are $2W$ samples per unit time, the capacity C per unit time can be written as

$$C' = W \ln\left(1 + \frac{S}{N}\right) \quad \text{(nats/s)} \tag{4-165}$$

The ideas developed thus far in this section have been somewhat abstract notions involving information sources and channels, channel capacity, and the various coding theorems. We now look more closely at conventional channels. Many aspects of these topics fall into the area often called *modulation theory*.

MODULATION

21. Modulation Theory. As discussed in Pars 4-1 to 4-3 and shown in Fig. 4-1, the central problem in most communication systems is the transfer of information originating in some source to a destination by means of a channel. It will be convenient in this section to call the sent message or intelligence $a(t)$ and to denote the received message by $a^*(t)$, a distorted or corrupted version of $a(t)$.

The message signals used in communication and control systems are usually limited in frequency range to some maximum frequency $f_m = \omega_m/2\pi$ Hz. This frequency is typically in the range of a few hertz for control systems and moves upward to a few megahertz for television video signals. In addition the bandwidth of the signal is often of the order of this maximum frequency so that the signal spectrum is approximately low-pass in character.

Such signals are often called *video signals* or *baseband signals*. It frequently happens that the transmission of such a spectrum through a given communication channel is inefficient or impossible. In this light, the problem may be looked upon as the one shown in Fig. 4-2, where an encoder E has been added between the source and the channel. However, in this case, the encoder acts to *modulate* the signal $a(t)$, producing at its output the *modulated wave* or signal $m(t)$.

Modulation may be defined as the modification of one signal, called the *carrier*, by another, called the *modulating signal*. The result of the modulation process is a modulated wave or signal. In most cases a frequency shift is one of the results. There are a number of reasons for producing modulated waves, of which the major ones are given.

  a. Frequency Translation for Efficient Antenna Design. It may be necessary to transmit the modulating signal through space as electromagnetic radiation. If the antenna used is to radiate an appreciable amount of power, it must be large compared to the signal wavelength. Thus translation to higher frequencies (and hence to smaller wavelengths) will permit antenna structures of reasonable size and cost at both the transmitter and receiver.

  b. Frequency Translation for Ease of Signal Processing. It may be easier to amplify and/or shape a signal in one frequency range than in another. For example, a dc signal may be converted to ac, amplified, and converted back again.

  c. Frequency Translation to Assigned Location. A signal may be translated to an assigned frequency band for transmission or radiation, e.g., in commercial radio broadcasting.

  d. Changing Bandwidth. The bandwidth of the original message signal may be increased or decreased by the modulation process. In general, decreased bandwidth will result in channel economies at the cost of fidelity. On the other hand, increased bandwidth will be accompanied by increased immunity to channel disturbances, as in wideband frequency modulation, for example.

  e. Multiplexing. It may be necessary or desirable to transmit several signals occupying the same frequency range or the same time range over a single channel. Various modulation techniques allow the signals to share the same channel and yet be recovered separately. Such techniques are given the generic name of *multiplexing*. As discussed later, multiplexing is possible in either the frequency domain (frequency-domain multiplexing FDM) or in the time domain (time-domain multiplexing TDM). As a simple example, the signals may be translated in frequency so that they occupy separate and distinct frequency ranges as mentioned in item *b* above.

Thus it can be seen that, in a general way, the process of modulation can be considered as a form of encoding used to match the message signal arising from the information source to the communication channel. At the same time it is generally true that the channel itself has certain undesirable characteristics resulting in distortion of the signal during transmission. A part of such distortion can frequently be accounted for by postulating noise disturbances in the channel. There noises may be additive and may also affect the modulated wave in a more complicated fashion, although it is usually sufficient (and much simpler) to assume additive noise only. Also, the received signal must be decoded (demodulated) to recover the original signal.

In view of this discussion, it is convenient to change the block diagram of Fig. 4-2 to that shown in Fig. 4-6. The waveform received at the demodulator (receiver) will be denoted by $r(t)$, where

$$r(t) = m^*[t, a(t), p(t)] + n(t) \qquad (4\text{-}166)$$

where $a(t)$ is the original message signal, $m[t,a(t)]$ is the modulated wave, $m^*[t,a(t),p(t)]$ is a corrupted version of $m[t,a(t)]$, and $p(t)$ and $n(t)$ are noises whose characteristics depend on the channel. Unless it is absolutely necessary for an accurate characterization of the channel, we will assume that $p(t) \equiv 0$ to avoid the otherwise complicated analysis that results.

The aim is to find modulators M and demodulators M^{-1} which make $a^*(t)$ a "good" estimate of the message signal $a(t)$. It should be emphasized that M^{-1} is not uniquely specified by M; for example, it is not intended to imply that $MM^{-1} = 1$. The form of the demodulator, for a given modulator, will depend on the characteristics of the message $a(t)$ and the channel as well as on the criterion of "goodness of estimation" used.

We now take up a study of the various forms of modulation and demodulation, their principal characteristics, their behavior in conjunction with noisy channels, and their advantages and disadvantages. We begin with some preliminary material on signals and their properties.

22. Elements of Signal Theory. The real time function $f(t)$ and its Fourier transform form a Fourier transform pair given by

$$F(\omega) = \int_{-\infty}^{\infty} f(t)e^{-j\omega t}\, dt \tag{4-167}$$

$$\text{and} \quad f(t) = \frac{1}{2\pi}\int_{-\infty}^{\infty} F(\omega)e^{j\omega t}\, d\omega \tag{4-168}$$

It follows directly from Eq. (4-167) that the transform $F(\omega)$ of a real time function has a real part which is even and an imaginary part which is odd.

Consider the function $f(t)$ shown in Fig 4-7a. This might be a pulsed signal or the

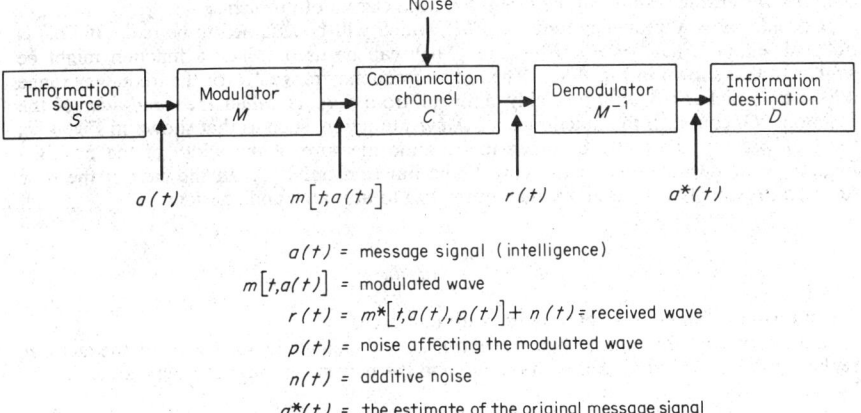

$a(t) =$ message signal (intelligence)

$m[t, a(t)] =$ modulated wave

$r(t) = m*[t, a(t), p(t)] + n(t) =$ received wave

$p(t) =$ noise affecting the modulated wave

$n(t) =$ additive noise

$a*(t) =$ the estimate of the original message signal

Fig. 4-6. Communication system involving modulation and demodulation.

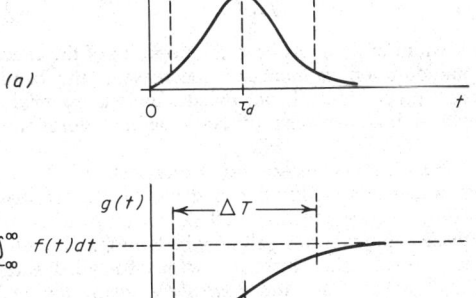

$$g(t) = \int_{-\infty}^{t} f(\tau)\, d\tau$$

Fig. 4-7. Duration and delay: (a) typical pulse; (b) integral of pulse.

impulse response of a linear system, for example. The time ΔT over which $f(t)$ is appreciably different from zero is called the *duration* of $f(t)$, and some measure, such as τ_d, of the center of the pulse is called the *delay* of $f(t)$. In system terms, the quantity ΔT is the system *response time* or *rise time*, and τ_d is the system delay. The integral of $f(t)$, shown in Fig. 4-7b, corresponds to the step-function response of a system with impulse response $f(t)$.

If the function $f(t)$ of Fig. 4-7 is nonnegative, the new function

$$\frac{f(t)}{\int_{-\infty}^{\infty} f(t)\,dt}$$

is nonnegative with unit area. We now seek measures of duration and delay which are both meaningful in terms of communication problems and also mathematically tractable. It will be clear that some of the results we obtain will not be universally applicable and, in particular, must be used with care when the function $f(t)$ can be negative for some values of t. However, the results will be useful for wide classes of problems.

Consider now a frequency function $F(\omega)$, which will be assumed to be real. If $F(\omega)$ is not real, either $|F(\omega)|^2 = F(\omega)F(-\omega)$ or $|F(\omega)|$ can be used. Such a function might be similar to that shown in Fig. 4-8a. The radian frequency range $\Delta\Omega$ (or the frequency range ΔF) over which $F(\omega)$ is appreciably different from zero is called the *bandwidth* of the function. Of course, if the function is a *bandpass* function, such as that shown in Fig. 4-8b, the bandwidth will usually be taken to be some measure of the width of the positive-frequency (or negative-frequency) part of the function only. As in the case of the time function previously discussed, we may normalize to unit area and consider

$$\frac{F(\omega)}{\int_{-\infty}^{\infty} F(\omega)\,d\omega}$$

Again this new function is nonnegative with unit area.

Let us consider the Fourier pair $f(t)$ and $F(\omega)$ and change the time scale by the factor a, replacing $f(t)$ by $af(at)$ so that both the old and the new signal have the same area, i.e.,

$$\int_{-\infty}^{\infty} f(t)\,dt = \int_{-\infty}^{\infty} af(at)\,dt \qquad (4\text{-}169)$$

For $a < 1$, the new signal $af(at)$ is stretched in time and reduced in height; its "duration" has been increased. For $a > 1$, $af(at)$ has been compressed in time and increased in height; its "duration" has been decreased. The transform of this new function is

$$\int_{-\infty}^{\infty} af(at)e^{-j\omega t}\,dt = \int_{-\infty}^{\infty} f(x)e^{-j(\omega/a)x}\,dx = F\left(\frac{\omega}{a}\right) \qquad (4\text{-}170)$$

The effect on the bandwidth of $F(\omega)$ has been the opposite of the effect on the duration of $f(t)$. When the signal duration is increased (decreased), the bandwidth is decreased (increased) in the same proportion. From the discussion, we might suspect that more fundamental relationships hold between properly defined durations and bandwidth of signals.

23. Duration and Bandwidth—Uncertainty Relationships. It is apparent from the discussion above that treatments of duration and bandwidth are mathematically similar although one is defined in the time domain and the other in the frequency domain. Several specific measures of these two quantities will now be found, and it will be shown that they are intimately related to each other through various so-called *uncertainty relationships*. The term "uncertainty" arises from the *Heisenberg uncertainty principle* of quantum mechanics, which states that it is not possible to determine simultaneously and exactly the position and momentum coordinates of a particle. More specifically, if Δx and Δp are the uncertainties in position and momentum, then

$$\Delta x\,\Delta p \geq h \qquad (4\text{-}171)$$

where h is a constant. A number of inequalities of the form of Eq. (4-171) can be developed relating the duration ΔT of a signal to its (radian) bandwidth $\Delta\Omega$. The value of the constant h will depend on the definitions of duration and bandwidth.

a. Equivalent Rectangular Bandwidth $\Delta\Omega_1$ and Duration ΔT_1. The *equivalent rectangular bandwidth* $\Delta\Omega_1$ of a frequency function $F(\omega)$ is defined as

$$\Delta\Omega_1 = \frac{\int_{-\infty}^{\infty} F(\omega)\,d\omega}{F(\omega_0)} \tag{4-172}$$

where ω_0 is some characteristic center frequency of the function $F(\omega)$. It is clear from this definition that the original function $F(\omega)$ has been replaced by a rectangular function of equal area, width $\Delta\Omega_1$, and height $F(\omega_0)$. For the low-pass case ($\omega_0 \equiv 0$), it follows from Eqs. (4-167) and (4-168) that Eq. (4-172) can be rewritten

$$\Delta\Omega_1 = \frac{2\pi f(0)}{\int_{-\infty}^{\infty} f(t)\,dt} \tag{4-173}$$

where $f(t)$ is the time function which is the inverse Fourier transform of $F(\omega)$.

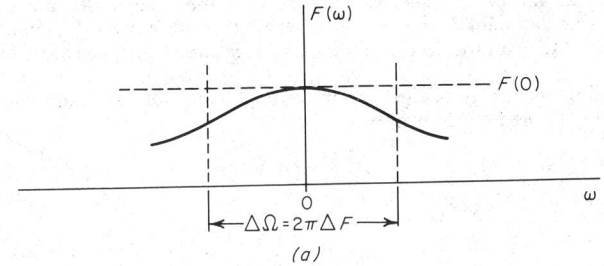

(a)

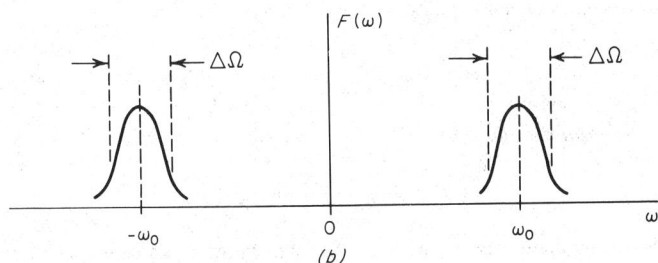

(b)

Fig. 4-8. Illustrations of bandwidth: (a) low-pass; (b) bandpass.

The same procedure may be followed in the time domain and the *equivalent rectangular duration* ΔT_1 of the signal $f(t)$ may be defined by

$$\Delta T_1 = \frac{\int_{-\infty}^{\infty} f(t)\,dt}{f(t_0)} \tag{4-174}$$

where t_0 is some characteristic time denoting the center of the pulse. For the case where $t_0 \equiv 0$, it is clear, then, from Eq. (4-173) and (4-174) that equivalent rectangular duration and bandwidth are connected by the uncertainty relationship

$$\Delta T_1 \Delta\Omega_1 = 2\pi \tag{4-175}$$

b. Second-Moment Bandwidth $\Delta\Omega_2$ and Duration ΔT_2. An alternative uncertainty relationship is based on the second-moment properties of the Fourier pair $F(\omega)$ and $f(t)$. The total energy $\mathcal{E}$ of the signal $f(t)$ may be defined by

$$\mathcal{E} = \int_{-\infty}^{\infty} f^2(t)\,dt = \frac{1}{2\pi}\int_{-\infty}^{\infty} |F(\omega)|^2\,d\omega \tag{4-176}$$

For an arbitrary real nonnegative function $g(x)$, the *mean* $\bar{x}$ of the function is given by

$$\bar{x} = \frac{\int_{-\infty}^{\infty} x g(x)\,dx}{\int_{-\infty}^{\infty} g(x)\,dx} \tag{4-177}$$

This quantity is just the center of gravity on the x axis of the mass distribution given by the function $g(x)$. If this function is pulselike, then the mean $\bar{x}$ of x is a measure of the pulse's location or point of concentration on the x axis. The *centered second moment*, or *variance*, $(\Delta x)^2$ of x is given by

$$(\Delta x)^2 = \frac{\int_{-\infty}^{\infty} (x - \bar{x})^2 g(x)\,dx}{\int_{-\infty}^{\infty} g(x)\,dx} \tag{4-178}$$

This quantity is a measure of the dispersion of the function $g(x)$ about the mean $\bar{x}$.

A small value for $(\Delta x)^2$ indicates that most of the area under the $g(x)$ curve is concentrated near the point $x = \bar{x}$. It is apparent that, at least for functions $g(x)$ which are nonnegative, the quantity $(\Delta x)^2$ can serve as a measure of the bandwidth or duration of $g(x)$.

This second moment may be used as a measure of bandwidth and duration by replacing the function $g(\cdot)$ in the time domain by

$$g(\cdot) \approx |f(t)|^2$$

and in the frequency domain by

$$g(\cdot) \approx \frac{1}{2\pi}|F(\omega)|^2$$

Thus a *second-moment bandwidth* $\Delta\Omega_2$ may be defined by

$$\mathcal{E}(\Delta\Omega_2)^2 = \frac{1}{2\pi}\int_{-\infty}^{\infty} (\omega - \bar{\omega})^2 |F(\omega)|^2\,d\omega \tag{4-179}$$

and a *second-moment duration* ΔT_2 by

$$\mathcal{E}(\Delta T_2)^2 = \int_{-\infty}^{\infty} (t - \bar{t})^2 |f(t)|^2\,dt \tag{4-180}$$

It is obvious that, without loss of generality, the total energy $\mathcal{E}$ can be set equal to unity; i.e.,

$$\mathcal{E} = \frac{1}{2\pi}\int_{-\infty}^{\infty} |F(\omega)|^2\,d\omega = \int_{-\infty}^{\infty} |f(t)|^2\,dt = 1$$

provided all results are divided by the actual value of $\mathcal{E}$. In other words, replace $g(\cdot)$ by $g(\cdot)/\mathcal{E}$.

In order to simplify Eq. (4-179) and (4-180) let us note that any t translation of $f(t)$, a linear change of variable, will not affect the value of $|F(\omega)|^2$ in Eq. (4-179). Similarly any ω translation in Eq. (4-179) will not affect the value of $|f(t)|^2$ in Eq. (4-180). Thus it can be assumed without loss of generality that $\bar{t}$ and $\bar{\omega}$ are zero. The derivative of $f(t)$ can be written from Eq. (4-168) as

$$f'(t) = \frac{1}{2\pi}\int_{-\infty}^{\infty} j\omega F(\omega)e^{j\omega t}\,d\omega \tag{4-181}$$

and we can form

$$\int_{-\infty}^{\infty} f'(t) f^{*\prime}(t)\,dt = \frac{1}{2\pi}\int_{-\infty}^{\infty} \omega^2 |F(\omega)|^2\,d\omega \tag{4-182}$$

From these results and the assumption that $\mathcal{E} = 1$, the bandwidth product $\Delta\Omega_2 \Delta T_2$ can be written as

$$(\Delta\Omega_2)^2(\Delta T_2)^2 = \int_{-\infty}^{\infty} tf(t)tf^*(t)\,dt \int_{-\infty}^{\infty} f'(t)f^{*\prime}(t)\,dt \qquad (4\text{-}183)$$

We now apply a form of the Schwarz inequality and, after some manipulation, obtain

$$\Delta\Omega_2 \Delta T_2 \geq \tfrac{1}{2} \qquad (4\text{-}184)$$

This expression is a second uncertainty relationship connecting the bandwidth and duration of a signal. Many other such inequalities can be obtained.

24. Continuous Modulation. As discussed in Par. 4-21, *modulation* may be defined as the modification of one signal, called the *carrier*, by another, called the *modulation*, *modulating signal*, or *message signal*. In this section we will be concerned with situations where the carrier and the modulation are both continuous functions of time. Later we will treat the cases where the carrier and/or the modulation have the form of pulse trains.

For our analysis, Fig. 4-6 may be modified to the system shown in Fig. 4-9. As shown in the latter, the message is sent through a modulator (or transmitter) to produce the modulated continuous signal $m[t,a(t)]$. This waveform is corrupted by additive noise $n(t)$ in transmission so that the received (continuous) waveform $r(t)$ may be written

$$r(t) = m[t,a(t)] + n(t) \qquad (4\text{-}185)$$

The purpose of the demodulator (or receiver) is to produce some best estimate $a^*(t)$ of the original message signal $a(t)$. As pointed out in Par. 4-21, a more general model of the transmission medium would allow corruption of the modulated waveform itself so that the received signal was of the form of Eq. (4-166). For example, in ionospheric radio-wave propagation, multiplicative disturbances might result due to multipath transmission or fading so that the received signal was of the form

$$r_1(t) = p(t)m[t,a(t)] + n(t) \qquad (4\text{-}186)$$

where both $p(t)$ and $n(t)$ are noises. However, we shall not treat such systems, confining ourselves to the simpler additive-noise model of Fig. 4-9.

25. Linear, or Amplitude, Modulation. In a general way, *linear (or amplitude) modulation* (AM) can be defined as a system where a *carrier wave* $c(t)$ has its amplitude varied linearly by some *message signal* $a(t)$. More precisely, a waveform is linearly modulated (or amplitude-modulated) by a given message $a(t)$ if the partial derivative of that waveform with respect to $a(t)$ is independent of $a(t)$. In other words, the modulated signal $m[t,a(t)]$ can be written in the form

$$m[t,a(t)] = a(t)c(t) + d(t) \qquad (4\text{-}187)$$

where $c(t)$ and $d(t)$ are independent of $a(t)$. Now we have

$$\frac{\partial m[t,a(t)]}{\partial a(t)} = c(t) \qquad (4\text{-}188)$$

and $c(t)$ will be called the *carrier*. In most of the cases we will treat, the waveform $d(t)$ will either be zero or will be linearly related to $c(t)$. It will be more convenient, therefore, to write Eq. (4-187) as

$$m[t,a(t)] = b(t)c(t) \qquad (4\text{-}189)$$

where $b(t)$ will be either

$$b_1(t) \equiv 1 + a(t) \qquad (4\text{-}190)$$

$$\text{or} \quad b_2(t) \equiv a(t) \qquad (4\text{-}191)$$

Also, at present it will be sufficient to allow the carrier $c(t)$ to be of the form

$$c(t) = C \cos(\omega_0 t + \theta) \qquad (4\text{-}192)$$

where C and ω_0 are constants and θ is either a constant or a random variable uniformly distributed on $(0, 2\pi)$.

Whenever Eq. (4-190) applies, it will be convenient to assume that $b_1(t)$ is nearly always nonnegative. This implies that if $a(t)$ is a deterministic signal, then

$$a(t) \geq -1 \tag{4-193}$$

It also implies that if $a(t)$ is a random process, the probability that $a(t)$ is less than -1 in any finite interval $(-T, T)$ is arbitrarily small; i.e.,

$$p[a(t) < -1] \leq \epsilon \ll 1 \quad -T \leq t \leq T \tag{4-194}$$

The purpose of these restrictions on $b_1(t)$ is to ensure that the carrier is not overmodulated and that the message signal $a(t)$ is easily recovered by simple receivers.

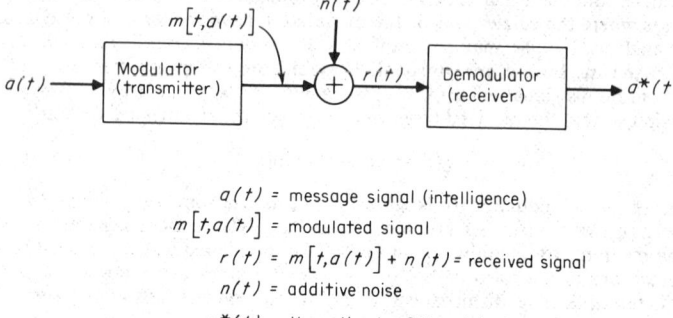

$$a(t) = \text{message signal (intelligence)}$$
$$m[t, a(t)] = \text{modulated signal}$$
$$r(t) = m[t, a(t)] + n(t) = \text{received signal}$$
$$n(t) = \text{additive noise}$$
$$a^*(t) = \text{the estimate of the original message signal}$$

Fig. 4-9. Communication-system model for continuous modulation and demodulation.

We may take the Fourier transform of both sides of Eq. (4-189) to obtain an expression for the frequency spectrum $M(\omega)$ of the general form of the linear modulated waveform $m[t, a(t)]$:

$$M(\omega) = \frac{1}{2\pi} \int_{-\infty}^{\infty} B(\omega - \nu) C(\nu) \, d\nu \tag{4-195}$$

where $B(\omega)$ and $C(\omega)$ are the Fourier transforms of $b(t)$ and $c(t)$ respectively. If the carrier $c(t)$ is the sinusoid of Eq. (4-192), then

$$C(\omega) = \pi C e^{j\theta} \delta(\omega - \omega_0) + \pi C e^{-j\theta} \delta(\omega + \omega_0) \tag{4-196}$$

and Eq. (4-195) becomes

$$M(\omega) = \frac{C}{2} e^{j\theta} B(\omega - \omega_0) + \frac{C}{2} e^{-j\theta} B(\omega + \omega_0) \tag{4-197}$$

Thus, for a sinusoidal carrier, linear modulation is essentially a symmetrical frequency translation of the message signal through an amount ω_0 rad/s, and no new signal components are generated. On the other hand, if $c(t)$ is not a sinusoid, so that the spectrum $C(\omega)$ has nonzero width, then $M(\omega)$ will represent a spreading and shaping as well as a translation of $B(\omega)$.

Suppose that the message signal $a(t)$ [and hence $b(t)$] is low-pass and strictly band-limited $(0, \omega_s)$ rad/s. Then the frequency spectrum of $b(t)$ obeys $B(\omega) = 0$, $|\omega| > \omega_s$, and $B(\omega - \omega_0)$ and $B(\omega + \omega_0)$ do not overlap. The spectrum $B(\omega - \omega_0)$ occupies only a range of positive frequencies while $B(\omega - \omega_0)$ occupies only negative frequencies.

The *envelope* of the modulated waveform $m[t, a(t)]$ is the magnitude $|b(t)|$. If, as mentioned earlier, the function $b(t)$ is restricted to be nonnegative almost always, then this envelope becomes just $b(t)$. In such cases, an *envelope detector* will uniquely recover $b(t)$ and hence $a(t)$. We now consider the common forms of simple amplitude, or linear, modulation.

26. Double-Sideband Amplitude Modulation (DSBAM). In this case the function $b(t)$ is given by Eq. (4-190) so that

$$m[t,a(t)] = C[1 + a(t)]\cos(\omega_0 t + \theta) \qquad (4\text{-}198)$$

The transform of $b_1(t)$ is just

$$B_1(\omega) = \mathcal{F}[1 + a(t)] = 2\pi\delta(\omega) + A(\omega) \qquad (4\text{-}199)$$

where $A(\omega)$ is the Fourier transform of $a(t)$ and $\delta(\omega)$ is the Dirac delta function. It follows, therefore, from Eq. (4-197) that the frequency spectrum of $m[t,a(t)]$ is given by

$$M(\omega) = \frac{C}{2}e^{j\theta}[2\pi\delta(\omega - \omega_0) + A(\omega - \omega_0)] + \frac{C}{2}e^{-j\theta}[2\pi\delta(\omega + \omega_0) + A(\omega + \omega_0)] \quad (4\text{-}200)$$

Depending on the form of $a(t)$, a number of special cases can be distinguished, but in any case the spectrum is given by Eq. (4-197).

The simplest case is where $a(t)$ is the *periodic function* given by

$$a(t) = \eta \cos \omega_s t \qquad \begin{matrix} \omega_s < \omega_0, \\ 0 \le \eta \le 1 \end{matrix} \qquad (4\text{-}201)$$

Then the frequency spectrum $A(\omega)$ is

$$A(\omega) = \pi\eta\delta(\omega - \omega_s) + \pi\eta\delta(\omega + \omega_s) \qquad (4\text{-}202)$$

For convenience, let us take the phase angle θ to be zero so that

$$M(\omega) = \pi C[\delta(\omega - \omega_0) + \delta(\omega + \omega_0) + \frac{\eta}{2}\delta(\omega - \omega_0 - \omega_s) \\ + \frac{\eta}{2}\delta(\omega - \omega_0 + \omega_s) + \frac{\eta}{2}\delta(\omega + \omega - \omega_s) + \frac{\eta}{2}\delta(\omega + \omega_0 + \omega_s)] \qquad (4\text{-}203)$$

This spectrum is illustrated in Fig. 4-10 together with the corresponding modulated waveform

$$m[t,a(t)] = C(1 + \eta \cos \omega_s t)\cos \omega_0 t \qquad (4\text{-}204)$$

Note that this last equation can also be written as

$$m[t,a(t)] = \underbrace{C \cos \omega_0 t}_{\text{carrier}} + \underbrace{\frac{C}{2}\eta \cos(\omega_0 - \omega_s)t}_{\text{lower sideband}} + \underbrace{\frac{C}{2}\eta \cos(\omega_0 + \omega_s)t}_{\text{upper sideband}} \qquad (4\text{-}205)$$

In this form it is easy to distinguish the *carrier*, the *lower sideband*, and the *upper sideband*.

Suppose that $a(t)$ is not a single sinusoid but is periodic with period P. Then it may be expanded in a Fourier series and each term of the series treated as in Eq. (4-203).

27. Suppressed Carrier DSBAM. If $b(t)$ is given by Eq. (4-191) so that

$$m[t,a(t)] = Ca(t)\cos(\omega_0 t + \theta) \qquad (4\text{-}206)$$

then the carrier is suppressed and Eq. (4-200) reduces to

$$M(\omega) = \frac{C}{2}e^{j\theta}A(\omega - \omega_0) + \frac{C}{2}e^{-j\theta}A(\omega + \omega_0) \qquad (4\text{-}207)$$

The principal advantage of DSBAM-SC over DSBAM is that the carrier is not transmitted in the former case, with a consequent saving in transmitted power. The principal disadvantages relate to problems in generating and demodulating the suppressed-carrier waveform.

28. Vestigial Sideband AM (VSBAM). It is apparent from Eq. (4-200) that the total information regarding the message signal $a(t)$ is contained in either the upper or lower sideband in conventional DSBAM. In principle it is only necessary to transmit one of these sidebands and the other could be eliminated, say by filtering, with a consequent reduction in bandwidth and transmitted power. Such a procedure is actually followed in single-sideband amplitude modulation (SSBAM), which will be discussed next. However, completely filtering out one sideband requires an ideal bandpass filter with infinitely sharp cutoff or an equivalent technique. A policy intermediate between the production of DSBAM and SSBAM is vestigial sideband amplitude modulation (VSBAM), where one sideband is attenuated much more than the other with some saving in bandwidth and transmitted power. The carrier may be present or suppressed.

The principal use of VSBAM has been in commercial television. In this case the video (picture) signal is transmitted by VSBAM with a consequent reduction in the total transmitted signal bandwidth and in the frequency difference that must be allowed between adjacent channels.

29. Single-Sideband AM (SSBAM). This important type of AM can be considered as a limiting form of VSBAM when the filter for the modulated waveform is an ideal filter with infinitely sharp cutoff so that one sideband, e.g., the lower, is completely eliminated. The modulation and demodulation of SSBAM are relatively complicated and are treated in more detail in the references, Par. **4-41.**

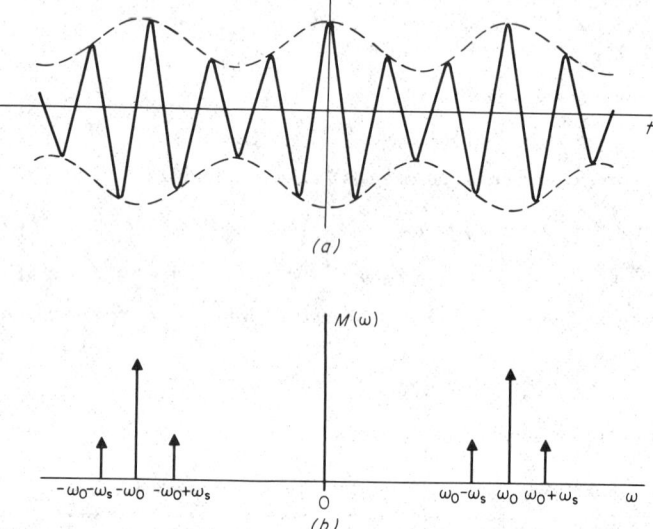

Fig. 4-10. (*a*) Double-sideband AM signal and (*b*) frequency spectrum.

30. Bandwidth and Power Relationships for AM. It is clear that the bandwidth of a given AM signal is related in a simple fashion to the bandwidth of the modulating signal since AM is essentially a frequency translation. If the modulating signal $a(t)$ is assumed to be low-pass and to have a bandwidth of W Hz, then the bandwidth ΔF of the modulated signal $m[t,a(t)]$ must satisfy

$$W \leq \Delta F \leq 2W \tag{4-208}$$

From the previous discussions, it is obvious that the upper limit of $2W$ Hz holds for double sideband modulation and the lower limit of W Hz for single-sideband. For the concept of bandwidth to be meaningful for a single-frequency sinusoidal modulating signal, it is assumed, of course, that a low-frequency sinusoid of frequency W Hz has a bandwidth W. Actually, what is really being assumed is that this sinusoid is the highest-frequency component in the low-pass modulating signal.

The only case where intermediate values of the inequality of Eq. (4-208) are encountered is in VSBAM. In principle any bandwidth between W and $2W$ Hz is possible. In this respect SSBAM can be considered as a limiting case of VSBAM. From a practical point of view, however, it is difficult (or expensive) to design a filter whose gain magnitude drops off sharply.

Power relationships are also simple and straightforward for AM signals. Consider Eq. (4-198), which is a general expression for the modulated AM waveform $m(t)$. Let the average power in this signal be denoted by P_{av}. The phase angle θ will be fixed if $a(t)$ is deterministic and will be taken to be a random variable uniformly distributed on $(0, 2\pi)$ if $a(t)$ is a random process. Also, it will be assumed that $a(t)_{av}$ is zero. It is clear that P_{av} is given by

$$P_{av} = \frac{C^2}{2}[1 + a^2(t)_{av}] \qquad (4\text{-}209)$$

where $a^2(t)_{av}$ is the average power in $a(t)$. The first term, $C^2/2$, is the carrier power, and the second term, $(C^2/2)a^2(t)_{av}$, is the signal power in the upper and lower sidebands. If $|a(t)| \leq 1$ to prevent overmodulation, then at least half the transmitted power is carrier power and the remainder is divided equally between the upper and lower sideband.

In double-sideband, suppressed-carrier operation, the carrier power is zero, and half the total power exists in each sideband. The fraction of information-bearing power is $\frac{1}{2}$. For single-sideband, suppressed-carrier systems, all the transmitted power is information-bearing, and, in this sense, SSB-SC has maximum transmission efficiency.

The disadvantages of both suppressed-carrier and single-sideband operation lie in the difficulties of generating the signals for transmission and in the more complicated receivers required. Demodulation of suppressed-carrier AM signals involves the reinsertion of the carrier or an equivalent operation. The local generation of a sinusoid at the exact frequency and phase of the missing carrier is either difficult or impossible unless a pilot tone of reduced magnitude is transmitted with the modulated signal for synchronization purposes or unless some nonlinear operation is performed on the suppressed-carrier signal to regenerate the carrier term at the receiver. In SSBAM, not only is the receiver more complicated, but transmission is considerably more difficult. It is usually necessary to generate the SSB signal at a low power level and then to amplify with a linear power amplifier to the proper level for transmission. On the other hand, DSB signals are easily generated at high power levels so that inefficient linear power amplifiers need not be used.

31. Angle (Frequency and Phase) Modulation. In angle modulation, the carrier $c(t)$ has either its phase angle or its frequency varied in accordance with the intelligence $a(t)$.

The result is not a simple frequency translation, as with AM, but involves both translation and the production of entirely new frequency components. In general, the new spectrum is much wider than that of the intelligence $a(t)$. The greater bandwidth may be used to improve the signal-to-noise performance of the receiver. This ability to *exchange* bandwidth for signal-to-noise enhancement is one of the outstanding characteristics and advantages of angle modulation.

In the form of angle modulation which will be called *phase modulation* (PM) the phase of the carrier is varied linearly with the intelligence $a(t)$. Thus the modulated signal is given by

$$m(t) = C \cos[\omega_0 t + \theta + k_p a(t)] \qquad (4\text{-}210)$$

where k_p is a constant and the *modulation index* $\varnothing_m$ is defined by

$$\varnothing_m = \max|k_p a(t)| \qquad (\text{radians}) \qquad (4\text{-}211)$$

In *frequency modulation*, the instantaneous frequency is made proportional to the intelligence $a(t)$. The modulated signal is given by

$$m(t) = C \cos\left[\omega_0 t + k_f \int_{-\infty}^{t} a(\tau)\, d\tau\right] \qquad (4\text{-}212)$$

where k_f is a constant. The *maximum deviation* $\Delta\omega$ is given by

$$\Delta\omega = \max|k_f a(t)| \qquad (\text{rad/s}) \qquad (4\text{-}213)$$

and, as before, a *modulation index* $\varnothing_m$ by

$$\varnothing_m = \max\left|k_f \int_{-\infty}^{t} a(\tau)\, d\tau\right| \qquad (4\text{-}214)$$

In general, the analysis of angle-modulated signals is difficult even for simple modulating intelligence. We will consider only the case where the modulating intelligence $a(t)$ is a sinusoid. Let $a(t)$ be given by

$$a(t) = \eta \cos \omega_s t \qquad \omega_s < \omega_0 \tag{4-215}$$

Then the corresponding PM signal is

$$m(t) = C \cos(\omega_0 t + k_p \eta \cos \omega_s t) \tag{4-216}$$

where the phase angle θ has been set equal to zero. In the same way, the FM signal is

$$m(t) = C \cos\left(\omega_0 t + \frac{k_f \eta}{\omega_s} \sin \omega_s t\right) \tag{4-217}$$

Let us now consider the FM signal of this last equation. Essentially the same results will be obtained for PM. The equation can be expanded to yield

$$m(t) = C \cos(\varnothing_m \sin \omega_s t) \cos \omega_0 t - C \sin(\varnothing_m \sin \omega_s t) \sin \omega_0 t \tag{4-218}$$

where the modulation index $\varnothing_m$ is given by $\varnothing_m = k_f \eta / \omega_s$. Sinusoids with sinusoidal arguments give rise to Bessel functions, and Eq. (4-218) can be expanded and rearranged to obtain

$$
\begin{aligned}
m(t) = \ &C J_0(\varnothing_m) \cos \omega_0 t \\[2pt]
&\underbrace{\hspace{3cm}}_{\text{carrier}} \\[6pt]
&+C \sum_{n=1}^{\infty} J_{2n}(\varnothing_m) [\cos(\omega_0 + 2n\omega_s)t + \cos(\omega_0 - 2n\omega_s)t] \\[-2pt]
&\hspace{2.6cm}\underbrace{\hspace{3cm}}_{\text{USB}} \quad \underbrace{\hspace{3cm}}_{\text{LSB}} \\[6pt]
&+C \sum_{n=1}^{\infty} J_{2n-1}(\varnothing_m) \{\cos[\omega_0 + (2n-1)\omega_s]t - \cos[\omega_0 - (2n-1)\omega_s]t\} \\[-2pt]
&\hspace{2.2cm}\underbrace{\hspace{3cm}}_{\text{USB}} \quad\;\; \underbrace{\hspace{3cm}}_{\text{LSB}}
\end{aligned}
\tag{4-219}
$$

where $J_m(x)$ is the Bessel function of the first kind and order m. This expression for $m(t)$ is relatively complicated even though the modulating intelligence is a simple sinusoid. In addition to the carrier, there are an infinite number of upper and lower sidebands separated from the carrier (and from each other) by integral multiples of the modulating frequency ω_s. Each sideband, and the carrier, has an amplitude determined by the appropriate Bessel function. When there is more than one modulating sinusoid, the complexity increases rapidly.

In the general case, an infinite number of sidebands exist dispersed throughout the whole frequency domain. In this sense, the bandwidth of an FM (or PM) signal is infinite. However, outside of some interval centered on ω_0, the magnitude of the sidebands will be negligible; this interval may be taken as a practical measure of the bandwidth. An approximation can be obtained by noting that $J_n(\varnothing_m)$, considered as a function of n, decreases rapidly when $n < \varnothing_m$. Therefore only the first $\varnothing_m$ sidebands are significant. If the highest-frequency sideband of significance is $\varnothing_m \omega_s$, then the bandwidth is given approximately by

$$BW \approx 2\omega_s \varnothing_m \qquad \text{(rad/s)} \tag{4-220}$$

A more accurate rule of thumb is the slightly revised expression

$$BW \approx 2\omega_s(\varnothing_m + 1) \qquad \text{(rad/s)} \tag{4-221}$$

which may be considered a good approximation when $\varnothing_m \geq 5$.

32. Pulse Modulation. In the preceding paragraphs, modulation schemes were considered which operated with sinusoidal carriers. In principle, any other continuous waveform could have been used, although analysis of the resulting modulated signal might have become very complicated. Next, systems are considered where the carrier is no longer continuous but consists of a pulse train, some parameter of which is suitably modified by the modulating intelligence. It will be seen that there are natural applications for such modulation schemes and that they may provide striking improvements in noise immunity.

33. Pulse Amplitude Modulation. In this modulation system, the amplitude of a pulsed carrier $p(t)$ is varied by the modulating signal $a(t)$. This form of modulation is called *pulse amplitude modulation* (PAM), with modulated signal given by

$$m(t) = m[t,a(t)] = a(t)\,p(t) \qquad (4\text{-}222)$$

Let us assume that the carrier $p(t)$ is a periodic pulse train with basic pulse shape $f(t)$. Since the pulse train is periodic, it can be expanded in a Fourier series as

$$p(t) = \sum_{k=-\infty}^{\infty} P_k e^{jk\omega_0 t} \qquad (4\text{-}223)$$

where $\omega_0 = 2\pi/T$ and the Fourier coefficient P_k is given by

$$P_k = \frac{1}{T}\int_{-T/2}^{T/2} p(t)e^{-jk\omega_0 t}\,dt \qquad (4\text{-}224)$$

The PAM signal of Eq. (4-222) can be rewritten with the aid of Eq. (4-223) as

$$m(t) = a(t)\,p(t) = \sum_{k=-\infty}^{\infty} P_k a(t)e^{jk\omega_0 t} \qquad (4\text{-}225)$$

with Fourier transform $M(\omega)$ given by

$$M(\omega) = \sum_{k=-\infty}^{\infty} P_k \mathscr{F}\{a(t)e^{jk\omega_0 t}\} = \sum_{k=-\infty}^{\infty} P_k A(\omega - k\omega_0) \qquad (4\text{-}226)$$

where $A(\omega)$ is the Fourier transform of $a(t)$. Suppose now that $a(t)$ is band-limited $(-B/2,B/2)$ rad/s where $B < \omega_0$. Then it is clear from Eq. (4-226) that the PAM signal $m(t)$ has a spectrum where the basic spectral shape is repeated periodically throughout the frequency domain but each spectral pulse $A(\omega - k\omega_0)$ is weighted by the appropriate Fourier coefficient P_k. The value of this coefficient depends on the amplitude, shape, and spacing of the carrier pulse train $p(t)$.

In practice, the width D of the basic pulse $p(t)$ is small compared to the pulse spacing T. The general effect of the modulating process is to sample the intelligence $a(t)$ at a fixed interval of T throughout the time domain. If $a(t)$ is strictly band-limited $(-B/2,B/2)$ rad/s, *then* the sampling theorem ensures that $a(t)$ can be reconstructed exactly from its samples spaced $2\pi/B$ s apart throughout the time domain. Thus a *necessary* condition for exact recovery of $a(t)$ from $a(t)\,p(t)$ is that

$$T \le 2\pi/B \quad \text{or} \quad BT \le 2\pi \qquad (4\text{-}227)$$

It is clear from Eq. (4-226) that this condition is the necessary restriction to prevent overlapping of the repeated spectra of the PAM signal $m(t)$. (Note that ω_0 has been defined as $2\pi/T$.)

In the demodulation of PAM signals, two cases must be distinguished. If the Fourier coefficient P_0 is not zero, the modulating intelligence $a(t)$ can be recovered by low-pass filtering. The resulting output is just the original intelligence $a(t)$ weighted by the constant P_0. When P_0 is zero but P_1 is not, the intelligence $a(t)$ can be recovered by some form of synchronous demodulation.

As a practical matter, the exact waveform given by Eq. (4-225) is not usually employed in PAM. More specifically, the tops of the pulse train are not usually shaped by the modulating signal. Instead they are kept flat, and the pulse height is determined by the value of $a(t)$ at some point in the pulse-length interval D. Thus the PAM signal is not exactly $a(t)\,p(t)$, and the resulting spectrum $M(\omega)$ does not yield $A(\omega)$ with a constant scale factor. Instead there is some distortion of the spectrum of $A(\omega)$ depending on the exact shape of the pulse train $p(t)$. In many cases this distortion can be kept negligibly small, and in other cases it can be substantially removed by a low-pass equalizing filter in the PAM receiver.

34. Quantizing and Quantizing Error. The PAM system just studied involved converting a time-continuous modulating intelligence $a(t)$ to a time-discrete form by *sampling*. The sampling theorem guaranteed that $a(t)$ could be reconstructed exactly from its samples $a(k/2W)$, $k = \cdots, -1, 0, 1, \ldots$, spaced throughout the time domain provided $a(t)$ had a spectrum $A(\omega)$ which was strictly band-limited $(-2\pi W, 2\pi W)$ rad/s. In a similar fashion, an amplitude-continuous signal is converted to an *amplitude-discrete*, or *digital*, signal by quantizing the amplitude domain into a finite number of fixed distinguishable amplitude levels, each a distance Q apart.

The process of quantizing is irreversible since regardless of how small the quantization level Q is taken to be, an unresolvable uncertainty of $\pm Q/2$ is associated after quantizing with each amplitude value. Thus a *quantization noise* N_q is inevitably associated with all quantized signals. This noise can be made as small as desired by choosing enough quantization levels or, equivalently, making each quantization level small enough, but it cannot be eliminated. In the pulsed-carrier systems to be discussed subsequently, it will turn out that the noise added in transmission can be almost completely eliminated at the receiver. In other words, the type of noise interference that is added externally (say in the channel) will be negligible, and the principal source of contamination will be the quantization noise.

The quantization levels are often taken to be equal; i.e., the spacing between amplitude levels is uniform. Nonuniform quantizing can be used to favor small amplitudes, where noise has a greater effect at the expense of large amplitudes. However, for this discussion, it will be assumed that the differences in amplitude levels are all equal to Q. Let us assume that the actual amplitude of the signal being sampled is equally likely to lie anywhere within the particular quantization level. In other words, the quantizing error in an amplitude interval is taken to be a uniformly distributed random variable q with density $p_q(x)$, given by

$$p_q(x) = \begin{cases} 0 & |x| > \dfrac{Q}{2} \\[2mm] \dfrac{1}{Q} & |x| \leq \dfrac{Q}{2} \end{cases} \tag{4-228}$$

The quantization noise N_q can be written

$$N_q = \int_{-Q/2}^{Q/2} x^2 p_q(x)\, dx = \frac{Q^2}{12} \tag{4-229}$$

The average value of the quantizing error is zero since the distribution of Eq. (4-228) has zero mean. In terms of 1-V quantization levels the rms error voltage is $1/\sqrt{12}$ V. Suppose that the peak-to-peak value of the signal to be quantized is limited to $\pm A$. Then the quantization level Q can be written

$$Q = \frac{2A}{d} \tag{4-230}$$

where d is the number of quantization levels. The signal-to-quantization noise power ratio for a sampled signal $a^*(t)$ can now be written

$$\frac{S}{N_q} = \frac{12[a^*(t)]^2 \mathrm{av}}{Q^2} = 3d^2 \frac{[a^*(t)]^2 \mathrm{av}}{A^2} \tag{4-231}$$

where the quantity $[a^*(t)]^2\mathrm{av}/A^2$ is the normalized sampled signal power.

35. Signal Encoding. The result of amplitude quantizing and time sampling a message signal is a sequence of numbers. We now consider advantageous ways to represent this sequence. Each number could be represented by a pulse of height proportional to that number. The result would be a quantized PAM pulse train. However, there can be great advantages in representing *each* number by a *sequence* of numbers which are allowed to assume fewer values than the original set. Such a system of representation was discussed in Par 4-7 and called a *code*. The principal advantages of encoding a message set are (1) that it is easier to discriminate in the presence of noise between a few possible numbers (pulse heights) than between many and (2) that an error in the value of a given code number will affect only part of the information contained in the original sample value. In general, an *m*-ary code will consist of *m* possible levels. Then a sequence of *n* symbols of this code can represent m^n possible message levels since there are $M = m^n$ distinguishable code groups of length *n*. As an example, suppose the original message is quantized at eight levels. The message sequence is to be encoded into a *binary* code consisting of two levels or symbols 0 and 1. In this case $m = 2$ and, since M must equal 8, we have

$$M = m^n \quad \text{or } 8 = 2^3$$

Thus each binary code word, representing one of the possible quantization levels, must consist of three binary digits. A logical encoding is

Quantization level	Code word	Quantization level	Code word
0	000	4	100
1	001	5	101
2	010	6	110
3	011	7	111

Of the possible *m*-ary codes that could be used, the binary code ($m = 2$) is the most common for at least two reasons: (1) the binary code offers the least number of choices, namely, two, in decoding; and (2) binary systems are most easily implemented electronically since the two possible states can be made to correspond to *on-off*, *open-closed*, or *conducting-nonconducting* conditions in circuits and systems.

It is clear from the previous discussion that the process of amplitude quantization, time sampling, and encoding can be used to convert a continuous band-limited message signal into a pulse train consisting of equally spaced pulses of two heights, e.g., zero and unity. The steps by which this is accomplished are as follows: (1) a low-frequency signal $a(t)$ band-limited $(-2\pi W, 2\pi W)$ is quantized into M amplitude levels and sampled at an interval of $1/2W$ s, resulting in a sequence of numbers; (2) each of these numbers is then encoded in a binary code; (3) the corresponding binary sequence becomes a pulse train consisting of pulses of height unity and zero to correspond to 1 and 0, respectively.

36. Pulse-Code Modulation. All the elements of pulse code modulation (PCM) have now been discussed. The steps are seen to be as follows:

1. A continuous signal $a(t)$ is sampled and quantized.
2. The quantized sample values are encoded into a pulse train $p(t)$.
3. The pulse train $p(t)$ is transmitted directly or used as the modulating signal in any appropriate modulation scheme.

The great advantage of PCM is its inherent resistance to external-noise corruption. If the external noise is not too large, it is clear that the original pulse $p(t)$ can be recovered *exactly* from the received pulse $p^*(t)$ since only a knowledge of the *presence* or *absence* of a pulse and not the pulse *shape* itself is needed at each pulse position to reproduce the original pulse train exactly. Thus, if the signal is large enough compared to corrupting noise, there will be *no error*. Furthermore, if it is desired to transmit the pulse train for a long distance, it can be regenerated exactly as often as desired at repeater stations along the way. In other words, no noise is added in transmission in the large-signal case. This situation is in striking contrast to the other modulation schemes already studied. Thus, in a properly designed large-signal PCM system, all error that results will arise from the original quantization noise discussed previously.

The PCM system can be considered a coded wideband system which trades complexity and bandwidth for noise immunity. In general, if there are d quantization levels, and if binary pulses are used, each code word must contain n pulses, where

$$d = 2^n \quad \text{or} \quad n = \log_2 d \tag{4-232}$$

Thus there must be $2nW$ pulses/s in the PCM pulse train, and any two adjacent pulses must be clearly distinguishable. If the bandwidth of the system transmitting the PCM pulse train is too small, the pulses will be broadened and will interfere with each other so that errors in identification will result. In fact, the uncertainty relationships previously developed bear directly on this problem. Thus Eq. (4-184) relates the bandwidth $\Delta\Omega$ and the duration ΔT of a pulse by

$$\Delta\Omega \, \Delta T \geq \tfrac{1}{2} \tag{4-233}$$

Since there must be $2nW$ pulses/s, each pulse cannot exceed a width $\Delta T \leq 1/2nW$ s.

It follows directly from Eq. (4-233) that the system bandwidth $\Delta\Omega$ is bounded from below by

$$\Delta\Omega \geq nW \quad \text{(Hz)} \tag{4-234}$$

It is clear that the PCM signal requires a transmission bandwidth which is at least n times that of the original intelligence, where n is determined by Eq. (4-232) and depends on the number of quantization levels desired. If the number of quantization levels is taken to be small (so that n is small), the quantization noise will increase as shown by Eqs. (4-229) and (4-230). The output signal-to-noise power ratio due to quantization noise is given by Eq. (4-231), where $[a^*(t)]_{av}^2$ is the power in the quantized signal and must be calculated. For reasonable signal distributions and uniform quantizers it can be shown that the signal-to-noise ratio increases approximately as 2^{2n}.

In many practical situations, equipment inadequacies and/or low-signal conditions will create situations where transmission noise as well as quantization noise is added. In other words, some pulses in the coded pulse train will be incorrectly identified. This problem is most conveniently formulated in terms of a *probability of error* which gives the average rate at which incorrect pulse identification occurs. This criterion was discussed in Par. **4-11** and is directly applicable here.

Thus far, no consideration has been given to the actual transmission of the PCM pulse train. It is clear that the pulse train cannot be radiated from any convenient antenna. As in other forms of modulation, some frequency-translation technique must be employed. A number of systems are used for such *digital* signaling. Some of the commonest are listed.

a. Binary On-Off Keying (OOK). This type of binary signaling is almost self-explanatory. A high-frequency sinusoidal signal is switched on and off so that on periods correspond to 1 and off periods to 0 in the PCM wave. In practice, the pulsed sinusoids do not start or stop abruptly but exhibit a transient buildup and decay.

b. Binary Frequency-Shift Keying (FSK). In this form of digital signaling, a continuous wave is sent which shifts between two frequencies, one representing the symbol 1 in the PCM wave and the other representing 0.

c. Binary Phase-Shift Keying (PSK). Here the frequency is kept constant, but the phase is shifted 180° whenever the basic signal changes from 0 to 1.

NOISE AND INTERFERENCE

37. General. In a general sense, *noise* and *interference* are used to describe any unwanted or undesirable signals in communication channels and systems. Since in many cases these signals are random or unpredictable, some study of random processes is a useful prelude to any consideration of noise and interference.

38. Random Processes. A random process $x(t)$ is often defined as an indexed family of random variables where the *index* or *parameter* t belongs to some set T; that is, $t \in T$.

The set T is called the *parameter set* or *index set* of the process. It may be finite or infinite, denumerable or nondenumerable; it may be an interval or set of intervals on the real line, or it may be the whole real line $-\infty < t < \infty$. In most applied problems, the index t will be time, and the underlying intuitive notion will be that of a random variable developing in time. However, other parameters such as position, temperature, etc., may also enter in a natural manner.

It follows from the definition that there are at least two ways to view an arbitrary random process: (1) as a set of random variables: this viewpoint follows from the definition; for each value of t, the random process reduces to a random variable; and (2) as a set of functions of time.   From this viewpoint, there is an underlying random variable each realization of which is a time function with domain T.   Each such time function is called a *sample function* or *realization* of the process.

From a physical point of view, it is the sample function which is important, since this is the quantity that will almost always be observed in dealing experimentally with the random process.   One of the important practical aspects of the study of random processes is the determination of properties of the random variable $x(t)$ for fixed t on the basis of measurements performed on a single sample function of the process $x(t)$.

A random process is said to be *stationary* if its statistical properties are invariant to time translation.   This invariance implies that the underlying physical mechanism producing the process is not changing with time.   Stationary processes are of great importance for two reasons: (1) they are frequently encountered in practice or approximated to a high degree of accuracy (actually, from the practical point of view, it is not necessary that a process be stationary for all time but only for some observation interval which is long enough to be suitable for a given problem); (2) many of the important properties of stationary processes commonly encountered are described by first and second moments.   Consequently, it is relatively easy to develop a simple but useful theory (*spectral theory*) to describe these processes.   Processes which are not stationary are called *nonstationary*, although they are also sometimes referred to as *evolutionary* processes.

The *mean* $m(t)$ of a random process $x(t)$ is defined by

$$m(t) = E\{x(t)\} \tag{4-235}$$

where $E\{\cdot\}$ is the mathematical expectation operator defined in Par. **4-4**.   In many practical problems, this mean is independent of time.   In any case, if it is known, it can be subtracted from $x(t)$ to form a new "centered" process $y(t) = x(t) - m(t)$ with zero mean.

The *autocorrelation function* $R_x(t_1,t_2)$ of a random process $x(t)$ is defined by

$$R_x(t_1,t_2) = E\{x(t_1)x(t_2)\} \tag{4-236}$$

In many cases, this function depends only on the time difference $t_2 - t_1$ and not on the absolute times t_1 and t_2.   In such cases, the process $x(t)$ is said to be *at least wide-sense stationary*, and by a linear change in variable Eq. (4-236) can be written

$$R_x(\tau) = E\{x(t)x(t + \tau)\} \tag{4-237}$$

If $R_x(\tau)$ possesses a Fourier transform $\varphi_x(\omega)$, this transform is called the *power spectral density* of the process and $R_x(\tau)$ and $\varphi_x(\omega)$ form the Fourier transform pair

$$\varphi_x(\omega) = \int_{-\infty}^{\infty} R_x(\tau)e^{-j\omega\tau}\,d\tau \tag{4-238}$$

and

$$R_x(\tau) = \frac{1}{2\pi}\int_{-\infty}^{\infty} \varphi_x(\omega)e^{j\omega\tau}\,d\omega \tag{4-239}$$

For processes which are at least wide-sense stationary, these last two equations afford a direct approach to the analysis of random signals and noises on a power ratio or mean-squared-error basis.   When $\tau = 0$, Eq. (4-239) becomes

$$R_x(0) = E\{x^2(t)\} = \frac{1}{2\pi}\int_{-\infty}^{\infty} \varphi_x(\omega)\,d\omega \tag{4-240}$$

an expression for the normalized power in the process $x(t)$.

As previously mentioned, in practical problems involving random processes, what will generally be available to the observer is not the random process but one of its sample functions or realizations.   In such cases, the quantities that are easily measured are various time averages, and an important question to answer is: Under what circumstances can these time averages be related to the statistical properties of the process?

We define the *time average* of the random process $x(t)$ by

$$A\{x(t)\} = \lim_{T\to\infty}\frac{1}{2T}\int_{-T}^{T} x(t)\,dt \tag{4-241}$$

The *time autocorrelation function* $\mathcal{R}_x(\tau)$ is defined by

$$\mathcal{R}_x(\tau) = A\{x(t)x(t+\tau)\} = \lim_{T\to\infty} \frac{1}{2T}\int_{-T}^{T} x(t)x(t+\tau)\,dt \qquad (4\text{-}242)$$

It is intuitively reasonable to suppose that, for stationary processes, time averages should be equal to expectations; e.g.,

$$E\{x(t)x(t+\tau)\} = A\{x(t)x(t+\tau)\} \qquad (4\text{-}243)$$

A heuristic argument to support this claim would go as follows. Divide the parameter t of the random process $v(t)$ into long intervals of T length. If these intervals are long enough (compared to the time scale of the underlying physical mechanism), the statistical properties of the process in one interval T should be very similar to those in any other interval. Furthermore, a new random process could be formed in the interval $(0,T)$ by using as sample functions the segments of length T from a single sample function of the original process. This new process should be statistically indistinguishable from the original process, and its ensemble averages would correspond to time averages of the sample function from the original process.

The foregoing is intended as a very crude justification of the condition of *ergodicity*. A random process is said to be *ergodic* if time averages of sample functions of the process can be used as approximations to the corresponding ensemble averages or expectations. A further discussion of ergodicity is beyond the scope of this treatment, but this condition can usually be assumed to exist for stationary processes. In this case, time averages and expectations can be interchanged at will, and, in particular,

$$E\{x(t)\} = A\{x(t)\} = u = \text{const} \qquad (4\text{-}244)$$

$$\text{and} \qquad R_x(\tau) = \mathcal{R}_x(\tau) \qquad (4\text{-}245)$$

39. Classification of Random Processes. A central problem in the study of random processes is their classification. From a mathematical point of view, a random process $x(t)$ is defined when all n-dimensional distribution functions of the random variables $x(t_1)$, $x(t_2)$, $\ldots, x(t_n)$ are defined for arbitrary n and arbitrary times $t_1, t_2, \ldots, t_n$. Thus classes of random processes can be defined by imposing suitable restrictions on their n-dimensional distribution functions. In this way, we can define the following (and many others):

a. Stationary processes, whose joint distribution functions are invariant to time translation.

b. Gaussian (or normal) process, whose joint distribution functions are multivariate normal.

c. Markov processes, where given the value of $x(t_1)$, the value of $x(t_2)$, $t_2 > t_1$, does not depend on the value of $x(t_0)$, $t_0 < t_1$; in other words, the future behavior of the process, given its present state, is not changed by additional knowledge about its past.

d. White noise, where the power spectral density given by Eq. (4-238) is assumed to be a constant N_0. Such a process is not realizable since its mean-squared value (normalized power) is not finite; i.e.,

$$R_x(0) = E\{x^2(t)\} = \frac{N_0}{2\pi}\int_{-\infty}^{\infty} d\omega = \infty \qquad (4\text{-}246)$$

On the other hand, this concept is of considerable usefulness in many types of analysis and can often be postulated where the actual process has an approximately constant power spectral density over a frequency range much greater than the system bandwidth.

Another way to classify random processes is on the basis of a model of the particular process. This method has the advantage of providing insights into the physical mechanisms producing the process. The principal disadvantage is the complexity that frequently results. On this basis, we may identify the following (natural) random processes.

e. Thermal noise is caused by the random motion of the electrons within a conductor of nonzero resistance. The mean-squared value of the thermal-noise voltage across a resistor of resistance $R\ \Omega$ is given by

$$\overline{v^2} = 4kTR\Delta f \text{ (volts}^2) \qquad (4\text{-}247)$$

where k is Boltzmann's constant, T is the absolute temperature in kelvins, and Δf is the bandwidth of the measuring equipment.

f. Shot noise is present in any electronic device (vacuum tube, transistor, etc.) where electrons move across a potential barrier in a random way. Shot noise is usually modeled as

$$x(t) = \sum_{i=-\infty}^{\infty} f(t - t_i) \tag{4-248}$$

where the t_i are random emission times and $f(t)$ is a basic pulse shape determined by the device geometry and potential distribution.

g. Partition noise results where the electrons in an electron beam can travel to two or more electrodes. Random fluctuations in the number of electrons reaching each electrode are the basis for this noise. It is frequently the predominant noise in multielectrode vacuum tubes such as the pentode.

h. Defect noise is a term used to describe a wide variety of related phenomena which manifest themselves as noise voltages across the terminals of various devices when dc currents are passed through them. Such noise is also called current noise, excess noise, flicker noise, contact noise, or $1/f$ noise. The power spectral density of this noise is given by

$$\varphi_x(\omega) = k\frac{I^\alpha}{\omega^\beta} \tag{4-249}$$

where I is the direct current through the device, ω is radian frequency, and k, α, and β are constants. The constant α is usually close to 2, and β is usually close to 1. At a low enough frequency this noise may predominate due to the $1/\omega$ dependence.

40. Man-Made Noise. The noises just discussed are more or less fundamental and are caused basically by the noncontinuous nature of the electronic charge. In contradistinction to these are a large class of noises and interferences which are more or less man-made and hence, in principle, under our control. The number and kinds here are too many to list and the physical mechanisms usually too complicated to describe. However, to some degree, they may be organized into three main classes.

a. Interchannel Interference. This includes the interference of one radio or television channel with another, which may be the result of inferior antenna or receiver design, variations in carrier frequency at the transmitter, or unexpectedly long-distance transmission via scatter of ionospheric reflection. It also includes crosstalk between channels in communication links and interference caused by multipath propagation or reflection. These types of noises can be removed, at least in principle, by better equipment design; e.g., by using a receiving antenna with a sufficiently narrow radiation pattern to eliminate reception from more than one transmitter.

b. Hum. This is a periodic and undesirable signal arising from the power lines. Usually it is predictable and can be eliminated by proper filtering and shielding.

c. Impulse Noise. Like defect noise, this term describes a wide variety of phenomena. Not all of them are man-made, but the majority probably are. This noise can often be modeled as a low-density shot process or, equivalently, as the superposition of a small number of large impulses. These impulses may occur more or less periodically, as in ignition noise from automobiles or corona noise from high-voltage transmission lines. On the other hand, they may occur randomly, as in switching noise in telephone systems or the atmospheric noise from thunderstorms. The latter type of noise is not necessarily man-made, of course. This impulse noise tends to have an amplitude distribution which is decidedly nongaussian, and it is frequently highly nonstationary. It is difficult to deal with in a systematic way because it is ill-defined. Signal processors which must handle this type of noise are often preceded by limiters of various kinds or by noise blankers which give zero transmission if a certain amplitude level is surpassed. The design philosophy behind the use of limiters and blankers is fairly clear. If the noise consists of large impulses of relatively low density, the best system performance is obtained if the system is limited or blanked during the noisy periods and behaves normally when the (impulse) noise is not present.

41. References. A large number of textbooks are listed on the following pages. The textbooks cited contain extensive references to the periodical literature.

General

1. LAWSON, J. W., and G. E. UHLENBECK   "Threshold Signals," McGraw-Hill, New York, 1950.

2. DAVENPORT, W. B., and W. L. ROOT   "An Introduction to Random Signals and Noise," McGraw-Hill, New York, 1958.

3. LEE, Y. W.   "Statistical Theory of Communication," Wiley, New York, 1960.

4. MIDDLETON, D.   "An Introduction to Statistical Communication Theory," McGraw-Hill, New York, 1960.

5. JAVID, M., and E. BRENNER   "Analysis, Transmission, and Filtering of Signals," McGraw-Hill, New York, 1963.

6. GOLOMB, S. W.   "Digital Communication with Space Applications," Prentice-Hall, Englewood Cliffs, N.J., 1964.

7. BENNETT, W. R., and J. R. DAVEY   "Data Transmission," McGraw-Hill, New York, 1965.

8. LATHI, B. P.   "Signals, Systems, and Communication," Wiley, New York, 1965.

9. PAPOULIS, A.   "Probability, Random Variables, and Stochastic Processes," McGraw-Hill, New York, 1965.

10. WOZENCRAFT, J. M., and I. M. JACOBS   "Principles of Communication Engineering," Wiley, New York, 1965.

11. SCHWARTZ, M., W. R. BENNETT, and S. STEIN   "Communication Systems and Techniques," McGraw-Hill, New York, 1966.

12. COOPER, G. R., and C. D. MCGILLEM   "Methods of Signal and System Analysis," Holt, Rinehart and Winston, New York, 1967.

13. LUCKY, R. W., J. SALZ, and E. J. WELDON   "Principles of Data Communication," McGraw-Hill, New York, 1968.

14. SAKRISON, D. J.   "Communication Theory: Transmission of Waveforms and Digital Information," Wiley, New York, 1968.

15. THOMAS, J. B.   "An Introduction to Statistical Communication Theory," Wiley, New York, 1969.

16. SCHWARTZ, M.   "Information Transmission, Modulation, and Noise," 2d ed., McGraw-Hill, New York, 1970.

Information theory

1. SHANNON, C. E., and W. WEAVER.   "The Mathematical Theory of Communication," University of Illinois Press, Urbana, 1949.

2. BELL, D. A.   "Information Theory and Its Engineering Applications," Pitman, London, 1952.

3. GOLDMAN, S.   "Information Theory," Prentice-Hall, Englewood Cliffs, N.J., 1953.

4. BRILLOUIN, L.   "Science and Information Theory," Academic, New York, 1956.

5. KHINCHIN, A. I.   "Mathematical Foundations of Information Theory," Dover, New York, 1957.

6. FEINSTEIN, A.   "Foundations of Information Theory," McGraw-Hill, New York, 1958.

7. KULLBACK, S.   "Information Theory and Statistics," Wiley, New York, 1959.

8. FANO, R. M.   "Transmission of Information," M.I.T. Press, Cambridge, Mass., 1961.

9. REZA, F. M.   "An Introduction to Information Theory," McGraw-Hill, New York, 1961.

10. WOLFOWITZ, J.   "Coding Theorems of Information Theory," Prentice-Hall, Englewood Cliffs, N.J., 1961.

11. ABRAMSON, N.   "Information Theory and Coding," McGraw-Hill, New York, 1963.

12. RAISBECK, G.   "Information Theory," M.I.T. Press, Cambridge, Mass., 1964.

13. ASH, R.   "Information Theory," Wiley-Interscience, New York, 1965.

14. BILLINGSLEY, P. B.   "Ergodic Theory and Information," Wiley, New York, 1965.

15. JELINEK, F.   "Probabilistic Information Theory," McGraw-Hill, New York, 1968.

16. GALLAGER, R.   "Information Theory and Reliable Communication," Wiley, New York, 1968.

Coding theory

1. PETERSON, W. W.   "Error-Correcting Codes," Wiley, New York, 1961.

2. BERLEKAMP, E. R.   "Algebraic Coding Theory," McGraw-Hill, New York, 1968.

Noise
1. Van der Ziel, A. "Noise," Prentice-Hall, Englewood Cliffs, N.J., 1954.
2. Smullin, L. D., and H. A. Haus "Noise in Electron Devices," Wiley, New York, 1959.
3. Bell, D. A. "Electrical Noise: Fundamentals and Physical Mechanism," Van Nostrand, London, 1960.
4. Bennett, W. R. "Electrical Noise," McGraw-Hill, New York, 1960.
5. MacDonald, D. K. C. "Noise and Fluctuations," Wiley, New York, 1962.

SECTION 5

SYSTEMS ENGINEERING

BY

C. NELSON DORNY Associate Professor, Moore School of Electrical Engineering, University of Pennsylvania; Senior Member, Institute of Electrical and Electronics Engineers

KENNETH A. FEGLEY Professor, Moore School of Electrical Engineering, University of Pennsylvania; Senior Member, Institute of Electrical and Electronics Engineers

EZRA S. KRENDEL Professor, Wharton School, University of Pennsylvania; Fellow, Institute of Electrical and Electronics Engineers

CONTENTS

Numbers refer to paragraphs

SECTION 5

SYSTEMS ENGINEERING

METHODS OF SYSTEMS ENGINEERING

1. Introduction. This section on systems engineering summarizes the state of the field and places it in perspective with respect to the general field of engineering: the application of science and technology in supplying human needs. Engineering takes such forms as the design of a valve for a process controller, the design of a new solid-state device, the design of a new radar circuit, or the design and development of a satellite communication system.

Systems engineering is distinguished from general engineering primarily in terms of the complexity of the system to be designed. Of the above-named four engineering tasks, the first three are relatively straightforward in terms of problem definition, if not in terms of technological difficulty. The fourth task involves considerable interaction among physical, social, economic, and political factors. Initially, the objectives may be unclear, although the motivation behind the task may be strong, e.g., the inadequacy of the present long-distance communication system. The objectives are then formulated as the project develops. Thus systems engineering is a term that describes activities near one extreme of the continuum of engineering work. It involves the design of relatively large scale, relatively complex systems.

2. Examples of Systems. A system is a set of objects and the relationships that tie them together. The following examples illustrate the complexity of systems typically met in systems engineering:

(a) Transportation systems. The design of a transportation system, whether for passengers or freight, interstate or urban, rail, air, or automobile, is a complex process. It involves traffic control (e.g., air traffic control or rail dispatching, signaling, and switching), obtaining routes and right-of-way, environmental considerations, and the question of compatibility with existing systems. The U.S. postal system (which is occasionally redesigned) can be considered a transportation system. In addition to the U.S. government vehicles and postal carriers, it involves assembly-line automation, problems of communication and control, and interaction with public carriers.

(b) Communications systems. The most notable communications systems have to do with telephones and the public radio and TV network. The cross-country microwave radio links, satellite communication system, and the direct-dial, automatic switching system are major subsystems. The communications requirements for the military, for fire and crime control, and for air traffic control also lead to major communications systems. Problems of highly variable loads and compatibility with existing systems are significant.

(c) Military weapons systems. A large naval ship is a 5,000-man city with all the systems and subsystems that exist in a city—power, communications, traffic control, etc. Such a floating city must have military capability and must function under extreme environmental conditions. Missile systems and civil defense warning systems also present major problems for the systems engineer.

It is evident that there is a long list of problems of complexity similar to the examples above.[35,40]* Such complex systems typically include important human, computer, control, and information-handling *components*.

The *terminal characteristics* of system components and the manner in which the components are *interconnected* are of more significance in system design than is the manner of *operation* of individual components. A focus on component interactions is central to systems engineering.

The term "systems" is often loosely applied to activities or subjects which emphasize

---

* Superior numbers correspond to numbered references in Par. 5-36.

component interactions. Thus control theory, circuit theory, and much of mathematics are in the "systems" category, whereas solid-state-device theory and much of physics are not.

3. Problem Definition. In systems endeavors of the complexity illustrated above, one of the major obstacles is the definition of the problem. There are usually many alternative systems which will fill the needs. Which system is best from the standpoint of economics, performance, acceptability to the user, environmental impact, etc.? Typically, the objectives to be met by the system must be reconsidered and redefined as information about alternatives comes to light. Since there can be no clearly defined criteria for decision making, the systems engineer is faced with a multifaceted optimization problem. In the methodology of systems engineering, therefore, the problem-definition phase permeates much of the process.

One man is not likely to be an expert in all aspects of a system as complex as those mentioned above. However, it is usually feasible to assemble a *team* of people with overlapping expertise that covers the necessary range of activities. Each team member can focus on one aspect or subsystem of the problem (e.g., technical design, management of the design process, development and testing) while keeping track of the efforts of the rest of the team in order to interface properly with them.

4. Operations Research. The term operations research is applied to activities similar in many respects to systems engineering. We distinguish between the two fields primarily as to the nature of the problems they address. Operations research teams are more likely to be involved in the design or optimization of a business operation than in the design of a new physical system; i.e., the team members have an interest and background in business or economics rather than in physical systems. There is, however, no reason to differentiate strongly between systems engineering and operations research. Both use the same systematic problem solving process in the design work. The terms "systems approach" and "systems science" are used loosely in reference to a rational systematic approach to problem solving.[17,43]

Fundamental to the systems approach is consideration of the larger setting, or *environment,* which surrounds and affects the problem, and it is necessary to avoid a myopic viewpoint during the processes of problem definition and goal setting. For example, a government planner should not try to solve the subproblem of poverty without considering the closely interacting subproblems of education, nutrition and health, housing and community development, etc.

5. Methods of Systems Engineering. Engineering, the application of science and technology in supplying human needs, satisfies a lack, a want, a demand, or a desire felt by an individual or a group. Thus the motivating force behind all problem solving is a state of tension or imbalance in the environment. From the economic, social, technological, and governmental environments opportunities arise to develop new systems to satisfy needs. The problem-solving process in systems engineering is thus usually the result of a social need, and the process of defining the problem consists to a great extent in analyzing the environment surrounding the motivating situation.

The significant initial step in the systems engineering process is the replacement of an ill-defined situation or need by a set of specific objectives, taking into account the tradeoffs between the cost and the performance of the system. Unfortunately, neither cost nor performance is easily measured for a complex system. System performance is a composite of the degree of alleviation of the need, the efficiency of operation of the system, the degree of environmental pollution, etc. The cost of a large system is measured partly in money, but partly in such social discomforts as condemned homes, congested streets, and increased pollution of the air.

While a good system design depends upon a correct and comprehensive definition of the problem, the problem often cannot be well defined until it is half solved and before many details of the design are explored. Some of the assumptions on which the problem definition is based can be verified only by detailed examination. Thus the design of a complex system is an *iterative process.*

We may summarize the systems engineering process in the following three steps.

1. Problem exploration. In the first phase of systems engineering we explore the motivating situation, gather data, define the problem, and set objectives. Then we propose various systems which we feel will meet these objectives. Each of these systems must be analyzed in sufficient detail to assess its feasibility and to determine its consequences in terms of performance, cost, market, environmental impact, etc. On the basis of these analyses we must decide whether any of the proposed systems is sufficiently likely to satisfy the needs and objectives to merit detailed development.

2. Project planning. We next formulate a specific detailed plan of development for a particular system. The development lists lower-level objectives and ways to meet them. The planning process requires that we repeat much of the problem-exploration process in greater detail in order that we may be able to forecast the development effort and recommend schedules. To this point, the effort is primarily on paper.

3. System development. System development ends with the testing of a prototype system which is acceptable to the customer. The design goals are detailed drawings and specifications for the manufacturers and instructions for installation, operation, and maintenance.

The primary systems engineering role during development is in appraising and supporting development activities and in coordinating the efforts of the many parties involved: device and subsystem development groups, manufacturers, and customers. As in phases 1 and 2 above, some adjustment of lower-level objectives may be necessary if, during construction and testing of the actual system, certain requirements are found to be unreasonable or unnecessary.

It is evident that iteration from design to setting of objectives and back to design occurs continually throughout all three phases. Hall[40] includes two additional phases in the process: an *initial phase* consisting in broad exploration of the environment—technological, economic, social, natural, and governmental—to create a background of information for use in future projects and to give perspective to the total program of work of the organization; and a final *evaluation phase* over the life of the system, which provides information helpful in future systems planning and development. Hall illustrates in detail the various phases of the systems engineering process by means of a case study—the design of the TD-2 microwave radio relay system. This system is the heart of our present long-distance television and telephone communication system. Other general references which deal with the methodology of systems engineering are given in Refs. 35, 38, and 54, Par. 5-36. The latter reference also contains good surveys of ancillary fields.

Listed below are a number of techniques which are particularly useful in various phases of the systems engineering process. These are described in more detail in this and other sections of this Handbook.

Techniques for problem definition
1. Trees
2. Utility functions
3. Human engineering

Techniques for systems analysis
1. Scheduling techniques (PERT)
2. Reliability theory
3. Queueing theory
4. Decision theory
5. Modeling and block diagrams
6. Simulation
7. Control theory
8. Optimization techniques

6. Military Air Traffic Control. As an example of the procedures of systems engineering, we discuss in Pars. **5-6** to **5-11** steps taken in the design of an improved system of military air traffic control. Tactical aircraft, fixed-wing and helicopter, form a central component of modern warfare. In a typical military operation, more than 100 friendly aircraft may occupy a 50×50 mile airspace. Even under good visual flight conditions, without enemy aircraft, ground fire, and missiles, congestion is a serious problem, endangering personnel, equipment, and the overall mission. At night and in extreme weather, operations are severely restricted by the weaknesses in air traffic control.

The focus of a typical military air traffic control system is a flight operations center (FOC) with which pilots file flight plans. The pilots report by voice to the FOC at regular intervals during their flights. The air traffic controllers in the FOC recognize potential congestion and collisions primarily by personally analyzing the flight plans and periodic pilot reports. There is little radar assistance in navigation. In battle areas, there is no traffic control; it is every man for himself.

Systems engineers, in attempting an improved design of this system, must have access to the tactical decision makers, pilots, and aircraft controllers who operate in this environment. They must also have the benefit of advice from designers and manufacturers of aircraft and detection and control equipment. By discussions with these sources of information, they can more clearly define the problems and gain understanding of the relative importance of the various factors and constraints. They must also be able to perform limited visual flight experiments to test new flight control policies.

Steps which might reduce the difficulty in air traffic control are (1) to develop better control policies, (2) to develop a system of air traffic control which is more automatic than the present system, and (3) to increase the accuracy and reliability of the equipment for automatic detection and control of aircraft.

A concept that arose from discussions of aircraft control is that of *formation flight*. In military operations, many aircraft fly between the same two points at approximately the same time. The flight of a group of aircraft as one unit reduces the effective number of units that need to be controlled. The military forces have used this concept extensively under good visual conditions; its use under poor visual conditions has been limited. Nonvisual formation flight requires a *station-keeping system* to maintain each of the follower aircraft in a given desired position with respect to the lead aircraft.

Systems engineering techniques useful in designing a station-keeping system (which will make formation flight feasible under nonvisual conditions) include modeling, control theory, simulation, probability and statistics, optimization techniques, PERT, queueing theory, and human factors. These techniques are touched on in the following paragraphs.

7. Computer-assisted Modeling. Evaluation of specific policies or equipment is difficult without being able to try them out, but the cost of extensive realistic testing under field conditions is prohibitive. In the case of proposed but nonexistent equipment, realistic trial is impossible. An alternative approach at this stage is computer-assisted modeling of, say, 100 fixed- and rotary-wing aircraft in an airspace of, say, 50×50 miles. This model can be used to simulate and evaluate various aircraft control systems and policies.

To evaluate the feasibility of the formation-flight concept, as a partial solution to the military air traffic control problem, the systems engineers develop a computer model of a group of 25 rotary-wing aircraft (helicopters) maintained in position in formation flight by a station-keeping system. With this model they simulate formation flight under conditions of wind gusts, rain, snow and ice, abrupt maneuvers, terrain avoidance, enemy aircraft, equipment failures, etc.

They also determine by simulation the functional requirements for nonvisual formation flight: What policies and equipment are necessary so that the pilots will be able to function and perform their overall mission? What data must be collected and transmitted between helicopters? What control automation is necessary for station-keeping in formation flight? What stress is placed upon the pilots? What maneuvers by and within the formation should be used for terrain avoidance and to minimize the effect of enemy action?

We cannot explore the whole model here or describe the results of simulation with the model, but details are available in the literature.[1] In the case cited here, formation flight does appear to be a feasible partial solution to the problem, and the concept merits taking the next systems engineering step, namely, planning the development of a system for formation flight.

8. Helicopter Model. A mathematical model of a helicopter is essential to the solution of the problem (we assume each of the helicopters in the formation flight is of the same design). The equations of motion for the helicopter are obtainable from a knowledge of the physical characteristics of the vehicle. To obtain accurate coefficients for the equations of motion, we use flight test data. A block diagram of the helicopter control system is shown in Fig. 5-1.

The helicopter is controlled by a pilot or by an autopilot. Under nonvisual conditions the autopilot would be used. The autopilot is in turn controlled by a set of signals whose values are functions of the position error of the helicopter and its present velocity and orientation. The block in Fig. 5-1 which generates the signals to control the autopilot is designated *station-keeping control laws*.

In the simulation of formation flight, the station-keeping control laws are simply

---

[1] The formation-flight feasibility study and the airspace simulation are described in Refs. 29 and 81, respectively, Par. 5-36.

equations whose independent variables are the position error, velocity, and orientation of the aircraft and whose dependent variables are control signals for the autopilot. In a physical station-keeping system used aboard helicopters that are to fly in formation, the block in Fig. 5-1 labeled "station-keeping control laws" would normally be a set of electric circuits which would generate the desired output signals in response to the given input signals.

The *pattern control computer* shown in Fig. 5-1 generates the desired position of each follower aircraft relative to the lead aircraft. This position may vary, depending upon the maneuver to be performed and upon aircraft velocity.

Hence each of the essential parts of the control system, including the helicopter, the autopilot, the station-keeping control laws, and the pattern control computer, are represented by mathematical models. Clearly, mathematical modeling (Pars. 5-27 to 5-31) is essential in designing a station-keeping system for formation flight.

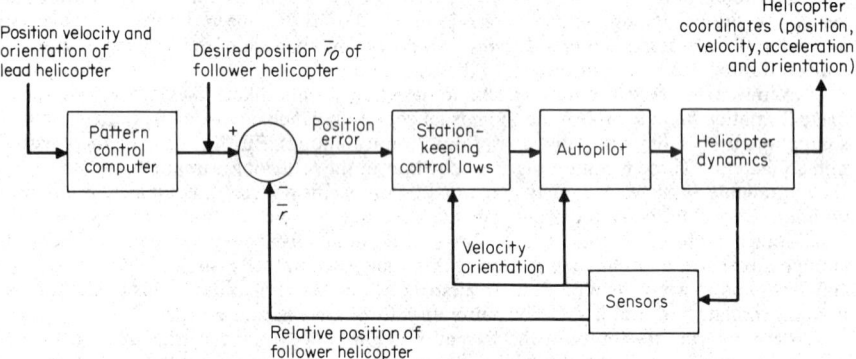

Fig. 5-1. A station-keeping system.

9. Design of a Station-keeping System. System design usually involves a trial-and-error procedure. That is, the system is designed conceptually, and then we perform an analysis to determine whether or not the system will meet the desired performance specifications. Modifications are made, and the effect of the modifications is determined. Because of the complexity of the system, these analyses usually require simulation (Par. 5-27).

In Fig. 5-1 there are three *feedback loops*. In one of the feedback loops, the relative position of the follower aircraft is compared with the desired relative position in order to generate signals to control the aerodynamic surfaces of the aircraft. The other two feedback loops provide aircraft orientation and velocity information to the station-keeping control laws and to the pilot-autopilot. In attempting to control the aircraft, we must be sure the system remains stable and does not have growing oscillations about the desired relative position. Hence the concepts of feedback control theory are important to us in this problem.

If helicopters are to be automatically maintained in a fixed formation, it is necessary to use *detection equipment* to determine their relative positions at all times. The measured relative positions of the vehicles are subject to some random errors. Also, the several vehicles are in an environment where they are subjected to random wind currents and gusts. Hence the systems engineer must consider the probabilistic aspects of the problem. Some of these concepts are introduced in the discussion of decision theory in Par. 5-26.

After the conceptual design of the station-keeping system is largely completed, we select the values of the parameters of the system to optimize the performance. It is desirable to make changes in the parameter values in a systematic manner so that near-optimal performance can be achieved with a minimum number of simulated flights. This process of selecting the values for parameters of the station-keeping system can be expedited if we use one or more of the techniques of optimization theory (Par. 5-31).

Before launching a project of designing a station-keeping system for formation flight of helicopters, we should prepare a schedule for the complete project, giving personnel requirements and completion dates for the several aspects of the project. This schedule may be accomplished with the assistance of PERT (program evaluation and review technique). A detailed description of this technique is given in Par. **5-20**.

A complete examination of the station-keeping system also includes the problems associated with takeoff and landing. If the landing zone is restricted in area and if the formation is large or is only one of several formations using the landing zone, then queueing theory (Par. **5-23**) is relevant to the problem.

Helicopter pilots agree that flying in close formation under adverse weather conditions requires considerable assistance from electronic equipment. On the other hand, pilots are unwilling to have the aircraft completely under the control of an automatic pilot unless they can intervene at any time they wish. The station-keeping system will not be used unless pilots are willing to use it; the system must accommodate to this and other requirements of the pilot. The pilot must be supplied with display equipment which is complete enough to assure him of his personal well-being, and he must be able to take control of the vehicle at his option. Hence, in designing a station-keeping system for the formation flight of helicopters, the systems engineer cannot ignore the important aspects of human engineering. Human factors are discussed in Pars. **5-13** to **5-18**.

10. Techniques for Problem Definition. Problem definition and the setting of objectives are at the core of systems engineering. The number of variables involved in a typical systems problem is immense, ranging from the availability of raw materials and electronic equipment to social, political, and economic pressure for change. There are only a few effective techniques available for measuring and organizing these variables. In short, we have few models for the problem-definition process.

11. Trees. A model which has wide applicability in showing relationships is the network, a set of points or vertices joined by line segments. A *tree* is a special type of network in which there is only one path between every pair of vertices. Figures 5-2 and 5-3 show the structure of a typical network and a tree, respectively. Figures 5-4 to 5-6 illustrate the use of trees in organizing the variables in an unstructured problem.

Relationship Trees. The *vertices* in the tree (Fig. 5-4) are items which must be considered, e.g., in an exploration of military air traffic control. Each vertex in the tree is related to the vertex which precedes it. As each vertex is developed, it becomes the focus for further exploration; this exploration leads to the definition of additional vertices in the tree. As the tree is examined in detail, it brings to mind different patterns of organization and different tree structures.

Objective Trees. In the tree of Fig. 5-5, each vertex is a specific *objective.* Each objective leads to subsidiary objectives. Development of an objective tree forces the systems engineer to develop well-defined objectives which build on the primary or initial objectives. For instance, both objectives which are subsidiary to "determine feasibility of formation flight" will probably require "development of mathematical models for aircraft" and "simulation of formation flight." The tree may be restructured to eliminate duplication of objectives.

Decision Trees. Each node of the tree in Fig. 5-6 constitutes a *decision.* Each possible outcome of a decision leads to a new point of decision. The decision tree is useful in helping to sort out the consequences of a sequence of decisions. Furthermore, if sufficient information is available to assign a priori probabilities to the outcome of each potential decision, the overall sequence of decisions can be, in some sense, optimized by the techniques of decision theory discussed later (see Par. **5-26**). Although the decision tree of Fig. 5-6 pertains to medical diagnosis, the same type of decision tree could find use many times over in the structuring of the problem of improving military air traffic control. Warfield and Hill [100] is a good general reference on the use of treelike models. See also Ref. 2 of Par. **5-36**.

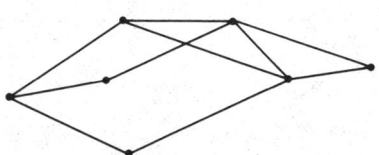

Fig. 5-2. Typical network.

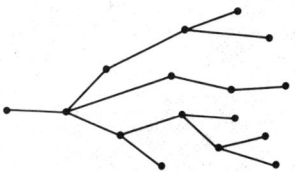

Fig. 5-3. Typical tree.

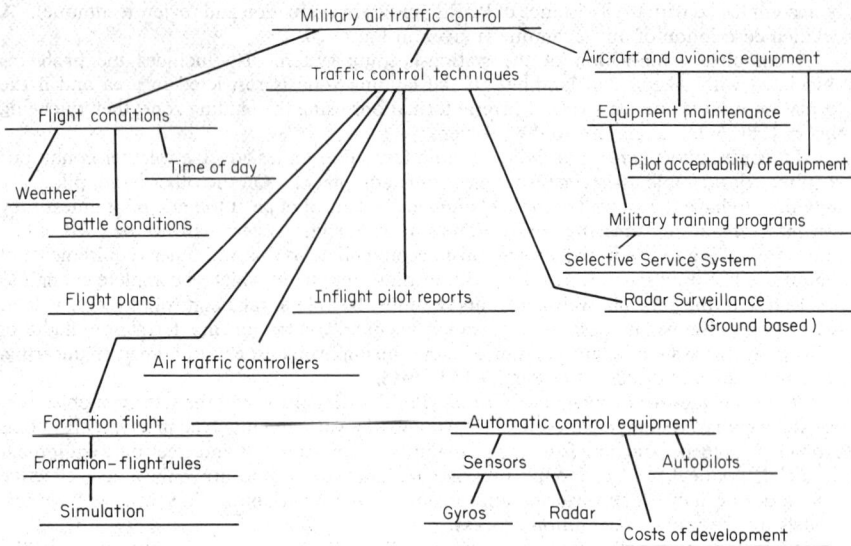

Fig. 5-4. Relationship tree.

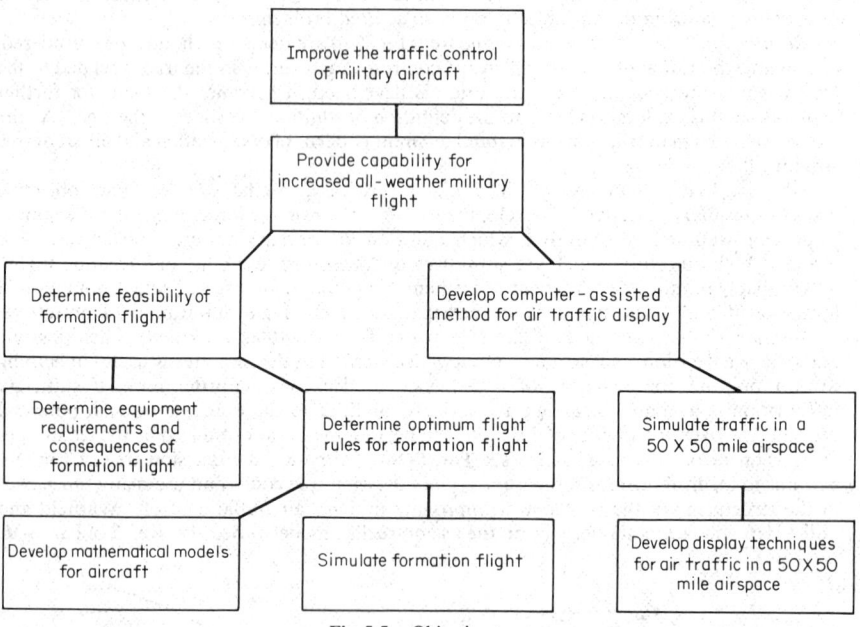

Fig. 5-5. Objective tree.

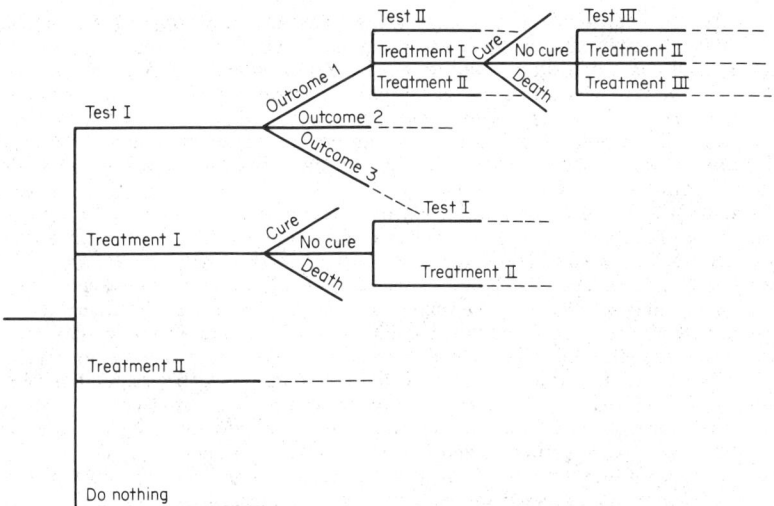

Fig. 5-6. Decision tree for medical diagnosis. *(From A. S. Ginsberg and F. L. Offensend, An Application of Decision Theory to a Medical Diagnosis-Treatment Problem, IEEE Trans. Systems Sci. Cybernetics, Vol. SSC-4, No. 3, September, 1968.)*

12. Utility Functions. Central to the setting of objectives is the concept of utility or value. A utility function expresses *preference;* that is, it is a measure of human satisfaction. The utility concept applies to any value—economic, political, moral, etc. The maximization of utility functions (or the minimization of cost functions) under various types of constraints is discussed in Pars. 5-31 to 5-35.

The concept of utility extends easily to the preference of an individual for several items jointly. The *joint utility function* indicates the relative preference of the individual for various combinations of the variables. The utility concept also applies to groups of people. We then refer to it as a *group utility function,* or *social utility function.*

To determine the utility function inherent in the feelings and opinions of the individuals and interest groups which are to be affected by a proposed system, measures must be found of the intensity of feeling or strength of desire of these individuals or interest groups. Von Neumann and Morgenstern have proposed a method for measuring and ordering individual preferences.[76] Suppose we are considering three items or variables, *A, B,* and *C.* These three items might be forms of entertainment: opera, theater, and movies.[1] First, the individual orders these three items according to his preferences, and we assign numbers to the ordering, for instance,

opera	theater	movies
5	2	1

Then we ask whether the individual prefers a choice of theater to a fifty-fifty combination of opera and movies; i.e., we ask if 2 is greater than (or is preferred to) $\frac{1}{2} \times 5 + \frac{1}{2} \times 1$. If the individual chooses theater, we must change the assigned numbers so that the preference is correct; we change the number assigned to theater to $3\frac{1}{2}$. We continue to ask similar preference questions and continue to change the number assigned to "theater" until we are satisfied that the three numbers are in the proper relation to each other. The utility function of the individual for these three forms of entertainment becomes

$$U(A,B,C) = 5A + NB + C$$

where N is the number which is eventually assigned to "theater." Then it makes sense to pick A, B, and C to maximize the utility subject to constraints (say, to stay within an entertainment budget of T dollars). Of course, we cannot assume the coefficient adjustment process will generate a utility function which precisely represents the preferences of the individual. In

---

[1] We here follow an example of Hall.[40]

some cases the individual may be comparing items that are not, in his mind, totally comparable. Furthermore, his preferences may not be constant in time.

In determining social utility for a group where all individuals may not have similar preferences, we are faced with the question of the relative importance of the preference of each individual in the group. Furthermore, no method of weighting the preferences of individuals will lead to a well-defined group-preference function unless "strength of preference" information is included. This fact is demonstrated by the following paradox, as stated by Arrow:*

"A natural way of arriving at the collective preference scale would be to say that one alternative is preferred to another if a majority of the community prefer the first alternative to the second, i.e., would choose the first over the second if those were the only two alternatives. Let A, B, and C be three alternatives, and 1, 2, and 3 three individuals. Suppose individual 1 prefers A to B and B to C (and therefore A to C); individual 2 prefers B to C and C to A (and therefore B to A); and individual 3 prefers C to A and A to B (therefore C to B). Then a majority prefer A to B, and a majority prefer B to C. We may therefore say that the community prefers A to B and B to C. If the community is to be regarded as behaving rationally, we are forced to say that A is preferred to C. But, in fact, a majority of the community prefers C to A. So the method just outlined fails to satisfy the condition of rationality as we ordinarily understand it."

The concept of utility applies to such values as pilot preference for different types of navigation equipment. Although utility seems a natural extension of the process of analytical modeling to the broad field of setting of objectives, its formal use is still infrequent. Many of the concepts of economics and psychology are helpful in defining and interpreting utility functions. References (Par. 5-36) which survey value theory and utility functions are 31, 32, 40, 53, 52, and 44. References of a more specialized nature are 22, 76, 42, 99, 5, 36, and 68. The latter three references pertain to group utilities. We summarize by noting that good objectives are tentative. The systems engineer must be free to change objectives as information becomes more complete.

HUMAN FACTORS

13. Introduction. The foregoing examples of air traffic control and of closely coordinated formation flight are typical of most major systems in that *human factors* enter into the setting of system objectives, as well as the design and implementation of the system itself. The setting of such objectives involves, either explicitly or implicitly, a tradeoff between human needs and/or preferences, on the one hand, and costs, on the other.

Human *needs* are most clearly evident in the design of life-support systems. The proper functioning of such support systems is intimately associated with the accomplishment of system objectives, e.g., performance of the lunar exploration mission and return to earth. *Preferences,* on the other hand, are often arrived at by either quantitative or qualitative techniques, which have many goals common to the techniques of market research.

Examples of incorporating user preferences in the design of a system can be seen in the design of a new transportation system (such as a Dial-a-Ride minibus system) or a new luxury-class high-speed rail system, where decisions must be made as to the functional relationship, as perceived by the consumer, of price to comfort, convenience, and time. Market research emphasizes direct preference surveys such as questionnaires or interviews. Traffic-demand analysts rely heavily on manifest preferences for various characteristics of an existing transportation system; these characteristics are related to the income and other attributes of the user and to characteristics of the region.

Preferences, whether measured or estimated, can be used to project the demand for certain characteristics of a proposed system. Such demand functions for transportation service have been based on assumptions and evidence about the value placed by customers on travel time and waiting time and on imperfectly defined measures of comfort and convenience. By comparing a synthesized demand function with observed demand functions for existing competitive systems, such as private automobiles, system design objectives concerning time, comfort, and convenience can be modified so as to result in an economically feasible configuration.

Economic feasibility is not the only criterion which must be traded against user preference.

*Arrow, Ref. 5, p. 3, Par. 5-36.

Social costs can be seen in the increased traffic congestion, noise, and pollution of the air attendant on the increased use of conventional automobiles. *Social equity* confronts planners as an issue when they lay out expressways and similar structures in urban areas. Thus social costs and social equity are both considerations in the design and implementation of most large systems.

One can list other systems in which design objectives are based on human preferences and demand curves. For example, user preferences concerning the maximum delay in obtaining a telephone dial tone or the minimum resolution permitted by a television raster may be measured either in laboratory tests or in a real situation.

14. Design Competition. The more conventional contributions of human-factors technology focus on the design and the implementation of systems whose objectives have been previously agreed upon. Most major systems arise out of a competitive design process. In this design competition, such criteria as cost, weight, and reliability augment the basic criteria established by system objectives and performance requirements. For example, in a competition between manned and unmanned operation of a deep-space exploration system, reliability is a major potential advantage for the manned system. The human, by his multimode capability, provides an advantage in both weight and reliability over an unmanned, redundant-component system. The human *modes* of obvious relevance are failure detection, replacement and repair, actuation and control, power generation, decision making, observation, and memory. The costs for the human component follow from the energy and weight requirements of the life support systems. In this example, and in others less obvious, the role of the human-factors specialist is to configure the man-machine interfaces and design the manual operations in such a way that, when human intervention is needed, this versatile component can maximally enhance the competitive advantage of the overall system. Systems for which human participation has been previously determined constitute the bulk of human-factors practice. Examples are the operation of automobiles, various industrial processes, computer consoles, any of a variety of military operations in which targets of opportunity may arise, and so forth.

15. Human Factors as a Discipline. Human factors as a technical discipline had its origin during World War II, but the activities in which human-factors specialists engage existed long before that time. In fact, Ulysses' adaptation of his transportation system to evade the Sirens is an example of the extensive past of human factors!

"Designed" is the word which distinguishes human factors from the field of industrial psychology with which it is occasionally confused. Industrial psychology evolved during the early twentieth century in response to the need to select industrial workers more effectively, to train them, and to maintain their morale so as to ensure good job performance. The industrial psychologists over the years developed batteries of tests to separate the more capable from the less capable. Industrial engineers concentrated on the man-production line, or man-tool interface, and developed methods for minimizing the time and effort required of a worker to accomplish his job. This effort established the field of time and motion studies.

During World War II, when it became necessary to expand the relatively small professional armies in Great Britain and in the United States and to utilize the new personnel in a variety of technically and physically demanding jobs, the batteries of selection tests were found to be ineffective. It was at this point that modern human factors came into being. The engineer of complex equipment was forced to examine his design and to accommodate it to the quantitatively expressed abilities and limits of the human operator.

16. Models of Human Behavior. The earliest work on *quantitative models* of human abilities was carried out in Great Britain under the auspices of military research agencies. The leading figure in this activity was Kenneth J. W. Craik.[19] The search for useful engineering models of human behavior, which Craik illustrates, is still the forefront of human factors. It is this sophisticated interest in human-operator plant dynamics, broadly viewed, which separates engineering psychology and human factors from the industrial engineering approach. A striking example of this process is described in a World War II report which demonstrated that secretaries, untrained in gunnery, could outperform trained gunners in the use of lead-computing gunsights when the sight dynamics were made appropriate for human needs.

Morale, motivation, social facilitation, and other questions of a more traditional psychological content have been bypassed, in the main, by human-factors specialists, because they do not lend themselves to the type of quantitative descriptions from which configuration competitions or equipment tradeoffs are feasible.

Individual differences, too, do not loom large in most human-factors studies because the populations examined are usually highly stratified, for example, military pilots or other highly selected personnel. Design techniques for maximizing human performance are more difficult when applied to highly diverse populations.

The role of human factors in setting system objectives is also an area which the main thrust of human-factors studies has avoided. This has been the bailiwick of political scientists and economists who have been concerned with the rules for governing group choices. Kenneth Arrow has been a particularly significant contributor.[5]

Finally, training, learning processes, and skill development have been underemphasized in much of human-factors work. There has been the dominating attitude that a well-coupled man-machine system minimizes most training problems. Representative references concerned with training are Bilodeau,[12] Kelley,[46] and Krendel-McRuer.[48]

Most of the human-factors principles which relate to the human as a component are derived from experimental psychology and from psychophysics. The bulk of the literature pertains to the *sensory inputs* (particularly vision and audition), the reception, storing, and transmitting of *information,* and the *physical outputs.* There is also a large store of human-operator information, ranging from general statements to handbook design data on such matters as vigilance, work cycles, and power output; man-machine task allocation; controls and displays; and the mass of detailed specifications on dials, knobs, work-space layout, and anthropometrics.

There are also descriptions of how the human component deteriorates, i.e., descriptions of the "graceful degradation" which characterizes human components in contrast to the catastrophic failure of inanimate components. Environmental effects—heat, vibration, noise, etc., fatigue, boredom, mental and physical efforts, drugs, etc., are considered. Finally, there is the literature on the performance of systems of men and machines, their communications needs, interfaces, and operating capacities.

17. Source References. Some useful source books are listed as Refs. 69, 97, 56, and 33 in Par. 5-36. These references present the core material for much of human-factors practice. More advanced thinking is found in Refs. 45 and 23. A good collection of early papers, many of which are difficult to obtain individually, is found in Ref. 92. References 25, 39, and 93 concentrate on human effort and work output. Several of these references deal with men in systems; Refs. 94 and 78 emphasize this area.

The most relevant scientific journals are *Ergonomics,* beginning in 1958; the IEEE publications which began in 1960 as the *IRE Transactions on Human Factors in Electronics,* and now after several changes is the *IEEE Transactions on Systems, Man and Cybernetics;* and *Human Factors,* beginning in 1961.

18. The Human as a Controller. The design of man-machine systems in which the human operator acts as a controller has a modern history beginning with World War II. An extensive literature has been developed on the subject, and some aspects have become routine engineering practice. McRuer and his coauthors[63,64] present tutorial discussions of models of a human operator as a component in a closed-loop man-machine system. For most system studies in which an operator performs compensatory tracking of a random appearing input, simple forms are adequate to describe the system low-frequency (0 to 1.5 Hz) characteristics of the operator (see Fig. 5-7).

Here the adaptive features—the human's prime virtue as a system component—are separated from the relatively unalterable characteristics. To estimate the operator's describing function, $Y_P(j\omega)$, in a given task, the adaptive equalizer characteristics are adjusted to values similar to those a servoengineer would select if he had the equalization $K(T_L j\omega + 1) / (T_I j\omega + 1)$ available to compensate for, and close the loop about, process dynamics comprising the unalterable operator characteristics in series with the controlled-element transfer function.

For a variety of plant dynamics, $Y_C(j\omega)$, this adjustment results in the open-loop system amplitude response $| Y_{OL} (j\omega) | = | Y_P(j\omega) Y_C(j\omega) |$ crossing the zero dB line with a slope of -20 dB decade. Transport delays and other lags arising from human operator dynamics can be represented near the crossover frequency, ω_C, by a single effective time delay, τ_e. Consequently, for a large range of circumstances, a two-parameter *crossover model,* $Y_{OL(j\omega)} = \omega_C e^{-j\omega\tau e}/j\omega$, occurs near ω_C. The phase margin is $\phi_M = \pi/2 - \tau_e\omega_C$.

Applications of the above general technique and its many elaborations require a reading of material in Refs. 62, 63, and 64. In particular, nonrandom appearing systems inputs, multiloop systems, environmental or physiological effects all impose different rules on the modeling process.

The adaptive model for random-input compensatory systems described above is useful for many different analytical purposes in manual control design. For example, it has been used:

1. To estimate human and overall man-machine system response and performance when the system forcing function and controlled-element dynamics are known.

2. To determine the barely uncontrollable controlled-element dynamics and controllability boundaries.

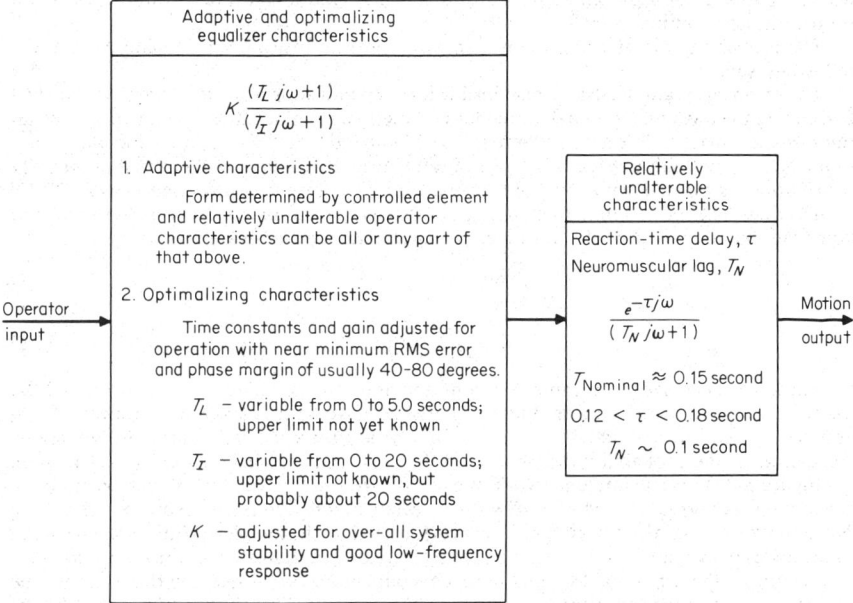

Fig. 5-7. Illustrative human operator model for compensatory operation; circa 1962.

3. To find the maximum forcing function bandwidth compatible with reasonable control action on the part of the human.

4. To indicate the type of additional system equalization (to be achieved via the display, in the control system, or by controlled-element modifications) desirable to achieve best manual control in the sense of minimum operator-equalization requirements.

5. To determine desirable machine outputs (for multidegree-of-freedom machines) for feedback to the display.

6. To delineate those features of the machine dynamics which are most important from a vehicle-handling-quality standpoint.

TECHNIQUES FOR SYSTEMS ANALYSIS

19. Introduction. The following paragraphs describe briefly a number of techniques of systems analysis. The principles of probability and statistics are of great importance for systems analysis. Feedback-control-theory principles are also valuable.

In the following discussion of techniques for systems analysis, several of the techniques (e.g., PERT, reliability and maintainability, theory of queues, decision theory) are based on the principles of probability and statistics, and other techniques (e.g., modeling, simulation) are closely related to the concepts of feedback control theory.

20. Program Evaluation and Review Technique (PERT).   PERT is a systematic way of scheduling and controlling a project. The method consists of three steps:

1. Developing the network of activities and events required to complete the project.
2. Estimating the time necessary to complete each activity in the network.
3. Analyzing the network to schedule effort effectively.

Each event in the network must be a specific accomplishment which occurs at a specific point in time. For instance, the initiation of the design of a subsystem for estimating the relative positions of 25 aircraft in formation flight constitutes an event. The design process itself is an activity which links two events.

The activities in the network include not only the design of physical equipment, but also design of operations, research, simulation and evaluation, testing, etc. Ideally, the time necessary to accomplish each activity in the network should be expressed in the form of a probability distribution.

The probability distributions are not known a priori. Instead, we estimate them in the following way.

The engineer or administrator responsible for accomplishing a particular activity is asked to estimate the most likely time required for completion, t_m, a pessimistic estimate t_p, and an optimistic estimate t_o. We assume there is little likelihood that the actual time for completion of the activity will fall outside this range of estimates. Regardless of the actual probability distribution associated with the time for each activity, extensive experience with PERT verifies that the expected time to complete an activity, t_e, and the variance σ^2 in the time to complete that activity are related to the above estimates approximately as

$$t_e = \frac{t_p + 4t_m + t_o}{6} \tag{5-1}$$

$$\sigma^2 = \left(\frac{t_o - t_p}{6}\right)^2 \tag{5-2}$$

The times necessary to complete various activities are essentially independent, and the expected values and variances are finite. Therefore, if we wish to add the times necessary to complete a sequence of activities, we can simply add the expected values and variances. Furthermore, the probability distribution of the sum tends toward a gaussian distribution.

Figure 5-8 shows a simple network of events. We use this network to demonstrate the scheduling technique. We work only with expected completion times (expressed in weeks). No activity in a chain can begin until the previous activity is completed (this is the meaning of the nodes or events). Under each node in the network we indicate the *earliest expected time* T_E, relative to the time of node 1, that the event could be accomplished. In this example, the earliest expected time for event 3 is 7 weeks after event 1. This time is determined by the longest or critical path, 1-4-3.

To find the earliest expected completion times, we start at the left of the network and, by adding expected activity times t_e, we work our way to the right. In all paths but the critical path, we could slow our efforts somewhat without delaying the 7-week expected completion time of the whole project.

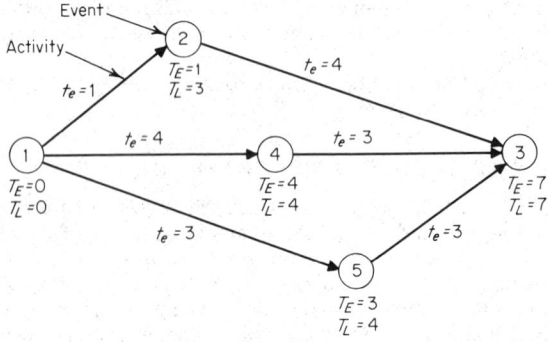

Fig. 5-8.   Simple event network.

To make explicit how much freedom (or slack) we have in these noncritical paths, we now start at the right of the network and move left; by subtracting expected activity times t_e, we find the *latest expected time* T_L, relative to the time of event 1, at which each event could be accomplished without delaying the 7-week expected completion time of the whole project. For some events, those not on the critical path, T_L will be later than T_E. The difference $T_L - T_E$ is the slack for that event. The manager can see quickly from the diagram where the critical paths are (by following events with zero slack) and how much slack or freedom he has in accomplishing various tasks. Some of this slack time might be better applied elsewhere, perhaps on a critical path.

PERT also yields an estimate of the probability of accomplishing any event in the network by any scheduled date. To obtain this information we need the probability distributions associated with each event. We assume these distributions are gaussian. The expected values of these distributions are the numbers T_E on the diagram. The variances are obtained by adding the variances of individual activities in a fashion similar to that used to obtain the numbers T_E. The probability of occurrence of an event by a particular time t is then equal to the integral of the appropriate gaussian probability density function up to the time t (see the shaded region in Fig. 5-9).

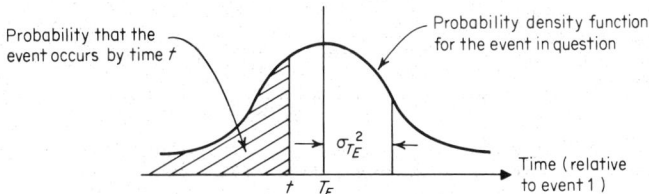

Fig. 5-9. Probability that event will occur by time t.

Computer programs are available which automatically compute critical paths, slack times, and probabilities of meeting schedules. Thus this tool is readily accessible to managers in efficient scheduling of projects as part of the second phase *(project planning)* of the systems engineering process. References that discuss PERT are 24, 77, and 2, Par. **5-36**. The latter reference also examines other scheduling methods.

21. Reliability and Maintainability. The reliability of a system is defined as the probability that it will perform its assigned functions for a specified period of time under given conditions. The maintainability of a system is the probability that, in the event of failure of the system, maintenance action under given conditions will restore the system within a specified time. Both these factors are extremely significant for complex systems. Low reliability or maintainability for a military air traffic control system, or even for a formation flight system, can have disastrous consequences in terms of human life and military operations. Estimates of reliability and maintainability must be included in the initial tradeoffs between performance, cost, and schedule.

The reliability of the system is affected not only by the choice of individual parts in the system, but also by manufacturing methods, quality of maintenance, type of user, etc. Most complex systems have significant electronic components. The communication, sensing, decision making, and control-stabilizing subsystems of a station-keeping system, for example, are primarily electronic in nature. Therefore we focus on the reliability of electronic equipment. The concepts which we discuss apply as well to other parts of the system.

22. Failure Behavior of Electronic Equipment. Electronic equipment exhibits a characteristic failure pattern (see Fig. 5-10). The failure rate is high but declining during the initial phase, as marginal components are discovered and eliminated. The second stage exhibits a relatively constant failure rate. Near the end of life of the system, the failure rate increases due to wearing out of components. The flat portion of the curve is the normal operating region. We estimate reliability only for operation during this normal operating period. Then failures can be considered random and independent; they are adequately represented by a Poisson process. Such a process with failure rate λ exhibits an exponential failure law:

$$R = e^{-\lambda t} \tag{5-3}$$

where R is the reliability or probability of successful operation for a length of time t. Notice that for a system to operate without failure for a given period with high reliability, the mean time between failures $1/\lambda$ must be much larger than the time period t.

The failure rate λ can be obtained for individual components (resistors, capacitors, transistors, small circuits, etc.) by testing under conditions similar to those expected in the final system. From the reliability of these individual components we determine the reliability of the overall electronic system. The simplest situation occurs if failure of one part is independent of failure of the others and if for successful operation every part must operate. Then the overall reliability of the system can be obtained by simply multiplying the reliabilities of the parts (or adding their failure rates).

$$R_{\text{system}} = e^{-\lambda_1 t} \times e^{-\lambda_2 t} \times \cdots \times e^{-\lambda_n t} \qquad (5\text{-}4)$$

The lack of accurate failure data for parts under appropriate operating conditions is one of the chief difficulties in reliability prediction. Yet the extreme shocks, vibrations, and temperatures to which components of many systems are subjected have a strong effect on their failure rate.

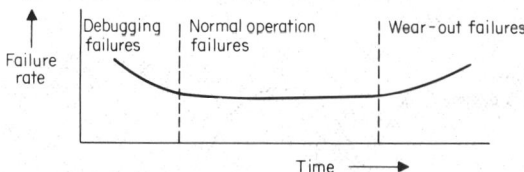

Fig. 5-10. Failure behavior for electronic equipment. *(From R. E. Machol (ed.), "Systems Engineering Handbook," Chap. 33, McGraw-Hill Book Company, New York, 1965.)*

There are several techniques for enhancing the reliability of a system. Each technique increases the cost of the system and usually degrades performance as well. One approach to reducing failures caused by a particularly hostile environment is to use particularly rugged (or derated) components. An alternative is to reduce the severity of the environment by shock mounting, temperature regulation, etc.

The basic factor in component reliability is that provided by *large-volume production* using unchanging processes. These conditions are also necessary to permit accurate prediction of the reliability of the components. Thus, to design a reliable system, we attempt to use standard parts—reliable parts are better than parts which are "optimum" but untested. Use of standardized circuitry is as important as use of standardized components in achieving reliability. Circuit proliferation also complicates manufacture, maintenance, and storage of spare parts in the field.

An important technique for increasing reliability is the use of *redundancy.* In a large continuously operating system, some form of redundancy is necessary to allow for periodic maintenance. In some cases, a completely redundant system can be kept in "standby" condition.

Information which is transmitted within the system can be coded redundantly in order that errors may be detected and corrected. Between these two extremes we have provision for redundant personnel for operating the system (e.g., a copilot), redundant circuits, redundant parts, etc. For example, a single diode can be replaced by a series-parallel combination of four diodes, as shown in Fig. 5-11, thereby protecting against an open or short circuit in a single diode. Of course, the use of such a redundancy technique results in reduced circuit speed and increased power dissipation. Furthermore, the determination of system reliability is complicated by the fact that certain combinations of part failures do not lead to a system failure.

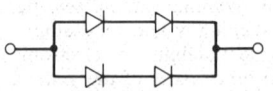

Fig. 5-11. Redundant circuit replacement for a single diode.

Determination of *system maintainability* is much like determination of reliability, but we test the *rates of repair* or replacement of units in the system. Experience has shown that repair times are adequately represented by a log-normal probability distribution (a normal distribution if plotted vs. the log of the repair time). By appropriate manipulation of failure rates and rates of repair, we can generate a curve indicating the probability of repairing the system within a given number of hours.

Because of the lack of appropriate data, the prediction of maintainability has not reached the same level of sophistication as has the prediction of reliability. However, certain design features are known to enhance maintainability. Equipment can be kept simple and accessible; modular construction adopted to facilitate quick repair; and test equipment can be made automatic and, in some cases, built into the equipment. Readily available parts can be used; the test equipment can be subjected to the same reliable design as the system itself; maintenance diagrams can be simply and logically arranged. References which deal with reliability and maintainability are 14, 87, 13, 7, and 86, Par. **5-36**.

23. Theory of Queues. The dynamic operation of a system is only partly described by the physical properties of the system. The operation is also affected by the fluctuating demands made on the system by human operators and external physical forces. These fluctuating demands are so complex that we classify them as random, and express them in terms of a random (stochastic) process, with a certain probability distribution. For example, the maneuvering required of an airplane or formation of airplanes in order to evade a missile is essentially a random process. The electrical signal produced by a position-sensing device and the number of aircraft waiting to land at a given airport (as a function of time) are also random processes.

We are concerned primarily with random processes which are stationary, processes for which the probability distribution does not vary with time. Furthermore, most processes of interest are markovian; the probability that the process takes on a certain value (or vector of values) at a given future time depends only on the value at the present time, but not on the past history of values. Each of the above examples is a Markov process.

24. Discrete-State Markov Processes. We focus here on Markov processes for which the set of possible values *(state space)* is discrete. Examples of discrete-state Markov processes, pertinent to a formation flight system, are the cumulative number of wind gusts which exceed a given threshold, the cumulative number of missiles directed at the formation, and the number of simultaneous events requiring pilot attention. Each of these numbers varies randomly throughout a particular flight. Almost every random counting process is a discrete-state Markov process. The example of waiting aircraft, mentioned earlier, is also of this type. Each of the following examples can be considered a random-service system: a sequence of units (aircraft, wind gusts, or events) arrives at random and is handled by a service facility (an airport, a flight formation, or a pilot). The time it takes for the server to handle an arrival (land and park a plane or handle a wind gust) is referred to as the service time. For obvious reasons, the theory of behavior of these service systems is called *queueing theory.*

Exploration of an air traffic control system would raise many questions concerning aircraft waiting to land. What is the expected wait before landing? The expected length of the queue? The probability that the queue will ever be longer than a given number? The probability of having no aircraft in the queue for a given length of time? Each of these questions must be explored under various assumptions concerning the characteristics of arrival of aircraft, characteristics of airport handling of arriving aircraft, availability of alternative airports, landing priorities for disabled aircraft, etc.

A fundamental characteristic of a stationary Markov process is that the probability of transition from a given state at time t_1 to another given state at time t_2 depends only on the time difference $t_2 - t_1$. For a service system to be of this type, arrival times and service times must be truly random. It can be shown that completely random arrivals must be Poisson arrivals; that is, they must obey

$$A(n) = \frac{(\lambda t)^n}{n!} e^{-\lambda t} \tag{5-5}$$

where $A(n)$ is the probability that precisely n arrivals occur in an arbitrary interval of length t, and λ is the mean rate of arrivals. For Poisson arrivals, the time between arrivals, $A(0)$, is exponentially distributed. If we consider the completion of service of one customer as a negative arrival, we recognize that for random (Poisson) completions of service, the service time (time between completions for a busy server) must be exponentially distributed; that is,

$$S(t) = e^{-\mu t} \tag{5-6}$$

where $S(t)$ is the probability that the service operation will take longer than t, and μ is the mean service rate. When arrivals or service times do not follow Eqs. (5-5) and (5-6), we can often add operations to the system model so that the model has Poisson arrivals and exponential service times. See Parzen[79], Par. 5-36.

We start with a service system which is discrete in time, one where we are interested in the state only at the end of discrete periods. This type of system is described by the equations

$$p_i(m+1) = \sum_j T_{ji}\, p_j(m) \qquad m = 0, 1, 2, \ldots \tag{5-7}$$

where $p_i(m)$ is the probability that the state is i at the end of the mth period, and T_{ji} is the probability of transition from state j to state i during one time period. Since $p_i(m)$ is a probability mass function, $p_i(m) \geq 0$ and $\sum_i p_i(m) = 1$. Similarly, $T_{ji} \geq 0$ and $\sum_i T_{ji} = 1$; T_{ji}, as a function of i, is a probability mass function.

Many of the interesting questions concerning a service system have to do with its steady-state behavior (its limiting behavior as $m \to \infty$). We denote the steady-state solution, if any, to Eq. (5-7) by the steady-state probabilities P_i, the limiting values of $p_i(m)$ as $m \to \infty$. When a steady-state occurs, P_i is independent of m and independent of the initial state. We obtain equations in the steady-state probabilities P_i from Eq. (5-7) and the properties of probability mass functions:

$$P_i = \sum_j T_{ji}\, P_j \qquad \sum_i P_i = 1 \tag{5-8}$$

As an example of a service system for which Eqs. (5-7) and (5-8) are the model, we take the following grossly simplified situation. An airport service shop performs regular aircraft maintenance in a fixed (unit) interval of time. Aircraft arrive for maintenance according to the Poisson distribution, Eq. (5-5). We observe the system only at unit intervals. No aircraft wait for service more than a unit interval of time; they go elsewhere for service. Thus there is no real queue. The shop has two states. State 0 is the idle state; state 1 is the busy state. The process is Markov; the state in the succeeding interval depends only upon the state and arrivals in the present interval.

We first determine the state transition probabilities T_{ji} for use in Eqs. (5-7) and (5-8). If the shop is idle at interval m, it will still be idle at interval $m+1$ if and only if there are no arrivals during the mth unit interval. Thus $T_{00} = A(0) = e^{-\lambda}$. Similarly, transition from the busy state to the idle state requires that there be no arrivals, and $T_{10} = e^{-\lambda}$. Since $T_{j0} + T_{j1} = 1$, $T_{01} = T_{11} = 1 - e^{-\lambda}$. Then Eq. (5-8) becomes

$$P_0 = T_{00}P_0 + T_{10}P_1 = e^{-\lambda}P_0 + e^{-\lambda}P_1$$
$$P_1 = T_{01}P_0 + T_{11}P_1 = (1-e^{-\lambda})P_0 + (1 - e^{-\lambda})P_1$$
$$1 = P_0 + P_1$$

In general, Eqs. (5-8) are either redundant or not solvable. In this instance the solution is

$$P_0 = e^{-\lambda} \qquad \text{and} \qquad P_1 = 1-e^{-\lambda}$$

Clearly, the fraction of idle time P_0 is never zero, even for large mean arrival rates λ.

In many operational systems, the changes in state can occur at any time, not just at the end of discrete periods. To use the Markov model, Eq. (5-7), for continuous-time operations, we subdivide time into infinitesimal intervals of length dt. We represent the transition probability T_{ji} by $R_{ji}\,dt$, the probability of transition from state j to state i during the small interval dt. Then R_{ji} is a rate of transition (with units equal to probability per unit time). We rewrite Eq. (5-7) as

$$p_i(t + dt) = \sum_{j \neq i} R_{ji}dt\, p_j(t) + T_{ii}p_i(t) \tag{5-9}$$

where $p_i(t+dt)$ is the probability that the system is in state i at time $t+dt$. We make use of the fact that $T_{ii} = 1 - \sum_{j \neq i} R_{ij}dt$ to rewrite Eq. (5-9) as a differential equation.

$$\lim_{dt \to 0} \frac{p_i(t+dt) - p_i(t)}{dt} = \lim_{dt \to 0}\left[\sum_{j \neq i} R_{ji}p_j(t) + \left(1 - \sum_{j \neq i} R_{ij}dt\right)\frac{p_i(t)}{dt} - \frac{p_i(t)}{dt}\right]$$

or

$$\frac{dp_i(t)}{dt} = \sum_j R_{ji}p_j(t) \qquad \text{where } R_{ji} \triangleq \sum_{j \neq i} R_{ij} \tag{5-10}$$

We are also interested in the steady-state solution for this continuous-time case. We obtain the steady-state equation corresponding to Eq. (5-8) by letting $\dfrac{dp_i(t)}{dt} \to 0$ and $p_i(t) \to P_i$ as t gets large.

$$\sum_j R_{ji}P_j = 0 \qquad \sum_i P_i = 1 \tag{5-11}$$

In effect, Eqs. (5-11) state that the average number of transitions out of state i equals the average number of transitions into state i.

25. Queueing in Air Traffic Control.   We illustrate the use of Eqs. (5-11) by the following example. Let the time required for an airport to service an aircraft (the time required for takeoff or for approach and landing) be exponentially distributed with mean service rate μ. Let the arrivals (in the air or on the ground) be Poisson with mean rate λ. We initially assume that the airport has two states, the idle state 0 and the busy state 1. Any arrivals during the busy state are turned away and there is no queue (a somewhat unlikely assumption for aircraft on the ground). We must determine the rates of transition R_{ji}. The probability of transition from state 0 during the interval dt is the probability of completion of service of an aircraft during the interval dt. Since the mean rate of completions is μ,

$$T_{10} = \mu \, dt = R_{10} \, dt$$

and $R_{10} = \mu$. Similarly, R_{01} is determined by the probability of an arrival at the idle airport during the interval dt. The mean arrival rate is λ; thus $T_{01} = \lambda \, dt = R_{01} \, dt$ and $R_{01} = \lambda$. Since $R_{ii} = -\sum\limits_{j \neq i} R_{ij}$, $R_{00} = -\lambda$ and $R_{11} = -\mu$.

The steady-state equations follow from Eqs. (5-11).

$$-\lambda P_0 + \mu P_1 = 0$$
$$\lambda P_0 - \mu P_1 = 0$$
$$P_0 + P_1 = 1$$

Thus the fraction of the time the airport will be busy is $P_1 = \lambda/(\lambda + \mu)$. The idle fraction will be $P_0 = \mu/(\lambda + \mu)$. Since aircraft arrive at random, the number of aircraft turned away by the system is equal to $\lambda/(\lambda + \mu)$, the fraction of time the airport is busy.

We modify the above problem to make it more realistic by requiring all arriving planes to wait for service; a queue is formed (including the planes waiting for takeoff, as well as those which are waiting to land). We define the state of the system as the length of the queue (including the single aircraft which is using the runway). We now have an infinite number of states, $0, 1, 2, \ldots$. We need the rates of transition for use in Eqs. (5-11). We found in the previous example that the probability of an arrival in the interval dt is $\lambda \, dt$; therefore $R_{i,i+1} = \lambda$. The probability of a completion of service in the interval dt is $\mu \, dt$; thus $R_{i,i-1} = \mu$. The probability of two or more events in the interval dt is zero; $R_{i,i-2} = R_{i,i+2} = 0$, etc. In order to have steady state, the average number of transitions out of a given state must equal the average number of transitions into that state. Therefore the steady-state equations are

$$\lambda P_0 = \mu P_1$$
$$(\lambda + \mu)P_k = \lambda P_{k-1} + \mu P_{k+1} \qquad k \geq 1 \tag{5-12}$$

We solve Eqs. (5-12) by means of generating functions (z transforms). Define the transform F of the set of steady-state probabilities P_k by

$$F(z) \triangleq \sum_{k=0}^{\infty} P_k z^k \tag{5-13}$$

Then

$$F(1) = \sum_{k=0}^{\infty} P_k = 1 \tag{5-14}$$

We can obtain P_k from $F(z)$ by using

$$P_k = \frac{d^k}{dz^k}F(z)\bigg|_{z=0} \tag{5-15}$$

We multiply the kth form of Eqs. (5-12) by z^k and sum from 1 to ∞ to obtain

$$(\lambda + \mu) \sum_{k=1}^{\infty} P_k z^k = \lambda \sum_{k=1}^{\infty} P_{k-1} z^k + \mu \sum_{k=1}^{\infty} P_{k+1} z^k$$

or

$$(\lambda+\mu)[F(z) - P_0] = \lambda z F(z) + (\mu/z)[F(z) - P_0 - P_1 z]$$

Then, letting $\rho = \lambda/\mu$, $F(z) = P_0/(1 - \rho z)$. Applying Eqs. (5-14) and (5-15), we find $P_0 = 1 - \rho$, $F(z) = (1 - \rho)/(1 - \rho z)$, and

$$P_k = \rho^k(1-\rho) \quad k = 0, 1, 2, \ldots \tag{5-16}$$

These are the probabilities that k aircraft are in the queue in the steady state.

The number $1 - P_0 = \rho = \lambda/\mu$ is the fraction of time that the airport is busy. The mean length of the queue is

$$L = \sum_{k=0}^{\infty} kP_k = F'(z)_{z=1} = \rho/1 - \rho \tag{5-17}$$

It should be intuitively clear that the mean queue length L equals the mean wait in the queue, W, divided by the arrival rate, λ, or

$$W = L/\lambda = \frac{1}{\mu - \lambda} \tag{5-18}$$

It is also apparent that these steady-state values apply only if $\lambda < \mu$. Otherwise, the queue grows without bound.

We see from Eqs. (5-17) and (5-18) that the mean length of the queue and mean wait in the queue are inversely proportional to the fraction of time the airport is idle, $P_0 = 1 - \rho$. The idle time allows for occasional peaks in arrivals without long waits for service. A general conclusion, which applies for many different arrival and service distributions, is that the queue will be very long unless the mean service rate is sufficiently greater than the mean arrival rate (and the idle time is appreciable).

Since it is not desirable to eliminate idle time, what can be done to allow the airport to handle more aircraft? Increasing the number of servers (runways) increases the mean service rate μ. Reducing the variability (variance) in service and arrivals can also significantly increase the number of aircraft handled by the airport without increasing the queue length or the wait in queue. The theory of queues is explored in detail in Refs. 79, 71, 70, 89, 98, and 85, Par. 5-36.

26. Decision Theory. During the design and testing of a physical system we find numerous occasions to make measurements, in order to evaluate the performance of system components. For example, we need to determine component reliability. A measurement constitutes a purchase of information. How much information (or how many measurements) should we purchase? The answer to this question depends upon the expense involved in taking each measurement and the consequences of an error in measured information.

Much measurement also takes place during operation of a complex system. The formation flight system discussed earlier will certainly require automatic measurement of velocities and accelerations of each aircraft relative to neighboring aircraft, the lead aircraft, and the ground. The information obtained from these measurements is needed to control the individual aircraft and the whole formation of aircraft. An aircraft radar system must decide, on the basis of electronic measurements, whether a target is present. We call this decision process *signal detection*, or *hypothesis testing*. If a target is present, the system must estimate the range and velocity of the target. We refer to this process as *parameter estimation*, or *signal extraction*.

A physical measurement generally includes a random error because of vibration of the transducer, electrical noise, finite resolution of the transducer, etc. How do we interpret our measurements so that decisions (e.g., the presence or absence of a target) and estimates (of velocities, distances, failure rates, etc.) are most likely to be accurate?

Suppose we take n independent measurements of a single quantity y, a random variable with mean m, and variance σ^2. The sample mean $\mathbf{y}$ is defined as

$$\mathbf{y} \triangleq \frac{1}{n} \sum_{k=1}^{n} y_k \tag{5-19}$$

The expected value of $\mathbf{y}$ is m. The variance V in the sample mean is

$$V(\mathbf{y}) = \sigma^2/n \tag{5-20}$$

Thus measurement of $\mathbf{y}$ (n measurements of y) provides a more reliable estimate of the parameter m than does a single measurement of y. A high value of σ^2 implies a low value of information per measurement. We can think of σ^2 as a measure of the *background noise,* or measurement error. As the sample size n increases, the total acquired information increases. Thus the reciprocal of $V(y)$ is analogous to the concept of signal-to-noise ratio which is used by communications engineers. We desire to obtain a low value of $V(y)$. By Eq. (5-20) we can obtain as low a value of $V(y)$ as we wish by taking enough measurements.

To avoid the expense of unnecessary measurements we should try to design the measurement process so that σ^2, the variance in y, is low (e.g., by use of an accurate, well-placed transducer). Furthermore, we take no more samples than necessary to obtain adequate statistical confidence in the sample mean. The sum of a set of independent measurements is generally a normally distributed random variable. Therefore the probability that a measurement of the sample mean y will deviate from the mean m by no more than k standard deviations is

$$P[\, | \, \mathbf{y} - m \, | \, < k\sqrt{V(\mathbf{y})}\,] = 2 \, \text{erf}(k) \qquad (5\text{-}21)$$

where the error function, erf, is the integral of the normal density function. Suppose we require that $P(\, | \, \mathbf{y} - m \, | \, < c) = 0.95$; that is, we select the number of measurements of y such that the probability is 0.95 that the average y of the measurements deviates from m by less than c. From a graph of the error function we find that $k = 2$ in order that $2 \, \text{erf}(k) = 0.95$. From Eqs. (5-20) and (5-21) we determine that $c = k \sqrt{V(\mathbf{y})} = 2 \sqrt{\sigma^2/n}$, or $n = 4\sigma^2/c^2$. Four measurements will be enough if we are satisfied with $c = \sigma$ and a confidence level of 0.95; it would take 400 measurements to yield $c = \sigma/10$ for the same confidence level.

Averaging of measurements is a suitable method for parameter estimation under conditions where repetitive sampling is possible. Control of aircraft velocity relative to a neighboring aircraft in a formation, however, requires almost continuous sampling of velocities. Furthermore, the samples in the sequence are correlated; we can obtain more information about the true velocity (as a function of time) by analyzing the whole sequence of measurements simultaneously than by estimating the velocity at each instant independently. We are faced with a multidimensional parameter estimation (or signal extraction) problem. By classical signal extraction techniques, we can devise an optimum estimate of the signal as a function of the sequence of measurements.[67] Like the function y of Eq. (5-19), this signal estimate is a random function. It specifies an estimate of the signal only when the measurements are given values. Such a random function is called an *estimator.* The classical estimators are optimum in a mean-square or minimum-variance sense. Of course, at each instant we want the expected value of the estimator to equal the expected value of the signal being estimated. Such an estimator is said to be *unbiased.* If the samples in the sequence are in fact uncorrelated, the minimum-variance unbiased estimator is just the average of the samples as in Eq. (5-19).

The problems of parameter estimation and signal detection can be approached through the optimization concepts of decision theory. Suppose that our noise-corrupted measurements M can be represented by

$$M = S + N \qquad (5\text{-}22)$$

where S is the signal we wish to measure, and N is the noise signal; i.e., we assume the measurement error or noise is additive. Suppose, further, that D is a decision variable; D may take on a range of values if we are estimating a parameter; it will take on only two values (say 1 or 0) if we are deciding between two hypotheses (e.g., the presence or absence of a target). We assume we know P_N, the probability density function for the noise. Then, because of the additive nature of the noise, we can determine $P_{M \, | \, S}$, the conditional probability of M given S.

$$P_{M \, | \, S}(M,S) = P_N(M - S) \qquad (5\text{-}23)$$

(Of course, if S and N are multidimensional, all probability density functions will be multidimensional.) We may or may not know a priori the probability density function for the signal, P_S. The basic problem is to determine a *decision function* (or estimator) $P_{D \, | \, M}(D,M)$ which assigns to each possible measurement M a unique decision D. Although D is usually a deterministic function of M, it is convenient to express it as the conditional probability density function $P_{D \, | \, M}$.

How do we select a decision function? We base our selection on a subjectively determined *cost function* $C(S,D)$ which indicates the cost of an incorrect decision. Typically, $C(S,D)$ is zero for a correct decision and increases progressively with the seriousness of the

SYSTEMS ENGINEERING

error. The decision function is chosen to minimize the expected value of the cost. Let $P_{M|S}$ and $P_{D|S}$ be the conditional probability densities for the measurement and the decision, respectively. Then, in terms of the decision function,

$$P_{D|S}(D,S) = \int P_{D|M}(D,M)\, P_{M|S}(M,S)\, dM \qquad (5\text{-}24)$$

We define the *conditional* risk $r(S)$ by

$$r(S) = \int P_{D|S}(D,S)\, C(S,D)\, dD \qquad (5\text{-}25)$$

Clearly, $r(S)$ is just the expected cost, given the signal S. Then the *average risk* R (or expected cost) is

$$R = \int r(S)\, P_S(S)\, dS \qquad (5\text{-}26)$$

A decision function which is selected to minimize Eq. (5-26) is called a *Bayes decision rule*.

We demonstrate the optimization process with a specific example. Suppose we wish to detect the presence or absence of a signal (of specified form but random amplitude) by analysis of noisy measurements. This is a *binary detection problem; S* and *D* each take on the value 1 (presence) or 0 (absence). Let q be the a priori probability that the signal is absent and $p = 1 - q$ the probability that it is present. Since the alternatives are discrete, we can express P_S in terms of the Dirac delta function:

$$P_S(S) = q\,\delta(S-0) + p\,\delta(S-1) \qquad (5\text{-}27)$$

We divide the space of measurements M into two regions, R_0 and R_1; R_0 is the region for which the decision is $D=0$ (signal absent), R_1 is the region for which the decision is $D=1$ (signal present). Our problem is to determine these regions. The decision function can be expressed in terms of these regions.

$$P_{D|M}(D,M) = \begin{cases} \delta(D-0) & \text{for } M \text{ in } R_0 \\ \delta(D-1) & \text{for } M \text{ in } R_1 \end{cases} \qquad (5\text{-}28)$$

We combine Eqs. (5-24) to (5-26), substitute Eqs. (5-27) and (5-28), and integrate with respect to S and D.

$$R = \int dS\, P_S(S) \int dD\, C(S,D) \int dM\, P_{D|M}(D,M)\, P_{M|S}(M,S)$$

$$= q \int dD\, C(0,D) \int dM\, P_{D|M}(D,M)\, P_{M|S}(M,0) + p \int dD\, C(1,D) \int dM\, P_{D|M}(D,M)\, P_{M|S}(M,1)$$

$$= q[C(0,0)\int_{R_0} P_{M|S}(M,0)\, dM + C(0,1)\int_{R_1} P_{M|S}(M,0)\, dM]$$

$$+ p\, C(1,0)\int_{R_0} P_{M|S}(M,1)\, dM + c(1,1)\int_{R_1} P_{M|S}(M,1)\, dM \qquad (5\text{-}29)$$

We note that Eq. (5-29) expresses the risk as a function of the probability P_f of false alarm (the probability that $D=1$ given $S=0$) and the probability P_m of missing the signal (the probability that $D=0$ given $S=1$):

$$P_f = \int_{R_1} P_{M|S}(M,0)\, dM = 1 - \int_{R_0} P_{M|S}(M,0)\, dM$$

$$P_m = \int_{R_0} P_{M|S}(M,1)\, dM = 1 - \int_{R_1} P_{M|S}(M,1)\, dM \qquad (5\text{-}30)$$

If we assign zero cost to each correct decision, $C(0,0) = C(1,1) = 0$, and the risk equation (5-29) becomes

$$R = qC(0,1)\, P_f + pC(1,0)\, P_m$$

$$= qC(0,1) + \int_{R_0} [pC(1,0)\, P_{M|S}(M,1) - qC(0,1)\, P_{M|S}(M,1)]\, dM \qquad (5\text{-}31)$$

The problem of picking the decision rule to minimize the risk reduces to picking the region R_0 to minimize Eq. (5-31). The minimum obviously occurs if R_0 contains all values of M for which the integrand is negative and no values of M for which it is positive. We express this condition in terms of the *likelihood ratio*

$$L(M) \triangleq \frac{pP_{M|S}(M,1)}{qP_{M|S}(M,0)} = \frac{pP_N(M,1)}{qP_N(M)} \tag{5-32}$$

[where we have substituted Eq. (5-23)] and the *threshold*

$$T \triangleq \frac{C(0,1)}{C(1,0)} \tag{5-33}$$

The boundary between the regions R_0 and R_1 in the measurement space is defined by

$$L(M) = T \tag{5-34}$$

A binary detection system which minimizes the risk would take a sequence of samples of the noisy signal from the measuring instrument (perhaps a radar). The sequence of samples constitutes a measurement vector M. The system would then compute the likelihood ratio, equation (5-32). If $L(M) \geq T$, the system would decide $D = 1$ (the signal is present); if $L(M) < T$, it would decide $D = 0$ (the signal is absent). The actual manipulations performed on the data to implement this decision rule depend upon the specific noise probability density function and upon the nature of the signal. The signal is called *coherent* if the time of its occurrence is known; otherwise it is *incoherent*. It can be shown that for gaussian noise and a coherent signal the operations required to implement the decision rule, equation (5-34), are linear; for incoherent signals, quadratic operations are required in order to correlate the measurements.[18]

Of course, the Bayes decision rule, Eq. (5-34), could not be implemented without the a priori signal probabilities p and q and the a priori noise density function P_N. Without this a priori information, we cannot hope to define a decision rule which is as accurate as Eq. (5-34). One approach which can be used if we do not have the a priori signal probabilities is to choose that decision rule which is optimum for the worst possible signal probabilities; that is, we can choose $P_{D|M}$ and P_S to yield the *minimax average risk*.

$$\min_{P_{D|M}} \max_{P_S} R \tag{5-35}$$

The minimax decision rule determined by relation (5-35) is the largest Bayes decision rule. Detailed examinations of decision theory and statistical communication theory may be found in Refs. 101, 67, and 82, Par. **5-36**.

27. Modeling and Simulation. See Refs. 15, 16, 27, 28, 37, 59, 60, 65, 73, 74, 90, 95, 96, and 104. Modeling and simulation play a paramount role in systems engineering. Through modeling and simulation we can evaluate plans, designs, or ideas and change them before implementing the real system. Models may be classified as mathematical or physical models; continuous or discrete models; stochastic or deterministic models; static or dynamic models.

Examples of physical models include scale models used in wind tunnels and water tanks to study aircraft and ship designs. Electric circuits are used as physical models of mechanical systems; electrical resistors, inductors, and capacitors are used as physical models of electrical transmission lines or of other distributed parameter systems.

In mathematical models the system is described by mathematical equations. For example, the voltage balance equation for the electric circuit of Fig. 5-12 is given by

$$L\ddot{q} + R\dot{q} + \frac{q}{C} = e(t) \tag{5-36}$$

where L, R, and C are the circuit inductance, resistance, and capacitance, respectively; q is the charge on capacitor; $\dot{q}$ is the time rate of change of the charge or the current i; $\ddot{q}$ is the second time derivative of the charge; and $e(t)$ is the source voltage. A mathematical model for the simple mechanical system of Fig. 5-13 is given by

$$M\ddot{x} + D\dot{x} + Kx = f(t) \tag{5-37}$$

where M, D, and K are the mass, dashpot damping constant, and spring constant, respectively; x, $\dot{x}$, and $\ddot{x}$ are, respectively, the displacement, velocity, and acceleration of the mass; and $f(t)$

is the external force applied to the system. Equations (5-36) and (5-37) have exactly the same form; only the symbols have been changed.

Equations (5-36) and (5-37) are mathematical models. They are also dynamic models. If $e(t)$ is given, Eq. (5-36) can be used to find the charge or the current $i = \dot{q}$ as a function of time t. If the voltage $e(t)$ is not a function of time but is a constant value E, then the charge on the capacitor will reach a fixed value and the rate of change of charge will be zero. Equation (5-36) then reduces to

$$\frac{q}{C} = E \tag{5-38}$$

Equation (5-38) is a static model for the circuit of Fig. 5-12. The static model for the mechanical system of Fig. 5-13 is $X = F/K$, where F is a constant external force.

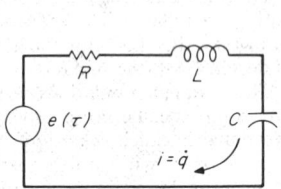
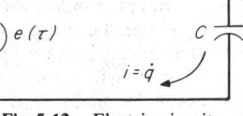
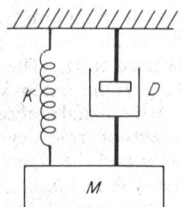

Fig. 5-12.   Electric circuit.   **Fig. 5-13.**   Mechanical system.

Equations (5-36) and (5-37) are continuous models in that they represent the electric circuit and the mechanical system at all points in time. If these equations are solved by analytic means or by an analog computer, the charge and displacement are given as continuous functions of time.

Consider the solution to Eq. (5-36) when $e(t)$ is a constant E and when $R/2L = 1/\sqrt{LC}$. The solution is then given by

$$q = CE[1 - e^{-\alpha t}(1 + \alpha t)] \tag{5-39}$$

where e is the base of the natural logarithm ($e = 2.71828+$) and $\alpha = R/2L$. Equation (5-39) is also a model for Fig. 5-13. Now suppose we use a digital computer to solve Eq. (5-36) and evaluate Eq. (5-39) only at the discrete times when $t = kT$, where T is a fixed value and $k = 0, 1, 2, \ldots$. The model is now a discrete mathematical model and is given by

$$q(kT) = CE[1 - e^{-\alpha kT}(1 + \alpha kT)] \quad k = 0, 1, 2, \ldots \tag{5-40}$$

The mathematical models given by Eqs. (5-36) through (5-40) are all deterministic, rather than stochastic, since the forcing function (the source voltage or the external force) is assumed to be a known value rather than a random variable.

It is now common practice to put mathematical models of dynamic systems in *state variable form*. In this form only first derivatives of variables appear explicitly. The standard state variable form is

$$\dot{\mathbf{x}} = A\mathbf{x} + B\mathbf{u} \tag{5-41}$$

where $\dot{\mathbf{x}}$, $\mathbf{x}$, and $\mathbf{u}$ are column vectors and A and B are matrices. The vector $\mathbf{x}$ is called the *state vector*, and the vector $\mathbf{u}$ is called the *control vector*. To write Eq. (5-36) in the form of Eq. (5-41), let

$$x_1 = q$$

$$x_2 = \dot{q} \qquad u_1 = e(t)$$

$$\dot{x}_2 = \ddot{q} \qquad u_2 = 0$$

Then

$$\dot{x}_1 = x_2$$

$$\dot{x}_2 = \frac{-1}{LC}x_1 - \frac{R}{L}x_2 + u_1 \qquad (5\text{-}42)$$

or

$$\dot{\mathbf{x}} = A\mathbf{x} + B\mathbf{u} \qquad (5\text{-}43)$$

where

$$\mathbf{x} = \begin{pmatrix} x_1 \\ x_2 \end{pmatrix} \qquad \mathbf{u} = \begin{pmatrix} e(t) \\ 0 \end{pmatrix}$$

$$A = \begin{pmatrix} 0 & 1 \\ -1/LC & -R/L \end{pmatrix} \qquad b = \begin{pmatrix} 0 & 0 \\ 1 & 0 \end{pmatrix}$$

The mathematical model of Fig. 5-12 as given by Eq. (5-43) is completely equivalent to the mathematical model as given by Eq. (5-36).

28. Analysis vs. Simulation.   Given a mathematical model of a system, it may be possible to find all the necessary information about the system by analytic means.   For the electric circuit of Fig. 5-12 or the mechanical system of Fig. 5-13, the mathematical model consists of a single second-order differential equation or two first-order differential equations.   These equations are readily solved analytically.   We may then analyze the solutions and determine the effects of changes in the several parameters.   This procedure is referred to as analysis.   If the system is complex, it may be desirable to use numerical computation to solve the equations, or it may be necessary to simulate the system.   Simulation is essentially a working analogy.   It involves the construction of a physical and/or mathematical model of a system and the use of this model in determining the dynamic behavior of the system as a function of time.

In constructing a mathematical model for use in a simulation, there is greater freedom than when an analytic solution is to be sought.   It may be impossible to deal analytically with a complex model; with simulation, while the time required to construct the model may increase with increased complexity, great complexity does not prevent simulation.

Simulation is an experimental procedure.   It provides a means for the systems engineer to evaluate a system experimentally and to determine the effect of modifications in the design of the system without the expense in time and money of constructing the actual system.

29. Computer Simulation—An Example.   Consider a simulation to determine the feasibility of formation flight of helicopters under adverse weather conditions.   The block diagram for the system is shown in Fig. 5-1.

The first step is to construct a mathematical model for the system.   The block diagram helps to structure this model.   In this example the major portions of the system are the aircraft dynamics, the autopilot, the station-keeping control laws, and the sensors.   Parallel efforts can begin to construct suitable models for each of these portions of the system.   The complexity of the mathematical model should be consistent with the desired objective of the simulation.

We briefly consider each of the blocks in Fig. 5-1.   The aircraft dynamics are complex, and to represent helicopter motion at all speeds adequately requires a set of nonlinear differential equations.   However, at the relatively constant speed at which a helicopter flies when in formation flight, the dynamics of a single rotor helicopter can be sufficiently represented by 13 first-order linear differential equations.   This change from nonlinear to linear equations greatly simplifies the model and reduces the time required for a simulated flight.   Since the principle of superposition applies to linear systems but not to nonlinear systems, a set of simulation runs for a linear system generally conveys more information than an equal number of simulation runs for a nonlinear system.

The *autopilot block* represents a pilot, an electromechanical autopilot, or a combination of a pilot and autopilot.   If this block is to represent an autopilot that has already been designed, mathematical equations can be written which will accurately represent the autopilot; or, as an alternative, the physical autopilot can be used directly in the simulation.   With the helicopter dynamics simulated by an analog computer, the analog outputs from the autopilot can be used directly as inputs to the analog computer; with helicopter dynamics simulated by a digital computer, an analog-to-digital converter must be used to tie the autopilot to the digital computer.

If a human pilot rather than an electromechanical autopilot is to be used in the system, there are two options: either the human pilot is modeled by mathematical equations or the human pilot is included in the simulation. If the human pilot is to be included in the simulation, he must be provided with a helicopter cockpit that includes the controls: the collective stick, the cyclic stick, and the rudder pedals. He must also be supplied with displays so that he knows how to move the controls. It is very difficult to adequately represent the human helicopter pilot by mathematical equations. Also, the use of a stationary cockpit and the human pilot has the shortcoming that the pilot does not feel the forces and motions he experiences when flying a helicopter. Use of a human pilot also requires that the simulation operate in real time so that the pilot will be required to react as quickly as he would have to react in the real system. For large systems this requirement for real-time simulation may make it very costly to include the human pilot. Clearly, it is easier to model an autopilot than to model a human pilot.

The block in Fig. 5-1 labeled "station-keeping control laws" is assumed to be undefined. That is, the systems engineer is to define the station-keeping control laws such that formation flight of helicopters under adverse weather conditions is feasible. Based on a reasonable, but perhaps limited, understanding of the dynamic behavior of the helicopter, the systems engineer defines station-keeping laws by mathematical equations. One possible set of station-keeping control laws is

$$\text{Longitudinal law:} \quad \theta = K_1 x_e + K_2 \dot{x}_r \tag{5-44}$$

$$\text{Lateral law:} \quad \phi = -K_3 y_e + K_4 \dot{y}_r \tag{5-45}$$

$$\text{Vertical law:} \quad H = H_l \tag{5-46}$$

where θ, ϕ, and H designate the desired pitch angle, roll angle, and elevation, respectively; K_1, K_2, K_3, and K_4 are constants; x_e and y_e are the longitudinal and lateral position errors, respectively, and $\dot{x}_r$ and $\dot{y}_r$ are the longitudinal and lateral velocities; H_l is the actual elevation of the lead helicopter. The longitudinal law calls for the follower helicopter to pitch downward in order to accelerate in the forward direction when the lateral separation between it and the lead helicopter becomes too great or when its speed relative to the lead helicopter is too small. The lateral law calls for the follower helicopter to roll in order to move laterally when the lateral position or lateral velocity needs correction. The vertical law calls for the follower helicopter to maintain the same elevation as the lead helicopter.

The block in Fig. 5-1 labeled "sensors" includes all the equipment necessary to determine the orientation, velocity, and relative position of the aircraft. Each helicopter will carry velocity-sensing equipment and gyros to determine orientation. It is also necessary to know the relative position of each aircraft with respect to the formation leader. The equipment required to determine the relative positions of the helicopters might be carried solely in the lead aircraft, or each aircraft might carry a portion of the total.

The *desired position* of each helicopter with respect to the lead helicopter may be a function of the maneuver being performed. For example, if the formation is to execute a sharp turn, it is generally desirable to have the followers trail the leader as ducklings trail their mother; but if the formation is flying straight and level, a more usual pattern is the Λ of a flock of geese.

30. Implementation of Computer Model. After completing the mathematical model, the model must be implemented on an analog, digital, or hybrid (combination of digital and analog) computer. While analog computers permit the continuous evaluation of variables, and digital computers permit the evaluation at only discrete points in time, digital computers are more widely used than analog or hybrid computers for the simulation of large systems. An analog computer simulation of formation flight with, say, seven aircraft in the formation would be extremely costly in analog computer equipment. Digital computation is more practical since the computations for the several aircraft are carried out in sequence rather than in parallel.

To implement the model on a digital computer, the systems engineer may either use a general-purpose language such as FORTRAN[20,51,55,58,66,103] or ALGOL or he may use a simulation language[83,84,90,104] such as GASP, GPSS, or SIMSCRIPT. Generally, the time required to execute a simulation run is greater if a simulation language is used than if a general-purpose language is used. On the other hand, the time required to write the simulation program in a general-purpose language is usually greater than the time required to write the program in a simulation language. Hence, where the system to be simulated is small and a small number of runs is required, it is usually most efficient to use a simulation language. Where a large system is to be simulated and a large number of runs is required, it is usually most efficient to use a general-purpose language. (See Sec. 23.)

In the implementation of the model of formation flight of helicopters with a *digital computer simulation,* it is convenient to use a relatively large number of subroutines which are called into play by a short main program. The aircraft dynamics may be represented by one subroutine, or perhaps by two subroutines, one subroutine to describe the linear motion along the rectangular axes of the vehicle and one subroutine to describe the angular motion about the axes. Other subroutines would (1) input or define all parameters necessary for a series of runs, (2) initialize and compute the desired trajectory of the lead aircraft, (3) simulate the station-keeping control laws, (4) simulate the functions of the autopilot, (5) integrate the time derivatives of the many variables, (6) compute the direction cosine matrices which relate aircraft axes to the local level reference frame, (7) compute the ground speed and air speed of the vehicles, and (8) print out the results of the computation.

The station-keeping control laws given by Eqs. (5-44) through (5-46) may be written in FORTRAN as

$$\text{THETA}(J) = K1 \cdot XE + K2 \cdot XRDOT \qquad (5\text{-}47)$$
$$\text{PHI}(J) = K3 \cdot YE + K4 \cdot YRDOT \qquad (5\text{-}48)$$
$$\text{H}(J) = \text{HL} \qquad (5\text{-}49)$$

Note that each left-hand side has a subscripted variable with subscript J; hence J refers to a particular helicopter. J takes on the successive values 1, 2, ... Hence these same three equations may be used for each of the aircraft. Since we do not store the values of the errors or velocities, it is not necessary to use subscripted values for XE, YE, XRDOT, and YRDOT. Note also that since THETA(J), PHI(J), H(J), XE, YE, etc., are functions of time, we must compute the values of these variables at discrete points in time. If the values of the variables are calculated at many closely spaced points, the cost of simulating a flight is high. If the points are not closely spaced, the accuracy of the simulation deteriorates. The spacing of the points is mainly a function of the dynamic behavior of the aircraft and the autopilot. For this simulation, calculating the values of the variables 100 s^{-1} gives good accuracy. That is, THETA(J) and all other variables are computed once for each helicopter for each 0.01-s interval. For a 15-min flight, THETA(J) would be computed 1,500 times for helicopter 2 and a like number of times for all other helicopters.

By making repeated simulated flights, it is possible to determine nearly optimum values for the parameters K_1, K_2, K_3, and K_4 in Eqs. (5-44) to (5-46) [the parameters K1, K2, K3, and K4 in Eqs. (5-47) to (5-49)]. By replacing these control laws by other station-keeping control laws, it can be established that the follower helicopters will maintain their desired positions quite accurately if the lateral control law, Eq. (5-45) or (5-48), is a function of the roll angle of the lead aircraft. The accuracy with which the follower helicopters can maintain their desired position is a function of the accuracy of the position and velocity data. The simulation permits the determination of the effect of data rate, data transmission delay, and data inaccuracy on the ability of the follower aircraft to maintain its position in the formation. By the addition of a subroutine which simulates wind gusts, the manner in which wind gusts affect the ability of the aircraft to maintain position can be determined.

31. Optimization Techniques[3,4,11,50,80,102]   Although a number of optimization techniques have been known to mathematicians for centuries, it was not until the electronic computer became generally available that it became feasible for engineers to apply optimization methods to a wide variety of practical problems. The increased application of optimization techniques has also led to the development of new methods of optimization. Today there are many optimization methods available to the systems engineer, and most large general-purpose computer installations provide software packages for much-used optimization techniques such as linear programming.

The systems engineer, to apply one of the optimization techniques that are available to him, must understand and adequately describe by a mathematical model the system he wishes to optimize. For example, to design an optimum station-keeping system for the formation flight of helicopters, the helicopter dynamics must be accurately represented by a mathematical model. To optimize the operation of an oil refinery, it is necessary to write flow balance equations for all the units in the refinery. For some systems there will be limits on money, time, raw materials, weight, or dimension which must be imposed. The equations of motion, the flow balance equations, and the equations which impose limits on particular variables form the constraint equations for the optimization problem.

To apply an optimization technique, the systems engineer must also choose the objective function he wishes to optimize. For example, in designing a station-keeping system he may wish to choose the parameters of the station-keeping control laws so that the maximum deviation of the helicopter from its desired position is minimized; or in designing a plant for the production of nitric acid he may wish to maximize profit or the production of nitric acid. For many systems the objective is well known and it is an easy task to form the objective function. For some systems, such as social systems, it may be impossible to form an objective function that meets the approval of all who are interested in the system.

Optimization techniques which are briefly discussed in the following pages include linear programming, quadratic programming, nonlinear programming, dynamic programming, and several methods of search. While this list of techniques is not complete, it does include the techniques which are most widely used.

The best optimization method to use depends upon the nature of the particular problem. If the function to be optimized is continuous and differentiable, with no constraints imposed on the variables, the calculus may be used. If the function to be optimized is linear and the variables are subject to only linear constraints, linear programming is applicable. In some cases there is no analytic expression for the quantity to be optimized. For example, we may be required to perform an experiment to determine the product yield in a chemical reaction corresponding to a particular temperature and pressure.

To find the temperature and pressure which give the maximum amount of product requires a series of experiments. In this case we wish to search optimally for the temperature-pressure combination which maximizes the product yield. Neither the calculus nor linear programming is suitable for this problem, but *simplex search* is applicable. Some techniques, such as *linear programming*, are easily applied to large-scale problems. Others, such as *dynamic programming* or *geometric programming*, require considerable effort to implement on a large scale.

Some methods of optimization, such as linear programming, consistently yield the *global optimum*. Others, such as the calculus, yield a *local optimum* which may or may not be the global optimum. While it is the global optimum that is normally sought, the nature of the problem dictates which optimization techniques are applicable and whether or not the optimum obtained is a global or a local optimum.

32. Linear Programming. In the past decade, linear programming[21] has been the optimization technique most used for systems problems. It has been widely used in problems of refinery scheduling, gasoline blending, and transportation scheduling. It has also been used in problems of contract award, plant scheduling, animal-feed mixing, plant expansion, structure design, control systems design, and in many other problems.

Linear programming deals with problems where it is desired to find the maximum or minimum of a linear function of a number of variables which are subject to linear constraints. Linear programming always locates the global optimum which always occurs on the boundary of the admissible region. The linear programming problem may be stated as follows:

Maximize (or minimize) the objective function y,

$$y = c_1 x_1 + c_2 x_2 + \cdots + c_n x_n$$

subject to the m constraints

$$a_{i1} x_1 + a_{i2} x_2 + \cdots + a_{in} x_n \begin{Bmatrix} \leq \\ = \\ \geq \end{Bmatrix} b_i \qquad i = 1, 2, \cdots, m$$

where $x_1, x_2, \ldots, x_n$ are nonnegative variables; $c_1, c_2, \ldots, c_n$ are constant coefficients and are usually referred to as costs; and $a_{i1}, a_{i2}, \ldots, a_{in}$ are constants.

For each of the constraints, only one of the three symbols ($\leq$, $\geq$, $=$) applies. The problem may, however, include both equality ($=$) constraints and inequality ($\leq$ and/or $\geq$)

constraints. The requirement that the variables $(x_1, x_2, \ldots, x_n)$ be nonnegative does not impose a real restriction; if a system variable is unrestricted in sign, it may be replaced by the difference of two nonnegative variables. That is, if x_1 is to be unrestricted in sign, let $x_1 = x_1^+ - x_1^-$, where x_1^+ and x_1^- are nonnegative.

Linear programming programs are generally available at large computer installations, so that the systems engineer need only write the objective function and the constraint in order to apply linear programming. A simple example will illustrate the formulation of a linear programming problem.

33. Example of Linear Programming. A golf course is to be fertilized to meet the following minimum per acre requirements:

Nitrogen . . . . 45 lb
Phosphoric acid . . . . 18 lb
Potash . . . . 12 lb

Two commercial fertilizers, designated A and B, are available. Costs and active-ingredient contents are:

Fertilizer	Cost/lb	Nitrogen	Phosphoric acid	Potash
A	$0.04	9%	6%	2%
B	$0.02	6%	2%	3%

The objective function y is selected as the total cost per acre for fertilizer. Let x_1 and x_2 represent the number of pounds required per acre of fertilizers A and B, respectively. The objective function is then to minimize y, where

$$y = 0.04x_1 + 0.02x_2 \tag{5-50}$$

The per acre requirements of nitrogen, phosphoric acid, and potash are given by the inequalities

Nitrogen: $\quad\quad\quad 0.09x_1 + 0.06x_2 \geq 45$
Phosphoric Acid: $\quad 0.06x_1 + 0.02x_2 \geq 18 \tag{5-51}$
Potash: $\quad\quad\quad\quad 0.02x_1 + 0.03x_2 \geq 12$

The objective function given by Eq. (5-50) together with the constraints given by Eq. (5-51) constitute the linear programming problem. The solution to this problem may be shown to be $x_1 = 100$ lb/acre, $x_2 = 600$ lb/acre, and minimum $y = \$16$/acre.

Computer programs are available which can handle linear programming problems with hundreds of variables and hundreds of constraints. Problems with thousands of variables and thousands of constraints have been solved, but generally require a large amount of computer time.

34. Other Mathematical Programming Methods.[1,8,9,10,47,61,75,105] Linear programming has the limitation that both the objective function and the constraints must be linear functions of the variables whose values are to be found. Quadratic programming requires linear constraints but permits an objective function that has quadratic terms. General nonlinear programming permits nonlinear constraints and a nonlinear objective function. The quadratic programming problem may be stated as follows:

Maximize (or minimize) the objective function y,

$$y = \sum_{i=1}^{n} c_i x_i + \frac{1}{2} \sum_{i=1}^{n} \sum_{j=1}^{n} c_{ij} x_i x_j$$

$$= c_1 x_1 + c_2 x_2 + \cdots + c_n x_n + c_{11} x_1^2 + c_{12} x_1 x_2 + \cdots + c_{nn} x_n^2$$

subject to the m linear constraints

$$a_{i1} x_1 + a_{i2} x_2 + \cdots + a_{in} x_n \begin{Bmatrix} \leq \\ = \\ \geq \end{Bmatrix} b_i \quad i = 1, 2, \cdots, m$$

where x_i $(i = 1, 2, \cdots, n)$ is nonnegative

As with linear programming, the systems engineer merely forms the constraints and the objective function and then applies a computer program to solve the problem.

The general nonlinear programming problem may be stated as follows:
Minimize y,

$$y = f(\mathbf{x})$$

subject to

$$g_j(\mathbf{x}) \ge 0 \qquad j = 1, 2, \cdots, l$$
$$g_j(\mathbf{x}) = 0 \qquad j = l + 1, \cdots, m$$

where $f(\mathbf{x})$ and $g_j(\mathbf{x})$ are, in general, nonlinear functions of the vector variables $\mathbf{x} = (x_1, x_2, \ldots, x_n)$. The problem statement given is quite general. If it is desired to maximize the function rather than minimize it, the sign of $f(\mathbf{x})$ is changed and the maximization problem becomes a minimization problem. If it is desired that $g_j(\mathbf{x}) \le 0$ rather than $g_j(\mathbf{x}) \ge 0$, merely change the sign of $g_j(\mathbf{x})$ and replace the symbol $\le$ by $\ge$. The technique most used for general nonlinear programming is SUMT (sequential unconstrained minimization technique).[4,11] In SUMT, the objective function is modified to include some influence by the constraints. If the constraints are not completely satisfied, the value of the objective function does not attain its true optimum. To use SUMT, the original problem is replaced by the following problem:
Minimize $z(\mathbf{x}, r_1)$, where

$$z(\mathbf{x}, r_1) = f(\mathbf{x}) + r_1 \sum_{j=1}^{l} 1/g_j(\mathbf{x}) + r_1^{-1/2} \sum_{j=l+1}^{m} g_j(\mathbf{x})$$

It is necessary first to locate a point $\mathbf{x} = (x_1, x_2, \ldots, x_n)$ which satisfies $g_j(\mathbf{x}) > 0, j = 1, 2, \ldots, l$. If such a point exists, then SUMT will seek a minimum of $z(\mathbf{x}, r_1)$ for a particular value of r_1. In the next iteration, r_1 is replaced by r_2, where $0 < r_2 < r_1$ and a minimum of $z(\mathbf{x}, r_2)$ is located. This process is continued with a sequence of values of r_k, each successive value being smaller than the last. For very small r_k, the point $\mathbf{x}^* = (x_1^*, x_2^*, \ldots, x_n^*)$ which minimizes $z(\mathbf{x}, r_k)$ approximately minimizes $f(\mathbf{x})$, subject to the constraints on $g_j(\mathbf{x})$ ($j = 1, \ldots, m$).

In linear, quadratic, and general nonlinear programming, the optimum values of the variables may be nonintegers. There are many cases where it is desirable to restrict the solution to integer values. If all the variables are restricted to integer values in an otherwise linear programming problem, we have an integer programming problem. If some, but not all, are restricted to integer values, we have a mixed-integer programming problem. If all the variables are restricted to values of zero or one in an otherwise linear programming problem, we have a binary programming problem. There are algorithms which have been implemented by computer programs to solve these integer, binary, and mixed-integer problems. In general, however, the solution time is much greater than for an equivalent linear programming problem. Even when integer solutions are required, it is often best to avoid integer programming. For example, suppose the optimal solution to a linear programming problem calls for production of 97.2, 178.4, and 24.7 finished items from three shops. It is probably not worthwhile to use integer programming to obtain an integer solution since the numerical results will be approximated the same.

Dynamic programming is a development of Richard Bellman.[9,10] It is most useful when the following conditions are met: (1) the problem can be broken into a large number of stages (where each stage represents an increment of time, space, equipment, etc.); (2) there are many decisions to be made; and (3) there are few variables. Dynamic programming is based upon the following *principle of optimality:* An optimal policy has the property that whatever the initial state and initial decisions are, the remaining decisions must constitute an optimal policy with regard to the state resulting from the first decision. Dynamic programming is a powerful technique for some problems. To use dynamic programming, however, the systems engineer usually cannot draw upon a computer program already prepared. Usually, a new program must be written to treat his particular problem.

35. Gradient Search and One-dimensional Search.[4,11,80,102] In climbing a mountain that is covered with trees so that the top of the mountain is not visible, the climber can select the steepest route in attempting to reach the top with the fewest steps. In optimizing a function, the gradient (steepest route) is frequently selected for reaching the maximum or minimum. There are a variety of gradient methods, which may be categorized as follows:

1. Methods which use only the functional values.
2. Methods which also use first-order derivatives.
3. Methods which use functional values, first-order derivatives, and second-order derivatives.

If the function to be optimized is a differentiable analytic function, the derivatives may be found analytically. If the function to be optimized is not given analytically, it is necessary to evaluate the function and its derivatives numerically if methods in class 2 or 3 are to be used. That is, suppose the product yield from a chemical reaction is to be optimized and that the yield is a function of temperature and pressure, but that the functional relationship between product yield, temperature, and pressure is not known. The approximate gradient can be evaluated at a temperature of, say, 150°C and a pressure of, say, 45 lb/in.^{2} by performing three experiments, one at 150°C, 45 lb/in.^{2}; one at, say, 155°C, 45 lb/in.^{2}; and one at 150°C, 50 lb/in.^{2} The gradient at the desired point can be calculated more exactly by taking the three points closer together. However, as the points get closer, the change in the product yield becomes smaller and measurement errors are more significant.

It is not always best to try to follow the true *gradient* (the steepest route). The mountain climber can follow the steepest route by continual observation. The systems engineer, however, must limit the number of points at which he evaluates the function and its derivatives. If a function of n variables $f(x_1, \ldots, x_n)$ and its gradient are evaluated numerically, the determination of the gradient requires evaluating the function $n + 1$ times. Hence, after evaluating the gradient, it is desirable to search along the gradient line for a maximum before evaluating the gradient again. But if the function exhibits a ridge, as shown in Fig. 5-14, the search may seesaw back and forth over the ridge as shown without progressing quickly toward the true maximum. To avoid this difficulty, the true

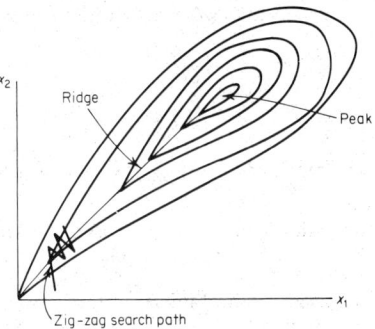

Fig. 5-14. Response function which exhibits a ridge.

gradient direction may be replaced by a direction that is dependent upon, say, the present gradient and the previous gradient.

Gradient methods do not directly provide for constraints, but merely provide a method to search for an optimum of a function whose variables may take on any values. Problems with constraints may frequently be framed as unconstrained problems. For example, see the discussion of SUMT given in Par. 5-34. Also, if the variables are to be constrained between given limits such as

$$x_{il} \leq x_i \leq x_{iu} \ (i=1, 2, \ldots, n)$$

where x_{il} and x_{iu} are numerical values for lower and upper limits on x_i, then a computer program can be written which will use a gradient method and will ensure that the constraints are not violated.

Gradient search methods that have received considerable use carry the names steepest ascent, parallel tangents, conjugate-direction method, conjugate-gradient method, Newton-Raphson method, Fletcher-Reeves method, and Davidon's method.

In gradient search methods we find the direction in which to move in order to achieve an optimum. It is then necessary to move in this direction to find the local optimum along the given path. A new direction in which to move is then determined, and another one-dimensional search is begun along the new path. If the objective function along this path can be expressed as an analytic and differentiable function, the calculus can be used to locate the local optimum along the path. Frequently, it is necessary to search along the one-dimensional path by a direct method such as dichotomous search, Fibonacci search, golden-section search, or polynomial approximation. Only one of these methods, golden-section search, is discussed here.

In *golden-section search*, each subsequent experiment (evaluation of the objective function) after the first experiment reduces the length of the one-dimensional region to be searched by a factor of 1.618. That is, if we let j be the number of experiments already run and L be the length of the region still to be searched, then

$$\frac{L_{j-1}}{L_j} = \frac{L_j}{L_{j+1}} = 1.618 = \tau$$

The points at which experiments are performed are selected so that this ratio is maintained. After n experiments, the length of the interval in which the optimum must lie is given by

$$L_n = \frac{T}{\tau^{n-1}}$$

where T is the total length of the path to be searched. The process is continued until the region in which the optimum lies is small enough for our purposes. As an example, suppose the region $0 \leq x \leq 1$ is to be searched for a maximum of $y(x)$ and that this maximum occurs at $x = 0.1$.

The locations of the first four experiments are shown in Fig. 5-15 at x equal to $1/\tau$, $1/\tau^2$, $1/\tau^3$, $1/\tau^4$, respectively. One experiment gives no information, but after two experiments we assume that the optimum lies to the left of $x = 1/\tau^2 = 0.382$, since this point gives a better result than the experiment at $x = 0.618$. Since the third experiment at $x = 1/\tau^3 = 0.236$ gives a better result than the second experiment, we assume after three experiments that the opti-

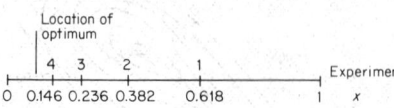

Fig. 5-15. Location of experiments in golden-section search.

mum lies to the left of $x = 0.382$. After four experiments, we assume that the optimum lies to the left of $x = 1/\tau^3 = 0.236$. Our assumptions concerning the position of the optimum apply if the function is unimodal, that is, if it has only one maximum in the region being searched. If the function is not unimodal, there is no assurance that this method will lead to an optimum.

36. References

1. ABADIE, J. (ed.)   "Nonlinear Programming"; Amsterdam, North-Holland Publishing Company, 1967.

2. ACKOFF, R. L., and SASIENI, M. S.   "Fundamentals of Operations Research"; New York, John Wiley & Sons, Inc., 1968.

3. AOKI, MASANAO   "Optimization of Stochastic Systems"; New York, Academic Press, Inc., 1967.

4. AOKI, M.   "Introduction to Optimization Techniques"; New York, The Macmillan Company, 1971.

5. ARROW, K. J.   "Social Choice and Individual Values"; New York, John Wiley & Sons, Inc., 1963.

6. ARROW, K. J.   'Public and Private Values' in Sidney Hook (ed.), "Human Values and Economic Policy"; New York, 1967.

7. BARLOW, R. E., and PROCHAN, F.   "Mathematical Theory of Reliability"; New York, John Wiley & Sons, Inc., 1965.

8. BEALE, E. M. L.   "Mathematical Programming in Practice"; New York, John Wiley & Sons, Inc., 1968.

9. BELLMAN, RICHARD   "Dynamic Programming"; Princeton, N.J., Princeton University Press, 1957.

10. BELLMAN, RICHARD, and DREYFUS, STUART E.   "Applied Dynamic Programming"; Princeton, N.J., Princeton University Press, 1962.

11. BEVERIDGE, GORDON S. G., and SCHECHTER, ROBERT S.   "Optimization: Theory and Practice"; New York, McGraw-Hill Book Company, 1969.

12. BILODEAU, E. A. (ed.)   "Acquisition of Skill"; New York, Academic Press, Inc., 1966.

13. BLANTON, H. E., and JACOBS, R. M.   A Survey of Techniques for Analysis and

Prediction of Equipment Reliability; *IRE Trans. Reliability Quality Control,* July, 1962, Vol. R QC-11, No. 2, pp. 18-35.

14. CALABRO, S. R. "Reliability Principles and Practices"; New York, McGraw-Hill Book Company, 1962.

15. CARNAHAN, BRICE "Numerical Methods, Optimization Techniques and Process Simulation for Engineers"; Ann Arbor, Mich., University of Michigan Press, 1967.

16. CHORAFAS, D. N. "Systems and Simulation"; New York, Academic Press, Inc., 1965.

17. CHURCHMAN, C. W. "The Systems Approach"; New York, Delacorte Press, Dell Publishing Co., 1968.

18. COOPER, GEORGE R. 'Decision Theory', Chap. 24 in Robert E. Machol (ed.), "Systems Engineering Handbook"; New York, McGraw-Hill Book Company, 1965.

19. CRAIK, KENNETH J. W. In Stephen L. Sherwood (ed.), "The Nature of Psychology: A Selection of Papers, Essays and Other Writings"; New York, Cambridge University Press, 1966.

20. CRESS, PAUL "FORTRAN IV with WATFOR"; Englewood Cliffs, N.J., Prentice-Hall, Inc., 1968.

21. DANTZIG, GEORGE B. "Linear Programming and Extensions"; Princeton, N.J., Princeton University Press, 1963.

22. DEBREU, G. "Theory of Value"; New York, John Wiley & Sons, Inc., 1959.

23. DE GREENE, K. B. (ed.) "Systems Psychology"; New York, McGraw-Hill Book Company, 1970.

24. DOD and NASA Guide, PERT COST, Office of the Secretary of Defense and NASA, June, 1962.

25. EDHOLM, O. G. "The Biology of Work"; New York, McGraw-Hill Book Company, 1970.

26. EKMAN, T. "Introduction to ALGOL Programming"; London, Oxford University Press, 1969.

27. EMSHOFF, J. "Design and Use of Computer Simulation Models"; New York, The Macmillan Company, 1970.

28. EVANS, G. W. "Simulation Using Digital Computers"; Englewood Cliffs, N.J., Prentice-Hall, Inc., 1967.

29. FEGLEY, KENNETH A. Formation Flight—Final Technical Report; *Res. Develop. Tech. Rept.,* June, 1971, ECOM-02411-21, Pennsylvania-Princeton Army Avionics Research Program.

30. FISHBURN, P. C. "Decision and Value Theory"; New York, John Wiley & Sons, Inc., 1964.

31. FISHBURN, P. C. On the Prospects of a Useful Unified Theory of Value for Engineering; *IEEE Trans. Systems Sci. Cybernetics,* August, 1966, Vol. SSC-2, pp. 27-35.

32. FISHBURN, P. C. Utility Theory; *Management Sci.,* January, 1968, Vol. 14, pp. 335-378.

33. FOGEL, L. J. "Biotechnology: Concepts and Applications"; Englewood Cliffs, N.J., Prentice-Hall, Inc., 1963.

34. GINSBURGH, A. S., and OFFENSEND, F. L. An Application of Decision Theory to a Medical Diagnosis-Treatment Problem; *IEEE Trans. Systems Sci. Cybernetics,* September, 1968, Vol. SSC-4, pp. 335-362.

35. GOODE, H. H., and MACHOL, R. E. "Systems Engineering"; New York, McGraw-Hill Book Company, 1957.

36. GOODMAN, L. A. On Methods of Amalgamation, pp. 39-48 in "Decision Processes"; R. M. Thrall, C. H. Coombs, and R. L. Davis (eds.), New York, John Wiley & Sons, Inc., 1954.

37. GORDON, GEOFFREY "Systems Simulation"; Englewood Cliffs, N.J., Prentice-Hall, Inc., 1969.

38. GOSLING, W. "The Design of Engineering Systems"; New York, John Wiley & Sons, Inc., 1962.

39. GRANDJEAN, E. "Fitting the Task to the Man"; London, Taylor & Francis, Ltd., 1969.

40. HALL, A. D. "A Methodology for Systems Engineering"; Princeton, N.J., D. Van Nostrand Company, Inc., 1962.

41. HAMMERSLEY, J. M.   "Monte Carlo Methods"; New York, John Wiley & Sons, Inc., 1964.

42. HAUSNER, M.   Multidimensional Utilities, pp. 167-180 in R. M. Thrall, C. H. Coombs, and R. L. Davis (eds.), "Decision Processes"; New York, John Wiley & Sons, Inc., 1954.

43. HOOS, IDA R.   A Realistic Look at the Systems Approach to Social Problems; *Datamation,* February, 1969.

44. HOUTHAKKER, H. S.   The Present State of Consumption Theory; *Econometrica,* October, 1961, Vol. 29, pp. 704-740.

45. HOWELL, W. C., and GOLDSTEIN, I. L. (eds.)   "Engineering Psychology: Current Perspectives in Research"; New York, Appleton Century Crofts, 1971.

46. KELLEY, C. R.   "Manual and Automatic Control"; New York, John Wiley & Sons, Inc., 1964.

47. KOWALIK, J., and OSBORNE, M. R.   "Methods for Unconstrained Optimization Problems"; New York, American Elsevier Publishing Company, Inc., 1968.

48. KRENDEL, E. S., and McRUER, D. T.   A Servomechanisms Approach to Skill Development; *Journ. Franklin Inst.,* 1960, Vol. 269, No. 1.

49. KRENDEL, E. S., and McRUER, D. T.   Psychological and Physiological Skill Development: A Control Engineering Model, in "Technical and Biological Problems of Control: A Cybernetic Review"; Pittsburgh, Pa., Instrument Society of America, 1970.

50. LAVI, ABRAHIM, and VOGL, THOMAS P. (eds.)   "Recent Advances in Optimization Techniques"; New York, John Wiley & Sons, Inc., 1966.

51. LEDLEY, R. S.   "FORTRAN IV Programming"; New York, McGraw-Hill Book Company, 1966.

52. LUCE, R. D., and RAIFFA, H.   "Games and Decisions: Introduction and Critical Survey"; New York, John Wiley & Sons, Inc., 1957.

53. LUCE, R. D., and SUPPES, P.   Preference, Utility and Subjective Probability, pp. 249-410, in R. D. Luce, R. R. Bush, and E. Galanter (eds.), "Handbook of Mathematical Psychology"; New York, John Wiley & Sons, Inc., 1965.

54. MACHOL, ROBERT E. (ed.)   "Systems Engineering Handbook"; New York, McGraw-Hill Book Company, 1965.

55. McCORMICK, J. M.   "Numerical Methods in FORTRAN"; Englewood Cliffs, N.J., Prentice-Hall, Inc., 1964.

56. McCORMICK, E. J.   "Human Factors Engineering," 3d ed.; New York, McGraw-Hill Book Company, 1970.

57. McCRACKEN, D. D.   "A Guide to ALGOL Programming"; New York, John Wiley & Sons, Inc., 1962.

58. McCRACKEN, D. D.   "A Guide to FORTRAN IV"; New York, John Wiley & Sons, Inc., 1965.

59. McLEOD, JOHN   "Simulation, the Dynamic Modeling of Ideas and Systems with Computers"; New York, McGraw-Hill Book Company, 1968.

60. McMILLAN, CLAUDE   "Systems Analysis: A Computer Approach to Decision Models"; Homewood, Ill., Richard D. Irwin, Inc., 1965.

61. McMILLAN, CLAUDE, JR.   "Mathematical Programming: An Introduction to the Design of Optimal Decision Machines"; New York, John Wiley & Sons, Inc., 1970.

62. McRUER, D. T., GRAHAM, D., and KRENDEL, E. S.   Manual Control of Single-Loop Systems"; *Journ. Franklin Inst.,* Parts 1 and 2, 1967, Vol. 283, Nos. 1 and 2.

63. McRUER, D. T., and KRENDEL, E. S.   The Man-Machine System Concept; *Proc. IEEE,* May 1962, Vol. 50, No. 5, pp. 1117-23.

64. McRUER, D. T., and WEIR, D. H.   Theory of Manual Vehicular Control; *IEEE Trans. Man-Machine Systems,* 1969, Vol. MMS-10, pp. 257-291.

65. MENDENHALL, W.   "Introduction to Linear Models and the Design and Analysis of Experiments"; Belmont, Calif., Wadsworth Publishing Company, Inc., 1968.

66. MERRILL, PAUL W.   "FORTRAN IV Programming for Engineers and Scientists"; Scranton, Pa., International Textbook Company, 1968.

67. MIDDLETON, D.   "An Introduction to Statistical Communication Theory"; New York, McGraw-Hill Book Company, 1960.

68. MINAS, J. S., and ACKOFF, R. L.   Individual and Collective Value Judgements, pp.

351–359 in M. W. Shelly and G. L. Bryan (eds.), "Human Judgements and Optimality"; New York, John Wiley & Sons, Inc., 1964.

69. Morgan, C. T., Chapanis, A., Cook, J. S., III, and Lund, M. W. (eds.) "Human Engineering Guide to Equipment Design"; New York, McGraw-Hill Book Company, 1963.

70. Morse, P. M. "Queues: Inventories and Maintenance"; New York, John Wiley & Sons, Inc., 1958.

71. Morse, Philip M. "Queues and Markov Processes"; Chap. 28 in Robert E. Machol (ed.), "Systems Engineering Handbook"; New York, McGraw-Hill Book Company, 1965.

72. Mueller, Eva Public Attitudes toward Fiscal Problems; Quart. Journ. Economics, 1963, Vol. 77, No. 2.

73. Naylor, T. H. "Computer Simulation Techniques"; New York, John Wiley & Sons, Inc., 1966.

74. Naylor, T. H. "Computer Simulation Experiments with Models of Economic Systems"; New York, John Wiley & Sons, Inc., 1971.

75. Nemhauser, George L. "Introduction to Dynamic Programming"; New York, John Wiley & Sons, Inc., 1966.

76. Von Neumann, J., and Morgenstern, O. "Theory of Games and Economic Behavior," 2d ed.; Princeton, N.J., Princeton University Press, 1947.

77. O'Neal, R. D., and Clayton, J. F. Management, Chap. 36 in Robert E. Machol (ed.), "Systems Engineering Handbook"; New York, McGraw-Hill Book Company, 1965.

78. Parsons, H. M. "Man-Machine System Experiments"; Baltimore, Md., The Johns Hopkins Press, 1972.

79. Parzen, E. S. "Stochastic Processes"; San Francisco, Calif., Holden-Day, Inc., Publisher, 1962.

80. Pierre, Donald A. "Optimization Theory with Applications"; New York, John Wiley & Sons, Inc., 1969.

81. Pikús, E. Simulation of a Computer-aided Flight Operation Center (CAFOC), Final Technical Report; June, 1970, Res. Develop. Tech. Rept., ECOM-02411-16, Pennsylvania-Princeton Army Avionics Research Program.

82. Pratt, J. W., Raiffa, H., and Schlaifer, R. "Introduction to Statistical Decision Theory"; New York, McGraw-Hill Book Company, 1965.

83. Pritsker, A. "Simulation with GASP II"; Englewood Cliffs, N.J., Prentice-Hall, Inc., 1969.

84. Pugh, A. L. "Dynamo User's Manual," 2d ed.; Cambridge, Mass., The M.I.T. Press, 1963.

85. Riordan, J. "Stochastic Service Systems"; New York, John Wiley & Sons, Inc., 1962.

86. Roberts, N. H. "Mathematical Methods in Reliability Engineering"; New York, McGraw-Hill Book Company, 1964.

87. Ross, Harold D., Jr. Reliability, Chap. 33 in R. E. Machol (ed.), "Systems Engineering Handbook"; New York, McGraw-Hill Book Company, 1965.

88. Saaty, Thomas L. "Optimization in Integers and Related Extremal Problems"; New York, McGraw-Hill Book Company, 1970.

89. Saaty, T. L. "Elements of Queueing Theory"; New York, McGraw-Hill Book Company, 1961.

90. Schriber, Thomas J. "General Purpose Simulation System"; Ann Arbor, Mich., Ulrich's Books, 1971.

91. Sherman, Philip M. "Techniques in Computer Programming"; Englewood Cliffs, N.J., Prentice-Hall, Inc., 1970.

92. Sinaiko, H. W. (ed.) "Selected Papers on Human Factors in the Design and Use of Control Systems"; New York, Dover Publications, Inc., 1961.

93. Singleton, W. T., Esterley, R. S., and Whitfield, D. C. (eds.) "The Human Operator in Complex Systems"; London, Taylor & Francis, Ltd., 1967.

94. Singleton, W. T., Fox, J. G., and Whitfield, D. C. (eds.) "Measurement of Man at Work"; London, Taylor & Francis, Ltd., 1971.

95. Smith, John U. M. "Computer Simulation Models"; New York, Hafner Publishing Company, 1968.

96. Stephenson, Robert E. "Computer Simulation for Engineers"; New York, Harcourt Brace Jovanovich, Inc., 1970.

97. STEVENS, S. S. (ed.) "Handbook of Experimental Psychology"; New York, John Wiley & Sons, Inc., 1951.

98. TAKACS, L. "Introduction to Theory of Queues"; Fair Lawn, N.J., Oxford University Press, 1962.

99. THRALL, R. M. Applications of Multidimensional Utility Theory, pp. 181–186, in R. M. Thrall, C. H. Coombs, and R. L. Davis (eds.), "Decision Processes"; New York, John Wiley & Sons, Inc., 1954.

100. WARFIELD, J. N., and HILL, J. D. "Systems Engineering Workshop Notebook"; Battelle, Seattle, Research Center, February, 1971.

101. WEISS, L. "Statistical Decision Theory"; New York, McGraw-Hill Book Company, 1961.

102. WILDE, DOUGLASS J., and BEIGHTLER, CHARLES S. "Foundations of Optimization"; Englewood Cliffs, N.J., Prentice-Hall, Inc., 1967.

103. WILF, H. S. "Programming for a Digital Computer in the FORTRAN Language"; Reading, Mass., Addison-Wesley Publishing Co., Inc., 1969.

104. WYMAN, FORREST P. "Simulation Modeling: A Guide to Using SIMSCRIPT"; New York, John Wiley & Sons, Inc., 1970.

105. ZANGWILL, WILLARD I. "Nonlinear Programming: A Unified Approach"; Englewood Cliffs, N.J., Prentice-Hall, Inc., 1967.

SECTION 6

PROPERTIES OF MATERIALS

BY

ILAN A. BLECH Associate Professor, Department of Materials Engineering, Technion-Israel Institute of Technology, Haifa

CONTENTS

Numbers refer to paragraphs

SECTION 6

PROPERTIES OF MATERIALS

CONDUCTIVE AND RESISTIVE MATERIALS

1. **Volume Electrical Conductivity.** The volume conductivity is the ratio between the electrical current density and the electric field strength,

$$\sigma = \frac{J}{E} \tag{6-1}$$

where σ = volume electrical conductivity (mhos/m, or $\mho$/m), J = electrical current density (A/m^2), and E = electric field strength (V/m). The conductivity of metals is independent of current density. Noncubic materials have an anisotropic electrical conductivity.

The volume conductivity is related to the conductance through

$$\sigma = \frac{Gl}{A} \tag{6-2}$$

where G = conductance ($\mho$ or Ω^{-1}), l = conductor length (m), and A = conductor cross section (m^2).

2. **Mass Conductivity.** Mass conductivity is defined as

$$\sigma_m = \frac{Gl^2}{m} \tag{6-3}$$

where σ_m = mass conductivity ($\mho \cdot m^2$/kg), G = conductance ($\mho$), l = conductor length (m), and m = conductor mass (kg).

3. **Volume Resistivity.** Volume resistivity is the electric field required to produce a unit current density.

$$\rho = \frac{E}{J} \tag{6-4}$$

where ρ = resistivity ($\Omega \cdot m$), J = current density (A/m^2), and E = electric field strength (V/m). The volume resistivity is the reciprocal of volume conductivity. The relation between the volume resistivity and electrical resistance is

$$\rho = \frac{RA}{l} \tag{6-5}$$

where R = resistance (Ω), A = conductor cross section (m^2), and l = conductor length (m). Table 6-1 lists electrical resistivities of elements. Table 6-2 and Fig. 6-1 show the electrical resistivity of alloys and compounds.

4. **Mass Resistivity.** The mass resistivity is defined as

$$\delta = \frac{Rm}{l^2} \tag{6-6}$$

where δ = mass resistivity ($\Omega \cdot kg/m^2$), m = conductor mass (kg), and l = conductor length (m).

5. **International Annealed Copper Standard (IACS).** The conductivity of annealed copper at 20°C has been chosen as a standard, and a value of 100% is assigned to it. The mass resistivity of annealed copper is 0.00015328 $\Omega \cdot kg/m^2$, and the volume resistivity is 1.7241 × $10^{-8} \Omega \cdot m$.

Table 6-1. Physical Properties of Pure Metals[114]

Metal	M.p., °C	B.p., °C	Density × 10⁻³, kg/m³, (20°C)	Thermal conductivity, Cal/s·cm·K* (0-100°C)	Mean specific heat, cal/g·°C† (0-100°C)	Resistivity × 10⁸, Ω·m, (20°C)	Temp. coeff. of resistivity × 10³, 1/°C, (0-100°C)	Coeff. of expansion × 10⁶, 1/°C, (0-100°C)
Aluminium	660.1	2450	2.70	0.57	0.219	2.69	4.2	23.5
Antimony	630.5	1440	6.68	0.042	0.050	42	5.1	8-11
Barium	710	1500	3.5	...	0.068	50	—	(18)
Beryllium	1284	2970	1.848	0.40	0.490	4-6	6.0	12
Bismuth	271	1560	9.80	0.019	0.0298	116	4.2	13.4
Cadmium	320.9	765	8.64	0.22	0.0557	7.4	4.3	31
Caesium	29.7	700	1.87	—	0.056	21	4.8	97
Calcium	850	1440	1.54	0.3	0.149	4.1 (soft) 4.37 (hard)	4.6	22
Cerium	804.44	3470	6.75	0.03	0.045	78	8.7	8
Chromium	1875	2682.7	7.1	0.165	0.110	12.9	2.14	6.5
Cobalt	1492	(2900)	8.9	0.165	0.102	6.24	6.04	12.5
Copper	1083	2590	8.96	0.94	0.0922	1.673	4.3	17.0
Gallium	29.8	2250	5.91	—	0.090	‡	—	18.3
Germanium	937	2830	5.32	0.14	0.074	46 × 10⁸	—	5.75
Gold	1063	(2950)	19.3	0.70	0.031	2.3	3.9	14.1
Hafnium	2220	5400	13.1	0.05	0.035	30.6	4.19	6.0
Indium	156.4	2075	7.3	0.196	0.058	9.0	4.7	24.8
Iridium	2443	4800	22.4	0.14	0.0312	5.3	3.9	6.8
Iron	1537	(3070)	7.87	0.17	0.109	9.71	6.51	12.1
Lead	327.4	1740	11.68	0.082	0.0310	20.6	3.36	29.0
Lithium	180	1329	0.534	0.17	0.84	9.35	4.75	56
Magnesium	650	1103	1.74	0.40	0.248	3.9	4.2	26.0
Manganese	1244	2060	7.4	—	0.116	160(α)	—	23
Mercury	-38.87	357	13.546	0.022	0.033	95.8	0.9	61
Molybdenum	2600	5560	10.2	0.34	0.062	5.7	4.23	5.1

Element								
Nickel	1453	2730	8.9	0.21	0.108	6.844	6.81	13.3
Niobium	2468	4927	8.6	0.13	0.064	14.5	3.95	7.2
Osmium	3010 ± 10	(5500)	22.5	—	0.031	9.5	4.2	4.57
Palladium	1552	(3900)	12.0	0.17	0.059	10.8	3.8	11.0
Platinum	1769	4240	21.45	0.17	0.0321	10.6	3.92	9.0
Potassium	63.6	775	0.86	0.22	0.180	6.86	5.8	83
Radium	960	1140	(5)					
Rhenium	3150	5900	21.0	0.17	0.033	19.1	3.11	6.6
Rhodium	1960	4500	12.4	0.20	0.060	4.7	4.57	8.5
Rubidium	38.8	680	1.53	—	0.085	12.5	5.53	90
Ruthenium	(2500)	(4900)	12.2	—	0.056	7.3	—	9.6
Silicon	1412	2600	2.34	0.2	0.174	23×10^{10}§	4.1	7.6
Silver	960.8	2210	10.5	1.00	0.054	1.6	5.0	19.1
Sodium	97.8	892	0.97	0.32	0.293	4.6	—	71
Strontium	770	1350	2.6	—	0.176	22.76	—	0.01%
Tantalum	2980	5429 ± 100	16.6	0.130	0.034	13.5	3.8	6.5
Tellurium	450	990	6.24	0.014	0.049	(400×10^3)	—	1.7
Thallium	303	1460	11.85	0.094	0.031	16.6	5.2	∥ c axis 27.5 / ⊥ c axis 30
Thorium	1850	3500 – 4200	11.5	0.09	0.024	18.62	3.8	11.2
Tin	231.9	2450	7.30	0.155	0.054	12.8	4.2	23.5
Titanium	1670	(3260)	4.5	0.041	0.126	55	4.1	8.9
Tungsten	3380	5900	19.3	0.394	0.033	5.5	4.6	4.5
Uranium	1130	3818	{19.05(α), 18.89(β)}	0.07	0.028	29(α)	3.4	¶
Vanadium	1860	(3350)	6.1	0.07	0.119	26	3.4	8.3
Zinc	419.5	907	7.14	0.265	0.094	5.92	4.2	31
Zirconium	1860	3580 – 3700	6.49	0.05	0.069	44.6	4.4	5.9

* To convert to W/m·K, multiply by 418.
† To convert to J/kg·K multiply by 4,180.
‡ 17.4 ∥ a axis, 8.1 ∥ b axis, 54.3 ∥ c axis.
§ Intrinsic single crystal.
¶ α-Uranium, 23 ∥ a axis, −3.5 ∥ b axis, 17 ∥ c axis, 25-300°C; β-uranium, 4.6 ∥ c axis, 23.0 ⊥ c axis, 20-720°C.

Table 6-2. **Electrical Resistivity of Some Alloys and Compounds**[115,116]

Material	Conductivity, % IACS	Resistivity × 10^8, $\Omega \cdot m$
92.5 Ag – 7.5 Cu.	85	2
60 Ag – 40 Pd	8	23
97 Ag – 3 Pt	50	3.5
90 Pt – 10 Ir	7	25
96 Pt – 4 W.	5	36
70 Pd – 30 Ag.	4.3	40
90 Au – 10 Cu.	16	10.8
75 Au – 25 Ag.	16	10.8
78.5 Ni – 20 Cr – 1.5 Si.	1.6	108.05
71 Ni – 29 Fe	9	19.95
80 Ni – 20 Cr	1.5	112.2
Carbon steel 0.65% C.	9.5	18
Electrical Si sheet steels	9.5 – 3	18 – 52
Stainless steel Type 302	3	72
Stainless steel Type 316	2.5	74
TaN	1.2	135
ZrN.	12.6	13.6
TiN	7.9	21.7
VN	2	85.9
TaC.	5.7	30
ZrC	2.7	63.4
SiC	1 – 1.7	100 – 200
WC	14	12
$MoSi_2$	4.5	37
$CbSi_2$.	26	6.5
ZrB_2	19	9
LaB_6	11.5	15
TiB_2.	8.6	20

6. Sheet Resistivity. The resistance of a conductor or resistor in sheet form is, according to Eq. (6-5),

$$R = \frac{\rho l}{db} = \frac{\rho}{d}\frac{l}{b} \qquad (6\text{-}7)$$

where d, l, and b are the conductor thickness, length, and width, respectively. ρ/d is called *sheet resistivity,* and is measured in ohms per square. l/b is the number of squares. The resistance of a sheet is its sheet resistivity multiplied by the number of squares.

7. Surface Resistivity. At high frequencies, electric current is mainly conducted near the surface of the conductor. The depth at which the current density falls to $1/\varepsilon$ of its value at the surface is called the *skin depth* δ. The surface resistivity R_s is the dc sheet resistivity of a conductor having a thickness of one skin depth.

$$R_s = \rho/\delta = 1/\sigma\delta \qquad (6\text{-}8)$$

where ρ = electrical resistivity ($\Omega \cdot m$), δ = thickness (m), and σ = electrical conductivity (Ω/m).

The skin depth, and therefore R_s, are functions of the ac frequency. Values of R_s are given in Table 6-3.

8. Temperature Coefficient of Electrical Resistivity. Over a narrow range of temperature the electrical resistivity changes approximately linearly with temperature.

$$\rho(t_2) = \rho(t_1)[1 + \alpha_{t_1}(t_2 - t_1)] \qquad (6\text{-}9)$$

where $\rho(t_1)$ = resistivity at temperature t_1 ($\Omega \cdot m$), $\rho(t_2)$ = resistivity at temperature t_2 ($\Omega \cdot m$), and α_{t_1} = temperature coefficient of electrical resistivity (1/°C). (See Table 6-1.)

9. Temperature Coefficient of Electrical Resistance. The electrical resistance changes with temperature similarly to the electrical resistivity.

$$R(t_2) = R(t_1)[1 + \alpha_{t_1}(t_2 - t_1)] \qquad (6\text{-}10)$$

where $R(t_2)$ = resistance at temperature t_2 (Ω), $R(t_1)$ = resistance at temperature t_1 (Ω), and α_{t_1} = temperature coefficient of electrical resistance (1/°C). It is customary to take t_1 = 20°C. Equation (6-10) holds in a range of temperatures of about 0 to 100°C. The resistance of most conductors increases with temperature. The resistance of carbon, semiconductors, several thin-film resistors, and electrolytes can decrease with temperature.

When the temperature of reference t_1 is changed to some other value t, the coefficient α_{t_1} will change to a new value α_t.

$$\alpha_t = \frac{\alpha_{t_1}}{1 + \alpha_{t_1}(t - t_1)} \qquad (6\text{-}11)$$

Equation (6-10) does not take into account changes in the dimensions of the conductor with temperature. Such changes depend on the temperature coefficient of linear expansion, which

Table 6-3. High-Frequency Resistivity Characteristics of Several Metals[117]

Material	Dc Resistivity ρ (relative to copper)	Skin depth δ at 2 GHz, μm	Surface resistivity, $R_s \times 10^7 / f^{1/2}$, Ω/square
Ag	0.95	1.4	2.5
Cu	1.0	1.5	2.6
Au	1.36	1.7	3.0
Al	1.6	1.9	3.3
W	3.2	2.6	4.7
Mo	3.3	2.1	4.7
Ni	5.1	0.31	55.0
Cr	7.6	4.0	7.2

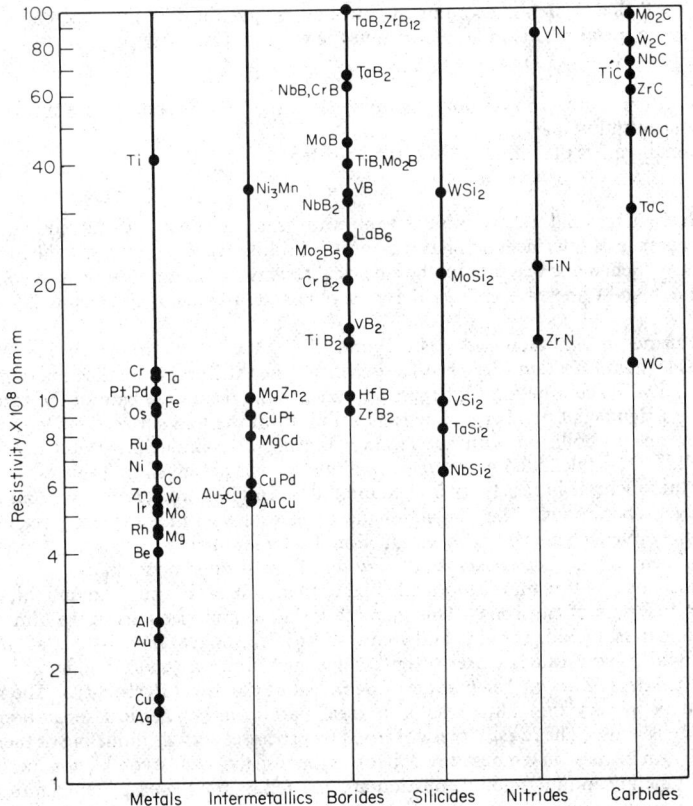

Fig. 6-1. Electrical resistivity of various materials.[61]

is generally much smaller than the temperature coefficient of electrical resistance. If dimensional changes are neglected, the coefficient of electrical resistance equals the coefficient of electrical resistivity.

10. Resistivity-Temperature Constant. The change of resistivity of a material per degree centigrade is called resistivity-temperature constant. For example, the resistivity of copper changes at the rate of 6.8×10^{-11} $\Omega \cdot$ m/°C, irrespective of the temperature.

11. Matthiessen's Rule. Electrical resistivity originates from two main sources: thermal scattering of the electrons *(thermal resistivity)* and scattering of electrons from imperfection in the lattice *(residual resistivity)*. According to Matthiessen's rule[*1], the total electrical resistivity ρ is the sum of thermal resistivity ρ_T and the residual resistivity ρ_R.

$$\rho = \rho_R + \rho_T \qquad (6\text{-}12)$$

At very low temperatures, $\rho_T \ll \rho_R$, and the electrical resistivity is mainly the residual resistivity. The residual resistivity arises from point defects: vacancies, solute atoms or interstitials; line defects: dislocations; surface defects: stacking faults or grain boundaries; and volume defects such as voids or second phase. The increase in resistivity due to vacancies is approximately 10^{-8} $(\Omega \cdot$ m)^2/at % of concentration[2]. The residual resistivity due to dislocations is, roughly,[3-6] $\rho_R = 10^{-25}C_D$ $\Omega \cdot$ m, where C_D is the dislocation density in meters per cubic meter. The effect of impurities on residual resistivity is discussed under Nordheim's rule (Par. **6-13**).

12. Resistivity Ratio. The resistivity ratio is the ratio of electrical resistivity at 298 K to that at 4.2 K. This ratio is a measure of the perfection of the conductor. Higher ratios are obtained for more perfect materials since their residual resistivity at 4.2 K is smaller. Ratios up to 100,000 have been obtained.

13. Nordheim's Rule. According to Nordheim's rule, the addition of solute of concentration X to a metal results in an increase in the residual resistivity of

$$\rho_R(X) = CX(1 - X) \qquad (6\text{-}13)$$

where $\rho_R(X)$ = increase in residual resistivity $(\Omega \cdot$ m), C = constant $(\Omega \cdot$ m/at %), and X = solute concentration (at %).

For small values of X, Eq. (6-13) can be written

$$\rho_R(X) = CX \qquad (6\text{-}14)$$

The increase in residual resistivity is due to electron scattering by the disordered solute atoms. The appearance of intermediate phases or ordered solutions generally lowers the resistivity of the alloy. Schematic diagrams of the electrical resistivity vs. composition are given in Fig. 6-2. The highest increase in resistivity is obtained in binary systems with complete miscibility.

For further information see Refs. 7 and 8. Values of C are given in Table 6-4.

14. Thickness Effect on Electrical Resistivity. The thickness of conducting films affects the resistivity. When the film thickness approaches the mean free path of the conduction electrons, a significant number of electrons collide with the film surfaces. The increase in resistivity due to collision with surfaces was treated by Sondheimer[9] and reviewed by Campbell.[10] The calculated mean free paths of electrons are given in Table 6-5.

A further increase in resistivity is encountered when films are very thin, so that they are physically discontinuous. The conduction mechanisms in very thin films have been recently reviewed by Neugebauer.[11] The conduction in these films proceeds by tunneling or thermionic emission. The resistance of very thin films is often nonohmic.

The resistivity of thin films is normally higher than bulk even when films are thicker than the mean free path of electrons. This increase is due to imperfections in the films such as grain boundaries, included oxides, and voids. Films deposited at higher substrate temperatures normally show values closer to bulk since such films are more structurally perfect.

15. Thickness Effect on Temperature Coefficient of Electrical Resistivity. The increase of resistivity in very thin films results in great part from geometrical separation of the conducting islands. The resistivity is governed by processes such as tunneling or thermionic emission. An increase in temperature will lower the resistivity since conduction mechanisms such as tunneling and thermionic emission are assisted by the increased temperature. The

---

* Superior numbers correspond to numbered references in Par. **6-257**.

Table 6-4. Maximum Solid Solubility c_{max} (at %) and the Atomic Resistivity Increase $C(10^{-8} \ \Omega \cdot m/at \ \%)$ for Cu, Ag, Au, and Al[8,63,69]

Solute	Base metal							
	Cu		Ag		Au		Al	
	c_{max}	C	c_{max}	C	c_{max}	C	c_{max}	C
Ag	4.9	0.14	. . .		Complete	0.36	23.8	1.15
Al	19.6	1.25	20.34	0.5	6	1.87		
As		6.8	8.8	8.5		8.00		
Au	Complete	0.55	Complete	0.36				
Be	16.4	0.62						
Bi			3	7.3		(6.5)		
Cd		0.30	42.2	0.38	32.5	0.63	0.11	0.057
Co	12.0	6.35			23.5	6.1		
Cr	0.8	(3.6)			(23)	4.25	0.40	7.7
Cu			14.1	0.077	Complete	0.45	2.48	0.785
Fe	4.5	9.3			75	7.9	0.025	5.30
Ga	19.9	1.42	18.7	2.36	13	2.2	→8.82	0.25
Ge	11.8	3.79	9.6	5.5	3.2	5.2		
In	11	1.06	20	1.78	12.6	1.39		
Ir		5.7						
Mg	7	0.65	29.3	1.95	(25)	1.30	16.26	0.49
Mn	Complete	2.90	47	1.60		2.41	0.90	5.97
Ni	Complete	1.25			Complete	0.79	0.023	1.77
P	3.5	6.7						
Pb			2.8	4.65		(3.9)		
Pd	Complete	0.89	Complete	0.44	Complete	0.41		
Pt	Complete	2.1	40.5	1.60	Complete	1.01		
Rh	(30)	4.40			(9.0)	4.15		
Sb	6	5.4	7.2	7.25	1.1	6.8		
Si	11.25	3.95					1.59	0.65
Sn	9.1	2.88	11.5	4.36	6.8	3.36		
Ti					(11)	12.9	→0.566	5.10
Tl			7.5	2.27	0.9	1.9		
V							0.32	6.90
Zn	38.3	0.32	40.2	0.64	31	0.95	66.4	0.211
Zr							0.085	5.85

Table 6-5. Electronic Mean Free Path of Several Metals[65]

Metal	Calculated mean free path, μm			Temp. coeff., ppm/°C (0–100)	Room temp. resistivity, × 10^8, $\Omega \cdot m$
	−200°C	0°C	100°C		
Li	0.0955	0.0113	0.0079	4,220	8.55
Na	0.1870	0.0335	0.0233	4,400	4.3
K	0.1330	0.0376	0.0240	5,500	6.1
Cu	0.2965	0.0421	0.0294	4,330	1.69
Ag	0.2425	0.0575	0.0405	4,100	1.47
Au	0.1530	0.0406	0.0290	4,000	2.44
Ni	. . .	0.0133	0.0080	6,750	7.24
Co		0.0130	0.0079	6,580	9.7
Fe	0.2785	0.0220	0.0156	4,110	8.85
Pt	0.0720	0.0110	0.0079	3,920	9.83

PROPERTIES OF MATERIALS

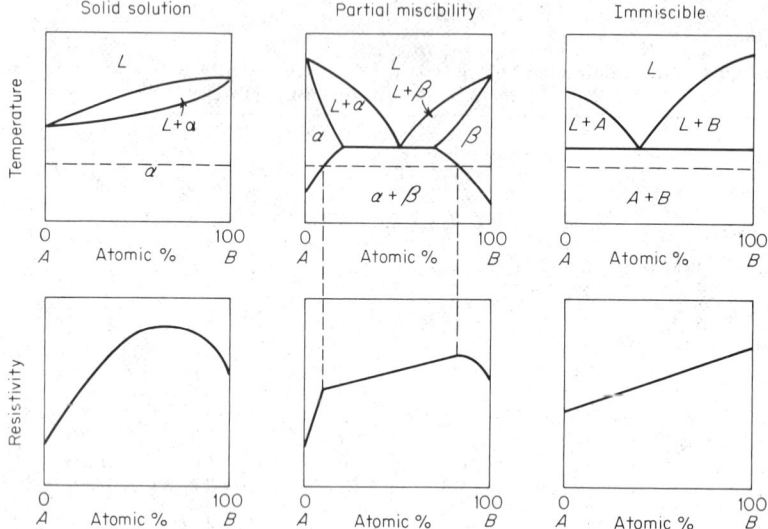

Fig. 6-2. Typical resistivity of binary alloys as a function of composition. α, β are solid solutions rich in A and B, respectively. L denotes liquid.[63]

temperature coefficient of electrical resistivity of very thin films will therefore be negative for most materials.

16. Thermal Conductivity. The heat flow rate along a rod is proportional to its cross section and to temperature gradient dT/dX along the rod (see Fig. 6-3).

$$q = - \lambda\, A\, \frac{dT}{dX} \qquad (6\text{-}15)$$

where dT/dX = temperature gradient (°C/m), A = cross section (m²), and λ = thermal conductivity (W/m · K). The thermal conductivity is a function of temperature (examples shown in Fig. 6-4). Values of λ are listed in Table 6-1.

17. Wiedemann-Frantz Ratio. The Wiedemann-Frantz ratio L is defined as

$$L = \frac{\lambda}{\sigma T} \qquad (6\text{-}16)$$

where λ = thermal conductivity (W/m · K), σ = electrical conductivity (℧/m), and T = temperature (K).

This ratio, approximately constant for most pure metals, is $2 - 3 \times 10^{-8}$(V/K)².

18. Temperature Coefficient of Linear Expansion. Materials change their dimensions with temperature. The temperature coefficient of linear expansion, α_L, is defined by

$$\alpha_L = \frac{1}{l}\frac{dl}{dT} \qquad (6\text{-}17)$$

where l = any linear dimension (length of a rod, for example) of the material (m), and T = temperature (°C).

If the length l_0 of a rod at temperature T_0 is known, the length of the rod at temperature T can be calculated from Eq. (6-17). When α_L is constant, then

$$l = l_0[1 + \alpha_L(T - T_0)] \qquad (6\text{-}18)$$

Values of α_L are given in Table 6-1.

A more exact relation between l and l_0 involves a power series of the temperature difference.

$$l = l_0[1 + \alpha_L(T - T_0) + \beta_L(T - T_0)^2 + \cdots] \qquad (6\text{-}19)$$

19. Volume Coefficient of Thermal Expansion. The volume coefficient of thermal expansion, α_V, is defined as

$$\alpha_\Gamma = \frac{1}{V}\frac{dV}{dT} \tag{6-20}$$

where V = volume (m^3) and T = temperature (°C). Since α_L and α_V are small, $\alpha_V \approx 3\alpha_L$.

20. Specific Heat. Specific heat at constant pressure, c_p (J/K · kg), is defined as the amount of heat needed to raise the temperature 1 kg of material by 1 K at constant pressure. Specific heat at constant volume, c_v (J/K · kg), is defined as the amount of heat needed to raise the temperature 1 kg of material by 1 K at constant volume. The difference between c_p and c_v is small in solids. Values of c_p are tabulated in Table 6-1.

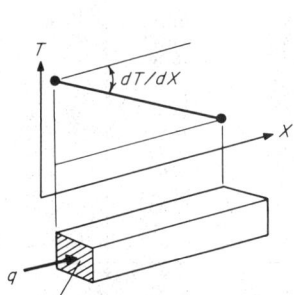

Fig. 6-3. Heat flow due to a temperature gradient.

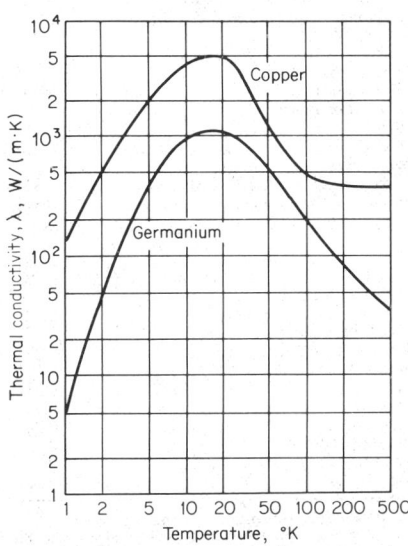

Fig. 6-4. Thermal conductivity of copper and germanium as a function of temperature.[66]

21. Tensile Stress. The engineering tensile stress is defined as the applied force per unit area.

$$\sigma = \frac{F}{A_0} \tag{6-21}$$

σ = stress (N/m^2), A_0 = cross section (m^2) (see Fig. 6-5), and F = force (N). Stress is often given in lb/in.^{2}, kg/mm^2, or kg/cm^2; 1 kg/cm^2 = 98,066.5 N/m^2; 1 lb/in.^{2} = 6,894.8 N/m^2. Tensile stress is defined to be positive; compression is negative.

22. Strain. The engineering strain ε_x is defined as the elongation of the material divided by the original length of the specimen.

$$\varepsilon_x = \frac{\Delta l}{l_0} = \frac{l - l_0}{l_0} \tag{6-22}$$

See Fig. (6-5) for l, l_0.

23. Young's Modulus. For small strains, the stress is proportional to the strain (Hooke's law). The proportionality constant is called *Young's modulus*,

$$\sigma = \varepsilon_x E \tag{6-23}$$

where σ = stress (N/m^2), ε_x = strain, and E = Young's modulus (N/m^2). Young's modulus depends on crystallographic direction, even for cubic materials. Values of Young's modulus are compiled in Table 6-6.

24. Poisson's Ratio. When a material is stressed in the x direction (Fig. 6-5), it deforms also in the transverse directions. The width b_0 will contract to a value b upon elongation from

Table 6-6. Elastic Constants for Polycrystalline Metals and Alloys at 20°C[114]

Metal	Poisson's ratio	Bulk modulus K	Young's modulus E	Rigidity modulus G
		In units of 10^{10} N/m²		
Aluminium..........................	0.345	7.52	7.06	2.62
Duralumin..........................	0.345	7.54	7.08	2.63
Brass (70% Cu, 30% Zn)	0.350	11.18	10.06	3.73
Chromium	0.210	16.02	27.90	11.53
Constantan.........................	0.327	15.64	16.24	6.12
Copper	0.343	13.78	12.98	4.83
Invar (36% Ni, 63.8% Fe, 0.2% C)....	0.259	9.94	14.40	5.72
Iron (soft).........................	0.293	16.98	21.14	8.16
Iron (cast)*	0.27	10.95	15.23	6.0
Lead*	0.44	4.58	1.61	0.559
Magnesium	0.291	3.56	4.47	1.73
Molybdenum.......................	0.293	26.12	32.48	12.56
Nickel, unmagnetized				
Soft*	0.312	17.73	19.95	7.60
Hard*..........................	0.306	18.76	21.92	8.39
Nickel silver				
(55% Cu, 18% Ni, 27% Zn)........	0.333	13.20	13.25	4.97
Niobium...........................	0.397	17.03	10.49	3.75
Silver..............................	0.367	10.36	8.27	3.03
Steel:				
Mild	0.291	16.92	21.19	8.22
¾% C	0.293	16.87	21.00	8.11
¾% C hardened	0.296	16.50	20.14	7.78
Tool†...........................	0.287	16.53	21.16	8.22
Tool, hardened†	0.295	16.52	20.32	7.85
Stainless‡	0.283	16.6	21.53	8.39
Tantalum	0.342	19.63	18.57	6.02
Tin................................	0.357	5.82	4.99	1.84
Titanium...........................	0.361	10.84	12.02	4.56
Tungsten...........................	0.280	31.10	41.10	16.06
Tungsten carbide	0.22	31.9	53.44	21.90
Vanadium..........................	0.365	15.80	12.76	4.67
Zinc...............................	0.249	6.94	10.45	4.19

NOTE: All compositions are percentage by weight.
* Approximate values for materials of variable composition.
† Oil hardening nondeforming tool steel, 0.98% C, 1.03% Mn, 0.65% Cr, 1.01% W, 0.1% V.
‡ Composition 0.2% C, 0.5% Si, 0.7% Mn, 2.0% Ni, 18.0% Cr.

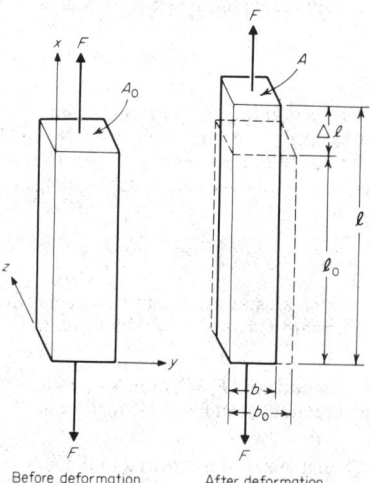

Fig. 6-5. Uniaxial tensile force and deformation Δl.

l to l_0. The transverse strain is $\varepsilon_y = (b - b_0)/b_0$. Poisson's ratio ν is defined as

$$\nu = -\frac{\varepsilon_y}{\varepsilon_x} \tag{6-24}$$

Values of Poisson's ratio are compiled in Table 6-6.

25. Shear. The shear stress is defined as the shear force F per unit area A_0 (see Fig. 6-6).

$$\tau = \frac{F}{A_0} \tag{6-25}$$

where τ = shear stress (N/m^2), F = shear force (N), and A_0 = area (m^2). The shear strain γ is defined from Fig. 6-6 as

$$\gamma = \frac{\Delta}{l_0} \tag{6-26}$$

26. Shear Modulus. For small strains the shear stress is proportional to the shear strain,

$$\tau = \gamma G \tag{6-27}$$

where τ = shear stress (N/m^2), γ = shear strain, and G = shear modulus (N/m^2). The proportionality constant G is called the shear modulus. (See also Table 6-6.)

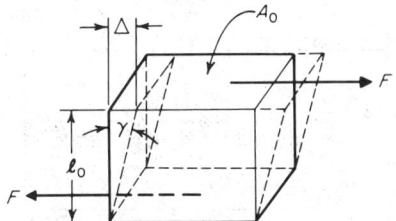

Fig. 6-6.   Shear force F and shear strain γ.

27. Compressibility. A material of volume V will contract by ΔV upon application of pressure Δp. The compressibility is the fractional volume change per unit of applied pressure,

$$-\frac{\Delta V}{V} = \beta \, \Delta p \tag{6-28}$$

where V = volume (m^3), ΔV = volume change (m^3), Δp = applied pressure (N/m^2), and β = compressibility (m^2/N). Values of $\beta = 1/K$ are compiled in Table 6-6.

28. Stress-Strain Diagrams. Typical stress-strain diagrams are shown in Fig. 6-7. Figure 6-7a, typical of low-carbon steels, displays a distinct *elastic region* up to a *sharp yield point* σ_Y. The elastic deformation is reversible; i.e., it can be removed by removal of the stress. Beyond the yield stress (or flow stress) the *plastic region* starts. The plastic deformation is not reversible; it remains after removal of the stress. The stress increases with deformation in the plastic region *(strain hardening)* until it reaches the *ultimate tensile strength* (UTS).

Figure 6-7b shows a typical diagram for ductile metals such as aluminum or copper. No distinct yield point is seen. It is customary to define an *offset yield strength* by plotting a line parallel to the slope at the origin at an offset of $\varepsilon = 0.2\%$ (see Fig. 6-7b) and noting its intersection with the stress-strain curve. Figure 6-7c shows a typical stress-strain diagram for a *brittle material*. The material breaks without any appreciable plastic deformation.

The ductility of materials is characterized by the final *elongation* of the sample after fracture.

$$\text{Elongation} = \frac{l - l_0}{l_0} \times 100 \tag{6-29}$$

where l_0 = original length (m), l = length after fraction (m), and by the final *reduction* in the cross section at fracture,

$$\text{Reduction} = \frac{A_0 - A}{A_0} \times 100 \tag{6-30}$$

where A_0 = original cross section (m^2) and A = cross section after fracture (m^2). See Table 6-7 for compilation of mechanical properties.

Table 6-7. Mechanical Properties of Metals and Alloys at Room Temperature[115,116]

Material	Composition	Condition	Yield point or 0.2% proof stress, $N/m^2 \times 10^7$	Ultimate tensile stress, $N/m^2 \times 10^7$	Elongation on 2 in., %	Hardness*
Ag	99.9	Annealed 600–650°C	0.76†	13.7	50	26 VHN
		Hard	...	38	4	90 VHN
Al	99.95	Rolled rod 0	...	5.5	61	17 B
Au	99.99	Soft, cast	0	12.1	30	33 B
		Hard, 60% red.	21.2	22.8	4	58 B
Co	99.9	Soft	19	24	4.8	124 B
		Hard	...	67.5	2–8	165 B
Cu	99.997	Annealed	34	35.1	60	
		Rod, cold-drawn	3.4	21.3	14	37 RB
Ni	>99.0	Annealed	13.8	48.2	40	100 B
		Cold-drawn	48.2	65.4	25	170 B
Pt	99.99	Annealed	...	12·13	25–40	38–40 VHN
		50% cold-rolled	18	20	3	92 VHN
Pd	99.9	Annealed	...	19	40	37 VHN
		50% cold-drawn	3.5	32	1.5	106 VHN
Ta	99.98	Annealed	18	20	36	90 VHN
	99.95	Cold-rolled	33	41	5	160 VHN
W	99.9	Swaged, recrys.	19.5	40.5	16	(200 VHN)
		Swaged	...	175	1–4	450–490 VHN
Aluminum alloys:						
1100	1% Si	0	2.75	6.9	45	19 B
1100		H18	12.4	13.1	15	35 B
3003	1–1.5% Mn	0	4.1	11	40	28 B
3003		H18	18.6	20	10	55 B
5056	4.5–5.6% Mg	0	15.2	29	35	65 B
5056		H38	34.5	41.5	15	100 B
7075	1.2–2% Cu, 2.1–2.9% Mg, 5.1–5.6% Z	0	10.4	22.7	...	60 B
7075		T6	50.5	57	...	150 B

Material	Composition	Condition				
Copper alloys:						
OHFC, copper		Wire, soft	...	24.1	35	46 RF
		Wire, hard	...	38	1.5	73 RB
Gilding metal	5% Zn	Annealed, strip	6.9	23.4	45	
		Extra hard	39	43.5	4	65 RB
Red brass	15% Zn	Annealed	7–13	28–32	48	
		Half hard	35	40.5	12	
		Extra hard	43.5	55.5	4	83 RB
Yellow brass	35% Zn	Annealed	10–15.7	32.6–37.6	54–65	58–78 RF
		Half hard	42.6	52.6	8	80 RB
		Extra hard	44	60.5	5	87 RB
Phosphor bronze	5% Sn	Annealed	15–22	34–39	57–48	33–46 RB
		Hard	56	57.5	8	89 RB
		Extra spring	71	72.5	3	98 RB
Beryllium copper	1.9% Be	Annealed	...	43–55	35	45–78 RB
		HT (cold worked and precipitation-hardened)	107	142	2	42 RC
Steels:						
C 1010	0.08–0.13% C	Hot-rolled	18.8	34	28	95 B
		Cold-drawn	31.8	38.3	20	105 B
C 1080	0.77–0.88% C	Hot-rolled	44	81	10	229 B
		Tempered 600°F	148	152	10	
12% manganese steel	12% Mn					
Stainless steel Type 304	9% Ni, 19% Cr	Annealed	34	59	60	160 B
		Cold-rolled	111	128	...	400 B

*VHN = Vickers hardness number; B = Brinell; RB = Rockwell B; RC = Rockwell C; RF = Rockwell F.
†0.01% proof stress.

29. Hardness. Hardness measures the resistance of the material to indentation. Hardness tests measure the plastic deformation (the size or depth) of an indentation. Brinell hardness tests use spheres as indenters; the Vickers test uses pyramids. Rockwell tests use cones or spheres. Microhardness tests for specimens are also available, using the Knoop method with miniature pyramid indenters. Another hardness scale is Mohs' scale, which lists the materials in order of their hardness, beginning with talc and ending with diamond. Hardness values are given in Table 6-7.

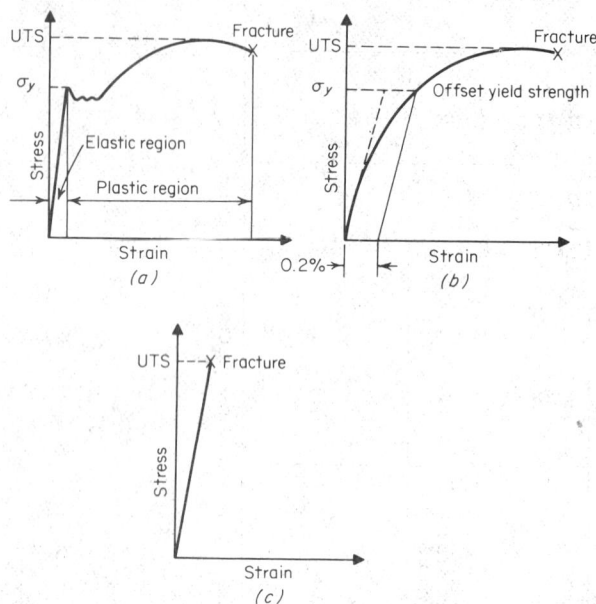

Fig. 6-7. Schematic stress-strain diagrams. (*a*) Sharp yield point; (*b*) ductile material without sharp yield point; (*c*) brittle material; σ_y = yield point, UTS = ultimate tensile stress.

30. Notch Toughness. Notch toughness is measured by the energy necessary to fracture a standard notched specimen. The notch toughness is a function of temperature. Many materials display a transition from high toughness (ductility) to low toughness (brittleness). The transition temperatures for low-carbon steels are in the region of -20 to $-70°C$.

31. Fatigue. Under cycling loading, materials tend to break after a large number of cycles even when stressed below their ultimate strength. Figure 6-8 shows the fracture stress vs. number-of-cycles curve (*S-N* curve) for steel. Ferrous metals tend to exhibit a lower limit of stress under which they do not fail due to fatigue. This limit is the *endurance limit*. Nonferrous metals do not show a definite endurance limit. *S-N* diagrams are usually reported for fully reversing stress.

Fatigue fracture starts with fine cracks that propagate from the surface through the specimen. The fatigue strength is therefore very sensitive to surface finish and tensile-stress risers such as keyholes and notches.

If a steady tensile stress is imposed on the alternating stress, the fatigue life will decrease. A compressive stress will increase fatigue life. Prestressing or hardening of the surface also results in an increased life. Fatigue strength is lower in corrosive environments (called *corrosion fatigue).*

32. Creep. Materials will continue to deform under a given stress at high temperatures. As function of time, three distinct stages occur. The first stage, primary creep, is characterized by a decelerating creep rate; the second stage is the steady creep region; during the third stage the creep rate accelerates and fracture finally results. The stress-rupture test records the rupture times at a given stress and temperature.

33. Cold Working. Plastic deformation at room temperature is called cold working.

Hardening and strengthening of a material by plastic deformation are called *work hardening*. Wire drawing or other cold-working processes raise the yield strength of metals. Plastic deformation may introduce a preferred crystallographic orientation, sometimes referred to as *texture*. Since some physical properties such as elastic constants and magnetic permeability are anisotropic, the texture strongly affects them.

34. Solute Effects on Mechanical Properties. A solute generally raises the yield and ultimate tensile strength of a material. Controlled precipitation of the solute from supersaturated solutions brings an even more drastic improvement in mechanical properties. The precipitation is achieved by first heat-treating above the solubility limit to dissolve all the solute (solute treatment), then quenching to room temperature, and finally heat-treating at 100 to 200°C to form very fine precipitates. The strengthening due to precipitation is often referred to as *age hardening*, or *precipitation hardening*. Figure 6-9 shows the change in mechanical properties with aging time for an Al alloy.

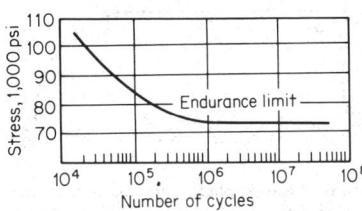

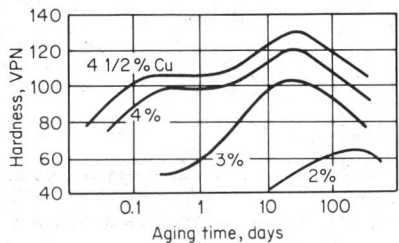

Fig. 6-8. Stress vs. number of cycles to failure for a hot-worked 4340 steel bar.[67]

Fig. 6-9. Hardness of various Al-Cu alloys aged at 130°C, as a function of time.[68]

35. Annealing. Softening of materials by exposure to high temperature for long times is referred to as annealing.

36. Stress Corrosion Cracking. Rapid corrosion of highly cold-worked materials is known as stress corrosion cracking. Examples are brass in mercury or steel in sodium hydroxide.

37. Melting Point. The temperature at which an element or compound transforms from solid to liquid under equilibrium conditions is the melting point (m.p.). The melting point is a function of the external pressure. See Table 6-1 for values of melting points.

38. Boiling Point. The temperature at which an element or compound transforms from liquid to vapor under equilibrium conditions is the boiling point (b.p.). The boiling point is a function of the external pressure. See Table 6-1.

39. Allotropic Forms. Different crystallographic forms of the same element or compound are called allotropic forms. For example, iron can exist in a body-centered cubic (bcc) form or a face-centered cubic (fcc) form.

40. Equilibrium Phase Diagrams. Phase diagrams show which phase or phases are most stable in a given system of components at a given temperature and composition. Most diagrams are binary (for two elements) or ternary. Sections through diagrams with more than three components are sometimes available.

Figure 6-10 is an example of a binary diagram with complete *miscibility*. Contrary to the pure materials, the alloy does not show a sharp melting point. The melting occurs over a wide temperature range. Figure 6-11 shows a system with partial miscibility. This is called a *eutectic* diagram. Except for the eutectic composition (38.1 wt % Pb) the alloys do not have a sharp melting point. Figure 6-12 shows a binary diagram of gold and tin. This diagram shows a series of eutectics, peritectics, and *intermetallic compounds*.

41. Diffusion. Diffusion is mass transport in a solid, liquid, or gas. The driving force for diffusion is generally chemical, originating from concentration gradients. The diffusion flux J in a binary alloy is proportional to the chemical gradient of the solute (Fick's law).

$$J = -D\frac{dC}{dX} \tag{6-31}$$

where C = concentration (atoms/m^3), X = direction of the material flux (m), J = flux (atoms/m$^2 \cdot$ s), and D = diffusion coefficient (m^2/s).

The diffusion coefficient is temperature-dependent, and can be approximated by

$$D = D_0 e^{-Q/kT} \tag{6-32}$$

where D_0 = preexponential constant (m²/s), Q = activation energy (J) (normally given in cal/mol or eV), k = Boltzmann's constant (J/K), and T = temperature (K). The diffusion coefficient is generally concentration-dependent. The diffusion coefficients are generally larger for structurally imperfect solvents. Thus heat-treatment, mechanical working, and radiation damage appreciably increase the diffusion coefficient. Diffusion in thin films, which normally are structurally imperfect, is much faster than in bulk materials.

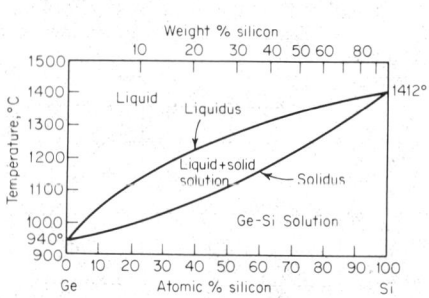

Fig. 6-10. Germanium-silicon phase diagram.[69]

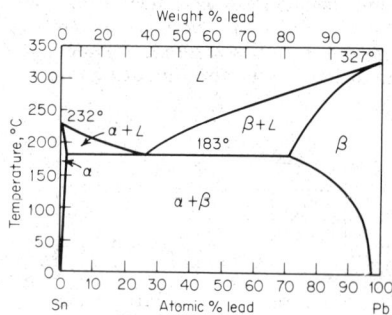

Fig. 6-11. Lead-tin phase diagram. α, β are the tin- and lead-rich solid solutions, respectively. L denotes liquid.[69]

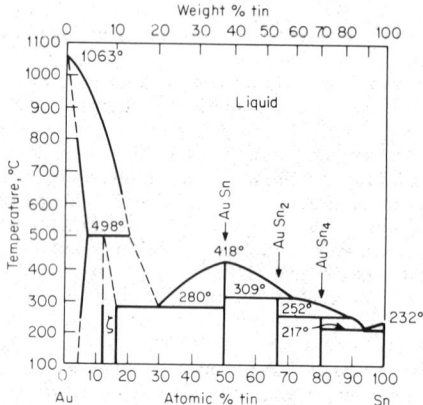

Fig. 6-12. Gold-tin phase diagram.[69]

Diffusion distance, or solute penetration, can be roughly estimated by the formula $X = \sqrt{Dt}$, where X = diffusion distance (m) and t = diffusion time (s).

42. Electromigration. An electric current density J induces in bulk conductors an atomic flux J_a (atoms/m²·s). The atomic flux is created by either simple electrostatic forces or by a momentum transfer from the electrons. Huntington[12] proposed the following relation:

$$J_a = \frac{ND}{kT} Z^* e \rho J \tag{6-33}$$

where N = density of metallic ions (m⁻³), D = diffusion coefficient of vacancies (m²/s), which may be expressed in terms of a constant D_0 and an activation energy Q,

$$D = D_0 e^{-Q/kT}$$

J = current density (A/m^2), k = Boltzmann's constant (J/K), ρ = resistivity of the conductor ($\Omega \cdot$ m), T = temperature (K), and Z^*e = *effective charge* of the metallic ions (C).

The effective charge on the metallic ions can be positive or negative, depending on whether the electrostatic field which creates an atomic flux toward the cathode is larger than the momentum transfer which creates a flux toward the anode.

Electromigration in thin films proceeds much faster than in bulk conductors since the diffusion processes in thin films are generally much faster than bulk values.[13,14]

43. Current-carrying Capacity. The current-carrying capacities of wires, cables, and bus bars are limited by heating effects produced by the current. The permissible temperature rise of bus bars is around 30°C above an ambient of 40°C. Current densities for copper buses are about 10^6 A/m^2 ($= 10^2$ A/cm^2). Aluminum current densities are 75% of those permitted in copper. Current-carrying capacities of various wires and cables are given in Table 6-8. Figure 6-13 shows the current-carrying capacity of copper conductors on printed-circuit boards.

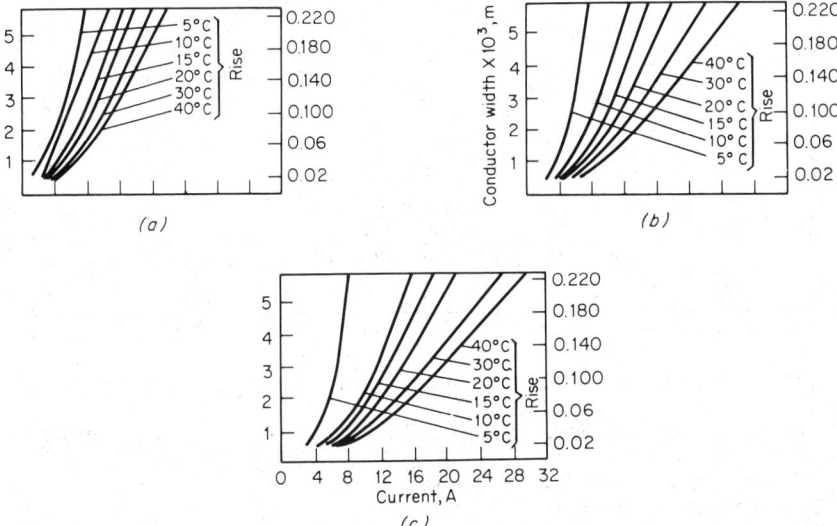

Fig. 6-13. Temperature rise vs. current for (*a*) 1-oz copper, (*b*) 2-oz copper, and (*c*) 3-oz copper.[71]

The current-carrying capacity of thin films is limited by electromigration (see Par. **6-42**). Due to the good thermal conduction from the films, it is possible to conduct currents in excess of 10^9 A/m^2. At higher current densities irregularities in the metal transport will create holes in the conductors, leading to their eventual failure.[14]

The lifetime of thin-film conductors depends strongly on their composition, structural perfection, width, thickness, length, temperature, and temperature gradients. There is no good set of tabulated lifetime data that encompasses all the above variables.

FORMS OF CONDUCTORS

44. Wire. Wires are the most common form of conductors for power transmission, smaller electrical signals, or as resistors. Wires are either solid or stranded. Most wires are round; occasionally square or rectangular conductors are used, such as integrated-circuit external leads.

Insulated wires are most often used in electronic circuits. The insulation provides electrical insulation and mechanical and chemical protection. There are several wire gages:

AWG (American B&S): This wire gage divides the range from 0.0050 in. (AWG No. 36) to 0.4600 in. (AWG No. 0000) into 39 intervals. Sizes progress in geometrical fashion with a ratio of 1.1229322 between adjacent gages.

PROPERTIES OF MATERIALS

Table 6-8. Maximum Current Capacity (Amperes) of Copper and Aluminum Conductors*[71]

Size, AWG	MIL-W-5088				National Electrical Code	Underwriters Laboratory		American Insurance Association	500 c mils/A
	Copper		Aluminum					
	Single-wire	Wire-bundled	Single-wire	Wire-bundled		60°C	80°C		
30	...	...	...	...	...	0.2	0.4	...	0.20
28	...	...	...	...	...	0.4	0.6	...	0.32
26	...	...	...	...	...	0.6	1.0	...	0.51
24	...	...	...	...	...	1.0	1.6	...	0.81
22	9	5	...	...	...	1.6	2.5	...	1.28
20	11	7.5	...	...	...	2.5	4.0	3	2.04
18	16	10	...	...	6	4.0	6.0	5	3.24
16	22	13	...	...	10	6.0	10.0	7	5.16
14	32	17	...	...	20	10.0	16.0	15	8.22
12	41	23	...	...	30	16.0	26.0	20	13.05
10	55	33	...	...	35	...	...	25	20.8
8	73	46	58	36	50	...	...	35	33.0
6	101	60	86	51	70	...	...	50	52.6
4	135	80	108	64	90	...	...	70	83.4
2	181	100	149	82	125	...	...	90	132.8
1	211	125	177	105	150	...	...	100	167.5
0	245	150	204	125	200	...	...	125	212.0
00	283	175	237	146	225	...	...	150	266.0
000	328	200	...	...	275	...	...	175	336.0
0000	380	225	...	...	325	...	...	225	424.0

* For further information consult Ref. 60.

BWG (Birmingham wire gage): An English designation based on the number of drawings necessary to produce the wire.

SWG (British Standard Wire Gauge): The legal standard in Great Britain. Similar to BWG.

Metric wire gage: 0.1 mm is No. 1, 0.2 mm is No. 2, etc.

Steel wire gage: Used exclusively for steel wire.

Wire size: A direct way to specify wire size is to give the diameter in mils (1 mil = 0.001 in.).

Circular mil: A unit of area that equals the cross-section area of a 1-mil-diam. wire. 1 cmil = 0.7854 mil^2.

See Table 6-9 for wire dimensions of the various gages. Table 6-10 shows the properties of copper wire.

45. Stranded Wire. A group of wires used as a single wire is called stranded wire. Due to their high flexibility, stranded wires are the most commonly used conductors.

Stranded conductors are based on a 1-wire or a 3-wire core. The total number of wires is $3n(n + 1) + 1$ for a 1-wire core (that is, 1,7,19, . . .) and $3n(n + 2) + 3$ for a 3-wire core (that is, 3, 12, . . .), $n = 0,1,2, ...$ There is no sharp distinction between stranded wire and cable. Tables of stranded wires are given in Table 6-11.

46. Cables. Cables are stranded conductors or a combination of stranded conductors. Cables can be made from one material or a combination of materials. Cables can be bare or insulated. There are various geometrical configurations for cables:

Concentric-lay cables: Successive layers of helically laid wires.

Bunch-stranded conductors: A group of conductors bunched together in no particular geometrical form. If conductors are parallel, they are referred to as *parallel strand,* and the strands are usually twisted together.

Rope lay: Made of helically laid successive layers of stranded conductors. The stranded conductors can be either bunch-stranded or concentric-lay-stranded.

Table 6-9.　Wire Gages—Diameter, Area, Copper Weight [118]

Size, AWG	Diameter at 20°C (68°F)		Area at 20°C (68°F)			Weight at 20°C (68°F), bare copper wire		
	mils	mm	mils²	c mils	mm²	lb/1,000 ft	lb/mile	kg/km
4/0	460.0	11.68	166,200	211,600	107.2	640.5	3,382	953.2
3/0	409.6	10.40	131,800	167,800	85.01	507.8	2,681	755.7
2/0	364.8	9.266	104,500	133,100	67.43	402.8	2,127	599.5
1/0	324.9	8.252	82,891	105,600	53.49	319.5	1,687	477.6
1	289.3	7.348	85,730	83,690	42.41	253.3	1,338	337.0
2	257.6	6.543	52,120	66,360	33.62	200.9	1,061	298.9
3	229.4	5.827	41,330	52,620	26.67	159.3	841.1	237.1
4	204.3	5.189	32,780	41,740	21.15	126.3	667.1	188.0
5	181.9	4.620	25,990	33,090	16.77	100.2	528.8	149.0
6	162.0	4.115	20,610	26,240	13.30	79.44	419.4	118.2
7	144.3	3.665	16,350	20,820	10.55	63.03	332.8	93.80
8	128.5	3.264	12,970	16,510	8.367	49.98	263.9	74.38
9	114.4	2.906	10,280	13,090	6.631	39.61	209.2	58.95
10	101.9	2.588	8,155	10,380	5.261	31.43	166.0	46.77
11	90.7	2.304	6,460	8,230	4.17	24.90	131	37.1
12	80	2.05	5,130	6,530	3.310	19.8	104	29.4
13	72.0	1.83	4,070	5,180	2.63	15.7	82.9	23.4
14	64.1	1.63	3,230	4,110	2.08	12.4	65.7	18.5
15	57.1	1.45	2,560	3,260	1.650	9.87	52.1	14.7
16	50.8	1.29	2,030	2,580	1.31	7.81	41.2	11.6
17	45.3	1.150	1,610	2,050	1.040	6.21	32.8	9.24
18	40.3	1.02	1,280	1,620	0.823	4.92	26.0	7.32
19	35.9	0.912	1,010	1,290	0.653	3.90	20.6	5.81
20	32.0	0.813	804.0	1,020	0.519	3.10	16.4	4.61
21	28.5	0.724	638.0	812.0	0.412	2.46	13.0	3.66
22	25.3	0.643	503.0	640.0	0.324	1.94	10.2	2.88
23	22.6	0.574	401.0	511.0	0.259	1.55	8.16	2.30
24	20.1	0.511	317.0	404.0	0.205	1.22	6.46	1.82
25	17.9	0.455	252.0	320.0	0.162	0.970	5.12	1.44
26	15.9	0.404	199.0	253.0	0.128	0.765	4.04	1.14
27	14.2	0.361	158.0	202.0	0.102	0.610	3.22	0.908
28	12.6	0.320	125	159.0	0.0804	0.481	2.54	0.715
29	11.3	0.287	100.0	128.0	0.0647	0.387	2.04	0.575
30	10.0	0.254	78.50	100.0	0.0507	0.303	1.60	0.450
31	8.9	0.226	62.20	79.2	0.0401	0.240	1.27	0.357
32	8.0	0.203	50.30	64.0	0.0324	0.194	1.02	0.288
33	7.1	0.180	39.60		0.0255	0.153	0.806	0.227
34	6.3	0.160	31.20	50.40	0.0201	0.120	0.634	0.179
35	5.6	0.142	24.60	39.7 / 31.4	0.0159	0.0940	0.501	0.141
36	5.0	0.127	19.60	25.00	0.0127	0.0757	0.400	0.113
37	4.50	0.114	15.90	20.20	0.0103	0.0613	0.324	0.0912
38	4.0	0.102	12.60	16.0	0.00811	0.0484	0.256	0.0721
39	3.5	0.0889	9.62	12.2	0.00621	0.0371	0.196	0.0552
40	3.1	0.0787	7.55	9.61	0.00487	0.0291	0.154	0.0433
41	2.8	0.0711	6.16	7.84	0.00397	0.0237	0.125	0.0353
42	2.5	0.0635	4.91	6.25	0.00317	0.0189	0.0999	0.0282
43	2.2	0.0559	3.80	4.84	0.00245	0.0147	0.0774	0.0218
44	2.0	0.0508	3.14	4.00	0.00203	0.0121	0.0639	0.0180
45	1.8	0.0457	2.54	3.24	0.00164	0.00981	0.0519	0.0146
46	1.6	0.0406	2.01	2.56	0.00130	0.00775	0.0409	0.0115
47	1.4	0.0356	1.54	1.96	0.000993	0.00593	0.0313	0.00883
48	1.2	0.0305	1.13	1.44	0.000730	0.00436	0.0230	0.00649
49	1.1	0.0279	0.950	1.21	0.000613	0.00366	0.0193	0.00545
50	1.0	0.0254	0.785	1.00	0.000507	0.00303	0.0160	0.00450

Table 6-10. Copper Wire—Weight, Breaking Strength, DC Resistance
(Based on ASTM Specifications B1-56, B2-52, B3-63)

Size, AWG	Area Diam., in.	Area c mils	Area in.²	Weight lb/1,000 ft	Weight lb/mile	Hard Breaking strength, min.* lb	Hard Dc resistance at 20°C (68°F), max.† Ω/1,000 ft	Medium Breaking strength, min.* lb	Medium Dc resistance at 20°C (68°F), max.† Ω/1,000 ft	Soft Breaking strength, max.‡ lb	Soft Dc resistance at 20°C (68°F), max.† Ω/1,000 ft
4/0	0.4600	211,600	0.1662	640.5	3382	8143	0.05045	6980	0.05019	5983	0.04901
3/0	0.4096	167,800	0.1318	507.8	2681	6720	0.06362	5666	0.06330	4744	0.06182
2/0	0.3648	133,100	0.1045	402.8	2127	5519	0.08021	4599	0.07980	3763	0.07793
1/0	0.3249	105,600	0.08291	319.5	1687	4518	0.1022	3731	0.1016	2985	0.09825
1	0.2893	83,690	0.06573	253.3	1338	3688	0.1289	3024	0.1282	2432	0.1239
2	0.2576	66,360	0.05212	200.9	1061	3002	0.1625	2450	0.1617	1928	0.1563
3	0.2294	52,620	0.04133	159.3	841.1	2439	0.2050	1984	0.2039	1529	0.1971
4	0.2043	41,740	0.03278	126.3	667.1	1970	0.2584	1584	0.2571	1213	0.2485
5	0.1819	33,090	0.02599	100.2	528.8	1590	0.3260	1265	0.3243	961.5	0.3135
6	0.1620	26,240	0.02061	79.44	419.4	1280	0.4110	1010	0.4088	762.6	0.3952
7	0.1443	20,820	0.01635	63.03	332.8	1030	0.5180	806.7	0.5153	605.1	0.4981
8	0.1285	16,510	0.01297	49.98	263.9	826.1	0.6532	644.0	0.6498	479.8	0.6281
9	0.1144	13,090	0.01028	39.61	209.2	660.9	0.8241	513.9	0.8199	380.3	0.7925
10	0.1019	10,380	0.008155	31.43	166.0	529.3	1.039	410.5	1.033	314.0	0.9988
11	0.0907	8,230	0.00646	24.9	131	423	1.31	327	1.30	249	1.26
12	0.0808	6,530	0.00513	19.8	104	337	1.65	262	1.64	197	1.59
13	0.0720	5,180	0.00407	15.7	82.9	268	2.08	209	2.07	157	2.00
14	0.0641	4,110	0.00323	12.4	65.7	214	2.63	167	2.61	124	2.52
15	0.0571	3,260	0.00256	9.87	52.1	170	3.31	133	3.29	98.6	3.18
16	0.0508	2,580	0.00203	7.81	41.2	135	4.18	106	4.16	78.0	4.02
17	0.0453	2,050	0.00161	6.21	32.8	108	5.26	84.9	5.23	62.1	5.05
18	0.0403	1,620	0.00128	4.92	26.0	85.5	6.64	67.6	6.61	49.1	6.39
19	0.0359	1,290	0.00101	3.90	20.6	68.0	8.37	54.0	8.33	39.0	8.05
20	0.0320	1,020	0.000804	3.10	16.4	54.2	10.5	43.2	10.5	31.0	10.1

AWG							B1-56		B2-52		B3-63
21	0.0285	812	0.000638	2.46	13.0	43.2	13.3	34.4	13.2	24.6	12.8
22	0.0253	640	0.000503	1.94	10.2	34.1	16.9	27.3	16.8	19.4	16.2
23	0.0226	511	0.000401	1.55	8.16	27.3	21.1	21.9	21.0	15.4	20.3
24	0.0201	404	0.000317	1.22	6.46	21.7	26.7	17.5	26.6	12.7	25.7
25	0.0179	320	0.000252	0.970	5.12	17.3	33.7	13.9	33.5	10.1	32.4
26	0.0159	253	0.000199	0.765	4.04	13.7	42.7	11.1	42.4	7.94	41.0
27	0.0142	202	0.000158	0.610	3.22	10.9	53.5	8.87	53.2	6.33	51.4
28	0.0126	159	0.000125	0.481	2.54	8.64	67.9	7.02	67.6	4.99	65.3
29	0.0113	128	0.000100	0.387	2.04	6.97	84.5	5.68	84.0	4.01	81.2
30	0.0100	100	0.0000785	0.303	1.60	5.47	108	4.48	107	3.14	104
31	0.0089	79.2	0.0000622	0.240	1.27	4.35	136	3.6	135	2.49	131
32	0.0080	64.0	0.0000503	0.194	1.02	3.53	169	2.90	168	2.01	162
33	0.0071	50.4	0.0000396	0.153	0.806	2.79	214	2.30	213	1.58	206
34	0.0063	39.7	0.0000312	0.120	0.634	2.20	272	1.82	270	1.25	261
35	0.0056	31.4	0.0000246	0.0949	0.501	1.75	344	1.44	342	0.985	331
36	0.0050	25.0	0.0000196	0.0757	0.400	1.40	431	1.16	429	0.785	415
37	0.0045	20.2	0.0000159	0.0613	0.324	1.13	533	0.944	530	0.636	512
38	0.0040	16.0	0.0000126	0.0484	0.256	0.898	674	0.750	671	0.503	648
39	0.0035	12.2	0.00000962	0.0371	0.196	0.691	880	0.577	876	0.385	847
40	0.0031	9.61	0.00000755	0.0291	0.154	0.543	1120	0.455	1120	0.302	1080
41	0.0028	7.84	0.00000616	0.0237	0.125	...	1380	...	1370	0.246	1320
42	0.0025	6.25	0.00000491	0.0189	0.0999	...	1730	...	1720	0.196	1660
43	0.0022	4.84	0.00000380	0.0147	0.0774	...	2230	...	2220	0.152	2140
44	0.0020	4.00	0.00000314	0.0121	0.0639	...	2700	...	2680	0.126	2590
ASTM Specification Designation..............							B1-56		B2-52		B3-63

* No. 19 AWG and smaller, based on Anaconda data.
† Based on nominal diameter and ASTM resistivities.
‡ No requirements for tensile strength are specified in ASTM B3-63. Values given here based on Anaconda data.

Table 6-11. Properties of Stranded Conductors[119]

Size desig- nation, AWG	Nominal conductor area, c mils	No. of strands	Allow- able no. of missing strands	Nominal diam. of individual strands, in.	Max. diam. of stranded conductor, in.	Max. resistance of finished wire at 20°C, Ω/1,000 ft			
						Tin- coated copper	Silver- plated copper	Nickel- plated copper	Silver- plated high- strength copper alloy
30	112	7	0	0.0040	0.013	107.0	101.0	109.0	116.0
28	175	7	0	0.0050	0.016	67.6	62.9	68.3	72.2
26	304	19	0	0.0040	0.021	39.3	36.2	40.1	41.5
24	475	19	0	0.0050	0.026	24.9	23.2	25.1	26.6
22	754	19	0	0.0063	0.033	15.5	14.6	15.5	16.8
20	1,216	19	0	0.0080	0.041	9.70	9.05	9.79	10.4
18	1,900	19	0	0.0100	0.052	6.08	5.80	6.08	6.65
16	2,426	19	0	0.0113	0.060	4.76	4.54	4.76	5.23
14	3,831	19	0	0.0142	0.074	2.99	2.87	3.00	3.30
12	7,474	37	0	0.0142	0.102	1.58	1.48	1.59	1.70
10	9,361	37	0	0.0159	0.118	1.27	1.20	1.27	1.38
8	16,983	133	0	0.0113	0.176	0.700	0.661	0.680	0.760
6	26,818	133	0	0.0142	0.218	0.436	0.419	0.428	0.483
4	42,615	133	0	0.0179	0.272	0.274	0.263	0.269	0.302
2	66,500	665	2	0.0100	0.345	0.179	0.169	0.174	0.194
0	104,500	1,045	3	0.0100	0.432	0.114	0.105	0.109	0.123

Annular conductors: Helically stranded wires over a central rope, copper helix or twisted I beam.

Expanded ACSR: Steel core covered with a filler material and helically laid hard-drawn aluminum wires.

Composite conductors: Made up from two different conductors to obtain desired ratios between mechanical and electrical properties.

47. Coaxial Cable. A wide variety of cables having a central conductor covered with a dielectric material, a metallic shield, and a jacket conductor are available. The shield is generally made of braided copper, metal tape, or solid shield. Properties of coaxial cables are given in Par. 9-3.

48. Bar. Bus bars of rectangular cross section are used in general for carrying high electric currents. Occasionally, tubular or angled bars are used. Tubular conductors are used for high-voltage applications. These conductors have a small skin-effect ratio, larger current-carrying capacity, and smaller corona losses.

49. Strip and Sheet. Conductors in strip (or foil) form can sometimes be found in interconnections of hybrid circuits. Flat flexible conductors are used for interconnections on printed-circuit boards. These wires are flat conductors laminated between layers of plastic insulation. The flexible wiring can be single- or double-layered. Shielded flat wiring has also been used.

Printed-circuit boards also contain flat conductors. The boards are produced by etching, electroplating, stamping, or molding the conductor (almost invariably copper) on an insulating rigid plastic board. Copper-cladding thicknesses are given in Table 6-12.

50. Thin-Film Conductors. Some of the interconnections in microelectronics are made of thin films. The films are generally vacuum-deposited and are about 1 μm thick.

In integrated circuits the most widely used conductor is aluminum. Other conductors such as Ti-Pt-Au or Ti-Pd-Au are also occasionally used.[15,16] In hybrid circuits, thicker films are used, up to 25 μm. These films are either vacuum-deposited, electroplated, or screen-printed and fired. Thin-film resistors, vacuum-deposited or directly diffused in the semiconductor, are also used.

51. Joining of Conductors. *Solders.* Solders are elements or alloys with low melting temperatures used for joining two or more conductors. Solders are available in various forms: wires, wires with a flux core, sheets (preforms), and balls.

Conductive Adhesives. Epoxies filled with conductive metals such as silver and gold are available for joining conductors. These glues are cured at relatively low temperatures.

Table 6-12. Copper-cladding Thickness Tolerances[71]

Nominal thickness		Nominal weight		Tolerance			Sheet resistivity $\times 10^3$ (based on $1.724 \times 10^{-8}/\Omega\cdot m$)
in.	mm	oz/ft²	g/mm²	By weight, %	By gauge		
					in.	mm	
0.0007	0.0178	1/2	1.5	± 10	± 0.0002	± 0.0051	0.9685
0.0014	0.0355	1	3.06	± 10	+ 0.0004	+ 0.0102	0.4843
					− 0.0002	− 0.0051	
0.0028	0.0715	2	6.12	± 10	+ 0.0007	+ 0.0178	0.2421
					− 0.0003	− 0.0076	
0.0042	0.1065	3	9.18	± 10	± 0.0006	± 0.0152	0.1614
0.0056	0.1432	4	12.24	± 10	± 0.0006	± 0.0152	0.1211
0.0070	0.1780	5	15.30	± 10	± 0.0007	± 0.0178	0.0968
0.0084	0.2130	6	18.36	± 10	± 0.0008	± 0.0204	0.0807
0.0098	0.2460	7	21.42	± 10	± 0.001	± 0.0254	0.0692
0.014	0.3530	10	30.6	± 10	± 0.0014	± 0.0355	0.0484
0.0196	0.4920	14	43.2	± 10	± 0.002	± 0.0508	0.0346

SPECIFIC CONDUCTOR MATERIALS

52. Copper. Copper is used very extensively for electrical conductors. It has a very high electrical and thermal conductivity and can easily be formed into wires, tubes, and sheet. Copper has a high resistance to corrosion, forming an oxide on its surface.

Copper is normally supplied as electrolytic tough pitch (ETP), containing 99.95% Cu, with about 0.04% oxygen and approximately 0.01% other impurities. This material is used for electrical wires, bus bars, switches, and terminals.

Other forms of commercially pure copper are deoxidized low- or high-residual phosphorus (DLP and DHP, respectively). The OFHC, oxygen-free high-conductivity copper (99.95% Cu) is an electrolytic copper free from cuprous oxide. This material is used for electrical conductors of various forms, bus bars, and waveguides.

For electrical connector and switch components, the free-machining copper is used. Both tellurium and lead additions are found. Additions of silver (zinc or chromium) to copper increases the resistance to creep and also raises the softening temperature after cold working.

When copper is used for contacts, the oxidation of copper is occasionally not desirable. The copper can be coated with a precious metal such as silver to prevent the oxidation. Nickel-plated copper is also used.

When high resistance to sticking, arcing, or welding is desired, copper-tungsten or copper-graphite mixture is used.

Copper alloys are also commonly used for plugs, connectors, and other contacts.

Yellow brass (65% Cu-35% Zn) has improved mechanical properties but reduced corrosion resistance and electrical conductivity (28% IACS).

Phosphor bronze is used where good resistance to wear is desired in electrical contacts (98.75% Cu-1.25% Sn). Bronzes with low additions of tin are desired since they have only a small loss of electrical conductivity.

Nickel silver (55% Cu-27% Zn-18% Ni) is used for telephone equipment, resistance wire, and contacts. It has 5.5% (IACS) conductivity.

Beryllium copper (97.9% Cu-1.9% Be-0.2% Ni or Co) alloy is used where a relatively high electrical conductivity (18% IACS cold-worked, 30% IACS annealed) and high mechanical strength are desired, as in spring contacts. This alloy also has good corrosion resistance.

Tin bronze (88% Cu-8% Sn-4% Zn) is used as collectors for electrical generators.

When copper, which contains cuprous oxides, is annealed in a hydrogen-bearing atmosphere (above about 500°C), the hydrogen diffused through the metal reduces the oxide and forms steam, which produces cracks in the metal. This cracking is termed *hydrogen embrittlement*. The embrittlement does not occur in OFHC copper due to the absence of oxides in this material.

Mechanical and electrical properties of copper wires and application of cables are given in Tables 6-11 and 6-13. Electrical properties of copper alloys are given in Table 6-14.

Table 6-13. Copper Cable—Stranding Classes and Applications (ASTM Specifications)

ASTM Designation	Construction	Class	Application		
B8-64	Concentric lay	AA	For bare conductors generally used in overhead lines		
		A	For weather-resistant (weatherproof), slow-burning conductors For bare conductors where greater flexibility than is afforded by Class AA is required		
		B	For conductors insulated with various materials such as rubber, paper, varnished cambric, etc. For the conductors indicated under Class A where greater flexibility is required		
		C D	For conductors where greater flexibility is required than is provided by Class B		
B173-64	Rope lay with concentric-stranded members	G	Conductor constructions having a range of areas from 5,000,000 c mils and employing 61 stranded members of 19 wires each down to No. 14 AWG containing 7 stranded members of 7 wires each (Typical uses are for rubber-sheathed conductors, apparatus conductors, portable conductors, and similar applications.)		
		H	Conductor constructions having a range of areas from 5,000,000 c mils and employing 91 stranded members of 19 wires each down to No. 9 AWG containing 19 stranded members of 7 wires each (Typical uses are for rubber-sheathed cords and conductors where greater flexibility is required, such as for use on takeup reels over sheaves and extra-flexible apparatus conductors.)		
B226-64	Annular-stranded		For bare conductors or covered with weather-resistant (weatherproof) materials or insulated with rubber, varnished cambric, or solid-type impregnated paper		

ASTM Designation	Construction	Class	Conductor size, AWG	Individual wire size		Application
				In.	AWG	
B174-64	Bunch-stranded	I	7, 8, 9, 10	0.0201	24	Rubber-covered, varnished-cambric, and paper-insulated conductors
		J	10, 12, 14, 16, 18, 20	0.0126	28	Fixture wire
		K	10, 12, 14, 16, 18, 20	0.0100	30	Fixture wire, flexible cord, and portable cord
		L	10, 12, 14, 16, 18, 20	0.0080	32	Fixture wire and portable cord with greater flexibility than Class K
		M	14, 16, 18, 20	0.0063	34	Heater cord and light portable cord
		O	16, 18, 20	0.0050	36	Heater cord with greater flexibility than Class M
		P	16, 18, 20	0.0040	38	More flexible conductors than provided in preceding classes
		Q	18, 20	0.0031	40	Oscillating fan cord; very great flexibility

Table 6-13. *(Continued)*

		Conductor size, c mils	Individual wire size		Application
			In.	AWG	
B172-64	Rope lay with bunched-stranded members	I	Up to 2,000,000	0.0201	24	Typical use, for special apparatus cable
		K	Up to 1,000,000	0.0100	30	Typical use, special portable cord and conductors
		M	Up to 1,000,000	0.0063	34	Typical use, for welding conductor

Table 6-14.   Electrical Properties of Copper Alloys[116]

Alloy	Electrical conductivity, % IACS	Electrical resistivity $\times 10^8$, $\Omega \cdot m$	Temperature coefficient of electrical resistivity, $1/°C$
Pure copper................	103.06	1.67	0.00404
Electrolytic copper (ETP)	101	1.71	0.00397
Oxygen-free copper (OF)	101	1.71	
Free-machining copper (1.0% Pb)........................	98	1.76	
Gilding metal, 5% Zn	56	3.10	0.00231
Red brass, 15% Zn	37	4.70	0.0016
Cartridge brass, 30% Zn	28	6.20	0.00148
Yellow brass, 35% Zn	27	6.40	
Phosphor bronze, Grade A, 5% Sn	15	11	
Cupro Nickel, 30% Ni........	4.6	37.00	0.00048
Beryllium copper	15-18	9.6-11.5 cold-worked 5.7-7.8 precipitation-hardened	

53. Aluminum.   Aluminum has a high electrical and heat conductivity.   It is malleable, very strong compared with its weight, and has good reflectivity and good corrosion resistance.

Aluminum is used for electrical conductors.   EC alloy (99.45% Al) is used for bus bars, wires, and stranded conductors.   For cables it is used with a central steel-core reinforcement. Aluminum-alloy conductors are also used for cables.   Aluminum-alloy bus bars (6061: 1% Ag, 0.6% Si, 0.25% Cu, 0.25% Cr) are also in use.   Alloying increases the mechanical strength and normally reduces electrical conductivity.   Some alloys can be age-hardened to obtain optimum mechanical properties.

In thin-film or fine-wire form it is used in the microelectronic industry.   In thin-film form it is prone to electrical failures due to electromigration (see Par. **6-42**).

Aluminum is used for heat sinks, radiators, and for reflective coatings.   Aluminum can be anodized to give it very good corrosion protection.

Properties of aluminum wires are listed in Table 6-15.   Resistivities of aluminum alloys are given in Table 6-16.

54. Silver.   Silver has the highest electrical conductivity at room temperature.   Silver also has excellent heat conduction.   Mechanically, it is malleable.   Silver has a good corrosion resistance but poor resistance to tarnishing.

Silver is used for electrical contacts, usually with lower current (up to 20 A) and voltages. Silver is used mainly in electrodeposited form for plugs, sockets, rotary switches, and occasionally slip rings.   For many applications silver alloys are used.   The alloys are harder and less prone to wear.   Silver-copper alloys from sterling silver (92.5% Ag-7.5% Cu) to the eutectic alloy (72% Ag-28% Cu) are used.   Ag-Cd are also used.

Silver can be used in screen conductors for hybrid circuits and as fillers in conducting low-

Table 6-15. Aluminum Wire—Dimensions, Weight, DC Resistance
(Based on ASTM Specifications B230-60, B262-61, and B323-61)

Conductor size, AWG	Diam. at 20°C (68°F), mils	Area at 20°C (68°F)		Dc resistance at 20°C (68°F),* Ω/1,000 ft	Weight at 20°C (68°F), † lb.		Length at 20°C (68°F), ft/Ω
		c mils	in.²		Per 1,000 ft	Per Ω	
2	257.6	66,360	0.05212	0.2562	61.07	238.4	3903
3	229.4	52,620	0.04133	0.3231	48.43	149.9	3095
4	204.3	41,740	0.03278	0.4074	38.41	94.30	2455
5	181.9	33,090	0.02599	0.5139	30.45	59.26	1946
6	162.0	26,240	0.02061	0.6479	24.15	37.28	1544
7	144.3	20,820	0.01635	0.8165	19.16	23.47	1225
8	128.5	16,510	0.01297	1.030	15.20	14.76	971.2
9	114.4	13,090	0.01028	1.299	12.04	9.272	769.7
10	101.9	10,380	0.008155	1.637	9.556	5.836	610.7
11	90.7	8,230	0.00646	2.07	7.57	3.66	484
12	80.8	6,530	0.00513	2.60	6.01	2.31	384
13	72.0	5,180	0.00407	3.28	4.77	1.45	305
14	64.1	4,110	0.00323	4.14	3.78	0.914	242
15	57.1	3,260	0.00256	5.21	3.00	0.575	192
16	50.8	2,580	0.00203	6.59	2.38	0.361	152
17	45.3	2,050	0.00161	8.29	1.89	0.228	121
18	40.3	1,620	0.00128	10.5	1.49	0.143	95.5
19	35.9	1,290	0.00101	13.2	1.19	0.0899	75.8
20	32.0	1,020	0.000804	16.6	0.942	0.0568	60.2
21	28.5	812	0.000638	20.9	0.748	0.0357	47.8
22	25.3	640	0.000503	26.6	0.589	0.0222	37.6
23	22.6	511	0.000401	33.3	0.470	0.0141	30.0
24	20.1	404	0.000317	42.1	0.372	0.00884	23.8
25	17.9	320	0.000252	53.1	0.295	0.00556	18.8
26	15.9	253	0.000199	67.3	0.233	0.00346	14.9
27	14.2	202	0.000158	84.3	0.186	0.00220	11.9
28	12.6	159	0.000125	107	0.146	0.00136	9.34
29	11.3	128	0.000100	133	0.118	0.000883	7.51
30	10.0	100	0.0000785	170	0.0920	0.000541	5.88

* Conductivity = 61.0% IACS.
† Density = 2.703 g/cm³ (0.09765 lb/in.³).

Table 6-16. Electrical Properties of Aluminum Alloys[115]

Alloy and heat-treatment*	Composition	Electrical conductivity, % IACS	Electrical resistivity × 10⁸, Ω·m
Aluminum	99.996% Al	64.94	2.655
EC (O and H 19)	99.45% Al	62	2.8
1100 (O)	1% Si,0.2% Cu,0.05% Mn,0.1% Zn	59	2.9
2011 (T3)	5-6% Cu,0.4% Si,0.7% Fe,0.3% Zn	36	4.8
3003 (O)	1.0-1.5% Mn,0.6% Si,0.7 Fe,0.2% Cu,0.1% Zn	50	3.4
5056 (H38)	4.5-5.6% Mg,0.3% Si,0.4% Fe,0.1% Zn,0.05-0.2% Mn, 0.05-0.2% Cr	27	6.4
6061 (T4 and T6)	0.8-1.2% Mg,0.4-0.8% Si,0.7% Fe,0.15% Mn,0.15-0.4% Cu, 0.15-0.35% Cr,0.25% Zn,0.15% Ti	40	4.31
7075 T6	5.1-6.1% Zn,2.1-2.9% Mg,1.2-2% Cu,0.5% Si,0.7% Fe, 0.3 Mn,0.18-0.4% Cr,0.2% Ti	30	5.74

* For heat-treatment designations see, for example, K. R. Van Horn (ed.), "Aluminum," Vol. 1, ASM The American Society for Metals, Metals Park, Ohio, 1967, p. 112.

curing-temperature adhesives. Silver is also used as a component in brazes: silver-copper or silver-copper-zinc alloys are extensively used as high-temperature brazes.

In the presence of humidity and electric fields, silver will migrate in the form of fine threads between silver conductors.[17] This ionic migration occurs between conductors on organic or ceramic surfaces and eventually causes electrical shorts.

55. Gold. Gold has excellent conductivity, similar to that of aluminum, as well as excellent corrosion and oxidation resistance. Gold is a very soft material: it can easily be fabricated to very small dimensions by cold working.

Gold is used in high-frequency conductors employed in corrosive environments and as a plated layer over plugs. Gold is used for fine-wire interconnections and integrated and hybrid circuits. Gold plating is extensively used on semiconductor packages, leads, and on circuit boards. Gold and glass frit are used as conductors on thick-film hybrid circuitry. Gold in thin-film form is used on thin-film hybrids and on beam-lead integrated circuits as the conductor material. Conductive adhesives occasionally contain gold.

Gold alloys are used for rotary switches and telephone relays. In particular, 60% Au-25% Ag-6% Pt is a useful alloy. Au-Ag alloys are also useful for low-current electrical contacts.

Au-Sn, Au-Si, and Au-Ge are some of the gold-base solders used whenever an intermediate melting temperature is desired. Au-Si is frequently used to solder integrated circuits to the gold-plated packages.

56. Platinum. Platinum is a precious metal with very good corrosion resistance and a high melting point: it is ductile and can be easily formed. Platinum is used for resistance thermometers and thermocouples.

Platinum is also used for electrical contacts, brushes, and precision potentiometer wires. Another use is in contacts to silicon and as part of the thin-film conductors on beam-lead integrated circuits.[15]

Platinum-palladium, rhodium, or ruthenium alloys are used as electrical contacts. Platinum and platinum-10% rhodium wires are frequently used for high-temperature thermocouples.

57. Palladium. Palladium is a precious metal with very good corrosion resistance and high melting point. It can be formed to various shapes and can be electroplated. It resembles platinum in appearance.

Palladium has wide use as telephone relay contacts because it is less prone to erosion, is relatively noise-free when used as contacts and is more economical than platinum.

A number of palladium alloys are used for relay applications: palladium-ruthenium, palladium-silver-nickel, palladium-copper. For sliding contacts, palladium-silver-gold-platinum and palladium-silver-copper-platinum-gold-zinc alloys are sometimes used.

Palladium has been suggested as a replacement for platinum in beam-lead circuits.

58. Nickel. Nickel-coated copper conductors can be used at temperatures up to 300°C. There is no enhanced corrosion at defective areas of the coating. Nickel can be electroplated or cladded. Nickel plating is frequently used as a coating under the final gold plating on conductors. Nickel is also an important constituent in resistance alloys.

59. Tungsten. Tungsten has a very high melting point and a relatively high electrical conductivity. Tungsten is a hard and brittle material. It tends to develop oxide films in air.

Since it has good resistance to erosion, arcing, and welding, it is used for vibrators, voltage regulators, and other low-current repetitive contacts. Tungsten has been suggested as a possible contact metal on integrated circuits. Tungsten is also used as heating wires and filaments for incandescent lamps.

60. Molybdenum. Molybdenum has a lower melting point than tungsten and a higher tendency for oxide formation. The electrical conductivity of molybdenum is similar to that of tungsten. It is not attacked but is wetted by mercury, and it is therefore used extensively in mercury switches.

Molybdenum has been used occasionally in combination with gold as the conductive film on integrated circuits.[18]

Molybdenum is also used as heating wires for high-temperature furnaces.

61. Sintered Materials. In electrical contacts sintered powders are sometimes used. The advantage of a composite powder is the added control over the physical properties of the conductors. Among the more common sintered powders are Ag-CdO; Ag-graphite; Ag-W; Ag-Mo; Ag-WC; Cu-W.

62. Rhodium. Rhodium is a hard, high-melting metal with relatively low electrical resistance and good corrosion resistance. Rhodium is used for sliding contacts and can be electroplated on silver or other base material. Rhodium is also used in thermocouple wires as an addition to platinum.

63. Contact Materials. See Ref. 81, Vol. 1, pp. 801-816, Par. **6-257.**

64. Materials for Wires and Cables. Aside from pure copper and aluminum, other wire and cable construction are available, as follows.

Copper-clad Steel. Steel wires covered with copper provide high strength with some conductivity loss (30-40% IACS). At high frequencies these wires have the same conductivity as solid-copper wires since most of the current is conducted near the surface.

Aluminum-clad Steel. This is used for communication or signal wires and cables. Again, the steel imparts the strength while the aluminum serves as the conductor.

Galvanized Steel. Galvanized-steel conductors are coated-steel wires with relatively low electrical conductivity and high mechanical strength. Galvanized-steel wires are not commonly used.

Copper Alloys. For high-strength wires, cadmium-copper, zirconium-copper, chromium-copper, and cadmium-chromium-copper alloys have been used.

Metal-coated Copper. Tin is used as a protective coating. The thickness of the coat can vary from 0.5 to 5 μm. In stranded cables the tin is prefused, overcoated, or top-coated on bare strands. Silver coating is used for high-temperature (up to 200°C) application. Silver-coated-copper wire is susceptible to enhanced corrosion at discontinuities, cracks, and pinholes in the coating. Nickel-coated copper can be used up to 300°C. Nickel has poor solderability.

65. Fusible Alloys. Alloys of low-melting-temperature materials such as lead, cadmium, bismuth, and tin have even lower melting temperatures than their pure components. Examples of low-melting-temperature alloys are given in Table 6-17.

66. Resistance Metals. Resistor materials have several requirements. The resistivity should be 50 to 150 $\times$ 10^{-8} $\Omega \cdot$ m. The thermoelectric potential against copper should be small. The wires should be solderable, and the temperature coefficient of electric resistance has to be very small. The wires have to be stable metallurgically and chemically. Stress relieving helps in achieving stability.

Most precision resistors are wire-wound nickel alloys: manganin or constantan. If high frequency and high stability are required, film-type resistors are used. These resistors are

Table 6-17. Compositions and Melting Temperatures of Eutectic Fusible Alloys[116]

Melting temp.		Composition				
°F	°C	Bi	Pb	Sn	Cd	Other
117	46.8	44.70	22.60	8.30	5.30	19.10 In
136	58	49.00	18.00	12.00	. . .	21.00 In
158	70	50.00	26.70	13.30	10.00	
197	91.5	51.60	40.20	. . .	8.20	
203	95	52.50	32.00	15.50		
217	102.5	54.00	. . .	26.00	20.00	
255	124	55.50	44.50			
281	138.5	58.00	. . .	42.00		
288	142	. . .	30.60	51.20	18.20	
291	144	60.00	. . .	. . .	40.00	
351	177	. . .	. . .	67.75	32.25	
362	183	. . .	38.14	61.86		
390	199	. . .	. . .	91.00	. . .	9.00 Zn
430	221.3	. . .	. . .	96.50	. . .	3.50 Ag
457	236	. . .	79.7	. . .	17.7	2.60 Sb
477	247	. . .	87.0	. . .	. . .	13.00 Sb

composed of a resistive film on a substrate. The least accurate resistors are the composition type, such as carbon resistors.

Resistor inks are used on thick-film hybrid circuits. These inks or pastes are screened on the surface of a ceramic substrate and subsequently fired. Thin-film hybrids or integrated circuits use different materials as resistors. Table 6-18 shows the composition and properties of thick- and thin-film resistors.

67. Carbon. Carbon occurs in two crystalline forms, graphite and diamond. Graphite is the stable form at room temperature. Carbon is also found in an amorphous (noncrystalline) form. Carbon or graphite is used with binders to form parts for electrical sliding contacts such as motor brushes. Graphite has a low shearing stress parallel to its basal plane which makes it a solid lubricant. Diamond is sometimes used as a substrate material for some unique applications where extremely good heat conduction is required.

Table 6-18. Thick- and Thin-Film Resistors

A. Thick-Film Resistors[*][123]

Trade name, ESL No.	Nominal sheet resistivity, Ω/square	Typical average TCR, ppm/°C (ref. to 25°C), −55 to 125°C	Trade name, ESL No.	Nominal sheet resistivity, Ω/square	Typical TCR, ppm/°C, when fired as recommended, +25 to 125°C
3810	1	+200 ± 100	7010	1	+300 to +400
3811	10	+50 ± 100	7011	10	+ 50 to +150
3812	100	0 ± 100	7012	100	+ 25 to +100
3813	1,000	0 ± 100	7013	1,000	+ 25 to +100
3814	10,000	0 ± 50	7014	10,000	+ 25 to +100
3815	100,000	0 ± 50	7024	20,000	+ 25 to +100
3816	1,000,000	−50 ± 150	7015	100,000	− 50 to +100
3817	10,000,000	−100 ± 150	7016 extended	1,000,000	−400 to −800
			7056 ranges	5,000,000	−500 to −1200
2831	30	150–250	7014B	10,000	+25 to +100
2812	100	150–250	7024B	20,000	+25 to +100
2813	1,000	150–250	7015B	100,000	− 50 to +100
2814	10,000	150–250	7016B	1,000,000	−400 to −800
2815	100,000	150–350	7056B	5,000,000	−500 to −1200

B. Thin-Film Resistors

Material	Sheet resistivity, Ω/square	Tolerance
Diffused silicon	50–250	± 10%
NiCr........................	40–400	± 10%
Ta, Ta-N	200	
Cr-SiO	100–10,000	± 20%
Si-Cr.......................	2,000–20,000	± 15 to ±30%
MoSi$_2$	200	± 10%

* Electro-Science Laboratories, Inc., Philadelphia, Pa.

68. Soldering Alloys. Soldering alloys are low-melting-temperature conducting alloys which join two conductors by wetting their surfaces and then solidifying to a mechanically strong solid.

Solders lose their mechanical strength above solidus temperature and become sufficiently fluid at about 50°C above the liquidus (see Fig. 6-11). The solder should be strong and creep- and corrosion-resistant and should be a reasonable electrical conductor.

Lead-tin solders are most commonly used for electronic applications. The eutectic composition (63% Sn–37% Pb) has good mechanical properties combined with high wettability. Electrical conductivity decreases from 13% IACS for tin to 7% IACS for lead. Antimony additions increase the strength of lead-tin alloys. To minimize scavenging of silver (when soldering onto thin silver films), silver is added to saturate the solders. Gold is unintentionally dissolved into the solder when soldering gold surfaces. Gold-tin intermetallics seem to impart brittleness to the joints. Aluminum and cadmium and zinc are detrimental to the solder. *Tin* melts at 232°C and has good wetting. Tin transforms into a gray brittle allotropic form at 13°C, although alloying additions will retard this occurrence. *Gold-silicon* or *Gold-germanium* solders are used for soldering semiconductor dice to gold-plated surfaces. *Tin-antimony* is an intermediate-temperature solder which is not, however, suitable for brass. *Lead-silver-(tin)* is used for higher temperatures, but has poor corrosion resistance. *Lead-5% indium* has good wetting at high temperatures. *Aluminum-zinc* is used for aluminum soldering.

69. Conductive Coatings and Films. Conductive films on printed-circuit boards are primarily of copper. The copper may be gold-plated or tinned.

On hybrid circuits conductors are formed by screen printing or electroplating (thick films) or by vapor deposition (thin films). A variety of conductive pastes are available (see Table 6-19). The conductive pastes containing the conductor and a binder are screen-printed on dielectric substrates (commonly alumina) and subsequently fired to remove the binder. The resulting conductor thickness is about 10 to 25 μm. Screened conductors are normally wider than 100 μm.

Another way of achieving a thick conductor is by electroplating. According to this method a thick conducting layer is electroformed through a photoresist pattern (pattern plating) on a previously deposited thin film. An alternative method is provided by electroplating on a previously deposited metallic film without use of a pattern, and subsequently etching the thick conductors. Gold is most commonly used as the conductor layer when electroforming is performed. Copper is occasionally used.

Thin conducting films are used on semiconductor devices. Al is the most commonly used film. The film thickness is about 1 to 2 μm (sheet resistivity of 0.015 to 0.03 Ω/square). Al thin films are susceptible to electromigration and corrosion. The reliability of thin films has been the subject of a number of studies.[16,19]

Alternative conductors are Mo-Au films. Mo serves as an adherence layer between the gold and the silicon or silicon oxide, and the gold is the main conductor. The Mo-Au films are not commonly used. Still another combination of films is Ti-Pt-Au. This system, developed at Bell Laboratories, consists of a titanium adherence layer, followed by a platinum diffusion barrier (to prevent gold diffusion into silicon), topped by the final gold conducting layer.[15] This system is relatively corrosion-free (electromigration in these films has not yet been extensively studied).

Recently, other conductive films have been reported as suitable for semiconductor devices: tungsten, tungsten-gold, tungsten-titanium-gold.[19]

70. Shielding. It is often desirable to shield part of the circuit from electromagnetic fields. The shields can *absorb, reflect,* or *degrade* (by multiple internal reflections) the electromagnetic energy. The most commonly used shields are braided copper. Tapes of copper or aluminum and copper or aluminum-coated *lossy* foils are also used. Solid shields can be found where complete shielding is necessary. Conductive plastics, yarns, and metal sprays also serve as shielding materials.

Magnetic shielding is obtained by high-permeability materials such as Mumetal.

71. Radar Absorbing Materials. Frequently, as in radar or antenna applications, it is necessary to absorb rf energy without reflection. This can be performed by *resonant* or *broadband* absorbing materials.[20] Examples of resonant absorbers are seen in Fig. 6-14a and b. Resonant absorbers have a certain effective bandwidth in which the reflections are strongly attenuated by interference, similar to light absorption in antireflection optical coatings.

Broadband absorbing materials are divided into several kinds. In $\mu = \varepsilon$ *absorbers,* the normal incidence reflection coefficient is zero when the relative magnetic permeability μ equals the relative electrical permittivity ε_r. Ferrites with $\mu_r \approx \varepsilon_r$ are used for this purpose (Fig. 6-14c). *Inhomogeneous layers* (Fig. 6-14d) are materials with an ε varying from a low value at the surface ($\varepsilon \approx 1$) to a high value of ε at the back of the layer. Geometric transition absorbers are pyramids or wedges of synthetic sponge rubber or plastic foam loaded with a material such as carbon (Fig. 6-14e). Sometimes these absorbers are constructed so that the absorption is increased toward the back of the layer. *Low-density absorbers* with $\varepsilon \approx \varepsilon_0$ have

Table 6-19. Thick-Film-Conductor Compositions[124]

Trade name	Material	Line resolution, × 1,000 in.	Thickness, μm	Sheet resistivity, Ω/square	Solderability 60 Sn/40/Pb, 215°C Initial	Solderability After refiring at 500°C	Adhesion lb. As fired	Adhesion lb. Aged 125°, 100/h	Die	Bondability Wire Au	Bondability Wire Al
8115	Au	10–15	15–17	.005–.010	U	U	…	…	Yes	Yes	Yes
DP-8237	Au	5–8	15–17	.005–.010	U	U	…	…	Yes	Yes	Yes
DP-8380	Au	2–5	25	.003	U	U	…	…	Yes	Yes	Yes
7553	Pt-Au	10–15	18	.080	E	E	2.4	1.0	*	Yes	†
DP-8351	Pt-Au	5–8	16	.100	E	E	3.5	3.0	*	Yes	†
8451	Pd-Au	5–8	7	.080	E	E	4.3	4.1	*	Yes	Yes
8227	Pd-Au	10–15	13	.100	E	E	3.7	3.0	*	Yes	Yes
DP-8267	Pd-Au	5–8	15	.050	E	E	3.7	3.0	*	Yes	Yes
8151	Pd-Ag	10–15	15	.050	E	V	4	1.5	*	Yes	Yes
8228	Pd-Ag	10–15	15	.100	E	V	4	1.5	*	Yes	Yes
DP-8263	Pd-Ag	5–8	15	.050	E	V	4	1.5	*	Yes	†
DP-8420	Pd-Ag	2–5	15	.060	E	E	6	2.5	*	Yes	Yes
DP-8363	Pd-Ag	5–8	15	.025	E	E	4	1.5	*	Yes	Yes
DP-8430	Pd-Ag	2–5	15	.030	E	E	6	2.5	*	Yes	Yes
DP-8463	Pd-Ag	5–8	15	.020	E	E	4	1.5	*	Yes	Yes
DP-8440	Pd-Ag	2–5	15	.025	E	E	6	2.5	*	Yes	†

*Nonbondable. †Under investigation.
E = excellent; V = variable; U = unsolderable.

6-33

a small reflection (Fig. 6-14*f*). Such materials are low-density plastic foams with some carbon particles as absorbers or lossy fibers.

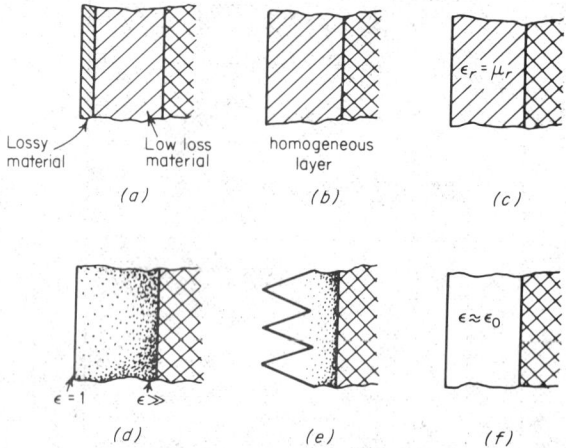

Fig. 6-14. Various configurations of radar absorbing materials. (*a*) Resonant absorber, Salisbury screen; (*b*) resonant absorber, homogeneous layer; (*c*) $\mu = \varepsilon$ absorber; (*d*) inhomogeneous layer; (*e*) geometric transition absorber; (*f*) low-density absorber.

72. Bibliography on Conductive and Resistive Materials

1. HARPER, C. A. (ed.) "Handbook of Materials and Processes for Electronics," New York, McGraw-Hill Book Company, 1970.

2. GERRITSEN, A. N. Metallic Conductivity, Experimental Part, in S. Flügge (ed.), "Handbuch der Physik," Vol. 19; Berlin, Springer Verlag OHG, pp. 137–226.

3. MEADEA, G. T. "Electrical Resistance of Metals," New York, Plenum Press, 1965.

4. PARKER, E. S. "Materials Data Book," New York, McGraw-Hill Book Company, 1967.

5. LYMAN, T. (ed.) "Metals Handbook," Cleveland, Ohio, The American Society for Metals, 1961.

DIELECTRIC AND INSULATING MATERIALS

Polarization of Dielectrics

73. Permittivity of Empty Space. The permittivity of empty space is defined through the force equation between two charges in empty space.

$$F = \frac{Q_1 Q_2}{4\pi\varepsilon_0 r^2} \tag{6-34}$$

where Q_1, Q_2 = charges (C), r = distance between the charges (m), F = force (N), ε_0 = permittivity of empty space, and $\varepsilon_0 = 8.85 \times 10^{-12}$ (F/m).

74. Permittivity. The force [Eq. (6-34)] is modified if the charges are separated by a dielectric medium.

$$F = \frac{Q_1 Q_2}{4\pi\varepsilon r^2} \tag{6-35}$$

where ε = permittivity of the dielectric (F/m).

75. Relative Permittivity (Dielectric Constant). The permittivity of a dielectric relative to that of empty space is called the relative permittivity,

$$\varepsilon_r = \frac{\varepsilon}{\varepsilon_0} \qquad (6\text{-}36)$$

where ε_r = relative permittivity (dielectric constant), ε = permittivity (F/m), and ε_0 = permittivity of empty space (F/m).

76. Electric Flux Density. The electric flux density in empty space is defined as

$$D_0 = \varepsilon_0 E \qquad (6\text{-}37)$$

where ε_0 = permittivity of empty space (F/m), E = electric field (V/m), and D_0 = electric flux density (C/m^2).

The electric flux density in a dielectric is

$$D = \varepsilon E \qquad (6\text{-}38)$$

where ε = permittivity of the dielectric.

77. Electric Polarization. The electric polarization P is the addition in electric flux density in a dielectric material to the density in free space.

$$P = D - \varepsilon_0 E \qquad (6\text{-}39)$$

where P = polarization (C/m^2), D = electric flux density (C/m^2), ε_0 = permittivity of free space (F/m), and E = electric field strength (V/m). The polarization is the total dipole moment induced in a unit volume of the dielectric.

78. Electric Susceptibility. The electric susceptibility x_e is the ratio between the polarization P and the electric flux density in empty space $\varepsilon_0 E$:

$$x_e = P/\varepsilon_0 E = \varepsilon/\varepsilon_0 - 1 \qquad (6\text{-}40)$$

79. Polarization Processes. The polarization within dielectric materials is determined by the displacement of charges. Four sources for polarization exist in the material: *electronic polarization,* due to displacement of electronic charges; *ionic polarization,* due to displacement of ions; *dipole polarization,* due to reorientation of permanent dipoles; and polarization by *space charges,* due to macroscopic displacement. Schematic diagrams of these four mechanisms are seen in Fig. 6-15.

Since the polarization process takes place in a finite time at any given temperature, it is expected that ε_r will be frequency- and temperature-dependent. A typical frequency dependence of the dielectric constant is seen in Fig. 6-16. The series of inflections in the curve occur at the relaxation times for the various polarization processes. For example, the dipoles of the material represented in Fig. 6-16 can follow audio frequencies but are incapable of following infrared frequencies. The dielectric constant at infrared frequencies is decreased by the dipole component and contains only the ionic and electronic components. Table 6-20 lists the dielectric constant at 1 MHz for several materials.

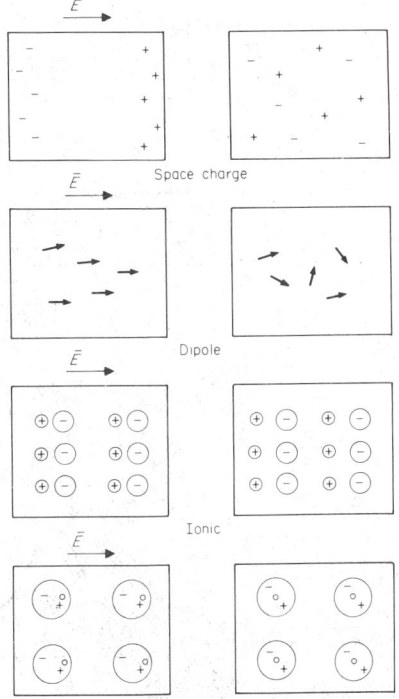

Fig. 6-15. Four types of polarization. On the right are schematic representations of the material without field; on the left, the material is under electric field.

Table 6-20. Dielectric Constant and Loss Factor for Several Dielectrics

Material	Dielectric constant at 10^6 Hz	Dissipation factor at 10^6 Hz
ABS resins	2.4 –3.2	0.005–0.016
Cellulose acetate	3.5 –7.5	0.01 –0.06
Fluorocarbons	2.1 –3.6	0.0003–0.0015
PTFE, FEP	2.0	0.0002–0.0003
Nylon 6 and nylon 10	3.5 –3.6	0.04
Polypropylene	2.20–2.28	0.0002–0.002
Polystyrene, model	2.45–4.0	0.0001–0.002
Silicones	3.4 –4.3	0.001–0.004
Polystyrene, foam	1.02–1.24	<0.0005
Hard rubber	2.95–4.80	0.007–0.028
Pyroceram 9606	5.58	0.0015
Cordierite	4.02–6.23	0.001–0.009
Forsterite	6.2 –6.5	0.0002–0.0004
Alumina	8.0–10.0	0.0001–0.0009
Beryllia		
Soda-lime glass	7.2	0.009
Borosilicate	4.1 –4.9	0.0006–0.005
Fused quartz	3.8	0.00001
Mica, muscovite	5.4 –8.7	0.0001–0.0004
Phlogopite	6.5	0.0001–0.0003
Glass-bonded mica	6.3 –9.3	0.0011–0.0025
Boron nitride	4.15	0.0002

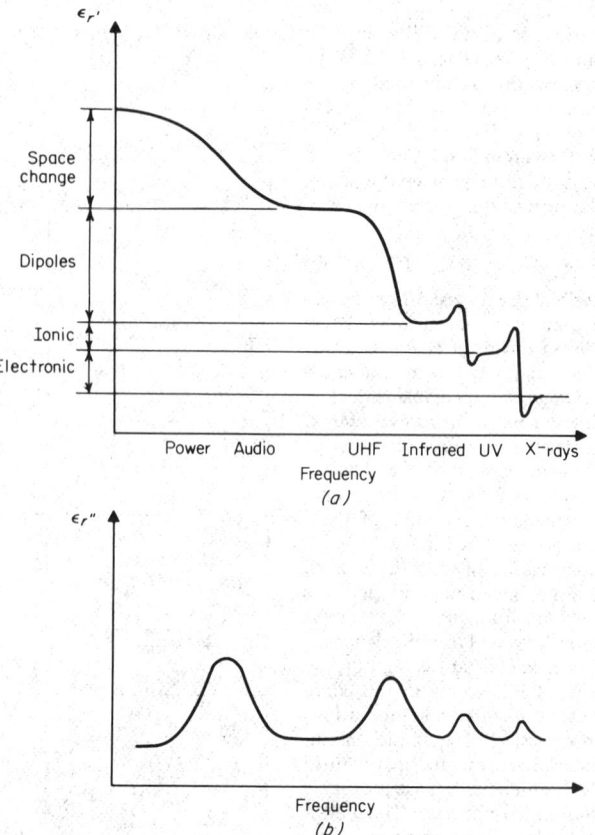

Fig. 6-16. Schematic diagram of the real (a) and imaginary part (b) of the dielectric constant as a function of frequency.[72]

80. Dielectric Constant.   The dielectric constant is generally described as a complex quantity,

$$\varepsilon_r = \varepsilon'_r + j\varepsilon''_r \qquad (6\text{-}41)$$

The imaginary part ε''_r is related to dielectric losses.

A typical frequency dependence of ε''_r is shown in Fig. 6-16, and a typical temperature dependence of ε''_r in Fig. 6-17.

In an ideal lossless dielectric, alternating current leads the voltage by 90°. In the presence of dielectric losses the current leads the voltage by 90° − δ, and the power losses are proportional to tan δ. Tan δ is called the *loss tangent,* or *dissipation factor.* The loss tangent is related to the dielectric constant

$$\tan \delta = \varepsilon''_r / \varepsilon'_r \qquad (6\text{-}42)$$

When, in addition to the polarization processes, Joule heating occurs by leakage currents, the total energy dissipated per unit volume is

$$W = \sigma E^2 + \varepsilon_0 \varepsilon_r \omega E^2 \tan \delta \qquad (6\text{-}43)$$

where W = losses per unit volume (W/m²), σ = conductivity of the dielectric (℧/m), ε_0 = permittivity of empty space (F/m), ε_r = dielectric constant, E = electric field (V/m), tan δ = loss tangent due to polarization processes, and ω = angular frequency (s⁻¹).

The Joule heating is often negligible, and the power losses are proportional to ε_r tan δ (often referred to as the *loss factor*). Table 6-20 shows the loss factor for various materials.

81. Electrical Conduction in Dielectrics.   The volume resistivity of dielectrics is defined as for conductors, i.e., as the electric field strength required to produce a unit current density. The resistivity is time-, temperature-, and field-dependent. The dc component of the

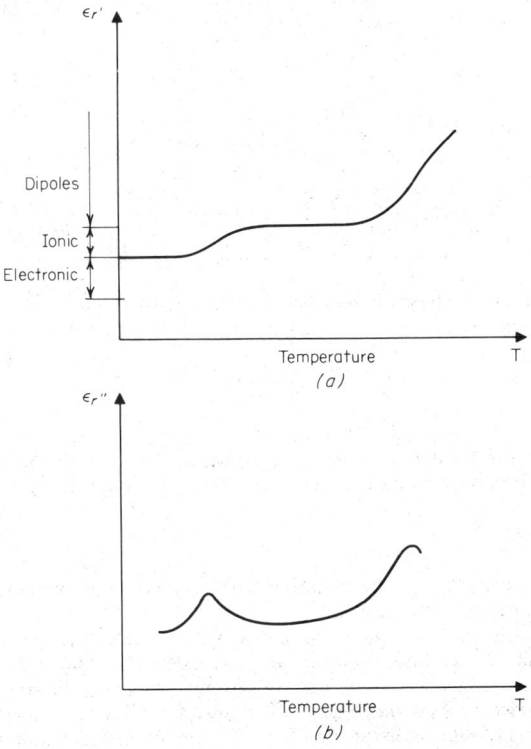

Fig. 6-17.   Schematic diagram of the real (*a*) and imaginary part (*b*) of the dielectric constant as a function of temperature.

conduction follows Ohm's law roughly up to fields of 10^7 V/m and then rises rapidly with increasing fields. The conduction in dielectrics is ascribed to electronic and ionic processes.

Electronic conduction can be electrode- or bulk-controlled. At high fields electrode-controlled processes (important in thin dielectric films) can be thermionic with Schottky lowering of the barrier, field emission, and temperature-aided field emission. The electronic conduction in the bulk is usually considered to proceed by hopping processes. The hopping can occur by tunneling between localized states or by thermal activation (Poole-Frenkel, or Poole, effect). As a consequence of such processes, the conductivity is found to increase as some exponential function of field and/or temperature. In addition, space charges produced by the carriers can determine the conductivity at higher current densities and increasing thicknesses. The ionic conductivity occurs by the motion of the charged ions in the electric field. The ionic conduction is a linear function of the field at low fields, but increases rapidly at high fields. The ionic conductivity increases exponentially with the inverse absolute temperature.

Low field resistivities of various dielectrics are given in Table 6-21. Tabulated values of resistivities should be treated with caution since the resistivity depends on fields, trapped space charges, defects, impurities, surface leakage currents, and time effects.

Table 6-21. Resistivities of Dielectrics

Material	Volume Resistivity, $\Omega \cdot$m
Cellulose acetate, molding	$10^8 - 10^{11}$
Epoxy, cast resin	$10^{14} - 10^{15}$
Methyl methacrylate, cast	$> 10^{13}$
Mica	$10^{12} - 10^{15}$
Nylon 6 and nylon 12	$10^{12} - 10^{13}$
Fluorocarbons	$10^{13} - > 2 \times 10^{16}$
Polyester, cast resin	10^{12}
Rubber, band	2×10^{13}
Silicones	$10^9 - 10^{11}$
Alumina	$10^9 - 10^{12}$
Beryllia	$> 10^{13}$
Forsterite	$> 10^{12}$
Cordierite	$> 10^{12}$
Boron nitride	10^{11}
Pyroceram 9606	2×10^{14}
Soda-lime glass	10^3
Borosilicate glass	$10^{14} - > 10^{15}$

82. Carrier Mobility. The mobility of charge carriers in an electric field is defined as their average drift velocity in a unit electric field,

$$\mu = \frac{V_{drift}}{E} \tag{6-44}$$

where μ = mobility (m²/V · s), V_{drift} = drift velocity (m/s), E = electric field strength (V/m).

If conduction is occurring by charges of one sign only, the conductivity σ is related to the mobility as

$$\sigma = \mu N q \tag{6-45}$$

where σ = conductivity ($\mho$/m), μ = mobility (m²/V · s), N = density of charge carriers (m^{-3}), and q = charge on a carrier (C).

83. Surface Resistivity. Surfaces of dielectrics can provide electrical conduction paths in parallel to the bulk. Such conduction can be pronounced when the dielectric is placed in a humid environment. The conduction can be electronic or ionic. The conductance of the surfaces is characterized by the *surface resistivity*. Figure 6-18 shows the surface resistivity of glass at various levels of relative humidity. Surface conduction plays an important role in semiconductor devices.[21] Charges drifting over dielectric surfaces have to be eliminated to prevent interference with device operation.

84. Breakdown Processes and Fields.[1] Electrical breakdown is a sequence of often rapid processes leading to a change from an insulating to a conducting state. Breakdown develops in a number of successive stages: *(a)* the initiating stage, increasing the conductivity, leading to *(b)* instability with current runaway; *(c)* voltage collapse with discharge of the electrostatic energy stored in the specimen through the breakdown channel; and *(d)* settling down to the low-voltage, high-current state.

Insulating media cannot sustain indefinite voltages, and breakdowns occur at electric fields varying from 10^5 to more than 10^9 V/m. It has been customary to assign to insulating media a dielectric strength which denotes the breakdown field. Dielectric strength is not a particularly useful concept. In most cases the breakdown field is found to be strongly dependent on such parameters as the time of voltage application to breakdown, thickness, temperature, electrode material, frequency, geometry, micro- and macroscopic defects, chemical changes, and the presence of interfaces.

Information on breakdown properties can be given in breakdown characteristics. For thin insulating films such characteristics show that the time to breakdown decreases as some exponential function of increasing field (Fig. 6-19) and that breakdowns occur over a wide range of fields. Increase in thickness generally decreases the breakdown field, while increase in temperature not only decreases, but also increases, the breakdown field. For further details see Refs. 22 to 29.

85. Breakdown in Gases. Breakdown in gases is produced by impact ionization of the molecules. Instability and current runaway arise due to the positive feedback of secondary ionization processes. Two kinds of conducting states after breakdown are observed: the glow discharge at voltages roughly above 70 V and arc discharge at much lower voltages. As sparking in air is possible above 330 V, electronic equipment can be easily exposed to the harmful effects of spark discharges. The breakdown field of gases can be enhanced by increase in pressure and by the use of gases such as sulfur hexafluoride or Freon. Breakdown in gases is observed from fields less than 10^5 to 10^8 V/m. High-intensity laser radiation can also cause gas breakdown.

86. Breakdown in Liquids. Several mechanisms were proposed for liquid breakdown, such as thermal breakdown or impact ionization in the liquid and field emission from the cathode. It has been suggested in recent years that breakdown can be initiated also by the formation of tiny gas bubbles in the high fields in the liquid, leading to completion of the event by gas discharges. A further mechanism was observed in impure liquids with conducting particles in suspension. Such particles coalesce into conducting bridges between the electrodes, causing breakdown. The breakdown field of liquids extends from less than 10^7 to 5×10^8 V/m. Lower values are found in impure liquids and in those containing dissolved gases.

87. Breakdown in Solids. Breakdown in solids is caused mainly by thermal, electronic, and internal and external discharge mechanisms. In a solid specimen that mechanism prevails which causes breakdown quicker or at a lower voltage. The discharge of the electrostatic energy stored in the specimen through the conducting channel determines mainly whether destruction arises. It is usually found that destruction occurs at applied fields larger than 10^7 V/m. At lower fields with no destruction, the processes lead to switching events.

Thermal breakdown is caused at low frequencies by the Joule heat of the leakage current and at very high frequencies, mainly by dielectric losses. At low frequencies the Joule heat leads to thermal instability and current runaway due to rapid, often exponential, rise of the conductivity with temperature. The breakdown fields decrease rapidly with increasing temperature and time of voltage application. Thermal instability has been observed from 10^5 to 10^9 V/m.

Electronic breakdown can be caused by a variety of processes such as impact ionization, field emission, double injection, and, possibly, by insulator-to-metal transition. Breakdown in insulators is usually ascribed to the interaction of impact ionization, charge carrier injection into the insulator, and the effect of space charges. In thin films breakdowns are found to occur randomly in space and in time, involving only a very small part of the insulator. The typical range of electronic breakdown in insulators is from 10^7 to more than

[1] The author thanks Dr. N. Klein for his valuable suggestions and for preparation of the sections on electrical breakdown.

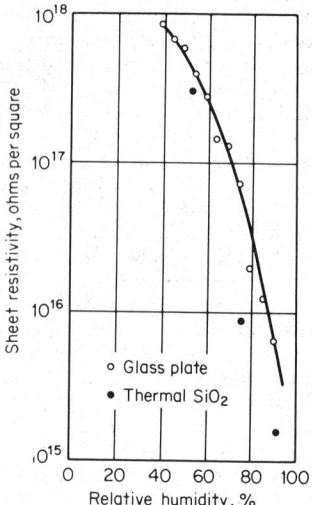

Fig. 6-18. Sheet resistivity of glass plate and SiO$_2$ (thermal oxidation of Si) vs. relative humidity.[94]

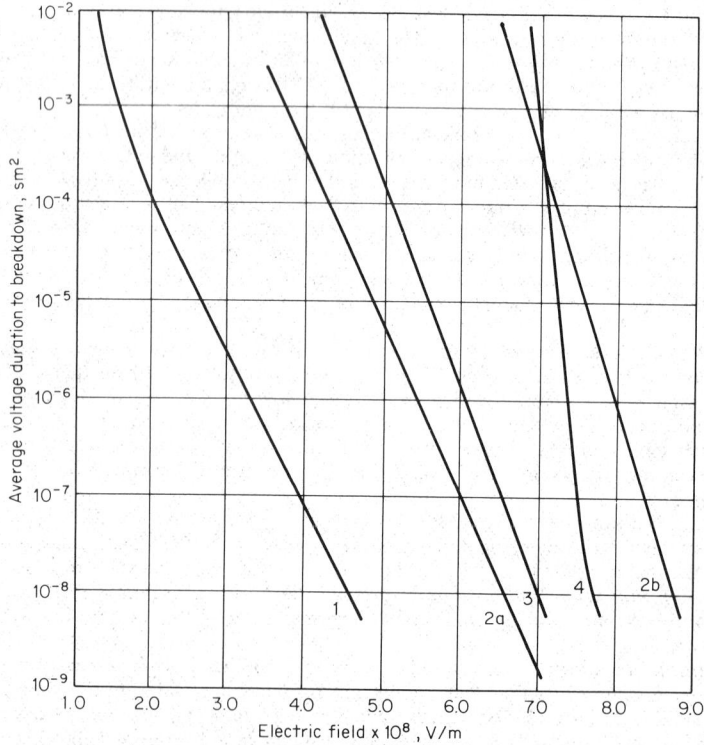

Fig. 6-19. Average voltage duration to breakdown as a function of breakdown in field.[73] (1) Hf - HfO$_2$ - Au, Hf neg, oxide 325 nm thick; (2) Al - Al$_2$O$_3$ - Au 170 nm Ta$_2$O$_5$ (*a*) Al neg, (*b*) Au neg; (3) Ta - Ta$_2$O$_5$ - Au, Ta neg, oxide 230 nm thick; (4) Si - SiO$_2$ = Al, Si pos, *n*-type 0.01 cm, oxide 150 nm thick, room temperature.

10^9 V/m. In semiconductors electronic breakdown is initiated either by impact ionization or by tunneling processes at fields of 10^6 to 10^8 V/m. It should be remarked that impact ionization is observed in semiconductors at liquid-helium temperatures at fields less than 10^3 V/m; these processes lead to switching but not to breakdown.

Breakdown by internal discharges is possible once the voltage is sufficient to cause sparking in the cavities in the solid. Thus, this kind of sparking does not occur in thin films. As sparking at atmospheric pressure can start at fields less than 10^6 V/m, the cavities are the weak links of thick insulators. Discharges in cavities can cause erosion by sputtering, chemical reactions, local melting, and evaporation and can promote breakdown in the adjacent solid. These local destructive processes can be fast at high voltages and high frequencies, but very slow at lower voltages and on direct current. The breakdown field is a decreasing function of increasing time, and breakdowns occurring after years of operation are known to be caused by internal discharges.

The effect of internal discharges is relatively small on dc operation, but the effect becomes marked in ac systems, increasing with frequency. When internal discharges are present, breakdown of thick insulators may arise readily above fields of 10^6 V/m, and the breakdown progresses through the insulator in branched channels in the form of *treeing*.

Breakdown fields are lowest when breakdown is produced by external discharges. Such discharges occur along the interface of an insulator with a gaseous or a liquid medium. These discharges, and also those occurring in cavities, are often denoted as *corona* discharges. The development of electrical discharges along surfaces is enhanced by the interaction of processes in the solid and in the gas or liquid. Favorable conditions for impact ionization and charge carrier emission produce sparking and lead to *arc tracking* on the surface of the solid. Tracking produces conducting channels in the form of treeing. Conditions for tracking are especially favorable on impure surfaces and in humid atmospheres, and the lowest design fields, less than 10^6 V/m, in electronic equipment are assigned to surfaces.

All the breakdown processes can be enhanced, and the lifetime of the insulator shortened, when the insulator is subject to *thermal aging* by chemical processes.

88. Thermal Properties. The thermal conductivity of dielectrics is defined as in conductors (see Par. **6-16**). Values of thermal conductivity are given in Table 6-22 and in Fig. 6-20. Dielectrics are in general poor heat conductors, although a few notable exceptions are diamond and BeO. Crystalline dielectrics are generally better heat conductors than amorphous dielectrics. Porous dielectrics have a smaller heat conduction than the dense materials.

Table 6-22. Thermal Conductivity of Dielectrics at Room Temperature

Material	Thermal Conductivity, J/ (m.K)
Diamond	658
BeO	167
Graphite	117–200
Bonded SiC	42
MgO	42
Al₂O₃	33
TiC	33
BN	28
Mullite	6.3
Titania	3.3–4.2
Porcelain	1.7
Fused silica glass	1.5
Soda-lime glass	0.8
Cast epoxies	0.17–1.3
Wood	0.5–2.5
Melamines	0.3–0.7
Mica	0.3–0.7
Phenolics (molded)	0.17–0.67
Polyethylenes	0.33
Nylons	0.17–0.25
TFE fluorocarbons	0.25
Elastomers	0.08–0.17
Foams	0.003–0.04

89. Potential Distribution in Dielectrics. *Uniform Dielectrics.* The electrostatic field in dielectrics is the voltage difference divided by the spacing. Nonuniform electrostatic fields can be calculated, a common case being the coaxial geometry:

$$E = \frac{V}{r \, \ln \, (R_2/R_1)} \qquad (6\text{-}46)$$

where E = electrostatic field (V/m), V = applied voltage (V), r = radius at which the value of E is sought (m), R_2 = outer radius of dielectric (m), and R_1 = inner radius of the dielectric (m). The maximum field occurs at $r = R_1$.

Composite Dielectrics. If the dielectric is composed of two substances parallel to the electrodes of thickness t_1, t_2, then on ac operation the electric fields are related as

$$\frac{E_1}{E_2} = \frac{\varepsilon_{r_2}}{\varepsilon_{r_1}} \qquad (6\text{-}47)$$

where E_1, E_2 = electric fields (V/m), and ε_{r_1}, ε_{r_2} = dielectric constants. The material with the lower dielectric constant will have the larger electric field.

In a static (dc) electric field,

$$\frac{E_1}{E_2} = \frac{\rho_1}{\rho_2} \qquad (6\text{-}48)$$

where ρ_1, ρ_2 = resistivities of the two dielectrics ($\Omega\cdot$m). In this case the larger field will occur in the dielectric with the higher resistivity.

If two dielectrics with different resistivities are placed in a field, space charges will accumulate in the interface between the layers. This space-charge accumulation has been used at high fields for memory application in MNOS (metal nitride oxide semiconductor) devices.[29] Here, accumulated charge was first produced in desired locations (writing) and then sensed at a later time (reading).

Nonuniform Dielectrics. Real dielectrics may have nonuniformities such as voids, changes in chemical composition, and also conducting inclusions. These nonuniformities act in a manner similar to that described in the preceding section. The accumulation of space charge at nonuniformities can cause additional dielectric losses which can reduce the breakdown field. Reduction of nonuniformities, such as the impregnation of porous dielectrics by oil, increases the dielectric strength. Intentional nonuniformities have been used in programmable memories where a semiconductor was buried in the insulator. A charge is injected through the dielectric into the buried layer. The stored charge can be subsequently sensed if desired.

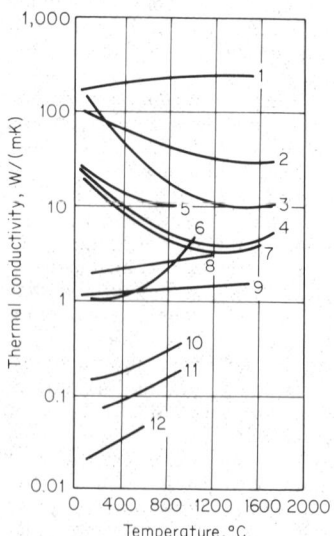

Fig. 6-20. Thermal conductivity of some ceramic materials as a function of temperature.[74] (1) Platinum; (2) graphite; (3) pure dense BeO; (4) pure dense MgO; (5) bonded SiC; (6) clear fused silica; (7) pure dense Al_2O_3; (8) fine-clay refractory; (9) dense stabilized ZrO_2; (10) 2800°F insulating firebrick; (11) 2000°F insulating firebrick; (12) powdered MgO.

SPECIFIC DIELECTRIC AND INSULATING MATERIALS

90. Gases. At low temperatures and low electric fields, gases have practically no electrical conduction. Ionization by electromagnetic radiation or particles can ionize gases and largely increase their conductivity. The dielectric constant of gases is slightly larger than unity. Breakdown occurs at high fields due to electron multiplication. The electric breakdown in uniform fields is a function of pressure × spacing (Fig. 6-21). The dielectric strength tends to increase with increasing molecular weight of the gases. Figure 6-21 shows that electrical breakdown can occur at voltages as low as a few hundred volts. Electrical breakdown can occur even in vacuum. This phenomenon, termed *vacuum breakdown,* is probably related to field emission from the cathode.

91. Liquids. *Oils.* Oils are used to provide electrical insulation by replacing the air in certain systems or to impregnate porous insulators. The properties of interest in the oils are breakdown voltage, electrical conductivity, heat conductivity, viscosity, flammability, and vapor pressure.

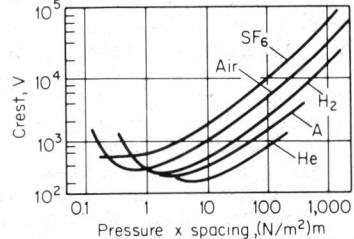

Fig. 6-21. Pressure-spacing dependence of the dielectric strength of gases (Pashen's curves).

The oils are usually mineral hydrocarbons derived from petroleum deposits. The oils contain aliphatic and aromatic compounds. The ratio of these compounds affects the properties of the oil. Properties of mineral oils are given in Table 6-23. Mineral oils can be chlorinated. The chlorination raises the viscosity and boiling point and reduces the dielectric constant.

Oils dissolve gases to 10 to 100% by volume, and water to 100 ppm at 100% relative humidity. Breakdown voltage of oils depends on the oil purity; i.e., water and particulate matter will reduce their dielectric strength. Oils can deteriorate with time due to oxidation leading to sludge and acidity. Electrode materials such as copper can enhance the oil oxidation. Oils are more stable in N_2 atmosphere. Oxidation inhibitors can also be used. The properties of oils are usually maintained by filtering. (See *IEEE Standards Publication* 64.)

Chlorinated Aromatics (Askarel Liquids). These are used at low frequencies. (These liquids are lossy at high frequency.) Two main types are used: (1) chlorinated diphenyl liquids for capacitors and (2) mixtures of highly chlorinated diphenyl with trichlorobenzene or tetrachlorobenzene for transformers. Additives are generally used in these liquids to avoid the effects of hydrogen chloride production in the liquid on discharge or arcing.

Fluorocarbon Liquids. Fluorocarbon liquids are nonflammable aliphatic compounds with low permittivities and conductivities and are chemically inert. They have been used for large transformers or filling electronic apparatus.

Silicon Fluids. These have high-stability, low-dielectric losses and high dielectric strength. They can be obtained at various levels of viscosity, from a fraction to millions of centistokes. Their working range is between 40 and 200°C.

Ester Fluids. These are used occasionally for high-frequency capacitors. They are prone to hydrolysis and have poor thermal stability.

Table 6-23. Characteristic Properties of Insulating Liquids[118]

Type of liquid	Mineral oil			Transformer Askarel
	Transformer	Cable and capacitor	Solid cable	
Specific gravity.	0.88	0.885	0.93	1.56–1.57
Viscosity, Saybolt sec at 37.8°C	57–59	0.100	100	52–56* / 41–45†
Flash point, °C	135	165	235	
Fire point, °C	148	185	280	None
Pour point, °C.	−45	−45	−5	$< -32°C*$ / $< -44°C†$
Specific heat.	0.425	0.412	. . .	0.251
Coefficient of expansion.	0.00070	. . .	0.00075	0.0007
Thermal conductivity, cal/cm·s·°C	0.39	. . .	. . .	0.30
Dielectric strength,‡ kV~	30	. . .	. . .	>35
Permittivity at 25°C	2.2	. . .	. . .	4.5–4.7
Resistivity, Ω·cm × 10^{12}	1–10	50–100	1–10	0.1

* A mixture of 60% hexachlorobiphenyl and 40% trichlorbenzene.
† A mixture of 45% hexachlorobiphenyl and 55% trichloro- and tetrachlorobenzenes.
‡ ASTM D877.

92. Plastics. Plastics are the most widely used dielectric materials in the electrical and electronics industry. There are numerous plastic materials available with a spectrum of electrical, mechanical, and chemical properties. Plastics can be divided into elastomers and thermosetting and thermoplastic materials.

Elastomers. Elastomers are natural or synthetic rubberlike materials which have outstanding elastic characteristics. Their elastic-deformation range is normally a few hundred percent. Elastomers are principally used where their outstanding mechanical properties are sought, i.e., for damping, sealing, or gasketing. Designation and applications of elastomers are given in Table 6-24, and properties are given in Table 6-25.

Thermosetting Plastics. These are materials which are cured and hardened to a desired form at room temperature or at a higher temperature. The chemical change on curing is permanent, and the material cannot be softened by reheating. Classification and properties of thermosetting plastics are given in Table 6-26, and properties of plastic materials are given in Table 6-27.

Thermoplastic Plastics. Thermoplastic materials do not cure or set upon heating. They soften and can be shaped by molding into any desired form. Thermoplastics can be repeatedly resoftened by heating. Table 6-28 lists the applications of several thermoplastics. Properties of thermoplastics are given in Table 6-29. Figures 6-22 and 6-23 show the dielectric constant, dissipation factor, and electrical breakdown fields of two fluorocarbons.

93. Fillers. Fillers in form of particles, fibers, or platelets are added mainly to thermosetting material to improve their properties. Fillers can improve the dimensional stability, stiffness, hardness, and tensile and impact strength, as well as improve chemical and heat resistance. Graphite and molybdenum disulfide are used to improve lubrication properties. Fillers can be organic, such as cotton, paper, or cellulose, or can be inorganic, such as glass, asbestos, mica, clay, carbonates, or oxides.

94. Laminates. Laminates are made by impregnating glass cloth, paper, cotton, or other fibers with various thermosetting compounds such as phenolics, melamines, silicones, or epoxies. The laminate is cured by heat and pressure. Laminates with thermoplastics are not common. Laminates find a very important use as the boards for printed circuits. Application characteristics of several copper-clad laminates are listed in Table 6-30.

95. Plastic Films, Sheets, and Tapes. Most thermoplastics are available in sheet, film, or tape form. Thin films are similar in most properties to bulk materials. The dielectric strength of thin films is higher, however, than that of bulk. Several tapes are available with adhesive backing. Properties of films are given in Table 6-31.

96. Plastic Foams. Plastic foams are cellular forms of urethanes, polystyrenes, vinyls, polyethylenes, polypropylenes, phenolics, epoxies, and a variety of other plastics. The foams are used for electrical, thermal, or acoustical insulators (see also radar absorbing materials, Par. 6-71). Properties of plastic foams are given in Table 6-32.

97. Plastic Coatings. Plastic coatings can be thermosetting, thermoplastic, or elastomeric. The coating properties are similar to the bulk properties. In addition, the coatings have characteristic properties such as coatability, adhesion, and wetting. In general, coatings can be applied by dipping, brushing, spraying, roller, screening, impregnation, fluidized beds, or vacuum deposition.

Varnishes are used for impregnation (such as transformer coils) or for coating electronic equipment. Varnishes may contain solvents for reducing the viscosity and then be air-dried or baked. Solventless coatings consist generally of a resin and a curing agent. The resins are most commonly epoxies, polyesters, or polyurethanes. Such coatings have, generally, a shorter working life than the solvent-based coatings.

Printed-circuit boards are coated with conformal coating by dipping the boards in thixotropic coatings. Such coatings are up to 1 mm thick. The coatings improve the protection against corrosion, electrical resistance degradation, and moisture. Coatings provide some mechanical protection. Coatings are sometimes filled to improve their properties. Flexible silicone coatings are used in packaging semiconductor devices. The flexible coating is applied directly on the semiconductor die surface after it is assembled on the package, and subsequently the entire assembly is covered with an outer shell by transfer molding or potting with silicone or epoxy.

The inner flexible coating, often referred to as junction coating, has to provide flexibility, electrical insulation, and some moisture resistance. The purity of this coating is essential to minimize the corrosion of the metallization and interconnecting fine wires which are used in this technology. Coatings are also used for wire and coil insulation. The coatings are generally high-molecular-weight thermoplastic or thermosetting. Some applications of coatings are shown in Table 6-33.

98. Paper. Paper is widely used as electrical insulation. Unimpregnated cellulose paper (kraft paper) is used for cable, small transformers, and capacitors. The dielectric constant, up to 6, and dissipation factor (as low as 0.001 for dry paper) are functions of humidity, temperature, and density. The dielectric strength of kraft paper is 6 to 12×10^6 V/m. Press boards of cellulose can also be found up to 0.5-in. thickness with dielectric strength of 2 to 12×10^6 V/m.

Table 6-24. Designations and Application Information for Elastomers[71]

ASTM D1418	Trade name or common name	Chemical type	Major application considerations
NR	Natural rubber	Natural polyisoprene	Excellent physical properties; good resistance to cutting, gouging, and abrasion; low heat, ozone, and oil resistance. The best electrical grades are excellent in most electrical properties at room temperature.
IR	Synthetic natural	Synthetic polyisoprene	Same general properties as natural rubber; requires less mastication in processing than natural rubber.
CR	Neoprene	Chloroprene	Excellent ozone, heat, and weathering resistance; good oil resistance; excellent flame resistance. Not so good electrically as NR or IR. However, the combination of generally good electricals for jacketing application, coupled with all the other good properties, gives this elastomer broad use for electrical wire and cable jackets.
SBR	GRS, Buna S	Styrene-butadiene	Good physical properties; excellent abrasion resistance; not oil-, ozone-, or weather-resistant. Electrical properties generally good but not specifically outstanding in any area.
NBR	Buna N, nitrile	Acrylonitrile-butadiene	Excellent resistance to vegetable, animal, and petroleum oils; poor low-temperature resistance. Electrical properties not outstanding; probably degraded by molecular polarity of acrylonitrile constituent.
IIR	Butyl	Isobutylene-isoprene	Excellent weathering resistance; low permeability to gases; good resistance to ozone and aging; low tensile strength and resilience. Electrical properties generally good but not outstanding in any area.
IIR	Chlorobutyl	Chloroisobutylene-isoprene	Same general properties as butyl.
BR	Cis-4	Polybutadiene	Excellent abrasion resistance and high resilience; used principally as a blend in other rubbers.
	Thiokol (PS) (Thiokol chemical)	Polysulfide	Outstanding solvent resistance; widely used for potting of electrical connectors.
R	EPR	Ethylene propylene	Good aging, abrasion, and heat resistance; not oil-resistant. Good general-purpose electrical properties.
R	EPT	Ethyl propylene terpolymer	Good aging, abrasion, and heat resistance; not oil-resistant. Good general-purpose electrical properties.
CSM	Hypalon (HYP) (Du Pont)	Chlorosulfonated polyethylene	Excellent ozone, weathering, and acid resistance; fair oil resistance; poor low-temperature resistance. Not outstanding electrically, but has some special-application uses based on other properties.
SIL	Silicone	Polysiloxane	Excellent high- and low-temperature resistance; low strength; high compression set. Among the best electrical properties in the elastomer grouping. Especially good stability of dielectric constant and dissipation factor at elevated temperatures.
	Urethane (PU)	Polyurethane diisocyanate	Exceptional abrasion, cut, and tear resistance; high modulus and hardness; poor moist-heat resistance. Generally good general-purpose electrical properties. Some special high-quality electrical grades available from formulators.
	Viton (FLU) (Du Pont)	Fluorinated hydrocarbon	Excellent high-temperature resistance, particularly in air and oil. Not outstanding electrically.
ABR	Acrylics	Polyacrylate	Excellent heat, oil, and ozone resistance; poor water resistance. Not outstanding for or widely used in electrical applications.

PROPERTIES OF MATERIALS

Table 6-25. Comparison of Properties of Rubbers and Elastomers[71]

Material	Di-electric constant, 10^6 Hz	Power factor $\times 10^2$, 10^6 Hz	Volume resistivity, $\Omega \cdot$cm	Surface resistivity, Ω	Di-electric strength, V/mil
Natural rubber	2.7-5	0.05-0.2	10^{15}-10^{17}	10^{14}-10^{15}	450-600
Styrene-butadiene rubber	2.8-4.2	0.5-3.5	10^{14}-10^{16}	10^{13}-10^{14}	450-600
Acrylonitrile-butadiene rubber	3.9-10.0	3-5	10^{12}-10^{15}	10^{12}-10^{15}	400-500
Butyl rubber	2.1-4.0	0.3-8.0	10^{14}-10^{16}	10^{13}-10^{14}	400-800
Polychloroprene	7.5-14.0	1.0-6.0	10^{11}-10^{12}	10^{11}-10^{12}	100-500
Polysulfide polymer*	7.0-9.5	0.1-0.5	10^{11}-10^{12}	...	250-325
Silicone	2.8-7.0	0.10-1.0	10^{13}-10^{17}	10^{13}	300-700
Chlorosulfonated polyethylene†	5.0-11.0	2.0-9.0	10^{13}-10^{17}	10^{14}	400-600
Polyvinylidene fluoride-copolymer-hexafluoropropylene‡	10.0-18.0	3.0-4.0	10^{13}	...	250-700
Polyurethane§	5.0-8.0	3.0-6.0	10^{10}-10^{11}	...	450-500
Ethylene propylene terpolymer¶	3.2-3.4	0.6-0.8	10^{15}-10^{17}	...	700-900

* Thiokol, Thiokol Corp.
† Hypalon, Du Pont, Inc.
‡ Viton A, Du Pont, Inc.
§ Adiprene, Du Pont, Inc.
¶ Nordel, Du Pont, Inc.

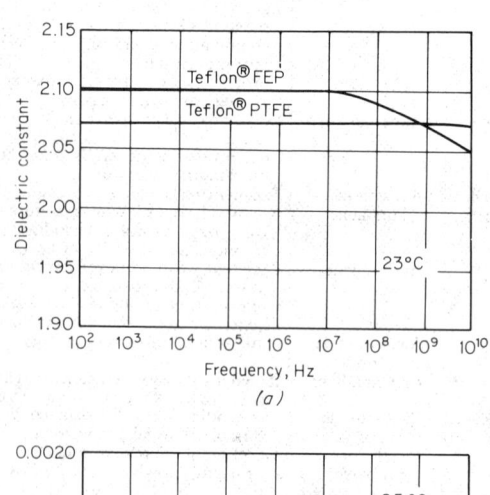

(a)

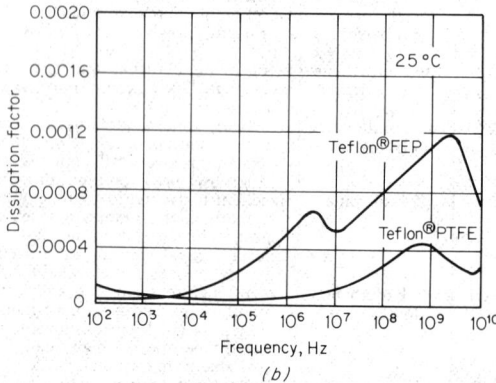

(b)

Fig. 6-22. Dielectric constant (*a*) and dissipation factor (*b*) for Teflon® FEP and Teflon® PTFE as a function of frequency.[76] (Teflon® is a registered trademark of E. I. Du Pont de Nemours, Inc., Wilmington, Del.)

Table 6-26.   Application Information for Thermosetting Plastics[71]

Material	Major application considerations	Common available forms
Alkyds	Excellent dielectric strength, arc resistance, and dry insulation resistance. Low dielectric constant and dissipation factor. Good dimensional stability. Easily molded.	Compression moldings, transfer moldings
Aminos (melamine formaldehyde and urea formaldehyde)	Available in an unlimited range of light-stable colors. Exhibit hard glossy molded surface and good general electrical properties, especially arc resistance. Excellent chemical resistance to organic solvents and cleaners and household-type cleaners.	Compression moldings, extrusions, transfer moldings, laminates, film
Diallyl phthalates (DAP) (allylics)	Unsurpassed among thermosets in retention of properties in high-humidity environments. Also, have among the highest volume and surface resistivities in thermosets. Low dissipation factor and heat resistance to 400°F or higher. Excellent dimensional stability. Easily molded.	Compression moldings, extrusions, injection moldings, transfer moldings, laminates
Epoxies	Good electrical properties, low shrinkage, excellent dimensional stability, and good to excellent adhesion. Extremely easy to compound, using nonpressure processes, for providing a wide variety of end properties. Useful over a wide range of environments. Bisphenol epoxies are most common, but several other varieties are available for providing special properties.	Castings, compression moldings, extrusions, injection moldings, transfer moldings, laminates, matched-die moldings, filament windings, foam
Phenolics	Among the lowest-cost, most widely used thermoset materials. Excellent thermal stability to over 300°F generally, and over 400°F in special formulations. Can be compounded to a broad choice of resins, fillers, and other additives.	Castings, compression moldings, extrusions, injection moldings, transfer moldings, laminates, matched-die moldings, stock shapes, foam
Polyesters	Excellent electrical properties and low cost. Extremely easy to compound using nonpressure processes. Like epoxies, can be formulated for either room-temperature or elevated-temperature use. Not equivalent to epoxies in environmental resistance.	Compression moldings, extrusions, injection moldings, transfer moldings, laminates, matched-die moldings, filament windings, stock shapes
Silicones (rigid)	Excellent electrical properties, especially low dielectric constant and dissipation factor, which change little up to 400°F and over. Nonrigid silicones are covered in elastomers and embedding-material sections.	Castings, compression moldings, transfer moldings, laminates
Urethanes (rigid foams)	Low-weight plastics. Excellent electrical properties, which are basically variable as a function of density. Easy to use for foam-in-place and embedding applications. Flexible urethane foams and nonrigid high-density urethanes are covered in sections on foams, elastomers, and embedding materials.	Castings, coatings

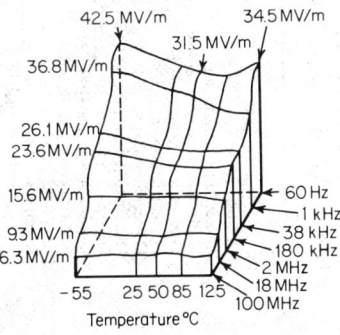

Fig. 6-23.   Effect of frequency and temperature on the short-term dielectric strength of Teflon® PTFE.[76] (Teflon® is a registered trademark of E. I. Du Pont de Nemours, Inc., Wilmington, Del.)

Table 6-27. Thermosetting Molding Materials[118]

Material		Diallyl phthalate			Epoxy		Melamine		
		SPI prefix: DAP					SPI prefix: MF		
Filler	ASTM No.	Glass fiber	Mineral	Synthetic fiber	Glass fiber	Mineral	Alpha cellulose	Asbestos	Glass fiber
Electrical properties:									
Volume resistivity, Ω-cm.	D257	10^{16}	10^{13}	10^{16}	10^{14}	10^{14}	10^{14}	10^{12}	10^{11}
Dielectric strength, short-time, V/mil	D149	450	420	400	400	400	400	430	300
Dielectric strength, step-by-step, V/mil	D149	400	400	410	400	400	300	320	240
Dielectric constant, 60 Hz	D150	4.3	5.2	5.0	5.0	5.0	9.5	10.2	11.1
Dielectric constant, 10^3	D150	4.4	5.3	3.9	5.0	5.0	9.2	9.0	..
Dielectric constant, 10^6	D150	4.5	4.0	3.6	5.0	5.0	8.4	6.7	7.5
Dissipation factor, 60 Hz	D150	0.01	0.03	0.026	0.01	0.01	0.030	0.07	0.14
Dissipation factor, 10^3	D150	0.004	0.03	0.004	0.01	0.01	0.015	0.07	..
Dissipation factor, 10^6	D150	0.009	0.02	0.012	0.01	0.01	0.027	0.041	0.013
Arc resistance, s	D495	180	190	130	180	190	180	180	180
Mechanical properties:									
Specific gravity	D792	1.78	1.68	1.39	2.0	2.0	1.52	2.0	2.0
Specific volume, in.^{3}/lb	D792	17.2	16.8	20.7	15.4	14.2	18.2	13.8	13.8
Tensile strength, lb/in.^{2}	D638,D651	11,000	8,700	6,800	30,000	15,000	13,000	7,000	10,000
Elongation, %	D638	..	..	..	4	..	0.9	0.45	..
Tensile modulus, lb/in.$^2 \times 10^5$	D638	22	22	6.0	30.4	..	14	19.5	24
Compressive strength, lb/in.^{2}	D695	35,000	32,000	30,000	40,000	40,000	45,000	30,000	35,000
Flexural strength, lb/in.^{2}	D790	19,000	11,000	19,000	60,000	15,000	16,000	11,000	23,000
Impact strength, ft-lb/in. of notcl.	D256	10.0	0.45	8.0	10.0	0.4	0.35	0.4	6.0
Hardness, Rockwell	D785	M108-110	M100-103	M108-115	M100-110	M100-110	M110-125	M110	
Thermal properties:									
Thermal conductivity, (cal/scm °C) × 10^{-4}	C177	5-10	7-25	5-6	4-10	4-30	7-10	13-17	11-0
Thermal expansion, °C$^{-1} \times 10^{-5}$	D696	3.6	4.2	6.0	3.5	5.0	4.0	2.0-4.5	1.5-1.7
Maximum-use temp., °F		350	350	300	300	300	210	250	300
Heat-distortion temp., °F at 264 lb/in.^{2}	D648	350	320	300	250	250	360	265	400

	ASTM								
Chemical properties:									
Water absorption, % in 24 h	D570	0.35	0.5	0.2	0.2	0.04	0.6	0.14	0.21
Burning rate	D635	Self-exting. to non-burning	Self-exting. to non-burning	Self-exting.	Self-exting.	Self-exting.	Self-exting.	Self-exting.	Self-exting.
Effect of sunlight	...	None	None	None	Slight	Slight	Slight color change	Slight color change	Slight
Effect of weak acids	D543	None	None	None	None	None	None	None to slight	None
Effect of strong acids	D543	Slight	Slight	Slight	Negligible	None	Decomposes	Decomposes	Decomposes
Effect of weak alkalies	D543	None to slight	None to slight	None	None	None	None	Very slight	None
Effect of strong alkalies	D543	Slight	Slight	Slight	None	Slight	Attacked	Slight attack	None to slight
Effect of organic solvents	D543	None	None	None	None	None	None	None	None
Trade names	...	Acme, Cosmic, Dapon, Diall, Poly-Dap, RX		Bakelite, Dri-Coat, Eccomold, EMC, Epiall, Eposet, Fibercore, Formitt, High Strength, Hysol, Plenco, Polyset, Scotchply, Smooth-on, Trevarno, Unipoxy			Amres, Cymel, Diaron, Melmac, Permelite, Plenco, Resimene, Resloom, Syr-U-Tex		

6-49

Table 6-27. Thermosetting Molding Materials *(concluded)*

Material		Phenolic SPI prefix PF			Polyester SPI prefix EA		Silicone SPI prefix S		Urea formaldehyde SPI prefix UF
Filler	ASTM No.	Woodflour and cotton flock	Asbestos	Glass fiber	Glass fiber	Mineral	Glass fiber	Mineral	Alpha cellulose
Electrical properties:									
Volume resistivity, Ω-cm	D257	10^{13}	10^{13}	10^{12}	10^{15}	10^{14}	10^{14}	10^{14}	10^{13}
Dielectric strength, short-time, V/mil	D149	400	350	400	420	450	400	400	400
Dielectric strength, step-by-step, V/mil	D149	375	300	270	390	350	300	380	300
Dielectric constant, 60 Hz	D150	13	50	7.1	7.3	7.5	5.2	3.6	9.5
Dielectric constant, 10^3	D150	9.0	30	6.9	4.68	6.2	5.0	...	7.5
Dielectric constant, 10^6	D150	6.0	10	6.6	6.4	5.5	4.7	6.3	6.8
Dissipation factor, 60 Hz	D150	0.05	0.1	0.05	0.011	0.009	0.004	0.004	0.035
Dissipation factor, 10^3	D150	0.04	0.1	0.02	...	0.01	0.0035	...	0.025
Dissipation factor, 10^6	D150	0.03	0.4	0.012–0.026	0.008	0.015	0.002	0.002	0.25
Arc resistance, s	D495	Tracks	120	120	180	150	250	420	150
Mechanical properties:									
Specific gravity	D792	1.45	1.9	1.95	2.3	2.30	2.0	2.82	1.52
Specific volume, in.^{3}/lb	D792	17.8	11.9	14.1	...	5.4	13.8	...	18.2
Tensile strength, lb/in.^{2}	D638,D651	10,000	7,500	18,000	10,000	8,000	5,000	3,500	13,000
Elongation, %	D638	0.8	0.50	0.2	...	...	...	...	1.0
Tensile modulus, lb/in.$^2 \times 10^5$	D638	17	30	33	25	26	...	...	15
Compressive strength, lb/in.^{2}	D695	36,000	35,000	70,000	30,000	25,000	15,000	18,000	45,000
Flexural strength, lb/in.^{2}	D790	12,000	14,000	60,000	20,000	10,000	14,000	8,000	18,000
Impact strength, ft-lb/in. of notch	D256	0.60	3.5	18	16.0	0.50	15	0.35	0.40
Hardness, Rockwell	D785	M96-120	M95-115	M95-100	...	...	M84	M71-95	M110-120
Thermal properties:									
Thermal conductivity, (cal/s·cm·°C) $\times 10^{-4}$	C177	4.7	8-22	9-14.5	10-16	15-25	7.51-7.54	11-13	7-10
Thermal expansion, °C$^{-1} \times 10^{-5}$	D696	3.0-4.5	0.8-4	0.8-1.6	2.5-3.3	3.5-5	0.8	2-4	2.2-3.6
Maximum-use temp., °F	...	350	350	350	300	300	>600	>600	170
Heat-distortion temp., °F at 264 lb/in.^{2}	D648	260	300	300	>400	350	>900	>900	260
Chemical properties:									
Water absorption, % in 24 h	D570	0.7	0.5	1.2	0.28	0.5	0.2	0.13	0.8

Property	ASTM							
Burning rate	D635	Self-exting.	Self-exting.	None to slow	None to slow	Slow to self-exting.	Slow to nonburning	Very low
Effect of sunlight	...	Pastels gray	None to slight	None to slight	None to slight	Depends on pigmentation	None	Darkens
Effect of weak acids	D543	None to slight	None	None to slight	None to slight	Slight	None	None
Effect of strong acids	D543	Decomposed if surface attacked	Slight	None to slight	Slight	Attacked	Attacked	Oxidizing acids
Effect of weak alkalies	D543	Slight to marked	None to slight	None to slight	None to slight	Slight to attacked	Attacked	Slight
Effect of strong alkalies	D543	Decomposes	Slight to marked	Slight to marked	Slight to marked	Slight to attacked	Decomposes	Attacked
Effect of organic solvents	D543	None to slight	Attacked by some	Attacked by some	Attacked by some	None	None	None
Trade names	...	Amres, Arodure, Beetle, Mouldrite, Nestorite, Plaskon, Resimene, Resloom, Sylplast, Syn-U-Flex, Synvarol, Tetra-Ria, Tybon	Dow Corning			Alpon, Co-Rezyn, Durez, Fibercore, Formadall, Glasdramatic, Glaskyd, Glasrin, Glastic, Haysite, Insulstruct, Mobaloy, Parr, Plaskon, Politen, Polyglas, Premix, Resistrac, Rosite, Trevarno		Amres, Arochem, Bakelite, Catalin, Celcron, Durez, Fiberite, Haveg, Mouldrite, Nestorite, Permelite, Plenco, Plyophen, Resinox, Snap-Cure, Synvaren, Synvorite, Tetra-Flex, Tybon, Varcum

Table 6-28. Application Information for Thermoplastics[71]

Material	Major application considerations	Common available forms
ABS (acrylonitrile-butadiene-styrene)	Extremely tough, with high impact resistance. Can be formulated over a wide range of hardness and toughness properties. Special grades available for plated surfaces with excellent pull-strength values. Good general electrical properties, but not outstanding for any specific electric applications.	Blow moldings, extrusions, injection moldings, thermoformed parts, laminates, stock shapes, foam
Acetals	Outstanding mechanical strength, stiffness, and toughness properties, combined with excellent dimensional stability. Good electrical properties at most frequencies, which are little changed in humid environments up to 125°C.	Blow moldings, extrusions, injection moldings, stock shapes
Acrylics (polymethyl methacrylate)	Outstanding properties are crystal clarity and resistance to outdoor weathering. Excellent resistance to arcing and electrical tracking.	Blow moldings, castings, extrusions, injection moldings, thermoformed parts, stock shapes, film, fiber
Cellulosics	There are several materials in the cellulosic family, such as cellulose acetate (CA), cellulose propionate (CAP), cellulose acetate butyrate (CAB), ethyl cellulose (EC), and cellulose nitrate (CN). Widely used plastics in general, but not outstanding for electronic applications.	Blow moldings, extrusions, injection moldings, thermoformed parts, film, fiber, stock shapes
Chlorinated polyethers	Good electrically, but most outstanding properties are corrosion resistance and good physical and thermal stability by thermoplastic standards.	Extrusions, injection moldings, stock shapes, film
Ethylene-vinyl acetates (EVA)	Excellent flexibility, toughness, clarity, and stress-crack resistance. Somewhat like a tough synthetic rubber or elastomer. Not widely used in electronics. Comparatively low resistance to heat and solvents.	
Fluorocarbons a. Chlorotrifluoroethylene (CTFE)	Excellent electrical properties and relatively good mechanical properties. Somewhat more stiff than TFE and FEP fluorocarbons, but does have some cold flow. Widely used in electronics, but not quite so widely as TFE and FEP. Useful to about 400°F.	Extrusions, isostatic moldings, injection moldings, film, stock shapes
b. Fluorinated ethylene propylene (FEP)	Very similar properties to those of TFE, except useful temperature limited to about 400°F. Easier to mold than TFE.	Extrusions, injection moldings, laminates, film
c. Polytetrafluoroethylene (TFE)	Electrically one of the most outstanding thermoplastic materials. Exhibits very low electrical losses and very high electrical resistivity. Useful to over 500°F and to below −300°F. Excellent high-frequency dielectric. Among the best combinations of mechanical and electrical properties, but relatively weak in cold-flow properties. Nearly inert chemically, as are most fluorocarbons. Very low coefficient of friction. Nonflammable.	Compression moldings, stock shapes, film
d. Polyvinyl fluoride	Mostly used as a weatherable, architectural facing sheet. Not widely used in electronics.	Extrusions, injection moldings, laminates, film
e. Polyvinylidine fluoride (PVF$_2$)	One of the easiest of the fluorocarbons to process. Stiffer and more resistant to cold flow than TFE. Good electrically. Useful to about 300°F. A major electronic application is wire jacketing.	Extrusions, injection moldings, laminates, film
Ionomers	Excellent combination of toughness, solvent resistance, transparency, colorability, abrasion resistance, and adhesion. Based on ethylene-acrylic copolymers with ionic bonds. Not widely used in electronics.	Film, coatings, injection moldings
Nylons (polyamides)	Good general-purpose for electrical and nonelectrical applications. Easily processed. Good mechanical strength, abrasion resistance, and low coefficient of friction. There are numerous types of nylons; nylon 6, nylon 6/6, and nylon 6/10 are most common. Some nylons have limited use due to moisture-absorption properties. Nylon 6/10 is best here.	Blow moldings, extrusions, injection moldings, laminates, rotational moldings, stock shapes, film, fiber

Table 6-28. *(concluded)*

Parylenes (polyparaxylylene)	Excellent dielectric properties and good dimensional stability. Low permeability to gases and moisture. Produced as a film on a substrate, from a vapor phase. Such vapor-phase polymerization is unique in polymer processing. Used primarily as thin films in capacitors and dielectric coatings. Numerous polymer modifications exist.	Film coatings
Phenoxies	Tough, rigid, high-impact plastic. Has low mold shrinkage, good dimensional stability, and very low coefficient of expansion for a thermoplastic. Useful for electronic applications below about 175°F. Useful in adhesive formulations.	Blow moldings, extrusions, injection moldings, film
Polyallomers	Thermoplastic polymers produced from two monomers. Somewhat similar to polyethylene and polypropylene, but with better dimensional stability, stress-crack resistance, and surface hardness than high-density polyethylene. Electronic application areas similar to polyethylene and polypropylene. One of the lightest commercially available plastics.	Blow moldings, extrusions, injection moldings, film
Polyamide-imides and polyimides	Among the highest-temperature thermoplastics available, having useful operating temperatures between about 400°F and about 700°F or higher. Excellent electrical properties, good rigidity, and excellent thermal stability. Low coefficient of friction. Polyamide-imides and polyimides are chemically similar but not identical in all properties. They are difficult to process, but are available in molded and block forms, and also as films and resin solutions.	Films, coatings, molded and/or machined parts, resin solutions
Polycarbonates	Excellent dimensional stability, low water absorption, low creep, and outstanding impact-resistance thermoplastics. Good electrical properties for general electronic packaging application. Available in transparent grades.	Blow moldings, extrusions, injection moldings, thermoformed parts, stock shapes, film
Polyethylenes and polypropylenes (polyolefins or polyalkenes)	Excellent electrical properties, especially low electrical losses. Tough and chemically resistant, but weak to varying degrees in creep and thermal resistance. There are three density grades of polyethylene: low (0.910-0.925), medium (0.926-0.940), and high (0.941-0.965). Thermal stability generally increases with density class. Polypropylenes are generally similar to polyethylenes, but offer about 50°F higher heat resistance.	Blow moldings, extrusions, injection molding, thermoformed parts, stock shapes, film, fiber, foam
Polyethylene terephthalates	Among the toughest of plastic films with outstanding dielectric strength properties. Excellent fatigue and tear strength and resistance to acids, greases, oils, solvents. Good humidity resistance. Stable to 135-150°C.	Film, sheet, fiber
Polyphenylene oxides (PPO)	Excellent electrical properties, especially loss properties to above 350°F, and over a wide frequency range. Good mechanical strength and toughness. A lower-cost grade (Noryl) exists, having somewhat similar properties to PPO, but with a 75-100°F reduction in heat resistance.	Extrusions, injection moldings, thermoformed parts, stock shapes, film
Polystyrenes	Excellent electrical properties, especially loss properties. Conventional polystyrene is temperature-limited, but high-temperature modifications exist, such as Rexolite or Polypenco cross-linked polystyrene, which are widely used in electronics, especially for high-frequency applications. Polystyrenes are also generally superior to fluorocarbons in resistance to most types of radiation.	Blow moldings, extrusions, injection moldings, rotational moldings, thermoformed parts, foam
Polysulfones	Excellent electrical properties and mechanical properties to over 300°F. Good dimensional stability and high creep resistance. Flame-resistant and chemical-resistant. Outstanding in retention of properties upon prolonged heat-aging, as compared with other tough thermoplastics.	Blow moldings, extrusions, injection-mold thermoformed parts, stock shapes, film sheet
Vinyls	Good low-cost, general-purpose thermoplastic materials, but not specifically outstanding electrical properties. Greatly influenced by plasticizers. Many variations available, including flexible and rigid types. Flexible vinyls, especially polyvinyl chloride (PVC), widely used for wire insulation and jacketing.	Blow moldings, extrusions, injection moldings, rotational moldings, film sheet

Table 6-29. Thermoplastic Molding Materials[118]

Material	ASTM No.	Acetal	ABS	Acrylic	Cellulose acetate	Cellulose acetate butyrate	Cellulose propionate	Chlorinated polyether	Chlorotri-fluoro-ethylene	Nylon (polyamide)	Poly-carbonate
SPI prefix		...	ABS	MM	CA	CAB	CP	...	HH	PA	...
Electrical properties:											
Arc resistance	D495	129	90	No track	200	...	180	...	>360	140	120
Dielectric constant	D150										
60 Hz		3.8	3.0	4.0	7.5	6.4	4.0	3.1	2.8	5.5	3.2
10^6 Hz		3.8	3.0	3.5	7.0	6.3	4.0	3.0	2.7	4.9	3.0
10^9 Hz		3.8	3.0	3.2	7.0	6.2	3.6	2.9	2.5	4.7	3.0
Dissipation factor	D150										
60 Hz		0.004	0.003	0.04	0.01	0.02	0.01	0.01	0.001	0.01	0.0009
10^6 Hz		0.004	0.003	0.03	0.01	0.02	0.01	0.01	0.023	0.01	0.0021
10^9 Hz		0.004	0.005	0.02	0.01	0.05	0.01	0.01	0.009	0.03	0.01
Dielectric strength, V/mil, step by step	D149	400	350	350	200	250	300	400	450	320	364
Volume resistivity, Ω-cm	D257	10^{14}	10^{16}	10^{14}	10^{13}	10^{14}	10^{15}	10^{15}	10^{18}	10^{15}	10^{16}
Mechanical properties:											
Tensile strength, lb/in.²	D651,D638	9,000	7,000	11,000	8,500	6,900	7,800	6,000	6,000	14,000	9,500
Tensile modulus, lb/in.² $\times 10^5$		4.1	3.5	4.5	4.0	2.0	2.2	1.6	3.0	3.8	3.5
Elongation, %		15		10	70	88	100	160	250	320	100
Compressive strength, lb/in.²	D695	18,000	7,000	18,000	36,000	22,000	22,000	...	7,400	13,000	12,500
Flexural strength, lb/in.²	D790	14,000	10,500	17,000	16,000	9,300	11,400	5,000	9,300	...	13,500
Impact strength, ft lb/in of notch	D256	1.4	7.0	0.5	5.2	6.3	11.5	0.4	2.7	4.0	16
Hardness, Rockwell	D785	R120	M110	M105	R125	R115	R122	R100	R95	R118	R118
Thermal properties:											
Heat-distortion temp. at 264 lb/in.²	D648	255	200	210	190	202	228	285	258	167	280
Maximum-use temp., °F		185	210	190	220	220	220	290	390	250	250
Coefficient of thermal expansion, °C$^{-1} \times 10^{-5}$	D696	8.1	11	9	16	17	17	8	7	8	7
Thermal conductivity, cal/S-cm °C	C177	5.5	3.0	0.0	8.0	8.0	8.0	3.1	5.3	5.9	4.6
Flammability, in./min	D635	1.1	...	1.2	...	...	1.3	No burn	No burn	No burn	No burn

Chemical properties:										
Water absorption, % in 24 h (D570 D543)	0.25	0.45	0.4	6.5	2.2	2.8	0.01	0.00	1.88	0.15
Not resistant to	Strong acids	Oxidizing acids, ketones, esters, chlorinated solvents	Ketones, esters, aromatic and chlorinated solvents	...	Strong bases, ketones, esters, aromatic and chlorinated solvents	...	Oxidizing acids	Chlorinated solvents	Strong acids, phenol	Alkalies, aromatic and chlorinated solvents
Trade names	Betalux Celcon Delrin Dielax Formaldafil Thermocomp	Abson Cycolac Cyclon Kralastic Lustran Royalite Sulivac Triform Tybrene	Acrapon Acrydass Acrylite Acrylux Aerysol Araset Bovick Glopaque Implex Interpole Kydex Lacite Lactrelite Oraglos Plexiglas Stsvol XT Polymer Zerlon	Cal-Stix Celanese Joda Kodacel Plasticel Yenite	Acelon Cabulite Joda Tenite Uvex	Forticel Tenite	Penton	Kel-F Halon	Capran Catalin Firestone Fosta-Nylon Glastil Moleculoy Monocast Nylafil Nylux Plaskon Spencer Thermocomp X-Tal Zytel	Dupilon Lexan Merlon Penntube IV Polycarbafil Thermocomp Zelux

6-55

Table 6-29. Thermoplastic Molding Materials (concluded)

Material	ASTM No.	Poly ethylene, low-density	Poly ethylene, med-density	Poly ethylene, high-density	Poly propylene	Poly styrene	Poly sulfone	Poly phenylene oxide	Phenoxy	Poly vinyl chloride	SAN	Tetra fluoro ethylene
SPI prefix	...	PE	PE	PE	...	PS	...	...	...	VC	...	HH
Electrical properties:												
Arc resistance	D495	140	200	200	185	100	122	75	...	80	150	>200
Dielectric constant	D150											
60 Hz		2.4	2.4	2.4	2.6	3.4	3.1	2.6	4.1	3.6	3.4	2.1
10^6 Hz		2.4	2.4	2.4	2.6	3.2	3.1	2.6	4.1	3.3	2.5	2.1
10^9 Hz		2.4	2.4	2.4	2.6	3.1	3.1	2.6	3.8	3.4	3.1	2.1
Dissipation factor	D150											
60 Hz		<0.0005	<0.0005	<0.0005	<0.0005	0.0004	0.0008	0.0004	0.001	0.007	0.004	<0.0002
10^6 Hz		<0.0005	<0.0005	<0.0005	<0.0005	0.0004	0.001		0.002	0.009	0.007	<0.0002
10^9 Hz		<0.0005	<0.0005	<0.0005	<0.0005	0.0004	0.005	0.0009	0.03	0.006	0.007	<0.0002
Dielectric strength, V/mil, step by step	D149	420	500	550	450	300	400	400	400	375	300	430
Volume resistivity, Ω·cm	D257	10^{16}	10^{16}	10^{16}	10^{16}	10^{16}	10^{17}	10^{13}	10^{13}	10^{16}	10^{16}	10^{18}
Mechanical properties:												
Tensile strength, lb/in.²	D651,D638	2,300	3,500	5,500	5,500	6,800	10,200	11,000	9,500	9,000	12,000	4,500
Tensile modulus, lb/in.², × 10^5		0.35	0.55	1.5	2.3	4.5	3.6	3.8	3.8	6.0	5.6	0.58
Elongation, %		800	600	100	700	80	100	80	100	40	3.5	400
Compressive strength, lb/in.²	D695	...	...	3,200	8,000	9,000	14,000	13,000	12,000	13,000	17,000	1,700
Flexural strength, lb/in.²	D790	...	7,000	1,000	8,000	10,000	15,400	15,000	14,000	16,000	19,000	
Impact strength, ft lb/in. of notch	D256	No break	>16	20	1.5	8	1.3	1.9	12	20	0.5	3.0
Hardness, Rockwell	D785	...	...	...	R110	R100	R120	R123	R123	...	M90	...
Thermal properties:												
Heat-distortion temp. at 264 lb/in.²	D648	105	120	120	145	205	...	375	...	164	215	>250
Maximum-use temp., °F		212	250	250	320	175	345	...	300	175	205	550
Coefficient of thermal expansion, °C⁻¹ × 10^{-5}	D696	18	16	13	10	21	6	3	6	18	8	10
Thermal conductivity, cal/S·cm·°C	C177	8.0	10.0	12.4	2.8	3.0	6.2	4.5	8.4	7.0	2.9	6
Flammability, in./min	D635	1.04	1.04	1.04	1.04	1.0	No burn	No burn	No burn	No burn	1.0	No burn
Chemical properties:												
Water absorption, % in 24 h	D570	<0.01	<0.01	<0.01	0.03	0.6	0.2	0.06	0.13	0.4	0.3	0.00

Not resistant to (ASTM D543)	Trade names
Oxidizing acids, aromatic and chlorinated solvents	Agilene, Althon, Bakelite, Cao X-L, Dylan, El-Rex, Epolene, Ethylux, Excelite, Fortiflex, Hi-Fax, Marlex, Modulene, Petrothene, Poly-Eith
...	Aetotut, Avison, Bakelite, Chevron, Escon, Marlex, Petrothene, Pro-Fax
Oxidizing acids, aromatic chlorinated solvents	Bakelite, Biax, Dylene, El-Rex, Exenglo, E-Z Flow, Fostarene, Gilco, Grace, Hypac, Kardel, Lustrex, Shell, Solar, Styrafil, Styroflex, Styroflux, Styron
Aromatic solvents	Bakelite
Aromatic solvents	PPO
Aromatic and chlorinated solvents	Bakelite
Ketones, esters, aromatic solvents	Bakelite, Biacar, Dacovin, Durelene, Esamgia, Ethyl, Excelon, Exon, Geon, Insular, Kenron, Kohinor, Marvinol, Nalgon, Opalon, Pliovic, Resinite, Rucoblend, Secron, Trulon, Tygon, Vicoa, Vygen, Vyran, Xyran
Acids, ketones, esters, chlorinated solvents	Acrylafl, Bakelite, Catalin, Kralac, Lustran, Plaxacrin, Tyril
	Halon, Teflon

Table 6-30. Properties of NEMA Copper-clad[95]

(See Table 6-30a)

Property	Conditioning procedure by ASTM Methods D618*	XXXP	XXXPC	FR-2	FR-3	FR-4	FR-5	G-10	G-11
Peel strength, min., lb/in. width									
1-oz copper:									
After solder dip	A	6	6	6	7	7	7	7	7
After elevated temp.	E-1/140†	6	6	6	7	7	7	7	7
2-oz copper:									
After solder dip	A	7	7	7	9	9	9	9	9
After elevated temp.	E-1/140†	7	7	7	9	9	9	9	9
Volume resistivity (min.), m$\Omega\cdot$cm	C-96/35/90	10,000	10,000	10,000	100,000	100,000	100,000	100,000	100,000
Surface resistance (min.), mΩ	C-96/35/90	1,000	1,000	1,000	1,000	1,000	1,000	5,000	5,000
Dielectric breakdown parallel to laminations, min., kV, step by step	D-48/50	15	15	15	30	30	30	30	30
Dielectric constant (avg. max.) at 1 MHz‡	D-48/50	5.3	5.3	5.3	5.0	5.8	5.8	5.8	5.8
Dissipation factor (avg. max.) at 1 MHz‡	D-48/50	0.05	0.05	0.05	0.045	0.045	0.045	0.045	0.045
Flexural strength (vg. min.), 16/in.²									
Lengthwise§	A	12,000	12,000	12,000	20,000	55,000	55,000	55,000	55,000
Crosswise§	A	10,500	10,500	10,500	16,000	45,000	45,000	45,000	45,000
Lengthwise¶						50,000	50,000	50,000	50,000
Crosswise¶						40,000	40,000	40,000	40,000
At elevated temp., % of condition A value retained, lengthwise:	E-1/150 T-150								
1/32-in. thickness							50		50
1/16- and 3/32-in. thicknesses							40		40
1/8- and 1/4-in. thicknesses							50		50
Flammability (avg. max.), seconds to extinguish	A		15	15	15	15	15		
Water absorption (avg. max.), %									
1/32-in. thickness		1.30	1.30	1.30		0.80	0.80	0.80	0.80
1/16-in. thickness		1.00	0.75	0.75		0.35	0.35	0.35	0.35
3/32-in. thickness		0.85	0.65	0.65	0.65	0.25	0.25	0.25	0.25
1/8-in. thickness		0.75	0.55	0.55	0.50	0.20	0.20	0.20	0.20

* Methods of Conditioning Plastics and Electrical Insulating Materials for Testing (ASTM Designation D618).
† For grades XXXP, XXXPC, FR-2, and FR-3, use condition E-1/120.
‡ Applies only to 3/32 and 1/8-in. thicknesses.
§ Applies only to 1/32 to 3/32-in. thicknesses.
¶ Applies only to 1/8 and 1/4-in. thicknesses.

Table 6-30a.   Description of NEMA grades listed in Table 6-30

NEMA grade	Resin	Reinforcement	Description
XXXP	Phenolic	Paper	Hot-punching grade for general use
XXXPC	Phenolic	Paper	Room-temperature punching for boards with small close holes
G-10	Epoxy	Glass cloth	General-purpose glass base; excellent electrical and physical properties; good moisture resistance
G-11	Epoxy	Glass cloth	Similar to G-10 but more thermally stable
FR-2	Phenolic	Paper	Similar to XXXPC but flame-retardant
FR-3	Epoxy	Paper	Room-temperature punching; flame-retardant; better physical properties than XXXPC
FR-4	Epoxy	Glass cloth	Similar to G-10 but flame-retardant
FR-5	Epoxy	Glass cloth	Similar to G-11 but flame-retardant

Synthetic-fiber papers are also used, having improved range of working temperatures and higher dielectric strengths.   Paper is also used with metallization for capacitors.

99. Impregnated Papers.   Liquid impregnation of papers considerably improves their dielectric strength and thermal stability.   The impregnating liquids are mineral oils or chlorinated hydrocarbons.   Solid impregnants such as wax or resins are also used.   Impregnated paper is widely used for capacitors and transformers.

100. Wood.   Wood is not commonly used as electrical insulation.   Some use remains as poles of electrical transmission lines.   Wood has a dielectric constant of 2.5 to 7.7 and resistivity of 10^{10} to 10^3 Ω·cm.   The dielectric strength is about 4×10^5 V/m.

101. Fabrics.   Both natural and synthetic organic fibers are used for insulation.   Inorganic fibers such as glass and asbestos are also used.   Fibers can be used in nonwoven form such as laminates.

Unimpregnated fibers are used in low-voltage applications such as thermocouple wires.   Impregnated fabrics are coated with elastomers and provide an improved dielectric strength and thermal endurance.   These flexible fabrics are used for cable, coil, or bus bars.

102. Mica.   Mica is available as muscovite or phlogopite.   The muscovite, or ruby mica, is $KAl_3Si_3O_{10}(OH)_2$, while the phlogopite, or amber mica, is $KMg_3AlSi_3O_{10}(OH)_2$.

Mica can be cleaved into very thin sheets.   Muscovite can be used up to about 500°C, and heat-treated phlogopite up to 800°C.   The resistivity of mica in the range 0 to 300°C is between 10^8 and 10^{14} Ω · m.   Dissipation factors for muscovite are 0.0001 to 0.0004 and for phlogopite 0.004 to 0.07 between 60 and 10^6 Hz.   The dielectric constant is about 6.5 to 8.7 for muscovite and 5 to 6 for phlogopite.   The dielectric strength is about 1 to 2×10^7 V/m for 0.0001- to 0.0003-in.-thick material in air.   Synthetic mica can be obtained as a compacted aggregate.   It has electrical properties similar to bulk muscovite, but can be heated to 1100 to 1200°C.   Mica paper, i.e., mica splittings, often oriented platelets, embedded in resins are very stable insulators.   Mica is used for capacitors interlayered with tin foil, silvered, or as mica paper.   Glass-bonded mica, Mycalex, for example, is a very stable insulator, possessing good arc resistance and high dielectric strength (2×10^7 V/m).

103. Ceramics.   Ceramics are extensively used as electric insulators or capacitors in the electrical and electronics industries.   The uses of ceramics are summarized in Table 6-34.   Melting points of some ceramics are given in Fig. 6-24.   A more detailed description of representative ceramics is given below.

Porcelain.   Most porcelain insulators consist of mullite and quartz embedded in a glassy matrix.   High-tension porcelain is completely vitrified, normally covered with a glaze to improve mechanical properties and corona resistance.

Porcelain can be formed in the soft state, then dried, glazed, and fired.   It is used for low-frequency application since it has a high dielectric loss factor due to its lossy glass content.   Porcelains have relatively low electrical resistivities due to their high content of mobile alkali ions.   Properties of porcelain are given in Table 6-35.

Steatite.   Steatite ceramic bodies are low-dielectric-loss porcelains.   Steatite is used for variable capacitors, switches, coil forms, resistor shafts, spacers, and bushings.   Steatite is mainly used for low-tension, high-frequency applications.   Natural steatite talc is often referred to as *lava*.   It is readily machinable in the soft state and can subsequently be fired to form a hard product.   Properties of steatite are given in Table 6-35.

Table 6-31. Insulating Films[118]

Film base	ASTM No.	Cellulose acetate	Cellulose triacetate	Polytrifluoro chloroethylene copolymers	Polyurethane elastomer	FEP-fluorocarbon	Polyethylene	Polyvinyl chloride	Ionomer	Polyimide
Method of processing	...	Cast, extruded	Cast	Extruded	Cast, calendered extruded	Extruded	Extruded	Cast, calendered extruded	Extruded	
Forms available	...	Rolls, sheets, tapes	Rolls, sheets	Rolls	Rolls, sheets	Rolls, sheets, tapes	Rolls, sheets, tapes	Rolls, sheets, tapes	Rolls, sheets, tapes	Rolls
Thickness range, in.	...	0.0005-0.250	0.0008-0.020	0.0005-0.030	0.005-0.030	0.0005-0.020	0.00075-0.010	0.00075-0.08	0.001-0.010	0.001-0.005
Maximum width, in.	...	60	45	54	52	48	240	84	60	18
Area factor in.²/lb·mil	...	22,000	21,400	13,000	22,000	12,900	29,500	23,000	29,400	19,400
Specific gravity $kg/m^3 \times 10^{-3}$	D792-60T	1.31	1.31	2.15	1.26	2.15	0.940	1.50	0.94	1.42
Tensile strength, lb/in.²	D882-61T	16,400	16,000	8,000	9,000	3,000	3,500	5,600	5,000	25,000
Elongation, %	D882-61T	70	40	150	650	300	650	500	450	70
Bursting strength, 1 mil thickness, Mullen points	D774-63T	40	70	42	...	11	...	20	...	75
Tearing strength, g	D689-62	10	10	26	...	125	300	1,400	70	...
Tearing strength, lb/in.	D689-62	415	395	900	600	600	...	490	...	8
Folding endurance, cycles	D643-63T	2,000	4,000	...	...	4,000	>100,000	...	...	10,000
Water absorption, % in 24 h	D570-63	9	2-4.5	0.00	0.55-0.77	<0.01	<0.01	...	0.4	...
Dielectric constant, 10³ Hz	D1531-62	3.6	3.2	2.7	7.5	2.0	2.2	6.0	2.4	2.9
Dielectric constant, 10⁶ Hz	D1531-62	3.2	3.3	2.4	7.1	2.0	2.2	4.0	2.4	3.5
Dielectric constant, 10⁹ Hz	D1531-62	3.2	3.2	2.3	...	...	2.2	...	2.4	3.4
Dissipation factor, 10³ Hz	D1531-62	0.10	0.10	0.027	0.060	0.0002	0.0003	0.16	0.007	0.003
Dissipation factor, 10⁶ Hz	D1531-62	0.10	0.10	0.017	...	0.007	0.0003	0.14	0.007	0.010
Dissipation factor, 10⁹ Hz	D1531-62	0.094	0.094	0.004	...	...	0.0003	...	0.007	...
Dielectric strength, V/mil	...	5,000	3,700	...	500	5,000	500	600	1,000	7,000
Volume resistivity, Ω-cm	...	10¹³	10¹⁵	10¹⁸	10¹¹	10¹⁹	10¹⁶	10¹⁴	10¹⁶	10¹⁸
Orientation possible	...	No	No	No	Yes	Yes	Yes	Yes	Yes	No

Trade names	...	Rexfilm Celanese Kodacel	Kel-F Polyfluoron	Adiprene Estane Perflex Texin	Teflon	Arnar Amerifilm Ampacet Auburn Boltaron Clysar Conolean Crown-Seal Dynafilm Ger-Pae Hypac Katharon Midlon Plicose Polython Prohi Seilon Vis-Queen Zee Zendel	Agilide Amerifilm Arnar Boltaflex Cadeo Clopane Colovin Dorn Fabray Fabtex Ger-Pac Katharon Koroseal Krene Monosol Pantex Randfilm Resinite Reynolon Ruccam Saplasco Touchstone Ultron Velon Vylene Wataseal	Surlyn	Kapton
	Auburn Acelon Bexfilm Bexoid Cadeo Campco Inceloid Kodacel Lumarith Midlon Saplasco Tenite Vuepak								

6-61

Table 6-31. Insulating Films (concluded)

Film base	ASTM No.	Polypropylene	Polyvinyl fluoride	Polyester	TFE tetra-fluoroethylene	Vinylidene chloride	Fluoro-halocarbon	Vinylidene fluoride	Polyamide
Method of processing	...	Extruded	Extruded	Cast	Skived, extruded	Extruded, cast	Extruded	Extruded	Extruded, cast
Forms available	...	Rolls, sheets, tapes	Rolls	Rolls, sheets, tapes	Sheets, tapes	Rolls, tapes	Rolls	Rolls	Rolls
Thickness range, in.	...	0.0005–0.020	0.0005–0.02	0.00015–0.014	0.0005–0.01	0.0004–0.02	0.005–0.02	0.001–0.010	0.0005–0.02
Maximum width, in.	...	60	138	120	38	40	54	16	54
Area factor in.²/lb·mil	...	31,300	17,200	22,600	12,800	16,800	13,000	15,000	24,500
Specific gravity kg/m³ × 10⁻³	D792-60T	0.9	1.38	1.4	2.2	1.64	2.2	1.76	1.13
Tensile strength, lb/in.²	D882-61T	10,000	18,000	30,000	4,000	20,000	11,000	6,500	11,500
Elongation, %	D882-61T	1,000	250	120	350	80	150	300	400
Bursting strength, 1 mil thickness, Mullen points	D774-63T	...	70	66	...	...	31	20	90
Tearing strength, g	D689-62	45	40	15	100	...	150	60	1,200
Tearing strength, lb/in.	D689-62	...	1,400	1,300	...	...	29	...	...
Folding endurance, cycles	D643-63T	>100,000	47,000	14,000	...	50,000	>100,000	75,000	>250,000
Water absorption, % in 24 h	D570-63	0.005	0.5	0.8	0.00	0.00	0.00	0.04	9.5
Dielectric constant, 10³ Hz	D1531-62	2.1	8.5	3.1	2.1	6.0	2.6	7.72	3.8
Dielectric constant, 10⁶ Hz	D1531-62	2.1	7.4	3.0	2.1	5.0	2.3	6.43	3.7
Dielectric constant, 10⁹ Hz	D1531-62	2.1	...	2.8	2.1	4.0	...	2.98	3.4
Dissipation factor, 10³ Hz	D1531-62	0.0003	0.009	0.0047	0.0002	0.045	0.039	0.019	0.010
Dissipation factor, 10⁶ Hz	D1531-62	0.0003	...	0.016	0.0002	0.075	0.037	0.159	0.016
Dissipation factor, 10⁹ Hz	D1531-62	0.0003	...	0.003	0.0002	0.08	...	0.11	0.025
Dielectric strength, V/mil	...	3,000	...	7,000	430	5,000	3,000	1,280	1,700
Volume resistivity, Ω·cm	...	10¹⁶	10¹³	10¹⁸	10¹⁸	10¹⁶	10¹⁷	10¹⁴	...
Orientation possible	...	Yes	No	Yes	No	Yes	No	No	No
Trade names	...	Bexphone Clysar Dynafilm Electro-film Hypac Midlon Olefane Profax Propylux Udel Vypro	Tedlar	Celenar Kodar Mylar Scotchpar Scotchpack Videne	Halon Teflon	Cryovac Saran	Aclar	Kynar	Cadco Califilm Capran Plastex Sapalaso

Cordierite.   Cordierite has a low thermal expansion coefficient and a high resistance to thermal shock.   Cordierite is used as electrical insulation where thermal stresses are important, as in heating apparatus, thermocouple insulators, etc.   See Table 6-35 for physical properties.

Forsterite.   The main advantages of forsterite are its ease of firing, low dielectric loss at high frequencies, and high resistivity.   Forsterite is used in small microwave tubes.   It has a high thermal expansion coefficient and poor thermal shock resistance.   See Table 6-35 for physical properties.

Alumina (Al_2O_3).   Alumina ceramics are composed of fine crystalline Al_2O_3 particles and a glass matrix.   High-purity alumina can be $>99.9\%$ Al_2O_3. Al_2O_3 can also be found as single crystals, termed sapphire.

Alumina ceramics have a high mechanical and dielectric strength, as well as a high resistivity and low dielectric loss.   Alumina ceramics are very stable chemically, retaining their properties over a wide temperature and frequency range.   The properties of alumina are affected by its purity.   For example, ionic impurities cause higher dielectric losses in impure alumina.   Porosity causes a pronounced degradation in thermal conductivity.   High-purity, high-density alumina can be obtained by sintering fine-grained powders at high temperatures.

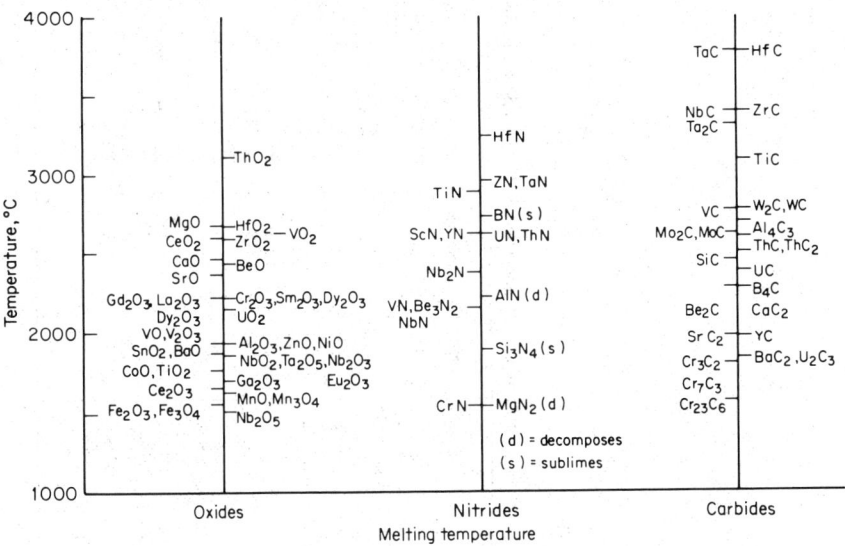

Fig. 6-24.   Melting points of oxides, nitrides, and carbides.[77]

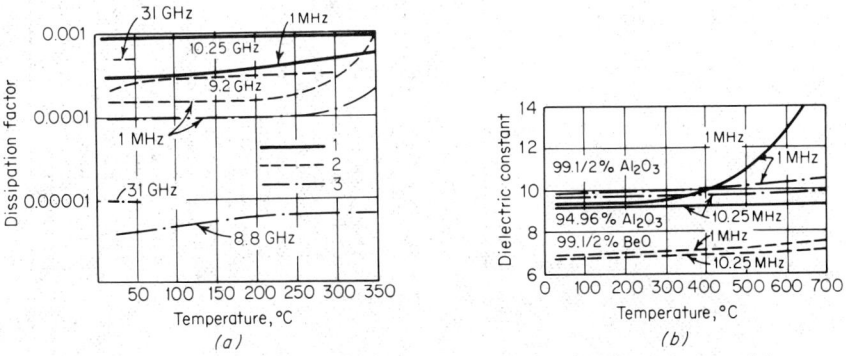

Fig. 6-25.   Dielectric constant (b) and dissipation factor (a) as a function of temperature and frequency for alumina and beryllia.[78]   (1) 94.9% alumina; (2) 99.5% alumina; (3) 99.5% beryllia.

Table 6-32. Properties of Rigid Plastic Foams

Type	ASTM No.	ABS	Cellulose acetate	Epoxy	Syntactic epoxy*		Phenolic	Polyethylene	Polypropylene
Density per cubic foot		31	6-8	5-8	36	42	2-4	34	5
Ther. cond. Btu/hr-ft2-°F-in.	177	0.58	0.31-0.32	0.24-0.28	4.56		0.20-0.22	0.92	0.27
Coef. of ther. exp. per °F × 10^{-5}	D696	9.7	2.5	2.3	4.5		0.5	4.18	
Water absorption, % vol	C272	0.6	13-17			1.5	<3	0.22	
Flammability, in./min	D1692	1.04	4.9				SE		1.6
Dielectric consta. at 10^6 Hz.		1.59	1.10-1.12	2.0	1.55			1.48	
Dissipation factor at 10^6 Hz		0.007	0.003	0.005	0.01			0.0003	
Max. rec. service temp. °F		200	350	400-500	300	300	270	195	230
Tensile strength, lb/in.2	D1623	1400	170	50-200	3300	4600	20-55	1000	170
Ultimate tension elong., %	D1623								
Mod. of elast. in tension, 1,000 lb/in.2	D1623	2.4				610			
Compr. strength lb/in.2 (10%)	D1621		125-150	60-90	9600	13,400	20-90	800	55
Mod. of elast. in compr., 1000 lb/in.2	D1621		5.5-13	2.1-6.5	373	480	25-65		1.2
Flex. strength, psi	D790	2.4	150	200-800	3800	6000		1900	230
Mod. of elast. in flex., 1,000 lb/in.2	D790	9	5.5	2.5-6				88	9.6
Shear strength	C273		140		3800	4400	15-30		
Mod. of elast. in shear, 1,000 lb/in.2	C273						0.4-0.75		
Hardness (Shore D)		60			80-85				

* Glass-microsphere-filled epoxy.

Type	ASTM No.	Polyvinyl chloride	Polystyrene (expanded)		Urea	Urethane		
Density per cubic foot		3	2	6	0.8–1.2	2–3	4–7	18–25
Ther. cond., Btu/h·ft.²·°F·in.	177	0.15–0.20	0.20–0.28	0.20–0.25	0.18–0.21	0.11–0.23	0.15–0.28	0.29–0.52
Coef. of ther. exp. per °F × 10⁻⁵	D696	2	2.7–4	2.7–4	…	3–4	4	4
Water absorption, % vol	C272	0.1	<0.1	<0.1	…	3–4	1.5–2	0.2
Flammability, in./min	D1692	SE-NB	2–8	2–8	SE	NB	NB	NB
Dielectric const. at 10^6 Hz	…	…	1.02–1.24	…	…	…	…	…
Dissipation factor at 10^6 Hz	…	…	<0.0005	…	…	…	…	…
Max. rec. service temp. °F	…	180	175	175	120	200–250	250–300	300–400
Tensile strength, lb/in.²	D1623	100–200	50–55	120	…	20–70	90–250	700–1300
Ultimate tension elong., %	D1623	5–20	5	2	…	…	…	…
Mod. of elast. in tension, 1,000 lb/in.²	D1623	3–4	740	6100	5	20–50	65–275	1200–2000
Compr. strength lb/in.² (10%)	D1621	70–100	25–30	100–150	…	0.3–0.6	1.5–4.5	10–40
Mod. of elast. in compr., 1,000 lb/in.²	D1621	3–4	0.55–2	3–6	17	60–100	200–350	700–2000
Flex. strength, lb/in.²	D790	120–160	55–75	200–300	0.7	0.8–0.9	0.8–15	12–100
Mod. of elast. in flex., 1,000 lb/in.²	D790	3–4	1.3–3.8	5–15	…	20–30	60–130	7600
Shear strength, lb/in.²	C273	60–80	35	150	…	20–30	…	…
Mod. of elast. in shear, 1,000 lb/in.²	C273	2–2.5	1.15–1.6	3	…	0.17–0.21	0.5–1.5	3–9

Alumina can be metallized by vacuum deposition or by screen printing metal oxides, and firing. Alumina is used for circuit breakers, spark plugs, resistor cores, substrates for hybrid microelectronic circuits, microwave integrated circuits, integrated circuits, and power transistors. See Table 6-35 and Fig. 6-25 for properties of alumina. Sapphires are used as substrates where high heat dissipation and a very flat surface are needed.

Beryllia (BeO). Beryllia is a unique material having very high heat conductivity with very low electrical conductivity. At room temperature BeO thermal conductivity is about that of aluminum (Fig. 6-20). BeO has good thermal shock resistance and good mechanical characteristics.

BeO is used where high heat dissipation is essential, as in high-power transistor bases, microwave windows, high-power klystrons, and lasers. Inhalation of BeO powder is hazardous. Properties are given in Fig. 6-25.

Magnesia (MgO). MgO has poor thermal shock resistance but shows little electrical conduction at high temperatures. MgO has found extensive use as a basic refractory in steelmaking, while in electronics it is used for heaters and thermocouple insulators.

Zirconia (ZrO_2). Zirconia has poor thermal conductivity and shock resistance. It is usually available with yttrium or calcium, which stabilize the zirconia. Changes in electrical conduction of zirconia have been used to measure oxygen partial pressure. ZrO_2-MgO combinations have been used for high-temperature heating elements.

Carbides. Silicon carbide is the most widely known carbide. SiC forms a protective-surface oxide layer which permits its use up to 1700°C in oxidizing and 2200°C in inert atmosphere. SiC has an electrical resistivity of about $10^{-3}\ \Omega \cdot m$ at room temperature. SiC semiconductor devices capable of performing up to 500°C have been made. The main uses of SiC are in cutting tools and heating elements.

Nitrides. Boron nitride is used as a high-temperature insulator. It is readily machinable and needs no firing. BN absorbs water and must therefore be kept dry. BN powder can be used up to 3000°C in nitrogen.

Silicon nitride is used in the integrated circuits primarily as a sodium diffusion barrier on the SiO_2. It has been used for double dielectric memories (MNOS). Si_3N_4 has an electrical resistivity of $10^{10}\ \Omega \cdot m$ at 200°C. Films can be formed by chemical vapor deposition, sputtering, and vapor deposition. Such films have been used as oxidation masks in integrated circuits.

104. Ferroelectric Ceramics. Ferroelectric materials have a **D-E** curve similar to the **B-H** hysteresis curve in ferromagnetic materials. This nonlinear nature of the induction originates from domain structure in the ferroelectrics. Each domain consists of parallel dipoles. When the field is applied, favorable domains grow at the expense of their neighbors, and very high polarization can be obtained with low fields. Ferroelectrics have a Curie temperature (1200°C for $BaTiO_3$) above which the ferroelectric behavior vanishes. The main materials used as ferroelectric ceramics are titanates, zirconates, niobates, tantalates, and stannates. Properties of some ferroelectrics are listed in Table 6-36.

Ceramic capacitors can be divided into two classes:

Class I. Capacitors which have a definite temperature coefficient of capacitance. The dielectric constants of such capacitors (usually containing TiO_2) are up to 500. The capacitors have loss factors up to 0.004. Class I capacitors are suited for resonator or oscillator circuits where good capacitance stability is required.

Class II. Capacitors which are highly nonlinear materials ($BaTiO_3$, for example) having very high dielectric constants (typically 500 to 10,000). These capacitors are not stable and have high losses. Class II capacitors are used in coupling, filtering, or bypass applications where capacitance stability is not important but miniaturization is essential.

105. Piezoelectricity. Piezoelectricity is the appearance of electric polarization with applied stress. Piezoelectric materials are noncentrosymmetric. Quartz, $BaTiO_3$, and rochelle salt are common piezoelectric materials. Piezoelectrics are used for ultrasonic cleaning, delay lines, strain gages, thickness monitors, accelerometers, etc. Application of electric fields causes various strains in crystals; such an effect is called *inverse piezoelectricity.*

The appearance of polarization can also come about by thermal expansion. Such an effect is called *pyroelectricity.*

106. Electrolytic Films. Electrolytic capacitors are formed by anodizing aluminum or tantalum foils to form Al_2O_3 or Ta_2O_5 dielectric films. Electrolytically formed oxide is between 0.01 and 1.0 μm thick. Tantalum oxide capacitors are used also in thin-film hybrid circuits where vapor-deposited tantalum is anodized to form the oxide on its surface.[30] The

Table 6-33.   Applications of Organic Coatings[125]

Coating	Distinguishing Characteristics (Max. Continuous-Use Temp., °F)
Thermosetting coatings:	
Acrylics	Excellent resistance to UV and weathering. (250)
Alkyds	Most widely used general-purpose coating. (200–250)
Epoxies	Excellent chemical resistance and good insulation coating. Widely used as circuit-board coatings. (400)
Phenolics	Good chemical resistance against alkalies and good insulation coating. (350–400)
Phenolic-oil varnishes	Electrical impregnating varnishes. (250)
Polybutadienes	Good electrical insulation properties. (450)
Polyesters	Good electrical insulation properties. (200)
Silicones	Excellent high-temperature electrical insulation properties. (500)
Thermoplastic coatings:	
Acrylics	Excellent resistance to UV and weathering. (180)
Cellulosics	Fast-dry commercial lacquers. (180)
Fluorocarbons	Excellent chemical resistance, excellent electrical insulation properties, even at high temperatures. (400–500)
Penton chlorinated polyethers	Primarily used as chemically resistant coatings for equipment. Coatings have good electrical properties, however. (250)
Phenoxies	Low coefficient of expansion coatings. (180)
Polyimides	High-temperature coating having excellent insulation properties. Widely used as high-temperature magnet wire enamel. (600–700)
Polyurethanes	Good electrical insulation coatings. Widely used as circuit-board coatings. (250)
Vinyls	There are many varieties of vinyl coatings. Some vinyls are outstanding in resistance to inorganic and plating chemicals. Vinyl plastisols are convenient for dip coating of electrical parts. (150)

* In addition to the above basic types of coatings, there are many specialty coatings, such as ablative coatings, thermal control coatings, flame-retardant coatings, fungus-resistant coatings, electrically conductive coatings, and magnetic coatings. In addition, many elastomers can be applied as coatings (e.g., neoprene). Most of the specialty coatings are specifically filled or otherwise modified variations of basic thermosetting, thermoplastic, or elastomeric coating polymers. Hence, the possible coatings available are as unlimited as the broad range of plastics and polymers.

Table 6-34.   Uses of Technical Ceramics[61]

Applications	Types of Ceramics
Electrical	
Low- and high-tension insulation	
Domestic fittings, power generation, and transmission	Electrical porcelain, zircon porcelain
Insulation at elevated temperatures	
Insulation for electric fires, ovens, and low-temperature kilns	Aluminous porcelain, cordierite
Insulation for fireproof cables	Magnesia
Insulation at high temperatures	
Thermocouple sheaths, furnace muffles, various furnace parts	Mullite, fused silica, alumina
Sparking plug insulators	Alumina
Electronic	
High-frequency insulation	
Rods, tubes, plates, coil formers, valve parts	Steatite, zircon porcelain, alumina
Capacitor dielectrics	
Trimmer capacitors	Steatite
High-permittivity dielectrics	Rutile, titanates
Transmitter capacitor dielectrics	Rutile, magnesium titanate
"Nonlinear" dielectrics	
Dielectric amplifiers, memory units, accelerometers, electromechanical transducers (e.g., gramophone pickup crystals, ultrasonic generators)	Barium titanate, lead zirconate titanate
Insulation and good heat conduction	
Transistors and integrated-circuit packages and hybrid-circuit substrates	Alumina, beryllia

Table 6-35. Some Typical Physical Properties of Ceramic Dielectrics[74]

Material	Vitrified products			
	High-voltage Porcelain	Alumina porcelain	Steatite	Forsterite
Typical applications	Power-line insulation	Spark plug cores thermocouple insulation, protection tubes	High-frequency insulation, electrical appliance insulation	High-frequency insulation, ceramic-to-metal seals
Specific gravity, g/cm³	2.3-2.5	3.1-3.9	2.5-2.7	2.7-2.9
Water absorption, %	0.0	0.0	0.0	0.0
Coefficient of linear thermal expansion/°C (20-700)	$5.0\text{-}6.8 \times 10^{-6}$	$5.5\text{-}8.1 \times 10^{-6}$	$8.6\text{-}10.5 \times 10^{-6}$	11×10^{-6}
Safe operating temperature, °C	1000	1350-1500	1000-1100	1000-1100
Thermal conductivity, cal/·cm² cm·s·°C	0.002-0.005	0.007-0.05	0.005-0.006	0.005-0.010
Tensile strength, lb/in.²	3000-8000	8000-30,000	8000-10,000	8000-10,000
Compressive strength, lb/in.²	25,000-50,000	80,000-250,000	65,000-130,000	60,000-100,000
Flexural strength, lb/in.²	9000-15,000	20,000-45,000	16,000-24,000	18,000-20,000
Impact strength, ft-lb; ½-in. rod	0.2-0.3	0.5-0.7	0.3-0.4	0.03-0.04
Modulus of elasticity, lb/in.²	$7\text{-}14 \times 10^6$	$15\text{-}52 \times 10^6$	$13\text{-}15 \times 10^6$	$13\text{-}15 \times 10^6$
Thermal shock resistance	Moderately good	Excellent	Moderate	Poor
Dielectric strength, V/mil; ¼-in.-thick specimen	250-400	250-400	200-350	200-300
Resistivity Ω·/cm³, at room temp	$10^{12}\text{-}10^{14}$	$10^{14}\text{-}10^{15}$	$10^{13}\text{-}10^{15}$	$10^{13}\text{-}10^{15}$
Te value, °C	200-500	500-800	450-1000	Above 1000
Power factor at 1 MHz	0.006-0.010	0.001-0.002	0.0008-0.0035	0.0003
Dielectric constant	6.0-7.0	8-9	5.5-7.5	6.2
L-grade (JAN Spec. T-10)	L-2	L-2-L-5	L-3-L-5	L-6

oxide forms at 16 Å/V. Formation voltages range between 30 and 250 V. The capacitance density is 0.004 to 0.0005 F/m², and the dielectric constant is 21.2. The temperature coefficient of capacitance can be up to 250 ppm/°C. Dissipation factor ranges from 0.002 to 0.01 at frequencies less than 10 kHz. Leakage currents are between 0.01 and 0.1 A/F. Manganese oxide (MnO_2) is used on Ta_2O_5 capacitors as a protective layer under the counterelectrodes. This layer permits large-area capacitors to be used. Electrolytic SiO_2 can be formed on silicon semiconductor devices as a dielectric. However, the anodic silicon oxide has not found a wide application in the microelectronic industry.

107. Glass. Glass is widely used in the electronic industry as a structural member, such as tube envelopes, hermetic seals to metal or ceramic, protective coating on hybrid and integrated circuits, insulating layers and crossovers in microelectronics, and capacitors. Semiconducting chalcogenide glasses have been reported as suitable for switching and memory application.

Glass is usually a noncrystalline solid having no long-range order, a state referred to as *vitreous*. Glasses have a very pronounced short-range order, for example, in vitreous SiO_2, where most silicon atoms have four oxygen neighboring atoms at distances typical to silicon-oxygen bond length.[31] Glasses can be devitrified (recrystallized) in order to modify their properties.

Most glasses are composed of a variety of oxides with hundreds of glass compositions available. Table 6-37 lists the composition of some typical glasses. Common additions to glass are alkali oxides to lower the melting point. These oxides add an undesirable ionic conductivity. In this respect, the potassium, with the larger ionic radius, is less detrimental than sodium. Alkali-earth oxides improve the weathering resistance of low-melting glasses. Aluminum is added to increase strength; PbO lowers the melting point; B_2O_3 lowers the melting point and results in chemical resistance without affecting the electrical resistivity. Volume resistivity of glass decreases with temperature (see Fig. 6-26). Surface conductivity of glass becomes pronounced above about 50% relative humidity (see Fig. 6-18). Chemical reduction of the lead in lead glasses can result in an electrically conductive layer on the surface of the glass. This occurrence has to be avoided when lead solder glasses are used as insulators or sealing glasses.

The dielectric strength of glasses is a function of thickness, temperature, time, frequency, and testing medium. The dielectric constants of glasses vary from 3.8 for fused silica to 2,000

Table 6-35. Some Typical Physical Properties of Ceramic Dielectrics[74] *(continued)*

			Semivitreous and refractory products			
Zircon porcelain	Lithia porcelain	Titania, titanate ceramics	Low-voltage porcelain	Cordierite refractories	Alumina, aluminum silicate refractories	Massive fired talc, pyrophyllite
Spark plug cores, High-voltage high-tempera- ture insulation	Temperature, stable induct- ances, heat- resistant insulation	Ceramic capacitors, piezoelectric ceramics	Switch bases, low-voltage wire holders, light receptacles	Resistor sup- ports, burner tips, heat insulation, arc chambers	Vacuum spacers, high- temperature insulation	High-frequency insulation, vacuum tube spacers, ceramic models
3.5-3.8	2.34	3.5-5.5	2.2-2.4	1.6-2.1	2.2-2.4	2.3-2.8
0.0	0.0	0.0	0.5-2.0	5.0-15.0	10.0-20.0	1.0-3.0
$3.5\text{-}5.5 \times 10^{-6}$	1×10^{-6}	$7.0\text{-}10.0 \times 10^{-6}$	$5.0\text{-}6.5 \times 10^{-6}$	$2.5\text{-}3.0 \times 10^{-6}$	$5.0\text{-}7.0 \times 10^{-6}$	11.5×10^{-6}
1000-1200	1000	...	900	1250	1300-1700	1200
0.010-0.015	...	0.008-0.01	0.004-0.005	0.003-0.004	0.004-0.005	0.003-0.005
10,000-15,000	...	4000-10,000	1500-2500	1000-3500	700-3000	2500
80,000-150,000	60,000	40,000-120,000	25,000-50,000	20,000-45,000	15,000-60,000	20,000-30,000
20,000-35,000	8000	10,000-22,000	3500-6000	1500-7000	1500-6000	7000-9000
0.4-0.5	0.3	0.3-0.5	0.2-0.3	0.2-0.25	0.17-0.25	0.2-0.3
$20\text{-}30 \times 10^6$	...	$10\text{-}15 \times 10^6$	$7\text{-}10 \times 10^6$	$2\text{-}5 \times 10^6$	$2\text{-}5 \times 10^6$	$4\text{-}5 \times 10^6$
Good	Excellent	Poor	Moderate	Excellent	Excellent	Good
250-350	200-300	50-300	40-100	40-100	40-100	80-100
$10^{13}\text{-}10^{15}$	...	$10^8\text{-}10^{15}$	$10^{12}\text{-}10^{14}$	$10^{12}\text{-}10^{14}$	$10^{12}\text{-}10^{14}$	$10^{12}\text{-}10^{15}$
700-900	...	200-400	300-400	400-700	400-700	600-900
0.0006-0.0020	0.05	0.0002-0.050	0.010-0.020	0.004-0.010	0.0002-0.010	0.0008-0.010
8.0-9.0	5.6	15-10,000	6.0-7.0	4.5-5.5	4.5-6.5	5.0-6.0
L-4	L-3					

for glasses containing ferroelectric materials. Typical dielectric properties of glasses are given in Table 6-38 and Fig. 6-27.

108. Fused Quartz and Fused Silica. Fused quartz and fused silica are vitreous forms of SiO_2. Fused quartz is prepared from rock crystal or white quartzite sand, and fused silica (translucent fused quartz) is made from white silica sand. Fused quartz and silica have very low expansion coefficients and poor heat conduction (see Fig. 6-20). They are mechanically strong and have a high softening point.

Vitreous SiO_2 is used extensively in thin-film form in the microelectronics industry. Thermally oxidized silicon provides a thin (1-μm) layer of SiO_2 which is used as insulation and as diffusion barriers. Sputtered SiO_2 has been used as dielectric layer for two-level metallizations. Chemically vapor deposited SiO_2 is also used as protective layers. Properties of fused silica are given in Table 6-38 and Fig. 6-27.

109. Glass Ceramics. Glass ceramics are glasses which are devitrified about 100°C below their softening point to form a very fine network of crystalline phase. The glass

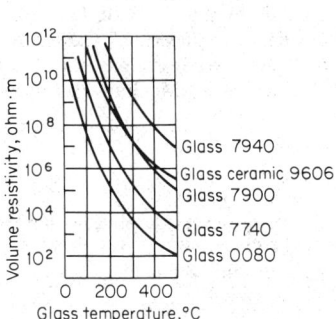

Fig. 6-26. Resistivity of several glasses as a function of temperature.[79]

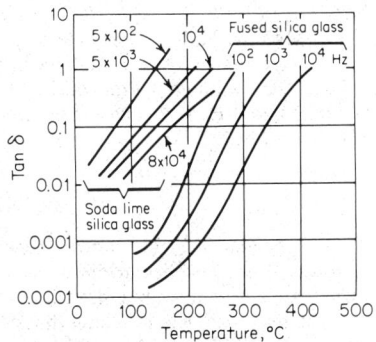

Fig. 6-27. Tan δ as a function of frequency and temperature for two glasses.[74]

Table 6-36.   The Curie Temperature T_c and the Spontaneous Polarization P_s of Several Ferroelectrics[127]*

		T_c, K	P_s in esu/cm², at T K	
KDP type	KH_2PO_4	123	16,000	[96]
	KD_2PO_4	213	13,500	
	RbH_2PO_4	147	16,800	[90]
	RbH_2AsO_4	111		
	KH_2AsO_4	96	15,000	[80]
	KD_2AsO_4	162		
	CsH_2AsO_4	143		
	CsD_2AsO_4	212		
TGS type.....	Triglycine sulfate	322	8,400	[293]
	Triglycine selenate	295	9,600	[273]
Perovskites ...	$BaTiO_3$	393	78,000	[296]
	$SrTiO_3$†	32	(9,000)	[4]
	WO_3	223		
	$KNbO_3$	712	90,000	[523]
	$PbTiO_3$	763	>150,000	[300]
	$LiTaO_3$	. . .	70,000	[720]
	$LiNbO_3$	1470	900,000	

* A table of 76 ferroelectric crystals (excluding solutions) given in F. Jona and G. Shirane, *"Ferroelectric crystals,"* Pergamon Press, New York, 1962, app. A. This reference is an excellent source of data on the crystal structure of ferroelectrics. For further data, see E. Nakumura, T. Mitsui, and J. Furuichi, *J. Phys. Soc. Japan,* Vol. 18, p. 1477, 1963.
† It is not at all certain that $SrTiO_3$ has a ferroelectric phase; it may be paraelectric down to at least 1 K. Possibly a ferroelectric phase can be induced by an electric field at low temperatures.

ceramics can be shaped at the soft state and then hardened by crystallization. They are mechanically strong and have a low thermal expansion coefficient, and therefore have a high thermal shock resistance. Glass ceramics retain their properties at higher temperatures than glasses. Properties of a glass ceramic, Pyroceram,[1] are given in Table 6-38.

110. Thick-Film Dielectrics. The dielectric films can be divided into crossover insulators, capacitors, overglazes, and solder-glass pastes. These films are formed as a screenable paste which are screened on substrates and subsequently fired to leave the desired dielectric layer. Table 6-39 lists some of the available pastes and their properties.

111. Bibliography on Dielectric Materials

1. HARPER, C. A. (ed.)   "Handbook of Electronic Packaging;" New York, McGraw-Hill Book Company, 1969.

2. KOHL, W. H.   "Handbook of Materials and Techniques for Vacuum Devices;" New York, Reinhold Book Corporation, 1967.

3. KINGERY, W. D.   "Introduction to Ceramics;" New York, John Wiley & Sons, Inc., 1960.

4. MEGAW, H. D.   "Ferroelectricity in Crystals;" London, Methuen & Co., Ltd., 1957.

5. VON HIPPEL, A. R.   "Molecular Science and Molecular Engineering;" Cambridge, Mass., The Technology Press of the M.I.T., and New York, John Wiley & Sons, Inc., 1959.

6. SHAND, E. B.   "Glass Engineering Handbook;" New York, McGraw-Hill Book Company, 1958.

MAGNETIC MATERIALS

112. Magnetic Field. Forces beyond electrostatic forces are exerted between moving charges. It is convenient to speak of a magnetic field set up by the moving charge, and forces which appear on a second moving charge when it traverses through the field.

The magnetic field in a long thin coil is

$$H = n \cdot I \tag{6-49}$$

---

[1] Trademark of Corning Glass Works, Corning, N.Y.

Table 6-37. Typical Composition of Some Glasses

Type	Corning No.	Constituent oxides, wt %									Thermal expansion coefficient 10^7, $1/°C$
		SiO_2	B_2O_3	Na_2O	K_2O	CaO	MgO	Al_2O_3	PbO	Others	
Fused silica	....	99.5									5.5
Silica glass	7900	96.3	2.9	0.2	0.2			0.4			8.0
Soda lime	0080	73.6		16.0	0.6	5.2	3.6	1.0			92
Lead silicate	0010	63.0	0.2	7.6	6.0	0.3	0.2	0.6	21.0	SO_3:0.2	93
Borosilicate	7740	80.5	12.9	3.8	0.4			2.2			33

6-71

Table 6-38. Comparison of Properties of Pyroceram, Glass, and Ceramic[128]

	Pyroceram				Glass				Ceramic		
	9605	9606	9607	9608	Fused silica 7940	Vycor 7900	Pyrex 7740	Lime glass 0080	High-purity alumina (93%+)	Steatites MgO-SiO$_2$	Forsterite 2MgO-SiO$_2$
Specific gravity 25°C	2.62	2.61	2.40	2.50	2.20	2.18	2.23	2.47	3.6	2.65–2.92	2.8
Water absorption, %	0.00	0.00	0.00	0.00	0.00	0.00	0.00	0.00	0.00	0–0.03	0–0.01
Gas permeability	0	0	0	0	0	0	0	0.00	0		0
Thermal											
Softening temp., °C	1350	1260	...	1250	1584	1500	820	696	1700	1349	1349
Specific heat, 25°	0.185	0.185	...	0.190	0.176	0.178	0.186	0.200	0.181		
Specific heat mean, 25–400°C	0.230	0.230	...	0.235	0.223	0.224	0.233	0.235	0.241		
Thermal conductivity, cgs, 25°C mean temp.	0.0100	0.0087	...	0.0047	0.0028		0.0026		0.052–0.058	0.0062–0.0065	0.010
Linear coeff. of thermal expansion × 10^7, 25–300°C	14	57	−7	4 to 20	5.5	8	32	92	73 (20–500°C)	81.5–99 (20–500°C)	(20–500°C)
Mechanical											
Modulus of elasticity × 10^{-6}, lb/in.^{2}	19.8	17.3	...	12.5	10.5	9.6	9.5	10.2	40	15	...
Poisson's ratio	...	0.245	...	0.25	0.17	0.17	0.20	0.24	0.32	...	...
Modulus of rupture (abraded) × 10^{-3}, lb/in.^{2}	...	20	...	16–23	...	5–9	6–10	...	40–50	20[3]	19[3]
Hardness: Knoop 100 g	..	698	...	703	...	532	481	...	1880	...	...
500 g	720	619	...	588	...	477	442	...	1530	...	...
Electrical											
Dielectric constant: 1 MHz 25°C	(6.1)	5.58	...	6.78	3.78	3.8	4.6	7.2	8.81	5.9	6.3
300°C	(6.3)	5.60	...			3.9	5.9		...	...	...
500°C		8.80	...						9.03	...	...
10 GHz 25°C	(6.1)	5.45	...	6.54	3.78	3.8	4.5	6.71	8.79	5.8	5.8
300°C	(6.1)	5.51	...	6.65	3.78	...	...	...	...	...	...
500°C	(6.1)	5.53	...	6.78	3.78	...	...	...	9.03	...	...

Property			C1	C2	C3	C4	C5	C6	C7	C8	C9	C10
Dissipation factor:												
1 MHz	25°C		(0.0017)	0.0015	0.0030	...	0.0005	0.0046	0.009	0.00035	0.0013	0.0003
	300°C		(0.014)	0.0154	...	...	0.0042	0.0130	...	0.012	...	...
	500°C		...	...	...	...	...	...	...	...	...	...
10 GHz	25°C		(0.0002)	0.00033	0.0068	...	0.0009	0.0085	0.017	0.0015	0.0014	0.0010
	300°C		(0.0008)	0.00075	0.0115	...	...	...	...	0.0021	...	...
	500°C		(0.0025)	0.00152	0.040	...	...	...	...	...	...	...
Loss factor:												
1 MHz	25°C		(0.010)	0.008	0.02	...	0.0019	0.0212	0.065	0.0031	0.0077	0.0019
	300°C		(0.078)	0.086	...	...	0.0164	0.0566	...	0.108	...	...
	500°C		...	...	...	...	...	...	...	...	...	...
10 GHz	25°C		(0.001)	0.002	0.045	...	0.0036	0.0282	0.114	0.0132	0.0082	0.0058
	300°C		(0.005)	0.004	0.077	...	...	...	...	0.019	...	...
	500°C		(0.015)	0.008	0.27	...	...	...	...	...	...	...
Volume resistivity $\log_{10}$, Ω·cm												
	250°C		10.1	10	8.1	12.0	9.7	8.1	6.4	14.0(100°C)	14 (20°C)	14 (20°C)
	350°C		8.7	8.6	6.8	9.7	8.1	6.6	5.1	12.95 (300°C)	...	...

NOTES: (1) Softening temperature; method of evaluation: (a) pyroceram; comparable with ASTMC 24-56; (b) glass, ASTMC 338-54T; (c) ceramics, ASTMC 24-56.
(2) Expansion coefficients depend on heat-treatment.
(3) Unabraded values.

PROPERTIES OF MATERIALS

where H = magnetic field strength (A/m), n = number of turns per meter (m^{-1}), and I = electric current through the coil (A).

113. Magnetic Pole Strength. The magnetic pole strength of a material is analogous to the charge in electrostatics. For example, the force on a magnetic pole of strength m in a magnetic field of strength H is

$$F = m \cdot H \qquad (6\text{-}50)$$

where F = force (N), m = magnetic pole strength, Webers (Wb), and H = magnetic field strength (A/m).

Table 6-39. Properties of Dielectric Thick-Film Screened Coatings[129]

	ESL No.	Dielectric constant	Properties	
Dielectrics— multilayer	4640	40 ± 5	Good adhesion to cofired conductors; also useful for low-value capacitors	
	4610	10–14	Excellent solderability of top conductors	
	4608	8–10	Alkali- and lead-free, low porosity, good for hermetic seals and capacitors	
	4608C	8–10	Similar properties to No. 4608, but fine definition for printing small vias in multilayer circuits	
	4608, black	. . .	For opaque overglazing of gold conductors for LED display panels	

	ESL No.	Dielectric constant	Dissipation factor	Capacitance, pF/in.-mil
Dielectrics— capacitor	4608	10 ± 2	0.1–0.5	2,000
	4110	50 ± 10	1.5	10,000
	4210	100 ± 20	1.5	20,000
	4310	250 ± 50	1.5	50,000
	4410	600 ± 100	1.5	120,000
	4510	1,000 ± 300 (Up to 2,000 with higher firing temp. and time)	2.5–4	200,000

	ESL No.	Firing temp.	Properties	
Overglazes	4770B	500–525°C	Resistor overglaze—transparent	
	4771B	525–540°C	Resistor overglaze—transparent	
	4770BC	500–525°C	Color-tinted resistor overglaze—translucent	
	4771BC	525–540°C		
	4770, black	500–550°C	Black opaque overglaze	

	ESL No.	Firing temp.	Properties	Applications
Solder Glass Pastes	4009	400–500°C Crystallizable, black color	Matches alumina: screen prints to give very thick coatings—low-temperature vehicle burnoff	Package sealing, designed for DIP package application
	4010	450–500°C Vitreous, black color	Matches soda lime glass: screen prints to give very thick coatings—low-temperature vehicle burnoff	Glass package sealing for flat display tubes
	4011	500–550°C Crystallizable, black color	Matches soda lime glass: screen prints to give very thick coatings—low-temperature vehicle burnoff	Glass package sealing for flat display tubes

* ESL = Electro-Science Laboratories, Inc., Philadelphia, Pa.

114. Magnetic Moment. In analogy to electrical dipoles, a magnetic dipole composed of two poles of strength $+m$ and $-m$ separated by distance l will have a magnetic moment

$$M = m \cdot l \tag{6-51}$$

where M = magnetic moment (Wb · m) and l = distance (m).

115. Magnetization. The magnetization I is defined as the magnetic moment per unit volume.

116. Magnetic Flux Density. In empty space the magnetic flux density (or induction) B is given by

$$B = \mu_0 H \tag{6-52}$$

where B = magnetic flux density (Wb/m^2, or tesla), and μ_0 = magnetic permeability of empty space = $4\pi \times 10^{-7}$ H/m.

In a material the flux density is

$$B = \mu \cdot H \tag{6-53}$$

where μ = magnetic permeability of the material (H/m).

The magnetic flux density can also be written as a sum of the flux in empty space $\mu_0 H$ and the flux density due to magnetization I.

$$B = \mu_0 H + I \tag{6-54}$$

117. Relative Magnetic Permeability. The ratio between the permeability of a material and that of empty space is the relative permeability,

$$\mu_r = \frac{\mu}{\mu_0} \tag{6-55}$$

where μ_r = relative permeability.

It follows that

$$B = \mu_0 \mu_r H \tag{6-56}$$

In analogy to the electrostatics, μ_r is written in a complex form for lossy materials.

$$\mu_r = \mu'_r - j\mu''_r \tag{6-57}$$

118. Magnetic Susceptibility. The magnetic susceptibility is the magnetization per unit magnetic field,

$$\chi = I/H \tag{6-58}$$

where χ = magnetic susceptibility (H/m). From Eq. (6-58) and Eqs. (6-53) and (6-54) it follows that

$$\mu = \mu_0 + \chi \tag{6-59}$$

and

$$\mu_r = 1 + \frac{\chi}{\mu_0} \tag{6-60}$$

The relative magnetic susceptibility $\bar{\chi}$ is defined as

$$\bar{\chi} = \frac{\chi}{\mu_0} \tag{6-61}$$

The values of $\bar{\chi}$ range from 10^{-5} to 10^6.

119. Magnetic Flux. The magnetic flux ϕ across a surface A is

$$\phi = \int B \cos \alpha \, dA \tag{6-62}$$

where ϕ = magnetic flux (Wb), α = between B and normal to the surface, B = magnetic flux density (Wb/m^2), and dA = surface element (m^2).

120. Magnetic Energy. The magnetic energy stored in a unit volume by increasing the magnetic flux density from B_1 to B_2 is

$$\text{Magnetic energy density} = \int_{B_1}^{B_2} H \, dB \quad \text{(J/m}^3\text{)} \tag{6-63}$$

121. Classification of Magnetic Materials. The atomic origin of magnetism stems from the orbital motion and spin of electrons. The main classes of magnetic materials (see Fig. 6-28) are as follows.

Diamagnetic Materials. These materials have a small negative susceptibility $\bar{\chi} \approx -10^{-5}$. The magnetism originates from induced currents opposing the external field.

Paramagnetic Materials. These materials have a $\bar{\chi} = +10^{-5}$ to 10^{-3}. The magnetism stems from partial alignment of existing spins, randomly oriented by thermal agitation in the absence of external fields.

Ferromagnetic Materials. Ferromagnetic materials have spins aligned parallel to each other. The material is divided into *magnetic domains*, each domain having a net magnetization even without an external field. This magnetization is called *spontaneous magnetization.* A bulk sample will generally not have a net magnetization since the spontaneous magnetization in the various domains will cancel each other.

Application of a small magnetic field will cause growth of favorable domains resulting in high magnetization and high $\bar{\chi}$ (up to 10^6). Above a critical temperature, Curie temperature, these materials become paramagnetic.

Ferrimagnetic Materials. These materials have two kinds of magnetic ions with unequal spins, oriented in an antiparallel fashion. The spontaneous magnetization can be regarded as the two opposing and unequal magnetizations of the ions on the two sublattices. Ferrimagnetic materials become paramagnetic above a Curie temperature.

Antiferromagnetic Materials. The materials have an antiparallel arrangement of equal spins resulting in very low $\bar{\chi}$ similar to paramagnetic materials. The spin arrangement of antiferromagnetic materials is not stable above a critical temperature (Neél temperature).

122. Magnetization of Ferromagnetic Materials. The magnetic induction of a ferromagnetic material is seen in Fig. 6-29 as a function of the applied magnetic field. The induction starts from 0 at zero field and reaches *maximum induction* B_m and *maximum field* H_m at saturation. The magnetization reaches an upper limit called the *saturation induction* B_s. When the field is decreased, the induction will follow a curve with higher

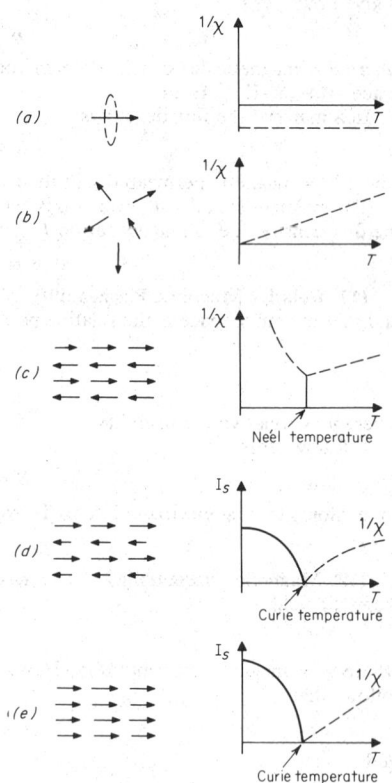

Fig. 6-28. Magnetic susceptibility and saturation magnetization for several magnetic materials. (*a*) Diamagnetic; (*b*) paramagnetic; (*c*) antiferromagnetic; (*d*) ferrimagnetic; (*e*) ferromagnetic. On the left are schematic representations of spin arrangements within the materials.

values than the original curve. At $H = 0$ there remains an induction B_r—*residual induction,* or *remanence.* The maximum residual induction (when materials are fully magnetized) is the *retentivity.* To remove the retentivity, a negative magnetic field is applied and the induction is completely removed at H_c, the *coercive force,* or its maximum, the *coercivity.* The process of removal of residual induction is often referred to as *demagnetization,* and the portion of curve between B_r and H_c is called the *demagnetization curve.* Application of higher negative fields will eventually saturate the material. Reversing the field again will complete the *B-H* curve and bring it again to the maximum field H_m and maximum induction B_m. The entire curve (Fig. 6-29) is called *hysteresis loop.*

123. Ferromagnetic Domains. Ferromagnetic materials are composed of domains which are spontaneously magnetized in a specific direction. The external magnetic field merely moves domain walls and at higher fields changes the direction of the magnetization in the domains. Domain theory explains the initial low slope of the hysteresis curve as a reversible wall movement, the second steep part of the curve as irreversible wall movement, and the final convergence toward saturation as magnetization rotation (see Fig. 6-30).

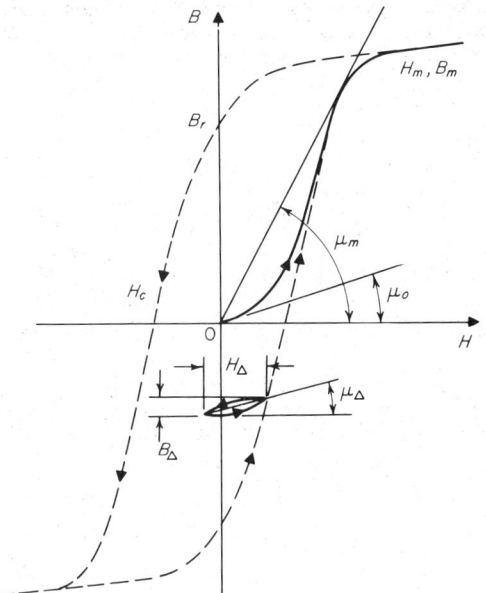

Fig. 6-29. Magnetization curve and hysteresis loop of a ferromagnetic material.

124. Energy Product Curve. The curve of $B \cdot H$ at any point on the demagnetization curve is termed the energy product curve. Figure 6-31 shows a typical demagnetization curve. The $B \cdot H$ product reaches a maximum value BH_{max}. This value is a good criterion for permanent magnetic materials; i.e., higher BH_{max} values are obtained with higher-quality magnets.

125. Hysteresis Losses. The work per unit volume required to magnetize a ferromagnetic material to saturation is $\int_0^{B_m} H \, dB$. In one cycle of the hysteresis loop there is a loss of energy per unit volume which equals the area within the hysteresis loop.

Materials are called *hard magnetic materials* when they have a high coercivity and a large hysteresis loop. *Soft magnetic materials* are low-loss, low-coercive-force, high-permeability materials.

126. Eddy Current Losses. The varying magnetic flux in a material induces an emf, which in turn creates eddy currents. The Joule heating due to the currents is an energy loss, termed *eddy current loss.* At low frequencies, when the flux penetration is complete, the eddy current losses are proportional to the square of the frequency, f^2, and inversely proportional to the electrical resistivity. At high frequency and constant flux amplitude, the losses become proportional to $f^{3/2}$. Eddy current losses are greatly reduced if cores are made from laminated sheets so as to confine the currents within the relatively high resistance sheets. Substantial reduction of eddy current losses can be achieved by using ferromagnetic material having very high electrical resistivities.

127. Total Core Losses. The sum of all losses in a core is termed total core loss. It should be noted that the losses are normally given for a sinusoidally varying magnetic field under symmetrically cycling conditions. Corrections have to be made for deviations from sinusoidal condition.

128. Demagnetizing Factor. If a material is magnetized, free poles are produced in it which create a field opposing the external applied field. This field is termed the demagnetizing field H_d,

$$H_d = \frac{NI}{\mu_0} \qquad (6\text{-}64)$$

where I = magnetization (Wb/m^2), μ_0 = permeability of empty space (H/m), and N = demagnetizing factor. For fields perpendicular to sheets or thin plates, $N = 1$, and for applied fields along the axis of an infinitely long rod, $N = 0$.

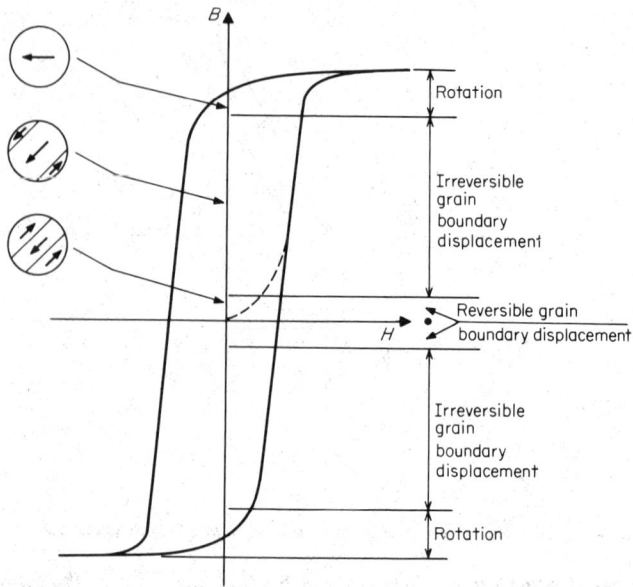

Fig. 6-30.   Ferromagnetic-domain configuration at various stages of magnetization.

129. Permeability of Ferromagnetic Materials. The permeability of a ferromagnetic material (B/H) is not constant (see Fig. 6-29). When the material is first magnetized, the initial slope of the B-H curve is called *initial permeability* μ_0. The maximum slope from the origin to the B-H curve is the *maximum permeability* μ_m. If a biasing field H_b is held constant and a small alternating field H_Δ is applied, then $B_\Delta/H_\Delta = \mu_\Delta$, the *incremental permeability*. *Differential permeability* is dB/dH, the local slope of the hysteresis curve. Permeability measured under ac excitation is termed *ac permeability*.

130. Magnetostriction. The change in a material's dimension due to magnetization is called magnetostriction. Changes in linear dimensions are called *linear* magnetostriction, while volume changes are referred to as *volume* magnetostriction. The change in length in the direction of the magnetization is called *Joule* magnetostriction. The symbol λ is generally used for the value of the fractional change in length of saturation. Figure 6-32 shows the magnetostriction field strength for iron, nickel, and cobalt.

131. Magnetic Anisotropy. The internal energy of a ferromagnetic or ferrimagnetic crystal depends on the direction of the spontaneous magnetization with respect to the crystal axes; i.e., it is anisotropic. This anisotropy is called *magnetocrystalline anisotropy*. It follows that the ease of magnetization of crystals depends on the direction of magnetization. For example, cobalt shows a uniaxial anisotropy, the easy and stable direction of spontaneous magnetization being its *c* axis. Iron has an easy axis of magnetization in the $<100>$ directions, while nickel has easy $<111>$ directions (Fig. 6-33). Polycrystalline materials with strong preferred orientation also show magnetic anisotropy.

In thin rods or thin films there is a tendency for the spontaneous magnetization to align itself in the direction of the rod axis or in the film plane, respectively. This form of anisotropy is called *shape anisotropy*.

Anisotropy can also be induced by strain due to the magnetostriction energy. In some alloy systems, that is, Ni-Fe (see Par. **6-141**), heat-treatment in the presence of a magnetic field results in a uniaxial magnetic anisotropy.

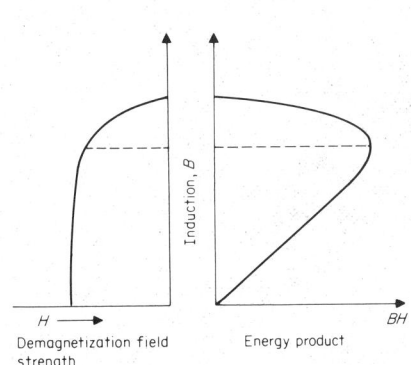

Fig. 6-31. Schematic demagnetization and energy product curve for a magnet.

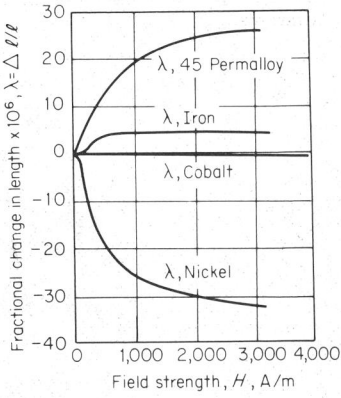

Fig. 6-32. Magnetostriction of some common materials.[80]

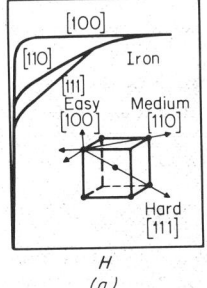

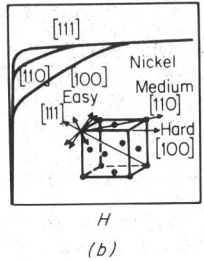

Fig. 6-33. Magnetization vs. applied field for field directions along the [100], [110], and [111] directions for (*a*) iron and (*b*) nickel single crystals.[82]

132. Aging Coefficient. The percentage change in magnetic properties of materials resulting from temperature aging is called aging coefficient. Typical aging treatments are 100 h at 150°C and 600 h at 100°C.

133. Magnetoresistance. The change in electrical resistance due to the application of a magnetic field is called magnetoresistance.

134. Magnetocaloric Effect. A change in temperature of the material due to a change in the magnetic field is called the magnetocaloric effect.

135. Magnetooptical Effects. Spontaneously magnetized material can affect the polarization of light transmitted through the materials (Faraday effect) or reflected from their surface (Kerr effect). These are called magnetooptical effects.

SPECIFIC MAGNETIC MATERIALS

136. Retentive and Nonretentive Materials. Magnetic materials can be classified into *nonretentive soft* materials and *retentive hard* materials (see Fig. 6-34). The nonretentive materials are low-loss materials used in transformers, motor or generator cores, electromagnetic apparatus, and memories. These materials can be further classified according to the frequency range for which they are used—from nonalloyed iron for low-frequency uses, silicon-iron, Permalloy, ferrites, and finally garnets for the highest-frequency use. Hard materials are high-loss, high-retentivity materials with a high-energy product used for permanent magnets. Hard magnetic alloys can be further classified into alloys undergoing martensitic transformation, precipitation-hardened alloys, ordered-hardened alloys, and powder magnets.

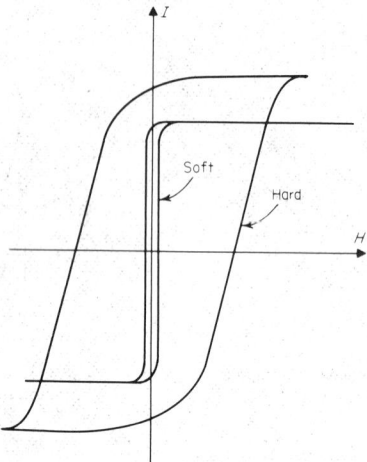

Fig. 6-34. Hysteresis loops for soft and hard magnetic materials.

Magnetic materials with special properties are also often desired, for example, square-loop materials for memory application and constant-permeability materials or alloys capable of operating at elevated temperatures.

Dc magnetization curves of various magnetic materials are given in Fig. 6-35.

NONRETENTIVE MAGNETIC MATERIALS

137. Iron. Iron is used for dc applications such as electromagnet cores and relays. Iron contains nonmetallic impurities which reduce its permeability and increase the hysteresis losses. *Hydrogen annealing* at high temperature can remove the impurities and greatly increases the permeability. Another common annealing procedure which increases permeability is an 800°C anneal followed by a slow (5°C/min) cool. Impurities such as carbon or nitrogen in iron cause magnetic aging, i.e., deterioration of magnetic properties with time.

Iron is available as electrolytic iron, which is a very pure form containing typically 0.006% C, 0.005% Si, 0.005% P, and 0.005% S. Very high permeabilities can be obtained in electrolytic iron by hydrogen annealing. The use of electrolytic iron is limited mainly by economic considerations. *Ingot iron* is highly refined iron of typical composition 0.08% C, 0.015% Mn, 0.16% Si, 0.06% P, 0.01% S, and 1.2 to 2.8% slag which is finely dispersed in the iron. Annealed ingot-iron bars are used for relay cores. *Low-carbon steels* have lower permeability and higher hysteresis losses and higher aging coefficient due to the carbon content. *Cast steels* are used where strength is required. *Gray cast iron* containing about 3% C and 1 to 3% Si can be cast to form complex shapes. *Ductile cast iron* is superior to gray cast iron in magnetic properties. Ductile cast iron contains the carbon in form of nodular graphite which imparts the ductility to the material. *Malleable cast iron* is magnetically superior to ductile cast iron.

138. Iron-Silicon Steels. The useful magnetic properties of iron-silicon alloys were reported at the end of the nineteenth century by Hadfield.[32] Silicon steels, known as electrical steels, are very widely used for low- and intermediate-frequency applications.

Their main advantages are low core losses, high maximum permeability, increase in electrical resistivity (lower eddy current loss), and reduction of magnetic aging. The effect of silicon additions on the coercive force, hysteresis and core losses, and maximum relative permeability is seen in Fig. 6-36.

The disadvantages of silicon additions are the decrease in saturation magnetization and the increase in brittleness. The increase in brittleness limits the amount of Si to 1 to 3% for rotating machinery and 3 to 4.5% for transformers. Si increases the yield and ultimate tensile strength of steel and improves the high-temperature mechanical properties. Table 6-40 lists the typical applications and properties of silicon steels. Conventional material is supplied in

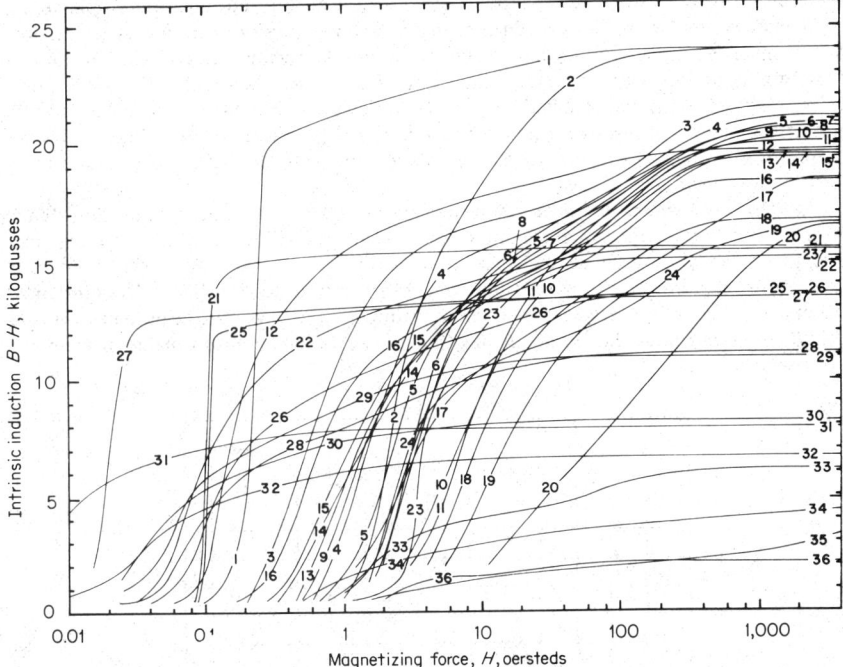

Fig. 6-35. Dc magnetization curves for various magnetic materials.[81] For identification of materials, see p. 6-88.

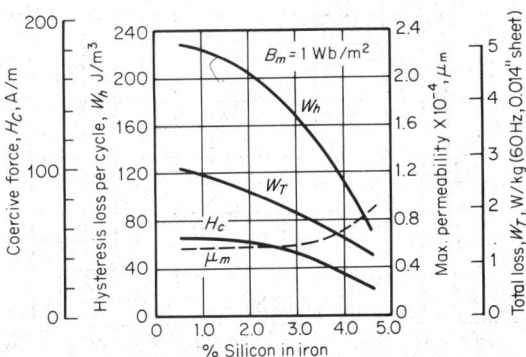

Fig. 6-36. Magnetic properties of hot-rolled commercial silicon-iron sheet as a function of silicon content. Some of the changes are due to higher annealing temperatures of the materials having higher silicon content.[80]

four classes: hot-rolled, fully processed; cold-rolled, fully processed; hot-rolled, semiprocessed; and cold-rolled, semiprocessed. Silicon steels can also be obtained in grain-oriented form. This product is rolled to obtain the easy [100] magnetization direction parallel to the rolling direction. Grain-oriented silicon steels have very low core losses. Very thin oriented silicon steels for higher-frequency applications are also available in tape form. Tapes with thickness of 0.001, 0.002, and 0.004 in. are available for pulse transformers, reactors, magnetic amplifiers, and small power transformers. Core laminations are frequently coated to provide electrical insulation between the laminations. These coatings can be organic or inorganic.

139. Iron-Cobalt Alloys. Iron-cobalt alloys are used when a high magnetic saturation is required. Figure 6-37 shows the curve for a 49% Co, 49% Fe, and 2% V (vanadium Permendur) alloy. Vanadium is added to the iron-cobalt alloys to permit easy processing. When annealed under a magnetic field, vanadium Permendur has a high maximum permeability and the highest residual induction of any other alloy with very low coercive force. A hysteresis curve of vanadium Permendur is seen in Fig. 6-37. Iron-cobalt alloy is one of the highest in positive magnetostriction ($\lambda = 60 - 70 \times 10^{-6}$). Other alloys are Hyperco: 35% Co, 0.5% Cr, and the balance Fe; and alloys with the addition of several percent Cr: stainless invar.

140. Iron-Aluminum and Iron-Aluminum-Silicon Alloys. Iron-aluminum alloys have high electrical resistivities, high maximum permeabilities, and low coercivity. Commercial alloys are available, such as the 16% Al, Alperm, and the 13% Al, Alfer (used as a magnetostriction material). Iron-aluminum alloys can be cold-rolled. Grain-oriented material can also be produced. Iron-aluminum-silicon alloys have good ac properties but are brittle. A commercial alloy of about 10% Si and 6% Al (Sendust) is available in sheet and powder form.

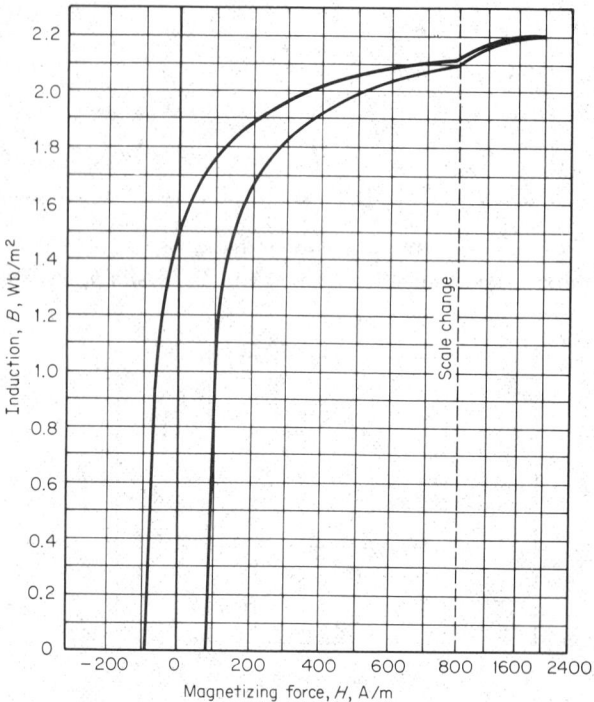

Fig. 6-37. Dc hysteresis loop of cold-rolled strip of vanadium Permendur. (Sample form: 2.5 in. OD × 1.9 in. ID stamped rings, annealed.)[82]

Table 6-40. Properties of Laminated or Wound Cores of Silicon Steel[82]

Grade	M-6	M-7	M-19	M-22	M-27	M-36	M-43
General applications	Power and distribution transformers, large turbogenerators, small power and audio transformers		Large rotating machines, communication transformers	Stators of induction motors and high-efficiency rotating equipment	High-efficiency motors, small transformers	Rotating machines, including ac-dc motors	Intermittent-duty rotating machines, pole pieces, relays
Nominal Si content, % cold-rolled	3.25	3.25	3.00	3.00	3.00	1.75	0.50
Core loss, W/lb, max.							
10,000B 0.014 in.	...	...	0.72	0.82	1.01	1.17	1.30
60 Hz 0.0185 in.	...	...	0.83	0.94	1.14	1.35	1.55
.025 in.	0.66	0.73	0.97	1.10	1.30	1.70	1.98
15,000B .014 in.	...	...	1.75	2.00	2.46	3.00	3.50
60 HZ 0.0185 in.	...	...	2.00	2.20	2.66	3.20	3.70
.025 in.	...	...	2.35	2.60	3.05	3.85	4.60
Saturation induction B_S							
Gauss	19,700	19,700	19,000	19,000	19,000	20,000	20,500
Webers per square meter	1.97	1.97	1.90	1.90	1.90	2.00	2.05
Specific gravity	7.65	7.65	7.65	7.65	7.65	7.75	7.75
Electrical resistivity $\times 10^8$, $\Omega\cdot$m	50	50	47	47	47	37	28

141. Iron-Nickel Alloys. Iron-silicon alloys have high permeabilities at high fields. However, for low-field applications which are common in higher-frequency operations, these alloys do not have a high enough permeability. The iron-nickel alloys have a high initial permeability which makes them suitable for low-field, high-frequency operations. Figure 6-38 shows the initial permeability of iron-nickel alloys as a function of nickel content. The maximum of the permeability occurs around 78% Ni, between the composition which corresponds to zero magnetocrystalline anisotropy (~75% Ni) and the one corresponding to $\lambda_{III} = 0$ (just below 80% Ni). Another permeability maximum occurs at about 45% Ni. At this composition the alloys have higher saturation magnetization, which permits the use of

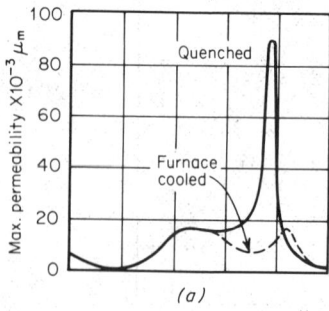 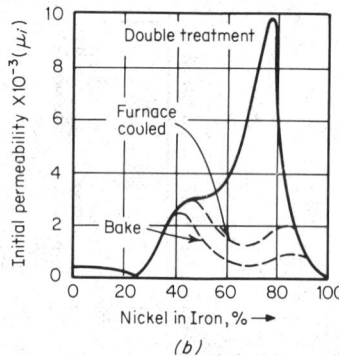

Fig. 6-38. (*a*) Maximum permeability and (*b*) initial permeability vs. composition of nickel-iron alloys.[83]

smaller volumes of material in core construction. Composition and properties of iron-nickel alloys are given in Table 6-41. *78 Permalloy* is a 78% Ni-22% Fe alloy. Permalloy has high initial and maximum permeabilities and low losses. High permeabilities are obtained by a high-temperature heat-treatment (preferably in hydrogen), cooling to 600°C, followed by a rapid quench. The quench prevents the formation of the low-permeability ordered phases.

The maximum and differential permeabilities and remanence (i.e., loop squareness) can be increased appreciably by inducing a uniaxial isotropy in iron-nickel alloys. This anisotropy is technically accomplished by rolling or by annealing in a magnetic field. Permalloys have high squareness and very low coercivities.

Permalloy has a higher saturation density than 78 Permalloy. This alloy is available under several trade names: 4750 Allegheny and Carpenter 5549 alloy. *Supermalloy* has very high initial and maximum permeabilities. Its composition is 79% Ni, 5% Mo, and 16% Fe. *Mo Permalloy* results from the addition of small amounts of Mo, Cr, or Cu to Permalloy (for example, 4% Mo Permalloy) to improve the permeability, increase the resistivity, and eliminate the necessity of quenching from 600°C. 2% Mo Permalloy is also pulverized and used as magnetic powder cores. Uniaxial anisotropy has been introduced in Mo Permalloy also by a radiomagnetic treatment, i.e., bombarding with high-energy electrons at 100°C under a magnetic field.[33] Iron-nickel alloys possessing uniaxial anisotropy are also used in thin-film form for high-speed memory application (in the 10^{-8}- to 10^{-9}-s range). These films are prepared by electroplating on wires or by vacuum deposition in the presence of a magnetic field. *Deltamax* is a recrystallized sheet with a (001) [100] texture of 50% iron-nickel alloy. Deltamax reaches saturation at very low fields. Deltamax is also available in tape form. *Isoperm* is a specially treated 50% Fe-Ni alloy with a constant permeability. Its linear magnetization curve is useful for circuit transformers. *Thermoprem* is a temperature-sensitive alloy containing 30% nickel with a Curie point just above room temperature. The alloy decreases in magnetization with temperature and is suitable for temperature-compensating elements.

Tape-wound cores of nickel-iron alloys are also available. Tapes range from 0.004 in. (0.1 mm) to 0.000125 in. (3.125 μm) in thickness. Typical properties of tape-wound and laminated cores are given in Tables 6-42 and 6-43.

Table 6-41. Magnetic Properties of Soft Magnetic Alloys[130]

Material	Composition	Heat-treatment,* °C	$\bar{\mu}_a$	$\bar{\mu}_{max}$	H_c A/m	H_c Oe	I_s Wb/m	I_s gauss	θ, °C	ρ, ×10^{-8} Ω·m	σ, g/cm³
Iron	0.2 (imp.)	950	150	5,000	80	1.0	2.15	1,710	770	10	7.88
Purified iron	0.05 (imp.)	1,480(H_2); 880	10,000	200,000	4	0.05	2.15	1,710	770	10	7.88
Silicon-iron	4 Si	800	500	7,000	40	0.5	1.97	1,570	690	60	7.65
Silicon-iron (oriented)	3 Si	800	1,500	40,000	8	0.1	2.00	1,590	740	47	7.67
Silicon-iron (cubic)	3 Si	...	...	116,000	5.6	0.07	2.00	1,590	740	47	7.67
Silicon-iron (cubic)	3 Si	...	...	65,000	6.4	0.08	2.00	1,590	750	47	7.67
Aluminum-iron	3.5 Al	1,100	500	19,000	24	0.3	1.90	1,510	...	90	7.5
Alfer	13 Al	...	700	3,700	53	0.66	1.20	955	400	140	6.7
Alperm	16 Al	600Q	3,000	55,000	3.2	0.04	0.80	637	...	153	6.5
16 Alfenol	16 Al	600Q	600→4,100	4,000→90,000	...	...	0.80	637	...	16	6.5
78 Permalloy	78.5 Ni	1,050; 600Q	8,000	100,000	4	0.05	1.08	860	600	60	8.6
Supermalloy	5 Mo, 79 Ni	1,300C	100,000	1,000,000	0.16	0.002	0.79	629	400	65	8.77
Cr Permalloy	3.8 Cr, 78 Ni	1,000	12,000	62,000	4	0.05	0.80	637	420	62	8.5
Mumetal	5 Cu, 2 Cr, 77 Ni	1,175(H_2)	20,000	100,000	4	0.05	0.65	517	...	45	8.58
Hipernik	50 Ni	1,200(H_2)	4,000	70,000	480	6	1.60	1,270	500	40	8.25
50 Isoperm	50 Ni	1,100CR	90	100	...	...	1.60	1,270	500	45	8.25
Deltamax	50 Ni	1,075	500	200,000	...	...	1.55	1,230	...	...	8.25
Thermoperm	30 Ni	1,000	...	...	...	...	0.20	159	...	...	...
Permendur	50 Co	800	800	5,000	160	2.0	2.45	1,950	980	7	8.3
45-25 Perminvar	25Co, 45 Ni	1,000; 400	400	2,000	95	1.2	1.55	1,230	715	19	...
7-70 Perminvar	7 Co, 70 Ni	1,000; 425	850	4,000	48	0.6	1.25	995	650	16	8.6

*Q-quenched; C-controlled cooling rate; CR-severely cold-rolled; H_2-annealed in pure hydrogen.

Table 6-42. High-Permeability Laminated or Wound Cores of Silectron† and Nickel Alloys[82]

Grade	Silectron cores, 0.012 in.	Silectron laminations, 0.014 in.	Mumetal, 0.014 in.	Molybdenum Permalloy, 0.014 in.	4750, 0.014 in.	Deltamax, 0.002 in.	Supermalloy, 0.004 in.
General applications	Power distribution transformers	Power and audio transformers	Low-induction filters and audio transformers		Servo and synchro motors, audio transformers	Magnetic amplifiers	Magnetic amplifiers, specialty transformers
Nominal composition, %:							
Molybdenum	...	...	...	4.00	...	...	5.00
Silicon	3.25	3.25	...	...	...	...	...
Nickel	...	...	77.0	79.00	48.00	50.00	79.00
Dc permeability:							
μ_{max}	50,000	...	100,000	200,000	80,000	100,000	700,000
B at μ_{max}	8,000	...	2,500	3,000	5,000	12,000	3,000
μ at 40B	4,000	...	25,000	30,000	8,000	500	75,000
μ at 100B	6,500	...	30,000	45,000	12,000	2,000	80,000
Ac permeability, 60 Hz:							
μ at 40B	3,500	...	20,000	26,000*	8,000*	500	70,000
μ at 200B	6,500	...	30,000	32,000*	13,500*	1,000	90,000
μ at 2,000B	15,000	...	40,000	54,000*	30,000*	20,000	160,000
Saturation induction B_S, gauss	19,700	19,700	7,500	8,000	15,500	16,000	7,800
Specific gravity	7.65	7.65	8.5	8.74	8.20	8.25	8.77
Electrical resistivity, $\mu\Omega\cdot$cm	50	50	60	55	45	45	65

* Minimum values.
† Allegheny Ludlum Steel Co.

Table 6-43. High-Frequency Silicon and Nickel Cores[82]

Grade	Monimax	Powder cores	Rotosil, 7 mils	Silectron, 4 mils	Silectron, 2 mils	Silectron, 1 mil	Ultrathin nickel irons		
							Square Permalloy	Delta-max	Super-malloy
General applications	Pulse transformers, high-frequency transformers	Loading coils, filters	High-frequency rotating machinery	Pulse transformers, high-frequency transformers, magnetic amplifiers			Magnetic amplifiers, computer cores		
Nominal composition, %:									
Nickel	47.00	81.00	...	...	...	...	79.00	50.00	79.00
Silicon	...	...	3.25	3.25	3.25	3.25	...	...	...
Molybdenum	3.00	2.00	...	...	...	...	4.00	...	5.00
Nominal frequency range	Audio range	Augio to low r.f.	400 Hz 800 Hz	400 Hz audio range	Pulse 0.5–10	Pulse under .50	Pulse at repetition rates up to 1 MHz		
Specific gravity	8.25	7.800	7.65	7.65	7.65	7.65	8.74	8.25	8.77
Saturation induction B_S, gauss	14,500	...	19,700	19,700	19,700	19,700	8,000	16,000	8,000
Electrical resistivity, $\mu\Omega \cdot cm$	65 min.	High	50	50	50	50	55	45	65

PROPERTIES OF MATERIALS

Other nickel-iron alloys are as follows:

Mumetal, containing 77% Ni, 4.8% Cu, 1.5% Cr, 14.9% Fe, has a high initial permeability, especially after hydrogen annealing. It is used as a magnetic shielding for electronic equipment and for magnetic amplifier cores. *Perminvar,* containing 25% Co, 45% Ni, 30% Fe, has a constant permeability over a wide range of inductions. This alloy must be annealed for long periods at 400 to 450°C. Perminvar can be made, by magnetic annealing, to have a square hysteresis loop. *Superinvar,* containing 31% Ni, 4 to 6% Co, is a magnetic alloy with a very small thermal expansion coefficient. *Elinvar,* containing 36% Ni, 12% Co, has constant elastic moduli, which are temperature-independent.

142. Powder Cores. For higher-frequency applications powdered ferromagnetic materials are used. These powdered materials are compacted with an insulating binder, thereby reducing considerably the eddy currents. The reduction in eddy current losses comes at the expense of the permeability, which is normally around 100. Powder grain size varies from 10 to 100 μm, approximately. Powder cores are made of pure carbonyl iron, Sendust (10% Si, 67% Al), or Permalloy. At low frequencies and for high stability, Permalloy cores are preferable. At audio frequencies all three powders are suitable, while at high frequencies iron and Sendust are useful. At very high frequency iron cores are used. Properties of powder cores are given in Table 6-44.

143. Ferrites. At higher frequencies nonconducting ferrimagnetic materials are used as core materials. The ferrimagnetic materials have smaller saturation inductions (up to 0.5 Wb/m^2) and very small core losses due to their high electrical resistivities. The ferrites are generally of the spinel, magnetoplumbite, or garnet types. The permeability of some common ferrites is shown in Fig. 6-39 as a function of frequency. It can be seen that the real part of the permeability decreases with increasing frequency, while the imaginary part, which is related to losses, reaches a maximum and then decreases. In garnets the losses at specific dc bias magnetic fields become important. Large losses are encountered at specific fields.

Table 6-44. Magnetic Properties of Pressed Powder and Ferrite Cores[130]

Material	Composition	Treat-ment	μ_a	H_c		I_s		Θ, °C	ρ, $\Omega \cdot m$
				A/m	Oe	Wb/m^2	gauss		
Carbonyl iron powder	100 Fe	Press	20	1200	15	1.56	1240	770	10^8
Mo Permalloy powder............	2 Mo, 81 Ni	Press, 650	125	...	...	0.70	560	480	10^4
Sendust powder......	5 Al, 10 Si	Press, 800	80	100	1.25	0.45	360	500	
Mn-Zn ferrite........	50 Mn, 50 Zn	1,150	2000	8	0.1	0.25	200	110	1
Cu-Zn ferrite........	40 Cu, 60 Zn	1,000	1100	40	0.5	...	...	90	10^3
Cu-Mn ferrite........	40 Cu, 60 Mn	1,250	...	80	1.0	0.29	230		
Ni-Zn ferrite.........	30 Ni, 70 Zn	1,050	80	...	...	0.40	320	130	10^{10}
Mg-Zn ferrite........	50 Mg, 50 Zn	...	500	80	1.0	0.26	207	120	
Mg-Mn ferrite	50 Mg, 50 Mn	1,400	...	40	0.5	0.27	215	130	

Dc magnetization curves for various magnetic materials:[81]

(1) Supermendur
(2) Vanadium Permendur (50% Co-2%V)
(3) Pure iron, annealed
(4) Ingot iron, annealed
(5) Hot-rolled low-carbon sheet steel
(6) M-50
(7) Cold-rolled low-carbon strip steel
(8) Cold-drawn carbon steel, annealed
(9) M-43 (0.5% Si) and M-36 (1.5% Si)
(10) Steel castings, as cast
(11) Carbon-steel forgings, annealed
(12) 3% Si strip, oriented
(13) M-27 (2.75% Si)
(14) M-22 (3.25% Si)
(15) M-19 (3.75% Si)
(16) M-14 (4.50% Si)
(17) Malleable iron castings, as cast
(18) Type 416 stainless steel, annealed
(19) Nodular cast iron
(20) Gray iron, as cast
(21) Deltamax, oriented
(22) 4750 alloy
(23) Perminvar (45% Ni-25% Co)
(24) Powdered iron, sintered and annealed
(25) Monimax, oriented
(26) Monimax, nonoriented
(27) 65 Permalloy, oriented
(28) Sinimax
(29) 78 Permalloy
(30) Mol-Permalloy
(31) Superalloy
(32) Mumetal
(33) Pure nickel, annealed
(34) Soft magnetic ferrite (Fe-Ni-Zn-V)
(35) 30% Ni-Fe temperature compensator alloy
(36) Monel, annealed

Ferrites can be classified also according to their use: high permeability, low frequency; low loss, high frequency; microwave materials; square-loop materials. Tables 6-45 and 6-46 list some properties of ferrites.

Impurities tend to decrease the initial permeability, and insoluble inclusions increase the coercive force. Pores at grain boundaries have little effect on the initial permeability. Pores within grains decrease the initial permeability. Pores both at grain boundaries and in grains can affect the maximum permeability, coercivity, and hysteresis losses.

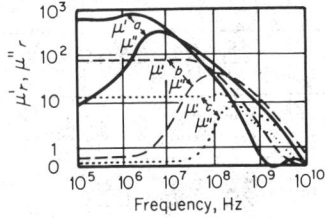

Fig. 6-39. Permeability spectra at room temperature of three commercial ferrites. (a) $Ni_{0.36}Zn_{0.64}Fe_2O_4$; (b) $Ni_{0.64}Zn_{0.36}Fe_2O_4$; (c) $NiFe_2O_4$. The μ scale is logarithmic for μ greater than 1 and linear for μ less than 1.[85]

144. Spinels. The ferrimagnetic spinel structure is obtained by replacing the Mg in $MgAl_2O_4$ with a divalent metal Me^{II} and Al with Fe^{III}. The divalent metal can be Mg, Ni, Co, Cu, Fe, Zn, Mn, or Cd. The spinel has 8 occupied tetrahedral sites and 16 occupied octahedral sites. The spins of the ions are oppositely directed on the two kinds of sites. The two sides can be occupied by Me^{II} and Fe^{III} in the following fashion:

$$Me_\delta^{II} Fe_{1-\delta}^{III} [Me_{1-\delta}^{II} Fe_{1-\delta}^{III}] O_4$$

$Me_\delta^{II}Fe_{1-\delta}^{III}$ ions are on the tetrahedral sites, while $Me^{II}_{1-\delta}Fe^{III}_{1-\delta}$ ions are on the octahedral site. Normal spinel has $\delta = 1$, and Me^{II} is found only on the tetrahedral sites. Inverse spinel has $\delta = 0$, and the Fe^{III} ions are equally divided between tetrahedral and octahedral sites, thus canceling their net magnetic divalent metal ions in the octahedral sites.

Ni-Zn ferrites are most commonly used in the frequency range 0.1 to 200 MHz. Figure 6-39 shows μ'_r, μ''_r for such ferrites as a function of frequency. Ni-Zn ferrites have high initial permeability and low coercive force and are used for recording tapes.

Mn-Zn ferrites have a very small coercive force and very high initial permeability, as well as a high-saturation magnetization. These ferrites are used in the 1-kHz to 1.5-MHz range and are lossy at higher frequencies.

Ni-Co ferrites have been used as magnetostrictive materials.

Square-loop ferrites—Mn-Cu, Li-Ni, and Mn-Mg ferrites—and doped YIG exhibit high permeability but low coercive forces. Such materials are used for high-speed memory cores.

145. Hexagonal Ferrites. The magnetoplumbites are the most common materials in this group. The mineral is of the formula $PbFe_{7.5}Mn_{3.5}Al_{0.5}Ti_{0.5}O_{19}$. The hexagonal ferrites show a considerable magnetocrystalline anisotropy, which makes them suitable for hard magnets.

The ferrites are denoted M, U, Z, Y, X, and W, where

$M = BaFe_{12}O_{19}$

$U = Ba_4Me_2Fe_{36}O_{60}$

$Z = Ba_6Me_4Fe_{48}O_{82}$

$Y = Ba_2Me_2Fe_{12}O_{22}$

$X = Ba_2Me_2Fe_{28}O_{46}$

$W = BaMe_2Fe_{16}O_{17}$

Me can be Ni, Mg, Co, Fe, Zn, Mn, or Cu.

Hexagonal ferrites can be used at very high frequencies.

146. Garnets. The highest-frequency ferrites available are garnets. Garnets are of the $Me_3^{II}Me_2^{III}Si_3^{IV}O_{12}$ type, where Me^{II} can be replaced by Ca, Fe, Mg, or Mn and the Me^{III}

Table 6-45. Summary of Properties of the Ferrites[88]

	Molecular wt.	Density, g/cm³	a_0, Å	ρ, Ω·cm (room temp.)	n_B (calc.)	n_B	σ_0	I_0 0°K	B_0	σ_s (room temp.)	I_s	B_s	θ_c, °C	K_1, K10⁻³	λ_s +10⁶	λ_{100}	λ_{111}	λ_{110} ×10⁶
Zn Fe₂O₄	241.1	5.33	8.44	10^2	Antiferromagnetic			$\theta_N = 9.5°$ K										
Mn Fe₂O₄	229.6	5.00	8.51	10^4	5	4.55	112	560	7000	80	400	5000	300	−40	−5	−25	+4.5	
Fe Fe₂O₄ (Fe₃O₄)	231.6	5.24	8.39	4×10^{-3}	4	4.1	98	510	6400	92	480	6000	585	−130	+40	−20	+78	57
Co Fe₂O₄	234.6	5.29	8.38	10^7	3	3.94	93.9	496	6230	80	425	5300	520	+2000	−110			
Ni Fe₂O₄	234.4	5.38	8.337	10^3–10^4	2	2.3	56	300	3800	50	270	3400	585	−69	−17	−46	+22	
Cu Fe₂O₄: Quenched	239.2	5.42	8.37	10^5	1	2.3	30	160	2000	25	135	1700	455	−63	−10			
Slow-cooled		5.35	8.70			1.3	31	140	1800	27	110	1500		−60				
Mg Fe₂O₄	200.0	4.52	8.36	10^7	0	1.1	69	330	4200	65	310	3900	440	−40	−6			
Li₀.₅Fe₂.₅O₄	207.1	4.75	8.33	10^2	2.5	2.6							670	−83	−8			
γFe₂O₃	159.7		8.34		2.5	2.3	81			73.5	417		575					
Mn Mn₂O₄ (Mn₃O₄)	278.8	4.84	b.c.t. $a_0 = 5.75$ $c_0 = 9.42$			1.85		218			185		42°K	−107				
Mg Mn₂O₄	198.2							25			85							

$n_B = 5\mu_B$

σ, emu/g⁻¹; K_1, erg/cm³; I, emu/cm³. a_0 = lattice parameter, ρ = electrical resistivity, n_B = Bohr magneton per molecule, λ_S = saturation magnetostriction. $\lambda_{100}\lambda_{111},\lambda_{110}$ = magnetostriction constants.

by Al, Cr, Fe, or Mn. Magnetic iron magnets are derived by substituting $Me^{III}Fe^{III}$ for $Me^{II}Si^{IV}$. Me^{III} can be Y, Sm, or Lu. The unit cell contains eight units of $Me_3^{III}Fe_5^{III}O_{12}$.

The garnet has 24 tetrahedral Fe^{III} sites (d sites), 16 octahedral Fe^{III} sites (a sites), and 24 dodecahedral Me^{III} sites (c sites).

Partial substitution of other trivalent ions such as aluminum is also possible. YIG, for example, is $Y_3Fe_5O_{12}$ and has a very low hysteresis loss at microwave frequencies. Saturation magnetization of $Gd_xY_{3-x}Fe_5O_{12}$ as a function of temperature and x, is given in Fig. 6-40. The peculiar shape of the saturation magnetization occurs because it is a sum of the magnetization due to the c sublattice and the a-d sublattice. The magnetization at high temperature is due to the Fe^{III} ions only. Properties of some microwave materials are given in Table 6-46.

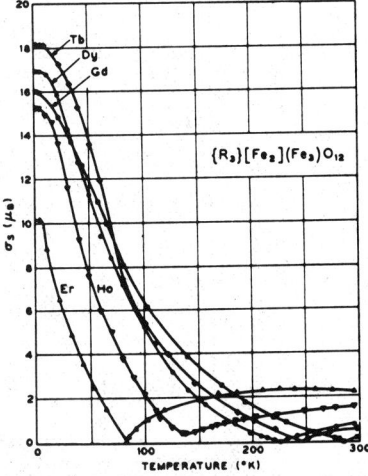

Fig. 6-40. Spontaneous magnetization in Bohr magnetons per formula unit vs. temperature for various rare-earth iron garnets.[86]

RETENTIVE MAGNETIC MATERIALS

147. Retentive materials are characterized by a high-energy product, making them suitable for permanent magnets. These materials have high remanences and coercive forces. Retentive materials are composed of a fine structure, which is responsible for the large coercive force.

There are four classes of retentive materials:

1. Martensitic lattice-transformation alloys.
2. Precipitation-hardened alloys.
3. Ordered alloys.
4. Fine-particle magnets.

148. Martensitic Lattice-Transformation Alloys (Quench- and Work-hardened Alloys). These are iron-rich alloys quenched from the fcc γ phase to form a martensitic structure. It is a very fine platelike structure which is created from the γ phase by a rapid shear transformation. This fine structure is both mechanically and magnetically hard. High-carbon steels containing W, Cr, Co, Al, or V are most commonly used. Properties and composition of some of these steels are given in Table 6-47, and the demagnetizing curves in Fig. 6-41. In some alloys the martensitic transformation can be induced by work hardening.

149. Precipitation-hardened Alloys. These alloys develop a very fine structure due to precipitation upon heat-treatment. This fine structure results in very high remanence, coercive force, and very high energy product.

Alnico is the best-known alloy in this group. Its maximum energy product can be increased by directional solidification in the <100> directions. Solution treatment at 1300°C followed by a 800°C heat-treatment will produce further improvements. The 800°C treatment will produce a network of fine α'-Fe, Co-rich precipitates with an α-Ni-Al-rich matrix. The α' has a higher magnetization. Annealing in the presence of a magnetic field results in elongated α' particles which improve the magnetic hardness due to shape anisotropy of the α'. Long heat-treatments at 580°C further increase the "hardness" of Alnico.

Alnico magnets can be cast to complex shapes. There are various kinds of Alnico alloys, and their magnetic properties are listed in Table 6-47. Demagnetization curves are given in Fig. 6-41. The remanence of a magnet is affected by temperature. Alnico 5 suffers a structural change at 550°C, but will lose part of the remanence at lower temperatures.

Alnico magnets are mechanically hard and brittle. To compromise these properties with the magnetic hardness, the ductile alloys were developed. *Cunife* is a ductile alloy, containing 20% Ni, 60% Cu, and 20% Fe, which has still a reasonable magnetic hardness. *Cunico* can be machined before heat-treatment but not after. *Vicalloy* is a Co-V-Fe alloy which is ductile before heat-treatment. *Silmanal* is a ductile Ag-Mn-Al alloy having a very high coercive

Table 6-46. Properties of Commercially Available Microwave Ferrites* (Measurements Made at X-Band)[131]

Supplier	Material code	Composition	Saturation magnetization $4\pi M_s$, (Ga)	Curie temp. T_C, (K)	Resonance line width ΔH, (Oe)	Landé g factor	Coercive force H_C, Oe	Dielectric constant ϵ	Dielectric loss factor $\tan \delta_x$	Density, g/cm³
						Garnets				
K	Y-1	Y	1800	575	60	2.00			<0.00025	5.06
S	G-54	Y	1780	555	45	2.01			<0.00025	
T	G-113	Y	1780	555	55	2.0	0.75	16.0	0.0005	
R	R-171	Y	1750	555	60	2.01		15		5.08
H	H-101	Y	1750	555	35			16.3		
M	MCL YIG	Y	1750	550	50			16.2	0.0005	5.1
A	C30P37	Y	1740	550	60	2.0				
T	G-1600	YGd	1600	540	60	2.0	1.1	16.0	<0.00025	
M	MCL1600	YAl	1540	550	50			14.8	0.0005	
S	G-57	YGd	1520	550	55	2.01			<0.00025	5.26
S	G-77	YAl	1470	525	40	2.01			<0.00025	5.04
S	G-59	YGd	1300	555	70	2.03			<0.00025	5.42
M	MCL1200	YAl	1220	515	55			14.6	0.0005	
R	R-132	YGd	1200	525	100	2.0	0.75	15	0.0005	
T	G-1200	YGd	1200	555	85	2.00	0.5	15	<0.00025	
M	MCL1110M	YAl	1200	535	60	1.99		15.5	<0.00025	
A	C32P37	YGdAl	1150	525	55			14.8	0.0005	
S	G-79	YAl	1120	500	60	2.0			<0.00025	5.0
S	G-37	YAl	1000	485	40	2.01	0.8		<0.00025	5.03
T	G-1002	YGd	1000	555	130	2.05	1.2	15.5	<0.00025	5.68
T	G-1000	YAl	1000	555	120	2.00		15.0	<0.00025	
M	MCL1000B	YAi	990	525	60	1.99		14.4	0.0005	
	G85	YAl	850	485	55				<0.00025	5.00
S	G-1003	YGd	850	455	40	2.01	1.0	15.5	<0.00025	
R	R-131	YGd	840	555	160	2.00		15	0.0005	
T	G-800	YAl	800	525	200	2.0	1.2	15.0	<0.00025	
M	MCL1118R	YGdAl	800	505	65	2.00		15.0	0.0005	
M	MCL800STA	YGdAl	750	525	110			15.0	0.0005	
S	G-52	YGd	750	525	80	2.08			<0.00025	5.85
T	G-1005	YGd	725	555	220	2.03	1.0	15.5	0.0005	

	Grade	Comp.								
T	G-600	YAl	680	475	70	2.00	1.2	15.0	<0.00025	5.1
T	G-610	YAl	680	455	45	2.00	1.0	14.5	<0.00025	5.73
A	C34P37	YGdAl	675	475	120	2.0				5.00
S	G-188	YGdAl	660	525	210	2.05		15	<0.00025	
S	G-91	YAl	650	445	40	2.02		14.3	<0.00025	
R	R-181	YAl	600	430	60	2.0		14.0	0.0005	
M	MCL1116FH	YAl	600	445	40			14.0	0.0005	5.63
T	G-500	YAl	550	435	75	2.00	1.2	14.0	<0.00025	
T	MCL500	YAl	475	430	50			13.8	0.0004	
M	G-400	YAl	400	405	45	2.01	1.0	14.0	<0.0004	
M	MCL400	YAl	400	400	45				0.0005	
S	MCL300LLS	YAl	370	400	40			14.0	0.0005	
T	G-189	YGdAl	340	475	310	2.10	0.4	13.4	<0.00025	
M	G-300	YAl	300	400	40	2.02		14.0	<0.00025	
T	MCL290TG	YAl	290	390	40		1.0	13.8	0.0005	
M	G-240	YAl	240	370	40	2.03			<0.00025	
M	MCL220	YAl	225	380	35				0.0004	

Nickel Ferrites

	Grade	Comp.								
T	TT2-111	NiZn	5000	650	135	2.08	0.9	12.5	0.001	5.1
A	C11P16	Ni	3000	850	350	2.3		13.0	0.0025	
T	TT2-101	NiCo	3000	860	350	2.21	12	13	0.005	
R	R-191	NiCo	3000	865	200	2.2		12	0.0005	
R	R-161	Ni	3000	826	500	2.3			>0.005	
S	54	NiCoAl	2440	825	260	2.33		12.0	0.002	5.05
H	H-300	NiTiAl	2220	835	500	2.45		12.8	0.001	4.80
T	TT2-125	NiAl	2000	820	450	2.31	4	9.0	0.0015	
T	TT2-120	NiCoAl	1800	775	1000	2.55	36	9.5	0.002	
T	TT2-118	NiCoAl	1800	755	800	2.55	21	11.3	0.0003	
K	AN-15-MW	NiAl	1600	725	450	2.53		12.0	0.001	4.7
T	TT2-115	NiCoAl	1600	850	330	2.45	6.5	9.4	0.0017	4.7
K	N-40-MW	Ni	1500	715	800	2.42				
A	C3P50	NiAl	1450	715	275	2.5		12.5	0.001	
A	C3P1	NiAl	1450	850	300	2.5	14		>0.005	
T	TT2-116	NiCoAl	1400	700	200	2.38		12.0	0.001	5.04
S	32	NiCoAl	1240	699	360	2.57	7		>0.0005	
T	TT2-130	NiAl	1000	675	300	2.7		9.0	0.0005	4.97
S	40	NiCoAl	800	642	600	2.94	5.5			
T	TT2-113	NiAl	500	435	155	1.54				
A	C6P37	NiAl	420	430	225	1.4				4.8
K	AN-50-MW	NiAl	350	555	225	1.46		8.2	0.0003	4.8

Table 6-46. Properties of Commercially Available Microwave Ferrites*
(Measurements made at X-Band) *(Concluded)*

Supplier	Material code	Composition	Saturation magnetization $4\pi M_s$ (Ga)	Curie temp. T_C (K)	Resonance line width ΔH (Oe)	Landé g factor	Coercive force H_C Oe	Dielectric constant ε	Dielectric loss factor $\tan \delta_\varepsilon$	Density, g/cm³
Magnesium Ferrites										
S	83	MgMn	2220	585	500	2.07	...	...	...	4.20
T	TT1-390	MgMn	2150	595	540	2.10	2.5	13.0	<0.00025	
R	R-151	MgMn	2100	575	425	2.0	...	12	<0.00025	
A	C8P35	MgMn	2000	555	350	2.0	...	...	0.0005	4.3
K	MGM-11-MW	MgMn	2000	575	400	2.07	...	12	0.0008	4.3
A	C9P28	MgMn	1950	515	150	2.0	...	...	...	4.3
I	R-4	MgMn	1780	595	380	2.06	1.4	12	0.0005	4.15
I	R-1	MgMnAl	1760	565	490	2.11	2.1	12	0.0005	4.10
T	TT1-105	MgMnAl	1700	500	225	2.00	1.0	12.0	<0.00025	
A	C7P12	MgMn	1650	575	650	2.0	...	...	...	4.3
A	C9P13	MgMnAl	1550	515	375	2.0	...	...	...	4.3
T	TT1-109	MgMnAl	1250	439	155	2.00	0.4	11.5	<0.00025	
A	C43P11	MgMnAl	1250	475	300	2.0	...	...	...	4.2
S	65	MgMnAl	1140	425	165	2.02	...	...	...	4.10
I	R-5	MgMnAl	1110	415	220	2.04	0.8	...	0.0003	3.95
I	R-6	MgMnAl	730	375	120	2.00	0.7	12	0.0004	3.90
T	TT1-414	MgMnAl	680	375	120	2.02	1.2	11.5	<0.00025	

* All data in this table are based on catalogs or specifications supplied by the manufacturers. Manufacturers are as follows:
A—Airtron, Division of Litton Industries, 200 East Hanover Ave., Morris Plains, N.J.
H—Hughes Aircraft Corp., Aerospace Group, Research & Development Div., Culver City, Calif.
I—Indiana General Corp., Electronics Div., Keasby, N.J.
K—Kearfott Div., General Precision, Inc., 1150 McBride Ave., Little Falls, N.J.
M—Microwave Chemicals Laboratory, Inc., 282 Seventh Ave., New York, N.Y.
R—Raytheon Co., Special Microwave Devices Operation, 130 Second Ave., Waltham, Mass.
S—Sperry Microwave Electronics Co., Clearwater, Fla.
T—Trans-Tech, Inc., 12 Meem Ave., Gaithersburg, Md.

force. *Remalloy* or *Comol* are Co or Mo ductile age-hardened iron alloys. Properties of age-hardened permanent magnets are listed in Table 6-47.

150. Ordered Alloys. These include the Fe-Pt and Co-Pt alloys which exhibit ordered superlattices on aging at 700°C. These alloys have very high coercive forces and the highest energy product. For properties see Table 6-47.

151. Fine-Particle Magnets. Fine-particle magnets are produced from compacted metallic powders or ceramic magnetic materials. Metallic-powder magnets with high-energy products have been produced from iron, 70% Fe-30% Co alloys, Alnico, Remalloy, and others. The properties of powder magnets have been further improved by the use of elongated-single-domain (ESD) powders which consist of elongated fine particles (formed electrolytically on a liquid-metal electrode) which impart very high retentivities to the magnets. MnBi powder (Bismanol) magnets have been produced with exceedingly high coercive forces (see Fig. 6-41).

Fig. 6-47. Properties of Permanent-Magnet Materials[87]

	Required magne-tizing force	Residual induc-tion	Coercive force	Peak energy product	Occurs at	
	H_S, Oe	B_r, G	H_c, Oe	$(B_d H_d)_{max}$, Mg·Oe	B	H
Cast Alnico 1	2,000	7,100	400	1.3	4,200	300
Cast Alnico 2	2,500	7,200	540	1.6	4,500	365
Cast Alnico 3	2,500	6,700	450	1.4	4,300	320
Cast Alnico 4	3,500	5,200	700	1.2	3,000	380
Cast Alnico 5	3,000	12,400	640	5.50	10,000	550
Cast Alnico 5DG..............	3,500	12,600	670	6.25	10,450	600
Cast Alnico 5-7	3,500	13,000	730	7.25	11,150	650
Cast Alnico 6	4,000	10,200	770	3.75	7,000	535
Cast Alnico 8A	8,000	8,500	1,600	5.0	5,000	1,000
Cast Alnico 8B...............	10,000	7,500	1,850	5.25	4,550	1,150
Sintered Alnico 2..............	2,500	6,800	520	1.50	4,300	350
Sintered Alnico 5..............	3,000	10,500	590	3.50	8,000	435
Sintered Alnico 6..............	4,000	8,600	790	3.0	6,000	500
Sintered Alnico 8..............	8,000	7,600	1,550	4.5	4,700	960
Platinum-cobalt	21,000	6,000	4,200	7.50	3,000	2,500
P-6 alloy*	300	14,000	60	0.55	10,500	52.4
Lodex 30	6,000	4,000	1,250	1.68	2,400	700
Lodex 31	6,000	6,250	1,140	3.4	4,250	800
Lodex 32	5,000	7,300	940	3.4	5,300	650
Lodex 33	5,000	8,000	860	3.2	5,800	550
1095 carbon steel	300	8,600	48	0.18	6,000	25
5% tungsten steel	300	10,300	70	0.32	7,400	50
3½% chromium steel	300	9,000	63	0.29	6,000	35
17% cobalt steel	750	9,000	170	0.65	5,900	110
36% cobalt steel	1,000	9,600	228	0.94	6,300	140
Cunife	2,400	5,400	550	1.50	4,000	375
Cunico........................	3,000	3,400	660	0.8	2,000	400
Barium ferrite (Isotropic)........	10,000	2,200	1,600	0.9	1,000	850
Oriented barium ferrite 1........	10,000	3,050	2,400	2.20	1,525	1,440
Oriented barium ferrite 2........	10,000	3,700	1,900	3.25	1,850	1,775
Remalloy (Comol)..............	1,000	10,000	230	1.1	6,900	160

NOTE: Properties are based on relatively simple geometry. In the case of antisotropic materials, these properties refer to the preferred direction of magnetization.
* When heat-treated at 600°C for 2 h.

An outstanding ceramic permanent-magnet material is the $BaFe_{12}O_{19}$ (Ferroxdure), belonging to the group of hexagonal ferrites. The ceramic magnets have a much higher ohmic resistance than metallic-powder magnets. They have high coercive forces, are chemically inert, and usually have a lower cast. Ceramic magnets can operate at high frequency and are sometimes used as resonance oscillators at the 50-GHz range. Ceramic magnets have been used with plastic or elastomeric binders. Such construction allows machining of the magnets to the desired shape. Properties of ceramic magnets are listed in Table 6-47.

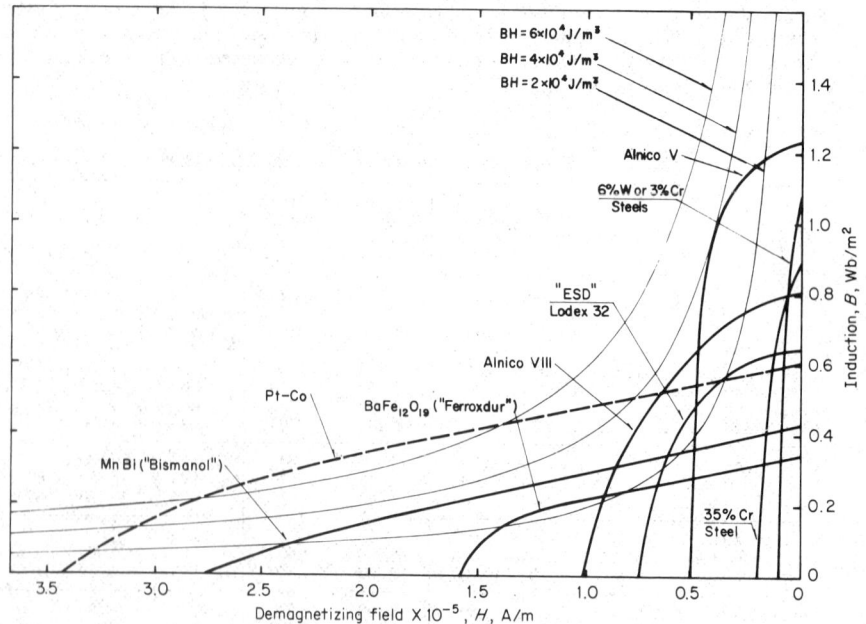

Fig. 6-41. Demagnetization curves for several hard magnetic materials.[87,88]

152. Bibliography on Magnetic Materials

1. CHIKAZUMI, S. "Physics of Magnetism;" New York, John Wiley & Sons, Inc., 1966.
2. BOZORTH, R. M. "Ferromagnetism;" Princeton, N.J., D. Van Nostrand Company, Inc., 1951.
3. TEBBLE, R. S., and CRAIK, D. J. "Magnetic Materials;" New York, John Wiley & Sons, Inc.-Interscience Publishers, 1969.
4. BERKOWITZ, A. E., and KNELLER, E. (eds.) "Magnetism and Metallurgy;" New York, Academic Press, Inc., 1969.
5. VON AULOCK, W. H. (ed.) "Handbook of Microwave Ferrite Materials;" New York, Academic Press, Inc., 1965.
6. SMIT, J. (ed.) "Magnetic Properties of Materials;" New York, McGraw-Hill Book Company, 1971.

SEMICONDUCTOR MATERIALS

153. Elemental and Compound Semiconductors. Semiconductors are usually materials which have energy-band gaps smaller than 2 eV. An important property of semiconductors is the ability to change their resistivity over several orders of magnitude by doping. Semiconductors have electrical resistivities between 10^{-5} and $10^7 \, \Omega \cdot m$. Semiconductors can be crystalline or amorphous.

Elemental semiconductors are single-element semiconductor materials (except for intentional minute additions) such as silicon or germanium. *Compound semiconductors* are materials containing more than one element, for example, III–V compounds such as GaAs, AlAs, or GaP.

154. Energy Bands and Gaps. According to band theory, discontinuities occur in energy levels at certain values of electronic wave vectors. The discontinuities are called energy gaps, or band gaps. Figure 6-42 shows the typical energy-band occupancy for metals, semiconductors, and insulators. In metals there is an incompletely filled conduction band (or overlapping bands) which allows free movement of electrons. Semiconductors have a filled *valence band* with a relatively small band gap separating it from the *conduction band.* At any finite temperature a small number of electrons are found at the bottom of the conduction band, and a number of unfilled states, or *holes,* are found in the valence band. Electrical conduction is possible by the motion of electrons or holes. Insulators have large energy gaps with filled valence bands and empty conduction bands, resulting in extremely small conductivity.

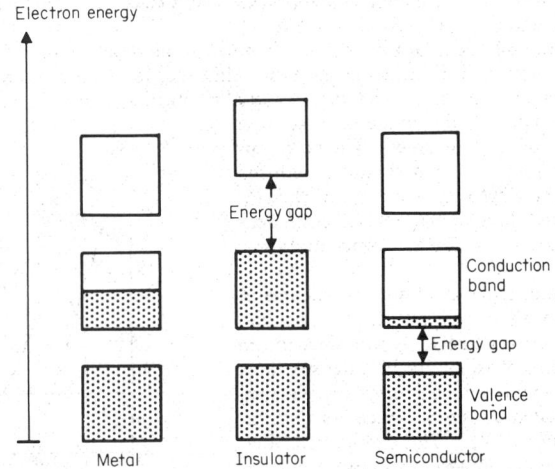

Fig. 6-42. Energy-band diagrams for metals, insulators, and semiconductors.

155. Intrinsic and Extrinsic Semiconductors. Semiconductors with a low impurity concentration having electrical properties characteristic of the pure semiconductor material are called *intrinsic* semiconductors. Semiconductors with electrical properties modified by impurities are called *extrinsic* semiconductors.

156. Charge Transport. A perfect semiconductor is, essentially, an insulator at absolute zero temperature, since its conduction band is vacant. As the temperature increases, electrons are thermally excited from the valence into the conduction band, and the conductivity increases. The free electrons in the conduction band and the holes in the valence band participate in electrical conduction. The electrons are negative charge carriers, while the holes are positive carriers. Actual crystals exhibit purely *intrinsic conduction* (i.e., independent of impurity concentration) only at high temperatures. At lower temperatures the impurity effect can be quite large.

157. Carrier Concentration in Intrinsic Semiconductors. The concentration of electrons, n, and of holes, p, in intrinsic semiconductors can be calculated based on a simple model of the band structure (assuming the distance between the Fermi level and the band edges to be large in comparison with kT).

$$n = N_c e^{-(E_c - E_f)/kT}$$

$$p = N_v e^{-(E_f - E_v)/kT}$$

(6-65)

6-97

where n and p = electron and hole densities (m^{-3}). $N_c = 2(\frac{2\pi m^* n kT}{h^2})^{3/2}$ can be thought of as an effective density of states at the bottom edge of the conduction band. $N_v = 2(\frac{2\pi m^* p kT}{h^2})^{3/2}$ is the effective density of states at the top edge of the valence band. E_f, E_c, E_v are the energies corresponding to the Fermi level, conduction-band edge, and valence-band edge, respectively. m^*_n, m^*_p are the effective masses of electrons and holes, respectively (kg), k = Boltzmann's constant (J/K), h = Planck's constant (J · s), and T = temperature (K).

The product of the electron and hole densities is

$$np = N_c N_v e^{-(E_c - E_v)/kT} = N_c N_v e^{-E_g/kT} \qquad (6\text{-}66)$$

where the energy gap $E_g = E_c - E_v$. The np product depends on the energy gap, and not on the Fermi level. For intrinsic semiconductors $n = p = n_i$ and

$$np = n^2_i \qquad (6\text{-}67)$$

$$n_i = (N_c N_v)^{1/2} e^{-E_g/kT} \qquad (6\text{-}68)$$

Figure 6-43 shows n_i as a function of T for Si, Ge, and GaAs. It should be noted that N_c and N_v are also slightly temperature-dependent.

158. Donor and Acceptor Impurities. Impurities can be added intentionally to semiconductors to modify their electrical properties. This addition is termed *doping*.

Silicon doped with pentavalent substitutional impurities such as P, As, or Sb will have a higher electron density due to the easily ionized fifth electrons of the impurities. The hole concentration in such a silicon crystal is reduced since the np product remains constant [Eq. (6-66)]. Dopants which increase the electron density are called *donors*. Dopant concentration is denoted N_D.

The addition of trivalent dopants such as B, Al, and Ga to silicon will attract electrons and reduce the electron density. Again, since the np product remains constant, the hole density must increase. Dopants which reduce the electron density are called *acceptors*. Acceptor concentration is denoted N_A. The energy-band diagrams for donor- and acceptor-doped semiconductors are shown in Fig. 6-44.

Donor-doped material is called *n*-type because the majority of carriers are electrons. The electrons in such semiconductors are the *majority carriers*, while the holes are the *minority carriers*. Acceptor-doped materials are called *p*-type. Here the holes are the majority carriers and electrons are the minority carriers. A highly doped *n*-type semiconductor is often referred to as degenerate and denoted by $n+$. Similarly, a highly doped *p*-type is denoted $p+$.

159. Carrier Concentration in Extrinsic Semiconductors. To calculate the carrier concentration in extrinsic semiconductors we use the electrical charge neutrality equation. For fully ionized impurities,

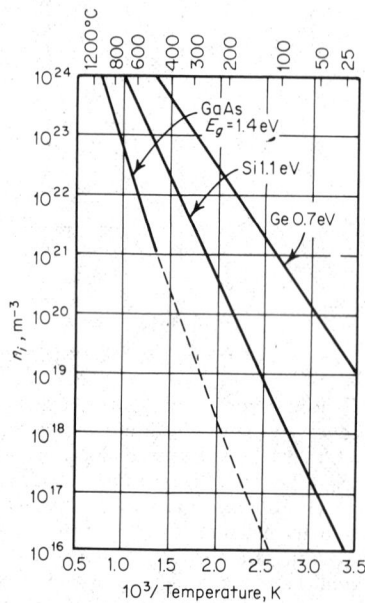

Fig. 6-43. Intrinsic carrier concentration in Si, Ge, and GaAs as a function of temperature.[90-92]

$$p_n + N_D = n_n + N_A$$
$$\qquad (6\text{-}69)$$
$$p_p + N_D = n_p + N_A$$

where p_n and n_n = hole and electron concentration in *n*-type semiconductors (m^{-3}), p_p and

n_p = hole and electron concentrations in a p-type semiconductor (m^{-3}), and N_D and N_A = donor and acceptor concentrations (m^{-3}).

Using Eqs. (6-69) and (6-67) and assuming $n_i << | N_D - N_A |$, the carrier concentration in semiconductors of

$$n\text{-type:} \begin{matrix} n_n = N_D - N_A \\ p_n = n_i^2/(N_D - N_A) \end{matrix}$$

$$p\text{-type:} \begin{matrix} p_p = N_A - N_D \\ p_n = n_i^2/(N_D - N_A) \end{matrix}$$

(6-70)

It can be seen from Eq. (6-70) that a reduction in charge carriers (and conductivity) can be obtained in materials with low $N_D - N_A$. This lowering can be done, for example, by intentionally adding donors into a p-type material. Such a material is called *compensated intrinsic.*

Electron and hole concentrations in n-type silicon as a function of inverse temperature are shown in Fig. 6-45. Partially ionized donors are the source of carriers at low temperature. At high temperature, impurities are completely ionized, and at still higher temperature, the thermally excited electrons from the valence band begin to dominate and the semiconductor becomes intrinsic.

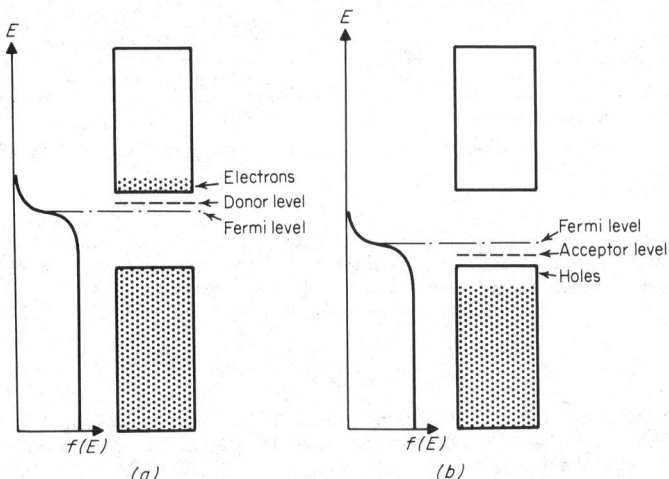

Fig. 6-44. Band diagrams for (*a*) n-type and (*b*) p-type semiconductors. The Fermi-Dirac distribution functions $f(E)$ are also shown.

The carrier concentration determines the position of the Fermi level. Figure 6-46 shows the Fermi level as a function of donor and acceptor concentrations.

160. Impurity Energy Levels. The doping impurities have characteristic energy states within the energy gap. Impurities with energy states close to the band edges are called *shallow states,* and those closer to the center of the gap are called *deep states.* Figure 6-47 shows the energy states of various impurities in Si, Ge, and GaAs. Some impurities have several energy states within the band gap.

161. Surface States. Oxidized semiconductors are of interest in most semiconductor devices. Such semiconductors have localized electronic states both in the oxide and at the semiconductor interface. The density of states on freshly cleaved surfaces is about 10^{19} m^{-2}. Oxidized surfaces show densities of 10^{15} to 10^{16} m^{-2}. The density of the states can be reduced by proper manufacturing techniques.

Surface states have been divided into slow and fast states, depending upon the speed of their interaction with the semiconductor space-charge region.

162. Conduction in Semiconductors. An electric field applied to a semiconductor will induce a drift on the charge carriers. At low fields, the drift velocity v_d is proportional to the field E:

$$v_d = \mu \cdot E \tag{6-71}$$

where the proportionality constant μ is the mobility.

The conductivity in semiconductors is given by

$$\sigma = e(n\mu_n + p\mu_p) \tag{6-72}$$

where σ = conductivity ($\mho \cdot$ m), n, p = electron and hole density, respectively (m^{-3}), μ_p, μ_n = mobilities of electrons and holes, respectively (m^2/V $\cdot$ s), and e = electronic charge (C).

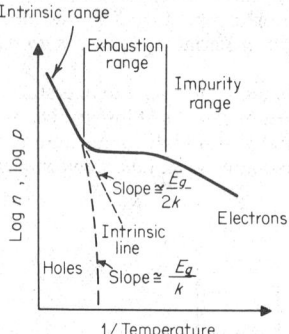

Fig. 6-45. Carrier concentration in an n-type semiconductor as a function of inverse temperature.[93]

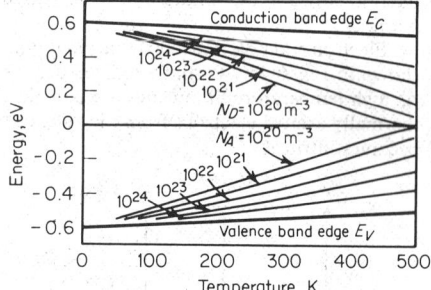

Fig. 6-46. The Fermi level in Si as a function of impurity concentration.[94]

The mobility of electrons and holes is affected by two main scattering mechanisms, *impurity* and *lattice scattering*. The mobility due to impurity scattering μ_I is proportional to $T^{3/2}$, while the mobility due to lattice scattering μ_L is proportional to $T^{-3/2}$. The total mobility μ is related to the individual components as

$$\frac{1}{\mu} = \frac{1}{\mu_I} + \frac{1}{\mu_L} \tag{6-73}$$

The mobility of electrons and holes in Si, Ge, and GaAs is shown as a function of impurity concentration in Fig. 6-48. The resistivity of these semiconductors as a function of impurity concentration is shown in Fig. 6-49.

Crystal imperfections also scatter electrons. This scattering is small and has only a slight temperature dependence. The mobilities near semiconductor surfaces are decreased to a fraction of the bulk mobilities. This decrease is in part due to extra scattering of the carriers at the surfaces.

163. Carrier Diffusion. Carrier concentration gradients cause a current flow in addition to the normal flow by electric fields. The current density for electrons or holes in a concentration gradient along the x axis is

$$J_n = eD_n \frac{dn}{dx}$$

$$\tag{6-74}$$

$$J_p = -eD_p \frac{dp}{dx}$$

where J_n, J_p = electron or hole current densities (A/m^2), e = electronic charge (C), D_n, D_p = diffusion coefficients for electrons or holes (m^2/s), and n, p = electron or hole concentrations (m^{-3}).

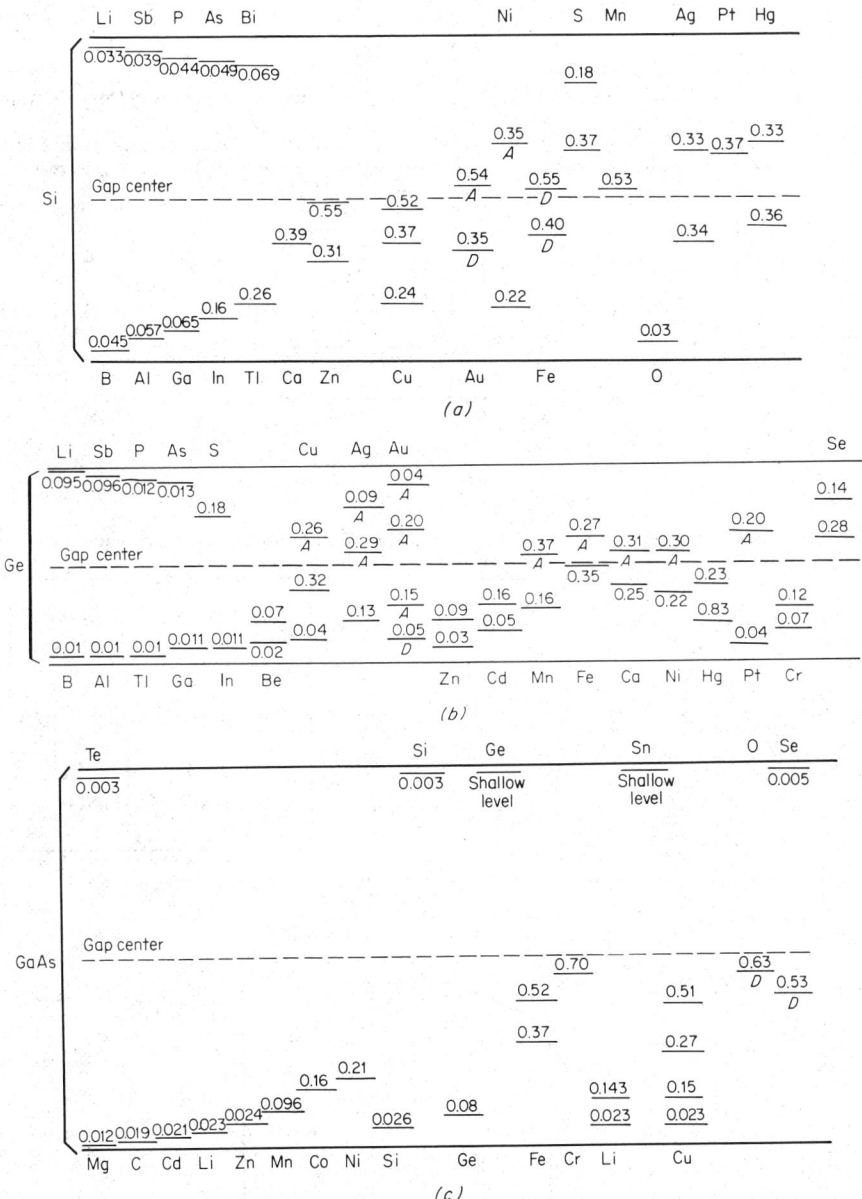

Fig. 6-47. Impurity energy levels in (*a*) Si, (*b*) Ge, and (*c*) GaAs.[95]

The diffusion coefficients are related to mobilities according to Einstein's relation,

$$D_n = \frac{kT}{e} \mu_n$$

$$D_p = \frac{kT}{e} \mu_p$$

(6-75)

where k = Boltzmann's constant (J/K), μ_n, μ_p = electron or hole mobilities (m²/V · s), and T = temperature (K).

164. Carrier Generation and Recombination. *Generation* of electron-hole pairs can be accomplished by means other than thermal, for example, by illuminating the semiconductor.

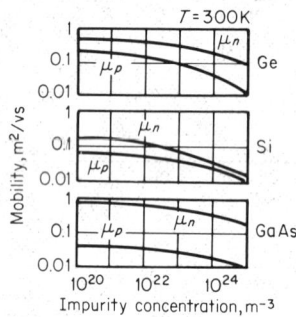

Fig. 6-48. Drift mobility of Ge and Si and Hall mobility of GaAs vs. impurity concentration at 300 K.[95]

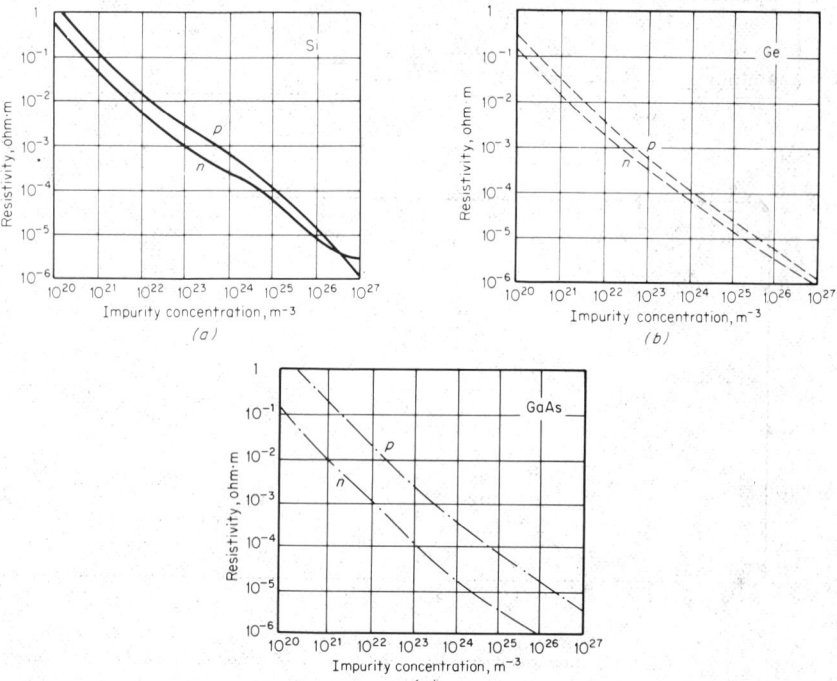

Fig. 6-49. Resistivity vs. carrier concentration at room temperature for (*a*) Si, (*b*) Ge, and (*c*) GaAs.[96-98]

A semiconductor with excess electron-hole pairs will return to equilibrium by *recombination* of holes and electrons. An n-type semiconductor with p_n minority carriers at equilibrium and p_{n0} carriers at time $t = 0$ (just after illumination removal) will have a concentration of p carriers at time t.

$$p = p_m + (p_{n_0} - p_n)e^{-t/\tau p} \qquad (6\text{-}76)$$

where τ_p = hole lifetime (s). (Similarly, τ_n is the excess electron lifetime in p-type semiconductors.)

Recombination can be accomplished by a band-to-band process or with the aid of impurities or imperfections. Impurities with deep levels near the center of the energy gap are efficient recombination-generation centers, for example, gold or platinum in silicon or copper in germanium. Defects produced by high-energy radiation can also serve as recombination centers and decrease the minority-carrier lifetime. τ_n, τ_p should not be confused with *relaxation time* for conduction, also denoted by τ.

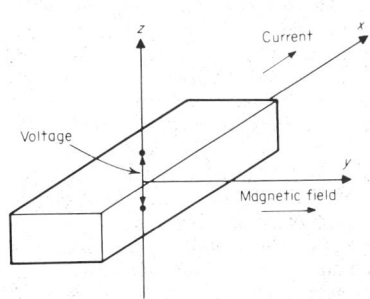

Fig. 6-50. Schematic diagram of Hall voltage developing in the z direction relative to an electric current in the x direction and a magnetic field in the y direction.

Surfaces are sites of enhanced recombination which causes a decrease in the minority-carrier concentration at the surface. This reduction in concentration produces strong concentration gradients which create hole and electron flow toward the surface. The diffusion flux toward the surface is balanced by the recombination at the surface. For example, the current density for holes is given by

$$J = e(p_{n_0} - p_n)S \qquad (6\text{-}77)$$

where J = electric current density normal to the surface (A/m^2), e = electronic charge (C), p_{n_0}, p_n = hole concentration (m^{-3}), and S = *surface recombination velocity* (m/s).

165. High-Energy Irradiation Effects. High-energy irradiation can cause crystal imperfections (such as vacancies and interstitialcies) in semiconductors which serve as recombination centers, and therefore reduce the minority-carrier lifetime. Irradiation damage of this sort is annealed out even at room temperature.

Irradiation with x-rays, γ-rays, electrons, or other ionizing radiation builds up positive space charges in SiO_2. The space charge originates from electron-hole pairs generated by the radiation. Application of an electric field on the oxide can separate the electrons and holes upon generation, accentuating the space-charge buildup.[21] Irradiation damage can be annealed at about 300°C.

166. Hall Effect. A magnetic field applied across a current-carrying material will force the moving carriers to crowd to one side of the conductor. An electric field will develop as a result of this crowding, shown schematically in Fig. 6-50, where the current density J_x is in the x direction and the magnetic induction B is in the y direction. The Hall coefficient R_H is for an n-type semiconductor, defined as

$$R_H = E_H/J_xB_y = \frac{1}{-en} \qquad (6\text{-}78)$$

where n = density of electrons (m^{-3}) and e = electronic charge (C). p-type semiconductors show a positive Hall coefficient.

167. Photoconductive Effects. Photons with high enough energies can generate electron-hole pairs. The excess carriers will cause an increase in conductivity, called *photoconductivity*. The photoconductivity can be utilized by using a reverse-biased pn junction in which the reverse current is small and the effect of illumination is very large. Illuminated nonbiased junctions can develop a voltage across the junction, called *photovoltaic effect*. Light emission from semiconductors is discussed in Par. **6-229.**

168. Thermoelectric Effects. Several semiconductors have large thermoelectric power.

The thermal and electrical properties of some semiconductors allow their use in thermoelectric heating and refrigeration.

169. Acoustoelectric Effect.   The acoustoelectric effect is an electric current generated by longitudinal acoustic waves.   The waves transmit some of their momentum to electrons and force a small current to flow.

JUNCTION PHENOMENA

170. pn Junctions.   When a p-type semiconductor is brought into contact with an n-type semiconductor, holes will diffuse from the p-type to the n-type material, while electrons will diffuse in the opposite direction.   The displacement of charges creates an electric field which opposes the diffusion.   The Fermi level stays constant across a junction in equilibrium.

Reverse biasing a pn junction (that is, p is negatively biased with respect to n) will tend to create an electron flow in the n-type material and a hole flow in the p-type material away from the junction.   Such currents resulting from electron-hole-pair generation are small.   Forward biasing a junction will tend to move the majority carriers toward the junction, where their recombination can create a large forward current.

The I-V characteristic of a pn junction is shown in Fig. 6-51.   Note that at high reverse bias the junction suffers an electrical *breakdown*.   The breakdown can be due to avalanche processes (multiplication of electron-hole pairs by collision of fast-moving charge carriers) or by *zener breakdown* (due to tunneling between the two sides of junction).   *Soft breakdown* is a very undesirable occurrence in semiconductor junctions.   This kind of breakdown is characterized by high-reverse-bias currents at voltages smaller than breakdown voltage. Copper and iron are notorious for producing "soft" junctions in silicon.   Junctions can be "hardened" by a gettering process with phosphorus or nickel.

Junctions cannot be operated above a temperature which causes the semiconductor to be intrinsic because the rectification properties vanish.

171. p^+n^+ Junctions.   Highly doped p^+n^+ junctions are used in tunnel diodes.   The energy-band diagram and I-V characteristic are shown in Fig. 6-52.   The forward biasing of

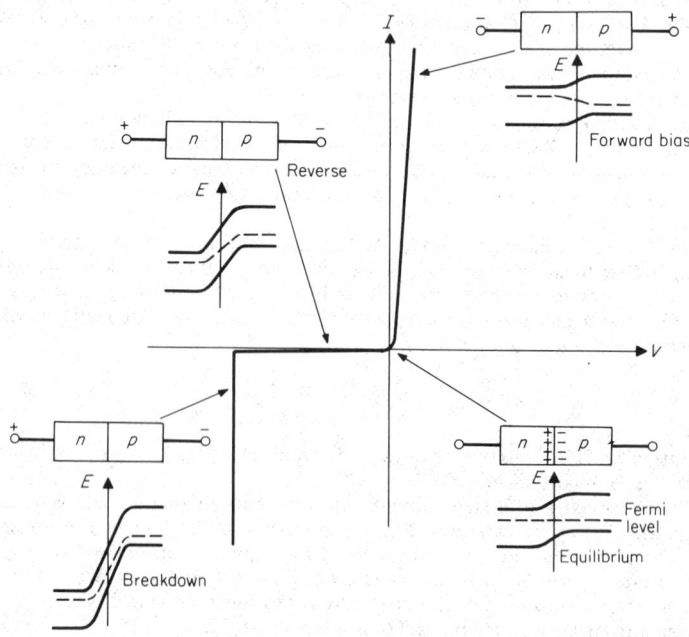

Fig. 6-51. Schematic diagram of pn junction current - voltage characteristics and related energy levels.

6-104

the diode will move the conduction states of the n^+ material close to the states of the p^+ valence band, and a large tunnel current will result. Higher forward bias will decrease the tunneling current since it will separate the n^+ and p^+ states.

172. Metal-Semiconductor Junctions. Metal semiconductor junctions can be either rectifying (nonohmic) or nonrectifying (ohmic) junctions. Low-resistance ohmic junctions are needed as contacts to the various p or n regions in semiconductors. It is customary to dope the semiconductor contacts heavily to achieve low-resistance ohmic junctions. Rectifying metal semiconductor junctions are often used as high-frequency diodes (Schottky barrier diodes).

173. Heterojunctions. Occasionally, junctions are produced between two kinds of semiconductors, for example, GaAs and Ge. Such junctions are called heterojunctions.

SPECIFIC SEMICONDUCTORS

Properties and applications of semiconductors are summarized in Tables 6-48 to 6-50 and Figs. 6-53 to 6-58.

174. Silicon. Silicon is the most common semiconductor material used today. It is used for diodes, transistors, integrated circuits, memories, infrared detection and lenses, light-emitting diodes (LED), photosensors, strain gages, solar cells, charge transfer devices, radiation detectors, and a variety of other devices.

Silicon belongs to group IV in the periodic table. It is a gray brittle material with a diamond cubic structure (lattice parameter = 5.43 Å). Silicon is available free of dislocations, but is generally used with dislocation densities of 10^6 to 10^{10} m/m^3. Additional dislocations are often generated in silicon during processing. The energy gap of silicon is 1.1 eV, this value permitting the operation of Si semiconductor devices at higher temperatures than germanium. Silicon is conventionally doped with P, As, and Sb donors and B, Al, and Ga acceptors.

Silicon is mainly processed by the planar technology. Silicon (n^+) single crystals are sawed into slices, and the slices polished mechanically and chemically. An epitaxial n layer is deposited onto the n^+ substrate; then a predeposition of p-type dopant is performed on the base region. Diffusion of the predeposited dopant drives the doping impurities into the silicon. The n-type emitter is next predeposited and diffused, and the desired npn structure is achieved. Reduction of minority-carrier lifetime is obtained by diffusing gold or platinum throughout the silicon slice. Phosphorus or nickel is also generally diffused into the silicon to obtain stable hard junctions. Silicon is also used in polycrystalline form as a thin-film conductor.

The metal oxide semiconductor (MOS) technology[21] is another common technology used for producing field effect transistors and large-scale memories.

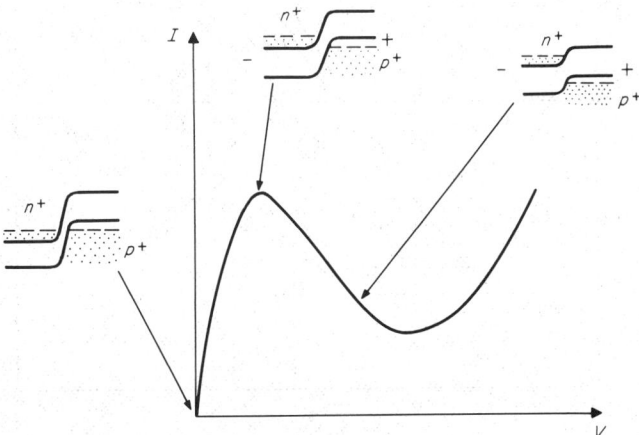

Fig. 6-52. Schematic diagram of p^+n^+ junction current - voltage characteristics and relative energy levels.

Silicon is sensitive to nuclear radiation which can introduce recombination centers, increase surface state density, and introduce space changes in biased silicon oxide gates.

Silicon is used for infrared lenses because it has good transmittance edge in the near-infrared region (Fig. 6-58).

175. Germanium. The use of germanium as a semiconductor is rather small compared with silicon. The transistor action was first observed in germanium. Germanium found uses in near-infrared detection and in nondispersive x-ray detectors.

Germanium is a group IV gray brittle material with a diamond cubic structure (lattice parameter = 5.657 Å). It is available with very low dislocation densities.

The energy gap of germanium, 0.67 eV, precludes its use at temperatures higher than 80°C. Germanium cannot be conveniently fabricated by the planar technology because its oxide is not stable.

Doped germanium crystals, cooled to liquid-helium temperature, serve as infrared detectors. Li-doped crystals are used at this temperature for nondispersive x-ray detection.

176. Tin. The gray cubic form of tin, stable below 13.2°C, is semiconducting. This allotropic form is diamond cubic (lattice parameter = 6.5041 Å). Gray tin has a very small energy gap and is very brittle. It has not been commercially used.

177. Diamond and Graphite. Diamond is a transparent, extremely hard cubic material (lattice parameter = 3.56 Å). Diamond has a very high band gap (~5.3 eV) at room temperature, making it essentially an insulator. Several p-type diamonds have been found, however, with resistivities of 0.1 to 1 Ω · m. Diamonds have not been used commercially as semiconductors, although they have found some use as heat sinks since they have the highest observed heat conductivity.

Graphite has a hexagonal structure ($a = 2.461$ Å, $c = 6.708$ Å). It has a high electrical resistivity perpendicular to the basal plane but is metallic along some directions. It can be made an n- or p-type semiconductor by doping.

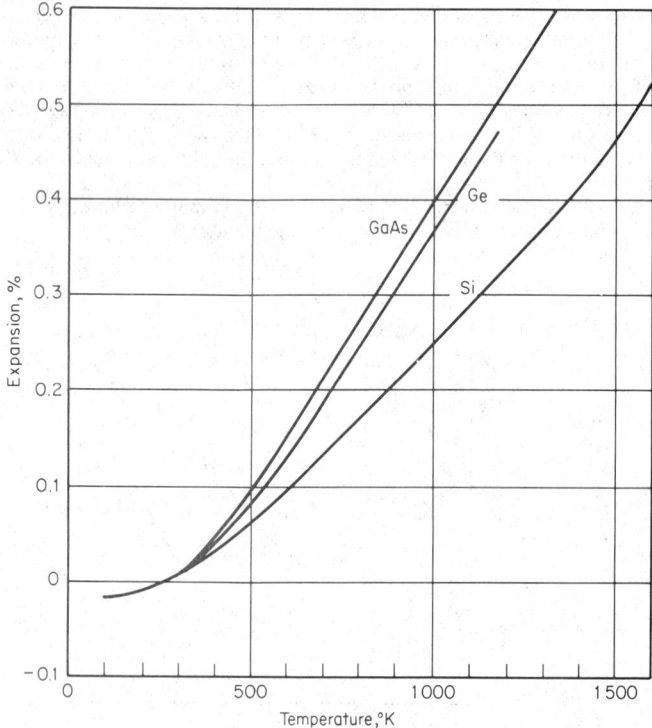

Fig. 6-53. Thermal expansion of Si, Ge, and GaAs.[61]

178. Selenium. Selenium is an element in the VI period group. It can be found in the monoclinic (α), monoclinic (β), and the hexagonal (γ) allotropic forms. Se can also be found in a noncrystalline form. All forms are semiconducting.

Selenium is used as a rectifier material, for photovoltaic cells, and for xerographic printing. As a rectifier it is prepared as a thin crystalline layer of doped Se on steel or nickel electrodes with a Sn or Cd counterelectrode. Se films are used also as photoelectric light meters. For xerography a vitreous Se layer is used on a plate. This layer can be electrostatically charged and then preferentially discharged by selective illumination. The electrostatic image can be transmitted to a paper by dusting the plate with fine powder and transferring the image to the paper.

179. Silicon Carbide. Silicon carbide has a large energy gap (~ 3.0 eV) and was consequently tried for use as a high-temperature semiconductor. SiC can be prepared by chemical vapor decomposition, from the melt, or by sublimation. SiC is a very hard material which decomposes at 2380°C. It is found in cubic form ($a = 4.36$ Å) and in a variety of

Table 6-48. Summary of Properties of Lightly Doped Semiconductors*[100]

Group	Semi-conductor	E_G, eV	μ_n, m³	μ_p, V·s	m_n/m_o	m_p/m_o	a, Å	T_m, °C	ε_s	d_s g/cm³
IV	C	5.3	0.1800	0.1600	...	...	3.56	3800	5.8	3.51
	Si	1.1	0.1350	0.0475	0.23	0.12	5.43	1417	11.7	2.33
	Ge	0.7	0.3900	0.1900	0.03	0.08	5.66	937	16.0	5.33
	SiC	2.8	0.0400	0.0050	0.60	1.20	4.36	2830	10.0	3.22
III-V	AlAs	2.2	.0180	...	...	...	5.66	1600	8.5	3.79
	AlP	3.0	.0080	...	...	...	5.46	1500	11.6	2.38
	AlSb	1.6	.0200	.0420	0.30	0.40	6.14	1050	10.1	4.26
	BN	4.6	...	...	...	...	3.62	3000	7.1	2.20
	BP	6.0	...	.0300	...	...	4.54	1250	11.6	2.97
	GaAs	1.4	.8500	.0400	0.07	0.09	5.65	1237	10.4	5.32
	GaP	2.3	.0110	.0075	0.12	0.50	5.45	1465	8.5	4.13
	GaSb	0.7	.4000	.1400	0.20	0.39	6.10	712	14.0	5.60
	InAs	0.4	3.3000	.0460	0.03	0.02	6.06	942	11.7	5.66
	InP	1.3	.4600	.0150	0.07	0.69	5.07	1070	10.3	4.78
	InSb	0.2	8.0000	.0750	0.01	0.18	6.48	525	15.6	5.77
II-VI	CdS	2.6	.0340	.0018	0.21	0.80	5.83	1750	5.4	4.84
	CdSe	1.7	.0600	...	0.13	0.45	6.05	1350	10.0	5.74
	CdTe	1.5	.0300	.0065	0.14	0.37	6.48	1098	11.0	5.86
	ZnS	3.6	.0120	.0005	0.40	...	5.41	1850	5.2	4.09
	ZnSe	2.7	.0530	.0016	0.10	0.60	5.67	1515	8.4	5.26
	ZnTe	2.3	.0530	.0900	0.10	0.60	6.09	1238	9.0	5.70
IV-VI	PbS	0.4	.0600	.0200	0.25	0.25	5.94	1077	17.0	7.50
	PbSe	0.3	.1400	.1400	0.33	0.34	6.15	1062	23.6	8.10
	PbTe	0.3	.6000	.4000	0.22	0.29	6.46	904	30.0	8.16
II-IV	Mg₂Ge	0.7	.0530	.0110	...	...	...	1115	...	...
	Mg₂Si	0.8	.0370	.0065	...	0.46	6.34	1102	...	1.94
	Mg₂Sn	0.4	.0210	.0150	...	...	6.75	778	...	3.66
II-V	Cd₃As₂	0.1	...	1.5000	0.05	...	...	721	...	6.21
	CdSb	0.5	.0300	.1000	0.16	0.10	6.47	456	...	6.92
	Zn₃As₂	0.9	...	.0010	...	...	...	1015	...	5.53
	ZnSb	0.5	.0010	.0350	0.15	...	...	546	...	6.33
V-VI	As₂Se₃	1.6	.0015	.0045	...	...	...	608	...	4.75
	As₂Te₃	1.0	.0170	.0080	0.36	...	14.40	360	...	6.00
	Bi₂Te₃	0.2	1.0000	.0400	0.32	0.21	10.45	580	...	7.70
	Sb₂Se₃	1.2	.0015	.0045	...	...	11.68	612	...	5.81
	Sb₂Te₃	0.3	...	.0270	...	0.34	...	620	...	6.50
V-VIII	PtSb₂	0.1	.0200	.1400	...	...	6.43	1240	...	...
III-VI	In₂Se₃	1.3	.0030	...	...	...	...	890	...	5.67
	In₂Te₃	1.0	.0340	...	0.70	1.23	6.15	667	...	5.78

* Explanation of symbols: E_a = energy gap; μ_n, μ_p = electron, hole mobility; m_n/m_o = effective electron mass; m_p/m_o = effective hole mass; a = lattice constant; T_m = melting point; ε_s = dielectric constant; d_s = density.

Table 6-49.   Applications of Semiconductors[100]

Effect	Cause	Application	Semiconductors
Transistor effect.....	Current multiplication	Amplifier, etc.	Si, Ge
Tunnel effect	*pn* junction in degenerate semiconductor	High-frequency switch and storage element, oscillator, amplifier	Si, Ge, GaAs
Avalanche effect	Carrier generation, hot electrons	Cryogenic switches, high-frequency generation, high-frequency amplification	Si, Ge, GaAs
Gunn effect.........	Hot electrons in semiconductors with two different band minima	High-frequency generation, high-frequency amplification	GaAs, InP, CdTe
Piezo effect	Polar cohesion in semiconductor	Electroacoustic amplifier	GaAs, CdS, CdSe
Piezoresistance......	Disformation of band structure by pressure	Pressure indicator	Si, Ge
Varactor effect	Voltage-dependent space charge and capacity at *pn* junction	Parametric amplification, frequency multiplication, tuning diode	Si, Ge, GaAs
Pair generation......	Carrier generation by light or irradiation at a *pn* junction	Photocell, solar cell, particle counter	Si, Ge, GaAs, Se
Electroluminescence.	Radiative carrier recombination at *pn* junction	Light displays, generation of incoherent light by injection	GaAs, GaP, InAs, InSb, SiC
Laser effect.........	Radiative carrier recombination by injection to degeneracy	Laser diode, generation of coherent light by injection	GaAs, InAs, InSb
Galvanomagnetic effect.............	Influence of magnetic field or carrier motion	Hall generator, field plate	Si, InAs, InSb
Plasma waves.......	Interaction between charge carriers and electromagnetic waves in a magnetic field	Gyrator	InSb

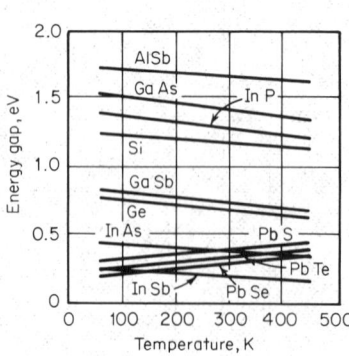

Fig. 6-54. Energy gap of several semiconductors as a function of temperature.[100]

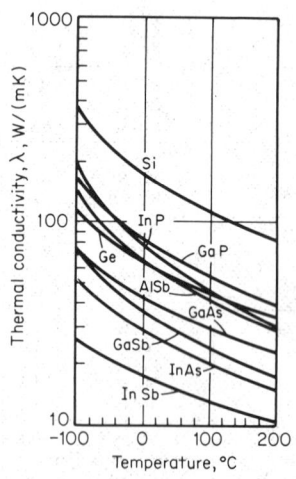

Fig. 6-55. Thermal conductivity of various near-intrinsic semiconductors as a function of temperature.[100]

Table 6-50.   Properties of Ge, Si, and GaAs (at 300 K, in Alphabetical Order)[132]

Property	Ge	Si	GaAs
Atoms/cm^3	4.42×10^{22}	5.0×10^{22}	2.21×10^{22}
Atomic weight	72.6	28.08	144.63
Breakdown field, V/cm	$\sim 10^5$	$\sim 3 \times 10^5$	$\sim 4 \times 10^5$
Crystal structure	Diamond	Diamond	Zincblende
Density, g/cm^3	5.3267	2.328	5.32
Dielectric constant	16	11.8	10.9
Effective density of states in conduction band, N_c, cm^{-3}	1.04×10^{19}	2.8×10^{19}	4.7×10^{17}
Effective density of states in valence band, N_v, cm^{-3}	6.1×10^{18}	1.02×10^{19}	7.0×10^{18}
Effective mass m^*/m_o:			
Electrons	$m_l^* = 1.6, m_t^* = 0.082$	$m_l^* = 0.97, m_t^* = 0.19$	0.068
Holes	$m_{lh}^* = 0.04, m_{hh}^* = 0.3$	$m_{lh}^* = 0.16, m_{hh}^* = 0.5$	0.12, 0.5
Electron affinity χ,V	4.0	4.05	4.07
Energy gap, eV, at 300 K	0.803	1.12	1.43
Intrinsic carrier concentration, cm^{-3}	2.5×10^{12}	1.6×10^{10}	1.1×10^7
Lattice constant, Å	5.65748	5.43086	5.6534
Linear coefficient of thermal expansion, $\Delta L/L\ \Delta T$, °C^{-1}	5.8×10^{-6}	2.6×10^{-6}	5.9×10^{-6}
Melting point, °C	937	1420	1238
Minority-carrier lifetime, s	10^{-3}	2.5×10^{-3}	$\sim 10^{-8}$
Mobility (drift), cm^2/V·s:			
μ_n (electrons)	3,900	1,500	8,500
μ_p (holes)	1,900	600	400
Raman phonon energy, eV	0.037	0.063	0.035
Specific heat, J/g·°C	0.31	0.7	0.35
Thermal conductivity at 300 K, W/cm·°C	0.64	1.45	0.46
Thermal diffusivity, cm^2/s	0.36	0.9	0.44
Vapor pressure, torr	10^{-3} at 1270°C 10^{-8} at 800°C	10^{-3} at 1600°C 10^{-8} at 930°C	1 at 1050°C 100 at 1220°C
Work function, V	4.4	4.8	4.7

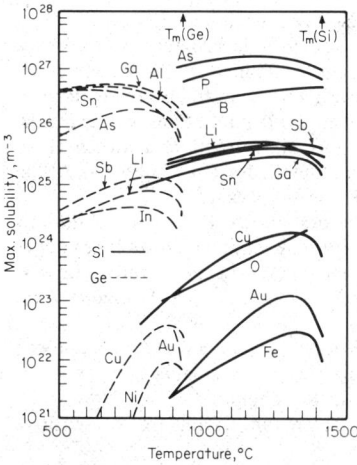

Fig. 6-56. Maximum solid solubility in Si and Ge as a function of temperature.[100]

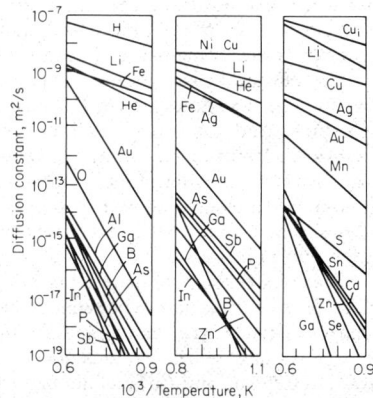

Fig. 6-57. Diffusion coefficients of various impurities in (left to right) Si, Ge, and GaAs.[101, 102]

hexagonal polytypes (a = 3.0806 Å, C = multiples of 2.52 Å). The band gap changes slightly with the polytypic form. SiC can be prepared both n-type (normally by nitrogen doping) or p-type (by aluminum doping). Mobilities are low (~0.01 m²/V · s). The acceptor level is 0.25 eV above the valence band, causing incomplete acceptor ionization at room temperature. Large variations in electrical properties occur in p-type SiC when the temperature is increased, due to changes in the degree of acceptor ionization.

180. Gallium Arsenide. Gallium arsenide is an important device material. Gallium arsenide has a large band gap (1.47 eV) at room temperature and high mobility; 0.85 (m²/V · s). It is used for Schottky barrier diodes, light-emitting diodes, Gunn diodes, and injection lasers. GaAs is a gray brittle material with the zincblende structure (a = 5.65 Å).

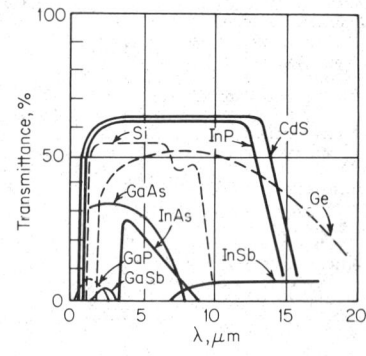

GaAs can be obtained with low dislocation density, although it often contains a large number of other defects. p-type material is normally obtained by Zn doping, while n-type is Te-doped. Deviation from stoichiometry has also the effect of adding donor or acceptor levels.

181. Indium Antimonide. Indium antimonide has a small band gap and a very high electron mobility. It is useful for infrared detectors, infrared filter material, and as magnetoresistance and Hall effect devices.

Fig. 6-58. Optical transmittance of several semiconductors as a function of wavelength.[100]

InSb has the zincblende structure with a lattice parameter of 6.48 Å. InSb becomes body-centered tetragonal above 23 kbars, and its properties become metallic. InSb crystals normally contain precipitates and dislocations.

The band gap is about 0.18 eV at 298 K and 0.23 eV at 80 K. Electron mobility is 7.7 m²/V · s at 290 K and larger than 50 m²/V · s at 77 K. Sn(Na), S, Se, and Te have been used as n-type dopants, while Zn, Cd, Mg, Hg, Ag, Au, Al, Ge, and Mn have been used as p-type dopants.

The magnetoresistance effect is quite large: 1 Wb/m² can raise the resistance of InSb (5 × 10¹² acceptors/m³) by 1.7%. The optical absorption edge for InSb is about 7×10^{-6} μm at 290 K. Donor doping shifts the absorption edge to lower wavelength.

Tunnel diodes, transistors, and laser diodes have also been made with InSb.

182. Gallium Phosphide. Gallium phosphide has a zincblende structure with a lattice parameter of 5.45 Å and 2.3-eV band gap. It is used for electroluminescent diodes which can emit either green or red light. Red light is obtained with zinc or cadmium oxide dopants acting as isoelectronic pairs. Green light is emitted when the dopant atoms are separated.

183. Isomorphous Systems. Isomorphous systems are solid solutions between III-V compounds. Examples are Ga(P, As), used for light-emitting diodes; (In, Ga)Sb, (In, Al)Sb, (Ga, In)As, and (Ga, Al)As, used for continuous-wave injection lasers.

184. Cadmium Compounds. *Cadmium sulfide* is the best-known II-VI compound. Its main use is in photodetectors. Cadmium sulfide is a brittle material which sublimes under atmospheric pressure. It has two allotropic forms: hexagonal (Wurtzite structure) with a = 4.1368 Å and C = 6.7163 Å, and cubic (ZnS structure) with a = 5.82 Å. The hexagonal form is the stable one.

The color of the pure material is pale yellow. CdS has an energy gap of 2.4 eV and can be prepared only as an n-type semiconductor. CdSe and CdTe have smaller band gaps. n-type and p-type CdTe can be obtained by Ga and Ag (or Sb) doping.

CdS is a most sensitive photoconductor in the 0.5- to 0.6-μm range, while CdSe is sensitive at 0.7 to 0.75 μm, and CdTe at about 0.85 μm.

185. Lead Compounds. Lead sulfide, selenide, and telluride have three possible applications: diodes and transistors at low temperatures, infrared detectors, or thermoelectric applications.

PbTe diodes have been operated at 4 K where the mobility may be greater than 10^2 m^2/V · s. PbS infrared detectors have a long-wavelength threshold of 3 μm with a maximum specific detectivity at 2 μm. PbTe has a threshold at 5 μm; PbSe at about 6 μm. These II-VI compounds are very sensitive to deviations from stoichiometry. For example, sulfur-deficient PbS is p-type, while excess sulfur makes PbS n-type.

186. Bismuth Telluride. Bi_2Te_3 is noted for its use as a thermoelectric element for refrigeration. It is used in a pure form or in solid solutions with Sb_2Te_3 or Be_2Sb_3.

187. Metal Oxides. Metal oxides normally have large energy gaps but can be conductive when they deviate from stoichiometry or when suitably doped. Mobilities are very small. Semiconductivity has been found in a variety of oxides: ZnO, MgO, transition metal oxides, and ferrites. Sintered mixtures of transition metal oxides have been used as thermistors (temperature-sensitive resistors).

188. Organic Semiconductors. One of the most studied organic semiconductors is the anthracene C_6H_4:(CH_2):C_6H_2. Semiconducting polymers have also been studied. Organic semiconductors have very low mobilities and have not been commercially used.

189. Amorphous Semiconductors.[1] The past few years have seen a great deal of research and development activity in the area of amorphous semiconductors. This has been motivated, in part, by the scientific interest in understanding such structurally disordered solids, which lack the crystalline periodicity underlying much of solid-state theory. However, the decisive factor in the present level of interest in these materials stems, in the traditional way, from technological considerations.

Uses of amorphous solids can be roughly categorized in two classes: applications in which they can function as well as crystalline solids but are much easier to use, and applications in which their potentially unique properties are exploited. The most obvious example of the first class is ordinary window glass, composed of amorphous oxide insulators (largely SiO_2). The existing technologies based on amorphous semiconductors also belong to this category. The largest-scale application is as the light-sensitive element in electrophotography,[37] as currently used in document-copying machines. The semiconducting chalcogenide glasses, Se and As_2Se_3, are employed here as large-area photoconductive films. Chalcogenide glasses (amorphous solids which contain one or more of the group VI elements S, Se, or Te) are also used as infrared-transmitting windows and lenses.

Applications of the second type, while not yet technologically significant, have provided much impetus to recent work. Glasses, while they can normally persist indefinitely in their amorphous state, are thermodynamically metastable. Thus, for example, the methods of preparing such solids typically employ rapid quenching to bypass crystallization: either from the melt, for bulk glasses relatively easy to prepare, such as SiO_2 or As_2S_3; or from the vapor, for amorphous films of more difficult glass formers such as Ge and Si. The possibility of reversible amorphous-crystalline transformation is a unique feature of semiconducting glasses which underlies some promising applications.[38] Such transformations could be induced electrically, optically, or thermally. Switching between a high- and a low-resistance condition in tellurium-based glasses such as $Te_{0.8}Ge_{0.2}$ probably involves such a mechanism, perhaps the phase separation of Te-rich conducting filaments.[39] Optically induced phase transitions may also find applications in image handling or optical mass memories.[40] This area, the investigation of structural transformations (catalyzed by various means) in amorphous semiconductors is a rapidly developing one; the scope of the eventual technological usefulness of such phenomena is very much an open question at this time.

190. Bibliography on Semiconductor Materials

1. SZE, S. M. "Physics of Semiconductor Devices;" New York, John Wiley & Sons, Inc.-Interscience Publishers, 1969.

2. AZAROFF, L. V., and BROPHY, J. J. "Electronic Processes in Materials;" New York, McGraw-Hill Book Company, 1963.

3. HANNAY, N. B. "Semiconductors;" New York, Reinhold Publishing Corporation, 1960.

4. WOLF, H. "Semiconductors;" New York, John Wiley & Sons, Inc.-Interscience Publishers, 1971.

---

[1] Dr. R. Zallen prepared the material on amorphous semiconductors; his contribution is gratefully acknowledged.

5. HOGARTH, C. A. (ed.) "Materials Used in Semiconductor Devices;" New York, Interscience Publishers, 1965.

6. GROVE, A. S.   "Physics and Technology of Semiconductor Devices;" New York, John Wiley & Sons, Inc., 1967.

7. MADELUNG, O.   "Grundlagen der Halbleiterphysik;" Berlin, Springer-Verlag OHG, 1970.

ELECTRON-EMITTING MATERIALS

191. Work Function and Surface-potential Barrier.   The potential energy of electrons increases when electrons are removed from the material.   This increase is due to the work needed to detach the electron from the positively charged surface.   The potential energy of the electrons far from the surface is called surface-potential-energy barrier.   The difference between the Fermi level (in the material) and the surface potential energy at absolute zero of temperature is called work function (see Fig. 6-59a).

192. Thermal Emission from Metals.   Thermal emission of electrons, *thermionic emission,* is an electron current leaving the surface of a material due to the thermal activation. Electrons with sufficient thermal energy to overcome the surface-potential barrier escape from the material's surface.   The thermally emitted electron current increases with temperature, since more electrons have the energy necessary to leave the material.

The current density of the thermal emission follows the Richardson-Dushman equation,

$$J_s = AT^2 \, e^{-e\phi/kT} \tag{6-79}$$

where J_s = emitted current density (A/m²), T = absolute temperature (K), e = electronic charge (C$_b$), $e\phi$ = work function (eV), A = Richardson's constant (theoretical value = 1.2 × 10⁶ A/m² · K).

The value of ϕ can be written

$$\phi = \phi_0 + \alpha T \tag{6-80}$$

where ϕ_0 and α are constants.

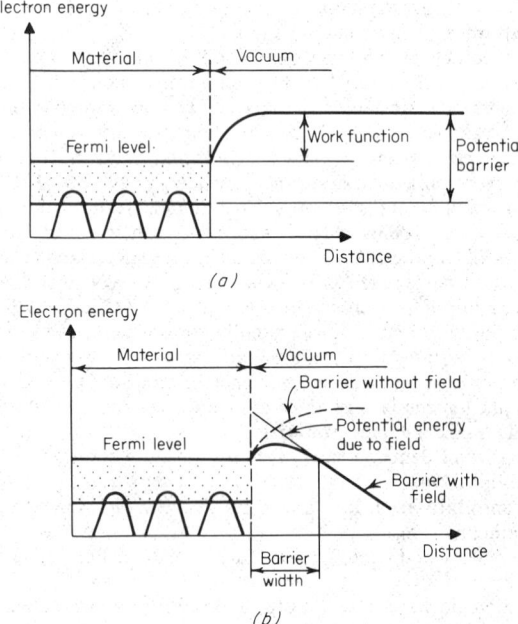

Fig. 6-59. Electron energy near a surface of a material (a) without electric field and (b) with an electric field normal to the surface.

Substituting Eq. (6-80) in (6-79),

$$J_s = A_0 T^2 e^{-e\phi_0/kT} \tag{6-81}$$

A plot of $\ln J_s/T^2$ versus $1/T$ yields the values of ϕ_0 and $A_0 e^{-e\phi_0/kT}$ by measuring the slope of the curve and its intercept at $1/T = 0$.

193. Schottky Effect. The thermal electron emission can be increased by applying an electric field to the cathode. The electric field lowers the surface-potential barrier, as seen in Fig. 6-59b, enabling more electrons to escape the material. This field-assisted emission is called Schottky effect.

194. Field Emission. When large electric fields ($\sim 10^9$ V/m) are applied to the material, the surface-barrier width decreases. When the barrier becomes very thin (~ 100 Å), electrons can quantum-mechanically tunnel through it. This phenomenon is called field emission. The current density of emitted electrons is almost unaffected by the temperature and is primarily field-dependent. The emitted current density is given by the Fowler-Nordheim equation, which as the form

$$J = C_1 E^2 e^{-C_2/E} \tag{6-82}$$

where C_1, C_2 = constants which depend on the material's work function, surface potential, and Fermi level; E = electric field (V/m); and J = emitted current density (A/m^2).

195. Secondary Emission. A material surface bombarded with energetic electrons will emit low-energy electrons called secondary electrons. The secondary yield δ is defined as the number of secondary electrons per primary incident electron. Secondary field as a function of primary-electron accelerating voltage is shown in Fig. 6-60. The yield rises rapidly, goes through a broad maximum, δ_{max}, and drops at higher energies, since high-energy electrons penetrate deeper in the material and the secondaries generated there are unable to reach the material surface with enough energy to be emitted.

196. Energy Distribution of Secondary Electrons. Figure 6-61 shows schematically the energy distribution of emitted electrons. There are two distinct energy regions which are populated: primary electrons with energies close to the primary energy, and secondary electrons with energies about 10 eV.

197. Effect of Surface Contamination. Since most secondary electrons have energies larger than the work function, their yield is relatively insensitive to monolayer-type absorbed films. However, films of few tens or hundreds of angstroms play an important role, since most electrons are emitted from this depth. Specifically, some oxide films may increase appreciably the electron yields, while absorbed hydrocarbons may reduce yields. This sensitivity to surface cleanliness accounts for some of the discrepancies of reported secondary-emission yields.

198. Effect of Angle of Incidence. The yield of secondary electrons increases with the increase of angle of incidence of the primary electrons. The angular distribution of the

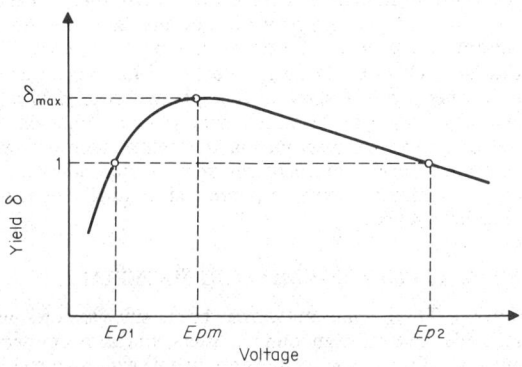

Fig. 6-60. Schematic diagram of the secondary electron yield as a function of the incident electron voltage.

secondary electrons is, roughly, a cosine distribution and is hardly affected by the primary incidence angle.

199. Photoemission. Electrons excited by photons can acquire enough energy to surmount the surface-potential barrier. Electron emission due to the illumination of surfaces is called photoemission. The maximum kinetic energy of electrons emitted by photons with frequency ν is, according to Einstein's equation,

$$KE_{max} = h\nu - e\phi \qquad (6\text{-}83)$$

where h = Planck's constant (J · s), ν = frequency of impinging radiation (s^{-1}), e = electronic charge (C), $e\phi$ = work function (eV), and KE_{max} = maximum kinetic energy of emitted electrons (J). The work function can be found experimentally by the use of Eq. (6-83).

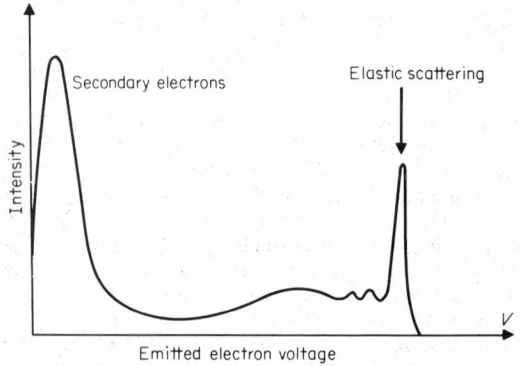

Fig. 6-61. Schematic diagram of the intensity of emitted electrons as a function of voltage.

Photoemission spectra from insulators or semiconductors are more complicated than those of metals. Electrons can be excited from different energy states and emitted. The photocurrent may have a series of maxima as a function of photon energy corresponding to excitation of electrons from various impurity levels. To emit electrons, the photons have first to produce energetic electrons, then the electrons have to travel through the material to the surface, and finally the electron has to overcome the surface-potential barrier.

In metals, the efficiency of energetic electron production is low due to the high reflectivity of light, and moreover, the motion of electrons through the solid is very inefficient, due to electron-electron scattering. In semiconductors and insulators the production and motion of energetic electrons are much easier.

Photoemission due to very energetic photons may produce secondary electrons (i.e., production of primary electrons, which in turn excite secondary electrons). Yields exceeding unity can be achieved by such a process. Electrons can be emitted by low-energy photons when 2 or 3 photons transmit their energy to a single electron.

200. Photoelectric Efficiency (Yield). The photoelectric yield, or quantum yield, is the number of emitted electrons per incident photon. The photoelectric yield is termed also photoelectric (or quantum) efficiency. It is also customary to use the photocurrent per watt incident radiation, mA/W, or microampere per lumen, μA/lm, the latter having reference to a tungsten lamp of a given temperature. The microampere per lumen values are useful when comparing photocathodes with similar spectral response and for applications where incandescent lamps are useful as light sources.

SPECIFIC ELECTRON-EMITTING MATERIALS

201. Thermionic Emitters. Thermionic emitters can be classified as pure-metal emitters, monolayer-type emitters, oxide emitters, compound emitters, and alloy emitters.

202. Pure-Metal Emitters. Thermionic emitters are metals with high melting temperatures, low vapor pressure, and preferably low work function. Tungsten and tantalum are the metals most suited for thermal emission. Thermal-emission characteristics of several metals are given in Table 6-51.

203. Monolayer-Type Emitters. Tungsten filaments containing small amounts of thoria have much better emission characteristics than pure tungsten. The work function of thoriated tungsten is reduced from 4.5 eV for pure W to 2.6 eV for the thoriated materials. The improved electron emission was found on addition of other electropositive elements such as Cs, Ba, and La. This lowering is probably due to formation of dipole monolayers of these elements on the tungsten, facilitating the escape of electrons. A monolayer of electronegative oxygen increases the work function considerably.

Additions of 0.5 to 2% thoria to tungsten (with other additions of alumina, SiO_2, Na_2O) improves the emission by several orders of magnitude.[41] Thoria prevents tungsten grain growth, and therefore improves its high-temperature mechanical properties. Barium monolayers reduce the work function to 1.6 eV. See Table 6-51 for properties of monolayer emitters. Carburization of thoriated tungsten filaments reduces considerably the rate of thorium evaporation and prolongs the filament lifetime. Carburized filaments have peak emission yields of about 3×10^4 A/m².

204. Oxide Emitters. Several types of oxide emitters are found. The dispenser-type cathodes are essentially porous tungsten or molybdenum-tungsten cathodes in which the emitting films are dispensed to the surface of the cathode from the interior. The *L-cathode* is a porous tungsten disk behind which barium-strontium carbonate is located.[43] A filament heats the carbonates, which decompose, and the barium and strontium migrate to the outer surface to form the emitting film. The "impregnated" cathode[44] uses a single pressed disk of barium calcium aluminates and tungsten rather than a tungsten bloc and a separate source of alkaline-earth carbonates. Such cathodes are capable of emitting up to 10^5 A/m² pulsed emission and can be operated up to 1150°C.

Another dispenser cathode is made by pressing Mo-W with alkaline-earth compounds.[45]

Dispenser cathodes can be poisoned by metal vapors; therefore care must be taken not to expose the cathodes to metal vapors.

Nickel-base pressed and sintered cathodes are widely used. The emitter is a disk of sintered Ni, ZrH, and Ba-Sr-Ca carbonate. The ZrH is used as a reducing agent. The top surface is again a monolayer of Ba continuously replenished from the sintered body.[46]

Another type of emitter is the oxide-coated cathodes[47] which are made of Ba-Sr carbonate layers on nickel. This layer is heated in a vacuum to form a BaO + SrO oxide layer. Apparently, the emitting surface of such cathodes is also a monolayer of Ba. These cathodes

Table 6-51. Thermal Emission of Different Cathodes[133]

Cathode form	Material	Work function, eV	Constant A (10^6 A/m²·K²)
Metal.............	W	4.5	0.60
	Zr	4.1	3.3
	Re	4.8	1.0
	Ta	4.3	1.2
	Mo	4.2	0.55
	Th	3.4	0.6
	Ba	2.5	0.6
	Cs	1.9	1.6
Film..............	W-Cs	1.5	0.03
	W-Ba	1.6	0.015
	W-Th	2.7	0.04
	Ta-Th	2.5	0.015
	W-O-Ba	1.3	0.015
Oxide.............	BaO, SrO	0.95	10^{-3}-10^{-4}
	Toria	2.5	0.03
	CsO	0.75	10^{-3}-10^{-4}
Compound.........	TiC	3.35	0.25
	SiC	3.5	0.64
	UC	2.9	0.33
	ThC_2	3.5	5.5
	LaB_6	2.7	0.29
	ThO_2	2.6	0.05

are also formed by triple carbonates containing Ba, Sr, and Ca carbonates. The carbonates are applied in paste form or by transfer tape to the cathode sleeve. Nickel or nickel-based alloys are almost exclusively used as sleeve materials. Properties of oxide emitters are given in Table 6-51.

205. Compound Emitters. Emitters such as uranium carbide and zirconium carbide have been studied. Lanthanum hexaboride has been recently introduced as emitters in several configurations.[48] Emission densities of up to 70 A/m^2 have been obtained with LaB$_6$. See Table 6-51 for properties of compound emitters.

206. Alloy Emitters. Emitters have also been produced by using alloys of W, Ta, and Mo with mobilizers such as Ti, Zr, or Hf. The alloys are impregnated with activators such as La, Ce, Gd, Th, or Ca to form a monolayer-type cathode.[49]

207. Secondary Electron Emitters. Secondary emitters can be classified according to the maximum yield δ and the accelerating voltage needed for maximum yield. Table 6-52 lists the maximum yield of various cathode materials. The listing includes metals, alloys, oxides, and glass. Large discrepancies in values of δ$_{max}$ can be found in the literature, mainly due to the variations in the emitting-surface condition.

Materials with high maximum yield are used for electron multiplier surfaces, called *dynodes*. Silver-magnesium alloys are used after oxidation. The surfaces are oxidized to form MgO. High secondary yields can be obtained with these emitters. Pure magnesium

Table 6-52. Secondary Electron Yields from Metals Insulators and Semiconductors[134]

		Metals			
Atomic no.	Element symbol	δ$_{max}$	E_{pm}	E_{p1}	E_{p2}
3	Li	0.5	85		
4	Be	0.5	200		
4	Be	0.60	800		
4	Be on Ag (26 layers)	0.61	300–1,500		
11	Na	0.82	300		
12	Mg	0.95	300		
12	Mg on Ta	8	~700		
13	Al	0.95	300		
19	K	0.7	200		
22	Ti	0.9	280		
26	Fe	1.3	(400)	120	1,400
27	Co	1.2	(500)	200	
28	Ni	1.35	550	150	1,750
28	Ni containing 1.5% Ba	2.8	800–900		
29	Cu	1.3	600	200	1,500
29	Cu containing 10% Mg	~13	~900		
37	Rb	0.9	350		
40	Zr	1.1	350	175	(600)
41	Cb	1.2	375	175	1,100
42	Mo	1.25	375	150	1,300
46	Pd	>1.3	>250	120	
47	Ag	1.47	800	150	>2,000
47	Ag	1.56	800		
48	Cd	1.14	450	300	700
49	In	1.3–1.4	~500		
50	Sn	1.35	500		
51	Sb	1.3	600	250	2,000
55	Cs	0.72	400		
56	Ba	0.82	400		
73	Ta	1.3	600	250	>2,000
74	W	1.35	650	250	1,500
74	W containing 25% Re	1.2	500–1,000		
75	Re	1.3	900		
78	Pt	1.5	750	350	3,000
79	Au	1.45	800	150	>2,000
80	Hg	1.3	600	350	>1,200
81	Tl	1.7	650	70	>1,500
82	Pb	1.1	500	250	1,000
82	Pb	1.1–1.3	600		
83	Bi	1.5	900	80	>2,000
90	Ta	1.1	800		

Table 6-52. Secondary Electron Yields from Metals Insulators and Semiconductors
concluded.

Semiconductors and insulators

Group	Substance	δ_{max}	E_{pm}
Semiconductive elements .	Ge (single crystal)	1.2-1.4	400
	Si (single crystal)	1.1	250
	Se (amorphous)	1.3	400
	Se (crystal)	1.35-1.40	400
	C (diamond)	2.8	750
	C (graphite)	1	250
	B	1.2	150
Semiconductive compounds	Cu_2O	1.19-1.25	400
	PbS	1.2	500
	MoS_2	1.10	
	MoO_2	1.09-1.33	
	WS_2	0.96-1.04	
	Ag_2O	0.98-1.18	
	ZnS	1.8	350
Intermetallic compounds .	$SbCs_3$	5-6.4	700
	SbCs		
	Initial	5.7	600
	After 50 h bombardment	4.0	400
	SbCs	1.9	550
	$BiCs_3$	6-7	1,000
	Bi_2Cs	1.9	1,000
	GeCs	7	700
	Rb_3Sb	7.1	450
Insulators..............	LiF (evaporated layer)	5.6	
	NaF (layer)	5.7	
	NaCl (layer)	6-6.8	600
	NaCl (single crystal)	14	1,200
	NaBr (layer)	6.2-6.5	
	NaBr (single crystal)	24	1,800
	NaI (layer)	5.5	
	KCl (layer)	7.5	1,200
	KCl (single crystal)	12	
	KI (layer)	5.5	
	KI (single crystal)	10.5	1,600
	RbCl (layer)	5.8	
	KBr (single crystal)	12-14.7	1,800
	BcO	3.4	2,000
	MgO (layer)	4	400
	MgO (single crystal)	23	1,200
	MgO film on Mg-Ag alloy	12	600
	BaO (layer)	4.8	400
	BaO	3-9	500-700
	BaO-SrO (layer)	5-12	1,400
	Common phosphors:		
	P-1	2.7	750
	P-2	3.4	750
	P-3	3.9	1,000
	P-4	3.7	700
Insulators..............	Al_2O_3 (layer)	1.5-9	350-1,300
	SiO_2	2.4	400
	Mica	2.4	300-384
Glasses................	Technical glasses	2-3	300-420
	Pyrex	2.3	340-400
	Quartz glass	2.9	420

NOTE: For E_{pm}, E_{p1}, E_{p2} see Fig. 6-60.

oxide has a very high yield (over 20). Nickel-beryllium and nickel-magnesium alloys can also be oxidized to form secondary emitters. Copper-beryllium is used extensively as a dynode material.

Occasionally, the secondary emission has to be surpressed. Materials with low secondary emission such as graphite can be used for this purpose.

208. Photoelectric Emitters. Photoelectric emitters are divided into metallic emitters and various types of semiconductor emitters.

209. Metallic Emitters. Metallic emitters have lost their importance since the advent of high-efficiency insulating and semiconducting emitters.

Photoelectric emission is confined to a layer of about 10 atomic layers at the surface of metals. Metallic surfaces are very sensitive to absorbed surface layers. These layers can form in a few seconds, even at pressures of 10^{-7} torr. The presence of surface layers casts doubt on many of the earlier experimental results. The alkali metals are the only metals with appreciable efficiency in the visible-light region. The spectral response of alkali metals is given in Fig. 6-62.

210. Semiconductor Emitters. Cs_3Sb. Cs_3Sb cathodes have a much higher efficiency than the alkali metals in the visible-light region. These cathodes are prepared by evaporating Sb on glass and exposing it to Cs vapor. The resulting structure is approximately Cs_3Sb. Additional sensitivity can be obtained by partially oxidizing the cathode. Both transparent and opaque cathodes can be fabricated. Transparent cathodes can be further improved by using a manganese oxide substrate. It is not possible to assign an exact spectral response curve for the Cs_3Sb cathodes. Instead, an S-number is assigned to various cathodes, and typical response curves have been agreed on to classify them. Table 6-53 describes some of these cathodes. The spectral response of Cs_3Sb can be seen in Fig. 6-63.

Changes in the performance of photoemitters with time are referred to as *fatigue*. Fatigue can be due to excessive exposure to the sun, temperatures above 100°C, chemical changes, nonhermetic sealing, or electron and ion bombardment.

K_3Sb, Rb_3Sb, Na_3Sb. K_3Sb photocathodes are prepared by evaporating Sb on glass and exposing it to glow discharge in oxygen followed by K vapor. Spectral responses of the K_3Sb, Rb_3Sb, and Na_3Sb cathodes can be seen in Fig. 6-63.

Multialkali Cathodes. Compounds of several alkali metals and antimonium have very high quantum yields. Na-K-Sb and Cs-Na-K-Sb cathodes (designated S-20) are examples of such cathodes. Na-K-Sb cathodes are formed by exposing K_3Sb films to Na. Sb and K are added to the film to obtain maximum sensitivity. The resulting compound is approximately Na_2KSb. Cs-Na-K-Sb cathodes are even more complicated. Superficial oxidation and manganese oxide substrates have no beneficial effect on these cathodes. Spectral response of Na_2KSb and $(Cs)Na_2KSb$ cathodes is seen in Fig. 6-64.

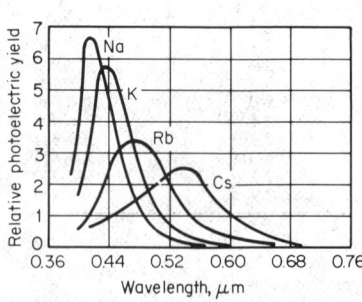

Fig. 6-62. Spectral response of alkali metals.[104]

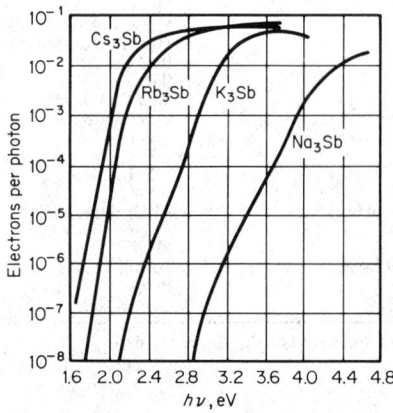

Fig. 6-63. Spectral response of Cs_3Sb, Rb_3Sb, K_3Sb, and Na_3Sb.[106]

211. Silver-Oxygen-Cesium Photocathodes. The Ag-O-Cs cathodes are fabricated by first evaporating, electroplating, or using other deposition methods for formation of a silver base. The silver is then oxidized, normally, by a glow discharge in oxygen. The Cs vapor is then deposited on the oxidized silver. The entire process requires care and skill. Occasionally, additional silvering and superficial oxidation are also performed. The resulting structure of Ag-O-Cs is probably $Ag-Cs_2O-Cs$. The spectral response of Ag-O-Cs is seen in Fig. 6-64. Another interesting cathode is the Bi-Ag-O-Cs combination, which has a panchromatic response.

212. Ultraviolet-sensitive Photocathodes. Cs_2Te and Rb_2Te are sensitive at the 0.2- to 0.35-μm-wavelength range. Cs_2Te cathodes are prepared by evaporating Te on a metallic or quartz substrate followed by exposure to Cs vapor at elevated temperature. In the 0.1- to 0.2-μm-wavelength range the alkali halides (with exception of some fluorides) reach a quantum yield of 0.1 electron per photon. In this range the silver halides and the oxides of the alkaline-earth metals (MgO, CaO, SrO, and BaO) have also high quantum yields. The alkali halides have high quantum yields also below 0.1 μm.

213. Special Cathodes. GaAs crystals with a Cs surface layer have potential infrared applications.[50] A variety of elemental and compound semiconductors have also been investigated, but none are used as photoemissive cathodes.

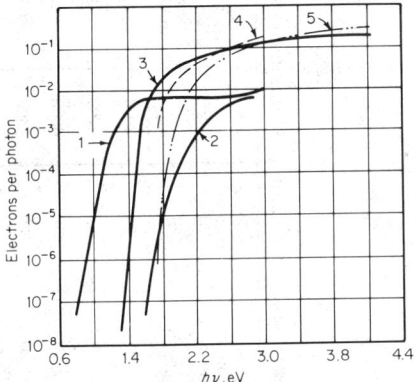

Fig. 6-64. Spectral response of (1) AgOCs, normal; (2) AgOCs, silver-deficient; (3) Cs(Na$_2$KSb); (4) K$_2$CsSb; (5) Na$_2$KSb.[107]

214. Bibliography on Electron-emitting Materials

1. KOHL, V. H. "Handbook of Materials and Techniques for Vacuum Devices;" New York, Reinhold Publishing Corporation, 1967.

2. BRUINING, H. "Physics and Applications of Secondary Electron Emission;" New York, McGraw-Hill Book Company, 1954.

3. SOMMER, A. H. "Photoemissive Materials;" New York, John Wiley & Sons, Inc., 1968.

RADIATION-EMITTING MATERIALS

215. Blackbody Radiation. Solids or liquids at finite temperatures radiate a continuous spectrum of electromagnetic energy. This radiation is called *thermal radiation*. The power radiated at different wavelengths is a function of temperature.

A body which completely absorbs all radiation falling on it (and, according to Kirchhoff's law, which also emits all the radiation) is called a *blackbody*. The power distribution of the energy radiated from blackbodies is given by Planck's formula

$$W_\lambda d\lambda = \frac{C_1 \lambda^{-5}}{\exp{(C_2/\lambda T)} - 1} \qquad (6\text{-}84)$$

where λ = wavelength (m), T = temperature (K), $W_\lambda d\lambda$ = energy emitted per unit area

Table 6-53. Characteristics of Various Photocathodes[136]

Device S-number[a]	Photocathode type and envelope	Conversion factor[b] k, lm/W	Typical luminous sensitivity[c] S_{typ}, μA/lm	Maximum luminous sensitivity[d] S_{max}, μA/lm	λ_{max}, Å	Typical radiant sensitivity[e] σ_{typ}, mA/W	Typical quantum efficiency,[f] %	Typical photocathode dark emission[g] at 25°C, A/cm²
S-1	Ag-O-Cs Lime-glass bulb	93.9	25	60	8000	2.35	0.36	900×10^{-15}
S-3	Ag-O-Rb Lime-glass bulb	286	6.5	20	4200	1.86	0.55	
S-4	Cs-Sb Lime-glass bulb	977	40	110	4000	39.1	12	0.2×10^{-15}
S-5[h]	Cs-Sb 9741 glass bulb	1252	40	80	3400	50.1[h]	18[h]	0.3×10^{-15}
S-8	Cs-Bi Lime-glass bulb	755	3	20	3650	2.26	0.77	0.13×10^{-15}
S-9	Cs-Sb Semitransparent, lime-glass bulb	683	30	110	4800	20.5	5.3	
S-10	Ag-Bi-O-Cs Semitransparent, lime-glass bulb	508	40	100	4500	20.3	5.6	70×10^{-15}
S-11	Cs-Sb Semitransparent, lime-glass bulb	804	60	110	4400	48.2	14	3×10^{-15}

S-number	Designation	k	s		λ_{max}	σ	Quantum eff.	Dark emission
S-13	Cs-Sb Semitransparent, fused-silica bulb	795	60	80	4400	47.7	13	4×10^{-15}
S-17	Cs-Sb Lime-glass bulb, reflecting substrate	664	125	160	4900	83	21	1.2×10^{-15}
S-19[i]	Cs-Sb		40	70		22[i]	11[i]	0.3×10^{-15}
S-20	Sb-K-Na-Cs Fused-silica bulb (Multialkali)	428	150	250	4200	64.2	18	0.3×10^{-15}
S-21	Cs-Sb Semitransparent, lime-glass bulb	779	30	60	4400	23.4	6.6	
	Cs-Sb Semitransparent, 9741 glass bulb							

[a] The S-number is the designation of the spectral-response characteristic of the device and includes the transmission of the device envelope.

[b] k is the conversion factor from amperes per lumen to amperes per watt at the wavelength of peak sensitivity.

[c] s is the luminous sensitivity for the photocathode for 2870 K color-temperature test lamp. In the case of a multiplier phototube, output sensitivity is μs, where μ is the amplification of the multiplier phototube.

[d] Care must be used in converting s_{max} to a σ_{max} figure. Photocathodes having maximum lumen sensitivity frequently have more red sensitivity than normal, and the formula cannot be applied without reevaluation of the spectral response for the particular maximum-sensitivity device.

[e] σ is the radiant sensitivity at the wavelength of maximum response.

[f] 100% quantum efficiency implies one photoelectron per incident quantum, or $e/h\nu = \lambda/12{,}395$, where λ is expressed in angstrom units. Quantum efficiency at λ_{max} is computed by comparing the radiant sensitivity at λ_{max} with the 100%-quantum-efficiency expression above.

[g] Most of these data are obtained from multiplier phototube characteristics. For tubes capable of operating at very high gain factors, the dark emission at the photocathode is taken as the output dark current divided by the gain (or the equivalent minimum anode-dark-current input multiplied by cathode sensitivity). On tubes where other dc dark-current sources are predominant, the dark noise figure may be used. In this case, if all the noise originates from the photocathode emission, it may be shown that the photocathode dark emission in amperes is approximately $0.4 \times 10^{18} \times$ (equivalent noise input in lumens times cathode sensitivity in amperes per lumen)^2. The data shown are all given per unit area of the photocathode.

[h] The S-5 spectral response is suspected to be in error. The data tabulated conform to the published curve, which is maximum at 3400 Å. Present indications are that the peak value should agree with that of the S-4 curve (4000 Å). Typical radiant sensitivity and quantum efficiency would then agree with those for S-4 responses.

[i] No value for k or λ_{max} is given because the spectral-response data are in question. The values quoted for σ and typical quantum efficiency are typical only of measurements made at the specific wavelength 2537 Å, and not at the wavelength of peak sensitivity as for the other data.

between λ and $\lambda + d\lambda$ (W/m^2), $C_1 = 3.740 \ 10^{-2} \ 10^{-16}$ (W $\cdot$ m^2), and $C_2 = 1.4385 \times 10^{-2}$ (m $\cdot$ K).

Figure 6-65 shows the emitted energy from a blackbody at various temperatures. Higher temperatures produce radiation with shorter wavelengths.

216. Relative Luminosity and Luminous Efficiency. The human eye has a brightness sensation which is wavelength-dependent. A standard luminosity curve which gives the relative brightness sensation for a standard observer of monochromatic light of different wavelengths is given in Fig. 20-1. It should be noted that this curve is not intended to predict retinal response under every lighting condition. At 555 nm, 1 W of radiant flux has a luminous flux of 685 lm.

The luminous efficiency is defined as (luminous flux)/(radiant flux). For a blackbody radiator the luminous efficiency can be calculated by dividing the total luminous flux to the total radiant flux. Figure 6-66 shows the luminous efficiency of a blackbody as a function of its temperature. The maximum efficiency is about 94 lm/W. Nonideally, blackbodies have smaller radiant emittance. They are characterized by their emissivity, which is the ratio of their emittance to that of a blackbody.

217. Coherent and Incoherent Radiation. A wavetrain which preserves its wavelength and phase is called *time-coherent.* Two time-coherent sources with identical wavelengths can cause interference since there is a definite relation between their phases. Wavetrains which have no phase relationship are called *incoherent waves.* Incoherent waves do not produce interference phenomena. Sources of incoherent radiation can produce interference phenomena by spatially separating the radiation to produce two (or more) wavetrains which are space-coherent. Interference occurs because there is a definite phase relationship between the separated beams originating from the same source. Most sources of light are time-incoherent. Coherent radiation persisting over small periods of time can be obtained from lasers.

218. Radiation from Heated Solids. Metal filaments heated below their melting point (m.p.) are used as radiation emitters in incandescent lamps. To obtain maximum luminous efficiency from such lamps, it is best to operate the filaments at temperatures of about 6000 K. Carbon has the highest melting point (3600°C), but it evaporates excessively above

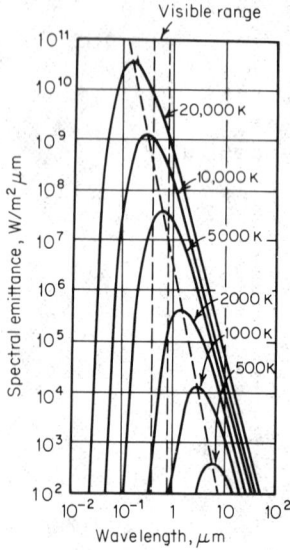

Fig. 6-65. Spectral emittance of a blackbody as a function of temperature.[108]

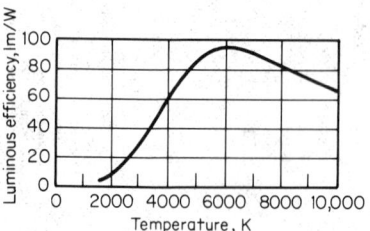

Fig. 6-66. Luminous efficiency of a blackbody as a function of temperature.[108]

1850°C. Incandescent lamps have been made from tungsten (3380°C m.p.), tantalum (2900°C m.p.), and osmium (2700°C m.p.). Another type of heated-solid-light source is the arc lamp. In open arc lamps the carbon electrode emits radiation due to its high temperature.

219. Radiation from Combustion. Radiation from flames produced by combustion arises mainly by heated fine carbon particles in the flame. Another source of radiation is the combustion of Al or Zr wires in photographic flashbulbs. The light originates from heated oxide particles obtained by the combustion of the metal.

220. Radiation from Excited Gases, Vapors, and Plasmas. An electric field applied to a gas can ionize or excite the gas atoms. The ionization (removal of electrons) or excitation (momentary raising of the energy of an electron) is produced by the impact of energetic electrons. Electromagnetic radiation is produced when the excited electrons return to lower energy levels.

Every gas has a series of possible energy levels which its electrons can occupy. Since there are specific energy changes possible for the electrons in a given gas, there will also be only a number of characteristic wavelengths emitted from the gas.

Discharge lamps use the light emitted from the characteristic lines of noble gases, mercury vapor, or sodium vapor at low pressures. Radiation can also be obtained from discharge in high pressure. The spectrum from high-pressure arcs does not show sharp distinct spectral lines but resembles the continuous blackbody radiation. The luminous part of the arc is confined to a central column of the gas, which may reach 6000 K.

221. Luminescence. Luminescence is the emission of light from a substance, above the thermal emission, due to external excitations. Luminescent materials are often classified according to the rise or decay time of the emitted light. Materials which emit light almost at the instant of excitation and have a rapid decay of emission with the removal of excitation are called *fluorescent* materials. Materials which have an afterglow, i.e., a long decay time for the emitted light, are called *phosphorescent* materials, or phosphors.

The emission of light from a substance can be excited by a variety of methods:

Chemical reactions. The slow oxidation of organic or inorganic materials can produce some light *(chemiluminescence).* Such reactions are responsible for the light emitted by fireflies and glowworms *(bioluminescence).*

Electromagnetic radiation. Light can be produced by illuminating a substance with visible or ultraviolet light *(photoluminescence).* Shorter wavelength, such as x-rays or γ radiation, can also produce light emission.

High-energy particles. Ions (protons, alpha particles) can produce visible radiation *(ionoluminescence).* Radioactive materials can excite light emission from solids by providing alpha particles, which in turn cause the light emission *(radioluminescence).*

Electrons or electric fields. Electrons moving in a gas can excite its atoms, which emit light upon return to their excited state (glow, or arc, discharge).

Electrons impinging on a substance can produce light *(cathodoluminescence),* for example, the cathode-ray-tube screen.

Electrons moving in a solid can excite its electrons, and light emission can occur upon return of electrons to the vacant energy states. Such a process is called *electroluminescence.* Light can be produced in such a fashion when an ac field is applied to phosphorus.

Light can be emitted from semiconductors due to recombination processes in a *pn* junction. Such electroluminescence is the basis of light-emitting diodes.

222. Stimulated Emission (Laser Radiation). Light can be emitted when an excited atom or molecule returns to a low-energy state. This light emission can occur spontaneously or can be triggered by radiation. A triggered emission is called stimulated emission. The stimulated emission is proportional to the incident radiation and is coherent with respect to it.

Normally, materials contain a small number of excited atoms, and consequently, incident radiation causes negligible stimulated emission. In lasers (*l*ight *a*mplification by *s*timulated *e*mission of *r*adiation) the population of excited states is inverted by pumping, i.e., by external excitation, for example by a flash lamp, and sizable stimulated emission can be obtained. Figure 11-18 shows a schematic diagram of a solid-state laser. A flash lamp excites the atoms in the material, and the emitted light from some excited atoms stimulates radiation. The two ends of the laser are silvered so that only stimulated radiation parallel to the rod axis is reflected back and forth within the laser cavity. This increases the radiation field within the cavity, thus further increasing the stimulation emission. When the losses in the cavity are overcome, a coherent beam emerges from the laser.

Laser action can also be obtained in semiconductors. The injection of minority carriers at a *pn* junction can invert the carrier concentration there. Stimulated emission from this inverted population can occur.

Laser operation can be divided into three modes: *continuous wave; pulses* of 10^{-6} to 10^{-3} s, a duration similar to that of the pumping pulses; and switched and *phase-locked pulses* in which the radiation emerges from the laser in short intense pulses (10^{-9} s).

SPECIFIC RADIATION-EMITTING MATERIALS

223. Metal Filaments. Incandescent lamps use a resistance-heated filament. The filament can be tungsten, tantalum, osmium, or platinum, tungsten being the most common filament material.

Tungsten has the highest melting point (3380°C) of all metals. Since incandescent lamps need to be operated at the highest possible temperature, tungsten is the best choice as the filament. A typical spectral emittance from a tungsten incandescent lamp is shown in Fig. 6-67.

Tungsten filaments are produced by first forming bars by powder metallurgy, swaging them to wire form, and finally drawing them to produce fine wires. Coiled filaments are produced by winding the fine wires and then etching the mandrel.

Tungsten filaments evaporate slowly, causing local hot spots at the thinner areas. These hot spots enhance further evaporation, followed by additional thinning, and finally failure. When the evaporated tungsten settles on the glass, it reduces light transmission. Filament evaporation can be reduced by filling the bulbs with argon gas. Normally, some nitrogen is added to the argon to reduce its tendency to ionization. Krypton has been used to increase the lamp efficiency since its heat conduction is smaller than that of argon.

At the high operating temperatures (2500°C) there is an excessive tungsten grain growth and a loss of strength. The grain

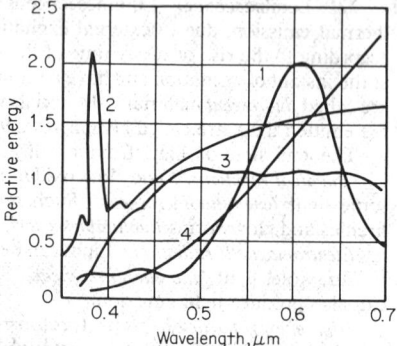

Fig. 6-67. Spectral energy distribution of emitted light from (1) white fluorescent lamp, (2) carbon arc, (3) sunlight, (4) 50-W incandescent lamp.[74]

growth can be reduced by addition of thoria (2%), which prevents grain boundary motion. Another approach is provided by additions of silica or alumina to enhance grain growth; a stable grain configuration is produced shortly after the filament is lit the first time.[51]

Failures due to tungsten evaporation have been drastically reduced by halogen-filled lamps. In the quartz iodine lamp the tungsten vapors reach the hot quartz envelope, and combine with iodine to form volatile tungsten compounds which are reduced to metallic tungsten on contact with the hot filament. The evaporated tungsten is continuously carried back from the envelope to the filament, reducing considerably the evaporation loss and envelope blackening.

224. Noble Gases. Noble gases are used in discharge lamps. The light is derived from either the typical resonance radiation or the continuum. The noble gases are He, Ne, Ar, Kr, Xe, and Rn.

Xenon. Xenon lamps are used for optical equipment, film projectors, high-power large-area lights, and electronic flashtubes. Xenon spectral distribution is close to sunlight distribution. Xenon emits a sizable power in continuous ultraviolet radiation. In addition, xenon lamps emit both characteristic and continuous radiation in the infrared. Xenon lamps normally have quartz envelopes because of their high operating temperatures. Xenon flashlight spectral distribution is more continuous.

Neon. Neon lamps are used as indicator lamps and advertising signs. The indicator lamps use the negative glow, while the advertising signs use a long positive glow. Neon lamps often contain some argon to permit lower striking voltage.

225. Sodium Vapor. Low-pressure sodium-vapor lamps are used as low cost sources for street lighting, for instance. The low-pressure lamps are filled with neon (at a few millimeters of mercury) and a small addition of argon and/or xenon. The neon serves both to start the discharge and also to enhance electron collision with sodium atoms. The small addition of argon serves to lower the starting voltage. Low-pressure sodium vapor has discrete resonance lines at 0.5890 and 0.5896 μm. At 270°C the sodium pressure is 3 to 4 mm Hg. High-pressure lamps operate at several hundred degrees centigrade. The spectrum of the high-pressure lamps tends to be continuous.

226. Mercury Vapor. Mercury-vapor lamps can be low, medium, high, and extra-high pressure. The low-pressure lamps are low-power, small sources for instrument-panel illumination or ultraviolet excitation of fluorescent materials. Another kind of low-pressure mercury lamp is the familiar fluorescent lamp. It has hot cathodes which are filled with low-pressure argon gas and a drop of mercury. The excitation of the mercury produces the 0.2537-μm resonant radiation in the ultraviolet. This radiation reaches the tube walls, which are phosphor-coated, and excites the phosphor to emit visible light. Medium-pressure mercury lamps (~ 1 atm) are not used to a large extent. High-pressure (2 to 10 atm) lamps are used for such applications as street illumination. The high-pressure lamps have a very high temperature at the central portion of the discharge. Lamps are double-walled, the inner envelope being quartz and the outer one clear or pearl-finished glass. The outer envelope can be coated with magnesium fluorogermanate phosphor for color correction. Extra-high pressure (20 atm) mercury lamps are also available.

Mercury-iodide discharge lamps are high-pressure mercury lamps with a metal iodide incorporated in the gas. The iodide decomposes in the high-temperature region, releasing the metal vapor, which emits light. The additional light emission improves the color balance of the lamp. The metal can never reach the lamp walls and damage them, since on approaching the cooler parts near the walls, the metal recombines and forms the iodides, which are not harmful to the lamp walls.

227. Phosphors. (See also Pars. **11-60** to **11-72a.**) Phosphors are inorganic or organic luminescent materials. Most phosphors are crystalline organic crystals (host crystals) with substitutional or interstitial impurities (activators) which increase the light emission of the host crystal (or of another impurity) or introduce new emission lines. Some impurities, such as Ni in ZnS or Ag, are "poisons" and tend to surpress light emission. Phosphors are compounds which have oxygen or fluorine anions (O-dominated), sulfur or selenium (S-dominated), mixed compounds, alkali halide crystals, or organic compounds.

The decay times of light emission of crystals may follow an exponential decay e^{-at}, a power law t^{-n}, or a mixed behavior.

Phosphors Excited by Photons. *Long-wave ultraviolet.* Little use is made of these phosphors, mainly due to the lack of economical sources in this wavelength region.

Short-wave ultraviolet. These phosphors are used in fluorescent lamps. Fluorescent lamps contain low-pressure mercury vapor which emits 2,537 μm radiation. This radiation is then exciting visible radiation from a phosphorescent coating inside the lamp envelope. Coating for fluorescent lamps are, for example: γ-$Ca_3(PO_4)_2$:Ce, Mg_2WO_5:[W] emits pale blue light; rhombohedral, Zn_2SiO_4:Mn emits green light; rhombohedral, $(Zn:Be)_2SiO_4$:Mn phosphors may emit pink, yellow, orange, gold, or red light.

X-rays. For x-ray fluoroscopy: hexagonal, (Zn:Cd)S:Ag, is used at voltages smaller than 100 kV and tetragonal, $CaWO_4$:[W]; rhombohedral, $BaSO_4$:Pb, or cubic, BaFCl for higher voltages. Short persistent luminescent crystals are used for scintillation counters, for example, cubic, NaI:Tl, hexagonal, ZnO:[Zn], or rhombohedral, $BaSO_4$:Pb. The microsecond delay time of such crystals allows high counting rates (see Table 6-54).

Phosphors Excited by Electrons. These phosphors are used for cathode-ray tubes, television screen, and electron microscopes. Table 11-13 lists the common phosphors. Short-persistent phosphors are P15, P5, P11, P3, and P1. Long-persistent phosphors are P12, P2, P14, and P7. (See Fig. 6-68.)

Phosphors Excited by Ions and Neutrons. The main use of these phosphors is in luminescent coating for dials. $RaBr_2$ or $RaSO_4$ is mixed with hexagonal ZnS:Cu. The decaying radium excites the ZnS to emit visible radiation. Neutrons can excite phosphors indirectly by mixing the phosphors with materials that emit charged particles when irradiated with neutrons. The neutrons cause charged-particle emission, which in turn excite the phosphors to emit light.

Phosphors Excited by Electric Fields. *ZnS phosphors* are mainly used for electroluminescent devices which operate by application of an ac field on the powder. Electroluminescent lamps are prepared by sandwiching the phosphor between two conducting layers which serve as the electrodes. The phosphor can be embedded in an organic material or in an enamel ("ceramic") lamps). The top electrodes can be conducting glass or plastic. The construction employs a bottom metal plate, a reflective conducting enamel followed by the phosphor, top electrode, and a transparent cover. A list of electroluminescent materials is given in Table 6-54.

Another type of electroluminescent lamp is the forward biased *pn semiconductor junction*. The GaAs, GaP light-emitting diodes operate on this principle. Radiation is emitted due to recombination of charge carriers in the semiconductor junction.

Avalanche in *reverse-biased pn junctions* causes hole-pair generation followed by light emission. Again GaAs is an example for this type of electroluminescence.

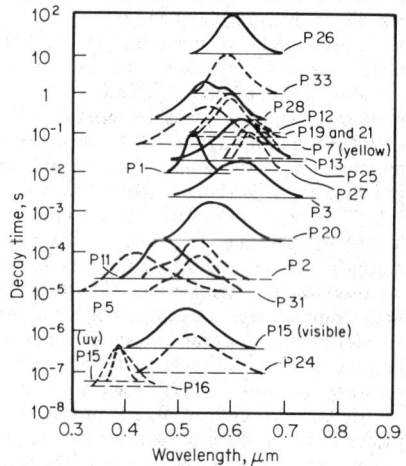

Fig. 6-68. Spectral emission curve of some commercial cathodoluminescent phosphors.[111]

Light emission can also occur due to electrons supplied by *tunneling* into the phosphor.

228. Solid-Laser Materials. Lasers can be classified according to the type of lasing material: solid lasers (actually, insulating solids incorporating ions), semiconductor lasers, liquid lasers, and gas lasers.

Solid lasers use solid materials which contain impurity ions that can be excited to high levels. Figure 6-69a shows a simplified energy diagram for ruby (Cr-doped Al_2O_3 crystals). This diagram is typical of three-level solid lasers. The Cr ions are excited to high energy level 3, and then return to level 2, which can be overpopulated with respect to level 1. The spontaneous radiation emitted by the return from level 2 to 1 stimulates more transfers from level 2 to 1 and further radiation.

Table 6-54. Characteristics of Some Scintillators in Current Use[111]

Phosphor	Density, g/cm³	Emission wavelength, max. Å	Absolute efficiency, %	Decay time
ZnS-Ag	4.1	4700	23	10 μs
NaI-Tl	3.7	4100	13	0.23 μs
CsI-Tl	4.5	4100–5800	6–7	1.5 μs
LiI–Eu	4.06	4750	4	0.94 μs
High-silica glass and Ce	2.5	3900	1	0.05 μs
Lithium silicate glass and Ce	2.5	3900	1	0.05 μs
BaPt(CN)₄4H₂O	2.1	5500	9	1.8 μs
CaWO₄	6.1	4300	2	3 μs
Anthracene	1.25	4450	4	30 μs
Naphthalene	1.15	3850	1	80 ns
trans-Stilbene	1.16	4100	2	6.4 ns
p-terphenyl in toluene	1	3200–4000	1.6	2–3 ns
p-terphenyl in xylene	1	3200–4000	1.1	2–3 ns
POPOP + p-terphenyl in toluene	1	4440	2.2	10 ns
POPOP + terphenyl in PVT	1	4400	2	10 ns

Population inversion is even easier to achieve in a four-level material (Fig. 6-69b). Most solid lasers have this type of energy-level diagram. These lasers contain transition metals, rare-earth metals, or actinide ions in glass, garnets, tungstate, or fluorides. Two impurities in one host crystal may affect each other. Sensitizing ions may absorb radiation and transmit it to the second ions, activators. Crystalline imperfections of laser crystals have an effect on the minimum power (threshold) needed to operate the laser because the scattering from imperfections must be overcome to obtain amplification. Lasers are also difficult to operate at high temperatures since it is harder to obtain population inversion. Table 11-2 lists some of the solid-laser materials.

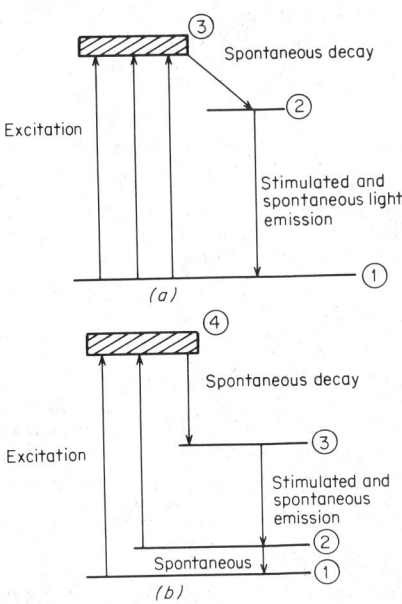

Ruby. The laser crystal is Al_2O_3 with addition of 0.05 wt % Cr_2O_3. The Cr addition is responsible for the pink color of the crystals. Ruby is a common solid-laser material. These lasers give powerful pulses which can be used for material processing, triggering chemical reactions, holography, and other applications.

Neodymium Crystal. 0.5 to 2% Nd ions in crystals such as $CaWO_4$, $SrWO_4$, $SrMoO_4$, $Ca(NbO_3)$, and $Y_3Al_5O_{12}$ (YAG) have been used for emission in the near-infrared region. Out of the above-mentioned materials, the YAG is a very useful material because it has better optical and mechanical properties. YAG lasers can be operated either in continuous or pulsed modes. They are useful for machining, communication, and ranging.

Fig. 6-69. Schematic energy-level diagram for (*a*) three-level and (*b*) four-level lasers.

229. Semiconductor-Laser Materials. Semiconductor lasers are efficient devices for converting electrical energy into light. The lasers, generally of linear dimensions smaller than 1 mm, are usually pumped by carrier injection at a *pn* junction. Other possible pumping modes are by electron bombardment by avalanche breakdown or optically.

Injection Lasers. GaAs is the most commonly used material. GaAs diodes are produced by preparing a *p*-type layer on an *n*-type substrate. The wafer is diced or cleaved to form individual lasers. The laser operates by application of sizable forward electric field for 10^{-6} s. Electrons from the *n*-region and holes from the *p*-region flow into the junction area, where they recombine and emit radiation. Injection-laser materials are listed in Table 6-55. Continuous operation at room temperature has been obtained in a heterostructure which consisted of *p*- and *n*-type layers of GaAs sandwiched between layers of $(Ga_{1-x}Al_x)As$.

Optically and Electron-beam-excited Semiconductor Lasers. The entire volume of the crystal can be used when semiconductors are optically or electron-beam-excited. The optical excitation is produced by another laser, while electron-beam excitation is produced by high-energy electrons. These pumping methods are important when junctions are difficult to obtain.

Excitation by Avalanche Breakdown. This type of laser operates by impact ionization due to avalanche breakdown in semiconductors. Low output powers were obtained when *n*-type GaAs was used for such lasers.

230. Liquid-Laser Materials. Liquid lasers have the advantage of eliminating the need for crystal growing. They conform easily to any shape, and their concentration can be easily changed. Their disadvantages are broad spectral lines and low efficiency. Three types are distinguishable: rare-earth chelate lasers, nonorganic neodymium–selenium oxychloride lasers, and the organic-dye lasers.

231. Gas-Laser Materials. A large variety of gas lasers have been constructed, although only a small number of these can be discussed in this available space.

PROPERTIES OF MATERIALS

Table 6-55. Injection-Laser Materials and Their Wavelengths[138]

Material	Wavelength, μm
GaAs	0.84
Ga(AsP)	0.64-0.84
(Ga,In)As	0.84-3.11
(Ga,Al)As	0.64-0.84
InAs	3.11
In(As,P)	0.90-3.11
InP	0.90
In(As,Sb)	3.11-5.18
InSb	5.18
GaSb	1.56
PbS	4.32
PbTe	6.5 (at 12 K)
PbSe	8.5 (at 12 K), 7.3 (at 77 K)
(Pb,Sn)Te	6-28
(Pb,Sn)Se	8-31

Noble-Gas Lasers. Helium-neon lasers are excited by glow discharge. The He is excited mainly by electron collisions and transfers its excitations to the Ne. The emitted radiations 0.6328, 1.15, and 3.39 μm are due to radiative transfers in Ne. He-Ne lasers are inexpensive and widely used in alignment, measurements, holography, communication, etc.

Helium lasers emit in the near infrared 1.9543 and 2.0603 μm and also in the far infrared 95.8 and 216 μm.

Neon lasers cover a spectral range from 0.59 to 133 μm; these lasers are not excited by another gas.

Argon-laser lines cover the range from 1.62 to 26.9 μm.

Krypton lasers cover the range from 1.68 to 7.06 μm, with two strong lines at 2.1165 and 2.1902 μm.

Xenon lasers extend from 2.02 to 18.5 μm.

Metal vapors such as Cd, Hg, Cs, Cu, Pb, and Mn have also been used for lasers.

Ion-Gas Lasers. Ion-gas lasers operate at powers several orders of magnitude higher than atomic lasers. Ionic lasers can be operated in the continuous mode, but are often operated in pulses to reduce power dissipation in the laser parts. Ionic-gas lasers can be used at the 0.26- to 0.7-μm-wavelength range with considerable power output.

Ionic lasers most frequently use ions of Ne, Ar, Kr, or Xe. Excitation of ionic lasers is accomplished by high-power gas discharges. Ion lasers are very convenient in photoelectric and photochemical work.

Noble-gas ions are chemically very reactive, similar to the halogens; therefore special precaution is necessary to protect the inner parts of such lasers. The noble-gas ion lasers are the best-known ion lasers. However, a large number of other ion lasers have been reported, among them Hg, halogens, O, N, and vaporized elements such as B, C, Si, Mn, Cu, Zn, Ge, As, Cd, In, Sn, and Pb.

Molecular Lasers. The best-known laser of this group is the CO_2 laser. This laser contains CO_2, N_2, and a large proportion of He, and sometimes also H_2O. The N_2 molecules are excited by electrons and transfer their excitation to the CO_2. The He improves the discharge characteristics of the gas mixture. H_2O gas helps to remove the energy of the CO_2 molecules in the terminal laser levels. CO_2 lasers have been constructed up to several meters long. These lasers can have very high output power and can be operated continuously or in a pulsed mode. These lasers have applications where a high-power output is required.

Another version of the pulsed CO_2 laser, the so-called TEA (transversely excited at atmospheric pressures), is capable of providing very high output pulses of the order of hundreds of joules.

The most powerful CO_2 laser is the gas-dynamic laser. The population inversion is obtained in this laser by letting a mixture of gases, including CO_2, expand from a region of high temperature and high pressure into a low-temperature and low-pressure region.

232. Chemical Lasers. In chemical lasers the population inversion is brought about by chemical means. An example is the population inversion created by mixing hydrogen and chlorine to produce excited HCl. Other types are photodissociated CS_2-O_2 mixtures.

233. Bibliography on Radiation-emitting Materials

1. LEVERENZ, H. W. "Luminescence of Solids;" New York, Dover Publications, Inc., 1968.

2. IVEY, H. F. Electroluminescence and Related Effects; *Advan. Electron. Electron Optics,* supplements, 1963.

3. GOLDBERG, P. (ed.) "Luminescence of Inorganic Solids;" New York, Academic Press, Inc., 1966.

4. HEWITT, H., and VAUSE, A. S. "Lamps and Lighting;" London, Edward Arnold (Publishers) Ltd., 1966.

5. LENGYEL, B. A. "Lasers;" New York, John Wiley & Sons, Inc.-Interscience Publishers, 1971.

OPTICAL AND PHOTOSENSITIVE MATERIALS

234. Index of Refraction. The index of refraction of a material is the ratio of the light velocity in empty space to the velocity in the material,

$$n = \frac{c}{v} \tag{6-85}$$

where c = velocity of light in empty space (m/s), v = velocity of light in the material (m/s), and n = index of refraction. The index of refraction for nonmagnetic materials is

$$n = \varepsilon_r^{1/2} \tag{6-86}$$

where ε_r = dielectric constant.

The index of refraction is a function of the wavelength. This effect is called *dispersion.* Figure 6-70 shows the dispersion in several glasses.

235. Dispersive Power. The dispersive power ω is defined by the indices of refraction in the blue (n_F), yellow (n_D), and red (n_C) portions of the spectrum,

$$\omega = \frac{n_F - n_C}{n_D - 1} \tag{6-87}$$

The dispersive power is a measure of the changes of the index of refraction of the material over the whole visible range, relative to the mean deviation of this index from unity.

236. Birefringence. Optically anisotropic crystals having two indices of refraction are called birefringent crystals, or doubly refracting crystals.

237. Dichroic Crystals. Doubly refracting materials with high absorption for one polarized component are called dichroic. Dichroic crystals can serve as light polarizers since they can be made to transmit only one polarization direction.

238. Refraction and Reflection. When light waves arrive at the boundary between two media, part of them enter the media, changing direction, while part are reflected. The angles of incidence and refracted and reflected rays are related.

$$n_1\sin\phi_1 = n_2\sin\phi_2 \tag{6-88}$$

$$\phi_1 = \phi_3$$

where n_1, n_2 = indices of refraction, and ϕ_1, ϕ_2, ϕ_3 = angles of incidence, refraction, and reflection, respectively, measured from the normal to the surface.

239. Transmittance and Reflectance. For a monochromatic light of wavelength λ the transmittance T_λ of a material is defined as

$$T_\lambda = \frac{P_T}{P_0} \tag{6-89}$$

where P_T = radiant flux transmitted through the material, and P_0 = radiant flux of the incident beam.

Reflectance of a monochromatic light of wavelength λ, R_λ, from a material is defined as

$$R_\lambda = \frac{P_R}{P_0} \tag{6-90}$$

where P_R = reflected radiant flux, and P_0 = radiant flux reflected from 100% reflecting (completely diffusing) sample.

Transmittance and reflectance can be defined also with respect to sensors for nonmonochromatic light. The transmittance of light with respect to a sensor with a given sensitivity is

$$T = \frac{\int P_{TS}\, d\lambda}{\int P_{0S}\, d\lambda} \qquad (6\text{-}91)$$

where T = transmittance and s = sensitivity of the sensor. In effect, the transmittance is the total transmitted luminous flux divided by the total incident luminous flux.

The reflectance, similarly, is

$$R = \frac{\int P_{RS}\, d\lambda}{\int P_{0S}\, d\lambda} \qquad (6\text{-}92)$$

where R = reflectance.

240. Transmission and Reflection Density. The transmission density D_T is defined as

$$D_T = \log_{10} 1/T = -\log_{10} T \qquad (6\text{-}93)$$

The reflection density D_R is defined as

$$D_R = \log_{10} 1/R = -\log_{10} R \qquad (6\text{-}94)$$

T, R are the transmittance and reflectance, respectively.

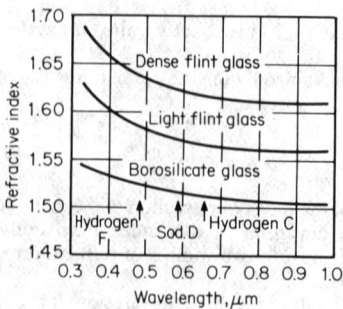

Fig. 6-70. Refractive index as a function of wavelength for various glasses.[74]

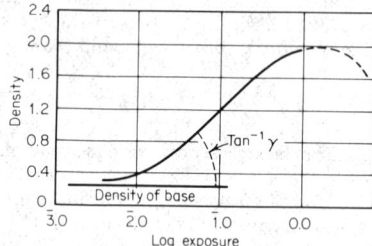

Fig. 6-71. Schematic H & D curve for a photographic emulsion.

241. Exposure. Exposure is defined as the amount of energy incident on a unit area of photographic emulsion.

242. Characteristic Curve of Emulsions. The density of the emulsion as a function of exposure is shown in Fig. 6-71. This curve is called *H-D curve* (Hurter and Driffield). The slope of this curve is called *gamma*. Emulsion speed can be defined by this curve. Exposures beyond the maximum of the curve begin to decrease the film density. This phenomenon is called *solarization,* or *reversal.*

243. Spectral Sensitivity of Photographic Emulsions. The sensitivity s of a photographic emulsion at a given wavelength λ is

$$s = \text{const} \cdot \frac{1}{U_\lambda} \qquad (6\text{-}95)$$

where U_λ = incident energy per unit emulsion area necessary to produce a given density D. Spectral-sensitivity distributions of various emulsions are given in Fig. 6-72.

244. Image Formation on Photographic Emulsions. Photographic emulsions are in most cases minute silver halide crystals dispersed in a gelatine carrier. The gelatine carrier separates the silver halide crystals and may contain additional materials such as sensitizers, desensitizers, and fogging or antifogging materials.

Although the theory of image formation is not yet completely founded, it is generally accepted that the incident light releases electrons from the halide ion and these electrons are captured by *sensitivity specks*. Interstitial silver ions can move toward the negatively charged specks and become neutralized. Sufficient neutral silver atoms render a grain developable. These minute chemical changes of the emulsion on exposure are called *latent image*. The latent image has not yet been detected directly, but can be made visible by the development process. The development process causes chemical decomposition almost exclusively in the grains in which silver atoms collected at sensitivity specks. The development process multiplies the chemical decomposition by a factor as high as 10^8.

245. Development and Fixing. During development, the developing agent differentially reduces the light-exposed silver halide grains to metallic silver. This development process is called *chemical developing*. Long development time will affect all the crystals in the emulsion. Most developers contain reducing agents such as hydroquinone or *p*-methylaminophenol sulfate (for black and white) and pyrogallol or derivatives of *p*-phenylenediamine (for color photography).

In addition, the developer contains sulfite, which serves as a preservative against oxidation and also a solvent for silver halide. Bromine is frequently added to the developer to promote adsorption onto silver halide grain surface. This retards the development of very weakly or unexposed grains, i.e., prevents fog. An alkali is added to the developer to control the proper pH of the solution. Other additions such as wetting agents are also common.

The fixing process dissolves the undeveloped silver halide grains. The photographic picture is then composed of the remaining dark silver particles.

246. Reversal Process. In this process the emulsion is first developed and then the silver is dissolved by a strong oxidizing agent. The silver remains only in the unexposed areas of the film. The film is now exposed and developed, and the result is a direct positive. Reversal process is used for 16-mm amateur motion picture films.

SPECIFIC OPTICAL AND PHOTOSENSITIVE MATERIALS

247. Optical Glasses. Glasses for optical instruments have to fulfill stringent requirements, those being that the glass have a specified refractive index and dispersion; it must have minimal variations in its optical properties; and it must be strain-free to minimize birefringence and have high dimensional stability and a very smooth surface finish.

Optical glasses are discussed in great detail by Morley.[52] The properties of some optical glasses are given in Fig. 6-73.

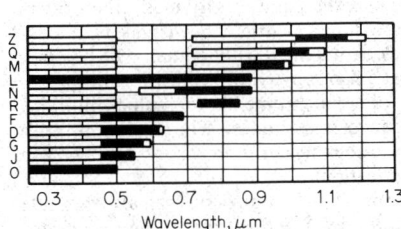

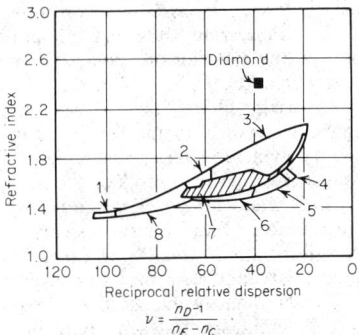

Fig. 6-72. Spectral sensitivity of various emulsion classes. The dark area denotes the useful working region. The letters indicate the various classes.[112,113]

Fig. 6-73. Range of optical properties obtained with ordinary glasses (shaded) and with Eastman Kodak fluoride and rare-earth glasses.[74] (1) E.K. fluoride glasses; (2) E.K. fluoborate glasses; (3) E.K. rare-element borate glasses; (4) E.K. fluogermanate glasses; (5) E.K. fluosilicate glasses; (6) titanium glasses; (7) ordinary optical glasses; (8) E.K. fluophosphate glasses.

248. Filters. Glasses can be obtained in different colors to serve as optical filters. The various colors result by compound additions:

Green—chromium oxide, iron oxide, vanadium oxide.

Blue—cobalt oxide, copper oxide.

Purple—manganese oxide.

Amber—carbon or sulfur.

Red—selenium and cadmium sulfide, cuprous oxide, elemental gold.

Opal glass—fluorine compounds (cryolide).

Filter glass is normally not of optical grade. A large selection of filters can be obtained with various transmittance spectra.

Filtering can also be obtained by multiple dielectric coating on glass. The multiple coating removes certain light wavelength by interference.

249. Mirrors. Reflecting of glass surfaces can be significantly improved by dense metallic coatings. Selective reflectance can be achieved by incorporation of gold or nickel in glass. For example, visible light can be transmitted; infrared radiation is reflected by certain mirrors. Interference coatings can also serve as selective reflectors.

250. Light-sensitive Glasses (UV, Visible, IR).[53-55] Light sensitive glasses can be divided into the following two types.

Photosensitive Glasses. The color of these glasses can be permanently affected by ultraviolet (UV) radiation. Such glasses are commonly lithium silicates. The glasses contain metallic ions such as silver and gold and small additions of cerium oxide and common reducing agents such as tin oxide.

The UV light is applied at room temperature, and the color changes are observable only after a heat-treatment above the glass-annealing range. The color is produced by elemental silver or gold precipitation and growth.

Photosensitive glasses can be made into photomachinable glasses by slight variations in composition. In these glasses lithium metasilicate is nucleated by the silver or gold particles, and crystals can be grown by heat-treatment. The lithium metasilicate phase is etched considerably faster in diluted hydrofluoric acid than the matrix. These chemically machined glasses are used, for example, for fluidic controlled systems and engravings.

The photomachined glasses may be further heated at slightly higher temperatures to produce lithium disilicate crystals. These crystals form an interlocking network of considerable mechanical strength similar to that of glass-ceramic bodies.

Photochromic Glasses.[54,55] These normally contain silver halides and change their transmittance reversibly with incident-light intensity. Their transmittance decreases upon strong illumination and is regained when the illumination is decreased. Such glass is used for windows, sunglasses, and for optical memories.

251. Polarizing Materials. Tourmaline is an example of a dichroic crystal which has a pale yellow color for one polarization direction and becomes almost entirely opaque to the other direction.

Polarizing filters, Polaroid, have been developed by E. H. Land.[56] Originally, these filters contained aligned crystallites of herapathite. Polaroid filters have later been made with long-chain polymers such as polyvinylene or polyvinyl alcohol treated with iodine. The polymer chains are aligned to produce the necessary optical anisotropy.

252. Monochrome Film Emulsions. A variety of monochrome photographic emulsions are available. Emulsion speeds vary over three orders of magnitudes. The resolution varies from less than 50 to 2,000 lines/mm. The contrast obtained in films also varies, and the appropriate emulsion must be chosen for each application.

The granularity changes from "coarse" to "extremely fine." Emulsions can also be made with different spectral sensitivity in the visible ultraviolet and infrared. Emulsions for nuclear, x-ray, and electron microscope work are also available.

The spectral sensitivities of unsensitized emulsions show some differences in response. For example, silver chloride is UV-sensitive, while silver bromide is blue-sensitive. The differences in these curves are due to differences in spectral absorption of those materials; i.e., yellow silver bromide absorbs blue light and therefore is sensitive to blue, while silver chloride absorbs only in the UV range, etc.

The emulsions can be either chemical-sensitized with sulfur-containing organic compounds, or optical-sensitized with dyes such as cyanines. The sensitizers increase the range of emulsion sensitivity. Supersensitizing of emulsion is obtained by addition of several dyes. Desensitizing of an emulsion can be used to decrease the sensitivity of a film to some desired wavelengths (such as yellow, to allow film developing in bright illumination).

Figure 6-72 shows the sensitive spectral regions for several film types. Orthochromatic emulsions are green-sensitive, panchromatic emulsions are sensitive to the entire visible range, and nonsensitized emulsions are blue-sensitive.

253. Photoresists. Photoresists are organic compounds whose structure can be changed upon exposure in the UV region. Photoresists can be preferentially dissolved after the exposure. Photoresists remaining in the areas which received the illumination are called *negative resists.* Resists remaining in the unilluminated areas are *positive resists.*

Negative resists contain polymers, sensitizers, and solvents. Such resists are kept from polymerizing. The light promotes the polymerization with the aid of the sensitizer. The polymerized photoresist remains insoluble during the development. Positive resists, when illuminated, render the illuminated area soluble.

Table 6-56 lists some of the common photoresists, and Fig. 6-74 shows the spectral sensitivity of several photoresists.

Photoresists have found use in chemical-machining metal foils, printed-circuit boards, and very extensively in microelectronics.

254. Other Recording Media. In addition to photoresists and photographic films, there are a variety of image-recording techniques. Electrophotographic recording includes techniques such as xerography, or elastrostatically produced deformations in plastic films. *Dielectric recording* (Electret) is another recording technique in which the image is transferred to paper as in xerography. Other techniques use metallic oxide reduction, electrolytic recording, diazo-recording photochromism, and photochemical recording. A review of recording techniques is found in Refs. 57 to 59, Par. **6-257.**

255. Color-Film Emulsions. Color-film emulsions are usually of the multilayer substractive type. The color films are prepared from several layers of emulsion. For example,

Table 6-56. Photoresist Materials[95]

Supplier	Name	Remarks
Du Pont.........	Riston	Supplied as a film; for plating and etching
Dynachem.......	DCR 3140	General-purpose resist
	DCR 3154	Improved adhesion to aluminum and improved resistance to alkaline etches
	DCR 3118, 3118H	Roller-coating formulation
	DCR 3116	Provides heavy resist layer
	DCR 3170	Microelectronic formulation
G.A.F.	...	Positive resist in field testing
Philip A. Hunt ...	Waycoat No. 10	General-purpose etching and plating resist
	Waycoat No. 20	Especially useful as a plating resist
Kodak..........	KPR	Used on copper and copper-based alloys; used on clear and light-colored anodized aluminum
	KPR2	Used on copper and copper-based alloys; electroplating resist
	KPR3	Formulated for dip-coating systems
	KPR4	Formulated for roller-coating systems; good for plated-through holes
	KOR	Similar to KPR 2; possesses greater spectral sensitivity than other products
	KMER	Used on all surfaces except copper and copper alloys and clear and light-colored anodized aluminum
	KTFR	Microelectronic formulation
	KPL	Used to increase viscosity of KPR; rarely used alone
Norland	Photoresist 30	One-part water-based resist for stainless regular steel, nickel, copper, brass
	Photoresist 22	Two-part water-based system for Kovar-type metals
Shipley (positive photoresists)....	AZ-111	General-purpose photoresist
	AZ-119	As for AZ-111, but formulated for roller coating
	AZ-340	Used for circuit boards and plated-through holes
	AZ-345	Higher solids and viscosity than AZ-340 to provide lands around plated-through holes
	AZ-1350	Used for microelectronic applications
	AZ-1350H	As for AZ-1350, but formulated for roller coating

Kodachrome has a top layer of blue-sensitive emulsion followed by a yellow filter. Underneath is an orthochromatic (mainly green-sensitive), and the bottom is a panchromatic, mainly red-sensitive, emulsion. The processing of such an emulsion is quite involved. It consists in developing, exposing to red light from the back side, then dye-coupler developing (i.e., developing which includes dyeing during the process). More development and dye coupling are necessary before the final image is obtained.

Ansco, Agfacolor, and Ektachrome are emulsions that can be developed in a simpler manner because the dye couplers are contained in the emulsion. Other emulsion types are Kodacolor, Ektacolor, and a Du Pont process.

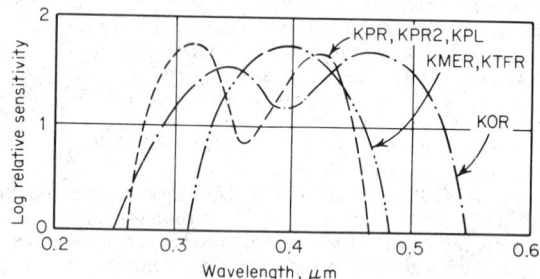

Fig. 6-74.   Spectral sensitivity of photoresists.[113]

256. Bibliography on Optical and Photosensitive Materials

1. MEES, C. E. KENNETH   "The Theory of Photographic Process;" New York, The Macmillan Company, 1942.

2. EVANS, R. M., HANSON, W. T., and BREWER, W. L.   "Principles of Color Photography;" New York, John Wiley & Sons, Inc., 1953.

257. References

1. MATTHIESSEN, A., and VOGT, C.   *Proc. Roy. Soc.,* 1863, Vol. 12, p. 652; *Ann. Phys. Chem.,* 1864, Vol. 122, p. 19.

2. THOMPSON, M. W.   "Defects and Radiation Damage in Metals"; New York, Cambridge University Press, 1969, p. 27.

3. CLAREBROUGH, L. M., HARGREAVES, M. E., and LORETTO, M. H.   *Phil. Mag.,* 1962, Vol. 1, p. 115.

4. PANSERI, C., and FEDERIGHI, T.   *Phil. Mag.,* 1958, Vol. 3, p. 1223.

5. YOSHIDA, S., KINO, T., KIRITANI, M., KABEMOTO, S., MAETA, H., and SHIMOMURA, Y. *J. Phys. Soc. Japan,* 1963, Vol. 18, Suppl. 2, p. 98.

6. COTTERILL, R. M. J.   *Phil. Mag.,* 1963, Vol. 8, p. 1937.

7. WASHBURN, E. W. (ed.)   "International Critical Tables of Numerical Data," Vol. 6; New York, McGraw-Hill Book Company, 1929, p. 135.

8. "Landolt-Bornstein," Zahlenwerte und funktioner aus naturwissenschaften und technik, Vol. 6; Berlin, Springer Verlag OHG, 1959.

9. SONDHEIMER, E. H.   The Mean Free Path of Electrons in Metals; *Advan. Phys.,* 1952, Vol. 1, p. 1.

10. CAMPBELL, D. S.   "The Use of Thin Films in Physical Investigations," New York, Academic Press, Inc., 1966, p. 299.

11. NEUGEBAUER, C. A.   Pages 191-220 in B. Schwartz and N. Schwartz (eds.), "Measurement Techniques for Thin Films"; New York, Electrochemical Society, 1967.

12. HUNTINGTON, H. B., and GRONE, A. R.   *J. Phys. Chem. Solids,* 1961, Vol. 20, p. 76.

13. ROSENBERG, R., and BERENBAUM, L.   *Appl. Phys. Letters,* 1968, Vol. 12, p. 201.

14. BLECH, I. A., and MEIERAN, E. S.   *J. Appl. Phys.,* 1968, Vol. 40, p. 485.

15. LEPSELTER, M. P.   *Bell Syst. Tech. J.,* 1966, Vol. 45, p. 233.

16. WILSON, R. W., and TERRY, L. E.   *Proc. IEEE,* 1969, Vol. 57, p. 1580.

17. KOHMAN, G. T., et al.   *Bell Syst. Tech. Jour.,* 1955, Vol. 34, p. 115.

18. CUNNINGHAM, J. A.   *Solid State Elec.,* 1965, Vol. 8, p. 735.

19. CUNNINGHAM, J. A., FULLER, C. R., and HAYWOOD, G. T. *IEE Trans.* Rel.-R-19, 1970, p. 182.

20. RUCK, G. T., BARRICK, D. E., STUART, W. D., and KRICHBAUM, C. K. "Radar Cross-section Handbook"; New York, Plenum Press, 1970.

21. GROVE, A. S. "Physics and Technology of Semiconductor Devices"; New York, John Wiley & Sons, Inc., 1967, pp. 346-350.

22. WHITEHEAD, S. "Dielectric Breakdown of Solids"; London, Oxford University Press, 1953.

23. O'DWYER, J. J. "The Theory of Dielectric Breakdown in Solids"; London, Oxford University Press, 1964.

24. KLEIN, N. Electrical Breakdown in Solids: *Advan. Elect. Electron Phys.,* 1969, Vol. 26, pp. 309-424.

25. MASON, J. H. Dielectric Breakdown in Solid Insulation; *Progr. Dielectrics* (London) 1959, Vol. 1, pp. 1-58.

26. ADAMCZEWSKI, I. "Ionization, Conductivity and Breakdown in Dielectric Liquids"; London, Taylor & Francis, 1969.

27. LLEWELLYN-JONES, F. "Ionization and Breakdown in Gases" and "Ionization Avalanches and Breakdown"; London, Methuen & Co., Ltd., 1957 and 1967.

28. VON ENGEL, A. "Ionized Gases," 2d ed.; London, Oxford University Press, 1965.

29. FROHMAN-BENTCHKOWSKI, D., and LENZLINER, M. *J. Appl. Phys.,* 1969, Vol. 40, p. 3307.

30. BERRY, R. W., HALL, P. M., and HARRIS, M. T. "Thin Film Technology"; Princeton, N.J., D. Van Nostrand Company, Inc., 1968.

31. WARREN, B. E. "X-ray Diffraction"; Reading, Mass., Addison-Wesley Publishing Company, Inc., 1969.

32. HADFIELD, R. A. "Metallurgy and Its Influence on Modern Progress"; Princeton, N.J., D. Van Nostrand Company, Inc., 1926.

33. SERY, R. S., and GORDON, D. I. *J. Appl. Phys.,* 1965, Vol. 36, p. 1221.

34. CZOCHRALSKI, J. *Z. Phys. Chem. Leipzig,* 1917, Vol. 92, p. 219.

35. MILLER, L. F. *IBM Tech. Rep.* 22.743, January, 1970.

36. HARPER, C. A. (ed.) "Handbook of Electronic Packaging"; New York, McGraw-Hill Book Company, 1969.

37. DESSAUER, J. H., and CLARK, E. H. (eds.) "Xerography and Related Processes"; New York, Focal Press, 1965.

38. OVSHINSKY, S. R., and FRITZCHE, H. *Met. Trans.,* 1971, Vol. 2, p. 641.

39. SIE, C., DUGAN, P., and MOSS, S. C. 4th International Conference on Amorphous and Liquid Semiconductors, Ann Arbor, Mich., August, 1971.

40. FEINLEIB, J., DE NEUTVILLE, J., MOSS, S. C., and OVSHINSKY, S. R. *Appl. Phys. Letters,* 1971, Vol. 18, p. 254.

41. LANGMUIR, I. *J. Franklin Inst.,* 1934, Vol. 217, p. 543.

42. AYER, R. B. *AIEE Trans.,* 1953, Vol. 72, p. 121.

43. LEMMENS, H. J., JANSEN, M. J., and LOOSJES, R. *Philips Tech. Rev.,* 1959, Vol. 11, p. 341.

44. LEVI, R. *J. Appl. Phys.,* 1952, Vol. 24, p. 233.

45. COPOLA, P. P., and HUGHES, R. C. *Proc. IRE,* 1956, Vol. 44, p. 351.

46. HADLEY, C. P., RUDY, W. G., and STOECKERT, A. J. *J. Electrochem. Soc.,* 1958, Vol. 105, p. 395.

47. HERMANN, G., and WAGENER, S. "The Oxide Coated Cathode"; London, Chapman & Hall, Ltd., 1951.

48. LAFFERTY, J. M. *J. Appl. Phys.,* 1951, Vol. 22, p. 299.

49. ALBERT, M. J., and ATTA, M. A. *Metals Mater.,* February, 1967, p. 43.

50. GOBELI, G. W., and ALLEN, F. G. *Phys. Rev.,* 1965, Vol. 137, p. 245A.

51. HEWITT, H., and VAUSE, A. S. "Lamps and Lighting"; London, Edward Arnold (Publishers) Ltd., 1966, p. 168.

52. MORLEY, G. W. "Properties of Glass," 2d ed.; *ACS Monograph* 77, 1954.

53. LILLIE, H. R. *Glass Technol.,* 1960, Vol. 1, p. 115.

54. ARAUJO, R. J., and STOOKEY, S. D. *Glass Ind.,* December, 1967, p. 687.

55. SMITH, G. P. *Jour. Photo. Sci.,* 1970, Vol. 18, No. 2.

56. LAND, E. H.   *J. Opt. Soc. Am.,* 1951, Vol. 41, p. 957.

57. ROBILLARD, J. J.   *Phot. Sci. Eng.,* 1964, Vol. 8, p. 18.

58. KOSAR, J., and CLARCK, W. (eds.)   "Wiley Series in Photographic Sciences and Technology"; New York, John Wiley & Sons, Inc., 1965.

59. SOULE, H. V.   "Electro-optical Photography at Low Illumination Levels"; New York, John Wiley & Sons, Inc., 1968.

60. FINK, D. G. (ed.)   "Standard Handbook for Electrical Engineers"; New York, McGraw-Hill Book Company, 1968, Secs. 4-144 and 4-145.

61. Compiled from several sources.

62. GOLDSMITH, A., et al. (eds.)   "Handbook of Thermophysical Properties of Solid Materials"; Oxford, Pergamon Press, 1961.

63. GERRITSEN, A. N.   Metallic Conductivity, Experimental Part, Vol. 19, p. 137, in S. Flugge (ed.), "Handbuch der Physik"; Berlin, Springer Verlag OHG, 1956.

64. SCOW, K. B., and THUN, R. E.   *Trans. 9th Vacuum Symp. Am. Vacuum Soc.,* New York, The Macmillan Company, 1962, pp. 151-155.

65. MOTT, N. F., and JONES, H.   "The Theory of the Properties of Metals and Alloys"; New York, Dover Publications, Inc., 1958.

66. GUY, A. G.   "Introduction to Materials Science"; New York, McGraw-Hill Book Company, 1971, p. 553.

67. GARWOOD, M. F., ZURBURG, H. H., and ERICKSON, M. A.   "Correlation of Laboratory Tests and Service Performance"; Cleveland, American Society for Metals, 1951.

68. HARDY, H. K., and HEAL, T. J.   *Prog. Metal Phys.,* 1954, Vol. 5, p. 195.

69. HANSEN, M.   "Constitution of Binary Diagrams"; New York, McGraw-Hill Book Company, 1958.

70. RAMO, S., WHINNERY, J. R., and DUZER, T. V.   "Fields and Waves in Communication Electronics"; New York, John Wiley & Sons, Inc., 1965, p. 297.

71. HARPER, C. A. (ed.)   "Handbook of Electronic Packaging"; New York, McGraw-Hill Book Company, 1969, pp. 1-52.

72. After MURPHY, E. J., and MORGAN, S. D.   *Bell Syst. Tech. Jour.,* 1935, Vol. 16, p. 493.

73. KLEIN, N.   *Thin Solid Films,* 1971, Vol. 7, p. 149.

74. KINGERY, W. D.   "Introduction to Ceramics"; New York, John Wiley & Sons, Inc., 1960.

76. "Design Data for Teflon"; Geneva, Du Pont de Nemours International S.A.

77. HAGUE, J. R., et al.   "Refractory Ceramics for Aerospace"; American Ceramic Society, Columbus, Ohio, 1964.

78. American Lava Corporation, Chattanooga, Tenn.

79. "Properties of Selected Commercial Glasses"; Corning, N.Y., Corning Glass Works, 1971.

80. BOZORTH, R. M.   "Ferromagnetism"; Princeton, N.J., D. Van Nostrand Company, Inc., 1951.

81. LYMAN, T. (ed.)   "Metals Handbook," 8th ed.; Cleveland, The American Society for Metals, 1961, p. 792.

82. "Electrical Materials Handbook"; Allegheny Ludlum Steel Co., Pittsburgh, Pa., 1961.

83. BOZORTH, R. M.   *Rev. Mod. Phys.,* 1953, Vol. 25, p. 42.

84. WILLIAMS, H. J., and GOERTZ, M.   *J. Appl. Phys.,* 1952, Vol. 23, p. 316.

85. SMIT, J. (ed.)   "Magnetic Properties of Materials"; New York, McGraw-Hill Book Company, 1971, p. 77.

86. GELLER, S., REMEIKA, J. P., SHERWOOD, R. C., WILLIAMS, H. J., and ESPINOSA, G. P.   *Phys. Rev.,* 1965, Vol. 137, p. A1034.

87. General Electric "Permanent Magnet Manual."

88. TEBBLE, R. S., and CRAIK, D. J.   "Magnetic Materials"; New York, Wiley-Interscience, 1969.

89. GOULD, J. E.   *Proc. IEE,* 1959, Vol. 106A, p. 493.

90. HALL, R. N., and RACETTE, J. H.   "Diffusion and Solubility of Copper in Extrinsic and Intrinsic Germanium, Silicon, and Gallium Arsenide"; *J. Appl. Phys.,* 1964, Vol. 35, p. 379.

91. MORIN, F. J., and MAITA, J. P.   Electrical Properties of Silicon Containing Arsenic and Boron; *Phys. Rev.,* 1954, Vol. 96, p. 28.

92. MORIN, F. J., and MAITA, J. P.   Conductivity and Hall Effect in the Intrinsic Range of Germanium, *Phys. Rev.,* 1954, Vol. 94, p. 1525.

93. AZAROFF, L. V., and BROPHY, J. J.   "Electronic Processes in Materials"; New York, McGraw-Hill Book Company, 1963.

94. GROVE, A. S.   "Physics and Technology of Semiconductor Devices"; New York, John Wiley & Sons, Inc., 1967.

95. HARPER, C. A. (ed.)   "Handbook of Materials and Processes for Electronics"; New York, McGraw-Hill Book Company, 1970.

96. SZE, S. M., and IRVIN, J. C.   *Solid State Electron.,* 1968, Vol. 11, p. 599.

97. CURTISS, D. B.   *Bell Syst. Tech. Jour.,* 1961, Vol. 40, p. 509.

98. IRVIN, J. C.   *Bell Syst. Tech. Jour.,* 1962, Vol. 41, p. 387.

99. Several sources.

100. WOLF, H.   "Semiconductors"; New York, John Wiley & Sons, Inc.-Interscience Publishers, 1971.

101. After BURGER, R. M., and DONOVAN, R. P. (eds.)   "Fundamentals of Silicon Integrated Device Technology," Vol. 1; Englewood Cliffs, N.J., Prentice-Hall Inc., 1967.

102. KENDALL, D. L.   Stanford, Calif., Dept. of Material Science, Rept. 65-29, Stanford University, August, 1965.

103. MÜLLER, H. O.   *Z. Phys.,* 1937, Vol. 104, p. 475.

104. ROSE, R. M., SHEPARD, L. A., and WULFF, J.   "The Structure and Properties of Materials," Vol. 4; New York, John Wiley & Sons, Inc., 1966.

105. SOMMER, A. H.   Fifth meeting of Conference of International Commission on Optics, August, 1959, p. 329.

106. SPICER, W. E.   *Phys. Rev.,* 1958, Vol. 112, p. 114, and *RCA Rev.,* 1958, Vol. 19, p. 555.

107. SOMMER, A. H.   "Photoemissive Materials"; New York, John Wiley & Sons, Inc., 1968.

108. SEARS, F. W.   "Optics"; Reading, Mass., Addison-Wesley Publishing Company, Inc., 1949.

109. HEWITT, H., and VAUSE, A. S.   "Lamps and Lighting"; London, Edward Arnold (Publishers) Ltd., 1966.

110. LEVERENZ, H. W.   "An Introduction to Luminescence of Solids"; New York, Dover Publications, Inc., 1968, p. 211.

111. GARLICK, G. F. J.   In P. Goldberg (ed.), "Luminescence of Inorganic Solids"; New York, Academic Press, Inc., 1966, p. 708.

112. "Kodak Data Book of Applied Photography," Vol. 4; London, Kodak Ltd.

113. Reprinted by permission from a copyrighted Kodak publication, Eastman Kodak Co., Rochester, N.Y., 1971.

114. SMITHELLS, C. J.   "Metals Reference Book"; London, Butterworth & Co. (Publishers) Ltd., 1967.

115. PARKER, E. (ed.)   "Material Data Book"; New York, McGraw-Hill Book Company, 1967.

116. LYMAN, T. (ed.)   "Metals Handbook," Vol. 1; Cleveland, The American Society for Metals, 1961.

117. CAULTON, M.   *Nerem Record,* November, 1968.

118. FINK, D. G., et al. (eds.)   "Standard Handbook for Electrical Engineers," 10th ed.; New York, McGraw-Hill Book Company, 1968.

119. MIL-W-81044.

120. ITT, Wire and Cable Division, Clinton, Mass.

121. Amphenol Corporation.

122. Sigmund Cohn Corp., Mt. Vernon, N.Y.

123. MANKO, H.   "Solders and Soldering"; New York, McGraw-Hill Book Company, 1964.

124. E. I. Du Pont de Nemours, Inc., Wilmington, Del.

125. LICARI, J. J., and BRANDS, E. R.   *Machine Design Mag.,* 1967.

126. Coors Porcelain Co.

127. KITTEL, C.   "Introduction to Solid State Physics," 4th ed.; New York, John Wiley & Sons, Inc., 1971.

128. Corning Glass Works, Corning, N.Y.

129. Electro-Science Laboratories, Philadelphia, Pa.
130. CHIKAZUMI, S. "Physics of Magnetism"; New York, John Wiley & Sons, Inc., 1964.
131. VON AULOCK, W. H. (ed.) "Handbook of Microwave Ferrite Materials"; New York, Academic Press, Inc., 1965.
132. SZE, S. M. "Physics of Semiconductor Devices"; New York, John Wiley & Sons, Inc.-Interscience Publishers, 1969.
133. KOHL, W. H. "Handbook of Materials and Techniques for Vacuum Devices"; New York, Reinhold Book Corporation, 1967.
134. GRAY, D. E. (ed.) "American Institute of Physics Handbook"; New York, McGraw-Hill Book Company, 1957.
135. ENGSTROM, E. W. *RCA Rev.,* 1960, Vol. 21, p. 184.
136. KAZAN, B., and KNOLL, M. "Electronic Image Storage"; New York, Academic Press, Inc., 1968.
137. IVEY, H. F. Electroluminescence and Related Effects, *Advan. Elect. Electron Phys.,* Supplement 1, 1963, and IVEY, H. F., private communication.
138. LENGYEL, B. A. "Lasers"; New York, John Wiley & Sons, Inc.-Interscience Publishers, 1971.

SECTION 7

DISCRETE CIRCUIT COMPONENTS

BY

E. A. GERBER Director, Electronic Components Laboratory (ret.), U.S. Army Electronics Command Fort Monmouth; Consultant, U.S. Army Electronics Technology and Devices Laboratory, Electronics Command, Fort Monmouth, New Jersey; Fellow, IEEE

THOMAS S. GORE, JR. Deputy Chief, Frequency Control and Signal Processing Devices Technical Area;* Member, IEEE

EMANUEL GIKOW Leader, Filter Devices Team (ret.); Senior Member, IEEE; Licensed Professional Engineer, New Jersey

GEORGE W. TAYLOR Supervisory Electronics Engineer, Beam, and Display Plasma Devices Technical Area; Member, IEEE

L. E. SCHARMANN Electronics Engineer, Beam, and Display Plasma Devices Technical Area (ret.); Member, IEEE

I. REINGOLD Leader, Display Devices Team; Senior Member, IEEE; Member, Society for Information Display; Licensed Professional Engineer, New Jersey

MUNSEY CROST Senior Scientist, Display Devices Team; Member, American Physical Society; Member, Society for Information Display

GREGORY J. MALINOWSKI Electronics Engineer, Semiconductor Devices and Integrated Electronics Technical Area

EDWARD B. HAKIM Solid State Physicist, Semiconductor Devices and Integrated Electronics Devices Technical Area; Member, IEEE

DAVID LINDEN Chief, Power Sources Technical Area; Fellow, American Institute of Chemists; Member, American Chemical Society; Member, Electrochemical Society

RICHARD BABBITT Physicist (Electromagnetics), Frequency Control and Signal Processing Devices Technical Area; Member, American Ceramic Society

JOSEPH M. GIANNOTTO Electronics Engineer, Frequency Control and Signal Processing Devices Technical Area; Senior Member, IEEE

GUNTER K. GUTTWEIN Deputy Chief, Semiconductor and Frequency Control Devices Technical Area (ret.); Senior Member, IEEE

JOSEPH M. KIRSHNER Chief, Fluidics Systems Research Branch, Harry Diamond Laboratories, Washington, D.C.; Member, American Society of Mechanical Engineers; Member, American Physical Society; Associate Member, IEEE

ROBERT A. GERHOLD Deputy Chief, Semiconductor Devices and Integrated Electronics Technical Area; Senior Member, IEEE

SAM DI VITA Supervisory Ceramic Engineer, Frequency Control and Signal Processing Devices Technical Area; Fellow, American Ceramic Society

JACK SPERGEL Chief Applications Engineer, General Cable Corporation, Colonia, N.J.; Senior Member, IEEE (Deceased)

---

*Unless stated otherwise, all authors are with the U.S. Army Electronics Technology and Devices Laboratory, Electronics Command, Fort Monmouth, N.J.

DISCRETE CIRCUIT COMPONENTS

CONTENTS

Numbers refer to paragraphs

SECTION 7

DISCRETE CIRCUIT COMPONENTS

RESISTORS

By Thomas S. Gore, Jr.

1. Fundamentals. A resistor is a device that introduces resistance into an electronic circuit and as such is used for setting biases, controlling gain, fixing time constants, matching and loading circuits, and for voltage division, heat generation, and other related functions. Resistance is a fundamental property of a conductor, as shown:

$$R = \rho l/A$$

where R = resistance in ohms, ρ = specific resistance or resistivity of the conductor material in ohm-centimeters, l = length of conductor in centimeters, and A = cross-section area in square centimeters. Resistance controls either the voltage or the flow of current in an electronic circuit and in so doing produces dissipation of power. Thus

$$\text{Resistance} = R = V/I \qquad \text{Ohm's law}$$
$$\text{Power dissipated in resistor} = P = V^2/R = I^2R$$

where V = voltage in volts, I = current in amperes, and P = power in watts.

Resistance-Temperature Characteristic. The magnitude of change in resistance due to temperature is usually expressed in percent per degree centigrade (%/°C) or parts per million per degree centigrade (ppm/°C). If the changes are linear over the operating temperature range, the parameter is known as *temperature coefficient;* if nonlinear, the parameter is known as *resistance-temperature characteristic.*

Hot-spot temperature is the maximum temperature measured on the resistor body due to both internal heating and the ambient operating temperature and is usually the maximum temperature at which the resistor is derated to zero power.

Noise is an unwanted voltage fluctuation generated within the fixed resistor. Total noise of a resistor always includes (1) Johnson noise, which is dependent only on resistance value and temperature of the resistance element, due to thermal agitation, and (2) noise caused by current flow, cracked bodies, and loose end caps or leads, depending on the type of resistor element and construction. For variable resistors, noise may also be caused by jumping of the moving contact over turns and by an imperfect electrical path between the contact and resistance element.

Maximum Working Voltage $(V = \sqrt{PR})$. This quantity represents the maximum voltage stress (dc or rms) that can be applied to the resistor. Its value is a function of the materials used, the physical dimensions, and the quality of performance desired.

High-Frequency Effects. For most resistors, the lower the resistance value, the less total impedance the resistor exhibits at high frequency. Resistors are not generally tested for total impedance at frequencies above 120 Hz; therefore this characteristic is not controlled. For the best high-frequency performance, the ratio of resistor length to the cross-section area should be a maximum. Dielectric losses are kept low by proper choice of the resistor base material. When dielectric binders are used, their total mass is kept to a minimum.

Film-type resistors have the best high-frequency performance (see Fig. 7-1). The effective dc resistance for most resistance values remains fairly constant up to 100 MHz and decreases at higher frequencies. In general, the higher the resistance value, the greater the effect of frequency.

2. Fixed Resistors. Typical characteristics of fixed resistors are listed in Table 7-1.

Precision Resistors. The precision resistor, available in metal-film or wire constructions, is designed for use in circuits requiring close tolerance, long-term resistance stability, low noise, and low temperature coefficient. The wire-wound variety is comparatively large and

Table 7-1.   Characteristics of Typical Fixed Resistors

Resistor types	Resistance range	Watt range (full rating at indicated temp.)*	Operating-temp. range, °C	Resistance-temp. characteristics, ppm/°C	Terminations
Precision					
Wire-wound	0.1 Ω–1.2 MΩ	⅛–¾ at 125°C	−55–+145	±10	Axial leads, printed circuit
Metal film ..	10 Ω–5 MΩ	1/20–½ at 125°C	−55–+125	±25	Axial leads
Semiprecision					
Metal oxide .	10 Ω–1.5 MΩ	¼–2 at 70°C	−55–+150	±200	Axial leads
Cermet	10 Ω–1.5 MΩ	1/20–½ at 125°C	−55–+175	±200	Axial leads
Deposited carbon	10 Ω–5 MΩ	⅛–1 at 70°C	−55–+165	+200, −500	Axial leads
General-purpose					
Composition	2.7 Ω–100 MΩ	⅛–2 at 70°C	−55–+130	±1500	Axial leads
Power					
Wire-wound:					
Tubular ..	0.1–180 kΩ	1–210 at 25°C	−55–+275	±260	Radial tab, axial leads
Chassis mount	1.0–38 kΩ	5–30 at 25°C	−55–+275	±50	Axial leads, radial tab
Precision .	0.1–40 kΩ	1–10 at 25°C	−55–+275	±20	Axial leads
Film	20 Ω–2 MΩ	7–1000 at 25°C	−55–+225	±500	Radial tab, ferrule
Chip resistor .	1–22 MΩ	...	−55–+125	±25 to ±200	Beam lead, tab, solder

* When resistors are operated above the temperature at which full wattage is listed, the wattage must be derated in accordance with applicable specifications.

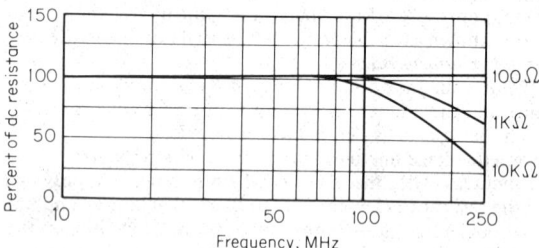

Fig. 7-1.   Resistance vs. frequency, film resistors, ½ W.

available in a limited resistance range only; however, it is the most stable of all resistors. The inductive L and capacitive C effects of the wire-wound units make them unsuitable for use above 50 kHz even when specially wound to reduce the associated inductance and capacitance. Wire-wound resistors usually exhibit an increase in resistance with higher frequencies because of skin effects. The units are usually low-power devices. The metal-film resistor element is normally nichrome, tin oxide, or tantalum nitride, and is not so stable as the wire-wound unit but is less inductive. Metal-film resistors are available either in hermetically sealed or molded-phenolic cases.

Semiprecision Resistors. The semiprecision resistor is designed for circuits requiring long-term temperature stability. This resistor is normally smaller than the precision unit and is less expensive. The units are used primarily for current-limiting or voltage-dropping functions in circuit applications.

General-Purpose Resistors. These types are small, of inexpensive composition (carbon with binder), most frequently used in electronic circuit applications (see Fig. 7-2). They are for use in circuits that are not critical of initial tolerances (e.g., 5% or more) or of long-term stability that may approach 20% under full rated power. These resistors should not be used where a low-temperature coefficient of resistance and low noise levels are desired. Composition resistors exhibit little change in effective dc resistance up to frequencies of about 100 kHz. Resistance values above 0.3 MΩ begin to show decreasing resistance at approximately 100 kHz; above frequencies of 1 MHz, all resistance values exhibit decreased resistance.

Power Resistors.   These resistors are available in both wire-wound and film constructions (see Fig. 7-3). The wire-wound units are designed for precision and general-purpose applications. The use of tapped resistors should generally be avoided because the insertion of taps weakens the resistors mechanically and lowers their effective power ratings. The power resistors are normally rated at a 25°C ambient. When necessary to operate these resistors above this temperature, a power-rating correction factor with reference to 25°C must

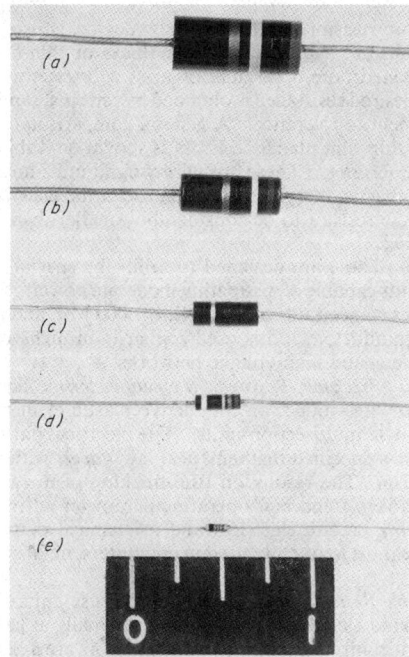

Fig. 7-2. General-purpose resistors (composition).   (a) 2 W; (b) 1 W; (c) ½ W; (d) ¼ W; (e) ¹/₁₀ W.

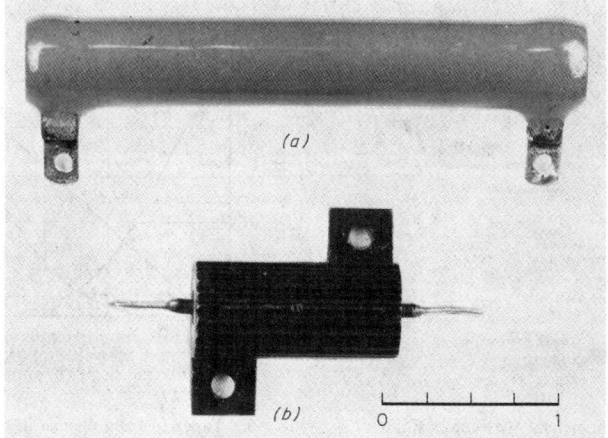

Fig. 7-3. Power resistors.   (a) 18-W wire-wound; (b) 25-W maximum, wire-wound, chassis mount.

be applied, as specified in applicable specification. The film types have the advantage of good performance at high frequencies, especially as dummy loads, and have higher resistance values than wire-wound types for a given size. The power resistors are used in power supplies, control circuits, and voltage dividers where an operational stability of 5% is satisfactory.

3. Special Resistors. *High-Megohm Resistors.* The high-megohm resistors (10^8 to 10^{15} Ω) are used in test instrumentation, photocell circuits, and other special applications. The resistance element is hermetically sealed within a glass or glazed-ceramic case. Extreme care must be taken in handling and mounting the resistor to prevent creation of leakage paths on the resistor surface.

Chip Resistor. The chip resistor, a relatively new device, is used in hybrid microelectronic circuits. These units are available in either thick- or thin-film construction. The thick-film element is screened onto a ceramic or glass substrate; the thin-film element is vacuum-deposited. The desired resistance is obtained by scribing, sandblasting, or otherwise adjusting the resistor element to tolerance. A general comparison of the characteristics of most common thick- and thin-film planar resistors is shown on Table 7-2.

Chassis-mounted Power Resistor. These are wire-wound units housed in metal cases for mounting directly on the chassis, thereby permitting use of the chassis as a heat sink. This usage permits the size of the resistor to be considerably smaller than conventional resistors of equivalent wattage rating.

High-Voltage Resistors. These are designed to fulfill the special requirement for high-voltage, high-resistance units capable of dissipating moderate power. Such resistors are rated from 5 to 20 kV, have a resistance range of 2,000 Ω to 1,000 MΩ, and are rated from 5 up to 20 W. The resistor is noninductive and is used primarily in high-voltage bleeder circuits, high-voltage voltage dividers, and high-voltage networks.

4. Variable Resistors. *Precision Resistors (Potentiometers).* The variable resistor is designed with either film, wire-wound, or conductive plastic elements and is available in single or multiturn and single multisection units. The electrical output in terms of applied voltage is linear or follows a specified mathematical law *(taper)* with respect to the angular position of the contact arm. The resolution (the tracking of the actual resistance to the theoretical resistance) is a very important operating parameter when the resistor is used in servo applications requiring precise electrical and mechanical output and quality performance. Such resistors are most frequently used in computers, flight control instrumentation, and radar-system circuitry.

General-Purpose Variable Resistors (Potentiometers). These units (Fig. 7-4) have composition, cermet, or wire-wound elements and are used principally as gain or volume controls, voltage dividers, or current controls in circuits. The resistors are available with linear taper *A,* clockwise taper *B,* or counterclockwise taper *C* (see Fig. 7-5). The resistors can be ganged, and a rear-mounted switch can be attached to the end resistor.

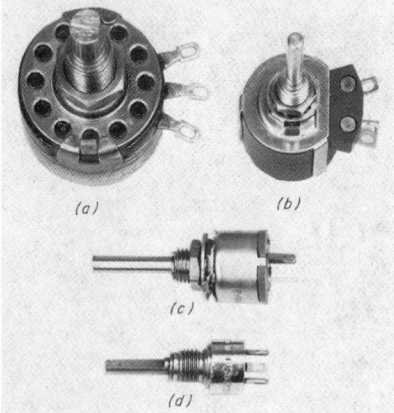

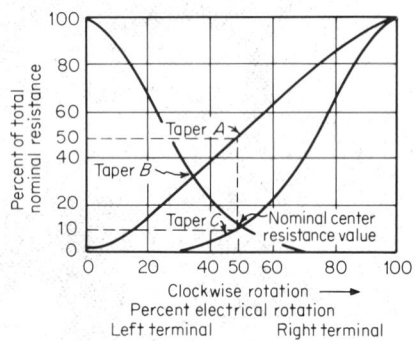

Fig. 7-4. Potentiometers. (*a*) composition ½ W; (*b*) wire-wound 1½ W; (*c*) composition ½ W; (*d*) cermet ¼ W.

Fig. 7-5. Taper (percent change in resistance vs. rotation).

Rheostat (Power). These are variable wire-wound resistors used as speed controls for motors, ovens, and heater controls and in applications where adjustments in voltage or current levels are required, such as voltage dividers and bleeder circuits.

Trimmer Resistors (Trim Pots). Screw-lead actuated trim pots (Fig. 7-6), in single or multiturn design, are available in composition, cermet, or wire-wound construction. These are principally used to control the low-current or bias voltages in transistor circuits and as matching, balancing, and adjusting circuit variables in critical circuit applications.

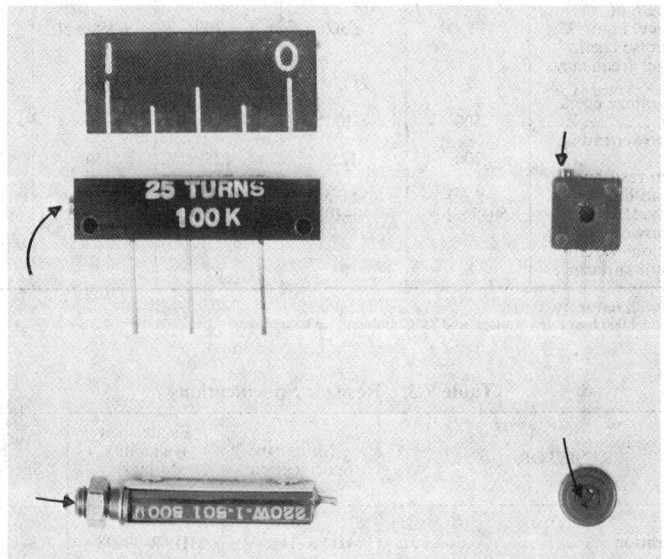

Fig. 7-6. Trimmer resistors (trim pots). Arrows point to adjustment screws.

5. Reference Specifications. Established Reliability (ER) Specifications are a new series of Military (MIL) Specifications that define the failure-rate levels for selected styles of resistors. Life failure-rate levels in percent per 1,000 h, with a confidence level of 60%, range from 1% (level M) to 0.001% (level S), M representing the level associated with the conventional MIL specification. The predicted failure rate is based on qualification-approval life testing and continuous-production life testing by the resistor manufacturer. Table 7-2 lists characteristics of film-type planar resistors. A list of the MIL and Electronic Industries Association resistor specifications is shown in Table 7-3.

CAPACITORS

By Thomas S. Gore, Jr.

6. Fundamentals. A capacitor consists basically of two conductors separated by a dielectric or vacuum so as to store a large electrical charge in a small volume. The capacitance is expressed as a ratio of electrical charge to the voltage applied $C = Q/V$, where Q = charge in coulombs, V = volts, C = capacitance in farads. A capacitor has a capacitance of one farad (F) when it receives a charge of one coulomb (C) at a potential of one volt (V). The electrostatic energy in watt-seconds (Ws) or joules (J) stored in the capacitor is given by

$$J = \frac{1}{2}CV^2$$

Dielectric. Depending on the application, the capacitor dielectric may be either air, gas, paper (impregnated), organic film, mica, glass, or ceramic, each having a different dielectric constant, temperature range, and thickness.

Equivalent Circuit. A practical capacitor possesses, in addition to capacitance, an

Table 7-2. General Comparison of Typical Thick- and Thin-Film Planar Resistor Types

Characteristic	Printed carbon	Cermet	Tin oxide	Nickel chromium	Tantalum nitride
Resistance range	4 Ω-20 MΩ	10 Ω-5 MΩ	10 Ω-100 kΩ	15 Ω-150 kΩ	15 Ω-150 kΩ
Resistance tolerance, %	±5, ±10	±1, ±5	±1, ±5	±2, ±5	±0.5, ±2
Temp. coeff. of resistance, ppm/°C.	−1500	±300	±100	±25, ±50	±25
Typical noise coeff., μV/V per frequency decade	2	<1.0*	<0.1	<0.1	<0.1
Typical voltage coeff., ppm/V	500	<100*	<25	5	5
Typical power rating, mW	200	125	125	50	50
Change in resistance after load life, %† ...	<−7	<0.5	<0.5	<0.25	<0.25
Resistivity, Ω/square .	50-5×10⁶	10-10⁵	20-400	25-300	25-300
Typical power dissipation, W/cm² of resistor surface ...	8	8	8	8	2

* Varies with resistivity.
† Based on 1,000 h at rated wattage and 85°C ambient, no heat sink.

Table 7-3. Resistor Specifications

RESISTOR	Military Specification	Established Reliability Military Specification	EIA Specification
Fixed:			
Composition	MIL-R-11	MIL-R-39008	RS-172A
Film:			
High stability	MIL-R-10509	MIL-R-55182	RS-196A
Power........................	MIL-R-11804		
Wire-wound	MIL-R-26	MIL-R-39007	RS-155B
			RS-344
Accurate	MIL-R-93	MIL-R-39005	RS-229A
Power chassis mounting	MIL-R-18546	MIL-R-39009	
Film, metal oxide	MIL-R-22684	MIL-R-39017	
Variable:			
Composition	MIL-R-94	...	RS-303
Wire-wound:			
Low operating temp.	MIL-R-19	...	RS-333
Power........................	MIL-R-22	...	RS-322
Precision	MIL-R-12934	...	
Lead-screw-actuated	MIL-R-27208	MIL-R-39015	RS-345
Non-wire-wound, lead-screw-actuated ...	MIL-R-22097	MIL-R-39035	RS-360

inductance and resistance. An equivalent circuit useful in determining the performance characteristics of the capacitor is shown in Fig. 7-7, where R_s = series resistance due to wire leads, contact terminations, and electrodes in ohms; R_p = shunt resistance due to resistivity of dielectric and case material in ohms, and to dielectric losses; and L = stray inductance due to leads and electrodes in henrys.

Equivalent Series Resistance (ESR).   The ESR is the ac resistance of a capacitor reflecting both the series resistance R_s and the parallel resistance R_p at a given frequency so that the loss of these elements can be expressed as a loss in a single resistor R in the equivalent circuit.

Capacitive Reactance.   The reactance of a capacitor is given by

$$X_c = \frac{1}{2\pi f C} = \frac{1}{\omega C} \quad \text{(ohms)}$$

where $f = \omega/2\pi$ is the frequency in hertz.

Impedance (Z). In practical capacitors operating at high frequency, the inductance of the leads must be considered in calculating the impedance. Specifically,

$$Z = \sqrt{R^2 + (X_L - X_C)^2}$$

where R is the ESR, and X_L reflects the inductive reactance.

The effects illustrated in Fig. 7-7 are particularly important at radio frequencies where a capacitor may exhibit spurious behavior due to these equivalent elements. For example, in many high-frequency tuned circuits the inductance of the leads may be sufficient to detune the circuit.

Power Factor (PF). The term PF defines the electrical losses in a capacitor operating under an ac voltage. In an ideal device the current will lead the applied voltage by 90°. A practical capacitor, due to its dielectric, electrode, and contact termination losses exhibits a phase angle of less than 90°. The PF is defined as the ratio of the effective series resistance R to the impedance Z of the capacitor and is usually expressed in percent.

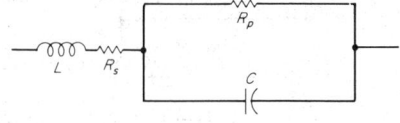

Fig. 7-7. Equivalent circuit of a capacitor.

Dissipation Factor (DF). The DF is the ratio of effective series resistance R to capacitive reactance X_c and is normally expressed in percent. The DF and PF are essentially equal when the PF is 10% or less.

Quality Factor Q. The Q is a figure of merit and is the reciprocal of the dissipation factor. It usually applies to capacitors used in tuned circuits.

Leakage Current, dc. Leakage current is the current flowing through the capacitor when a dc voltage is applied.

Insulation Resistance. The insulation resistance is the ratio of the applied voltage to the leakage current and is normally expressed in megohms. For electrolytic capacitors, the maximum leakage current is normally specified.

Ripple Current/Voltage. The ripple current/voltage is the rms value of the maximum allowable alternating current/voltage (superimposed on any dc level) at a specified frequency at which the capacitor may be operated continuously at a specified temperature.

Surge Voltage. The surge voltage applicable to electrolytic capacitors is a voltage in excess of the rated voltage which the capacitor will withstand for a specified limited period at any temperature.

7. Fixed Capacitors. Table 7-4 shows the types of capacitors which would be generally used over the frequency range from direct current to 10 GHz.

Precision Capacitors. Capacitors falling into the precision category (see Table 7-5) are generally those having exceptional capacitance stability with respect to temperature, voltage, frequency, and life. They are available in close capacitance tolerances and have low-loss (high-Q) dielectric properties. These capacitors are generally used at radio frequencies in tuner, rf filter, coupling, bypass, and temperature-compensation applications. Typical capacitor types in this category are mica, ceramic, glass, and polystyrene. The polystyrene capacitor has exceptionally high insulation resistance, low losses, low dielectric absorption, and a controlled temperature coefficient for film capacitors.

Semiprecision Units. Paper- and plastic-film capacitors listed in Table 7-5 with foil or metallized dielectric constitute a large portion of the present applications. These capacitors are nonpolar and generally fall between the low-capacitance precision types, such as mica and ceramic, and the high-capacitance electrolytics.

General-Purpose. Electrolytic aluminum and tantalum capacitors and the large-usage general-purpose (hi-K) ceramic capacitors are so grouped in Table 7-5 because both have broad capacitance tolerances, are temperature-sensitive, and have high volumetric efficiencies (capacitance-volume ratio). They are primarily used as bypass and filter capacitors where high capacitance is needed in small volumes, and with guaranteed minimum values. These applications do not require low dissipation factors, stability, or high insulation resistance found in precision and semiprecision capacitors. On a performance-vs.-cost basis, the general-purpose capacitors are the least expensive of the groups. High-capacitance aluminum electrolytic capacitors have been designed for computer applications featuring low equivalent series resistance and long life.

Table 7-4.   Useful Frequency Range of Capacitors
(Dashed Lines Indicate Fringe Areas of Application)

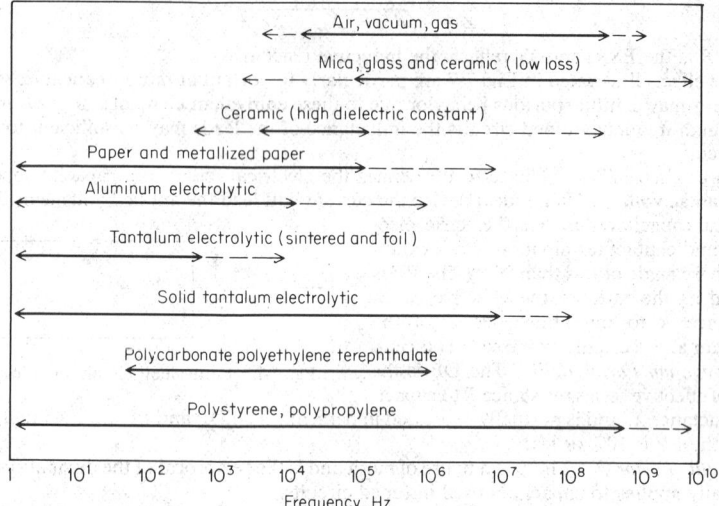

SOURCE: Adapted from G. W. Dummer and Harold M. Nordenberg, "Fixed and Variable Capacitors," McGraw-Hill Book Company, New York, 1960.

Suppression Capacitors. The feed-through capacitors listed in Table 7-5 are three-terminal devices designed to minimize effective inductance and to suppress rf interference over a wide frequency range. For heavy feed-through currents, applications in 60- and 400-Hz power supplies, paper or film dielectrics are normally used. For small low-capacitance, low-current units, the ceramic and button-mica feed-through high-frequency styles are used.

Capacitors for Microelectronic Circuits. Table 7-6 lists representative styles of discrete miniature capacitors electrically and physically suitable for microelectronic circuit usage (filtering, coupling, tuning, bypass, etc.). The chip capacitor, widely used in hybrid circuits, is available in single-wafer or multilayer (monolithic) ceramic, or in tantalum constructions, both offering reliable performance, very high volumetric efficiency, and a wide variety of capacitance ranges at moderate cost. Temperature-compensating ceramic chips are used where maximum capacitance stability or predictable changes in capacitance with temperature are required.

General-purpose ceramic and solid electrolytic tantalum chips are used for coupling and bypass applications where very high capacitance and small size are necessary. The ceramic chips are unencapsulated and leadless, with suitable electrodes for soldering in microcircuits. The beam-leaded tantalum chips are attached by pressure bonding. The tantalum chip is also available in a multiple-unit assembly in the dual-in-line package for use on printed-circuit boards.

Transmitter Capacitors. The principal requirements for transmitter capacitors are high rf power-handling capability, high rf current and voltage rating, high Q, low internal inductance, and very low effective series resistance. Mica, glass, ceramic, and vacuum- or gas-filled capacitors are used primarily for transmitter applications. Glass dielectric transmitter capacitors have a higher self-resonating frequency and current rating than comparable mica styles. The ceramic capacitors offer moderately high rf current ratings, operating temperatures to 105°C, and high self-resonant frequencies. The gas or vacuum capacitor is available in a wide range of capacitance and power ratings and is used where very-high-power, high-voltage, and high-frequency circuit conditions exist. These units are also smaller than other transmitting types for comparable ratings. The circuit performance of the transmitter capacitor is highly dependent on the method of mounting, lead connections, operating temperatures, air circulation, and cooling, which must be considered for specific applications. Typical ratings are listed in Table 7-7, and typical fixed capacitors shown in Fig. 7-8.

Table 7-5. Characteristics of Typical Fixed Capacitors

Capacitor type	Typical capacitance range	Typical voltage-rating range, V, dc	Operating temp. range, °C	Q, 1 MHz, min.	% dissipation factor, 1 kHz, max.	Min. insulation resistance, MΩ at 25°C	Typical nominal temp. coeff., ppm/°C	Max. capacitance change over temp. range, %	Terminations
Precision									
Mica	1–91,000 pF	100–2,500	−55–+125	1,200	...	7,500	−20, +100, ±200	...	Axial leads, Radial leads, tab
Glass	1–10,000 pF	300–500	−55–+125	1,500	...	10,000	+140	...	Axial leads
Ceramic	1–1,100 pF	150–500	−55–+85	1,000	...	10,000	+100 to −750*	...	Axial, radial leads
Semiprecision									
Film polystyrene	1,000–222,000 pF	200–600	−55–+85	...	0.1	100,000	...	±3	Axial leads
Plastic Film	1,000 pF–10 µF	30–1,000	−55–+125	...	0.5	10,000	...	+15 −10	Axial leads
Paper-plastic film (metallized)	4,700 pF–10 µF	50–400	−55–+125	...	1	10,000	...	+20 −10	Axial leads
General-purpose									
Ceramic (hi-K)	10–100,000 pF	50–200	−55–+125	...	2	10,000	...	±15	Axial, radial leads
Tantalum oxide, sintered, solid electrolyte, polar	4,700 pF–330 µF	6–100	−55–+125	...	6†	...	...	+8 −10	Axial leads
Foil polar	1–580 µF	10–300	−55–+125	...	...	...	...	+50 −40	Axial leads
Sintered (wet), polar	3.6–560 µF	4.0–85	−55–+125	...	...	...	...	+20 −64	Axial leads
Aluminum oxide, dry, polar	3.3–1,000 µF	7–250	−55–+125	...	20†	...	...	+25 −30	Axial
Computer-grade polar	150–120,000 µF	5–450	−40–+85	...	...	...	...	...	Solder lugs
Suppression									
Feed-through									
Paper	10,000 pF–3.0 µF	100–600	−55–+125	...	...	...	...	±15	Axial
Mica, button	5–2,400 pF	500	−55–+125	1,000	...	100,000	...	±100	Special
Ceramic	100–1,500 pF	500–1,500	−55–+125	...	2	10,000	...	±15	Axial leads, solder lugs screw type
Bypass paper	10,000–25,000 pF	100–500‡	−55–+85	...	...	3,000	...	±25	Solder lugs

* +100, 0, −330, −750, preferred values.
† pF at 120 Hz.
‡ ac/dc.

7-11

Table 7-6. Characteristics of Typical Capacitors for Microelectronic Circuits

Type of capacitor	Typical capacitance range	Typical working voltage range, V, dc	Temperature range, °C	Dissipation factor, %	Minimum insulation resistance, MΩ at 25°C	Temp. coeff., ppm/°C	Max.-capacitance change over temp. range, %	Termination
Chip:								
Ceramic, temperature-compensating	1 pF–0.027 µF	50–200	−55–+125	0.1 at 1 MHz	50,000	0 to −750	...	Metallized
Ceramic, general-purpose	390 pF–0.47 µF	25–200	−55–+125	3.0 at 1 kHz	10,000	...	±15	Metallized
Tantalum oxide (beam-lead), dry	100–3,000 pF	12–50	−55–+85	0.6 at 1 kHz	10,000	+200	...	Beam lead
Tantalum oxide Polar, solid electrolyte	0.1–47 µF	3–35	−55–+85	4.0 at 120 Hz	...	...	±10	Metallized
Metallized, film (metal case)	0.1–10 µF	50–100	−55–+85	2.5 at 1 kHz	5,000	...	±10	Axial
Variable ceramic trimmer	Min. 1–3 pF, max. 5–25 pF	25	−55–+125	0.2 at 1 MHz	10,000	...	±5–±15	Printed circuit

Table 7-7.   Transmitter Capacitors (Fixed)

Type of capacitor	Typical capacitance range, pF	Peak voltage range, kV, ac	Current rating, A
Mica.............	47–100,000	0.25–35	0.51–75 at 1 MHz
Glass	22–150,000	0.5–6	0.33–25 at 1 MHz
Gas-filled..........	500–2,000	5–14	120–150 at 16 MHz
Vacuum-filled	500–1,000	2–35	30–200 at 16 MHz

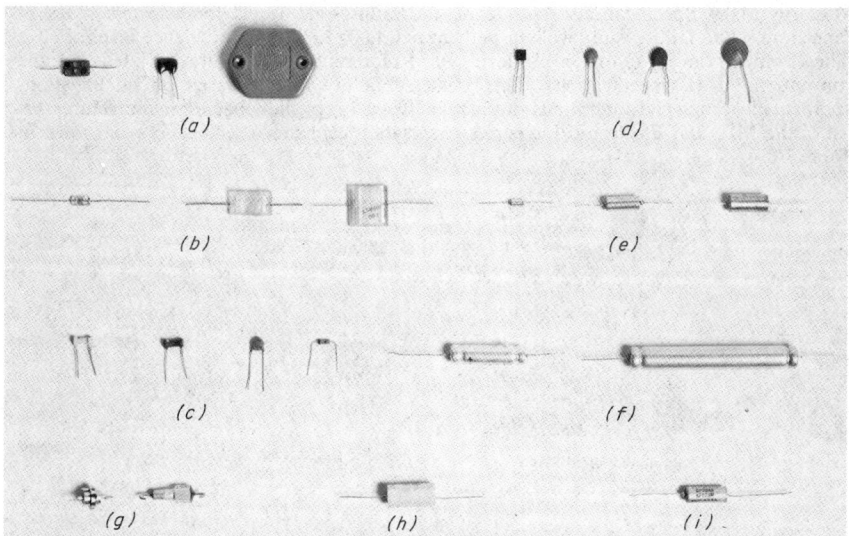

Fig. 7-8. Typical fixed capacitors: (a) Mica; (b) glass; (c) temperature-compensation ceramic; (d) general-purpose ceramic; (e) solid electrolyte tantalum; (f) foil tantalum; (g) feed-through button mica and ceramic; (h) general-purpose plastic film; (i) general-purpose paper.

8. Variable Capacitors.   Variable capacitors are used for tuning and for trimming or adjusting circuit capacitance to a desired value.

Tuning Capacitor (Air-, Vacuum-, or Gas-filled).   The parallel plate (single or multisection style) is used for tuning receivers and transmitters.   In receiver applications, one section of a multisection capacitor is normally used as the oscillator section, which must be temperature-stable and follow prescribed capacitance-rotation characteristics.   The remaining capacitor sections are mechanically coupled and must track to the oscillator section.   The three most commonly used plate shapes are semicircular (straight-line capacitance with rotation), midline (between straight-line capacity and straight-line frequency), and straight-line frequency (which are logarithmically shaped).   For transmitter applications, variable vacuum- or gas-filled capacitors of cylindrical construction are used.   These capacitors are available in assorted sizes and mounting methods.   Typical ratings are shown below in the table.

	Capacitance, pF	Peak rf voltage, kV	Rf current, A
Gas-filled:			
Min.............	7–50	3.5	35 at 50 MHz
Max.............	500–3,000	14.0	150 at 16 MHz
Vacuum:			
Min.	3–30	7.5	36 at 16 MHz
Max.	100–5,000	15	125 at 16 MHz

Typical variable capacitors are shown in Figs. 7-9 and 7-10.

9. Special Capacitors. *High-Energy Storage Capacitors.* Oil-impregnated paper and/or film dielectric capacitors have been designed for voltages of 1,000 V or higher for pulse-forming networks. For lower voltages, special electrolytic capacitors can be used which have a low inductance and equivalent-series resistance.

Commutation Capacitors. The widespread use of SCR devices (see Par. 7-51) has led to the development of a family of oil-impregnated paper and film dielectric capacitors for use in triggering circuits. The capacitor is subjected to a very fast rise time (0.1 ms) and high current transients and peak voltages associated with the switching.

Reference Specifications. Established Reliability (ER) Specifications are a new series of Military (MIL) Specifications that define the failure-rate levels for selected styles of capacitors. Life failure-rate levels in percent per 1,000 h, with a confidence level of 60%, range from 1% (level *M*) to 0.001% (level *S*), *M* representing the level associated with the conventional MIL specification. The predicted failure rate is based on qualification-approval life testing and continuous-production life testing by the capacitor manufacturer. A list of the MIL and Electronic Industries Association capacitor specifications is shown in Table 7-8.

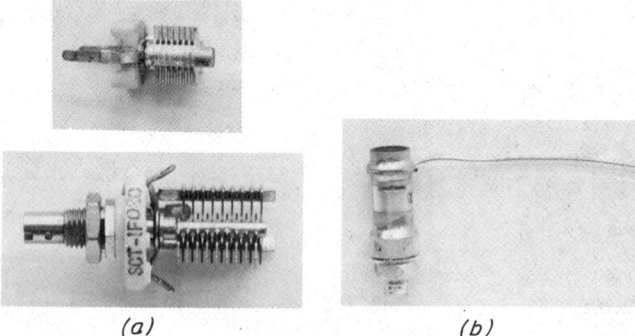

(a)   *(b)*

Fig. 7-9. Trimmer capacitors. (*a*) air dielectric; (*b*) glass or quartz dielectric.

Fig. 7-10. Three-gang air tuning capacitor.

Table 7-8.   Capacitor Specifications

Capacitor	Military Specification	Established-Reliability Military Specification	EIA Specification
Fixed, mica dielectric	MIL-C-5	MIL-C-39001	RS-153A TR-109
Fixed, ceramic, temperature-compensating	MIL-C-20	...	RS-198A
Fixed, glass dielectric	...	MIL-C-23269	
Fixed, metallized paper or plastic film, hermetically sealed	...	MIL-C-39032	RS-377
Fixed, plastic or metallized plastic dielectric, dc in nonmetallic cases	...	MIL-C-55514	RS-377 RS-164A
Fixed, paper or plastic dielectric (hermetically sealed)...	...	MIL-C-14157	RS-376
Fixed, paper dielectric in metal cases for dc applications	...	...	RS-218A
Fixed, ceramic dielectric (general-purpose)	...	MIL-C-39014	RS-198A
Fixed, electrolytic, tantalum, solid electrolyte	...	MIL-C-39003	RS-228A
Fixed, ceramic dielectric (1,000-7,500 V)	...	...	RS-165A
Fixed, electrolytic (nonsolid electrolyte)	...	MIL-C-39006	RS-228A
Fixed, electrolytic (aluminum oxide)	MIL-C-39018		
Fixed, aluminum dc dry electrolyte	MIL-C-62	...	RS-154B
Feed-through, radio interference reduction (ac and dc)	MIL-C-11693	...	RS-361
Fixed, mica dielectric, button styles	MIL-C-10950		
Fixed, Bypass, radio interference reduction, paper Dielectric, ac and dc	MIL-C-12889		
Variable, ceramic dielectric	MIL-C-81		
Variable, air dielectric (trimmer)	MIL-C-92		
Variable (piston type, tubular trimmer)	MIL-C-14409		
Fixed or Variable, vacuum dielectric	MIL-C-23183		

INDUCTORS AND TRANSFORMERS

By Emanuel Gikow

10. Introduction. Among the many families of electronic parts, inductive components are generally unique in that they must be designed for a specific application. Whereas resistors, capacitors, meters, switches, etc., are available as standard or stock items, inductors and transformers have not usually been available as off-the-shelf items with established characteristics. With recent emphasis on miniaturization, however, wide varieties of chokes have become available as stock items. Low inductance values are wound on a nonmagnetic form, powdered iron cores are used for the intermediate values of inductance, and ferrites are used for the higher-inductance chokes. High-value inductors use both a ferrite core and sleeve of the same magnetic material. Wide varieties of inductors are made in tubular, disk, and rectangular shapes. Direct-current ratings are limited by the wire size or magnetic saturation of the core material.

11. Solenoids. Inductance can be calculated readily from the geometry of the inductors. The nomograph shown in Fig. 7-11 can be used as a design guide for calculating the inductance of single-layer air-core solenoids. As shown by the "Key," if three known values are connected, the fourth unknown can be read off at the intersection. For example: If $L = 100 \mu H$, $d = 2$ cm, shape factor $= 0.5$, turns/cm $= 28$, total turns $= 112$.

12. Distributed Capacitance. The distributed capacitance between turns and windings has an important effect upon the characteristics of the coil at high frequencies. At a frequency determined by the inductance and distributed capacitance, the coil becomes a parallel resonant circuit. At frequencies above self-resonance, the coil is predominantly capacitive. For large-valued inductors the distributed capacitance may be reduced by winding the coil in several sections so that the distributed capacitances of the several sections are in series.

13. Toroids. The toroidal inductor using magnetic-core material has a number of advantages. Using ungapped core material, the maximum inductance per unit volume can be obtained with the toroid. It has the additional advantages that leakage flux tends to be minimized and shielding requirements are reduced. The inductance of the toroid and its temperature stability are directly related to the average incremental permeability of the core material. For a single-layer toroid of rectangular cross section without air gap,

$$L = 0.0020\ N^2 b / \mu_d \ln r_2 / r_1$$

where L = inductance in microhenrys, b = core width in centimeters, μ_d = average incremental permeability, r_1 = inside radius in centimeters, r_2 = outside radius in centimeters, and N = total number of turns.

Incremental permeability is the apparent ac permeability of a magnetic core in the presence of a dc magnetizing field. As the dc magnetization is increased to the point of saturating the magnetic core, the effective permeability decreases. This effect is commonly used to control inductance in electronically variable inductors.

Inductors with magnetic cores are often gapped to control certain characteristics. In magnetic-cored coils the gap can reduce nonlinearity, prevent saturation, lower the temperature coefficient, and increase the Q.

14. Adjustable Inductors. The variable inductor is used in a number of circuits: timing, tuning, calibration, etc. The most common form of adjustment involves the variation of the effective permeability of the magnetic path. In a circuit comprised of an inductor and capacitor, the slug-tuned coil may be found to be preferred over capacitive tuning in an environment with high humidity. Whereas the conventional variable capacitor is difficult and expensive to seal against the entry of humidity, the inductor lends itself to sealing without seriously inhibiting its adjustability. For example, a slug-tuned solenoid can be encapsulated, with a cavity left in the center to permit adjustment of the inductance with a magnetic core. Moisture will have little or no effect on the electrical performance of the core. The

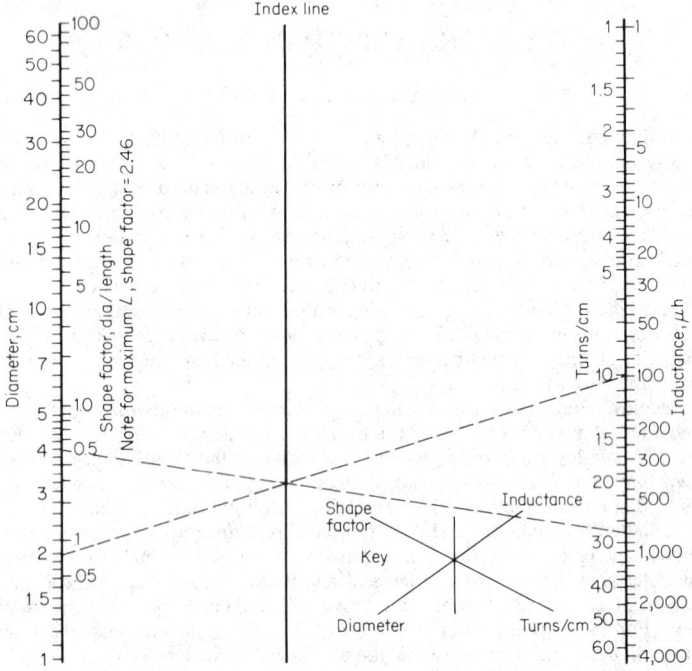

Fig. 7-11. Single-layer solenoid design chart. *(Adapted from K. Henney, "Radio Engineering Handbook," 5th ed., McGraw-Hill Book Company, New York, 1959.)*

solenoid inductor using a variable magnetic core is the most common form of adjustable inductor. The closer the magnetic material is to the coil (the thinner the coil form), the greater will be the adjustment range. Simultaneous adjustment of both a magnetic core and magnetic sleeve will increase the tuning range and provide magnetic shielding. In another form, the air gap at the center post of a cup core can be varied to control inductance. Not only is the result a variable inductor, but also the introduction of the air gap provides a means for improving the temperature coefficient of the magnetic core by reducing its effective permeability.

15. Power Transformers. Electronic power transformers normally operate at a fixed frequency. The most popular frequencies are 50, 60, and 400 Hz. Characteristics and designs are usually determined by the voltampere (VA) rating and the load. For example, when used with a rectifier, the peak inverse voltage across the rectifier will dictate the insulation requirements. With capacitor input filters, the secondary must be capable of carrying higher currents than with choke input filters. With the increased use of semiconductors, and their small size, there is considerable interest in a concomitant reduction in the size and weight of all other component parts. There are a number of ways by which the size and weight of power transformers can be reduced, as follows:

Operating frequency. For a given volt-ampere rating, size and weight can be reduced as some inverse function of the operating frequency.

Maximum operating temperature. By the use of high-temperature materials and at a given VA rating and ambient temperature, considerable size reduction can be realized by designing for an increased temperature rise.

Ambient temperature. With a given temperature rise and for a fixed VA rating, transformer size can be reduced if the ambient temperature is lowered.

Regulation. If the regulation requirements are made less stringent, the wire size of the windings can be reduced, with a consequent reduction in the size of the transformer.

16. Audio Transformers. These transformers are used for voltage, current, and impedance transformation over a nominal frequency range of 20 to 20,000 Hz. The equivalent circuit is the starting point for the basic analysis of the transformer frequency response.

Figure 7-12 is the equivalent circuit of a transformer where E_g = generator voltage, R_g = generator impedance, C_p = primary shunt and distributed capacitance, L_p = primary leakage inductance, L_s = secondary leakage inductance, C_s = secondary shunt and distributed capacitance, R_p = primary winding resistance, R_e = equivalent resistance corresponding to core losses, L_e = equivalent magnetizing (open-circuit) inductance of primary, n = ideal transformer primary-to-secondary turns ratio, C_{ps} = primary-to-secondary capacitance (interwinding capacitance), R_s = secondary winding resistance, R_L = load impedance.

A prime consideration in the design, size, and cost of audio transformers is the span of the frequency response. In wideband audio transformers the frequency coverage can be separated into three nearly independent ranges for the purpose of analysis. Thus, in the high-frequency range (Fig. 7-13) leakage inductance and distributed capacitance are most significant. In the low-frequency region (Fig. 7-14) the open-circuit inductance is important. At the medium-frequency range, approximately 1,000 Hz, the effect of the transformer on the

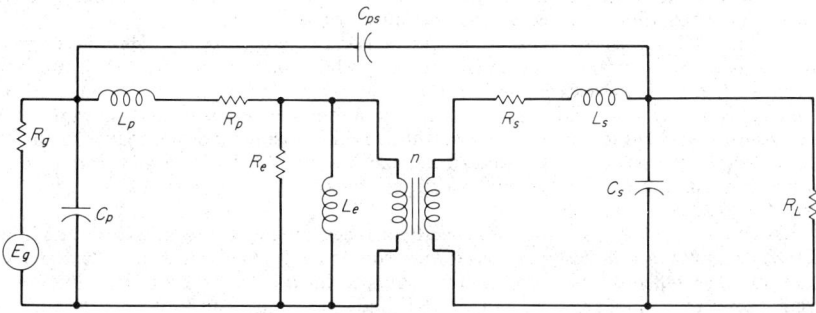

Fig. 7-12. Equivalent circuit of a broadband transformer.

frequency response can be neglected. In the above discussion, the transformation is assumed to be that of an ideal transformer.

17. Miniaturized Audio Transformers. *Frequency Response.* The high-frequency response (Fig. 7-13) is dependent on the magnitude of the leakage inductance and distributed capacitance of the windings. These parameters decrease as the transformer size decreases. Consequently, miniaturized audio transformers generally have an excellent high-frequency response. On the other hand, the small size of these transformers results in increased loss and in a degradation of the low-frequency response, which is dependent on the primary open-circuit inductance.

Air Gap. The primary open-circuit inductance is proportional to the product of the square of the turns and the core area, and inversely proportional to the width of the gap. As the transformer size is reduced, both the core area and number of turns must be reduced. Consequently, if the open-circuit inductance is to be maintained, the air gap must be reduced. Table 7-9 shows the reduction in open-circuit inductance (OCL) as the air gap increases. A butt joint has of necessity a finite air gap; as a result the full ungapped inductance is not realized. An interlaced structure of the core, where the butt joints for each lamination are staggered, most closely approximates an ungapped core. The substantially higher inductance, 120 H, is shown in Table 7-9. The problem with taking advantage of this effect is that, as the air gap is reduced, the allowable amount of unbalanced direct current flowing in the transformer winding must be lowered to prevent core saturation.

Table 7-9. Effect of Air Gap on Inductance

Lamination Assembly	Primary OCL, H
Interlace 3 × 3	120
Butt joint:	
No gap	32
0.001-in. gap	20
0.002-in. gap	12

Data of test transformer: lamination size and material, EE 24-25, alloy 4750; number of turns, 1,000; core area $\frac{1}{16}$ in[2].

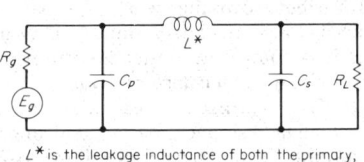

L^* is the leakage inductance of both the primary, and of the secondary referred to the primary

Fig. 7-13. Equivalent circuit of audio transformer at high frequency (approximately 15,000 Hz).

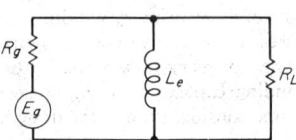

Fig. 7-14. Equivalent circuit of audio transformer at low frequency (approximately 300 Hz).

Operating Voltage Level. To avoid core saturation, which will result in distortion, the voltage level of operation must be lowered as core size is reduced. Existence of a dc magnetizing field will further reduce the operating voltage.

18. Pulse Transformers. Pulse transformers used in high voltage on high-power applications, above 300 W peak, are generally used in modulators for radar sets. Their function is to provide impedance matching between the pulse-forming network and the magnetron. Prime concern is transformation of the pulse with a minimum of distortion. Lower-power pulse transformers fall into two categories: those used for coupling or impedance matching similar to the high-power pulse transformers, and blocking oscillator transformers used in pulse-generating circuits. Pulse widths for such transformers most commonly range from about 0.1 to 20 μs.

Assuming the pulse transformer is properly matched and the source is delivering an ideal rectangular pulse, a well-designed transformer should have small values of leakage inductance and distributed capacitance. Within limits dictated by pulse decay time the open-circuit inductance should be high. Figure 7-15 shows the pulse waveform with the various types of distortions that may be introduced by the transformer.

19. Broadband RF Transformers. At the higher frequencies, transformers provide a simple, low-cost, and compact means for impedance transformation. Bifilar windings and powdered-iron or ferrite cores provide optimum coupling. The use of cores with high permeabilities at the lower frequencies reduces the number of turns and distributed capacitance. At the upper frequencies the reactance increases, even though the permeability of the core may fall off.

20. Transmission-Line Transformers. Where dc isolation is not a factor, transmission-line transformers can be made to provide polarity reversal, balanced to unbalanced, 4:1 impedance transformation. By hooking these transformers in tandem, higher impedance transformations are possible. Very broad bandwidths have been attained with a single transformer, as high as 0.1 to 1,000 MHz within ± 1 dB. Successful transformers were made using ferrite toroidal cores.

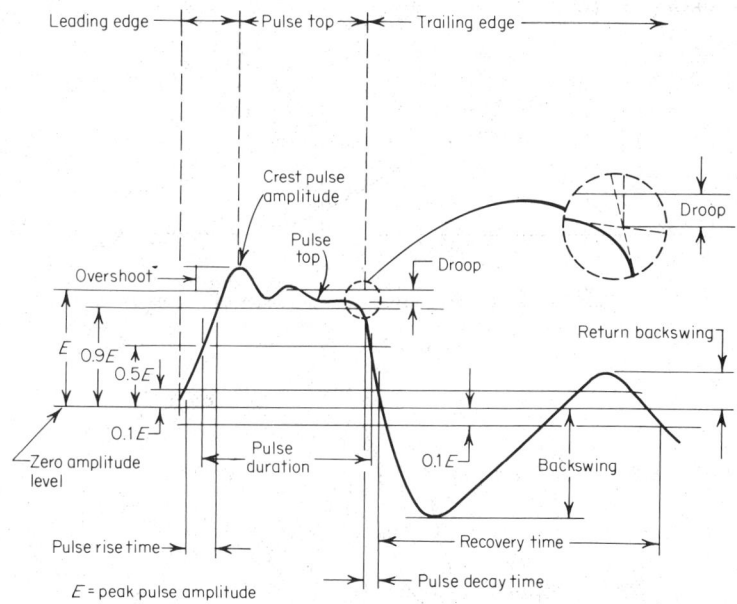

Fig. 7-15. Pulse waveform.

21. Inductive Coupling. There are a variety of ways to use inductive elements for impedance-matching or coupling one circuit to another. Autotransformers and multiwinding transformers which have no common metallic connections are a common method of inductive coupling. In a unity-coupled transformer, N_1 = number of primary turns, N_2 = number of secondary turns, k = coefficient of coupling, M = mutual inductance, n = turns ratio, L_1 = primary open-circuit inductance, L_2 = secondary open-circuit inductance, I_1 = primary current, I_2 = secondary current, E_1 = primary voltage, E_2 = secondary voltage, Z_1 = primary impedance with matched secondary, Z_2 = secondary impedance with matched primary. Transformer relationships for unity-coupled transformer, $k = 1$, assuming losses are negligible:

$$n = N_2/N_1 = E_2/E_1 = I_1/I_2 = Z_2/Z_1 \qquad M = \sqrt{L_1 L_2}$$

22. Single-tuned Circuits. Single-tuned circuits are most commonly used in both wide- and narrowband amplifiers. Multiple stages which are cascaded and tuned to the same frequency are synchronously tuned. The result is that the overall bandwidth of the cascaded amplifiers is always narrower than the single-stage bandwidth. The shrinkage of bandwidth can be avoided by stagger tuning. A stagger-tuned system is a grouping of single-tuned circuits where each circuit is tuned to a different frequency. For a flat-topped response the individual stages are geometrically balanced from the center frequency. Figure 7-16 illustrates the response of a staggered pair: f_0 = center frequency of overall response, f_1 and f_2 = resonant frequency of each stage. The frequencies are related as follows:

$$f_0/f_1 = f_2/f_0 = \alpha.$$

23. Double-tuned Transformers. One of the most widely used circuit configurations for if systems in the frequency range of 250 kHz to 50 MHz is the double-tuned transformer. It consists of a primary and secondary tuned to the same frequency and coupled inductively to a degree dependent on the desired shape of the selectivity curve. Figure 7-17 shows the variation of secondary current vs. frequency for the undercoupled (1), critically coupled (2), and overcoupled (3) cases.

24. Bandwidth. A comparison of the relative 3-dB bandwidth of multistage single- and double-tuned circuits is shown in Table 7-10. Most significant is the lower skirt ratio of the double-tuned circuit, i.e., relative value of the ratio of the bandwidth at 60 dB (BW_{60}) to the bandwidth at 6 dB (BW_6).

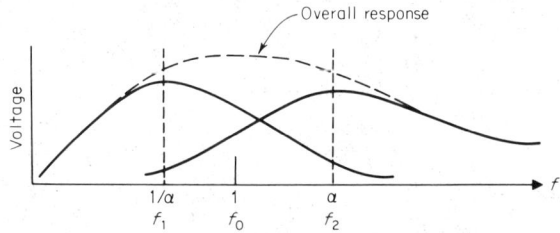

Fig. 7-16. Response of a staggered pair, geometric symmetry.

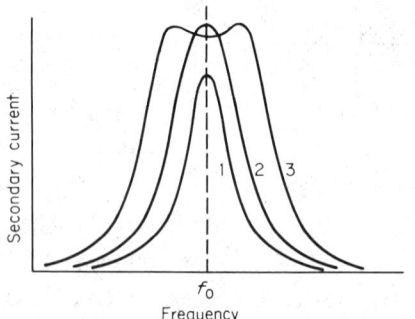

Fig. 7-17. Variation of secondary current and gain with frequency and with degree of coupling.

Table 7-10.   Relative Values of 3-dB Bandwidth for Single- and Double-tuned Circuits

No. of stages	Relative values of 3-dB bandwidth		Relative values of BW_{60}/BW_6	
	Single-tuned	Double-tuned*	Single-tuned	Double-tuned
1	1.00	1.00	577	23.9
2	0.64	0.80	33	5.65
3	0.51	0.71	13	3.59
4	0.44	0.66	8.6	2.94
6	0.35	0.59	5.9	2.43
8	0.30	0.55	5.0	
10	0.27	0.52	4.5	

* Based upon identical primary and secondary circuits critically coupled.

25. Fabrication Techniques. *Control of Inductance.* In a single-layer winding, if the end turns are spaced, a finer adjustment of inductance can be made by repositioning the final turns. In a universal winding, the slight sponginess of the winding permits a final adjustment of inductance by squeezing it parallel to the coil axis or perpendicular to the axis.

Q Adjustment. It is sometimes difficult and uneconomical to design a coil to specific *Q* values. Accordingly, if the coil is designed to a somewhat higher *Q* than is needed, a shunt resistor can be used to reduce the effective *Q* of the coil.

POWER AND RECEIVING TUBES

By G. W. Taylor and L. E. Scharmann

26. Introduction. Power and receiving tubes are active devices utilizing either the flow of free electrons in a vacuum or electrons and ions combined in a gas medium. The vast majority of uses for low-power electron tubes are found in the area of high vacuum with controlled free electrons. The source of free electrons is a heated material that is a thermionic emitter (cathode). The control element regulating the flow of electrons is called a *grid.* The collector element for the electron flow is the *anode,* or *plate.* Power tubes, in contrast to receiving-type tubes, handle relatively large amounts of power, and for reasons of economy, a major emphasis is placed on efficiency. The traditional division between the two tube categories is at the 25-W plate dissipation rating level.

Receiving tubes (and transistors, see Pars. 7-52 to 7-61) provide the essential active-device function in electronic applications. The general uses of receiving tubes cover nearly all functions of radio and television receivers, low-power transmitters, telephone systems, industrial control, and measurement devices. The majority of the large number of receiving-type tubes produced at the present time is for replacement use in existing equipments. Most of the electronic functions of the new equipment designs are being handled by solid-state devices.

Power tubes are widely used as high-power-level generators and converters in radio and television transmitters, radar, sonar, manufacturing-process operations, and medical and industrial x-ray equipment.

27. Classification of Types. Power and receiving-type tubes may be separated into groups according to their circuit function or by the number of electrodes they contain. Table 7-11 illustrates these factors and compares some of the related features. The physical shape and location of the grid relative to the plate and cathode are the main factors that determine the amplification factor, or μ, of the triode. The μ values of triodes generally range from about 5 to over 200. The mathematical relstionships between the three important dynamic tube factors are:

$$\text{Amplification factor } \mu = \Delta e_b / \Delta e_{c1} \qquad i_b = \text{const}$$
$$\text{Dynamic plate resistance } r_p = \Delta e_b / \Delta i_b \qquad e_{c1} = \text{const}$$
$$\text{Transconductance } Sm \text{ or } Gm = \Delta i_b / \Delta e_{c1} \qquad e_b = \text{const}$$

where e_b = total instantaneous plate voltage, e_{c1} = total instantaneous control grid voltage, and i_b = total instantaneous plate current. Note that $\mu = Gm \times r_p$. Figure 7-18 shows the curves of plate and grid current as a function of plate voltage at various grid voltages for a typical triode with a μ value of 30.

The tetrode, a four-element electron tube, is formed when a second grid (screen grid) is mounted between grid 1 (control grid) and the anode (plate). The plate current is almost independent of plate voltage. Figure 7-19 shows the curves of plate current as a function of plate voltage at a fixed screen voltage and various grid voltages for a typical power tetrode. Table 7-12 lists typical receiving tubes.

28. Basic Construction. *Cathodes* used in most power and receiving tubes obtain the energy required for electron emission from heat. The two types of cathodes based upon the method of heating are directly heated (filament types) and indirectly heated. The two basic types of emitting surfaces most commonly used are thoriated tungsten and alkaline-earth oxides. Most receiving tubes and many power tubes use the oxide cathode because of its high efficiency, i.e., high emission current per watt of heating power. Thoriated tungsten is more

DISCRETE CIRCUIT COMPONENTS

tolerant to ion bombardment and is used in many high-power, high-voltage tubes. The characteristics of these two emitting surfaces are given in Table 7-13.

Grids. The most widely used grid construction in receiving-type tubes involves helically wound fine wire of nickel alloy supported on two side rods. Much of the mechanical strength of this construction is derived from the strength of the lateral grid wires themselves. Grids of high electrical performance require finer wires of low mechanical strength and are not good for use as a support. A structure that is used in modern, close-spaced, high-performance receiving tubes is the frame grid. Extremely fine lateral wire is wound under tension on a rigid open frame with close spacing between grid wires and between tube electrodes, thereby improving tube performance. Grid wire of 12 μm diameter or less is practical using this construction technique.

Table 7-11. Tube Classification by Construction and Use

Tube type	No. of active electrodes	Typical use	Relative features and advantages
Diode	2	Rectifier	High back resistance
Triode	3	Low- and high-power amplifier, oscillator, and pulse modulator	Low cost, circuit simplicity
Tetrode	4	High-power amplifier and pulse modulator	Low drive power, low feedback
Pentode	5	Low-power amplifier	High gain, low drive, low feedback, low anode voltage
Hexode, etc. ...	6 or more	Special applications	Multiple-input mixers, converters

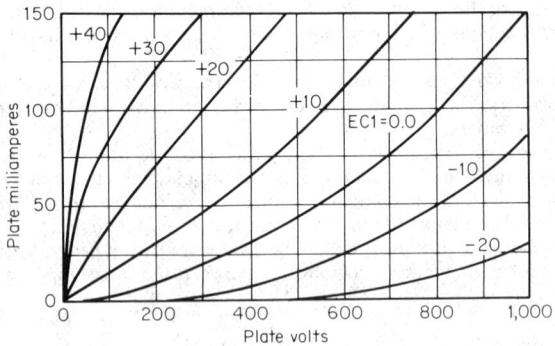

Fig. 7-18. Typical triode plate characteristics.

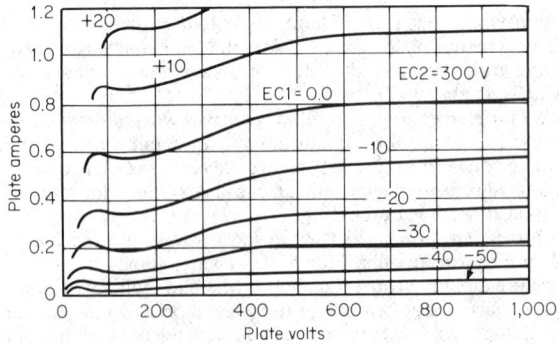

Fig. 7-19. Typical tetrode plate characteristics.

7-22

The grids of power tubes are made to very close tolerances and must maintain their shape and spacing at elevated temperatures. They must also withstand shock and vibration. For this reason, most grids in power tubes are made with tungsten or molybdenum, which have good hot strength. The grids are also required not to emit either primary or secondary electrons. Gold plating of the wires is the most widely used technique to inhibit primary emission in tubes with oxide cathodes. The maximum safe operating temperature for gold plating is on the order of 550°C. Many special coatings have been developed for specific applications that are effective in reducing grid emission.

Anodes. The anode (plate) usually encloses the other electrodes, and in most cases it is basically cylindrical in cross section. The major exception is the planar construction used in high-frequency types. The anode of most receiving tubes and many older medium-power tubes is completely enclosed by the tube envelope. Stamped or formed sheet nickel is generally used in receiving tubes, although some tubes have been made with aluminum-clad steel. The most commonly used material for power tubes with enclosed internal anodes is zirconium-clad molybdenum. In most power tubes, the anode is a part of the envelope, and since the outer surface is external to the vacuum, it can be cooled directly (see Fig. 7-20). The material normally used for external anodes is copper.

Envelopes of most receiving tubes and power tubes of older design are glass. Glass is easy to form, readily available, and low in cost. Ceramic material is used as the envelope material in power tubes of recent design and in some receiving tubes for special applications. Ceramic envelopes are more expensive than glass; however, higher-temperature processing improves the vacuum, resulting in longer tube life. Metal envelopes have been used in the manufacture of receiving tubes, such as types 6L6, 12SQ7, 12SK7. Receiving tubes made with iron (or steel) envelopes possessed advantages with respect to electrostatic shielding and were less subject to mechanical damage. However, this is an expensive construction and not widely used today. An exception is

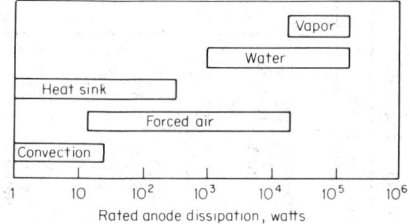

Fig. 7-20. Anode cooling methods.

the *nuvistor* construction, which uses a ceramic-metal envelope to house a small cantilever-supported cylindrical structure. Examples of this construction are the types 6DS4 and 6CW4.

29. Life Expectancy. Life expectancy is one of the most important factors to be considered in the use of tubes. Whereas it is impossible to predict the life for an individual tube, accurate average life expectancy can be determined for a group of tubes run under closely controlled conditions. Two types of failures are encountered.

Catastrophic failure is the premature demise of the electron tube long before its statistical life expectancy. Most of these failures are in the mechanical category, i.e., broken filament, poor weld, cracked insulators, etc. These failures generally occur early in life and can be reduced by careful design, selection of materials, and rigorous control of manufacturing techniques. These types of failures have been dramatically reduced in modern tubes.

Wear-out failure results from gradual deterioration of electrical characteristics. Modern tube technology has virtually overcome manufacturing problems that result in fast wear-out. The one remaining consideration for achieving ultimate reliability, i.e., the proper application of the tube, is the responsibility of the user. A modern tube, properly used, can be reasonably expected to perform for many thousands of hours.

The cathode is the heart of any tube, and cathode temperature, together with current loading, is of major importance in determining tube life. Oxide cathodes are used in almost all receiving tubes and in the majority of modern power tubes. Figure 7-21 is a plot of the tube wear-out life capability as a function of cathode current density. This curve assumes the tubes are properly cooled, the cathode temperature is adjusted to the optimum value required by the current demand, the heater voltage is regulated to 1%, and the tube is protected from arc damage. Thoriated tungsten emitters, which are used in some high-voltage power tubes, have very long life. This type of emitter, operating at 1650°C with a filament-voltage variation of less than 1%, has a life capability in excess of 25,000 h. An increase of 5% in filament voltage will reduce life approximately 50%.

Table 7-12. Typical Receiving Tubes

A. Standard Commercial Types

Type	Function description	Base pins*	Bulb	Heater, V/A	Max. plate/screen, V	Plate current, mA	Grid voltage or cathode resistance	Load resistance, Ω, or plate resistance, MΩ	Transconductance, μmho	Output, W	Max. plate dissipation, W
1B3GT	Diode H-V rectifier	8+TC	T-9	1.25/0.2	22 kV	0.5					
3AT2	Diode H-V rectifier	12+TC	T-9	3.15/0.22	30 kV	1.7					P-5
6AF11	Duotriode-pentode	12	T-9	6.3/1.05	330/180	P24; T7.9	100; 220	0.068; 0.010	11,000; 5,500-4,400		T-1.2; 12
6AQ5A	Audio-beam pentode	7	T-5½	6.3/0.45	275/275	45	-12.5	5000	4100	4.5	12
6AR11	IF twin pentode	12	T-9	6.3/0.8	330/180	11; 11	56; 56	0.2; 0.2	10,500; 10,500		3.1; 3.1
6AX3	TV horiz. damper diode	12	T-9	6.3/1.2	5 kV	165					5.3
6BQ5 (EL 84†)	Audio pentode	9	T-6½	6.3/0.76	300/300	48	135	5,200	11,300	6.0	12
6CL6	Video pentode	9	T-6½	6.3/0.65	300/300	30	-3.0	7,500	11,000	2.8	7.5
6CW4	Nuvistor triode	5+S	MT-4	6.3/0.135	110	7	130	0.0066	9,800		1.5
6DJ8	Twin triode, frame grid	9	T-6½	6.3/0.365	220	12	120	0.003	11,500		1.5
6EJ7 (EF 184†)	Sharp C.O. frame-grid pentode	9	T-6½	6.3/0.3	250/250	10	-2.5	0.350	15,000		2.5
6JE6	TV horiz. sweep-beam pentode	Novar+TC	T-12	6.3/2.5	7.0 kV	115	-25	0.0055	10,500		24
6L6GC	Audio-beam pentode	8	T-12	6.3/0.9	500/300	54	-18	4200	5,200	10.8	30
6T10	Audio detector and output pentode-tetrode	12	T-9	6.3/0.95	275/275	35; 1.3	-8.0; 560	0.10; 0.15	6,500; 1,000		T-10; P-1.7
6U8A	Triode-pentode	9	T-6½	6.3/0.45	330/180	9.5; 13.5	-1.0; -1.0	0.200; 0.005	5,000; 7,500		P-3.0; T-2.5
12AT7 (ECC 81†)	Twin triode	9	T-6½	6.3/0.3; 12.6/0.15	300	10; 10	200; 200	0.0011; 0.0011	5,500; 5,500		2.2; 2.2
12AU7A (ECC 82†)	Twin triode	9	T-6½	6.3/0.3; 12.6/0.15	300	10.5; 10.5	-8.5; -8.5	0.0077; 0.0077	2,200; 2,000		2.75; 2.75
12AV6	Duodiode triode	7	T-5½	12.6/0.15	330	1.2	-2.0	0.0625	1,600		0.55
12AX7 (ECC 83†)	Twin triode	9	T-6½	6.3/0.3; 12.6/0.15	330	1.2; 1.2	-2.0; -2.0	0.0625; 0.0625	1,600; 1,600		1.2; 1.2
12BA6	RF pentode	7	T-5½	12.6/0.15	330	11	68	1.0	4,400		3.4
12BE6	Mixer heptode	7	T-5½	12.6/0.15	330/110	2.9	-1.5		425Sc		1.1
12HG7	Video pentode	9	T-9	12.6/0.26; 6.3/0.52	400/165	31	47	0.06	32,000		10

38HE7	TV horiz. diode-tetrode	12	T-12	38/0.45	5 kV / 4.2 kV	60	−22	0.0062	8,800	…	10-T
50C5	Audio-beam pentode	7	T-5½	50/0.15	150	200	−8.0	2,500	7,500	…	4-D
7077	Grounded grid ceramic triode	2+3	0.3 in.	6.3/0.24	250	49	82	0.010	10,000	2.3	7
7587	Nuvistor tetrode	5+S	MT-4	6.3/0.15	125/50	6.5 / 10	68	0.2	10,600	450 MHz	1.1 / 2.2
7768	Ceramic triode	2+3	0.55 in.	6.3/0.4	330	22	22	0.0045	50,000	…	5.5
B. Military and Industrial Types											
5654/6AK5W	Sharp C.O. pentode	7	T-5½	6.3/0.175	180/180	7.7	180	0.5	5,100	…	1.7
5702WA	Sharp C.O. pentode	7-pin, sub min.	T-3	6.3/0.2	165/155	7.5	200	0.34	5,000	…	1.1
5725/6AS6W	Cathode resistance pentode, dual control			6.3/0.175	200/155	5.2	−2.0	…	3,200	…	1.65
5749/6BA6W	Remote C.O. pentode	7	T-5½	6.3/0.3	300/300	11	68	1.0	4,400	…	3.3
5750/6BE6W	Mixer heptode	7	T-5½	6.3/0.3	300/100	2.6	−1.5	1.0	475Sc	…	1.0
5751	Twin triode, high-μ	9	T-6½	6.3/0.35, 12.6/0.175	330, 330	1.0, 1.0	−3.0, −3.0	0.058, 0.058	1,200, 1,200	…	0.8, 0.8
5814WA	Twin triode, med.-μ	9	T-6½	6.3/0.35, 12.6/0.175	330	10.5, 10.5	−8.5, −8.5	0.0077, 0.0077	2,200, 2,200	…	3.0, 3.0
5840	Sharp C.O. pentode	8-pin, sub min.	T-3	6.3/0.15	165/155	7.5	150	0.26	5,000	…	1.1
5902	Audio-beam pentode	8-pin, sub min.	T-3	6.3/0.45	165/155	30	270	3000	4,200	1.0	3.7
6005/6AQ5W	Audio-beam pentode	7	T-5½	6.3/0.45	250/250	45	−12.5	5000	4,100	4.5	12
6021	Twin triode, med.-μ	8-pin, sub min.	T-3	6.3/0.3	165	6.5	150	0.0065	5,400	…	0.7
6111	Twin triode, med.-μ	8-pin, sub min.	T-3	6.3/0.3	165	8.5	220	0.004	5,000	0.95	
6112	Twin triode, high-μ	8-pin, sub min.	T-3	6.3/0.3	165	1.75	820	0.028	2,500	…	0.3
7788 (E810F†)	RF pentode, wideband amplifier	9	T-6½	6.3/0.34	250/200	35	47	0.05	50,000	…	5

* TC = top cap; S = metal shield.
† European equivalents.
SOURCE: A portion of the material in this table is from Donald G. Fink and John M. Carroll, "Standard Handbook for Electrical Engineers," 10th ed., McGraw-Hill Book Company, New York, 1968.

Table 7-13.　Typical Characteristics of Emitters

	Oxide	Thoriated tungsten
Method of heating	Direct and indirect	Direct
Operating temperature	700–820°C	1600–1700°C
Emission per watt of heating:		
Average current	250 mA at 800°C	8–10 mA
Peak current	10 A for 10 μs or less	80–100 mA

30. Cooling Methods.　Cooling of the tube envelope, seals, and anode, if external, is a major factor affecting tube life.　The data sheets provided by tube manufacturers include with the cooling requirements a maximum safe temperature for the various external surfaces.　The temperature of these surfaces should be measured in the operating equipment.　The temperature can be measured with thermocouples, optical pyrometers, a temperature-sensitive paint such as Tempilaq, and temperature-sensitive tapes.

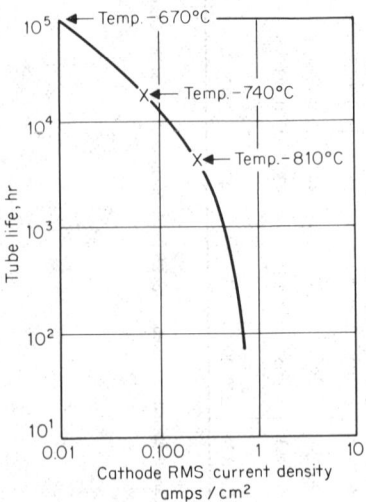

Fig. 7-21.　Tube life vs. current density.

The envelopes and seals of most tubes are cooled by convection of air around the tube or by using forced air.　The four principal methods used for cooling external anodes of tubes are by air, water, vapor, and heat sinks.　Other cooling methods occasionally used are oil, heat pipes, refrigerants, such as Freon, and gases, such as sulfahexafluoride.　Figure 7-20 shows the range of anode dissipation generally used for the four principal methods used for anode cooling.

31. Protective Circuits.　Arcs can damage or destroy electron tubes, which may significantly increase the equipment operating cost.　In low- or medium-power operation, the energy stored in the circuit is relatively small and the series impedance is high, which limits the tube damage.　In these circuits, the tube will usually continue to work after the fault is removed; however, the life of the tube will be reduced.　Since these tubes are normally low in cost, economics dictate that inexpensive slow-acting devices, such as circuit breakers and common fuses, be used to protect the circuit components.

High-power equipment has large stored energy, and the series impedance is usually low.　Arcs in this type of equipment will often destroy expensive tubes, and slow-acting protective devices offer insufficient protection.　The two basic techniques used to protect tubes are *energy diverters* and special *fast-blow fuses.*　The term *crowbar* is commonly used for energy diverters.　The typical circuit for a crowbar is shown in Fig. 7-22.　In the event of a tube arc, the trigger-fault sensor unit "fires" to a crowbar, which is a very-low-impedance gas-discharge device.　The firing time can be less than 2 μs.　The low impedance in the crowbar arm

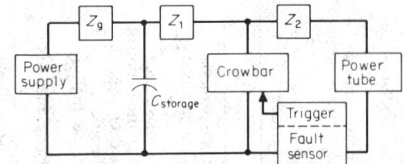

Fig. 7-22.　Energy-diverter circuit.

is in shunt with the Z_2 and tube arm, and diverts current from the tube during arcs.　The impedance Z_2, which is in series with the tube, is used to ensure that most of the current is diverted through the crowbar.　The value of Z_2 is primarily limited by efficiency considerations during normal operation.　The impedance Z_1 is required to limit the fault current in the storage condenser to the maximum current rating of the condenser.　The impedance Z_g is the internal power-supply impedance.　Devices used as crowbars are thyratrons, ignitrons, triggered spark gaps, and plasmoid triggered vacuum gaps.

32. Fast Fuses. Two types of fast-blow fuses are used to protect power tubes. They are *exploding-wire* and *exothermic* fuses. Exploding-wire fuses require milliseconds to operate and are limited in their ability to protect the tube. Exothermic fuses, although faster-acting than exploding wires, are significantly slower than crowbars in clearing the fault current in the tube. A second disadvantage of fuses is that after each fault the power supply must be turned off and the fuse replaced. For this reason, fuses are limited to applications where the tubes seldom arc. The major advantage of fuses is low cost.

33. Special Application Considerations. *Noise.* The electronic noise generated with an electron tube is a basic limitation to the magnitude of amplification achievable. Resistor and tube noise can never be avoided; however, much can be done by proper circuit design and choice of tube types to approach the minimum attainable noise. Random division of current between electrodes contributes to the noise in multielectrode tubes and makes pentodes three to five times as noisy as the same tube connected as a triode. Mixer tubes are also noisy, the triode converter being the lowest in noise. Mechanical noise results from some aspect of tube construction or use. Vibration of elements may cause microphonic noise. The design of the control grid is the most important factor in microphonic effects. Hum can come from poor heater design or construction. Modern receiving tubes properly applied have a minimum of mechanical-noise problems.

Frequency. *Transit time* is a large factor in considering the upper frequency limitation of electron tubes. A finite time is taken by the electron to traverse the space from the cathode, through the grid, and on to the plate. As the frequency of operation is raised, a frequency is finally reached at which electron transit-time effects become significant. The point of onset is dependent on accelerating voltages to the grid and anode planes and the respective spacings. Closer-spaced tubes, in particular, with close spacing in the grid-to-cathode region, have reduced transit-time effects.

Lead inductance and *capacitance* act to reduce the upper frequency limit of electron tubes. Modern high-frequency electron tubes have short structures, short lead-electrode connections, and a minimum of unnecessary or parasitic capacitance.

Power limitations are interrelated with cathode emission capability, electrode sizes, and design-form factors, materials, and methods of cooling. There is also a power-frequency relationship in that, as the frequency of required operation is increased, closer spacing and smaller-sizes electrodes are used. This reduces the power-handling capability for tube designs having the same form factor. Figure 7-23 illustrates the maximum continuous-wave power-output capability as a function of frequency for high-power gridded tubes.

CATHODE-RAY STORAGE AND CONVERSION DEVICES

By I. Reingold and M. Crost

34. Introduction. Electronic charge-storage tubes are divided into four broad classes: electrical-input, electrical-output types; electrical-input, visual-output types; visual-input, electrical-output types; and visual-input, visual-output types. An example of each class is cited in Table 7-14. Tubes under these classes in which storage is merely incidental, such as camera tubes and image converters, are classed under Conversion Devices. See also Sec. 11.

35. Electrical-Input Devices. *Electrical-Output Types.* The *radechon, or barrier-grid storage tube,* is a single-electron-gun storage tube with a fine-mesh metal screen in contact

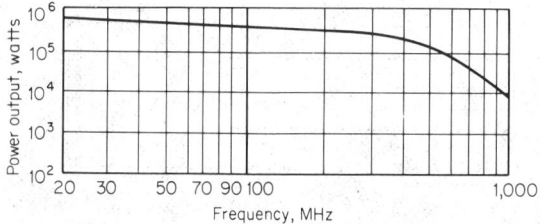

Fig. 7-23. Continuous-wave output power capability.

Table 7-14.   Storage-Tube Classes

Type of input	Type of output	Subclass	Representative example
Electrical	Electrical	Single-gun Multiple-gun	Radechon Graphechon
	Visual	Bistable Halftone	Memotron Tonotron
Visual	Electrical	Nonscanned	Correlatron
	Visual	Time-sharing	Storage image tube

Table 7-15.   Conversion Devices

Type of input	Type of output	Subclass	Representative example
Visual	Electrical	Photoemissive Photoconductive	Image orthicon Vidicon
	Visual	Photoemissive	Image amplifier

with a mica storage surface. The metal screen, called the barrier grid, acts as a very-close-spaced collector electrode, and essentially confines the secondary electrons to the apertures of the grid in which they were generated. The very thin mica sheet is pressed in contact with a solid metal backplate. A later model was developed with a concave copper bowl-shaped backplate and a fritted-glass dielectric storage surface. A similarly shaped fine-mesh barrier grid was welded to the partially melted glass layer. The tube is operated with backplate modulation; i.e., the input electrical signal is applied to the backplate while the constant-current electron beam scans the mica storage surface. The capacitively induced voltage on the beam side of the mica dielectric is neutralized to the barrier-grid potential at each scanned elemental area by collection of electrons from the beam and/or by secondary electron emission (equilibrium writing). The current involved in recharging the storage surface generates the output signal.

In a single-electron-gun *image-recording storage tube* (Fig. 7-24) information can be recorded and can be read out at a later time. The intended application is the storage of complete halftone images, such as a television picture, for later readout. In this application, write-in of information is accomplished by electron-beam modulation, in time-sharing sequence with the readout mode. Reading is accomplished nondestructively, i.e., without intentionally changing the stored pattern, by not permitting the beam to impinge upon the storage surface. The electron-gun potentials are readjusted for readout so that none of the electrons in the constant-current beam can reach the storage surface, but rather divide their current between the storage mesh and the collector in proportion to the stored charge. This process is called *signal division.*

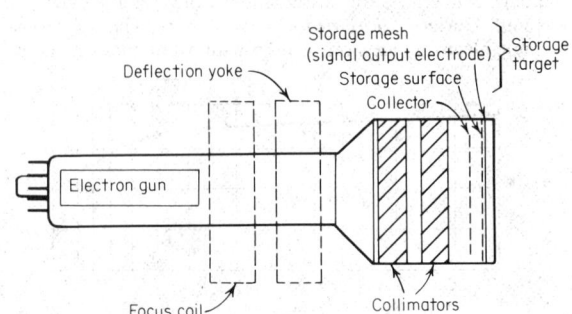

Fig. 7-24. Electrical signal storage tube, basic structure. *(Hughes Aircraft Corp.)*

One type of double-ended, multiple-electron-gun *recording storage tube* with nondestructive readout operates by recording halftone information on an insulating coating on the bars of a metal mesh grid, which can be penetrated in both directions by the appropriate electron beams. Very high resolution and long storage time with multiple readouts, including simultaneous writing and reading, are available with this type of tube. Radio-frequency separation or signal cancellation must be used to suppress the writing signal in the readout during simultaneous operation.

Figure 7-25 is a representative schematic drawing of a double-ended, multiple-electron-gun *membrane-target storage tube* with nondestructive readout, in which the writing beam and the reading beam are separated by the thin insulating film of the storage target. In this case there is a minimal interference of the writing signal with the readout signal in simultaneous operation, except for capacitive feed-through, which can readily be canceled. Writing is accomplished by beam modulation, while reading is accomplished by signal division. Very high resolution and long storage time with multiple readouts are available with this group of tubes.

The *graphechon* differs from the two groups of tubes just described in that its storage target operates by means of electron-bombardment-induced conduction (EBIC). Halftones are not available from tubes of this type in normal operation. The readout is of the destructive type, but since a large quantity of charge is transferred through the storage insulator during writing, a multitude of readout copies can be made before the signal displays noticeable degradation. In simultaneous writing and reading, signal cancellation of the writing signal in the readout is generally accomplished at video frequencies.

Visual-Output Types. This class comprises the *display storage tubes (DSTs)* or *direct-view storage tubes (DVSTs)*. Figure 7-26 shows a schematic diagram of a typical DVST with one electrostatic focus-and-deflection writing gun (other types may have electromagnetic focus and/or deflection) and one flood gun, which is used to provide a continuously bright display and may also establish bistable equilibrium levels. The storage surface is an insulating layer

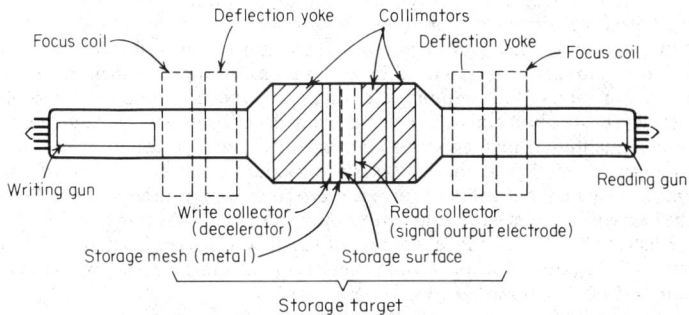

Fig. 7-25. Typical double-ended scan converter, basic structure. *(Hughes Aircraft Corp.)*

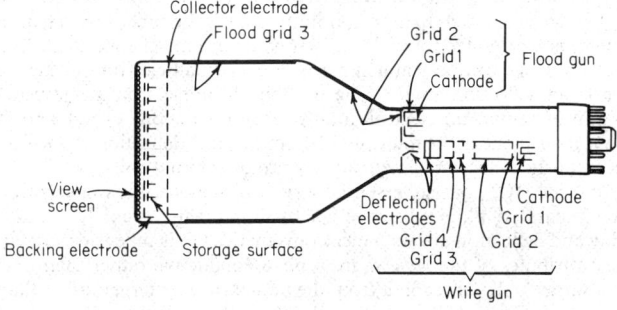

Fig. 7-26. Cross-section view, direct-view storage tube. *(Westinghouse Electric Corp.)*

deposited on the bars of the metal-mesh backing electrode. The view screen is an aluminum-film-backed phosphor layer.

The *memotron* is a DST that operates in the bistable mode; i.e., areas of the storage surface may be in either the cutoff condition (flood-gun-cathode potential) or in the transmitting condition (collector potential), either of which is stable. The focusing and deflection are electrostatic. Normally, the phosphor remains in the unexcited condition until a trace to be displayed is written into storage; then this trace is displayed continuously until erased, so long as the flood beam is maintained in operation.

The *tonotron* is typical of a large number of halftone DSTs. These operate in the nondestructive readout mode, with the storage surface at or below flood-cathode potential. Writing is accomplished by depositing halftone charge patterns by electron-beam modulation.

36. Visual-Input Devices. *Electrical-Output Types.* The *correlatron* is a storage tube that receives a visual input to a photoemissive film, focuses the photoelectron image upon a transmission-grid type of storage-target, where it is stored, and later compares the original image with a similar image. A single total output current is read out for the entire image, with no positional reference. The purpose of this comparison is to ascertain whether the first and second images correlate.

Visual-Output Types. In the *storage image tube,* a positive electron image may be stored upon the insulated bars of the storage mesh. Then, if the photocathode is uniformly flooded with light to produce a flood electron cloud, a continuously bright image of the stored charge pattern may be obtained upon the phosphor screen in the following section of the tube. A high degree of brightness gain may be achieved with this type of tube, or a single snapshot picture of a continuous action may be "frozen" for protracted study.

CONVERSION DEVICES

37. Principles of Conversion Devices. The conversion devices discussed receive images in visible or infrared radiation and convert these images by internal electronic processes into a sequence of electrical signals or into a visible output image. Some of these devices may employ an internal storage mechanism, but this is generally not the primary function in their operation. These tubes are characterized by a photosensitive layer at their input ends that converts a certain region of the quantum electromagnetic spectrum into electron-hole pairs. Some of these layers are photoemissive; i.e., they emit electrons into the vacuum if the energy of the incoming quantum is high enough to impart at least enough energy to the electron to overcome the work function at the photosurface-vacuum interface. Others do not emit electrons into the vacuum but conduct current between the opposite surfaces of the layer by means of the electron-hole pairs; i.e., they are photoconductive. The transmission characteristics of the material of the entrance window of the tube can greatly modify the effective characteristics of any photosurface. If the material is photoemissive, the total active area is called a *photocathode.* See also Sec. 11.

The types of conversion devices discussed are divided into visual-input devices, electrical-output and visual-output types. An example of each type is cited in Table 7-15.

38. Visual-Input Conversion Devices (see also Sec. 11). *Electrical-Output Types.* This class covers the large group of devices designated camera tubes. In the *image dissector,* the incoming visual image is focused upon the photocathode of the tube. The resulting electron analog of the image is accelerated toward and focused upon a metal plate some distance away, so that the electron image can be scanned across a very small aperture in the center of the plate without a large deflection angle. The portion of the electron image penetrating the aperture at any instant enters an electron multiplier in the region beyond, and the signal is obtained from the final anode of the multiplier. Focusing and deflection of the electron image are usually accomplished electromagnetically by use of external coils.

The *image orthicon* (IO) consists essentially of two sections, the image section and the orthicon section, separated in the tube envelope by the storage target. In the most common version, focusing and deflection are both electromagnetic. This tube is described in Sec. 20.

The defining attributes of the *vidicon* are a photoconductive rather than photoemissive image surface or target, a direct readout from the photosensitive target rather than by means of a return beam, and a much smaller size than the above camera tubes. The original vidicons employed coincident electromagnetic deflection and focusing. Many later versions employ either or both electrostatic focusing and deflection.

A very important group of tubes is the *semiconductor diode-array vidicons*. In place of the usual photoconductive target, these tubes include a very thin monolithic wafer of the semiconductor, usually single-crystal silicon. On the beam side of this wafer a dense array of junction photodiodes has been generated by semiconductor diffusion technology. These targets are very sensitive compared with the photoconductors, and they have very low leakage, low image lag, and low blooming from saturation.

Visual-Output Types. The *image tube,* or *image amplifier,* with input in the visible spectrum is used principally to increase the light level and dynamic range of a very low light-level image to a level and contrast acceptable to a human observer, a photographic plate, or a camera tube. The image tube consists basically of a photoemissive cathode, a focusing and accelerating electron-optical system, and a phosphor screen.

Since the photocathode would be illuminated by light returning from the phosphor screen, the internal surface of the phosphor is covered with a very thin film of aluminum that can be penetrated by the high-energy image electrons. The aluminum film also serves as the tube anode and as a reflector for the light that would otherwise be emitted back into the tube.

SEMICONDUCTOR DIODES AND CONTROLLED RECTIFIERS

By G. J. Malinowski

39. Semiconductor Materials and Junctions. Transistors and diodes are fabricated from semiconductor materials, a form of matter situated between metals and insulators in their ability to conduct electricity. Typical values of electrical resistivity of conductors, semiconductors, and insulators are 10^{-6} to 10^{-5}, 10 to 10^4, and 10^{12} to 10^{16} Ω·cm, respectively. By far the most widely used semiconducting materials are germanium and silicon. Other semiconductor materials such as gallium arsenide, selenium, cadmium sulfide, and copper oxide have electrical properties that make them useful in special applications.

Unlike the vacuum tube, where current flow arises from the motion of charge carriers within a vacuum, semiconductor devices develop current flow from the motion of *charge carriers* within a crystalline solid. The conduction process in semiconductors is most easily visualized in terms of silicon and germanium. The atoms of each of these elements have four electrons in the outer shell (valence shell). These electrons are normally bound in the crystalline lattice structure. Some of these valence electrons are free at room temperature, and hence can move through the crystal; the higher the temperature, the more electrons are free to move. Each vacancy, or hole, left in the lattice can be filled by an adjacent valence electron. Since a hole moves in a direction opposite to that of an electron, a hole may be considered as a positive-charge carrier. Electrical conduction is due to the motion of holes and electrons under the influence of an applied field.

Intrinsic (pure) semiconductors exhibit a negative coefficient of resistivity, since the number of carriers increases with temperature. Conduction due to thermally generated carriers, however, is usually an undesirable effect, because it limits the operating temperature of the semiconductor device.

At a given temperature, the concentration of thermally generated carriers is related to the energy gap of the material. This is the minimum energy (stated in electron volts) required to free a valence electron (1.1 eV for silicon and 0.7 eV for germanium). Silicon devices perform at higher temperatures because of the wider energy gap.

The conductivity of the semiconductor material can be altered radically by doping with minute quantities of *donor* or *acceptor impurities.* Donor (*n*-type) impurity atoms have five valence electrons, whereas only four are accommodated in the lattice structure of the semiconductor. The extra electron is free to conduct at normal operating temperatures. Commonly used donor impurities include phosphorus, arsenic, and antimony. Conversely, acceptor (*p*-type) atoms have three valence electrons; a surplus of holes is created when a semiconductor is doped with them. Typical acceptor dopants include boron, gallium, and indium.

In an *extrinsic (doped) semiconductor,* the current-carrier type introduced by doping predominates. These carriers, electrons in *n*-type material and holes in *p*-type, are called *majority carriers.* Thermally generated carriers of the opposite type are also present in small quantities and are referred to as *minority carriers.* Resistivity is determined by the concentration of majority carriers.

Lifetime is the average time required for excess minority carriers to recombine with majority carriers. Recombination occurs at "traps" caused by impurities and imperfections in the semiconductor crystal. Semiconductor junctions are formed in material grown as a single continuous crystal to obtain the lattice perfection required, and extreme precautions are taken to ensure exclusion of unwanted impurities during processing. However, in some applications short lifetime is desired, and in such cases gold doping is used to achieve this.

Carrier mobility is the property of a charge carrier which determines its velocity in an electric field. Mobility also determines the velocity of a minority carrier in the diffusion process. High mobility yields a short transit time and good frequency response.

40. pn Junctions. If *p*- and *n*-type materials are formed together, a unique interaction takes place between the two materials, and a *pn* junction is formed. In the immediate vicinity of this junction (in the *depletion region*), some of the excess electrons of the *n*-type material diffuse into the *p* region, and likewise holes diffuse into the *n*-type region. During this process of recombination of holes and electrons, the *n* material in the depletion region acquires a slightly positive charge and the *p* material becomes slightly negative. The space-charged region thus formed repels further flow of electrons and holes, and the system comes into equilibrium. Figure 7-27 shows a typical *pn* junction.

To keep the system in equilibrium, two related phenomena constantly occur. Due to thermal energy, electrons and holes diffuse from one side of the *pn* junction to the other side. This flow of carriers is called *diffusion current.* When a current flows between two points, a potential gradient is produced. This potential gradient across the depletion region causes a flow of charge carriers, drift current, in the opposite direction to the diffusion current. As a result, the two currents cancel at equilibrium and the net current flow is zero through the region. An energy barrier is erected such that further diffusion of charge carriers becomes impossible without the addition of some external energy source. This energy barrier formed at the interface of the *p*- and *n*-type materials provides the basic characteristics of all junction semiconductor devices.

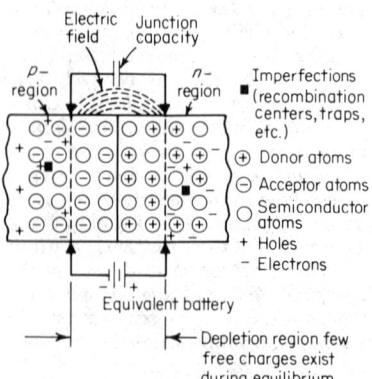

Fig. 7-27. *pn* junction. *(General Electric Co.)*

41. Fabrication. Silicon diodes can be made by using any of the various technologies employed in the fabrication of transistors, including alloying, growing, or diffusion (see Par. 7-54). The planar epitaxial process is widely employed. See also Pars. 8-4 to 8-32.

42. pn Junction Characteristics. When a dc power source is connected across a *pn* junction, the quantity of current flowing in the circuit is determined by the polarity of the applied voltage and its effect upon the depletion layer of the diode. Figure 7-28 shows the classical condition for a reverse-biased *pn* junction. The negative terminal of the power supply is connected to the *p*-type material, and the positive terminal to the *n*-type material. When a *pn* junction is reverse-biased, the free electrons in the *n*-type material are attracted toward the positive terminal of the power supply and away from the junction. At the same time, holes from the *p*-type material are attracted toward the negative terminal of the supply, and therefore away from the junction. As a result, the depletion layer becomes effectively wider, and the potential gradient increases to the value of the supply. Under these conditions the current flow is very small because no electric field exists across either the *p* or *n* region.

Figure 7-29 shows the positive terminal of the supply connected to the *p* region and the negative terminal to the *n* region. In this arrangement, electrons in the *p* region near the positive terminal of the supply break their electron-pair bonds and enter the supply, thereby creating new holes. Concurrently, electrons from the negative terminal of the supply enter the *n* region and diffuse toward the junction. This condition effectively decreases the

depletion layer, and the energy barrier decreases to a small value. Free electrons from the
n region can then penetrate the depletion layer, flow across the junction, and move by way
of the holes in the *p* region toward the positive terminal of the supply. Under these
conditions, the *pn* junction is said to be *forward-biased.*

A general plot of voltage and current for a *pn* junction is shown in Fig. 7-30. Here both
the forward- and reverse-biased conditions are shown. In the forward-biased region, current
rises rapidly as the voltage is increased and is quite high. Current in the reverse-biased
region is usually much smaller, and remains low until the breakdown voltage of the diode is
reached. Thereupon the current increases rapidly. If the current is not limited, it will
increase until the device is destroyed.

Junction Capacitance. Since each side of the depletion layer is at an opposite charge with
respect to each other, each side can be viewed as the plate of a capacitor. Therefore a *pn*
junction possesses capacitance. As shown in Fig. 7-31, junction capacitance changes with
applied voltage.

DC Parameters of Diodes. The most important of these parameters are as follows:

Forward voltage V_F is the voltage drop at a particular current level across a diode when
it is forward-biased.

Breakdown voltage *BV* is the voltage drop across the diode at a particular current level
when the device is reverse-biased to such an extent that heavy current flows. This is known
as *avalanche.*

Reverse current I_R is the current specified at a voltage less than *BV* when the diode is
reverse-biased.

AC Parameters of Diodes. The most important of these parameters are as follows:

Capacitance C_0 is the total capacitance of the device, which includes junction and package
capacitance. It is measured at a particular frequency and bias level.

Rectification efficiency R_E is defined as the ratio of dc output (load) voltage to the peak of
the input voltage, in a detector circuit. This provides an indication of the capabilities of the
device as a high-frequency detector.

Forward recovery time t_{fr} is the time required for the diode voltage to drop to a specified
value after the application of a given forward current.

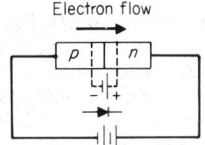

Fig. 7-28. Reverse-biased diode.

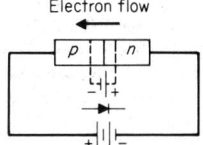

Fig. 7-29. Forward-biased diode.

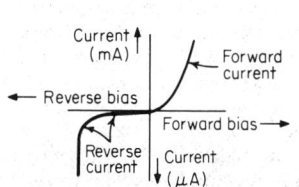

Fig. 7-30. Voltage-current characteristics for a *pn* junction.

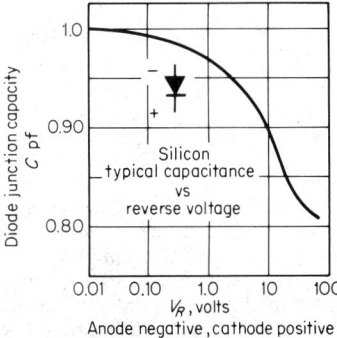

Fig. 7-31. Diode junction capacitance vs. reverse voltage. *(General Electric Co.)*

7-33

Reverse recovery time t_{rr} is specified as the time between the application of reverse voltage and the point where the reverse current has dropped to a specified value. It is further necessary to have the diode forward-biased prior to the application of the reverse voltage pulse.

Stored charge C_s can be used to characterize reverse recovery time of diodes.

Transient thermal resistance provides data as to the instantaneous junction temperature as a function of time with constant applied power. This parameter is valuable in ensuring reliable operation of diodes in pulse circuits.

43. Small-Signal Diodes. Small-signal diodes are the most widely used semiconductor devices. The capabilities of the general-purpose diode as a switch, demodulator, rectifier, limiter, capacitor, and nonlinear resistor suit it to many applications. One of the highest-volume uses of the semiconductor diode is in computers. In this application, the computer diode must have low forward resistance, high reverse resistance, low intrinsic capacitance and inductance, and fast recovery time. The computer diode is manufactured specifically for digital applications; it is usually gold-doped and is packaged as small as practical.

The most important characteristics of all small-signal diodes are forward voltage, reverse breakdown voltage, reverse leakage current, junction capacitance, and recovery time.

44. Silicon Rectifier Diodes. Silicon rectifier diodes are *pn* junction devices that have up to several hundred amperes forward-current-carrying capability. An ideal rectifier has an infinite reverse resistance, infinite breakdown voltage, and zero forward resistance. The silicon rectifier approaches these ideal specifications in that the forward resistance is only a few tenths of an ohm, while the reverse resistance is in the megohm range.

Since silicon rectifiers are primarily used in power supplies, thermal dissipation must be adequate. To avoid excessive heating of the junction, the heat generated must be efficiently transferred to a heat sink. The relative efficiency of this heat transfer is expressed in terms of the thermal resistance of the device. For stud-mounted rectifiers, the thermal-resistance range is typically 0.1 to 1°C/W.

45. Zener Diodes. Externally, the zener diode is similar in appearance to other silicon rectifiers, and electrically, it is capable of rectifying alternating current. While it serves in a variety of applications, its primary use is as a voltage reference or regulator element. Its ability to maintain a desired operating voltage is limited by the temperature coefficient and the impedance of the zener device.

The voltage-reference and regulation performance of the zener diode is based on the avalanche characteristics of the *pn* junction. When a source of voltage is applied to the diode in the reverse direction (anode negative), a reverse current I_r is observed. As the reverse potential is increased beyond the knee of the current-voltage curve, avalanche-breakdown current becomes well developed. This occurs at the zener voltage V_z. Since the resistance of the device drastically drops at this point, it is necessary to limit the current flow by means of an external resistor. Avalanche breakdown of the operating zener diode is not destructive as long as the rated power dissipation of the junction is not exceeded.

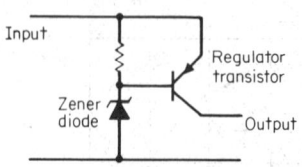

Fig. 7-32. Transistor voltage regulator using a zener diode for reference.

The zener diode can be used to control the reference voltage of a transistor power supply, as shown in Fig. 7-32. The zener diode also finds use in audio or rf applications where a source of stable reference voltage is required. The design of zener diodes permits them to absorb overload surges, and thereby serves the function of protecting circuitry from transients.

46. Varactor Diodes. The varactor diode is a *pn* junction device that has useful nonlinear voltage-dependent variable-capacitance characteristics. Varactor diodes are useful in microwave oscillators when employed with the proper filter and impedance-matching circuitry. Varactor diodes are also used for sensitive microwave amplification in parametric amplifiers. The voltage-dependent capacitance effect in the diode permits its use as an electrically controlled tuning capacitor in radio and television receivers, to replace conventional mechanical variable capacitors.

47. The Tunnel Diode. The tunnel diode is a semiconductor device whose primary use arises from its negative conductance characteristic. In a *pn* junction, a *tunnel* effect is obtained when the depletion layer is made extremely thin. Such a depletion layer is obtained by heavily doping both the *p* and *n* regions of the device. In this situation it is possible for an electron in the conduction band on the *n* side to penetrate, or tunnel, into the valence band of the *p* side. This gives rise to an additional current in the diode at a very small forward bias, which disappears when the bias is increased. It is this additional current that produces the negative resistance of the tunnel diode.

Commercial tunnel diodes are fabricated from germanium, silicon, and gallium arsenide. Typical applications of tunnel diodes include oscillators, amplifiers, converters, and detectors.

48. The Schottky Barrier Diode. This diode (also known as the surface-barrier diode, metal-semiconductor diode, and hot-carrier diode) consists of a rectifying metal-semiconductor junction in which majority carriers carry the current flow. When the diode is forward-biased, the carriers are injected into the metal side of the junction, where they remain majority carriers at some energy greater than the Fermi energy in the metal; this gives rise to the name *hot carriers*. The diode can be switched to the OFF state in an extremely short time (in the order of picoseconds). No stored minority-carrier charge exists.

The reverse dc current-voltage characteristics of the device are very similar to those of conventional *pn* junction diodes. The reverse leakage current increases with reverse voltage gradually, until avalanche breakdown is reached.

Schottky barrier diodes usually consist of silicon or gallium arsenide semiconductor material onto which gold, platinum, palladium, or silver is deposited by evaporation techniques.

Schottky barrier diodes utilized in detector applications have several advantages over conventional *pn* junction diodes. They have lower noise and better conversion efficiency, and hence have greater overall detection sensitivity.

49. Diode Light Sensors (Photodiodes). When a semiconductor junction is exposed to light, photons generate hole-electron pairs. When these charges diffuse across the junction, they constitute a photocurrent. Junction light sensors are normally operated with a load resistance and a battery which reverse-biases the junction. The device acts as a source of current which increases with light intensity.

Silicon sensors are used for sensing light in the visible and near-infrared spectra. These may be fabricated as phototransistors in which the collector-base junction is light-sensitive. Phototransistors are more sensitive than photodiodes because the photon-generated current is amplified by the current gain of the transistor.

50. Light-emitting Diodes (LEDs). These devices have found wide use in visual displays, isolators, and as digital storage elements.

LEDs are principally manufactured from gallium arsenide. While it has been known for some time that *pn* junctions emit visible light when biased into the avalanche-breakdown region, useful light levels have been achieved at low power levels only in recent years. LEDs are capable of providing light of different wavelengths by varying their construction. They can be used individually or in matrix arrays.

51. Silicon Controlled Rectifiers (SCR). A silicon controlled rectifier is basically a four-layer *pnpn* device that has three electrodes (a cathode, an anode, and a control electrode called the *gate*). Figure 7-33 shows the junction diagram and voltage-current characteristics for an SCR.

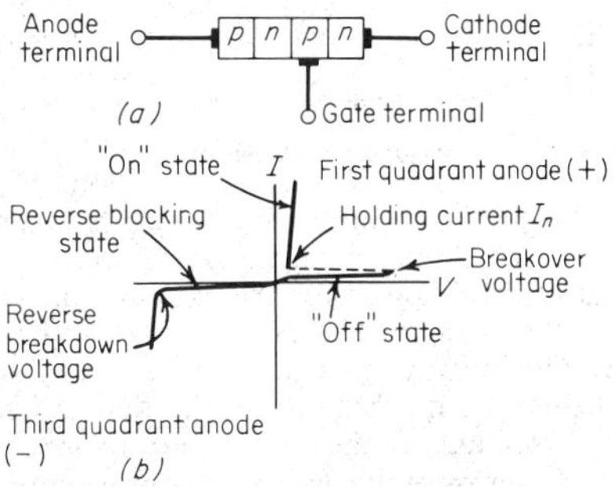

When an SCR is reverse-biased (anode negative with respect to the cathode), it is similar in characteristics to that of a reverse-biased silicon rectifier or other semiconductor diode. In this bias mode, the SCR exhibits a very high internal impedance, and only a very low reverse current flows through the *pnpn* device. This current remains small until the

Fig. 7-33. (*a*) SCR junction diagram; (*b*) typical SCR characteristics.

reverse voltage exceeds the reverse breakdown voltage; beyond this point, the reverse current increases rapidly.

During forward-bias operation (anode positive with respect to the cathode), the *pnpn* structure of the SCR is electrically bistable and may exhibit either a very high impedance (OFF *state*) or a very low impedance (ON *state*). In the forward-blocking state (OFF), a small forward current, called the *forward OFF-state current*, flows through the SCR. The magnitude of this current is approximately the same as that of the reverse-blocking current that flows under reverse-bias conditions. As the forward bias is increased, a voltage point is reached at which the forward current increases rapidly, and the SCR switches to the ON state. This voltage is called the *forward breakover voltage*.

When the forward voltage exceeds the breakover value, the voltage drop across the SCR abruptly decreases to a very low value, the forward ON-state voltage. When the SCR is in the ON state, the forward current is limited primarily by the load impedance. The SCR will remain in this state until the current through the SCR decreases below the holding current and then reverts back to the OFF state.

The breakover voltage of an SCR can be varied, or controlled, by injection of a signal at the gate. When the gate current is zero, the principal voltage must reach the breakover value of the device before breakover occurs. As the gate current is increased, however, the value of breakover voltage becomes less until the devices go to the ON state. This enables an SCR to control a high-power load with a very-low-power signal.

Silicon controlled switches (SCS) are basically SCRs with a second gate. This second gate can be used to turn the device on or off, depending upon the circuitry employed and the polarity of the applied signal. SCSs are basically low-current devices and have found application in low-power digital circuits.

TRANSISTORS

By E. B. Hakim

TRANSISTOR JUNCTIONS

52. Semiconductor *pn* Junctions. See Pars. 7-40 to 7-42.

53. Transistor Action. A transistor consists of two junctions in close proximity within a single crystal. An *npn* transistor is shown in Fig. 7-34. In normal bias conditions, the emitter-base junction is forward-biased and the collector-base junction is reverse-biased.

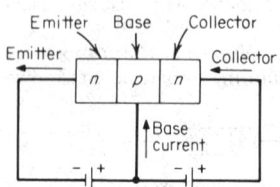

Fig. 7-34.　*npn* junction transistor.

Forward bias of the emitter-base junction causes electrons to be injected into the base region, producing an excess concentration of minority carriers there. These carriers move by diffusion to the collector junction, where they are accelerated into the collector region by the field in the depletion region of the reverse-biased collector junction. Some of the electrons recombine before reaching the collector. Current flows from the base terminal to supply the holes for this recombination process. Another component of current flows in the emitter-base circuit because of the injection of holes from the base into the emitter.

Practical transistors have narrow bases and high lifetimes in the base to minimize recombination. Injection of holes from the base into the emitter is made negligible by doping the emitter much more heavily than the base. Thus the collector current is less than, but almost equal to, the emitter current.

In terms of the emitter current I_E, the collector current I_C is

$$I_C = \alpha I_E + I_{CBO}$$

where α is the fraction of the emitter current that is collected, and I_{CBO} is due to the reverse-current characteristic of the collector-base junction. Increase of I_{CBO} with temperature sets the maximum temperature of operation.

High-frequency transistors are fabricated with very narrow bases to minimize the transit time of minority carriers across the base region. Germanium transistors have a better high-frequency response than silicon devices because of the higher carrier mobility of germanium.

54. Transistor Fabrication. Today the most widely used technique for transistor fabrication is the diffusion process. Without the development of this technique of junction formation, the rapid growth made in all fields of transistor electronics would have been impossible. The details of the diffusion process appear in Par. 7-56.

55. Alloy Process. The alloy technique preceded the development of the diffusion process. Alloy transistors were among the first types manufactured and are still widely used, because the process is inexpensive and performance of the devices is good at low frequencies. Both germanium and silicon-alloy transistors, for example, are used in applications such as audio stages of transistor radios, where high power at low frequency is required.

In alloy transistors, donor or acceptor impurities are applied directly to the top and bottom surfaces of a carefully prepared wafer of semiconductor material. The impurities on what will eventually become the collector are made to cover more area than on the emitter side to improve the current gain characteristics. The assembly is brought up to such a temperature that the impurities alloy into the semiconductor wafer, whereupon a saturated liquid solution of both materials is formed on both sides. This assembly is then permitted to cool. The resulting alloy fronts form abrupt (step) junctions with the base. The final electrical characteristics of the alloy transistor are determined in part by the area the impurities cover and their depth of penetration into the wafer.

56. Diffusion Process. The diffusion process has many advantages over the alloy process, i.e., precise control of junction areas and layer width, nonuniform-resistivity regions to provide for a variety of electrical characteristics, graded junctions instead of abrupt junctions, and a variety of geometries to optimize current handling and frequency response. There are many variations of this process, but it basically involves exposing a semiconductor wafer of predetermined resistivity to a gaseous flow of impurities in a furnace. The gaseous-impurity atoms thus diffuse into the semiconductor surface, forming a *pn* junction. The time, temperature, gas flow, and concentration (all of which must be accurately controlled) determine the junction depth, width, resistivity, impurity concentration, and general electrical characteristics of the *pn* junction.

An innovation in the diffusion process which has led to the development of integrated circuits, large-scale integration, and other advanced technologies in semiconductor devices is the *planar process*. The term planar refers to a device in which each of the junctions, emitter-base and collector-base, are brought to a common plane surface. This structure is distinguished from the mesa structure, in which the junctions are terminated at the edge of the layers constituting the device. The planar structure is illustrated in Fig. 7-35. The significance of the planar process is that the *pn* junctions are terminated and protected beneath a silicon oxide layer. Thus many of the surface problems associated with other types of transistor fabrication techniques, i.e., high leakage currents and poor low-current dc gain, are eliminated.

To improve switching speed, operating frequency, dc characteristics, collector voltage ratings, power dissipation, and reliability, the *epitaxial* collector was introduced to the planar transistor devices, as shown in Fig. 7-35. The epitaxial process provides a means of growing a very thin high-purity single crystal layer of semiconductor material on a very heavily doped crystal wafer of the same type. The epitaxial process can be used for epitaxial-base devices, as well as collector fabrication. The topography and geometry of transistors take many shapes. In power transistors, the geometry is chosen to favor current-handling capability, whereas in small-signal transistors, high-speed operation is the design goal.

TRANSISTOR OPERATION

57. Circuit Models of the Transistor. Performance of the transistor as an active circuit element is analyzed in terms of various small-signal equivalent circuits. The low-frequency T-equivalent circuit (Fig. 7-36) is closely related to the physical structure. This circuit model is used here to illustrate the principle of transistor action. Carriers are injected into the base region by forward current through the emitter-base junction. A fraction α (near unity) of this current is collected. The incremental change in collector current is determined essentially by the current generator αi_e, that is, where i_e is the incremental change of emitter current. The collector resistance r_c in parallel with the current generator accounts for the finite resistance of the reverse-biased collector-base junction. The input impedance is due to the dynamic resistance r_e of the forward-biased emitter-base junction and the ohmic resistance r_b of the base region.

The room-temperature value of r_e is about $26/I_E$ Ω, where I_E is the dc value of emitter current in milliamperes. Typical ranges of other parameters are as follows: r_b varies from tens of ohms to several hundred ohms; α varies from 0.9 to 0.999; and r_c ranges from a few hundred ohms to several megohms. The symbolic representations of an *npn* and *pnp* transistor are shown in Fig. 7-37. The direction of conventional current flow and terminal voltage for normal operation as an active device are indicated for each. The voltage polarities and current for the *pnp* are reversed from those of the *npn*, since the conductivity types are interchanged.

Transistors may be operated with any one of the three terminals as the common, or grounded, element, i.e., common base, common emitter, and common collector. These configurations are shown in Fig. 7-38 for an *npn* transistor.

Common Base. The transistor action shown at the left in Fig. 7-38 is that of the common-base connection whose current gain (approximately equal to α) is slightly less than 1. Even with less than unity current gain, voltage and power amplification can be achieved, since the output impedance is much higher than the input impedance.

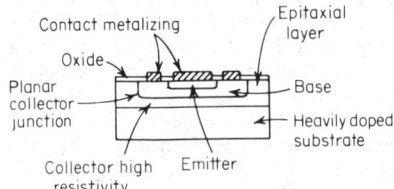

Fig. 7-35. Double-diffused epitaxial planar transistor structure.

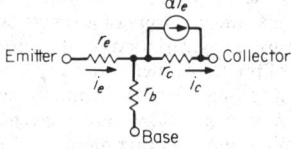

Fig. 7-36. Common-base T-equivalent circuit.

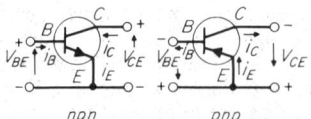

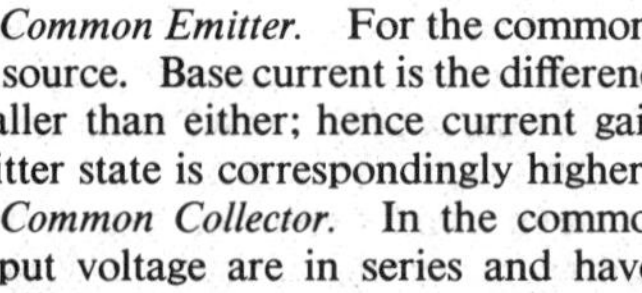

Fig. 7-37. Transistor symbols.

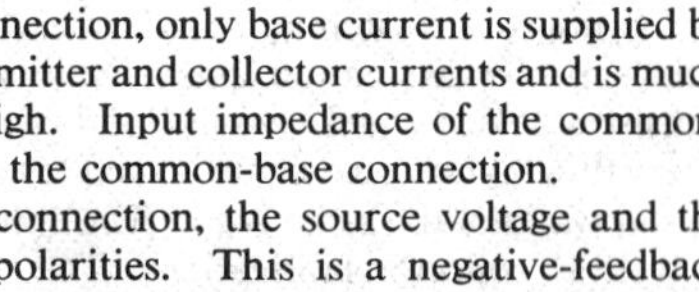

Fig. 7-38. Circuit connections for *npn* transistor.

Common Emitter. For the common-emitter connection, only base current is supplied by the source. Base current is the difference between emitter and collector currents and is much smaller than either; hence current gain I_c/I_b is high. Input impedance of the common-emitter state is correspondingly higher than it is in the common-base connection.

Common Collector. In the common-collector connection, the source voltage and the output voltage are in series and have opposing polarities. This is a negative-feedback arrangement, which gives a high input impedance and approximately unity voltage gain. Current gain is about the same as that of the common-emitter connection. The common-base, common-emitter, and common-collector connections are roughly analogous to the grounded-grid, grounded-cathode, and grounded-plate (cathode-follower) connections, respectively, of the vacuum tube.

h Parameters. Low-frequency performance of transistors is commonly specified in terms of the small-signal h parameters listed in Table 7-16. In the notation system used, the second subscript designates the circuit connection (b for common-base and e for common-emitter). The forward-transfer parameters (h_{fb} and h_{fe}) are current gains measured with the output short-circuited. The current gains for practical load conditions are not greatly different. The input parameters h_{ib} and h_{ie}, although measured for short-circuit load, approximate the

Table 7-16. Transistor Small-Signal h Parameters

	Input parameter	Transfer parameter	Output parameter
Common-base	$h_{ib} \approx r_e + (1-\alpha)r_b$	$h_{fb} \approx \alpha$	$h_{ob} \approx 1/r_c$
Common-emitter	$h_{ie} \approx r_e/(1-\alpha)+r_b$	$h_{fe} \approx \alpha/(1-\alpha)$	$h_{oe} \approx 1/r_c(1-\alpha)$

Table 7-17. h-Parameter Nomenclature

Parameter	Nomenclature	Unit
h_{ib}	Input impedance (common-base)	Ohms (Ω)
h_{ie}	Input impedance (common-emitter)	Ohms
h_{fb}	Forward-current transfer ratio (common-base)	Dimensionless
h_{fe}	Forward-current transfer ratio (common-emitter)	Dimensionless
h_{ob}	Output admittance (common-base)	mhos ($\mho$)
h_{oe}	Output admittance (common-emitter)	mhos

input impedances of practical circuits. The output parameters h_{ob} and h_{oe} are the output admittances.

The current gain of the common-base stage is slightly less than unity; common-emitter current gains may vary from ten to several hundred. Input impedance and output admittance of the common-emitter stage are higher than those of the common-base circuit by approximately h_{fe}. Nomenclature and units for h parameters are given in Table 7-17.

Although matched power gains of the common-base and common-emitter connections are about the same, the higher input impedance and lower output impedance of the common-emitter stage are desirable for most applications. For these reasons, the common-emitter stage is more commonly used. For example, the voltage gain of cascaded common-base stages cannot exceed unity unless transformer coupling is used.

The common-collector circuit has a higher input impedance and lower output impedance than either of the other connections. It is used primarily for impedance transformation.

High-Frequency Limit. The current gain of a transistor decreases with frequency, principally because of the transit time of minority carriers across the base region. The frequency f_T at which h_{fe} decreases to unity is a measure of high-frequency performance. Parasitic capacitances of junctions and leads also limit high-frequency capabilities. These high-frequency effects are shown in the modified equivalent circuit of Fig. 7-39. The maximum frequency $f_{\max}$ at which the device can amplify power is limited by f_T and the time constant $r'_b C_c$, where r'_b is the ohmic base resistance and C_c is that portion of the collector-base junction capacitance which is under the emitter stripe. Values of f_T greater than 2 GHz and $f_{\max}$ exceeding 10 GHz are obtained by maintaining very thin bases ($< 3 \times 10^{-5}$ cm) and narrow emitters ($< 3 \times 10^{-4}$ cm).

58. Transistor Voltampere Characteristics. The performance of a transistor over wide ranges of current and voltage is determined from static characteristic curves, such as the common-emitter output characteristics of Fig. 7-40. Collector current I_C is plotted as a

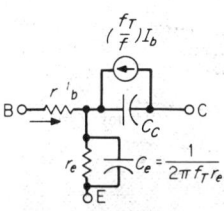

Fig. 7-39. High-frequency common-emitter equivalent circuit.

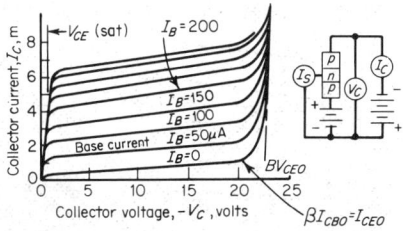

Fig. 7-40. $V_c - I_c$ characteristic for grounded-emitter junction transistor.

function of collector-to-emitter voltage V_C for constant values of base current I_B. Maximum collector voltage for grounded emitter is limited by either punch-through or avalanche breakdown, whichever is lower, depending on the base resistivity and thickness. When a critical electric field is reached, avalanche occurs due to intensive current multiplication. At this point current increases rapidly with little increase in voltage. The common-emitter breakdown voltage BV_{CEO} is always less than the collector-junction breakdown voltage BV_{CBO}. Another characteristic evident from Fig. 7-40 is the grounded-emitter saturation voltage $V_{CE.sat}$. This parameter is especially important in grounded-emitter switching applications.

Two additional parameters, both related to the emitter junction, are BV_{EBO} and V_{BE}. The breakdown voltage emitter-to-base with the collector open-circuited, BV_{EBO}, is the avalanche-breakdown voltage of the emitter junction. The base-to-emitter forward voltage of the emitter junction, V_{BE}, is simply the junction voltage necessary to maintain the forward-bias emitter current.

The leakage current I_{CBO} in the common-base connection is the reverse current of the collector-base junction; common-emitter leakage is higher by the factor $1/(1 - \alpha)$ because of transistor amplification. In either case, the leakage current increases exponentially with temperature. Maximum junction temperatures are limited to about 100°C in germanium and 250°C in silicon. The locus of maximum power dissipation is a hyperbola on the voltampere characteristic curve. Power dissipation must be decreased when higher ambient temperatures exist. Large-area devices and physical heat sinks of high thermal dissipation are used to extend power ratings (see also Par. 7-60).

Dynamic variations of voltage and current are analyzed by a load line on the characteristic curves, as in vacuum tubes. For a linear transistor amplifier with load resistance R_L, the output varies along a load line of slope $- 1/R_L$ about the dc operating point (Fig. 7-41). Since the minimum voltage $V_{CE.sat}$ is quite low, good efficiencies can be obtained with low values of supply voltage. The operating point on the V_{CE}-I_C coordinates is established by a dc bias current in the input circuit. Transistor circuits should be biased for a fixed emitter current rather than a fixed base current to maintain a stable operating point, since the lines of constant base current are variable between devices of a given type and with temperature.

The common-emitter circuit can also be used as an effective switch, as shown by the load line of Fig. 7-42. When the base current is zero, the collector circuit is effectively open-circuited and only leakage current flows in the collector circuit. The device is turned on by applying base current I_{BI}, which decreases the collector voltage to the saturation value.

59. Transistor Applications.   Representative circuit functions are discussed in Secs. 13 to 17. *RC*-coupled amplifiers are used for audio and video applications. Bandwidths in excess of 250 MHz can be achieved with high-frequency transistors. Transistors are used in Class B push-pull amplifiers for high-power linear applications. Since the transistor is a high-current, low-voltage device, low-impedance loads such as speakers and servomotors can be driven without a matching transformer. Use of complementary stages provides versatility of source and load connections not possible with vacuum tubes. The *npn* and *pnp* transistors conduct during alternate half cycles, since they are forward-biased for opposite polarities of the input signal.

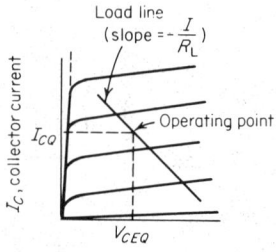

Fig. 7-41. Load line for linear transistor amplifier circuit.

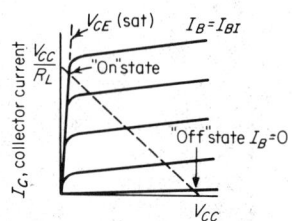

Fig. 7-42. Switching states for common-emitter circuit.

High-frequency applications of transistors include amplifiers, oscillators, and mixers in communications systems. They provide useful power gains at frequencies as high as 5 GHz, with noise performance superior to that of vacuum tubes. High-frequency power transistors used in mobile transmitters supply as much as 20 W at 500 MHz.

Indicated in Fig. 7-43 is a frequency-power curve of silicon rf power transistors available in the early 1970s. Of significance is the rapid falloff of power above 1.0 GHz. Also shown is the theoretical limit for silicon transistors.

60. Transistor Reliability. The life of a transistor in any system should be inherently greater than other electronic components used in the system. However, due to manufacturing defects, poor circuit design, hostile environments, or improper troubleshooting, the life of a transistor may be limited to hours. In the early days of semiconductors, reliability was determined by putting a number of devices on various life tests. Today, this is impractical, due to the hundreds of thousands of hours required to induce failure in high-quality parts. Various accelerated test programs are used in conjunction with long-term life tests to determine failure rates.

Most test programs involve temperature as a stress, to predict the behavior of transistors over a period of time. Temperature is used since the failure rate is related to an exponential function involving the inverse of temperature. Figure 7-44 is an example of data obtained from an accelerated test. The points *A, B,* and *C* are obtained at three different temperatures (junction temperature if the devices were operated, and ambient temperature if tested at elevated storage conditions only). If there are no new failure modes introduced to change the slope of the curve, the line can be extrapolated to anticipated equipment-use conditions, point *D.* See also Pars. **8-108** to **8-116.**

A typical military Group B, Mechanical-Environmental specification requirement, is

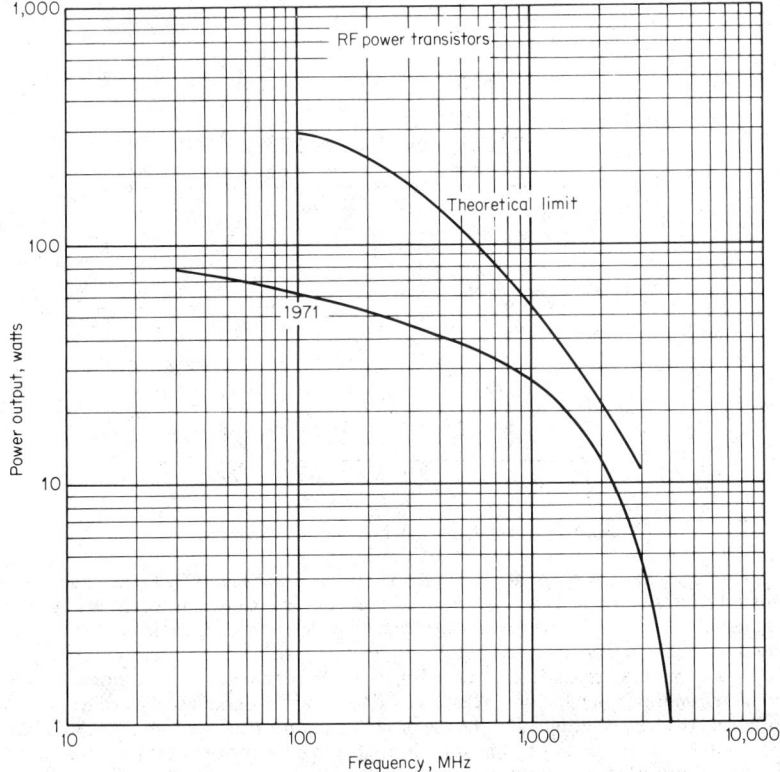

Fig. 7-43. Frequency power curve of available silicon power transistors.

shown in Table 7-18. LTPD, lot tolerance percent defective, assures that a lot will not be accepted that has a proportion defective equal to or greater than the specified LTPD.

Figure 7-45 indicates some of the common packages used today. Packages 1, 2, and 3 are hermetically sealed, having a void between the semiconductor and the package and using seals which are usually metal-to-metal or glass-to-metal. Packages 4 and 5 are constructed of plastic. This type of encapsulation is used, literally, by the hundreds of millions, because of the lower cost. In long-term high-reliability systems, designers have not accepted plastic encapsulation, due to insufficient field reliability data.

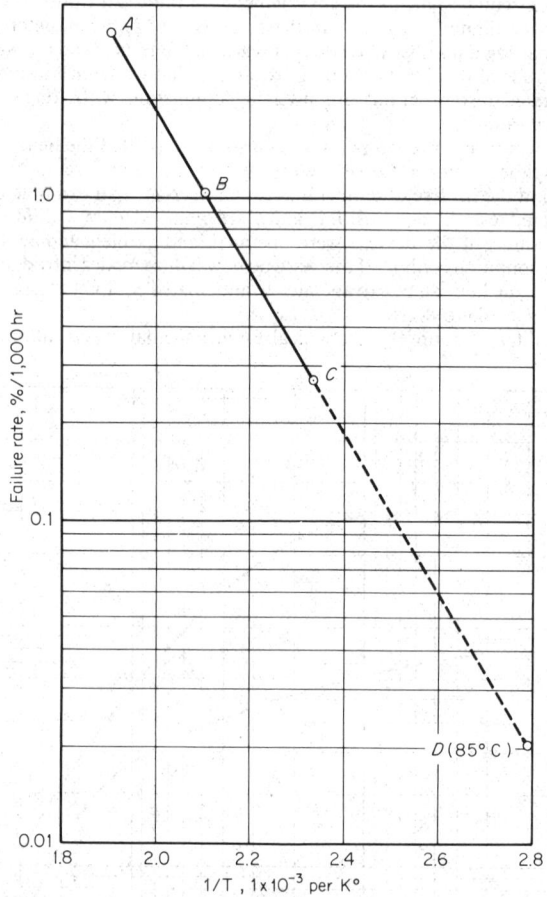

Fig. 7-44. Failure-rate plot as a function of temperature.

Second breakdown, a phenomenon common in power transistors (both audio and rf), is a potentially destructive mode of operation. This condition is caused by localized hot-spotting and thermal runaway. To prevent devices from being operated at conditions which can cause second breakdown, manufacturers specify on their data sheets "safe operating areas." One type of safe operating area is illustrated in Fig. 7-46. In conjunction with safe area, the device *thermal resistance* (θ_{JC}), junction-to-case, is used to avoid hot-spotting. The value of thermal resistance is usually determined by the use of a temperature-sensitive parameter, that is, V_{BE}, whose value as a function of temperature is predictable. However, this technique gives an average indication of the junction temperature. For greater reliability assurance *hot-spot thermal resistance* θ_{HS} is becoming popular. This technique requires measurement of the surface temperature under various operating conditions. The most common tools used for these measurements are the infrared radiometer, thermal graphic phosphors, and liquid crystals.

FIELD EFFECT AND UNIJUNCTION TRANSISTORS

61. Field Effect (Unipolar) Transistors. There are two general types of field effect transistors (FET): junction FETs and metal oxide semiconductor (MOS) transistors. See also Pars. **8-32** to **8-49**.

The cross section of a p-channel junction FET is shown in Fig. 7-47. Channel current is controlled by reverse-biasing the gate-to-channel junction so that the depletion region reduces the effective channel width.

The input impedance of these devices is high because of the reverse-biased diode in the

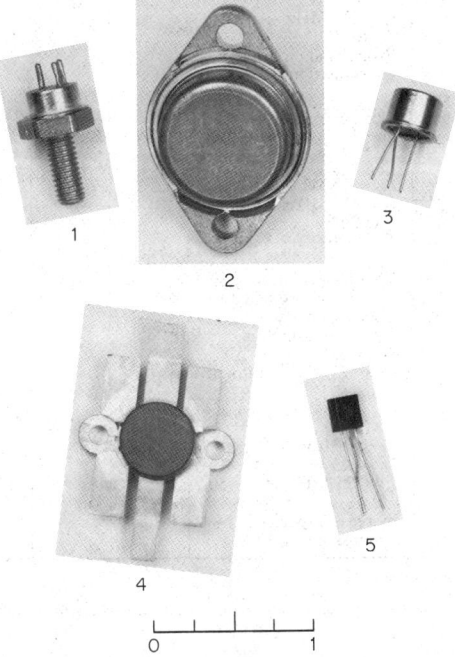

Fig. 7-45. Common transistor packages.

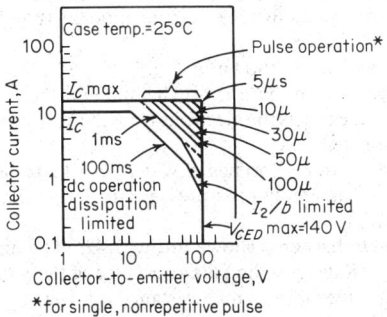

Fig. 7-46. Safe-operating-area chart.

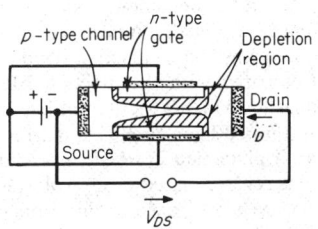

Fig. 7-47. p-Channel junction field effect transistor.

Table 7-18.   Typical Military Group B, Mechanical-Environmental Specification Requirement

(Group B Inspection)

Examination or test	Conditions	LTPD*	Symbol	Min.	Max.	Unit
Subgroup 1 .... Physical dimensions ...	... MIL-STD-750	20				
Subgroup 2 .... Soldering temperature cycling	... -65 to $+200°C$; 5 cycles†	10				
Subgroup 3 .... Shock	... Nonoperating 5 blows in each orientation X_1, Y_1, Y_2, and Z_1 (total of 20 blows)	10				
Subgroup 4 .... Rf hot-spot thermal ... resistance‡	$V_{CC} = 35$ V $f = 225$ MHz T_C 100°C Broadband circuit; 50-Ω termination	10 ...	Q_{Rf}	...	5.0	°C/W
Subgroup 5 .... Storage life	... Tstg = 200°C 1,000 h	10				
Subgroup 6 .... Rf operation life	... $T_c = 85°C$ $V_{CE} = 25\ ^{.3}_{-0}$ V $P_0 = 25$ W, cw $f = 225$ MHz, $t = 1,000$ h $R_L = 50\ \Omega$	10				
Subgroup 7 .... VSWR test	... VSWR = 10:1 Phase angle adjusted to achieve max. device power dissipation $T_C = 55°C$, $t = 1$ h $P_0 = 25$ W, cw, into 50 Ω $V_C = 28$ V dc, $f = 225$ MHz	20				
Peak rf voltage capability	$V_{CC} = 30$ V dc $P_0 = 25$ W $f = 225$ MHz, $B_L = 50$ Ω Increase V(peak), rf by increasing rf drive	...	$V_{(peak)rf}$	120	...	Volts

* LTPD = lot tolerance percent defective.
† MIL-STD-202, Method 102A.
‡ θ_{rf} is defined as the thermal resistance under rf operating conditions between the hottest spot on the chip to the case.

input circuit.   In fact, the voltampere characteristics are quite similar to those of a vacuum tube.   Another important feature of the junction FET is the excellent low-frequency noise characteristics, which surpass those of either the vacuum tube or conventional (bipolar) transistor.

The cross section of the p-channel MOS transistor is shown in Fig. 7-48.   This device operates in the depletion mode.   For zero gate voltage there is no channel and the drain current is very small.   A negative voltage on the gate repels the electrons from the surface and produces a p-type conduction region under the gate.

Compared with the junction FET, the MOS transistor has a wider gain-bandwidth product and a higher input impedance ($> 10^{11}\ \Omega$).   Its simple structure is particularly suitable for integrated logic circuits and arrays.   See Sec. **8**.

62. Unijunction Transistors.   A unijunction transistor is shown in Fig. 7-49.   The input diode is reverse-biased at low voltages owing to IR drop in the bulk resistance of the n-type region.   When V_E exceeds this drop, carriers are injected and the resistance is lowered.   As a result, the IR drop and V_E decrease abruptly.   The negative-resistance characteristic is useful in such applications as oscillators and as trigger devices for silicon controlled rectifiers.

BATTERIES AND FUEL CELLS

By D. Linden

PRINCIPLES OF OPERATION

63. Electrochemical Principles and Reactions. A battery is a device which converts the chemical energy contained in its active materials directly into electrical energy by means of an oxidation-reduction electrochemical reaction. This type of reaction involves the transfer of electrons from one material to another. In a nonelectrochemical reaction this transfer of electrons occurs directly and only heat is involved. In a battery (Fig. 7-50) the negative electrode or anode is the component capable of giving up electrons, being oxidized during the reaction. It is separated from the oxidizing material, which is the positive electrode or cathode, the component capable of accepting electrons. The transfer of electrons takes place in the external electric circuit, connecting the two materials, and in the electrolyte, which serves to complete the electric circuit in the battery by providing an ionic medium for the electron transfer.

64. Fuel Cells. The operation of the fuel cell is similar to that of a battery except that one or both of the reactants are not permanently contained in the electrochemical cell, but are fed into it from an external supply when power is desired. The fuels are usually gaseous or liquid (compared with the metal anodes generally used in batteries), and oxygen or air is the oxidant.

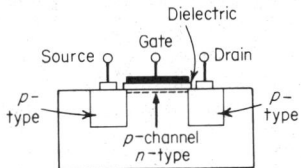

Fig. 7-48. *p*-channel MOS transistor.

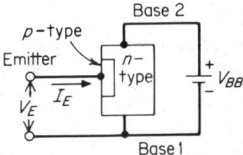

Fig. 7-49. Unijunction transistor.

65. Theoretical Cell Voltage and Capacity. The theoretical capacity (ampere-hours) of a battery system is determined by its active materials; the maximum electrical energy (watthours) corresponds to the free-energy change of the reaction. The theoretical voltage and ampere-hour capacities of a number of electrochemical systems are given in Table 7-19. The voltage is determined by the active materials selected; the ampere-hour capacity is determined by the amount (weight) of available reactants. One gram-equivalent weight of material will supply 96,500 C, or 26.8 Ah, of electrical energy.

66. Factors Influencing Battery Voltage and Capacity. In actual battery practice, only a small fraction of the theoretical capacity is realized. This is due not only to the presence of nonreactive components (containers, separators, electrolyte) that add to the weight of the battery, but to many other factors that prevent the battery from performing at its theoretical level. Factors influencing the voltage and capacity of a battery are as follows.

Voltage Level. When a battery is discharged in use, its voltage is lower than the theoretical voltage. The difference is caused by *IR* losses due to cell resistance and polarization of the active materials during discharge. This is illustrated in Fig. 7-51. The theoretical discharge curve of a battery is shown as curve 1. In this case, the discharge of the battery proceeds at the theoretical voltage until the active materials are consumed and the capacity fully utilized. The voltage then drops to zero. Under load conditions, the discharge curve is similar to curve 2. The initial voltage is lower than theoretical, and it drops off as the discharge progresses.

The Current Drain of the Discharge. As the current drain of the battery is increased, the *IR* loss increases, the discharge is at a lower voltage, and the service life of the battery is reduced (curve 5). At extremely low current drains it is possible to approach the theoretical capacities (in the direction of curve 3). In a very long discharge period, chemical self-deterioration during the discharge becomes a factor and causes a reduction of capacity.

Table 7-19. Characteristics of Batteries and Fuel Cells

System	Anode	Cathode	Theoretical battery		Practical battery		
			Voltage, V	Capacity, Ah/kg	Typical voltage, V	Capacity[a]	
						Wh/kg	Wh/dm³
Primary:							
Leclanche	Zn	MnO_2	1.6	230	1.2	65	175
Magnesium	Mg	MnO_2	2.0	270	1.5	100	195
Organic cathode	Mg	m-DNB	1.8	1,400	1.15	130	180
Alkaline MnO_2 .	Zn	MnO_2	1.5	230	1.15	65	200
Mercury	Zn	HgO	1.34	185	1.2	80	370
Mercad	Cd	HgO	0.9	165	0.85	45	175
Silver oxide	Zn	AgO	1.85	285	1.5	130	310
Zinc-air	Zn	Air (O_2)	1.6	815	1.1	200	190
Li-organic electrolyte	Li	b	2.1-5.4	130-660	1.8-3.2	250	400
Secondary:							
Lead-acid	Pb	PbO_2	2.1	55	2.0	37	70
Edison	Fe	Ni oxides	1.5	195	1.2	29	65
Nickel-cadmium	Cd	Ni oxides	1.35	165	1.2	33	60
Silver-zinc	Zn	AgO	1.85	285	1.5	100	170
Silver-cadmium .	Cd	AgO	1.4	230	1.05	55	120
Zinc-nickel oxide	Zn	Ni oxides	1.75	185	1.6	55	110
Zinc-air[c]	Zn	Air (O_2)	1.6	815	1.1	150	155
Cadmium-air[c]	Cd	Air (O_2)	1.2	475	0.8	90	90
Zinc-O_2	Zn	O_2	1.6	610	1.1	130	120
H_2-O_2	H_2	O_2	1.23	3,000	0.8	45	65
Reserve:							
Cuprous chloride	Mg	CuCl	1.5	240	1.4	45	65[d]
Silver-chloride ..	Mg	AgCl	1.6	170	1.5	60	95[d]
Zinc-silver oxide	Zn	AgO	1.85	285	1.5	30	75[e]
Thermal	Ca	b	2.8	240	2.6	10	20[f]
Ammonia-activated	Mg	m-DNB	2.2	1,400	1.7	22	60[g]
High-	Na	S	2.1	685	1.8	200	h
temperature ..	Li	S	2.2	1,150	1.8	200	i
Solid electrolyte	Ag	Polyiodide	0.66	...	0.6	180	75[j]
Fuel cell:							
Hydrogen[c]	H_2	Air	1.23	26,000	0.7		
Hydrazine[c]	N_2H_4	Air	1.5	2,100	0.7	800[b]	l
Methanol[c]	CH_2OH	Air	1.3	1,400	0.9	175[b]	185[m]

[a] Delivered capacity when discharged at normal temperatures (20°C) at normal discharge rates.
[b] Based on fuel consumption only.
[c] Weight of air not considered in computation of watt hours.
[d] Water-activated.
[e] Automatically activated; high rate discharge; 2- to 10-min rate.
[f] Fused salt; heat-activated; high rate discharge; 2- to 10-min rate.
[g] Four-minute discharge rate.
[h] β-alumina electrolyte, 300°C operation.
[i] Fused salt; 350°C operation.
[j] Solid $RbAg_4I_5$ electrolyte.
[k] Several different cathode materials used.
[l] Fuel consumption is based on source of H_2.
[m] Based on methanol battery-fuel supply in situ.

Voltage Regulation.   The voltage regulation required by the equipment is most important. As is apparent by the curves in Fig. 7-51, design of equipment to operate to the lowest possible end voltage results in the highest capacity and longest service life. Similarly, the upper voltage limit of the equipment should be established to take full advantage of the battery characteristics. In some applications, where only a narrow voltage range can be tolerated, voltage regulators may have to be employed to take full advantage of the battery's capacity. If a storage battery is used in conjunction with another energy source, which is permanently connected in the operating circuit, allowances must be made for the voltage required to charge the battery, as illustrated in curve 7, Fig. 7-51. The maximum voltage must just exceed the maximum voltage on charge.

The Type of Discharge (Continuous, Intermittent, etc.).   When a battery stands idle after a discharge, certain chemical and physical changes take place which can result in voltage recovery. Thus the voltage of a battery which has dropped during a heavy discharge will rise after a rest period, giving a sawtooth-shaped discharge. Curve 6 of Fig. 7-51 shows the characteristic of a battery discharged intermittently at the same drain as curve 2. The improvement resulting from the intermittent discharge depends on the current drain, length of the recovery period, discharge temperature, end voltage, and the particular battery system and design employed. Some battery systems, during inactive stand, develop a protective film on the active-material surface. These batteries, instead of showing a recovery voltage, may momentarily demonstrate a lower voltage after stand until this film is broken by the discharge. This is known as a *voltage delay*.

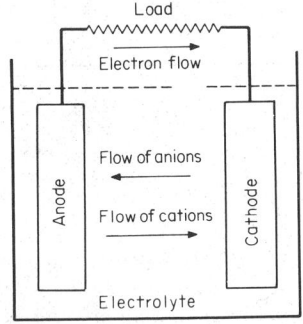

Fig. 7-50. Electrochemical operation of a battery.

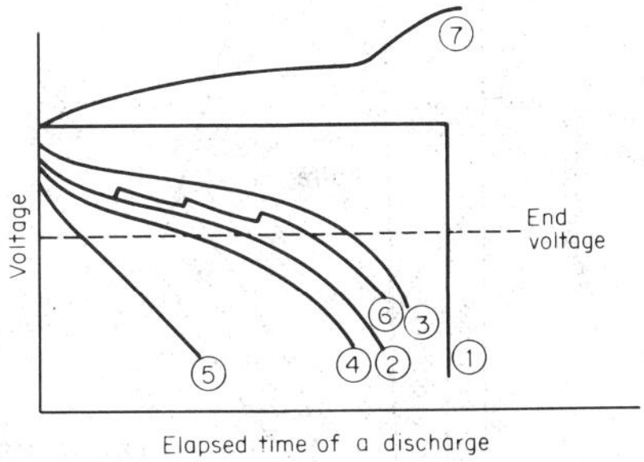

Fig. 7-51. Battery discharge characteristics.

Temperature of the Battery During Discharge. The temperature at which the battery is discharged has a pronounced effect on its service and voltage characteristics. This is due to the reduction in chemical activity and the increase in battery internal resistance at lower temperatures. This is illustrated by curves 3, 2, 4, and 5 of Fig. 7-51, representing discharges at the same current drawn at progressively reduced temperatures. The specific characteristics vary for each battery system and discharge rate, but generally best perfor-mance is obtained between 20 and 40°C. At higher temperatures, chemical deterioration may be rapid enough during discharge to cause a loss in capacity.

Effect of Size on Capacity. Battery size influences the voltage characteristics by its effect on current density. A given current drain may be a severe load on a small battery, giving a discharge similar to curve 4 or 5 (Fig. 7-51), but be a mild load to a larger battery, and give a discharge similar to curve 3. It is possible often to obtain more than a proportional increase in the service life by increasing the size of the battery. The absolute value of current, therefore, is not the key influence, although its relation to the size of the battery, i.e., the current density, is important.

The Age and Storage Condition of the Battery. Batteries are a perishable product and deteriorate as a result of chemical action that proceeds during storage. The type of battery, design, temperature, and length of storage period are factors which affect the shelf life of the battery. Since self-discharge proceeds at a lower rate at reduced temperatures, refrigerated storage extends the shelf life. Refrigerated batteries should be warmed before discharge to obtain maximum capacity.

CLASSIFICATION OF CELLS AND BATTERIES

67. General Characteristics. The many and varied requirements for battery power and the multitude of environmental and electrical conditions under which they must operate necessitate the use of a number of different battery types and designs, each having superior performance under certain discharge conditions. Figure 7-52 lists the various batteries and identifies the power level and operational time in which each excels. The key theoretical and performance characteristics of the major primary and secondary batteries and fuel cells are listed in Table 7-19.

68. Types of Batteries. Batteries are generally identified as primary or secondary.

Primary batteries are not capable of being easily recharged electrically, and hence are used or discharged a single time and discarded. Many of the primary batteries, in which the electrolyte is contained by absorbent or separator materials (i.e. there is no free or liquid electrolyte), are termed *dry cells.*

Secondary batteries are those which are capable of being recharged electrically, after discharge, to their original condition by passing current through them in the opposite direction to that of the discharge current. They are electrical-energy storage devices and are known also as storage batteries or accumulators.

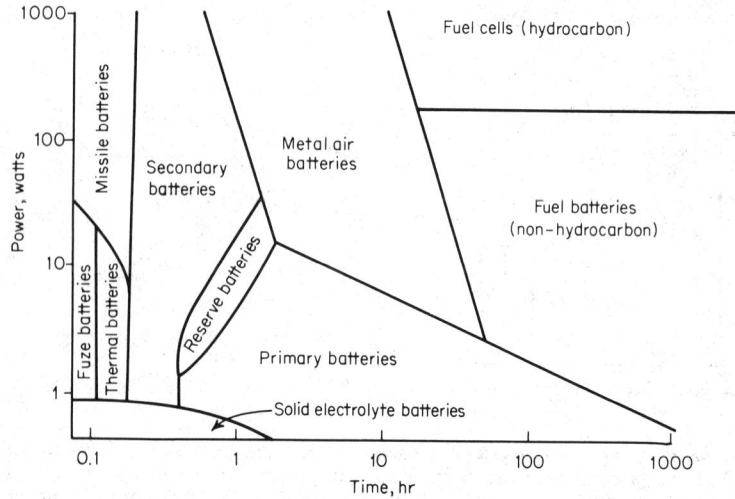

Fig. 7-52. Optimum power level and operational time of batteries.

Reserve batteries are primary types in which a key component is separated from the rest of the battery prior to activation. In the inert condition, the battery is capable of long-term storage. Usually, the electrolyte is the component that is isolated; in other systems, such as the thermal battery, the battery is inactive until it is heated, melting a solid electrolyte, which then becomes conductive.

PRIMARY BATTERIES

69. Leclanche Battery (Zn-MnO$_2$). The Leclanche dry cell, known for over a hundred years, is still the most widely known and used of all the dry-cell batteries.

The most common construction (Fig. 7-53) uses a cylindrical zinc container as the negative electrode, and manganese dioxide as the positive element, made externally available through a carbon electrode. A paste or paper separator separates the two electrodes. Another common design is the flat cell, used only in multicell batteries, which offers better volume utilization. The Leclanche cell is fabricated in a number of sizes of varying diameter (or cross section) and height.

Service Life. Typical discharge curves for the Leclanche dry cell are shown in Fig. 7-54.

Shelf Life. Figure 7-55 shows the capacity retention (shelf life) of the Leclanche battery at different temperatures. These data point out the advantage of storing batteries at low temperatures for preserving their capacity.

Low Temperature. The capacity of the standard Leclanche cell falls off rapidly with decreasing temperature and is essentially inoperative below $-20°C$. A low-temperature battery, using a low-freezing-point electrolyte and a battery design minimizing cell resistance, has been developed, but has not achieved popularity because its overall characteristics are not satisfactory. To obtain useful service at low ambient temperatures, batteries should be kept warm. The vest battery has proved effective, for example, in military applications. This

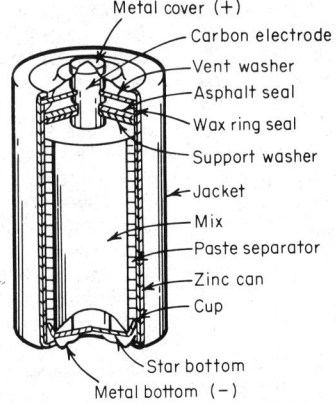

Fig. 7-53. Cross section of a Leclanche cylindrical cell.

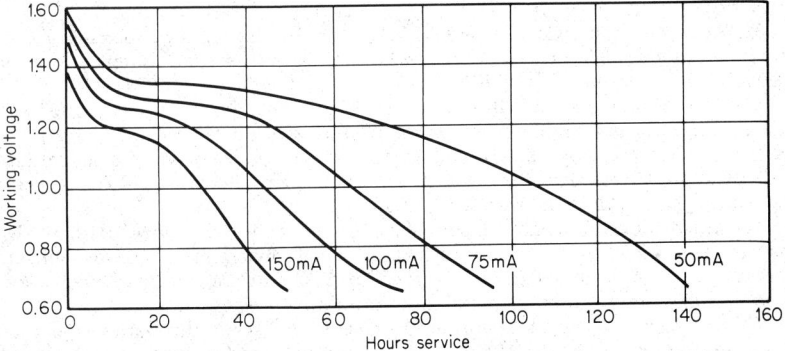

Fig. 7-54. Typical discharge curves for Leclanche dry cell (size D).

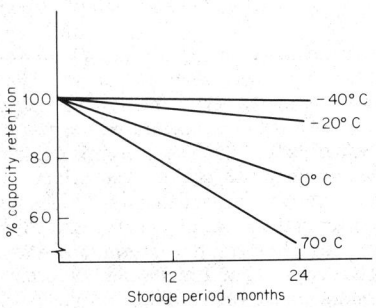

Fig. 7-55. Capacity retention of Leclanche cells.

battery is worn under the user's clothing and utilizes body heat to maintain the battery at a satisfactory operating temperature.

70. Magnesium Dry-Cell Batteries ($Mg-MnO_2$). The magnesium battery was developed for military use and has two principal advantages over the zinc dry cell: (1) it has twice the capacity or service life of an equivalently sized zinc cell, and (2) it has the ability to retain this capacity during storage, even at elevated temperatures (Fig. 7-56). The construction of the magnesium dry cell is similar to that of the cylindrical zinc cell, except that a magnesium can is used instead of the zinc container. The magnesium cell has a mechanical vent for the escape of hydrogen gas, which forms as a result of a parasitic reaction during the discharge of the battery. Magnesium batteries have not been fabricated successfully in flat-cell designs.

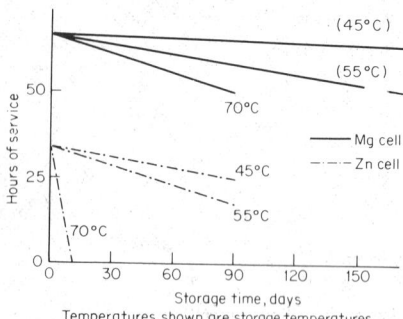

Fig. 7-56. Comparison of magnesium vs. Leclanche (zinc) dry cell.

The good shelf life of the magnesium battery results from a film which forms on the inside of the magnesium can, preventing corrosion. This film, however, is responsible for a delay action, a delay in the battery's ability to deliver full output voltage after it is placed under load. The delay is usually less than 0.3 s, but can be longer at low temperatures and high current drains.

71. Zinc-Mercuric Oxide Battery (Zn-HgO). The zinc-alkaline-mercuric oxide battery is noted for its high capacity per unit volume, a relatively constant output voltage during its discharge, and good storageability.

The zinc-mercuric oxide cell is constructed in a sealed but vented structure, with the active materials balanced to prevent formation of hydrogen in a discharged battery. Three basic structures are used: the wound-anode, the flat-pressed powdered-anode, and the cylindrical-pressed powdered-electrode types. The cell is available in a number of sizes, from the miniature 16-mAh to the largest 14-Ah size.

The general discharge characteristics of the mercury battery are shown in Fig. 7-57. The mercury battery is suited to use at normal and elevated temperatures. The low-temperature performance is poor, particularly under heavy loads, and its use at temperatures below 0°C is not recommended, except in special low-temperature designs.

The mercury cell has good storage characteristics. While this varies with cell size, capacity retention is about 90% after 2 years storage at 20°C and over 80% after 1 year at 45°C.

72. Alkaline-MnO_2 Battery (Zn-MnO_2). This battery uses the same electrochemically active materials as the Leclanche cell, zinc and manganese dioxide. It differs in construction and in the electrolyte, and thus has a lower internal resistance. Its advantage on low-rate or intermittent discharge is marginal, but at high- or continuous-drain conditions, it can deliver from 2 to 10 times more ampere-hours of service than the Leclanche cell. Its operation at low temperature is superior to other available commercial dry batteries, giving satisfactory performance to −25°C.

73. Cadmium-Mercuric Oxide (Cd-HgO). The cadmium-alkaline-mercuric oxide battery is similar in design to the zinc-mercuric oxide battery. The substitution of cadmium for zinc lowers the cell voltage but gives a very stable system, with a predicted shelf life of up to 10 years, as well as performance at low temperatures. Its watthour capacity, because of the lower voltage, is about 60% of that of the zinc-mercuric oxide battery.

74. Zinc-Silver Oxide Battery (Zn-AgO). The primary zinc-silver oxide battery is similar in design to the zinc-mercuric oxide battery, but uses silver oxide in place of mercuric oxide. This gives it a higher cell voltage and energy capacity, which makes it a desirable battery for use in hearing aids, photographic applications, and electronic watches.

75. Lithium Primary Batteries. The lithium primary battery is a new development using nonaqueous organic or inorganic solvents rather than aqueous electrolytes. Cell voltages range between 2 and 3 V, depending on the cathode material, and energy densities up to 200 Wh/kg can be obtained at 20°C with performance extending down to −30°C. The most advanced cell, which is in production, uses the lithium/organic electrolyte (SO_2) electrochemical system. Cells using inorganic solvents, for example, thionyl chloride (which acts as the solvent as well as the cathode material), have yielded capacities in excess of 400 Wh/kg on an experimental basis.

76. Air Batteries. The use of air as a depolarizer (oxidizer) results in a primary battery which possesses high energy density. This system, using bulky zinc anodes, a carbon-air electrode, and a potassium hydroxide electrolyte in a glass jar, has been used successfully for railway signals and similar applications. Miniature designs, employing thin Teflon-coated fuel-cell-type electrodes, are now being developed. These batteries can give up to 220 Wh/kg, but moisture loss requires careful packaging during storage and limits service time after the battery is put into service.

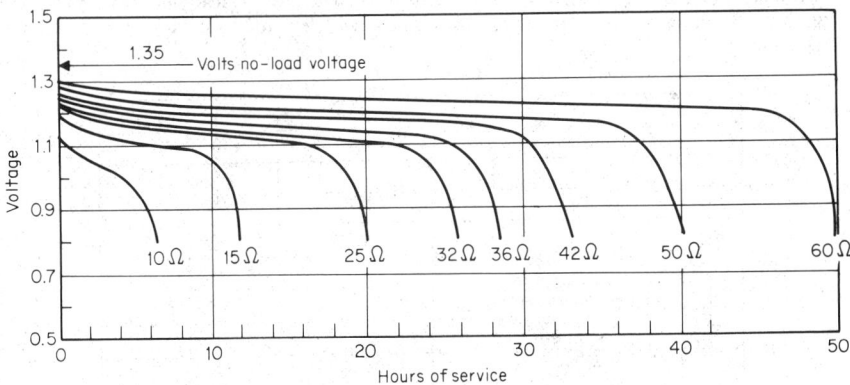

Fig. 7-57.   Discharge curves for the zinc-HgO cell.

77. Other Primary Batteries. Many other electrochemical systems have been used as primary batteries to obtain special performance characteristics. The more prominent ones are listed in Table 7-19.

78. Hybrid Configuration. In the hybrid configuration a high-energy-density primary battery is combined with a high-power-density secondary battery to improve the overall battery performance. The secondary battery handles the high-power-pulse and low-temperature-performance requirements (where the primary battery is most inefficient) and is recharged by the primary battery. The primary battery is discharged under a light load, i.e., to provide its highest energy output. This design is of particular interest to obtain high efficiency under adverse environmental and load conditions.

79. Recharging Primary Batteries. Most primary batteries can be recharged for a small number of cycles under controlled conditions. With Leclanche cells the charging must be done at a slow rate and on freshly discharged cells which have not been discharged below 1.0 V per cell. The cells must be returned to service soon after recharging, since recharged cells have poor shelf life.

Recharging dry cells is usually impractical and can be hazardous, particularly with cells that are not properly vented to eliminate the gases that form during the charge.

SECONDARY (STORAGE) BATTERIES

80. General. For many applications, it is more economical or practical to employ a secondary battery that can be recharged to its original condition rather than continuously replace primary batteries. The storage battery also is ideal for emergency or no-break power because it can be kept fully charged, in a state of readiness, during periods of normal operation.

81. Lead-Acid Battery. This is the most widely used and economical secondary battery. It uses sponge lead for the negative electrode, lead oxide for the positive, and a sulfuric acid solution for the electrolyte. As the cell discharges, the active material is converted into lead sulfate, and the sulfuric acid solution is diluted; that is, its specific gravity decreases. On charge the reverse actions take place. The state of charge of the battery can be determined by measuring the specific gravity.

The lead-acid battery is manufactured in many sizes, ranging from small plastic encased batteries of less than 1 Ah capacity to large automotive and stationary units with hundreds of ampere-hours of capacity. The most common construction is the *pasted-plate design*, where the lead oxides are pasted onto a flat antimonial lead grid. Maintenance-free designs, in which the electrolyte is immobilized and loss of water is minimized, have been introduced and are particularly advantageous in portable applications.

The general performance characteristics of the lead-acid battery are given in Fig. 7-58. Several limitations of the lead-acid battery are its poor low-temperature characteristics (particularly its inability to accept a charge at low temperatures), its loss of capacity on stand (self-discharge), and its relatively weak mechanical structure. A lead-acid battery, left in a discharged condition for more than six months, will become "sulfated" and difficult to recharge.

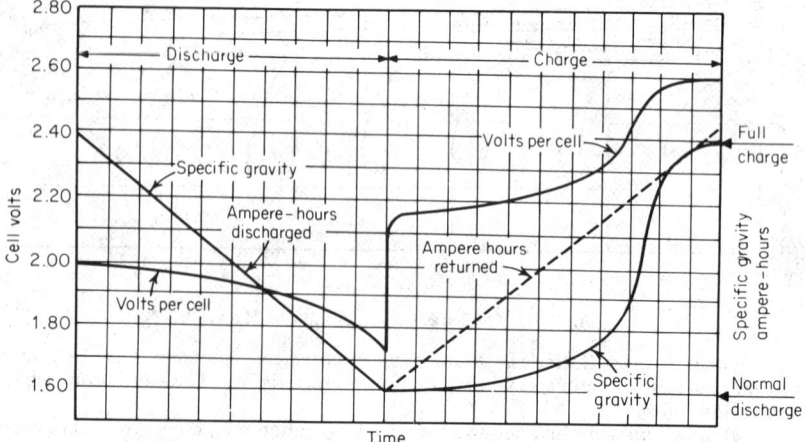

Fig. 7-58. Typical voltage and gravity characteristics of a lead-acid battery (constant rate discharge and charge).

Lead-acid batteries, since they cannot be maintained indefinitely in a charged condition, are usually stored *dry-charged*. In this condition, they can retain their charge for as long as 2 years and can be put into service by adding the acid electrolyte. Recently, a water-activated battery, which contains the sulfuric acid in dry form and only requires the addition of water, has been introduced into the market.

A lead-acid battery can be charged at any rate that does not produce excessive gasing or high temperatures. The most common practice for recharging a fully discharged battery is to start the charge at the $C/5$ rate (amperes, corresponding to 1/5 the rated capacity of the battery) for 5 h, which will return about 80 to 85% of the battery's capacity. The charging current is then tapered to about one-half to complete the charge. In an emergency, "fast" or "boost" charging is used. In this type of charge, the current should not exceed the C rate because the battery can suffer damage if this rate is exceeded.

The battery can also be *float-* or *trickle*-charged when it is continuously connected to an electrical system. The current should be regulated at a low level to maintain the battery in a charged condition (sufficient just to replace capacity loss due to stand) and to prevent overcharging. The battery manufacturers can supply detailed charging instructions.

82. Nickel-Cadmium Batteries. The major alkaline secondary battery is the nickel-cadmium battery, which is noted for high power capability, long cycle life, good low-temperature performance, ruggedness, and reliability. It uses nickel oxide for the positive electrode, cadmium for the negative, and a solution of potassium hydroxide for the electrolyte.

The *pocket-type vented batteries* are used primarily for standby service or emergency lighting, with the battery maintained in a fully charged condition by trickle charging. Their long life and low cost make them ideally suited for these applications.

The *sintered-plate vented battery* is a more recent development. It is more expensive than the pocket type, but gives better performance at high discharge rates and at low temperatures due to its lower resistance. The characteristics of the vented sintered-plate nickel-cadmium battery are given in Fig. 7-59.

The nickel-cadmium battery requires little maintenance and can take considerable abuse. Unlike the lead-acid battery, it can be stored in either the charged or discharged condition without damage. The self-discharge rate is very low, tapering off to about 3%/month, with a capacity loss of less than 50% during a year's storage at normal temperatures. Cycle life is excellent—over 1,000 cycles on deep discharges; many more on partial discharges. The battery can be rapidly recharged. Reasonable overcharging has no detrimental effect, and float charging for extended periods of time is permissible, although overheating the battery should be avoided.

In the sealed construction, the nickel-cadmium battery is completely sealed so that water cannot be lost through evaporation or as a result of gasing during charge.

83. Zinc-Silver Oxide Battery. The zinc-potassium hydroxide–silver oxide battery is noted for its high capacity per unit weight and volume and the ability to deliver this capacity at high current drains (Fig. 7-60). It is useful in applications where high energy density is a prime requisite. The battery is not recommended for general storage battery use because its cost is high, its performance at low temperatures falls off more markedly than other storage batteries, and its cycle life is limited. Charging can be accomplished efficiently by normal methods, but it is necessary to limit overcharging due to the growth of zinc dendrites which can short-circuit the cell internally.

84. Other Alkaline Secondary Batteries. There are a large number of electrochemical systems which can perform as secondary batteries, other than those described above. Of these, the iron–nickel oxide Edison cell was an important battery for traction and standby applications. The cadmium–silver oxide battery, while expensive, has better cycle-life and low-temperature performance than the zinc–silver oxide battery, but has a lower energy density. The zinc–nickel oxide battery now under development should have an energy

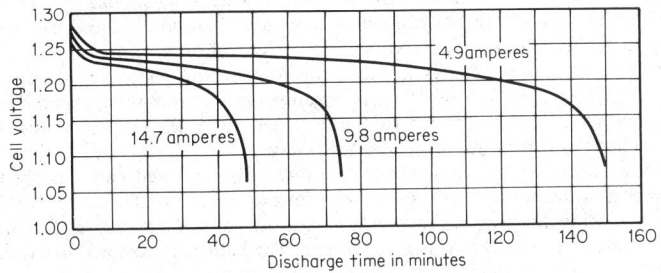

Fig. 7-59. Discharge curves for vented sintered-plate nickel-cadmium battery (12.5 Ah).

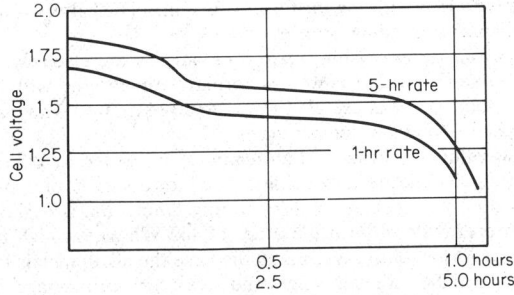

Fig. 7-60. Discharge curves for the zinc-AgO battery.

7-53

DISCRETE CIRCUIT COMPONENTS

density twice that of the nickel-cadmium battery and a lower cost; however, it will be difficult to achieve long cycle life with this system.

85. Mechanically Rechargeable Batteries. The mechanically rechargeable battery is a new type of power source using air cathodes and a replaceable metal anode. The discharged or spent anode is replaced, and the battery thus "recharged." The key advantages of this system are (1) high capacity—200 Wh/kg and (2) rapid (nonelectrical) recharging.

Each cell of the battery consists of an air-permeable, watertight container formed by two air electrodes. The anode (zinc), containing the KOH electrolyte in solid form, is inserted into the cell container after water is added. Figure 7-61 presents the discharge characteristics of this battery at various continuous rates.

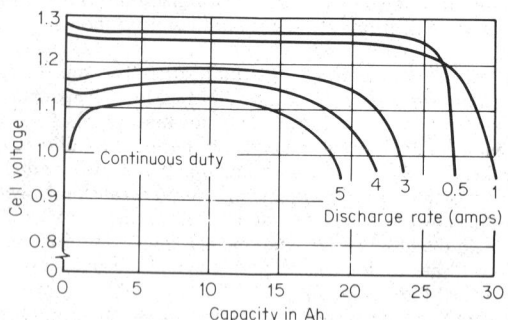

Fig. 7-61. Discharge curves for the zinc-air battery.

RESERVE AND SPECIAL BATTERIES

86. Reserve Batteries. Many batteries use highly active materials to obtain the high-energy content and power levels that are required, and are designed in a reserve construction to withstand long or severe environmental storage conditions. These batteries are used primarily to deliver high power for very short periods of time.

An important reserve battery is the *automatically activated zinc-silver oxide battery* used for missiles. The electrolyte of this battery is contained in a copper tube, sealed by a metal-foil diaphragm and activated by firing a gas squib which forces the electrolyte into the battery. The battery is fully discharged within 10 to 20 min.

The most prominent water or electrolyte manually activated batteries are the magnesium-silver chloride and magnesium-cuprous chloride systems. These are used for weather balloons or marine applications (sonobuoys, radios) to take advantage of their good low-temperature capability and high energy content. The battery is used within a few hours after activation.

The *thermal, or heat-activated, battery* employs a solid nonconducting electrolyte when the battery is inactive. The cell is activated by heating it sufficiently high to melt the electrolyte, making it conductive and permitting the flow of current. The battery is used in applications requiring its full capacity in minutes or less.

The *liquid-ammonia battery,* using liquid ammonia as the electrolyte, is an interesting system, being operable in the cold as well as normal temperatures, with little change in cell voltage and energy output. It is useful at the 1- to 20-min rate. The major problem is the volatilization of gas at the higher temperatures.

87. High-Temperature Systems. High-temperature batteries, operating at fused-salt temperatures (300 to 500°C), are now being studied for a variety of applications, including electric-vehicle propulsion. At the high operating temperatures it should be possible to achieve both high energy density (in the order of 400 Wh/kg) at high power densities (in excess of 400 W/kg). Most of the systems studied use the alkali metals lithium and sodium for the negative because they are lightweight and give a high-cell voltage. Similarly, the best positive materials are the halogens (F_2, Cl_2) or the chalcogens (Te, Se, S). The electrolytes are nonaqueous because of the reactivity of the alkali metals with water and are either fused salts or conducting glasses, such as β-alumina.

88. Solid Electrolyte Batteries. Most batteries depend on the ionic conductivity of liquid electrolytes for their operation. The solid electrolyte batteries depend on the ionic conductivity of an electronic nonconducting insulative salt in the solid state, for example, Ag^+ ion mobility in silver iodide. Cells designed to use this solid electrolyte have a long shelf life and the capability of operating over a wide temperature range. Early battery designs were limited to low current densities in the microampere range, but the discovery of new highly conductive electrolytes has stimulated investigation in higher-power and higher-capacity systems, in both primary and rechargeable modes.

FUEL CELLS

89. General. The potentially most efficient of the newer approaches to electric-power generation is the fuel cell. It permits conversion of liquid or gaseous fuels electrochemically, without the thermal losses typical of the internal-combustion engine and the *Carnot cycle* limitations. It combines low fuel consumption and silent operation.

The operation of a fuel cell is shown graphically in Fig. 7-62. The basic fuel cell consists of an anode ($-$) and cathode ($+$) electrode separated by a conducting electrolyte, such as potassium hydroxide. A fuel, e.g., hydrogen, is fed to the negative electrode, where it is oxidized, releasing electrons to the load. Air or oxygen is fed to the cathode, where it collects the electrons and is reduced. The circuit is completed by ionic conduction through the electrolyte. Water is the reaction product in the hydrogen-air cell.

The ultimate objective is a direct fuel cell using conventional hydrocarbon fuels which would have a conversion efficiency in excess of 65%. A number of technological developments must be achieved, however, before such a device will be practical. Typical approaches to fuel-cell design are listed in Fig. 7-63.

90. Direct-Fuel-Cell Systems. Hydrazine is a particularly effective fuel for portable applications, since it lends itself readily to direct oxidation. Its energy content is only about 18,000 Btu/kg, and its cost is relatively high. The 60-W hydrazine fuel cell contains three components (the fuel-cell module, the power conditioner, and the hydrazine fuel tank) weighs 7 kg; and occupies 0.01 cm^3. A secondary battery is mated to the fuel cell in a hybrid

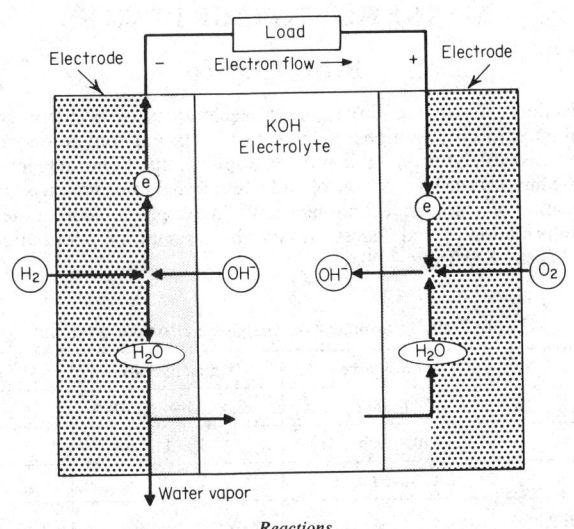

Reactions

Anode
$$H_2 + 2OH^- \rightarrow 2H_2O + 2e$$

Cathode
$$2e + \tfrac{1}{2}O_2 + H_2O \rightarrow 2OH^-$$

Fig. 7-62. Operation of a fuel cell.

configuration to handle peak power loads. The hydrazine fuel cell will deliver over 800 Wh/kg of fuel; the advantage of the performance over battery systems for continuous operation in excess of 50 h is shown in Fig. 7-64. Fuel-cell life is in excess of 1,000 h.

91. Indirect-Fuel-Cell Systems. The indirect systems are those using chemicals which can provide a source of hydrogen, which in turn is used in a hydrogen fuel cell. Metal hydrides, for example, react with the water to produce H_2; 1 lb of of hydride will produce about 800 Wh in a fuel cell operating at 50% efficiency.

The indirect fuel cells using low-cost conventional fuels, having a heat content ranging from 35,000 to 44,000 Btu/kg, are most practical for higher-power applications. Steam reforming and thermal decomposition appear to be the best approaches for conversion of the fuel to hydrogen. The current state-of-art design of a 10-kW indirect fuel cell is 0.1 cm³, 200 kg, and an overall thermal efficiency of 40%. Considerable reduction in size and weight is expected in future designs.

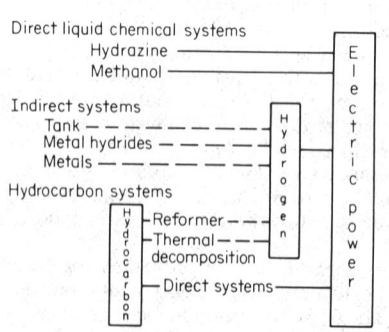

Fig. 7-63. Approaches to fuel cells.

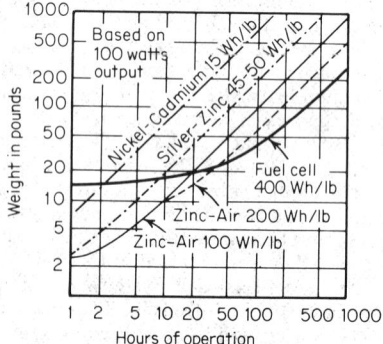

Fig. 7-64. Comparison of power source systems, weight vs. service.

MAGNETIC STORAGE DEVICES

By R. Babbitt

92. Introduction. There are three general techniques of magnetic storage: random access, sequential, and combinations of these two. The random-access and combination techniques are used solely in digital-computer applications. The sequential storage technique (magnetic tape) is used for storage of audio and visual information as well as computer data. The selection of a storage technique usually involves a tradeoff of the time it takes to retrieve the information stored (access time), the amount of information to be stored (capacity), and cost. See Table 7-20.

Table 7-20.   Summary of Magnetic Storage Systems

Type	Access time	Capacity, bits	Cost/bit
Core	200–500 ns	2×10^4–4×10^7	1¢–5¢
Plated wire	100–300 ns	1×10^5–1×10^8	5¢–8¢
Tape	10–60 s	3×10^8	<0.001¢*
Fixed-head drum and disk	5 ms	2×10^7	0.01¢
Moving-head disk ...	40 ms	2×10^8–4×10^9	0.006¢*

* Does not include the cost of the transport and associated circuitry, since these are changeable items.

93. Random Access. Random access has the characteristic that its data location (address) has a nonfixed sequence; that is, any address may be selected at random, and access to information stored is direct. This method is generally faster than other storage techniques, and more costly.

Cores. Cores (toroids), made of ferrite, are currently the main computer storage element in random-access memories because of their high operating speeds and moderate cost per bit. Each magnetic core stores one bit of digital information. This information is put into memory (write) or retrieved from memory (read) by a coincident-current selection technique, shown in Fig. 7-65. To illustrate the write function, assume all cores are in the 0 state. A 1 is to be written into the $X_1 Y_3$ core. When a current sufficient to produce a magnetic force of $+ H_m/2$ is applied to each of the X_1 and Y_3 drive wires, the $X_1 Y_3$ core is thereby switched to the 1 position, and this digit is then stored in the core. The read function involves a similar operation; that is, the sense wire is used to determine if a 0 or 1 is stored at a specific address. In a coincident system it is necessary that the core have a high squareness ratio (B_a/B_m) so that the half-amplitude drive pulse $(H_m/2)$ produces a minimum flux change on the other cores.

Magnetic Film (Plated Wire). A plated-wire memory is made by plating a nonmagnetic wire with a thin magnetic Permalloy film, with its magnetic easy-axis circumferential as shown in Fig. 7-66. A 1 is written by the coincidence of a current on the word strap (word current), and a current on the plated conductor (bit current). When the bit current is in the 1 direction, the magnetization vector is left in the 1 stored position after the currents are released. To read, a word current is applied which produces a field along the axis of the wire. This field tilts the magnetization vector toward the axis of the wire, producing a voltage of either polarity, depending on whether a 1 or 0 is stored, which is sensed by the plated wire. Unlike cores, which have to be electrically reset after a read operation, plated wire returns to its initial magnetization state after the word current is released.

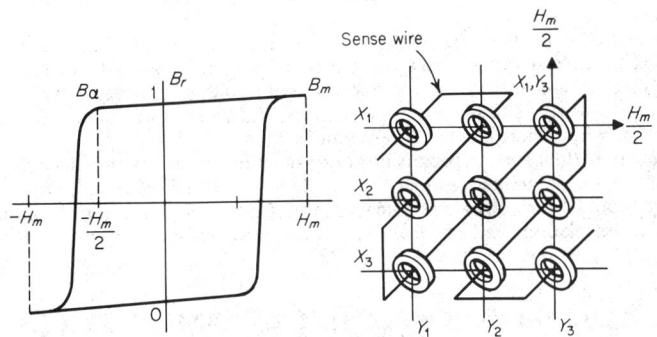

Fig. 7-65. Magnetic-core storage.

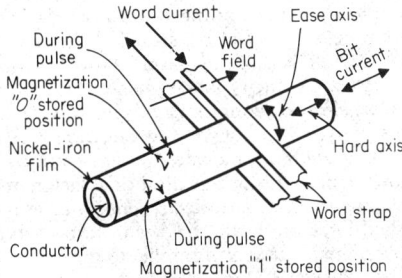

Fig. 7-66. Plated-wire storage.

94. Sequential Access (Magnetic Tape). Sequential access requires that all information stored before the desired information must be sensed in sequence. Because of this and the physical motion required, rather than electronic means, the access time is several orders of magnitude greater than for random access. Because of its low cost, magnetic tape is a popular medium for storing large quantities of information.

Magnetic tape is generally a plastic tape with a coating of magnetic material. Information is written or retrieved by use of a magnetic head, shown in Fig. 7-67. Magnetic head magnetizes a small area of film, producing a magnetic flux in one of two directions, depending on whether a 0 or 1 is to be stored. The information is retrieved when the tape is passed under the head, and the flux lines cut by the head induce a voltage which indicates whether a 0 or 1 is stored.

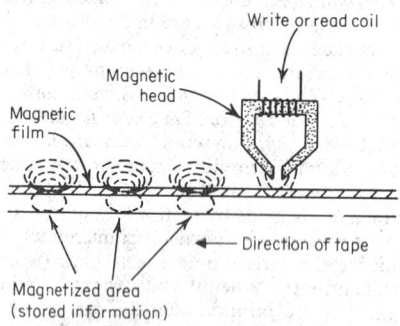

Fig. 7-67. Magnetic-tape storage.

95. Combinations. Magnetic storage which uses a mixture of random-access and sequential techniques is less expensive than random-access memories and has a shorter access time than sequential memories. In a combination storage device the files (tracks) are selected randomly, while the individual bits within a file are sequentially sensed. The two popular types of combination storage devices are magnetic drums and magnetic disks.

Magnetic Drum (Fixed-Head). A magnetic drum is a rotating cylinder coated with a magnetic material. Similar to magnetic tape, magnetic heads are used for read and write operations. A drum consists of several tracks, each with its own head (fixed head). The tracks are random-access electrically, while a specific bit within a track is access by rotating the drum (sequentially).

Magnetic Disks. A magnetic disk, like a drum, is composed of several tracks for bit storage. Magnetic disks are of two types, fixed-head and moving-head. In a fixed-head disk the track selection is electrical; the operation is similar to the fixed-head drum. In a moving-head disk, i.e., where one head covers many tracks, it is necessary to move the magnetic head mechanically, which increases the access time.

96. Magnetic Bubbles. A promising new magnetic storage system is the bubble memory. It is similar to a disk memory, in that the bubble moves to a fixed point in order to be read or to write. Bubbles are magnetic domains in thin magnetic films. Information storage is based on the association of a 1 or a 0 with the presence or absence of a domain at any specific point.

POLED FERROELECTRIC CERAMIC DEVICES

BY J. M. GIANNOTTO

97. Introduction. The utility of ferroelectrics rests on two important characteristics, asymmetry and high dielectric constant. Poled ferroelectric devices are capable of doing electric work when driven mechanically, or mechanical work when driven electrically. In poled ferroelectrics, the piezoelectric effect is particularly strong. From the design standpoint, they are especially versatile because they can be used in a variety of ceramic shapes.

98. Basic Properties. Piezoelectricity is the phenomenon of coupling between elastic and dielectric energy. Piezoelectric ceramics have gained wide use in the low-frequency range up to a few megahertz over the strongly piezoelectric nonferroelectric single crystals such as quartz, lithium sulfate, lithium niobate, lithium tantalate, and zinc oxide. High dielectric strength and low manufacturing cost are prime factors for their utility.

The magnitude and character of the piezoelectric effect in a ferroelectric material depend upon orientation of applied force or electric field with respect to the axis of the material. With piezoelectric ceramics, the polar axis is parallel to the original dc polarizing field. In all cases the deformations are small, when amplification by mechanical resonance is not involved. Maximum strains with the best piezoelectric ceramics are in the range of 10^{-3}.

Figure 7-68 illustrates the basic deformations of piezoelectric ceramics and typical applications.

99. Transducers. The use of barium titanate as a piezoelectric transducer material has been increasingly replaced by lead titanate zirconate solid-solution ceramics since the latter offer higher piezoelectric coupling, wider operating temperature range, and a choice of useful variations in engineering parameters. Table 7-21 reveals the characteristics of different compositions.

The high piezoelectric coupling and permittivity of PZT-5H have led to its use in acoustic devices such as phonograph pickups, where its high electric and dielectric losses can be tolerated. For hydrophones or instrument applications PZT-5A is a better choice, since its higher Curie point leads to better temperature stability. The low elastic and dielectric losses of the PZT-8 composition at high drive level point to its use in high-power sonic or ultrasonic transducers. The very low mechanical Q of lead metaniobate has encouraged its use in ultrasonic flaw detection, where the low Q helps the suppression of ringing. The high acoustic velocity of sodium potassium niobate is of advantage in high-frequency thickness-

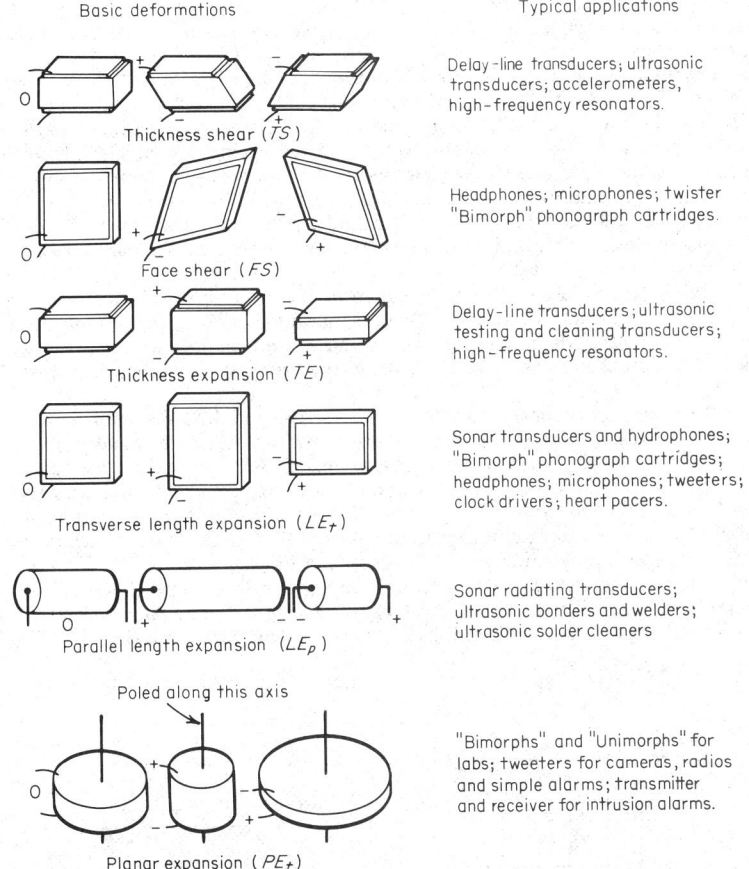

Basic deformations

Thickness shear (TS)

Face shear (FS)

Thickness expansion (TE)

Transverse length expansion (LE_t)

Parallel length expansion (LE_p)

Poled along this axis

Planar expansion (PE_t)

Typical applications

Delay-line transducers; ultrasonic transducers; accelerometers, high-frequency resonators.

Headphones; microphones; twister "Bimorph" phonograph cartridges.

Delay-line transducers; ultrasonic testing and cleaning transducers; high-frequency resonators.

Sonar transducers and hydrophones; "Bimorph" phonograph cartridges; headphones; microphones; tweeters; clock drivers; heart pacers.

Sonar radiating transducers; ultrasonic bonders and welders; ultrasonic solder cleaners

"Bimorphs" and "Unimorphs" for labs; tweeters for cameras, radios and simple alarms; transmitter and receiver for intrusion alarms.

Fig. 7-68. Basic piezoelectric action depends upon the type of material used and the geometry. Generally, two or more of these actions are present simultaneously. TE and TS are high-frequency (>1 MHz) modes and FS, LE_t, LE_p, and PE_t are low-frequency (<1 MHz) modes. The thickness controls the resonant frequency for the TE and TS modes, the diameter for PE_t, and the length for the LE_t and LE_p modes. Typical applications illustrate selections based on performance characteristics and cost. *(Electro-Technology.)*

Table 7-21. Ceramic Compositions (Courtesy of IEEE)

	k_{33}	k_p	$\varepsilon_{33}^{T}/\varepsilon_0$	Change in N_1 −60 to +85°C, %	Change in N_1 per time decade, %
PZT-4	0.70	0.58	1,300	4.8	+1.5
PZT-5A	0.705	0.60	1,700	2.6	+0.2
PZT-5H	0.75	0.65	3,400	9.0	+0.25
PZT-6A	0.54	0.42	1,050	<0.2	<0.1
PZT-8	0.62	0.50	1,000	2.0	+1.0
$Na_{0.5}K_{0.5}NbO_3$	0.605	0.46	500	?	?
$PbNb_2O_6$	0.38	0.07	225	3.3	?

k_{33} = coupling constant for longitudinal mode.
k_p = coupling constant for radial mode.
ε_{33}^{T} = permittivity parallel to poling field, stress-free condition.
N_1 = frequency constant (resonance frequency × length).

extensional and thickness-shear transducers, since this allows greater thickness, and therefore lower capacitance.

Since ceramic materials can be fabricated in a wide range of sizes and shapes, they lend themselves to designs for applications which would be difficult to achieve with single crystals. Figure 7-69 illustrates the use of simple piezoelectric elements in a high-voltage source capable of generating an open-circuit voltage of approximately 40 kV. Piezoelectric accelerometers suitable for measuring vibrating accelerations over a wide frequency range are readily available in numerous shapes and sizes. Figure 7-70 shows a typical underwater transducer which utilizes a hollow ceramic cylinder polarized through the wall thickness. Flexing-type piezoelectric elements exhibit the capability of handling larger motions and smaller forces than single plates.

100. Resonators and Filters. The development of temperature-stable filter ceramics has spurred the development of ceramic resonators and filters. These devices include simple resonators, multielectrode resonators and cascaded combinations thereof, mechanically coupled pairs of resonators, and ceramic ladder filters, covering a frequency range from 50 Hz to 10 MHz.

Two lead titanate-zirconate compositions, PZT-6A and PZT-6B ceramics, are most widely used for resonator and filter applications. PZT-6A, having high electromechanical coupling coefficient (45%) and moderate mechanical Q (400), is used for medium-to-wide bandwidth applications, while PZT-6B, with moderate coupling (21%) and higher Q (1,500), is used for narrow bandwidths. The compositions exhibit a frequency constant stable to within ±0.1% over a temperature range from −40 to +85°C. The frequency characteristics increase slowly with time at less than 0.1%/decade of time.

101. Ceramic Resonators. A thin ceramic disk with fully electroded faces, polarized in its thickness direction, has its lowest excitable resonance in the fundamental radial mode. The impedance response of such a disk and its equivalent circuit are shown in Fig. 7-71.

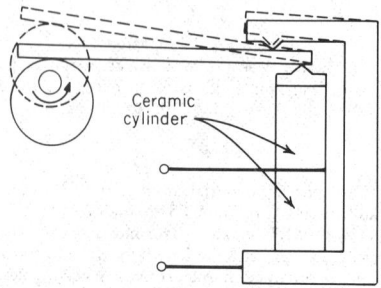

Fig. 7-69. High-voltage generator.

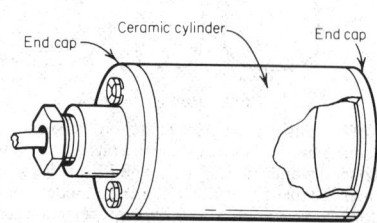

Fig. 7-70. Underwater sound transducer.

Ceramic resonators can be used in various configurations for single-frequency applications or combined in basic L-sections to form complete bandpass filter networks (see Fig. 7-72). Table 7-22 presents typical performance characteristics for ferroelectric ceramic filter elements.

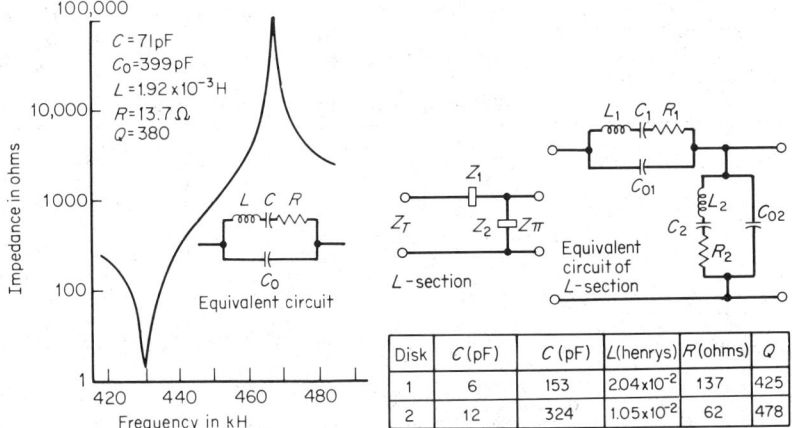

Fig. 7-71. Impedance response of a fundamental radial resonator.

Fig. 7-72. Ceramic L-section.

Disk	C (pF)	C (pF)	L (henrys)	R (ohms)	Q
1	6	153	2.04×10^{-2}	137	425
2	12	324	1.05×10^{-2}	62	478

Table 7-22. Ferroelectric Ceramic Filter Elements

Center frequency	50 Hz to 10 MHz
Bandwidth	$\frac{1}{2}$ to 10% of center frequency
Impedance	A few ohms to 50,000 ohms
Temperature stability	$\pm 0.1\%$ from -55 to $+85°C$
Operating temperature	To 125°C
Aging stability	$+0.06\%$ center frequency/decade of time

PIEZOELECTRIC CRYSTAL DEVICES

By G. K. Guttwein

102. Introduction. Piezoelectric devices provide for precise frequency control and selection in electronic equipment. The piezoelectric material used for most applications is quartz. A quartz crystal acts as a stable mechanical vibrator which, by its piezoelectric behavior, determines the frequency generated in an oscillator circuit. Crystals are available in the frequency range from about 2 kHz to 200 MHz. To cover this wide range, different modes of vibration are used. Table 7-23 lists the commonly used modes, their designations and normal frequency ranges. In the manufacture of different types of vibrators, quartz wafers are cut from the mother crystal in different directions with regard to the crystallographic axes. After lapping to required dimensions, metal electrodes are applied to the quartz blank, which is then mounted in a holder structure. The assembly is called a *crystal unit.*

103. The Equivalent Circuit. The circuit designer treats the behavior of a quartz crystal unit by considering its equivalent circuit, shown in Fig. 7-73. The mechanical resonance in the crystal vibrator is represented by L_1, C_1, R_1. Because it is a dielectric with electrodes attached, the quartz also displays an electrical capacitance C_0. The parallel combination of C_0 and the *motional arm* $L_1 C_1 R_1$ represents the equivalent circuit.

As shown at the left of Fig. 7-73, the equivalent series reactance X_e and equivalent series resistance R_e of this circuit vary with frequency. Below the resonance frequency f_r, the reactance X_e is capacitive; at f_r it goes to zero, and then becomes inductive, until a parallel resonance f_a with C_0 occurs. At this frequency R_e reaches a maximum.

The Q values $(Q = 2\pi f_r L_1/R_1)$ of quartz-crystal units are much higher than those attainable with other circuit elements. In general-purpose units the Q is usually between 10,000 and 100,000. The intrinsic Q of quartz is limited by internal losses. For AT-cut resonators, the intrinsic Q has been experimentally determined as 15×10^6 at 1 MHz; it is inversely proportional to frequency. At 5 MHz, for example, the intrinsic Q is 3×10^6. The best units made at this frequency approach a limiting Q value of 2.5×10^6.

Table 7-23. Modes of Vibration and Frequency Ranges

Usual description	Mode of vibration	Usual frequency range	Approximate capacitance ratio, $r = C_0/C_1$
X-Y bar	Flexure	2–20 kHz	500
NT	Length-width flexure	8–100 kHz	900
+5°X	Extensional	40–300 kHz	130
DT	Face shear	100–500 kHz	400
CT	Face shear	200–800 kHz	350
AT	Thickness shear (fundamental)	800 kHz–30 MHz	250
AT	Thickness shear (nth overtone)	2.5–200 MHz	$250n^2$
BT	Thickness shear (fundamental)	3–40 MHz	650
BT	Thickness shear (nth overtone)	15–150 MHz	$650n^2$

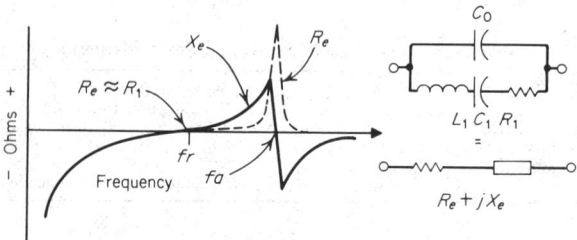

Fig. 7-73. Equivalent circuit and frequency-reactance relation.

APPLICATION OF CRYSTAL UNITS

104. Oscillators. The commonly used crystal oscillator circuits fall into two main groups. In the first group, called *series resonance oscillators,* the crystal operates at zero reactance (frequency f_r). In the second group, called *antiresonance,* or *positive-reactance oscillators,* the crystal is used as a positive reactance (frequency between f_r and f_a). In this latter mode of operation the oscillator circuit provides a *load capacity* to the crystal unit. The crystal then operates at that frequency at which its positive reactance cancels the reactance of the load capacitance. If the load capacitance is changed, the oscillator frequency will change. An important parameter in this connection is the capacitance ratio $r = C_0/C_1$. Numerical values of r for the different crystal cuts are listed in Table 7-23.

The frequency changes encountered are usually referred to the resonance frequency of the motional arm $f_s = 1/(2\pi\sqrt{L_1 C_1})$; f_s is approximately equal to f_r, and the following relation holds:

$$(f_a - f_s)/f_s = \tfrac{1}{2} r$$

In an oscillator with a load capacitance C_L the operating frequency f_0 is determined by

$$\frac{f_0 - f_s}{f_s} = \frac{C_0}{2r(C_0 + C_L)} = \frac{C_1}{2(C_0 + C_L)}$$

For a typical AT-cut crystal unit with $r = 250$, $C_0 = 5$ pF, $C_L = 30$ pF ($C_1 = 20$ fF), this results in a frequency shift of 286 parts per million (ppm) relative to f_s. If this crystal unit were erroneously placed in a series resonance oscillator, it would operate 286 ppm below its nominal frequency. An error of only 1 pF in the load capacitance results in a frequency error of approximately 10 ppm.

105. Filters. Quartz crystals are used as selective components in crystal filters. With the constraint imposed by the equivalent circuit of Fig. 7-73, existing filter design techniques are applicable to obtain bandpass or bandstop filters with prescribed characteristics. Crystal filters exhibit low insertion loss, high selectivity, and excellent temperature stability.

The main difference between a filter and oscillator crystal is given by the requirement that the filter crystal should have only one strong resonance in its region of operation, with all other responses *(unwanted modes)* attenuated as much as possible. This posed a serious problem for the so-called *thickness resonators* (Table 7-23), which was solved by the conception and application of the *energy-trapping theory.* If electrode size and electrode thickness are selected in accordance with that theory, the energy of the main response is "trapped" between the electrodes, whereas the unwanted modes are "untrapped" and propagate toward the edge of the crystal resonator, where their energy is dissipated. It is possible to manufacture AT-cut filter crystals with 30- to 40-dB attenuation of the unwanted modes relative to the main response.

106. Crystal-Unit Standardization. Adequate standardization exists concerning dimensions and performance characteristics of crystal units. The principal documents are the U.S. Military Standards, the standards issued by the Electronics Industry Association (EIA), and the International Electrotechnical Commission (IEC). The generally used designations are those of the Military Standard, that is, HC-6 and HC-18 for commonly used crystal enclosures. In addition, transistor-type holders and flat-pack and other configurations are used as enclosures for quartz crystals.

FACTORS AFFECTING FREQUENCY STABILITY

107. Temperature. Most crystal cuts have a parabolic frequency-temperature characteristic, as illustrated in Fig. 7-74 for the BT, CT, and DT cuts. The vertex of the parabola is the point of zero temperature coefficient, arbitrarily placed at 20°C in the figure. By a change of the angle of cut with respect to the crystallographic axes, this point can be shifted within the useful temperature range.

The frequency-temperature characteristic of the AT cut follows a third-order law, as shown in Fig. 7-75. The frequency stability of the AT cut over a wide temperature range is greatly superior to that of the other cuts discussed. A slight change in the orientation angle (7′ in the example shown in the figure) greatly changes the frequency-temperature characteristic. Curve 1 is optimal for a wide temperature range (−55 to 105°C). Curve 2 gives minimum frequency deviation over a narrower temperature range. The AT characteristic has two points of zero-temperature coefficient, called the *lower* and *upper turnover points.*

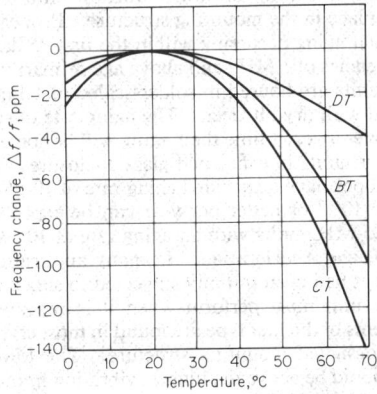

Fig. 7-74. Parabolic frequency-temperature characteristic of some crystal cuts.

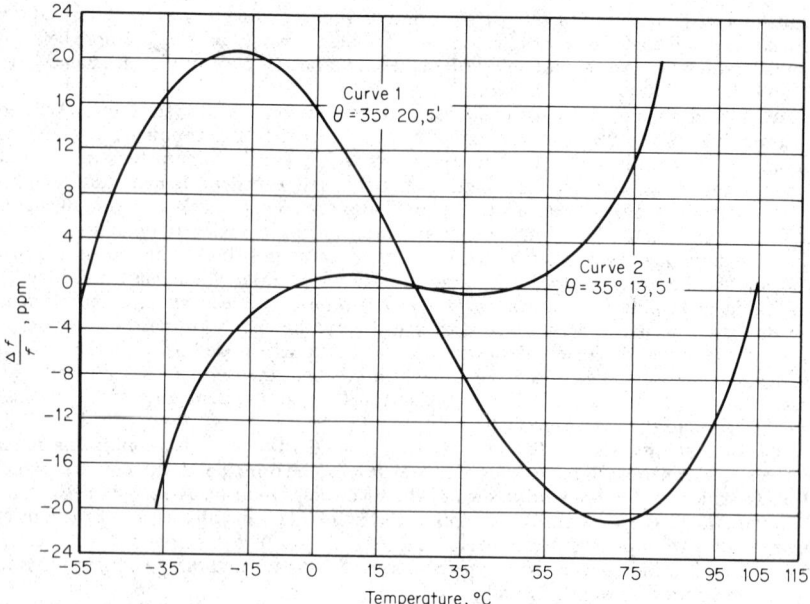

Fig. 7-75. Frequency-temperature characteristic of AT-cut crystals.

108. Drive Level. The frequency of a crystal vibrator depends on its motional amplitude. As a rule of thumb it can be assumed that an AT-cut crystal unit will increase its frequency by 1 ppm if the drive level is raised by 1 mW. Because of this effect, the drive level must be specified. General-purpose units usually operate around 2 mW. In precision units the drive level is kept as low as possible, usually less than 1 μW.

109. Aging. This term denotes the change of frequency with time, experienced in all crystal units. The main causes for aging are mass transfer to or from the resonator surface and strain relief within the mounting structure or at the interface between quartz and electrodes.

The aging of a properly made crystal unit will be greatest when it is new. As time elapses, stabilization occurs within the unit and the aging rate decreases. A final aging rate will be reached after a stabilization period of from several weeks to as long as 2 years.

The paramount reason for aging of low-frequency units (below 800 kHz) is that mechanical changes take place in the mounting structure. Properly made units may age 10 to 20 ppm/year, half of that aging occurring within the first 35 days.

Crystal units for frequencies of 1 MHz and above age primarily because of mass transfer. General-purpose crystal units are housed in solder-sealed metal enclosures of the HC-6 or HC-18 configuration filled with dry nitrogen. The aging rate of such units is specified as 5 ppm for the first month; over a year's time their aging will be somewhat better than 60 ppm. If lower aging is important, units in evacuated glass enclosures should be used. The best types in HC-6 glass envelopes have a specified aging rate of 10^{-8} per week after 30 days of operation. Aging of 2×10^{-7} or better per year can be expected thereafter. Advanced methods have resulted in 5-MHz units with an aging rate of 10^{-10}/day or better.

110. Shock, Vibration, and Acceleration. In many applications the crystal unit must operate satisfactorily after it had been initially subjected to shock and vibration conditions. In other applications the unit must perform when it is repeatedly exposed to such an environment. Requirements of the first type are found in most crystal specifications and can be met with a properly designed mounting structure. The lowest mechanical-structure resonance, for example, should be above the highest vibration frequency. Crystal units with mounting resonances above 3,000 Hz are available. Small crystal units capable of withstanding impact shocks of 15,000 G have also been developed.

For the second requirement, that the crystal unit must operate within tolerances in a shock, vibration, or acceleration environment, an additional problem is imposed by the fact that all crystal vibrators suffer a frequency change when subjected to external mechanical forces. A crystal unit changes its frequency, for example, when accelerated in a centrifuge. The best units tested under this condition have shown a frequency-acceleration coefficient of approximately $10^{-9}/G$, when acceleration forces will be encountered in all directions, and $10^{-10}/G$, when the crystal can be oriented for an acceleration force occurring only in one direction.

METHODS FOR TEMPERATURE CONTROL

111. Crystal Ovens. A crystal oven consists of a thermally isolated cavity designed to accept one or several crystal units, appropriate heating elements, and control circuitry suitable for regulating the heater current so that the temperature is maintained in a specified narrow range within the cavity. Bimetallic thermostats are used today as controlling elements only in inexpensive applications. For most other purposes proportional control ovens are preferred. In this type of oven the output of a bridge circuit (containing temperature-sensitive components) is amplified and powers the heater, resulting in power application proportional to the difference between the nominal and actual cavity temperature. With this arrangement continuous adjustment of the operating temperature can also be obtained. Because of crystal-manufacturing tolerances, the nominal turnover temperature will usually show a spread of $\pm 5°C$. For optimum stability the oven temperature should be adjusted to the actual turnover point.

112. Temperature Compensated Oscillator (TCXO). These devices avoid two disadvantages of crystal ovens, namely, warmup time and additional power consumption, at the expense of somewhat greater circuit complexity. The frequency-temperature behavior of the crystal is compensated by applying a dc correction voltage to a voltage variable capacitor placed in series with the crystal. The correction voltage is derived from a thermistor-resistor network. TCXOs have been developed for a wide variety of applications. The following is an example of a commercially available high-stability type: frequency range 3 to 5 MHz; frequency stability $\pm 5 \times 10^{-7}$ from -40 to $75°C$; total power consumption 50 mW, volume 2 in^3. The best TCXOs developed thus far for this temperature range have reached $\pm 1 \times 10^{-7}$ stability.

FLUIDIC DEVICES

By J. M. Kirshner

113. Introduction. Fluidics, the use of fluids to perform the functions of sensing, computation, and/or control, is a widely applied technique. Fluidic devices are used in preference to electronic methods for reasons of safety (as in explosive environments), for immunity to high temperatures and corrosive environments, and for special capabilities in sensing applications (e.g., thread, the presence of double sheets of paper, and the thickness of frost in refrigerator coils). They combine convenience, reliability, and ease of maintenance.

Fluidics as a technology dates from 1959, when B. M. Horton, R. E. Bowles, and R. W. Warren of the Harry Diamond Laboratories began work on no-moving-part fluid devices.

Fluidic systems comprising sensors, bistable and proportional amplifiers, diodes, oscillators, resistors, and capacitors are operable and reliable in environments in which electronic systems cannot survive. They are especially advantageous as control devices in fluid-powered systems such as jet engines and on vehicles such as ships, aircraft, and missiles.

114. Bistable Amplifier. As shown in Fig. 7-76, fluid flow (a two-dimensional ribbon-shaped turbulent jet) is initiated from a slitlike nozzle. The jet entrains fluid on both sides. Some of the molecules are evacuated by entrainment from the regions between the jet and the walls, leaving low-pressure regions in which counterflows develop. Because of minor fluctuations, the jet may momentarily bend toward one wall, thereby restricting the counterflow, lowering the pressure between the jet and that wall, and resulting in a pressure difference across the jet that causes the jet to attach to the wall after a very short time, as shown in Fig. 7-77.

The wall-attachment effect is used to make a bistable device by adding controls and a splitter (Fig. 7-78). The jet, at random, attaches to one wall or the other. It can be switched by closing the control on the unattached side or increasing the flow through the control on the attached side. If one outlet is blocked, the fluid flows out of the other outlet. To prevent this, bleeds (or vents), shown in Fig. 7-79, are provided.

Commercial fluidic digital amplifiers (binary switches) have a fan-out of three or more, pressure recoveries of 35 to 50%, flow recoveries of 80% or more, and switching times of 1 ms or slightly less. *Fan-out* is the number of similar devices that can be actuated by one device. *Pressure recovery* is the percentage of power jet pressure recovered at the output. *Flow recovery* is the percentage of the power jet flow recovered at the output. Fan-outs of 32 have been achieved, but usually at the expense of stability.

115. Logic Devices. Various logic devices (AND, NAND, OR, NOR) are available. An important type of NOR unit known as the turbulence amplifier is based on transition from laminar to turbulent flow and depends on the fact that there is a range of Reynolds numbers where a normally laminar jet is very easily disturbed to turbulence.

A laminar jet from a small pipe enters another small pipe in line with the first but an appreciable distance downstream. If the jet is disturbed by a control jet, it quickly becomes turbulent and spreads, and very little of the flow reaches the downstream pipe. If there are two or more control jets, this device is a NOR unit (Fig. 7-80). Because there is almost no effect on the controls when the outputs are loaded, this device is easy to operate in successive stages.

The turbulence amplifier consumes little power (20 to 30 mW), but amplification is required before the signal can be used to actuate a mechanical device. In addition, its switching time (about 10 ms) is greater than that of the higher-power devices. It has a high fan-out (around 16) and a pressure recovery of about 30%.

116. Vortex Diode. Consider fluid tangentially driven into a cylinder by a suddenly

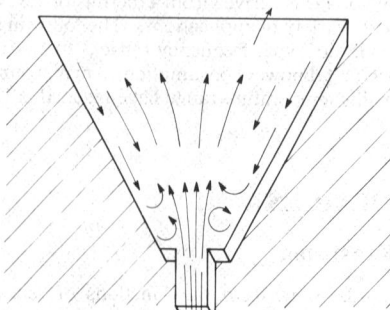

Fig. 7-76. Jet flow in initial stages.

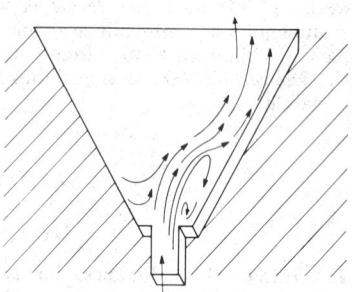

Fig. 7-77. Jet attached to wall.

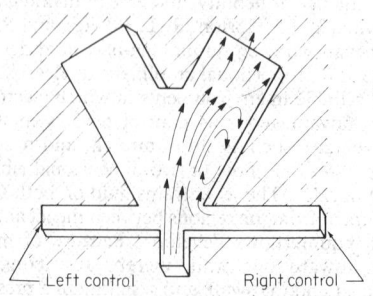

Fig. 7-78. Elementary bistable device.

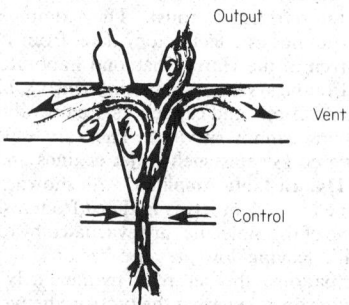

Fig. 7-79. Bistable device with vents and output orifices.

applied pressure p_c and leaving through an axial drain, as shown in Fig. 7-81. Because of the conservation of angular momentum, the tangential velocity of the fluid (and the corresponding centrifugal force) increases as it spirals in toward the drain. If the drain is sufficiently small, the centrifugal force of the inner layers of rotating fluid is sufficient to balance the applied pressure forces and cut off the input flow.

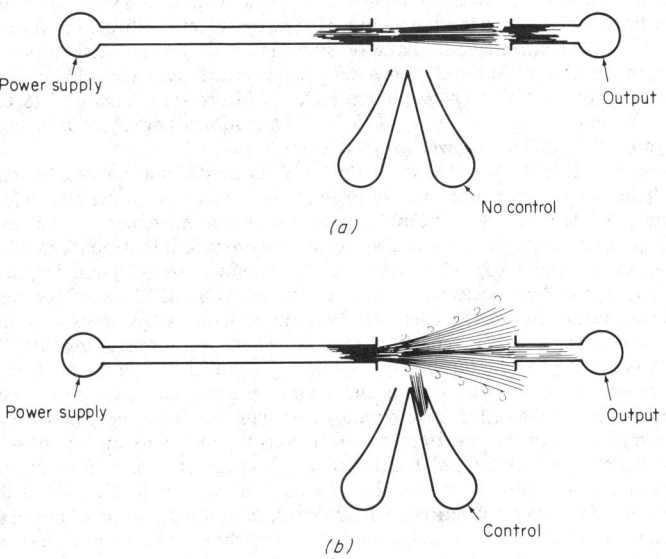

Fig. 7-80. Turbulence amplifier; (*a*) laminar jet (high recovery); (*b*) turbulent jet (low recovery).

It follows that nonviscid fluid would be trapped and merely continue whirling, without leaving the drain; but flow in the other direction (from the drain toward the tangential arm) is not restricted by a similar phenomenon. The result is a fluid diode. Unfortunately, viscosity effects are important in this device, so that vortex diodes cannot really completely cut off the flow. Because of nonlinearity, the ratio of forward to backward pressure drop is a function of flow. A typical value of this ratio is 50.

117. Vortex Throttle. If another source of flow is added, as shown in Fig. 7-81, a vortex throttle, or fluidic triode, is obtained. Fluid from the power source p_j ordinarily proceeds straight in toward the drain. The addition of flow from the control p_c causes the resultant flow to spiral in toward the center, resulting in a decreased output because of the centrifugal-force effect. Since the change in net flow is greater than the control flow causing it, the device has gain.

The vortex throttle, or amplifier, is a flow-controlled device. It requires control pressures ranging from 20 to 75% greater than the supply pressure and has a turndown ratio of about 10; that is, the output varies from full flow to about 10% of full flow as the control is increased. Maximum pressure recovery is 95%, resulting in power recoveries of 50 to 75%, with bandwidths of 50 to 100 Hz. The primary advantage of the vortex amplifier is that it is a throttling device rather than a diverting device, but it is relatively noisy and the control pressure must be higher than the power jet

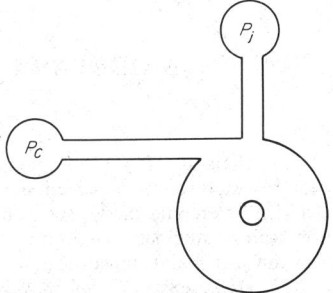

Fig. 7-81. Vortex triode.

pressure. Because of these factors the vortex device is more useful as a power output device than for fine control, or where staging of many elements is involved.

118. Beam Deflection Amplifier. The beam deflection amplifier obtains proportional control in a device similar to that shown in Fig. 7-79 by minimizing the effect of the walls. To do this, the interaction region is opened to the atmosphere or to a reservoir through vents, or the splitter is moved up closer to the nozzle, or both methods are used. A center vent to prevent instability under load is often included. The power jet is deflected by the momentum and pressure of the control jets, so that a pressure difference appears across the outlets that is proportional to the pressure difference across the control over a specific pressure range.

Commercial beam deflection amplifiers have listed pressure gains of 4 to 6, dynamic ranges of 200, and bandwidths up to 4,000 Hz. Operational amplifiers are available with pressure gains up to 200 and power gains to 2,500.

119. Angular Rate Sensor. Most of the effort in fluidic rate sensing has utilized the velocity change that occurs as a radially injected fluid moves from the circumference of a vortex device to the drain; the tangential forces are due to movement of the device. The sensitivity, threshold, and response of a vortex rate sensor are interdependent, and tradeoffs are consequently employed, depending on the intended use. Examples of available characteristics for a 10-cm vortex rate sensor are sensitivity, 0.35 kP/deg·s; response, 10 Hz.

120. Temperature Sensors. There are two fluidic temperature-measuring techniques. One type depends on a pressure change across a capillary resistor resulting from the change in viscosity of the gas flowing through the resistor. The other type of fluidic thermometer is an oscillator that changes in frequency as the temperature changes because the speed of wave propagation in the oscillator feedback circuit varies with temperature. The sensor has been used as a temperature probe in extreme environments because of its small size and ability to operate with a fast response at high temperatures. The characteristics of the sensor depend on the gas and temperature of the gas being sensed, as well as on the size of the device.

121. Fluidic-Systems Applications. There are several important problems in systems of fluidic components. These are concerned with matching (i.e., minimizing reflections), minimizing noise, ease of fabrication, size and weight of the system, and power consumption. It is possible to build systems without detailed knowledge of fluid dynamics, by techniques that characterize fluidic components as *black boxes,* so that they can be put together purely on the basis of their inputs and outputs, with no concern for internal operation.

The simplest items to work with are sets of compatible modular units that can be easily connected (by plugging into a board or by stacking). A number of modular components of both the stacking and the plug-in variety are presently on the market.

Typical applications include boiler control on ships, control of bottle-filling machines, machine-tool controls, coil winders, air gaging, determination of proper filling of candy boxes, actuating trimming knives of commercial sewing machines, the control of high-speed glass presses, papermaking machinery, and bottle-casing machines and in liquid-level controls of various types.

MODULAR COMPONENT ASSEMBLIES

By R. A. Gerhold

122. General. Physical implementation of a circuit schematic into a functional component assembly requires selection of the discrete component parts, mechanical support, electrical interconnection, provision for dissipation of heat generated, and adequate protection against anticipated environmental factors. Design of such a functional assembly must also consider maintenance, cost, and reliability.

123. Definitions. *Circuit packaging:* The branch of electronic design concerned with the physical assembly and electrical interconnection of elements into circuits, and with location, mounting, assembly, protection, and interconnection of circuits into subassemblies, assemblies, and equipments to meet performance, operational, and maintenance requirements.

Module, equipment: An assembly or subassembly of an equipment which displays dimensional regularity and separable repetition within a given equipment. It is generally designed to be handled as a single unit to facilitate supply and installation, operations and/or maintenance. It can be either repairable or nonrepairable.

Printed wiring: A pattern of interconnecting wiring formed on a common insulating base.

124. Modular Assembly. There are four specific approaches to modular assembly: cordwood, planar, stacked-wafer, and embedment.

125. Cordwood Assembly. This concept derives its name from the ordered array of cut logs of firewood stacked side by side in a cordwood pile; the logs may vary in diameter without upsetting the efficiency of the stacked pile. Similarly, cylindrical axial-leaded discrete components (resistors, capacitors, diodes, etc.) may be "bundled" together and their leads interconnected to form a cordwood component assembly (see Fig. 7-82) which is compact and rugged. In the example shown, the component-part leads are held in position by the end plates and welded to nickel ribbons to accomplish the module intraconnections. Similarly, printed-wiring boards soldered to the component leads may be used at each end of the cordwood assembly. Plating techniques have also been employed to interconnect the leads at the ends of the module.

126. Planar Assembly. Printed wiring on rigid laminated plastic boards provides a planar interconnection wiring array. Axial-leaded components are mounted on the printed-wiring boards; their leads are bent at right angles and inserted into predrilled or prepunched holes in the printed-wiring boards (see Fig. 7-83). Soldering of the leads to the printed wiring provides both electrical connection and mechanical support of the components into the completed printed-wiring assembly.

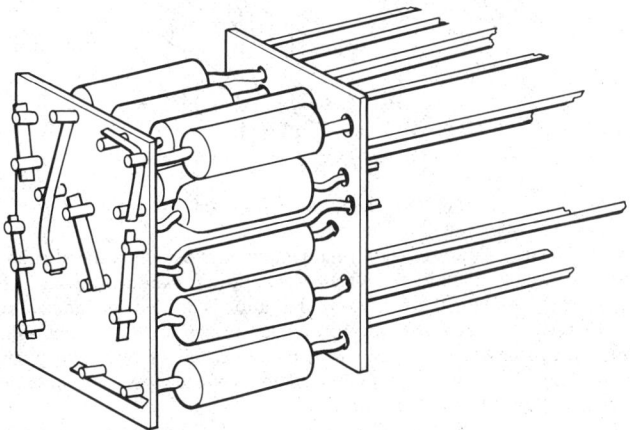

Fig. 7-82. Cordwood module.

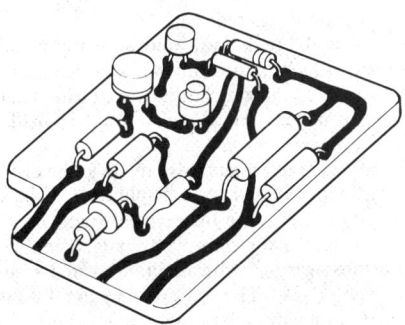

Fig. 7-83. Printed-wiring assembly.

127. Stacked-wafer assembly permits elements and circuits to be fabricated on individual wafers. The wafers are then assembled into a stack which provides mechanical and electrical interconnection (see Fig. 7-84). The side connections, or risers, may be round wire or ribbon, and connections may be made by soldering, welding, or plating techniques.

128. Embedment as a modular-assembly concept may be applied to any of the various approaches described above and, in addition, may be applied to assemblies of components in free form which are electrically interconnected but which require encapsulating resin to provide mechanical support and environmental protection. Encapsulation or embedment material should be applied to individual components only as part of a controlled production process for that component. Module design should be completely verified for the particular encapsulation or embedment materials and processes to be employed. In particular, extreme temperature, aging, and temperature-cycling tests are required to verify adequacy of the design.

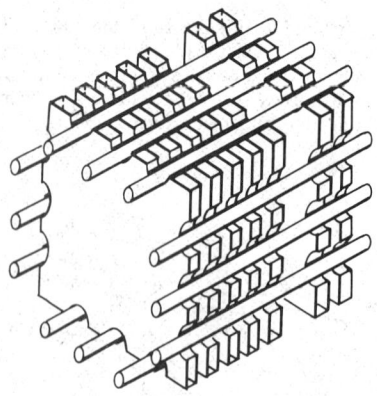

Fig. 7-84. Stacked-wafer assembly.

129. Printed Wiring. The printed-wiring conductor is usually a metal such as copper, and the *insulator* is usually a laminated plastic sheet. In the *subtractive process*, the starting point is a copper-clad laminate; the desired configuration is printed with an acid resist on the foil surface. Etching then removes the unwanted surface areas, leaving a foil pattern.

130. In **additive processes** the printed-wiring pattern is built up by plating on the insulator surface. Plating techniques are also employed to produce *through connections* between circuits on opposite sides of a plastic board.

Thickness of the conductor foil and *width* of individual printed-wiring lines are determined by the *current-carrying capacity* required, as indicated in Fig. 7-85.

131. Printed-wiring-board material and *thickness* are determined mainly by mechanical considerations of strength and stiffness vs. cost and fabricability. Some thin materials are used where flexible printed wiring is desired.

132. Insulation resistance and breakdown voltage between adjacent conductors are severely degraded by surface contamination in the presence of high humidity. *Spacing requirements* between conductors for various classes of application and voltages are indicated in Table 7-24. As noted in this table, applications for military field equipment require that all printed-wiring boards be *conformally coated.*

133. Connection. The above requirements also relate to the distinction between the two broad classes of connectors for printed-wiring assemblies. In the card-edge, or one-part, connector, the plug is printed on the end of the printed-wiring board as part of the printed wiring. The surface may be plated for improved contact performance. However, this printed-plug portion of a printed-wiring card cannot be treated with the protective coating, as indicated in the note to Table 7-24. The solution for humid military environments is to require the use of a two-part connector where the plug portion is formed separately and is mounted as another component on the printed-wiring board. Appropriate receptacles are provided for the two types of connector plugs.

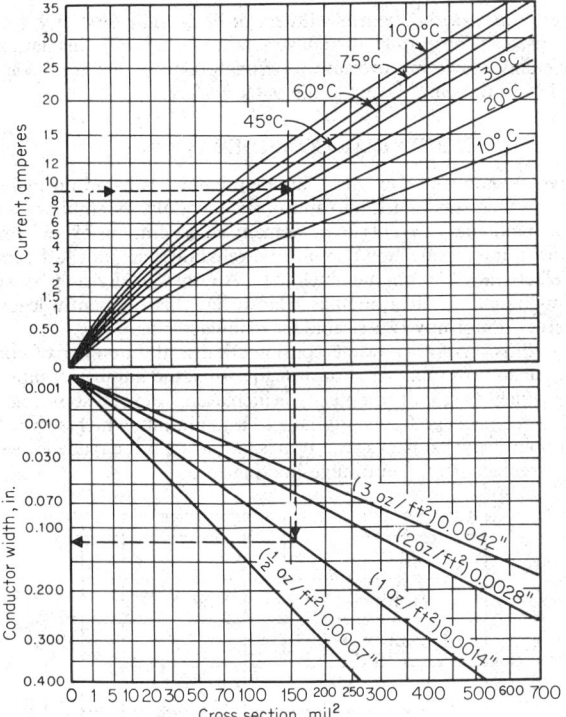

For four standard thicknesses of copper.

(For use in determining current carrying capacity and sizes of etched copper
conductors for various temperature rises above ambient.)

Fig. 7-85. Printed-wiring conductor thickness and width.

Table 7-24. Spacing Requirements

Voltage between conductors, dc or ac peak, V	Minimum spacing
Conductor spacing, uncoated printed-wiring boards, sea level through 10,000 ft	
0–150	0.025 in.
151–300	0.050 in.
301–500	0.100 in.
Greater than 500	0.0002 in./V
Conductor spacing, uncoated printed-wiring boards, over 10,000 ft	
0–50	0.025 in.
51–100	0.060 in.
101–170	0.125 in.
171–250	0.250 in.
251–500	0.500 in.
Greater than 500	0.001 in./V
Conductor spacing, conformal coated printed-wiring boards, applicable to all altitudes	
0–30	0.010 in.
31–50	0.015 in.
51–100	0.020 in.
101–300	0.030 in.
301–500	0.060 in.
Greater than 500	0.00012 in./V

NOTE: For the Department of the Army, all printed-wiring boards shall be conformally coated and the minimum
distance between uncoated conductive areas shall be 0.080 in. and the potential between such areas shall not exceed 50 V.

134. Multilayer printed-wiring boards with six or eight individual layers of conductive patterns are also used. These individual layers, alternating with insulating layers, are laminated together under heat and pressure. Through connections are made between the various layers and from the inner layers to the outer surfaces.

ELECTRICAL CONSIDERATIONS

135. Functional Assemblies. Design of large equipment is based on major equipment assemblies, which are then subdivided into appropriate modular assemblies for convenience in manufacture and maintenance. This involves many considerations of performance, failure location, replacement, interconnection complexity, reliability, subunit test capability, standardization, and electrical isolation and shielding. An obvious division between modules correlates with functional circuit groupings. However, this frequently leads to unequal complexities, whereas uniformity is desirable for manufacturing.

136. Connector Pins. Another prime consideration is the number of connector pins required, per module and overall. As shown in Fig. 7-86, the number of pins required per package may vary widely as circuit complexity is increased. Beyond some maximum point, the number of pins per package frequently falls to a minimum, which serves to indicate a desirable division point between packages. Provision must also be made for ease of testing, installation, repair, replacement, and troubleshooting.

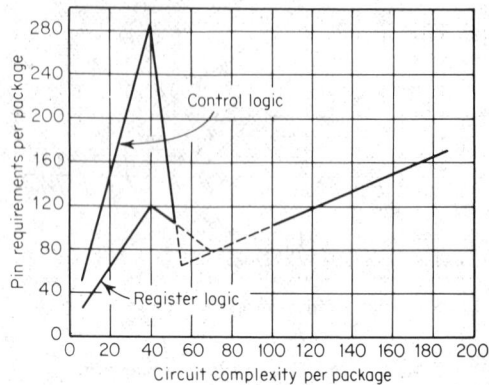

Fig. 7-86.   Relationship of pin requirements to circuit complexity.

137. Standard Modules. Standard modules, of specific design, are available for use throughout an entire system. Thus a variety of systems can be built and changes incorporated simply by changing interconnecting wiring. However, modules built specifically for the individual purposes usually provide simplified wiring, smaller size, and improved electrical performance. These advantages must be weighed against the cost, reliability, and availability which may accrue from the use of the standard modules.

Consideration must also be given to the need to minimize interference between stages, excess propagation time, pickup on signal lines, surface-leakage currents under humidity, changes in components by heat of soldering, resistance of joints in series, and effect of component hot spots on operation. Voltage distribution and grounding systems can be series, tree, loop, and point types. Multilayered printed-wiring boards facilitate use of multiplaned grounding systems and of several layers of voltage distribution. Additional electrostatic and electromagnetic shielding may be used.

ENVIRONMENTAL CONSIDERATIONS

138. Humidity. Consideration must be given to the effect of volumetric absorption of moisture, to surface absorption or condensation of moisture, and to possibilities of moisture entrapment within the case or container or between closely spaced surfaces. Aside from performance changes, moisture greatly accelerates corrosion of metals, particularly where voltage differences are present. Even in sealed containers, temperature drops (such as occur in high-altitude flight) may increase the relative humidity to the dew point and result in condensed moisture on the circuitry. Localized heating from power-dissipating components may aid in drying off surfaces.

139. Temperature. Temperature considerations involve expected environmental temperatures, plus additional heat rise within the equipment due to power dissipation. Hot-spot temperatures of critical components must be maintained within acceptable limits. Low temperature, particularly with thermal cycling, can cause severe mechanical degradation due to cracking. Figure 7-87 indicates the wide range of coefficients of thermal expansion of commonly used materials.

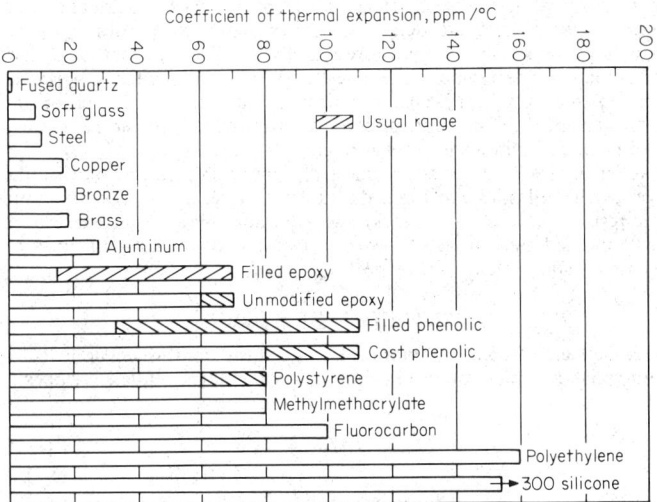

Fig. 7-87. Coefficients of thermal expansion. *(After C. A. Harper, ed.)*

LIFE-CYCLE CONSIDERATIONS

140. Initial design and production cost is but one aspect of the overall life-cycle of a piece of equipment. For complex equipment, the cost of locating faults during the operating life of the equipment, of replacing failed modular assemblies, and of repairing such replaced assemblies can add significantly to total life-cycle cost. For military applications, this additional cost may run several times the initial cost of the equipment.

141. Reliability. The failure rate of a particular equipment is directly related to the sum of the failure rates of its individual component parts. This failure-rate total is also affected by any redundancy of circuitry and by the degree of criticality of application (i.e., the severity of the failure criteria) of the parts in the individual circuits.

142. Repair vs. Discard. In each application, there is a cost value for the modular-component assembly below which it is more economic to discard a failed assembly and to replace it with a new one, rather than to locate and repair the failure. Figure 7-88 indicates the type of decision curve that applies to such an analysis. The curve itself is a function of the cost of repair time, repair parts, test equipment, transportation, failure analysis, and quantities involved. Such a curve varies widely with the specific type of equipment and its manner of application. Entering Fig. 7-88 at the lower left, the known or anticipated failure rate, when applied to the total number of units in the system, leads to the total number of expected failures per year. The decision to repair or discard after failure is determined by the intersection point between this total number of expected failures and the card or assembly purchase price. If this intersection falls above the decision curve (in the white area), it indicates that sufficient failures are anticipated to warrant setting up a cost-effective procedure for repairing such failures when they occur. If the intersection point falls below the decision curve (in the shaded area), it indicates that an insufficient total number of failures is expected to support a repair procedure that would be cost-effective compared with the cost of discarding failed assemblies.

INSULATORS

By Sam Di Vita

143. General. Ceramics and plastics are the principal materials for electronics insulation and mounting parts. Ceramic materials are outstanding in their resistance to high temperature, mechanical deformation, abrasion, chemical attack, electrical arc, and fungus attack. Ceramics also possess excellent electrical insulating properties and good thermal conductivity and are impervious to moisture and gases. These properties of ceramics are retained throughout a wide temperature range, and are of particular importance in high-power applications such as vacuum-tube envelopes and spacers, rotor and end-plate supports for variable air capacitors, rf coil forms, cores for wire-wound resistors, ceramic-to-metal seals, and feed-through bushings for transformers.

The properties of plastics differ rather markedly from ceramics over a broad range. In a number of properties, plastics are more desirable than ceramics. These include lighter weight; better resistance to impact, shock, and vibration; higher transparency; and easier fabrication with molded-metal inserts (however, glass-bonded-mica ceramic material may be comparable with plastic in this latter respect).

CERAMIC INSULATORS

144. Linear Dielectric Radio Insulators. Ceramic insulators are linear dielectrics having low loss characteristics, which are used primarily for coil forms, tube envelopes and bases,

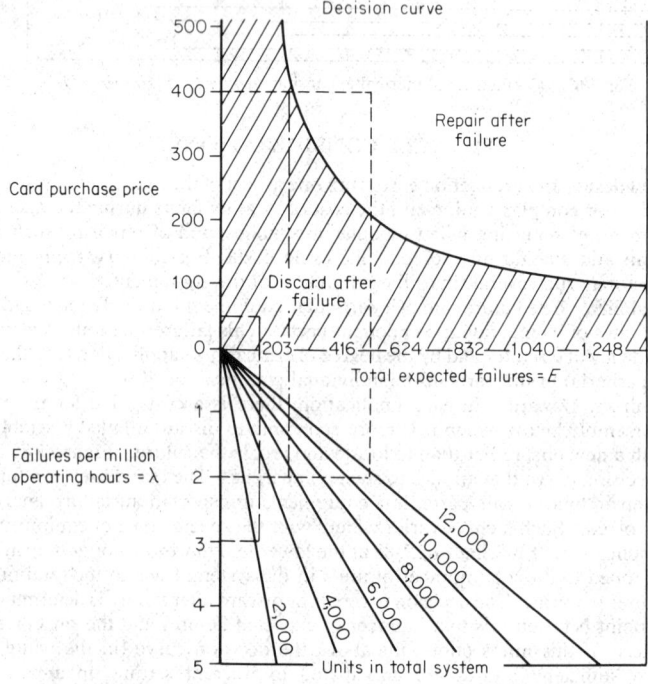

Fig. 7-88. Repair vs. discard decision curve.

and bushings, which all require loss factors less than 0.035 when measured at standard laboratory conditions. Dielectric loss factor is the product of power factor and dielectric constant of a given ceramic. Military Specification MIL-I-10B, Insulating Compound Electrical, Ceramic Class L, covers low-dielectric-constant (12 or under) ceramic electrical insulating materials, for use over the spectrum of radio frequencies used in electronic communications and in allied electronic equipments. In this specification the "grade designators" are identified by three numbers, the first representing dielectric loss factor at 1 MHz, the second dielectric strength, and the third flexural strength (modulus of rupture).

Table 7-25 lists the various types of ceramics and their grade designators approved for use in the fabrication of Military Standard ceramic radio insulators specified in MIL-I-23264A, Insulators—Ceramic, Electrical and Electronic, General Specification For. This specification covers only those insulators characterized by combining the specific designators required for the appropriate Military Standard insulators used as standoff, feed-through, bushing, bowl, strain, pin, spreader, and other types of insulators. As a typical example of one military standard for insulators: feed-through is shown in Fig. 7-89, and Fig. 7-90 shows other commercial electronic ceramic parts.

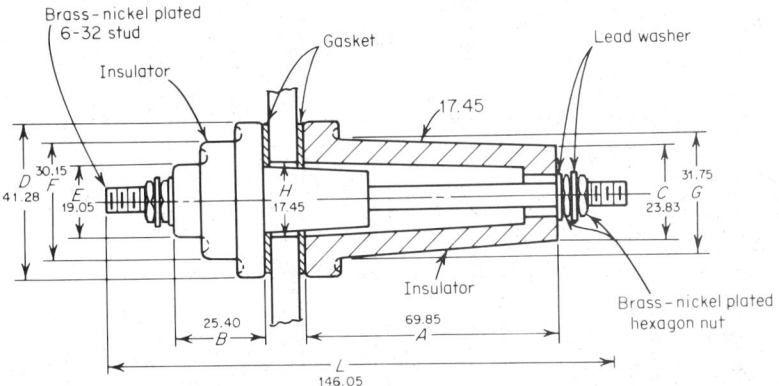

Fig. 7-89. Feed-through insulator as per MIL-23264/13A, Type-NL 422B34-046 (dimensions in millimeters).

Currently, Grade L-242 is typical of porcelain, L-422 of steatite, L-523 of glass, L-746 of alumina, L-442 of glass-bonded mica, and L-834 of beryllia.

145. Good Practice Design. The data compiled in Table 7-26 represent good design practice for ceramic radio insulators but do not necessarily imply that closer tolerances or any important dimension or special designs cannot be produced by special handling.

146. Surface Finishes. Surface finish is the deviation of the heights and depths of surface irregularities from a central reference line.

An "applied" finish results from a modification of the "as fired" condition by grinding, lapping, polishing, tumbling, or glazing. Glazed (vitreous coating) surfaces provide the smoothest surfaces, finishes of 1 μin. (25 nm) or better being fairly common. Composition specifically designed to have smooth as-fired finishes can be made having surface finishes of 5 μin. (125 nm) or smoother, whereas composition designed for other applications may run up to 60 μin. (1,500 nm) surface finish.

147. High-Thermal-Shock-resistant Ceramics. Lithia porcelain is the best thermal-shock-resistant ceramic because of its low (close to zero) coefficient of thermal expansion. It is followed in order by fused quartz, cordierite, high-silica glass, porcelain, steatite beryllium oxide, alumina, and glass-bonded mica. Those materials find wide use for rf coil forms, cores for wire-wound resistors, stator supports for air dielectric capacitors, coaxial cable insulators, standoff insulators, capacitor trimmer bases, tube sockets, relay spacers, and base plates for printed radio circuits.

148. High Thermal Conductivity. High-purity beryllium oxide is unique among homo-

Table 7-25. Property Chart of Insulating Ceramic Materials Qualified under MIL-I-10

(Various Types of Ceramics and Their Electrical and Mechanical Properties Which Qualified Them under MIL-I-10 Specifications) MIL-I-Government

Class L ceramics	Grade designators	Power factor, 1 MHz	Dielectric constant, 1 MHz	Dielectric loss factor, 1 MHz	Dielectric strength, V/mil	Flexural strength, lb/in.^{2}	Applications, electronic and electrical
Steatite:							Electronic and electrical
Unglazed	L-523	0.00069-0.0010	6.42-6.03	0.0041-0.0063	230	21,200	Bushings
Glazed	L-543	0.0008-0.0014	5.73-6.14	0.005-0.008	330	28,600	Coil forms
Porcelain:							Capacitor leads
Glazed	L-232	0.0076-0.0099	5.42-6.01	0.041-0.059	249	13,600	Electronic packages
Zircon:							Envelopes, tubes
Unglazed	L-433	0.0012-0.0014	8.14-8.22	0.010-0.012	259	22,900	Insulators, antenna
Glazed	L-413	0.00119	8.92	0.011	191	29,800	Insulators, cyclotron
Alumina:							Insulators, spark plug
Unglazed	L-746	0.0001-0.0008	8.14-	0.0009	500	68,000	Insulators, thermocouple
Glass:							Insulators, tube element
Borosilicate (Pyrex) .	L-622	0.00074	4.19	0.0031	226	16,000	Housing lamp
High silica (Vycor) .	L-541	0.0017	3.78	0.0065	363	8,520	Magnetron parts
Glass-bonded mica:							Printed-circuit boards
Unglazed	L-442	0.0017-0.0018	7.08-7.44	0.012-0.013	382	18,600	Radomes
Forsterite:							Resistor bases
Glazed	L-723	0.0003	6.37	0.002	200	20,000	Supports, tube element
Cordierite:							Shafts, condenser
Unglazed	L-321	0.0049	4.57	0.022	245	13,000	Substrates
Wallastonite:							Terminals
Unglazed	L-621	0.0004	6.49	0.003	293	13,700	Tube windows
Beryllia	L-834	0.00015	6	0.0009	295	29,000	Transformer bushings
							Tuner-coupling arms

Table 7-26. Good Design Practice for Ceramic Radio Insulator: *Dimension Tolerances*
(1 in. = 2.54 cm)

Material	Type of dimensions	Dimensions up to and including 12½ Inch	Dimensions over 12½ Inch	Cylindrical shapes (OD and ID) Inch	Hole center* Inch	Hole diameter† Inch
Glass or glass-bonded mica‡	*Noncritical and critical*	±0.010	±0.015	±0.015	12½ diam. and less, ±0.005; over 12½ diam., ±0.007	0.500 diam. and less, ±0.005; over 0.500 diam. ±1%
	Wall	Not applicable	Not applicable	±0.005	Not applicable	Not applicable
Porcelain	*Noncritical*	±3% with ±³⁄₁₆ max., ±¹⁄₃₂ min.	±1.5%	Not applicable	Not applicable	Not applicable
	Critical Unglazed	±1.5% with ±0.015 min.	±1.5%	±1.5% with ±0.015 min.	±1.5% with ±0.015 min.	±1.5% with ±0.010 min.
	Thickness	±0.015	Not applicable	±0.015 ` + 0.015 with ±0.020 min.	Not applicable	Not applicable
	Glazed	±2% with ±0.1875 max.	±1.5%	±1.5% + 0.015 with ±0.020 min.	±1.5% with ±0.015 min.	±1.5% with 0.020 min.
Steatite and other fired ceramics (except porcelain)	*Noncritical*	±2% with ±⅛ max., ±¹⁄₆₄ min.	±1%	Not applicable	Not applicable	Not applicable
	Critical Unglazed	±1% with ±0.005 min.	±1%	±1%	Less than 0.500 diam. ±0.005; 0.500 diam. and greater ±1%	Less than 0.500 diam. ±0.005; 0.500 diam. and greater ±1%
	Thickness	±0.010	Not applicable	±0.010 (nominal wall)	Not applicable	Not applicable
	Glazed	±2% + 0.012	±2% + 0.012	1 in. diam. or less, ±2% with ±0.012 min.; over 1 in. diam. ±1.5% + 0.010	Less than 0.500 diam. ±0.005; 0.500 diam. and greater ±1%	Less than 0.500 diam. ±0.015; 0.500 diam. and greater ±3%
	Thickness	±1% + 0.012	Not applicable	±0.010 (nominal wall)	Not applicable	Not applicable

* If pin gages are used for inspecting hole-center spacings, the design of the gage should be such as to meet the specific requirements. If pin gages are used, consideration should be given to the tolerance between hole centers as well as hole-diameter tolerance.

† Holes leading from the glazed surfaces should have the tolerances for glazed surfaces applied except when specified that the glaze should be removed from the hole, in which case the tolerances for unglazed surfaces should apply. Holes which are not perfect circles should have the same tolerances applied to the minor axis only.

‡ The flatness tolerance should be ±0.0015 in./in in. When angles or V cuts are required in the edges of flat pieces, the tolerance on such angular dimensions should be ±1 deg.

geneous materials in possessing high thermal conductivity comparable with metal, together with excellent electrical insulating properties.

Care must be exercised in the use of beryllium oxide because its dust is highly toxic. Although it is completely safe in dense ceramic form, any operation that generates dust, fumes, or vapors is potentially very dangerous.

Some typical uses of beryllium oxide are:
1. Heat sinks for high-power rf amplifying tubes, transistors, and other semiconductors.
2. Printed-circuit bases.
3. Antenna windows and tube envelopes.
4. Substrates for vapor deposition of metals.
5. Heat sinks for electronic chassis or subassemblies.

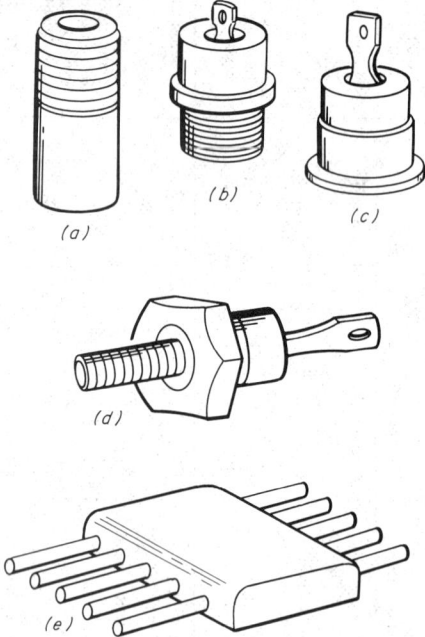

Fig. 7-90. Some typical metallized ceramic parts. (*a*) threaded ceramic bushing; (*b*) threaded metallized bushing; (*c*) metallized terminal bushing; (*d*) terminal stud feed-through; (*e*) hermetic-sealed microcircuit package.

149. Mounting of Ceramic Parts. Parts to be mounted on a flat surface should be designed with one, two, or three mounting bosses that can be ground flat after firing to prevent breakage during the mounting operation. If two bosses are used, they should be spaced 180 deg, and if three bosses are used they should be spaced 120 deg. When screws are used, they should be properly secured with corrosion-resistant lock washers or, if this is not practical, secured with weatherproof cement. Ceramic insulators should be cushioned with fungus-proof resilient gasket, fish paper, cork, lead, or other shock-absorbing material.

PLASTIC INSULATORS

150. General. The term plastics usually refers to a class of synthetic organic materials (resins) which are solid in finished form but at some stage in their processing are fluid enough to be shaped by application of heat and pressure. The two basic types of plastic are *thermoplastic resins,* which, like wax or tar, may be softened and resoftened repeatedly without undergoing a change in chemical composition, and *thermosetting resins,* which undergo a chemical change with application of heat and pressure and cannot be resoftened.

151. Choice of Plastics. Some of the differences between plastics that should be considered when defining specific needs are degree of stiffness or flexibility; useful temperature range; tensile, flexural, and impact strength; intensity, frequency, and duration of loads; electrical strength and dielectric losses; color retention under environment; stress-crack resistance over time; wear and scratch resistance; moisture and chemical resistance at high temperature; gas permeability; weather and sunlight resistance over time; order, taste, and toxicity; and long-term creep properties under critical loads.

152. Reinforced Plastics. These comprise a distinct family of plastic materials which consist of superimposed layers of a synthetic resin-impregnated or resin-coated filler. Fillers such as paper, cotton fabric, glass fabric or fiber, glass mats, felted asbestos, nylon fabric— either in the form of sheets or macerated—are impregnated with a thermosetting resin (phenolic, melamine, polyester, epoxy, silicone). Heat and pressure fuse these materials into a dense, insoluble solid and nearly homogeneous mass, which may be fabricated in the form of sheets or rods or in molded form.

153. Plastic Insulators. Table 7-27 shows typical properties of plastic (glass-reinforced-resin) materials for electrostructural devices and gives pertinent data on physical and electrical properties and their electronic applications.

High-pressure reinforced laminates seem destined to play a big role in industrial electronics, electronic home appliances (transistor radios, TV sets, hi-fi amplifiers), and in electronic business-computer applications, and particularly in the control circuits for guided missiles.

154. Mounting Plastic Parts. Choice of assembly method depends on the strength needed, contour to the parts, appearance demands, and mold design. Parts may be attached to each other or to other materials by bolts or screws, but care must be exercised to assure a tight fit without crushing or damaging the plastic. Bolts with large heads and washers should be used to distribute the damping force. Parts must fit snugly; loose fit allows movement, causing rapid wear and deterioration of the part by abrasion. Load-bearing parts are often made with metal inserts to hold the bolts and distribute the stress.

RELAYS, SWITCHES, CONNECTORS

By J. Spergel*

155. Introduction. The primary function of electromechanical components is the transmission and control of electric current accomplished by mechanical contacting and actuating devices. In recent years, solid-state (nonmechanical) switching devices have come into wide use and their applications are extending rapidly.

156. Relay Types. The simplified diagram of a relay shown in Fig. 7-91 illustrates the basic elements that constitute an electromagnetic relay (EMR). The most common relay types are as follows:

General-Purpose. Design, construction, operational characteristics, and ratings are adaptable to a wide variety of uses.

Latch-in. Contacts lock in either the energized or deenergized position until reset either manually or electrically.

Polarized (or Polar). Operation is dependent upon the polarity of the energizing current. A permanent magnet provides the magnetic bias.

Differential. Functions when the voltage, current, or power difference between its multiple windings reaches a predetermined value.

Telephone. An armature relay with an end-mounted coil and spring-pickup contacts mounted parallel to the long axis of the relay coil. Ferreeds are also widely used for telephone cross-point switches.

Stepping. Contacts are stepped to successive positions as the coil is energized in pulses; they may be stepped in either direction.

Interlock. Coils, armature, and contact assemblies are arranged so that the movement of one armature is dependent upon the position of the other.

Sequence. Operates two or more sets of contacts in a predetermined sequence. (Motor-driven cams are used to open and close the contacts.)

---

* Deceased.

Table 7-27. Typical Properties of Plastic (Glass-reinforced-Resin) Materials for Electronic Applications

Material	Dielectric constant, at 1 MHz	Power factor, at 1 MHz	Dielectric strength, V/mil	Volume resistivity, Ω-cm	Impact strength, ft.lb/in. of notch, Izod	Flexural strength, lb/in.²	Resistance to heat, °F, continuous
Epoxies.........	5	0.01	400	10^{14}	30	40,000	500
Phenolics	6	0.04	400–900	10^{13}	18	60,000	550
Polyesters	6	0.008	420	10^{15}	16	20,000	350
Silicones	4	0.004	400	10^{15}	15	14,000	900

Plastic Insulator Applications

Loop-antenna mountings
Switching-device mountings
Focus-coil insulation
Rotary-switch and shaft parts
Flyback insulators
Rotary-switch parts
Picture-tube socket and cable supports
Delay-cable insulators
High-voltage-transformer coil forms
Turner shafts

Printed-circuit and copper-clad boards
Fixed-capacitor insulation
Terminal plates
Coil forms
Standoff insulators
Resistor strips
Waveguides
Radomes
UHF antennas

Jack and plug insulators
Panelboards
Fuse link casings

Bank insulators
Cord-terminal-plug handles

Time-Delay. A synchronous motor is used for accurate long time delay in opening and closing contacts. Armature-type relay uses a conducting slug or sleeve on the core to obtain delay.

Marginal. Operation is based on a predetermined value of coil current or voltage.

157. Performance Criteria. The design or selection of a relay should be based on the following circuit-performance criteria:

Operating Frequency. Electrical operating frequency of relay coil.

Rated Coil Voltage. Nominal operating voltage of relay coil.

Rated Coil Current. Nominal operating current for relay.

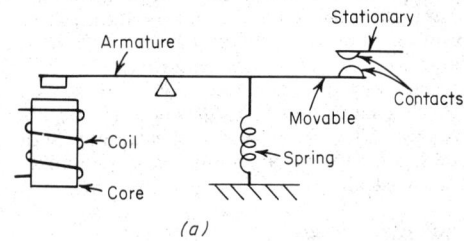

(a)

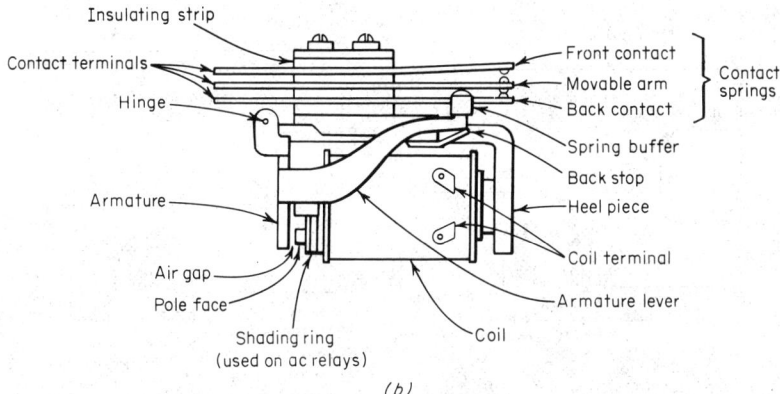

(b)

Fig. 7-91. Simplified diagram of a relay and typical relay structure. (*a*) Single-pole, single-throw, normally open relay; (*b*) structure of conventional relay. *(From "Electronics Components Handbook," McGraw-Hill Book Company, New York, 1958.)*

Nonoperate Current (or Voltage). Maximum value of coil current (or voltage) at which relay will not operate.

Operate Voltage (or Current). Minimum value of coil voltage (or current) at which switching function is completed.

Release Voltage (or Current). Value of coil voltage (or current) at which contacts return to the deenergized position.

Operate Time. Time interval between application of power to coil and completion of relay-switching function.

Release Time. Time interval between removal of power from coil and return of contacts to deenergized position.

Contact Bounce. Uncontrolled opening and closing of contacts due to forces within the relay.

Contact Chatter. Uncontrolled opening and closing of contacts due to external forces such as shock or vibration.

Contact Rating. Electrical load on the contacts in terms of closing surge current, steady-state voltage and current, and induced breaking voltage.

Figure 7-92 illustrates some of the contacting characteristics during energization and deenergization. Table 7-28 provides characteristic data on a variety of relay types. Figure 7-93 illustrates the effect on relay operation of changing coil current. Detailed discussion and analysis of relay design parameters, as well as data on magnetic-core materials, winding coils, and general formulas for temperature of electromagnets, are provided by Peek and Wagar (see reference Par. 7-181).

158. General Design and Application Considerations. The dynamic characteristics of the moving system, i.e., armature and contact assembly, are primarily determined by the mass of the armature and depend upon the magnet design and flux linkage. Typical armature configurations shown in Fig. 7-94 are clapper or balanced armature, hinged or pivoted lever about a fixed fulcrum; rotary armature; solenoid armature; and reed armature. Contact and restoring-force springs are attached or linked to the armature to achieve the desired make and/or break characteristics. The types of springs used for the contact assembly and restoring force are generally of the cantilever, coil, or helically wound spring type. Primary characteristics for spring materials are modulus of elasticity, fatigue strength, conductivity, and corrosion resistance. They should also lend themselves to ease of manufacture and low cost. Typical materials for springs are spring brass, phosphor bronze, beryllium copper, nickel silver, and spring steel.

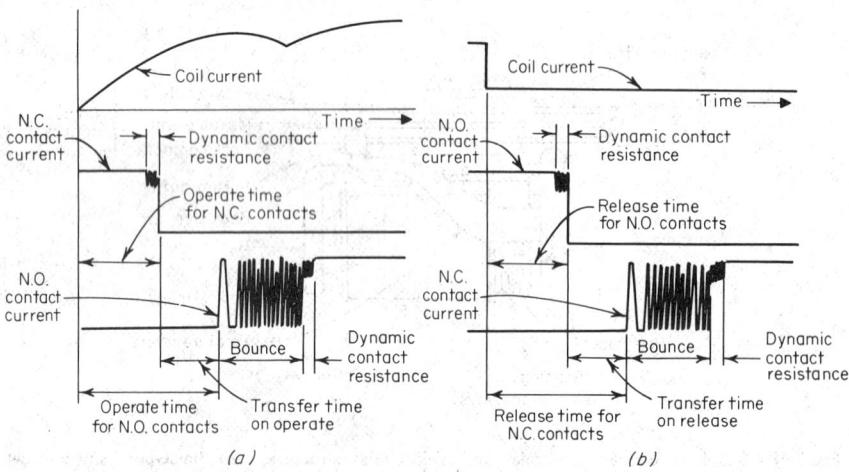

Fig. 7-92. Typical oscillograph pictures of contacting characteristics during (*a*) energization and (*b*) deenergization of a relay. *(Automatic Electric, Northlake, Ill.)*

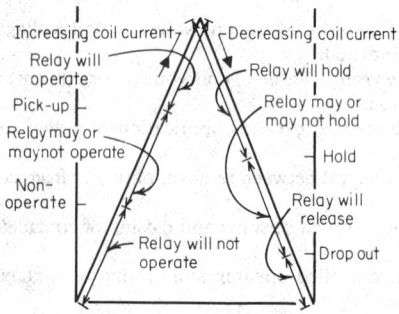

Fig. 7-93. Illustration of the effect of changing coil current. *(Automatic Electric, Northlake, Ill.)*

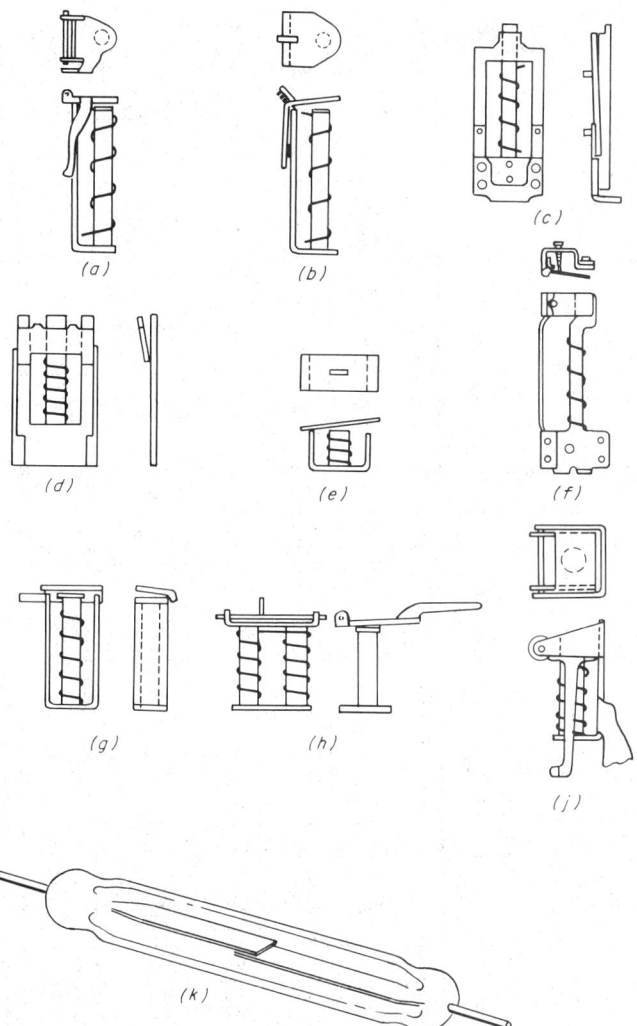

Fig. 7-94. Typical examples of armature designs (*a–j* are clapper types; *k* is a reed type). *(Copyright 1955, Litton Educational Publishing, Inc. Reprinted by permission of Van Nostrand Reinhold Publishing Company, New York, from "Switching Relay Design.")*

Contacts. These include stationary and moving conducting surfaces that make and/or break the electric circuit. The materials used depend on the application; the most common are palladium, silver, gold, mercury, and various alloys. Plated and overlaid surfaces of other metals such as nickel or rhodium are used to impart special characteristics such as long wear and arc resistance or to limit corrosion.

The heart of the relay is the contact system that is typically required to make and/or break millions of times and provide a low, stable electrical resistance. The mechanical design of the relay is aimed principally at achieving good contact performance. Because of the numerous operations and arcing often occurring during operation, the contacts are subject to a wide variety of hazards that may cause failure, such as:

Table 7-28. Comparison of Performance Characteristics of Various Relay Types
(Common ranges for general-purpose characteristics*)

Relay types	Contact rating			Sensitivity				Size	Environment			Notes
	Contact form†	Amperes	Volts	Operations × 10⁶	mW (or W), dc	VA, ac	Speed, ms	in.³	Vib., g	Shock, g	Temp. °C	
Telephone, standard	2C to 8C	0.01-2.0	28 dc or 115 ac	1.0-100	100-500	...	5-30	3-10	1-5	1-20	-10 to +70°	For clean areas free of vibration or shock
Telephone, wire-contact	2C to 18C	0-0.25	28 dc or 115 ac	50-500	1-8 W	...	2-10	5-15	1-10	1-20	-10 to +70°	Sensitivity is that required for rated speed
General-purpose	2C to 4C	5-10	28 dc or 230 ac	0.1-10	0.25-2 W	1-3	...	2-5	5-10	1-30	-10 to +70°	Suitable for adverse industrial environments
Telephone, miniature	2C to 6C	2-5	28 dc or 115 ac	0.1-10	500‡-1,000	...	10-25	1	10	20-30	-55 to +85°	Suitable for most mobile service
Reed relay, ≈2.5 in.	1A, 1B, and 1C	10 mA-1A	250 V and 25 VA max.§	5-20	100-500	...	1-5	0.5-2	10-20	10-50	-55 to +125°	Four or more switches can be mounted in one coil; unsuitable for reactive loads
Reed relay, ≈1 in.	1A, 1B, and 1C	10 mA-500 mA	150 V and 10 VA max.§	5-20	250-400	...	0.5-3	0.1-0.5	10-30	10-100	-65 to +125°	

Mercury-wetted contact	1A and 1C	2-5	300 V to 100 VA	1,000	10-100	...	1-5	1-5	...	...	-39 to +105°	Normally mounted within 30° of vert.
Crystal can, standard	2C	2	28 dc or 115 ac	0.1-1.0	250-400	...	4-11	0.3	20-30	50-100	-65 to +125°	All aircraft and missile requirements
Telegraph	1C	0.06-0.5	120 dc inductive	100-1,000	0.5-50	...	1-3	5-15	...	...	-10 to +50°	Several watts coil drive req. for rated speed

* All relays are capable of use or special adjustments outside the common ranges, but gain in one characteristic is usually at the expense of some other.
† Form A-SPNO, form B-SPNC, form C-SPDT.
‡ Up to 50 mW sensitivity can be achieved when load and environmental requirements permit.
§ Special switches are available up to 5,000 V.
SOURCE: W. Holcombe, Relay Facts, *Electromech. Design,* February, 1969.

Film formation: Effect of inorganic and organic corrosion, causing excessive resistance, particularly at dry-circuit conditions.

Wear erosion: Particles in contact area which can cause bridging between small contact gaps.

Gap erosion: Metal transfer and welding of contacts.

Surface contamination: Dirt and dust particles on contact surfaces can prevent achievement of low resistance between contacts and may actually cause an open circuit.

Cold welding: Clean contacts in a dry environment will self-adhere or cold-weld.

One of the major factors in determining relay-contact life is the arcing that occurs at the contact surface during the period in which the contacts are breaking the circuit. Contact life (and hence relay reliability) can be greatly enhanced by the addition of appropriate contact-protection circuits. Such circuitry can reduce the effects of load transients, which are especially deleterious. A variety of circuits may be employed using bifilar coil windings, RC networks, diodes, varistors, etc. As a rule of thumb, for compact relays with operating speeds of from 5 to 10 ms the approximate parameters for suitable RC networks are approximately R, as low as possible, but sufficient to limit the capacitor discharge current to the resistive load rating of the contacts; and C, a value in microfarads, approximately equal to the steady-state load current in amperes.

Details on the effects of various methods of suppression are given in the 19th Relay Conference Proceedings (see reference, Par. 7-181).

159. Packaging. Relays are packaged in a wide variety of contact arrangements (see Table 7-29) and in many package configurations ranging from T05 cans and crystal cans to relatively large enclosures, as well as plastic encapsulation, and open construction with plastic dust covers. The packaging adopted depends on the environment and reliability requirements. If the environment is controlled, as in a telephone exchange, a relatively inexpensive and simple package may be used. In a military or aerospace environment, hermetically sealed or plastic-encapsulated packages are essential, to prevent corrosion of contacts. In this regard the reed switch has had great impact on relay design since it provides a sealed-contact enclosure in a relatively simple and inexpensive package. It has become widely used in the telephone and electronics industry because it has been able to extend relay life to millions of cycles. A comparison between the reed and typical conventional relays is given in Table 7-30.

With the growth of the microelectronic technology, the development of solid-state relays (SSR) has been achieved and is making an impact in the relay industry. Table 7-31 highlights the important differences in performance of SSR and EMR.

SWITCHES

160. Introduction. Switches are electromechanical devices used to make or break an electric circuit by manual or mechanical operation. Switches are available in many types for many functions. Figure 7-95 illustrates some typical configurations.

161. Selection Criteria. Some of the more important criteria are as follows:

Switching Speed. Duration of contact travel during *make* or *break* function. It is generally desirable to have high speed during make to minimize the duration of an arc or flashover. If the duration is excessive, the contact surface will deteriorate and welding of the contacts may occur. Arc-suppression techniques should be used if higher currents are anticipated. During break, it is generally desirable to have a slower speed to minimize transients, particularly for dc and inductive circuits.

Electrical Noise. Electromagnetic radiations may occur during make and break of switches, causing interference in sensitive circuits or high-gain amplifiers. Suppression of arc may be necessary to reduce such noise to acceptable levels.

Capacitance. In some circuits switches may look like a capacitor and its capacitance may be sufficient to cause complications, particularly in frequencies of 60 to 100 MHz.

Frequency. Switch ratings are generally given for direct current, 60 and 400 Hz, and would not apply at higher frequencies. Such applications may necessitate a special switch design to meet specified electrical requirements.

Contact Snap-over and Bounce Time. Snap-over time is the time a contact separates from a normally closed position and travels to a normally open position and makes contact with the circuit. Bounce time is the interval between initial contact and steady contact during

Table 7-29. Nomenclature for Basic Contact Forms

Form	Description	Symbol	Form	Description	Symbol
A	Make or SPSTNO		J	Make, make, break or SPST (M-M-B)	
B	Break or SPSTNC		K	Single pole, double throw, center off or SPDTNO	
C	Break, make or SPDT (B-M)		L	Break, make, make or SPST (B-M-M)	
D	Make, break or SPDT (M-B)		U	Double make, contact on arm or SPSTNODM	
E	Break, make, break or SPDT (B-M-B)		V	Double break, contact on arm or SPSTNCDB	
F	Make, make or SPST (M-M)		W	Double break, double make, contact on arm or STDTNC-NO(DB-DM)	
G	Break, break or SPST (B-B)		X	Double make or SPSTNODM	
H	Break, break, make or SPST (B-B-M)		Y	Double break or SPSTNCDB	
I	Make, break, make or SPST (M-B-M)		Z	Double break double make or SPDTNC-NO(DB-DM)	

Poles, single (SP), double (DP); throws, single (ST), double (DT); normal position, open (NO), closed (NC); double make (DM), double break (DB).

Table 7-30. Estimated Load-Life Capability of Two-Pole Miniature Relays

	Dry reed, 0.125 A	Conventional crystal can, 5 A	Miniature power, 10 A
λ at 1×10^6 (%/10^4)	0.002	1.2	0.80
$R_{(.999)}$ (10^3 operation)	700	2.8	6.0
$R_{(.90)}$ (10^3 operation)	10,000	60	120.0

λ = failure rate in %/10,000 h operation; $R_{(.999)}$ = operating life with 99.9% probability; $R_{(.90)}$ = operating life with 90% probability.

Toggle switch			L	D	H	h	Bushing thread
Standard	– SPDT···		1.3	0.625	1.0	0.480	15/32 – 32
	DPDT···		1.3	0.750	1.0	0.480	15/32 – 32
Miniature	– SPDT···		0.562	0.350	0.749	0.170	1/4 – 40
	DPDT···		0.562	0.560	0.749	0.170	1/4 – 40
Subminiature	–SPDT···		0.492	0.250	0.515	0.354	1/4 – 40
	DPDT···		0.492	0.450	0.515	0.354	1/4 – 40

Snap switch (momentary contact)		L	D	H
	Standard········	1.94	0.68	0.91
	Miniature········	1.09	0.41	0.63
	Subminiature······	0.78	0.25	0.38
	Subsubminiature···	0.50	0.25	0.38

Rotary switch (available in "make – before – break" and "break – before – make")

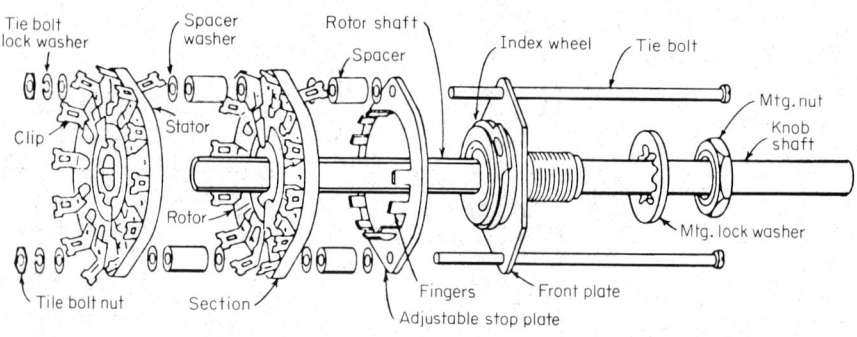

	Section dia.	Bushing thread	Shaft dia.		Positions per section
Standard····	1.875	3/8 – 32	1/4	0.625	2 to 24
	1.500	3/8 – 32	1/4	0.558	2 to 24
Miniature ····	1.280	3/8 – 32	1/4	0.558	2 to 24
Subminiature	0.940	1/4 – 32	1/8	0.125	2 to 12

Fig. 7-95. Typical switch configurations. *(From "Handbook of Electronic Packaging," McGraw-Hill Book Company, New York, 1969.)*

which the contact bounces as a result of impact of a moving contact on a stationary contact. These times should be kept to a minimum because the time intervals and/or the contacting instability may influence critical or sensitive circuits.

162. Switch Configurations. Switches are available in numerous configurations and packages to meet special equipment designs. These are often *ganged* or assembled in matrices. A variety of unique switching capabilities can be achieved by using a permanent magnet in combination with reed capsules. In addition, many pushbutton switches are available with illuminated faces. Mil-Std-454, Standard General Requirements for Electronic Equipment, and MIL-Std-1132, Selection and Use of Switch and Associated Hardware, are useful guides for procurement and application.

CONNECTORS

163. Introduction. Electrical connectors are electromechanical devices which provide the capability of interconnecting discrete or remotely located subassemblies or equipment components of a major electronic system. This capability simplifies the manufacture, installation, and maintenance of electronic equipment. However, the introduction of connectors into the circuit adds a discontinuity in both the conducting path and the insulation, which may cause deterioration of the circuit due to the effects of the environment. Accordingly, connectors are required to provide minimum electrical contact resistance, ruggedness, and environmental protection for stable circuit performance. The references (Par. 7-181) should be consulted for comprehensive data on coaxial and multicontact connectors, as well as the various manufacturers' catalogs and Military Specifications.

164. Wire Terminations. There are several methods of terminating wire directly to a circuit or component. The most popular are solder, weld, crimp, or wire-post terminations. The techniques of soldering wire to electrical contacts or other terminals are described in NASA Handbook SP-5002, "Soldering Electrical Connections."

To achieve a good weld joint, a prescribed weld cycle must be determined for each combination of materials. The parameters of welding also depend upon the techniques and specific equipment. In view of the variety of methods available, care must be taken in following equipment manufacturers' instructions.

Crimping of contacts to wire should be performed with crimp tools of specification MIL-T-22520. Wire-wrap joints which utilize solid wire are coming into wide use for back-plane (or "mother"-board) wiring and should satisfy requirements of MIL-Std-1130.

165. Contacts. The heart of any connector is the contact. The design of the connector is aimed at providing a stable contact which can be easily engaged and disconnected with good durability and can be protected from the environment and mechanical hazards. The materials of the contacts are generally a copper-base alloy and are coated with a precious metal, as indicated in Table 7-32. There is no one optimum plating; selection of material should be based on the specific application. The use of selective plating techniques and/or lubricants may be desirable to minimize cost and improve performance. The contact may be either fixed in the connector where the wire can be soldered or welded to the contact, or it can be removable where the contact is crimped to the wire (Fig. 7-96). Table 7-33 summarizes contact requirements in many of the current connector specifications.

166. Design Features. A wide variety of connective devices are generally available to satisfy various application and installation requirements. MIL-Std-454 lists the different types of connectors available and their general application. The following features should be incorporated in connectors:

Polarization of mated halves, which should take place before contact engagement and provide positive-contact alignment.

Closed-entry contacts and chamfered socket cavities in hard insert to assure proper contact engagement with minimum pin bending or insulator damage.

Quick and rugged coupling designs to withstand field abuse and maintain positive mating for stable performance. Bayonet or double-coarse-thread designs are preferred.

Scoop-proof shell to prevent accidental pin damage. This feature locates contact pins within the shell so that the mating connector shell cannot come into contact with the pins.

Cable housing should provide *adequate space for assembly, sealing, and strain relief.*

Table 7-31. Relative Comparison of Electromagnetic Relays (EMR) vs. Solid-State Relays (SSR)

Characteristic	EMR	SSR	Advantage
Life	From 100,000 to millions of cycles. Reed contacts are outstanding.	No moving parts. When properly designed should last life of equipment.	SSR
Isolation	Infinite dielectric isolation.	Not dielectrically isolated; however, several techniques are available to achieve up to 10 $k M\Omega$.	EMR
EMI (RFI)	Can generate EMI by switching of its coil, thereby requiring special isolation (i.e., shielding).	Noise generated is negligible compared with EMR.	SSR
Speed	Order of milliseconds.	Up to nanoseconds.	SSR
Operate power	Uses more power than SSR.	Lower power requirements but requires continuous standby power.	SSR
Contact voltage drop	Relatively low voltage drop because of low contact resistance.	High voltage drop which is dissipated into heat.	EMR
Thermal power dissipation	Primarily concerned with dissipating coil power.	Higher voltage drop develops appreciable heat to be dissipated.	EMR

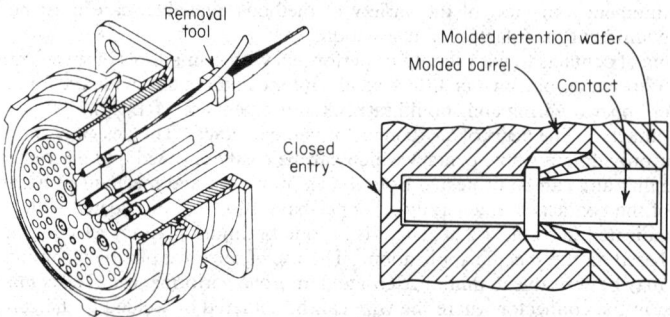

Fig. 7-96. Typical connector with rear-release removable contacts. Contacts are inserted from rear of connector and held by cone-shaped retainers. Removal is accomplished by use of a simple plastic removal tool that fits around terminal end of contact, expanding retaining cone in insert, and thus allowing contact to be removed.

Table 7-32. Plating for Connector Contacts

Material and Thickness, In.	Remarks
0.0003 hard or soft gold over 0.0002 (min.) silver	Suitable for most crimped-contact applications. Provides low-contact resistance for signal circuits and good wear resistance. Corrosion resistance is good unless excessive wear exposes the silver underplating, which is then sensitive to sulfide atmosphere. Porous gold or thickness less than the specified 0.0003 will also result in sulfide contamination.
0.00005 hard or soft gold over 0.0002 (min.) silver	Same as above, but much more resistant to wear, with subsequent higher resistance to chemical deterioration.
0.00005 gold over 0.0002 nickel	Excellent for wear resistance and hostile environments.
0.00005 gold over copper flash	Excellent for low-level circuit applications with low to medium insertion and withdrawal requirements, as in many data processing and computer applications.
0.0005 to 0.001 silver	Suitable for power contacts with relatively high contact forces.
0.0003 electrotin	For use in low-cost applications where few disconnects are anticipated.

Connector design should be *compatible with the cable*. This is particularly important because many failures occur at the cable-connector junction.

167. References on Resistors

JOHNSON, J. B. Thermal Agitation of Electricity in Conductors; *Phys. Rev.*, July, 1928, Vol. 32, pp. 97–109.

FINK, DONALD G., and CARROLL, JOHN M. "Standard Handbook for Electrical Engineers"; New York, McGraw-Hill Book Company, 1968.

HENNEY, KEITH "Radio Engineering Handbook"; New York, McGraw-Hill Book Company, 1959.

VARIABLE RESISTIVE COMPONENTS INSTITUTE "Wirewound and Non-wirewound Precision Potentiometers"; Chicago, Ill.

ELECTRONIC INDUSTRY STANDARDS AND ENGINEERING PUBLICATIONS Washington, D.C. (See Table 7-3.), May 1971.

MILITARY SPECIFICATIONS ON RESISTORS. (See Table 7-3.)

168. References on Capacitors

DUMMER, G. W. A., and NORDENBERG, HAROLD M. "Fixed and Variable Capacitors"; New York, McGraw-Hill Book Company, 1960.

FINK, DONALD G., and CARROLL, JOHN M. (eds.) "Standard Handbook for Electrical Engineers"; New York, McGraw-Hill Book Company, 1968.

169. References on Inductors and Transformers

SCHLICKE, H. M. "Dielectromagnetic Engineering"; New York, John Wiley & Sons, Inc., 1961.

DEAN, C. E. Bandwidth Factors for Cascade Tuned Circuits, Electronic Reference Sheet, *Electronics*, July, 1941, pp. 41–42.

HENNEY, K. (ed.) "Radio Engineering Handbook"; New York, McGraw-Hill Book Company, 1959.

RUTHROFF, C. L. Some Broad-band Transformers; *Proc. IRE*, August, 1959, Vol. 47, No. 8, pp. 1337–1342.

KAJIHARA, H. H. Miniaturized Audio Transformer Design for Transistor Applications; *IRE Trans.—Audio*, January–February, 1956.

HOWE, J. G., et al. "Final Report for High Power, High Voltage, Audio Frequency Transformer Design Manual"; August, 1964, Contract No. BSR 87721, Project Serial No. SR008-03-02, Task 9599, U.S. Navy.

GROVER, F. W. "Inductance Calculations"; Princeton, N.J., D. Van Nostrand Company, Inc., 1946.

U.S. DEPARTMENT OF COMMERCE Circular of the National Bureau of Standards C74, 1952.

Table 7-33. Mechanical and Electrical Characteristics of Contacts

Connector specification	Applicable contact specification	Contact plating	Contact size	Accept wire gage (AWG)	Contact terminal type	Contact rating, A†
MIL-C-5015		Gold over silver	16		Solder*	22
			12			41
			8			73
			4			135
MIL-C-22992		Gold	0			245
MIL-C-26482		Gold over silver	20		Solder	7.5
			16			13.0
			12			
	MIL-C-23216 MIL-C-39029	Gold over silver	20	24	Crimp	3.0
				22		
				20		7.5
			16	20		7.5
				18		
				16		13.0
			12	14		17.0
				12		23.0
MIL-C-26500	MIL-C-26636	Rhodium over silver or nickel	20	24	Crimp	3.0
				22		5.0
				20		7.5
			16	20		7.5
				18		16.0
				16		22.0
			12	14		32.0
				12		41.0
NAS 1599	NAS 1600	Gold or silver or nickel	20	24	Crimp	3.0
				22		5.0
				20		7.5
			16	20		7.5
				18		15.0
				16		20.0
			12	14		25.0
				12		35.0
MIL-C-25955		Gold over silver	20	30	Crimp	
				24		
				22		
				20		7.5
MIL-C-27599		Gold over silver	16		Solder	13.0
			20			7.5‡
			22			5.0
			22M			3.0
MIL-C-38999		Gold over silver	16	16	Crimp	13.0
				18		10.0
				20		7.5
			20	20		7.5
				22		5.0
				24		3.0
			22	22		5.0
				24		3.0
				26		2.0
			22M	24		3.0
				26		2.0
				28		1.5
MIL-C-81511	MIL-C-39029	Gold over copper	22	26	Solder	2.0
			20	24	Crimp	3.0
			16	22		7.5
				20		10.0
				18		13.0
			12	16		17.0
				14		23.0
				12		

* Available with crimp/removable contacts under manufacturer's part numbers.
† Maximum contact rating for individual contact.
‡ Hermetic contacts rated at 5.0 and 10.0 A, respectively.

170. References on Vacuum Tubes

DOOLITTLE, H. D.   Vacuum Power Tubes for Pulse Modulation; *Machlett Cathode Press,* 1964, Vol. 21, No. 1.

KOHL, W. H.   "Handbook of Materials and Techniques for Vacuum Devices"; New York, Reinhold Publishing Corporation, 1967.

MILLMAN, J., and SEELY, S.   "Electronics"; New York, McGraw-Hill Book Company, 1951.

SCHNEIDER, S., and TAYLOR, G. W.   Transients in High-power Modulators; *IEEE Trans. Electron Devices,* 1966, Vol. ED-13, No. 12.

SPANGENBERG, K. R.   "Vacuum Tubes"; New York, McGraw-Hill Book Company, 1948.

TERMAN, F. E.   "Electronic and Radio Engineering," 4th ed.; New York, McGraw-Hill Book Company, 1955.

EIMAC DIVISION OF VARIAN   "Care and Feeding of Power Grid Tubes"; San Carlos, Calif., 1967.

"Receiving Tube Manual"; Harrison, N.J., RCA, 1973.

171. References on Storage and Conversion Devices

KAZAN, B., and KNOLL, M.   "Electronic Image Storage"; New York, Academic Press, Inc., 1968.

IEEE Trans. Electron Devices; September, 1971, Vol. ED-18, No. 9.

IRE STANDARDS ON ELECTRON TUBES   "Methods of Testing," Part 8, Camera Tubes; Part 10, Cathode-Ray Charge Storage Tubes; New York, Institute of Electrical and Electronic Engineers, 1962.

CROWELL, M. H. and LABUDA, T. M.   Silicon Diode Array Camera Tube; *Bell Syst. Tech. J.,* May-June, 1969, Vol. 48, No. 5, pp. 1481-1528.

172. References on Semiconductor Diodes and SCRs

CLEARY, J. F.   "G.E. Transistor Manual"; General Electric Company, 1964.

GUTZWILLER, F. W.   "G.E. Silicon Controlled Rectifier Manual"; General Electric Company, 1967.

EVERITT, W. L.   "Semiconductor Controlled Rectifiers"; Englewood Cliffs, N.J., Prentice-Hall, Inc., 1964.

"RCA Tunnel Diode Manual"; Radio Corporation of America, 1963.

"RCA Silicon Power Circuits Manual"; Radio Corporation of America, 1969.

LINDMAYER, J., and WRIGLEY, C. Y.   "Fundamentals of Semiconductor Devices"; Princeton, N.J., D. Van Nostrand Company, Inc., 1965.

173. References on Transistors

PHILLIPS, A. B.   "Transistor Engineering"; New York, McGraw-Hill Book Company, 1962.

GARTNER, W. W.   "Transistors: Principles, Design and Applications"; Princeton, N.J., D. Van Nostrand Company, 1960.

JOYCE, M., and CLARKE, K. K.   "Transistor Circuit Analysis"; Reading, Mass., Addison-Wesley Publishing Company, Inc., 1961.

CLEARY, J. F.   "G.E. Transistor Manual"; General Electric Company, 1964.

174. References on Batteries and Fuel Cells

FALK, S. UNO, and SALKIND, ALVIN J.   "Alkaline Storage Batteries"; New York, John Wiley & Sons, Inc., 1969.

MANTELL, C. L.   "Batteries and Energy Systems"; New York, McGraw-Hill Book Company, 1970.

HEISE, GEORGE W., and CAHOON, N. COREY   "The Primary Battery," Vol. 1; New York, John Wiley & Sons, Inc., 1971.

FLEISCHER, ARTHUR, and LANDER, JOHN L.   "Zinc–Silver Oxide Batteries"; New York, John Wiley & Sons, Inc., 1971.

Proc. Power Sources Conf., 1956-1970; Red Bank, N.J., PSC Publications Committee.

"Application Engineering Handbook"; Gainesville, Fla., General Electric Company, 1971.

Eveready Battery Applications Engineering Data, Union Cabride Corp., 1971.

FINK, D. G., and CARROLL, JOHN M. (eds.) "Standard Handbook for Electrical Engineers"; New York, McGraw-Hill Book Company, 1968.

175. References on Magnetic Memory Devices

BARTEE, THOMAS C. "Digital Computer Fundamentals," 2d ed.; New York, McGraw-Hill Book Company, 1966.

BOBECH, ANDREW H. Magnetic Bubbles; *Sci. Am.,* June, 1971, Vol. 224, No. 6.

176. References on Ferroelectric Devices

JAFFE, HANS Piezoelectric Applications of Ferroelectrics; *IEEE Trans. Electron Devices,* June, 1969, Vol. ED-1G, No. 6, pp. 557-561.

BERLINCOURT, DON A. Piezoelectric Transducers; *Electro-Technol.,* January, 1970, pp. 33-38.

177. References on Piezoelectric Devices

BUCHANAN, J. P. "Handbook of Piezoelectric Crystals for Radio Equipment Designers," WADC Tech. Rep. 54-243; Springfield, Va., National Technical Information Service, Oct. 1956.

Proc. Ann. Symp. Frequency Control, 1956-1968, National Technical Information Service, Springfield, Va.; 1969-1971, Electronics Industries Association (EIA), Washington, D.C.

GERBER, E. A., and SYKES, R. A. State of the Art; Quartz Crystal Units and Oscillators; *Proc. IEEE,* February, 1966, Vol. 54, pp. 103-116.

IEEE Standards Nos. 176, 177, and 178, Military Standards MIL-C-3098, EIA Publication R 192A.

178. References on Fluidics

KIRSHNER, J. M. (ed.) "Fluid Amplifiers"; New York, McGraw-Hill Book Company, 1966.

FOSTER, K., and PARKER, G. A. "Fluidics Components and Circuits"; New York, John Wiley & Sons, Inc.-Interscience Publishers, 1971.

BELSTERLING, C. "Fluidic Systems Design"; New York, John Wiley & Sons, Inc.-Interscience Publishers, 1971.

Fluids Quarterly, published by Fluid Amplifier Associates, P.O. Box 1244, Ann Arbor, Mich.

Fluidics Feedback (devoted primarily to current fluidic bibliographical references and abstracts), published by The British Hydromechanics Research Association, Cranfield, Bedford, England.

"Fluerics," a bibliography published by the Harry Diamond Laboratories, available from the National Technical Information Service, Sills Bldg., Springfield, Va.

179. References on Modular Assemblies

HARPER, C. A. (ed.) "Handbook of Electronic Packaging"; New York, McGraw-Hill Book Company, 1969.

KEONJIAN, E. "Microelectronics"; New York, McGraw-Hill Book Company, 1963.

MIL-STD-275C, Printed Wiring for Electronic Equipment, Jan. 9, 1970.

180. References on Insulators and Mountings

"Inorganic Dielectrics Research," a history of 23 years of ceramic dielectric research sponsored by the U.S. Army Electronics Command, Fort Monmouth, N.J., at Rutgers University; *Eng. Res. Bull.* 50, New Brunswick, N.J., Dec. 1969.

Military Specification MIL-I-10, Insulating Compound, Electrical Ceramic, Class L; U.S. Army Electronics Command, ATTN:AMSEL-PPEM-2, Fort Monmouth, N.J., Dec. 1966.

Military Specification MIL-L-23264A, Insulators, Ceramic, Electrical and Electronic General Specification for use by the Dept. of the Army, Navy, and Air Force, July 1968.

Military Specification MIL-S-55620, Substrates Ceramic for Deposition Thin Film Microcircuits; Commanding General, U.S. Army Electronic Command, ATTN:AMSEL-PPEM-2, Fort Monmouth, N.J., Jan. 1969.

American Standards Association B46.1-1962 Standard for Surface Finishes.

Standards of the Alumina and Steatite Ceramic Manufacturers Association, New York, 1964.

"Modern Plastics Encyclopedia," 1970-1971, Vol. 47, No. 10a; New York, McGraw-Hill Publications.

181. References on Relays, Switches, and Connectors

PEEK, R. L., and WAGAR, H. N. "Switching Relay Design"; Princeton, N.J., D. Van Nostrand Company, Inc., 1955.

HENNEY, K., WALSH, C., and MILEAF, H. "Electronics Components Handbook," Vol. 1, Chap. 5, Relays, and Vol. 3, Chap. 2, Connectors; New York, McGraw-Hill Book Company, 1958.

HARPER, C. A. (ed.) "Handbook of Electronic Packaging," Chap. 6, Connectors, and Chap. 8, Packaging with Conventional Components; New York, McGraw-Hill Book Company, 1969.

Proceedings, Relay Conferences, 1953–1971; Scottsdale, Ariz., National Association Relay Manufacturers, and Oklahoma University.

Proceedings, Holm Seminars on Electrical Contact Phenomena, 1955–1971; Chicago, Illinois Institute of Technology and ITT Research Institute.

HARPER, C. A. (ed.) "Handbook of Wiring, Cabling and Interconnecting for Electronics," Chap. 3, Hook-up Wiring and Connector Systems, and Chap. 4, Coaxial Cable and Connector Systems; New York, McGraw-Hill Book Company.

SECTION 8

INTEGRATED CIRCUITS

BY

ALAN B. GREBENE Vice President, Engineering, Exar Integrated Systems, Inc., Sunnyvale, California; Member, Institute of Electrical and Electronics Engineers

HANS H. STELLRECHT Signetics Corporation, Sunnyvale, California; Member, Institute of Electrical and Electronics Engineers

CONTENTS

Numbers refer to paragraphs

SECTION 8

INTEGRATED CIRCUITS

INTRODUCTION

1. Overview of Integrated Circuits. The field of integrated circuits is a relatively new branch of electronics. Its history starts with the development of the *planar process*[1] in the late 1950s and early 1960s. Since then integrated circuits have taken over a significant segment of the electronics field from their discrete counterparts. In most applications, integrated circuits offer several advantages over conventional circuitry: lower cost, higher reliability, smaller size and weight, and improved performance. Integrated circuits and microelectronics represent an overlap of a wide variety of technical disciplines, from physical chemistry and metallurgy to solid-state device and circuit theory.

Integrated circuits can be classified in different categories, depending on the fabrication technology and their functional applications. In terms of *structure*, integrated circuits fall into three classes: monolithic circuits, thin- or thick-film circuits, and hybrid circuits.

The *monolithic circuits* are the most fundamental form of integrated circuits where the entire circuit function is formed in a monolithic body of semiconductor material. All the active and passive components are physically an integral part of the substrate, and are embedded in it.

Thin- or thick-film circuits are fabricated by depositing conductive or resistive films on an insulating substrate and by imposing patterns on them to form an electrical network.

Hybrids form a natural extension of the first two classes of integrated circuits. They contain passive and active devices, as well as monolithic integrated circuits assembled and interconnected on an insulating ceramic substrate common to all.

In terms of their *functional applications,* integrated circuits can be classified in three general categories: linear, digital, and microwave. Included in the linear category are the circuit functions of amplifiers, regulators, multipliers, modulators, and other classes of analog signal-processing circuitry. *Digital integrated circuits* cover various types and classes of logic circuits, as well as semiconductor memories. *Microwave integrated circuits* cover the frequency range from 0.5 to 15 GHz and rely heavily on hybrid circuit technology.

INTEGRATED CIRCUIT FABRICATION TECHNOLOGY

2. Introduction. The planar technology is the principal method used in the fabrication of semiconductor devices for hybrid and monolithic integrated circuits. As applied to semiconductor devices, the term planar means that the devices and components fabricated by this process extend below the surface of one plane of a silicon substrate; the surface structure of the semiconductors remains essentially unaltered during the fabrication process. Silicon is especially well suited for this process because of the relative ease with which an insulating oxide layer can be formed at the surface. This oxide layer can act as a barrier to the diffusion of dopants used for the device fabrication of integrated circuits. Thus, by selectively etching openings into the oxide layer, components can be formed by localized diffusions of impurities.

Fabrication of an integrated circuit requires a sequence of several independent processing steps, as follows: (1) material preparation, (2) epitaxial growth, (3) surface passivation, (4) photolithography, (5) junction formation, and (6) film deposition. The sequence of these basic processing steps is shown in Fig. 8-1.

3. Material Preparation. The basic material preparation starts with the growing of the single-crystal semiconductor material. The grown crystal is sliced into thin *wafers,* which are lapped, polished, and chemically etched to provide a smooth, defect-free semiconductor surface.

---

[1] J. A. Hoerni, U.S. Patent 3,025,589, assigned to Fairchild Camera and Instrument Corp., New York, 1960.

Crystal Growing. The most common method used for the growth of single crystals for integrated circuit fabrication is the Czochralsky *pulling* technique. In this method an induction-heated melt of high-purity silicon, with the desired doping impurity, is prepared in an inert atmosphere. A small seed crystal of silicon is then dipped into the melt, rotated, and very slowly withdrawn from the melt. If temperature conditions of the melt are properly maintained as the seed is withdrawn, it grows as a single, oriented crystal bar.

Wafer Preparation. After the single crystal has been grown, it is sliced into wafers. The crystal is first oriented according to crystallographic planes by x-ray techniques. A diamond saw is used to slice the oriented crystal into wafers. The wafers are then lapped to the required thickness and polished with fine-graded diamond polish. Often a chemical polish is used as final preparation step to remove any mechanical damage at the surface. The final dimensions of the prepared silicon wafer are 1 to 3 in. in diameter and approximately 250 to 400 μm in thickness.

4. Epitaxial Growth. Epitaxy, derived from the Greek word meaning "arranged upon," is a growth technique where the crystal structure of a silicon substrate is extended by arranging new layers of atoms in precisely the same lattice structure as the substrate. The additional layers are formed by vapor-phase deposition in an epitaxial reactor. By controlling the deposition rates and introducing controlled amounts of impurities into the carrier-gas stream, the thickness and the resistivity of the deposited layer can be controlled.

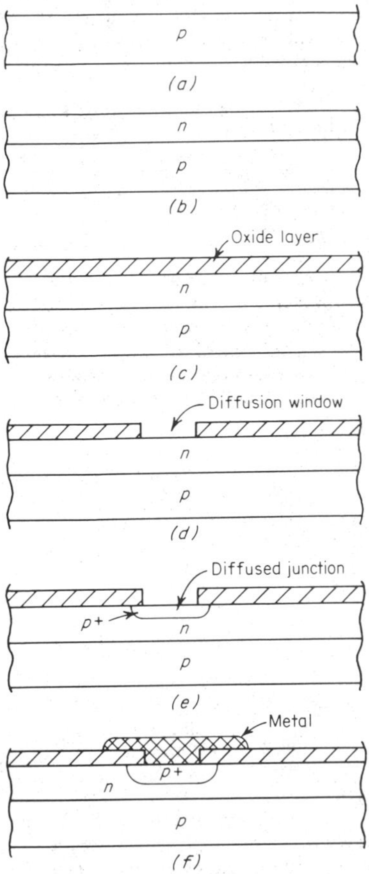

Different systems have been developed for the growth of epitaxial layers. The most commonly used systems presently are open-tube systems, where silicon is deposited on the surface of heated substrates by the reaction[1] of hydrogen with silicon tetrachloride ($SiCl_4$) or by the decomposition[2] of silane (SiH_4). Figure 8-2 shows a simplified diagram of a typical epitaxial system using both horizontal and vertical reactors. The resistivity of the deposited layers can be controlled by introducing controlled amounts of *p*- or *n*-type impurities into the carrier gas.

Fig. 8-1. Basic processing steps in planar technology. (*a*) *p*-type starting material after material preparation; (*b*) epitaxial growth; (*c*) surface passivation; (*d*) photolithography; (*e*) junction formation; (*f*) metallization.

$SiCl_4$ is an inexpensive source of silicon; it is easy to purify and is nontoxic, but requires relatively high temperatures, which causes considerable redistribution of previously diffused impurities.

An alternative epitaxial growth process which has gained considerable importance is based on the *pyrolytic decomposition of silane* (SiH_4) into Si and H_2. The disadvantage of this process is that SiH_4 is highly toxic and difficult to handle. However, this disadvantage is compensated by the advantage of lower-temperature processing.

Since the purity and crystal perfection of the epitaxial film are strongly dependent on the

---

*Superior numbers correspond to the numbered references in Par. 8-117.

surface condition of the substrate, final cleaning of the substrate wafers is very important. This cleaning is usually done in the reaction chambers with vapor-phase hydrochloric acid.

Typical epitaxial-layer thicknesses and resistivity ranges for integrated circuit applications are given in Table 8-1. Unlike the diffusion process, epitaxial growth proceeds by uniform addition of atomic layers onto the substrate. Thus the dopant impurities are relatively uniformly distributed through the epitaxial layer, and do not show a significant concentration gradient.

Table 8-1. Typical Epitaxial-Layer Thickness and Sheet Resistivity

Circuit type	Thickness, μm	Resistivity range, $\Omega \cdot cm$
Linear, low frequency	8–15	1–6
Linear, high frequency	6–10	0.2–1
Digital	5–8	0.22

The epitaxial growth process takes place at temperatures comparable with those encountered during the diffusion steps. Consequently, when the epitaxial-layer resistivity is different from the background resistivity, some impurity redistribution can take place across the epitaxial-layer-substrate interface, due to the exchange of impurities by diffusion.[3]

Figure 8-3 shows the resistivity of uniformly doped p- or n-type silicon substrate or epitaxial layers as a function of impurity concentration.[4]

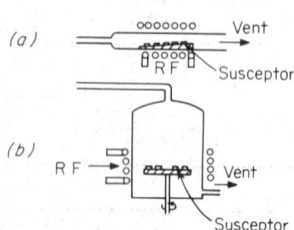

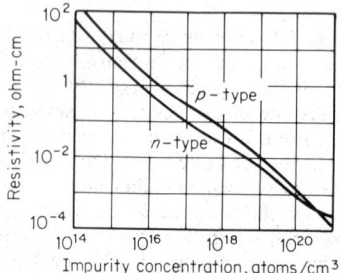

Fig. 8-2. Typical epitaxial reactor systems. (a) Horizontal reactor; (b) vertical reactor.

Fig. 8-3. Resistivity of silicon as a function of uniform impurity concentration at 300 K (Ref. 4).

5. Surface Passivation. The passivation of the silicon surface by an insulating dielectric layer is one of the principal features of the planar process. Passivation is normally achieved by thermal oxidation of silicon (Par. **8-6**). The thin layer of silicon dioxide obtained in this manner serves four functions:

1. It serves as a diffusion mask and allows selective diffusion into the silicon through the openings etched into the silicon dioxide.
2. It protects the diffused junctions from impurity contamination.
3. It serves as a surface insulator which separates devices and metal interconnections.
4. It can serve as a dielectric film for monolithic capacitors.

6. Thermal Oxidation. Thermal oxidation[5,6] of the silicon wafers is usually carried out in a diffusion furnace at temperatures between 900 and 1200°C. If a pure oxygen ambient is used, the oxide layer is formed on the silicon surface through the basic chemical reaction

$$Si + O_2 \rightarrow SiO_2$$

This reaction is significantly accelerated in the presence of water vapor, and proceeds in accordance with the reaction

$$Si + 2H_2O \rightarrow SiO_2 + 2H_2$$

Figure 8-4 shows typical growth rates of SiO_2 as a function of temperature for the cases of dry and wet oxygen (steam).[7]

7. Masking Properties of SiO_2. The diffusion coefficients of most dopants in SiO_2 are between two and four orders of magnitude smaller than in silicon. Therefore an SiO_2 layer of proper thickness can serve as a diffusion barrier against these dopants. Figure 8-5 shows the typical oxide thickness required to mask against boron or phosphorus diffusion, as a function of diffusion times and temperatures.[8]

Ionic Contamination. Some positively charged ions such as sodium (Na^+) or hydrogen (H^+) have relatively large diffusion coefficients in SiO_2 at low temperatures. This can be especially detrimental to integrated circuit devices containing lightly doped *p*-type regions.

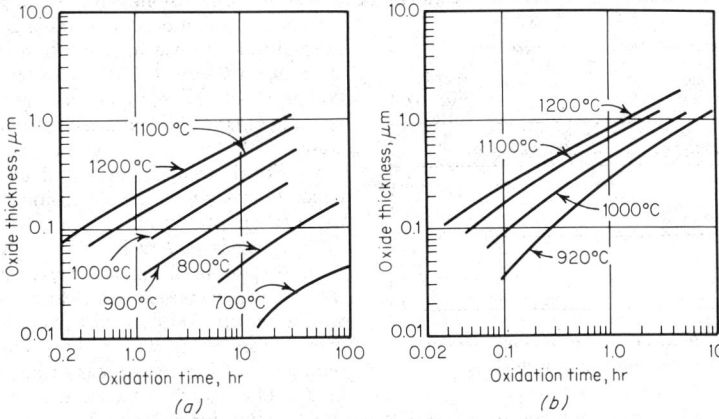

Fig. 8-4. Typical growth rates of SiO_2 as a function of temperature. (*a*) Dry O_2; (*b*) oxygen plus water vapor (adapted from Ref. 7).

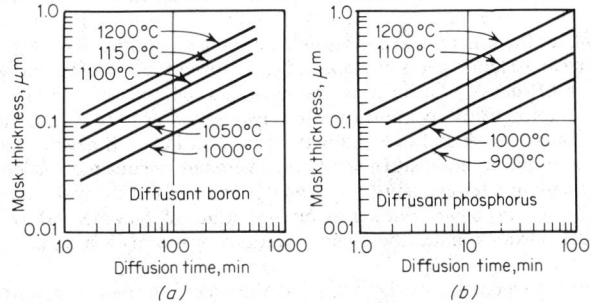

Fig. 8-5. Masking oxide thickness for (*a*) boron, and (*b*) phosphorus (Ref. 8).

8. Silicon Nitride Passivation. Silicon nitride (Si_3N_4) is more resistant to ionic contamination than SiO_2. It can be used as an additional passivating layer to prevent surface contaminants from reaching the SiO_2 layer. This is especially true of bipolar circuits designed to operate at low current levels and for devices such as MOS transistors which rely heavily on surface phenomena. The silicon nitride layer can be formed by pyrolytic deposition at temperatures of 800 to 1000°C, or by rf or reactive-sputtering techniques. Often an additional silicon dioxide layer is deposited over the silicon nitride to facilitate photomasking.

Deposited Oxides. The high temperatures needed for the growth of thermal oxides cause redistribution of impurities within the previously diffused layers. To avoid this, passivating oxide layers can be formed by low-temperature deposition processes such as pyrolytic decomposition of oxysilane, vapor-phase reactions, anodizing, sputtering, or plasma oxidation.[9]

9. Postmetal Passivation. Sometimes it is desirable to form a protective dielectric coating over the integrated circuit surface after the metal interconnections have been formed. This can be done by pyrolytic deposition of SiO_2 or silicon glass, commonly referred to as *s*-

glass. This postmetal passivation involves additional processing steps, but in return it provides surface protection for the interconnections and for the thin-film devices on the surface. In addition to this it allows the use of multiple-layer metallization, which in turn greatly simplifies the circuit layout. A structural diagram of a monolithic circuit chip using postmetal passivation is shown in Fig. 8-6.

10. Photolithography. Initial design and layout of an integrated circuit are carried out on a scale several hundred times larger than the final desired dimensions of the circuit pattern.

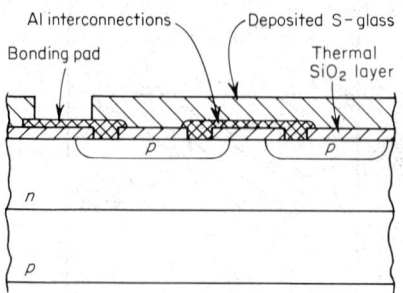

Fig. 8-6. Postmetal passivation of wafer surface by pyrolytic deposition of silicon glass.

The initial layout of an integrated circuit is normally done on a magnification scale in the range of 250:1 to 500:1. This layout is a composite of different mask patterns corresponding to the different masking steps associated with the fabrication process. The dimensional accuracy of the final circuit depends strongly on the dimensional stability of this master layout.

Each mask pattern is cut into a dimensionally stable plastic laminate layer, called a *rubylith,* which consists of a clear mylar base with a peelable opaque ruby overlay. The overlay can be cut with a sharp knife and removed to form clear areas in an opaque field. Photographic techniques are then used to reduce each of these mask layers to the final circuit dimensions.

The reduced form of the mask is then reproduced many times on a transparent glass slide using a photographic step-and-repeat process. The end result of this process is a final mask which has multiple images of the circuit pattern to cover the entire surface of the silicon wafer to be masked. Therefore an array of a large number of identical masks can be applied simultaneously over the wafer surface in a single masking operation. The larger the number of circuits which can be fabricated simultaneously, the lower the cost of the individual circuit.

11. Photoresist Process. During the masking operation, the wafer surface to be masked is coated with a photosensitive coating known as *photoresist,* or *resist.* The masking plate is then used as a contact mask and exposed with ultraviolet light. If negative-acting resist is used, the portions of the photosensitive resist not covered by the opaque portions of the mask polymerize and harden after exposure. The unexposed parts of the resist can be dissolved and washed away, leaving a *photoresist mask* on the surface of the wafer. If a positive-acting resist is used, the portions under the opaque sections of the mask are left to perform the masking function.

After the photoresist coating has been applied, the wafer is heated for a short time to drive out solvent traces and to densify the resist layer. Then the circuit mask is placed over the wafer and aligned to its desired position by two-directional micrometer adjustments. This alignment process requires the aid of a microscope.

12. Etching. The photomasking step is followed by an etching step. In this step, the parts of the SiO_2 layer which are not covered by photoresist are etched away, thereby exposing the bare silicon to form diffusion or contact windows in the oxide layer. A buffered hydrofluoric acid solution is used to etch the SiO_2 at a controlled rate. After the etching process, the remaining photoresist is stripped in a special solvent. This photomasking procedure is repeated for all the remaining diffusion steps, as well as in forming the metal-interconnection pattern. A flow chart of standard mask making and photolithography steps is shown in Fig. 8-7.

13. Photolithography Defects and Limitations. The three fundamental limitations of the photolithography process are the alignment and resolution of the mask patterns and etching defects.[10]

Mask Alignment. The pattern of each mask layer applied to the silicon-wafer surface must align properly with the previously applied patterns.

Resolution. The resolution of the photomasking step may be measured by the minimum line width needed to resolve two parallel lines spaced one line width apart. The main limitations of mask resolution are due to the finite grain size in the photographic emulsions and to the diffraction of light at the mask edges.

Etching Defects. Etching defects come about due to irregular thickness or adherence properties of the photoresist, or the nonuniform etching properties of the oxide. These defects can result in poorly defined oxide edges and may increase the size of resulting diffusion windows.

14. Junction Formation. To form a *pn* junction in an integrated circuit, a controlled number of impurities must be introduced into the silicon substrate. This can be done by either *diffusion* or by *ion-implantation techniques,* as described below.

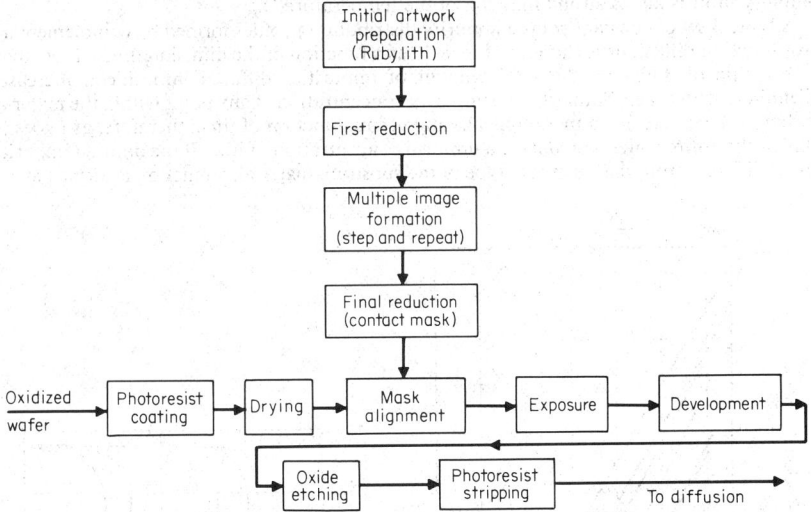

Fig. 8-7. Flow chart of mask-making and photolithography steps.

15. Diffusion. The diffusion process is presently the most widely used method of introducing controlled amounts of impurities into the silicon substrate. It is a relatively well understood and highly reproducible process which readily lends itself to the batch-processing advantages of the planar technology, where many silicon wafers can be processed in a single operation.

In the crystalline silicon lattice, the impurities can move by two different diffusion mechanisms: *substitutional* and *interstitial diffusion.*[11] In substitutional diffusion, the impurity atoms propagate through the crystal lattice by jumping from one lattice site to the next, thus substituting for the original host atom. The number of available lattice sites to which the impurity atom can jump is given by the number of vacancies in the lattice. In the case of interstitial diffusion, the impurity atoms move through the crystal lattice by jumping from one interstitial site to the next. Since there are five interstitial voids in a unit cell of the silicon lattice, and only a few are occupied by point defects, an interstitial impurity can diffuse through the lattice at a much faster rate.

The impurities which are diffused into silicon to determine the type and the resistivity of various regions of the semiconductor material are elements from group III or V of the periodic table, for *p*- or *n*-type doped regions, respectively. Some of these dopants are listed in Table 8-2. For integrated circuit fabrication, B, P, As, and Sb are the most commonly used dopants. All these dopants are substitutional impurities. Figure 8-8 shows the diffusion coefficients of silicon dopants as a function of temperature.[12]

16. Constant-Source Diffusion. In this type of diffusion, the impurity concentration at the semiconductor surface is maintained at a constant level throughout the diffusion cycle. The constant impurity level on the surface is determined by the temperature and the carrier-gas flow rate of the diffusion furnace. In most constant-source diffusion systems, it is

Table 8-2.   p- and n-Type Dopants for Silicon-Device Fabrication

p-Type	n-Type
Boron (B) Aluminum (Al) Gallium (Ga) Indium (In)	Phosphorus (P) Arsenic (As) Antimony (Sb)

convenient to let the surface concentration N_0 be determined by the solid-solubility concentration limit of the particular dopant in silicon. As shown in Fig. 8-9, the solid-solubility limit is also a strong function of the temperature.[13]

A typical set of constant-source impurity distribution profiles formed by complementary error function diffusions is shown in Fig. 8-10 as a function of the diffusion time. Note that, in this type of diffusion, the total amount of impurities diffused into silicon increases indefinitely with time. Similarly, the impurity concentration at any point within the material (except at the surface) is a monotonically increasing function of time; therefore, as t goes to infinity, the entire wafer would have a uniform concentration of N_0. If the diffused impurity type is different from the resistivity type of the substrate material, a junction is formed at the

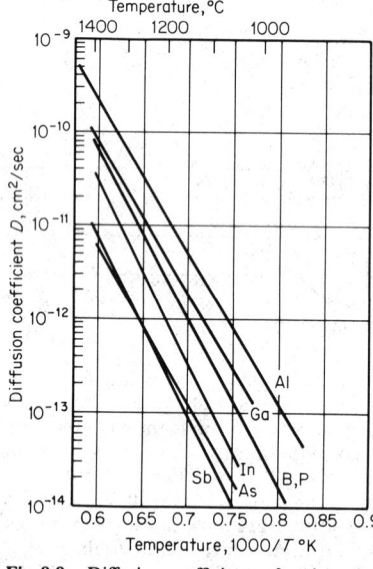

Fig. 8-8.   Diffusion coefficients of various dopants in silicon.

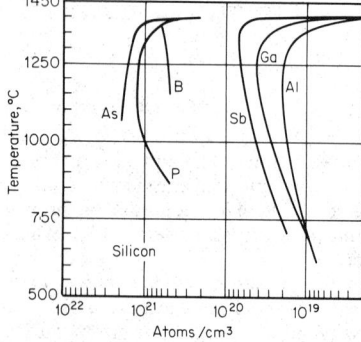

Fig. 8-9.   Solid solubilities of various dopants in silicon (Ref. 16).

points where the diffused impurity concentration is equal to the background concentration already present in the substrate. These junction depths are shown as points x_1, x_2, and x_3, respectively, for the diffusion profiles of the figure. In the fabrication of monolithic circuits, constant-source diffusion is commonly used for the isolation and the emitter diffusion steps.

17. Limited-Source (Gaussian) Diffusion. In this type of diffusion, the total amount of impurity introduced into the semiconductor during the diffusion step is limited. This is achieved by depositing on the silicon surface a fixed amount of impurity atoms per unit of exposed surface area, during a short "predeposition" step prior to the actual diffusion. This predeposition is then followed by a "drive-in" cycle where the impurity already deposited during the predeposition step is diffused into the silicon substrate.

Figure 8-10b shows a sketch of the limited-source diffusion profile, for increasing values of time. Note that in this type of diffusion, the surface concentration N_0 is inversely proportional to the square root of the diffusion time. In integrated circuit fabrication, the limited-source diffusion is commonly used in forming the transistor base regions.

18. Basic Properties of the Diffusion Processes. In the design and layout of monolithic integrated circuits, the following three fundamental properties of the diffusion process must be considered:

   a. All diffusions proceed simultaneously. The impurities introduced in an earlier diffusion step continue to diffuse during the subsequent diffusion cycles. Therefore, when calculating the total effective diffusion time for a given impurity profile, one must often consider the effects of the subsequent diffusion cycles. The effects of the subsequent diffusions on a given impurity profile can be estimated by defining an effective product equal to the sum of the products of the diffusion coefficients and the respective times of diffusion.

$$Dt \approx D_1 t_1 + D_2 t_2 + D_2 t_3 + \cdots$$

Thus, for example, in the planar device fabrication, the emitter region of a bipolar transistor is formed by a diffusion process which succeeds the base diffusion step. Therefore the effective Dt product of the base region contains a finite contribution from the emitter diffusion step.

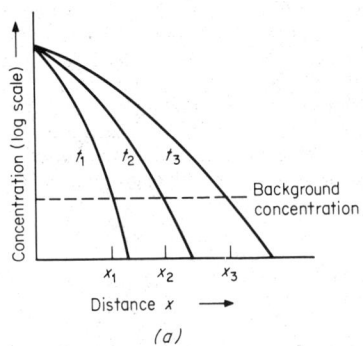 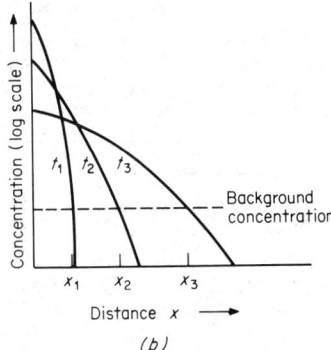

Fig. 8-10   Constant source (*a*) and limited sources (*b*) diffusion profiles as a function of time.

   b. For a given surface and background concentration, the *junction depths x* associated with the two separate diffusions having different diffusion times and temperature are related as the square root of the ratio of the respective products.

$$x_1/x_2 = \sqrt{D_1 t_1 / D_2 t_2}$$

   c. The diffusion proceeds *sideways from a diffusion window* as well as downward. In considering the lateral dimensions of the planar devices, these lateral diffusion effects must be taken into account. Figure 8-11 shows the lateral diffusion effects for various concentration ratios.[14] Figure 8-11a gives the constant concentration contours for the case of a constant-source diffusion. The diffusion contours of Fig. 8-11b correspond to the limited-source case. Typical side diffusion effects are about 75 to 85% of the vertical penetration for most impurity concentration levels encountered in integrated circuit fabrication.

   19. Ion Implantation.   Ion implantation provides an alternative method for introducing impurities into a semiconductor.[15,16] In contrast to diffusion processes, the number of implanted ions is controlled by the external system, rather than by the physical properties of the substrate. This allows a precise monitoring of the dose of the implanted layers, even at temperatures at which normal diffusion is insignificant. The implanted dopant concentration is not limited by solubility considerations; and a much wider variety of dopant elements may be used. An important side benefit of this is that implanted layers can be formed without affecting previously diffused device structures.

   In the ion-implantation process, the desired impurities are introduced into the semiconductor lattice by bombarding the surface with high-energy impurity ions. The implantation operation takes place in vacuum. Impurity ions are accelerated from an ion source, and a mass spectrometer is used to separate the undesired impurities from the beam. The ion beam is then focused to a small area (typically smaller than a $\frac{1}{4}$ in.²) and is scanned across the semiconductor wafer which serves as the target.

   The impurity ions penetrate the target lattice at typical energy levels used for implantation applications, in the 50- to 150-keV range. The depth of penetration is a function of the kinetic energy of the impurity ions and the mass of the target atoms, the lattice spacing, and the crystal plane facing the incident ion beam.

Because of better reproducibility, ions are usually implanted with the beam at a small angle with respect to the $\langle 111 \rangle$ crystal plane of the semiconductor target. This prevents channeling because the semiconductor appears amorphous to the incident beam, thereby resulting in better reproducibility.[16]

The semiconductor surface can be masked against the implanted ions by using a metal layer (such as aluminum) or a thick oxide layer as a mask. Thus ion-implanted regions can be readily patterned at the silicon surface in the same manner as diffused regions, using photomasking techniques.

20. Range Distribution. The range is the total distance the ion travels within the target before coming to rest. The projection of this distance on the direction of incidence is called the *projected range* X_p. A typical range distribution in an amorphous substrate is approximately gaussian in shape, and can be characterized by a mean range, and a finite range distribution about this mean value, as shown in Fig. 8-12. Typical value of the mean range for 100-keV ions is approximately $0.1 \ \mu m$. Figure 8-13 shows X_p as a function of the implantation energy E_{I0} for different impurity ions. The standard deviation σ_R associated with the range distribution is also shown in the figure.[17]

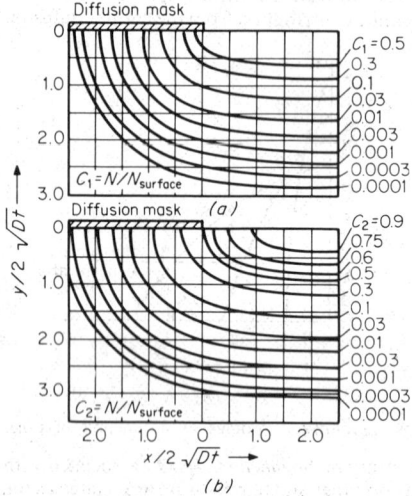

Fig. 8-11. Lateral diffusion profiles for (*a*) constant-source and (*b*) limited-source diffusions (Ref. 14).

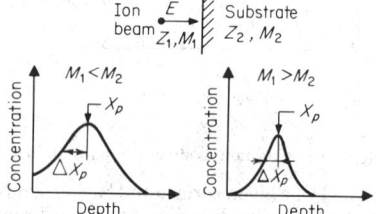

Fig. 8-12. Depth distribution of implanted atoms in an amorphous target for the case in which the ion mass is less or greater than the mass of the substrate atoms. To a first approximation the mean depth X_p depends on ion mass M_1 and incident energy E, whereas the relative width $\Delta X_p / X_p$ of the distribution depends primarily on the ratio of the mass of the ions to the mass of the substrate atoms M_2 (Ref. 16).

Table 8-3 shows the projected range and standard deviation of some important dopants in silicon. The impurity profile in a semiconductor after ion implantation is determined by random scattering of ions. Unlike the diffusion processes, the highest impurity concentration is not found at the semiconductor surface, but at a distance X_p from the surface, as shown in Fig. 8-12. Figure 8-14 shows a typical gaussian impurity profile obtained by implantation with different ion doses.

Table 8-3. Projected Range and Standard Deviation of Selected Ions as a Function of Initial Kinetic Energy E_{I0} in Silicon for Nonchanneling Conditions at 300 K (After Ref. 17)

Ion type	Projected range, Å			Standard deviation, Å		
	$E_{I0} = 10$	10^2	10^3 keV	$E_{I0} = 10$	10^2	10^3 keV
Al	160	1,450	13,020	65	420	1,610
As	95	585	5,730	25	125	820
B	385	3,980	23,230	190	940	1,810
Ga	100	610	6,130	25	135	890
In	90	465	3,820	15	75	490
P	145	1,230	11,760	55	355	1,530
Sb	85	455	3,680	15	75	465

21. Crystal Damage. Ion bombardment of a crystalline semiconductor usually results in structural damage to the crystal lattice of the target material. After an incident ion has dissipated its kinetic energy upon entering the target, it typically lodges in an interstitial position of the target lattice. During its penetration, the incident ion displaces many target atoms, and creates a large number of interstitials and vacancies. The extent of this damage depends on the properties of the impurity ions and the target, the ion energy, the impurity concentration, and the ambient conditions. At sufficiently high doses, a noncrystalline or amorphous layer is formed. The crystal damage can be almost completely removed by a high-temperature annealing heat-treatment. If the damage is confined to isolated disordered regions, anneal temperatures of 200 to 300°C are sufficient to remove most of the damage.

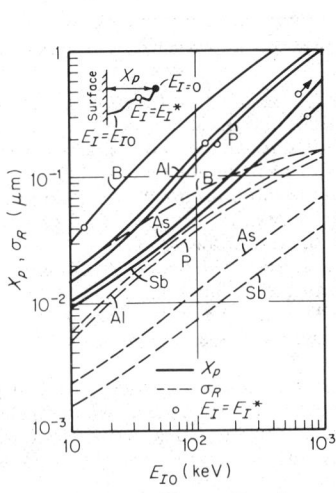

Fig. 8-13. Projected range X_p and standard deviation σ_p versus ion energy at semiconductor surface E_{I0} for various impurities in silicon at 300 K (after Ref. 17).

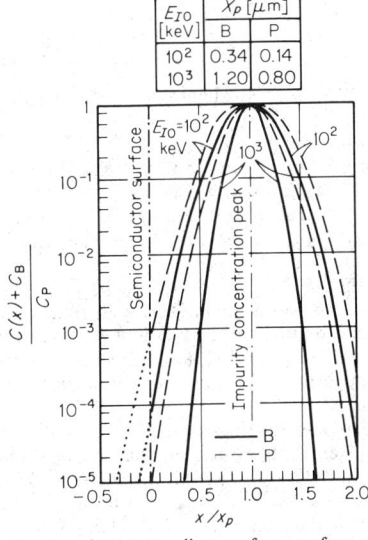

E_{I0} [keV]	X_p [μm]	
	B	P
10^2	0.34	0.14
10^3	1.20	0.80

Fig. 8-14. $C(x)$ at a distance from surface x and location of impurity concentration peak, X_p, for boron and phosphorus in silicon at 300 K (after Ref. 17).

22. Applications. The spatial distribution of implanted impurities can be determined by channeling-effect measurements, and their electrical behavior can be measured using Hall effect techniques. After the implantation step, only a small fraction of the implanted atoms are electrically active.

Ion implantation offers a wide variety of applications in the fabrication[19] of monolithic devices. Some of its most important applications are in fabricating complementary bipolar or MOS transistors, implanted resistors, and MOS transistors with self-aligned gate structures and in adjusting the threshold voltage of MOS transistors.

23. Isolation Techniques. In monolithic integrated circuits, electrical isolation between the devices on the same substrate is achieved by fabricating them within electrically isolated regions of the substrate known as isolation *pockets*, or *tubs*. The electrical separation of these pockets can be achieved by one of three methods: reverse-biased junctions (junction isolation), dielectric barrier layers (dielectric isolation), or beam-lead-connected components (air isolation).

24. Junction Isolation. Junction-isolation technique is the most common and economical method used in monolithic circuits. Electrical isolation is achieved by separating the individual components by reverse-biased *pn* junctions. This method requires few additional process steps and wastes relatively little surface area. However, parasitic effects are created which can affect circuit performance. Figure 8-15 shows the cross section of a junction-isolated pocket containing an *npn* bipolar transistor. In fabricating this structure the basic planar process is used. Prior to the epitaxy step, a selective n^+ diffusion is made into the *p*-type substrate. This buried layer provides a low-resistivity current path from the collector contact to the active collector area directly below the base region.

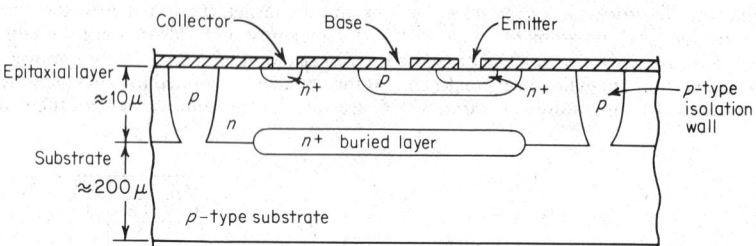

Fig. 8-15. Structural diagram of a *pn* junction-isolated pocket containing an *npn* transistor.

Following the epitaxy and surface oxidation steps, an *isolation mask* is applied to the wafer surface, which opens the diffusion windows outlining the isolation pockets. The *p*-type isolation walls are then diffused from the wafer surface, through the *n*-type epitaxial layer into the *p*-type substrate. This isolation diffusion forms a continuous *p*-type wall around a selected region of the *n*-type epitaxial layer. When the substrate and the isolation walls are at negative dc potential with respect to the *n*-type pocket, the reverse-biased *pn* junction surrounding the pocket electrically isolates it from the rest of the wafer.

Isolation diffusion is a relatively noncritical diffusion step, with the basic requirement that the final depth of the isolation wall be greater than the epitaxial layer thickness. After completion of the isolation diffusion, the transistor base and the emitter regions are formed by respective diffusions into the *n*-type pocket, which serves as the collector of the *npn* transistor.

25. Dielectric Isolation. In certain applications, the parasitic junction capacitances or leakage currents associated with the junction isolation methods may not be acceptable. In such cases, a superior electrical isolation is obtained by insulating each pocket with a dielectric layer, as shown in Fig. 8-16. Normally, thermally grown SiO_2 is used as the dielectric material.

In forming the dielectrically isolated pockets on the wafer surface, a number of alternative fabrication techniques can be utilized. Figure 8-17 shows a typical sequence of fabrication steps in forming the dielectrically isolated single-crystal silicon pockets or islands. Starting with an *n*-type substrate, a nonselective n^+ layer is diffused into the wafer surface. Following the initial n^+ diffusion, the wafer surface is oxidized, and a mirror-image mask of the desired isolation grid pattern is applied to the wafer, to re-

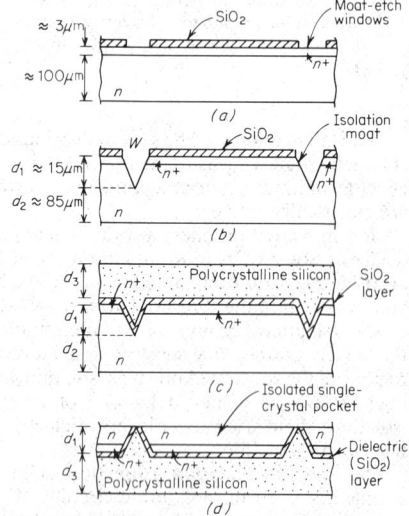

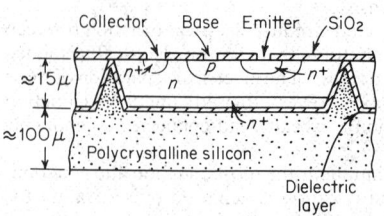

Fig. 8-16. Structural diagram of a dielectrically isolated pocket containing an *npn* transistor.

Fig. 8-17. Sequence of processing steps in dielectric isolation.

move the oxide along the isolation grid. The exposed silicon surface is then etched by a potassium hydroxide (KOH)–based etch. If a $<111>$ oriented silicon crystal is used as the substrate, this preferential etching results in the formation of a V-shaped isolation groove, or moat, on the wafer surface, as shown in Fig. 8-17b. After the preferential etching step, the exposed silicon is reoxidized, and a thick layer of polycrystalline silicon is deposited over the oxide layer, as shown in Fig. 8-17c.

Finally, the original wafer is flipped around and the bottom surface of Fig. 8-17c now corresponds to the top of the device structure. Then the single-crystal n layer of thickness d_2 is backlapped until the tips of the isolation moats forming the isolation grid appear on the wafer surface. This results in an isolated n-type pocket as shown in Fig. 8-17d.

After the isolated pockets are formed, the fabrication of the integrated devices within the pockets is completed by a sequence of conventional masking and diffusion steps, resulting in the isolated-device structure of Fig. 8-16.

26. Air Isolation (Beam-Lead Technology). The two isolation techniques described above use silicon and silicon dioxide as a support medium to give the integrated circuit structural integrity. In air isolation thick metallic interconnections are used as supporting "beams" to provide structural support for the circuit, as well as serve as electrical interconnections.[21] These beams are cantilevered over the edge of the devices and also serve as external bonding connections. Besides providing superior isolation, this type of structure greatly simplifies the interconnection and packaging of integrated circuits.

Figure 8-18 shows a cross section of an air-isolated beam-lead structure. The preliminary fabrication steps for beam-lead devices are the same as those described earlier in connection with the planar process. After contact holes have been etched, a special metallization process is used which makes it possible to realize good ohmic contacts, as well as provide good adherence to the SiO_2 layer. Normally, multiple metal layers are required for this operation. Initially, platinum silicide ohmic contacts are formed; then titanium and platinum layers

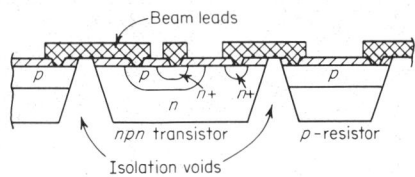

Fig. 8-18. Cross section of an air-isolated integrated circuit with beam-lead interconnections.

are sputtered on the SiO_2 surface; and finally the heavy gold beam leads are electroformed on the platinum base. Electroforming is a special form of electroplating where material is built up in selective areas only.

After the metallization pattern has been applied, the wafer is turned over and the isolation-etch-masking pattern is applied in registry with the metallization pattern on the other side of the wafer. The unmasked areas are then etched away, leaving discrete silicon mesas containing active and passive devices interconnected by the beam leads.

The beam-lead technology has several advantages over conventional processes: Higher device yields can be realized because of chemical separation rather than the scribe-and-break technique normally used; mounting and bonding of the circuit are greatly simplified; and higher reliability is obtained. The major advantage is the improvement in the electrical performance due to the reduction of the parasitic capacitances between the active device areas.

27. Thin- and Thick-Film Deposition. The electrical interconnection of integrated components is achieved by evaporation of conductive thin films on the wafer surface. Resistor and capacitor structures can also be formed on the passivated silicon surface by deposition of resistive or dielectric thin-film layers. A wide variety of thick and thin films are available in integrated circuits. The term *thin film* is used to describe approximate film thickness of 1 μm or less, as compared with the larger geometry and thicker films associated with hybrid integrated circuits. Tables 8-4 and 8-5 list some of the commonly used thin-film materials and their electrical properties.[22]

High-conductivity metallic films are used for circuit interconnections. Table 8-6 lists some of the films used in forming the circuit interconnections. Among these, aluminum (Al) is the most often used thin-film material because of its high electrical conductivity and good adherence to the SiO_2 surface. In forming the resistor patterns, resistive thin films such as tantalum (Ta), nickel-chromium (Ni-Cr) alloys, and tin oxide (SnO_2) are the most commonly used materials.

Table 8-4. Thin-Film Materials in Microelectronics

Interconnections/ Terminals	Insulation/ Passivation	Encapsulation	Resistive	Capacitive	Semiconducting		Processing	Development
Al	$SiO_2 \cdot P_2O_5$	SiO_2	Ta	SiO_2	Si	InAs	SiO_2	Si_3N_4-SiO_2
Al alloys (Cu, Si)	SiO_2	$SiO_2 \cdot P_2O_5$	TaN_2	Ta_2O_5	Se	InSb	Si	PbO
Cu	Si_3N_4	Al_2O_3	Cr-SiO	HfO_2	Te	PbS	Photopolymer	Nb_2O_5
Mo-Au	Al_2O_3	Parylene	NiCr	ZrO_2	SiC	PbTe	Mo	Ge-Si-Te
Ti-Ag	BN	$PbO \cdot B_2O_3 \cdot SiO_2$	SnO_2	$PbTiO_3$	GaAs	CdS	Cr	ZnO
Cr-Ag	SiO		Kanthal		GaP	CdSe		Polymer
Pt-Au					AlN	ZnSe		
Cr-CuAu								
Pb-Sn								

8-14

Table 8-5.   Classification of Typical Microelectronic Thin-Film Material by Resistivity

Type	Example	Resistivity, Ω·cm
Dielectric	SiO_2 $SiTiO_3$	10^{18} 10^{14}
Resistor	NiCr $CrSiO_2$	10
Semiconductor	CdS PbTe	10^{-3}
Conductor	Al Mo Pt	10^{-6}

Table 8-6.   Properties of Metals Employed as Thin Films

Metal	Limiting current density J_1, A/cm^2 $\times$ 10^5	Resistivity, $\times$ 10^{-6} Ω·cm	Remarks
Silver	4.0	1.59	Poor adhesion
Copper	4.0	1.67	Poor adhesion, corrosion
Gold................	7.0	2.35	Silicon eutectic 370°C, poor adhesion
Aluminum	0.5	2.65	Silicon eutectic 577°C, electromigration
Aluminum + Cu	2.0		
Magnesium		4.45	Extremely reactive
Rhodium		4.51	Poor adhesion
Iridium		5.3	Poor adhesion
Tungsten	20.0	5.6	Difficult etching
Molybdenum	10.0	5.7	Corrosion susceptibility
Platinum		9.8	Poor adhesion
Titanium		55	

Figure 8-19 shows the cross-section diagrams of some of the thin-film components used in various types of integrated circuits.

28. Deposition Techniques.   A variety of deposition techniques are available for forming thick- or thin-film layers on a dielectric substrate.   Some of these processes are briefly outlined below.

Vacuum Evaporation.   The passivated substrate, together with the source of the material to be evaporated, is placed in a bell jar under high-vacuum conditions (10^{-5} to 10^{-6} torr). The material to be evaporated is heated by an electrical element until it vaporizes.   Under the high-vacuum conditions used, the mean free path of the vaporized molecules is comparable with the dimensions of the bell jar; therefore the vaporized material radiates in all directions within the bell jar.   Some of the vaporized material then deposits on the substrate, which is placed some distance from the source to ensure uniformity of deposition. The substrate is also maintained at an elevated temperature to provide a good adhesion of the deposited film.

Both conductive and resistive films can be deposited by vacuum evaporation.   Aluminum, gold, and silver are among the conductive films formed in this manner.   Nickel-chromium resistors can also be deposited by vacuum-evaporation techniques, except that in this case, due to high power densities required to vaporize the source, electron-beam bombardment, rather than thermal heating of the source material, is used.

Cathode Sputtering.   Sputtering process takes place in a low-pressure gas atmosphere.   A glow discharge is formed by applying a high voltage (typically 5,000 V) between the cathode and the anode sections of the sputtering apparatus.   The cathode is coated with the material to be evaporated, and the substrate is attached to the anode, or placed within the glow-discharge region.   Normally, an inert gas, such as argon (A), is used as the sputtering medium.   The A$^+$ ions generated by the glow discharge accelerate toward the cathode, due to the negative cathode potential.

When these high-energy ions impinge on the cathode, they cause the atoms or the molecules of the cathode to break away, or sputter, from the cathode surface. Then a part of these cathode particles which float away are intercepted by the substrate, and are deposited in the form of a thin layer. Under the low-vacuum conditions used in sputtering, the mean free path of the source atoms is much shorter than the source-to-substrate spacing. Therefore the deposition rates in the sputtering process are much slower than in vacuum evaporation.

Vapor-Phase Deposition.   In vapor- or gas-phase deposition, halide compounds of the material to be deposited are chemically reduced, and the resulting metal atoms are deposited on the substrate. This basic deposition process very closely resembles the epitaxial-growth

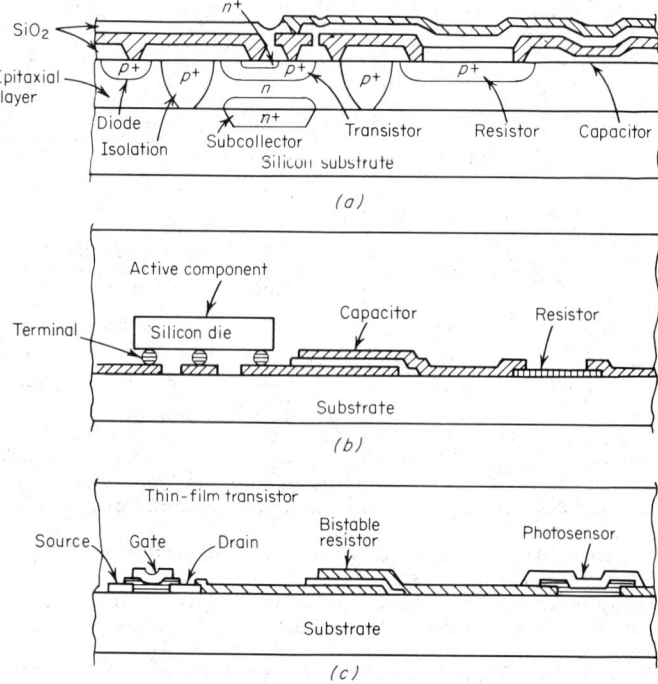

Fig. 8-19.   Cross-sectional diagram of the three types of microelectronic circuits using thin films. (*a*) Monolithic silicon integrated circuit; (*b*) hybrid silicon thin-film circuit; (*c*) all-thin-film integrated circuit.

step of the planar process. Vapor deposition is particularly useful for obtaining thick layers of deposited films (up to 20 μm). It is commonly utilized for forming aluminum oxide ($Al_2O_3 \cdot SiO_2$) dielectric layers or tin oxide (SnO_2) resistive films. The sheet resistance of SnO_2 films can be controlled by introducing group III or group V ions (such as In or Sb) to increase or reduce the sheet resistance. In this manner, a sheet resistivity range of 100 to 5,000 Ω/square can be obtained.

Plating Techniques.   Two basic kinds of plating techniques are used in forming metallic films on semiconductor substrate: electroplating and electroless plating. In electroplating, the substrate to be plated is placed at the cathode terminal of the plating apparatus and is immersed in an electrolytic solution. An electrode made up of the metal to be plated serves as the anode. When a dc current is passed through the solution, the positive-charged metal ions which dissolve into the solution from the anode migrate and plate at the cathode. This method of plating is often used for forming conductive films of gold or copper.

In electroless plating, simultaneous reduction and oxidation of a chemical agent is used in forming a free metal atom or molecule. Since this method does not require electrical conduction during the plating process, it can be used with insulating substrates. Nickel, copper, and gold are among the most common metals that can be deposited in this manner.

Anodization. Anodization is an electrochemical process for converting a metal to its oxide. In the fabrication of thin-film devices, anodization can be used to replace deposited dielectrics. This can be done by forming an anodized layer of metal oxide to serve as the dielectric. Two basic metals commonly used in the anodization process are tantalum and titanium. Anodization techniques are often used in forming dielectric layers of Ta_2O_5 on tantalum films to fabricate capacitors. Since Ta_2O_5 has a higher dielectric constant than SiO_2, a larger capacitance value can be obtained per unit area. Another application of anodization methods is in trimming tantalum resistors to obtain a stable resistor with a tight absolute-value tolerance.

29. Patterning and Etching of Thin Films. With minor modifications, the basic photo-masking and etching techniques can be utilized in patterning the thin-film components. One significant exception is the case of multiple thin-film layers (such as aluminum interconnections over thin-film resistors), where additional care must be taken in the choice of the etchant to ensure that the bottom film is not damaged by the patterning of the top layer.

In the case of Al, SnO_2, Ta, or $Al_2O_3 \cdot SiO_2$ layers, patterning and etching can be achieved by direct photoresist techniques. In the case of very thin (300 to 500 Å) nickel-chromium films, an *inverse metal-masking* technique can be used. In this process, a thin layer of metal film (typically copper) is deposited and etched by the inverse of the desired final metal pattern. Then the desired thin-film layer is deposited on this inverse metal pattern. In the final etching step, the inverse metal pattern of the initial metal film is etched away, taking with it the layer of desired metal deposited on it, and only the portion of the desired metal layer which directly adheres to the substrate is left behind.

In a simpler version of the inverse metal-masking technique, a photoresist layer may be used in place of the first metal layer, on which the desired metal can be deposited. Then the photoresist can be etched and cleared away to leave behind the desired metal pattern, which adheres directly to the substrate. Since photoresist is an organic polymer, it cannot withstand exposure to high temperatures. Therefore inverse photoresist techniques are limited to thin-film processes where the substrate temperature is maintained below 250°C.

30. Thick Films. No clear-cut definition exists for separating the thick and the thin films; generally, film layers of greater than 10-μm thickness are referred to as thick films. The structure and the dimensions of most thick films are not compatible with monolithic circuits. Therefore their use is mostly limited to hybrid structures.

The resistive thick films are normally deposited and patterned on a ceramic substrate using silk-screening techniques. The silk-screening process used in thick-film circuit fabrication is an adaptation of that used in the graphic arts industry. This technique can be used to make the metal-interconnection pattern for hybrid circuits, or in other applications where close dimensional tolerances are not required. A screen made of silk or fine stainless-steel mesh is mounted on a rigid frame and coated with a photographic emulsion or a resist. The screen is then exposed through a mask and developed, thus leaving clear areas where the film is to be deposited, and opaque areas elsewhere.

The screen is then placed on the circuit substrate, and a thin layer of the film material in viscous suspension with powdered glass is squeezed through the open areas of the screen onto the ceramic substrate. The substrate is then fired in a furnace to drive off liquids and to form the bond to the substrate. The temperature and time of the firing cycle depend on the type of ceramic used and on the combination of metals used in the thin-film material. Some of the materials which can be applied by this method are carbon, molybdenum-manganese, gold-palladium, silver-palladium, and tin-antimony.

Resistive thick films are composed of a suspension of conductive metal particles in a ceramic matrix, with an organic resin as the filler. Following the silk-screening operation, the substrate is fired at 1400 to 1600°C to ensure a permanent adhesion of the resistive film. Because of their ceramic and metal base, these resistors are known as *cermets.* Typical cermet structures are composed of Cr, Ag, or PbO in an SiO or SiO_2 matrix. By proper control of the original resistor composition, sheet resistivities as high as 10,000 Ω/square can be obtained.

This process in general does not lend itself to tight control of sheet resistance. Therefore, where resistor accuracy is desired, the deposited resistor values need to be adjusted by trimming techniques. For this purpose either abrasive or laser trimming methods can be used.

The electrical contacts to the resistive thick films are normally made by gold or nickel interconnections which are deposited by electroless plating techniques and patterned by photographic etching.

Figure 8-20 is a flow chart of process steps associated with the fabrication of a thick-film hybrid integrated circuit.[23]

31. Interconnections. *Ohmic Contacts.* The basic requirement for the conductive films used for interconnections is that they should make good ohmic contact with the diffused components or other metallic films deposited on the device surface. A good ohmic contact is defined as one that exhibits a linear voltage-current relationship and introduces a negligible amount of series resistance.

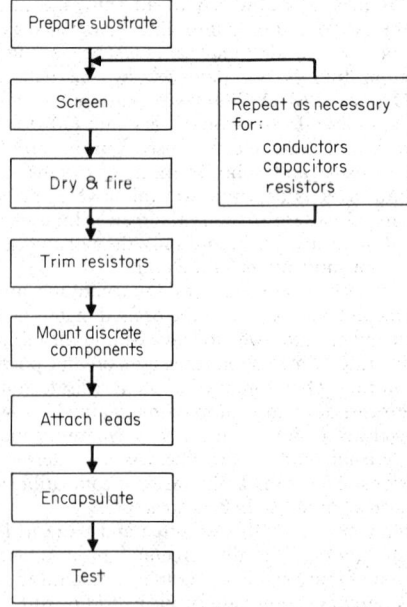

Fig. 8-20. Typical thick-film process sequence.

The exposure of contact areas to the ambient atmosphere often results in the formation of parasitic oxide layers over the chip areas to be interconnected. Therefore, to provide good ohmic contact, the interconnecting metal must be chemically active so that it can be alloyed or sintered through these parasitic oxide layers. The most commonly used interconnection metal is aluminum, which can be readily alloyed into the silicon substrate, to form ohmic contacts.

In conventional monolithic circuit fabrication, the alloying of the Al interconnections into silicon is the last step of the planar process. It is normally accomplished by a short heat-treatment in an inert atmosphere, typically about 10 min at 500°C.

A troublesome metal-interconnection problem can occur in devices which employ two active dissimilar metals in their interconnection scheme. At the interface of two dissimilar metals, parasitic intermetallic compounds and oxides can form. A typical example of this is the intermetallic gold-aluminum compounds forming between the aluminum bonding pads and the gold wires which may be used to connect the chip to the package terminals.

Interconnections in Monolithic Circuits. Aluminum is the most commonly used interconnection metal in monolithic integrated circuits. In other types of integrated circuits, such as hybrid or beam-lead devices, another conductive metal layer, such as gold or molybdenum, may be used for interconnections. The deposition and patterning techniques for these metal film layers are discussed in Pars. **8-28** and **8-29**.

Crossovers. If the interconnection cannot be accomplished in one plane, different methods of producing crossovers are available without additional processing steps. For example, the area over a diffused resistor can be used for a metal crossover, as shown in Fig. 8-21a. The emitter diffusion can be used if a low-value resistance is permissible for the interconnection. This is shown in Fig. 8-21b. Figure 8-21c shows how the geometry of an *npn* transistor can be modified to facilitate a crossover near the device. By use of these crossover methods, it is possible to interconnect most integrated circuits in one plane. However, if very high packing densities of devices are required, or if stray capacitances must be minimized, multilayer metallization can be used.

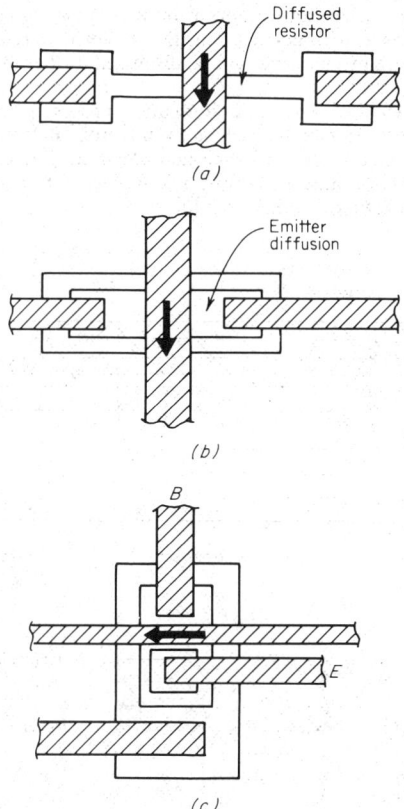

Fig. 8-21. Crossovers in monolithic integrated circuits. (*a*) Interconnection crossing diffused resistor; (*b*) crossover obtained by using emitter diffusion; (*c*) interconnection passing over base region of *npn* transistor.

Multilayer Metallization. Multilayer interconnections are formed by depositing alternate layers of patterned interconnections separated by dielectric layers, as shown in Fig. 8-22a. The dielectrics used in this application must have sufficient dielectric strength, volume resistivity, and thermal and mechanical stability to withstand the processing operations used to form the interconnect metallization. The first dielectric layer is usually the thermal oxide used to passivate the silicon surface. Subsequent dielectric layers must be formed by deposition techniques. Most multilayer metal processes use aluminum as the interconnection metal. The multiple metal-interconnection layers are then interconnected by means of the *via holes* cut in the separating dielectric layer, as shown in Fig. 8-22b.

Reliability of Interconnections. The reliability of interconnections is a serious problem in microcircuits. This is mostly due to the sharp oxide steps which must be covered by the deposited metal layers. In many cases, the step in the oxide thickness may be significantly greater than the thickness of the metal-interconnection layer. This is particularly true in the case of multilayer metal systems, and can lead to the formation of *microcracks* in the metal layer at the edges of contact windows or at sharp oxide steps.

Figure 8-22c shows a structural diagram indicating the possible location of microcracks in the vicinity of a contact window. It can be shown that if the step height is an appreciable part of the film thickness, cracking due to geometrical shadowing effects can be expected from any source configuration. The most satisfactory solution to this problem appears to be a reduction of the height of the oxide step or tapering of the oxide, as shown in Fig. 8-22d.

Various other failure mechanisms can occur in interconnection systems. Some of these are *electromigration* effects at high current densities, or formation of parasitic intermetallic compounds. These failure mechanisms will be discussed further, under Reliability, Par. **8-114**.

32. Testing and Assembly. The basic fabrication processes covered in the preceding sections deal with the fabrication of the *intrinsic* circuit only. Before the circuit is functional, it has to be assembled in a circuit package and electrically tested to guarantee certain performance requirements. Figure 8-23 shows a flow chart of the complete integrated circuit processes, including the testing and the assembly steps.

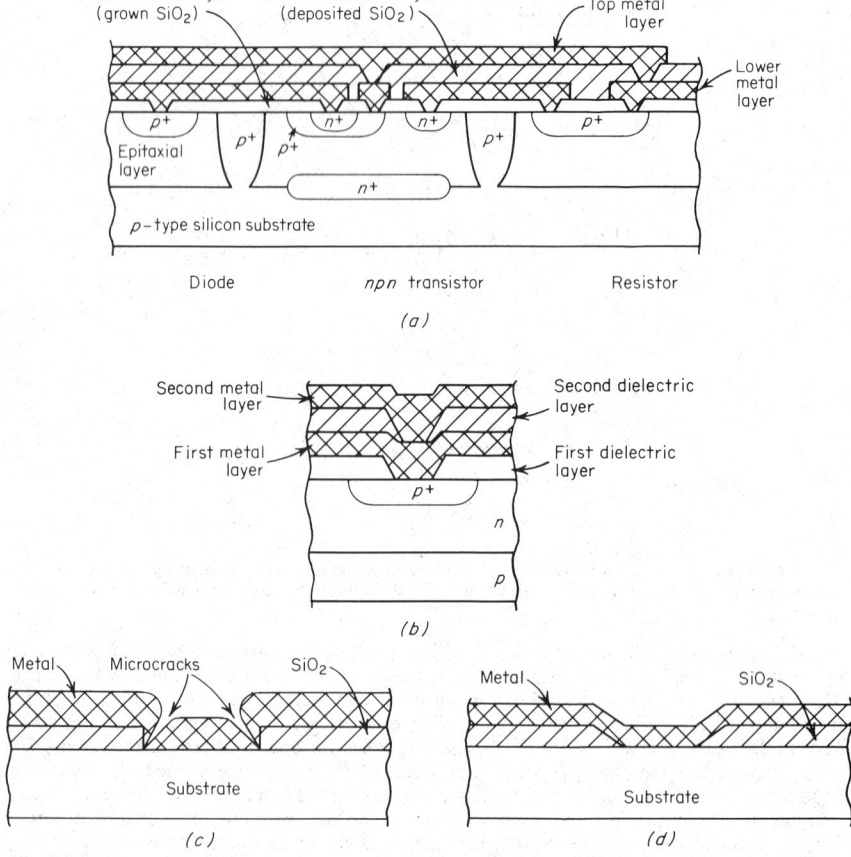

Fig. 8-22. (*a*) *pn* junction-isolated integrated circuit structure using double-layer metallization. (*b*) Double layer via connection. (*c*) Microcrack formation at oxide cut. (*d*) Good step coverage with tapered oxide steps.

The basic fabrication steps outlined in the figure can be briefly described as follows: After the wafer fabrication is completed, each finished wafer undergoes a *wafer-sort* process where each and every circuit on the wafer is contacted and electrically tested using miniaturized probes. Units that do not meet the desired electrical specifications are marked on the wafer, to be discarded. The tested wafer is then *scribed* by using a diamond-tipped cutting tool, and the individual circuit chips, or dice, are separated. Those chips which have passed the wafer-sort step are assembled in packages, where the circuit terminals are brought out to the package leads. After the assembly operation, the packaged circuits undergo a final-test step, which completes the fabrication cycle.

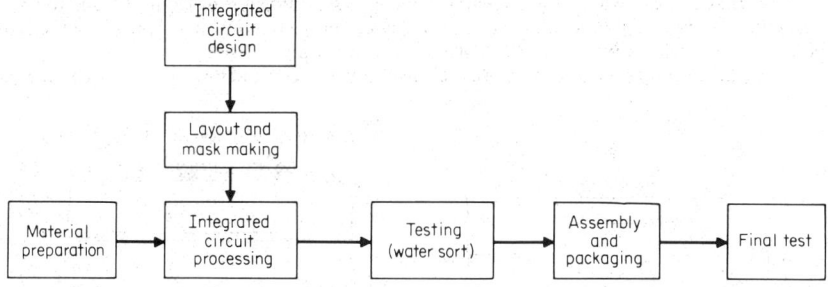

Fig. 8-23. Flow chart showing testing and assembly steps.

Testing and assembly steps do not enjoy the batch-processing advantages of the basic fabrication steps because each and every circuit needs to be individually tested and assembled. For this purpose a larger number of automated test and assembly techniques have been developed. These will be covered in detail in Pars. **8-108** to **8-111.**

MICROELECTRONIC COMPONENTS AND DEVICES

33. Introduction. The following paragraphs present a brief survey of active and passive components available in integrated circuits. In multichip hybrid circuits, it is possible to combine a number of dissimilar devices on a ceramic substrate. Therefore the device and component constraints of these circuits in general do not differ significantly from discrete (nonintegrated) circuits. The same is not true for monolithic integrated circuits (ICs). Since all the components on a monolithic substrate are fabricated simultaneously and with the same set of process steps (see Par. **3-31**) the available device structures are somewhat restricted. In many cases, the electrical characteristics of these devices also differ significantly from their discrete counterparts.

BIPOLAR TRANSISTORS

34. *npn* Transistor. The *npn* transistor is the workhorse of bipolar integrated circuits. In monolithic bipolar circuits, the choice of the *npn* transistor structure and impurity profile serves as the starting point of the design.

Figure 8-24 shows a comparison of the discrete and integrated *npn* bipolar transistor structures. In the case of the *discrete device,* a heavily doped (low-resistivity) *n*-type substrate is used as the starting material, on which an *n*-type epitaxial layer of desired resistivity and thickness is grown to serve as the collector region. The entire substrate region of this device serves as a low-resistivity collector contact.

In the case of the *integrated npn bipolar transistor,* the collector region is accessible only from the top surface of the wafer. This results in a higher value of collector series resistance r_{cs} for the integrated device. A low-resistivity n^+ layer (normally called a *buried layer*) is used along the collector-substrate interface to reduce this additional series resistance introduced by the topside collector contact. The presence of a *pn* junction isolation pocket around the device also introduces two additional parasitics. These parasitics are the collector-substrate diode D_{ss} and its associated junction capacitance C_{ss}, as shown in Fig. 8-25.

Different applications of *npn* transistors often require different device characteristics.[24] In linear-circuit applications, the devices are used in a nonsaturating mode and are in general required to have high current gain and high breakdown voltages. A device structure satisfying this requirement is one which uses a relatively thick epitaxial layer (typically 8 to 12 μm) in the resistivity range of 1 to 5 Ω·cm. Figure 8-26a shows the typical impurity profile for a monolithic *npn* transistor suitable for linear IC applications.[25] In general, the base diffusion can be approximated by a *gaussian* impurity profile, and the emitter by a complementary error function type of profile. The isolation diffusion does not directly contribute to the electrical characteristics of the device; therefore it is not shown separately in the figure.

Note that the base width W_B is defined by the intersection of the base-emitter and the base-collector impurity contours. For typical *npn* devices utilized in linear ICs, W_B is of the order of 0.5 to 1 μm, with a typical control tolerance of ± 0.1 μm.

In digital circuit applications where high switching speeds and low saturation voltages are

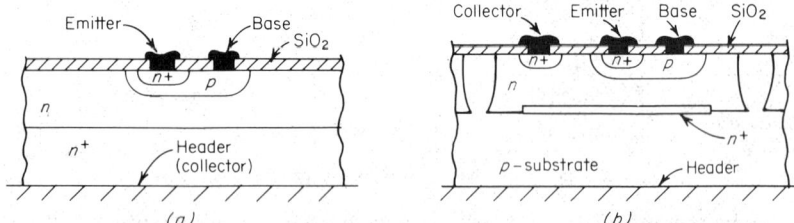

(a)   (b)

Fig. 8-24. Comparison of (*a*) discrete and (*b*) integrated *npn* bipolar transistors.

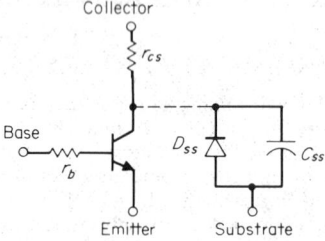

Fig. 8-25. Inherent parasitics associated with a junction-isolated transistor.

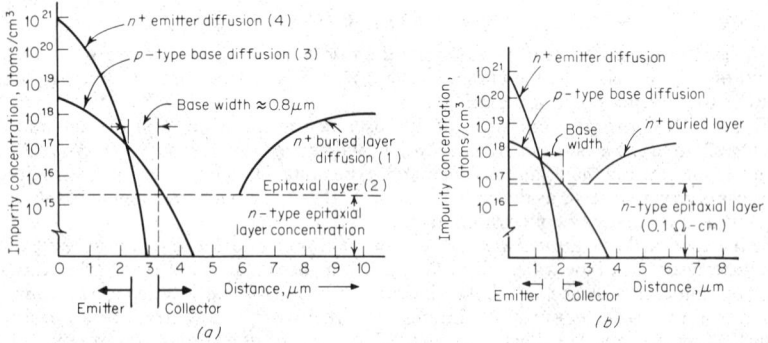

Fig. 8-26. Typical impurity profile for (*a*) linear integrated circuit transistor and (*b*) digital integrated circuit transistor.

needed, and where low breakdown voltages (≤ 20 V) can be tolerated, a somewhat different impurity profile is used.[26] A typical example of this is shown in Fig. 8-26b. A high-speed digital circuit transistor normally uses a thinner epitaxial collector region and shallow base and emitter diffusions. This choice is made to reduce the parasitic *sidewall* capacitances associated with the side portions of diffused regions.

In saturating logic circuits, a small quantity of gold is also diffused into the device structure. Gold atoms diffuse very rapidly through the entire silicon lattice and serve as recombination centers for free carriers in silicon. The net effect of gold doping is the reduction of minority-carrier lifetime, which in turn reduces the *storage time* or the *turn-off* delay associated with switching an *npn* transistor from saturated operation to "OFF" condition.

35. *npn* **Device Layout.** Figure 8-27 shows the lateral geometry and the structural cross section of a typical small-signal *npn* transistor, having the impurity profile shown in Fig. 8-26a. Note that the vertical dimensions of the structural cross section are not drawn to scale.

The common-emitter current gain β is a function of collector I_c. Figure 8-28 shows the β versus I_c characteristics for a typical small-signal *npn* transistor, having the geometry of Fig. 8-27 and the impurity profile of Fig. 8-26a. Current-gain characteristics of various other types of monolithic bipolar transistors of comparable size and geometry are also shown in the figure for comparison purposes.

The degradation of β at low current levels is due primarily to parasitic surface-recombination effects. Low-current β can be improved by additional surface-passivation steps (such as nitride passivation) and by minimizing the emitter periphery-to-area ratio.[27]

At high current levels, transistor current gain degrades, due to two separate effects: (1) decrease of emitter efficiency and (2) current crowding at the emitter periphery. The emitter efficiency can be improved by increasing the emitter area, and the current-crowding effects can be minimized by increasing emitter periphery-to-area ratio. This can be done using multiple emitter *stripes* in the device layout. Figure 8-29 shows the practical layout for an integrated power transistor. Note that the device has three emitter stripes, surrounded by interdigitated base contacts and a low-resistivity collector contact that surrounds the active-device region.

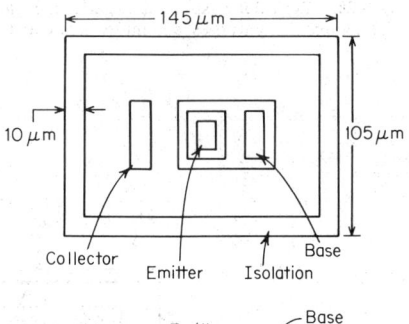

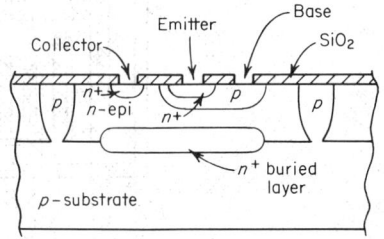

Fig. 8-27. Lateral geometry and the structural cross section of a small-signal *npn* transistor (vertical dimension not to scale).

36. Frequency-Response Parameters. For small-signal applications, the frequency response of bipolar transistors can be closely approximated by the hybrid-π model shown in Fig. 8-30. In the circuit model, the parameters r_b, C_c, and g_0 are the parasitics inherent in any bipolar transistor structure, either discrete or integrated. The base spreading resistance r_b is the physical resistance of the path traversed by the base current from the base contact to the active base region, directly below the emitter. C_c is the collector-base junction capacitance, which is directly proportional to collector-base junction area. The finite output conductance g_0 comes about as a result of base-width modulation.

Collector substrate capacitance C_{ss} and the collector series resistance r_{cs} are the two additional parasitics particularly associated with integrated transistor structures (see Figs. 8-24 and 8-25).

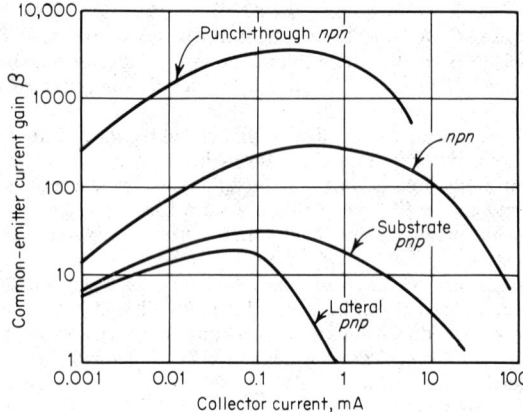

Fig. 8-28. Typical β versus I_c characteristics of various integrated circuit transistors of equal emitter area. (For comparison purposes, an emitter area equal to that shown in Fig. 8-27 is assumed.)

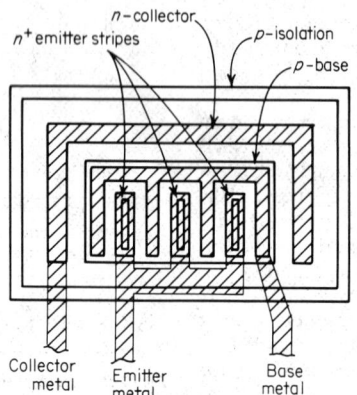

Fig. 8-29. Lateral geometry of a high-current *npn* transistor.

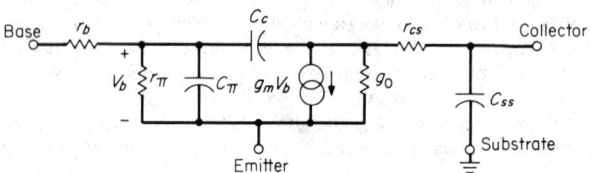

Fig. 8-30. Small-signal equivalent circuit for an integrated bipolar transistor.

The device transconductance g_m and the dynamic resistance r_π of the base emitter junction can be expressed as

$$g_m = \frac{q\,I_E}{kT} = \frac{I_E}{V_T} \quad \text{and} \quad r_\pi = \beta_0/g_m$$

where I_E is the emitter bias current, q is the electronic charge ($= 1.6 \times 10^{-19}$ C), k is Boltzmann's constant ($= 1.38 \times 10^{-16}$ ergs/K), and T is the temperature in degrees Kelvin. V_T ($= kT/q$) is the thermal voltage, and β_0 is the low-frequency current gain of the device in common-emitter configuration.

The key design parameters describing the high-frequency capability of bipolar transistors are the unity current-gain-bandwidth frequency f_t and the maximum frequency of oscillation f_{max}. These can be related to hybrid-model parameters as

$$2\pi f_t = \frac{g_m}{C_\pi + C_c} \quad \text{and} \quad f_{max} = \sqrt{\frac{f_t}{8\pi r_b C_c}}$$

37. Electrical Characteristics. The electrical characteristics of integrated circuit transistors vary greatly, depending on the particular choice of device geometry and impurity profile. Table 8-7 lists some of the typical electrical parameters of small-geometry *npn* bipolar transistors encountered in monolithic circuits. For comparison purposes a device geometry similar to that shown in Fig. 8-27 is assumed. The devices listed in the first and last columns of the table correspond, respectively, to linear- and digital-type device structures having impurity profiles similar to those shown in Fig. 8-26.

Table 8-7.   Typical Electrical Parameters of Integrated *npn* Bipolar Transistors*

	2.5 Ω·cm collector (no gold)	0.5 Ω·cm collector (no gold)	0.1 Ω·cm collector (gold-doped)	Measurement condition
Low-frequency current gain β	150	50	25	$I_C = 0.1$ mA
	150	60	50	$I_C = 1$ mA
	80	60	50	$I_C = 10$ mA
LV_{CEO}, V	60	35	22	
BV_{CBO}, V	85	45	25	$I_0 = 10$ μA
BV_{EBO}, V	7.0	6.7	6.7	$I_0 = 10$ μA
V_{BE}, V	0.65	0.69	0.73	$I_E = 1$ mA
$V_{CE,\text{sat}}$, V	1.2	0.5	0.26	$I_C = 5$ mA
Emitter transition cap.	4	6	9	Forward-bias condition
C_{TE}, pF				5 V reverse bias
	0.5	0.8	1.5	
C_{ss}, pF	0.6	1	1.5	5 V reverse bias
C_{ES}, pF	1.8	2.9	4.6	5 V reverse bias
r_b, ohms	100	80	60	
r_{cs}, ohms	200	75	15	
f_T, MHz	400	440	550	$I_C = 5$ mA, $V_{CE} = 5$ V

* For device geometry and impurity profiles refer to Figs. 8-26 and 8-27.

Some of the process tolerances and temperature coefficients associated with the key *npn* device parameters are listed separately in Table 8-8. Note that even though the absolute-value tolerances of these device parameters are relatively poor by discrete-device standards, their matching and thermal-tracking properties are far superior to those of discrete devices.

38. *pnp* Transistors. In monolithic IC structures, high-performance *pnp* and *npn* devices are not readily compatible. To fabricate a high-performance *pnp* transistor on the same substrate with an *npn* bipolar transistor usually requires additional critical diffusion steps. In certain applications where the *pnp* performance is not critical, it is possible to fabricate functional *pnp* transistors using the same process steps as for *npn* devices. The two types of *pnp* transistors which can be fabricated in this manner are the lateral-*pnp* and the substrate-*pnp* transistors.

39. Lateral-*pnp* Transistors. The simplest *pnp* transistor structure which can be fabricated simultaneously with the *npn* bipolars is the lateral-*pnp* transistor, which requires no additional masking or diffusion steps.[28] Figure 8-31 shows the plane view and the structural diagram of a lateral-*pnp* transistor. The base region of the device is formed by the *n*-type

Table 8-8. Typical Design Tolerances for Integrated *npn* Transistors

Device parameter	Typical value*	Typical design tolerance	Temperature coefficient	Matching tolerance	Temperature tracking tolerance
V_{BE}	0.65 V	± 20 mV	− 2 mV/°C	± 1 mV	± 10 μV/°C
β	150	± 25%	+ 3000 ppm/°C	± 8%	± 100 ppm/°C
BV_{EBO}	7.0 V	± 0.3 V	+ 3 mV/°C	± 25 mV	± 100 μV/°C
BV_{CBO}	85 V	± 15 V	+ 60 mV/°C	± 2 V	± 3 mV/°C
r_b	100 Ω	± 20%	+ 1500 ppm/°C	± 3%	± 100 ppm/°C
r_{cs}	200 Ω	± 20%	+ 2500 ppm/°C	± 3%	± 200 ppm/°C

* An impurity profile similar to that of Fig. 8-26*a* is assumed.

epitaxial layer, which serves as the collector of the *npn* transistors. The *p*-type base diffusion of the *npn* is used to form the emitter and the collector regions of the lateral *pnp;* the n^+ emitter diffusion of the *npn* is used to form the n^+ contact region for the *pnp* base.

In such a device structure, the transistor action takes place in the lateral direction, i.e., parallel to the device surface. The minority carriers injected into the base diffuse laterally toward the collector region. The carrier transport across the base region is the most efficient at or near the surface of the device, where the separation between the collector and the emitter is minimum. This minimum spacing is the effective base width W_B for the device. Due to masking-tolerance and voltage-breakdown requirements, W_B is constrained to be of the order of 6 to 12 μm. This value for the effective base width is more than an order of magnitude larger than that of a vertical-*npn* transistor. Consequently, the current gain and the frequency characteristics of lateral *pnp*s are inferior to those of *npn* devices.

As shown in Fig. 8-31, the lateral-*pnp* structure uses the n^+ buried layer at the episubstrate interface. In the case of a lateral *pnp*, this n^+ layer serves a dual function: it avoids parasitic *pnp* action between the lateral-*pnp* emitter and the substrate, and it provides a low-resistivity path for *pnp* base current.

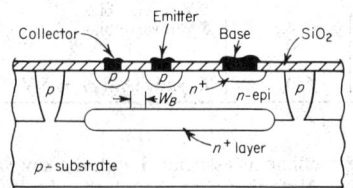

Fig. 8-31. Plane view and the structural diagram of a lateral-*pnp* transistor.

The low-frequency current gain β of the lateral *pnp* is limited by poor emitter efficiency, wide base width, and surface recombination effects. Typical values of β are in the range of 5 to 20. Typical dependence of β on the collector current of the lateral *pnp* is shown in Fig. 8-28.

Frequency-response characteristics of the lateral *pnp* are dominated by the base transit time. Its cutoff frequency can be approximated as[29]

$$f_t \approx 2.43 \, D_p / (W_B)^2$$

where D_p is the diffusion constant of holes in the base region. Typical values of f_t for lateral-*pnp* structures are of the order of 2 to 5 MHz.

The low current gain of a lateral *pnp* can be improved by combining it with a monolithic *npn* transistor, to form the composite transistor structure shown in Fig. 8-32. The polarity and the current gain of such a composite device are equivalent to those of a single *pnp* device having an effective current gain $\beta_t = \beta_p \beta_n$, where β_p and β_n are the current gains of the *pnp* and the *npn* devices. However, the frequency response of the composite device is determined by the lateral *pnp*.

40. Substrate *pnp* Transistors. A functional *pnp* transistor can also be formed by using the base region of the *npn* transistor as the emitter, the *n*-type epitaxial layer as the base, and the *p*-type substrate as the collector. Such a device is called a *substrate pnp* and has the typical layout and the cross section shown in Fig. 8-33. The n^+ buried layer is not present in this device structure. Since the epitaxial-layer thickness is now directly related to the effective base width W_B of the *pnp*, a tighter control of the epitaxial-layer thickness is necessary (typically to $\pm 1 \mu m$).

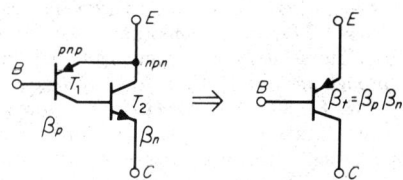

Fig. 8-32. Composite connection of complementary transistors.

The collector of the substrate *pnp* is formed by the *p*-type substrate, which is common to the rest of the circuit and is at all times ac-grounded. Therefore the substrate *pnp* is available only in the grounded-collector configuration.

The current gain and the frequency response of the substrate *pnp* are also limited by its relatively large base width ($W_B \approx 6$ to 10 μm) and poor emitter efficiency. The typical values of β_0 for the device are in the range of 5 to 30. Since the entire bottom surface of the emitter is electrically active, the substrate *pnp* can handle a higher amount of current than the lateral *pnp* of comparable geometry. Typical values of f_t are in the 10- to 30-MHz range.

41. Punch-through Transistors. In certain analog circuits, such as the input stages of operational amplifiers, it is necessary to have very high input impedance and low input bias currents. For such an application, β_0 of a typical integrated *npn* transistor is not high enough, since the device design requires a compromise between the current-gain and the voltage-breakdown requirements.

The current gain can be greatly increased at the expense of the collector-emitter breakdown voltage by using a very narrow base structure ($W_B \approx 2000$ Å). Such a transistor structure can provide current gains of 1000 to 3000 at collector current levels of 10 to 20 μA, with a collector-emitter voltage of 0.5 V (see Fig. 8-28). However, the collector-emitter

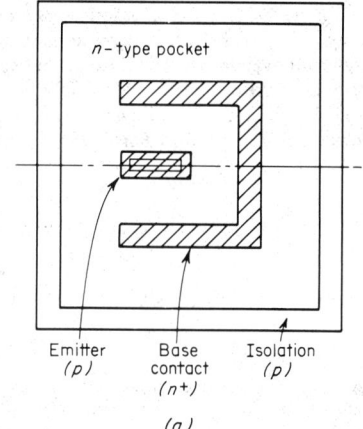

(a)

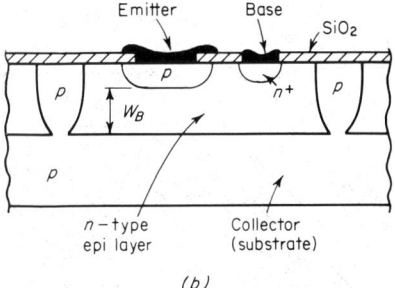

(b)

Fig. 8-33. Structural diagram of a substrate-*pnp* transistor. (*a*) Plane view; (*b*) cross section.

voltage breakdown of such a device structure is limited to a 2- to 3-V range because of the collector-base depletion layer punching through the active base region into the emitter. The term *punch-through transistor* is derived from this phenomenon.

Punch-through transistors can be fabricated simultaneously with the conventional *npn* bipolar transistors, with the addition of one extra masking and diffusion step. The emitter of the punch-through transistor is partially diffused prior to the emitter diffusion cycle for conventional *npn* transistors. Due to this initial diffusion step, the emitter of the punch-through transistor is diffused to a slightly greater depth than the conventional *npn* device. Thus, if both devices have the same base diffusion depth, the resulting base width W_B of the punch-through device is significantly smaller (2500 versus 7000 Å) than that of a conventional *npn* transistor on the same silicon substrate.

In circuit applications, punch-through transistors are often used together with a conventional integrated circuit transistor which can provide the necessary voltage protection for it. Figure 8-34 shows a composite connection of a punch-through transistor T_1 with a lateral-*pnp* transistor T_2 to form an equivalent *npn* transistor T_{eq} which has the effective β of the punch-through *npn* but the breakdown voltage of the lateral *pnp*.

42. High-Performance Complementary Transistors. In the design of high-performance analog circuits, particularly for operation within a nuclear radiation environment, the performance characteristics of the lateral- or the substrate-*pnp* transistors are not acceptable. For these specialized applications, a number of high-performance complementary bipolar structures have been developed. Each of these structures requires additional processing steps beyond those required for the basic *npn* transistor. Their use is therefore limited to design applications where the added fabrication cost or complexity can be justified.

Figure 8-35 shows a practical *pnp-npn* device structure using dielectric isolation techniques.[30] In fabricating complementary bipolar devices, the use of dielectric isolation is preferred over conventional isolation because it provides access to the backside of the device structure during a portion of the fabrication steps. In fabricating the device structure of Fig. 8-35, the process steps follow the basic sequence of steps associated with dielectrically isolated devices, as discussed in Par. **8-25** (see Fig. **8-17**).

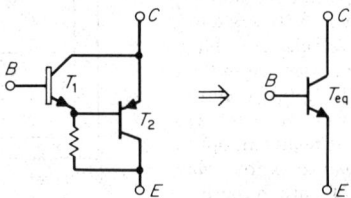

Fig. 8-34. Composite connection of a punch-through *npn* and a lateral *pnp* to simulate a high-gain *npn* transistor with high breakdown voltage.

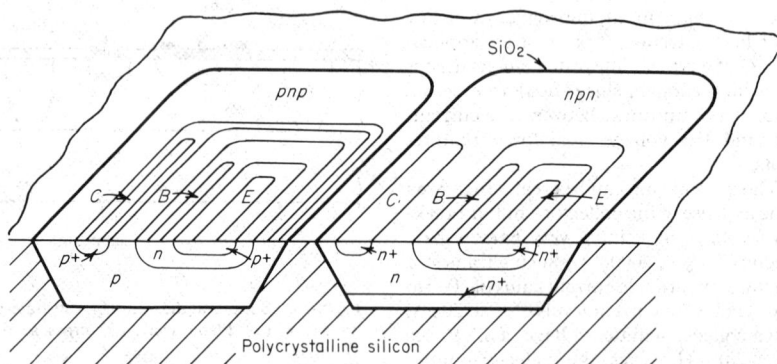

Fig. 8-35. A high-performance complementary bipolar transistor structure using dielectric isolation.

The resulting *pnp* device shows significant improvement over the lateral- or the substrate-type *pnp* transistors and has the following typical electrical characteristics:

$$\beta \approx 50 \text{ to } 80 \qquad LV_{CEO} \approx 60 \quad \text{(volts)}$$
$$f_t \approx 150 \text{ (MHz)} \qquad BV_{EBO} \approx 9 \quad \text{(volts)}$$

FIELD-EFFECT TRANSISTORS

43. Terminology. The field-effect transistor (FET) is a voltage-controlled device in which the current conduction between the *source* and the *drain* regions is controlled or modulated by means of a control voltage applied to the *gate* terminal. Depending on the physical structure of the gate region, the FETs are classified as junction-gate (JFET) or insulated-gate (IGFET) devices.

44. Junction-Gate Field-Effect Transistor (JFET). *Principle of Operation.* Figure 8-36 shows the cross-section diagram of an integrated JFET with an n-type channel region. In such a device structure, reverse bias is applied to the gate-channel depletion layer and causes the layer to extend into the channel region. This modulates the effective width of the conductive path between the source and the drain. Assuming that the impurity concentration in the gate region is much higher than that of the channel and that the channel has a uniform n-type impurity concentration N_D, the thickness d of the gate-channel depletion layer can be related to the reverse bias V_{GC} across the gate-channel junction as

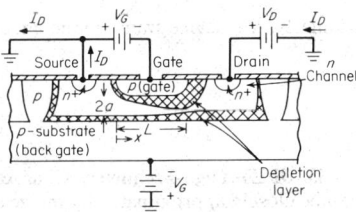

Fig. 8-36. Structural diagram of an n-channel JFET. (Crosshatched areas denote the gate-channel depletion layer.)

$$d \approx \left(\frac{2\epsilon V_{GC}}{q N_D}\right)^{\frac{1}{2}}$$

where ϵ is the dielectric constant of silicon and q is the electronic charge.

The reverse bias across the gate-channel junction, which would cause d to extend into the channel to deplete, or *pinch*, the entire channel, is known as the *pinch-off voltage V_p*. For a uniformly doped channel having a half-width a (see Fig. 8-36), V_p can be written

$$V_p = \frac{q N_D a^2}{2\epsilon}$$

At low values of drain voltage, the JFET operates as a voltage controlled resistor, with an effective source-drain resistance R given as

$$R = R_0 \left(1 - \frac{V_G}{V_p}\right)^{-\frac{1}{2}}$$

where R_0 is the bulk resistance of the channel, with zero depletion layer.

At any given gate bias $V_G > V_p$, as the drain voltage is increased, the drain current also increases, thus increasing the ohmic drop along the channel. At any point along the channel, this voltage drop adds to the net bias across the gate-channel interface, and thus causes the depletion region to extend further into the channel, in the vicinity of the drain.

Consequently, for drain voltages in excess of the pinch-off voltage, a space-charge region is formed near the drain end of the channel. This space-charge layer then causes I_D to reach a saturation level and be relatively insensitive to the further increase of the drain potential. This is known as the pinched operation of the FET, where the device functions as a voltage-controlled current source. For operation below the drain-current saturation (that is, $V_D > V_p - V_G$), the drain-current–voltage characteristics can be approximated as

$$I_D = \frac{2azV_p}{3\rho L}\left(3V_D/V_p - 2\frac{V_D + V_G}{V_p} + 2V_G/V_p\right)^{3/2}$$

where z is the dimension of the channel measured normal to a and L.

Figure 8-37 shows the typical I_D versus V_D characteristics of an n-channel JFET, with the gate bias V_G as a parameter. The region of validity of the equation for I_D corresponds to the area to the left of the dotted line, where the net voltage across the gate-drain junction is less than the pinch-off voltage.[31] For higher values of the drain voltage, the first-order theory predicts a total saturation of I_D. However, in practical devices, I_D still exhibits a slight increase with increasing V_D, thus leading to a nonzero value of dynamic drain conductance g_d. This effect comes about due to the modulation of the effective channel length L by the space-charge region near the drain, in a manner analogous to the base-width modulation effect in bipolar transistors.[32]

The saturation value of the drain current I_{DS}, can be expressed as

$$I_{DS} \approx \frac{V_p}{3R_0}\left[1 - \frac{V_G}{V_p} + 2\left(\frac{V_G}{V_p}\right)^{3/2}\right]$$

Similarly, the device transconductance g_m for pinched operation can be written

$$g_m = \frac{\partial I_D}{\partial V_G} = \frac{1}{R} = \frac{1 - \sqrt{\frac{V_G}{V_p}}}{R_0}$$

45. JFET High-Frequency Characteristics. The frequency performance of the JFET can be closely approximated by the ac equivalent circuit of Fig. 8-38 for device operation in its pinched region. In the figure, R_s represents the parasitic bulk resistance in series with the source contact. For typical integrated JFET structures (see Fig. 8-39) R_s is of the order of 50 to 80 Ω. C_{gs} and C_{gd} are the gate-source and gate-drain capacitances; g_d is the dynamic output conductance due to channel-length modulation effects. In the device layout, the drain area is made as small as possible to minimize the C_{gd}, because this capacitance provides parasitic coupling between the drain and the gate terminals and reduces the frequency capability of the FET in a manner similar to C_c of bipolar transistors.

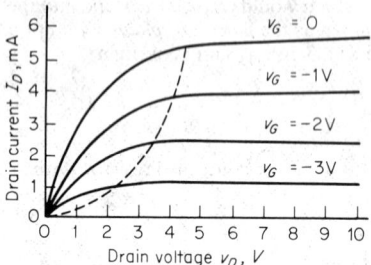

Fig. 8-37. Current-voltage characteristics for a typical JFET.

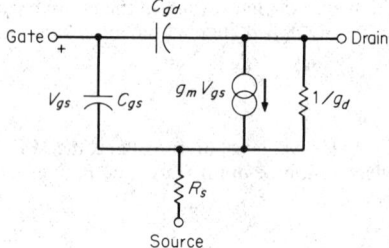

Fig. 8-38. Ac small-signal equivalent circuit for a JFET.

A useful figure of merit for high-frequency capability of an FET is its transconductance cutoff frequency f_c, which is inversely proportional to the total capacitance C_g, seen looking into the gate terminal. For a uniform-channel device with a channel resistivity ρ_c, f_c can be related to the device dimensions as

$$f_c = \frac{g_m}{2\pi C_g} = \frac{\rho_c}{2\pi}\,(a/L)^2$$

In a typical integrated JFET structure with a 5 $\Omega\cdot$cm channel resistivity, having a channel half-width of 2.5 μm and a length of 10 μm, f_c is of the order of 150 to 200 MHz.

46. JFET Device Layout. For monolithic integrated circuits, the most useful JFET structures are those which are compatible with the *npn* bipolar technology and can be fabricated simultaneously with *npn* transistors. Figure 8-39 shows the layout and the structural diagram of an *n*-channel JFET which uses the *n*-type epitaxial collector region of the *npn* transistor as the channel of the FET. Similarly, the *p*-type base diffusion of the *npn* transistor is used to form the control gate with the source and the drain contacts formed by the *n*+ emitter diffusions for the *npn* transistor.

To obtain a narrow channel width without degrading the *npn* bipolar breakdown characteristics, a *p*+ subepitaxial layer is diffused under the FET-gate region, as well as under the isolation walls. Then, during the isolation diffusion, this subepitaxial p^+ layer outdiffuses into the epitaxial layer and reduces the effective channel width of the n-channel FET. In such a structure, the p-type substrate is also a part of the FET gate. However, since the substrate is common to the rest of the circuit, it cannot be used as a control terminal. Therefore only the top gate functions as the control electrode, and the bottom gate is at all times connected to a fixed negative potential, with respect to the rest of the device.

Figure 8-40 shows two additional JFET structures which are also compatible with the npn bipolar technology. The structure of Fig. 8-40a is a p-channel device with a diffused channel region. The p-type channel region can be formed either by npn base diffusion or by an additional p-diffusion step, resulting in a deeper junction structure with higher gate-channel breakdown voltage.

The device structure of Fig. 8-40b is particularly suitable for high-frequency applications and uses dielectric isolation. The gate and the source and drain contacts are again formed by the npn base and emitter diffusions. The channel is formed by a wedge-shaped groove etched under the gate region.

One of the major drawbacks of JFET devices is the strong dependence of device parameters on channel geometry, and particularly on the channel half width a. Table 8-9 lists the dependence of some of the significant JFET parameters on channel dimensions. Due to

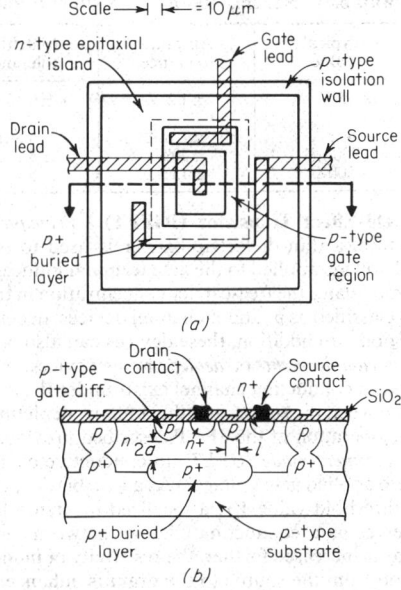

Fig. 8-39. Layout and structural diagram of a typical integrated n-channel JFET.

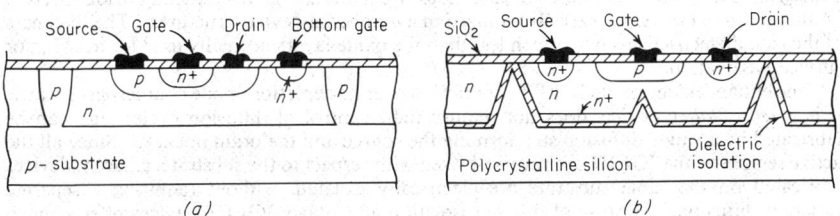

Fig. 8-40. Other possible JFET structures compatible with monolithic integrated circuits.

tolerances associated with the epitaxial-growth and diffusion steps, the absolute tolerance associated with the channel width is of the order of $\pm 15\%$. This can result in absolute tolerances of $\pm 25\%$ and $\pm 40\%$ for V_p and I_{DS}.

Table 8-10 lists some of the typical parameter values and tolerances associated with the JFET structure of Fig. 8-39, assuming a channel resistivity of 5 Ω·cm and a channel half-width of 2.5 μm and a substrate bias of -6 V.

Table 8-9. Dependence of Junction-Gate FET Parameters on Channel Dimensions

Device parameter	Dependence on channel dimensions	
	Half-width a	Length L
Saturation current I_{DS}	a^3	$1/L$
Transconductance g_m	a	$1/L$
Cutoff frequency	a^2	$1/L^2$
Drain conductance beyond pinch-off, g_d	a^3	$1/L^2$
Pinch-off voltage V_p	a^2	
"On" resistance R_{on}	$1/a$	L
Input capacitance C_{in}		L

Table 8-10. Typical Device Parameters and Tolerances Associated with the JFET Structure of Fig. 8-39, with $\rho_c = 5$ Ω·cm, and $a = 2.5$ μm and Substrate Bias of -6 V

Device parameter	Typical value	Absolute tolerance	Matching tolerance	Temperature coefficient
Pinch-off voltage V_p ..	5 V	± 1.2 V	± 0.15 V	-600 ppm/°C
Drain saturation current I_{DS}	4 mA	± 2 mA	± 0.5 mA	-4000 ppm/°C
Transconductance g_m .	900 $\mu \mho$	$\pm 20\%$	$\pm 8\%$	-4000 ppm/°C
"On" resistance R_{on} ..	500 Ω	$\pm 15\%$	$\pm 5\%$	$+4000$ ppm/°C

47. Insulated-Gate Field-Effect Transistor (IGFET). *Principle of Operation.* In the insulated-gate FET structure, a thin dielectric barrier is used to isolate the gate and the channel.[33,34] The control voltage applied to the gate terminal induces an electric field across the dielectric barrier and modulates the free-carrier concentration in the channel region. The IGFET structures can be classified as p- and n-channel devices, depending on the conductivity type of the channel region. In addition, these devices can also be classified according to their mode of operation as *enhancement-* or *depletion-type* devices.

In a *depletion-mode FET,* a conducting channel exists under the gate, and the gate voltage controls the current flow between the source and the drain by depleting a part of this channel. This is very similar to the operation of the JFET described previously.

In the case of the *enhancement-mode IGFET,* no conductive channel exists between the source and the drain at zero applied gate voltage. As a gate bias of proper polarity is applied and increased beyond a threshold value V_T, a localized inversion layer is formed directly below the gate, which serves as a conducting channel between the source and the drain electrodes. If the gate bias is increased further, the resistivity of induced channel is reduced, and the current conduction from the source to the drain is enhanced.

Figure 8-41 shows a cross-section diagram of a p-channel enhancement-mode IGFET. The particular polarities of the gate and the drain bias for the proper operation of the device are also identified on the figure. Silicon dioxide (SiO_2) is the most commonly used gate dielectric. However, other types of gate-dielectric materials such as silicon nitride (Si_3N_4) or aluminum oxide (Al_2O_3) are also utilized in a number of device structures. The thickness of the gate dielectric is usually much less than the oxide layers normally used for masking or surface passivation.

The enhancement-mode IGFET is preferred over its depletion-mode counterpart because it is a *self-isolating* device, does not require tight control of diffusion cycles, and can be fabricated by a single diffusion step forming the source and the drain pockets. Since all the active regions of the IGFET are reverse-biased with respect to the substrate, adjacent devices fabricated on the same substrate are electrically isolated, without requiring a separate isolation diffusion. Because of this self-isolation advantage, IGFET devices offer a much higher packing density per unit area of silicon surface than the bipolar transistors.

If the drain voltage is increased in the polarity shown in Fig. 8-41, a finite drain current I_D flows through the induced channel. This current also causes an ohmic drop along the channel which substracts from the net gate voltage $(V_G - V_T)$. Since the apparent sheet conductance at any point along the channel is proportional to this net gate voltage, the voltage gradient due to I_D causes the channel to deplete along its length. Thus an increase of V_D causes the drain current to increase, which in turn causes the channel to deplete, or pinch off, near the drain. For values of the drain voltage in excess of the net gate bias, a space-charge layer is formed at the drain end of the channel, since the net gate bias (i.e., the applied gate voltage minus the voltage drop along the channel) at this point is no longer sufficient to maintain an induced channel. This leads to a saturation of drain current similar to the case of JFET, for values of $V_D \geq (V_G - V_T)$, and results in a set of drain-current versus drain-voltage characteristics shown in Fig. 8-42.

The threshold voltage V_T is one of the key design parameters in IGFET structures, since it appears in most of the device equations. V_T is a complex function of the gate capacitance, the Fermi level in the silicon substrate, the work-function difference between the gate conductor and silicon, and excess charge built up at the silicon-dielectric interface. The excess charge is created by the presence of surface states at the semiconductor-dielectric interface.[34] Figure 8-43 shows a plot of the threshold voltage as a function of impurity concentration for various values of surface state concentration N_{ss}. For illustration purposes, a gate oxide thickness of 1000 Å with aluminum gate electrode is assumed.[35] The threshold voltage decreases linearly with decrease of work-function difference between the gate electrode and silicon. For example, V_T would be reduced by approximately 1.5 V if polycrystalline silicon, rather than aluminum, were used as the gate electrode. This has led

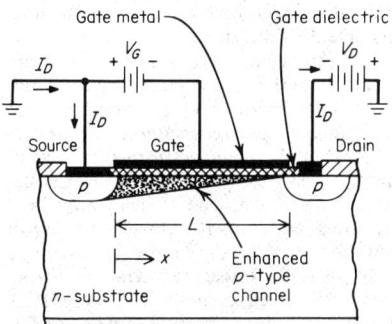

Fig. 8-41. Cross section of a p-channel enhancement-mode IGFET.

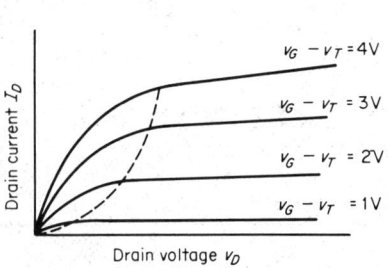

Fig. 8-42. Typical current-voltage characteristics of a p-channel enhancement-mode IGFET.

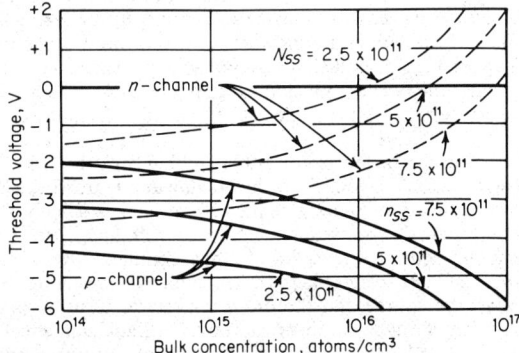

Fig. 8-43. Threshold voltage vs. bulk doping level and surface state density N_{ss} (Ref. 12).

to the development of so-called *silicon-gate* IGFET structures, which offer significantly lower threshold (≈ 1.2 V) compared with conventional devices ($V_T \approx 3$ V).

48. IGFET High-Frequency Characteristics.   For small-signal operation, the frequency performance of an IGFET can be closely approximated by the equivalent circuit of Fig. 8-44 for $V_D > V_G - V_T$.   It should be noted that, with the exception of the source-substrate and the drain-substrate capacitances C_{ss} and C_{ds}, this equivalent circuit is identical with that of Fig. 8-38.   Normally, the IGFET is used in the grounded-source configuration, with the substrate either ac-grounded or short-circuited to the source.   In this latter case, C_{ss} is short-circuited out, and C_{gs} appears between the drain and the ground terminal.   As in the JFET case, C_{gd} and C_{gs} refer to the components of the gate capacitance associated with the source and the drain electrodes.   The drain output conductance g_d is due to the modulation of the effective channel length L by the drain space-charge layer width.   Although not shown explicitly in Fig. 8-44, the parasitic bulk resistances R_S and R_D are still present in the IGFET devices and have the same degradation effects on the device performance as discussed earlier in connection with junction-gate FET transistors.

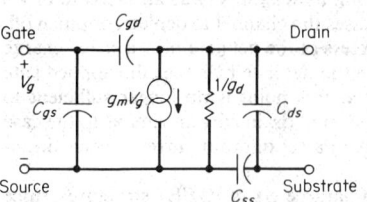

Fig. 8-44.   Ac equivalent circuit for an IGFET.

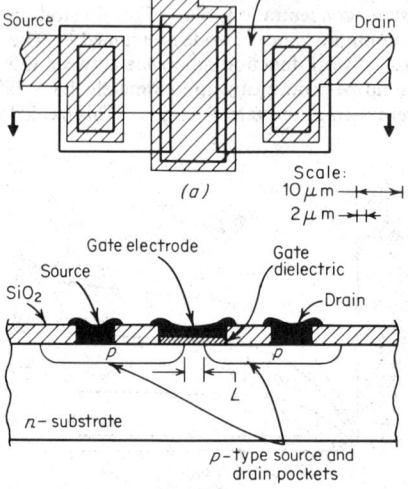

Fig. 8-45.   Lateral geometry and cross section of a *p*-channel enchancement-mode IGFET. (Vertical dimensions not to scale.)

To optimize the high-frequency performance of an IGFET, it is necessary to reduce the channel length L and to minimize the parasitic *gate overlap capacitance* due to physical overlap of gate dielectric over the source and drain regions.   In practical MOS structures, $L \geq 5$ µm, due to gate alignment tolerances and punch-through breakdown limitations between the source and the drain.

49. IGFET Integrated Device Structures.   The most commonly used IGFET structure is the *p*-channel enhancement-mode device.   Figure 8-45 shows a typical layout and structural cross section for such a device.   The resistivity of the *n*-type substrate is in the range of 5 to 8 Ω·cm.   In the fabrication of the device, *p*-type beds, forming the source and the drain, are diffused first.   The separation between these beds defines the effective channel length L.   In considering the final dimensions of the channel, the side diffusion of *p*-type source and drain pockets must be taken into consideration.   After the pocket diffusion, a relatively thick oxide layer (≈ 1 to 1.4 µm) is grown over the device surface.   This oxide is later etched to form the contact areas to the device.

The thin gate dielectric can be formed by either etching back the oxide layer on the gate to the desired thickness (≈ 1000 to 1500 Å) or by etching and regrowing a new layer of gate dielectric.   In certain applications other gate dielectrics such as silicon nitride (Si_3N_4) or aluminum oxide (Al_2O_3) can also be used in conjunction with SiO_2 to form a multilayer gate structure.   The device structure of Fig. 8-45 is then completed by depositing the metal interconnections and the gate electrode.   In an enhancement-mode IGFET, the gate electrode must overlap the source and the drain pockets to form a continuous channel.

The crystal orientation of *n*-type starting material has a significant effect on the threshold voltage of the device.   A $<100>$ oriented silicon crystal exposes a smaller number of incomplete interatomic bonds at the crystal surface than the $<111>$ oriented crystal.   This reduces the charge accumulation at the silicon-dielectric interface and reduces the threshold voltage V_T. Typical ranges of values of V_T for different device structures are listed in Table 8-11.

Compared with bipolar transistors, enhancement-mode IGFET structures have two distinct advantages for monolithic fabrications: (1) They are self-isolating; i.e., the adjacent devices on the same substrate are electrically isolated without requiring separate isolation pockets. (2) They can be fabricated without requiring any critical diffusion step.

Table 8-11.   Typical Threshold-Voltage Ranges for Various IGFET Structures

Device type	Crystal orientation	Threshold voltage range, V
Conventional p-channel	$\langle 111 \rangle$ $\langle 100 \rangle$	3.5-5 2.0-3.0
Silicon-gate p-channel	$\langle 111 \rangle$ $\langle 100 \rangle$	1.5-2.0 0.4-1.2

Complementary IGFET devices can be fabricated on the same monolithic substrate, with only a moderate increase in process complexity.[35,36] Figure 8-46 shows the structural cross section of a pair of n- and p-channel IGFET devices on the same silicon substrate. In fabricating complementary IGFET structures, a relatively deep (≈ 5 to 7 μm) p-type island, or pot, is diffused into the n-substrate. This p-type pot forms the isolated island on which n-channel IGFET is later formed. Following the pot diffusion, the p- and n-type beds are diffused to form the source and drain regions of respective devices; and finally, the gate dielectric layer is grown or deposited, and the aluminum interconnections are deposited to complete the structure.

An alternative device structure which offers a low-threshold voltage and higher packing density than conventional IGFETs is the *silicon-gate* IGFET. In such a device structure, p-type doped polycrystalline silicon layer is used as the gate electrode, instead of aluminum. This reduces the work-function difference between the gate electrode and the silicon surface and decreases V_T.[37] The sequence of process steps in fabricating the silicon-gate FET is shown in Fig. 8-47. The polycrystalline silicon layer forming the gate electrode has very low sheet resistivity. Therefore it also works as an additional layer of interconnection between the devices. This allows

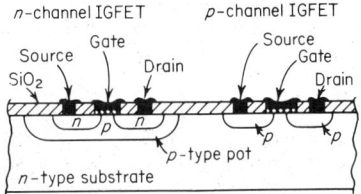

Fig. 8-46. Complementary IGFET structures on a monolithic substrate.

the active devices to be placed close to each other on the chip, without metal-crossover problems.

Ion implantation also provides an added degree of flexibility to IGFET fabrication.[38,39] Figure 8-48 shows the cross section of an ion-implanted-device structure. In this type of structure, the source and drain beds are located relatively far apart, and the actual channel length L is defined at a later step, by ion implantation. After the gate dielectric is grown, aluminum gate electrode is deposited and etched to define the gate region. Then a p-type layer is implanted through the thin gate dielectric. During this implantation step, the aluminum gate electrode works as a mask; therefore the region directly below the gate is not implanted. In this manner, the channel is self-defined by the outline of the gate metal. Since no mask alignment is necessary in defining the channel, this is known as a *self-aligning* structure.

OTHER ACTIVE DEVICES

50. Unijunction Transistors (UJT). The unijunction transistor (UJT) is a single-junction device with two ohmic base contacts. Figure 8-49 shows a simplified diagram of a UJT, along with its integrated circuit realization. The device exhibits a voltage-controlled negative-resistance characteristic as a result of the conductivity modulation within the high-resistivity base region. For a given base bias V_B, negligible anode current I_j flows until the anode terminal is raised to a sufficiently positive voltage level V_p to cause the junction to be forward-biased. The minority carriers injected into the base region cause the effective value of R_A to decrease, due to conductivity modulation, thus further forward-biasing the junction. This results in a negative-resistance characteristic, where I_j increases rapidly, without requiring V_j to rise, and the device switches to its "on" state. Similarly, it can be returned back to its stable "off" state by reducing the anode voltage to below V_B. The voltage-controlled negative-resistance characteristics of the UJT make it useful for relaxation-oscillator application. However, due to excessive charge-storage effects in the high-resistivity base region, the switching speed of the UJT is several orders of magnitude lower than that of the bipolar transistor.

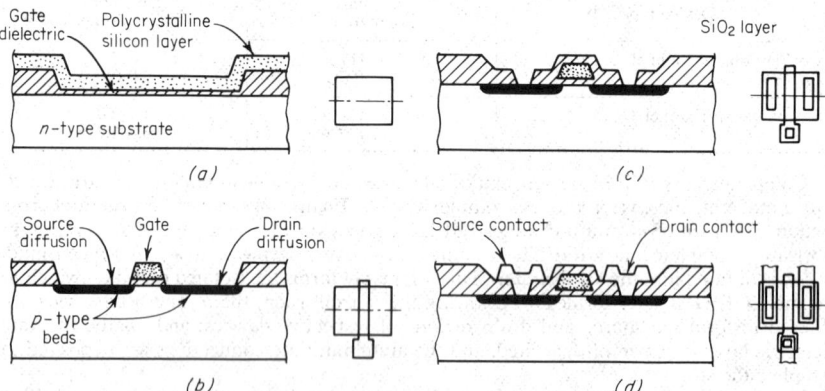

Fig. 8-47. Sequence of process steps in the fabrication of a silicon-gate insulated-gate FET. (*a*) Gate oxide and polycrystalline silicon are grown; (*b*) source and drain beds are diffused; (*c*) source and drain contacts are defined; (*d*) aluminum interconnections are deposited and etched.

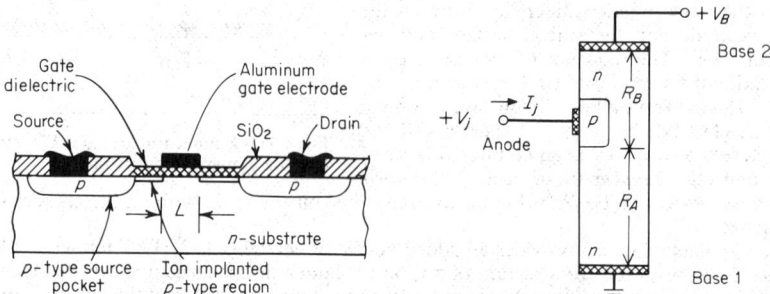

Fig. 8-48. Ion-implanted insulated-gate field-effect transistor.

Fig. 8-49. Schematic representation of a unijunction transistor (UJT).

Since the UJT depends on the conductivity-modulation effects for its operation, its I-V characteristics are strongly temperature-dependent. This effect, along with its slow speed, severely limits the use of UJT in integrated circuits.

51. Four-Layer Diodes. The four-layer diode *(pnpn)* and its three-terminal version, the thyristor, each contains four semiconductor regions. As shown in Fig. 8-50, for analysis purposes a *pnpn* diode can be decomposed into a set of cross-coupled *pnp* and *npn* transistors. It can be shown that if one gradually raises the anode voltage, the device initially remains in a nonconductive state, until a turn-on voltage V_{BO} is reached. Then the device suddenly switches to its "on" state, where it can carry a large amount of current with very little voltage drop across it. The device can be turned off by reducing the current to a level below the critical current level, known as the *holding current*. This is shown as the current level I_H in Fig. 8-50.

An additional control of the turn-on characteristics can be obtained by connecting a *gate* electrode to any one of the two center regions. By injecting a small amount of current into this gate, the turn-on voltage can be kept at a value below V_{BO}, giving the device an additional degree of control. Such a three-terminal *pnpn* device is known as a *thyristor*, or *controlled rectifier*. See Par. **7-51**.

52. Integrated Diodes. Any one of the semiconductor junctions forming the monolithic circuit structure can be used as a diode. Figure 8-51 shows the basic diodes associated with an integrated *npn* transistor. D_{BE} and D_{BC} represent the diodes formed by the base-emitter and the base-collector junctions; D_{CS} is the collector-substrate diode in junction-isolated

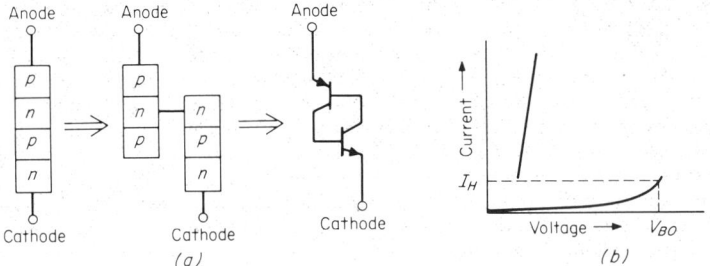

Fig. 8-50. The four-layer diode. (*a*) Transistor equivalent circuit; (*b*) current-voltage characteristics.

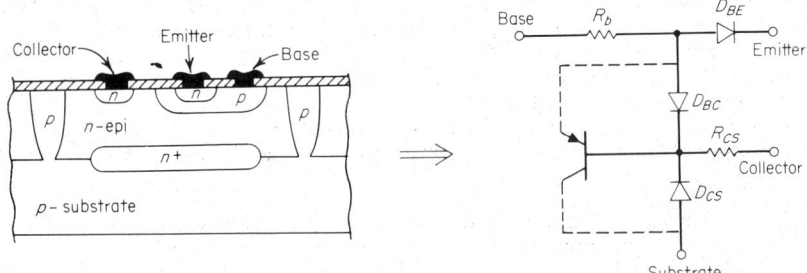

Fig. 8-51. Possible diodes available in a monolithic *npn* structure.

circuits. The resistors R_b and R_{cs} represent the parasitic bulk resistances between the device terminals and the actual diode junctions. A parasitic *pnp* transistor is formed when the diode D_{BC} is forward-biased and D_{CS} is reverse-biased. The current gain of this parasitic *pnp* can be greatly reduced by using an n^+ buried layer between the diode junction and the *p*-type substrate.

The diode current I_D is exponentially related to the voltage V_D applied across the diode as

$$I_D = I_0 \left(e^{qV_D/kT} - 1\right)$$

where I_0 is the reverse saturation current and is proportional to the junction area. For a typical integrated diode the forward (I_D, V_D) characteristics predicted by this equation are valid over six orders of magnitude in current. For a 1-mil^2 base-emitter junction area, this range covers current values of 10 nA to 10 mA.

Under forward bias, the junction diode exhibits a differential conductance g_d given as

$$g_d = \partial I_D / \partial V_D = qI_D/kT$$

The diode forward voltage V_D shows a strong negative temperature coefficient. This thermal drift is a highly predictable and repeatable effect. For the case of the base-emitter diode, which is the most commonly utilized diode connection, the temperature coefficient of V_D falls within the narrow range of -1.9 to -2.1 mV/°C.

Only two of the diodes, D_{BE} and D_{BC}, associated with the *npn* structure of Fig. 8-51 are readily suitable for circuit applications. The collector-substrate diode D_{CS} is not as useful, since its cathode, the substrate, is common to the rest of the circuit. The semiconductor junctions which make up the npn bipolar transistor can be interconnected as a diode in any one of the five possible configurations shown in Table 8-12. The series bulk resistances associated with each diode are in general the most significant parasitics. In this table these parasitic resistances are expressed in terms of the parameters of the npn bipolar equivalent circuit of Fig. 8-30. For most circuit applications where the low reverse breakdown is not a problem, the diode connection is preferred since this configuration has the least series resistance and the lowest storage time.

53. Avalanche Diodes. The avalanche-breakdown characteristics of junctions can be used for voltage reference or dc level-shift purposes. Typical breakdown voltages associated with each of the five diode connections are listed in Table 8-12. The base-emitter breakdown, which falls within the 6- to 9-V range, is the most commonly used avalanche diode because its breakdown voltage is compatible with the voltage levels available in most integrated circuits. For this purpose, any of the diode configurations (1) or (2) can be utilized. Each of these diodes has a reverse-breakdown resistance approximately equal to r_b of Fig. 8-30.

The avalanche-breakdown voltage BV_{EB} associated with the base-emitter junction shows a typical positive temperature coefficient in the range of 2 to 5 mV/°C. Since the thermal drifts of the diode forward voltage V_D and BV_{EB} are in opposite directions, it is possible to partially compensate the thermal drift of an avalanche-breakdown diode by connecting a forward-biased diode in series with it.

54. Schottky Barrier Diodes. When a metal and a semiconductor material are brought into contact, an electrostatic barrier is formed at the interface which causes the metal-semiconductor interface to have rectifying properties.[40] Figure 8-52 shows the thermal-equilibrium diagrams of Schottky barriers on n- and p-type semiconductors, where E_c and E_v are the conduction and the valence band edges, E_g is the band gap, and E_F is the Fermi level. The bending of the energy bands at the interface occurs in such a manner as to retard the flow of majority carriers into the metal. Thus, in the case of the n-type semiconductor, an energy barrier of height ϕ_{Bn} is present at the contact. It is found that the barrier height is essentially independent of semiconductor doping level.

A metal-semiconductor junction having an energy diagram similar to that of Fig. 8-52 behaves as a rectifying junction, with the metal as the anode. When a forward bias V_F is

Table 8-12. Typical Characteristics of Integrated Circuit Diodes

Diode connection	Series resistance	Reverse breakdown	Storage time (I=2 mA)	Parasitic pnp
	Low ($\approx r_b$)	Low ($\approx 6 - 9$V)	High (≈ 60 ns)	No
	Low ($\approx r_{cs} + r_b$)	Low ($\approx 6 - 9$V)	Low (≈ 15 ns)	No
	High ($\approx r_b + r_{cs}$)	High > 30 V	High (≈ 80 ns)	Yes
	High ($\approx r_b + r_{cs}$)	High > 30 V	High (≈ 50 ns)	Yes
	High ($\approx r_b + r_{cs}$)	Low ($\approx 6 - 9$V)	High (≈ 100 ns)	Yes

applied to the interface, the barrier height is lowered, and current flows freely. If a reverse voltage V_R is applied, the barrier height increases, and the current flow is blocked.

Compared with conventional *pn* junction diodes, Schottky barrier diode characteristics have the following significant differences:

1. In Schottky barrier diodes current flow is by majority carriers, rather than by minority-carrier diffusion. Thus switching speeds of Schottky diodes are not limited by storage-time delays.

2. For a given diode area and forward-current level, Schottky diodes have a smaller forward-voltage drop V_F. (For a typical forward current of 10 μA, $V_F \approx 0.25$ V for a Schottky diode, and ≈ 0.55 V for a comparable *pn* junction diode.)

These two properties of Schottky barrier diodes make them well suited for various switching or clamping applications.

Figure 8-53 shows typical Schottky diode structures suitable for integrated circuits.[41] The structure of Fig. 8-53a is the simplest diode structure, formed by placing a metal such as aluminum in contact with lightly doped *n*-type silicon ($N_D < 10^{17}$ atoms/cm^3). The positive fixed surface charge at the silicon surface causes the space-charge region to narrow along the

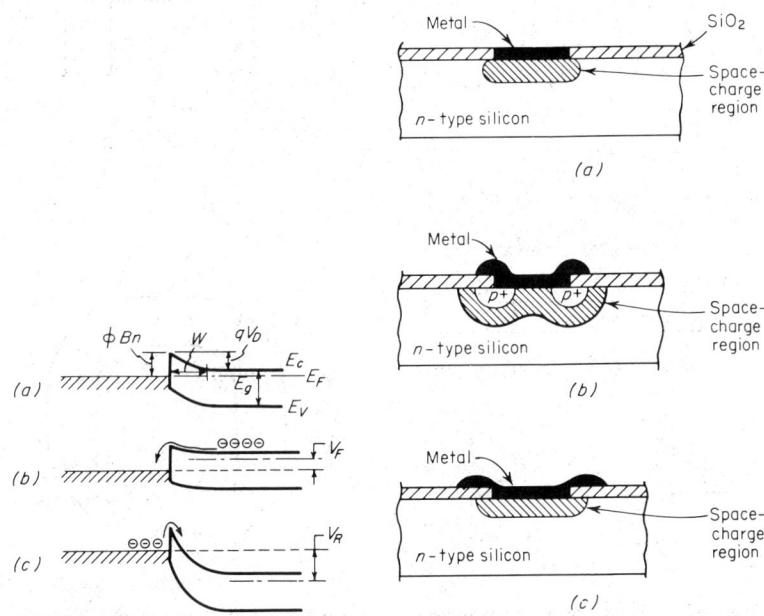

Fig. 8-52. Energy-band diagrams for a Schottky barrier junction under various bias conditions. (*a*) Thermal equilibrium; (*b*) forward bias; (*c*) reverse bias.

Fig. 8-53. Three basic Schottky barrier structures. (*a*) Without metal overlap; (*b*) with p^+ guard ring; (*c*) with metal overlap.

periphery of the diode. This results in leaky, or "soft," reverse characteristics for the diode. In the diode structure (Fig. 8-53b) this problem is avoided by putting a p^+ guard ring around the diode. However, this guard ring adds additional shunt capacitance to the diode.

Schottky barrier diodes can be fabricated simultaneously with ohmic contacts by using conventional aluminum evaporation or sputtering techniques. The nature of the contact (i.e., ohmic or rectifying) is determined by the resistivity of the semiconductor region directly below the metal.

The most important application of Schottky diodes in integrated circuits is to perform clamping functions for *npn* bipolar transistors. In this application, the Schottky barrier diode is connected in parallel with the collector-base junction of an *npn* transistor to prevent the transistor from going into heavy saturation and thereby reducing the switching time. Figure

8-54 shows the structural diagram and the electrical equivalent of a Schottky clamped *npn* transistor. The Schottky diode is incorporated in the structure with little increase in device size or processing complexity.

RESISTORS IN INTEGRATED CIRCUITS

55. Integrated Circuit Resistors. Integrated circuit resistors can be classified into two general categories: (1) semiconductor resistors and (2) deposited (film) resistors. The semiconductor resistors make use of the bulk resistivity of doped semiconductor regions to obtain a desired resistance value. Depending on their structure, semiconductor resistors can also be subdivided into four groups: diffused, bulk, pinched, and ion-implanted resistors.

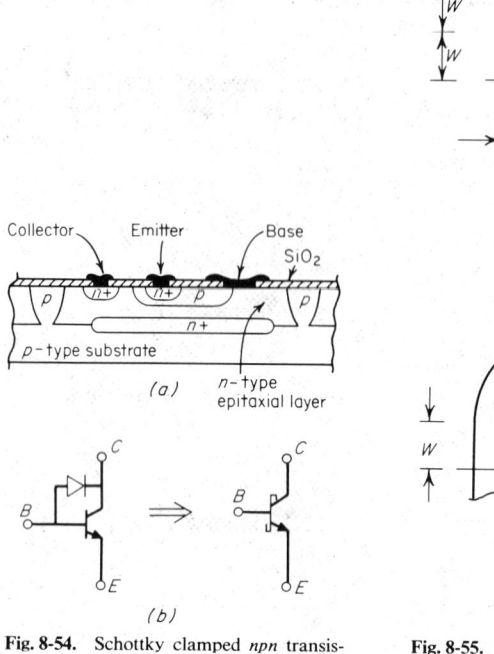

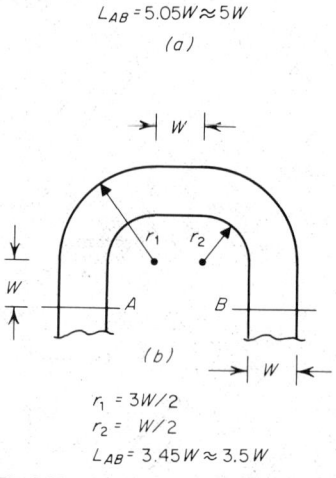

Fig. 8-54. Schottky clamped *npn* transistor. (*a*) Structural diagram; (*b*) equivalent circuit.

Fig. 8-55. Calculation of effective resistor length around corners. (*a*) Square corners; (*b*) round corners.

Deposited, or film, resistors are formed by depositing resistive films on an insulating substrate. These films are then etched and patterned to form the desired resistive network. Depending on the thickness and the dimensions of the deposited films, film resistors are classified into two groups: thin- and thick-film resistors.

To obtain a given resistor L/W ratio with efficient use of the chip area often requires that the resistor follow a folded or zigzag structure. This can be achieved by making square or round bends in the lateral layout of the resistor. The effective length of a resistor around such a bend can be readily calculated by numerical integration techniques. Figure 8-55 shows the effective resistor length around round and square bends for some commonly used resistor layouts.

56. Diffused Semiconductor Resistors. The diffused resistor is formed by the bulk resistivity of a diffused region in a semiconducting substrate. In monolithic circuits based on the *npn* bipolar technology, both the *p*-type base and the *n*-type emitter diffusions can be utilized to form a diffused resistor.   Because of its higher breakdown and R_s values, p-type base diffusion is almost always preferred over the emitter diffusion.

Figure 8-56 shows the typical geometry and cross section of a p-type base-diffused resistor. The sheet resistance R_s is normally dictated by the npn transistor base-diffusion specifications, and is typically in the range of 80 to 250 Ω/square.

The lower limit of the resistor width, W, is normally set by the photomasking tolerances, and is approximately 5 to 7 μm for production circuits.   Best matching is obtained for the values of W in the range of 20 to 50 μm.   Table 8-13 shows the typical properties of base-diffused resistors, along with other types of integrated resistor structures.   The matching tolerance shown in the table corresponds to a 10-μm-wide resistor structure.

The diffused resistor structure has a distributed capacitance associated with it, due to the reverse-biased junction which outlines the resistor.   For a diffused resistor of total value R_1, the high-frequency equivalent circuit looks as shown in Fig. 8-57, where C_1 is the total

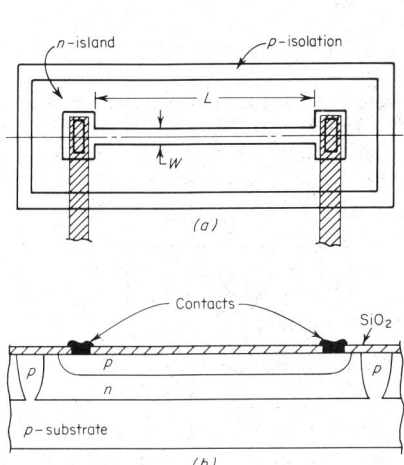

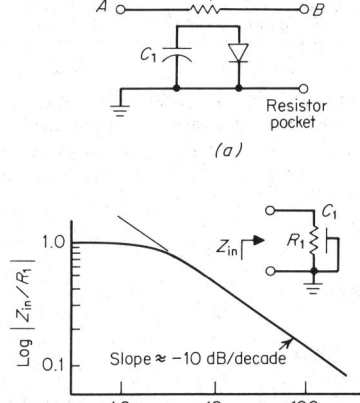

Fig. 8-56. Basic diffused resistor structure. (*a*) Lateral geometry; (*b*) cross section.

Fig. 8-57. High-frequency characteristics of diffused resistors. (*a*) Equivalent circuit; (*b*) typical frequency response (Ref. 43).

distributed capacitance associated with R_1.   The effect of C_1 is to shunt the ac signal to ground and cause excessive phase lag.   Figure 8-57b shows the typical frequency response of a diffused resistor in terms of the magnitude of its driving-point impedance.[43]   Assuming that the capacitance C_1 is uniformly distributed along R_1, one can show that the magnitude of the driving-point impedance z_{in} is down by approximately 3 dB at the frequency f_1 given as

$$f_1 \approx \frac{1}{3R_1C_1} \approx \frac{1}{3R_sC_0L^2}$$

where C_0 is the capacitance per unit area of the resistor junction.

Bulk Resistors.   The bulk resistance of the n-type epitaxial layer can be used in some applications to form a noncritical high-value resistor.   This can be done by using a device structure as shown in Fig. 8-58.   The resulting structure, known as a *bulk resistor,* has a sheet resistivity $R_s = \rho_e/d$.

Pinched Resistors.   The sheet resistivity of a semiconductor region can be increased by reducing its effective cross-section area.   In a pinched-resistor structure, this technique is used to obtain a high-value sheet resistance from an ordinary diffused or bulk resistor. Figure 8-59 shows the structural diagram of a base-pinch resistor formed by placing an n^+-type emitter diffusion over the p-type diffused resistor.

The current-voltage characteristics of a pinched resistor are similar to those of a JFET since the resistor body is analogous to the FET channel, with the surrounding junction area

Table 8-13. Typical Characteristics of Integrated Circuit Resistors

Resistor type	Sheet resistivity, Ω/square	Temp. coef., ppm/°C	Absolute tolerance	Matching tolerance
Semiconductor:				
Diffused	80–250	1200 to 2000	±12%	±1.5%
Pinched	3–10 kΩ	3000 to 5000	+80 to −50%	±10%
Bulk	1–10 kΩ	3500 to 6000	±30%	±5%
Ion-implanted	500–20 kΩ	100 to 1200	±6%	±1%
Thin-film:				
Tantalum	10–1,000	±100	±5%	±1%
Ni-Cr	40–400	±100	±5%	±1%
SnO$_2$	80–4,000	0 to −1500	±8%	±2%
Cermet (Cr-SiO) ...	30–2,500	±50 to ±150	±10%	±2%
Thick-film:				
Palladium-silver	100–100 kΩ	±150 to −500	±10%	±2%
Ruthenium-silver ...	10–10 MΩ	±200	±10%	±2%

functioning as the gate. Consequently, for increasing voltages, the pinched-resistor current I_p saturates at a value I_0, as shown in Fig. 8-59. In the case of base-pinch resistors, the resistor breakdown voltage is quite low, i.e., equal to transistor base-emitter breakdown BV_{EBO} (6 to 8 V). Typical matching and tracking tolerances of base-pinch resistors are listed in Table 8-13.

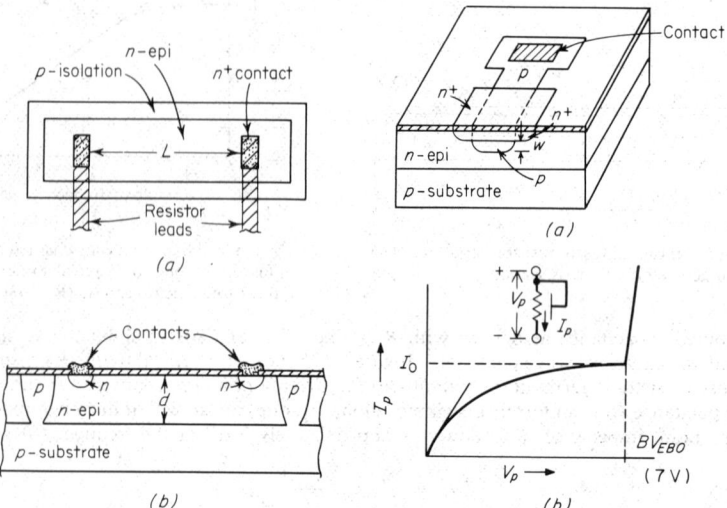

Fig. 8-58. Bulk resistor. (*a*) Lateral geometry; (*b*) cross section.

Fig. 8-59. Structural diagram and electrical characteristics of a base-pinch resistor.

Ion-implanted Resistors. Ion-implantation techniques can be utilized to form resistor structures on the semiconductor surface. With this technique the impurities are introduced into the silicon lattice by bombarding the wafer surface with high-energy ions. The implanted ions lie within a very shallow layer (typically of the order of 0.1 to 0.8 µm) along the silicon surface. Thus, for the similar doping levels, the implanted layers yield a sheet resistivity which is roughly 20 times higher than a correspondingly doped diffused layer of 2- to 4-µm thickness.[39,44]

57. Thin-Film Resistors. Resistive films can be deposited and patterned on a dielectric substrate by using the deposition and photomasking techniques outlined in Par. 3-31. In some cases, in addition to the basic deposition and patterning steps, an additional dielectric deposition is necessary to stabilize the resistor structure by sealing it off from the ambient atmosphere.[45]

Compared with the diffused resistors, thin films offer the following advantages:

1. Low temperature coefficient.
2. Tighter absolute-value control.
3. Lesser parasitics.
4. Higher sheet resistivity.

Table 8-13 and Fig. 8-60 summarize the basic properties of some of the thin-film resistors used in integrated circuits. Among those listed, tantalum (Ta), nickel-chromium (Ni-Cr), and tin oxide (SnO_2) are by far the most commonly used.

58. Thick-Film Resistors. Thick-film resistors are deposited on an insulating substrate by screening, firing, or pyrolytic deposition techniques. The most commonly used deposition technique is screening, where the resistive paste is squeezed onto the substrate through a fine silk or stainless-steel mesh. This part is then dried, etched, and fired (typically ≈ 30 min at 600 to 900°C) to stabilize the resistivity value. After firing, typical resistor absolute-value tolerances are in the range of $\pm 10\%$. When closer tolerances are required, resistors are often designed for about 80% of their target value, and then are trimmed to the desired value. Compositions of most of the available thick-film pastes are held proprietary. Some of the commonly used thick films are mixtures of palladium or ruthenium with conductive metals such as gold or silver.[45] Typical electrical characteristics of these films are listed in Table 8-13.

59. Trimming of Film Resistors. After deposition, the absolute-value tolerances of film resistors can be trimmed to within 1 to 0.01% of the desired value. For this purpose four methods are available.

Oxidation. By heating resistor films in an oxidizing atmosphere, some of the resistive material can be oxidized and becomes nonconductive, and thus increases the total resistance value.

Annealing. Annealing causes the resistor grain structure to reorient itself in a more dense fashion, and causes the resistivity to be lowered.

Laser Trimming. By selectively evaporating a small portion of a film resistor, its effective resistance can be increased. A commonly used technique for this purpose is the L-shaped groove cut into the resistor, as shown in Fig. 8-61. A focused laser beam (spot size $\approx 1\ \mu m$) is first moved perpendicular to the resistor for coarse trimming; as the resistance approaches its final value, the beam is moved parallel to resistor length for fine adjustment of resistor value.

Abrasion Techniques (Thick Films Only). A part of the resistive film can be removed by sand- or air-blasting techniques to trim the resistor value.

CAPACITORS IN INTEGRATED CIRCUITS

60. Integrated Capacitors. The most fundamental limitation to integrated capacitors is size. A general expression for the capacitance of a parallel-plate capacitor can be written $C = C_0 A$, where C_0 is the capacitance per unit area and A is the area of one of the plates. The value of C_0 is usually restricted to a narrow range (typically, of the order of 0.05 to 1 pF/mil²)

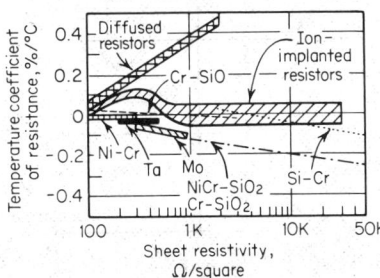

Fig. 8-60. Temperature coefficient vs. sheet resistivity for various semiconductor and film resistors.

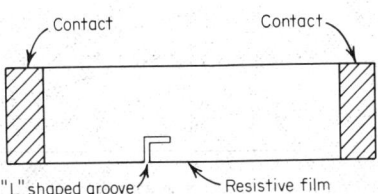

Fig. 8-61. Adjustment of a film resistor by laser trimming.

due to the type of dielectric materials available in integrated circuits and their voltage-breakdown properties. Thus the area requirement increases quite rapidly with the required capacitor value.

There are two basic classes of capacitor structures available in integrated circuits: *junction* and *thin-film capacitors.*

61. Junction Capacitors. Application of a reverse bias across a semiconductor junction forces the mobile carriers to move away from the immediate vicinity of the junction. This creates a depletion layer of thickness x_d across the junction, and causes the junction to behave as a parallel-plate capacitor with a plate separation x_d.[46]

In general, the voltage dependence of most junction capacitances that can be fabricated with the integrated circuit processes can be described by an expression of the form

$$C_0 = K_1 (1/V_R)^n$$

where the exponent n is within the narrow range of 0.33 to 0.5, K_1 is a constant of proportionality depending on the impurity concentration levels in the immediate vicinity of the junction, and V_R is the total reverse bias. The $n = 0.5$ case corresponds to the step junction structure, and $n = 0.33$ corresponds to a linearly graded junction. A detailed family of capacitance vs. voltage curves is available in the literature.[47]

Figure 8-62 shows the family of C_0 versus V_R curves for a step junction, for various values of N. Figure 8-63 shows the three separate sets of junction capacitors available in bipolar integrated circuits. Also shown in the figure is the resultant interconnection of these three capacitors, along with the ideal diodes in shunt with them to indicate the bias polarity requirements for each capacitor. Note that each capacitance also has a finite bulk resistance in series with it. The collector-substrate capacitance C_{ss} has only a very limited application since one of its terminals, the substrate, is common to the rest of the circuit, and represents an ac ground point in any junction-isolated-device structure. The remaining capacitances C_{EB} and C_{BC} can be eliminated when not needed by omitting the emitter or the base diffusion steps. Typical values of C_0 associated with each of these junctions are listed in Table 8-14 for a device impurity profile similar to that shown in Fig. 8-26a.

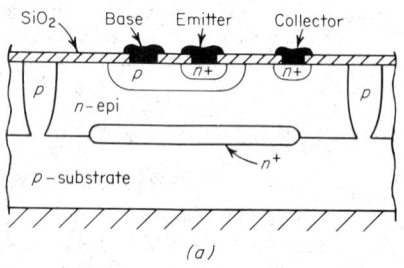

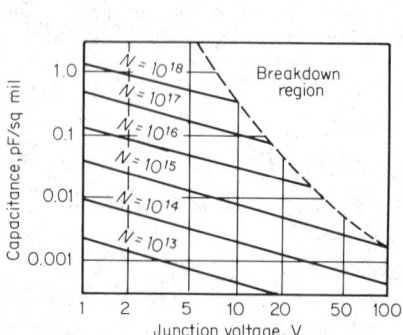

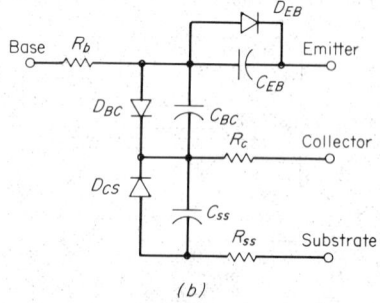

Fig. 8-62. Capacitance per unit area vs. junction voltage for a step junction (N = concentration of the lighter-doped side).

Fig. 8-63. Junction capacitances in bipolar integrated circuits. (a) Physical structure; (b) equivalent circuit of junction capacitances.

Table 8-14. Typical Values of Capacitance per Unit Area Associated with Integrated Transistor Junctions

| Applied voltage, V | Typical junction capacitance, pF/mil² | | | |
	C_{EB}	C_{BC}	C_{cs} without n^- layer	C_{cs} with n^- layer
0	0.9	0.26	0.12	0.17
5	0.65	0.11	0.04	0.06
10		0.08	0.025	0.035

62. Thin-Film Capacitors. The thin-film capacitor is a direct miniaturization of the conventional parallel-plate capacitor. It is comprised of two conductive layers separated by a dielectric. In integrated circuits, it can be fabricated in either one of the two forms shown in Fig. 8-64: *(a)* using a metal oxide semiconductor (MOS) structure; *(b)* using a thin dielectric film between two conducting metal layers. The MOS structure is the most commonly employed thin-film capacitor in monolithic circuits because it is readily compatible with the conventional processing technology and does not require multiple metallization layers.

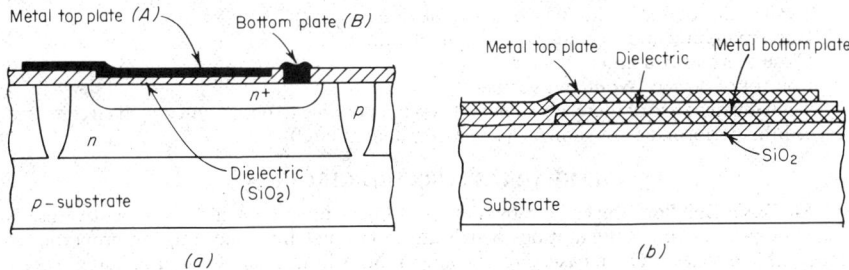

Fig. 8-64. Thin-film capacitor structures for integrated circuits. (*a*) Metal oxide semiconductor (MOS) capacitor; (*b*) thin-film capacitor using multiple metal layers.

The capacitance per unit area of a thin-film capacitor is equal to the ratio of the permittivity ε_x to the thickness T_x of the dielectric layer. In MOS capacitors either SiO_2 or silicon nitride (Si_3N_4) can be used as the dielectric layer. Although SiO_2 is more readily available, Si_3N_4 is preferred whenever the extra processing steps can be justified, since it offers a higher dielectric constant (see Table 8-15). The minimum thickness of the dielectric layer is set by the yield and process control and breakdown requirements, and typically, about 500 to 2500 Å is preferred.

Unlike their junction counterparts, thin-film capacitors are not a function of the magnitude or the polarity of the voltage applied across them, and can offer higher capacitance per unit area with lesser parasitics. However, thin-film capacitors fail by the breakdown of dielectric layer when their voltage rating is exceeded. This is a destructive irreversible failure mechanism; therefore additional care should be taken in their current application to provide overvoltage protection.

Table 8-15. Thin-Film Capacitor Characteristics

| | Dielectric material | | | |
	SiO_2	Si_3N_4	Al_2O_3	Ta_2O_5
Capacitance, pF/mil²	0.25–0.4	0.5–1.0	0.3–0.5	2–3.5
Relative dielectric constant	2.7–4.2	3.5–9	4–8.5	24–28
Breakdown voltage, V	50	50	20–40	20
Absolute tolerance, %	±20	±20	±20	±20
Matching tolerance, %	±3	±3	±5	±5
Temperature coef., ppm/°C	15	4–10	300	200–500
Q, at 10 MHz	25–80	20–100	10–100	10–100

Table 8-15 gives a summary of the electrical characteristics of thin-film capacitors available in integrated circuits. Note that the first two on the list are the MOS type of structures, and the last two columns correspond to the multilayer metal structure of Fig. 8-64b.

INTEGRATED CIRCUIT DESIGN

63. Discrete vs. Integrated Circuit Design. For a circuit designer trained in the use of discrete circuits, the constraints and the limitations of integrated circuit technology may pose a difficult challenge. The constraints are the most severe in the case of monolithic circuits. Some of the basic limitations of monolithic components are:

Poor absolute-value tolerances.
Poor temperature coefficients.
Limitations on component values.
Lack of integrated inductors.
Limited choice of compatible active devices.
Limited power-handling capability.

On the other hand, integrated circuit fabrication methods offer a number of advantages to the circuit designer:

Availability of a large number of active devices.
Good matching and tracking of component values.
Close thermal coupling.
Control of device layout and geometry.

Through efficient use of these advantages it is often possible to design integrated circuits that exceed the performance of similar discrete-component circuits.

LINEAR INTEGRATED CIRCUITS

64. Basic Building Blocks. In the design of linear integrated circuits, repetitive use is made of basic circuit configurations which utilize the matching and tracking properties of monolithic devices. These basic circuit configurations form a set of useful building blocks, which in turn serve as the starting points for the design of larger and more complex functional circuits. Several of these are described in the following paragraphs.

65. Constant-Current Stages. In the design of monolithic circuits, constant-current stages or constant-current sources are often utilized for biasing purposes. They can also be used as *active loads*, to simulate large-value resistors in high-gain amplifier stages.[48] Figure 8-65 shows some commonly used constant-current stages, utilizing *npn* bipolar transistors. In each case, a reference current I_1 applied to one branch of the circuit generates an equal current I_2 at the output of the stage, over a wide range of temperature or bias conditions.

The circuit of Fig. 8-65a relies solely on the close matching of base-emitter voltage V_{BE} between two monolithic transistors. The emitter current I_E of a transistor is related to V_{BE} as

$$V_{BE} = V_T \ln (I_E/I_0)$$

where $V_T (= kT/q)$ is the thermal voltage* and I_0 is the base-emitter reverse leakage current

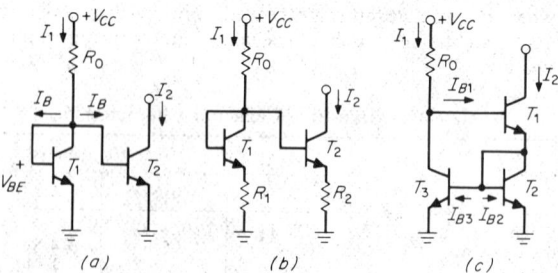

Fig. 8-65. *npn* constant-current stages. (*a*) Diode-biased; (*b*) diode-resistor-biased; (*c*) internal-feedback constant-current stages.

---

* At 300 K, $V_T = 0.0259$ V.

which is proportional to emitter area. For transistor current gain much greater than unity, I_E can be replaced by collector current I_C. Then, for a pair of transistors T_1 and T_2, biased as shown in Fig. 8-65a, the collector currents I_{C1} and I_{C2} of the respective transistors are related as $I_{C1}/A_1 = I_{C2}/A_2$, where A_1 and A_2 are the effective emitter areas of T_1 and T_2. If $A_1 = A_2$, the input and output currents can be related as

$$I_2 = I_1 - 2I_B \approx I_1$$

Figure 8-65b shows an alternative constant-current stage, where the current level is determined by a resistor ratio, given as

$$I_2 = I_1 R_1/R_2 + V_T/R_2 \ln(I_1/I_2)$$

For $I_1 R_1 \gg V_T$, only the first term of the equation is dominant, and $I_2 \approx I_1 (R_1/R_2)$. By setting $R_1 = 0$ in this circuit, one can also obtain a constant-current stage for low current levels.[49]

An improved version of the diode-biased constant-current stage is shown in Fig. 8-65c. This circuit uses internal feedback through the connection between T_2 and T_3 to reduce the effects of transistor base currents.[50] The input and output currents are related as

$$I_2 = I_1 + (I_{B3} + I_{B2} - 2I_{B1})$$

If transistor current gains are well matched, the last term is negligible and independent of the absolute value of β.

Lateral-pnp transistors can also be used in designing constant-current stages. Figure 8-66 shows a composite connection of npn and lateral-pnp transistors for such an application.

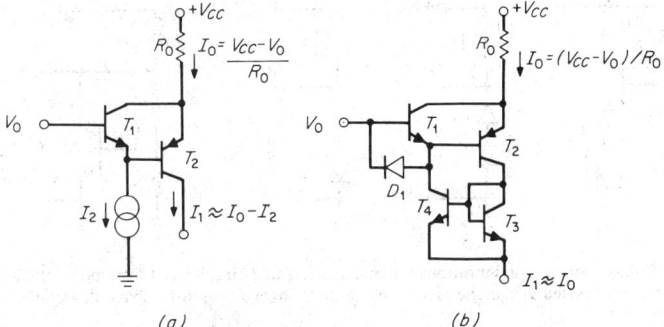

Fig. 8-66. Current-source circuits using composite connection of lateral-pnp and conventional-npn transistors. (a) Basic current-source circuit; (b) improved version.

Note that, in each case, the high-β characteristics of the npn devices are utilized to minimize the effects of low β values associated with the lateral pnp. Assuming the npn β is sufficiently high, the circuit performance becomes independent of pnp gain characteristics. The simplified circuit in Fig. 8-66a shows a method of generating a constant current I_1 independent of pnp characteristics. The circuit shown in Fig. 8-66b is an improved version of this circuit in which the current source I_2 of the previous circuit is replaced by a diode-biased npn constant-current stage formed by T_3 and T_4.[51]

66. Voltage Sources. In many circuit applications, it is necessary to establish a low-impedance point within the circuit which can serve as an internal voltage supply. Such a voltage reference point should have both a very low ac impedance and a very stable dc voltage level, insensitive to power-supply and temperature variations. Usually, however, either the low impedance or the dc voltage stability is of prime importance. The circuits which primarily fulfill the low-impedance requirements are known as *voltage sources*, whereas those specifically designed to provide a constant voltage independent of the supply or the temperature changes are called *voltage references*.

Figure 8-67 shows typical voltage-source configurations often used in integrated circuit design. In the circuit of Fig. 8-67a the low output impedance of an emitter follower is used to simulate a low-impedance source, the diode D_1 providing the temperature compensation for the V_{BE} drift of T_1. The circuits of Fig. 8-67b and c use a temperature-compensated avalanche diode and a diode string, respectively, to provide a low-impedance voltage output. The circuit of Fig. 8-67d uses a shunt-feedback transistor amplifier stage to amplify the transistor V_{BE} drop. This circuit provides an efficient substitute for the diode string of the circuit of Fig. 8-67c if a large number of diodes are involved.[52]

67. Voltage References. In the design of analog circuits, such as the low-drift amplifiers or voltage regulators, it is often necessary to establish an internal voltage reference within the circuit. Unlike the case of voltage sources, the main emphasis in a voltage reference stage is on the thermal stability of the reference voltage. Temperature-stability requirements for a reference voltage are typically ≤ 100 ppm/°C. Temperature coefficients of most monolithic circuit components or devices are much greater than this. However, by making use of the matching and tracking properties of integrated components, and the close thermal coupling on the chip, it is possible to compensate the thermal drifts to a few parts per million. Figure 8-68 shows two circuit configurations which can be used for such an application.

68. DC-Level-Shift Stages. Since large-value coupling capacitors are not available in monolithic circuits, all broadband gain stages need to be dc-coupled. In a bipolar gain stage using an *npn* transistor, the output dc level is always higher than the dc level of the input. Therefore, if a number of such gain stages are cascaded, the output dc level rapidly builds up toward the positive supply voltage. This in turn limits amplitude and the linearity of the available output swing. Ideally, such a dc-level buildup can be avoided by using complementary *pnp-npn* gain stages. However, the *pnp* transistors available in monolithic form have relatively poor frequency-response and current-gain characteristics.

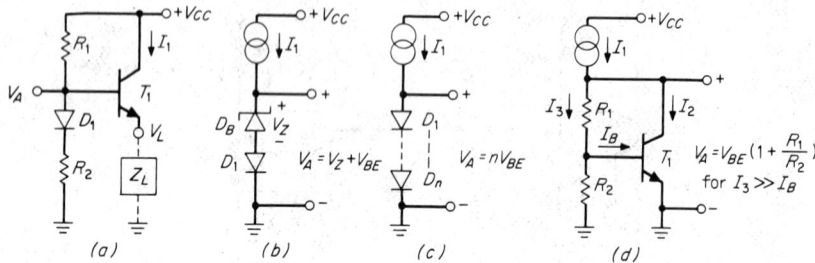

Fig. 8-67. Voltage-source configurations for linear integrated circuits. (a) Common-collector stage; (b) temperature-compensated avalanche diode; (c) diode string; (d) shunt-feedback diode amplifier.

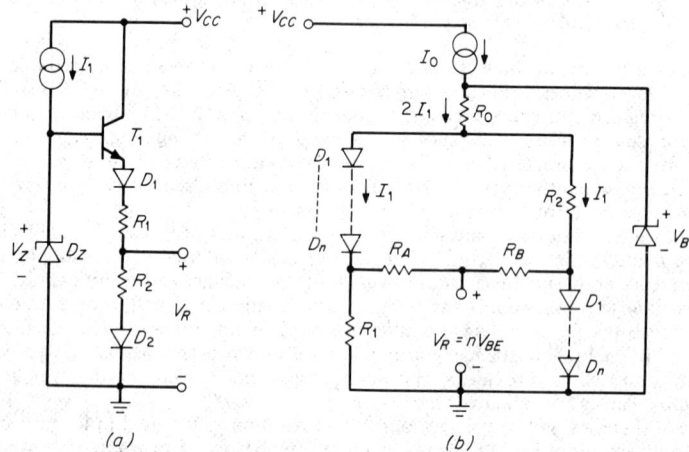

Fig. 8-68. Circuit configurations for generating a temperature-independent voltage reference V_R.

Figure 8-69 shows some practical dc-level-shift stages commonly used in monolithic design. A level-shift stage also serves as a unilateral buffer between the successive gain stages; therefore it is required to have a high input impedance and a relatively low output impedance to prevent interstage loading. A possible exception to this is the resistor–current-source level-shift circuit of Fig. 8-69d, which has an output impedance approximately equal to R_1.

In certain low-frequency applications, the composite *npn*-lateral-*pnp* connection of Fig. 8-70 can also be used as a level-shift stage. As described earlier (see Fig. 8-32), this composite connection of the *npn-pnp* transistors is equivalent to a single transistor which has the polarity of the *pnp* and the β of the *npn*. Hence, a level-shift stage such as shown in Fig. 8-70 can also provide a voltage gain $A_V \approx -(R_2/R_1)$.

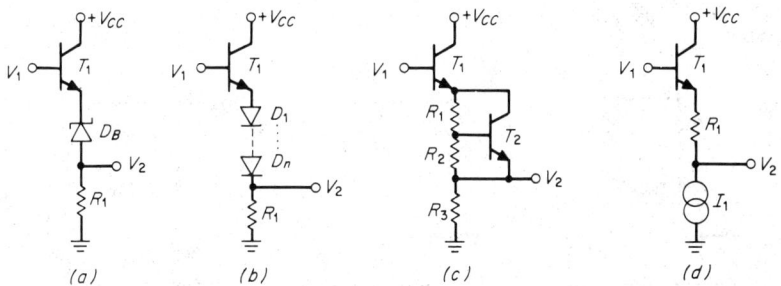

Fig. 8-69. Dc-level-shift stages. (*a*) Avalanche diode; (*b*) diode string; (*c*) diode amplifier; (*d*) resistor–current-source combination.

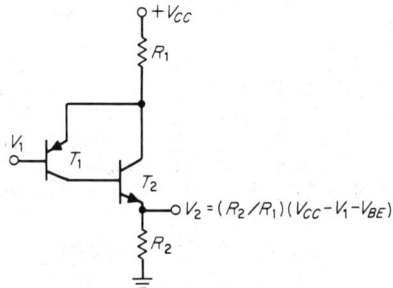

Fig. 8-70. *pnp-npn* level-shift stage.

69. Differential Gain Stages. The differential gain stage is a balanced amplifier circuit, designed to amplify only the *difference* between the two input signals.[53,54] Figure 8-71 shows the basic differential gain stage in its simplest form. The types of inputs which are applied simultaneously to both inputs of the circuit are called *common-mode* (CM) inputs. Due to the symmetry of the circuit, CM inputs cause the current and voltage levels within both branches of the circuit to vary in an identical level. Thus, for common-mode inputs, the differential gain stage has no voltage gain, and can be described by the equivalent circuit of Fig. 8-71b. On the other hand, if a set of *differential-mode* (DM) signals are applied to the inputs, the current and voltage levels in the circuit vary in a differential (e.g., antisymmetric) manner, and a net output voltage is produced. For differential input signals, the response of the amplifier can be described by the equivalent circuit of Fig. 8-71c.

Figure 8-72 shows some of the differential gain stage configurations commonly used in integrated circuits. The expression for differential voltage gain, A_v ($=V_{out}/V_{in}$), is also shown in the figure for each circuit configuration.

In spite of the inherent matching advantages of monolithic devices, small differences exist between two identical transistors on the same chip. These mismatches result in a finite *offset*

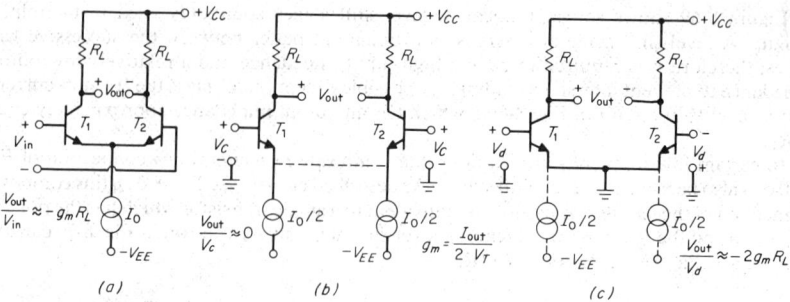

Fig. 8-71. Differential gain stages. (*a*) Basic circuit; (*b*) common-mode equivalent; (*c*) differential-mode equivalent.

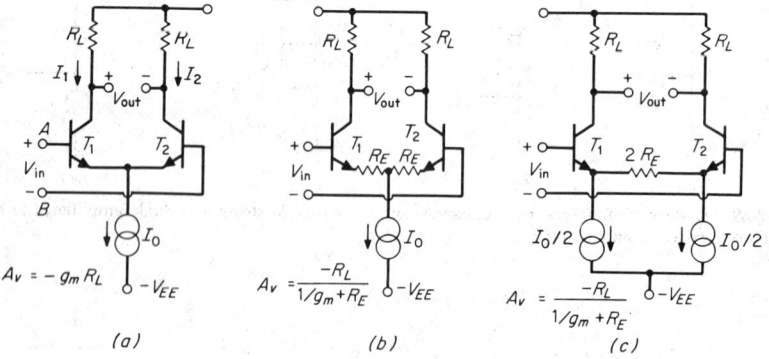

Fig. 8-72. Commonly used differential gain stages.

between the currents in the two branches of the circuit, under zero input condition. In a generalized differential amplifier stage, as shown in Fig. 8-73, the V_{BE} and β mismatches of active devices can be related to the collector-current offset ΔI_c as[52]

$$\frac{\Delta I_c}{I_1} = \frac{\Delta V_{BE} + \dfrac{R_B I_1}{\beta_1}(1 - \beta_1/\beta_2)}{V_T + R_E I_1 + R_B I_1/\beta_1} \qquad (8\text{-}1)$$

where $\Delta I_c = I_1 - I_2$ and $\Delta V_{BE} = V_{BE1} - V_{BE2}$.

If significant resistor mismatches were present in the circuit, the collector-current mismatch expression of Eq. (8-1) could be made to include these by replacing the ΔV_{BE} term in the equation by an equivalent offset voltage ΔV_{io}, given as

$$\Delta V_{io} = V_{BE} + (\Delta R_E) I_1 + (\Delta R_B) I_1/\beta \qquad (8\text{-}2)$$

where ΔR_E and ΔR_B are the emitter and the base resistor mismatches. Even though it is fairly complicated, Eq. (8-2) gives a good insight into the circuit parameters which contribute to the current unbalance in a differential stage.

70. Active Loads. In a number of circuit applications, it is necessary to obtain very high voltage gain from an amplifier stage. This can often be achieved by using active devices to simulate high-value load resistors. Figure 8-74 shows a simplified circuit schematic of an *npn* common-emitter amplifier stage, using a lateral-*pnp* transistor as an active load. In this case, the lateral *pnp* is biased as a constant-current stage. The voltage gain of the stage can be expressed as

$$A_v = -g_m r_0$$

where g_m is the transconductance of T_1, and r_0 is the combined collector impedance of T_1 and T_2.

71. Class B Output Stages. In most high-gain amplifier circuits it is necessary to have a sizable output-current capability without high quiescent-power dissipation. This requires the use of class AB or class B type of output stages. Figure 8-75 shows some output circuit configurations commonly used in integrated circuit design.

The circuit of Fig. 8-75a is an all-*npn* class B amplifier stage. This circuit configuration, often referred to as the *totem-pole* topology, was initially developed for digital integrated circuits, but is readily adaptable to analog integrated circuits. The operation of the circuit can be briefly described as follows: For positive swing of the input voltage, T_2 is in its active region, and the load current flows toward the negative supply through D_2 and T_2. The diode drop across D_2 assures that T_1 is off during this half cycle. For negative-going input signals, T_2 is cut off and the load current is supplied from T_1. For standby condition, with no ac signal at the input, T_1 stays off, and T_2 is biased near cutoff with a small standby current I_1 equal to V_{CC}/R_1.

Figure 8-75b shows a more conventional class B output stage using a substrate-*pnp* transistor, along with a conventional *npn*. The substrate *pnp* has a relatively low β but a good high-current capability. The diodes D_1 and D_2 are diode-connected transistors whose V_{BE} drops match those of T_1 and T_2, respectively. Therefore, in the quiescent state, a bias current I_2 flows through both T_1 and T_2 such that $I_2 = mI_1$, where m is the emitter area ratio of T_1 to D_1 or T_2 to D_2. Therefore this circuit operates as a class AB amplifier since both T_1 and T_2 can be on in standby condition.

72. Operational Amplifiers. An "ideal" operational amplifier can be defined as a voltage-controlled voltage amplifier circuit which offers infinite voltage gain with an infinite input impedance and zero output impedance. The advantage of such an idealized block of gain is that it is possible to perform a large number of mathematical "operations," or generate

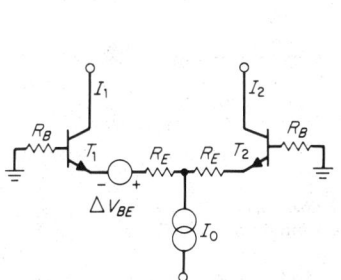

Fig. 8-73. Offset sources in a current-biased differential amplifier.

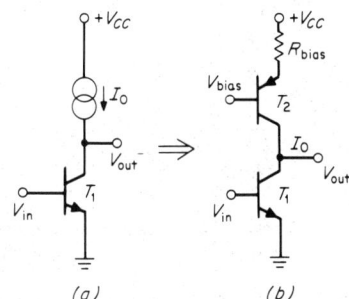

Fig. 8-74. Active loads. (a) Equivalent circuit; (b) actual implementation.

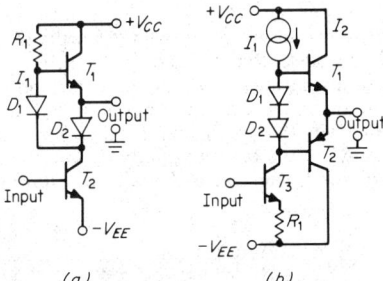

Fig. 8-75. Class B output stages. (a) All-*npn* totem-pole output stage; (b) complementary output stage using substrate *pnp*.

a number of circuit functions, by applying passive feedback around the amplifier. Figure 8-76 shows some of the basic functional applications of an operational amplifier. If the input and the output impedance levels of the amplifier are respectively high and low as compared with the source and feedback impedances, and if the voltage gain is sufficiently high, the resulting amplifier performance becomes determined solely by the external feedback components.

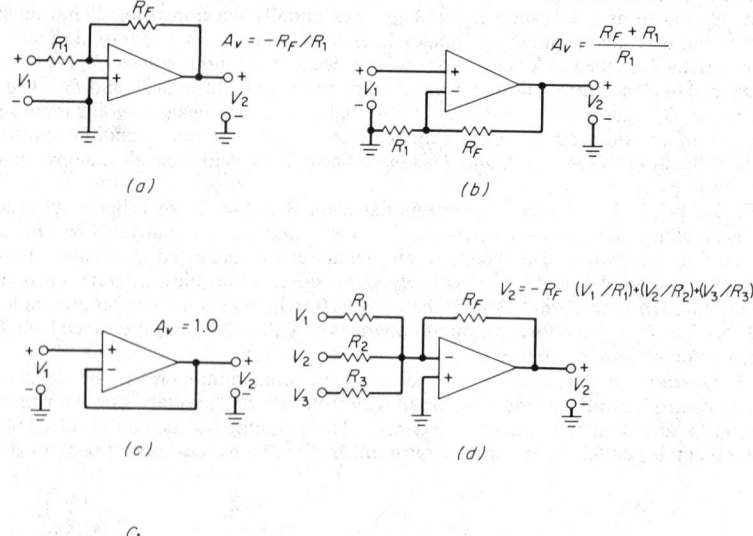

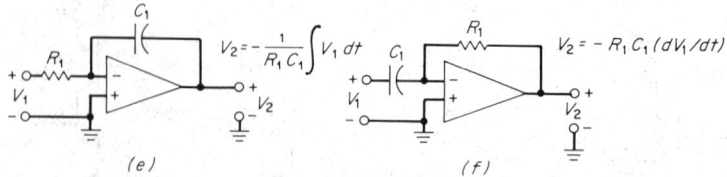

Fig. 8-76. Functional applications of an operational amplifier. (*a*) Inverting amplifier; (*b*) noninverting amplifier; (*c*) voltage follower; (*d*) summing amplifier; (*e*) integrator; (*f*) differentiator.

Because of its versatility as a general-purpose building block, the operational amplifier is the most widely accepted class of linear integrated circuits. Other widely used classes of linear integrated circuits closely related to the operational amplifiers are the *voltage comparators* and the *sense amplifiers.* Voltage comparators compare the amplitude and the polarity of an input signal, and produce a large dc level shift at the output if the input exceeds the reference level.

Operational amplifiers possess finite voltage gain (typically of the order of 80 to 100 dB) and large but finite input impedance (typically above 400 kΩ).

The circuit model of an actual integrated operational amplifier is shown in Fig. 8-77, where I_B indicates the finite input bias currents, dc sources V_{io} and I_{io} represent the voltage and current offsets associated with the circuit, and A is the voltage amplification factor. Due to nonzero values of V_{io} and I_{io}, the output voltage $V_{out} \neq 0$ for input voltage $V_{in} = 0$.

Neglecting the nonzero output impedance, the overall voltage gain $A_V (= V_{out}/V_{in})$ can be expressed as

$$A_r = V_{out}/V_{in} = -\frac{R_F}{R_S} \frac{1}{1 + (1/A)(1 + R_F/R_S + R_F/R_{in})}$$

where R_{in} is the input impedance of the amplifier. For large values of R_{in}, as $A \to \infty$, A_V becomes determined solely by the external components, as $A_V = -R_F/R_S$.

73. Operational-Amplifier Terminology. Since the operational amplifier has become a universal building block for circuit and system design, a number of widely accepted design terms have evolved which describe the comparative merits of various integrated operational amplifier circuits. Some of these terms are:

Input offset voltage: The input voltage which must be applied across the input terminals to obtain zero output voltage.

Input offset current: The difference of the currents into the two input terminals with the output at zero volts.

Common-mode range: Maximum range of input voltage that can be simultaneously applied to both inputs without causing cutoff or saturation of amplifier gain stages.

Common-mode rejection ratio: Ratio of the differential open-loop gain to the common-mode open-loop gain.

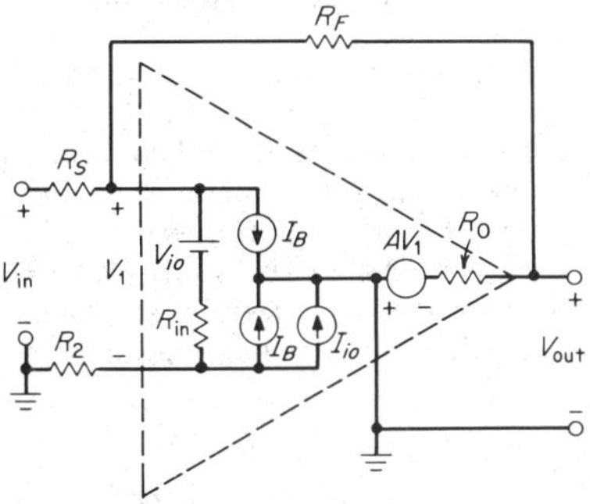

Fig. 8-77. Equivalent circuit of an operational amplifier showing the effect of finite input impedance, bias currents, and current and voltage offsets.

Supply-voltage rejection ratio: Input offset voltage change per volt of supply voltage change.

Slew rate: Maximum rate of change of output voltage, for a step input. It is normally measured with unity gain, at the zero crossing point of the output waveform.

Full-power bandwidth: Maximum frequency over which the full output voltage swing can be obtained.

Unity-gain bandwidth: Small-signal 3-dB bandwidth, with unity-gain closed-loop operation.

Overload recovery time: Time required for the output stage to return to the active region, when driven into hard saturation.

74. Monolithic Operational Amplifiers. In the design of a monolithic general-purpose operational amplifier, it is not possible to optimize all the performance characteristics associated with the device. For example, the requirement of high input impedance may not be compatible with the requirement of low input offset voltage; or the requirements of high slew rate or unity-gain bandwidth may make the frequency compensation of the circuit difficult. Therefore a large number of design compromises are necessary in the design of a monolithic operational amplifier.

In the design of integrated operational amplifiers, the input stage is often the most critical stage. In a high-performance operational amplifier circuit, the input stage must fulfill the following requirements:

High input impedance (above 500 kΩ).

Low input bias current (below 500 nA).

Small input voltage and current offset (V_{io} below 5 mV).

High common-mode rejection ratio (above 60 dB).

High common-mode range (greater than $V_{CC}/2$).

High differential input range (greater than $V_{CC}/2$).

High voltage gain (≥ 40 dB).

Figure 8-78 shows the simplified circuit diagram of an operational-amplifier input stage which satisfies most of the above requirements. In the circuit the composite connection of transistors (T_1,T_3) and (T_2,T_4) effectively have the polarity of a *pnp* but the high β of an *npn* transistor. They produce a high-voltage amplification (typically >60 dB) by working against the active loads formed by current sources T_8 and T_9. A common-mode feedback loop formed by T_5, T_6 and current source I_B is used to set the operating point of the *pnp* transistors.[55]

The high-input-impedance requirements of operational-amplifier input stages can also be satisfied by using field-effect transistors (FET) or punch-through bipolar transistors.[56,57] Since the matching of FET characteristics is poorer than bipolar transistors, integrated operational amplifiers using FET inputs have somewhat poorer offset and drift characteristics than all-bipolar circuits.

Figure 8-79 shows the complete circuit diagram for an integrated operational-amplifier circuit.[55] The circuit uses the input-stage configuration shown in Fig. 8-78. It is a two-stage amplifier circuit. The second gain stage is formed by the transistor pair (T_{16}, T_{17}) with the lateral *pnp*, T_{13}, serving as an active load. The circuit can be internally frequency-compensated by connecting a 30-pF capacitor across this gain stage. The output circuit uses a class B stage (see Fig. 8-75b), with T_{20} designed as a substrate-*pnp* transistor.

75. Voltage Regulators. The function of a voltage regulator is to maintain a constant output-voltage level, irrespective of the changes of the input voltage or the output current. Voltage regulator circuits suitable for monolithic integration are *series*-type regulators which are connected in series between the load and the unregulated supply line. The key design parameters for a voltage regulator are its *line*- and *load*-regulation characteristics. In

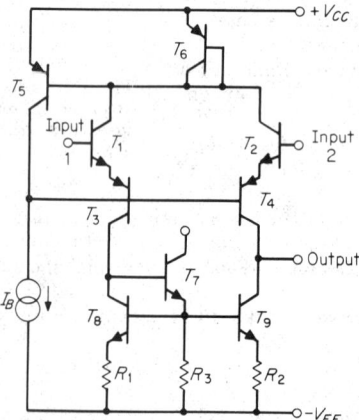

Fig. 8-78. Operational-amplifier input-stage configuration using active loads.

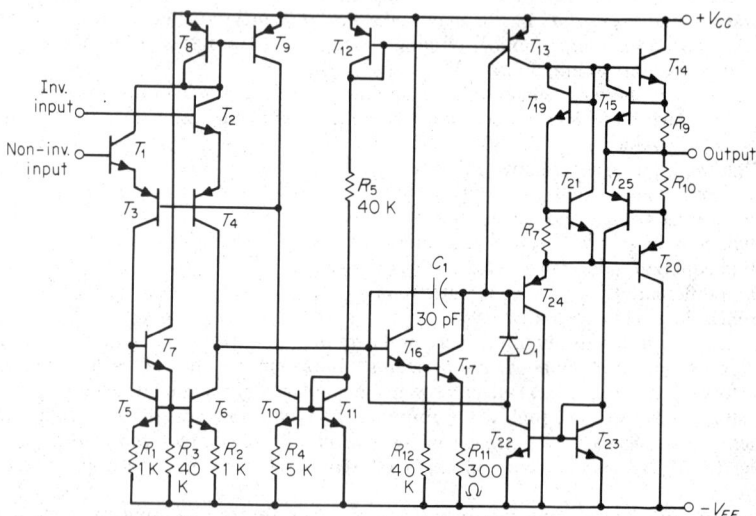

Fig. 8-79. Circuit of a monolithic operational amplifier using internal compensation and overload protection.

addition, the output voltage level is required to have very low temperature coefficient, and to be protected from burning out under accidental output short-circuit conditions.

An integrated voltage regulator circuit in general consists of three main parts: the reference voltage element, the error amplifier, and the series pass element. Figure 8-80 shows a simplified circuit diagram for an integrated regulator circuit where each of these basic sections is identified. The voltage reference V_R must be insensitive to supply voltage or temperature changes. It can be either external or internal to the monolithic design (see Fig. 8-68). For most applications the open-loop voltage gain A for the error amplifier is of the order of 60 to 70 dB, which can be obtained from a single high-gain differential stage using active (current-source) loads. The pass element is normally a power transistor connected in common-collector configuration. In most cases a Darlington connection of two transistors can be used as the pass device to reduce the output impedance and to provide adequate buffering of the error amplifier from the load.

76. Wideband Amplifiers. Lack of large-value coupling and bypass capacitors and matching transformers makes the design of wideband integrated amplifiers a difficult problem. In the integrated design, the ac design and performance of the circuit cannot be considered as a separate problem from the dc bias considerations. The availability of a large number of well-matched active devices again provides a distinct advantage for the designer. For example, one can now use compound connections of two or more devices to replace a single transistor for improved high-frequency performance, or since the devices are well matched, their parasitics can be neutralized by proper circuit layout or interconnections.[59] Among the monolithic devices, the *npn* bipolar transistor has the best high-frequency capability. For this reason almost all wideband amplifiers use *npn* devices in the signal path; the lateral- or substrate-*pnp* devices, when used, are confined to the biasing applications.

Compound Devices. The gain-bandwidth product of an amplifier stage can be significantly improved by using a multiple number of active devices as a direct replacement for a single transistor.

Figure 8-81*a* shows the simplified circuit diagram of a common-collector/common-emitter stage. In this compound device configuration, T_1 effectively buffers the gain stage T_2 from the input. Therefore capacitive loading at the input due to the Miller capacitance of T_2 is greatly reduced. This results in an improved gain-bandwidth product for the compound device over a single common-emitter stage, particularly for large values of source resistance R_s.

The cascode, or common-emitter/common-base stage of Fig. 8-85*b*, also forms a useful compound device topology suitable for wideband amplifier design. The improvement in the high-frequency performance is obtained by the impedance mismatch between the transistors T_1 and T_2. The collector load of T_1 is formed by the input impedance of T_2. Since T_2 is operated in common-base configuration, its input impedance is very low. This in turn keeps the voltage gain across T_1 low, and minimizes the capacitive loading at the input due to the base-collector capacitance of T_1. The advantage of the cascode configuration over a simple common-emitter stage is particularly significant at large values of R_L.

Feedback Stages. Feedback is one of the most commonly used broadbanding techniques. It allows one to exchange gain for bandwidth in amplifier performance. Feedback can be

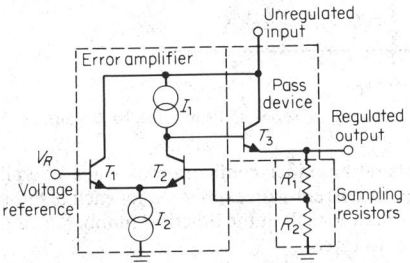

Fig. 8-80. Simplified diagram for a series-type voltage regulator.

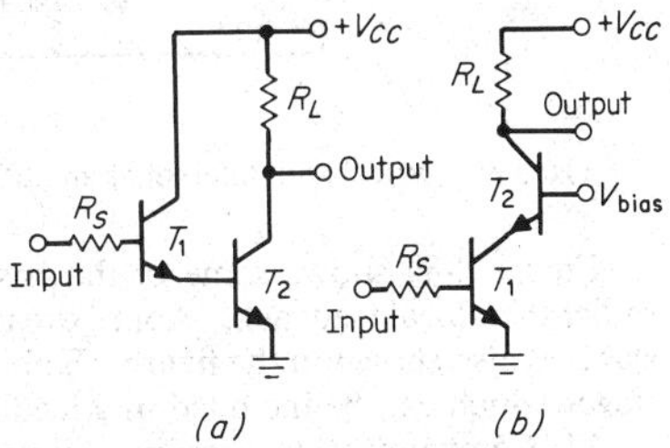

Fig. 8-81. Wideband amplifier stages using compound transistors. (*a*) Common-collector/common-emitter cascade; (*b*) common-emitter/common-base cascade (cascode connection).

either applied locally (around each stage) or it can be applied around the overall amplifier. Figure 8-82 shows the two basic single-stage amplifier configurations using local feedback. Series feedback stage offers high input and low output impedance; therefore it operates best as a voltage amplifier. Shunt feedback stage is best suited to current amplification because of its low input and high output impedance. In cascading feedback stages, one normally alternates between series and shunt feedback. In this manner the shunt feedback stage can provide the necessary low-impedance drive for the series feedback stage; and conversely, the series feedback output can provide the high-impedance drive for the shunt feedback stage succeeding it.

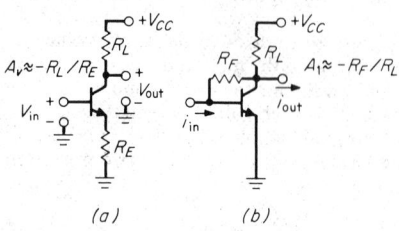

Fig. 8-82. Basic feedback stages suitable for integration. (*a*) Series feedback; (*b*) shunt feedback.

Figure 8-83 shows the circuit diagram of a practical wideband amplifier circuit which uses a cascade of differential series and shunt feedback stages to obtain broadband performance. In this circuit, the output emitter followers Q_5 and Q_6 also function as buffer stages and provide low output impedance in spite of shunt feedback applied around the second gain stage. The circuit gain can be adjusted by interconnecting the taps A and B and A', B' at the input stage. The monolithic circuit offers a bandwidth of 70 MHz with a voltage gain of 35 dB.

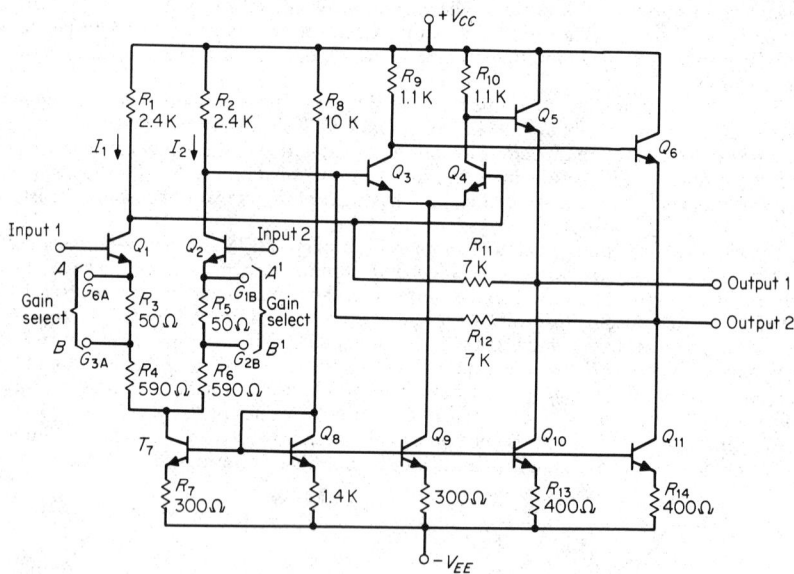

Fig. 8-83. Wideband differential amplifier using a cascade of series and shunt feedback stages.

Figure 8-84 shows some of the basic wideband amplifier configurations using overall rather than local feedback. Approximate voltage or current-gain expression for each of these stages is also shown in the figure. Stability considerations limit the practical number of gain stages which can be included in a feedback loop to three.

77. Communication Circuits. *Balanced Modulators.* A circuit configuration which is often used in modulator or mixer applications is the cross-coupled differential stage shown in Fig. 8-85. The differential configuration of the circuit makes it particularly suited to integration.[64] Because of the differential symmetry of the circuit, it provides an output signal only if both of the inputs are present simultaneously. Because of this property, it is called

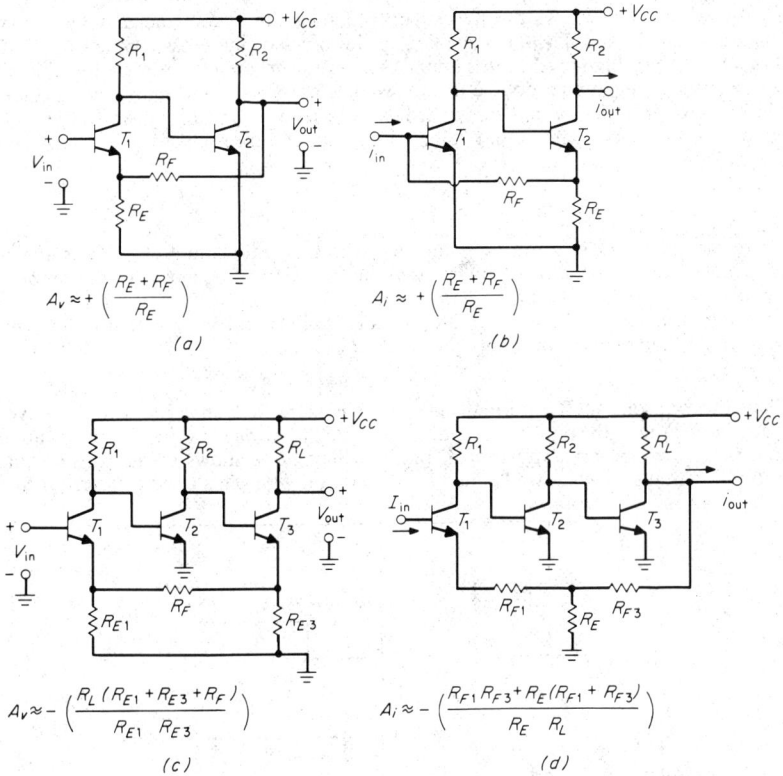

Fig. 8-84.   Basic feedback pairs and triples for wideband amplifier design. (*a*) Series-shunt pair; (*b*) shunt-series pair; (*c*) series-series triple; (*d*) shunt-shunt triple.

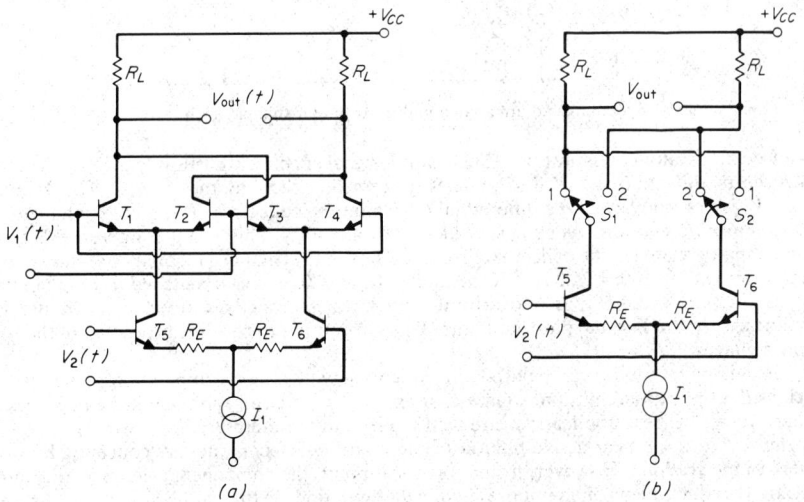

Fig. 8-85.   Balanced modulator circuit (*a*) and its equivalent circuit model (*b*).

a balanced modulator. In the normal operation of the circuit, the high-level constant-amplitude carrier signal $V_1(t)$ is applied to the bases of the cross-coupled transistors T_1 through T_4. For high-level inputs [$V_1(t) > 25$ mV, rms] these transistors function as a set of synchronous single-pole, double-throw switches, as shown in the equivalent circuit model. The low-level modulation input is applied to the bases of T_5 and T_6 and is effectively *chopped* by the carrier signal. The output voltage $V_0(t)$ can be expressed as

$$V_0(t) = \frac{R_L}{R_E} v_2(t)\, S(t)$$

where $S(t)$ is a square-wave switching waveform of unit amplitude. Thus the output corresponds to a symmetrical square wave at the carrier frequency, whose amplitude is modulated by $V_2(t)$.

Analog Multipliers. In analog-multiplier applications, an analog output V_0 is required to be related to the two sets of input voltages V_x, V_y as

$$V_0 = KV_x V_y$$

where the gain constant K is normally chosen to be 1/10. If the polarity of the input voltages is also conserved at the output, the multiplier is said to be *four-quadrant,* since it can handle both positive and negative values of V_x and V_y. Figure 8-86 shows a practical four-quadrant multiplier circuit which is well suited for integration.[65,66] This circuit is derived from the

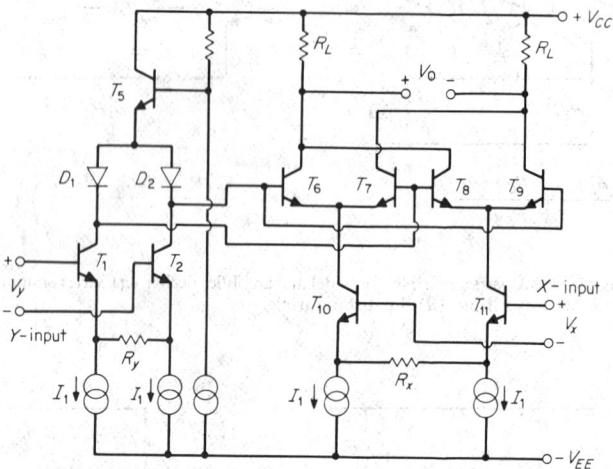

Fig. 8-86. Four-quadrant analog multiplier circuit suitable for integration.

balanced modulator configuration. The X-input signal effectively corresponds to $V_2(t)$ of the balanced modulator (see Fig. 8-85), and is processed linearly by the circuit. The Y-input signal V_y is first converted to a differential current at the collectors of T_1 and T_2, and causes a *logarithmic* voltage difference across diodes D_1 and D_2. This voltage difference is then exponentially related to the collector currents of the doubly balanced modulator section of the circuit formed by T_6 through T_9. If the diodes D_1 and D_2 are well matched to base-emitter diodes of T_6 through T_9, the logarithmic nonlinearity exactly cancels the exponential nonlinearity in the Y-signal path, and thus allows the output to be a linear product of the two input voltages.

The analog multiplier can be used in a wide variety of analog computation applications, such as multiplication, division, square-root extraction, squaring, and obtaining root-mean-square values. These applications are well covered in the literature.[65,66]

Active Filters. Linear-active-filter design and synthesis techniques are conceptually well suited to integration. However, from a practical point, the component tolerance and gain-sensitivity requirements often impose significant limitations in their application to integrated circuits. The requirement of tight component absolute-value tolerance for linear active filters stems from one fundamental fact: The performance characteristics are a very strong function of the system natural frequencies of "poles" which are determined by the resistor-capacitor *(RC)* products and the overall loop gain in the feedback circuit. The absolute value of the loop gain can be desensitized at low frequencies by using local feedback around gain stages. However, the absolute-value control of a product of two dissimilar circuit elements such as a resistor and a capacitor requires a tight control of the absolute value of each element, and does not benefit from the matching and tracking between "similar" monolithic components. In many cases, this drawback restricts the linear active filters to hybrid rather than monolithic integrated circuits where additional trimming of component values is possible after circuit fabrication. Unfortunately, this also sacrifices some of the inherent batch-processing advantages of integrated circuits.

A large number of design and synthesis techniques have been developed for designing active filters. These are well covered in the literature.[67-69]

Phase-locked Loop Circuits. The phase-locked loop (PLL) is one of the fundamental circuit blocks often used in communication systems. It is a frequency-selective circuit comprised of a phase comparator, a low-pass filter, and an amplifier, interconnected to form a feedback system as shown in Fig. 8-87. Some of the basic applications of a PLL are frequency-selective AM or FM demodulation, signal conditioning, frequency synchronization, and frequency synthesis.[70]

The basic principle of operation of a PLL can be briefly explained as follows: With no signal input to the system, the error voltage $V_d(t)$ of Fig. 8-87a is equal to zero. Then the voltage controlled oscillator (VCO) operates at a set frequency, $\omega_0 = 2\pi f_0$, known as its *free-running* frequency. When an input signal $V_s(t)$ is applied to the system, the phase comparator compares the phase and frequency of the input with the VCO frequency and generates an error voltage, related to the frequency and phase difference between the two signals.

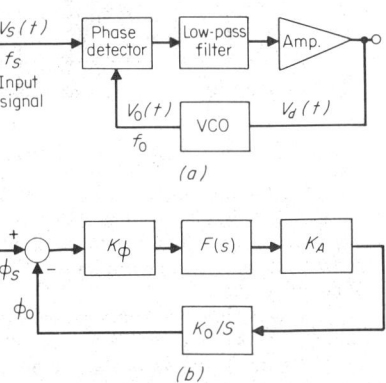

(a)

(b)

Fig. 8-87. The phase-locked loop. (*a*) Functional block diagram; (*b*) linearized model as a negative-feedback system.

This error voltage is filtered, amplified, and applied to the control terminals of the VCO. In this manner, the control voltage $V_d(t)$ forces the VCO frequency to vary in a direction which reduces the frequency difference between f_0 and the input signal. If the input frequency f_s is sufficiently close to f_0, the feedback nature of the PLL causes the VCO to synchronize, or *lock*, with the incoming signal. Once in lock, the VCO frequency is identical with the input signal, except for a finite phase difference. This net phase difference ϕ_0 is necessary to generate the corrective error voltage V_d to shift the VCO frequency from its free-running value to the input signal frequency f_s, and thus keep the PLL in lock. This self-correcting ability of the system also allows the PLL to *track* the frequency changes of the input signal once it is locked.

The range of frequencies over which the PLL can *maintain* lock with an input signal is defined as the *lock range* of the system. This is always larger than the band of frequencies over which the PLL can acquire lock with an incoming signal. This latter range of frequencies is known as the *capture range* of the system. Thus the PLL has a high degree of frequency selectivity since it responds only to a narrow band of frequencies falling within the lock and capture ranges of the system. When the PLL is in lock, it can be approximated and analyzed as a linear feedback system, as shown in Fig. 8-87b, where $F(s)$ is the low-pass-filter transfer function.

The basic building blocks forming the PLL are well suited for integration. Figure 8-88 shows the circuit diagram of a monolithic phase-locked loop circuit, designed for FM demodulation and frequency synthesis applications.[71] The circuit uses a balanced modulator type of phase comparator (see Fig. 8-85).

The VCO section is designed as an emitter-coupled multivibrator. The feedback path of the PLL is closed externally by connecting the VCO outputs to the *reference input* terminals of the phase comparator section. The idling frequency of the VCO is determined by a single external capacitor connected across oscillator terminals 5 and 6 in the figure. The low-pass-filter section can be implemented by connecting a capacitor across the phase comparator outputs (terminals 13 and 14).

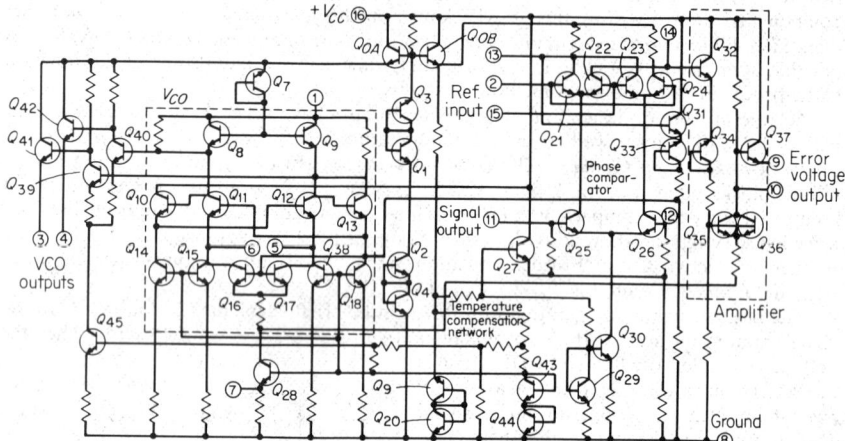

Fig. 8-88.   Circuit schematic for a monolithic PLL circuit (Signetics NE562).

DIGITAL CIRCUITS

78. Digital Integrated Circuits.   The greatest impact of integrated circuits has been in the area of digital circuits, for the following reasons: (1) digital systems employ large quantities of relatively simple but repetitive circuit functions derived from a few basic and well-standardized circuit configurations; (2) digital circuits show low sensitivity to tolerances of component values.

The logic gate is the fundamental circuit block in digital integrated circuits, and it is customary to use the basic gate circuit for comparison between various logic families. Standard symbols for various gate types are shown in Fig. 8-89.

Seven major circuit families are used in digital integrated circuits.[72-74] The basic characteristics and applications of each of these are briefly described below.

79. Resistor Transistor Logic (RTL).   As the name implies, this logic family contains only resistors and transistors. The schematic diagram of a three-input RTL gate is shown in Fig. 8-90 together with its logic symbols.

The operation of the circuit in Fig. 8-90a is as follows: When input A goes to a high level, current will flow through R_1 and turns T_1 "on." The input voltage must be high enough to supply sufficient base current to drive T_1 into saturation. The base resistors must be large enough to prevent current hogging, which could be caused by V_{BE} mismatches between the transistors T_1, T_2, and T_3.

If any one or combination of the inputs goes to a high level, the respective transistors conduct, and current flow through R_L causes the output to go to a low state. For *positive logic*, the circuit performs the NOR function; i.e., if one or more inputs are high, the output is low. For *negative logic*, the circuit performs the NAND function; i.e., when all inputs are at a low level, each of the transistors is turned off, and the output voltage is at high level.

The output dc level at the high state of the RTL gate depends on the effective load resistance seen at the output port. This load resistance is in turn determined by the fan-out requirements. *Fan-out* is defined as the number of gate inputs which are driven by the output. Figure 8-90c shows the output voltage as a function of fan-out for a typical RTL gate with a load resistance of 640 Ω and R_B = 450 Ω, as indicated by the equivalent circuit.

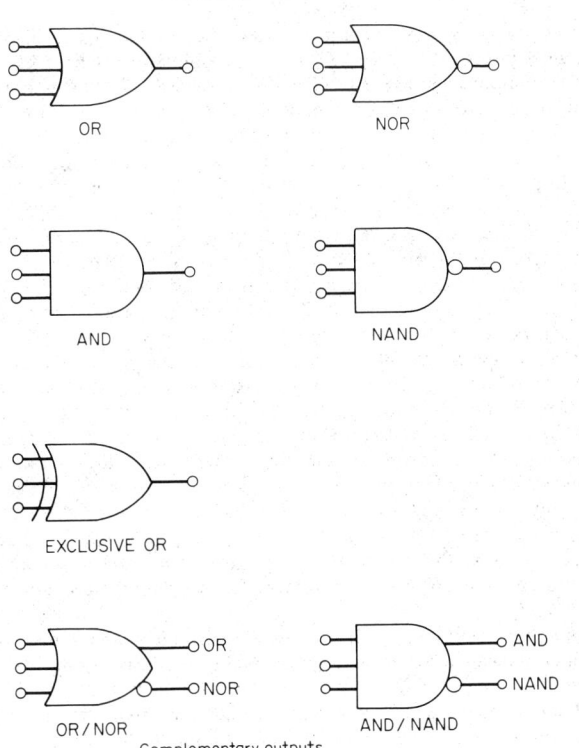

Fig. 8-89. Standard symbols for logic gates.

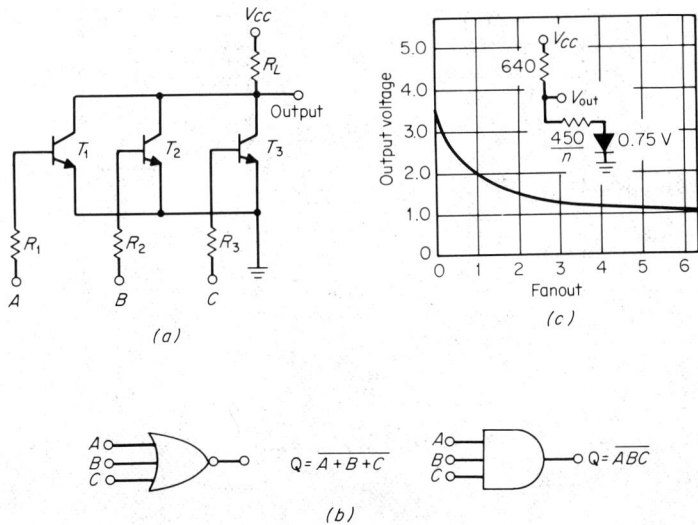

Fig. 8-90. Resistor-transistor logic (RTL). (*a*) Three-input RTL gate design; (*b*) positive and negative logic symbols; (*c*) output voltage vs. fan-out and equivalent circuit for a typical RTL device (*n*-fan-out).

Temperature dependence of device parameters and the absolute-value tolerances associated with resistor values limit the "worst-case" fan-out capability of practical RTL circuits to five.

80. Diode-Transistor Logic (DTL).   The diode-transistor logic gate circuit of Fig. 8-91 avoids some of the inherent shortcomings of RTL, such as current hogging or poor noise immunity, by using diodes in series with each of the input terminals.   In the circuit of Fig. 8-91a, if one or more of the inputs is pulled low, the base drive to the output transistor is shunted to ground and T_1 stays in the "off" state. T_1 is turned "on" only when all the logic inputs are high.   The transistor acts simply as an inverting amplifier, while the logic function is performed by the diode network.   The output voltage for the low level is $V_{CE.\text{sat}}$ (normally 0.2 V).   Additional diodes may be added at the X input for increasing fan-in capability.   The input threshold voltage and noise margin is determined by the saturation voltage of the transistors driving the input diodes and the forward voltage across diodes D_4 and D_5.   Thus noise margin can be selected by the number of diodes in series with D_4.   The pulldown resistor R_2 provides a discharge path for stored charge in the output transistor in saturation, thus reducing turn-off delay.   This pulldown resistor also improves the noise immunity by keeping the transistor turned off during short positive-input transients.

A modified version of the basic DTL gate is obtained if one of the two series diodes is replaced by a transistor, as shown in Fig. 8-91b.   In this design the gain of the transistor T_2 is used to reduce the required input power and to enhance the performance characteristics of the basic circuit.   In the modified circuit the input current drawn through an input diode is reduced by approximately one-third.   Since the transistor T_2 acts as an emitter follower, the available base drive for the output transistor T_1 is increased by almost a factor of 2.   This reduces the minimum current-gain requirement for the output transistor and doubles the output fan-out capability.

DTL circuits lend themselves well to integrated circuit fabrication.   For example, the input diodes can be placed into a single isolation pocket, because they share a common anode. This conserves area, especially in circuits with a large number of inputs.   Another desirable feature of DTL logic is that the input impedance is high, for a high-level input, thus effectively unloading the driving circuit for positive signals.   Figure 8-91c gives the logic symbols for DTL logic, for both positive and negative conventions.

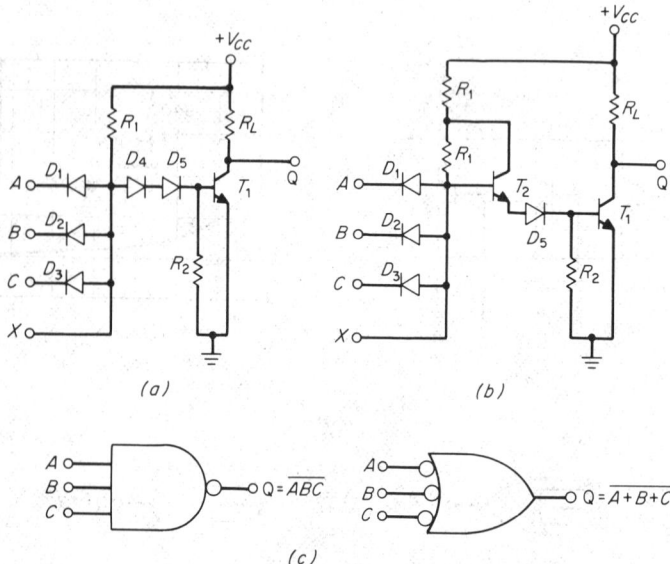

Fig. 8-91.   Diode-transistor logic (DTL). (*a*) Basic DTL gate design; (*b*) modified DTL gate design; (*c*) DTL positive- and negative-logic symbols.

81. High-Threshold Logic (HTL). Many digital logic applications in noisy environments require circuitry with considerably greater noise immunity than that available from the standard logic families. High-threshold logic (HTL) has been designed for such applications. HTL designs are characterized by higher supply voltages (approximately 15 V), higher noise immunity, and higher thresholds than other logic families. These characteristics are usually obtained by adding a zener-diode voltage drop to the normal diode voltage drops in DTL designs. The circuit form is basically the same as DTL except for the zener diode, which replaces a conventional diode, as shown in Fig. 8-92a. To prevent excessive power dissipation at the higher supply voltages, the resistor values in the circuit are increased several fold over DTL circuits.

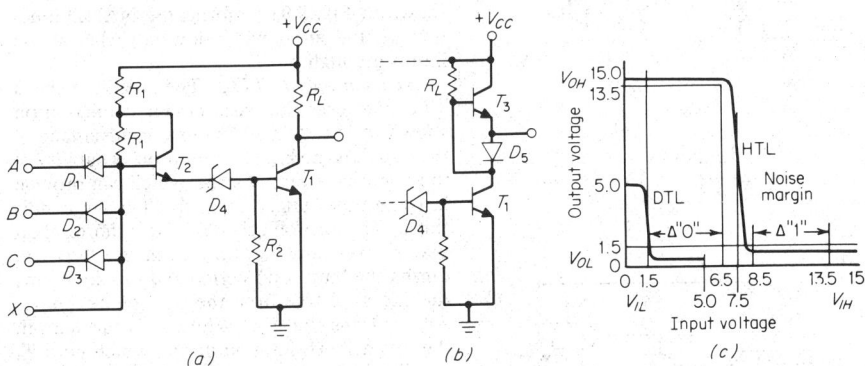

Fig. 8-92. High-threshold logic (HTL). (a) HTL gate design with passive pull-up; (b) active pull-up for HTL gate of part a; (c) worst-case noise margins HTL and DTL, $V_{CC} = 15$ V.

To reduce the output impedance of the HTL gate, an active pull-up configuration as shown in Fig. 8-92b can be used. With transistor T_1 turned off, current flows through R_L into the base of the pull-up transistor T_3. This makes a large amount of current available to move the output voltage close to V_{CC}. When transistor T_1 is turned on, the load current flows through the diode D_5, thereby turning off the emitter-base junction of T_3. A disadvantage of the active pull-up circuit is that the implied AND function can no longer be used. Except for the higher voltage levels, the basic HTL gate operation is the same as the DTL.

The threshold voltage is dependent on the zener voltage of D_4 and on the base-emitter voltage of the output transistor T_1. The zener diode is a reverse-biased base-emitter junction and has a process-dependent voltage drop of approximately 6.9 V. This, together with the typical V_{BE} turn-on voltage of the output transistors, gives a 7.5-V gate threshold. Figure 8-92c shows the typical worst-case noise margins for HTL as compared with DTL.

The design shown in Fig. 8-92a and b allows a worst-case voltage noise immunity of 5 V. The thresholds are fairly insensitive to temperature changes since the positive temperature coefficient of the zener diode is approximately compensated by the negative temperature coefficient of the base-emitter diode drop.

82. Transistor-Transistor Logic (TTL). TTL is one of the most commonly used logic families, because of its versatility and high-speed capability. TTL is basically an extension of the DTL logic family, but it has higher noise immunity and output current capability. TTL logic circuits are classified as medium-speed and high-speed types.

Figure 8-93 shows the operation of the basic TTL gate. The input transistor T_1 performs the same function as the input diodes in a DTL circuit. For normal operation the clamp diodes D_1, D_2, and D_3 are reverse-biased and can be neglected. If any one of the inputs is at a low level, current flows through R_1, causes T_1 to conduct and keep collector of T_1 at a low voltage level, and thus prevents T_2 from conducting. Thus T_2 stays "off" and the output reads high.

When all the inputs are high, the base-emitter junction of T_1 is reverse-biased. Under this condition T_1 is "off"; yet its base-collector junction becomes conducting and provides the base current for T_2. Thus T_2 becomes conducting, and the output level reads low.

The output circuitry of Fig. 8-93 is known as the *totem-pole* configuration. It provides a higher output current drive capability than the RTL or DTL type of circuit. The output drive transistor T_2 is known as a *phase-splitter* stage because it provides a simultaneous in-and-out-of-phase drive to bases of T_4 and T_3. When T_2 is "off," T_3 functions as an emitter follower and the output level is two V_{BE} below V_{CC}, corresponding to a high reading. If T_2 is "on" or saturated, T_4 conducts and the output level is equal to the saturation voltage $V_{CE,sat}$ of T_4 (≈ 0.2 V). The basic three-input TTL gate shown in Fig. 8-93 performs the NAND function; i.e., the output will be low only when all the inputs are high.

Medium-Speed TTL. The medium-speed TTL gate uses the basic circuit configuration shown in Fig. 8-93a. One major advantage of the circuit is the high current drive capability of the totem-pole output stage, which can provide fast rise times under capacitive loading conditions. The turn-off delay of TTL is shorter than that of a comparable DTL gate, since T_1 turns on during the "turn-off" period and rapidly drains the excess charge from the base of the phase-splitter transistor T_2. While T_2 is turning off, there is a short duration during which both T_3 and T_4 are simultaneously "on." This results in a brief low-impedance state across the power supply lines, which can cause a transient current spike. Total propagation delays associated with a medium-speed TTL gate are of the order of 10 to 15 ns.

High-Speed TTL. The basic circuit configuration for a high-speed TTL gate is shown in Figure 8-93c. This circuit uses a lower resistance value for the base pulldown resistors to minimize parasitic RC time constants. The lower totem-pole transistor T_6 also has a lower base drive impedance provided by T_3, which increases its current-handling capability. The addition of the active base-pulldown transistor T_3 also reduces the temperature dependence of the switching characteristics by compensating for the temperature drift of the V_{BE} drop of T_6. The Darlington connection of T_4 and T_5 on the upper part of the output totem pole reduces the transient current spike by reducing the turn-off delay. Typical values of propagation delay for high-speed TTL circuits are in the range of 6 to 10 ns.

Schottky Clamped TTL. In saturated logic circuits, storage-time delay associated with saturated transistors is the most significant limitation on switching speed. Storage-time delay can be eliminated by preventing the transistors from saturating. This can be accomplished by placing a clamping diode across the base-collector junction. If the clamping diode has a lower turn-on voltage than that of the collector-base junction diode, it will keep the collector-base junction from being forward-biased. Such a low turn-on voltage can be obtained by using a Schottky barrier diode as described in Par. **8-54** (see Figs. 8-52 to 8-54). The clamped *npn* transistor is formed by placing an integrated Schottky barrier diode in parallel with the base-collector junction, as shown in Fig. 8-54. A metal electrode is then connected across the *p*-type base and to the lightly doped *n*-type collector region, where it forms a rectifying contact. The Schottky diode turns on at a forward bias of about 0.4 V and keeps the transistor in its active region.

Figure 8-94 shows the schematic diagram of a high-speed TTL circuit using Schottky clamped transistors. The Schottky clamped transistors are identified with the electrical symbol defined in Fig. 8-54. Typical propagation delay for Schottky clamped TTL circuits

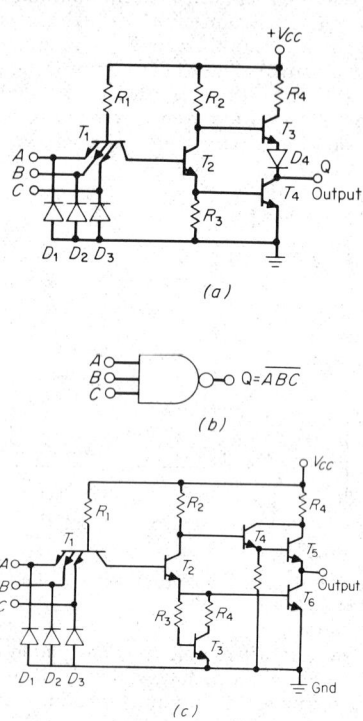

Fig. 8-93. Transistor-transistor logic (TTL). (*a*) Typical medium-speed TTL gate; (*b*) TTL NAND gate for positive-logic convention; (*c*) high-speed TTL circuit.

is 3.5 to 6 ns. Figure 8-95 gives a comparison of speed and power-dissipation properties of various types of TTL circuits.[74]

83. Emitter-coupled Logic (ECL). In emitter-coupled logic (also called current-mode logic) the transistors are switched between "off" and "active" states, without going into saturation. The transistors are kept out of saturation by limiting the total current flow through them. Since the transistors never go into saturation, storage-time delays are eliminated, and the overall propagation delay can be significantly reduced. Figure 8-96 shows the basic circuit configuration and the logic symbol for a three-input ECL gate. If all three inputs are low (approximately -1.6 V), the emitter voltage V_3 will be one diode drop below V_{BB} (approximately -1.9 V). Under this condition, none of the input transistors

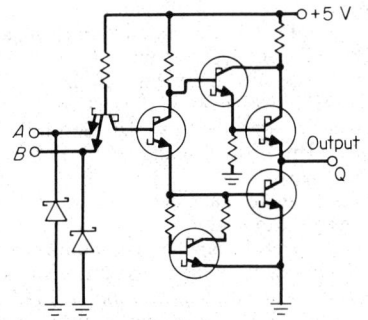

Fig. 8-94. Schematic diagram of active-pulldown TTL circuit employing Schottky diode clamping technique.

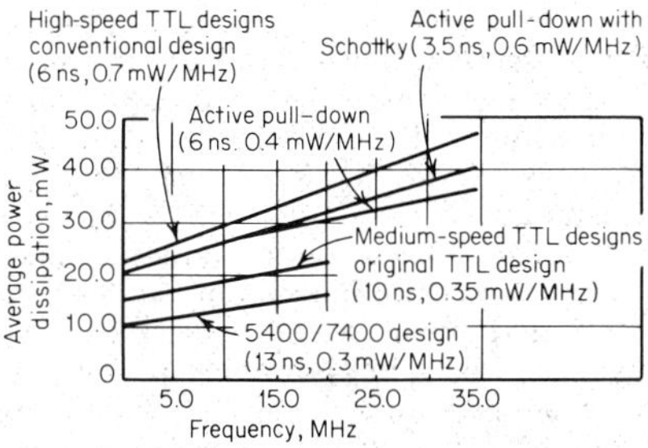

Fig. 8-95. Average TTL power dissipation vs. frequency for typical gate, with 50% duty cycle.

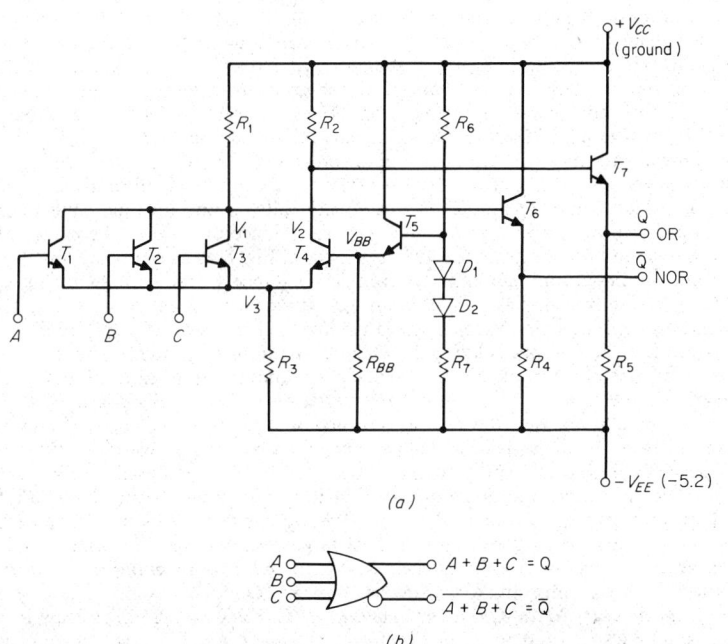

(a)

(b)

Fig. 8-96. Emitter-coupled logic (ECL). (*a*) Emitter-coupled logic gate with complementary outputs; (*b*) ECL gate symbol.

conduct; thus the base of T_6 is at ground potential, and the NOR output at the emitter of T_6 is one V_{BE} below ground.

If one or more of the inputs is high (-0.75 V), current will flow through R_1, causing the output of emitter follower T_6 to drop to about -1.65 V. This corresponds to a NOR function, since the output is *not* high if one or more of the inputs is high. The output at the emitter of T_7 is 180° out of phase with the NOR output; thus it corresponds to a logical OR function. The internal reference voltage V_{BB} determines the switching level of the gate. The emitter-follower outputs provide a low output impedance of about 15 Ω. The input circuit of the gate is similar to that used in most operational amplifiers. Its input impedance is of the order of 100 kΩ.

Recent versions of ECL circuits exhibit the best speed-power products and have the shortest propagation delay of any logic form. In contrast to other logic families, the current drain is essentially constant regardless of frequency. The ECL circuit is capable of very-high-frequency operation, due to the use of nonsaturating transistors and small voltage swings. Propagation delays of 1 to 2 ns can be achieved using ECL circuits. To utilize the maximum-speed advantages of ECL, it is necessary to use special circuit-board layout and termination techniques. The main disadvantages of ECL are high power dissipation and low noise immunity. Its low swing levels make interfacing with other logic families difficult.

84. Metal Oxide Semiconductor (MOS) Logic. The small size and relatively simple device structure of metal oxide semiconductor (MOS) transistors are very attractive for digital circuit design. These devices are also known as insulated-gate field-effect transistors (IGFET) as discussed in detail in Par. **8-47** (see Figs. 8-42 to 8-48).

MOS logic circuits offer three significant advantages over the bipolar logic families: high component density, low power dissipation, and high fan-out capability. The last advantage comes about because of the high input impedance associated with the gate terminal of MOS devices. MOS logic circuits also have the disadvantages of low operating speeds and low current drive capability. In addition, MOS logic circuits often require two power supplies for proper operation.

Because of its simplicity and ease of fabrication, the p-channel enhancement-mode MOS transistor is one of the most widely used devices in MOS logic circuits. In integrated circuit terminology, this device is often referred to as the PMOS transistor. For PMOS devices the threshold voltage V_T (i.e., the gate-source bias at which a conductive channel is formed between the source and the drain) is in the range of 2 to 5 V. This threshold voltage is controlled by the thickness of the gate dielectric, the crystal orientation of the semiconductor, and the impurity concentration in the channel region (see Fig. 8-43 and Table 8-11).

The "on" resistance of MOS devices depends on the applied gate voltage and on the channel dimensions. For conventional small-geometry PMOS devices, the "on" resistance is of the order of several kilohms. Device performance can be significantly improved by using advanced fabrication techniques such as silicon-gate or ion-implantation technologies. Silicon-gate-device structures shown in Fig. 8-47 can reduce the threshold voltage to below 1 V. Ion-implanted MOS transistors such as shown in Fig. 8-48 also provide low threshold voltages and minimize the gate-drain overlap capacitance to increase switching speeds.

Figure 8-97a shows the basic three-input NAND gate using MOS transistors and the corresponding symbols for positive and negative logic. Transistors T_1, T_2, and T_3 are used as input gates, and the output is low only when all the inputs are high. The lower device, T_4, acts as a constant current source. The input impedance is capacitive and can be considered a dc open circuit. The output characteristic is resistive. The resistance to ground is approximately 2 kΩ for each device that is turned on for a device geometry similar to that shown in Fig. 8-45. With the devices off, the output sees an impedance of about 25 kΩ to V_{DD}. The load MOS device is kept turned on with a negative potential applied to V_{GG}.

Typical power-supply requirements and other pertinent electrical characteristics of PMOS logic circuits are listed in Table 8-16. PMOS logic permits high packing density of circuit functions and is therefore well suited for large complex repetitive functions such as shift registers and memories. Computer-aided design and layout techniques are often used to minimize the design time and expense for complex MOS logic circuits.

85. Complementary Metal Oxide Semiconductor (CMOS) Logic. Significant improvements in switching speed and power dissipation can be made if both n- and p-channel devices are used in a logic circuit. CMOS transistors can be fabricated on a monolithic substrate using the device structure of Fig. 8-46.

Figure 8-98a shows a typical circuit connection for a CMOS three-input NOR gate. The circuit symbols for p- and n-channel MOS devices are defined in Fig. 8-98b. Note that the

direction of the arrow from channel to substrate corresponds to the polarity of the diode formed by the channel-substrate interface. The logic symbol for the circuit is given in Fig. 8-98c.

In CMOS logic circuits, the complementary devices are connected in series between the power supplies V_{DD} and V_{SS}, with PMOS devices located adjacent to the positive supply voltage V_{DD}, as shown in Fig. 8-98. At any given time, only one device is turned "on," and its complementary counterpart is "off." Thus, at steady-state conditions, either the p- or the n-channel devices are "off" under all logic conditions, and negligible current flows from V_{DD} to V_{SS}. This results in extremely low power dissipation. The only substantial power dissipation occurs during switching when both p- and n-channel devices may be "on" simultaneously, for a short duration.

In CMOS logic circuits, the devices also function as "active loads" for each other during switching, where one tends to turn "on" while its complement is turning "off." This creates an internal positive-feedback effect and sharpens the transfer characteristics between logic states.

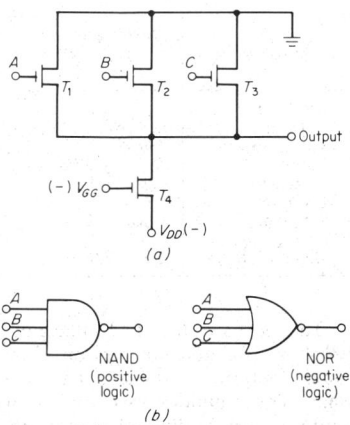

Fig. 8-97. p-channel MOS logic. (a) Basic three-input PMOS NAND gate; (b) logic symbols for basic PMOS gate.

The MOS devices in general have a symmetrical structure, where the source and the drain terminals can be interchanged. Thus the circuit of Fig. 8-98a can be operated with either a positive or negative supply. For most applications, the nominal value of the supply voltage V_{DD} is in the range of 3 to 20 V (see Table 8-16).

86. Comparison of MOS and Bipolar Logic Families. MOS transistors are inherently slower than bipolar devices in logic applications, due to high impedance levels and parasitic capacitances associated with them. The speed-power product of bipolar circuits is normally 10 to 100 times greater than MOS circuits.

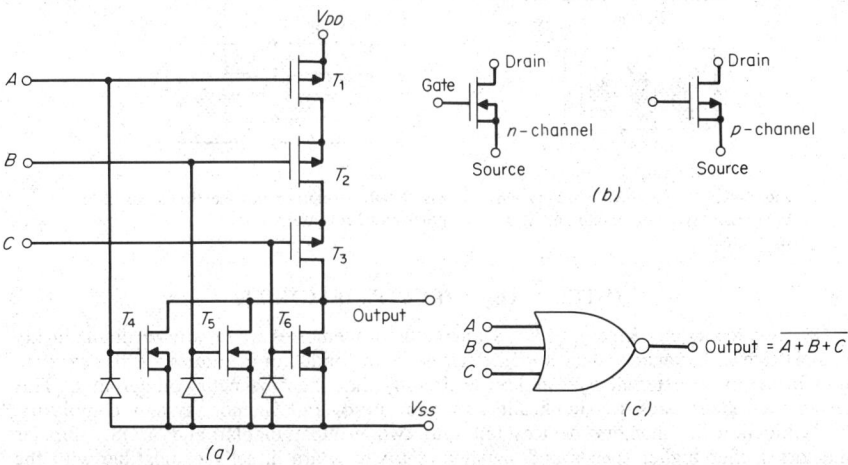

Fig. 8-98. Complementary MOS logic circuit. (a) Three-input NOR gate using CMOS transistors; (b) circuit symbols for n- and p-channel MOS transistors; (c) logic symbol for three-input NOR gate.

Table 8-16. Typical Supply-Voltage Requirements and Performance Characteristics of MOS Logic Circuits for Various Fabrication Processes

Process	*p*-channel MOS (PMOS)			Complementary MOS (CMOS)	
	Metal gate, high threshold	Metal gate, low threshold	Silicon gate, low threshold	Metal gate	Silicon gate
Threshold voltage	-3 to -5 V	-1.5 to -2.5 V	-1.7 to -2.5 V	1.5 to 2.5 V	0.5 to 1.0 V
Typical power-supply voltages	$V_{DD} = -13$ V $V_{GG} = -27$ V $V_{SS} = 0$ V	$V_{DD} = -5$ V $V_{GG} = -12$ V $V_{SS} = +5$ V	$V_{DD} = -5$ V $V_{GG} = -12$ V $V_{SS} = +5$ V	3 to 18 V	1.5 to 16 V
Maximum frequency	2.0 MHz	3.0 MHz	8 MHz	1 to 15 MHz	3 to 25 MHz
Power dissipation	1.5 mW/gate	0.7 mW/gate	0.7 mW/gate	10 nW/gate	1 nW/gate

The most significant advantage of MOS devices is their small size. Figures 8-99 and 8-100 show a relative comparison of small-geometry bipolar and MOS transistors for digital circuit applications, the layout and the cross section of each device being drawn to the same scale. The minimum size of each device is determined by masking tolerances. In Fig. 8-100, Δ is the minimum dimension or the narrowest acceptable line width for a contact opening or the width of an aluminum conductor. As shown by the relative area requirements, the MOS transistor structure offers approximately a factor-of-5 improvement in packing density over the minimum-geometry bipolar transistor.[75]

Table 8-17 gives a comparison of the fundamental IC logic families.

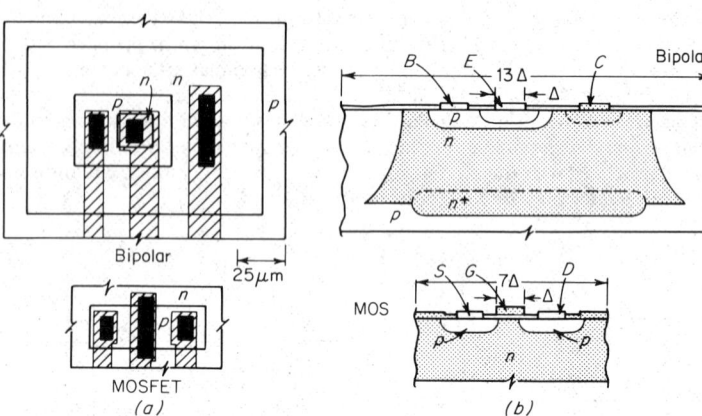

Fig. 8-99. Size comparison between the planar layout of bipolar and MOS transistor.

Fig. 8-100. Relative cross-section areas of bipolar and MOS transistors.

INTEGRATED CIRCUIT MEMORIES

87. Semiconductor Memories[76]. Semiconductor memories are rapidly becoming highly competitive with magnetic data storage devices. Semiconductor memories are also easier to interface with the external drive and sense circuitry than their magnetic counterparts. This advantage often leads to significant reduction in overall memory-system complexity.

Semiconductor memory devices fall into two groups: bipolar and MOS. Bipolar memories offer higher speeds and signal levels, which are directly compatible with the remaining bipolar logic circuits most often used in computer design. MOS memories, on the other hand, are less expensive to fabricate and offer higher bit densities. In addition, the

Table 8-17. Comparison of the Main Integrated Circuit Digital Logic Families

Parameter	RTL	Low-power RTL	DTL	HTL	12-ns TTL	6-ns TTL	4-ns ECL	1-ns ECL	PMOS	CMOS
Circuit form	resistor-transistor	resistor-transistor	diode-transistor	diode-zener transistor	transistor-transistor	transistor-transistor	emitter-coupled current mode	emitter-coupled current mode	p-channel MOS	Complementary MOS
Positive logic function of basic gate	NOR	NOR	NAND	NAND	AND-OR-INVERT	AND-OR-INVERT	OR/NOR	OR/NOR	NAND	NOR or NAND
Wired positive logic function	implied AND (some functions)	implied AND (some functions)	implied AND	implied AND (A-O-I)			implied OR (all functions)	implied OR (all functions)		none
Typical high-level Z_o	640 Ω	3.6 kΩ	6 or 2 kΩ	15 or 1.5 kΩ	70 Ω	10 Ω	15 Ω	6 Ω	2 kΩ	1.5 kΩ
Typical low-level Z_o	R_{sat}	R_{sat}	R_{sat}	R_{sat}	R_{sat}	R_{sat}	15 Ω or 2.7 mA	6 Ω or 21 mA	25 kΩ	1.5 kΩ
Fan-out	5	4	8	10	10	10	25	10 low-Z inputs or 50 Ω	20	50 or higher
Specified temperature range, °C	-55 to 125 0 to 75 15 to 55	-55 to 125 0 to 75 15 to 55	-55 to 125 0 to 75	-30 to 75	-55 to 125 0 to 70	-55 to 125 0 to 75	-55 to 125 0 to 75	0 to 75	-55 to 125 0 to 75	-55 to 125
Supply voltage	3.0 V ± 10% 3.6 V ± 10%	3.0 V ± 10% 3.6 V ± 10%	5.0 V ± 10%	15 ± 1 V	5.0 V ± 10% 5.0 V ± 5%	5.0 V ± 10% 5.0 V ± 5%	-5.2 V + 20% - 10%	-5.2 V ± 10%	-27 ± 2 V -13 ± 1 V	4.5 to 16 V
Typical power dissipation per gate	12 mW	2.5 mW	8 or 12 mW	55 mW	12 mW	22 mW	40 mW	55 mW plus load	0.2 to 10 mW	0.01 mW static ≈ 1 mW at 1 MHz
Immunity to external noise	Nominal	Fair	Good	Excellent	Very good	Very good	Good	Good	Nominal	Very good
Noise generation	Medium	Low medium	Medium	Medium	Medium high	High	Low	Medium	Medium	Low medium
Propagation delay per gate, ns	12	27	30	90	12	6	4	1	300	70
Typical clock rate for flip-flops, MHz	8	2.5	12 to 30	4	15 to 30	30 to 60	60 to 120	400	2	5

Source: After Ref. 74, Par. 8-117.

8-69

power consumption of MOS circuits is usually lower, particularly in the case of CMOS circuitry. The high impedance of MOS devices also makes dynamic circuits possible, in which the information is temporarily stored as a charge on a capacitor, and is replenished periodically. The dynamic approach often allows a higher functional density than the static approach.

Depending on their functional use, semiconductor memories can also be classified as random-access read/write memories (RAM) or the read-only memories (ROM). The basic difference between these two memory types is that the bit pattern of the stored information is fixed in the read-only memory, while it can be changed during normal operation in the random-access memory.

88. Bipolar Memory Cells. The bistable flip-flop, made up of two cross-coupled-inverter stages, is the most widely used basic memory cell for both bipolar and static MOS memory circuits.[77,78] This cell is inherently fast, simple to design, and insensitive to process variations. Figure 8-101a shows a basic bipolar memory cell made up of *npn* transistors and collector load resistors. In this circuit one transistor is normally on, which keeps the other transistor turned off. When an external signal is used to force the "off" transistor into the "on" state, the "on" transistor turns off.

Thus the flip-flop can have two stable states, and will remain in either of these states until an external signal is used to change its state. These two stable states can be interpreted as stored logic 1 and 0. This type of memory cell is generally used for random-access read/write memories. To get information in and out of the circuit—writing and reading—a gating arrangement is used. This gating can be achieved with dual-emitter transistors, as shown in Fig. 8-101b. One of the emitters of each transistor is tied to a common word line, while the other two emitters are each tied to one of the bit lines.

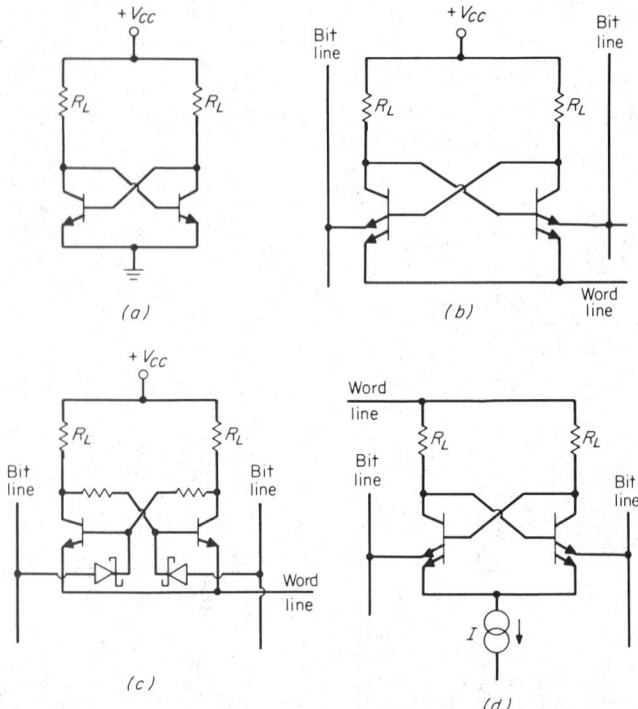

Fig. 8-101. Basic-bipolar-memory-cell structures. (*a*) The basic bipolar flip-flop memory cell; (*b*) bipolar memory cell with multiemitter transistors; (*c*) bipolar memory cell with Schottky diode gating; (*d*) emitter-coupled memory cell.

A large memory array can be formed by interconnecting many such flip-flops. Any particular memory cell may be selected or addressed, for writing or reading. To read the contents of a cell, the word-line voltage is raised. This causes the flip-flop current, which normally flows through the word line, to transfer to one of the bit lines. The signal current is then detected by a current-sensing amplifier.

For writing, a cell is similarly selected by the word line; then, unbalancing the voltage at the two bit lines forces the memory cell into the desired state. For unselected cells, the word-line voltage is low, forcing the cell current through the word line. For this condition, no signal current flows through the bit line, and the cell content is not sensed. A change in voltage on the bit line will have no effect on the state of the cell. Because of its simplicity and small area, this cell structure is used in many bipolar memory circuits for read/write applications.

A Schottky diode (see Figs. 8-53 and 8-54) may also be used to address the basic flip-flop, as shown in the cell configuration of Fig. 8-101c. This cell may be selected for reading by lowering the word-line voltage. Signals can then be detected on the bit lines through the Schottky diode. For writing, a large current through the Schottky diode simultaneously turns on the "off" transistor and forces the previously "on" transistor to turn off by increasing the load current and decreasing the base drive. This cell has the advantage of small size, low power operation, and high speed, because the collector resistor R_C can be made large without degrading the access speed of the memory cell.

Figure 8-101d shows a memory-cell structure based on the emitter-coupled logic (ECL) configuration. In this type of memory cell, bit selection is achieved by raising the word-line voltage. The write/read operation of the ECL memory cell is much as in the dual emitter cell shown in Fig. 8-101d. In the ECL memory cell, the voltage across the selected cell is higher than across the unselected cell, making a large sense current available. Since the voltage across the supply terminals of unselected cells can be made quite low, standby power dissipation can also be kept low.

89. MOS Memory Cells. Because of the self-isolating feature of MOS devices, MOS memory cells can be packed much closer together than comparable bipolar memory cells. The relative sizes of MOS and bipolar devices are shown in Fig. 8-99.

Static Memory. In static-memory cells, information can be stored indefinitely, provided that the bias supplies are not turned off. The basic bipolar memory cells of Fig. 8-101 all fall into this category. In the case of MOS circuits, the basic cross-coupled flip-flop configuration can also be used for static-memory applications.

A circuit configuration suitable for this application is shown in Fig. 8-102a. In the figure, transistors T_1, T_2 and T_3, T_4 make up the basic bistable circuit. Transistors T_5 and T_6 provide the gating function for the cell. To read the cell, T_5 and T_6 are turned on by the word line, and the data on T_1 and T_3 are transferred to the bit lines.

Writing can be achieved by again turning on T_5 and T_6 and then forcing the cell into the desired logic state by applying the proper voltage to the bit lines.[79] If p-channel MOS transistors are used, V_{DD} and V_{GG} are negative supply voltages. The basic MOS flip-flop storage cell using six transistors per bit is used in many read/write MOS memories.

Dynamic (Charge-Storage) Memory Cells. The dynamic MOS memory cells overcome the two major disadvantages of bistable flip-flop cells, namely, high-power-dissipation and large-chip-area requirements. In these types of memory cells, one makes use of the high-impedance and low-leakage-current properties of MOS devices, which permit using the parasitic device capacitances for temporary charge storage. For example, the gate of a MOS transistor may be used as storage node. For a p-channel device, the presence of a sufficient amount of negative charge on its gate will turn the device on. If this charge is removed, the device is off. This can be interpreted as a logic 1 or 0 condition. If provisions are made to supply and remove charge from the gate, and for sensing the presence or absence of charge, a memory cell can be obtained.[80,81]

Figure 8-102b shows a circuit configuration of such a charge-storage memory cell. Transistor T_1 is used as the charge-storage device. Transistor T_2 provides a connection between T_1 and the "read data" terminal, when the "read select" line is activated. This permits interrogation of the cell for its content. Transistor T_3 permits access to the charge-storage node for write purposes. This is accomplished by activating the "write select" line. Since charge can be stored for a limited time only, a means for periodic refreshing of the memory data must be provided. This is also done through T_3 by reinforcing the charge condition of the storage node from a refresh amplifier.

Read and write select functions or data lines may be combined as shown in Fig. 8-102c and d, respectively. These differences can save interconnections at the expense of more complex drive circuitry. In general, dynamic read/write memory cells achieve higher performance (less than 100 μW/bit power dissipation in the active mode) and higher packing densities than static memory cells. As a result of these advantages, dynamic memory structures can be competitive with either core or plated-wire memories.

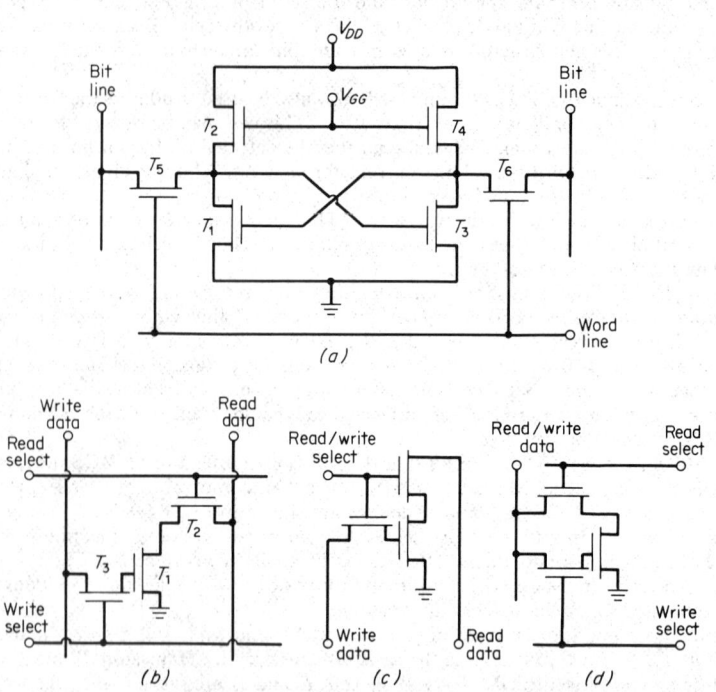

Fig. 8-102. Basic PMOS memory cells. (a) Static memory cell; (b) basic dynamic read/write memory cell; (c and d) simplified cell structures using combined read/write and data select lines.

90. Functional Classification of Semiconductor Memories. Depending on their functional application, semiconductor memories can be divided into three basic categories: read-only memories, random-access memories, and shift registers.

91. Read-Only Memories (ROM). This type of memory contains a "permanent" set of information written into the memory. Under ordinary operation, the contents of the memory remains unchanged. Semiconductor ROMs can be divided into three groups:

1. Mask programmed ROMs.

2. Fusible-link ROMs, in which a permanent and irreversible change in the memory interconnection pattern is caused by either an electrical pulse or by mechanical means.

3. Alterable ROMs, in which a reversible change in active-device characteristics is induced electrically.

In both the mask programmed memories and the fusible-link memories, the data in the individual memory cells are "written" at the time the memory is fabricated. Since the ROM cells are very small, such a memory takes up much less chip area than a RAM of comparable bit count.

Mask programmed ROMs are frequently used to store microprograms for code conversion, process control, or for character generation. In these cases a custom bit pattern is specified by the customer for a special application. Typically, the customer specifies the input-output requirements of the custom ROM on a coding sheet. This information is then used in the fabrication process to generate custom interconnection masks for the integrated circuit.

Small size and high functional density of MOS devices make them particularly useful for read-only memories. In most cases, the additional circuitry required for address decoding and input buffering is included on the same chip with the memory matrix.

Figure 8-103 shows a block diagram of an MOS read-only memory which is organized as a 16-word matrix, with 2 bits per word.[82] The basic storage elements are the MOS transistor cells. Since the memory has 32 transistor locations, it can be coded with 32 bits of information by inserting or deleting these transistors at the desired locations.

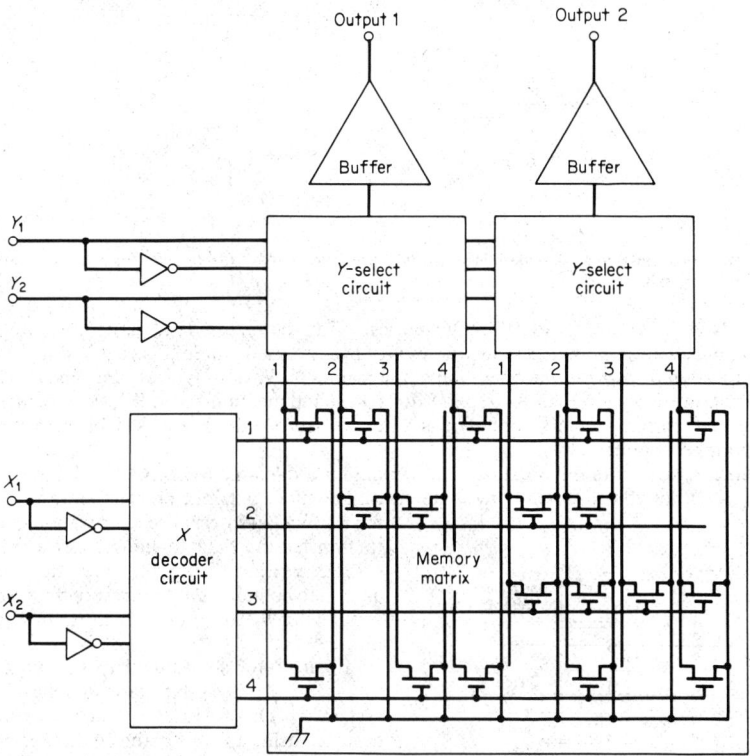

Fig. 8-103. Block diagram of a 16-word, 2 bits/word MOS ROM. Logic 1 in a cell is represented by a transistor at the intersection of the decoded X and Y address lines (Ref. 82).

Fusible-link ROM circuits are programmed after the monolithic circuit fabrication and packaging process is completed. This programming is achieved by fusing, or burning out certain critical interconnections at the desired storage locations. It can be done by the customer with the help of special programming equipment. These memories are also called *field-programmable ROMs* (FROM). One disadvantage of this type of memory is that they cannot be reprogrammed, which prevents complete functional testing before delivery and also makes it impossible to modify the content of the memory in case of an error.

Alterable ROM Circuits. In these circuits the memory contents can be reprogrammed by inducing a change in the active-device characteristics. An example of such an "alterable" active device is the floating-gate avalanche injection metal oxide semiconductor (FAMOS) structure, shown in Fig. 8-104a.[83] This device is essentially a p-channel silicon-gate MOS transistor (see Fig. 8-47) with no electrical contact made to the gate. The operation of this memory cell depends on the charge transport of electrons to the floating gate by avalanche injection from either the source or the drain pn junctions. An applied high-voltage pulse causes an accumulation of charge on the gate, which induces a conductive inversion layer from source to drain. When the applied voltage is removed, no discharge path is available to the accumulated electrons, since the gate is surrounded by thermal oxide. Thus this injected charge becomes "trapped" at the gate electrode, and maintains an induced channel under the gate until a high-voltage pulse of opposite polarity is applied to alter the memory setting. The circuit diagram of a FAMOS memory cell is shown in Fig. 8-104. The programming of a memory bit is accomplished by simultaneously selecting the X- and Y-select lines.

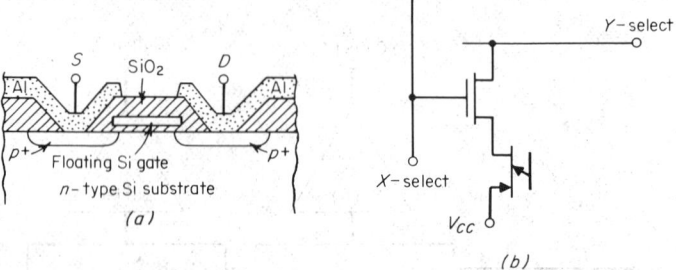

Fig. 8-104. Floating-gate avalanche injection MOS transistor (FAMOS). (*a*) Device cross section; (*b*) typical memory cell.

92. Random-Access Read/Write Memories. Random-access read/write memories, often simply called random-access memories (RAM), are used where it is required to change the bit location during routine operation of the memory. Various typical static and dynamic RAM cells are shown in Figs. 8-101 and 8-102. Fast and medium-speed RAMs are generally used for scratch-pad memory applications, while low-speed RAMs are used for main storage or in peripheral buffers.

Figure 8-105 shows the organization diagram for a dynamic MOS RAM. Reading and writing occur for all cells in one row simultaneously. Since only one bit at a time is available for writing, an internal read operation is used to transfer the data to the refresh amplifier prior to writing. In this manner, the refresh amplifier contains data corresponding to the contents of the row into which writing takes place.[79]

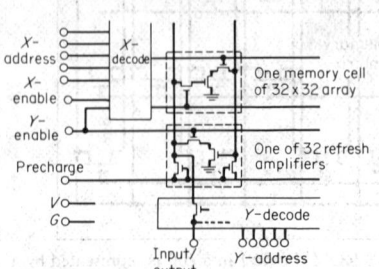

Fig. 8-105. Organization diagram of a dynamic MOS random-access memory system.

Figure 8-106 shows the basic organization for a 16-word by 4-bit static bipolar RAM system.[84] The basic cell structures can be designed using any one of the bipolar memory-cell circuits shown in Fig. 8-101. Reading the contents of a particular word is accomplished by applying a signal to the appropriate word-select line and detecting the outputs of the sense amplifiers (SA) at terminals D_0 through D_3. Writing is accomplished by enabling word and write lines and applying a signal to the proper write amplifier.

93. Shift Registers. A shift register is an arrangement of an arbitrary number of storage cells in a row and is used primarily for temporary storage of digital information. Some common applications of shift registers are in serial-data entry and serial-data output, as well as in serial-to-parallel converters. The serial in–serial out shift register can perform similarly to a high-speed drum memory; however, unlike the mechanical drum memory, it may be stopped instantaneously. Serial-to-parallel converters are often used in accumulators, where the data are entered in serial fashion, for example, from a keyboard, and then acted upon in parallel, as by an adder.

The shift registers can be designed for either static or dynamic operation. Static shift registers make use of the basic flip-flop circuits of Figs. 8-101 and 8-102 for data-storage purposes. Dynamic shift registers operate in the same manner as the dynamic RAM circuits, where each bit of information is constantly refreshed and recycled. Figure 8-107 shows the circuit diagram of a section of a dynamic MOS shift-register circuit. The circuit operates with two-phase clock pulses, $V_{\phi1}$ and $V_{\phi2}$. After each clock pulse, one bit of information is inverted and transferred half a cell to the right; thus, after two clock pulses, the content of each cell is shifted over to the next one.

94. Economic Considerations. The trend in integrated circuits is toward increasing complexity of circuit functions available in a single microelectronic package. Integrated

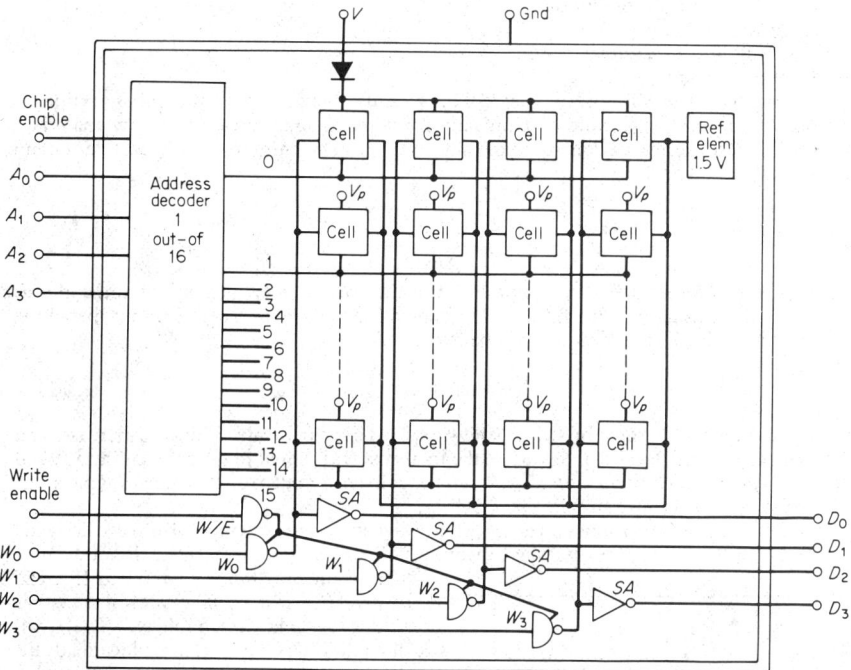

Fig. 8-106. Typical organization diagram of a 16-word by 4-bit (64-bit total) bipolar random-access memory (Ref. 84).

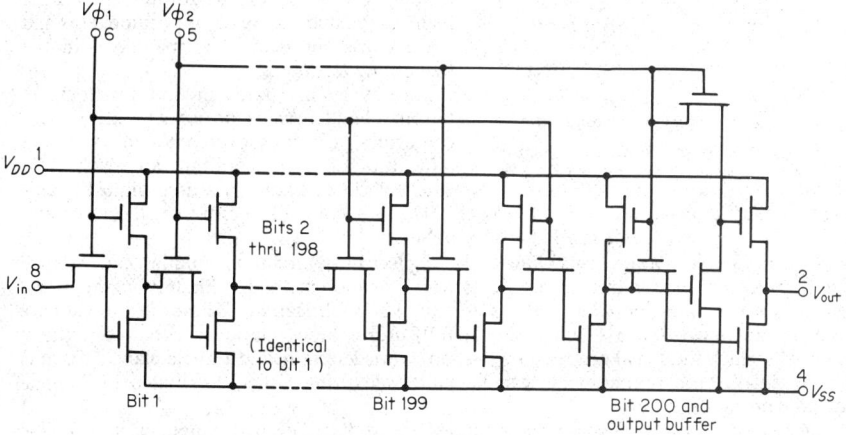

Fig. 8-107. Circuit diagram of one section of a 200-bit dynamic shift register.

circuits of a high degree of functional complexity are classified as large-scale integration (LSI) circuits. Typically, monolithic circuits having a functional complexity equal to or greater than 100 simple gate circuits are classified as LSI. It is possible to build large subsystems (such as computer memories) or even complete systems (such as desk calculators) on a single chip or a wafer of silicon.

The practical limitations as to the extent of integration of systems are set by any one or any combination of the following factors:

1. Fundamental limitation of silicon processing technology.
2. Package and pin limitations.
3. Power dissipation.
4. Constraints on systems organization.
5. Economic factors.

A possible criterion for the degree of systems integration is the minimization of manufacturing costs per unit. Manufacturing costs include the cost of raw materials, processing, testing, and packaging of the silicon chip. The chip-processing cost per circuit may be expressed as[85]

$$C = \frac{pa}{y(n)}$$

where p is the processing cost per unit area, a is the area of an elementary circuit, and $y(n)$ is the chip yield as a function of the level of integration (n is the number of circuits per chip).

The packaging cost per circuit can be expressed as

$$P = \alpha a + \frac{\gamma}{n}$$

where αa accounts for batch processes, and γ for fixed costs per wafer. Since C increases and P decreases with the level of integration, the total cost C_T is equal to the sum of C and P, and has a minimum which occurs, roughly, when the chip costs approximate the packaging costs. This relationship is illustrated in Fig. 8-108.

95. Microwave Integrated Circuits. Microwave integrated circuits (MICs) are designed to operate at frequencies beyond the capabilities of conventional integrated circuits. They cover the frequency range from 0.5 to 15 GHz. Microwave ICs can be fabricated with either monolithic or hybrid technology. The small size of monolithic circuits is a significant advantage for microwave applications because it minimizes the lead inductance problem associated with discrete circuits. However, the use of monolithic IC technology for microwave applications has been somewhat limited because many of the microwave circuit components and devices are not readily compatible with the monolithic technology.

The hybrid integrated circuit technology, on the other hand, permits the use of a wider variety of devices, and thus overcomes many of the difficulties of the monolithic approach. To date, hybrid technology is used almost exclusively in the frequency range from 1 to 15 GHz.[86] Details of microwave structures and semiconductors are provided in Pars. 9-50 to 9-62.

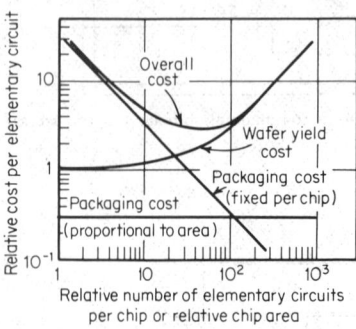

Fig. 8-108. Trends in manufacturing costs with increasing levels of integration (Ref. 85).

96. Microwave Circuit Techniques. Microelectronic circuit techniques can provide significant cost reduction over the conventional fabrication methods. Photolithography and screening are the most popular methods used in hybrid fabrication. These technologies are discussed in Pars. 8-10 to 8-13 and 8-30. To fully utilize these techniques, circuit forms must be used in which the signal propagation properties are determined in a single plane. Circuits which satisfy these requirements can be fabricated using either distributed or lumped components.

Distributed Circuits. The most commonly used form of distributed circuits is the microstrip transmission line[87] as illustrated in Fig. 8-109a. The line consists of a strip conductor which is separated from the ground plane by a dielectric layer. The circuit properties are determined by the impedance and the length of the lines. The main part of the propagation field is confined to the region of the dielectric below the strip conductor, and the propagation approximates a TEM mode. An alternative approach to distributed hybrid MICs is the use of the suspended substrate line.

The cross section of a suspended substrate line is shown in Fig. 8-109b, where the metal shield surrounding the system acts as a ground plane, and the ceramic substrate serves as a mechanical support for the suspended line. Two other types of distributed lines used in hybrid MICs are also shown in the figure. The finite slot line of Fig. 8-109c is a geometrical and electrical dual of the coplanar waveguide line of Fig. 8-109d. In the case of both these lines, the wave propagation mode is not TEM, and there are longitudinal as well as transverse magnetic rf fields.

Lumped Circuits. This class of MICs use lumped circuit elements such as resistors, capacitors, or inductors whose values are independent of frequency, within the frequency band of interest. In order for electrical components to behave as lumped elements, it is necessary that their physical dimensions be much smaller than the wavelength of the electrical signal they are supposed to handle. Small size of integrated circuit components is a distinct advantage in this application since these components can maintain their lumped characteristics up to much higher frequencies than their discrete counterparts.

Some of the lumped inductor, conductor, and capacitor structures used in fabricating hybrid microwave ICs[87] are shown in Fig. 8-110a. Typical lumped LC circuits[88] which can resonate in the frequency range of 4 to 12 GHz are shown in Fig. 8-110b and c.

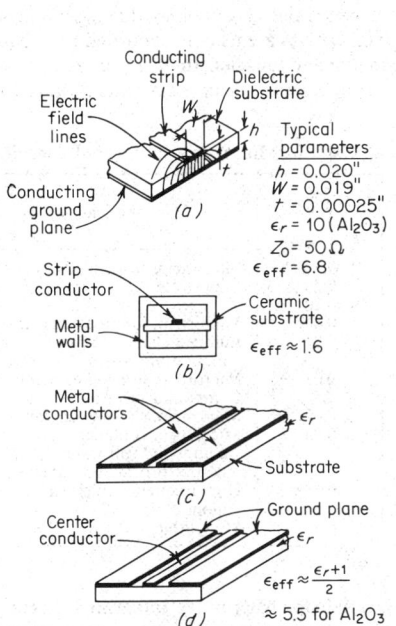

Fig. 8-109. Commonly used MIC transmission lines. (a) Microstrip transmission line; (b) suspended substrate line; (c) slot line; (d) coplanar waveguide.

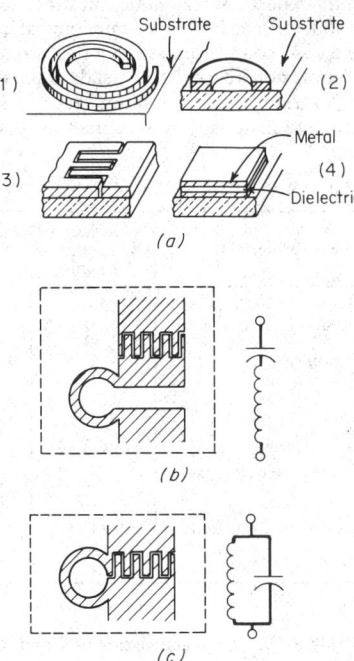

Fig. 8-110. Lumped components for microwave integrated circuits. (a) Typical structures: (1) spiral inductor, (2) strip inductor, (3) interdigitated capacitor, (4) metal-oxide-metal capacitor; (b) series LC circuit using strip inductor and interdigitated capacitor; (c) parallel LC circuit using strip inductor and interdigitated capacitor.

Most hybrid MIC structures require a three-layer sandwich of metal-dielectric-metal on a dielectric substrate. Figure 8-111 shows the cross section and the equivalent circuit of such a multilayer circuit structure.[89] First a layer of chromium-copper-chromium is deposited on the substrate. The thin layers of chromium (or titanium) on either side of the copper (or gold) conductive layer are necessary to ensure proper adhesion to the dielectric surfaces. After the bottom metal layer is deposited and patterned, an SiO_2 film is deposited as the dielectric layer; and finally, the top metal layer is deposited and etched.

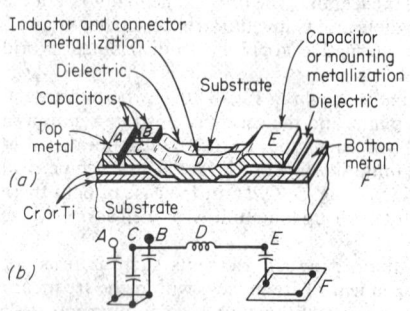

Fig. 8-111. Multilayer metal-dielectric films for MICs. (*a*) Cross section of integrated structure; (*b*) its equivalent circuit contacts (Ref. 89).

97. Materials for Microwave ICs. *Substrates.* Substrates for MICs should have low dielectric loss. The surface finish is important because it determines the definition of the circuit pattern, the yield in thin-film MOS capacitors, and rf conductor loss. The relative dielectric constant ε_r should be in the range of 8 to 16. Heat conductivity is important where high-power devices are used. Some properties and applications[87] of substrates that are commonly used for MIC applications are listed in Table 8-18.

Conductors. Important considerations for MIC conductors are the rf resistance and skin depth, deposition technique, substrate adherence, and thermal expansion during processing. Conductors can be divided into four categories, as shown in Table 8-19. The categories range from good conductors with poor substrate adherence to relatively poor conductors with good substrate adherence. The metals of the first two categories are usually deposited by vacuum evaporation, or electron-beam heating. Molybdenum and tungsten, in the third category, are refractive materials, and vacuum evaporation using electron-beam heating is required for

Table 8-18. Properties of Substrate Materials Used in Microwave Integrated Circuits

Material	Surface roughness Δ, μm	ε_r	K, W/cm²-C°	MIC applications
Alumina: 99.5%	2–8	10	0.3	Microstrip, suspended
96%	20	9	0.28	substrate
85%	50	8	0.20	
Sapphire	1	9.3–11.7	0.4	Microstrip. lumped element
Glass	1	5	0.01	Lumped element quasi-monolithic MICs
Quartz (fused)	1	3.8	0.01	Microstrip, lumped element
Beryllia	2–50	6.6	2.5	Compound substrates
Rutile	10–100	100	0.02	Microstrip, slot-line coplanar
Ferrite/garnet	10	13–16	0.03	Microstrip, coplanar compound substrates, nonreciprocal components
GaAs (high resistivity)	1	13	0.3	High-frequency microstrip, monolithic MICs
Si (high resistivity) ...	1	12	0.9	Monolithic MICs

Table 8-19. Characteristics of Conductor Materials for Microwave Integrated Circuits

Material	Dc resistivity ρ (relative to Cu)	Skin depth δ at 2 GHz, μm	α_r thermal expansion, °C⁻¹ × 10⁶	Adherence to dielectric
I (Ag, Cu, Au, Al) ...	0.95–1.6	1.4–1.9	15–26	Poor
II (Cr, Ta, Ti)	7.6–48	4.0–10.5	8.5–9.0	Good
III (Mo, W)	3.3	2.6	6.0, 4.6	Fair
IV (Pt, Pd)	6.2	3.6	9–11	

deposition. Sputtering works with all these conductors and is especially useful for category III materials. The resistivity of the metals listed in the table is normalized to that of copper ($\approx 1.7 \times 10^6$ Ω·cm).

Dielectrics and Resistors. Isolation and capacitor dielectrics for MIC applications must be reproducible, withstand high voltages, and be able to undergo processing without developing pinholes. SiO, SiO_2, and tantalum pentoxide are the most widely used. Resistive films should have resistivities in the range of 10 to 500 Ω/square, a low-temperature coefficient of resistance, and good stability. The most widely used resistive films are nichrome and tantalum.

INTEGRATED CIRCUIT ASSEMBLY AND PACKAGING

98. Assembly Techniques. The batch processing of integrated circuits on semiconductor wafers represents the most efficient part of the IC manufacturing process. After processing is complete, the wafer must be separated into individual circuit chips, and each circuit must be handled and packaged individually. This assembly-and-package processing is inherently less efficient, since it does not enjoy the advantage of batch processing. Packaging and assembly are one of the most critical steps in the integrated circuit fabrication process, both from the point of view of cost as well as of reliability. Packaging costs can easily exceed the cost of the integrated circuit chip; and the majority of device failures are either assembly- or packaging-oriented.

99. Scribing and Die Separation. After the completion of the wafer fabrication process, monolithic chips are electrically tested while still on the wafer. In this step, called the *wafer-sort* step, the chips which do not meet the electrical specifications are automatically marked with an ink drop. This wafer-sort step is followed by a scribing and die-separation step, where the individual dice are cut apart and physically separated. A number of methods have been designed for separating the individual chips. Diamond scribing and breaking are almost universally employed in integrated circuits. A diamond tool is used to scribe lines into the surface of the wafer in a grid pattern. The wafer is then fractured into individual dice along these scribe lines. Recently, the use of a laser beam for scribing purposes has also gained acceptance. Other die-separation methods, such as etch-cutting, ultrasonic dicing, and diamond sawing, have found relatively little use in integrated circuits due to their excessive waste. If the wafers are diced using the scribe-and-break process, the dice must be cleaned to remove scribing dust before they are attached to the circuit package.

100. Die Attachment. The attachment of the die to the package is generally accomplished by using a eutectic brazing alloy such as gold-silicon or gold-germanium.[90] In the gold-silicon system the eutectic point is at approximately 31 at % silicon and 370°C. In practice, several methods can be used to form the eutectic bond: (1) firm contact is made between the gold-plated silicon chip and gold-plated metal substrate and then heat is applied; (2) a thin piece of gold foil is sandwiched between the unmetallized silicon die and the gold-plated substrate; (3) a gold-silicon eutectic preform is inserted between the silicon die and the gold-plated metal substrate. During the die-attach step, the chip is held in position on the substrate with the preform, and then heated to above the gold-silicon eutectic temperature. This procedure is normally performed in a nonoxidizing atmosphere in either a continuous belt furnace or on a special die-mounting apparatus.

101. Lead Bonding. In this step, electrical connections are provided between the integrated circuit and the package leads. Normally, this is accomplished by using either thermocompression or ultrasonic bonding techniques to attach gold or aluminum wires between the contact areas on the silicon chip and the package leads. Other bonding methods combine the electrical connection and die attachment in one step. Flip-chip bonding and beam leads are examples of these methods. All the different lead bonding and welding processes rely on obtaining intimate contact between two metallic surfaces.

Thermocompression Bonding. Thermocompression bonding depends on the simultaneous application of pressure and heat to the two metals to be joined together. In the region to which pressure is applied, interface films are broken and diffusion occurs between the metals, resulting in a metallic bond. Two major types of thermocompression-bonding methods are ball bonding and wedge bonding.

In *ball bonding* (also called *nailhead bonding*) a small ball is formed at the end of the bonding wire and deformed under pressure against the pad area of the integrated circuit chip, giving a bond. Only gold wire is used in this method. The bonding tool consists of a capillary tube through which the bonding wire is fed. A hydrogen flame is used to melt off the end of the wire to break continuity. This melting process then forms a new ball at the end of the capillary tube, which is used to initiate the next bonding cycle.

In *wedge-bonding* operation, a wedge- or chisel-shaped bonding tool is used to apply pressure to the lead wire which has been located on the bonding pad. The pad must be heated to the bonding temperature. Two precise positioning operations are required in the wedge-bonding process. The package with the silicon chip attached is first positioned under a microscope. This position corresponds to the terminal position of the wedge. Then the wire must be accurately positioned over the bonding pad. The bonding operation is performed in an inert-gas atmosphere.

Ultrasonic Bonding. Unlike the thermocompression technique, ultrasonic bonding uses mechanical rather than thermal energy to form a metallic bond. In this technique, normally, Al wires are used for interconnections, since aluminum is a soft metal which can be easily deformed under low pressures.

In the ultrasonic bonding device, an elastic vibration is created by the rapid expansion and contraction of a magnetostrictive transducer which is driven by high-frequency alternating current. A bonding tip serves to transfer the vibrations to the material to be welded. The bonding tip vibrates in a direction parallel to the interface of the bond. Ultrasonic bonding is often preferred to thermocompression bonding since it does not require the circuit chip to be heated during the bonding operation.

102. Flip-Chip Bonding. The flip-chip bonding technique combines die attachment and die bonding in a single fabrication step. To achieve this, the bonding pads are processed to give them raised contact bumps. In the bonding operation, the chip is "flipped over" and bonded face down on a matching lead pattern. The raised contacts, or bumps, are usually solder, aluminum, or a silver-tin combination, chosen to be compatible with commonly used substrate materials. The bumps must provide reliable electrical and mechanical connection and must be capable of sufficient deformation to compensate for nonplanar surfaces and bump heights.

Two different techniques for deforming the bumps are used. The first is based on application of pressure, as in thermocompression or ultrasonic bonding, to cause plastic flow of the bump. Typical materials used for this process are gold, silver, aluminum, and copper. In the second technique, the bump material becomes liquid at some point in the bonding cycle. Solder reflow is often used to implement this method. Figure 8-112 shows a cross section of a flip-chip-bonded die on a package substrate.

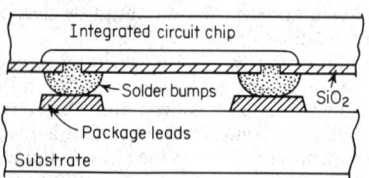

Fig. 8-112. Cross section of a flip-chip-bonded die on a package substrate.

103. Beam-Lead Bonding. Beam-lead bonding is similar to the flip-chip method in that the integrated circuit die is attached face down and bonded in a single operation. The beam leads thus serve as mechanical mounts and as electrical connections and are easily visible since they extend beyond the edge of the chip. The chips are bonded either by thermocompression or ultrasonically. The beam-lead contacts are usually electroformed, as discussed in Par. **8-1**.

104. Spider Bonding. In the spider-bonding process an accurately stamped lead frame (spider) is bonded directly to the bonding pads on a standard integrated circuit chip. The lead frame can either be bonded to a standard dual in-line lead frame for packaging or it can be cut to leave ends extending over the edge of the chip, similar to beam-lead devices.

Typical materials used for the spider-lead frames are aluminum and copper. If copper is used, the leads can be soldered or thermocompression-bonded directly to printed-circuit boards or ceramic substrates. The bonding process requires high-precision equipment. First the chip is automatically aligned, and then transferred to the spider-lead frame. Then the leads are bonded simultaneously by ultrasonic techniques. After the spider is bonded to the chip, the die may be eutectically bonded to a header.

105. Integrated Circuit Packaging. The purpose of an integrated circuit package is to meet the electrical, thermal, chemical, and physical requirements associated with the ultimate application of the IC. Some of the basic requirements for an integrated circuit package can be outlined as follows.[91]

Chip protection (environment isolation). The construction of the package must be such that it protects the electronic device from the surrounding environment. Common tests to ensure adequate protection are:

  a. Hermeticity.
  b. Thermal shock and cycling.
  c. Mechanical shock and centrifuge.
  d. Humidity tests.

Compatibility with system requirements. The basic properties of the materials used in the package manufacture and the design can greatly affect the circuit performance. Of particular interest in this area are such things as:

  a. Heat-transfer capability.
  b. Radiation hardening.
  c. Electrical properties, such as the lead resistance and capacitance.

Mechanical configuration. This category contains the requirements of size, weight, shape, number of leads, type of leads, and lead location. Particular points to consider are:

  a. Effect on size of finished equipment.
  b. Usable device space per size of package.
  c. Ease of device assembly and test.

Interfacing between the die and the electronic system. The package supports the chip and provides the electrical connection to the outside world in a convenient and reliable manner. Points to consider are:

  a. Minimum voltage drop within connection elements.
  b. Isolation between the circuits.
  c. Ease of assembly of packages into equipment.

Cost objectives. The cost of the package is usually a major part of the device. One must consider not only the manufacturing cost of the package, but also the cost of handling the package during assembly and test of the device, and assembly of the final equipment. The most economical design may not use the cheapest package. The ability to use automatic handling during the processing may well justify a package several times the cost of a cheaper package.

106. Standard IC Packages. The most widely used package types are the radial-lead type, the flat pack, and the dual-in-line package. The radial-lead type of packages are mostly made of kovar (an alloy of iron, nickel, and cobalt) with hard glass seals and kovar leads. Since this material is a relatively poor conductor of heat and electricity, the substrate and leads are usually gold-plated. Other packages use metal-lead frames combined with ceramic, glass, and metal components. Dual-in-line packages often use injection-molded plastics. Construction details of these various packages are discussed below.

Radial Lead (TO-Type) Packages. The small TO-5, TO-18, and TO-47 headers which were originally developed for transistors have been adapted to accommodate the larger number of leads required for integrated circuits. The various-size packages have been standardized into a series of TO packages. Each of the packages in the series, TO-1 to TO-51, is produced to a standard set of dimensions by all manufacturers. The most common radial-lead packages are the TO-5 and TO-18 packages. Typical dimensions of a TO-5 package are shown in Fig. 8-113. TO-18 is a smaller version of TO-5, with a 0.1-in.-diameter pin circle. TO packages have a very high reliability rating and good thermal characteristics. The TO-type packages have two main disadvantages: (1) available number of leads is limited (12 leads maximum); and (2) leads bend easily and are difficult to insert into sockets.

Flat Packs. The flat package was developed to improve on the volume, weight, and pin-count limitations of the TO-type packages. The flat pack has about one-fifth the volume and one-tenth the weight of a TO-5 package. Flat packs with up to 22 leads are commercially available. They can be produced in both round and rectangular shapes. Typical dimensions of a 14-pin rectangular flat pack are shown in Fig. 8-114. The flat package has the following advantages: (1) light weight and small volume and (2) large die area compared with package size. Its major disadvantages are high cost and difficulty in handling.

Dual-in-Line Package (DIP). The dual-in-line package represents a compromise between the pin-limited transistor-type packages and the costly and difficult-to-handle flat packs. The

DIP packages are available in 8-, 14-, 16-, 18-, or 24-pin versions. Figure 8-115 shows the typical dimensions for a 14-pin DIP package.

Dual-in-line packages can be fabricated using side-brazed metal leads on a ceramic base, as well as glass-to-metal seals or plastic injection molding techniques. This last type of DIP package is the least expensive type; however, it is nonhermetic and has lower resistance to moisture and relatively poor thermal conductivity.

Face-down-Bonded Packages. Face-down-bonded packages are used in connection with flip-chip or beam-lead assembly techniques. They require very close spacing of the lead pattern (typically 4-mil spacing is used between adjacent leads). The flatness of the lead pattern also needs to be very critically controlled to ensure electrical contact to all the lead terminals.

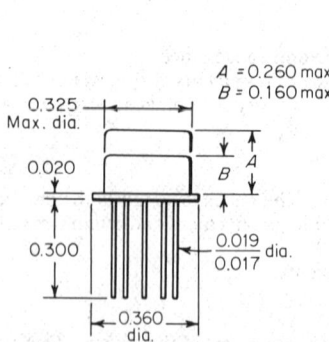

Fig. 8-113. Dimensions of a TO-5 package (all dimensions in inches).

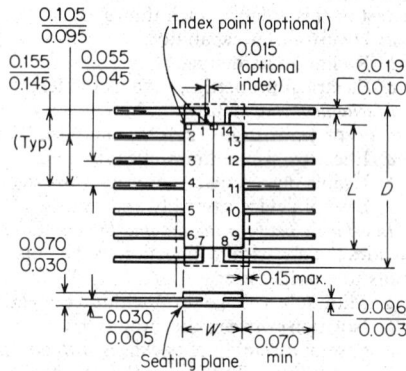

Fig. 8-114. Typical dimensions of a 14-pin rectangular flat pack (all dimensions in inches).

In these types of packages, the contact to the chip is made only through the raised contacts. Therefore the thermal resistance from the chip to the package is significantly higher.

Special-Purpose Packages. In addition to the standard classes of integrated circuit packages described above, a variety of other integrated circuit packaging techniques also find limited use in specialized applications. Some of these are the *power packages,* designed to handle high-power dissipation circuits, and multichip packages, designed to house two or more monolithic chips in one package.

107. Thermal Considerations. One of the basic limitations of integrated circuits is the dissipation of heat produced during operation of the circuit. This heat must be transferred to some sink without causing excessive temperature rise in the circuit elements or interconnections.

Heat transfer may take place by any one or the combination of three mechanisms: conduction, convection, and thermal radiation. In integrated circuits, most of the heat produced by an element is removed by conduction through the substrate. Radiation, convection, and conduction through the leads are usually negligible by comparison. Certain material properties govern the rate of heat transfer for any given temperature difference. For heat transfer by conduction, the thermal conductivity of materials is an important property. Thermal conductivities of substrate and heat-sink materials commonly used in integrated circuits are listed in Table 8-20.

Since the rate of heat transfer depends on the thermal conductivity of each body and each interface in the thermal path, it is desirable to minimize thermal resistance. In the case of a packaged integrated circuit, a model as shown in Fig. 8-116 may be used. The heat-transfer equation for this system may be written

$$Q = \frac{T_j - T_c}{R_{jc}} + \frac{T_c - T_A}{R_{cA}}$$

where Q is the amount of heat flow, and $T_j, T_c,$ and T_A are the junction, case, and the ambient temperatures, respectively. R_{jc} and R_{cA} are the thermal resistances from junction to case, and

Table 8-20. **Thermal Conductivities of Substrate and Heat-Sink Materials Commonly Used in Integrated Circuits**

Material	Thermal conductivity, $\dfrac{W \cdot cm}{°C \cdot cm^2}$	
	At 25°C	At 100°C
Silicon	1.45	1.05
Alumina (Al$_2$O$_3$)	0.29	0.23
Beryllia (BeO)	2.27	1.87
Glass (Pyrex)	0.01	0.01
Copper	3.87	3.76
Aluminum	2.02	2.06
Steel	0.45	0.43

from case to ambient. Thermal resistance is normally measured in degrees centigrade per watt.

The thermal resistance from junction to case is specified by the device manufacturer; the case to ambient thermal resistance depends on the application. The maximum ambient temperature is part of the system specification, and the maximum junction temperature is usually given for certain IC devices. Typical upper temperature limit for silicon devices is 175°C. However, since reliability is inversely proportional to the operating temperature, the actual operating points may have to be significantly lower than 175°C.

Some typical values of total thermal resistance $R_T (= R_{jc} + R_{cA})$ are in the range of 200 and 95°C/W for ceramic flat packs and TO-5 packages, respectively.

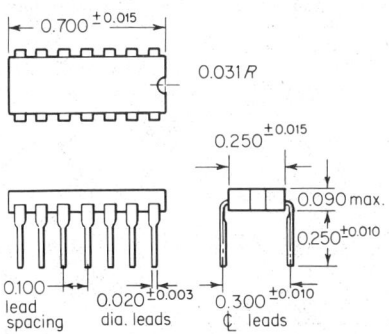

Fig. 8-115. Typical dimensions for a dual-in-line package (all dimensions in inches).

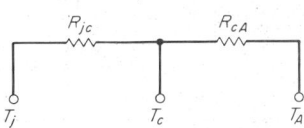

Fig. 8-116. Electrical analog of thermal resistance.

TESTING AND RELIABILITY

108. Testing of Integrated Circuits. During the fabrication and the assembly steps, integrated circuits undergo three separate test cycles: (1) in-process testing, such as continuous monitoring of sheet resistivities, junction depths, and other pertinent device parameters, such as current gain and voltage breakdown; (2) a preliminary electrical testing called the wafer-probe or the wafer-sort test performed prior to the scribing and the die-separation steps; (3) a detailed testing of all pertinent circuit parameters after the completion of the assembly and the packaging operations. In this discussion, these last two types of electrical testing will be discussed.

109. Wafer-probe Testing. The wafer-probe test, also called the wafer-sort operation, takes place after the wafer processing is completed and the final metal interconnections are applied to the circuit. It is then followed by the die-separation (scribing) and packaging operations. These tests are usually made with multiprobe instruments which have adjustable probes to contact the bonding pads of the circuit to be tested. Each individual circuit undergoes a number of basic electrical tests. If found defective, it is marked with an ink marker for identification, to be discarded after the die-separation step. Wafer-probe test is normally performed using an automatic test station where approximately one hundred separate tests can be performed on each chip in a time duration of less than one second. At the completion of the test sequence, the probe assembly is automatically lifted up and the probes are indexed over to the next chip to be tested on the wafer. In this manner, each and every die on the wafer gets tested prior to the start of the assembly operation.

110. Final Testing. The large number of tests which are often necessary to fully characterize an integrated circuit has made testing one of the more expensive and time-consuming parts of the manufacturing process. Automatic high-speed testing is practically mandatory for the final testing of modern ICs because a large number of complex tests are required to check even the simplest types of circuits. Integrated circuit testing can be divided into three general categories:[92]

Dc testing. This includes the static parameters.

Dynamic testing. This includes pulse-amplitude and time measurements as well as complex waveforms.

Ac testing. This includes sinusoidal measurements.

Usually, digital ICs require both dc and dynamic testing while linear ICs require both dc and ac testing. Table 8-21 lists some of the typical circuit parameters which may be tested during the final test cycle.

Because of the large number and the complexity of tests required, the computer has become a major element in IC testing. A generalized block diagram of a computer-controlled test system is shown in Fig. 8-117. The test program can be loaded into the computer by

Table 8-21.   Typical Parameters Measured in Integrated Circuit Evaluation

Input voltage	Voltage gain
Output voltage	Offset voltage
Preset voltage	Delay time
Dc gain	Turn-on time
Leakage current	Turn-off time
Input current	Propagation delay
Output current	Preset time
Pulse amplitude	Commutation time
Pulse overshoot	Saturation voltage
Noise feed-through	Pulse-period width
Noise immunity	Reverse recovery time
Rise time	Step response
Fall time	Phase response
Storage time	Bandwidth
Output distortion	Common-mode rejection

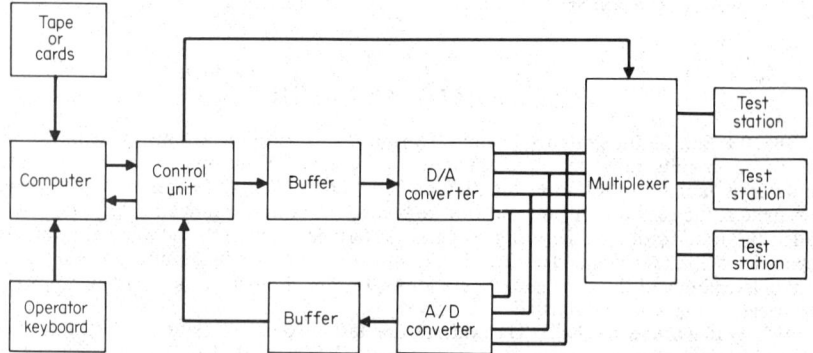

Fig. 8-117.   Functional block diagram of a computer-controlled IC test system (Ref. 92).

punched cards, paper tape, or magnetic tape. Instructions from the computer are then sent to an interface or control unit which controls the various elements of the system. Stimuli instructions for the integrated circuits are buffered, converted into analog voltages, and delivered to the pins of multiplexed test stations or wafer probes, which are time-shared under computer control. Analog-digital (A/D) converters convert the output functions of the IC into digital form.

This information is then buffered, and returned to the computer for processing. The computer makes a "go" or "no go" decision based on the test results and automatically sorts the device into the appropriate container. Normally, such a tester would also have a *data-logging* capability such that preliminary modes of failure can be recorded for failure analysis and yield-improvement purposes.

111. Digital IC Testing. *Bipolar ICs.* Since bipolar digital ICs represent the largest segment of IC production, relatively standard testing techniques have evolved. Usually, testing includes:

Functional testing to find catastrophic failures caused by improper packaging, bonding, metallization, etc.

Dc and pulse parametric testing to find failures due to surface defects such as channels, pinholes, etc.

While virtually all ICs are tested functionally and for dc characteristics, pulse and dynamic testing are performed mainly on high-speed logic families such as the TTL and the ECL circuits.

Functional testing of digital ICs consists in applying certain input codes to a device and comparing the resulting output codes with the desired output codes. Figure 8-118 shows a block diagram of a simple comparison-type tester. Here a binary or random pattern is applied simultaneously to the device under test as well as to a reference unit having the same truth table. The outputs are compared, and a "reject" signal is generated if they differ. More complex systems use algorithmically generated test patterns designed to detect all the failure modes intrinsic to the logic to be tested.

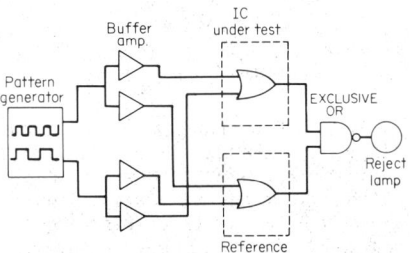

Fig. 8-118. Block diagram of a comparison-type functional system.

MOS ICs. MOS digital circuits are tested in a manner similar to their bipolar counterparts. Functional testing of MOS circuits is conducted near the maximum frequency of the devices (at clock rates of 5 to 10 MHz). Also, for dynamic circuits, measurement of the "stay-alive" time for the minimum operating frequency of the device near 1 Hz may be included. The MOS tester must be able to supply high-frequency test signals as well as the different phases, each precisely settable with respect to the others. For two- or four-phase clock inputs, a phase resolution of better than 1 ns is required. The drivers of an MOS test system must be able to swing 30 V at a slope of 1 ns/V or better, with minimum overshoot, ringing, or crosstalk.

112. Testing of Semiconductor Memories. Semiconductor memories are tested both functionally and parametrically, similarly to other ICs. Various test patterns are used to ensure that the reading or writing of a memory bit will not affect the state of an adjacent cell.

Commonly used routines are checkerboard patterns of 1s and 0s or floating of a 1 or 0 from cell to cell while the adjacent cells are maintained in the opposite state. For larger memories, the generation of these test patterns requires a large number of functional tests. On-line pattern generators allow a test system to produce patterns as the test proceeds, saving memory space in the control computer.

113. Testing of Linear ICs. Until recently, operational amplifiers made up almost all linear ICs, and linear IC testing meant the measurement of offset current, offset voltage, open-loop gain, power capability, common-mode rejection, frequency response, etc. However, with the volume introduction of other specialized linear ICs such as voltage regulators, phase-locked loops, D/A and A/D converters, as well as a variety of consumer circuits, linear IC testing has become a complex procedure, which cannot be treated generally.

114. Reliability. One of the most significant attributes of monolithic integrated circuits is high reliability. Integrated circuits are far more reliable than their discrete-component counterparts, and their reliability is improving rapidly with increased knowledge of the processing techniques and the understanding of the possible failure mechanisms.

115. Reliability Measurements. It is customary to assume the exponential failure distribution for integrated circuits.[93] In its simplest form, this distribution is known as the reliability function, and can be expressed as

$$R(t) = e^{-\lambda t}$$

where $R(t)$ is the probability that a component will not fail for the required operating time t. λ is a constant related to the failure rate of the component; it is known as the *reliability estimator*. The reciprocal of λ is equal to the mean time between failures (MTBF).

The difficulty of demonstrating a given failure rate or MTBF becomes apparent from the requirement that testing time must be at least as long as the MTBF. Generally, the MTBF for ICs is greater than 10^7 h. To demonstrate this reliability with 90% confidence, approximately 2.3×10^7 h of operational life testing with no failures is required.

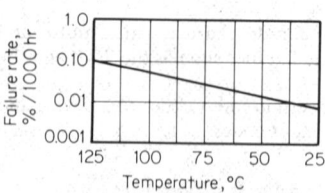

Fig. 8-119. Integrated circuit failure rate vs. junction temperature (Ref. 93).

To reduce this testing time, accelerated life tests may be used. This can be done by aging the IC at accelerated stress conditions such as elevated temperatures. As shown in Fig. 8-119, the reliability of the circuit decreases with increased junction temperature. This effect can be utilized to accelerate the testing time.

Step-stress testing is an alternative approach to reducing the length of IC life testing. In this form of life test ICs are subjected to a succession of discrete increases in stress levels until all or a large portion of the sample fails. This method is particularly useful in obtaining reliability information in a relatively short period of time. However, it does not provide failure data descriptive of ICs under normal use conditions.

Storage Life Tests. Storage life tests are the basic reliability tests for IC manufacturing. Circuits under test are stored at elevated ambient temperatures for 1,000 h and up. Periodically, their characteristics are measured after cooling to room temperature. Usually, storage-life-test temperatures are between 125 and 350°C.

Operating Life Tests. The operating life test is another basic reliability test, which is carried out at 25° and/or 125°C. Operating conditions such as supply-voltage application and circuit operation may vary widely. These tests simulate actual use conditions more closely.

Accelerated Operating Life Tests. The failure modes encountered under operating-life-test conditions can be accelerated by enhancing the stresses responsible for some of the basic failure modes. Failure modes due to bulk imperfections are accelerated by increasing current flow. High-temperature stress will cause failure due to other mechanisms before bulk failure occurs. Reverse-bias high-temperature operation is best suited to the acceleration of surface-failure mechanisms, and the metallization failures are accelerated by high-current, high-temperature operation.

116. Failure Modes and Mechanisms. Most commonly encountered failure mechanisms in integrated circuits can be attributed to one of the following three sources: (1) bulk failures, (2) surface-related failures, and (3) failures of metallization or interconnections.

Bulk Failure. Bulk failure is a relatively unimportant failure mode in integrated circuits. Good starting material is essential in the fabrication of reliable ICs. Crystallographic defects such as dislocations, stacking faults, and growth strains enhance long-term degradation mechanisms, and therefore contribute to the unreliability of ICs. Failure modes associated with bulk silicon include die breakage, shorts due to secondary breakdown, uncontrolled *pnpn* switching, and degradation of electrical characteristics. Bulk-failure mechanisms are accelerated at high operating current densities, due to localized heating effects.

Surface-related Failures. These failures are statistically second only to failures in the interconnection system. Typically, 35% of all IC failures result from surface effects. Surface effects significantly influence *pn* junction characteristics and tend to control transistor gains, junction breakdown voltages, and leakage currents. Charge migration along the silicon surface, especially in the vicinity of *pn* junction, is a major mechanism of surface instability. This instability is often caused by ionic contaminants in the oxide on the surface or near the silicon-oxide interface. The charge buildup due to ionic contamination may be high enough to cause the formation of an *inversion* layer along the surface, where the resistivity type of the underlying silicon may be reversed. Failures due to surface effects are accelerated by increasing the temperature and reverse-biasing the *pn* junctions. Both conditions tend to increase the ionic charge mobility and enable ionic contaminants to induce inversion layers near the junctions.

Metallization and Interconnection Failures. The most common failure modes in ICs are open or short circuits in the circuit metallization and in the bonding. These conditions contribute more failures than all other failure types combined (typically between 50 and 60%).

Under high current densities ($\geq 5 \times 10^5$ A/cm^2), *electromigration* effects become a dominant failure mode. Electromigration is a mass-transport effect which causes the atoms of the interconnection metal to migrate gradually toward the more positive end of the conductor. This mass-transport phenomenon takes place along the grain boundaries of the metal interconnections, and results in the formation of voids in the interconnection pattern which may eventually lead to an open circuit. Electromigration effects are enhanced at elevated temperatures. An additional failure mechanism associated with the metal interconnections is the formation of *microcracks,* as discussed in Par. **8-31.** In the case of aluminum interconnections, chemical reactions between silicon or SiO$_2$ and the aluminum layer can also result in breaks in the conductive interconnection pattern.

Bond Failures. The most serious reliability problems in ICs are associated with the bonding of the lead wires between the package and the chip. Various bonding techniques utilized for this purpose have been discussed in Pars. **8-101** to **8-104.** One of the most serious failure mechanisms associated with gold-wire bonds on aluminum is known as *purple plague,* and is due to the formation of gold-rich intermetallic compounds such as Au$_2$Al, Au$_4$Al, and Au$_5$Al$_2$. These compounds create porous regions in the bond which are mechanically weak and electrically nonconductive. As these intermetallic compounds are formed, the differences in their structure and thermal expansion can stress the porous interface layer to the point of rupture. However, the "plague" formations are a serious problem only at elevated temperatures (typically $\geq 200°$C) and are more likely to occur in step stressing than in actual use. Therefore gold-wire–aluminum metallization and bonding systems are still considered to be the most reliable under normal operating conditions.

Package Testing and Reliability. Most packages use a number of different materials, such as metal, glasses, ceramics, or plastics, to isolate the integrated circuit chip from its environment. Special testing procedures have been developed to test the actual sealing ability and the hermeticity of these packages. One of the tests commonly used for this purpose is the *helium-leak test,* where the package is immersed into a helium atmosphere under pressure, for extended periods of time (usually 1 h). The package is then transferred to a mass-spectrometer chamber and tested for helium leaks. *Radioactive tracer methods* can also be used to detect or trace leaks in the package seals.

Thermal stresses introduced between leak tests can point out losses in the package integrity and the cracking of seals. Thermal shock tests typically consist of cycling the package between the temperature extremes of -55 and $125°$C for 10 to 20 times. Other structural tests include *lead fatigue tests,* where the leads are bent back and forth for a given number of times; *soldering tests,* where the device must withstand typical soldering temperatures applied to the leads; *acceleration* and *shock tests,* where the integrity of the package and the leads is examined under centrifugal or inertial shock conditions. In specialized applications, other parameters of the circuit and the package may have to be measured. An example of such a parameter is the *radiation resistance test,* which tests the package and circuit operation during and after irradiation by neutrons, x-rays, and gamma rays.

117. References

Fabrication technology

1. H. C. THEURER Epitaxial Silicon Films by Hydrogen Reduction of SiCl$_4$, *J. Electrochem. Soc.,* Vol. 108, pp. 649–653, 1961.

2. RESEARCH TRIANGLE INSTITUTE "Integrated Silicon Device Technology: Epitaxy," Vol. 9, No. ASD-TDR 63-316, pp. 42–47, 1965.

3. A. S. Grove, A. Roder, and C. T. Sah   Impurity Distribution in Epitaxial Growth, *J. Appl. Phys.*, Vol. 36, p. 802, 1965.

4. J. C. Irvin   Resistivity of Bulk Silicon and of Diffused Layers in Silicon, *Bell System Tech. J.*, Vol. 41, pp. 387–410, 1962.

5. B. E. Deal   The Oxidation of Silicon in Dry Oxygen, Wet Oxygen and Steam, *J. Electrochem. Soc.*, Vol. 110, p. 527, 1963.

6. A. S. Grove   "Physics and Technology of Semiconductor Devices," Chap. 2, Wiley, New York, 1967.

7. B. E. Deal and A. S. Grove   General Relationships for the Thermal Oxidation of Silicon, *J. Appl. Phys.*, Vol. 36, pp. 3370–3378, 1965.

8. S. K. Ghandhi   "The Theory and Practice of Microelectronics," Chap. 6, Wiley, New York, 1968.

9. Research Triangle Institute   "Integrated Silicon Device Technology: Oxidation," Vol. 7, No. ASD-TDR 63-316, p. 76, 1965.

10. Research Triangle Institute   "Integrated Silicon Device Technology: Photoengraving," Vol. 3, No. ASD-TDR 63-316, 1964.

11. C. S. Fuller and J. A. Ditzenberger   Diffusion of Donor and Acceptor Elements in Silicon, *J. Appl. Phys.*, Vol. 27, pp. 544–553, 1956.

12. Research Triangle Institute   "Integrated Silicon Device Technology: Diffusion," Vol. 4, No. ASD-TDR 63-316, 1964.

13. F. A. Trumbore   Solid Solubilities of Impurity Elements in Germanium and Silicon, *Bell System Tech. J.*, Vol. 39, pp. 205–234, 1960.

14. D. P. Kennedy and R. R. O'Brian   Analysis of the Impurity Atom Distribution near the Diffusion Mask for a Planar P-N Junction, *IBM J. Res. Dev.*, Vol. 9, No. 3, pp. 179–186, 1965.

15. J. T. Burrill, W. J. Kind, S. Harrison, and P. McNally   Ion Implantation as a Production Technique, *IEEE Trans. Electron Devices*, Vol. ED-14, No. 1, pp. 10–17, January, 1967.

16. J. W. Mayer, L. Ericksson, and J. A. Davies   "Ion Implantation in Semiconductors," Academic, New York, 1970.

17. H. T. Wolf   "Semiconductors," Chaps. 2–5, Wiley-Interscience, New York, 1971.

18. H. H. Stellrecht, D. S. Perloff, and J. T. Kerr   Precision Ladder Networks Using Ion Implanted Resistors, *WESCON Tech. Program Record*, 1971.

19. J. D. MacDougall and K. E. Manchester   Implanted Components in Microcircuits, *Proc. Nat. Electron. Conf.*, Vol. 25, pp. 140–145, December, 1969.

20. H. A. Waggener, R. C. Krogness, and A. L. Tyler   Anisotropic Etching for Forming Isolation Slots in Silicon Beam-leaded Integrated Circuits, *Proc. IEEE Int. Electron Devices Conf.*, Washington, D.C., October, 1967.

21. M. P. Lepselter   Beam Lead Technology, *Bell System Tech. J.*, Vol. 45, No. 2, February, 1966.

22. L. V. Gregor   Thin Film Processes for Microelectronic Application, *Proc. IEEE*, Vol. 59, pp. 1390–1403, October, 1971.

23. T. C. Reissing   An Overview of Today's Thick Film Technology, *Proc. IEEE*, Vol. 59, pp. 1448–1454, October, 1971.

Components and devices

24. A. B. Phillips   "Transistor Engineering," Chaps. 8–17, McGraw-Hill, New York, 1962.

25. A. B. Grebene   "Analog Integrated Circuit Design," Chap. 2, Van Nostrand, New York, 1972.

26. Motorola Semiconductor Products Div., Engineering Staff   "Integrated Circuits: Design Principles and Fabrication," Chap. 7, McGraw-Hill, New York, 1965.

27. C. A. Bittmann, G. H. Wilson, R. J. Whittier, and R. K. Waits   Technology for the Design of Low-Power Circuits, *IEEE J. Solid State Circ.*, Vol. SC-5, No. 1, pp. 29–37, February, 1970.

28. H. C. Lin, T. B. Tan, G. Y. Chang, and B. Van der Leest   Lateral Complementary Transistor Structure for Simultaneous Fabrication of Functional Blocks, *Proc. IEEE*, Vol. 49, pp. 1491–1495, 1964.

29. H. C. Lin   "Integrated Electronics," Chap. 8, Holden-Day, New York, 1967.

30. B. Polata   Compatible High Performance Complementary Bipolar Transistors for Integrated Circuits, *IEEE Int. Electron Devices Meeting,* Washington, D.C., 1969.

31. R. Cobbold   "Theory and Applications of Field-Effect Transistors," Wiley, New York, 1970.

32. A. B. Grebene and S. K. Ghandhi   General Theory for Pinched Operation of Junction-Gate FET, *Solid State Electron.,* Vol. 12, pp. 573-589, 1969.

33. H. K. J. Ihantola and J. L. Moll   Design Theory of a Surface Field-Effect Transistor, *Solid State Electron.,* Vol. 7, pp. 423-430, 1964.

34. A. S. Grove   "Physics and Technology of Semiconductor Devices," Chaps. 8 and 9, Wiley, New York, 1967.

35. H. R. Camenzind   "Designing Integrated Systems," Chap. 11, Van Nostrand, New York, 1972.

36. M. H. White and T. R. Cricchi   Complementary MOS Transistors, *Solid State Electron.,* Vol. 9, p. 991, 1966.

37. L. Vadasz, A. S. Grove, T. A. Rowe, and G. E. Moore   Silicon-Gate Technology, *IEEE Spectrum,* October, 1969, pp. 28-35.

38. R. W. Bower, H. G. Dill, K. G. Aubuchon, and S. A. Thompson   MOS Field-Effect Transistors Formed by Gate-masked Ion Implantation, *IEEE Trans. Electron Devices,* Vol. ED-15, pp. 757-761, 1968.

39. J. D. MacDougall and K. E. Manchester   Implanted Components in Microcircuits, *Proc. Nat. Electron. Conf.,* Vol. 25, pp. 140-145, December, 1969.

40. D. Khang and M. P. Lepselter   Planar Epitaxial Schottky Barrier Diodes, *Bell System Tech. J.,* Vol. 44, p. 1525, 1965.

41. A. Y. C. Yu   The Metal-Semiconductor Contact: An Old Device with a New Future, *IEEE Spectrum,* March, 1970, pp. 83-89.

42. J. C. Irvin   Resistivity of Bulk Silicon and Diffused Layers in Silicon, *Bell System Tech. J.,* Vol. 41, pp. 287-410, 1962.

43. A. B. Grebene   A Practical Method for Reducing the Effects of Parasitic Capacitances in Integrated Circuits, *Proc. IEEE,* Vol. 55, No. 2, pp. 235-236, 1967.

44. H. H. Stellrecht, D. S. Perloff, and J. T. Kerr   Precision Ladder Networks Using Ion Implanted Resistors, *WESCON Tech. Program Record,* 1971.

45. M. Fogiel   "Microelectronics," Chaps. 3 and 4, Research and Education Association, New York, 1968.

46. A. B. Phillips   "Transistor Engineering," Chap. 5, McGraw-Hill, New York, 1962.

47. H. Lawrence and R. M. Warner   Diffused Junction Depletion Layer Calculation, *Bell System Tech. J.,* Vol. 39, pp. 389-404, 1960.

Integrated circuit design

48. H. R. Camenzind and A. B. Grebene   An Outline of Design Techniques for Linear Integrated Circuits, *IEEE J. Solid State Circ.,* Vol. SC-4, pp. 110-122, June, 1969.

49. R. J. Widlar   Some Circuit Design Techniques for Linear Integrated Circuits, *IEEE Trans. Circ. Theory,* Vol. CT-12, pp. 586-590, December, 1965.

50. G. R. Wilson   A Monolithic Junction FET-NPN Operational Amplifier, *Int. Solid State Circ. Conf., Digest Tech. Papers,* Vol. 11, pp. 20-21, February, 1968.

51. J. A. Mattis and H. R. Camenzind   A New Phase-locked Loop with High Stability and Accuracy, *Signetics Corp. Application Note,* September, 1970.

52. A. B. Grebene   "Analog Integrated Circuit Design," Chap. 4, Van Nostrand-Reinhold, New York, 1972.

53. R. D. Middlebrook   "Differential Amplifiers," Wiley, New York, 1963.

54. L. J. Giacoletto   "Differential Amplifiers," Wiley, New York, 1970.

55. D. Fullagar   A New High Performance Monolithic Operational Amplifier, *Fairchild Semiconductor Application Brief,* May, 1968.

56. J. E. Solomon, W. R. Davis, and P. L. Lee   A Self-compensated Monolithic Operational Amplifier with Low Input Current and High Slew Rate, *IEEE Int. Solid State Circ. Conf., Digest Tech. Papers,* Vol. 12, pp. 14-15, February, 1969.

57. R. J. Widlar   Design Techniques for Monolithic Operational Amplifiers, *IEEE J. Solid State Circ.,* Vol. SC-4, pp. 184-191, August, 1969.

58. R. J. Widlar   New Developments in IC Voltage Regulators, *IEEE J. Solid State Cir.,* Vol. SC-6, pp. 2-7, February, 1971.

59. G. W. Haines   C_c Compensated Transistors, *IEEE Int. Electron Devices Meeting,* Washington, D.C., October, 1966.

60. S. W. Director and R. A. Rohrer   Automated Network Design: The Frequency Domain Case, *IEEE Trans. Circ. Theory,* Vol. CT-16, pp. 318-323, August, 1969.

61. G. C. Temes and D. A. Callahan   Computer-aided Network Optimization: The State-of-the-Art, *Proc. IEEE,* Vol. 55, pp. 1832-1863, November, 1970.

62. B. A. Wooley   Automated Design of DC-coupled Monolithic Broadband Amplifiers, *IEEE J. Solid State Cir.,* Vol. SC-6, pp. 24-34, February, 1971.

63. R. A. Rohrer, L. W. Nagel, and R. Meyer   CANCER: Computer Analysis of Nonlinear Circuits Excluding Radiation, *IEEE Int. Solid State Circ. Conf.,* Vol. 14, pp. 124-125, February, 1971.

64. A. Bilotti   Applications of a Monolithic Analog Multiplier, *IEEE J. Solid State Circ.,* Vol. SC-3, pp. 373-380, December, 1968.

65. B. Gilbert   A Precise Four-Quadrant Multiplier with Subnanosecond Response, *IEEE J. Solid State Circ.,* Vol. SC-3, No. 4, pp. 365-373.

66. E. Renschler   Theory and Application of a Linear Four-Quadrant Monolithic Multiplier, *IEEE Mag.,* Vol. 17, No. 5, May, 1969.

67. S. K. Mitra (ed.)   "Active Inductorless Filters," IEEE Press, New York, November, 1971.

68. R. W. Newcomb   "Active Integrated Circuit Synthesis," Prentice-Hall, Englewood Cliffs, N.J., 1968.

69. L. P. Huelsman (ed.)   "Active Filters: Lumped, Distributed, Integrated, Digital, and Parametric," McGraw-Hill, New York, 1970.

70. A. B. Grebene and H. R. Camenzind   Frequency Selective Integrated Circuits Using Phase-lock Techniques, *IEEE J. Solid State Circ.,* Vol. SC-4, pp. 216-225, August, 1969.

71. A. B. Grebene   The Monolithic Phase-locked Loop: A Versatile Building Block, *IEEE Spectrum,* March, 1971, pp. 38-49.

72. L. S. Garrett   Integrated Circuit Digital Logic Families, Part I, *IEEE Spectrum,* October, 1970, pp. 46-58.

73. L. S. Garrett   Integrated Circuit Digital Logic Families, Part II, *IEEE Spectrum,* November, 1970, pp. 63-72.

74. L. S. Garrett   Integrated Circuit Digital Logic Families, Part III, *IEEE Spectrum,* December, 1970, pp. 30-42.

75. R. M. Warner, Jr.   Comparing MOS and Bipolar Integrated Circuits, *IEEE Spectrum,* June, 1967, pp. 50-58.

76. D. A. Hodges   "Semiconductor Memories," IEEE Press, New York, 1971.

77. M. E. Hoff, Jr.   Application Considerations for Semiconductor Memories, *IEEE Mag.,* June, 1970, pp. 62-69.

78. L. L. Vadasz, H. T. Chua, and A. S. Grove   Semiconductor Random Access Memories, *IEEE Spectrum,* pp. 40-48, May, 1971.

79. J. S. Schmidt   Integrated MOS Random-Access Memory, *Solid State Design,* January, 1965, pp. 21-25.

80. L. Boysel, W. Chan, and J. Faith   Random-Access MOS Memory Packs More Bits to the Chip, *Electronics,* Vol. 43, pp. 109-115, February, 1970.

81. W. M. Regitz and J. A. Garp   Three-Transistor-Cell 1020-Bit 500-ns MOS RAM, *IEEE J. Solid State Circ.,* Vol. SC-5, pp. 182-186, October, 1970.

82. M. R. McCoy   Semiconductor Memories, pp. 47-58, Chap. 6 in J. Eimbinder (ed.), Wiley-Interscience, New York, 1971.

83. D. Frohman-Bentchkowsky   A fully-decoded 2048-Bit Electrically Programmable FAMOS Read-Only Memory, *IEEE J. Solid State Circ.,* Vol. SC-6, No. 5, October, 1971.

84. M. G. Snyder   Semiconductor Memories, pp. 159-168, Chap. 6 in J. Eimbinder (ed.), Wiley-Interscience, New York, 1971.

85. H. Johnson   The Anatomy of Integrated Circuit Technology, *IEEE Spectrum,* February, 1970, pp. 56-66.

86. H. Sobol   Applications of Integrated Circuit Technology to Microwave Frequencies, *Proc. IEEE,* Vol. 59, No. 8, pp. 1200-1211, August, 1971.

87. M. Caulton   Film Technology in Microwave Integrated Circuits, *Proc. IEEE,* Vol. 59, No. 10, pp. 1481-1489, October, 1971.

88. C. S. AITCHISON et al.   Lumped Circuit Elements at Microwave Frequencies, *IEEE Trans. Microwave Theory Techniques,* Vol. MTT-19, p. 928, December, 1971.

89. M. CAULTON et al.   Status of Lumped Elements in Microwave Integrated Circuits: Present and Future, *IEEE Trans. Microwave Theory Techniques,* Vol. MTT-19, No. 7, pp. 588–599, July, 1971.

Assembly and packaging

90. RESEARCH TRIANGLE INSTITUTE   "Integrated Silicon Device Technology," "Interconnections and Encapsulation," Vol. 14, Report No. ASD-TDR 63-319, May, 1967.

91. G. FEHR   A Survey of Today's Microcircuit Packaging, *1970 NEP/CON Proc.,* February, 1970, pp. 5.12–5.20.

Testing and reliability

92. F. VAN VEEN   An Introduction to IC Testing, *IEEE Spectrum,* December, 1971, pp. 28–37.

93. RESEARCH TRIANGLE INSTITUTE   "Integrated Silicon Device Technology," "Reliability," Vol. 15, Report No. ASD-TDR 63-316, May, 1967.

SECTION 9

UHF AND MICROWAVE DEVICES

BY

JOSEPH FEINSTEIN Varian Associates, Palo Alto, CA; Fellow, Institute of Electrical and Electronics Engineers

RICHARD B. NELSON Varian Associates, Palo Alto, CA; Fellow, Institute of Electrical and Electronics Engineers

GEORGE K. FARNEY Varian Associates, Beverly, MA; Fellow, Institute of Electrical and Electronics Engineers

DONALD A. PRIEST Varian Associates, Carlos, CA; Fellow, Institute of Electrical and Electronics Engineers

CONTENTS

Numbers refer to paragraphs

SECTION 9

UHF AND MICROWAVE DEVICES

PASSIVE MICROWAVE COMPONENTS

By Joseph Feinstein

1. Introduction. While the physical concepts and mathematical theory underlying electromagnetic wave propagation in confined structures were developed at the end of the nineteenth century, the practical utilization of wavelengths shorter than one meter began during World War II. A wide variety of devices is now available for the transmission, sampling, filtering, and impedance matching of uhf and microwave power. Because of the short wavelengths at these frequencies (10 cm at 3 GHz, 1 cm at 30 GHz), most of these components employ distributed elements, and obtain specific reactances by judicious use of short-circuited lengths of transmission line. However, the trend toward microcircuitry has led recently to the introduction of lumped elements in some low-power, low-Q applications. In addition, the use of strip and microstrip transmission lines marks a return to open-wire media, with the attendant radiation loss kept low by close spacing and the presence of the dielectric filler.

Except for special applications (such as rotating joints for antenna feeds where a cylindrical member is essential), the use of rectangular waveguide, dimensioned to transmit in the dominant (lowest-order) mode, is standard for high-power transmission. Coaxial cable is employed for short-distance runs where the advantage of its flexibility outweighs its higher attenuation. Ridged waveguide is useful for designing matching sections and for providing very-wide-bandwidth single-mode transmission. Oversize cylindrical guide operated in the circular electric mode is finding use in millimeter-wave-carrier telephony where its extremely low attenuation justifies the special precautions which must be taken to avoid mode conversions.

Reciprocal and Nonreciprocal Components. Of the components which have become standard in this field, perhaps the most novel are ferrite devices. When biased with the proper magnetic field, ferrites act as nonreciprocal elements with respect to microwave transmission in an appropriate frequency band. This behavior allows isolation of a signal source from reflections and the separation of incident from reflected power along the same transmission line (by means of a ferrite device called a *circulator*). It is also possible to vary the phase of a transmitted wave by adjusting the magnetizing field on the ferrite. Such phase shifters are capable of handling high powers with low loss.

All other types of microwave components, such as hybrid junctions and directional couplers, are reciprocal in their action. The latter are employed for power division and for signal sampling. A wide variety of transmission components such as variable attenuators, matched-load terminations, and slotted lines are used in microwave measurements and design.

High-Q resonators are formed from completely enclosed short-circuited sections of waveguide, with slit or loop coupling. Resonators employing lengths of strip line or coaxial line provide smaller size, but at lower Q. Finally, lumped elements such as varactor diodes or YIG spheres provide electrically tunable resonators for microwave oscillators and receivers.

2. Transmission-Line Relationships. The basic transmission-line equations are derived for distributed parameters R (series resistance), G (shunt conductance), L (series inductance), and C (shunt capacitance), all defined per unit length of line. Some useful relations are shown in Table 9-1. For zero losses (R, $G = 0$) one obtains the ideal line expressions shown

Table 9-1. Summary of Transmission-Line Equations

Quantity	General line expression	Ideal line expression
Propagation constant	$\gamma = \alpha + j\beta = \sqrt{(R + j\omega L)\,(G + j\omega C)}$	$\gamma = j\omega\sqrt{LC}$
Phase constant β	Imaginary part of γ	$\beta = \omega\sqrt{LC} = \dfrac{2\pi}{\lambda}$
Attenuation constant α	Real part of γ	0
Characteristic impedance	$Z_0 = \sqrt{\dfrac{R + j\omega L}{G + j\omega C}}$	$Z_0 = \sqrt{\dfrac{L}{C}}$
Input impedance	$Z_{-l} = Z_0 \dfrac{Z_r + Z_0 \tanh \gamma l}{Z_0 + Z_r \tanh \gamma l}$	$Z_{-l} = Z_0 \dfrac{Z_r + jZ_0 \tan \beta l}{Z_0 + jZ_r \tan \beta l}$
Impedance of short-circuited line, $Z_r = 0$	$Z_{sc} = Z_0 \tanh \gamma l$	$Z_{sc} = jZ_0 \tan \beta l$
Impedance of open-circuited line, $Z_r = \infty$	$Z_{oc} = Z_0 \coth \gamma l$	$Z_{oc} = -jZ_0 \cot \beta l$
Impedance of line an odd number of quarter wave-lengths long	$Z = Z_0 \dfrac{Z_r + Z_0 \coth \alpha l}{Z_0 + Z_r \coth \alpha l}$	$Z = \dfrac{Z_0^2}{Z_r}$
Impedance of line an integral number of half wave-lengths long	$Z = Z_0 \dfrac{Z_r + Z_0 \tanh \alpha l}{Z_0 + Z_r \tanh \alpha l}$	$Z = Z_r$
Voltage along line	$V_{-l} = V_i(1 + \Gamma_0 e^{-2\gamma l})$	$V_{-l} = V_i(1 + \Gamma_0 e^{-2j\beta l})$
Current along line	$I_{-l} = I_i(1 - \Gamma_0 e^{-2\gamma l})$	$I_{-l} = I_i(1 - \Gamma_0 e^{-2j\beta l})$
Voltage reflection coefficient	$\Gamma = \dfrac{Z_r - Z_0}{Z_r + Z_0}$	$\Gamma = \dfrac{Z_r - Z_0}{Z_r + Z_0}$

Note: l = length of transmission line.

on the right. Table 9-2 gives some equations that are useful for relating the measured voltage standing wave ratio (VSWR) to wave transmission and reflection.

The nomographs in Figs. 9-1 and 9-2 provide a convenient means for determining these quantities. The transmission-line impedance transformations of Table 9-1 may be seen graphically on the Smith Chart, shown in Fig. 9-3. Circles of constant reflection coefficient are concentric with the center of the chart ($R/Z_0 = 1.0$). The VSWR magnitudes are given by their horizontal intercepts. Radial lines from the center are loci of constant phase angle, read against the degree or fractional wavelength scales at the circumference of the chart. Normalized values of impedance seen as one moves along the line can be read at each radial intercept from the resistance and reactance grids on which the circle of constant VSWR has been drawn. This chart is also of value in matching reactances, by providing a graphical determination of the range of impedance transformations available from a stub tuner by changes in line length.

Although L and C for a waveguide cannot be defined uniquely, because its electric and magnetic field patterns are not purely transverse, it is possible to construct a useful equivalent circuit for the dominant mode. This is based on a capacitance defined from the total electric field energy storage across the guide cross section and on an impedance which is related to power flow down the guide. The relations given in Tables 9-1 and 9-2 may then be employed for the dominant-mode guide. When nondominant-mode transmission is possible, the situation becomes far more complex, since mode conversion can occur at any discontinuity.

3. Coaxial Cable Data. The attenuation and power-handling capability of flexible 50-Ω polyethylene coaxial cable are shown in Fig. 9-4 as a function of frequency. Figure 9-5 gives these quantities for rigid coaxial of various outer diameters as marked on the curves. The frequency scale on this figure extends only to 3 GHz because such high-power transmitting coaxial is generally not used at higher frequencies. Figure 9-6 gives the variation of

Table 9-2.   Some Miscellaneous Relations in Low-Loss Transmission Lines

Equation	Explanation
$r = \dfrac{1 + \mid \Gamma \mid}{1 - \mid \Gamma \mid}$	$r = $ VSWR
$\mid \Gamma \mid = \dfrac{r - 1}{r + 1}$	$\mid \Gamma \mid = $ magnitude of reflection coefficient
$\Gamma = \dfrac{R - Z_0}{R + Z_0}$	$\Gamma = $ reflection coefficient (real) at a point in a line where impedance is real (R)
$r = \dfrac{R}{Z_0}$	$R > Z_0$ (at voltage maximum)
$r = \dfrac{Z_0}{R}$	$R < Z_0$ (at voltage minimum)
$\dfrac{P_r}{P_i} = \mid \Gamma \mid^2 = \left(\dfrac{r - 1}{r + 1}\right)^2$	$P_r = $ reflected power $P_i = $ incident power
$\dfrac{P_t}{P_i} = 1 - \mid \Gamma \mid^2 = \dfrac{4r}{(r + 1)^2}$	$P_t = $ transmitted power
$\dfrac{P_b}{P_m} = \dfrac{1}{r}$	$P_b = $ net power transmitted to load at onset of breakdown in a line where VSWR $= r$ exists $P_m = $ same when line is matched, $r = 1$
$\dfrac{\alpha_r}{\alpha_m} = \dfrac{1 + \Gamma^2}{1 - \Gamma^2} = \dfrac{r^2 + 1}{2r}$	$\alpha_m = $ Attenuation constant when $r = 1$, matched line $\alpha_r = $ attenuation constant allowing for increased ohmic loss caused by standing waves
$r_{\max} = r_1 r_2$	$r_{\max} = $ maximum VSWR when r_1 and r_2 combine in worst phase
$r_{\min} = \dfrac{r_2}{r_1} \qquad r_2 > r_1$	$r_{\min} = $ minimum VSWR when r_1 and r_2 are in best phase
$\mid \Gamma \mid = \dfrac{\mid X \mid}{\sqrt{X^2 + 4}}$ $\mid X \mid = \dfrac{r - 1}{\sqrt{r}}$	Relations for a normalized reactance X in series with resistance Z_0
$\mid \Gamma \mid = \dfrac{\mid B \mid}{\sqrt{B^2 + 4}}$ $\mid B \mid = \dfrac{r - 1}{\sqrt{r}}$	Relations for a normalized susceptance B in shunt with admittance Y_0

9-4

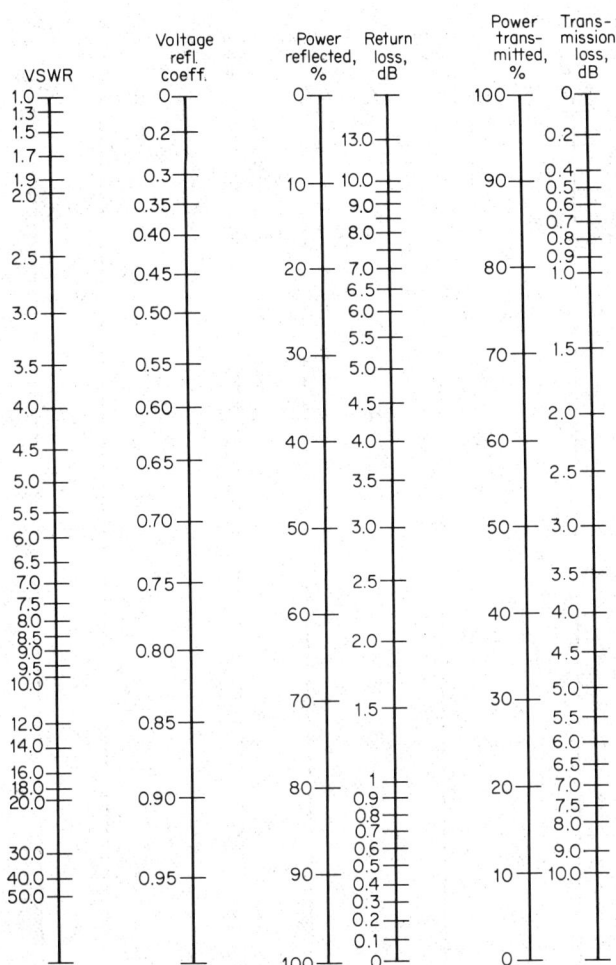

Fig. 9-1. Nomograph for transmission and reflection of power at high voltage standing wave ratios (VSWR).

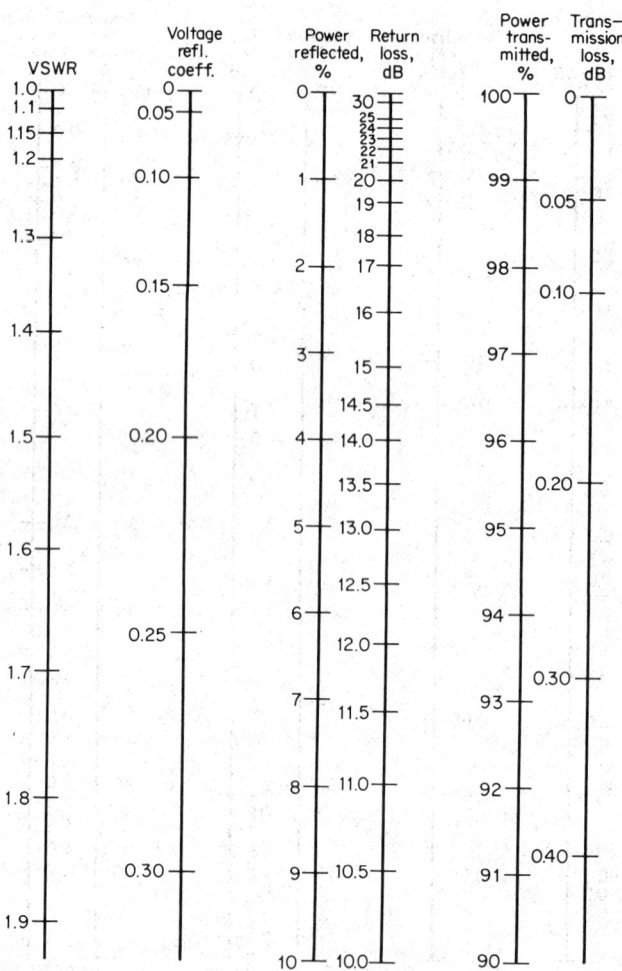

Fig. 9-2. Nomograph for low VSWR.

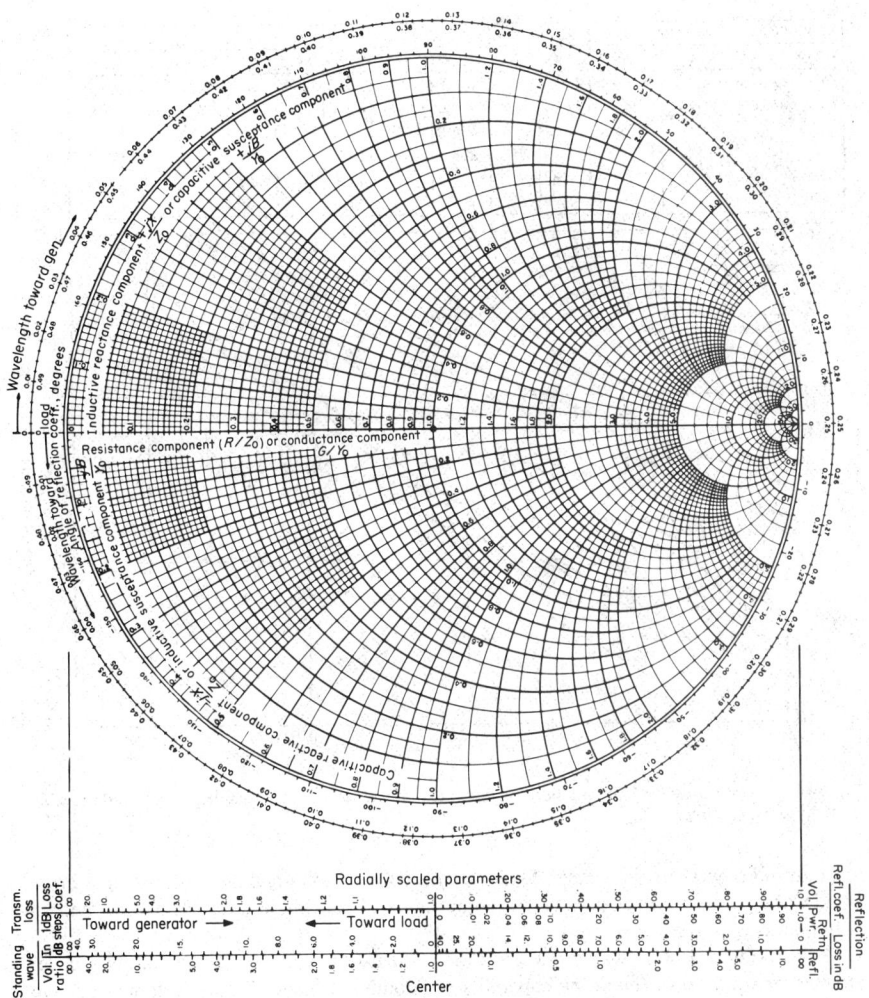

Fig. 9-3. The Smith Chart.

impedance with dimensions for a standard coaxial line (curve 1) as well as for several other forms of inner-conductor TEM-mode configurations.

Care must be taken to avoid operation of a coaxial line at wavelengths where it becomes possible for additional modes to propagate. This occurs when the mean circumference between the inner and outer cylinders forming the transmission path equals a full wavelength; a stable standing wave is then possible in a circumferential direction. The higher attenuation

USASI C83.2 cable group	1	2	3	4	5	6	7	8	9	10	11
Approx size O.D. inches	0.080	0.110	0.160	0.200	—	0.330	0.415	0.415	0.550	0.725	0.870
RG-( )/U Cable designation	178,A,B 196,A (Teflon)	174 188,A	122	55,A,B 58,A,C 142A,B 223	75Ω cables	5,A,B 21A,B 143,A 212 222 304	115,A	8,A 9,A,B 213 214 225	14,A 217	Teflon cables	17,A 218

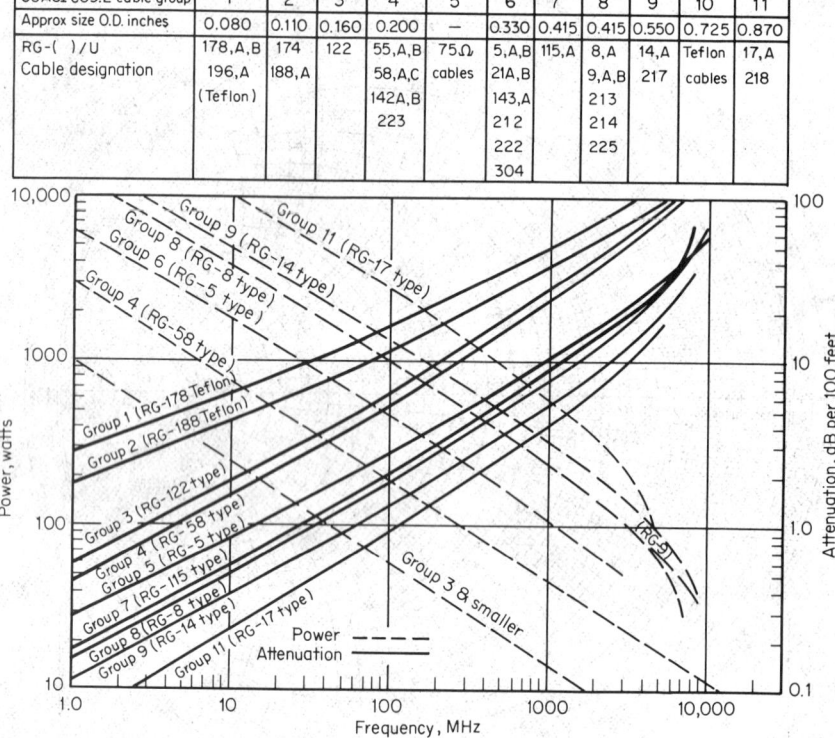

Fig. 9-4. Power rating and attenuation of flexible polyethylene coaxial cables. For Teflon cables multiply power ratings by 5.

and reduced power-carrying capability which accompany the small dimensions necessary to avoid higher-order modes generally lead to the choice of waveguide as a transmission medium at frequencies above 3 GHz.

4. Waveguide Data. Table 9-3 gives the standard dimensions, flange coupling codes, attenuation, and power-handling capability for dominant-mode (TE_{10}) rectangular wave-guide. The electric field pattern of this mode is a half sinusoid across the transverse-guide dimension with its maximum at the center of the broad wall. The frequency range shown for each guide size follows conventional practice in avoiding operation too close to the cutoff wavelength λ_c (= $2a$, where a is the broad dimension of the guide), on the low-frequency side, or too close to the next higher mode (usually, λ'_c = a), on the high-frequency side.

When higher power must be transmitted than is possible because of breakdown in air with these restrictions on dimensions, pressurization of the air or a sulfur hexafluoride gas fill is generally employed. The increase in power rating obtained is shown in Fig. 9-7.

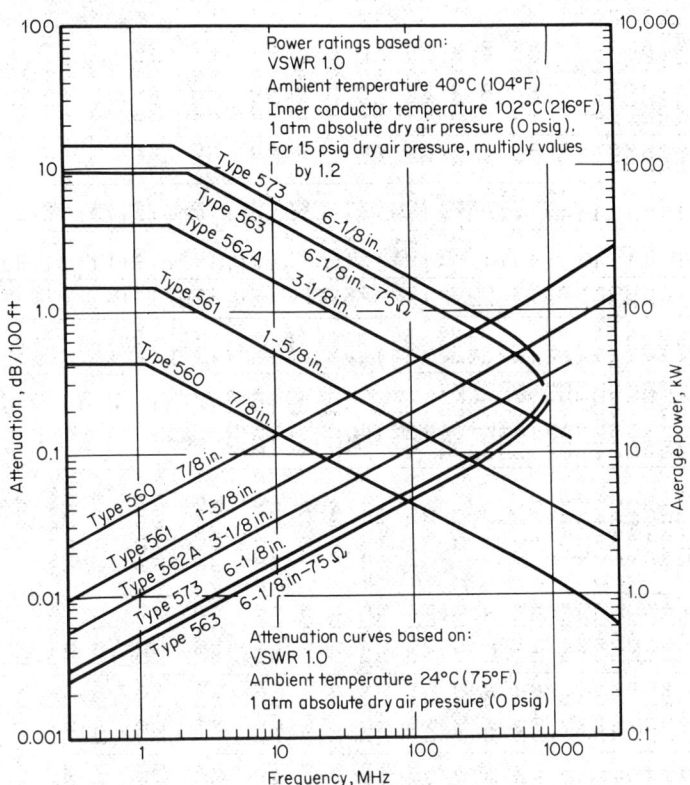

Fig. 9-5. Power rating and attenuation for rigid coaxial cables.

Table 9-3. Rectangular Waveguide Data and Fittings

EIA WG Desig. WR ( )	Recommended operating range for TE$_{10}$ mode — Frequency, GHz	Recommended operating range — Wavelength, cm	Cutoff for TE$_{10}$ mode — Frequency, GHz	Cutoff — Wavelength, cm	Range in $\frac{2a}{\lambda}$	Range in $\frac{\lambda_g}{\lambda}$	Theoretical cw power rating, lowest to highest frequency, MW	Theoretical attenuation, lowest to highest frequency, dB/100 ft	Material alloy	JAN WG Desig. RG ( ) /U	JAN flange — Choke UG ( ) /U	JAN flange — Cover UG ( ) /U	EIA WG Desig. WR ( )	Dimension Inside, in.	Inside Tolerance	Outside	Outside Tolerance	Wall thickness, nominal
2300	0.32–0.49	93.68–61.18	0.256	116.84	1.60–1.05	1.68–1.17	153.0–212.0	0.051–0.031	Alum.	—	—	—	2300	23.000–11.500	±.020	23.250–11.750	±.020	0.125
2100	0.35–0.53	85.65–56.56	0.281	106.68	1.62–1.06	1.68–1.18	120.0–173.0	0.054–0.034	Alum.	—	—	—	2100	21.000–10.500	±.020	21.250–10.750	±.020	0.125
1800	0.41–0.625	73.11–47.96	0.328	91.44	1.60–1.05	1.67–1.18	93.4–131.9	0.056–0.038	Alum.	201	—	—	1800	18.000–9.000	±.020	18.250–9.250	±.020	0.125
1500	0.49–0.75	61.18–39.97	0.393	76.20	1.61–1.05	1.62–1.18	67.6–93.3	0.069–0.050	Alum.	202	—	—	1500	15.000–7.500	±.015	15.250–7.750	±.015	0.125
1150	0.64–0.96	46.84–31.23	0.513	58.42	1.60–1.07	1.82–1.18	35.0–53.8	0.128–0.075	Alum.	203	—	—	1150	11.500–5.750	±.015	11.750–6.000	±.015	0.125
975	0.75–1.12	39.95–26.76	0.605	49.53	1.61–1.08	1.70–1.19	27.0–38.5	0.137–0.095	Alum.	204	—	—	975	9.750–4.875	±.010	10.000–5.125	±.010	0.125
770	0.96–1.45	31.23–20.67	0.766	39.12	1.60–1.05	1.66–1.18	17.2–24.1	0.201–0.136	Alum.	205	—	—	770	7.700–3.850	±.005	7.950–4.100	±.005	0.125
650	1.12–1.70	26.76–17.63	0.908	33.02	1.62–1.07	1.70–1.18	11.9–17.2	0.317–0.212 / 0.269–0.178	Brass / Alum.	69, 103	—	417A, 418A	650	6.500–3.250	±.005	6.660–3.410	±.005	0.080
510	1.45–2.20	20.67–13.62	1.157	25.91	1.60–1.05	1.67–1.18	7.5–10.7	0.588–0.385 / 0.501–0.330	Brass / Alum.	104, 105	—	435A, 437A	510	5.100–2.550	±.005	5.260–2.710	±.005	0.080
430	1.70–2.60	17.63–11.53	1.372	21.84	1.61–1.06	1.70–1.18	5.2–7.5	0.877–0.572 / 0.751–0.492	Brass / Alum.	112, 113	—	—	430	4.300–2.150	±.005	4.460–2.310	±.005	0.080
340	2.20–3.30	13.63–9.08	1.736	17.27	1.58–1.05	1.78–1.22	3.1–4.5	—	Brass / Alum.	—	—	553, 554	340	3.400–1.700	±.005	3.560–1.860	±.005	0.080
284	2.60–3.95	11.53–7.59	2.078	14.43	1.60–1.05	1.67–1.17	2.2–3.2	1.102–0.752 / 0.940–0.641	Brass / Alum.	48, 75	54A, 585	53, 584	284	2.840–1.340	±.005	3.000–1.500	±.005	0.080
229	3.30–4.90	9.08–6.12	2.577	11.63	1.56–1.05	1.62–1.17	1.6–2.2	—	Brass / Alum.	—	—	—	229	2.290–1.145	±.005	2.418–1.273	±.005	0.064
187	3.95–5.85	7.59–5.12	3.152	9.510	1.60–1.08	1.67–1.19	1.4–2.0	2.08–1.44 / 1.77–1.12	Brass / Alum.	49, 95	148B, 406A	149A, 407	187	1.872–0.872	±.005	2.000–1.000	±.005	0.064
159	4.90–7.05	6.12–4.25	3.711	8.078	1.51–1.05	1.52–1.19	0.79–1.0	—	Brass / Alum.	—	—	—	159	1.590–0.795	±.004	1.718–0.923	±.004	0.064
137	5.85–8.20	5.12–3.66	4.301	6.970	1.47–1.05	1.48–1.17	0.56–0.71	2.87–2.30 / 2.45–1.94	Brass / Alum.	50, 106	343A, 440A	344, 441	137	1.372–0.622	±.004	1.500–0.750	±.004	0.064
112	7.05–10.00	4.25–2.99	5.259	5.700	1.49–1.05	1.51–1.17	0.35–0.46	4.12–3.21 / 3.50–2.77	Brass / Alum.	51, 68	52A, 137A	51, 138	112	1.122–0.497	±.004	1.250–0.625	±.004	0.064
90	8.20–12.40	3.66–2.42	6.557	4.572	1.60–1.06	1.68–1.18	0.20–0.29	6.45–4.48 / 5.49–3.83	Brass / Alum.	52, 67	40A, 136A	39, 135	90	0.900–0.400	±.003	1.000–0.500	±.003	0.050
75	10.00–15.00	2.99–2.00	7.868	3.810	1.57–1.05	1.64–1.17	0.17–0.23	—	Brass / Alum.	—	—	—	75	0.750–0.375	±.003	0.850–0.475	±.003	0.050
62	12.4–18.00	2.42–1.66	9.486	3.160	1.53–1.05	1.55–1.18	0.12–0.16	9.51–8.31 / 6.14–5.36	Brass / Alum. / Silver	91, 107	541	419	62	0.622–0.311	±.0025	0.702–0.391	±.003	0.040
51	15.00–22.00	2.00–1.36	11.574	2.590	1.54–1.05	1.58–1.18	0.080–0.107	—	Brass / Alum. / Silver	—	—	—	51	0.510–0.255	±.0025	0.590–0.335	±.003	0.040
42	18.00–26.50	1.66–1.13	14.047	2.134	1.56–1.06	1.60–1.18	0.043–0.058	20.7–14.8 / 17.6–12.6 / 13.3–9.5	Brass / Alum. / Silver	53, 121, 66	596, 598	595, 597	42	0.420–0.170	±.0020	0.500–0.250	±.003	0.040
34	22.00–33.00	1.36–0.91	17.328	1.730	1.57–1.05	1.62–1.18	0.034–0.048	—	Brass / Silver	—	—	—	34	0.340–0.170	±.0020	0.420–0.250	±.003	0.040
28	26.50–40.00	1.13–0.75	21.081	1.422	1.59–1.05	1.65–1.17	0.022–0.031	21.9–15.0	Brass / Silver	96	600	599	28	0.280–0.140	±.0015	0.360–0.220	±.002	0.040
22	33.00–50.00	0.91–0.60	26.342	1.138	1.60–1.05	1.67–1.17	0.014–0.020	31.0–20.9	Brass / Silver	97	—	383	22	0.224–0.112	±.0010	0.304–0.192	±.002	0.040
19	40.00–60.00	0.75–0.50	31.357	0.956	1.57–1.05	1.63–1.16	0.011–0.015	—	Brass / Silver	—	—	—	19	0.188–0.094	±.0010	0.268–0.174	±.002	0.040
15	50.00–75.00	0.60–0.40	39.863	0.752	1.60–1.06	1.67–1.17	0.0063–0.0090	52.9–39.1	Brass / Silver	98	—	385	15	0.148–0.074	±.0010	0.228–0.154	±.002	0.040
12	60.00–90.00	0.50–0.33	48.350	0.620	1.61–1.06	1.68–1.18	0.0042–0.0060	93.3–52.2	Brass / Silver	99	—	387	12	0.122–0.061	±.0005	0.202–0.141	±.002	0.040
10	75.00–110.00	0.40–0.27	59.010	0.508	1.57–1.06	1.61–1.18	0.0030–0.0041	152–99	Silver	138	—	—	10	0.100–0.050	±.0005	0.180–0.130	±.002	0.040
8	90.00–140.00	0.333–0.214	73.840	0.406	1.64–1.05	1.75–1.17	0.0018–0.0026	163–137	Silver	136	—	—	8	0.080–0.040	±.0003	0.156 dia	±.001	—
7	110.00–170.00	0.272–0.176	90.840	0.330	1.65–1.05	1.77–1.18	0.0012–0.0017	308–193	Silver	135	—	—	7	0.065–0.0325	±.00025	0.156 dia	±.001	—
5	140.00–220.00	0.214–0.136	115.750	0.259	1.61–1.05	1.78–1.17	0.00071–0.00107	384–254	Silver	137	—	—	5	0.051–0.0255	±.00025	0.156 dia	±.001	—
4	170.00–260.00	0.176–0.115	137.520	0.218	1.61–1.05	1.69–1.17	0.00052–0.00075	512–348	Silver	139	—	—	4	0.043–0.0215	±.00020	0.156 dia	±.001	—
3	220.00–325.00	0.136–0.092	173.280	0.173	1.57–1.06	1.62–1.18	0.00035–0.00047	—	Silver	—	—	—	3	0.034–0.0170	±.00020	0.156 dia	±.001	—

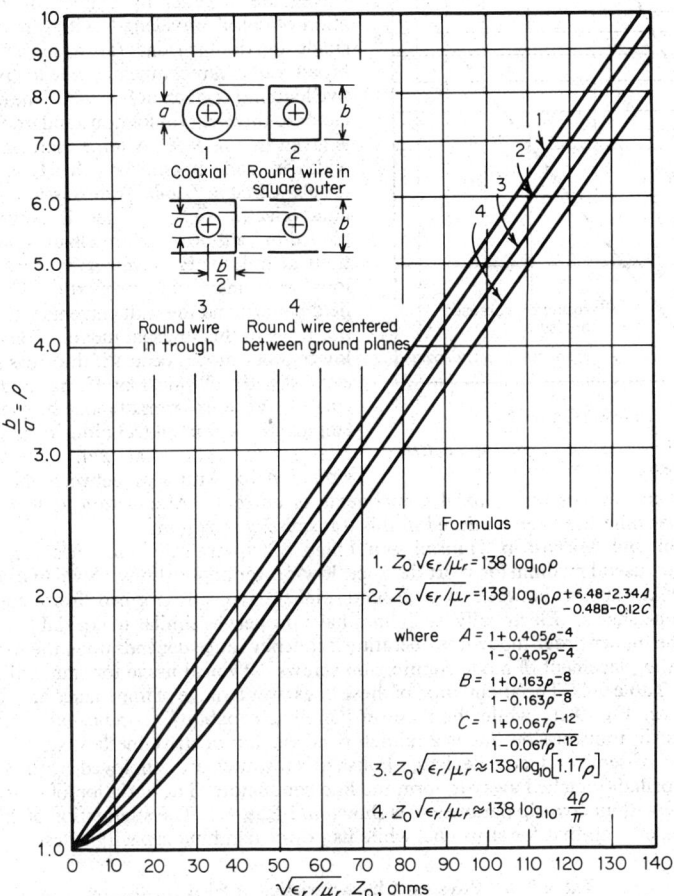

Fig. 9-6. Characteristic impedance of coaxial lines.

Formulas

1. $Z_0\sqrt{\epsilon_r/\mu_r} = 138\log_{10}\rho$

2. $Z_0\sqrt{\epsilon_r/\mu_r} = 138\log_{10}\rho + 6.48 - 2.34A - 0.48B - 0.12C$

    where $A = \dfrac{1+0.405\rho^{-4}}{1-0.405\rho^{-4}}$

    $B = \dfrac{1+0.163\rho^{-8}}{1-0.163\rho^{-8}}$

    $C = \dfrac{1+0.067\rho^{-12}}{1-0.067\rho^{-12}}$

3. $Z_0\sqrt{\epsilon_r/\mu_r} \approx 138\log_{10}\left[1.17\rho\right]$

4. $Z_0\sqrt{\epsilon_r/\mu_r} \approx 138\log_{10}\dfrac{4\rho}{\pi}$

Unlike the case for the TEM mode of a coaxial or parallel-wire line, the guide wavelength λ_g departs from the free-space wavelength λ_0 and varies with frequency, producing phase distortion in a wideband signal. The relationship is

$$\lambda_g = \lambda_o / \sqrt{1 - (\lambda_o/\lambda_c)^2}$$

5a. Ridge Waveguide. To obtain broader bandwidth in a single mode, as well as increased flexibility in the choice of impedance, ridge waveguide is generally employed. The variation of cutoff wavelength with dimensions for single and double ridges (always inserted on the broad wall of a rectangular guide to give capacitive loading) is given in Fig. 9-8. The correction factor to these curves for nonstandard (b/a) ratio is given in Fig. 9-9. A large increase in bandwidth is made possible by this type of guide.

5b. Circular Mode Transmission. The circular electric (TE$_{01}$) mode is currently employed in long-distance broadband communications at millimeter wavelengths because of its low-loss transmission property. The electric field pattern and the wall currents form concentric rings in this unusual mode. Conversion to lower-order modes occurs if the cross section is even slightly elliptical or if the guide axis is curved. Various forms of mode suppression are employed; a common technique consists in fabricating the cylindrical wall from a tightly wound helix with loss between the turns to damp out modes possessing axial components of current. Attenuation as low as a few decibels per mile has been achieved in this transmission medium.

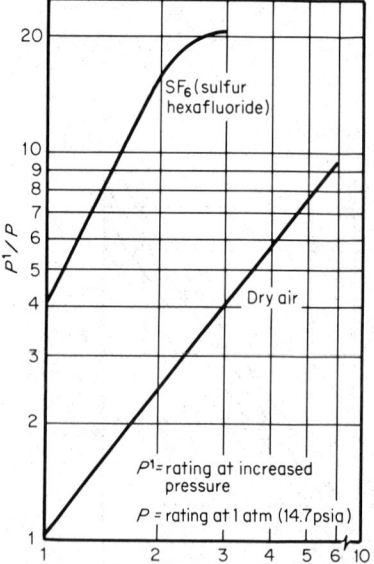

Fig. 9-7. Power rating of pressurized transmission lines.

6. Strip and Microstrip Transmission Lines. Strip transmission lines constitute an increasingly useful medium for short-distance, low-power applications. Strip line consists of a printed conductor between two ground planes, typically formed from copper-clad polyethylene sheets. Electrically, such lines have properties similar to coaxial transmission lines. The higher-order-mode-free operating frequency range depends upon the width of the strip and the placement of mode-suppression screws. Typical usage for standard 50-Ω line is given in Table 9-4. The attenuation of these lines over a range of impedance and frequency is shown on Fig. 9-10, while their power-handling capability is indicated in Fig. 9-11.

Microstrip transmission line is a miniaturized version of strip line best suited to circuit integration of semiconductor devices. Polished substrates are employed, with a precious metal deposited or etched away to form the line conductor. The variation of characteristic impedance with microstrip parameters is shown in Fig. 9-12. The attenuation of this type of line is greater than that for strip line, while its power-handling capacity is less.

Table 9-4. Physical Characteristics of 50-Ω Strip Line

Standard Designation	Material thickness,* in.	Copper weight,† oz.	Strip width, in.	Screw spacing, in.		Upper frequency limit,¶ GHz
				Long.‡	Lat.§	
MPC-062-2	0.062	2	0.083 ± 0.0015	0.375	0.200	7.5
MPC-125-2	0.125	2	0.182 ± 0.003	0.375	0.300	5.0
MPC-187-2	0.187	2	0.280 ± 0.004	0.375	0.400	3.6
MPC-250-2	0.250	2	0.380 ± 0.005	0.375	0.500	2.8

* Material thickness measured over copper cladding.
† Copper thickness: 1 oz, 0.0014 in.; 2 oz, 0.0028 in.
‡ Type 4-40 screws recommended.
§ Lateral screw spacing measured from edge of strip to center of screw.
¶ Recommended for this lateral spacing from circuits. For higher-frequency limits, the specified spacing should be reduced.

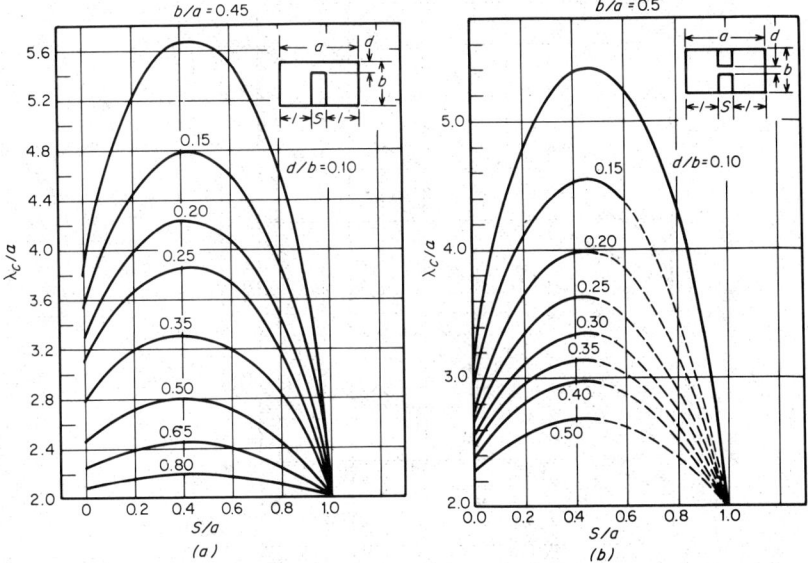

Fig. 9-8. Cutoff wavelengths of ridge waveguide. (*a*) Single ridge; (*b*) double ridge.

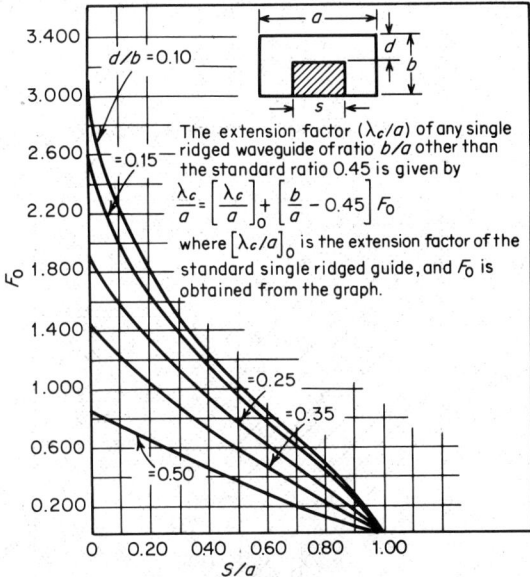

Fig. 9-9. Variation of cutoff wavelength for nonstandard *b/a* ratios.

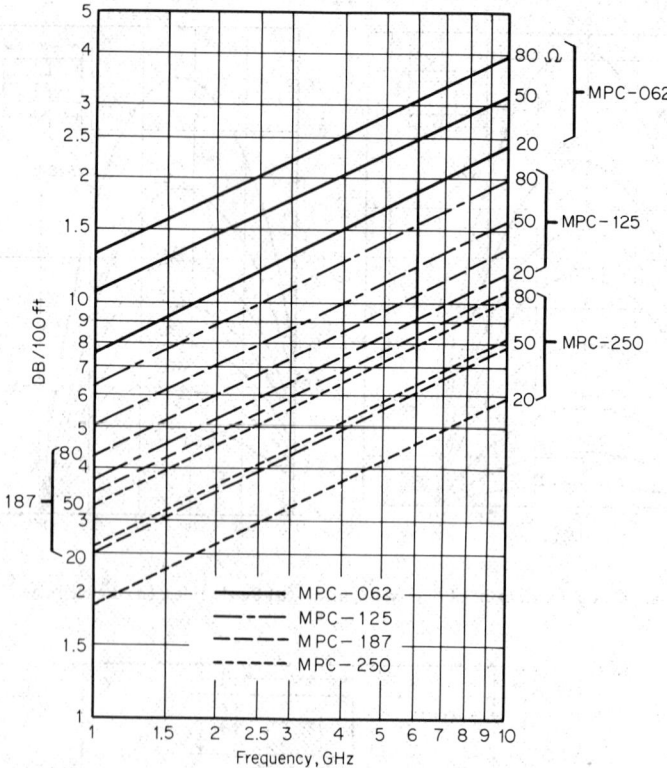

Fig. 9-10. Attenuation characteristics of strip line.

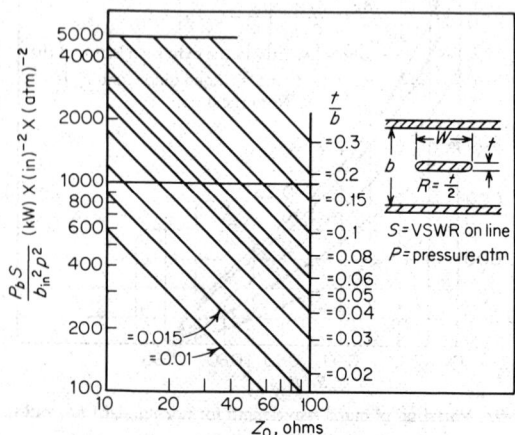

Fig. 9-11. Power-carrying capacity of strip line.

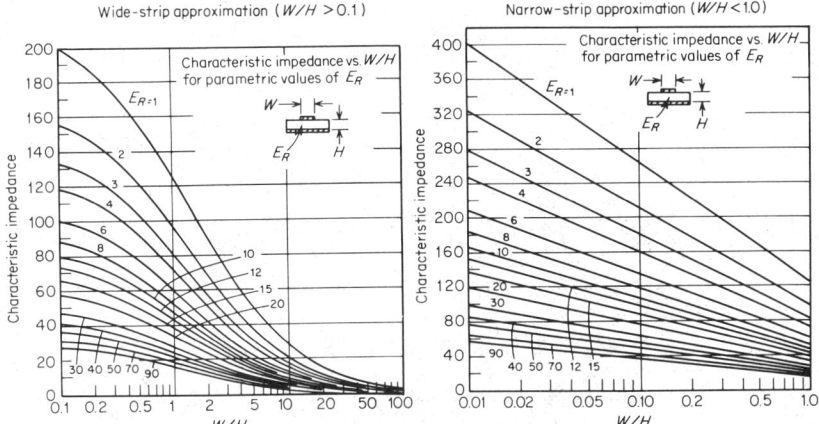

Fig. 9-12. Characteristic impedance of microstrip line: wide-strip approximation (left); narrow-strip approximation (right).

7. Reciprocal Circuit Components. The basic circuit theory for microwaves is identical with that for lower frequencies, but the physical form of the components is usually quite different from conventional lumped elements because of the distributed nature of microwave electric and magnetic field patterns.

Resonant circuits are generally formed from short-circuited lengths of transmission line. The dimensions for such a resonator are selected from a mode chart of the type shown in Fig. 9-13 for the right circular cylinder. The Q which is obtained for some typical cavity resonators in hollow-cylinder, coaxial, and rectangular configurations is shown in Fig. 9-14.

A metal post, screw, or dielectric discontinuity inserted in a transmission line acts as a lumped circuit element. Typical examples of capacitive and inductive equivalent circuits in waveguide are shown in Fig. 9-15.

Directional Couplers. Directional couplers are employed to sample power for measurement purposes. By spacing a series of holes at quarter-wavelength intervals one obtains phase addition of the propagation field coupled from one guide into another for the forward direction of power flow, and phase cancellation for the reverse direction. The coupling ratios obtained for holes in the broadwall of rectangular guide are shown in Fig. 9-16. The cross-guide coupler, another popular type, has the characteristics shown in Fig. 9-17.

Impedance Matching. The smooth flow of power from one type of transmission medium into another requires a matching of the field patterns across the boundary so as to launch the wave into the second medium with a minimum of reflection. Coaxial line is generally matched into rectangular waveguide by extending the center conductor of the coaxial line through the broadwall of the guide, parallel to the electric field lines across the guide. Alternatively, the center conductor may be formed into a loop and oriented to couple the magnetic field of the guide mode.

The higher-order-mode field patterns, which are necessary for analytical reasons to completely match the transition, represent lumped reactances provided no propagation of these modes is possible, i.e., that they are cut off. Compensating susceptances can then be introduced in the form of tunable stubs designed by means of the Smith Chart (Fig. 9-3) but with an empirical fine adjustment to cancel reflections. If an impedance transformation is required in addition to the pattern match described, quarter-wave transformers are generally employed. These consist of $\lambda_g/4$ lengths of line designed for an intermediate impedance equal to the square root of the product of the two impedances to be matched.

For broader bandwidth, multiple quarter-wave sections are called for, leading in the limit to a smooth taper of the dimension(s) so that the fractional change in impedance with distance is constant. The resultant reflection coefficients for various forms of tapers are compared as a function of taper length in Fig. 9-18. Figure 9-19 shows the bandwidth obtainable from a two-section (quarter-wave) transformer for given VSWR (denoted by S on the curves).

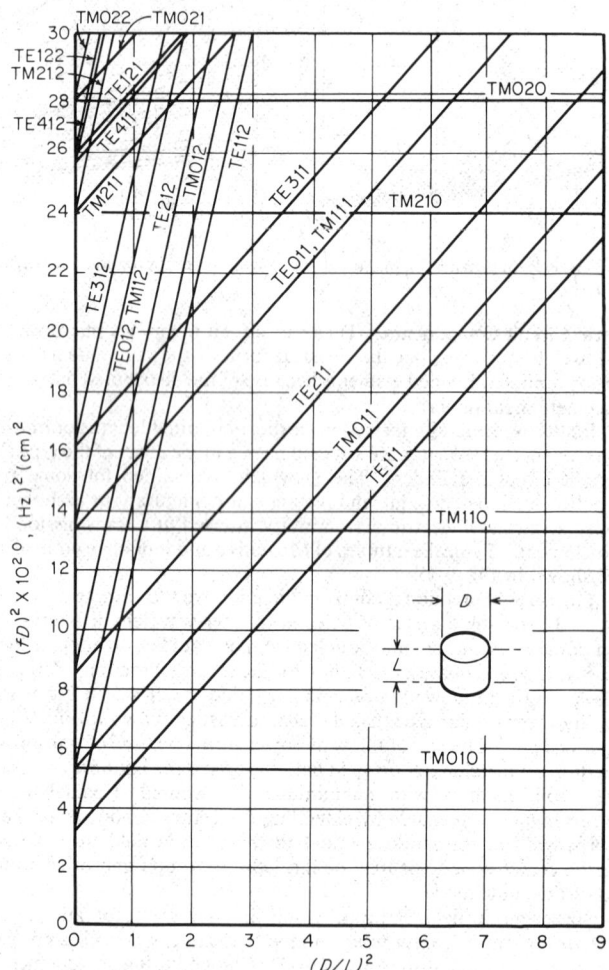

Fig. 9-13. Mode chart for right circular cylinder.

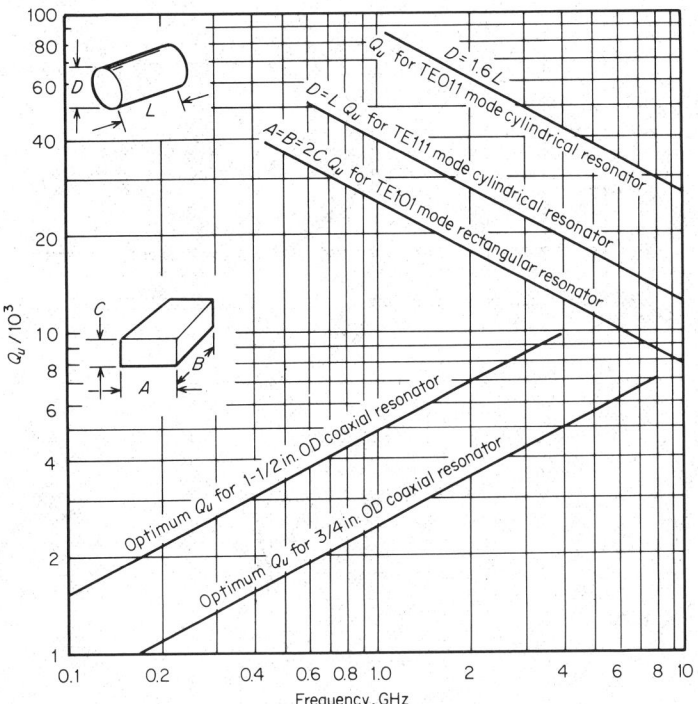

Fig. 9-14. Q for typical cavity resonators.

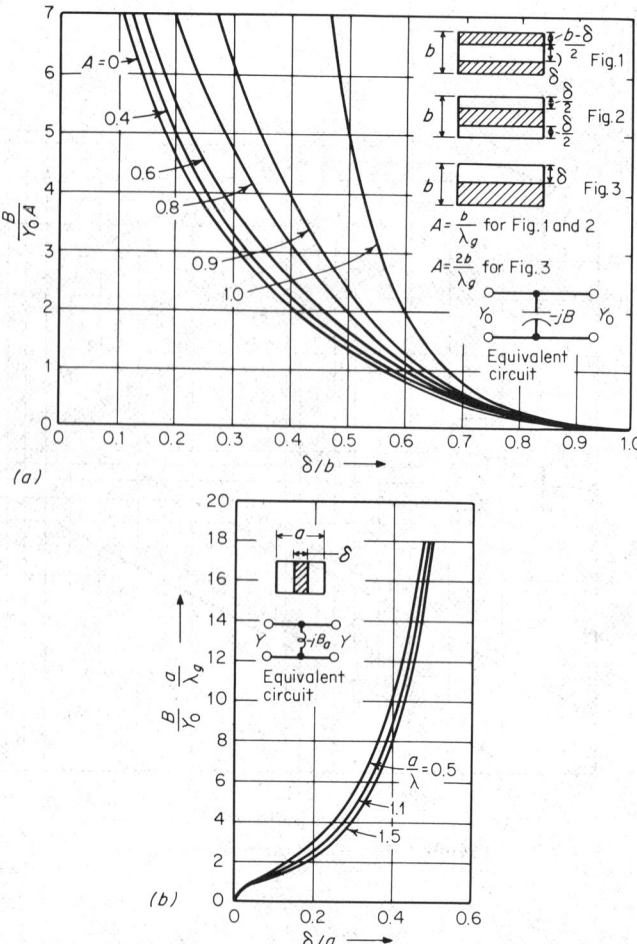

Fig. 9-15. Susceptance of waveguide elements. (*a*) Capacitive irises; (*b*) inductive, centered thin vane.

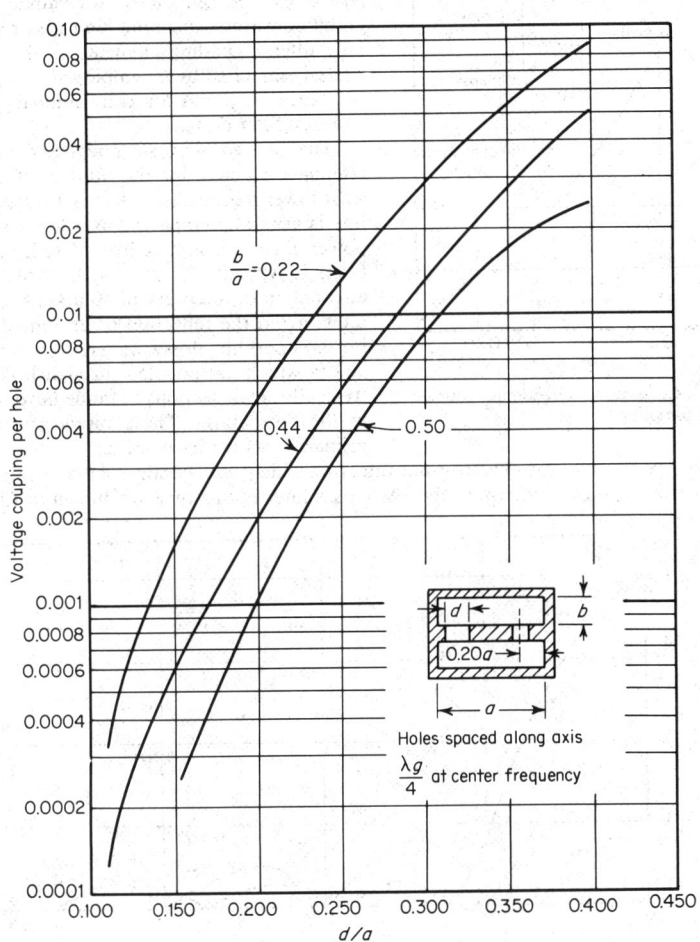

Fig. 9-16. Broadwall directional coupler ratios.

Filters. Frequency filters are employed to separate the components of a composite waveform for signal-processing purposes or to suppress radio-frequency interference (RFI)

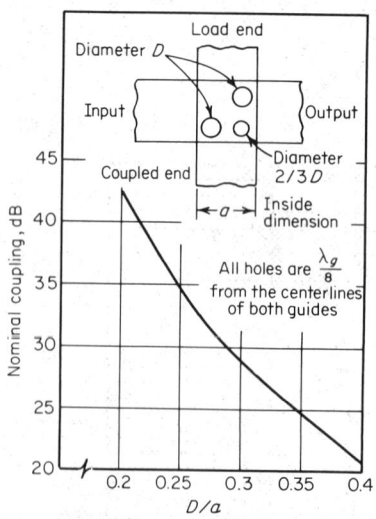

Fig. 9-17. Cross-guide directional coupler for signal sampling.

which results from the spurious output of transmitters. The latter problem has only recently become serious at microwave frequencies as this area of the spectrum has become congested. High-power capability is required for such filters, leading generally to the use of waveguide structures. A section of waveguide beyond cutoff constitutes a simple high-pass reflective-type filter. Loading elements in the form of posts, irises, or stubs are employed to supply the reactances required for conventional lumped-constant-filter design.

The desired skirt steepness and stopband attenuation determine the number of sections as at lower frequencies. A disk-loaded coaxial line is generally employed as a low-pass high-power filter. Insertion loss of reflective-type filters is typically 0.1 to 0.2 dB, with stopband attenuation of the order of 50 dB. Absorption filters avoid the reflection of unwanted energy by incorporating lossy material in secondary guides which are coupled through leaky walls (typically small sections of guide beyond cutoff in the passband). These filters are effective primarily against harmonics.

For low-power applications, strip-line filters are widely used because of their compact size and low cost. Typical dimensions for a low-pass filter of this type are shown in Fig. 9-20.

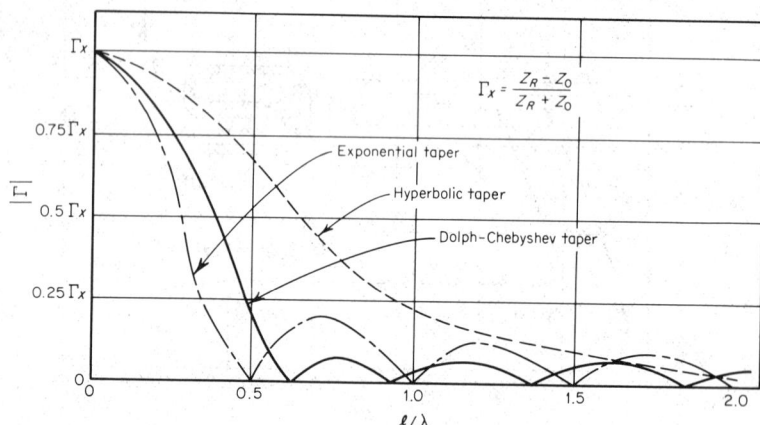

Fig. 9-18. Use of tapers for impedance matching.

Among the components useful for measurement purposes are wavemeters, attenuators, and matched loads. Mechanically tuned resonant cavities are generally employed for frequency determination. For a transmission wavemeter such a cavity is coupled in series into the transmission path, while for an absorptive indication it is coupled in shunt. A dominant mode (TE_{111}) cylindrical resonator is most widely used for this purpose, but for highest selectivity a circular electric mode (TE_{011}) resonator is employed. Variable attenuators take the form of thin absorptive material introduced tangential to the electric field typically through a slot in the broadwall of rectangular guide so as to produce minimum

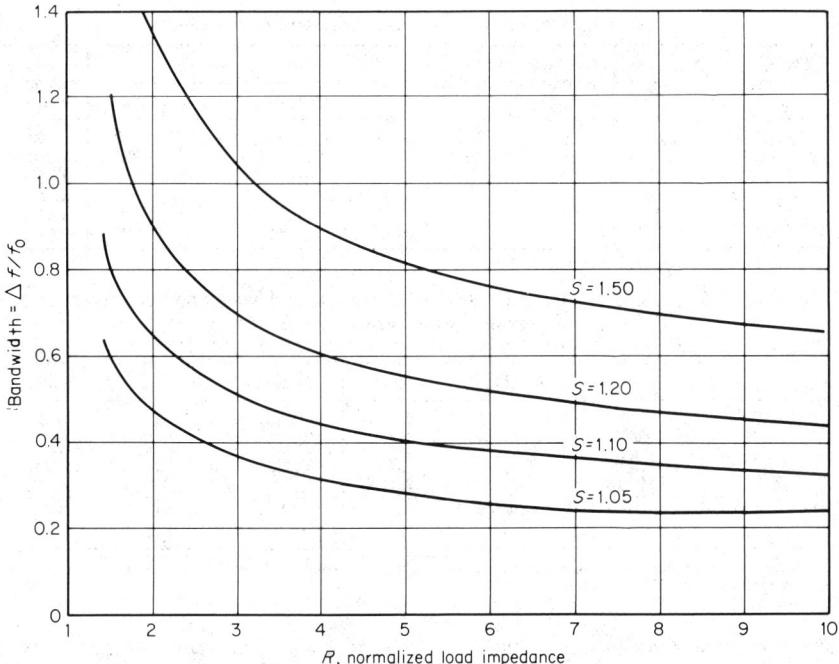

Fig. 9-19. Bandwidth and VSWR of quarter-wave transformers.

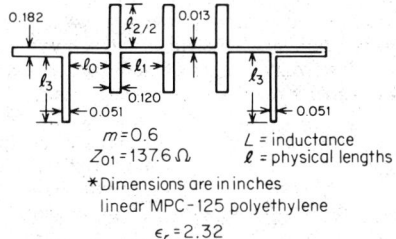

$m = 0.6$
$Z_{01} = 137.6\ \Omega$

L = inductance
ℓ = physical lengths

*Dimensions are in inches
linear MPC-125 polyethylene
$\epsilon_r = 2.32$

Frequency	Dimensions			
f_c	ℓ_0	ℓ_1	$\ell_{2/2}$	ℓ_3
	$0.084\lambda_{\epsilon_r}$	$0.100\lambda_{\epsilon_r}$	$0.144\lambda_{\epsilon_r}$	$0.200\lambda_{\epsilon_r}$
1 GHz	0.6653	0.7920	1.093	1.548
2 GHz	0.3326	0.3960	0.533	0.756
3 GHz	0.2220	0.2640	0.333	0.493

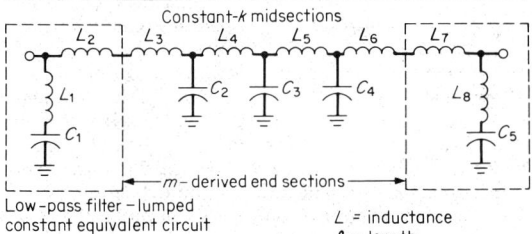

Constant-k midsections

← m- derived end sections →

Low-pass filter – lumped
constant equivalent circuit

L = inductance
ℓ = length

Fig. 9-20. Strip-line-filter design.

9-21

Table 9-5. Performance of Field Displacement Isolators

Frequency range, MHz	Maximum forward loss, dB	Minimum reverse loss, dB	Minimum return loss, dB	Ferrite description		Applied dc magnetic field, Oe
				Composition	Gauss	
3,700–4,200	0.41	19.2	30	$Mg_{.9} Cu_{.1} Al_{.35} Fe_{1.4} Mn_{.04}$	1,200	590
5,925–6,425	0.22	30	30	$Mg_{.9} Mn_{.1} Fe_{1.2}$	1,960	660
10,700–11,700:						
Single slab	0.25	28	32	$Ni_{.8} Cu_{.1} Zn_{.1} Fe_{1.9} Mn_{.02}$	3,800	1,150
double slab	0.99	65	27		3,800	1,150
23,500–24,470	0.60	34	...	$Ni_{.6} Zn_{.4} Fe_{1.9} Mn_{.02}$	4,900	3,670

Table 9-6. Scaled and Actual Values of Geometric and Magnetic Parameters in Isolators at Various Frequency Bands

(6-GHz Isolator is used as basis. Quantities in parentheses are scaled values.)

Frequency, MHz	δ, mils	b, mils	h, mils	L, mils	$4\pi Ms$, G	H, applied, Oe
5,925–6,425	180	70	550	1,590	1,960	660
3,700–4,200	280.5	120	828	2,290	1,200	590
	(288)	(112)	(880)	(2,544)	(1,225)	(412)
10,700–11,700	101	27.5	303	900	3,800	1,150
	(100)	(39)	(305)	(883)	(3,528)	(1,188)
23,500–24,470	45	19	137	420	4,900	3,670
	(45)	(17)	(137)	(397)	(7,840)	(2,640)

SOURCE: S. Weisbaum and H. Boyet, reprinted from *IRE Trans. Microwave Theory Techniques,* Professional Group on Microwave Theory and Techniques, July, 1959.

reflection. Load terminations are tapered attenuators designed to produce at least 40 dB of return loss while maintaining a good match (maximum VSWR less than 1.2) through the specified band. For high power, water cooling is provided either around a loaded ceramic or Teflon absorber or by introducing a tube of water directly into the guide to act as the absorber. Calorimetric determination of power is possible with such water loads.

8. Ferrite Components. *Isolators.* Ferrite isolators owe their usefulness to the large ratio of reverse loss to forward loss which occurs when the proper dc magnetic field bias is applied to material chosen for a specific microwave frequency range. The original concept was that of a Faraday rotation, which required positioning a ferrite rod in regions of circular polarization of the field pattern. The difference in phase velocity for the two directions of propagation is then a function of the applied magnetic field. The resultant differential phase shift can be utilized in conjunction with 3-dB power splitters to provide nonreciprocal attenuation. The nonreciprocal change in field pattern obtained when a ferrite slab is positioned transversely across a rectangular guide has led to several new configurations.

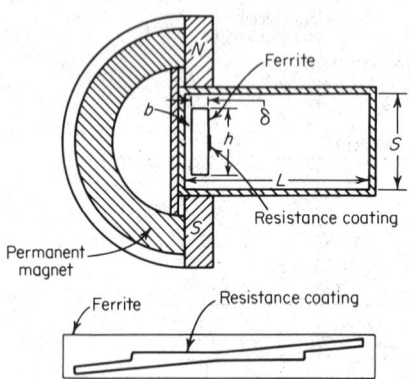

Fig. 9-21. Ferrite field-displacement isolator.

Circulators. Typical construction and operating characteristics of field-displacement isolators, once widely employed in low-power applications, are shown in Fig. 9-21 and Tables 9-5 and 9-6. For higher powers the absorption function is separated from the nonreciprocal reflection so that the ferrite is physically removed from the power dissipation. In the ferrite circulator, a three-port device, the reflected power is brought out from a separate port from the incident and transmitted power ports, where it may be monitored, absorbed in a load, or employed as useful output in the case of a reflective-type amplifier.

Circulators have also found use as duplexers, providing greater than 20 dB isolation between transmitter and antenna with insertion loss of the order of 1/4 dB. Ferrite components have been designed which can handle power levels of the order of megawatts peak and many kilowatts average over the uhf and microwave frequency bands.

The most widely employed ferrite component today is the Y-junction circulator. In this device, which has been developed for both strip line and waveguide, the ferrite material is placed at the junction of three equiangularly spaced transmission paths, and the dc magnetic field applied perpendicular to the plane of these paths. A complex coupling of dielectric resonator modes takes place in the ferrite, leading to excellent circulator characteristics (low insertion loss, low VSWR, and high port-to-port isolation). Octave-bandwidth strip-line versions and full-band waveguide models are commercially available.

Phase Shifters. Reciprocal phase shifters, useful in phased arrays, are obtained by centering a thin ferrite rod longitudinally in a rectangular guide and applying an axially directed magnetic field to it. A full 180° phase variation at 9 GHz requires a change of only a few hundred gauss in a 2-in.-long × 1/4-in.-diameter ferrite rod.

Digital Latching. Latching ferrites, named for the remanent state of magnetization on a hysteresis loop into which the ferrite is switched by the application of coil current pulses, are widely used for digital phase shifting. Such remanent states require only small biasing fields.

Limiters. The nonlinear behavior of ferrites at high-power levels is utilized in ferrite limiters. Such limiters are replacing TR gas discharge tubes in radar today. Peak powers of tens of kilowatts at X-band can be handled, but with considerable spike energy leakage which requires a follow-on solid-state (*pin*) diode stage. Typical insertion loss is about 0.5 dB, with 30 dB of flat limiting and about 6 dB of spike limiting for the ferrite alone. The diode stage increases the insertion loss to about 1 dB, but reduces all leakage an additional 30 dB. The recovery time is very short, typically 100 ns, and is determined primarily by the diode section. Characteristics of such diodes are discussed further in Pars. **9.50** and **9.59**.

8a. Bibliography on Passive Microwave Components

RAMO, S., and WHINNERY, J. R. "Fields and Waves in Modern Radio," Wiley, New York, 1953.

REICH, H. J., SKALNIK, J. G., ORDUNG, P. F., and KRAUSS, H. L. "Microwave Principles," Van Nostrand, New York, 1957.

GINZTON, E. L. "Microwave Measurements," McGraw-Hill, New York, 1957.

SAAD, T. S. (ed.) "Microwave Engineers' Handbook," Artech House, 1971.

YOUNG, L. (ed.) "Microwave Filters, Impedance-matching Networks, and Coupling Structures," McGraw-Hill, New York, 1972.

PLANAR MICROWAVE TUBES AND CIRCUITS

By Donald A. Priest

9. Development of Planar Tubes. *Triodes* for rf power generation at frequencies above about 1 GHz were developed just before World War II in Europe and the United States for both continuous-wave (cw) and pulsed operation. Cylindrical and planar electrode configurations were used with oxide-coated cathodes in Britain. In the United States the "lighthouse tube" used planar geometry and an internal anode. All used metal and glass envelopes. Continuous-wave rf powers of the order of 1 to 10 W at 1 GHz and peak pulse powers of the order of kilowatts were obtained in coaxial-line circuits in several industrial laboratories.

The need for centimeter-wave equipment for military purposes stimulated development of microwave tubes and associated circuitry during World War II, particularly in the United States, and by 1946 the 2C39 type triode with external anode emerged as the most popular design. The version with an aluminum oxide ceramic envelope (3CX100A5) was produced in very large quantities and with minor internal modifications is still in use today. Before the emergence of the power Klystron in the 1950s, these tubes and combinations of them (e.g., in an *annular circuit*) produced the highest rf power outputs obtainable above 1 GHz.

The power gain of rf amplifiers with planar triodes is sufficient to make practical the use of multistage amplifiers driven by crystal controlled oscillators with high-frequency stability. For example, the TD2 microwave radio relay system operating at 4 GHz, in service since 1951, uses a planar triode of advanced design (type 416A/1553) in broadband circuitry for television and multichannel telephone transmission.[1] Planar triodes are preferred today in some applications (airborne radar and navigation equipment and space communications) because of their small size and/or low cost relative to other technical approaches. They have been improved considerably in performance over the years, and are manufactured in Europe, the United States, and Japan. Triodes of small size using cylindrical electrodes ("pencil triodes") are available for similar applications.

Tetrodes (4X150A, 4X150G, 1949, and later models; 6816, 7213, and others) also give useful performance at 1 GHz and above, especially in pulse service, but are less widely used than the triodes because of greater difficulty in designing and fabricating the associated cavity circuitry and the tube socket. These difficulties increase rapidly with frequency, and the improvement in power gain over the planar triode decreases rapidly. In general, therefore, tetrodes are preferred below 1 GHz and triodes above 1 GHz today.

10. Performance vs. Frequency. Power output and power gain of amplifiers with a single tube fall off with increasing frequency, for several basic reasons. Physical size imposes upper limit related directly to wavelength and inversely to frequency. Ideally, the rf voltages between electrodes should be uniform, but this cannot be realized unless the major electrode dimensions are much less than one-quarter wavelength (radius of planar triodes, axial length of cylindrical electrode triodes or tetrodes). The radius cannot be increased indefinitely because of the tendency to introduce circumferential variations of electric field which can exist when circumference exceeds about one wavelength. These factors cause maximum tube electrode area, and therefore power output, to fall off inversely with the square of the frequency.

In addition, *electron transit time* effects cause a further fall-off unless interelectrode spacings, mainly grid to cathode, and grid-wire diameters are scaled inversely with frequency. At 1 GHz, the grid cathode spacing must not exceed a few mils, or transit time will result in excessive loading of the drive source, significantly reducing power gain of amplifiers, causing back-heating of the cathode by electron bombardment and reduction of conversion efficiency due to increase of the *conduction angle* (fraction of the rf cycle during which electrons arrive at anode).

Another cause of reduced performance with increasing frequency is the increase in rf displacement currents drawn by interelectrode capacitances, resulting in increased i^2R heating of the grid wires, cathodes, and connecting leads, including the seal areas between metal leads and the insulating envelope.

For these reasons, the performance of a given tube falls off very rapidly above a specified frequency, and in practice an upper limit of frequency (sometimes called f_1) is designated by manufacturers above which the dc power input to the anode circuit must be reduced to prevent overheating. Below this frequency, performance stays roughly constant as the frequency is reduced, since the basic limiting factors are then cathode emission density and voltage breakdown between electrodes, which are not functions of frequency.

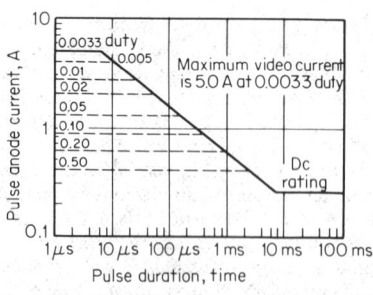

Fig. 9-22. Derating of peak pulse current with increasing pulse length. *(EIMAC Division of Varian Associates.)*

11. Pulsed Operation. In pulsed operation higher frequencies can be reached than with cw operation because the higher voltages, used intermittently, result in increased electron velocities and decreased transit times with higher conversion efficiencies. At a given frequency the rf power output during a short pulse can be as much as 100 times the cw output. This is partly due to the higher voltages, but mainly due to the greatly increased emission density available from the cathodes during short pulses. The usable cathode current depends on the pulse length and the duty factor (pulse duration divided by pulse-to-pulse time interval, see Fig. 9-22). The cathode temperature, and therefore heater voltage, may also be different for pulsed operation. Lifetime of tubes depends on other

---

[1] Superior numbers correspond to numbered references, Par. **9-19a.**

operating conditions besides cathode temperature, and tube manufacturers should advise about these conditions if maximum life is to be obtained, whether in cw or pulsed operation.

Examples of performance vs. frequency from available tubes are given in Fig. 9-23 for cw and pulsed operation. The curves show maximum rf power output, pulsed and cw, considered to be obtainable from tubes, using present technology if designed for narrowband operation at the frequency chosen, and assuming zero circuit losses and a tube life of 1,000 h or more. Pulse lengths of a few microseconds are assumed.

Higher peak power can be obtained if the plate voltage is pulsed (rather than if dc plate voltage is used with grid pulsing) because the higher probability of internal arcing requires a lower dc voltage than the allowable pulse voltage. For longer pulses, cathode current must be reduced, and therefore less rf power output can be obtained. Typical operating conditions for a few planar triodes are given in Table 9-7.

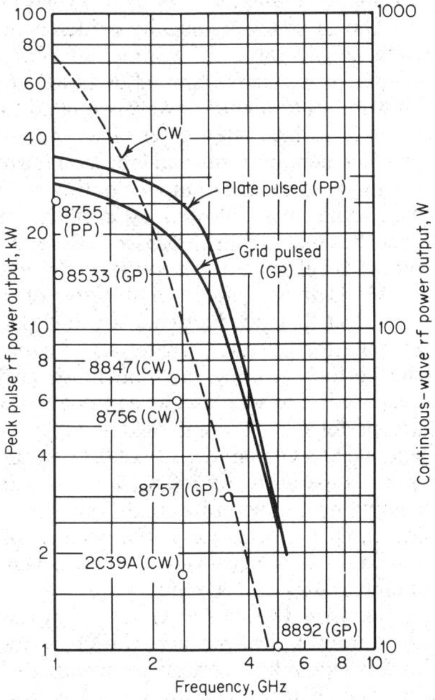

12. Multitube Arrangements. To obtain more power output, several tubes can be used together if care is taken to suppress *moding*, which tends to occur when some dimension of the combining circuitry used becomes large in terms of the wavelength. For example, using an annular circuit of 14 planar triodes can produce 500 W cw output at 1 GHz.[2] This approach to higher power has lost popularity at microwave frequencies with the advent of the Klystron, traveling-wave tube, and crossed-field amplifier, all of which are capable of very much higher power and are competitive in price with multiple triodes because of the high cost of the combining circuitry required.

13. Efficiency and Gain. Typical plate conversion efficiencies, assuming zero circuit losses, narrowband operation, and optimum conditions for planar triodes, fall from about 60% at 1 GHz to about 30% at 5 GHz when

Fig. 9-23. Power output vs. frequency, of typical planar triode and manufacturer's ratings of typical tubes.

pulsed, or to about 5% at 5 GHz, cw. Circuit losses reduce these numbers, depending on the circuit design.

The power gain obtainable from amplifiers depends on circuit design (see below), but seldom exceeds 12 dB at 1 to 2 GHz, falling off with increasing frequency. For this reason self-excited oscillators are used at the higher frequencies. The bandwidth of such amplifiers also depends on circuit design, but seldom exceeds a few percent of the center frequency.

Table 9-7. Planar Tubes

	3CX100A5, 7289	8892, 18651	8756	416B*
Frequency, GHz	3.0	5.0	2.35	4.0
Service	Plate pulsed oscillator	Grid pulsed oscillator	Cw amplifier	Cw amplifier
Plate voltage, kV	3.5	2.0	1.25	0.3
Plate current, A	3.0	2.0	0.15	0.03
Rf output	2.0 kW	1.0 kW	60 W	0.5 W
Rf drive, W	...	...	3.0	...
Plate efficiency, %	19	25	33	...
Gain, dB	...	...	13	6–9
Heater power, W	6.0	4.5	5.1	...

*See Ref. 4, Par. 9-19a.

14. Design and Construction of Planar Tubes.   The envelopes of planar tubes are almost always of ceramic insulation with penetrating metal members supporting the electrodes. High-purity alumina (above 99% pure) is most commonly used, but beryllia is used when the added performance made possible by its increased thermal conductivity (10 times) and lower dielectric constant (2/3) justifies the increased cost (5 to 10 times).   The metal members are usually of a material (such as kovar) with thermal expansion coefficient close to that of the insulator, to minimize thermal stresses in fabrication and operation.   They are coated with a highly conductive metal (e.g., copper or silver) to reduce i^2R losses.   The metal members are shaped either as disks or disks with cylindrical projections.

The cathodes are usually oxide-coated (Ba, Sr, Ca on Ni), indirectly heated.   There has been considerable developmental effort to increase emission density, to increase power output (since size is limited as noted above).   Increased electric field strength, which also contributes to higher power, is related to cathode performance because the condition of the cathode surface often determines the voltage at which internal arcing occurs.   Because rf displacement currents and emission currents can cause significant cathode heating, attempts are made to reduce cathode-coating resistance.   Mixing Ni powder with carbonates of Ba, etc., sometimes followed by application of pressure, is beneficial.   Other successful approaches include the *dispenser cathode,* a metal sponge impregnated with Ba compounds.   A recent development uses osmium doping.

15. Cathode Life.   An objective of cathode development is to improve emission density and electrical conductivity without decreasing tube life and without increasing rate of evaporation of emitter material, which deposits on other electrodes and may actually block the perforations in very fine mesh grids, if excessive.   The life tends to decrease, and evaporation to increase, as cathode temperature is raised, so that low-temperature emitters are preferred.   Typical oxide-coated cathodes of good design, carefully processed, may operate at a current density of 0.25 A/cm^2 (averaged over one rf cycle) under cw conditions, at a temperature of 750°C, and the corresponding life may be from 2,000 to 10,000 h, depending on operating conditions.   It is nearly always possible, with a given tube, to obtain increased performance at the expense of life by increasing cathode temperature.   In pulse operation it is necessary to reduce peak current density as pulse length is increased, to maintain a given tube life.

16. Grid Design.[3,4]   Grid design is very important in planar triodes, and it presents a major mechanical problem if high performance is required.   Some electron current, which ideally would flow entirely from cathode to anode, is intercepted by the grid, causing heating, in addition to heating by radiation from the cathode and heating of the wires by circuit currents.   Since the periphery of the grid is colder than the center, thermal stresses tend to cause buckling or departure from planeness, if not counteracted.   In some designs the counteracting means is by prestressing, in others by a rigid cellular supporting structure.

Increasing the grid temperature not only tends toward deformation, but also leads to emission of primary electrons, which adversely affects performance.   Grids are usually made of wires of a refractory metal (W, Mo) for high tensile strength and thermal conductivity, coated with gold or other material to inhibit grid emission.

To obtain high performance, the grid must have the thinnest possible wires, spaced as close as possible to the cathode.   This not only minimizes electron transit time losses, but also maximizes amplifier power gain per stage, which is limited to approximately the tube amplification factor because of the *grounded-grid* circuitry invariably used.   Wires are typically of 1 to 0.4 mil in diameter, and closest grid-to-cathode spacing is 2 to 3 mils.

17. Anode Design.   Anodes are usually of Cu, the main function being to conduct the heat of electron bombardment to an external heat sink, usually air-cooled.   A design problem exists because thermal expansion may change anode-to-grid spacing and capacitance, detuning the resonant circuit needed for efficient energy transfer to the load.

The internal structural details of the type 7289 planar triode are shown in Fig. 9-24.

18. Circuits for Planar Triodes.   Planar triodes for 1 GHz and above are used in resonant-cavity-type circuits, of waveguide, coaxial line, or strip line.   These confine the rf fields to the interior, minimizing radiation loss, and provide convenient and efficient means of presenting the correct impedance between tube terminals.   Since the tube itself usually provides the major part of the shunt capacitance reactance of the resonant circuit, and also has substantial series inductive reactance, it is fundamental that the tube and associated circuits must be considered together during the design process.

The basic amplifier circuit used is the *grounded-grid (common grid, grid separation,* or *cathode-driven)* type. At low frequency the equivalent circuit is shown in Fig. 9-25a. It is characterized by relatively low power gain, large input bandwidth, and high stability, resulting from the inherent large negative feedback due to the loading of the rf drive source by the in-phase rf component of the pulsating anode current.

At microwave frequencies the equivalent circuit is made more complicated (Fig. 9-25b) by (1) the dominating influence of circuit elements inside the tube envelope and the capacitance of the envelope itself; (2) the effect of contact fingers between tube and circuit, which may have significant reactance; (3) the distributed reactance of cavity resonators and the tube itself; and (4) electron transit time, which causes resistive loading and phase shifts of considerable magnitude.

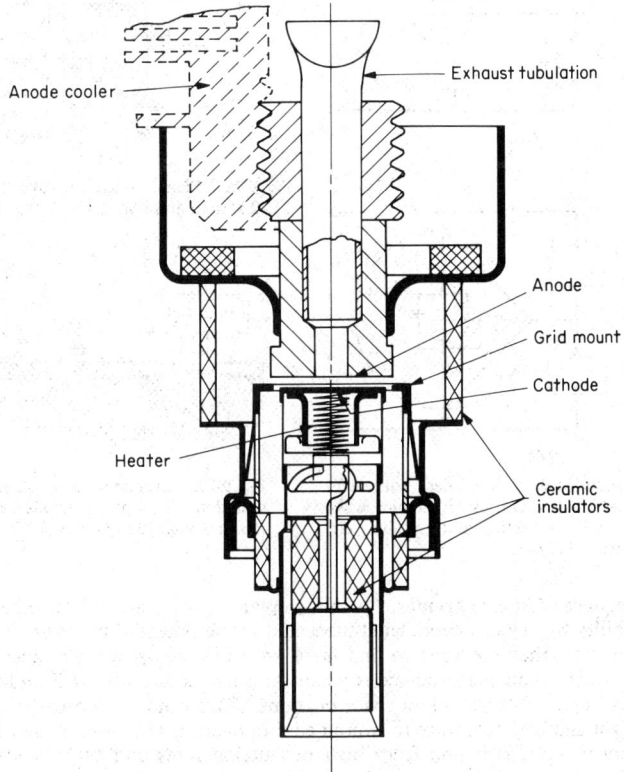

Fig. 9-24. Section through type 7289 planar triode. *(EIMAC.)*

An example of the physical arrangement of tube, cavities, and coupling elements is shown in Fig. 9-26. This is an untuned amplifier. Tuned amplifiers require variation of cavity inductance with sliding contacts or noncontacting capacitive plungers, both of which introduce added circuit losses. The design of planar triode amplifiers has always been difficult and highly specialized, and is more so today because of demands for large bandwidth and untuned circuits, which stem from similar characteristics obtained from solid-state amplifiers used to drive the tube amplifiers.

In general, a single stage will provide power gain of 5 to 10 dB, depending on bandwidth, frequency, and drive level. Stages may be cascaded to obtain more gain. Interstage coupling may use either waveguide or coaxial-line elements. A good example of waveguide technique is used in the TD2 transmitter operating at 4 GHz. Three stages have 18 dB gain over a 20-MHz bandwidth between points 0.1 dB down.[1]

The use of double- and triple-tuned circuits to improve bandwidth is discussed in Refs. 3 and 5, Par. **9-19a.** In some applications where reduced frequency stability can be tolerated and small size is paramount, oscillators are used, for example in airborne altimeters. Useful rf power can be obtained from tubes as oscillators at frequencies higher than the maximum frequency for amplification with useful power gain. Circuits used are derived from the basic amplifier form in coaxial or strip-line construction, with the addition of positive feedback. An example is shown in Fig. 9-27.

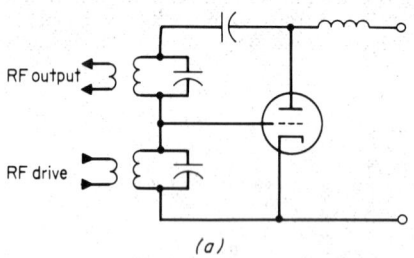

(a)

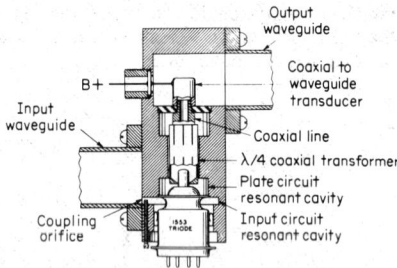

Fig. 9-26. Planar triode in waveguide cavity for TD-2 system amplifier (Ref. 1, Par. **9-19a**).

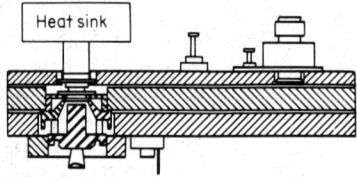

(b)

Fig. 9-25. Grounded-grid amplifier equivalent circuits. (*a*) At low frequencies; (*b*) at microwave frequencies. Cathode heating and grid bias circuits not shown. *(EIMAC.)*

Fig. 9-27. Oscillator using strip-line construction with a planar triode. *(From U.S. Patent 3,596,130 to Melvin D. Clark.)*

19. Economics of Planar Triodes. Although planar triodes are outclassed technically in power capability by velocity-modulated tubes and in life potential by transistors and other solid-state devices, they continue to find favor with new-equipment designers for special applications where their performance is adequate, such as airborne IFF and navigational aids, both civil and military, and for space missions. This is not only because of their small size and weight and high tolerance to aircraft environment, but because of their low first cost and replacement cost, resulting from high production rates and high yields in modern manufacturing plants and price competition in the industry. For example, type 7289 triodes were sold in quantities in 1971 for under $10 each. It is to be expected that triodes will be preferred for the power-frequency range shown in Fig. 9-23 in the foreseeable future.

19a. References on Planar Tubes

1. ROETKEN, A. A., SMITH, K. D., and FRIIS, R. W. The TD-2 Microwave Radio Relay System, *Bell System Tech. J.,* October, 1951, p. 1041.

2. PREIST, D. Annular Circuits for High Power Multiple Tube RF Generators at VHF and UHF, *Proc. IRE,* May, 1950.

3. MORTON, J. A., and RYDER, R. M. Design Factors of the B.T.L. 1553 Triode, *Bell System Tech. J.,* October, 1950, p. 496.

4. BOWEN, A. E., and MUMFORD, W. W. A New Microwave Triode: Its Performance as a Modulator and as an Amplifier, *Bell System Tech. J.,* October, 1950, p. 531.

5. BEGGS, J. E., and LAVOO, N. T. High Performance Experimental Power Triodes, *IEEE Trans.,* ED-13 No. 5, p. 502, May, 1966.

6. GUREWITSCH, A. M., and WHINNERY, J. R.   Microwave Oscillators Using Disk Seal Tubes, *Proc. IRE,* May, 1947.

7. "Klystrons and Microwave Triodes," Radiation Laboratory Series, Vol. 7, McGraw-Hill, New York, 1948.

KLYSTRONS

BY RICHARD B. NELSON

20. Introduction.   For high frequencies, linear-beam tubes overcome the transit-time limitations of grid-controlled tubes by accelerating the electron stream to high velocity before it is modulated. Modulation is accomplished by varying the velocity, with consequent drifting of electrons into bunches to produce rf space current. The rf circuits for coupling signals to and from the electron beam are generally integral parts of the tube. Two basic types are important today, Klystrons and traveling-wave tubes (see Pars. **9-20** to **9-28**). Different versions are used as oscillators and amplifiers.

In a Klystron, the rf circuits are resonant cavities which act as transformers to couple the high-impedance beam to low-impedance transmission lines. The frequency response is limited by the impedance-bandwidth product of the cavities, but can be increased by stagger tuning and by multiple-resonance filter-type cavities.

Table 9-8 lists the principal types of linear-beam tubes, with typical power and bandwidth values.

Table 9-8.   Linear-Beam Tubes

	Klystrons		Traveling-wave tubes		
	Type	Watts	Type	Watts	Freq. range
Oscillators	Reflex Two-cavity Extended interaction	0.01–1 1–100 0.1–1 (mm wave)	Helix BWO Power BWO	0.01–0.1 0.1–1.0	2:1 1.4:1
Amplifiers	Two-cavity Multicavity	1–10 10^3–10^5 cw 10^3–10^7 pulse	Low-noise helix Medium-power helix Ring-bar Coupled-cavity	0.001–0.01 1–1,000 10^3–10^5 pulse 10^3–10^4 cw 10^3–10^6 pulse	2:1 2:1 1.3:1 1.1:1 1.1:1
	Hybrid Twystron			10^6–10^7 pulse	1.1:1

21. Reflex Klystrons.   In the reflex Klystron a single resonator is used to modulate the beam and extract rf energy from it, making the tube simple and easy to tune. The beam passes through the cavity and is reflected by a negative-charged electrode to pass through again in the reverse direction. With proper phasing determined by applied voltages, oscillating modes occur for n + three-quarters-cycle transit time between passes through the cavity. The frequency may be modulated by varying voltage on the reflector (which draws no current). Figure 9-28 shows a schematic of a reflex Klystron.

Most reflex tubes have two grids to concentrate the electric field of the cavity in the hole through which the beam passes. Some are tuned by deforming the cavity envelope to vary the spacing between grids, and hence the capacity loading. A tuning range up to 1.4:1 is thus obtainable.

Reflex tubes requiring stable output, even under shock and vibration, are tuned by a second resonant cavity outside the vacuum envelope, which is tightly coupled to the evacuated cavity through an iris. A capacitive post in the outer cavity tunes the circuit. The tuning range is typically 1.1:1.

Figure 9-29 shows the power-supply circuit, and Fig. 9-30 the operating parameters of a typical reflex Klystron, the VA-244E.

UHF AND MICROWAVE DEVICES

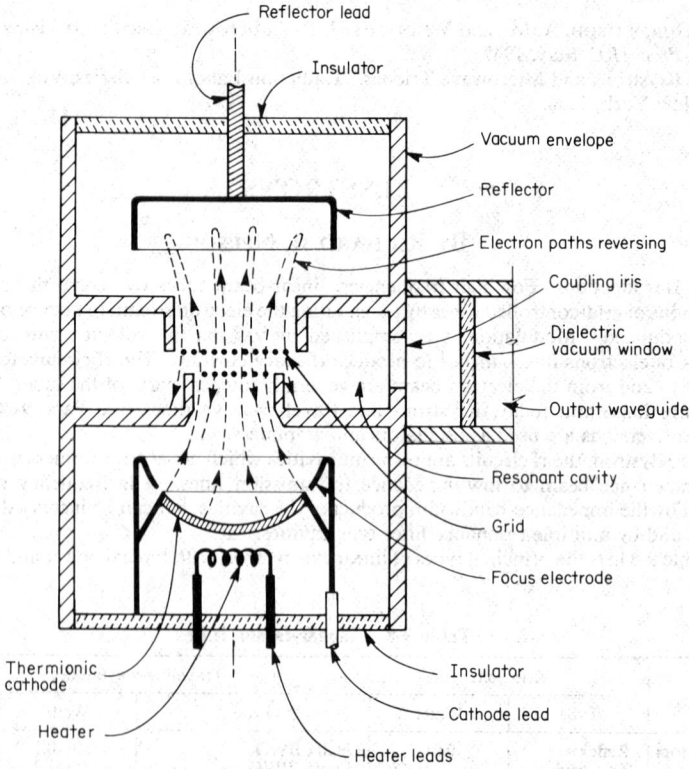

Fig. 9-28. Schematic cross section of reflex oscillator. *(Varian Associates.)*

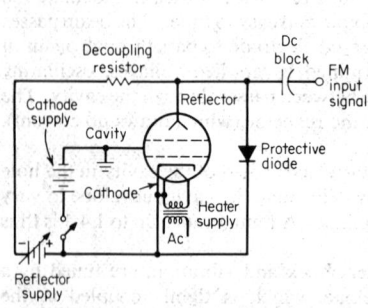

Fig. 9-29. Supply circuits for a reflex Klystron. The reflector bias should be turned on before the cathode voltage. The protective diode prevents transient positive excursions of the reflector.

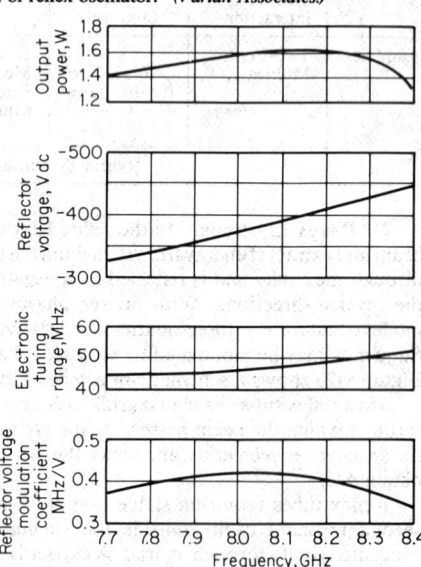

Fig. 9-30. Operating parameters of reflex Klystron VA-244E: beam voltage, 750 V dc; heater voltage, 6.3 V; beam current, 78 mA dc; heater current, 0.75 A.

Reflex Klystrons are used as test signal sources, receiver local oscillators, pump sources for parametric amplifiers, and low-power transmitters for FM line-of-sight relays. Among microwave devices they are cheap and highly reliable. Reflex-tube frequencies cover the entire microwave range from 1 to 100 GHz. Current usage is mainly from 4 to 40 GHz.

22. Two-Cavity Klystron Oscillators. In all Klystrons except the reflex, the beam goes through each cavity in succession, and so there is no feedback. The tube is a buffered amplifier, with each stage isolated from those upstream. Electromagnetic feedback may be provided to make an oscillator.

The specialized two-cavity oscillator has a coupling iris in the wall between the cavities. This tube is more efficient and more powerful than the reflex Klystron. It may be frequency-modulated by varying the cathode voltage around the center of the oscillating mode, but requires more modulator power than a reflex Klystron.

Two-cavity oscillators are used where moderate power, stable frequency, and low sideband noise are needed. Examples are the transmitter source in Doppler navigators, pumps and parametic amplifiers, and master oscillators for cw Doppler radar illuminators. To improve stability, the tubes are usually fixed-frequency, or have at most a few percent tuning range.

Figure 9-31 shows the power and frequency of cataloged reflex Klystron tubes. Figure 9-32 shows operating characteristics of a typical high-power, low-noise tube.

23. Extended-Interaction Oscillators. At millimeter wave frequencies the losses in Klystron cavities make it hard to build up the impedance necessary to oscillate with the very small low-current beams which are required.

If a series of cavities are coupled together and interact sequentially with the beam in the proper phase, the total interaction impedance increases directly with the number of cavities. The circuits of extended-interaction oscillators resemble those of traveling-wave tubes. They operate with a complete standing wave (at the cutoff of the traveling-wave passband), and hence the tubes may be classed as Klystrons. Various names are used for tubes of this type. The Laddertron uses a ladder-shaped periodic circuit and a flat-ribbon electron beam. *Multicavity Klystron oscillators* use coupled cavities and cylindrical beams. Most of the extended-interaction tubes are still experimental.

24. Two-Cavity Amplifiers. In the simplest Klystron the driving signal is coupled through a transmission line to the input cavity. The cavity voltage produces velocity modulation of the beam. After a single drift space, the resultant density modulation induces current in the output resonator, from which power is extracted through another transmission line. The beam is usually focused electrostatically.

The gain of a two-cavity Klystron is about 10 dB. Usage is limited because more gain is desired in high-power tubes and solid-state amplifiers are available at low powers.

25. Multicavity Amplifiers. Downstream from the input cavity, cascaded intermediate cavities are inserted. These have no external coupling; they are driven by the rf beam current and in turn remodulate the beam velocity. Figure 9-33 shows the structure schematically.

Each cavity tuned to the signal frequency adds about 20 dB of gain to the 10 dB of a two-cavity Klystron. Net gain up to 60 dB is practical. The penultimate cavity is usually detuned to a higher frequency to improve efficiency by about 5%. Other intermediate cavities are often detuned, or *staggered,* to increase bandwidth, at the expense of gain. Up to eight cavities have been used.

Focusing. In multicavity amplifiers the electron beam is long and requires focusing forces to keep it uniformly small. A series of electrostatic lenses is used in a few tubes where light weight or stray magnetic fields are important. Most Klystrons use a uniform magnetic field parallel to the beam. Permanent magnets are used for rf power below 5 kW. At higher power the tubes are inserted interchangeably in electromagnets.

Tuning. The cavities are tuned mechanically in several ways, depending on the range required. Some examples follow:

Fixed-frequency: Set on frequency by permanent deformation of a cavity wall.

3% tuning: One wall of the cavity is a thin diaphragm which is forced in and out by a mechanism outside the vacuum.

10% tuning: A movable interior cavity wall is not part of the vacuum wall but is moved through a flexible bellows.

(The above three methods vary the inductance of the cavity.)

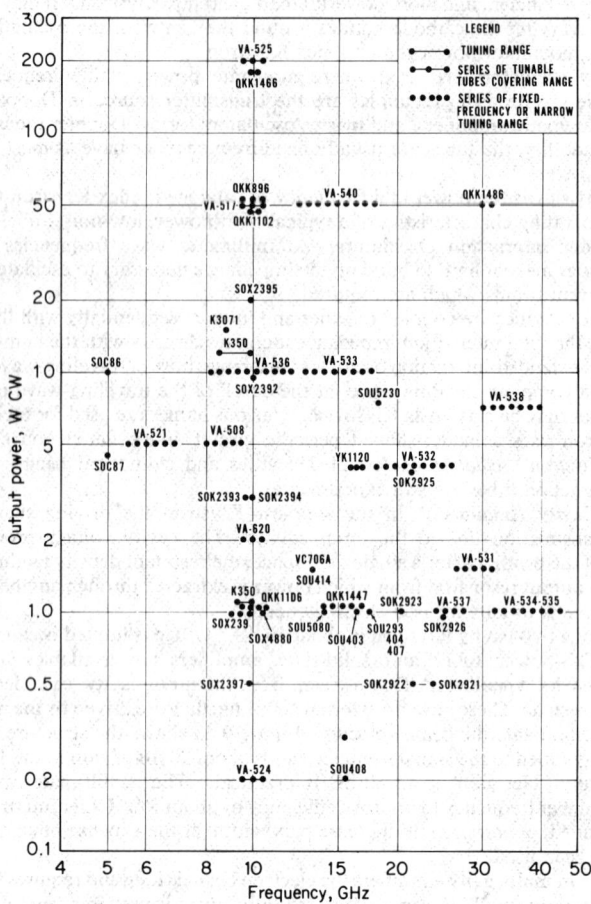

Fig. 9-31. Power and frequency of cataloged Klystron tubes.

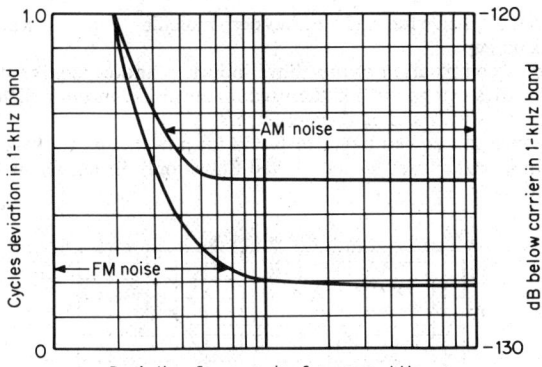

Fig. 9-32. Two-cavity oscillator noise characteristics, type VA-517:

Frequency .... 10.000 ±0.125 GHz
Output power .... 75 W
Beam voltage .... 10 kV dc
Beam current .... 90 mA dc
Temperature coefficient ..... 200 kHz/°C
Beam voltage coefficient .... 1.8 kHz/V
Coolant flow, water .... 0.5 gal/min

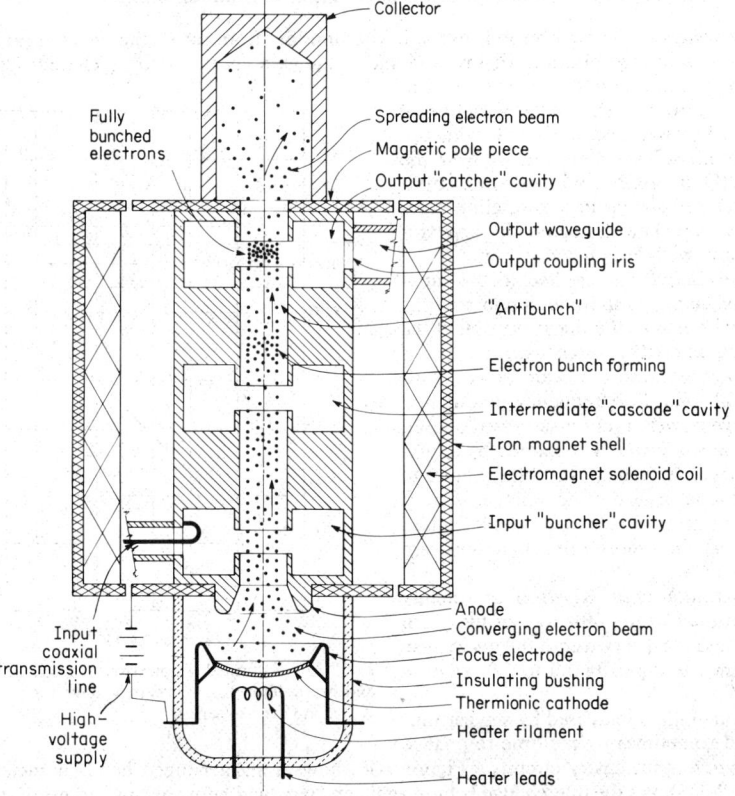

Fig. 9-33. Cross section of cascade Klystron amplifier. *(Varian Associates.)*

25% tuning: A paddle inside the cavity moves perpendicular to the beam and adds capacity across the interaction gap.

35% tuning: (a) A combination of the above inductive and capacitive tuners inside the cavities. (b) The cavities extend outside the vacuum envelope, through dielectric windows, and have movable walls with sliding-contact fingers.

Cooling. Klystrons are cooled by air or liquid for powers up to 5 kW. Higher-power cooling is by boiling water (used in the United States only in television transmitters) or recirculating liquid.

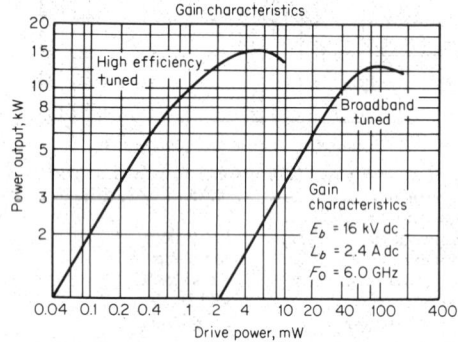

Fig. 9-34. Typical gain, output-power and drive-power characteristics.

Performance. The Klystron is a true linear amplifier from zero signal level up to 2 or 3 dB below saturated output. Figure 9-34 shows the gain characteristic. The efficiency at maximum output is in the range from 35 to 55%. Figure 9-35 shows the variation of saturated power and efficiency with applied beam voltage. The data are from the VA-884D, a 14-kW cw broadband amplifier designed for ground-to-satellite communication. The characteristics are typical of any well-designed Klystron.

Rf modulation is applied to the input drive signal. Amplitude modulation is usually limited to the linear portion of the gain characteristic, with consequent loss of efficiency, because the beam power is always full on. For frequency modulation the drive power is set for saturated output.

Pulse modulation is obtained by applying a negative rectangular voltage pulse to the cathode instead of dc voltage. The rf drive (saturation value) is usually pulsed on for a slightly shorter time than the beam pulse.

Bandwidth of a Klystron is roughly proportional to the fifth root of the beam power and the 4/5 power of the perveance. Perveance is typically 1.0 to 2.0 × 10⁻⁶ A/V³ᐟ².

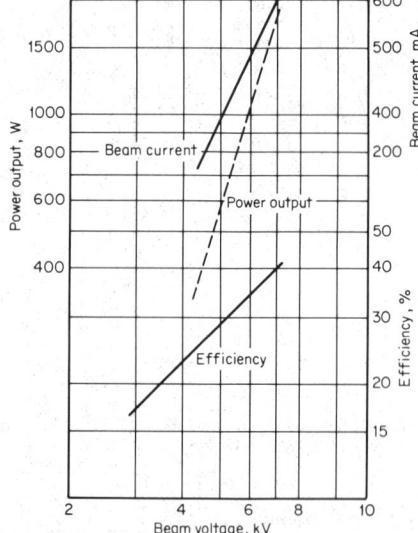

Fig. 9-35. Saturated rf power output, beam current and beam efficiency vs. beam voltage for a 15-kW Klystron (VA-884 series).

Bandwidth is increased by stagger tuning and sometimes by multiple-resonance filter-type output cavity circuits. Figure 9-36 shows a stagger-tuned band characteristic. Figure 9-37 shows the interchange between gain and bandwidth for various degrees of stagger tuning.

Distortion. *Amplitude distortion* is reasonably predictable because the nonlinearity of the

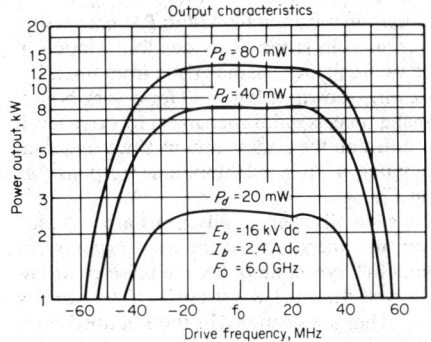

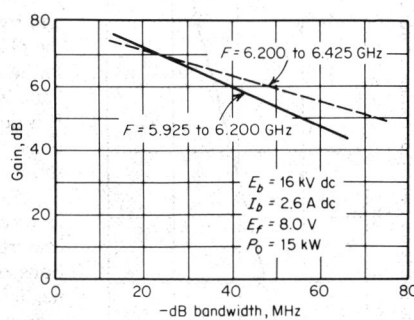

Fig. 9-36. Gain and output vs. frequency characteristics under saturated and unsaturated rf drive
conditions.

Fig. 9-37. Type VA-884 series Klystron gain-
bandwidth characteristics.

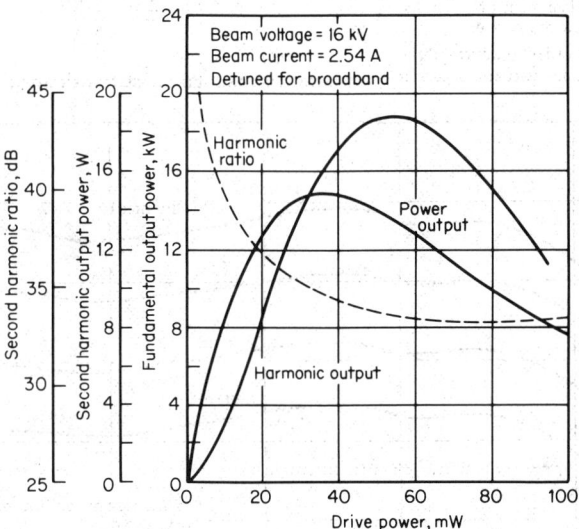

Fig. 9-38. Harmonic output of typical Klystron.

gain characteristic is quite similar for all Klystrons. Figure 9-38 shows the increase in output

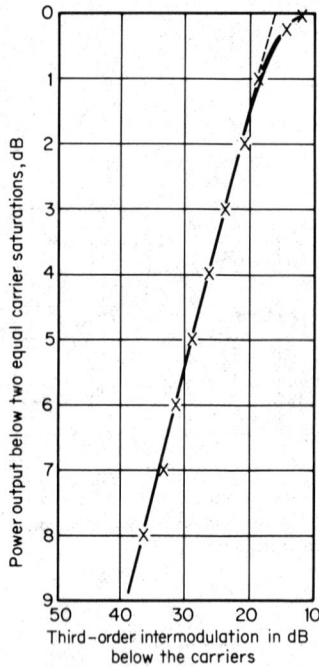

of second harmonic with drive power. Other harmonics are similar, but with rapidly decreasing amplitude above the third harmonic. When multiple carriers are combined in the signal, the second-order intermodulation products are always outside the band. The most important intermodulation is the third-order $(2f_2 - f_1)$, shown in Fig. 9-39.

FM distortion is caused by AM to PM conversion and by variation in group delay due to deviation of the phase vs. frequency characteristic from linearity. *Phase linearity,* shown in Fig. 9-40 for a very broadband X-band tube, is quite unchanged by drive level. It is dependent on the stagger tuning, since optimum phase requires a rounded amplitude response as shown on Fig. 9-41.

26. Noise in Klystrons. Klystrons are not used as receiver amplifiers, so that the *noise figure* is not important. In a power amplifier the noise contributions of a well-designed Klystron itself are usually negligible. They are swamped by the amplified noise output of the master oscillator and by modulation due to power-supply ripple or noise. Amplitude ripple in percent of carrier power is 5/2 times the percentage beam-voltage ripple. Phase ripple is given by

$$\frac{\Delta\phi}{\phi} = \frac{1}{2}\frac{\Delta V}{V}$$

where ϕ is the total phase transit angle through the rf

Fig. 9-39. Third-order intermodulation distortion under two equal carrier conditions.

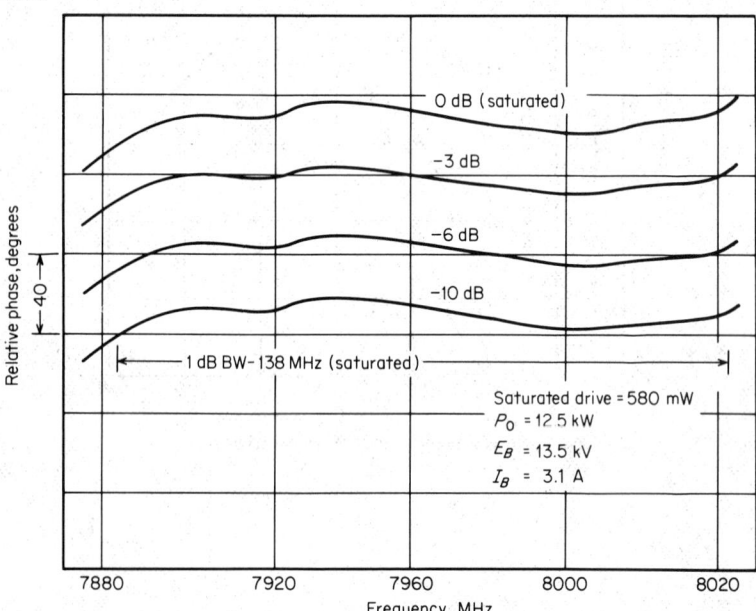

Fig. 9-40. Phase linearity of type VKX-7753 Klystron. The deviation from linearity is unaffected by the drive level.

section of the Klystron. Typically, $\phi \approx 10(N-1)$ rad, where N is the number of cavities, and so, in a four-cavity tube.

$$\Delta\phi \approx 15\frac{\Delta V}{V}$$

27. Applications and Circuits. The principal uses of Klystron amplifiers include microwave heating; transmitters for radar; television; tropospheric scatter; ground to satellite; earth to spacecraft; and illuminators for guided missiles.

A typical power-supply schematic for a cw amplifier is shown in Fig. 9-42. Protective devices often required for Klystron amplifiers include sensors to monitor cooling air velocity, cooling water flow, collector overtemperature, cathode-heating time delay, cathode overcurrent, body overcurrent, and output waveguide arcing.

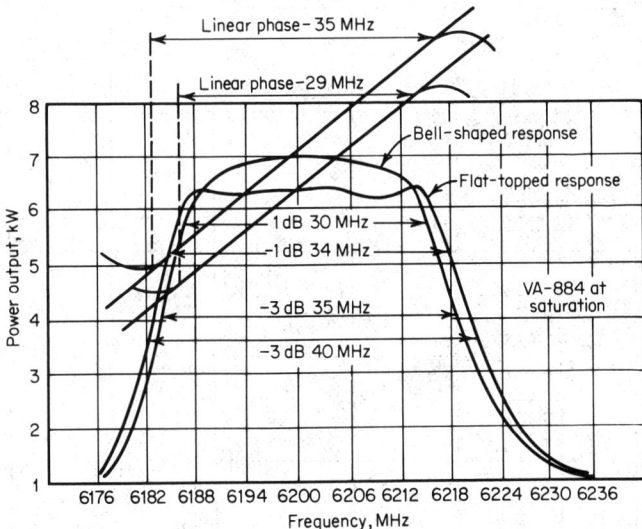

Fig. 9-41. Response curves for bell-shaped and flat-topped tuned amplifiers.

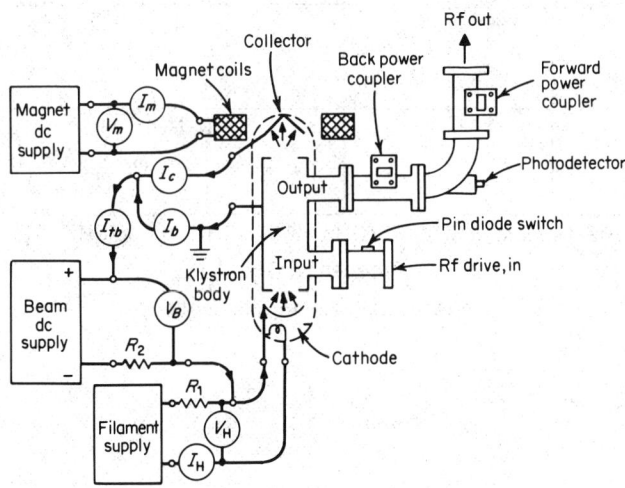

Fig. 9-42. Circuits for a cw Klystron amplifier.

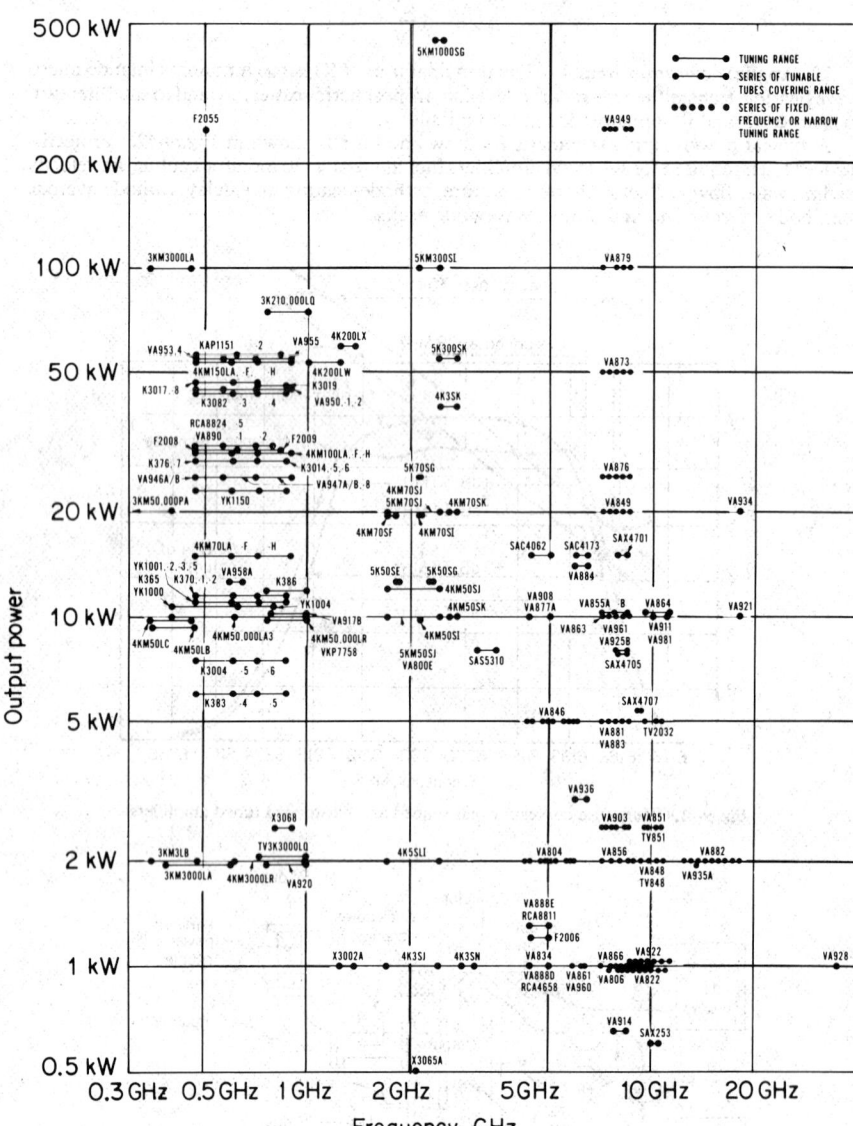

Fig. 9-43. Cataloged cw Klystron amplifiers.

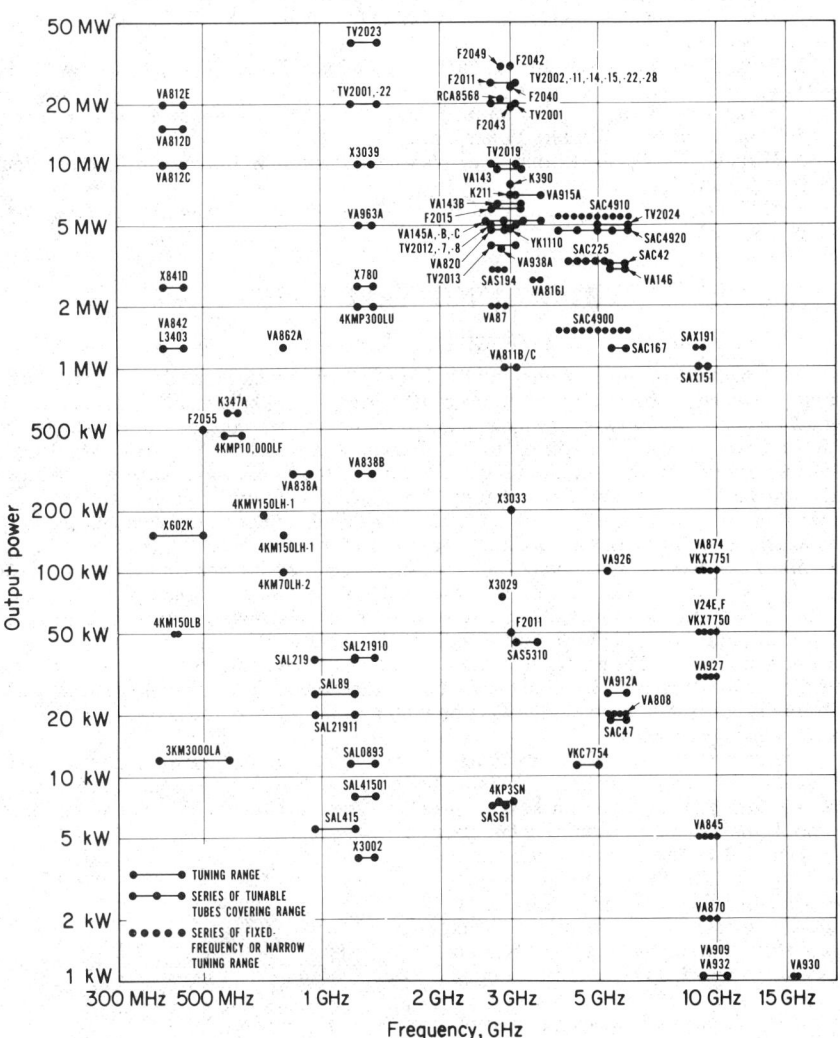

Fig. 9-44. Cataloged pulsed Klystron amplifiers.

28. Klystron Types.   In Figs. 9-43 and 9-44 catalog cw and pulsed klystrons are arranged by power and frequency coverage.

28a. References on Klystrons

1. HARRISON, A. E.   "Klystron Tubes," McGraw-Hill, New York, 1947.
2. PIERCE, J. R., and SHEPHERD, W. C.   "Reflex Oscillators," *Bell System Tech. J.,* Vol. 26, No. 3, p. 460, July, 1947.
3. BECK, A. H. W.   "Velocity Modulated Thermionic Tubes," Macmillan, New York, 1948.
4. WARNECKE, R. R., and GUENARD, P. R.   "Le tube electronique à commande par modulation de vitesse," Gauthier-Villars, Paris, 1951.
5. HUTTER, R. G. E.   "Beam and Wave Electronics in Microwave Tubes," Van Nostrand, New York, 1960.
6. MORENO, T.   "High Power Axial Beam Tubes," *Advan. Electron. Electron Phys.,* Vol. 14, p. 299, 1961.

TRAVELING-WAVE TUBES

BY RICHARD B. NELSON

29. Introduction.   In a traveling-wave tube (TWT) a linear electron beam passes through a circuit which propagates an electromagnetic wave with a phase velocity approximately equal to the electron velocity.   A signal injected on the circuit produces velocity modulation of the beam, which is converted by the bunching process to rf current modulation. Interaction of the space-charge wave on the beam with the circuit wave causes both to grow synchronously as they travel down the tube.   The rf energy comes from a reduction in average velocity of the electrons.   The mismatch between the high-impedance beam and the low-impedance circuit is overcome because they interact for many cycles.

Many characteristics common to all linear-beam tubes are described in Pars. **9-23** to **9-28** and will only be referred to here.   The types of TWTs are listed in Table 9-8.

30. Low-Power Backward-Wave Oscillators.   A traveling-wave tube used as a signal generator employs a helix slow-wave circuit and hollow electron beam, focused by a permanent magnet.   The beam interacts with a backward-wave space harmonic of the circuit field which has a phase velocity in the direction of the beam.   The power flow is upstream, i.e., in the opposite direction to the electron motion.   Such devices are known as *backward-wave oscillators* (BWOs).

The downstream end of the circuit has a nonreflecting termination; the upstream end is matched to the output transmission line.   The tube oscillates at a frequency such that the beam velocity is equal to the phase velocity of the dispersive circuit.   The beam voltage thus controls the frequency, permitting fast electronic sweeping.

Figure 9-45 shows a schematic of a helix BWO.   The electron beam is hollow because the rf electric field of the useful space harmonic is maximum near the helix and falls to zero on the axis.   The beam is focused by an axial magnetic field from a permanent magnet.   Most of the spent beam is collected on the helix, an acceptable method at low-power level.   An anode at fixed voltage and a grid are used to regulate the beam current, while the variable tuning voltage is applied to the helix.

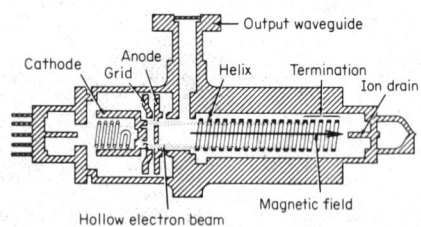

Fig. 9-45.   Schematic of a helix backward-wave oscillator.

Figure 9-46 shows a typical voltage-tuning characteristic, and Fig. 9-47 the variation of output with tuning voltage.   The latter is often flattened by programming the grid bias or by servo control of an electronic attenuator in the output line.   In such devices, the FM noise has a typical total excursion of 10^{-6} of the carrier frequency.   It is usually masked by power-supply noise.   The AM noise is some 90 dB below carrier in a 1-MHz-band distant from the carrier.   There are also coherent noise signals 70 to 75 dB below carrier at a few megahertz from it.   Total noise at 30 MHz from the carrier in a 1-MHz band is 90 to 100 dB down.   Harmonics are 20 to 25 dB down.

BWOs are used as electronically tunable signal sources in test signal generators, search receivers, and some transmitters with wideband FM. Power output is in the range 10 to 100 mW. The operating frequencies cover the entire microwave band. At S-band and below, the tuning range is typically one octave. At C-band and above, a waveguide band (one-half octave) is typical.

31. Power Backward-Wave Oscillators. High-power linear-beam backward-wave oscillators are used for generating millimeter waves. The helix is not suitable as a slow-wave circuit at high power, and so all-metal bandpass periodic structures are used. One such structure is a series of Klystron-like cavities coupled by inductive irises (see Par. **9-35**).

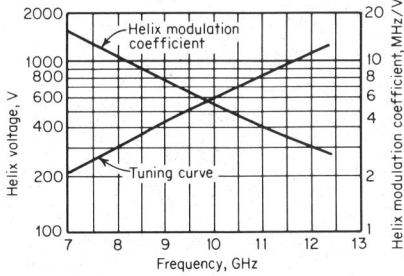

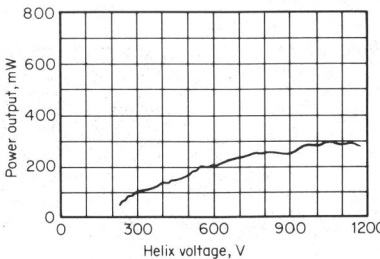

Fig. 9-46. Tuning and modulation characteristic of a BWO.   **Fig. 9-47.** Power vs. helix voltage of a BWO.

Another is a row of parallel vanes projecting from the wall of a hollow surrounding pipe. The fundamental mode of these circuits is a backward wave, and so they interact strongly with the electron beam. The efficiency is less than 1%. The electronic interaction is not inherently efficient when the beam sees an attenuating rather than a growing wave. Also, these tubes are used at frequencies where circuit losses are high. Such oscillators have been built at frequencies from 25 to 300 GHz, with output powers ranging from 40 W to 1 mW.

32. Low-Noise Traveling-Wave Amplifiers. TWTs are used as input amplifiers of moderately low-noise receivers. Their advantages are broad bandwidth; room-temperature operation; wide dynamic range; and freedom from damage by overloading. Tubes for this service use helix circuits and low-perveance beams, magnetically focused. The noise figure improves with increased magnetic field and with decreased beam current (hence low power output). The magnets used, in order of decreasing weight and increasing noise figure, are solenoids, permanent magnets with straight field, straight field with direction reversals, and periodic permanent magnets.

Table 9-9 shows the noise figure vs. power of typical tubes. Higher noise figures than these often result from compromises with low weight, extended bandwidth, etc. Attainable noise figures also increase with frequency.

33. Medium-Power Helix Traveling-Wave Tubes. The slow-wave circuit formed by a close-spaced metal helix supported between dielectric rods has the widest bandwidth of all. It is used universally in TWTs below 1-kW output. Figure 9-48 shows the slow-wave structure. The conductor is usually a ribbon, but may be a round wire.

The commonly attained bandwidth is one octave (2:1), except at high frequency and power, where waveguides limit the band to 1.4:1. Two octaves (4:1) are available in some

Table 9-9. Low-Noise TWTs
(Typical Minimum Noise Figures of TWT Amplifiers)

Saturated Power Out, dBm	Noise Figure, dB
−10	5–6
0	6–7
10	7–10
20	10–16
35	22–25

tubes. The efficiency, typically in the range 10 to 20%, can sometimes be increased to over 50% by depressing the potential of the collector below that of the circuit so that the electrons are slowed down before they strike the collector. Figure 9-49 shows the power-supply connections for depressed collector operation. Figure 9-50 gives operating characteristics of a typical tube.

The gain is usually in the range of 30 to 60 dB. To make the amplifier stable against regeneration, the circuit has an attenuator midway along its length to absorb reflected power, while forward power is carried through by the electron beam. For higher gain the circuit is cut into sections, each terminated by attenuators, with no rf connections between them. Thus the total power transfer is by the one-way electron beam.

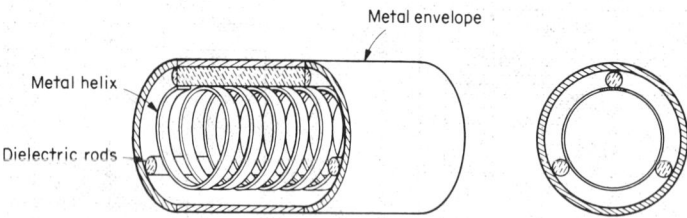

Fig. 9-48. Helix slow-wave circuit. The helix ribbon and outer envelope are metal, the support rods dielectric.

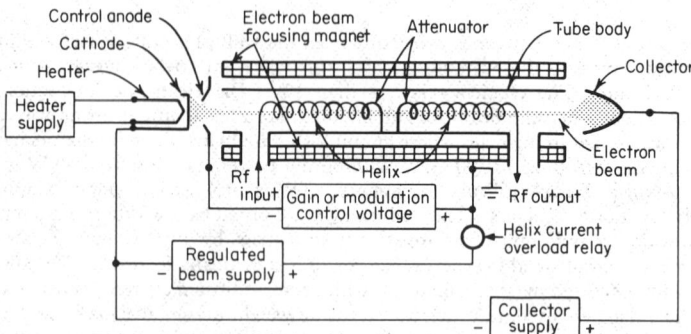

Fig. 9-49. Traveling-wave-tube circuit functions.

The beam in these tubes is focused by straight field solenoids or periodic permanent magnets. Focusing by periodically reversing magnetic field requires that the pole pieces be close to the beam. This is easily done because the helix circuit has a small diameter compared with a wavelength. Such tubes are cooled by convection at power levels around 1 W average output, by conduction to a heat sink for 10 W, by forced air for 100 W, and by circulating liquid for 1 kW.

Helix tubes are used as transmitters for line-of-sight communications, low-power radar, satellite-borne relays, electronic-countermeasure transmitters, and as drivers for high-power amplifiers. A very large number of models are available.

34. Ring-Bar Medium-Power TWTs. For peak powers much above 1 kW, the helix TWT tends to oscillate in a backward-wave mode. Other low-pass circuits are used, called *helix-derived* because they are formed from one or more wire conductors insulated from the envelope. They have slow phase velocity and are periodic. The most common circuit is formed of two helices of opposite pitch connected at their crossovers. This is often deformed into the electrically equivalent ring-bar circuit shown in Fig. 9-51.

Ring-bar TWTs are used for peak powers of a few kilowatts to a few hundred kilowatts. With no metallic connections to the envelope, the circuit is not capable of high heat dissipation, and so these tubes are generally not designed for continuous-wave (cw) operation.

Bandwidth is around 15%. Ring-bar tubes are frequently used as drivers for high-power transmitters.

35. Coupled-Cavity High-Power TWTs. For high average or peak power there are a variety of all-metal slow-wave circuits which have low rf losses and high heat dissipation by thermal conduction. The circuits are equivalent to a series of cavities, similar to Klystron cavities but mutually coupled to propagate a traveling wave.

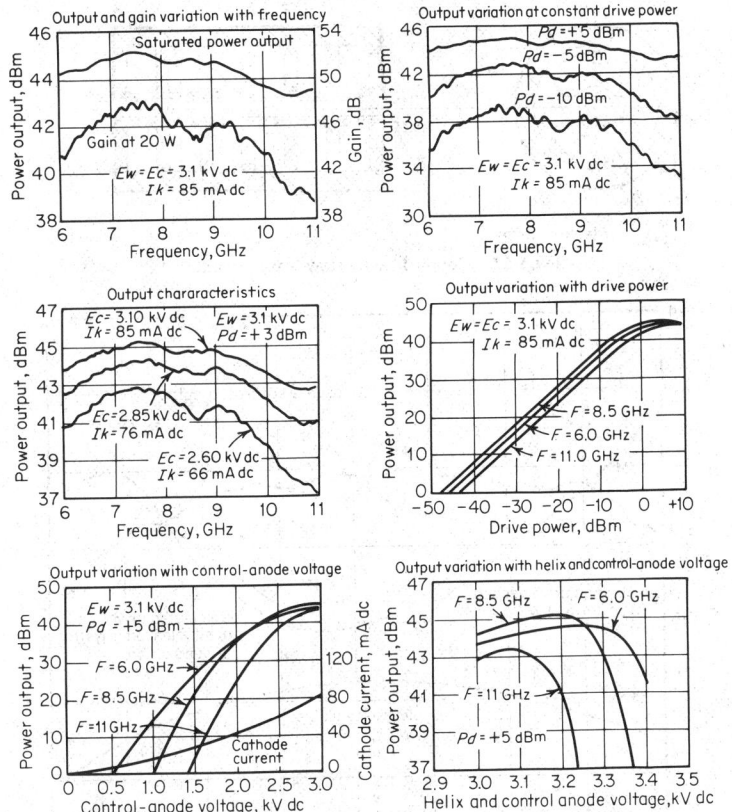

Fig. 9-50. Typical performance of a helix TWT, type VTH-6170A1. Ripples in gain, due to mismatch between the helix and the transmission lines, smooth out at saturation drive.

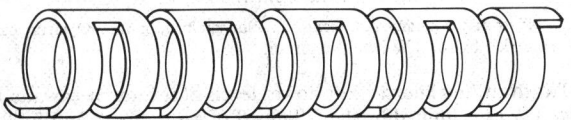

Fig. 9-51. The ring-bar circuit, electrically equivalent to two crossed helices.

In the range from 1 to 100 kW the cavities are usually simple pillboxes coupled by irises which, being shorter than a half wavelength, act as mutual-inductance coupling elements. The fundamental mode of this circuit is a backward wave, and so the useful beam interaction is with a forward space harmonic. Figure 9-52 shows such a circuit.

For very-high-power applications, from 1 to 10 MW, circuits are used in which the phase of the inductive coupling is artificially reversed, producing a forward fundamental wave. The *cloverleaf* circuit has the cavity walls deformed to reverse the current across the coupling slots in successive cavities. The *centipede* uses a multiplicity of coupling loops reversed in S shapes. TWTs of this class have been largely replaced by hybrid Twystrons. (See Par. **9-36.**)

Coupled-cavity TWTs are used in transmitters for radar, electronic countermeasures, ground-to-satellite communication, tropospheric scatter transmitters, and in guided-missile illuminators.

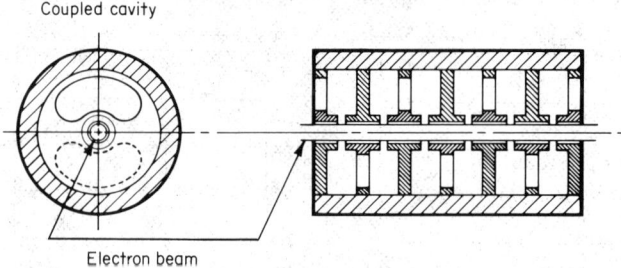

Fig. 9-52. Coupled-cavity slow-wave circuit.

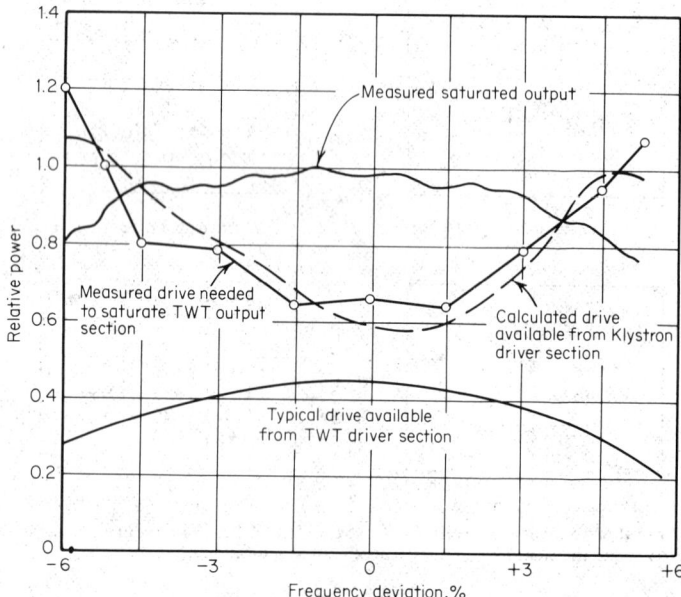

Fig. 9-53. Hybrid Twystron frequency equalization. Stagger-tuning the Klystron cavities gives the double-peaked drive needed by the TWT output circuit.

36. Hybrid Twystron Amplifiers. For power levels of several megawatts, the bandwidth of coupled-cavity TWTs is limited by the falling off of gain at both ends of the band. This can be corrected by passing the electron beam, before it enters the output circuit of the TWT, through a series of Klystron cavities. These are stagger-tuned to provide a double-peaked frequency characteristic with higher gain at the edges of the band. Figure 9-53 illustrates the resultant bandpass. Other characteristics are similar to a coupled-cavity TWT. Twystrons are used principally in radar transmitters.

37. Modulation of TWTs. The slow-wave structure of a TWT is normally at ground potential. Rf modulation, AM or FM, may be applied by (1) the cathode biased at ground or slightly positive and pulsed negative to draw current; (2) the cathode held negative and a *modulating anode* between cathode and circuit biased slightly negative to the cathode and pulsed positive to ground potential; or (3) a high-μ control grid between cathode and anode biased negative to the cathode and pulsed positive. The last method requires less modulator power than others, but limits the power and performance of the tubes. The three modulation methods are shown in Fig. 9-54.

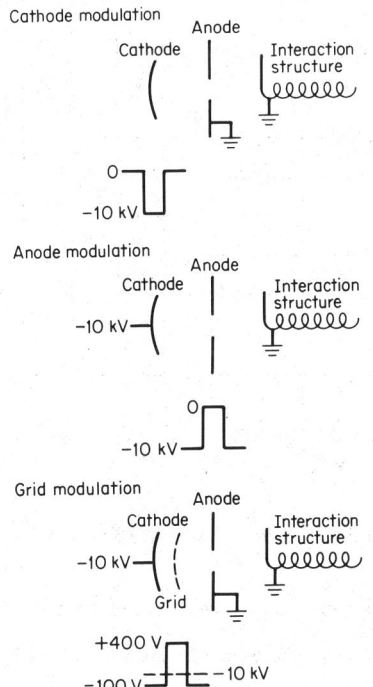

38. Spurious Outputs of TWTs. *AM Distortion.* Like the Klystron, the TWT is a linear amplifier of small signals which slowly saturates with higher inputs (Fig. 9-50). The spurious outputs generated are harmonics; also intermodulation products when multiple carriers are present. The exact values of these are similar to the Klystron's, but not as predictable, because the shape of the saturation characteristic is somewhat dependent on velocity and impedance variations in the TWT circuit. An important exception occurs in helix tubes with more than one-octave bandwidth where second-harmonic and second-order intermodulation frequencies may be in the

Fig. 9-54. Methods of modulation.

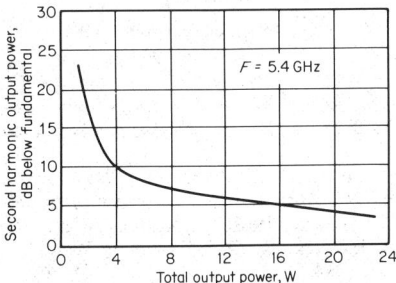

Fig. 9-55. Typical TWT harmonic output.

band and be amplified to an appreciable fraction of the total output. Figure 9-55 shows harmonics of a cw coupled-cavity TWT, and Fig. 9-56 the intermodulation.

AM-PM Conversion. The phase of the output advances with increased drive, reaching a maximum beyond the saturation point. Figure 9-57 shows a typical characteristic.

FM Distortion. With the wideband distributed circuit of a TWT it is possible to get a very linear phase characteristic. In practice, mismatches at the ends of the circuit often cause ripples in the phase response, with a frequency periodicity inversely proportional to the total electrical length. Thus there is no "typical" phase-frequency characteristic.

Shot Noise. This is the limiting factor in low-noise receiving TWTs. It is caused by the finite lumps of charge on the electrons; it appears as velocity and current noise waves on the beam, augmented by partition noise when part of the beam is intercepted and by positive-ion effects. The basic shot-noise current is proportional to the square root of the beam current. The best noise figures are thus obtained in low-current, low-power tubes, and there is consequently a tradeoff between noise figure and dynamic range.

Beam Supply Noise. Amplitude modulation due to beam voltage ripple is not a simple function, as in the Klystron, but depends on the beam-circuit synchronization as well as the beam power fluctuation. The graph at lower right in Fig. 9-50 shows typical values. Frequency-modulated ripple is also a complex effect. In a TWT, about half the energy flows in the circuit and half in the beam, and so the phase shift per unit length due to beam voltage is less than in a Klystron, where the beam velocity alone controls it. On the other hand, a TWT is about twice as long as a Klystron with equal gain; so the phase vs. voltage change is comparable.

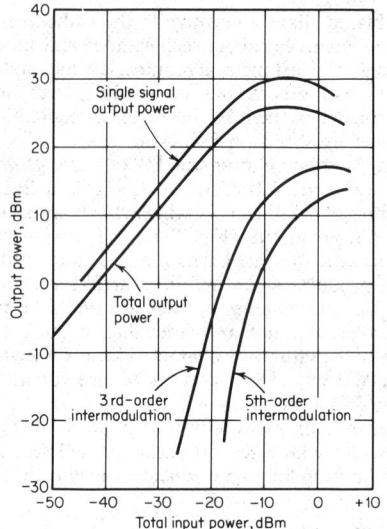

Fig. 9-56. Intermodulation in a TWT.

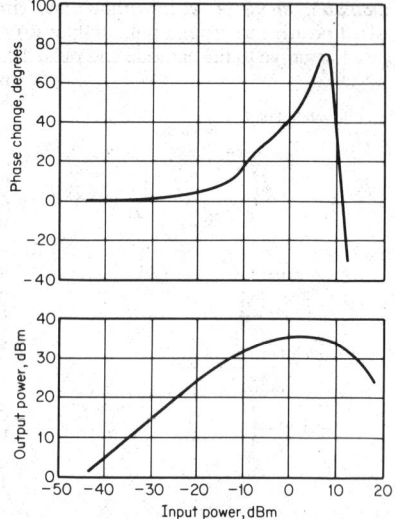

Fig. 9-57. Typical phase change and output vs. input.

38a. References on Traveling-Wave Tubes

1. BECK, A. H. W. "Velocity Modulated Thermionic Tubes," Macmillan, New York, 1948.

2. PIERCE, J. R. "Traveling-Wave Tubes," Van Nostrand, New York, 1950.

3. WARNECKE, R. R., and GUENARD, P. R. "Le tube electronique à commande par modulation de vitesse," Gauthier-Villars, Paris, 1951.

4. HUTTER, R. G. "Traveling-Wave Tubes," *Advan. Electron. Electron Phys.,* Vol. 6, p. 371, 1954.

5. GITTINS, J. "Power Traveling-Wave Tubes," Elsevier, New York, 1965.

CROSSED-FIELD TUBES

BY GEORGE K. FARNEY

39. Crossed-Field Interaction Mechanism. A crossed-field microwave tube is a device that converts dc electrical power to microwave power using an electronic energy conversion process similar to that utilized in a magnetron oscillator. These devices differ from beam tubes in that they are potential-energy converters rather than kinetic-energy converters. The term *crossed field* is derived from the orthogonality of the dc electric field supplied by the source of dc electrical power and the magnetic field required for beam focusing in the interaction region. These tubes are sometimes called *M Tubes.*

Typically, the magnetic field is supplied by a permanent-magnet structure. The electronic interaction is illustrated schematically in Fig. 9-58. Electrons moving to the right in the figure experience electric field deflection forces ($f_e = -\epsilon E$) toward the electrically positive anode, while the magnetic deflection forces ($f_m = -\epsilon \bar{v} \times \bar{B}$) resulting from the motion of the negatively charged electron in the orthogonal magnetic field cause deflection toward the negative electrode. This electrode is also called the *sole.*

The forces are balanced when an electron is traveling in a parallel direction between the electrodes with a velocity numerically equal to the ratio of the dc electric field to the magnetic field ($v_e = E/B$). Any alteration of the electron velocity leads to an unbalanced condition. Reduction of the electron forward motion causes the magnetic deflection force to become less, and the electron trajectory is deviated toward the positive electrode. Conversely, an increase of velocity causes a greater magnetic deflection force, which causes trajectory deviation toward the negative electrode.

Electronic interaction with a traveling wave occurs when the positive electrode is an rf-guiding slow-wave circuit whose phase velocity for the traveling wave is numerically equal to the ratio of the dc electric field to the magnetic field ($v_p = E/B$). Under these conditions synchronous interaction occurs between the rf fields on the slow-wave circuit and the stream of electrons traveling in the interaction region.

Two general kinds of motion result, as illustrated schematically in Fig. 9-58. In this figure a moving frame of reference is shown traveling from left to right at a velocity equal to the phase velocity of the circuit wave, so that the instantaneous rf fields are seen as stationary. The electronic motion resulting from interaction with the tangential components of the

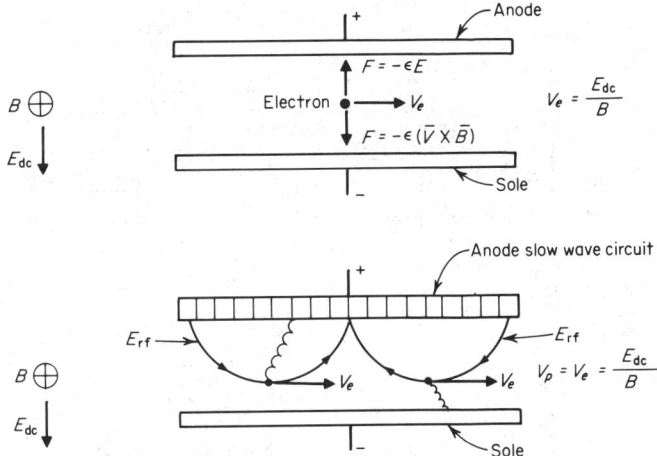

Fig. 9-58.   Forces exerted on a moving electron in a crossed-field environment.

additional rf electric fields depends on the location of the electron relative to the phase of the rf fields of the slow-wave circuit. Those so located that their forward motion is retarded by the rf electric field are slowed, and the energy they lose is transferred to the rf wave on the circuit. These slower-moving electrons are subsequently accelerated toward the anode by the dc electric field, and their velocity is increased to the synchronous condition. The energy exchange cycle can then be repeated. Electrons moving in this phase of the rf field pattern transfer energy to the rf wave on the circuit while maintaining nearly constant kinetic energy. The energy transfer results from the loss of potential energy of the electrons as they move to the anode.

Electrons located in the alternate phase of the rf field pattern are accelerated by the rf field and move away from the anode. The intensity of the slow wave decreases exponentially with distance away from the slow-wave circuit so that the magnitude of this interaction decreases. The result is the transfer of dc electrical power to microwave power on the slow-wave circuit, with the phase-sorted electrons in the crossed-field interaction region providing the necessary coupling mechanism. Electron current thus flows to the anode only in the region of suitably phased rf electric fields.

The components of rf field which are perpendicular to the forward motion of the electrons exert forces which phase-lock the electron near the center of the pattern. These regions are called *spokes* because of the similarity, in a magnetron oscillator, to the spokes in a rotating wheel. The phase locking of the sorted space charge relative to the traveling rf wave on the slow-wave circuit reduces the effect of power-supply variations on the electron trajectories. The details of the electron trajectories are extremely complex and have been calculated only approximately, using sophisticated computer techniques.

It is an important fundamental of crossed-field interaction that very high electronic conversion efficiency can be obtained because the kinetic energy of the electrons lost as heat upon ultimate impact with the slow-wave circuit can be designed to be a small fraction of the total potential energy transferred from the power supply. The ideal electronic conversion efficiency is given by $\eta = 1 - V_0 / V$, where η is efficiency, V_0 is the synchronous voltage, and V is the cathode-to-anode voltage. Large ratios of V/V_0 lead to high efficiencies.

Many microwave devices have been investigated which use variations of this basic concept, but a majority have not proceeded beyond laboratory evaluation. Devices in practical use fall into two broad categories, which differ primarily in the method by which the electron current is delivered to the interaction region. These are illustrated schematically in Fig. 9-59. In one class, called the *injected-beam crossed-field tubes,* the electron stream is

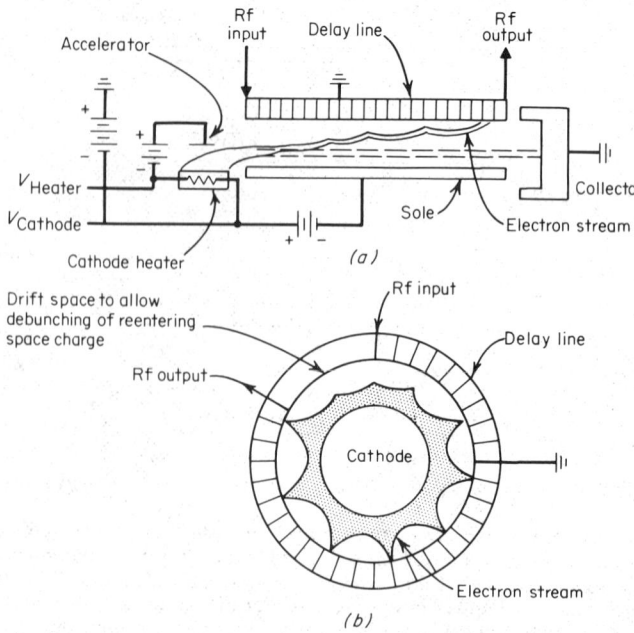

Fig. 9-59. Linear injected beam (*a*) and reentrant emitting-sole crossed-field amplifier (*b*).

produced by an electron gun located external to the interaction region in a manner somewhat similar to a traveling-wave tube. In the second class, the *emitting-sole tubes,* the electron current for interaction is produced directly within the interaction region by secondary electron emission, which results when some electrons are driven to the negative electrode and allowed to strike it. In the latter devices, this electrode is formed from a material capable of delivering copious secondary-emission electrons.

40. Slow-Wave Circuits for Crossed-Field Tubes. Electron current in crossed-field interaction moves toward the slow-wave circuit rather than through the circuit as in beam tubes. This leads to the use of open circuits that present an rf wave-guiding surface to the electron stream. Maximum energy conversion efficiency is usually obtained when the current is intercepted on the slow-wave circuit; so the structures must withstand the thermal stress associated with electron bombardment. Electronic interaction can occur using either forward-wave or backward-wave traveling-wave circuits, as well as with circuits supporting a standing wave. Examples of circuits suitable for use in forward-wave interaction are various *meander lines, helix-derived structures, bar* and *vane structures,* which are capacitively loaded by ground planes, and *capacitively-strapped bar circuits.*

A helix-coupled vane circuit and a ceramic-mounted meander line are shown in Fig. 9-60a. The most common backward-wave circuits are derivatives of the interdigital line and strapped-bar and vane circuits. Examples of a choke-supported interdigital line and a strapped-bar circuit are shown in Fig. 9-60b. Traveling-wave circuits are used mostly in amplifiers. Standing-wave circuits are resonant and used typically in magnetron oscillators. The most commonly used standing-wave circuits are composed of arrays of quarter-wave resonators that may or may not be strapped for improved oscillating-mode stability.

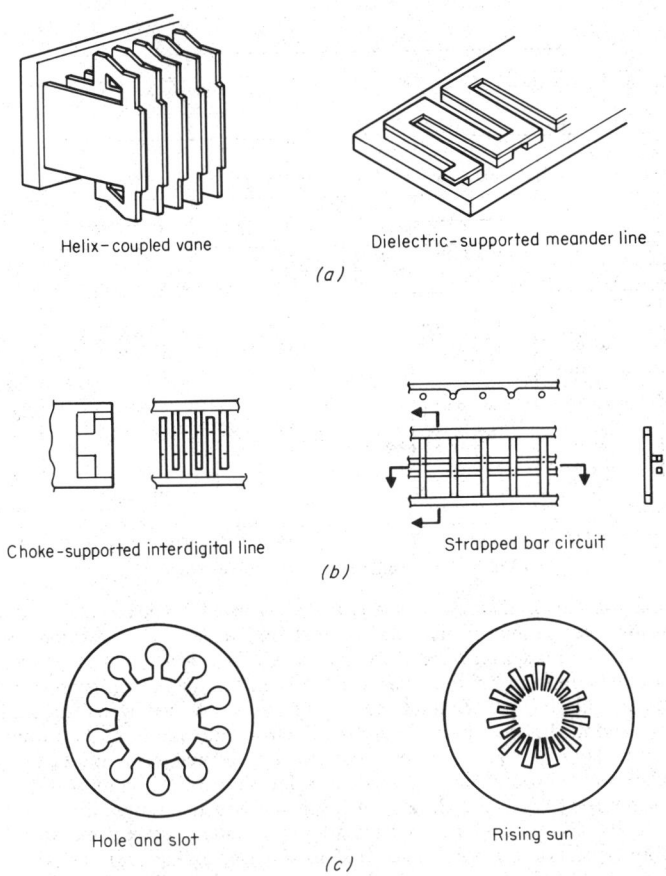

Helix-coupled vane   Dielectric-supported meander line

(a)

Choke-supported interdigital line   Strapped bar circuit

(b)

Hole and slot   Rising sun

(c)

Fig. 9-60. Slow-wave circuits for crossed-field tubes.

Variations of these circuits include *hole* and *slot resonators* and *rising-sun* anodes. Examples are shown in Fig. 9-60c. Cooling of vane structures for high average power is obtained by heat conduction along the vanes to the back wall of the anode to a heat sink which may be liquid- or forced-air-cooled. Bar structures are cooled by passage of liquid coolant through the tubular bars of the slow-wave circuit.

41. Crossed-Field Family Tree. The current crossed-field tube types can be classified in a family tree (Fig. 9-61), divided into two major divisions as *injected-beam* and *emitting-sole* tubes. Subdivisions relate to *linear* and *circular format*, *reentrant* and *nonreentrant* electron stream, and as to whether the rf-wave electronic interaction takes place with a *forward-traveling wave*, a *backward-traveling wave*, or a *standing wave* on the circuit.

Magnetron oscillators are single-port devices. Both the slow-wave circuit and the electron stream are reentrant; i.e., the circular geometry is always used. Traveling-wave crossed-field oscillators are single-port devices but use a nonreentrant electron stream. They use either the linear or circular format.

Injected-beam and emitting-sole amplifiers are two-port devices with rf input and output ports. They are fabricated in both linear and circular format. Linear tubes must use a nonreentrant electron stream. Some circular-format amplifiers use a reentrant electron stream and some do not. Both forward-wave and backward-wave amplifiers have been developed.

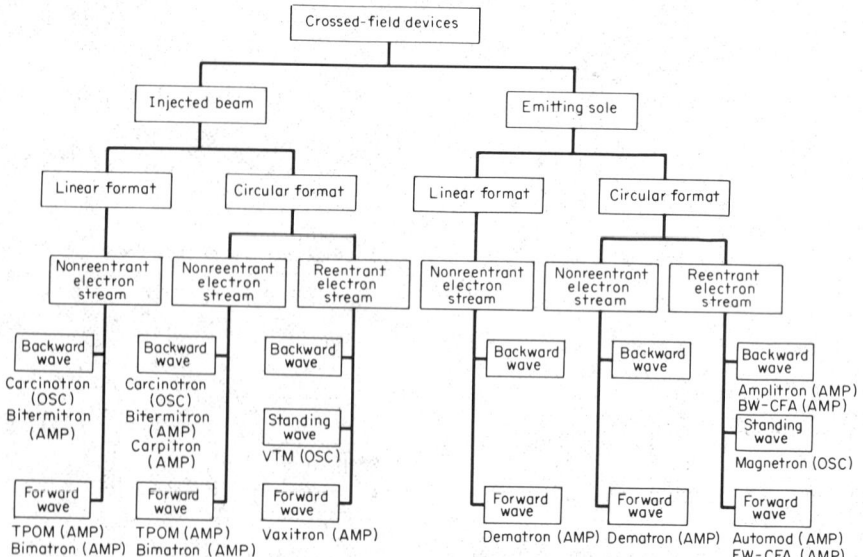

Fig. 9-61. Family tree of crossed-field tubes.

42. Crossed-Field Oscillators. *Conventional Magnetrons.* The conventional magnetron is an emitting-sole, circular-format, reentrant-stream device with electronic interaction between the circulating current and a π-mode, rf standing wave on the slow-wave circuit. Oscillation builds up from noise contained initially in thermionic-emission current from a heated cathode. During operation interaction current is obtained from a circulating hub of space charge supplied primarily by secondary electron emission from the cathode surface. This is illustrated in Fig. 9-62. Large peak currents are obtainable, permitting the generation of high peak power at lower voltages than are used for beam tubes of comparable peak power.

Pulsed magnetrons have been developed covering frequency ranges from a few hundred megahertz to 100 GHz. Peak power from a few kilowatts to several megawatts has been obtained with typical overall efficiencies of 30 to 40%, depending upon the power level and frequency range. Continuous-wave magnetrons have also been developed with power levels of a few hundred watts, in tunable tubes, at an efficiency of 30%. As much as 25 kW cw has been obtained for a 915-MHz fixed-frequency magnetron at efficiency greater than 70%. Figure 9-63 illustrates typical performance.

Pulsed magnetrons are used primarily in radar applications as sources of high peak power. Low-power pulsed magnetrons find applications as beacons. Magnetrons operate electrically as a simple diode, and pulsed modulation is obtained by applying a negative rectangular voltage pulse to the cathode with the anode at ground potential. Voltage values are less critical than for beam tubes, and line-type modulators are often used to supply pulsed electrical power. Tunable cw magnetrons are used in electronic countermeasure applications. Fixed-frequency magnetrons are used as microwave heating sources.

Mechanical tuning of conventional magnetrons is accomplished by moving capacitive tuners, near the anode straps or capacitive regions of the quarter-wave resonators, or by inserting symmetrical arrays of plungers into the inductive portions. Tuner motion is produced by a mechanical connection through flexible bellows in the vacuum wall. Tuning ranges of 10 to 12% bandwidth are obtained for pulsed tubes, and as much as 20% for cw tubes.

43. Coaxial Magnetrons. The frequency stability of conventional magnetrons is affected by variations in the microwave load impedance (frequency pulling) and by cathode current fluctuations (frequency pushing). When the mode control becomes marginal, the tube may occasionally fail to produce a pulse. The coaxial magnetron minimizes these effects by using

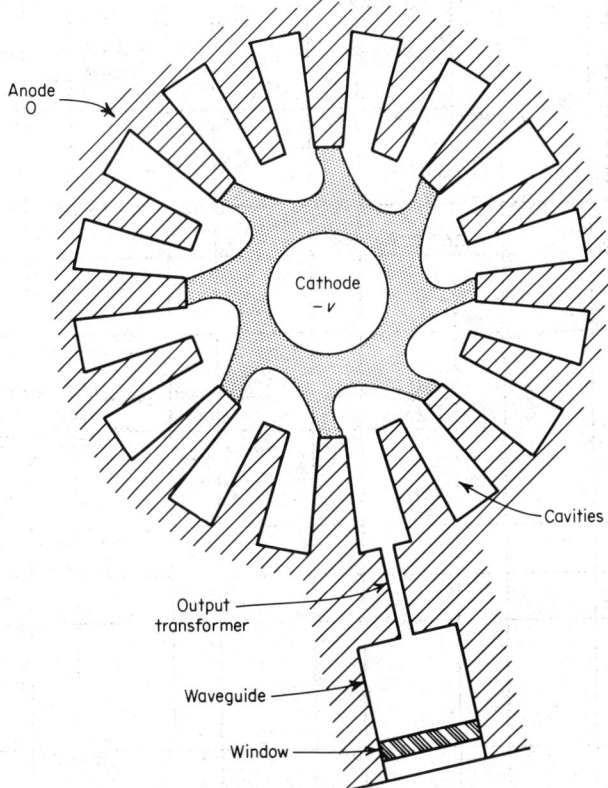

Fig. 9-62. Conventional magnetron structure.

the anode geometry shown in Fig. 9-64. Alternate cavities are slotted to provide coupling to a surrounding coaxial cavity. π-mode operation of the vane structure provides in-phase currents at the coupling slots which excite the TE_{011} circular electric coaxial mode. The unique rf field pattern of the circular electric mode permits effective damping of all other cavity modes with little effect on the TE_{011} mode, and oscillation in other cavity modes is thereby prevented. Additional resistive damping is used adjacent to the slots, but removed from the vanes to prevent oscillations in unwanted modes associated with rf energy stored in the vanes and slots that does not couple to the coaxial cavity.

The oscillation frequency is controlled by the combined vane system and resonant cavity. Sufficient energy is stored in the TE_{011} cavity to provide a marked frequency-stabilizing effect on the oscillation frequency. Hence the coaxial magnetron is much less subject to frequency pushing and pulling than conventional magnetrons, and it exhibits fewer missed pulses. Tunable versions of this tube type are tuned by a movable end plate in the coaxial cavity similar to a tunable coaxial wavemeter. This is illustrated in Fig. 9-65. The larger resonant volume for energy storage leads to a slower buildup time for oscillation than in conventional magnetrons. This causes greater statistical variation in the starting time for oscillation (leading-edge jitter).

UHF AND MICROWAVE DEVICES

These factors are compared in Table 9-10 for the SFD-349 coaxial magnetron, which was designed as an improved retrofit for the 7008 conventional magnetron. The operating efficiency of the SFD-349 was deliberately degraded to meet retrofit requirements. Typically, coaxial magnetrons operate with an efficiency of 40 to 50% or higher.

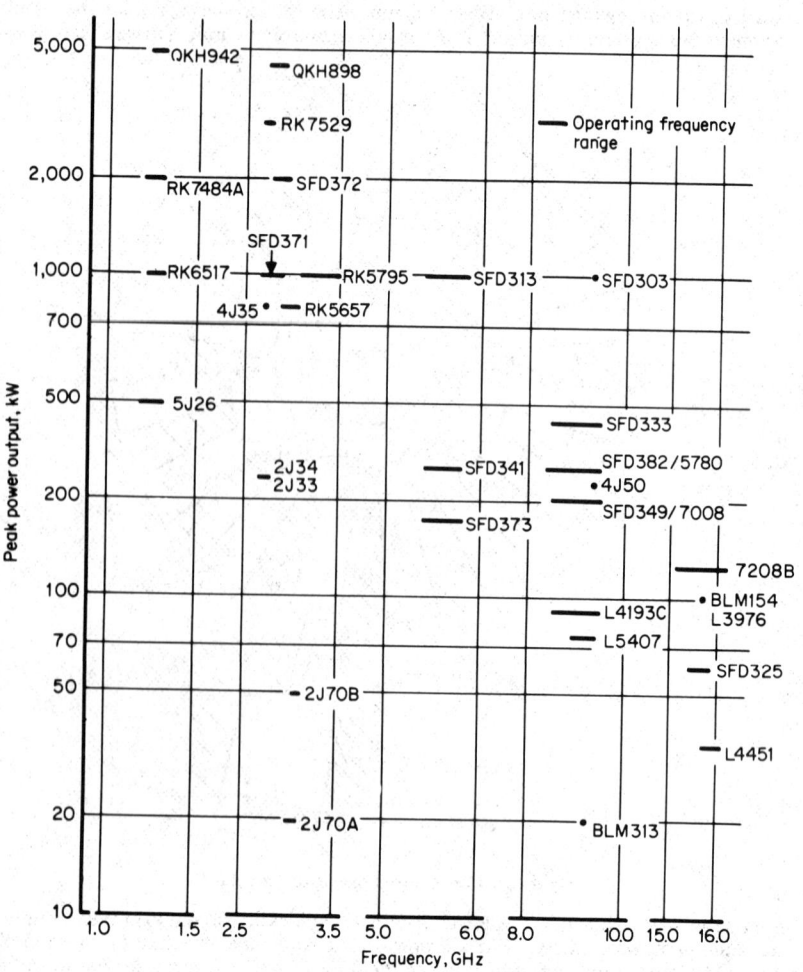

Fig. 9-63. Performance of typical conventional magnetrons.

Table 9-10. Comparison of the 7008 Magnetron and the SFD-349 Coaxial Magnetron

	7008	SFD-349
Efficiency	38%	38%
Leading-edge jitter, rms	1.2 ns	1.5 ns
Pushing factor: Specified	500 kHz/A	100 kHz/A
Typical	200 kHz/A	50 kHz/A
Pulling factor (VSWR 1.5)	15 MHz	5 MHz
Spectra side lobes	8–9 dB	12–13 dB
Missing pulses	1%	0.01%
Pulse-frequency jitter, rms	60 kHz	5 kHz
Life: Specified	500 h	1,250 h
Typical	700–800 h	3,000–3,500 h

Spurious Noise. The circulating space charge in the hub of both conventional and coaxial magnetrons contains wideband noise-frequency components that can couple to the output. In conventional magnetrons this spurious noise can couple directly to the output wave-guide. Spurious noise power measured in a 1-MHz bandwidth is typically greater than 40 to 50 dB below the carrier. The coaxial cavity in the coaxial magnetron provides some isolation between the spurious noise coupled to the vanes and the output waveguide. The spurious-noise power from coaxial magnetrons is typically 10 to 20 dB lower than conventional magnetrons of comparable peak power level.

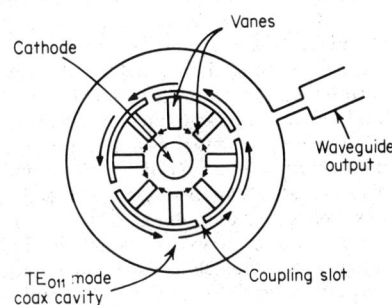

Fig. 9-64. Coaxial magnetron coupling.

44. Frequency-agile Magnetrons. To improve radar signal detection and electronic countermeasures, rapid frequency-changing signal sources have been developed. Frequency-agile conventional magnetrons are available with rapidly rotating capacitive tuners *(spin-tuned magnetrons)* or hydraulic-driven, *mechanically tuned* tubes. The operational advantages of the coaxial magnetron are preserved in frequency-agile dither-tuned and gyro-tuned coaxial magnetrons.

Dither-tuned magnetrons employ a mechanically tuned coaxial magnetron with an integral motor and *resolver* to provide high-speed, narrowband frequency-agile operation. Mechanical linkage between the rotating motor and the tuning plunger provides approximately sinusoidal tuning of the magnetron frequency. Mechanical limitations imposed by acceleration forces determine the attainable tuning range and tuning rates. A voltage output from the resolver is made proportional to the magnetron frequency and is used to adjust the receiver local oscillator to track the rapidly tuned frequency of the magnetron. X-band tubes, 200

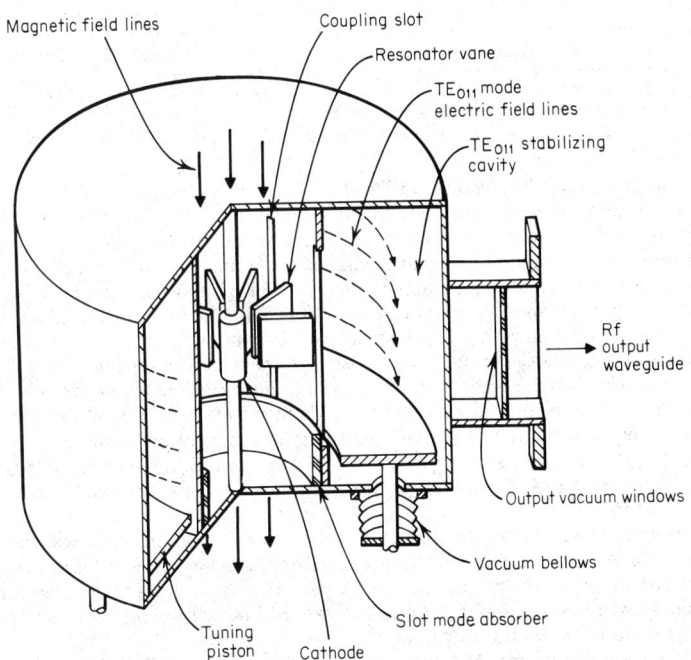

Fig. 9-65. Schematic of coaxial magnetron.

kW, with narrowband dither-tuned frequency ranges of 30 to 50 MHz, are tuned at rates of 200 Hz. Wider-band frequency excursions of 250 to 500 MHz are dithered at rates of 25 to 40 Hz.

Gyro-tuned coaxial magnetrons employ several rotating dielectric ceramic paddles in the stabilizing coaxial cavity, which cause frequency variation as they are rotated in a plane normal to the rf electric field of the TE_{011} mode. The anode vane system of the tube is surrounded by a ceramic cylinder bonded to the ends of the coaxial cavity to form the vacuum wall for the electronic interaction region. The stabilizing cavity, containing the tuning, is outside of this vacuum wall and is pressurized with sulfur hexafluoride, to inhibit arcing or corona caused by high rf fields. The tuner drive motor and frequency readout generator are also located within the pressurized section of the magnetron. The symmetry and inherently low rotational mass of the dielectric paddles result in a mechanism in which tuning speed and rf frequency excursion are essentially independent. It is therefore possible to attain higher tuning rates and relatively broader frequency excursion simultaneously than is achieved currently with dither tuning. Ku-band, 60 kW peak power, gyro-tuned magnetrons obtain frequency excursions of 300 MHz at 200-Hz tuning rates.

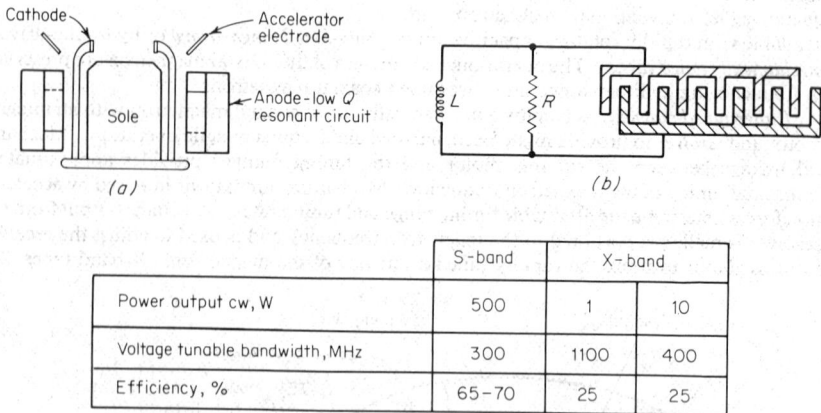

	S-band	X-band	
Power output cw, W	500	1	10
Voltage tunable bandwidth, MHz	300	1100	400
Efficiency, %	65–70	25	25

Fig. 9-66. Schematic of voltage-tunable magnetron. (*a*) Structure; (*b*) equivalent circuit.

45. Voltage-tunable Magnetrons (VTMs). The voltage-tunable magnetron uses a circular-format, reentrant-stream injected beam which interacts with a standing wave on a low-Q resonant structure. A hollow electron beam is launched axially from an electron gun into the interaction region. The hollow beam begins to rotate in the annular region between the anode and sole as it enters the interaction region. This is shown schematically in Fig. 9-66. The anode consists of an interdigital line which capacitively loads the resonant cavity in a manner similar to the grids in a Klystron cavity. Oscillating energy in the resonant cavity provides electric fields of opposite polarity between the fingers of the interdigital line, which creates a π-mode standing-wave pattern. Varying the voltage between the anode and sole changes the circumferential velocity of the circulating electron stream by changing the ratio E/B. The oscillation frequency is determined by the circumferential velocity of the electrons interacting with the π-mode field pattern, and voltage tuning is obtained. Voltage-tuning ranges of more than 10% bandwidth have been obtained. Some achieved power levels are indicated in Fig. 9-66.

Low-power VTMs (milliwatts) have low-noise power output that is at least 80 dB/MHz below the carrier power level. These tubes find application as receiver local oscillators and as signal sources in swept frequency generators. High-power VTMs (watts) are used in electronic countermeasure applications to generate FM noise by applying a noise-modulated voltage between the sole and anode.

46. M-Carcinotrons and M-backward-Wave Oscillators (MBWOs). M-carcinotron and M-backward-wave oscillators use a circular- or linear-format, nonreentrant injected beam which interacts with a traveling wave on a dispersive backward-wave circuit. Voltage-tunable oscillations are generated in a manner similar to that of linear-beam backward-wave oscillators. Voltage tuning is obtained by varying the anode-to-sole voltage. Tubes have been developed in frequency ranges between P- and Ku-band. Continuous-wave power greater than 100 W has been generated at X-band with 30% voltage-tunable frequency range. Several hundred watts of cw power has been generated at lower frequencies with comparable tuning range. The primary use for these tubes is for noise-jamming signal sources in electronic countermeasure equipment.

47. Crossed-Field Amplifiers (CFAs). *Injected-Beam Types.* Nonreentrant injected-beam crossed-field amplifiers have been developed using linear- and circular-format geometry. Current for interaction is obtained from a heated, emissive cathode located external to the interaction region as in a TWT (see Pars. **9-29** to **9-38**). The power-supply requirements for these tubes are greater than for emitting-sole oscillators and amplifiers, both in the number of voltages and stability required. An accelerator electrode, positive with respect to the cathode, draws current from the cathode surface. Magnetic forces deflect the current toward the interaction region over the surface of the cathode.

The presence of nonuniform space charge between the accelerator electrode and the cathode surface introduces nonuniformities in the intervening electric field. This can cause nonuniform cathode emission densities and perturbation of the electron launch trajectories. Noise generation and beam instabilities can result from an inadequate electron gun. Variation of the voltage between anode and sole causes variation in beam velocity. Low-current beams in low-power, high-gain amplifiers require power-supply stability comparable with TWTs to minimize output signal power and phase variation. Low-gain, high-power, high-current amplifiers are less sensitive to beam voltage variation by a factor of 3 to 5.

Backward-wave amplifiers *(Bitermitron, Carpitron)* are voltage-tunable, narrow-band-width devices that are essentially injected frequency-locked M-carcinotrons and MBWOs. The achievable power levels and frequency ranges are comparable. Amplification of 15 to 20 dB has been obtained at 30 to 35% efficiency. Noise power measured far from the carrier in Carpitron tubes is more than 65 dB/MHZ below the carrier.

Extensive development efforts have been devoted to nonreentrant forward-wave amplifiers *(TPOM, Bimatron)* for both cw and pulsed application. Continuous-wave amplifiers with a few hundred watts output have been developed with gain of 20 to 30 dB and with efficiencies of 20 to 35%. Proper control of the electron stream at large values of gain requires the beam to be physically close to the anode at synchronism. Operation at low values of V/V_0 (3 to 6), together with increased insertion loss for tubes with greater circuit length for greater gain, leads to lower efficiency values. Like TWTs, the attainable bandwidth is dependent upon the electron-beam optics and the dispersiveness of the slow-wave circuit. Half-octave bandwidth has been obtained at constant-voltage settings, and full-octave bandwidth has been obtained with adjustment of the anode-to-sole voltage. Stable operation at gain in excess of 20 dB requires the utilization of circuit severs or distributed attenuation as in TWTs. Useful gain in excess of 30 dB is difficult because of excessive noise buildup in the electron stream.

A major source of noise comes from modulation of the potential minimum in front of the cathode emissive surface. Closely spaced grids located in front of the cathode have been employed to minimize this effect by electrostatic shielding of the potential minimum region. The grids are sets of parallel wires perpendicular to the magnetic field. The magnetic field deflection of the electrons minimizes beam interception. Grids are also used to modulate the emitted cathode current, permitting a cw amplifier to be pulsed to higher-peak-power output while maintaining the same average power level. Dual-mode operation is complicated by the necessary change in electron-beam optics. Peak-to-cw power ratios as great as 10:1 may be feasible by using multiple-element depressed collectors to preserve efficiency in the two operational modes.

Narrowband (10%), pulsed, high-peak-power (5 MW) injected-beam amplifiers, which operate with efficiencies greater than 50%, have been developed for use in radar application. These tubes have gain of 11 to 15 dB. The lower gain values (greater input signal level) permit the use of less critical beam optics, shorter slow-wave circuits, and greater ratios of V/V_0.

Reentrant-stream, forward-wave injected-beam amplifiers *(Vaxitron)* use an interaction geometry similar to the voltage-tunable magnetron, with the low-Q resonant structure replaced by a broadband, two-port, nondispersive forward-wave circuit. These tubes have not yet been fully developed, but experimental versions have demonstrated 30-dB gain with efficiencies in excess of 50% when constructed with multiple-element depressed collectors.

By 1972 no type of injected-beam CFA had been extensively used in the United States.

48. Emitting-Sole Crossed-Field Amplifiers. Electron current for emitting-sole crossed-field amplifiers is obtained from the sole electrode in the interaction space by electron-beam back bombardment, as in the magnetron oscillator. Unlike the magnetron, thermionic emission is not required to initiate current flow. Current flow can be started in an emitting-sole CFA by the admission of an rf signal to the input of the slow-wave circuit when the proper magnetic field and anode-cathode voltage are present in the interaction region. Amplifiers with rf-induced current flow are called *cold*-cathode amplifiers, regardless of cathode temperature, provided there is no thermionic emission from the cathode. In the absence of an rf input signal, these amplifiers remain quiescent even with full operational voltage applied. The details of the starting mechanism of rf-induced current flow are not fully understood, but the phenomenon is reliable in a properly designed amplifier.

Radio-frequency-induced current flow permits several modulation techniques for pulsed emitting-sole CFAs. These include cathode-pulsed CFAs, dc-operated CFAs with combination of dc voltage and a pulsed turn-off voltage, and dc-operated CFAs with only dc voltages applied. Cathode-pulsed amplifiers, like magnetron oscillators, use pulse modulators to supply the required dc electrical power during amplification. Direct-current-operated amplifiers obtain electrical input power for amplification from a dc power supply. Electron current flow is initiated by an rf input signal and is terminated at the end of the rf input signal either by a voltage pulse or a dc bias voltage applied to a quench electrode.

Cold-cathode starting for cathode-pulsed amplifiers is assured by correct temporal alignment so that the rf input pulse bridges the cathode voltage pulse. The rf signal is present on the slow-wave circuit as the applied voltage pulse increases to synchronous value. Current flow is initiated, and amplification occurs during the voltage pulse and ceases upon removal.

Both forward- and backward-wave cathode-pulsed CFAs are available. They use the circular-format, reentrant-stream geometry. This is illustrated in Fig. 9-67a. A reentrant electron stream is advantageous because electrons which have delivered only part of their available potential energy to the circuit wave as they exit the interaction region can reenter for further participation, thereby leading to higher electronic conversion efficiency. (Overall amplifier efficiencies of 45 to 50% or more are common.) The exiting phase-sorted electrons contain rf modulation. Some reentrant-stream amplifiers use a circuit geometry with the rf input and output ports spatially separated by a sufficiently long circuit-free region (called the *drift space*) so that internal space-charge forces cause dispersal of the electron spokes. This removes the rf modulation from the reentrant stream while preserving the reentrant-stream

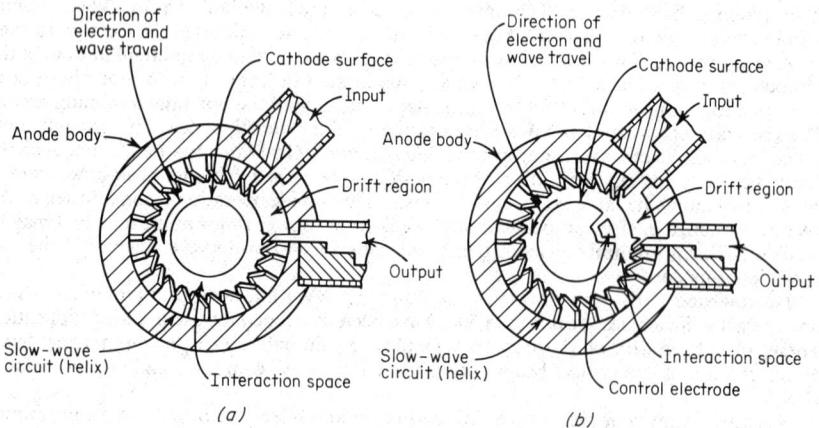

Fig. 9-67. Diagrams of reentrant-stream crossed-field amplifiers. (*a*) Cathode-pulsed; (*b*) with control electrode turn-off.

efficiency advantage. Other reentrant backward-wave amplifiers *(Amplitrons)* employ the modulated electron stream to create a regenerative amplifier. A minimal drift space is used, so that the electron spokes reenter the interaction region before the modulation is dispersed. By suitable design of the shorter drift-space dimensions, the modulated stream reenters with positive phase to enhance the interaction.

Greater electronic conversion efficiency (amplifier efficiency in excess of 70% has been obtained) can be obtained at the expense of lower gain-bandwidth product than can be obtained with amplifiers which remove the reentrant modulation. The use of regenerative amplification is not feasible with a forward-wave amplifier because, at a fixed operating voltage, the simultaneous regenerative gain at frequencies other than the drive signal could lead to unwanted auxiliary oscillations or selective peaks in spurious-noise output power. This is avoided with a backward-wave amplifier because the dispersive slow-wave circuits require different voltages to obtain adequate amplification at separated frequencies.

Spurious Noise. Dispersal by space-charge forces of the non-phase-locked electrons in a drifting spoke can lead to a rapid buildup of broadband spurious-noise components in the reentering electron stream. This is prevented from becoming severe by use of a sufficiently large rf input signal to lock out noise-signal growth. CFA power output as a function of rf drive signal is illustrated in Fig. 9-68. Reduction of the rf input signal leads to a reduced amplifier output signal at higher values of gain, but is accompanied by an increased relative amount of broadband noise power output. (Reentrant emitting-sole CFAs with terminated input have been used for high-efficiency broadband noise generators.) Ade-

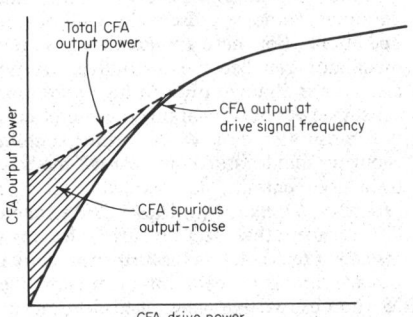

Fig. 9-68. Emitting-sole crossed-field power output.

quate lockout of the noise power is obtained at reduced signal gain when the amplifier is driven well into saturation. The rapid growth of noise at small drive signals precludes the use of distributed attenuation and of circuit severs for emitting-sole amplifiers.

Relatively short slow-wave circuits are used with the minimum attainable insertion loss between the rf input and output connections. These amplifiers are called *transparent tubes.* Reflected signals from a load mismatch travel backward through the amplifier to the rf input, where they can be reflected, possibly leading to oscillation. Judicious use of ferrite isolators and circulators at the rf input and/or output minimizes this effect. The overall stable gain of emitting-sole CFAs is limited to 20 dB or less (typically 13 to 15 dB) because of the requirement for a sufficiently large rf input signal for adequate lockout of spurious-noise power and the need to limit gain to avoid oscillations caused by multiply reflected signals. Transparency is often used to advantage in radar systems employing the final amplifier in the transmitter chain in a nonoperating feed-through mode to provide coarse programming of the output power level.

Bandwidth Characteristics. Cathode-pulsed forward-wave amplifiers offer 10 to 15% instantaneous bandwidth at a fixed value of pulsed voltage. Backward-wave amplifiers provide only 1 to 2% instantaneous bandwidth under comparable conditions, but can accommodate 10% bandwidth by adjustment of cathode voltage. The static impedance of both tube types varies as a function of frequency. The constant voltage vs. frequency characteristics of forward-wave amplifiers is readily accommodated by a hard-tube cathode modulator, providing nearly constant power output across the frequency band of the amplifier. Constant power output vs. frequency for a variable-voltage backward-wave amplifier is nearly achieved with a constant-current modulator. For restricted bandwidth (4 to 6%) this condition is approximated by using a line-type modulator.

Modulation Requirements. A simplification in modulator requirements is obtained with a broadband, dc-operated, rf-triggered CFA. The termination of the rf input signal after rf turn-on leaves uncontrolled circulating space charge that can generate cw spurious output signals of magnitude as large as half the amplified signal output. To avoid this, a *control electrode* isolated from the cathode is mounted as part of the cathode structure and is located in the drift space. This location minimizes interference with amplifier performance. This is illustrated in Fig. 9-67*b*. This electrode is pulsed positive with respect to the cathode, coincident in time with removal of the rf input signal. The circulating space charge in the interaction space is collected upon the control electrode, and the cathode current flow is terminated. Modulator requirements are simplified because the pulse voltage required for the turn-off electrode is typically one-quarter to two-thirds of the anode-cathode voltage, and the peak collected current is less than one-half of the peak cathode current flow during amplification. The duration of the collection time for the circulating current is approximately one transit time for electron flow around the interaction region. Typically, this is a small fraction of the time duration of the amplified signal. Consequently, the modulator energy required for the control electrode per amplifier pulse is much less than that required from the modulator for full-cathode-pulsed amplifiers.

Modern radar systems often employ high-pulse repetition frequencies, very-short-pulse durations, complex-pulse codes with possible variable-pulse widths, and/or burst modes of operation. For these applications even the simpler requirements for the control electrode modulator can become restrictive. A typical interelectrode capacitance for a turn-off electrode is 50 to 70 pF. At high repetition rates or very short amplifier pulses the energy consumed per pulse charging the interelectrode capacity becomes a significant factor affecting the transmitter design. A further simplification is obtained by employing an amplifier geometry that leads to complete self-modulation by the rf input signal. This is accomplished by a nonreentrant, dc-operated, emitting-sole CFA of either linear- or circular-format geometry *(Dematron)*. These CFAs function similarly to reentrant-stream, emitting-sole CFAs, except that electron current leaving the interaction region passes to a beam collector instead of recirculating for further participation. Obtainable gross-performance characteristics are similar to reentrant-stream amplifiers, but the overall efficiency is limited to about 35% because partially interacted electron current exiting from the interaction region is usually collected at full beam voltage. Complete rf-controlled modulation of the beam current is provided because removal of the rf input signal permits the electron current to pass unidirectionally through the interaction region to exit for removal from further participation by the beam collector.

Self-Modulation. Reentrant-stream amplifiers that provide complete self-modulation by the rf input signal have been demonstrated in the laboratory. These dc-operated amplifiers *(AutoMod)* use an isolated turn-off electrode in the cathode structure located within the interaction region rather than in the drift space. A dc bias voltage, positive with respect to the cathode, is applied to the turn-off electrode. This electrode behaves as a beam collector similarly to a normal control electrode located in the drift space when a positive turn-off voltage pulse is applied. During rf turn-on and normal amplification the fringing rf electric fields from the slow-wave circuit sufficiently perturb the circulating electron trajectories in the vicinity of the dc-biased electrode so that enough electrons pass through the collector region to provide adequate reentrancy for normal amplification. Upon removal of the rf input signal the bias electrode becomes totally effective for collecting circulating current, and the amplifier turns off automatically. Complete rf modulation is obtained at the expense of some power loss due to partial beam collection by the bias electrode during normal amplification. Overall efficiency of 40 to 45% has been demonstrated, which is midway between that achievable by Dematrons and control electrode modulated amplifiers.

Self-emitting Cathodes. A variety of materials are used for secondary-electron-emission cathodes in rf-triggered amplifiers. The selection is based on the amplifier drive signal level, peak power output, and the intended operating voltage for the tube. Materials used include pure metals, such as aluminum, beryllium, and platinum, as well as a variety of composite materials, such as dispenser cathodes and cermets. Dispenser cathodes and pure-platinum cathodes are suitable for drive signal levels in excess of 10 kW. Amplifiers with drive signals from a few hundred watts to 10 kW are better accommodated with metals supporting oxide surface layers such as aluminum and beryllium. Oxide layers are susceptible to erosion under electron-beam bombardment, and some CFAs employ a low-level background pressure of pure oxygen supplied from a suitable reservoir to rejuvenate and extend the active cathode life.

Frequency Range. Emitting-sole amplifiers have been developed at frequency ranges extending from vhf to Ku-band, with experimental models at lower and higher frequencies.

Examples of peak power levels available include 100 kW at Ku-band, 1 MW at X-band, and 3 MW at S-band. Average power levels vary from 200 W at Ku-band to several kilowatts at lower frequencies. Laboratory models have demonstrated as much as 400 kW of cw power at S-band. The performance of typical cataloged amplifiers is shown in Fig. 9-69.

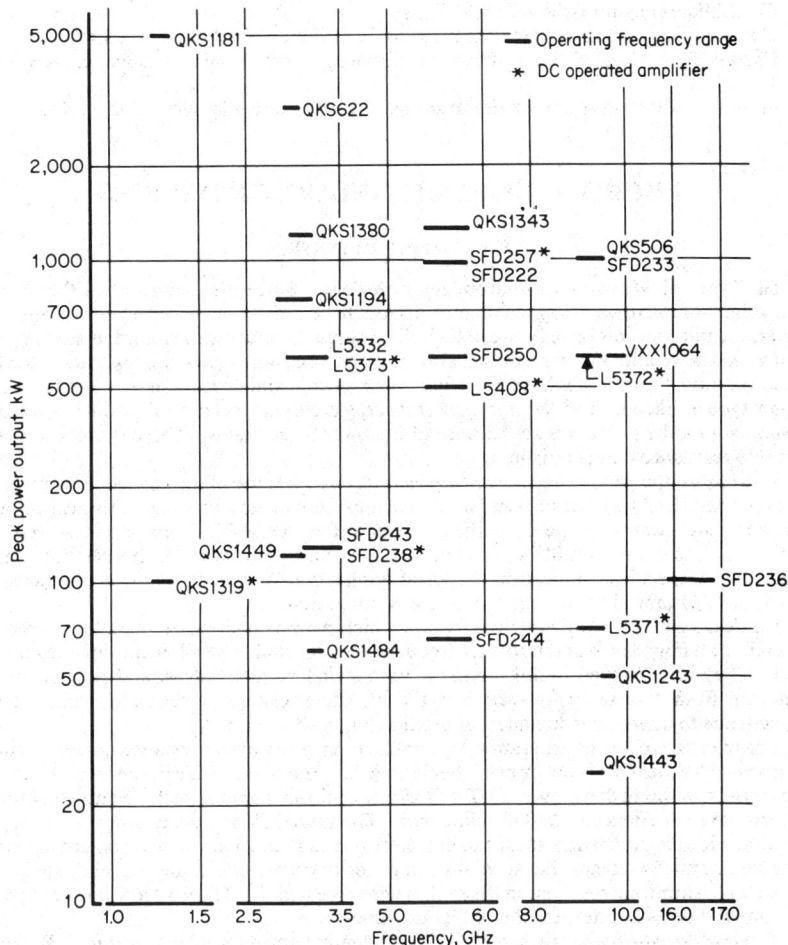

Fig. 9-69. Performance of cataloged emitting-sole crossed-field amplifiers.

Noise Power. Noise-power output from a reentrant emitting-sole CFA with a space-charge-dispersing drift space measured in a 1-MHz bandwidth far from the carrier is typically greater than 30 dB below the carrier power level. Noise levels greater than 40 dB below the carrier are not uncommon. Noise power close to the carrier frequency tends to be less, due to the phase-lockout phenomenon associated with high drive-signal levels. The maximum power density of the main lobe of the spectrum of a pulsed CFA is typically greater than 40 dB above the radiated interspectral line noise when both are normalized to a 1-MHz bandwidth. Ratios of more than 50 dB are not uncommon. Phase locking of the space charge by the drive signal also minimizes phase variation due to the voltage change and drive-signal variation. Saturated amplifiers with 13 to 15 dB gain have output phase variations of 3 to 8° for a 1% change in anode-cathode voltage. A comparable phase change occurs with a 1-dB variation in drive-signal level.

Applications.   The primary use for emitting-sole CFAs is for transmitter tubes in coherent radars.   CFAs have been used in pulse compression radars and pulse-coded and phased-coded radars, as well as phased-array radars.   Light weight, low-voltage cathode-pulsed amplifiers are attractive for airborne applications.   High-power dc-operated CFAs are used in ground-based radars.

49. Bibliography on Crossed-Field Tubes

COLLINS, G. B.   "Microwave Magnetrons," McGraw-Hill, New York, 1948.

OKRESS, E.   "Crossed-Field Microwave Devices," Vols. 1 and 2, Academic, New York, 1961.

———   "Microwave Power Engineering," Vol. 1, Academic, New York, 1968.

MICROWAVE SEMICONDUCTOR DEVICES

BY JOSEPH FEINSTEIN

50. Types of Microwave Semiconductor Devices.   Since the middle 1950s the number and variety of microwave semiconductor devices have greatly increased as new techniques, materials, and concepts have been applied.   The oldest structure is the tungsten-silicon *point-contact diode,* employing a metal-whisker contact, used for signal mixing and detection. More recently, these diodes have been fabricated by epitaxial deposition of a thin layer of *p*-on-*p*$^+$ type of silicon.   The *Schottky barrier diode,* a rectifying metal-semiconductor junction, is supplanting the point contact because of its lower noise figure.   These devices display a variable-resistance characteristic.

In contrast, the *variable-reactance (varactor) diode* makes use of the change in capacitance of a reverse-biased *pn* junction as a function of applied voltage.   Physically, this capacitance change results from widening the depletion layer as the reverse-bias voltage is increased.   By controlling the doping profile at the junction, the functional forms of this relation can be tailored to a specified application.   Typical applications of varactor diodes are harmonic generation, parametric amplification, and electronic tuning.

pin diodes employ a wide intrinsic region which permits high-power-handling capability and offers an impedance at microwave frequencies controllable by a lower frequency (or dc) bias.   They have proved useful for microwave switches, modulators, and protectors.   In changing from reverse to forward bias the *pin* diode changes electrically from a small capacitance to a large conductance, approximating a short circuit.

For microwave power generation or amplification, a negative-resistance characteristic at microwave frequencies is necessary.   Beginning with the *tunnel diode* in the early 1960s and progressing to the higher-power *IMPATT diodes* and *Gunn diodes,* such negative-resistance devices have experienced rapid development.   The tunnel diode utilizes a heavily doped *pn* junction which is sufficiently abrupt so that electrons can tunnel through the potential barrier near zero applied voltage.   Because this is a majority-carrier effect, the tunnel diode is very fast-acting, permitting response in the millimeter-wave region.   The very low power at which the tunnel diode saturates has limited its usefulness.

The *transferred-electron oscillator* (TEO), originally named for its discoverer, J. B. Gunn, depends on a specific form of quantum-mechanical band structure for its negative resistance. This band structure is found in gallium arsenide, the semiconductor material generally associated with this class of device, and in a few other III–V compounds.   *Gunn* oscillators have power output in the tens to hundreds of milliwatts at low dc operating voltage (9 to 28 V).   They have a wide range of tunability and reasonably low AM and FM noise.

The IMPATT (*imp*act *a*valanche *tr*ansit *ti*me) diode owes its negative resistance to the classical effects of phase shift introduced by the time lag between maximum field and maximum avalanche current multiplication and by the transit time through the device of this current.   These effects can occur in all semiconductor *pn* junctions under sufficient reverse bias to initiate breakdown.   While IMPATT diodes were originally developed as silicon devices (still the predominant type), gallium arsenide IMPATTs have recently come into prominence because of their higher power (several watts) and superior noise characteristics. IMPATT diodes find applications in point-to-point microwave communications transmitters and as pumps for parametric amplification.

Bipolar transistors have penetrated the microwave region through the refinement of fabrication techniques based on the planar photolithography technology. By reducing the emitter width and base thickness to micron dimensions (and by paralleling stripes of structure to maintain power capability), transit times and charging times (resistance-capacitance product) can be kept low enough to provide useful devices to frequencies of about 5 GHz. Such microwave transistors have displaced other devices for low-noise receiver stages and for amplifiers of moderate power.

Field-effect transistors (FETs) also have the capability of reaching microwave frequencies, provided sufficiently precise fabrication techniques are employed. To date the maximum frequency (10 GHz) has been attained by utilizing the higher mobility of GaAs, with the lower charging time of a Schottky barrier gate. Because of the low-power limitation of FETs, their application is limited to low-noise receiver input stages.

The *ruby maser* (acronym for *m*icrowave *a*mplification through *s*timulated *e*mission of *r*adiation), while historically very important in demonstrating the feasibility of the inverted population principle, has been replaced by the *cooled parametric amplifier* in applications requiring ultralow-noise reception. This obsolescence has been a consequence of the low power saturation and narrow bandwidth of the maser, together with the liquid-helium temperature necessary for its operation. It is used only in a few radio astronomy installations.

Acoustic-wave amplifiers came into prominence in electronics through the discovery that amplification could be obtained by coupling the charge carriers in a semiconductor to an acoustic wave propagating in a piezoelectric material. Although a great deal of experimental work continues on such amplifiers, the practical applications of acoustic waves have been in passive devices utilizing the slow propagation velocity (relative to electromagnetic waves) and compressed wavelengths of acoustic waves. *Microwave delay lines* and *signal-processing devices* making use of these properties have been developed.

51. Microwave Mixer and Detector Diodes.   Crystal rectifiers used as mixers have good noise performance in comparison with thermionic tubes at microwave frequencies. Noise figures lower than 6 dB are now possible at Ku-band using Schottky barrier diodes. These improvements have resulted from better control of semiconductor purity, epitaxial material to achieve low series resistance, and photolithographic techniques to achieve small-area Schottky diodes.

Mixing is defined as the conversion of a low-power-level signal from one frequency to another by combining it with a higher-power (local-oscillator) signal in a nonlinear device. In general, mixing produces a large number of sum and difference frequencies. Usually, the difference frequency between signal and local oscillator (the intermediate frequency) is of interest and is at a low power level.

In *microwave detection,* solid-state devices are used as nonlinear elements to accomplish direct rectification of an applied rf signal. The sensitivity of a detector is usually much less than that of a superheterodyne receiver, but the detector circuit is simple and easy to adjust. Recently this sensitivity has been improved considerably by the use of a wide-band low-noise microwave preamplifier preceding the detector.

Tunnel diodes, back diodes, point-contact diodes, and Schottky barrier diodes are majority-carrier devices that have nonlinear resistive characteristics and are useful for mixing and detecting. Because of greater susceptibility to rf burnout, circuit complications, and fabrication difficulties, tunnel and back diodes have not found as wide acceptance as mixers and detectors at microwave frequencies. The point-contact (pressure contact between metal and semiconductor) and Schottky barrier diodes (formed by deposition of metal on semiconductor surface) are the primary mixer and detector microwave devices. However, back diodes are used in selected applications such as broadband low-level detectors and Doppler mixers.

Point contact, Schottky barrier, and back diodes are used as mixers and detectors from uhf to millimeter frequencies. The construction and current-voltage characteristics of these diodes are shown in Fig. 9-70. The point contact is fabricated by a metal whisker forming a rectifying junction in contact with the semiconductor. The Schottky is formed generally by an evaporated metal contact, and the back diode by an alloyed junction.

52. Point Contact Diodes.   The point contact diode is the oldest structure. Until 1965 point contact diodes were fabricated utilizing moderately low resistivity material with the rectifying contact established by contacting the semiconductor surface with a metal whisker (normally, tungsten for silicon and phosphorus bronze for germanium and GaAs).

Since that time the semiconductor epitaxial deposition has been applied to point contact diodes (as well as to Schottky diodes) to maximize the frequency cutoff (minimize the $R_s C_j$ product). This is a significant consideration since conversion loss is directly proportional to the product of diode series resistance R_s and junction capacitance C_j, as shown in Fig. 9-71. Further, the conversion loss L can be shown to be related to semiconductor properties as

$$L \approx R_s C_j \approx \frac{W \varepsilon^{\frac{1}{2}}}{N^{\frac{1}{2}} \mu}$$

where ε is the semiconductor dielectric constant, and W, N, and μ are the epitaxial-layer thickness, carrier concentration, and mobility, respectively. Present-day silicon point contact diodes are mainly fabricated from p-on-p^+ epitaxial silicon (p layer 0.25 to 0.5 μm thick).

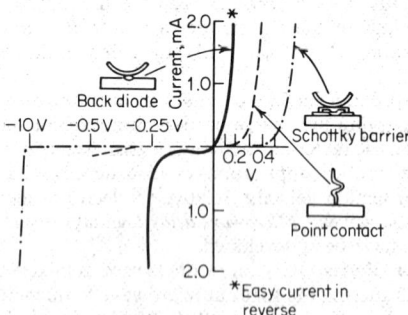

Fig. 9-70. Current-voltage characteristics of back, point-contact, and Schottky barrier diodes. *(Proc. IEEE, August, 1971.)*

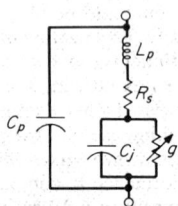

Fig. 9-71. Equivalent circuit of varistor diode.

53. Schottky Barrier Diodes. The Schottky barrier diode is a rectifying metal-semiconductor junction formed by plating, evaporating, or sputtering a variety of metals on n- or p-type semiconductor materials. Schottky diodes are fabricated from n-on-n^+ epitaxial material (n layer 0.5 to 1.0 μm thick), where the p and n layers are optimized in thickness and carrier concentration for minimum conversion loss and maximum rf burnout or power-handling capability. Generally, n-type silicon and n-type GaAs are used. Due to their higher cutoff frequency, GaAs devices are preferred in applications above X-band frequencies. This results from the higher mobility of electrons in GaAs than in silicon. Although in practice this advantage is not as significant as predicted, a conversion-loss improvement of 0.5 dB at Ku-band is readily obtainable with GaAs, compared with silicon.

Schottky diodes are fabricated by a planar technique. A SiO_2 layer (10,000 Å thick) is thermally grown or deposited on the semiconductor wafer, and windows are etched in the SiO_2 by photolithography techniques. Schottky junctions are formed by evaporation, sputtering, or plating techniques. Metal on the oxide is removed by a second photo step. Junction diameters as small as 5 μm are made by this technique.

54. Mixer-Diode Parameters. Figure 9-72 gives the noise figure for a variety of mixer diodes as a function of local oscillator power at 16 GHz. Below this frequency, silicon Schottky diodes exhibiting a minimum noise figure of 5.5 dB are generally used, while GaAs is preferred at higher frequencies.

The *tangential signal sensitivity* (TSS) is the most widely used criterion for detector performance. It indicates the ability of the detector to detect a signal against a noise background and also includes the noise properties of the diode and video amplifier. Figure 9-73 gives the TSS as a function of microwave frequency for good-quality detector diodes. This quantity varies with the square root of the amplifier bandwidth, since square-law response is obtained at low signal levels. As a rough basis of comparison, the TSS rating corresponds to a signal-to-noise ratio of about 2.5.

Typical mixer- and detector-diode mounts are shown in Fig. 9-74. These are designed to be compatible with a particular type of microwave circuitry (i.e., waveguide, coaxial, or strip-line). The encapsulations shown in Fig. 9-74a and b are generally used up to about 12 GHz. The coaxial package in Fig. 9-74c permits operation up to about 30 GHz, while the waveguide wafer in Fig. 9-74d is employed at millimeter wavelengths. Diodes for strip-line circuitry are generally unencapsulated, employing oxide protection and gold beam leads for attachment.

55. Varactor Diodes. The active element of a varactor diode consists of a semiconductor

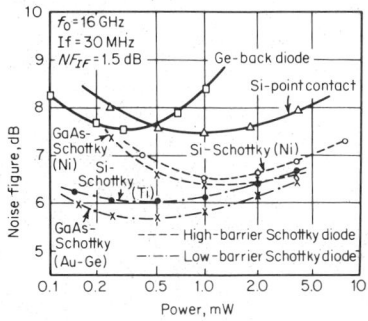

Fig. 9-72. Noise figure vs. local oscillator power for point contact diodes and n-type Schottky barrier diodes.

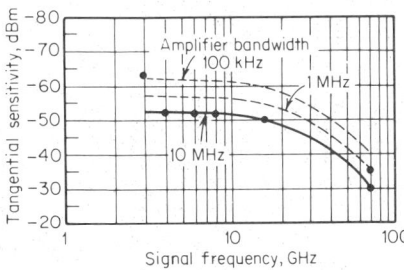

Fig. 9-73. Tangential sensitivity vs. signal frequency of detector diodes in 1971. *(After Watson, Par. 9-64.)*

wafer containing a junction of well-defined geometry, usually formed by diffusion. Figure 9-75 shows a cutaway view of a typical varactor wafer and its equivalent circuit. The equivalent circuit of the wafer includes:

C_j, junction capacitance, which is a function of the applied voltage.

R_j, junction resistance, which is in shunt with C_j and is also a function of bias.

R_s, series resistance, which may be a function of bias, including the resistance of the semiconductor on either side of the junction through which the conduction current passes and the resistance of the ohmic electrical contacts to the wafer.

Also shown in Fig. 9-75 are typical values for wafer dimensions and equivalent circuit parameters for two types of varactors. The harmonic-generator varactor characteristics listed could be obtained by using an epitaxial silicon wafer with an arsenic-doped substrate of 0.004 Ω·cm resistivity and a 1 Ω·cm epitaxial layer, 9 μm thick. Boron is diffused into the epitaxial layer to a depth of 4 μm. An ohmic contact is made to a small circular area on top of the wafer, and most of the epitaxial layer is etched away, except that which is under the contact. In this way a mesa of the desired diameter can be formed. Such a device is used in a 1 to 4 GHz quadrupler, with 500 mW output and 50% conversion efficiency. The parametric amplifier diode is of similar construction but is used at higher frequencies (6 to 12 GHz).

56. Reverse-Biased Operation of Varactor Diodes. Varactor diodes are normally operated under reverse bias, where the junction resistance (ordinarily 10 MΩ or more) is negligible in comparison with the microwave capacitive reactance of the junction. Therefore the equivalent circuit of the reverse-biased varactor wafer at microwave frequencies is simply a capacitance and resistance in series. The equivalent circuit of a forward-biased varactor at microwave frequencies is generally more complicated, since it must include the diffusion capacitance of the injected carriers, as well as the effect of these carriers on the conductance of the semiconductor material.

For forward bias the diode current increases exponentially with the applied voltage, and for reverse bias a small saturation current I_s flows. When the reverse bias is increased to the avalanche breakdown voltage V_B, the diode reverse current increases very rapidly, since it is limited only by the small diode resistance and any external resistance that is present in the circuit.

Figure 9-76 shows typical dc current-voltage characteristics, microwave series resistance, and 1-MHz C-V characteristics for the harmonic-generator diode described in Fig. 9-75.

Charge-Storage Effects. In some applications the injected charge-storage capacitance is more important than the capacitance variation associated with the varying width of the depletion region. Charge-storage capacitance is produced by the injection of minority carriers during the forward-biased excursion of the varactor pump voltage and the with-

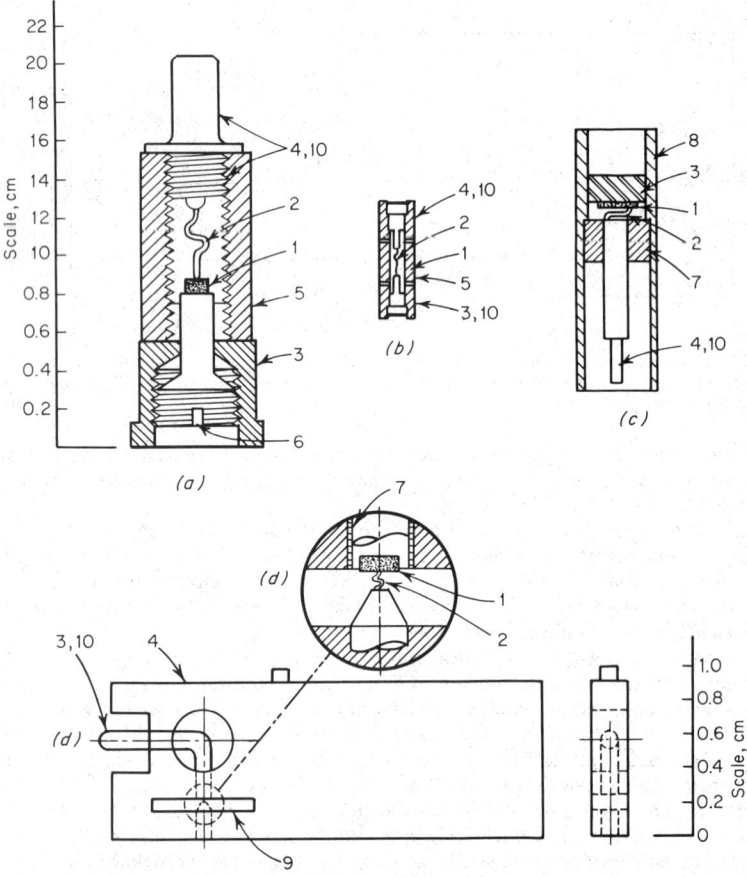

Fig. 9-74. Mixer and detector diode encapsulations. (1) Semiconductor; (2) whisker; (3) wafer contact; (4) external whisker contact; (5) ceramic case; (6) adjustment screw; (7) insulating spacer; (8) outer conductor; (9) waveguide rf input port; (10) connection to if amplifier or detector. *(After Watson, Par. 9-64.)*

drawal of this charge during the reverse-biased portion of the cycle. The resultant waveform is shown in Fig. 9-77. Efforts to maximize charge-storage effects in microwave diodes have led to a class of devices known as snapback, or step-recovery, diodes. These feature steep doping profiles and narrow junctions so as to give fast recovery of injected charge, typically in a transition period of a few tenths of a nanosecond, yielding a high harmonic content. However, this design results in lower breakdown voltage, reducing the power capability of the diode. Frequency multipliers use these diodes where high-order multiplication (above eight) and circuit simplicity are desired, at the expense of power output.

57. Varactor Frequency Multipliers. Figure 9-78 gives the characteristics of diffused epitaxial GaAs and Si varactor diodes employed in a conventional tripler (4 to 12 GHz) frequency multiplier. Because of its higher cutoff frequency, GaAs gives higher efficiency but lower maximum output power than Si.

58. Varactor Parametric Amplifiers (Paramps). The most commonly used parametric amplifier is the one-port nondegenerate configuration in which a circulator is employed to separate output from input signal. Typical bandwidths of practical amplifiers with single-tuned signal and idler circuits at 20 dB gain are of the order of 2 or 3%. To achieve wider

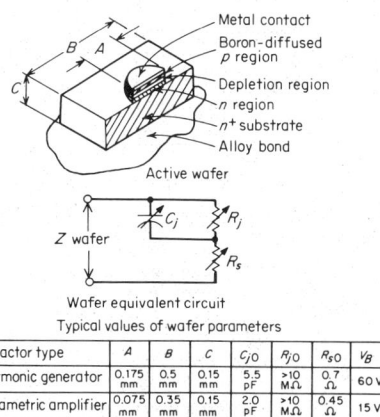

Typical values of wafer parameters

Varactor type	A	B	C	C_{j0}	R_{j0}	R_{s0}	V_B
Harmonic generator	0.175 mm	0.5 mm	0.15 mm	5.5 pF	>10 MΩ	0.7 Ω	60 V
Parametric amplifier	0.075 mm	0.35 mm	0.15 mm	2.0 pF	>10 MΩ	0.45 Ω	15 V

Fig. 9-75. Typical varactor wafer and equivalent circuit. *(After Watson, Par. 9-64.)*

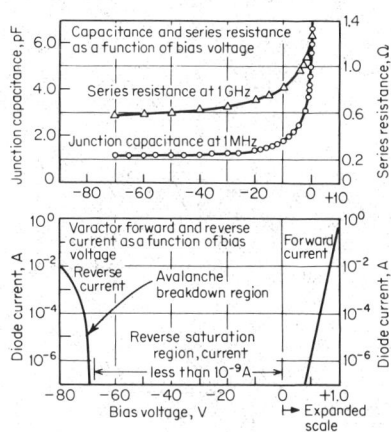

Fig. 9-76. Varactor junction capacitance, series resistance, forward current, and reverse current as functions of bias voltage. *(After Watson, Par. 9-64.)*

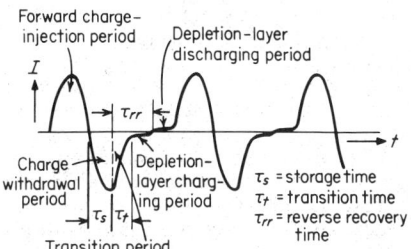

Fig. 9-77. Current waveform of sinusoidally switched step-recovery diode. *(After Watson, Par. 9-64.)*

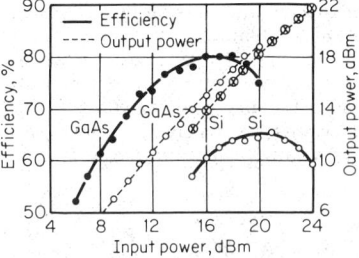

Fig. 9-78. Conversion efficiency and output power of varactors used as frequency triplers, with bias adjusted to maximum output at each point. *(After Watson, Par. 9-64.)*

bandwidth, gain is lowered, and an additional tuning resonator is added to broaden and flatten the passband. The low gain necessitates cascading paramp stages to make the noise contribution of subsequent devices negligible. For example, at 4 GHz, three double-tuned 10-dB stages are cascaded to give a 500-MHz 30-dB-gain low-noise receiver.

Most varactors currently used in paramps are made of gallium arsenide, in the form of epitaxial diffused junction diodes, with a dynamic cutoff frequency of about 100 GHz. These give, typically, noise temperatures of 120°K uncooled in the 7-GHz satellite communications band and 80°K at 3 to 4 GHz. When cooled by a 17°K cryogenic refrigerator, these noise temperatures drop to 30 and 20°K, respectively. GaAs Schottky barrier type varactor diodes have recently appeared with a cutoff frequency above 500 GHz. When pumped in the 60-GHz range, amplifiers employing such diodes have exhibited uncooled noise temperatures of the order of 60°K at 7 GHz.

Mounting. Typical varactor diodes are mounted in hollow dielectric (usually alumina) cylinders with kovar or copper end caps, as shown in Fig. 9-79. A flexible connection is made to the diode mesa from one end cap by means of a thin gold strap. The equivalent circuit shown represents the coupling between the diode junction region and the package surface.

59. *pin* Diodes. *pin* diodes consist of heavily doped *p* and *n* regions separated by a layer of high-resistivity intrinsic material. Typical construction of such a diode is shown in Fig. 9-80, and Fig. 9-81 illustrates a metal-ceramic package employed at microwave frequencies.

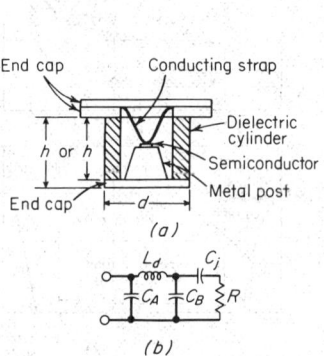

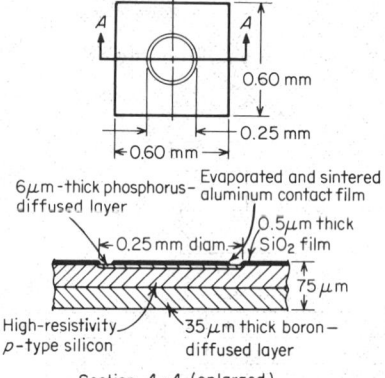

Fig. 9-79. Diode construction and equivalent circuit. *(Microwave J., November, 1970.)*

Fig. 9-80. Typical planar *pin* microwave wafer.

Under zero and reverse bias this type of diode has a very high impedance, whereas at moderate forward current it has a very low impedance. This permits its use as a switch in microwave transmission lines. Generally, the diode is placed in shunt across a strip line, allowing unimpeded transmission when reverse-biased, but short-circuiting the line to produce almost total reflection when forward-biased by as little as one volt. The wide intrinsic layer permits high microwave peak power to be controlled since the breakdown voltage is of the order of a kilovolt. Very little power is dissipated by the diode itself because reflection-type switching is employed.

pin diodes can be utilized as limiters, replacing TR tubes for peak powers smaller than 100 kW. At higher peak power, these diodes are useful, following the TR box to eliminate any spike leakage, although if fast response is required (less than one microsecond), a varactor diode is used. In attenuator applications the diode behaves as a current-controlled resistance in parallel with the capacitance of the intrinsic region.

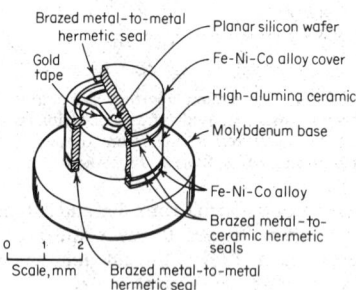

Fig. 9-81. Typical *pin* diode in microwave package.

Electrically controllable, rapid-acting microwave phase shifters are finding increasing use in phased-array systems. *pin* diodes are employed to switch lengths of transmission line, providing digital increments of phase in individual transmission paths, each capable of carrying many kilowatts of peak power.

60. Microwave Bipolar Transistors. To achieve operation at microwave frequencies, individual transistor dimensions must be reduced to the micrometer range. To maintain current and power capability, various forms of internal paralleling on the chip are employed. These geometries fall into three general types, as shown on Fig. 9-82, interdigitated fingers forming emitter and base, overlay groupings of emitter and base stripes, and a mesh or matrix of emitter and base spots. All microwave transistors are now planar in form, and almost all are of the silicon *npn* type.

Construction and Fabrication. High-quality microwave transistors are built up on an *n*-type epitaxial layer of the order of 1 Ω·cm resistivity (deposited on lower-resistivity silicon) to keep the collector depletion layer narrow. Following a first thermal oxidation of the silicon, the base area is opened by masking and photoetching operations, and diffusion of *p*-type dopant (typically boron) proceeds.

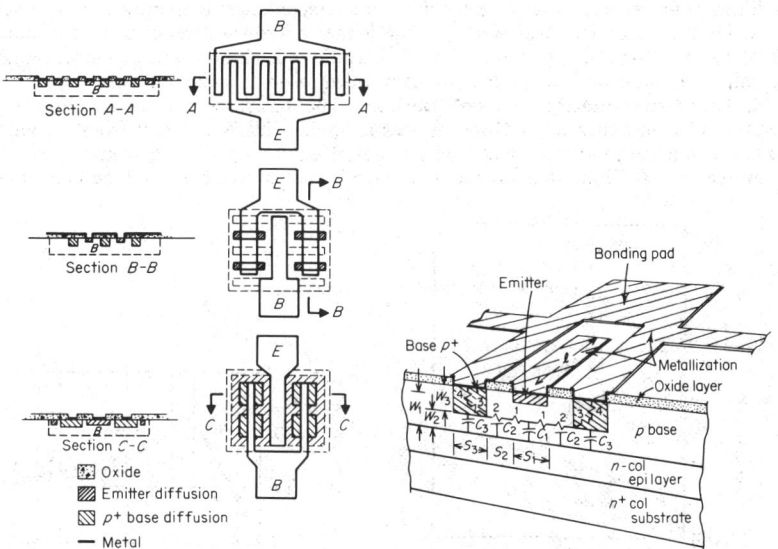

Fig. 9-82. Typical geometries of bipolar microwave transistors.

Fig. 9-83. Cross section of planar transistor, showing (1-3) r'_b and (4) contact resistance R_c. *(Proc. IEEE, August, 1971.)*

Following base formation, a second oxide is deposited on the slice, generally by thermal decomposition of a gas or liquid in a low-temperature process designed to minimize the possibility of further base diffusion. The emitter is then defined through photomasks, and the appropriate dopant (phosphorus, or preferably arsenic) diffused in. The emitter diffusion is quite critical, since too shallow or too deep a penetration can result in no transistor action at all. Metal contacts are then evaporated or sputtered to interconnect the elements of the device.

Figure 9-83 illustrates the steps in the fabrication process. Silicon nitride is replacing silicon oxide as a passivating agent when extremely shallow junctions are to be protected, because of its superior properties.

Power Capability. Most power microwave transistors include some form of integral emitter resistors to aid in equalizing the current over the distributed emitter structure. The *overlay transistor* utilizes an integral diffused resistor as part of each emitter stripe, while *thin-film resistors* are deposited as part of the contacts on interdigitated devices.

A figure of merit can be defined for the transistor in terms of r_b', the base resistance, C, the collector capacitance, and τ_{ec}, the emitter-to-collector signal delay time:

$$\text{(Power gain)}^{1/2} \text{(bandwidth)} \approx \frac{1}{4\pi (r_b' C \tau_{ec})^{1/2}}$$

τ_{ec} is composed of the transit time of carriers across the base and collector depletion layer, plus the charging time of the emitter-base junction capacitance and the collector capacitance. The maximum frequency of oscillation is obtained by setting the gain equal to unity in this expression.

Noise. The principal sources of noise in a microwave transistor are shot noise associated with the emitter and collector current and thermal noise in the base resistance. It is increase in the latter and reduction in current gain which are responsible for the deterioration of noise figure with frequency.

The current state of the art in transistor noise figure and power output in the microwave region is shown in Fig. 9-84. The input-output isolation of the transistor, as well as its unconditional stability in properly designed circuits, makes it the component of choice in those applications where its power output or noise is competitive. Most circuit configurations require that the collector be isolated from the ground plane. A ceramic (commonly beryllium oxide because of its superior thermal characteristics) is employed in the package to provide the necessary insulation. At the higher microwave frequencies, conventional packaging results in a loss of gain, primarily because of lead inductance. Direct mounting of a chip in a microwave integrated circuit is called for in this application.

61. Transferred Electron (Gunn) Devices. The basic mechanism involved in the operation of bulk materials in Gunn devices is fundamentally different from the junction effects which occur in semiconductor devices described in the preceding paragraphs. It was discovered by J. B. Gunn that when a dc voltage is applied to a bar of *n*-type GaAs or InP,

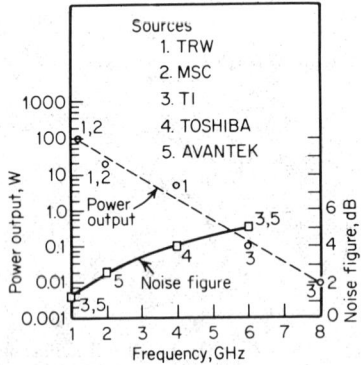

Fig. 9-84. Noise figure and power output of microwave transistors in 1970. *(Proc. IEEE, August, 1971.)*

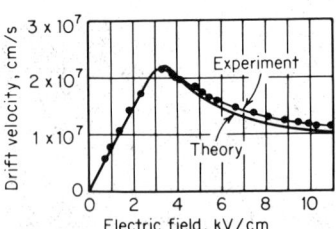

Fig. 9-85. Theoretical and experimental velocity vs. electric field of GaAs, as used in Gunn devices. *(IEEE Trans. MTT, November, 1970.)*

the current first increases linearly with voltage and then oscillates when the average electric field increases beyond a threshold field of several kilovolts per centimeter. The time period of this oscillation was found to be approximately equal to that of the transit time of the carriers from the cathode to the anode. Further experiments by Gunn revealed that these current oscillations were associated with the transit of high-field dipole *domains* nucleating cyclically at the cathode and being collected at the anode. Figure 9-85 shows the theoretical and experimental velocity–electric field (v-E) characteristic for GaAs. It will be seen that for electric fields exceeding the threshold field of about 3 kV/cm the electrons have a negative differential mobility. For the transferred electron effect to occur, the energy gap ΔE must be smaller than the band gap E_g of the semiconductor. This condition is met in GaAs, where $E_g = 1.4$ eV.

Under constant-current conditions, the negative resistance associated with a bar of negative-mobility material is always stable. Under constant-voltage conditions, however, the negative resistance becomes unstable if $n_0 L$ exceeds a certain critical value. If the negative resistance can be stabilized, the bar can be used as the active element in stable rf amplifiers. When the negative resistance is not stabilized, the bar will generate rf oscillations.

To visualize the behavior of space charge in an oscillating device, consider a bar of uniformly doped *n*-type GaAs biased to negative mobility. The cathode contact forms a large discontinuity in doping, and consequently space-charge buildup starts there. The exponential increase in space charge causes the associated space-charge electric field to increase, and consequently the voltage across the space-charge region increases. This space-charge growth continues until the voltage drop across the region is so large that the field outside the space-charge region falls below the threshold field; this space-charge region is then called a *mature domain*. Since this space-charge readjustment takes place in a moving electron stream, the high-field domain is also in motion. It nucleates at the cathode contact and moves to the anode, where it is collected. The cycle then repeats.

In the limited space-charge accumulation (LSA) mode the formation of traveling domains is suppressed by an rf voltage whose amplitude is large enough to drive the device below threshold during every rf cycle, and whose frequency is too high to allow enough time for domain formation during the part of the rf cycle when the device voltage is above threshold. The frequency of oscillation in this mode is independent of the transit time of the carriers, but instead is determined by the circuit configuration. As a result, a device designed for LSA-mode operation can be made much thicker than one operating in a domain mode, leading to higher impedance and efficiency. Many practical devices operate in a hybrid mode in which partial domains are formed, so that sensitivity to load and to rf circuit is less than for a true LSA mode.

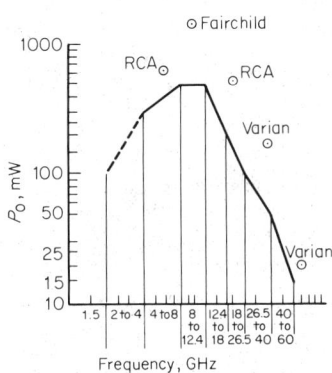

Fig. 9-86. Continuous-wave Gunn diodes and oscillators. *(After Fank, WESCON, 1971.)*

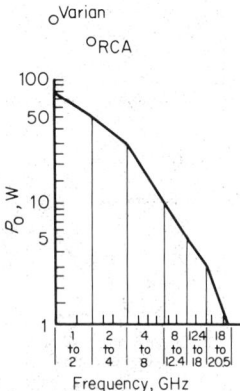

Fig. 9-87. Pulsed Gunn diodes and oscillators. *(After Fank, WESCON, 1971.)*

Fabrication of Gunn Devices. The fabrication of these devices requires stringent control over the GaAs material employed. Good electrical performance results from extremely pure and uniform material with a minimum of deep donor levels and traps and the use of very-low-loss contacts. Modern devices use an n-type epitaxial layer of GaAs grown from the liquid or vapor phase on n^+ bulk material. Typical carrier concentrations range from 10^{14} to 10^{16} cm^{-3}, while device lengths range from a few micrometers to several hundred micrometers.

The power levels commercially available from Gunn diode oscillators are shown in Figs. 9-86 and 9-87 for cw and pulsed operation, respectively. The points shown above the curves are for laboratory devices, but still utilizing single chips. Multiple diodes in specially designed circuits would allow higher total power per package. Efficiencies range from 2 to 4% for cw devices to 10 to 20% for pulsed operation. The indications are that the 10% range can be attained in cw operation.

Noise. One of the more important operating characteristics of the Gunn type of oscillator is its low-noise performance as compared with other types of microwave sources. Figure 9-88 gives such a comparison for FM noise. Injection locking with a low-power source which has been cavity-stabilized (since FM noise is inversely proportional to loaded Q) can be employed to reduce noise further.

Electronic Tuning. Electronically tuned Gunn oscillators are available employing YIG spheres, varactor diodes, and *pin* diodes as the tuning element, in addition to mechanically tuned devices. Typical ranges covered are 8 to 12 and 12 to 18 GHz at the rather slow YIG magnetic tuning speed of 100 Hz and of the order of 10% bandwidth at the much faster varactor rate of up to 10 MHz.

Reflection Amplification. By utilizing a circulator to separate input from output, negative-resistance, reflection-type amplification can be achieved. A single-stage gain of 10 dB over bandwidths of 1 to 2 GHz at X-band with a noise figure of 15 dB represents the state of the art.

62. Microwave Avalanche (IMPATT) Diodes. The power obtainable from IMPATT devices is greater than that available from the Gunn diodes described in Par. **9-61**, but at the expense of higher noise and higher operating voltage. There are two fundamental physical processes pertinent to the operation of avalanche diodes: the drift velocity at which carriers travel under a reverse-biased electric field and the avalanche multiplication which occurs at sufficiently high fields.

Read Diode. An understanding of the dynamic operating characteristics of IMPATT diodes can be best obtained by considering the operation of the Read diode. The structure, field distribution, and current waveforms for a reverse-biased Read diode are shown in Fig. 9-89. The Read diode consists of two regions: a narrow avalanche region (p region), in which

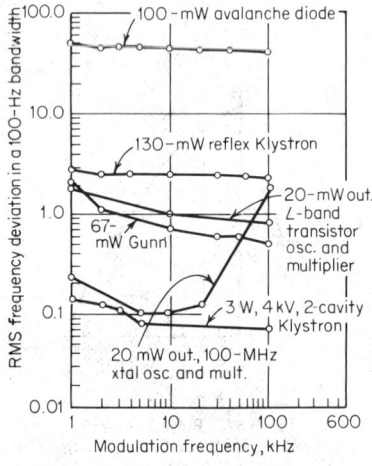

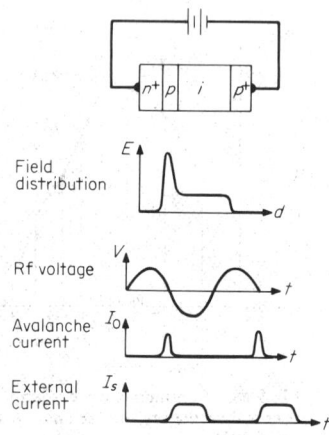

Fig. 9-88. Comparison of rms frequency deviation in Klystrons and semiconductor microwave devices. *(After Fank, WESCON, 1971.)*

Fig. 9-89. Waveforms of the Read diode. *(IEEE Trans. MTT, November, 1970.)*

carrier multiplication by impact ionization occurs, and a drift region (I region), in which the carriers drift at saturated or field-independent velocities and where no impact ionization occurs.

The negative resistance or conductance of a Read diode is attributed to a phase shift between the current through the diode and the voltage across it. This phase shift consists of two components. There is a phase delay of the current caused by the avalanche multiplication process and by the finite transit time of the holes drifting through the drift region. The phase delay caused by the avalanche process results from the condition that the rate of generation of electron-hole pairs in the avalanche region is proportional to both the electric field and the density of electrons and holes that are already present.

If a small rf voltage of sufficiently high frequency is assumed to be superimposed along with a dc voltage near breakdown across the diode, the rate of generation of electron-hole pairs will exceed the rate at which the pairs leave the avalanche region. Thus, both the density of carriers and the current will grow exponentially with time as long as the rf and dc voltages add to give a total field above breakdown. When the rf voltage changes sign and subtracts from the dc voltage, causing the field to fall below breakdown, the generation rate starts decreasing, causing the current to decrease.

Thus the current generated by the avalanche process has its maximum when the rf field goes through zero. That is, as shown in Fig. 9-89, the avalanche process contributes a 90° inductive phase lag to the current generated in the avalanche region. This current is then injected into the drift region of the diode. The current induced in the external circuit by this drifting charge is shown at the bottom of Fig. 9-89. It is obvious that the fundamental component of the external current is more than 90° out of phase with the rf voltage, causing the resistance or conductance of the diode to be negative. If the diode is to operate as a stable oscillator, the negative conductance of the diode must decrease with increasing rf voltage. The rf voltage across the diode will grow until the admittance of the diode is balanced by the admittance of the diode mount and microwave circuit.

The effect of the transit time on the frequency range of operation can be understood by assuming that the current generated in the avalanche region is generated in a sharp pulse. The precise time in the cycle that this pulse is generated is a function of the sum of the dc current and the rf voltage, as discussed previously. The phase delay caused by the transit time is proportional to $\omega\tau$, where τ is the transit time in seconds. For a given diode the transit time τ remains constant but the phase delay decreases with decreasing frequency. At a sufficiently low frequency the phase delay of the transit time plus the phase lag of the avalanche process are not sufficient to have the fundamental component of the external current lag the rf voltage by more than 90°. Therefore, below a certain cutoff frequency, the conductance of the diode becomes positive.

IMPATT Diode. At present the semiconductor materials employed in commercial IMPATTs are silicon and gallium arsenide. The latter is operated in this case at much higher electric fields than corresponds to the region of negative differential mobility utilized in the Gunn effect. The basic structure of a typical *pn* junction silicon IMPATT is shown in Fig. 9-90.

Fabrication. This structure is fabricated by first growing a thin epitaxial layer of *n*-type silicon on a heavily doped *n*-type (n^+) substrate and then diffusing a *p*-type dopant, or growing a heavily doped p^+ layer upon it. In operation the device is reverse-biased past the point of avalanche breakdown, so that a dc current of typically 50 to 250 mA flows through the diode. A coaxial circuit configuration for IMPATT operation is shown in Fig. 9-91. Typical values for X-band diode equivalent circuit parameters are $R_D = -2\,\Omega$, $X_D = 0.5$ pF ($-j30\,\Omega$), and $L_p = 0.6$ nH ($+j40\,\Omega$).

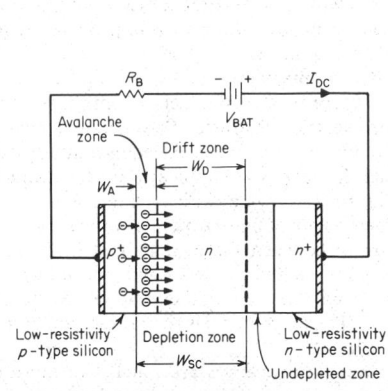

Fig. 9-90. *pn* junction of IMPATT diode under reverse bias. *(After Cowley, WESCON, 1971.)*

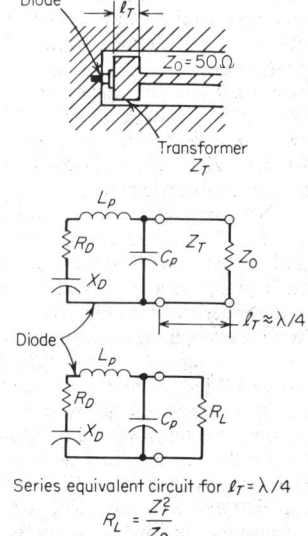

Series equivalent circuit for $\ell_T = \lambda/4$

$$R_L = \frac{Z_T^2}{Z_0}$$

Fig. 9-91. Low-Q coaxial circuit for use with IMPATT diodes as oscillators or amplifiers. *(After Cowley, WESCON, 1971.)*

Cooling. A difficult technological problem in IMPATT diode packaging is efficient heat removal from the active portion of the device. These diodes operate at high dc power densities, typically 10^4 to 10^5 W/cm², and since only a fraction of this power (6 to 12%) is converted to rf power, the remainder must be removed as heat. Inverted mesa thermocompression bonding to a copper heat sink has been employed for this purpose. A better approach electroplates the heat sink at the wafer stage before individual diodes have been fabricated. Diamond has also been employed as a heat sink in experimental devices utilizing inverted chips.

Performance. Figure 9-92 shows the power performance attained in multiple-mesa (but single-package) devices. Ring and stripe geometries are also used to provide efficient heat removal in larger-area structures. In addition, 1 W cw has been obtained at 50 GHz; that is, the IMPATT diode has millimeter-wave capability.

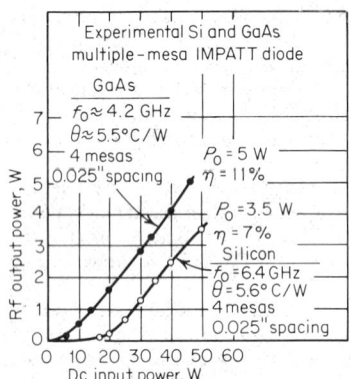

Fig. 9-92. Comparison of Si and GaAs IMPATT diodes. *(After Cowley, WESCON, 1971.)*

The noise performances of various types of IMPATTs are compared in Fig. 9-93. The high-noise characteristic of these devices precludes their use in receiver front ends, but does not inhibit higher-power applications.

63. Acoustic-Wave Devices. Although interest in acoustic waves at uhf and microwave frequencies was triggered by the discovery of the acoustic-wave amplifier, which is a device employing a piezoelectric semiconductor such as cadmium sulfide to provide interactive gain between its drifting charge carriers and the electric field component of its acoustic wave, practical applications of these waves has thus far been restricted to passive delay lines and signal processors. Acoustic waves owe their usefulness to their relatively low velocity, typically 10^{-5} of the electromagnetic velocity, permitting relatively long electrical signal delay times to be obtained in a physically small space. Both bulk-mode propagation and surface waves have been employed, the latter gaining in popularity because of the relative ease of access to intermediate points along the propagation path for tapping.

Transducers. The transducers designed for the two types of acoustic modes are physically quite different, but both contain electrodes spaced either a quarter or half an acoustic wavelength in a piezoelectric material. Microwave bulk-wave transducers consist of multiple films of ZnO or CdS, each approximately $\lambda/4$ thick, separated by similar films of gold or other nonpiezoelectric material deposited on the bulk medium so as to provide electric to acoustic-wave coupling and impedance transformation.

Acoustic surface waves are generally excited by means of interdigital transducers. An interdigital transducer, illustrated in Fig. 9-94, consists of two sets of interleaved metal electrodes, called *fingers*, deposited on the piezoelectric substrate. To generate a wave an rf potential is applied between the adjacent sets of fingers, which are spaced by a distance equal to one-half wavelength at the transducer design frequency. A typical 100-MHz transducer on LiNbO₃ has aluminum fingers 0.2 μm thick by 9 μm wide, with 9-μm gaps.

The wave excited by the rf potential between a pair of fingers travels at the surface-wave velocity. By the time the wave arrives midway between the next pair of fingers, the rf excitation potential has reversed sign, and the wave excited by the second pair of fingers will be in phase with the wave from the first pair. Thus the excitation due to the second pair is added to the excitation from the first, and so on. The mechanism is reciprocal, and hence the transducer that excites a wave will also detect it.

The transducer bandwidth is limited because its properties are much like those of its electromagnetic counterpart, the end-fired antenna array. If there are a large number of fingers, and thus the interdigital line is many wavelengths long, even a slight change from synchronism frequency causes a large mismatch in phase, and net excitation is relatively small. Therefore a long transducer has a relatively narrow bandwidth.

A surface-wave transducer is a three-port device, i.e., with one electric and two elastic ports. Figure 9-95 shows the conversion loss as a function of frequency from a 50-Ω source to one of the two acoustic outputs for three different transducers on a lithium niobate surface. These calculated curves show that for this particular case the use of five finger pairs provides the widest bandwidth and smallest conversion loss. The attenuation of the wave, once it has been launched, is given in Fig. 9-96 for lithium niobate. This quantity generally varies as the square of the frequency. Surface-wave amplification, obtained by coupling to the carriers in

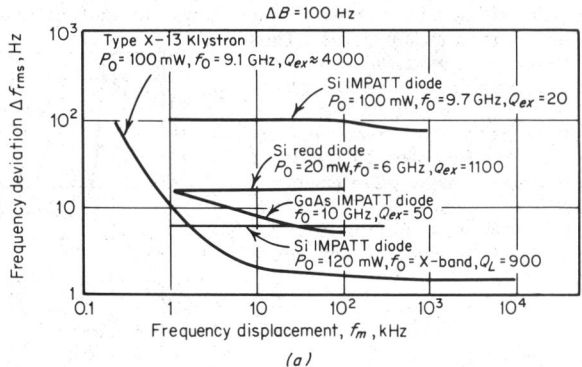

(a)

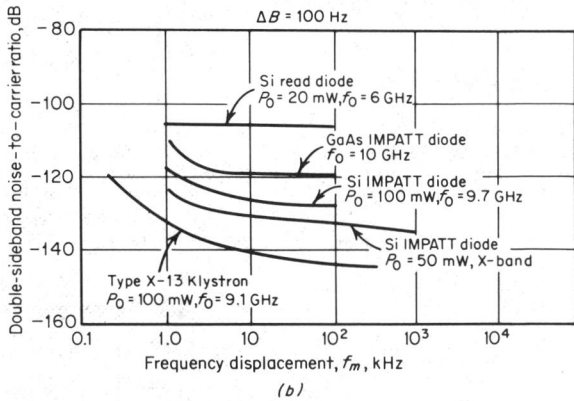

(b)

Fig. 9-93. Noise data for IMPATT diodes. (*a*) FM noise; (*b*) AM noise. *(IEEE Trans. MTT, November, 1970.)*

a thin film of semiconducting material mounted adjacent to the surface, has been employed to compensate this attenuation, but is still in an experimental stage.

The lowest operating frequency of acoustic surface-wave devices is limited by the allowable size. At present the upper operating frequency is limited by fabrication difficulties to about 1 GHz. A typical 500-MHz transducer on quartz has interleaved metal fingers 1.5 μm wide by 3 mm long, separated by 1.5-μm gaps. Usual photolithography techniques can be employed to make transducers up to 600 MHz routinely. Transducers with an operating frequency up to 3.5 GHz have been made by using a scanning electron microscope to expose the photoresist in the photolithographic process.

Tapped Delay Lines. The surface-wave devices now being readied for use in radar, sonar, communication, and computer apparatus are tapped delay lines. The interdigital transducers are employed for input and output and as taps on the surface-wave delay line. The required signal-processing function is achieved through proper choice of the location, the frequency response, and the coupling of each tap.

Pulse Compression. Coded waveforms are being increasingly used in both radar and communication systems; surface-wave delay-line devices are very well suited to this application and are being actively developed. Frequency-coded waveforms have been used for many years in sophisticated radars to increase range resolution and discrimination against certain kinds of interference. In the most common form a *chirped* (i.e., linear FM) pulsed carrier signal is radiated and returning echoes are processed in a pulse-compression filter. Much work has been done to develop surface-wave devices for this application.

The device shown schematically in Fig. 9-97a has a broadband input transducer on the left. The finger-pair spacing in the output transducer is monotonically decreased, and so the pairs on the far left are most widely spaced. Thus the left part of the output transducer responds to lower frequencies, and the right to higher frequencies. A pulse of constant-amplitude, chirped rf voltage, in which the frequency decreases linearly with time (Fig. 9-97b), is applied to the input transducer. As the surface-wave pulse enters the output transducer from the left-hand side, the low-frequency part of the output transducer does not respond well to the leading high-frequency part of

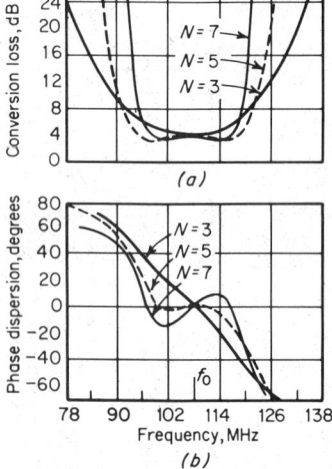

(a)

(b)

Fig. 9-95. Coupling between electric port and one acoustic port for transducers with 3, 5, and 7 interdigital periods. (*a*) Theoretical conversion loss; (*b*) phase dispersion.

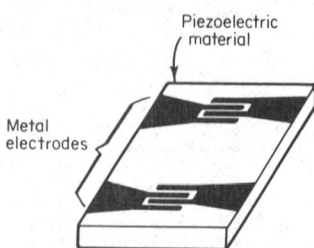

Fig. 9-94. Piezoelectric surface-wave microwave device, with interdigital electrodes. *(IEEE Spectrum, August, 1971.)*

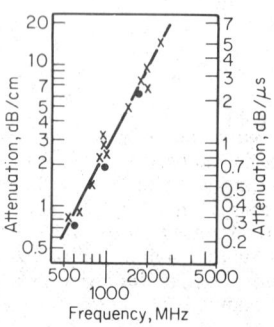

Fig. 9-96. Surface-wave attenuation of Y-cut lithium niobate. *(After Armstrong, WESCON, 1971.)*

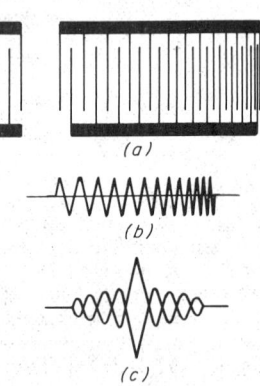

Fig. 9-97. Pulse compression by an acoustic-wave filter. (*a*) Structure of filter; (*b*) linear frequency-modulated input signal; (*c*) output voltage, with peak at center frequency of input.

the surface-wave pulse. When the pulse exactly fills the transducer, each finger pair is located at the correct position and has the correct spacing to respond well to the frequency component of the pulse just under it. As the pulse leaves to the right, the high-frequency half of the output transducer does not respond well to the trailing low-frequency part of the surface-wave pulse. The actual output voltage has the input-pulse center frequency with a $(\sin \alpha t)/t$ amplitude modulation, as shown in Fig. 9-97c. The filter compresses the chirped input pulse to the narrower central lobe of the output. Pulse compressions of 100:1 or greater can be obtained by this technique.

64. Bibliography on Microwave Semiconductor Devices

WATSON, H. A. "Microwave Semiconductor Devices and Their Circuit Applications," McGraw-Hill, New York, 1969.

IEEE Trans. Microwave Theory Techn., Vol. MTT-18, No. 11, November, 1970 (Special Issue on Bulk Devices).

Proc. IEEE, Vol. 59, No. 8, August, 1971 (Special Issue on Microwave Semiconductors).

KINO, G. S., and MATTHEWS, H. Signal Processing in Acoustic Surface-Wave Devices, *IEEE Spectrum,* Vol. 8, No. 8, pp. 22–35, August, 1971.

YOUNG, L. (ed.) *Advan. Microwaves,* Vols. 1–6, 1965–1971.

FANK, B. Review of Gunn Effect Devices, Session 26, 1971 Wescon Technical Paper.

SECTION 10

TRANSDUCERS

BY

HARRY N. NORTON Member, Technical Staff, Jet Propulsion Laboratory, California Institute of Technology; Author, "Handbook of Transducers for Electronic Measurement Systems"

CONTENTS

Numbers refer to paragraphs

SECTION 10

TRANSDUCERS

TRANSDUCER CHARACTERISTICS

1. Nomenclature. The term transducer has been applied to a variety of devices, including measuring instruments, acoustic-energy transmitters, signal converters, and phonograph cartridges. With the recent vast increase in the development and use of electronic measuring systems, however, instrumentation engineers found it necessary to devise a more limited definition of transducer as a device used for measurement purposes.

The Instrument Society of America (ISA) published its Standard S37.1 in 1969. This standard, "Electrical Transducer Nomenclature and Terminology," defines a *transducer* as "a device which provides a usable output in response to a specified measurand." The *measurand* is "a physical quantity, property or condition which is measured." The *output* is "the electrical quantity, produced by a transducer, which is a function of the applied measurand." Only the last of these three definitions applies specifically to electrical transducers. It could apply equally well to transducers with pneumatic output if the word "electrical" in the definition were omitted. However, only electrical transducers are covered in this Handbook, as well as in the ISA Standard.

ISA S37.1-1969 also applies to the construction of transducer nomenclature (see Table 10-1). When used in titles or for indices, the sequence shown in the table should be used, e.g., "transducer, acceleration, potentiometric, ±5 g." When the nomenclature is used in the text, the opposite of the sequence shown in the table should be used, e.g., "A 0- to 8-cm dc output reluctive displacement transducer was installed on the actuator."

2. Transducer Elements. In most transducers, the measurand acts upon a *sensing element*. The response of the sensing element causes the output of the transducer to be originated within the *transduction element*. An example of this is a Bourdon tube type of potentiometric pressure transducer. A pressure increase causes the Bourdon tube to deflect. This deflection is used to move a wiper arm over a resistance winding across which a voltage is applied. In some transducers, e.g., resistive temperature transducers, the transduction of temperature changes into resistance changes occurs within the sensing element itself. Most transducer types require *excitation* power since their transduction element is "passive" (e.g., a strain-gage bridge). Others, the "self-generating" types (e.g., thermocouples, piezoelectric transducers), require no excitation.

3. Integral Conditioning. Any transducer type can usually be equipped with excitation- and output-conditioning circuitry packaged integrally within the case (housing) of the transducer. An example of this is the "dc-to-dc" reluctive transducer intended to operate from an unregulated 12-V or 28-V-dc source and capable of providing a 0- to 5-V-dc full-scale output. Such a transducer contains provision for regulating the dc excitation and converting it into the ac voltage needed by the transduction element. It further contains a demodulator and an amplifier to convert the relatively low-level ac output of the transduction element into an amplified dc signal.

4. Basic Characteristics. Every transducer design can be characterized by an ideal or theoretical output-measurand relationship ("transfer function"). This relationship is capable of being described exactly by a prescribed or known *theoretical curve* (see Fig. 10-1), stated in terms of an equation, a table of values, or a graphical representation. This applies primarily to the *static characteristics* of a transducer, i.e., the output-measurand relationship for a steady-state or very slowly varying measurand. It can also apply to the transducer *dynamic characteristics*, i.e., the output-measurand characteristics for a relatively rapidly fluctuating measurand. However, this dynamic behavior is described by relationships other than the transducer's theoretical curve.

Table 10-1. Construction of Typical Transducer Nomenclature and Examples of Modifiers

Main noun	First modifier, measurand, examples	Second modifier, restricts measurand, examples	Third modifier, electrical transduction principle, examples	Fourth modifier,[3] sensing element, special features or provisions, examples	Range, examples	Unit, examples
Transducer	Acceleration	Absolute	Capacitive	Ac output	0 to 1,000	A
	Air speed	Angular	Electromagnetic	Amplifying	±5	°C
	Attitude	Differential	Inductive	Bellows	-100 to +500	cm
	Attitude rate	Gage	Ionizing	Bondable	-430 to -415	cm/s
	Current	Infrared	Photoconductive	Bonded		deg
	Displacement	Intensity	Photovoltaic	Bourdon tube		°F
	Flow rate	Linear	Piezoelectric	Capsule[4]		ft/s
	Force	Mass	Potentiometric	Dc output		g
	Heat flux	Radiant	Reluctive	Diaphragm		Hz
	Humidity	Relative	Resistive	Digital output		in./s
	Jerk	Surface	Strain gage	Discrete Increment		in.
	Light	Total	Thermoelectric	Dual-output		K
	Liquid level	Volumetric		Exposed element		kg
	Mach No.			Frequency output		lb/min.
	Nuclear radiation			Gyro		m
	Pressure			Integrating		mmHg
	Speed[1]			Self-generating		N
	Sound pressure			Semiconductor		% RH
	Strain			Servo[5,6]		lb/in.[2]
	Temperature			Switch		kPa
	Torque			Toothed-rotor		mbar
	Velocity[2]			Triaxial		rad/s
				Turbine		
				Ultrasonic		
				Unbonded		
				Vibrating-element[7]		
				Weldable		

1 Scalar quantity.
2 Vector quantity.
3 Nomenclature may include two of these terms.
4 Preferred to "aneroid."
5 Preferred to "force balance" or "null balance."
6 When this modifier is used, the third modifier ("transduction principle") may be omitted.
7 When this modifier is used together with "frequency output," the third modifier may be omitted.

5. Transducer Errors. Because of a variety of factors, the behavior of a real transducer is nonideal. These factors include production variations, as well as the use of nonideal materials, production methods, ambient conditions during manufacture, and testing methods. It must also be recognized that many tradeoffs enter into the design of a marketable transducer, and that man's knowledge (the *state of the art*) is limited with regard to producing an ideal transducer design and then compensating it perfectly for aging effects and a variety of environmental conditions the transducer may be subjected to during its operation.

Hence the measurand value indicated by the transducer usually differs from the true measurand value or the specified theoretical value. The algebraic difference between the indicated and the true value is the *error*. The most common errors attributable to the transducer and causing deviations from the theoretical curve are listed in Table 10-2.

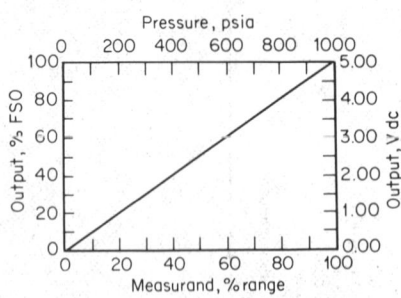

Fig. 10-1. Output-measurand relationship of ideal linear-output transducer as exemplified for a dc-output pressure transducer. *(Prentice-Hall, Inc., by permission.)*

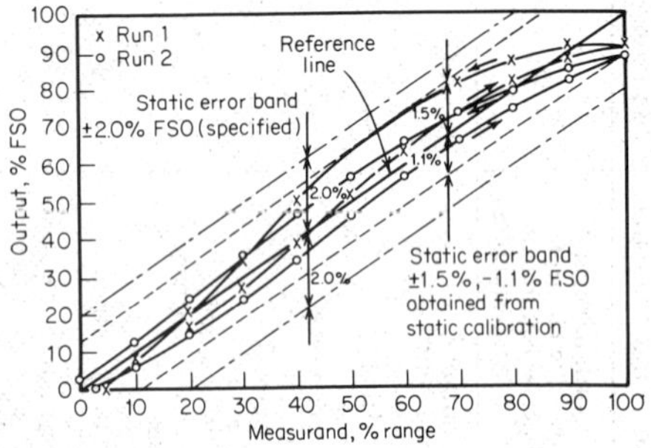

Fig. 10-2. Static error band referred to terminal line (error scale 10:1). *(Prentice-Hall, Inc., by permission.)*

6. Error Band. A convenient manner of determining or specifying transducer errors is to state them in terms of the band of maximum (or maximum allowable, for a specification) deviations from a specified reference line or curve. This band is defined as the error band. The *static error band* (see Fig. 10-2) is that error band obtained (or obtainable) by means of a *static calibration*, which is performed under "room conditions" (controlled room temperature, humidity, and atmospheric pressure) and in the absence of any vibration, shock, or acceleration (unless one of these is the measurand) by applying known values of measurand to the transducer and recording corresponding output readings. Other types of error band are applicable under somewhat different (and rigorously specified) conditions.

Table 10-2. The Most Common Transducer Errors
(Definitions According to ISA Standard S37.1-1969)

A. Static Errors

   Creep: A change in output occurring over a specific time period while the measurand and all environmental conditions are held constant.

   Friction error: The maximum change in output, at any measurand value within the specified range, before and after minimizing friction within the transducer by dithering.

   Dithering: The application of intermittent or oscillatory forces just sufficient to minimize static friction within the transducer.

   Hysteresis: The maximum difference in output, at any measurand value within the specified range, when the value is approached first with increasing, and then with decreasing, measurand.

   Linearity: The closeness of the calibration curve to a specified straight line.

   Linearity, end-point: Linearity referred to the end-point line.

   Linearity, independent: Linearity referred to the "best straight line."

   Linearity, least-squares: Linearity referred to the least-squares line.

   Linearity, terminal: Linearity referred to the terminal line.

   Linearity, theoretical-slope: Linearity referred to the theoretical slope.

   "Best straight line": A line midway between the two parallel straight lines closest together and enclosing all output vs. measurand values on a calibration curve.

   End-point line: The straight line between the end points.

   End points: The outputs at the specified upper and lower limits of the range.

   Least-squares line: The straight line for which the sum of the squares of the residuals (deviations) is minimized.

   Terminal line: A theoretical slope for which the theoretical end points are 0 and 100% of both measurand and output.

Table 10-2.—Continued

Theoretical end points: The specified points between which the theoretical curve is established and to which no end-point tolerances apply.

Theoretical slope: The straight line between the theoretical end points.

Repeatability: The ability of a transducer to reproduce output readings when the same measurand value is applied to it consecutively, under the same conditions, and in the same direction.

Sensitivity shift: A change in the slope of the calibration curve due to a change in sensitivity.

Sensitivity: The ratio of the change in transducer output to a change in the value of the measurand.

Zero shift: A change in the zero-measurand output over a specified period of time and at room conditions.

Zero-measurand output: The output of a transducer, under room conditions unless otherwise specified, with nominal excitation and zero measurand applied.

B. Environmental Errors

Acceleration error: The maximum difference, at any measurand value within the specified range, between output readings taken with and without the application of specified constant acceleration along specified axes.

Ambient pressure error: The maximum change in output, at any measurand value within the specified range, when the ambient pressure is changed between specified values.

Attitude error: The error due to the orientation of the transducer relative to the direction in which gravity acts upon the transducer.

Temperature error: The maximum change in output, at any measurand value within the specified range, when the transducer temperature is changed from room temperature to specified temperature extremes.

Temperature gradient error: The transient deviation in output of a transducer at a given measurand value when the ambient temperature or the measured fluid temperature changes at a specified rate between specified magnitudes.

Thermal sensitivity shift: The sensitivity shift due to changes of the ambient temperature from room temperature to the specified limits of the operating temperature range.

Thermal zero shift: The zero shift due to changes of the ambient temperature from room temperature to the specified limits of the operating temperature range.

Vibration error: The maximum change in output, at any measurand value within the specified range, when vibration levels of specified amplitude and range of frequencies are applied to the transducer along specified axes.

C. Other Errors

Conduction error: The error in a temperature transducer due to heat conduction between the sensing element and the mounting of the transducer.

Drift: An undesired change in output over a period of time, which change is not a function of the measurand.

Loading error: An error due to the effect of the load impedance on the transducer output.

Mounting error: The error resulting from mechanical deformation of the transducer caused by mounting the transducer and making all measurand and electrical connections.

Output noise: The rms, peak, or peak-to-peak (as specified) ac component of a transducer's dc output in the absence of measurand variations.

Output regulation: The change in output due to a change in excitation.

Reference pressure error: The error resulting from variations of a differential-pressure-transducer's reference pressure within the applicable reference-pressure range.

Reference pressure: The pressure relative to which a differential-pressure transducer measures pressure.

Stability: The ability of a transducer to retain its performance characteristics for a relatively long period of time.

Strain error: The error resulting from a strain imposed on a surface to which the transducer is mounted.

Transverse sensitivity: The sensitivity of a transducer to transverse acceleration or other transverse measurand.

Transverse acceleration: An acceleration perpendicular to the sensitive axis of the transducer.

---

7. Dynamic Characteristics. When a step change in measurand is applied to a transducer, the transducer output does not instantaneously indicate the new measurand level. Examples of such step changes are mechanical shock, a sudden pressure rise when a solenoid valve opens, or a temperature transducer rapidly immersed in a very cold liquid. The lag between the time the measurand reaches its new level and the corresponding steady (final) transducer output reading is defined in various ways. The time required for the output change to reach 63% of its final value is the *time constant* of the transducer. The time required to reach a different specified percentage of this final value (e.g., 90 or 98%) is the *response time*. The time in which the output changes from a small to a large specified percentage of the final value (usually from 10 to 90%) is the *rise time*. The output may rise beyond the final value before it stabilizes at that value. This *overshoot* depends on the *damping* characteristics of the transducer.

When the measurand fluctuates (sinusoidally) over a stated frequency range, the transducer output may not be able to indicate the correct amplitude of the measurand over these excursions. An example of this is the mechanical vibration (vibratory acceleration) of an engine housing. The output may be somewhat higher at certain measurand frequencies but usually drops off as the frequency increases until the output is essentially zero. The change with measurand frequency of the output-measurand amplitude ratio is the *frequency response* of the transducer, always stated for a specified frequency range. The above characteristics, as well as other transducer dynamic characteristics, are defined in the terminology section of ISA Standard S37.1-1969.

8. Environmental Characteristics. In most applications transducers are used only under the controlled room conditions of the facility where they are calibrated and where various static performance characteristics are determined. The external conditions to which a transducer is exposed not only while operating but also during shipping, storage, and handling can contribute additional errors, such as temperature error, acceleration error, or attitude error. Such environmental conditions (which can also include corrosive atmosphere, salt-water immersion, or nuclear radiation) may even cause a permanent deterioration or malfunction in the transducer.

9. Criteria for Transducer Selection. The basic considerations involved in the selection of a transducer for a given application are listed in Table 10-3.

10. Transduction Principles. The most essential determinant of any one transducer type is its transduction principle. The manner in which the electrical output is originated affects

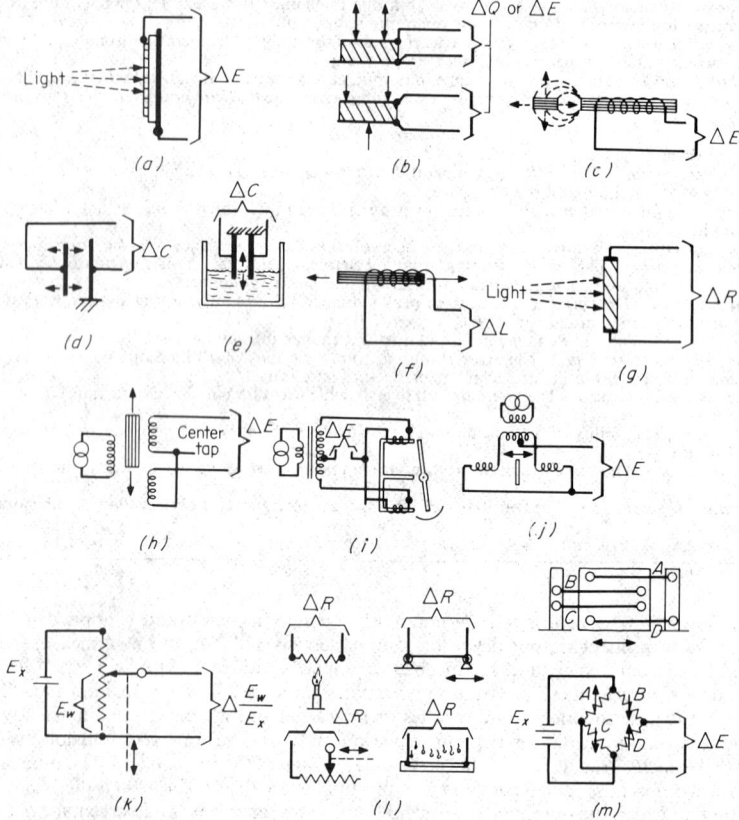

Fig. 10-3. Transduction principles. (*a*) Photovoltaic; (*b*) piezoelectric; (*c*) electromagnetic; (*d-e*) capacitive; (*f*) inductive; (*g*) photoconductive; (*h-j*) reluctive; (*k*) potentiometric; (*l*) resistive; (*m*) strain gage. *(Prentice-Hall, Inc., by permission.)*

very strongly most other characteristics of the transducer. The most frequently utilized transduction principles are described below and illustrated in Fig. 10-3. It should be noted that photovoltaic, piezoelectric, and electromagnetic transduction are used in *self-generating* transducers, whereas all other transduction methods illustrated require some sort of external excitation power.

Photovoltaic Transduction. The measurand is converted into a change in the voltage generated when a junction between certain dissimilar materials is illuminated. Used primarily in optical sensors, this principle has also been employed in transducers incorporating mechanical-displacement shutters to vary the intensity of a light beam between a built-in light source and the transduction element.

Piezoelectric Transduction. The measurand is converted into a change in the voltage E or

Table 10-3. General criteria for transducer selection*

A. Measurement
   1. What is the real purpose of the measurement?
   2. What is the measurand?
   3. What range of the measurand must be displayed in final data?
   4. What measurand overloads may occur before and during the time data are required?
   5. With what accuracy must the measurement be presented in final data?
   6. What are the lower and upper limits of flat frequency response (or the response time) needed in final data?
   7. What is the nature of the fluid to be measured?
   8. Where will the transducer be installed?
   9. What ambient environmental conditions will exist around a transducer for this measurement?

B. Data System Capability
   1. What data transmission system is used?
   2. What data processing system is used?
   3. What data display system is used?
   4. What are the accuracy and frequency response capabilities of the transmission, processing, and display systems?
   5. What transducer output will the transmission system accept with minimum signal conditioning?
   6. What transducer excitation voltage is most readily available?
   7. How much current can be drawn from the excitation power supply?
   8. What load will the transmission circuit present to a transducer?
   9. Does the transmission system provide sufficient limiting for abnormal transducer outputs?
   10. Is filtering of the transducer output required, and can this be adequately provided by the transmission or data processing system?
   11. To what extent does the transmission system allow for detection of, or provide compensation for, errors in the transducer output?

C. Transducer Design
   1. What are the configurational limitations?
   2. What maximum error can be tolerated during static conditions and during and after exposure to known environmental conditions?
   3. What are the limitations on excitation and output?
   4. What power drain can be tolerated?
   5. Which transduction principle (reluctive, potentiometric, strain-gage, etc.) should be utilized?
   6. What are the effects of the measured fluid on the transducer?
   7. Will the transducer affect the measurand so that erroneous data are obtained?
   8. What cycling or operating life is required?
   9. What test methods will be used to prove performance? Are these methods adequate? Are they sufficiently simple? Are they firmly established?
   10. What are the failure modes of the transducer? What hazards would a failure present to the component or system in which it is installed, to adjacent components or systems, or to other portions of the data system?
   11. What is the lowest level of technical competence of all personnel expected to handle, install, and use the transducer? What human engineering requirements should affect the transducer design?

D. Transducer Availability
   1. Is a transducer which fulfills all requirements now available?
   2. What manufacturer has successfully demonstrated his ability to produce a transducer similar to the required item?
   3. What experience history exists in dealing with a proposed manufacturer?
   4. Will minor redesign of an existing transducer be sufficient or will a major development effort be required?
   5. Is the cost of the transducer compatible with the necessity for the measurement?
   6. Can the transducer be delivered in time to meet installation schedules?
   7. What is the cost and length of time of all minimum-performance verification testing?

* Prentice-Hall, Inc., by permission.

electrostatic charge Q generated by certain crystals when mechanically stressed by compression or tension forces or by bending forces. Either natural or synthetic crystals (usually ceramic mixtures) are used in such transduction elements.

Electromagnetic Transduction. The measurand is converted into a voltage (electromotive force) induced in a conductor by a change in magnetic flux, usually due to a relative motion between a magnetic material and a coil having a ferrous core (electromagnet).

Capacitive Transduction. The measurand is converted into a change of capacitance. This change occurs typically either by having a moving electrode move to or from a stationary electrode or by a change in the dielectric between two fixed electrodes.

Inductive Transduction. The measurand is converted into a change of the self-inductance of a single coil.

Photoconductive Transduction. The measurand is converted into a change in conductance (resistance change) of a semiconductive material due to a change in the illumination incident on the material. This transduction is implemented in a manner similar to that explained for the case of photovoltaic transduction, above.

Reluctive Transduction. The measurand is converted into an ac voltage change by a change in the reluctance path between two or more coils while ac excitation is applied to the coil system. This transduction principle applies to a variety of circuits, including the differential transformer and the inductance bridge.

Potentiometric Transduction. The measurand is converted into a change in the position of a movable contact on a resistance element. The displacement of the contact (wiper arm) causes a change in the ratio between the resistance from one end of the element to the wiper arm and the end-to-end resistance of the element. In its most common applications the resistance ratio is used in the form of a voltage ratio when excitation is applied across the resistance element.

Resistive Transduction. The measurand is converted into a change of resistance. This change is typically effected in a conductor or semiconductor by heating or cooling, by the application of mechanical stresses, by sliding a wiper arm across a rheostat-connected resistive element, or by drying or wetting electrolytic salts.

Strain-Gage Transduction. The measurand is converted into a resistance change, due to strain, usually in two or four arms of a Wheatstone bridge. This principle is a special version of resistive transduction. However, the output is always given by the bridge-output voltage change. In the typical configuration illustrated in Fig. 10-3 the upward arrows indicate increasing resistance, and the downward arrows decreasing resistance, in the respective bridge arms for sensing link motion toward the left.

TRANSDUCERS FOR SOLID-MECHANICAL QUANTITIES

11. Terminology. *Acceleration,* a vector quantity, is the time rate of change of velocity with respect to a reference system. When the term acceleration is used alone, it usually refers to *linear acceleration a,* which is then related to linear (translational) velocity v, and time t by $a = dv/dt$. *Angular acceleration* α is related to angular (rotational) velocity ω and time t by $\alpha = d\omega/dt$. Mechanical *vibration* is an oscillation wherein the quantity, varying in magnitude with time so that this variation is characterized by a number of reversals of direction, is mechanical in nature. This quantity can be stress, force, displacement, or acceleration; however, in measurement technology the term vibration is usually applied to *vibratory acceleration* and sometimes to *vibratory velocity.* Mechanical *shock* is a sudden nonperiodic or transient excitation of a mechanical system.

12. Acceleration transducers (accelerometers) are used to measure acceleration as well as shock and vibration. Their sensing element is the *seismic mass,* restrained by a spring. The motion of the seismic mass in this acceleration sensing arrangement is usually damped (see Fig. 10-4a). Acceleration applied to the transducer case causes motion of the mass relative to the case. When the acceleration stops, the mass is returned to its original position by the spring (see Fig. 10-4b). This displacement of the mass is then converted into an electrical output by various types of transduction elements in *steady-state acceleration transducers* whose frequency response extends down to essentially zero hertz. In piezoelectric accelerometers the mass is restrained from motion by the crystal transduction element, which is thereby mechanically stressed when acceleration is applied to the transducer. Such *dynamic* *acceleration transducers* do not respond appreciably to acceleration fluctuating at a rate of less than 5 Hz. They are normally used for vibration and shock measurements.

Capacitive and photoelectric accelerometers have been produced at various times, and vibrating-element accelerometers (in which the mass, as it tends to move, applies tension to a wire or ribbon, thereby changing the frequency at which the wire can oscillate) have been used in some aerospace programs. However, the most commonly used steady-state acceleration transducers are the potentiometric, reluctive, strain-gage, and servo types. For vibration and shock measurement the piezoelectric accelerometers are most frequently used because of their inherently high frequency-response capability; some miniature semiconductor-strain-gage accelerometers are also used for these measurements since they can respond to fairly high acceleration frequencies.

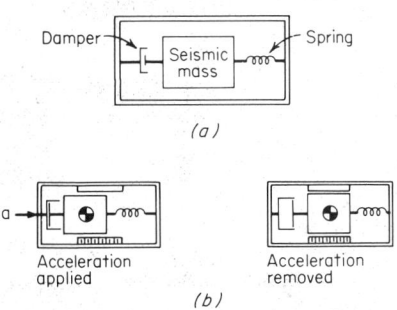

Fig. 10-4. Basic operating principle of an acceleration transducer. (*a*) Spring-mass system; (*b*) displacement of seismic mass. *(Prentice-Hall, Inc., by permission.)*

13. Potentiometric accelerometers usually employ a mechanical linkage to amplify the motion of the seismic mass so as to produce the necessary extent of wiper-arm travel over the resistance element. The mass is supported by flexural springs or a cantilever spring in some models. In others it slides on a central coaxial shaft, restrained by calibrated coil springs. Magnetic, viscous, or gas damping is normally used in potentiometric accelerometers, primarily to reduce output noise due to wiper-arm whipping and transient wiper-contact resistance changes. Overload stops keep the wiper arm from moving beyond the resistance-element ends in the presence of acceleration beyond the range of the accelerometer.

14. Reluctive accelerometers require ac excitation power having a frequency greater than the upper limit of the transducer's frequency response. When moderately high frequency response is needed, the inductance-bridge version has been found most suitable. In a typical design the seismic mass is attached to a spring-restrained ferromagnetic armature plate, pivoted at its middle and placed above two coils so that the small seesaw motion of the plate, due to acceleration acting on the mass, causes a decrease of inductance in one coil and an increase of inductance in the other. Since the coils are in opposite bridge arms, these inductance changes are additive and produce a bridge output voltage double that obtainable from having only one coil change its inductance. When a relatively low frequency response is needed, a differential-transformer synchro or microsyn transduction circuit can be employed to convert the seismic-mass displacement into the required electrical output.

15. Strain-gage accelerometers are very popular and exist in several design versions. Some use unbonded metal wire stretched between the seismic mass and a stationary frame or between posts on a cross-shaped spring to whose center the seismic mass is attached and whose four tips are attached to a stationary frame. Other designs use bonded-metal wire, metal foil, or semiconductor gages bonded to one or two elastic members deflected by the displacement of the seismic mass. The transducer shown in Fig. 10-5 has two semiconductor strain gages bonded to the upper surface and two to the lower surface of a flat cantilever beam to whose tip the seismic mass is attached. The gages are usually connected as a four-active-arm bridge.

16. Servo accelerometers are closed-loop force-balance, torque-balance, or null-balance transducers. The displacement of the seismic mass is detected by a position-sensing element, usually reluctive or capacitive, whose output is the error signal in the servo system. This signal is amplified and fed back to a torquer or restoring coil so that the restoring force is equal and opposite to the acceleration-induced force. The coil or torquer is attached to the seismic mass and returns the mass to its original position, when the feedback current is sufficient, so that the position error signal is reduced to zero. The current, which is proportional to acceleration, passes through a resistor. The *IR* drop across the resistor is the accelerometer output voltage, proportional to the acceleration.

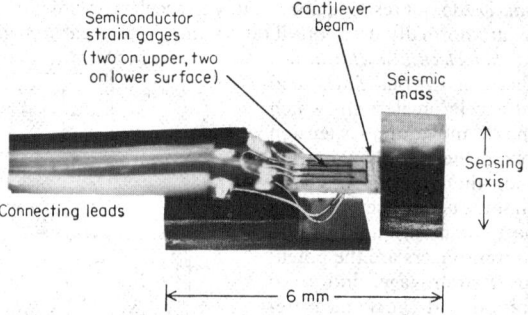

Fig. 10-5. Semiconductor-strain-gage acceleration transducer. *(Entran Devices, Inc.)*

17. Piezoelectric accelerometers exist in several design versions, two of which are illustrated in Fig. 10-6. Both contain a seismic mass which applies a force, due to acceleration, to a piezoelectric crystal. With acceleration acting perpendicular to the base, an output is generated by the crystal due to compression force in one design and to shear force in the other. Crystal materials include quartz and several ceramic mixtures such as titanates, niobates, and zirconates.

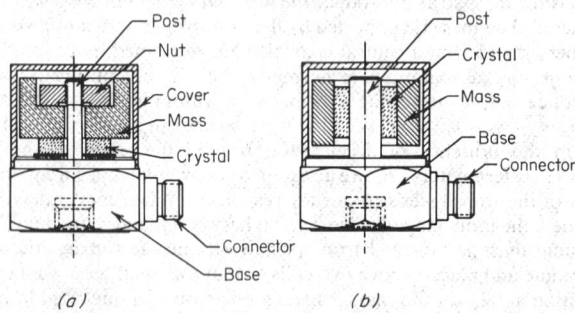

Fig. 10-6. Piezoelectric acceleration transducers. (*a*) Single-ended compression type; (*b*) shear type. *(Endevco, Division of Becton, Dickinson & Co.)*

Ceramic crystals are used more frequently than natural crystals. They gain their piezoelectric characteristics by exposure to an orienting electric field during cooling after they are fired at a high temperature. If they are subsequently heated, as during transducer operation at elevated temperature, they can lose their piezoelectric qualities if that temperature is above the *Curie point,* which varies between about 100 and 600°C, depending on the materials used in the crystal. Piezoelectric accelerometers almost invariably require some signal-conditioning circuitry to provide a usable output since they have a relatively low output amplitude and a very high output impedance. In some designs, the necessary conditioning circuitry is included in the transducer. For most such accelerometers, a separate charge or voltage amplifier is needed, connected to the transducer by a thin shielded coaxial cable of special low-noise construction to avoid noise pickup from within the cable itself.

18. Criteria for selection of an acceleration transducer are primarily the required acceleration range and frequency response. They are mutually dependent, e.g., a typical ±2-g potentiometric accelerometer design will have an upper frequency limit for flat response of about 12 Hz, whereas a ±20-g accelerometer of the same design can have an upper frequency-response limit of about 40 Hz. As frequency-response requirements increase, the reluctive, servo, metal-strain-gage, semiconductor-strain-gage, and piezoelectric transducers successively become candidates for selection. The best accuracy characteristics are provided by servo accelerometers, which are also most suitable for low-range (±0.2 g or lower) applications.

19. Attitude and Attitude-Rate Transducers. Attitude is the relative orientation of a vehicle or an object represented by its angles of inclination to three orthogonal reference axes. Attitude rate is the time rate of change of attitude.

The sensing methods employed by attitude transducers are best categorized by the kind of reference system to which the orientation to be measured is related. The *inertial* reference system is provided by a *gyroscope (gyro)* in which a rotating member will continue turning about a fixed axis as long as no forces are exerted on the member, and the member is not accelerated. *Gravity* reference is used to establish a vertical reference axis. This principle is applied in *pendulum-type transducers* in which a weight is attached to a wiper arm and a potentiometric element is attached to the case, so that an output change is obtained when the object, to which the case is mounted, deflects from a vertical position.

A *magnetic reference axis* may be established by the poles of a magnetic field which remains fixed in position. This reference system is employed by certain navigational transducers related to the compass. *Flow-stream reference* refers to the direction of fluid flow past an object moving within that fluid, a reference system employed in *angle-of-attack* transducers mounted well forward of the nose of high-speed aircraft and rockets, so that the flow stream sensed is not altered in direction by the vehicle itself.

Optical reference systems are utilized by electrooptical transducers mounted (in a known attitude) so as to sense a remote light source, or a light-dark interface, whose position is known. This establishes a reference axis between the object on which the transducer is mounted and the target sensed by the transducer. *Optically referenced* transducers include such aerospace (primarily spacecraft) devices as the sun sensor, Canopus sensor, star tracker, and horizon sensor, as well as military target-locating equipment.

20. Gyros are the most widely used attitude and attitude-rate transducers. The operating principle of the gyro is illustrated in Fig. 10-7. A fast-revolving rotor turns about the *spin axis* of the gyro. This axis, which remains fixed in space as long as the rotor revolves, establishes the inertial reference axis. The rotor shaft ends are supported by a *gimbal* frame which is free to pivot about the gimbal axis. The pivot points are part of the gyro housing structure, which is attached to the object whose changes in attitude about the gimbal axis are to be measured. An angular-displacement transduction element (pick-off) is then used to provide an output proportional to attitude. A simple example of such an element is a wiper arm, attached to the gimbal frame at the pivot point, wiping over a ring-shaped potentiometric resistance element attached to the inside of the case. Potentiometric transduction as well as reluctive (especially synchro) and, occasionally, capacitive and photoconductive transduction are used in most gyros.

Gyro attitude transducers (free gyros) are often designed as two-degree-of-freedom gyros, i.e., those providing an output for each of two of a vehicle's three attitude planes (pitch, yaw, and roll, or *x*, *y*, and *z* axes). The design illustrated in Fig. 10-8 provides an inner gimbal for one axis and an outer gimbal for the other axis, with a separate pick-off for each axis. The caging mechanism (symbolized by the hand) is used to lock the inner gimbal to a reference position until the time is reached when the spin axis is to start serving as inertial reference axis. At this point the gyro is uncaged (after the rotor has come up to speed). Ac or dc motors are commonly used to turn the

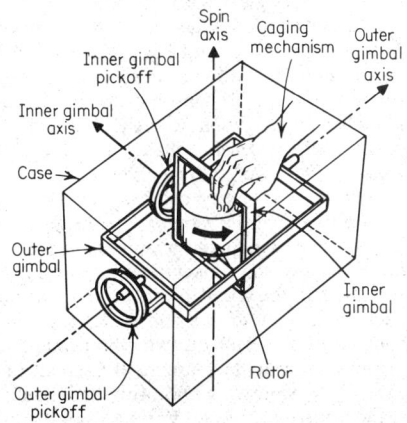

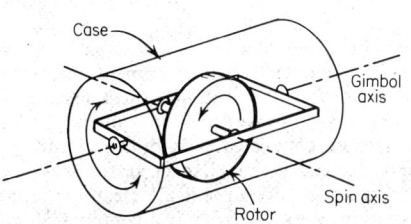

Fig. 10-7. Basic single-degree-of-freedom gyro. *(Prentice-Hall, Inc., by permission.)*

Fig. 10-8. Two-degree-of-freedom gyro. *(Conrac Corp.; Prentice-Hall, Inc., by permission.)*

rotor. Some gyros use a clock spring, wound before each use, or a pyrotechnic charge which, when activated, forces a stream of combustion gases into a small turbine.

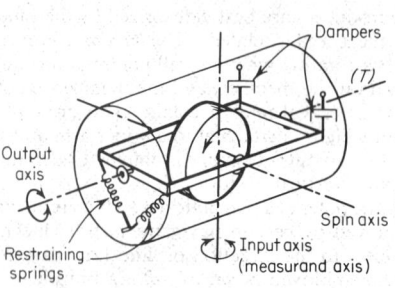

Fig. 10-9. Basic-rate gyro. *(Prentice-Hall, Inc., by permission.)*

Rate gyros are attitude-rate transducers. They provide an output proportional to angular velocity (time rate of change of attitude). The operating principle of the rate gyro (see Fig. 10-9) is similar to that of the single-degree-of-freedom free gyro, except that the gimbal is elastically restrained and its motion is damped. The output is representative of gimbal deflection about the output axis in response to attitude-rate changes about the input axis. The deflection of the gimbal (precession) is caused by the torque T applied to it. The applied torque is the product of the instantaneous attitude rate about the input axis and the angular momentum of the gyro.

When selecting a gyro, attention must be paid not only to the usual characteristics (weight, size, range, linearity, repeatability, threshold, etc., and dynamic characteristics for rate gyros) but also to *drift*, the amount of precession of the spin axis from its intended position due to internal unwanted torques, and the time period (after spin-motor runup or after uncaging) during which measurements must be obtained continuously.

21. Displacement and Position Transducers. *Position* is the spatial location of a body or point with respect to a reference point. *Displacement* is the vector representing a change in position of a body or point with respect to a reference point. Displacement transducers are used to measure linear and angular displacements, as well as to establish position from a displacement measurement.

The sensing element of most displacement transducers is the *sensing shaft* with its coupling device, which must be of a design suitable to make the motion of the sensing shaft truly representative of the motion of the measured point (driving point). A spring-loaded sensing shaft (without coupling device) is used for some applications. A number of *contactless-sensing* transducer designs are also in use. These require no coupling or sensing shaft. Various transduction principles are employed in displacement transducers.

22. Capacitive Displacement Transducers. In these devices a linear or angular motion of the sensing shaft causes a change in capacitance either by relative motion between one or more moving *(rotor)* electrodes and one or more stationary *(stator)* electrodes or by moving a sleeve of insulating material, having a dielectric constant different from that of air, between two stationary electrodes.

23. Inductive displacement transducers can be of the coupled or the noncontacting types. Coupled designs contain a coil whose self-inductance is varied as a nonmagnetic sensing shaft moves a magnetically permeable core gradually into or out of the central hollow portion of the coil. Some designs incorporate an additional coil (balancing coil) having a fixed inductance value equal to the inductance of the transduction coil at a predetermined "zero" position of the sensing shaft. The two coils are connected as two arms of an inductance bridge. This two-coil principle is used in some noncontacting displacement transducers in which the transduction coil has a stationary core but changes its inductance with the distance between itself and a moving ferromagnetic object.

24. Photoconductive displacement transduction is employed in at least one coupled displacement transducer design and in several noncontacting displacement measuring systems. In the coupled version the sensing shaft moves a shutter (a plate with a small slit) between a light source and either a potentiometric arrangement of photoconductive and conductive material or two photoconductive sensors connected in a bridge circuit and mounted so as to decrease the light incident on one sensor as the other sensor receives more light. Noncontacting photoconductive sensors usually require an optical reflector mounted to the measured object. Various optical configurations are used to obtain an output from the photoconductive element as the intensity, the phase, or the position of the reflected light beam changes. The *laser interferometer* is included in this group of displacement sensing devices.

25. Potentiometric displacement transducers are widely used because of their relative simplicity of construction and their ability to provide a high-level output. All these designs use a sensing shaft. The wiper arm is either attached directly to the shaft (but insulated from it) or mechanically connected to it through an amplification linkage. Straight potentiometric resistance elements are used in linear displacement transducers; circular or arc-shaped elements in angular displacement transducers. The elements are usually wire-wound, but conductive plastic, carbon film, metal film, or ceramic-metal mixtures (cermets) are also used. Some transducers have two or more wiper-element combinations moved by the same sensing shaft. A good sliding seal is needed at the point where the sensing shaft enters the transducer case, to protect the internally exposed resistance elements from atmospheric contaminants and moisture.

Reluctive displacement transducers are as commonly used as the potentiometric types. The reluctive transduction circuits employed in linear- and angular-displacement transducers are illustrated in Fig. 10-10. Only the linear-variable differential transformer (LVDT) and the inductance-bridge circuits are used for linear-displacement measurements. Many winding configurations exist for the LVDT transducers; one manufacturer offers 12 different "off-the-shelf" configurations, including several with two separate secondary windings. Alternating-current excitation is required for all reluctive transducers. However, some designs are available with integral ac/dc output conversion and, in some cases, also integral dc/ac excitation conversion. Synchro-type transducers are often connected to a synchro-type receiver, which indicates the measured angle directly, e.g., on a dial.

A few strain-gage displacement transducers have been designed for the measurement of small linear and angular displacement. The gages are usually attached to the top and bottom surfaces of a cantilevered or end-supported beam which is deflected by the displacement.

26. Digital-output displacement transducers (encoders) are frequently referred to as *linear encoders* and *angular* or *shaft-angle encoders*, respectively. These consist essentially of a strip (for linear displacements) or a disk (for angular displacements), coded so as to provide a digital readout for discrete (sometimes very small) displacement increments and a reading head. Three types of encoders are in common use:

1. Brush-type encoders, in which the reading head is a sliding contact (brush) which wipes over a partly conductive, partly nonconductive coded pattern.

2. Photoelectric encoders, in which the reading head consists of a light-source assembly on one side of the disk or strip and a corresponding light-sensor assembly facing it on the other side of the disk or strip; the coded pattern is partly translucent, partly opaque.

3. Magnetic encoders, with a magnetic reading head and a partly magnetized, partly nonmagnetized coded pattern.

Incremental encoders have a simple, alternately "on" and "off" coded pattern. They provide an output in the form of number of *counts* between the start and end of the displacement. Hence the start position must be known if the end position is to be determined in absolute terms. *Absolute encoders* have a code pattern such that a unique digital word is formed for each discrete displacement increment. Various codes are used for this purpose, such as the binary, binary-coded-decimal (BCD), and the Gray code.

Among displacement-transducer *selection criteria*, the most critical are range, resolution, starting force, overtravel, and type and magnitude of full-scale output. Accuracy and dynamic characteristics, type of available excitation supply, and freedom from contamination by the ambient atmosphere or other fluids need to be considered as well, for all transducer applications.

27. Force, Torque, Mass, and Weight Transducers. *Force* is the vector quantity necessary to cause a change in momentum. *Mass* is the inertial property of a body, a measure of the quantity of matter in the body and of its resistance to change in its motion. *Weight* is the gravitational force of attraction; where gravity exists, it is equal to mass times acceleration due to gravity. *Torque* is the moment of force, the product of force and the perpendicular distance from the line of action of the force to the axis of rotation (lever arm).

28. Force transducers (load cells) are used for force measurements (compression, tension, or both) as well as for weight determinations in any locality where gravity exists and the gravitational acceleration g is known. The standard g (on Earth) is 9.80665 m/s^2. Mass can be determined from weight, which is expressed in force units. A mass of one kilogram, for example, "weighs" 2.205 lb$_f$ (pounds-force) on Earth. Torque is measured by *torque transducers.*

The sensing elements of force and torque transducers usually convert the measurand into a mechanical deformation of an elastic element. This deformation, in terms of either local strains or gross deflection, is then converted into a usable output by a suitable transduction

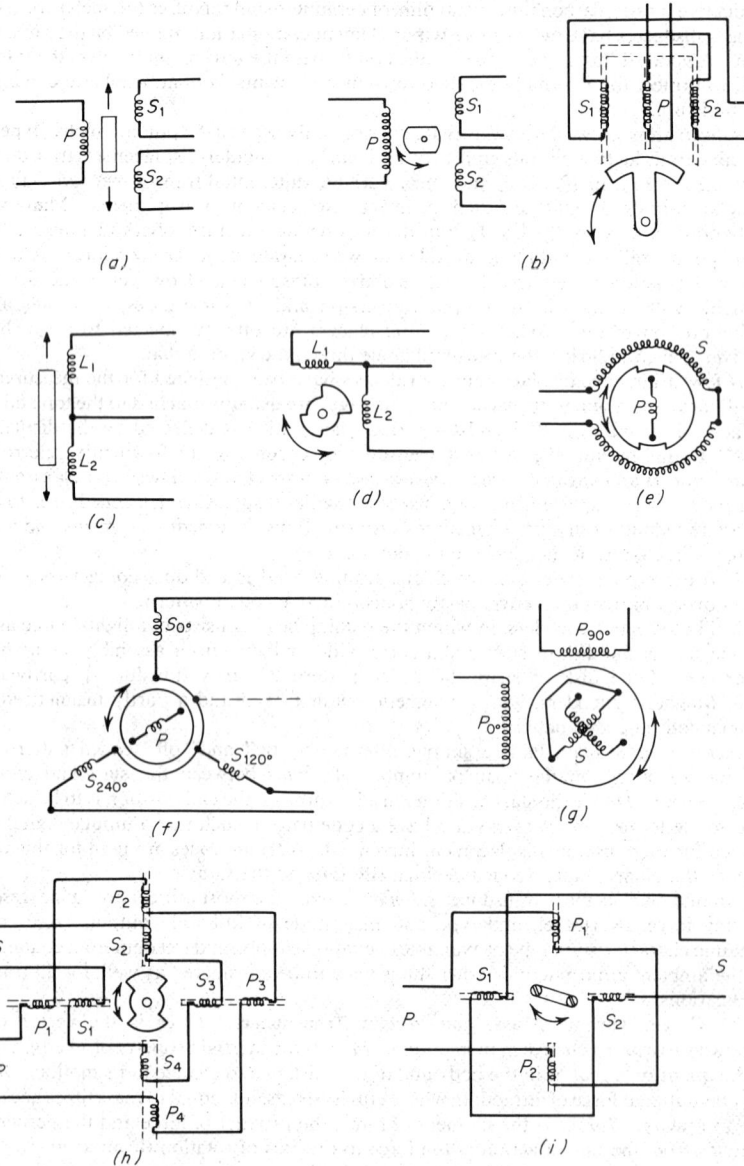

Fig. 10-10. Transduction circuits of reluctive displacement transducers. (*a*) Linear differential transformer; (*b*) angular differential transformer; (*c*) linear inductance bridge; (*d*) angular inductance bridge; (*e*) induction potentiometer; (*f*) synchro; (*g*) resolver; (*h*) microsyn; (*i*) shorted-turn signal generator. *(Prentice-Hall, Inc., by permission.)*

element. Bending beams (cantilever, end-supported, or end-restrained), solid rings or frames *(proving rings)*, and solid or hollow rectangular or cylindrical columns are the most commonly used force sensing elements. Special solid or notched shafts are used as torque sensing elements.

29. Piezoelectric force transducers are used for dynamic compression-force measurements. A typical design has the shape of a thick washer. The annular piezoelectric crystal segments are sandwiched between two hollow cylindrical columns. Bidirectional force measurements can be obtained by preloading this *force washer*. An amplifier is used to boost the low-level output signals.

30. Reluctive force transducers use proving-ring sensing elements in most design versions. The deflection of the proving ring is converted into an ac output by an inductance-bridge or differential-transformer transduction element. An entirely different design uses the permeability changes due to stresses in a laminated column to vary the voltage induced by a primary winding in a secondary winding.

31. Strain-gage force transducers are the most widely used type. Bonded-metal foil and metal wire gages predominate, but unbonded wire gages and bonded semiconductor gages are

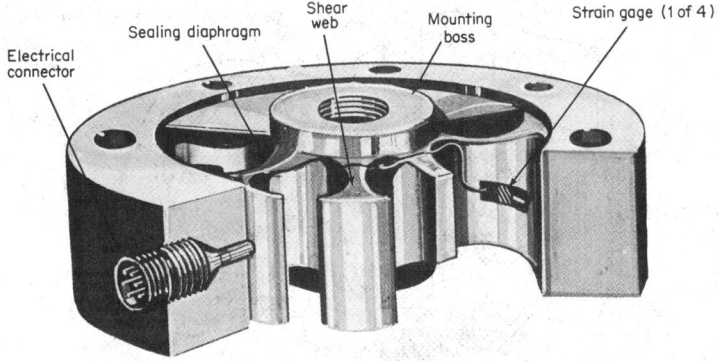

Fig. 10-11. Strain-gage force transducer. *(Interface, Inc.)*

used in some designs. Columns and proving rings are the usual sensing elements. The shear-web sensing element of the force transducer shown in Fig. 10-11 is related to the column, but is reported to offer greater transduction efficiency.

32. Torque transducers are mostly of the reluctive, photoelectric, or strain-gage type. The latter is more widely used than the two former types. The metal-foil strain gages in the transducer shown in Fig. 10-12 are located on the sensing shaft, which is enclosed within a cylindrical *torque sensor* housing. The leads from the gages are carried through the shaft to slip rings. Brushes ride on the slip rings to provide stationary external connections. The brush assembly can be lifted off the slip rings to increase brush life during periods when torque is not monitored. The speed sensing provisions are described in Par. **10-36**. In some strain-gage torque transducers the slip rings and brushes are replaced by a rotary transformer.

33. Reluctive torque transducers utilize changes in shaft permeability, due to torque-induced stresses in the shaft, to change the voltage coupled from a primary winding to two secondary windings. Photoelectric torque transducers use two incremental-encoder disks, one on each end of the shaft, to change the illumination on a light sensor when one disk undergoes a small angular deflection, due to torque, relative to the other disk.

Selection criteria include the usual range, accuracy, excitation, and output characteristics, case configuration and dimensional constraints, overload rating, the thermal environment, and, for torque transducers, maximum shaft speed and proximity of any magnetic fields that may cause reading errors. A frequent application of force transducers is in automatic weighing systems.

34. Speed and Velocity Transducers. *Speed* (a scalar quantity) is the magnitude of the time rate of change in displacement. *Velocity* (a vector quantity—magnitude and direction) is the time rate of change of displacement with respect to a reference system. *Velocity transducers* are almost invariably linear-velocity transducers, whereas speed transducers are normally angular-speed transducers *(tachometers)*.

35. Velocity transducers are usually of the electromagnetic type, exemplified by a coil in which a permanent-magnet core moves freely. The core has a sensing-shaft extension, and the shaft is attached to the object whose (usually oscillatory) velocity is to ` . measured. The . ate at which lines of magnetic flux from the core are cut by the coil turns determines the amount of electromotive force generated in the coil; hence the output is proportional to the velocity of the measured point. In some designs the coil moves within a fixed magnetic field instead.

36. Tachometers are also predominantly of the electromagnetic type. Such angular-speed transducers as the *dc tachometer generator*, the *ac induction tachometer*, and the *ac*

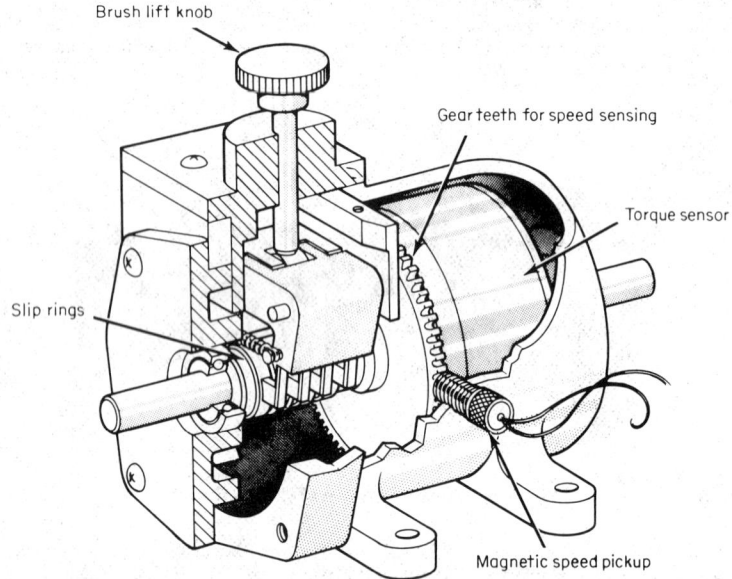

Brush lift knob

Gear teeth for speed sensing

Torque sensor

Slip rings

Magnetic speed pickup

Fig. 10-12. Strain-gage torque transducer. *(Lebow Associates, Inc.)*

permanent-magnet tachometers (ac magneto) are electrical generators. Their output amplitude increases with angular (rotational) speed. In the case of the ac magneto, the output frequency also increases with speed. The output of a *toothed-rotor tachometer* also varies in both amplitude and frequency, but the frequency variation is much greater than the amplitude variation and represents the angular speed much more accurately.

The speed-sensing gear teeth and sensing coil (pickup) incorporated in the torque transducer of Fig. 10-12 constitute a toothed-rotor tachometer. A pulse is generated in the electromagnetic sensing coil every time a ferromagnetic tooth passes by it. Since there are 60 teeth on the gear shown in the illustration, 60 pulses per revolution are provided by the sensing coil. By counting the pulses over a fixed time interval the angular speed can be determined with very close accuracy. *Photoelectric tachometers* provide the same degree of accuracy, typically by chopping a beam between a light source and a light sensor into equidistant pulses by an incremental-encoder disk attached to the sensing shaft. The pulse-frequency output can also be converted into a dc output voltage if the degraded accuracy, resulting from the conversion, can be tolerated.

Selection criteria include, besides range and accuracy characteristics, the mounting position and required frequency response for velocity transducer, and the type of available readout or signal-conditioning and telemetry equipment in the case of tachometers.

37. Strain Transducers. *Strain* is the deformation of a solid resulting from *stress,* the force acting on a unit area in a solid. Strain is measured as the ratio of dimensional change to the total value of the dimension in which the change occurs. Essentially, all strain transducers are resistive and are referred to as *strain gages.* Their most essential characteristic is their sensitivity *(gage factor),* the ratio of the unit change in resistance to the unit change in dimension (length).

38. Strain gages employ either a conductor or semiconductor, of small cross-sectional area, suitable for mounting to the measured surface so that it elongates or contracts with that surface and changes its resistance accordingly. A typical metal-wire gage is shown in Fig. 10-13. Other types of metal gages are made of thin metal foil, die-cut or etched into the required pattern and deposited on an insulating substrate through a pattern mask by bombardment or evaporative methods. The metals used in strain gages are usually copper-nickel alloys; other alloys such as nickel-chromium, platinum-tungsten, and platinum-iridium are also used. Semiconductor gages are usually made from doped-silicon wafers or blocks.

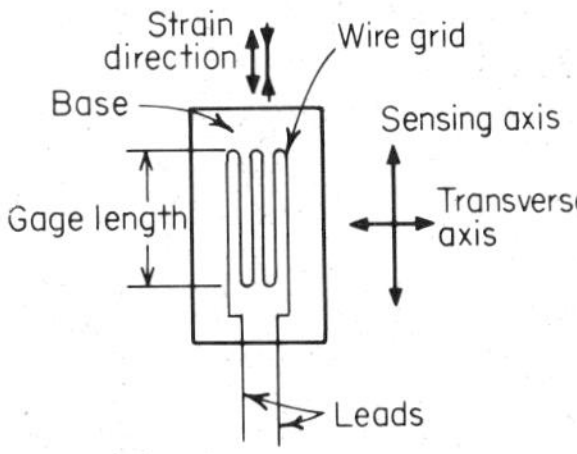

Fig. 10-13. Bondable wire-grid strain gage. *(Prentice-Hall, Inc., by permission.)*

Strain gages can be *bare (surface-transferable,* free-filament), bonded to an insulating carrier sheet on one side only, or completely *encapsulated* in a bondable (usually plastic) or weldable (metal) carrier, the latter insulated internally from the gage. Bare gages are normally supplied with a strippable insulating substrate *(carrier).* Since two or more gages are normally used to obtain a strain measurement, for temperature-compensation, linearity-compensation, and output-multiplication purposes, strain-gage *rosettes* are sometimes used, combining two, three, or four gages, mutually aligned as to their strain-sensing axes, on one carrier.

Selection criteria involve the desired type and size (always including gage length and width), type, and material of connecting leads and spacing between them on the gage itself, type of carrier or encapsulation, gage resistance, gage factor, transverse sensitivity tolerances, allowable overload *(strain limit),* and maximum excitation current for a given application. Semiconductor gages must often be shielded from illumination, which can cause reading errors. Proper methods of attachment and of connection into a Wheatstone bridge circuit are very critical for strain gages.

TRANSDUCERS FOR FLUID-MECHANICAL QUANTITIES

39. Density Transducers. *Density* is the ratio of the mass of a homogeneous substance to a unit volume of that substance. Density transducers *(densitometers)* are used for the determination of the density of fluids (gases, liquids, and slurries). They are, however, not related to densitometers used to measure optical density, as of a photographic image, nor to equipment used to determine spectral density (e.g., power spectral density).

Three methods are primarily used for density sensing.

Sonic density sensing is achieved by an arrangement of piezoelectric sound (usually ultrasound) transmitters and receivers producing outputs proportional to the speed of sound in the fluid and to the acoustic impedance of the fluid. Since acoustic impedance varies with the product of speed of sound and density, a signal proportional to density can be derived from the transducer and signal conditioning system.

Radiation density sensing relies upon the attenuation, due to density, of the radiation passing from a radioisotope source, on one side of the fluid-carrying pipe or vessel, to a radiation detector on the opposite side.

Vibrating-element density sensing employs a simple mechanical structure, such as a cylinder or a plate, electromagnetically set into vibration at its resonant frequency. This frequency changes with density, and an output is produced, proportional to density, which is related directly to the square of the period of vibration. The transducer illustrated in Fig. 10-14 uses a vibrating plate in its sensing head. The plate is installed as the end-supported beam in a circular support structure containing a crystal detector close to one of the attachment points of the beam. A magnetostrictive drive sets the beam into vibration at its resonant frequency. This frequency and its variations due to density change are converted into a usable output by the crystal detector and built-in preamplifier. Additional methods, used to infer density from other measurements, have also been employed in measurement systems.

40. Flow Transducers. *Flow* is the motion of a fluid. *Flow rate* is the time rate of motion expressed either as fluid volume per unit time *(volumetric flow rate)* or as fluid mass per unit time *(mass flow rate)*. Transducers used for flow measurement *(flowmeters)* generally measure flow rate. Most flowmeters measure volumetric flow rate, which may be converted to mass flow rate by simultaneously measuring density and computing mass flow rate from the two measurements. Some flowmeters measure mass flow rate directly. Flow sensing elements can be categorized as follows:

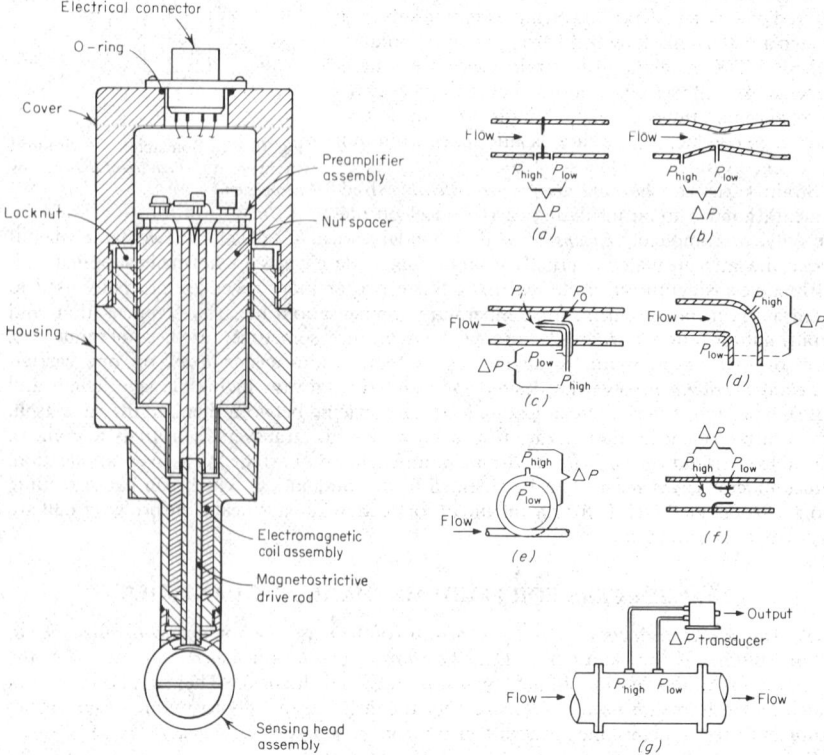

Fig. 10-14. Vibrating-beam density transducer. *(ITT Barton.)*

Fig. 10-15. Flow measurement using differential pressure sensing elements. (*a*) Orifice plate; (*b*) venturi tube; (*c*) pitot tube; (*d*) centrifugal section (elbow); (*e*) centrifugal section (loop); (*f*) nozzle; (*g*) measurement of differential pressure due to flow rate. *(Prentice-Hall, Inc., by permission.)*

(a) Differential-pressure flow sensing elements: Sections of pipe provided with a restriction or curvature which produces a pressure differential (ΔP), proportional to flow rate, across two points of the device (see Fig. 10-15). The output of a differential pressure transducer whose input ports are connected to these two points is representative of flow rate through the sensing element. Known relationships of ΔP versus flow rate exist for each type of element.

(b) Mechanical flow sensing elements: Freely moving elements (e.g., turbine or propeller) or mechanically restrained elements (e.g., a float in a vertical tapered tube, a spring-restrained plug, a hinged or cantilevered vane) whose displacement, deflection, or angular speed is proportional to flow rate.

(c) Flow sensing by fluid characteristics: Certain transduction elements can be so designed and installed that they will interact with the moving fluid itself and produce an output relative to flow rate. The heated wire of a *hot-wire anemometer* transfers more of its heat to the fluid as the flow rate increases, thereby causing the resistance of the heated wire to decrease. When small amounts of radioisotope tracer material are added to the fluid, a radiation detector close to the moving fluid will respond with increasing output as the flow rate increases *(nucleonic flowmeter).*

In the *fluid-conductor magnetic flowmeter* an increasing electromotive force is induced in an electrically conductive fluid, flowing through a transverse-magnetic field, as the flow rate increases. In the *thermal flowmeter* two thermocouple junctions are immersed into the moving fluid, one upstream, the other downstream, from an electric heater immersed in the same fluid, and the two junctions are connected as a differential thermocouple, the output of the latter increasing with mass flow rate. In a similar device, the *boundary-layer flowmeter,* only the portion of the fluid immediately adjacent to the inside wall of the pipe is heated and thermally sensed.

41. Turbine Flowmeter. The turbine flowmeter (see Fig. 10-16) is among the most

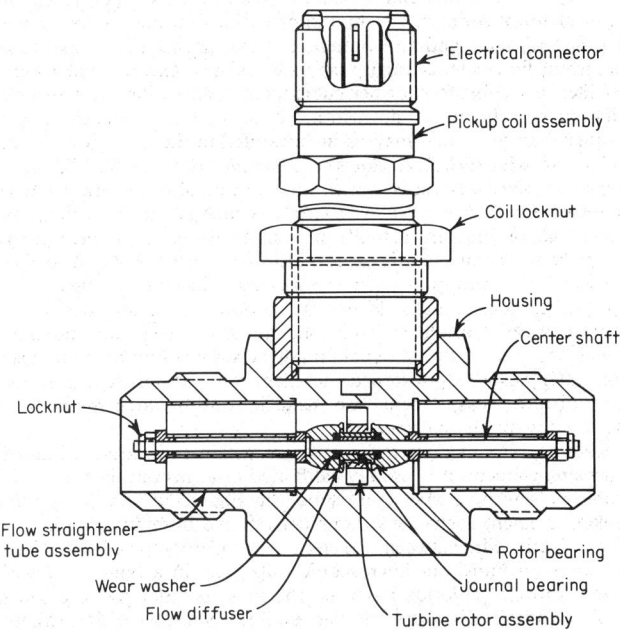

Fig. 10-16. Turbine flowmeter. *(ITT Barton.)*

widely used flow-rate transducers. Its operating principle is similar to that of the toothed-rotor tachometer described in Par. **10-36.** The bladed rotor (turbine) rotates at an angular speed proportional to volumetric flow rate. Rotational friction is reduced as much as possible by special bearing design. As each magnetic rotor blade cuts the magnetic flux of the pickup coil's pole piece, a pulse is induced in the pickup coil (sensing coil). A frequency meter is used to display the frequency output of the flowmeter, or a frequency-to-dc converter can be utilized to provide a dc voltage increase with flow rate. The rotor blades can be so machined that the variable-frequency ac voltage across the sensing coil terminals is virtually sinusoidal. This permits use of an FM demodulator as frequency-to-dc converter. The number of turbine blades, the pitch of the blades, and the internal geometry of the flowmeter determine the range of output frequencies for a given flow-rate range.

42. Oscillating-Fluid Flowmeter. In this device the fluid is first forced into a swirling motion, then passes through a venturi-like cavity at a point of which the flow oscillates about the axis of the flowmeter. A fast-response resistive temperature transducer at that point provides an output in terms of frequency of resistance changes. This frequency, proportional to flow rate and converted into voltage variations, can then be displayed on a counter after it has been amplified, filtered, and wave-shaped.

43. Other flowmeter designs include the *ultrasonic flowmeter,* typically using pairs of piezoelectric transducers to establish sonic paths. Changes in flow rate produce corresponding changes in the propagation velocity of sound along the path. *Strain-gage flowmeters* use cantilevered vanes or beam-supported drag bodies which deflect or displace due to fluid flow. The strain in the deflecting beam is then transduced by strain gages. A few types of *angular-momentum mass flowmeters* have been developed in which the fluid either imparts angular momentum to a circular tube through which it flows or receives angular momentum by a rotating impeller. The angular momentum is then used to cause an angular displacement or a torque in a mechanical member, either of which can be transduced to provide an output proportional to mass flow.

Selection criteria involve, first, a choice of either a flowmeter alone or a complete flow-rate or flow measuring system which can include signal conditioning and display equipment and, when required, a flow totalizer. Among the essential flowmeter characteristics are the (mass or volumetric) flow-rate range, the properties and type(s) of the measured fluid (gas, liquid, mixed-phase, slurry), the nominal and maximum pressure and temperature of the fluid, the configuration, mechanical support, weight and provisions for connection of the flowmeter, the required time constant, and the output, as well as accuracy, specifications. The sensitivity of a turbine flowmeter is usually expressed as the *K factor,* stated in hertz (or cycles) per gallon, per liter, per cubic foot, or per cubic meter. Attention must also be paid to the length of straight pipe upstream and downstream of the flowmeter and the necessity for flow straighteners other than those that may be incorporated in the transducer itself.

44. Humidity and Moisture Transducers. *Humidity* is a measure of the water vapor present in a gas. It is usually measured as relative humidity or dew-point temperature, sometimes as absolute humidity. *Relative humidity,* which is temperature-dependent, is the ratio of the water-vapor pressure actually present to water-vapor pressure required for saturation at a given temperature; it is expressed in percent (% RH). The *dew point* is the temperature at which the saturation water-vapor pressure is equal to the partial pressure of the water vapor in the atmosphere. Hence any cooling of the atmosphere, even a slight amount below the dew point, produces water condensation. The relative humidity at the dew point is 100% RH. *Moisture* is the amount of liquid adsorbed or absorbed by a solid; it is also the amount of water adsorbed, absorbed, or chemically bound in a nonaqueous liquid. Humidity and moisture measurements are made by one of three methods: hygrometry, psychrometry, and dew-point determination.

45. Hygrometers. The hygrometer is a device which can measure humidity directly, with a single sensing element; it is usually calibrated in terms of relative humidity. Three types of hygrometric sensing elements are shown in Fig. 10-17. In the *resistive* humidity-transducer sensing element a change in ambient relative humidity produces a change in resistance of a conductive film between two electrodes. Carbon powder in a binder material has been used for such films, but hygroscopic salts, also in a binder material, are more commonly used. Lithium chloride has been the most popular hygroscopic salt in such applications. The *mechanical* hygrometric element is the oldest type. It uses a material, such as human hair or animal membrane, which changes its dimension with humidity. The resulting displacement on an attaching point on the material is then transduced into an output proportional to humidity. The *oscillating-crystal* hygrometric element consists of a quartz crystal with a hygroscopic coating, so that the total crystal mass changes as water is adsorbed on, or desorbed from, the coating. When the crystal is connected into an oscillator circuit, the oscillator output frequency will change with changes in humidity.

Several other types of hygrometric sensing elements have also been developed. In the *aluminum oxide element* an impedance (resistance and capacitive reactance) change occurs with changes in humidity. The *Brady array* also provides an ac output when excited with alternating current (at about 1 kHz). However, it differs from other devices in that it consists of an array of semiconducting crystal matrices which look electrically neutral to the water molecule. Vapor pressure then allows the molecules to drift in and out of the interstices, creating an exchange of energy within the structure. The *porous-glass-disk* hygrometric element has electrodes plated on the two surfaces of the disk. When water vapor permeates the pores in the glass, it is decomposed electrolytically when a voltage is applied across the electrodes. The current necessary to decompose the water is then a measure of relative humidity.

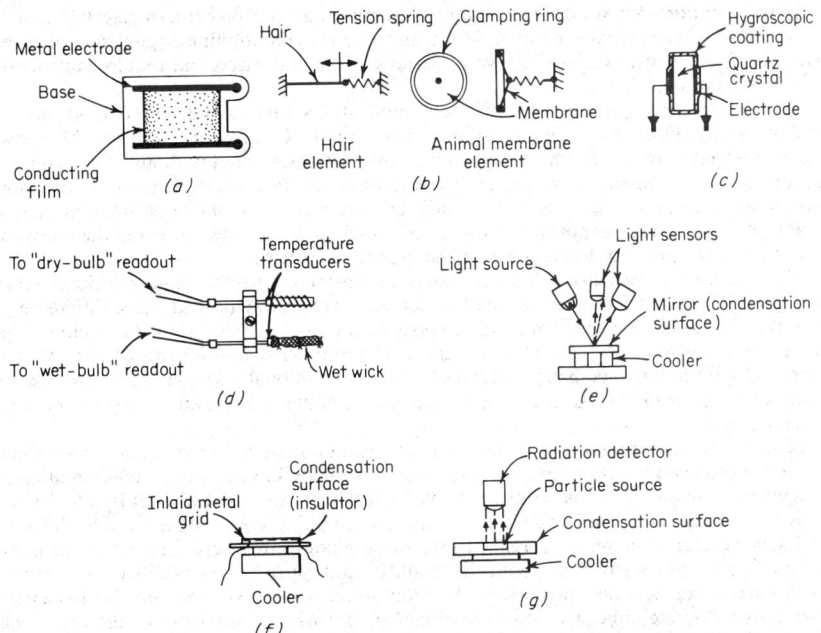

Fig. 10-17. Sensing elements of humidity transducers. (*a*) Resistive; (*b*) mechanical; (*c*) oscillating crystal; (*d*) psychrometric; (*e*) photoelectric; (*f*) resistive; (*g*) nucleonic; (*e*), (*f*), and (*g*) are dew-point sensors. *(Prentice-Hall, Inc., by permission.)*

46. Psychrometers. Psychrometers use two temperature sensing elements (see Fig. 10-17). One element, *dry bulb,* measures ambient temperature, the other, *wet bulb,* covered with a water-saturated wick or similar device, measures temperature reduction due to evaporative cooling. Relative humidity can be determined from the dry-bulb temperature reading, the differential temperature between dry-bulb and wet-bulb readings, and knowledge of the barometric pressure by referring to a *psychrometric table* of numbers. Such tables are available from government agencies (e.g., weather service), as well as from manufacturers. The temperature sensing elements are usually resistive (platinum- or nickel-wire windings or thermistors), sometimes thermoelectric (see Pars. **10-58 to 10-65** for further description).

47. Dew-point sensing elements are dual elements. The condensation-detection element senses the first occurrence of dew on a surface whose temperature is being lowered. The temperature sensing element measures the temperature of this surface so that the dew point (the temperature at which condensation first occurs as the temperature is lowered) can be determined by monitoring the output of both elements simultaneously. Typical condensation detectors (see Fig. 10-17) include a photoelectric device in which light sensors detect the difference in light, reflected from a mirror that serves as the condensation surface, when the dew point is reached; a resistive element in which a change in conductivity occurs in an inlaid metal grid at the condensation surface when condensation occurs; and a nucleonic device in which drop in particle flux, emitted from a radiation source at the condensation surface, indicates the dew point.

48. Auxiliaries. Resistive humidity transducers are generally more popular than other types when a transducer, rather than a complete measurement system, is required. Almost all types require ac excitation. The electrodes are spiral, helical, or loop-shaped to obtain as large a resistance change as feasible for a given element size. Other hygrometric transducers usually require at least an excitation and signal conditioning unit.

Psychrometric transducers are typically complemented by a signal conditioning and readout system. A small blower *(aspirator)* is often included to blow the ambient air over the two sensing elements so that a faster response can be obtained.

Dew-point humidity transducers require, as a minimum, a cooler (thermoelectric coolers are often used), its associated control circuit, and a power conditioning circuit, as well as the two sensing elements. However, several designs are miniaturized and require sufficiently little signal conditioning.

Selection Criteria. Humidity transducer applications should first be examined to see whether relative humidity or dew point is to be measured. Relative humidity can, of course, also be inferred from psychrometric and dew-point readings, but not without a look-up table or calculations. Among performance characteristics the measurement range is the most important; measurement accuracy can usually be improved when only a partial range needs to be measured. Other important characteristics include the temperature and the chemical properties of the ambient atmosphere or the measured material.

49. Liquid-Level Sensing. A large variety of sensing approaches and transducer types have been developed for the determination of the level of liquids and quasi-liquids (e.g., slurries and powdered or granular solids) in open or enclosed vessels (tanks, ducts, etc.). Not only is the knowledge of the *level* itself important, but other measurements can be inferred from level. If the tank geometry and dimensions are additionally known, the *volume* of the liquid can be determined. If, additionally, the density of the liquid is known, its *mass* can be calculated.

Level is generally sensed by one of two methods: obtaining a discrete indication when a predetermined level has been reached *(point sensing)* or obtaining an analog representation of the level as it changes *(continuous sensing)*. Point sensing is also used when it is only desired to establish whether a liquid or a gas exists at a certain point (e.g., in a pipe). The different level sensing methods can be classified into those lending themselves primarily to point sensing, to continuous sensing, or both. It should be understood, of course, that point sensing systems are usually simpler and cheaper than continuous sensing systems and should be used when only a discrete indication has to be obtained. Even when two or more discrete levels must be established in one vessel, the use of two or more point sensors may be preferable to a continuous sensing system. On the other hand, electronic circuitry can be used to provide one or more discrete level indications from a continuous sensing system.

50. Point level sensing methods are usually aimed at indicating the interface between a liquid and a gas, sometimes the interface between two different liquids. Three methods are illustrated in Fig. 10-18. *Heat-transfer sensing* is employed by two types of sensors: the resistive sensor (wire-wound or thermistor) is heated to some degree by the current passing through it so that its resistance changes due to cooling when contacted by the liquid; the thermoelectric sensor detects the cooling, upon liquid contact, of a wire-wound heater it is in thermal contact with. *Optical sensing* relies either on the presence or absence of reflection of a light beam from the interface between a prism surface in contact with gas (reflection) or liquid (no reflection) or on the greater attenuation of a light beam when it passes through liquid on its way to a light sensor. In *damped-oscillation sensing* the mechanical vibration of an element, excited into such vibration electrically, is either stopped (in a magnetostrictive or piezoelectric element) or reduced in amplitude (e.g., in an oscillating-paddle element) due to acoustic damping or viscous damping, respectively, when the measured fluid changes to a liquid.

51. Continuous Level Sensors. Three classic continuous level sensing methods are illustrated in Fig. 10-19. The level, volume, or mass of a liquid in a tank of known geometry can be determined by *weighing* the tank continuously, as by means of a load cell (force transducer), and substracting the tare weight of the tank or compensating for the tare weight. *Pressure sensing* relies on the pressure *(head)* developed at the base of a liquid column. This pressure increases with the column height, and hence with level above the point at which pressure is sensed. The differential pressure P_D, measured by the differential-pressure transducer, on the tank shown in the illustration, is equal to the difference in pressures between the bottom and top of the tank $(P_L - P_H)$. The level h of the liquid above the bottom sensing point is then given by $h = (P_L - P_H)/w$, where w is the specific weight of the liquid. Pressure acting at the bottom of the tank can also be sensed by a diaphragm built into the tank bottom, used as pressure sensing element. A third method uses the increasing pressure near the bottom of a tank, as the level rises, to compress electrically conductive strips against a resistive element, gradually shorting out the resistance (and therefore decreasing this resistance) as the level increases. The *level-sensing float* mechanically actuates a transduction

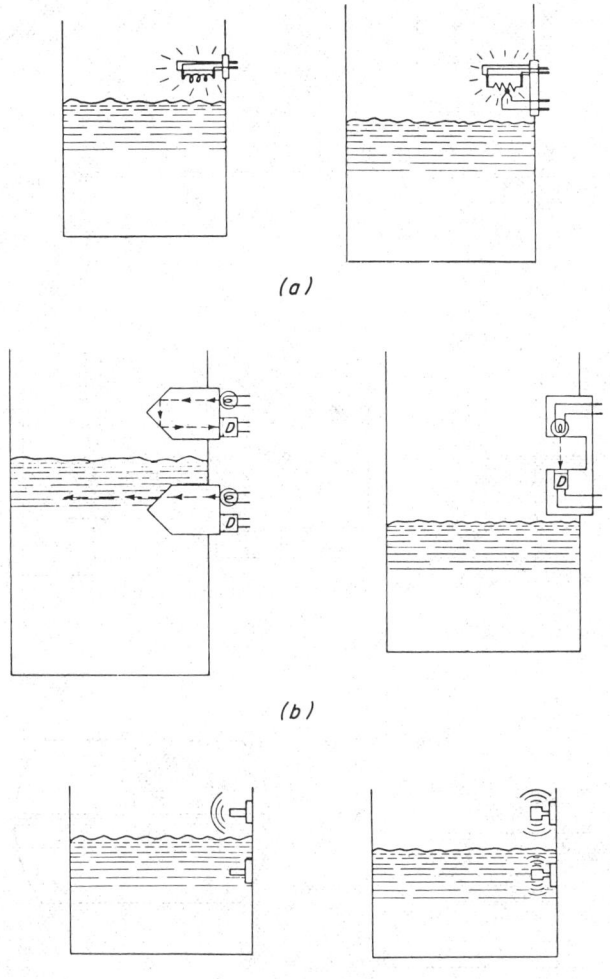

Fig. 10-18. Point-level sensing methods. (*a*) By heat-transfer rate; (*b*) by optical means; (*c*) by oscillation damping. *(Prentice-Hall, Inc., by permission.)*

element, usually a potentiometer, sometimes a reluctive element or one or more magnetic reed switches. A radically different method (not illustrated) is *cavity-resonance sensing,* where electromagnetic oscillations are excited (from a coupling element at the tank top) within the gaseous cavity enclosed by the liquid surface and the upper tank walls, and the change in resonant frequency, as the liquid surface changes in location, becomes a measure of liquid level.

Several methods are equally useful for point and continuous level sensing (see Fig. 10-20). *Conductivity level sensing* is usable with even mildly conductive liquids. The resistance between two electrodes (the tank wall may serve as one of the two) changes continuously (or suddenly, in the case of the point-sensor version) as the liquid level rises or falls.

Dielectric-variation sensing is used primarily for nonconductive liquids, which then play the role of dielectric materials between two (sometimes four) concentric electrodes which are used (and electrically connected) as plates of a capacitor. The capacitance changes continuously (or suddenly, for the point sensor) as the vertical distance h of the level changes.

10-23

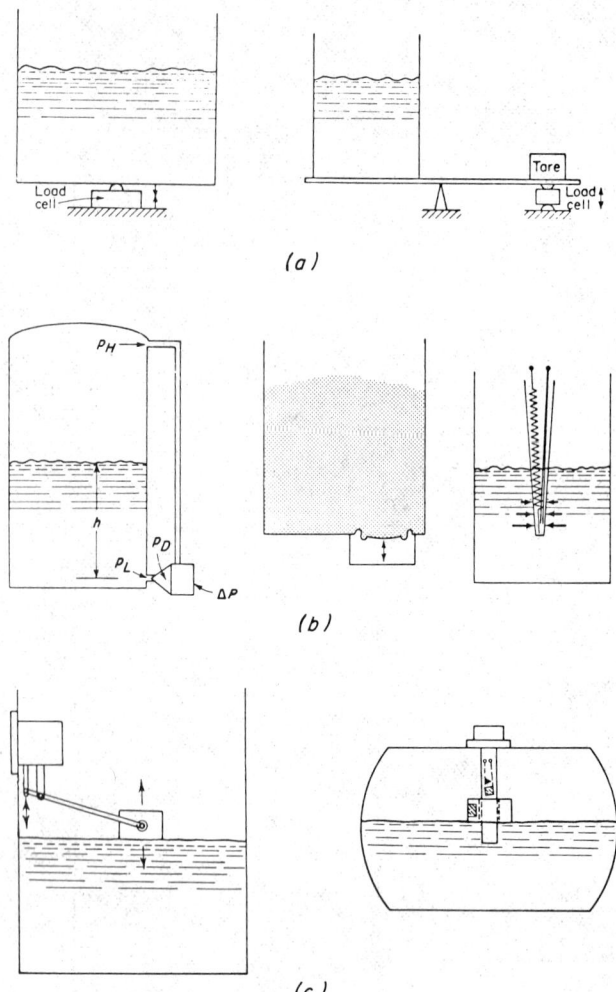

Fig. 10-19. Continuous-level sensing methods. (*a*) By weighing; (*b*) by pressure sensing; (*c*) by float. *(Prentice-Hall, Inc., by permission.)*

If it is necessary to compensate for changes in the liquid's characteristics during measurement, a reference capacitor, always submerged, can be employed so that the ratio of the capacitance change equals the ratio of the measured level to the vertical dimension of the reference capacitor ($\Delta C/\Delta C_R = h/h_R$).

Sonic level sensing uses ultrasound either emitted from a sound projector and detected by a sound receiver or emitted and detected by a single sound transceiver operating alternately in the "transmit" and "receive" mode. An echo-ranging technique is commonly used, the liquid-gas interface (the liquid level) acting as the target. The difference in attenuation or travel time of the beam of sound between liquid or gas in its path can also be used for sonic level sensing, especially for point sensing.

Radiation sensing is a nucleonic sensing method employing usually one, sometimes two or more, radioisotope sources and radiation detectors to indicate level changes by virtue of the changes in attenuation of the radiation due to level changes. The attenuation in the liquid

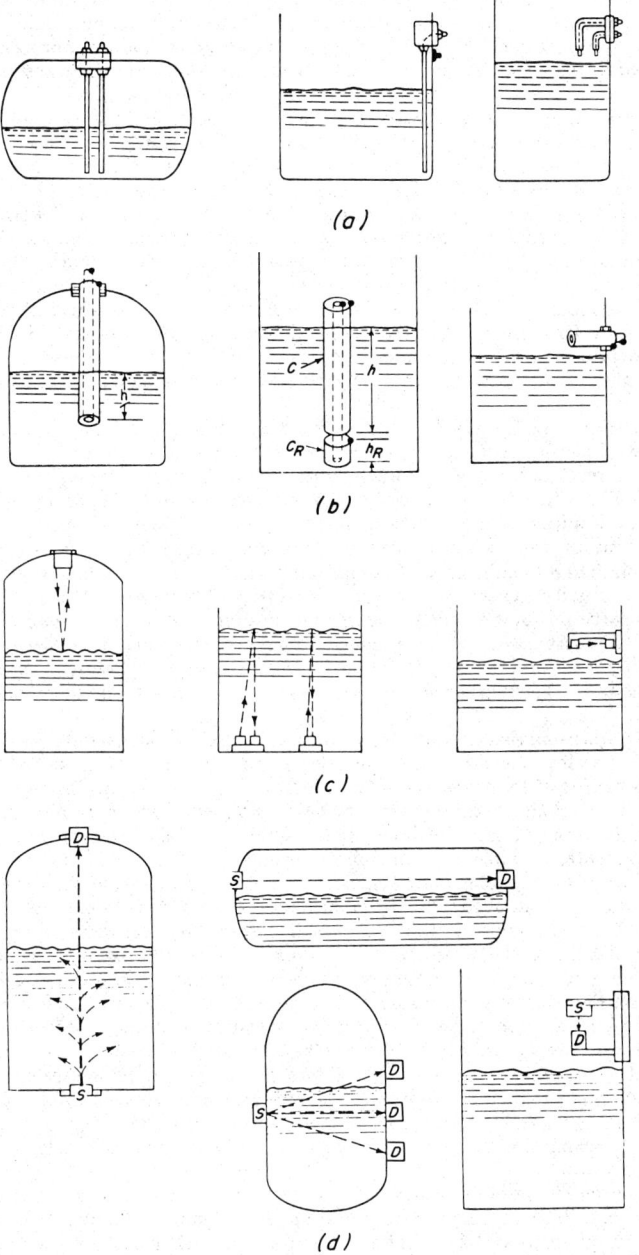

Fig. 10-20. Continuous- and point-level sensing. (*a*) Conductivity; (*b*) dielectric variation; (*c*) sonic sensing; (*d*) radiation sensing. *(Prentice-Hall, Inc., by permission.)*

10-25

is caused mainly by absorption. Such nucleonic methods have also been used to study density profiles and the location and extent of vortices in tanks and of gas bubbles in pipes.

52. Liquid-level transducers, in their most common configuration, are probes, flange- or boss-mounted through the tank or duct wall. Some pipe-wall-mounted transducers are so designed that their sensing end is flush with the inside of the wall, to prevent obstructions to flow. Nucleonic transducer systems are attached to the outside of the wall.

The transduction principle of liquid-level transducers is given by the sensing technique employed. Dielectric-variation sensing demands capacitive transducers, using ac excitation having a frequency between 400 Hz and 200 kHz. Magnetostrictive and piezoelectric transducers, whose probe tip oscillates at a frequency in the vicinity of 40 kHz, find their application in the sonic, as well as the damped-oscillation, sensing techniques. Ionization-type, as well as solid-state, transducers are used in nucleonic systems. Photoelectric transducers are utilized in optical sensing systems. Potentiometric and reluctive transduction elements are found in float-actuated liquid-level transducers. Resistive transducers are used for heat-transfer sensing and, in a somewhat different form, for conductivity and pressure sensing. Thermoelectric elements are found in some heat-transfer sensors. Vibrating-element (notably vibrating-paddle) transducers find their use in damped-oscillation sensing systems.

Selection criteria involve, first of all, the choice of one or more point-level sensors or a continuous level sensor. After this choice has been made, together with an evaluation of end-to-end system requirements, the characteristics of the measured liquid are of primary importance. These include its conductivity, viscosity, temperature, chemical properties and, for installation in pipes or ducts, its flow rate and pressure. The transducer must also be designed and installed in such a manner as to prevent false level indications due to slosh, spray, and splash and to adherence of liquid to the transducer with falling level.

53. Pressure and Vacuum Transducers. *Pressure* is force acting on a surface; it is measured as force per unit area, exerted at a given point. *Absolute pressure* is measured relative to zero pressure, *gage pressure* relative to ambient pressure, and *differential pressure* relative to a *reference pressure* or a range of reference pressures. A perfect *vacuum* is zero absolute pressure. Vacuum measurement, however, is the measurement of very low pressures.

54. Pressure sensing elements are almost invariably mechanical in nature (see Fig. 10-21). They can be described generally as thin-walled elastic members which deflect when the pressure on one side of their wall is not balanced by a pressure on the opposite side. The former pressure is the measured pressure; the latter is either a vacuum or near vacuum (for absolute-pressure transducers), the ambient atmosphere (for gage-pressure transducers), or some other pressure (for differential-pressure transducers).

The *diaphragm* is a circular plate fastened around its periphery so that its center will deflect when pressure is applied to it. It can be flat or, when a greater deflection is required, contain a number of concentric corrugations which increase the effective area upon which the force (pressure) can act. Two corrugated diaphragms, welded, brazed, or soldered together around their periphery, form a *capsule* sensing element (aneroid). Two or more capsules can be fastened together so that the pressure acts on all of them. The displacement obtainable at the end of such a multiple-capsule element nearly equals the displacement of one capsule multiplied by the number of capsules in the assembly. The *bellows* sensing element is typically made from a thin-walled tube formed into deep convolutions and sealed at one end, whose displacement can then be made to act on a transduction element. In the *straight-tube* sensing element, again sealed at one end, applied pressure causes an expansion of the tube diameter. This expansion, though slight, can be converted into a usable output by a transduction element.

The *Bourdon tube* is one of the most widely used sensing elements, particularly for pressure ranges higher than 20 bars (about 300 lb/in.^{2}). The Bourdon tube, elliptical in cross section and sealed at its tip, tends to straighten from its curved, twisted, helical, or spiral shape, thus causing the tip to deflect sufficiently to act upon a transduction element. The number of turns or twists in a Bourdon tube tends to multiply the tip travel.

55. Pressure transducers, using the sensing elements described above, provide their outputs by means of a large variety of transduction elements (see Table 10-4). Many designs are available with integrally packaged output- and excitation-conditioning circuitry. Certain designs, notably potentiometric, reluctive, and strain-gage transducers, are more prevalent

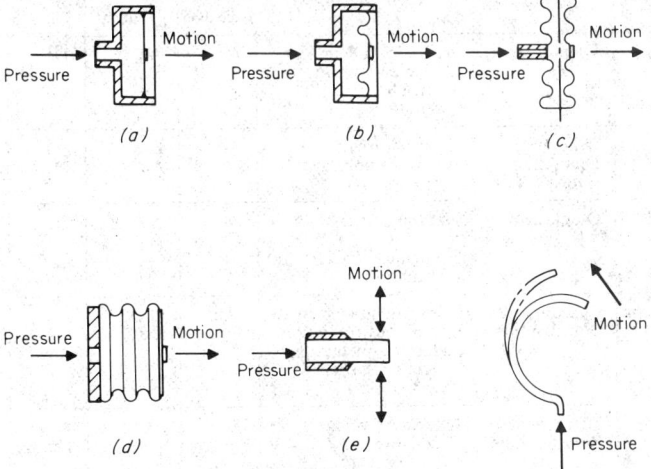

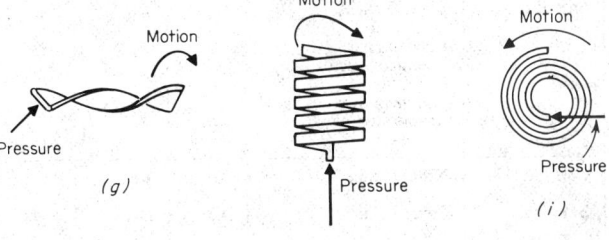

Fig. 10-21. Pressure sensing elements. (*a*) Flat diaphragm; (*b*) corrugated diaphragm; (*c*) capsule; (*d*) bellows; (*e*) straight tube; (*f*) C-shaped Bourdon tube; (*g*) twisted Bourdon tube; (*h*) helical Bourdon tube; (*i*) spiral Bourdon tube. *(Prentice-Hall, Inc., by permission.)*

than other types. Piezoelectric transducers are usable only for dynamic pressure measurements. Inductive transducers are subject to severe temperature effects and are not used extensively.

A *potentiometric pressure transducer* is illustrated in Fig. 10-22. The dual-capsule sensing element transfers its displacement to a lever-type wiper arm by means of a pushrod. The wiper then slides over the curved resistance element. Capsule elements are commonly used in such transducers for pressure ranges up to 25 bars (about 360 lb/in.^{2}).

Reluctive pressure transducers use either the inductance bridge circuit or, primarily when only the normal ac output is required, the differential-transformer circuit. When inductance bridge transducers use a diaphragm sensing element, the magnetic diaphragm itself, positioned between two coils, acts as the armature which increases the inductance of one coil while decreasing the inductance of the other coil. When inductance bridge transducers use a Bourdon tube sensing element, a flat armature plate, positioned over two coils, tilts more toward one coil than toward the other as the Bourdon tube tip rotates slightly with applied pressure. In differential-transformer transducers the sensing-element displacement is used to move a magnetic core within the transformer.

Most *strain-gage pressure transducers* use a diaphragm sensing element, although at least one good design uses a straight tube. Most designs have a four-active-arm strain-gage bridge, with the gages either on the diaphragm or on a beam actuated by the diaphragm.

Table 10-4. Pressure Transducers

Transduction	Sensing elements	Type variations	Normal		Optional	
			Excitation	Output	Excitation	Output
Capacitive	Diaphragm	Ac bridge unbalance Variable ionization Rf-tank-circuit detuning	Ac Ac Ac	Ac Dc Freq.	Dc	Dc
Inductive	Bellows Diaphragm–Bourdon tube	LC tank circuit Ac bridge unbalance	Ac Ac	Freq. Ac		
Piezoelectric	Diaphragm	Natural crystal Synthetic crystal Ceramic	None	Ac	Dc	Amplified ac
Potentiometric	Capsule Bourdon tube	Wire-wound element Continuous-resolution element (metal film, cermet, plastic, carbon film)	(Ac) Dc	(Ac) Dc		
Reluctive	Diaphragm Bourdon tube	Inductance bridge Differential transformer	Ac	Ac	Ac Dc	Dc Dc
Strain gage	Diaphragm Straight tube	Unbonded gages, metal wire Bonded gages, metal wire Bonded gages, metal foil Bonded gages, semiconductor Diffused semiconductor gages Evaporated metal gages	(Ac) Dc	(Ac) Dc	Dc	Amplified dc
Servo type	Capsule Bellows	Null balance Force balance	Ac Ac Dc	Ac Dc Dc		Encoder (digital) Synchro Potentiometric
Vibrating element	Diaphragm Straight tube	Vibrating wire Vibrating cylinder Vibrating diaphragm	Ac Dc	Freq. Freq.		

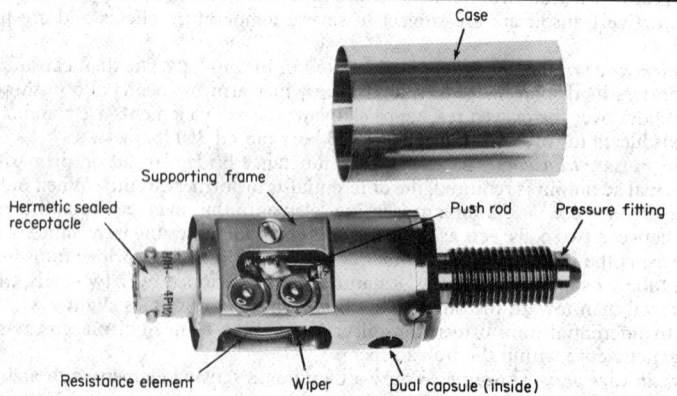

Fig. 10-22. Potentiometric pressure transducer. *(Bourns, Inc.)*

When the sensing-element displacement is not sufficient for a given transduction element, a mechanical amplification linkage can be inserted between the two elements. Special design considerations apply to differential-pressure transducers when the measured fluid (at one of the two pressure ports) must not come in contact with the transduction element. One solution to this problem has been to fill the affected inside portion of the transducer with a *transfer fluid,* sealed off by a thin *membrane* to which the measured fluid can be applied safely. Gage-pressure transducers have the inside of their case (which usually acts as the *reference cavity*) vented to the outside through a small hole *(gage vent),* equipped with a fine-mesh screen, a porous plug, or another filter to prevent internal contamination.

Flush-diaphragm transducers are designed for high-frequency-response applications where use of tubing, or even the cavity formed by a mounting boss, may reduce response; these transducers are so designed that the diaphragm is flush (when installed) with the inside surface of the pipe wall (or other wall) through which they are mechanically fastened.

Specification characteristics of pressure transducers deserve particular attention since pressure is one of the two most common measurands (the other is temperature). Table 10-5 lists those characteristics which should be considered when preparing a specification for a pressure transducer. Not all these characteristics need always be specified; some can be omitted when sufficient knowledge of the application permits.

56. Vacuum transducers are an important subgroup of pressure transducers, though bearing little resemblance to them with regard to design and operation. The pressure constituting a practical dividing line between pressure and vacuum measurement is not too well defined. Some pressure transducers are usable for very-low-pressure measurement. Generally, however, pressure measurements extending substantially below 1.3 mbars (= 1 torr = 0.02 lb/in.^{2}) can be considered as vacuum measurements.

Vacuum transducers (see Table 10-6) exist in two major categories, given by their transduction principles.

Thermoconductive vacuum transducers measure pressure as a function of heat transfer by the measured gas. As the number of gas molecules within the transducer decreases, the quantity of heat transferred from a heated filament, through the gas, and to the case of the transducer, will decrease proportionally. The *Pirani gage,* as well as the *thermocouple gage* (which may use a thermopile instead of a single junction), both use this principle. A basic thermocouple gage is illustrated in Fig. 10-23.

Ionizing vacuum transducers measure pressure as a function of gas density by measuring ion current. Since different gases have different densities, the calibration of such a transducer will usually differ as well. The gas is usually ionized by electrons, except in one type using alpha particles for this purpose.

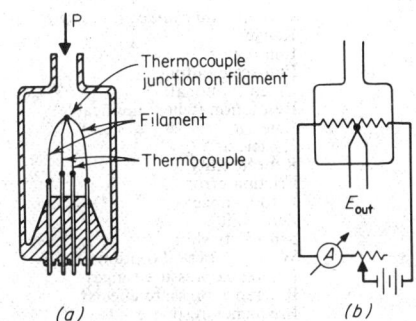

Fig. 10-23. Thermoelectric thermoconductive vacuum transducer. (*a*) Transducer; (*b*) typical circuit. *(Prentice-Hall, Inc., by permission.)*

In thermionic vacuum transducers the electrons are emitted by a filamentary cathode, and positive ions are collected at the anode. Various modifications of the original triode type have helped to extend its lower range limit from 10^{-8} to 10^{-10} torr (*Bayard-Alpert gage,* by reducing internal x-ray effects) to 10^{-11} torr (*Nottingham gage,* by reducing electrostatic-charge effects) and to 10^{-12} torr (*Schuemann modification,* by virtually eliminating x-ray effects). The ion current, representative of pressure, is in the microampere region.

Several ionizing vacuum transducer types, whose electrons are emitted from either hot or ion-bombarded cold cathodes, use a magnetic field to increase the electron path length by forcing this path to be helical so that the probability of electron collisions with gas molecules is increased *(magnetron gages).* The hot-cathode versions include the *Lafferty gage.* The *Philips* (or *Penning) gage* and the *Redhead gage* are examples of the cold-cathode versions.

Selection criteria for vacuum transducers (and any necessary ancillary equipment for them) are primarily the required measuring range; secondarily, size, weight, ruggedness, and

Table 10-5. Specification Characteristics for Pressure Transducers*

Mechanical Design Characteristics

a. Specified by user and manufacturer

Configuration and dimensions (shown on drawing)
Mountings (shown on drawing)
Type and location of pressure ports (shown on drawing)
Type and location of electrical connections (shown on drawing)
Nature of pressure to be measured, including range
Measured fluids
Case sealing (explosionproof, burstproof, or waterproof enclosure)
Isolation of transduction element
Mounting and coupling torque
Weight
Identification
Nameplate location (shown on drawing)

b. Stated by manufacturer

Sensing element
Transduction-element details
Materials in contact with measured fluids
Dead volume
Type of damping (including type of damping oil if used)

Electrical Design Characteristics

Excitation (nominal and limits)
Power rating (optional)
Input impedance (or element resistance)
Output impedance
Insulation resistance (or breakdown-voltage rating)
Wiper noise (in potentiometric transducers)
Output noise (in dc output transducers)
Electrical connections and wiring diagram
Integral provisions for simulated calibration (optional)

Performance Characteristics

a. Individual characteristics spec.	*b. Error-band spec.*
Range	Range
End points	Full-scale output (nominal)
Full-scale output	End points (defining reference line)
Creep (optional)	Resolution (where applicable)
Resolution (where applicable)	Static error band
Linearity	Reference-pressure range†
Hysteresis	Warm-up period (optional)
Repeatability	Frequency response
Friction error	
Zero balance	
Zero shift	
Sensitivity shift	
Warm-up period (optional)	
Reference pressure range†	
Reference pressure effects†	
Frequency response	
Operating temperature range	Operating temperature range
Temperature error *or*	Temperature error band
Thermal zero shift *and*	
Thermal sensitivity shift	
Temperature-gradient error	Temperature-gradient error
Ambient-pressure error	Ambient-pressure error band
Acceleration error	Acceleration error band
Vibration error	Vibration error band

Performance after exposure to:

Shock (triaxial)
Humidity
Salt spray or salt atmosphere

Performance during and after exposure to:

High sound-pressure levels
Sand and dust
Ozone
Nuclear radiation
High-intensity magnetic fields
Etc.

* By permission of Prentice-Hall, Inc.
† Applies to differential-pressure transducers only.

Table 10-6. Vacuum Transducers

Sensing element	Common name	Output	Nominal range, mbars or torr
Thermoconductive transduction			
Heated filament	Pirani gage	Resistance change	10^{-3}–1
	Thermocouple gage	Thermoelectric emf	10^{-3}–1
Ionizing transduction			
Thermionic (triode)	Bayard-Albert gage	Direct current	10^{-10}–10^{-3}
Thermionic (photo-multiplier)		Direct current	10^{-18}–10^{-3}
Hot cathode, magnetic field	Hot-cathode magnetron gage (Lafferty gage)	Direct current	10^{-14}–10^{-5}
Cold cathode, magnetic field	Cold-cathode magnetron gage (Philips gage, Penning gage)	Direct current	10^{-7}–10^{-3}
Cold cathode, magnetic field, with flash filament		Direct current	10^{-12}–10^{-3}
Cold cathode, magnetic field, with auxiliary cathode	Redhead gage	Direct current	10^{-13}–10^{-4}
Radioactive, alpha particles	Alphatron	Dc voltage (from amplifier)	10^{-5}–1,000
Radioactive, beta particles		Dc voltage (from amplifier)	10^{-8}–1,000

complexity. Considerations for the selection of a pressure transducer are primarily range, type of excitation and output, accuracy and frequency response; secondarily, the properties of the measured fluid and environmental conditions.

57. Ranges of Pressure Transducers. Some of the pressure transducers described in this section are available for ranges up to 1,000 bars (about 15,000 lb/in.^{2}). Special sensing devices have been designed for pressures up to 70 kbars (about 1,000,000 lb/in.^{2}). Pressure transducers are also used to measure altitude (a known nonlinear relationship exists between atmospheric pressure and altitude above sea level), water depth [pressure increases at the rate of approximately 10 mbars/m (0.44 lb/in.^{2} ft) when descending below the water surface], and airspeed (by measuring the difference between impact pressure, obtained from a pitot tube, and static pressure, while in flight).

TRANSDUCERS FOR THERMAL QUANTITIES

58. Temperature Transduction. The *temperature* of a body or substance is (*a*) its potential of heat flow, (*b*) a measure of the mean kinetic energy of its molecules, and (*c*) its thermal state considered with reference to its power of communicating heat to other bodies or substances. *Heat* is energy in transfer, due to a difference in temperature, between a system and its surroundings or between two systems, substances, or bodies. Heat energy is transferred by one or more of the following methods of *heat transfer:* (*a*) *conduction,* by diffusion through solid material or stagnant liquids or gases; (*b*) *convection,* by the movement of a liquid or gas between two points; and (*c*) *radiation,* by electromagnetic waves.

The sensing elements of temperature transducers typically act as transduction elements as well. The two most commonly used sensing-transduction elements are the *thermoelectric* element *(thermocouple)* and the *resistive* element *(resistance thermometer).* Among other sensing-transduction elements the only one which has found commercial acceptance is the *oscillating-crystal* element, essentially a quartz crystal (connected into an oscillator circuit) which has a substantial and highly linear temperature coefficient of frequency.

59. Thermocouples. A thermocouple is an electric circuit consisting of a pair of wires of different metals joined together at one end *(sensing junction)* and terminated at their other end in such a manner that the terminals *(reference junction)* are both at the same and known temperature *(reference temperature).* Connecting leads from the reference junction to some sort of load resistance (an indicating meter or the input impedance of other readout or signal-conditioning equipment) complete the thermocouple circuit. Both these connecting leads can be of copper or some other metals different from the metals joined at the sensing junction. Due to the *thermoelectric effect (Seebeck effect)*, a current is caused to flow through the circuit whenever the sensing junction and the reference junction are at different temperatures. In practice, the reference junction is either held at a known constant temperature (e.g., at 0°C) or is electrically compensated for variations from a preselected temperature.

The electromotive force *(thermoelectric emf)*, which causes current flow through the circuit, is dependent in its magnitude on the sensing-junction wire materials, as well as on the temperature difference between the two junctions. Commonly used wire materials are Chromel (CR) and Alumel (AL) (both registered trade names of Hoskins Mfg. Co., Detroit, Mich.), Constantan (CN, an alloy of 53% copper and 45% nickel), copper (Cu), iron (Fe), platinum (Pt), an alloy of platinum and (either 10 or 13%) rhodium (Rh), tungsten (W), tungsten-rhenium (Rh) alloys (5 or 26% rhenium content is typical), nickel (Ni), and ferrous nickel alloys.

The characteristics of certain combinations of wire materials, such as their thermoelectric emf vs. temperature characteristics, their accuracy tolerances, and wire-insulation color

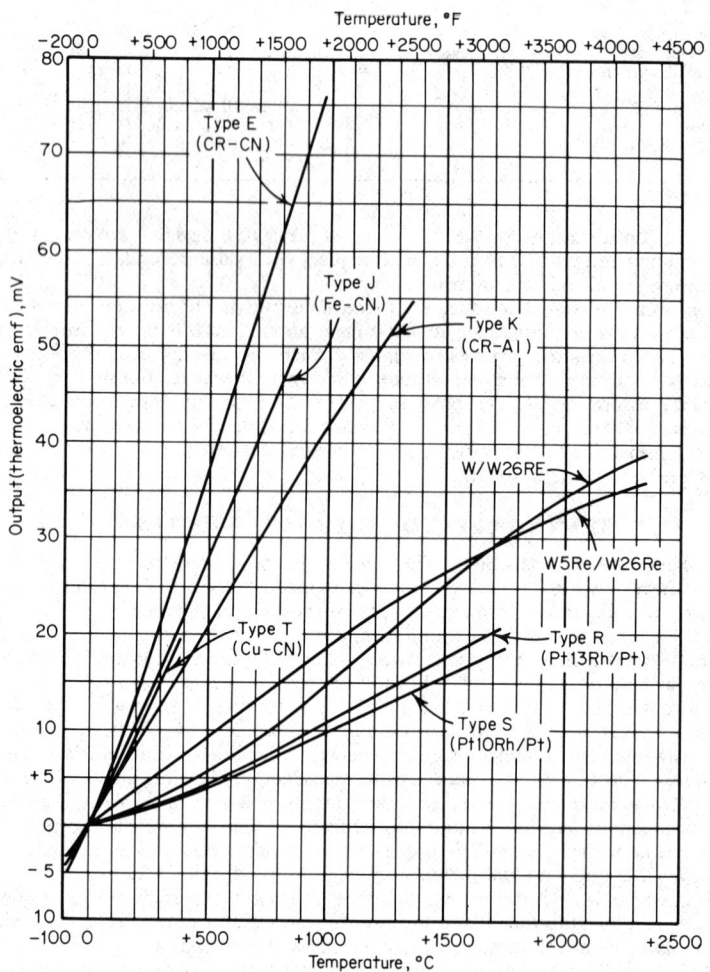

Fig. 10-24. Thermocouple output vs. temperature characteristics. (Reference junction at 0°C.)

codes, were standardized by ANSI Standard C96.1 (which is based on ISA Recommended Practice RP1) in such a manner that materials of different brand names can be used as long as the characteristics assigned to a specific type of thermocouple are maintained.

The names of the wire materials constituting, in their combination, a thermocouple sensing junction are now listed only as typical examples. Thus typical materials of a *type K* thermocouple are Chromel and Alumel. The ANSI Standard favors the use of type-letter designations in lieu of the names of the two metals used. Fig. 10-24 shows the thermoelectric emf obtainable from various types of thermocouples when the reference temperature is held at 0°C.

Thermopiles (see Fig. 10-25) consist of several sensing junctions of the same material pairs, in close proximity to each other and connected in series so as to multiply the output obtainable from a single sensing junction. The isothermal reference junctions are usually also in close proximity to each other to assure an equal temperature for each reference junction.

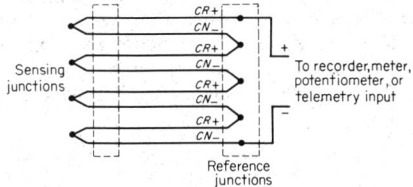

Fig. 10-25. Thermopile, schematic diagram; CR-CN combination shown as example. *(Prentice-Hall, Inc., by permission.)*

60. Resistive temperature sensing elements are either conductive or semiconductive. Conductive elements are usually wire-wound, sometimes made of metal foil or film. Elements wound with high-purity annealed platinum wire are best suited for most applications. Other metal-wire elements are wound of nickel or nickel alloy. Copper-wire elements are rarely used any more. Tungsten-wire elements have shown some promising characteristics but are generally considered too difficult to manufacture and too brittle to stay reliable.

A platinum-wire element has been used to define the International Practical Temperature Scale from -183 to $+630°C$, and it is expected that this upper limit will be extended to the melting point of gold ($+1063°C$). The resistance vs. temperature curve of such an element follows a well-defined theoretical relationship, thus making most points on the curve calculable within very close tolerances when only a few measured points have been established. Repeatabilities within about 0.01°C have been obtained at temperatures up to the gold point. Semiconductive resistive temperature sensing elements include thermistors, germanium, and silicon crystals, carbon resistors, and gallium arsenide diodes. Thermistors have a nonlinear and negative temperature coefficient of resistance and an empirical resistance vs. temperature relationship.

61. Temperature transducers are classified into two general categories: surface-temperature transducers, which are cemented, welded, bolted, or clamped to a surface whose temperature is to be measured, and immersion probes, which are immersed into stagnant or moving fluids to measure their temperature. The fluid can be in a pipe, a duct, a tank, or other enclosed vessel, where the immersion probe is mounted through a pressure-sealed opening. It can also be freely moving, even at almost imperceptible rates of motion, e.g., an open body of water, an outdoor or indoor atmosphere.

Thermoelectric temperature transducers have the same sort of sensing junction, whether they are intended for surface temperature measurement or as immersion probes. The junctions between the two dissimilar-metal wire pairs are made by butt-welding the wire ends, by crossing them and welding them, by coiling one wire end around the other, or twisting the two ends about each other, then welding, brazing, or soldering the junction, or by welding both wire ends, in very close proximity to each other, to a metallic surface or to the metallic inside of an immersion-probe tip.

For surface measurements, the junctions are soldered, brazed, or welded to a surface (if it is metallic) or cemented to it (if it is not). If it is cemented, care must be taken to have the junction in solid thermal contact with the measured surface. Taping a junction to a surface is poor practice, since even a very small gap between junction and surface can introduce considerable errors. For immersion measurements, thermocouples are often produced with an integral sheath, or inserted into a sealed immersion sheath *(thermowell).*

Junctions for thermoelectric immersion probes can be grounded (metallic contact from junction to sheath or thermowell) or isolated (ungrounded). In some cases, exposed junctions, at the tip of a probe, are immersed into the fluid without use of an integral sheath or thermowell. If terminals or connectors must be used between the sensing junction and the reference junctions, the terminals as well as the *extension wires* must be made of the same types of metals as used for the junction.

Thermocouples are usually made from two-conductor insulated cable, rarely from reels of individual bare-wire materials. The cables have a variety of insulation, over each conductor as well as over the conductor pair, and can be shielded or unshielded. Useful for many applications is thin (2 to 10 mm outside diameter) metal-sheathed, ceramic-insulated thermocouple cable.

Differential thermocouples can be used when the measurement objective is to measure the temperature difference between two points. In this case the sensing junction at the other measured point replaces the reference junction. The first wire of the first junction and the second wire of the second junction must still be brought to isothermal terminals; however, it is not necessary that the temperature of these terminals be known.

62. Resistive Temperature Transducers. Electrically conductive surface temperature transducers are usually small and flat enough not to be influenced by convective heat transfer, but only by conductive transfer from the measured surface. After installation they may be coated or covered to minimize any radiative heat transfer to them. The sensing element is usually a metal wire either wound around a thin insulating "card" or a coiled wire cemented to the base (see Fig. 10-26). Some metal-foil transducers (encapsulated or *free-grid*) are in the shape of a zigzag pattern. All designs are aimed at exposing the maximum of sensing surface to the conductive heat transfer in an area of minimum size.

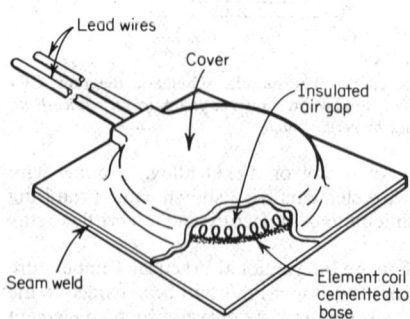

Fig. 10-26. Platinum-wire resistive surface temperature transducer. *(Rosemount Engineering Co.)*

Resistive metal-wire *immersion probes,* most commonly with a platinum-wire element but sometimes with elements of nickel or nickel-alloy wire, are widely used for industrial, as well as scientific, fluid-temperature measurements. The probe-type transducer, illustrated in Fig. 10-27, has a ceramic encapsulated (coated) element within a perforated protective sheath so as to be usable for a variety of measured fluids over a wide temperature range. For applications in relatively stagnant fluids an unencapsulated *(exposed)* element is used to provide a shorter time constant. Some fluids require an element completely *enclosed* within a metallic well, but with good thermal contact between well and element. The threaded mounting allows for compression sealing by means of a gasket or O-ring between the housing and the mounting boss.

63. Thermistors are used for surface-temperature as well as fluid measurements. Because of their nonlinear (essentially negative exponential) resistance vs. temperature characteristics, they are particularly useful when a large resistance change is needed for a narrow range of temperature. Where a short time constant is required, a glass-coated thermistor bead, as small as 0.3 mm in diameter, can be suspended on its 0.03-mm-diameter precious-metal-alloy leads. Where somewhat more ruggedness is required, a glass-encapsulated bead about 1.5 mm around the tip and 4 mm long can be employed. Excitation power must be kept low to avoid errors due to self-heating. Thermistor-type temperature transducers are available in a large variety of configurations, some of which are illustrated in Fig. 10-28.

64. Germanium thermometers are made of germanium crystals with highly selected and controlled impurities (dopants). They are intended primarily for cryogenic temperature measurements (below $-195°C$). Carbon resistors have also been used for such applications, as have gallium-arsenide junction diodes, which are additionally usable to somewhat higher temperatures. Silicon-wafer transducers have been used for surface temperature measurements in the range -50 to $275°C$, where their resistance vs. temperature characteristics are similar to those of some metal wires.

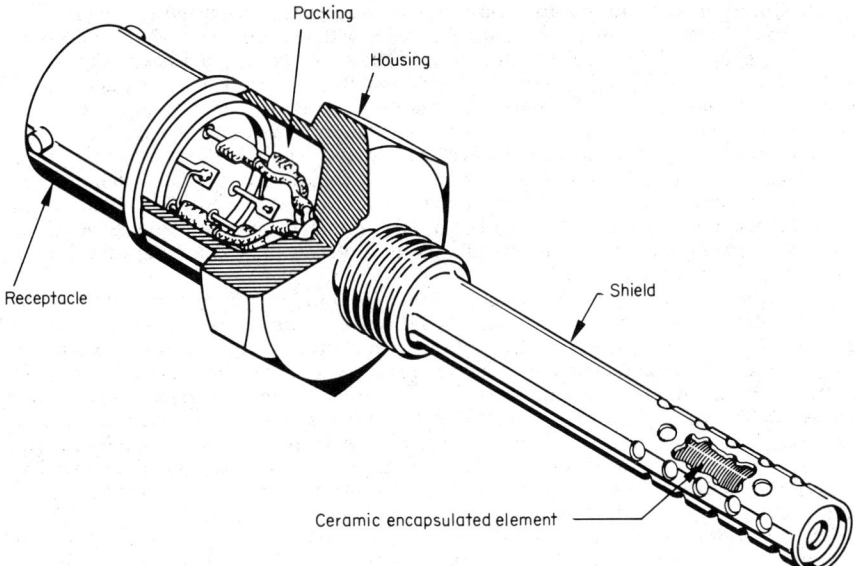

Fig. 10-27. Platinum-wire resistive immersion-probe temperature transducer. *(Rosemount Engineering Co.)*

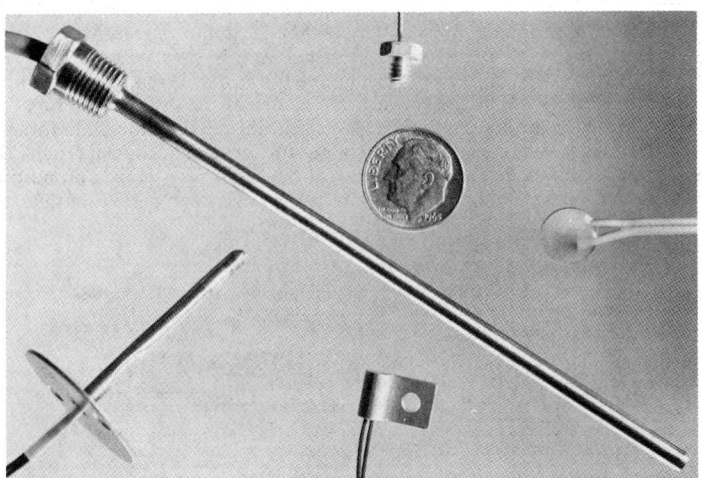

Fig. 10-28. Typical thermistor-transducer configurations. *(Fenwal Electronics, Inc.)*

65. Quartz-crystal temperature transducers use oscillating-crystal sensing elements in such a manner that the change of oscillator frequency with temperature is nearly linear over a range from about -50 to $250°C$. They are usually furnished with associated electronics and readout equipment. This tends to limit their usability for general telemetry application without, however, detracting from their advantages in laboratory applications.

The selection of a temperature transducer requires somewhat more complex considerations than the selection of most other types of transducers. The objective is to select a design whose sensing element will attain the temperature of the measured material within the time available to make the measurement. Among primary selection criteria are, then, the characteristics and properties of the measured solid or fluid, the measuring-range limits, the required response time (time constant), and the type of excitation and signal conditioning available or intended to be used.

66. Radiation pyrometers are noncontacting temperature transducers which respond to radiative heat transfer from the measured surface or material. This radiation occurs primarily in the infrared portion of the electromagnetic spectrum (wavelengths between 0.75 and 1,000 μm). Typical radiation pyrometers resemble a motion-picture camera in appearance. They employ an optical lens or mirror system (sensitive in the infrared region) which focuses the radiation on a thermoelectric or resistive (usually photoconductive) sensing element. The output of the sensing element can be correlated, by calibration, to the temperature of the measured surface. Radiation pyrometers are used primarily for high-temperature measurements (up to about 3500°C), but have also been found useful for noncontacting measurements in the medium temperature range (down to about $-50°C$).

67. Heat-Flux Transducers. Two basic types of transducers have been developed to measure *heat flux,* heat transfer in terms of the total amount of thermal energy (heat flux is commonly expressed in W/cm^2 or $Btu/ft^2 \cdot s$). The *calorimeter* provides an output proportional to convective as well as radiant thermal energy *(total heat flux).* The *radiometer* responds to radiant thermal energy *(radiant heat flux)* only. Virtually all heat-flux transducers have thermoelectric sensing elements.

68. Calorimeters. The *foil calorimeter (membrane calorimeter,* Gardon gage) acts as a copper-Constantan differential thermocouple. When heat flux is received by the thin Constantan sensing disk (Fig. 10-29a) which is metallurgically bonded around its rim to a copper heat sink, the heat absorbed by the membrane is transferred radially to the heat sink. This causes a temperature difference between the center of the disk and its rim. A thin copper wire is attached to the bottom surface of the disk, at its exact center, thus forming one copper-to-Constantan sensing junction. The copper-to-Constantan contact around the rim of the disk forms the other junction. The output of the calorimeter is then proportional to the energy absorbed. When heat flux must be measured over long periods of time, the foil

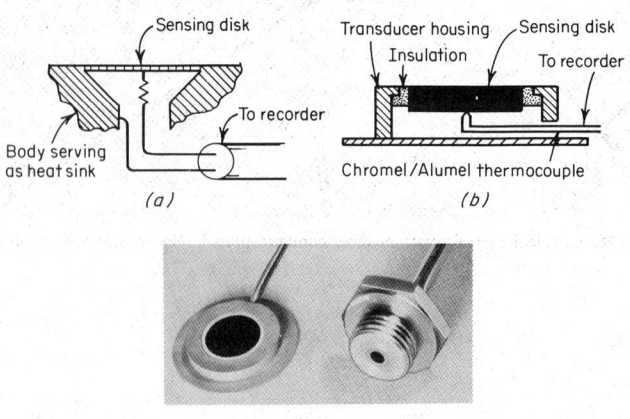

Fig. 10-29. Calorimeters. (*a*) Slug; (*b*) foil; (*c*) typical appearance. *(Hy-Cal Engineering.)*

10-36

calorimeter can be provided with tubing and an internal flow path so that it can be water-cooled.

The *slug calorimeter (slope calorimeter)* uses a relatively thick thermal-mass sensing disk with an external high-emissivity (black) coating, which is thermally insulated from the transducer housing (Fig. 10-29*b*). A thin-wire thermocouple is attached to the bottom of the disk (slug), at its center. When heat flux is received by the slug, an output signal is produced by the thermocouple. The signal is proportional to the temperature rise of the slug. Two of the many available slug-calorimeter configurations are illustrated in Fig. 10-29*c*.

69. Radiometers. A typical *radiometer* is, essentially, a foil calorimeter with a *window* (usually of quartz or synthetic sapphire) mounted over the sensing disk so that the disk can receive radiant heat flux but no convective heat flux. The cavity formed by window, transducer housing, and sensing disk is usually sealed, but provisions for gas purging of this cavity can be made to prevent window clouding when the radiometer is to be used in a contaminating atmosphere. Radiometers can also be water-cooled. The sensitivity of a radiometer can be increased by using a differential (multijunction) thermopile instead of the two-junction differential thermocouple.

TRANSDUCERS FOR ACOUSTIC QUANTITIES

70. Terminology. *Sound* is an oscillation in pressure, stress, particle displacement, etc., in an elastic or viscous medium. *Sound sensation* is the auditory sensation evoked by the oscillations associated with sound. *Sound pressure* is the total instantaneous pressure at a given point, in the presence of a sound wave, minus the static pressure at that point. *Sound pressure level* (SPL or L_p) is 10 times the logarithmic ratio of the mean-square sound pressure p to a mean-square reference pressure p_{ref}. It is normally expressed as $SPL = 20 \log_{10}(p_{rms}/p_{ref,rms})$. See Fig. 10-30. The reference pressure is usually specified as 2×10^{-4} μbar, sometimes as 1 μbar (1 dyn/cm^2). *Sound level* is a weighted sound-pressure-level reading obtained with a meter complying with ANSI Standard S1.4, "Specification for General-Purpose Sound Level Meters."

71. Sound-Pressure Transducers. The sensing element of a sound-pressure transducer is almost invariably a diaphragm (see Pars. **10-53** to **10-55**). The reference cavity behind the diaphragm is vented to the ambient atmosphere, by means of a small hole in the transducer case, so that static pressures on both sides of the diaphragm are equalized and only sound pressure is sensed.

A perforated cap over the diaphragm protects the diaphragm mechanically and, by its shape and geometry of perforations, provides some control over the transducer's directivity characteristics.

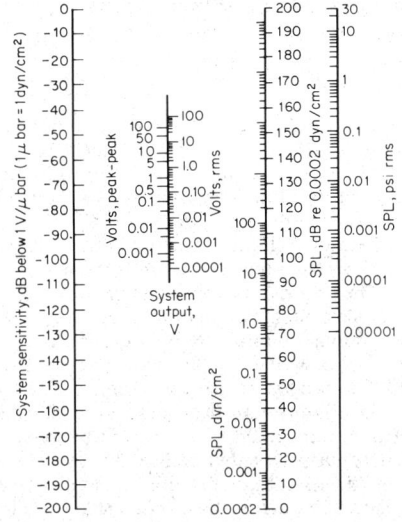

Fig. 10-30. Nomograph for sound-pressure-level calculations. *(Prentice-Hall, Inc., by permission.)*

Sound-pressure transducers can be described, essentially, as special-purpose gage-pressure transducers. *Capacitive sound-pressure transducers* (sometimes called *condenser microphones*) use the sensing diaphragm as one electrode of a capacitor and a rigidly supported back plate, insulated from the rest of the structure but provided with a connecting lead or terminal, as the other electrode. A dc polarization voltage, applied across the two electrodes through a high-series resistance, maintains a constant charge on them. Capacitance changes due to diaphragm deflection cause changes in the voltage across the electrodes. The transducer output is first fed to an emitter follower so as to reduce the output impedance to a workable value. The output is then amplified. The emitter-follower (or cathode-follower) circuitry is sometimes built into the transducer case to keep the coupling path short. A shielded cable connects the transducer to the amplifier.

Piezoelectric and, to a limited extent, *inductive pressure transducers* have also been designed as sound-pressure transducers. Some piezoelectric designs have sealed cases, primarily to protect the internal components from atmospheric moisture and contaminants. The absence of a gage vent, however, necessitates correction of output readings when the transducer is used at low ambient pressures (e.g., high altitudes). Piezoelectric transducers do not require an excitation power supply. They require an amplifier, however.

The primary performance characteristics of sound-pressure transducers are range, output, frequency response, and directivity (directional response). Output is usually expressed as sensitivity or sensitivity level, sometimes as full-scale output for a stated range of sound pressures or sound-pressure levels. The nomograph shown in Fig. 10-30 can be used to correlate output characteristics stated in various ways.

72. Sound-level meters are complete, self-contained measuring systems, typically battery-operated and portable. A sound-level meter consists of a sound-pressure transducer (microphone), amplifier, standardized weighting networks, a calibrated attenuator, and an indicating meter. The sound-level range is always referred to a sound pressure of 10^{-4} μbar. The weightings denote different frequency-response characteristics of the measuring system. Referred to merely as *A, B,* or *C,* they are defined in a national standard as, for the United States, ANSI Standard S1.4.

73. Underwater sound detectors are used either for listening *(hydrophone)* or, in conjunction with an *underwater sound projector,* in sonar (*s*ound *n*avigation *a*nd *r*anging) systems. The transmitting and receiving functions in a sonar system are frequently combined in a single device (sound transceiver). In the sonar field, underwater sound detectors, as well as projectors and transceivers, are commonly referred to as *transducers.*

TRANSDUCERS FOR OPTICAL QUANTITIES

74. Terminology. *Light* is a form of radiant energy, an electromagnetic radiation whose wavelength is between approximately 10^{-2} cm (100 μm) and 10^{-6} cm (0.01 μm). By strict definition, only visible light (0.4 to 0.76 μm wavelength) can be considered as light, and infrared or ultraviolet light is then termed *radiation.* The light spectrum, in terms of wavelength (the unit μm is called *micrometer*), frequency, photon energy, and blackbody temperature (all interrelated by physical laws), is illustrated in Fig. 10-31, with the visible-light spectrum (color spectrum) brought out in detail.

The transduction elements of light transducers (light sensors, photocells, photosensors, photodetectors, light detectors) also act as sensing elements since they convert electromagnetic radiation into a usable electrical output. Four transduction principles are commonly used: photovoltaic, photoconductive, photoconductive junction, and photoemissive (see Fig. 10-32). Section 11 provides detailed data on optical sensors and systems.

75. Photovoltaic light sensors are self-generating in that their output voltage is a function of the illumination of a junction between two dissimilar materials. These materials are semiconductive, either nonmetallic or metal compounds (Fig. 10-32a). Photons (particles of light) first pass through a thin conductive layer, and then impinge on the junction, causing an electron flow across the junction area in such a manner that the conductive layer becomes the negative terminal of the sensor. Various materials constitute the conductive and semiconductive portions of a photovoltaic light sensor.

The *selenium cell* consists of an iron base (conductive, positive terminal) with a very thin selenium coating (semiconductive), which is in contact with cadmium oxide (semiconductive), which is part of a cadmium film (conductive, negative terminal). A silver ring on the outside surface of the cadmium film forms the connecting terminal for the cadmium electrode. The cadmium film and its oxide layer are so thin as to be essentially transparent to light. This type of sensor has also been referred to as *barrier-layer photocell* because the cadmium oxide acts as barrier to any reverse flow of electrons. The spectral response of a selenium cell peaks around 0.57 μm.

The *silicon cell* (silicon photovoltaic cell, silicon solar cell) uses an arsenic-doped *n*-type silicon wafer. Boron is diffused into the upper (light-receiving) surface to create a thin *p*-type silicon layer. The *pn* junction between the layer and the wafer acts as a permanent electric

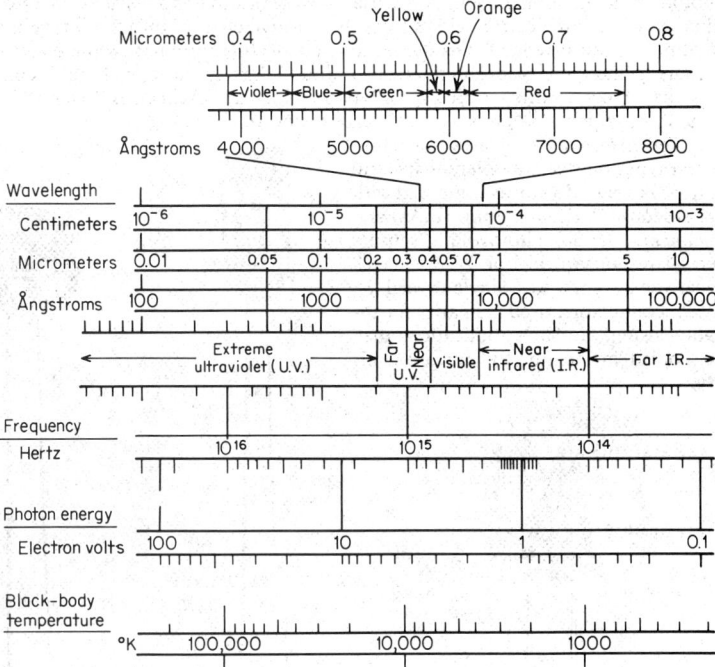

Fig. 10-31. The light spectrum. *(Prentice-Hall, Inc., by permission.)*

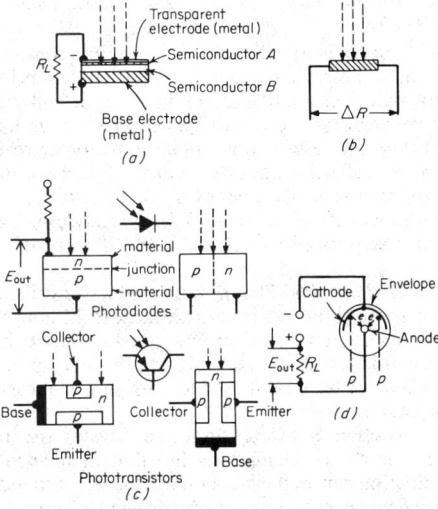

Fig. 10-32. Basic methods of light transduction. (*a*) Photovoltaic; (*b*) photoconductive; (*c*) photoconductive junction semiconductor; (*d*) photoemissive. *(Prentice-Hall, Inc., by permission.)*

field. Photons incident upon the junction cause a flow of positive and negative charges. The *pn* junction, acting as an electric field, directs the positive charges into the *p*-type material (nickel plating around its edge forms the positive connecting terminal) while directing the negative charges into the *n*-type material (solder around the bottom edge of the silicon wafer forms the negative connecting terminal). A typical silicon photosensor is illustrated in Fig. 10-33. The third connecting pin serves as a case connection.

Germanium photovoltaic light sensors are similar to silicon types but have a different spectral response (a peak near 1.55 μm, as compared with 0.8 μm for silicon). *Indium-arsenide* (InAs) and *indium-antimonide* (InSb) photovoltaic sensors have spectral response peaks near 3.2 and 6.8 μm, respectively. They are single-crystal *pn* junction semiconductors, used primarily for infrared-light sensing. In such applications they are often cooled artificially to increase their sensitivity.

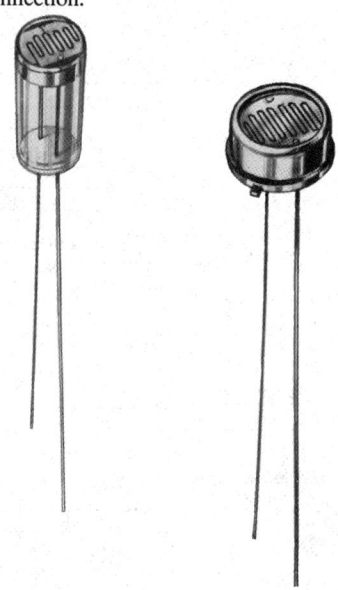

Fig. 10-33. Silicon photovoltaic cell. *(Solar Systems Div., Tyco.)*

Fig. 10-34. Photoconductive (CdS or CdSe) cells. *(Clairex Corp.)*

76. Photoconductive light sensors are widely used for control functions, such as automatic exposure control in cameras, in addition to their photometric (light-measurement) applications. The photoconductors are polycrystalline films or bulk single-crystal materials which, when contained between two conductive electrodes, act as light-sensitive resistors (Fig. 10-32b) whose resistance decreases as incident illumination increases. *Cadmium sulfide* (CdS) and *cadmium selenide* (CdSe) are popular because of their spectral response peaks in the visible-light region (approximately 0.6 μm for CdS, 0.72 μm for CdSe) and because of their relatively high output without artificial cooling. Two typical designs are shown in Fig. 10-34. Some photoconductive sensors employ mixed CdS-CdSe crystals to obtain a response peak around 0.66 μm.

Lead sulfide (PbS) and *lead selenide* (PbSe) photoconductive cells are used for infrared-light sensing because their spectral response peaks are close to 2.2 μm. The spectral response curve of PbSe, however, is shallow enough to provide good sensitivity between about 1.8 and 3.6 μm, and its time constant is less than one-tenth that of PbS. *Mercury-doped-germanium* (HgGe) photoconductive sensors have been used for far-infrared measurements while being cooled to cryogenic temperatures.

77. Photoconductive Junction Sensors. In these devices the resistance across the junction in a semiconductor device changes as function of incident light (Fig. 10-32c). Increasing incident illumination causes the junction photocurrent to increase. This category of photosensors includes *photodiodes* and *npn* as well as *pnp phototransistors*. They are made of silicon, with a spectral peak near 1.0 μm, except when germanium must be used to raise the spectral peak to about 1.6 μm. Time constants of photodiodes and phototransistors are less than 1 μs (as compared with 10 μs for PbSe photoconductive sensors; 100 to 700 μs for PbS photoconductive sensors; 10 ms for CdSe photoconductive cells; and 100 ms for CdS cells; but around 20 μs for silicon photovoltaic cells). Transistors also provide some amplification of the light-induced signal. Both types are usually sealed into a standard transistor can, sometimes furnished with a lens or window. A special silicon photodiode design has also been developed for ultraviolet measurements (0.06- to 0.25-μm region).

78. Photoemissive Sensors. The earliest *photoemissive light sensor* was the phototube, in which electrons are emitted by the cathode of a vacuum (or gas-filled) diode tube when photons impinge on the cathode surface (Fig. 10-32d). When a closely spaced anode is at a positive potential with respect to the cathode, the anode collects some of these electrons, and the resulting anode current can produce an output voltage as an *IR* drop across a suitable load resistor (R_L).

The photoemissive principle is now employed mostly by *photomultiplier* tubes in which additional electrodes *(dynodes)* are placed between cathode *(photocathode)* and anode to amplify the electron current by means of secondary emission. An additional dynode is located behind the anode. A voltage divider network is used to apply a successively higher voltage to each of the dynodes as they approach the anode in proximity. Various spectral response peaks can be obtained, depending on the photocathode material and material of that portion of the sealed envelope directly in front of the cathode (the *window*). Photomultiplier tubes are particularly useful in the visible-light and ultraviolet regions. They can provide very high sensitivities without artificial cooling.

Selection criteria for light sensors are, primarily, spectral response characteristics and sensitivity and, secondarily, operating temperatures, relative complexity of associated circuitry, ruggedness, and cost. Where the spectral response must be limited to only a portion of a sensor's basic response capabilities, a window can be placed in front of the sensing surface. Windows are optical filters which have spectral response characteristics of their own, depending on their material. Typical window spectral responses are 0.2 to 1.4 μm (quartz crystal or fused silica), 0.4 to 1.2 μm (borosilicate glass), 0.15 to 1.6 μm (cultured sapphire), 0.11 to 1.8 μm (lithium fluoride), 0.12 to 11 μm (calcium fluoride), and 0.25 to 70 μm (cesium iodide).

79. Spectrometers and Colorimeters. Light sensors in conjunction with light sources are used in a number of measuring devices other than for photometry. In optical *spectrometers* the incident light is passed through a *monochromator*, a grating or prism whose angular displacement relative to the incoming light beam can be closely correlated with the single wavelength of the light beam it sends on to the light sensor, whose spectral response is additionally known and selected for a specific range of wavelengths. Some monochromators consist simply of a set of windows, each having a different (and narrow) spectral response. The output of the light sensor then shows the light intensity in each of the sampled portions (groups of wavelengths) of the spectrum. The motion of the monochromator can be mechanized so that a given spectrum is scanned at a known rate over a known time interval. Wavelength can then be determined from the time counted from the start of the scan.

The *colorimeter* is similar to an optical spectrometer except that it responds to light of known characteristics (emanating from a built-in or associated light source) which is reflected by the surface whose color is to be determined.

80. Turbidity and Opacity Meters. In *turbidimeters, nephelometers,* and *opacity meters* a fixed amount of light from a light source is made to pass through the measured fluid and on to a light sensor whose output, then, varies with changes in the transmittance of the measured liquid or gaseous medium.

TRANSDUCERS FOR NUCLEAR RADIATION

81. Terminology. *Nuclear radiation* is the emission of charged and uncharged particles and of electromagnetic radiation from atomic nuclei. *Charged particles* include (a) *alpha particles,* helium-atom nuclei consisting of two protons and two neutrons and having a double positive charge; (b) *beta particles,* negative electrons or positive electrons (positrons) emitted when *beta decay* occurs in a nucleus, a radioactive transformation by which the atomic number is changed by $+1$ or -1, while the mass number remains unchanged; (c) *protons,* positively charged elementary particles of mass number 1. *Uncharged particles* are typified by the *neutron,* an uncharged elementary particle of mass number 1. Nuclear *electromagnetic radiation* includes (a) *gamma rays,* electromagnetic radiation quanta resulting from quantum transitions between two energy levels of a nucleus; (b) *x-rays,* quanta of electromagnetic radiation originating in the extranuclear part of the atom.

82. Ionization and Scintillation. Two basic transduction methods are used to convert nuclear radiation into a usable electrical output. When *ionizing* transduction is employed, the transduction element acts also as sensing element. This method relies upon the production of ion pairs in a gas or solid by incidence of nuclear radiation. The two ions of an ion pair are subatomic particles, one with a positive charge, the other with a negative charge. An electric field is then used to separate the positive and negative charges to produce an electromotive force. When *photoelectric* transduction is employed, a *scintillator* material is used as sensing element which converts the nuclear radiation into light. A light sensor is then utilized as transduction element.

Ionization occurs in gases and solids when charged particles pass atoms at a high enough velocity to separate one of the outer electrons from them. The resulting ion pair consists of an ion and a secondary electron. The same incoming particle can cause such ion pairs to be produced a number of times before its energy is expended. Uncharged particles can produce ion pairs by collision with atomic nuclei. Nuclear electromagnetic radiation can produce ion pairs by removing secondary electrons from atoms with which the radiation interacts.

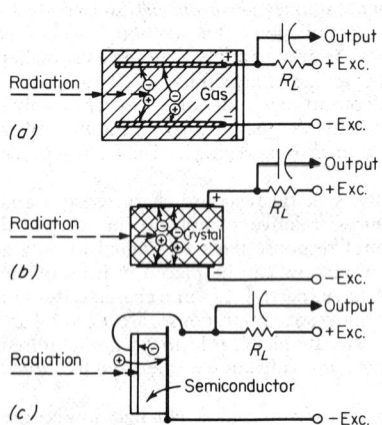

Fig. 10-35. Transduction of nuclear radiation. (*a*) Ionization in gas; (*b*) ionization in a solid crystal; (*c*) ionization in a solid semiconductor. *(Prentice-Hall, Inc., by permission.)*

When ion pairs are produced in a gas (e.g., argon, neon, xenon, hydrogen) the electric field between a cathode and an anode in a gas-filled envelope, with a dc potential applied across the two electrodes, causes charge separation (see Fig. 10-35a). The *ionization current,* the current due to flow of charges to electrodes of opposite polarities, can then be monitored as the average IR drop across the load resistor R_L. This principle is used in three types of nuclear radiation sensors.

83. Ionization Chambers. The ionization chamber consists typically of a metallic outer cylinder (cathode) with a metallic rod or wire (anode) mounted along the axis of the cylinder. One end of the cylinder is the sealed base of the sensor which provides the cathode terminal and a well-insulated anode terminal. The other end of the cylinder is also sealed, either by a thin mica, nylon, or metal window, or by a metal disk when a window is provided in a portion of the cathode. In windowless ionization chambers the thin-walled cathode itself acts as a window. The gases or gas mixtures, contained within the ionization chamber at a pressure below, above, or equal to atmospheric pressure, are selected on the basis of ionization potential and energy per ion pair produced. The material and thickness of the window control the types of radiation to which the ionization chamber will respond with the necessary output. Tungsten is the most commonly used anode material.

84. Proportional counters are ionization chambers whose anode potential is above a certain limiting value. Above this voltage value, different for different fill gases, *gas amplification* occurs within the chamber, the production of additional ions by the *avalanche effect.* The incident radiation produces an ion pair, including a secondary electron. Each secondary electron produces more ion pairs, including more secondary electrons, and they, in turn, produce additional ion pairs, etc. The gas amplification can be up to about 10,000. For each incident particle, the total amount of ionization, which dictates the magnitude of the output current pulse, is proportional to the energy of radiation.

85. Geiger Counter (Geiger-Mueller Tube). The Geiger counter is a proportional counter to whose fill gas a *quenching agent* has been added and whose anode voltage is higher than it would be for a proportional counter of similar design. An incident radiation particle (or other ionizing event related to radiation) causes a discharge within the fill gas that spreads along the entire length of the anode. The quenching vapor (a halogen such as chlorine or bromine, or alcohol) quenches the discharge by preventing the production of secondary electrons by positive ions at the cathode.

Because of the discharge phenomenon, the magnitude of the output pulse is independent of the type and energy of the radiation measured. After the anode voltage is reached at which the ionization counter starts operating as a Geiger counter, a further increase causes no change in operation until, finally, a voltage is attained at which the quenching action begins to fail. The anode voltage is therefore chosen somewhere around the center of this "plateau." Since no such plateau characteristic exists for ionization chambers or proportional counters, a well-regulated power supply must be used for these sensors to avoid inadvertent calibration changes. In its typical operation a Geiger counter produces an output pulse with a rise time of less than 1 μs and a duration of several microseconds. The pulse then decays because of quenching. After the decay the counter is inoperative over a *dead time* of 50 to 150 μs and recovers gradually over an additional, but shorter, *recovery time*.

86. Ionization in Solid Crystals (See Fig. 10-35*b*). Crystals have been used for the transduction of nuclear radiation in some cases. Such crystals must have a near-perfect crystal structure. Charge separation and the resulting ionization current flow are effected by the electric field established by two electrodes of opposite polarity on opposite crystal surfaces.

Ionization in solid semiconductors (see Fig. 10-35*c*), such as silicon or germanium, has been utilized in many useful nuclear radiation sensors. The positive-negative charge pair, resulting from an ionizing event due to incident radiation, is separated by the permanent electric field established by the junction between the *p* and *n* materials. Although *intrinsic* semiconductors (those having negligible impurity concentrations) have been used in some radiation-detection applications, *extrinsic* semiconductors (those having significant controlled impurity concentrations) make up the bulk of semiconductor radiation sensor designs because of their more favorable transduction characteristics. Two major types of extrinsic-semiconductor radiation sensors have been developed in various designs: the surface-barrier type and the diffused-junction type.

Surface-barrier radiation sensors (see Fig. 10-36) usually consist of a *p*-type layer of silicon dioxide formed on one surface of a wafer of *n*-type single-crystal silicon. The very thin oxide layer is then covered with a vacuum-evaporated film to serve as electrical contact.

Diffused-junction radiation sensors typically consist of a *p*-type single-crystal silicon wafer into one surface of which a shallow diffusion of *n*-type material (e.g., phosphorus) has been made. Some such sensors use a *p*-type diffusion into an *n*-type base instead.

In a *pin junction* radiation sensor the *n*- and *p*-type semiconductors are separated by an intrinsic semiconductor. The foremost example of such a device is the *lithium-drifted* radiation detector, in which lithium ions are made to form an intrinsic region slightly below the sensor surface.

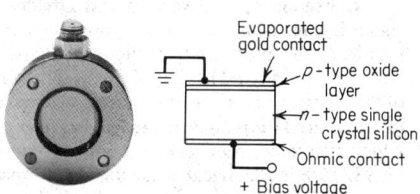

Fig. 10-36. Surface-barrier-type semiconductor radiation sensor. *(ORTEC, Inc.)*

87. Photoelectric transduction of nuclear radiation is used in *scintillation counters.* The scintillator material—inorganic or organic crystals, solid plastics, or plastics in liquid solution—is placed in front of a light sensor, usually a photomultiplier tube, in a single enclosure which is lightproof (except at the scintillator's sensing surface) and acts as a magnetic shield. The scintillator converts nuclear radiation into light (photons), which is then converted into an electrical output by the light sensor.

88. Neutron detection involves the interaction of neutrons with a *conversion material* which emits particles other than neutrons when neutrons impinge on them. In gas-filled sensors the conversion material (a stable isotope of one of several selected elements) can be applied either as a fill gas or as internal coating. In semiconductor sensors the material is usually in the form of a film or foil placed in front of the sensing surface. In scintillation counters the material can be a constituent of the scintillator material.

Selection criteria of nuclear radiation transducers involve primarily the type of radiation to be measured and its expected energy levels; secondarily, the presence of other nuclear or light radiation which should not influence the measurement. Special circuitry has been developed to improve the performance of radiation transducers. Since the transducer output is almost invariably in pulse form, logic circuitry has been used extensively to isolate the measured radiation from other prevailing but unwanted events also capable of causing ionization or scintillation.

TRANSDUCERS FOR ELECTRICAL QUANTITIES

89. Terminology. *Current measurements* can be made by various devices acting as transducers. A *series resistance,* inserted into the conductor where current is to be measured, provides a usable voltage across it due to the *IR* drop caused by the resistance. When the series resistance is relatively low, it is referred to as a *shunt.*

If it is necessary to keep the measurement circuit electrically isolated from the measured circuit (which is always desirable), a differential amplifier can be used, with its input terminals closely coupled across the series resistance so that its output terminals can be isolated from the measured circuit, and so that a signal sufficient for telemetry can be provided without inserting too high a series resistance into the measured circuit. This method is usable for both ac and dc currents. Other isolating current transducers are the *saturable reactor,* an adjustable inductor in which the input current vs. output voltage relationship is adjusted by controlled magnetomotive forces applied to the core, and the *Hall effect* current transducer in which an output-voltage change is produced by measured-current-originated electromagnetic effects on a semiconductor placed in a magnetic field. The *current transformer* can be used to convert the measured current, with circuit isolation, into an output current or voltage.

90. Electrometers. Small currents (down to 10^{-15} A) can be measured by means of an *electrometer tube* or by special semiconductor devices such as an amplifier with varactor diodes, metal oxide semiconductor field-effect transistors (MOSFET), or junction field-effect transistors. All these devices require output amplification to provide signals suitable for remote measurement. When current must be measured on the high-voltage secondary side of a transformer, the current in the low-voltage primary can often be measured instead, and a suitable calibration used to correlate the two currents.

91. Voltage Monitors. Dc and ac voltages can be monitored by means of a voltage divider connected across the two terminals to be measured. A voltage divider consists of two resistors in series, with the output taken across only one of the two resistors when a signal lower than the actual voltage is required by the measurement system. Ac voltages can also be measured by use of a transformer.

92. Frequency Converters and Dividers. Frequency can be converted into a voltage signal by use of a tuned discriminator or of integrating circuitry. When a digital signal is required, the measured frequency can be passed through an electronic "gate" which is "opened" for a fixed period of time. The pulse count over the gated period is then indicative of the frequency. If the measuring system can accept a frequency, but one much lower than the measured frequency, a *frequency divider* circuit can be used to provide an output frequency which is a fixed fraction of the input frequency.

93. Other Electrical Transducers. *Power measurements* are usually derived from simultaneous but separate voltage and current measurements. Power (especially at microwave frequencies) is also measured by using a portion of the power to raise the temperature of a resistive temperature transducer (e.g., a thermistor), then measuring the temperature change, which can be correlated to measured power by a suitable calibration.

Conductivity, typically of a liquid material, is measured by a *conductivity cell,* a chamber containing two electrodes of selected materials, connected into a bridge circuit.

Inductance, capacitance, and resistance characteristics are converted into a usable electrical output in the manner described for inductive, capacitive, and resistive transduction elements in Par. **10-10** and Table 10-3.

94. Bibliography

ARONSON, M. H. (ed.) "Temperature Measurement and Control Handbook," Instruments Publishing Co., Pittsburgh, Pa., 1961.

BOLLINGER, L. E. Transducers for Measurement, Part IV, Fluid Flow, *ISA Jo.,* Vol. 11, No. 11, pp. 64–69, November, 1964.

BRADSPIES, R. W. Bourdon Tubes, *Giannini Controls Tech. Note,* Conrac Corp., Duarte, Calif., 1961.

BROCK, T. E., and MOON, C. J. (eds.) "A Bibliography on Hot-Wire Anemometry," British Hydromechanics Research Association, Cranfield, Bedford, England, October, 1965.

CANFIELD, E. B. "Electromechanical Control Systems and Devices," Wiley, New York, 1965.

CERNI, R. H., and FOSTER, L. E. "Instrumentation for Engineering Measurement," Wiley, New York, 1962.

CORRUCCINI, R. J.   Interpolation of Platinum Resistance Thermometers, 20 to 373.15°K, *Rev. Sci. Instrum.,* Vol. 31, pp. 637-640, 1960.

DEAN, M., III (ed.)   "Semiconductor and Conventional Strain Gages," Academic, New York, 1962.

DEARNALEY, G., and NORTHROP, D. C.   "Semiconductor Counters for Nuclear Radiation," 2d ed., Barnes and Noble, New York, 1966.

DOEBELIN, E. A.   "Measurement Systems: Application and Design," McGraw-Hill, New York, 1966.

DOZER, B. E.   Liquid Level Measurement for Hostile Environments, *Instrum. Technol.,* Vol. 14, No. 2, pp. 55-58, February, 1967.

EISENMANN, W. L.   "Properties of Photodetectors," Naval Ordnance Laboratory, Photodetector Series, NOLC Report 637, Corona, Calif., Feb. 15, 1966.

FRAZINE, D. F.   "The Design and Construction of Thin Film Radiation Thermopiles," Arnold Engineering Development Center, Report AEDC-TR-66-38, USAF-AFSC, Arnold Air Force Sta., Tenn., 1966.

FREDERICK, J. R.   "Ultrasonic Engineering," Wiley, New York, 1965.

HARRIS, C. M., and CREDE, C. E. (eds.)   "Shock and Vibration Handbook," 3 vols., McGraw-Hill, New York, 1961.

HARRISON, T. R.   "Radiation Pyrometry and Its Underlying Principles of Radiant Heat Transfer," Wiley, New York, 1960.

HOLZBOCK, W. G.   "Instruments for Measurement and Control," 2d ed., Reinhold, New York, 1962.

HUGHES, W. G.   "Attitude Control: Gyros as Sensors," Royal Aircraft Establishment, Report ESRO-TM-32, Farnborough, England, 1966.

KEAST, D. N.   "Measurements in Mechanical Dynamics," McGraw-Hill, New York, 1967.

LLOYD, C., and GIARDINI, A. A.   Measurement of Very High Pressures, *Acta IMEKO 1964, Paper* 21-USA-267, Budapest, Hungary, 1964.

MERRIAM, J. D., EISENMAN, W. L., and NAUGLE, A. B.   "Properties of Photodetectors," Naval Ordnance Laboratory, NOLC Report 621, Corona, Calif., Apr. 1, 1965.

MERRILL, J. J. (ed.)   "Light and Heat Sensing," Macmillan, New York, 1963.

NEUBERT, H. K. P.   "Instrument Transducers," Oxford, London, 1963.

NOKES, M. C.   "Radioactivity Measuring Instruments: A Guide to Their Construction and Use," Philosophical Library, New York, 1959.

NORTON, H. N.   Error Band Concept Defines Transducer Performance, *Ground Support Equip.,* January, 1963.

———Specification Characteristics of Pressure Transducers, *Instrum. Control Systems,* Vol. 36, December, 1963.

———"Handbook of Transducers for Electronic Measuring Systems," Prentice-Hall, Englewood Cliffs, N.J., 1969.

OSTROVSKIJ, L. A.   "Elektrische Messtechnik" ("Electrical Measurement Technology"), (German transl. by D. Hoffman), VEB Verlag Technik, Berlin, 1969.

PITMAN, G. R., JR. (ed.)   "Inertial Guidance," Wiley, New York, 1962.

ROEHRIG, J. R.   "High Vacuum Measuring Instrumentation and Methodology," U.S. Dept. of Commerce, Office of Technical Services, Report FDL-TDR-64-68 (AFSC-R&TD-AFFDL), 1964.

RUSKIN, J. M., JR.   Thermistors as Temperature Transducers, *Data Systems Eng.,* Vol. 19, No. 2, pp. 24-27, February, 1964.

SAVET, P. H.   "Gyroscopes: Theory and Design," McGraw-Hill, New York, 1961.

SCHWEPPE, J. L., et al.   "Methods for the Dynamic Calibration of Pressure Transducers," National Bureau of Standards Monograph 67, Washington, D.C., 1963.

SHARPE, J.   "Nuclear Radiation Measurement," Temple Press, London, 1960.

SPAULDING, CARL P.   "How to Use Shaft Encoders," DATEX Division of Conrac Corp., Monrovia, Calif., 1965.

SPINK, L. K.   "Principles and Practice of Flow Meter Engineering," 9th ed., The Foxboro Co., Foxboro, Mass., 1967.

STEMPEL, F. C., and RALL, D. L.   "Applications and Advancements in the Field of Direct Heat Transfer Measurements," Instrument Society of America, Preprint 8.1.63, Pittsburgh, Pa., 1963.

STILTZ, H. L. (ed.) "Aerospace Telemetry," Prentice-Hall, Englewood Cliffs, N.J., 1961.

STREETER, V. L. (ed.) "Fluid Dynamics," McGraw-Hill, New York, 1948.

TAYLOR, J. M. "Semiconductor Particle Detectors," Butterworth, London, 1963.

TYSON, F. C. "Industrial Instrumentation," Prentice-Hall, Englewood Cliffs, N.J., 1961.

VAN DER PYL, L. M. Bibliography on Diaphragms and Aneroids, *ASME Paper* 60-WA-122, 1960.

VOLK, J. A. Gyroscopes, *Data Systems Eng.,* Vol. 19, No. 2, pp. 28–31, February, 1964.

WERNER, F. D. Time Constant and Self-heating Effect for Temperature Probes in Moving Fluids, *Rosemount Engineering Co. Bull.* 106017, Minneapolis, Minn., 1960.

WEXLER, A. (ed.) "Humidity and Moisture Measurement and Control in Science and Industry," Vols. 1–3, Reinhold, New York, 1965.

WOLF, W. L. "Handbook of Military Infrared Technology," Office of Naval Research, Washington, D.C., 1966.

"Temperature: Its Measurement and Control in Science and Industry," Vol. 2, Reinhold, New York, 1955.

"IRE Standards on Nuclear Techniques: Definitions for the Scintillation Counter Field, 1960," Institute of Electrical and Electronics Engineers, 60 IRE 13.S1, New York, 1960.

"Telemetry Transducer Handbook," WADD Technical Report 61-67, Vols. I and II, ASD, Air Force Systems Command, USAF, Wright-Patterson AFB, Ohio, 1961.

"Strain Gage Handbook," BLH Electronics, *Bull.* 4311A, Waltham, Mass., 1962.

"Temperature: Its Measurement and Control in Science and Industry," Vol. 3, Parts 1–3, Reinhold, New York, 1962.

"Synchro and Resolver Evaluation and Test Equipment" (editorial survey), *Electro-Technol.,* Vol. 71, No. 5, pp. 201–210, May, 1963.

"Thermistor Definitions and Test Methods," Electronic Industries Association, EIA Standard RD-275, Washington, D.C., 1963.

"Specifications and Tests for Piezoelectric Acceleration Transducers," Instrument Society of America, RP37.2, Pittsburgh, Pa., 1964.

"Standard Gyro Terminology" (rev.), Aerospace Industries Association, Washington, D.C., September, 1964.

"Strain-Gages, Bonded Resistance," Aerospace Industries Association, National Aerospace Standard NAS 942, Washington, D.C., 1964.

"Temperature Measurement Thermocouples," American National Standards Institute, Am. Nat. Std. C96.1, New York, 1964.

"Load Cells, Bonded Strain Gage, General Specification for," U.S. Naval Weapons Center, Specification NOTS-PD-101, China Lake, Calif., 1966.

"Photoconductive Cell Manual," Clairex Corp., New York, 1966.

"Specifications and Tests of Potentiometric Pressure Transducers for Aerospace Testing," Instrument Society of America, ISA Standard S37.6, Pittsburgh, Pa., 1967.

"Dynamic Response Testing of Process Control Instrumentation," Instrument Society of America, ISA Standard S26, Pittsburgh, Pa., 1968.

"Electrical Transducer Nomenclature and Terminology," Instrument Society of America, ISA Standard S37.1, Pittsburgh, Pa., 1969.

"Specifications and Tests for Piezoelectric Pressure and Sound-Pressure Transducers," Instrument Society of America, ISA Standard S37.10, Pittsburgh, Pa., 1969.

"Specifications and Tests for Strain-Gage Pressure Transducers," Instrument Society of America, ISA Standard S37.3, Pittsburgh, Pa., 1970.

SECTION 11

SOURCES AND SENSORS OF INFRARED, VISIBLE, AND ULTRAVIOLET ENERGY

BY

M. W. KLEIN Associate Director for R&D, Night Vision Laboratory, U.S. Army Electronics Command

ASSISTED BY

C. S. FOX Systems Development Technical Area, Night Vision Laboratory, U.S. Army Electronics Command; Member: Sigma Xi, Scientific Research Society of North America, Illuminating Engineering Society

D. J. HOROWITZ Image Intensification Technical Area, Night Vision Laboratory, U.S. Army Electronics Command

E. J. SHARP Image Intensification Technical Area, Night Vision Laboratory, U.S. Army Electronics Command

P. R. MANZO Optical Radiation Technical Area, Night Vision Laboratory, U.S. Army Electronics Command; Member: American Physical Society

G. M. JANNEY Laser Department, Hughes Research Laboratory, Hughes Aircraft Company; Member: Optical Society of America, Institute of Electrical and Electronics Engineers

L. F. GILLESPIE Visionics Technical Area, Night Vision Laboratory, U.S. Army Electronics Command; Member: Army Laser Advisory Group

R. R. SHURTZ II Image Intensification Technical Area, Night Vision Laboratory, U.S. Army Electronics Command; Member: Research Society of America

S. B. GIBSON Chief, Optical Radiation Team, Systems Development Technical Area, Night Vision Laboratory, U.S. Army Electronics Command; Member: Research Society of America, Illuminating Engineering Society, American Ordnance Society

N. A. DIAKIDES Image Intensification Technical Area, Night Vision Laboratory, U.S. Army Electronics Command; Member: American Chemical Society, Electrochemical Society, Institute of Electrical and Electronics Engineers

M. E. CROST Beam, Plasma and Display Technical Area, Electronics Technology and Devices Laboratory, U.S. Army Electronics Command; Member: Society for Information Display, American Physical Society, American Association for the Advancement of Science, American Institute of Physics

I. REINGOLD Leader, Display Devices Team, Beam, Plasma and Display Technical Area, Electronics Technology and Devices Laboratory, U.S. Army Electronics Command; Senior member: Institute of Electrical and Electronics Engineers; Fellow: Society for Information Display, Licensed Professional Engineer, State of New Jersey

W. KLEIN Chief, Negative Affinity Photocathode Team, Image Intensification Technical Area, Night Vision Laboratory, U.S. Army Electronics Command; Member: American Physical Society, German Physical Society

C. A. JOHNSON Systems Development Technical Area, Night Vision Laboratory, U.S. Army Electronics Command; Member: Institute of Electrical and Electronics Engineers, Research Society of America

RADIANT ENERGY DEVICES

R. D. GRAFT Systems Development Technical Area, Night Vision Laboratory, U.S. Army Electronics Command; Member: Optical Society of America, Ohio Academy of Science

R. E. FRANSEEN Chief, Remote View Tube Team, Systems Development Technical Area, Night Vision Laboratory, U.S. Army Electronics Command; Member: Institute of Electrical and Electronics Engineers Subcommittee on Diode Type Vidicons; Associate Member: American Society of Mechanical Engineers

W. A. GUTIERREZ Image Intensification Technical Area, Night Vision Laboratory, U.S. Army Electronics Command

J. R. PREDHAM Systems Development Technical Area, Night Vision Laboratory, U.S. Army Electronics Command; Member: American Institute of Physics

S. P. RODAK Far Infrared Technical Area, Night Vision Laboratory, U.S. Army Electronics Command

E. GADDY Engineering Physics Division, Goddard Space Flight Center, National Aeronautics and Space Administration

H. K. POLLEHN Chief, Tube Technology Team, Image Intensification Technical Area, Night Vision Laboratory, U.S. Army Electronics Command; Member: German Physical Society

REVIEWED BY

B. GOLDBERG Former Director, Night Vision Laboratory, U.S. Army Electronics Command; Member: Optical Society of America, American Association for the Advancement of Science, Research Society of America, Philosophical Society of Washington, D.C.

R. S. WISEMAN Director, Research, Development, and Engineering, and Director of Laboratories, U.S. Army Electronics Command; Fellow: Institute of Electrical and Electronics Engineers, Illuminating Engineering Society; Member: Optical Society of America, American Society for Engineering Education, Armed Forces Communication and Electronics Association, Association of the U.S. Army

CONTENTS

Numbers refer to paragraphs

RADIANT ENERGY DEVICES

SECTION 11

SOURCES AND SENSORS OF INFRARED, VISIBLE, AND ULTRAVIOLET ENERGY

LAMPS, LUMINOUS TUBES, AND OTHER NONCOHERENT ELECTRIC RADIATION SOURCES

By Clifton S. Fox

1. Generation of Light. Light is produced by the transitions of electrons from states of higher energies to states of lower energies. The law of conservation of energy is satisfied in these transition processes[1,2]* by the emission of a photon or quantum of light whose energy corresponds to the difference in energy of the initial and final energy states of the electron.

GLOSSARY

2. Blackbody Radiation. A blackbody is defined as a body which, if it existed, would absorb all and reflect none of the radiation incident upon it. It is thus a perfect absorber and a perfect emitter. The blackbody curves for several values of T are plotted on a logarithmic scale in Fig. 11-1.

3. The total emissivity ε of a thermal radiator at a given temperature is the ratio of the total radiation output of that radiator to that of a blackbody of the same temperature.

4. The spectral emissivity $\varepsilon(\lambda)$ of a thermal radiator is defined as the ratio of the output of the source at the wavelength λ to that of a blackbody at the same wavelength and operating temperature.

5. Graybody Radiation. If the emissivity of a thermal radiator is a constant less than 1 for all wavelengths, the radiator is called a graybody.

6. Selective Radiation. A thermal radiator whose spectral emissivity is not constant but is a function of wavelength is called a selective radiator.

7. The color temperature of a thermal radiator is the temperature of a blackbody chosen such that its output is the closest possible approximation to a perfect color match with the thermal radiator. Figure 11-2 shows the spectral distribution of a tungsten filament operating at a color temperature of 3000°K as compared with a blackbody of the same temperature and a graybody whose emissivity is the same as tungsten in the visible.

8. The candela is the unit of luminous intensity. Luminous intensity is the amount of luminous flux per unit solid angle in a given direction. This is measured as the luminous flux on a target normal to the direction divided by the solid angle (measured in steradians, abbreviated sr) subtended by the target as viewed from the source.

9. The lumen is the unit of luminous flux. It is equal to the flux in a unit solid angle from a uniform point source of one candela (1 cd) intensity.

10. The luminous efficacy of a light source is the measure of light-producing efficiency of the source. It is the ratio of the total luminous flux output to the total input power of the source. Luminous efficacy is measured in lumens per watt (lm/W).

11. Radiative efficiency of a light source is the ratio (in percent) of total output power of the source measured in watts to the input power to the source.

12. The most commonly used light sources are the tungsten filament, electric discharge, electroluminescent, and solid-state or light-emitting diodes. The first is incandescent; the others are luminescent.

---

*Superior numbers correspond to numbered references, Par. 11-23a.

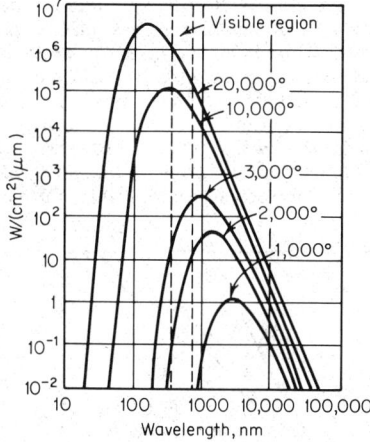

Fig. 11-1. Blackbody distribution curves for several values of temperature in degrees Kelvin.

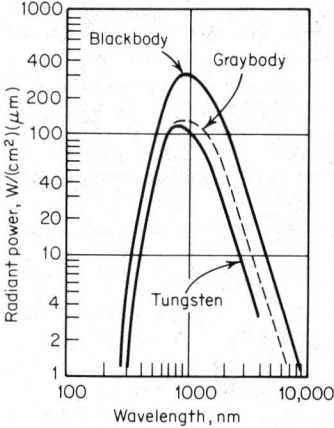

Fig. 11-2. Spectral distribution of tungsten at a color temperature compared with blackbody and graybody of the same temperature.

TUNGSTEN FILAMENT LAMPS

13. Filament. The higher the operating temperature of a solid filament, the higher will be the percentage of its radiation which falls in the visible portion of the electromagnetic spectrum. Tungsten, with its high melting point (3653°K), low vapor pressure, and other favorable characteristics, is the most frequently used filament material. In higher-power incandescent lamps (generally above 40 W) an inert gas instead of vacuum surrounds the filament to reduce the evaporation rate of the tungsten.

14. Lamp Types. Tungsten filament lamps[4-6] are divided into the following categories: general-service lamps; high- and low-voltage lamps; series burning lamps; projector and reflector lamps; showcase lamps; spotlight, floodlight, and projection lamps; halogen cycle lamps; and infrared lamps.

Tungsten halogen-cycle lamps have a quartz envelope and use a halogen fill, usually iodine, to keep the bulb clean by chemical reaction with sublimated tungsten. This reaction provides a high-lumen maintenance throughout the life of the lamp by redepositing evaporated tungsten on the filament instead of on the bulb. *Infrared lamps* are tungsten filament lamps which operate at low filament temperature.

15. Interrelationship of Lamp Parameters. The quantities voltage, current, resistance, temperature, watts, light output, efficacy, and life of a filament lamp are interrelated, and one cannot be changed without changing the others. Figure 11-3 shows how these quantities change typically as a function of the voltage for large gas-filled lamps.

Some useful exponential relations frequently applied to incandescent filament lamps are (capital letters indicate normal rated values)

$$\frac{\text{life}}{\text{LIFE}} = \left(\frac{\text{VOLTS}}{\text{volts}}\right)^d \qquad \frac{\text{lumens}}{\text{LUMENS}} = \left(\frac{\text{volts}}{\text{VOLTS}}\right)^k$$

$$\frac{\text{LM/W}}{\text{lm/w}} = \left(\frac{\text{VOLTS}}{\text{volts}}\right)^g \qquad \frac{\text{watts}}{\text{WATTS}} = \left(\frac{\text{volts}}{\text{VOLTS}}\right)^n$$

For approximate calculations the following average exponents may be used: $d = 13$, $g = 1.9$, $k = 3.4$, and $n = 1.6$.

ELECTRIC DISCHARGE LAMPS

16. Fluorescent lamps[4,5,7] are electric discharge lamps in which light is produced through the excitation of phosphors by the ultraviolet energy from a mercury arc. The lamp usually consists of a phosphor-coated tubular bulb with electrodes sealed into each end and containing mercury vapor at low pressure along with an inert starting gas such as argon or an argon-neon mixture. Various colors, grades of white light, and even black light[8] (near ultraviolet) are obtained by the choice of available phosphors. Cool white is the most widely used white-light fluorescent tube (Fig. 11-4). The de luxe fluorescent lamp has extra red output which enhances the natural appearance of red objects.

Lamp Types. Fluorescent lamps are classified as *hot-cathode* or *cold-cathode* type. There are three classes of hot-cathode-type fluorescent lamps: *preheat, instant-start,* and *rapid-start.* Preheat lamps allow preheating of the cathodes for a few seconds before striking the mercury

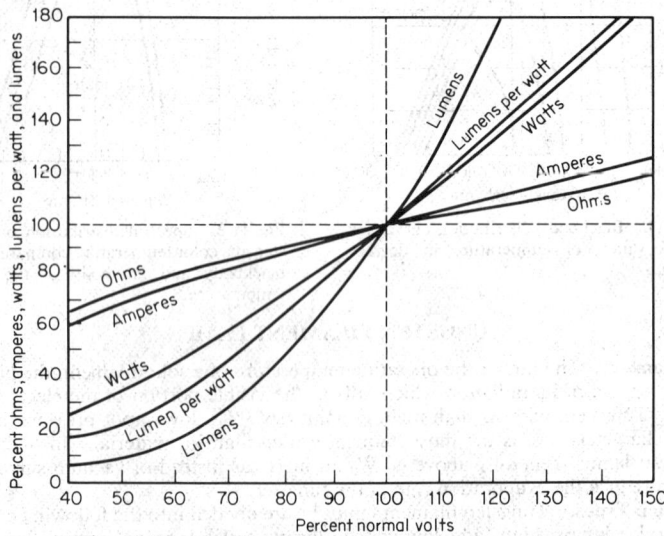

Fig. 11-3. Interrelation of lamp parameters for large tungsten filament lamps. (*From GE Technical Pamphlet TP 110, GE Large Lamp Dept., Nela Park, Cleveland, Ohio, 1969.*)

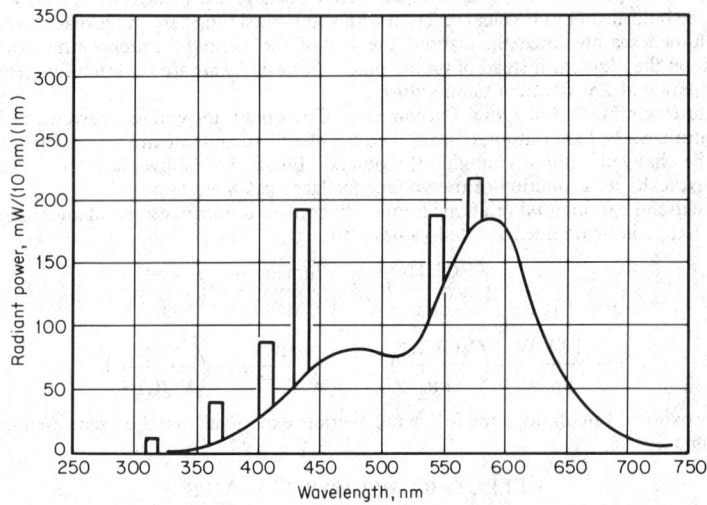

Fig. 11-4. Spectrum of cool-white fluorescent lamp. (*From GE Technical Pamphlet TP 111, GE Large Lamp Dept., Nela Park, Cleveland, Ohio, 1969.*)

arc. Instant-start lamps require no preheat because sufficient voltage is applied between the electrodes to strike the arc very quickly. Rapid-start lamps have continuously heated cathodes requiring a lower voltage than instant-start lamps. This feature also allows dimming and flashing of the lamp.

Life and Efficiency. The main advantages of fluorescent tubes are high efficiency and long life. A typical cool-white fluorescent tube renders 22% of its input energy in the form of visible light. Typical rated life for a 40-W fluorescent tube operated 3 h per start is 6,000 to 12,000 h.

17. Mercury Lamps.[4,5,9] Mercury-vapor discharge lamps of the wall-stabilized variety are used primarily for general lighting. Mercury lamps with additives such as the metal halide lamp and sodium vapor lamp and mercury compact arcs are treated in separate succeeding paragraphs. Most mercury vapor lamps have two bulbs. The inner bulb, called the *arc tube,* contains the arc and the electrodes between which the arc burns. The outer bulb protects the arc tube from drafts and stabilizes the operating temperature. Mercury lamps for general lighting are available in input power sizes from 50 to 3,000 W.

In addition to mercury in the arc tube, an easily ionized inert gas such as argon is present to facilitate starting. The arc is generally ignited through use of a starting electrode and current-limiting starting resistor. An arc is first struck between the starting electrode and the adjacent main electrode. The heating and additional ionization resulting from this arc allow the large arc to form between the main electrodes.

Electrical and Radiation Characteristics. Figure 11-5 shows the spectrum of a typical clear mercury lamp. The color-rendering properties of a clear mercury lamp are only fair, due to the line structure in the blue end of the spectrum. A clear mercury lamp of 1,000 W input power has a typical initial luminous efficacy of 56 lm/W.

Life. General lighting lamps of 100 to 1,000 W input power have mean lifetimes in excess of 24,000 h, based on 5 h burning time per start and operation from the correct ballast.

18. Metal halide mercury vapor lamps[4,5,9] are very nearly the same as regular mercury lamps except that additives such as the iodides of sodium, thallium, and indium are contained in the arc tube for the purpose of improving color rendition and efficiency (Fig. 11-6). The typical initial luminous efficacy of a metal halide mercury lamp of 1,000 W input power is 90 lm/W.

19. High-pressure sodium vapor lamps[4,9] are presently the most efficient source of man-

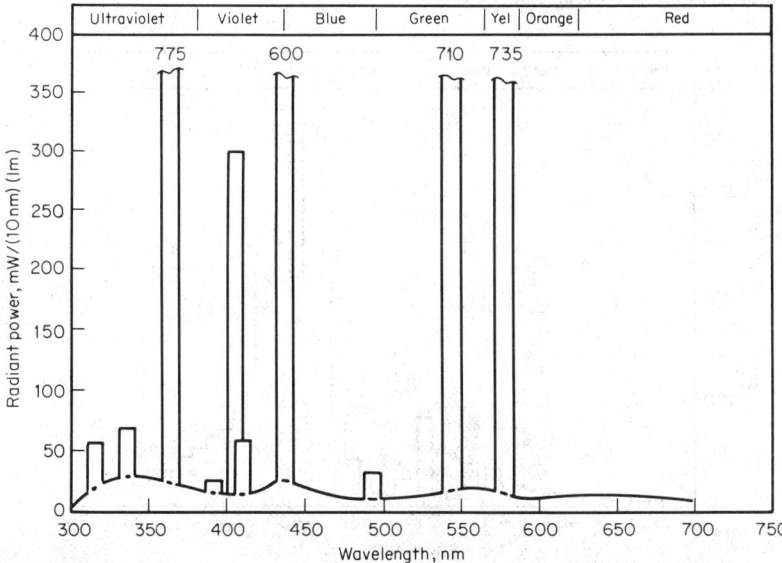

Fig. 11-5. Spectrum of typical clear mercury lamp. (*From GE Technical Pamphlet TP 109, GE Large Lamp Dept., Nela Park, Cleveland, Ohio, 1969.*)

made light (Fig. 11-7).   A typical 400-W high-pressure sodium lamp has an initial luminous efficacy of 115 lm/W.   The theoretical efficiency of white light in the visible, assuming 100% conversion of power into a continuum output, is 220 lm/W.

High-pressure sodium lamps require a high-transmission ceramic envelope such as alumina to contain the alkali metal at high temperature and an alkali-resistant high-temperature metal seal.   The corrosive effects of sodium at high operating temperature prohibit the use of quartz and other glasses as an arc-tube material.   Xenon is used as the readily ionized starting gas.   When the xenon starting gas is ionized, the arc is struck, producing heat, and the vapor pressure starts to rise.

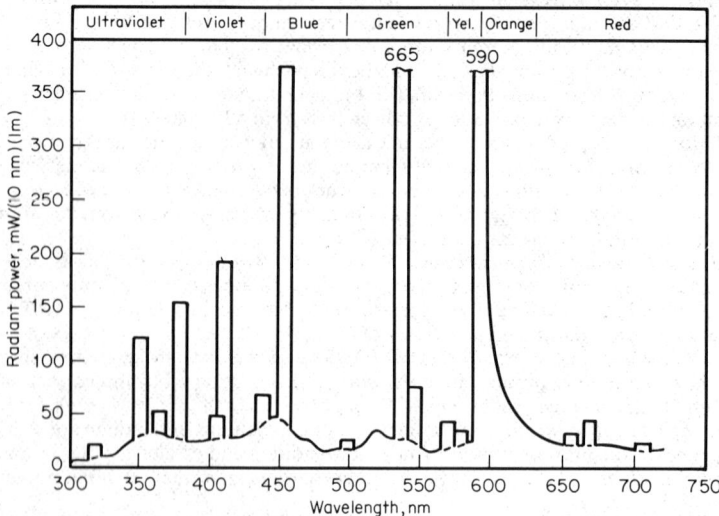

Fig. 11-6. Spectrum of a metal halide mercury lamp. (*From GE Technical Pamphlet TP 109, GE Large Lamp Dept., Nela Park, Cleveland, Ohio, 1969.*)

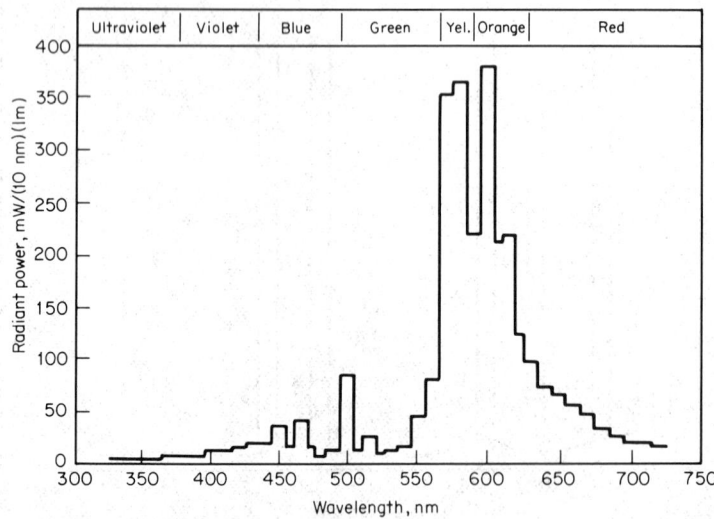

Fig. 11-7. Spectrum of a high-pressure sodium lamp. (*From GE Technical Pamphlet TP 109, GE Large Lamp Dept., Nela Park, Cleveland, Ohio, 1969.*)

11-8

20. Arc-Light Sources. *a. Short-arc or compact-arc lamps*[4,5] are of the enclosed-bulb type. Most are made for dc operation, which results in long life and good arc stability. They have high operating pressure, a comparatively short arc gap, and very high luminance (photometric brightness). The arc length may vary from about $\frac{1}{3}$ mm to about 17 mm. Since these lamps have the highest luminance of any continuous-operation light source and provide maintenance-free, clean operation, they are replacing the carbon arc in many of its applications.

Most short-arc lamps have a quartz bulb with a spherical or ellipsoidal center section. The electrodes are generally made of tungsten, and the cathode is thoriated for high-current-density operation.

Mercury short-arc lamps contain a low pressure of inert gas such as argon for starting. After the arc is struck, the lamp warms up and the mercury vapor reaches full operating pressure within a few minutes. Warmup time is reduced to approximately one-half of that of the mercury arc if xenon at one atmosphere or more pressure is added. The spectra of mercury and mercury-xenon lamps (Fig. 11-8) are essentially the same. The luminous efficacy of these lamps is about 50 lm/W for a 1,000-W lamp. These lamps are available in a power range from 30 to 5,000 W.

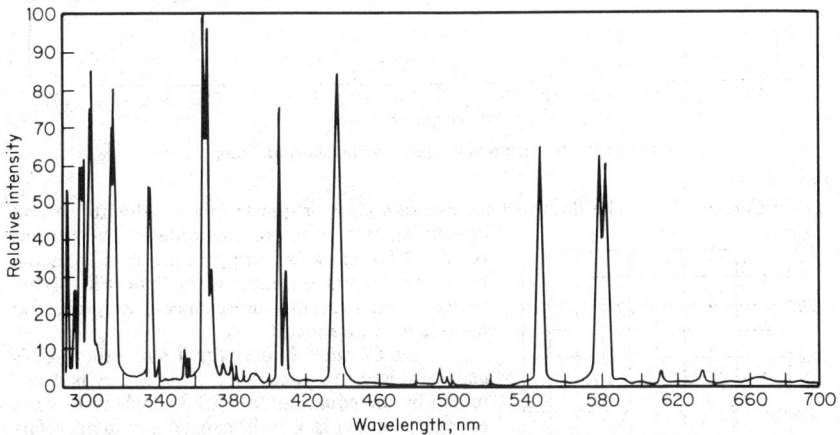

Fig. 11-8. Spectrum of a mercury xenon short-arc lamp.

Xenon short-arc lamps are filled to a cold pressure of several atmospheres with high-purity xenon gas. Operating pressure is roughly double the cold pressure. These lamps do not have as long a warmup time as the metal-vapor types. Eighty percent of the light output is obtained within a second of startup. Xenon has excellent color-rendering characteristics due to its continuous spectrum (Fig. 11-9) in the visible. Luminous efficacy at 5,000 W is approximately 45 lm/W.

b. Carbon arcs are of three basic types: the low-intensity arc, the flame arc, and the high-intensity arc.

The *low-intensity arc* has as its source of light the incandescent tip of the positive carbon which is maintained near the sublimation point of carbon (3700°C). The heat supplied to the positive carbon is from high-current-density electron bombardment originating from the negative carbon.

The *flame arc* is obtained by enlarging the core of the electrodes of a low-intensity arc and replacing the removed material with compounds of rare-earth elements such as cerium.

The *high-intensity arc* is obtained from the flame arc by increasing the core size and the current density so that the anode spot spreads over the entire tip of the carbon. A crater is formed, and this becomes the primary source of light.

21. Flashtubes[4] are designed to produce high-luminance flashes of light of very short duration. They are used in optical pumping of lasers (see Par. 11-35), stroboscopic work, photographic applications, and for many other purposes requiring flashing lights.

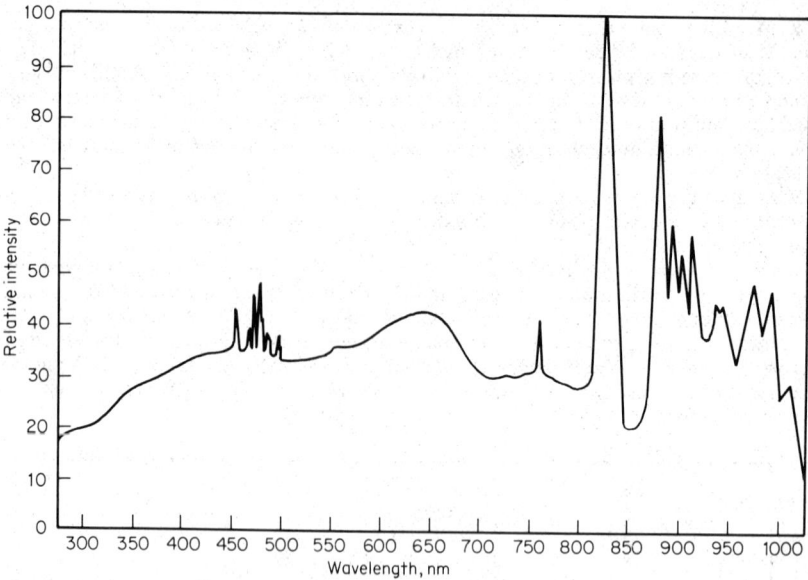

Fig. 11-9. Spectrum of a typical xenon short-arc lamp.

Lamp Construction. The flashtube consists of a glass or quartz tube filled with gas and containing two or more electrodes. The fill gas preferred for most flashtube applications is xenon, because of its high output of white light (Fig. 11-10). Other gases, including argon, neon, krypton, and hydrogen, are frequently used.

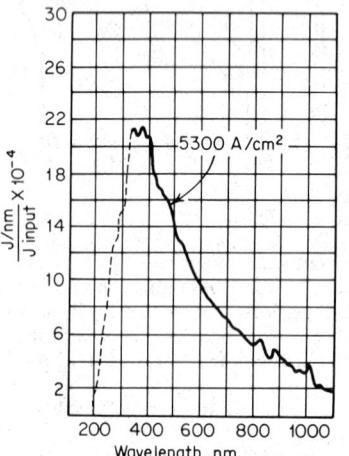

Fig. 11-10. Typical spectrum of a xenon flashtube.

Driving Circuit. Energy for flashing the tube is usually stored in a capacitor. This energy is determined by the equation $E = CV^2/2$, where E is the energy in joules (J), C is the capacitance in microfarads (μF), and V is the voltage on the capacitor in kilovolts (kV). The duration of the flash usually depends on the resistance of the discharge and the capacitance of the storage capacitor, the duration being approximately $3RC$, with R in ohms (Ω) and C in farads. For short pulses of one microsecond or less duration, frequently the inductance of the tube or circuit is the dominant factor over the resistance in determining pulse length.

Many circuits have been used to flash lamps. One basic method is to hold off a voltage higher than the self-breakdown voltage of the lamp with an electronic switch such as a thyratron, silicon controlled rectifier, or spark gap, and trigger the switch when a flash is desired.

ELECTROLUMINESCENT LAMPS

22. An electroluminescent lamp[4] is a thin source in which light is produced through excitation of a phosphor by an alternating electric field. The lamp is essentially a flat parallel plate capacitor with a phosphor embedded in the dielectric material sandwiched between two conducting plates. The front plate is a transparent sheet of either plastic, glass, or ceramic with a thin transparent conductive film on it. The back conductive plate may be a metal sheet or film or may be a transparent material like the front plate.

23. Light-emitting diodes (LEDs) (solid-state lamps) are semiconductor *pn* junction radiation sources. For theory of operation see Semiconductor Lasers, Par. **11-48**. The LED emits spontaneous rather than stimulated radiation. The spectral wavelength bandwidth is on the order of 400 to 1000 Å, and the radiance is much lower than that of the laser diode. These noncoherent emitters selectively produce radiation throughout the visible and near-infrared spectral ranges, as shown in Table 11-1. The LED has gained wide acceptance in the electronics field for photoelectric systems, visual displays, and electrooptical components.[10]

Table 11-1. Summary of Light-emitting Diodes Showing Color Selectivity

Type	Bandwidth at 1/2 power, Å	Spectral peak, Å	Color	Brightness or power output
Gallium arsenide	30	5,400	Green	80 ft.-lm
Coated with phosphor	600	10,000	Infrared	500 μW
Gallium arsenide phosphide	400	5,600	Green	300 ft.-lm
Silicon carbide	800	5,900	Yellow	150 ft.-lm
Gallium arsenide phosphide	400	6,100	Amber	200 ft.-lm
Gallium arsenide phosphide	400	6,800	Red	450 ft.-lm
Gallium aluminum arsenide	400	8,000	Red	1 mW
Gallium aluminum arsenide	400	8,500	Infrared	5 mW
Gallium arsenide	400	9,000	Infrared	10 mW
Gallium arsenide	600	9,800	Infrared	500 mW

23a. References on Noncoherent Sources

1. McNally, J. Rand Atomic Spectra Including Zeeman Effect and Stack Effect, in E. U. Condon and Hugh Odishaw (eds.), "Handbook of Physics," 2d ed., McGraw-Hill, New York, 1967.

2. Ditchburn, R. W. "Light," 2d ed., Interscience, New York, 1963.

3. Pivovonsky, Mark, and Nagel, Max R. "Tables of Blackbody Radiation Functions," Macmillan, New York, 1961.

4. Kaufman, John E. "IES Lighting Handbook," 4th ed., Illuminating Engineering Society, New York, 1966.

5. "Westinghouse Lighting Handbook," Westinghouse Electric Corporation, Bloomfield, N.J., 1969.

6. "Incandescent Lamps," General Electric Company, Pamphlet TP-110, Cleveland, Ohio.

7. "Fluorescent Lamps," General Electric Company, Pamphlet TP-111, Cleveland, Ohio, 1970.

8. "Black Light," General Electric Company, Pamphlet TP-215, Cleveland, Ohio, 1969.

9. "High Intensity Discharge Lamps," General Electric Company, Pamphlet TP-109, Cleveland, Ohio, 1968.

10. "Directory of GaAs Lite Products," Monsanto Co., Electronics Special Products Pamphlet, Cupertino, Calif.

11. Opto-Electronics Products Directory, *Electro-Optical Systems Design,* Vol. 3, No. 5, May, 1971.

LASERS (COHERENT SOURCES)

By Daniel J. Horowitz

24. Laser Light Compared with Nonlaser Light. The main characteristic of laser light [1-6]* is coherence, although laser light is usually more intense, more monochromatic, and more highly collimated than light from other sources.

Coherence is the property wherein corresponding points on the wavefront are in phase. A coherent beam can be visualized as an ideal wave whose spatial and time properties are

---

*Superior numbers correspond to numbered references, Par. 11-28a.

clearly defined and predictable. Ordinary noncoherent light consists of random and discontinuous phases of varying amplitudes. The noncoherent beam has an average intensity and a predominant wavelength, but it is basically a superposition of different waves. The characteristic grainy appearance of laser light is due to interference effects which result from coherence.

Intensity of laser light can be very high. For example, power densities of over 1,000 MW/cm^2 can be obtained. A beam of such intensity can cut through and vaporize materials.

A laser beam is often highly *monochromatic* and highly *collimated*, both varying with the type of laser.

25. Stimulated vs. Spontaneous Emission. The laser operates on the principle of stimulated emission, an effect which is rarely observed except in connection with lasers.

Spontaneous emission is the usual method whereby light is emitted from excited atoms or molecules. Assume that the laser material has energy levels which can be occupied by electrons, that the lowest level or ground state is occupied, and that the next upper level is unoccupied. An excitation process can then raise an electron from the ground state to this upper state. The electron, after a variable time interval, returns to the ground state and emits a photon whose direction and phase of the associated wave are random and whose energy corresponds to the energy difference between the states. The upper-level lifetime may be comparatively short (less than 10^{-11} s) or it may be long (greater than 10^{-6} s). In the latter case the level is referred to as a *metastable level* or *state,* and the light emission is referred to as *fluorescence.*

Stimulated Emission. When the electron is in the upper level, if a light wave of precisely the wavelength corresponding to the energy difference strikes the electron in the excited state, the light stimulates the electron to transfer down to the lower level and emit a photon. This photon is emitted precisely in the same direction, and its associated wave is in the same phase as that of the incident photon. Thus a traveling wave of the proper frequency is produced, passing through the excited material and growing in amplitude as it stimulates emission.

26. Pumping and Population Inversion. The process of exciting the laser material (raising electrons to excited states) is referred to as *pumping.* Pumping can be done optically using a lamp of some kind, by an electrical discharge, a chemical reaction, or in the case of the semiconductor laser, by injecting electrons into an upper energy level by means of an electric current.

A *population inversion* is necessary to initiate and sustain laser action. Normally, the ground state is almost entirely occupied and the upper level or levels, assuming they are more than a few tenths of an electron volt above the ground state at room temperature, are essentially unoccupied. When the upper level has a greater electron population than the lower level, a population inversion is said to exist. This inverted population can support lasing, since a traveling wave of the proper frequency can stimulate downward transitions and be amplified.

27. Optical Resonators. The addition of a positive-feedback mechanism to a lasing medium permits it to serve as an oscillator.

The Fabry-Perot resonator, which provides optical feedback, consists of two parallel mirrors, the rear mirror fully reflecting and the front mirror partly reflecting and partly transmitting at the laser wavelength. The light reflected back from the front and rear mirrors serves as positive feedback to sustain oscillation, and the light transmitted through the front mirror serves as the laser output. Laser action is started by spontaneously emitted light with the proper direction to travel down the axis of the laser rod and be reflected on itself from the end mirrors. The two mirrors form an optical cavity which can be tuned by varying the spacing of the mirrors. The laser can operate only at wavelengths for which a standing-wave pattern can be set up in the cavity, i.e., for which the length of the cavity is an integral number of half wavelengths. Mirrors can be separate from the laser rod, or they can be deposited on its end faces.

Spectral Modes. The output usually consists of a series of evenly separated wavelength spikes within the fluorescent line width of the material (Fig. 11-11). The fluorescent line width is the natural spectral line width of the transition from the metastable state to the terminal level. If only one output wavelength is desired, a subsidiary wavelength-selecting mechanism such as a prism can be used.

28. Q Switching. The typical output of an optical laser consists of a series of spikes occurring during the major portion of the time that the laser is pumped (Fig. 11-12). These spikes result because the inverted population is being alternately built up and depleted. *Q switching* (Q spoiling) is a means of obtaining all the energy in a single spike of very high peak power. As an example, an ordinary laser might generate 100 mJ over a time interval of 100 μs for a peak power (averaged over this time interval) of 1,000 W. The same laser Q-switched might emit 80 mJ in a single 10-ns pulse for a peak power of 8 MW. The term Q switching is used by analogy to the Q of an electric circuit. By lowering the Q of the optical cavity, the laser cannot oscillate, and a large inverted population builds up. When the cavity Q is restored, a single "giant pulse" (see Fig. 11-12) is generated. This high-peak-power pulse is useful in optical ranging and communication and in producing nonlinear effects in materials. (See Par. 11-36.)

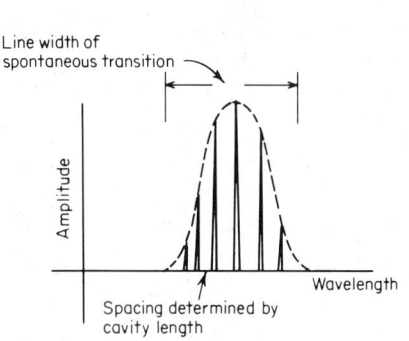

Fig. 11-11. Typical spectral modes of an optically pumped laser.

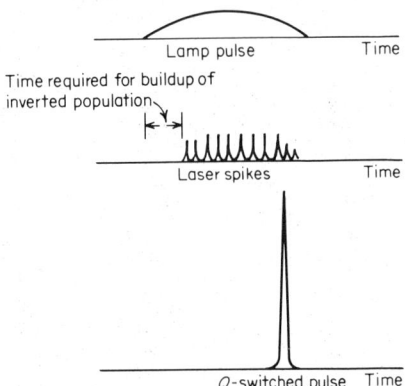

Fig. 11-12. Typical optically pumped laser output for Q-switched and non-Q-switched operation.

28a. References on Laser Fundamentals

1. SMITH, W. V., and SOROKIN, P. P. "The Laser," McGraw-Hill, New York, 1966.
2. STEELE, E. "Optical Lasers in Electronics," Wiley, New York, 1968.
3. MARSHALL, S. L. "Laser Technology and Applications," McGraw-Hill, New York, 1968.
4. LENGYEL, B. A. "Introduction to Laser Physics," Wiley, New York, 1966.
5. YARIV, A. "Quantum Electronics," Wiley, New York, 1967.
6. HEAVENS, O. A. "Optical Lasers," Wiley, New York, 1964.

SOLID OPTICALLY PUMPED LASER MATERIALS

BY EDWARD J. SHARP

29. Laser Ions and Hosts. The current list of ions which can be incorporated into a solid host and made to lase number about 20. The total number of different wavelengths available from these ions is approximately 150. (See Ref. 1, Par. **11-33a**, for a complete listing.) The solid host is any crystal or glass which can accommodate trivalent or divalent rare-earth and iron-group ions. The iron-group ions in which laser oscillation has been achieved are divalent nickel and cobalt and trivalent chromium and vanadium. The rare-earth ions in which laser oscillation has been achieved are divalent samarium and dysprosium and trivalent europium, praseodymium, ytterbium, neodymium, erbium, thulium, and uranium. Table 11-2 gives a brief list of these ions and the emission wavelength of the ion and host, as well as the mode and temperature of operation.

Table 11-2. Laser Ions and Hosts

λ nm	Ion	Host	Mode and temperature °K	Transition	Reference*
551.2	Ho^{3+}	CaF_2	P 77	$^5S_2 \to {}^5I_8$	2
598.5	Pr^{3+}	LaF_3	P 77	$^3P_0 \to {}^3H_6$	3
611.3	Eu^{3+}	Y_2O_3	P 220	$^5D_0 \to {}^7F_2$	4
694.3	Cr^{3+}	Al_2O_3	P CW 300	$^2E(A) \to {}^4A_2$	5
854.8	Er^{3+}	CaF_2	P 77	$^4S_{3/2} \to {}^4I_{13/2}$	6
946.0	Nd^{3+}	$Y_3Al_5O_{12}$	P 230	$^4F_{3/2} \to {}^4I_{9/2}$	7
1029.6	Yb^{3+}	$Y_3Al_5O_{12}$	P 77	$^2F_{5/2} \to {}^2F_{7/2}$	8
1040.0	Pr^{3+}	$Ca(NbO_3)_2$	P 77	$^1G_4 \to {}^3H_4$	9
1053.0	Nd^{3+}	$LiYF_4$	P 300	$^4F_{3/2} \to {}^4I_{11/2}$	10
1054.0	Nd^{3+}	$Ba_2MgGe_2O_7$	P 300	$^4F_{3/2} \to {}^4I_{11/2}$	11
1058.0	Nd^{3+}	$CaWO_4$	CW 300	$^4F_{3/2} \to {}^4I_{11/2}$	12
1059.7	Nd^{3+}	$NaLa(MoO_4)_2$	P 300	$^4F_{3/2} \to {}^4I_{11/2}$	13
1061.0	Nd^{3+}	$CaMoO_4$	CW 300	$^4F_{3/2} \to {}^4I_{11/2}$	14
1062.9	Nd^{3+}	$Ca_5(PO_4)_3F$	P 300	$^4F_{3/2} \to {}^4I_{11/2}$	15
1063.3	Nd^{3+}	$Y_3Ga_5O_{12}$	P 300	$^4F_{3/2} \to {}^4I_{11/2}$	16
1064.1	Nd^{3+}	YVO_4	P 300	$^4F_{3/2} \to {}^4I_{11/2}$	17
1064.5	Nd^{3+}	$YAlO_3$	P 300	$^4F_{3/2} \to {}^4I_{11/2}$	18
1064.8	Nd^{3+}	$Y_3Al_5O_{12}$	P CW 300	$^4F_{3/2} \to {}^4I_{11/2}$	16
1076.0	Nd^{3+}	La_2O_2S	P 300	$^4F_{3/2} \to {}^4I_{11/2}$	19
1079.5	Nd^{3+}	$YAlO_3$	P 300	$^4F_{3/2} \to {}^4I_{11/2}$	18, 20
1111.0	Nd^{3+}	$Y_3Al_5O_{12}$	P 300	$^4F_{3/2} \to {}^4I_{11/2}$	21
1116.0	Tm^{2+}	CaF_2	CW 27	$^2F_{5/2} \to {}^2F_{7/2}$	22
1121.3	V^{3+}	MgF_2	P 77	$^4T_2 \to {}^4T_1$	23
1314.4	Ni^{2+}	MgO	P 77	$^3T_2 \to {}^3A_2$	24
1358.0	Nd^{3+}	$Y_3Al_5O_{12}$	P 300	$^4F_{3/2} \to {}^4I_{13/2}$	25
1623.0	Ni^{2+}	MgF_2	P 77	$^3T_2 \to {}^3A_2$	24
1660.2	Er^{3+}	$Y_3Al_5O_{12}$	P 77	$^4I_{13/2} \to {}^4I_{15/2}$	8
1663.0	Er^{3+}	$YAlO_3$	P 300	$^4S_{3/2} \to {}^4I_{9/2}$	26
1750.0	Co^{2+}	MgF_2	P 77	$^4T_2 \to {}^4T_1$	23
1821.0	Co^{2+}	$KMgF_3$	P 77	$^4T_2 \to {}^4T_1$	23
1911.0	Tm^{3+}	$CaWO_4$	P 77	$^3H_4 \to {}^3H_6$	27
2013.2	Tm^{3+}	$Y_3Al_5O_{12}$	P 300	$^3H_4 \to {}^3H_6$	8
2059.0	Ho^{3+}	$CaWO_4$	P 77	$^5I_7 \to {}^5I_8$	28
2097.5	Ho^{3+}	$Y_3Al_5O_{12}$	CW 77-P 300	$^5I_7 \to {}^5I_8$	8
2165.0	Co^{2+}	ZnF_2	P 77	$^4T_2 \to {}^4T_1$	29
2358.8	Dy^{2+}	CaF_2	CW 77-P 145	$^5I_7 \to {}^5I_8$	30
2407.0	U^{3+}	SrF_2	P 90	$^4I_{11/2} \to {}^4I_{9/2}$	31
2556.0	U^{3+}	BaF_2	P 20	$^4I_{11/2} \to {}^4I_{9/2}$	32
2613.0	U^{3+}	CaF_2	P 300	$^4I_{11/2} \to {}^4I_{9/2}$	33
2691.0	Er^{3+}	CaF_2	P 300	$^4I_{11/2} \to {}^4I_{13/2}$	34

* See Par. 11-33a.

30. Characteristics of the Hosts. The host material to which the active ion (dopant) is incorporated can be either crystalline or glass and should exhibit:

a. *High thermal conductivity.*

b. *Ease of fabrication.*

c. *Hardness* (to prevent degradation of optical finishes).

d. *Resistance to solarization* or radiation-induced color centers.

e. *Chemical inertness* (i.e., not water-soluble).

f. *High optical quality,* which implies uniformity of refractive index and absence of voids or inclusions or other scattering centers.

31. Crystalline hosts offer as advantages in most cases their hardness, high thermal conductivity, narrow fluorescence line width, and, for some applications, their optical anisotropy. Crystalline hosts usually have as disadvantages their poor optical quality, inhomogeneity of doping, and generally narrower absorption lines.

32. Glass laser hosts are optically isotropic and easy to fabricate, possess excellent optical quality, and are hard enough to accept and retain optical finishes. In most cases glasses may be more heavily and more homogeneously doped than crystals, and in general, glasses possess broader absorption bands and exhibit longer fluorescence decay times. The primary disadvantages of glass are its broad fluorescence line widths (leading to higher thresholds), its significantly lower thermal conductivity (a factor of 10, leading to thermally induced birefringence and distortion when operated at high pulse repetition rates or high average powers), and its susceptibility to solarization [darkening due to color centers which are formed in the glass as a result of the ultraviolet (UV) radiation from the flashlamps]. These disadvantages limit the use of glass laser rods for cw (see Par. **11-34**) and high-repetition-rate lasers. Table 11-3 gives a brief listing of glass lasers.

Table 11-3.   Glass Lasers

Dopant	Glass	λ nm	Mode and temperature	Transition	Reference*
Gd^{3+}	Li-Mg-Al-Si	312.5	P 77	$^6P_{7/2} \to {}^8S_{7/2}$	55
Ho^{3+}	Li-Mg-Al-Si	1950.0	P 77	$^5I_7 \to {}^8I_8$	56
Nd^{3+}	K-Ba-Si	1060.0	P 300	$^4F_{3/2} \to {}^4I_{11/2}$	57
Nd^{3+}	Barium crown	1060.0	CW 300	$^4F_{3/2} \to {}^4I_{11/2}$	58
Nd^{3+}	Ba-Cs-Si	920.0	P 300	$^4F_{3/2} \to {}^4I_{9/2}$	59
Nd^{3+}	La-Ba-Th-B	1370.0	P 300	$^4F_{3/2} \to {}^4I_{13/2}$	60
Yb^{3+}	Li-Mg-Al-Si	1015.0	P 77	$^5F_{5/2} \to {}^5F_{7/2}$	61

* See Par. 11-33a.

33. Sensitized Lasers.   Laser performance and efficiency can be enhanced through the technique of energy transfer. A second ion (sensitizer) is incorporated into the host in addition to the laser ion (activator) to accomplish this effect. The sensitizer may be a color center. Pump energy is absorbed by the sensitizer, and is transferred to the activator, which then emits this energy at the laser wavelength. A list of sensitized lasers is given in Table 11-4.

33a. References on Solid Laser Materials

1. WEBER, M. J.   Insulating Crystal Lasers, in R. J. Pressley (ed.), "Handbook of Lasers," Chemical Rubber Co., Cleveland, Ohio, 1971.
2. VORONKO, U. K., KAMINSKII, A. A., OSIKO, V. V., and PROKHOROV, A. N.   *JETP Letters,* 1:3 (1965).
3. SOLOMON, R., and MUELLER, L.   *Appl. Phys. Letters,* 3:135 (1963).
4. CHANG, N. C.   *J. Appl. Phys.,* 34:3500 (1963).
5. MAIMAN, T. H.   *Brit. Commun. Electron.,* 7:674 (1960).
6. VORONKO, Y. K., and SYCHUGOV, V. A.   *Phys. Stat. Sol.,* 25:K119 (1968).
7. WALLACE, R. W., and HARRIS, S. E.   *Appl. Phys. Letters,* 28A:111 (1969).
8. JOHNSON, L. F., GEUSIC, J. E., and VAN UITERT, C. G.   *Appl. Phys. Letters,* 7:127 (1965).
9. BALLMAN, A. A., PORTO, S. P. S., and YARIV, A.   *J. Appl. Phys.,* 34:3155 (1963).
10. HARMER, A. L., LINZ, A., and GABBLE, D. R.   *J. Phys. Chem. Solids,* 30:1438 (1969).
11. ALAM, M., GOOEN, K. H., DI BARTOLO, B., LINZ, A., SHARP, E., GILLESPIE, L., and JANNEY, G.   *J. Appl. Phys.,* 39:4738 (1968).
12. JOHNSON, L. F., BOYD, G. D., NASSAU, K., and SODEN, R. R.   *Proc. IRE,* 50:213 (1962).
13. MOROZOV, A. M., TOLSTOI, M. N., FEOTILOV, P. P., and SHAPOVALOV, V. N.   *Opt. Spectrom.,* 22:224 (1967).
14. DUNCAN, R. C.   *J. Appl. Phys.,* 36:874 (1965).
15. OHLMANN, R. C., STEINBRUEGGE, K. B., and MAZELSKY, R.   *Appl. Opt.,* 7:905 (1968).
16. GEUSIC, J. E., MARCOS, H. M., and VAN UITERT, L. G.   *Appl. Phys. Letters,* 4:182 (1964).
17. BAGDASAROV, K. S., BUGOMOLOVA, G. A., KAMINSKII, A. A., and POPOV, V. I.   *Sov. Phys. Dok.,* 13:516 (1968).
18. WEBER, M. J., BASS, M., AMDRINGA, K., MANCHAMP, R. R., and CAMPERCHIO, E.   *Appl. Phys. Letters,* 15:342 (1969).
19. ALVES, R. V., BUCHANAN, R. A., WICKERSHEIM, K. A., and YATES, E. A. C.   *J. Appl. Phys.,* 42:3043 (1971).
20. BAGDASAROV, K. S., and KAMINSKII, A. A.   *JETP Letters,* 9:303 (1969).
21. SMITH, R. G.   *IEEE J. Quantum Electron.,* QE-4:505 (1968).

Table 11-4

Active ion	Sensitizing ion	Host	λ, nm	Transition	Mode and temp., °K	Reference*
Cr³⁺(Pr)	Cr³⁺	Al₂O₃	704.0	$\cdots$	P 77	35, 36, 37
Er³⁺	Yb³⁺	Silicate glass	1542.6	$^4I_{13/2} \rightarrow {}^4I_{15/2}$	P 300	38
Er³⁺	Color center	CaF₂	1530.8		P 4	39
Ho³⁺	Er³⁺	Er₂O₃	2121.0	$^5I_7 \rightarrow {}^5I_8$	CW P 77	40
Ho³⁺	Er³⁺	CaMoO₄	2070.0	$^5I_7 \rightarrow {}^5I_8$	P 77	41
Ho³⁺	Cr³⁺	Y₃Al₅O₁₂	2097.5	$^5I_7 \rightarrow {}^5I_8$	P 77	42
Ho³⁺	Er³⁺, Tm³⁺	Y₃Fe₅O₁₂	2086.0	$^5I_7 \rightarrow {}^5I_8$	P 295	43
Ho³⁺	Er³⁺, Tm³⁺, Yb³⁺	Y₃Al₅O₁₂	2128.8	$^5I_7 \rightarrow {}^5I_8$	CW 85	44
Ho³⁺	Er³⁺, Tm³⁺, Yb³⁺	Y₃Al₅O₁₂	2122.7	$^5I_7 \rightarrow {}^5I_8$	P 298	44
Ho³⁺	Er³⁺, Tm³⁺, Yb³⁺	CaF₂	2060.0	$^5I_7 \rightarrow {}^5I_8$	P 80	45
Ho³⁺	Yb³⁺	Silicate glass	2080.0	$^5I_7 \rightarrow {}^5I_8$	P 77	46
Ho³⁺	Er³⁺	Y₃Al₅O₁₂	2066.0	$^5I_7 \rightarrow {}^5I_8$	P CW 300	47
Nd³⁺	Cr³⁺	Y₃Al₅O₁₂	1061.2	$^4F_{3/2} \rightarrow {}^4I_{11/2}$	P 300	48
Nd³⁺	Cr³⁺	YAlO₃	1064.0	$^4F_{3/2} \rightarrow {}^4I_{11/2}$	P 90	49
Nd³⁺	Ce³⁺	CeF₃	1060.0	$^4F_{3/2} \rightarrow {}^4I_{11/2}$	P	50
Nd³⁺	Mn³⁺	Phosphate glass	1060.0	$^4F_{3/2} \rightarrow {}^4I_{11/2}$		51
Nd³⁺	UO₂²⁺	Barium crown glass	1060.0	$^4F_{3/2} \rightarrow {}^4I_{11/2}$		52
Tm³⁺	Er³⁺	Er₂O₃	1934.0	$^3H_4 \rightarrow {}^3H_5$	P 77	53
Tm³⁺	Er³⁺	CaMoO₄	1911.5	$^3H_4 \rightarrow {}^3H_5$	P 77	41
Tm³⁺	Cr³⁺	Y₃Al₅O₁₂	2019.0	$^3H_4 \rightarrow {}^3H_5$	P 295	42
Yb³⁺	Nd³⁺	Borate glass	1018.0	$^2F_{5/2} \rightarrow {}^2F_{7/2}$	P 77	54

* See Par. 11-33a.

22. KISS, Z. J., and DUNCAN, R. C.   Appl. Phys. Letters, 3:23 (1963).
23. JOHNSON, L. F., GUGGENHEIM, H. J., and THOMAS, R. A.   Phys. Rev., 149:179 (1966).
24. JOHNSON, L. F., DIETZ, R. E., and GUGGENHEIM, H. J.   Phys. Rev. Letters, 11:318 (1963).
25. DESERNO, U., ROSS, D., and ZEIDLER, G.   Phys. Letters, 28A:422 (1968).
26. WEBER, M. J., BASS, M., DE MARS, G. A., and ANDRINGA, K.   IEEE J. Quantum Electron., QE-6:654 (1970).
27. JOHNSON, L. F., BOYD, G. D., and NASSAU, K.   Proc. IRE, 50:86 (1962).
28. JOHNSON, L. F., BOYD, G. D., and NASSAU, K.   Proc. IRE, 50:87 (1962).
29. JOHNSON, L. F., DIETZ, R. E., and GUGGENHEIM, H. J.   Appl. Phys. Letters, 5:21 (1964).
30. KISS, Z. J., and DUNCAN, R. C.   Proc. IRE, 50:1531 (1962).
31. PORTO, S. P. S., and YARIV, A.   Proc. IRE, 50:153 (1962).
32. PORTO, S. P. S., and YARIV, A.   Proc. IRE, 50:1542 (1962).
33. BOYD, G. D., COLLINS, R. J., PORTO, S. P. S., YARIV, A., and HARGRAVES, G. W. Phys. Rev. Letters, 8:269 (1962).
34. ROBINSON, M., and DEVOR, D. P.   Appl. Phys. Letters, 10:167 (1967).
35. SOFFER, M. B., and HOSKINS, R. H.   Appl. Phys. Letters, 6:200 (1965).
36. SCHAWLOW, A. L., and DEVLIN, G. E.   Phys. Rev. Letters, 6:96 (1961).
37. POWELL, R. C., DI BARTOLO, B., BIRANG, B., and NAIMAN, C. S.   Phys. Rev., 155:296 (1967).
38. SNITZER, E., and WOODCOCK, R.   Appl. Phys. Letters, 6:45 (1965).
39. FORRESTER, P. A., and SAMPSON, D. F.   Proc. Phys. Soc., 88:199 (1966).
40. SOFFER, B. H., and HOSKINS, R. H.   IEEE J. Quantum Electron., QE-2:253 (1966).
41. JOHNSON, L. F., VAN UITERT, L. G., RUBIN, J. J., and THOMAS, R. A.   Phys. Rev., 133:A494 (1964).
42. JOHNSON, L. F., GEUSIC, J. E., and VAN UITERT, L. G.   Appl. Phys. Letters, 7:127 (1965).
43. JOHNSON, L. F., REMEIKA, J. P., and DILLON, J. F., JR.   Phys. Letters, 21:37 (1966).
44. JOHNSON, L. F., GEUSIC, J. E., and VAN UITERT, L. G.   Appl. Phys. Letters, 8:200 (1966).
45. ROBINSON, M., and DEVOR, D. P.   Appl. Phys. Letters, 10:167 (1967).
46. GANDY, H. W., GINTHER, R. J., and WELLER, J. F.   Appl. Phys. Letters, 6:237 (1965).
47. REMSKI, R. L., JAMES, L. T., GOOEN, K. H., DI BARTOLO, B., and LINTZ, A.   IEEE J. Quantum Electron., 5:214 (1969).
48. KISS, Z. J., and DUNCAN, R. C.   Appl. Phys. Letters, 5:200 (1964).
49. BASS, M., and WEBER, M. J.   Appl. Phys. Letters, 17:395 (1970).
50. O'CONNOR, J. R., and HARGRAVES, W. A.   Appl. Phys. Letters, 4:208 (1964).
51. MELAMED, N. T., HIRAYAMA, C., and DAVIS, E. K.   Appl. Phys. Letters, 7:170 (1965).
52. MELAMED, N. T., and HIRAYAMA, C.   Appl. Phys. Letters, 6:431 (1965).
53. SOFFER, B. H., and HOSKINS, R. H.   Appl. Phys. Letters, 6:200 (1965).
54. PEARSON, A. D., and PORTO, S. P. S.   Appl. Phys. Letters, 4:202 (1964).
55. GANDY, H. W., and GINTHER, R. J.   Appl. Phys. Letters, 1:25 (1962).
56. GANDY, H. W., and GINTHER, R. J.   Proc. IRE, 50:2113 (1962).
57. SNITZER, E.   Phys. Rev. Letters, 7:444 (1961).
58. YOUNG, C. G.   Appl. Phys. Letters, 2:151 (1963).
59. ROBINSON, C. C., SHAW, R., and WOODCOCK, R. F.   American Optical Corp., Interim Report, Contract No. DAAK02-70-C-0009, USAMERDC, Ft. Belvoir, Va., Aug. 1970.
60. MAURER, P. B.   Appl. Opt., 3:153 (1964).
61. ETZEL, H. W., GANDY, H. W., and GINTHER, R. J.   Appl. Opt., 1:534 (1962).

OPTICALLY PUMPED LASER DEVICES

By DANIEL J. HOROWITZ

34. Currently Used Laser Materials. The major optically pumped laser materials are ruby, neodymium-doped glass, and neodymium-doped yttrium aluminum garnet (YAG). Yttrium orthoaluminate appears to be a promising replacement for YAG. Ruby is the choice for visible light because it radiates at 6943 Å. YAG and glass, doped with neodymium, radiate at approximately 1.06 μm, in the near infrared. The properties of the materials are shown in Table 11-5.

Table 11-5. Properties of Common Laser Materials

	Output wavelength	Doping level	Chemical formula	Thermal conductivity, W/cm·°C
Ruby	6943 Å	≈0.04% chromium	Al_2O_3 : Cr	0.34
YAG	1.06 μm	≈1% Nd	$Y_3Al_5O_{12}$: Nd	0.11
YALO	1.06 or 1.08 μm depending on crystal orientation	≈1% Nd	$YAlO_3$: Nd	0.11
Glass Nd	1.06 μm	≈3% Nd	Varies with manufacturer	0.012

35. Pulsed Operation. *Flash Lamps.* Most common is flash-lamp pumping, using a linear, low-pressure lamp (see Par. **11-21**). Linear flash lamps have tungsten electrodes and are filled to a few atmospheres pressure with xenon or krypton. Xenon has been more commonly used, but recently krypton has been reported to give higher efficiency for pumping neodymium-doped materials at low input levels (about 10 J into a 2-in. flash lamp) because the emission lines of krypton provide a better match to the neodymium absorption bands. At higher input levels, the blackbody spectrum becomes stronger than the emission lines, and this advantage is lost.

Pumping Reflectors. The reflectors are usually elliptical in cross section, with the rod at one focus and the lamp at the other. Sometimes two lamps are used, in which case a double elliptical cavity is used. Often a round cross section is used, resulting in an afocal system. In general, the smallest cross section is best, and the highest efficiency is obtained with a *close-coupled* arrangement, where the rod and lamp are almost in contact and the reflector closely encloses them.

Coolants. Lasers that operate at a high repetition rate need to be cooled, since most of the energy expended in the flash lamp is converted into heat and would quickly cause overheating. Distilled or deionized water is often used as a coolant, sometimes with a deionizer in the coolant circuit. Where operation below 0°C is required, a mixture of water and ethylene glycol can be used. On occasion other liquids such as alcohol or certain fluorocarbons are used, as well as gases under pressure. Air cooling is used where cooling requirements are not too severe. The coolant circuit includes a pump for circulation, and a radiator, often with a fan, to dissipate the heat.

Firing and Pulsing Electronics. The energy to be discharged into the lamp is stored in a capacitor and discharged through the lamp via a choke whose value is selected[1*] to give the desired pulse width with minimum ringing. The selection of the C and L values is discussed by Emmett and Markiewicz.[1] Sometimes more complicated pulse-forming networks (PFNs) are used. The firing of the lamp is accomplished by a high-voltage (approximately 20 kV) discharge from an ignition coil.

Mirrors. Mirrors may be deposited on the laser rod-end surfaces. These mirrors are multilayer dielectric types, designed for the specified reflectivity at the wavelength of operation. Depositing the mirrors on the rod eliminates the problem of alignment, provided the rod ends are originally finished flat and parallel. The reflectivity of the rear mirror is normally desired to be 100%, but the reflectivity of the front mirror is selected for best performance, considering the laser-material properties, rod size, and operating power level. The mirrors must be capable of withstanding the laser optical power, and must be kept clean, because dirt or dust may char and cause damage. Where separate mirrors are used, the rod ends should be AR (antireflection)-coated or else cut at Brewster's angle. The separate mirrors must be mounted in holders that permit fine adjustment about two axes.

36. Q-switching Operation. *a. Basic Technique.* Q switching is done by effectively blocking the optical path to one of the mirrors for the majority of the time during which the

---

* See reference, Par. 11-37a.

rod is being pumped, causing the rod to store energy. The Q switch then quickly restores the optical path to the mirror and a "giant pulse" (Fig. 11-12), results (see Par. **11-27**).

b. Practical Q Switches. The four main types of Q switches are the electrooptical, the rotating prism or mirror, the acoustooptical, and the saturable absorber.

Electrooptical Q switches operate by changing the polarization of the light going through them, and thereby require an auxiliary polarizer, usually an air-spaced calcite Glan prism to block light transmission.

The materials usually employed are deuterated KDP, referred to as KD*P, and lithium niobate (see Par. **11-46**). Of the two, KD*P gives better performance. It is, however, less convenient than the niobate, because it needs to be shielded from the air, due to its being hygroscopic. With ruby, KD*P must be used, because niobate is damaged by intense visible light. Figure 11-13 shows the optical arrangement of a typical Q-switched laser.

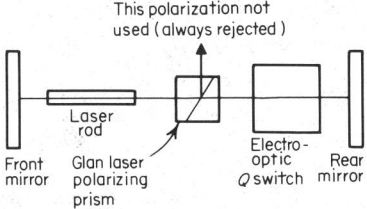

Fig. 11-13. Optical arrangement of typical optically pumped Q-switched laser.

The *rotating prism or mirror* requires careful mechanical design to provide rotation at the speeds required (30,000 r/min is common) and maintain alignment. The extinction ratio is infinite and the insertion loss negligible, enabling this type of Q switch to attain high efficiency.

The *acoustooptical Q switch* operates by using an acoustic wave in a material to diffract an optical wave and thereby change its direction. This device is useful mainly in cw applications.

The *saturable absorber Q* switch, used primarily in research, of which the most common is the bleachable dye, does not require any electric or mechanical control. The dye is opaque until it suddenly is bleached by fluorescence from the rod and permits generation of a Q-switched pulse.

37. Cw Operation. Continuous-wave (cw) operation attains efficiencies comparable with pulse operation. The pumping reflectors are similar, and the light sources used include tungsten-halogen lamps and mercury, krypton, and xenon arc lamps. These lamps are more rugged in design than those used for pulsed operation, since they must withstand continuous use.

37a. Reference on Optically Pumped Lasers

1. EMMETT, J. L., and MARKIEWICZ, J. P. *IEEE J. Quantum Electron.*, QE-2:707–711, November, 1966.

LIQUID LASERS

By Patrick R. Manzo

38. Liquid lasers employ a liquid as the laser medium in place of a large single crystal or a gas (Fig. 11-14). Their properties are intermediate between those of gaseous and solid lasers. They are easy to prepare in large samples with excellent optical quality, and for certain types their energy output can be as high as 10^8 W peak and several tens of watts average. The wavelength coverage available to liquid lasers is considerably greater than that of both the solid and the gaseous lasers. There are two types of liquid lasers in common usage:

The aprotic liquid laser consists of a rare-earth salt dissolved in an inorganic solvent which does not contain hydrogen. Energy from the excitation source is absorbed by the solvent and then transferred to the rare-earth ion, which lases. The absence of hydrogen in solution greatly increases the efficiency of this energy transfer process due to the fact that it lessens the possibility that this energy could be transferred into vibrations of the molecule. The output wavelength is that of the rare-earth salt (see Table 11-2). In system gain, in output power levels, and in overall efficiencies, the aprotic liquid lasers are comparable with the solid-state lasers. They are capable of extremely high peak powers (10^8 W) and sustained high average powers (50 to 100 W). The principal aprotic liquid laser materials are Nd^{3+} in $SeOCl_3$ with $SnCl_4$ or in $POCl_3$ with $SnCl_4$.

The dye laser uses highly fluorescent organic molecules as the laser medium, and unlike the aprotic liquid, these molecules do not contain the rare-earth salts. The lasing transition is extremely short-lived—approximately 10 ns compared with several hundreds of microseconds for the rare-earth ions. This requires the utilization of quite different excitation techniques to achieve laser action. The first organic dye lasers required Q-switched lasers or extremely fast rise-time flash lamps in order to invert the population fast enough to achieve laser action. However, there are techniques[4][*] now available which alleviate the necessity for such fast excitation sources. Recently, Rhodamine 6G and several other dyes have been made

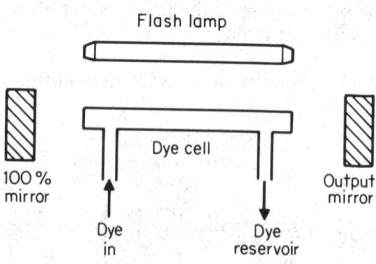

Fig. 11-14. Typical dye-laser configuration.

to operate in a continuous mode[3] with average powers as high as 10 watts.

The organic dyes exhibit excellent optical quality in solution, and they are extremely easy to prepare and handle. The absorption and fluorescent bands of the molecule are extremely broad (i.e., several hundred angstroms) due to the large number of vibrational and rotational energy levels associated with each electronic energy level. Thus laser output can be broadband (300 Å wide) or very narrow (0.1 Å). Table 11-6 illustrates the broad wavelength coverage possible by selecting the appropriate dye. In each case the wavelength listed is at approximately the center of the emission bandwidth.

[*] Superior numbers correspond to numbered references, Par. **11-38a**.

Table 11-6.   Most Commonly Used Organic Dye-Laser Materials

Dye	Solvent	Lasing wavelength, Å
p-Terphenyl	Cyclohexane	3410
p-Quaterphenyl	Dimethyl sulfoxide	3710
p,p'-Diphenylstilbene	Benzene	4080
9,10-Diphenylanthracene	Cyclohexane	4325
Acridone	Ethanol	4370
9-Aminoacridine hydrochloride	Ethanol	4585
4-Methyl-7-hydroxycoumarin	H_2O	4500
7-Diethylamino-4-methyl coumarin	Ethanol	4600
7-Hydroxycoumarin	H_2O	4600
Trypaflavin	Ethanol	5050
Acriflavin hydrochloride	Ethanol	5100
Fluorescein	Aqueous alkaline	5180
Na-fluorescein	H_2O	5270
Eosin	Ethanol	5400
Rhodamine 6G	Ethanol	5900
Uranine 6	Ethanol	5600
Rhodamine B	Ethanol	6200
Acridine red	Ethanol	6150
3,3'-Diethyloxadicarbocyanine iodide	Methanol	6580
3,3'-Diethyl-2,2'-thiadicarbocyanine iodide	Acetone	7110
3,3'-Diethyloxytricarbocyanine iodide	Ethanol	7085
Cryptocyanine	Glycerin	7450
Naphthalene green	Glycerin	7560
Malachite green	Isoamyl alcohol	7600
Chloro-aluminum phthalocyanine	Dimethyl sulfoxide	7615
3,3'-Diethylthiatricarbocyanine iodide	Ethanol	8075
	Methanol	8350
Methylene green	Sulfuric acid	8230
Toluidine blue	Sulfuric acid	8480
Phthalocyanine (metal-free)	Sulfuric acid	8630
1,1'-Diethyl-2,2'-quinotricarbocyanine iodide	Acetone	8980
1,1'-Diethyl-4,4'-quinotricarbocyanine iodide	Acetone	10,000

38a. References on Liquid Lasers

1. LEMPICKI, A., SAMELSON, H., and BRECHER, C. *Appl. Opt., Suppl.,* **2**:205 (1965).
2. LEMPICKI, A., and SAMELSON, H. Organic Laser Systems, in A. K. Levine (ed.), "Lasers," Vol. 1, pp. 181–252, Marcel Dekker, Inc., New York, 1966.
3. PETERSON, O. G., TUCCIO, S. A., and SNAVELY, B. B. *Appl. Phys. Letters,* **17**, 245 (1970).
4. SNAVELY, B. B. Flashlamp Excited Organic Dye Lasers, *Proc. IEEE,* **57**(8):1374–1390 (August, 1969).
5. SOROKIN, P. P. Organic Lasers, *Sci. Amer.,* **220**:30–40 (February, 1969).

GAS LASERS

BY GARETH M. JANNEY

39. General Characteristics. Gas lasers can best be characterized by their variety. The laser medium may be a very pure, single-component gas or mixture of gases. It may be a permanent gas or a vaporized solid or liquid. The active species in a gas laser may be a neutral atom, an ionized atom, or a molecule. The operating pressures range from a fraction of a torr to atmospheric pressure, and the operating temperature from -196 to $1600°C$. Excitation methods include electric discharges (glow, arc, pulsed, rf, dc), chemical reactions, supersonic expansion of heated gases (gas dynamic), and optical pumping. The average output power of useful gas lasers ranges from a few microwatts to tens of kilowatts (ten orders of magnitude), and the peak power ranges from a fraction of a watt to 100 MW. The range of output wavelengths extends from 0.16 to 774 μm at discrete wavelengths.

40. Available Gas Lasers. Table 11-7 is an abbreviated listing of selected gas lasers, arranged by output wavelength.

41. Multiple Wavelengths. Most gas-laser materials have a number of distinct laser transitions (i.e., different wavelengths). For example, the neon atom has more than 100, and the argon ion has more than 30. Lasers using these materials can operate with a multiwavelength output, or one of the transitions at a time can be selected by a simple adjustment (rotating a prism or diffraction grating in the optical cavity). Table 11-8 shows several commercially available lasers which provide more than one wavelength of operation.

42. Electric-Discharge Gas Lasers. Most gas lasers are excited by electric discharges. Electrons which have been accelerated by an electric field transfer energy to the gas atoms and molecules by collisions. These collisions may excite the upper laser level directly. Indirect excitation is also possible by cascading from higher-energy levels of the same atom (or molecule) or resonant energy transfer from one atom (or molecule) to another by collision.

Typical Configuration (Fig. 11-15). The gas is contained in a glass tube having an electrode near either end. The ends are sealed by windows mounted at Brewster's angle to minimize reflections at the windows (for one plane of polarization). An optical cavity is formed by two mirrors (usually both are concave), at least one of which is partially transmitting. When an electric discharge is produced in the tube between the electrodes, the gas atoms or molecules are excited and laser action commences. Listed below are some typical parameters for several common types of electric-discharge gas lasers.

Laser species	Gas mixture and pressure, torr		Current density, A/cm²
Ne (neutral atom)	He,	1.0	0.05–0.5
	Ne,	0.1	
Ar (ion)	Ar,	0.3	100–2,000
CO₂	He,	5–10	0.01–0.1
	N₂,	1.5	
	CO₂,	1.0	

The transverse-electric-discharge configuration (Fig. 11-16) is used for some gas lasers, especially for high-average-power, fast-flowing gas lasers. It provides high electric fields at practical voltages while retaining long paths of excited gas.

A pair of long electrodes (or linear arrays of electrodes) are located parallel to the optical

Table 11-7. Representative Gas Laser Transitions

Wavelength, μm	Laser species	Power, W	Mode — Pulsed	Mode — cw	Comments
0.1523–0.1613	H_2	$>10^5$ (e)	X	...	Special high-voltage, traveling-wave discharge circuit
0.2358	Ne IV	...	X	...	Low power
0.3250	Cd II	15×10^{-3} (c)	...	X	
0.3371	N_2	10^5 (c), 2×10^6 (e)	...	X	Up to 0.5 W average power (c) (short pulses 10^{-8} s)
0.3511–0.3638	Ar III	5 total (e)	...	X	
0.4416	Cd II	50×10^{-3} (c)	...	X	
0.4880	Ar II	2 (c), >100 (e)	...	X	
0.5105	Cu I	4×10^4 (e)	X	...	Metal vapor; tube heated to 1600°C; very high gain
0.5145	Ar II	3 (c), >100 (e)	...	X	
0.5401	Ne I	10^4 (c)	X	...	Short pulses (3 × 10^{-9} s)
0.6328	Ne I	10^{-1} (c)	...	X	Most widely used gas laser
0.7229	Pb I	2×10^3 (e)	X	...	Very high gain
1.15	Ne I	10^{-1} (c)	...	X	
2.6–2.9	HF	640 (e)	...	X	Fast-flowing chemical laser
3.39	Ne I	10^{-1} (c)	...	X	
3.6–4.1	DF	400 (e)	...	X	Fast-flowing chemical laser
4.9–5.6	CO	10 (c), 2×10^3 (e)	...	X	
10.6	CO_2	2×10^3 (c), cw; 6×10^4 (e), cw; 2×10^6 (c), pulsed	X	X	Highest-power cw laser; wide variety of cw and pulsed configurations; much development in progress
27–374	H_2O HCN, etc.	1–10 (c), pulsed; 10^{-3}–10^{-2} (c), cw	X	X	
774	ICN	5×10^{-1}	X	...	

NOTES: Roman numerals indicate state of ionization—neutral, singly ionized, etc., for I, II, ...
Commercially available or experimental are indicated by (c) and (e), respectively.
Most cw lasers can also be operated pulsed, with slightly higher peak powers.

Table 11-8. Multiple-Wavelength Gas Lasers

Laser type	Wavelength, μm	Remarks
Helium-neon	0.5939 1.080 0.6046 1.084 0.6118 1.152 0.6294 1.162 0.6328 1.177 0.6352 1.199 0.6401 3.39 0.7305	Rotate prism to select one wavelength at a time.
Noble-gas ions	Argon Krypton Xenon 0.3511 0.4619 0.4955 0.3638 0.4762 0.5007 0.3795 0.5208 0.5160 0.4579 0.5682 0.5260 0.4765 0.6471 0.5353 0.4965 0.6764 0.5395 0.5017 0.5956 0.5145 0.5287	Rotate prism to select lines from one gas. Change gas in tube to change set of wavelengths. Mixtures of gases can be used to extend the range. Simultaneous oscillation on many lines is possible.
Carbon dioxide	9.1–11.3	Several groups of closely spaced lines in this wavelength region. Line separations within groups are approximately 0.02 μm. Rotate diffraction grating to select lines one at a time.

11-22

axis, within the envelope which contains the gas. The discharge current flows transverse to the optical axis. Lasers employing this configuration include high peak power, pulsed N_2, Ne, and CO_2 and high-average-power, fast-flowing CO_2 lasers.

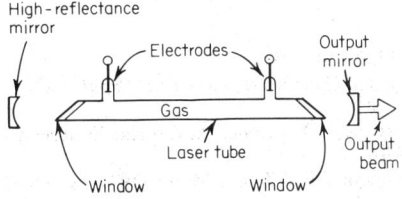

Fig. 11-15. Electric-discharge gas laser.

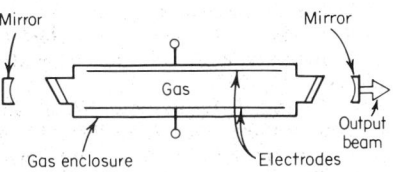

Fig. 11-16. Transverse-electric-discharge gas laser.

43. Chemical Lasers. Chemical lasers derive their energy from the free-energy change of a chemical reaction. The chemical reaction may be initiated by some other source of energy, such as light or electric discharges. Since the chemical reaction consumes the reactants, a flowing system is necessary for repetitive-pulsed or for cw operation. Figure 11-17 shows an arrangement used to produce high-power, cw laser output from hydrofluoride (HF) and other molecules. A series of chemical reactions is required to sustain the laser operation in practice, but in simplified form the reaction can be expressed as

$$F_2 + H_2 \rightarrow 2HF + \Delta E$$

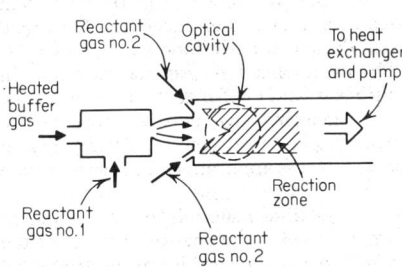

Fig. 11-17. Fast-flowing chemical laser.

where F_2 is reactant 1, H_2 is reactant 2, and ΔE is the free-energy change, some of which is in the form of vibrationally excited HF molecules. Laser action occurs on vibrational transitions of the HF molecule in the wavelength region 2.6 to 2.9 μm. Chemical lasers of this general type are potential sources of high average power (multikilowatt).

44. Gas Dynamic Lasers. The expansion of a hot, high-pressure mixture of CO_2 and N_2 through a supersonic nozzle results in a lowering of the gas temperature in a time which is short compared with the vibrational relaxation time of the CO_2 molecule. A differential relaxation time between the upper and lower laser levels results in a population inversion for a short distance downstream from the supersonic nozzle, and laser operation in this region is possible. Average output power of 60 kW at 10.6 μm has been obtained from such a device.

45. Properties of the Gas-Laser Output Beam. *Wavelength and Frequency.* It is customary in laser terminology to refer to a laser transition by its wavelength (in micrometers or angstroms) and to discuss the fine structure of the transition in terms of frequency. Wavelength λ and frequency ν are related by $\lambda\nu = c$, where c is the velocity of light. The wavelength of a laser transition, as well as the wavelength interval or width of the transition over which optical gain exists, is a property of the laser atom or molecule. The transition width is also related to gas temperature and pressure. Typical line widths of common gas lasers are:

Laser	Wavelength, μm	Line width, MHz
HeNe	0.6328	1,700
	1.15	920
	3.39	310
Ion lasers	0.5	2,500–3,000
CO_2	10.6	60

The fine structure or frequency of the laser is determined by the optical cavity. Oscillation can occur at discrete frequencies (cavity modes) at frequency spacings of $\Delta\nu =$

$c/2L$, where L is the separation of cavity mirrors. Depending on the nature of the laser medium and the geometry of the optical cavity, oscillation may occur at a number of different modes, within the line width of the transition. For example, for helium-neon at 0.6328 μm, with a cavity-mirror spacing of 100 cm, the mode spacing is 150 MHz, and 11 (axial) modes could oscillate.

45a. References on Gas Lasers

1. BLOOM, A. L.   Gas Lasers, *Appl. Opt.,* 5:1500-1514 (1966).

2. BENNETT, W. R., JR.   Inversion Mechanisms in Gas Lasers, *Appl. Opt., Suppl. 2, Chem. Lasers,* p. 34, 1965.

3. GEUSIC, J. E., BRIDGES, W. B., and PANKOVE, J. I.   Coherent Optical Sources for Communications, *Proc. IEEE,* 58:1419-1439 (1970).

4. EMMETT, J. L.   Frontiers of Laser Development, *Phys. Today,* March, 1971, pp. 24-33.

   Texts

5. BLOOM, A. L.   "Gas Lasers," Wiley, New York, 1968.

6. SINCLAIR, D. C., and BELL, W. E.   "Gas Laser Technology," Holt, New York, 1969.

ELECTROOPTICS AND NONLINEAR OPTICS

By LESTER F. GILLESPIE

ELECTROOPTICAL EFFECTS

46. Basic Principles.   The optical properties of many materials can be altered by a strong electric field. If these alterable properties can be made to interact properly with light propagating through the material, the material can serve as a transducer from electrical signals to optical signals; in short, an electrooptical device. In electrooptical practice, the index of refraction of the material used is altered by the application of an electric field. In this manner isotropic materials can be made birefringent, and birefringent materials can be made more or less so, or their axes may be changed. Since birefringent materials can change the nature of polarized light, an electrooptical material placed between a polarizer and an analyzer is capable of modulating a transmitted beam[1]* by the application of the appropriate electric field to the electrooptical material.[2]

a. Linear Electrooptical Effect, or Pockel's Effect.   Crystalline materials such as potassium dihydrogen phosphate (KDP) and lithium niobate are used. The change in index of the material is linearly related to the applied electric field; hence the term linear electrooptical effect. Metal electrodes are applied to rectangular blocks of these crystals, and the resultant structures are referred to as *Pockel cells.* When the electrodes are applied so as to produce a field parallel to the direction of light propagation, the field is called *longitudinal.* In such cases the electrodes, if metal, must have holes to permit passage of the light. Conductive (Nesa) coatings and transparent metallic films may be used, but they do not withstand high optical peak power.

b. The quadratic electrooptical effect, or Kerr effect, requires a higher voltage. The most frequently used material is nitrobenzene, a liquid. The voltage is applied transversely, and the cell used in much the same way as Pockel cells. Kerr cells are not used for high-frequency modulation because rf heating would take place.

c. EOMs as Light Modulators.   The Pockel and the Kerr cells described above are called *electrooptical modulators,* or EOMs. Space communication, terrestrial communication through optical waveguides, line-of-sight communication, video displays, and high-density photographic data storage and retrieval are examples of present and anticipated applications of EOMs. One specialized application is Q switching (see Par. 11-28). Light beams modulated by EOMs can carry a large amount of information because of the high frequency of the carrier wave.

NONLINEAR OPTICS

47. Basic Principles.   In electronics it is well known that any circuit element operating outside its linear region can produce the second harmonic of the applied signal. Such a nonlinear device can mix two applied signals to produce their sum and difference, separate a driving signal into two output frequencies whose sum equals the driving signal frequency

---

* Superior numbers correspond to numbered references, Par. **11-47a.**

(called *parametric oscillation*), and other nonlinear effects. Effects analogous to these effects can be observed at optical frequencies in nonlinear optical materials.[2] In general, such a material is one in which an optical property (such as index of refraction) is changed by the electromagnetic field of light propagating through the material. One would expect that any electrooptical material, being susceptible to dc electric fields, would tend to satisfy this requirement, and this is indeed the case. Most electrooptical materials can also be used as nonlinear materials.

Second-harmonic Generation.[3] Barium sodium niobate, $Ba_2NaNb_5O_{15}$, colloquially known as "bananas," and lithium iodate are commonly used for frequency doubling of 1.06-μm radiation from neodymium-doped laser materials, producing 0.53-μm light, which is green. These materials are of great practical significance because of two factors. First, the neodymium-doped materials such as YAG are highly efficient in producing 1.06-μm radiation, and second, the frequency-doubling process can be almost 100% efficient in converting this to coherent green light. Frequency doubling takes place many orders of magnitude more efficiently if the doubled frequency passes through the material with the same velocity as the fundamental frequency.

Parametric Oscillators and Amplifiers.[2,4] The optical parametric oscillator uses a nonlinear material to convert pump light into two signals, the sum of whose frequencies equals the frequency of the pump light. These frequencies can be tuned by index matching, within practical limits. This is normally done by controlling the temperature of the crystal. Since a parametric oscillator has gain at its two output frequencies, it can be used as an amplifier of either of these frequencies. Parametric "up-conversion," or frequency conversion, permits pumping a nonlinear crystal with a laser source at the difference frequency between a signal and an output.

47a. References on Electrooptics and Nonlinear Optics

1. SHURCLIFF, W. A. "Polarized Light," Harvard University Press, Cambridge, Mass., 1962.
2. YARIV, A. "Quantum Electronics," Wiley, New York, 1967.
3. LENGYEL, B. A. "Introduction to Laser Physics," Wiley, New York, 1966.
4. SMITH, W. V., and SOROKIN, P. P. "The Laser," McGraw-Hill, New York, 1966.

SEMICONDUCTOR LASERS

By RICHARD R. SHURTZ II

48. Semiconductor lasers range in wavelength from 0.33 to 31.2 μm, the most efficient materials lasing in the 0.6- to 0.95-μm range. The fundamental light-producing mechanism in the semiconductor is the recombination of electrons and holes when a conduction-band electron is captured by a valence-band hole. The wavelength of the emitted radiation is related to the band gap of the material by $\lambda = 1.24/Eg$, with λ in micrometers and Eg in electron volts. The most efficient method of pumping the semiconductor is the injection of electrons across a *pn* junction.

A typical semiconductor *pn* junction laser is shown in Fig. 11-18. The Fabry-Perot surfaces are formed by natural cleavage planes of the crystal, whose reflectance is typically 0.36. The sides of the laser are rough-sawed to inhibit lasing in the wrong direction. Ohmic contacts are placed on the top and bottom of the laser. The size is typically $14 \times 5 \times 3$ mils. The back Fabry-Perot surface is coated with an insulator, then a metallic reflector with a reflectance close to 1, so that all the coherent radiation emerges from the front face. In actual applications the laser chip is mounted on a heat sink, traditionally a transistor header for discrete diode applications, or on beryllium oxide for array application.

In use the junction laser is positively biased, *p*-side-positive. Electrons are injected from the *n* side of the junction (where there are excess electrons in the conduction band) to the *p* side, where they recombine with excess holes. For efficient lasing action, it has been found that the *p* side of the junction must be compensated (both *n*- and *p*-type impurities introduced), with the overall electrical properties *p*-type. Both sides of the junction must be degenerately doped. In practice this means that a carrier concentration of greater than 1×10^{18} cm^{-3} for the *n* and 1×10^{19} cm^{-3} for the *p* is needed.

49. Threshold Conditions. At low current levels, the electrons recombine with holes spontaneously emitting radiation in all directions. At higher current levels enough minority carriers are injected across the junction so that it is possible to achieve an inverted population, yielding a positive gain in the lasing region. The lasing threshold is reached when a light pulse can traverse a round trip in the Fabry-Perot cavity without attenuation, i.e.,

$$R_2 R_1 \exp[(g-a)2L] = 1$$

where R_1 and R_2 are the reflectances at the cavity ends, g is the gain per unit length, a is the absorption per unit length, and L is the cavity length. The prime cause of absorption is defect scattering and free-carrier absorption.

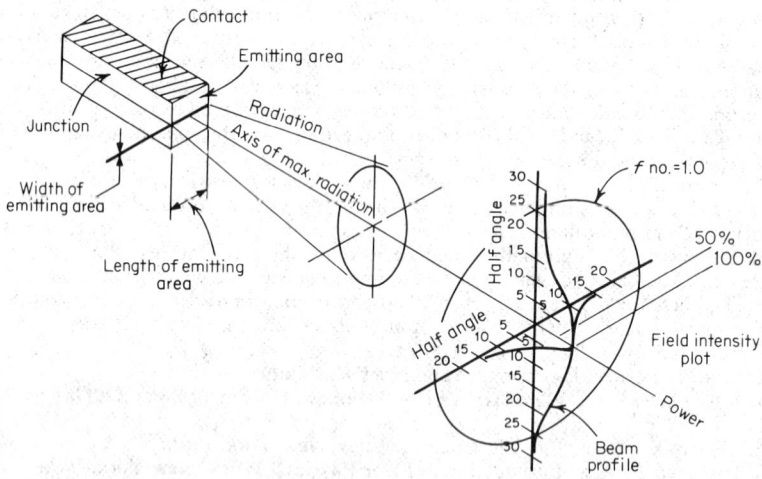

Fig. 11-18. A typical semiconductor laser. The beam-spread pattern is for a single heterostructure design. (*From "RCA Injection Lasers and Infrared Emitting Diodes," RCA Solid State Division, Somerville, N.J., 1971.*)

Threshold current density is a strong function of temperature. Because of the very fast transition lifetime in the direct materials, very large current densities are required to achieve inverted population. At room temperature current densities in the range of 8,000 to 40,000 A/cm² are required. Above 100 K the threshold current density varies[2]* approximately as the third power of T. To minimize the problem of Joule heating (and hence thermal quenching), the lasers must be pulsed with pulses typically about 100 to 400 ns long at a repetition rate of 1 to 10 kHz.

50. Materials. *Variable-band-gap ternary compounds* are formed by combining two binary compounds which have one constituent in common (like GaAs and AlAs). The band gap of the mixture varies between the gaps of the two individual compounds.

GaAlAs varies between 1.4 and 2.2 eV as the aluminum arsenide concentration is increased.[3] Another commonly used ternary compound is the mixture of GaAs and GaP.[4]

Room-Temperature Operation. To date the only materials which have lased at room temperature in a junction laser configuration are GaAs, GaAlAs, and GaAsP. With these materials a spectral output in the 0.61- to 0.91-μm range can be obtained. All other materials require some type of cooling.

Spectral dependence upon temperature is an important factor to be considered. The spectral output of GaAs, for instance, changes at a rate of approximately 2.5 Å/K. At 77 K it lases at 8500 Å, while at 300 K it lases at 9050 Å. Other laser materials change in a similar way.

Selection of the proper material depends upon the required spectral peak and the available temperature of operation. To date spectral outputs between 0.55 and 31.2 μm have been obtained using electron injection across a *pn* junction. The most efficient of these are in the 0.61- to 0.95-μm region. Generally, the further the output differs from this region, the less efficient the diode.

---

* Superior numbers correspond to numbered references, Par. **11-51a.**

A list of lasing semiconductors[5] is shown in Table 11-9. A spectral range is included for all variable-band-gap ternaries.

Table 11-9. Semiconductor Injection Lasers

Compound	λ, μm	$h\nu$, eV
GaAlAs*	0.63–0.90	2.0–1.4
GaAsP*	0.61–0.90	2.0–1.4
GaAs*	0.91	1.50–1.38
InAlP	0.55–0.91	1.36–2.3
InP	0.91	1.36
GaAsSb	0.9–1.5	1.4–0.83
InAsP	0.9–3.2	1.4–3.9
GaSb	1.55	0.80
InGaAs	0.85–3.1	1.45–0.4
InAs	3.1	0.39
InAsSb	3.1–5.4	0.39–0.23
InSb	5.2	0.236
PbTe	6.5	0.19
PbSSe	3.9–8.5	0.32–0.146
PbSe	8.5	0.146
PbSnTe	28	0.045
PbSnSe	8–31.2	0.155–0.040

* Lase at 300 K; all others require cooling.

51. Semiconductor Laser Configurations. *Homojunction Laser.* This standard laser structure is formed by diffusing zinc into a thin *n*-type crystal layer grown on an oriented substrate (see Fig. 11-19*a*). These homojunction lasers have room-temperature-threshold current densities which vary from 40,000 to 100,000 A/cm² and a beam spread of about 12° × 15° full angle. Room-temperature efficiencies are less than 1%, and efficiencies at 77 K are as high as 60%. This structure is primarily used when cooled.

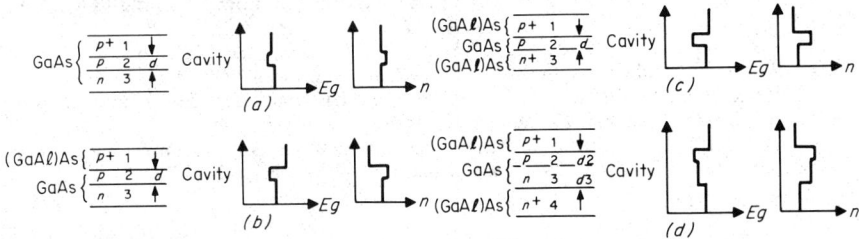

Fig. 11-19. (*a*) Homojunction laser; (*b*) single-heterojunction laser; (*c*) double-heterojunction laser; (*d*) large-optical-cavity laser.

Single-heterostructure lasers, developed to reduce the room-temperature threshold, have a GaAlAs layer located about 2 μm from the GaAs homojunction on the *p* side[6,7] (see Fig. 11-19*b*). Because the GaAlAs has a larger band gap and a lower index of refraction than the GaAs, the injected electrons are confined to a narrow recombination region, and the optical modes to an asymmetrical waveguide formed by the *pn* junction and the GaAs-GaAlAs heterojunction. This structure has a typical threshold current density in the 8,000 to 12,000 A/cm² range and has made room-temperature operation of the laser practical. The beam spread of this device is about 12° × 30° full angle.

The double-heterostructure laser has a second GaAlAs layer added at the *pn* junction to provide a symmetrical waveguide[8] (see Fig. 11-19*c*). The fraction of modes confined to the active region is much higher for this structure. Threshold currents can be reduced to arbitrarily low values by reducing the heterojunction separation and increasing the aluminum concentration in the outside layers for improved waveguiding. The first *cw room-temperature laser* was fabricated this way.[9]

Large-Optical-Cavity (LOC) Laser. Both the single- and double-heterostructure lasers undergo *catastrophic degradation* at low current levels, where the reflecting surfaces of the

Fabry-Perot cavity are destroyed. The destruction is caused by the high-optical-flux densities in the waveguide, which must be no wider than 2 μm for effective carrier confinement.[10] The large-optical-cavity laser (LOC) was developed to solve this problem[11] (see Fig. 11-19d). The optical cavity here is allowed to be wider than the recombination region, thus lowering the flux densities and raising the threshold for catastrophic degradation. By optically coating the front surface, this threshold can be raised even higher. The LOC lasers have power outputs as high as 7 W/mil emitting-junction length, as opposed to about 3 for the single and double heterostructures with 2-μm waveguides. Typical threshold current densities can be obtained in the 4,000 to 12,000 A/cm^2 region. This device[12,13] emits two narrow beams of light separated by 50 to 90°.

51a. References on Semiconductor Lasers

1. SZE, S. M. "Physics of Semiconductor Devices," Wiley, New York, 1969.

2. STERN, F. Effect of Band Tails on Stimulated Emission of Light in Semiconductors, *Phys. Rev.,* **148**:186 (1966).

3. NELSON, H., and KRESSEL, H. Improved Red and Infrared Light Emitting Al$_x$-Ga$_{1-x}$As Laser Diodes Using the Close-Confinement Structure, *Appl. Phys. Letters,* **15**:7 (1969).

4. HOLONYAK, N., JR., and BEVACQUA, S. F. Coherent (Visible) Light Emission from Ga(As$_{1-x}$P$_x$) Junction, *Appl. Phys. Letters,* **1**:82 (1962).

5. GEUSIC, J. E., BRIDGES, W. B., and PANKOVE, J. I. *Proc. IEEE,* **58**:1419 (1970).

6. KRESSEL, H., and NELSON, H. Close-Confinement GaAs *pn* Junction Lasers with Reduced Optical Loss at Room Temperature, *RCA Rev.,* **30**:106 (1969).

7. HAYASKI, I., PANISH, M. B., and FOY, P. W. A Low-Threshold Room Temperature Injection Laser, *IEEE J. Quantum Electron.,* **5**:211 (1969).

8. PANISH, M. B., HAYASKI, I., and SUMSKI, S. Double-Heterostructure Injection Lasers with Room Temperature Thresholds as Low as 2300 A/cm^2, *Appl. Phys. Letters,* **16**:326 (1970).

9. HAYASKI, I., PANISH, M. B., FOY, P. W., and SUMSKI, S. Junction Lasers Which Operate Continuously at Room Temperature, *Appl. Phys. Letters,* **17**:109 (1970).

10. KRESSEL, H., LOCKWOOD, H. F., and HAWRYLO, F. Z. Low Threshold LOC GaAs Injection Lasers, *Appl. Phys. Letters,* **18**:43 (1971).

11. LOCKWOOD, H. F., KRESSEL, H., SOMMERS, H. S., JR., and HAWRYLO, F. Z. An Efficient Large Optical Cavity Injection Laser, *Appl. Phys. Letters,* **17**:499 (1970).

12. BYER, N. E., and BUTLER, J. K. Optical Field Distribution in Close Confined Laser Structures, *IEEE J. Quantum Electron.,* **6**:291 (1970).

13. BUTLER, J. K. "Theory of Transverse Cavity Mode Selection in Homojunction and Heterojunction Semiconductor Laser Diodes," *J. Appl. Phys.,* **42**:4447 (1971).

APPLICATION OF SEMICONDUCTOR LASERS

BY STEVEN B. GIBSON

52. Applications of semiconductor radiation sources[1*] may be divided into two major categories, *signaling* and *illumination*. The choices include laser diodes, laser diode stacks, and laser diode arrays, as well as noncoherent light-emitting diodes (LEDs) (see Par. **11-23**). As shown in Table 11-10, the *laser source* becomes the choice when high radiance, high peak power, and/or high average radiant power are required in conjunction with narrow projected beam angles.

53. The wavelength selectivity of laser diodes ranges from 0.8 to 0.92 μm. Gallium aluminum arsenide and gallium arsenide phosphide laser diodes cover the spectral range from 0.8 to 0.89 μm. Gallium arsenide lasers cryogenically cooled to 77 K emit at 0.855 μm. At room temperature these laser diodes emit at 0.905 μm. The shift in spectral output with temperature is approximately 2.5 Å/°C. The best sensors available with spectral sensitivity in this wavelength region are the S-25 photoemissive detector and the silicon and gallium arsenide solid-state photodetectors (see Table 11-15, Par. **11-98**).

---

* Superior numbers correspond to numbered references, Par. **11-59a**.

Table 11-10. Diode-Source Application Selection Chart

Optics	Application	Type of operation	Room temperature devices				Cryogenic laser array
			Emitter	Laser diode	Laser diode stack	Laser array	
	Signaling applications						
None	Paper tape reader	Cw	X				
	Card reader	Cw	X				
	Shaft encoder	Cw	X				
	Keyboard	Cw or coded	X				
	Circuit isolator coupler—"dc transformer"	Mod	X				
Fiber optics	Data transmission	Mod	X	X			
	Line finder/edge sensor	Cw or pulse	X	X			
	Intrusion alarm	Mod or pulse	X	X	X		
Single-lens	Remote control signaling	Mod	…	X	X		
	Voice communications	Pulse	…	…			
	Ranging	Pulse	…	X	X	X	
	Illumination systems						
Single-lens	Illuminator for gated viewers	Pulse	…	…	…	X	X
	Target designator	Pulse or cw	…	X	X		

11-29

54. The electronic and optical-emission characteristics, as shown in Table 11-11, vary widely for operating temperatures of 77 K and room temperature. The average radiant power output and power dissipation must be calculated based on peak power output, duty factor, and power efficiency. In selecting a laser diode source it is of utmost importance to consider the effect of operating temperature on diode performance. Other important parameters shown in Table 11-11 are the width of the diode and the peak power output per unit of junction width. These two parameters are useful in determining the number of lasers in a diode stack or array and operating currents to obtain required radiant power outputs.

Table 11-11. Comparison of Laser Performance at Cryogenic and Room Temperatures

	Cryogenic temp., 77°K	Room temp., 27°C
Driving current	4 A	25 A
Threshold current	0.7 A	7 A
Radiant power output (peak)	2.5 W	6 W
Pulse duration	2 μs	0.2 μs
Duty factor	2%	0.1%
Power efficiency	40%	4%
Driving voltage	1.6 V	9 V
Wavelength	8450 Å	9050 Å
Diode width	6 mils	6 mils
Power output per unit width	0.4 W/mil	1 W/mil

55. Designing a semiconductor laser system requires the characterization of four subelements:

   a. Laser diode source.
   b. Thermal dissipation element or heat sink.
   c. Collection and projection optics.
   d. Electronic power supply.

56. Laser Diode Source. In signaling applications the peak radiant power output is the major parameter, while average power is more important in illumination systems. In room-temperature pulsed applications, where the peak power output exceeds 15 W or the average power is above 15 mW, it is advantageous to employ diode stacks or arrays. A diode stack is composed of from two to five lasers placed one upon another on the same heat sink. An array is composed of many laser diodes ranging from 5 to 1,000 in number.[3] The size of such a 200-laser diode array source using straightforward arraying techniques would be approximately 0.25 × 0.25 in. The average power of this source operating at room temperature with a duty factor of 0.01% is 200 mW. The power dissipation is 3.8 W, based on an external power efficiency of 5%. This heat can be dissipated by an air-cooled radiator or a thermoelectric cooling unit.

57. Thermal Dissipation. Temperature rise at the *pn* junction causes shift in spectral output, increase in threshold current, and reduction in radiant power output. An electronic design which minimizes losses due to ohmic contacts and impedances is the first step in reducing the temperature-rise problem. Single-diode-laser sources are usually placed on copper heat sinks designed to limit the temperature rise to less than 20°C. Closely packed multiple-diode arrays and stacks present severe thermal problems because of the high thermal flux density. To increase the area of the heat sink, a three-dimensional design in the shape of an inverted V is employed.[4] This design compromises between the effective optical size of the source and the heat sink.

The present approach to improved thermal heat sinking is the use of fiber optics in arraying multiple-diode lasers. As shown in the cross-sectional view of Fig. 11-20, the laser diodes are mounted on individual heat sinks and the radiant output is piped through the fibers into a small emitting area. The top view of Fig. 11-20 shows the large separation distances between stacks of diodes, which permit much larger heat sinks to reduce the thermal flux density. Other cooling systems include forced-air or liquid-cooled radiators, thermoelectric coolers, cryogenic refrigerators, and liquid-nitrogen dewars.

58. Collection and Projection Optics for Diode Lasers. An image projected into the far field by a focused lens is characterized by dark and bright spots, making it necessary to use light-integrating optics, such as condensing lenses with an aperture or light pipes[5] (Fig. 11-21) to obtain uniform beam distribution for target illumination. The second optical problem is the relatively wide intrinsic radiation emission angle from diode lasers. This angular spread to one-tenth of the peak intensity is typically $\pm 22°$, and the intensity distribution curve is approximated by $\cos^2 4\theta$, where θ is the half-angle ($22°$). This characteristic requires that efficient collection-projection optics have an f-number of <1.25, which results in large lens systems. Fiber optics–diode-coupling techniques (Pars. **11-57** and **11-146**) are of great importance because the source area of an array can be reduced by a factor of 20 to 30.

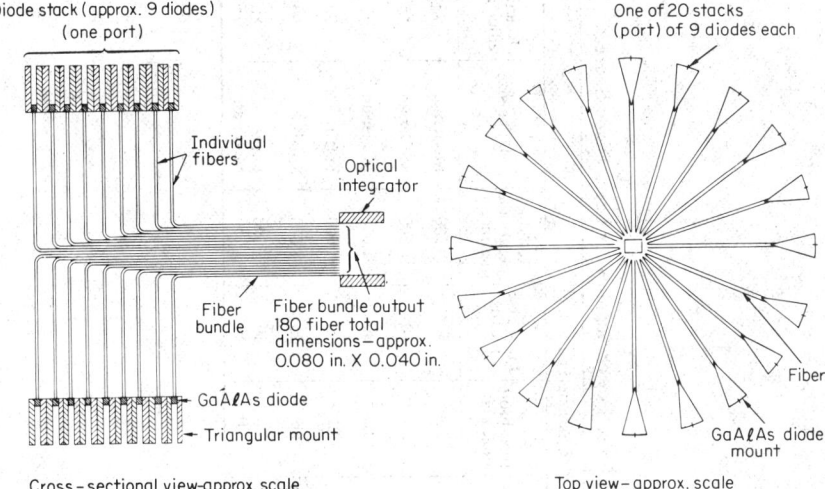

Fig. 11-20. Fiber-optic-coupled laser array design.

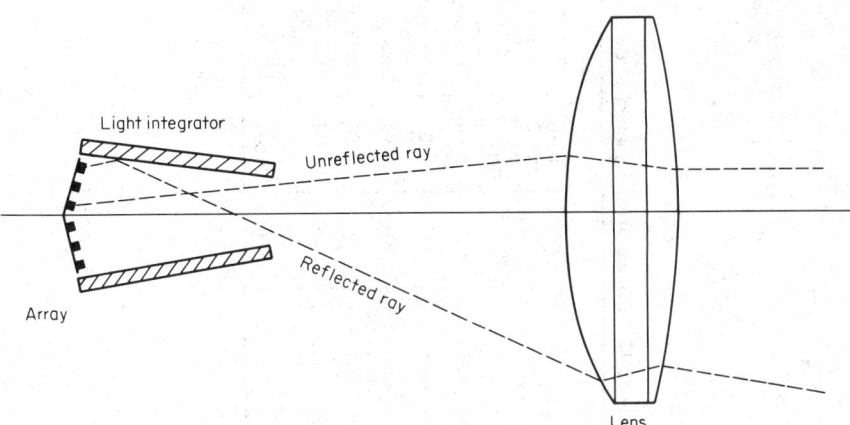

Fig. 11-21. Light integrator and projection lens.

59. Electronic Power Supply for Diode Lasers. A variety of pulse power supplies (Table 11-12) has been developed for driving laser diodes and diode arrays. The major problem areas are impedance matching due to the nonlinear characteristics of the diodes, the high threshold voltages required by series-connected diodes, and the protection of the diode from reverse-current surges. Each approach requires a trigger circuit, electronic switch, protective circuits, and a primary power source.[4] The switching elements employed are usually SCRs and transistors; however, gas tubes and electromechanical devices are also available.

The limitations on voltage and/or current of the SCR and transistor switching elements require that the laser diodes in a multiple-diode array be electrically connected in series-

Table 11-12. Comparative Summary of the Types of Pulse Power Supplies for Laser Diode Sources

Type	Peak current range, A	Peak supply voltage range, V	Pulse width control	Characteristics Pulse repetition rate, kHz	Pulse rise time, ns	Reliable operational lifetime, h	Relative size	Relative efficiency
Delay line—SCR	100	1,000	Fixed step adjust.	10	50	500	Moderate	75
Capacitor—SCR	100	1,000	Fixed step adjust.	10	10	500	Moderate	65
Pulse transformer—SCR	300	1,000	Fixed	10	40	1,000	Moderate	70
Avalanche transistor ..	30	100	Continuously variable	10	2	1,000	Small	90
Transistor	60	100	Continuously variable	30	25	1,000	Small	90
Mechanical relay	100	1,000	Fixed	1	10	100	Large	80
Gas tube	1,000	1,000	Fixed	1	10	50	Moderate	75

parallel circuits. As the number of diodes connected in series increases, the operating voltage and ohmic resistance rapidly increase to values causing impedance mismatching problems. As shown in Table 11-12, the transistor switching elements have the advantage of higher efficiency and continuous control over pulse width and pulse repetition rate.

59a. References on Applications of Semiconductor Lasers

1. Optoelectronic Product Directory, *Electro-Opt. Systems Design,* Vol. 3, No. 5, May, 1971.

2. VALLESE, L. M. Temperature Dependence of Semiconductor Laser Characteristics, *Solid State Technol.,* Vol. 14, No. 1, January, 1971.

3. GLICKSMAN, R. Technology and Design of GaAs Laser and Non-coherent IR Emitting Diodes, Part II, *Solid State Technol.,* Vol. 13, No. 10, October, 1970.

4. GIBSON, S. B., SMATHERS, S. E., and REPASY, A. J. Optics and Pulse Power Supply for Laser Diode Array Sources, *1970 Proc. Electro-Opt. Design Conf.,* Industrial and Scientific Conference Management, Inc., Chicago, Ill., 1970.

5. SMATHERS, S. E. Computer Designed Optical Integrating Devices and Projection Systems for Semiconductor Laser Arrays, *1971 Proc. Electro-Opt. Design Conf.,* Industrial and Scientific Conference Management, Inc., Chicago, Ill., 1971.

6. MILLMAN, J., and TAUB, H. "Pulse, Digital and Switching Waveforms: Devices and Circuits for Their Generation and Processing," McGraw-Hill, New York, 1965.

7. REPASY, A. J., and PEELER, J. L. A Microminiature Gallium Aluminum Arsenide Laser Diode Pulser, *1970 Govt. Microcirc. Appl. Conf.,* October, 1970.

PHOSPHOR SCREENS

By NICHOLAS A. DIAKIDES

60. Phosphor screens are used to convert electron energy to radiant energy in image tubes, cathode-ray tubes, and storage cathode-ray tubes. These screens are comprised of a thin layer of luminescent crystals, phosphors, which emit light when bombarded by electrons (cathodoluminescence).

THEORY OF OPERATION

61. Cathodoluminescence occurs when the energy of the electron beam is transferred to electrons in the phosphor crystal, raising their energy levels. When the electron returns to its initial state (ground state) following the excitation, it releases a quantum of light energy. The property of light emission during excitation is termed *fluorescence,* and that immediately after excitation is removed is *phosphorescence.* The latter emission process is also referred to as *persistence,* or *decay characteristic.*

62. Phosphor materials are highly purified inorganic crystals containing traces of other elements which serve as activators and, in combination with the host crystals, promote the phenomenon of luminescence.[1-4*] The activator determines the luminous efficiency of the phosphor, and may also affect other phosphor characteristics such as persistence and spectral emission. The most commonly used activators are metals such as copper, silver, manganese, and chromium. Typical characteristics of various standard phosphors which are used in image tubes and cathode-ray tubes are shown in Table 11-13.

63. Standardization of phosphor types has been accomplished by registration of the various phosphors with the Joint Electron Device Engineering Council (JEDEC) of the Electronic Industries Association.[5] Registered phosphors are designated by a number series known as P-numbers, such as P1, P2, etc., to P45 (see Table 11-13).

64. Phosphor-Screen Deposition Methods. Phosphor screens are usually fabricated by depositing very small crystal grains on a glass faceplate to form a thin phosphor layer. The most common method used by industry for fabrication of phosphor screens is sedimentation,[6,7] known as *phosphor settling.* This method allows the phosphor crystals to settle from a liquid suspension under the influence of gravity.

---

*Superior numbers correspond to numbered references, Par. 11-72a.

Table 11-13. Phosphor Characteristics

Type	Color* Fluorescent	Color* Phosphorescent	Persistence†	Intended use
P1	YG	YG	M	Oscillography; radar
P2	YG	YG	M	Oscillography
P3	YO	YO	M	Oscillography
P4	W	W	MS	Direct-view TV
P5	B	B	MS	Photographic
P6	W	W	S	
P7	B	Y	MS(B), L(Y)	Radar
P8		Replaced by P7		
P9		Not registered		
P10		Dark-trace screen		
P11	B	B	VL	Radar
P12	O	O	MS	Photographic
P13	RO	RO	L	Radar
P14	B	YO	M	Radar
P15	UV	G	MS(B), M(YO)	Flying-spot scanners
P16	UV	UV	UV(VS), G(S)	Flying-spot scanners; photographic
P17	B	Y	VS	Oscillography; radar
P18	W	W	S(B), L(Y)	Projection TV
P19	O	O	M-MS	Radar

	Fluorescence	Phosphorescence	Persistence†	Application
P20	YG	YG	M-MS	Storage tubes
P21	RO	RO	M	Radar
P22	W(R, B, G)	W(R, B, G)	MS	Tricolor TV
P23	W	W(R, B, G)	MS	Direct-view TV
P24	G	G	S	Flying-spot scanner
P25	O	O	M	Radar
P26	O	O	VL	Radar
P27	RO	RO	M	Color TV monitor
P28	YG	YG	L	Radar; indicators
P29	P2 and P25 stripes			
P30	Canceled			
P31	G	G	MS	Oscillography: bright TV
P32	PB	YG	L	Radar
P33	O	O	VL	Radar
P34	BG	YG	VL	Radar; oscillography
P35	G	B	MS	Oscillography
P36	YG	YG	VS	Flying-spot scanner
P37	B	B	VS	Flying-spot scanner; photographic
P38	O	O	VL	Radar
P39	YG	YG	L	Radar
P40	B	B	MS(B), L(YG)	Low repetition rate (P12 and P16)
P41	UV	O	VS(UV), L(O)	Radar with light trigger

*Colors: B = blue; P = purple; G = green; O = orange; Y = yellow; R = red; W = white; UV = ultraviolet.
†Persistence to 10% level: VS = <1 μs; S = 1 to 10 μs; MS = 10 μs to 1 ms; M = 1 to 100 ms; L = 100 ms to 1 s; VL = >1 s.
Source: M. I. Skolnik (ed.), "Radar Handbook," pp. 6-8, McGraw-Hill Book Company, New York, 1970, by permission.

65. Electrophoretic Screens. Another method for depositing phosphor screens makes use of the effect of particle motion under the influence of an electric field and is termed *electrophoretic*, or *cataphoretic, screening*.[8,9] Here the phosphor powder is first fractionated to select the desired particle-size range. The phosphor is then placed in suspension in an electrolytic solution. Under suitable conditions of electrolyte, phosphor suspension, electrode geometry, and electrical uniformity of the substrate, the resulting phosphor deposit is very uniform and compact and has higher resolution characteristics than settled screens.

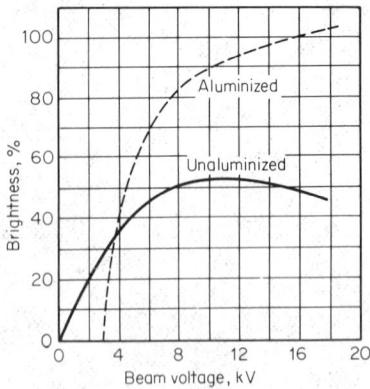

Fig. 11-22. Light output of aluminized and unaluminized phosphor screens. (*From D. G. Fink, "Television Engineering Handbook," McGraw-Hill, New York, 1957.*)

66. Transparent Screens. For certain applications where optimum image resolution is required, transparent phosphor screens prepared by vacuum evaporation are sometimes employed.[13,14] Activators may be coevaporated or diffused into the host material subsequent to the deposition. This is accomplished usually by heat-treatment following the deposition.

Thin-film phosphor screens are also fabricated by the vapor reaction method.[15] These are converted to the vapor state during processing and react with the gaseous atmosphere to provide film growth of the desired phosphor material.

67. Aluminizing,[16,17] or coating the phosphor-screen surface with a thin (1000 to 1400 Å) metallic film (usually aluminum), prevents the charging of the phosphor and permits accurate control of the primary electron energy. The reflecting aluminum surface redirects the backward emission of light from the phosphor toward the observer and almost doubles (see Fig. 11-22) the effective light output of the phosphor screen. Another important function of the aluminum backing is protection of the phosphor screen from ion-bombardment damage.

PHOSPHOR-SCREEN CHARACTERISTICS

68. Spectral Emission. Phosphors are commercially available with cathodoluminescent emission over the entire visible band, including ultraviolet and near-infrared. Typical absolute spectral characteristics of commercial phosphors are shown in Fig. 11-23.

The visible effectiveness of a phosphor is measured by comparing its spectral-emission curve with a standard visibility function (eye-response curve). This efficiency is usually stated in lumens per radiated watt and is referred to as *lumen equivalent*. It can be calculated[18] using the eye response per wavelength interval v and the phosphor output per wavelength interval P_λ:

$$\text{Lumen equivalent} = 680 \int_0^\infty vP_\lambda d\lambda \tag{11-1}$$

The constant 680 is the lumen content for 1 W of radiation at 555 nm (the wavelength of maximum eye response).

69. Brightness depends on various factors,[19] such as the type of phosphor used, accelerating voltage, electron-beam current, duration of excitation, and screen deposition method and is determined by the following expression:[20]

$$B = KI(V - V_0) \tag{11-2}$$

where B is the output brightness in lumens; K is a constant defined by both the *luminous efficiency*, in lumens per watt of radiated energy, and the *efficiency factor* of the specific phosphor; I and V are the average beam current and voltage, respectively; and V_0 is the voltage drop across the aluminized screen, which is a function of the thickness of the aluminum film. Phosphor brightness as a function of voltage for a P20 phosphor screen is shown in Fig. 11-24.

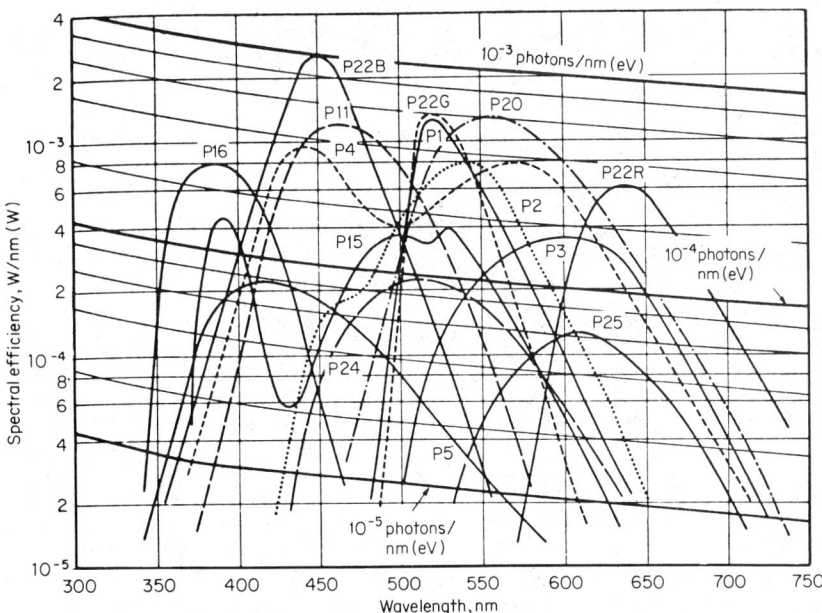

Fig. 11-23. Spectral characteristics of commercial phosphors. (*From "Reference Data for Radio Engineers" (5th ed.), Howard W. Sams & Co., Inc., Indianapolis, Ind., 1968.*)

Luminous efficiency (LE) is usually measured with a calibrated eye-corrected photometer which measures the phosphor brightness of a known area on the screen. Both the electron beam current and voltage are controlled. The luminous efficiency may be computed from the expression

$$LE = \frac{1.076 \times 10^{-3} B}{VI_d} \qquad (11\text{-}3)$$

where B is the phosphor brightness in foot-lamberts, I_d is the beam current density in amperes per square centimeter, V is the electron voltage in volts, and 1.076×10^{-3} is a conversion factor required to change the brightness measurement from foot-lamberts to lumens per square centimeter.

Conversion efficiency is a measure of the ability of the phosphor screen to convert input electrical energy into emitted radiant energy. The conversion-efficiency percentage of the phosphor is simply luminous efficiency [Eq. (11-3)] divided by the lumen equivalent [Eq. (11-1)].

70. Persistence of phosphors is generally characterized by approximately exponential decay of the form $\varepsilon^{-\alpha t}$ or of the power law t^{-n} or combinations of these forms. The decay characteristic of a phosphor is a function of numerous

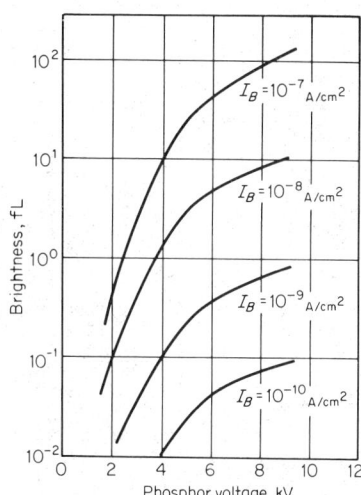

Fig. 11-24. Phosphor brightness as a function of voltage.

variables such as anode voltage, current density of the electron beam, duration of excitation, and pulse repetition rate. For this reason, empirical evaluation is required for the selection of phosphor persistence in visual display devices.

Table 11-14. Special Phosphor Characteristics

Phosphor	P1	P7A	P7N	P19	P19	P26	P31
Refresh rate for flicker-free display, Hz	32	29	27	18	28	17	55
Low-level persistence, s	1	15	25	500	160	700	0.6
Burn resistivity relative to P1 with an assigned weight of 100 (raster)	100	62	50	18	50	19	47
Burn resistivity relative to P1 with an assigned weight of 100 (spot)	100	27	22	9	42	12	30
Buildup	1.35	1.63	1.66	2.83	2.00	3.10	1.45
Light output relative to P1 with an assigned weight of 100 (5 kV, 10 mA)	100	58	32	1	24	34	100
Light output relative to P1 with an assigned weight of 100 (10 kV, 6.3 mA)	100	115	44	32	16	21	185
Spot size, mm	0.254	0.433	0.416	0.345	0.330	0.330	0.254
Apparent resolution relative to P1 with an assigned weight of 100	100	74	90	52	100	78	100
Short-time excitation magnitude of required pulse	28	25	30	65	50	60	20
Rise time, ms	20	20	20	120	60	100	0.2
Fall time, ms	28	60	55	230	80	350	1.2

SOURCE: H. R. Luxenberg and R. L. Kuehn, "Display Systems Engineering," p. 264, McGraw-Hill Book Company, New York, 1968, by permission.

Persistence requirements for flicker-reduced displays are shown in Table 11-14. Phosphors are rated according to the persistence of phosphorescence as *short* (<1 s), medium (<2 s), and long (>1 min).

71. Long-Persistence Phosphors. *Cascade* (P7) and *dark-trace* (P10) phosphor screens are used where very long persistence is desired. Cascade phosphors are composed of two layers, the top layer emitting ultraviolet radiation. During operation, the electron beam excites this layer, which in turn excites the second layer (long-persistence photoluminescent phosphor). The dark-trace phosphor consists of a layer of potassium chloride crystals which exhibit the phenomenon of induced absorption bands. In operation, the screen is viewed by white light and shows darkening of the surface under electron bombardment.

72. Rise Time. The rise of luminescence depends on the composition of the luminescent material, manufacturing process, crystal size, impurity content, and method of excitation. The period of luminescence from the beginning of the exciting pulse to the time the phosphor light output reaches a value of 90% of maximum brightness is defined as rise time.

72a. References on Phosphors

1. CURIE, D., and GARLICK, G. F. J. "Luminescence in Crystals," Wiley, New York, 1963.

2. LEVERENZ, H. W. An Introduction to Luminescence of Solids, Dover, New York, 1968.

3. GOLDBERG, P. "Luminescence of Inorganic Solids," Academic, New York, 1962.

4. KALLMAN, H. P., and SPRUCH, G. M. "Luminescence of Organic and Inorganic Materials," Wiley, New York, 1962.

5. Optical Characteristics of Cathode-Ray Tube Screens, *JEDEC Electron Tube Council Publ.* 16B, 1971.

6. SADOWSKI, M. *RCA Rev.,* **95**:112 (1957).

7. PAKSWER, S., and INTISO, P. J. *J. Electro Chem. Soc.,* **99**:146 (1952).

8. CERULLI, N. F. Method of Electrophoretic Deposition of Luminescent Materials and Product Resulting Therefrom, U.S. Patent 2,851,408, Sept. 9, 1958.

9. LINDEU, B. R. *Advan. Electron. Electron Phys.,* **16**:311, 1962.

10. KOLLER, L. R. *J. Opt. Soc. Amer.,* **43**:620 (1953).

11. BEESLEY, J., and NORMAN, D. J. *Advan. Electron. Electron Phys.,* **22A**:551, 1966.

12. LEHMANN, W. Method of Forming a Uniform Layer of Luminescent Material on a Surface, U.S. Patent 2,798,821, July 9, 1957.

13. FELDMAN, C., and O'HARA, M. *J. Opt. Soc. Amer.,* **47**:300, 1957.

14. KOLLER, L. R. Thin Film Phosphors, *Electrochem. Soc. Meeting,* Washington, D.C., May 13, 1957.

15. STUDER, F. J., CUSANO, D. A., and YOUND, A. H. *J. Opt. Soc. Amer.,* **41**:559, 1951.

16. McGEE, J. D., AIREY, R. W., and ASLAN, M. *Advan. Electron. Electron Phys.,* **22A**:571-581, 1966.

17. SADOWSKY, M. J. *J. Soc. Motion Picture Television Eng.,* **70**:81 (February, 1961).

18. MOON, P. "The Scientific Basis of Illuminating Engineering," Dover, New York, 1961.

19. POOLE, H. H. "Fundamentals of Display Systems," pp. 335-338, Spartan Books, Washington, D.C., 1966.

20. SHERR, S. "Fundamentals of Display System Design," p. 68, Wiley, New York, 1970.

CATHODE-RAY TUBES (CRTs)

By MUNSEY E. CROST AND IRVING REINGOLD

73. Introduction. The cathode-ray tube (CRT) produces visible or ultraviolet radiation by bombardment of a thin layer of a phosphor material by an energetic beam of electrons. The great preponderance of applications involves the use of a sharply focused electron beam directed time-sequentially toward relevant locations on the phosphor layer by means of externally controlled electrostatic or electromagnetic fields. In addition, the current in the electron beam can be controlled or modulated in response to an externally applied varying electrical signal.

DESIGN FEATURES

74. General Principles. The generalized modern cathode-ray tube consists of an electron-beam forming system, electron-beam deflecting system, phosphor screen, and evacuated envelope (see Figs. 11-25 and 26).

The electron beam is formed in the electron gun, where it is modulated and focused. The electron beam then travels through the deflection region, where it is directed toward a specific spot or sequence of spots on the phosphor screen. At the phosphor screen the electron beam gives up some of the energy of the electrons in producing light or other radiation, some in generating secondary electrons, and the remainder in producing heat.

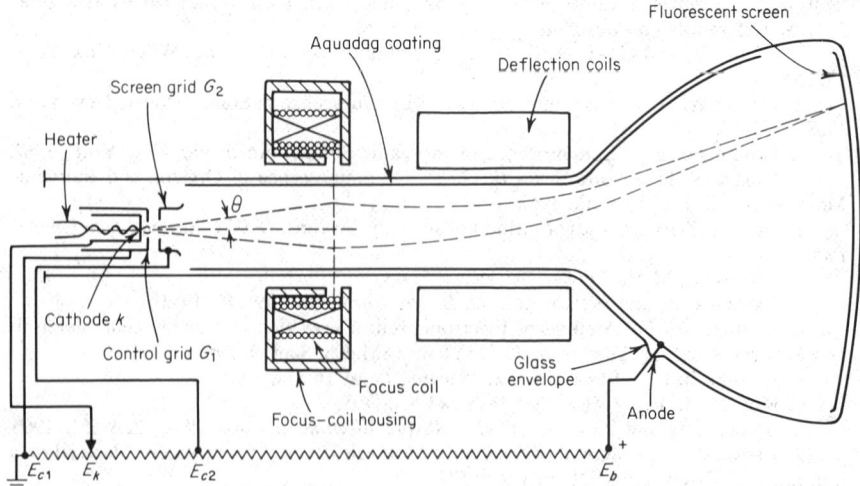

Fig. 11-25. Generalized schematic of cathode-ray tube with electromagnetic focus and deflection. (*From "Cathode-Ray Tubes," Radiation Laboratory Series, Vol. 22, p. 47, McGraw-Hill, New York, 1948.*)

75. Electron-Gun Heater-Cathode. Almost all the CRT electron guns now available have indirectly heated cathodes in the form of a small capped nickel sleeve or cylinder with an insulated coiled tungsten heater inserted from the back end. Most present-day heaters operate at 6.3 V ac with 600 mA current. Since this is rather wasteful for the emission current required, some tubes with 300-mA current are available, and one series of very-low-power heaters operates with 1.5 V and 140 mA.

76. Modulating Grid. The cathode assembly is mounted on the axis of the modulating or control grid cylinder, or simply grid, which is a metal cup of low-permeability steel or stainless steel about ½ in. in diameter and ⅜ to ½ in. long. A small aperture on the order of 10 mils diameter is punched or drilled in the cap.

77. Anodes, Accelerators, and Electrostatic Lenses. To obtain any electron current from the cathode through the grid aperture, there must be another electrode beyond the aperture at a positive potential sufficiently great so that its electrostatic field penetrates the aperture to the cathode surface. Since there are a multitude of electrode-gun designs, this next electrode may have many designs, a wide range of voltages, and several different names. In a simple accelerating lens, in which successive electrodes have progressively higher voltages, this electrode may also be used for focusing the electron beam upon the phosphor, in which case it may be designated the *focusing, or first, anode* a_1. This is usually a cylinder, longer than its diameter and probably containing one or more disk apertures.

Another type of gun employs a screen grid g_2, usually in the form of a short cup with an aperture facing the grid aperture. The voltage is usually maintained unadjusted between 200 and 400 V positive. In an electrostatically focused electron gun, the screen grid is usually followed by the focusing anode. In a magnetically focused electron gun the screen grid may be followed directly by the final anode (Fig. 11-25).

In another type of electrostatically focused electron gun in widespread use (Fig. 11-26), the grid is followed immediately by a long apertured cylinder at the voltage of the principal anode a_2. This is called the *accelerator,* or *preaccelerator.* It is followed in sequence by either two short cylindrical electrodes or apertured disks.

The last electrode and preaccelerator are connected within the tube. The set of three electrodes constitutes an *einzel lens.* By proper design of the einzel lens, the focal condition may be made to occur when the voltage on the central element is zero or a small positive voltage with reference to the cathode.

78. Electromagnetic Focusing Lenses. The focusing systems previously described use electrostatic electron lenses, but a large and important class of CRTs use external magnetic components for focusing. The magnetic field imparts no kinetic energy to the electron, since it always acts in a direction perpendicular to the velocity.

The common method of magnetic focusing of CRTs employs a short magnetic lens which operates by means of the radial inhomogeneity of the magnetic field and can have both the object and image points distant from the lens. The typical short magnetic lens or focus coil for a CRT (see Fig. 11-25) consists of a large number of turns of fine wire with a total resistance of several hundred ohms, wound on a bobbin of insulating paper or plastic. The bobbin and coil are almost totally enclosed in a soft iron shell, except for an annular gap of about ⅜ in. at one end of the core tubing.

79. Deflection of the Electron Beam. There are two basic methods of deflection of the electron beam in a CRT, by a transverse electrostatic field and by a transverse electromagnetic field.

80. Electrostatic Deflection. In electrostatic deflection, metallic deflection plates are used in pairs within the neck of the CRT (see Fig. 11-26).

The simplest deflection plates are merely flat rectangular plates parallel to and facing each other, with the electron beam directed along the central plane between them. The deflection plates are located in the field-free space within the second-anode region, and the plates are essentially at second-anode voltage when no deflection signal is applied. Deflection of the electron beam is accomplished by establishing an electrostatic field between the plates.

The well-made modern electrostatic deflection CRT does not exhibit excessive deflection defocusing until the beam deflection angle off axis exceeds the neighborhood of about 20°. Most electrostatic-deflection CRTs are used to display electrical waveforms as a function of time.

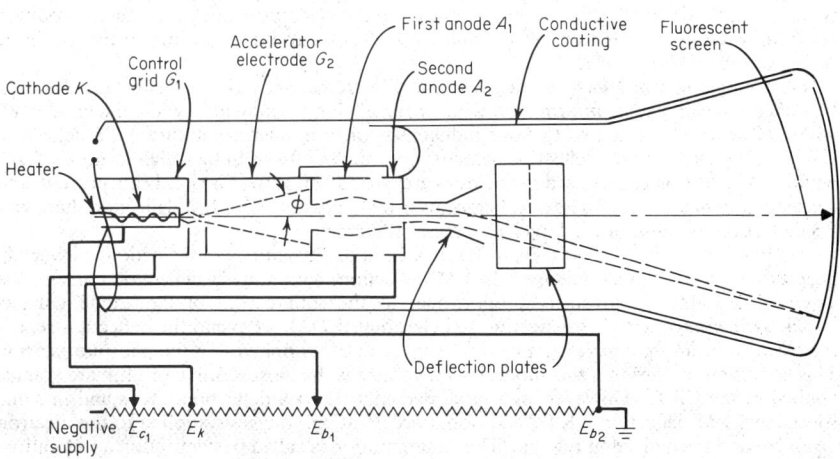

Fig. 11-26. Generalized schematic of cathode-ray tube with electrostatic focus and deflection. An *einzel,* or singlet, focusing lens is depicted. (*From "Cathode-Ray Tubes," Radiation Laboratory Series, Vol. 22, p. 47, McGraw-Hill, New York, 1948.*)

RADIANT ENERGY DEVICES

To display the electrical waveform it is necessary to generate a sweep representing passage of time and to superimpose on this an orthogonal deflection representing signal amplitude. This is most readily accomplished by the use of two pairs of deflection plates. The second pair of deflection plates must have an entrance window large enough to accept the maximum deflection of the beam produced by the first pair. This requires that although the plates may be close enough together at the entrance to afford high deflection sensitivity, they must also have an appreciable width, which results in high capacitance. The plates must also diverge, to accommodate their own deflection of the beam.

To obtain an acceptable deflection sensitivity, the plates must be made long, and consequently the capacitance is increased.

Electrostatic deflection CRTs are particularly suited for the display of arbitrary waveforms, as opposed to electromagnetic deflection CRTs, because the deflection plates generally have capacitances with reference to each other and to all other electrodes of the order of 10 pF or less.

81. Electromagnetic Deflection Systems. In contrast to electrostatic deflection systems, in such systems the deflection components are almost universally disposed outside the tube envelope, rather than inside the vacuum. Since the neck of the CRT beyond the electron gun is free of obstructions, a larger-diameter electron beam may be used in the magnetic deflection CRTs than in the electrostatic deflection CRTs, which permits a much greater beam current to the phosphor screen and, consequently, a much brighter picture than if electrostatic deflection were used. In fact, included deflection angles of 110° (55° off axis) are commonly used in television picture tubes without excessive spot defocusing. As is apparent, large deflection angles permit CRTs to be made with shorter bulb sections for any given screen size.

82. The electromagnetic deflection yoke is most suitable for repetitive types of scan, such as raster scans (parallel-line scans sweeping out a rectangular area), or plan-position-indicator (PPI) scans (radial line scan directed outward from the center to cover a circular area). In recent years, however, electromagnetic deflection has been used more and more frequently for random address. The principal problem with random deflection is the inductance of the deflection coils. For any specific field strength an ampere-turns product must be established. Therefore a low inductance implies a high current, which may be difficult to supply, especially with large bandwidth. Normally, for each axis the yoke includes two coils, each bent into a saddle shape and extending halfway around the CRT neck.

83. PPI Deflection. For PPI deflection one common arrangement is to have the single-axis yoke rotated physically by an external motor or self-synchronous repeater driven by the radar antenna. In this arrangement, a constant-amplitude triggered linear-sawtooth wave of current is supplied to the yoke. A second common method employs a stationary yoke with two orthogonal deflection axes. One axis receives a current waveform of the linear sawtooth with its amplitude coefficient varying according to the algebraic sine of the antenna rotation angle, and the other axis receives a similar waveform, varying according to the algebraic cosine of the rotation angle.

84. Comparison of Electrostatic (ES) and Electromagnetic (EM) Deflection Systems. ES deflection can display position information with extremely large bandwidth, while the bandwidth with EM deflection is limited by yoke inductance and amplifier capabilities. ES deflection is limited by unacceptable deflection defocusing to angles off axis in the neighborhood of 20°, while EM deflection can be used to angles as great as 55° off axis. Greater beam current, and consequently higher brightness of the trace, is usually possible with EM deflection than with ES deflection, because of the larger usable beam diameter.

With a given deflection voltage, in ES deflection the magnitude of deflection is inversely proportional to the anode voltage. In EM deflection, with a given deflection current, the angle of deflection is inversely proportional to the square root of the anode voltage.

85. Drift-Space and Postdeflection Acceleration (PDA). Beyond the deflection region the electron beam may travel in a field-free space until it impinges on the phosphor screen. This condition implies that the anode, or a_2, voltage is the most positive or ultimate voltage applied in the CRT. However, in a large percentage of ES deflection CRTs and in some specialized EM deflection CRTs, additional acceleration voltage is applied to the electron beam beyond the deflection region. This configuration is called postdeflection acceleration, or PDA.

86. Postdeflection Accelerator. In ES deflection tubes the postdeflection accelerator electrode (also called the *postaccelerator*, or *third anode* a_3) generally takes the form of a wide graphite band around the inside of the envelope funnel just behind the faceplate and connected to the aluminum film if the phosphor is aluminized.

Another type of postdeflection accelerator that may be used with a considerably higher a_3/a_2 voltage ratio without detrimental effects caused by the localized electron lens is the spiral accelerator. This accelerator consists of a high-resistance narrow circumferential spiral stripe of graphite of low screw pitch painted over a substantial length of the inside of the envelope funnel. The ends of the spiral are electrically connected to the a_2 and a_3 terminals. In operation the spiral accelerator requires a small continuous direct current to establish a nearly uniform potential variation from the a_2 to the a_3 voltage. In effect, this constitutes a thick electron lens, and since there are no abrupt changes in potential, the trace distortion effects are much smaller than in the thin lens.

87. Cathode-Ray-Tube Envelopes. The cathode-ray-tube envelope consists of the faceplate, bulb, funnel, neck, base press, base, faceplate safety panels, shielding, and potting. Not all CRTs will incorporate each of these components, of course.

The *faceplate* is the most critical component of the envelope, since the display on the phosphor must be viewed through it. Most faceplates are now pressed in molds from molten glass and are trimmed and annealed before further processing. Some specialized CRTs for photographic recording or flying-spot scanning use optical-quality glass faceplates sealed to the bulb section in such a way as to produce minimum distortion.

To minimize the return scattering of ambient light from the white phosphor, many CRT types, especially for television applications, use a neutral-gray tinted faceplate. While the display information will be attenuated as it makes a single pass through this glass, ambient light will be attenuated both going in and coming out, thus squaring the attenuation ratio and increasing contrast.

Certain specialized CRTs have faceplates made wholly or partially of fiber optics, which may have extraordinary characteristics, such as high ultraviolet transmission. A fiber optic region in the faceplate permits direct-contact exposure of photographic or other sensitive film without the necessity for external lenses or space for optical projection.

88. Bulb Section. The bulb section of the CRT is the transition section necessary to enclose the full deflection volume of the electron beam between the deflection region and the phosphor screen on the faceplate. In most CRTs, this is a roughly cone-shaped molded-glass component.

Instead of an ordinary glass bulb, many of the larger TV-type CRTs have metal cone sections made of a glass–sealing iron alloy. The metal cones are generally lighter in weight than the corresponding glass sections.

89. Funnel Section. The junction region of the bulb of a CRT with the neck section is critical as to geometry, since tubes made with these separate sections are intended for electromagnetic deflection, and this region is just where the deflection yoke is intended to be located.

90. Tube Neck. The neck diameter of a CRT depends to a great extent upon the type of deflection used and the intended application of the CRT. In general, the ES deflection CRTs have large neck diameters, while the EM types have small diameters.

91. Implosion Protection. As an additional safety factor, many of the larger TV-type CRTs are manufactured with a separate glass *implosion panel,* contoured to match the faceplate curvature, permanently bonded to the outside surface of the faceplate by means of a transparent rubbery silicone adhesive filler. This combination offers protection to the viewer in much the same manner as laminated safety glass in automobile windows. The implosion panel may be etched on its front surface to produce a ground-glass effect to avoid specular reflections from ambient light, and/or it may include neutral-gray attenuation.

CATEGORIES OF CATHODE-RAY TUBES

92. Oscilloscope Tubes. For oscilloscopic applications the general requirements on a CRT include a sharp, bright, rapidly deflectable, single-line trace with a minimum of deflection defocusing or astigmatism. The rapidity of deflection and the fact that arbitrary waveforms must be displayed dictate the use of ES deflection, at least for the vertical direction. For general use, both horizontal and vertical axes employ ES deflection. Since the included deflection angle must be small, usually less than 45°, to preserve good spot size and shape, these CRTs are relatively long compared with the face diameter.

The phosphors generally used for oscilloscope CRTs are P1 (green, medium persistence) or P2 (yellow-green, medium persistence, but with a much longer, low-level "tail" than P1, see Par. 11-63).

93. Radar Display Tubes. Except for the *A-scope* radar display, which is essentially the same as an oscilloscope display, most radar displays consist of a two-dimensional coordinate display with beam intensity modulation. Since the coordinate scans are mathematically regular and at preselected rates, EM deflection is generally used, inasmuch as this permits greater deflection angles and, consequently, shorter tubes to be used for a given face diameter.

Especially in filtered radar displays, it is often necessary to include alphanumeric characters, symbols, and vectors in the display along with the radar information. Shaped-beam tubes, such as the Charactron, or a multiple-beam tube, in which one beam is devoted to the tracing of the characters or symbols and the other to the plan-position-indicator (PPI) display, are used for this purpose.

Long-persistence phosphors are generally used in CRTs for radar displays, since it is desirable to be able to see the radar situation in the entire area covered at any given time.

94. Television Picture Tubes. *Monochrome Tubes.* Since the standards for television transmission in the United States call for 30 frames of two interlaced fields each per second, producing the effect of 60 pictures per second, which is above the flicker fusion frequency for all light levels, there is no stringent limitation on phosphor persistence for monochrome TV picture tubes, so long as the persistence does not cause picture smearing. The white luminescence used for most applications is achieved by a mixture of phosphors rather than any single component. Several white-luminescing combinations, all designated P4, have been in common use: the all-silicates, the silicate-sulfide mixture, and the all-sulfides.

Color Tubes. Many types of full-color CRTs have been developed for television use, but the shadow-mask tube is in most widespread use. This type of CRT uses a cluster of three electron guns in a wide neck, one gun for each of the colors red, green, and blue. All the guns are aimed at the same point at the center of the shadow mask, which is a metal grid with an array of perforations in triangular arrangement, spaced 0.025 in. between centers. Phosphor dots on the faceplate just beyond the shadow mask are arranged so that the electron beam from each gun, after passing through the perforations, can strike only the dots emitting one color. All three beams are deflected by a single large-diameter deflection yoke. The group of three phosphors together are designated P22, with individual phosphors denoted by the numbers P22R, P22G, and P22B.

Two other classes of multicolor CRTs are of current interest, those with parallel-stripe phosphors and those with voltage penetration phosphors. In the *parallel-stripe class* of CRTs, such as the Japanese *Trinitron,* sets of very fine stripes of red, green, and blue emitting phosphors are deposited continuously across the faceplate, generally in a vertical orientation. The Trinitron, unlike conventional color CRTs, has a single electron gun which emits three electron beams. Each beam is directed to the proper color stripe by means of the internal tube-deflecting structure and a slitted grille. The *Lawrence tube* is another example of the parallel-stripe phosphor class of color CRT. It uses a single electron beam, and color selection is accomplished solely by control voltages applied to the grille itself.

In the *voltage penetration type* of phosphor screens, two or three unstructured layers of phosphors are deposited one upon another for each color, sometimes with an inert, transparent barrier layer between them for better color differentiation. A single electron beam is used, and the resultant color is determined by preselected beam accelerating voltages, which are changed to control the depth of beam penetration into the phosphor layers.

95. Recording Tubes. Cathode-ray tubes for recording or transcribing information on photographic or otherwise sensitized film are usually of the very-high-resolution (vhr) or ultrahigh resolution (uhr) types. The great majority of these types have faceplate diameters of nominally 4 or 5 in. The spot diameters of the vhr and uhr tubes range from approximately 0.0015 in. down to 0.00033 in.

96. Computer Terminal Display Tubes. CRTs for computer display are very similar to tubes used in high-resolution video monitors, but since the display is principally alphanumeric and vector-graphical, the linearity of the beam modulation characteristics is less important. Well-focused round spots with minimum spot growth or deflection aberrations from the center to the useful edges of the display area are required. High legibility is of primary importance, implying high contrast. White-emitting phosphors are not necessary, so that highly efficient, high-visual-response phosphors emitting in the yellow or green spectral regions are applicable. Most of these CRTs are made with rectangular faceplates.

96a. References on Cathode-Ray Tubes

1. "Cathode-Ray Tube Displays," MIT Radiation Laboratory Series, Vol. 22, McGraw-Hill, New York, 1953.

2. LUXENBERG and KUEHN, "Display Systems Engineering," McGraw-Hill, New York, 1968.

3. POOLE, H. "Fundamentals of Display Systems," Spartan Books, Washington, D.C., 1966.

4. "IRE Standards on Electron Tubes, Methods of Testing (1962), Part 2, Cathode-Ray Tubes," Institute of Electrical and Electronics Engineers, New York, 1962.

5. *IEEE Trans. Electron Devices, Spec. Issue Info. Display,* Vol. ED 18, No. 9, September, 1971.

6. CROST, MUNSEY E. Display Devices and the Human Observer, *Proc. Interlab. Seminars Component Technology,* Part 1, *R&D Tech. Rept.* ECOM-2865, U.S. Army Electronics Command, Fort Monmouth, N.J., August, 1967, pp. 365–429.

7. SHERR, S. "Fundamentals of Display System Design," Wiley, New York, 1970.

PHOTOEMISSIVE ELECTRON TUBES, IMAGE CONVERTERS, AND INTENSIFIERS

By WOLFGANG KLEIN AND CLARENCE A. JOHNSON

97. General Characteristics. In their various configurations, photoemissive electron tubes are important devices for sensing, detecting, imaging, processing, amplifying, and displaying photon information. Devices are available with spectral sensitivity in the far-ultraviolet, visible, or near-infrared regions. They have linear characteristics over wide signal input ranges. Their good signal-to-noise characteristics make them very useful in low-photon-level detection.

98. Photoelectron and Secondary Electron Emitters. The front element of the photoemissive electron tube is a photocathode which converts photons into electrons. Photons are absorbed in the photocathode material and excite mobile photoelectrons which diffuse toward the photocathode-vacuum interface and escape into the vacuum if the work function of the surface is low enough. Once in the vacuum, the photoelectrons can be collected, amplified, and/or displayed.

Photocathodes for practical applications are formed by evaporation and oxidation of metals. The lack of control over these processes, along with an insufficient understanding of their operation, has led to many recipes for their preparation. Solid-state theory, environmental control over the photocathode, and good-quality semiconductor material have led to introduction of the zero and negative electron affinity photoemitters (Fig. 11-27). Recently, opaque and transparent photocathodes of the III-V semiconductors have been made, up to 2,000 and 400 μA/lm, respectively. Photocathode energy-band diagrams are shown in Fig. 11-27.

Photocathodes can operate either in a transmission or reflection mode. For the transmission mode a transparent substrate is required because free-standing films of the required thickness are not obtainable, with the exception of the silicon photocathode. The substrate can be an insulator like glass, sapphire, LiF, or a wide-band-gap semiconductor like GaP, AlAs, or GaAs.

For reviews and comprehensive literature, see Refs. 1 to 7, Par. **11-105a**. The limitation in the long-wavelength threshold is presently around 1.3 μm. In the future, longer-wavelength thresholds will be obtainable from tunnel emitters and heterojunction emitters.[15*] Typical photocathodes and their properties are listed in Table 11-15. Their cathode sensitivity is given in microamperes per lumen, measured with a standardized tungsten light source of 2854°K color temperature. As the evaluation with this light source tends to favor near-infrared-sensitive photocathodes, the quantum efficiencies at a specified wavelength are also given. Complete spectral-yield data for several selected photocathodes are shown in Fig. 11-28.

---

*Superior numbers correspond to numbered references, Par. **11-105a**.

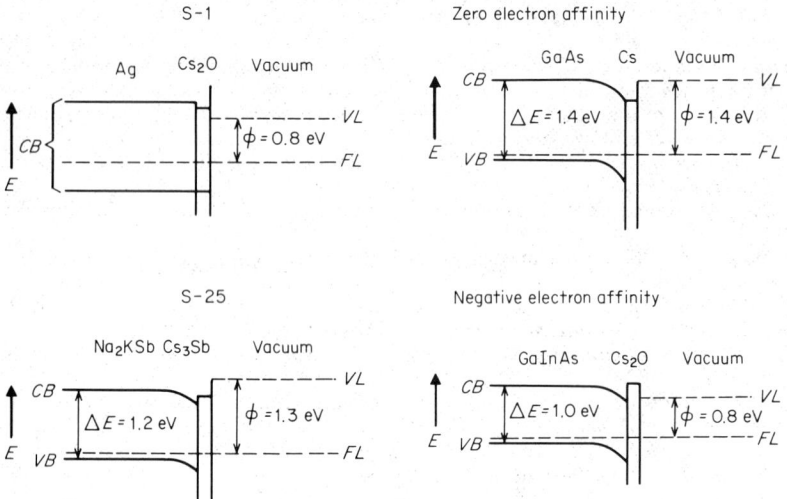

Fig. 11-27. Simplified energy-level diagram for the photocathode. Top left, S-1; bottom left, S-25; top right, GaAs cathode (zero-electron-affinity case); bottom right, GaInAs cathode with optimized 1.06-μm photoresponse (negative-electron-affinity case). E = energy; CB = conduction band; VB = valence band; VL = vacuum level; FL = fermi level. R (eV) = 1.2397/λ (μm), electron affinity EA = CB in the bulk − VL.

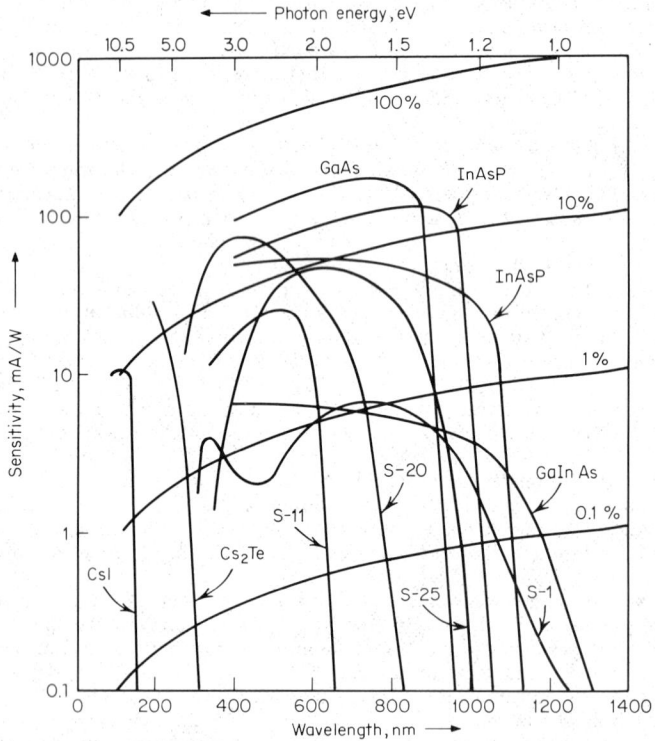

Fig. 11-28. Spectral response curves for various photocathodes. Sensitivity to incident light in milliamperes per watt is plotted on the ordinate. The curved lines with percentage numbers are quantum-efficiency lines.

Table 11-15. Typical Photocathodes and Their Properties

Cathode Material	Operation (transmission or reflection)	Threshold, nm	White-light sensitivity, µA/lm	Quantum, %	Eff. at λ, nm	Dark current, A/cm²
CsI	T	145	...	11	120	10^{-16}–10^{-17}
Cs$_2$Te	T	330	...	8	250	10^{-16}–10^{-17}
K$_3$Sb	T	550	60	24	380	10^{-14}–10^{-15}
Rb$_3$Sb	T	580	25	8	380	10^{-15}
Cs$_3$Sb (S-11)	T	670	60	15	440	10^{-15}
Na$_2$KSb+Cs (S-20)	T	870	300	20	400	10^{-15}
Na$_2$KSb+Cs$_3$Sb (S-25)	T	950	450	9	550	10^{-15}
Ag+Cs$_2$O (S-1)	R	1,150	80	0.1	1,060	10^{-10}–10^{-13}
GaAs+Cs$_2$O	R	930	2,062	35	550	10^{-16}
InAsP+Cs$_2$O	R	960	1,200	12	940	10^{-15}
GaInAsP+Cs$_2$O	R	1,100	1,200	7	1,060	10^{-14}
GaInAs+Cs$_2$O	R	1,300	150	0.3	1,060	10^{-12}
Si+Cs$_2$O	R	1,150	500	0.5	1,060	10^{-9}

99. Secondary Emission.[8-14] Every good photoemitter is a potentially good secondary electron emitter. Secondary electron emitters are operated in reflection mode, except for porous potassium chloride layers and silicon films. The processes involved are the same except for the excitation. For a compilation of secondary emitters, see Table 11-16. The values are for normal incidence of the primary electron beam. For a good signal-to-noise ratio the multiplication factor of the first dynode of the phototube must be as high as possible. This requires the use of $GaP + Cs_2O$ or $Si + Cs_2O$ as dynode material. Time lag is a problem in the zero- and negative-electron-affinity secondary emitter because of the long travel time of the excited electrons in the emitter. This effect may be offset by the use of fewer dynodes for a given amplification.

Table 11-16. Secondary Emitters

Material	Secondary-emission yield δ_{max}	Primary energy E_{max}, V, at δ_{max}
Pt	1.8	700
KCl	7.5	1,200
BeO	12	700
MgO	24	1,200
GaP+Cs₂O	240	9,000
Si+Cs₂O	950	20,000

100. Phototubes and Photomultipliers. Phototubes and photomultipliers are glass, ceramic, or metal vacuum bottles with a transmission or reflection photocathode in end-on or side-window configuration and an anode (in the case of the phototube) or discrete or continuous dynodes in front of an anode (in the case of photomultipliers). Discrete dynode structures, listed in order of improving time response, include venetian-blind, box-and-grid, circular-cage, and linear multiplier (Fig. 11-29). Collection efficiency considerations determine location and shape of the photocathode-to-first-dynode region. The typical photocathode diameter for phototubes and photomultipliers is $\frac{1}{4}$ to 2 in. The number of dynodes in photomultipliers varies between 5 and 16. A typical current amplification value is 1×10^6. The dc supply voltage lies between 1,000 and 3,000 V, with maximum values of 300 V between cathode and first dynode and 200 V between consecutive dynodes. Ruggedized versions resistant to shock, vibration, acceleration, and high temperature are available.

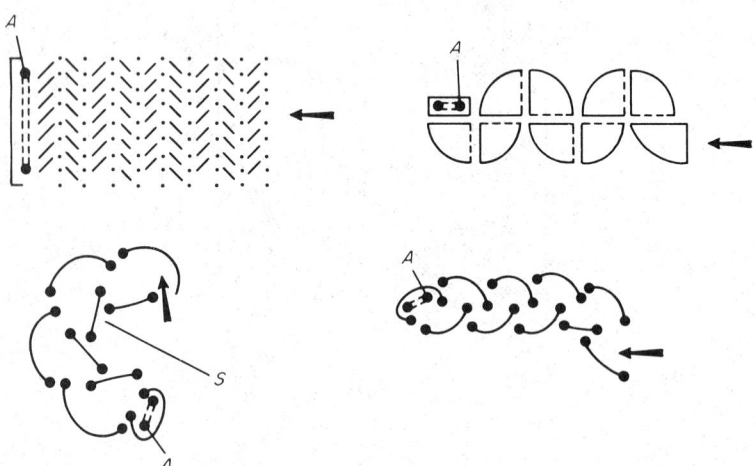

Fig. 11-29. Multiplier structures. Venetian-blind (upper left), box-and-grid (upper right), circular-cage (lower left), and linear (lower right). The arrow shows the direction of the incident electrons. A = anode; S = shield.

101. Microchannel electron multipliers (multipliers) are continuous dynodes consisting of a glass tubing coated on the inside with a secondary-electron-emitting material. The glass tube has electrodes at the entrance and exit. The entrance can be conical or straight, while the main section is straight, bent, or spiraled. The output current has to be a tenth or less of the strip current; otherwise the multiplier operates in a saturated mode. The amplification depends on the length-to-diameter ratio of the multiplier, the axial field strength, and the secondary-electron-emitter material.

IMAGE CONVERTERS AND INTENSIFIERS

102. Principle of Operation. An image tube converts an image in one spectral region directly into an image in another spectral band, usually with an increase in intensity. If the primary purpose is to convert the spectral region of the image (example, near infrared or ultraviolet to visible), the tube is called an *image converter* tube (Fig. 11-30). If the primary purpose is to intensify the image without regard to the spectral conversion, it is called an *image intensifier* tube (Fig. 11-31). Normally, a tube designed to do both is categorized as an image intensifier tube.

Direct-view image tubes such as shown in Fig. 11-30 utilize a photoemissive input surface (Par. **11-98**) to form an electron image of the scene being viewed, an electron lens system biased by a high voltage ranging from 4 to 20 kV to accelerate and focus the image, and a phosphor screen (Par. **11-60**) to display the image, all contained in a single vacuum envelope.[16] The exception to this arrangement is the solid-state image converter, which has no vacuum envelope (Fig. 11-32). The images formed by these devices are two-dimensional and are converted or amplified at all points simultaneously rather than being scanned as in a television system.

Image tubes may be obtained with many types of photocathodes, depending upon the particular application and the spectral region[18] desired (Table 11-17). Typical examples are the infrared image converter (Fig. 11-30), ultraviolet image converter, and infrared or visible-image intensifier (Fig. 11-31). Image tubes may also be obtained with a wide variety of phosphor viewing screens, depending on the persistence and/or the spectral characteristics required (Table 11-13).

103. Electron Optics. Image tubes may be categorized by the characteristics of their electron optics, which make them suitable for particular applications.

a. Proximity-focused Tubes[16,20] (Fig. 11-33). When the photocathode and phosphor screen are spaced a few millimeters apart and a potential difference of several kilovolts is maintained between them, the resulting electrostatic field focuses the electron image from the photocathode onto the phosphor screen with increased brightness and with a resolution of up to 40 line pairs/mm (a line pair consists of one black line and one white line of equal width).

A more sophisticated version of this tube type is the microchannel wafer intensifier (Fig. 11-34), which utilizes a microchannel array as the electron multiplier. These tubes are used in devices such as night-vision goggles, hand-held viewing devices, and night observation systems.

Since these tubes do not invert the images formed, fiber-optic twists or inverters (fiber bundles twisted 180° to invert the images) are used where an inverted image is necessary. Typical performance characteristics of this tube type are listed in Table 11-17.

b. Electrostatic-focus tubes are by far the most common:

1. *Infrared image converter.* The single-stage electrostatic-focus infrared converter tubes (Fig. 11-30) are usually diode, fixed-focus types with curved-glass photocathode and flat-glass–phosphor screen faceplates.[30] However, there are also triode tubes which require a focusing voltage. They are made in photocathode sizes from 18 to 25 mm, and their magnification ranges from 0.50 to 0.75. These tubes have been used mainly in military-weapon sights, driving binoculars, and armored-vehicle night-vision systems, as well as for photographic-film inspection and in scientific apparatus such as ultraviolet (UV) and infrared (IR) microscopes. Figure 11-35 and Table 11-17 contain typical characteristics of this tube type.

2. *Image intensifiers.*[18,19] The preponderance of the imaging tubes in use today fall into this category. Most utilize modular construction and have fiber-optic faceplates on input and/or output ends (Fig. 11-31). The electron lenses in the electrostatic-focus types invert

Table 11-17. Typical Characteristics of the Most Common Image Converters and Intensifiers

Tube type	Useful diam. photocathode, mm	Useful diam. screen, mm	Center magnification	Center resolution	Operating voltage	Gain[1]	EBI,[2] cm^-2
Image Converters							
8598	15.9	19.0	1.0	72	12	15 CI[3]	2.5×10^{-7}
6929	19.0	14.5	0.75	50	12	LT CI	3.3×10^{-7}
6914	25.0	21.8	0.76	50	16	15 CI	2.5×10^{-7}
8857	18.0	18.0	0.97	64	12	65 B	2×10^{-11}
8585	25.0	25.0	0.92	63	15	65 B	2×10^{-11}
8605	40.0	40.0	0.95	57	15	65 B	2×10^{-11}
Three-stage-assemblies, electrostatic focus							
8858[4]	18.0	18	0.84	32	2,655[5]	30 K	2×10^{-11}
8586[4]	25.0	25	0.82	28	2,800[6]	35 K	2×10^{-11}
8606[4]	40.0	40	0.82	28	2,800	35 K	2×10^{-11}
Microchannel intensifier assemblies							
Wafer MCP[4]	18.0	18	1.0	28	2,655[5]	10 K	2×10^{-11}
Electrostatic inverter MCP[4]	25.0	25	0.97	28	2,655[5]	25 K	2×10^{-11}
Magnetic-focus image converters and intensifiers							
Single-stage converter	40.0	40	1.0	70	12 kV	75	5×10^{-12}
Two-stage intensifier	40.0	40	1.0	50	120 kV	3.5 K	5×10^{-12}
Three-stage intensifier	40.0	40	1.0	35	36 kV	200 K	5×10^{-12}

[1] Unless otherwise noted, gain listed is luminous gain.
[2] EBI = equivalent background input.
[3] CI = conversion index, the ratio of luminous flux from the fluorescent screen to the product of the luminous flux incident on a Corning 2540 infrared filter and the filter factor of 10.8%.
[4] Packaged assemblies include supplies.
[5] Refers to dc input voltage.
[6] Refers to ac input voltage from external oscillator. The ABC (automatic brightness control) versions of these assemblies have built-in oscillators with 6.5 V dc input voltages.

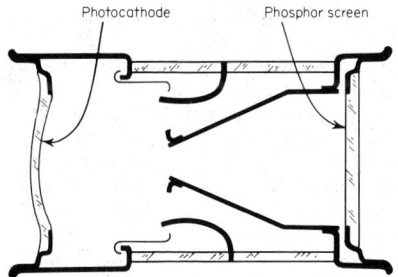

Fig. 11-30. Image converter tube.

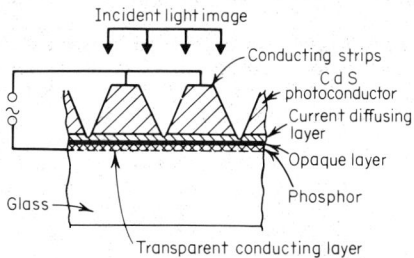

Fig. 11-32. Solid-state image converter.

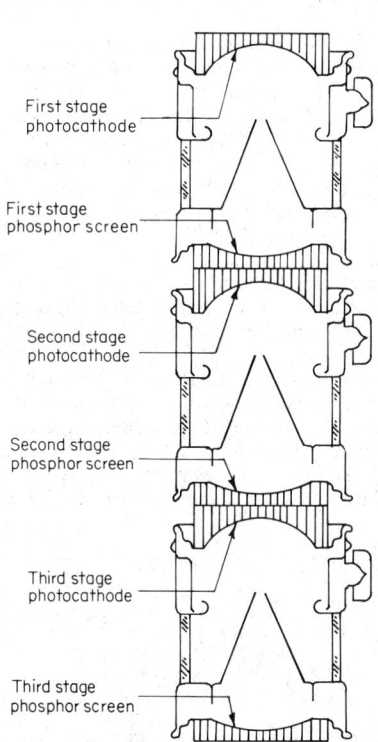

Fig. 11-31. Typical three-stage electrostatic-focus image intensifier.

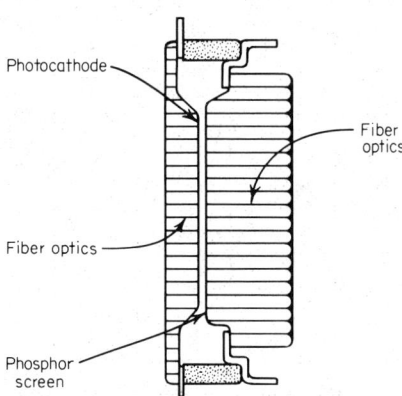

Fig. 11-33. Wafer-type proximity-focus tube with fiber-optics output.

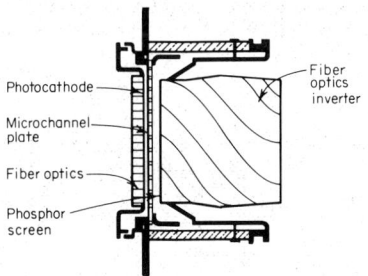

Fig. 11-34. Wafer-type microchannel tube with 180° fiber-optics twist for image inversion.

the image, and both the object plane and image plane of electrostatic-focus electron optics are curved.

Fiber optics often are used in electrostatic-focus tubes. They characteristically have magnifications on the order of 0.75 to 1.00; however, there are versions available with variable magnification *(zoom)* (Fig. 11-36). In addition, there are versions made with fixed magnifications other than unity, i.e., demagnification down to 0.3X and magnification up to 10X.

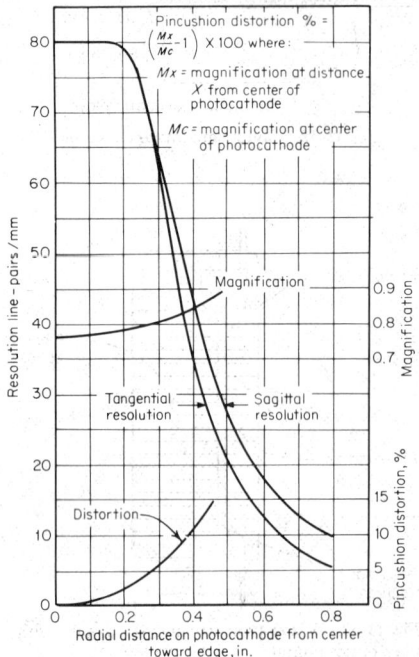

Fig. 11-35. Resolution, magnification, and distortion curves for a typical image converter tube.

Included in this grouping are the multistage tubes[19] (Fig. 11-31), which are cascaded via fiber optics for increasing useful gain.

Some versions have automatic brightness control (ABC) (see Figs. 11-37 and 11-38 for typical performance characteristics of regular and ABC-type multistage intensifiers), which extends their useful range to light levels two orders of magnitude higher than conventional versions. Typical characteristics of this tube type are listed in Table 11-17.

3. *The electrostatic-focus microchannel intensifier* (Fig. 11-39) utilizes electrostatic input focusing and proximity output focusing. This tube, like the proximity-focus wafer tube, utilizes a microchannel array (to be described) as an amplifier. It is used in military night sights because of its compact size, high resolution, and high gain.

c. Magnetic-Focus Image Intensifier.[16,20] Magnetically focused intensifiers (Fig. 11-40) are furnished as two-, three-, or four-stage tubes. These multistage devices are usually built into one vacuum envelope[21] and are focused with toroidal permanent magnets or electromagnets. These tubes are characterized by their high resolution, lack of distortion, low background, and high gain in the three- and four-stage types.

The bulk and weight of their focusing magnets limit the portability of systems utilizing this tube type, but it has found great favor for use on astronomical telescopes, where its low

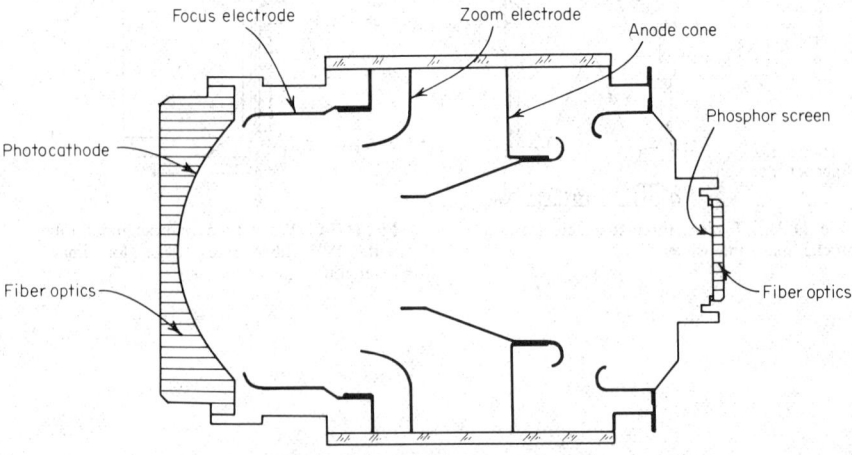

Fig. 11-36. Electrostatic-focus variable-magnification tube.

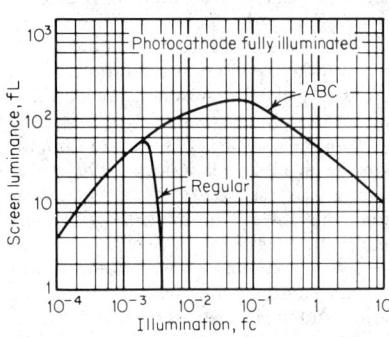

Fig. 11-37. Screen luminance vs. incident illumination for a three-stage assembly with and without automatic brightness control.

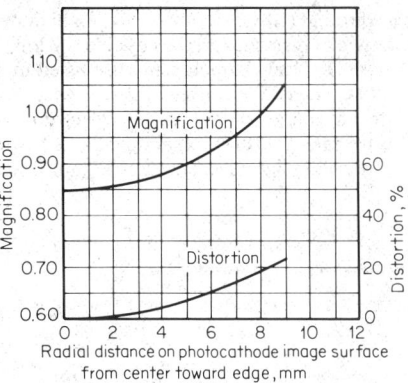

Fig. 11-38. Typical magnification and distortion characteristics of a three-stage assembly.

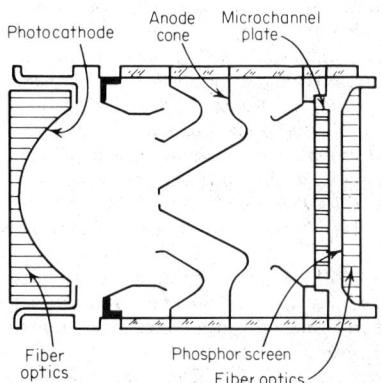

Fig. 11-39. Typical electrostatic-focus microchannel inverter tube.

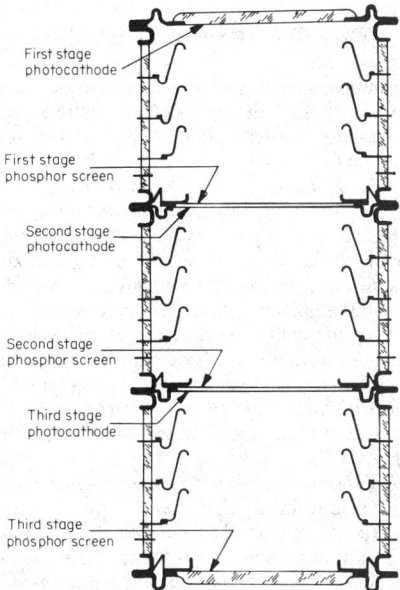

Fig. 11-40. Three-stage magnetic-focus image intensifier.

background (EBI, 5×10^{-12} cm^{-2}), high resolution (45 line pairs/mm for three stages), and absence of distortion make possible the long exposures necessary in astronomical photography.[17] Typical characteristics are listed in Table 11-17.

104. Gain Mechanism. *Electron Acceleration.* Electrons emitted from a photocathode with an energy of only a few electron volts are given tremendous acceleration by an electrostatic field of several kilovolts. These electrons then strike the phosphor screen, which emits visible light; i.e., substantial gain is available.

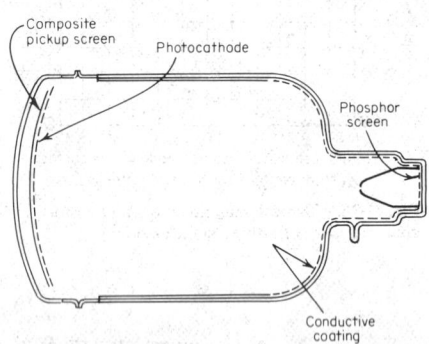

Fig. 11-41. X-ray image intensifier tube.

Demagnification. The image brightness in an electrostatic-focus tube varies inversely as the square of the magnification; the lower the magnification, the higher the brightness. The electron acceleration in combination with demagnification concentrates the electron flux at the screen and thus increases the gain. The x-ray intensifier image tube (Fig. 11-41) falls into this category.

Cascading. When several image-tube modules are placed in tandem or optically coupled to increase the gain, the method is called cascading. The cascading is accomplished by coupling several modules having fiber-optic faceplates, as in Fig. 11-31. Several stages may also be built into one envelope, as in Fig. 11-40. The cascaded assembly is called an *image intensifier,* and contains three stages in the case of electrostatic focus, because three stages yield the desired gain (in the range of 30,000 to 150,000) while providing reasonable resolution (25 to 40 time pairs/mm) and an erect image.

Microchannel Array. A microchannel array is a mosaic of several million long thin tubes of glass or ceramic with a high-resistance coating on the inner surface to provide continuous-channel electron multipliers. When a potential difference of 500 to 1,000 V is applied between the metalized ends, a uniform electric field is established within each channel. A single electron entering a channel creates one or more secondary electrons with a transverse velocity equivalent to about one electron volt of energy. This transverse velocity carries it across the channel, while the electric field accelerates it along the channel. A sufficient amount of energy is imparted to the typical electron so that, on the average, it will generate more than one secondary upon collision with the opposite wall. Thus a cascading action is started which can produce electron gains in excess of 100,000. The microchannel array is used as the gain mechanism for the wafer tube (Fig. 11-34) and the inverter-type intensifier (Fig. 11-39).

Transmission Secondary-Electron Multiplication (TSEM). When electrons are accelerated into a thin foil at high energy and under proper conditions of film thickness and electron energy, secondary electrons can be collected on the opposite, or transmission, side. Figure 11-42 illustrates a tube utilizing this principle, with several such films or dynodes.

105. Special Tube Types. *X-Ray Image Intensifiers* (Fig. 11-41). The composite target (composed of an x-ray phosphor deposited on an aluminum supporting substrate) of the x-ray image intensifier converts an x-ray image into a visible image which is intensified by demagnification and displayed on a phosphor screen for direct viewing. X-ray image intensifiers may also be coupled to television camera tubes by the use of fiber optics for remote viewing. X-ray image intensifiers are used mainly for medical diagnosis and structural inspection.

Image Magnifier (Fig. 11-43). The image magnifier is an electrostatic-focus image converter with a magnification usually of 5X and 10X. The gain in such tubes is low because, as previously explained, the gain varies inversely as the square of the magnification; thus there is lower gain in the image magnifier. Magnifier tubes have typical gains of 6 to 8 for the 5X type and 2 to 3 for the 10X type. To increase the gain in magnifier tubes, additional unity magnification stages are coupled to the input by means of fiber optics.

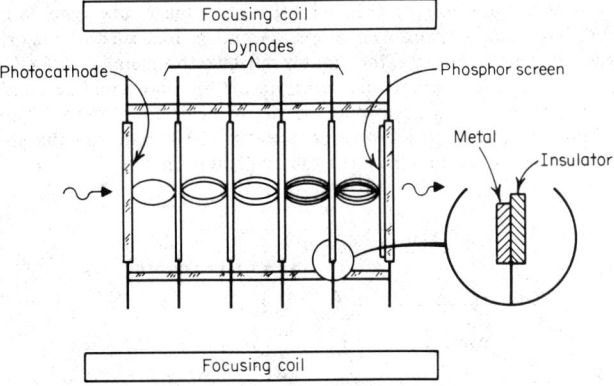

Fig. 11-42. **Transmission secondary-electron multiplication tube.**

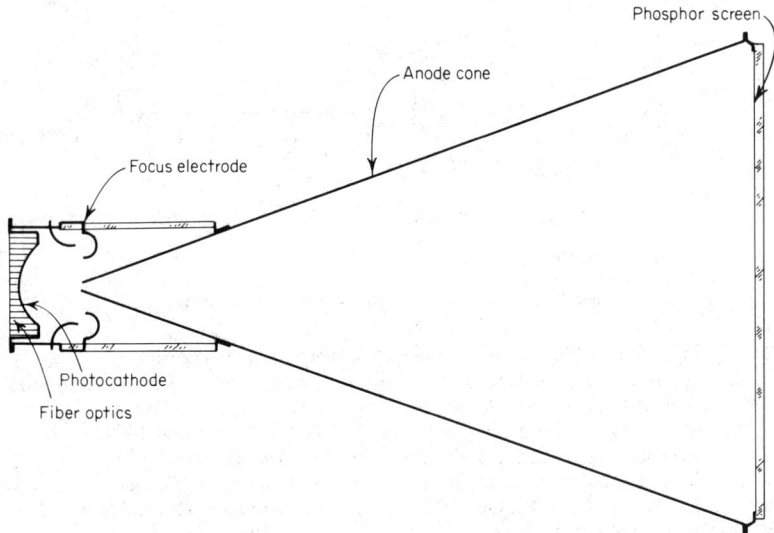

Fig. 11-43. **Image magnifier tube.**

Photoreflectronic Image Converter (Fig. 11-44). This image-tube type derives its name from light optics built into its vacuum envelope. An image focused onto the primary mirror through the glass entrance window at the input is reflected to a metallic surface (which serves both as the secondary mirror and as the substrate for the photocathode), and the emitted electron image is focused onto the phosphor screen by the electron optics. This image-tube type is used when the photocathode is opaque (see Par. **11-98**), requiring that the light image enter and the electrons be emitted from the same vacuum side.

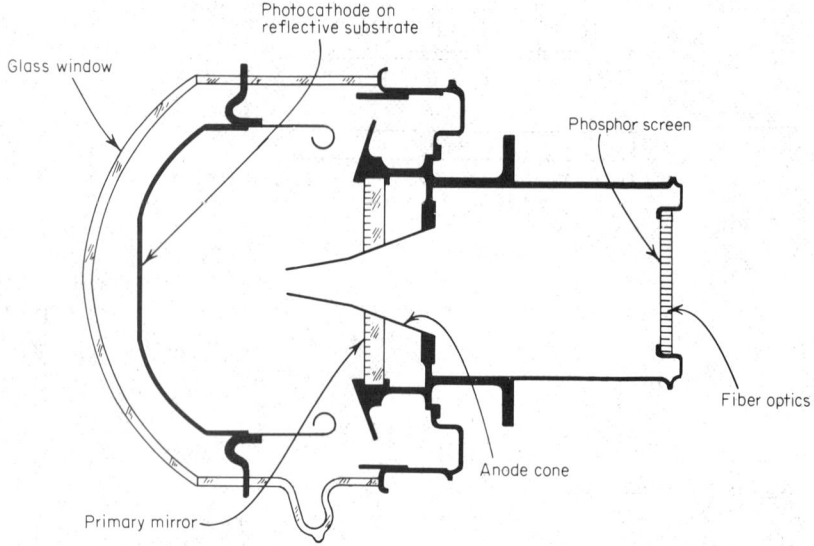

Fig. 11-44. Photoreflectronic image converter tube.

105a. References on Photoemissive Tubes, Converters, and Intensifiers

1. GOERLICH, P. Recent Advances in Photoemission, *Advan. Electron. Electron Phys.*, **11**:1 (1959); and BERNDT, M., and GORLICH, P. *Phys. Stat. Sol.*, **3**:963 (1963).

2. SOMMER, A. H. "Photoemissive Materials," Wiley, New York, 1968.

3. VAN LAAR, J., and SCHEER, J. J. *Philips Techn. Rev.*, **29**:54 (1970).

4. SCHAGEN, P., and TURNBULL, A. A. New Approaches to Photoemission at Long Wavelength, and SYMS, C. H. A., GaAs Thin Film Photocathodes, *Advan. Electron. Electron Phys.*, **28A**:93 and 399 (1969).

5. BELL, R. L., and SPICER, W. E. *Proc. IEEE*, **58**:1788 (1970).

6. ROME, M. Photoemissive Cathodes, I, and SYMS, C. H. A., Photoemissive Cathodes, II, in Biberman, L. M., and Nudelman, S. (eds.), "Photoelectronic Imaging Devices," Vol. 1, pp. 147 and 161, Plenum, New York, 1971.

7. BELL, R. L. "Negative Electron Affinity Devices," Clarendon Press, Oxford, 1973.

8. MARTINELLI, R. U. *Appl. Phys. Letters*, **17**:313 (1970) and *Conf. Photoelectric Secondary Electron Emission, Paper* D1, Minneapolis, Minn., Aug. 18-19, 1971.

9. DEKKER, A. J. Secondary Electron Emission, *Solid State Phys.*, **6**:251 (1958), and HACHENBERG, O., and BRAUER, W., Secondary Electron Emission from Solids, *Advan. Electron. Electron Phys.*, **11**:413 (1959).

10. BRUINING, H. Physics and Applications of Secondary Electron Emission, McGraw-Hill, New York, 1954.

11. "RCA Photomultiplier Manual," RCA Electronic Components, Technical Series PT-61, Harrison, N.Y., 1970.

12. MARK, D. Basics of Phototubes and Photocells, J. F. Rider, New York, 1956.

13. KOLLATH, R. Sekundaereliktronen-Emission fester Koerper bei Bestrahlung mit Elektronen, in S. Fluegge (ed.), "Handbuch der Physik," Vol. 21, p. 232, Springer, Berlin, 1956.

14. SIMON, R., and WILLIAMS, B. F. *IEEE Trans. Nucl. Sci.,* **15**:167 (1968).

15. MILNES, A. G., and FEUCHT, D. L. "Heterojunctions and Metal Semi-Conductor Junctions," Academic, New York, 1972.

16. GRIVET, P. "Electron Optics," Pergamon, New York, 1965.

17. SOULE, HAROLD V. "Electro-optical Photography at Low Illumination Levels," Wiley, New York, 1968.

18. MORTON, G. A. Image Intensifiers and the Scotoscope, *Appl. Optics,* June, 1964.

19. *Proc. Image Intensifier Symp.,* Office of Scientific and Technical Information NASA, NASA SP-2, Oct. 1961.

20. SPANGENBERG, KARL R. "Vacuum Tubes," McGraw-Hill, New York, 1948.

21. KOHL, WALTER H. "Materials and Techniques for Electron Tubes," Reinhold, New York, 1960.

TELEVISION CAMERA TUBES

BY RONALD D. GRAFT AND RICHARD E. FRANSEEN

106. Types of Television Camera Tubes. Modern television camera tubes are available in a wide variety of sizes and types to meet a multitude of applications. *Antimony trisulfide vidicons, lead oxide vidicons,* and *image orthicons* are the workhorse television tubes, accounting for perhaps 70% of all applications. Other tube types such as the *SEC vidicon, image isocon, silicon vidicon,* and *silicon intensifier vidicon* provide special capabilities not generally required in routine applications.

107. Theory of Operation. Television camera tubes can be conveniently discussed in terms of the three major subassemblies shown in Fig. 11-45. The *image section* uses a photoemissive surface and electron optics to convert an optical image to an electron image which is focused upon the surface of the storage target to create a corresponding electrical charge image. The *storage target* integrates, or stores, the focused electrical charge prior to readout and erasure by an electron beam generated in the *scan section.* The low-velocity electron beam repetitively scans the back surface of the target, thereby generating a time-varying electrical signal proportional to the magnitude of the spatial charge distribution. Several vidicon-tube types forgo an image section, using a photoconductive target to perform the transducing function.

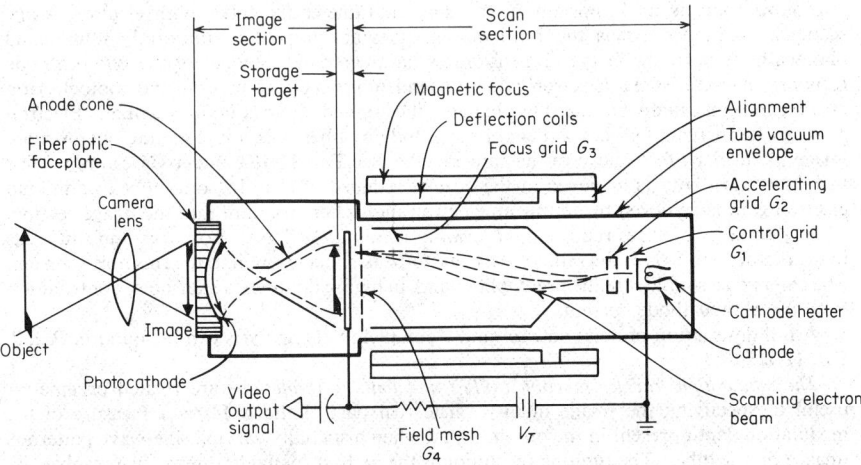

Fig. 11-45. Television camera tube schematic.

108. Performance. The major tube performance parameters are *sensitivity, modulation transfer function* (MTF), *limiting resolution, dark current,* and *lag.* Excluding dark current and lag, performance parameters are determined by limitations and conditions existing in each of the three major subassemblies. Dark current is exclusively determined by the electrical properties of the storage target; lag, or incomplete erasure, is dependent upon the charge transport characteristics and capacitance of the target and the electron velocity distribution in the scanning beam.

The sensitivity of a camera tube is the ratio of output signal current to uniform tube faceplate illumination. Some tubes, such as the antimony trisulfide vidicon, do not have a unique sensitivity for all illuminations; they are nonlinear or subunity gamma tubes. *Gamma* is the slope of a log-log plot of the light transfer characteristic, as shown in Fig. 11-46. Sensitivity is dependent on the spectral energy distribution of the illumination; therefore, the nature of the illumination must always be stated as part of the sensitivity parameter.

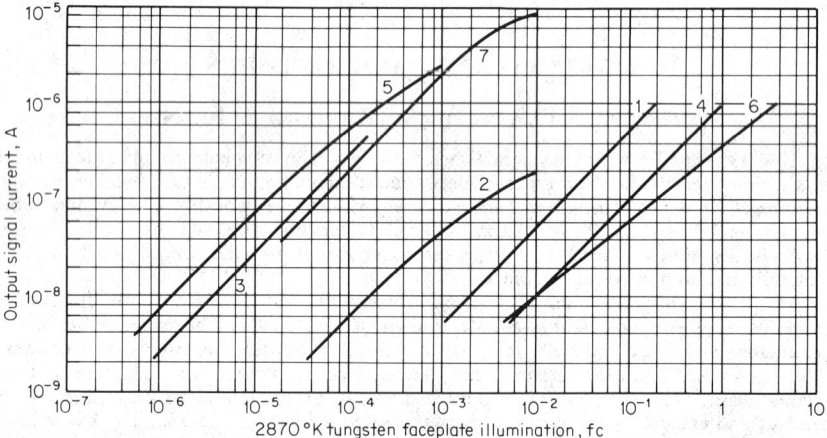

Fig. 11-46. Light transfer characteristics: (1) Tivicon silicon vidicon; (trademark of Texas Instruments Inc.); (2) WL-30893 SEC vidicon; (3) C21130 silicon intensifier vidicon; (4) one-inch-diameter plumbicon (trademark of N. V. Phillips, Holland); (5) 7967 image orthicon; (6) 8507 vidicon; (7) C21095C image isocon.

Electron gain is an important performance parameter for tubes with electron image sections. It is defined as the ratio of output signal current to uniformly illuminated photocathode current. It is a dimensionless parameter independent of the photocathode response characteristics. Electron gain is dependent upon target material and photoelectron energy, and can usually be varied by changing the applied photocathode potential. Electron gain can be increased several orders of magnitude by fiber optically coupling one or more image intensifier tubes ahead of the camera tube (see Par. 11-104, under Cascading). The increased sensitivity is achieved at the cost of reduced MTF. In some tubes, additional electron gain is achieved by adding an electron multiplier structure into the image section.

High-quality pictures require high *signal-to-noise ratios* (S/N). For large uniform areas in the picture, the S/N is the ratio of the peak-to-peak signal current to the rms noise current. The chief noise sources are the preamplifier and, in high-gain tubes, a shot-noise contribution from the photocathode current.

A full description of noise sources and a complete S/N analysis can be found in Ref. 2, Par. **11-115a.**

The modulation transfer function (MTF) and limiting resolution[2] are related parameters useful in specifying the image quality. The real part of the MTF is a measure of the modulation depth present in the output signal when a spatially varying sine-wave pattern is imaged on the tube. The limiting resolution is the highest spatial frequency just resolved by the camera tube-monitor-eye combination. In practice, *amplitude-response* data are commonly quoted for square-wave patterns (black-and-white bar pattern), as in Fig. 11-47. If desired, these data may be converted to sine-wave response via a well-known transformation.[1]* It is common television practice to state the spatial frequency in TV lines/raster height. The MTF of several cascaded elements is the product of the MTF of each individual element. As a result, the MTF of a multielement tube can be discussed in terms of the MTF of each element. The limiting resolution is commonly used to evaluate the image capabilities at reduced light levels; see Fig. 11-48.

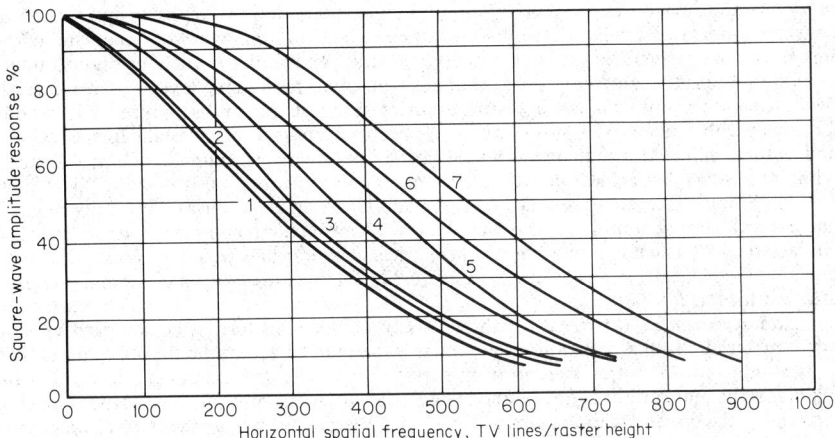

Fig. 11-47. Square-wave amplitude response characteristics. (See Fig. **11-46** for tube identification.)

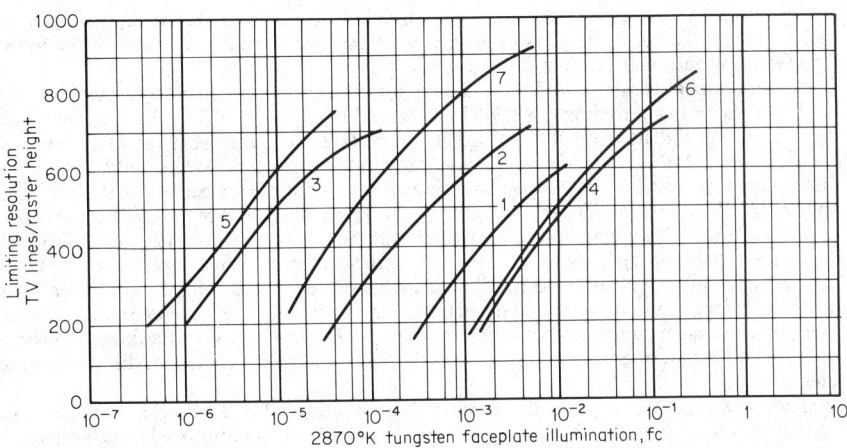

Fig. 11-48. Limiting resolution as a function of faceplate illumination. (See Fig. **11-46** for tube identification.)

Dark current is an apparent signal current resulting from target leakage currents between successive scans by the electron beam. In the dark, the rear surface of an ideal vidicon-type target is clamped to near cathode potential; the front surface is maintained at some fixed positive potential by an applied voltage. If target current flows between successive scans, beam current is subsequently deposited on the rear surface, creating an apparent signal current (dark current).

---

*Superior numbers correspond to number references, Par. 11-115a.

11-59

Lag arises from the failure of the beam to return the rear surface of a charged target to cathode potential after each scan. In viewing a dynamic scene, objectionable image smear will be noted.

109. Image Sections. Most camera tubes employ an image section identical with the direct-view image intensifiers discussed in Par. **11-102**, except that the photoelectrons are imaged on the charge storage target instead of a light-emitting phosphor. Image sections may be either magnetically focused, as generally employed for image orthicons, or electrostatically focused, as generally employed for silicon intensifier vidicons and SEC vidicons. Image orthicons and image isocons employ an additional mesh element in the image section to collect secondary electrons emitted by the target. The light input window (faceplate) of the image section or vidicon is used to restrict the spectral bandwidth to which the photoconductive layer is sensitive and may serve to support substrates for sensing layers. The materials being used are mainly glasses, high-silica glasses, fused quartz, and sapphire. Fiber-optic faceplates with concave back surfaces serve as faceplates for most electrostatic image sections and as the faceplate for vidicons to which image intensifiers are coupled. Optical systems with very short back focal length may also require the use of fiber-optic faceplates.

110. Storage Target Materials. Target-material parameters must be carefully selected and controlled to obtain high image quality, low blemish count, and reproducible electrical characteristics. The target must store an electrical charge image for at least $\frac{1}{30}$ s with small lateral charge leakage to ensure high image quality. Target materials most commonly used are described as follows.

Antimony trisulfide (Sb_2S_3), a high-resistivity photoconductor, is the standard vidicon target material. In the dark, a fixed potential is maintained across the thin-film target by a voltage applied to a transparent front-surface electrode; the scanning electron beam maintains the rear surface at near cathode potential. Light from the imaged scene is focused directly on the target, creating electron-hole pairs, thereby increasing the bulk conductivity. Charge is then free to migrate through the target between successive scans, allowing the rear surface to approach some positive voltage whose magnitude is dependent upon the leakage current.

At the next beam scan, the original rear-surface potential is reestablished by deposited beam current. The deposited charge generates a video signal via capacitive coupling to the target front surface and external circuitry.

Under normal operating conditions, the photon conversion efficiency (light sensitivity) of the Sb_2S_3 layer is approximately 20% (5500 Å) and nonlinear.[3,9] The efficiency and dark current are dependent upon the applied target voltage, both increasing at high voltages.

Lead oxide (PbO) is a photoconductive target,[6,10] in many respects similar to Sb_2S_3, discussed above. The target consists of a three-layered structure: a SnO_2 signal plate layer and two PbO layers of different conductivity. A nearly intrinsic PbO layer is sandwiched between the SnO_2 layer and a *p*-type polycrystalline PbO layer. The surface of the intrinsic layer nearest the SnO_2 is *n*-type, so that the structure forms a *pin* layer. The SnO_2 is maintained at a fixed positive potential, and the target rear surface driven to near cathode potential by the scanning beam. The structure then acts like a reverse-biased diode with small reverse current and high collection efficiency. The image is stored as positive charge (holes) in the *p*-type crystals, which are about 1 $\mu m^2 \times 0.1$ μm long, acting as discrete storage sites.

The PbO target circumvents three major shortcomings of the Sb_2S_3 vidicon: high dark current, low sensitivity, and high lag. The low dark current and improved sensitivity are a result of the *pin* layer structure, since reverse leakage currents are low and the intrinsic layer provides a strong field region for efficient carrier collection. Low lag is due to the absence of photoconductive lag and the low target capacitance. The current transfer characteristics are nearly linear, with a gamma between 0.8 and 0.9. The light sensitivity is target-voltage-dependent up to saturation near 20 V.

The silicon diode array target is a promising recent development[4,5] (see Fig. 11-49). The diodes are back-biased by an applied positive voltage and the scanning action of the electron beam. Incident photons or electrons generate electron-hole pairs which diffuse to the diode depletion region. The holes are swept into the *p* region, discharging the diode and reducing the back bias. On a subsequent scan, the electron beam returns the diode side to cathode potential, generating a video signal.

Potassium chloride (KCl) targets are used in secondary-electron conductivity (SEC) tubes. A highly porous KCl layer is deposited on a thin aluminum conducting plate. When an electron image from a photoemissive surface is focused on the target, the electrons penetrate the conducting coating, creating secondary electrons within the KCl layer. The secondary electrons are drawn through the porous structure by an applied electric field, creating a stored charge analog of the optical image. Reading by the scanning electron beam is identical with that discussed previously.

Thin-film target materials include ionic and electronic conducting glasses and MgO. The targets are commonly used in the image orthicon and image isocon. Although differing in detail, operation of the three target types is similar. A scanning electron beam drives the thin-film target to near cathode potential. An electron image from the photocathode is focused on the target. The energy of the incident electrons is selected to provide a net secondary-electron-emission ratio greater than unity. A closely spaced mesh, located on the target write side, collects the secondary electrons emitted by the target.

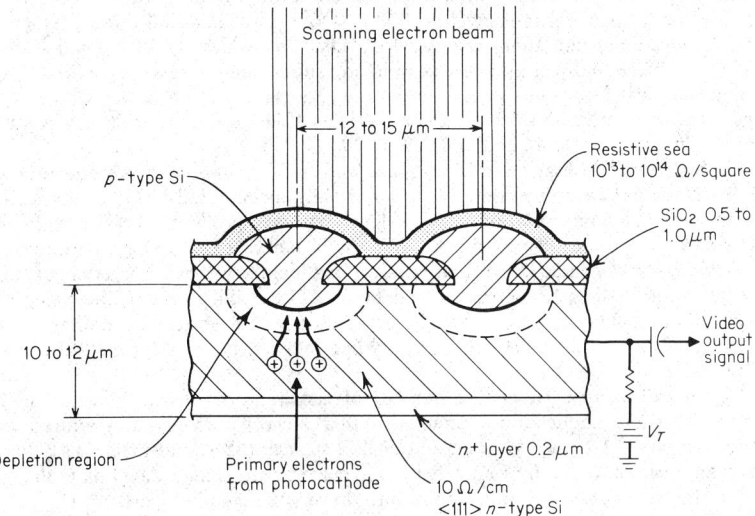

Fig. 11-49. Silicon diode array target.

The target becomes positively charged in proportion to the light intensity, the maximum excursion being determined by the mesh potential. The positive-potential pattern is then erased from the opposite reading side of the target by the scanning beam, and the video signal appears as modulation in the scattered and reflected components of the electron beam. A multiplier structure provides an electron gain of about 500 to the return beam prior to coupling to an external circuit.

111. Scanning Sections. The function of the scanning section is to generate, focus, and deflect an electron beam. The electron gun is similar in design and operation to the CRT gun described in Par. 11-74. Magnetic, electrostatic, or combined magnetic-electrostatic fields are used in focusing and deflection. All-magnetic tubes generally have the highest center and corner resolution but require more electrical power. Magnetic-focus electrostatic deflection tubes have resolution capabilities approaching, and sometimes exceeding, all-magnetic tubes, but require significantly less overall tube length. Power requirements are comparable. All-electrostatic tubes require the least electrical power, but also have reduced resolution capabilities. Electrostatic-focus magnetic deflection tubes have intermediate power requirements and provide intermediate resolution. Thus cameras designed for minimum size and power frequently employ all-electrostatic tubes. Cameras designed for maximum performance employ all-magnetic or magnetic-focus electrostatic deflection tubes.

Return-beam camera tubes such as the image orthicon and image isocon generate an output signal via modulation of the return beam from the storage target. The return electron beam is composed of scattered and reflected components, both of which are influenced by the electrostatic field and the magnetic-focusing field. The image orthicon collects and amplifies both components via an electron multiplier structure. The image isocon employs special steering plates to separate the scattered and reflected components. Only the scattered component enters the multiplier and is used for signal generation. Since the scattered component is a minimum in the dark areas of the scene, the effects of beam shot noise are reduced. Return-beam tubes are generally large high-performance devices.

112. Broadcast Service Tubes. Television camera tubes suitable for broadcast service must have good sensitivity, high reliability, extremely low blemish count, low lag, and high image quality. For color applications using three-tube cameras, careful matching of the spectral response of the individual tubes is required. The introduction of lead oxide vidicons in the 1960s resulted in the gradual replacement of image orthicons, until now approximately 90% of live color pickup cameras employ the lead oxide vidicon. Conventional Sb_2S_3 vidicons are also useful in broadcast service, particularly for film pickup applications. A representative listing of broadcast television camera tubes is given in Table 11-18. A limited amount of performance data is given in Figs. 11-46 to 11-48 and in Table 11-19.

113. Educational and Industrial Service Tubes (see Table 11-18). Single-tube color cameras and black-and-white cameras employing simple and less expensive Sb_2S_3 vidicons account for nearly all applications.

Performance parameters for a number of educational and industrial service tubes are given in Figs. 11-46 to 11-48 and in Table 11-19.

114. Space and Military Service. Space and military applications frequently require tube operation under extreme conditions not normally encountered in studio or closed-circuit applications. Wide temperature and light-level extremes and high vibration and corrosive environments are frequently specified. In many instances, extraordinary performance and compact size are demanded. Typical applications include airborne target acquisition and surveillance, ground-based perimeter and battlefield surveillance, satellite mapping, and tactical missile guidance. A listing of tube types useful in space and military service is provided in Table 11-18. Some tube performance data are shown in Figs. 11-46 to 11-48 and in Table 11-19.

115. Special-Service Tubes. A number of camera tubes, listed in Table 11-18, are designed for special applications. Included in this category are x-ray- and infrared-sensitive devices (see Par. **11-102**), slow-scan and charge storage tubes, and image dissector tubes. Slow-scan tubes are useful for remotely located cameras requiring a data link to the monitor wherein the bandwidth reduction significantly lowers the data-link cost.

Table 11-18.   Television Camera Tubes*

	Focusing and deflection†	Comments‡
	Broadcast service tubes	
Image orthicons:		
7629A	M, M	MgO target; good sensitivity; 3-in. tube
27888	M, M	Electronic glass target; field mesh; 3-in. tube
8685	M, M	Portable broadcast; electronic glass target; 2-in. tube
8092A	M, M	Low-light-level color or black-and-white; semiconductor target; 3-in. tube
Lead oxide vidicons:		
XQ1020	M, M	Black-and-white and color pickup; 30-mm target; separate mesh
XQ1070	M, M	1-in. version of XQ1020
4592	M, M	Color pickup; 30-mm target; separate mesh
TD1391	M, M	Color pickup; integral mesh
Vidicons (Sb_2S_3):		
7038, 7038V	M, M	Color and black-and-white cameras; 1-in. gun
8572	M, M	Separate-mesh version of 7038; 1-in. gun
8480	E, M	Film pickup; high resolution; 1½-in. gun

Table 11-18.—*Continued*

Educational and industrial service tubes

Vidicons (Sb$_2$S$_3$):		
7735	M, M	Low price; integral mesh; 1-in. gun
4427	M, M	Small size; $\frac{1}{2}$-in. gun
7325	M, M	High sensitivity; 1-in. gun
Z7873	M, E	High resolution; compact design; 1-in. gun
Z7940	M, E	High resolution; $1\frac{1}{2}$-in. gun
TD-1343-001	E, E	Ruggedized, low power; 1-in. gun
Undesignated	E, E	Wide variety of photoconductors; low power
Silicon diode array vidicons:		
Tivicon	M, E	16-mm target; 1-in. gun
Z7927	M, M	16-mm target; 1-in. gun
4532	M, M	16-mm target; 1-in. gun
S10X0	M, M	16-mm target; 1-in. gun

Space and military-service tubes

Silicon intensifier vidicons:		
C21130	M, M	25-mm photocathode; 16-mm target; fiber-optic input
C21125A	M, M	16-mm photocathode; 16-mm target; fiber-optic input
C21117C	M, M	40-mm photocathode; 16-mm target; fiber-optic input
WX31793	M, M	25-mm photocathode; 20-mm target; fiber-optic input
S50XQ	M, M	25-mm photocathode; 20-mm target; fiber-optic input
SEC vidicons:		
WL-30893	M, M	25-mm photocathode; 25-mm target; fiber-optic input
WL-30654	M, M	40-mm photocathode; 25-mm target; fiber-optic input
WL-30691	M, M	25-mm photocathode; 18-mm target; fiber-optic input
Image orthicons:		
Z5395	M, M	S-1 photocathode; MgO target; 3-in. tube
Z7882	M, M	S-20 photocathode; antihalation; fiber-optic input; 3-in. tube
Z7987	M, M	S-25 photocathode; MgO target; 3-in. tube
7967	M, M	S-20 photocathode; 3-in. tube; high sensitivity
Image isocons:		
C21095C	M, M	40-mm photocathode; 3-in. gun; fiber-optic input; easy setup
C21093A	M, M	40-mm photocathode; 3-in. gun
Vidicons (Sb$_2$S$_3$):		
4514	E, E	Ruggedized; government end use only; 1-in. gun
Z7917	M, E	Extremely short; high resolution; 1-in. gun
FPS-V	M, E	Ultrahigh resolution; 3-in. gun

Special-service tubes

Storage tubes (optical input):		
WX-5123	M, M	30 min storage; continuous readout; also useful for slow-scan applications
TD-1362	M, M	Up to 15 min storage; continuous nondestructive readout
Slow-scan vidicons:		
4542	M, M	Weather radar and signal storage application
TD-1342-010	M, M	Highly ruggedized; 1-in. gun; space-qualified
TD-1343-001	E, E	Highly ruggedized; 1-in. gun
X-ray-sensitive tubes:		
WL-5157	M, M	X-ray-sensitive vidicon
TD-1306-011	M, M	X-ray-sensitive vidicon
8541X	M, M	X-ray-sensitive vidicon
Image dissectors:		
WL-23111	. . .	High resolution; 3-in. tube; variable aperture sizes available from several manufacturers
Miscellaneous:		
C74137A	M, M	Return-beam vidicon, ultrahigh resolution
WL-5140	M, M	High resolution; 2-in. vidicon

* This table is intended to be a guide to the range of available tube types; it should not be used to select an optimum tube for a specific application.

† The letters M and E denote magnetic and electrostatic. The first letter is the focusing field, the second the deflection field.

‡ Additional information may be solicited from the manufacturer.

Table 11-19. Representative Camera Tube Performance

Tube type	Lag, %*	Dark current, A†	Sensitivity or electron gain	Gamma
3-in. image orthocon	7	0‡	1,500	1.0
3-in. image isocon	7	0	1,500	1.0
1-in. Sb_2S_3 vidicon	20	20×10^{-9}	250 $\mu A/lm$§	0.65
1-in. PbO vidicon	4	3×10^{-9}	400 $\mu A/lm$	0.95
1-in. silicon vidicon	12	8×10^{-9}	3,000 $\mu A/lm$	1.0
25-mm SEC vidicon	5	0	100	1.0
40-mm silicon intensifier vidicon	12	8×10^{-9}	2,000	1.0

　* Lag, or residual signal, is measured as a percentage of the original signal, three fields after exposure. Measured under typical operating conditions.
　† Measured under typical operating conditions, 25°C.
　‡ The image orthocon signal is inverted with maximum current in blacks, minimum in highlights, but after signal inversion the dark current can be assumed to be zero if the rms shot-noise current in the reading beam is not considered to be dark current.
　§ Measured at 0.1-footcandle (fc) illumination.

115a. References on Television Camera Tubes

　1. COLTMAN, J. W.　*J. Opt. Soc. Amer.,* 44:468 (1954).
　2. BIBERMAN, L. M., and NUDELMAN, S. (eds.)　"Photoelectronic Imaging Devices," Vols. 1 and 2, Plenum, New York, 1971.
　3. WEIMER, P. K., FORGUE, S. V., and GOODRICH, R. R.　*RCA Rev.,* 3:306 (1951).
　4. CROWELL, M. H., and LABUDA, E. F.　*Bell System Tech. J.,* 48:1481 (May–June, 1969).
　5. GORDON, E. I., and CROWELL, M. H.　*Bell System Tech. J.,* 47:1855 (November, 1968).
　6. DE HAAN, E. F., VAN DER DRIFT, A., and SCHAMPERS, P. P. M.　*Phillips Tech. Rev.,* 25:133 (1963/1964).
　7. KAZAN, B., and KNOLL, M.　"Electronic Image Storage," Academic, New York, 1968.
　8. ZWORYKIN, V. K., and MORTON, G. A.　"Television," 2d ed., Wiley, New York, 1954.
　9. WEIMER, P. K., FORGUE, J. V., and GOODRICH, R. R.　The Vidicon Photoconductive Camera Tube, *RCA Rev.,* 12(1):306–313 (1951).
　10. DE HAAN, E. F., VAN DER DRIFT, A., and SCHAMPERS, P. P. M.　The Plumbicon: A New Television Camera Tube, *Phillips Tech. Rev.,* 25(6,7):133–155 (1963–1964).

PHOTOCONDUCTIVE AND SEMICONDUCTOR JUNCTION DETECTORS

BY WILLIAM A. GUTIERREZ

116. Photoconductors and junction devices constitute an important class of solid-state photodetectors for the 0.2- to 2-μm spectral region. These detectors convert electromagnetic energy directly into electrical energy via the photoconductivity effect that occurs in semiconductors.

117. Figures of merit[1,2*] may be found in Par. 11-130.

PHOTOCONDUCTORS

118. Operation. The simplest photoconductor detector is a bar of relatively low conductivity n- or p-type semiconductor (in bulk or thin-film form) with ohmic contacts at its ends (Fig. 11-50). The photoconductor varies its electrical resistance in accordance with the light wavelength and intensity it receives. Its operation depends on the photoconductivity[3,4] which occurs in semiconducting materials. Electrons in bound states in the valence band (intrinsic) or in forbidden-gap levels (extrinsic) absorb the energy of the incident photons and are excited into the free states in the conduction band, where they remain for a characteristic lifetime. Electrical conduction may take place either by the electrons in the conduction band or by the positive holes vacated in the valence band. The electrical resistance of the material thus decreases on illumination, and this resistance change can be translated into a change in the current that flows through the output circuit.

---

　* Superior numbers correspond to numbered references, Par. 11-126a.

119. The performance of photoconductor detectors is measured not only in terms of D^* (Par. 11-130), but also in terms of photoconductivity gain, response time, dark current, spectral response, and temperature coefficient. For a photoconductor in which the conductivity is dominated by one carrier (either holes or electrons) the gain is given by the ratio of free-carrier lifetime to the transit time of this carrier. It can also be expressed as

$$\text{Gain} = \frac{\tau \mu V}{L^2}$$

where τ is the free-carrier lifetime, μ is the mobility, V is the applied voltage, and L is the spacing between ohmic contacts, as shown in Fig. 11-50. The maximum D^* and spectral dependence of commercially available CdS and CdSe photodetectors are shown in Fig. 11-51. Table 11-20 lists typical gains, response times, and dark current.

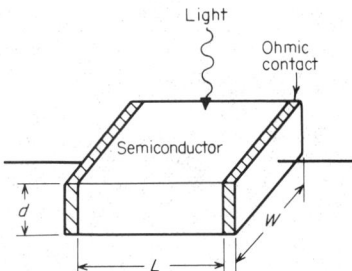

Fig. 11-50. Diagram of a photoconductor detector.

Table 11-20. Parameters of Various Photoconductive Detectors

Photodetector	Gain	Response time, s	Dark current
Photoconductor	10^5	10^{-3}	1-10 mA
pn junction	1	10^{-11}	1-10 μA
Metal-semiconductor...........................	1	10^{-11}	
Avalanche diode...............................	10^4	10^{-10}	
Point contact.................................	1	...	1-3 mA
Heterojunction photodiode	1	...	High
Phototransistor	10^2	10^{-8}	1 nA
Photofet	10^2	10^{-7}	1 μA

120. Properties of Specific Photoconductors. The long-wavelength threshold for photoconductivity is usually determined by the band gap of the material according to the relationship

$$\lambda_c = \frac{1.24}{Eg}$$

where λc is the threshold wavelength in micrometers (μm), and Eg is the band gap in electron volts (eV).

Band-gap values of materials commonly used as photodetectors in the 0.2- to 2-μm region are given in Table 11-21. The normalized spectral response of some of them is shown in Fig. 11-52.

Table 11-21. Band-Gap Values for Photoconductor Materials

Material	Threshold λ_c, μm	Band gap, eV
CdS	0.52	2.4
CdSe	0.73	1.7
ZnS	0.33	3.7
ZnSe	0.48	2.6
GaAs..	0.89	1.4
InP	1.03	1.2
Ge	1.77	0.7
Si	1.13	1.1

121. Semiconductor junction detectors (photodiodes and phototransistors) differ from photoconductive detectors in that their operation depends essentially on a reverse-biased diode whose leakage current is varied by electron-hole pairs generated near or at the depletion region by light absorption. Their response time is characteristically short.[5-7]

11-65

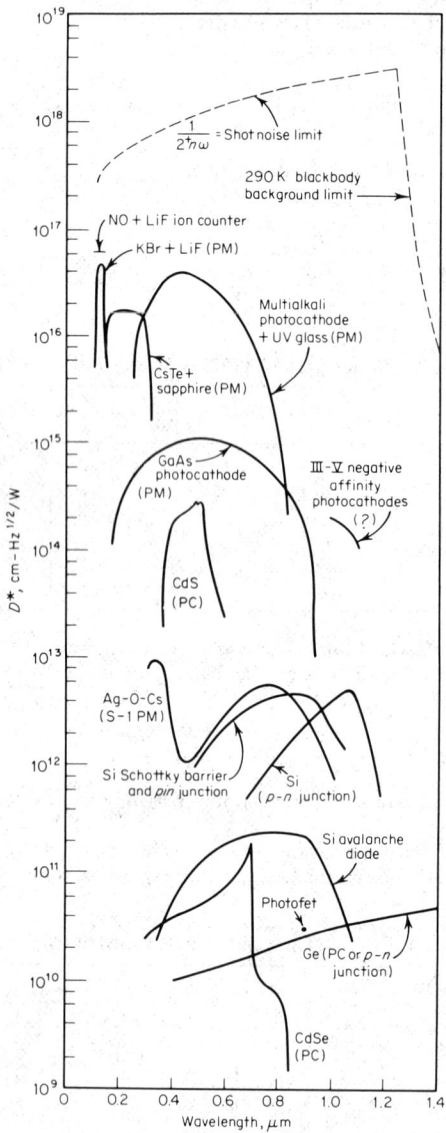

Fig. 11-51. Detectivity vs. wavelength for various photodetectors. PC indicates photoconductive mode. PM indicates photomultiplier mode.

122. Basic Classes of Photodiode Detectors. Photodiodes fall into two general categories, the *depletion-layer type* and the *avalanche type*. The distinguishing feature between them is the existence of a gain mechanism.

a. The depletion-layer photodiode[8] family includes the *pn* junction diode, the *pin* diode, the Schottky barrier (metal-semiconductor) diodes, the point contact diode, and the heterojunction diode.

1. *pn junction diode.* Figure 11-53 is a diagram of a *pn* junction diode. The junction is reverse-biased,[9] and the diode is illuminated either at the *n* or *p* region, away from the depletion region (Fig. 11-53*a*), or it is illuminated right at the depletion region (Fig. 11-53*b*). Their built-in field enables them to be operated in the photovoltaic mode (i.e., no externally applied bias); however, the photoconductive mode, with a fairly large reverse bias, is usually the more common mode of operation.

2. *pin photodiode.* Figure 11-54 shows a cross-sectional diagram of a typical *pin* photodiode.[10] The sensitivity range and frequency response of this type of diode depend principally on the thickness of the intrinsic layer (which defines the depletion layer). Light passes through the *p* region before it arrives at the

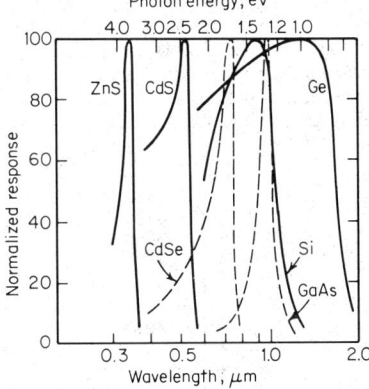

Fig. 11-52. Normalized spectral response of some photoconductors.

depletion region, where it excites hole-electron pairs that are very quickly swept out by the large electric field present.

3. *Metal-Semiconductor photodiode.* Figure 11-55 is a cross-sectional diagram of a metal-semiconductor (Schottky barrier) photodiode.[11] In this case, light passes through a thin (≈ 100 Å) metal film with a suitable antireflection coating to minimize large absorption and

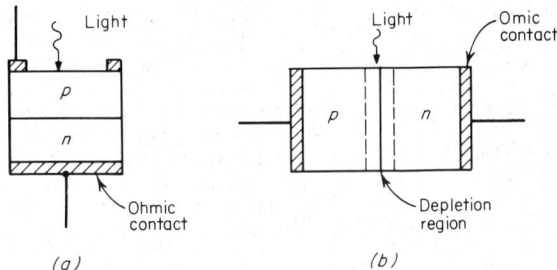

Fig. 11-53. Diagram of *pn* junction photodiode. It is illuminated (*a*) away from depletion region or (*b*) at depletion region.

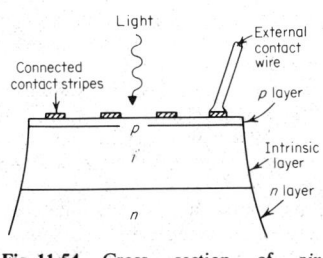

Fig. 11-54. Cross section of *pin* photodiode.

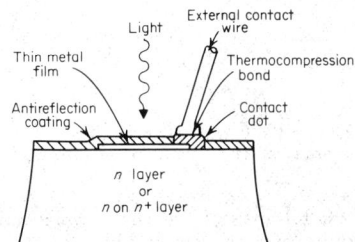

Fig. 11-55. Cross section of metal-semiconductor photodiode.

reflection losses. As with the *pn* and *pin* diodes, the photogenerated electron-hole pairs in the semiconductor give rise to an output signal current.

4. *Point contact photodiode.* Figure 11-56 shows a diagram of a point contact detector.[12] Light is incident onto the Schottky barrier through an etched cavity in the semiconductor. This detector is extremely fast because of small dimensions and low capacitance.

5. *Heterojunction photodiode.* A depletion-layer photodiode can be constructed by forming a junction between two semiconductors of different band gaps. Figure 11-57 shows a photodiode[13] made up of *n*⁻ GaAs and *p*⁻ Ge. Light is absorbed almost completely in the low-band-gap material. Large dark currents could arise due to spontaneous electron-hole generation in the depletion region from a large density of interface states. This could have deleterious effects on the signal-to-noise ratio at low light levels.

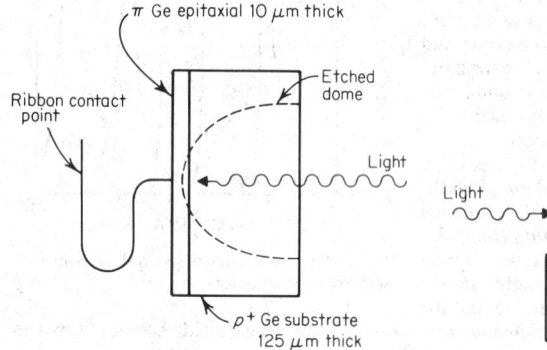

Fig. 11-56. Diagram of point contact photodiode.   **Fig. 11-57.** Diagram of heterojunction photodiode with applied reverse bias.

b. Avalanche Photodiodes. Depletion-layer photodiodes, when operated at higher reverse-biased voltages, give an increased output signal. This is due to internal carrier multiplication via the avalanche effect. If the field in the depletion region can impart an energy equal to or greater than the band-gap energy to an electron, this electron can create another hole-electron pair by collision, and this pair can be accelerated to create an additional pair, and so on. This gives rise to carrier multiplication and to internal gain in the photodiode. The avalanche photodiode is therefore the counterpart of the photomultiplier tube, and its multiplication factor M is given by

$$M = K(1 - V/V_B)^{-1}$$

where K is a constant, and V_B is the breakdown voltage.

Figure 11-58 shows two types of avalanche photodiodes[14-16] with guard rings. The guard ring prevents a high field breakdown region from reaching the surface.

123. Performance of Photodiodes. The spectral dependence of D^*, gain, speed of response, and dark current of the various photodiodes is shown in Table 11-20 and Fig. 11-51.

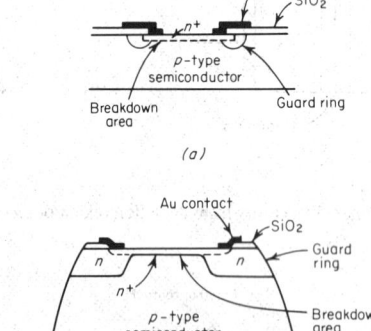

Fig. 11-58. Cross section of avalanche photodiode with guard ring. (*a*) Planar type; (*b*) mesa type.

124. Phototransistor. A *pnp* or *npn* junction transistor can act as a photodetector, with the possibility of large internal gain.[17,18] A *npn* structure, for example, is usually operated as a two-terminal device with the base floating and the collector positively biased.

Phototransistors are generally fabricated of Ge or Si in the same manner as conventional transistors, except that a lens or window is provided in the transistor to admit light at the base or base-collector junction. Response time of 10^{-8} and peak sensitivities of 30 A/lm are possible. Gains of several hundred have been attained with dark currents as low as nanoamperes.

125. Photofet, SMS Photodetector, and *pnpn* **Device.** The *photofet*, or *photosensitive field-effect transistor*, combines a photodiode and high-impedance amplifier in one device to achieve photodetection with large gain.[19]

A *semiconductor-metal-semiconductor (SMS) device* is essentially a Schottky barrier device with gain.[20]

A silicon controlled rectifier *(SCR),*[21] or *pnpn device* can be used as a photosensitive switch, with photogenerated current taking the place of the usual gate current.

126. D^* Comparison of Photoconductive Devices with Photoemissive Devices. Representative D^* values for various commercial types of radiation detectors in the 0.2- to 2-μm range are shown in Fig. 11-51. The D^* values given for the III-V negative-electron-affinity photoemitter represent an estimate, since these detectors are not yet fully developed. Because of rapid improvements being made in III-V photoemitters, the D^* values can be expected to increase significantly.

126a. References on Photoconductive and Semiconductor Devices

1. Ross, M. "Laser Receivers: Devices, Techniques, Systems," Wiley, New York, 1966.

2. Kruse, P. W., McGlauchlin, L. D., and McQuistan, R. B. "Elements of Infrared Technology," Wiley, New York, 1962.

3. Rose, A. "Concepts in Photoconductivity and Applied Problems," Interscience, New York, 1963.

4. Bube, R. H. "Photoconductivity of Solids," Wiley, New York, 1960.

5. Riesz, R. P. *Rev. Sci. Instrum.,* 33:994 (1962).

6. Gartner, W. W. *Phys. Rev.,* 116:84 (1959).

7. Schneider, M. V. *Bell System Tech. J.,* 45:1611 (1966).

8. Gartner, W. W. *Phys. Rev.,* 116:84 (1959).

9. Sawyer, D. E., and Rediker, R. H. *Proc. IRE,* 46:1122 (1958).

10. Reitz, R. P. *Rev. Sci. Instrum.,* 33:994 (1962).

11. Schneider, M. V. *Bell System Tech. J.,* 45:1611 (1966).

12. Sharpless, W. M. *Proc. IEEE,* 52:207 (1964).

13. Rediker, R. H., Quist, T. M., and Lax, B. *Proc. IEEE,* 51:218 (1963).

14. Anderson, L. K., McMullin, P. G., D'Asaro, L. A., and Goetzberger, A. *Appl. Phys. Letters,* 6:62 (1965).

15. Baertsch, R. D. *IEEE Trans. Electron Devices,* Ed-13:987 (1966).

16. Emmons, R. B., and Lucovsky, G. *IEEE Trans. Electron Devices,* Vol. Ed-13, 1966.

17. Hunter, L. P. (ed.) "Handbook of Semiconductor Electronics," 2d ed., pp. 4-16, 16-17, McGraw-Hill, New York, 1970.

18. Schuldt, S. B., and Kruse, P. W. *J. Appl. Phys.,* 39:5573 (1968).

19. Shipley, M. *Solid State Design,* 5:28 (1964).

20. Reynolds, J. H. *Trans. Metal. Soc. AIME,* 239:326 (1967).

21. Gentry, F. E., Gutzwiller, F. W., Holonyak, N. H., and Van Zastrow, E. E. "Semiconductor Controlled Rectifiers," Prentice Hall, Englewood Cliffs, N.J., 1964.

INFRARED DETECTORS AND ASSOCIATED CRYOGENICS

By James R. Predham and Stanley P. Rodak

127. Infrared detectors[1-4*] provide an electrical output which is a useful measure of the incident infrared radiation. It is usually necessary to cool these detectors to cryogenic temperatures to reduce the thermal noise inherent in an electrical transducer. Infrared detectors can be divided into two categories, *thermal detectors* and *quantum detectors*.

128. Thermal detectors are of two types: *(a)* the *bolometric detector* reacts to changes in temperature by a change in its electrical conductivity; *(b)* the *thermovoltaic detector* is a

---

*Superior numbers correspond to numbered references, Par. 11-138a.

junction of two dissimilar metals (a thermocouple). As the temperature of the junction changes, the voltage generated at the junction will change.

129. Quantum detectors are of four types:

The *photovoltaic detector* is a *pn* semiconductor junction. Fluctuations in incident photon flux cause variations in the voltage produced by this junction.

A *photoconductive detector* is a semiconductor in which the fluctuations in incident photons cause fluctuations in the number of free charge-carriers, thus causing a change in conductivity.

A *photoelectromagnetic detector* is a semiconductor in which photon-generated charge carriers created at the surface diffuse through the bulk and are separated by a magnetic field. Fluctuations in photon flux produce fluctuations in the voltage caused by this separation.

A *photoemissive detector* is one in which incident photons impart sufficient energy to surface electrons to free them from the detector surface. The topic of photoemissive surfaces is treated in Par. 11-98.

130. Detector Parameters. The parameters most often used in the description of infrared detectors are as follows:

Responsivity R is the ratio between the rms signal voltage (or current) and the rms incident signal power, referred to an infinite load impedance and to the terminals of the detector ($R = V_{s,\text{rms}}/P_{s,\text{rms}}$). Spectral responsivity R_λ refers to a monochromatic input signal, and blackbody responsivity R_{BB} refers to an input signal having a blackbody spectrum. The units of responsivity are volts per watt (or amperes per watt). Responsivity is a function of wavelength λ, signal frequency f, operating temperature T, and bias voltage V.

Noise equivalent power (NEP) is that value of incident rms signal power required to produce an rms signal-to-noise ratio of unity (NEP $= V_{n,\text{rms}}/R$). Spectral-noise equivalent power NEP$_\lambda$ refers to a monochromatic input signal, and blackbody-noise equivalent power NEP$_{BB}$ refers to an input signal having a blackbody spectrum. The unit of NEP is the watt. NEP is a function of wavelength λ, detector area A, signal frequency f, electrical bandwidth ΔF, temperature of blackbody T_{BB}, field of view Ω, and background temperature T_B.

D-star (D^*) (sometimes referred to as detectivity) is a normalization of the reciprocal of the noise equivalent power to take into account the area and electrical-bandwidth dependence ($D^* = A\Delta f/\text{NEP}$). Spectral D-star is written D^*_λ ($\lambda, f, \Delta f$), and blackbody D-star is written D^* ($T_{BB}, f, \Delta f$), where λ is the wavelength of interest, f is the signal frequency, Δf is the electrical bandwidth, and T_{BB} is the blackbody temperature. This figure of merit may be used to compare detectors of different types since, when the terms in the parentheses are identical, the detector with the greater value of D^* is a better detector under those conditions. An exception is those detectors limited by fluctuations in background photon flux. For such detectors, field of view and background temperature must be specified. The units of D^* are cm/Hz$^{1/2} \cdot$ W^{-1}.

Quantum efficiency QE, or η, is the ratio of countable output events to the number of incident photons.

Time constant τ is a measure of the speed of response of a detector. It is usually defined as $\tau = (2\pi f_c)^{-1}$, where f_c is that signal frequency at which the responsivity has fallen to 0.707 of its maximum value.

131. Background-limited infrared photodetectors (BLIP detectors) are detectors whose D^* is limited by the noise produced by random fluctuations in the arrival rate of background photons, and are said to be background-limited. This value of D^* is the theoretical limit of detectivity for a photon detector. Although this limit of detectivity has not been attained by existing detectors, several detectors are available which exhibit near BLIP performance.

Figure 11-59 is a plot of relative improvement in D^* versus angular field of view for a BLIP detector. Cooled spectral filters also improve the spectral D^* of a BLIP or near-BLIP detector by attenuating radiation at wavelengths which are not of interest (e.g., where atmospheric attenuation degrades the target signal to below background levels).

132. Noise Mechanisms in Photodetectors. In addition to background noise, already discussed, there are four other noise mechanisms.

Johnson Noise (Also called Nyquist, or thermal, noise). The random motion of charge carriers in a resistive element at thermal equilibrium generates a random electrical voltage across the element.

Generation-Recombination Noise (GR). Variations in the rate of generation and recombination of charge carriers in the detector create electrical noise.

Shot Noise. Since the electric charge is discrete, there is a noise current flowing through the detector which is a result of current pulses produced by individual charge carriers.

1/f Noise. The mechanism involved in this type of noise is not well understood. It is characterized by a $1/f^n$ noise power spectrum, where n varies from 0.8 to 2.

133. Detector Data. Tables 11-22 and 11-23 list commercially available detectors. They include the operating temperature, cutoff wavelength, and peak spectral D^*.

Figures 11-60 and 11-61 are plots of spectral D^* versus wavelength for selected detectors. Since detector noise is a function of operating temperature, D^* will also vary with temperature. Figure 11-62 is a plot of peak D^* versus temperature for selected detectors. All values tabulated are data from above-average single-element detectors.

134. Detector Selection. Though no absolute guidelines for the choice of an infrared detector for a specific application are possible, certain criteria may be used to eliminate many of the available detectors. Among these criteria are spectral region of interest, maximum signal frequency, required sensitivity, and available cooling.

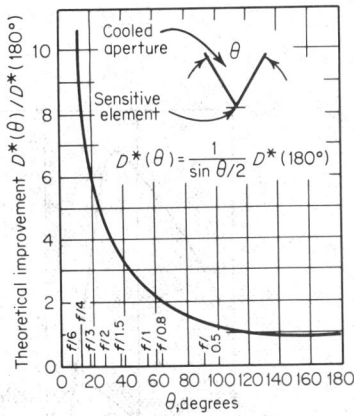

Fig. 11-59. Relative increase in D^* for BLIP detectors obtained by using cold shielding.

$$D^*(\theta) = \frac{1}{\sin \theta/2} D^*(180°)$$

Table 11-22.   Thermal Detectors

Type	Operating temp., °K	Detectivity D^*	Wavelength region, μm	Response time, ms	Resistance
TGS pyroelectric bolometer	300	$>6 \times 10^8$	>2	1	>10 MΩ
Thermistor bolometer .	300	2×10^8	1–40	1–10	2 MΩ
Semiconductor *pin* thermistor	300	3×10^9	1–30	30	10–100 Ω
Golay cell	300	3×10^9 (NEP)$^{-1}$	1–2,000	15	
Tin bolometer (superconducting) ..	<3.7	3×10^{11} (NEP)$^{-1}$	>10	1,000	100 Ω
Carbon bolometer	<2.1	3×10^{10} (NEP)$^{-1}$	>10	10	100 kΩ
Germanium bolometer	<2.1	3×10^{11} (NEP)$^{-1}$	>10	10	100 kΩ

135. Cryogenic Cooling. Temperatures of 200°K or lower are required for BLIP performance of intrinsic infrared detectors; extrinsic, or doped-type, photoconductive detectors require temperatures of 80°K or less (Table 11-23).

136. Cooling Specifications. The mission definition and type detector used will help define cryogenic requirements. Variables used for cooler specification include refrigeration load, cooling temperature, cool-down time, temperature stability, life, reliability, duty cycle, environment, weight, configuration, noise (acoustical, electromagnetic, mechanical), power (ac, dc, limits), and detector housing geometry. Table 11-24 shows a summary of cooling features and limitations.

137. Cooling Methods. *Leidenfrost Transfer.* The simplest cryogenic cooling method is two-phase liquid transfer known as Leidenfrost transfer. One liquid-transfer system[5] introduces stored liquid nitrogen (LN2) at 1.5 lb/in.^{2} gage into the supply line, which introduces the LN2 to a shock heat load, vaporizing a fraction of the LN2. Vaporized LN2 rapidly moves down the supply line, carrying small beads of LN2. Just enough heat is added via a variable thermal element to keep the transfer line frost-free and flexible at all times. At the delivery end of the line vaporized LN2 is bled off and the beads of LN2 are collected in

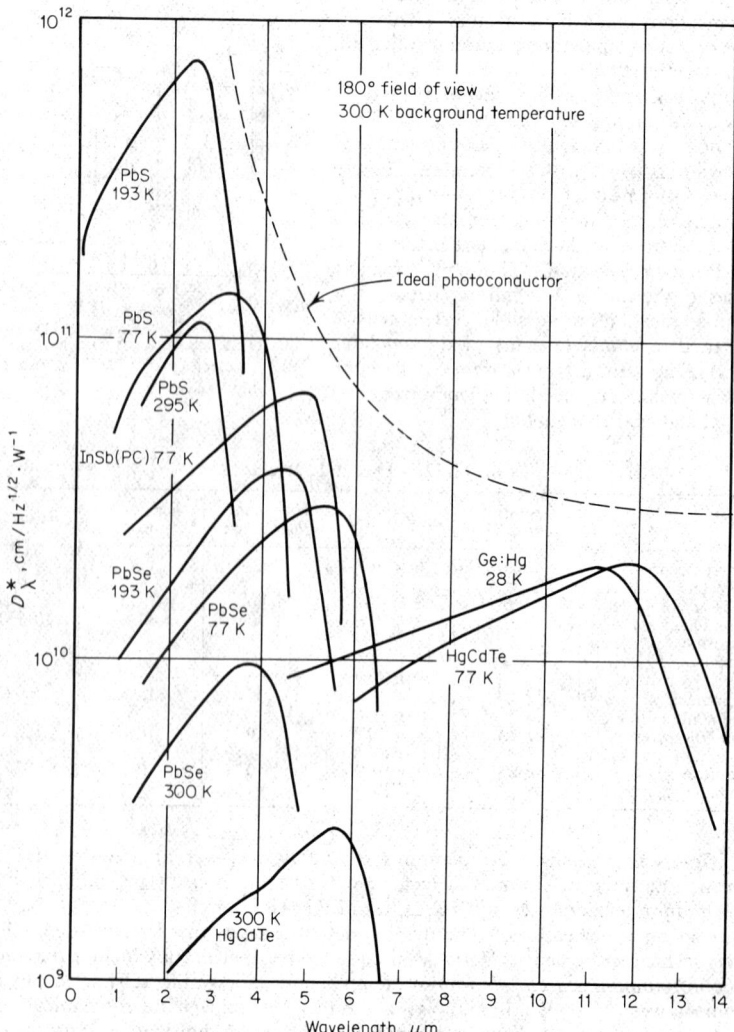

Fig. 11-60. Spectral D^* of photoconductive detectors.

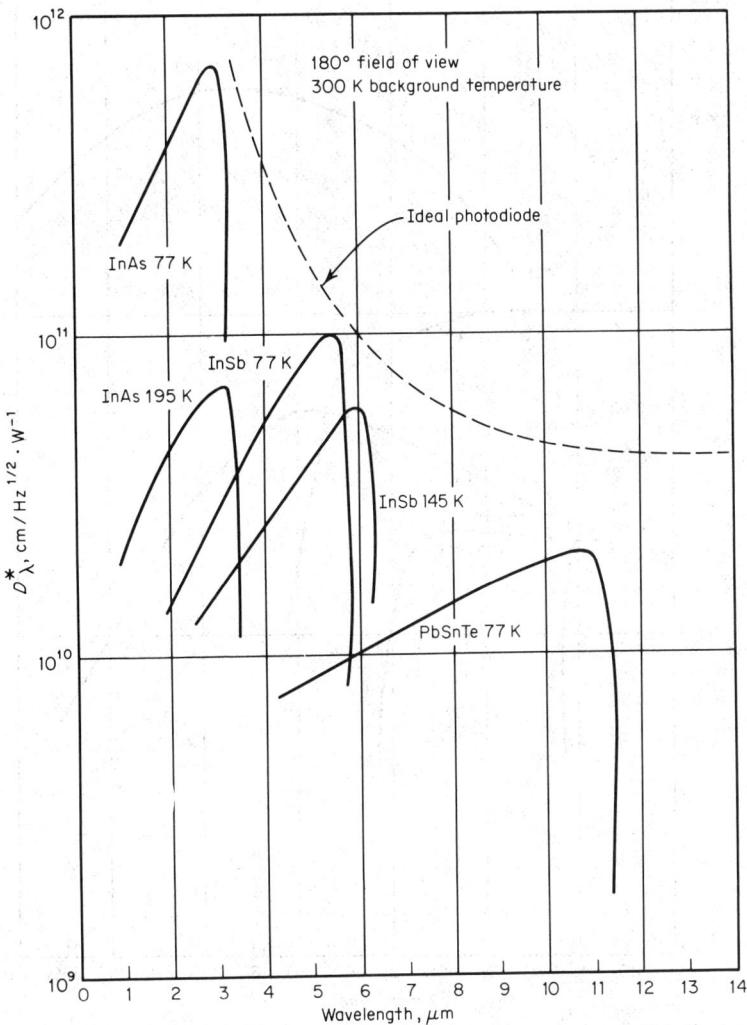

Fig. 11-61. Spectral D^* of photovoltaic detectors.

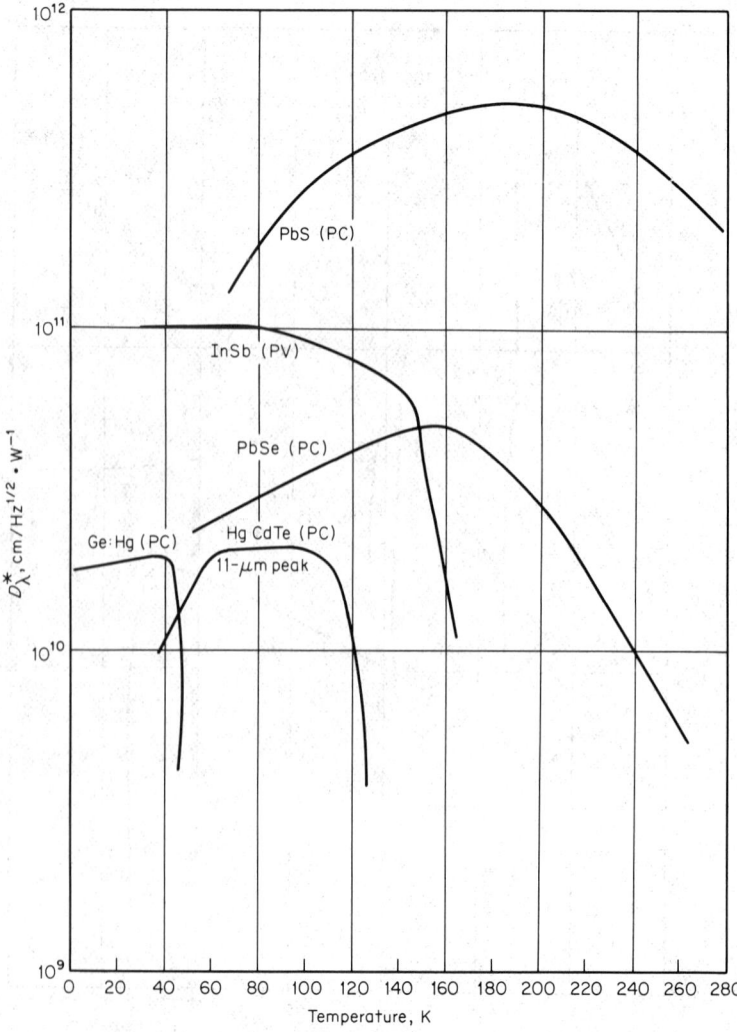

Fig. 11-62. Peak D^* versus temperature for selected detectors. PC = photoconductive mode; PV = photovoltaic mode.

Table 11-23. Quantum Detectors

Type	Operating temp., °K	Peak wavelength, μm	Detectivity at peak wavelength	Response time, μs	Resistance or impedance
Si-PV	300	.9	2.5×10^{12}	1,000	500 kΩ
Ge-PV	300	1.6	4×10^{11}	1,000	
PbS-PC	300	2.5	8×10^{10}	300	500 kΩ
PbS-PC	195	2.6	7×10^{11}	3,000	
PbSe-PC	300	3.5	1×10^{10}	2	500 kΩ
PbSe-PC	195	4.3	4×10^{10}	30	2 MΩ
PbSe-PC	77	5	3×10^{10}	100	>20 MΩ
InAs-PV	300	3.4	7×10^{9}	1	20 Ω
InAs-PV	195	3.2	7×10^{10}	1	50 Ω
InAs-PV	77	3.0	7×10^{11}	1	10 MΩ
PbTe-PV	77	5	6×10^{10}	1	
InSb-PV	77	5.4	1×10^{11}	1	>710 MΩ
InSb-PEM	300	6.2	3×10^{8}	<1	20 Ω
HgCdTe-PC	300	4.3	3×10^{9}	1	
HgCdTe-PC	77	11.5	2×10^{10}	1	40 Ω
PbSnTe-PV	77	11.5	2×10^{10}	<1	100–500 Ω
Ge:Au-PC	60	5	3×10^{10}	$\leq$1	1 MΩ
Ge:Hg-PC	27	12	2×10^{10}	$\leq$1	>1 MΩ
Ge:Cu-PC	15	25	3×10^{10}	0.1	>3 kΩ

PV = photovoltaic; PC = photoconductive; PEM = photoelectromagnetic.

a receiver dewar. A regulated cooling capacity up to 15 W with operating efficiency up to 70% has been proved for transfer lines up to 12 ft.

Joule-Thomson.[6,7] The Joule-Thomson (JT) refrigeration process exploits the cooling effect obtained by the adiabatic expansion of nonideal gases through an orifice under certain conditions. Figure 11-63 is a JT cryostat able to liquefy high-purity gas. A small temperature-sensitive, gas-filled, hermetically sealed bellows, fixed near the orifice, can be made to move a needle into and out of the orifice. This throttling of gas flow to meet heat-load demands and to maintain temperature reduces excessive liquid production. By judicious combinations of cascade staging, gas flow rate, and pressure above liquid, temperatures throughout the cryogenic span can be reached and maintained. Figure 11-64 shows the coefficient of performance (COP) for commercial closed-cycle systems.

138. Expansion-Engine Coolers. Closed-cycle, expansion-engine coolers described below often are used for infrared component refrigeration. Cold production by the refrigerator is based on expansion of a gas from a high to a low pressure, with consequent reduction of working-gas temperature. Note that a cooler cannot have a coefficient of performance (COP) greater than that for an ideal Carnot engine operating between the same absolute temperatures T_a and T_c; that is, COP $= Q_c/W = T_c/(T_a - T_c)$.

Stirling. This refrigeration cycle is well developed. Commercial models are able to provide up to 20 W usable cooling at 20°K. Cooling is obtained by cyclic out-of-phase motion of a compression piston and a displacer-regenerator (Fig. 11-65). The working gas is compressed while occupying the ambient space, temperature T_a, by an upward motion of compression piston, reducing gas volume. Heat of compression is rejected to ambient. COP for ideal working gas is equal to that of the Carnot engine. Stirling cycle refrigerators have best COP in practice (Fig. 11-64) and best ratio of total weight per watt refrigeration when compared with other refrigerators.[11]

Vuilleumier.[8] This refrigeration cycle is exploited for its long life and low vibration, due in part to inherently very low dynamic forces on moving parts. Coolers have been built that provide refrigeration at less than 20°K. Cooling is obtained by cyclic out-of-phase motion of two displacer-regenerators (Fig. 11-66). The working gas throughout the entire cooler is compressed by downward motion of the hot displacer-regenerator as it transfers part of the gas at the ambient end into the heated end. COP is equal to that for two Carnot heat engines in series: COP $= (T_c/T_h)(T_h - T_a)/(T_a - T_c)$.

Gifford-McMahon[9] *and Solvay.*[10] By separating the expander from the compressor, a refrigeration system can be constructed that consists of a simple lightweight cooling unit and a compressor which can be located remotely, connected to the expander with pressure lines.

Table 11-24. Detector Cooling Features and Limitations [14]

Cooler	Design feature	Design limitation
Direct-contact/ Leidenfrost	Simple in design; reliable, remote cooling; light in weight.	Operating time limited by liquid capacity, dewar heat load, and system loss.
Joule-Thomson	Remote cooling, light and small cooling head, mechanically and thermally quiet.	Requires high-pressure, high-purity gas for nonclogging; operating time limited by pressure-vessel capacity.
Stirling	Most efficient cooler; continuous-duty.	Not free of acoustic noise and mechanical vibration. Mean time between failure is on the order of 500 h.
Vuilleumier	Continuous-duty; low acoustic noise and mechanical vibration. Can use any high-temperature thermal energy to run engine (i.e., electrical, exothermic, nuclear, or solar).	Less efficient than Stirling. Mean time between failures is on the order of 1000 h.
Gifford-McMahon Solvay	Continuous-duty; remote, lightweight cooling head; long life (mean time between failures is on the order of 3000 h).	Compressor noise and mechanical vibration may be problem. Requires large amount of electric power.
Thermoelectric	Reliable; long life; small size; lightweight; silent in operation.	Very low cooling capacity. Cannot be exposed to temperatures greater than limit of eutectic solders used to join stages and couples (presently about 100°C).

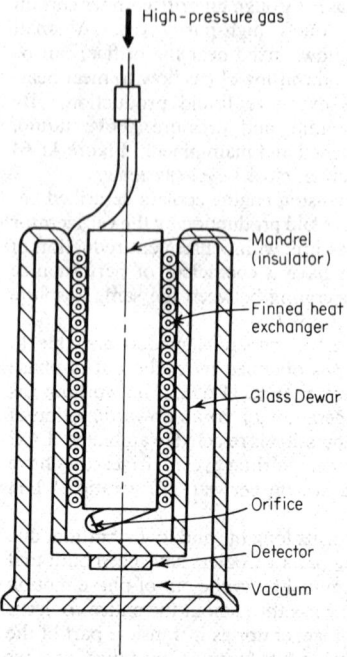

Fig. 11-63. Single-stage, open-cycle Joule-Thomson cooling system.

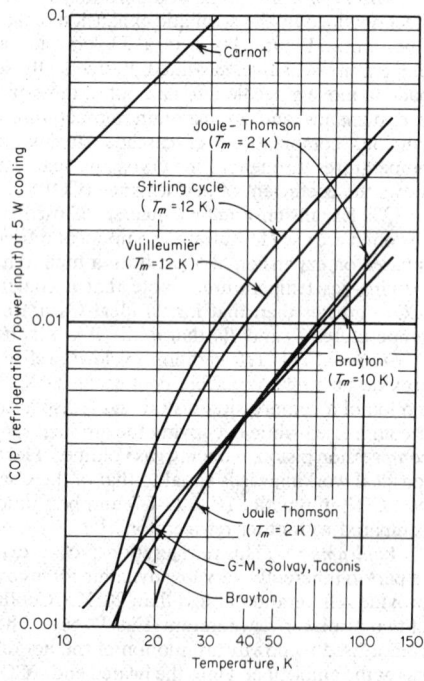

Fig. 11-64. Coefficient of performance (COP) for several refrigeration processes at 5 W cooling vs. refrigeration temperature.

Expander types include Solvay and Gifford-McMahon cryorefrigerators. Figure 11-67 shows a GM expander. The piston displacer-regenerator is pneumatically moved up and down by timed valving of the high and low working-gas pressure. Cyclic charging and discharging of the expander working-gas pressures with timed piston motion will pump heat from the cold to the ambient end. Heat pumped from the cold end using the GM cycle is rejected at the ambient end of the expander. Heat pumped from the cold end using modified Solvay cycle is rejected along pressure lines and at the compressor.

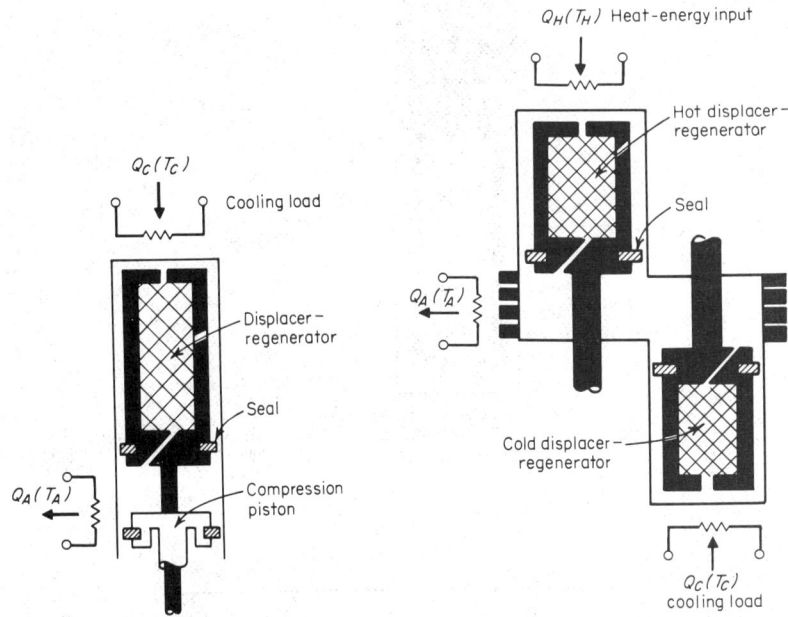

Fig. 11-65.   Stirling cycle refrigerator.   Fig. 11-66.   Vuilleumier cycle refrigerator.

Thermoelectric.[17,18]   The basic operating principle of the thermoelectric cooler is the Peltier cooling effect,[12] caused by absorption or generation of heat as a current I passes through a junction of two dissimilar materials (Fig. 11-68). Electrons passing across the junction absorb or give up an amount of energy equal to the transport energy and the energy difference between the dissimilar-materials conduction bands.

Cryogenic temperatures are reached using heat rejected from one thermoelectric cooler stage to supply thermal input to the adjacent stage. Nine stage devices have reached 145°K using less than 50 W power. Cooling loads of 0.2 W or less for 195°K operation are common. Table 11-25 summarizes data for these miniature devices.

Table 11-25.   COP Performance for Multistage Thermoelectric Refrigerators

T_h, °K	T_c, °K	COP		Wt., g
		Practical	Potential	
300	223	0.1	0.15	5
300	195	0.02	0.03	10
300	145	0.004	0.01	25

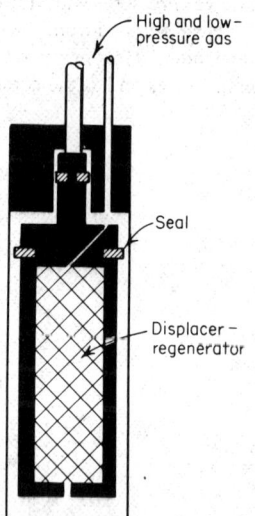

Fig. 11-67. Expander for Gifford-McMahon cryorefrigerator.

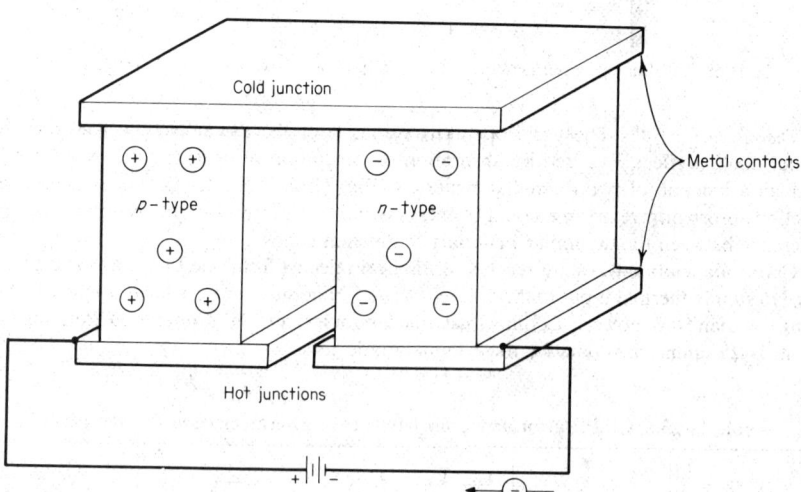

Fig. 11-68. Single-stage thermoelectric refrigerator.

138a. References on Infrared Detectors and Cooling

1. HOLTER, M., ET AL. "Fundamentals of Infrared Technology," Macmillan, New York, 1962.

2. KRUSE, P. W., McGLAUCHLIN, L. D., and McQUISTON, R. B. "Elements of Infrared Technology," Wiley, New York, 1962.

3. WOLFE, W. L. "Handbook of Military Infrared Technology," U.S. Government Printing Office, Washington, 1965.

4. PUTLEY, E. H. Solid State Devices for Infrared Detection, *J. Sci. Instrum.*, **43**:857–68, 1966.

5. Oxytronics, Inc., Buffalo, N.Y., technical correspondence and sales sheets.

6. BEAN, J. W., and MANN, D. B. The Joule-Thomson Process in Cryogenic Refrigeration Systems, *NBS Tech. Note* 227, February, 1965.

7. BULLER, J. S. "The Miniature Self-Regulation Rapid Cooling Joule-Thomson Cryostat," Santa Barbara Research Center, Goleta, Calif., June, 1970.

8. TIMMERHAUS, K. D. *Advan. Cryogenic Eng.*, **14**:332–377 (1968).

9. ACKERMANN, R. A. An Investigation of Gifford-McMahon Cycle and Pulse-Tube Refrigerators, *U.S. Army ECOM Tech. Rept.* 3245, March, 1970.

10. LONGSWORTH, R. C. A Modified Solvay Cycle Cryogenic Refrigerator, *1970 Cryogenics Conf., Paper* K-5, Boulder, Colo.

11. JENSON, H. L., ET AL. "Investigation of External Refrigeration Systems for Long Term Cryogenic Storage," Lockheed Missiles and Space Co., Report LMSC-A981632, February, 1971.

12. IOFFE, A. F. "Semiconductor Thermoelements and Thermoelectric Cooling," Tri-Litho Offset Ltd., Great Britain, 1957.

13. MUHLENHAUPT, R. C., and STROBRIDGE, T. R. An Analysis of the Brayton Cycle as a Cryogenic Refrigerator, *NBS Tech. Note* 366, 1968.

14. BARRON, R. B. "Cryogenic Systems," McGraw-Hill, New York, 1966.

15. FOWLE, A. A. Cooling with Solid Cryogens: A Review, *Advan. Cryogenic Eng.*, **2**:198–201, 1965.

16. ZEMANSKY, M. W. "Heat and Thermodynamics," pp. 157–66, McGraw-Hill, New York, 1957.

17. Borg-Warner Corp., R. C. Ingersoll Research Center, Des Plaines, Ill., technical correspondence.

18. Nuclear Systems, Inc., Garland, Tex., technical correspondence.

SOLAR CELLS AND SOLAR ARRAYS

By EDWARD GADDY

139. Theory of Operation. A solar cell is a device that converts electromagnetic radiation into electricity. Solar cells may be divided into two classes, *thin-film* and *single-crystal*. The theory of operation of thin-film cells is very complicated and not fully understood;[1*] it will not be discussed here.

Single-crystal solar cells are actually diodes, modified so that light can reach the junction. Therefore the same theory that applies to diodes also applies to solar cells. Light impinging on a solar cell forms electron-hole pairs. The electrons are attracted to the positive charge in the n-type material, and the holes are attracted to the negative charge in the p-type material. These changes dilute the equilibrium charge concentrations in the junction, and if enough electron-hole pairs are formed, the voltage normally across the junction becomes significantly less than in the unlighted situation. This in turn causes a voltage to appear across the solar cell's external contacts. (It should be remembered that in the unlighted diode there is a voltage across the junction; this voltage is equal and opposite to the contact potential; thus the dark diode has no voltage across its electrodes. Therefore, if the voltage across the junction becomes smaller, a voltage across the diode will appear.[2])

140. Fabrication. The only solar cells consistently made in production quantities are

---

* Superior numbers correspond to numbered references, Par. 11-142a.

single-crystal silicon cells, although germanium and gallium arsenide may also be used.[4] The thin-film cadmium sulfide cells are produced in small production lots sporadically.

A *single crystal* of silicon is doped and grown, usually in a Czochralski furnace,[3] then cut to size, cleaned, and polished. The junction of the cell is then formed by a diffusion process. The junction between the *n* and *p* layers is formed over the entire surface of the cell. This junction is etched off the back and sides of the cell with acid. Next, electrical contacts, usually made by depositing first titanium and then silver, are formed on the cell. The contacts are generally covered with solder to protect them from humidity damage. The finished cell is shown in Fig. 11-69.

Thin-film solar-cell construction is not as advanced as it is with single-crystal silicon solar cells. Thin-film solar cells can be fabricated from gallium arsenide, indium phosphide, cadmium telluride, cadmium sulfide, and similar compounds. The cadmium sulfide solar cell[1] has a thin layer of cadmium sulfide deposited on a metal or metallized plastic substrate which serves as the negative electrode. Copper sulfide is then deposited on top of the cadmium sulfide. The cell's junction is formed by the union of these two materials. A metal grid for

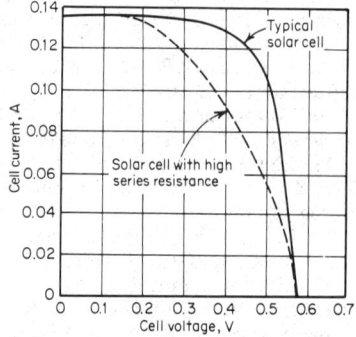

Fig. 11-70. Performance characteristics for a typical solar cell and a solar cell with high internal series resistance. Both cells have 140 mW/cm² of sunlight incident on them and are nominally 2 × 2 cm.

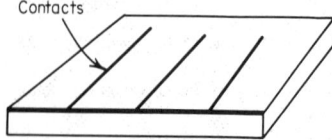

Fig. 11-69. Completed single-crystal silicon solar cell.

the positive electrode is pressed on top of the copper sulfide along with a transparent cover. Cadmium sulfide solar cells are roughly 0.01 cm thick, very pliable, and extremely lightweight.

141. Performance Characteristics. Since the only cells in constant production are made of single-crystal silicon, engineering data are supplied only for these cells. Figure 11-70 shows the effect of internal series resistance on such cells.

Figure 11-71 shows the effect of varying incident-light intensity on single-crystal silicon solar cells. For a solar cell that has a low internal series resistance, the short-circuit current is directly proportional to the intensity striking the cell.

The effect of temperature on a solar cell is shown in Fig. 11-72.

Single-crystal silicon solar cells respond only to wavelengths of light from approximately 4000 to approximately 12,000 Å; their peak response can be varied slightly by changing doping techniques.

Single-crystal silicon solar cells have a density of about 2.22 g/cm³. They are made in sizes from approximately 2 by 1 cm to 2 by 6 cm and from 0.015 to 0.045 cm thick. (The thicker cells are slightly more efficient than the thinner ones.)

The best thin-film cadmium sulfide cells run, in sunlight, at about one-half the 11% efficiency typical of single-crystal silicon solar cells and are prone to degrade.

142. Application. Virtually all solar cells produced are fabricated for spacecraft or directly related uses; and the vast majority of spacecraft launched with lifetimes over six months are powered by solar cells. Since proton or electron flux degrades solar cells, and since these fluencies are encountered in space, solar cells used on spacecraft are protected by a cover glass, usually made of fused silica.

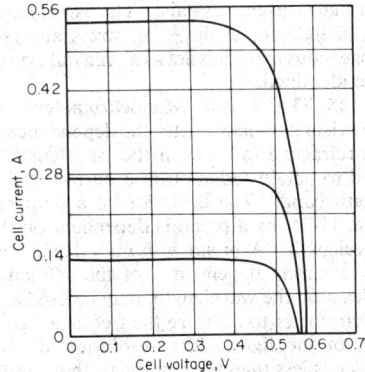

Fig. 11-71. Performance characteristics for a typical 2- × 2-cm single-crystal silicon solar cell at varying intensities. The spectrum is that of unfiltered sunlight.

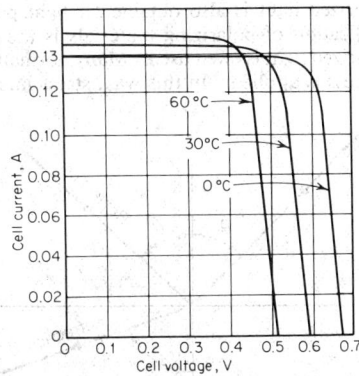

Fig. 11-72. Performance characteristics of a typical 2- × 2-cm single-crystal silicon solar cell at different temperatures.

142a. References on Solar Cells and Arrays

1. BOER, K. W. "Research Study of the Photovoltaic Effect in Cadmium Sulfide," Final Report Jet Propulsion Laboratory Contract 952666 (under NAS7-100), 1970.

2. ALTMAN, MANFRED "Elements of Solid State Energy Conversion," pp. 240-263, Van Nostrand, New York, 1969.

3. RUNYAN, W. R. "Silicon Semiconductor Technology," p. 34, McGraw-Hill, New York, 1965.

4. ANGRIST, STANLEY W. "Direct Energy Conversion," pp. 197-200, Allyn and Bacon, Boston, Mass., 1965.

OPTICAL ACCESSORIES

BY HERBERT K. POLLEHN

143. Geometrical Optics.[1*]   In geometrical optics it is assumed that all dimensions of obstacles and openings in diaphragms are large compared with the wavelength of the light, and that the field is completely incoherent (see Par. **11-24**).   The electromagnetic waves can then be represented as light rays.   The planes normal to these rays are called *geometrical wavefronts.*

The laws of reflection and refraction explain most geometrical optical phenomena.   In Fig. 11-73, light rays traveling through a homogeneous medium with an index of refraction n_1 impinge on a homogeneous medium with an index n_2.   The index of refraction is a constant of the materials and is dependent on the wavelength of the light.   (The variation of n with wavelength is called *dispersion.*)   From the conditions described in Fig. 11-73, the following formulas are obtained:

$$\text{Reflection: } \sin \theta = \sin \theta' \tag{11-4}$$

$$\text{Refraction: } \frac{\sin \theta}{\sin \theta''} = \frac{n_2}{n_1} = n \tag{11-5}$$

SPECIAL APPLICATIONS

144. Linear polarized light[1] is obtained if $\theta = \theta''$ (except for normal incidence).   For all other angles, the reflected and refracted light are partially polarized (see Par. **11-46**).

* Superior numbers correspond to numbered references, Par. **11-159**.

Polarized light is also obtained if light passes through certain crystals. The best-known application of polarizing materials is the control of light intensities.[2] In stress analysis, polarized light is often used. Many normally homogeneous materials act as a uniaxial crystal if stress is applied. In this way, stress lines can be identified.

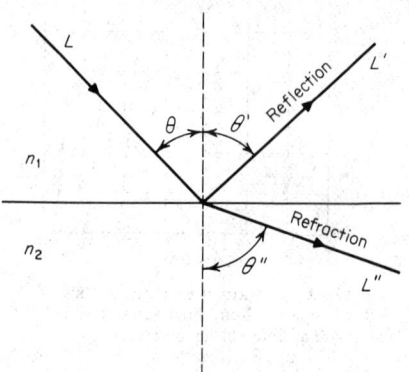

Fig. 11-73. Reflection and refraction. L = incoming ray; L' = reflected ray; L'' = refracted ray.

145. The Prism Monochrometer[2] In the prism monochrometer the dependence of the refractive index n on the wavelength is used to obtain radiation in a narrow wavelength band. The deviation of a light ray (Fig. 11-74) by a prism is dependent on the wavelength. A prism with high dispersion, i.e., a strong dependence of the refractive index n on the wavelength, may be used as a spectrometer to measure the frequency content of the light. The resolution of this device is less than the resolution that can be obtained with a grating spectrometer or an interferometer (see Par. 11-155).

146. Fiber optics[3] are based on the phenomena of total reflection (see Fig. 11-75). Total reflection occurs for $\theta'' > 90°$ [Eq. (11-5)]. Transparent materials such as glass or plastic with a high refractive index are used for the fiber, while a material having a low refractive index is used for the surrounding (cladding) material. As a result, total reflection at the boundary of the fiber and the outer (cladding) occurs even for large angles of incidence. The numerical aperture (NA), defined

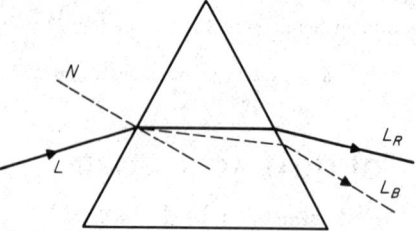

Fig. 11-74. Refraction and dispersion by a prism. L = incident light ray; L_R = refracted ray—red light; L_B = refracted ray—blue light; N = normals to prism surface.

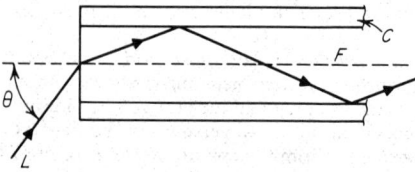

Fig. 11-75. Fiber-optics geometry. F = fiber; C = cladding material.

(cladding) occurs even for large angles of incidence. The numerical aperture (NA), defined as in lens systems (Par. **11-152**), can be made even higher than 1.0, exhibiting very high light-collecting properties. Fibers can be made with a diameter of only a few micrometers, and many individual fibers can be fused together to form a rigid or flexible bundle. A 100-m-long bundle can transmit more than 90% of the input light. If the fibers are carefully arranged, they can be used for image transfer, as in fiberscopes and image-tube faceplates (see Par. **11-103b**) or field flatteners in electrooptical devices.[4] In recent years it has been possible to vary the refractive index through the diameter of a single fiber to give the individual fiber-imaging properties. These are known commercially as Selfoc.[5]

147. Properties of Lens Systems.[3] *Aberrations.* In any lens system an optical axis and entrance and exit pupils, P_1 and P_2, can be defined (Fig. 11-76). The emerging wavefronts S from the object point P are spherical. If the outgoing wavefronts at the exit pupil were also spherical, all the light rays would converge to one point P', giving a *perfect lens system*. It is not possible.to construct such a perfect lens system, and deviations A from the sphere S are called aberrations.[1,2] In some optical systems, specially shaped glass plates, often aspheric, can be placed into the light beam to correct at least partially the deviations from the sphere *(Schmidt optics)*.

Chromatic Aberration.[2] Due to the dependence of the refractive index on the wavelength, the wavefronts are different for different wavelengths of light, and therefore converge to different points. Chromatic aberration is usually corrected by the incorporation into the lens of glasses having different indices of refraction.

148. Gaussian optics (first-order)[2] is the term commonly used to describe and specify the properties of a lens system. In gaussian optics, aberrations are neglected. This can be done strictly only for paraxial rays, for which Eq. (11-5) can be written

$$\frac{\theta}{\theta''} = \frac{n_2}{n_1}$$

(11-6)

A good measure for the deviations from a perfect lens caused by aberrations and diffraction is given by the *modulation transfer function*.

149. Thin Lens (Fig. 11-77). In a thin lens the thickness t of the lens is much less than the distance s or s' of the object h or the image h' from the lens: $t \ll s$ and $t \ll s'$. Assuming that the refractive index of the lens is n and the refractive index for the object and the image

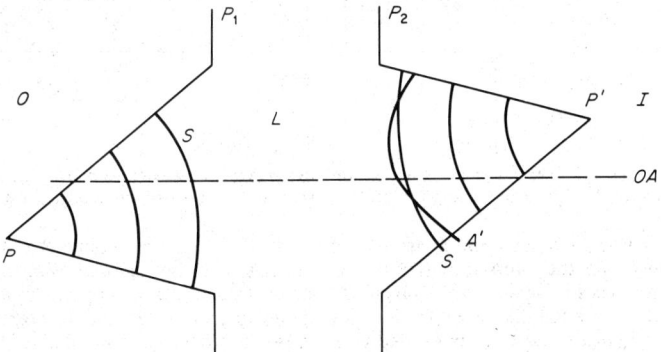

Fig. 11-76. Wavefronts in optical system. OA = optical axis; O = object space; I = image space.

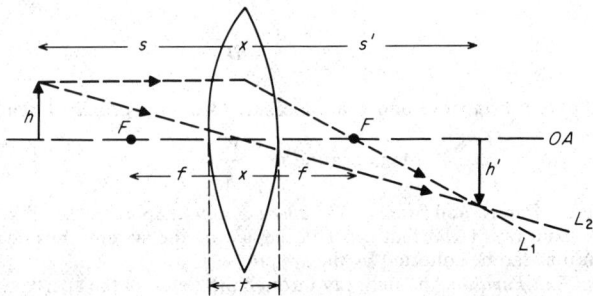

Fig. 11-77. Thin-lens geometry. h = object; h' = image; F = focal points; f = focal length; L_1, L_2 = light rays to construct image; OA = optical axis.

space is 1 and that the surfaces of the lens are spherical with radii R_1 and R_2, then, using Eq. (11-6), it can be shown that

$$\frac{1}{s} + \frac{1}{s'} = (n-1)\,\frac{1}{R_1} - \frac{1}{R_2} = \frac{1}{f} \tag{11-7}$$

where f is the *focal length*.

The *power* of the lens is expressed in diopters as $D = 1/f$ (in meters). The *magnification* m is given by

$$m = \frac{h'}{h} = \frac{-s'}{s} \tag{11-8}$$

All rays parallel to the optical axis converge to the focal points.

150. Thick-Lens System (Fig. 11-78). In a thick-lens system the focal points F and F' are found where the light rays parallel to the optical axis cross the optical axis on the other side

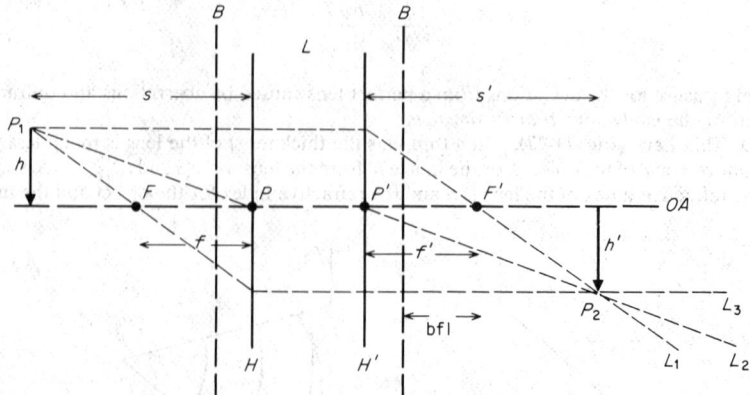

Fig. 11-78. Thick lens or lens system. L = lens or lens system; h = object height; h' = image height; s = object distance; s' = image distance; bfl = back focal length; L_1, L_2, L_3 = light rays to construct image.

of the lens. Focal lengths f and f' are measured from the principal points P and P'. The *principal points* are the intersections of the optical axis with the *principal planes* H and H'. They are not necessarily within the physical boundaries B of the lens or lens system. The principal planes are found easily if the location of one object point P_1 and the corresponding image point P_2 are known. H' is located where a ray through P_1 and parallel to the optical axis crosses a ray through P_2 and F_1. H is located where a ray through P_2 and parallel to the optical axis crosses a ray through P_1 and F. The focal lengths f and f' are equal if the refractive index is the same in the image and in the object space. With these definitions, the formula is

$$\frac{1}{s} + \frac{1}{s'} = \frac{1}{f} = D$$

The image and object distances s and s' are measured from the principal planes.

$$\text{Magnification } m = \frac{h'}{h} = \frac{s'}{s}$$

151. Apertures, Pupils, and Stops. *The aperture stop* (diaphragm D_1, Fig. 11-79) limits the cone of the axial rays (AR) that can be accepted by the system, thus determining the amount of light that can be collected by the system.

Entrance and Exit Pupils. Any light ray through the center of the aperture stop is called a *chief ray* (CR). Entrance pupils, EnP, and exit pupils, ExP, are located where the chief ray passes the optical axis (OA). The cone of accepted axial rays determines the diameters of the pupils.

The field stop D_3 limits the field of view and is determined by the chief ray with the largest angle u that can be accepted by the system. In Fig. 11-79, D_2 is the field stop. Often the field stop is in the image plane D_3. In cameras, the field stop is the film size. In image tubes, it is the photocathode surface.

152. Other Optical Parameters. The *entrance window* (EW) is the image of the field stop in object space. It does not necessarily coincide with the object.

The *field of view* is determined by the ray from the edge of the entrance window to the center of the entrance pupil (CR in Fig. 11-79). Twice the angle u, formed by this ray and the optical axis, is called the *angular field of view*.

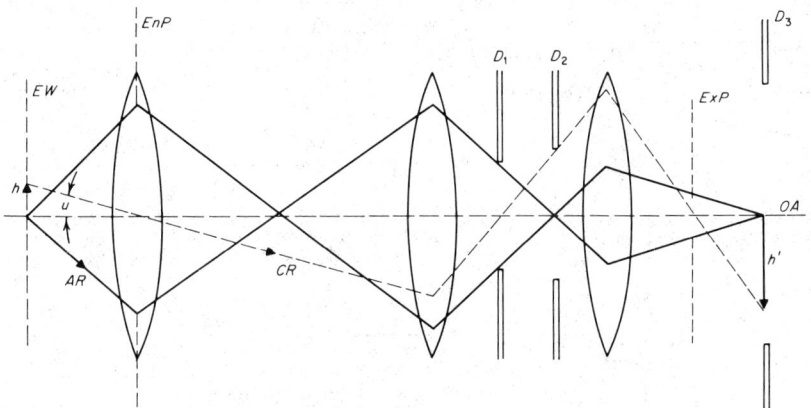

Fig. 11-79.   Pupils, stops, and windows.

The *f number* (f/no) is the ratio of the focal length f to the clear aperture D (diameter of the entrance pupil if the object is at infinity).

$$f/no = f/D \qquad (11\text{-}9)$$

The *numerical aperture* (NA) is the index of refraction of the object space times the sine of the half angle ϕ of the cone of axial rays that can be accepted by the system.

$$NA = n \cdot \sin \phi \qquad (11\text{-}10)$$

The irradiance E (illuminance)[6] in an optical image is given by the following equations. For very *distant objects* (telescopes, camera objectives, or image intensifiers):

$$E = \frac{\pi L}{4} \frac{1}{(f/no)^2} \qquad (11\text{-}11)$$

where L is the object radiance (brightness).

For *nearby objects* (especially microscopes and magnifiers) the principal plane (Par. 11-150):

$$H = \frac{\pi N}{n^2 m^2} (NA)^2 \qquad E = \frac{\pi L}{n^2 m^2} (NA)^2 \qquad (11\text{-}12)$$

where n is the refractive index in object space, and m is the magnification of the lens system.

These equations do not include light losses due to scattering and absorption in the atmosphere or lens system, and it is assumed that the objects radiate as a lambertian surface.

153. Special Lens Systems[2] *The Lupe (Magnifier)* (Fig. 11-80). The magnification is

$$m = \frac{u'}{u} \qquad (11\text{-}13)$$

m can also be written

$$m = \frac{d}{f} \qquad (11\text{-}14)$$

where $d = 25$ cm is the near point, or distance of most distinct vision.

The Microscope (Fig. 11-81). The objective forms a real image at P_2. The magnification $m_1 = s'/s$. The object is very close to the focal point of the objective $s \approx f_1$. The eyepiece acts as a lupe, with a magnification $m_2 = d/f_2 = 25/f_2$. The total magnification is

$$m = m_1 \cdot m_2 = \frac{-s'}{f_1} \cdot \frac{25}{f_2} \qquad (11\text{-}15)$$

The Astronomical Telescope (Fig. 11-82). The light rays from a far point in the object space are practically parallel on entering the objective. An image is formed in the focal plane

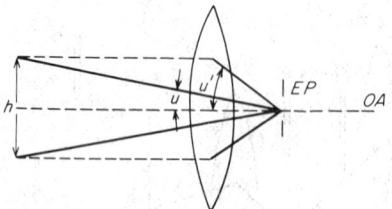

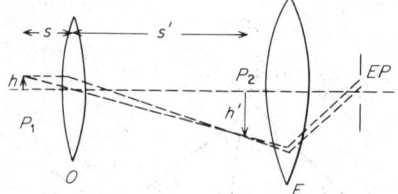

Fig. 11-80. Lupe. h = object height; OA = optical axis; EP = eye pupil.

Fig. 11-81. Microscope. O = objective; E = eyepiece; EP = eye pupil; h = object; h' = intermediate image.

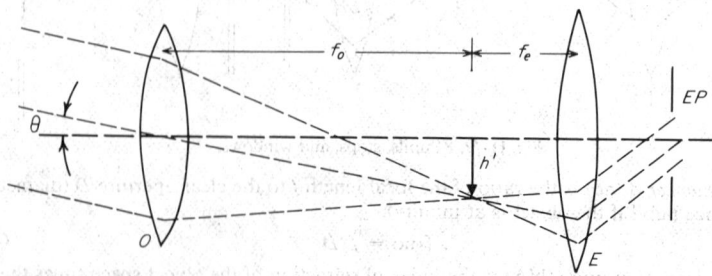

Fig. 11-82. Telescope. C = objective; E = eyepiece; EP = eye pupil; f_o = focal length of objective; f_e = focal length of eyepiece; h' = intermediate image.

of the objective, which coincides with the focal plane of the eyepiece. The eyepiece acts as a lupe. The magnification is

$$m = \frac{\theta'}{\theta} \qquad (11\text{-}16)$$

or

$$m = \frac{f_o}{f_e} \qquad (11\text{-}17)$$

Collimators[2] are used in many optical experiments and instruments where parallel (collimated) light is needed, as in monochromers, spectrometers, searchlights, or the alignment of optical systems. Each point of a light source, placed in the focal plane of a lens system, will generate a beam of parallel light. Different points will generate beams of parallel light which are divergent to each other. The angle θ, enclosed by two parallel beams originating from two points separated by a distance x, is approximately x/f, where f is the focal length of the collimator lens.

Reflecting mirrors are used with light sources when light is desired in only certain directions. The light source can then be operated at lower intensities, maintaining the required output in the desired direction (Fig. 11-83).

Radiometers (photometers) are used to measure the light power.[2] With synchronous detection systems, or photomultipliers that allow single-photon counting (see Par. **11-100**), measurements at very low light levels can be performed. Variable power from a light source may be obtained with *neutral density filters*.[3] For these filters, the attenuation of the radiant flux is independent of the wavelength over a wide spectral range. The attenuation is determined by the density

$$D = \log \frac{\phi_0}{\phi_t} \qquad (11\text{-}18)$$

where ϕ_0 = radiant flux incident on the filter, and ϕ_t = radiant flux transmitted by the filter.

COHERENT OPTICS AND DIFFRACTION[1,8,9]

154. Diffraction phenomena are observed if small particles or obstacles (on the order of the wavelength) or diaphragms with small openings (pinholes) are placed in the light beam. In such cases, light is observed in the geometrical shadow of the objects. *Coherent light* (see Par. **11-24**) must be used to obtain usable diffraction effects, as in the instruments to be described. Light consists of wave groups with a finite length ΔX and always contains a band of frequencies. The length ΔX is inversely proportional to the bandwidth Δf. To obtain coherent waves and to observe a time-invariant interference pattern, the following conditions have to be satisfied: the wave groups have to overlap within their coherence length ΔX, and the wave groups have to originate from the same "point" of a light source. A point, in this sense, is defined by the *coherence condition* $2y \sin u \ll \lambda/2$, where $2y$ is the size of the light source, and u is the angle in which light is emitted by the source.

To satisfy the coherence condition, a small aperture is placed in front of the light source. Excellent sources for coherent light

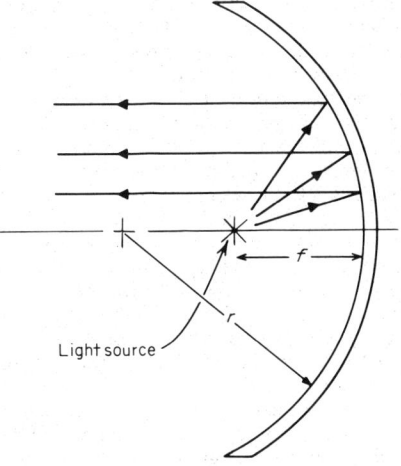

Fig. 11-83. Searchlight reflector.

of very high intensity are *lasers* (see Par. **11-24**). The angle u is very small, and a coherence length ΔX of several meters can be obtained (small Δf).

155. Division of wavefronts, caused by diffraction, is obtained if small obstacles (or a diaphragm with small openings) are placed in the original light wave. The *grating spectrometer* is used to measure the frequency content of the light. The grating consists of small, very long slits or grooves separated by a distance d on the order of the wavelength. Most high-resolution gratings contain grooves, often on a parabolic surface, and are used in the reflection mode. The parabolic surface serves the same purpose as the lens used with the transmission grating. The transmission grating is placed in the path of a collimated light wave, coming from a light source that satisfies the coherence conditions (Fig. 11-84). Each

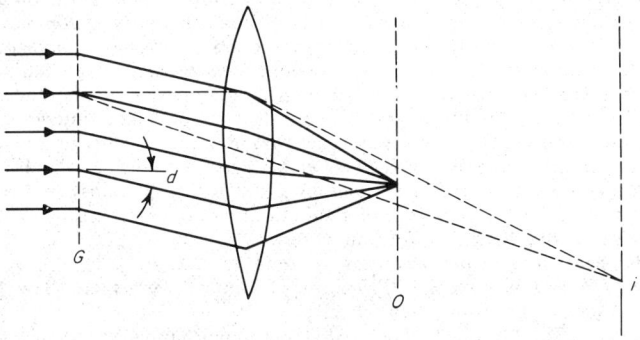

Fig. 11-84. Grating. G = grating = object; O = observation plane = focal plane; i = image of grating.

slit point is the source of a spherical wave. For infinitely long slits, cylindrical waves are obtained. In Fig. 11-84 some normals to the wavefronts are drawn (arrows) which will combine at infinity or at the focal plane (observation plane) of the lens. In the observation plane, a maximum of intensity (phase difference $x \times 2\pi$) is obtained for

$$d \cdot \sin \alpha = m \cdot \lambda \qquad m = 1,2,3, \ldots \qquad (11\text{-}19)$$

The maxima for different wavelengths appear at different angles α or different locations in the observation plane.

The intensity distribution in the observation plane of a grating is called an *interference pattern*. In Fig. 11-84 the light source and the observation plane are at infinity. In this case, the diffraction is called *Fraunhofer diffraction* and the diffraction pattern is the complex—amplitude and phase—*Fourier transform* of the object. The frequency is in cycles per unit length, depending on λ. If either light source or observation plane, or both, are not at infinity, the diffraction is called *Fresnel diffraction*.

156. Spatial Filtering.[8] In the Fraunhofer diffraction pattern, the different spatial frequencies appear at different locations in the plane. By blocking special frequencies or changing the phase, the image can be changed in a specific way. If the zero order is blocked or attenuated, a low-contrast scene (high dc level) will be imaged with higher contrast, and is commonly called *contrast enhancement*. If the zero order and several low frequencies are blocked, the image contains only the edges of the objects; this is referred to as *differentiation*. *Integration* is obtained by blocking the high frequencies. Very often a special object has to be found in a scene. Knowing the spatial-frequency content of the object, a filter can be constructed to enhance the image of the object and to depress all scene information—*signal enhancement*. By operating on the phase in the diffraction pattern so as to filter with a different refractive index, and thus change the optical path length $n \cdot s$, objects that differ from their environments only by their refractive indices (i.e., biological objects, gas in air, acid in water) can be made visible. This is called the *Schlieren method* and is used in the *phase contrast microscope*.

157. Image Resolution.[1,2,9] The intensity $s(r)$ of an image from a point object is always spread over some finite area even if aberrations are neglected. The spread is caused by diffraction in the apertures of the lens system.

The intensity distribution of this point image is called the *Airy disk*, which gives the final resolution limit of any optical system. Aberrations increase the spreading of the image, thus decreasing the resolution. The image of any object can be calculated by convoluting the normalized image of a point object (spread function) with the object.

The Fourier transform of the spread function—normalized at zero spatial frequency—is called the *optical transfer function*.[2] The concept of the optical transfer function is extensively used to specify and test optical systems.

Division of amplitude is obtained by dividing the original wave into two or more subwaves, mostly by means of reflection and refraction (beam splitter). Instruments based on this principle are known as *interferometers*,[10] such as the Michelson and Fabry-Perrot, which are very accurate devices used as spectrometers and to determine minute differences in length and optical path.

158. Holography utilizes both a division of wavefronts and a division of amplitude. From the Fraunhofer diffraction pattern, the image of the object is obtained by a complex *(amplitude and phase)* Fourier transform. By recording the diffraction pattern with a photographic plate, the phase information is lost and the image cannot be reconstructed. If the diffracted waves interfere with a coherent background wave obtained by a division of amplitude, the phase information is recorded on film as an intensity distribution. This interference pattern between diffracted waves and a coherent background wave is called a *hologram*. The reconstruction of the image is obtained, after proper processing of the hologram, by illumination with the same (or different, but always coherent) illumination. Holographic images appear three-dimensional if the original object was three-dimensional. Holograms are also obtained with a Fresnel diffraction pattern.

159. References on Optical Accessories

1. BORN, M., and WOLFE, E. "Principles of Optics," Pergamon, New York, and Sommerfeld, "Optics," Academic, New York.

2. KINGSLAKE, R. "Applied Optics and Optical Engineering," Academic, New York, 1967.

3. KAPANY, N.   "Fiber Optics," Academic, New York, 1967.

4. *Fiber Opt. Seminar Proc.,* 1968 and 1970, Society of Photo-Optic Instrumentation Engineers, Redonda Beach, Calif.

5. UCHIDA, TEIJI, and ICHIRO, KILANO   "Selfoc: A New Light Focusing Fiber Guide," *Japan Electron.,* February, 1969, p. 22.

6. U.S.A. Standard Nomenclature and Definitions for Illuminating Engineering, Illumination Engineering Society, New York.

7. WOLFE, E.   "Progress in Optics," Vol. 4, North Holland Publishing Co., Amsterdam, 1965.

8. ALWARD, J.   Spatial Frequency Filtering in "Handbook of Military Infrared Technology," Department of the Navy, Office of Naval Research, Washington, 1965.

9. O'NEILL, E.   "Introduction to Statistical Optics," Addison-Wesley, Reading, Mass., 1963.

10. FRANCON, M.   "Optical Interferometry," Academic, New York, 1966.

For suppliers of optical instruments, systems, components, and materials see:

"The Optical Industry and Systems Directory," Optical Publishing Co., Pittsfield, Mass.

SECTION 12

FILTERS, COUPLING NETWORKS, AND ATTENUATORS*

BY

MILTON DISHAL   Senior Scientist, International Telephone and Telegraph Corporation Laboratories; Fellow, Institute of Electrical and Electronics Engineers; former Chairman of the Radio Receivers Committee, IEEE; Administrative Committee of the Professional Group on Circuit Theory; IEEE Editorial Review Committee; former Adjunct Professor of Electrical Engineering, Polytechnic Institute of Brooklyn

CONTENTS

Numbers refer to paragraphs

* Adapted from material generated by M. Dishal for "Reference Data for Radio Engineers," H. P. Westman and M. Karsh (eds.), 5th ed., Chaps. 8 to 10, by arrangement with International Telephone and Telegraph Corporation; copyright 1968 Howard W. Sams & Co., Indianapolis, Ind.

SECTION 12

FILTERS, COUPLING NETWORKS, AND ATTENUATORS

The design information in this section results from the application of modern network theory to electric wave filters and coupling networks. Only design results are supplied, and a careful study of the cited references is required for an understanding of the synthesis procedures that underlie these results.

FILTER DESIGN

1. **Image-Parameter Design.** Consider the *simple low-pass ladder network of* Fig. 12-1A. Two simultaneous design equations are provided by classical image-parameter theory:

$$(Z_1/4Z_2)_{f=f_c} = -1 \text{ and } 0 \tag{12-1}$$

$$Z_{0T} = (Z_1 Z_2)^{1/2}[1 + (Z_1/4Z_2)]^{1/2} \tag{12-2}$$

Z_1 and Z_2, the full series- and shunt-arm impedances, respectively, must be suitably related to make Eq. (12-1) true at the desired cutoff frequencies, and the generator and load impedances must satisfy Eq. (12-2). Under the image-parameter theory, the resulting attenuation for the low-pass case is

$$V_p/V = \begin{cases} 1.0 & (\omega/\omega_c) < 1 \\ \exp[(n-1)\cosh^{-1}(\omega/\omega_c)] & (\omega/\omega_c) > 1 \end{cases} \tag{12-3}$$

where n is the number of arms in the network of Fig. 12-1, and V_p/V and ω are as in Fig. 12-3.

Equation (12-1) offers no problems. The application of Eq. (12-2) to Fig. 12-1A demands *terminating impedances that are physically impossible with a finite number of elements.* The generator and load impedances for Fig. 12-1A must be pure resistances of $(L/C)^{1/2}$ Ω at zero frequency. As frequency increases, the value of resistance must decrease to a short circuit at the cutoff frequency, and with further increase in frequency must behave like a pure inductance starting at zero value at the cutoff frequency and increasing to $L/2$ at infinite frequency.

The physical impracticability of devising such terminating impedances explains why element values obtained by Eq. (12-1) cannot simultaneously satisfy Eq. (12-2). The relative attenuation indicated by Eq. (12-3) is similarly incorrect and cannot be realized in practice.

Lattice-configuration filters also require impractical terminating impedances when designed by image-parameter theory. (Constant-resistance lattices are an exception but, for practical reasons, are seldom used.) The practical use of resistive terminations automatically makes element values computed on the basis of ideal impedance terminations incorrect.

For more than four decades, filters have been designed according to the image-parameter theory. Their commercial acceptance is due in no small part to the highly approximate requirements for most filters. Where more exact characteristics are required, shifting of element values in the actual filter has usually resulted in an acceptable design. For precise amplitude and phase response in the passband, the simple and approximate solutions obtained through image-parameter theory must give way to equations based on modern network theory.

2. **Modern Design.** A typical low-pass filter with resistive generator and load is shown in Fig. 12-1B. It is composed of lumped inductors, capacitors, and the resistive elements

unavoidably associated therewith. The circuit equations for the complete network can be written by applying Kirchhoff's laws. Modern network theory does just this, and then solves the equations to find the network parameters that will produce optimum performance in some desired respect.

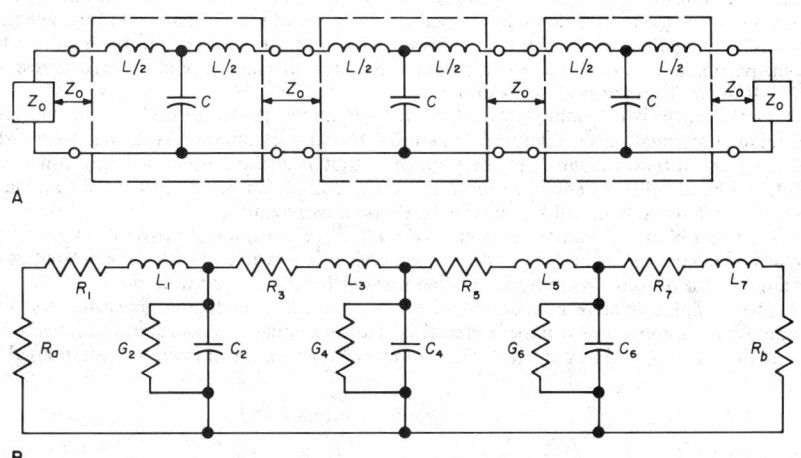

A

B

Fig. 12-1. A seven-element low-pass filter considered on the basis of image-parameter theory at *A*, and of modern network theory at *B*.

A block diagram of a generalized filter is illustrated in Fig. 12-2. This may be low-pass, high-pass, bandpass, band-rejection, phase-compensating, or other type. The elements of the filter include resistors, capacitors, self-inductors and mutual inductors, and possibly coupling elements such as electron tubes or transistors, all according to the design. The terminations shown are a constant-voltage generator (the same voltage at all frequencies) with a series resistor at the input and a resistive load. (Frequently it is preferable to stipulate a constant-current generator with a shunt conductance.) The generator and load resistors need not be equal, and they can be assigned any value between zero and infinity. Characteristic impedance plays no part in the modern network theory of filters.

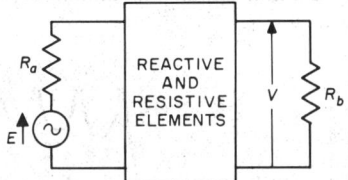

Fig. 12-2. Block diagram of a filter. The generator and load must be considered part of the filter.

Either or both the generator and load can be reactive, in which case the reactances are absorbed inside the block of Fig. 12-2 as specified parts of the filter. Either, but not both, R_a or R_b can be zero or infinite.

The term *bandwidth* (BW) as used here has two different meanings, according to the type of filter. For low- or high-pass filters, it is synonymous with the actual frequency of the point in question, or equivalent to the number of hertz in a band terminated on one side by zero frequency and on the other by the actual frequency. The actual frequency can be anywhere in the pass or the reject region. For symmetrical bandpass (Fig. 12-4) and band-reject filters, it is the difference in hertz between two particular frequencies (anywhere in the pass or reject regions) with the requirement that their geometric mean be equal to the geometric midfrequency f_0 of the pass or reject band.

Typical filter characteristics are plotted in Fig. 12-3 for a low-pass filter. In Fig. 12-3A, the magnitude of the output voltage V is plotted against radian bandwidth ω. Several specific points are indicated on the diagram. V_p is the peak voltage output, and V_m is the maximum voltage that could be developed across the load were it matched to the generator through an ideal network. Symbol ω_β designates a specified frequency or bandwidth where some particular characteristic is exhibited by the filter, such as the point where the response is 3 dB down from the peak, for example.

The characteristic of major interest to the filter engineer is the plot, shown in Fig. 12-3B, of relative attenuation vs. relative bandwidth. Relative attenuation is defined as the ratio of the peak output voltage V_p to the voltage output V at the frequency being considered. Relative bandwidth is defined as the ratio of the bandwidth being considered to a clearly specified reference bandwidth (e.g., the 3-dB-down bandwidth).

It should be noted that the elements of a filter are not uniquely fixed if only a certain relative attenuation shape is specified; in general, it is possible also to demand that at one frequency the absolute magnitude of some transfer function be optimized.

The complex relative attenuation of a complete filter (including generator and load) composed of lumped linear passive elements is always equal to a constant multiplied by the ratio of two polynomials in $j\omega$. The complex roots of the numerator polynomial are

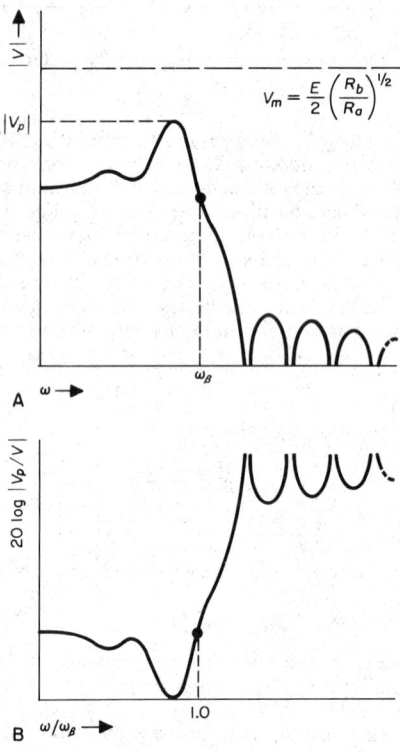

Fig. 12-3. Low-pass-filter output voltage vs. frequency at *A;* attenuation vs. normalized frequency at *B.* *A* is the actual voltage across the load as a function of frequency and is for the low-pass case. *B* uses the information in *A* to produce a plot of *relative* attenuation against *relative* bandwidth.

commonly called *attenuation poles*, or *transfer-function zeros*; the complex roots of the denominator polynomial are commonly called *attenuation zeros*, or *transfer-function poles*. Modern filter theory has derived various expressions for optimum relative attenuation shapes that can be physically realized from these complex expressions. The shapes are optimum in that they give the maximum possible rate of cutoff between the accept and reject bands for a given number of filter components, with a specified allowable equal ripple in the accept band and a specified required equal ripple in the reject band. See Fig. 12-4 for typical shapes of attenuation characteristic for bandpass filters.

3. **Chebyshev and Butterworth Shapes.** The attenuation-curve shapes illustrated in Fig. 12-4A and B are termed Chebyshev, and that in Fig. 12-4C is termed Butterworth. The equations for these shapes are Eq. (12-4) and Eq. (12-5), respectively. The Butterworth shape is the same as the limiting case of the Chebyshev shape when we set $V_p/V = 1.0$.

4. **Chebyshev and Butterworth Equations**

Chebyshev:

$$(V_p/V)^2 = 1 + [(V_p/V_v)^2 - 1] \times \cosh^2[n \cosh^{-1}(x/x_v)] \qquad (12\text{-}4)$$

Butterworth:

$$(V_p/V)^2 = 1 + (x/x_{3\text{dB}})^{2n} \qquad (12\text{-}5)$$

where V = output voltage at point x, V_p = peak output voltage in the passband, V_v = valley output voltage in the passband, and n = number of poles, equal to the number of arms in the ladder network being used. For low-pass and high-pass filters, n = number of reactances in the filter. For bandpass and band-reject, n = total number of resonators in the filter, x = a variable found in the following tabulations, x_v = value of x at the point on the skirt where attenuation equals valley attenuation, and $x_{3\text{dB}}$ = value of x at the point on the skirt where attenuation is 3 dB below V_p.

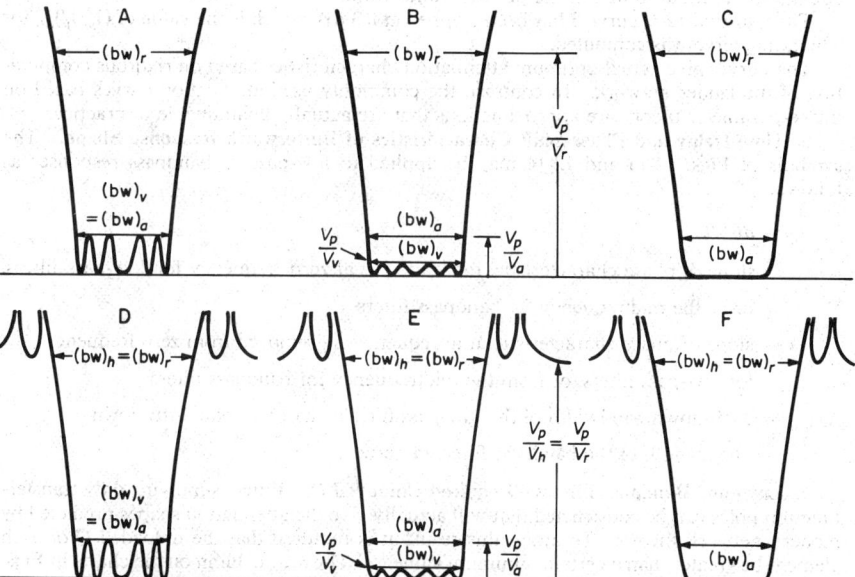

Fig. 12-4. *A* to *C* are the optimum relative attenuation shapes of Eqs. (12-4) and (12-5) that can be produced by networks supplying only transfer-function poles. *D* to *F* are the optimum relative attenuation shapes of Eqs. (12-8), (12-12), (12-13), and (12-16) that can be produced by networks supplying both transfer-function poles and zeros.

5. Significance of x.
Low-pass filters:

$$x = \omega = 2\pi f$$

High-pass filters:

$$x = -1/\omega = -1/2\pi f$$

Symmetrical bandpass filters:

$$x = \omega/\omega_0 - \omega_0/\omega = (f_2 - f_1)/f_0 = \text{BW}/f_0$$

Symmetrical band-reject filters:

$$x = -1/(\omega/\omega_0 - \omega_0/\omega) = -f_0/\text{BW}$$

where $f_0 = (f_1 f_2)^{1/2} = $ midfrequency of the pass or reject band, and $f_1, f_2 = $ two frequencies where the characteristic exhibits the same attenuation.

Working charts for these filters, derived from Eq. (12-4) and Eq. (12-5), are presented in Figs. 12-5 to 12-12 for values of n from 1 to 8, respectively. These curves give

$$(V_p/V)_{\text{dB}} = 20 \log(V_p/V)$$

versus x/x_{3dB}.

For low-pass and bandpass filters

$$x/x_{\text{3dB}} = \text{BW}/\text{BW}_{\text{3dB}}$$

For high-pass and band-reject filters, the scale of the abscissa gives $\text{BW}_{\text{3dB}}/\text{BW}$.

In Figs. 12-5 to 12-12, the family of curves toward the right side gives the attenuation shape for points where it is less than 3 dB, while those toward the left are for the reject band (greater than 3 dB). Each curve of the former family has been stopped where the attenuation is equal to that of the peak-to-valley ratio.

Thus, in Fig. 12-6, curve 3 has been stopped at 0.3 dB, which is the value of $(V_p/V_v)_{\text{dB}}$ for which the curve was computed.

The curves give actual optimum attenuation characteristics based on rigorous computation of the ladder network. In contrast, the commonly used attenuation curves based on image-parameter theory are approximations that are actually unattainable in practice.

6. Time-Delay and Phase-Shift Characteristics of Butterworth Response Shape. The symbols of Figs. 12-13 and 12-14 may be applied to low-pass or bandpass responses as follows:

$t_0 = d\theta/d\omega$

= slope of phase characteristic, rad/rad · s^{-1} at zero frequency for low-pass filters, or at the midfrequency for bandpass filters

t = slope of phase characteristic at a frequency Δf removed from zero frequency for low-pass filters, or from the midfrequency for bandpass filters

Δf_{3dB} = 3-dB-down bandwidth of the low-pass filter or half the total 3-dB-down bandwidth of the bandpass filter, in hertz

7. Low- and Bandpass Filters—Required Unloaded Q. Filters supplying only transfer-function poles can be constructed that will actually give the attenuation shapes predicted by modern network theory. To attain this result, it is required that the unloaded Q of each element be greater than a certain minimum value.[1-3*] The $q_{\min}$ column on the charts in Figs. 12-5 to 12-12 is used in the following manner to obtain this minimum allowable value: For the internal reactances of low-pass circuits,

$$Q_{\min} = q_{\min}$$

---

* Superior numbers correspond to numbered references, Par. **12-30.**

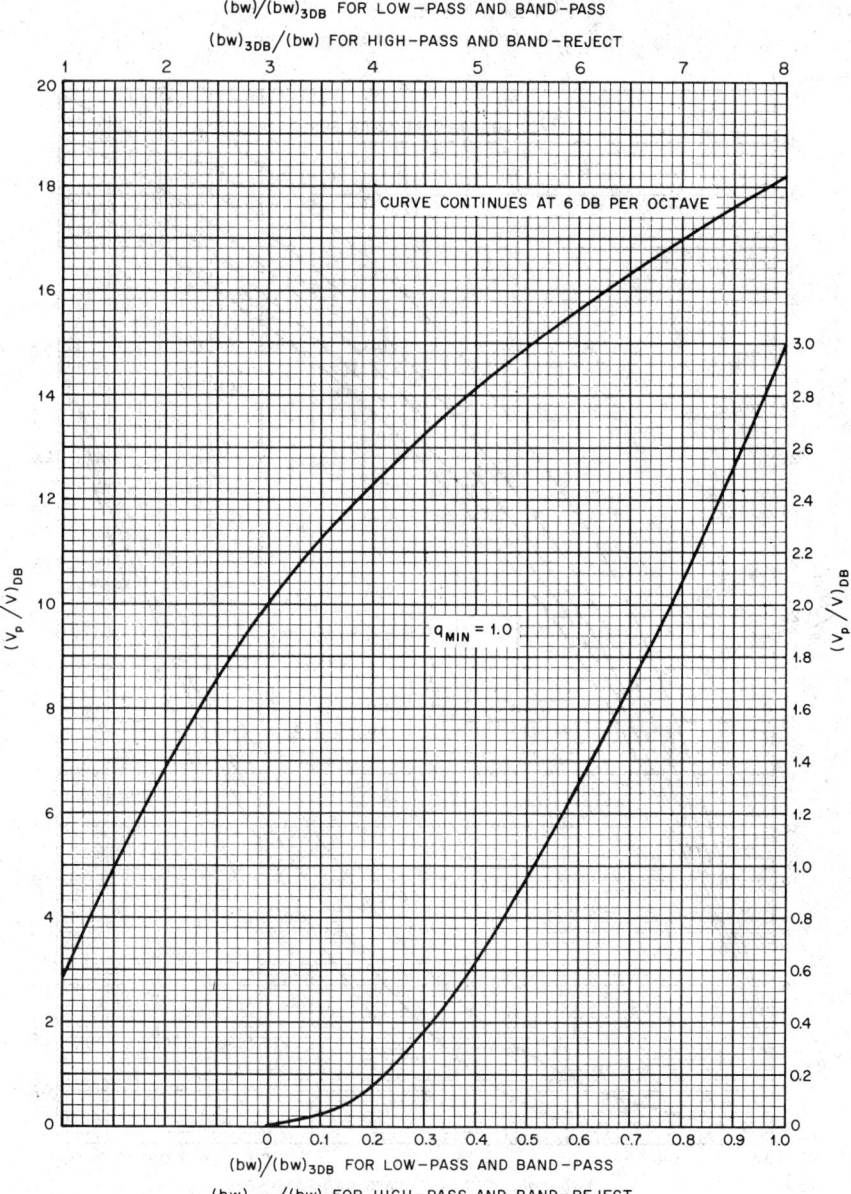

Fig. 12-5. Relative attenuation for a 1-pole network.

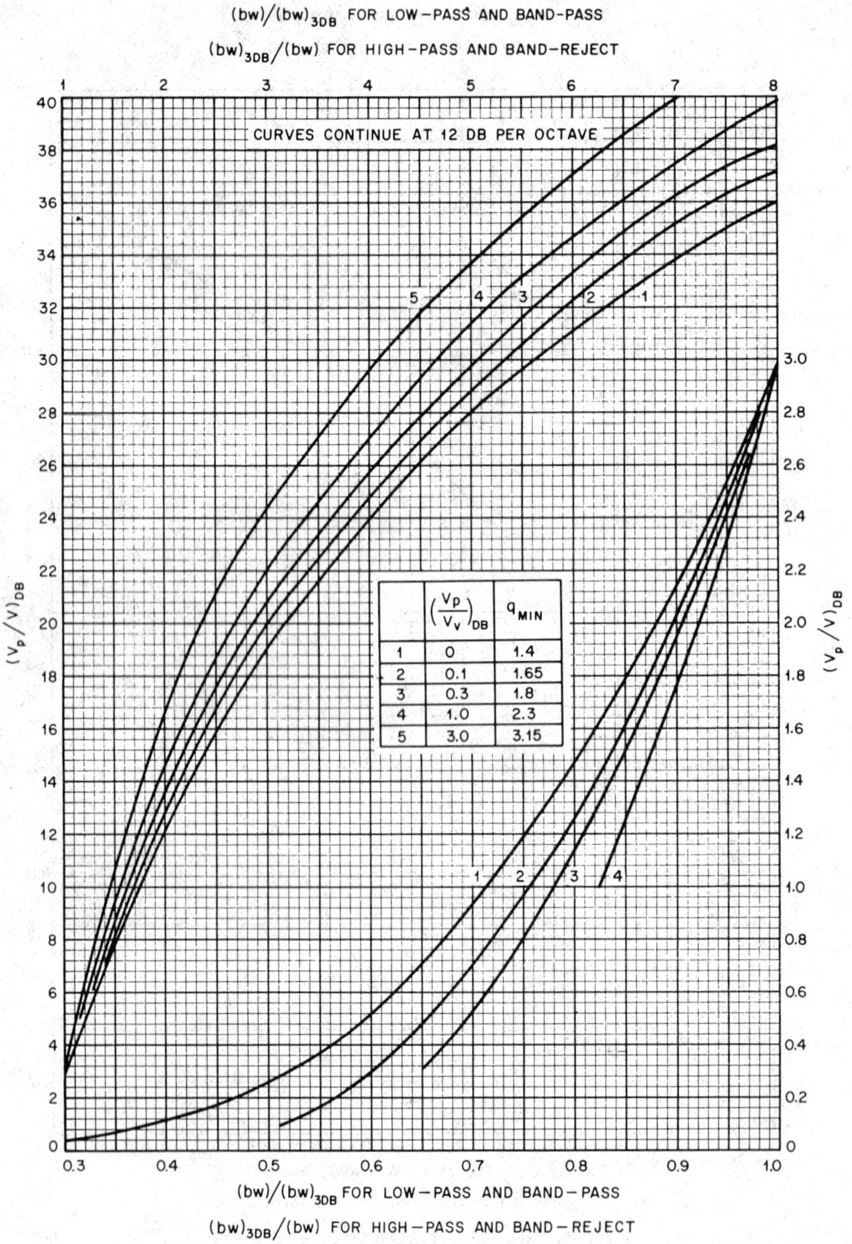

Fig. 12-6. Relative attenuation for a 2-pole network.

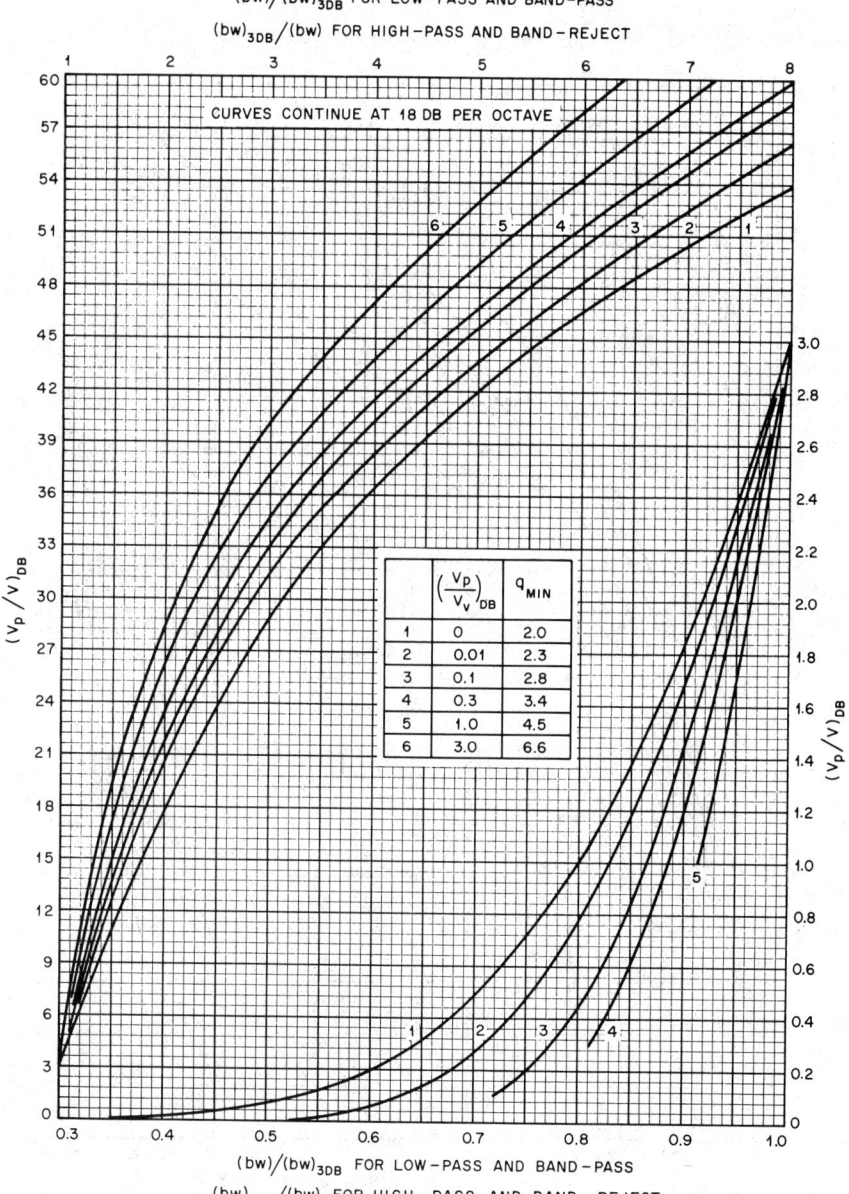

Fig. 12-7. Relative attenuation for a 3-pole network.

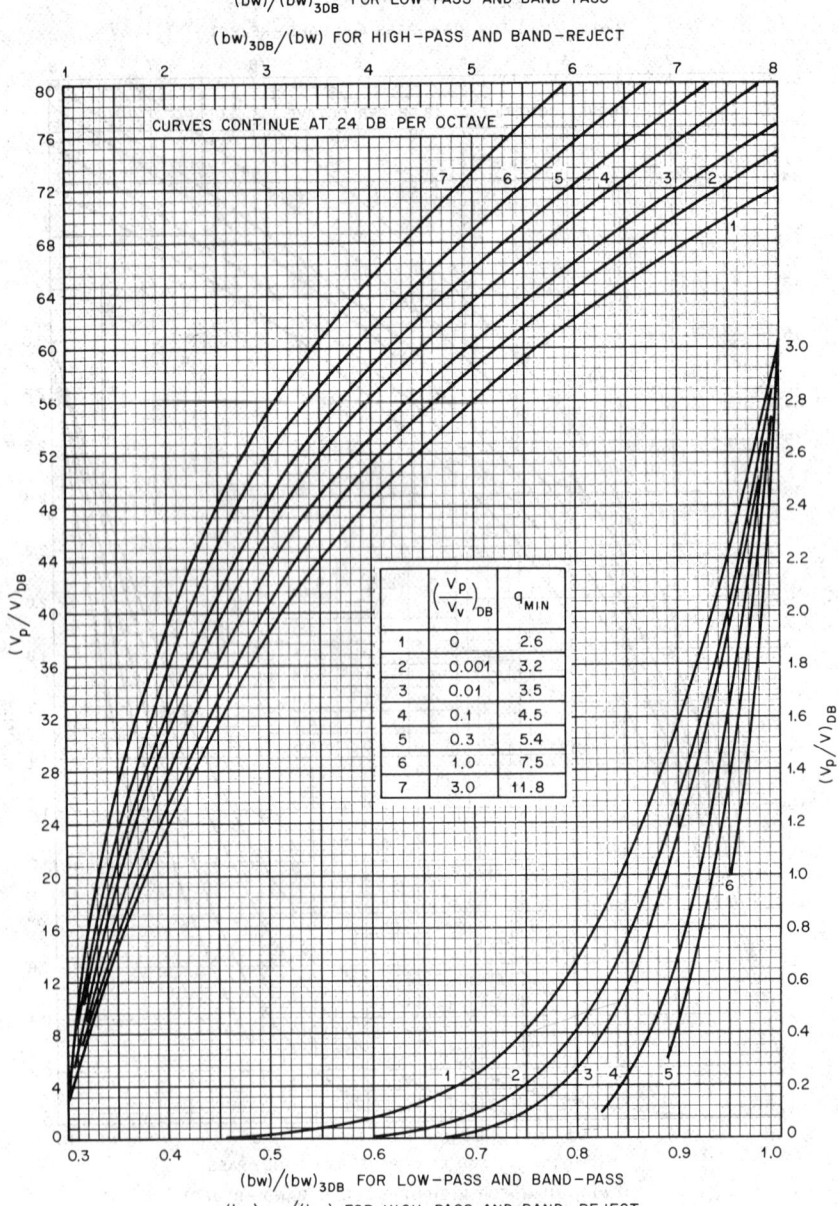

Fig. 12-8.   Relative attenuation for a 4-pole network.

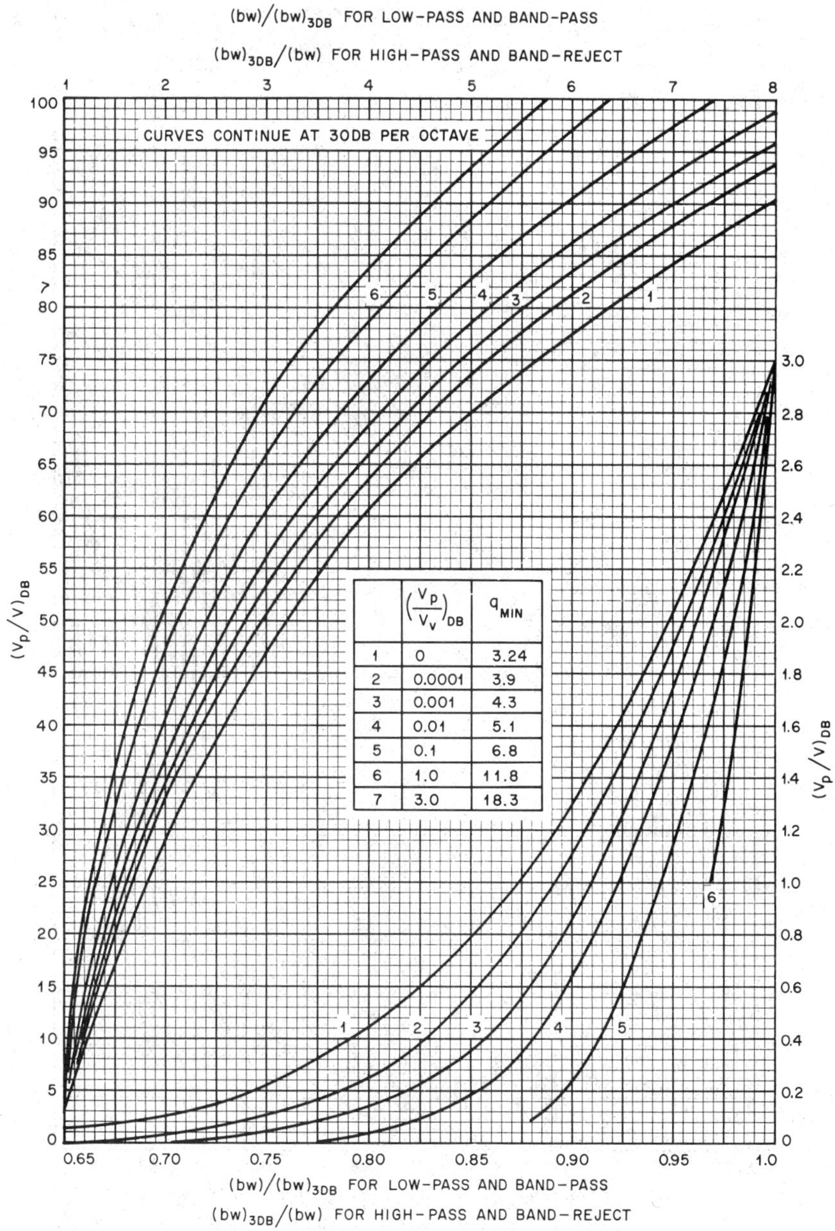
(bw)/(bw)3DB FOR LOW-PASS AND BAND-PASS
(bw)3DB/(bw) FOR HIGH-PASS AND BAND-REJECT
1 2 3 4 5 6 7 8
100 95 90 85 80 75 70 65 60 55 50 45 40 35 30 25 20 15 10 5 0
CURVES CONTINUE AT 30DB PER OCTAVE
7 6 5 4 3 2 1
(Vp/V)DB
3.0 2.8 2.6 2.4 2.2 2.0 1.8 1.6 1.4 1.2 1.0 0.8 0.6 0.4 0.2 0
(Vp/V)DB
6
1 2 3 4 5
0.65 0.70 0.75 0.80 0.85 0.90 0.95 1.0
(bw)/(bw)3DB FOR LOW-PASS AND BAND-PASS
(bw)3DB/(bw) FOR HIGH-PASS AND BAND-REJECT

	$\left(\frac{V_p}{V_v}\right)_{DB}$	q_{MIN}
1	0	3.24
2	0.0001	3.9
3	0.001	4.3
4	0.01	5.1
5	0.1	6.8
6	1.0	11.8
7	3.0	18.3

Fig. 12-9. Relative attenuation for a 5-pole network.

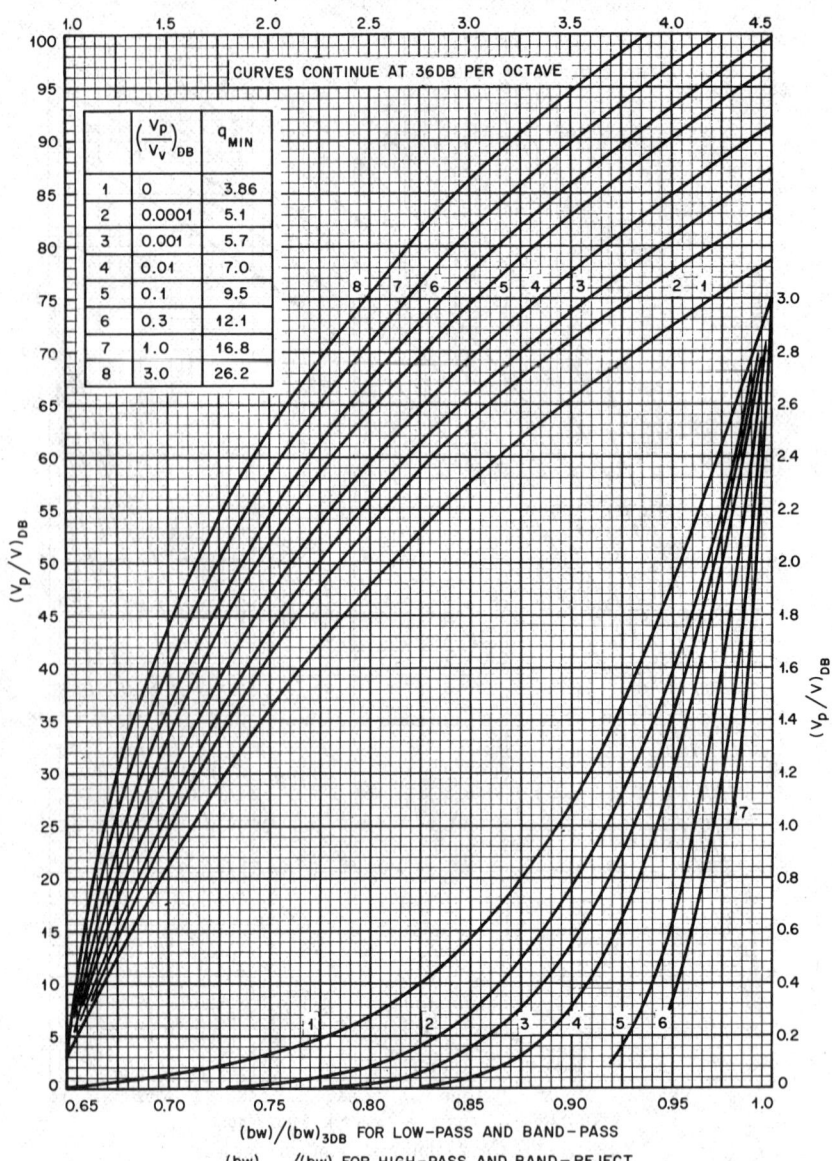

Fig. 12-10. Relative attenuation for a 6-pole network.

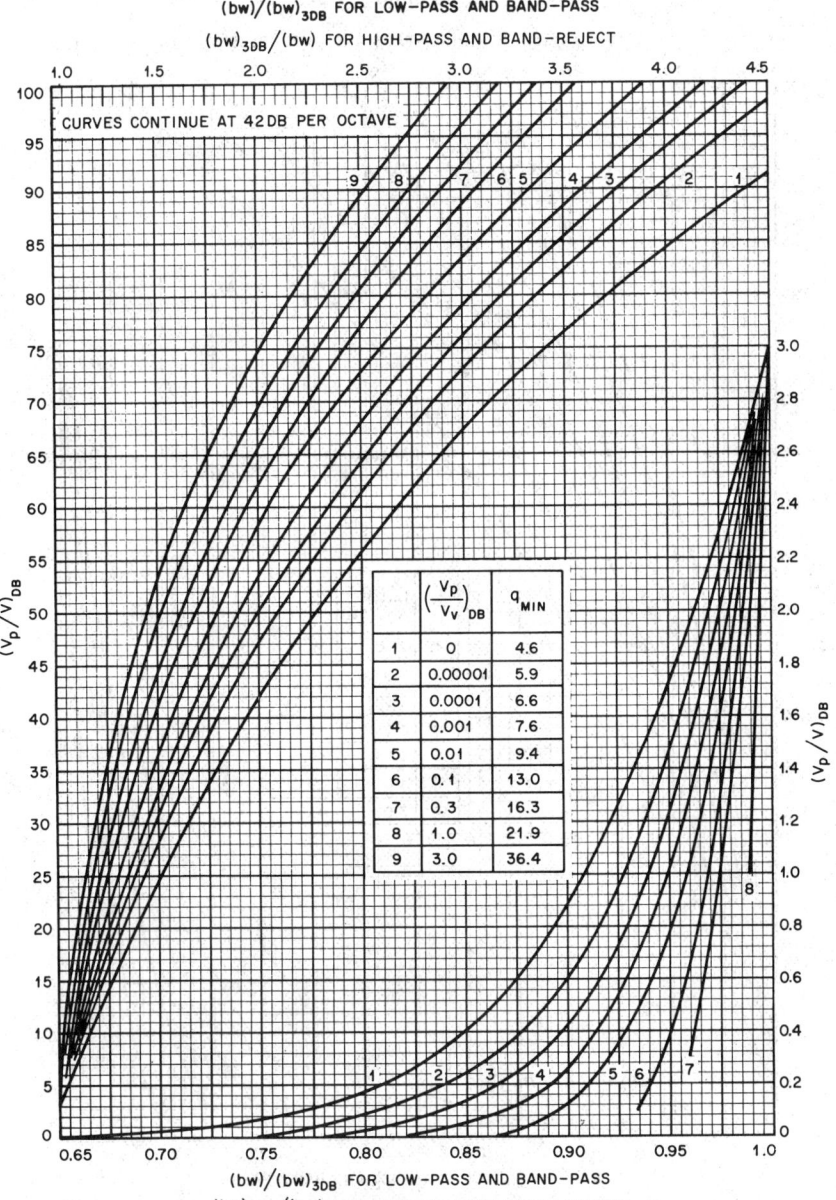

Fig. 12-11.   Relative attenuation for a 7-pole network.

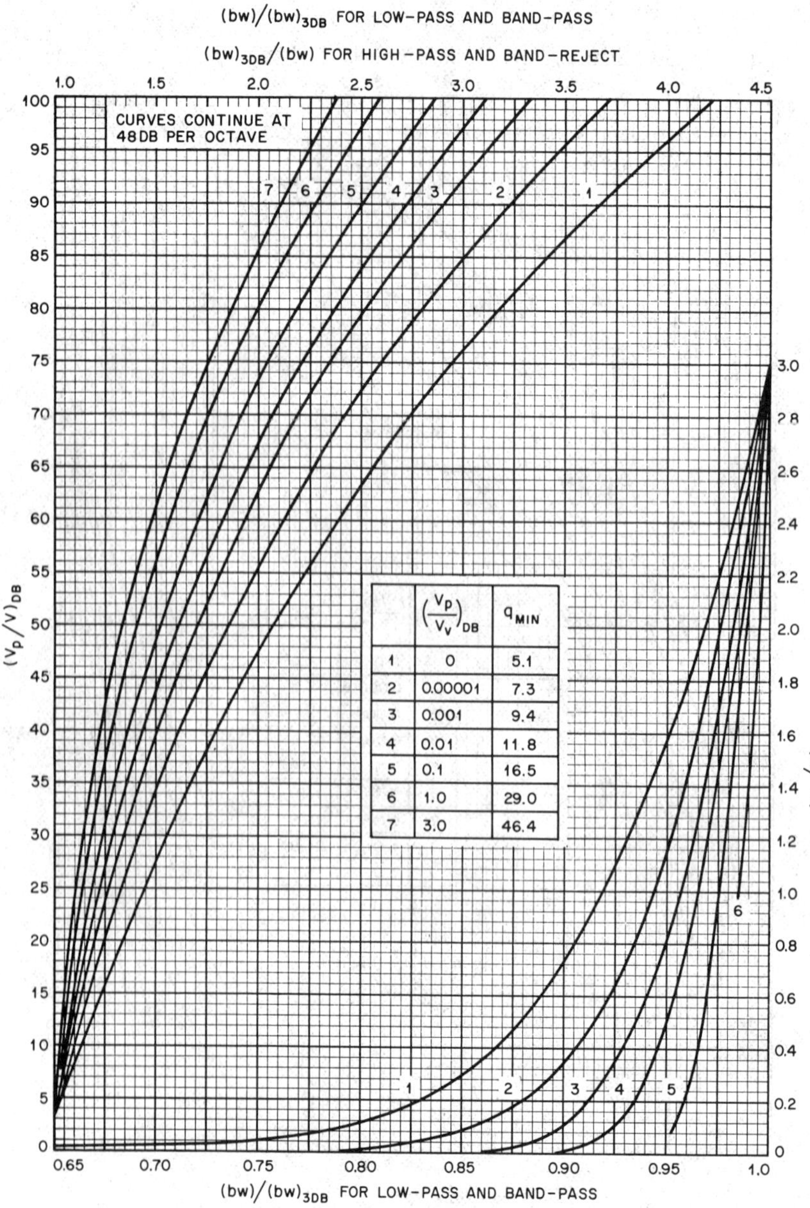

	$\left(\frac{V_p}{V_v}\right)_{DB}$	q_{MIN}
1	0	5.1
2	0.00001	7.3
3	0.001	9.4
4	0.01	11.8
5	0.1	16.5
6	1.0	29.0
7	3.0	46.4

Fig. 12-12. Relative attenuation for an 8-pole network.

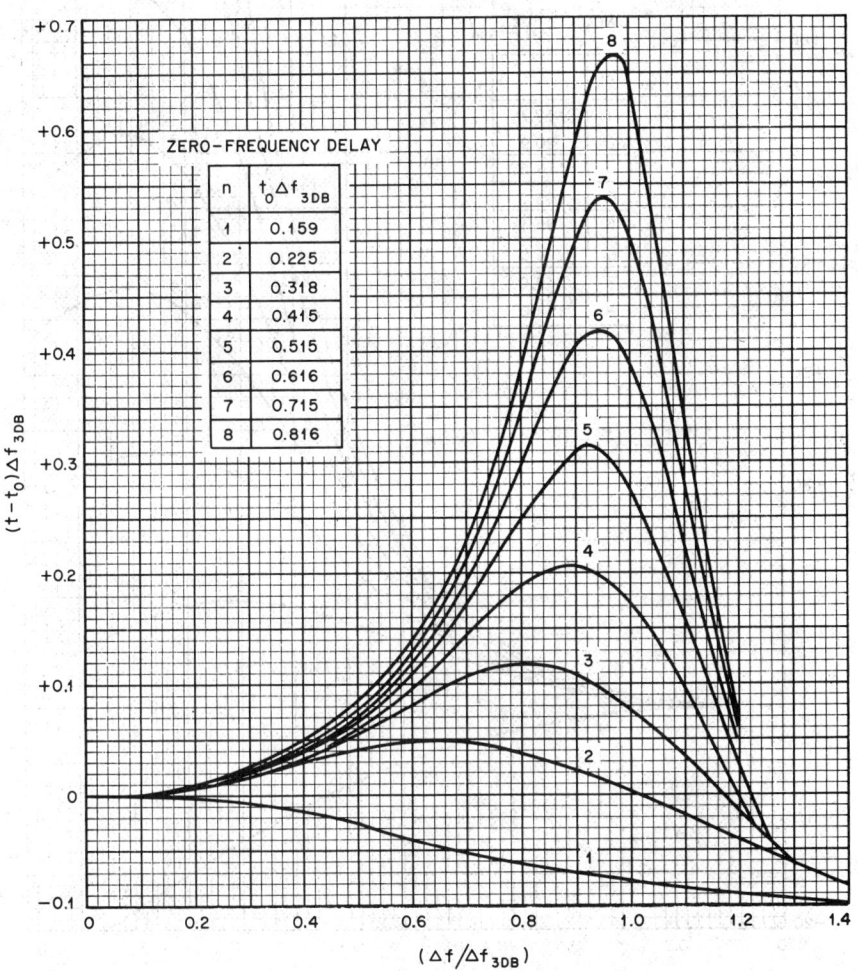

Fig. 12-13. Delay distortion of Butterworth response shape.

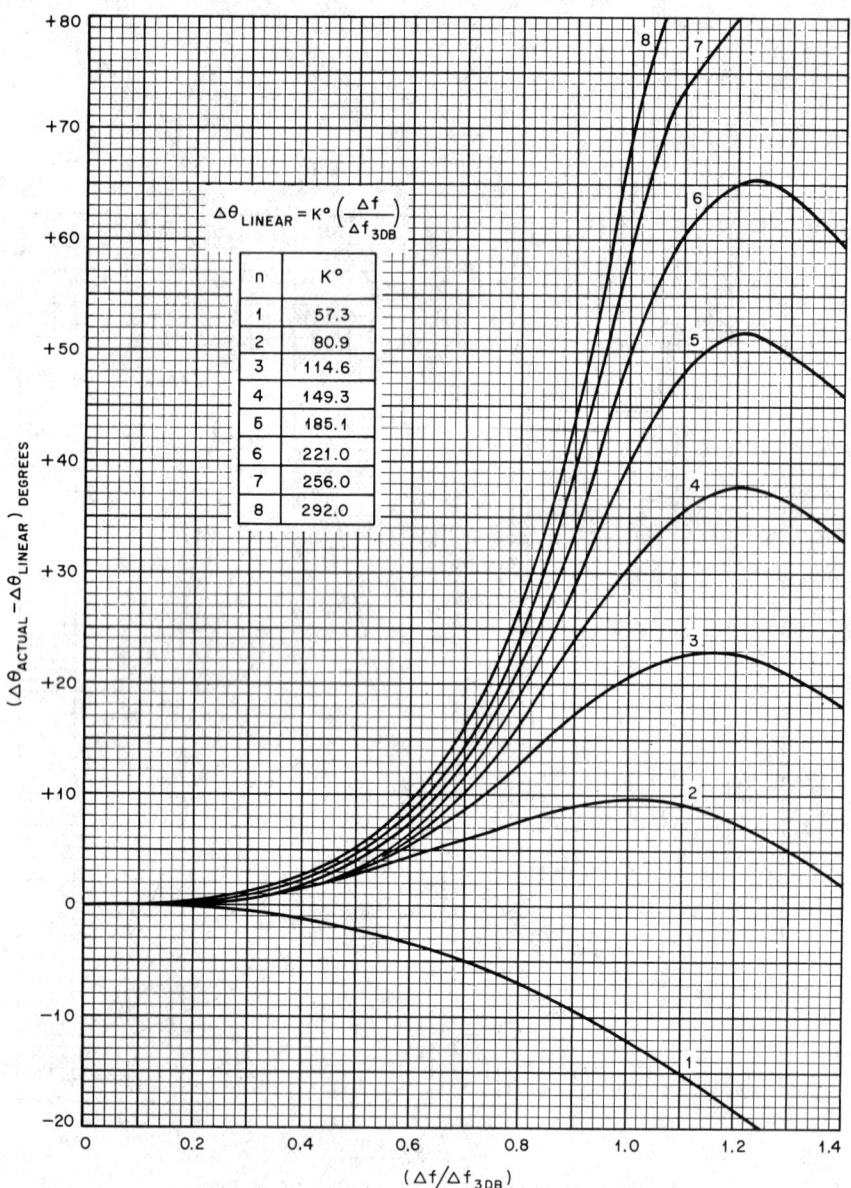

Fig. 12-14. Phase-shift distortion of Butterworth response shape.

For the internal resonators of bandpass circuits,

$$Q_{min} = q_{min}(f_0/BW_{3dB})$$

8. Examples of Shape Calculations. (*a*) In a low-pass filter without any peaks of infinite attenuation at a finite frequency, how few elements are required to satisfy the following specifications, and what minimum Q must they have? Response to be 1 dB down at 30 kHz and 50 dB down at not more than 75 kHz, compared with the peak response.

The allowable ripple is 1 dB in the passband. Then

$$BW_{50dB}/BW_{1dB} < 75/30 = 2.5$$

$$(V_p/V_v)_{dB} \leq 1.0 \text{ dB}$$

Since BW_{1dB} will be slightly less than BW_{3dB}, we must have BW_{50dB}/BW_{3dB} a little less than 2.5 when $(V_p/V)_{dB} = 50_{dB}$. Consulting Figs. 12-5 to 12-12 and examining curves for $(V_p/V_v)_{dB} = 1.0$, it is found that a 5-pole network (Fig. 12-9) is the least that will meet the requirements. Here, curve 6 gives

$$BW_{50dB}/BW_{3dB} = 2.14$$

while

$$BW_{1dB}/BW_{3dB} = 0.97$$

Then

$$BW_{50dB}/BW_{1dB} = 2.14/0.97 = 2.20$$

The 3-dB-frequency will be

$$30 \ BW_{3dB}/BW_{1dB} = 30/0.97 = 31 \text{ kHz}$$

At this frequency, the Q of each capacitor and inductor must be at least equal to $q_{min} = 11.8$, as shown in Fig. 12-9.

(*b*) Consider a bandpass filter with requirements similar to the above: bandwidth 1 dB down to be 30 kHz, 50 dB down at 75-kHz bandwidth, and 1-dB allowable ripple. Further, let the midfrequency be $f_0 = 500$ kHz. The solution at first is the same as above, and a 5-pole network is required.

The 3-dB bandwidth is 31 kHz, and the Q of each resonator must be at least

$$11.8 f_0/BW_{3dB} = 11.8 \times 500/31 = 190$$

where 11.8 is q_{min} as read from the table in Fig. 12-9. If a Q of 190 is not practical to attain, a greater number of resonators can be used. Suppose 7 resonators or poles are tried, as in Fig. 12-11. Then curve 2 gives

$$BW_{50dB}/BW_{1dB} = 2.10/0.93 = 2.26$$

The table shows the peak-to-valley ratio of 10^{-5} dB and $q_{min} = 5.9$. The 3-dB bandwidth is $30/0.93 = 32.2$ kHz. Then the minimum Q of each resonator can be 5.9 $\times$ 500/32.2 = 92, which is less than half that required if 5 resonators are used.

(*c*) In the bandpass filter, suppose the filter is subdivided into N identical stages in cascade, isolated by active devices or decoupling capacitors or resistors. For each stage the response requirements are the original number of decibels divided by N. For $N = 2$ stages,

$$BW_{25dB}/BW_{0.5dB} < 2.5$$

$$(V_p/V_v)_{dB} \leq 0.5 \text{ dB}$$

Proceeding as before, it is found that a 3-pole network (Fig. 12-7) for each stage will just suffice, curve 4 giving

$$(V_p/V_v)_{dB} = 0.3$$

and $BW_{25dB}/BW_{0.5dB} = 2.1/0.84 = 2.5$

To find the required minimum Q of each of the 6 resonators, the 3-dB bandwidth of each stage is

$$30/0.84 = 35.8 \text{ kHz}$$

For curve 4, $q_{min} = 3.4$, and so the minimum allowable Q for each resonator is

$$3.4 \times 500/35.8 = 47.5$$

9. Maximally Linear Phase Response.   In the design of filters where the linearity of the phase characteristic inside the passband is important, certain changes in design are necessary compared with the previously considered cases.   For filters supplying only transfer-function poles, rate of change of phase with frequency becomes more and more linear as the number of arms is increased, provided the design produces a complex relative attenuation characteristic given by the polynomial of the equation.[4]

$$\frac{V_p}{V} = \frac{n!}{(2n)!} \sum_{r=0}^{n} \frac{2^r(2n-r)!}{r!(n-r)!}\left[j\left(\frac{x}{x_\beta}\right)^r\right] \tag{12-6}$$

where r = a series of integers.   The magnitude of Eq. (12-6) is plotted in Figs. 12-5 and 12-6 for several values of n.   The former is for the relative attenuation inside the 3-dB points, and the latter for the response outside these points.   The curves for $n = \infty$ are plotted from the following equation, which is the gaussian shape that the attenuation characteristic approaches as n approaches infinity

$$10 \log(V_p/V)^2 = 3(x/x_{3dB})^2 \tag{12-7}$$

With a filter supplying only transfer-function poles, a maximally linear phase response can be produced only at the limitation of a rounded attenuation shape in the passband, as illustrated in Figs. 12-15 and 12-16.

The column labeled q_{min} in Fig. 12-15 gives the minimum allowable Q, measured at the 3-dB-down frequency, of the inductors and capacitors of a low-pass filter.   For bandpass filters, the minimum allowable unloaded Q at the midfrequency f_0 is $q_{min}f_0/BW_{3dB}$.   For the phase response figures in Fig. 12-15, the symbols are as follows.

   (a)   Low-pass filter:

$$t_0 = d\theta/d\omega$$

$$= \text{slope of phase characteristic at zero frequency, rad/rad} \cdot \text{s}^{-1}$$

$$t_{3dB} = \text{slope at } f_{3dB}$$

$$f_{3dB} = \text{frequency of 3-dB-down response}$$

$$BW_{3dB} = 2f_{3dB}$$

   (b)   Bandpass filter:

$$t_0 = \text{slope at midfrequency}$$

$$t_{3dB} = \text{slope at 3-dB-down bandwidth}$$

$$BW_{3dB} = \text{total 3-dB bandwidth}$$

The column $(t_0 - t_{3dB})\,BW_{3dB}$ shows the *group-delay* distortion over the passband. It shows numerically that the phase slope becomes much more constant as the number of elements is increased, in a filter designed for this purpose.

FILTERS SUPPLYING BOTH TRANSFER-FUNCTION POLES AND ZEROS

10. Elliptic Function Shapes   for these filters are shown in Figs. 12-4D to F.   The modern network theory of these filters has been treated by Norton and by Darlington.[1]   The attenuation shapes produced may be called *elliptic* and *inverse-hyperbolic* and are optimum

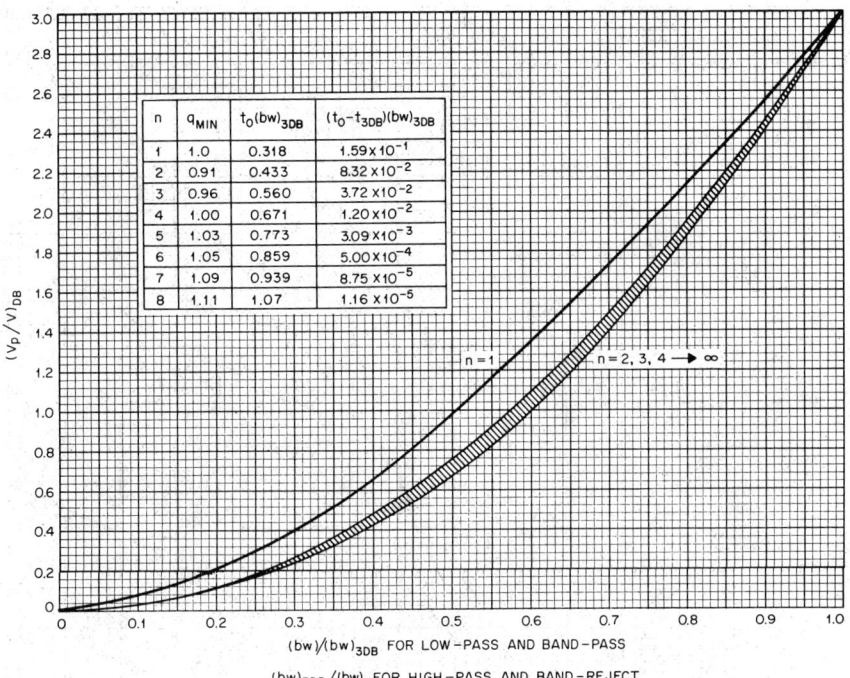

The table within the figure:

n	q_{MIN}	$t_0 (bw)_{3DB}$	$(t_0 - t_{3DB})(bw)_{3DB}$
1	1.0	0.318	1.59×10^{-1}
2	0.91	0.433	8.32×10^{-2}
3	0.96	0.560	3.72×10^{-2}
4	1.00	0.671	1.20×10^{-2}
5	1.03	0.773	3.09×10^{-3}
6	1.05	0.859	5.00×10^{-4}
7	1.09	0.939	8.75×10^{-5}
8	1.11	1.07	1.16×10^{-5}

Axis labels: vertical $(V_p/V)_{DB}$; horizontal $(bw)/(bw)_{3DB}$ FOR LOW–PASS AND BAND–PASS; $(bw)_{3DB}/(bw)$ FOR HIGH–PASS AND BAND–REJECT.

Fig. 12-15. Attenuation shape within the 3-dB-down passband for n-pole maximally flat time-delay filters.

in the sense that the rate of cutoff between the accept and reject bands is a maximum. The following equation gives the elliptic function shape:

$$(V_p/V)^2 = 1 + [(V_p/V_v)^2 - 1] \times cd_v^2[n(K_v/K_f)cd_f^{-1}(x/x_v)] \qquad (12\text{-}8)$$

where $cd = cn/dn$, the ratio of the two elliptic functions cn and dn (Ref. 5), $n = $ number of poles supplied by the filter, $x = $ a bandwidth variable described under Eq. (12-5), and K_v, $K_f = $ complete elliptic integrals of the first kind, evaluated for the modulus value given by the respective subscript.

Referring to the symbols in Fig. 12-4, the moduli v and f are given in

$$v = \{[(V_p/V_v)^2 - 1]/[(V_p/V_h)^2 - 1]\}^{1/2} \qquad (12\text{-}9)$$

$$f = x_v/x_n = BW_v/BW_h \qquad (12\text{-}10)$$

These are not independent, but must satisfy the equation

$$\log q_v = n \log q_f \qquad (12\text{-}11)$$

where q_k is called the *modular constant* of the modulus value k, the latter being equal to v or f, respectively. A tabulation of $\log q$ is available in the literature.[6]

In the limit, when $V_p/V_v = 1.0$ or zero decibels (Fig. 12-4F), the ripples in the accept band vanish. Then Eq. (12-8) reduces to the inverse hyperbolic shape of

$$\left(\frac{V_p}{V}\right)^2 = 1 + \frac{(V_p/V_h)^2 - 1}{\cosh^2[n \cosh^{-1}(x_h/x)]} \qquad (12\text{-}12)$$

12-19

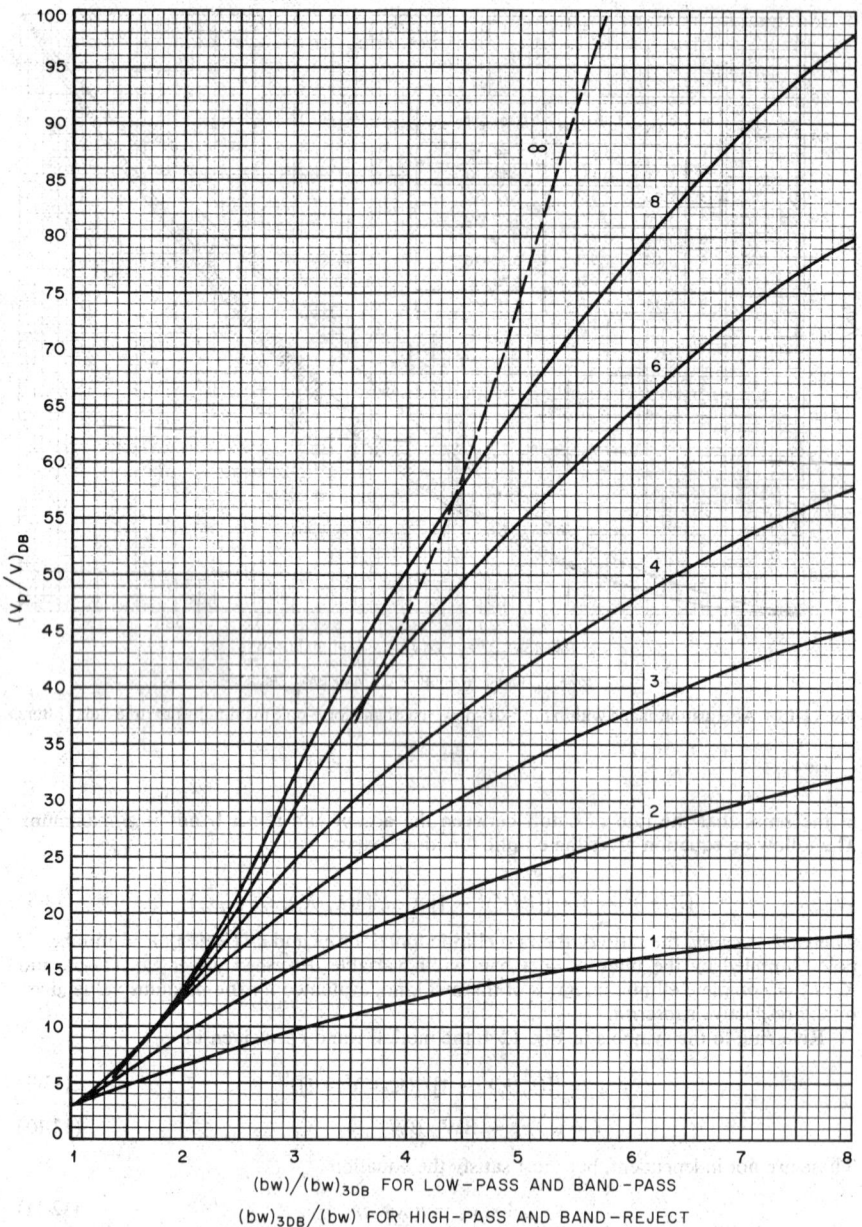

Fig. 12-16.   Attenuation shape beyond 3-dB-down passband for *n*-pole maximally flat time-delay filters.

Curves plotted from Eqs. (12-8) and (12-12) are presented in Figs. 12-17 to 12-22. Those labeled $V_p/V_v = 0$ dB, for n poles, m zeros, are plotted from Eq. (12-12), while the others are from Eq. (12-8). For all these shapes, n = number of poles = number of arms in the ladder network. When n is an even number, the number of zeros $m = n$. When n is odd, $m = n - 1$. The following description of Fig. 12-17 can be extended to cover the entire group of figures mentioned above.

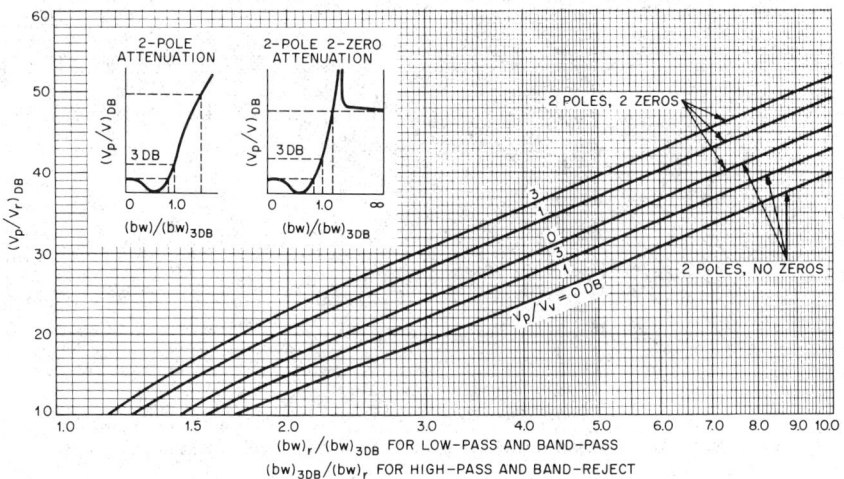

Fig. 12-17. Maximum rate of cutoff for 2-pole and for 2-pole 2-zero filters.

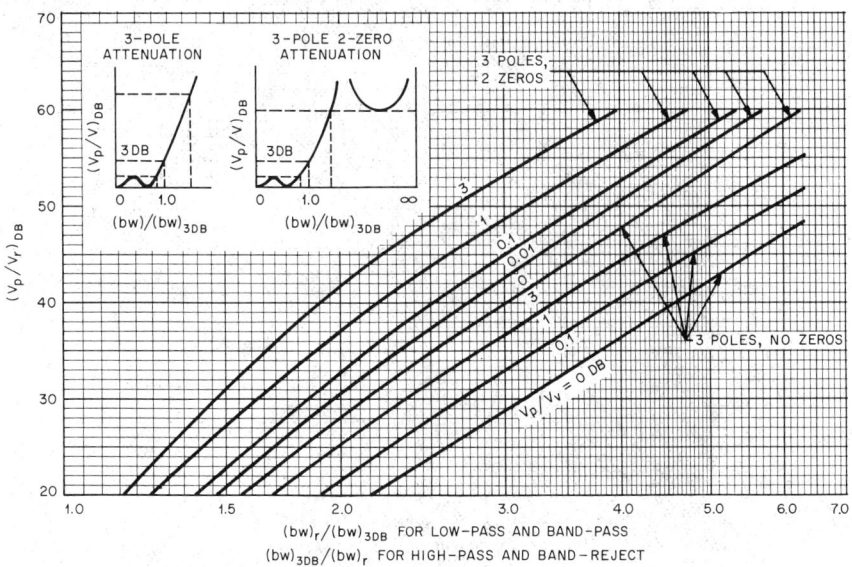

Fig. 12-18. Maximum rate of cutoff for 3-pole and for 3-pole 2-zero filters.

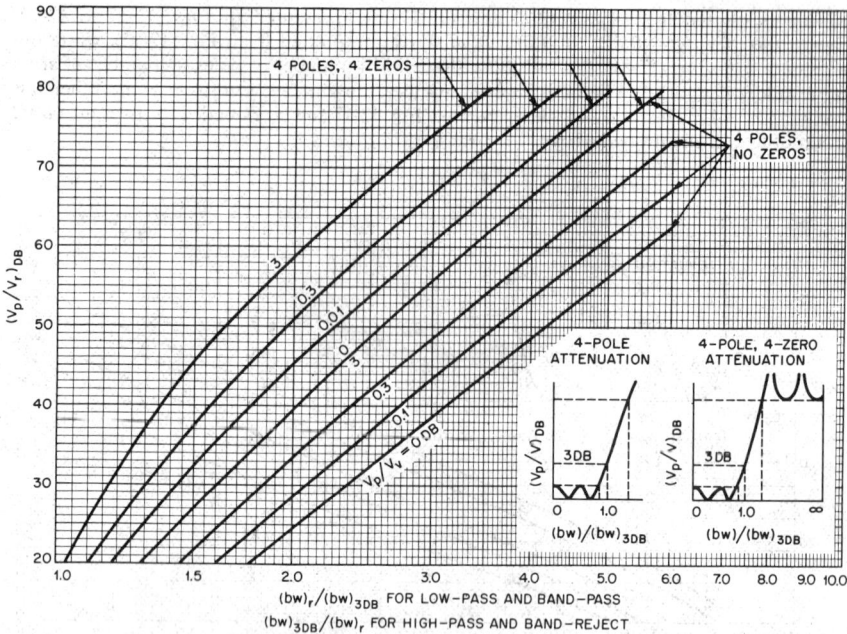

Fig. 12-19. Maximum rate of cutoff for 4-pole and for 4-pole 4-zero filters.

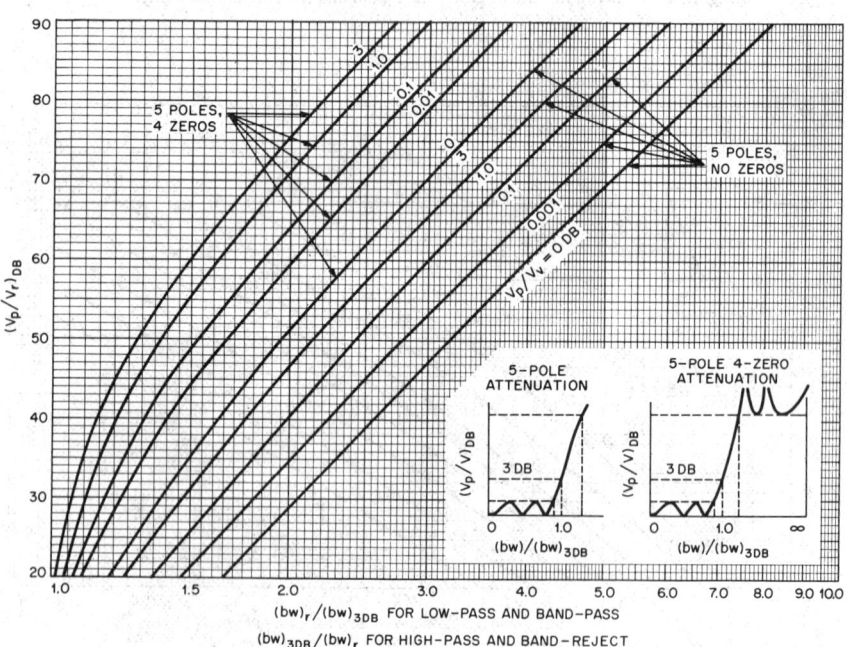

Fig. 12-20. Maximum rate of cutoff for 5-pole and for 5-pole 4-zero filters.

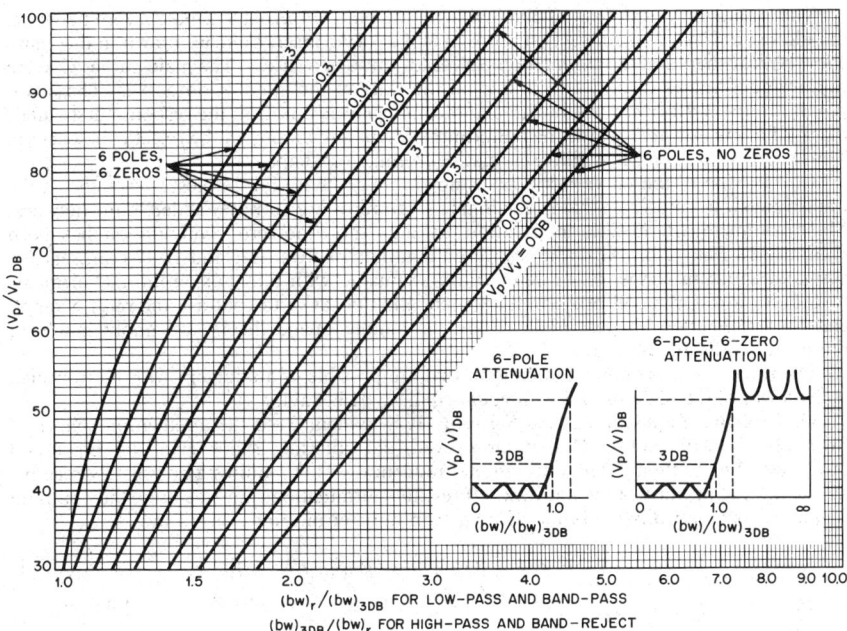

Fig. 12-21. Maximum rate of cutoff for 6-pole and for 6-pole 6-zero filters.

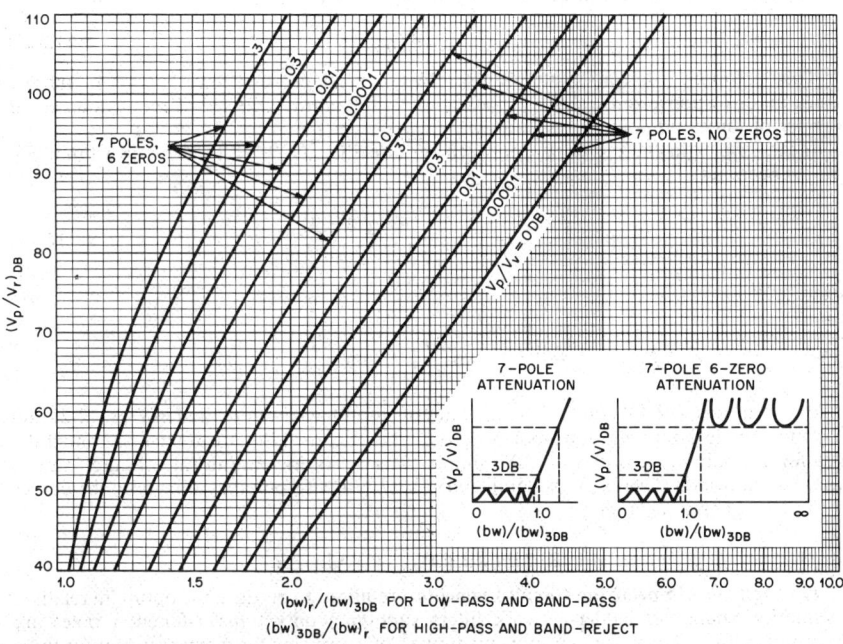

Fig. 12-22. Maximum rate of cutoff for 7-pole and for 7-pole 6-zero filters.

The maximum rates of cutoff obtainable with 2-pole no-zero and 2-pole 2-zero networks are plotted in Fig. 12-17 for several ratios of V_p/V_v. Two insert sketches drawn in the figure show typical shapes of the attenuation curves for these two cases. The main curves give the relative coordinates of only two points on the skirt of the attenuation curve. These two points are the 3-dB-down bandwidth and the *hill bandwidth* (where the response first equals that of the *response hills*, where the uniform minimum attenuation in the reject band occurs). Thus each point specifies a different relative attenuation shape.

Comparison of the curves for 2-poles no-zero with those for 2-poles 2-zeros shows the improvement in cutoff rate that is obtainable when zeros are correctly added to the network. More complete attenuation information on the 2-pole no-zero configuration has been presented in Fig. 12-6. Again, it is stressed that data of Figs. 12-6 and 12-17 represent the actual attenuation shapes and rate of cutoff attainable with filters using finite-Q elements (except for a rounding off of the infinite attenuation peaks). In contrast, the rates of cutoff and the attenuation shapes predicted by the simple image theory are unobtainable in physically realizable networks.

The rates of cutoff shown are the best that are possible of attainment with the specified number of poles and zeros and with equal-ripple-type behavior.

11. Resistive Terminations and n Even. It is evident from the attenuation shapes of Figs. 12-17, 12-19, and 12-21 that for n even, the optimum shape given by Eq. (12-8) produces a finite attenuation at an infinite frequency. This requires a completely reactive termination at one end of the network. If resistive terminations must be used, the optimum shape that is practically realizable with an even number of arms is given by

$$(V_p/V)^2 = 1 + [(V_p/V_v)^2 - 1]cd_v^2[n(K_v u/K_f)] \tag{12-13}$$

where

$$u = sc_f^{-1}\{[(x_v/x)^2 - 1]^{1/2}[dn_f(K_f/n)]/f'\} \tag{12-14}$$

The modulus v is given by Eq. (12-9), and the modulus f by Eq. (12-10). Solving Eq. (12-13) then gives the ratio of hill-to-valley bandwidth as

$$x_h/x_v = [fcd_f(K_f/n)]^{-1} \tag{12-15}$$

This optimum attenuation shape, Eq. (12-13), produces two fewer points of infinite rejection, or response zeros, than response poles. In contrast, Eq. (12-8) requires an equal number of zeros and poles.

If the ripples in the passband approach zero decibels ($V_p/V_v = 1$), then, as a limit, Eq. (12-13) becomes

$$\left(\frac{V_p}{V}\right)^2 = 1 + \frac{(V_p/V_h)^2 - 1}{\cosh^2(n \cosh^{-1}y)} \tag{12-16}$$

where

$$y = \left[\left(\frac{x_h}{x}\cos\frac{90}{n}\right)^2 + \sin^2\frac{90}{n}\right]^{1/2}$$

Based on Eqs. (12-13) and (12-16), the rates of cutoff have been plotted in Figs. 12-23 and 12-24 for 4-pole 2-zero and for 6-pole 4-zero filters. Figure 12-6 already has presented the data for a 2-pole no-zero network, the simplest case. An increase in rate of cutoff results when $n - 2$ response zeros are suitably added to n response poles, as shown by the curves in Figs. 12-23 and 12-24.

CIRCUIT-ELEMENT VALUES

12. Methods of Specifying Circuit Elements   required to produce the optimum relative-attenuation shapes of ladder-network filters supplying only transfer-function poles are discussed below. There are two convenient ways of expressing the element values for these ladder networks:

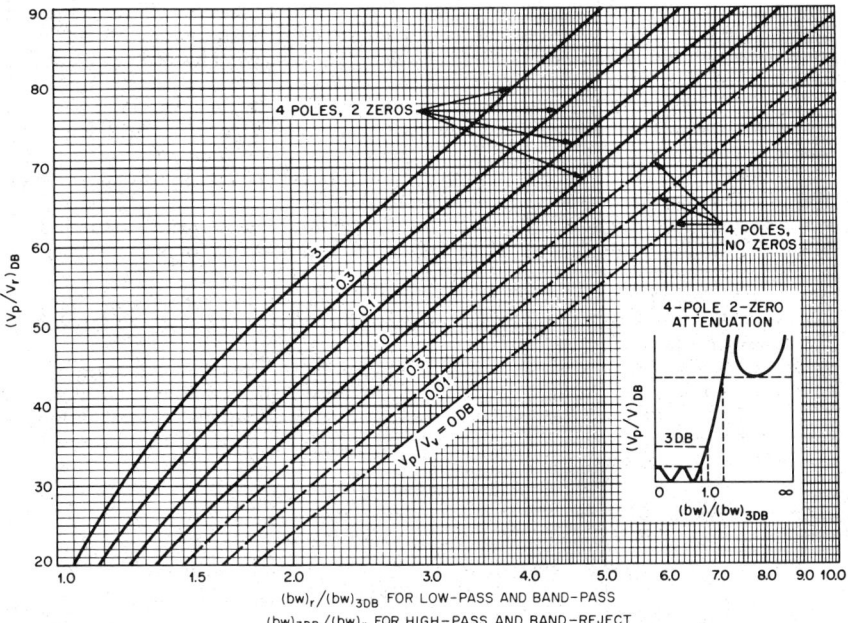

Fig. 12-23. Maximum rate of cutoff for 4-pole and for 4-pole 2-zero filters.

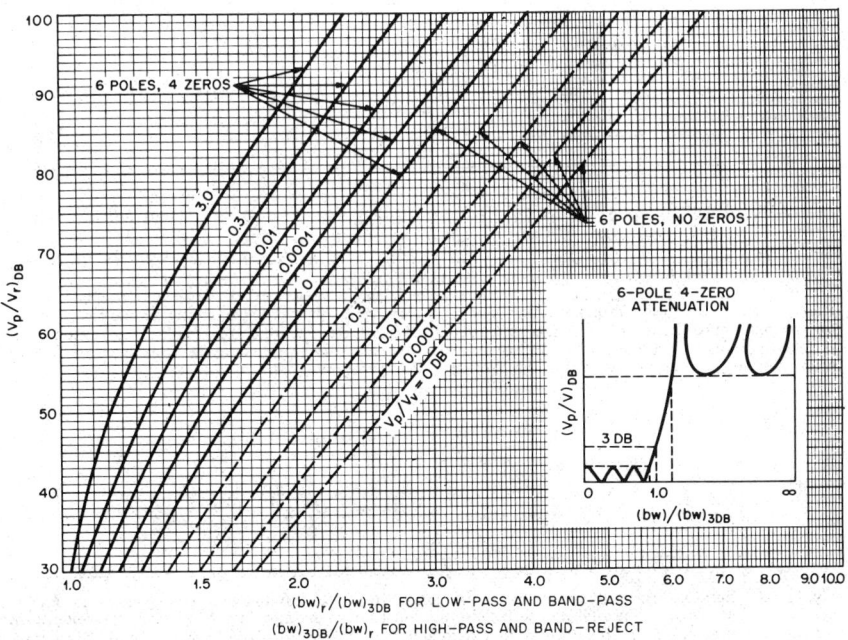

Fig. 12-24. Maximum rate of cutoff for 6-pole and for 6-pole 4-zero filters.

1. The reactive and resistive components of each element may be related to one of the terminating resistances (or to a completely arbitrary normalizing resistance R_0) and also to a definite bandwidth, usually the 3-dB-down value. The numerical results are called *ladder-network coefficients*, or *singly loaded Q's*.

2. The reactive component of each element may be related to the reactive part of the immediately preceding element, and to a definite bandwidth such as the 3-dB-down value. These numerical results are called the *normalized coefficients of coupling*. The resistive component of each element is related to its reactive part, and the numerical values are called *normalized decrements*, or, when inverted, *normalized Q's*.

The latter form of normalized coefficients of coupling k and normalized Q's (= q) will be used because the numerical values may be applied directly to the adjustment and checking of actual filters.[5]

Figures 12-25 to 12-28 relate the normalized k and q to the inductance, capacitance, and resistance values for low-pass and bandpass filters. Figs. 12-29 to 12-31 do this for high-pass and band-reject filters.

For low-pass filters, Fig. 12-25 shows that k gives the ratio of resonant frequency of two immediately adjacent elements to the overall 3-dB-down frequency. The resonant frequency of C_1 and L_2 in this example must be k_{12} times the required overall 3-dB-down bandwidth.

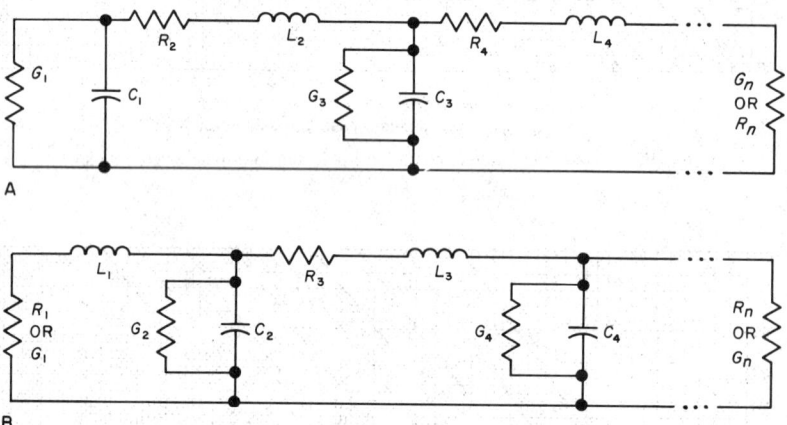

Fig. 12-25.   Relations among normalized k and q and values of inductance, capacitance, and resistance for low-pass and large-percentage bandpass circuits. (*A*) Shunt arm at one end. $1/(C_1L_2)^{1/2} = k_{12\omega3dB}$, $1/(L_2C_3)^{1/2} = k_{23\omega3dB}$, $1/(C_3L_4)^{1/2} = k_{34\omega3dB}$, etc. $G_1/C_1 = (1/g_1)\omega_{3dB}$, $q_2 = (\omega_{3dB}L_2)/R_2$, $q_3 = (\omega_{3dB} C_3)/G_3$, $q_4 = (\omega_{3dB} L_4)/R_4$, etc. (*B*) Series arm at one end. $1/(L_1C_2)^{1/2} = k_{12\omega3dB}$, $1/(C_2L_3)^{1/2} = k_{23\omega3dB}$, $1/(L_3C_4)^{1/2} = k_{34\omega3dB}$, etc. $R_1/L_1 = (1/q_1)\omega_{3dB}$, $q_2 = (\omega_{3dB} C_2)/G_2$, $q_3 = (\omega_{3dB} L_3)/R_3$, $q_4 = (\omega_{3dB} C_4)/G_4$, etc. To design a bandpass circuit that supplies a frequency response having geometric symmetry, for any percentage bandwidth the total required 3-dB-down bandwidth should replace ω_{3dB}, an inductor should be connected across each shunt capacitor, and a capacitor put in series with each series inductor, each such circuit being resonated to the geometric mean frequency $f_0 = (f_1 f_2)^{1/2}$.

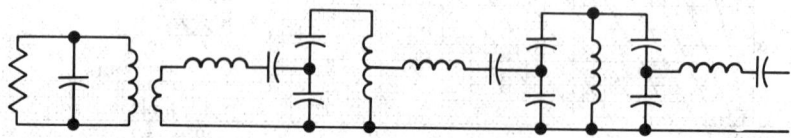

Fig. 12-26.   Bandpass circuit supplying a frequency response having geometric symmetry for any percentage bandwidth. Parallel and series circuits must alternate, and Fig. 12-27 gives the relationship between element values and resulting actual coefficient of coupling $K\{ = k(BW)_{3dB}/f_0\}$. Any adjacent pair of resonators may be coupled by any of the methods shown.

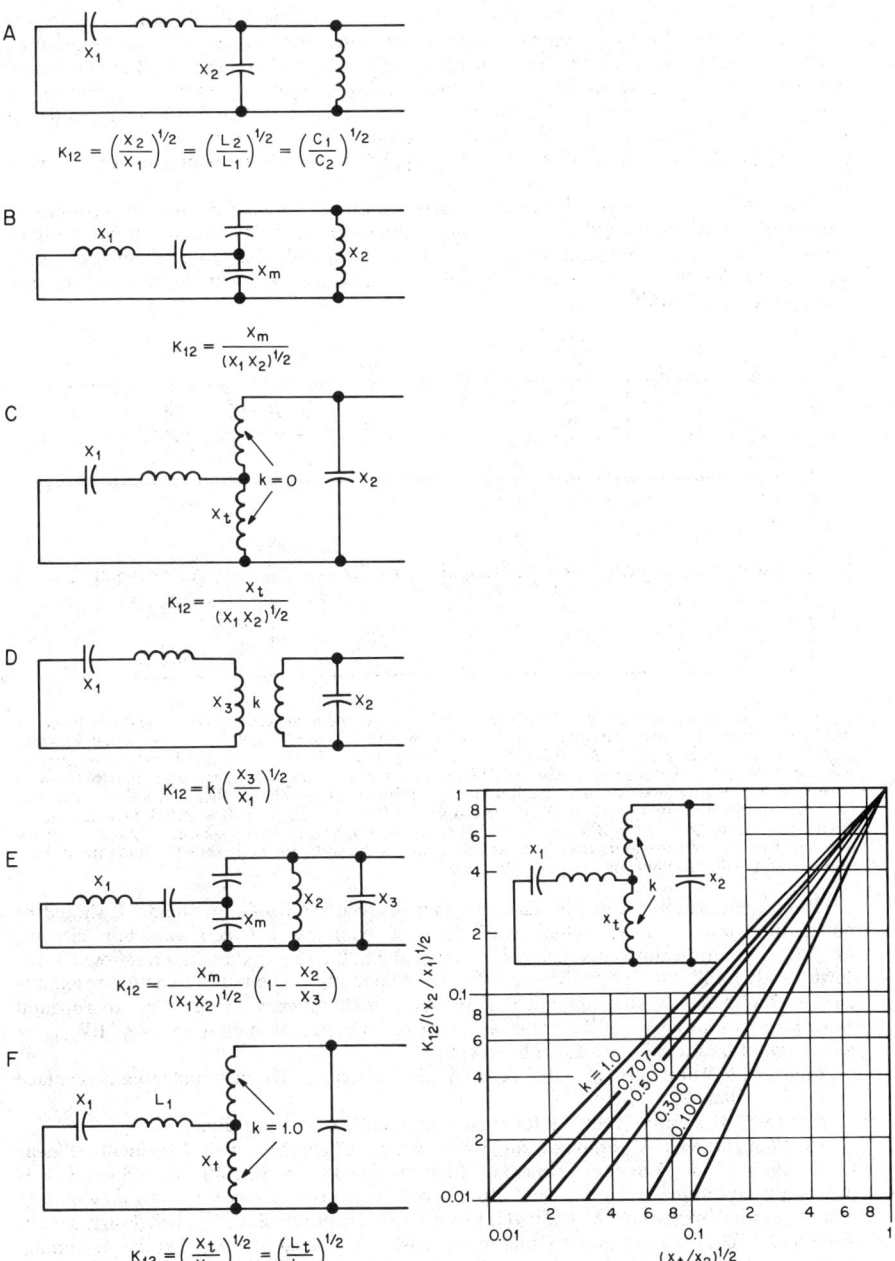

Fig. 12-27. Coefficient-of-coupling configurations for the bandpass circuit of Fig. 12-26. The node resonator is tuned to the desired midfrequency with the mesh resonator open-circuited, or the mesh resonator is tuned to this midfrequency with the node resonator short-circuited.

Figure 12-25 also gives, as the inverse of q, the ratio of the 3-dB-down bandwidth of a single element resulting from the resistive load and losses associated with it to the required 3-dB-down bandwidth of the overall filter. Thus $1/R_1C_1$ is the 3-dB-down radian bandwidth of C_1 and the conductance G_1 that must be shunted across it. If C_1 and G_1 are properly chosen, the measured bandwidth of these elements at their 3-dB-down point will be $1/q_1$ times the required overall 3-dB-down bandwidth of the filter.

The legend of Fig. 12-25 shows how it is applicable also to large-percentage bandpass filters.

When the procedure of Fig. 12-25 is used to design bandpass circuits of medium- or small-percentage bandwidths, the resulting element value ratios become impractical to obtain. The circuit configuration shown in Figs. 12-26 and 12-27 give practical element-value ratios for these two cases while still providing true geometric symmetry for any percentage of bandwidth.

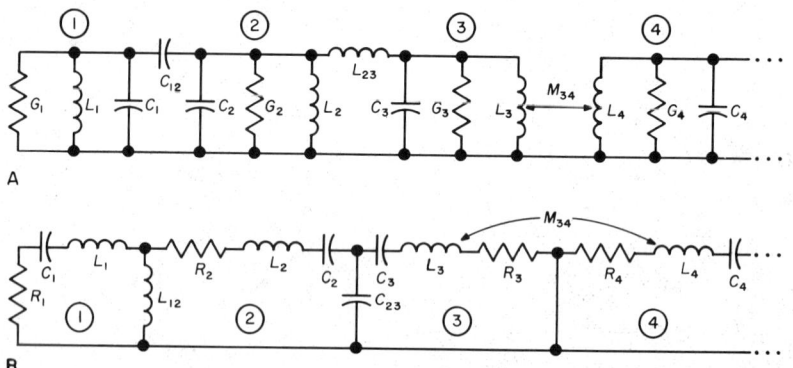

Fig. 12-28. Relations among normalized k and q and values of inductance, capacitance, and resistance for small-percentage bandpass circuits. (*A*) Parallel-resonant circuits, $C_{12}/(C_1C_2)^{1/2} = k_{12}(\text{BW}_{3\text{dB}}/f_0)$, $(L_2L_3)^{1/2}/L_{23} = k_{23}(\text{BW}_{3\text{dB}}/f_0)$, $M_{34}/(L_3L_4)^{1/2} = k_{34}(\text{BW}_{3\text{dB}}/f_0)$, etc. $Q_1 = q_1(f_0/\text{BW}_{3\text{dB}})$, $q_2 = Q_2/(f_0/\text{BW}_{3\text{dB}})$, $q_3 = Q_3/(f_0/\text{BW}_{3\text{dB}})$, $q_4 = Q_4/(f_0/\text{BW}_{3\text{dB}})$, etc. Any adjacent pair of resonators may be coupled by any of the three methods shown. Each node must resonate at f_0 with all other nodes short-circuited. (*B*) Series-resonant circuits. $L_{12}/(L_1L_2)^{1/2} = k_{12}(\text{BW}_{3\text{dB}}/f_0)$, $(C_2C_3)^{1/2}/C_{23} = k_{23}(\text{BW}_{3\text{dB}}/f_0)$, $M_{34}/(L_3L_4)^{1/2} = k_{34}(\text{BW}_{3\text{dB}}/f_0)$, etc. $Q_1 = q_1(f_0/\text{BW}_{3\text{dB}})$, $q_2 = Q_2/(f_0/\text{BW}_{3\text{dB}})$, $q_3 = Q_3/(f_0/\text{BW}_{3\text{dB}})$, $q_4 = Q_4/(f_0/\text{BW}_{3\text{dB}})$. Any adjacent pair of resonators may be coupled by any of the three methods shown. Each mesh must resonate at f_0 with all other meshes open-circuited.

Similar data are given in Fig. 12-28 for small-percentage bandpass filters. It should be noted that the required actual coefficient of coupling between resonant circuits, $M_{ab}/(L_aL_b)^{1/2}$ for example, is obtained by multiplying the required overall fractional 3-dB-down bandwidth by the normalized coefficient of coupling. The required actual resonant-circuit Q results from multiplying the fractional midfrequency by q. An experimental procedure for checking k and q values is available.[7] Fractional midfrequency $f_0/\text{BW}_{3\text{dB}} =$ reciprocal of fractional 3-dB-down bandwidth.

Figures 12-29 and 12-30 give the required information for high-pass and large-percentage band-reject filters.

Figure 12-31 supplies the data for small-percentage band-reject filters.

13. Element Values Required for Butterworth, Chebyshev, and Maximally Linear Phase Shapes Using Lossless Elements. Elegant closed-form equations for k and q values producing optimum Chebyshev and Butterworth response shapes for filters having any number of total arms may be obtained if lossless reactances are used.[8-11] The design data in Tables 12-1 to 12-7 are based on such equations. The k and q values for the maximally linear phase shape result from the Darlington synthesis procedure applied to Eq. (12-6). The tables provide data for two limiting cases of terminations: maximum-power-transfer loading at the two ends of the filter and resistive loading at only one end.

For Tables 12-1 to 12-7, the $(V_p/V_v)_{\text{dB}}$ column gives the ripple in decibels in the passband, and the corresponding curves on Figs. 12-5 to 12-12 give the complete attenuation shape.

For low-pass circuits, $q_{2,3, \ldots (n-1)}$, is the required unloaded Q, measured at the required 3-dB-down frequency, of the internal inductors and capacitors to be used. For bandpass circuits, the unloaded resonator Q required in the internal resonators is obtained by multiplying the required 3-dB fractional midfrequency $(f_0/\text{BW}_{3\text{dB}})$ by $q_{2,3,\ldots(n-1)}$.

For the detailed way in which the q and k columns fix the required element values, see Figs. 12-25 to 12-31 and the related discussion.

To be exactly correct, the design values given in the tables require that (as shown in the second column of each table except Table 12-1) infinite unloaded Q's be available for the internal elements of the filter. It should be realized that designs can be made using elements having finite unloaded Q's; these designs are given in Figs. 12-33 to 12-58.

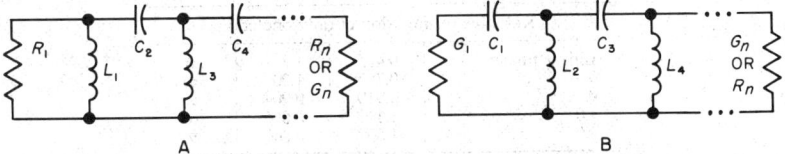

A   B

Fig. 12-29. Relations among normalized k and q and values of inductance, capacitance, and resistance for high-pass and large-percentage band-reject circuits. (A) Shunt arm at one end. $1/(L_1C_2)^{1/2} = (1/k_{12})\omega_{3\text{dB}}$, $1/(C_2L_3)^{1/2} = (1/k_{23})\omega_{3\text{dB}}$, $1/(L_3C_4)^{1/2} = (1/k_{34})\omega_{3\text{dB}}$, etc. $(R_1/L_1) = q_1\omega_{3\text{dB}}$. All reactances are assumed to be lossless. (B) Series arm at one end. $1/(C_1L_2)^{1/2} = (1/k_{12})\omega_{3\text{dB}}$, $1/(L_2C_3)^{1/2} = (1/k_{23})\omega_{3\text{dB}}$, $1/(C_3L_4)^{1/2} = (1/k_{34})\omega_{3\text{dB}}$, etc. $(G_1/C_1) = q_1\omega_{3\text{dB}}$. All reactances are assumed to be lossless. To design a band-reject circuit supplying a frequency response having geometric symmetry for any percentage bandwidth, the total required 3-dB-down bandwidth should replace $\omega_{3\text{dB}}$, a capacitor should be placed in series with each shunt inductor, and an inductor in shunt of each series capacitor, each such circuit being resonated to the geometric mean frequency $f_0 = (f_1 f_2)^{1/2}$.

$$f_0 \rightarrow \boxed{\;\;\overset{X_0}{\underset{X_0}{\quad\quad}}\;\;} \;\doteq\; \boxed{\;\;\overset{\leftarrow f_0}{\underset{(X_0/k^2)}{k\doteq 1.0}}\;\;}$$

Fig. 12-30. A useful approximate equivalence. After the band-reject procedure given in Fig. 12-29 is applied, then at frequencies where iron cores enable high-Q, tightly coupled transformers to be built, this equivalence allows the parallel resonant circuits to be embodied using more convenient component values.

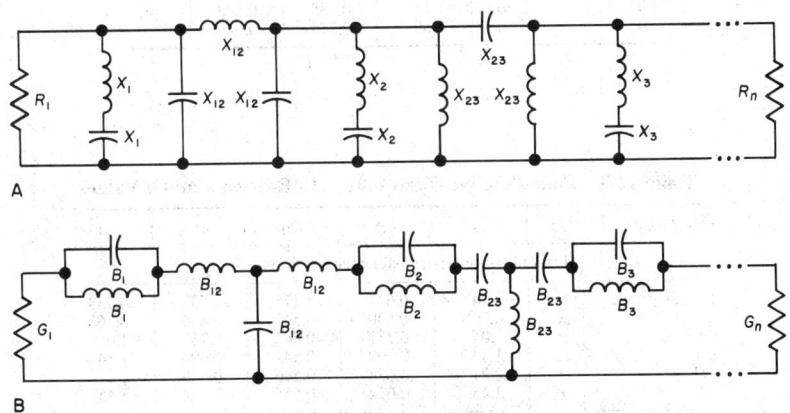

A

B

Fig. 12-31. Relations among normalized k and q and values of inductance, capacitance, and resistance for small-percentage band-reject circuits. (A) Series-resonant circuits. $X_{12}/(X_1X_2)^{1/2} = (1/k_{12})$ $(\text{BW}_{3\text{dB}}/f_0)$, $X_{23}/(X_2X_3)^{1/2} = (1/k_{23})(\text{BW}_{3\text{dB}}/f_0)$, etc. $X_1/R_1 = (1/q_1)\,[\,(f_0\text{BW}_{3\text{dB}})\,]$, $X_n/R_n = (1/q_n)\,[\,f_0/(\text{BW}_{3\text{dB}})\,]$. All resonant circuits are assumed to be lossless. Any adjacent pair of resonators may be coupled by either of the two π (or their dual-T) couplings shown. The reactances X are measured at the midfrequency of the reject band. (B) Parallel-resonant circuits. $B_{12}/(B_1B_2)^{1/2} = (1/k_{12})$ $(\text{BW}_{3\text{dB}}/f_0)$, $B_{23}/(B_2B_3)^{1/2} = (1/k_{23})$ $(\text{BW}_{3\text{dB}}/f_0)$, etc. $B_1/G_1 = (1/q_1)\,(f_0/\text{BW}_{3\text{dB}})$, $B_n/G_n = (1/q_n)\,(f_0/\text{BW}_{3\text{dB}})$. All resonant circuits are assumed to be lossless. Any adjacent pair of resonators may be coupled by either of the two T (or their dual-π) couplings shown. The susceptances B are measured at the midfrequency of the reject band.

Table 12-1.   Two-Pole No-Zero Filter, 3-dB-down k and q Values

$(V_p/V_v)_{dB}$	q_1	k_{12}	q_2
Maximum-power-transfer terminations			
Linear phase	0.576	0.899	2.15
0	1.414	0.707	1.414
0.3	1.82	0.717	1.82
1.0	2.21	0.739	2.21
3.0	3.13	0.779	3.13
Resistive termination at only one end			
Linear phase	0.455	1.27	∞
0	0.707	1.00	∞
0.3	0.910	0.904	∞
1.0	1.11	0.866	∞
3.0	1.56	0.840	∞

Table 12-2.   Three-Pole No-Zero Filter, 3-dB-down k and q Values

$(V_p/V_v)_{dB}$	q_2	q_1	k_{12}	k_{23}	q_3
Maximum-power-transfer terminations					
Linear phase	∞	0.338	1.74	0.682	2.21
0	∞	1.00	0.707	0.707	1.00
0.1	∞	1.43	0.665	0.665	1.43
1.0	∞	2.21	0.645	0.645	2.21
3.0	∞	3.36	0.647	0.647	3.36
Resistive termination at only one end					
Linear phase	∞	0.293	2.01	0.899	∞
0	∞	0.500	1.22	0.707	∞
0.1	∞	0.714	0.961	0.661	∞
1.0	∞	1.11	0.785	0.645	∞
3.0	∞	1.68	0.714	0.649	∞

Table 12-3.   Four-Pole No-Zero Filter, 3-dB-down k and q Values

$(V_p/V_v)_{dB}$	$q_{2,3}$	q_1	k_{12}	k_{23}	k_{34}	q_4
Maximum-power-transfer terminations						
Linear phase	∞	2.24	0.644	1.175	2.53	0.233
0	∞	0.766	0.840	0.542	0.840	0.766
0.01	∞	1.05	0.737	0.541	0.737	1.05
0.1	∞	1.34	0.690	0.542	0.690	1.34
1.0	∞	2.21	0.638	0.546	0.638	2.21
3.0	∞	3.45	0.624	0.555	0.624	3.45
Resistive termination at only one end						
Linear phase	∞	0.211	2.78	1.29	0.828	∞
0	∞	1.383	1.56	0.765	0.644	∞
0.01	∞	0.524	1.20	0.666	0.621	∞
0.1	∞	0.667	1.01	0.626	0.618	∞
1.0	∞	1.10	0.781	0.578	0.614	∞
3.0	∞	1.72	0.692	0.567	0.609	∞

Table 12-4. Five-Pole No-Zero Filter, 3-dB-down k and q Values

$(V_p/V_v)_{dB}$	$q_{2,3,4}$	q_1	k_{12}	k_{23}	k_{34}	k_{45}	q_5
Maximum-power-transfer terminations							
Linear phase	∞	0.175	3.36	1.56	1.06	0.631	2.26
0	∞	0.618	1.0	0.556	0.556	1.0	0.618
0.001	∞	0.822	0.845	0.545	0.545	0.845	0.822
0.1	∞	1.29	0.703	0.535	0.535	0.703	1.29
1.0	∞	2.21	0.633	0.538	0.538	0.633	2.21
3.0	∞	3.47	0.614	0.538	0.538	0.614	3.47
Resistive termination at only one end							
Linear phase	∞	0.162	3.62	1.68	1.14	0.804	∞
0	∞	0.309	1.90	0.900	0.655	0.619	∞
0.001	∞	0.412	1.48	0.760	0.603	0.606	∞
0.1	∞	0.649	1.044	0.634	0.560	0.595	∞
1.0	∞	1.105	0.779	0.570	0.544	0.595	∞
3.0	∞	1.74	0.679	0.554	0.542	0.597	∞

Proceeding across each table, figuratively from the left end of the filter, the next column gives q_1, from which, with the aid of Figs. 12-25 to 12-31, the relation between the terminating resistance R_1 and the first reactance element is obtained. The next column for k_{12} (with Figs. 12-25 to 12-31) provides for the relation between the first and second reactances. Continuing across each table, all relations between adjacent elements are obtained, including that of the right-hand terminating resistance.

Example: Reverting to the previous example, a filter is required having BW_{50dB}/BW_{1dB} = 2.5, and $V_p/V_v < 1$ dB. The 5-pole no-zero response with a passband peak-to-valley ratio of 1 dB in Fig. 12-9 satisfied the requirement.

Table 12-4 is for 5-pole networks and, if the terminations are to be maximum-power-transfer loads, the upper part of the table should be used. If a shunt capacitance is to appear at one end of the low-pass filter, Fig. 12-25A will apply.

Reading along the row for $(V_p/V_v)_{dB}$ = 1, the second column gives normalized unloaded Q's of infinity at the overall 3-dB-down frequency, which, for this example, is 31 kHz. It should be understood that much lower unloaded-Q designs can be accomplished.

The required value of $q_1 = 2.21$ is found in the third column. From Fig. 12-25A, $1/R_1 C_1 = \omega_{3dB}/2.21 = 0.451\omega_{3dB}$, from which R_1 or C_1 may be obtained. Experimentally, the 3-dB-down bandwidth of $R_1 C_1$ must measure 0.451 times the required 3-dB-down bandwidth, or $31 \times 0.451 = 14$ kHz.

Table 12-5. Six-Pole No-Zero Filter, 3-dB-down k and q Values

$(V_p/V_v)_{dB}$	$q_{2,3,4,5}$	q_1	k_{12}	k_{23}	k_{34}	k_{45}	k_{56}	q_6
Maximum-power-transfer terminations								
0	∞	0.518	1.17	0.606	0.518	0.606	1.17	0.518
0.0001	∞	0.679	0.967	0.573	0.518	0.573	0.967	0.679
0.01	∞	0.936	0.810	0.550	0.518	0.550	0.810	0.936
0.1	∞	1.27	0.716	0.539	0.518	0.539	0.716	1.27
1.0	∞	2.21	0.633	0.531	0.520	0.531	0.633	2.21
3.0	∞	3.51	0.610	0.532	0.524	0.532	0.610	3.51
Resistive termination at only one end								
Linear phase	∞	0.129	4.55	2.09	1.42	1.09	0.803	∞
0	∞	0.259	2.26	1.05	0.732	0.606	0.606	∞
0.0001	∞	0.340	1.76	0.869	0.650	0.573	0.596	∞
0.01	∞	0.468	1.34	0.725	0.591	0.550	0.591	∞
0.1	∞	0.637	1.06	0.642	0.560	0.539	0.589	∞
1.0	∞	1.12	0.771	0.566	0.533	0.531	0.589	∞
3.0	∞	1.75	0.673	0.546	0.529	0.531	0.591	∞

Table 12-6. Seven-Pole No-Zero Filter, 3-dB-down k and q Values

$(V_p/V_v)_{dB}$	$q_{2,3,4,5,6}$	q_1	k_{12}	k_{23}	k_{34}	k_{45}	k_{56}	k_{67}	q_7
Maximum-power-transfer terminations									
0	∞	0.445	1.34	0.669	0.528	0.528	0.669	1.34	0.445
0.00001	∞	0.580	1.10	0.611	0.521	0.521	0.611	1.10	0.580
0.001	∞	0.741	0.930	0.579	0.519	0.519	0.579	0.930	0.741
0.01	∞	0.912	0.830	0.560	0.519	0.519	0.560	0.830	0.912
0.1	∞	1.26	0.723	0.541	0.517	0.517	0.541	0.723	1.26
1.0	∞	2.25	0.631	0.530	0.517	0.517	0.530	0.631	2.25
3.0	∞	3.52	0.607	0.529	0.519	0.519	0.529	0.607	3.52
Resistive termination at only one end									
Linear phase	∞	0.105	5.53	2.53	1.72	1.33	1.08	0.804	∞
0	∞	0.223	2.62	1.20	0.824	0.659	0.579	0.598	∞
0.00001	∞	0.290	2.05	0.981	0.710	0.601	0.552	0.589	∞
0.001	∞	0.370	1.64	0.830	0.642	0.570	0.541	0.588	∞
0.01	∞	0.456	1.38	0.744	0.602	0.551	0.538	0.588	∞
0.1	∞	0.629	1.08	0.648	0.560	0.531	0.530	0.587	∞
1.0	∞	1.12	0.770	0.564	0.530	0.521	0.527	0.587	∞
3.0	∞	1.76	0.669	0.542	0.523	0.520	0.528	0.588	∞

Table 12-7. Eight-Pole No-Zero Filter, 3-dB-down k and q Values

(V_p/V_v)dB	$q_{2,3,4,5,6,7}$	q_1	k_{12}	k_{23}	k_{34}	k_{45}	k_{56}	k_{67}	k_{78}	q_8
				Maximum-power-transfer terminations						
0	∞	0.391	1.52	0.734	0.551	0.510	0.551	0.734	1.52	0.391
0.00001	∞	0.545	1.16	0.640	0.534	0.510	0.534	0.640	1.16	0.545
0.001	∞	0.717	0.960	0.592	0.524	0.510	0.524	0.592	0.960	0.717
0.01	∞	0.896	0.843	0.567	0.520	0.510	0.520	0.567	0.843	0.896
0.1	∞	1.25	0.727	0.545	0.516	0.510	0.516	0.545	0.727	1.25
1.0	∞	2.20	0.633	0.530	0.514	0.511	0.514	0.530	0.633	2.20
3.0	∞	3.53	0.605	0.527	0.515	0.513	0.515	0.527	0.605	3.53
				Resistive termination at only one end						
0	∞	0.199	2.98	1.36	0.920	0.721	0.615	0.562	0.591	∞
0.00001	∞	0.272	2.17	1.04	0.749	0.627	0.567	0.543	0.587	∞
0.001	∞	0.358	1.69	0.859	0.660	0.580	0.544	0.535	0.584	∞
0.01	∞	0.448	1.40	0.755	0.610	0.556	0.533	0.529	0.583	∞
0.1	∞	0.627	1.08	0.651	0.564	0.534	0.523	0.525	0.582	∞
1.0	∞	1.10	0.779	0.567	0.531	0.519	0.516	0.523	0.583	∞
3.0	∞	1.76	0.668	0.541	0.522	0.516	0.516	0.524	0.585	∞

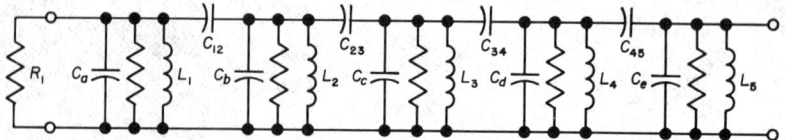

Fig. 12-32. A 5-resonator filter with high-side capacitance coupling.

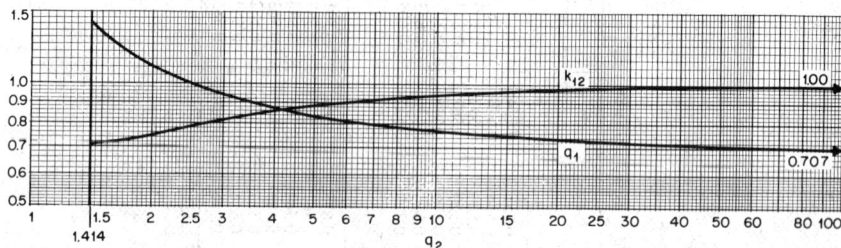

Fig. 12-33. (Loaded on one side only.) 2-pole filter of finite-Q elements producing a maximally flat amplitude shape. See curve 1 of Fig. 12-6.

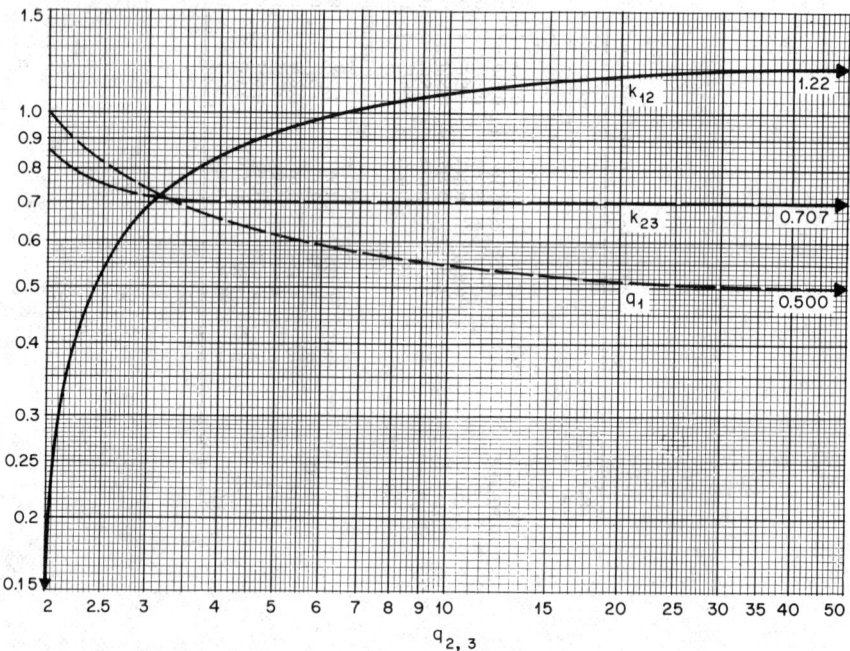

Fig. 12-34. (Loaded on one side only.) 3-pole filter of finite-Q elements producing a maximally flat amplitude shape. See curve 1 of Fig. 12-7.

12-34

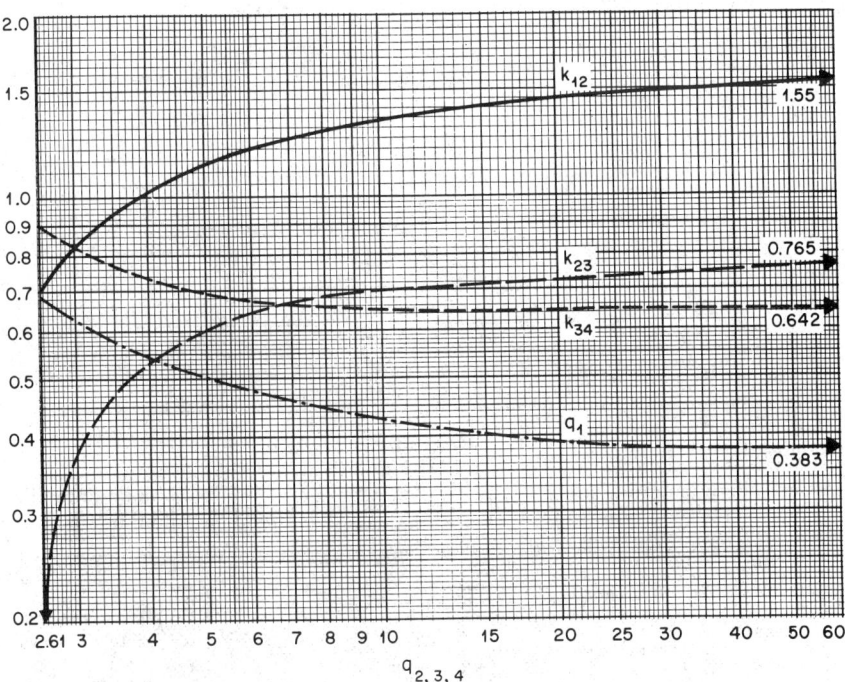

Fig. 12-35. (Loaded on one side only.) 4-pole filter of finite-Q elements producing a maximally flat amplitude shape. See curve 1 of Fig. 12-8.

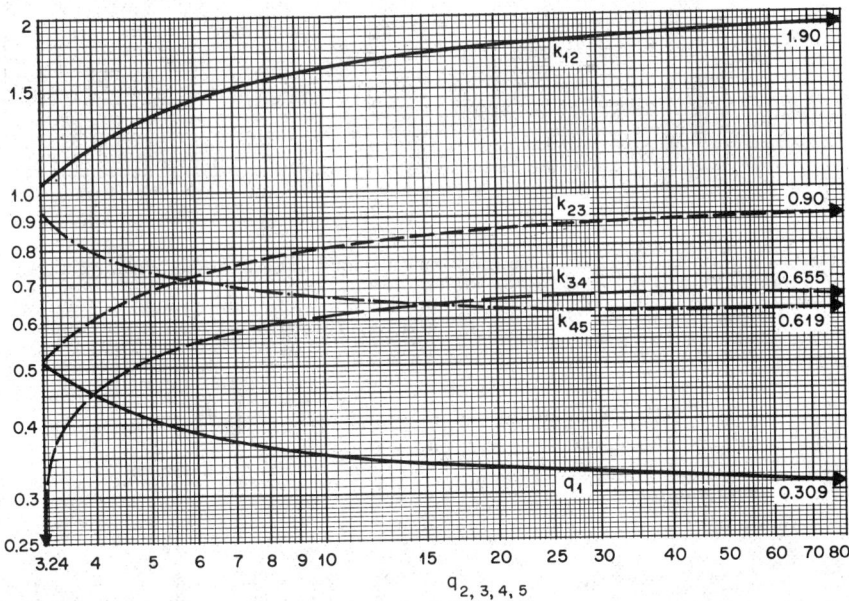

Fig. 12-36. (Loaded on one side only.) 5-pole filter of finite-Q elements producing a maximally flat amplitude shape. See curve 1 of Fig. 12-9.

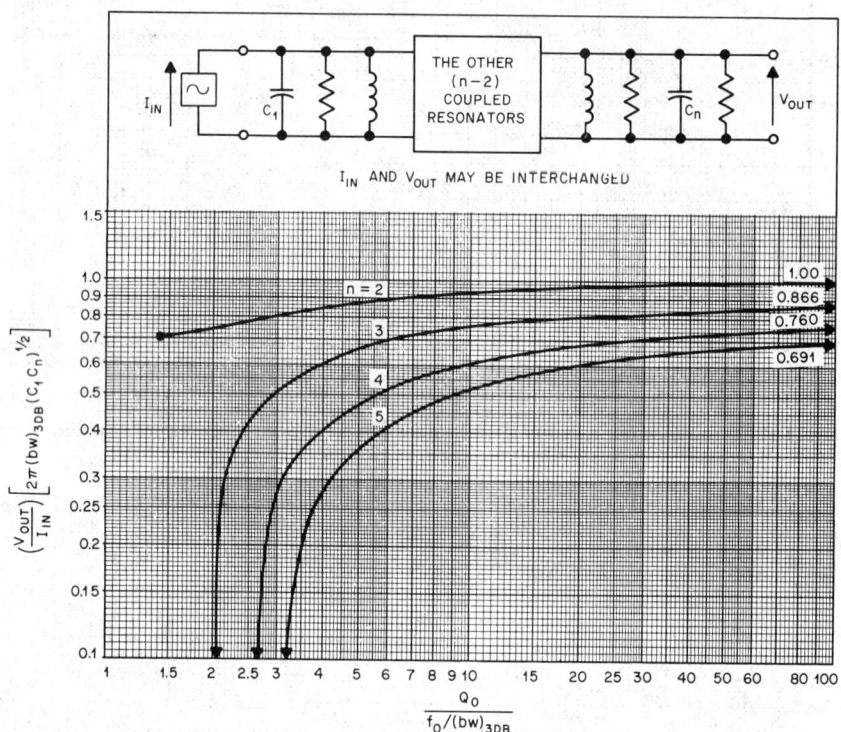

Fig. 12-37. Midfrequency transfer-impedance magnitude for the maximally flat-amplitude-shape designs of Figs. 12-33 to 12-36 with direct resistive loading on one side only.

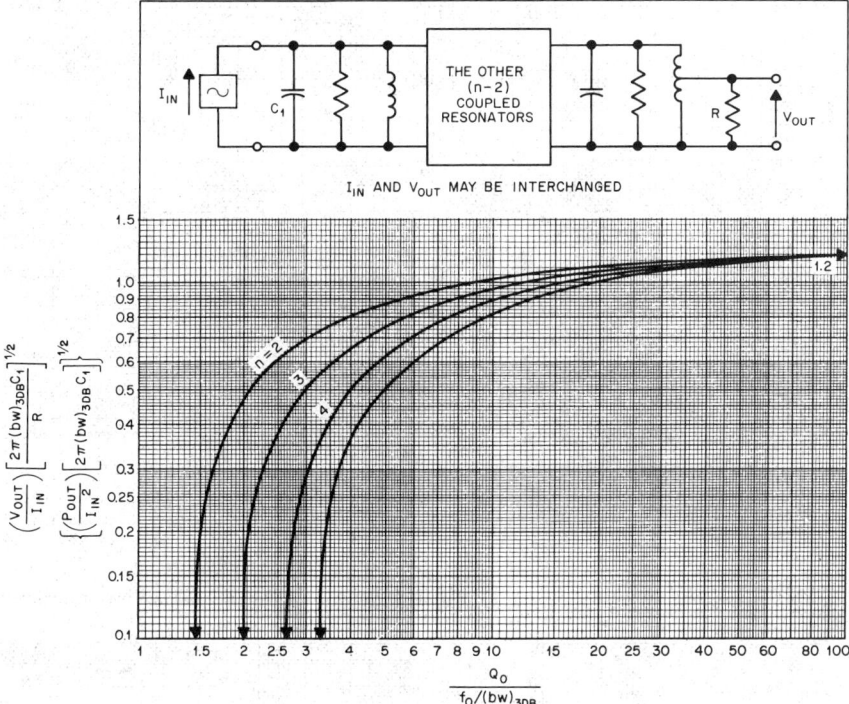

Fig. 12-38. Midfrequency transfer-impedance magnitude for the maximally flat-amplitude-shape designs of Figs. 12-33 to 12-36 with transformed resistive loading on one side only.

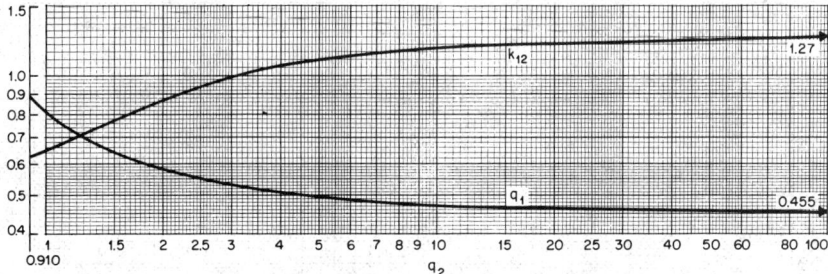

Fig. 12-39. (Loaded on one side only.) 2-pole filter of finite-Q elements producing a maximally linear phase shape. See Figs. 12-15 and 12-16.

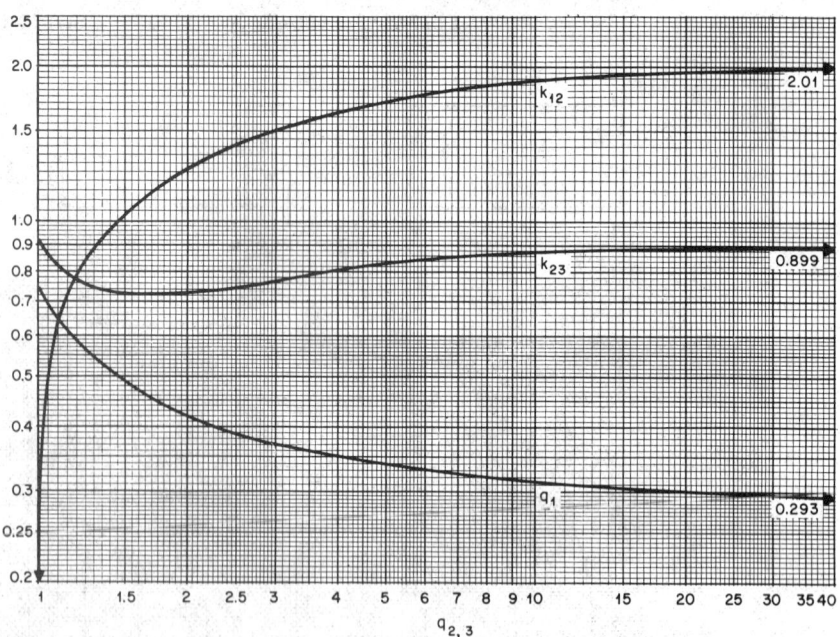

Fig. 12-40. (Loaded on one side only.) 3-pole filter of finite-Q elements producing a maximally linear phase shape. See Figs. 12-15 and 12-16.

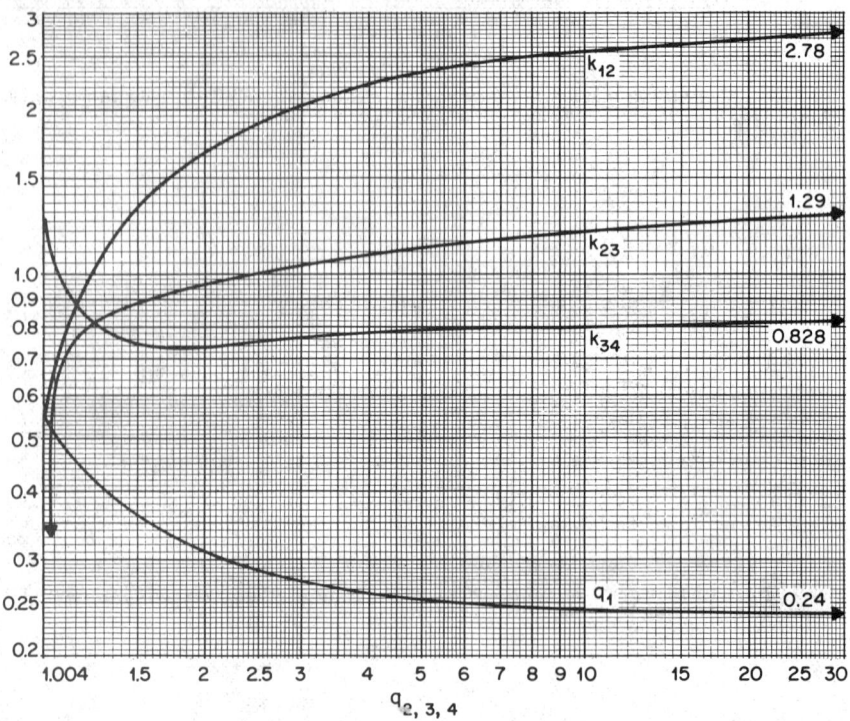

Fig. 12-41. (Loaded on one side only.) 4-pole filter of finite-Q elements producing a maximally linear phase shape. See Figs. 12-15 and 12-16.

12-38

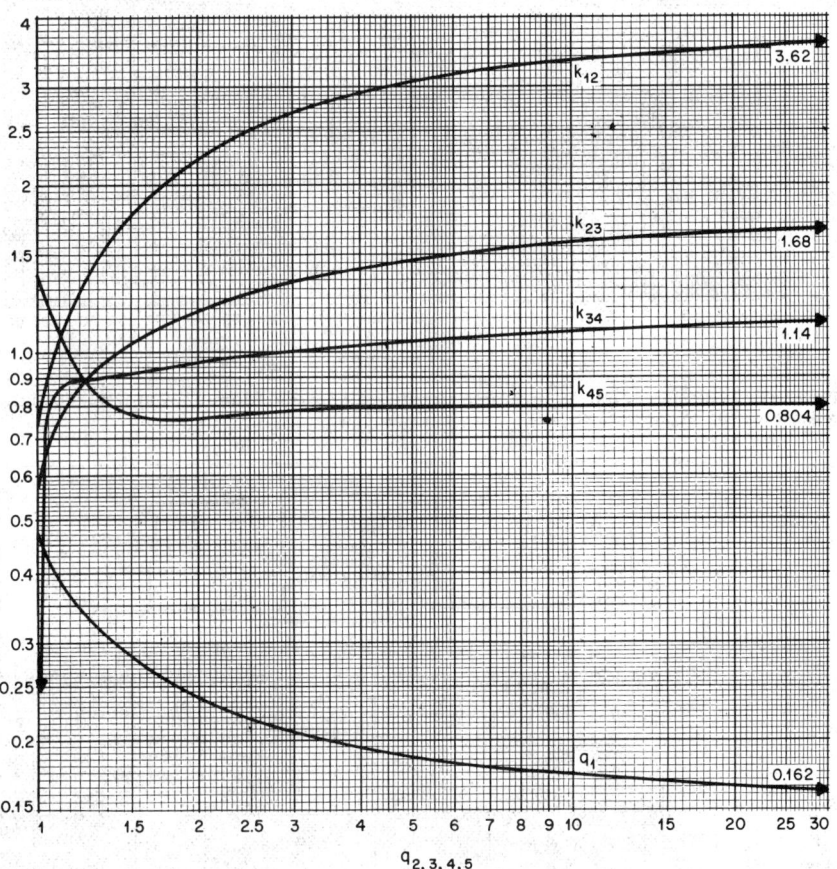

Fig. 12-42. (Loaded on one side only.) 5-pole filter of finite-Q elements producing a maximally linear phase shape. See Figs. 12-15 and 12-16.

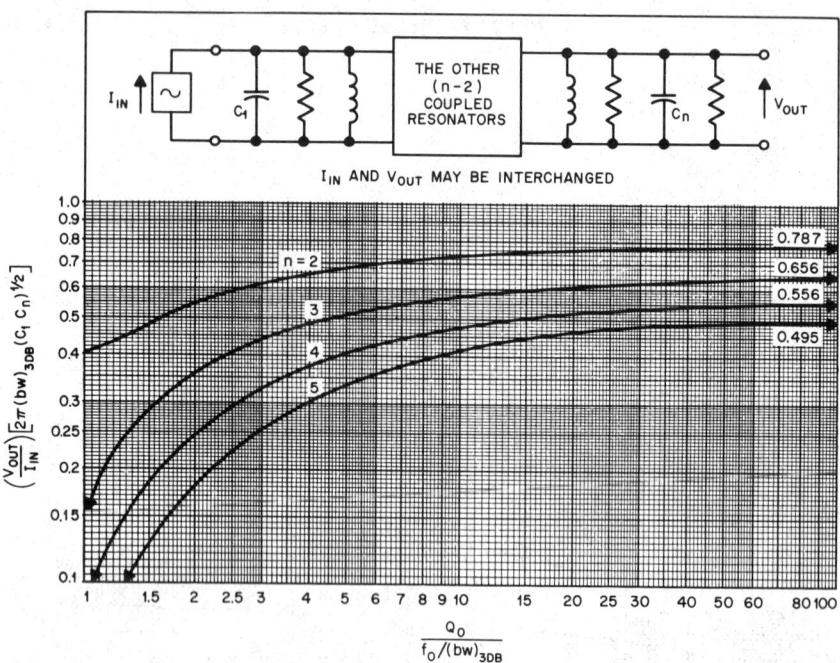

Fig. 12-43. Midfrequency transfer-impedance magnitude for designs of Figs. 12-39 to 12-42 with direct resistive loading on one side only.

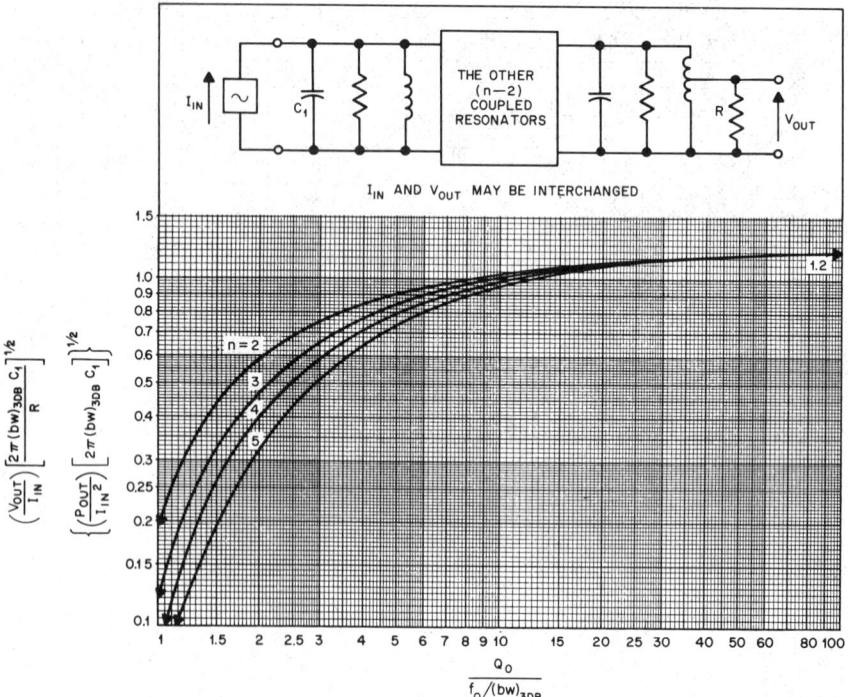

Fig. 12-44. Midfrequency transfer-impedance magnitude for designs of Figs. 12-39 to 12-42 with transformed resistive loading on one side only.

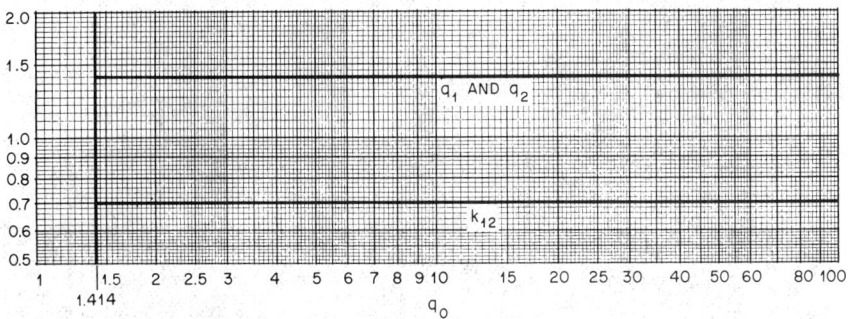

Fig. 12-45. Resistive generator–resistive load 2-pole filter of finite-Q elements producing a maximally flat amplitude shape. See curve 1 of Fig. 12-6.

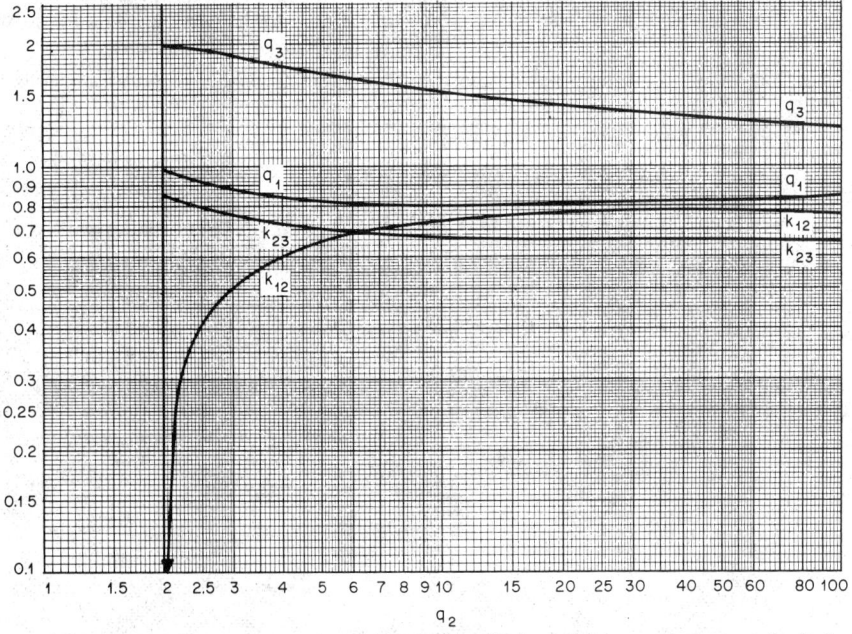

Fig. 12-46. Resistive generator–resistive load 3-pole filter of finite-Q elements producing a maximally flat amplitude shape. See curve 1 of Fig. 12-7.

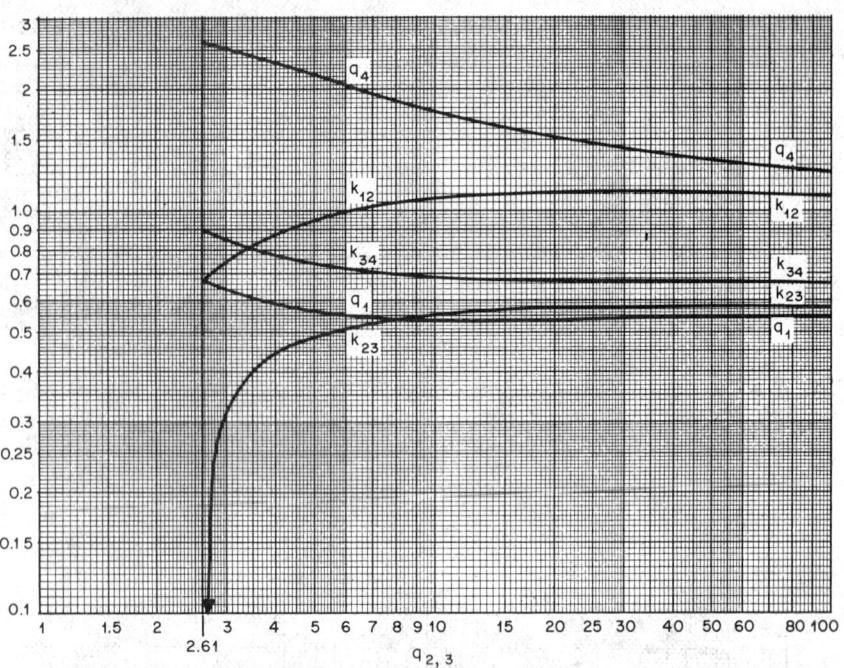

Fig. 12-47. Resistive generator–resistive load 4-pole filter of finite-Q elements producing a maximally flat amplitude shape. See curve 1 of Fig. 12-8.

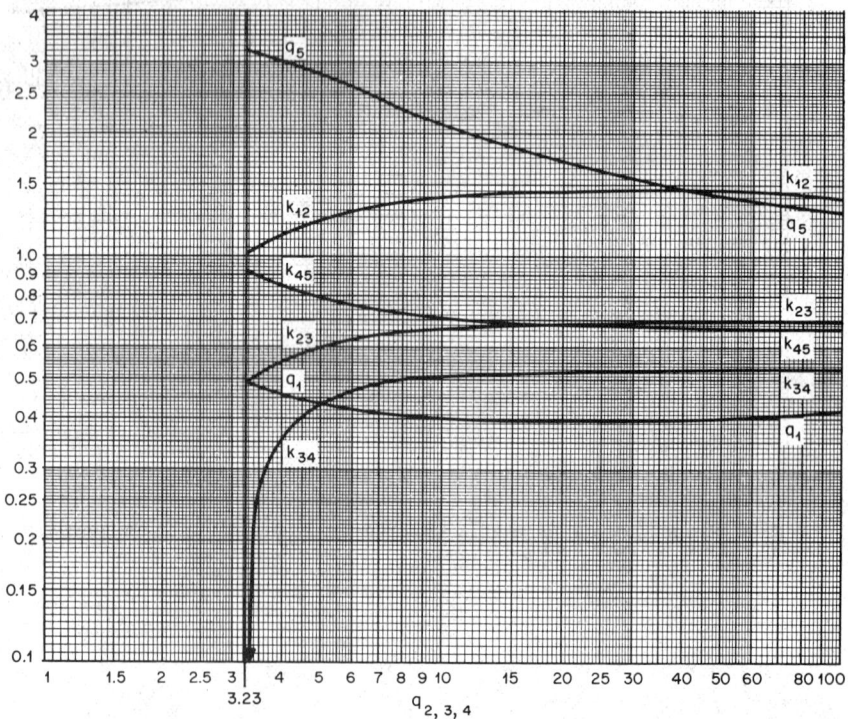

Fig. 12-48. Resistive generator–resistive load 5-pole filter of finite-Q elements producing a maximally flat amplitude shape. See curve 1 of Fig. 12-9.

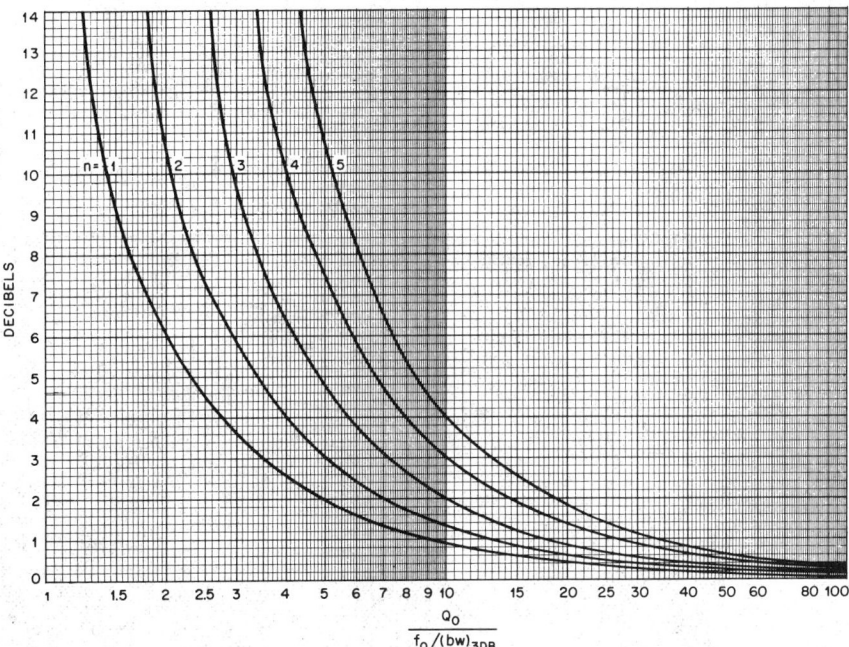

Fig. 12-49. Ratio of power available from generator to power delivered to load for designs of Figs. 12-45 to 12-48.

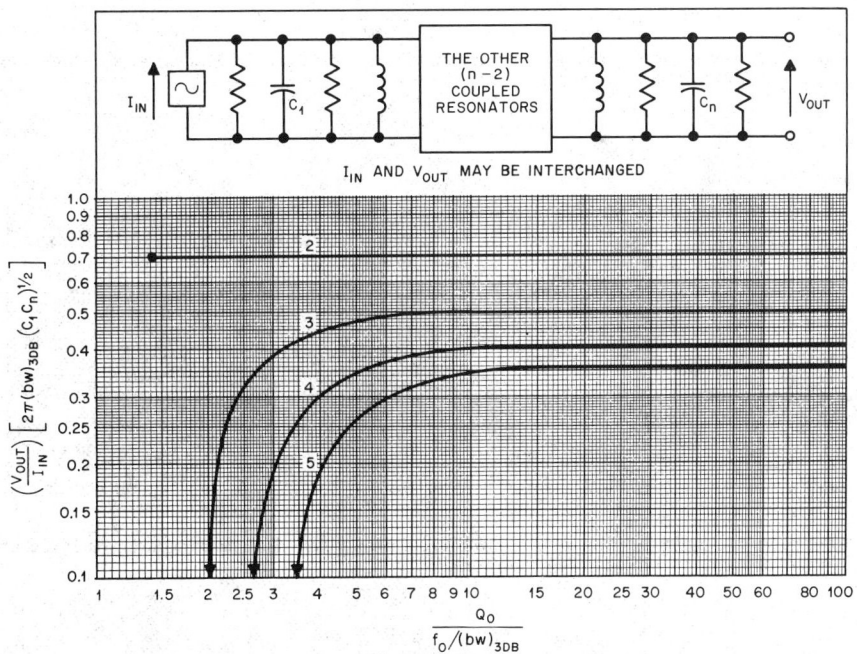

Fig. 12-50. Midfrequency transfer-impedance magnitude for designs of Figs. 12-45 to 12-48 with direct resistive loading on both sides.

12-43

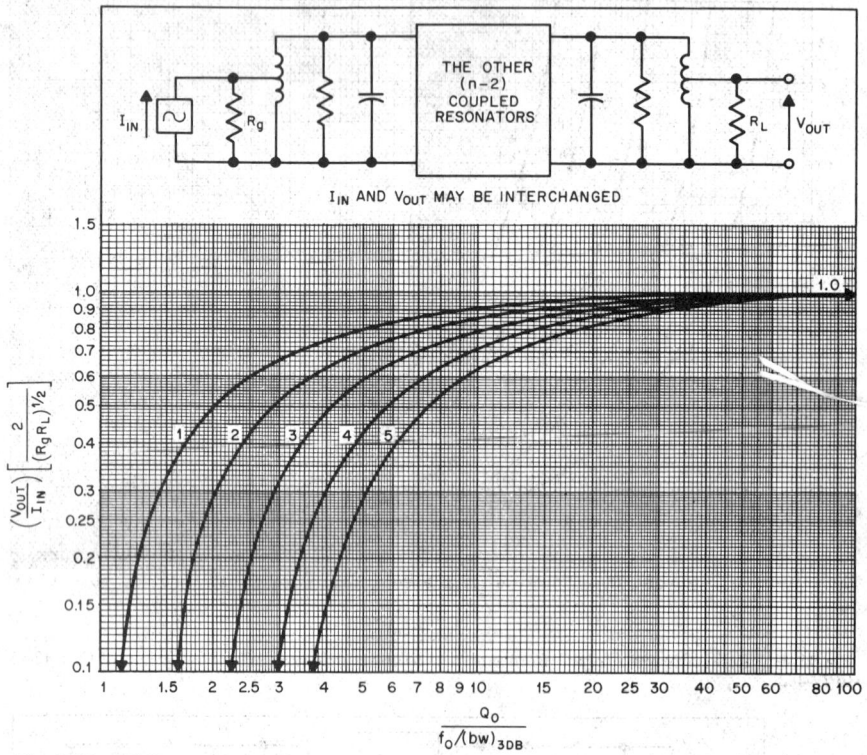

Fig. 12-51. Midfrequency transfer-impedance magnitude for designs of Figs. 12-45 to 12-48 with transformed resistive loading on both sides.

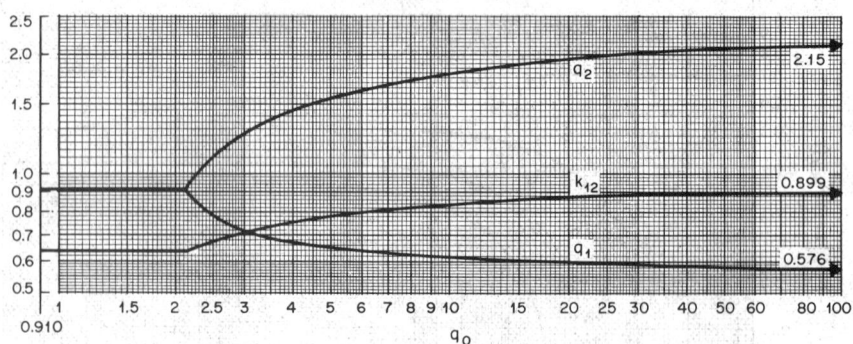

Fig. 12-52. Resistive generator–resistive load 2-pole filter of finite-Q elements producing a maximally linear phase shape. See Figs. 12-15 and 12-16.

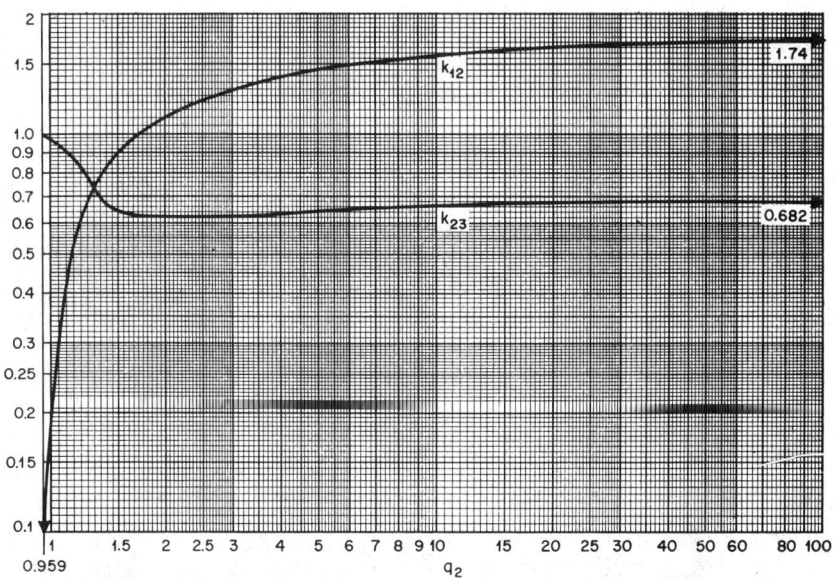

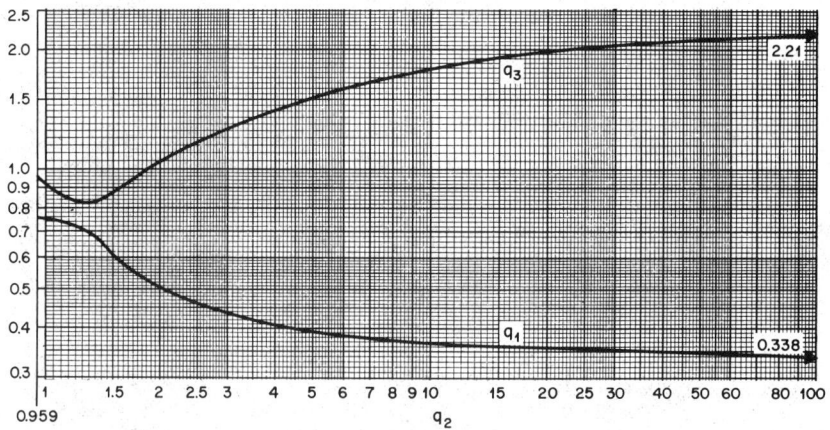

Fig. 12-53. Resistive generator–resistive load 3-pole filter of finite-Q elements producing a maximally linear phase shape. See Figs. 12-15 and 12-16.

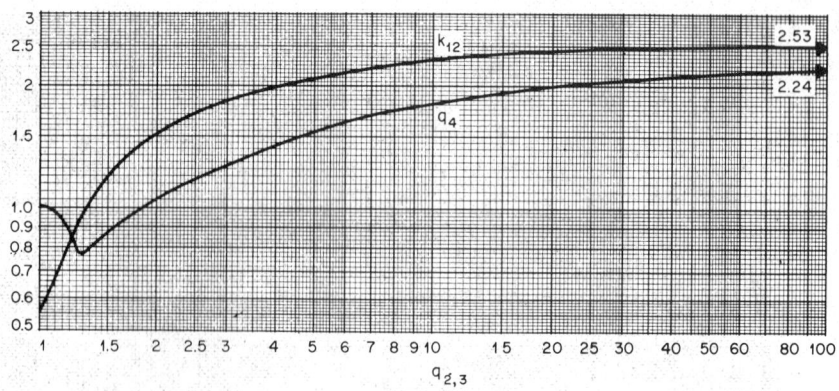

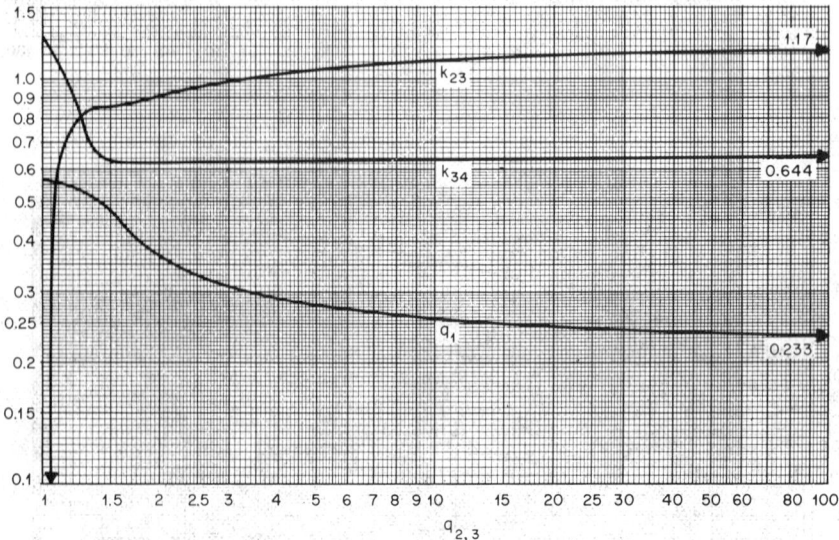

Fig. 12-54. Resistive generator-resistive load 4-pole filter of finite-Q elements producing a maximally linear phase shape. See Figs. 12-15 and 12-16.

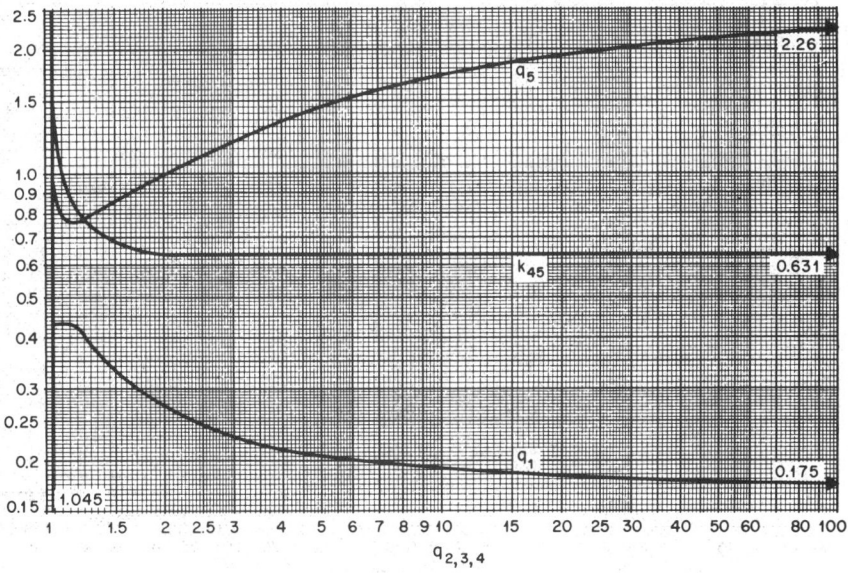

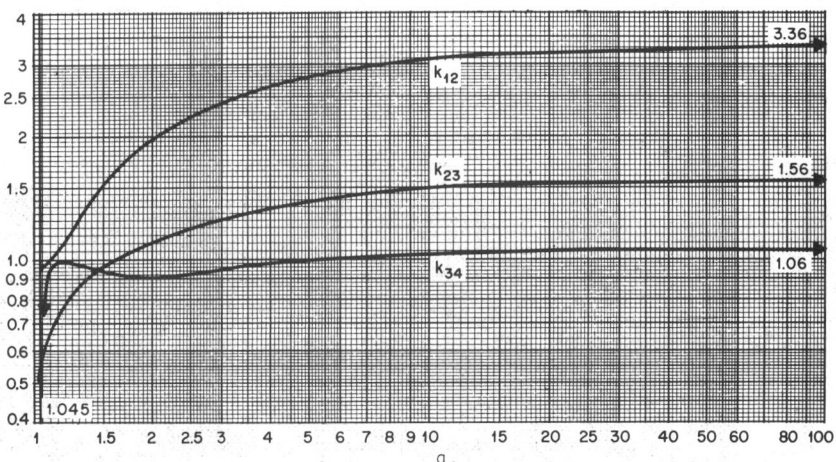

Fig. 12-55. Resistive generator-resistive load 5-pole filter of finite-Q elements producing a maximally linear phase shape. See Figs. 12-15 and 12-16.

12-47

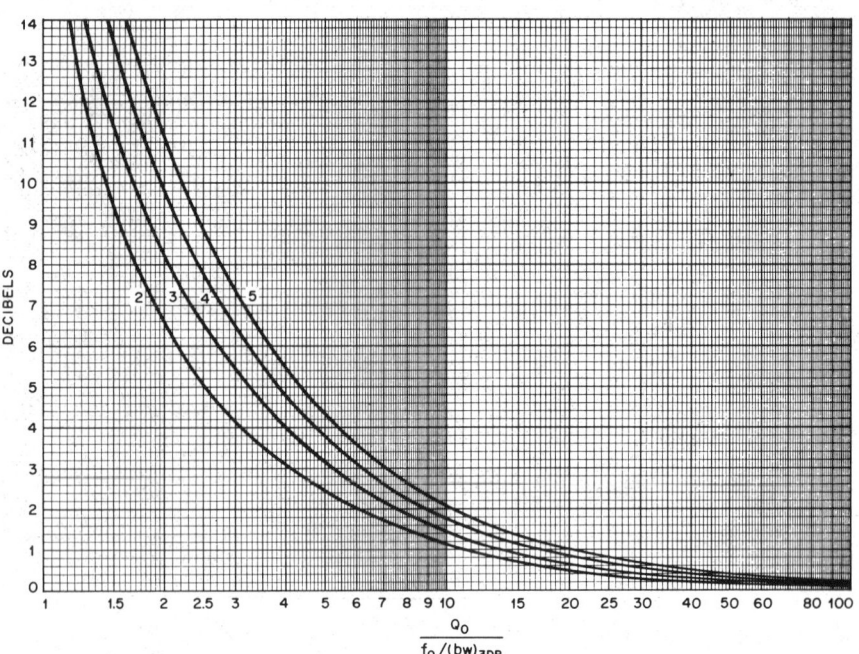

Fig. 12-56. Ratio of power available from generator to power delivered to load for designs of Figs. 12-52 to 12-55.

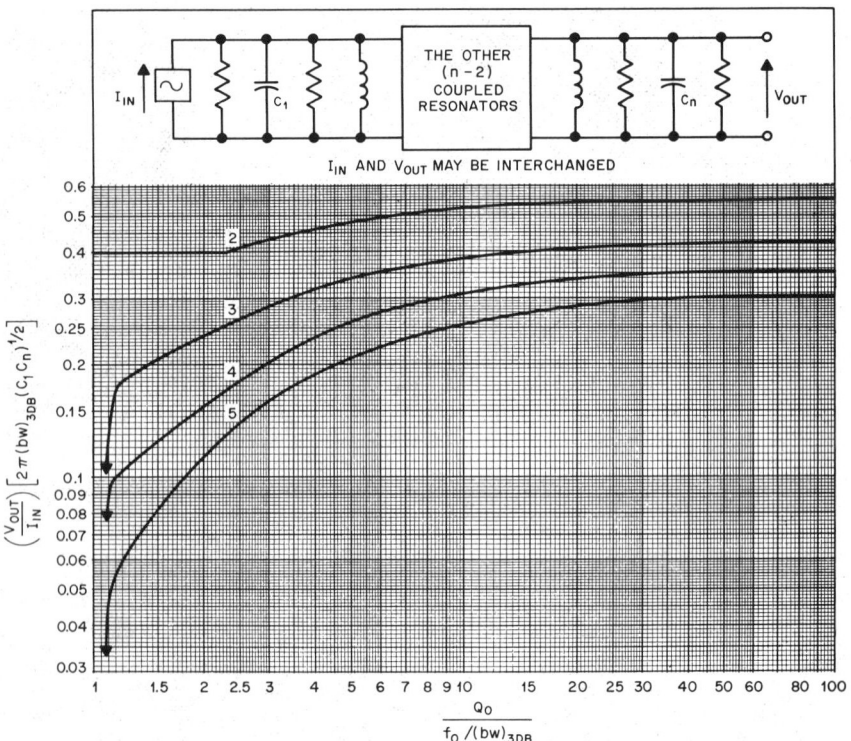

Fig. 12-57. Midfrequency transfer-impedance magnitude for designs of Figs. 12-52 to 12-55 with direct resistive loading on both sides.

12–48

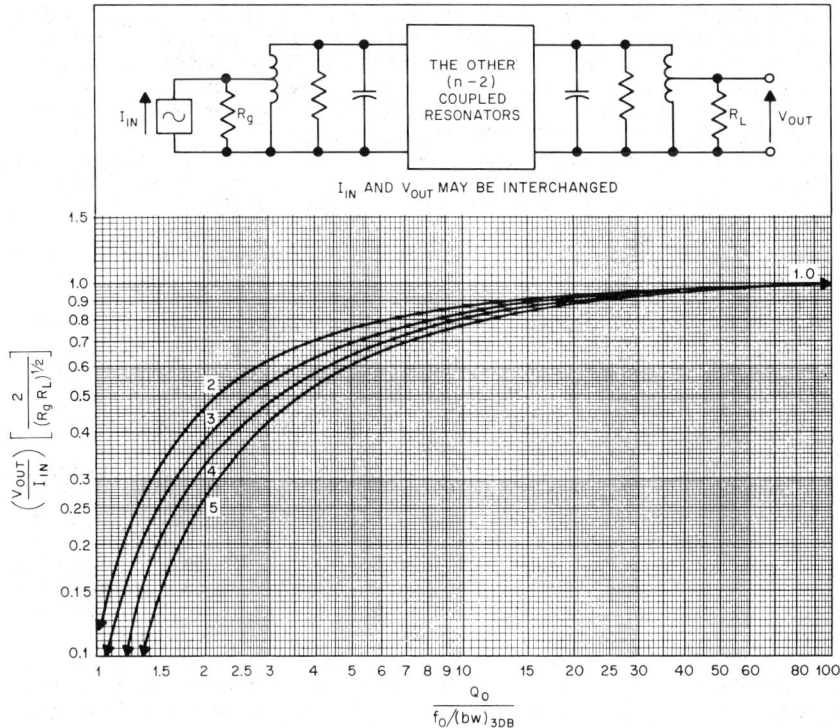

Fig. 12-58. Midfrequency transfer-impedance magnitude for designs of Figs. 12-52 to 12-55 with transformed resistive loading on both sides.

From the table a value of 0.633 is obtained for k_{12} and from Fig. 12-25A it is found that $1/(C_1 L_2)^{1/2} = 0.633\omega_{3dB}$. This means that a resonant circuit made up of C_1 and L_2 must tune to 0.633 times the required 3-dB-down bandwidth, or $31 \times 0.633 = 19.7$ kHz.

In this fashion, all the remaining elements are determined. Any one of them may be set arbitrarily (for instance, the input load resistance R_1), but once it has been set, all other values are rigidly determined by the k and q factors.

Designs may be accomplished for elements having finite unloaded Q's. These designs are necessary for small-percentage bandpass filters. As is evident from Fig. 12-28, the Q of the internal resonators measured at the midfrequency must be the normalized q multiplied by the fractional midfrequency f_0/BW_{3dB}. If the bandwidth percentage is small, the fractional midfrequency, and therefore the actually required Q, will be large.

Practical values of end q's and all k's will result if the internal elements have finite q's above the minimum values given in Figs. 12-5 to 12-12. For a required response shape, the resulting data can be expressed as in Figs. 12-30 to 12-58. These curves are for zero-decibel ripple (Butterworth) and for the maximally linear phase shape.

14. Finite-Q, Singly Loaded, Maximally Flat Amplitude Designs Using Lossy Elements. These designs, given in Figs. 12-33 to 12-38, produce the maximally flat amplitude (Butterworth) response shape for 2-, 3-, 4-, and 5-pole filters using finite-Q elements.

15. Finite-Q, Singly Loaded, Maximally Linear Phase Designs Using Lossy Elements. These designs, given in Figs. 12-39 to 12-44, produce the maximally linear phase response shape for 2-, 3-, 4-, and 5-pole filters using finite-Q elements.

16. Finite-Q, Doubly Loaded, Maximally Flat Amplitude Designs Using Lossy Elements. These designs, given in Figs. 12-45 to 12-51, produce the maximally flat amplitude (Butterworth) response shape for 2-, 3-, 4-, and 5-pole filters using finite-Q elements.

17. Finite-Q, Doubly Loaded, Maximally Linear Phase Designs Using Lossy Elements. These designs, given in Figs. 12-52 to 12-58, produce the maximally linear phase response shape for 2-, 3-, 4-, and 5-pole filters using finite-Q elements.

18. Finite-Q Element Considerations. There are three possible generator and load conditions.

1. Reactive generator and load. The transfer impedance or admittance is the significant factor, and a loading resistance must be added to either or both end resonators.

2. Resistive generator and reactive load or vice versa. The function to be considered here is the transfer impedance or admittance. The resistive impedance must be transformed onto the associated resonator to produce the required q_1 or q_n.

3. Resistive generator and resistive load. It is usually desirable to maximize the ratio of the power delivered to the load to that available from the generator. The generator resistance and the load resistance must be transformed onto their associated resonators to produce the required q_1 and q_n.

Figures 12-33 to 12-44 give optimum design information for cases (1) and (2). Figures 12-45 to 12-58 give optimum design information for case (3).

In Figs. 12-33 to 12-58, the abscissa is the normalized unloaded Q of the resonators being used, that is, $Q_0/[f_0/\mathrm{BW}_{3\mathrm{dB}}]$. q_1 and q_n given by the graphs are the required resultant normalized Q's of the end resonators due to their unloaded Q's, plus the coupled loading from the resistive generator and/or resistive load.

Example: (a) The filter to be designed must have a $\mathrm{BW}_{3\mathrm{dB}} = 4$ kHz, and $f_0 = 80$ kHz, and a relative attenuation of $\mathrm{BW}_{70\mathrm{dB}}/\mathrm{BW}_{3\mathrm{dB}} = 5$, and there must be no ripple in the passband. Curve 1 of Fig. 12-9 satisfies these conditions and calls for a 5-pole network.

(b) The specified fractional midfrequency is 20 (passband = 5% of the midfrequency), the $Q_{\min}$ from Fig. 12-9 becomes $3.24 \times 20 = 65$. Assume further that resonators with unloaded midfrequency Q's of 100 are available. As the normalized unloaded q is the actual unloaded Q divided by the fractional midfrequency, the filter must produce a Butterworth shape with 5 resonators having normalized unloaded q's of $100/20 = 5$.

(c) Assuming a high-resonant impedance filter to be required, the node network of Fig. 12-28A could be used.

It should here be noted that, when node resonators are used, capacitive coupling causes the rate of cutoff on the high-frequency side to be slower, and the low-frequency skirt to fall more rapidly, than arithmetic symmetry would indicate. The opposite result is produced when inductive coupling is used. The magnitude of this resulting nonsymmetry is a function of the percentage passband involved, and the correct mixture of capacitive and inductive couplings can be used in the filters of Fig. 12-27 to obtain approximate arithmetic symmetry. However, in this simplified example, all high-side capacitance couplings will be used in Fig. 12-32. The element values will be obtained from Fig. 12-28A and Fig. 12-36.

(d) The q_1 curve of Fig. 12-36 intersects the abscissa value of 5 at 0.405. By tapping a resistive generator or load onto it, or placing a resistor across it, the resonator $C_1 L_1$ must be loaded to produce an actual Q_1 of $0.405 f_0/\mathrm{BW}_{3\mathrm{dB}} = 8.1$ (see Fig. 12-28A).

(e) As a convenience, the same size of inductor may be used for resonating each node, say 4 mH. For a required midfrequency of 80 kHz, for this example, each node total capacitance will be 1,000 pF.

(f) Again from Fig. 12-36, we get k_{12} of 1.35 for an abscissa value of 5. From Fig. 12-28A,

$$C_{12} = 1.35(\mathrm{BW}_{3\mathrm{dB}}/f_0)(C_1 C_2)^{1/2}$$

$$= 1.35 \times 0.05 \times 1,000$$

$$= 67.5 \text{ pF}.$$

At the midfrequency of 80 kHz, node 1 must be resonant when all other nodes are short-circuited. To produce the required capacitance in shunt of L_1, C_a must be $1,000 - 67.5 = 932.5$ pF.

(g) From Fig. 12-36, a value of 0.67 is obtained for k_{23}, and $C_{23} = 0.67 \times 0.05 \times 1,000$ = 33.5 pF. To resonate node 2 at the midfrequency with all other nodes short-circuited, $C_b = 1,000 - 33.5 - 67.5 = 899$ pF.

(h) Additional computations give values for C_{34} of $0.53 \times 0.05 \times 1,000 = 26.5$ pF.

$$C_c = 1,000 - 33.5 - 26.5 = 940$$
$$C_{45} = 0.73 \times 0.05 \times 1,000 = 36.5$$
$$C_d = 1,000 - 36.5 - 26.5 = 937$$

and

$$C_6 = 1,000 - 36.5 = 963.5 \text{ pF}$$

All inductances will be identical and of 4 mH, and there will be no inductive coupling among them.

(i) If, for example, direct resistive loading is used to produce the required Q of step d, then Fig. 12-37 shows that the midfrequency transfer-impedance magnitude will be

$$\frac{V_{\text{out}}}{I_{\text{in}}} = 0.36 \frac{1}{2\pi \times 4 \times 10^3 \times 10^{-9}} = 14,000 \ \Omega$$

ATTENUATOR NETWORK DESIGN

An attenuator is a network designed to introduce a known loss when working between resistive impedances Z_1 and Z_2 to which the input and output impedances of the attenuator are matched. Either Z_1 or Z_2 may be the source and the other the load. The attenuation of such networks expressed as a power ratio is the same regardless of the direction of working.

Three forms of resistance network may be conveniently used to realize these conditions. These are the T section, the π section, and the bridged-T section. Equivalent balanced sections also are shown. Methods are given for the computation of attenuator networks, the hyperbolic expressions giving rapid solutions with the aid of tables of hyperbolic functions. Tables of the various types of attenuators are given in the following paragraphs.

19. Symbols

Z_1 and Z_2 are the terminal impedances (resistive) to which the attenuator is matched.

N is the ratio of the power absorbed by the attenuator from the source to the power delivered to the load.

K is the ratio of the attenuator input current to the output current into the load. When $Z_1 = Z_2$, $K = N^{1/2}$. Otherwise K is different in the two directions.

Attenuation in decibels $= 10 \log_{10} N$.

Attenuation in nepers $= \theta = 1/2 \log_e N$.

20. Notes on Error Equations. The equations and figures for errors, given in Tables 12-8 to 12-11, are based on the assumption that the attenuator is terminated approximately by its proper terminal impedances Z_1 and Z_2. They hold for deviations of the attenuator arms and load impedances up to $\pm 20\%$ or somewhat more. The error due to each element is proportional to the deviation of the element, and the total error of the attenuator is the sum of the errors due to each of the several elements.

When any element or arm R has a reactive component ΔX in addition to a resistive error ΔR, the errors in input impedance and output current are

$$\Delta Z = A(\Delta R + j\Delta X)$$
$$\Delta i/i = B[(\Delta R + j\Delta X)/R]$$

where A and B are constants of proportionality for the elements in question. These constants can be determined in each case from the figures given for errors due to a resistive deviation ΔR.

The reactive component ΔX produces a quadrature component in the output current, resulting in a phase shift. However, for small values of ΔX, the error in insertion loss is negligibly small.

Description	Unbalanced Configuration	Balanced Configuration	Hyperbolic Equations*
Unbalanced T and balanced H	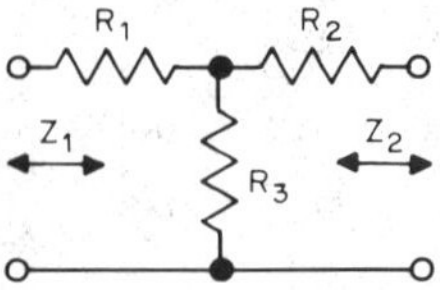	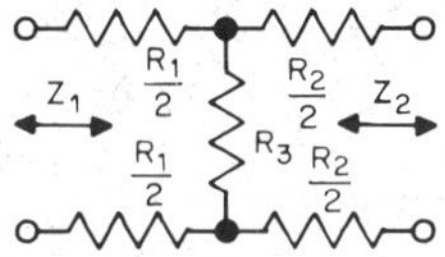	$R_3 = (Z_1 Z_3)^{1/2}/\sinh\theta$ $R_1 = (Z_1/\tanh\theta) - R_3$ $R_2 = (Z_2/\tanh\theta) - R_3$
Symmetrical T and H $(Z_1 = Z_2 = Z)$ (refer to Table 12-8)	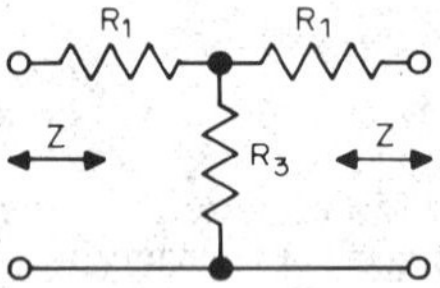	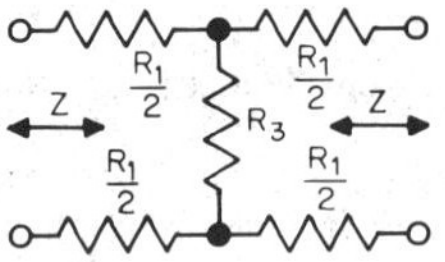	$R_3 = Z/\sinh\theta$ $R_1 = Z \tanh(\theta/2)$
Minimum-loss pad matching Z_1 and Z_2 $(Z_1 > Z_2)$ (refer to Table 12-11)	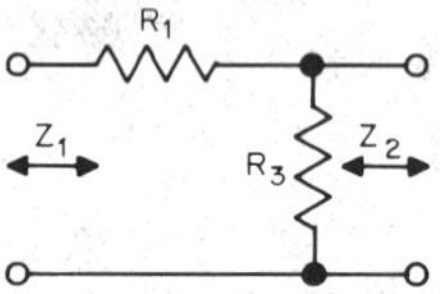	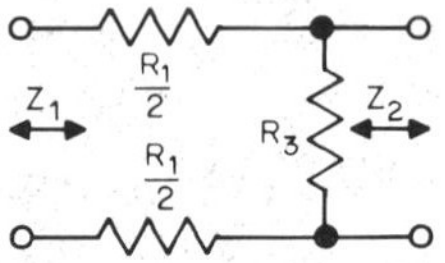	$\cosh\theta = (Z_1/Z_2)^{1/2}$ $\cosh 2\theta = 2(Z_1/Z_2) - 1$
Unbalanced π and balanced 0	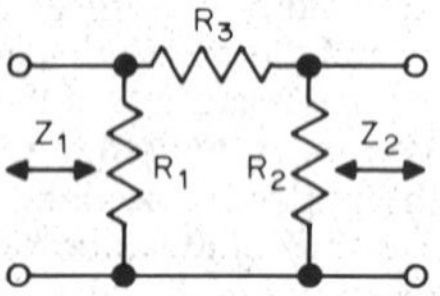	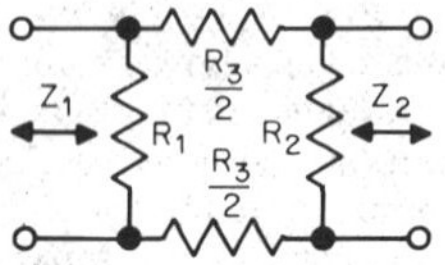	$R_3 = (Z_1 Z_2)^{1/2} \sinh\theta$ $1/R_1 = (1/Z_1 \tanh\theta) - (1/$ $1/R_2 = (1/Z_2 \tanh\theta) - (1/$
Symmetrical π and 0 $(Z_1 = Z_2 = Z)$ (refer to Table 12-9)	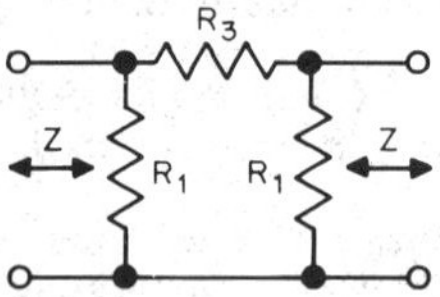	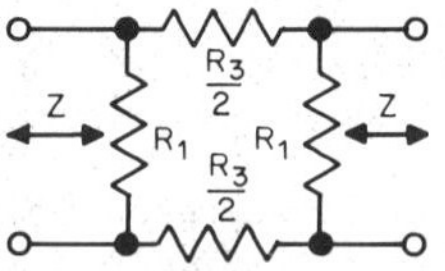	$R_3 = Z \sinh\theta$ $R_1 = Z/\tanh(\theta/2)$
Bridged T and bridged H (refer to Table 12-10)	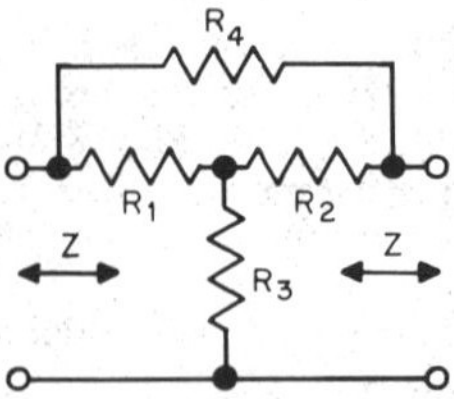	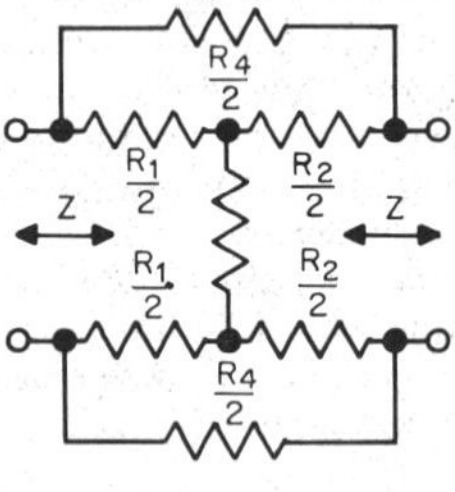	

*Four-terminal networks: The hyperbolic equations above are valid for passive linear 4-terminal networks in gen working between input and output impedances matching the respective image impedances. In this case: Z_1 and Z_2

Fig. 12-59. Attenuator network design.

Arithmetic Equations	Checking Equations
$R_3 = 2(NZ_1Z_2)^{1/2}/(N-1)$ $R_1 = Z_1[(N+1)/(N-1)] - R_3$ $R_2 = Z_2[(N+1)/(N-1)] - R_3$	
$R_3 = \dfrac{2Z(N)^{1/2}}{N-1} = \dfrac{2ZK}{K^2-1}$ $= \dfrac{2Z}{K - 1/K}$ $R_1 = Z[(N^{1/2}-1)/(N^{1/2}+1)] = Z[(K-1)/(K+1)]$ $= Z[1 - 2/(K+1)]$	$R_1R_3 = \dfrac{Z^2}{1+\cosh\theta} = Z^2\dfrac{2K}{(K+1)^2}$ $R_1/R_3 = \cosh\theta - 1 = 2\sinh^2\theta/2)$ $= (K-1)^2/2K$ $Z = R_1[1 + 2(R_3/R_1)]^{1/2}$
$R_1 = Z_1[1 - (Z_2/Z_1)]^{1/2}$ $R_3 = Z_2/[1 - (Z_2/Z_1)]^{1/2}$	$R_1R_3 = Z_1Z_2$ $R_1/R_3 = (Z_1/Z_2) - 1$ $N = \{(Z_1/Z_2)^{1/2} + [(Z_1/Z_2) - 1]^{1/2}\}^2$
$R_3 = \frac{1}{2}(N-1)(Z_1Z_2/N)^{1/2}$ $1/R_1 = (1/Z_1)[(N+1)/(N-1)] - (1/R_3)$ $1/R_2 = (1/Z_2)[(N+1)/(N-1)] - (1/R_3)$	
$R_3 = Z[(N-1)/2(N)^{1/2}] = Z[(K^2-1)/2K]$ $= Z(K - 1/K)/2$ $R_1 = Z[(N^{1/2}+1)/(N^{1/2}-1)] = Z[(K+1)/(K-1)]$ $= Z[1 + 2/(K-1)]$	$R_1R_3 = Z^2(1 + \cosh\theta) = Z^2[(K+1)^2/2K]$ $R_3/R_1 = \cosh\theta - 1 = (K-1)^2/2K$ $Z = R_1/[1 + 2(R_1/R_3)]^{1/2}$
$R_1 = R_2 = Z$ $R_4 = Z(K - 1)$ $R_3 = Z/(K - 1)$	$R_3R_4 = Z^2$ $R_4/R_3 = (K-1)^2$

the image impedances; R_1, R_2, and R_3 become complex impedances; and θ is the image transfer constant. $\theta = \alpha + j\beta$, where α is the image attenuation constant and β is the image phase constant.

Table 12-8. Symmetrical T and H Attenuator Values
$Z = 500 \ \Omega$ Resistive (Diagram in Fig. 12-59)

Attenuation, dB	Series arm R_1, Ω	Shunt arm R_3, Ω	$1,000/R_3$	log R_3
0.0	0.0	∞	0.0000	
0.2	5.8	21,700	0.0461	
0.4	11.5	10,850	0.0921	
0.6	17.3	7,230	0.1383	
0.8	23.0	5,420	0.1845	
1.0	28.8	4,330	0.2308	
2.0	57.3	2,152	0.465	
3.0	85.5	1,419	0.705	
4.0	113.1	1,048	0.954	
5.0	140.1	822	1.216	
6.0	166.1	669	1.494	2.826
7.0	191.2	558	. . .	2.747
8.0	215.3	473.1	. . .	2.675
9.0	238.1	405.9	. . .	2.608
10.0	259.7	351.4	. . .	2.546
12.0	299.2	268.1	. . .	2.428
14.0	333.7	207.8	. . .	2.318
16.0	363.2	162.6	. . .	2.211
18.0	388.2	127.9	. . .	2.107
20.0	409.1	101.0	. . .	2.004
22.0	426.4	79.94	. . .	1.903
24.0	440.7	63.35	. . .	1.802
26.0	452.3	50.24	. . .	1.701
28.0	461.8	39.87	. . .	1.601
30.0	469.3	31.65	. . .	1.500
35.0	482.5	17.79	. . .	1.250
40.0	490.1	10.00	. . .	1.000
50.0	496.8	3.162	. . .	0.500
60.0	499.0	1.000	. . .	0.000
80.0	499.9	0.1000	. . .	-1.000
100.0	500.0	0.01000	. . .	-2.000

SYMMETRICAL T OR H ATTENUATORS

21. Interpolation of Symmetrical T or H Attenuators (Table 12-8). Column R_1 may be interpolated linearly. Do not interpolate the R_3 column. For 0 to 6 dB interpolate the $1,000/R_3$ column. Above 6 dB, interpolate the column log R_3 and determine R_3 from the result.

22. Errors in Symmetrical T or H Attenuators. *Series Arms R_1 and R_2 in error.* Error in input impedances:

$$\Delta Z_1 = \Delta R_1 + (1/K^2)\Delta R_2$$

and

$$\Delta Z_2 = \Delta R_2 + (1/K^2)\Delta R_1$$

Error in insertion loss, in decibels:

$$\text{Decibels} = 4[(\Delta R_1/Z_1) + (\Delta R_2/Z_2)] \text{ approx.}$$

Shunt arm R_3 in error:
Error in input impedance:

$$\frac{\Delta Z}{Z} = 2\frac{K-1}{K(K+1)}\frac{\Delta R_3}{R_3}$$

Error in output current:

$$\frac{\Delta i}{i} = \frac{K-1}{K+1}\frac{\Delta R_3}{R_3}$$

See notes under Par. **12-20**.

Shunt Arm R_3 in Error (10% High)

Designed loss, dB	Error in insertion loss, dB	Error in input impedance, $100 (\Delta Z/Z)$ %
0.2	−0.01	0.2
1	−0.05	1.0
6	−0.3	3.3
12	−0.5	3.0
20	−0.7	1.6
40	−0.8	0.2
100	−0.8	0.0

SYMMETRICAL π AND 0 ATTENUATORS

23. Interpolation of Symmetrical π and 0 Attenuators (Table 12-9). Column R_1 may be interpolated linearly above 16 dB, and R_3 up to 20 dB. Otherwise interpolate the $1,000/R_1$ and log R_3 columns, respectively.

24. Errors in Symmetrical π and 0 Attenuators.

Error in input impedance:

$$\frac{\Delta Z'}{Z'} = \frac{K-1}{K+1}\left(\frac{\Delta R_1}{R_1} + \frac{1}{K^2}\frac{\Delta R_2}{R_2} + \frac{2}{K}\frac{\Delta R_3}{R_3}\right)$$

Error in insertion loss, in decibels: Decibels = $-8(\Delta i_2/i_2)$ approx.

$$= 4\frac{K-1}{K+1}\left(-\frac{\Delta R_1}{R_1} - \frac{\Delta R_2}{R_2} + 2\frac{\Delta R_3}{R_3}\right)$$

See notes under Par. **12-20.**

Table 12-9.   Symmetrical π and 0 Attenuators
$Z = 500$ Ω Resistive (Diagram in Fig. 12-59)

Attenuation, dB	Shunt arm R_1, Ω	$1,000/R_1$	Series arm R_3, Ω	log R_3
0.0	∞	0.000	0.0	
0.2	43,400	0.023	11.5	
0.4	21,700	0.046	23.0	
0.6	14,500	0.069	34.6	
0.8	10,870	0.092	46.1	
1.0	8,700	0.115	57.7	
2.0	4,362	0.229	116.1	
3.0	2,924	0.342	176.1	
4.0	2,210	0.453	238.5	
5.0	1,785	0.560	304.0	
6.0	1,505	0.665	373.5	
7.0	1,307	0.765	448.0	
8.0	1,161.4	0.861	528.4	
9.0	1,049.9	0.952	615.9	
10.0	962.5	1.039	711.5	
12.0	835.4	1.197	932.5	
14.0	749.3	1.335	1,203.1	
16.0	688.3	1.453	1,538	
18.0	644.0	...	1,954	
20.0	611.1	...	2,475	3.394
22.0	586.3	...	3,127	3.495
24.0	567.3	...	3,946	3.596
26.0	552.8	...	4,976	3.697
28.0	541.5	...	6,270	3.797
30.0	532.7	...	7,900	3.898
35.0	518.1	...	14,050	4.148
40.0	510.1	...	25,000	4.398
50.0	503.2	...	79,100	4.898
60.0	501.0	...	2.50×10^5	5.398
80.0	500.1	...	2.50×10^6	6.398
100.0	500.0	...	2.50×10^7	7.398

BRIDGED-T OR BRIDGED-H ATTENUATORS

25. Interpolation of Bridged-T or Bridged-H Attenuators (Table 12-10).

Bridge Arm R_4.   Use the formula log $(R_4 + 500) = 2.699 + $ decibels/20 for $Z = 500$ Ω. However, if preferred, the tabular values of R_4 may be interpolated linearly, between 0 and 10 dB only.

Shunt Arm R_3.   Do not interpolate R_3 column.   Compute R_3 by $R_3 = 10^6/4R_4$, for $Z = 500$ Ω.

Table 12-10.   Values for Bridged-T or Bridged-H Attenuators
$Z = 500$ Ω Resistive, $R_1 = R_2 = 500$ Ω (Diagram in Fig. 12-59)

Attenuation, dB	Bridge arm R_4, Ω	Shunt arm R_3, Ω
0.0	0.0	∞
0.2	11.6	21,500
0.4	23.6	10,610
0.6	35.8	6,990
0.8	48.2	5,180
1.0	61.0	4,100
2.0	129.5	1,931
3.0	206.3	1,212
4.0	292.4	855
5.0	389.1	642
6.0	498	502
7.0	619	404
8.0	756	331
9.0	909	275.0
10.0	1,081	231.2
12.0	1,491	167.7
14.0	2,006	124.6
16.0	2,655	94.2
18.0	3,472	72.0
20.0	4,500	55.6
25.0	8,390	29.8
30.0	15,310	16.33
40.0	49,500	5.05
50.0	157,600	1.586
60.0	499,500	0.501
80.0	5.00×10^6	0.0500
100.0	50.0×10^6	0.00500

NOTE: For attenuators of 60 dB and over, the bridge arm R_4 may be omitted provided a shunt arm is used having twice the resistance tabulated in the R_3 column. (This makes the input impedance 0.1 of 1% high at 60 dB.)

26. Errors in Bridged-T or Bridged-H Attenuators.   Resistance of any one arm 10% higher than correct value:

Element in error (10% high)	Error in loss	Error in terminal impedance
Series arm R_1 (analogous for arm R_2)	Zero	B, for adjacent terminals*
Shunt arm R_3	$-A$†	C
Bridge arm R_4	A‡	C

NOTE: A, B, and C are given in tabulation below.
* Error in impedance at opposite terminals is zero.
† Loss is lower than designed loss.
‡ Loss is higher than designed loss.

Designed loss, dB	A, dB	B, %	C, %
0.2	0.01	0.005	0.2
1	0.05	0.1	1.0
6	0.2	2.5	2.5
12	0.3	5.6	1.9
20	0.4	8.1	0.9
40	0.4	10	0.1
100	0.4	10	0.0

Error in input impedance:

$$\frac{\Delta Z_1}{Z_1} = \left(\frac{K-1}{K}\right)^2 \frac{\Delta R_1}{R_1} + \frac{K-1}{K^2}\left(\frac{\Delta R_3}{R_3} + \frac{\Delta R_4}{R_4}\right)$$

For $\Delta Z_2/Z_2$ use subscript 2 in the equation in place of subscript 1.
Error in output current:

$$\frac{\Delta i}{i} = \frac{K-1}{2K}\left(\frac{\Delta R_3}{R_3} - \frac{\Delta R_4}{R_4}\right)$$

See notes under Par. **12-20**.

MINIMUM-LOSS PADS

27. Interpolation of Minimum-Loss Pads. Table 12-11 may be interpolated linearly with respect to Z_1, Z_2, or Z_1/Z_2 except when Z_1/Z_2 is between 2.0 and 1.2. The accuracy of the interpolated value becomes poorer as Z_1/Z_2 passes below 2.0 toward 1.2, especially for R_3.

Table 12-11. Values for Minimum-Loss Pads Matching
Z_1 and Z_2, Both Resistive (Diagram in Fig. 12-59)

Z_1, Ω	Z_2, Ω	Z_1/Z_2	Loss, dB	Series arm R_1, Ω	Shunt arm R_2, Ω
10,000	500	20.00	18.92	9,747	513.0
8,000	500	16.00	17.92	7,746	516.4
6,000	500	12.00	16.63	5,745	522.2
5,000	500	10.00	15.79	4,743	527.0
4,000	500	8.00	14.77	3,742	534.5
3,000	500	6.00	13.42	2,739	547.7
2,500	500	5.00	12.54	2,236	559.0
2,000	500	4.00	11.44	1,732	577.4
1,500	500	3.00	9.96	1,224.7	612.4
1,200	500	2.40	8.73	916.5	654.7
1,000	500	2.00	7.66	707.1	707.1
800	500	1.60	6.19	489.9	816.5
600	500	1.20	3.77	244.9	1,224.7
500	400	1.25	4.18	223.6	894.4
500	300	1.667	6.48	316.2	474.3
500	250	2.00	7.66	353.6	353.6
500	200	2.50	8.96	387.3	258.2
500	160	3.125	10.17	412.3	194.0
500	125	4.00	11.44	433.0	144.3
500	100	5.00	12.54	447.2	111.80
500	80	6.25	13.61	458.3	87.29
500	65	7.692	14.58	466.4	69.69
500	50	10.00	15.79	474.3	52.70
500	40	12.50	16.81	479.6	41.70
500	30	16.67	18.11	484.8	30.94
500	25	20.00	18.92	487.3	25.65

28. For Other Terminations. If the terminating resistances are to be Z_A and Z_B instead of Z_1 and Z_2, respectively, the procedure is as follows. Enter the table at $Z_1/Z_2 = Z_A/Z_B$ and read the loss and the tabular values of R_1 and R_3. Then the series and shunt arms are, respectively, MR_1 and MR_3, where $M = Z_A/Z_1 = Z_B/Z_2$.

29. Errors in Minimum-Loss Pads.

Series Arm R_1 10% high: Loss is increased by D dB from the table below. Input impedance Z_1 is increased by E percent. Input impedance Z_2 is increased by F percent.

Shunt Arm R_3 10% high. Loss is decreased by D dB from the table below. Input impedance Z_2 is increased by E percent. Input impedance Z_1 is increased by F percent.

Impedance ratio Z_1/Z_2	D, dB	E, %	F, %
1.2	0.2	+4.1	+1.7
2.0	0.3	7.1	1.2
4.0	0.35	8.6	0.6
10.0	0.4	9.5	0.25
20.0	0.4	9.7	0.12

Errors in input impedance:

$$\Delta Z_1/Z_1 = (1 - Z_2/Z_1)^{1/2}(\Delta R_1/R_1 + N^{-1}\Delta R_3/R_3)$$

$$\Delta Z_2/Z_2 = (1 - Z_2/Z_1)^{1/2}(\Delta R_3/R_3 + N^{-1}\Delta R_1/R_1)$$

Error in output current, working either direction:

$$\Delta i/i = \tfrac{1}{2}(1 - Z_2/Z_1)^{1/2}(\Delta R_3/R_3 - \Delta R_1/R_1)$$

See notes under Par. **12-20**.

30. References

1. S. DARLINGTON   Synthesis of Reactance 4-Poles, *J. Math Phys.*, vol. 18, pp. 257-353, September, 1939.

2. M. DISHAL   Design of Dissipative Band-Pass Filters Producing Desired Exact Amplitude-Frequency Characteristics, *Proc. IRE*, vol. 37, pp. 1050-1069, September, 1949; also, *Elect. Commun.*, vol. 27, pp. 56-81, March, 1950.

3. M. DISHAL   Concerning the Minimum Number of Resonators and the Minimum Unloaded Q Needed in a Filter, *Trans. IRE Professional Group on Vehicular Commun., vol. PGVC-3*, pp. 85-117, June, 1953; also, *Electr. Commun.*, vol. 31, pp. 257-277, December, 1954. Institution; Publ. 3863, Washington, D.C., 1947.

4. W. E. THOMSON   Networks with Maximally Flat Delay, *Wireless Eng.*, vol. 29, pp. 256-263, October, 1952.

5. G. W. and R. M. SPENCELY   Smithsonian Elliptic Function Tables, Smithsonian Institution, Publ. 3863, Washington, D.C., 1947.

6. E. JAHNKE and F. EMDE   "Table of Functions with Formulas and Curves." 4th ed., pp. 49-51, Dover Publications, New York, 1945.

7. M. DISHAL   Alignment and Adjustment of Synchronously Tuned Multiple-resonant-Circuit Filters, *Proc. IRE*, vol. 39, pp. 1448-1455, November, 1951; Also, *Electr. Commun.*, vol. 29, pp. 154-164, June, 1952.

8. V. BELEVITCH   Tchebyshev Filters and Amplifier Networks, *Wireless Eng.*, vol. 29, pp. 106-110, April, 1952.

9. M. DISHAL   Two New Equations for the Design of Filters, *Electr. Commun.*, vol. 30, pp. 324-337, December, 1952.

10. H. J. ORCHARD   Formulas for Ladder Filters, *Wireless Eng.*, vol. 30, pp. 3-5, January, 1953.

11. E. GREEN   Exact Amplitude-Frequency Characteristics of Ladder Networks, *Marconi Rev.*, vol. 16, no. 108, pp. 25-68, 1953.

31. Bibliography

Image-parameter filter design

NOTE: The bibliographies in these references will direct the reader to other useful publications.

T. E. SHEA   "Transmission Networks and Wave Filters," D. Van Nostrand Co., Inc., New York, 1929.

F. E. TERMAN   "Radio Engineers' Handbook," pp. 226-244, McGraw-Hill Book Co., New York, 1943.

W. P. MASON   "Electromechanical Transducers and Wave Filters," D. Van Nostrand Company, Inc., New York, 1948.

"Reference Data for Radio Engineers," 5th ed., Chap. 7, Howard W. Sams & Company, Inc., New York, 1968.

Other modern network theory design information

E. GREEN   "Amplitude-Frequency Characteristics of Ladder Networks," Marconi's Wireless Telegraph Company, Essex, England, 1954.

R. SAAL   "The Design of Filters by a Catalog of Normalized Lowpass Filters," Telefunken GmbH, Backnang/Württenberg, West Germany, 1961.

P. GEFFE   "Simplified Modern Filter Design," Hayden Book Co., New York, 1963.

J. K. SKWIRZYNSKI   "Design Theory and Data for Electrical Filters," D. Van Nostrand Co., New York, 1965.

E. CHRISTIAN and E. EISENMANN "Filter Design Tables and Graphs," John Wiley & Sons, Inc., New York, 1966.

A. I. ZVEREV "Handbook of Filter Synthesis," John Wiley & Sons, Inc., New York, 1967.

Computer-aided design of filters

Special Issue on Computer Aided Circuit Design, *IEEE Trans. Circuit Theory,* vol. CT-18, pp. 3–163, January, 1971.

Papers on Network Design by Computers, *IRE Trans. Circuit Theory,* vol. CT-9, pp. 184–237, September, 1961.

SECTION 13

AMPLIFIERS AND OSCILLATORS

BY

G. BURTON HARROLD   Senior Engineer, Member Institute of Electrical and Electronics Engineers; JOSEPH P. HESLER   Consulting Engineer; SAMUEL M. KORZEKWA   Senior Engineer; JOHN W. LUNDEN   Senior Engineer, Member IEEE; CHANG S. KIM   Consulting Engineer, Member IEEE; ROBERT J. McFADYEN   Consulting Engineer, Member IEEE; STEPHEN W. TEHON   Consulting Engineer, Fellow IEEE, all of the Electronics Laboratory, General Electric Company.

RICHARD W. FRENCH   ELEMEK Inc., Member IEEE

HAROLD W. LORD   Consulting Engineer, Fellow IEEE

CONRAD E. NELSON   Senior Consulting Engineer, Heavy Military Electronic Systems, Senior Member IEEE

GUNTER K. WESSEL   Professor, Physics Department, Syracuse University

CONTENTS

Numbers refer to paragraphs

SECTION 13

AMPLIFIERS AND OSCILLATORS

PRINCIPLES OF OPERATION—AMPLIFIERS

By G. B. Harrold

1. Gain. In most amplifier applications the prime concern is gain. A generalized amplifier is shown in Fig. 13-1; the most widely applied definitions of gain using the quantities defined there are:

Voltage gain:
$$A_v = \frac{e_{22}}{e_{11}}$$

Current gain:
$$A_i = \frac{i_2}{i_1}$$

Available power from source:
$$P_{avs} = \frac{|e_s|^2}{4 \, \mathrm{RE}(Z_s)}$$

where $\mathrm{RE}(\cdot)$ = real part of complex impedance $(\cdot)$.

Output load power:
$$P_L = \frac{|e_{22}|^2}{\mathrm{RE}(Z_L)}$$

Input power:
$$P_I = \frac{|e_{11}|^2}{\mathrm{RE}(Z_{in})}$$

Available power at output:
$$P_{avo} = \frac{|e_{22}|^2}{4 \, \mathrm{RE}(Z_{out})}$$

Transducer gain:
$$G_T = \frac{P_L}{P_{avs}}$$

Available power gain:
$$G_A = \frac{P_{avo}}{P_{avs}}$$

Power gain:
$$G = \frac{P_L}{P_I}$$

Insertion power gain:
$$G_I = \frac{\text{power into load with network inserted}}{\text{power into load with source connect to load}}$$

2. Bandwidth and Gain-Bandwidth Product. Bandwidth is a measure of the range of frequencies within which an amplifier will respond. The frequency range (passband) is usually measured between the half-power (3-dB) points on the output response vs. frequency curve, for constant input. In some cases it is defined at the quarter-power points (6 dB). See Fig. 13-2.

The gain-bandwidth product of a device is a commonly used figure of merit. It is defined for a bandpass amplifier as

$$F_a = A_r B$$

where F_a = figure of merit in radians per second; A_r = reference gain, either the maximum

13-3

gain or the gain at the frequency where the gain is purely real or purely imaginary, and $B =$ 3-dB bandwidth in radians per second.

For low-pass amplifiers

$$F_a = A_r W_H$$

where F_a = figure of merit in radians per second, A_r = reference gain, and W_H = upper cutoff frequency in radians per second.

In the case of vacuum tubes and certain other active devices this definition is reduced to

$$F_a = g_m/C_T$$

where F_a = figure of merit in radians per second, g_m = transconductance of the active device, and C_T = total output capacitance, plus input capacitance of the subsequent stage.

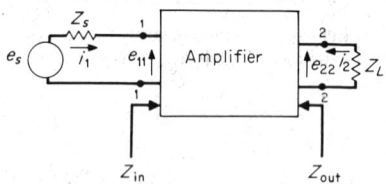

Fig. 13-1.   Input and output quantities of generalized amplifier.

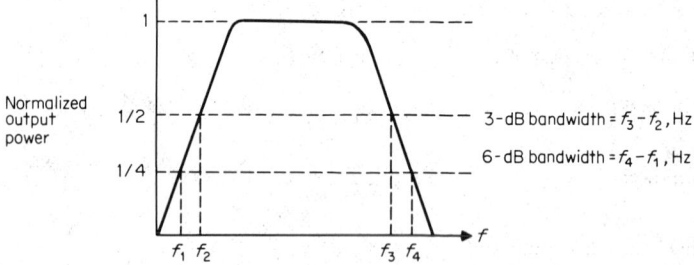

Fig. 13-2.   Amplifier response and bandwidth.

3. Noise.   The major types of noise are illustrated in Fig. 13-3.   Important relations and definitions in noise computations are

Noise factor:

$$F = \frac{S_i/N_i}{S_o/N_o}$$

where S_i = signal power available at input, S_o = signal power available at output, N_i = noise power available at input at $T = 290$ K, and N_o = noise power available at output.

Available noise power:

$$P_{n,\text{av}} = \frac{e_n^2}{4R} = KTB \qquad \text{for thermal noise}$$

where the quantities are as defined in Fig. 13-3.

Excess noise factor:

$$F - 1 = N_e/N_i$$

where $F - 1$ = excess noise factor, N_e = total equivalent device noise referred to input, and N_i = thermal noise of source at standard temperature.

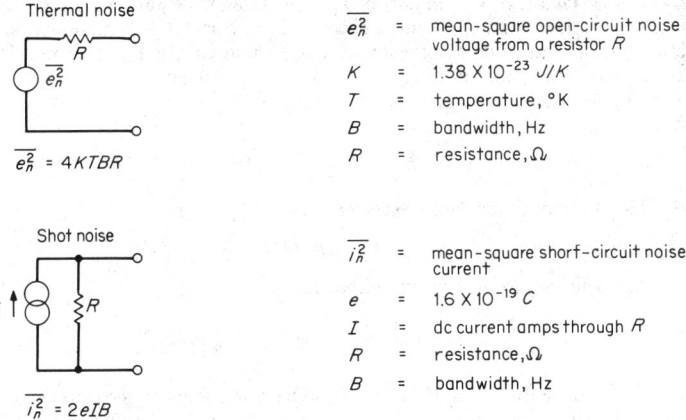

Thermal noise

$\overline{e_n^2} = 4KTBR$

$\overline{e_n^2}$	=	mean-square open-circuit noise voltage from a resistor R
K	=	1.38×10^{-23} J/K
T	=	temperature, °K
B	=	bandwidth, Hz
R	=	resistance, Ω

Shot noise

$\overline{i_n^2} = 2eIB$

$\overline{i_n^2}$	=	mean-square short-circuit noise current
e	=	1.6×10^{-19} C
I	=	dc current amps through R
R	=	resistance, Ω
B	=	bandwidth, Hz

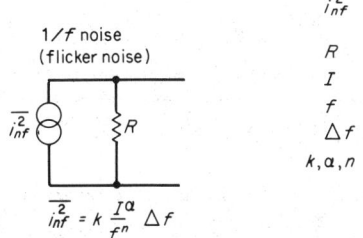

1/f noise
(flicker noise)

$\overline{i_{nf}^2} = k \, \dfrac{I^a}{f^n} \, \Delta f$

$\overline{i_{nf}^2}$	=	mean-square short-circuit flicker noise current
R	=	resistance, Ω
I	=	dc current
f	=	frequency, Hz
Δf	=	frequency interval
k, a, n	=	empirical constants depending on device and mode of operation

Fig. 13-3.   Noise equivalent circuits.

Noise temperature:

$$T = P_{n,\text{av}}/KB$$

where $P_{n,\text{av}}$ = average noise power available.

Effective input noise temperature: at a single input/output frequency in a two-port,

$$T_e = 290(F - 1)$$

Cascaded noise factor:

$$F_T = F_1 + \frac{F_2 - 1}{G_A}$$

where F_T = overall noise factor, F_1 = noise factor of first stage, F_2 = noise factor of second stage, and G_A = available gain of first stage.

Generalized Noise Factor. It can be shown[1]* that a general representation of noise performances can be expressed in terms of Fig. 13-4. This is the representation of a noisy two-port in terms of external voltage and current noise sources with a correlation admittance. In this case the noise factor becomes

$$F = 1 + \frac{G_u}{G_s} + \frac{R_N}{G_s}[(G_s + G_\gamma)^2 + (B_s + B_\gamma)^2]$$

---

* Superior numbers correspond to numbered references, Par. 13-17.

where F = noise factor, G_s = real part of Y_s, B_s = imaginary part of Y_s, G_u = conductance due to the uncorrelated part of the noise current, Y_γ = correlation admittance between cross product of current and voltage noise sources, G_γ = real part of Y_γ, B_γ = imaginary part of Y_γ, and R_N = equivalent noise resistance of the noise voltage.

The optimum source admittance is $Y_{\text{opt}} = G_{\text{opt}} + jB_{\text{opt}}$

$$G_{\text{opt}} = \left(\frac{G_u + R_N G_\gamma^2}{R_N} \right)^{1/2} \qquad \text{where } B_{\text{opt}} = -B_\gamma$$

and the value of the optimum noise factor F_{opt} is

$$F_{\text{opt}} = 1 + 2R_N(G\gamma + G_0)$$

The noise factor for an arbitrary source impedance is

$$F = F_{\text{opt}} + \frac{R_N}{G_s}[(G_s - G_0)^2 + (B_s - B_0)^2]$$

The values of the parameters of Fig. 13-4 can be determined by measurement of (1) noise figure versus B_s with G_s constant and (2) noise figure versus G_s with B_s at its optimum value.

4. Dynamic Characteristic, Load Lines and Class of Operation. Most active devices have two considerations involved in their operation. The first is the dc bias condition that establishes the operating point (the *quiescent point*). The choice of operating point is determined by such considerations as signal level, uniformity of the device, temperature of operation, etc.

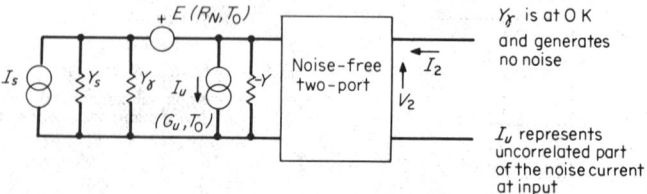

Fig. 13-4. Noise representation using correlation admittance.

The second consideration is the ac operating performance, related to the slope of the dc characteristic and to the parasitic reactances of the device. These ac variations give rise to the *small-signal parameters*. The ac parameters may also influence the choice of dc bias point when basic constraints, such as gain and noise performance, are considered.

For frequencies of operation where these parasitics are not significant, the use of a load line is valuable. The class of amplifier operation is dependent upon its quiescent point, its load line, and input signal level. The types of operation are shown in Fig. 13-5.

5. Distortion. Distortion takes many forms, most of them undesirable. The basic causes of distortion are nonlinearity in amplitude response and nonuniformity of phase response. The most commonly encountered types of distortion are:

Harmonic distortion is due to nonlinearities in the amplitude transfer characteristics. The typical output contains not only the fundamental frequency but integer multiples of the fundamental frequency.

Crossover distortion is due to the nonlinear characteristics of a device when changing operating modes (e.g., in a push-pull amplifier). It occurs when one device is cut off and the second turned on, if the "crossover" is not smooth between the two modes.

Intermodulation distortion is a spurious output resulting from the mixing of two or more signals of different frequencies. The spurious output occurs at the sum or difference of integer multiples of the original frequencies.

Cross-modulation distortion occurs when two signals pass through an amplifier and the modulation of one is transferred to the other.

Phase distortion results from the deviation from a constant slope of the output phase vs. frequency response of an amplifier. This deviation gives rise to echo responses in the output that precede and follow the main response, and a distortion of the output signal when an input signal having a large number of frequency components is applied.

6. Feedback Amplifiers. Feedback amplifiers fall into two categories: those that have positive feedback (usually oscillators) and those having negative feedback. The positive-feedback case is discussed under oscillators, Par. **13-12.** The following discussion is concerned with negative-feedback amplifiers.

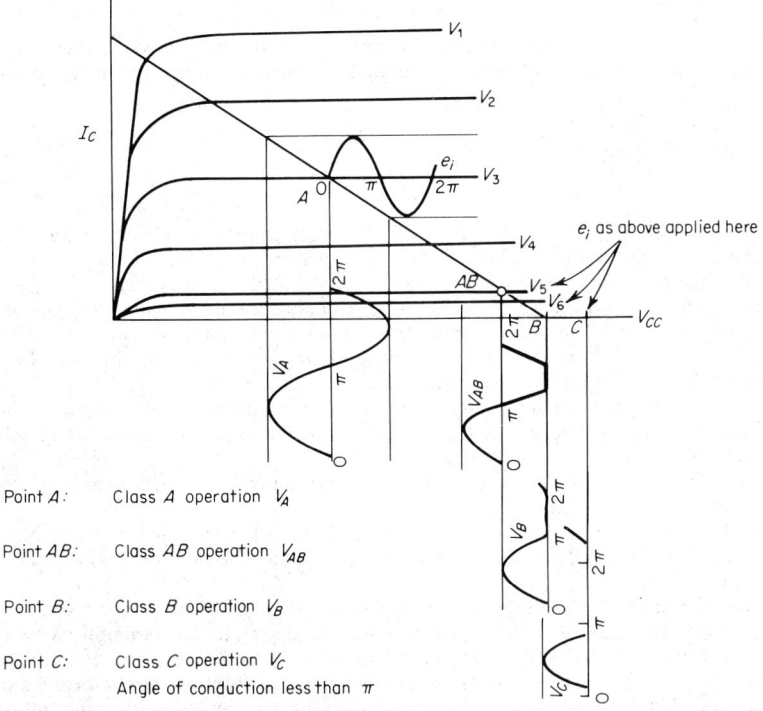

Point A: Class A operation V_A

Point AB: Class AB operation V_{AB}

Point B: Class B operation V_B

Point C: Class C operation V_C
              Angle of conduction less than π

Fig. 13-5. Classes of amplifier operation. Class S operation is a switching mode in which a square-wave output is produced by a sinewave input.

7. Negative Feedback. A simple representation of a feedback network is shown in Fig. 13-6. The closed-loop gain is given by

$$\frac{e_2}{e_1} = \frac{A}{1 - BA}$$

where A = forward gain with feedback removed, and B = the fraction of the output returned to the input.

For negative feedback, A provides a 180° phase shift in midband, so that

$$1 - AB > 1 \qquad \text{in this frequency range}$$

The quantity $1 - AB$ is called the *feedback factor,* and if the circuit is cut at any X point in Fig. 13-6, the open-loop gain is AB.

It can be shown that for large loop gain AB the closed-loop transfer function reduces to

$$\frac{e_2}{e_1} \approx \frac{1}{B}$$

The gain then becomes essentially independent of variations in A. In particular, if B is passive, the closed-loop gain is controlled only by passive components. It can also be shown[2] that feedback has no beneficial effect in reducing unwanted signals at the input of the amplifier, e.g., input noise, but does reduce unwanted signals generated in the amplifier chain (e.g., output distortion).

The return ratio can be found if the circuit is opened at any point X (Fig. 13-6) and a unit signal P is injected at that X point. The return signal P' is equal to the return ratio, since the input P is unity. In this case the return ratio T is the same at any point X and is

$$T = -AB$$

The minus sign is chosen because the typical amplifier has an odd number of phase reversals and T is then a positive quantity. The return difference is by definition

$$F = 1 + T$$

It has been shown by Bode that

$$F = \Delta/\Delta^0$$

where Δ = the network determinant with XX point connected, and Δ^0 = the network determinant of the amplifier when the gain of the active device is set to zero.

8. Stability. The stability of the network may be analyzed by several techniques. Of prime interest are the Nyquist, Bode, Routh, and root locus techniques of analyzing stability.

Nyquist Method. The basic technique of Nyquist involves the plotting of T on a polar plot as shown in Fig. 13-7 for all values $s = j\omega$ for ω between minus and plus infinity. Stability is then determined by the following method:

a. Draw a vector from the $-1 + j0$ point to the plotted curve and observe the rotation of this vector as ω varies from $-\infty$ to $+\infty$. Let R be the net number of counterclockwise revolutions of this vector.

b. Determine the number of roots of the denominator of $T = -AB$ which have positive real parts. Call this number P.

c. The system is stable if and only if $P = R$. Note that in many systems A and B are stable by themselves, so that P becomes zero and the net counterclockwise revolution N becomes zero for stability.[3]

Bode's Technique. A technique that has found wide usage in determining stability and performance, especially in control systems, is the Bode diagram. The assumptions used here for this method are that $T = -AB$, where A and B are stable when the system is open-circuited, and consists of minimum-phase networks. It is also necessary to define a phase margin γ such that $\gamma = 180 + \phi$, where ϕ is the phase angle of T and is positive when measured counterclockwise from zero, and γ, the phase margin, is positive when measured counterclockwise from the 180° line (Fig. 13-7). The stability criterion under these conditions becomes: Systems having a positive phase margin when their return ratio equal to 20 log|T|

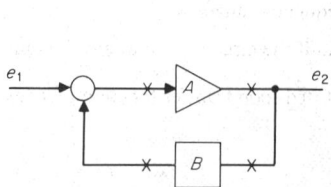

Fig. 13-6. Amplifier with feedback loop.

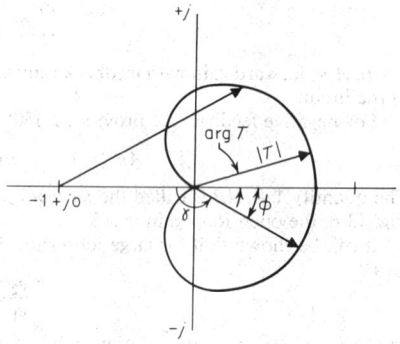

Fig. 13-7. Nyquist diagram for determining stability.

goes through 0 dB (i.e., where $|T|$ crosses the unit circle in the Nyquist plot) are stable; if a negative γ exists at 0 dB, the system is unstable.

The usefulness of this technique lies in the fact that Bode's theorems show that the phase angle of a system is related to the attenuation or gain characteristic as a function of frequency. It is shown in Ref. 3 that the use of Bode's technique relies heavily on straight-line approximation. Figure 13-8 shows the use of these straight-line approximations applied to the factored form of T. It can also be shown that the rate of change of 20 log $|D|$ as it crosses the 0 dB axis must be less than 40 dB/decade in order for the system to be stable. Complicated systems may be analyzed by means of this technique by breaking them down into the products of their various components.

Routh's Criterion for Stability. Routh's method is used to test the characteristic equation or return difference $F = 1 + T = 0$, to determine whether it has any roots that are real and positive or complex with positive real parts that will give rise to growing exponential responses and hence instability.

Root Locus Method. The root locus method of analysis is a means of finding the variations of the poles of a closed-loop response as some network parameter is varied. The most convenient and commonly used parameter is that of the gain K. The basic equation then used is

$$F = 1 + KT(s) = 1 - K\frac{(S - S_2)(S - S_4)\cdots}{(S - S_1)(S - S_3)\cdots} = 0$$

This is a useful technique in feedback and control systems, but it has not found wide application in amplifier design. A detailed exposition of the technique is found in Ref. 4.

9. Active Devices Used in Amplifiers. There are numerous ways of representing active devices and their properties. Several common equivalent circuits are shown in Fig. 13-9. Active devices are best analyzed in terms of the *immittance* or *hybrid matrices.* Figures 13-10 and 13-11 show the definition of the commonly used matrices, and their interconnections are shown in Fig. 13-12. The requirements at the bottom of Fig. 13-12 must be met before the interconnection of two matrices is allowed.

The matrix that is becoming increasingly important at higher frequencies is the S matrix. Here the network is embedded in a transmission-line structure, and the incident and reflected powers are measured and reflected coefficients and transmission coefficients are defined.

10. Cascaded and Distributed Amplifiers. Most amplifiers are cascaded (i.e., connected to a second amplifier). The two techniques commonly used are shown in Fig. 13-13. In the cascade structure the overall response is the product of the individual responses; in the distributed structure the response is one-half the sum of the individual responses, since each stage's output is propagated in both directions. In cascaded amplifiers the frequency response and gain are determined by the active device as well as the interstage networks. In simple audio amplifiers these interstage networks may become simple RC combinations, while in rf amplifiers they may become critically coupled double-tuned circuits. Interstage coupling networks are discussed in subsequent sections.

In distributed structures (Fig. 13-13b), actual transmission lines are used for the input to the amplifier, while the output is taken at one end of the upper transmission line. The propagation time along the input line must be the same as that along the output line, or distortion will result. This type of amplifier is noted for its wide frequency response, and is discussed in a later section, Par. 13-62.

PRINCIPLES OF OPERATION—OSCILLATORS

By G. B. Harrold

11. Introduction. An oscillator can be considered as a circuit that converts a dc input to a time-varying output. This discussion deals with oscillators whose output is sinusoidal, as opposed to the relaxation oscillator whose output exhibits abrupt transitions (see Sec. 16). Oscillators often have a circuit element that can be varied to produce different frequencies.

An oscillator's frequency is sensitive to the stability of the frequency-determining elements as well as the variation in the active-device parameters (e.g., effects of temperature,

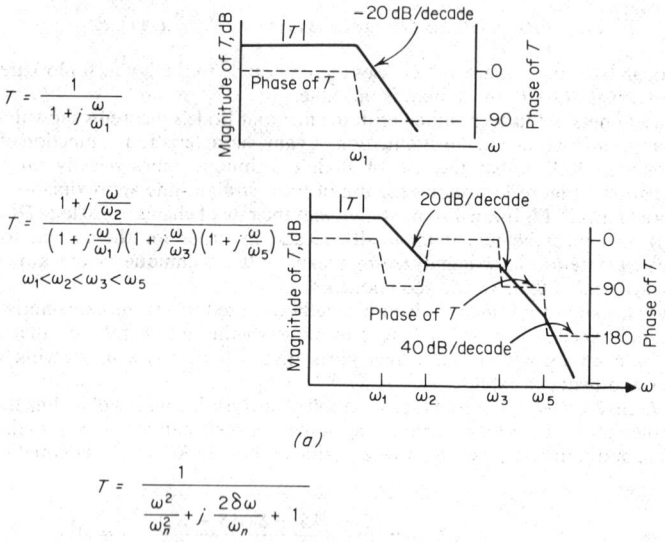

$$T = \frac{1}{1 + j\frac{\omega}{\omega_1}}$$

$$T = \frac{1 + j\frac{\omega}{\omega_2}}{\left(1 + j\frac{\omega}{\omega_1}\right)\left(1 + j\frac{\omega}{\omega_3}\right)\left(1 + j\frac{\omega}{\omega_5}\right)}$$

$$\omega_1 < \omega_2 < \omega_3 < \omega_5$$

(a)

$$T = \frac{1}{\frac{\omega^2}{\omega_n^2} + j\frac{2\delta\omega}{\omega_n} + 1}$$

(Second-order function with damping)

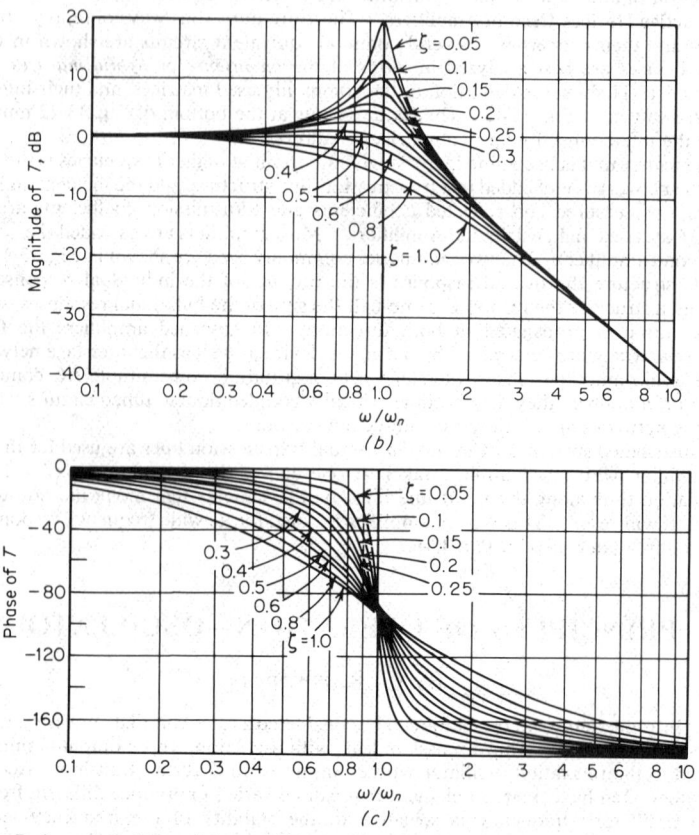

Fig. 13-8. Bode method of determining stability. (*a*) Typical straight-line approximations; (*b*) magnitude of T versus frequency; (*c*) phase of T versus frequency.

13-10

bias point, and aging). In many instances the oscillator is followed by a second stage serving as a buffer, so that there is isolation between the oscillator and its load. The amplitude of the oscillation can be controlled by automatic gain control (AGC) circuits, but the nonlinearity of the active element usually determines the amplitude. Variations in bias, temperature, and component aging will have a direct effect on amplitude stability.

12. Requirements for Oscillation. Oscillators can be considered from two viewpoints: as using positive feedback around an amplifier or as a one-port network in which the real component of the input immittance is negative. An oscillator must have frequency-determining elements (generally passive components), an amplitude-limiting mechanism, and sufficient closed-loop gain to make up for the losses in the circuit. It is possible to predict the operating frequency and conditions needed to produce oscillation from a Nyquist or Bode analysis. The prediction of output amplitude requires the use of nonlinear analysis, commonly in the form of graphical techniques.

13. Oscillator Circuits. Typical oscillator circuits applicable up to uhf frequencies are

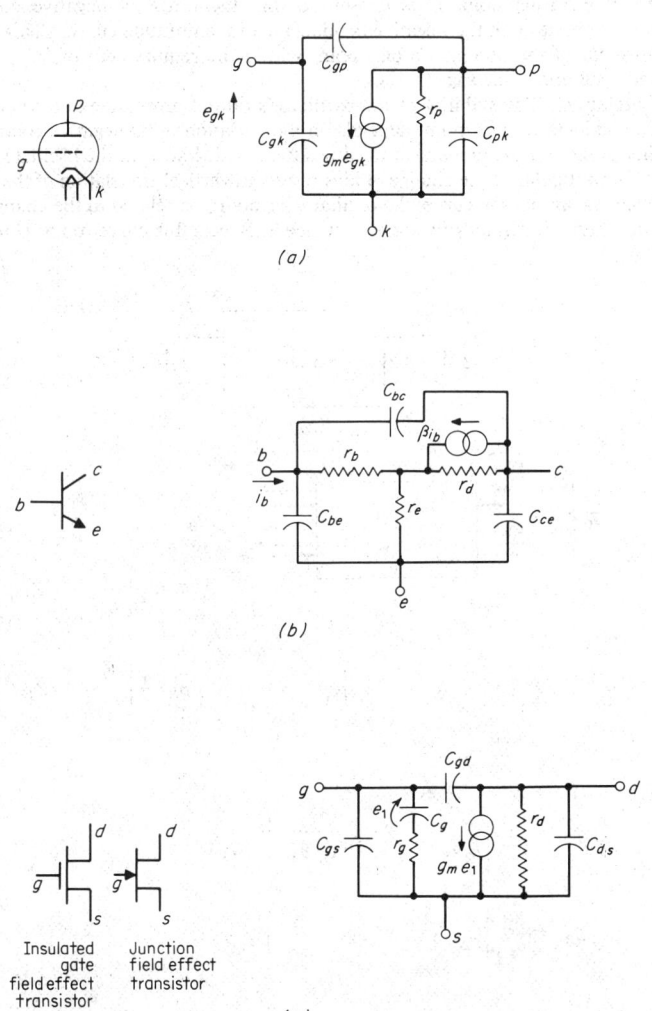

(a)

(b)

(c)

Fig. 13-9. Equivalent circuits of active devices, (*a*) vacuum tube; (*b*) bipolar transistor; (*c*) field-effect transistor (FET).

shown in Fig. 13-14. These are discussed in detail in the following subsections. Also of interest are crystal oscillators. In this case the crystal is used as the passive frequency-determining element. The frequency range of crystal oscillators extends from a few hundred hertz to over 200 MHz by use of overtone crystals. The analysis of crystal oscillators is best done using the equivalent circuit of the crystal (see also Sec. 7).

14. Synchronization. Synchronization of oscillators is accomplished by use of phase-locked loops or by direct low-level injection of a reference frequency into the main oscillator. The diagram of a phase-locked loop is shown in Fig. 13-15, and that of an injection-locked oscillator in Fig. 13-16. Detailed discussions are contained in References 5 to 7, Par. **13-17**.

15. Harmonic Content. The harmonic content of the oscillator output is related to the amount of oscillator output power at frequencies other than the fundamental. From the viewpoint of a negative-conductance (resistance) oscillator,[8] better results are obtained if the curve of the negative conductance (or resistance) vs. amplitude of oscillation is smooth and without an inflection point over the operating range. Harmonic content is also reduced if the oscillator's operating point Q is chosen so that the range of negative conductance is symmetrical about Q on the negative conductance vs. amplitude curve. This may be done by adjustment of the oscillator's bias point within the requirement of $|G_C| = |G_D|$ for sustained oscillation (see Fig. 13-17).

16. Stability. The stability of the oscillator's output amplitude and frequency from a negative-conductance viewpoint depends on the variation of its negative conductance with operating point and the amount of fixed positive conductance in the oscillator's associated circuit. In particular, if the change of bias results in vertical translation of the conductance (resistance) vs. amplitude curve, the oscillator's stability is related to the change of slope at the point where the circuit's fixed conductance intersects this curve (point Q in Fig. 13-17).

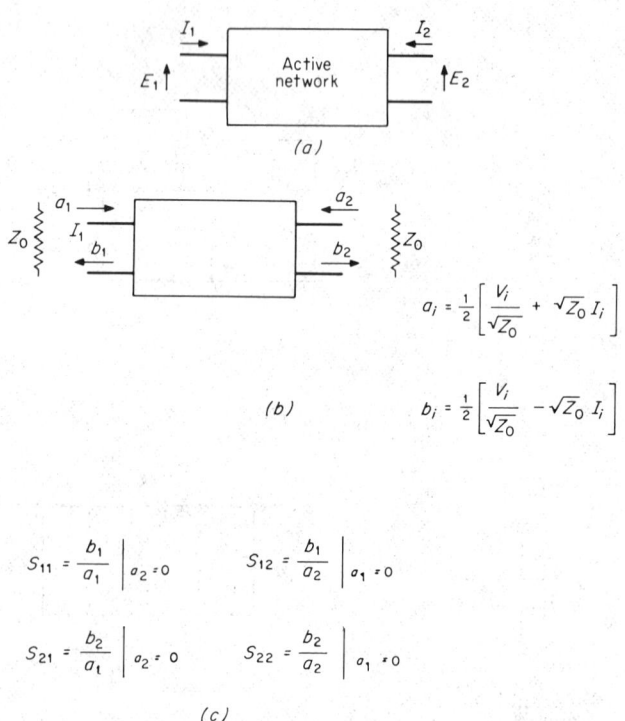

$$a_i = \frac{1}{2}\left[\frac{V_i}{\sqrt{Z_0}} + \sqrt{Z_0}\, I_i\right]$$

$$b_i = \frac{1}{2}\left[\frac{V_i}{\sqrt{Z_0}} - \sqrt{Z_0}\, I_i\right]$$

$$S_{11} = \frac{b_1}{a_1}\bigg|_{a_2=0} \qquad S_{12} = \frac{b_1}{a_2}\bigg|_{a_1=0}$$

$$S_{21} = \frac{b_2}{a_1}\bigg|_{a_2=0} \qquad S_{22} = \frac{b_2}{a_2}\bigg|_{a_1=0}$$

(c)

Fig. 13-10. Definitions of active-network parameters. (a) General network; (b) ratios a_i and b_i of incident and reflected waves (square root of power); (c) s parameters.

$$\begin{bmatrix} E_1 \\ E_2 \end{bmatrix} = \begin{bmatrix} z_{11} & z_{12} \\ z_{21} & z_{22} \end{bmatrix} \times \begin{bmatrix} I_1 \\ I_2 \end{bmatrix}$$

$$z_{11} = \left(\frac{E_1}{I_1}\right)_{I_2=0} \qquad z_{12} = \left(\frac{E_1}{I_2}\right)_{I_1=0}$$

$$z_{21} = \left(\frac{E_2}{I_1}\right)_{I_2=0} \qquad z_{22} = \left(\frac{E_2}{I_2}\right)_{I_1=0}$$

$$\begin{bmatrix} I_1 \\ I_2 \end{bmatrix} = \begin{bmatrix} y_{11} & y_{12} \\ y_{21} & y_{22} \end{bmatrix} \times \begin{bmatrix} E_1 \\ E_2 \end{bmatrix}$$

$$y_{11} = \left(\frac{I_1}{E_1}\right)_{E_2=0} \qquad y_{12} = \left(\frac{I_1}{E_2}\right)_{E_1=0}$$

$$y_{21} = \left(\frac{I_2}{E_1}\right)_{E_2=0} \qquad y_{22} = \left(\frac{I_2}{E_2}\right)_{E_1=0}$$

$$\begin{bmatrix} E_1 \\ I_2 \end{bmatrix} = \begin{bmatrix} h_{11} & h_{12} \\ h_{21} & h_{22} \end{bmatrix} \times \begin{bmatrix} I_1 \\ E_2 \end{bmatrix}$$

$$h_{11} = \frac{1}{y_{11}} \qquad h_{12} = \left(\frac{E_1}{E_2}\right)_{I_1=0}$$

$$h_{21} = \left(\frac{I_2}{I_1}\right)_{E_2=0} \qquad h_{22} = \frac{1}{z_{22}}$$

$$\begin{bmatrix} I_1 \\ E_2 \end{bmatrix} = \begin{bmatrix} l_{11} & l_{12} \\ l_{21} & l_{22} \end{bmatrix} \times \begin{bmatrix} E_1 \\ I_2 \end{bmatrix}$$

$$l_{11} = \frac{1}{z_{11}} \qquad l_{12} = \left(\frac{I_1}{I_2}\right)_{E_1=0}$$

$$l_{21} = \left(\frac{E_2}{E_1}\right)_{I_2=0} \qquad l_{22} = \frac{1}{y_{22}}$$

$$\begin{bmatrix} E_1 \\ I_1 \end{bmatrix} = \begin{bmatrix} a_{11} & a_{12} \\ a_{21} & a_{22} \end{bmatrix} \times \begin{bmatrix} E_2 \\ -I_2 \end{bmatrix}$$

$$a_{11} = \frac{1}{l_{12}} \qquad a_{12} = -\frac{1}{y_{21}}$$

$$a_{21} = \frac{1}{z_{21}} \qquad a_{22} = -\frac{1}{h_{21}}$$

$$\begin{bmatrix} E_2 \\ I_2 \end{bmatrix} = \begin{bmatrix} b_{11} & b_{12} \\ b_{21} & b_{22} \end{bmatrix} \times \begin{bmatrix} E_1 \\ -I_1 \end{bmatrix}$$

$$b_{11} = \frac{1}{h_{12}} \qquad b_{12} = -\frac{1}{y_{12}}$$

$$b_{21} = \frac{1}{z_{12}} \qquad b_{22} = -\frac{1}{l_{12}}$$

$$\begin{bmatrix} b_1 \\ b_2 \end{bmatrix} = \begin{bmatrix} S_{11} & S_{12} \\ S_{21} & S_{22} \end{bmatrix} \times \begin{bmatrix} a_1 \\ a_2 \end{bmatrix}$$

$$S_{11} = \left(\frac{b_1}{a_1}\right)_{a_2=0} \qquad S_{12} = \left(\frac{b_1}{a_2}\right)_{a_1=0}$$

$$S_{21} = \left(\frac{b_2}{a_1}\right)_{a_2=0} \qquad S_{22} = \left(\frac{b_2}{a_2}\right)_{a_1=0}$$

Fig. 13-11. Network matrix terms.

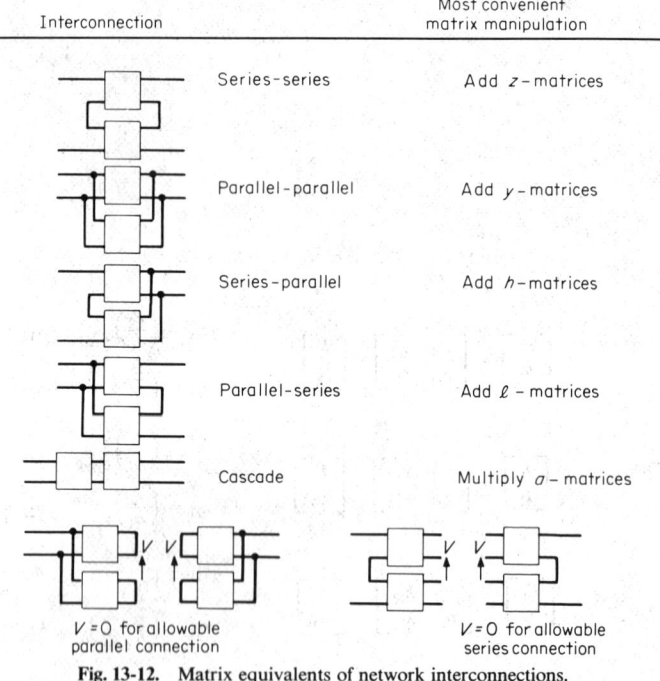

Interconnection	Most convenient matrix manipulation
Series–series	Add z–matrices
Parallel–parallel	Add y–matrices
Series–parallel	Add h–matrices
Parallel–series	Add ℓ–matrices
Cascade	Multiply a–matrices

$V = 0$ for allowable parallel connection

$V = 0$ for allowable series connection

Fig. 13-12. Matrix equivalents of network interconnections.

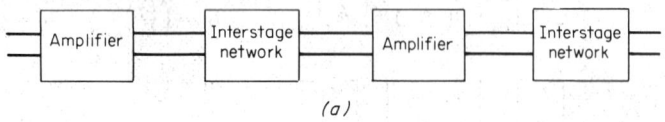

(a)

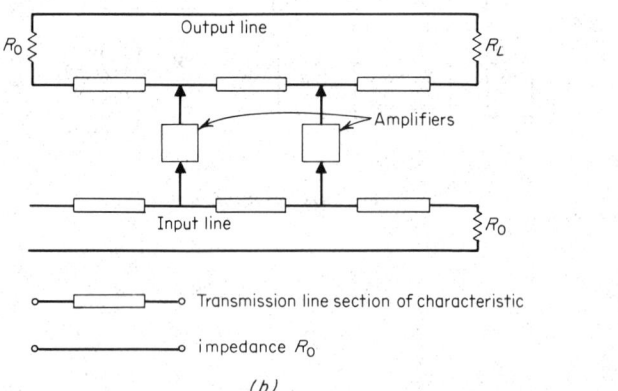

(b)

Fig. 13-13. Multiamplifier structures. (*a*) Cascade; (*b*) distributed.

13-14

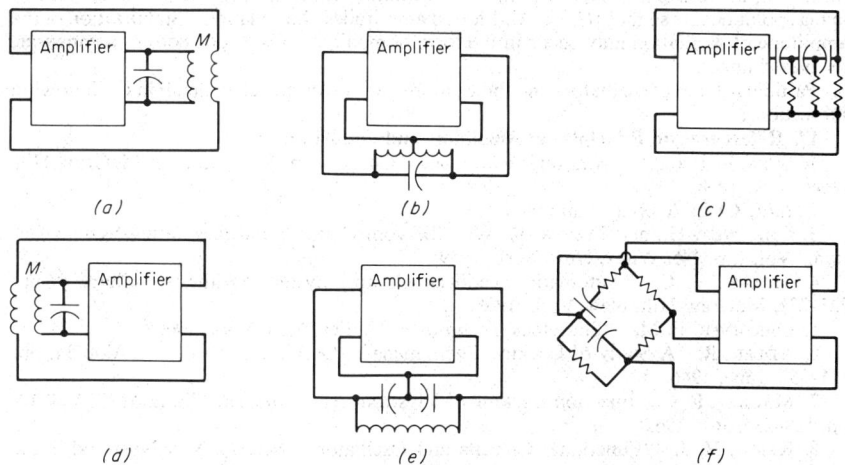

Fig. 13-14. Types of oscillators. (*a*) Tuned-output; (*b*) Hartley; (*c*) phase-shift; (*d*) tuned-input; (*e*) Colpitts; (*f*) Wien bridge.

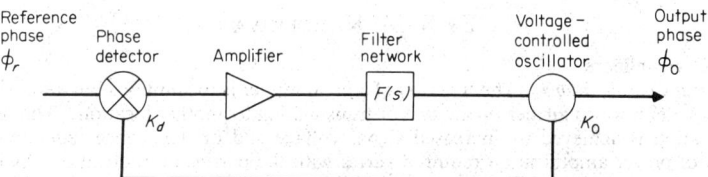

Fig. 13-15. Phase-locked loop oscillator.

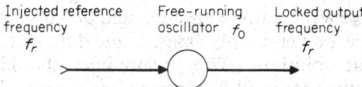

Fig. 13-16. Injection-locked oscillator.

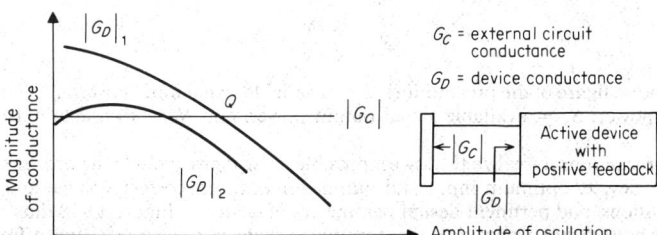

Fig. 13-17. Device conductance vs. amplitude of oscillation.

If the $|G_D|$ curve is of the shape of $|G_D|_2$, the oscillation can stop when a large enough change in bias point occurs so that $|G_D| < |G_C|$ for all amplitudes of oscillation. Stabilization of the amplitude of oscillation may occur in the form of modifying G_C, G_D or both to compensate for bias changes.

Particular types of oscillators and their parameters are discussed in detail in the following subsections.

17. References on Principles of Amplifiers and Oscillators

1. SHEA, R. F. (ed.) "Amplifier Handbook," Chap. 7, pp. 7-11 and 7-19, McGraw-Hill, New York, 1968.

2. *Ibid.*, Chap. 6, pp. 6-4 and 6-5.

3. CHESTNUT, H., and MAYER, R. W. "Servomechanism and Regulating System Design," Vol. I, p. 145, Wiley, New York, 1959.

4. TRUXAL, J. C. "Automatic Feedback Control System Synthesis," Chap. 4, pp. 221-277, McGraw-Hill, New York, 1955.

5. GARDINER, F. M. "Phaselock Techniques," Wiley, New York, 1966.

6. ADLER, R. A Study of Locking Phenomena in Oscillators, *Proc. IRE*, Vol. 34, pp. 351-357, June, 1946.

7. MACKEY, R. C. Injection Locking of Klystron Oscillators, *IRE Trans. MTT*, Vol. 10, pp. 228-235, July 1962.

8. REICH, H. J. "Functional Circuits and Oscillators," Sec. 74, Van Nostrand, New York, 1961.

AUDIO-FREQUENCY AMPLIFIERS

By S. M. KORZEKWA

18. Preamplifiers

General Considerations. The function of a preamplifier is to amplify a low-level signal to a higher level prior to further processing or transmission to another location. The required amplification is achieved by increased signal voltage and/or impedance reduction. The amount of power amplification required varies with the particular application. A general guideline is to provide sufficient preamplification so that further signal handling adds minimal (or acceptable) signal-to-noise degradation.

Signal-to-Noise Considerations. The design of a preamplifier must consider all potential signal degradation from sources of noise, whether generated externally or within the preamplifier itself.

Examples of externally generated noise are hum and pickup, which may be introduced by the input-signal lines or the power-supply lines. Shielding of the input-signal lines often proves to be an acceptable solution. The preamplifier should be located close to the transmitting source, and the preamplifier power gain must be sufficient to override interference that remains after these steps are taken.

A second major source of noise is that internally generated in the preamplifier itself. The noise figure specified in decibels for a preamplifier, which serves as a figure of merit, is defined as the ratio of the available input-to-output signal-to-noise power ratios:

$$F = \frac{S_i/N_i}{S_o/N_o}$$

where F = noise figure of the preamplifier, S_i = available signal input power, N_i = available noise input power, S_o = available signal output power, and N_o = available noise output power.

Design precautions to realize the lowest possible noise figure include the proper selection of the active device, optimum input and output impedances, correct voltage and current biasing conditions, and pertinent design parameters of devices. Figure 13-18 illustrates the effects on the noise figure of specific transistors due to changes in source resistance, frequency, and current bias.[5]*

Sensor/System Characteristics. A preamplifier can be used to compensate the sensor's signal characteristics to realize a specific effect, as in phonograph pickups. In a magnetic

---

* Superior numbers refer to corresponding numbered references, Par. 13-33.

sensor, when the stylus motion causes an output signal proportional to stylus velocity, the output signal level vs. frequency increases at 20 dB/decade for a constant input-signal level. If the loading is varied, the signal output will be further modified.

In contrast, a ceramic phonograph pickup is basically capacitive; i.e., its high impedance is susceptible to loading and the output signals are proportional to the amplitude of the stylus movement. The signals derived from magnetic tapes are often processed via the preamplifier used for phonograph records (see Sec. 19). The standard RIAA recording characteristic (Fig. 19-139) indicates the proper system frequency response when using an inductive pickup. The specified frequency shaping takes into account both the low-frequency limitations of the original recording and also the higher-frequency performance of the sensor. The resultant frequency compensation is ultimately obtained via preamplifier design, pickup loading, or both. Note that what is being specified via these standards is essentially the system transfer function, which takes into account specific processing and sensor limitations and/or attributes.

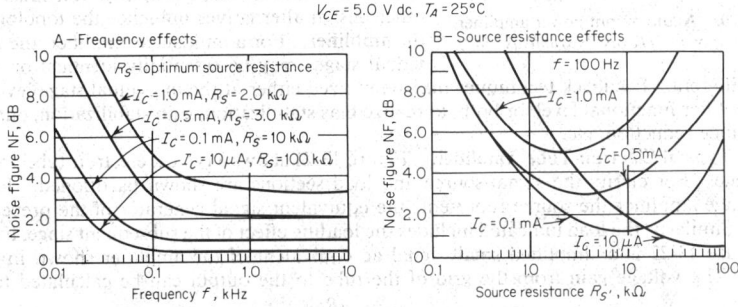

Fig. 13-18. Noise-figure data for the 2N5088 and 2N5089 silicon *npn* transistors (Ref. 5).

Figure 13-19 represents a phonograph preamplifier designed for a 5,000- to 10,000-pF-capacitance ceramic cartridge. With an astatic Model 137 cartridge, the output reference level of 1 V is 13 dB below maximum output and 69 dB above the unweighted noise level. The total harmonic distortion is less than 0.6% at the reference level. This preamplifier is properly equalized in accordance with the RIAA standard.[3]

19. Low-Level Amplifiers. The low-level designation applies to amplifiers operated below maximum permissible power-dissipation, current, and voltage limits. Thus many low-level amplifiers are purposely designed to realize specific attributes other than delivering the maximum attainable power to the load, such as gain stability, bandwidth, optimum noise figure, low cost, etc.

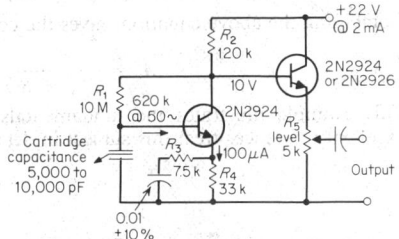

Fig. 13-19. Phonograph preamplifier for ceramic cartridge, with RIAA equalization (Ref. 3).

In an amplifier designed to be operated with a 24-V power supply and a specified load termination, for example, the operating conditions may be such that the active devices are just within their allowable limits. If operated at these maximum limits, this is not a low-level amplifier. However, if this amplifier also fulfills its performance requirements at a reduced power-supply voltage of 6 V, with resulting much lower internal dissipation levels, this amplifier becomes a low-level amplifier.

20. Medium-Level and Power Amplifiers. The medium-power designation for an amplifier implies that some active devices are operated near their maximum dissipation limits, and precautions must be taken to protect these devices. Taking power-handling capability as the criterion, the 5- to 100-W power range is a current demarcation line. As higher-power-handling devices come into use, this range will tend to shift to higher power levels.

The amount of power that can be safely handled by an amplifier is usually dictated by the dissipation limits of the active devices in the output stages, the efficiency of the circuit, and the means used to extract heat to maintain devices within their maximum permissible temperature limits. The classes of operation (A, B, AB, C) are discussed in Par. 13-29. In case single active devices do not suffice, multiple series or parallel configurations may be used to achieve higher voltage or power operation. Figure 13-20 illustrates an audio output power stage capable of providing 70 W to a load without the use of an output transformer.[6]

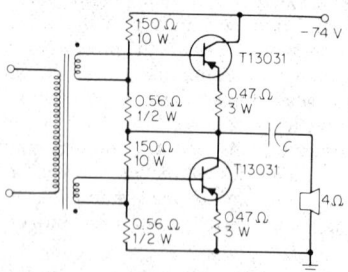

Fig. 13-20. Audio output power amplifier of 70 W power. *(Texas Instruments, Inc.)*

21. Multistage Amplifiers. An amplifier may take the form of a single stage or a complex single stage or it may employ an interconnection of several stages. Various biasing, coupling, feedback, and other design alternatives influence the topology of the amplifier. For a multistage amplifier, the individual stages may be essentially identical or radically different. Feedback techniques may be utilized either at the individual stage level, at the amplifier functional level, or both, to realize bias stabilization, gain stabilization, output-impedance reduction, etc.

22. Typical Electron-Tube Amplifier. Figure 13-21 shows a typical electron-tube amplifier stage. For clarity the signal-source and load sections are shown partitioned. For a multistage amplifier the source represents the equivalent signal generator of the preceding stage. Similarly, the load indicated includes the loading effect of the subsequent stage, if any.

Figure 13-22 is a simplified small-signal ac equivalent of the amplifier shown in Fig. 13-21. The voltage gain from the grid of the tube to the output can be calculated to be

$$A_{v1} = -\frac{\mu R_l}{r_p + R_l}$$

Similarly, the voltage gain from the source to the tube grid is

$$A_{v2} = \frac{R_1}{(R_1 + R_g) + 1/j\omega C}$$

Combining the above equations gives the composite amplifier voltage gain:

$$A_v = -\frac{\mu R_1 R_l}{(r_p + R_l)[(R_1 + R_g) + 1/j\omega C]}$$

This example illustrates the fundamentals of an electron-tube amplifier stage. Many excellent references treat this subject in detail.[7,8]

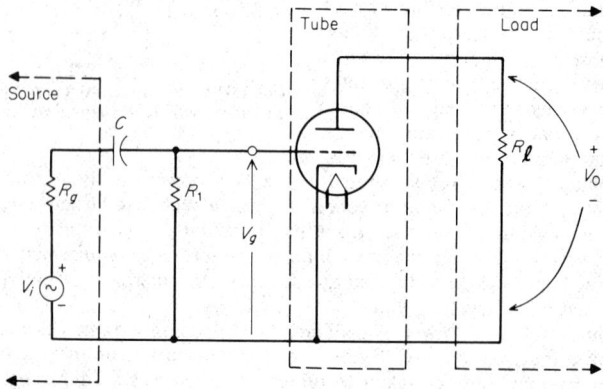

Fig. 13-21. Typical triode electron-tube amplifier stage (biasing not shown).

23. Typical Transistor Amplifier. The analysis techniques used for electron-tube amplifier stages generally apply to transistorized amplifier stages. The principal difference is that different active-device models are used.

The typical transistor stage shown in Fig. 13-23 illustrates a possible form of biasing and coupling. The source section is partitioned and includes the preceding-stage equivalent generator, and the load includes subsequent stage loading effects. Figure 13-24 shows the generalized *h*-equivalent circuit representation for transistors. Table 13-1 lists the *h*-parameter transformations for the common-base, common-emitter, and common-collector configurations.

Table 13-1. *h* Parameters of the Three Configurations[2]

	Common-base	Common-emitter	Common-collector
h_{11}	h_{ib}	$h_{ib}(h_{fe} + 1)$	$h_{ib}(h_{fe} + 1)$
h_{12}	h_{rb}	$h_{ib}h_{ob}(h_{fe} + 1) - h_{rb}$	1
h_{21}	h_{fb}	h_{fe}	$-(h_{fe} + 1)$
h_{22}	h_{ob}	$h_{ob}(h_{fe} + 1)$	$h_{ob}(h_{fe} + 1)$

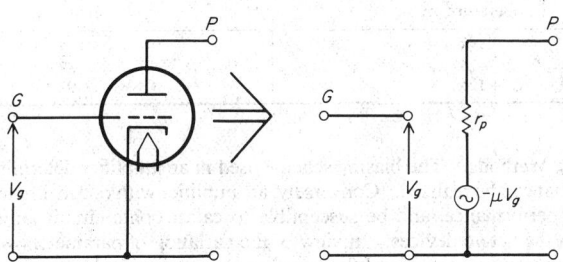

Fig. 13-22. Equivalent circuit (right) of electron tube (left).

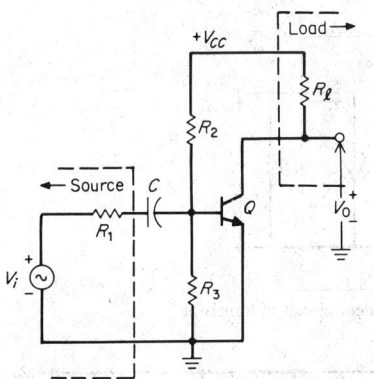

Fig. 13-23. Typical bipolar transistor amplifier stage.

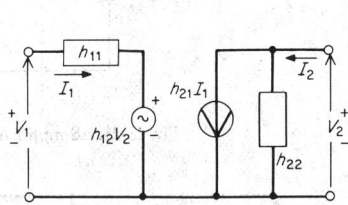

Fig. 13-24. Equivalent circuit of transistor, based on *h* parameters.

While these parameters are complex and frequency-dependent, it is often feasible to use simplifications. Most transistors have their parameters specified by their manufacturers, but it may be necessary to determine additional parameters by test.

Figure 13-25 illustrates a simplified model of the transistor amplifier stage of Fig. 13-23. The common-emitter *h* parameters are used to represent the equivalent transistor. The voltage gain for this stage is

$$A_v = \frac{V_0}{V_i} = -\frac{h_{fe}R_l}{R_g + h_{ie}}$$

The complexity of analysis depends on the accuracy needed. Currently, most of the more complex analysis is performed with the aid of computers. Several transistor-amplifier-analysis references treat this subject in detail.[2-4]

24. Typical Multistage Transistor Amplifier. Figure 13-26 is an example of a capacitively coupled three-stage transistor amplifier. It has a broad frequency response, illustrating the fact that an audio amplifier can be useful in other applications. The component values marked on the diagram are as follows:

$$R_1 = 16,000\ \Omega \qquad R_2 = 6,200\ \Omega \qquad\qquad R_3 = 1,600\ \Omega$$
$$R_4 = 1,000\ \Omega \qquad R_L = 560\ \Omega \qquad Q_1, Q_2, Q_3 = 2N1565$$
$$C_1 = 10\ \mu F \qquad\quad C_2 = 100\ \mu F$$

This amplifier is designed to operate over a range of -55 to $+125°$ C, with an output voltage swing of 2 V peak to peak and frequency response down 3 dB at approximately 200 Hz and 2 MHz. The overall gain at 1000 Hz was as follows:

Temperature, °C	Gain, dB
-55	83
$+25$	88
$+125$	91

25. Biasing Methods. The biasing scheme used in an amplifier determines the ultimate performance that can be realized. Conversely, an amplifier with poorly implemented biasing may suffer in performance, and be susceptible to catastrophic circuit failure due to high stresses within the active devices. In view of the variation of parameters within the active devices, it is important that the amplifier function properly even when the initial and/or end-of-life parameters of the devices vary.

26. Electron-Tube Biasing. Biasing is intended to maintain the quiescent currents and

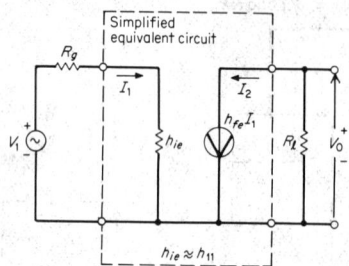

Fig. 13-25. Simplified equivalent circuit of transistor.

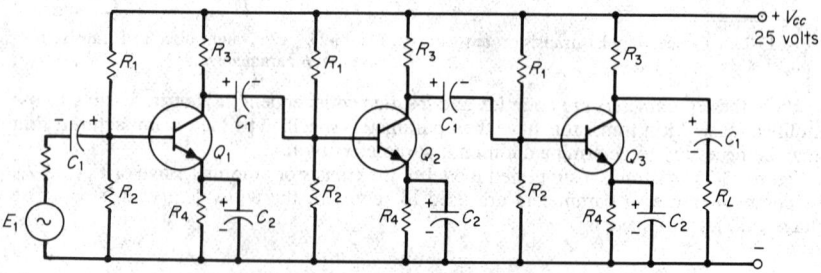

Fig. 13-26. Typical three-stage transistor amplifier (Ref. 2).

voltages of the electron tube at the prescribed levels. The tube-plate characteristics represent the biasing relations between the tube parameters.

The single-stage amplifier shown in Fig. 13-27 illustrates self-biasing (cathode biasing) via the cathode feedback components R_k and C_k. The cathode current tends to increase until the correct quiescent positive cathode bias results.

The principal bias parameters (steady-state plate and grid voltages) can be readily identified by the construction of a load line on the plate characteristic, as illustrated in Fig. 13-28. The operating point Q is located at the intersection of the selected plate characteristic with the load line. The load-line end points are determined by

$$i_b = 0, \qquad E_b = E_{bb}$$

$$i_b = \frac{E_{bb}}{R_L}, \qquad e_b = 0$$

It is evident that there are many possible grid-voltage vs. anode-current combinations, each of which results in a different quiescent I_b current. When a particular grid bias voltage is

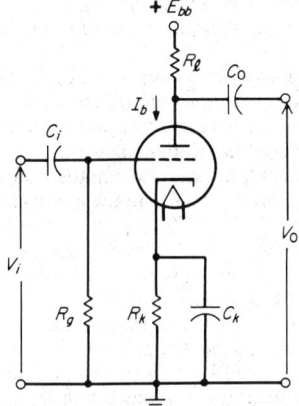

Fig. 13-27. Self-biased electron-tube stage.

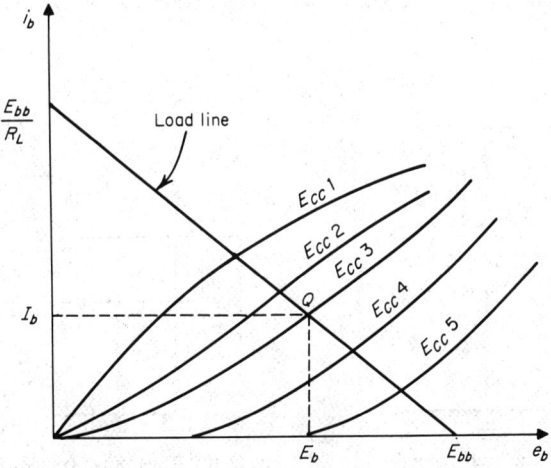

Fig. 13-28. Load-line determination of operating point Q.

13-21

selected, the resulting I_b current directly determines the value of cathode resistor R_k. The capacitor C_k is selected to give the required frequency performance.

27. Transistor Biasing. The methods of biasing a transistor amplifier stage are in many respects similar to those of an electron-tube amplifier. However, there are many different types of transistors, each characterized by different curves. Bipolar transistors are generally characterized by their collector and emitter families, while field-effect-type transistors have different characterizations. The *npn*-type transistor requires a positive base bias voltage and current (with respect to its emitter) for proper operation; the converse is true for a *pnp*-type transistor.

Figure 13-29 illustrates a commonly used biasing technique. A single power supply is used, and the transistor is self-biased with the unbypassed emitter resistor R_e. Although a graphical solution of the value of R_e could be found by referring to the collector-emitter curves, an iterative solution, described below, is also commonly used.

Because the performance of the transistors depends on the collector current and collector-to-emitter voltage, these are often selected as starting conditions for biasing design. The unbypassed emitter resistor R_e and collector resistor R_c are the primary voltage-gain-determining components; these are determined next, taking into account other considerations such as the anticipated maximum signal level, available power supply V_{cc}, etc. The last step is to determine the R_1 and R_2 values.

28. Coupling Methods. Transformer coupling and capacitance coupling are commonly used in transistor and electron-tube audio amplifiers. Direct coupling is also used in transistor stages, and particularly in integrated transistor amplifiers. Capacitance coupling, referred to as RC coupling, is the most common method of coupling stages of an audio amplifier. The discrete-component transistorized amplifier stage shown in Fig. 13-29 serves as an example of RC coupling, where C_i and C_o are the input and output coupling capacitors, respectively.

Transformer coupling is commonly used to match the input and output impedances of electron-tube amplifier stages. Since the input impedance of an electron tube is very high at audio frequencies, the design of an electron-tube stage is primarily dependent on the transformer parameters. The much lower input impedances of transistors demand that many other factors be taken into account, and the design becomes more complex. The output-stage transformer coupling to a specific load is often the optimum method of realizing the best power match. Figure 13-30 illustrates a typical transformer-coupled transistor audio amplifier stage.

The direct-coupling approach is now also used for discrete-component-type transistorized amplifiers, and particularly in integrated amplifier versions. The level-shifting requirement is realized by selection from the host of available components, such as *npn* and *pnp* transistors

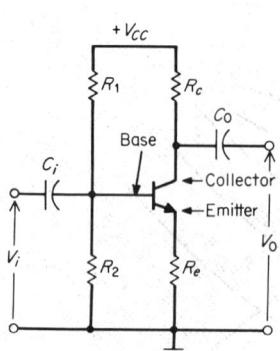

Fig. 13-29. Capacitively coupled *npn* transistor amplifier stage.

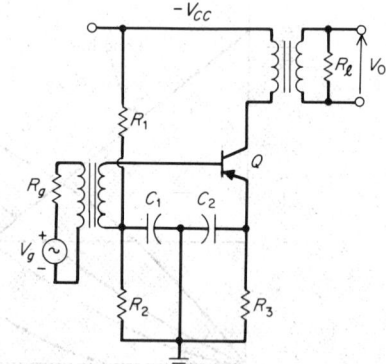

Fig. 13-30. Transformer-coupled *pnp* transistor amplifier stage.

and Zener diodes. Since it is difficult to realize large-size capacitors via integrated circuit techniques, special methods have been developed to direct-couple integrated amplifiers.

29. Classes A, B, AB, and C Operation. The output or power stage of an amplifier is usually classified as operating class A, B, AB, or C, depending on the conduction characteristics of the active devices (see Fig. 13-5). These definitions can also apply to any intermediate amplifier stage. Figure 13-31 illustrates the class of operation vs. conduction relations using transistor parameters.[2] This figure would be essentially the same for an electron-tube amplifier with the tube plate current and grid voltage as the equivalent device parameters.

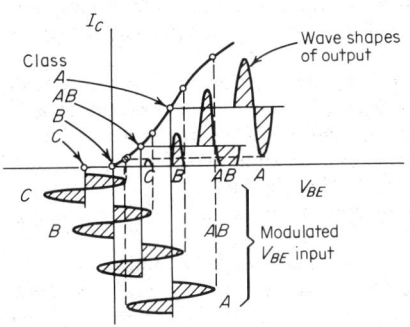

Fig. 13-31.   Classes of amplifier operation, based on transistor characteristics.

Subscripts may be used to denote additional conduction characteristics of the device. For example, the electron-tube grid conduction can also be further classified as A_1 to show that no grid current flows, or A_2 to show that grid current conduction exists during some portion of the cycle.

30. Push-Pull Amplifiers. In a single-ended amplifier the active devices continuously conduct. The single-ended configuration is generally used in low-power applications, operated in class A. For example, preamplifiers and low-level amplifiers are generally operated single-ended, unless the output power levels necessitate the more efficient power handling of the push-pull circuit.

In a push-pull configuration there are at least two active devices that alternately amplify the negative and positive cycles of the input waveform. The output connection to the load is most often transformer-coupled. An example of a transformer input and output in a push-pull amplifier is illustrated in Fig. 13-32. Direct-coupled push-pull amplifiers and capacitively coupled push-pull amplifiers are also feasible, as illustrated in Fig. 13-33 push-pull emitter-follower examples.

The active devices in push-pull are usually operated either in class B or AB because of the high power-conversion efficiency. Feedback techniques can be used to stabilize gain, stabilize biasing or operating points, minimize distortion, etc.

31. Output Amplifiers. The function of an audio output amplifier is to interface with the preceding amplifier stages and to provide the necessary drive to the load. Thus the output-amplifier designation does not uniquely identify a particular amplifier class. When several different types of amplifiers are cascaded between the signal source and its load, e.g., a high-power speaker, the last-stage amplifier is designated as the output amplifier. Because of the

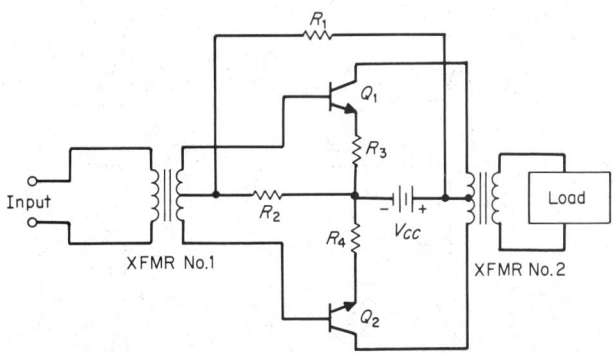

Fig. 13-32.   Transformer-coupled push-pull transistor stage.

13-23

high-power requirements, this amplifier is usually a push-pull type operating either in class B or AB.

32. Stereo Amplifiers. A stereo amplifier provides two separate audio channels properly phased with respect to each other. The objective of this two-channel technique is to enhance the audio reproduction process, thereby making it more realistic and lifelike. It is also feasible to extend the system to contain more than two channels of information (e.g., four-channel, see Sec. 19). A stereo amplifier is a complete system that contains its power supply and other commonly required control functions.

Each channel has its own preamplifier, medium-level stages, and output power stage, with different gain and frequency responses for each mode of operation, e.g., for tape, phonograph, etc. The input signal is selected from the phonograph input connection, tape input, or a tuner output. Special-purpose trims and controls are also used to optimize performance on each mode. The bandwidth of the amplifier extends to 20 kHz or higher.

The block-diagram representation of a high-quality stereo amplifier is shown in Fig. 13-34. The performance specifications[2] for this amplifier are:

Power output: Stereophonic, each channel 40 W music power (36 W continuous, 72 W peak). Monophonic, 80 W music power (72 W continuous, 144 W peak), at 1½% intermodulation distortion (60 Hz − 7 kHz 4:1).

Outputs: 16-, 8-, and 4-Ω left and right speakers; two recording, third channel.

Inverse feedback: 16 dB.

Damping factor: 5:1.

Frequency response (36 W): 20 Hz to 20 kHz ± 1/2 dB.

Tone-control range: 15 kHz, 17-dB boost or cut. 40 Hz, 16-dB boost, 19-dB cut.

Rumble filter: 27 Hz, 17-dB rejection; 70 Hz less than 1 dB down.

Sensitivity: Radio 0.25 V, tape 1.4 mV, phonograph 1.2 mV; all inputs are adjustable with level control.

Maximum input: Phonograph, 200 mV for less than 1% distortion. Radio, adjustable with level control.

Maximum hum and noise: Volume control, minimum 100 dB (weighted) below rated output. Radio input (controls maximum), 90 dB (weighted) below rated output. Phonograph input (controls flat), 60 dB below rated output, 72 dB below 10 mV (equivalent to 1/2 µV referred to input grid).

Interchannel crosstalk: Less than −50 dB at 1 kHz.

Power consumption: 110 to 120 V, 60 Hz, 150 W, 1.3A.

Tube complement: Four 7868, six 12AX7/ECC83, four silicon rectifiers.

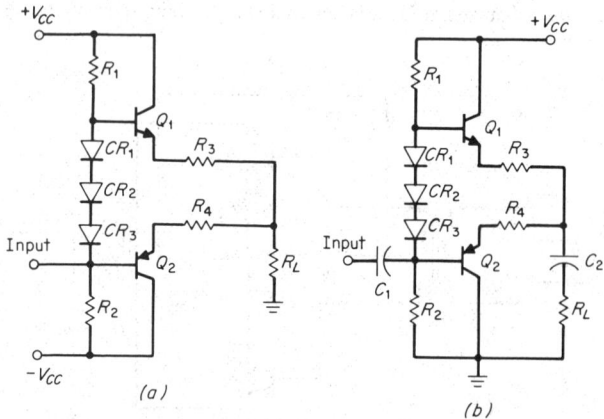

Fig. 13-33. Direct (*a*) and capacitively (*b*) coupled push-pull stages.

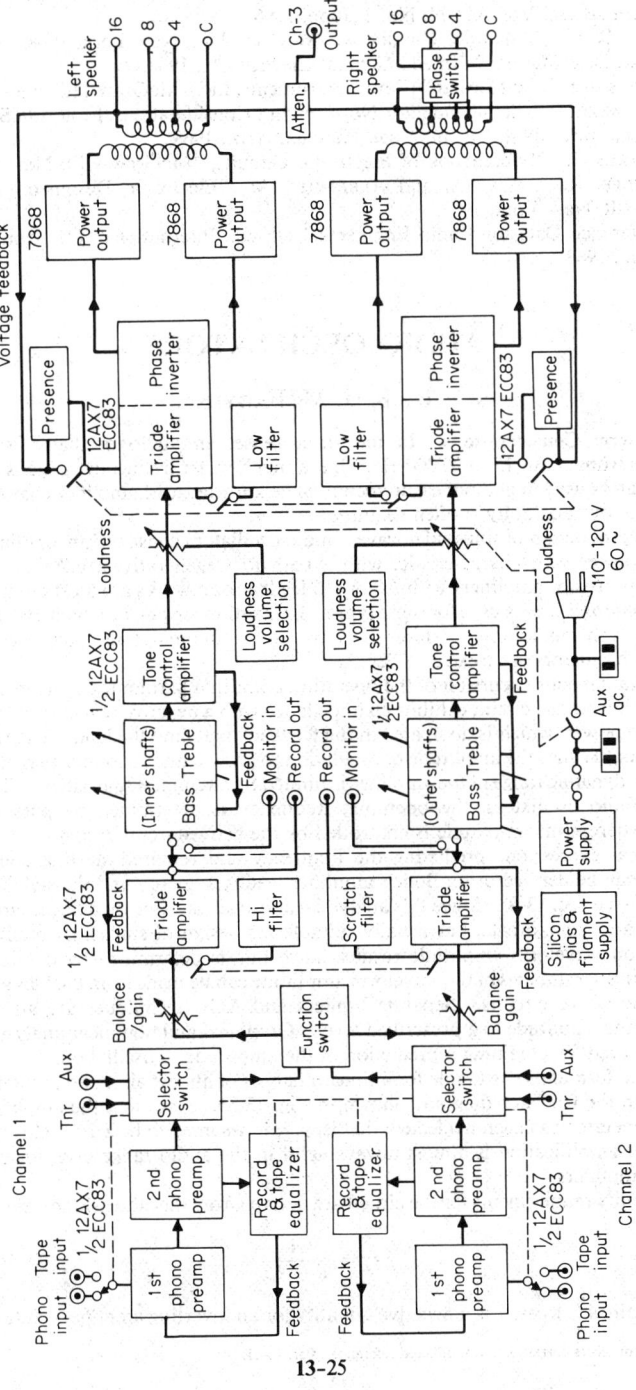

Fig. 13-34. Block diagram of high-fidelity stereo amplifier. *(Sherwood Electronic Laboratories, Inc.)*

33. References on Audio Amplifiers.
1. IEEE Standard on Definitions of Terms for Audio and Electroacoustics, *IEEE Trans. Audio Electroacoust.,* Vol. AU-14, No. 2, June, 1966.
2. SHEA, R. F. "Amplifier Handbook," McGraw-Hill, New York, 1968.
3. "Transistor Manual," 7th ed., General Electric Co., 1964.
4. "Transistor Circuit Design" (Texas Instruments, Inc.), McGraw-Hill, New York, 1963.
5. BRUBAKER, R. Semiconductor Noise Figure Considerations, Motorola Semiconductor Products, Inc., AN-421 Appl. Note, Phoenix, Ariz., 1968.
6. MARKUS, J. "Sourcebook of Electronic Circuits," McGraw-Hill, New York, 1968.
7. LANDEE, R., DAVIS, D., and ALBRECHT, A. "Electronic Designers' Handbook," McGraw-Hill, New York, 1957.
8. "Reference Data for Radio Engineers," 5th ed., International Telephone and Telegraph Corp., 1968.

AUDIO OSCILLATORS

By R. J. McFadyen

34. General Considerations. In the strict sense, an audio oscillator is limited to frequencies from about 15 to 20,000 Hz,[1]* but a much wider frequency range is included in most oscillators used in audio measurements, since knowledge of amplifier characteristics in the region above audibility is often required.

For the production of sinusoidal waves, audio oscillators consist of an amplifier having a nonlinear power gain characteristic, with a path for regenerative feedback. Single and multistage transistor amplifiers with *LC* or *RC* feedback networks are most often used. The term *harmonic oscillator* is used for these types. Relaxation *oscillators,* which may be designed to oscillate in the audio range, exhibit sharp transitions in the output voltages and currents. Relaxation oscillators are treated in Sec. **16.**

The instantaneous excursion of the operating point in a harmonic oscillator is restricted to the range where the circuit exhibits an impedance with a negative real part. The amplifier supplies the power, which is dissipated in the feedback path and the load. The regenerative feedback would cause the amplitude of oscillation to grow without bound, were it not for the fact that the dynamic range of the amplifier is limited by circuit nonlinearities. Thus, in most sinewave audio oscillators, the operating frequency is determined by passive-feedback elements, whereas the amplitude is controlled by the active-circuit design.

Analytical expressions predicting the frequency and required starting conditions for oscillation can be derived using Bode's amplifier feedback theory and the stability theorem of Nyquist[2] (see Sec. **13-8**). Since this analytical approach is based on a linear circuit model, the results are approximate, but usually suitable for design of sinusoidal oscillators. No prediction on waveform amplitude results, since this is determined by nonlinear circuit characteristics. Estimates of the waveform amplitude can be made from the bias and limiting levels of the active circuits. Separate limiters and AGC techniques are also useful for controlling the amplitude to a prescribed level. Graphical and nonlinear analysis methods[3] can also be used for obtaining a prediction of the amplitude of oscillation.

A general formulation suitable for a linear analysis of almost all audio oscillators can be derived from the feedback diagram[4] shown in Fig. 13-35. Note that the amplifier internal feedback generator has been neglected; that is, y_{12A} is assumed to be zero. This assumption of unilateral amplification is almost always valid in the audio range even for single-stage transistor amplifiers.

The stability requirements for the circuit are derived from the closed-loop-gain expression

$$A_c = \frac{A}{1 - A\beta} \tag{13-1}$$

where the gain A is treated as a negative quantity for an inverting amplifier. Infinite closed-

---

* Superior numbers correspond to numbered references, Par. **13-41.**

loop gain occurs when $A\beta$ is equal to unity, and this defines the oscillatory condition. In terms of the equivalent circuit parameters used in Fig. 13-35,

$$1 - A\beta = 1 - y_{21A}\frac{y_{12\beta}}{(y_{11A} + y_{11\beta})(y_{22A} + y_{22\beta}) - y_{12\beta}y_{21\beta}} \quad (13\text{-}2)$$

In the audio range, y_{21A} remains real, but the fractional portion of the function is complex, because β is frequency-sensitive. Therefore, the open-loop gain $A\beta$ may be expressed in the general form

$$A\beta = y_{21A}\frac{A_r + jA_i}{B_r + jB_i} \quad (13\text{-}3)$$

It follows from Nyquist's stability theorem that this feedback system will be unstable if, first, the phase shift of $A\beta$ is zero and, second, the magnitude is equal to or greater than unity. Applying this criterion to Eq. (13-3) yields the following two conditions for oscillation:

$$A_iB_r - A_rB_i = 0 \quad (13\text{-}4)$$

$$y_{21}^2 \geq \frac{B_r^2 + B_i^2}{A_r^2 + A_i^2} \quad (13\text{-}5)$$

Equation (13-4) results from the phase condition and determines the frequency of oscillation. The inequality in Eq. (13-5) is the consequence of the magnitude constraint and defines the necessary condition for sustained oscillation. Equation (13-5) is evaluated at the oscillation frequency determined from Eq. (13-4).

A large number of single-stage oscillators have been developed in both vacuum-tube and transistor versions. The transistor circuits followed by direct analogy from the earlier vacuum-tube circuits. In the following examples, transistor versions are illustrated, but the y-parameter equations apply to other devices as well.

35. LC Oscillators.[4-6] The *Hartley oscillator* circuit is one of the oldest forms; the transistor version is shown in Fig. 13-36. With the collector and base at opposite ends of the tuned circuit, the 180° phase relation is secured, and feedback occurs through mutual inductance between the two parts of the coil. In the figure, the frequency and condition for oscillation are expressed in terms of the transistor y parameters and feedback inductance L, inductor coupling coefficient k, inductance ratio n, and tuning capacitance C. The admittance parameters of the bias network R_1, R_2, and R_3, as well as the reactance of bypass capacitor C and coupling capacitor C_2, have been neglected. These admittances could be included in the amplifier y parameters in cases where their effect is not negligible. If

$$\frac{C}{L} \gg \frac{n(1 - k^2)(y_{11A}y_{22A})}{1 + 2k\sqrt{n} + n} \quad (13\text{-}6)$$

the frequency of oscillation will be essentially independent of transistor parameters.

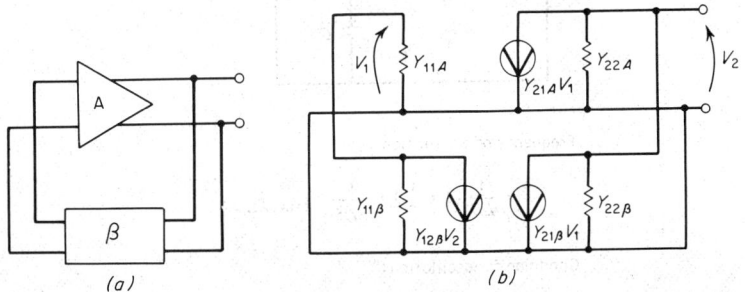

Fig. 13-35. Oscillator representations. (*a*) Generalized feedback circuit; (*b*) equivalent-Y-parameter circuit.

The transistor version of the *Colpitts oscillator* is shown in Fig. 13-37. Capacitors C and NC in combination with the inductance L determine the resonant frequency of the circuit. A fraction of the current flowing in the tank circuit is regeneratively fed back to the base through the coupling capacitor C_2. Bias resistors R_1, R_2, R_3, and R_L, as well as capacitors C_1 and C_2, are chosen so as not to affect the frequency or conditions for oscillation. Alternatively, the bias element admittances may be included in the amplifier y parameters.

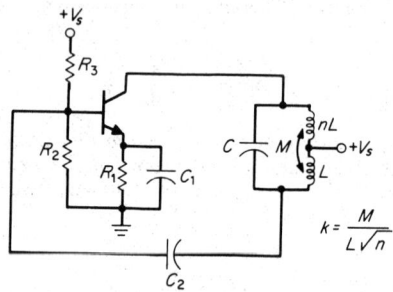

Frequency of oscillation:

$$\omega^2 = \frac{1}{LC\,(1 + 2k\sqrt{n} + n) + nL^2\,(1-k^2)(y_{11A}\,y_{22A})}$$

Condition for oscillation:

$$y_{21A} \geq \frac{y_{11A} + ny_{22A} + n\omega^2\,LC\,(1-k^2)(y_{11A}\,y_{22A})}{k\sqrt{n} + n\omega^2 LC\,(1-k^2)}$$

Fig. 13-36. Hartley oscillator circuit.

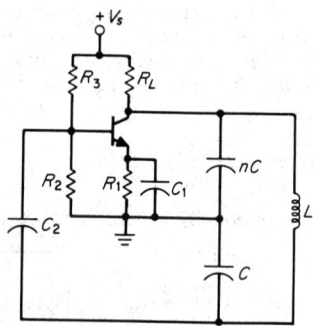

Frequency of oscillation:

$$\omega^2 = \frac{1}{LC}\left(1 + \frac{1}{n}\right) + \frac{1}{nC^2}\,(y_{11A}\,y_{22A})$$

Condition for oscillation:

$$y_{21A} \geq \omega^2\,LC\left(ny_{11A} + y_{22A}\right) - (y_{11A} + y_{22A})$$

Fig. 13-37. Colpitts oscillator circuit.

In the Colpitts circuit, if the ratio of C/L is chosen so that

$$\frac{C}{L} \gg \frac{y_{11A} y_{22A}}{1+n} \tag{13-7}$$

the frequency of oscillation is essentially determined by the tuned-circuit parameters.

Another oscillator configuration useful in the audio-frequency range is the tuned collector circuit shown in Fig. 13-38. Here regenerative feedback is furnished via the transformer turns ratio n from the collector to base. If the ratio of C/L is such that

$$\frac{C}{L} \gg N^2 y_{11A} y_{22A}(1-k^2) \tag{13-8}$$

the frequency of oscillation is specified by $\omega^2 = 1/LC$. This circuit may be tuned over a wide range by varying the capacitor C and is compatible with simple biasing techniques.

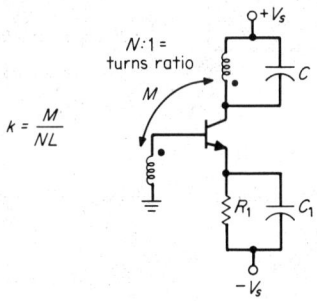

Frequency of oscillation:

$$\omega^2 = \frac{1}{LC + N^2 L^2\, y_{11A}\, y_{22A}\,(1-k^2)}$$

Condition for oscillation:

$$y_{21A} \geq \frac{1}{Nk}(N^2 Y_{11A} + Y_{22A}) - \frac{\omega^2 NLCY_{11A}}{k}\,(1-k^2)$$

Fig. 13-38. Tuned collector oscillator.

36. RC Oscillators.[4,5] Audio sinusoidal oscillators may be designed using an RC ladder network (of three or more sections) as a feedback path in an amplifier. This scheme originally appeared in vacuum-tube circuits, but the principles have been directly extended to transistor design. RC phase shift oscillators may be distinguished from tuned oscillators in that the feedback network has a relatively broad frequency response characteristic.

Typically, the phase shift network has three RC sections of either a high- or a low-pass nature. Oscillation occurs at the frequency where the total phase shift is 180° when used with an inverting amplifier. Figures 13-39 and 13-40 show examples of high-pass and low-pass feedback-connection schemes. The amplifier is a differential pair with a transistor current source, a configuration which is common in integrated circuit amplifiers. The output is obtained at the opposite collector from the feedback connection, since this minimizes external loading on the phase shift network. The conditions for, and the frequency of, oscillation are derived, assuming that the input resistance of the amplifier, which loads the phase shift network, has been adjusted to equal the resistance R. The load resistor R_L is considered to be part of the amplifier output resistance, and it is included in y_{22A}.

37. Null Network Oscillators.[4,5,7] In almost all respects null network oscillators are superior to the RC phase shift circuits described above (Par. 13-36). While many null network configurations are useful (including the bridged-T and twin-T), the Wien bridge design predominates.

The general form for the Wien bridge oscillator is shown in Fig. 13-41. In the figure, an ideal differential voltage amplifier is assumed, i.e., one with infinite input impedance and zero output impedance. An integrated circuit operational amplifier that has a differential input stage is a practical approximation to this type of amplifier and is often used in bridge oscillator designs.

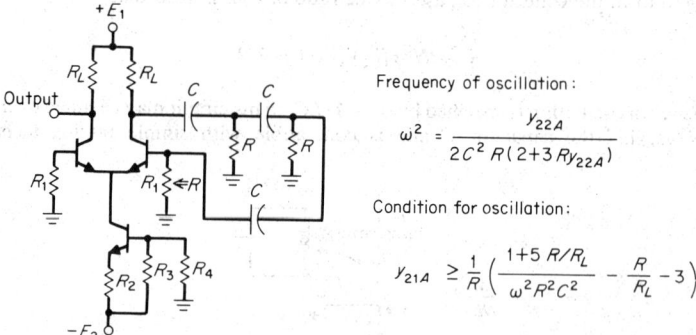

Frequency of oscillation:

$$\omega^2 = \frac{y_{22A}}{2C^2 R(2+3Ry_{22A})}$$

Condition for oscillation:

$$y_{21A} \geq \frac{1}{R}\left(\frac{1+5R/R_L}{\omega^2 R^2 C^2} - \frac{R}{R_L} - 3\right)$$

Fig. 13-39. *RC* oscillator with high-pass feedback network.

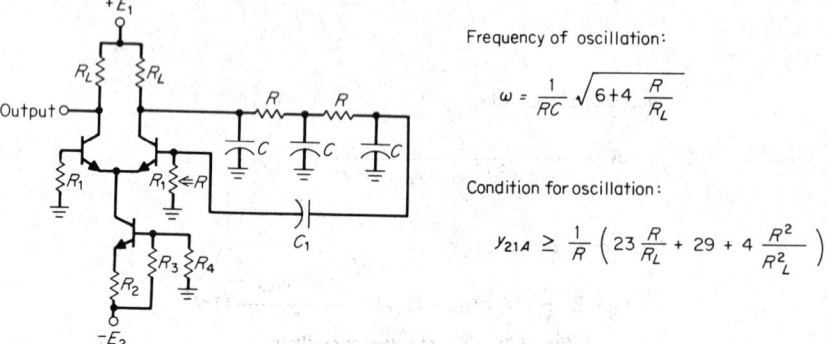

Frequency of oscillation:

$$\omega = \frac{1}{RC}\sqrt{6+4\frac{R}{R_L}}$$

Condition for oscillation:

$$y_{21A} \geq \frac{1}{R}\left(23\frac{R}{R_L} + 29 + 4\frac{R^2}{R_L^2}\right)$$

Fig. 13-40. *RC* oscillator with low-pass feedback network.

The Wien bridge is used as the feedback network, with positive feedback provided through the *RC* branches for regeneration and negative feedback through the resistor divider. Usually the resistor divider network includes an amplitude-sensitive device in one or both arms which provides automatic correction for variation of the amplifier gain. Circuit elements such as a tungsten lamp, thermistor, and field-effect transistor used as the voltage-sensitive resistance element maintain a constant output level with a high degree of stability. Amplitude variations of less than $\pm 1\%$ over the band from 10 to 100,000 Hz are realizable.[8] In addition, since the amplifier is never driven into the nonlinear region, harmonic distortion in the output waveform is minimized. For the connection shown in Fig. 13-41, an increase in V will cause a decrease in R_2, restoring V to the original level.

The lamp or thermistor have thermal time constants that set at a lower frequency limit on this method of amplitude control. When the period is comparable with the thermal time constant, the change in resistence over an individual cycle distorts the output waveform. There is an additional degree of freedom with the field-effect transistor, since the control voltage must be derived by a separate detector from the amplifier output. The time constant of the detector, and hence the resistor, are set by a capacitor, which can be chosen commensurate with the lowest oscillation frequency desired.

At ω_o the positive feedback predominates, but at harmonics of ω_o the net negative feedback reduces the distortion components. Typically, the output waveform exhibits less than 1% total harmonic distortion. Distortion components well below 0.1% in the mid-audio-frequency range are also achieved.[8]

Unlike *LC* oscillators, in which the frequency is inversely proportional to the square root of *L* and *C*, in the Wien bridge ω_o varies as $1/RC$. Thus, a tuning range in excess of 10:1 is easily achieved. Continuous tuning within one decade is usually accomplished by varying both capacitors in the reactive feedback branch. Decade changes are normally accomplished by switching both resistors in the resistive arm. Component tracking problems are eased when the resistors and capacitors are chosen to be equal.

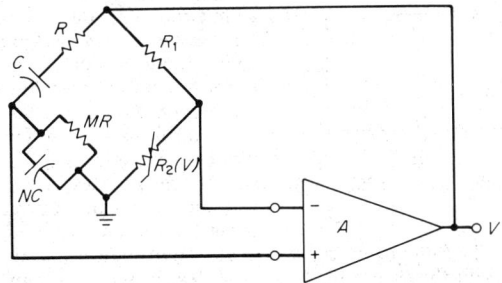

Frequency of oscillation ($M = N = 1$):

$$\omega_0 = \frac{1}{RC}$$

Condition for oscillation:

$$A \geq \delta = \frac{3(R_1 + R_2)}{R_1 - 2R_2}$$

Fig. 13-41. Wien bridge oscillator circuit.

Almost any three-terminal null network can be used for the reactive branch in the bridge; the resistor divider network adjusts the degree of unbalance in the manner described. Many of these networks lack the simplicity of the Wien bridge since they may require the tracking of three components for frequency tuning. For this reason networks such as the bridged-T and twin-T are usually restricted to fixed-tuned applications.

38. Low-Frequency Crystal Oscillators.[9-11] Quartz-crystal resonators are used where frequency stability is a primary concern. The frequency variations with both time and temperature are several orders of magnitude lower than obtainable in *LC* or *RC* oscillator circuits. The very high stiffness and elasticity of piezoelectric quartz make it possible to produce resonators extending from approximately 1 kHz to 200 MHz. The performance characteristics of a crystal are dependent on both the particular cut and the mode of vibration (see Sec. 7). For convenience, each "cut-mode" combination is considered as a separate piezoelectric element, and the more commonly used elements have been designated with letter symbols. The audio-frequency range (above 1 kHz) is covered by elements *J, H, N,* and *XY*, as shown in Table 13-2.

The temperature coefficients vary with frequency, i.e., with the crystal dimensions, and except for the *H* element, a parabolic frequency variation with temperature is observed. The *H* element is characterized by a negative temperature coefficient on the order of -10 ppm/°C. The other elements have lower temperature coefficients, which at some temperatures are zero because of the parabolic nature of the frequency deviation curve. The point where the zero temperature coefficient occurs is adjustable and varies with frequency. At

Table 13-2.   Low-Frequency Crystal Elements

Symbol	Cut	Mode of vibration	Frequency range, kHz
J	Duplex 5°X	Length-thickness flexure	0.9–10
H	5°X	Length-width flexure	10–50
N	NT	Length-width flexure	4–200
XY	XY	XY flexure	8–40

temperatures below this point the coefficient is positive, and at higher temperatures it is negative. On the slope of the curves the temperature coefficients for the N and XY elements are on the order of 2 ppm/°C, whereas the J element is about double at 4 ppm/°C.

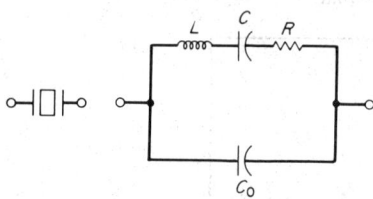

Fig. 13-42. Symbol and equivalent circuit of a quartz crystal.

Although the various elements differ in both cut and mode of vibration, the electrical equivalent circuit remains invariant. The schematic representation and the lumped constant equivalent circuit are shown in Fig. 13-42. As is characteristic of most mechanical resonators, the motional inductance L resulting from the mechanical mass in motion is large relative to that obtainable from coils. The extreme stiffness of quartz makes for very small values of motional capacitance C, and the very high order of elasticity allows the motional resistance R to be relatively low. The shunt capacitance C_0 is the electrostatic capacitance existing between crystal electrodes with the quartz plate as the dielectric and is present whether or not the crystal is in mechanical motion. Some typical values for these equivalent circuit parameters are shown in Table 13-3.

The H element can have a high Q value when mounted in a vacuum enclosure; however, it then has the poorest temperature coefficient. The N element exhibits an excellent temperature characteristic, but the piezoelectric activity is rather low, so that special care is required when it is used in oscillator circuits. The J and XY elements operate well in low-frequency oscillator designs, the latter having lower temperature drift. For the same frequency the XY crystal is about 40% longer than the J element. Where extreme frequency stability is required, the crystals are usually controlled to a constant temperature.

The reactance curve of a quartz resonator is shown in Fig. 13-43. The zero occurs at the frequency f_s, which corresponds to series resonance of the mechanical L and C equivalences.

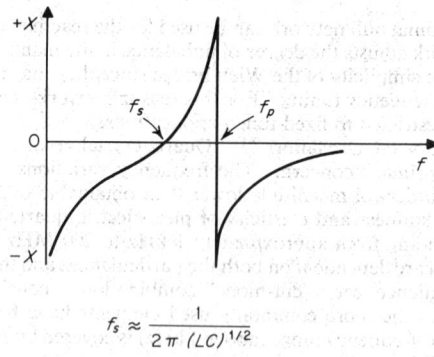

$$f_s \approx \frac{1}{2\pi (LC)^{1/2}}$$

$$f_p \approx \frac{1}{2\pi \left(\dfrac{LC}{1 + C/C_0}\right)^{1/2}}$$

Fig. 13-43. Quartz-crystal reactance curve.

Table 13-3. Typical Crystal Parameter Values

Element	Frequency, kHz	L, H	C, pF	R, kΩ	C_0, pF	Q, approx.
J	10	8,000	0.03	50	6	2×10^4
H	10	2,500	0.1	10	75	2×10^4
N	10	8,000	0.03	75	30	10^4
XY	10	12,000	0.02	30	20	3×10^4

The antiresonant frequency f_p is dependent on the interelectrode capacitance C_0. Between f_s and f_p the crystal is inductive, and this frequency range is normally referred to as the *crystal bandwidth*:

$$BW = \frac{f_s}{2(C_0/C)} \tag{13-9}$$

In oscillator circuits the crystal can be used as either a series or a parallel resonator. At series resonance the crystal impedance is purely resistive, but in the parallel mode the crystal is operated between f_s and f_p and is therefore inductive. For oscillator applications the circuit capacitance shunting the crystal must also be included when specifying the crystal, since it is part of the resonant circuit. If a capacitor C_L, that is, a negative reactance, is placed in series with a crystal, the combination will series-resonate at the frequency f_R of zero reactance for the combination.

$$f_R = f_s \left[1 + \frac{1}{2\dfrac{C_0}{C}\left(1 + \dfrac{C_L}{C_0}\right)} \right] \tag{13-10}$$

The operating frequency can vary in value due to changes in the load capacitance, and this variation is prescribed by

$$\Delta f_R = \frac{f_s}{2C_0/C} \frac{\Delta C_L/C_0}{(1 + C_L/C_0)^2} \tag{13-11}$$

This effect can be used to "pull" the crystal for initial alignment, or if the external capacitor is a voltage-controllable device, a VCO with a range of about $\pm$ 0.01% can be constructed. Phase changes in the amplifier will also give rise to frequency shifts since the total phase around the loop must remain at zero degrees in order to maintain oscillation.

Although single-stage transistor designs are possible, more flexibility is available in the circuit of Fig. 13-44, which uses an integrated circuit operational amplifier for the gain

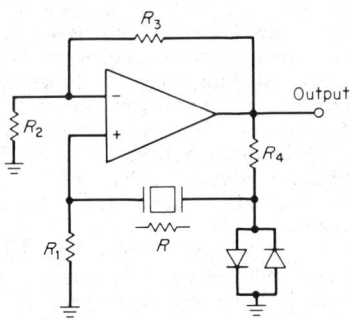

Condition for oscillation:

$$\left(\frac{R_1}{R_1 + R_4 + R} \right) \left(1 + \frac{R_3}{R_2} \right) \geq 1$$

Fig. 13-44. Crystal oscillator using an integrated circuit operational amplifier.

element. The crystal is operated in the series mode, and the amplifier gain is precisely controlled by the negative-feedback divider R_2 and R_3. The output will be sinusoidal if

$$\frac{V_D R_1}{R_1 + R}\left(1 + \frac{R_3}{R_2}\right) < V_{\lim} \qquad (13\text{-}12)$$

where V_D = limiting diode forward voltage drop, and V_{lim} = limiting level of amplifier output.

39. Frequency Stability. Many factors contribute to the ability of an oscillator to hold a constant output frequency over a period of time. These may range from short-term effects, caused by random noise, to longer-term variations, caused by circuit parameter dependence on temperature, bias voltage, and the like. In addition to the temperature and aging effects of the frequency-determining elements, nonlinearities, impedance loading, and amplifier phase variations also contribute to the instability problem.

Harmonics generated by circuit nonlinearities are passed through the feedback network, with various phase shifts, to the input of the amplifier.[12] Intermodulation of the harmonic frequencies produces a fundamental frequency component that differs in phase from the amplifier output. Since the condition $A\beta = 1$ must be satisfied, the frequency of oscillation will shift so that the network phase shift cancels the phase perturbation caused by the nonlinearity. Therefore the frequency of oscillation is influenced by an unpredictable amplifier characteristic, viz., the saturation nonlinearity. This effect is negligible in the Wien bridge oscillator, where automatic level control keeps harmonic distortion to a minimum.

The relationships shown in Fig. 13-41 were derived assuming that the amplifier does not load the bridge circuit on either the input or output sides. In the practical sense this is never true, and changes in the input and output impedances will load the bridge and cause frequency variations to occur.

Another source of frequency instability are small phase changes in the amplifier. This effect is minimized by using a network with a large stability factor defined by

$$S = \frac{d\phi}{d\omega/\omega_o}\bigg|_{\omega=\omega_o} \qquad (13\text{-}13)$$

For the Wien bridge oscillator, which has amplitude-sensitive resistive feedback, the RC impedances can be optimized to provide a maximum stability factor value.[13,14] As shown in Fig. 13-41, this amounts to choosing proper values for M and N. The maximum stability factor value is $A/4$, and it occurs for $N = \frac{1}{2}$ and $M = 2$. Most often the bridge is used with equal resistor and capacitor values; that is, $M = N = 1$, in which case the stability factor is $2A/9$. This represents only a slight degradation from the optimum.

40. Synchronization. It is often desirable to lock the oscillator frequency to an input reference. Usually this is done by injecting sufficient energy at the reference frequency into the oscillator circuit. When the oscillator is tuned sufficiently close to the reference, natural oscillations cease and the synchronization signal is amplified to the output. Thus the circuit appears to oscillate at the injected signal frequency. The injected reference is amplitude-stabilized by the AGC or limiting circuit in the same manner as the natural oscillation. The frequency range over which locking can occur is a linear function of the amplitude of the injected signal. Thus, as the synchronization frequency is moved away from the natural oscillator frequency, the amplitude threshold to maintain lock increases. The phase error between the input reference and the oscillator output will also deviate as the input frequency varies from the natural frequency.

Methods for injecting the lock signal vary and depend on the type of oscillator under consideration. For example, LC oscillators may have signals coupled directly to the tank circuit, whereas the lock signal for the Wien network is usually coupled into the center of the resistive side of the bridge, i.e., the junction of R_1 and R_2 in Fig. 13-41.

If the natural frequency of oscillation can be voltage-controlled, synchronization can be accomplished with a phase-locked loop.[15] Replacing both R's with field-effect transistors, or alternatively shunting both C's with varicaps, provides an effective means for voltage-controlling the frequency of the Wien bridge oscillator. Although more complicated in structure, the phase-locked loop is more versatile, and has many diverse applications (see Pars. 13-14 and 13-17).

41. References on Audio Oscillators.

1. IEEE Standard on Definitions of Terms for Audio and Electroacoustics, *IEEE Trans. Audio Electroacoust.,* Vol. AU-14, No. 2, June, 1966.

2. Bode, H. W. "Network Analysis and Feedback Amplifier Design," Van Nostrand, New York, 1945.

3. Cunningham, W. J. "Nonlinear Analysis," McGraw-Hill, New York, 1958.

4. Hakim, S. S. "Junction Transistor Circuit Analysis," Iliffe Books, Ltd, London, and Wiley, New York, 1962.

5. Reich, H. J. "Functional Circuits and Oscillators," Van Nostrand, New York, 1961.

6. Millman, J. "Vacuum Tube and Semiconductor Electronics," McGraw-Hill, New York, 1958.

7. Strauss, L. "Wave Generation and Shaping," 2d ed., McGraw-Hill, New York, 1970.

8. Owen, R. E. Solid State RC Oscillator Design for Audio Use, *J. Audio Eng. Soc.,* Vol. 14, No. 1, Jan., 1966.

9. Silver, J. F. "Design Notes for Quartz Crystals in Oscillators and Filter Applications," CTS Knight Co., Sandwich, Ill., 1962.

10. Buchanan, J. P. "Handbook of Piezoelectric Crystals for Radio Equipment Designers," Wright Air Development Center, WADC Tech. Rept. 54-248, 1954.

11. Firth, D. "Quartz Crystal Oscillator Circuits Design Handbook," Magnavox Co., Fort Wayne, Ind., 1965.

12. Groszkowski, J. The Interdependence of Frequency Variation and Harmonic Content and the Problem of Constant Frequency Oscillators, *Proc. IRE,* Vol. 21, pp. 958–981, July, 1933.

13. Stevens, B. L., and Manning, R. P. Improvements in the Theory and Design of RC Oscillators, *IEEE Trans. Circuit Theory,* Vol. CT-18, No. 6, pp. 636–643, November, 1971.

14. Mehta, V. B. Comparison of RC Networks for Frequency Stability in Oscillators, *Proc. IEEE,* Vol. 112, pp. 296–300, February, 1965.

15. Gardner, F. M. "Phaselock Techniques," Wiley, New York, 1967.

RADIO-FREQUENCY AMPLIFIERS

By G. B. Harrold

42. Small-Signal RF Amplifiers. The prime considerations in the design of first-stage rf amplifiers are gain and noise figure. As a rule, the gain of the first rf stage should be greater than 10 dB, so that subsequent stages contribute little to the overall amplifier noise figure. The tradeoff between amplifier cost and noise figure is an important design consideration. For example, if the environment in which the rf amplifier operates is noisy, it is uneconomic to demand the ultimate in noise performance. Conversely, where a direct tradeoff exists in transmitter power vs. amplifier noise performance, as it does in many space applications, money spent to obtain the best possible noise figure is fully justified.

Another consideration in many systems is the input/output impedance match of the rf amplifier. For example, TV cable distribution systems require an amplifier whose input and output match produce little or no line reflections. The performance of many rf amplifiers is also specified in handling large signals, to minimize cross- and intermodulation products in the output. The wide acceptance of transistors has placed an additional constraint on first-stage rf amplifiers, since many rf transistors having low noise, high gain, and high frequency response are susceptible to burnout and must be protected to prevent destruction in the presence of high-level input signals.

It is also common to require that first rf stages be gain-controlled by automatic gain control (AGC) voltage. The amount of gain control and the linearity of control are system parameters. Many rf amplifiers have the additional requirement that they be tuned over a range of frequencies. In most receivers, regardless of configuration, local-oscillator leakage back to the input is strictly controlled by government regulation. Finally, the rf amplifier must be stable under all conditions of operation.

43. Device Evaluation for RF Amplifiers. An important consideration in an rf amplifier is the choice of active device. This information on device parameters can often be found in published data sheets. If parameter data are not available, or not at a suitable operating point, the following characterization techniques can be used.

*Rx Meter.** This measurement technique is usually employed at frequencies below 200 MHz, for active devices that have high input and output impedances. The technique[1,2†] is summarized in Fig. 13-45 with assumptions tacit in these measurements. The biasing

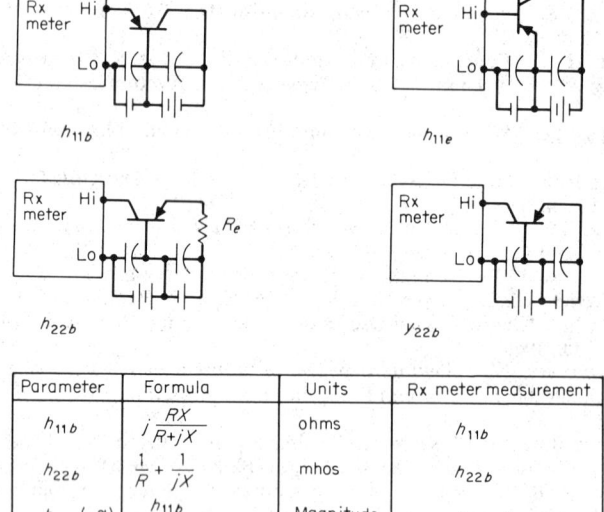

Parameter	Formula	Units	Rx meter measurement
h_{11b}	$j\dfrac{RX}{R+jX}$	ohms	h_{11b}
h_{22b}	$\dfrac{1}{R} + \dfrac{1}{jX}$	mhos	h_{22b}
$h_{21b}(-\alpha)$	$\dfrac{h_{11b}}{h_{11e}} - 1$	Magnitude angle	h_{11b}, h_{11e}
h_{12b}	$(y_{22b} - h_{22b})\dfrac{h_{11b}}{-h_{21b}}$	Magnitude angle	h_{11b}, h_{11e}, y_{22b}, h_{22b}

Assumes: Determinate of $|h| \ll h_{21}$ and $h_{12} \ll 1$

R and X are Rx meter's reading of parallel resistance and reactance

Fig. 13-45. Use of the Rx meter in device characterization.

techniques are shown. In particular, the measurement of h_{22b} requires a very large resistor (R_e) to be inserted in the emitter, and this may cause difficulty in achieving the proper biasing. Care should be taken to prevent burnout of the bridge when a large dc bias is applied. The bridge's drive to the active device may be reduced for more accurate measurement by varying the B+ voltage applied to the internal oscillator.

*Vector Voltmeter.** This characterization technique measures the S parameters defined in Par. 13-9 and Fig. 13-10. The measurement consists in inserting the device in a transmission line, usually 50 Ω characteristic impedance, and measuring the incident and reflected voltages at the two ports of the device. The setup is shown in Fig. 13-46 and discussed in Ref. 3.

This approach is used between frequencies of 100 and 1,000 MHz, on devices whose impedance levels are compatible with 50 Ω. The vector voltmeter measures the voltages present at A and B and gives the phase of voltage B referenced to that of voltage A. The initial calibration is performed by placing a short-circuited section in place of the transistor jig, the probe B at B_1, and adjusting the line stretcher to give zero-degree phase shift between input A and B. This calibration can usually be held across a reasonable frequency band. The short-circuited section is then replaced by the transistor in its jig, and bias is applied. S_{11} is then found by taking the ratio of the vector voltages V_A and V_{B_1}.

$$S_{11} = \frac{V_{B_1}}{V_A} \Big/ \arg V_{B_1} - \arg V_A$$

---

* Trademark of the Hewlett Packard Co., Palo Alto, Calif.
† Superior numbers correspond to numbered references, Par. **13-56.**

The calculation is simplified if V_A is taken as one unit of voltage, with argument zero. Care must be taken not to drive the transistor into saturation, or the measurement of the small-signal parameters is not valid.

Next, with a through-section replacing the transistor jig and the B probe in the B_2 position, the phase difference between V_A and V_B is again adjusted to zero by means of the line stretcher. Replacing the through-section with the transistor in its jig and applying bias, S_{21} can then be measured as the ratio between the vector voltages V_A and V_B.

$$S_{21} = \frac{V_{B_2}}{V_A} \Big/ \underline{\text{arg } V_{B2} - \text{arg } V_A}$$

The measurements of S_{12} and S_{22} are achieved by using the above calibration technique but reversing the transistor jig.

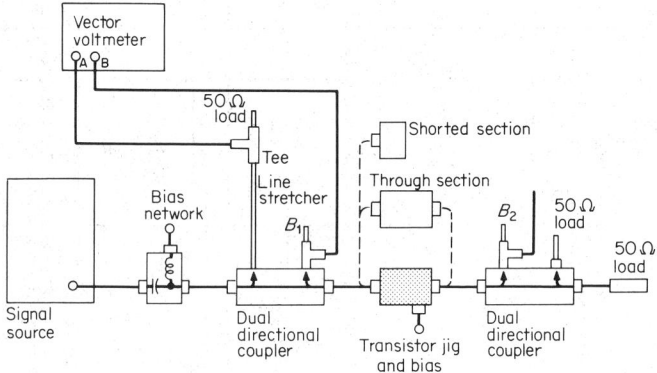

Fig. 13-46. Use of the vector voltmeter and S parameters in device characterization.

Use of 8743 Reflectometer.* This modification of the S-parameter technique can be used above 1 GHz and up to 12.4 GHz. This technique is described in Ref. 4.

Additional Techniques. Several additional evaluation techniques have been used. The Wayne-Kerr three-terminal measurement is discussed in Ref. 5. The General Radio Bridge GR 1607 measurement of transistor parameters between 30 and 1,000 MHz is described in Ref. 6, and the Rhode-Schwartz Diagraph technique is discussed in Ref. 7.

A new technique employs the Type 8750A Automatic Network Analyzer,* which has computer control to measure device parameters automatically from 100 MHz to 12 GHz, with CRT display and teletypewriter printout of the desired parameters.

44. Noise in RF Amplifiers. A common technique employing a noise source to measure the noise performance of an rf amplifier is shown in Fig. 13-47. Initially the external noise source (a temperature-limited diode) is turned off, the 3-dB pad short-circuited, and the reading on the output power meter recorded. The 3-dB pad is then inserted, the noise source is turned on, and its output increased until a reading equal to the previous one is obtained. The noise figure can then be read directly from the noise source, or calculated from 1 plus the added noise per unit bandwidth divided by the standard noise power available, KT_0 where $T_0 = 290°$ Kelvin and K = Boltzmann's constant = 1.38×10^{-23} $J/°K$. See Ref. 8.

At higher frequencies, the use of a temperature-limited diode is not practical, and a gas discharge tube or a hot-cold noise source is employed. The Y-factor technique of measurement is used.[9] The output from the device to be measured is put into a mixer, and the noise output converted to a 30- or 60-MHz center frequency if output. A precision attenuator is then inserted between this if output and the power-measuring device. The attenuator is adjusted to give the same power reading for two different conditions of noise power output represented by effective temperatures T_1 and T_2. The Y factor is the difference

---

* Manufactured by the Hewlett-Packard Co.

in decibels between the two precision attenuator values needed to maintain the same power-meter reading. See Refs. 8 and 9. The noise factor is

$$F = \left[\left(\frac{T_2}{290} - \frac{T_1 Y}{290}\right)\Big/(Y - 1)\right] + 1$$

where T_1 = effective temperature at reference condition 1, T_2 = effective temperature of reference condition 2, and Y = decibel reading defined in the text, converted to a numerical ratio.

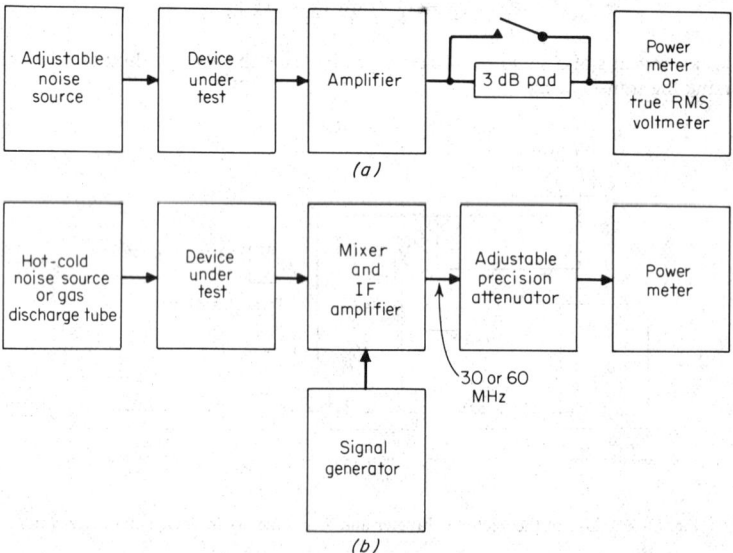

(a)

(b)

Fig. 13-47. Noise-measurement techniques. (*a*) At low frequencies; (*b*) at high frequencies.

In applying this technique it is often necessary to correct for the second-stage noise. This is done by use of the cascade formula

$$F_1 = F_T - \frac{F_2 - 1}{G_1}$$

where F_1 = noise factor of the first stage, F_T = overall noise factor measured, F_2 = noise factor of the second stage mixer and if amplifier, and G_1 = available gain of the first stage.

It is often necessary to determine the optimum source impedance and best noise figure for a particular device. Since many devices have essentially the same first-order noise representation, a generalized analysis can be used. For example, an analysis based on Ref. 10 shows that, for a typical transistor, a first-order assessment of its noise performance, neglecting $1/f$ noise, may be made using the equivalent circuit of Fig. 13-48. The analysis technique is to determine noise-power output from all sources, internal as well as external, and then find the output noise power due to the external source. The noise figure is the ratio of the first divided by the second.

By applying the usual minimization procedures, the optimum source resistance $R_{s,\text{opt}}$ becomes

$$R_{s,\text{opt}} = \frac{1}{g_m}\sqrt{h_{FE}} \cdot \sqrt{1 + 2g_m r_b}$$

and the minimum noise figure $F_{0,\text{min}}$ becomes

$$F_{0,\text{min}} = 1 + \frac{1}{\sqrt{h_{FE}}} \cdot \sqrt{1 + 2g_m r_b}$$

13-38

where h_{FE} is the dc beta, and the other terms are as defined in Fig. 13-48. The above analysis is not valid at the high-frequency end of the transistor's performance, and the optimum source impedance for best noise performance is not necessarily that source impedance for optimum power gain.

The typical noise figure vs. frequency characteristic of an active device is relatively flat in the middle of its frequency range, and the optimum source impedance in this region may be varied within limits without seriously affecting the noise performance.

The question of input match is determined by the various required performance characteristics. If best noise performance is required, the input matching network must transform the source impedance to the optimum-noise source impedance for the active device. On the other hand, the matching network may be designed for optimum power gain, or to present an input impedance that is constant across a wide band of frequencies required by coaxial signal distribution systems.[11]

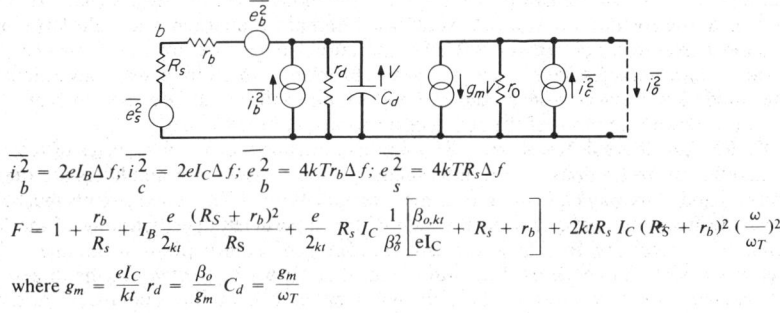

$$\overline{i_b^2} = 2eI_B\Delta f; \; \overline{i_c^2} = 2eI_C\Delta f; \; \overline{e_b^2} = 4kTr_b\Delta f; \; \overline{e_s^2} = 4kTR_s\Delta f$$

$$F = 1 + \frac{r_b}{R_s} + I_B\frac{e}{2_{kt}}\frac{(R_S + r_b)^2}{R_S} + \frac{e}{2_{kt}}R_s I_C\frac{1}{\beta_o^2}\left[\frac{\beta_{o,kt}}{eI_C} + R_s + r_b\right] + 2ktR_s \, I_C \, (R_S + r_b)^2 \, (\frac{\omega}{\omega_T})^2$$

$$\text{where } g_m = \frac{eI_C}{kt} \quad r_d = \frac{\beta_o}{g_m} \quad C_d = \frac{g_m}{\omega_T}$$

  e = electronic charge
  β_o = low frequencies common emitter current gain
  ω_T = radian frequency where $h_{fe} = 1$
  Δf = bandwidth
  k = Boltzmann's constant (1.38×10^{-23} joules/°K)
  R_s = source resistance
  T = Kelvin temperature (290°K)
  r_b = intrinsic base resistance

Fig. 13-48.   Noise equivalent circuit of a typical transistor.

45. Large-Signal Performance of Rf Amplifiers. The large-signal performance of an rf amplifier can be specified in many ways. A common technique is to specify the input where the departure from a straight-line input/output characteristic is 1 dB. This point is commonly called the *1-dB compression point*. The greater the input before this compression point is reached, the better the large-signal performance.

Another method of rating an rf amplifier is in terms of its third-order intermodulation performance. Here two different frequencies, f_1 and f_2, of equal powers, p_1 and p_2, are inserted into the rf amplifier, and the third frequency, internally generated, $2f_1 - f_2$ or $2f_2 - f_1$, has its power p_{12} measured. All three frequencies must be in the amplifier passband. With the intermodulation power p_{12} referred to the output, the following equation may be written:

$$P_{12} = 2P_1 + P_2 + K_{12}$$

where P_{12} = intermodulation output power at $2f_1 - f_2$ or $2f_2 - f_1$, P_1 = output power at input frequency f_1, P_2 = output power at input frequency f_2, all in decibels referred to 1mW(dBm); and K_{12} = constant associated with the particular device.

The value of K_{12} in the above formula can be used to rate the performance of various device choices. Higher orders of intermodulation products can also be used.[12]

A third measure of large-signal performance commonly used is that of cross-modulation. In this instance, a carrier at f_D with no modulation is inserted into the amplifier. A receiver is then placed at the output and tuned to this unmodulated carrier. A second carrier at f_I with amplitude modulation index M_I is then added. The power P_I of f_I is increased, and its modulation is partially transferred to f_D. The equation becomes

$$10 \log \frac{M_K}{M_I} = P_I + K$$

where MK = cross-modulation index of originally unmodulated signal at f_D, M_I = modulation index of signal F_I, P_I = output power of signal at f_I, all in decibels referred to 1 mW(dBm), and K = cross-modulation constant.

46. Maximum Input Power. In addition to the large-signal performance, the maximum power of voltage input into an rf amplifier is specified, with a requirement that device burnout must not occur at this input. There are two ways of specifying this input: by a stated pulse of energy or by a requirement to withstand a continuously applied large signal. It is also common to specify the time required to unblock the amplifier after removal of the large input. With the increased use of transistors (FETs especially) having good noise performance, these overload characteristics have become a severe problem. In many cases, conventional or zener diodes, in a back-to-back configuration shunting the input, are used to reduce the amount of power the input of the active devices must dissipate.

47. RF Amplifiers in Receivers. Rf amplifiers intended for the first stages of receivers have additional restrictions placed upon them. In most cases, such amplifiers are tunable across a band of frequencies with one or more tuned circuits. The tuned circuits must track across the frequency band, and in the case of the superheterodyne, tracking of the local oscillator is necessary so that a constant frequency difference (if) is maintained. The receiver's rf section may be tracked with the local oscillator by the two- or the three-point method, i.e., with zero error in the tracking at either two or three points. The design technique is discussed in Ref. 13.

A second consideration peculiar to rf amplifiers used for receivers is the automatic gain control (AGC). This requirement is often stated by specifying a low-level rf input to the receiver and noting the power out. The rf signal input is then increased with the AGC applied until the output power has increased a predetermined amount. This becomes a measure of the AGC effectiveness. The AGC performance can also be measured by plotting a curve of rf input vs. AGC voltage needed to maintain constant output, compared with the desired performance.

A third consideration in superheterodynes is the leakage of the local oscillator in the receiver to the outside. This spurious radiation is carefully specified[14] by the Federal Communications Commission (FCC) in the United States, to be according to Table 13-4. Reference 14 states the various radiation levels for community antenna television systems and suggests techniques of measurement.

Table 13-4. Allowable Spurious Radiation from Receivers

Frequency of radiation	Signal strength, μV/m at 100 ft or more from receiver
0.45 up to and including* 25 MHz	*For TV receivers* Less than 100 μV: Between 0.45 and 25 MHz *All other receivers* Less than 100 μV: Up to 9 MHz Less than 1,000 μV: 10 MHz up to 25 MHz; linear increase from 100 to 1,000 μV between 9 and 10 MHz
Over 25 up to and including 70 MHz	Less than 32 μV
Over 70 up to 130 MHz	Less than 50 μV
130 to 174 MHz	Less than 50 to less than 150 μV; linear interpolation
174 to 260 MHz	Less than 150 μV
260 to 1,000 MHz	Less than 150 to less than 500 μV; linear interpolation
470 to 1,000 MHz	Less than 500 μV

* Measurement in this frequency band of rf voltage is between each power line and ground for receivers so powered.

48. Design Using Immittance and Hybrid Parameters. The general gain and input/output impedance of an amplifier can be formulated, in terms of the Z or Y parameters, to be

$$Y_{in} = y_1 - \frac{y_{12}y_{21}}{y_{22} + y_L}$$

$$Y_{out} = y_{22} - \frac{y_{12}y_{21}}{y_{11} + y_s}$$

where y_L = load admittance, y_s = source admittance, Y_{in} = input admittance, Y_{out} = output admittance, G_T = transducer gain, and the transducer gain is

$$G_T = \frac{4 \operatorname{Re}(y_s)\operatorname{Re}(y_L)|y_{21}|^2}{|(y_{11} + y_s)(y_{22} + y_L) - y_{12}y_{21}|^2}$$

for the y parameters, and interchange of z for y is allowed.

The stability of the circuit can be determined by either Linvill's C or Stern's k factor as defined below. Using the y parameters, $y_{ij} = g_{ik} + jB_{ik}$, these are
Linvill:

$$C = \frac{|y_{12}y_{21}|}{2g_{11}g_{22} - \operatorname{Re}(y_{12}y_{21})}$$

where $C < 1$ for stability; does not include effects of load and source admittance.
Stern:

$$k = \frac{2(g_{11} + g_s)(g_{22} + g_L)}{|y_{12}y_{21}| + \operatorname{Re}(y_{12}y_{21})}$$

where $k > 1$ for stability, g_L = load conductance, and g_s = source conductance.

The preceding C factor defines only unconditional stability; i.e., no combination of load and source impedance will give instability. Rollett[15] has shown that there is an invariant quantity K defined as

$$K = \frac{2 \operatorname{Re}(\gamma_{11})\operatorname{Re}(\gamma_{22}) - \operatorname{Re}(\gamma_{12}\gamma_{21})}{|\gamma_{21}\gamma_{12}|} \qquad \begin{array}{l} \operatorname{Re}(\gamma_{11}) = > 0 \\ \operatorname{Re}(\gamma_{22}) = > 0 \end{array}$$

where γ represents either the y, z, g, or h parameters, and $K > 1$ denotes stability.

This quantity K has then been used to define maximum available power gain G_{max} (only if $K > 1$).

$$G_{max} = \left|\frac{\gamma_{21}}{\gamma_{12}}\right|(K - \sqrt{K^2 - 1})$$

To obtain this gain, the source and load immittance are found to be ($K > 1$)

$$\gamma_s = \frac{\gamma_{12}\gamma_{21} + |\gamma_{12}\gamma_{21}|(K + \sqrt{K^2 - 1})}{2 \operatorname{Re}(\gamma_{22})} - \gamma_{11} \qquad \gamma_s = \text{source immittance}$$

$$\gamma_L = \frac{\gamma_{12}\gamma_{21} + |\gamma_{12}\gamma_{21}|(K + \sqrt{K^2 - 1})}{2 \operatorname{Re}(\gamma_{11})} - \gamma_{22} \qquad \gamma_L = \text{load immittance}$$

The procedure is to calculate the K factor, and if $K > 1$, calculate G_{max}, γ_s, and γ_L. If $K < 1$, the circuit can be modified either by use of feedback or by adding immittances to the input/output.

49. Design Using S Parameters. The advent of automatic test equipment and the extension of vacuum tubes and transistors to the gigahertz frequency range have led to design procedures using the S parameters. These parameters were described in the introduction to this section. Following the previous discussion, the input and output reflection coefficient can be defined as

$$p_{in} = S_{11} + p_L \frac{S_{12}S_{21}}{1 - p_L S_{22}} \qquad p_L = \frac{Z_L - Z_0}{Z_L + Z_0}$$

$$p_{out} = S_{22} + p \frac{S_{12}S_{21}}{1 - pS_{11}} \qquad p_s = \frac{Z - Z_0}{Z + Z_0}$$

$$Z_0 = \text{characteristic impedance}$$

where p_{in} = input reflection coefficient, and p_{out} = output reflection coefficient. The transducer gain can be written

$$G_{transducer} = \frac{|S_{21}|^2(1 - |p_s|^2)(1 - |p_L|^2)}{|(1 - S_{11}p_s)(1 - S_{22}p_L) - S_{21}S_{12}p_s p_L|^2}$$

The unconditional stability of the amplifier can be defined by requiring the input (output) impedance to have a positive real part for any load (source) impedance having a positive real part.[16]

This requirement gives the following criterion:

$$|S_{11}|^2 + |S_{12}S_{21}| < 1$$

$$|S_{22}|^2 + |S_{12}S_{11}| < 1$$

and

$$\eta = \frac{1 + |\Delta_s|^2 - |S_{11}|^2 - |S_{22}|^2}{2|S_{12}S_{21}|} > 1$$

$$\Delta_s = S_{11}S_{22} - S_{12}S_{21}$$

Similarly, the maximum transducer gain, for $\eta > 1$, becomes

$$G_{max \ transducer} = \left|\frac{S_{21}}{S_{12}}\right|(\eta \pm \sqrt{\eta^2 - 1})$$

(positive sign when $|S_{22}|^2 - |S_{11}|^2 - 1 + |\Delta_s|^2 > 0$) for conditions listed above.

The source and load to provide conjugate match to the amplifier, when $\eta > 1$, are the solutions of the following equations, which give $|p_s|$, and $|p_L|$ less than 1.

$$p_{ms} = C_1^* \left[\frac{B_1 \pm \sqrt{B_1^2 - 4|C_1|^2}}{2|C_1|^2}\right]$$

$$p_{mL} = C_2^* \left[\frac{B_2 \pm \sqrt{B_2^2 - 4|C_2|^2}}{2|C_2|^2}\right]$$

where $\quad B_1 = 1 + |S_{11}|^2 - |S_{22}| - |\Delta_s|^2$

$\qquad\qquad B_2 = 1 + |S_{22}|^2 - |S_{11}|^2 - |\Delta_s|^2$

$\qquad\qquad C_1 = S_{11} - \Delta_s S_{22}^*$

$\qquad\qquad C_2 = S_{22} - \Delta_s S_{11}^*$

the star (*) denoting conjugate.

If $|\eta| > 1$ but η is negative or $|\eta| < 1$, it is not possible to simultaneously match the two-port with real source and load admittances.

50. Intermediate-Frequency Amplifiers. Intermediate-frequency amplifiers consist of a cascade of a number of stages whose frequency response is determined either by a filter or by tuned interstages. The design of the individual active stages follows the techniques discussed earlier, but the interstages become important for frequency shaping. There are various forms of interstage networks; several important cases are discussed below.

Synchronous-tuned Interstages. The simplest forms of tuned interstages are synchronously tuned circuits. The two common types are the single- and double-tuned interstage. The governing equations are

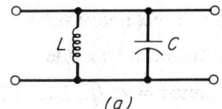

(a)

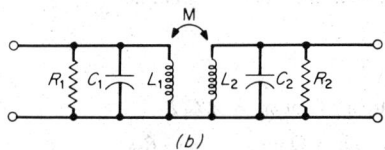

(b)

Fig. 13-49. Interstage coupling circuits. (a) Single-tuned; (b) double-tuned.

(a) Single-tuned interstage (Fig. 13-49a):

$$A(j\omega) = -A_r \frac{1}{1 + jQ_L(\omega/\omega_0 - \omega_0/\omega)}$$

where Q_L = loaded Q of the tuned circuit greater than 10, ω_0 = resonance frequency of the tuned circuit = $1\sqrt{LC}$, ω = frequency variable, A_r = midband gain equal to g_m times the midband impedance level. For an n-stage amplifier with n interstages.

$$A_T = A^n(j\omega) = A_r^n \left[1 + \left(\frac{\omega^2 - \omega_0^2}{B\omega}\right)^2\right]^{-n/2}$$

where $B = \omega_0/Q_L$ = single-stage bandwidth, n = number of stages, ω_0 = center frequencies, and Q_L = loaded Q. $B_n = B\sqrt{2^{1/n} - 1}$ is the overall bandwidth reduction due to n cascade.

(b) Double-tuned interstage (Fig. 13-49b):

$$A(j\omega) = \frac{g_m k}{C_1 C_2 (1 - k^2)\sqrt{L_1 L_2}} \frac{j\omega}{\omega^4 - ja_1\omega^3 - a_2\omega^2 + ja_3\omega + a_4}$$

(for a single double-tuned stage)

where
$$a_1 = \omega_r\left(\frac{1}{Q_1} + \frac{1}{Q_2}\right)$$

$$a_2 = \frac{\omega_r^2}{Q_1 Q_2} + \frac{1}{1 - k^2}(\omega_1^2 + \omega_2^2)$$

$$a_3 = \frac{\omega_r}{1 - k^2}\left(\frac{\omega_2^2}{Q_1} + \frac{\omega_1^2}{Q_2}\right)$$

$$a_4 = \frac{\omega_1^2\omega_2^2}{1 - k^2}$$

13-43

The circuit parameters are:

R_1 = total resistance primary side

C_1 = total capacitance primary side

L_1 = total inductance primary side

R_2 = total resistance secondary side

C_2 = total capacitance secondary side

L_2 = total inductance secondary side

M = mutual inductance = $k\sqrt{L_1 L_2}$

k = coefficient of coupling

ω_r = resonant frequency of amplifier

$\omega_1 = 1/\sqrt{L_1 C_1}$

$\omega_2 = 1/\sqrt{L_2 C_2}$

Q_1 = primary Q at $\omega_r = \omega_r C_1 R_1$

Q_2 = secondary Q at $\omega_r = \omega_r C_2 R_2$

g_m = transconductance of active device at midband frequency

Simplification. If $\omega_1 = \omega_2 = \omega_0$, that is, primary and secondary tuned to the same frequency, then

$$\omega_r = \frac{\omega_0}{\sqrt{1 - k^2}}$$

is the resonant frequency of the amplifier and

$$A(j\omega_r) = \frac{+jkg_m\sqrt{R_1 R_2}}{\sqrt{Q_1 Q_2}(k^2 + 1/Q_1 Q_2)}$$

is the gain at this resonant frequency.

For maximum gain,

$$k_c = \frac{1}{\sqrt{Q_1 Q_2}} = \text{critical coupling}$$

and for maximum flatness,

$$k_T = \sqrt{\frac{1}{2}\left(\frac{1}{Q_1^2} + \frac{1}{Q_2^2}\right)} = \text{transitional coupling}$$

If k is increased beyond k_T, a double-humped response is obtained.

Overall bandwidth of an n-stage amplifier, having equal Q circuits with transitional coupled interstages whose bandwidth is B, is

$$B_n = B(2^{1/n} - 1)^{1/4}$$

The governing equations for the double-tuned-interstage case are shown above. The response for various degrees of coupling related to $k_T = k_C$ in the equal-coil-Q case is shown in Fig. 13-50.

51. Maximally Flat Staggered Interstage Coupling. This type of coupling consists of n single-tuned interstages that are cascaded and adjusted so that the overall gain function is maximally flat. The overall cascade bandwidth is B_n; the center frequency of the cascade is ω_c; and each stage is a single-tuned circuit whose bandwidth B and center frequency are determined from Table 13-5. The gain of each stage at cascade center frequency is $A(j\omega_c) =$

Table 13-5.   Design Data for Maximally Flat Staggered *n*-tuples

n	Name of circuit	No. of stages	Center frequency of stage	Stage bandwidth
2	Staggered pair	2	$\omega_c \pm 0.35B_n$	$0.71B_n$
3	Staggered triple	2	$\omega_c \pm 0.43B_n$	$0.50B_n$
		1	ω_c	$1.00B_n$
4	Staggered quadruple	2	$\omega_c \pm 0.46B_n$	$0.38B_n$
		2	$\omega_c \pm 0.19B_n$	$0.92B_n$
5	Staggered quintuple	2	$\omega_c \pm 0.29B_n$	$0.81B_n$
		2	$\omega_c \pm 0.48B_n$	$0.26B_n$
		1	ω_c	$1.00B_n$
6	Staggered sextuple	2	$\omega_c \pm 0.48B_n$	$0.26B_n$
		2	$\omega_c \pm 0.35B_n$	$0.71B_n$
		2	$\omega_c \pm 0.13B_n$	$0.97B_n$
7	Staggered septuple	2	$\omega_c \pm 0.49B_n$	$0.22B_n$
		2	$\omega_c \pm 0.39B_n$	$0.62B_n$
		2	$\omega_c \pm 0.22B_n$	$0.90B_n$
		1	ω_c	$1.00B_n$

For $Q_L > 20$

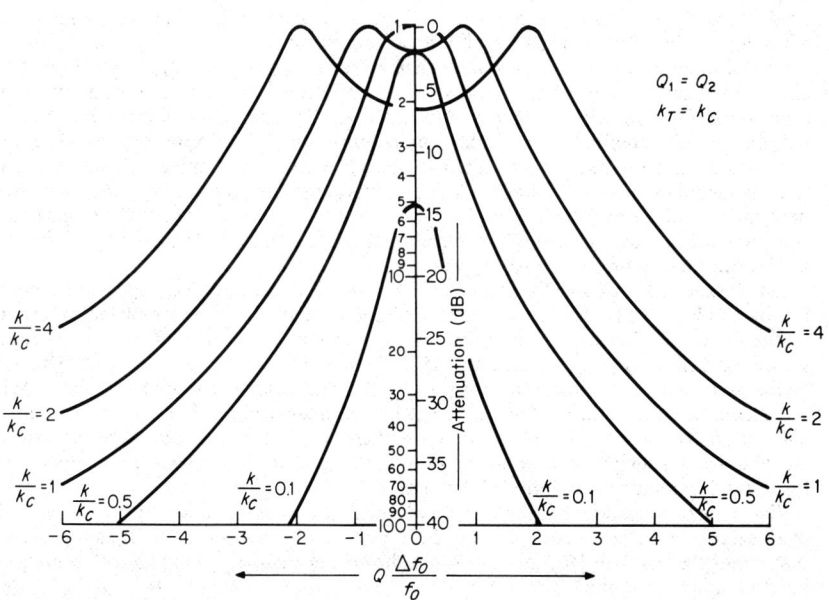

Fig. 13-50. Selectivity curves for two identical circuits in a double-tuned interstage circuit, at various values of k/k_c.

13-45

$-g_m/C_T[B + j(\omega_c^2 - \omega_0^2)/\omega_c]$, where C_T is the sum of the output capacitance/input capacitance to the next stage and wiring capacitance of the cascade; B is the stage bandwidth; ω_0 is the center frequency of the stage; and ω_c is the center frequency of the cascade.[19]

52. Other IF Interstage Coupling Systems. There are several other methods of obtaining the desired passband response of an if amplifier. The design of cascaded single-tuned circuits with active devices in between can also achieve a Chebyshev ripple passband response.[19] Also coming into vogue, with the use of integrated circuits, is the use of lumped gain followed by a passive filter and additional gain. The amount of gain before and after the filter is a system compromise between noise performance and the overloading of the wideband stages that are used.

RADIO-FREQUENCY OSCILLATORS

BY G. B. HARROLD

53. General Considerations. Oscillators at rf frequencies are usually of the class A sinewave output type.

RF oscillators (in common with audio oscillators, see Pars. **13-34** to **13-41**) may be considered either as one-port networks that exhibit a negative real component at the input or as two-port-type networks consisting of an amplifier and a frequency-sensitive passive network that couples back to the input port of the amplifier. It can be shown that the latter type of feedback oscillator also has a negative resistance at one port. This negative resistance is of a dynamic nature and is best defined as the ratio between the fundamental components of voltage and current.

The sensitivity of the oscillator's frequency is directly dependent upon the effective Q of the frequency-determining element and the sensitivity of the amplifier to variations in temperature, voltage variation, and aging. For example, the effective Q of the frequency-determining element is important because the percentage change in frequency required to produce the compensating phase shift in a feedback oscillator is inversely proportional to the circuit Q; thus the larger the effective Q, the greater the frequency stability. The load on an oscillator is also critical to the frequency stability since it affects the effective Q and, in many cases, the oscillator is followed by a buffer stage for isolation.

It is also desirable to provide some means of stabilizing the oscillator's operating point, either by means of a regulated supply, dc feedback for bias stabilization, or the use of oscillator self-biasing schemes such as grid-leak bias. This not only stabilizes the frequency, but the output amplitude, by tending to compensate any drift in the active device's parameters. It is also necessary to eliminate the harmonics in the output since these give rise to cross-modulation products that produce currents at the fundamental frequency that are not necessarily in phase with the dominant oscillating mode. The use of high-Q circuits and the control of the nonlinearity helps in controlling harmonic output. The basic block diagrams for feedback oscillators are shown in Fig. 13-14.

54. Negative-Resistance Oscillators. The analysis of the negative-impedance oscillator is shown in Fig. 13-51. The frequency of oscillation at buildup is not completely determined by the LC circuit, but has a component that is dependent upon the circuit resistance. At steady state, the frequency of oscillation is a function of $1 + R/R_{iv}$ or $1 + R_{ic}/R$, depending on the particular circuit where the ratios R/R_{iv}, R_{ic}/R are usually chosen to be small. While R is a fixed function of the loading, R_{ic} or R_{iv}/R must change with amplitude during oscillator buildup, so that the condition of $\alpha = 0$ can be reached. Thus R_{iv}, R_{ic} cannot be constant, but are dynamic impedances defined as the ratio of the fundamental voltage across the element to the fundamental current into the element.

The type of dc load for biasing and the resonant circuit required for the proper operation of a negative-resistance oscillator depend on the type of active element. It can be shown[20] that R must be less than $|R_{iv}|$ or that R must be greater than $|R_{ic}|$ in order for oscillation to build up and be sustained.

The detailed analysis of the steady-state oscillator amplitude and frequency can be undertaken by graphical techniques. The magnitude of G_i or R_i is expressed in terms of its voltage dependence. Care must be taken with this representation, since the shape of the G_i or R_i curve is dependent upon the initial bias point.

The analysis of negative-resistance oscillators may now be performed by means of admittance diagrams. The assumption for oscillation to be sustaining is that the negative-resistance element, having admittance y_i, must equal $-y_c$, the external circuit admittance. This may be summarized by $G_i = -G_c$ and $B_i = -B_c$. A typical set of admittance curves

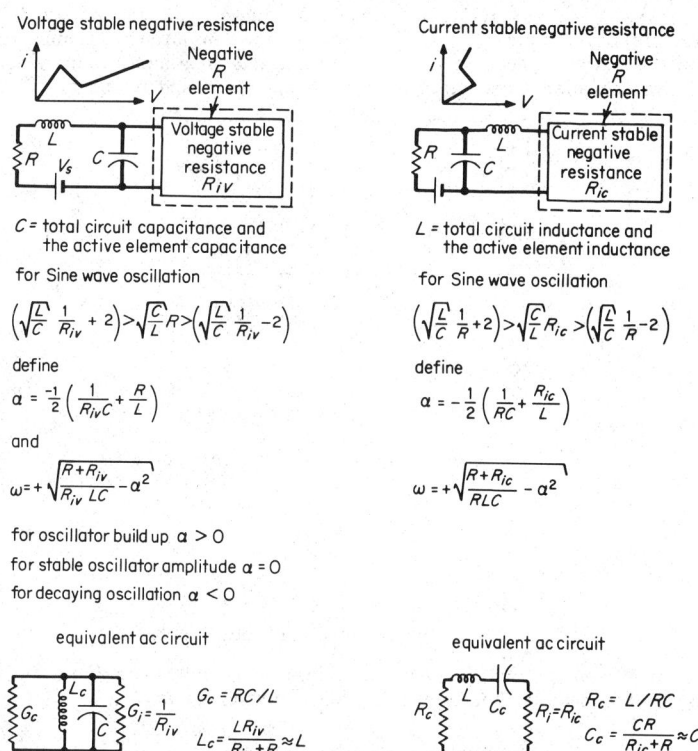

Fig. 13-51. General analysis of negative-resistance oscillators.

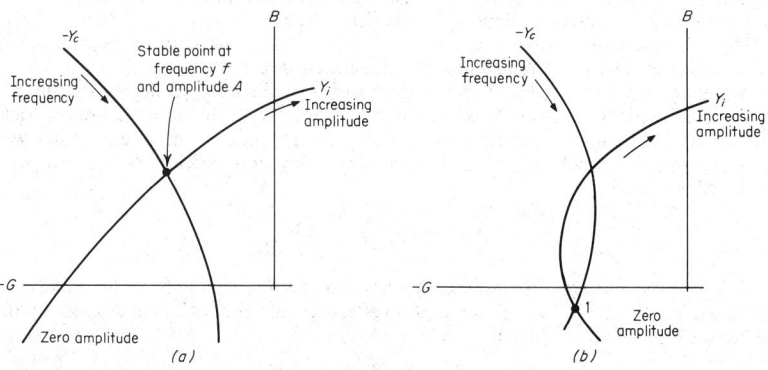

Fig. 13-52. Admittance diagrams of voltage-stable negative-resistance oscillators. (a) Self-starting case, $\alpha > 0$; (b) circuit starts oscillating only if externally excited beyond point 1.

is shown in Fig. 13-52. In this construction, it is assumed that $B_i = -B_c$, even during the oscillator build-up. Also shown is the fact that G_i at zero amplitude must be larger than G_c so that the oscillator can be started, that is, $\alpha > 0$, and that it may be possible to have two or more stable modes of oscillation.

55. Feedback Oscillators. Several techniques exist for the analysis of feedback oscillators.[21-23] In the generalized treatment, the active element is represented by its y parameters whose element values are at the frequency of interest, having magnitudes defined by the ratio of the fundamental current divided by fundamental voltage. The general block diagram and equations are shown in Fig. 13-53. Solution of the equations given yields information as to the oscillator's performance. In particular, equating the real and imaginary parts of the characteristic equation gives information as to amplitude and frequency of oscillation.

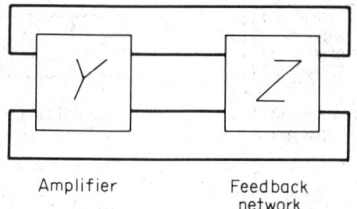

Characteristic equation:

$$y_{21}\,z_{21} + y_{11}\,z_{22} + y_{22}\,z_{11} + y_{12}\,z_{12} + \Delta_y\,\Delta_z + 1 = 0$$

$$\Delta_y = y_{11}\,y_{22} - y_{12}\,y_{21} \qquad \Delta_z = z_{11}\,z_{22} - z_{12}\,z_{21}$$

$$\text{If } y_{21} \gg y_{12} \qquad y_{12} \approx 0 \qquad \text{and } [z] \text{ passive } z_{12} = z_{21}$$

Then:

$$y_{21}\,z_{12} + y_{11}\,z_{22} + y_{22}\,z_{11} + \Delta_y\,\Delta_z + 1 = 0$$

Fig. 13-53. General analysis of feedback oscillators.

In many instances, many simplifications to these equations can be made. For example, if y_{11} and y_{12} are made small (as in vacuum-tube amplifiers), then

$$y_{21} = -\frac{1}{z_{21}}(y_{22}z_{11} + 1) = -\frac{1}{Z}$$

This equation can be solved by equating the real and imaginary terms to zero to find the frequency and the criterion for oscillation of constant amplitude. This equation can also be used to draw an admittance diagram for oscillator analysis.

These admittance diagrams are similar to those discussed under negative-resistance oscillators (Sec. **13-54**). The technique is illustrated in Fig. 13-54.

At higher frequencies, the S parameters may also be used to design oscillators (Fig. 13-55). The basis for the oscillator is that the magnitude of the input reflection coefficient must be greater than unity, causing the circuit to be potentially unstable (in other words, it has a negative real part for the input impedance). The input reflection coefficient with a Γ_L output termination is

$$S_{11}' = S_{11} + \frac{S_{12}\,S_{21}\,\Gamma_L}{1 - S_{22}\Gamma_L}$$

Either by means of additional feedback or adjustment of Γ_L it is possible to make $|S_{11}'| > 1$. Next, establishing a load Γ_s such that it reflects all the energy incident upon it will cause the circuit to oscillate. This criterion is stated as

$$\Gamma_s S_{11}' = 1$$

at the frequency of oscillation.

This technique can be applied graphically, using a Smith Chart as before. Here the reciprocal of S'_{11} is plotted as a function of frequency since $S'_{11} > 1$. Now choose either a parallel-or a series-tuned circuit and plot its Γ_s. If f_1 is the frequency common to $1/S'_{11}$ and Γ_s and satisfies the above criterion, the circuit will oscillate at this point.

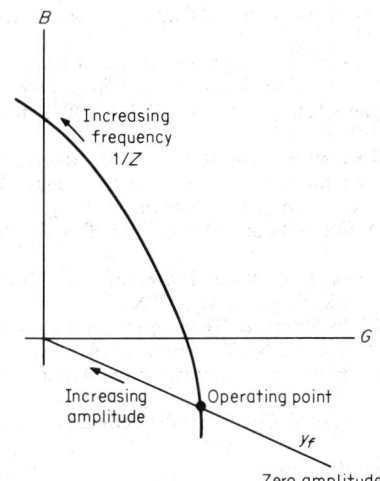

Fig. 13-54. Admittance diagram of feedback oscillator.

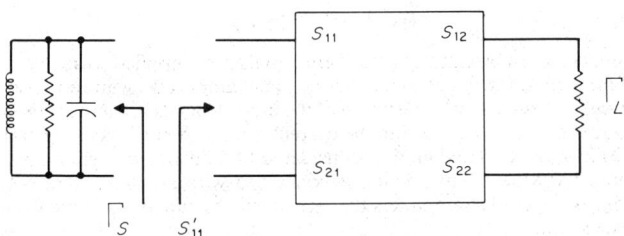

Fig. 13-55. *S*-parameter analysis of oscillators.

56. References on Radio-Frequency Amplifiers and Oscillators

1. McCASLAND, G. P. Transistor Measurements with the HF-VHF Bridge, *Boonton Radio Notebook* 19, pp. 1-6, 1958.

2. McCASLAND, G. P. More about Transistor Measurements, *Boonton Radio Notebook* 20, 1959.

3. Transistor Parameter Measurements, *Hewlett Packard, Appl., Note* 77-1, Palo Alto, Calif.

4. An Advanced New Network Analyzer for Sweep Measuring Application and Phase from 0.1 to 12.4 GHz, *Hewlett Packard J.*, Vol. 18, No. 6, February, 1967.

5. SCARLETT, R. M. Measuring Transistor Parameters with Wayne Kerr R. F. Bridges, *Electron. Design*, Vol. 15, No. 7, pp. 38-41, July 22, 1959.

6. THURSTON, W. R., and SODERMAN, R. A. Type 1607-A Transfer Function and Admittance Bridge, *Gen. Radio Exp.*, Vol. 33, No. 5, pp. 3-11, May, 1959.

7. ABRAHAM, R. P., and KIRKPATRICK, R. J. Transistor Characterization at VHF, *Bell Telephone System, Monogr*, 3007 Tech. Publ., 1957.

8. "Instruction Book for VHF-UHF Noise Generator PRD Type 904-A," PRD Electronics (Div. of Harris-Intertype Corp.), Westbury, N.J., 1967.

9. "Instruction Book for AIL Type 70 Hot-Cold Body Standard Noise Generator," Cutler-Hammer, Deer Park, N.Y., 1968.

10. CHENETTE, E. R.   Low Noise Transistor Amplifiers, *Solid State Design,* February, 1964, pp. 27-30.

11. LO, A. W., ENDRES, R. O., et al.   "Transistor Electronics," pp. 318-326, Prentice-Hall, Englewood, N.J., 1955.

12. Interference Analysis of New Components and Circuits, Rome Air Development Center *Tech. Doc. Rept.* RADC-TDR-64-161, May, 1964, AD 601850.

13. "Radiotron Designer's Handbook," Sec. 25.3, pp. 1002-1017, Amalgamated Wireless Co., Sydney, Australia, 1953.

14. FCC Rules and Regulations, Vol. II, Sec. 15.61, August, 1969.

15. ROLLETT, J. M.   Stability and Power Gain Invariants of Linear Two Ports, *IRE Trans. Circuit Theory,* Vol. CT-9, pp. 29-32, March, 1962.

16. HAYKIN, S. S.   "Active Network Theory," pp. 272-276, Addison-Wesley, Reading, Mass., 1970.

17. BODWAY, G. E.   Two Port Power Flow Analysis Using Generalized Scattering Parameters, *Microwave J.,* Vol. 10, No. 6, May, 1967.

18. ANDERSON, R. W.   S-Parameter Techniques for Faster, More Accurate Network Design, *Hewlett Packard J.,* Vol. 18, No. 6, February, 1967.

19. MARTIN, T. L., JR.   Electronic Circuits, Chaps. 4 and 5, Prentice-Hall, Englewood, N.J., 1955.

20. REICH, H. J.   "Functional Circuits and Oscillators," p. 317, Van Nostrand, New York, 1961.

21. Reference 19, Chap. 10.

22. SEELY, S.   "Electronic Tube Circuits," Chap. 12, McGraw-Hill, New York, 1950.

23. Reference 20, *loc. cit.*

BROADBAND AMPLIFIERS

By J. W. LUNDEN

57. Introduction.   In broadband amplifiers signals are amplified so as to preserve over a wide band of frequencies such characteristics as signal amplitude, gain response, phase shift, delay, distortion, and efficiency.   The width of the band depends upon the active device used, the frequency range, and power level in the current state of the art.   As a general rule, above 100 MHz, a 20% or greater bandwidth is considered broadband, whereas an octave or more is typical below 100 MHz.   This section is concerned with amplifiers that cover the band without tuning, as opposed to amplifiers that are tuned as a function of time to cover a broad band of the spectrum.

Whether one should use the frequency domain or the equivalent time domain in analysis and measurement techniques depends on many factors, including instrumentation, experience in interpretation, and the specific application.   Historically, frequency-domain techniques have dominated, but recent developments in time domain reflectometry, wide-band-sampling oscillography, and Fourier transform analysis have turned attention to the time domain.

58. Low-, Mid-, and High-Frequency Performance.   Consider the basic common-cathode and common-emitter-broadband *RC* coupled configurations shown in Fig. 13-56. Simplified low-frequency equivalent circuits are shown in Fig. 13-57.

The voltage gain of the tube amplifier stage under the condition that all reactances are negligibly small is the midband value (at frequency f)

$$[A_{\text{mid}}]_{\text{tube}} = \frac{-g_m}{1/r_p + 1/R_L + 1/R_g}$$

$$= \frac{-\mu[R_L R_g/(R_L + R_g)]}{r_p + R_L R_g/(R_L + R_g)} \approx -g_m R_L$$

If the low-frequency effects are included, this becomes

$$[A_{\text{low}}]_{\text{tube}} = \frac{g_m R_L}{1 + 1/j\omega R g C_g}\left[\frac{1 + 1/j\omega R_K C_K}{1 + (1 + g_m R_K)/j\omega C_K R_K}\right]$$

The low-frequency cutoff is due principally to two basic time constants, $R_g C_g$ and $R_K C_K$. For C_K values large enough so that its time constant is much longer than that associated with C_g, a low-frequency cutoff or half-power point may be determined as

$$[f_1]_{\text{tube}} = \frac{1}{2\pi C_g [R_g + r_p R_L / (r_p + R_L)]}$$

If the coupling capacitor is very large, the low-frequency cutoff is due to C_K. The slope of the actual roll-off is a function of the relative effect of these two time constants. Evidently the design of coupling and bypass circuits to achieve very-low-frequency response requires very large values of capacitance.

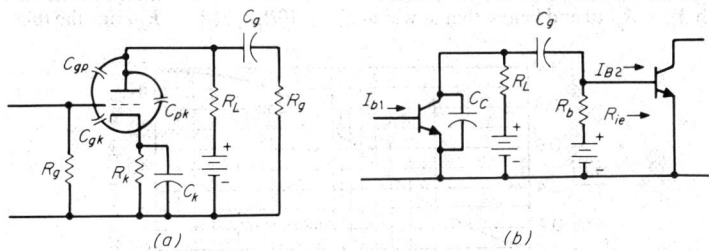

Fig. 13-56. *RC*-coupled stages. (*a*) Electron-tube form; (*b*) transistor form.

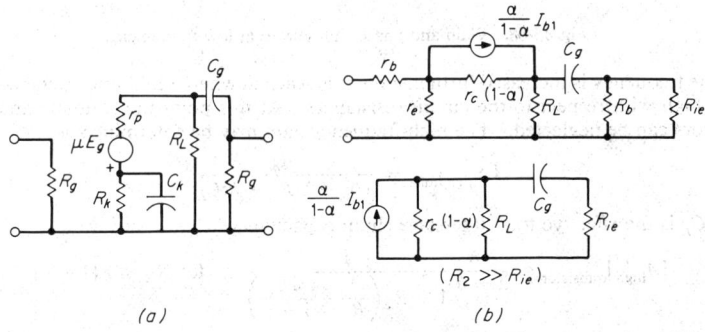

Fig. 13-57. Equivalent circuits of the stages shown in Fig. 13-56. (*a*) Electron-tube form; (*b*) transistor forms.

Similarly, for a transistor stage, the midband current gain may be determined as

$$[A_{\text{mid}}]_{\text{transistor}} = \frac{-\alpha r_c R_L}{[R_L + r_c(1 - \alpha)]\left[R_{ie} + \dfrac{R_L r_c(1 - \alpha)}{R_L + r_c(1 - \alpha)}\right]} \approx \frac{-\alpha}{1 - \alpha}\frac{R_L}{R_L + R_{ie}}$$

where $R_{ie} = r_b + \dfrac{r_e}{1 - \alpha}$

Including low-frequency effects, this becomes

$$[A_{\text{low}}]_{\text{transistor}} \approx \frac{-\alpha}{1 - \alpha}\frac{R_L}{R_L + R_{ie} - j/\omega C_g} \qquad \text{for } R_L \ll r_c(1 - \alpha)$$

and

$$[f_1]_{\text{transistor}} = \frac{1}{2\pi C_g\left[R_{ie} + \dfrac{R_L r_c(1 - \alpha)}{R_L + r_c(1 - \alpha)}\right]} \approx \frac{1}{2\pi C_g}\frac{1}{R_{ie} + R_L}$$

13-51

If the ratio of low- to midfrequency voltage or current gain is taken, its reactive term goes to unity at $f = f_1$, that is, the cutoff frequency.

$$\frac{A_{low}}{A_{mid}} = \frac{1}{1 - j(f_1/f)} \qquad \phi_{low} = \tan^{-1}\frac{f_1}{f}$$

These quantities are plotted in Fig. 13-58 for a single time-constant roll-off.

Caution should be exercised in assuming that interelectrode reactances are negligible. Although this is generally the case, gain multiplicative effects can result in input or output values greater than the values assumed above, e.g., by the Miller effect:

$$C_{in} = C_{gK} + C_{gP}(1 + g_m R_L')$$

Typically, the midfrequency gain equation may be employed for frequencies above that at which $X_c = R_g/10$ and below that at which $X_{cg} = 10R_g R_L/(R_g + R_L)$ (for the tube circuit).

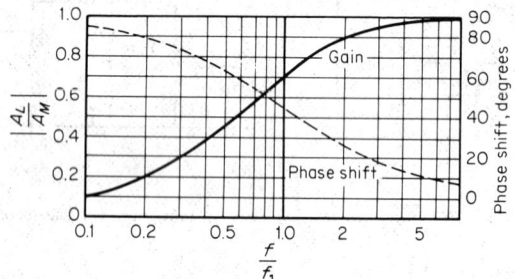

Fig. 13-58. Gain and phase-shift curves at low frequencies.

If the frequency is increased further, a point is reached where the shunt reactances are no longer high with respect to the circuit resistances. At this point the coupling and bypass capacitors can be neglected. The high-frequency gain may be determined as

$$[A_{high}]_{tube} = \frac{-g_m}{1/r_p + 1/R_L + j\omega C_L}$$

where C_L is the effective total interstage shunt capacitance.

$$[A_{high}]_{transistor} \approx \frac{-\alpha}{1 - \alpha} \frac{1}{1 + R_{ie}\left(\dfrac{1}{R_L} + \dfrac{j\omega C_c}{1 - \alpha}\right)} \qquad \text{for } R_L \ll r_c(1 - \alpha)$$

The ratio of high- to midfrequency gains may be taken, and upper cutoff frequencies determined.

$$\left[\frac{A_{high}}{A_{mid}}\right]_{tube} = \frac{1}{1 + j\omega C_g \dfrac{1}{\dfrac{1}{r_p} + \dfrac{1}{R_L} + \dfrac{1}{R_g}}}$$

$$[f_2]_{tube} = \frac{1}{2\pi C_g}\frac{1}{r_p} + \frac{1}{R_L} + \frac{1}{R_g}$$

$$\left[\frac{A_{high}}{A_{mid}}\right]_{transistor} = \frac{1}{1 + \dfrac{j\omega C_c r_c R_L R_{ie}}{R_{ie}[R_L + r_c(1 - \alpha)] + R_L r_c(1 - \alpha)}}$$

$$[f_2]_{transistor} \approx \frac{1 - \alpha}{2\pi C_c}\left(\frac{1}{R_L} + \frac{1}{R_{ie}}\right)$$

and

$$\phi_{high} = -\tan^{-1}(f/f_2)$$

Dimensionless curves for these gain ratios and phase responses are plotted in Fig. 13-59.

59. Compensation Techniques. To extend the cutoff frequencies f_1 and f_2 to lower or higher values, respectively, compensation techniques may be employed.

Figure 13-60 illustrates two techniques for low-frequency compensation. If the condition $R_g C_g = C_X R_X R_L /(R_X + R_L)$ is fulfilled (in circuit a or b), the gain relative to the midband gain is:

$$\frac{A_{\text{low}}}{A_{\text{mid}}} = \frac{1}{1 - j(1/\omega R_g C_g)[R_L/(R_L + R_X)]}$$

and

$$f_1 = \frac{1}{2\pi R_g C_g} \frac{R_L}{R_L + R_X}$$

Hence, improved low-frequency response is obtained with increased values of R_X. This value is related to R_L and restricted by active-device operating considerations. Also, R_L is

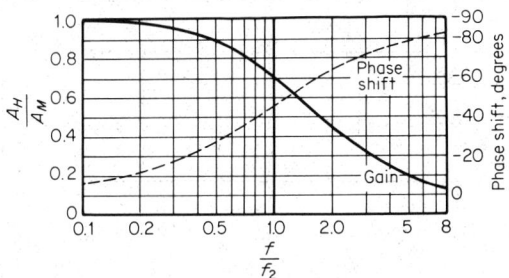

Fig. 13-59. Gain and phase-shift curves at high frequencies.

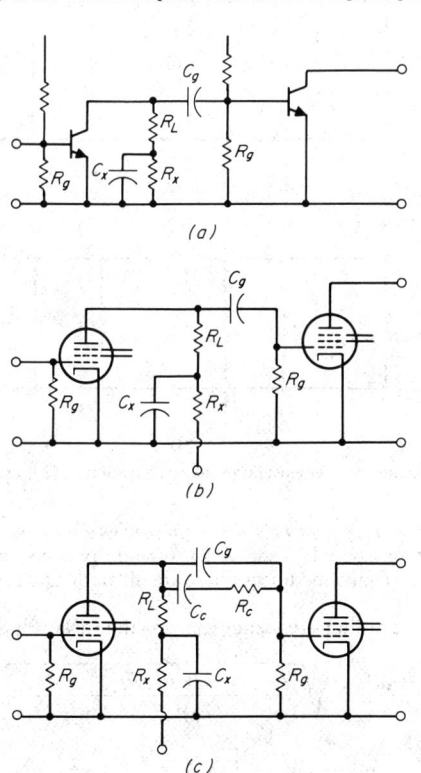

Fig. 13-60. Low-frequency compensation networks. (a) Transistor version; (b,c) tube versions.

dependent on the desired high-frequency response. It can be shown that equality of time constants $R_L C_X = R_g C_g$ will produce zero phase shift in the coupling circuit (for $R_X > 1/\omega C_X$). The circuit shown in Fig. 13-60c is more critical. It is used with element ratios set to $R_L/R_X = R_g/R_c$ and $C_X/C_g = R_c/R_X$.

Various compensation circuits are also available for high-frequency-response extension. Two of the most common, the series- and shunt-compensation cases, are shown in Fig. 13-61. The high-frequency-gain expressions of these configurations can be written

$$\left|\frac{A_{\text{high}}}{A_{\text{mid}}}\right| = \sqrt{\frac{1 + a_1(f/f_2)^2 + a_2(f/f_2)^4 + \cdots}{1 + b_1(f/f_2)^2 + b_2(f/f_2)^4 + b_3(f/f_2)^6 + \cdots}}$$

The coefficients of the terms decrease rapidly for the higher-order terms, so that if $a_1 = b_1$, $a_2 = b_2$, etc., to as high an order of the f/f_2 ratio as possible, a maximally flat response curve is obtained.

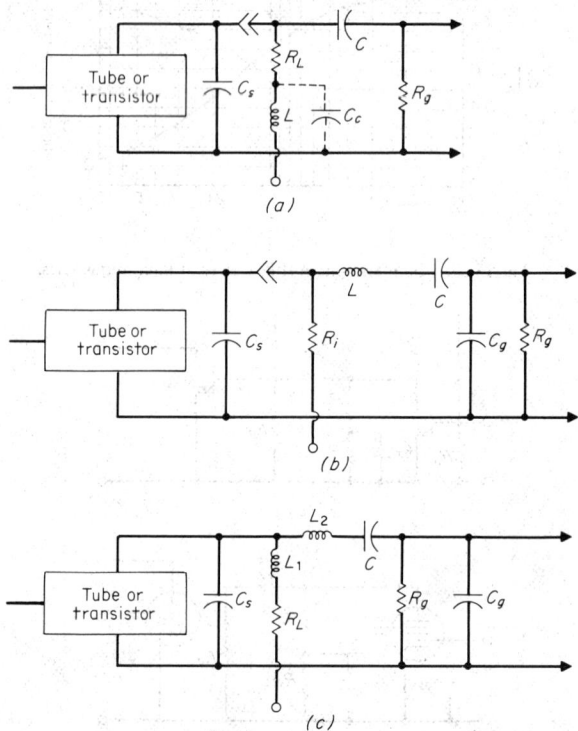

Fig. 13-61. High-frequency compensation schemes. (*a*) Shunt; (*b*) series; (*c*) shunt-series.

For the phase response, $d\phi/d\omega$ may also be expressed as a ratio of two polynomials in f/f_2 and a similar procedure may be followed. A flat time-delay curve results. Unfortunately, the sets of conditions for flat gain and linear phase are different, and compromise values must be used.

Shunt Compensation. The high-frequency gain and time delay for the shunt-compensated stage are

$$\left|\frac{A_{\text{high}}}{A_{\text{mid}}}\right| = \sqrt{\frac{1 + \alpha^2(f/f_2)^2}{1 + (1 - 2\alpha)(f/f_2)^2 + \alpha^2(f/f_2)^4}}$$

$$\phi = -\tan^{-1}\frac{f}{f_2}\left[1 - \alpha + \left(\frac{f}{f_2}\right)^2\alpha^2\right]$$

where

$$\alpha = L/C_g R_L^2 \quad \text{and} \quad R_g \gg R_L$$

Generalized amplitude and phase (delay) response curves for the shunt-compensated amplifiers are shown in Fig. 13-62 for several values of α.

A case when R_g cannot be assumed to be high, such as the input of a following transistor stage, is considerably more complex, depending on the transistor equivalent circuit utilized. This is particularly true when operating near the transistor f_T and/or above the vhf band.

Series-Compensation. In the series-compensated circuit, the ratio of C_s to C_g is an additional parameter. If this can be optimized, the circuit performance is better than in the shunt-compensated case. Typically, however, control of this parameter is not available due to physical and active-device constraints.

The gain and phase response for series compensation are

$$\left| \frac{A_{high}}{A_{mid}} \right| = \sqrt{\cfrac{1}{1 + \left(\frac{f}{f_2}\right)^2 \left[1 - \frac{2LC_g}{R_L^2(C_s + C_g)^2} \right]}}$$

$$+ \left(\frac{f}{f_2}\right)^4 \left[\frac{L^2 C_g^2}{R_L^4(C_s + C_g)^4} - \frac{2 LC_s C_g}{R_L^2(C_s + C_g)^3} \right]$$

$$+ \left(\frac{f}{f_2}\right)^6 \frac{L^2 C_s^2 C_g^2}{R_L^4(C_s + C_g)^6}$$

$$\phi = -\tan^{-1} \frac{f}{f_2} \cfrac{\left(\frac{f}{f_2}\right)^2 \frac{LC_s C_g}{R_L^2(C_s + C_g)^3} - 1}{1 - \left(\frac{f}{f_2}\right)^2 \frac{LC_g}{R_L^2(C_s + C_g)^2}}$$

For maximal flatness, $C_2/(C_1 + C_2) = 0.75 = K_2$ and $L = mR_L^2(C_s + C_g)$, with $m = 0.667$. The maximum improvement in rise time or bandwidth with this technique is about a factor of 4. The response curves for series compensation are given in Fig. 13-63 for several values of m.

These two basic techniques may be combined to improve the response at the expense of complexity. The shunt-series-compensation case (Fig. 13-64) and the so-called "modified" case are examples. The latter involves a capacitance added in shunt with the inductance L or placing L between C_s and R_L.

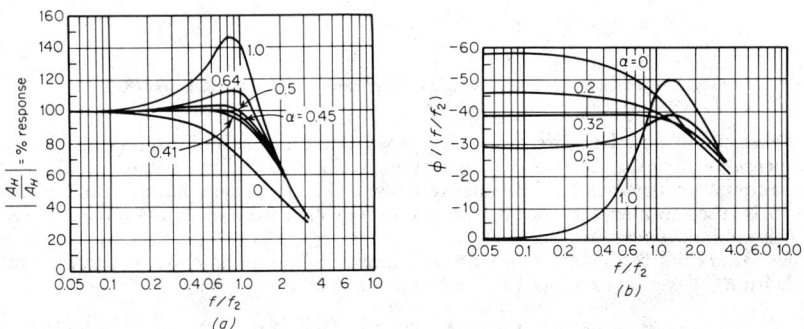

Fig. 13-62. Gain (*a*) and phase-shift (*b*) curves for shunt compensation.

Response curves for the latter case are given in Fig. 13-65. For the modified-shunt case, the added capacitance C_c permits an additional degree of freedom and associated parameter, $K_1 = C_c/C_s$.

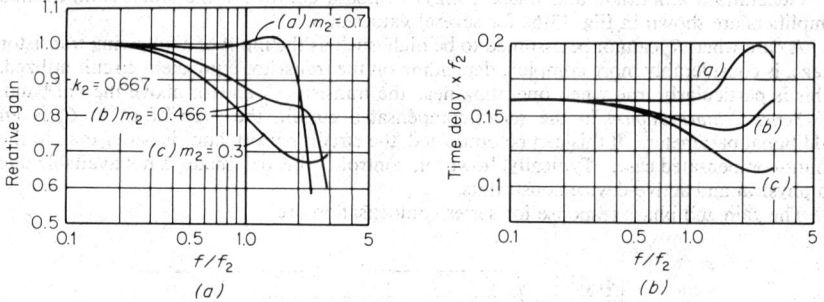

Fig. 13-63. Gain (a) and time-delay (b) curves for series compensation (Ref. 4).

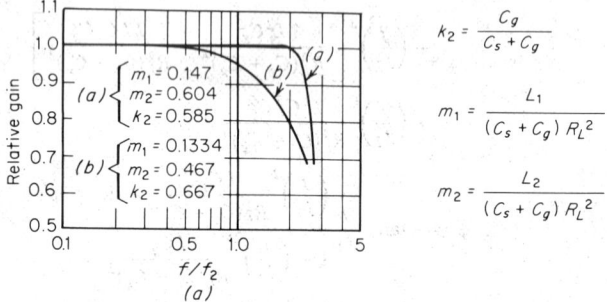

$$k_2 = \frac{C_g}{C_s + C_g}$$

$$m_1 = \frac{L_1}{(C_s + C_g) R_L^2}$$

$$m_2 = \frac{L_2}{(C_s + C_g) R_L^2}$$

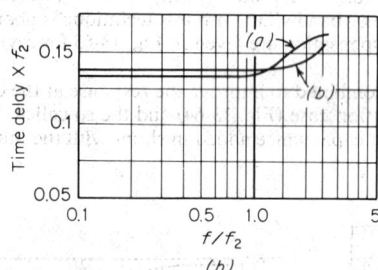

Fig. 13-64. Gain (a) and time delay (b) of shunt-series compensation (Ref. 4).

Other circuit variations exist for specific broadband compensation requirements. Phase compensation, for example, may be necessary as a result of cascading a number of minimum-phase circuits designed for flat frequency response. Circuits such as the lattice and bridged-T can be used to alter the system response by reducing the overshoot without severely increasing the overall rise time.

60. Bandwidth of Cascaded Broadband Stages. When an amplifier is made up of n cascaded RC stages, not necessarily identical, the overall gain A_n may be written

$$\left| \frac{A_n}{A_{\text{mid}}} \right| = \left[\frac{1}{1 + (f/f_a)^2} \right]^{\frac{1}{2}} \left[\frac{1}{1 + (f/f_b)^2} \right]^{\frac{1}{2}} \cdots \left[\frac{1}{1 + (f/f_n)^2} \right]^{\frac{1}{2}}$$

where $f_a, f_b, \ldots f_n$ are the f_1 or f_2 values for the respective stages, depending on whether the overall low- or high-frequency gain ratio is being determined. The phase angle is the sum of the individual phase angles. If the stages are identical, $f_a = f_b = f_x$ for all, and

$$\left| \frac{A_n}{A_{\text{mid}}} \right| = \left[\frac{1}{1 + (f/f_x)^2} \right]^{n/2}$$

Stagger Peaking. The principle of stagger tuning has been well known for some time. A number of individual bandpass amplifier stages are cascaded with frequencies skewed according to some predetermined criteria. The most straightforward is with the center frequencies adjusted so that the f_2 of one stage coincides with the f_1 of the succeeding stage, etc. The overall gain-bandwidth then becomes

$$(G \cdot BW)_n = \sum_{n=1}^{N} (G \cdot BW)_n$$

A significant simplifying criterion of this technique is stage isolation. Isolation, in transistor stages particularly, is not generally high, except at low frequencies. Hence the overall design equations and subsequent overall alignment can be significantly complicated due to the interactions. Complex device models and computer-aided design greatly facilitate the implementation of this type of compensation. The simple shunt-compensated stage has found extensive use in stagger-tuned pulse-amplifier applications.

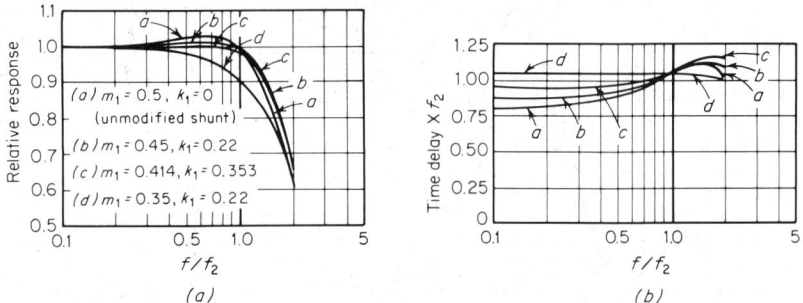

Fig. 13-65. Gain (a) and time delay (b) of modified shunt compensation (Ref. 4).

61. Transient Response. Time-domain analysis is particularly useful for broadband applications. Extensive theoretical studies have been made of the separate effects of nonlinearities of amplitude and phase response.[1-3] These effects can be investigated starting with a normalized low-pass response function.

$$\frac{A(jw)}{A(0)} = \exp(a^m w^m - jb^n w^n)$$

where a and m = constants describing the amplitude-frequency response, and b and n = constants describing the phase-frequency response. Figure 13-66 illustrates the time response to an impulse and a unit step forcing function for various values of m, with $n = 0$. Rapid change of amplitude with frequency (large m) results in overshoot. Nonzero, but linear, phase-frequency characteristics ($n = 1$) result in a delay of these responses, without introducing distortion. Further increase in n results in increased ringing and asymmetry of the time function.

An empirical relationship between rise time (10 to 90%) and bandwidth (3 dB) can be expressed as

$$(t_r)BW = K$$

where K varies from about 0.35 for circuits with little or no overshoot to 0.45 for circuits with about 5% overshoot. K is 0.51 for the ideal rectangular low-pass response with 9% overshoot; for the gaussian amplitude response with no overshoot, $K = 0.41$.

The effect on rise time of cascading a number of networks (n) depends on the individual network pole-zero configurations. Some general rules are:

(a) For individual circuits having little or no overshoot, the overall rise time is

$$t_{rt} = (t_{r1}^2 + t_{r2}^2 + t_{r3}^2 + \ldots)^{\frac{1}{2}}$$

(b) If $t_{r1} = t_{r2} = t_{rm}$,

$$t_{rt} = 1.1 \sqrt{n}\, t_{r1}$$

(c) For individual stage overshoots of 5 to 10%, total overshoot increases as $\sqrt{n}$.

(d) For circuits with low overshoot ($\sim 1\%$), the total overshoot is essentially that of one stage.

The effect of insufficient low-frequency response of an amplifier is sag of the time response. A small amount of sag ($< 10\%$) can be described by the formula

$$\frac{E_{sag}}{E_{total}} = \frac{T}{2.75\,RC}$$

where T is one-half period of a square-wave voltage.

Figure 13-67 illustrates the response to a step function of n capacitor-coupled stages with the same time constant. The effect of arithmetic addition of individual stage initial slopes determining the net total initial transition slope can be seen.

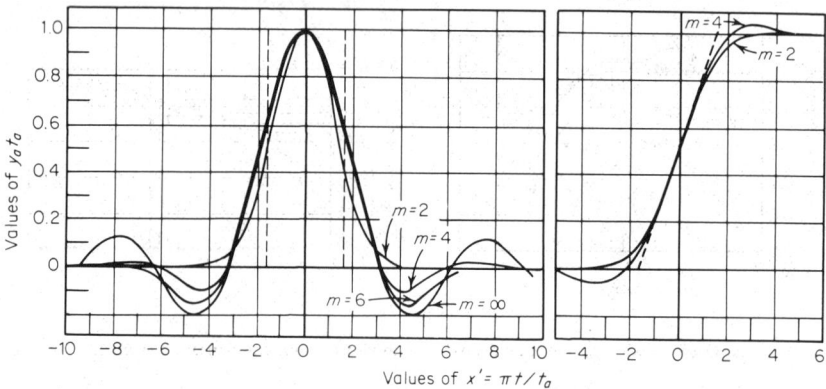

Fig. 13-66.   Transient responses to unit impulse (left) and unit step (right) for various values of m (Ref. 2).

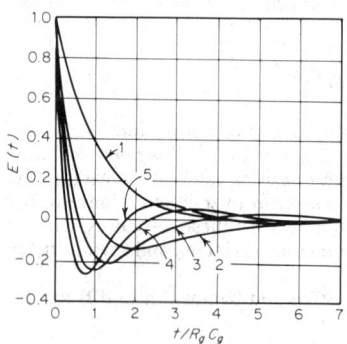

Fig. 13-67. Response to a unit step of n capacitively coupled stages of the same time constant (Ref. 5).

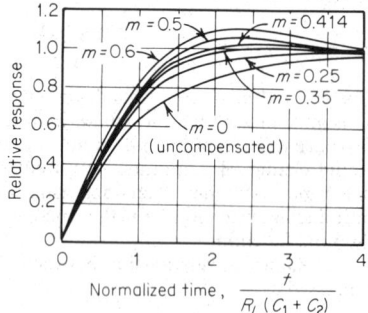

Fig. 13-68. Response to a unit step of a shunt-compensated stage (Ref. 4).

Figure 13-68 shows the effect of high-frequency shunt compensation on the time response to a step function. The series peaked time response is shown in Fig. 13-69. Similar curves for the many other compensation circuits and variations are given in the literature.[4-6]

62. Distributed Amplifiers. This technique is useful for operation of devices at or near the high-frequency limitation of gain-bandwidth. While the individual stage gain is low, the stages are cascaded so that the gain response is additive instead of multiplicative. The basic principle is to allow the input and output capacitances to form the shunt elements of two delay lines. This is shown in Fig. 13-70.

If the delay times per section in the two lines are the same, the output signals traveling forward add together without relative delay. Care must be taken to ensure proper terminating conditions. The gain produced by n tubes, each of mutual conductance g_m, has a value of $G = ng_m Z_{02}/2$. Performance to very low frequencies may be achieved. The high-frequency limit is determined by the cutoff frequencies of the input and output lines or effects within the active devices themselves other than parasitic shunt capacities (e.g., transit time or alpha-fall-off effects).

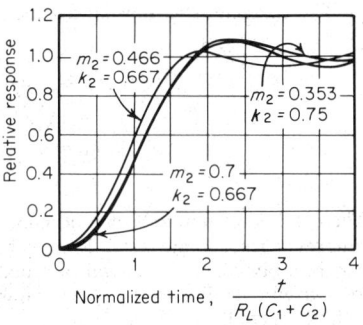

Fig. 13-69. Response to a unit step of a series-compensated stage (Ref. 4).

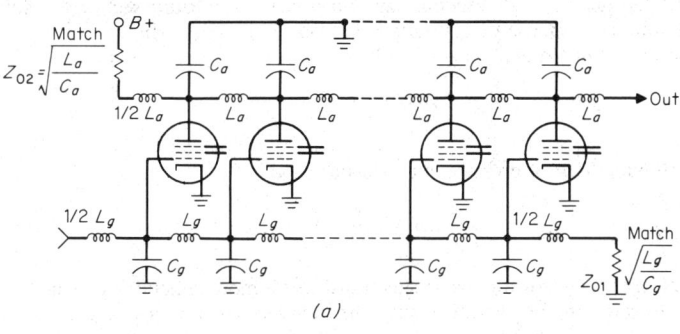

(a)

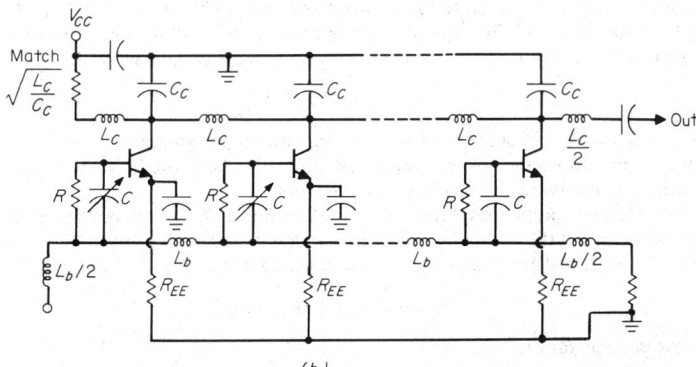

(b)

Fig. 13-70. Distributed amplifier circuits. (a) Tube version; (b) transistor version.

For input and output lines of different characteristic impedances, the overall gain is given by

$$G = \frac{2Z_{01}}{Z_{01} + Z_{02}} \frac{n g_m Z_{02}}{2}$$

The characteristic impedances and cutoff frequencies are given as

$$Z_{01} = \frac{L_g}{\sqrt{C_g}} \qquad Z_{02} = \frac{L_a}{\sqrt{C_a}}$$

$$f_{c1} = \pi \frac{1}{\sqrt{L_g C_g}} \qquad f_{c2} = \pi \frac{1}{\sqrt{L_a C_a}}$$

There does not seem to be an optimum choice for Z_{02}; a device with a high figure of merit is simply chosen, and the rest follows. There is, however, an optimum way in which N devices may be grouped, in m identical cascaded stages, each with n devices, so that N is a minimum for a given overall gain A. N is a minimum when $m = \ln A$. Consequently, the optimum gain per stage $G = e = 2.72$.

Various techniques are utilized to determine the characteristics of the lumped transmission lines. Constant-K and m-derived filter sections are most common, augmented by several compensation circuit variations.[7,8] The latter include paired grid-plate (base-collector), bridged-T, and resistive loading connections. The constant-K lumped line has several limitations, including the fact that the termination is not a constant resistance and that impedance increases with frequency and time delay also changes with frequency.

m-derived terminating half sections and time-delay equalizing sections[9] result in a frequency-amplitude response that is quite flat at very high frequencies.

The effects of input loading and/or line loss modify the gain expression to

$$G = \frac{2(GB)}{f_c} \frac{e^{-\alpha}(1 - e^{-n\alpha})}{1 - e^{-\alpha}}$$

where α = the real part of propagation constant γ and

$$GB = \frac{g_m}{2\pi C_{eff}}\bigg|_{tube}$$

The design of distributed amplifiers using transistors is more difficult due to the low input impedance. In addition, the intrinsic gain of the transistor (β or α) is a decreasing function of frequency.

The simplest approach in overcoming this problem is to connect a parallel RC in series with the base, as shown in Fig. 13-70b. By setting $RC = \beta_0/2\pi f_T$, the gain is essentially independent of frequency up to about $f_T/2$, with an overall voltage gain of

$$A_v = (n\beta_0 Z_0/2R)^m$$

where m = the number of stages and n = the transistors per stage.

The increasing impedance of the constant-K line helps keep the frequency response flat, compensating for the increased loading of the transistor.

Another transistor approach[10] involves the division of the frequency range into three regions to account for the losses. Each region is associated with a linear-Q variation with frequency of the transistor input/output circuits, approximated by shunt RC elements.

$$Q_{in} = w(C_{in} + C)/(G_{in} + G_k)$$

The lossless voltage gain is

$$A_L = f_T Z_0 / [2f(1 - (f/f_c)^2)^{\frac{1}{2}}]$$

In region I, the input impedance is high and the transistor supplies high voltage gain. In regions II and III, the voltage gains are given by

$$A_{II} = A_L n\{1 - nK_{11} Z_0 (f/f_c)^2/[4r_b'(1 - (f/f_c)^2)^{\frac{1}{2}}]\}$$

$$A_{III} = A_L n\{1 - nK_{111} Z_0/[4r_b'(1 - (f/f_c)^2)^{\frac{1}{2}}]\}$$

where $K_{11} = (f_c/f_Q)^2$ $K_{111} = 1$, f_Q = frequency of min. Q

63. Broadband Matching Circuits. Broadband impedance transformation interstage-coupling matching can be achieved with balun transformers,[11] quarter-wave transmission-line sections,[12] lumped reactances in configurations other than discussed above, and short lengths of transmission lines.[13]

Balun Transformers. In conventional coupling transformers, the interwinding capacitance resonates with the leakage inductance, producing a loss peak. This limits the high-frequency response. A solution is to utilize transmission-line transformers in which the turns are arranged physically to include the interwinding capacitance as a component of the characteristic impedance of a transmission line. With this technique, bandwidths of hundreds of megahertz can be achieved. Good coupling can be realized without resonances, leading to the use of these transformers in power dividers, couplers, hybrids, etc.

Typically, the lines take the form of twisted wire pairs, although coaxial lines can also be used. In some configurations, the length of the line determines the upper cutoff frequency. The low-frequency limit is determined by the primary inductance. The larger the core permeability, the fewer the turns required for a given low-frequency response. Ferrite toroids have been found to be satisfactory with widely varying core-material characteristics. The decreasing permeability with increasing frequency is offset by the increasing reactance of the wire itself, causing a wide-band, flat-frequency response.

Quarter-Wave Transformers. The quarter-wavelength line transformer is another well-known element. It is simply a transmission line one-quarter-wavelength long, with a characteristic impedance

$$Z_{line} = \sqrt{Z_{in} \cdot Z_{out}}$$

where Z_{in} and Z_{out} are the terminating impedances. The insertion loss of this line section is

$$10 \log\left[1 + \frac{(r - 1)^2}{4r} \cos^2\theta\right] \quad (dB)$$

where

$$r = Z_{in}/Z_{out} \quad \text{and} \quad \theta = \frac{2\pi L}{\lambda} = 90° \quad \text{at } f_0$$

Figure 13-71 shows the bandwidth performance of the quarter-wave line for several matching ratios. Such lines may be cascaded to achieve broader-band matching by reducing the matching ratio required of each individual section.

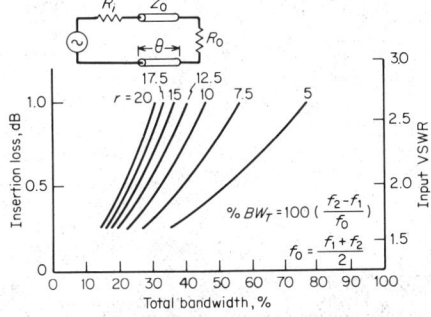

Fig. 13-71. Bandwidth of a quarter-wave matching transformer.

Lumped and Pseudo-lumped Transformations. The familiar lumped-element ladder network, depending on the element values and realization, approximates a short-step transmission-line transformer or a tapered transmission line.[14] Convenient tables of element values as a function of bandwidth, ripple, transformation ratio, and number of elements are available.[15, 16]

64. Feedback and Gain Compensation. *Feedback.* The bandwidth of amplifiers can be increased by the application of inverse feedback. A multiplicity of feedback combinations and analysis techniques are available and extensively treated in the literature.[18-20] In addition, the related concept of feed-forward has been investigated and successfully implemented.[21] Figure 13-72 shows four feed-back arrangements and formulas describing their performance. A major consideration, particularly for rf applications, is the control of impedance and loop-delay characteristics to insure stability.

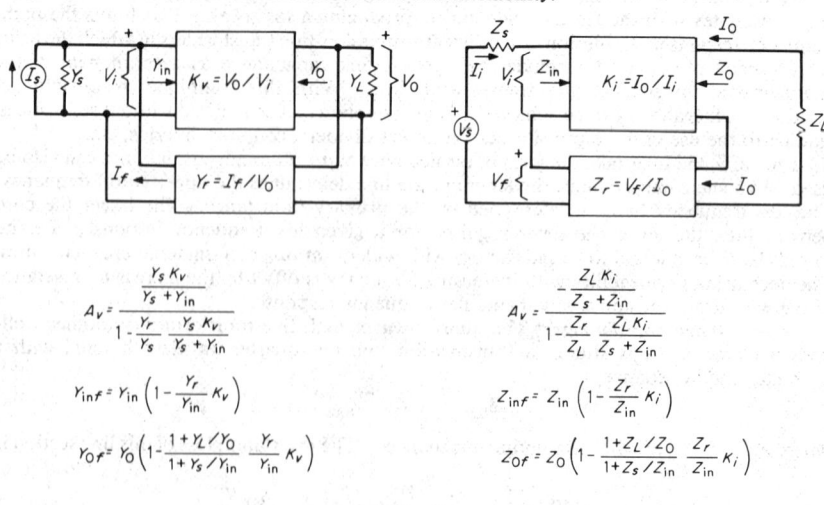

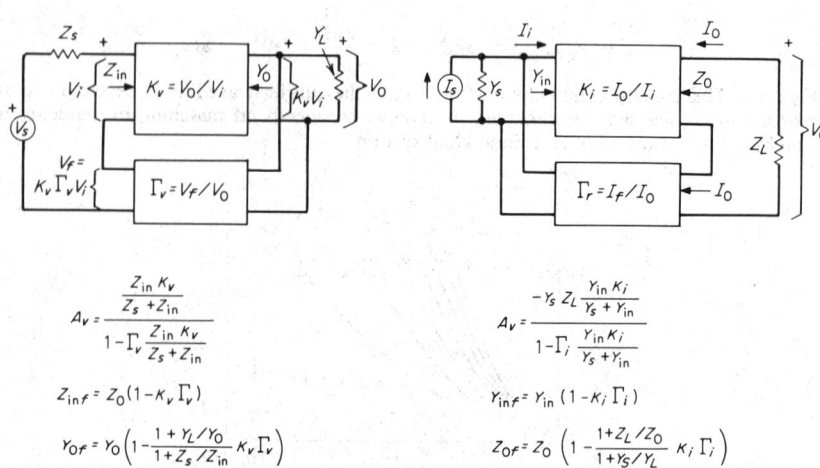

Fig. 13-72. Four methods of applying feedback, showing influence on gain and input and output immittances. *(From Hakim, "Junction Transistor Circuit Analysis," by permission of John Wiley & Sons, Inc., New York)*

Gain Compensation. The power gain of a transistor amplifier typically falls 6dB/octave with increasing frequency above the f_β value. The gain can be leveled (bandwidth-widened) by exact matching of the source impedance to the device at the upper frequency only, causing increasing mismatch with decreasing frequency and the associated gain loss. The overall flat gain is the value at the high best-match frequency. Sufficient resistive loss is usually required in this interstage to prevent instabilities in either driver or driven stage due to the mismatch unloading.

65. Power-combining Amplifiers. Many circuit techniques have been developed to obtain relatively high-output powers with given modest-power devices. Two approaches are the direct-paralleling and hybrid splitting/combining techniques. The direct-paralleling approach is limited by device-to-device variations and the difficulties in providing balanced conditions due to the physical wavelength restrictions. A technique commonly used at uhf and microwave frequencies to obtain multi-octave response incorporates matched stages driven from hybrid couplers in a balanced configuration. The coupler offers a constant-source impedance (equal to the driving-port impedance) to the two stages connected to its output ports. Power reflected due to the identical mismatches is coupled to the difference port of the hybrid and dissipated in the idler load. The hybrid splitting/combining approach has proved quite effective in implementing high-output-level requirements (e.g., kilowatts at 1.4 GHz with transistors).

The hybrid splitting/combining approach enhances circuit operation. In particular, quadrature hybrids effect a VSWR-canceling phenomenon which results in extremely well matched power-amplifier inputs and outputs that can be broadbanded upon proper selection of particular hybrid types. Also, the excellent isolation between devices enables reliable power-amplifier service.

Several hybrid/directional-coupler configurations are possible, including the split-tee, branch-line, magic-T, backward-wave, lumped, etc. Important factors in the choice of hybrid for this application are coupling bandwidth, isolation, fabrication ease, and form. The equiamplitude, quadrature-phase, reverse-coupled TEM $\lambda/4$ coupler is a particularly attractive implementation due to its bandwidth and amenability to strip-transmission-line circuits. Figure 13-73 illustrates the coupler type and balanced configuration.

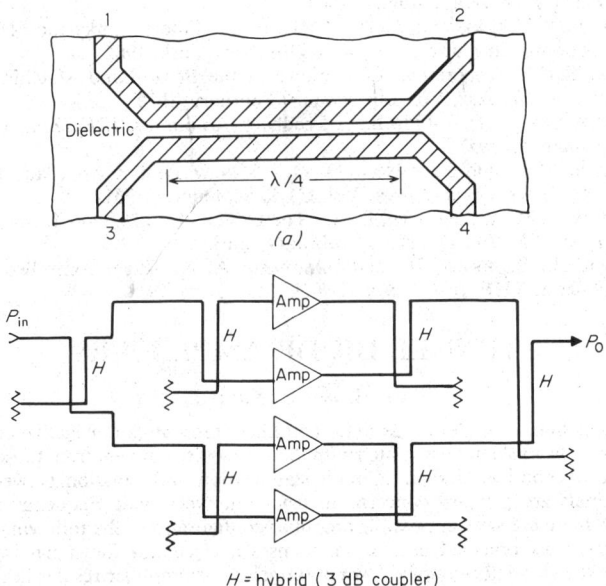

H = hybrid (3 dB coupler)

(b)

Fig. 13-73. Wide-band combining techniques. (*a*) Wide-band quarter-wave coupler; (*b*) balanced combining amplifier configuration.

66. References on Broadband Amplifiers.

1. WHEELER, H. A.　The Interpretation of Amplitude and Phase Distortion in Terms of Paired Echoes, *Proc. IRE,* Vol. 27, June, 1939.

2. DiTORO, M. J.　Phase and Amplitude Distortion in Linear Networks, *Proc. IRE,* Vol. 36, January, 1948.

3. BANGERT, J. T.　Practical Applications of Time Domain Theory, *IRE WESCON Conv. Record,* 1959, Pt. 3.

4. GLASFORD, G. M.　"Fundamentals of Television Engineering," McGraw-Hill, New York, 1955.

5. VALLEY, G. E., JR., and WALLMAN, H.　"Vacuum Tube Amplifiers," McGraw-Hill, New York, 1948.

6. MULLER, F. A.　High Frequency Compensation of RC Amplifiers, *Proc. IRE,* August, 1954.

7. LEWIS, I. A. D., and WELLS, F. H.　"Millimicrosecond Pulse Techniques," Pergamon, New York, 1959.

8. MOORE, A. D.　Synthesis of Distributed Amplifiers for Prescribed Amplitude Response, *Stanford Univ. Tech. Rept.* 53, August, 1952.

9. GINZTON, E. L., HEWLETT, W. R., JASBERG, J. H., and NOE, J. D.　Distributed Amplification, *Proc. IRE,* August, 1948.

10. ROESHOT, L. F.　UHF Broadband Transistor Amplifiers, *Elec. Design News,* January, 1963–March, 1963.

11. RUTHROFF, C. L.　Some Broad-band Transformers, *Proc. IRE,* August, 1959, pp. 1337–1342.

12. YOUNG, LEO　Tables of Cascaded Homogeneous Quarter Wave Transformers, *IRE Trans. GMTT,* Vol. MTT-7, April, 1959.

13. COHN, S. B.　Optimum Design of Stepped Transmission Line Transformers, *IRE Trans. GMTT,* Vol. MTT-3, April, 1955.

14. WOMACK, C. P.　The Use of Exponential Transmission Lines in Microwave Components, *IRE Trans.,* Vol. MTT-10, March, 1962.

15. MATTHAEI, G. L.　Tables of Chebyshev Impedance Transforming Networks of Low Pass Filter Form, *Proc. IEEE,* August, 1964.

16. MATTHAEI, YOUNG, and JONES　"Microwave Filters, Impedance Matching Networks, and Coupling Structures," McGraw-Hill, New York, 1964.

17. FANO, R. M.　Theoretical Limitations on the Broad-Band Matching of Abitrary Impedance, *J. Franklin Inst.,* Vol. 249, January-February, 1950.

18. WALDHAUER, F. D.　Wide Band Feedback Amplifiers, *IRE Trans. Circuit Theory,* Vol. CT-4, September, 1957.

19. GHAUSI, M.S., and PEDERSON, D. O.　A New Design Approach for Feedback Amplifiers, *IRE Trans. Circuit Theory,* Vol. CT-9, September, 1961.

20. CHERRY, E. M., and HOOPER, D. E.　The Design of Wide-Band Transistor Feedback Amplifiers, *Proc. IEE,* Vol. 110, No. 2, February, 1963.

21. SEIDEL, H., BEURRIER, H., and FRIEDMAN, A. N.　Error-controlled High Power Linear Amplifier at VHF, *Bell System Tech. J.,* May–June, 1968.

TUNNEL DIODE AMPLIFIERS

By C. S. KIM

67. Introduction.　Tunnel diode (TD) amplifiers are one-port negative-conductive devices.[1]* Hence the problems associated with them are quite different from those encountered in conventional amplifier design. Circuit stabilization and isolation between input and output terminals are primary concerns in using this very wide frequency range device.

Although there are several possible amplifier configurations, the following discussion is limited to the most practical design, which uses a circulator for signal isolation. The advantage of the circulator-coupled form of tunnel diode amplifier resides in the fact that it is thereby possible to convert a bilateral one-port amplifier into an ideal unilateral two-port amplifier. Tunnel diodes can provide amplification at microwave frequencies with a relatively simple structure and at a low noise figure.

---

*Superior numbers correspond to numbered references, Par. **13-72.**

68. Tunnel Diodes. Three kinds of tunnel diodes are available, namely, those using Ge, GaAs, and GaSb. V-I characteristics and corresponding small-signal conductance-voltage relationships are shown in Figs. 13-74 and 13-75, respectively. A typical small-signal tunnel diode equivalent circuit[2] is shown in Fig. 13-76. Here g_j, C_j, r_s, L, and C_p are the small-signal conductance, junction capacitance, series resistance, series inductance, and shunt capacitance. Noise generators e_s and i_j are included for subsequent discussion.

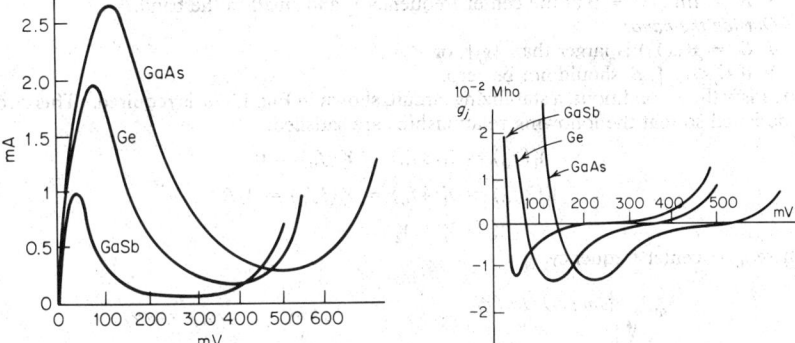

Fig. 13-74. Characteristics of tunnel diodes.

Fig. 13-75. Voltage versus g_j characteristics of tunnel diodes of Fig. 13-74.

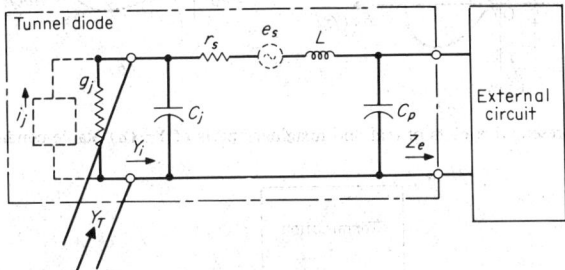

Fig. 13-76. Small-signal equivalent circuit of tunnel diode amplifier.

69. Stability. Several studies of the stability conditions of tunnel diode amplifiers have been reported.[3,4] The stability criteria are derived from the immittance expression across the diode terminals and are quite complicated.

Using the short-circuit stable condition,[5] the stability criteria can be simplified considerably. With reference to Fig. 13-76, with the external circuit connected, the total admittance Y_T across g_j can be expressed as

$$Y_T(s) = Y_i(s) - g_j = p(s)/q(s)$$

where $Y_i(s)$, the admittance facing g_j, is a positive real function.[6] Since Y_i is connected in parallel with a short-circuit stable device with negative conductance g_i, Y_T is short-circuit stable.[5,7] This implies that, since $q(s)$ is always a Hurwitz polynomial, the stability condition is that $p(s)$ must also be a Hurwitz polynomial.[6]

A simple graphical interpretation of this stability condition is as follows: The plot* of $Y_T(\omega)$ can be obtained from the plot of $Y_i(\omega)$ by shifting the imaginary axis by $|g_j|$ along the real axis, as shown in Fig. 13-77. Since $q(s)$ has no roots in the right half plane, any encirclement of the origin of $Y_T(\omega)$ must come from the right-half-plane roots of $p(s)$ only. Therefore, the circuit will be stable if, and only if, $Y_T(\omega)$ does not encircle the origin (Fig. 13-77a). If g_j becomes large so that the origin is encircled by $Y_T(\omega)$ (Fig. 13-77b) the circuit is unstable.

---

* Here, the plot of $Y_T(\omega)$ represents the case for $Z_e(\omega)$ short-circuited. However, similar plots can be obtained for more general cases.

70. Tunnel Diode Amplifier Design. A simplified block diagram of a circulator-coupled TD amplifier is shown in Fig. 13-78. The three basic circuit parts are the tunnel diode, the stabilizing circuit, and the tuning circuit, which includes a four-port circulator. The following conditions are imposed on the amplifier design:

In the band:
1. $G_i = \text{Re}(Y_i)$ is slightly larger than $|g_j|$.
2. G_i is contributed by the tuning circuit only.
3. $B_i = \text{Im}(Y_i) = 0$ at the center frequency f_0 and small in the band.

Outside the band:
4. $G_i = \text{Re}(Y_i)$ is larger than $|g_j|$, or
5. If $G_i \leq |g_j|$, B_i should not be zero.

To satisfy these conditions, a stabilizing circuit, shown in Fig. 13-78, is required. This circuit is designed so that the following relationships are satisfied:

$$Y_1(f_0) = Y_1(3f_0) = Y_2(f_0) = 0$$

$$Y_1(2f_0) = Y_1(4f_0) = Y_2(3f_0) = 1/R$$

$$Y_s = Y_1 + Y_2$$

where f_0 = center frequency.

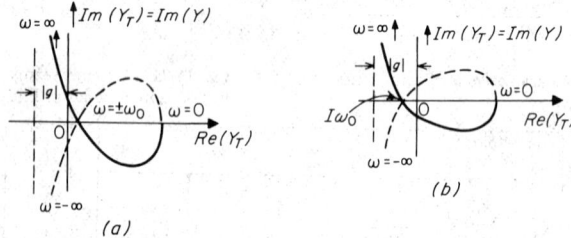

(a)

(b)

Fig. 13-77. Representative plots of real and imaginary parts of Y_T. (*a*) Stable condition; (*b*) unstable condition.

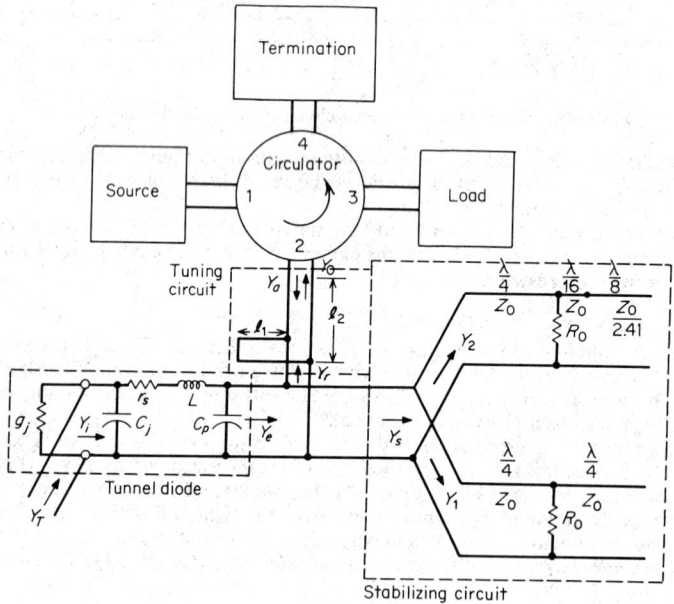

Fig. 13-78. Tunnel diode amplifier using circulator.

The equivalent circuit of Fig. 13-76 can be transformed into the parallel equivalent circuit of Fig. 13-79. The following identities relate the parameters of Figs. 13-76 and 13-79:

$$g_{jp} = 1/|Z_s|^2 g_j X$$

$$g_{sp} = r_s/|Z_s|^2$$

$$B = \omega C_p - \left[\omega L - (1 - \tfrac{1}{X})\frac{1}{\omega C_j}\right]/|Z_s|^2$$

where $X = 1 + \omega^2 c_j^2/g_j^2$

$$|Z_s|^2 = \gamma_s - 1/|g_j|X + [\omega L - (1 - 1/X)1/\omega C_j]^2$$

The gain can be expressed by

$$|\Gamma|^2 = \left|\frac{Y_e - Y_d}{Y_e + Y_d}\right|^2$$

$$|\Gamma|^2_{f=f_0} = \left|\frac{G_e - g_{sp} + |g_{jp}|}{G_e + g_{sp} - |g_{jp}|}\right|^2$$

where $B_e - B = 0$ at $f = f_0$.

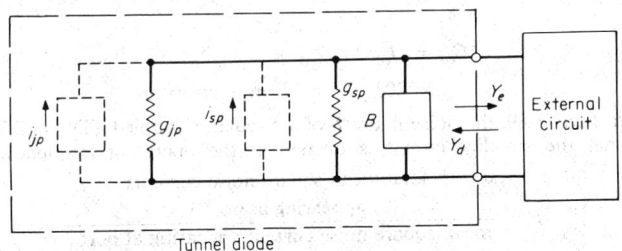

Fig. 13-79. Parallel version of equivalent circuit of Fig. 13-76.

In this case, the tuning circuit is a more general matching circuit using combinations of parallel and series transmission lines. It should be noted that $|\Gamma|$ becomes 1, or $g_{sp} = |g_{jp}|$, as the operating frequency increases to f_r, the resistive cutoff frequency, which is defined by

$$f_r = \frac{|g_j|}{2\pi C_j}\sqrt{1 - \frac{1}{r_s|g_j|}}$$

It is desirable to have a device with f_r several times (at least three) larger than f_0. Furthermore, it is desirable to make the self-resonance frequency f_x

$$f_x = \frac{1}{2\pi}\sqrt{\frac{1}{LC_j} - \left(\frac{g_j}{C_j}\right)^2}$$

as high as possible (higher than f_c) to improve the stability margin.

The gain expression can be modified to include the input amplifier admittance Y_a of Fig. 13-78 as follows:

$$|\Gamma|^2 = \left|\frac{Y_0 - Y_a}{Y_0 + Y_a}\right|^2$$

Similarly, the bandwidth can be determined from the expression for Γ^2 and f_x, above. Typical germanium tunnel diode parameters pertinent to S, C, and X-band amplifiers with approximate gains of 10 dB are shown in Table 13-6.

Table 13-6.   Typical Germanium Tunnel Diode Parameters

			S-band	C-band	X-band
$\lvert r_j \rvert = 1/g_j$	Ω		70	70	70
C_j	pF	$<$	1	0.5	0.2
L	H	$<$	0.1	0.1	0.1
r_s	Ω	$<$	$0.1 \cdot \lvert r_j \rvert$	$0.1 \cdot \lvert r_j \rvert$	$0.1 \cdot \lvert r_j \rvert$

71. Noise Figure in TD Amplifiers. A noise-equivalent circuit can be completed[9] by inserting a current generator, $\overline{i_f^2}$ and a voltage generator $\overline{e_s^2}$ as shown in Fig. 13-76 (dotted lines). The mean-square values are determined from

$$\overline{i_j^2} = 2eI_{eq}\Delta f$$

$$\overline{e_s^2} = 2KTr_s\Delta f$$

where I_{eq} = equivalent shot noise current = I_{dc} = dc current in the negative-conductance region.

The noise-equivalent circuit of Fig. 13-76 can be transformed into a parallel equivalent circuit of Fig. 13-79 having current generators $\overline{i_{jp}^2}$ and $\overline{i_{sp}^2}$ (dotted lines). $\overline{i_{jp}^2}$ and $\overline{i_{sp}^2}$ can be derived from the two equivalent circuits to be

$$\overline{i_{jp}^2} = 4KT(G_{eq}/\,\lvert g_j \rvert)\,\lvert g_{jp} \rvert \Delta f$$

$$\overline{i_{sp}^2} = 4KTg_{sp}\,\Delta f$$

where

$$G_{eq} = eI_{eq}/2KT$$

$$= 20I_{eq}\qquad \text{at room temperature}$$

Referring to Fig. 13-80, the noise figure F of a circulator-coupled TD amplifier is derived, assuming that the stability circuit is opened at the operating frequency, as follows:

$$F = \frac{\text{total mean-square noise currents}}{\text{mean-square noise current appearing at port 3}}$$
$$\text{appearing at port 3}$$
$$\text{contributed by the source}$$

$$= \frac{\overline{i_s^2}\lvert\Gamma\rvert^2 + \overline{i_2^2}\lvert\Gamma - 1\rvert^2}{\overline{i_s^2}\lvert\Gamma\rvert^2}\left(1 + \frac{\overline{i_2^2}\lvert\Gamma - 1\rvert^2}{\overline{i_s^2}\lvert\Gamma\rvert^2}\right)$$

where

$$\overline{i_s^2} = 4KT_0 G_0\Delta f$$

$$\overline{i_2^2} = \overline{i_{sp}^2} + \overline{i_{jp}^2}$$

$$\Gamma = \frac{Y_0 - Y_a}{Y_0 + Y_a}$$

If $\lvert\Gamma\rvert$ approaches infinity, F becomes

$$F = 1 + \frac{\overline{i_2^2}}{\overline{i_s^2}} = 1 + \frac{(g_{sp} + G_{eq}\lvert g_{jp} \rvert\,/\,\lvert g_j \rvert)T}{G_s T_0}$$

$$= \frac{1 + G_{eq}/\lvert g_j \rvert}{(1 - r_s\lvert g_j \rvert)\,[1 - (f/f_r)^2]}$$

For low F_0, $r_s\lvert g_j \rvert$, f/f_r, and $G_{eq}/\lvert g_j \rvert$ should be made small. If I_{eq} could be approximated by I_{dc} in the negative conductive region, $I_{dc}/\lvert g_j \rvert$ would be determined by the material used. $I_{dc}/\lvert g_j \rvert$ is minimum at the voltage where $\lvert g_j \rvert$ becomes maximum.

A typical $r_s|g_j|$ value is 0.1. Therefore, if $f/f_r = 1/3$, then for $I_{eq}/|g_j| = 0.06$(Ge) and $I_{eq}/|g_j| = 0.04$ (GaSb), F_0 becomes, in each case,

$$F_0 = \frac{1 + 20 \times 0.06}{(1 - 0.1)\left(1 - \dfrac{1}{3^2}\right)} = 4.4 \text{ dB for Ge}$$

$$F_0 = \frac{1 + 20 \times 0.04}{(1 - 0.1)\left(1 - \dfrac{1}{3^2}\right)} = 3.5 \text{ dB for GaSb}$$

Tunnel diodes provide good noise performance with a relatively simple amplifier structure requiring only a dc source for bias. It is therefore a useful low-noise, small-signal amplifier for microwave applications.

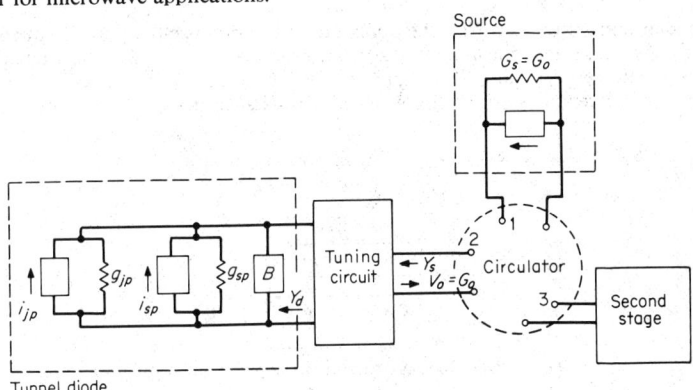

Fig. 13-80. Noise equivalent circuit of circulator-coupled tunnel diode amplifier.

72. References on Tunnel Diode Amplifiers.

1. SHEA, R. F. "Amplifier Handbook," Chap. 12, McGraw-Hill, New York, 1968.

2. KIM, C. S., and LEE, C. W. Microwave Measurements on Tunnel Diode Parameters, *Microwave*, November, 1964.

3. SNIDEN, L. I., and YONTA, D. C. Stability Criteria for Tunnel Diodes, *Proc. IRE*, Vol. 49, pp. 1206–1207, July, 1961.

4. HINES, M. E. High-Frequency Negative-Resistance Circuit Principles for Esaki Diodes Application, *Bell System Tech. J.*, Vol. 39, May, 1960.

5. BODE, H. W. "Network Analysis and Feedback Amplifier Design," Van Nostrand, Princeton, N.J., 1945.

6. GUILLEMIN, E. A. "Synthesis of Passive Networks," Wiley, New York, 1957.

7. KIM, C. S., and BRANDLI, A. High Frequency High Power Operation of Tunnel Diodes, *IRE Trans. Circuit Theory*, December, 1962, pp. 416–426.

8. KUH, E. S., and PATTERSON, J. D. "Design Theory of Optimum Negative Resistance Amplifier," University of California, Electronics Research Laboratory, December 6, 1960.

9. TIEMANN, J. J. Shot Noise in Tunnel Diode Amplifier, *Proc. IRE*, Vol. 48, No. 8, p. 1418, August, 1960.

PARAMETRIC AMPLIFIERS

By C. E. NELSON

73. Introduction. The term parametric amplifier (paramp) refers to an amplifier (with or without frequency conversion) utilizing a nonlinear or time-varying reactance. Development of low-loss variable-capacitance (varactor) diodes resulted in the development of varactor-diode parametric amplifiers with low noise figure in the microwave frequency region. Types of paramps include one-port, two-port (traveling-wave), degenerate (pump frequency twice the signal frequency), nondegenerate, multiple pumps, and multiple idlers. The most widely used amplifier is the nondegenerate one-port paramp with a circulator, because it achieves very good noise figures without undue circuit complexity.

74. One-Port Paramp with Circulator. The one-port paramp with circulator is illustrated in Fig. 13-81, and the simplified circuit diagram is shown in Fig. 13-82. The input signal and the amplified output signal (at the same frequency) are separated by the circulator. The cw pump source is coupled to a back-biased varactor to drive the nonlinear junction capacitance at the pump frequency. The signal and pump currents mix in the nonlinear varactor to produce voltages at many frequencies. The additional (idler) filter allows only the difference or idler current to flow at the idler frequency; i.e.,

$$f_i = f_p - f_s$$

The idler current remixes with the pump current to produce the signal frequency again. The phasing of this signal due to nonlinear reactance mixing is such that the original incident signal is reenforced (i.e., amplified) and reflected back to the circulator. The one-port paramp at band center is essentially a negative-resistance device at the signal frequency.

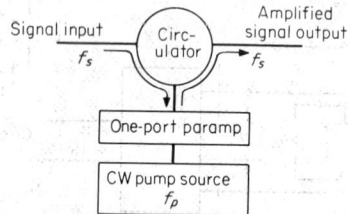

Fig. 13-81.   One-port parametric amplifier and circulator.

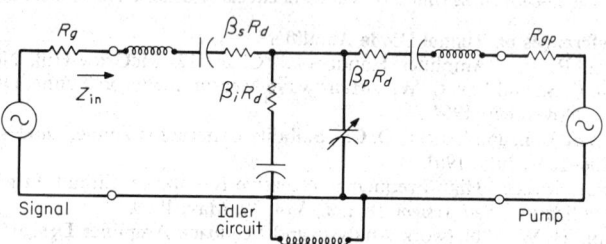

Fig. 13-82.   Circuit of one-port paramp using single-tuned resonators.

75. Power Gain and Impedance Effects. The one-port paramp power gain at the signal frequency is

$$G = \left| \frac{Z_{\text{in}} - R_g}{Z_{\text{in}} + R_g} \right|^2$$

where R_g = signal circuit equivalent generator resistance, and Z_{in} = input impedance at the signal frequency.

For a single-tuned signal circuit and a single-tuned idler circuit the input impedance at the signal frequency is

$$Z_{\text{in}} = \beta_s R_d + jX_s - \frac{\sigma \beta_s \beta_i R_d^2}{\beta_i R_d - jX_i}$$

where

$$\sigma = \frac{m_1^2 f_c^2}{\beta_s \beta_i f_s f_i}$$

The diode loss resistance R_d has been modified at the signal and idler frequencies to include circuit losses (that is, $\beta \geq 1$). The signal and idler reactances are X_s and X_i, respectively, and must include the varactor junction capacitance. (In this simplified circuit the diode package capacitance has been neglected.) For a fully pumped varactor the cutoff frequency is

$$f_c = \frac{1}{2\pi R_d C_R}$$

where $C_R =$ the junction capacitance at reverse breakdown. The modulation ratio m_1 for an abrupt-junction diode is 0.25. (See Ref. 3.*)

76. Bandwidth and Noise. Factors that determine the overall paramp bandwidth include the varactor characteristics (cutoff frequency, junction and package capacitance, and lead inductance), choice of idler (pump) frequency, the nature of the signal and idler resonant circuits, and the choice of band-center gain. Multiple-tuned signal and idler circuits are often used to increase the overall paramp bandwidth.

At band center the effective noise temperature of the one-port paramp (due to the diode and circuit loss resistances at the signal and idler frequencies) is

$$T_e = T_d \left[\frac{G-1}{G} \right] \left[\frac{f_p}{f_i} \left(\frac{\sigma}{\sigma - 1} \right) - 1 \right]$$

where $T_d =$ temperature of the varactor junction and loss resistances. The effective noise temperature can be reduced by cooling the paramp below room temperature and/or by proper choice of pump frequency. Circulator losses must be included when determining the overall paramp noise figure.

77. Pump Power. The pump power required at the diode to fully drive a single varactor is

$$P_p = k_1 \beta_p R_d [\omega_p C_R (\phi - V_R)]^2 \qquad \text{(watts)}$$

where the pump circuit loss resistance and diode resistance $= \beta_p R_d$, and $k_1 = 0.5$ for an abrupt-junction varactor.

78. Gain and Phase Stability. The stability of the cw pump source (amplitude and frequency) is a significant factor in the overall paramp gain and phase stability. For small pump-power changes an approximate expression for the paramp power-gain variation is

$$\frac{\Delta G}{G} \approx \frac{G-1}{\sqrt{G}} \frac{\Delta P_p}{P_p}$$

Environmental temperature changes often require a temperature-regulated enclosure for the paramp and pump source.

At low levels of nonlinear distortion, the amplifier third-order relative power intermodulation product at band center is a function of the one-port amplifier gain and the incident signal power level; i.e.,

$$\Delta IMP \propto \frac{(G-1)^4 P_s^2}{G} \approx G^3 P_s^2 \approx GP_{\text{out}}^2$$

Thus low-band center amplifier gains reduce this nonlinear distortion.

79. References on Parametric Amplifiers.

1. MANLEY, J. M., and ROWE, H. E. Some General Properties of Nonlinear Elements, Part 1, General Energy Relations, *Proc. IRE,* Vol. 44, pp. 904–913, July, 1956.

2. BLACKWELL, L. A., and KOTZEBUE, K. L. "Semiconductor-Diode Parametric Amplifiers," Prentice-Hall, Englewood Cliffs, N.J., 1961.

3. PENFIELD, P., JR., and RAFUSE, R. P. "Varactor Applications," The M.I.T. Press, Cambridge, Mass., 1962.

4. CHANG, K. K. N. "Parametric and Tunnel Diodes," Prentice-Hall, Englewood Cliffs, N.J., 1964.

---

*For references, see Par. 13-79.

5. Aitchison, C. S., Davies, R., and Gibson, P. J. "A Simple Diode Parametric Amplifier Design for the Use at S, C, and X Band," *IEEE Trans.*, Vol. MTT-15, pp. 22-31, January, 1967.

6. Porra, V., and Somervuo, P. "Broadband Matching of a Parametric Amplifier by Using Fano's Method," *IEEE Trans.*, Vol. MTT-16, pp. 880-882, October, 1968.

7. Takahashi, S., Nojima, M., Fukuda, T., and Yamada, A. K-Band Cryogenically Cooled Wide-Band Non-degenerate Parametric Amplifier, *IEEE Trans.*, Vol. MTT-18, pp. 1176-1178, Dec., 1970.

8. Okean, H. C., Allen, C. M., Sard, E. W., and Weingart, H. Integrated Parametric Amplifier Module with Self-contained Solid-State Pump Source, *IEEE Trans.*, Vol. MTT-19, pp. 491-493, May, 1971.

MASER AMPLIFIERS

By G. K. Wessel

80. Introduction. A maser is a microwave active device whose name is an acronym derived from *m*icrowave *a*mplification by *s*timulated *e*mission of *r*adiation. A laser is an amplifier or oscillator also based on stimulated emission but operating in the optical part of the spectrum. The expressions microwave, submillimeter, infrared, optical, and uv maser are also in common use. Here the word maser is used for the whole frequency range of devices, the *m* standing for *m*olecular.

A description of the laser principle, some of its properties and means of achieving laser operation, is given in Sec. 11 of this Handbook. A complete treatment of the subject is available in monographs[1-4] and review articles.[5,6]

In these devices, no electronic tubes or transistors are employed, but the emission properties of atoms or molecules—either as gases, liquids, or solids—serve for the amplification of the signals. Oscillators require the addition of a feedback mechanism. The interaction of the electromagnetic radiation with the maser material occurs in a suitable cavity or resonance structure. This structure often serves the additional purpose of generating a desired phase relationship between different spatial parts of the signal. The application of external energy, required for the amplification or oscillation process, is referred to as *pumping*. The pumping power consists, in many cases, of external electromagnetic radiation of a different frequency from that for the signal.

Microwave masers are used as low-noise preamplifiers and as time and frequency standards. The stimulated emission properties of atoms and molecules at optical frequencies, with their relatively high noise content, make the laser more useful for high-power light amplification and oscillation.

81. Maser Principles. In the processes of emission and absorption, atoms or molecules interact with electromagnetic radiation. It is assumed that the atoms possess either sharp internal-energy states or broader energy bands. A change of internal energy from one state to another is accompanied by absorption or emission of electromagnetic radiation, depending on the direction of the process. The difference in energy from the original to the final state is proportional to the frequency of the radiation, the proportionality constant being Planck's constant h.

The processes of the spontaneous emission and absorption were defined by Einstein:[7]

Spontaneous emission: $$-\frac{dN_2}{dt} = A_{21} N_2 \qquad (13\text{-}14)$$

Absorption: $$-\frac{dN_1}{dt} = B_{12}(\nu_{12}, T) N_1 \qquad (13\text{-}15)$$

where N_1 and N_2 = population densities of atoms or molecules of material having energy states E_1 and E_2 and an energy separation $E_2 - E_1 = h\nu_{12}$ between them.

At high temperature, an inconsistency in the equilibrium system arises unless a second emission process, stimulated emission, takes place. The rate of stimulated emission is defined

---

*Superior numbers correspond to numbered references, Par. 13-91.

very similarly to that of absorption (the stimulated emission is sometimes called *negative absorption*):

$$\frac{dN_2}{dt} = B_{21}\,\mu(\nu_{12}, T)\,N_2 \tag{13-16}$$

In temperature equilibrium, the rates of emission must be equal to the rate of absorption.

$$A_{21}N_2 + B_{21}\mu N_2 = B_{12}\mu N_1 \tag{13-17}$$

Assuming that the ratio of the population densities is equal to the Boltzmann factor,*

$$\frac{N_1}{N_2} = \exp\frac{h\nu_{12}}{kT} \tag{13-18}$$

Equation (13-15) yields, for the radiation density,

$$\mu(\nu_{12}, T) = \frac{A_{21}/B_{21}}{B_{12}/B_{21}\exp(h\nu_{12}/kT) - 1} \tag{13-19}$$

Equation (13-19) is identical with Planck's radiation law if we set

$$B_{12} = B_{21} \tag{13-20a}$$

and

$$\frac{A_{21}}{B_{21}} = \frac{8\pi h\nu_{12}^3}{c^3} \tag{13-20b}$$

where c = velocity of light.

Equation (13-20a) shows the close relationship between stimulated emission and absorption. The rates of population decrease, for these two processes [Eqs. (13-15) and (13-16)] depend only on their respective population densities. Neglecting the spontaneous emission, the net absorption or the net stimulated emission of an incoming radiation depends, therefore, only on the difference in population density: for absorption, if $(N_1 - N_2) > 0$; for amplification, if $(N_1 - N_2) < 0$. In particular, if a system is in quasi-equilibrium such that the upper energy state, E_2, is more populated than the lower one, E_1, it is capable of amplifying electromagnetic radiation.

To create and maintain the quasi-equilibrium requires application of external energy since the natural equilibrium has the opposite population excess $(N_1 > N_2)$. Different masers differ widely in the methods of how to accomplish the reverse in population density.

82. Properties of Masers. The properties of masers can be understood by analyzing Eqs. (13-14) to (13-20b), as follows:

Signal-to-Noise Ratio. From Eqs. (13-14) and (13-16) it follows that the Einstein coefficients A and B have different dimensions, but the ratio B_{21}/A_{21} is dimensionless. Here μ is proportional to the strength of the incoming signal and B_{21} is proportional to the amplified signal strength. A_{21} is proportional to the noise contribution due to spontaneous emission. After rewriting Eq. (13-20b),

$$\frac{B_{21}\mu}{A_{21}} = \frac{c^3\mu}{8\pi h\nu_{12}^3} \tag{13-21}$$

the ratio is found to be proportional to the signal-to-noise ratio of the amplifier. It thus becomes clear that the noise contribution (thermal or Johnson noise) is very small at microwave frequencies, whereas it becomes very large at optical frequencies (an increase of 15 orders of magnitude for an increase of 5 orders for the frequency). Microwave masers are therefore very useful as low-noise microwave preamplifiers.

Optical, infrared, and ultraviolet masers (lasers), on the other hand, are commonly used for power amplification and as powerful light sources (oscillators). The high content of spontaneous-emission noise does not make them easily applicable for low-noise amplification or sensitive detection of light, except in a few special cases.

83. Linearity and Line Width. The proportionality between the incoming signal

---

‡It is assumed that the statistical weights are equal to 1.

strength, the radiation density μ, and the amplified signal strength means that a maser is a linear amplifier as long as it is not driven into saturation. The latter occurs when the excess population becomes $N_2 - N_1 \approx N_2$.

Line Width. In the absence of a strong interaction of the individual atoms of the material, either with the environment or among themselves, the line width is determined by the average length of time of interaction of an atom with the radiation field. However, if regeneration due to feedback is taking place, the line width may be much narrower because of the following effect. It is assumed that the incoming radiation has a frequency distribution as shown in Fig. 13-83. The amplified signal is proportional to the rate of stimulated emission dN_2/dt. According to Eq. (13-16), the latter is proportional to $\mu(\nu_{12})$. Thus the center portions of the line will be more amplified than the wings, leading to increasingly narrower line widths. In an oscillator the final width is ultimately limited by statistical fluctuations due to the quantum nature of the radiation.

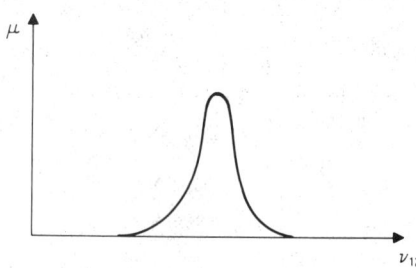

Fig. 13-83. Frequency distribution of radiation incoming to maser.

The narrow line width coupled with the low-noise property of a microwave maser makes the latter useful as a time and frequency standard device. To reduce interaction, gases are used as microwave maser materials.

In low-noise microwave amplifiers one is usually interested in a relatively large bandwidth (typically 50 MHz or more). For such amplifiers solid-state materials are used, with strong spin-spin interactions of atoms (for example, Cr^{3+} in ruby).

84. Tuning. Solid-state microwave masers can be tuned over a wide range of frequencies by adjustment of an external magnetic field, while the relative tuning range by a magnetic field is smaller by a factor of 10^{-4} to 10^{-5} in optical masers. Generally, in lasers, one has to depend on fixed frequencies wherever a suitable spectral line occurs in the material. However, modern liquid-dye lasers can be tuned over a relatively wide range in the visible spectrum.

85. Power Output. The power output due to stimulated emission depends on the number of excess atoms $(N_2 - N_1)$ available for the transition. Assuming p to be the probability of a transition per second, the power output is given by

$$P = (N_2 - N_1)h\nu_{12}p \tag{13-22}$$

Eq. (13-22) shows that the power output of a microwave gas maser must be very low (for example, 10^{-10} to 10^{-12}W) because the population density and the frequency are small. On the other hand, solid-state lasers and some gas lasers with very high efficiencies may have cw power outputs of 1,000 W and more.

86. Coherence. The similarity between absorption and stimulated emission [as expressed in Eqs. (13-15), (13-16), and 13-20a)] means that the stimulated emission is a coherent process. In practice, this means that a beam of radiation will be amplified only in the direction of propagation of the incoming beam and when all atoms participate coherently in the amplification. This is very different from the noise producing spontaneous emission, where the radiation is emitted isotropically and no phase relationship exists between the radiation coming from different atoms.

The coherence of the stimulated emission enables one to produce coherent light and radiation patterns similar to those which are well known and applied in the radio and microwave parts of the spectrum. In particular, it is possible by a suitable geometry of the device to produce a plane-parallel light beam whose divergence angle is limited only by diffraction. The divergence angle is given by

$$\alpha \approx \frac{\lambda}{d}$$ (13-23)

87. Time and Frequency Standards. The ammonia, as well as hydrogen, atomic-beam masers are used as time and frequency standards with operating frequencies of about 24 GHz and about 1.4 GHz, respectively. The reverse of population difference is achieved by focusing atoms in the excited state into a suitable microwave cavity, whereas the atoms in the lower state are defocused by a special focuser and these atoms do not reach the interior of the cavity. The microwave field in the cavity causes stimulated emission and amplification. If the cavity losses can be overcome, oscillations will set in. Time and frequency accuracy of 1 part in 10^{12} or higher can be obtained with modern maser standard devices.

88. Three- and Four-Level Devices. In many of the masers of all frequency regions, population reverse is obtained by a *three-level-maser* method first described by Bloembergen.[8] Radiation from an external power source at the pumping frequency v_p (Fig. 13-84) is applied to the material. E_3 may be a bond to make the pumping more economical. The material is contained in a suitable cavity, a slow-wave structure, or other resonance structure which is resonant at the signal frequency v_s. Under favorable conditions of the pumping power, the frequency ratio v_s/v_p, and the involved relaxation times, an excess population in the excited state E_2 over the ground state E_1 can be obtained. The interaction with the electromagnetic field of frequency v_s causes stimulated emission. Thus amplification (and if required, oscillation) is produced.

The four-level maser (Fig. 13-85) has the additional advantage that the population of the ground state does not have to be reduced to less than 50% of its original equilibrium value and the level E_4 can remain a relatively wide band. The transfer of energy from E_4 to E_3, and E_1, is by spontaneous decay or some other means like spin-lattice relaxation. Most solid-state microwave and optical masers are of the three- or four-level variety.

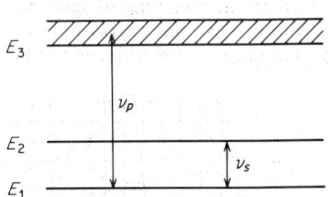

Fig. 13-84. Three-level maser.

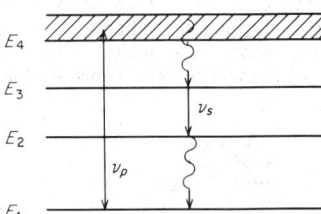

Fig. 13-85. Four-level maser.

89. Semiconductor Lasers. Stimulated light emission can be obtained by carrier injection in certain semiconductor materials. Upon application of an external dc potential, recombination light emission can take place at the interface of an N- and P-type semiconductor (for example, Zn and Te in GaAs). The semiconductor laser requires only a very simple power source (a few volts of direct current). However, the laser material has to be cooled to liquid-nitrogen temperature and the power output is relatively low.

90. Applications. The particular properties of devices based on stimulated emission have resulted in numerous applications which often surpass the capabilities of standard devices. In some instances, as in the case of holography, new fields have opened up through the emergence of lasers and masers.

Solid-state microwave preamplifiers for communication, radar, and radio astronomy have by far the best signal-to-noise ratio of all microwave amplifiers. The hydrogen-beam maser is the most accurate of all frequency and time standards.

Lasers have revolutionized the field of optical instrumentation and spectroscopy. They have made possible the new field of nonlinear optics. The laser is or may be used in many applications where large energy densities are required, as in microwelding, medical surgery, or even in the cracking of rocks in tunnel building. Lasers may be used as communication media where extremely large bandwidths are required. Optical radar, with its high precision due to the short wavelength, employs lasers as powerful, well-collimated light sources. Their range extends as far as the moon.

91. References on Maser Amplifiers

1. SINGER, J. R. "Masers," Wiley, New York, 1959.
2. UNGER, H. G. "Introduction to Quantum Electronics," Pergamon, New York, 1970.
3. ROSS, D. "Lasers, Light Amplifiers and Oscillators," Academic, New York, 1969.
4. LENGYEL, B. A. "Lasers," Wiley-Interscience, New York, 1971.
5. Optical Pumping and Masers, *Appl. Opt.*, Vol. 1, No. 1, January, 1962.
6. *Proc. IEEE*, Issue on Optical Electronics, Vol. 54, No. 10, October, 1966.
7. EINSTEIN, A. Zur Quantumtheorie der Strahlung, *Z. Phys.*, Vol. 18, No. 121, 1917.
8. BLOEMBERGEN, N. *Phys. Rev.*, Vol. 104, No. 324, 1956.

ACOUSTIC AMPLIFIERS

BY S. W. TEHON

92. Acoustoelectric Interaction. The acoustic amplifier stems from the announcement by Hutson, McFee, and White,[1]* in 1961, that they had observed a sizable influence on acoustic waves in single crystals of CdS caused by a bias current of charge carriers. CdS is both a semiconductor and a piezoelectric crystal, and the interaction was found to involve an energy transfer via traveling electric fields generated by acoustic waves, producing, in turn, bunching of the drift carriers. Quantitative analyses showing that either loss or gain in wave propagation could be selected by controlling the drift field were published by White[2] and Blotekjaer and Quate.[3]

Figure 13-86 illustrates the nature of the interaction. As an acoustic wave propagates in the crystal, its stresses induce a similar pattern of electric field through piezoelectric coupling. Since the coupling is linear, compressive stresses in half the wave induce a forward-directed field, and tensile stresses in the remainder of the wave induce a backward-directed field. Drifting charge carriers tend to bunch at points of zero field to which they are forced by surrounding fields. When a charge carrier loses velocity in bunching, it gives its excess kinetic energy to the acoustic wave; when it gains velocity, it extracts energy from the wave. Therefore the drift field is effective for determining attenuation or amplification in the range

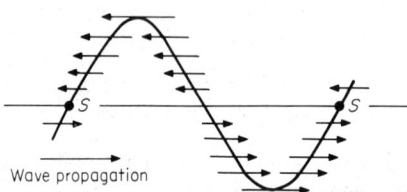

Fig. 13-86. Traveling wave of electric field intensity induced by traveling stress wave.

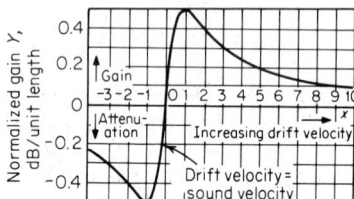

Fig. 13-87. Normalized gain as a function of bias field. *(After R. F. Shea (ed.), "Amplifier Handbook," Chap. 30, McGraw-Hill, New York, 1968.)*

near which carriers move at the speed of sound. Figure 13-87 illustrates the gain characteristic as a function of drift velocity. At the zero-crossover point, the carrier velocity is just equal to the velocity of propagation for the acoustic wave; the gain ranges from maximum attenuation up to maximum amplification over a relatively small change in bias. Beyond this range of drift velocity, bunching becomes decreasingly effective, and the interaction has less effect on the acoustic wave.

93. Piezoelectric Materials. Piezoelectricity is the linear, reversible coupling between mechanical and electrical energy due to displacement of charges bound in molecular structure. Pressure applied to a piezoelectric material produces a change in observed surface density of charge, and conversely, charge applied over the surfaces produces internal stress and strain. If S is the strain, T the stress, E the electric field intensity, and D the dielectric

---

*Superior numbers correspond to numbered references, see Par. 13-97.

displacement, then the piezoelectric effect at a point in a medium is described by the pair of linear equations

$$S = sT + dE$$

$$D = dT + \epsilon E$$

where constant s = elastic compliance, ϵ = permittivity, and d = piezoelectric constant. The quantity $s\epsilon - d^2$ equal to the determinant of the array of these material constants is defined as k^2, where k is the coefficient of electromechanical coupling. As a consequence of conservation of energy, it can be shown that k is a number less than unity, and that a fraction k^2 of applied energy (mechanical or electrical) is stored in the other form (electrical or mechanical.) Since D and E are vectors, and S and T are tensors, the piezoelectric equations are tensor equations, and are equivalent to nine algebraic equations, describing three vector and six tensor components.

Materials that are appreciably piezoelectric are either crystals with anisotropic properties or ceramics with ferroelectric properties which can be given permanent charge polarization through dielectric hysteresis. Ferroelectric ceramics, principally barium titanate and lead zirconate titanate, are characterized by relative dielectric constants ranging from several hundred to several thousand, by coupling coefficients as high as 0.7, and by polycrystalline grain structure which will propagate acoustic waves with moderate attenuation at frequencies extending up to the low-megahertz range.

Single crystals are generally suited for much higher acoustic frequencies, and in quartz, acoustic-wave propagation has been observed at 125 GHz. Quartz has low loss, but a low coupling coefficient. Lithium niobate is a synthetic crystal, ferroelectric and highly piezoelectric; lithium tantalate is somewhat similar. The semiconductors cadmium sulfide and zinc oxide are moderately low in loss and show appreciable coupling.

94. Stress Waves in Solids. Acoustic waves propagate with low loss and high velocity (5,000 m/s) in solids. Solids also have shear strength, whereas sound waves in gases and fluids are simple pressure waves, manifested as traveling disturbances measurable by pressure and longitudinal motion.

Sound waves in solids involve either longitudinal or transverse particle motion. The transverse waves may be propagation of simple shear strains, or may involve bending in flexural waves. Since different modes of waves travel with different velocities, and since both reflections and mode changes can occur at material discontinuities, the general pattern of wave propagation in a bounded solid medium is quite complicated.

Bulk waves are longitudinal or transverse waves traveling through solids essentially without boundaries; e.g., the wave-fronts extend over many wavelengths in all directions. A solid body supporting bulk waves undergoes motion and stress throughout its volume. A *surface wave* follows a smooth boundary plane, with elliptical particle motion which is greatest at the surface and drops off so rapidly with depth that almost all the energy is carried in a one-wavelength layer at the surface.

Ideally, the wave medium for surface-wave propagation is regarded as infinitely deep; practically if it is many wavelengths deep, its properties are equivalent. A surface wave following a surface free from forces is a Rayleigh wave; if an upper material with different elastic properties bounds the surface, the motion may be a Stonely wave.

95. Bulk-Wave Devices. Most acoustic amplifiers utilizing bulk waves have the components shown in Fig. 13-88. The input and output transducers are piezoelectric crystals or deposited thin layers of piezoelectric material, used for energy conversion at the terminals. The amplifier crystal is generally CdS, which is not only piezoelectric but also an N-type

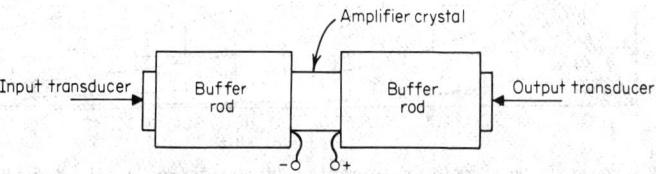

Fig. 13-88. Typical structure for acoustic amplification measurements. *(After R. F. Shea (ed.), "Amplifier Handbook," Chap. 30, McGraw-Hill, New York, 1968.)*

semiconductor. Electrodes are attached at the input and output surfaces of the CdS crystal, for bias current. Since the mobility of negative-charge carriers in CdS is only about 250

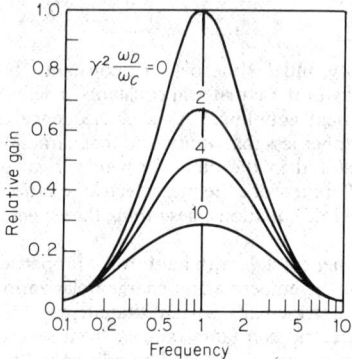

cm²/V · s, large bias fields are required to provide a drift velocity equal to acoustic velocity: in CdS, 4,500 m/s, for longitudinal waves and 1,800 m/s for shear waves. The buffer rods shown in Fig. 13-88 are therefore added for electrical isolation of the bias supply. Furthermore, the transducers and amplifying crystal are cut in the desired orientation, to couple to either longitudinal or shear waves.

In the analysis of bulk-wave amplification [2,3] the crystal properties are characterized by:

f_D = diffusion frequency = $\omega_D/2\pi$

f_C = dielectric relaxation, or conductivity frequency = $\omega_C/2\pi$

$\gamma = (1 - f)$ times the drift velocity divided by the acoustic phase velocity, where f is the fraction of space charge removed from the conduction band by trapping

k = coefficient of electromechanical coupling

$\bar{\omega}$ = radian frequency of maximum gain = $\sqrt{\omega_C \omega_D}$

Fig. 13-89. Normalized plots of gain vs. frequency. Note symmetry of frequency response and effects of bias and critical frequency values. *(After R. F. Shea, (ed.), "Amplifier Handbook," Chap. 30, McGraw-Hill, New York, 1968.)*

Figure 13-89 shows the computed gain curves, using these constants, for four values of drift velocity. The frequency of maximum gain can be selected by control of crystal conductivity, which is easily accomplished in CdS by application of light. The amount of gain, and to some extent the bandwidth, are controlled by the bias field.

Figure 13-90 shows characteristics specifically calculated for CdS with shear acoustic-wave amplification. Large bias voltages (800 to 1,300 V/cm) are required, and extremely high gains are possible. Generally, heat dissipation due to the power supplied by bias is so high (up to 13 W/cm³ at $\gamma = 0.76$) that only pulsed operation is feasible.

96. Surface-Wave Devices. The disadvantages in bulk-wave amplifiers, evident as excessive bias requirements, could be alleviated if materials with high mobility were available. However, the amplifier crystal must show low acoustic loss, high piezoelectric coupling, and high mobility; a suitable material combining these properties has not been

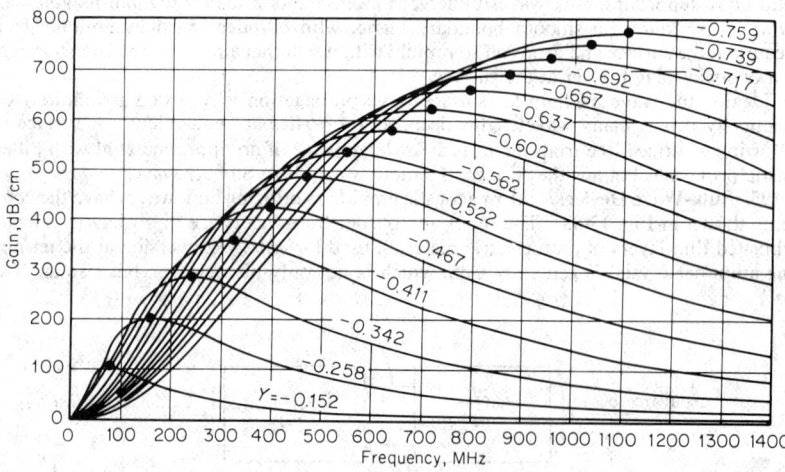

Fig. 13-90. Gain vs. frequency for CdS shear-wave amplification at $f_D = 796$ MHz. *(After R. F. Shea (ed.), "Amplifier Handbook," Chap. 30, McGraw-Hill, New York, 1968.)*

found. Figure 13-91 shows an acoustic amplifier structure which operates with acoustic surface waves and provides charge carriers in a thin film of silicon placed adjacent to the insulating piezoelectric crystal. Coupling for the interaction takes place in the electric field across the very small gap between piezoelectric and semiconductor surfaces. Transducers are formed by metallic fingers, interlaced to provide field for piezoelectric coupling in the regions between parallel fingers. This interdigital-array technique is flexible, providing means for complicated transducer designs.

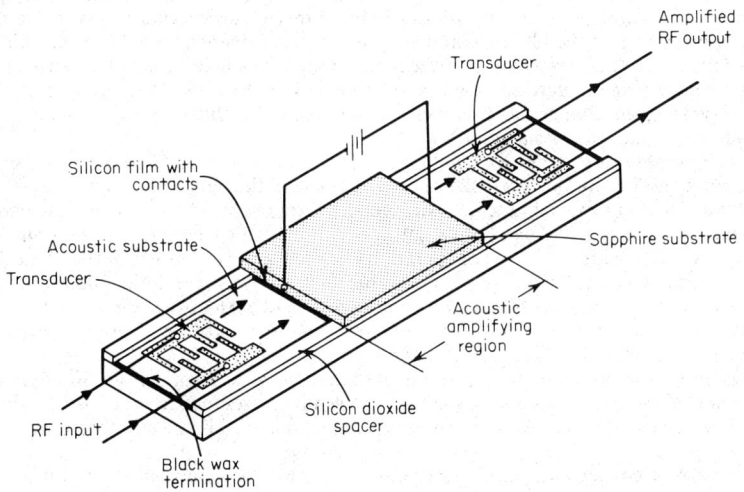

Fig. 13-91. Structure of a surface-wave acoustic amplifier. *(After J. H. Collins and P. J. Hagon, Electronics, Dec. 8, 1969.)*

Typically, gain in an amplifier of this type can be as much as 100 dB in a crystal less than 1 cm long, operating with as much as 50% bandwidth at frequencies up to several hundred megahertz. The amount and direction of gain are controlled by the bias field, and the bandwidth is determined by geometry of the interdigital arrays. The silicon film operates at much lower bias field than required in bulk-wave amplifiers, and can be deposited on a sapphire substrate in narrow strips to permit parallel excitation at low voltage. The upper frequency limits are set by resolution in the photolithographic processes used to form the fingers, and have been extended to more than 3 GHz by using electron-beam processing to secure high resolution.

97. References on Acoustic Amplifiers

1. HUTSON, A. R., McFEE, J. H., and WHITE, D. L.   Ultrasonic Amplification in CdS, *Phys Rev. Letters,* Vol. 7, No. 6, pp. 237–239, Sept. 15, 1961.

2. WHITE, D. L.   Amplification of Ultrasonic Waves in Piezoelectric Semiconductors, *J. Appl. Phys.,* Vol. 33, No. 8, pp. 2547–2554, August, 1962.

3. BLOTEKJAER, K., and QUATE, C. F.   The Coupled Modes of Acoustic Waves and Drifting Carriers in Piezoelectric Crystals, *Proc. IEEE,* Vol. 52, No. 4, pp. 360–377, April, 1965.

MAGNETIC AMPLIFIERS

By H. W. LORD

98. Static magnetic amplifiers may be divided into two classes, identified by the terms saturable reactor and self-saturating magnetic amplifier. A *saturable reactor* is defined as "an adjustable inductor in which the current versus voltage relationship is adjusted by control magneto-motive forces applied to the core."[1] * A *magnetic amplifier* is defined as "a device using saturable reactors either alone or in combination with other circuit elements to secure amplification or control."[1] A *simple magnetic amplifier* is defined as "a magnetic amplifier consisting only of saturable reactors."[1] The abbreviation SR is used in this section to denote a *saturable reactor* and/or simple magnetic amplifier.

A self-saturating magnetic amplifier is a magnetic amplifier in which half-wave rectifying circuit elements are connected in series with the output windings of saturable reactors. It has been shown[2] that saturable reactors can be considered to have negative feedback. Half-wave rectifiers in series with the load windings will block this intrinsic feedback. A self-saturating magnetic amplifier is therefore a parallel-connected saturable reactor with blocked intrinsic feedback. This latter term avoids the term self-saturation, which, although extensively used, does not have a sound physical basis. The abbreviation MA is used here to denote this type of high-performance magnetic amplifier. Trade names for this type of magnetic amplifier include Amplistat, Magnestat, and Mag-Amp.

99. Saturable Reactors (SR) Amplifiers. The SR can be considered to have a very high impedance throughout one part (the *exciting interval*) of the half-cycle of alternating supply voltage and to abruptly change to a low impedance throughout the remainder of the half-cycle (the *saturation interval*). The phase angle at which the impedance changes is controlled by a direct current. This is the type of operation obtained when the core material of the SR has a highly rectangular hysteresis loop. Two types of operation, representing limiting cases, are discussed here, namely, *free even-harmonic currents* and *suppressed even-harmonic currents*. Intermediate cases are very complex, but use of one or the other of the two extremes is sufficiently accurate for most practical applications.

The present treatment is limited to resistive loads, the most usual type for SR applications. Reference 3 discusses inductive dc loads and Ref. 4 discusses inductive ac loads. The basic principles of operation of SRs and MAs are given in more detail in Ref. 5 than space permits here.

100. Series-connected Saturable Reactor Amplifiers. Basically, an SR circuit consists of the equivalent of two identical single-phase transformers. Figure 13-92 shows two transformers, SR_A and SR_B, interconnected to form a rudimentary series-connected SR circuit.

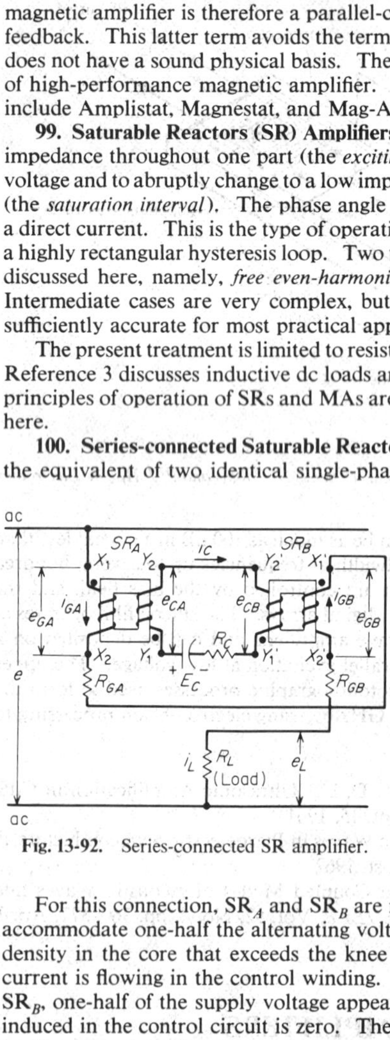

Fig. 13-92.   Series-connected SR amplifier.

The two series-connected windings in series with the load are called *gate windings*, and the other two series-connected windings are called *control windings*. Note, from the dots that indicate relative polarity, that the gate windings are connected in series additive and the control windings are connected in series subtractive. By reason of these connections, the fundamental power-frequency and all *odd* harmonics thereof will not appear across the total of the two control windings, but any *even* harmonic induced in one control winding will be additive with respect to a corresponding even harmonic induced in the other control winding.

For this connection, SR_A and SR_B are normally so designed that each gate winding will accommodate one-half the alternating voltage of the supply without producing a peak flux density in the core that exceeds the knee of the magnetization curve, assuming no direct current is flowing in the control winding. Under these conditions, if SR_A is identical with SR_B, one-half of the supply voltage appears across each gate winding, and the net voltage induced in the control circuit is zero. The two SRs operate as transformers over the entire portion of each half-cycle.

When a direct current is supplied to the control circuit, each SR will have a saturation interval during a part of each cycle, SR_A during half-cycles of one polarity and SR_B during half-cycles of the opposite polarity. The ratio of the saturation interval to the exciting interval can be controlled by varying the direct current in the control winding.

---

* Superior numbers correspond to numbered references, Par. **13-109**.

When an SR core has a saturation interval during part of half-cycles of one polarity, and there is a load or other impedance in series with the gate winding, even-harmonic voltages are induced in all other windings on that core. In the series connection, one SR gate winding can be the series impedance for the SR, and large even-harmonic voltages will appear across the individual gate windings and control windings when direct current flows in the control windings. The amount of even-harmonic current which flows in the control circuit of Fig. 13-92 will depend upon the impedance of the control circuit to the harmonic voltages. If the control circuit impedance between terminals Y_1 and Y'_1 is low with respect to the induced harmonic voltages of the control windings (usually referred to as a *relatively low* control circuit impedance), the harmonic currents can flow freely in this circuit, and the SR circuit is identified by the term *free even-harmonic currents*. If the control circuit impedance is high with respect to the induced harmonic voltages (a *relatively high* control circuit impedance), the harmonic current flow is suppressed and the SR circuit is identified by the term *suppressed even harmonic currents*.

If R_C in Fig. 13-92 is relatively low and the source impedance of the dc control current is low, the circuit is of the free even harmonic currents type and the two SRs are tightly coupled together by the control circuit. If one SR is in its saturation interval, it will reflect a low impedance to the gate winding of the other SR, even though it is then operating in its exciting interval. Thus, during the saturation interval of any core, transformer action causes both gate windings to have low impedances and current can flow from the ac source to the load, with correspondingly high harmonic circulating currents in the control circuit. When both cores are in their exciting intervals, only the low core exciting current can flow and the current circulating in the control circuit is substantially zero.

If the exciting current is so low as to be negligible, and if N_G are the turns in each gate winding, N_C are the turns in each control winding, I_C is the rectified average of the load current, and I_C is the average value of the control current, then applying the law of equal ampere-turns for transformers provides the following expression for the circuit in Fig. 13-92: $I_C N_C = I_G N_G$. This is the law of equal ampere-turns for the series-connected SR with resistive load. It applies to operation in the so-called *proportional region*, the upper limit of this region being that point in the control characteristic where the load current is limited solely by the load circuit resistance. Figure 13-93 shows the gate current circuit and control circuit current at one operating point in the proportional region. An increase in control current causes a decrease in the angle α, thereby increasing I_L, the rectified average of the current to the load.

101. Parallel-connected Saturable Reactor Amplifiers. Figure 13-94 shows the circuit diagram for the parallel-connected SR. Each gate winding is connected directly between the ac supply and the load resistance, thus providing two parallel paths through the SR. There is therefore a low-impedance path for the free flow of even-harmonic currents, even though the impedance of the control circuit happens to be relatively high. As a result, the parallel-connected SR is always of the *free flow of even-harmonic currents* type, and the cores operate in the same manner as described in Par. **13-100**, and the waveshapes of currents to the *load*

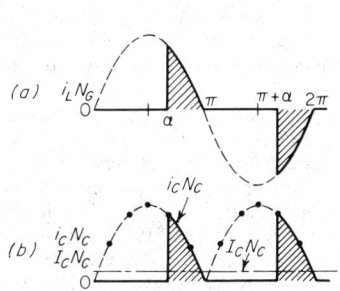

Fig. 13-93. Currents in circuit of Fig. 13-92, for free even-harmonic current conditions. (*a*) Gate current; (*b*) control-circuit current.

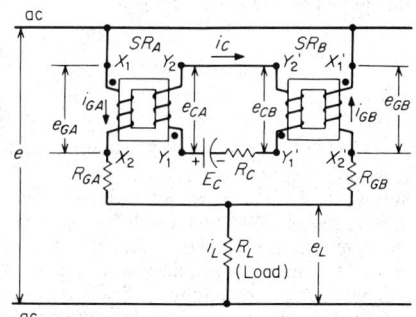

Fig. 13-94. Parallel-connected SR amplifier.

are as shown in Fig. 13-93. However, the gate winding currents are different, as shown in Fig. 13-95.

It is obvious from Fig. 13-95 that $I_L' = -I_L/2$. Since, except for the low-excitation requirements for the core, the net ampere-turns acting upon the core must be zero during the exciting interval, then $I_L' N_G + I_C N_C = 0$. Using the above equation to eliminate I_L', the result is $I_C N_C = I_L N_G/2$. Figure 13-96 shows the control characteristic of saturable core amplifiers applicable both to free and suppressed even-harmonic modes of operation.

102. Series-connected SR Circuit with Suppressed Even Harmonics. This circuit is usually analyzed by assuming operation into a short circuit and the control *current* from a current source. Figure 13-97 shows idealized operating minor hysteresis loops for this circuit, and Fig. 13-98 shows pertinent current and voltage waveshapes lettered to correspond to Fig. 13-97. Note that this circuit supplies a squarewave of current to the load so long as it is operating in the proportional region. Energy is interchanged between the power-supply circuit and the control circuit so as to accomplish this type of operation. If a large inductor is used to maintain substantially ripple-free current in the control circuit, the energy interchanged between the supply voltage and control circuit is alternately stored in and given up by the control-circuit inductor.

103. Gain and Speed of Response of Saturable Reactors. It is obvious from the generalized control characteristic that the current gain is directly proportional to the ratio of the control winding turns to the gate winding turns. Thus $I_L/I_C = N_C/N_G$. If R_C is the control-circuit resistance and R_L is the load resistance, the power gain GP is $N_C^2 R_L^2 / N_G^2 R_C^2$.

The response time of saturable reactors is the combination of the time constants of the control circuit and of the gate-winding circuit. Expressed in terms of the number of cycles of the supply frequency, the time constant of the control winding of a series-connected SR is[6] $\tau_C = R_0 N_C^2 / 4 R_C N_C^2$ cycles. If the SR is operating underexcited, a transportation lag may cause an additional delay in response.[7]

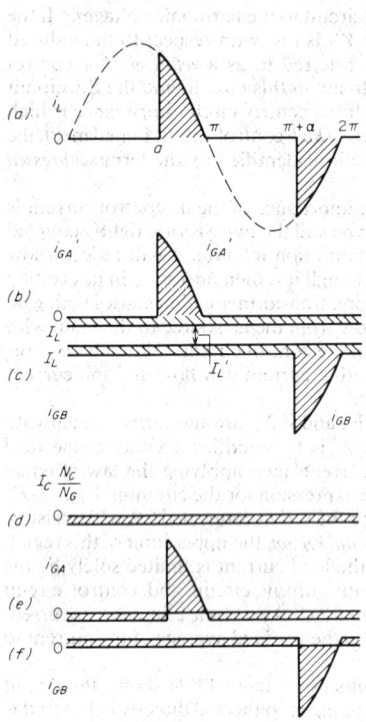

Fig. 13-95. Currents in the circuit of Fig. 13-94 for free even-harmonic current conditions. (*a*) Load current; (*b*) components of load current in SR$_A$ gate winding; (*c*) same in SR$_B$; (*d*) control-circuit current; (*e*) gate current in SR$_A$; (*f*) same in SR$_B$.

104. High-Performance Magnetic Amplifier (MA). If a rectifier is placed in series with each gate winding of a parallel-connected SR and the rectifiers are poled to provide an ac output, it becomes a type of high-gain magnetic amplifier circuit called the *doubler circuit*. When this is done, the law of equal ampere-turns no longer applies, and the transfer characteristic is mainly determined by the magnetic properties of the SR cores. The design of an MA therefore requires more core-materials data than is required for SRs in simple magnetic amplifier circuits. The text of this section assumes that the designer has the required magnetic-core-materials data on hand, since such data are readily available from the manufacturers of cores for use in high-performance magnetic amplifiers.

Descriptions of the operation of MAs require several terms which are defined here for reference.[8]

Firing: In a magnetic amplifier, the transition from the unsaturated to the saturated state of the saturable reactor during the conducting or gating alternation. Firing is also used as an adjective modifying phase or time to designate when the firing occurs.

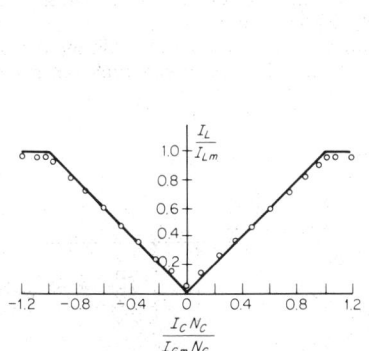

Fig. 13-96. Control characteristic of SR amplifier, applying both to free and suppressed even-harmonic conditions.

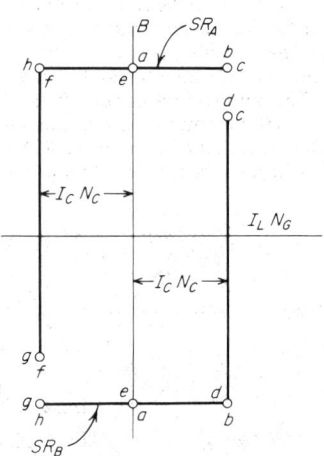

Fig. 13-97. Idealized operating hysteresis loops for suppressed even-harmonic current condition.

Gate angle (firing angle): The angle at which the gate impedance changes from a high to a low value.

Gating: The function or operation of a saturable reactor or magnetic amplifier that causes it, during the first portion of the conducting alternation of the ac supply voltage, to block substantially all the supply voltage from the load; and during a later portion allows substantially all the supply voltage to appear across the load. The "gate" is said to be virtually closed before firing and substantially open after firing.

Reset, degree of: The reset flux level expressed as a percentage or fraction of the reset flux level required to just prevent firing of the reactor in the subsequent gating alternation under given conditions.

Reset flux level: The difference in saturable-reactor core flux level between the saturation level and the level attained at the end of the resetting alternation.

Resetting (presetting): The action of changing saturable-reactor core flux level to a controlled ultimate reset level, which determines the gating action of the reactor during the subsequent gating alternation. The terms resetting and presetting are synonymous in common usage.

Resetting half-cycle: The half-cycle of the magnetic amplifier ac supply voltage at which resetting of the saturable reactor may take place.

105. Half-wave MA Circuits. Figure 13-99 shows a half-wave MA circuit with control from a source of controllable direct *current*. Figure 13-100 shows a half-wave MA circuit with control from a controllable source of direct *voltage*. The following sequence of operation is the same for both circuits and is described in connection with Fig. 13-101:

a. During the gating half-cycle (diode REC or REC_1 conducting) the core flux increases toward a saturation level (3') from some reset flux level (0) and the current to the load R_L is very low.

b. When saturation of the SR occurs at 2', firing occurs and current flows to the load for the rest of the gating half-cycle, leaving the core flux at 4.

c. During the resetting half-cycle (diode REC or REC_1 blocking) the SR core is reset from 4 through 5′ to a value of reset flux level corresponding to B′ and 0.

The waveshape of the current to the load is the same for both circuits, being a portion of one polarity of sinewave (phase-controlled half-wave). The two types of control circuits differ as follows:

a. The curve on the reset portion of the hysteresis loop between 4 and 0 is as shown in Fig. 13-101 for the current control type of control circuit (Fig. 13-99), but for the circuit of Fig. 13-100, the resetting portion of the curve in the region of 5′ is not a vertical line and may coincide with the outer major hysteresis loop for a substantial portion of the resetting period.

b. The output of the circuit of Fig. 13-99 is a maximum for zero dc control current, but for Fig. 13-100 the output is a minimum for zero dc control voltage.

The operation of a half-wave MA may be summarized as follows: (*a*) During the gating half-cycle, the core acts to withhold current from the load for a portion of each half-cycle, the

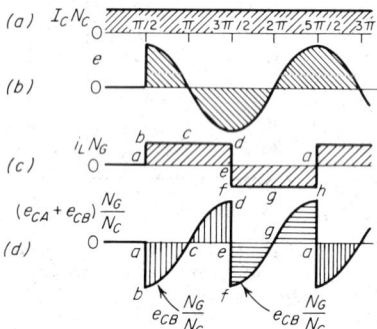

Fig. 13-98. Series-connected suppressed even-harmonic conditions. (*a*) Control-circuit current; (*b*) applied input voltage; (*c*) gate winding and load current; (*d*) voltage across series-connected control coils.

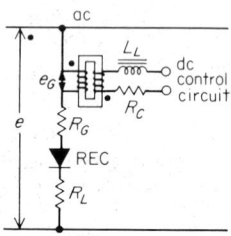

Fig. 13-99. Half-wave MA circuit.

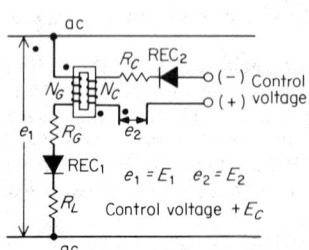

Fig. 13-100. Voltage-controlled MA circuit.

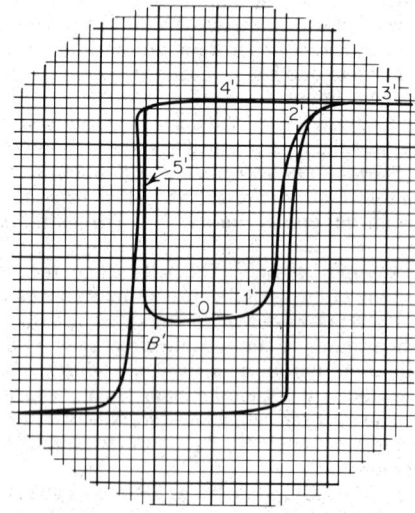

Fig. 13-101. Major and operating minor dynamic hysteresis loops for circuit of Fig. 13-99.

13-84

length of the withholding period being determined by the degree of reset of core flux provided during the immediately preceding resetting half-cycle. (*b*) During the resetting half-cycle, the degree of reset can be controlled by varying the amount of direct current of proper polarity in the control winding, or by varying the amount of average voltage of proper polarity applied to the control winding. (*c*) The amount of current and power required to reset the core is primarily a function of the excitation requirements of the core and bears no direct relation to the power delivered to the load.

106. Full-wave MA Circuits. Two half-wave MA circuits can be combined to provide either full-wave ac output (Fig. 13-102) or full-wave dc output (Fig. 13-103). During the gating half-cycle, all gate windings act in the same manner as described for the half-wave MA circuit. Because of an interaction of the common load resistor voltage upon the availability of inverse voltage across a rectifier during its resetting half-cycle, each half-wave part of Fig. 13-102 cannot be assumed to operate independently of its mate. The differences in control characteristics are shown by Fig. 13-104 where curve *A* applies to the circuit of Fig. 13-99 and curve *B* applies to the circuit of Fig. 13-102.

107. Equations for Magnetic Amplifier Design. For SRs in simple magnetic amplifier circuits, since the ampere-turn balance rules apply, their design is so similar to transformer design that ordinary transformer design procedures are applicable. For SRs in MA circuits, the gate windings must carry rms load currents that are related to the ac and dc load currents in the same manner that rectifier transformer secondary rms currents are related to the load current. Gate windings must be able to withstand a full half-cycle of ac supply voltage, without saturation, for a total flux swing from $-B\mu_m$ to $+B\mu_m$, where $B\mu_m$ is the flux density at which the permeability of the core material is a maximum.

Using the cgs system of units:

Maximum ac supply voltage, rms: $E_e = 4.44 N_G B_{\mu m} A f \times 10^{-8}$, where N_G is gate winding turns, A is core area, and f is supply frequency.

Maximum load current for firing angle $\alpha = 0°$ (rectified average): $I_{LM} = E_e/1.11(R_G + R_L)$, where R_G is total gate winding resistance, including diode drops, and R_L is the effective load resistance.

For current-controlled MA circuits the following equations apply:

Minimum load current ($\alpha = 180°$); $I_{LX} = 2I_X = 2H_{csf}l/0.4\pi N_G$, where I_X is the exciting current in amperes of one SR core, H_{csf} is the sine flux coercive force of the core material in oersteds, and l is the mean length of magnetic circuit of one core.

Control current for upper end of control current: $I_C = Hcl/0.4\pi N_C$, where H_C is the dc coercive force of the core material.

Control current for minimum load-current point: $I_C = H_{csf}l/0.4\pi N_C$. Ampere-turn gain $G_{AT} = (I_{LM}/2I_X)\pi$.

Time constant $\tau_C = 0.9\pi E_e N_C^2/4fI_X R_C N_G^2$ seconds, where R_C is the control circuit resistance and assuming $\alpha = 90°$. This equation does not take into account a transportation lag or any underexcited effects which increase the delay time.

Power gain $G_p = G_{AT}^2 (N_C/N_G)^2 (R_L/R_C)$ where k_f, is the form factor (ratio of rms to average values). If only average values are used, k_f is unity. At $\alpha = 90°$, dynamic power gain $G_D = G_P/\tau_C = \pi fI_{LM}R_L/I_X(R_G + R_L)$ per second.

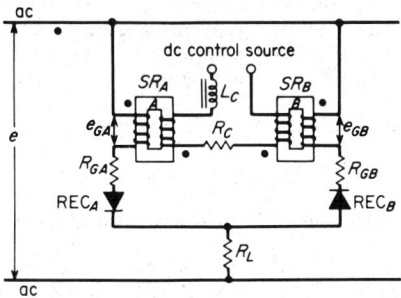

Fig. 13-102. Full-wave ac output MA circuit with high-impedance control circuit.

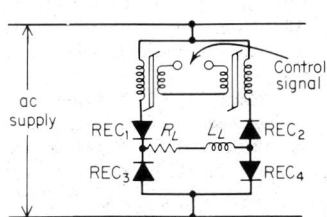

Fig. 13-103. Full-wave bridge MA circuit with inductive load.

For voltage-controlled MA circuits, the following equations apply to Fig. 13-100: AC bias voltage $E_2 > E_1 N_C/N_G$ and of indicated relative polarities. A value of $E_2 = 1.2E_1 N_C/N_G$ is usually adequate. Since the dc control voltage E_C acts to inhibit the resetting effect of E_2, a firing angle $\alpha = 0°$ occurs when $E_C \geq \sqrt{2E_2}$.

A single-stage amplifier has a half-cycle transport lag and no other time delay unless the resetting is inhibited by common-load interactions with other half-wave-amplifier elements.

Maximum average load current $I_{LM} = 0.9E_1/(R_G + R_L)$.

Minimum average load current $I_{LX} = I_X = H_{csf}l/0.4\pi N_G$.

The calculation of the power gain of voltage-controlled MA circuits largely depends upon the control circuit. The dc control voltage of Fig. 13-100 actually must *absorb* power from the ac bias circuit during much of its control range. Assuming that $E_2 = 1.2E_1 N_C/N_G$ and a resistor R'_C is connected across the control source, which has a drop of $0.2E_2$ at cutoff, then $R'_C = 0.24E_1 N_C^2/N_{GX}^2$.

The maximum dc control voltage is then $1.2\sqrt{2}E_1 N_C/N_G$.

When the dc control voltage $E_C > \sqrt{2E_2}$, rectifier REC_2 is blocking all the time, and so the only load on the control source is R'_C. Therefore the maximum power from the control source $P_C = E_C^2/R'_C = 5\sqrt{2E_1 I_X}$.

The maximum power output is $P_o = I_{LM}^2 R_L (0.9E_1)^2/(R_G + R_L)^2$.

The power gain is then $G_P = P_o/P_C$ if the gain is assumed to be linear over most of the control characteristic. This can also be written $G_P = I_{LM}^2 G_p - I_{LM}^2 R_L/5\sqrt{2E_1 I_X}$.

108. Core Configurations. Figures 13-105 and 13-106 show two coil-and-core configurations commonly used for SRs. The best core-and-coil geometry for the SRs and MAs is the

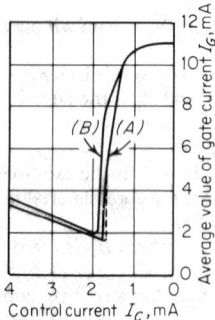

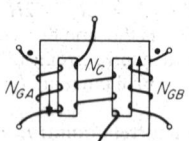

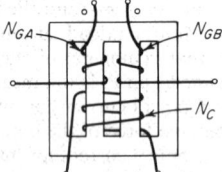

Fig. 13-104. MA circuit control characteristics. (*a*) Half-wave circuit; (*b*) full-wave ac output circuit.

Fig. 13-105. Three-legged SR core and coil configuration.

Fig. 13-106. Four-legged SR core and coil configuration.

toroid-shaped core with the gate winding uniformly wound over the full 360° of the core. Full-wave operation requires two such cores. After winding a gate winding on each core of a matched pair, the two cores and coils may be stacked together coaxially and the required control coils wound over the stack. The gate coils are so connected into the load circuit that no fundamental or odd-harmonic voltage is induced in the control windings.

109. References on Magnetic Amplifiers

1. "The International Dictionary of Physics and Electronics," Van Nostrand, Princeton, N.J., 1956.

2. STORM, H. F. "Magnetic Amplifiers," Chap. 5, J. Wiley, New York, 1955.

3. Reference 2, Chaps. 11 and 12.

4. WILSON, T. G. Series-connected Magnetic Amplifier with Inductive Loading, *Trans. AIEE*, Vol. 71, Pt. I, pp. 101–110, 1952.

5. SHEA, R. F. "Amplifier handbook," Chaps. 8 and 21, McGraw-Hill, New York, 1968.

6. Reference 2, p. 148.

7. Reference 2, p. 150.

8. Terms and Definitions for Magnetic Amplifiers, AIEE Committee Report, *Trans. AIEE*, Vol. 73, Pt. I, pp. 265–270, 1954.

DIRECT-COUPLED, OPERATIONAL, AND SERVO AMPLIFIERS

By S. M. Korzekwa

110. Direct-coupled Amplifiers. A direct-coupled amplifier has frequency response that starts at zero frequency (dc) and extends to some specified upper limit. To obtain the zero-frequency capability, such amplifiers are normally direct-coupled throughout; i.e., they do not use capacitive or transformer coupling (except for auxiliary higher-frequency compensation or signal transmission).

The primary sources of error of a direct-coupled amplifier are initial offset, drift, and gain variations, errors usually dependent on temperature, aging, etc. The gain-variation problem can be minimized by feedback gain-stabilization techniques, but offset and drift errors are usually not handled so effectively by feedback techniques. A bias shift or drift error cannot be distinguished from a signal response because their output responses are identical. Various techniques are available to minimize this drift problem, and several different methods may be used in an amplifier.

One method of minimizing drift is to use a balanced topology, as in differential amplifiers, whereby the drift errors tend to cancel out. A more complicated but effective method is the modulated-carrier-amplifier approach: the signal is first converted to a carrier signal, amplified using an ac-coupled amplifier, and then demodulated to a baseband signal. The function performed by the modulated-carrier amplifier is identical with that of a direct-coupled amplifier, but the signal processing technique is different.[1,2]*

An example of a direct-coupled transistor amplifier is illustrated in Fig. 13-107. The primary signal path, shown in heavy lines, directly couples the input signal to the amplifier

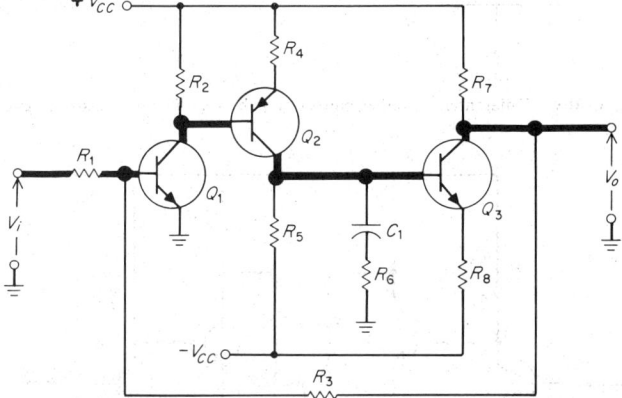

Fig. 13-107. Direct-coupled transistor amplifier.

output. This configuration also provides gain and bias stabilization via the direct-coupled R_3 feedback path. Usually the low-frequency amplifier voltage gain is primarily determined via the R_1 and R_3 components, while the function of the C_1 capacitor and R_6 resistor is to provide high-frequency compensation or stabilization. Sources of error for this topology include the initial V_{BE} offset of the Q_1 input transistor and the subsequent drift caused by the temperature dependence of this offset.

111. Differential Amplifiers. A differential amplifier is a dual-input amplifier that amplifies the difference between its two signal inputs. This amplifier may have an output that is single-ended (one output) or it may have a differential output.

The differential amplifier eliminates or greatly minimizes many common sources of error. The drift problem encountered in direct-coupled amplifiers can be handled more effectively by the differential approach. A second major advantage of a differential amplifier is its ability

---

* Superior numbers correspond to numbered references, Par. **13-116**.

to reject common-mode signals, that is, unwanted signals present at both of the amplifier inputs or other common points. Common-mode performance is usually a critical requirement in instrumentation amplifiers.

The basic differential active-device circuits commonly used in differential amplifiers are shown in Fig. 13-108. Such differential pairs may be constructed using separate devices or in integrated circuit form. The integrated package yields additional advantages since the parameter differences between the units of the integral differential pair are usually much less than if separate devices are used. Thus the units of such integral pairs tend to track differentially more closely, even though their individual parameters may vary in absolute value. Also, many of the passive components in the integrated amplifiers track better. Figure 13-109 shows a typical differential transistor amplifier.[3] For further detailed information, such as analysis procedures, design techniques, and application data, refer to the literature.[4,5]

112. Chopper (Modulated-Carrier) Amplifiers. The chopper amplifier performs the function of a direct-coupled amplifier by using a carrier frequency and an ac-coupled amplifier. The modulated-carrier approach is used specifically to minimize drift and offset types of errors. Although this approach is often more complex to implement, the performance improvement makes this technique desirable for applications demanding low-drift performance.

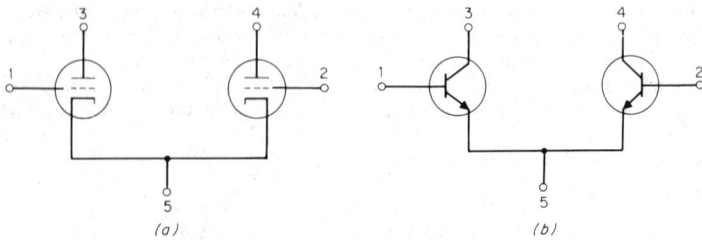

Fig. 13-108. Differential amplifier pairs. (*a*) Tube version; (*b*) transistor version.

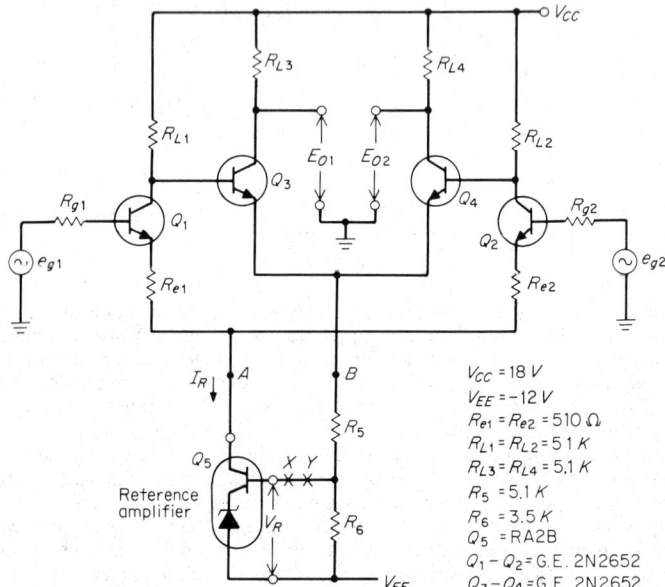

$V_{CC} = 18\ V$
$V_{EE} = -12\ V$
$R_{e1} = R_{e2} = 510\ \Omega$
$R_{L1} = R_{L2} = 51\ K$
$R_{L3} = R_{L4} = 5.1\ K$
$R_5 = 5.1\ K$
$R_6 = 3.5\ K$
$Q_5 = RA2B$
$Q_1 - Q_2 = G.E.\ 2N2652$
$Q_3 - Q_4 = G.E.\ 2N2652$

Fig. 13-109. Two-stage differential amplifier with common-mode feedback. *(General Electric Company.)*

Figure 13-110 shows the block diagram of a modulated-carrier amplifier. While the input and output signals are dc-coupled, the interstage coupling between the modulator, amplifier, and demodulator may be either a capacitor or a transformer.

Since the modulator and demodulator are usually operated synchronously, the carrier delay must be maintained at acceptable levels. Low pass filters are generally included in the modulator and demodulator. The choice of carrier frequency depends on the application. Typical examples of low-frequency carriers are 60 and 400 Hz. Carrier frequencies above 10 kHz are also feasible. Since the chopper and demodulator generate noise at their carrier frequency and its harmonics, the carrier frequency should normally be chosen above the baseband-frequency range of interest. Early chopper modulators and demodulators were mechanical devices, but present-day types employ solid-state electronics.

A variation of the modulated-carrier technique uses the chopper-stabilized approach illustrated in Fig. 13-111, which includes an additional parallel ac-coupled high-frequency signal path. The two signal paths have gains and bandwidths tailored to a crossover frequency; e.g., when the high-frequency signal path becomes dominant, the modulated-carrier signal path ceases to contribute to the sum total. This variation offers the low-drift advantages of the carrier-modulated system with much higher bandwidth.

In a typical example of a commercially available chopper-stabilized amplifier (Analog Devices Model 210/211 chopper-stabilized operational amplifier) the following characteristics are obtained:

Drift	0.5 μV/°C
	1 pA/°C
Slewing rate	100 V/μs
Bandwidth	20 MHz
Overload recovery	0.2 μs
Long-term stability	1 μV/day
Voltage gain	10^8
Noise	5 μV(p−p)

113. Operational Amplifiers. An operational amplifier is intended to realize specific signal processing functions. For example, the same operational amplifier (depending on the externally added components) can be used as an integrating amplifier, a differentiating amplifier, an active filter, or an oscillator, among others. Applications of operational amplifiers also include such functions as impedance transformers, regulators, and signal conditioning. They are versatile "building blocks" that can also be used in nonlinear applications to realize functions such as logarithmic amplifiers, comparators, ideal rectifiers, etc.

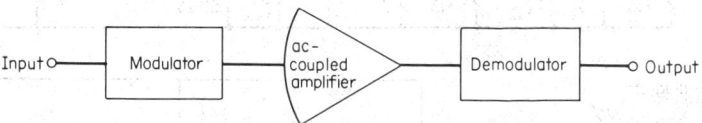

Fig. 13-110. Modulated-carrier (chopper) amplifier.

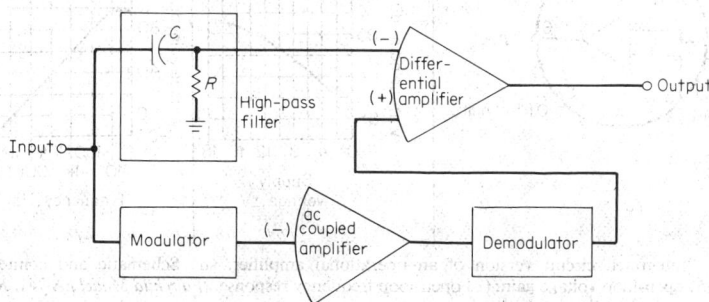

Fig. 13-111. Chopper-stabilized amplifier.

The early application of operational amplifiers was largely in the area of analog computations. The requirements placed upon the amplifiers were severe, and the cost was high. Currently, however, these high-performance functions are available at low cost, and they are widely used.

An operational amplifier can be either direct-coupled or ac-coupled. Most operational amplifiers have differential inputs and consequently realize common-mode rejection; however, many operational amplifiers are single-ended. Their power capability covers a wide range, and they can function as a power driver. The open-loop bandwidth can range from below 1 kHz up to the megahertz range. Voltage gain can be designed from unity to above 100,000.

To design or use the many excellent models now available commercially, it is necessary to understand the attributes and limitations of these versatile building blocks. There are many excellent references[1,4] that include design approaches, analysis techniques, and application data.

An example of a currently available and widely used integrated transistorized operational amplifier is shown in Fig 13-112 (see Ref. 7). Its electrical characteristics are listed in Table 13-7.

114. Application of Operational Amplifiers. Figure 13-113 shows the block diagram for an operational amplifier with associated external elements. A_1 represents the transfer

Schematic diagram

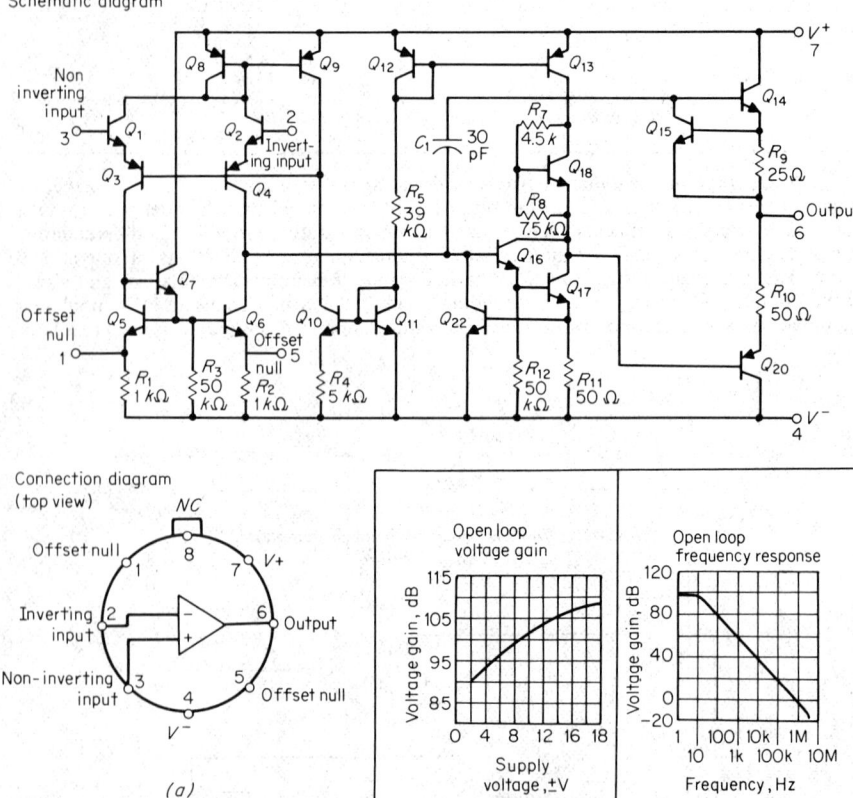

Fig. 13-112. Integrated circuit version of an operational amplifier. (*a*) Schematic and connection diagrams; (*b*) open-loop voltage gain; (*c*) open-loop frequency response. (*Fairchild Model μA 741, Ref. 7; Fairchild Semiconductor Corporation.*)

Table 13-7. Characteristics of Operational Amplifier Shown in Fig. 13-112*

Parameter	Conditions	Min.	Typical	Max.	Unit
Input offset voltage	$R_S \leq 10$ kΩ	...	1.0	5.0	mV
Input offset current	...	...	30	200	nA
Input bias current	...	...	200	500	nA
Input resistance	...	0.3	1.0	...	MΩ
Large-signal voltage-gain	$R_L \geq 2$ kΩ, $V_{out} = \pm 10$ V	50,000	200,000	...	V
Output voltage swing	$R_L \geq 10$ kΩ	±12	±14	...	V
	$R_L \geq 2$ kΩ	±10	±13	...	V
Input voltage range		±12	±13	...	V
Common-mode rejection ratio	$R_S \leq 10$ kΩ	70	90	...	dB
Supply-voltage rejection ratio	$R_S \leq 10$ kΩ	...	30	150	μV/V
Power consumption	...	...	50	85	mW
Transient response (unity gain):	$V_{in} = 20$ mV, $R_L = 2$ kΩ, $C_L \leq 100$ pF				
Rise time		...	0.3	...	μs
Overshoot		...	5.0	...	%
Slew rate (unity gain)	$R_L \geq 2$ kΩ	...	0.5	...	V/μs
Specifications for −55°C ≤ T_A ≤ +125°C:					
Input offset voltage	$R_S \leq 10$ kΩ	...	...	6.0	mV
Input offset current	...	...	...	500	nA
Input bias current	...	...	...	1.5	μA
Large-signal voltage gain	$R_L \geq 2$ kΩ, $V_{out} = \pm 10$ V	25,000	...	...	V
Output voltage swing	$R_L \geq 2$ kΩ	±10	...	...	V

*($V_S = \pm 15$ V, $T_A = 25$°C unless otherwise specified.)

function of the amplifier, either current gain or voltage gain, which is generally frequency-dependent. Z_i and Z_f are the primary elements that normally determine the closed-loop transfer function for this operational amplifier. The indicated Z_{in} and Z_l are equivalent summing node and load impedances, respectively, which are usually factored into the error terms of the composite transfer function. All the elements are general impedances that may be real or complex. Thus the simplified transfer function of this operational amplifier can be written

$$e_o/e_i \approx -Z_f/Z_i \qquad (13\text{-}24)$$

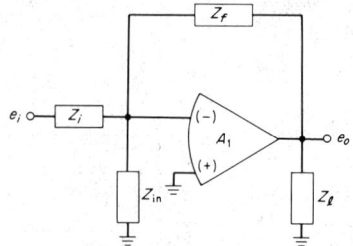

Fig. 13-113. Operational amplifier A_1 with external impedances which determine its functional application.

The error terms of the complete transfer function equation are not shown in Eq. (13-24); however, the complete equation with error terms can be readily derived or obtained from the literature.[1,2,5] Thus, if the required constraints are adhered to, Eq. (13-24) can be used to generate various transfer functions, as illustrated in the examples given below.

Integrating Amplifier. An example of an integrating amplifier is shown in Fig. 13-114. Assume that the external-component values are as indicated and that the A_1 amplifier is that previously described in Fig. 13-112. Using Laplace transform notation, the integrator transfer function becomes

$$\frac{e_0}{e_i} S = \frac{(-)1}{RCS} \qquad (13\text{-}25)$$

Using the component values specified in the figure, the integrator transfer function can then be written

$$\frac{e_0}{e_i}(S) = \frac{(-)5,000}{S} \qquad (13\text{-}26)$$

An alternative frequency-domain representation of the above integrator transfer function can also be used.

$$\frac{e_0}{e_i}(f) = \frac{(-)800}{jf} \qquad (13\text{-}27)$$

The closed- and open-loop frequency responses of this operational integrator amplifier are illustrated in Fig. 13-114b.

The frequency range of application depends primarily on the accuracy desired. In this example, 1-Hz to 10-kHz frequency is considered a realistic range of operation. For very low frequencies the error tends to increase due to inadequate excess gain within the operational

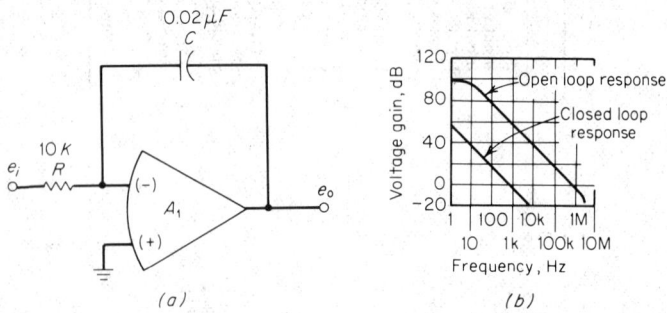

Fig. 13-114. Operational amplifier of Fig. 13-112 used for integration. (*a*) Basic schematic; (*b*) frequency response.

amplifier, whereas for high frequencies the closed-loop gain becomes low, and drift and offset errors may then become important.

Differentiating Amplifier. An example of a differentiating amplifier is shown in Fig. 13-115. The following simplified closed-loop equations are applicable:

$$\frac{e_0}{e_i}(S) \approx (-)RCS \tag{13-28}$$

$$\frac{e_0}{e_i}S \approx (-)0.165S \tag{13-29}$$

$$\frac{e_0}{e_i}(f) \approx (-)jf \tag{13-30}$$

The frequency responses of this operational amplifier differentiator are shown in Fig. 13-115b. The open-loop response of the operational amplifier of Fig. 13-112 is used, and the differentiator closed-loop response is superimposed onto it.

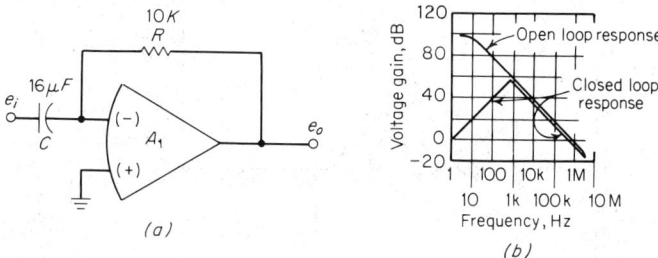

Fig. 13-115. Operation amplifier of Fig. 13-112 used for differentiation. (*a*) Basic schematic; (*b*) frequency response.

The frequency range of this operational amplifier differentiator depends primarily on the closed-loop differentiation accuracy required. The difference between open- and closed-loop transfer functions (usually referred to as *excess gain*) can be used to predict the accuracy of the function being generated. A realistic frequency range is from 0.01 to 100 Hz. At very low frequencies the closed-loop gain becomes very small, and consequently errors such as drift and offset may become critical. At high frequencies the accuracy degrades, and ultimately an integration function is generated rather than the differentiation function intended.

115. Servo Amplifiers. The function of a servo amplifier, one of the principal components in a control feedback system, is to amplify the input (usually low-level) error signals and to provide rated drive power to the load (the servo actuator). Servo amplifiers can be dc- or ac-operated, and can be linear or nonlinear.

Direct-coupled Servo Amplifiers. A direct-coupled servo amplifier can operate on dc error signals, i.e., zero frequency signals. The bandwidth of a dc servo amplifier is often quite large, and thus it has the capability of dynamic operation. In practice it is often operated in the dynamic mode, and the transient-response characterization is often used.

A servo amplifier is intended to drive or control its load to some prescribed reference level. It is the drift and offset errors associated with this control function that are the main concern in this dc mode of operation. Thus, if the actuator is positioned at its exact reference level, the amplifier should provide zero drive power. However, due to initial offsets or subsequent temperature or aging effects, the amplifier may still provide unwanted drive to the load. The usual solution is to minimize these problems by circuit design, e.g., tracking or balanced configurations, or by use of an intermediate-carrier modulated by the dc error input signals, subsequently demodulated for use as the direct coupled-drive to the actuator.

Alternating-Current Servo Amplifiers. An ac servo amplifier operates at a selected fixed frequency and is consequently a carrier-system amplifier. The most commonly used carrier frequencies are 60 and 400 Hz, but ac servo systems can be designed to operate at almost any frequency compatible with the servo actuator used.

The principal advantage of ac servo amplifiers is that the previously discussed drift and offset problems present in dc servo amplifiers are virtually eliminated. Another advantage is that the prime power from the actuator now can be supplied directly from the line, for example, 60-Hz 115-V source.

For applications wherein the servo actuator or load is a servomotor, the load is often tuned to the carrier frequency via the addition of a capacitor in parallel with the motor. Note that the servomotor impedance is a function of the motor speed; however, the losses (usually resistive) change, whereas the parallel inductance remains virtually constant. Thus tuning the load causes it to become real (resistive), and consequently the efficiency of the amplifier output stage improves significantly.

The output stage of an ac servo amplifier is usually operated class B or AB and in a push-pull configuration to obtain the best possible drive efficiency. Figure 13-116 shows an example of a push-pull tuned-load ac servo amplifier.[3]

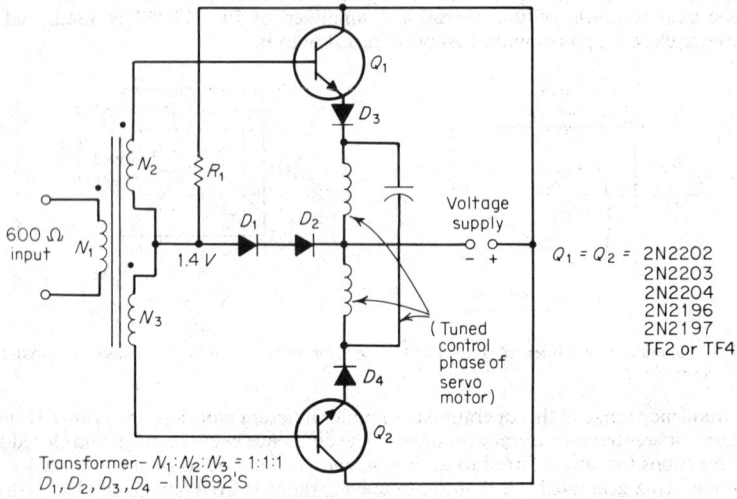

Fig. 13-116. Servo amplifier used as a motor drive, in the power range of 1 to 4 W. *(General Electric Company.)*

Nonlinear Servo Amplifiers. This class of servo amplifier uses the load to filter the highly nonlinear drive signals resulting in improved drive efficiencies and higher realizable driver-power capabilities. The drivers essentially act as switches wherein the dissipation losses are low when the switches are either on or off. Since the loads or servo actuators are usually highly reactive, they can be advantageously used in this manner. The drive power can be readily derived from a dc source via a pulse-width modulated scheme. In addition, an ac source of power can also be used via a phase-modulation technique. Some examples of high-power switching devices used for ac nonlinear servo amplifiers are thyratrons and silicon controlled rectifiers (SCRs).

116. References on Direct-coupled, Operational, and Servo Amplifiers

1. SHEA, R. F. "Amplifier Handbook," McGraw-Hill, 1968.

2. HUNTER, L. P. "Handbook of Semiconductor Electronics," McGraw-Hill, New York, 1970.

3. "Transistor Manual," 7th ed., General Electric Co., 1964.

4. TOBEY, G. E., HUELSMAN, L. P., and GRAEME, J. G. "Operational Amplifiers," McGraw-Hill, New York, 1971.

5. MIDDLEBROOK, R. D. "Differential Amplifiers," Wiley, New York, 1963.

6. "Model 210/211 Chopper Stabilized Operational Amplifiers," Analog Devices, Tech. Data Pub., Cambridge, Mass.

7. "Fairchild Semiconductor Integrated Circuit Data Catalog 1970," Fairchild Semiconductor, Mountain View, Calif., 1969.

OPERATIONAL AMPLIFIERS FOR ANALOG ARITHMETIC

By J. P. Hesler

117. Analog Multiplier Circuits. Circuits used for analog multiplication fall into three categories: transconductance multipliers, averaging-type multipliers, and exponential multipliers.

118. Transconductance Multipliers. The most prominent type of transconductance multiplier employs the property of the bipolar transistor, that is, its collector current and transconductance are linearly related. The balanced transistor differential amplifier used in these circuits offers good accuracy, wide bandwidth, and low cost.

In the differential amplifier as shown in Fig 13-117, the collector currents I_{c_1} and I_{c_2} are functions of the differential input voltage $V_{b_1} - V_{b_2}$. It is assumed that the current transfer ratios of the two transistors (Q_1 and Q_2), the junction ambient temperatures, and the base-emitter junction saturation currents are the same, as is probable in integrated circuit fabrication.

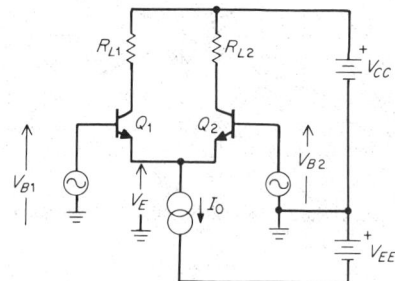

Fig. 13-117. Basic differential amplifier for analog multiplication.

The diode equation governs the relationship between the transistor emitter currents I_{E_i} and the base-to-emitter voltages V_{BE_i}.

$$I_E = I_S\left(\exp\frac{V_{BE}\,q}{kT} - 1\right)$$

where I_s = junction saturation current, q = electron charge, 1.6×10^{19}C, k = Boltzmann's constant 1.38×10^{-23}W/s · K, and T = junction temperature, K.

The relationships of collector currents, shown graphically in Fig. 13-118, are

$$I_{c_1} = \frac{\alpha_1 I_0}{1 + \exp\left[\dfrac{(V_{BE_2} - V_{BE_1})q}{kT}\right]} \qquad I_{c_2} = \frac{\alpha_2 I_0}{1 + \exp\left[\dfrac{(V_{BE_1} - V_{BE_2})q}{kT}\right]}$$

For zero differential input, $V_{BE_1} = V_{BE_2}$, and $\alpha_1 = \alpha_2$, and so

$$I_{c_1} = I_{c_2} = \frac{\alpha I_0}{2}$$

At zero input, $\Delta V_{BE} = 0$, the maximum single-ended transconductance occurs at

$$gm_{\max} = \frac{\alpha I_0}{4kt/q}$$

For a differential output connection the maximum transconductance is twice this value. The linear range of transconductance extends over a differential input signal range of about 50 mV peak to peak at room temperature.

   Linear multiplication occurs when one input varies the I_o term and the other input is used to vary ΔV_{BE}.

$$\Delta I_c = gm\Delta V_{BE} = \frac{\alpha I_o}{2kT/q}\Delta V_{BE}$$

   The linear range of input voltage may be extended by inserting resistance in series with each emitter. This increase in allowable signal is accompanied by a corresponding reduction in transconductance. For I_o equal to 2 mA, the addition of 50 Ω in series with each emitter increases the linear input swing by a factor of 3 and reduces the transconductance to one-third. Optimization of the transconductance multiplier operation requires a tradeoff between linearity and error sources due to offset voltages and thermal noise.

   A typical four-quadrant multiplier using the differential amplifier as building blocks is shown in Fig. 13-119. This circuit is generally used with differential inputs to obtain maximum linearity and common-mode signal rejection.

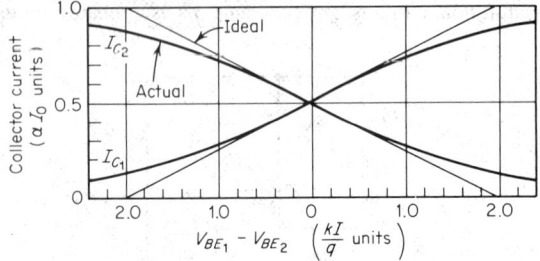

Fig. 13-118. Transfer curves on which analog multiplication is based.

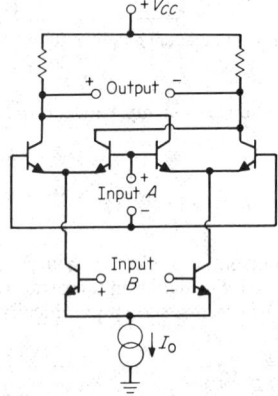

Fig. 13-119. Four-quadrant transconductance multiplier.

   119. Averaging-Type Multipliers. Two types of averaging multipliers have found extensive use. The first type, the pulse height/width multiplier, uses one of the input variables to modulate the height or amplitude of a pulse train, while the pulse width is modulated with the other input variable. The pulse area, averaged over a suitable period, is proportional to the product of the input variables. High pulse rates permit short averaging intervals and fast response. These multipliers offer very good accuracy and stability, but are more expensive than the transconductance multipliers. A block diagram of a typical pulse height/width averaging multiplier is shown in Fig. 13-120.

   A second version of the averaging multiplier is the triangle averaging multiplier. In this type a high-frequency triangular waveform is generated and combined with the input variables to form an averaged output proportional to the product of the input signals. These multipliers are less accurate than the pulse height/width multipliers and are being displaced by transconductance multipliers.

A third type of averaging multiplier is the time-base multiplier. This approach uses a comparator to sense the time interval required for the integral of a reference input voltage to equal the amplitude of a sample of one input variable. During the same interval, 0 to T, the other input variable is integrated to produce an output V. This type of circuit can be built with a few components to provide moderate accuracy, 1 to 5%, at low cost.

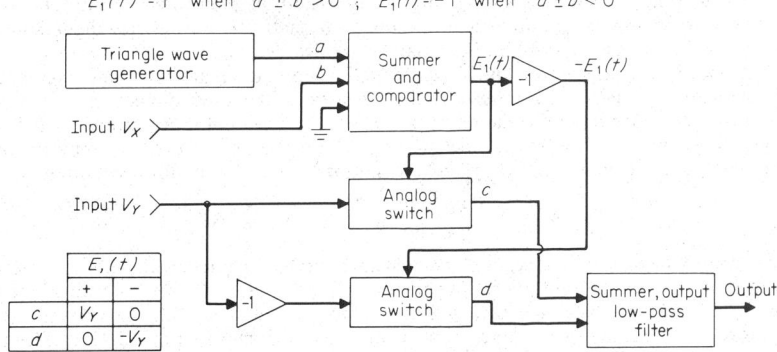

$E_1(t) = 1$ when $a \pm b > 0$; $E_1(t) = -1$ when $a \pm b < 0$

Fig. 13-120.   Block diagram of an averaging-type multiplier.

120. Exponential Multipliers. The first electronic multipliers were of the exponential type. In these circuits, resistor-diode networks are designed to provide a current or voltage output approximating the square of the input. These multipliers are also called *quarter-square* multipliers based on the identity

$$XY = \tfrac{1}{4}[(X + Y)^2 - (X - Y)^2]$$

The piecewise-linear approximations to the squared response result in "lumpy" error characteristics. Also, the amount of circuitry required to compute the quarter-square algorithm is expensive. Although these multipliers are capable of good accuracy and bandwidth, they are becoming obsolete.

A second type of exponential multiplier uses logarithmic amplifiers, a summer, and an antilog amplifier to implement the relationship

$$XY = \mathrm{antilog}_a (\log_a X + \log_a Y)$$

Semiconductor diodes are available with excellent logging characteristics over many decades of bias current. These diodes are used with operational amplifiers to realize the necessary functions. The circuits can provide moderate to good accuracy (with thermal compensation) and good bandwidth. Applications are restricted by unipolar input requirements and differential drift due to thermal effects.

121. Multiplier Error Sources. *Offset Error.* There are two subclasses of offset error. The first is static offset caused by variances in component parameters such as saturation current I_s, in transistors and diodes. The second error is due to drift in component parameters. While the initial static offset can be trimmed with external adjustments, the drift components must be compensated by introducing complementary temperature coefficients within the circuit. A common source of drift is local heating of diode junctions and resistive components having nonzero temperature coefficients.

Feed-through Error. Feed-through errors result from nonideal transfer characteristics. Two feed-through error contributions can exist. The first, E_{FY}, is defined as the output due to the input variable Y when the X input is zero. The second, E_{FX}, is the complementary function due to an X input. Feed-through errors may result in dc, fundamental, and harmonic components of the contributing input signal.

Gain and Nonlinearity Errors. Gain errors produce output deviations from the expected scale factor of the multipliers. Nonlinearity of the transfer functions of the multiplier can produce additional error contributions, as previously discussed. Gain is most apt to vary as a function of the combined input-signal values because the internal components are operated over a range of bias conditions.

In certain multiplier applications, required transient responses may overtax the bandwidth capabilities of the circuits. In these instances additional error terms may appear as a result of limited slew rate of the circuits and differential phase response between the input channels.

122. Squaring Circuits. Squaring circuits are readily implemented by introducing the variable to both inputs of a multiplier circuit. Alternatively, the resistor-diode squaring circuits used in exponential multipliers may be applied directly.

123. Dividing and Square-Root Circuits. Division and square-root functions may be implemented with basic multiplier circuits by altering the interconnections. A typical dividing-circuit connection is shown in Fig. 13-121. The multiplier output is fed back through a summing amplifier to one of the multiplier inputs. The summing amplifier maintains an equivalence between the numerator Y and the multiplier output XE_0/K.

$$\frac{Y}{R} = \frac{-XE_0}{KR} \quad \text{or} \quad E_0 = \frac{-KY}{X}$$

Square-root circuits can be implemented in a similar method with another feedback connection involving a multiplier. In this instance, the output is used as each of the two multiplier input variables, and the multiplier output is summed with the variable whose square root is desired. The square-root connection is shown in Fig. 13-122. The non-negative input limitations should be stated for division and square root connections.

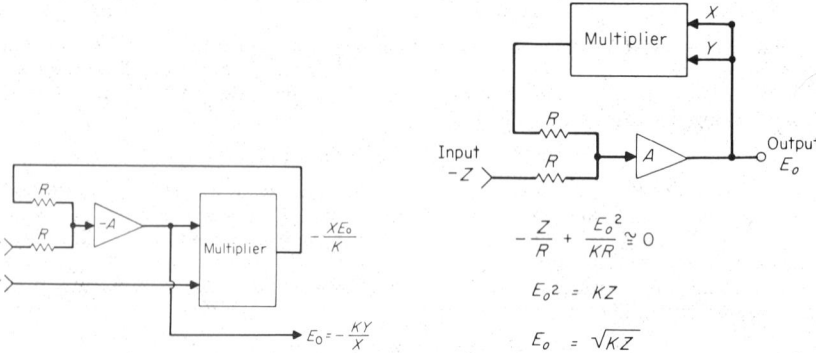

Fig. 13-121. Dividing circuit.   **Fig. 13-122.** Square-root circuit.

$$-\frac{Z}{R} + \frac{E_0^2}{KR} \cong 0$$

$$E_0^2 = KZ$$

$$E_0 = \sqrt{KZ}$$

124. References on Analog Multiplier Circuits

1. "Evaluating, Selecting, and Using Multiplier Circuit Modules for Signal Manipulation and Function Generation," Analog Devices, Cambridge, Mass., 1970.

2. Programmable Multiplier/Divider IC Includes a Variable Scale Factor, *Elect. Design News,* Nov. 15, 1972, p. 72.

3. Cate, T. Modern Techniques of Analog Multiplication, *Electron. Eng.,* April, 1970, pp. 75-79.

4. Abbott, H. W., and Mathis, V. P. Elapsed Time Computation, *Proc. Nat. Electron. Conf.,* Chicago, Ill., Vol. 25, Oct. 12-14, 1959.

HIGH-POWER AMPLIFIERS

By R. W. French

125. Thermal Considerations. A problem common to all high-power amplifier and oscillator equipment is that of removal of the excess thermal energy produced in the active devices and other circuit components so that operating temperatures consistent with reliable performance are maintained. Available cooling methods are radiation, natural-convection, forced-convection, liquid, evaporative, and conduction. Radiation and evaporative cooling are dependent for suitable operation upon a rather high temperature for the device being cooled and are thus generally restricted to use with vacuum tubes. The remaining methods are suitable for use in both solid-state and vacuum-tube systems.

126. High-Power Broadcast-Service Amplifiers. Transmitters for amplitude modulation broadcast service may employ high-level (plate) modulation or low-level (grid or screen) modulation of the output stage, or modulation of an intermediate stage, followed by linear amplification in the following stage(s). Generally, the latter approach is employed in very-high-power transmitters, because of the difficulty and expense in designing and building a modulation transformer to handle the high audio modulating power required. High-level modulation has been used in am transmitters up to at least 250 kW carrier power with a modulator output power requirement of 125 kW. By a unique design in which the positive terminal of the modulator plate supply is grounded, an autotransformer is utilized as the modulation transformer, with a significant reduction in size and cost.

127. Class B Linear Rf Amplifiers. The conventional means for achieving linear amplification of an am signal is the class B radio-frequency amplifier circuit, often referred to simply as a *linear amplifier*. The plate efficiency, plate dissipation, and output power are highly dependent upon the drive level. It is convenient, therefore, to define a *drive ratio*, or normalized drive level, k.

$$k = \frac{E_{pm}}{E_{bb}} \tag{13-31}$$

where E_{bb} = dc plate supply voltage, and E_{pm} = peak ac plate signal voltage.

In an ideal class B amplifier with sinusoidal drive, the dc plate current is

$$I_b = \frac{k i_{bm}}{\pi} \tag{13-32}$$

where i_{bm} = peak value of the plate current at full output power level.

The dc plate input power is thus

$$P_{DC} = E_{bb} I_b = \frac{k E_{bb} i_{bm}}{\pi} \tag{13-33a}$$

The output power is

$$P_o = \frac{(E_{pm})(k i_{bm})}{4} = \frac{k^2 E_{bb} i_{bm}}{4} \tag{13-33b}$$

The plate efficiency is

$$\eta_P = \frac{P_o}{P_{DC}} = \frac{k^2 E_{bb} i_{bm}/4}{k E_{bb} i_{bm}/\pi} = k\frac{\pi}{4} \tag{13-34}$$

The plate dissipation is

$$P_D = P_{DC}(1 - \eta_P) = \frac{k E_{bb} i_{bm}}{\pi}\left(1 - \frac{k\pi}{4}\right) \tag{13-35}$$

The maximum output power level occurs at full drive, i.e., for $k = 1$. Substituting $k = 1$ into Eq. (13-33b) gives

$$P_{o,\max} = \frac{E_{bb} i_{bm}}{4} \tag{13-36}$$

Normalizing P_{DC}, P_o, and P_D given by Eqs (13-33a), (13-33b), and (13-35), respectively, by the maximum output power, $P_{o,\max}$, from Eq. (13-36), gives

$$\frac{P_{DC}}{P_{o,\max}} = k\frac{4}{\pi} \tag{13-37}$$

$$\frac{P_o}{P_{o,\max}} = k^2 \tag{13-38}$$

$$\frac{P_D}{P_{o,\max}} = \frac{4k}{\pi} - k^2 \tag{13-39}$$

Equations (13-37) to (13-39) and (13-34) are plotted in Fig. 13-123 vs. the normalized drive level k defined in Eq. (13-31). The necessity for a high drive level to obtain a high efficiency is plainly evident from the curves in this figure.

The greatest single limitation on the plate efficiency obtainable from a practical class B amplifier is the inability to achieve a value of unity for the normalized drive level k, that is, the inability to achieve a peak ac plate voltage swing equal to the dc plate voltage. The minimum instantaneous plate voltage must never fall below the instantaneous peak positive grid voltage in a triode and should preferably be two to three times the latter; hence k must always be less than unity. A reasonable value for k in the best presently available triodes is

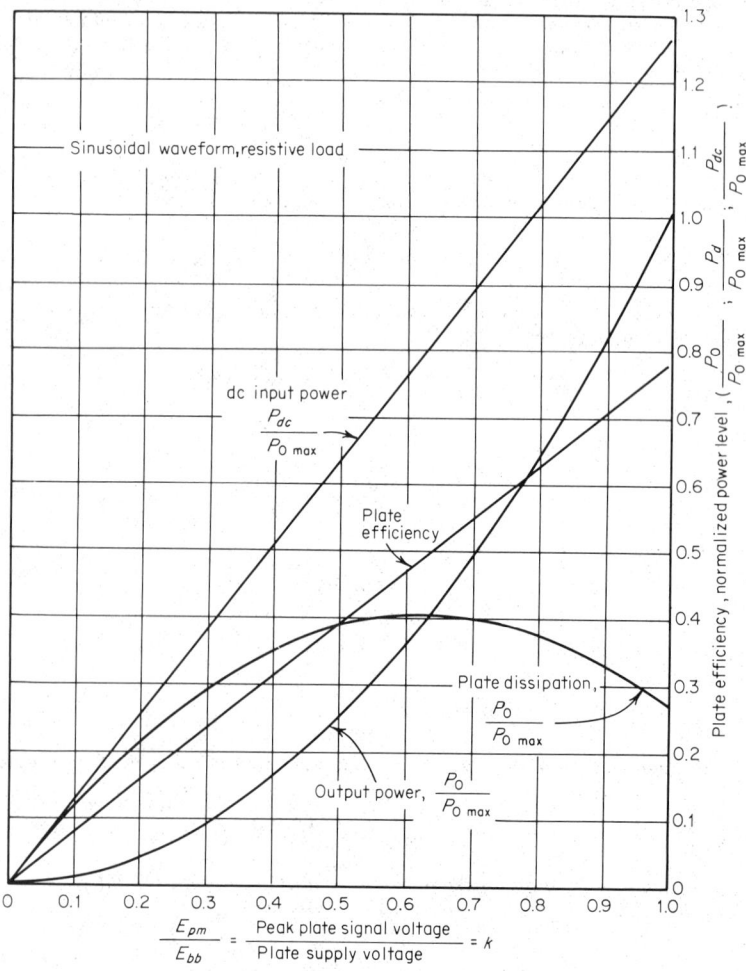

Fig. 13-123. Ideal class B amplifier characteristics.

about 0.9. Thus from Fig. 13-123, a maximum plate efficiency of 70.6% could be realized. Assuming an efficiency of 0.9 for the output-coupling circuitry, an overall plate circuit efficiency of 63.5% could be expected.

Class B amplifiers employing transistors can approach the theoretical maximum collector efficiency of $\pi/4$, or 78.54%, quite closely, as a consequence of the very low collector saturation voltage of transistors, which gives a correspondingly high value for k.

If an ideal class B amplifier is used to amplify an am signal, the efficiency is less than that attainable with a cw signal, because the average value of the drive ratio k is less for the am signal. With no modulation (carrier condition), the amplifier drive is adjusted to produce a plate voltage swing equal to one-half the maximum, so that, with 100% modulation, the full maximum swing is utilized on positive-modulation peaks. The carrier condition corresponds to a k of 0.5 in the ideal amplifier, and from Fig. 13-23, it is seen that the theoretical plate efficiency for this condition is 39.3%, or half the theoretical maximum. It is evident from the curve for dc plate input power in Fig. 13-123 that there will be no change in average input power with modulation, assuming a modulating waveform having zero average value. However, since output power varies as k^2, the output power, and thus the efficiency, should increase with modulation depth.

By making use of the relationship between the modulation index m of the am signal and the normalized drive ratio k, expressions for efficiency, plate output power, and dc plate input power can be derived for the ideal linear amplifier. The required equation relating m and k is

$$k = \frac{1}{2}(1 + m \cos \omega_a t) \tag{13-40}$$

where ω_a = modulating frequency. The derived equations are
Plate dissipation:

$$P_D = \frac{E_{bb} i_{bm}}{\pi}\left[\frac{1}{2} - \frac{\pi}{16}\left(1 + \frac{m^2}{2}\right)\right] \tag{13-41}$$

Output power:

$$P_o = \frac{E_{bb} u_{bm}}{16}\left(1 + \frac{m^2}{2}\right) \tag{13-42}$$

Direct-current plate input power (sum of P_D and P_o):

$$P_{DC} = \frac{E_{bb} i_{bm}}{2\pi} \tag{13-43}$$

Plate efficiency:

$$\eta_P = \frac{\pi}{8}\left(1 + \frac{m^2}{2}\right) \tag{13-44}$$

P_D and P_o, normalized by P_{DC}, are shown plotted in Fig. 13-124, along with η_p, versus the modulation index.

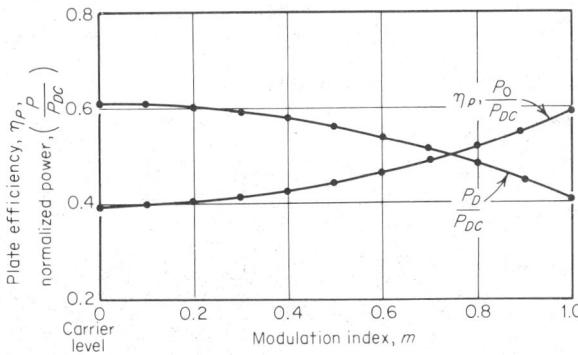

Fig. 13-124. Performance of ideal linear amplifier with sinusoidal amplitude modulation.

13-101

128. High-Efficiency Power Amplifiers. The plate efficiency at carrier level in a practical linear amplifier circuit is about 33%, while that of a high-level (plate) modulated stage is 65 to 80%. When the efficiency of the modulator is taken into account the net efficiency of the high-level modulated stage is still higher than that of the linear amplifier, since the output power level of the modulator is at most one-third of the total output power of the modulated stage.

Because of the difficulty and expense involved in the design and manufacture of very large modulation transformers and chokes, several linear amplifier circuits having much greater efficiency than the class B amplifier, while amplifying an amplitude modulated waveform, have been developed. Low-level modulation can be employed, and the need for the large, expensive modulator components is circumvented. Several such high-efficiency amplifier circuits are described in the following paragraphs.

129. Chireix Outphasing Modulated Amplifier. The Chireix outphasing modulation system permits generation of a high-level amplitude-modulated signal at good efficiency by driving the grids of two output-stage tubes with signals whose relative phase varies with the modulating signal, the output signals of the two tubes being applied by suitable networks to a common load.

Figure 13-125 shows the basic circuit arrangement of the output stage. If the control grids of the two tubes are driven with signals which are 180° out of phase, the output power will be zero, which would correspond to a negative-modulation peak of a 100% modulated am signal. By causing the angle θ to vary with modulating signal amplitude, the output power from the stage is made to vary.

Because of the outphasing method employed to produce the modulated output signal, the power factor associated with the output stage is of special significance. The basic modulation scheme would produce a unity power factor at zero output level which would decrease with increasing output level. To modify this undesirable characteristic the output plate circuits are detuned, one above and one below resonance, to produce an offset or dephasing of $\theta = \theta_0$. For $\theta_0 = 18°$, the power factor is zero for zero output ($\theta = 0$), rises rapidly to unity at $\theta = \theta_0$, and remains at 0.9 or higher for $\theta \geq \theta_0$. The low power factor at low output level does not have a significant effect on overall efficiency. The angle θ will have a value of approximately 25° at carrier level and 45 to 50° at a 100% positive-modulation peak. The

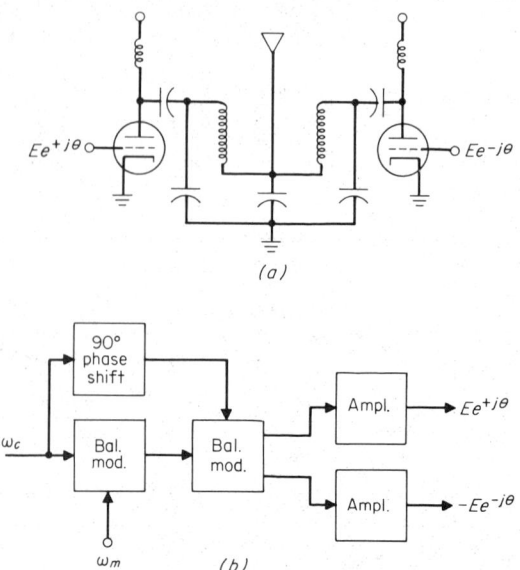

Fig. 13-125. Chireix outphasing modulated amplifier. (*a*) Output circuit; (*b*) drive signal generation.

linearity of the modulation characteristic is good, and efficiency is reasonably high, 60% being typical at carrier level, decreasing slightly with modulation.

The phase-modulated carrier-frequency driving signals for the output stage are generated at low level and amplified to the power level required. A suppressed-carrier am signal is first produced by combining the carrier and modulating signals in a balanced modulator. To this suppressed-carrier signal is added a carrier having a quadrature phase relative to the original (suppressed) carrier. Two such composite signals are generated having opposite sense of phase rotation. The amplitude variation of the composite signal is only about 10% for a phase variation of 50°.

An important difficulty in implementing the Chireix outphasing modulation circuit is maintaining the very stable phase characteristics of the circuits producing the two driving signals for the output stage, since a relative phase of ±45° between the two signals produces 100% modulation.

130. Dome High-Efficiency Modulated Amplifier. In the Dome high-efficiency modulated circuit, shown schematically in Fig. 13-126, modulation is achieved by load-line modification during positive-modulation swings and by linear amplification during negative-modulation swings. Load-line modification is achieved by absorption of a portion of the generated radio-frequency power; however, a major portion of the absorbed power is returned to the plate power supply, rather than being dissipated, and high-power efficiency results.

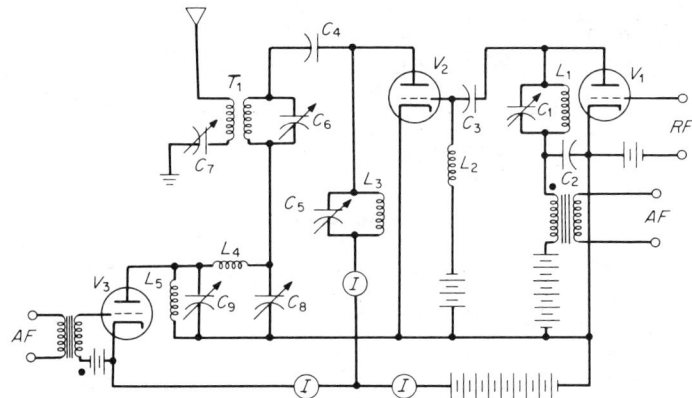

Fig. 13-126. Dome high-efficiency modulated amplifier.

The operation of the circuit shown in Fig. 13-126 is as follows: Tube $V1$ is used in a plate-modulated driver stage supplying power to the grid circuit of the power amplifier tube $V2$. The total load impedance in the plate circuit of $V2$ is that load impedance reflected into the primary of rf transformer $T1$ (from the antenna) in series with the impedance appearing across the $C8$ terminals of the 90° phase-shift network consisting of $C8$, $C9$, and $L4$. The impedance appearing across the $C8$ terminals of the 90° network is inversely related to the impedance across the other terminals of the network, i.e., the effective ac impedance of tube $V3$. Thus, with tube $V3$ cutoff, a short circuit is reflected across the $C8$ terminals of the network. Tube $V3$ is a modulated rectifier, the audio modulating signal voltage being applied to its control grid. Dome calls this tube a *modifier*.

With no modulation (carrier condition), tube $V3$ is biased to cut off and the drive to $V2$ is adjusted until the output power is equal to four times carrier power (corresponding to a positive-modulation peak). Note that all the plate signal voltage of $V2$ appears across the primary of the transformer $T1$ for this condition. The bias on tube $V3$ is then reduced, lowering the ac impedance of the tube, and therefore reflecting an increasing impedance at the $C8$ terminals of the 90° network. The bias on $V3$ is reduced until carrier power level is being delivered to the antenna. An amount of power equal to the carrier power is then being rectified by $V3$ and returned to the plate power supply, except for that portion dissipated on the $V3$ plate. The drive level to $V2$ is now adjusted so that the tube is just out of saturation.

For positive-modulation swings, tube $V3$ grid voltage is driven negative, reaching cutoff for a positive-modulation peak (100%). For negative-modulation swings, tube $V2$ acts as a linear amplifier. $V3$ does not conduct during the negative-modulation swing since the peak rf voltage on its plate is less than dc supply voltage on its cathode.

Because of the serial loss incurred in the power amplifier tube and in the modifier tube for energy returned to the power supply, the circuit is not as efficient as the Doherty circuit, 55 to 60% efficiency at carrier level being typical. The efficiency does not vary appreciably with modulation.

131. Doherty High-Efficiency Amplifier. The Doherty amplifier circuit is perhaps the most widely applied of the several types of high-efficiency circuits. This circuit is a high-efficiency linear amplifier circuit as opposed to high-efficiency circuits that achieve modulation as well as amplification. Doherty was the first to employ the 90° network as an impedance inverter to achieve load-line modification as a function of the power level in an amplifier stage.

In the circuit shown in Fig. 13-127a, the carrier tube $V1$ is biased class B and its loading and drive are such that it is operating at maximum linear voltage swing at carrier level. Tube

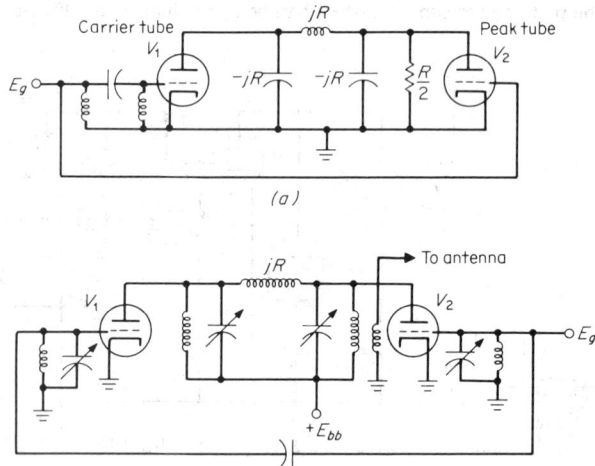

Fig. 13-127. Doherty high-efficiency amplifier.

$V2$ is biased class C such that at carrier level it is just beginning to conduct plate current. Each tube delivers an output power equal to twice carrier power when working into a load impedance of R Ω. At carrier level the reflected load impedance at the plate of $V1$ is $2R$ Ω, which is the correct value of load impedance for carrier-level output power at full plate voltage swing. The impedance-inverting property of the 90° network is like that of a quarter-wave transmission line; i.e.,

$$Z_{\text{in}} = \frac{Z_0^2}{Z_{\text{load}}}$$

where Z_0 = the characteristic impedance of the line. If the three reactances in the 90° network are each equal to R Ω, the input impedance is

$$R_{\text{in}} = \frac{R^2}{R_{\text{load}}}$$

Thus, for $R_{\text{load}} = R/2$,

$$R_{\text{in}} = \frac{R^2}{R/2} = 2R$$

The 90° phase lead network in the grid circuit of $V1$ is required to compensate for the 90° phase lag produced by the 90° network (impedance-inverting network) in the plate circuit. For negative-modulation swings the carrier tube performs as a linear amplifier with load impedance of $2R\ \Omega$, and the peak tube is inoperative.

On positive-modulation swings the peak tube conducts and contributes power to the $R/2$ load resistance. This is equivalent to connecting a negative resistance in shunt with the $R/2$ load resistance, so that the load resistance seen at the load end of the 90° network increases. This increase is reflected through the network as a decrease in load resistance at the plate of $V1$, causing an increase in its output current, and hence in its output power. The drive levels on the tube are adjusted so that each contributes the same power (equal to twice carrier power) at a positive-modulation peak. For this condition a load of $R\ \Omega$ is presented to each tube.

The important aspect of the circuit operation is the change in load impedance on tube $V1$ with modulation which enables it to deliver increased output power at constant plate voltage swing, and thus at high efficiency and good linearity. Some distortion of the modulation envelope will occur at carrier level as the peak tube comes into operation. The distortion can be reduced by a reduction in bias on the peak tube, but this causes a reduction in efficiency. Envelope feedback is normally used to improve the linearity. The efficiency of the Doherty circuit is 60 to 65%, essentially independent of modulation depth.

A more practical circuit for the Doherty amplifier is shown in Fig. 13-127b, where the shunt reactances of the phase-shift networks are supplied by detuning the related tuned circuits, the tuned circuits in the grid circuits being tuned above the operating frequency, while those in the plate circuits are tuned below the operating frequency.

132. Terman-Woodyard High-Efficiency Modulated Amplifier. The Terman-Woodyard high-efficiency grid-modulated amplifier uses the basic scheme of Doherty for achieving high efficiency, i.e., the impedance-inverting property of a 90° phase-shift network. However, the Terman-Woodyard circuit employs grid modulation of both the carrier tube and the peak tube, which allows both tubes to be operated class C, with the result that an increase in efficiency over the Doherty circuit is obtained. It has the highest efficiency of the systems currently available.

Referring to Fig. 13-128, $V1$ is the carrier tube and $V2$ is the peak tube. In the absence of modulation, $V1$ is operating as a class C amplifier, supplying the carrier power, and $V2$ is

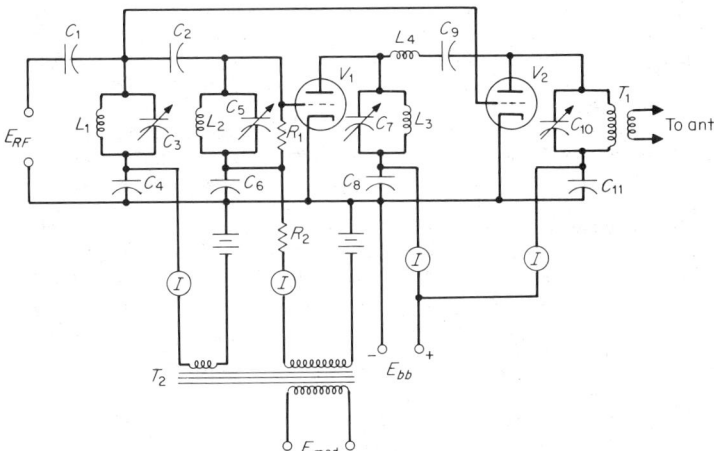

Fig. 13-128. Terman-Woodyard high-efficiency modulated amplifier.

biased so that it is just starting conduction. The efficiency at carrier level is thus essentially that of a class C amplifier.

The shunt reactances of the 90° phase-lead network in the grid circuit of $V1$ and the 90° phase-lag network (impedance-inverting network) in the plate circuit of $V1$ are provided by detuning the tuned circuits $L1$-$C3$, $L2$-$C5$, $L3$-$C7$, and $C10$ and the primary inductance of $T1$. $C2$ is the series element of the phase-lead network, and $L4$ is the series element of the phase-lag network. Capacitor $C9$ enables the metering of the individual plate currents of $V1$ and $V2$.

Resistor $R1$ is included to prevent overdriving of the grid of $V1$ on positive-modulation swings. The relative modulating voltage applied to the two tubes is controlled by adjustment of $R2$.

During a positive-modulation swing, tube $V2$ conducts, and at a (100%) modulation peak the two tubes are supplying equal amounts of power to the load, similar to the Doherty amplifier operation. During a negative-modulation swing, $V2$ is cut off and $V1$ performs as a standard grid-modulated amplifier.

Terman and Woodyard show curves of plate efficiency as a function of modulation percentage (sinusoidal modulation) for both circuits based upon a peak plate voltage swing equal to 80% of the dc plate supply voltage. The curves indicate an efficiency for the Doherty amplifier of 62.8% at carrier level, dropping to a minimum of 57% at 50% modulation and rising back to 62.8% at 100% modulation. The curve for the Terman-Woodyard circuit shows an efficiency of 80% at carrier level and a minimum of 68% at 50% modulation and 73.4% at 100% modulation. These efficiency values do not take into account output-circuit losses.

The adjustments in rf drive level, modulation voltage amplitude, and tuning and loading to obtain proper operation of the Terman-Woodyard circuit are tedious. The application of envelope feedback is very desirable in order to reduce distortion and minimize the effects of misadjustment of the operating conditions on performance. Large amounts of feedback can normally be applied in this circuit because the feedback loop includes only the one rf stage.

133. Induction Heating Circuits. Induction heating is achieved by placing a coil carrying alternating current adjacent to a metal workpiece so that the magnetic flux produced by the current in the coil induces a voltage in the workpiece, which produces the necessary current flow.

Power sources for induction heating, in addition to direct use of commercial power, include spark-gap converters, motor-generator sets, vacuum-tube oscillators, and inverters. Motor-generator sets generally provide outputs from 1 kW to more than 1MW and from 1 to 10 kHz. Spark-gap converters are generally used for the frequency range from 20 to 400 kHz and provide output power levels up to 20 kW. Vacuum-tube oscillators operate at frequencies from 3 kHz to several MHz and provide output levels from less than 1 kW to hundreds of kW. Inverters using mercury-arc tubes have been used up to about 3 kHz. Solid-state inverters have been developed in recent years which operate at frequencies up to about 10 kHz and at power levels of several megawatts. These solid-state inverters generally employ thyristors (silicon controlled rectifiers) and are replacing motor-generator sets and mercury-arc inverters.

Induction heaters employing vacuum tubes generally operate at frequencies of 300 kHz and higher and are available for output power levels from about 5 to 200 kW in single equipments. A single power tube is usually employed in an oscillator circuit. A simplified circuit diagram of a typical 20-kW, 450-kHz induction heater is shown in Fig. 13-129.

134. Dielectric Heating. Whereas induction heating is used to heat materials which are electrical conductors, dielectric heating is used to heat nonconductors or dielectric materials. The basic arrangement for a dielectric heating system is that of a capacitor in which the material to be heated forms the dielectric or insulator. The heat generated in the material is proportional to the loss factor, which is the product of the dielectric constant and the power factor. Because the power factor of most dielectric materials is quite low at low frequencies, the range of frequencies employed for dielectric heating is higher than for induction heating, extending from a few megahertz to a few gigahertz.

The power generated in a material is given by

$$P = 141 V^2 f \frac{K \cos \phi}{t} \times 10^{-6} \quad \text{watts}$$

where V = voltage across material in volts, f = frequency of the power source in megahertz,

A = area of the material in square inches, K = dielectric constant, t = material thickness in inches, and cos ϕ = power factor of the dielectric material.

The voltage that may be applied to a particular material is limited by the insulation properties of the material at the required process temperature. The frequency that may be used is limited by voltage standing waves on the electrodes, which will be appreciable when the electrode dimensions are comparable with one-eighth wavelength (10% voltage variation).

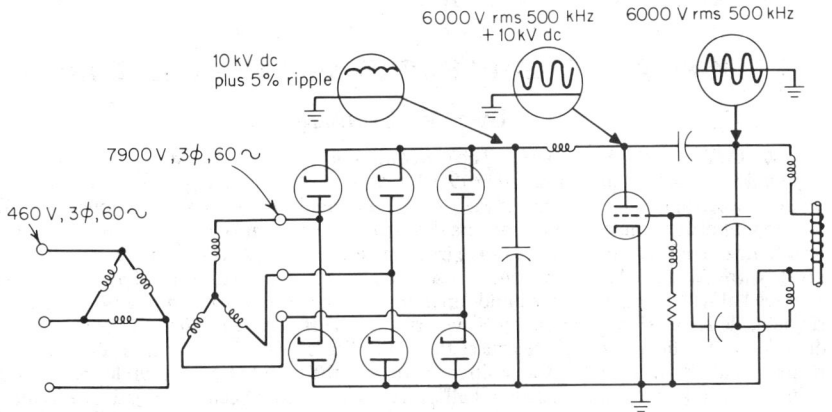

Fig. 13-129. Induction heating circuit of 20-kW power.

135. References on High-Power Amplifiers

1. CHIREIX, H.   High Power Outphasing Modulation, *Proc. IRE*, Vol. 23, pp. 1370-1392, November, 1935.

2. DOME, R. B.   High-Efficiency Modulation System, *Proc. IRE*, Vol 26, pp. 963-982, August, 1938.

3. DOHERTY, W. H.   A New High Efficiency Power Amplifier for Modulated Waves, *Proc. IRE*, Vol. 24, pp. 1163-1182, September, 1936.

4. TERMAN, F. E., and WOODYARD, J. R.   A High-Efficiency Grid-modulated Amplifier, *Proc. IRE*, Vol. 26, pp. 929-945, August, 1938.

5. FISHER, S. T.   A New Method of Amplifying with High Efficiency a Carrier Wave Modulated in Amplitude by a Voice Wave, *Proc. IRE*, Vol. 34, pp. 3p-13p, January, 1946.

6. SAINTON, J. B.   A 500 Kilowatt Medium Frequency Broadcast Transmitter, *Machlett Cathode Press*, Vol. 22, No. 4, 1965.

7. Greenville, U.S.A., the World's Largest Broadcast Facility 4.8 MW, *Machlett Cathode Press*, Vol. 22, No. 4, 1965.

8. CURTIS, F. W.   "High-Frequency Induction Heating," McGraw-Hill, New York, 1964.

9. BROWN, G. H., HOYLER, C. N., and BIERWIRTH, R. A.   "Theory and Application of Radio-Frequency Heating," Van Nostrand, New York, 1947.

10. STANSEL, N. R.   "Induction Heating," McGraw-Hill, New York, 1949.

11. SPASH, D. I.   "High Frequency Heating," Pt 3 of Vol. 3 in A. H. Beck (ed.), "Handbook of Vacuum Physics," Macmillan-Pergamon, New York, 1964.

12. CABLE, J. WESLEY   "Induction and Dielectric Heating," Reinhold, New York, 1954.

13. HUGHES, L. E. G., and HOLLAND, F. W. (eds.)   "Electronic Engineer's Reference Book," 3d. ed., Heywood Books, London, 1967.

14. FRANK, W. E.   New Developments in High-Frequency Power Sources, *IEEE Trans. Ind. Gen. Appl.*, Vol. IGA-6, No. 1, pp. 29-35.

15. ROSS, N. V.   A System for Induction Heating of Large Steel Slabs, *IEEE Trans. Ind. Gen. Appl.*, Vol. IGA-6, No. 5, pp. 449-454.

16. DEWAN, S. B., and HAVAS, G.   A Solid-State Supply for Induction Heating and Melting, *IEEE Trans. Ind. Gen. Appl.*, Vol. IGA-5, No. 6, pp. 686-692.

17. Hatchard, D. G.   Induction Heating of Bars and Semifinished Steel, *IEEE Trans. Ind. Gen. Appl.,* Vol. IGA-2, No. 5, pp. 346–352.

18. Shea, R. F. (ed.)   "Amplifier Handbook," Chap. 20, McGraw-Hill, New York, 1968.

19. Cockrell, W. D. (ed.)   "Industrial Electronics Handbook," Chap. 5c, McGraw-Hill, New York, 1958.

20. A New Hardboard Production System Uses a 600 Kilowatt Dielectric Heater, *Machlett Cathode Press,* Vol. 25, No. 2, 1968.

MICROWAVE AMPLIFIERS AND OSCILLATORS

By J. W. Lunden

136. IMPATT Diode Circuits.   The generation of microwave power in a reverse-biased *pn* junction was originally suggested in 1958 by W. T. Read.[1]*   Read proposed that the finite delay between applied rf voltage and the current generated by avalanche breakdown, with the subsequent drift of the generated carriers through the depletion layer of the junction, would lead to negative resistance at microwave frequencies.   The diode is biased in the avalanche-breakdown region.   As the rf voltage rises above the dc breakdown voltage during the positive half-cycle, excess charge builds up in the avalanche region, reaching a peak when the rf voltage is zero.   Hence this charge waveform lags the rf voltage by 90°.   Subsequently, the direction of the field in the diode causes the multiplied carriers to drift across the depletion region.   This, in turn, induces a positive current in the external circuit while the diode rf voltage is going through its negative half-cycle.   This is equivalent to negative resistance, which is maximum when the transit angle is approximately 0.74π.   See also Sec. **9.**

A simplified equivalent circuit of the IMPATT diode circuit is shown in Fig. 13-130.   The resistance R_D includes both the parasitic positive resistance due to contacts, bulk material,

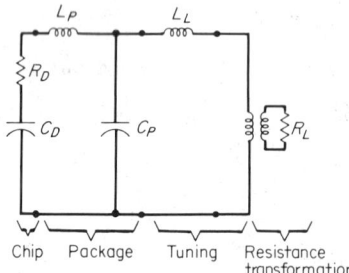

Fig. 13-130.   Approximate rf equivalent circuit of IMPATT oscillator.

etc., and the dynamic negative resistance.   The net magnitude is typically in the range -0.5 to -4.0 Ω and varies with current.   The capacitance C_D is the voltage-sensitive depletion-layer capacitance and can be approximated sufficiently accurately with the value at breakdown.   The diode resistance variation results in a stable operating point for any positive load resistance equal to or less than the diode peak negative value.

Oscillations will build up and be maintained at the frequency for which the net inductive reactance of the package parasitics and the external load equals the capacitive reactance of C_D.   The values for L_P and C_P vary with package or mounting style.   Typical values range from 0.3 to 0.6 nH and 0.2 to 0.4 pF, respectively.   It is important to minimize these parasitics since they limit the operating frequency and bandwidth.

IMPATT diodes may be utilized in several mounting configurations, including coaxial, waveguide, strip-line, or microstrip.   It is important, to ensure that a good heat flow path is provided[2] (due to the typically low efficiency) and that low-electrical-resistance contacts be made to both anode and cathode.

In avalanche breakdown, the diode tends to look like a voltage source.   Hence a current source is desirable for dc bias.   Several circuits are possible.   The *RC* bias circuit (Fig. 13-

---

* Superior numbers correspond to numbered references, Par. **13-146.**

131*a*) is the simplest, but is inefficient, and the transistor current regulator (Fig. 13-131*b*) may be more desirable. In either case, the loading of the diode with shunt capacitance or a resonance path to ground (at some frequency) must be avoided to prevent instabilities (noise and/or spurious frequencies).

Two broadly tunable diode loading circuits are the multiple-slug cavity[3] and the variable-package-inductance types. Coaxial implementations of these two techniques are shown in the cross sections of Fig. 13-132. In Fig. 13-132*a*, slugs $\lambda/8$ and/or $\lambda/4$ long at the desired center frequency of operation with characteristic impedance of between 10 and 20 Ω are adjusted in position along the centerlines to provide a load reactance equal to the negative of the diode reactance and a load resistance equal to or less than the magnitude of the diode net negative resistance. Circuit *b* tunes the diode by recessing it into the holder, effectively decreasing the net series inductance, which is resonant with the diode capacitance. Single- or multisection transformers can also be included for resistive matching to the load R_L. Similar waveguide-circuit implementations can be used with typically narrower bandwidths, better frequency stability, and lower fm noise.

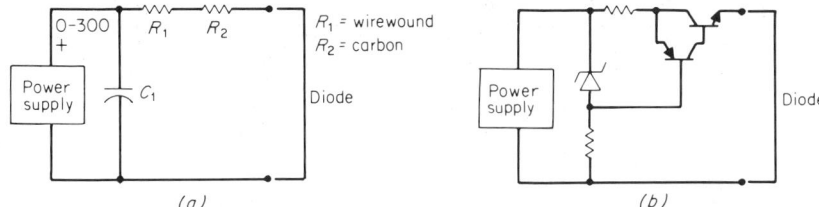

Fig. 13-131.   Bias circuits for IMPATT diodes. (*a*) *RC* type; (*b*) transistor-regulated type.

The IMPATT diode can function as an amplifier if the load resistance presented to it is larger in magnitude than the negative resistance of the diode. Typically, a circulator is used in conjunction with the tuned diode circuit to separate input and output signals. At the center frequency, the power gain of the amplifier is given by

$$G_0 = \left(\frac{R_D - R_L}{R_D + R_L}\right)^2$$

where R_D and R_L are as shown in Fig. 13-130.

In general, amplifier operation for a given diode requires a smaller shunt tuning capacitor or a larger transformer characteristic impedance. An estimate of the diode rf current and voltage can be obtained[3] from

$$V_D \approx \frac{\sqrt{2\,P_o R_L}}{\omega C_j(V_b)}\left(1 + \frac{1}{\sqrt{G_0}}\right)$$

$$I_D \approx \omega C_j(V_b)V_D$$

where V_b = diode breakdown voltage, C_j = junction capacitance, P_o = power output and G_0 = gain. Higher output powers have been achieved by multiple-diode series-parallel configurations and/or hybrid combining techniques.

In general, properly designed IMPATT oscillator circuits can have noise performance comparable with reflex Klystron or Gunn oscillators. The noise performance of both Si and GaAs IMPATT diode amplifiers and oscillators has been extensively treated in the literature.[4-7]

137. TRAPATT Diode Circuits.   The TRAPATT (*tr*apped *p*lasma *a*valanche *tr*iggered *tr*ansit) mode of operation is characterized by high efficiency, operation at frequencies well below the transit-time frequency, and a significant change in the dc operating point when the diode switches into the mode. The basic understanding of this high-efficiency mode of operation has proceeded from consideration of IMPATT diode behavior with large signals.[9,10] Two saturation mechanisms in the IMPATT diode, space-charge suppression of the avalanche and carrier trapping, reduce the power generated at the transit-time frequency, but play an important role in establishing the "trapped-plasma" states for high-efficiency operation.

To manifest the high efficiency of the TRAPATT mode, four important circuit conditions must be met: Large IMPATT-generated voltage swings must be obtained by trapping the IMPATT oscillation in a high-Q-cavity circuit; selective reflection and/or rejection of all subharmonics of the IMPATT frequency except the desired subharmonic must be realized (typically with a low-pass filter); sufficient capacitance must be provided near the diode to sustain the high-current state: and tuning or matching to the load must be provided at the TRAPATT frequency.   Figure 13-133 illustrates a widely used circuit that achieves the foregoing conditions.

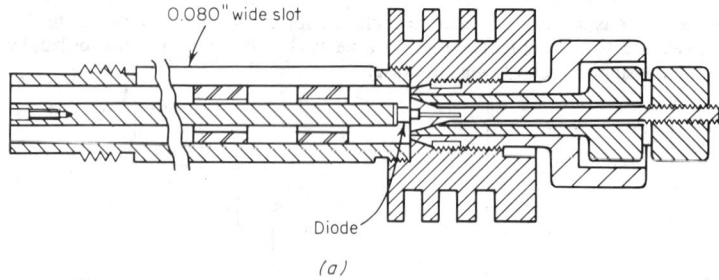

(a)

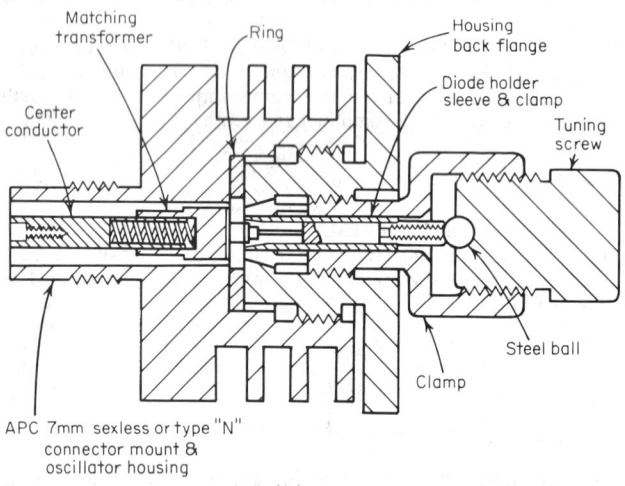

(b)

Fig. 13-132.   Mechanical tuning of IMPATT circuits. (*a*) Multiple-slug type; (*b*) variable-inductance (diode-recess) type. *(Hewlett Packard Company.)*

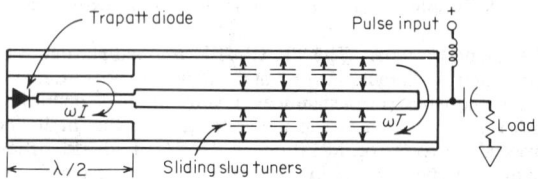

Fig. 13-133.   Typical TRAPATT diode circuit configuration.

The TRAPATT diode is typically represented by a current pulse generator and the diode's depletion-layer capacitance, as shown in the simplified schematic of Fig. 13-134. This permits a simple interpretation of operation. Initially, a large voltage pulse reaches the diode, triggering the traveling avalanche zone in the diode. This initiates a high-current low-voltage state. The drop in voltage propagates down the line *l*, is inverted (due to the -1 reflection coefficient of the low-pass filter), and travels back toward the diode. The process then repeats. Consequently, the frequency of operation is inversely proportional to the length of line *l*. The period of oscillation is slightly modified by the finite time required for the diode voltage to drop to zero. The inductance L is due to the bond lead and is helpful in driving the diode voltage high enough to initiate the TRAPATT plasma state. The capacitance C is provided to supply the current required by the diode to the extent that the transmission line is insufficient. The total capacitance which can be discharged during the short interval of high conduction current in the diode is

$$C_T = C + \tau_P/Z_0$$

where τ_P = time that the high-current or trapped plasma state exists. τ_P must be at least as large as the transit time of carriers through the diode, thus putting an upper limit on C. This is contrary to the high-current requirements to initially drive the avalanche zone through the diode (10,000 to 20,000 A/cm^2) in large-area devices.

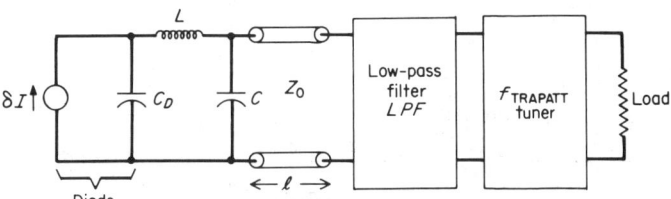

Fig. 13-134. Simplified schematic diagram of TRAPATT circuit.

Since the TRAPATT frequency is generally an integral submultiple of the diode transit-time frequency, the line length l ($l = \lambda/2$ at $f_{TRAPATT}$) presents the low-pass-filter short circuit to the diode as a series resonance at the IMPATT frequency. This net series resonance, however, includes the diode series-reactive elements. Further, this circuit should have a high Q at the transit-time frequency to reduce the buildup time to TRAPATT initiation.

Several TRAPATT circuit configurations are shown in Fig. 13-135. The coaxial cavity circuit in (*a*) places the diode in the reentrant gap of the half-wave cavity resonator. The output coupling loop passes only the fundamental to a triple stub tuner to match to the load. Proper dc biasing and bypassing are included. This circuit is good into lower L-band. The lumped circuit (*b*) is compact and very useful for vhf and uhf. The series capacitor is resonated with the inductance of a copper bar and the self-inductance of the diode. The trimmer controls the resonance of the third and fifth harmonics. The circuit of (*c*) is a variation on the circuit of Fig. 13-133. The use of additional filter sections and/or lumped elements provides better harmonic filtering and results in higher efficiencies. Circuits analogous to those of Fig. 13-133 can be implemented in waveguide for the higher frequencies. In all these circuits, the presence of the third and fifth harmonics has been found essential for stable and high-efficiency performance. Higher power levels can be achieved by operating multiple diodes in series and/or parallel configurations.[12]

Another useful technique for extending both power and frequency is the antiparallel diode configuration. The circuit consists of two diodes, placed with opposite polarity approximately one-half fundamental wavelength apart in a transmission line.[13] Operation is similar to a free-running multivibrator. Output may be extracted with a transmission line connected to the midpoint of the diodes, followed by the usual low-pass filter. The position of this filter should be adjusted so that the round-trip delays from midpoint to diodes and the filter are equal. A microstrip circuit realization for antiparallel operation is given in Fig. 13-136.

Operation of the anomalous avalanche diode for microwave amplification has also been established.[15] A 10-dB dynamic range is typical, with a low-level threshold gain decreasing with increasing power level to a saturated condition. The pulsed bias may be replaced by a simple dc source and storage capacitor. Unlike the *locked-oscillator* mode of operation, only a small residual output is present without the input signal. The locked oscillator will typically display only a 3:1 power change between locked and unlocked cases.

TRAPATT operation has yielded output power levels of 10 to 500 W with efficiencies of 20 to 75% in the frequency range from 0.5 to 10 GHz.

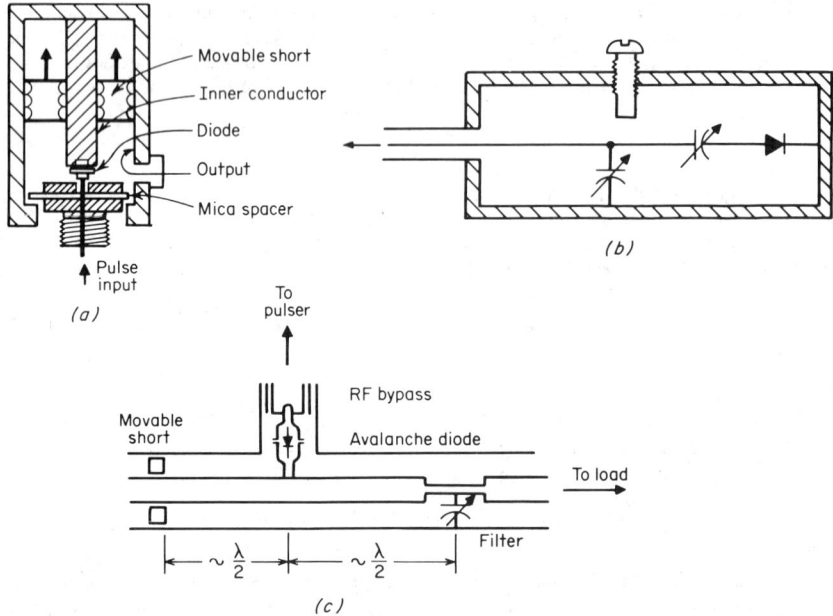

Fig. 13-135. TRAPATT diode circuit arrangements. (*a*) Tuned coaxial cavity; (*b*) lumped circuit; (*c*) coaxial circuit.

138. Transferred Electron Effect Device (TED) Circuits. This class of circuit, using both the Gunn and LSA (limited-space-charge accumulation), devices depends on the internal negative resistance due to carrier motion in the semiconductor at high electric fields.[16,17] When the material is biased above the critical threshold field, a negative dielectric relaxation time is exhibited, which results in amplification of any carrier concentration fluctuations, causing a deviation from space-charge neutrality. The resultant *domain* drifts toward the anode and is extinguished, and a new domain is formed at the cathode. The current through the sample consists of a series of narrow spikes with a period equal to the transit time of the domain.

When an rf voltage is superimposed on the bias, in a given period of time, the terminal voltage can be below both the threshold voltage V_{th} and the domain-sustaining voltage V_s. The domain is quenched at any place in the sample when the latter occurs, and the nucleation of a new domain is delayed until the voltage again exceeds V_{th}. Therefore the frequency of oscillation is determined by the resonant circuit, including the impedance of the sample. Experimental results and computer modeling have shown that the device may be tuned over greater than an octave bandwidth by the external circuit cavity. Although other modes of operation are possible, depending on the characteristics of the external circuit, the LSA mode appears to be the most important.

An approximate equivalent circuit is given in Fig. 13-137, with values dependent on frequency, bias, and power level. The capacitance includes the diode static capacitance, in addition to that due to traveling high field domains.

One of the simplest tuned circuits is the coaxial-line cavity as shown in Fig. 13-138a. The diode is mounted concentric with the line, at one end to facilitate heat sinking. The frequency of oscillation is determined primarily by the length of the cavity, the position of the output coupling loop (or plate) determining the load impedance.

A rectangular waveguide cavity configuration (Fig. 13-138b) is more widely used due to its higher Q and better performance at X-band and higher frequencies. The diode post acts as a large inductive susceptance, which, with the inductive iris, produces the resonant frequency for which the length l is $\lambda/2$. The tuning rod lowers the frequency as its insertion length increases.

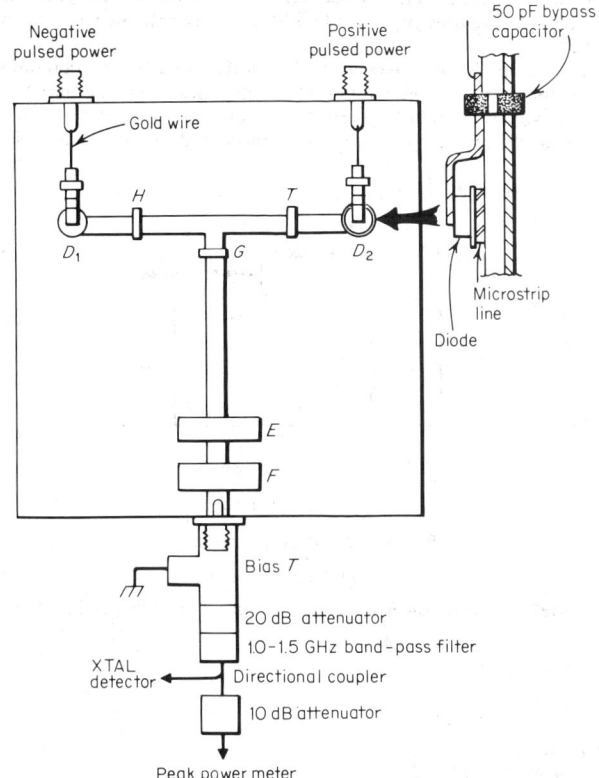

Fig. 13-136.   Microstrip version of antiparallel TRAPATT circuit.

In addition to mechnical tuning, both YIG[18] and varactor[19] tuning techniques are applicable. YIG tuning has the potential for the widest tuning range but is limited in tuning speed, hysteresis, and physical bulk. Figure 13-139 illustrates a typical varactor-tuned Gunn oscillator equivalent circuit and characteristics. The varactor diode Q is highest at the maximum reverse voltage and decreases with decreasing voltage due to an increase in R_{vs}. Figure 13-140 illustrates two varactor-tuned implementation techniques.

The noise characteristic of a GaAs TED is comparable with that of a Klystron. Various noise-source models and measuring equipments are discussed in the literature. Several methods can be employed to reduce the AM and FM noise of a Gunn oscillator, including increasing the loaded Q of the cavity circuit; biasing at or near the frequency and/or power turnover points (i.e., bias at which $df/dv = 0$ and $dP_o/dv = 0$); minimizing power-supply ripple; and diode selection.

The TED can also be operated as an amplifier, typically using circulator or hybrid techniques[20,21] similar to the IMPATT circuits. Parameters of importance are saturation characteristics, bandwidth, gain and phase tracking, linearity, efficiency, and dynamic range. The block diagram of a four-stage chain is shown in Fig. 13-141*a* with operating characteristics in Fig. 13-141*b*. Hybrid coupling was found to provide greater linear power output. Gains in excess of 20 dB can be realized. Efficiency and bandwidth are typically less than 10% and greater than 35%, respectively.

TEDs may be operated with a pulsed bias, allowing extremely high power density without damage to the device. Over 1,000 W (see ref. 22) at 10 GHz has been achieved in the LSA mode with short low-duty cycle pulses. Reactive termination and unloading of the circuit at the harmonic frequencies can significantly improve the efficiency. A problem associated with pulsed oscillators is the significant frequency change during each pulse caused by rapid temperature rise. Starting-time jitter can be alleviated by *priming,* i.e., injecting a weak cw signal into the circuit.

139. Transistor Amplifier and Oscillator Microwave Circuits. Silicon bipolar and GaAs field-effect transistors are available with cutoff frequencies extending into X- and K-bands, respectively. Equivalent circuits for these transistor types are given in Fig. 13-142. The intrinsic chip element values, with variations considered at low frequencies, are further

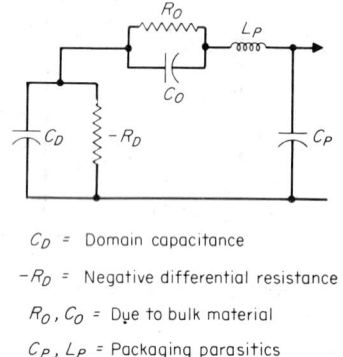

C_D = Domain capacitance

$-R_D$ = Negative differential resistance

R_O , C_O = Due to bulk material

C_P , L_P = Packaging parasitics

Fig. 13-137. Approximate equivalent circuit of a TED device and its package.

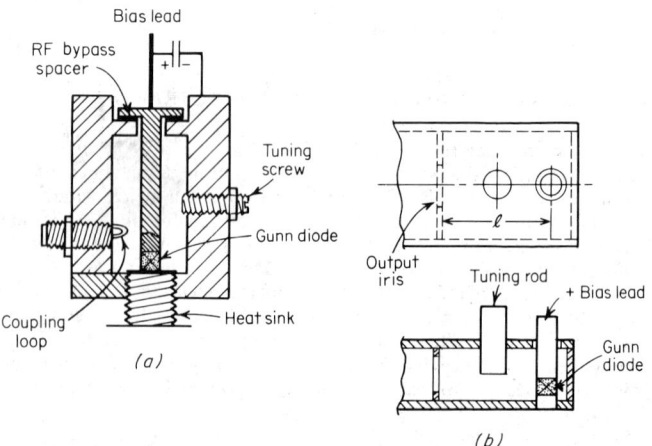

Fig. 13-138. Gunn diode cavity circuits. (*a*) Coaxial form; (*b*) waveguide cavity form.

modified by high-frequency effects. A small-signal figure of merit has been defined in terms of the contributing time constraints.

$$K = (\text{power gain})^{1/2}(\text{bandwidth}) = \frac{1}{4\pi(r'_b C_c \tau_{ec})^{1/2}}$$

where $\tau_{ec} = \tau_e + \tau_b + \tau_l + \tau_c$, τ_e = emitter barrier charging time, τ_b = base transit time, τ_l

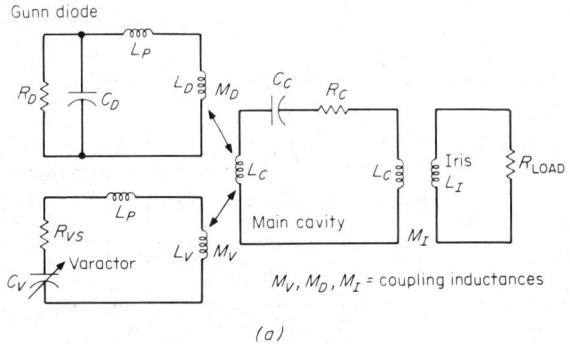

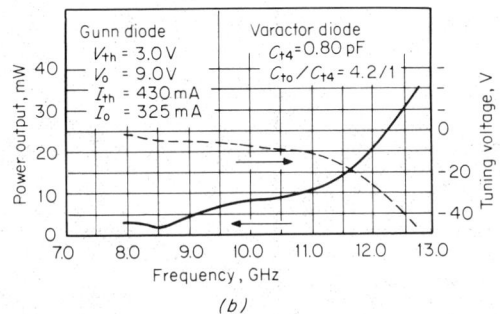

Fig. 13-139. Varactor tuning of Gunn oscillator. (a) Simplified equivalent circuit; (b) tuning characteristic.

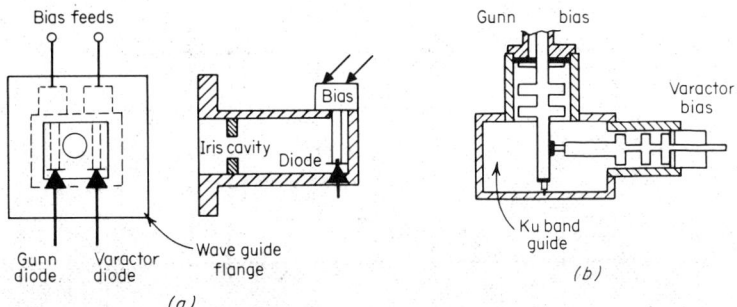

Fig. 13-140. Varactor tuning techniques for Gunn diodes. (a) Front and side views of double-port waveguide type; (b) type used at K_n-band.

13-115

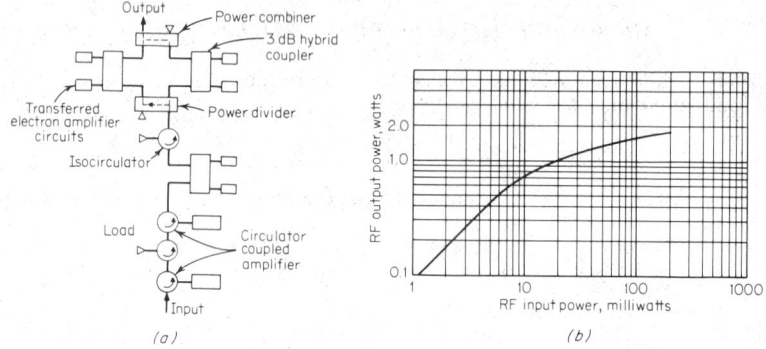

Fig. 13-141. Four-stage ganged TED amplifier chain for 5.7 GHz. (*a*) Block diagram; (*b*) power transfer curve.

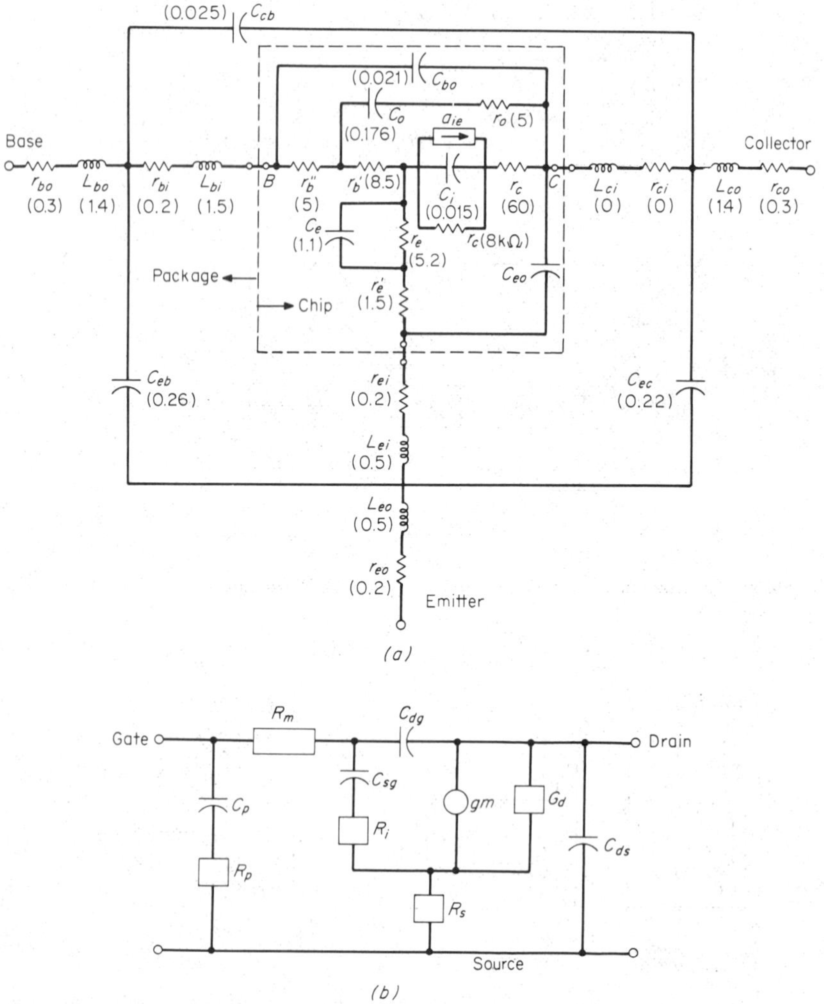

(a)

Fig. 13-142. Equivalent circuits of microwave solid-state devices. (*a*) Bipolar transistor; (*b*) Schottky barrier field-effect transistor (chip only).

13-116

= collector transit time, and τ_c = collector depletion-layer charging time. The maximum frequency of oscillation is defined as the frequency for which the power gain is unity.

$$f_{max} \approx \frac{1}{4\pi(r_b' C_c \tau_{ec})^{1/2}}$$

The application of transistors at microwavelengths demands that considerable attention be given to packaging, fixturing, and impedance characterization. Historically, characterization has taken the form of f_T, $r_b' C_c$ specification and/or h-y-parameter techniques. A more desirable method is by scattering parameters. Scattering parameters describe the relationship between the incident and reflected power waves in any N-port network.[23,24] As such, this technique offers substantial advantages, including remote measurement, broadband (no tuning), stability (no short-circuited or open terminations), accuracy, and ease of measurement.

From Fig. 13-143, the scattering equations describing the two-port network can be written.

$$b_1 = S_{11} a_1 + S_{12} a_2$$

$$b_2 = S_{21} a_1 + S_{22} a_2$$

Solving for the S parameters yields

$$S_{11} = \frac{b_1}{a_1}\bigg|_{a_2=0} = \text{input reflection coefficient with } Z_L = Z_0$$

$$S_{12} = \frac{b_1}{a_2}\bigg|_{a_1=0} = \text{reverse transmission gain with } Z_L = Z_0$$

$$S_{21} = \frac{b_2}{a_1}\bigg|_{a_2=0} = \text{forward transmission gain with } Z_L = Z_0$$

$$S_{22} = \frac{b_2}{a_2}\bigg|_{a_1=0} = \text{output reflection coefficient with } Z_L = Z_0$$

Other linear two-port parameters may be calculated from S parameters (for example, y parameters for feedback analysis). Either manual or complex automatic measurement techniques can be used. In either case, the transistor chip package is embedded in a system with a given reference impedance Z_0 and well-defined reference wave planes.

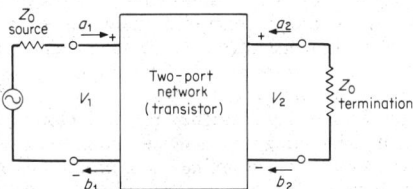

Fig. 13-143. Basic two-port configuration for a microwave transistor.

A number of useful relationships can be calculated from the S parameters and used in amplifier-oscillator design.

Reflection coefficient-impedance relationship:

$$S_{11} = \frac{Z - Z_0}{Z + Z_0} \qquad Z_0 = \text{reference impedance}$$

Input reflection coefficient with arbitrary Z_L:

$$S_{11}' = S_{11} + \frac{S_{12} S_{21} \Gamma_L}{1 - S_{22} \Gamma_L}$$

Output reflection coefficient with arbitrary Z_S:

$$S_{22}' = S_{22} + \frac{S_{12} S_{21} \Gamma_S}{1 - S_{11} \Gamma_S}$$

13-117

Stability factor:

$$K = \frac{1 + |D|^2 - |S_{11}|^2 - |S_{22}|^2}{2(S_{12}S_{21})}$$

Transducer power gain:

$$G_T = \frac{|S_{21}|^2(1 - |\Gamma_S|^2)(1 - |\Gamma_L|^2)}{|(1 - S_{11}\Gamma_S)(1 - S_{22}\Gamma_L) - S_{12}S_{21}\Gamma_L\Gamma_S|^2}$$

Maximum available power gain:

$$G_{max} = \left| \frac{S_{21}}{S_{12}}(k \pm \sqrt{k^2 - 1}) \right| \qquad \text{for } k > 1$$

Source and load reflection coefficients for simultaneous match:

$$\Gamma_{ms} = M^* \left[\frac{B_1 \pm \sqrt{B_1^2 - 4|M|^2}}{2|M|^2} \right]$$

$$\Gamma_{mL} = N^* \left[\frac{B_2 \pm \sqrt{B_2^2 - 4|N|^2}}{2|N|^2} \right]$$

where Γ_S, Γ_L = source and load reflection coefficients

$$M = S_{11} - DS_{22}^*$$
$$N = S_{22} - DS_{11}^*$$
$$D = S_{11}S_{22} - S_{12}S_{21}$$
$$B_1 = 1 + |S_{11}|^2 - |S_{22}|^2 - |D|^2$$
$$B_2 = 1 + |S_{22}|^2 - |S_{11}|^2 - |D|^2$$

The maximum power gain is obtained only if the transistor is terminated with the Γ_{ms} and Γ_{mL} resultant impedances. A lossless transforming network is placed between the source and load to realize this transformation.

Generally, the embedding circuits utilized take the form of simple ladder networks: series-shunt combinations of L's and C's or their transmission-line equivalents. These elements can be determined by moving on the Smith Chart from the value of the terminating impedance to the center of the chart along constant resistance-conductance, impedance-susceptance contours for series-shunt elements, respectively. The circuits are typically implemented in strip-line or lumped form, as shown in Fig. 13-144.

Optimum design at more than one frequency requires plotting of gain circles at each frequency, with subsequent terminating impedance iteration to obtain the best compromise across the band. Several computer-aided optimization programs[25] are available to simplify this routine. A plot of gain circles and impedance loci for a typical S-band design is shown in Fig. 13-145, which corresponds to the collector circuit of Fig. 13-144a.

140. Noise Performance of Microwave Transistor Circuits. The noise performance of a well-designed amplifier depends almost entirely upon the noise figure of the first transistor. It has been shown[26] that the noise factor of a transistor amplifier is related to its equivalent circuit parameters and external circuit by

$$NF = 1 + \frac{r_{bb}'}{R_g} + \frac{(r_e)}{2R_g} + \frac{(r_{bb}' + r_e + R_g)^2}{2r_e R_g h_{feo}}\left[1 + \left(\frac{f}{f_\alpha}\right)^2(1 + h_{feo})\right]$$

As a result, r_{bb}' should be as low as possible, and the alpha cutoff frequency should be high. The source resistance providing minimum noise figure is

$$R_g\Big|_{F_{min}} = \left[(r_{bb}' + r_e)^2 + \frac{(r_{bb}' + 0.5r_e)(2h_{feo}r_e)}{1 + (f/f_\alpha)^2(1 + h_{feo})}\right]$$

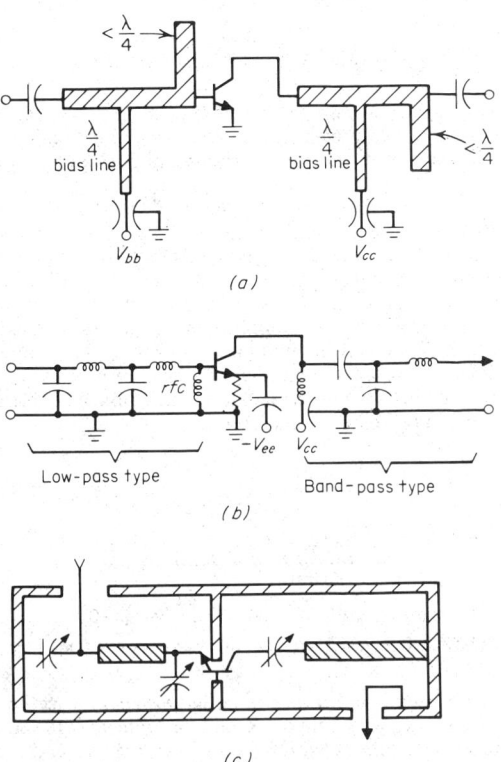

Fig. 13-144. Circuit realizations for microwave transistor circuits. (a) strip-line; (b) lumped design; (c) coaxial design.

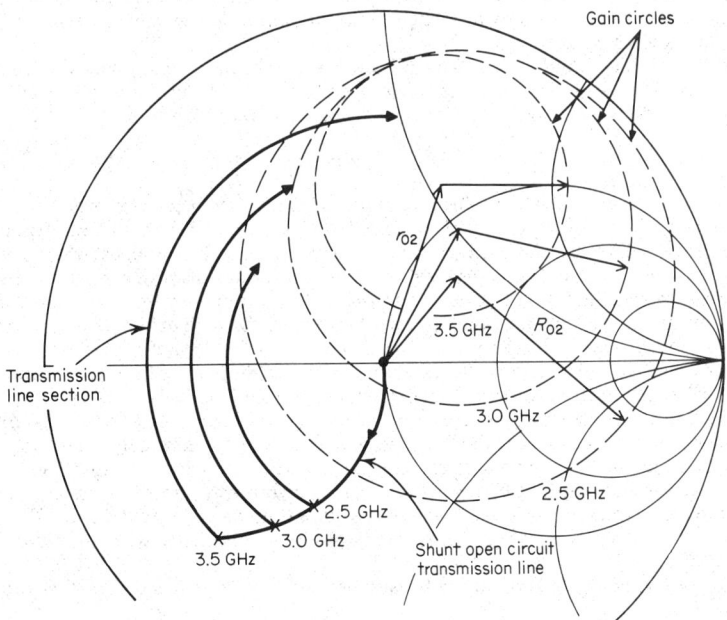

Fig. 13-145. Use of Smith Chart in microwave transistor circuit design.

13-119

This value is typically close to the value providing maximum power gain for the common-emitter configuration. Care must be taken in the matching-circuit implementation to minimize any losses in the signal path, e.g., by using high-Q elements and isolated bias resistances.

141. High-Power Microwave Transistor Amplifiers. The difficulties arising in power-amplifier operation are due to the nonlinear variation of device parameters as a function of time and to bias conditions. Saturation, junction capacitance-voltage dependence, h_{fe} current (and voltage) level dependence, and charge storage effects are the prime contributors to this situation. Class A operation is normally not used, implying collector current conduction angles less than $360°$, resulting in further complication of the time-averaging effects. With these qualifications, the basic equivalent circuit given in Fig. 13-142 applies. Figure 13-146 shows a greatly simplified equivalent circuit useful for first-order design. Of particular note is the low input resistance, high output capacitance, and a nonnegligible feedback element causing bandwidth, gain, and stability limitations.

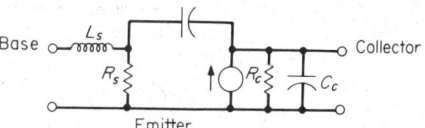

$$L_S \approx 1\text{nH} \quad ; \quad R_S = 0.5 \text{ to } 5\Omega \quad ; \quad C_c = 2 \text{ to } 20\,\text{pF}$$

Fig. 13-146. Equivalent circuit of microwave power transistor.

Several methods have been used to determine large-signal-device characteristics with varying degree of success. One method involves the measurement of the embedding circuitry at the plane of the transistor terminals, with the transistor removed and the source and load properly terminated.[27] The circuit is tuned for maximum power gain before the transistor is removed to make the measurement. An average or effective device impedance is then the complex conjugate of the measured circuit impedance at that frequency. The maximum power obtainable from a particular transistor is determined by thermal considerations (cw operation), avalanche-breakdown voltages, current gain falloff at high current levels and second-breakdown effects.

Bandwidth. The input impedance has been found to be the primary bandwidth-limiting element. R_S varies inversely with the area of the transistor. Hence, for a given package L_S, the Q increases and bandwidth decreases with higher-power transistors.

$$Q = \frac{\omega_0 L_S}{R_S} \qquad BW \bigg|_{3dB} \approx \frac{f_0}{Q}$$

Impedance Matching. The problem of matching complex impedances over a wide band of frequencies has been treated by Fano and others (see Ref. 17, Par. 13-66). Essentially, high-order networks may be used to achieve nearly rectangular bandpass characteristics. However, without mismatching, the 3-dB bandwidth determined above cannot be extended. In fact, the greater the ratio of generator resistance R_g to transistor input resistance R_S the greater will be the reduction of the intrinsic bandwidth, for a given ripple and number of circuit elements. Hence the external circuit design must consider both the transistor-input-circuit Q value and the impedance level relative to the driving source for a given bandwidth.

Either lumped- or transmission-line-element networks of relative simplicity are typically used to achieve the necessary input/output matching. Although the bandpass type yields somewhat better performance, the low-pass configuration is more convenient to realize physically. Quarter-wave line sections are particularly useful for broadband impedance transformations and bias feed-bypassing functions. Eighth-wave transformers are useful to match the small complex impedances directly without tuning-out mechanisms. The input impedance to a $\lambda/8$ section is real if it is terminated in an impedance with magnitude equal to the Z_0 of the line.

Load Resistance. The desired load resistance may be determined to a first order by

$$R_L \approx \frac{(V_{CC} - V_{CE,\text{sat}})^2}{2P_o}$$

where P_o = desired fundamental power output, and V_{CC} = collector supply voltage. This expression is altered by several factors, including circuit Q, harmonic frequencies, leakage, current conduction angle, etc. Assuming only the presence of the fundamental frequency, V_{CC} is limited to $\frac{1}{2} BV_{CBO}$ by resonant effects. Recently, it has been shown that BV_{CBO} may be exceeded for short pulses without causing avalanche.

Power Gain. The power gain depends on the dynamic f_T or large-signal current gain-transient capability, the dynamic input impedance, and the collector load impedance. A simple expression for power gain is

$$PG = \frac{(f_T/f)^2 R_L}{4 R_e(Z_{in})}$$

The high current at f_T is of particular importance. The effect of parasitic common-terminal inductance is to reduce this gain in the common-emitter and to cause regeneration in the common-base connection. The latter configuration is more commonly used at frequencies near or above the f_T value, whereas the former generally results in a more stable circuit below f_T. This situation is highly dependent on the parasitic-element situation with respect to the specific frequency. Various forms of instabilities, such as hysteresis (jump modes), parametric, low-frequency, and thermal, can occur due to the parameter values changing with time-varying high-signal levels. Usually, most of these difficulties can be eliminated or minimized by careful design of bias circuits (including fewer elements), ground returns, parasitics, and out-of-band terminating impedances.

The *collector efficiency* of a transistor amplifier is the ratio of rf power output to dc power input. High efficiency implies low circuit losses, high ratio of output resistance to load resistance, high f_T, and low collector saturation voltage. In addition, experiments and calculations show that for high efficiency, the impedance presented to the collector by the output network should be inductive for the favored generation of second harmonic. If the phase is correct, the amplitude of the fundamental is raised beyond the limit otherwise set by the difference between the supply voltage and $V_{C,sat}$. Figure 13-147 illustrates this effect.

A high value of f_T relative to the operating frequency improves efficiency by causing the operating point to spend less time (per cycle) in the high-dissipation active region between cut-off and saturation. The integrity of the transistor die bond has been found to have a substantial effect on efficiency due to the effects of low intrinsic bulk collector resistance and thermal gradient.

142. Transistor Oscillators. Transistor oscillators may be designed by choosing the source (or load) terminating impedances such that $S'_{11} \Gamma_S \geq 1 (S'_{22} \Gamma_L \geq 1)$. The design is facilitated by plotting stability circles[28] on the Smith Chart.

A number of circuit configurations are appropriate, including the standard Colpitts, Hartley, Clapp, coupled-hybrid, etc. The major difference, however, is the proper inclusion of device parasitic reactances into the intended configuration. The frequency-determining element(s) may be in the input, output, or feedback circuit, but should have a high Q for good

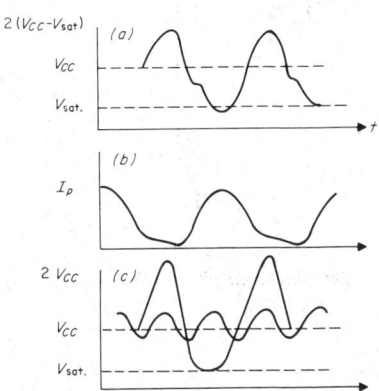

Fig. 13-147. Typical voltage-current waveforms in power applications. (*a*) Collector voltage; (*b*) collector current; (*c*) second-harmonic enhancement.

frequency stability. Care must be taken to decouple or resistively load the circuit at frequencies outside the band where oscillatory conditions are satisfied. This is particularly true for lower frequencies where the current gain is much greater.

The common-base and common-collector transistor connections are generally more unstable than common-emitter and most commonly used, depending on the power requirements and frequency of oscillation.

Several methods are utilized to vary the frequency dynamically. Bias variation is effective only over a narrow band and results in substantial power variation. The YIG sphere provides very high Q but is somewhat bulky and of limited tuning speed. The varactor diode requires very-low-power octave-tuning bandwidth and fast tuning in a form compatible with hybrid integration. Figure 13-148 shows the schematic of an S-band varactor-tuned wideband oscillator implemented in hybrid thin-film form. The frequency-determining elements are connected in the high-Q common-base input circuit. This also allows maximum isolation from load mismatch effects (pushing) and permits the balance of the circuitry to be low-Q (broadband).

Both high- and low-power oscillators are large-signal, that is, the output power is limited primarily by beta falloff, with increasing current at a given frequency and collector voltage.

143. Traveling-Wave-Tube Circuits. The traveling-wave tube[29] is a unique structure capable of providing high amplification of rf signals varying in frequency over several octaves without the need for any tuning or voltage adjustment (see Sec. 9). Figure 13-149 shows the principal components of an amplifier using this tube. Electrons emitted by the electron-gun assembly are sharply focused, drawn through the length of the slow-wave rf structure, and eventually dissipated in the collector. Synchronism between the rf electro-magnetic wave and the beam electrons results in a cumulative interaction which transfers energy from the dc beam to the rf wave. For details on this type of amplifier see Pars. **9**-29 to **9**-38a.

Figure 13-150 shows the gain characteristics of a typical broadband TWT amplifier. As more energy is extracted from the electron beam, it slows down. This loss of synchronism results in lower gains at higher power levels. One advantage of this overdrive characteristic is the protection of following stages against strong signals.

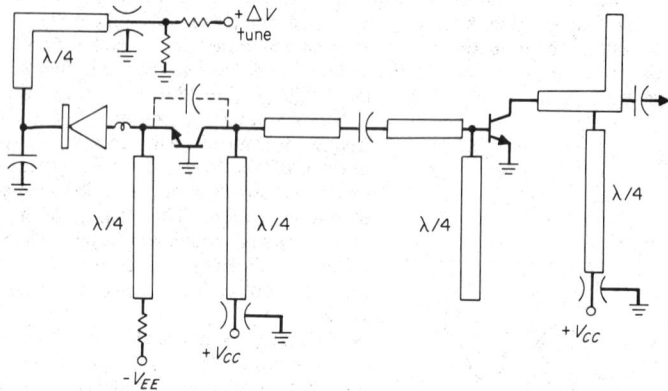

Fig. 13-148. Varactor-tuned microstrip oscillator-buffer circuit.

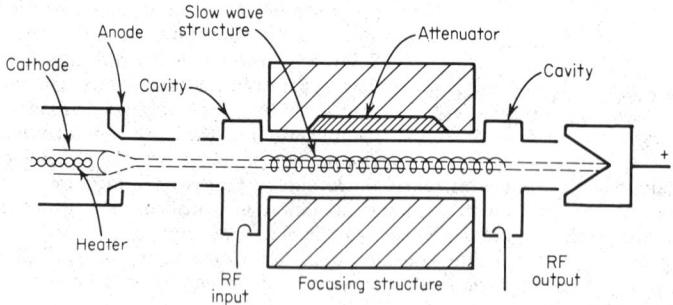

Fig. 13-149. Traveling-wave amplifier circuit.

13-122

The backward-wave oscillator[30] is essentially a TWT device making use of the interaction of the electron stream with an electromagnetic wave whose phase and group velocities are 180° apart. At a sufficient beam current, oscillations are produced as discussed above, without a reverse-wave attenuator. These devices are voltage-tunable. Frequency is proportional to the $\frac{1}{2}$ power of the cathode-helix voltage and the dimensions of the structure. Multioctave tuning is possible, depending on output-power-variation requirements. These oscillators have low pulling figures and, typically, high pushing figures. Frequency stability is excellent, usually dependent on power-supply variations. These devices are typically low-power (< 1 W).

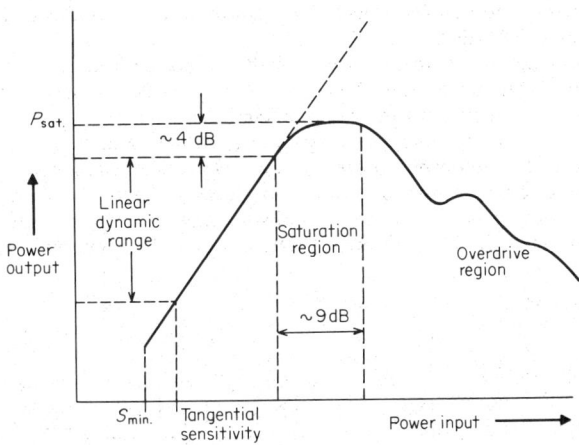

Fig. 13-150. Typical gain characteristic of a broadband TWT amplifier.

144. Klystron Oscillators and Amplifiers. The chief advantages of the Klystron amplifier oscillator are that it is capable of large stable output power (10 MW) with good efficiency (40%) and high gain (70 dB). Basically, the mechanism involves modulation of the velocity of electrons in a beam by an input rf signal. This is converted into a density-modulated (bunching) beam from which a resonant cavity extracts the rf energy and transforms it to a useful load.

Klystron amplifiers (see Pars. **9-20** to **9-28a**) may be conveniently divided into three categories: (1) two resonator single-stage high- and low-noise voltage amplifiers, (2) two resonator single-stage (*optimum bunching*) power amplifiers, and (3) multiresonator cascade-stage voltage and power amplifiers. The power gain of a two-cavity voltage amplifier can be given by[8]

$$G = \frac{M_1^2 M_2^2}{240\beta} \left(\frac{\pi a}{\lambda} \right)^2 \frac{G_0}{(G_{BR})^2}$$

where M_1, M_2 = beam coupling factors, a = beam radius, G_0 = beam conductance, G_{BR} = cavity shunt conductance contributions due to beam loading and ohmic losses, and β = electron velocity/velocity of light.

A simplified schematic representation of a Klystron amplifier is shown in Fig. 13-151. Multicavity tubes are typically used for high pulse power and cw applications. The intermediate cavities serve to remodulate the beam, causing additional bunching and higher gain-power output. Optimum power output is obtained with the second cavity slightly detuned. Further, loading of this cavity serves to increase the bandwidth (at the expense of gain). These tubes typically use magnetically focused high-perveance beams.

The broadbanding of a multicavity Klystron is accomplished in a manner analogous to that of multistage if amplifiers. A common technique is stagger tuning, which is modified somewhat, due to nonadjacent cavity interactions. The gain and bandwidth of multicavity Klystrons have been calculated by Kreuchen.[32]

The two-cavity amplifier can be made to oscillate by providing a feedback loop from output to input with proper phase relationship. Klystrons may also be used for frequency multiplication, using the high harmonic content of the bunched-beam current waveforms.

Reflex Klystron. A simple Klystron oscillator results if the electron-beam direction is reversed by a negative electrode, termed the *reflector.* A schematic diagram of such a structure is given in Fig. 13-152. Performance data for a reflex Klystron are usually given in the form of a reflector-characteristic chart. This chart displays power output and frequency deviation as a function of reflector voltage. Two distinct classes of reflex Klystrons are low-power tubes for oscillator, pump, and test applications and higher-power tubes (10 W) for frequency-modulator applications. Operating voltage varies from 300 to 2,000 V with bandwidths up to $\sim$ 200 MHz.

145. Crossed-Field-Tube Circuits.[34] Practically all crossed-field tubes have integrally attached distributed constant circuits. Operation within a critical range of beam current is necessary to maintain the proper bandpass characteristics.

Magnetron.[35] The original microwave tube was a magnetron diode switch with oscillations due to the cyclotron resonance frequency. Several oscillator circuits were utilized until the standard cavity resonator magnetron was introduced in 1940. See Pars. **9-41** to **9-45**.

The relations between power, frequency, and voltage vs. the load admittance are shown in the Rieke diagram (Fig. 13-153). Such charts illustrate the compromises necessary to

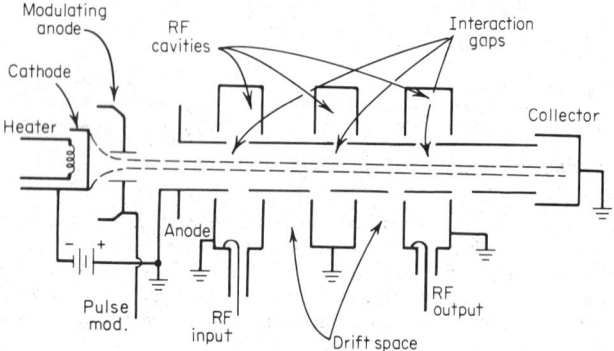

Fig. 13-151. Klystron amplifier structure.

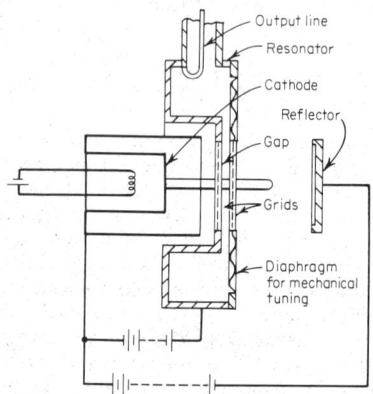

Fig. 13-152. Reflex Klystron with coaxial-line loop output.

obtain desired operating conditions. In general, good efficiency results from increasing anode current and magnetic field strength.

Various load effects may be considered with the Rieke diagram. The pulling figure (measure of frequency change for a defined load mismatch—SWR = 1.5 at all phase angles) and long-line effect are examples.

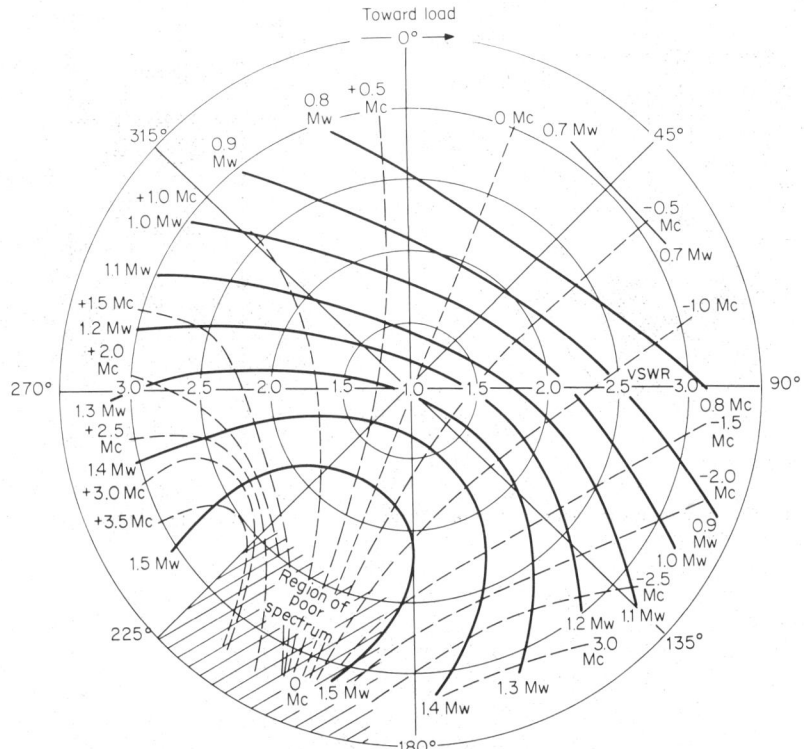

Fig. 13-153. Rieke diagram for an L-band magnetron at 1,250 MHz. Solid curves are contours of constant power; dashed curves of constant frequency. *(Raytheon Company.)*

The area of highest power on the Rieke diagram is called the *sink* and represents the tightest load coupling to the tube. The highest efficiency results in this region. However, a poor spectrum or instability typically results. The buildup of oscillations in the antisink region is closer to ideal. However, this lightly loaded condition may also result in instability. Stability is a measure of the percentage of missing pulses, usually defined at 30% energy loss.

Tuning. Methods used to tune a magnetron are classified as mechanical, electronic, and voltage tuning. In the mechanical method, the frequency of oscillation is changed by the motion of an element in the resonant circuit. Two types are shown in Fig. 13-154. An electron beam injected into the cavities will change the effective dielectric constant, and hence the frequency. Utilizing the frequency pushing effect, it has been possible to tune the magnetron over a frequency range of 4 to 1 (typically, 0.1 to 2 MHz/V) using voltage tuning. The frequency change is usually linear, but the power output is not constant over the tuning range. Magnetrons with power outputs greater than 5 MW and efficiencies 50% or more are available.

Amplitron. The Amplitron is essentially a magnetron with two external couplings, enabling amplifier operation. It is characterized by high power, broad bandwidth, very high efficiency, and low gain. The output is independent of rf input, but dependent on dc input.

It acts as a low-loss passive transmission line in the absence of high voltage. A typical plot of power for an Amplitron is shown in Fig. 13-155. Conversion efficiency of an Amplitron can be as high as 85%. The gain can be increased by inserting mismatches into the input and output transmission lines.

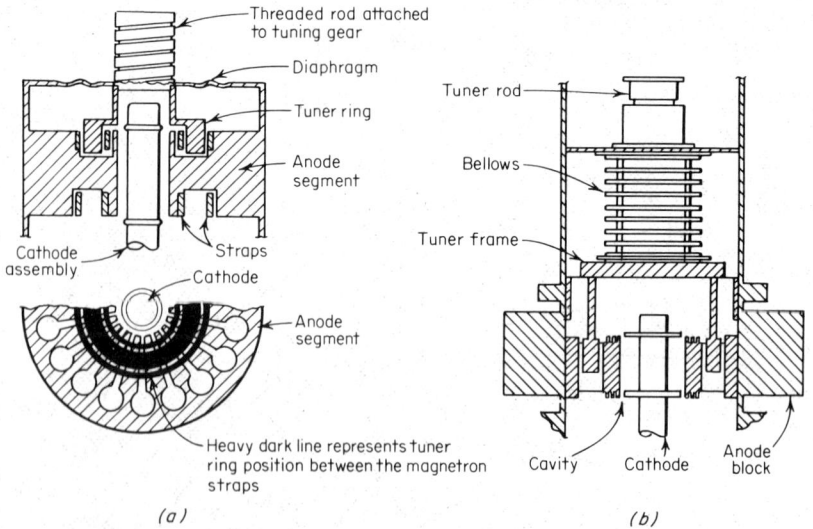

(a)   *(b)*

Fig. 13-154. Mechanical magnetron tuning mechanisms. (*a*) Capacitance type; (*b*) inductance type. *(Raytheon Company.)*

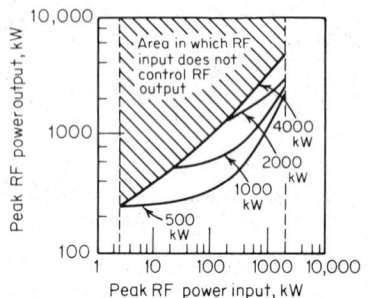

Fig. 13-155. Gain characteristic of Model QK520 L-band amplifier.

Stabilitron. Feedback with a stabilized cavity, resulting in highly stable oscillator performance, is used in the Stabilitron. Steady oscillations occur when the total phase shift from an output reflection to the cavity reflection and return is an integral multiple of 2π and the loop gain is greater than 1. Frequency stability can be 5 to 100 times greater and pulling figure 5 to 20 less than that of a magnetron oscillator.

146. References on Microwave Amplifiers and Oscillators

1. READ, W. T. "A Proposed High-Frequency Negative Resistance Diode," *Bell System Tech. J.,* Vol. 37, 1958.

2. HAITZ, R. H., et al. A Method for Heat Flow Resistance Measurements in Avalanche Diodes, *IEEE Trans. Electron Devices,* Vol. ED-16, p. 438, May, 1969.

3. Microwave Power Generation and Amplification Using IMPATT Diodes, *Hewlett Packard Appl. Note* 935, June, 1971.

4. GUMMEL, H. K., and BLUE, J. L.   A Small Signal Theory of Avalanche Noise in IMPATT Diodes, *IEEE Trans. Electron Devices,* Vol. ED-14, 1967.

5. HINES, M. E.   Noise Theory for Read Type Avalanche Diodes, *IEEE Trans. Electron Devices,* Vol. ED-13, 1966.

6. SCHERER, E. F.   Investigations of the Noise Spectra of Avalanche Oscillators, *IEEE Trans.,* Vol. GMTT-16, September, 1966.

7. KUVAS, R. L.   Noise in IMPATT Diode Oscillators and Amplifiers, *Proc. 3d Biennial Cornell Elec. Eng. Conf.,* High Frequency Generation and Amplification: Devices and Applications, 1971.

8. CHAN, V. W., and LEVINE, P. A.   A Comparative Study of IMPATT Diode Noise Properties, *Proc. 3d Biennial Cornell Elec. Eng. Conf.,* High Frequency Generation and Amplification: Devices and Applications, 1971.

9. PRAGER, H. J., CHANG, K. K. N., and WEISBROD, S.   High Power, High Efficiency Silicon Avalanche Diodes at Ultra High Frequencies, *Proc. IEEE (Letters),* Vol. 55, April, 1967.

10. EVANS, W. J.   Circuits for High Efficiency Avalanche-Diode Oscillators, *IEEE Trans.,* Vol. MTT-17, No. 12, December, 1969.

11. *Ibid.*

12. LIU, S. G., and RISKO, J. J.   Fabrication and Performance of Kilowatt L-Band Avalanche Diodes, *RCA Review,* Vol. 31, March, 1970.

13. KAWAMOTTO, H. Anti-parallel Operation of Four High Efficiency Avalanche Diodes, *IEEE ISSCC Digest,* February, 1971.

14. DELOACH, B. C., JR., and SCHARFETTER, D. L.   Device Physics of TRAPATT Oscillators, *IEEE Trans. Electron Devices,* Vol. ED-17, January, 1970.

15. LIU, S. G., PRAGER, H. J., CHANG, K. K. N., RISKO, J. J., and WEISBROD, S.   High Power Harmonic Extraction and Triggered Amplification with High Efficiency Avalanche Diodes, *IEEE ISSCC Digest,* 1971.

16. GUNN, J. B.   Microwave Oscillators of Current in III-IV Semiconductors, *Solid State Commun.,* Vol. 1, 1963.

17. McCUMBER, D. E., and CHYNOWETH, A. G.   Theory of Negative-Conductance Amplification and of Gunn Instabilities in "Two-Valley" Semiconductors, *IEEE Trans. Electron Devices,* Vol. ED-13, No. 1, January, 1966.

18. HANSON, D. C.   YIG Tuned TED Using Thin Film Microcircuits, *IEEE ISSCC Digest Tech. Papers,* February, 1969.

19. LARGE, D.   Octave Band Varactor-tuned Gunn Diode Sources, *Microwave J.,* Vol. 13, No. 10, October, 1970.

20. PERLMAN, B. S., UPADHYAYULA, L. C., and MARX, R. E.   Wide-Band Reflection Type Transferred Electron Amplification, *IEEE Trans.,* Vol. MTT-8, No. 11, November, 1970.

21. SIEKANOWICZ, N. W., PERLMAN, B. S., BERSON, B. E., MARX, R. E., and KLATSKIN, W. E.   Performance of Medium Power, High Gain, C. W. Transferred Electron Amplifiers at C-Band, *Proc. 3d Biennial Cornell Elect. Eng. Conf.,* Cornell University, Ithaca, N.Y., 1971.

22. CAMP, W. O., BRAVMAN, J. S., and WOODARD, D. W.   The Operation of Very High Power LSA Transmitters, *Proc. 3rd Biennial Cornell Elect. Eng. Conf.,* Cornell University, Ithaca, N.Y., 1971.

23. KUROKAWA, K.   Power Waves and the Scattering Matrix, *IEEE Trans. GMTT,* Vol. MTT-13, No. 2, March, 1965.

24. BODWAY, G.   Two Part Power Flow Analysis Using Generalized Scattering Parameters, *Microwave J.,* Vol. 10, No. 6, May, 1967.

25. GELNOVATCH, V. G., CHASE, I. L., and ARELL, T.   A 2-4 GHz Integrated Transistor Amplifier Designed by an Optimal Seeking Computer Program, *IEEE ISSCC Digest,* February, 1970.

26. HUNTER, L. P.   "Handbook of Semiconductor Electronics," 3d ed., McGraw-Hill, New York, 1969.

27. LEE, H. C.   Microwave Power Transistors, *Microwave J.,* February, 1969.

28. FROEHNER, W. H.   Quick Amplifier Design Using Scattering Parameters, *Electronics,* 1967.

29. PIERCE, J. R.   "Traveling Wave Tubes," D. Van Nostrand, New York, 1950.

30. HEFFNER, H. Analysis of the Backward Wave Traveling Wave Tube, *Proc. IRE,* Vol. 42, June, 1954.

31. HAMILTON, D. R., KNIPP, J. K., KUPER, J. B. H. "Klystrons and Microwave Triodes," McGraw-Hill, New York, 1948.

32. KREUCHEN, K. H., AULD, B. A., and DIXON, N. E. A Study of the Broad Band Frequency Response of the Multi-Cavity Klystron Amplifier, *J. Electron.,* Vol. 2, May, 1957.

33. DAIN, J. Ultra High Frequency Power Amplifiers, *Proc. IEEE,* Vol. 105, Pt. B, November, 1958.

34. OKRESS, E. "Crossed Field Microwave Devices," Vols. 1 and 2, Academic, New York, 1961.

35. COLLINS, G. B. "Microwave Magnetrons," MIT Radiation Laboratory Series, Vol. 6, McGraw-Hill, New York, 1948.

SECTION 14

MODULATORS, DEMODULATORS, AND CONVERTERS

BY

JOSEPH L. CHOVAN Senior Engineer

MYRON D. EGTVEDT Senior Engineer, Member IEEE

JOSEPH P. HESLER Consulting Engineer

GEORGE F. PFEIFER Senior Engineer, Member IEEE

NOBLE R. POWELL Consulting Engineer, Member IEEE; all of the Electronics Laboratory, General Electric Company

GLENN B. GAWLER Senior Engineer, Barker Manufacturing Company, Member IEEE

CONTENTS

Numbers refer to paragraphs

MODULATORS, DEMODULATORS, AND CONVERTERS

SECTION 14

MODULATORS, DEMODULATORS, AND CONVERTERS

AMPLITUDE MODULATORS AND DEMODULATORS

By Joseph P. Hesler

1. Amplitude Modulation. Frequency translation is the key to radio communications in that it produces signal energy, in proportion to variations of an information source, at frequencies that have desirable transmission characteristics, such as antenna size, freedom of interference from similar information sources, line-of-sight to long-range propagation, and freedom of interference from particular noise sources. Frequency translation permits the efficient utilization of open and closed propagation media by many simultaneous users and/or signals.

One of the most used forms of frequency translation is linear modulation, the most common of which is amplitude modulation. In general, amplitude modulation consists in varying the magnitude of a carrier signal in direct correspondence to the instantaneous fluctuations of a modulating signal source, as illustrated in Fig. 14-1.

Variations of the basic amplitude modulation process have been developed to achieve more efficient spectrum utilization and to reduce transmitter power requirements. These

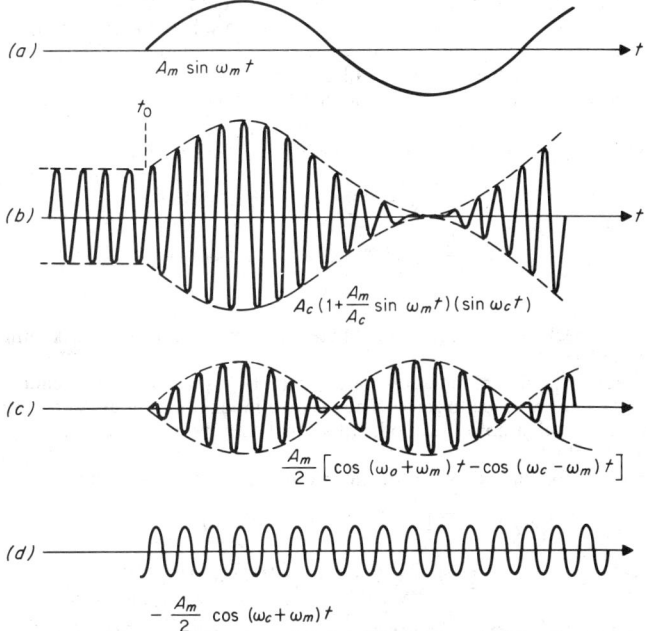

Fig. 14-1. Amplitude modulation. (*a*) Modulating signal; (*b*) double-sideband amplitude-modulated signal; (*c*) double-sideband suppressed carrier amplitude-modulated signal; (*d*) single-sideband suppressed carrier amplitude modulated signal.

include suppressed-carrier systems such as vestigial-sideband, single-sideband, and double-sideband modulation systems.   The companion form of frequency translation used in amplitude modulation systems is *detection*.   This is the process whereby the originally translated information is recovered as a baseband signal.   Linear amplitude modulation has been the most widely used form of frequency translation in general communications for three reasons: relative ease of implementation, efficient utilization of bandwidth, and availability of devices to implement a simple detection procedure.

It can be argued that amplitude modulation includes such modulation methods as pulse code keying, pulse amplitude modulation, frequency shift keying (the sequential keying of multiple carrier signals), and variations of these, such as pulse position modulation and pulse width modulation.   This sub-section (Pars. **14-1** to **14-15**) is restricted to the amplitude modulation by signal sources whose outputs are continuous time functions.   The special types of modulation mentioned above are discussed in Pars. **14-26** to **14-41**.

The general expression for the output of a linear amplitude modulator with a sinusoidal modulation input is

$$E = E_0(1 + m \sin \omega_m t)[\sin(\omega_c t + \phi)] \tag{14-1}$$

where E_0 — peak amplitude of the carrier signal, ω_m = modulating signal frequency in radians per second, ω_c = carrier frequency in radians per second, m — modulation index, ϕ = arbitrary carrier phase angle in radians, and t = time in seconds.

Expansion of the Eq. (14-1) provides

$$E = E_0 \sin(\omega_c t + \phi) + \frac{mE_0}{2} \cos[(\omega_c - \omega_m)t + \phi]$$
$$- \frac{mE_0}{2} \cos[(\omega_c + \omega_m)t + \phi] \tag{14-2}$$

Note that the carrier signal is reproduced exactly as if it carried no modulation.   The carrier in itself does not carry any information.   The second and third terms in Eq. (14-2) represent sideband signals produced in the modulation process.   These signals are displaced from the carrier signal in the frequency spectrum, on each side of the carrier, by a frequency difference equal to the modulating signal frequency.   The magnitudes of the sideband signals are equal and are proportional to the modulating index m.

Both positive and negative amplitude modulation can be produced in an unsymmetrical manner.

For each case the amplitude modulation index m is defined as

$$m = \begin{cases} \dfrac{E_{max} - E_0}{E_0} & \text{positive modulation} \\ \dfrac{E_0 - E_{min}}{E_0} & \text{negative modulation} \end{cases} \tag{14-3}$$

where E_{max} = peak amplitude of modulated carrier, and E_0 = peak amplitude of unmodulated carrier.

The maximum negative-modulation index of unity results from the reduction of the instantaneous carrier envelope to zero.   The positive-modulation index is not limited.   The maximum symmetrical amplitude modulation that can be produced corresponds to a modulation index of unity.

For a complex modulating signal $G(t)$ the modulated carrier spectrum is

$$F(\omega) = \mathcal{F}\{E_0[1 + mG(t)][\sin(\omega_c t + \phi)]\}$$
$$= \frac{E_0}{2\pi} \int_{-\infty}^{\infty} [1 + mG(t)]\sin(\omega_c t + \phi)e^{-j\omega t}\, dt \tag{14-4}$$

where $F(\omega)$ = Fourier transform of the time function $\mathcal{E}$ $[f(t)]$.

2. Types of Amplitude Modulation.   Generation of an amplitude-modulated waveform requires the multiplication of signals, $f_1(t)f_2(t)$.   Both signals need not be time-variant; for example, a carbon microphone modulates a dc potential with voice signals.   The main applications, however, concern the modulation of *carrier* signals to exploit desirable transmission characteristics, that is, $f_1(t)[A \sin (\omega_c t + \phi)]$. Two classes of circuits are used as modulators, *square-law* devices and *linear* modulators.[16-18]*

Square-Law Modulation. Any device having a nonlinear transfer function may be expressed in a power series form, for example, the current in a diode, $i(e) = a_0 + a_1 e + a_2 e^2 + a_3 e^3 + \cdots$.

When the diode characteristic and bias conditions are chosen so as to enhance the coefficient a_2 with respect to the other coefficients, the device is considered to be a squarelaw nonlinear element.[2]

Under these conditions, when two signal imputs $f_1(t)$ and $f_2(t)$ are summed and used as the driving function e, a significant portion of the output power will exist as $a_2 e^2$.

$$e = f_1(t) + f_2(t) \tag{14-5}$$

$$e^2 = 2f_1(t)f_2(t) + [f_1(t)]^2 + [f_2(t)]^2 \tag{14-6}$$

Suitable nonlinear characteristics are exhibited by various types of rectifiers, triodes, and transistors.

Linear Modulation. Linear modulators are devices with transfer functions that are linearly related to a control parameter. Examples include the outputs of class C rf amplifiers as a function of the B+ supply voltage, the transconductance gain of transistor differential amplifiers vs. emitter current-source magnitude, and Hall effect devices whose transconductance is proportional to the applied magnetic field.[6]

3. Methods of Amplitude Modulation. *Square-Law Amplitude Modulators.*[2,17-19] A square-law modulator requires three features: a method of summing the two input signals $f_1(t)$ and $f_2(t)$, a device with a nonlinear transfer function, and a tuned circuit and coupling network for extracting the desired modulation products. Voltage summing or current summing are used depending on the transfer characteristic of interest. The nonlinear device is biased in a region that enhances the second-order coefficient of the power series that represents the nonlinear transfer function. The most common devices used for this type of modulation are semiconductor diodes and vacuum-tube triodes. An example of a typical circuit is shown in Fig. 14-2.

The efficiency of this type of amplitude modulation is generally low, and all the output energy is supplied by the driving functions.

Consider two input signals:

$$e_m = E_m \cos \omega_m t \qquad \text{modulating signal} \tag{14-7}$$

$$e_c = E_c \cos \omega_c t \qquad \text{carrier signal} \tag{14-8}$$

The input applied to the nonlinear device is

$$e_s = E_m \cos \omega_m t + E_c \cos \omega_c t \tag{14-9}$$

If the transfer function is represented by the two terms of interest from a Taylor series,

$$e_0 = a_1 e_s + a_2 e_s^2 \tag{14-10}$$

The output components resulting are

$$\begin{aligned}
e_0 &= \frac{a_2}{2}(E_m^2 + e_c^2) & \text{dc rectified component} \\
&+ a_1 E_m \cos \omega_m t & \text{modulating signal} \\
&+ a_1 E_c \cos \omega_c t & \text{carrier} \\
&+ \frac{a_2}{2} E_m^2 \cos^2 2\omega_m t & \text{second harmonic of modulation} \\
&+ \frac{a_2}{2} E_c^2 \cos^2 2\omega_c t & \text{second harmonic of carrier} \\
&+ a_2 E_c E_m \cos(\omega_c - \omega_m)t & \text{lower sideband} \\
&+ a_2 E_c E_m \cos(\omega_c + \omega_m)t & \text{upper sideband}
\end{aligned} \tag{14-11}$$

---

*Superior numbers correspond to numbered references, Par. **14-15**.

There would be other terms, also, from the higher-order coefficients of the Taylor series. The degree of modulation is expressed as

$$\text{Modulation index} = 2\frac{a_2}{a_1}E_m \qquad (14\text{-}12)$$

The desired outputs for double-sideband amplitude modulation are

$$e_0' = a_1 E_c \cos \omega_c t + a_2 E_c E_m \cos(\omega_c \pm \omega_m)t \qquad (14\text{-}13)$$

The square-law devices are reciprocal in that the modulating frequency will appear as an output if a modulated signal is applied as the input. Thus the square-law device may also be used as a demodulator or detector.

4. Low- and Medium-Power Linear Modulators. (See Refs. 1, 3, 6–8, 19, 21, 23, and 25.) Many applications exist in mobile equipment for amplitude modulators with output powers from milliwatts to tens of watts. Transistor circuits are used almost exclusively for these circuits. Carrier frequencies above 1 GHz can be used in the lower-power transistor circuits. The most common methods of amplitude modulation used are class C collector modulated stages with the rf applied in the common-emitter or common-base configuration, as shown in Fig. 14-3. The common-emitter configuration provides the maximum power gain and excellent efficiency. Common-base stages are used to increase the upward modulation capabilities, where the maximum modulation indices are important. Increased linearity can be achieved at the expense of efficiency by biasing the amplifier class B so that a nominal collector current flows under no-modulation conditions.

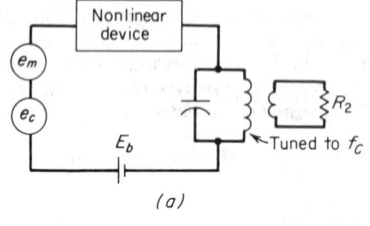

(a)

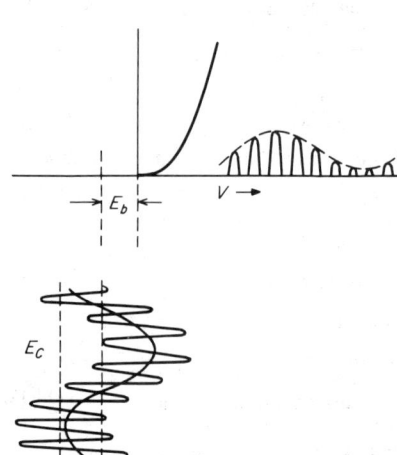

(b)

Fig. 14-2. Square-law modulators.

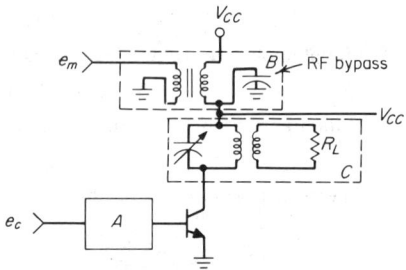

A = input impedance matching network

B = modulator circuit for producing V_{CC}' which follows the modulating signal, e_m, and provides a low source impedance at f_c

C = Frequency selective network, $f_c \pm f_m$, and output impedance matching network

Fig. 14-3. Collector-modulated transistor (class C rf amplifier).

For class C operation the transistor transfer characteristics of the modulated amplifier are determined from the large-signal input and output parallel equivalent impedance data.[12,15] These are determined experimentally or provided on device specification sheets. The transistors are operated in a very nonlinear manner as class C amplifiers; therefore the experimental data should be representative of the expected operating point, because the small-signal transistor parameters are not adequate.

5. Power Relationships. The instantaneous rms output voltage E varies about the unmodulated carrier rms level E_0. The maximum rms output is $E_0(1 + m)$, where m is the modulation index which can vary from zero to unity for symmetrical modulation. The minimum rms output is $E_0(1 - m)$. The unmodulated power into the load, R, is

$$P_o = \frac{E_0^2}{R} \tag{14-14}$$

The peak power into the load is

$$P_{max} = \frac{[E_0(1 + m)]^2}{R} \tag{14-15}$$

The minimum is

$$P_{min} = \frac{[E_0(1 - m)]^2}{R} \tag{14-16}$$

The average power into the load for sinusoidal modulation is

$$P_{av} = P_o\left(1 + \frac{m^2}{2}\right) = \frac{E_0^2}{R}\left(1 + \frac{m^2}{2}\right) \tag{14-17}$$

The unmodulated output power E_0^2/R is supplied by the class C amplifier, and the sideband energy $(E_0^2/R)(m^2/2)$ is supplied by the modulator. The class C amplifier can be biased very close to the peak modulated output envelope swing, $V_{CC} \approx \sqrt{2} m\, E_0$. For 100% modulation, $V_{CC} \approx \sqrt{2} E_0$. The dissipation in the output voltage stage is the difference between the total input power, consisting of the dc collector bias and the input rf drive, and the output rf power. The input rf drive for an amplifier with power gain A is

$$P_{drive} = \frac{E_0^2}{RA} \tag{14-18}$$

The input dc bias, unmodulated, is slightly greater than

$$P_{dc} = \sqrt{2} E_0 \cdot \bar{I}_C \tag{14-19}$$

where $\bar{I}_C$ = average dc collector current.

The output transistor dissipation unmodulated is

$$P_{TR_o} \approx \sqrt{2} E_0 \cdot \bar{I}_C + P_{drive} - \frac{E_0^2}{R} \tag{14-20}$$

$$P_{TR_o} \approx \sqrt{2} E_0 \cdot \bar{I}_C - \frac{E_0^2}{R}\frac{A - 1}{A} \tag{14-21}$$

In higher-power systems the output transistors can be paralleled. At higher frequencies the gain of the output transistors may be such that the output power is limited by the dissipation in the driver. In these instances the driver may also be collector-modulated to achieve adequate upward modulation and to reduce the power dissipation in the driver.[9]

The input and output impedances of the transistor class C amplifiers are characteristically low. Typical parallel equivalent input impedances are

$$2 < R_{in} < 50\Omega$$

$$30 < C_{in} < 5,000\text{pF}$$

The collector load impedance resistance component is determined from the bias voltage and output power

$$R'_L = \frac{(V_{CC})^2}{2 P_o}$$ (14-22)

where V_{CC} = dc collector bias voltage and the output voltage swing is $2V_{CC}$ peak to peak, and P_0 = unmodulated output power.

The collector parallel output capacitance is dependent on the device geometry and may range from a few picofarads for very-high-frequency lower-power devices to a few hundred picofarads for large-geometry high-power devices.

Interstage and output matching networks are used to obtain conjugate matches for maximum power gain.

Care must be exercised in the selection of components, especially capacitors, used in the input matching networks. The low input impedance and high ratio of reactive to resistive components of the input impedance can result in very high circulating currents in these networks.

The class C transistor amplifiers have the advantage over tube circuits that a zero bias condition at the base-to-emitter junction will reduce the collector current to the value of collector leakage current; hence loss of drive will not result in destructive device dissipation, as may be the case with grid-leak-biased tube circuits.

Other types of linear transistor modulators may be used where efficiency is less critical or where very-wide-band operation precludes effective output filtering for the elimination of harmonics. A differential amplifier circuit, as shown in Fig. 14-4, will have a transconduc-

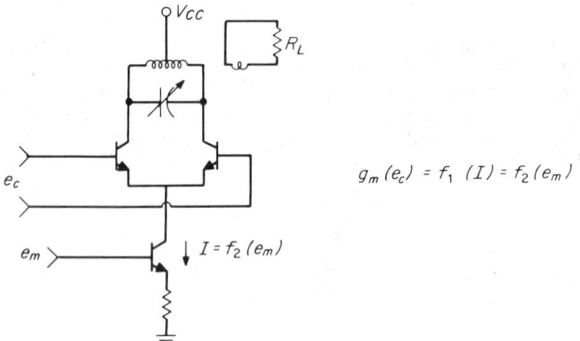

$$g_m (e_c) = f_1 \; (I) = f_2 (e_m)$$

$$I = f_2 (e_m)$$

Fig. 14-4. Differential amplifier amplitude modulator (without dc bias details).

tance gain that is very nearly proportional to the emitter current-source magnitude. Modulation of the current source will produce a nearly ideal multiplication of the rf input signal and the modulation signal for a wide range of low current levels. The differential amplifier must be biased class A with minimum dc offset at the base inputs. A single transistor multiplier may also be used, as shown in Fig. 14-5 for very-low-level outputs. An emitter bypass capacitor for the rf signal is used in place of the additional transistor for the rf return.

6. High-Power Linear Modulators. High-power linear modulators, 50 W and up, are generally constructed using class C plate-modulation vacuum-tube circuits (Fig. 14-6). Some of the intermediate power and frequency applications use paralleled transistor configurations with class C collector modulation.[14]

Triodes, tetrodes, and pentodes are employed in the vacuum-tube circuits. Multigrid tubes require screen-grid modulation in conjunction with the control-grid modulation to achieve space-charge modulation and to minimize screen current and screen dissipation. The two methods of screen modulation commonly used are:

1. Self-bias of the screen grid with a bypassed dropping resistor or inductor from the screen to the plate supply or the screen-grid supply (Fig. 14-7a).

2. Screen modulation via a separate winding on the modulation transformer (Fig. 14-7b).

7. Grid Modulation. The amplitude of a class C rf amplifier output can also be modulated by changing the grid bias with the modulating signal. The modulating signal is added to the rf input signal. The effect is to change the magnitude of the plate current pulses, and hence the fundamental component of the plate current (Fig. 14-8).

The disadvantage associated with grid modulation is that the fixed plate supply voltage

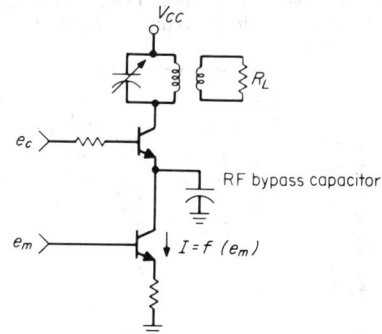

Fig. 14-5.   Two-transistor transconductance modulator.

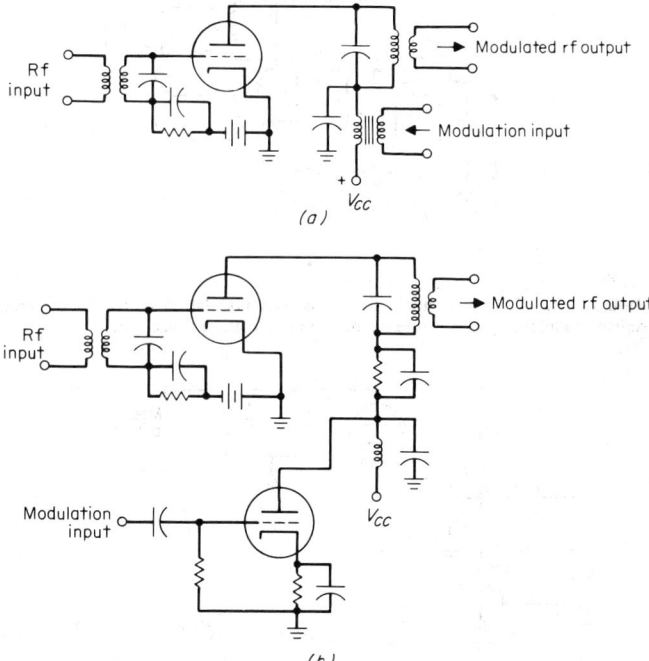

Fig. 14-6.   Typical class C plate modulators. (*a*) Modulation transformer; (*b*) class A modulating driver.

must be twice the peak rf voltage without modulation. This causes high plate dissipation and lowers the plate efficiency to the range of 35 to 45% when unmodulated.

The carrier power obtained from a plate-modulated class C amplifier is about three times that available from a grid-modulated circuit using the same tube.

The principal advantage of grid modulation is the reduction in modulator voltage and power. Grid modulation is used in systems where plate modulation transformers cannot provide adequate bandwidth and a class A modulator is required. Linearity in grid-modulated class C amplifiers is more difficult to obtain at a high modulation index while maintaining maximum efficiency.

8. Cathode Modulation of Class C RF Amplifiers. Cathode modulation can be used with a class C rf amplifier, as shown in Fig. 14-9. The modulation transformer output varies the grid-cathode as well as the plate-cathode voltages. The ratio of grid and plate modulation

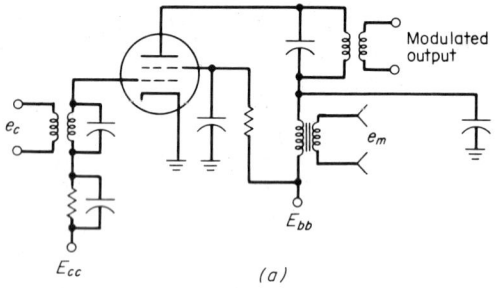

(a)

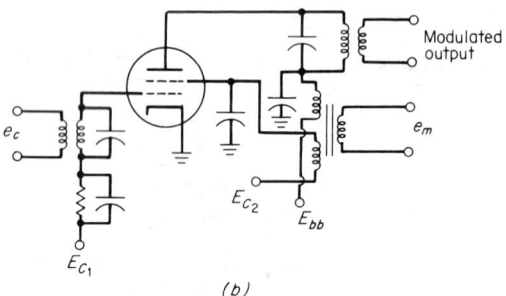

(b)

Fig. 14-7. Multigrid class C modulation circuits. (a) Screen modulation via RC from unmodulated plate supply; (b) screen modulation via separate winding on modulation transformer.

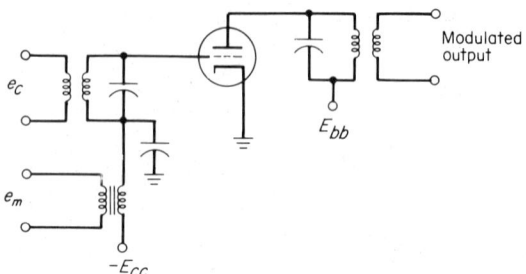

Fig. 14-8. Grid-modulated class C rf amplifiers.

can be selected by varying the tap; thus the circuit provides a means of producing varying combinations of grid and plate modulation. Some grid leak bias is normally used to improve linearity.

9. Modified Amplitude Modulation Methods. The information transmitted by an amplitude-modulated carrier is contained wholly in the modulation sidebands. The transmission of the carrier energy simplifies the receiver detector implementation but adds no information. In addition, each sideband contains the same information, and only one is required to transmit the intelligence. Elimination of the carrier and/or one sideband can effect a substantial transmitter power saving. For 100% amplitude modulation the carrier power is two-thirds of the transmitter power, and each sideband one-sixth. Elimination of the carrier only results in double-sideband suppressed-carrier modulation. Elimination of one sideband while retaining the carrier or a substantial portion of the carrier results in vestigial-sideband transmission. Elimination of the carrier and one sideband is called *single-sideband suppressed-carrier modulation.*

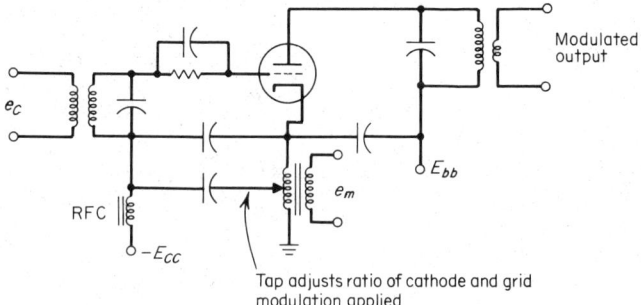

Fig. 14-9. Cathode-modulated class C rf amplifier.

The easiest of these modulation schemes to implement is the vestigial-sideband transmission, both from a transmitter and a receiver viewpoint. The unwanted sideband is generally filtered out at low levels in the transmitter chain, and is known as *transmitter attenuation* (TA). One sideband can also be eliminated in the receiver by selective filtering, and is called *receiver attenuation* (RA). The latter scheme is not practical from a spectrum-utilization sense; hence it is normally used only as in television broadcast to complete unwanted sideband rejection that is performed primarily at the transmitter.

A vector notation may be used to illustrate the phenomena of various types of amplitude modulation. The vector is a complex function. The sinusoidal function of time that is of interest is the real part or the vector projection on the real axis of the complex plane. Thus the real part of $Ae^{j\omega t}$ is $A \cos \omega t$, as shown in Fig. 14-10a. The projection on the real axis of the vector $Ae^{j\omega t}$ can be considered as the carrier signal in subsequent amplitude modulation discussions. Since the unmodulated carrier signal in amplitude modulation processes is a fixed peak amplitude and fixed frequency function of time, a modified vector representation can be used to describe the envelope of the modulated waveform. The modified vector diagram maintains the carrier vector as a fixed, nonrotating vector. By this means subsequent illustrations are referenced to a complex plane rotating at the carrier angular rate, and the projection on the real axis corresponds to the *modulation envelope variations with time.*

For an amplitude modulation system using a sinusoidal modulating function at 100% modulation, $m = 1$, the addition of two vectors to the basic vector diagram is required. The two additional vectors represent the sideband signals produced in the modulation process. The two sideband signals are displaced on either side of the carrier signal in the modulated signal spectrum by a frequency equal to the modulating frequency. Therefore, with respect to the modified reference system, the lower sideband vector will rotate clockwise at the modulation-signal angular rate, and the upper sideband signal will rotate counterclockwise at the same rate.

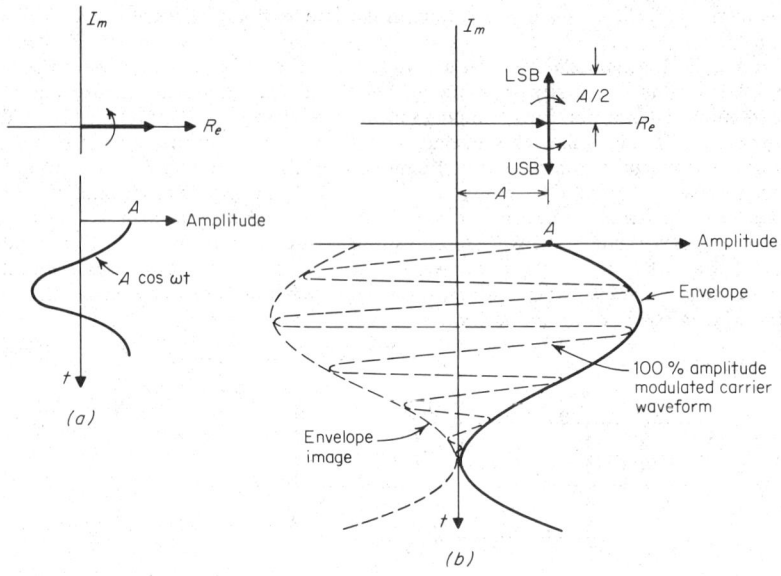

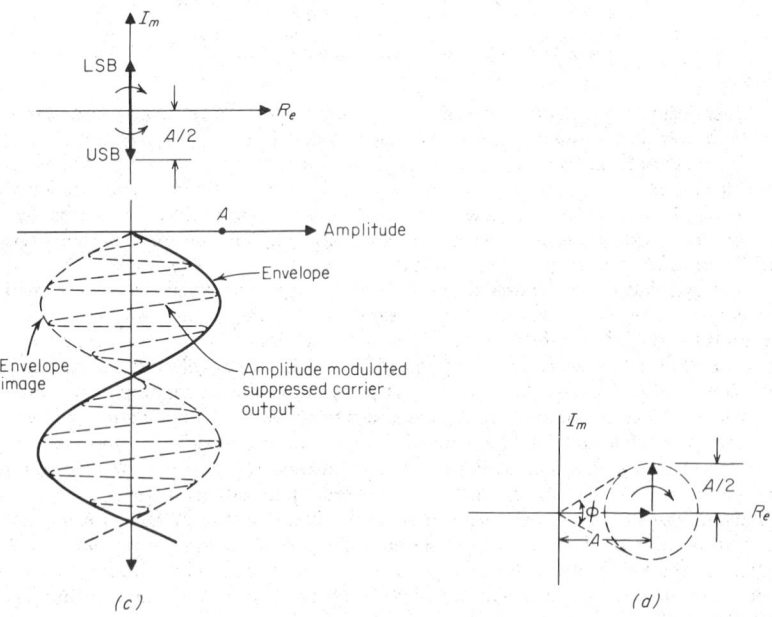

Fig. 14-10. Vector representation of amplitude modulation. (*a*) Rotating carrier vector and its real projection; (*b*) vector representation with sideband vectors resulting from 100% amplitude modulation. Envelope function is locus of projection on the real axis with a non-rotating carrier vector; (*c*) double sideband suppressed carrier vector representation. Envelope function is locus of projection on the real axis with coordinate rotation as in (*b*); (*d*) vector representation of a single sideband amplitude modulated waveform showing peak to peak carrier phase modulation, φ, resulting from the absence of the other sideband signal.

The relative phases of the two sideband vectors are such that no charge in carrier phase angle is introduced when they are both present; i.e., the imaginary parts of the sideband contributions cancel. The initial angles are set by the phase of the modulating function. As the sideband vectors rotate in the modified complex plane, the vector sum of the carrier plus sidebands describes sinusoidal projection on the real axis at the modulating frequency rate.

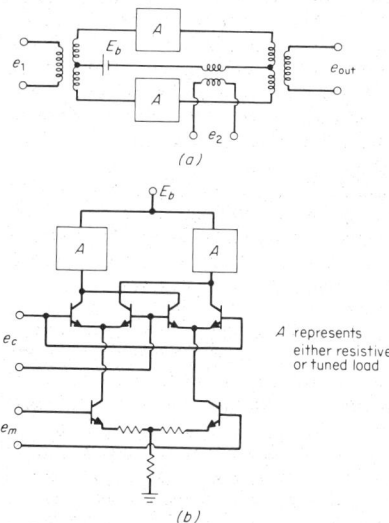

(a)

(b)

Fig. 14-11. Double-sideband suppressed-carrier modulators. (*a*) General form of a balanced modulator for suppressed carrier output using non linear elements, A and bias voltage E_b; (*b*) four-quadrant multiplier.

A represents either resistive or tuned load

Note that the unmodulated carrier projection is constant and equal to the peak carrier magnitude. The 100% modulated carrier projection is nonnegative. If the original complex plane were used, the entire vector system would rotate at the carrier angular rate, and the projection on the real axis would be the actual time function produced by the modulator. This function would have negative portions and would fill the envelope function and its image with amplitude-modified sinusoids at the carrier frequency, as shown in Fig. 14-10*b* with the dashed lines within the envelope.

10. Vestigial-Sideband Systems. Vestigial-sideband transmission introduces angle modulation on the carrier because the symmetry of the contrarotating sideband vectors is lost, as illustrated in Fig. 14-10*d*. A standard envelope detector can still be used to recover the modulating signal, however. The primary objective in the application of vestigial-sideband transmission is to conserve spectrum in the transmission medium.

11. Suppressed-Carrier Systems. Suppressed-carrier systems for AM transmission and reception require modifications to the receiver. The double-sideband suppressed-carrier signal cannot be envelope-detected without the reinsertion of a carrier signal. The frequency and the phase of the reinserted carrier are critical. This type of transmission is used with special phase-locked receivers or with the transmission of low-level carrier to permit the reconstitution of the carrier frequency and phase at the receiver.

A double-sideband suppressed-carrier signal can be generated with a balanced modulator, as shown in Fig. 14-11*a*.

For a sinusoidal modulating signal the outputs of the two modulators are

$$e_1 = E_c \cos \omega_c t + \frac{E_m}{2} \cos(\omega_c + \omega_m)t + \frac{E_m}{2} \cos(\omega_c - \omega_m)t \qquad (14\text{-}23)$$

$$e_2 = E_c \cos \omega_c t \frac{E_m}{2} \cos[(\omega_c + \omega_m)t + \pi] + \frac{E_m}{2} \cos[(\omega_c - \omega_m)t + \pi] \qquad (14\text{-}24)$$

When these two signals are combined in push-pull, the output becomes

$$e_0 = e_1 + e_2 = E_m\cos(\omega_c + \omega_m)t + E_m\cos(\omega_c - \omega_m)t \qquad (14\text{-}25)$$

The nonlinear devices can be diodes or modulated class C rf amplifiers. The balanced modulator simplifies the tuned-output-circuit design because all even harmonics of the modulating process tend to be canceled along with the carrier.

Single-sideband suppressed-carrier modulation simplifies the receiver design in some applications. The phase of the reinserted carrier at the receiver is not critical as in double-sideband suppressed-carrier systems. Also, a frequency error in the reconstituted carrier will result only in a corresponding shift in the demodulated signal frequencies, which may be tolerable. If accurate modulation-frequency preservation is required, a low-level carrier may be transmitted to aid in the reconstruction of the carrier signal at the receiver. The reinserted carrier amplitude with respect to the received sideband signal is generally made large to minimize angle modulation of the carrier at the detector.

Two methods are usually employed to generate suppressed-carrier signal-sideband transmissions. The most direct method is to filter out the undesired sideband and carrier at low levels in the transmitter chain. This filtering problem is difficult when the modulation sidebands of interest are close to the carrier and the carrier frequency is high. This problem can be eased by using successive modulators where the first carrier frequency is low. The single-sideband signal is filtered from the output of a low-frequency-carrier balanced modulator to eliminate the lower sideband. This signal is then used to modulate a second balanced modulator. The balanced modulators eliminate the carriers and cause the second-modulation sidebands to be separated by twice the first carrier frequency to ease the second filtering.

An alternative method of suppressed-carrier single-sideband modulation is by unwanted sideband cancellation. Two balanced modulators are used, with their outputs combined in push-pull. One modulator is driven with carrier and modulating signals that have been shifted 90° with respect to the inputs to the other modulator. Special wide-band phase-shift networks are required to handle the modulation input because orthogonality must be maintained across the bandwidth of the modulating signal.

12. Modulated Oscillators. A direct modulated class C oscillator can be used as an AM transmitter. The linearity of such circuits is generally as good as or better than the plate-modulated class C amplifiers. The main disadvantage is the tendency for carrier-frequency pulling which results from the changes in the oscillator operating point as the modulation signal varies.

A very useful circuit for the generation of low-level double-sideband suppressed-carrier signals is the four-quadrant multiplier. The same circuit will also operate as a double-sideband modulator with carrier and as low-level demodulator in phase-locked receiver systems. Multipliers are available in integrated circuit form, and they can be used at frequencies from dc to beyond 100 MHz (Fig. 14-11*b*).

AMPLITUDE DEMODULATORS

13. Detectors. The most commonly used amplitude modulation detector or demodulator is a diode rectifier. The ideal diode detector passes current in only one direction and will essentially follow the envelope of an amplitude-modulated waveform when used in a circuit as shown in Fig. 14-12 (see also Pars. **14-60** to **14-68**).

The charging time constant R_sC must be short, so that the capacitor voltage follows the input signal E_{rf} when the diode is forward-biased or conducting. Conversely, the discharge

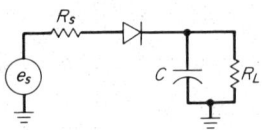

Fig. 14-12. Diode envelope detector.

time constant R_LC must be long enough to retain most of the rectified voltage between cycles of the carrier signal, but not so large that the capacitor voltage will not discharge at the maximum rate of change of the input signal envelope. The envelope detector is essentially insensitive to residual angle modulation of the carrier, and it is therefore usable in single-sideband receivers.

Practical diode rectifiers have nonlinear resistance characteristics in the conduction bias region, and are therefore operated at fairly high input signal levels, on the order of 2 to 10 V peak for semiconductor diodes.

14. Product Detectors. Another type of amplitude demodulator is the product detector, or multiplier circuit (Fig. 14-13). This circuit has distinct advantages and disadvantages. The advantages include the ability to detect lower-level signals with a linear response; the ability to differentiate phrase reversals in the modulated waveform, resulting from balanced amplitude modulation with suppressed carrier; and the ability to produce, in some designs, error signals for automatic frequency control systems in receivers.[20]

The analytical expression for the output of a product detector is

$$e(t) = E\underbrace{[1 + m\,\sin(\omega_m t + \phi_m)][\cos(\omega_c t + \phi_c)]}_{\text{amplitude-modulated signal}}\underbrace{[\cos(\omega_c t + \phi_d)]}_{\substack{\text{local-oscillator}\\\text{signal}}} \qquad (14\text{-}26)$$

Expansion of Eq. (14-26) shows that the product of the incoming carrier signal, $\cos(\omega_c t + \phi_c)$, and the local-oscillator signal, $\cos(\omega_c t + \phi_c)$, produces a dc term, except when these two inputs are in quadrature phase. The output dc term is proportional to the cosine of the relative phase angle of the carrier and local-oscillator signals. A four-quadrant multiplier circuit capable of performing this type of detection is shown in Fig. 14-13.

The details of the dc bias network are not included in the elementary schematic of the multiplier circuit. The input signals, rf and local-oscillator, are applied to either input port. Balanced-differential or single-ended inputs can be used, although the balanced inputs give the added performance of increased linearity and common-mode signal rejection. Two outputs of opposite polarity are available at the collectors of the upper-rank transistors. Design options are available to increase the efficiency and linearity of the circuit. Normally, an overdrive is applied to the local-oscillator port to make that section of the multiplier operate in a switching mode. The effect is to multiply the rf signal with a square wave at the carrier frequency instead of a sine-wave. This type of operation also produces additional outputs at the higher harmonics of the carrier frequency.

For balanced linear in-phase inputs at both ports, the outputs consist of

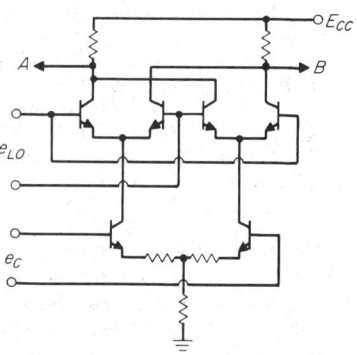

Fig. 14-13. Four-quadrant multiplier used as a product detector. The local oscillator input e_{LO} must be phase coherent with the carrier of the modulated input signal e_c. Complementary outputs are obtained at A and B.

$$e_0 = \tfrac{1}{2}(A_{LO} + A_{LO}\cos 2\omega_c t)E[1 + m\sin(\omega_m t + \phi_m)]$$
$$= \tfrac{1}{2}A_{LO}E[1 + m\sin(\omega_m t + \phi_m)] \qquad (14\text{-}27)$$
$$+ \tfrac{1}{2}A_{LO}E\cos 2\omega_c t$$

The linearity versus dc offset of the multiplier can be improved by using emitter degeneration at the rf signal port. Without degeneration the maximum peak-to-peak differential drive that will not cause distortion is on the order of 50mV. At this signal input level the dc offsets in the circuit may cause undesirable output voltage shifts. A compromise can be made to minimize the ratio of output dc offset to peak signal output by degenerating the rf port input and applying a larger input signal level.

The disadvantages of the product detector are relative circuit complexity and the need for a phase-coherent local-oscillator signal. The circuit complexity can be circumvented for input signal frequencies up to 100 MHz by using integrated circuit multipliers. This approach also minimizes the possibility of serious dc offset problems due to device mismatch. The generation of the coherent local-oscillator signal can be achieved in two ways. For simple DSB amplitude-modulated signals, the carrier can be stripped from the input rf signal, with a parallel limiting amplifier with narrow bandwidth. This approach cannot be used in suppressed-carrier AM systems where the carrier phase reversals are introduced in the modulation process.

The more general approach is to use an additional multiplier circuit as a phase detector. When the rf and local-oscillator signals are equal in frequency and in phase quadrature, the multiplier output has no dc component. Any relative phase shift from quadrature will produce an odd function error signal at dc. This signal can be used with a voltage controlled oscillator to correct the phase of the local oscillator or input rf signal. The system described is a phase-locked loop. The bandwidth of this loop can be controlled independently from the rf bandwidth for optimum acquisition and noise-suppression characteristics. This type of system is capable of producing a stable and noise-free local-oscillator signal that tracks any variations in the frequency and phase of the input rf signal.

With additional modifications the phase-locked receiver system is capable of reinserting the desired carrier in suppressed-carrier double- and vestigial-sideband systems.

14-15

15. References on Amplitude Modulators and Demodulators.

1. WILSON, J. P.   A Simple High-Speed Analogue Multiplier, *Electron. Eng.* (Great Britain), Vol. 39, 11-14, January, 1967.

2. LEENOV, D.   PIN Diode Microwave Switches and Modulators, *Solid State Design,* Vol. 6, pp. 37-40, April, 1965.

3. DEKOLD, RONALD   Amplitude Modular Is Highly Linear, *Electronics,* June 5, 1972, pp. 101-102.

4. MINTON, ROBERT   Design Trade-offs for RF Transistor Power Amplifiers, *Electron. Eng.,* March, 1967, pp. 68-73.

5. HEJHALL, ROY   For High-Frequency Communications Equipment Use Balanced Modulators, *EDN/EEE,* Feb. 15, 1972, pp. 28-32.

6. OPPENHEIMER, MICHAEL   In IC Form, Hall-Effect Devices Can Take On Many New Applications, *Electronics,* Aug. 2, 1971, pp. 46-49.

7. COTE, T.   Modern Techniques of Analog Multiplication, *Electron. Eng.,* April, 1970, pp. 75-79.

8. GILBERT, B.   A Precise Four Quadrant Multiplier with Subnanosecond Response, *IEEE J. Solid State Circuits,* Vol. SC-3, No. 4, pp. 365-373, December, 1968.

9. RHEINFELDER, W. A.   Modulation of Driver Stage to Increase Power Output of AM Transmitter, *Motorola Semiconductor Appl. Note* AN-114, reprint from *Semiconductor Products Mag.,* March, 1962.

10. HEJHALL, ROY   Getting Transistors into Single-Sideband Amplifiers, *Motorola Semiconductor Products Appl. Note* AN-150, reprinted from *Electronics.*

11. Principles and Techniques of Single-Sideband Modulation, *Electro-Technology,* July, 1962.

12. BRUBAKER, RICHARD J.   An All-Solid-State Marine Band Transmitter, *Motorola Semiconductor Products Appl. Note* AN-156, reprinted from *SSD/CDE.*

13. HEJHALL, ROY C.   A 50 Watt 50 MHz Solid State Transmitter, *Motorola Semiconductor Products Appl. Note* AN-246.

14. BRUBAKER, R.   A Broadband 4-Watt Aircraft Transmitter, *Motorola Semiconductor Products Appl. Note* AN-481.

15. MARTENS, CARL   A 40-W, 50-MHz Transmitter for 12.5 Volt Operation, *Motorola Semiconductor Products Appl. Note* AN-502.

16. TERMAN, FREDERICK E.   "Electronic and Radio Engineering," Chap. 15, McGraw-Hill, New York, 1955.

17. LANDEE, R. W., DAVIS, D. C., and ALBRECHT, A. P.   "Electronic Designers' Handbook," Secs. 5.1-5.4, McGraw-Hill, New York, 1957.

18. GRAY, TRUMAN S.   "Applied Electronics," Chap. 12, Arts. 1-14, Wiley, copyright 1943, 1954, Second Printing, March, 1955.

19. BILOTTI, ALBERTO   Application of a Monolithic Analog Multiplier, *IEEE J. Solid State Circuits,* Vol. SC-3, No. 4, December, 1968.

20. COSTAS, J. P.   Synchronous Communications, *Proc. IRE,* Vol. 44, pp. 1713-1718, December, 1956.

21. SHAPIRO, G. R.   Analog Multipliers Offer Solutions to Video Modulation Problems, *EDN,* Sept. 1, 1972, pp. 40-41.

22. Microwave Power Transistor Brochure MPT-700, *RCA Solid State Dev.*

23. Analog Multiplier Principles, *Analog Devices Tech. Bull.* AD530.

24. A New Linear Power Transistor for SSB Equipment, *RCA Commercial Eng., Appl. Note* AN-4591, Harrison, N. J.

25. Wideband Analog Multipliers 424/5. Analog Devices, Cambridge, Mass.

26. CARSON, J. R.   The Equivalent Circuit of the Vacuum-Tube Modulator, *IRE Proc.,* Vol. 9, pp. 243-249, 1921.

27. NEILSEN, J. R.   Tranformerless Ring Modulator, *EEE,* February, 1970, p. 116.

28. SONDE, B. S.   (Correspondence), Micropower Amplitude Modulator, *Proc. IEEE,* July, 1971, pp. 114-116.

29. KELLY, R. G.   Linear Modulator Has Excellent Temperature Stability, *EEE,* July, 1968, p. 102.

30. PICHARD, A.   100% Amplitude Modulation with Two Transistors, *Electronics,* June 12, 1967, pp. 104-105.

31. McDermott, C. Suppressed Carrier Modulator with Noncritical Components, *Electronics,* Oct. 31, 1966, p. 70.

32. Rockwell, R. J. Cathanode Modulation System, *IEEE Trans. Broadcasting,* Vol. BC-13, p. 19, January, 1967.

FREQUENCY AND PHASE (ANGLE) MODULATORS

By N. R. Powell

16. Angle Modulation. The representation of angle modulation is conveniently made in terms of the notion of the analytic function.[1]* For real continuous functions of time $x(t)$, consider

$$m(t) = x(t) + jHx(t)$$

where $x(t) =$ a real continuous function

$$j = (-1)^{1/2}$$

$Hx =$ Hilbert transformation of x

$$Hx(t) = \pi^{-1} \int_{-\infty}^{\infty} x(\tau)(t - \tau)^{-1} d\tau$$

The angle of $m(t)$ is said to be the angle θ defined by the relation

$$\theta = \tan^{-1}(Hx/x)$$

Angle modulation may be considered as the change in θ with time; or angle modulation may be considered that portion of the total change in θ which can be associated with the phenomenon of interest. If the relationship between the changes in θ and the effect of interest is direct, the modulation is called *phase modulation* (*PM*) and the devices producing this relationship *phase modulators*. If the relationship between the changes in the derivative $d\theta/dt$ and the effect of interest is direct, the modulation is called *frequency modulation* (*FM*), and the devices producing this relationship are called *frequency modulators*.

The derivative $d\theta/dt = \dot\theta$ is related to the components of the analytic function $m(t)$ by

$$\dot\theta = \begin{vmatrix} x & Hx \\ \dot x & H\dot x \end{vmatrix} \Big/ [x^2 + (Hx)^2] \qquad (14\text{-}28)$$

for functions $x(t)$ for which the differential and Hilbert operators commute.

Thus in angle modulators, whether implemented functionally in the general form (as with a general-purpose computer) suggested by the representations of θ and $\dot\theta$ or as some special combination of electronic networks, a direct or proportional relationship is established within the device between either of these two functions and an effect, call it the input $v(t)$, of interest. Diagramatically, a linear angle modulator may be considered to be a device that transforms $v(t)$, as shown in Fig. 14-14.

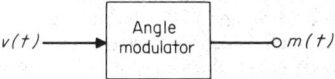

Fig. 14-14. Linear angle modulator.

Demodulators simply perform the inverse of this operation, providing an output function proportional to $\theta(t)$ from a function proportional to the angle of $m(t)$ as an input.

17. Angle Modulation Spectra. The spectral distribution of power for angle-modulated waveforms varies widely with the nature of the input function $v(t)$.

Random Modulation. For a random process having sample functions

$$x(t) = b \sin(2\pi ft + \phi), \qquad b \text{ constant}$$

---

*Superior numbers correspond to numbered references, Par. **14-25**.

where f and ϕ = independent random variables, ϕ being uniformly distributed over $-\pi \leq \phi \leq \pi$, and f having a symmetric probability density $p(f)$, this stationary random process has a spectral-density function

$$S_{xx}(f) = (b^2/2)p(f)$$

Note that if $p(f)$ is not a discrete distribution, the process is not in general periodic, and $S_{xx}(f)$ must be considered continuous.

Periodic Modulation. For a random process having sample functions

$$x(t) = b \cos[\omega_c t - \phi(t) + \theta] \qquad b, \omega_c \text{ constants}$$

θ uniformly distributed over $-\pi \leq \theta \leq \pi$, $\phi(t)$ is a stationary process independent of θ; that is,

$$\phi(t) = d \cos(\omega_m t + \theta') \qquad d, \omega_m \text{ constants}$$

θ' uniformly distributed over $-\pi \leq \theta' \leq \pi$, the spectral-density function is

$$S_{xx}(f) = (b^2/4)(J_0^2(d)[\delta(f - f_c) + \delta(f + f_c)]$$
$$+ \sum_{n=1}^{\infty} J_n^2(d)\{\delta[f - (f_c \pm nf_m)] + \delta[f + (f_c \pm nf_m)]\}) \qquad (14\text{-}29)$$

where $J_n(d)$ = the nth-order Bessel function of the first kind evaluated at d, δ = the Kronecker delta function, and $S_{xx}(f)$ = the Fourier transformation of the autocorrelation function of x.

Deterministic Modulation. For a function of a completely specified type, e.g.,

$$x(t) = b \sin(\omega_c t + d \sin \omega_m t)$$

frequently it is possible to re-express $x(t)$ in terms of $J_n(d)$ as

$$x(t) = \sum_{n=-\infty}^{\infty} J_n(d)\sin(\omega_c t + n\omega_m t) \qquad (14\text{-}30)$$

Examination of Eqs. (14-28) and (14-29) and tables of Bessel functions[2] permits the construction of Fig. 14-15, showing[3] the implied increase in bandwidth vs. modulation index for FM. Modulation index for sinusoidal modulation may be defined as

$$d = |\Delta f|/f_m$$

where $|\Delta f| = df_m$ = amount of instantaneous

frequency change, $\dot{\theta}$, to be

associated with the input $v(t)$.

Such a set of curves can be readily constructed for most deterministic modulation waveforms, since $J_n(d)$ decreases monotonically and rapidly with n for $n > d > 1$. Using the criteria for n indicated for curves A, B, and C, the frequency range (or bandwidth) centered about ω_c, which contains all such spectral components, is indicated. These are the components that are found to be below the value of nf_m for which $J_n(d)$ is monotonically decreasing and equal to the value for each case. Thus the bandwidth (BW) required at the frequency ω_c is

$$\text{BW} = 2(df_m)(1 + I) \qquad (14\text{-}31)$$

Such criteria should be used with care, since the relationship which these measures bear to distortion of the input $v(t)$, when carried at frequency ω_c through linear-tuned circuits as angle modulation, is rather indirect. An approximation for the signal-to-distortion ratio (SDR) for gaussian baseband modulation, uniform in $(-B,B)$ of modulation index d, which is passed through a single-pole bandpass filter with half-bandwidth f_c, is[9]

$$\text{SDR} \approx 15/2B^2 d^4 \qquad B/f_c < 0.3 \qquad (14\text{-}32)$$

18. Angle Modulation Signal-to-Noise Improvement. One of the principal reasons for using frequency modulation in communications and telemetry systems is that it provides a convenient and power-efficient method of trading power for bandwidth while providing high-quality transmission of the input. This is expressed in phase modulation by the relationship

$$\left.\frac{S}{N}\right|_{f_m} = ad^2 \left.\frac{C}{N}\right|_{f_m}$$

where $\left.\dfrac{S}{N}\right|_{f_m}$ = demodulated output signal-to-noise ratio measured in a bandwidth f_m

$\left.\dfrac{C}{N}\right|_{f_m}$ = demodulator input signal-to-noise ratio measured in a bandwidth f_m (14-33)

d = modulation index

a = a constant of proportionality

The constant of proportionality a is unity for sinusoidal phase modulation of constant-amplitude sinusoidal carrier. The constant varies between 0.5 and 3.0 with class of modulation waveforms, type of network compensation, and noise spectrum; however, Eq. (14-33) may be conservatively applied to FM single-channel voice systems for $a = 3/2$ and preemphasis and deemphasis networks which preshape the spectrum of the modulation to match the sloped noise spectrum and to restore the original spectrum after demodulation.

The tradeoff between the signal-to-noise improvement and the required bandwidth corresponding to curve A in Fig. 14-15 is shown for FM in Fig. 14-16. These curves have been prepared for a conventional demodulator operating at an input carrier-to-noise ratio 1 dB above the threshold (13 dB) measured in the input noise bandwidth to the demodulator, with a constant of proportionality a equal to 0.5. Output signal-to-noise ratio referred to the output information bandwidth $(S/N)_{fm}$ is

$$(S/N)|_{fm} = 10 + G \quad (dB) \tag{14-34}$$

where G is obtained from the figure along with r, the ratio of premodulator bandwidth to output information bandwidth. The carrier-to-noise ratio in a noise bandwidth equal to the output information bandwidth is

$$(C/N)|_{fm} = 13 + 10 \log r \quad (dB) \tag{14-35}$$

19. Noise Threshold Properties of Angle Modulation. The signal-to-noise improvement represented by Eq. (14-33) is achievable only when the input carrier-to-noise ratio is above certain minimum levels. These levels depend upon the type of modulating waveforms, the type of noise interference prevalent, and the type of demodulator employed. As the foregoing discussion indicates, whenever a demodulator without phase or frequency feedback is employed, an input carrier-to-noise ratio (measured in the premodulator bandwidth) of roughly 12 dB is required. Unless this condition is met, a small decrease in carrier-to-noise ratio will result in a sharp decrease in output signal-to-noise ratio, accompanied by undesirable noise effects, such as loud clicking sounds in the case of voice modulation.

For properly designed feedback demodulators, the noise threshold is not determined by the condition prevalent in the premodulator bandwidth as much as by the closed-loop noise bandwidth of the demodulator and the match between internal loop filtering, modulation waveforms, and noise properties.

A set of curves illustrating the effect of thresholding in properly designed feedback demodulators is shown in Fig. 14-17. Curve 1 is an information-theoretic limit based upon gaussian modulation, infinite predemodulator bandwidth, and the Shannon upper bound on information flow. The Wiener-Hopf filter limit corresponds to the use of optimal feedback-demodulator filter design and is shown as curve 2. Curve 3 corresponds to the use of a phase-

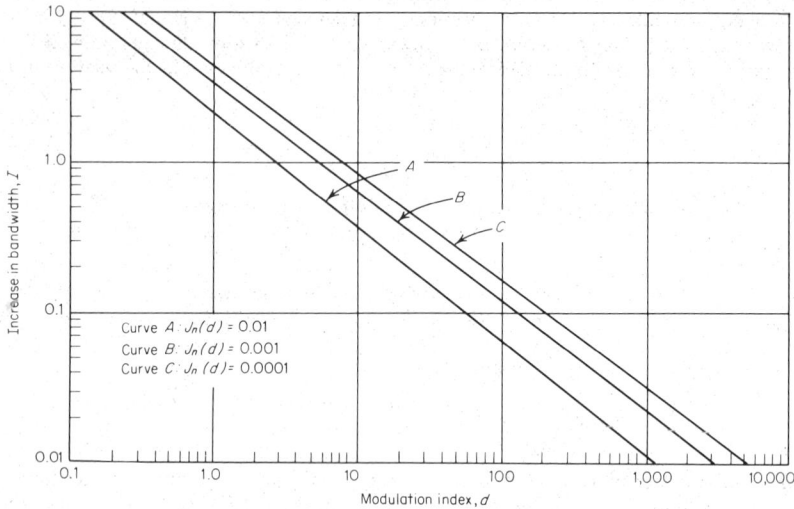

Fig. 14-15. Bandwidth increase vs. modulation index.

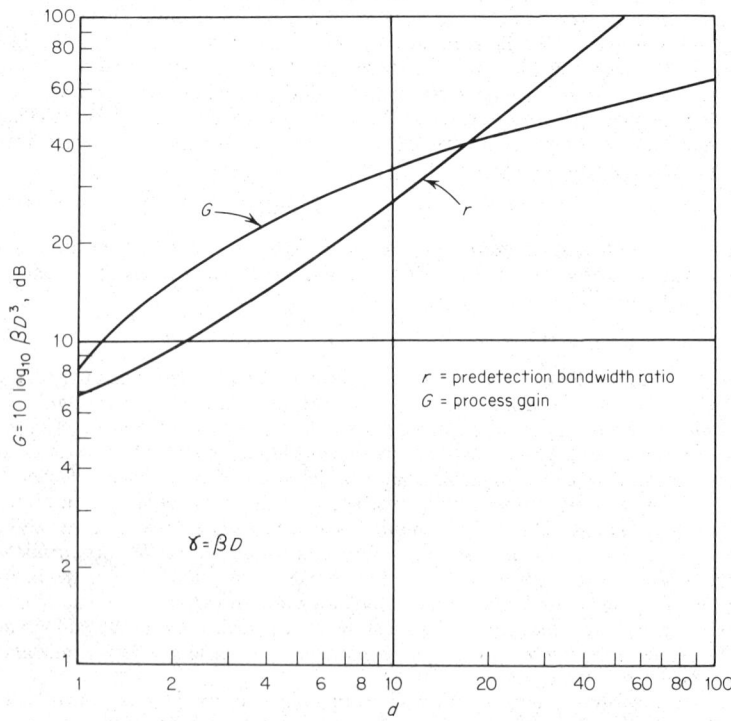

Fig. 14-16. Process gain and rf bandwidth vs. deviation ratio.

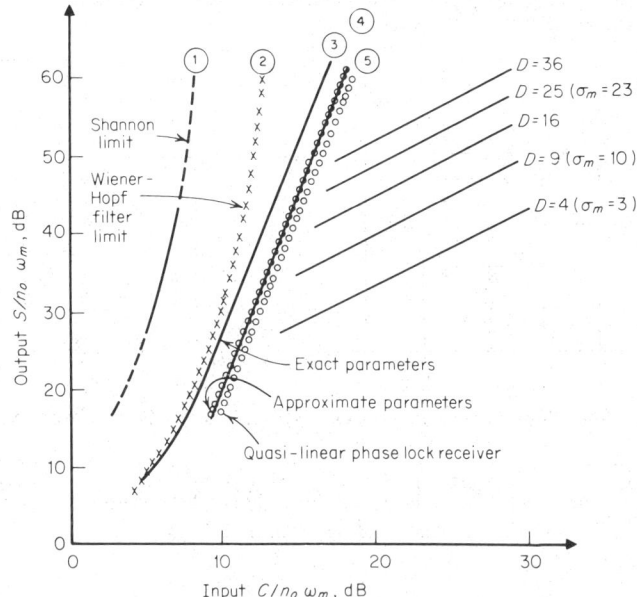

Note : Curves 1, 2, 5 correspond to gaussian modulation, σ_m

Curves 3, 4 correspond to voice modulation, D

Fig. 14-17. Threshold properties of FM.

locked demodulator employing frequency feedback,[4] with a simple proportional plus integral filter, the time constants for which are selected according to the prevalent signal-to-noise ratio and the modulation parameters. Curves 4 and 5 represent threshold performance achievable with a fixed-parameter design using the same linear filter approximation to the optimum filter. The noise threshold for curve 4 is, approximately,

$$(S/N)|_{fm} = 0.48 \times 10^{-2}(C/N)^5|_{fm} \qquad (14\text{-}36)$$

The modulation indices shown for operation above threshold are for single-channel voice and single-channel gaussian[5] modulation.

20. Angle Modulators. Angle modulators for communications and telemetry purposes generally fall into the category of what may be termed "hard" oscillators having relatively high-Q frequency-determining networks; or they fall into the category of "soft" oscillators having supply and bias sources as the frequency-determining networks. Examples of each are shown in Fig. 14-18.

Control of the hard oscillator is executed by symmetrical incremental variation of the reactive components. For the case of a Hartley oscillator, Z_1 and Z_2 are inductors and Z_3 is a capacitor, allowing the use of varicaps paralleling Z_3 as the voltage-controllable reactance. In such a case

$$f_0 = (2\pi)^{-1}[C_3(L_1 + L_2)]^{-1/2}$$

where C_3 = total capacitance of the varicaps and fixed capacitor of Z_3. Note that the bandwidth of this modulator is determined by the frequency-determining impedances of the network, i.e., the overall Q and center frequency. In view of the need for certain minimum bandwidth requirements from Fig. 14-15 and the need for good oscillator stability, the total frequency deviation required is sometimes obtained by following the oscillator with a series of frequency multipliers and frequency translators, as shown in Fig. 14-19. This configuration permits the attainment of good oscillator stability, constant proportionality between output frequency change and input voltage change, and the necessary modulator bandwidth to achieve wideband FM.

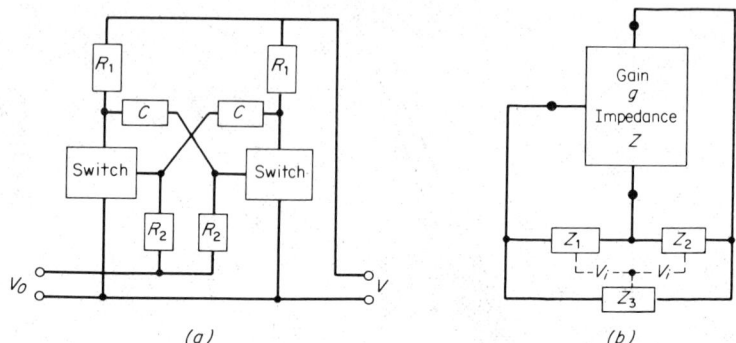

(a)   (b)

Fig. 14-18. Voltage controlled oscillators. (*a*) Soft oscillator; (*b*) hard oscillator.

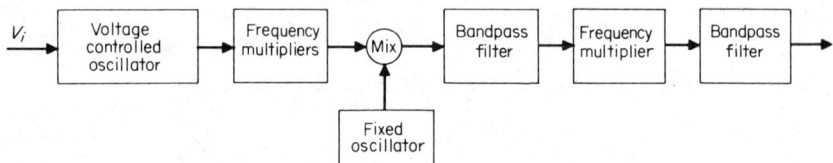

Fig. 14-19. Frequency modulator configuration.

Control for the soft oscillator is introduced as a change in the switching level of the active-device switches. The frequency of a transistor version of this oscillator is, roughly,

$$f = [2R_1 C \ln(1 + V/V_j)]^{-1}$$

Since this type of oscillator is a relaxation oscillator, the rate at which the frequency of oscillation can be changed is limited only by the rate at which the switching points can be altered by voltage control. Modulators of this kind can be designed with bandwidths greater than the frequency of oscillation. The disadvantage of such networks is the relatively poor frequency stability compared with the high-Q hard oscillators.

ANGLE DEMODULATORS

By N. R. Powell

21. Discriminators. Basic angle demodulators may be designed using balanced tuned networks with suitably connected nonlinearities (such as diode switches); or angle demodulators may be designed using voltage controlled oscillators (VCO), multipliers, and appropriate feedback networks.

The former are simpler to implement than the latter, but require much higher input signal-to-noise ratios for operation above the noise threshold at which the angle modulation produces signal process gain. Two popular versions of the discriminator type of frequency demodulator are shown in Fig. 14-20. The diode discriminator is designed so that one diode conducts more with increases, the other with decreases, in frequency. For greatest linearity, the mutual coupling between the tuned circuits is generally greater than unity. The other basic type of conventional demodulator simply implements a version of the definition [Eq. (14-29)] of FM for the case of a constant-amplitude sinusoidal waveform. The mixing operation can be replaced by a simple phase shifter that provides a 90° phase relationship between x and Hx. Frequently, the reference oscillator will be phase-locked to the average value of the $J_0(d)$ component indicated by Eqs. (14-29) and (14-30).

22. Feedback Demodulators (FMFB). These angle demodulators have the advantage of lower distortion, lower noise threshold, and little or no drift in center frequency of operation and can be designed to be less sensitive to interference. No limiter is required for good performance, and there is no requirement to maintain a minimum predemodulator input carrier-to-noise ratio to avoid noise threshold.

A block diagram of a basic synchronous-filtering demodulator is shown in Fig. 14-21. The synchronous filter is indicated by the dashed lines which enclose a phase-locked loop designed to follow the instantaneous excursions in the phase of the angle-modulated signal ϕ_2. The mixer, if amplifier, discriminator, filter 2, and the voltage controlled oscillator VCO-2 form a frequency feedback loop for the demodulator.

Aside from the synchronous filter, the basic configuration may be considered that of a conventional FMFB demodulator[6] which compresses the wide-band FM input signal so that it may be passed through a relatively narrow-bandwidth fixed-tuned filter and to a discriminator for detection. It should be noted that the configuration shown here may also be considered simply as a phase-locked loop (PLL) with frequency feedback around it.

The significant advantages of each of these techniques (FMFB and PLL) may be combined. It is important to consider the design from both the synchronous and frequency-feedback viewpoints. Note in this regard Figs. 14-22 and 14-23 in which $F_1(S)$ and $F_2(S)$ have been assigned. Observe that Fig. 14-24 is an FMFB equivalent linear form obtained by substituting the closed-loop transfer function for the synchronous filter; however, by retaining the inner loop and combining phase comparators, we obtain the synchronous phase-locked loop form as shown in Fig. 14-25. To the extent that linear analysis and quasi-linear substitutions may be made in these block diagrams, the remarks which follow are thus

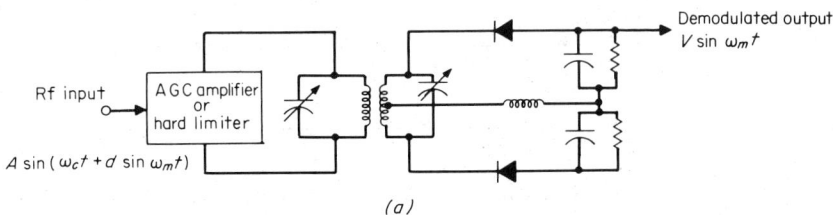

(a)

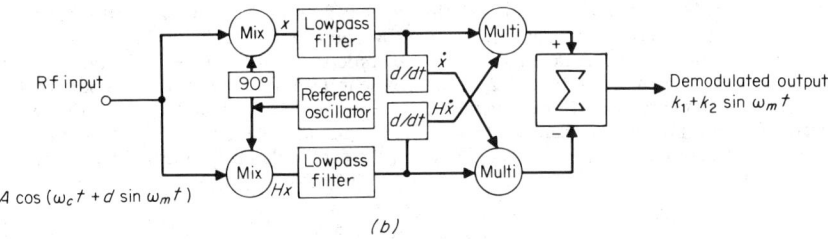

(b)

Fig. 14-20. Conventional demodulators. (*a*) Diode discriminator; (*b*) phase shift discriminator.

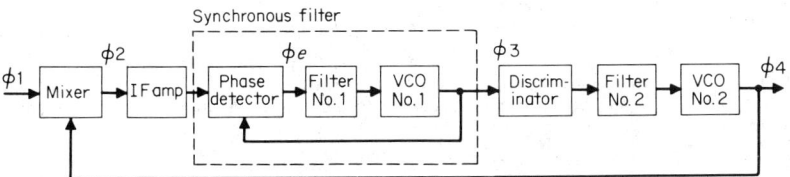

Fig. 14-21. Block diagram of basic feedback demodulator.

relevant to both PLL and FMFB forms implemented with synchronous filters or with broadband intermediate-frequency amplifiers in cascade with single-tuned filters.

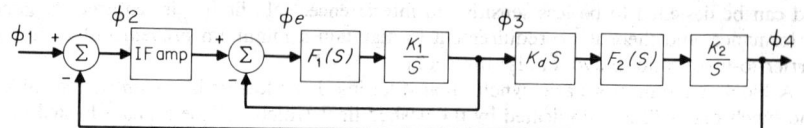

Fig. 14-22.   Equivalent network of basic demodulator shown in Fig. 14-21.

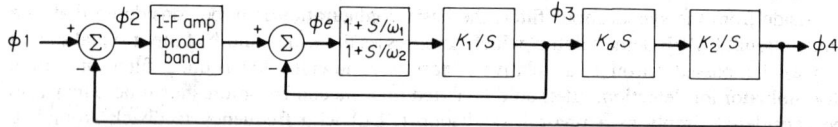

Fig. 14-23.   Direct linear equivalent form of basic demodulator.

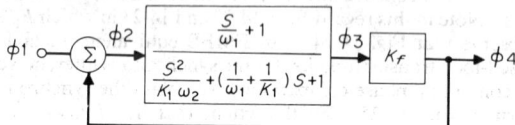

Fig. 14-24.   FMFB demodulator.

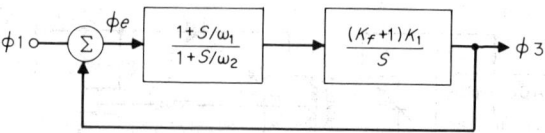

Fig. 14-25.   PLL demodulator.

23. Phase-locked Loop Demodulators. The introduction of the phase-locked loop between the if amplifier and discriminator may be viewed simply as a means by which the if signal may be tracked, limited, and filtered regardless of Doppler shift or oscillator drift. The synchronous filter is employed in conjunction with the relatively wide-bandwidth if amplifier shown on the block diagram to perform this critical filtering function, as well as to take advantage of the phase coherence between the FM signal sidebands.

The operation of the demodulator may be understood from an examination of the equivalent network shown in Fig. 14-22 in which the transfer functions relate to phase as the input and output variables. If the bandwidth of the if amplifier is broad by comparison with both the synchronous-filter bandwidth and significant modulation sidebands, it may be ignored in the equivalent linear representation of the demodulator. Since the synchronous-filter function is

$$\frac{\phi_3(s)}{\phi_2(s)} = \frac{(K_1/s)F_1(s)}{1 + (K_1/s)F_1(s)} \tag{14-37}$$

the signal ϕ_2 is related to the input by

$$\frac{\phi_2(s)}{\phi_1(s)} = \frac{1}{1 + \dfrac{K_1 K_2 K_d F_1(s) F_2(s)}{s[1 + (K_1/s)F_1(s)]}} \tag{14-38}$$

For frequency components of $\phi_1(s)$ lying well within the bandwidths of $F_2(s)$ and the synchronous filter, this transfer function reduces to the familiar form of a type-zero feedback network; i.e.,

$$\frac{\phi_2(s)}{\phi_1(s)} = \frac{1}{1 + K_2 K_d} = \frac{s\phi_2(s)}{s\phi_1(s)} \tag{14-39}$$

Since this is also the transfer function with respect to frequency variations, the compression of frequency excursions is evident. If

$$\phi_1(t) = \frac{\Delta\omega}{\omega_m} \sin \omega_m t = D \sin \omega_m t \qquad (14\text{-}40)$$

where $\phi_1(t)$ represents the instantaneous variation of the phase of the input signal relative to some reference carrier phase, and if the synchronous filter follows this instantaneous variation, the effective phase excursion to the discriminator is reduced to

$$\phi_3(t) = \frac{D}{1 + K_f} \sin \omega_m t \qquad K_f = K_2 K_d \qquad (14\text{-}41)$$

if ω_m is well within the passband of $F_2(s)$ and the phase-locked loop. This reduction in deviation ratio by the use of frequency feedback gain K_f suggests that an optimum gain and synchronous-filter bandwidth combination should be sought for given modulation and noise characteristics, just as the proper if amplifier, and K_f must be chosen in a conventional FMFB demodulator.

Figure 14-24 shows the closed-loop transfer function of the synchronous filter along with the frequency-feedback loop. It has been assumed that the if amplifier bandwidth is very broad compared with the bandwidth occupied by the significant portions of the signal spectrum as it appears at if frequencies. This amounts to assuming that the dispersive effect produced by the if amplifier is negligible. If the phase-locked loop gain may be considered large compared with the filter zero ($K_1 \gg \omega_1$ is usually satisfied in practice), the following relationships between the synchronous-filter and the complete feedback demodulator parameters can be written

$$\zeta_f^2 = (K_f + 1)\zeta_\phi^2$$

$$\omega_{n_f}^2 = (K_f + 1)\omega_{n_\phi}^2 \qquad (14\text{-}42)$$

$$B_{n_f} = B_{n_\phi}\left(1 + K_f \frac{1}{1 + 1/4\zeta_\phi^2}\right)$$

where ζ_f = damping of demodulator, ω_{n_f} = natural frequency of demodulator, ζ_ϕ = damping of synchronous filter, ω_{n_ϕ} = natural frequency of synchronous filter, B_{n_f} = noise bandwidth of demodulator, B_{n_ϕ} = noise bandwidth of synchronous filter, and in terms of the actual network parameters,

$$\zeta_\phi^2 = \frac{K_1 \omega_2}{4\omega_1^2}$$

$$\omega_{n_\phi}^2 = K_1 \omega_2 \qquad (14\text{-}43)$$

$$B_{n_\phi} = \frac{1}{2}\left(\frac{K_1 \omega_2}{\omega_1} + \omega_1\right) \qquad \text{(Hz)}$$

Note that all the demodulator response variables can be controlled by simple RC adjustments.

If, as previously, $\phi_1(t) = D \sin \omega_m t$,

$$\phi_2(t) = \frac{D}{K_f + 1}\left\{\sin \omega_m t + \frac{K_f \omega_m^2/\omega_{n_f}^2}{\left[\left(1 - \frac{\omega_m^2}{\omega_{n_f}^2}\right)^2 + \left(\frac{\omega_m}{\omega_1}\right)^2\right]^{1/2}} \sin(\omega_m t + \psi_1)\right\} \qquad (14\text{-}44)$$

$$\text{where} \qquad \psi_1 = \pi - \tan^{-1}\frac{K_1}{\omega_1} \frac{1}{1 - \left(\frac{\omega_m}{\omega_{n_f}}\right)^2}$$

$$\approx 90° \qquad \text{for } \frac{K_1}{\omega_1} \gg 1 \text{ and } \left(\frac{\omega_m}{\omega_{n_f}}\right)^2 \ll 1$$

$$\phi_4(t) = \frac{K_f D}{K_f + 1} \left\{ \sin \omega_m t - \frac{(\omega_m/\omega_{nf})^2 \sin(\omega_m t + \psi_1)}{\left[\left(1 - \frac{\omega_m^2}{\omega_{nf}^2}\right)^2 + \left(\frac{\omega_m}{\omega_1}\right)^2 \right]^{1/2}} \right\}$$

(14-45)

$$\phi_3(t) = \frac{D}{K_f + 1} \left\{ \sin \omega_m t - \frac{(\omega_m/\omega_{nf})^2 \sin(\omega_m t + \psi_1)}{\left[\left(1 - \frac{\omega_m^2}{\omega_{nf}^2}\right)^2 + \left(\frac{\omega_m}{\omega_1}\right)^2 \right]^{1/2}} \right\}$$

(14-46)

$$\phi_e(t) = \frac{(D\omega_m^2/\omega_{nf}^2) \sin(\omega_m t + \psi_1)}{\left[\left(1 - \frac{\omega_m^2}{\omega_{nf}^2}\right)^2 + \left(\frac{\omega_m}{\omega_1}\right)^2 \right]^{1/2}}$$

(14-47)

These four equations describe the instantaneous time relationships which exist throughout the demodulator. Examination of them along with relationships (14-42) indicates the manner in which the demodulator parameters may be employed to improve the response characteristics of the demodulator and to maintain the phase error in Eq. (14-47) within the proper bounds.

24. FM Feedback Demodulator Design Formulas.

Acquisition: The rate at which the VCO in a phase-locked demodulator can be swept through the frequency and phase at which pull-in and phase lock will occur is indicated by the representative curves[7]* of Fig. 14-26.

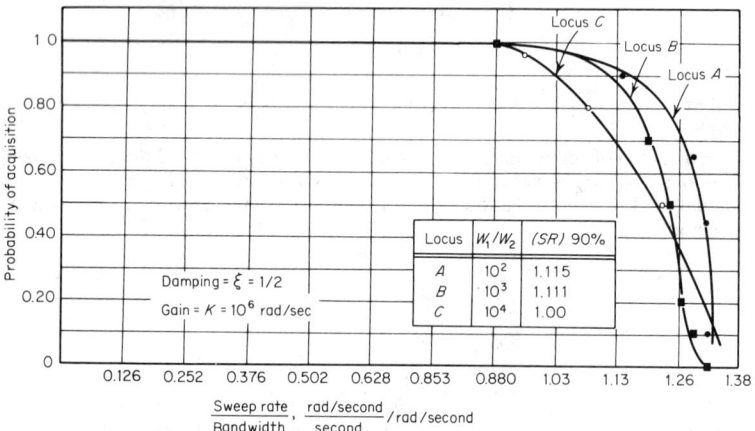

Fig. 14-26. Probability of acquisition vs. sweep rate/bandwidth.

Pull-in: The frequency difference $\Delta\omega$ (between an unmodulated carrier and the VCO) within which a phase-locked loop will pull into phase synchronism is, roughly,[8]

$$|\Delta\omega|_p = (2\rho K \omega_n)^{1/2} \qquad \text{for } K/\omega_n \gg 1$$

where K = total open-loop gain in radians per second, ω_n = loop natural frequency in radians per second, and ρ = a dimensionless constant, approximately unity.

The time required for pull-in is given by[9]

$$T = 4(\Delta f)^2 / B_n^3 \qquad \text{(seconds)}$$

for loop damping of 0.5 and $|\Delta f| < 0.8 |\Delta f|_p$, where Δf = difference frequency in hertz, and B_n = closed-loop noise bandwidth in hertz.

*Superior numbers correspond to numbered references, Par. **14-25.**

Stability: The condition of sustained beat-note stability without the loop capacity to be swept into lock, which is exhibited by high-grain, narrow-bandwidth, phase-locked demodulators, is a condition of loop oscillation arising from the presence of the phase detector nonlinearity and extraneous memory, such as that in the tuned circuits in phase detectors and in voltage controlled oscillators. This condition can be predicted from the approximate condition

$$\frac{K^2|G(\omega_0)|}{2|\omega_0|^2} \cos[\phi(\omega_0)] + 1 = 0$$

where K = total open-loop gain, G = complete transfer function between phase detector and VCO, and ϕ = angle of G. This condition can be used to predict the nonlinear network oscillations preventing lockup.

Parameter Variations: The rate at which second-order demodulators damped for minimum-noise bandwidth can have the loop natural frequency changed at constant damping is given by[4]

$$\frac{d\omega_n/dt}{\omega_n^2} \leq 0.1$$

for peak phase errors less than 5°, where ω_n is the natural frequency of the loop.

Minimum-Threshold Parameters: The parametric relationships for minimum-noise threshold for the demodulator in Fig. 14-23 are shown in Fig. 14-27.

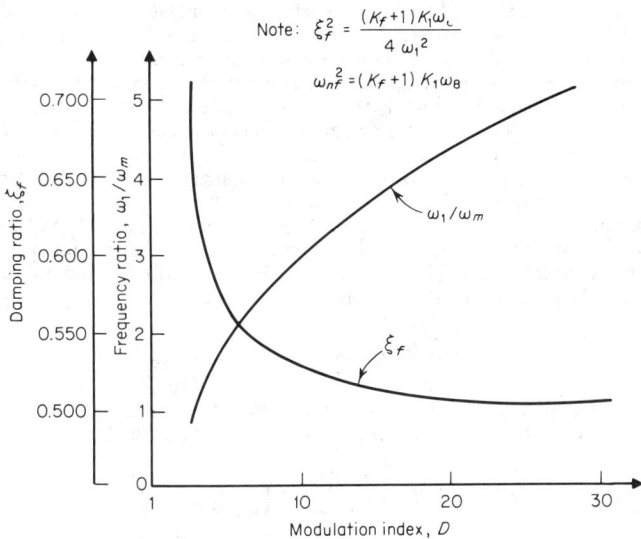

Note: $\xi_f^2 = \dfrac{(K_F+1)K_1\omega_L}{4\,\omega_1^2}$

$\omega_{nF}^2 = (K_F+1)K_1\omega_B$

Fig. 14-27. Parametric relationships for minimum threshold (voice).

25. References on Angle Modulators and Demodulators

1. BEDROSIAN, E. The Analytic Signal Representation of Modulated Waveforms, *Proc. IRE*, Vol. 50, p. 2071, October, 1962.

2. JANHKE, E., and EMDE, F. "Tables of Functions with Formulae and Curves," Dover, New York.

3. CORRINGTON, M. S. Variation of Bandwidth with Modulation Index in Frequency Modulation, *Proc. IRE*, October, 1947.

4. POWELL, N. R. Controlled Parameter Phase-Feedback FM Demodulation, *Seventh Ann. Internatl. Space Electron. Symp. Proc.*, 6-b-1, October, 1964.

5. DEVELET, J. A., JR. A Threshold Criterion for Phase-Lock Demodulation, *Proc. IEEE*, Vol. 51, No. 2, pp. 349-356, 1963.

6. CHAFFEE, H. C.   The Application of Negative Feedback to F. M. Systems, *Bell System Tech. J.*, 1939.

7. POWELL, N. R., and WOODS, C. R.   A Study of Acquisition Capabilities of Phase-locked Oscillators, *G. E. Rept.* ASER 28-60, March, 1960.

8. GRUEN, W. J.   Theory of AFC Synchronization, *Proc., IRE,* Vol. 53, pp. 1043–1048, August, 1953.

9. BEDROSIAN, E., and RICE, S. O.   Distortion and Crosstalk of Linearly Filtered Anglemodulated Signals, *Proc. IEE,* Vol. 56, No. 1, pp. 2–13, January, 1968.

PULSE MODULATORS AND DEMODULATORS

By G. F. PFEIFER

26. Pulse modulation is, in general, the encoding of information by means of varying one or more pulse parameters.   It finds application in both the communication and the control fields.   The control applications are usually confined to the use of pulse time modulation (PTM) and pulse frequency modulation (PFM), where on-off control power can be used to minimize device dissipation.   All pulse modulation schemes require sampling analog signals, and some, such as pulse code modulation (PCM) and delta modulation, require the additional quantization of the analog signals.

In communications, the chief application of pulse modulation is found where it is desired to time-multiplex by interleaving a number of single-channel, low-duty-cycle pulse trains. The pulse trains may, in turn, be used for compound modulation by amplitude or angle modulation of a continuous carrier.   In usual applications, subcarriers are pulsed, time-division-multiplexed, and then used to frequency-modulate a carrier.

Since noise is present in all systems, a prime consideration in modulation selection is the choice of a waveform based upon its signal-to-noise efficiency.   For instance, PTM is more efficient than PAM, which offers no improvement over continuous AM; however, PTM is less efficient than PCM or delta modulation.   A chief advantage of pulsed systems such as PTM, PCM, and delta is improved signal-to-noise ratio in exchange for increased bandwidth, in the same manner as continuous FM improves over AM.

27. Sampling and Smoothing.   An ideal impulse sampler can be considered as the multiplication of an impulse train, period T seconds, with the continuous signal $f(t)$. Following the notation[1] of Ref. 6,* this is shown in Fig. 14-28a for the impulse train defined as $p_T(t) = \text{rep}_T[\delta(t)]$ where $\delta(t)$ is an impulse at $t = 0$ and

$$\text{rep}_T[u(t)] = \sum_{n=-\infty}^{\infty} u(t - nT) \tag{14-48}$$

The output spectrum function is the convolution of $F(f)$, the Fourier transform of the input signal, and the transform of $p_T(t)$, which is $(1/T)\,\text{comb}_{1/T}(1)$, defined as

$$\text{comb}_{1/T}[U(f)] = \sum_{n=-\infty}^{\infty} U\left(\frac{n}{T}\right)\delta\left(f - \frac{n}{T}\right) \tag{14-49}$$

Thus the transform $R(f) = F(f)*(1/T)\,\text{comb}_{1/T}(1)$ with spectrum is as shown in Fig. 14-28b. The result of ideal impulse sampling has been to repeat the original signal spectrum, assumed to be band-limited, each $1/T$ Hz, and multiply each by a $1/T$ scale factor.   Since all the signal information is present in each lobe of Fig. 14-28b, it is only necessary to recover a single lobe through filtering in order to recover the signal function reduced by a scale factor.

Consider an ideal low-pass rectangular filter of bandwidth f_f Hz defined as $T(jf) = A(f)e^{-j\theta(jf)}$

$$\text{where } A(f) = \begin{cases} 1 & |f| < f_f \\ 0 & |f| > f_f \end{cases}$$

$$\theta(jf) = 2\pi\alpha f \quad \text{for all } f$$

---

*Numbered references are listed in Par. **14-41.**

The cutoff frequency f_f is adjusted to select the output spectral lobe about zero $f_c \nearrow f_f < 1/T - f_c$ and will fall in the guard band between lobes. That portion of $R(f)$ selected is

$$R_0(f) = \frac{1}{T}F(f)e^{-j2\pi\alpha f} \qquad (14\text{-}50)$$

with inverse transform

$$r_0(t) = \frac{1}{T}f(t - \alpha) \qquad (14\text{-}51)$$

which is identical with the signal function, with the amplitude reduced by a scale factor and function shifted by α seconds. If $\alpha = 0$, signifying no delay, the filter is termed a "cardinal data hold"; otherwise, it is an "ideal low-pass filter." Unfortunately, these filters cannot be realized in practice, since they are required to respond before they are excited.[2]

Examination of Fig. 14-28b gives rise to the sampling theorem accredited to Shannon and/or Nyquist which states that, when a continuous time function with band-limited spectrum $-f_c < f < f_c$ is sampled at twice the highest frequency, $f_s = 2f_c$, the original time function can be recovered. This corresponds to the point where the sampling frequency $f_s = 1/T$ is decreased so that the spectral lobes of Fig. 14-28b are just touching. To decrease

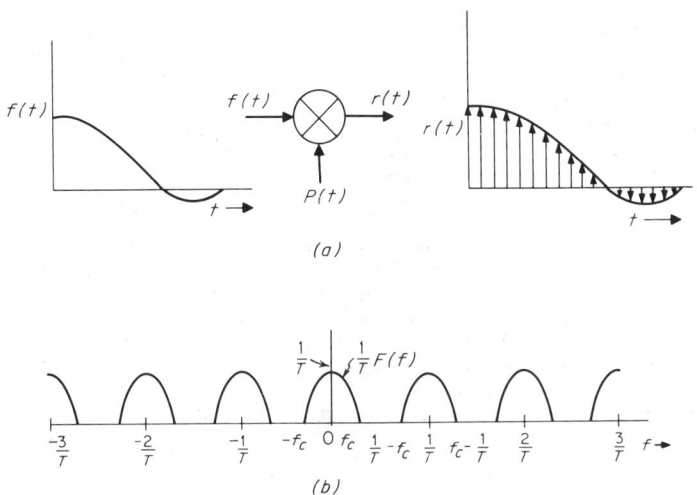

Fig. 14-28. Pulse modulation. (*a*) Output spectrum; (*b*) sampling configuration.

f_s beyond the value of $2f_c$ would cause spectral overlap and make recovery with an ideal filter impossible. A more general form of the sampling theorem states that any $2f_c$ independent samples per second will completely describe a band-limited signal, thus removing the restriction of uniform sampling, as long as independent samples are used.[2] In general, for a time-limited signal of T seconds, band-limited to f_c Hz, only $2f_cT$ samples are needed to specify the signal completely.

In practice, the signal is not completely band-limited, so that it is common to allow for a greater separation of spectral lobes, called the *guard band.* This guard band is generated simply by sampling at greater than $2f_c$, as in the case for Fig. 14-28b. Although the actual tolerable overlap depends on the signal spectral slope, setting the sampling rate at about $3f_c = f_s$ is usually adequate to recover the signal.

In practice, narrow but finite-width pulse trains are used in place of the idealized impulse sampling train. To determine the effect of finite-width pulses, consider the pulse train made up of τ duration pulses repeating at a $1/T$ rate represented by

$$s(t) = \mathrm{rep}_T\left[\mathrm{rect}\left(\frac{t}{\tau}\right)\right] \qquad (14\text{-}52)$$

14-29

where

$$\text{rect}\left(\frac{t}{\tau}\right) = \begin{cases} 1 & |t| < \tau/2 \\ 0 & |t| > \tau/2 \end{cases}$$

The corresponding Fourier transform is

$$S(f) = \frac{\tau}{T}\,\text{comb}_{1/T}[\text{sinc}(\tau f)] \qquad (14\text{-}53)$$

where

$$\text{sinc}(\tau f) = \frac{\sin \pi\tau f}{\pi\tau f}$$

Since the transform of the sampler output function is the convolution of the signal transform and the pulse train, the resulting response $R_s(f)$ is given by

$$R_s(f) = F(f)^* S(f)$$
$$= F(f)^* \frac{\tau}{T}\,\text{comb}_{1/T}[\text{sinc}(\tau f)] \qquad (14\text{-}54)$$

The pulse width is usually much smaller than the period $\tau \ll T$, so that the comb function is an impulse train in frequency with an envelope that follows a $\sin x/x$ function. The convolution with the signal $F(f)$ thus yields a spectrum that is almost the same as for the impulse sampling train, except that the lobes decrease in amplitude as $\text{sinc}(\tau f)$ determined by the pulse width τ. The ideal reconstruction filter can be approximated by commonly used hold circuits, the characteristics of which are shown[10] in Fig. 14-29.

28. Pulse Amplitude Modulation (PAM). Pulse amplitude modulation is essentially a sampled-data type of encoding where the information is encoded into the amplitude of a train of finite-width pulses. The pulse train can be looked upon as the carrier in much the same way as the sinewave is for continuous amplitude modulation. There is no improvement in signal-to-noise when using PAM, and furthermore, PAM is not considered wideband in the sense of FM or pulse time modulation (PTM). Thus PAM would correspond to continuous AM, or PTM corresponds to FM. Generally, PAM is used chiefly for time multiplex systems employing a number of channels sampled, consistent with the sampling theorem.

There are a number of ways of encoding information as the amplitude of a pulse train. These ways include both bipolar and unipolar pulse trains for both instantaneous or square-topped sampling and for exact or top sampling. In top sampling, the magnitude of the individual pulses follows the modulating signal during the pulse duration, while for square-topped sampling, the individual pulses assume a constant value, depending on the particular exact sampling point that occurs somewhere during the pulse time. These various waveforms are shown in Fig. 14-30.

The spectrum for the top-modulation bipolar sampling case is given by Eq. (14-54), since this type of modulation, shown in Fig. 14-30c, is simply sampling with a finite-pulse-width train. Carrying out the convolution indicated yields

$$R_{STB}(f) = \frac{\tau}{T} \sum_{n=-\infty}^{\infty} \text{sinc}\left(\frac{\tau n}{T}\right) F\left(f - \frac{n}{T}\right) \qquad (14\text{-}55)$$

Using a square-topped rectangular spectrum for the original signal spectrum, the spectrum for top-modulation bipolar sampling is shown in Fig. 14-31a. The signal spectrum repeats with a $\sin x/x$ scale factor determined by the sampling pulse width, with each repetition a replica of $F(f)$.

Unipolar sampling can be implemented by adding a constant bias A to $f(t)$, the signal, to produce $f(t) + A$, where A is large enough to keep the sum positive; that is, $A > |f(t)|$. Sampling the new sum signal by multiplication with the pulse train results in the unipolar top-modulated waveform of Fig. 14-30e. The spectrum can be found by substituting the new signal into Eq. (14-54) and results in

$$R_{STU}(f) = \frac{\tau}{T} \sum_{n=-\infty}^{\infty} \text{sinc}\left(\frac{\tau n}{T}\right)\left[F\left(f - \frac{n}{T}\right) + A\delta\left(f - \frac{n}{T}\right)\right] \qquad (14\text{-}56)$$

The cutoff frequency f_f is adjusted to select the output spectral lobe about zero $f_c \smallfrown f_f < 1/T$ $- f_c$ and will fall in the guard band between lobes. That portion of $R(f)$ selected is

$$R_0(f) = \frac{1}{T}F(f)e^{-j2\pi\alpha f} \qquad (14\text{-}50)$$

with inverse transform

$$r_0(t) = \frac{1}{T}f(t - \alpha) \qquad (14\text{-}51)$$

which is identical with the signal function, with the amplitude reduced by a scale factor and function shifted by α seconds. If $\alpha = 0$, signifying no delay, the filter is termed a "cardinal data hold"; otherwise, it is an "ideal low-pass filter." Unfortunately, these filters cannot be realized in practice, since they are required to respond before they are excited.[2]

Examination of Fig. 14-28b gives rise to the sampling theorem accredited to Shannon and/or Nyquist which states that, when a continuous time function with band-limited spectrum $-f_c < f < f_c$ is sampled at twice the highest frequency, $f_s = 2f_c$, the original time function can be recovered. This corresponds to the point where the sampling frequency $f_s = 1/T$ is decreased so that the spectral lobes of Fig. 14-28b are just touching. To decrease

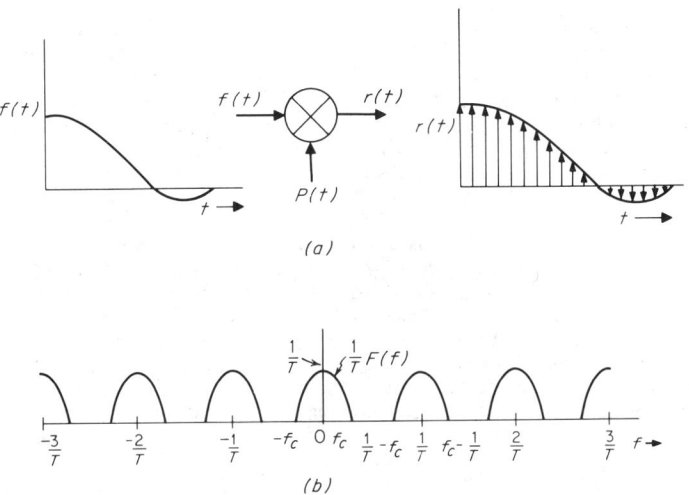

Fig. 14-28. Pulse modulation. (*a*) Output spectrum; (*b*) sampling configuration.

f_s beyond the value of $2f_c$ would cause spectral overlap and make recovery with an ideal filter impossible. A more general form of the sampling theorem states that any $2f_c$ independent samples per second will completely describe a band-limited signal, thus removing the restriction of uniform sampling, as long as independent samples are used.[2] In general, for a time-limited signal of T seconds, band-limited to f_c Hz, only $2f_cT$ samples are needed to specify the signal completely.

In practice, the signal is not completely band-limited, so that it is common to allow for a greater separation of spectral lobes, called the *guard band*. This guard band is generated simply by sampling at greater than $2f_c$, as in the case for Fig. 14-28b. Although the actual tolerable overlap depends on the signal spectral slope, setting the sampling rate at about $3f_c = f_s$ is usually adequate to recover the signal.

In practice, narrow but finite-width pulse trains are used in place of the idealized impulse sampling train. To determine the effect of finite-width pulses, consider the pulse train made up of τ duration pulses repeating at a $1/T$ rate represented by

$$s(t) = \text{rep}_T\left[\text{rect}\left(\frac{t}{\tau}\right)\right] \qquad (14\text{-}52)$$

where

$$\text{rect}\left(\frac{t}{\tau}\right) = \begin{cases} 1 & |t| < \tau/2 \\ 0 & |t| > \tau/2 \end{cases}$$

The corresponding Fourier transform is

$$S(f) = \frac{\tau}{T}\text{comb}_{1/T}[\text{sinc}(\tau f)] \tag{14-53}$$

where

$$\text{sinc}(\tau f) = \frac{\sin \pi\tau f}{\pi\tau f}$$

Since the transform of the sampler output function is the convolution of the signal transform and the pulse train, the resulting response $R_s(f)$ is given by

$$R_s(f) = F(f)^* S(f)$$
$$= F(f)^* \frac{\tau}{T}\text{comb}_{1/T}[\text{sinc}(\tau f)] \tag{14-54}$$

The pulse width is usually much smaller than the period $\tau \ll T$, so that the comb function is an impulse train in frequency with an envelope that follows a $\sin x/x$ function. The convolution with the signal $F(f)$ thus yields a spectrum that is almost the same as for the impulse sampling train, except that the lobes decrease in amplitude as sinc (τf) determined by the pulse width τ. The ideal reconstruction filter can be approximated by commonly used hold circuits, the characteristics of which are shown[10] in Fig. 14-29.

28. Pulse Amplitude Modulation (PAM). Pulse amplitude modulation is essentially a sampled-data type of encoding where the information is encoded into the amplitude of a train of finite-width pulses. The pulse train can be looked upon as the carrier in much the same way as the sinewave is for continuous amplitude modulation. There is no improvement in signal-to-noise when using PAM, and furthermore, PAM is not considered wideband in the sense of FM or pulse time modulation (PTM). Thus PAM would correspond to continuous AM, or PTM corresponds to FM. Generally, PAM is used chiefly for time multiplex systems employing a number of channels sampled, consistent with the sampling theorem.

There are a number of ways of encoding information as the amplitude of a pulse train. These ways include both bipolar and unipolar pulse trains for both instantaneous or square-topped sampling and for exact or top sampling. In top sampling, the magnitude of the individual pulses follows the modulating signal during the pulse duration, while for square-topped sampling, the individual pulses assume a constant value, depending on the particular exact sampling point that occurs somewhere during the pulse time. These various waveforms are shown in Fig. 14-30.

The spectrum for the top-modulation bipolar sampling case is given by Eq. (14-54), since this type of modulation, shown in Fig. 14-30c, is simply sampling with a finite-pulse-width train. Carrying out the convolution indicated yields

$$R_{STB}(f) = \frac{\tau}{T}\sum_{n=-\infty}^{\infty}\text{sinc}\left(\frac{\tau n}{T}\right)F\left(f - \frac{n}{T}\right) \tag{14-55}$$

Using a square-topped rectangular spectrum for the original signal spectrum, the spectrum for top-modulation bipolar sampling is shown in Fig. 14-31a. The signal spectrum repeats with a $\sin x/x$ scale factor determined by the sampling pulse width, with each repetition a replica of $F(f)$.

Unipolar sampling can be implemented by adding a constant bias A to $f(t)$, the signal, to produce $f(t) + A$, where A is large enough to keep the sum positive; that is, $A > |f(t)|$. Sampling the new sum signal by multiplication with the pulse train results in the unipolar top-modulated waveform of Fig. 14-30e. The spectrum can be found by substituting the new signal into Eq. (14-54) and results in

$$R_{STU}(f) = \frac{\tau}{T}\sum_{n=-\infty}^{\infty}\text{sinc}\left(\frac{\tau n}{T}\right)\left[F\left(f - \frac{n}{T}\right) + A\delta\left(f - \frac{n}{T}\right)\right] \tag{14-56}$$

14-30

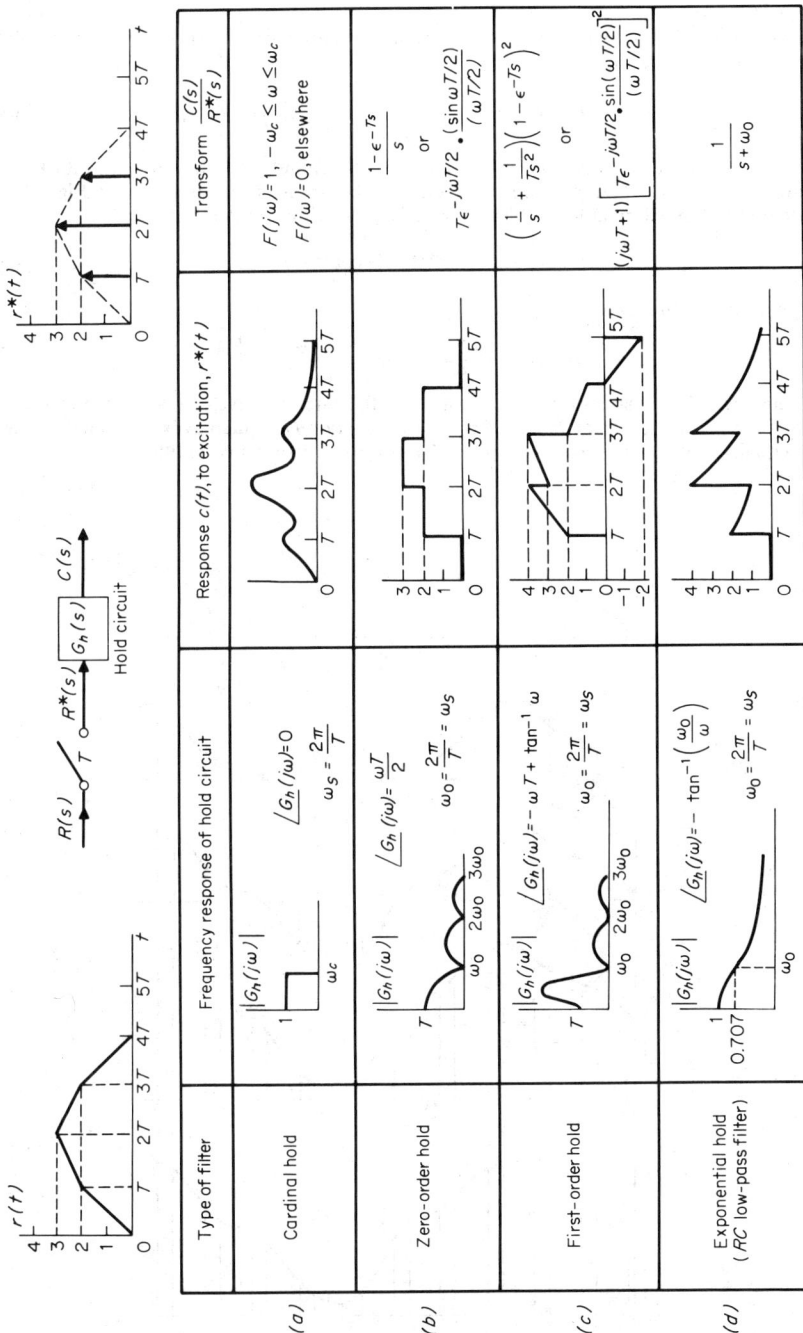

Fig. 14-29. Characteristics of various filter circuits (Ref. 10).

14-31

However, the delta-function part of the summation reduces to the spectrum function of the pulse train $S(f)$ by Eq. (14-53). Thus

$$R_{\text{STU}}(f) = AS(f) + \frac{\tau}{T} \sum_{n=-\infty}^{\infty} \text{sinc}\left(\frac{\tau n}{T}\right) F\left(f - \frac{n}{T}\right) \tag{14-57}$$

The resulting spectrum of top-modulation unipolar sampling is the same as with bipolar sampling plus the impulse spectrum of the sampling pulse train, as shown in Fig. 14-31b. For square-topped-modulation bipolar sampling, the time domain result is

$$r_{\text{SSB}}(t) = \text{rect}\left(\frac{t}{\tau}\right) * \text{comb}_T[f(t)] \tag{14-58}$$

with spectrum function

$$R_{\text{SSB}}(f) = \frac{\tau}{T} \text{sinc}(f\tau) \sum_{n=-\infty}^{\infty} F\left(f - \frac{n}{T}\right) \tag{14-59}$$

In this case, the signal spectrum is distorted by the sinc $(f\tau)$ envelope, as shown in Fig. 14-30c. This frequency distortion is referred to as *aperture effect* and may be corrected by use of an equalizer of sinc $(f\tau)$ form, following the low-pass reconstruction filter.

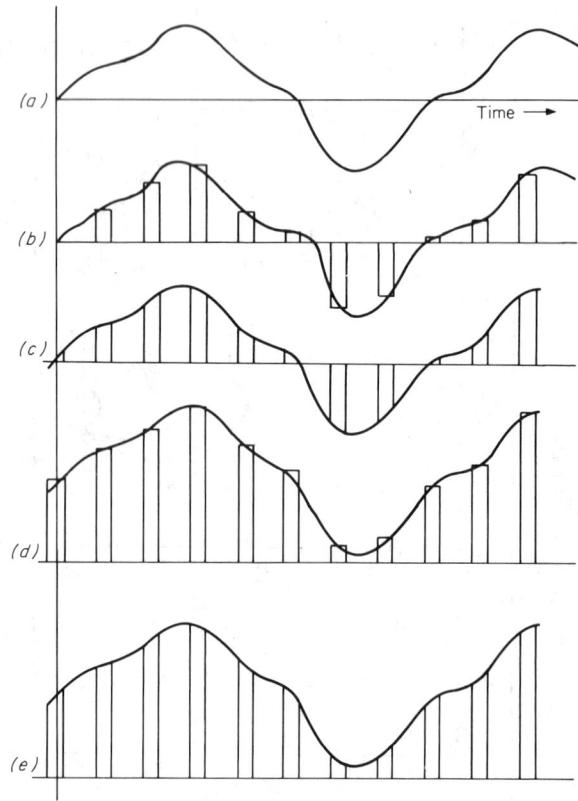

Fig. 14-30.   PAM waveforms. (a) Modulation; (b) square topped sampling, bipolar pulse train; (c) top sampling, bipolar pulse train; (d) square topped sampling, unipolar pulse train; (e) top sampling, unipolar pulse train.

As in the previous case of unipolar sampling, the resulting spectrum for square-topped modulation will contain the pulse train spectrum, as shown in Fig. 14-30d. The expression is

$$R_{SSU}(f) = AS(f) + \frac{\tau}{T}\text{sinc}(f\tau) \sum_{n=-\infty}^{\infty} F\left(f - \frac{n}{T}\right) \quad (14\text{-}60)$$

The signal information is generally recovered, in PAM systems, by use of a low-pass filter which acts on the reduced signal energy around zero frequency, as shown in Fig. 14-31.

29. Pulse Time (PTM), Pulse Position (PPM), and Pulse Width (PWM) Modulation. In PTM the information is encoded into the time parameter instead of, for instance, the

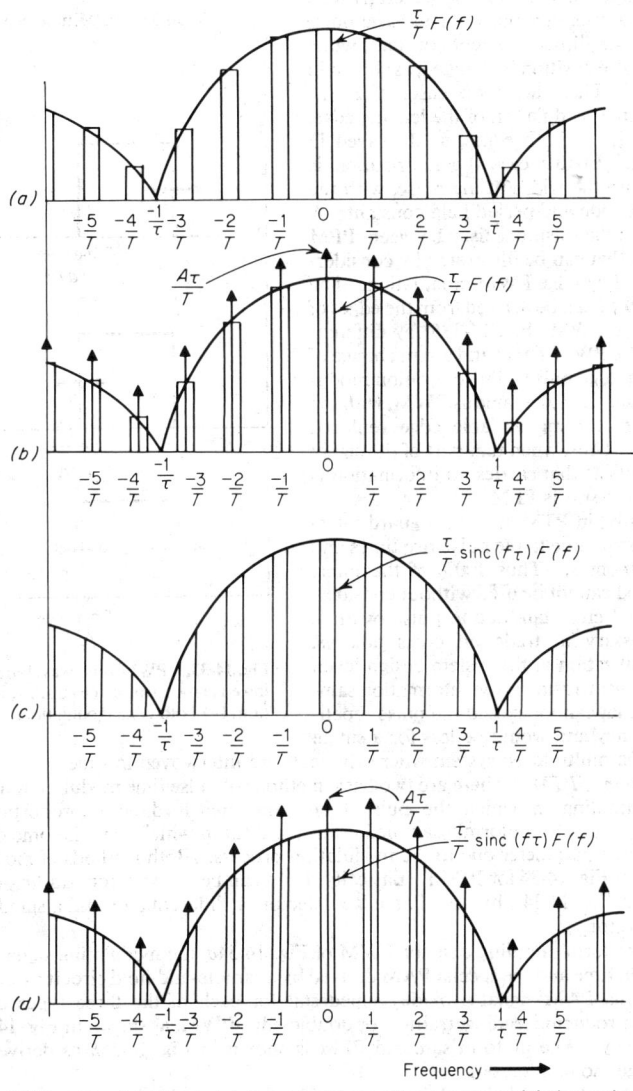

Fig. 14-31. PAM spectra. (a) Top modulation, bipolar sampling; (b) top modulation, unipolar sampling; (c) square top modulation, bipolar sampling; (d) square top modulation, unipolar sampling.

amplitude, as in PAM. There are two basic types of PTM: pulse position modulation (PPM) and pulse width modulation (PWM), which is also known as pulse duration (PDM) or pulse length modulation (PLM). The PTM allows the power driver circuitry to operate at peak saturation amplitude level, thus conserving power loss. Operating at saturation level, full on, full off, is especially important for heavy-duty high-load control applications, as well as for communication applications.

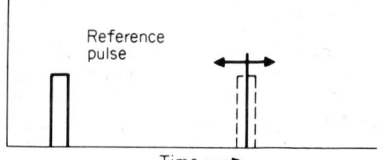

Fig. 14-32.   PPM time waveform.

In PPM the information is encoded into the time position of a narrow pulse, generally with respect to a reference pulse. The basic pulse width and amplitude are kept constant, while only the pulse position is changed, as shown in Fig. 14-32. There are three cases of PWM which are the modulation of the leading edge, trailing edge, or both edges, as displayed in Fig. 14-33. In this case the information is encoded into the width of the pulse, with the pulse amplitude and period held constant. A derivative relationship exists between PPM and PWM that can be illustrated by consideration of trailing-edge PWM modulation. The pulses of PPM can be derived from the edges of trailing-edge PWM (Fig. 14-33b) by differentiation of the PWM signal and a sign change of the trailing-edge pulse. Pulse position modulation is essentially the same as PWM, with the information-carrying variable edge replaced by a pulse. Thus, when that part of the signal power of PWM that carries no information is deleted, the result is PPM.

Generally, in PTM systems a guard interval is necessary due to the pulse rise times and system responses. Thus 100% of the interpulse period cannot be used without considerable channel cross talk due to pulse overlap. It is necessary to trade off cross talk vs. channel utilization at the system design level. Another consideration is the information sampling rate that cannot exceed the pulse repetition frequency and would be less for a single channel of a multiplexed system where channels are interwoven in time.

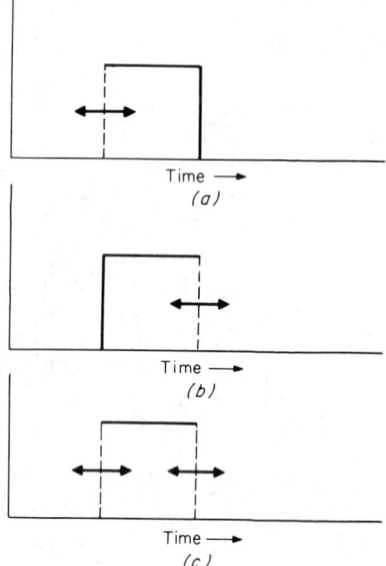

Fig. 14-33.   PWM time waveforms. (*a*) Leading edge modulation; (*b*) trailing edge modulation; (*c*) both edge modulation.

Generation of PTM.   There are two basic methods of pulse time modulation: (1) based on uniform sampling in which the pulse time parameter is directly proportional to the modulating signal at uniformly sampled points, and (2) in which there is some distortion of the pulse time parameter due to the modulation process. Both methods of modulation are illustrated in Fig. 14-34 for PWM. Basically, PPM can be derived from trailing-edge PWM, as shown in Fig. 14-34c, by use of an edge detector or differentiator and a standard narrow pulse generator.

In the uniform sampling case for PWM of Fig. 14-34a, the modulating signal is sampled uniformly in time and the special PAM derived by a sample-and-hold circuit as shown in Fig. 14-35a. This PAM signal provides a pedestal for each of the three types of sawtooth waveforms producing leading, trailing, or double-edge PWM, as shown in Fig. 14-35c, e, and g, respectively. The uniform sampled PPM is shown in Fig. 14-35h as derived from the trailing-edge modulation of g.

Nonuniformly sampled modulation, termed *natural sampling* by some authors, is shown in Fig. 14-36, and results from the method of Fig. 14-34b, where the sawtooth is added directly to the modulating signal. In this case the modulating waveform influences the time when the

14-34

samples are actually taken. This distortion is small when the modulating amplitude is small compared with the interpulse period T. The distortion is caused by the modulating signal distorting the sawtooth waveform when they are added, as indicated in Fig. 14-34b. The information in the PPM waveform is similarly distorted because it is derived from the PWM waveform, as shown in Fig. 14-35h.

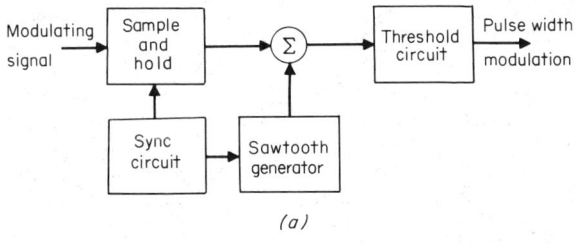

(a)

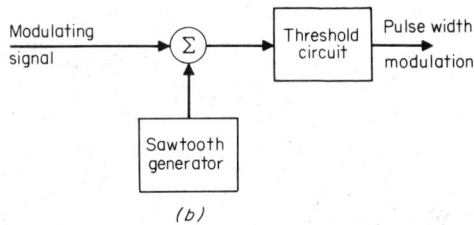

(b)

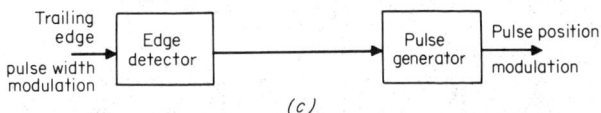

(c)

Fig. 14-34. PTM generation. (a) Pulse width modulation generation, uniform sampling; (b) pulse width modulation generation, non-uniform sampling; (c) pulse position modulation generation.

30. Pulse Time Modulation Spectra. The spectra are smeared in general, for most modulating signals, and are difficult to derive; however, some idea of what happens to the spectra with modulating is possible by considering a sinusoidal modulation of form

$$A \cos 2\pi f_s t \qquad (14\text{-}61)$$

The amplitude $A < T/2$, where T is the interpulse period, assuming no guard band.
For PPM with uniform sampling and unity pulse amplitude, the spectrum is given by

$$x(t) = \frac{\tau}{T} + \frac{2\tau}{T} \sum_{m=1}^{\infty} \text{sinc}(mf_0) J_0(2\pi A m f_0) \cos 2\pi m f_0 t$$

$$+ \frac{2\tau}{T} \sum_{n=1}^{\infty} \text{sinc}(nf_s) J_n(2\pi A n f_s) \cos\left(2\pi n f_s t - \frac{n\pi}{2}\right)$$

$$+ \frac{2\tau}{T} \sum_{m=1}^{\infty} \sum_{n=1}^{\infty} \left\{ \text{sinc}(mf_0 + nf_s) J_n[2\pi A(mf_0 + nf_s)] \cos\left[2\pi(mf_0 + nf_s)t - \frac{n\pi}{2}\right] \right.$$

$$\left. + \text{sinc}(nf_s - mf_0) J_n[2\pi A(nf_s - mf_0)] \cos\left[2\pi(nf_s - mf_0)t - \frac{n\pi}{2}\right] \right\}$$

$$(14\text{-}62)$$

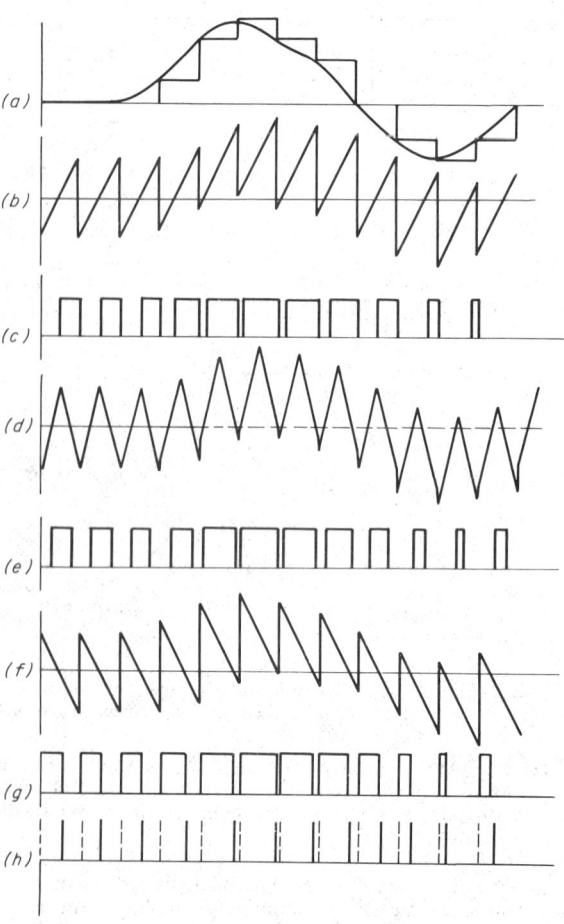

Fig. 14-35. Pulse time modulation, uniform sampling. (*a*) Modulating signal and sample and hold waveform; (*b*) sawtooth added to sample and hold waveform; (*c*) leading edge modulation; (*d*) sawtooth added to sample and hold waveform; (*e*) double edge modulation; (*f*) sawtooth added to sample and hold waveform; (*g*) trailing edge modulation; (*h*) pulse position modulation (reference pulse dotted).

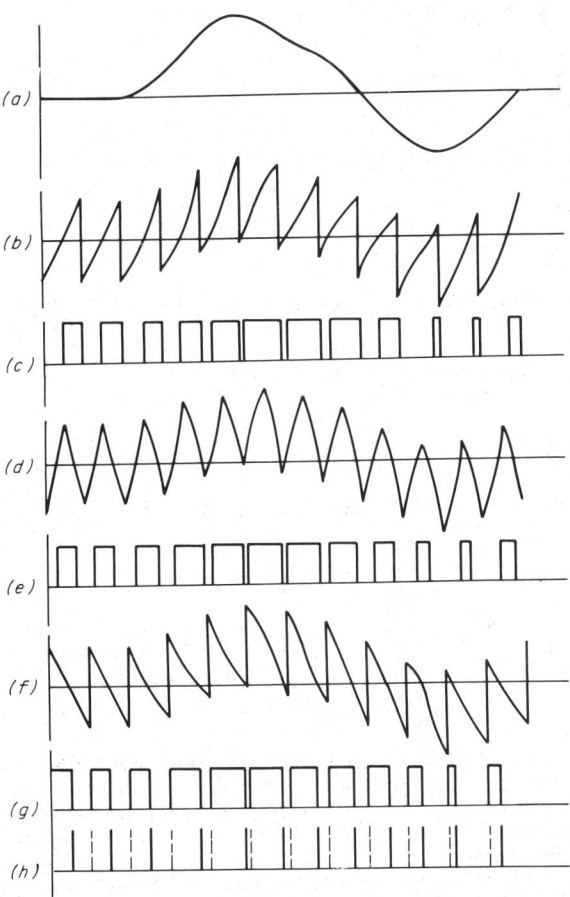

Fig. 14-36. (*a*) Modulating signal; (*b*) sawtooth added to modulation; (*c*) leading edge modulation; (*d*) sawtooth added to modulation; (*e*) double edge modulation; (*f*) sawtooth added to modulation; (*g*) trailing edge modulation; (*h*) pulse position modulation.

where τ = pulse width, T = pulse period, f_s = modulation frequency, J_n = Bessel function of first kind, nth order, and $f_0 = 1/T$.

As is apparent, all the harmonics of the pulse repetition frequency and the modulation frequency are present, as well as all possible sums and differences. The dc level is τ/T, with the harmonics carrying the modulation. The pulse shape affects the line amplitudes as a sinc function, reducing the spectra for higher frequencies.

The spectrum for PWM is similar to that of PPM, and for uniformly sampled trailing-edge sinusoidal modulation is given by

$$x(t) = \frac{1}{2} + \frac{1}{\pi T} \sum_{m=1}^{\infty} \frac{1}{mf_0} \cos\left[2\pi mf_0 t + \frac{\pi}{2}(2m - 1)\right]$$

$$+ \frac{1}{\pi T} \sum_{m=1}^{\infty} \frac{1}{mf_0} J_0(2\pi Amf_0) \cos\left(2\pi mf_0 t - \frac{\pi}{2}\right)$$

$$+ \frac{1}{\pi T} \sum_{n=1}^{\infty} \frac{1}{nf_s} J_n(2\pi Anf_s) \cos\left[2\pi nf_s t - (n + 1)\frac{\pi}{2}\right] \tag{14-63}$$

$$+ \frac{1}{\pi T} \sum_{\substack{m=1 \\ n=1}}^{\infty} \left\{ \frac{1}{mf_0 + nf_s} J_n[2\pi A(mf_0 + nf_s)]\cos\left[2\pi(mf_0 + nf_s)t - (n + 1)\frac{\pi}{2}\right] \right.$$

$$\left. + \frac{1}{nf_s - mf_0} J_n[2\pi A(nf_s - mf_0)]\cos\left[2\pi(nf_s - mf_0)t - (n + 1)\frac{\pi}{2}\right] \right\}$$

The same comments apply for PWM as for PPM.

A more compact form is given in Ref. 7 (Par. 14-41) for PPM and PWM, respectively, as

$$x(t) = \frac{1}{T} \sum_{\substack{m=\infty \\ n=\infty}}^{\infty} (-j)^n J_n[2\pi A(mf_0 + nf_s)]P(mf_0 + nf_s)e^{j2\pi(mf_0 + nf_s)t} \tag{14-64}$$

where $P(f)$ is Fourier transform of the pulse shape $p(t)$, and

$$x(t) = \frac{1}{2} + \frac{1}{T} \sum_{\substack{m=-\infty \\ m\neq 0}}^{\infty} j^{2m-1} \frac{e^{j2\pi mf_0 t}}{2\pi mf_0}$$

$$- \frac{1}{T} \sum_{\substack{m=-\infty \\ n=-\infty \\ |m|+|n|\neq 0}}^{\infty} (-j)^{n+1} \frac{J_n[2\pi A(mf_0 + nf_s)]}{2\pi(mf_0 + nf_s)} e^{j2\pi(mf_0 + nf_s)t} \tag{14-65}$$

31. Demodulation of PTM. Demodulation of PWM or PPM can be accomplished by low-pass filtering if the modulation is small compared with the interpulse period. However, in general, it is best to demodulate on a pulse-to-pulse basis that usually requires some form of synchronization with the pulses. The distortion introduced by nonuniform sampling cannot be eliminated and will be present in the demodulated waveform. However, if the modulation is small compared with the interpulse period T, the distortion will be minimized.

To demodulate PWM each pulse can be integrated and the maximum value sampled and held and low-pass-filtered, as shown in Fig. 14-37a. To sample and reset the integrator, it is necessary to derive sync from the PWM waveform, in this case trailing-edge-modulated.

Generally, PPM is demodulated by conversion to PWM and then demodulated as PWM. Although in some demodulation schemes the actual PWM waveform may not exist as such, the general demodulation scheme is the same. PPM can be converted to PWM by the configuration of Fig. 14-37b. The PPM signal is applied to an amplitude threshold, usually termed a *slicer*, that rejects noise except near the pulses. The pulses are applied to a flip-flop synchronized to one particular state by the reference pulse, and it generates the PWM as its output. More detailed information on PTM is available in Refs. 1 to 5, 7, and 8.

32. Pulse Frequency Modulation (PFM). In PFM the information is contained in the frequency of the pulse train, which is composed of narrow pulses. The highest frequency possible ideally occurs when there is no more interpulse spacing left for finite-width pulses. This frequency, given by $1/\tau$, where τ is the pulse width, will not be achieved in practice, due to the pulse rise time. The lowest frequency is determined by the modulator, usually a voltage controlled oscillator (VCO), in which in practice a 100 to 1 ratio of high to low frequency is easily achievable Examination of Fig. 14-38 indicates why PFM is used mostly for control purposes rather than communications. The wide variation, coupled with the uncertainty, of pulse position does not lend itself to time multiplexing, which requires the interweaving of channels in time. Since one of the chief motivations of pulse modulation in communication systems is to be able to time-multiplex a number of channels, PFM is not used. On the other hand, PFM is a good choice for on-off control applications, especially where fine control is required. A classic example of PFM control is for the attitude control of near-earth satellites that have on-off gas thrusters where a very close approximation to a linear system response is achievable.

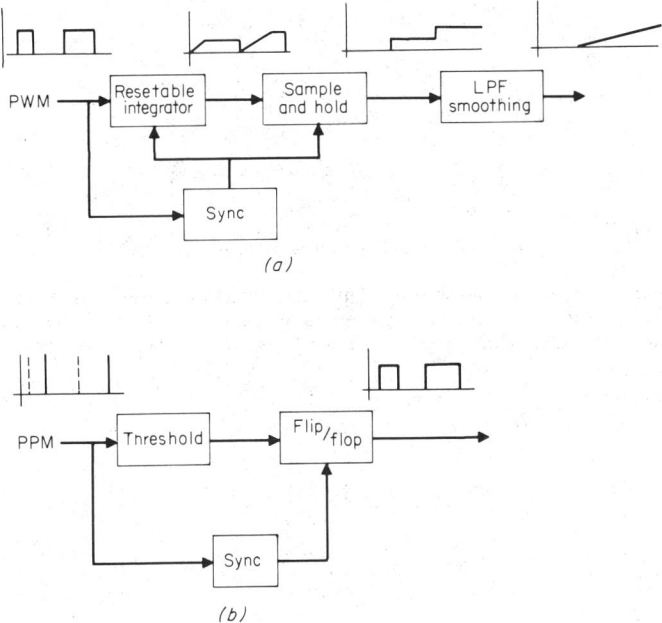

Fig. 14-37. Pulse time demodulation. (*a*) PWM demodulation; (*b*) PPM to PWM for demodulation.

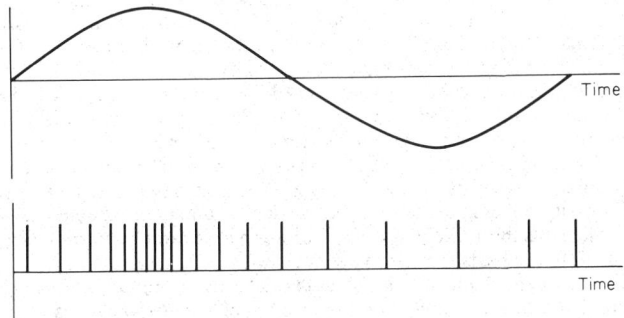

Fig. 14-38. PFM modulation.

Generation of PFM.   Basically, PFM is generated by modulation of a VCO as shown in Fig. 14-39a. A constant reference voltage is added to the modulation so that the frequency can swing above and below the reference-determined value. For control applications it is usually required that the frequency follow the magnitude of the modulation, its sign determining which actuators are to be turned on, as shown in Fig. 14-39b.

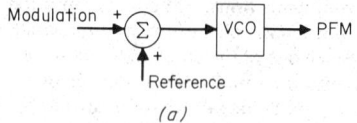

(a)

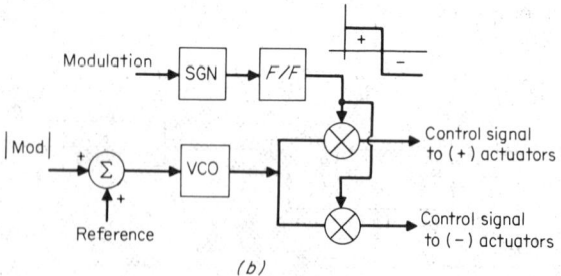

(b)

Fig. 14-39.   Generation of PFM. (*a*) PFM modulation; (*b*) PFM for control.

Spectrum for PFM.   The spectrum can be determined from consideration of the following expression,[1] for the case of sinusoidal modulation of the form $\beta \sin(\omega_m t + \phi)$, where $\beta = \Delta\omega/\omega_m$ and the pulse train is of amplitude A with τ width pulses and spacing T:

$$x(t) = \frac{A\tau}{T} + \sum_{k=1}^{\infty} \text{sinc}\left(\frac{k\tau}{T}\right)\left(J_0(k\beta)\cos\frac{2\pi kt}{T}\right)$$

$$+ \sum_{n=1}^{\infty} J_n(k\beta)\left\{\cos\left[\left(\frac{2\pi k}{T} + 2\pi n f_m\right)t + n\phi\right] + (-1)^n\right. \qquad (14\text{-}66)$$

$$\left.\cos\left[\left(\frac{2\pi k}{T} - 2\pi n f_m\right)t - n\phi\right]\right\}$$

The kth harmonic of the pulse repetition frequency is frequency-modulated with modulation index $k\beta$. Hence it could be demodulated with no harmonic distortion by a bandpass filter set to extract one of the harmonics and a frequency discriminator. Also note that the spectral amplitude decreases as a sinc function due to its relationship with a rectangular pulse.

33. Pulse Code Modulation (PCM).   In PCM the signal is encoded into a stream of digits. This differs from the other forms of pulse modulation by requiring that the sample values of the signal be quantized into a number of levels and subsequently coded as a series of pulses for transmission. By selecting enough levels, the quantized signal can be made to approximate closely the original continuous signal at the expense of transmitting more bits per sample. The PCM scheme lends itself readily to time multiplexing of channels and will allow widely different types of signals; however, synchronization is strictly required. This synchronization of the system can be on a single-sample or code-group basis. The synchronizing signal is most likely inserted with a group of samples from different channels, on a frame or subframe basis to conserve space.

The motivation behind modern PCM is that improved implementation techniques of solid-state circuitry allow extremely fast quantization of samples and translation to complex codes with reasonable equipment constraints. PCM is an attractive way to trade bandwidth for signal-to-noise and has the additional advantage of transmission through regenerative repeaters with a signal-to-noise that is substantially independent of the number of repeaters. The only requirement is that the noise, interference, and other disturbances be less than one-half a quantum step at each repeater. Also, systems can be designed that have error-detecting and error-correcting features.

34. PCM Coding and Decoding. Coding is the generation of a PCM waveform from an input signal, and decoding is the reverse process. There are many ways to code and many code groups to use; hence standardization is necessary when more than one user is considered. Each sample value of the signal waveform is quantized and represented to sufficient accuracy by an appropriate code character. Each code character is composed of a specified number of code elements. The code elements can be chosen as two-level, or binary; three-level, or ternary; or n-ary. However, general practice is to use binary, since it is not affected as much by interference introduced by the required increased bandwidth. An example of binary coding is shown in Fig. 14-40 for three-bit or eight levels of quantization. Each code group is composed of three pulses, with the pulse trains shown for on-off pulses in Fig. 14-40b and bipolar pulses in Fig. 14-40c.

A generic diagram of a complete system is shown in Fig. 14-41. The recovered signal is a delayed copy of the input signal degraded by noise due to sources such as sampling, quantization, and interference. For this type of system to be efficient, both sending and receiving terminals are required to be synchronized. This synchronism is required to be monitored continuously and be capable of establishing initial synchronism when the system is out of frame. The synchronization is usually accomplished by use of special sync pulses that establish frame, subframe, or word sync.

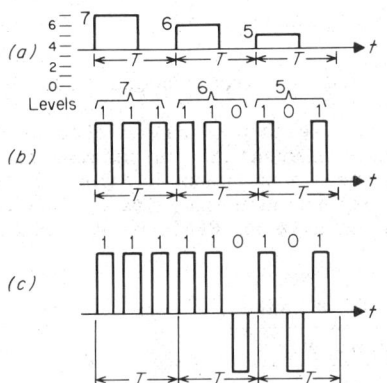

Fig. 14-40. Binary pulse coding. (a) Quantized samples; (b) on-off coded pulses; (c) bipolar coded pulses. (Ref. 3)

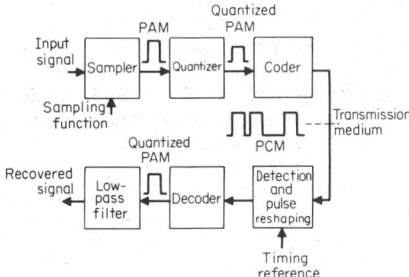

Fig. 14-41. Basic operations of a PCM system. (Ref. 3)

There are three basic ways to code, namely, feedback and subtractions, pulse counting, and parallel comparison.[5] In *feedback subtraction* the sample value is compared with the most significant code-element value and that value subtracted from the sample value if the element value is less. This process of comparison and subtraction is repeated for each code-element value down to the least significant bit. At each subtraction the appropriate code element or bit is selected to complete the coding. In *pulse counting* a gate is established by using the PWM pulse corresponding to a sample value. Clock pulses are gated using the PWM gate and are counted in a counter. The output of a decoding network attached to the counter is read out as the PCM. *Parallel comparison* is the fastest method since the sampled value is applied to a number of different threshold values. The thresholders are read out as the PCM.

35. System Considerations for PCM. Quantization introduces an irremovable error into the system, referred to as *quantization noise*. This kind of noise is characterized by the fact that its magnitude is always less than one half a quantum step, and it can be treated as uniformly distributed additive noise with zero mean value and rms value equal to $1/\sqrt{12}$

times the total height of a quantum step.[3]   Using as a measure of fidelity the ratio of signal power to quantization noise power at the quantizer output, the improvement with quantizer levels is shown in Fig. 14-42 for different kinds of signals.

In general, using an n-ary code with m pulses allows transmission of n^m values.   For the binary code this reduces the 2^m values which approximate the signal to 1 part in 2^m-1 levels. Encoding into plus/minus pulses, assuming either pulse is equally likely, results in an average power of $A^2/4$, which is half the on-off power of $A^2/2$, where the total pulse amplitude, peak to peak, is A.   The channel capacity for a system sampled at the Nyquist rate of $2f_m$ and quantized into s levels is

$$C = 2f_m \log_2 s \qquad \text{(bits/s)} \qquad (14\text{-}67)$$

or for m pulses of n values each.

$$C = mf_m \log_2 n^2 \qquad \text{(bits/s)} \qquad (14\text{-}68)$$

Since the encoding process squeezes one sample into m pulses, the pulse widths are effectively reduced by $1/m$; thus the transmission bandwidth is increased by a factor of m, or $B = mf_m$.

The maximum possible ideal rate of transmission of binary bits is

$$C = B \log_2\left(1 + \frac{S}{N}\right) \qquad \text{(bits/s)} \qquad (14\text{-}69)$$

according to Shannon.[2]   For a system sampling at the Nyquist rate, quantizing to $K\sigma$ per level and using the plus/minus pulses, the channel capacity is

$$C = B \log_2\left(1 + \frac{12S}{K^2 N}\right) \qquad \text{(bits/s)} \qquad N = \sigma^2 \qquad (14\text{-}70)$$

where S = average power over a large time interval, and σ = rms noise voltage at the decoder input.

There exists in PCM a fairly definite threshold, as indicated in Table 14-1, at about 20 dB (ratio of 9.2) peak signal pulse to rms noise voltage, where the error rate decreases quite rapidly.   Using the threshold ratio of 9.2 for an average error rate of about 10^{-6} yields a PCM binary system that requires 7 times (8.5 dB) the power of the ideal binary system.   Although PCM is much more efficient than uncoded systems such as FM and PPM, it is still 8.5 dB less efficient than the ideal system.

Table 14-1.   Probability of Error

S/N, dB	Probability of Error
13	10^{-2}
17	10^{-4}
20	10^{-6}
21	10^{-8}
22	10^{-10}
23	10^{-12}

The effect of regenerative repeaters is illustrated in Fig. 14-43, where the input carrier signal-rms noise ratio is expressed as a function of the number of repeaters for a 60-dB output signal-to-noise.   Note that for regenerative repeaters the 60-dB output is achievable relatively independently of the number of repeaters used.

36. Delta Modulation (DM).   Delta modulation is basically a one-digit PCM system where the analog waveform has been encoded in a differential form.   In contrast to the use of n digits in PCM, simple DM uses only one digit to indicate the changes in the sample values.   This is equivalent to sending an approximation to the signal derivative.   At the receiver the pulses are integrated to obtain the original signal.   Although DM can be simply implemented in circuitry, it requires a sampling rate much higher than the Nyquist rate of $2f_m$ and a wider bandwidth than a comparable PCM system.   Most of the other characteristics of PCM apply to DM.

Delta modulation differs from differential PCM in which the difference in successive signal samples is transmitted.[11] In DM only 1 bit is used to express and transmit the difference. Thus DM transmits the sign of successive slopes.

Coding and Decoding DM. There are a number of coding and decoding variations in DM, such as single-integration, double-integration, mixed-integration, delta-sigma, and high-information DM (HIDM). In addition, companding the signal which is compressing the signal at transmission and expanding it at reception is also used to extend the limited dynamic range. The simple single-integration DM of the coding-decoding scheme is shown in Fig. 14-44. In the encoder the modulator produces positive pulses when the sign of the difference signal $\epsilon(t)$ is positive, and negative pulses otherwise; and the output pulse train is integrated and compared with the input signal to provide an error signal $\epsilon(t)$, thus closing the encoder feedback loop. At the receiver the pulse train is integrated and filtered to produce a delayed

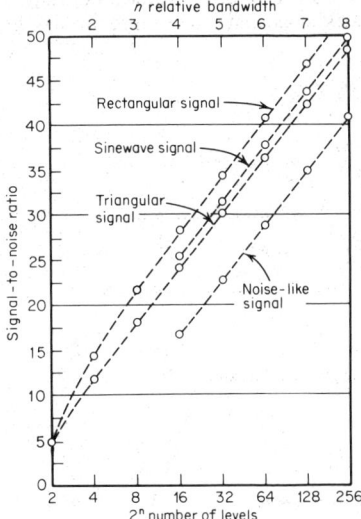

Fig. 14-42. PCM signal-noise improvement with number of quantization levels. (Ref. 3)

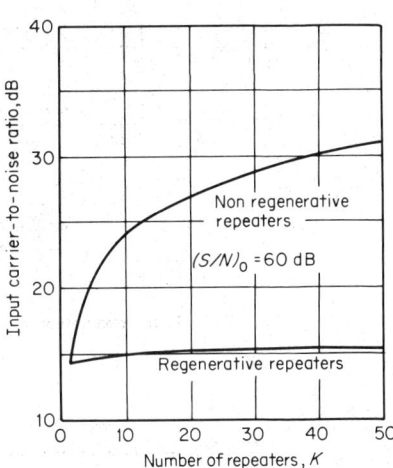

Fig. 14-43. Input carrier-to-noise ratio vs. number of repeaters for constant-output signal-to-noise ratio. (Ref. 1)

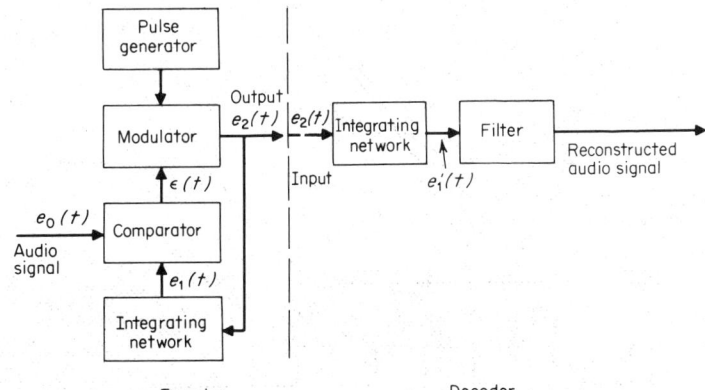

Fig. 14-44. Basic coding-decoding diagram for DM. (Ref. 1)

14-43

approximation to the signal, as shown in Fig. 14-45. The actual circuit implementation with operational amplifiers and logic circuits is very simple.

By changing to a double-integration network in the decoder, a smoother replica of the signal is provided. This decoder has the disadvantage, however, of not recognizing changes in the slope of the signal. This gave rise to a scheme to encode differences in slope instead of amplitude. This led to coders with double integration; however, systems of this type are marginally stable and can oscillate under certain conditions.[11] Waveforms of a double-integrating delta coder are shown in Fig. 14-46. Single and double integration can be combined to give improved performance while avoiding the stability problem. These mixed systems are often referred to in the literature as *delta modulators* with double integration.[11] A comparison of waveforms is shown in Fig. 14-47.

System Considerations for DM. The synthesized waveform can change only one level each clock pulse; thus DM overloads when the slope of the signal is large. The maximum signal power will depend on the type of signal, since the greatest slope that can be reproduced is the

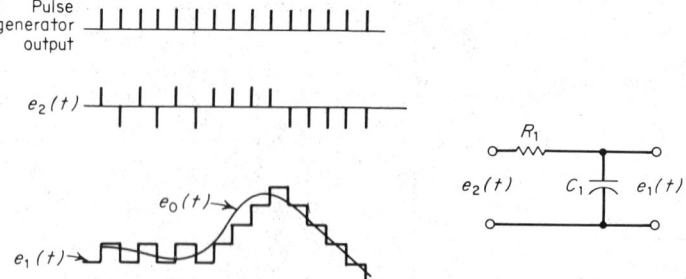

Fig. 14-45.   Delta modulation waveforms using single integration. (Ref. 1)

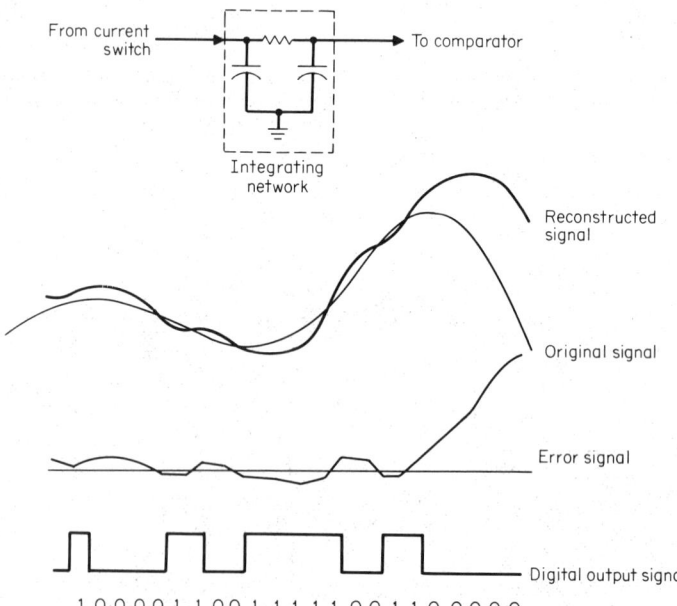

1 0 0 0 0 1 1 0 0 1 1 1 1 1 0 0 1 1 0 0 0 0 0

Fig. 14-46.   Waveforms for delta coder with double integration.

14-44

integration of one level in one pulse period. For a sinewave of frequency f, the maximum amplitude signal is

$$A_{max} = \frac{f_s \sigma}{2\pi f} \qquad (14\text{-}71)$$

where f_s = sampling frequency, and σ = one quantum step.

It has been observed that a DM system will transmit a speech signal without overloading if the amplitude of the signal does not exceed the maximum permissible amplitude of an 800-Hz sinewave.[1] The DM coder overload characteristic is shown in Fig. 14-48 along with the spectrum of a human voice. Notice that they decrease in frequency together, indicating that DM can be used effectively with speech transmission. Generally speaking, transmission of speech is the chief application of DM, although various modifications and improvements are being studied to extend DM to higher frequencies and transmission of the dc lost component.

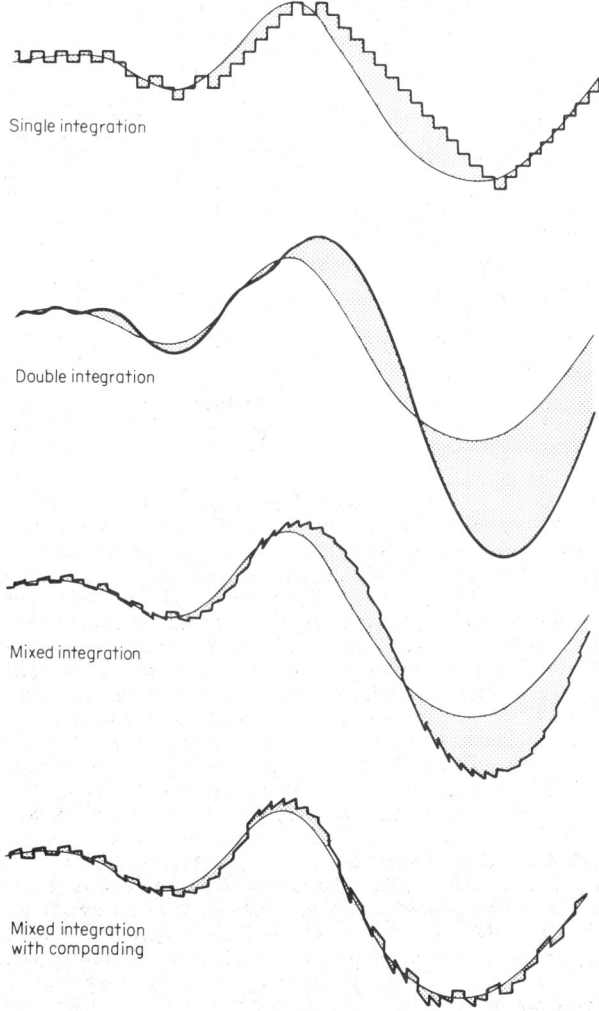

Single integration

Double integration

Mixed integration

Mixed integration
with companding

Fig. 14-47. Waveforms for various integrating systems. (Ref. 11)

14-45

Among these techniques is delta-sigma modulation, where the signal is integrated and compared with an integrated approximation to form the error signal similar to $\varepsilon(t)$ of Fig. 14-44. The decoding is accomplished with a low-pass filter and requires no integration. The signal-to-quantization noise ratio for single-integration DM is given by[11]

$$\frac{S}{N} = \frac{0.2\, f_s^{3/2}}{f f_0^{1/2}} \tag{14-72}$$

where f_s = sampling frequency, f = signal frequency, f_0 = signal bandwidth. For double or mixed DM,[11]

$$\frac{S}{N} = \frac{0.026\, f_s^{5/2}}{f f_0^{3/2}} \tag{14-73}$$

A comparison of signal-to-noise for DM and PCM is shown in Fig. 14-49, along with an experimental DM system for voice application. Note that DM at $40K$ pulses per second sampling rate is equal in performance with a 5-bit PCM system.

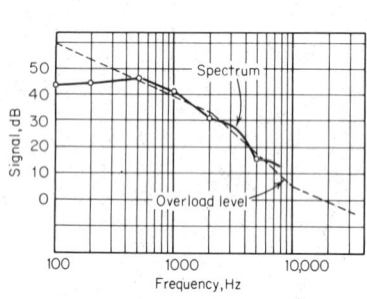

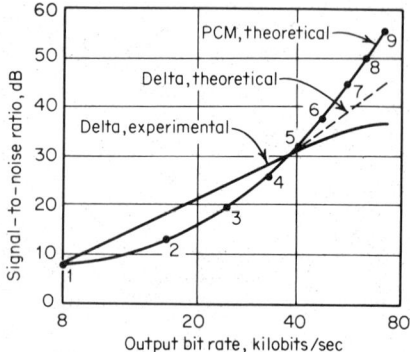

Fig. 14-48. Spectrum of the human voice compared with delta-coder overload level. (Ref. 11)

Fig. 14-49. Signal-to-noise ratio for delta modulation and PCM. (Ref. 1)

Extended-Range DM. A system termed *high-information* DM (HIDM, developed by M. R. Winkler in 1963) falls in the category of companded systems and encodes more information in the binary sequence than normal DM. Basically, the method doubles the size of the quantization step when two identical, consecutive binary values appear, and takes one-half of the step after each transition of the binary train. The HIDM system is capable of reproducing the signal with smaller quantization and overload errors. This technique also increases the dynamic range. The response of HIDM compared with that of DM is shown in Fig. 14-50. For a more extensive discussion of recent companding schemes see Ref. 11.

Implementation of HIDM is similar to that of DM, as shown in Figs. 14-51 and 14-52, with the difference only in the demodulator. The flip-flop of Fig. 14-52 changes state on the polarity of the input pulses. While the impulse generator initializes the experimental generators each pulse time, the flip-flop selects either the positive or negative one. The integrator adds and smooths the exponential waveforms to form the output signal. The scheme has a dynamic range with slope limiting of 11.1 levels per pulse period, which is much greater than DM and is equivalent to a 7-bit linear-quantized PCM system.[1]

37. Digital Modulation. Digital modulation is concerned with the transmission of a binary pulse train over some media. The output of, say, a PCM coder would be used to modulate a carrier for transmission. This modulation is treated here. In PCM systems, for instance, the high-quality reproduction of the analog signal is a function only of the probability of correct reception of the pulse sequences. Thus the measure of digital modulation is the probability of error, resulting from the digital modulation. The three basic types of digital modulation, namely; amplitude shift keying (ASK), frequently shift keying (FSK), and phase shift keying (PSK), are treated below.

38. Amplitude Shift Keying (ASK). In ASK the carrier amplitude is turned on or off, generating the waveform of Fig. 14-53 for rectangular pulses. Pulse shaping such as raised cosine, etc., is sometimes used to conserve bandwidth. The elements of a binary ASK receiver are shown in Fig. 14-54. The detection can be either coherent or noncoherent. However, if the added complexity of coherent methods is to be applied, a higher performance can be achieved by using one of the other methods of digital modulation.

The error rate of ASK with noncoherent detection is given in Fig. 14-55. Note that the curves approach constant values of error for high signal-to-noise ratios.

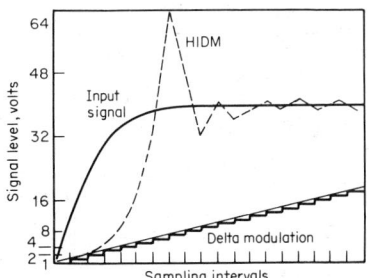

Fig. 14-50. Step response for a high-information delta modulation. (Ref. 11)

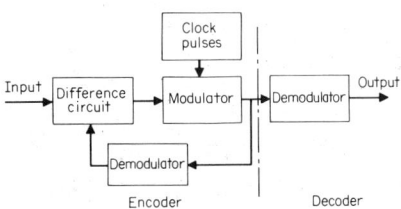

Fig. 14-51. Block diagram of HIDM system. (Ref. 1)

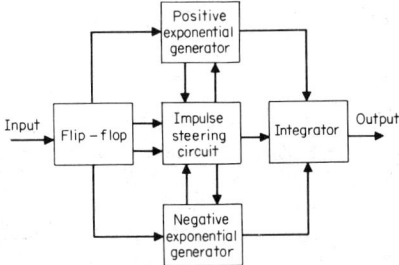

Fig. 14-52. Block diagram of HIDM demodulator. (Ref. 1)

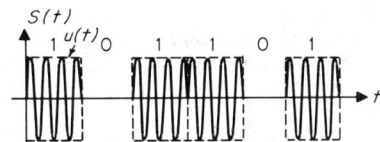

Fig. 14-53. ASK modulation. (Ref. 3)

The probability of error for the coherent detection scheme of Fig. 14-54c is shown in Fig. 14-56. The coherent-detection operation is equivalent to bandpass filtering of the received signal plus noise, followed by synchronous detection, as shown. At the optimum threshold shown in Fig. 14-56, the probability of error of marks and spaces is the same. The curves also tend toward a constant false-alarm rate, as in the noncoherent case.

39. Frequency Shift Keying (FSK). In FSK the frequency is shifted rapidly between one of two frequencies. Generally, two filters are used in favor of a conventional FM detector to discriminate between the marks and spaces, as illustrated in Fig. 14-57. As with ASK, either noncoherent or coherent detection can be used, although in practice coherent detection is not often used. This is because it is just as easy to use PSK with coherent detection and achieve superior performance.

In the noncoherent FSK system shown in Fig. 14-58a, the largest of the output of the two envelope detectors determines the mark-space decision. Using this system results in the curve for noncoherent FSK in Fig. 14-59. Comparison of the noncoherent FSK error with that of the noncoherent ASK results in the conclusion that both achieve an equivalent error rate at the same average SNR at low error rates.[3] FSK requires twice the bandwidth of ASK because of the use of two tones. In ASK, in order to achieve this performance, it is required to optimize the detection threshold at each SNR. The FSK system threshold is independent of SNR, and thus is preferred in practical systems where fading is encountered.

By synchronous detection of FSK (Fig. 14-58b) is meant the availability of an exact replica of each possible transmission at the receiver. The coherent-detection process has the effect of rejecting a portion of the bandpass noise. Coherent FSK involves the same difficulties as phase shift keying but achieves poorer performance. Also, coherent FSK is significantly advantageous over noncoherent FSK only at high error rates. The probability of error is shown in Fig. 14-59.

40. Phase Shift Keying (PSK). Phase shift keying is optimum in the minimum-error-rate sense from a decision-theory point of view. The PSK of a constant-amplitude carrier is

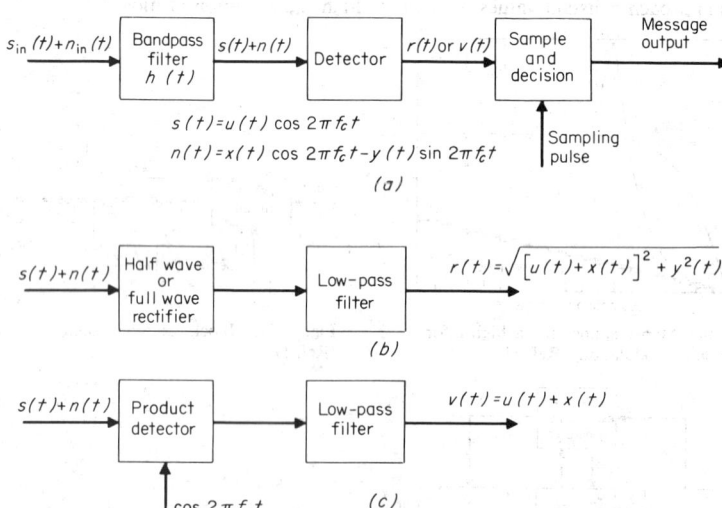

(a)

(b)

(c)

Fig. 14-54. Elements of a binary digital receiver. (*a*) Elements of a simple receiver; (*b*) noncoherent (envelope) detector; (*c*) coherent (synchronous) detector. (Ref. 3)

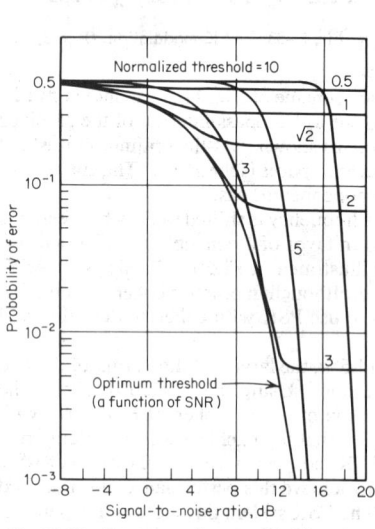

Fig. 14-55. Error rate for on-off keying, non-coherent detection. (Ref. 3)

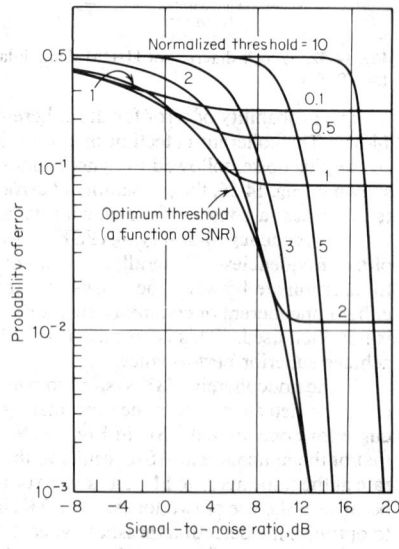

Fig. 14-56. Error rate for on-off keying, coherent detection. (Ref. 3)

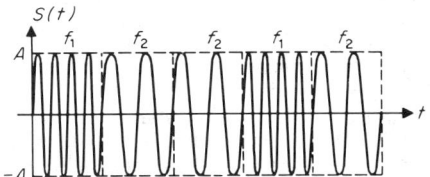

Fig. 14-57. FSK waveform, rectangular pulses. (Ref. 3)

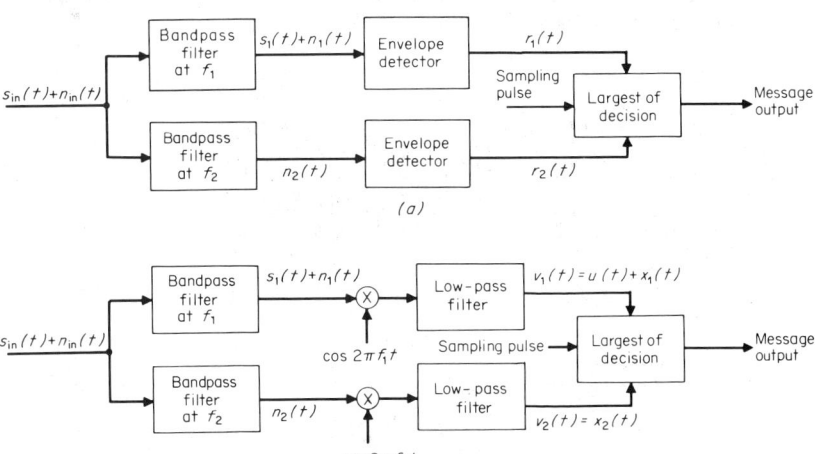

Fig. 14-58. Dual-filter detection of binary FSK signals. (*a*) Noncoherent detection tone f_1 signaled; (*b*) coherent detection tone f_1 signaled. (Ref. 3)

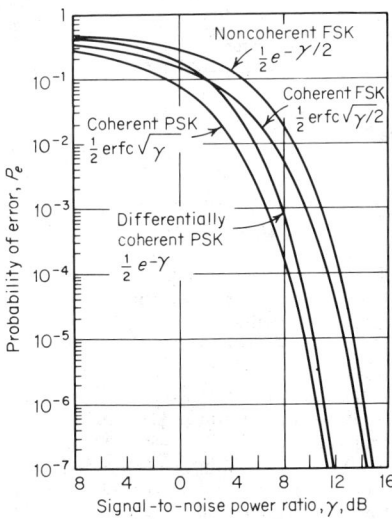

Fig. 14-59. Error rates for several binary systems. (Ref. 3)

shown in Fig. 14-60, where the two states are represented by a phase difference of π rad. Thus PSK has the form of a sequence of plus/minus rectangular pulses of a continuous sinusoidal carrier. It can be generated by double-sideband suppressed-carrier modulation by a bipolar rectangular waveform or by direct phase modulation. It is also possible to phase-modulate more complex signals than a sinusoid.

There is no performance difference in binary PSK between the coherent detector and the normal phase detector, both of which are shown in Fig. 14-61. Reference to Fig. 14-59 shows that there is a 3-dB design advantage for ideal coherent PSK over ideal coherent FSK, with about the same equipment requirements. Practically, PSK could suffer if very much phase error, $\Delta\phi$, is present in the system, since the signal is reduced by $\cos \Delta\phi$. This phase error may be introduced by relative drifts in the master oscillators at transmitter or receiver or be due to phase drift or fluctuation in the propagation path. In most cases this phase error can be compensated at the expense of requiring long-term smoothing.

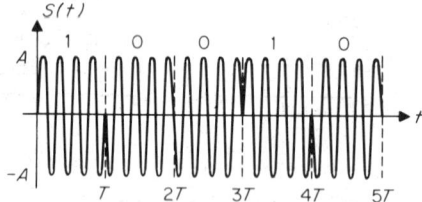

Fig. 14-60.   PSK signal, rectangular pulses. (Ref. 3)

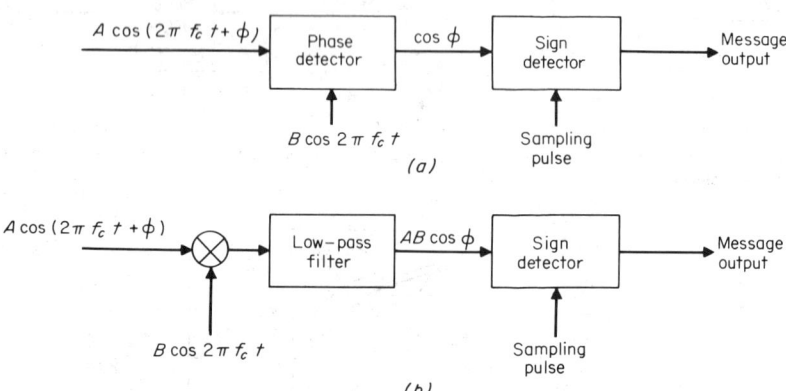

Fig. 14-61.   Two detection schemes for ideal coherent PSK. (*a*) Phase detection; (*b*) coherent detection. (Ref. 3)

An alternative to PSK is differential phase shift keying (DPSK), where it is required that there be enough stability in the oscillators and transmission path to allow negligible phase change from one information pulse to the next. Information is encoded differentially in terms of phase change between two successive pulses. For instance, if the phase remains the same from one pulse to the next (0° phase shift), a mark would be indicated. However, a phase shift of π from the previous pulse to the next would indicate a space. A coherent detector is still required where one input is the current pulse, with the other input the previous pulse.

The probability of error is shown in Fig. 14-59. At all error rates DPSK requires 3 dB less SNR than noncoherent FSK for the same error rate. Also, at high SNR, DPSK performs almost as well as ideal coherent PSK at the same keying rate and power level.

41. References on Pulse Modulators and Demodulators.

1. PANTER, P. F.   Modulation, Noise, and Spectral Analysis, McGraw-Hill, New York, 1965.

2. Schwartz M. "Information Transmission, Modulation and Noise," 2d ed., McGraw-Hill, New York, 1970.

3. Stein, S., and Jones, J. J. "Modern Communication Principles," McGraw-Hill, New York, 1968.

4. Landee, R., Davis, D., and Albrecht, A. "Electronic Designer's Handbook," McGraw-Hill, New York, 1957.

5. Black, H. S. "Modulation Theory," Van Nostrand, New York, 1953.

6. Woodward, P. M. "Probability and Information Theory, with Applications to Radar," Pergamon, New York, 1953.

7. Rowe, H. E. "Signals and Noise in Communication Systems," Van Nostrand, New York, 1965.

8. Nichols, M. H., and Rauch, L. L. "Radio Telemetry," Wiley, New York, 1954.

9. Truxal, J. G. "Automatic Feedback Control System Synthesis," McGraw-Hill, New York, 1955.

10. Mishkin, E., and Braun, L. "Adaptive Control Systems," McGraw-Hill, New York, 1961.

11. Schindler, H. R. Delta Modulation, *IEEE Spectrum,* October, 1970, p. 69.

SPREAD-SPECTRUM MODULATION

By M. D. Egtvedt

42. Spread-Signal Modulation. In a receiver designed exactly for a specified set of possible transmitted waveforms (in the presence of white noise and in the absence of such propagation defects as multipath and dispersion), the performance of a matched filter or cross-correlation detector depends only on the ratio of signal energy to noise power density E/n_0, where E is the received energy in one information symbol, and $n_0/2$ is the rf noise density at the receiver input. Since signal bandwidth has no effect on performance in white noise, it is interesting to examine the effect of spreading the signal bandwidth in situations involving jamming, message and traffic-density security, and transmission security. Other applications include random-multiple-access communication channels,[11]* multipath propagation analysis,[4] and ranging.

The information-symbol waveform may be characterized by its time-bandwidth (TW) product. Consider a binary system with the information symbol defined as a *bit* (of time duration T), while the fundamental component of the binary waveform is called a *chip*. For this *direct-sequence* system, the ratio (chips per bit) is equal to the TW product. An additional requirement on the symbol waveforms is that their cross-correlation with each other and the noise or extraneous signals be minimal.

Spread-spectrum systems occupy a signal bandwidth much larger (>10) than the information bandwidth, while the conventional systems have a TW of well under 10. FM with a high modulation index might slightly exceed 10, but is not optimally detectable and has a processing gain only above a predetection signal-noise threshold.

43. Nomenclature of Secure Systems. While terminology is not subject to rigorous definition the following terms apply to the following material:

Security and privacy: Relate to the protection of the signal from an unauthorized receiver. They are differentiated by the sophistication required. Privacy protects against a casual listener with little or no analytical equipment, while security implies an interceptor familiar with the principles and using an analytical approach to learn the *key.* Protection requirements must be defined in terms of the interceptor's applied capability *and* the time value of the message. Various forms of *protection* include:

Crypto security: Protects the information content, generally without increasing the TW product.

Antijamming (AJ) security: Spreads the signal spectrum so as to provide discrimination against energy-limited interference by using cross-correlation or matched-filter detectors. The interference may be natural (impulse noise), inadvertent (as in amateur radio or aircraft

---

*Superior numbers correspond to numbered references, Par. **14-49**.

channels), or deliberate (where the jammer may transmit continuous or burst cw, swept cw, narrow-band noise, wide-band noise, or replica or deception waveforms).

Traffic-density security: Involves capability to switch data rates without altering the apparent characteristics of the spread-spectrum waveform. The TW product (processing gain) is varied inversely with the data rates.

Transmission security: Involves spreading the bandwidth so that, beyond some range from the transmitter, the transmitted signal is buried in the natural background noise. The process gain (TW) controls the reduction in detectable range vis-à-vis a "clear" signal.

Use in Radar. It is usual to view radar applications as a variation on communication; i.e., the return waveforms are known except with respect to noise, Doppler shift, and delay. Spectrum spreading is applicable to both cw and pulse[2] radars. The major differentiation is in the choice of cross-correlation or matched-filter detector. The TW product is the key performance parameter, but the covariance function properties must frequently be determined to resolve Doppler shifts as well as range delays.

44. Classification of Spread-Spectrum Signals. Spread-spectrum signals can be classified on the basis of their spectral occupancy vs. time characteristics, as sketched in Fig. 14-62. Direct sequence (DS) and pseudo-noise (PN) waveforms provide continuous full

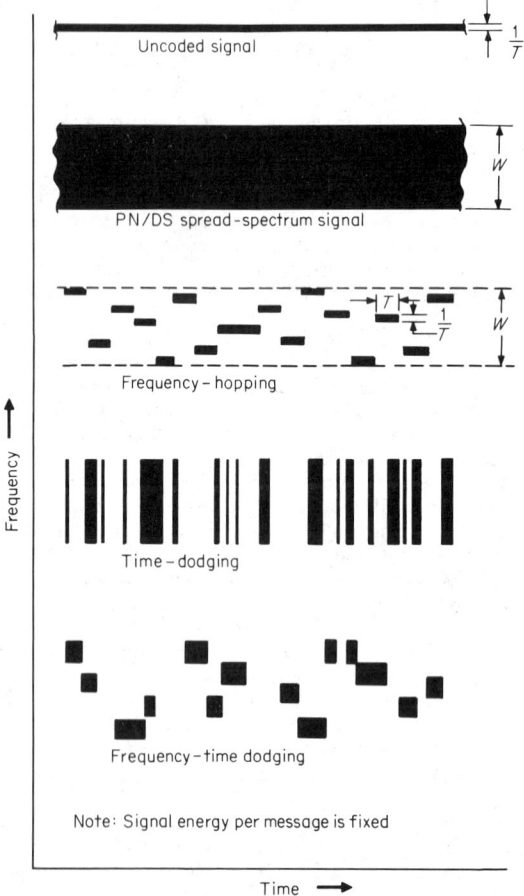

Fig. 14-62. Spectral occupancy vs. time characteristics of spread-spectrum signals.

coverage, while frequency hopping (FH), time-dodging, and frequency-time dodging (F-TD), fill the F-T plane only in a long-term averaging sense.

DS waveforms are pseudo-random digital streams generated by digital techniques and transmitted without significant spectral filtering. If heavy filtering is used, the signal amplitude statistics become quite noiselike, and this is called a *PN waveform.* In either case correlation detection is generally used because the waveform is dimensionally too large to implement a practical matched filter, and the sequence generator is relatively simple and capable of changing codes.

In FM schemes the spectrum is divided into subchannels spaced orthogonally at $1/T$ separations. One or more (e.g., two for FSK) are selected to pseudorandom techniques for each data bit. In time-dodging schemes the signal burst time is controlled by pulse repetition methods, while F-TD combines both selections. In each case a jammer must either jam the total spectrum continuously or accept a much lower effectiveness (approaching $1/TW$).

45. Correlation Detection Systems. The basic components of a typical direct sequence (DS) type of link are shown in Fig. 14-63. The data are used to select the appropriate waveform, which is shifted to the desired rf spectrum by suppressed-carrier frequency-conversion techniques, and transmitted. At the receiver identical locally generated waveforms multiply with the incoming signal. The stored reference signals are often modulated

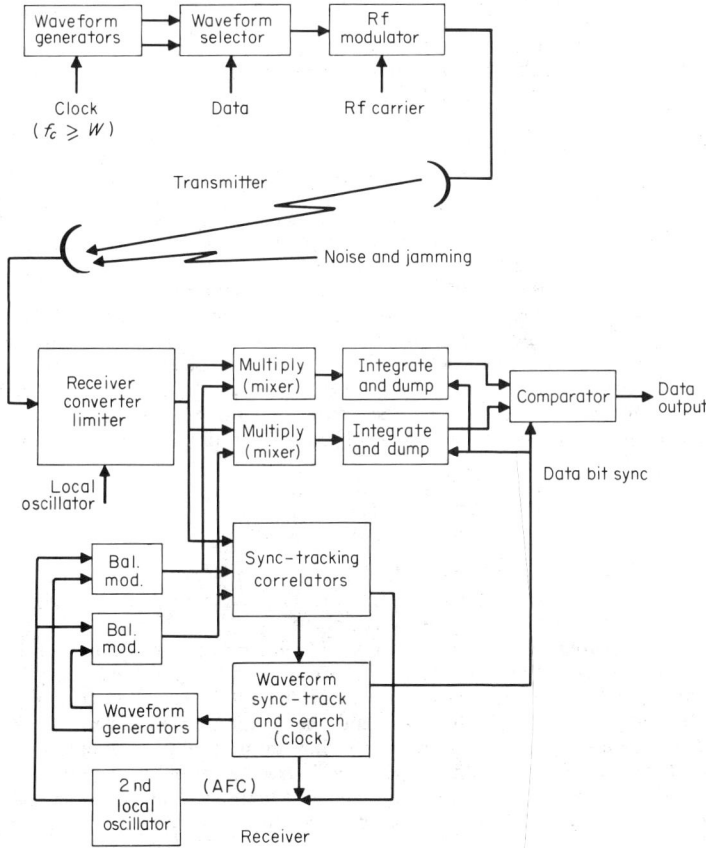

Fig. 14-63. Direct-sequence link for spread-spectrum system.

14-53

onto a local oscillator, and the incoming rf may be converted to an intermediate frequency, usually with rf/if limiters.

The mixing detectors are followed by linear integrate-and-dump filters, with a "greatest of" decision at the end of each period. The integrator is either a low-pass or bandpass quenchable narrow-band filter. Digital techniques are increasingly being used.

Synchronization is a major design and operational problem. Given a priori knowledge of the transmitted sequences, the receiver must bring its stored reference timing to within $\pm 1/(2W)$ of the width of the received signal, and hold it at that value. In a system having a 19-stage PN generator, a 1-MHz PN clock, and a 1-kHz data rate, the width of the correlation function is $\pm 1/2 \mu s$, repeating 1/2 s separations, corresponding to 524,287 clock periods. In the worst case, it would be necessary to inspect each sequence position for a millisecond, i.e., 524 s would be required to acquire sync. If oscillator tolerances and/or Doppler lead to frequency uncertainties equal to or greater than the 1-kHz data rate, then parallel receivers or multiple searches are required.

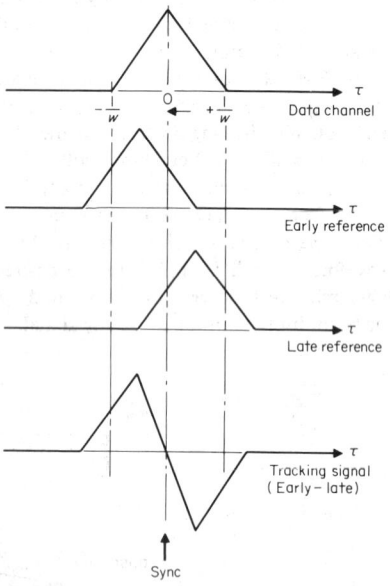

Fig. 14-64. Sync tracking by early-late correlators.

Ways to reduce the sync acquisition time include using jointly available timing references to start the PN generators, using shorter sequences for acquisition only; "clear" sync triggers; and paralleling detectors. Titsworth[3] discusses composite sequences which allow acquiring each component sequentially, searching $N_1 + N_2 + N_3$ delays, while the composite sequence has length $N_1 \cdot N_2 \cdot N_3$. These methods have advantages for space-vehicle ranging applications, but have reduced security to jamming.

Sync tracking is usually performed by measuring the correlation at early and late times, $\pm \tau$, where $\tau \leq 1/W$, as shown in Fig. 14-64. Subtracting the two provides a useful time discrimination function, which controls the PN clock. The displaced values can be obtained by two tracking-loop correlators, or by time-sharing a single unit. "Dithering" the reference signal to the signal correlator may also be used, but with performance compromises.

The tracking function can also be obtained by taking the time derivative of one of the inputs:

$$\frac{d\varphi_{XY}(\tau)}{d\tau} = \overline{\frac{dX(t)}{dt} \cdot Y(t + \tau)}$$

A third approach has been to add by modulo-2 methods the clock to the transmitted PN waveform. The spectral envelope is altered, but very accurate peak tracking can be accomplished by phase locking to the recovered clock.

46. Limiters in Spread-Spectrum Receivers. Limiters are frequently used in spread-spectrum receivers to avoid overload saturation effects, such as circuit recovery time, and incidental phase modulation. In the usual low-input signal-noise range, the limiter tends to normalize the output noise level, which simplifies the decision circuit design. In repeater applications (e.g., satellite), a limiter is desirable to allow the transmitter to be fully modulated regardless of the input-signal strength. When AGC is used, the receiver is highly vulnerable to pulse jamming, while the limiter causes a slight reduction of the instantaneous signal-to-jamming ratio and a proportional reduction of transmitter power allocated to the desired signal.

The signal-to-jamming ratios (SJRs) in and out of a limiter can be expressed by

$$\text{SJR}_{out} + \alpha \cdot \text{SJR}_{in}$$

where α is a function of the signal and jamming and the time and spectral characteristics of each. This problem has been analyzed[5,6,7,13,16,17] for various cases. Jones[5] covers the case of cw signal against a cw jammer plus noise, or noise alone (including noise jamming). The problem of gaussian signals has been treated also, for gaussian noise, by Price[13] and others. As an approximation, α can be taken as 1- to 2-dB loss for noise jamming and up to 6-dB loss for cw jamming.

Additional discrimination against narrow-band jamming can be obtained by fixed or adaptive notch filters, or by the Kirbar fix,[18] in which the spectrum is divided into several contiguous bands, each of which is limited, and the outputs combined. This procedure limits the cw energy to $1/n$th of the total, where n is the number of contiguous bands.

47. Deltic-aided Search. The sync search may be accelerated by use of deltic-aided (delay line time compression) circuits, if logic speeds permit.[19] The basic deltic consists of a recirculating shift register (or a delay line) which stores M samples, as shown in Fig. 14-65. The incoming spread-spectrum signal must be sampled at a rate above W (W = bandwidth). During each intersample period the shift register is clocked through $M + 1$ shifts before accepting the next sample. If $M \geq 2W$, a signal period at least equal to the data integration period is stored, and is read out at M different delays during each period T, permitting many high-speed correlations against a similarly accelerated (but not time-advancing) reference.

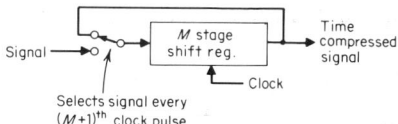

Fig. 14-65. Delay line time compression (deltic) configuration.

For a serial deltic and shift-register delay-line clock rate is at least $4TW^2$. Using a deltic with K parallel interleaved delay lines, the internal delay lines are clocked at $4TW^2/K^2$ and the demultiplexed output has a $4TW^2/K$ bit rate, providing only M/K discrete delays. This technique is device-limited to moderate signal bandwidths, primarily in the acoustic range up to about 10 kHz.

48. Waveforms. The desired properties of a spread-spectrum signal include:

An autocorrelation function which is unity at $\tau = 0$ and zero elsewhere.

A zero cross-correlation coefficient with noise and other signals.

A large library of orthogonal waveforms is available.

Maximal-Length Linear Sequences. A widely used class of waveforms is the maximal-length sequence (MLS) generated by a tapped re-fed shift register,[9,12,14] as shown in Fig. 14-66a and as a one-tap unit in Fig. 14-66b. The modulo-2 half adder ($\oplus$) and EXCLUSIVE-OR logic gate are identical for 1-bit binary signals. Analyses of this mode of operation are given by Birdsall and Ristenbatt[1] and by Golomb.[10]

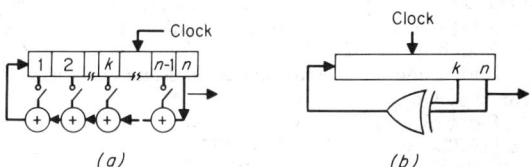

Fig. 14-66. Maximal-length-sequence (MLS) system.

If analog levels $+1$ and -1 are substituted, respectively, for 0 and 1 logic levels, the circuit is observed to function as a 1-bit multiplier.

Pertinent properties of the MLS are as follows: Its length, for an n-stage shift register, is $2^n - 1$ bits. During $2^n - 1$ successive clock pulses, all n-bit binary numbers (except all zeros) will have been present. The autocorrelation function is unity at $\tau = 0$, and at each $2^n - 1$ clock pulses displacement, and $1/(2^n-1)$ at all other displacements. This assumes that the sequences repeat cyclically; i.e., the last bit is closed onto the first. The autocorrelation function of a single (noncyclic) MLS shows significant time side lobes (Frank[8]). Titsworth[3] has analyzed the self-noise of incomplete integration over p chips, obtaining for MLSs,

$$\sigma^2(t) = \frac{(p - t)(p^2 - 1)}{p^3 t}$$

which approaches $1/t$ for the usual case of $p \gg t$. Since $t \approx TW$, the self-noise component is usually negligible.

Another self-noise component is frequently present due to amplitude and dispersion differences, caused by filtering, propagation effects, and circuit nonlinearities. In addition to intentional clipping, the correlation multiplier is frequently a balanced modulator, which is linear only to the smaller signal, unless deliberately operated in a bilinear range.[13,15] The power spectrum is shown in Fig. 14-67. The envelope has a $\sin^2(X)/X^2$ shape ($X = \pi/\omega_{clock}$), while the individual lines are separated by $\omega_{clock}/(2^n - 1)$.

An upper bound on the number of MLS for an n-stage shift register is given in terms of the Euler ϕ function:

$$N_u = \phi(2^n - 1)/n \leq 2^{(n - \log_2 n)}$$

where $\phi(k) = $ number of positive integers less than k, including 1, which are relatively prime to k.

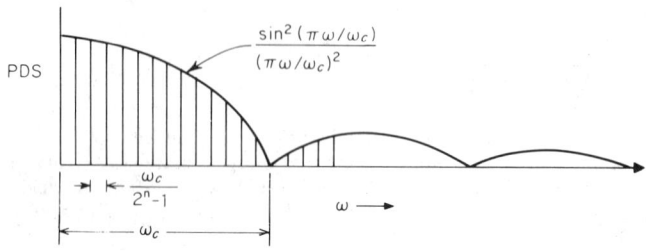

Fig. 14-67. Spectrum of MLS system.

49. References on Spread-Spectrum Modulation

1. BIRDSALL, T. G., and RISTENBATT, M. P. Introduction to Linear Shift-Register Generated Sequences, *Univ. Michigan Res. Inst. Tech. Rept. 90*, Ann Arbor, Mich., October, 1958.

2. COOK, CHARLES E. Pulse Compression: Key to More Efficient Radar Transmission, *Proc. IRE*, March, 1960, p. 310.

3. TITSWORTH, ROBERT C. Correlation Properties of Cyclic Sequences, *California Inst. Technol., Jet Propulsion Lab., Tech. Rept. 32-388*, Pasadena, Calif., July, 1963.

4. PRICE, R., and GREEN, P. E., JR. A Communication Technique for Multipath Channels, *Proc. IRE*, March, 1958, pp. 555-570.

5. JONES, J. J. Hard Limiting of Two Signals in Random Noise, *IEEE Trans. Inf. Theory*, Vol. IT-9, January, 1963.

6. MANASSE, R., PRICE, R., and LERNER, R. Loss of Signal Detectability in Bandpass Limiters, *IRE Trans. Inf. Theory*, Vol. IT-4, May, 1958.

7. DAVENPORT, W. B., JR. Signal-to-Noise Ratios in Bandpass Limiters, *J. Appl. Phys.*, Vol. 24, pp. 720-727, June, 1953.

8. FRANK, R. L.   Polyphase Codes with Good Nonperiodic Correlation Properties, *IEEE Trans. Inf. Theory.* Vol. IT-9, January, 1963.

9. MARSH, RICHARD W.   "Table of Irreducible Polynomials over GF(2) through Degree 19," NSA, distributed by Commerce Dept., Office of Technical Services, Washington, D.C., October, 1957.

10. GOLOMB, S. W.   "Shift Register Sequences," Holden-Day, San Francisco, Calif., 1967.

11. COSTAS, J. P.   Poisson, Shannon, and the Radio Amateur, *Proc. IRE,* December, 1959, p. 2058.

12. PETERSON, W. W.   "Error Correcting Codes," Wiley-M.I.T. Technical Press, New York, 1961.

13. PRICE, ROBERT   A Useful Theorem for Nonlinear Devices Having Gaussian Inputs, *IRE Trans. Inf. Theory,* June, 1958.

14. NIKIFORUK, P. N., and GUPTA, M. M.   A Bibliography on the Properties, Generation and Control System Applications of Shift Register Sequences, *Int. J. Control,* Vol. 9, No. 2, pp. 217-234, 1969.

15. GREEN, P. E., JR.   The Output Signal-to-Noise Ratio of Correlation Detectors, *IRE Trans. Inf. Theory,* March, 1957.

16. BUSSGANG, J. J.   Cross-correlation Functions of Amplitude-distorted Gaussian Signals, *M.I.T. Res. Lab. Elec. Tech. Rept.* 216, March, 1952.

17. BAUM, R. F.   The Correlation Function of Smoothly Limited Gaussian Noise, *IRE Trans. Inf. Theory,* September, 1957.

18. KIRKPATRICK, G. M.   "Signal Processing Arrangement with Filters in Plural Channels Minimizing Underdesirable Interference to Narrow and Wide Pass Bands," U.S. Patent 3,112,452, Nov. 1963.

19. ALLEN, W. B., and WESTERFIELD, E. C.   Digital Compressed-Time Correlators and Matched Filters for Active Sonars, *J. Acoust. Soc. Amer.,* January, 1964, pp. 121-139.

OPTICAL MODULATORS AND DEMODULATORS

BY J. L. CHOVAN

50. Modulation of Beams of Radiation.   This discussion of optical modulators is restricted to devices that operate on a directed beam of optical energy to control its intensity, phase, or frequency, according to some time-varying modulating signal.   Devices that deflect a light beam, or devices that spatially modulate a light beam, such as light-valve projectors, are treated in Sec. **20.**

Phase or frequency modulation requires a coherent light source, such as a laser.   Optical heterodyning is then used to shift the received signal to lower frequencies where conventional FM demodulation techniques can be applied.

Intensity modulation can be used on incoherent as well as coherent light sources. However, the properties of some types of intensity modulators are wavelength-dependent. Such modulators are restricted to monochromatic operation, but not limited to the extremely narrow laser line widths required for frequency modulation.

Optical modulation depends on either perturbing the optical properties of some material with a modulating signal or mechanical motion to interact with the light beam.   Modulation bandwidths of mechanical modulators are limited by the inertia of the moving masses. Optical index modulators generally have a greater modulation bandwidth, but typically require critical and expensive optical materials.

Optical index modulation can be achieved with electric or magnetic fields or by mechanical stress.   Typical modulator configurations are presented below, as in heterodyning, which is often useful in demodulation.   Optical modulation can also be achieved using semiconductor junctions.   This approach, which is comparatively new and presently under development, is discussed in Ref. 4.*

---

*Numbered references are listed in Par. **14-59.**

51. Optical Index Modulation: Electric Field Modulation.   *Pockels and Kerr Effects.* In some materials, an electric field vector $\overline{E}$ can produce a displacement vector $\overline{D}$ whose direction and magnitude depend on the orientation of the material.   Reference 1 shows that such a material can be completely characterized in terms of three independent dielectric constants associated with three mutually perpendicular natural directions of the material.   If all three dielectric constants are equal, the material is *isotropic.*   If two are equal and one is not, the material is *uniaxial.*   If all three are unequal, the material is *biaxial.*

The optical properties of such a material can be described in terms of the *ellipsoid of wave normals* (Fig. 14-68).   This is an ellipsoid whose semiaxes are the square roots of the associated dielectric constants.   The behavior of any plane monochromatic wave through the

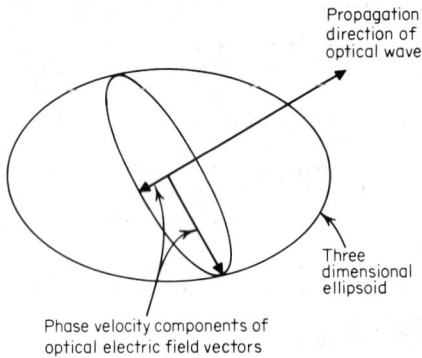

Fig. 14-68.   Ellipsoid of wave normals.

medium can be determined from the ellipse formed by the intersection of the ellipsoid with a plane through the center of the ellipsoid and perpendicular to the direction of wave travel. The instantaneous electric field vector $\overline{E}$ associated with the optical wave has components along the two axes of this ellipse.   Each of these components travels with a phase velocity that is inversely proportional to the length of the associated ellipse axis.

Thus there is a differential phase shift between the two orthogonal components of the electric field vector after it has traveled some distance through such a birefringent medium. The two orthogonal components of the vector vary sinusoidally with time but have a phase difference between them, which results in a vector whose magnitude and direction vary to trace out an ellipse once during each optical cycle.   Thus linear polarization is converted to elliptical polarization in a birefringent medium.

In some materials it is possible to induce a perturbation in one or more of the ellipsoid axes by applying an external electric field.   This is the electrooptical effect (Par. **14-54**).   The electrooptical effect is most commonly used in optical modulators presently available.   More detailed configurations using these effects are discussed in Par. **14-54**.   Reference 12 presents design considerations for various configurations and tabulates material properties.

Stark Effect.   Materials absorb and emit optical energy at frequencies which depend on molecular or atomic resonances characteristic of the material.   In some materials an externally applied electric field can perturb these natural resonances.   This is known as the Stark effect.

Reference 3 discusses a modulator for the CO_2 laser in the 3- to 22-μm region.   The laser output is passed through an absorption cell whose natural absorption frequency is varied by the modulating signal, using the Stark effect.   Since the laser frequency remains fixed, the amount of absorption depends on how closely the absorption cell is tuned to the laser frequency.   Intensity modulation results.

52. Magnetic Field Modulation.   *Faraday Effect.*   Two equal-length vectors circularly rotating at equal rates in opposite directions in space combine to give a nonrotating resultant whose direction in space depends on the relative phase between the counterrotating components. Thus any linearly polarized light wave can be considered to consist of equal right and left circularly polarized waves.

In a material which exhibits the Faraday effect, an externally applied magnetic field causes a difference in the phase velocities of right and left circularly polarized waves traveling along the direction of the applied magnetic field. This results in a rotation of the electric field vector of the optical wave as it travels through the material. The amount of the rotation is controlled by the strength of a modulating current producing the magnetic field. Reference 4 discusses an infrared modulator (1.2 to 4.5 μm) that uses the Faraday effect in yttrium-iron-garnet (YIG).

Zeeman Effect. In some materials the natural resonance frequencies at which the material emits or absorbs optical energy can be perturbed by an externally applied magnetic field. This is known as the Zeeman effect.

Intensity modulation can be achieved using an absorption cell modulated by a magnetizing current in much the same manner as the Stark effect absorption cell is used. The Zeeman effect has also been used to tune the frequency at which the active material in a laser emits.

53. Mechanical Stress Modulation. In some materials the ellipsoid of optical-wave normals can be perturbed by mechanical stress. An acoustic wave traveling through such a medium is a propagating stress wave which produces a propagating wave of perturbation in the optical index.

When a sinusoidal acoustic wave produces a sinusoidal variation in the optical index of a thin isotropic medium, the medium can be considered, at any instant of time, as a simple phase grating. Such a grating diffracts a collimated beam of coherent light into discrete angles whose separation is inversely proportional to the spatial period of the grating.

This situation is analogous to an rf carrier phase-modulated by a sinewave. A series of sidebands results which correspond to the various orders of diffracted light. The amplitude of the mth order is given by an mth-order Bessel function whose argument depends on the peak phase deviation produced by the modulating signal. The phases of the sidebands are the appropriate integral multiples of the phase of the modulating signal.

The mth order of diffracted light has its optical frequency shifted by m times the acoustic frequency. The frequency is increased for positive orders and decreased for negative orders.

Similarly, a thick acoustic grating refracts light mainly at discrete input angles. This condition is known as *Bragg reflection* and is the basis for the *Bragg modulator* (Fig. 14-69). In the Bragg modulator, essentially all the incident light can be refracted into the desired order, and the optical frequency is shifted by the appropriate integral multiple of the acoustic frequency.

Figure 14-69 shows the geometry of a typical Bragg modulator. The input angles for which Bragg modulation occurs are given by

$$\sin \theta = m\lambda/2\Lambda$$

where θ = angle between the propagation direction of the input optical beam and the planar acoustic wavefronts, λ = optical wavelength in the medium, Λ = acoustic wavelength in the medium, $m = \pm 1, \pm 2, \pm 3, \ldots$, and $m\theta$ = angle between the propagation direction of the output optical beam and the planar acoustic wavefronts.

The ratio of optical to acoustic wavelength is typically quite small, and m is a low integer, so that the angle θ is very small. Critical alignment is thus required between the acoustic wavefronts and the input light beam.

If the modulation bandwidth of the acoustic signal is broad, the acoustic wavelength varies, so that there is a corresponding variation in the angle θ for which Bragg reflection occurs. To overcome this problem, a phased array of acoustic transducers is often used to steer the angle of the acoustic wave as a function of frequency in the desired manner.

A limitation on bandwidth is the acoustic transit time across the optical beam. Since the phase grating in the optical beam at any instant of time must be essentially constant frequency if all the light is to be diffracted at the same angle, the bandwidth is limited so that only small changes can occur in this time interval. Reference 5 presents a figure of merit from Bragg modulator materials. References 6 and 7 review acoustooptical devices and compare several materials used in such devices. Lithium niobate is a material commonly used in commercially available Bragg modulators.

54. Modulator Configurations: Intensity Modulation. *Polarization Changes.* Linearly polarized light may be passed through a medium exhibiting an electrooptical effect, and the output beam passed through another polarizer. The modulating electric field controls the eccentricity and orientation of the elliptical polarization, and hence the magnitude of the component in the direction of the output polarizer. Typically, the input linear polarization is oriented to have equal components along the fast and slow axes of the birefringent medium, and the output polarizer is orthogonal to the input polarizer. The modulating field causes a phase differential varying from 0 to π rads. This causes the polarization to change from linear (at 0) to circular (at $\pi/2$) to linear normal to the input polarization (at π). Thus the intensity passing through the output polarizer varies from 0 to 100% as the phase differential varies from 0 to π rads.

Figure 14-70 shows this typical configuration. The following equations relate the optical intensity transmission of this configuration to the modulation.

$$\frac{I_o}{I_i} = \frac{1}{2}(1 - \cos \phi)$$

where I_o = output optical intensity, I_i = input optical intensity, and ϕ = differential phase shift between the fast and slow axes.

In the Pockels effect the differential phase shift is linearly related to applied voltage; in the Kerr effect it is related to the voltage squared.

$$\text{Pockels effect:} \qquad \phi = \pi \frac{v}{V}$$

$$\text{Kerr effect:} \qquad \phi = \pi \left(\frac{v}{V} \right)^2$$

where v = modulation voltage, and V = voltage to produce π rads differential phase shift.

Figure 14-71 shows the intensity transmission given by the above expression. The most linear part of the modulation curve is at $\phi = \pi/2$. Often a quarter-wave plate is added in series with the electrooptical material to provide this fixed bias at $\pi/2$. A fixed-bias voltage on the electrooptical material can also be used.

This arrangement is probably the most commonly used broadband intensity modulator. Early modulators of this type used a uniaxal Pockels cell with the electric field in the direction of optical propagation. In this arrangement, the induced phase differential is directly proportional to the optical path length, but the electric field is inversely proportional to this path length (at a fixed voltage). Thus the phase differential is independent of the path length and depends only on applied voltage. Typical materials require several kilovolts for a differential phase shift of π in the visible-light region.

Since the Pockels cell is essentially a capacitor, the energy stored in it is $\frac{1}{2}CV^2$ (where C is the capacitance and V is the voltage). This capacitor must be discharged and charged during each modulation cycle. Discharge is typically done through a load resistor, where this energy is dissipated. Due to the high voltages involved, the dissipated power at high modulation rates is appreciable.

The high-voltage problem can be overcome by passing light through the medium in a direction normal to the applied electric field. This permits a short distance between the electrodes, so that a high-E field is obtained from a low voltage, and a long optical path in the orthogonal direction, so that the cumulative phase differential is experienced.

Unfortunately, materials available are typically uniaxial, having a high eccentricity in the absence of electric fields. When oriented in a direction that permits the modulating electric field to be orthogonal to the propagation direction, the material has an inherent phase differential which is orders of magnitude greater than that induced by the modulating field. Furthermore, minor temperature variations cause perturbations in this phase differential which are large compared with those caused by modulation.

This difficulty is overcome by cascading two crystals which are carefully oriented so that temperature effects in one are compensated for by temperature effects in the other. The modulation electrodes are then connected so that their effects add. This approach is discussed in Ref. 8. Commercially available electrooptical modulators are of this type.

The Kerr effect is often used in a similar arrangement. Kerr cells containing nitrobenzene are commonly used as high-speed optical shutters.

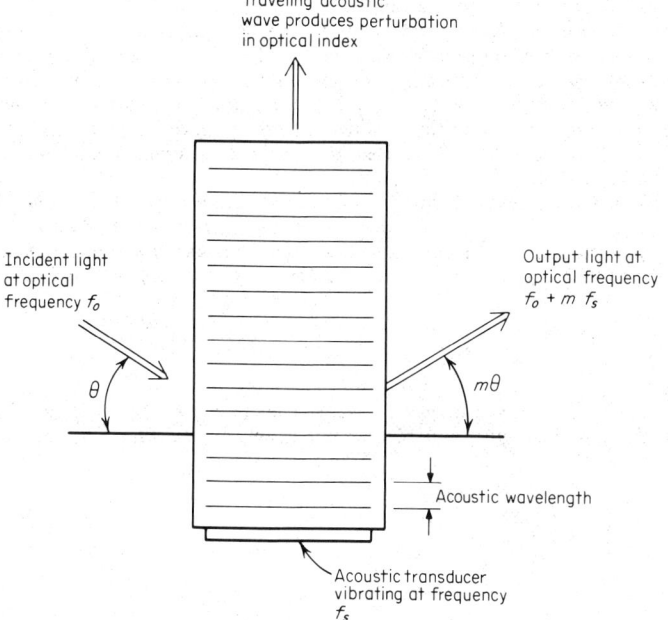

Fig. 14-69. The Bragg modulator.

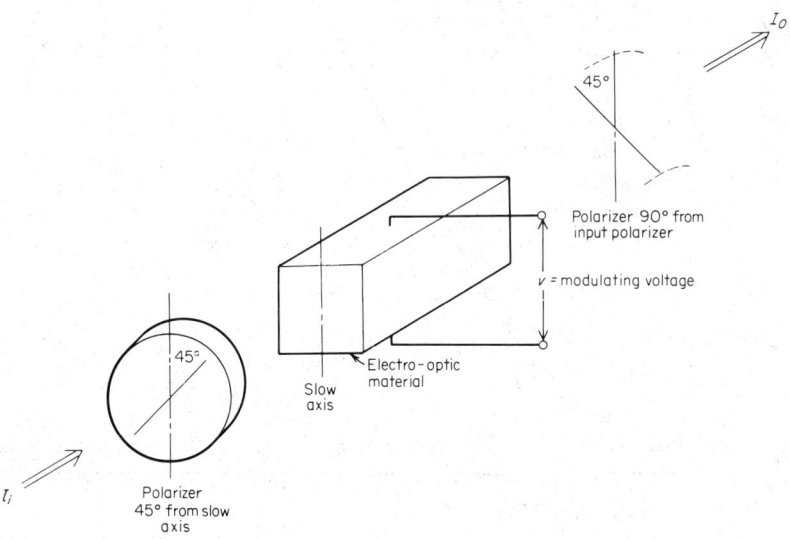

Fig. 14-70. Electrooptical intensity modulator.

Polarization rotation produced by the Faraday effect is also used in intensity modulation by passing through an output polarizer in a manner similar to that discussed above. The Faraday effect is more commonly used at wavelengths where materials exhibiting the electrooptical effect are not readily available.

Controlled Absorption.   As noted above, the frequency at which a material absorbs energy due to molecular or atomic resonances can be tuned over some small range in materials exhibiting the Stark or Zeeman effect. Laser spectral widths are typically narrow compared with such an absorption line width. Thus the absorption of the narrow laser line can be modulated by tuning the absorption frequency over a range near the laser frequency. Although such modulators have been used, they are not as common as the electrooptical modulators discussed above.

55. Phase and Frequency Modulation of Beams.   *Laser-Cavity Modulation.*   The distance between mirrors in a laser cavity must be an integral number of wavelengths. If this distance is changed by a slight amount, the laser frequency changes to maintain an integral number. The following equation relates the change in cavity length to the change in frequency:

$$\Delta f = \frac{C}{L} \frac{\Delta L}{\lambda}$$

where Δf = change in optical frequency, ΔL = change in laser-cavity length, L = laser-cavity length, λ = optical wavelength of laser output, and C = velocity of light in laser cavity.

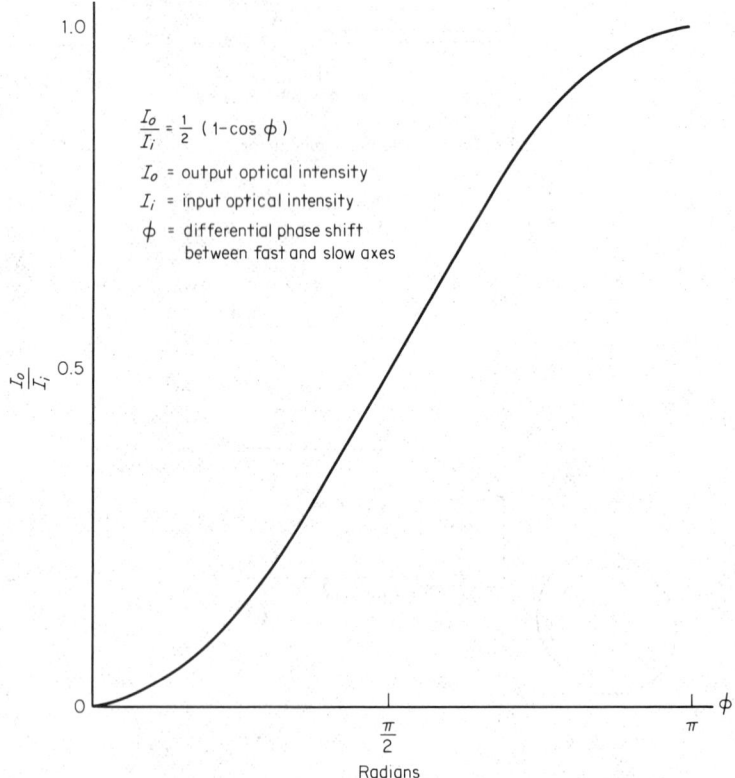

$$\frac{I_o}{I_i} = \frac{1}{2}(1 - \cos \phi)$$

I_o = output optical intensity

I_i = input optical intensity

ϕ = differential phase shift between fast and slow axes

Fig. 14-71.   Transmission of electrooptical intensity modulator.

14-62

In a cavity 1 m long, a change in mirror position of one optical wavelength produces about 300 MHz frequency shift. Thus a laser can be frequency-modulated by moving one of its mirrors with an acoustic transducer, but the mass of the transducer and mirror limit the modulation bandwidths that can be achieved.

An electrooptical cell can be used in a laser optical cavity to provide changes in the optical path length. The polarization is oriented so that it lies entirely along the axis of the modulated electrooptical material. This produces the same effect as moving the mirror, but without the inertial restrictions of the mirror's mass.

Under such conditions, the ultimate modulation bandwidth is limited by the Q of the laser cavity. A light beam undergoes several reflections across the cavity, depending on the Q, before an appreciable portion of it is coupled out. The laser frequency must remain essentially constant during the transit time required for these multiple reflections. This limits the upper modulation frequency.

Modulation of the laser-cavity length produces a conventional FM signal with modulating signal directly proportional to change in laser-cavity length. Demodulation is conveniently accomplished by optical heterodyning to lower rf frequencies where conventional FM demodulation techniques can be used.

56. External Modulation. The Bragg modulator (Fig. 14-69) is commonly used to modulate the optical frequency. As such it produces a single-sideband suppressed-carrier type of modulation.

Demodulation can be achieved by optical heterodyning to lower rf frequencies, where conventional techniques can be employed for this type of modulation. It is also possible to reinsert the carrier at the transmitter for a frequency reference. This is done by using optical-beam splitters to combine a portion of the unmodulated laser beam with the Bragg modulator output.

Conventional double-sideband amplitude modulation has also been achieved by simultaneously modulating two laser beams (derived from the same source) with a common Bragg modulator to obtain signals shifted up and down. Optical-beam splitters are used to combine both signals with an unmodulated carrier.[9] Conventional power detection demodulates such a signal.

Optical phase modulation is commonly accomplished by passing the laser output beam through an electrooptical material, with the polarization vector oriented along the modulated ellipsoid axis of the material. Demodulation is conveniently achieved by optical heterodyning to rf frequencies, FM demodulation, and integrating to recover the phase modulation in the usual manner.

For low modulation bandwidths, the electrooptical material can be replaced by a mechanically driven mirror. The light reflected from the mirror is phase-modulated by the changes in the mirror position. This effect is often described in terms of the Doppler frequency shift, which is directly proportional to the mirror velocity and inversely proportional to the optical wavelength.

57. Traveling-Wave Modulation. In the electrooptical and magnetooptical modulators described thus far, it is assumed that the modulating signal is essentially constant during the optical transit time through the material. This sets a basic limit on the highest modulating frequency that can be used in a lumped modulator.

This problem is overcome in a traveling-wave modulator. The optical wave and the modulation signal propagate with equal phase velocities through the modulating medium. This allows the modulating fields to act on the optical wave over a long path, regardless of how rapidly the modulating fields are changing. The degree to which the two phase velocities can be matched determines the maximum interaction length possible.

Reference 10 describes such a traveling-wave optical modulator using microwave modulation frequencies in a carbon disulfide Kerr cell.

58. Optical Heterodyning. Two collimated optical beams, derived from the same laser source and illuminating a common surface, produce straight-line interference fringes. The distance between fringes is inversely proportional to the angle between the beams. Shifting the phase of one of the beams results in a translation of the interference pattern, such that 2π rad phase shift translates the pattern by a complete cycle. An optical detector having a sensing area small compared with the fringe spacing has a sinusoidal output as the sinusoidal intensity of the interference pattern translates across the detector.

A frequency difference between the two optical beams produces a phase difference between the beams that changes at a constant rate with time. This causes the fringe pattern to translate across the detector at a constant rate, producing an output at the difference frequency. This technique is known as *optical heterodyning* in which one of the beams is the signal beam, the other the local oscillator.

The effect of the optical alignment between the beams is evident. As the angle between the two collimated beams is reduced, the spacing between the interference fringes increases, until the spacing becomes large compared with the overall beam size. This permits a large detector which uses all the light in the beam. If converging or diverging beams are used instead of collimated beams, the situation is similar, except that the interference fringes are curved instead of straight. Making the image of the local-oscillator point coincide with the image of the signal-beam point causes the desired infinite fringe spacing.

Optical heterodyning provides a convenient solution to several possible problems in optical demodulation. In systems where a technique other than simple amplitude modulation has been used (e.g., single-sideband, frequency, or phase modulation), optical heterodyning permits shifting to frequencies where established demodulation techniques are readily available.

In systems where background radiation, such as from the sun, is a problem, heterodyning permits shifting to lower frequencies, so that filtering to the modulation bandwidth removes most of the broadband background radiation. The required phase front alignment also eliminates background radiation from spatial positions other than that of the signal source.

Many systems are limited by thermal noise in the detector and/or front-end amplifier. Cooled detectors and elaborate amplifiers are often used to reduce this noise to the point that photon noise in the signal itself dominates. This limit also can be achieved in an optical heterodyne system with noncooled detector and normal amplifiers by increasing the local-oscillator power to the point where photon noise in the local oscillator is the dominant noise source.[11] Under these conditions, the signal-to-noise power ratio is given by the following equation:

$$\frac{s}{n} = \frac{\eta \lambda P}{2hBC}$$

where s/n = signal power/noise power, η = quantum efficiency of the photo detector, λ = optical wavelength, h = Planck's constant, C = velocity of light, B = bandwidth over which s/n is evaluated, and P = optical signal power received by the detector.

59. References on Optical Modulators and Demodulators.

1. BORN, M., and WOLF, E. "Principles of Optics," Chap. 14, Pergamon, New York, 1959.

2. *Ibid.*, Chap. 12.

3. FANDMAN, A., MARANTZ, H., and EARLY, V. Light Modulation by Means of Stark Effect in Molecular Gases-Application to CO_2 Lasers, *Appl. Phys. Letters,* Dec. 1, 1969, pp. 357-360.

4. NELSON, D. F. The Modulation of Laser Light, *Sci. Amer.,* June, 1968.

5. GORDON, E. I. Figure of Merit for Acoustic-optical Deflection and Modulation Devices, *IEEE J. Quantum Electron.* (correspondence), May, 1966, pp. 104-105.

6. GORDON, E. I. A Review of Acousto-optical Deflection and Modulation Devices, *Proc. IEEE,* October, 1966, pp. 1391-1401.

7. ADLER, R. Interaction between Light and Sound, *IEEE Spectrum,* May, 1967, pp. 42-54.

8. FEY, J. M., and WEBB, R. J. Low Voltage Light-Amplitude Modulation, *Electron. Letters,* May 31, 1968, pp. 213-215.

9. DIXON, R. W., and GORDON, E. I. Acoustic Light Modulator Using Optical Heterodyne Mixing, *Bell System Tech. J.,* Vol. 46, p. 367, 1967.

10. CHENAWETH, A. J., GADELY, O. L., and HOLSHOUSER, D. F. Carbon Disulfide Traveling Wave Kerr Cells, *Proc. IEEE,* October, 1966, pp. 1414-1418.

11. PRATT, W. K. "Laser Communication System," Chap. 10, Wiley, New York, 1969.

12. KAMINOW, I. P., and TURNER, E. H. Electro-optic Light Modulators, *Proc. IEEE,* October, 1966, p. 1374.

FREQUENCY CONVERTERS AND DETECTORS

By G. B. Gawler

60. General Considerations of Frequency Converters. A frequency converter usually consists of an oscillator (called a *local oscillator*, or LO) and a device used as a mixer. The mixing device is either nonlinear or its transfer parameter can be made to vary in synchronism with the local oscillator. A signal voltage with information in a frequency band centered at frequency f_s enters the frequency converter, and the information is reproduced in the intermediate-frequency (if) voltage leaving the converter. If the local-oscillator frequency is designated f_{LO}, then the if voltage information is centered about a frequency $f_{if} = f_{LO} \pm f_s$. The situation is shown pictorially in Fig. 14-72. Characteristics of interest for design in systems using frequency converters are gain, noise figure, image rejection, spurious responses, intermodulation and cross-modulation capability, desensitization, local-oscillator rf/if isolation. These characteristics will be discussed at length in the descriptions of different types of frequency-converter mixers and their uses in various systems. First, explanations are in order for the above terms.

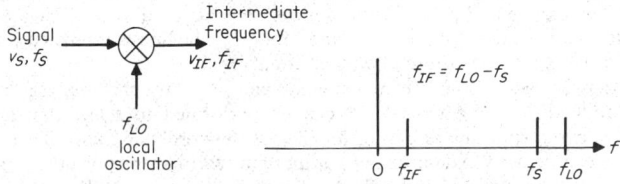

Fig. 14-72.   Frequency-converter terminals and spectrum.

Frequency-Converter Gain. The available power gain of a frequency converter is the ratio of power available from the if port to the power available at the signal port. Similar definitions apply for transducer gain and power gain.

Noise Figure of Frequency Converter. The noise factor is the ratio of noise power available at the if port to the noise power available at the if port due to the source alone at the signal port.

Image Rejection. For difference mixing $f_{if} = f_{LO} - f_s$, and the image is $2f_{LO} - f_s$. For sum mixing $f_{if} = f_{LO} + f_s$, and the image is $2f_{LO} + f_s$. An undesired signal at the difference mixing frequency $2f_{LO} - f_s$ results in energy at the if port. This condition is called *image response* and attenuation of the image response is image rejection, measured in decibels.

Spurious Responses. Spurious external signals reach the mixer and result in generation of undesired frequencies that may fall into the intermediate-frequency band.[26] The condition for an interference in the if band is

$$ mf_s' \pm nf_1 = \pm f_{if} $$

where m and n are integers, and f_s' represents spurious frequencies at the signal port of the mixer.

Example: There is a strong local station in the broadcast band at 810 kHz and a weak distant station 580 kHz. A receiver is tuned to the distant station, and a whistle, or beat, at 5kHz is heard on the receiver (refer to Fig. 14-73).

An analysis shows that the second harmonic of the local oscillator interacts with the second harmonic of the 810-kHz signal to produce a mixer output at 450 kHz in the if band of the receiver:

$$ 580 + 455 = 1,035\text{kHz} = \text{LO frequency} $$

$$ 2 \times 1,035 - 2 \times 810 = 450\text{kHz} = \text{if interference frequency} $$

The interference at 450 kHz then mixes with the 455-kHz desired signal in the second detector

---

*Superior numbers correspond to numbered references, Par. 14-69.

to produce the 5-kHz whistle. Notice that if the receiver is slightly detuned upward by 5 kHz, the whistle will zero-beat. Further upward detuning will create a whistle of increasing frequency.

Karpen and Mohr[26] have shown that a mixer spurious response of signal order m is of the following form:

$$E(m, n) = k(m,n)E_S^m$$

or in terms of decibels,

$$P(m,n) = mP_S' + K(m,n)$$

where n is the order of the LO and

$$mf_s' \pm nf_{LO} = \pm f_{if}$$

and P denotes power in dBm or dBw.

Significance of the Formulation of $P(m,n)$. Extensive measurements show that larger values of numbers m and n give smaller and smaller values of $P(m,n)$. Therefore, for practical purposes, a limited range of m and n can be selected for any mixer evaluation, and the numbers $K(m,n)$ can be found from a set of measurements on a mixer. Figure 14-74 shows some sample measurements from Karpen and Mohr.[26] The lines are labeled $n \times m$. Notice that higher orders give less output at a given input level.

As an example, take a double conversion system. A spur chart analysis such as that in the ITT Handbook[27] or by Hoigaard[28] can be performed to select first and second if frequencies to avoid spurious responses for the first few orders of m,n. Then the numbers $K(m,n)$ above can be used to determine filtering required to keep spurious responses $P(m,n)$ below system threshold sensitivity. Charts of the $K(m,n)$ values for Schottky diode doubly balanced mixers are available.

For a simple numerical example (refer to Fig. 14-75), suppose the first if is 100 MHz and a spurious signal is due to the first LO at $f_s' = 200$ MHz. Let $K(3,1) = -35$, and require that spurious responses be below -110 dBm. What is the required filtering at 200 MHz in the 100-MHz if strip?

$$-110 \text{ dBm} = P(3,1) = 3P_S' + K(3,1) = 3P_S' - 35$$

or

$$P_S' = \tfrac{1}{3}(-110 + 35) = -25 \text{ dBm}$$

That is, the level of the 200-MHz leakage from the first LO at the second mixer input must be no higher than -25 dBm.

61. Intermodulation. Intermodulation is particularly troublesome because a pair of strong signals that pass through a receiver preselector can cause interference in the if passband, even though the strong signals themselves do not enter the passband. Ebstein et al.[29] show typical generating mechanisms for intermodulation and cross-modulation in mixers.

Consider two undesired signals at 97 and 94 MHz passing through a superheterodyne receiver tuned to 100 MHz. Suppose, further, that the if is sufficiently selective that a perfect mixer allows no response to the signals (see Fig. 14-77). Third-order intermodulation in a physically realizable mixer will result in interfering signals at the if frequency and 9 MHz away (corresponding to 100 and 91 MHz rf frequencies, respectively). Fifth-order intermodulation will produce interferences 3 and 12 MHz from the intermediate frequency (103 and 88 MHz rf frequencies).

There is a formula for variation of intermodulation products that is quite useful. Figure 14-76 shows typical variations of desired output and intermodulation with input power level. Desired output increases 1 dB for each 1-dB increase of input level, whereas third-order intermodulation increases 3 dB for each 1-dB increase of input level. At some point the mixer saturates and the above behavior no longer exists. Since the interference of the intermodulation product is primarily of interest near the system sensitivity limit (usually

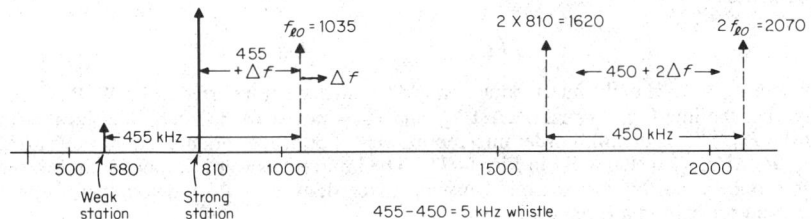

Fig. 14-73.   Spurious response in AM receiver.

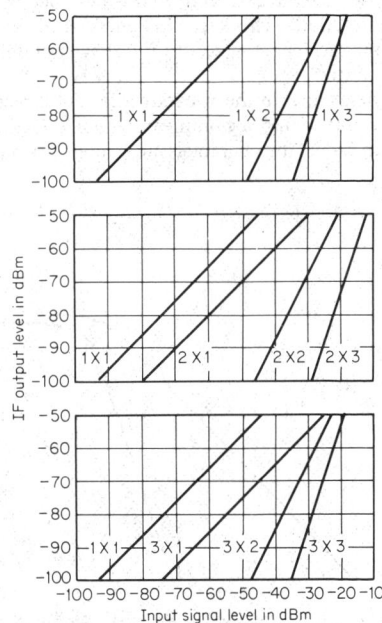

Fig. 14-74.   Spurious-response behavior of a balanced mixer. (Ref. 26)

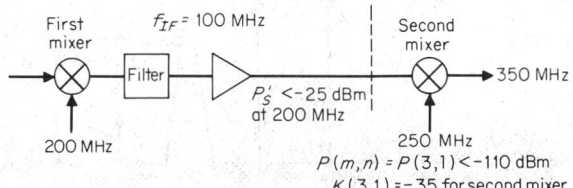

Fig. 14-75.   Spurious-response analysis.

14-67

somewhere below -20 dBm), the 1 dB/1 dB and 3 dB/1 dB patterns hold. The formula can be written

$$P_{21} = 2P_N + P_F - 2P_{I_{21}},$$

where P_{21} = level of the intermodulation product in decibels referred to 1 mW, P_N = power level of the interfering signal nearest P_{21}, and P_F = power of the interfering signal farthest from P_{21}. $P_{I_{21}}$ is the third-order intercept power. For proper orientation, $f_N = 97$ MHz, $f_F = 94$ MHz, $f_{21} = 100$ MHz in Fig. 14-77. The intercept power is a function of frequency. It can be used for comparisons between mixer designs and for determining allowable preselector gain in a receiving system.

62. Frequency-Converter Isolation. There are two paths in a mixer where isolation is important. The so-called *balanced mixers* give some isolation of the local-oscillator energy at the rf port. This keeps the superheterodyne receiver from radiating excessively. The doubly balanced mixers also give rf-to-if isolation. This keeps interference in the receiver rf environment from penetrating the mixer directly at the if frequency. Less important, but still significant, is the LO-to-if isolation. This keeps LO energy from overloading the if amplifier. Also, in multiple-conversion receivers low LO-to-if leakage minimizes spurious responses in subsequent frequency converters.

Desensitization. A strong signal in the rf bandwidth, not directly converted to if, drives the operating point of the mixer into a nonlinear region. The mixer gain is then either decreased or increased. In radar, the characteristic of concern is pulse desensitization. In

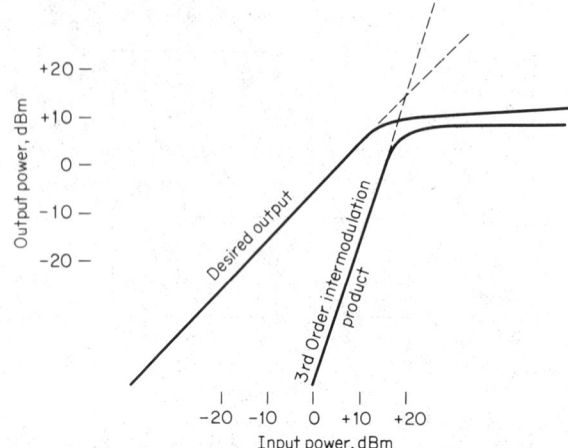

Fig. 14-76. Third-order intermodulation intercept power.

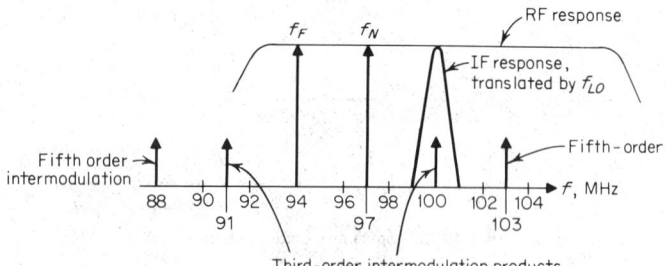

Fig. 14-77. Intermodulation in a superheterodyne receiver.

14-68

television receivers the characteristic is called cross-modulation. Here the strong undesired adjacent TV station modulates the mixer gain, especially during synchronization intervals, where the signal is strongest. The result appears in the desired signal as a contrast modulation of picture with the pattern of the undesired sync periods, corresponding to mixer gain *pumping* by the strong adjacent channel.

63. Schottky Diode Mixers. The Schottky barrier diode, a relatively new device, is an improvement over the point-contact diode. The Schottky diode has two features that make it very valuable in high-frequency mixers. First, it has low series resistance and virtually no charge storage, which results in low conversion loss. Second, it has a noise-temperature ratio very close to unity. The noise factor of a mixer-if amplifier cascade is[1]

$$F = L_M(t_D + F_{if} - 1)$$

where L_M = mixer loss, t_D = diode noise-temperature ratio, and F_{if} = if noise factor. Since t_D is near unity and L_M is in the range[2] of 2.4 to 6 dB, overall noise factor is quite good, with F_{if} near 1.5 dB in well-designed systems.

Basic theory for diode mixer operation is given in Refs. 5 to 7 (par. **14-69**). Torrey and Whitmer[5] show that the complete conversion matrix involves LO harmonic sums and differences, as well as signal, if, and image frequencies. They restrict their treatment of crystal rectifiers to the following third-order matrix:

$$\begin{bmatrix} I_1 \\ I_2 \\ I_3^* \end{bmatrix} = [Y] \begin{bmatrix} V_1 \\ V_2 \\ V_3^* \end{bmatrix} \quad Y = \begin{bmatrix} y_{11} & y_{12} & y_{13} \\ y_{21} & y_{22} & y_{23} \\ y_{31} & y_{32} & y_{33} \end{bmatrix}$$

where the subscripts denote: 1, signal port; 2, if port; 3 image port.

With point-contact diodes, the series resistance is so large that not much improvement is realized by terminating the image frequency, and terminating the other frequencies involved is less significant.

With the advent of Schottky barrier diodes, which have much smaller series resistances, proper termination of pertinent frequencies, other than signal and if frequencies, results in a minimizing of conversion loss. This, in turn, leads to a minimizing of noise figure.

An outline of the mathematical approach to conversion-loss minimization involves the maximum available gain concept discussed by Fukui.[30] First, let the conversion-loss matrix be reduced to a two-port matrix for signal and if frequencies by termination of all other pertinent frequencies:

$$\begin{bmatrix} I_1 \\ I_2 \end{bmatrix} = \begin{bmatrix} y_{11} & y_{12} \\ y_{21} & y_{22} \end{bmatrix} \begin{bmatrix} V_1 \\ V_2 \end{bmatrix}$$

where subscripts 1 = signal port and 2 = if port.

Then the maximum available gain is

$$\left| \frac{y_{21}}{y_{12}} \right| \frac{1}{k + \sqrt{k^2 - 1}}$$

where

$$k = \frac{2g_{11}g_{22} - \mathrm{Re}(y_{12}y_{21})}{|y_{12}y_{21}|}$$

Although maximum available gain corresponds to minimum available loss, the process is not so simple. There are constraints on how pertinent frequencies (such as the image) are terminated, because a diode has only one physical port, and it is difficult to control the impedances presented to all these frequencies. Nevertheless, conceptually the effort should be made to terminate the pertinent frequencies in such a manner that signal-if two-port having the largest value of maximum available gain is obtained.

Herold et al.[6] discuss termination of the image frequency, while Johnson[4] includes termination of the sum frequency, as well as termination at harmonic sums and differences.

Katoh and Akaiwa[2] report a 4.1-dB noise figure (including if) for an image-terminated mixer at 4 GHz. Johnson[4] achieved a 6.7-dB overall noise figure at X-band with his image-terminated mixer.

Several different configurations are used with Schottky mixers. Figure 14-78 shows an image-rejection mixer, which is used for low if frequency systems where rf filtering of the image is impractical. Kurpis and Taub[3] report 20-dB image rejection, 11-dB noise figure (including if), and 8- to 12-GHz frequency range. Their image-rejection mixer is balanced and has a 200-Mhz if.

There is a general rule of thumb for obtaining good intermodulation, cross-modulation, and desensitizable performance in mixers. It has been found experimentally that pumping a mixer harder extends its range of linear operation. The point-contact diode had a rapidly increasing noise figure with high LO power level and could easily burn out with too much power. The Schottky diode however, degrades in noise figure relatively slowly with increasing LO power, and it can tolerate quite large amounts of power without burnout.

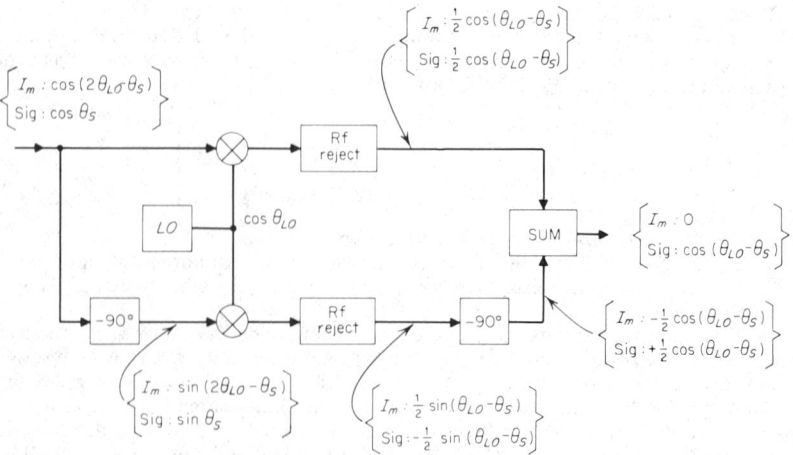

Fig. 14-78. Mixer designed for image rejection.

There is a limit to this process of increasing LO power, and that is that the Schottky diode series resistance begins to appear nonlinear. This leads to another rule of thumb: Pump the diode between two linear regions, and spend as little time as possible in the transition region. Application of the above two rules of thumb leads to the doubly balanced Schottky mixer. The reason for this is that one pair of diodes conducts hard and holds the other pair off. Hence large LO power is required, and one diode pair is conducting well into its linear region while the other diode pair is held in its linear nonconducting region.

64. Doubly Balanced Mixers. The diode doubly balanced mixer, or ring modulator, is shown in Fig. 14-79. This configuration has been analyzed by Caruthers,[8] Tucker,[9] and Belevitch.[10] The doubly balanced mixer is used up to and beyond 1 GHz in this configuration. The noise-figure optimization process previously discussed applies to this type of mixer. It exhibits good LO-to-rf and-if isolation, as shown in Fig. 14-80. Typical published data on mixers quote 30 dB rf-to-if isolation below 50 MHz and 20 dB isolation from 50 to 500 MHz.

Figure 14-81 gives a list of spurious responses for 50-MHz operation of an available mixer. With reference to the spurious-response formula (Par. **14-60**) relating $P(m,n)$, P_S', *and* $K(m,n)$, the numbers shown in Fig. 14-81 are $P_S' - P(m,n)$ in decibels for $P_S' = -10$dBm. From these values an array of $K(m,n)$ can be calculated. Then other sets of $K(m,n)$ values can be found by measurement for other frequencies. The information for this mixer can be compared with similar data for other mixers to determine optimum mixer designs for spurious-response performance. Of course, the data can also be used to determine filtering requirements in design of superheterodyne receiving systems. Finally, they quote an intermodulation ratio of − 60 dB for − 10 dBm applied. This corresponds to a third-order intercept power of + 20 dBm.

Another feature of balanced mixers is their local-oscillator noise-suppression capability. The noise-suppression power ratio is[15]

$$S = \frac{(\sqrt{R_2 L_2} + \sqrt{R_1 L_1})^2}{(\sqrt{R_2 L_2} - \sqrt{R_1 L_1})^2}$$

where R_1 and R_2 are the if impedances of the diodes, and L_1 and L_2 are their conversion losses.

Balanced mixers can take forms other than the ring modulator. Rafuse[18] has described a doubly balanced mixer using MOS field-effect transistors to achieve high dynamic range. With a local-oscillator power of 3 W at 40 MHz, he measures a third-order intercept power of + 60 dBm. The MOSFET mixer is shown in Fig. 14-82.

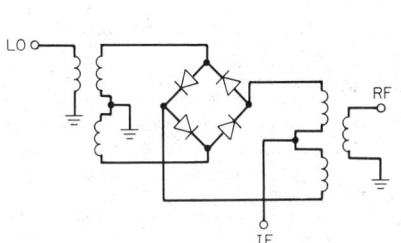

Fig. 14-79. Doubly balanced mixer.

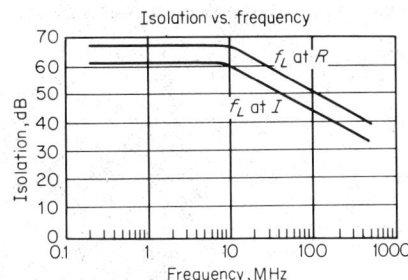

Fig. 14-80. Rf-to-if isolation in doubly balanced mixer. *(Model M1, RELCOM Division of Watkins Johnson.)*

Harmonic intermodulation signals

7 f_R	98	93	>100	90	94	84	>100	83	95
6 f_R	97	89	96	90	97	95	>100	92	>100
5 f_R	83	73	86	70	92	67	92	67	82
4 f_R	85	82	96	85	95	84	96	85	91
3 f_R	66	61	64	55	65	51	69	44	72
2 f_R	70	65	79	65	82	66	80	67	77
f_R	24	0	37	13	40	29	43	32	45
0		27	37	26	53	34	54	38	55
	0	f_L	2f_L	3f_L	4f_L	5f_L	6f_L	7f_L	8f_L

(Left axis label: Harmonics of f_R)

Harmonics of f_L

Values of $P_S' - P(m,n)$ for $P_S' = -10$ dBm

Fig. 14-81. Spurious-response chart of mixer shown in Fig. 14-80. *(RELCOM Division of Watkins Johnson.)*

Another balanced mixer uses a pair of high-gain junction FETs at 250 MHz. The reported third-order intercept is + 32 dBm, with noise figure at 6.5 dB and gain at 3 dB.

65. Parametric Converters. Parametric converters make use of time-varying energy-storage elements. Their operation is in many ways similar to that of parametric amplifiers, which is covered in Sec. 13 of this Handbook (Pars. 13-73 to 13-79). The difference is that output and input frequencies are the same in parametric amplifiers, while the frequencies differ in parametric converters. The device most widely used for microwave parametric converters today is the varactor diode, which has a voltage-dependent junction capacitance. The time variation of varactor capacitance is provided by a local oscillator, usually called the *pump.*

Attainable gain of a parametric converter is limited by the ratio of output to input frequencies. Therefore up-conversion is generally used to achieve some gain. Because lower-

14-71

sideband up-conversion results in negative resistance,[12] the upper sideband is generally used. This results in simpler circuit elements to achieve stability.

Gemulla[13] and Maninger[14] mention achievable gain of 2 to 3 dB and noise figure of about 3 dB in the X-band region. Maninger quotes a third-order intermodulation intercept of $+10$ dBm, while Gemulla gives $+5$ dBm at 100 mW pump power. Maninger reports these spurious responses, $mf_2 - nf_{LO} = \pm f_{if}$:

m	n	Interfering input level, dBm	Distortion product level referred to input, dBm
2	1	-4	-95
1	2	-75	-90

There is a distinct advantage to up-conversion; image rejection is easily achievable by a simple low-pass filter. Finally, the above values of intercept power are quite good, compared with values achievable with doubly balanced Schottky mixers at lower frequencies.

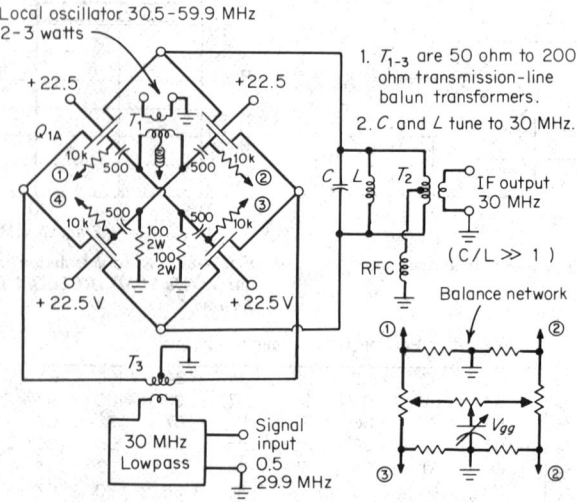

Fig. 14-82. Broadband MOSFET mixer of high dynamic range.

66. Transistor Mixers. One of the original concerns in transistor mixers was their noise performance.[19,20] The base spreading resistance r_b is very important in noise performance. The reason is that mixing occurs across the base-emitter junction; then the if signal is amplified by transistor action. However, r_b is a lossy part of the terminations at the if, signal, image, and all other frequencies present in the mixing process. Hence r_b dissipates some energy at each of the frequencies present, and all these contributions add to appear as a loss in the signal-to-if conversion. This loss, in turn, degrades noise figure. Vogel and Strutt[20] have measured as low as 3 dB noise figure at 1 MHz for an OC45 germanium unit. Other values range as high as 15 dB. Recently, a 6-dB noise figure at 200 MHz has been reported.

Manufacturers do not promote transistors used as mixers, probably because of their intermodulation and spurious-response performance. Estimates of intermodulation intercept power go as high as $+12$ dBm, while one measurement gave $+5$ dBm at 200 MHz. However, a cascode transistor mixer is used in a commercial vhf television tuner.

Characterization of Linearity in TV Tuners. The television tuner is a good example for discussing nonlinearities in mixers. Table 14-2 shows the various types of interferences generated, and they all result from the third-order, or cubic, nonlinearity. This explains the appearance of $P_{I_{21}}$ in each formula. Cross-modulation is the nonlinearity observed for characterizing the tuner, while the other three types actually cause interferences observed on

Table 14-2 **Interference Mechanisms in TV Tuners**

Interference	Mechanism	Formula
Intermodulation	$(v_N + v_F)^3 \rightarrow 3v_N^2 v_F$	$P_{21} = 2P_N + P_F - 2P_{I_{21}}$
Cross-modulation	$[v_U(1 + m_U \cos \theta_M) + v_D]^3 \rightarrow 6m_U \cos \theta_M v_U^2 v_D$	$20 \log \dfrac{m_X}{m_U} = 2(P_U - P_{I_{21}}) + 12 \text{ dB}$
Desensitization	$(v_U + v_D)^3 \rightarrow 3v_U^2 v_D$	$20 \log \left\|\dfrac{\Delta G}{G}\right\| = 2(P_U - P_{I_{21}}) + 6 \text{ dB}$
920-kHz beat	$(v_S + v_P + v_C)^3 \rightarrow 6v_S v_P v_C$	$P_{\text{beat}} = P_S + P_P + P_C - 2P_{I_{21}} + 6 \text{ dB}$

All power in decibels referred to 1 mW.
U = undesired
D = desired
S = sound carrier
P = picture carrier
C = chroma
m_X = cross-modulation index
m_U = modulation index on undesired signal

the picture tube. Figure 14-83 shows the frequencies involved in the 920-kHz beat generation.

Investigators who measure cross-modulation performance in TV tuners usually apply the undesired signal at 6 MHz or so from band center of the desired signal. This allows the rf-mixer interstage filter to provide attenuation and does not give a true measure of mixer linearity. The measurement does provide information on performance in the presence of a strong adjacent channel, because there is a correspondence between cross-modulation and the densensitization due to the sync tips of the adjacent channel. However, the 920-kHz beat that occurs due to the mixer is not controlled by results of the above measurement. The latter phenomenon results from mixing of the picture chrominance and sound carriers to produce a distortion product 920 kHz away from the picture carrier. The three carriers and the distortion product are all within the desired passband. Hence the cross-modulation measurement taken with the undesired signal carrier 6 MHz away is not adequate. Nevertheless, the 920-kHz beat results from a third-order phenomenon, and is therefore related to both the cross-modulation and intermodulation phenomena. An extra set of measurements taken near band center would suffice to predict performance for the 920-kHz beat.

67. Measurement of Spurious Responses. Figure 14-84 shows an arrangement for measuring mixer spurious responses. The filter following the signal generator implies that generator harmonics are down, say 40 dB. This ensures that frequency-multiplying action

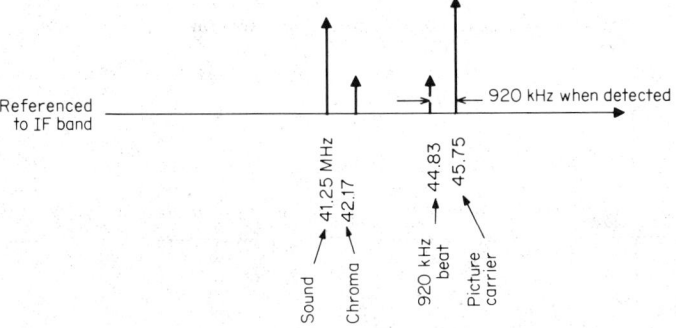

Fig. 14-83. Generation of 920-kHz beat in TV tuners.

14-73

is due only to the mixer under test. The attenuator following the mixer can be used to be sure that a spurious response of the receiver is not being measured. That is, a 6-dB change in attenuator setting should be accompanied by a 6-dB change on the indicator.

Generally the most convenient way of performing the spurious-response test is to first obtain an indication on the indicator. Then tune the signal generator to the desired frequency and record the level required to obtain the original response. This should be repeated at one or two more levels of the undesired signal to ensure that the spur follows the appropriate laws. For example, if the response is fourth-order (4 times the signal frequency $\pm$ *n* times the LO frequency), then the measured value should change 4 dB for 1 dB change in undesired frequency level. The order of the spurious response can be determined by either of two methods. The first method is simply by knowing with some accuracy the undesired signal frequency and the local-oscillator frequency, and then determining the harmonic numbers required to obtain the if frequency. The other technique entails observing the incremental changes of the if frequency with known changes in the undesired signal frequency and the local-oscillator signal frequency.

This completes the measurement for one spurious response. The procedure should be repeated for each of the spurious responses it is desired to measure.

The intermodulation test setup is shown in Fig. 14-85. In general, a diplexer is preferable to a directional coupler for keeping generator 1 signal out of generator 2. This is necessary so that the measurement is not limited by the test setup. A good idea would be to establish that no third-order intermodulation occurs due to the setup alone. To do this, initially remove the mixer/LO circuit. Then tune generator 1 off from center frequency to about 10 or 20 dB down on the skirt of the receiver preselector. Tune generator 2 twice this amount from the receiver center frequency. Set generator levels equal and at some initial value, say − 30 dBm. Then vary one generator frequency slightly and look for a response peak on the indicator. If none is noticed, increase the generator level to − 20 dBm and repeat the procedure. Usually, except for very good receivers, the third-order intermodulation response is found. Vary the attentuator by 6 dB, and look for a 6-dB variation in the indicator reading. If the latter is not 6 dB but 18 dB, then intermodulation is occurring in the receiver. If the indicator variation is between 6 and 18 dB, then intermodulation is occurring in the circuitry preceding the attenuator and in the receiver. To obtain trustworthy measurements with a mixer in the test position, the indicator should read at least 20 dB greater than without the mixer, while the generator levels should be lower by mixer gain + 10 dB than they were without the mixer. This ensures that the test setup is contributing an insignificant amount to the intermodulation measurement.

With the mixer in test position and the above conditions satisfied, obtain a reading on the indicator and let the power referred to the mixer input be denoted by *P* (dBm). Turn down both generator levels, and retune generator 1 to center frequency. Adjust generator 1 level to obtain the previous indicator reading. This essentially calibrates the measurement setup.

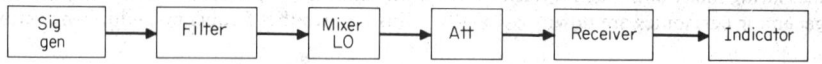

Fig. 14-84. Test equipment for measuring mixer spurious responses.

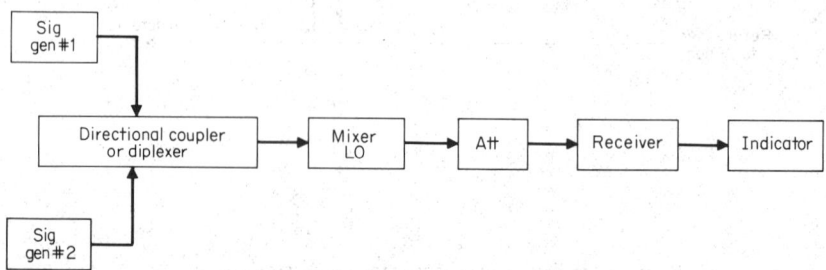

Fig. 14-85. Test equipment for measurement of mixer intermodulation.

Denote generator 1 level referred to the mixer input by P_{21} (dBm). Then the intermodulation intercept power is given by

$$P_{I_{21}} \text{ dBm} = \frac{3P - P_{21}}{2}$$

The subscripts on intercept power $P_{I_{21}}$ refer to second order for the near frequency and first order for the far frequency (see Fig. 14-76).

The procedure should be repeated for one or two lower values of P. The corresponding values of $P_{I_{21}}$ should asymptotically approach a constant value. The constant value of $P_{I_{21}}$ so obtained is then a valid number for predicting behavior of the mixer near its sensitivity limit.

68. Detectors (Frequency Deconverters) Detectors have become more complex and versatile since the advent of integrated circuits. Up to the mid-1950s most radio receivers used the standard single-diode envelope detector for AM and a Foster-Seeley discriminator or ratio detector for FM. Today, integrated circuits are available with if amplifier, detector, and audio amplifier functions in a single package.

Figure 14-86 shows three conventional AM detectors. In Fig. 14-86a an envelope detector is shown. In order for the detected output to follow the modulation envelope faithfully, the RC time constant must be chosen so that $RC < 1/\omega_m$, where ω_m is the maximum angular modulation frequency in the envelope. Figure 14-86b shows a peak detector. Here the RC time constant is chosen large, so that C stays charged to the peak voltage. Usually, the time constant depends on the application. In a television field strength meter, the charge on C should not decay significantly between horizontal sync pulses separated by 62.5 μs. Hence a time constant of 1 to 6 ms should suffice. On the other hand, an AGC detector for single-sideband use should have a time constant of 1 s or longer.

Figure 14-86c shows a product (synchronous) detector. This type of detector has been used since the advent of single-sideband transmission. The product detector multiplies the signal with the LO, or beat frequency oscillator (BFO), to produce outputs at sum and difference frequencies. Then the low-pass filter passes only the difference frequency. The result is a clean demodulation with a minimum of distortion for single-sideband signals.

The two classical FM detectors widely used up to the present are the Foster-Seeley discriminator and the ratio detector. Figure 14-87 shows the Foster-Seeley discriminator and its phasor diagrams. The circuit consists of a double-tuned transformer, with primary and secondary voltages series-connected. The diode connected to point A detects the peak value of $V_1 + V_2/2$, and the diode at B detects the peak value of $V_1 - V_2/2$. The audio output is then the difference between the detected voltages. When the incoming frequency is in the center of the passband, V_2 is in quadrature with V_1; the detected voltages are equal; and audio output is zero. Below the center frequency the detected voltage from B decreases, while that from A increases, and the audio output is positive. By similar reasoning, an incoming

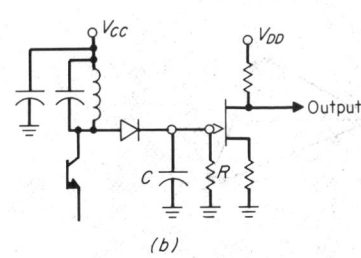

(a)

$\tau = RC \dfrac{1}{\omega_m}$

To audio amplifier

(b)

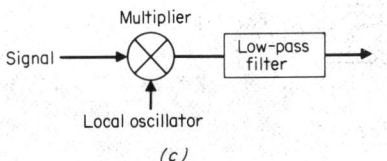

(c)

Fig. 14-86. Amplitude modulation detectors. (a) Am envelope detector; (b) peak detector; (c) product detector.

frequency above band center produces a negative audio output. Optimum linearity requires that $KQ = 2$, where K is the transformer coupling and Q is the primary and secondary quality factor.

Figure 14-87c shows a ratio detector, which has an advantage over the Foster-Seeley discriminator in being relatively insensitive to AM. The ratio detector uses a tertiary winding (winding 3) instead of the primary voltage, and one diode is reversed. However, the phasor diagrams also apply to the ratio detector. The AM rejection feature results from choosing the $(R_1 + R_2)C$ time constant large compared with the lowest frequency to be faithfully reproduced. The voltages E_{OA} and E_{OB} represent the detected values of rf voltages across OA and OB, respectively. With the large time constant above, voltage on C changes slowly with AM and the conduction angles of the diodes vary, loading the tuned circuit so as to keep the rf amplitudes relatively constant.

Capacitor C_0 is chosen to be an rf short circuit, but small enough to follow the required audio variations. In the AM rejection process, AF voltage on C_0 does not follow the AM because the charge put on by one diode is removed by the other diode. With FM variations on the rf, voltage on C_0 changes to reach the condition, again, that charge put on C_0 by one diode is removed by the other diode. The ratio detector is generally used with little or no previous limiting of the rf, while the Foster-Seeley discriminator must be preceded by limiters to provide AM rejection.

With the recent trend toward integrated circuits, there has been increased interest in using phase-locked loops and product detectors. These techniques have been selected because they do not require inductors, which are not readily available in integrated form. Figure 14-88 shows a phase-locked loop (PLL) as an FM detector. The phase comparator merely provides a dc voltage proportional to the difference in phase between signals represented by f_M and f.

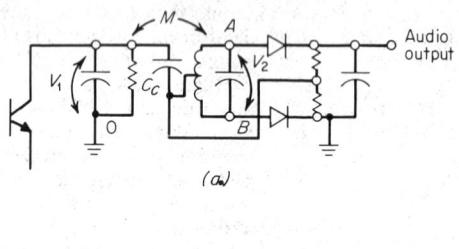

(a)

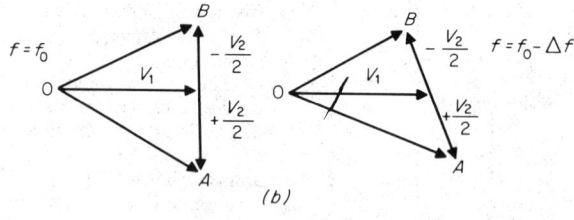

(b)

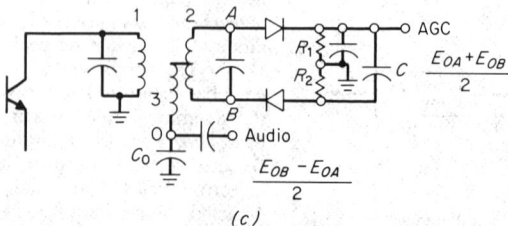

(c)

Fig. 14-87. Frequency modulation detectors. (*a*) Foster-Seeley fm discriminator; (*b*) phasor diagrams; (*c*) ratio detector.

Initially, f and f_M are unequal, but because of high loop gain, $GH \gg 1$, f and f_M quickly become locked and stay locked. Then as f_M varies, f follows exactly. But because of the high loop gain, response is essentially $1/H$, which is the voltage controlled oscillator (VCO) characteristic. Hence the PLL serves as an FM detector. Grebene and Camenzind[24,25] give more detail on integrated circuit implementation of the PLL technique.

AM product detectors also make use of the PLL to provide a carrier locked to the incoming signal carrier. The output of the VCO is used to drive the product detector. Grebene[25] describes such an arrangement in integrated circuit form. Probably one of the most stringent uses of the product detector is in an FM stereo decoder. The *left minus right* $(L - R)$ subcarrier is located at 38 kHz with sidebands from 23 to 53 kHz. There may also be an SCA signal centered about 67 kHz which is used to provide music service for restaurants and commercial offices. The $L - R$ product detector is driven by a 38-kHz VCO, the output of which also goes to a 2-to-1 counter. The counter output is compared with the 19-kHz pilot signal in a phase comparator, and the phase-comparator output then controls the VCO. Because of the relatively small pilot signal and the presence of $L + R$, $L - R$, and SCA information, the requirement for phase locking is stringent.

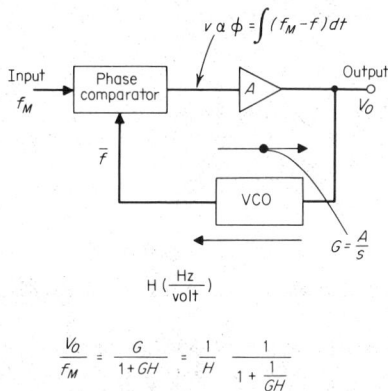

$$\frac{V_O}{f_M} = \frac{G}{1+GH} = \frac{1}{H} \frac{1}{1+\frac{1}{GH}}$$

Fig. 14-88. FM detector using phase-locked loop.

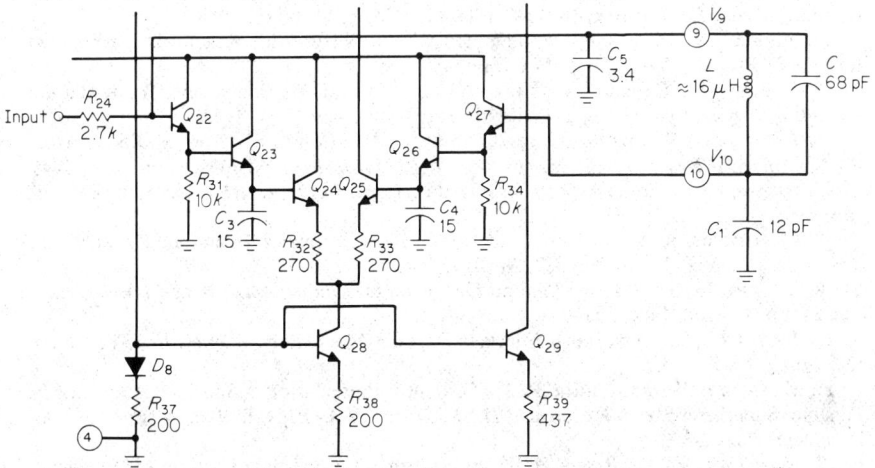

Fig. 14-89. Differential peak detector for TV FM sound, a portion of the CA3065 integrated circuit. *(RCA Corporation.)*

An interesting new FM detector is the *differential peak detector*. It is included in the RCA CA3065 integrated circuit, which serves as a complete TV sound system (excluding audio output stage). The FM detector portion is shown in Fig. 14-89. Transistors $Q23$ and $Q26$ are peak detectors, and the difference of their detected voltages divided by $(R32 + R33)$ is the dc differential current at the $Q25$ collector. Transistors $Q22$ and $Q27$ are buffers for rf voltages. Figure 14-90 gives an idea of detector operation. Starting from low frequencies, voltages at $Q23$ and $Q26$ detectors are near equal, and differential output current is near zero. As frequency increases, the LC parallel circuit series-resonates with C_1, so that a short circuit appears at V_9 (PIN 9). This is a maximum at V_{10} (PIN 10). At higher frequencies LC approaches parallel resonance, where $V_{10} \approx 0$ and V_9 reaches maximum. This, along with the differential operation, results in the S shape required for FM detection.

$$V_9 = F(\omega)\left[1-\omega^2 L(C+C_1)\right] \qquad C_1 = 12\,pF$$
$$V_{10} = F(\omega)\,(1-\omega^2\,LC) \qquad\quad C = 68\,pF$$
$$F(\omega) = \text{three pole transfer function}$$

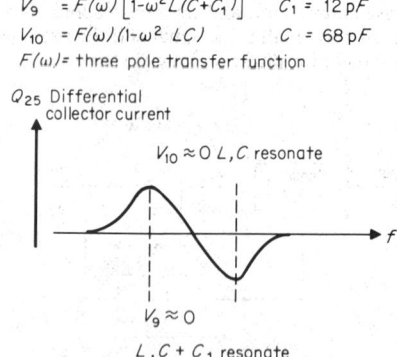

Fig. 14-90. Characteristic of differential peak FM detector.

69. References on Frequency Converters and Detectors.

1. Osborne, T. L., Kibler, L. U., and Snell, W. W. Low-Noise Receiving Down Converter, *Bell System Tech. J.,* Vol. 48, No. 6, pp. 1651-1663, July-August, 1969.

2. Katoh, M., and Akaiwa, Y. 4 GHz Integrated-Circuit Mixer, *IEEE Trans.,* Vol. MTT-19, No. 7, pp. 634-637, 1971.

3. Kurpis, G. P., and Taub, J. J. Wideband X-Band Microstrip Image Rejection Balanced Mixer, *IEEE Trans.,* Vol. MTT-18, No. 12, pp. 1181-1182, 1970.

4. Johnson, K. M. X-Band Integrated Circuit Mixer with Reactively Terminated Image, *IEEE Trans.,* Vol. ED-15, No. 7, pp. 450-459, 1968.

5. Torrey, H. C., and Whitmer, A. C. "Crystal Rectifiers." M.I.T. Radiation Laboratory Series, Vol. 15, pp. 111-178, McGraw-Hill, New York, 1948.

6. Herold, E. W., Bush, R. R., and Ferris, W. R. Conversion Loss of Diode Mixers Having Image-Frequency Impedance, *Proc. IRE,* Vol. 33, pp. 603-609, September, 1945.

7. Strum, P. D. Some Aspects of Crystal Performance, *Proc. IRE,* Vol. 41, pp. 875-889, July, 1953.

8. Caruthers, R. S. Copper Oxide Modulators in Carrier Telephone Systems, *Bell System Tech. J.,* Vol. 18, pp. 315-337, April, 1939.

9. Tucker, D. G. Intermodulation Distortion in Rectifier Modulators, *Wireless Eng.,* Vol. 31, pp. 145-152, June, 1954.

10. Belevitch, V. Non-linear Effects in Rectifier Modulators, *Wireless Eng.,* Vol. 27, p. 130, 1950.

11. Karpen, E. W., and Mohr, R. J. Graphical Presentation of Spurious Responses in Tunable Superheterodyne Receivers, *IEEE Trans.,* Vol. EMC-8, No. 4, pp. 192-196, December, 1966.

12. Manley, J. M., and Rowe, H. E. Some General Properties of Non-linear Elements, Pt. I, General Energy Relations, *Proc. IRE,* Vol. 44, pp. 904-913, July, 1956; Pt. II, Small Signal Theory, Vol. 46, pp. 850-860, May, 1958.

13. GEMULLA, W. J.   Parametric Up-Converters for Low-Noise Broadband Microwave Receivers, *WESCON Tech. Papers,* Vol.14, Session 9, Paper 3, 1970.

14. MANINGER, L.   Wideband Receivers Using Varactor Diode Frequency Upconverters, *Microwave J.,* Vol. 11, pp. 49-55, August, 1968.

15. POUND, R. V.   "Microwave Mixers," M.I.T. Radiation Laboratory Series, Vol. 16, Chap. 6, McGraw-Hill, New York, 1948.

16. OHTOMO, M.   Experimental Evaluation of Noise Parameters in Gunn and Avalanche Oscillators, *IEEE Trans.,* Vol. MTT-20, No. 7, pp. 425-437, July, 1972.

17. TAUB, J. J., and GIORDANO, P. J.   Use of Crystals in Balanced Mixers, *IRE Trans.,* Vol. MTT-2, pp. 26-38, July, 1954.

18. RAFUSE, R. P.   Symmetric MOSFET Mixers of High Dynamic Range, *ISSCC Digest Tech. Papers,* pp. 122-123, 1968.

19. WEBSTER, R.   The Noise Figure of Transistor Converters, *IRE Trans.,* Vol. BTR-7, pp. 50-65, November, 1961.

20. VOGEL, J. S., and STRUTT, M. J. O.   Noise in Transistor Mixers, *Proc. IEEE,* Vol. 51, No. 2, pp. 340-349, February, 1963.

21. VAN DER ZIEL, A.   "Noise," Prentice-Hall, Englewood Cliffs, N. J., 1954.

22. KNIGHT, M. B.   A New Miniature Beam-Deflection Tube, *RCA Rev.,* pp. 266-289, June, 1960.

23. SCHLESINGER, K.   The Synchrotector: A Sampling Detector for Television Sound, *IRE Trans.,* Vol BTR-2, pp. 34-42, July, 1956.

24. GREBENE, A. B., and CAMENZIND, H. R.   Frequency-selective Integrated Circuits Using Phase-Lock Techniques, *IEEE J. Solid State Circuits,* Vol. SC-4, No. 4, pp. 216-225, August, 1969.

25. GREBENE, A. B.   An Integrated Frequency-selective AM/FM Demodulator, *IEEE Trans. Broadcast TV Receivers,* Vol. BTR-17, pp. 71-80, May, 1971.

26. KARPEN, E. W., and MOHR, R. J.   Graphical Presentation of Spurious Responses in Tunable Superheterodyne Receivers, *IEEE Trans. Electromagn. Compatibility,* Vol. EMC-8, No. 4, pp. 192-196, December, 1966.

27. "Reference Data for Radio Engineers," 4th ed., p. 774, International Telephone and Telegraph Corp., 1962.

28. HOIGAARD, J. C.   Spurious Frequency Generation in Frequency Converters, *Microwave J.,* July, 1967, pp. 61-64; pp. 78-82, August, 1967.

29. EBSTEIN, B., HUENEMANN, R., and SEA, R.   The Correspondence of Intermodulation and Cross Modulation in Amplifiers and Mixers, *Proc. IEEE,* Vol. 55, No. 8, pp. 1514-1516, August, 1967.

30. FUKUI, H.   Available Power Gain, Noise Figure, and Noise Measure of Two-Ports and Their Graphical Representations, *IEEE Trans. Circuit Theory,* Vol. CT-13, No. 2, pp. 137-142, June, 1966.

SECTION 15

POWER ELECTRONICS

BY

W. E. NEWELL, Senior Member, Institute of Electrical and Electronics Engineers; P. F. PITTMAN, Senior Member, IEEE; J. C. ENGEL; J. W. MOTTO, Member, IEEE; B. R. PELLEY; T. M. HEINRICH, Member, IEEE; R. M. OATES; and L. GYUGYI; members, technical staff, Westinghouse Research Laboratories

CONTENTS

Numbers refer to paragraphs

SECTION 15

POWER ELECTRONICS*

INTRODUCTION

By W. E. Newell

1. Power Electronics Defined. Power electronics deals with the application of electronic devices and associated components to the conversion, control, and conditioning of electric power. The primary characteristics of electric power which are subject to control include its basic form (ac or dc), its effective voltage or current (including the limiting cases of initiation and interruption of conduction), and its frequency and power factor (if ac). The control of electric power is frequently desired as a means for achieving control or regulation of one or more nonelectrical parameters, e.g., the speed of a motor, the temperature of an oven, the rate of an electrochemical process, or the intensity of lighting.

2. Efficiency Requirements. Aside from the obvious difference in function, power-electronics technology differs markedly from the technology of low-level electronics for information processing in that much greater emphasis is required on achieving high power efficiency. Few low-level circuits exceed a power efficiency of 15%, but few power circuits can tolerate a power efficiency less than 85%. High efficiency is vital, first, because of the economic value of the wasted power and, second, because of the cost of dissipating the heat it generates. This high efficiency cannot be achieved by simply scaling up low-level circuits; a different approach must be adopted.

Variable-Resistance Approach. In theory, the desired control of power could be achieved by means of a high-speed rheostat connected in series with the load. Variation of the rheostat resistance from zero to infinity permits continuous and smooth control of the load voltage and power from their maximum values to zero. However, the power efficiency is proportional to the load voltage and is only 50% when the load voltage is equal to half of the source voltage. At its intermediate settings, the rheostat must be capable of dissipating up to 25% of the maximum load power. The consequences of this dissipation are trivial in the milliwatt power region. They can be tolerated into the power region of hundreds of watts, where other approaches may cause more severe problems for certain applications. But in the kilowatt power region and above, this approach becomes completely impracticable.

On-Off Approach. The alternative is based on the fact that the rheostat dissipates very little power in its two extreme positions, corresponding to a closed or open switch. In this case, control is achieved by means of the timing of repetitive switching action. Because of wear and limited switching speed, mechanical switches are ordinarily not suitable, but electronic switches have made this approach feasible into the megawatt power region while maintaining high power efficiencies over wide ranges of control. However, the inherent nonlinearity of the switching action leads to the generation of transients and spurious frequencies that must be considered in the design process.

3. Types of Switching Devices. The origins of modern power electronics can be traced to the technology of rectifiers and inverters developed many years ago (Rissik, 1939)[1] to utilize mercury-arc devices. Today solid-state power-switching devices have won nearly universal acceptance in these applications and are making many new applications feasible because of their greater reliability, faster speed, higher efficiency, smaller size, and lower cost. As in microwave technology, much of the technology of power electronics is devoted to the

---

*Overall coordination of this section was by W. E. Newell. In addition to the other authors listed, he wishes to acknowledge the important contributions made by R. P. Putkovich, P. Wood, J. Rosa, and A. H. B. Walker.
[1] Author and date references in parentheses refer to the Bibliography, Par. 15-52.

problems of utilizing the full capability of state-of-the-art solid-state devices while minimizing the effects of their inherent limitations.

For example, most power converters depend on the ability to connect any of two or more input lines to any of two or more output lines, a function served admirably by a multipole selector switch. The lack of electronic switching devices capable of fulfilling this function directly leads to a large (and to the beginner, often bewildering) variety of device and circuit configurations for performing the function indirectly by means of unilateral, single-pole single-throw switches, which can be turned on but which cannot interrupt current to turn themselves off.

Because of the key role played by solid-state switching devices, it is advisable to summarize their main characteristics and distinguish between their main types before proceeding further. Most solid-state devices are inherently *unilateral;* i.e., they are designed to carry current in only one direction, known as the *forward direction.** Appreciable current flow in the reverse direction is blocked, although there is a brief transient of reverse current when reverse voltage is applied immediately following forward conduction. In brief, most solid-state switches open automatically when reverse polarity is applied across their terminals.

The first type of solid-state device which should be distinguished is a *two-terminal switch,* which closes automatically when forward polarity is applied. Such devices are commonly known as *rectifier diodes* or simply diodes.

The second type of solid-state switch has an additional *control terminal* and will block the flow of forward current until an appropriate turn-on pulse is applied to the control terminal. Thereafter the device continues to conduct as long as forward current flows. The control electrode cannot interrupt forward current. This type of device is known officially as a *reverse-blocking triode thyristor* but is commonly called simply a thyristor or SCR (for *silicon controlled rectifier*).

The third type of device is similar to the second type except that it also has *turnoff ability.* Forward current can be interrupted by means of the control electrode. *Switching transistors* belong to this type because collector current ceases when the control signal is removed from the base. Bistable devices of this type are known as *turnoff thyristors* or *gate-controlled switches.*

4. Commutation. Commutation refers to the process by which forward current is interrupted or transferred from one switching device to another. In most circuits where power is supplied from an ac source, turn-on control is adequate and turnoff occurs naturally when the ac cycle causes the polarity of voltage across a given device to reverse.† Such circuits, whether supplied from an ac or a dc source, are said to have *natural commutation.*

Most circuits supplied from a dc source and some ac circuits must be able to interrupt a nonzero current. Because of the limited power capacity of presently available devices with inherent turnoff capability, it is often necessary to achieve turnoff by indirect means. These include auxiliary components which momentarily reverse bias and divert the current from a turn-on thyristor until it reestablishes its blocking state. This artificial turnoff of devices which are capable only of turn-on is known as *forced commutation,* and the development of turnoff techniques is currently extending the usefulness of power electronics into many new functions and new applications.

For an overview of power electronics, the following references are recommended: Gutzwiller, 1967; Storm, 1969; Hoft, 1972; Kusko, 1972; Bates and Colyer, 1973; and Newell, 1974.

SOLID-STATE POWER DEVICES

By P. F. Pittman, J. C. Engel, and J. W. Motto

5. Introduction. Power circuits utilize a wide variety of solid-state devices, including low-level devices for processing the signals which control the switching devices. Figure 15-1 summarizes the junction structures, circuit symbols, and main-terminal *VI* characteristics of the most common power devices. These devices are discussed in subsequent paragraphs.

---

* One exception is the *triac,* which is an integration of two unilateral devices in parallel opposition into a single structure.
† Care should be taken to distinguish between the reversal of device voltage and the reversal of source voltage. Neglecting transient effects, the device voltage reverses as its current passes through zero. This may or may not occur in synchronism with the reversal of the source voltage.

Device name	Structure	Symbol	V-I characteristic	Principal use
Conventional diode	A — [P | N] — K	A — ▷|— K		Rectifier
Avalanche diode	A — [P | N] — K	A — ▷|— K		Rectifier
Thyristor	A — [P | N | P | N] — K, G	A — ▷|— K, G		Controlled turn-on switch
Triac	MT2 — [P | N | P | N] — MT1, G	MT1, MT2, G		Bidirectional controlled turn-on switch
Transistor	E — [N | P | N] — C, B / E — [P | N | P] — C, B	C, B, E (NPN / PNP)		Controlled turn-on and turn-off switch

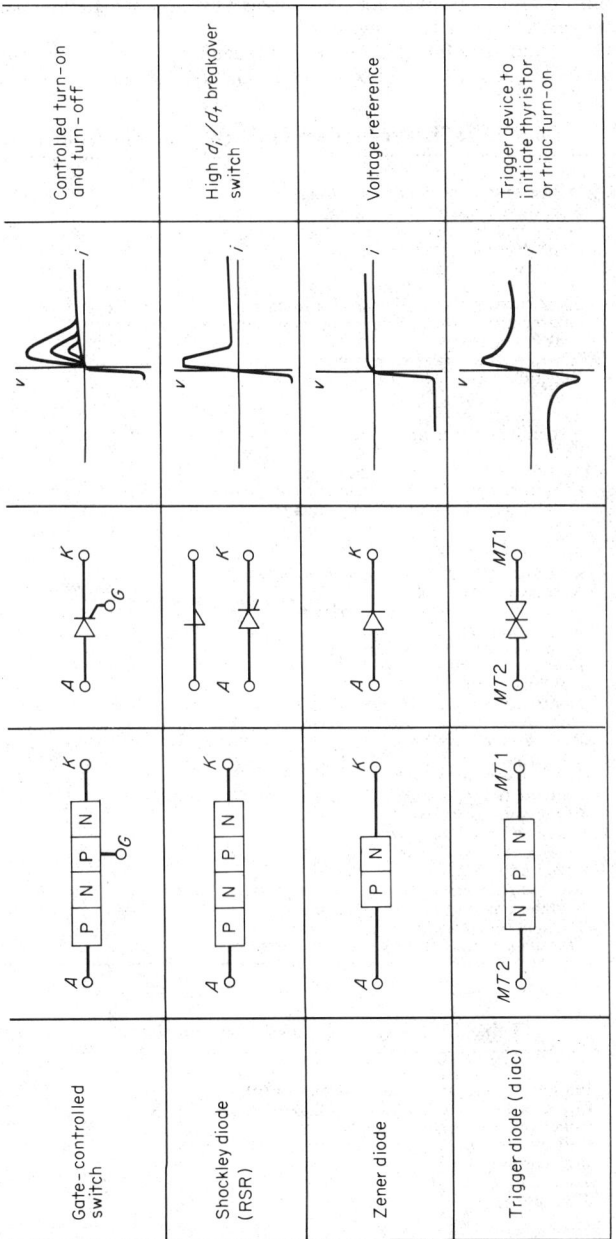

Fig. 15-1. Devices commonly used in power circuits.

Tables 15-1 and 15-2 list the standard letter symbols used to designate the more important ratings and characteristics of power devices.

Further information on solid-state devices is available in manufacturers' handbooks, in textbooks (Gentry, et al., 1964; Hibberd, 1968; Delhom, 1968), and in several IEEE

Table 15-1. Letter Symbols for Diodes and Thyristors

Ratings:

V_{DSM}	Nonrepetitive peak off-state voltage
V_{DRM}	Repetitive peak off-state voltage
V_{RSM}	Nonrepetitive peak reverse voltage
V_{RRM}	Repetitive peak reverse voltage
$I_{T,rms}$	Rms on-state current
$I_{T,av}$	Average on-state current
I_{TSM}	Surge (nonrepetitive) on-state current
$I^2 t$	Nonrepetitive (rms) ampere^2-seconds overcurrent capability
I_{GM}	Peak gate current
V_{GRM}	Repetitive peak reverse gate voltage
I_{GRM}	Repetitive peak reverse gate current
P_{GM}	Peak gate power
$P_{G,av}$	Average gate power
T_J	Virtual junction temperature
T_{stg}	Storage temperature
$P_{av,max}$	Maximum average device dissipation
di/dt	Critical rate of rise of on-state current
dv/dt	Critical rate of rise of off-state voltage

Characteristics:

v_{BO}	Breakover voltage
V_{TM}	Maximum (steady-state) on-state voltage
$V_{T,av}$	Average on-state voltage
V_{TO}	Dynamic on-state voltage (during turn-on)
I_{DRM}	Repetitive peak off-state current
I_{RRM}	Repetitive peak reverse current
I_L	Latching current
I_H	Holding current
I_{DSM}	Nonrepetitive peak off-state current
I_{RSM}	Nonrepetitive peak reverse current
I_{RQM}	Peak reverse recovery current
V_{GT}	Gate trigger voltage
I_{GT}	Gate trigger current
V_{GD}	Gate nontrigger voltage
I_{GD}	Gate nontrigger current
t_d	Gate-controlled delay time
t_r	Gate-controlled rise time
t_f	Fall time
t_{rr}	Reverse recovery time
t_q	Circuit-commutated turn-off time
$R_{\theta JC}$	Thermal resistance, junction to case
$Z_{\theta JC(t)}$	Transient thermal impedance, junction to case

Table 15-2. Letter Symbols for Transistors

Ratings:

V_{EBO}	Emitter-to-base dc voltage (collector open)
$V_{CEO,sus}$	Collector-to-emitter sustaining voltage (base open)
V_{CEO}	Collector-to-emitter voltage (base open)
V_{CEV}	Collector-to-emitter voltage (base at specified voltage)
I_B	Base current (dc)
I_C	Collector current (dc)
P_T	Total power dissipation

Characteristics:

h_{FE}	Dc current gain (common emitter)
$V_{CE,sat}$	Collector-to-emitter saturation voltage
$V_{BE,sat}$	Base-to-emitter saturation voltage
I_{EBO}	Emitter-to-base cutoff current (collector open)
t_s	Storage time
C_{ob}	Output capacitance
f_T	Gain bandwidth
SOA	Safe operating area

publications on power devices (IEEE, 1967, 1970). Pertinent standards include USAS C34.2-1968 and EIA-1972.

RECTIFIER DIODES

6. Conventional Diodes. A conventional semiconductor diode employs a single *pn* junction, as shown in Fig. 15-1. The diode *VI* characteristic is such that current can easily flow in one direction while its flow in the other direction is blocked by the junction. The physical theory of the action of a *pn* junction under conditions of forward and reverse bias is discussed in Sec. 7, Par. 7-42.

Silicon diodes have increased in power-handling capability at least as rapidly as triode power-control devices. Diodes with current and voltage ratings of 1500 A or 3000 V repetitive reverse voltage are readily available, although not in the same device. Larger devices are in development, and as a result, diode power-handling capabilities will continue to increase.

Silicon-diode ratings include voltage, current, and junction temperature. The device current rating I_T is primarily determined by the geometric design of the silicon die and the method of heat sinking, while the spread of voltage ratings V_{RRM} results from the yield distribution of the manufacturing process.

Reverse voltage ratings are designated as repetitive V_{RRM} and nonrepetitive V_{RSM}. The repetitive value pertains to steady-state operating conditions, while the nonrepetitive peak value applies to occasional transient or fault conditions. Care must be exercised when applying a device to ensure that the voltage rating is never exceeded, even momentarily. When the blocking capability of a conventional diode is exceeded, leakage currents flow through localized areas at the edge of the crystal. The resulting localized heating can easily cause instant device failure.

Although even low-energy reverse overvoltage transients are likely to be destructive, the silicon diode is remarkably rugged with respect to *forward current transients*. This property is demonstrated by the I_{TSM} rating which permits a one-half cycle peak surge current of nearly 10 times the I_T rating. For shorter current pulses, less than 4 ms, the surge current is specified by an I^2t rating similar to that of a fuse.

Proper circuit design must ensure that the *maximum average junction temperature* will never exceed its design limit of typically 200°C. Good design practice for high reliability, however, limits the maximum junction temperature to a lower value. The average junction-temperature rise above ambient is calculated by multiplying the average power dissipation, given approximately by the product of V_T and I_T, by the thermal resistance. Transient junction temperatures can be computed from the transient thermal-impedance curve.

Device ratings are normally specified at a given *case temperature* and *operating frequency*. The proper use of a device at other operating conditions requires an appreciation of certain basic device characteristics. This is especially true in applications where the operating conditions of a number of devices are interdependent, as in series and parallel operation. For example, the forward voltage drop of a silicon diode has a negative temperature coefficient of 2 mV/°C for currents below the rated value. This variation in forward drop must be considered when devices are to be operated in parallel.

The reverse blocking voltage of a diode, at a specified reverse current, can either decrease or increase with temperature. The tendency to decrease comes from the fact that the reverse leakage current of a junction increases with temperature, thereby decreasing the voltage attained at a given measuring-current level. If the leakage current is very low, the maximum reverse voltage will be determined by avalanche breakdown in the silicon, which has a coefficient of approximately 0.1%/°C. It should be noted that the reverse blocking voltage of a conventional diode is usually determined by imperfections at the edge of the die, and thus an ideal avalanche breakdown is usually not observed.

The *reverse recovery time* of a diode causes its performance to degrade with increasing frequency. Because of this effect, the rectification efficiency of a conventional diode used in a power circuit at high frequency is poor. In order to serve this application, a family of fast-recovery diodes has been developed and is presently available at a premium price. The stored charge of these devices is low, with the result that the amplitude and duration of the sweep-out current are greatly reduced compared with those of a conventional diode.

However, improved turnoff characteristics of the fast-recovery diodes are obtained at some sacrifice in blocking voltage and forward drop compared with a conventional diode.

7. Avalanche Diodes. An avalanche diode differs from a conventional diode in its ability to repeatedly absorb large reverse energy pulses of a joule or more. This capability results from the fact that avalanche breakdown is a bulk crystal phenomenon, whereas the edge breakdown of the conventional diode occurs due to defects in the crystal lattice. To ensure that bulk breakdown occurs first, the edges of an avalanche diode are tapered or beveled to reduce the local field intensity. The positive temperature coefficient of the avalanche voltage (approximately 0.1%/°C), coupled with the heating that results from localized reverse currents, causes the area of reverse conduction to spread throughout the entire diode. The reverse energy is thus absorbed by the complete wafer, not a small localized spot.

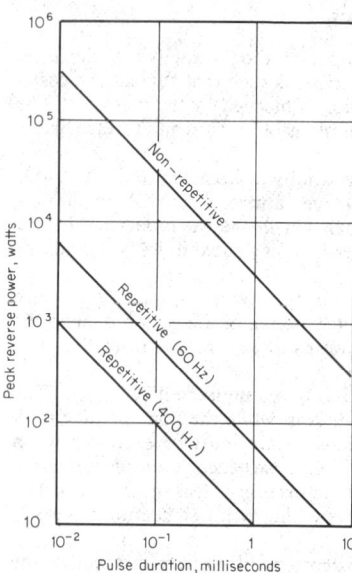

Fig. 15-2. Power-surge rating for a typical 250-A avalanche diode.

The ratings of current, forward voltage drop, reverse blocking voltage, and operating and storage temperatures for an avalanche diode are similar to those of a conventional diode of the same wafer size. The difference in reverse characteristics results in the repetitive and nonrepetitive peak-voltage ratings being replaced by reverse-pulse power ratings. These are usually given in curve form, as shown in Fig. 15-2 for a 250-A device. As can be seen, the device can absorb nearly 100 kW for 10 μs (1 J), which makes it relatively insensitive to ordinary transient or fault conditions in a circuit.

Following forward conduction, several microseconds are required for an avalanche diode to regain its reverse blocking voltage. If several diodes are connected in series, the variation in the reverse recovery times of individual diodes forces the first diodes that recover to pass the high reverse recovery current of the remaining diodes while supporting their full avalanche voltage. Such a condition would be instantly fatal to an ordinary diode but is not harmful to an avalanche diode. This characteristic can therefore greatly simplify high-voltage rectifier circuitry where series operation is required. Because of the extra steps involved in its manufacture, an avalanche diode is ordinarily more expensive than a conventional diode and thus is usually not employed unless required.

TURN-ON DEVICES

8. Three-Terminal Devices. Power at a high level can be controlled by a signal at a much lower level when a three-terminal, or triode, device is used. In general, load power flows through two of the device terminals when a control signal, referenced to one of the power terminals, is applied to the third, or control, terminal. The control signal may bear a linear relationship to the load current, as in the transistor; it may serve only to initiate conduction, as in the thyristor and triac; or it may serve both to turn on and turn off load current, as in the gate-controlled switch.

9. Thyristors (SCRs). The thyristor is a triode semiconductor device composed of four layers of silicon, as discussed briefly in Sec. 7, Par. 7-51. In contrast to the linear relation which exists between load and control currents in a transistor, the thyristor is bistable.

The four-layer structure of the thyristor is shown in Fig. 15-1. The anode and cathode terminals, whose names were carried over from vapor-rectifier-tube terminology, are connected in series with the load to which power is to be controlled. The thyristor is turned on by application of a low-power control signal between the third terminal, or gate, and the cathode.

The anode-to-cathode VI characteristic of a thyristor is also shown in Fig. 15-1 for several fixed values of gate current. The reverse characteristic is determined by the outer two junctions, which are reverse-biased in this region. With zero gate current, the forward characteristic in the off, or blocking, state is determined by the center junction, which is reverse-biased. However, if the applied voltage exceeds the forward blocking voltage, the thyristor switches to its on, or conducting, state. The effect of gate current is to lower the blocking voltage at which switching takes place.

This behavior can be explained in terms of the two-transistor analog shown in Fig. 15-3. The two transistors are regeneratively coupled so that if the sum of their current gains (α's) exceeds unity, each drives the other into saturation. In the forward blocking state, the leakage current is small, both α's are small, and their sum is less than unity. Gate current increases the current in both transistors, increasing their α's. When the sum of the two α's equals 1, the thyristor switches to its on state.

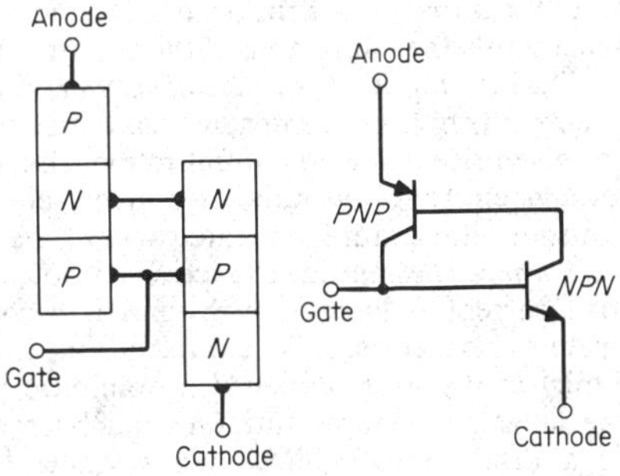

The form of the *gate-to-cathode VI characteristic* of the thyristor is similar to that of a diode. With positive gate bias, the gate-cathode junction is forward-biased and permits the flow of a large current in the presence of a low voltage drop. When negative gate voltage is applied, the gate-cathode junction is reverse-biased and prevents the flow of current until

Fig. 15-3. Two-transistor analog of a thyristor.

the avalanche breakdown voltage is reached. Recently, minute gate-cathode shorts have been intentionally introduced into the thyristor during fabrication to reduce the sensitivity of the device to unwanted dv/dt triggering. One result of the presence of the shorts is increased reverse gate-leakage current.

A summary is provided in Table 15-1 of some of the ratings which must be considered when choosing a thyristor for a given application. Both forward and reverse repetitive and nonrepetitive voltage ratings must be considered, and a properly rated device must be chosen so that the maximum voltage ratings are never exceeded. In most cases, either forward or reverse voltage transients in excess of the nonrepetitive maximum ratings result in destruction of the device.

The *maximum rms* or *average current ratings* given are usually those which cause the junction to reach its maximum rated temperature. Because the maximum current will depend upon the current waveform and upon thermal conditions external to the device, the rating is usually shown as a function of case temperature and conduction angle. The peak single half-cycle surge-current rating must be considered, and in applications where the thyristor must be protected from damage by overloads, a fuse with a I^2t rating smaller than the maximum rated value for the device must be used. Maximum ratings for both forward and reverse gate voltage, current, and power also must not be exceeded.

The maximum rated *operating junction temperature* must not be exceeded, since device performance, in particular voltage-blocking capability, will be degraded. Junction temperature cannot be measured directly but must be calculated from a knowledge of steady-state thermal resistance and the average power dissipation. For transients or surges, the transient thermal impedance must be used.

The *maximum average power dissipation* is related to the maximum rated operating junction temperature and the case temperature by the steady-state thermal resistance. In general, both the maximum dissipation and its derating with increasing case temperature are provided.

The number of thyristor characteristics specified varies widely from one manufacturer to another—whether only typical values are given or minima and maxima are also provided and whether charts or graphs showing their variations are included. Table 15-1 summarizes some of the characteristics provided.

Included are the values of conducting drop under both steady-state and turn-on transient conditions. Junction leakage and saturation currents are usually given for the maximum rated operating junction temperature. Gate conditions of both voltage and current to ensure either nontriggered or triggered device operation are included.

The *turn-on* and *turnoff transients* of the thyristor are characterized by switching times listed in Table 15-1. The turn-on transient can be divided into three intervals: gate-delay interval, turn on of initial area, and spreading interval.

The *gate-delay interval* is simply the time between application of a turn-on pulse at the gate and the time the initial area turns on. This delay decreases with increasing gate drive current and is of the order of a few microseconds.

The second interval, the time required for *turn-on of the initial area*, is quite short, less than 1 μs. In general, the initial area turned on is a small percentage of the total useful device area. After the initial area turns on, conduction spreads throughout the device by diffusion and in general fills the entire area of the thyristor in about 50 μs.

During the third, or *spreading, interval,* load current flows, but the forward drop is abnormally high until spreading has been completed. The presence of this spreading interval has given rise to a load current rate-of-rise, or di/dt, rating for a thyristor. The di/dt rating for a given device operating at a given set of conditions is governed by the maximum local hot-spot temperature generated within the device at the restricted initial area turned on.

Localized heating due to excessive load current rate of rise can be a problem at 60 Hz if the load current is large enough, and it becomes increasingly difficult as the frequency of operation is increased. Localized heating from high di/dt applications has led to a whole new family of thyristors designed to minimize this problem. These new devices, which have special gate structures, turn on a much larger initial area than conventional thyristors, and, as a result, greater di/dt can be tolerated without increasing the localized hot-spot temperature. Rated values of di/dt for this new family of thyristors are of the order of 500 A/μs, compared with about 50 A/μs for conventional thyristors.

Thyristor turnoff times range from 10 to 50 μs. Turnoff time is generally important when force-commutated turnoff is used. A thyristor which has been carrying current is force-commutated off by suddenly reducing the current to zero at a rapid rate and then applying a reverse bias. Before the device will block reverse current, a *reverse sweep-out current* will flow until all stored charge is removed. When the flow of sweep-out current stops, the device will block reverse voltage. A short time later, the reverse voltage can be reduced to zero and forward voltage reapplied at a fixed rate, or dv/dt. If forward voltage is reapplied too soon or at too rapid a rate, the thyristor will turn on immediately, causing the circuit to malfunction. Values of dv/dt ranging from 50 to 100 V/μs are acceptable for the presently available thyristors which include distributed shorts in the cathode-gate junction to bypass capacitive currents.

Both *steady-state thermal resistance* and *transient thermal impedance* are specified. In general, $R_{\theta JC}$ is given as a number, while $Z_{\theta JC}$ is presented in the form of a graph. The double-sided flat package which makes heat removal possible from both sides of the device is becoming widely accepted and is replacing the stud-mounted package in many applications. On some data sheets, vendors provide data for both single- and double-sided cooling. Care must be taken to use the appropriate data which apply to the intended method of use.

10. Triacs. The triac is equivalent to two thyristors connected together in inverse parallel. When a single thyristor is used to control power in an ac circuit, either by off-on switching or phase control, other devices, such as diodes in a bridge arrangement, must be connected around the thyristor to accommodate both directions of current flow. On the other hand, the diodes can be eliminated if a second thyristor is connected in inverse parallel across the first one.

The triac was developed to satisfy the need for ac power control with a single device. The structure of the device is shown in Fig. 15-1. The triac is basically a five-layer silicon device of the form *npnpn*, as shown. The *n* regions at the extreme ends are shorted to the adjacent *p* regions by the contact metallization. The overall result of these shorts is to minimize the effect of the *n* region, which is positively biased. As a result, the triac looks like a *pnpn* configuration independent of the polarity of the bias voltage and thus performs functionally as though two thyristors were connected in inverse parallel.

The symbol used for the triac (Fig. 15-1) is a combination of the symbols for two thyristors. Because the triac is a relatively new device, the terminology used to designate the device terminals differs from one vendor to another. Some use the terms high, common, and gate for the three terminals, while others use main terminal 1 (MT1), main terminal 2 (MT2), and gate. For convenience, the latter terminology will be used in this discussion.

A family of *triac characteristics* is shown in Fig. 15-1 with gate current as a variable. Load current may be triggered on by any combination of MT2 voltage and gate-current polarities, but triggering levels differ depending upon the combination used. If unidirectional gate-current pulses are used for applications where the MT2 voltage alternates in polarity, a preferred gate-current polarity is recommended by the vendor and should be used.

The types of ratings specified for a triac are quite similar to those used for a thyristor with a few exceptions. A list of parameters used in the rating of a thyristor is shown in Table 15-1. In the case of the triac, the rated blocking voltage V_{DRM} applies to either polarity of MT2 voltage. In addition, because the gate characteristic is symmetrical, the reverse gate-voltage ratings do not apply.

The thyristor characteristics listed in Table 15-1 apply to the triac as well, with the following exceptions. Breakover-voltage and conducting-voltage drop apply to both polarities of MT2 voltage and current instead of only one polarity, as for the thyristor. All load-current ratings must be changed to accommodate the flow of bidirectional load current. In addition, gate characteristics for both triggering and nontriggering must be changed to accommodate both polarities of gate and MT2 voltage.

The *turn-on transient characteristics* must be enlarged to include turn-on with both polarities of MT2 voltage. The turnoff transient must be characterized differently because circuit-commutated turnoff does not apply to the triac. During the turnoff transient of a thyristor, reverse bias is applied to the device following sweep-out. Such a condition is impossible to establish in a triac because it will turn on with potential of either polarity. In general, values for fall time, reverse-recovery time, and turnoff time are not specified because they are of little value in most triac applications. Owing to the special structure of the triac, the rate of reapplication of blocking voltage following termination of load current is much lower than that allowable with the thyristor. In general, the reapplied voltage of the triac must be held to 5 V/μs.

TURNOFF DEVICES

11. Power Transistors. The thyristor and triac discussed in the previous section can initiate the flow of load current by turning on as the result of the application of a low-level gating signal. Once load current flows, however, the switching device loses control and current must be made to go to zero by external means. The transistor and gate-controlled switch are capable not only of initiating the flow of current but of interrupting it also under the control of a low-level gating signal.

The transistor is a triode device composed of three layers of semiconductor material, either *npn* and *pnp*, as shown in Fig. 15-1. Transistor operation and fabrication are discussed in Sec. 7, Pars. **7-52** to **7-60**.

Most power transistor circuits, other than linear regulators, do not use the linear relationship between collector and base currents. The transistor is driven from cutoff to saturation to operate as an off-on switch, avoiding the high dissipation encountered with sustained operation in the linear region.

The rating which defines the collector-to-emitter voltage blocking capability is the *sustaining voltage*, $V_{CEO,\text{sus}}$. This rating should never be exceeded even on a transient basis, except in very special circumstances, if reliable operation is desired. A similar maximum rating applies to reverse base-to-emitter voltage V_{BEO}, which, if exceeded, will cause the emitter to fail.

Ratings for maximum peak and continuous *collector currents* are given for the most favorable conditions of ambient temperature. A rating for maximum power dissipation at 25°C case temperature is also provided, as is its variation with increasing case temperature. One of these ratings will determine the maximum allowable collector current for a given application. A similar maximum rating for *base current* is specified and must be observed.

The number and types of characteristics specified on data sheets, and whether only typical values or minima or maxima are given, vary widely between vendors. Values for reverse-biased junction *leakage currents* (collector cutoff currents) are sometimes given at room and at elevated temperatures (I_{CEX}, I_{CEO}, and I_{EBO}). In general, typical values for switching times comprising delay t_d, rise t_r, storage t_s, and fall times t_f are provided although some data sheets provide maximum limits also. Other characteristics include thermal resistance $R_{\theta JC}$, collector-to-base saturation voltage $V_{CE,\text{sat}}$, base-to-emitter saturation voltage $V_{BE,\text{sat}}$, and small-signal current gain h_{FE}.

Of special importance in switching circuits is the *safe operating area* (SOA) *characteristic* usually supplied in graphical form. If the device is not applied within the limits of the SOA curves, it may latch up during a switching transient, causing the circuit to malfunction and possibly causing device failure. For a more complete discussion of transistor SOA characteristics, see Sec. 7, Par. **7-60**, and RCA, 1971, pp. 134-145, referred to on page 15-60.

12. Gate-controlled Switch (GCS). A gate-controlled switch or *turnoff thyristor* is a four-layer triode silicon semiconductor device which resembles the thyristor in performance and structure but has the additional attribute that anode current can be turned off by control of the gate. It is effectively a semiconductor analog to an electromechanical toggle switch.

The basic four-layer structure of the GCS, shown in Fig. 15-1, is identical to that of the thyristor. The difference in performance arises from the differences in geometry of the cathode and gate designs on the upper surface of the device. In the design of the GCS, the gate structure is distributed throughout the cathode to permit the gate to exert maximum influence on all parts of it. This is not true of the thyristor, where only a small area of the cathode is in intimate contact with the gate.

The form of the *VI* characteristic of the GCS is the same as that of the thyristor, as shown in Fig. 15-1. The parameters used to rate a GCS are similar to those used for the thyristor with several exceptions. All those shown in Table 15-1 apply to the GCS also. For the GCS, however, it is also necessary to rate the maximum peak anode current which can be interrupted by a signal applied to the gate. When this maximum anode current is exceeded, gate current no longer can cause anode current to cease and the device may be destroyed.

All the characteristics shown in Table 15-1 for thyristors apply to the GCS except reverse recovery time and circuit-commutated turnoff time. For the GCS, turnoff is initiated by the device itself, and, in general, anode voltage does not reverse but builds up in the forward direction as anode current goes to zero.

To the characteristics listed, a *turnoff gain* must be added. This characteristic is the ratio of anode current to required gate current when the turnoff transient begins. Because the value of this characteristic depends upon many factors, including the level of anode current and the length of time reverse gate current has been applied, the conditions under which it is measured must be specified.

NATURALLY COMMUTATED CIRCUITS

AC-DC CONVERTERS

By B. R. Pelly

13. General Considerations. The basic feature common to all the ac-dc converters described in the following paragraphs is that they are connected to a source of ac voltage which causes natural commutation. In most cases, power flow is from the ac terminals to a dc load, and the process is known as *rectification*. However, some members of this general family of converters can be controlled so that power flow occurs from the dc terminals back into the ac line. This process is known as *synchronous inversion* to distinguish it from inversion into a passive ac load. In the latter case, forced commutation is usually required.

The simplest member of this family of converters is the well known *half-wave single-phase rectifier*. Although widely used for low-current dc power supplies, this type of rectifier is not used for higher power because of the large ripple voltage in its output and because the unidirectional current causes dc magnetization of transformer cores.

The number of different converter circuits is very large. Two basic configurations should first be distinguished, the *bridge circuit* (also known as double-way circuit) and the *midpoint circuit* (also known by the names center tap and single way).

Converter circuits are also distinguished according to whether the ac line is *single-phase* or *3-phase*. The effective number of phases of a 3-phase line can be further increased by connecting transformer windings to give intermediate phase shifts. Increasing the number of phases increases the *ripple frequency* of both the dc output voltage and the ac line current, making filtering easier. The ratio of the fundamental ripple frequency of the dc voltage to the ac line frequency is known as the *pulse number*.

If all the switching devices are diodes, the converter can operate only as an *uncontrolled rectifier* with the average dc output voltage fixed by the input ac voltage and by the circuit configuration. If half of the diodes in a bridge are replaced by thyristors, the average dc output voltage can be controlled by changing the phase angle at which the thyristors are fired, but the circuit is still capable only of rectifier action with power flow from the ac terminals to the dc terminals. Such circuits, known as *half-controlled converters* or *semiconverters,* belong to the category of *one-quadrant converters* since only one polarity of dc voltage and one polarity of dc current are possible.

Replacement of the diodes in a rectifier by thyristors produces a fully controlled converter, or *two-quadrant converter.* This type permits dc current to flow in only one direction, but the dc voltage may have either polarity. With one polarity, power flows from the ac to the dc terminals, and the converter acts as a *rectifier.* With the opposite polarity of dc voltage, net power flows from the dc terminals to the ac terminals, and the circuit acts as a *synchronous inverter.*

In some applications both polarities of both dc current and dc voltage must be permitted. This *four-quadrant action* can be achieved by interconnecting two similar two-quadrant converters, the combination being known as a *dual converter.*

The applications for this family of naturally commutated converters embrace a very wide range, including dc power supplies for electronic equipment, battery chargers, and speed controllers for fractional-horsepower motors, as well as dc supplies delivering many thousands of amperes for electrochemical and other industrial processes, high-performance reversing drives for dc machines rated at hundreds of horsepower, and high-voltage dc transmission in the megawatt power region.

Throughout the discussion which follows, unless otherwise stated, the following simplifying assumptions are made:

1. The voltage drop across switching devices is neglected while they are conducting, and the leakage current is neglected while they are blocking. Stray resistances are neglected.

2. Device turn-on and turnoff occur instantaneously.

3. The dc terminals are connected to an ideal filter (an infinite inductance), which suppresses all ripple current.

Ac-dc converters are treated in greater detail in Pelly (1971) and Schaefer (1965).

14. Two-Quadrant Converters. The circuit configurations, waveforms, and design relationships for various one- and two-quadrant converters are tabulated in subsequent paragraphs. In this paragraph, the operation of several typical circuits is discussed.

Two-Pulse Midpoint Circuit. Figure 15-4 shows a single-phase two-pulse midpoint converter and the associated source and load waveforms for various values of the firing delay angle α. For $\alpha = 0$, the converter is equivalent to an uncontrolled rectifier, and the thyristors could be replaced by diodes. During the positive half cycle of the supply voltage, thyristor Th1 and transformer secondary S1 carry the load current, the voltage across the load is v_{s_1}, and thyristor Th2 is reverse-biased. During the negative half cycle, Th2 and S2 carry the load current, the load voltage is v_{s_2}, and Th1 is reverse-biased. The load-voltage waveform consists of a direct component V_{d0} plus a superimposed ac ripple having a fundamental frequency which is twice the supply frequency (hence the name, two-pulse). The fundamental component of the supply current is in phase with the supply voltage.

As α is increased in the range $0 < \alpha < 90°$, the delay in firing causes the average load voltage to decrease, as shown in Fig. 15-4b. Note that the assumption of smooth load current, i.e., constant throughout the cycle, implies a highly inductive load.[1] When the instantaneous load voltage goes negative, the reactive emf of the inductance forces power back into the source in order to maintain the current constant. However, over a half cycle, the net power flow is from the ac source to the dc load, and the supply current has a lagging power factor.

When α becomes equal to 90°, the instantaneous load voltage is negative for as long as it is positive, so that the average dc component of load voltage is zero (see Fig. 15-4c), and the supply current lags the supply voltage by 90°.

---

[1] Under this assumption, Th1 cannot cease conduction until Th2 is fired. Therefore each thyristor conducts for 180°. If, however, the load is purely resistive, each thyristor will cease conduction when its half of the supply voltage goes negative, and the current will pulsate.

If α is increased beyond 90°, the continuous current flow can be maintained only if an external negative dc source is connected to the dc terminals. Net power flow is from the dc terminals to the ac terminals, and the converter is performing synchronous inversion (see Fig. 15-4d). Since the polarity of the average dc voltage has reversed, operation has shifted from quadrant I to quadrant IV. But because the current cannot reverse, quadrants II and III are forbidden. Hence this is a two-quadrant converter.

Finally, in Fig. 15-4e, α is nearly equal to 180°, and the dc voltage approaches its maximum negative value. In practice, α must be limited to about 160° or less to permit the thyristor which is being commutated off to regain its blocking ability before forward voltage is reapplied to it. Otherwise a commutation failure occurs. Operation in the inversion region is frequently described in terms of the advance angle $\beta = 180° - \alpha$. The margin of safety from commutation failure is described by the recovery angle δ between the completion of commutation and the next zero crossing at which forward voltage is reapplied.

Three-Pulse Midpoint Circuit. The simplest type of phase-controlled converter which operates from a 3-phase supply is the three-pulse midpoint circuit, shown in Fig. 15-5 with

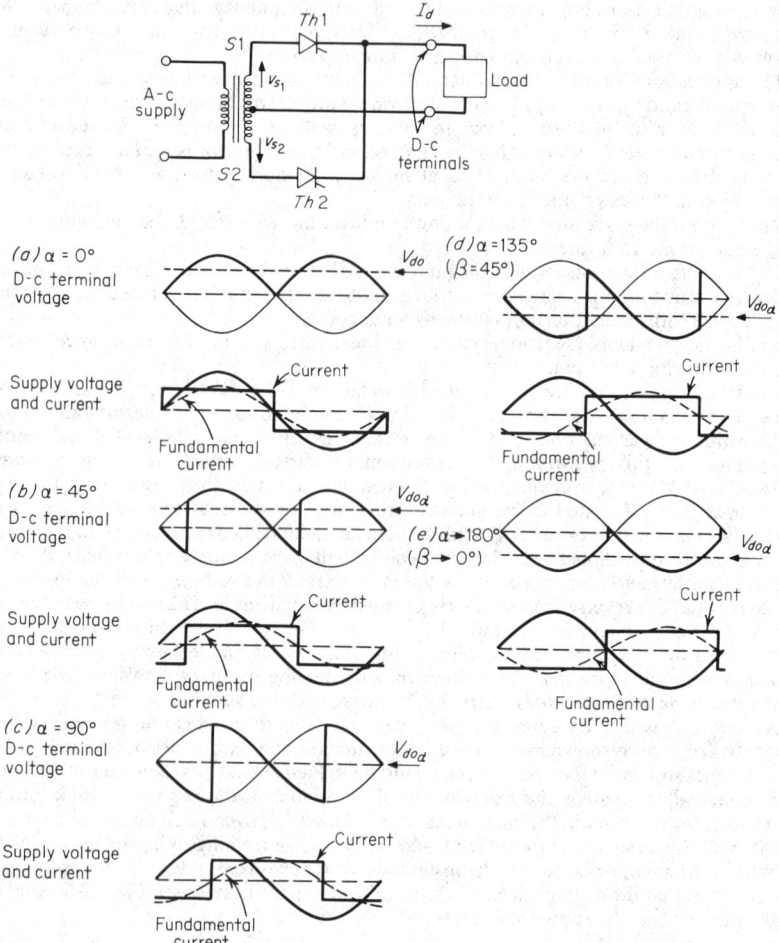

Fig. 15-4. Two-pulse midpoint converter circuit and associated waveforms (smooth direct current assumed).

idealized waveforms. The zigzag connection of the transformer secondary windings prevents dc magnetization of the transformer core by permitting equal and opposite currents to flow in the two secondary windings in each phase.

The waveforms illustrate that this circuit has the same basic operating characteristics as the two-pulse circuit of Fig. 15-4. That is, continuous control of the mean dc terminal voltage from maximum positive to maximum negative is achieved by controlling the phase of the thyristor firing pulses through a theoretical range of 180°. This is accompanied by a continuously increasing shift in the phase of the input current from 0 to 180° lagging. In fact, these characteristics are common to all two-quadrant converters.

Six-Pulse Midpoint Circuit. The outputs of two three-pulse converters having mutually displaced input voltages can be combined in parallel through an interphase reactor as shown in Fig. 15-6. Each three-pulse converter operates independently of the other. In the ideal case, the load current is shared equally between the two groups, and there is no dc magnetization of the core of the interphase reactor. In practice, some relatively small unbalance of currents may occur.

Because of the phase displacement between the ac ripple voltages at the dc terminals of

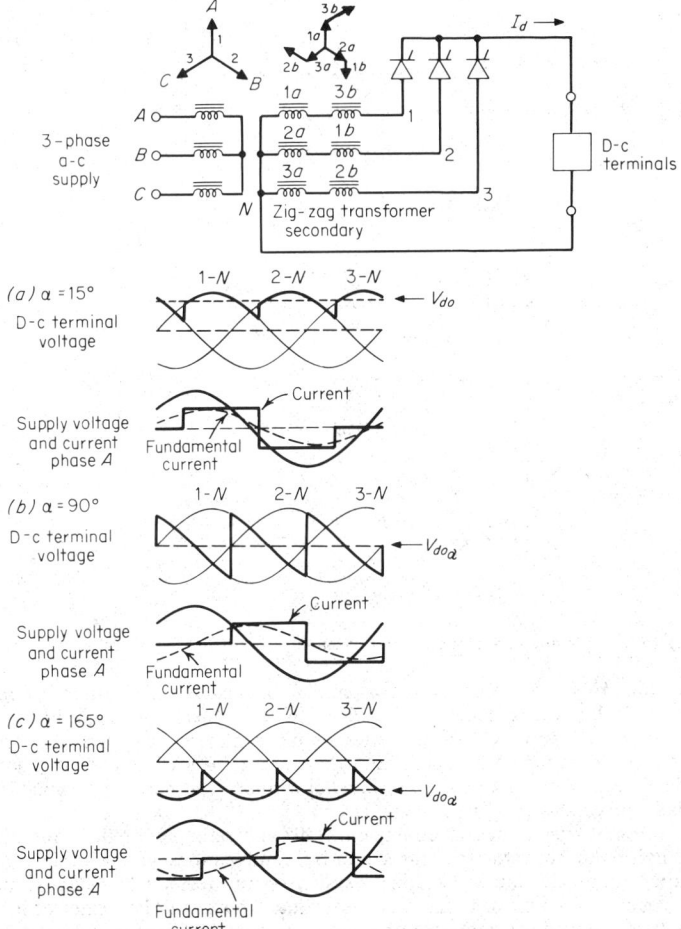

Fig. 15-5. Three-pulse midpoint converter circuit and associated theoretical waveforms.

the individual groups, the fundamental frequency of the ripple in the output voltage is 6 times the input frequency. The fundamental frequency of the ripple voltage across the interphase reactor is 3 times the input frequency.

If the interphase reactor is eliminated by making a solid connection between the dc terminals of the three-pulse groups, a six-pulse voltage waveform is still obtained at the output. However, the utilization factor of the circuit decreases because each thyristor then conducts for only 60° instead of the previous 120°.

Six-Pulse Bridge Circuit. The dc terminals of two three-pulse groups can be connected in series with one another to give an overall six-pulse operation. The resulting bridge, one of the most commonly used converter circuits, is shown in Fig. 15-7. So far as the ac lines are

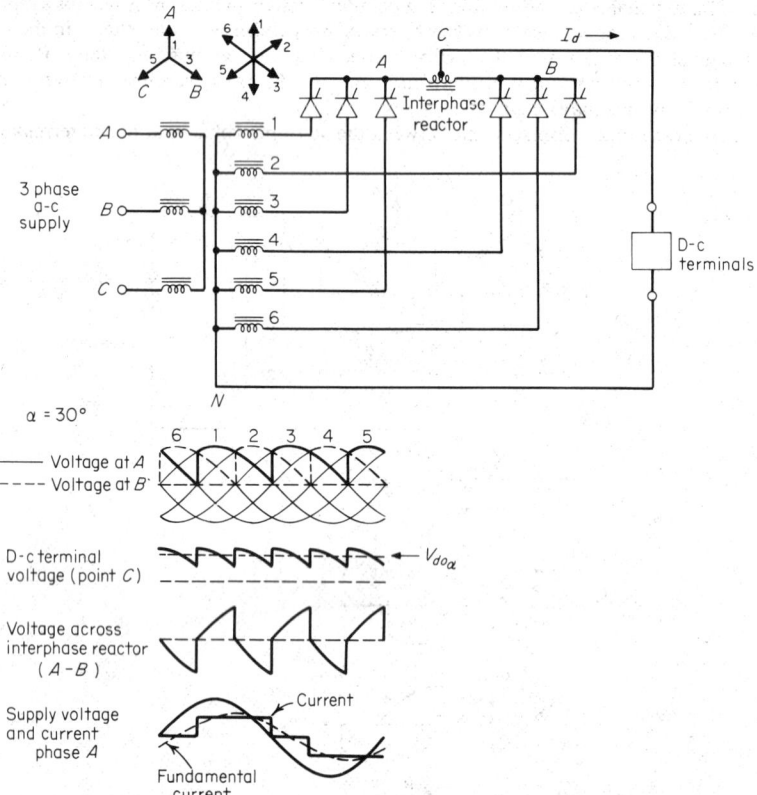

Fig. 15-6. Six-pulse midpoint converter circuit and associated theoretical waveforms.

concerned, the bridge circuit contains two similar oppositely poled groups of rectifying devices; thus, it draws a balanced current from the line, ideally with no dc component.

Higher Pulse Numbers. Other converter-circuit configurations having higher pulse numbers can be constructed by connecting the dc terminals of individual groups, with suitably displaced ac voltages, in series or parallel with one another, or by combining series and parallel connections into one system.

In practice, a thyristor conduction angle of 120° is greatly preferred. Thus, almost all practical multipulse converter circuits comprise combinations of the basic three-pulse commutating groups. Each group within the system operates essentially independently of all the other groups. When the dc terminals of individual groups must be connected in parallel, the connections are made through interphase reactors to maintain independent operation of the groups. On the other hand, groups can be connected in series with "solid" connections at the dc terminals. Series connections of bridges, however, require isolation between the transformer secondaries connected to the individual bridges.

15. One-Quadrant Converters. Many applications require operation with only one polarity of dc output voltage; i.e., they operate only in the rectifying mode. In this case, it is generally advantageous to connect uncontrolled diodes into certain parts of the circuit.

In bridge-connected circuits (but not midpoint circuits) uncontrolled diodes can be used in place of half of the thyristors. With this *half-controlled converter,* it is possible to control the mean dc terminal voltage continuously from maximum to virtually zero, but reversal of the mean voltage is not possible.

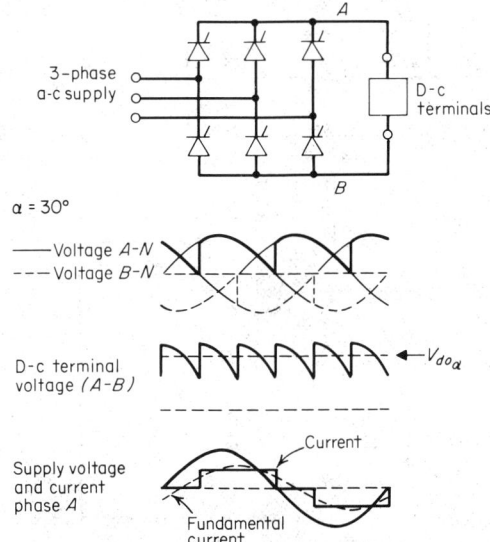

Fig. 15-7. Six-pulse bridge converter and associated theoretical waveforms.

The half-controlled bridge has economic advantages over the fully controlled circuit because diodes are less expensive than equivalent thyristors. In addition, the input-power factor at relatively low levels of output voltage is improved over that of a fully controlled converter. Except for a single-phase bridge circuit, however, this advantage is obtained at the expense of a 2:1 reduction in ripple frequency at the dc terminals.

Either bridge or midpoint two-quadrant converters can be limited to one-quadrant operation by connecting a *freewheeling diode* across the dc terminals to conduct when the terminal voltage instantaneously tends to go negative. The diode reduces the ripple and improves the input-power factor for low dc output voltages. A further feature of the freewheel diode is that it provides a bypass path for inductive load currents if the supply lines become disconnected, thereby preventing reverse-voltage surges.

Both the half-controlled converter and the fully controlled converter with freewheeling diode have the advantage that the ratio of input current to dc output current decreases as the output voltage is reduced toward zero. In an ideal two-quadrant converter this ratio remains constant.

16. Commutation Overlap. The preceding discussion and the idealized waveforms that have been shown assume instantaneous commutation of current from one thyristor which is turning off to the next which is being turned on. In practice, circuit inductance causes conduction in the two devices to overlap for a time that is usually not negligible. The process is known as commutation overlap, and its duration relative to the period of a cycle is expressed in terms of the overlap angle u.

The physical explanation of commutation overlap depends upon the fundamental voltage-current relationship of an inductor, $\Delta i = (1/L) \int v\,dt$, which states that the change in current is equal to the voltage-time area, i.e., integral, divided by the inductance.[1]   Hence transformer-leakage inductance and inductance in the ac line introduce a delay until the voltage-time area is sufficient to bring about the necessary redistribution of currents.   During this delay, the current in the thyristor being turned on increases, and that in the thyristor turning off decreases at the same rate, since the total current is constant.   If each thyristor has an equal series inductance, during the overlap the dc terminal voltage will be the average of the two source voltages to which the thyristors are connected.

The effects of commutation overlap are illustrated in Fig. 15-8 for a three-pulse group having inductance L in series with each thyristor.   The voltage at the dc terminal is shown

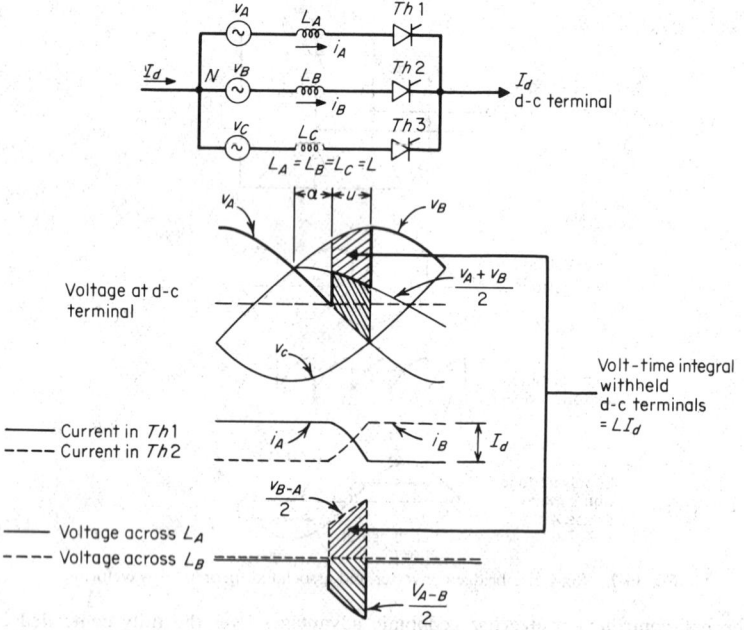

Fig. 15-8.   The commutation process for a three-pulse group.

as the current commutates from Th1 to Th2 after a phase delay α.   The voltage-time area required to change the current in L_B from zero to I_d is shown shaded and is subtracted from the ideal output-voltage waveform.   The average value of voltage withheld from the dc terminals is directly proportional to the product of the direct current and the inductance, and it is independent of the firing-delay angle.   Thus during the overlap, the output voltage follows the curve $(v_A + v_B)/2$.

In general, the relationship between the firing-delay angle α and the overlap angle u for a three-pulse commutating group is

$$\cos \alpha - \cos(u + \alpha) = \sqrt{\frac{2}{3}} \frac{X_c I_d}{V_s}$$

For inverter operation, the corresponding relationship between the advance angle β and the recovery angle $\delta = \beta - u$ is

$$\cos \delta - \cos \beta = \sqrt{\frac{2}{3}} \frac{X_c I_d}{V_s}$$

---

[1] This relationship is widely useful in analyzing the smoothing action of an interphase or filter reactor and other aspects of converter operation.

Table 15-3. Letter Symbols Used in the Analysis of Converter Circuits

V_S	Rms value of phase-to-neutral voltage at converter input terminals
$\hat{V}_s$	Peak value of V_s
V_n	Rms value of phase-to-neutral voltage at primary converter transformer
h	Ratio of V_n to V_S
V_d	Average value of voltage at dc terminals of converter under load, at any firing angle
$V_{d_{0\alpha}}$	Average value of voltage at dc terminals of converter at firing angle α, with no commutation overlap
$V_{d_{max}}$	Maximum possible average value of voltage at dc terminals of converter, obtained at $\alpha = 0°$, with no commutation overlap
V_{FB}	Maximum instantaneous value of forward blocking voltage applied across thyristor
V_{RB}	Maximum instantaneous value of reverse blocking voltage applied across thyristor
V_D	Maximum instantaneous value of reverse blocking voltage applied across diode
r	Ratio of V_d to $V_{d_{max}}$
I_d	Direct current at output of converter
I_1	Rms value of the fundamental component of converter input line current
I_{1P}	Rms value of the "in-phase" or "power" component of I_1
I_{1Q}	Rms value of the "quadrature" or "reactive" component of I_1
$I_{av,Th}\ (I_{av,D})$	Average value of thyristor (diode) current
$I_{rms,Th}\ (I_{rms,D})$	Rms value of thyristor (diode) current
P_d	Average power at output of converter
P_{d_0}	Theoretical average power at output of converter, at $\alpha = 0°$ with no commutation overlap
VA_0	Theoretical rms volt-amperes of transformer windings at $\alpha = 0°$ with no commutation overlap
L_s	Line-to-neutral commutating inductance at transformer secondary
L_p	Line-to-neutral commutating inductance at transformer primary
X_c	Commutating reactance at input frequency, referred to transformer secondary
α	Converter firing-delay angle, measured from the point at which the converter operates as if it were an uncontrolled rectifier circuit
β	Inverter advance angle; angle in advance of the zero crossing of the line-to-line commutating voltage at which the commutation is initiated: $\beta = 180° - \alpha$
δ	Inverter recovery angle; angle in advance of the zero crossing of the line-to-line commutating voltage at which the commutation is completed: $\delta = 180° - (\alpha + u)$
u	Commutation overlap angle
u^*	Overlap angle for commutation of current into freewheeling path
ϕ	Displacement angle between fundamental component of converter input current and associated line-to-neutral voltage
$\cos \phi$	Displacement factor of fundamental component of converter input current: $\cos \phi = I_{1P}/\sqrt{I_{1P}^2 + I_{1Q}^2}$
λ	Power factor at a given point in the converter input circuit; ratio of the average power to the rms volt-amperes
μ	Distortion factor of the current at a given point in the converter input circuit; ratio of the rms value of the fundamental component to the total rms value $\mu = \lambda/(\cos \phi)$
p	Pulse number of converter = ratio of fundamental output ripple frequency to ac supply frequency (with steady delay angle)
ω	Angular frequency of input supply

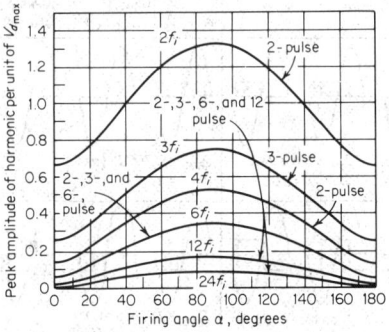

Fig. 15-9. Variation with firing angle of the principal harmonic components present in the dc terminal voltage of various two-quadrant converters with continuous conduction and no commutation overlap. *(From B. R. Pelly, "Thyristor Phase-controlled Converters and Cycloconverters," Wiley-Interscience, New York, 1971. Used by permission.)*

POWER ELECTRONICS

For definitions of X_c, I_d, and V_s, see Table 15-3. These relationships are also valid for multipulse converters consisting of noninteracting three-pulse groups.

At the input side of a converter, the effect of commutation overlap is to cause rounding of the edges of the waveforms of line current. This means that the amplitudes of the higher-order harmonic terms are progressively reduced, compared with the theoretical amplitudes of these components with no overlap. In addition, the duration of each segment of the waveform of the input current is stretched by the overlap angle, resulting in a slight additional lagging phase shift of the fundamental component of current.

17. Waveforms and Data for Converter Circuits. Table 15-3 lists the letter symbols most

Table 15-4. Waveforms and Data for Various Rectifier and Converter Circuits

	2-pulse midpoint rectifier	2-pulse midpoint converter (2-quadrant)	2-pulse midpoint converter with freewheel diode (1-quadrant)	2-pulse bridge rectifier	2-pulse bridge converter (2-quadrant)
Circuit	Circuit same as converter, with diodes instead of thyristors			Circuit same as converter, with diodes instead of thyristors	
D-c terminal voltage		$V_d = \frac{2}{\pi}\hat{V}_s \cos\alpha - I_d \frac{X_c}{\pi}$ $X_c = \omega(\frac{2L_p}{h^2}+L_S)$ For harmonic distortion, see Fig. 9	$V_d = \frac{2}{\pi}\hat{V}_s(\frac{1+\cos\alpha}{2}) - I_d\frac{X_c}{\pi}$ $X_c = \omega(\frac{2L_p}{h^2}+L_S)$ For harmonic distortion, see Fig.10		$V_d = \frac{2}{\pi}\hat{V}_s \cos\alpha - I_d\frac{2X_c}{\pi}$ $X_c = \omega(\frac{L_p}{h^2}+L_S)$ For harmonic distortion, see Fig
Device voltage and current		$V_{FB}=2.0\hat{V}_s$ $I_{AV_{Th}}=0.5I_d$ $V_{RB}=2.0\hat{V}_s$ $I_{RMS_{Th}}=0.707I_d$	$V_{FB}=\hat{V}_s$ $\binom{I_{AV_{Th}}=0.5I_d}{I_{RMS_{Th}}=0.707I_d}\binom{I_{AV_D}=I_d}{I_{RMS_D}=I_d}$ $V_{RB}=2\hat{V}_s$ $V_D=2\hat{V}_s$ $(\alpha=0)$ $(\alpha=\pi)$		$V_{FB}=\hat{V}_s$ $I_{AV_{Th}}=0.5I_d$ $V_{RB}=\hat{V}_s$ $I_{RMS_{Th}}=0.707I_d$
Transformer secondary voltage and current		$I_{RMS}=0.707I_d$ $\cos\phi=\cos\alpha=r$ $VA_o=1.57P_{do}$ $\lambda=0.637r$	$I_{RMS}=0.707I_d\ (\alpha=0)$ $VA_o=1.57P_{do}$ $\cos\phi=\cos\frac{\alpha}{2}=\sqrt{r}$ λ see Fig.12		$I_{RMS}=I_d$ $\cos\phi=\cos\alpha=$ $VA_o=1.11P_{do}$ $\lambda=0.9r$
Transformer primary voltage and current		$I_{RMS}=I_d/h$ $\cos\phi=\cos\alpha=r$ $VA_o=1.11P_{do}$ $\lambda=0.9r$ For harmonic distortion, see Fig.13	$I_{RMS}=I_d/h\ (\alpha=0)$ $VA_o=1.11P_{do}$ $\cos\phi=\cos\frac{\alpha}{2}=\sqrt{r}$ λ-see Fig.12 For harmonic distortion, see Fig.13		$I_{RMS}=I_d/h$ $\cos\phi=\cos\alpha$ $VA_o=1.11P_{do}$ $\lambda=0.9r$ For harmonic distortion, see Fig

frequently used in the analysis of converter circuits. Tables 15-4 to 15-6 summarize the idealized waveforms and design relationships for the more common single-phase and 3-phase, one- and two-quadrant converters.

The relationships between firing angle and the principal harmonic components in the dc terminal voltage of these converters are shown in Figs. 15-9 and 15-10.

For all two-quadrant converters, the input-displacement factor is equal to the dc voltage ratio; and for all half-controlled converters, the input-displacement factor is equal to the square root of the dc voltage ratio. These relationships are indicated in the tables. The corresponding relationships for the converter circuits with freewheel diodes are illustrated in Fig. 15-11.

2-pulse half-controlled bridge converter (1-quadrant)	3-pulse midpoint rectifier	3-pulse midpoint converter (2-quadrant)	3-pulse midpoint converter with freewheel diode (1-quadrant)
$V_d = \frac{2}{\pi}\hat{V}_S - I_d \frac{2X_c}{\pi}$ (α=0) $= \frac{2}{\pi}\hat{V}_S \frac{(1+\cos\alpha)}{2} - I_d \frac{X_c}{\pi}$ (α ≥ u") $X_c = \omega(\frac{L_p}{h2} + L_S)$ For harmonic distortion, see Fig.10	Circuit same as converter, with diodes instead of thyristors	$V_d = \frac{3\sqrt{3}}{2\pi}\hat{V}_S \cos\alpha - I_d \frac{3X_c}{2\pi}$ $X_c = \omega(\frac{L_p}{h2} + L_S)$ For harmonic distortion, see Fig.9	$V_d = \frac{3\sqrt{3}}{2\pi}\hat{V}_S \cos\alpha - I_d \frac{3X_c}{2\pi}$ (0≤α≤$\frac{\pi}{6}$) $= \frac{3\sqrt{3}}{2\pi}\hat{V}_S \frac{[1+\cos(\alpha+\frac{\pi}{6})]}{\sqrt{3}} - I_d \frac{3X_c}{2\pi}$ $X_c = \omega(\frac{L_p}{h2} L_S)$ | ($\frac{\pi}{6} \le \alpha \le \frac{5\pi}{6}$) For harmonic distortion, see Fig.10
$V_{FB} = \hat{V}_S$ ($I_{AVTh}=0.5I_d$) ($I_{AV}=I_d$) $V_{RB} = \hat{V}_S$ ($I_{RMSTh}=0.707I_d$) ($I_{RMS}=I_d$) $V_D = \hat{V}_S$ ⌐←(α=0) ←(α=π)		$V_{FB}=1.732\hat{V}_S$ $I_{AVTh}=0.333I_d$ $V_{RB}=1.732\hat{V}_S$ $I_{RMSTh}=0.577I_d$	$V_{FB}=\hat{V}_S$ ($I_{AVTh}=0.333I_d$) ($I_{AVD}=I_d$) $V_{RB}=1.732\hat{V}_S$ ($I_{RMSTh}=0.577I_d$) ($I_{RMSD}=I_d$) $V_D=1.732\hat{V}_S$ ←(0≤α≤$\frac{\pi}{6}$) ←(α=π)
$I_{RMS}=I_d$ (α=0) $VA_0 = 1.11 P_{do}$ $\cos\phi=\cos\frac{\alpha}{2}=\sqrt{r}$ λ—see Fig.12		$I_{RMS} = 0.577I_d$ $\cos\phi=\cos\alpha=r$ $VA_0 = 1.71P_{do}$ λ =0.585r	$I_{RMS}=0.577I_d$ (0≤α≤$\frac{\pi}{6}$) $VA_0 = 1.71P_{do}$ cosφ,λ,see Fig.11,12
$I_{RMS}=I_d/h$ (α=0) $VA_0 = 1.11P_{do}$ $\cos\phi=\cos\frac{\alpha}{2}=\sqrt{r}$ λ- see Fig.12 For harmonic distortion, see Fig.13		$I_{RMS}=0.272 I_d/h$ $\cos\phi=\cos\alpha=r$ $VA_0 = 1.21P_{do}$ λ=0.827r For harmonic distortion, see Fig.13	$I_{RMS}=0.272 I_d/h$ (0≤α≤$\frac{\pi}{6}$) $VA_0 = 1.21P_{do}$ cosφ,λ see Fig.10,11 For harmonic distortion, see Fig.13

Table 15-5.   Waveforms and Data for Various Rectifier and Converter Circuits

	6-pulse midpoint rectifier	6-pulse midpoint converter (2-quadrant)	6-pulse midpoint converter with freewheel diode(1-quadrant)	6-pulse bridge rectifier
Circuit	Circuit same as converter, with diodes instead of thyristors	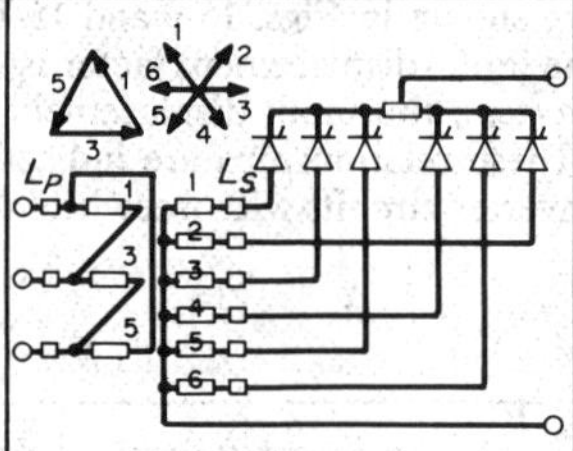	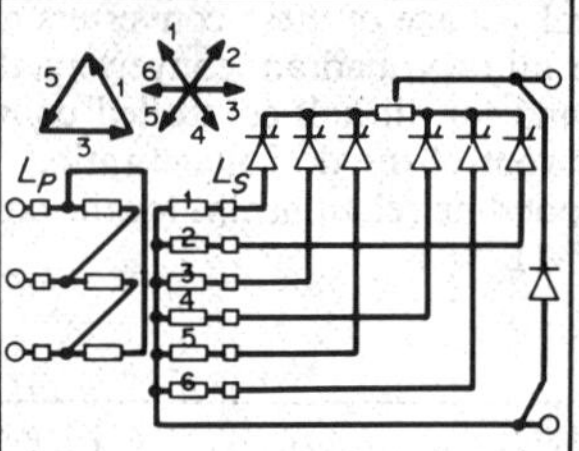	Circuit same as converter, with diodes instead of thyristors
D-c terminal voltage	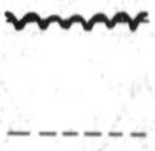	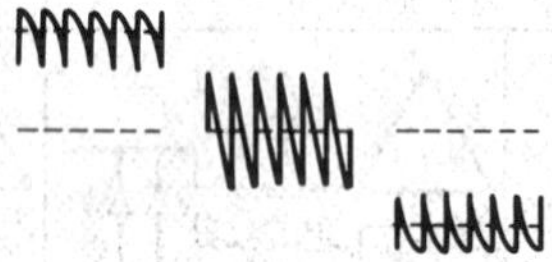 $V_d = \dfrac{3\sqrt{3}}{2\pi}\hat{V_S}\cos\alpha - I_d\dfrac{3X_c}{4\pi}$ $X_c = \omega\left(\dfrac{L_p}{h^2} + L_s\right)$ For harmonic distortion, see Fig. 9	 $V_d = \dfrac{3\sqrt{3}}{2\pi}\hat{V_S}\cos\alpha - I_d\dfrac{3X_c}{4\pi}\ \left(0\leqslant\alpha\leqslant\dfrac{\pi}{3}\right)$ $= \dfrac{3\sqrt{3}}{2\pi}\hat{V_S}\left[1+\cos\left(\alpha+\dfrac{\pi}{3}\right)\right] - I_d\dfrac{3X_c}{2\pi}$ $X_c = \omega\left(\dfrac{L_p}{h^2}+L_s\right)\ \left\|\ \left(u^* + \dfrac{\pi}{3}\leqslant\alpha\leqslant 2\dfrac{\pi}{3}\right)\right.$ For harmonic distortion, see Fig.10	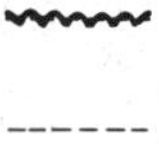
Device voltage and current	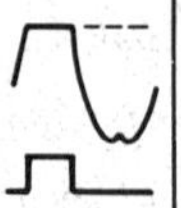	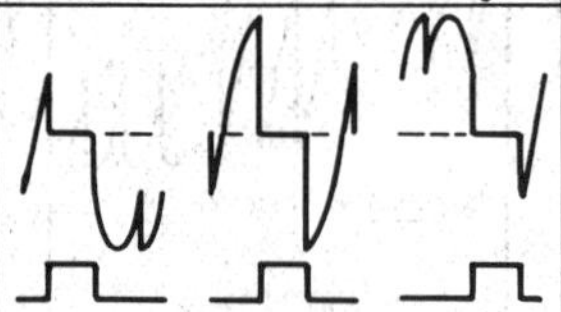 $V_{FB} = 1.732\,\hat{V_S}$ $I_{AVTh} = 0.167\,I_d$ $V_{RB} = 1.732\,\hat{V_S}$ $I_{RMSTh} = 0.288\,I_d$	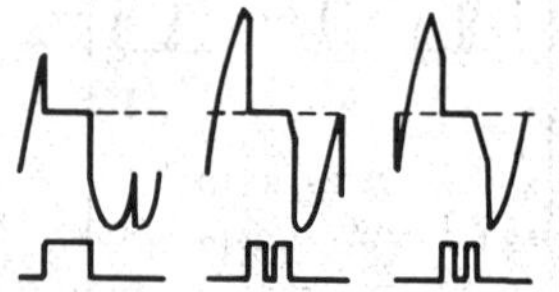 $V_{FB} = 1.5\,\hat{V_S}$ $\{I_{AVTh} = 0.167\,I_d\ \{I_{AVD} = I_d$ $V_{RB} = 1.732\,\hat{V_S}\ [I_{RMSTh} = 0.288\,I_d\ \{I_{RMSD} = I_d$ $(0\leqslant\alpha\leqslant\dfrac{\pi}{3})\quad (\alpha=\pi)$	
Transformer secondary voltage and current		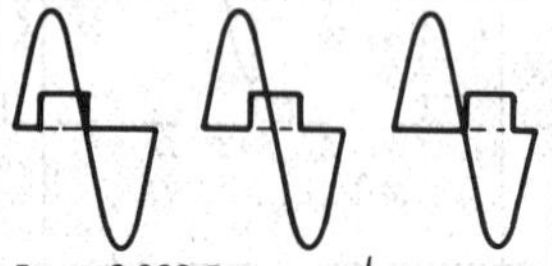 $I_{RMS} = 0.288\,I_d$ $\cos\phi = \cos\alpha = r$ $VA_o = 1.48\,P_{do}$ $\lambda = 0.675\,r$	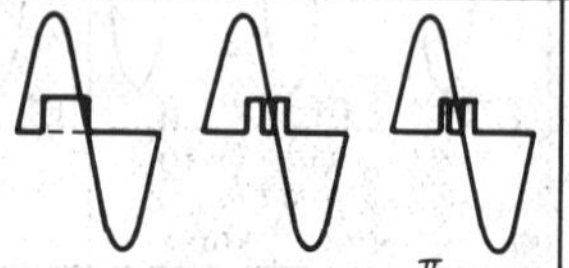 $I_{RMS} = 0.288\,I_d$ $\left(0\leqslant\alpha\leqslant\dfrac{\pi}{3}\right)$ $VA_o = 1.48\,P_{do}$ $\cos\phi,\lambda$,see Fig.11,12	
Transformer primary voltage and current		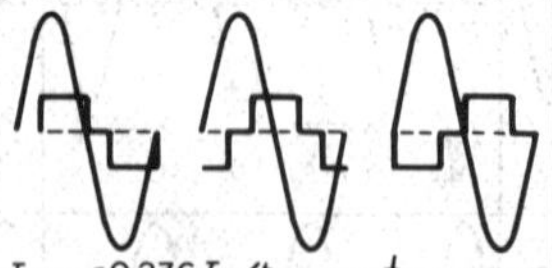 $I_{RMS} = 0.236\,I_d/h$ $\cos\phi = \cos\alpha = r$ $VA_o = 1.05\,P_{do}$ $\lambda = 0.955\,r$ For harmonic distortion, see Fig.13	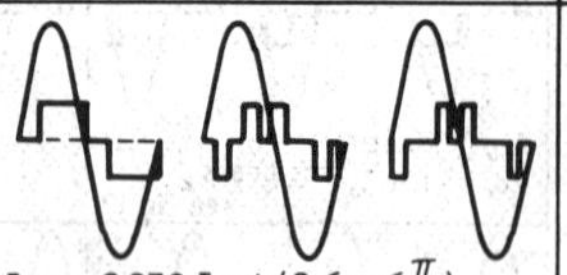 $I_{RMS} = 0.236\,I_d/h$ $\left(0\leqslant\alpha\leqslant\dfrac{\pi}{3}\right)$ $VA_o = 1.05\,P_{do}$ $\cos\phi,\lambda$,see Fig. 11,12 For harmonic distortion, see Fig 13.	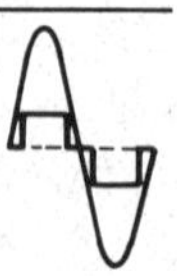

6-pulse bridge converter (2-quadrant)	6-pulse bridge converter with freewheel diode (1-quadrant)	3-pulse half-controlled bridge converter (1-quadrant)
 $V_d = \dfrac{3\sqrt{3}}{\pi}\,\hat{V}_S\cos\alpha - I_d\dfrac{3X_c}{\pi}$ $X_c = \omega\left(\dfrac{L_p}{h^2}+L_s\right)$ For harmonic distortion, see Fig. 9	 $V_d = \dfrac{3\sqrt{3}}{\pi}\,\hat{V}_S - I_d\dfrac{3X_c}{\pi}$ $(0\leqslant\alpha\leqslant\frac{\pi}{3})$ $= \dfrac{3\sqrt{3}}{\pi}\,\hat{V}_S[1+\cos(\alpha+\frac{\pi}{3})]-I_d\dfrac{6X_c}{\pi}$ $X_c=\omega\left(\dfrac{L_p}{h^2}+L_s\right)$ $(u^*+\frac{\pi}{3}\leqslant\alpha\leqslant 2\frac{\pi}{3})$ For harmonic distortion, see Fig. 10	 $V_d = \dfrac{3\sqrt{3}}{\pi}\,\hat{V}_S\dfrac{(1+\cos\alpha)}{2} - I_d\dfrac{3X_c}{\pi}$ $X_c=\omega\left(\dfrac{L_p}{h^2}+L_s\right)$ For harmonic distortion, see Fig. 10
 $V_{FB}=1.732\,V_S$ $I_{AV_{Th}}=0.333\,I_d$ $V_{RB}=1.732\,V_S$ $I_{RMS_{Th}}=0.577\,I_d$	 $V_{FB}=1.5\,\hat{V}_S$ $\left(\begin{array}{l}I_{AV_{Th}}=0.333\,I_d\\[2pt]I_{RMS_{Th}}=0.577\,I_d\\[2pt]\scriptstyle(0\leqslant\alpha\leqslant\frac{\pi}{3})\end{array}\right.$ $\left.\begin{array}{l}I_{AV_D}=I_d\\[2pt]I_{RMS_D}=I_d\\[2pt]\scriptstyle(\alpha-\pi)\end{array}\right\}$ $V_{RB}=0.732\,\hat{V}_S$	 $V_{FB}=1.732\,V_S$ $I_{AV_{Th}}=0.333\,I_d$ $I_{AV_D}=0.333\,I_d$ $V_{RB}=1.732\,V_S$ $I_{RMS_{Th}}=0.577\,I_d$ $I_{RMS_D}=0.577\,I_d$
 $I_{RMS}=0.817\,I_d$ $\cos\phi=\cos\alpha=r$ $VA_o=1.05\,P_{do}$ $\lambda=0.955\,r$	 $I_{RMS}=0.817\,I_d$ $(0\leqslant\alpha\leqslant\frac{\pi}{3})$ $VA_o=1.05\,P_{do}$ $\cos\phi,\lambda$, see Fig. 11,12	 $(0\leqslant\alpha\leqslant\frac{\pi}{3})$ $I_{RMS}=0.817\,I_d$ $\cos\phi=\cos\dfrac{\alpha}{2}=\sqrt{r}$ $VA_o=1.05\,P_{do}$ λ-see Fig. 12
 $I_{RMS}=0.471\,I_d/h$ $\cos\phi=\cos\alpha=r$ $VA_o=1.05\,P_{do}$ $\lambda=0.955\,r$ For harmonic distortion, see Fig.13	 $I_{RMS}=0.471\,I_d/h$ $(0\leqslant\alpha\leqslant\frac{\pi}{3})$ $VA_o=1.05\,P_{do}$ $\cos\phi,\lambda$, see Fig.11,12 For harmonic distortion, see Fig.13	 $(0\leqslant\alpha\leqslant\frac{\pi}{3})$ $I_{RMS}=0.471\,I_d/h$ $\cos\phi=\cos\dfrac{\alpha}{2}=\sqrt{r}$ $VA_o=1.05\,P_{do}$ λ, see Fig.12 For harmonic distortion, see Fig.13

POWER ELECTRONICS

Table 15-6. Waveforms and Data for Various Rectifier and Converter Circuits

	12-pulse midpoint rectifier	12-pulse midpoint converter (2-quadrant)	12-pulse bridge rectifier
Circuit	Circuit same as converter, with diodes instead of thyristors	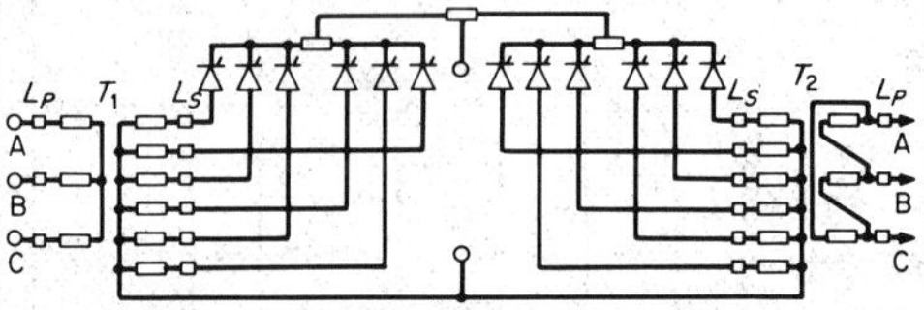	Circuit same as converter, with diodes instead of thyristors
D-c terminal voltage	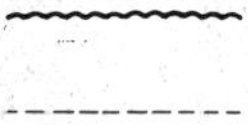	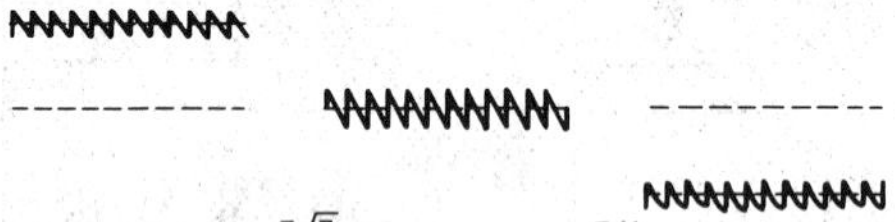 $$V_d = \frac{3\sqrt{3}}{2\pi} \hat{V}_S \cos\alpha - I_d \frac{3X_C}{8\pi}$$ $$X_C = \omega\left(\frac{L_p}{h^2} + L_S\right)$$ For harmonic distortion, see Fig. 9	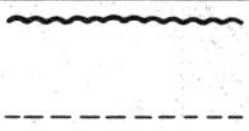
Device voltage and current	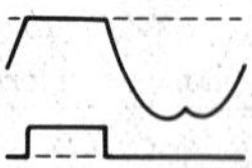	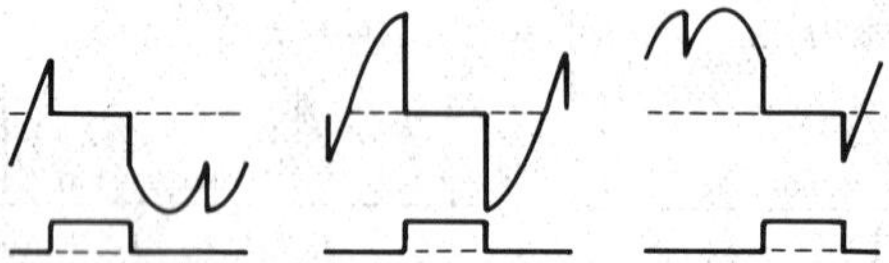 $V_{FB} = 1.732\,\hat{V}_S$ $V_{RB} = 1.732\,\hat{V}_S$ $I_{AV\,Th} = 0.083\,I_d$ $I_{RMS\,Th} = 0.144\,I_d$	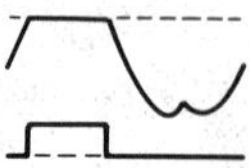
Transformer secondary voltage and current	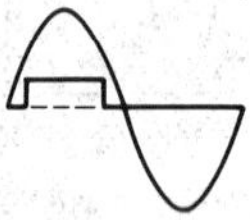	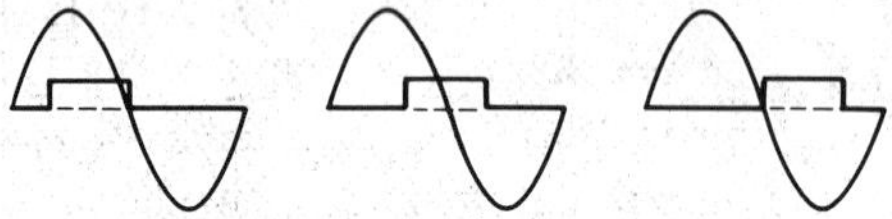 $I_{RMS} = 0.144\,I_d$ $VA_o = 1.48\,P_{do}$ $\cos\phi = \cos\alpha = r$ $\lambda = 0.675\,r$	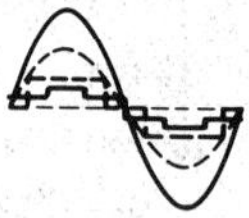
Transformer primary voltage and current	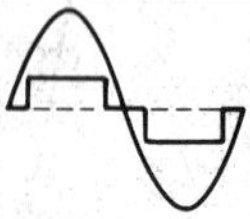	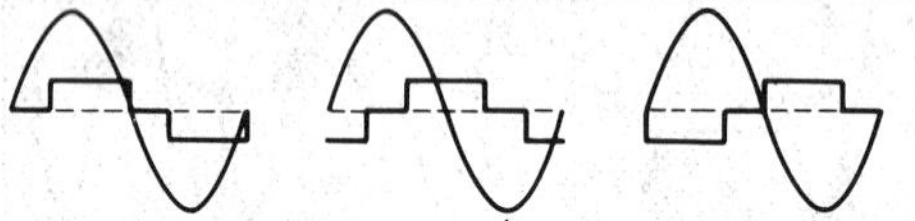 $I_{RMS} = 0.204\,I_d/h\,(T1)$ $\quad= 0.118\,I_d/h\,(T2)$ $\cos\phi = \cos\alpha = r$ $\quad VA = 1.05\,P_{do}$ $\lambda = 0.955\,r$ For harmonic distortion, see Fig. 13	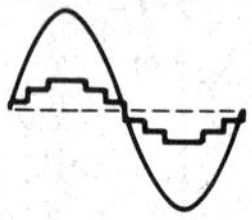

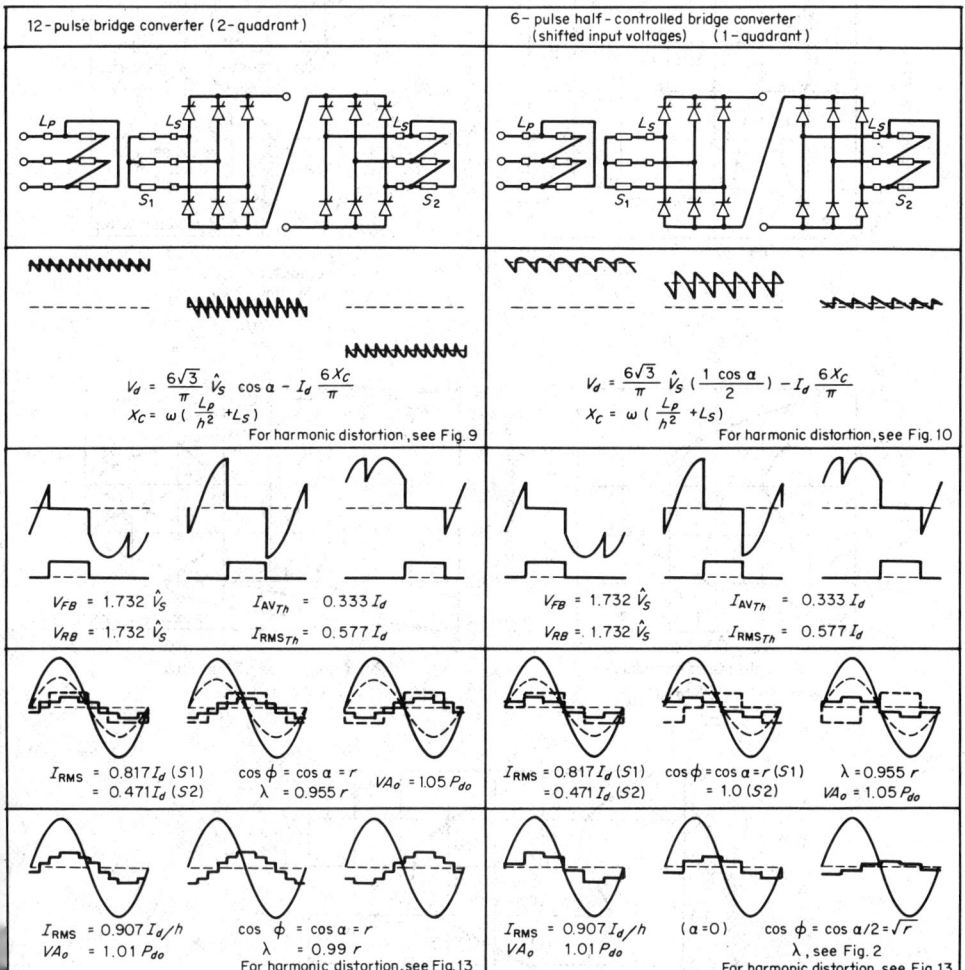

12-pulse bridge converter (2-quadrant)	6-pulse half-controlled bridge converter (shifted input voltages) (1-quadrant)

$$V_d = \frac{6\sqrt{3}}{\pi} \hat{V}_S \cos \alpha - I_d \frac{6X_c}{\pi}$$
$$X_c = \omega \left(\frac{L_p}{h^2} + L_S \right)$$
For harmonic distortion, see Fig. 9

$$V_d = \frac{6\sqrt{3}}{\pi} \hat{V}_S \left(\frac{1 \cos \alpha}{2} \right) - I_d \frac{6X_c}{\pi}$$
$$X_c = \omega \left(\frac{L_p}{h^2} + L_S \right)$$
For harmonic distortion, see Fig. 10

$V_{FB} = 1.732 \hat{V}_S$ $I_{AV_{Th}} = 0.333 I_d$
$V_{RB} = 1.732 \hat{V}_S$ $I_{RMS_{Th}} = 0.577 I_d$

$V_{FB} = 1.732 \hat{V}_S$ $I_{AV_{Th}} = 0.333 I_d$
$V_{RB} = 1.732 \hat{V}_S$ $I_{RMS_{Th}} = 0.577 I_d$

$I_{RMS} = 0.817 I_d \ (S1)$ $\cos \phi = \cos \alpha = r$ $VA_o = 1.05 P_{do}$
$\quad\quad = 0.471 I_d \ (S2)$ $\lambda = 0.955 \, r$

$I_{RMS} = 0.817 I_d \ (S1)$ $\cos \phi = \cos \alpha = r \ (S1)$ $\lambda = 0.955 \, r$
$\quad\quad = 0.471 I_d \ (S2)$ $\quad\quad = 1.0 \ (S2)$ $VA_o = 1.05 P_{do}$

$I_{RMS} = 0.907 I_d / h$ $\cos \phi = \cos \alpha = r$
$VA_o = 1.01 P_{do}$ $\lambda = 0.99 \, r$
For harmonic distortion, see Fig.13

$I_{RMS} = 0.907 I_d / h$ $(\alpha = 0)$ $\cos \phi = \cos \alpha / 2 = \sqrt{r}$
VA_o $1.01 P_{do}$ λ, see Fig. 2
For harmonic distortion, see Fig.13

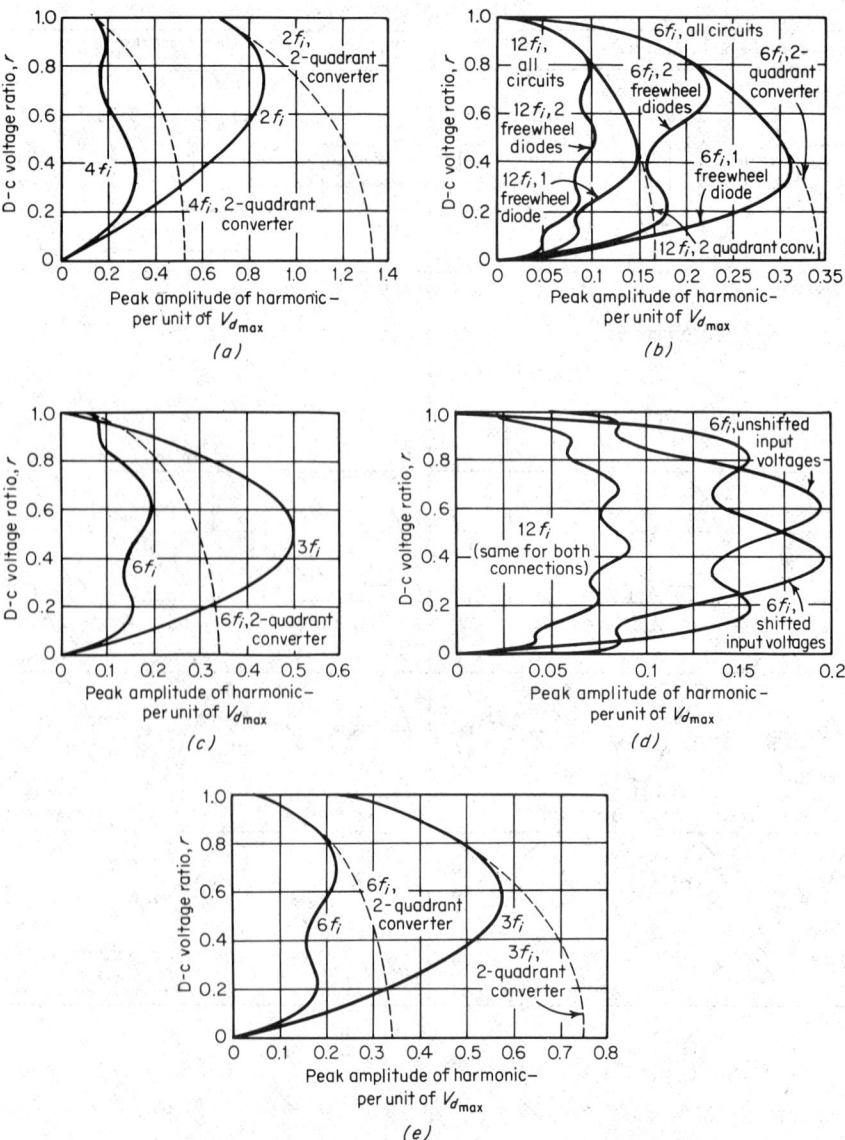

Fig. 15-10. Principal harmonic components in the dc terminal voltage of various one-quadrant converters with no commutation overlap. Curves for corresponding two-quadrant converters are shown for comparison. (*a*) 2-pulse half-controlled bridge circuit, and 2-pulse midpoint circuit with freewheel diode, (*b*) 6-pulse circuit with one and two freewheel diodes, (*c*) 3-pulse half-controlled bridge circuit, (*d*) 6-pulse half-controlled bridge circuit, with 30° "shifted" and "unshifted" input voltages for the two bridges, (*e*) 3-pulse circuit with freewheel diode. *(From B. R. Pelly, "Thyristor Phase-controlled Converters and Cycloconverters," Wiley-Interscience, New York, 1971. Used by permission.)*

For all two-quadrant converters, the input-power factor and the dc voltage ratio are also directly proportional to one another, as indicated in the tables. The corresponding relationships for the one-quadrant converters are illustrated in Fig. 15-12.

Figure 15-13 shows the principal-harmonic components present in the input line current of each of the converters shown in Tables 15-4 to 15-6.

All the above theoretical relationships assume ripple-free current at the dc terminals of the converter, with no commutation overlap.

18. Four-Quadrant Converters. A four-quadrant converter, or *dual converter,* can operate with both polarities of both voltage and current at the dc terminals. Such converters permit, for example, dc motors to be driven and regeneratively braked in both forward and reverse directions. A six-pulse bridge four-quadrant converter formed by paralleling two oppositely polarized two-quadrant converters is shown in Fig. 15-14.

If both converters are active simultaneously, in principle one operates as a rectifier while the other operates as an inverter with the same average voltage. In practice, the instantaneous difference between the voltages of the two converters tends to cause a large circulating current. One solution is to parallel the two converters through a circulating-current reactor, as shown in Fig. 15-14. Another solution, which is usually preferable, involves deactivating the idle converter either by removing its firing pulses or by appropriately adjusting its relative delay angle. This control can be achieved automatically in various ways (Pelly, 1971).

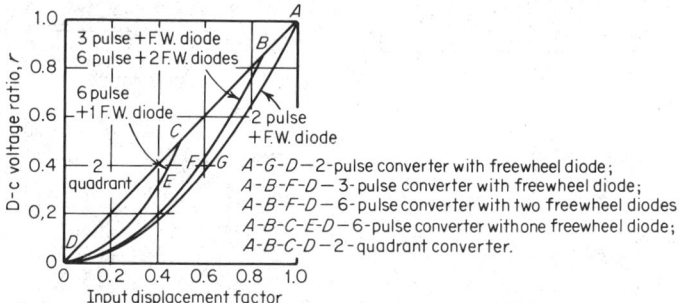

Fig. 15-11. Relationships between the dc terminal voltage ratio and the input-displacement factor for two-, three-, and six-pulse converters with freewheel diodes and no commutation overlap. *A-G-D,* two-pulse with freewheel diode; *A-B-F-D,* three-pulse with freewheel diode; *A-B-F-D,* six-pulse with two freewheel diodes; *A-B-C-E-D,* six-pulse with one freewheel diode; *A-B-C-D,* two-quadrant converter.

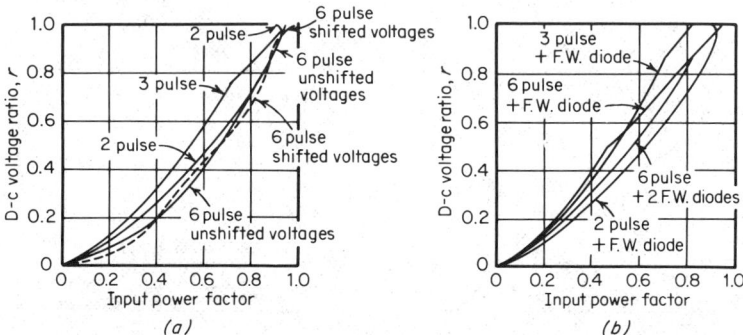

Fig. 15-12. Relationships between the dc voltage ratio and the input-power factor for various one-quadrant converters with no commutation overlap. (*a*) Two-, three-, and six-pulse half-controlled circuits; (*b*) two-, three-, and six-pulse circuits with freewheel diodes. These curves apply to transformer primary and secondary for bridge circuits and to primary for midpoint circuits. For transformer-secondary power factor of midpoint circuits, multiply by 0.707.

OTHER NATURALLY COMMUTATED CIRCUITS

19. AC Switches and Regulators. Various applications require ac power to be regulated or switched on and off without converting it to dc or to ac of a different frequency. Naturally commutated solid-state switches can also perform these functions.

Such switches and regulators require bilateral conduction. An obvious way to achieve single-phase bilateral conduction is the antiparallel connection of two unilateral thyristors,

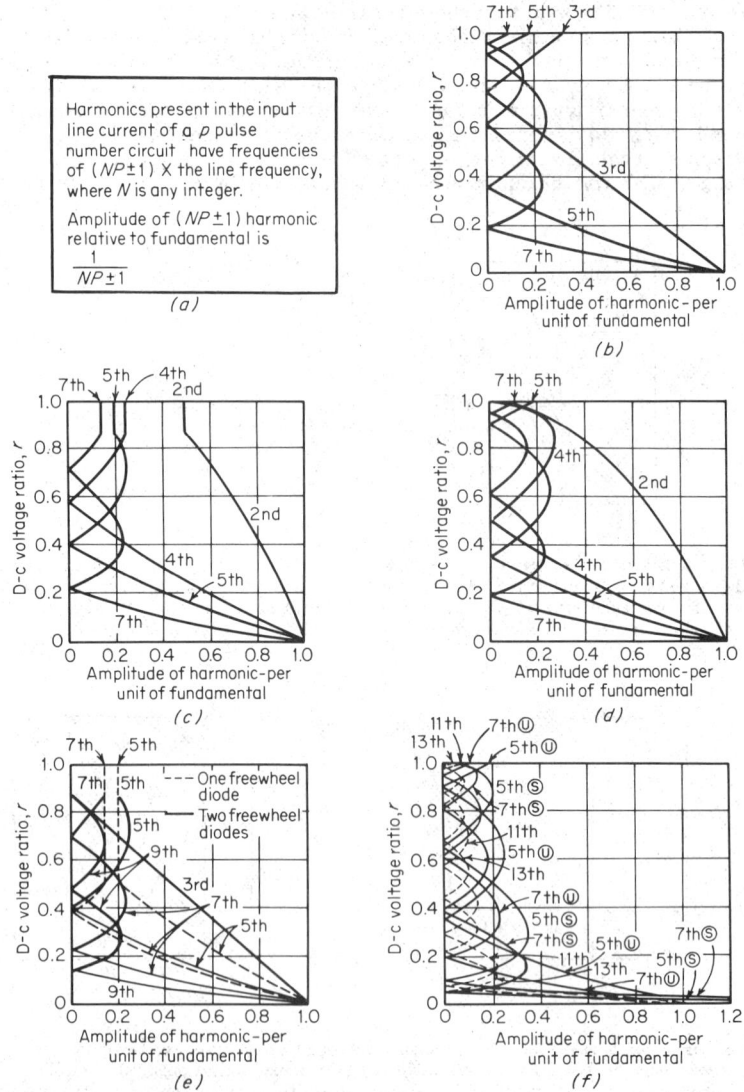

Fig. 15-13. Principal-harmonic components in the input line current of various converter circuits (with smooth direct current and no commutation overlap).

as shown in Fig. 15-15a. The triac, whose symbol is shown in Fig. 15-15b, integrates both thyristors into a single device structure and is widely used where its power and frequency ratings permit. Two other configurations for a bilateral switch are shown in Fig. 15-15c and d.

For 3-phase applications, single-phase switches may be placed in two or all three of the ac supply lines. Two alternate connections which allow full 3-phase control with only three thyristors are shown in Fig. 15-16. The first circuit can be used only in balanced 3-wire systems; connection of the neutral causes loss of control on the negative half cycle and a corresponding dc component. The second circuit can be used only if the load can be split into three balanced ungrounded single-phase loads or if the load is coupled through Y-connected transformer windings with the thyristor delta-connected into the common point.

20. On-Off Control and Protection. Solid-state ac switches can be used to replace mechanical relays and circuit breakers for on-off control and protection of electric circuits.

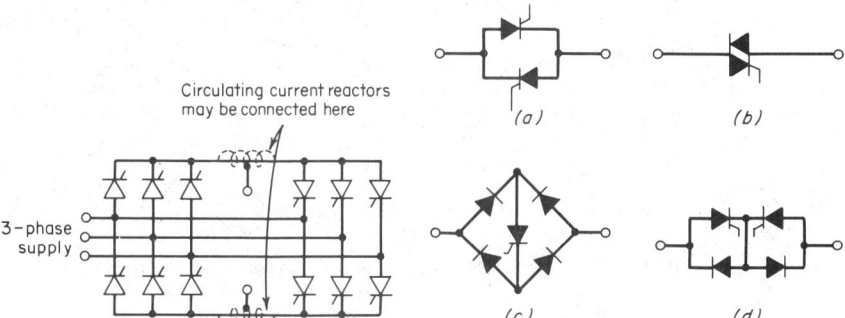

Fig. 15-14. Six-pulse bridge dual-converter circuit.

Fig. 15-15. Single-phase ac switches.

Since there are no moving contacts to wear or arc, useful life and reliability can be very long. Electromagnetic noise and transient surges can be minimized by delaying turn-on until the next voltage zero. Normally ac switches are switched off by removing the gate-firing signals, allowing the conducting devices to commutate naturally at the next current zero. In certain applications, a hybrid combination of solid-state and mechanical switches is advantageous. The solid-state switch eliminates arcing and allows precise control of the instant of switching, whereas parallel mechanical contacts carry the continuous load current and avoid the heat-sink requirements of the solid-state devices.

The fast response of solid-state switches can also be utilized in circuit breakers which offer protection before a surge reaches its peak value. In some applications, a thyristor across the power line is fired when a fault is detected, thereby blowing a fuse in series with the line. Although it is the fuse which clears the circuit, the thyristor "crowbar" is able to limit the volt.ge surge seen by a protected circuit much more rapidly than a fuse acting alone. A thyristor in series with the power line can also be used as a current-limiting circuit breaker. However, subcycle interruption of a fault current before it reaches its peak requires the use of forced commutation, discussed in subsequent paragraphs.

21. Regulation of AC Power. With appropriate firing signals applied to the gates, the same circuits can also be used for the regulation of ac power. Such regulation may be occasional, as in *soft starting* to limit the inrush current of a motor, or continuous, as in a lamp dimmer or a speed control for a universal motor. Regulators may also be categorized as shown in Fig. 15-17. The operation of a phase-delay regulator is similar to that of a controlled rectifier in that it permits full control of the average ac voltage from zero to maximum. It also has the same potential problems: harmonics and electrical noise resulting from the nonsinusoidal waveforms and lagging power factor from the delayed conduction.

Integral-cycle control, in which firing pulses are simply omitted for the required proportion of cycles, reduces these problems but introduces subharmonic frequencies that can cause noticeable flicker in lighting fed from the same supply line.

If control over a wide range of voltage is unnecessary, the waveform distortion and attendant problems can be reduced by using *differential control,* in which the output voltage is switched between the minimum and maximum values rather than between zero and the maximum value. Either phase-delay or integral-cycle control can be used for intermediate values.

22. Naturally Commutated Cycloconverters. The frequency of ac power can be converted from one value to another in two steps by a *rectifier* which supplies power to an *inverter.* Frequency changing can also be performed in a single step. Classical power-frequency changers utilize natural commutation and are known as cycloconverters. Cycloconverters synthesize the output waveform from segments of several half cycles of the input waveform. Therefore they are limited to decreasing the frequency of the input power (Rissik, 1939, Chaps. 11 and 12).

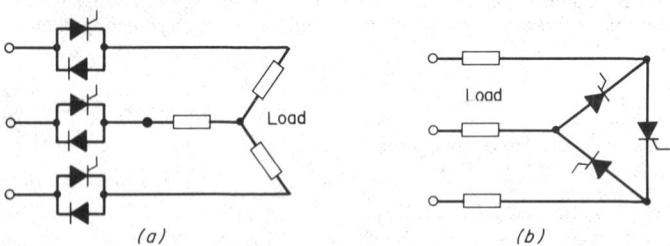

(a)   *(b)*

Fig. 15-16.   Special 3-phase ac switches.

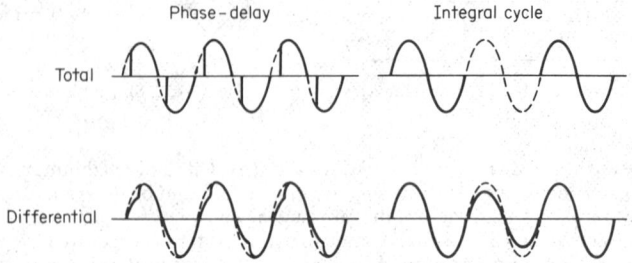

Fig. 15-17.   Waveforms of various types of ac regulators.

More recently, forced commutation has made it feasible to build power-frequency changers which are not restricted to decreasing the frequency and which have other useful properties. Since naturally commutated cycloconverters are a special case of these more generalized circuits, the theory of their operation is discussed under force-commutated power-frequency changers (see Par. 15-40).

CIRCUITS USING FORCED COMMUTATION

23. General Considerations. Many of the new applications of power electronics depend on the ability to interrupt a nonzero current. These applications were not feasible with mercury-arc devices because their long turnoff times required large energy storage to achieve forced commutation. *Forced commutation* is the general name given to a variety of techniques which permit current to be interrupted by devices without inherent turnoff ability.

One type of forced commutation uses *resonance* to generate an alternation which brings the current in a conducting thyristor to zero. Such circuits typically require large reactive elements with considerable energy storage, but the functions of commutation and filtering for waveform improvement can often both be performed by the same reactive elements (Rice, 1970, pp. 8-47 to 8-54).

Another type of forced commutation, known as *impulse commutation,* momentarily diverts the load current while reverse-biasing the thyristor until it regains its blocking state. Some circuit configurations include a capacitor which achieves an automatic transfer of current when the next load-carrying thyristor is fired. These circuits are said to be self-commutated. Other circuits use an auxiliary thyristor in series with a capacitor to divert load current briefly and then cease conduction.

The *energy storage* required to achieve impulse commutation is considerably less than that required for resonant commutation. Because of the crucial role played by stored energy in achieving forced commutation, the fault current which can be safely interrupted is limited by the commutation circuit rather than by the thyristors.

The *power capacity* required by many applications of the circuits to be described can be obtained only by using conventional thyristors. Achieving reliable and economical forced commutation is then a major factor in the design process. For applications within the available ratings of power transistors or gate-controlled switches, these devices can be used to avoid the costs of the additional components needed for forced commutation.

In its on state, a transistor is generally operated in or near saturation, where its dissipation is minimized. If the collector current appreciably exceeds the product of the base current and the gain, the transistor comes out of saturation and is subject to very high dissipation, which can cause permanent damage very quickly. The transistor should never be exposed to voltage and current outside its safe operating area (SOA) (see Sec. 7, Par. **7-60**).

All circuits which are required to interrupt current in an inductive load must provide a safe path for absorbing the energy stored in the inductance. Otherwise the inductive voltage spikes are likely to destroy the long-term reliability, if not the immediate operation of the circuit. A freewheeling diode across the load is a common technique which permits the energy to be dissipated in the load itself. In circuits where the load is resonated, this energy is transferred to a capacitor, where it is available for reuse.

24. DC Regulators (Choppers). The average power supplied from a dc source to a dc load can be regulated with high efficiency by a series switch that repetitively opens and closes, thereby "chopping" the current which flows between the two. Hence dc regulators are frequently called choppers. Control is achieved by varying the relative on time or duty cycle. Obviously the duty cycle can be changed by changing the on time, the off time, or both, and all three control modes are used in practical choppers (Bedford and Hoft, 1964, Chap. 10).

25. Transistor Choppers. A typical transistor chopper is shown in Fig. 15-18. The transistor is switched by a drive circuit connected to its base. The duty cycle of the drive circuit is determined by a circuit which compares the load voltage to a reference voltage. The capacitor across the source protects the transistor from inductive spikes when the source current is interrupted, and a freewheeling diode shunts inductive spikes from the load. An *LC* filter between the chopper and the load reduces the ripple in the output voltage.

The frequency at which the switch is operated is determined by the desired efficiency and size. As the frequency is increased, the size of the filter decreases, but the switching losses in the transistor increase, reducing efficiency and requiring a larger heat sink. Transistor choppers are typically operated at frequencies in the range of 1 to 4 kHz. However, in some specialized applications where the power requirements are low and minimum size and weight are important, frequencies above 20 kHz can be used.

26. Thyristor Choppers. The principles of operation of thyristor choppers are basically the same as for transistor choppers except that forced commutation is required with thyristors. Of the variety of commutation circuits in use, two typical circuits are shown enclosed in dashes in Fig. 15-19. In the first circuit *(a),* capacitor *C* is first charged by switching on Q1. Resonant charging through *L* causes the voltage across *C* to rise to a peak value which is greater than the supply voltage, after which Q1 ceases to conduct. After *C* is

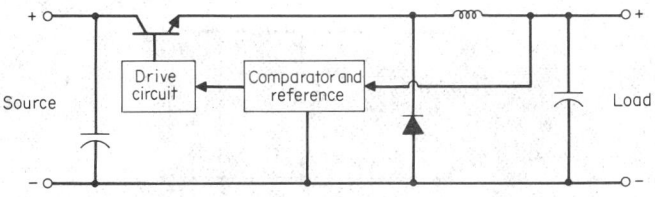

Fig. 15-18. Chopper using a transistor switch.

charged, main thyristor Q2 is switched on and current is delivered to the load. To turn Q2 off, Q3 is fired. The voltage on C reverse-biases Q2 until it is able to block forward voltage. When C is discharged, Q3 ceases to conduct.

In the second circuit *(b)*, capacitor C is charged to the supply voltage by switching Q1 on. After C has charged, Q2 is switched on to deliver a load pulse and to allow the voltage on C to reverse by resonantly discharging through the path provided by D1 and L. After the voltage on C has reversed, Q1 may be switched on to commutate Q2 off.

The maximum current which can be reliably controlled in a thyristor chopper depends upon the commutating capability designed into the circuit. In many applications, if the current exceeds that which the commutating circuit can switch off, the main thyristor will remain on and the full supply voltage will be delivered to the load. Normal operation can then be restored only by opening the source or load circuit. For this reason, many thyristor choppers have automatic current limiting built into their control circuits. When the load current approaches a preset limit (determined either by the commutating circuit or by the thyristor rating), the control signal from the limiter overrides the signal from the output comparator and the chopper operates in a current-limited mode.

The *frequency of operation* of thyristor choppers is lower than that obtained with transistors because of the combined effects of device switching and recovery times and the di/dt and dv/dt limitations of thyristors. Frequencies under 2 kHz generally provide the most favorable compromise between efficiency, size, and cost. If necessary, higher frequencies can be achieved by replacing the main thyristor by a number of parallel thyristors. The duty cycle of each thyristor is reduced by time-shared conduction on sequential cycles.

Thyristor choppers find wide application for speed control of dc motors used in traction applications. They provide smooth, efficient control from standstill to full vehicle speed. Choppers are used both with series- and shunt-wound motors and frequently incorporate controlled dynamic or regenerative braking of the wheels. The range of traction applications varies from battery-powered forklift trucks, using controllers operating at 36 V and several hundred amperes, to high-speed train drives using series thyristors operating at several thousand volts and several hundred amperes.

27. Voltage Step-up Choppers. Choppers can also be used to produce a dc output voltage which is higher than the supply voltage. This is accomplished by means of the circuit shown in Fig. 15-20. Initially Q is turned on, and the source stores energy in inductor L. When Q is turned off, the energy in L discharges into the load through diode D. During this

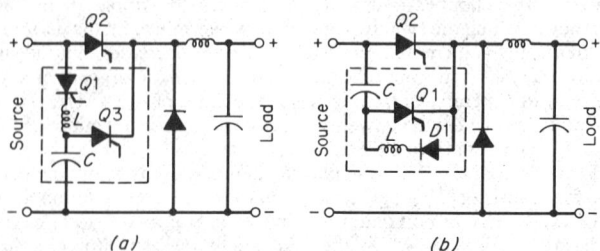

Fig. 15-19. Two typical thyristor choppers showing circuits for forced commutation.

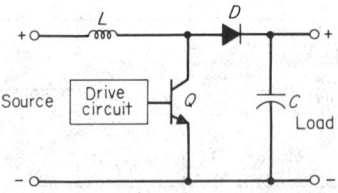

Fig. 15-20. Voltage step-up chopper.

discharge period, the voltage across L adds to the source voltage. Capacitor C across the load smooths the voltage pulsations. This circuit is equivalent to a dc step-up transformer.

INVERTERS

BY T. M. HEINRICH and R. M. OATES

28. General Considerations. An inverter is a power converter in which the normal direction of power flow is from a dc source to an ac load. In contrast to naturally commutated converters, which may operate as synchronous inverters, as described previously (Pars. **15-13** to **15-18**), the thyristor inverters to be discussed here must be force-commutated unless the load happens to have a leading power factor. However, the power flow is still reversible. By properly phasing the control signals, the dc source can be made to absorb power from an active ac load, such as a synchronous motor which is being dynamically braked.

Table 15-7 shows the four basic inverter circuits, corresponding to a center-tapped load, a center-tapped source, and single- and 3-phase bridges. The relationships given were derived with the aid of the following simplifying assumptions: the switches operate instantaneously and have no voltage drop when closed or leakage current when open; the ideal filter removes all harmonics from the output voltage without attenuation or phase shift to the fundamental; and the load is resistive. As illustrated, the switches must block only one polarity of voltage, but they must be capable of conducting both polarities of current. In practice these switches are implemented by shunting a transistor or thyristor by a diode which carries the reverse current.

29. Self-commutated Inverters. The half-bridge thyristor inverter shown in Fig. 15-21 illustrates the principles of self-commutation. R_0 is the load resistor, L_c and C_c are the commutating components, and L_F and C_F constitute a simple filter to smooth the load-voltage waveform. When Th1 is turned on to initiate the first positive half cycle, C_c charges to a voltage of $+E/2$. When the first negative half cycle is to be initiated, Th2 is turned on, causing the lower half of the source voltage to add to the voltage stored on C_c across the lower half of L_c.

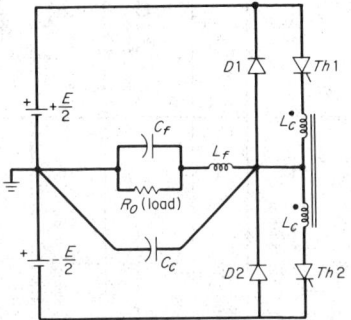

Fig. 15-21. Half-bridge thyristor inverter with forced commutation.

Because of the mutual coupling between the two halves of the commutating inductor, a voltage equal to E is induced in the upper half. This voltage causes Th1 to become reverse-biased so that it begins to turn off. C_c and L_c are designed to hold a reverse bias on Th1 long enough for it to recover its blocking ability. Thus the firing of Th2 automatically transfers current from Th1 to Th2, and, following the commutation transient, the polarity of the load voltage reverses, causing C_c to charge to $-E/2$. When Th1 is again fired, Th2 is turned off in a complementary manner, and the first full cycle of operation is complete.

During the time that C_c is recharging toward $-E/2$, the current in the lower half of L_c approximately doubles. Because of diode D2, the voltage across inductor L_c cannot reverse by more than the combined forward drops of D2 and Th2. The excess current so trapped

Table 15-7. Basic Inverter Circuits

	Center-tap	Half-bridge	Full-bridge	Three-phase bridge
Circuit diagram	(transformer circuit, η:1:1 Turns ratio, SW1, SW2, E_{DC}, Ideal filter, Z_L, V_{UNF}, V_{out}, I_L, I_{DC})	($E_{DC}/2$ dual supply, 1A, 1B, Ideal filter, V_{UNF}, V_{out}, Z_L, I_L, I_{DC})	(E_{DC}, 1A, 1B, 2A, 2B, Ideal filter, V_{UNF}, Z_L, V_{out}, I_L, I_{DC})	(E_{DC}, 1A, 2A, 3A, 1B, 2B, 3B, Ideal filter, V_{UNF}, V_{out}, Z_L, I_L, I_{DC})
Circuit name	Center-tap	Half-bridge	Full-bridge	Three-phase bridge
Output voltage V_{out} — Unfiltered voltage waveform	square wave $+\eta E_{DC}$, $-\eta E_{DC}$ (at π, 2π)	square wave $+\tfrac{1}{2}E_{DC}$, $-\tfrac{1}{2}E_{DC}$ (at π, 2π)	square wave $+E_{DC}$, $-E_{DC}$ (at π, 2π)	stepped wave E_{DC} at $\tfrac{2}{3}\pi$, $-E_{DC}$ at $\tfrac{5}{3}$; Contains no third harmonic (at π, 2π)
Output voltage V_{out} — RMS value of V_{out} (fundamental component only)	$\dfrac{2\sqrt{2}}{\pi}\,\eta\,E_{DC}$	$\dfrac{\sqrt{2}}{\pi}\,E_{DC}$	$\dfrac{2\sqrt{2}}{\pi}\,E_{DC}$	$\dfrac{\sqrt{6}}{\pi}\,E_{DC}$
Input current I_{DC} — Waveform	rectified sine pulses, peak $I_L\sqrt{2}$, phase φ (at π, 2π)	rectified sine pulse, peak $I_L\sqrt{2}$, phase φ (at π, 2π)	rectified sine pulses, peak $I_L\sqrt{2}$, phase φ (at π, 2π)	ripple waveform (at π, 2π)

15-34

	Input current I_{dc}			Switch stress			
	I_{DC} (avg value)	$\dfrac{I_{PK}}{I_{DC}}$ (avg)	$\dfrac{f_{ripple}}{f_{inverter}}$	Voltage waveform	Current waveform	RMS value of reverse current I_{REV}	RMS value of forward current as a function of I_{REV}
	$\dfrac{2\sqrt{2}}{\pi}\,\eta I_L \cos\varphi$	$\dfrac{\pi}{2\cos\varphi}$	2	$2E_{DC}$	$\eta I_L\sqrt{2}$; $\eta I_L\sqrt{2}\cos\varphi$	$\dfrac{1}{2}\eta I_L\sqrt{\dfrac{2\varphi-\sin 2\varphi}{\pi}}$	$\sqrt{\dfrac{\eta^2 I_L^2}{2}-(I_{REV})^2}$
	$\dfrac{\sqrt{2}}{\pi}\,I_L \cos\varphi$	$\dfrac{\pi}{\cos\varphi}$	1	E_{DC}	$I_L\sqrt{2}$; $I_L\sqrt{2}\cos\varphi$	$\dfrac{1}{2}I_L\sqrt{\dfrac{2\varphi-\sin 2\varphi}{\pi}}$	$\sqrt{\dfrac{I_L^2}{2}-(I_{REV})^2}$
	$\dfrac{2\sqrt{2}}{\pi}\,I_L \cos\varphi$	$\dfrac{\pi}{2\cos\varphi}$	2	E_{DC}	$I_L\sqrt{2}$; $I_L\sqrt{2}\cos\varphi$	$\dfrac{1}{2}I_L\sqrt{\dfrac{2\varphi-\sin 2\varphi}{\pi}}$	$\sqrt{\dfrac{I_L^2}{2}-(I_{REV})^2}$
	$\dfrac{3\sqrt{2}}{\pi}\,I_L \cos\varphi$	$\dfrac{\pi}{3\cos\varphi}$ $\left(0 \le \omega \le \dfrac{\pi}{6}\right)$	6	E_{DC}	$I_L\sqrt{2}$; $I_L\sqrt{2}\cos\varphi$	$\dfrac{1}{2}I_L\sqrt{\dfrac{2\varphi-\sin 2\varphi}{\pi}}$	$\sqrt{\dfrac{I_L^2}{2}-(I_{REV})^2}$

15-35

continues to circulate through D2 and Th2 until all the trapped inductor energy is dissipated. Aside from the extra dissipation in the devices, the trapped energy is bothersome in that it hinders the commutation process. If it is not removed, the inductor current for subsequent commutations will become progressively greater until C_c can no longer supply sufficient commutating energy.

A resistor may be inserted in series with diodes D1 and D2 to dissipate the trapped energy, but for frequencies up to about 400 Hz, most of this energy can be recovered by employing a tapped transformer primary (McMurray and Shattuck, 1961). A practical circuit with trapped-energy-recovery transformer is shown in Fig. 15-22. The tap at n provides an additional voltage in the discharge loop to absorb energy from L_c. The energy absorbed by the transformer is passed along to the load or returned to the dc source if the load is unreceptive. The tap n is generally placed at 10 to 20% of the primary turns, tending toward 20% if the dc input voltage is low and inverter frequency is high. Note that in Fig. 15-22 the commutating capacitor C_c has been split between the $+$ dc and $-$ dc supplies. With this arrangement a center-tapped dc supply is unnecessary for commutation, and two half bridges can be combined into a full bridge, or three half bridges can be combined into a 3-phase bridge, as shown in Table 15-7.

The values of the commutating capacitor and commutating inductor are given by

$$C_c = \frac{t_r \hat{I}}{0.425\,E} \quad \text{farads} \quad \text{and} \quad L_c = \frac{t_r E}{0.425\,\hat{I}} \quad \text{henrys}$$

where t_r is the turnoff time required by thyristor, E is the total dc supply voltage, and $\hat{I}$ is the maximum thyristor anode current to be commutated.

30. Auxiliary Commutation. Above 400 Hz, the trapped-energy problem of the McMurray-Shattuck circuit degrades circuit efficiency to such an extent that more complex circuits are justified. A half-bridge circuit which uses auxiliary thyristors for commutation is shown in Fig. 15-23. This circuit was suggested by W. McMurray. It has better voltage-regulation characteristics than the self-commutated circuit and can be used at frequencies up to about 5 kHz.

Operation of the circuit is initiated by firing Th1 and Th2A, thereby applying $+E/2$ to the load and charging the commutating capacitor C. When C is fully charged, the current in Th2A goes to zero and it ceases conduction. To end the first half cycle, auxiliary thyristor Th1A is fired. Inductor L limits the rate of current increase in D1 and Th1A. As the current in L increases, the load current is diverted from Th1 to Th1A and C. After a delay of about $2.4\sqrt{LC}$ s, the forward drop across D1 reverse biases Th1 and turns it off. Then Th2 is fired to begin the negative half cycle. In the meantime, C charges to the opposite polarity for the next commutation before Th1A ceases conduction. Th2 is turned off by Th2A in the same way that Th1 was turned off by Th1A.

The values of the commutating components are given by (Bedford and Hoft, 1964, p. 180):

$$C = 0.893\frac{\hat{I}\,t_r}{E} \quad \text{farads} \qquad L = 0.397\frac{E t_r}{\hat{I}} \quad \text{henrys}$$

where E, $\hat{I}$, and t_r are as defined previously.

The circuit as shown generates severe dv/dt transients on all the thyristors, which require snubber circuits for protection (see Par. **15-46** on the protection of thyristors).

31. Output-Voltage Waveform. For some applications, such as motor drives and dc-to-dc converters, a square-wave output from an inverter may be acceptable. Much of the time, however, sinusoidal voltage waveforms with limited total harmonic distortion are desired. A typical limit in equipment specifications would be 5% total harmonics relative to the magnitude of the fundamental frequency.

Various second- and third-order filter networks are commonly used to eliminate undesirable harmonics from the inverter output, but all tend to be large, heavy, costly, and, in general, highly load-dependent. For this reason, it is desirable to provide an inverter waveform which is inherently devoid of low-order harmonics. Higher-order harmonics can then be filtered with a relatively small network, producing an output waveform which is nearly sinusoidal. Common methods for producing such waveforms from square-wave inverters can be placed in two main categories: harmonic neutralization and pulse-width modulation. Harmonic neutralization involves a combination of several phase-shifted square-wave inverters, each switching at the fundamental frequency, whereas pulse-width modulation involves switching a single inverter at a frequency higher than the fundamental (Kernick et al., 1962). Both schemes give satisfactory results, and actual selection of a method would depend on many factors such as the output-power level, the fundamental frequency, the speed of the switching devices, and the type of commutation circuit. Harmonic neutralization is especially suited for 3-phase outputs.

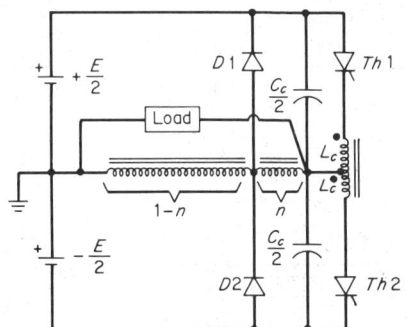

Fig. 15-22. Half-bridge thyristor inverter with forced commutation and energy-recovery transformer.

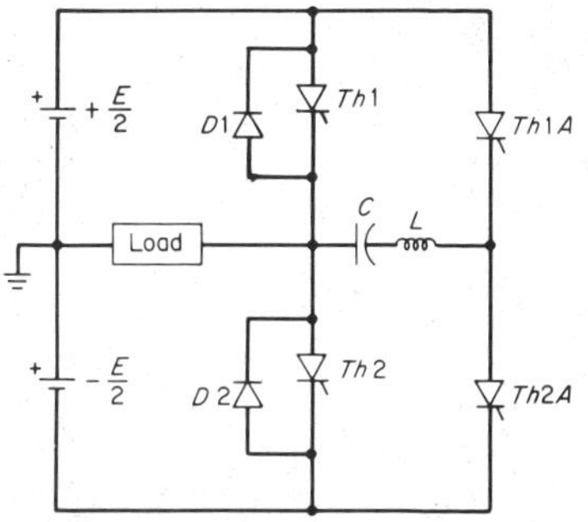

Fig. 15-23. Auxiliary impulse-commutated inverter.

32. Harmonic Neutralization. A harmonic-neutralized inverter consists of N square-wave inverter stages which are sequentially phase shifted by $180/N$ electrical degrees (Kernick and Heinrich, 1964; Heinrich, 1967). In general, for a polyphase ac output, each inverter stage contributes to the output of each phase by means of a process of phasor addition performed by transformer windings. In place of an overall square-wave output containing all odd harmonics of the fundamental frequency, the output voltage is a stepped approximation to a sine wave in which most of the harmonics have been neutralized. The remaining harmonics occur in pairs and have frequencies of $2kN \pm 1$, where $k = 1, 2, 3, \ldots$ The amplitudes of the harmonics which remain, relative to the fundamental, are inversely proportional to their frequencies, as in a square wave.

Each stage of the inverter will share the total output power equally if the load is balanced, but the voltage contributed to each phase by each stage must be properly adjusted. In general, these voltages will not be equal but are given by $(\pi V_{rms} \cos \Psi_{MW})/\sqrt{2}N$, where V_{rms} is the desired line-to-neutral output voltage and Ψ_{MW} is the phase angle between stage M and phase W.

As an example, consider the 3-phase six-stage inverter shown in Fig. 15-24. The firing angles of the respective stages are separated by $180°/6 = 30°$, giving the phasor diagrams shown for the fundamental components of the individual stages and phases. It is assumed that stage A and phase X are each arbitrarily assigned a 0° phase angle and the line-to-neutral voltage is to be 120 V. Using the relationship above, the transformer-turns ratios in the various stages should be chosen as follows for phase X:

$$V_{AX} = (\pi \times 120 \cos 0°)/\sqrt{2} \times 6 = 44 \text{ V}$$
$$V_{BX} = (\pi \times 120 \cos 30°)/\sqrt{2} \times 6 = 38 \text{ V}$$

Similarly, $V_{CX} = 22$ V, $V_{DX} = 0$, $V_{EX} = -22$ V, $V_{FX} = -38$ V. Since $V_{DX} = 0$, no winding is needed, and phase X is formed by the series connection of the other five windings.

The individual square waves and the corresponding output-voltage waveform for phase X are shown in Fig. 15-25. In a similar way, the contribution of each stage to the other two phases can be calculated. The only harmonics present in the output waveform and their amplitudes relative to the fundamental are the eleventh ($\frac{1}{11}$) and thirteenth ($\frac{1}{13}$), twenty-third ($\frac{1}{23}$) and twenty-fifth ($\frac{1}{25}$), etc.

The ripple frequency of the current into the inverter is $2N$, or 12 times the line frequency,

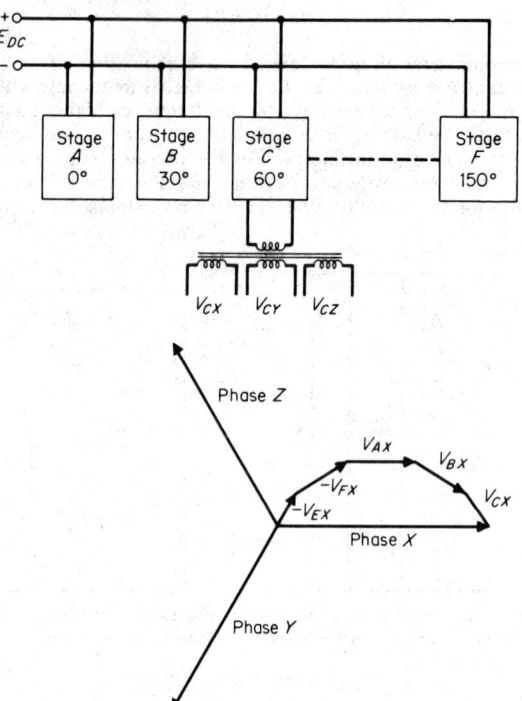

Fig. 15-24. Six-stage harmonic-neutralized inverter.

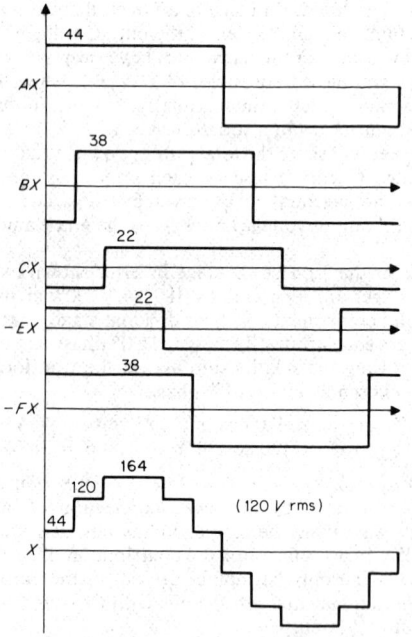

Fig. 15-25. Waveforms from individual inverter stages summed to form phase X.

15-38

thereby reducing the size of the input filter. The combined rating of the transformers is about 1.4 times the rating of the inverter.

Although the inverter described synthesizes the output from isolated single-phase stages, many variations are possible depending on the particular application. For instance, the same result could be achieved using a pair of 3 phase bridge inverters and 3-phase transformers (Oates, 1970).

33. Pulse-Width Modulation (PWM). Pulse-width-modulated inverters approximate sine-wave outputs by switching the power elements at a rate higher than the fundamental frequency. These inverters, categorized by the manner in which this switching takes place, fall into three basic groups: programmed waveform, modulated carrier, and optimum response (bang-bang) (Kernick and Haque, 1969). Although the harmonic content of the output varies in the three methods, the total harmonic distortion is always inversely proportional to the switching rate.

In a *programmed-waveform PWM inverter*, the power stage or inverter is given a fixed switching pattern which is periodic. This pattern is designed to produce the best possible waveform for the number of switching operations permitted per cycle. Center-tap and half-bridge inverters are capable of providing positive or negative but not zero instantaneous output. The output from such an inverter is called a *noncommutated waveform*. The full-bridge inverter can also produce a zero instantaneous voltage, and its output is called *commutated*. In general, the commutated waveforms give lower distortion for the same number of switching operations per cycle. Figure 15-26 presents a summary of useful programmed waveforms of both types along with their harmonic content.

Carrier-modulated PWM is usually accomplished by comparing a reference sine wave at the fundamental frequency to a sawtooth signal having a fixed frequency higher than the fundamental (Ravas et al., 1967). The power elements are switched at the zero crossing of these two signals, as shown in Fig. 15-27. Distortion of the output waveform occurs at the carrier frequency and its sidebands and at multiples of the carrier frequency and their associated sidebands. This distortion may or may not be harmonic, depending on whether or not the carrier frequency is synchronized with the fundamental reference. The magnitude of the distortion depends on the degree of modulation (relative magnitude of the sine-wave peak compared to the carrier peak) and is lowest at 100% modulation.

Another type of PWM, known as *optimum-response switching*, is shown in Fig. 15-28 (Geyer and Kernick, 1971). This scheme, unlike the others, must operate with an output filter, and it must have closed-loop control. Hysteresis in the feedback path sets the allowable deviation of the output from a sinusoidal reference. The switching rate varies throughout the cycle and is determined by the amount of hysteresis and by the characteristics of the load and filter. Very high switching rates are generally required to keep the error small.

The control for such an inverter is very simple, and voltage regulation is automatically accomplished. However, many applications will not permit optimum-response PWM because of the inherent voltage ripple and the asynchronous output waveform.

34. Voltage Control. Most inverter applications require direct control of the output voltage. For motor-drive inverters, it must be continuously variable from zero to full value, depending upon torque and speed requirements. For ac power-supply inverters, the voltage must be held nearly constant over a certain load and input range. In addition, many inverters are required to provide a specified amount of current into a short circuit, making it necessary to cut back the output voltage to nearly zero. A typical load profile is shown in Fig. 15-29.

Varying the dc input voltage and internal pulse-width control are the most common methods of controlling the output voltage where this control is not inherent, as it is in carrier-modulated PWM and optimum-response inverters.

DC-Input Control. Control of the dc input is the most straightforward method. If an inverter's switching pattern remains invariant, the output voltage is directly proportional to the dc-input voltage for all types of inverters. If the power source is ac, a phase-controlled rectifier can be used to control the dc input to the inverter. If the power source is dc, it is necessary to use a dc regulator, i.e., chopper.

The main advantages of using dc-input control are that the switching requirements and control complexity are not increased. In addition, the harmonic content of the output-voltage waveform does not vary with the input voltage. However, it has the disadvantage that the power must often pass through an extra stage of conditioning, thus reducing overall efficiency. Also, it is often impossible to use input-voltage control when the control range is large because reliable forced commutation depends on input voltage.

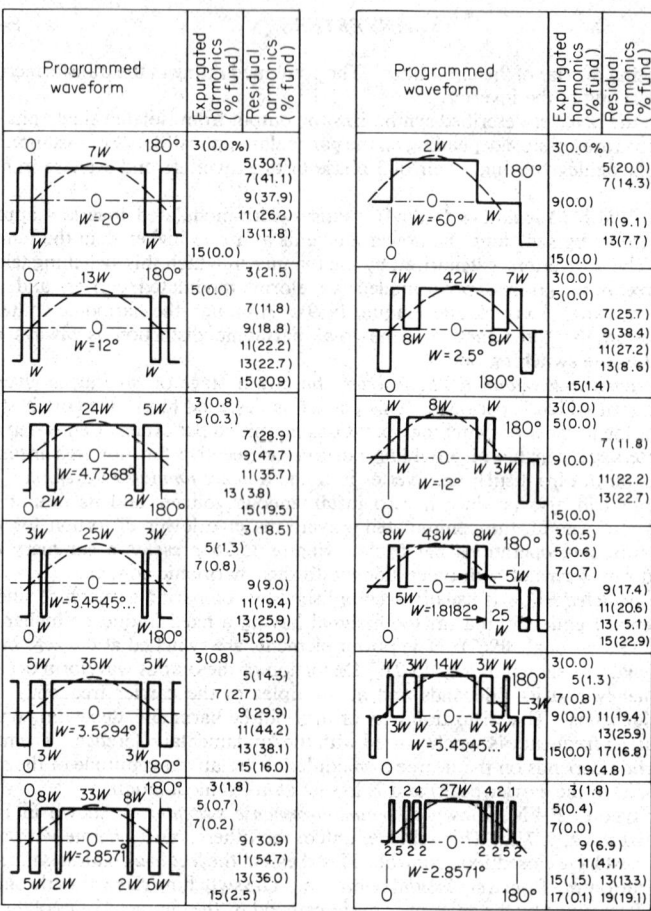

Fig. 15-26. Summary of programmed waveforms and their harmonic content. W is the unit increment of time for each waveform in degrees.

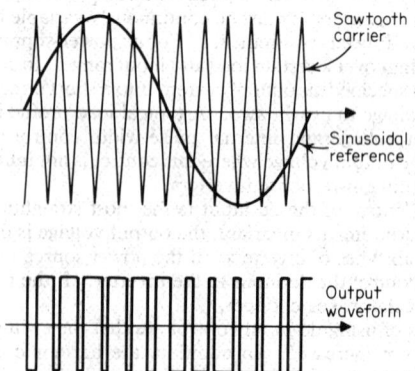

Fig. 15-27. Carrier-modulated pulse-width-modulation waveforms.

Pulse-Width Control. Inverter output voltage can also be controlled by varying the conduction time of the power switches. The pulse width of a full-bridge inverter can be controlled by introducing a delay between the turnoff of each pair of switches and the turn-on of the other pair to produce the waveform shown in Fig. 15-30. The rms value of the fundamental component of this waveform varies as the cosine of half of the delay angle θ. The fundamental frequency remains unchanged. All the odd harmonics are present, but their magnitudes change with θ. Figure 15-31 shows the variation of the third, fifth, and seventh harmonics, expressed as a percentage of the fundamental voltage. Notice that as θ approaches 180°, the harmonics become as large as the fundamental.

Center-tap and half-bridge circuits cannot be modulated in this simple way but require more complex switching at a higher frequency.

Pulse-width techniques can be used to control the output voltage of harmonic-neutralized inverters by controlling each individual inverter stage. The output continues to obey the

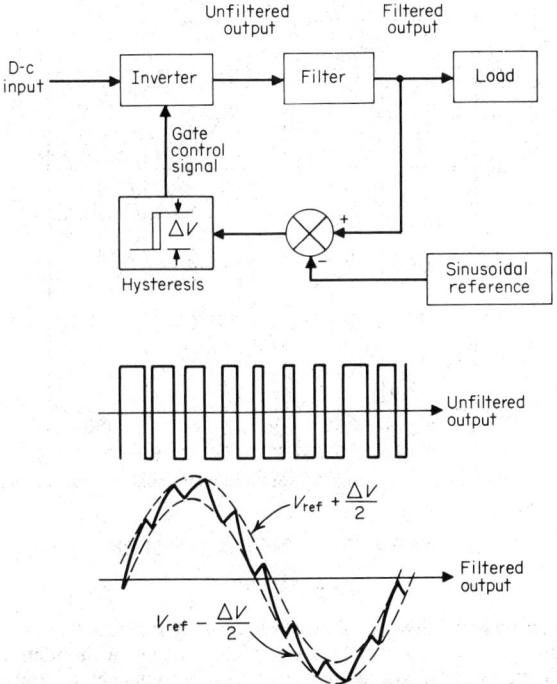

Fig. 15-28. Optimum-response pulse-width modulation.

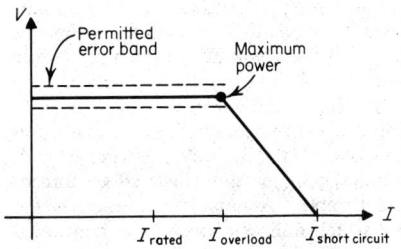

Fig. 15-29. Typical inverter load profile.

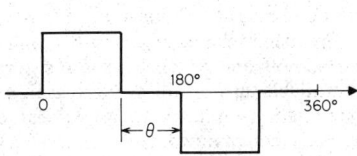

Fig. 15-30. Pulse-width voltage control of a full-bridge inverter.

cosine dependence on θ, and all neutralized harmonics remain neutralized. The remaining harmonics vary with θ.

Pulse-width control of programmed-waveform inverters is more difficult. To preserve the harmonic cancellation, each conduction period must be reduced by the same proportion while maintaining its relative position within the cycle constant. The width of each pulse *(picket)* must be reduced from both directions about the center of that picket.

For carrier-modulated and optimum-response PWM inverters, pulse-width control is accomplished by simply reducing the width of each pulse. This reduction occurs automatically as the amplitude of the sinusoidal reference is decreased.

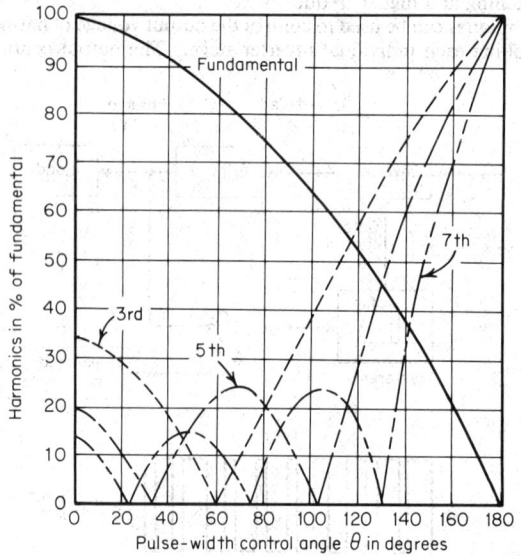

Fig. 15-31. Variation of harmonics for a pulse-width-controlled square wave.

POWER FREQUENCY CHANGERS

By L. Gyugyi

35. Basic Principles and Circuits. Power frequency changers are static systems usually employing solid-state switching devices, capable of directly, i.e., without an intermediate dc link, converting single or polyphase ac power of a given frequency to single or polyphase power of a chosen frequency. They may be used to link two ac power systems of different frequencies, to provide power at controllable frequency for variable-speed ac motor drives, or to convert the output of variable-speed ac generators to constant frequency.

Functionally, frequency changers are wave synthesizers. They fabricate the output-voltage wave(s) of desired amplitude and frequency by sequentially applying appropriate segments of the input-voltage wave(s) to the output. This is accomplished by arrays of static switches arranged to make *bilateral connections,* for controlled time intervals, between the input and output terminals, i.e., between the supply voltages and loads.

Frequency changers generally require controllable power switches with intrinsic turn-on and turnoff ability (such as transistors and gate-controlled switches) or switches with controllable turn-on ability (such as thyristors and triacs) complemented by auxiliary forced-commutation circuitry to implement controllable turnoff. A notable exception is the naturally commutated cycloconverter (Par. 15-40), which utilizes conventional controlled rectifiers.

The basic circuit configurations of static frequency changers are identical with polyphase converters characterized by their pulse number (see Tables 15-4 to 15-6) except that each unidirectional thyristor is replaced by a bidirectional ac switch. Typical bilateral solid-state switch configurations, applicable in frequency-changer circuits, are shown in Fig. 15-15. As in converters, increased pulse number leads to reduced distortion of the output-voltage and input-current waves. In practical applications, frequency changers are often required to produce 3-phase output; in this case three identical converter circuits, one for each output phase, are employed. Three-pulse frequency changers with single-phase and 3-phase output are shown in Fig. 15-32a and b, respectively. The bilateral-switch symbols represent any one of the previously described bidirectional solid-state switch arrangements.

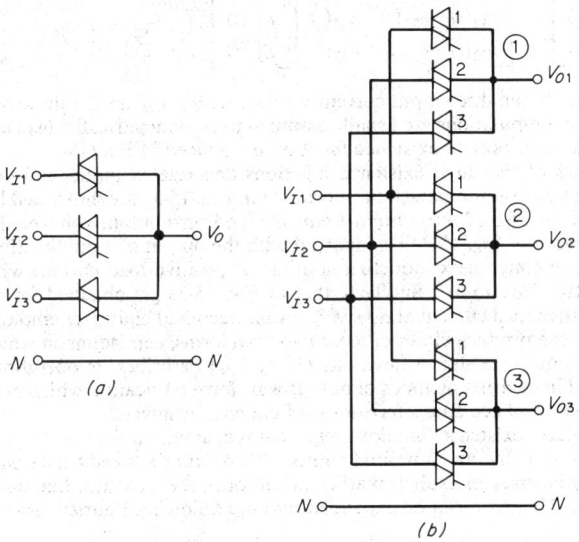

Fig. 15-32. Three-pulse frequency changers (a) with single-phase output and (b) with 3-phase output.

Frequency changers fabricate the output-voltage wave with the desired (or "wanted") frequency and amplitude by sequentially connecting the input voltages to the output(s) for appropriate time intervals. The output-voltage wave(s) are thus composed of segments of the input-voltage waves. The length of each segment is determined by the duration of closure of the corresponding switch. However, an output-voltage wave of given frequency and amplitude can be obtained in several distinctly different ways (Gyugyi, 1970) characterized by the control (modulation) of the repetition rate and/or duration of switch closures. The method of output-waveform fabrication uniquely determines the external performance characteristics of the frequency changer, the most important of which are the distortion of the output-voltage and input-current waves and the input-displacement and power factors.

36. Fundamental Principles. Consider the simple three-pulse frequency-changer circuits shown in Fig. 15-32. These circuits convert 3-phase input power of frequency f_I into a single- or 3-phase output power of frequency f_O. The relationship between the input and the generated output *voltage waves* can be described by the matrix equation

$$[v_O(t)] = [H(t)][v_I(t)]$$

or
$$\begin{bmatrix} v_{O1}(t) \\ v_{O2}(t) \\ v_{O3}(t) \end{bmatrix} = \begin{bmatrix} h_{11}(t) & h_{12}(t) & h_{13}(t) \\ h_{21}(t) & h_{22}(t) & h_{23}(t) \\ h_{31}(t) & h_{32}(t) & h_{33}(t) \end{bmatrix} \begin{bmatrix} v_{I1}(t) \\ v_{I2}(t) \\ v_{I3}(t) \end{bmatrix}$$

(15-1)

where v_{O1}, v_{O2}, v_{O3} are the time functions of generated voltage waves; v_{I1}, v_{I2}, v_{I3} are the three input-voltage waves, which are usually sinusoids, that is,

$$v_{I1} = V_I \sin \omega_I t \qquad v_{I2} = V_I \sin\left(\omega_I t - \frac{2\pi}{3}\right) \qquad v_{I3} = V_I \sin\left(\omega_I t - \frac{4\pi}{3}\right)$$

and each h_{ij} ($i = 1, 2, 3; j = 1, 2, 3$) is a time-varying existence function which defines whether a given switch h_{ij}, connecting output terminal i to input terminal j, is open ($h_{ij} = 0$) or closed ($h_{ij} = 1$) at a given time t.

The input *current waves* drawn from the supply by a three-pulse frequency changer can be similarly expressed in terms of the output (load) currents:

$$[i_I(t)] = [H(t)]^T [i_O(t)]$$

or

$$\begin{bmatrix} i_{I1}(t) \\ i_{I2}(t) \\ i_{I3}(t) \end{bmatrix} = \begin{bmatrix} h_{11}(t) & h_{21}(t) & h_{31}(t) \\ h_{12}(t) & h_{22}(t) & h_{32}(t) \\ h_{13}(t) & h_{23}(t) & h_{33}(t) \end{bmatrix} \begin{bmatrix} i_{O1}(t) \\ i_{O2}(t) \\ i_{O3}(t) \end{bmatrix} \tag{15-2}$$

where i_{I1}, i_{I2}, i_{I3} are the three input current waves, i_{O1}, i_{O2}, i_{O3} are the three output current waves, which for computations are usually assumed to be symmetrically displaced sinusoids, and each h_{ij} is an appropriate existence function introduced in Eq. (15-1).

The basic sets of the three existence functions and related input- and output voltage waveforms of the three-pulse power circuit shown in Fig. 15-32 are illustrated in Fig. 15-33a and b for the trivial case of zero output frequency and zero output voltage. Note that the output-voltage wave of Fig. 15-33a is identical with the output of a unidirectional naturally commutated ac-dc converter conducting continuous positive load current when the delay angle α is 90° (see Fig. 15-5). Similarly that of Fig. 15-33b is obtained from a converter conducting negative load current at $\alpha = 90°$. A bidirectional converter employing bilateral turnoff switches can produce either of these two waveforms, depending on which of the two sets of complementary existence functions (h_{ij} or $h_{ij\pi}$) describes its operation. This free option is utilized in devising methods of output-waveform fabrication which provide desired operating and performance characteristics for frequency changers.

The unmodulated existence functions represent rectangular pulses with repetition period $1/f_I$, pulse duration $1/3f_I$, and amplitude unity. To obtain the steady-state output voltages of the frequency changer in explicit mathematical form, the existence functions h_{ij} and $h_{ij\pi}$ shown in Fig. 15-33a and b can be expanded into the following Fourier series:

$$h_{ij}(\omega_I t) = \frac{1}{3} - \frac{2}{\pi} \sum_{n=0}^{\infty} \frac{\sin n \frac{2\pi}{3}}{n} \cos \left\{ n \left[\omega_I t - (j-1)\frac{2\pi}{3} \right] \right\} \tag{15-3}$$

and

$$h_{ij\pi}(\omega_I t) = \frac{1}{3} + \frac{2}{\pi} \sum_{n=0}^{\infty} \frac{\sin n \frac{\pi}{3}}{n} \cos \left\{ n \left[\omega_I t - (j-1)\frac{2\pi}{3} \right] \right\} \tag{15-4}$$

where $i = 1, 2, 3$, $j = 1, 2, 3$, and subscript π indicates that the second set is displaced by π rad with respect to the first.

The existence functions defined by Eqs. (15-3) and (15-4) describe two complementary but otherwise equivalent modes of operation of the switches in the power converter which result in zero desired output frequency and voltage. (The output waveform consists entirely of "unwanted" components.)

A desired output frequency differing from zero is obtained by appropriately modulating the repetition frequency of the basic existence functions, which is equivalent to varying the commencement and/or duration of the conduction intervals of the corresponding switches. Mathematically this means that a modulating function $M(\omega_0 t)$ ($\omega_0 = 2\pi f_0$; f_0 is the "wanted" output frequency), is added to the arguments of the basic existence functions given by Eqs. (15-3) and (15-4). To keep the wanted (fundamental) component identical in the two output-voltage waves obtainable by the use of the two complementary sets of existence functions, the modulating functions must also be mutually displaced by π. Similarly, for balanced 3-phase output, the modulating functions used to generate the three output voltage waves must also be mutually displaced by $2\pi/3$. Thus

$$h_{ij}(\omega_I t, \omega_0 t) = \frac{1}{3} - \frac{2}{\pi} \sum_{n=0}^{\infty} \frac{\sin n \frac{2\pi}{3}}{n} \cos \left\{ n \left[\omega_I t + M \left[\omega_0 t - (i-1)\frac{2\pi}{3} \right] - (j-1)\frac{2\pi}{3} \right] \right\} \tag{15-5}$$

and

$$h_{ij\pi}(\omega_I t, \omega_0 t) = \frac{1}{3} + \frac{2}{\pi} \sum_{n=0}^{\infty} \frac{\sin n\frac{\pi}{3}}{n} \cos\left\{ n\left[\omega_I t + M\left[\omega_0 t - (i-1)\frac{2\pi}{3} - \pi \right] - (j-1)\frac{2\pi}{3} \right] \right\}$$

(15-6)

The modulating function in Eqs. (15-5) and (15-6) can be visualized as a time-varying angle which effectively changes (modulates) the repetition frequency of the existence functions from or about their quiescent frequency f_I.

Mathematical expressions for the resulting output-voltage waveforms of a general P-pulse frequency changer can be obtained from Eqs. (15-1) and (15-5) and (15-1) and (15-6) in terms of the modulating functions $M(\omega_0 t)$ and $M(\omega_0 t - \pi)$, respectively (Gyugyi, 1970):

$$v_{Oi} = \frac{3\sqrt{3}}{2\pi} V_I \sin\left[M\left(\omega_0 t - (i-1)\frac{2\pi}{3} \right) \right]$$

$$+ \frac{3\sqrt{3}}{2\pi} V_I \sum_{k=1}^{\infty} \left(\frac{\sin\left\{ Pk\omega_I t + (Pk-1)M\left[\omega_0 t - (i-1)\frac{2\pi}{3} \right] \right\}}{Pk-1} \right.$$

$$\left. + \frac{\sin\left\{ Pk\omega_I t + (Pk+1)M\left[\omega_0 t - (i-1)\frac{2\pi}{3} \right] \right\}}{Pk+1} \right)$$

(15-7)

and

$$v_{Oi\pi} = -\frac{3\sqrt{3}}{2\pi} V_I \sin\left[M\left(\omega_0 t - (i-1)\frac{2\pi}{3} - \pi \right) \right]$$

$$- \frac{3\sqrt{3}}{2\pi} V_I \sum_{k=1}^{\infty} (-1)^k \left(\frac{\sin\left\{ Pk\omega_I t + (Pk-1)M\left[\omega_0 t - (i-1)\frac{2\pi}{3} - \pi \right] \right\}}{Pk-1} \right.$$

$$\left. + \frac{\sin\left\{ Pk\omega_I t + (Pk+1)M\left[\omega_0 t - (i-1)\frac{2\pi}{3} - \pi \right] \right\}}{Pk+1} \right)$$

(15-8)

where $i = 1, 2, 3$, P is the pulse number, and $M(\omega_0 t)$ specifies the modulation of the basic repetition frequency f_I of the existence functions.

Similar equations can be written for the input-current wave, and after laborious computation the performance characteristics can be numerically obtained.

Equations (15-7) and (15-8) indicate that the modulating function entirely determines the operation and performance characteristics of a frequency-changer circuit defined by its pulse number P. The modulating function is actually a mathematical description of the control defining the operation of the power switches and thereby the method of output-waveform fabrication. Various control methods (modulating functions) can be applied to the same power circuit to generate output-voltage (or input-current) waveforms of widely differing characteristics to meet practical requirements. In the following, the five most important operation modes, resulting in practically desirable output waveforms and performance characteristics, will be summarized and illustrated for the case of the single-phase output, three-pulse circuit shown in Fig. 15-32a. All the output waveforms considered can be derived from the two complementary output-voltage waveforms obtained from Eqs. (15-7) and (15-8) by the substitution of the modulating function

$$M(\omega_0 t) = \sin^{-1}(r \sin \omega_0 t)$$

(15-9)

where r is the output-voltage ratio, that is, $r = V_O/V_{Omax}$ ($r \leq 1$). The practical derivation of this modulating function and the subsequent generation of the two complementary waveforms v_O and $v_{O\pi}$ are graphically illustrated for $r = 1$ in Fig. 15-33c to j.

Figures 15-33c and d illustrate that the modulating function $M(\omega_0 t)$ of Eq. (15-9) is a mathematical expression for the well-known sine-wave crossing technique widely used to control the firing angle of thyristor converters (Pelly, 1971, Chap. 9). Using this technique, the magnitude of $M(\omega_0 t)$, and thus the modulation of each existence function h_{ij} about its quiescent point (zero output), is determined by the crossing point of a corresponding timing wave v_T with a reference sinusoid v_R. Derivation of the complementary sets of $M(\omega_0 t)_\pi$ and

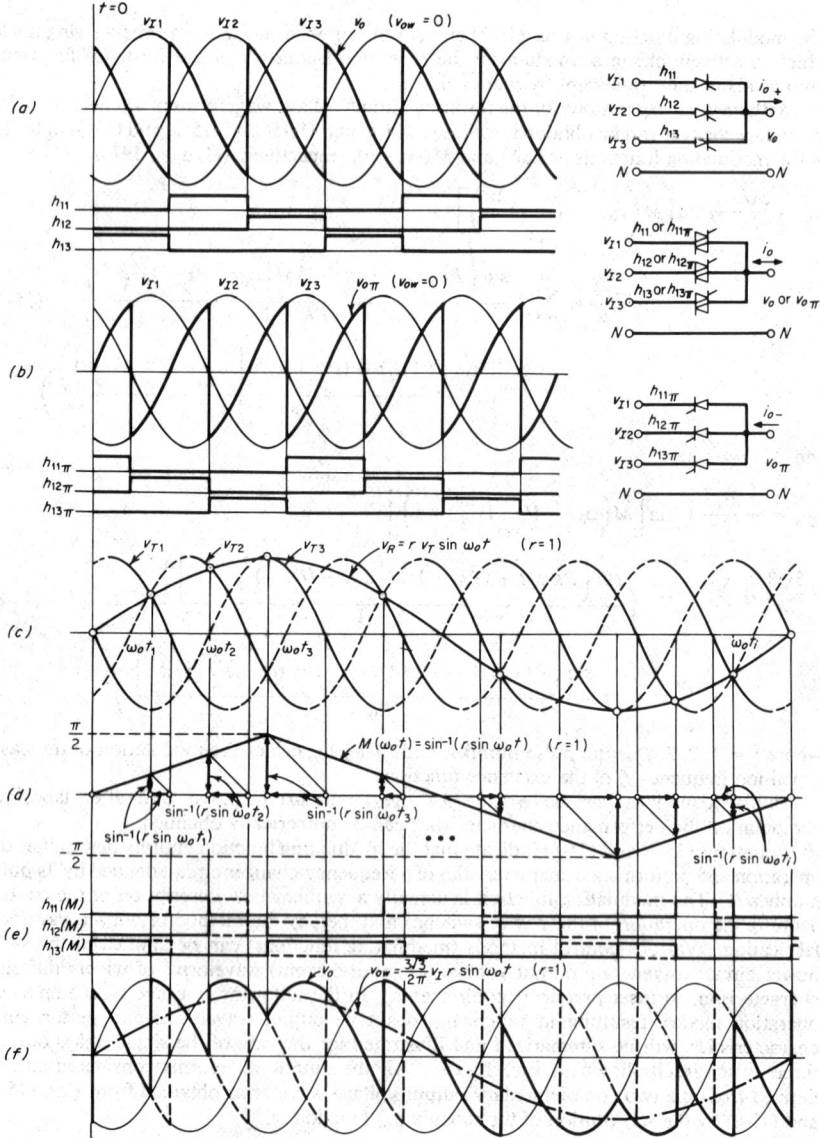

Fig. 15-33. Waveforms illustrating the generation of the two complementary waveforms V_0 and $V_{0\pi}$.

15-46

$h_{ij\pi}$ are shown in Fig. 15-33g to i. The timing waves v_{T1}, v_{T2}, v_{T3}, $v_{T1\pi}$, $v_{T2\pi}$, and $v_{T3\pi}$ are opposite half-period sections of sine waves synchronized to the source voltages with a phase relationship such that at zero reference the mean of the output voltages v_O and $v_{O\pi}$ is zero.

Figures 15-33e, f, i, and j show that the output waveforms v_O and $v_{O\pi}$, at $r = 1.0$, are generated by the act of periodically stepping up and down (v_O) and down and up ($v_{O\pi}$) the original f_I repetition frequency of the existence functions, and thus that of the power switches, to $f_I + f_O$ and $f_I - f_O$, respectively.

The spectral characteristics (frequency and amplitude) of the two complementary output-voltage waveforms v_O and $v_{O\pi}$ shown in Fig. 15-33f and j are identical, as are those of the corresponding input-current waves. Since, however, the two complementary waveforms have a mutually complementing internal relationship (certain characteristics of the output cycle of v_O, observable during a given *half* period, are identical to those of $v_{O\pi}$, observable during the following output half cycle and vice versa), it is possible to fabricate new output waveshapes from the two complementary waveforms which satisfy given output- and/or input-performance requirements. In the following section, synthesis of the four important frequency-changer output waveforms, using the two basic complementary waves, is described, and the pertinent operating conditions and performance characteristics are summarized.

37. Practical Frequency Changers. Utilizing the properties of the complementary output-voltage waveforms derived in the preceding paragraphs, frequency changers having the following special, *mutually exclusive* characteristics can be devised:

1. Unity or controllable-input displacement factor.
2. Natural (input-line) commutation of the power switches.
3. Unity input-power factor and minimum output-voltage distortion.
4. Unrestricted output-to-input frequency ratio.

To establish the necessary operating conditions for the above characteristics, consider Fig. 15-34a and b where the two three-pulse complementary output waveforms v_O and $v_{O\pi}$, together with the voltage waves (v_{I1}, v_{I2}, and v_{I3}) of the 3-phase supply and an assumed sinusoidal load current i_O having an arbitrary phase angle ϕ_0, are shown. The input-current

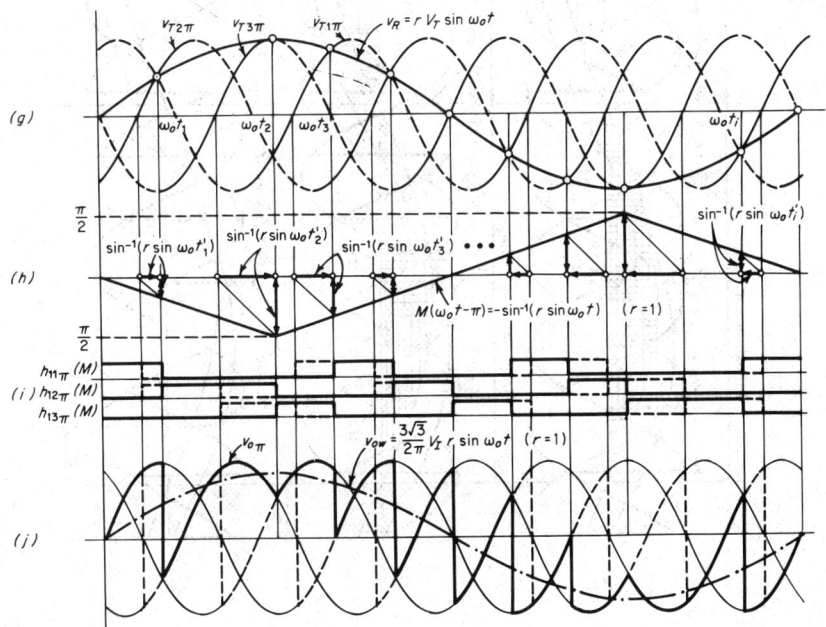

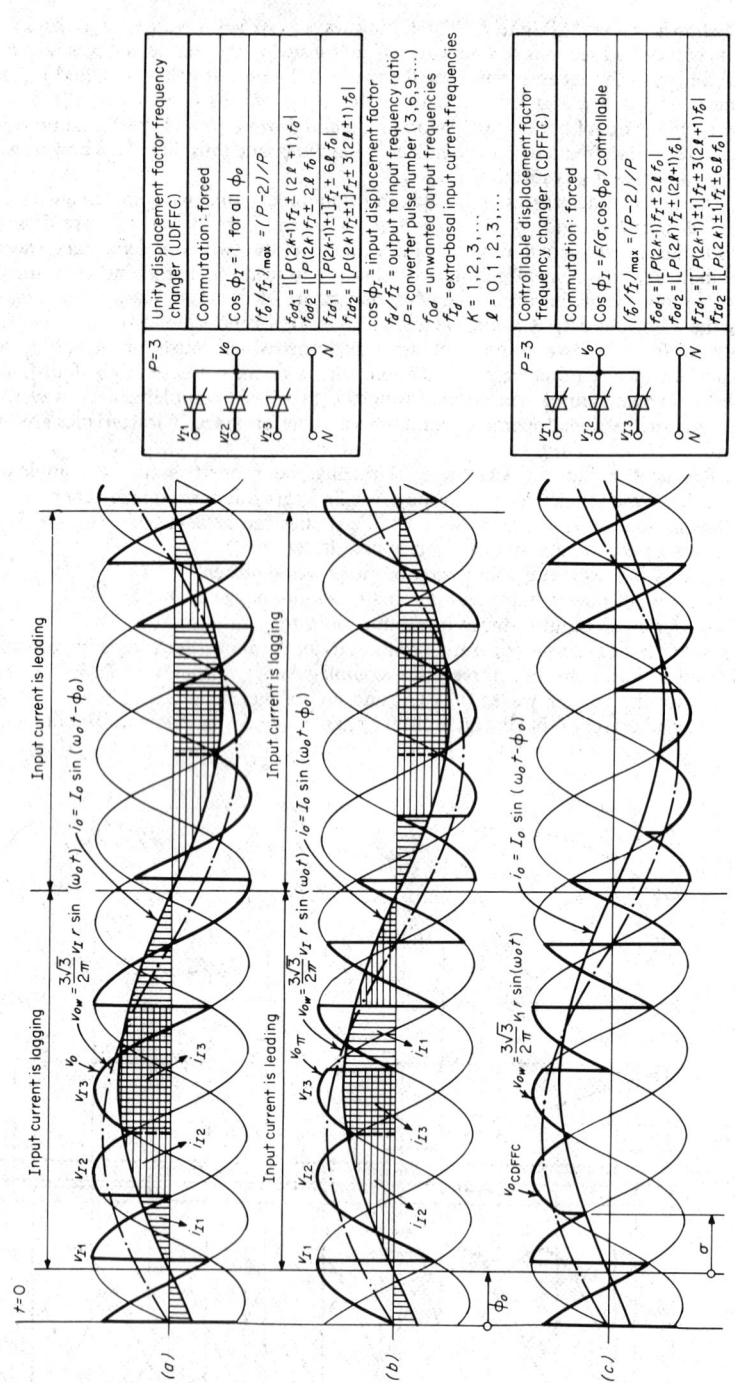

	P=3	Unity displacement factor frequency changer (UDFFC)
		Commutation: forced
		$\cos \phi_I = 1$ for all ϕ_o
		$(f_o/f_I)_{max} = (P-2)/P$
		$f_{od_1} = \lvert[P(2k+1)f_I \pm (2\ell+1)f_o]\rvert$
		$f_{od_2} = \lvert[P(2k)f_I \pm 2\ell f_o]\rvert$
		$f_{Id_1} = \lvert[P(2k+1)\pm1]f_I \pm 6\ell f_o\rvert$
		$f_{Id_2} = \lvert[P(2k)f_I\pm1]f_I \pm 3(2\ell\pm1)f_o\rvert$

$\cos \phi_I$ = input displacement factor
f_o/f_I = output to input frequency ratio
P = converter pulse number $(3,6,9,....)$
f_{od} = unwanted output frequencies
f_{Id} = extra-basal input current frequencies
$K = 1,2,3,....$
$\ell = 0,1,2,3,....$

	P=3	Controllable displacement factor frequency changer (CDFFC)
		Commutation: forced
		$\cos \phi_I = F(\sigma, \cos \phi_o)$ controllable
		$(f_o/f_I)_{max} = (P-2)/P$
		$f_{od_1} = \lvert[P(2k-1)f_I \pm 2\ell f_o]\rvert$
		$f_{od_2} = \lvert[P(2k)f_I \pm (2\ell+1)f_o]\rvert$
		$f_{Id_1} = \lvert[P(2k-1)\pm1]f_I \pm 3(2\ell+1)f_o\rvert$
		$f_{Id_2} = \lvert[P(2k)\pm1]f_I \pm 6\ell f_o\rvert$

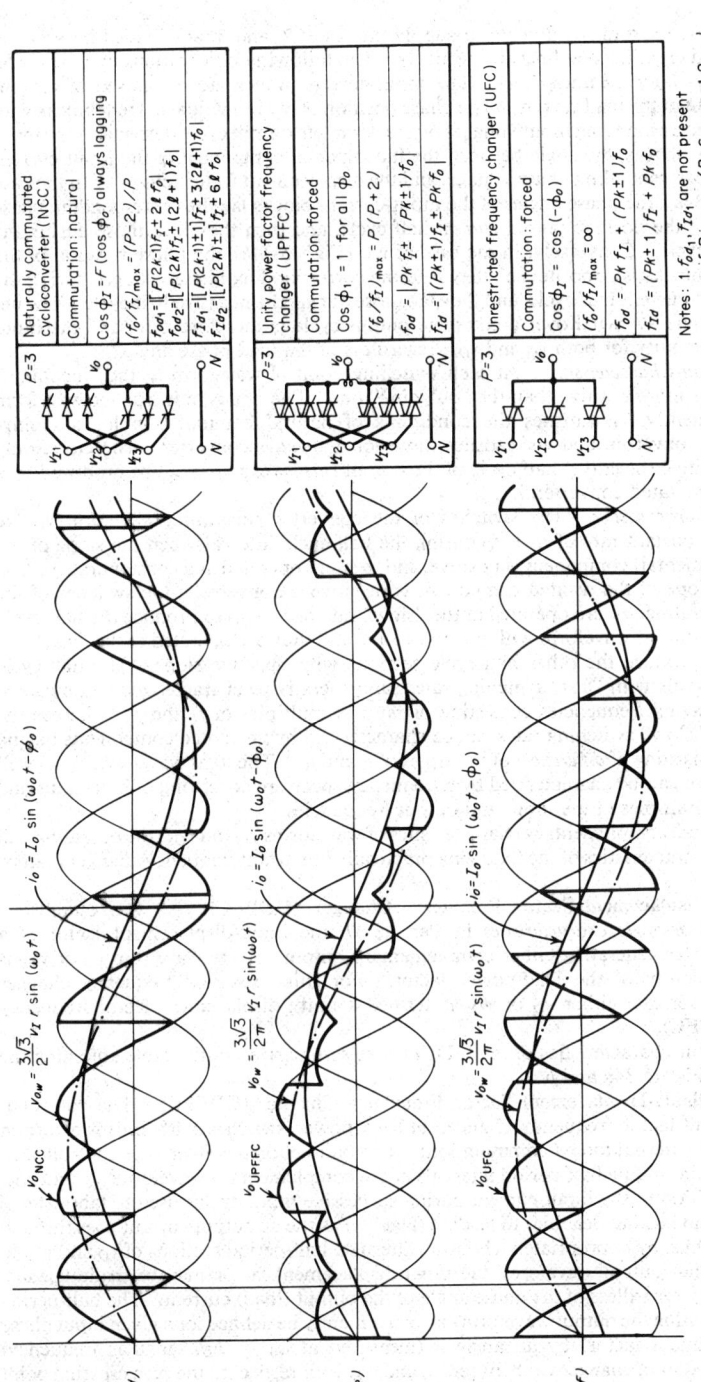

Fig. 15-34. Waveforms illustrating the derivation of practical frequency-changer output waveforms from the two complementary waves.

15–49

waveshapes, i_{I1}, i_{I2}, and i_{I3}, flowing in supply phases 1, 2, and 3, are shaded by vertical, horizontal, and crosshatched lines, respectively. The following observations may be made.

Input-Phase-Angle Characteristics. The input-current waves are composed of current "blocks" cut out of the load current. The phase position of the individual current blocks with respect to the corresponding input voltages over a complete output cycle determines the input phase angle ϕ_I, that is, the angle between the fundamental component of the input-current wave and the corresponding phase voltage, and the displacement factor (cos ϕ_I). In the case of v_O (Fig. 15-34a), the phase angles of the input-current blocks lag the corresponding phase voltages during the positive output-*current* half cycle, and then they lead those during the negative half cycle. Conversely, in the case of $v_{O\pi}$ (Fig. 15-34b), the phase position of the input-current blocks is opposite; i.e., they lead the corresponding phase voltages during the positive output-current half cycle, and they lag those during the negative half cycle. The net input phase angle averaged over a complete output cycle is therefore zero, and the input displacement is unity for both v_O and $v_{O\pi}$, regardless of the load phase angle.

Commutation Characteristics. At each switching point of waveform v_O the "incoming" input voltage is more positive than the "outgoing" one. The opposite is true for waveform $v_{O\pi}$. Consequently, v_O satisfies the conditions of natural commutation for a *positive* unidirectional converter, and $v_{O\pi}$ satisfies those for a *negative* converter. Therefore, only during the positive (negative) half cycle of the output *current* can v_O ($v_{O\pi}$) be produced by a naturally commutated converter.

Spectral Characteristics. The switches of the converter generating waveform v_O are operated at a constant rate of $f_I + f_O$ during the half-cycle interval when the slope of the wanted (fundamental) component is positive, and they are operated at a constant rate of $f_I - f_O$ when the slope of the wanted component is negative. Conversely, the switches of the converter generating $v_{O\pi}$ are operated at the "slow" rate of $f_I - f_O$ to produce the half-cycle sections of the output waveform with positive slope, and they are operated at the "fast" rate of $f_I + f_O$ to produce the other half-cycle sections with negative slope. The half-cycle waveform intervals with "fast" switching rates can generally be characterized by unwanted components having frequencies consisting of *sums* of multiples of f_I and f_O, whereas the intervals with "slow" switching rates can be characterized by unwanted components having frequencies consisting of *differences* of multiples of f_I and f_O. The total waveform, v_O as well as $v_{O\pi}$, therefore, can be characterized by a frequency spectrum consisting of both sums and differences of multiples of the input and output frequencies.

Utilizing these complementary characteristics of waveforms v_O and $v_{O\pi}$, the operating and performance characteristics of the following practically important frequency changers can be established.

38. Unity-Displacement-Factor Frequency Changer (UDFFC). As was established under *input-phase-angle characteristics* in Par. 15-37, the input-displacement factor of a bilateral converter generating either complementary output-voltage waveform v_O or $v_{O\pi}$ is unity independently of the load-power factor. For this reason, a frequency changer controlled to fabricate either v_O or $v_{O\pi}$ is termed a unity displacement factor frequency changer (UDFFC).

The pertinent characteristics of the UDFFC are summarized in the table adjoining the waveforms of Fig. 15-34a and b.

39. Controllable-Displacement-Factor Frequency Changer (CDFFC). The controllable-displacement-factor frequency changer utilizes power switches with intrinsic turnoff ability (or with external forced commutation). The bidirectional converter is controlled so as to generate alternating half-period intervals of the complementary waveforms v_O and $v_{O\pi}$. The phase position of the input current during successive input cycles used to fabricate v_O ($v_{O\pi}$) varies from lagging (leading) to leading (lagging) as the ac output current goes through a full cycle. Thus, by appropriately choosing alternate half-period sections of v_O and $v_{O\pi}$ to fabricate the final output waveform, the input-displacement factor may be made lagging, leading, or unity regardless of the phase angle of the output (load) current. The half-period sections constituting the output waveform can conveniently be defined for a given input phase angle (displacement factor) by an angle σ (measured at $\omega_0 = 2\pi f_O$ angular frequency) specifying the point of changeover between v_O and $v_{O\pi}$ with respect to the zero crossing point of the output current, as illustrated in Fig. 15-34c. The relationship between the input phase angle and angle σ is shown in Fig. 15-35 for various ϕ_0 load phase angles.

The intervals during which the converter switches are operated at high and low rates (f_I

$+ f_O$ and $f_I - f_O$) depend upon the angle σ, that is, the output-to-input displacement-factor transfer. The frequencies of the dominant unwanted components may thus be either the sums or differences, or both, of multiples of f_I and f_O. The frequency spectrum is therefore a function of σ.

The characteristics of the CDFFC are summarized in the table adjoining Fig. 15-34c. The unique input characteristic of the CDFFC offers a number of intriguing application possibilities. One of them is a power-generating system which utilizes a squirrel-cage induction machine and a CDFFC. In this system the reactive-excitation requirement of the induction generator and the reactive kilovolt-ampere demand of the load are provided by the static frequency changer itself. The frequency changer thus has two basic functions: (1) it converts the generally variable generator frequency to precisely regulated output frequency, and (2) it provides a controllable excitation for the induction machine.

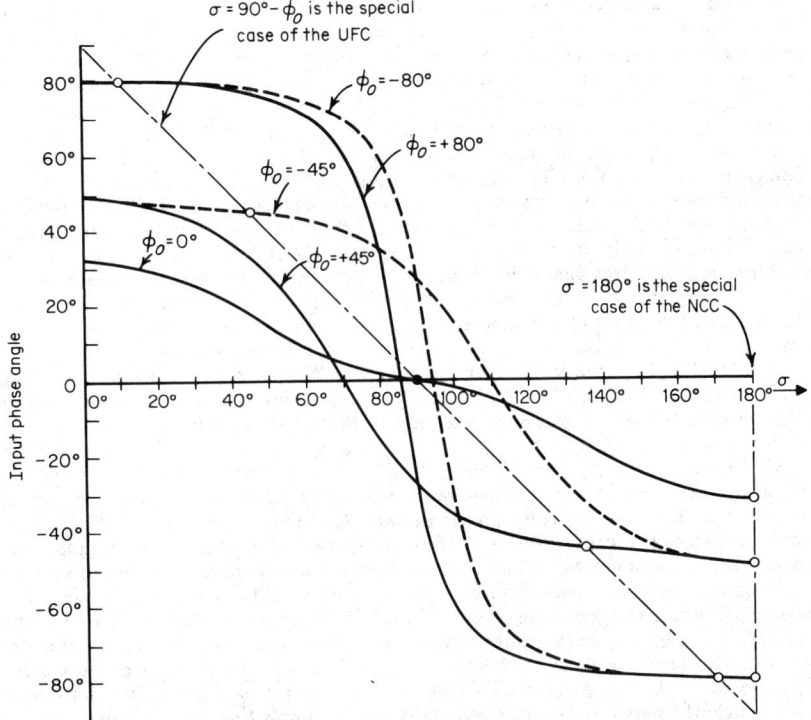

Fig. 15-35. Relationship between the load phase angle ϕ_0, angle σ, and the input phase angle.

40. Naturally Commutated Cycloconverter (NCC). The naturally commutated cyclo-converter, in compliance with the conditions outlined under *commutation characteristics* in **Par. 15-37**, consists of two unidirectional inverse-parallel connected converter circuits (dual converter). The positive and negative converters are controlled to produce output wave-forms v_O and $v_{O\pi}$, respectively. However, the positive converter is gated on only during the positive half cycles of the output *current*, and the negative converter is operated only during the negative half cycles. The output waveform is therefore composed of half-period segments of the complementary waveforms with the changeover taking place between v_O and $v_{O\pi}$ at the zero crossing points of the ac output current, as shown in Fig. 15-34d. This mode of operation ensures that the switches of the converter can operate by natural commutation.

The output voltage waveform v_O results in a lagging input phase angle for positive output current, and $v_{O\pi}$ results in the same for negative output current. Consequently, the input-displacement factor of the NCC is always lagging regardless of the load-power factor; this characteristic is inherent for all naturally commutated phase-controlled converters.

The intervals during which the converter switches are operated at high and low rates ($f_I + f_O$ and $f_I - f_O$) depend upon the load-power factor. The frequencies of the dominant unwanted components may thus be either the sums (leading load-power factor) or differences (lagging load-power factor), or both (unity load-power factor), of multiples of f_I and f_O; therefore the frequency spectrum depends on the load-power factor.

Note that the characteristics of the NCC (frequency spectra, input-displacement factor, etc.) are identical to those of the CDFFC operated at fixed $\sigma \equiv 180°$ (see Fig. 15-35 and tables adjoining Fig. 15-34c and d).

The practical significance of the NCC is due to the fact that presently the voltage and current ratings of thyristors are considerably higher than those of other semiconductor switches having internal turnoff ability. For this reason, the NCC currently offers the most economical if not the only feasible solution to very high power frequency-changer applications.

41. Unity-Power-Factor Frequency Changer (UPFFC). The unity-power-factor frequency changer is the combination of two bilateral converter circuits operated from a common ac source and supplying the same load. One converter is controlled to generate v_O, the other $v_{O\pi}$. The final output waveform is produced by summing, or generating the arithmetic mean of the two complementary waveforms (see Fig. 15-34e). Note that the output rating of the combined system is the sum of the ratings of the constituent converters.

The input-displacement factor of each constituent converter, and thus that of the combined system, is unity. The advantage of this arrangement is that in addition to unity input-displacement factor (regardless of load-power factor), certain groups of unwanted components present in the output-voltage (input-current) waves of the constituent converters cancel out, resulting in greatly improved frequency spectra, increased f_O/f_I ratio, and rms distortion decreased by a factor of $\sqrt{2}$ (see table adjoining Fig. 15-34e). The reduction in the distortion of the input-current wave results in a "near unity" input-power factor λ, which is the product of the input-displacement and current-distortion factors ($\lambda = 0.9$ for a three-pulse, $\lambda = 0.977$ for a six-pulse, and $\lambda = 0.995$ for a twelve-pulse system).

The UPFFC is particularly advantageous in applications where the required output power is higher than that obtainable from a single converter, in which case multiple converters can be used advantageously to increase the power rating as well as to improve the performance of the system.

42. Unrestricted Frequency Changer (UFC). The unrestricted frequency changer utilizes a single bilateral converter whose switches are operated at the constant "fast" rate of $f_I + f_O$. Therefore, as discussed under *spectral characteristics* in Par. **15-37**, the output waveform of the UFC can be synthesized from half-period sections of the two complementary waveforms, v_O providing the output when the slope of the wanted component is positive and $v_{O\pi}$ when it is negative. The changeover points between v_O and $v_{O\pi}$ thus coincide with the peaks of the wanted voltage component (see Fig. 15-34f). Because of the constant switching rate of $f_I + f_O$, the frequencies of the unwanted components are only sums of multiples of f_I and f_O and therefore are always higher than f_O, regardless of the f_O/f_I ratio (see table adjoining Fig. 15-34f). Consequently, the UFC can generate a high-quality output waveform having a frequency which may be lower or higher than the ac supply frequency or equal to it. The described operation of the converter switches also results in a unique output- to input-power-factor transfer characteristic; i.e., the UFC reflects the negative of the load phase angle back to the source (Fig. 15-35) and therefore the input- and output-displacement factors are mirror images of each other (an inductive load is seen capacitive and vice versa).

The UFC is an ideal system to provide ac output power over a wide frequency range (which may extend from zero to well above the ac supply frequency) to control the speed of ac motors.

43. Control of the Output Voltage. In the previously described operation modes, frequency changers supply ac power at the maximum output voltage obtainable from the given supply voltages. In many practical applications, e.g., speed control of ac motors and regulated ac supplies, the effective value of the output-voltage waveform, i.e., the amplitude of the wanted component, has to be controllable independently of the input source. This can be achieved by varying either the depth of modulation used to generate the output waveform or the conduction intervals of the switches while maintaining their repetition rate as required for maximum output voltage (pulse-width modulation). The first type of voltage control can be accomplished simply through sine-wave crossing control by varying the amplitude of the reference wave. This method is compatible, i.e., does not significantly affect the output and input performance characteristics obtained at maximum voltage, with the UDFFC, NCC, CDFFC, and UPFFC (Gyugyi, 1970). Typical three-pulse NCC and UPFFC output waveforms with a relative amplitude of 0.7 ($v_O/v_{O,max}$ = 0.7) are shown in Figs. 15-36a and b, respectively.

Pulse-width-modulation voltage control is the only type which is completely compatible with the UFC (Gyugyi, 1970). Its essence is to subdivide the conduction periods into *active* and *passive* intervals. During the active interval, the switches are operated in the usual manner. During the passive interval, the load is reconnected to the input phase used for the preceding active interval, as illustrated by the three-pulse UFC waveform in Fig. 15-36c.

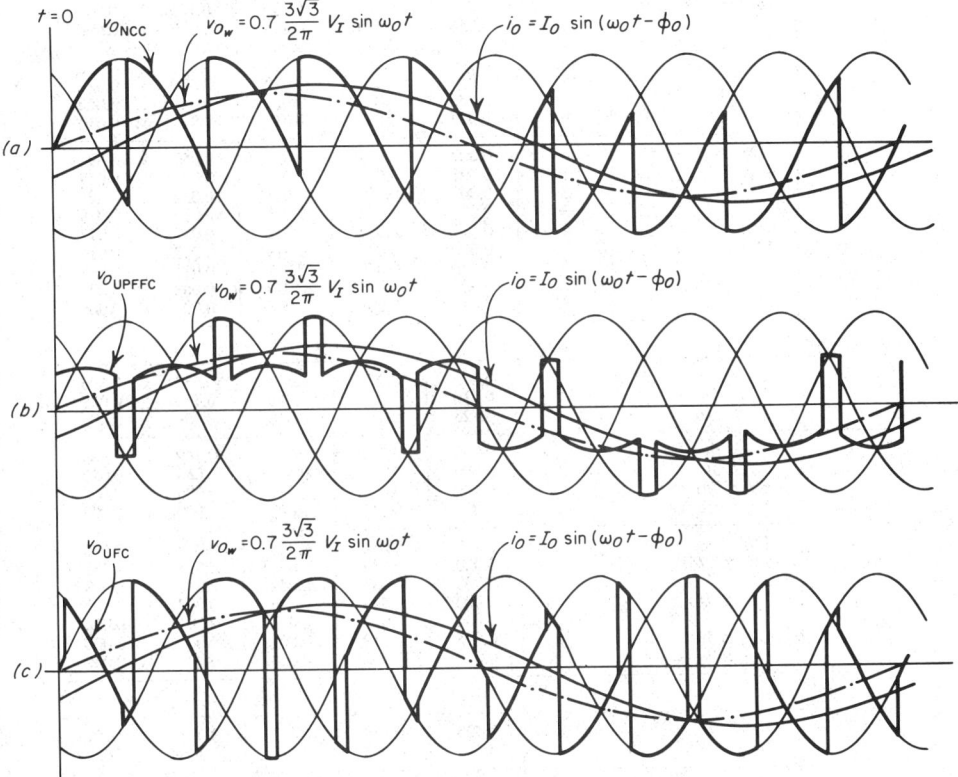

Fig. 15-36. Waveforms illustrating amplitude control of the output voltage by reducing the depth of modulation (*a* and *b*, NCC and UPFFC) and employing pulse-width modulation (*c*, UFC).

[Note that for even-pulse-number converters (six, twelve, etc.), the output voltage is zero and the load is actually shorted during the passive intervals.] By controlling the relative duration of the active and passive intervals within the original conduction period, the mean output voltage can be continuously varied from maximum to zero at any given output frequency.

The amplitudes of the dominant unwanted components present in the output voltage waves of three-, six-, and twelve-pulse NCC, CDFFC, UPFFC, and UFC are given as per unit values of $v_{O,max}$ in Table 15-8 for five discrete values of the output-voltage ratio $r = v_O/v_{O,max}$. The data presented are pertinent to the sine-wave crossing control for the NCC (at unity load-power factor, that is, ϕ_0 = 0 or 180°), the CDFFC (at $\sigma + \phi_0$ = 0 or 180°), and the UPFFC. For the UFC, the data are relevant to PWM control.

Table 15-8. Amplitudes of the Dominant Unwanted Frequency Components of the Output Voltage for Various Voltage Ratios

Number of Pulses		l	r				
			1.0	0.8	0.6	0.4	0.2
	NCC at $\phi_0 = 0$ and CDFFC at $\sigma + \phi_0 = \{\begin{smallmatrix}0\\180°\end{smallmatrix}$						
3	$3f_1 \pm 2lf_0$	0	0.318	0.470	0.596	0.682	0.734
		1	0.272	0.276	0.230	0.163	0.084
		2	0.164	0.112	0.071	0.040	0.018
		3	0.064	0.038	0.025	0.015	0.007
		4	0.021	0.018	0.013	0.008	0.004
6	$6f_1 \pm (2l+1)f_0$	0	0.006	0.069	0.130	0.178	0.208
		1	0.055	0.123	0.138	0.120	0.090
		2	0.105	0.111	0.084	0.060	0.047
		3	0.100	0.065	0.045	0.036	0.032
		4	0.055	0.035	0.029	0.026	0.025
		5	0.026	0.023	0.022	0.021	0.020
		6	0.022	0.021	0.018	0.017	0.017
12	$12f_1 \pm (2l+1)f_0$	0	0.002	0.022	0.035	0.057	0.091
		1	0.001	0.021	0.045	0.071	0.059
		2	0.007	0.029	0.061	0.056	0.029
		3	0.005	0.052	0.051	0.030	0.018
		4	0.027	0.053	0.032	0.017	0.013
		5	0.050	0.036	0.018	0.012	0.010
		6	0.049	0.020	0.012	0.010	0.009
		7	0.027	0.013	0.009	0.008	0.007
		8	0.011	0.009	0.008	0.007	0.006
	UPFCC						
3	$3f_1$		0.000	0.099	0.310	0.530	0.690
	$3f_1 \pm 2f_0$		0.250	0.275	0.210	0.110	0.028
	$3f_1 \pm 4f_0$		0.125	0.050	0.018	0.004	0.000
6	$6f_1 \pm f_0$		0.000	0.061	0.025	0.169	0.163
	$6f_1 \pm 3f_0$		0.000	0.098	0.133	0.068	0.012
	$6f_1 \pm 5f_0$		0.100	0.089	0.031	0.006	0.000
	$6f_1 \pm 7f_0$		0.071	0.013	0.003	0.000	0.000
12	$12f_1 \pm f_0$		0.300	0.014	0.020	0.047	0.086
	$12f_1 \pm 3f_0$		0.000	0.019	0.035	0.063	0.033
	$12f_1 \pm 5f_0$		0.000	0.027	0.047	0.040	0.003
	$12f_1 \pm 7f_0$		0.000	0.026	0.046	0.007	0.000
	$12f_1 \pm 9f_0$		0.000	0.052	0.012	0.001	0.000
	$12f_1 \pm 11f_0$		0.045	0.019	0.001	0.000	0.000
	$12f_1 \pm 13f_0$		0.038	0.002	0.000	0.000	0.000
	UFC						
3	$3f_1 + 2f_0$		0.500	0.678	0.822	0.922	0.981
	$3f_1 + 4f_0$		0.250	0.038	0.171	0.346	0.464
6	$6f_1 + 5f_0$		0.200	0.350	0.399	0.335	0.196
	$6f_1 + 7f_0$		0.143	0.071	0.235	0.229	0.179
12	$12f_1 + 11f_0$		0.091	0.176	0.037	0.141	0.162
	$12f_1 + 13f_0$		0.077	0.121	0.111	0.075	0.148

OTHER CONSIDERATIONS

44. Thermal Considerations. The internal or junction temperature of a solid-state device has considerable influence on the terminal characteristics of the device. The reverse leakage current of a pn junction, for example, increases rapidly with temperature. In a thyristor, this leakage current in the forward blocking state plays a role which is similar to gate current. Hence if the junction temperature of a thyristor exceeds its rated maximum (usually 125°C) even instantaneously, the thyristor cannot be depended upon to block forward voltage until the temperature drops. As a result, the circuit may temporarily malfunction or be permanently faulted.

Consequently, precautions must be taken to ensure that both average and instantaneous peak junction temperatures are within the limits of good design. The average junction-temperature rise can be calculated by multiplying the dissipation (averaged over a cycle) by the total steady-state thermal resistance from the junction to the cooling medium, which is assumed to have a known constant temperature. The total thermal resistance is the sum of several components, including the junction-to-case thermal resistance of the device itself, the thermal resistance of the heat sink to the coolant, and the thermal resistances of any interfaces through which the heat must flow.

In normal 60-Hz operation, the average dissipation will be determined primarily by the forward current and voltage drop and by the proportion of the cycle during which conduction takes place. Manufacturers' data sheets usually show this dissipation plotted against average current for various conduction angles. For operation at higher frequencies, the switching dissipation during turnoff and turn-on must be included and may become predominant (Golden, 1971).

The junction-to-case thermal resistance is determined by the internal construction of the device and is specified on the manufacturer's data sheet. Typical values range from about 1°C/W for a device with a ¼-in. stud to 0.06°C/W for double-sided cooling of a flat device. To minimize the interface resistance, the contacting surfaces must be smooth and flat and should be properly covered by a thermal lubricant. Stud nuts should be tightened to the recommended torque to minimize thermal resistance without causing internal damage from excessive mechanical stresses. Typical case-to-sink thermal resistances range from 0.14 to 0.05°C/W.

The heat sink-to-ambient thermal resistance varies widely, depending on the type of heat sink and on the type and flow rate of the coolant. The thermal resistance of several typical air-cooled heat sinks is plotted in Fig. 15-37 as a function of air velocity. Water-cooled heat sinks are usually designed for a thermal resistance of 0.1°C/W or less.

45. Transient Thermal Impedance. The calculation of instantaneous junction temperature involves thermal capacity (specific heat and mass) as well as thermal resistance. Because solid-state power devices are very small compared to other electrical components, such as transformers and motors, with comparable power ratings, their thermal capacities are also much smaller. Consequently, the thermal response of a solid-state device to transient, overload, and fault conditions requires close attention if long life and reliable operation are to be achieved.

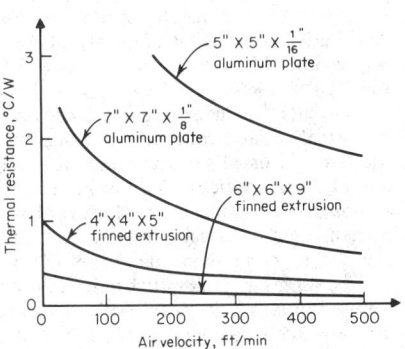

Fig. 15-37. Steady-state thermal resistance of typical air-cooled heat sinks.

Manufacturers' data sheets for power devices normally contain a graph of transient thermal impedance plotted as a function of time. Physically this plot shows how the junction temperature would rise as a function of time following the application of a 1-W step function of dissipation. In practice, the transient temperature rise resulting from a rectangular pulse of power of duration t is obtained by multiplying the transient thermal impedance at t by the amplitude of the power pulse. If the power during the pulse is not constant, a stepped approximation and linear superposition may be used to calculate the temperature response because heat conduction is a linear process (Gutzwiller and Sylvan, 1961). The same curve of transient thermal impedance may also be used to calculate the rate of cooling of the junction following a power pulse.

This method of finding the instantaneous junction temperature is valid once the total area of the device is turned on. During the first 20 μs or so of the turn-on transient, the conducting area is increasing rapidly, and the hot-spot temperature is held to a safe value by observing the turn-on di/dt rating (Mapham, 1964).

For long pulse durations, internal heat flow approaches equilibrium, and the transient thermal impedance approaches the steady-state thermal resistance as an upper limit. The time at which the transient thermal impedance becomes nearly equal to its limit is the thermal time constant of the device and its package. If the duration of a transient overload exceeds this time, the transient thermal impedance of the heat sink must be added to that of the device to obtain the thermal response. For shorter times, the transient temperature rise is determined solely by the internal construction of the device.

46. Protection of Thyristors. In many applications, if a thyristor or other solid-state device is specified only on the basis of its steady-state ratings, it is likely to fail immediately or after very short service. To achieve the reliability and long life for which these devices are well known, the surrounding circuit must be carefully designed to protect them from transients and overloads which exceed their ratings. Of particular importance are both the maximum dv/dt and peak voltage of transients when a device is in its off state, the maximum di/dt when it is being turned on, and the peak current once it is fully on.

A capacitor (typically 0.1 to 1 μF) is frequently connected in parallel with a thyristor to act as a "snubber" which limits dv/dt to prevent unintentional firing and also absorbs energy from voltage spikes. A resistor of 10 to 50 Ω is usually required in series with this capacitor to prevent excessive di/dt when the thyristor is turned on (McMurray, 1972; Rice and Nickels, 1968; von Zastrow and Galloway, 1965).

Excessive voltage transients may also be limited by nonlinear voltage suppressors, which are matched to the maximum voltage rating of the thyristor and have sufficient energy capacity to dissipate the transient (Gutzwiller, 1959; IEEE Comm. Rep., 1970; Harnden, 1972; Lawatsch and Weisshaar, 1972).

The di/dt at turn-on is usually limited by the inductance inherent in the circuit or by an inductor added for this purpose (Paice and Wood, 1967). High-frequency inverters and other applications requiring high di/dt can utilize fast turn-on thyristors developed especially for this purpose. The gate construction of these devices causes conduction to be initiated over a larger area than in conventional thyristors, thereby increasing the di/dt rating. The magnitude and rise time of the signal applied to the gate also influence the di/dt capability of a device. Manufacturers' recommendations should be followed for dependable performance (Dyer, 1966).

Currents which exceed the steady-state rating of a device fall into two categories. Abnormal or fault conditions presumably occur infrequently. Under these conditions the equipment is usually made temporarily inoperative if necessary to prevent the destruction of the solid-state devices. The surge rating of a device determines when this must be done. Although the terminal characteristics of a device should not be noticeably different after it has been stressed to its surge rating, this is a "nonrepetitive" rating. A device should not be exposed to more than 100 surges of this intensity during the entire expected lifetime.

The surge current through a device is normally limited by a fast-acting series fuse. In the past, I^2t ratings have been used to achieve proper coordination between the fuse and the device which it protects (Gutzwiller, 1958). However, this rating is not constant but varies with the peak current and the clearing time of the fuse. Recently $I^2\sqrt{t}$ has been proposed as a rating which is more nearly constant and hence more valid (Motto, 1971). Another approach to protection which may sometimes be used in place of fuses involves modification of the gate circuit such that subsequent firing pulses are disabled when a fault is detected.

In some applications, such as motor controls, the solid-state devices are repetitively loaded beyond their steady-state ratings for short periods of time. This type of overload is "normal" in these applications and obviously must not blow a fuse each time. The capacity of a device to withstand this type of overload is determined by the allowable peak junction temperature and the transient thermal impedance. Applications of this type may require that oversize solid-state devices be chosen on the basis of their overload capacity rather than their steady-state ratings.

High-voltage or high-current applications which exceed the available ratings of individual solid-state devices require that multiple devices be combined in series or parallel arrays. In theory, matched devices could be used in an array, but in practice typical spreads of device characteristics frequently necessitate equalizing networks to avoid stress concentrations under both steady-state and transient conditions (Rice, 1970; Grafham and Hey, 1972).

Parallel devices tend to share current unequally, with the highest current flowing in the device with the lowest forward drop. An impedance in series with each device helps to equalize the steady-state current sharing. Fortunately at high current levels the temperature coefficient of forward voltage becomes positive, so that surge currents tend to be self-equalizing.

Because of the variation in leakage currents, series devices share blocking voltages unequally in steady state. Also, the differences in junction capacitance and switching times cause unequal voltage distributions during transients, turn-on, or turnoff. An *RC* network in parallel with each device can be designed to equalize both the steady-state and transient voltage distribution.

POWER FILTERS

47. General Considerations. Of necessity, high-power controls and converters utilize switching devices because these permit high power efficiency to be achieved. But the switching action generates transients and spurious frequencies (usually harmonics of the fundamental frequency) which may have intolerable effects on the power source, the load, and/or other nearby equipment, by way of electromagnetic interference (EMI) radiated or conducted through the supply line. Frequently the internal design of a power converter, such as a harmonic-neutralized inverter, is chosen to minimize the most troublesome spurious frequencies. In addition it is often necessary to filter the input, the output, or both.

Because all power filters must handle substantial volt-amperes, their component dissipation, cost, size, and weight are all important design factors, although these are usually not optimized by conventional small-signal filter-synthesis procedures. Basically, a low-loss filter operates by storing energy during an unwanted peak in a waveform and then discharging the energy back into the circuit during an unwanted trough. The cost of the filter obviously increases as the required energy storage increases.

Power filters may be classified according to whether their main purpose is to improve the power waveform or to remove EMI. Filters for waveform improvement usually deal with frequencies in the audio range. EMI filters are usually concerned with frequencies of 455 kHz or higher, although coupling to telephone lines or interference with low and very low frequency communications can be a problem at much lower frequencies.

48. Input Filters. Input filters for waveform improvement normally consist of three to five series-resonant traps across the input power lines. These traps provide a low-impedance path in which the dominant low-order harmonic currents required by the converter can circulate. Without a filter, these currents would have to circulate through the source impedance of the supply line, thereby deteriorating the voltage waveform.

$L_1 C_1 R_1$ in Fig. 15-38 is a single-tuned trap of this type. The Q of these traps is relatively high, with only enough damping to accommodate variations in line frequency and changes in component values due to initial tolerances, aging, and temperature variation. Although the ease of adjustment is decreased, the cost of filter components can be reduced by combining two single-tuned traps into a double-tuned trap, as shown in the middle of Fig. 15-38.

In contrast to the low-order harmonics which must be individually suppressed by high-Q traps, higher-order harmonics can usually be adequately suppressed by a single damped filter section like that shown at the right of Fig. 15-38.

It should be noted that all these input filters draw leading current at the fundamental fre-

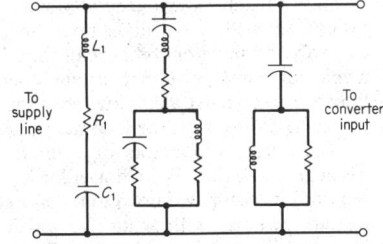

Fig. 15-38. Typical input filter for waveform improvement.

quency, a property which is often useful for power-factor correction. It should also be observed that poles of impedance interleave the zeros which suppress the dominant harmonics. Care must be taken to assure that residual harmonics at other frequencies do not excite undesired resonance at these poles.

If the output frequency is variable, either intentionally (as in a variable-frequency cycloconverter) or because of appreciable variation in the supply frequency, high-Q traps are unsatisfactory and broadband filters must be used.

Design procedures for input-power filters are given in Cory (1965, Chap. 7) and Kimbark (1971, Chap. 8).

Input filters for EMI suppression usually consist of one or more low-pass *LC* L sections between the converter and the supply lines. The input and output terminals of the filter are transposed from the usual low-pass section, as shown in Fig. 15-39, because it is desired to minimize the current-transfer ratio rather than the voltage-transfer ratio. With this transposition, treating the converter as a current source and the supply line as a zero-impedance load, conventional design tables can be used (Geffe, 1964, App. 3). Second-order filters are usually critically damped, i.e., Butterworth response, while filters of higher order can be made to have a steeper edge to the stop band by using the Chebyshev design criterion.

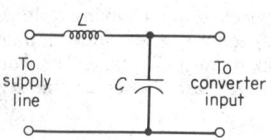

Fig. 15-39. Transposed low-pass input filter to prevent EMI from entering the supply line.

49. Output Filters. Output filters for waveform improvement may be divided into two categories: dc filters for rectifiers and choppers and ac filters for inverters and cycloconverters.

Conventional single-section low-pass *LC* filters are used almost universally for dc applications. The inductor is usually chosen to be larger than the critical value which will maintain continuous current for the worst-case ripple-load combination (see Distler and Munshi, 1965). The capacitor is then chosen to obtain the desired reduction in ripple voltage. However, the resonant frequency must also be chosen so that it does not coincide with a residual harmonic below the fundamental ripple frequency. Although in theory these harmonics of the supply frequency are canceled, the cancellation is never perfect in practice.

The amplitudes of the unfiltered harmonics are summarized in Figs. 15-9 and 15-10, and design data for rectifier filters are presented in Terman (1943, Sec. 8) and Langford-Smith (1953, Chap. 31).

Simple second-order low-pass sections are ordinarily not used to filter the ac output of inverters and cycloconverters because of their insertion loss at the fundamental frequency. However, the series arm can be resonated to minimize this loss in fixed-frequency equipment, as in the Ott filter, which also supplies commutating capacitance (Ott, 1963; Rice, 1970, pp. 8-47 to 8-54). The shunt arms can be series-resonated at the dominant low-order harmonics and/or parallel-resonated at the fundamental frequency.

Passive filters are not suitable for use with variable-frequency converters unless the frequency range is restricted so that the lowest harmonic of the lowest fundamental frequency is considerably higher than the highest fundamental frequency.

50. Filter Components. The capacitors used in power filters are required to pass high currents. Therefore they should always be of extended-foil construction with a dielectric having a low loss over the required frequency range. Paper-oil, plastic-film, polycarbonate, or mica capacitors are generally suitable but must be chosen to have adequate transient ratings as well as steady-state ratings. Filters for dc applications nearly always use electrolytic capacitors because of their lower cost and smaller volume, even if their limited ripple-current capacity necessitates overdesign of the filter to stay within their ratings. The filter designer must also remember that in certain applications, e.g., a rectifier feeding dc power to an inverter, the load may cause significant additional ripple currents.

The design of inductors for dc filters is well established (Langford-Smith, 1953, Chap. 5.6; Terman, 1943, Sec. 2). Chokes for ac filters may be air- or iron-cored, depending on the required inductance, kilovolt-ampere rating, and frequency range. Even iron-cored inductors are designed with an air gap which determines the inductance. The design of low-loss inductors is complicated by skin effect and winding capacitance, which often dictate the use of strip, tubular, or Litz conductors, and by fringing flux at the gap, which may require using powdered iron to distribute the gap.

51. Active Filters. Conventional passive filters frequently represent a substantial part of the total cost, weight, and size of power electronics equipment. For this reason, improved types of filters are constantly being sought. Active filters represent a new approach still in its infancy about which practically nothing has been published. The approach is related to the theory of low-level active-feedback filters but requires considerable adaptation to achieve power efficiency within the capability of available devices. In effect, a power operational amplifier in a feedback loop with a single energy-storage element (an inductor or capacitor) serves to minimize the difference between the actual waveform and the desired waveform.

The "filter" may become an integral part of the converter itself (see the optimum-response inverter described in Fig. 15-28). Although the cost of active filters tends to be high at present, their adaptability permits problems to be solved that could not be solved otherwise, as in a variable-frequency inverter, for instance.

BIBLIOGRAPHY

BATES, J. J., and R. E. COLYER (1973) "The Impact of Semiconductor Devices on Electrical Power Engineering," *Radio and Electronic Engineer*, Vol. 43, No. 1/2, pp. 115-124.

BEDFORD, B. D., and R. G. HOFT (1964) "Principles of Inverter Circuits," Wiley, New York.

CORY, B. J. (ed.) (1965) "High Voltage Direct Current Convertors and Systems," Macdonald, London.

DELHOM, L. A. (1968) "Design and Application of Transistor Switching Circuits," McGraw-Hill, New York.

DISTLER, R. J., and S. G. MUNSHI (1965) "Critical Inductance and Controlled Rectifiers," *IEEE Trans.*, Vol. IECI-12, pp. 34-37.

DYER, R. F. (1966) The Rating and Application of SCR's Designed for Power Switching at High Frequencies, *IEEE Trans.*, Vol. IGA-2, pp. 5-15.

EIA (1971) Recommended Standards for Thyristors, *EIA Stand. RS*-397.

GEFFE, P. (1964) "Simplified Modern Filter Design," Rider, New York.

GENTRY, F. E., et al. (1964) "Semiconductor Controlled Rectifiers: Principles and Applications of PNPN Devices," Prentice-Hall, Englewood Cliffs, N.J.

GEYER, M. A., and A. KERNICK (1971) Time Optimal Response Control of a Two-Pole Single-Phase Inverter, *Power Cond. Spec. Conf. Rec.*, pp. 101-109, *IEEE Pub.* 71C15-AES.

GOLDEN, F. B. (1971) Thyristor Switching Losses and Their Measurement, *Direct Curr. Power Electron.*, Vol. 2, pp. 112-120.

GRAFHAM, D. R., and J. C. HEY (eds.) (1972) *SCR Manual*, 5th ed., General Electric Semiconductor Products Dept., Syracuse, N. Y.

GUTZWILLER, F. W. (1958) The Current-limiting Fuse as Fault Protection for Semiconductor Rectifiers, *Trans. AIEE*, Vol. 77, pt. 1, pp. 751-754.

GUTZWILLER, F. W. (1967) "Thyristors and Rectifier Diodes—The Semiconductor Workhorses," *IEEE Spectrum*, August 1967; pp. 102-111.

——— (ed.) (1967) "General Electric SCR Manual," 4th ed., General Electric Semiconductor Products Department, Syracuse, N.Y.

——— and T. P. SYLVAN (1961) Power Semiconductor Ratings under Transient and Intermittent Loads, *Trans. AIEE*, Vol. 79, Pt. 1, pp. 699-705.

GYUGYI, L. (1970) Generalized Theory of Static Power Frequency Changers, Ph.D. thesis, University of Salford, England.

HARNDEN, J. D., et al (1972) "Metal-Oxide Varistor: A New Way to Suppress Transients," *Electronics*, Oct. 9; pp. 91-95.

HEINRICH, T. M. (1967) Static Inverter with Neutralization of Harmonics, M.S. dissertation, University of Pittsburgh, Pa.

HIBBERD, R. G. (1968) "Solid State Electronics," McGraw-Hill, New York.

HNATEK, E. R. (1971) "Design of Solid-State Power Supplies," Van Nostrand Reinhold, New York.

HOFT, R. G. (1972) "Static Power Converters in the USA," *Conf. Rec. IEEE Intnl. Semiconductor Power Converter Conf.*, pp. 2-8-1 to 8.

IEEE (1967) Special Issue on High-Power Semiconductor Devices, *Proc. IEEE*, Vol. 55, August.

——— (1970) Special Issue on High-Power Semiconductor Devices, *IEEE Trans.*, Vol. ED-17, September.

——— (1972) "Applications of High Power Semiconductor Devices," IEEE Press, New York.

IEEE Comm. Rep. (1970) Bibliography on Surge Voltages in AC Power Circuits Rated 600 Volts and Less, *IEEE Trans.*, Vol. PAS-89, pp. 1056-1061.

KERNICK, A., et al. (1962) Static Inverter with Neutralization of Harmonics, *Trans. AIEE*, Vol. 81, Pt. 2, pp. 59-68.

—— and T. M. HEINRICH (1964)   Controlled Current Feedback in a Static Inverter with Neutralization of Harmonics, *IEEE Trans. Aerosp.*, Vol. 2, pp. 985-992.

—— and I. U. HAQUE (1969)   Programmed Waveform Static Inverter, *Proc. 23d Annu. Power Sources Conf.*, pp. 59-63, sponsored by U.S. Army Electronics Command, Fort Monmouth, N.J.

KIMBARK, E. W. (1963)   A Chart Showing the Relations between Electrical Quantities on the AC and DC Sides of a Converter, *IEEE Trans.*, Vol. PAS-82, pp. 1050-1054.

—— (1971)   "Direct Current Transmission," Vol. I, Wiley-Interscience, New York.

KUSKO, A. (1969)   "Solid-State DC Motor Drives," M.I.T. Press, Cambridge, Mass.

KUSKO, A. (1972)   "Solid-State Motor-Speed Controls," *IEEE Spectrum*, Oct. 1972; pp. 50-55.

LANGFORD-SMITH, F. (ed.) (1953)   "Radiotron Designer's Handbook," RCA, Harrison, N.J.

LAWATSCH, H., and E. WEISSHAAR (1972)   "A Silicon Voltage Limiter for Power Thyristor Circuits," *Brown Boveri Review*, pp. 476-482.

MAPHAM, N. (1964)   The Rating of Silicon Controlled Rectifiers when Switching into High Currents, *Trans. AIEE*, Vol. 83, Pt. 1, pp. 515-519.

McMURRAY, W. (1972)   Optimum Snubbers for Power Semiconductors, *IEEE Trans.*, Vol. IA-8, pp. 593-600.

—— and D. P. SHATTUCK (1961)   A Silicon Controlled Rectifier Inverter with Improved Commutation, *Trans. AIEE*, Vol. 80, Pt. 1, pp. 531-542.

MOTTO, J. W. (1971)   A New Quantity to Describe Power Semiconductor Subcycle Current Ratings, *IEEE Trans.*, Vol. IGA-7, pp. 510-517.

NEWELL, W. E. (1974)   "Power Electronics—Emerging from Limbo," *IEEE Trans.*, Vol. IA-10, Jan./Feb. 1974.

OATES, R. M. (1970)   Inverter Harmonic Neutralization Using Interphase Reactors, M.S. dissertation, University of Pittsburgh, Pa.

OTT, R. R. (1963)   A Filter for SCR Commutation and Harmonic Attenuation in High-Power Inverters, *IEEE Trans. Commun. Electron.*, Vol. 82, pp. 259-262.

PAICE, D. A., and P. WOOD (1967)   Nonlinear Reactors as Protective Elements for Thyristor Circuits, *IEEE Trans.*, Vol. MAG-3, pp. 228-232.

PELLY, B. R. (1971)   "Thyristor Phase-controlled Converters and Cycloconverters," Wiley-Interscience, New York.

RAVAS, R. J., et al. (1967)   Staggered Phase Carrier Cancellation: A New Circuit Concept for Lightweight Static Inverters, *EASTCON Tech. Conv. Rec., Suppl. IEEE Trans.*, Vol. AES-3, No. 6, pp. 432-444; *IEEE Pub.* 10-C-57.

RCA (1971)   "Solid-State Power Circuits (SP-52)," RCA Solid State Division, Somerville, N.J.

RICE, J. B., and L. E. NICKELS (1968)   Commutation *dv/dt* Effects in Thyristor Three-Phase Bridge Converters, *IEEE Trans.*, Vol. IGA-4, pp. 665-672.

RICE, L. R. (ed.) (1970)   "Westinghouse Silicon Controlled Rectifier Designers Handbook," 2d ed., Westinghouse Semiconductor Division, Youngwood, Pa.

RISSIK, H. (1939)   "The Fundamental Theory of Arc Converters," Chapman & Hall, London (contains extensive bibliography).

SCHAEFER, J. (1965)   "Rectifier Circuits: Theory and Design," Wiley, New York.

STORM, H. F. (1969)   "Solid-State Power Electronics in the USA," *IEEE Spectrum*, Oct. 1969; pp. 49-59.

TERMAN, F. E. (1943)   "Radio Engineers' Handbook," McGraw-Hill, New York.

USAS C34.2-1968   "USA Standard Practices and Requirements for Semiconductor Power Rectifiers," American National Standards Institute, New York.

ZASTROW, E. E. VON and J. H. GALLOWAY (1965)   Commutation Behavior of Diffused High Current Rectifier Diodes, *IEEE Trans.*, Vol. IGA-1, pp. 157-166.

SECTION 16

PULSED CIRCUITS AND WAVEFORM GENERATORS

BY

PAUL G. A. JESPERS Professor of Electrical Engineering, Institut d'Electricite, Batiment Maxwell, Catholic University of Louvain, Louvain-la-Neuve, Belgium; Senior Member, Institute of Electrical and Electronics Engineers

CONTENTS

Numbers refer to paragraphs

SECTION 16

PULSED CIRCUITS AND WAVEFORM GENERATORS

PASSIVE WAVEFORM SHAPING

1. Linear Passive Networks. Waveform generation is customarily performed in active nonlinear circuits. Since passive networks, linear as well as nonlinear, enter into the design of pulse-forming circuits, this survey starts with the study of the transient behavior of passive circuits.

Among linear passive networks, the single-pole RC and RL networks are the most widely used. Their transient behavior in fact has a broad field of applications since the responses of many complex higher-order networks are dominated by a single pole; i.e., their response to a step function is very similar to that of a first-order system.

2. Transient Analysis of the RC Integrator. The step-function response of the RC circuit (shown in Fig. 16-1a), after closing of the switch S, is given by

$$V(t) = E\left[1 - \exp\left(-\frac{t}{T}\right)\right] \tag{16-1}$$

where T, the time constant, is equal to the product RC. The inverse of T is called the cutoff pulsation ω_0 of the circuit.

The Taylor series expansion of Eq. (16-1) yields

$$V(t) = E\frac{t}{T}\left(1 - \frac{t}{2!\,T} + \frac{t^2}{3!\,T^2} - \cdots\right) \tag{16-2}$$

When the values of t are small compared to T, a first-order approximation of Eq. (16-2) is

$$V(t) \simeq E\frac{t}{T} \tag{16-3}$$

In other words, the RC circuit of Fig. 16-1 behaves like an imperfect integrator. The relative error ϵ with respect to the true integral response is given by

$$\epsilon = \frac{t}{T}\left(-\frac{t}{2!\,T} + \frac{t^2}{3!\,T^2} - \frac{t^3}{4!\,T^3} + \cdots\right)$$

The experimental step-function response of (16-1) and the ideal-integrator output of (16-3) are represented in Fig. 16-1b.

Small values of t with respect to T correspond in the frequency domain (Fig. 16-1c) to frequency components situated above ω_0, that is, the transient signal whose spectrum lies to the right of ω_0 in the figure. In that case, the difference is small between the response curve of the RC filter and that of an ideal integrator (represented by the -6 dB/octave line in the figure). The circuit shown in Fig. 16-1a thus approximates an integrator, provided either of the following conditions is satisfied: (1) the time under consideration is much smaller than T or (2) the spectrum of the signal lies almost entirely above ω_0.

3. Transient Analysis of the RC Differentiator. When the resistor and the capacitor of the integrator are interchanged, the circuit (Fig. 16-2a) is able to differentiate signals. The step-function response (Fig. 16-2a) is able to differentiate signals. The step-function response (Fig. 16-2b) of the RC differentiator is given by

$$v(t) = E\exp\left(-\frac{t}{T}\right) \tag{16-4}$$

16-2

The time constant T is equal to the product RC, and its inverse ω_0 represents the cutoff of the frequency response of the circuit. As the values of t become large compared to T, the step-function response becomes more like a sharp spike; i.e., it increasingly resembles the delta function (Par. 16-8).

The response differs from the ideal delta function, however, because both its amplitude and its duration are always finite quantities. The area under the exponential pulse, equal to ET, is the important quantity in applications where such a signal is generated to simulate a delta function, as in the measurement of the impulse response of a system. These considerations may be transposed in the frequency domain (Fig. 16-2c).

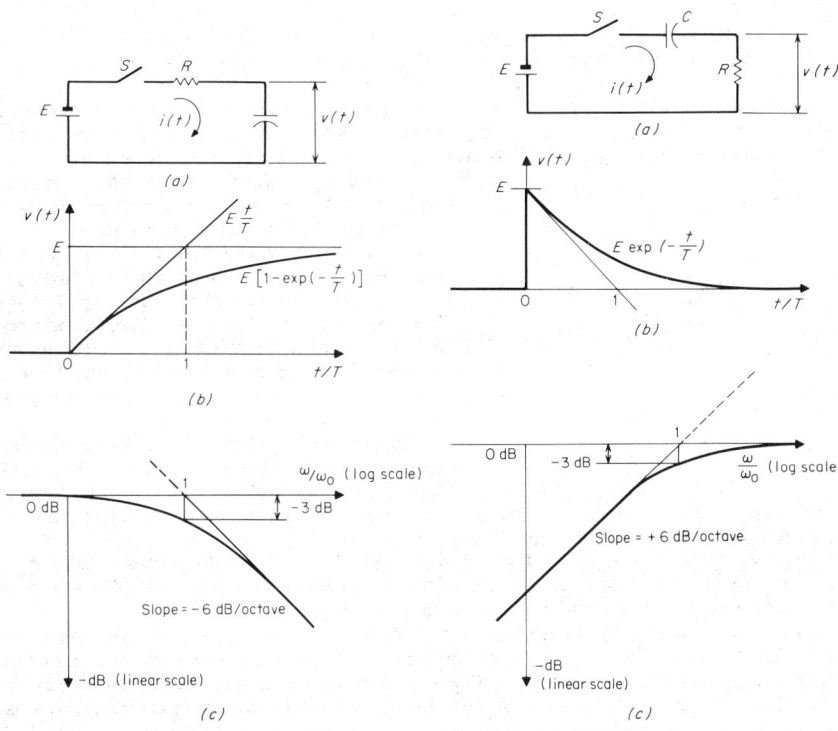

Fig. 16-1. (*a*) *RC* integrator circuit; (*b*) voltage vs. time across capacitor; (*c*) attenuation vs. angular frequency.

Fig. 16-2. (*a*) *RC* differentiator circuit; (*b*) voltage across resistor vs. time; (*c*) attenuation vs. angular frequency.

4. Transient Analysis of RL Networks. Circuits involving a resistor and an inductor are also often used in pulse formation. Since integration and differentiation are related to the functional properties of first-order systems rather than to the topology of actual circuits, *RL* networks may perform the same functions as *RC* networks. The duals of the circuits represented in Figs. 16-1 and 16-2, respectively, are shown in Figs. 16-3 and 16-4 and exhibit identical functional properties. In the first case, the current in the inductor increases

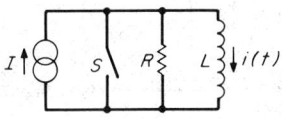

Fig. 16-3. *RL* current integrator circuit, the dual of the circuit in Fig. 16-1*a*.

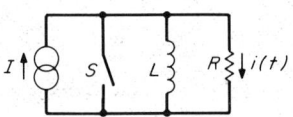

Fig. 16-4. *RL* current differentiator circuit, the dual of the circuit in Fig. 16-2*a*.

exponentially from zero to I with a time constant equal to L/R, while in the second case it drops exponentially from the initial value I to zero, with the same time constant. Similar behavior can be obtained regarding voltage instead of current by changing the circuit from Fig. 16-3 to that of Fig. 16-5 and from Fig. 16-4 to Fig. 16-6, respectively. This duality applies also to the RC case.

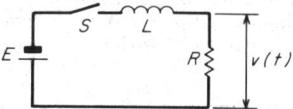

Fig. 16-5. RL voltage integrator.

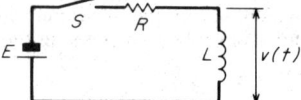

Fig. 16-6. RL voltage differentiator.

5. Compensated Attenuator. The compensated attenuator is a widely used network, e.g., as an attenuator probe used in conjunction with cathode-ray oscilloscopes. The compensated attenuator (Fig. 16-7) is designed to perform the following functions:

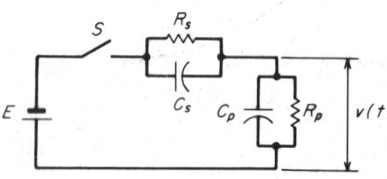

Fig. 16-7. Compensated attenuator circuit.

1. To provide remote sensing with a very high input impedance, thus producing a minimum perturbation to the circuit under test.

2. To deliver a signal to the receiving end (usually the input of a wide-band oscilloscope) which is an accurate replica of the signal at the input of the attenuator probe. These conditions can be met only by introducing substantial attenuation to the signal being measured, but this is a minor drawback since adequate gain to compensate the loss is usually available.

Diagrams of two types of cathode-ray-oscilloscope attenuator probes are given in Fig. 16-8, similar to the circuit of Fig. 16-7. In both cases, the coaxial-cable capacitance parallels the input capacitance of the receiver end; C_p represents the sum of both capacitances.

The shunt resistor R_p has a high value, usually 1 MΩ, while the series resistor R_s is typically 9MΩ. The dc attenuation ratio of the attenuator probe therefore is 1/10, while the input impedance of the probe is 10 times that of the receiver.

At high frequencies the parallel and series capacitors C_p and C_s play the same role as the resistive attenuator. Ideally these capacitors should be kept as low as possible to achieve a high input impedance even at high frequencies. Since it is impossible to reduce C_p below the capacitance of the coaxial cable, there is no alternative other than to insert the appropriate

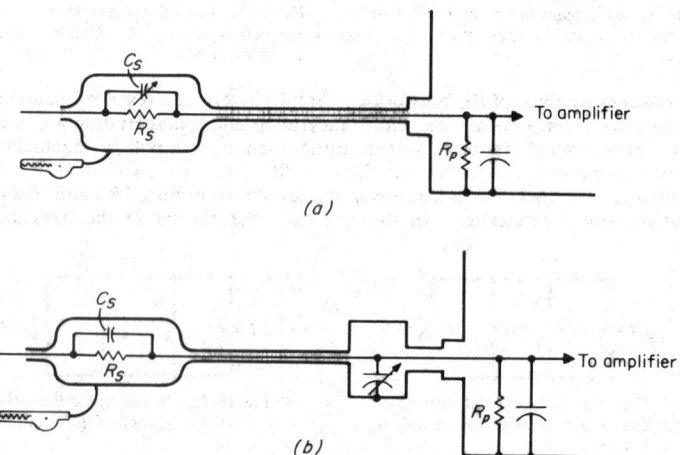

Fig. 16-8. Coaxial-cable type of attenuator circuit: (*a*) series adjustment; (*b*) shunt adjustment.

value of C_s to achieve a constant attenuation ratio over the required frequency band. In consequence, as the frequency increases the nature of the attenuator changes from resistive to capacitive. However, the attenuation ratio remains unaffected, and no signal distortion is produced. The condition that ensures constant attenuation ratio is given by

$$R_p C_p = R_s C_s \qquad (16\text{-}5)$$

The step-function response of the compensated attenuator illustrates clearly how distortion may occur when the above condition is not met. The output voltage $V(t)$ of the attenuator is given by

$$V(t) = \frac{C_s}{C_s + C_p}\left\{1 - (1 - K)\left[1 - \exp\left(-\frac{t}{T}\right)\right]\right\} \qquad (16\text{-}6)$$

where K represents the ratio of the resistive attenuation factor to that of the capacitive attenuation factor

$$K = \left(\frac{R_p}{R_p + R_s}\right)\left(\frac{C_s}{C_p + C_s}\right)$$

and

$$T = (R_p \parallel R_s)(C_s + C_p) \qquad (16\text{-}7)$$

The $\parallel$ sign stands for the parallel combination of two elements, e.g., in the present case $R_p \parallel R_s = R_p R_s/(R_p + R_s)$. Only when K is equal to 1, in other words when Eq. (16-5) is satisfied, will no distortion occur, as shown in Fig. 16-9.

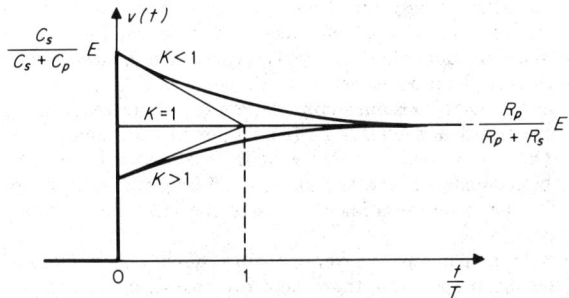

Fig. 16-9. Voltage vs. time responses of attenuator, showing correctly compensated condition at $K = 1$.

In all other cases, there is a difference between the initial amplitude of the step-function response (which is controlled by the attenuation ratio of the capacitive divider) and the steady-state response (which depends on the resistive divider only).

A simple adjustment to compensate an attenuator consists in trimming one capacitor, either C_p and C_s, to obtain the proper step-function response. Adjustments of this kind are provided in attenuators like those shown in Fig. 16-8.

6. Variable Step Attenuators. A variable, step, attenuator is a set of fixed compensated attenuators that may either be interchanged or placed in cascade to achieve desired levels of attenuation. The conditions imposed on a variable step attenuator are like those enumerated in Par. 16-5, but an additional requirement is introduced, namely, the requirement for constant input impedance

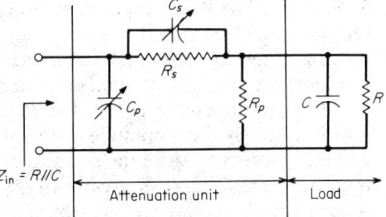

Fig. 16-10. Compensated attenuator suitable for use in cascaded circuits.

when cells are cascaded. This introduces a different structure compared to the compensated attenuator, as shown in Fig. 16-10. The resistances R_p and R_s must be chosen so that the impedance R is kept constant. The capacitor C_s is adjusted to compensate the attenuator, while C_p provides the required additional capacitance to make the input susceptance equal to that of the load.

7. Periodic Input Signals. Repetitive transients are typical input signals to the majority of pulsed circuits. In linear networks there is no difficulty in predicting the response of circuits to a succession of periodic step functions, alternatively positive and negative, since in linear circuits the principle of superposition holds. There is no need for transient solution of linear circuits when the period of the excitation lasts for a time longer than that needed to attain the steady state. Hence we restrict our attention here to two simple cases, the square-wave response of an RC integrator and an RC differentiator.

Figure 16-11 represents, at the left, the buildup of the response of the RC integrator, assuming that the period τ of the input square wave is smaller than the time constant of the circuit T. On the right in the figure the steady-state response is shown. The triangular waveshape represents a fair approximation to the integral of the input square wave. The triangular wave is superimposed on a dc pedestal of amplitude $E/2$. Higher repetition rates of the input reduce the amplitude of the triangular wave without affecting the dc pedestal. When the frequency of the input square wave is high enough, the dc component is the only remaining signal; i.e., the RC integrator then acts like an ideal low-pass filter.

A similar presentation of the behavior of the RC differentiator is shown in Fig. 16-12a and b. The steady-state output in this case is symmetrical with respect to the zero axis because no dc component can flow through the series capacitor. When, as shown in Fig. 16-12b, no overlapping of the pulses occurs, the steady-state solution is obtained from the first step.

8. Delta (Dirac) Function and Corresponding Physical Impulse. The step function and the delta function (Dirac function) are widely used to determine the dynamic behavior of physical systems. Theoretically the delta function is a pulse of infinite amplitude and infinitesimal duration but having a finite area (product of amplitude and time). In practice the question of the equivalent physical impulse arises. The answer to this question involves the system under consideration as well as the impulse itself.

The delta function has a constant amplitude over the whole frequency spectrum from zero to infinity. Other transient signals of finite area (amplitude × time) have different spectral distributions. On a logarithmic scale of frequency, the spectrum of any finite-area transient signal tends to be constant between zero and a cutoff frequency which depends on the shape of the signal. The shorter the duration of the signal, the wider the constant-amplitude portion of the spectrum.

If such a signal is used in a system whose useful frequency band is located within the cutoff frequency of the signal spectrum, the system response is indistinguishable from its delta impulse response. Any transient signal with a finite area, whatever its shape, can thus be considered as a delta function relative to the given system, provided that the flat portion of its spectrum embraces all or a part of the system's useful frequency range. A measure of the effectiveness of a pulse to serve as a delta function is given by the approximation of useful spectrum bandwidth $B = 1/\tau$, where τ represents the mid-height duration of the pulse.

9. Pulse Generators. Very short pulses are used in various applications in order to measure the delta-function response of systems. In the field of radio-frequency interference, for instance, the basic response curve of the receiver is defined in terms of its response to regularly repeated pulses. In this case, the amplitude of the uniform portion of the pulse spectrum must be calibrated, i.e., the area under the pulse must be a known constant which is a function of a limited number of circuit parameters.

The step-function response of an RC differentiator provides such a convenient signal. Its area is given by the amplitude of the input step multiplied by the time constant RC of the circuit. Moreover, since the signal is exponential in shape, its -3 dB spectrum bandwidth is equal to $1/RC$. In the circuit of Fig. 16-13, R_1 is much larger than R; when the switch S is open, the capacitor charges to the voltage E of the dc source. When the switch is suddenly closed, the capacitor discharges through R, producing an exponential signal of known amplitude and duration (known area).

A circuit based on the same principle is shown in Fig. 16-14. Here the coaxial line plays the role of energy storage source. If the line is lossless, its characteristics impedance is given

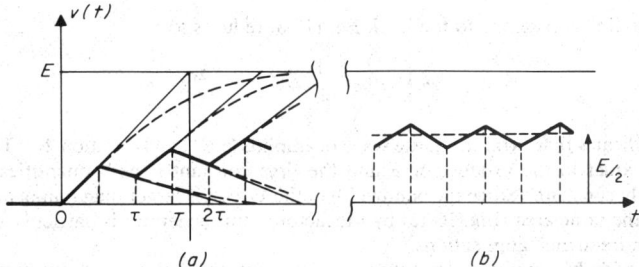

Fig. 16-11. *RC* integrator with square-wave input of period smaller than *RC:* (*a*) initial buildup; (*b*) steady state.

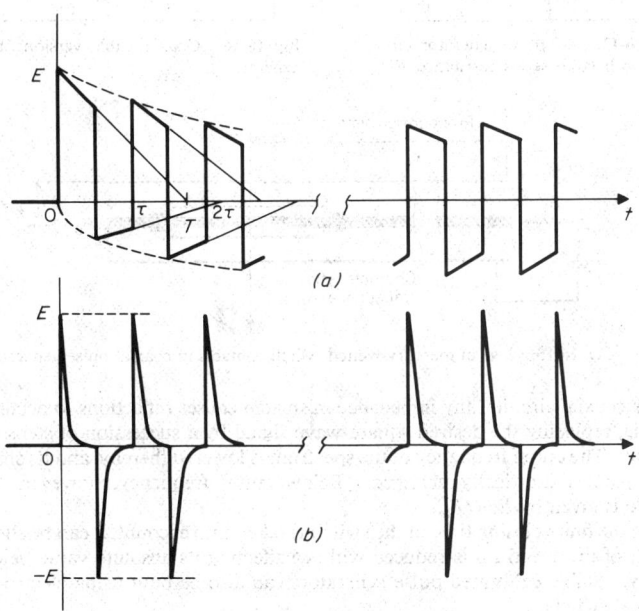

Fig. 16-12. *RC* differentiator with square-wave input: (*a*) period of input signal smaller than *RC;* (*b*) input period longer than *RC.*

by R_0, the propagation delay is equal to τ, and the Laplace transform of the voltage drop across R is

$$V(p) = \frac{1}{p}E\left(1 + \frac{R_0}{R}\coth p\tau\right)^{-1} \tag{16-8}$$

When the line is matched to the load, Eq. (16-8) reduces to

$$V(p) = \frac{1}{2p}E(1 - e^{-p2\tau}) \tag{16-9}$$

which indicates that $V(t)$ is a square wave of amplitude $E/2$ and duration 2τ. The area of the pulse is equal to the product of E and the time constant r; both quantities can be kept reasonably constant. Since the bandwidth of this circuit is larger than that of an exponential pulse of the same area (Fig. 16-13) by the factor π, this generator is particularly suitable for very high frequency applications.

Very wide bandwidth pulse generators based on this principle use a coaxial mercury-wetted switch built into the line (Fig. 16-15) to achieve low standing-wave ratios. A bandwidth of several gigahertz can be obtained in this manner.

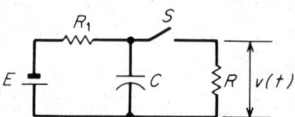

Fig. 16-13. *RC* pulse-generator circuit with large series resistance R_1.

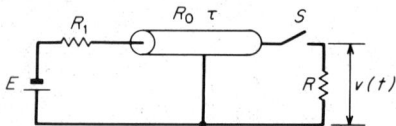

Fig. 16-14. Coaxial-cable version of *RC* pulse generator.

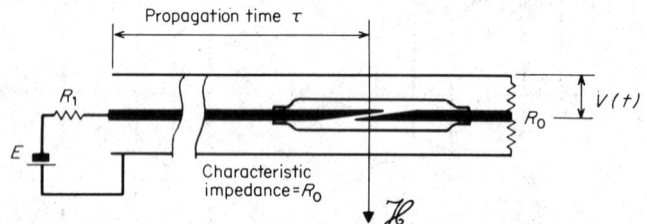

Fig. 16-15. Use of mercury-wetted switch contacts in coaxial pulse generator.

In the coaxial circuits, any impedance mismatch causes reflections to occur at both ends of the line, replacing the desired square-wave signal by a succession of steps of decreasing amplitude. The cutoff frequency of the spectrum is lowered thereby, and its shape above the uniform part is drastically changed. Below cutoff frequency, however, the spectrum amplitude is given by $E\tau R/R_0$.

When the finite closing time of the switch is taken into account, it can be shown that only the width of the spectrum is reduced without affecting its absolute value below the cutoff frequency. Stable calibrated pulse generators can also be built using electronic instead of mechanical switches.

10. Nonlinear-Passive-Network Waveshaping. Nonlinear passive networks offer wider possibilities for waveshaping than linear networks, especially when energy-storage elements such as capacitors or inductors are used with nonlinear devices. Since the analysis of the behavior of such circuits is difficult, we first consider purely resistive nonlinear circuits.

11. Resistive Clamping Circuits. Diodes provide a simple means for clamping a voltage to a constant value. Both forward conduction and avalanche (zener) breakdown are well suited for this purpose. Avalanche breakdown usually offers sharper nonlinearity than forward biasing, but the latter consumes less power.

Clamping action may be obtained in many different ways. The distinction between series and parallel clamping is shown in Fig. 16-16. Clamping occurs in the first case when the diode conducts, in the second it is blocked.

Since the diode is not an ideal device, it is useful to introduce an equivalent network that takes into account some of its imperfections. The complexity of the equivalent network is a trade-off between accuracy and ease of manipulation.

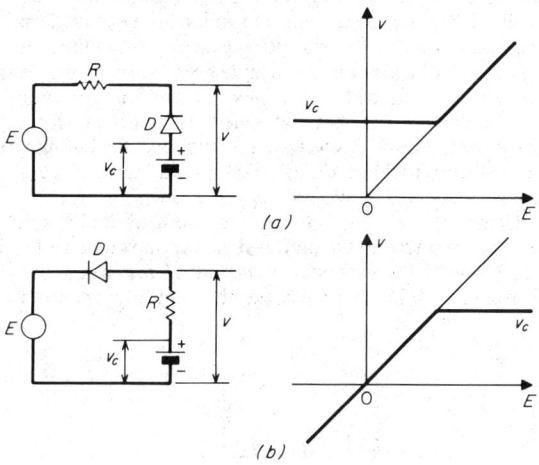

Fig. 16-16. Diode clamping circuits and voltage vs. time responses: (*a*) shunt diode; (*b*) series diode.

The physical diode is characterized by

$$I = I_S\left(\exp\frac{V}{V_T} - 1\right) \qquad (16\text{-}10)$$

where I_S represents the leakage current and $V_T = kT/q$, typically 26 mV at room temperature.

The leakage current is usually quite small, typically 100 pA or less. Therefore, V must be at least several hundred millivolts, typically 600 mV or more, to attain values of forward current I in the range of milliamperes. A first approximation of the forward-biased real diode consists therefore of a series combination of the ideal diode and a small emf (Fig. 16-17). Moreover, to take into account the finite slope of the forward characteristic, a better approximation is obtained by inserting a small resistance in series. If the leakage current under reverse-bias conditions also is important, an additional improvement consists in introducing a current source in parallel.

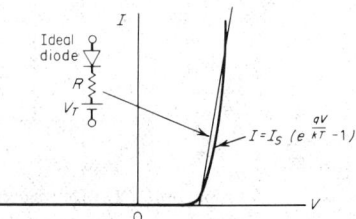

Fig. 16-17. Actual and approximate current-voltage characteristics of ideal and real diodes.

12. Nonlinear Passive Networks with Storage Elements. There is no simple theory of the behavior of nonlinear circuits with storage elements, such as capacitors or inductances. Acceptable solutions can be found, however, by breaking the analysis of the circuit under investigation into a series of linear solutions. A typical example is the dc restorer circuit.

13. DC Restorer Circuit. The circuit shown in Fig. 16-18 resembles the *RC* differentiator but exhibits properties which differ substantially from those examined previously. The diode *D* is assumed to be ideal to simplify the analysis of the circuit, which is carried out in two steps, i.e., with the diode forward- and reverse-biased. In the first step, the output of the circuit is short-circuited; in the second, the diode has no effect, and the circuit is identical to the linear *RC* differentiator.

When a series of alternatively positive and negative steps is applied at the input, after the first positive step is applied, no output voltage is obtained. The first positive step causes a large transient current to flow through the diode and charges the capacitor. Since D is assumed to be an ideal short circuit, the current will be close to a delta function (Par. **16-8**) as long as the internal impedance of the generator connected at the input is small.

In practice, the finite series resistance of the diode must be added to the generator internal impedance, but this does not affect the load time constant significantly, since it is smaller than the time between the first positive step and the following negative step. This allows the circuit to attain the steady-state conditions between steps. When the input voltage suddenly returns to zero, the output voltage undergoes a large negative swing whose amplitude is equal to that of the input step. The diode is then blocked, and the capacitor discharges slowly through the resistor. If the time constant is assumed to be much larger than the period of the input wave, when the second positive voltage step is applied, the output voltage swings back to zero and only a small current flows through the forward-biased diode to restore the charge lost when the diode was under the reverse-bias condition.

If the finite resistance of the diode is taken into consideration, a series of short positive exponential pulses must be added to the output signal, as shown in the lower part of Fig. 16-18. The first pulse, which corresponds to the initial full charge on the capacitor, is substantially higher than the following pulse, but this is of little importance in the operation of the circuit.

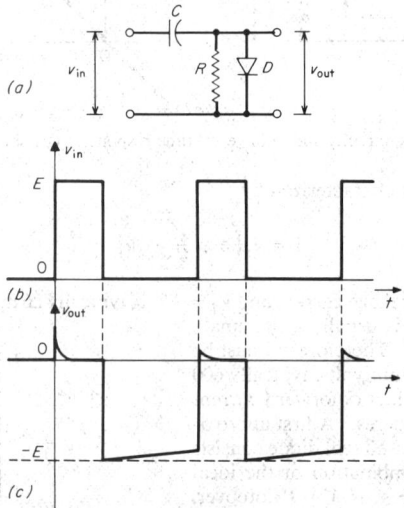

Fig. 16-18. (*a*) Dc restorer circuit; (*b*) input signal; (*c*) output signal.

An interesting feature of the dc restorer circuit lies in the fact that although no dc component can flow from input to output, the output signal has a well-defined level, although determined only by the amplitude of the negative steps (assuming of course that the lost charge between two steps is negligible). This circuit is used extensively in video systems to prevent the average brightness level of the image from being affected by its varying video content. In this case, the reference steps are the line-synchronizing pulses (see Par. **20-12**).

PASSIVE AND ACTIVE ELEMENTS USED AS SWITCHES

14. The Ideal Switch. An ideal switch is a two-pole device that satisfies the following conditions:

1. *Closed-Switch Condition.* The voltage drop across the switch is zero whatever the current flowing through the switch may be.

2. *Open-Switch Condition.* The current through the switch is zero whatever the voltage across the switch may be.

Mechanical switches are usually electrically ideal, but they suffer from other drawbacks; e.g., their switching rate is low, and they exhibit jitter. Moreover bouncing of the contacts may be experienced after closing, unless mercury-wetted contacts are used. Electronic switches do not exhibit these effects, but they are less ideal in their electrical characteristics.

15. Bipolar Transistor Switches (Static Characteristics). A blocked transistor approximates an open switch, since the leakage current I_{CEO} between collector and emitter is typically of the order of magnitude of 1 nA in silicon transistors. Under forward-bias conditions, the voltage drop $V_{CE,\text{sat}}$ across a saturated transistor is small, typically 100 mV. This drop may be considered negligible in many applications.

A rigorous approach to the transistor static switching characteristics is based on the Ebers and Moll equations

$$\begin{bmatrix} I_E \\ I_C \end{bmatrix} = \frac{1}{1 - \alpha_N \alpha_I} \begin{bmatrix} -I_{EBO} & \alpha_I I_{CBO} \\ \alpha_N I_{EBO} & -I_{CBO} \end{bmatrix} \begin{bmatrix} \exp \dfrac{qV_E}{kT} - 1 \\ \exp \dfrac{qV_C}{kT} - 1 \end{bmatrix} \tag{16-11}$$

where I_E and I_C represent the emitter and collector currents, respectively; V_E and V_C represent the voltage drops across the emitter and collector junctions, respectively; positive voltages stand for forward bias, negative for reverse bias; α_N and α_I represent the common-base dc current gains for the transistor connected in normal and inverse mode, respectively (inverse mode means that the collector becomes the emitter and vice versa); I_{CBO} and I_{EBO} represent the leakage current of the collector and emitter junctions, respectively, with the third terminal left open. From the physics of the transistor, it can be shown that

$$\alpha_N I_{EBO} = \alpha_I I_{CBO} \tag{16-12}$$

When the transistor is blocked, both the emitter and the collector junctions are reverse-biased. Assuming that $V_E < -kT/q$ and $V_c < -kT/q$, the collector current is given by

$$I_C = \frac{1 - \alpha_I}{1 - \alpha_N \alpha_I} I_{CBO} \tag{16-13}$$

and the emitter current by

$$I_E = \frac{1 - \alpha_N}{1 - \alpha_N \alpha_I} I_{EBO} \tag{16-14}$$

In a typical example, $\alpha_N = 0.99$, $\alpha_I = 0.10$, $I_{CBO} = 0.1$ nA, and, by Eq. (16-12), $I_{EBO} = 0.01$ nA. The collector leakage current I_C is approximately equal to I_{CBO}, since both junctions are reverse-biased.

The saturated transistor is more difficult to analyze. The curves representing I_C vs. V_{CE} for constant values of I_B are superimposed and closely parallel to the collector-current axis. The fact that all the curves have nearly the same vertical shape means that the series resistance of the saturated transistor is quite small, which is a desired consideration for use as a potential switch. Since the characteristics do not coincide with the vertical coordinate axis, however, there is an equivalent voltage in series with the transistor. To evaluate the magnitude of this voltage, we may determine the intersection point between the saturated transistor characteristic and the blocked transistor leakage-current curve computed by Eq. (16-13). This determines a point A shown in Fig. 16-19, whose coordinates $V_{CE,\text{sat}}$ and I_{CBO} represent the limitations of the transistor used as a switch, in the blocked and saturated conditions, respectively.

At point A, $I_C \simeq I_{CBO}$ and $I_B \gg I_C$. It follows that $|I_E| \simeq |I_B|$. When this approximation is introduced in the Ebers-Moll equations we find

$$V_{CE,\text{sat}} = \frac{kT}{q} \log_n \frac{1}{\alpha_I} \tag{16-15}$$

In the previous example, this expression yields $V_{CE,\text{sat}} = 60$ mV, a value close to that obtained in practice. The Ebers-Moll model of the transistor does not include the inevitable collector series resistance r_C (usually 30 Ω for an integrated transistor), which contributes an increase in $V_{CE,\text{sat}}$ by a few tens of millivolts per milliampere of collector current.

Similar reasoning for the inverted transistor leads to the symmetrical expression

$$V_{EC,\text{sat}} = \frac{kT}{q} \log_n \frac{1}{\alpha_N} \qquad (16\text{-}16)$$

A notable feature of Eq. (16-16) is that because a_N is close to $1, V_{EC,\text{sat}}$ may be much smaller that $V_{CE,\text{sat}}$. In the example, $V_{EC,\text{sat}}$ drops to 0.26 mV; that is, the inverted transistor is better than the normal transistor so far as the static residual-voltage drop under saturated conditions is concerned.

16. IGFET Transistor Switches (Static Characteristics). The isolated-gate field-effect transistor (IGFET) is often used as a switch. Its static behavior is simpler to analyze than that of the bipolar transistor. When the IGFET transistor is blocked, the residual current flowing through the drain terminal represents the leakage current between this junction and the substrate. The typical drain leakage current lies in the same range as the junction leakage current, i.e., from tenths of a nanoampere to a few nanoamperes.

When the IGFET transistor is turned on, its characteristics differ substantially from those of the bipolar switch. Since there is no residual emf in series, the transistor is comparable to a resistor whose value lies in the range of a few kilohms. This can be shown from the nonsaturated IGFET equation

$$I_{DS} = \mu C_0 \frac{W}{L} \left[(V_{GS} - V_T) V_{DS} - \frac{V_{DS}^2}{2} \right] \qquad (16\text{-}17)$$

where μ represents the minority carriers mobility in the vicinity of the surface, C_0 is the gate isolation-layer capacitance per unit area, W is the width of the induced channel, L is the length of the induced channel, V_{GS} is the voltage applied between the gate and the source, V_T is the gate threshold voltage, V_{DS} is the voltage applied between the drain and the source, and I_{DS} is the source-to-drain current.

Usually the factor μC_0 is constant for metal oxide-silicon transistors (MOST) and approximately equal to $6 \times 10^{-6} \text{A/V}^2$.

Assuming that

$$K = \mu C_0 \frac{W}{L} \simeq 6 \times 10^{-6} \frac{W}{L} \qquad (\text{A/V}^2)$$

and considering only small values of V_{DS} compared to $V_{GS} - V_T$, the output admittance of the MOST switch [given by the partial derivative of Eq. (16-17) with respect to V_{DS}] yields the value

$$K(V_{GS} - V_T) \qquad (16\text{-}18)$$

The conductance of the MOST switch thus varies linearly with the voltage applied to the gate. For instance, a transistor having a channel characterized by a ratio W/L of 10 and a

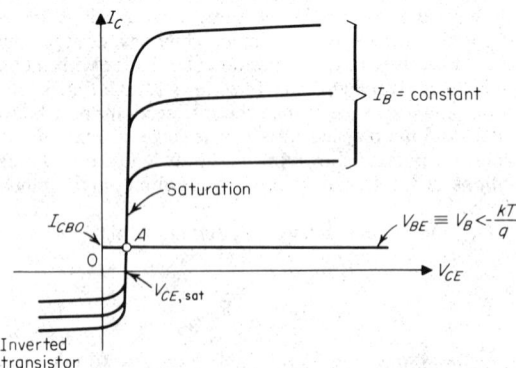

Fig. 16-19. Static characteristics curves and parameters of a bipolar transistor switch.

threshold voltage of 3 V, exhibits a conductance of 6×10^{-5} mhos for V_{GS} equal to 4 V and 6×10^{-4} mhos for V_{GS} equal to 14 V.

17. Dynamic Behavior of Bipolar Transistor Switches. The imperfections of transistors with respect to their dynamic behavior have a more limiting effect than those in the static case. This is due to the inevitable reactive contributions associated with the diffusion mechanism of minority carriers in the base region and with the junction depletion capacitances. These reactive contributions are nonlinear in essence, the first being a function of the current flowing through the base, the second depending on the voltage across the junctions. Since switching involves rapid changes from the blocked condition to saturation and vice versa, the nonlinearities must be taken into consideration to obtain accurate predictions of circuit behavior.

The fundamentals of transistor switching are embodied in the so-called charge-control-model theory of the transistor. The following assumptions are introduced to simplify the analysis:

1. Effects associated with depletion capacitances are neglected initially and reintroduced later.

2. Base-width modulation (known as the *Early effect*) is neglected since it is dominated by other effects.

3. For simplicity, a uniformly doped base is assumed. This assumption, although far from the actual case, does not affect the validity of the analysis because the important quantity is the total charge in the base, rather than the profile of the minority carriers.

Because the circuit to which the transistor is connected controls the switching of the transistor as well as the physical operation of the device, a classical inverter network is considered (Fig. 16-20). The transistor is considered to be blocked initially. When a current step of amplitude I_{B1} is suddenly applied to the base contact, the current drives the transistor into the active region (or into saturation). The boundary between the two regions is defined in terms of the base current as

$$I_{B0} = \frac{1}{h_{FE}} \frac{E_{cc}}{R} = \frac{1}{h_{FE}} I_{C,\,max} \qquad (16\text{-}19)$$

where h_{FE} represents the dc current gain of the transistor.

As long as the transistor operates in the active region, minority carriers tend to build up in the base according to a straight-line profile which rotates around point B of Fig. 16-21 since the junction is reverse-biased. The slope of the profile determines the collector current because only diffusion current is considered (the base is assumed to be uniformly doped). When the collector current I_C reaches its maximum possible value E_{cc}/R (determined by the circuit and not by the transistor), saturation occurs and the collector junction becomes forward-biased. If more charge is then injected into the base, the minority-carrier profile moves up, remaining parallel to itself, to maintain I_C constant.

The minority-carrier charge in the base (which can be regarded as the excess charge with respect to the minority-carrier charge under thermal-equilibrium conditions) is balanced, moreover, by an equal amount of excess charge of majority carriers whose spatial distribution

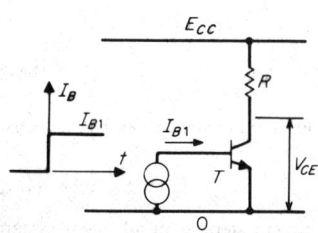

Fig. 16-20. Bipolar transistor as an inverter.

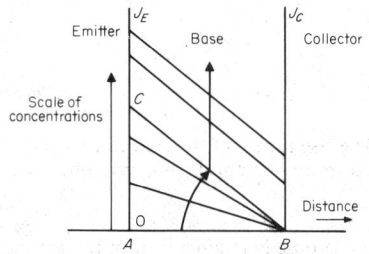

Fig. 16-21. Minority-carrier profiles in the base of a bipolar transistor during switching from cut-off condition to saturation.

closely approximates that of the minority carriers, thus preserving electrical neutrality in the base.

The total minority-carrier charge Q_N in the base may be written

$$Q_N = \tfrac{1}{2} q A p_{nb} e^{qV_E/kT} \; W \qquad (16\text{-}20)$$

where q represents the electron charge, p_{nb} is the minority-carrier concentration in the base under thermal-equilibrium conditions, V_E is the voltage drop over the forward-biased emitter junction, W represents the base width, and A represents the base section. The collector current I_C is given by

$$I_C = -qA \text{ grad } p = qAD_{pb}p_{nb}\frac{e^{qV_E/kT}}{W} \qquad (16\text{-}21)$$

where D_{pb} is the diffusion constant of the minority carriers in the base.

From Eqs. (16-20) and (16-21), it follows that

$$I_C = \frac{Q_N}{\tau_N} \qquad (16\text{-}22)$$

where the constant τ_N represents the base transit time.

Under steady-state linear conditions

$$I_B = \frac{1}{h_{FE}} I_C = \frac{Q_N}{\tau_{BN}} \qquad (16\text{-}23)$$

where

$$\tau_{BN} = h_{FE}\tau_N \qquad (16\text{-}24)$$

If the emitter efficiency is assumed to be equal to 1, the steady-state base current supplies the charge necessary to balance the recombination in the base. However, if a change ΔI_C in collector current is required, additional base current is needed in order to supply the corresponding change in base charge ΔQ_N. Therefore Eq. (16-23) should be modified to take into account transients

$$I_B = \frac{Q_N}{\tau_{BN}} + \frac{dQ_N}{dt} \qquad (16\text{-}25)$$

This first-order equation in fact considers the base as a lumped capacitor. In reality, the base acts like a distributed line, but this effect may be neglected because the width of the base must be small compared to the diffusion length of the minority carriers to achieve high current gain.

Solving Eq. (16-25) with a step function I_{B1} for input function gives

$$Q_N = \tau_{BN} I_{B1}(1 - e^{-t/\tau_{BN}}), \qquad (16\text{-}26)$$

or

$$I_C = h_{FE} I_{B1}(1 - e^{-t/\tau_{BN}}) \qquad (16\text{-}27)$$

The rise time t_r of the collector current, defined as the time between 10 and 90% levels of the collector current, is thus given by

$$t_r = 2.\,2\tau_{BN} \qquad (16\text{-}28)$$

This rise time is independent of the amplitude of the current step, as in all linear systems. However, if the base-current step exceeds I_{B0}, saturation of the transistor occurs and the rise time may be drastically shortened since the collector current is then at its maximum.

When a saturation factor n is defined as the ratio of I_{B1} to I_{B0}, the rise time t_r is given by

$$t_r = \tau_{BN} \log_n \frac{1 - 0.1/n}{1 - 0.9/n} \tag{16-29}$$

When $n \gg 1$, the expression can be simplified to

$$t_r \simeq \tau_{BN} \frac{0.8}{n} \tag{16-30}$$

When the collector current has become saturated, no further modification of collector or base current can be expected, but this does not mean that the transistor has reached the steady-state condition. In fact, the process of charge increase in the base continues till another equilibrium like Eq. (16-22) is obtained. But the charge then becomes

$$Q_N = Q_{N0} + Q_S \tag{16-31}$$

where Q_{N0} is the minority-carrier charge in the base, obtained when I_C is equal to E_{cc}/R (Q_{N0} is proportional to the area of the triangle ABC of Fig. 16-21), and Q_S is the minority-carrier charge in excess of Q_{N0}. This charge is proportional to the area of the parallelogram above the limit BC associated with I_{B0} in Fig. 16-21.

The following relationship must hold when the base has reached the steady-state condition

$$I_{B1} = \frac{Q_{N0}}{\tau_{BN}} + \frac{Q_S}{\tau_{BS}} \tag{16-32}$$

τ_{BS}, which differs from τ_{BN}, is known as the saturation time constant. It corresponds to a completely different functioning of the transistor. Charge may indeed be injected from the collector as well as from the emitter since both junctions are now forward-biased, in contrast to the previous hypothesis of Eq. (16-23).

Fortunately, the new problem is not difficult because the relation between base charges and current is linear. This results from the physics of semiconductors and is true as long as high injection is not considered. The problem therefore is divisible into two separate subproblems by the principle of superposition. Thus, dividing Q_S into two equal parts, according to Fig. 16-22, yields

$$Q_S = (\tau_N + \tau_I)I_{CS} \tag{16-33}$$

where I_{CS} is the amount of collector current which corresponds to Q_S and τ_I pertains to the transistor functioning in the inverse mode. The time constant τ_I (defined as the ratio of base minority-carrier charge to the current collected by the emitter) may be substantially different from τ_N, because the emitter now acts as a collector. Figure 16-23 reveals that due to the geometry of planar transistors, quite different situations occur regarding charges in the base, according as the emitter or the collector injects. Therefore, the normal and inverse current gains are different because the recombination effects differ. Thus

$$I_{BN} = \frac{1}{h_{FEN}} I_C \tag{16-34}$$

and

$$I_{BI} = \frac{1}{h_{FEI}} I_E \tag{16-35}$$

If I_{BS}, defined as $I_{BS} = I_{B1} - I_{B0}$, is written as the sum of I_{BN} and I_{BI}, and if Eqs. (16-33) to (16-35) are combined, the value of τ_{BS} is

$$\tau_{BS} = \frac{Q_S}{I_{BS}} = \frac{\tau_N + \tau_I}{1/h_{FEN} + 1/h_{FEI}} \tag{16-36}$$

This relationship leads to the differential equation describing the transistor while it is saturated

$$I_{B1} = \frac{Q_{N0}}{\tau_{BN}} + \frac{Q_S}{\tau_{BS}} + \frac{dQ_S}{dt} \tag{16-37}$$

which determines how the charges in the base reach their steady-state value.

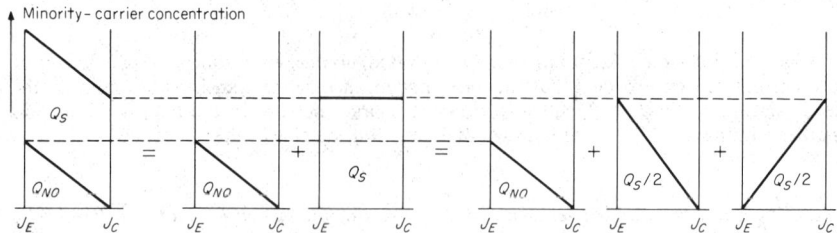

Fig. 16-22.   Equal division of Qs between normal-mode and inverse-mode transistors.

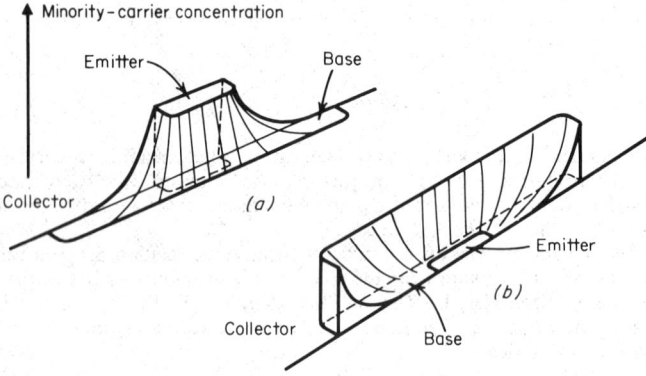

Fig. 16-23.   Minority-carrier charge-concentration patterns: (*a*) normally biased transistor; (*b*) inverted transistor.

Of greater interest is another problem, the reverse change from saturation to the active region and to the blocked conditions. We assume that an additional opposite-polarity step of base current I_{B2} is applied and define a desaturation factor *m*, such as

$$m = -\frac{I_{B2}}{I_{B0}} \tag{16-38}$$

(*m* is a positive quantity). First, the extra charge Q_S is removed from the base. This happens while I_C is still at maximum; i.e., no change will occur in the collector current. This situation is depicted in Fig. 16-24.

During this time, the extra charge drops from its initial steady-state value given by Eq. (16-32) to zero. The time required, the desaturation time t_s, is given by solving the equation

$$Q_S + \tau_{BS}(I_{B2} - I_{B1})(1 - e^{-t_s/\tau_{BS}}) = 0 \tag{16-39}$$

where

$$Q_S = \tau_{BS}(I_{B1} - I_{B0}) \tag{16-40}$$

and

$$t_s = \tau_{BS} \log_n \frac{1 + n/m}{1 + 1/m} \tag{16-41}$$

When both n and m are much larger than 1, t_s may be given by

$$t_s \simeq \tau_{BS} \log_n \left(1 + \frac{n}{m}\right) \tag{16-42}$$

Once Q_S is reduced to zero, the transistor is brought back into the active region. The problem then is treated as in Eqs. (16-25) to (16-28).

The decay time t_d of the collector current is

$$t_d = \tau_{BN} \log_n \frac{1 + 0.9/m}{1 + 0.1/m} \tag{16-43}$$

or

$$t_d = \tau_{BN} \frac{0.8}{m} \quad \text{if } m \gg 1 \tag{16-44}$$

A numerical example for a transistor having $h_{FEN} = 100$, $h_{FEI} = 10$, $\tau_N = 2$ ns, $\tau_I = 5$ ns, $n = 5$, and $m = 10$ is

$$\tau_{BN} = 200 \text{ ns} \quad \text{and} \quad t_r = 35.6 \text{ ns}$$
$$\tau_{BI} = 50 \text{ ns} \quad\quad\quad t_s = 19.7 \text{ ns}$$
$$\tau_{BS} = 63.6 \text{ ns} \quad\quad\quad t_d = 15.2 \text{ ns}$$

The 10 to 90% rise time of this transistor, operating in the active region exclusively, is 440 ns. This result shows clearly how effective overdriving may be.

Also of special interest is the desaturation time, 20 ns in the present example, during which the transistor is insensitive to the input signal. This extra delay, which would not exist if the transistor had not been driven into saturation, plays an important role in bipolar transistor logic and explains why unsaturated logic, like emitter-coupled logic, is faster than saturated-logic circuits.

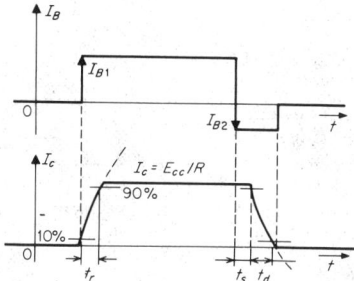

Fig. 16-24. Base (input) and collector (output) currents vs. time.

It should be noticed that in Fig. 16-24, since I_{B2} is larger than I_{B1}, a constant inverse current flows through the base contact as long as the transistor is not blocked. Usually current steps are obtained by means of square-wave voltage sources with large series resistors. Once the transistor is cut off, the load impedance of course becomes very large, and the step generator at the input turns out to be more like a voltage source than a current source. This explains why the base current in Fig. 16-24 drops to zero at the same time that I_C vanishes.

We have thus far neglected the effects of junction capacitances on the switching times. The charge-control method offers a convenient way of introducing capacitances by adding additional terms Q_E and Q_C to Eq. (16-37), representing the total charges stored in the emitter and collector transition regions, respectively.

18. Dynamic Switching of Diodes. The inverse base current, due to the mechanism of minority-carrier storage, also exists in diodes. This effect causes the reverse-biased conventional diode to act like a short circuit for a very short time after forward conduction. This time is a monotonic function of the amplitudes of the forward and reverse currents. Desaturation times of junction diodes usually range in nanoseconds.

Schottky barrier diodes, on the other hand, operate by field-emission effects, and their storage times are exceedingly small. An interesting application of Schottky barrier diodes is shown in Fig. 16-25. In this circuit, the Schottky barrier diode is reverse-biased at all times except when the transistor tends to become saturated. When this happens, the collector-to-emitter voltage drops below the base-to-emitter voltage by at least a few tenths of a volt, causing the diode to conduct. In this manner a negative-feedback loop is established between collector and base which prevents the transistor from going further into saturation.

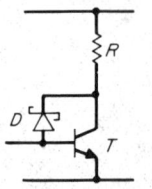

In fact, in the Schottky barrier diode the excess base current above I_{B0} is directly derived, preventing excess charge Q_s from accumulating in the base. The consequence is that desaturation time of the diode is not a problem, even if the base is heavily overdriven to achieve low rise and decay times. Figure 16-26 shows that the Schottky diode can be implemented, in the case of a planar *npn* transistor, by extending the base contact over the collector region to take advantage of the particular nature of the contact between the aluminum and the lightly doped *n* type silicon.

Fig. 16-25. Schottky barrier diode switching circuit.

19. Dynamic Behavior of IGFET Transistor Switches. Large-scale integration technology takes advantage of the possibilities offered by IGFET (isolated-gate field-effect) transistors, such as MOS transistors or silicon gate transistors (see Pars. 8-47 to 8-49). Their small size and ease of fabrication make them very attractive for producing complex digital systems on a single chip. However, their slow speed, compared with bipolar transistors, is a drawback. In examining the dynamics of IGFET transistor switches we again consider an inverter stage. The output load is formed by another IGFET transistor, as shown in Fig. 16-27.

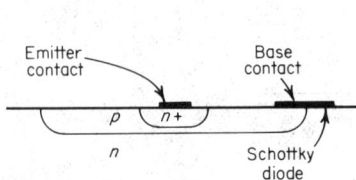

Fig. 16-26. Planar *npn* transistor and Schottky diode in integrated-circuit form.

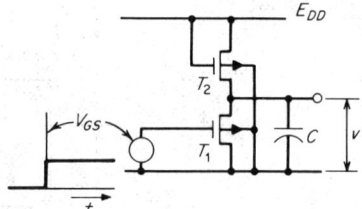

Fig. 16-27. IGFET transistors in switching circuit.

Field-effect integrated transistors provide an easy way of realizing high-value resistances that are physically much smaller than diffused resistors. A 20-kΩ resistor, for instance, made from a typical 200 Ω/square base diffusion requires 100 squares; i.e., if the section of the diffused ribbon is 15 μm, a total length of 1.5mm is required, whereas a 25-μm square is sufficient for an IGFET transistor having the same resistance.

Since the load transistor of Fig. 16-27 has the same gate and drain voltages, it can be shown that this transistor is always in saturation, whatever the source-to-substrate voltage may be. Therefore, its series resistance can be computed from

$$I_{DS} = \frac{K_L}{2}(V_{GS} - V_T)^2 \qquad (16\text{-}45)$$

where K_L is the value the constant K takes for the load transistor. This expression is obtained by so taking V_{DS} in Eq. (16-17) as to maximize I_{DS}. Equation (16-45) extends to the whole region of saturation, neglecting the output admittance of the common source transistor. When the transistor is used as a load, as shown in Fig. 16-27, its series conductance G is equal to its transconductance:

$$G = K_L(V_{GS} - V_T) = \sqrt{2 K_L I_{DS}} \qquad (16\text{-}46)$$

The voltage gain of the inverter stage is given by

$$A_{vf} = -SG^{-1} \qquad (16\text{-}47)$$

16-18

where S represents the transconductance of T_1. As long as T_1 is in saturation, Eq. (16-47) can be simplified to

$$A_{vf} = -\sqrt{\frac{K}{K_L}} \qquad (16\text{-}48)$$

If the channel length L is the same for both T_1 and T_2

$$A_{vf} = -\sqrt{\frac{W}{W_L}} \qquad (16\text{-}49)$$

In other words, the inverter-stage voltage gain is directly related to the ratio of the respective channel widths.

In analyzing switching by this stage, capacitors must be added to the ideal dc transadmittance which represents the IGFET transistor under static conditions. The source and collector are in fact reverse-biased junctions exhibiting a capacitive coupling with respect to the substrate. The gate also is capacitively linked to the substrate and to the source and the drain also. The gate-drain capacitance (known as the *Miller capacitance*) is particularly important in respect to the switching speed. None of these capacitances involves storage mechanisms, however, because the propagation time in the channel is very short. Their effect is essentially negligible compared to the time constants associated with depletion and gate capacitances.

The switching time of the inverter stage of Fig. 16-27 is thus wholly controlled by the total capacitance C existing between its output terminal and the substrate. The drain and source-to-substrate capacitance of T_1 and T_2, plus any additional parasitic capacitance which is related to the same output port, must be combined to find C.

To investigate the switching mechanism, we assume that a voltage step is applied to the gate of T_1. Since an inverter stage is usually driven by the output of another inverter stage, we assume that the amplitude of the input step is equal to the output dynamic range of the inverter itself, that is, $E_{DD} - V_T$.

No sudden change of output voltage is experienced after the input step has been applied to the blocked transistor T_1, since this would require an infinite current step. The trajectory of the operating point of transistor T_1 thus jumps abruptly along a vertical line from I to II in the characteristic curves of this transistor (Fig. 16-28) until it reaches the V_G characteristic, which corresponds to the amplitude of the input signal ($E_{DD} - V_T$). This first phase is very rapid. Only the internal impedance of the step-function generator and the input capacitance of the transistor determine the rate at which V_{GS} increases. Since this time constant is small, it is legitimate to ignore it compared to the other time constants involved.

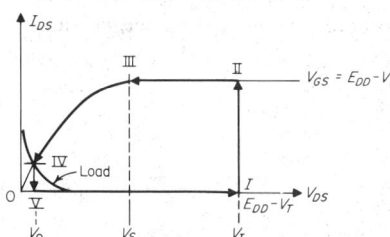

Fig. 16-28. Operating path of IGFET switch.

The operating point then shifts toward its steady-state value, corresponding to the logical 0, located at the intersection of the particular curve considered above and the load curve represented in Fig. 16-28. This sequence may be divided into two parts; the first corresponds to the period during which T_1 is still saturated, and the second is related to the unsaturated mode of operation. The first part comprises the horizontal motion to point III, and the second corresponds to the bent part of the trajectory, from III to IV. As long as the capacitive load sees a saturated transistor, e.g., from II to III, the circuit simply resumes the discharge of C by means of a constant current source, neglecting the contribution of T_2, which is usually small. The discharge thus follows a linear relationship

$$\frac{V}{E_{DD} - V_T} = \frac{t}{\tau_1} \qquad (16\text{-}50)$$

with

$$\tau_1 = \frac{2C}{K(E_{DD} - V_T)} \qquad (16\text{-}51)$$

During the next period, the problem is a little more difficult since T_1 then introduces a nonlinear relation between current and voltage. When Eq. (16-17) is written as

$$I_{DS} = f(V_{DS}) = f(V) \quad \text{and} \quad I_{DS} = C\frac{dV}{dt}$$

the following differential equation is obtained

$$\frac{C}{K}\frac{dV}{(E_{DD} - V_T)V - V^2/2} = dt \tag{16-52}$$

which leads to the implicit equation relating time to output voltage

$$\frac{V}{E_{DD} - V_T} = 1 - th\frac{t}{\tau} \tag{16-53}$$

V tends asymptotically toward V_0, as shown in Fig. 16-29. The rate at which V decreases from V_s (which corresponds to III), toward V_0 is relatively constant at the beginning and extends continuously from the first part. Therefore a reasonable approximation of the total switching time may be obtained by linearizing the response curve over the entire dynamic range of the output signal from V_1 to V_0. The response time t_r thus becomes proportional to $0.8\tau_1$, and

$$t_r = 0.8\tau_1 \tag{16-54}$$

assuming the voltage drop at point IV is negligible.

We next consider the trajectory of the operating point of T_1 after cutoff. T_1 is again assumed to be cut off almost instantaneously so that the operating point jumps abruptly from IV to V. Thereafter, a very slow recharging process takes place from V to I. This effect is explained by the high resistance of the saturated transistor T_2 recharging the capacitor C. Since T_2 again introduces a nonlinear relationship between current and voltage, we must write a new differential equation

$$\frac{2C}{K_L}\frac{dV}{(E_{DD} - V_T - V)^2} = dt \tag{16-55}$$

the solution of which is given by

$$\frac{V}{E_{DD} - V_T} = \frac{t/\tau_2}{1 + t/\tau_2} \tag{16-56}$$

with

$$\tau_2 = K_L\frac{2c}{(E_{DD} - V_T)} \tag{16-57}$$

The time t_d between the 10 and 90% steady-state amplitude levels is equal to $8.9\tau_2$. This time is substantially longer than t_r. Furthermore, t_d can be rewritten as

$$t_d = 8.9\frac{K}{K_L}\tau_1 = 8.9A_{vf}^2\tau_1 \tag{16-58}$$

Thus, to achieve fast switching, one must aim at low voltage gain, but a gain less than 2 should not be accepted since this would cause deterioration of the transfer characteristic of the inverter stage. For a gain of 2, t_d is 36 times longer than t_r.

In a typical example an MOS transistor with W/L equal to 4 is used as the common-source amplifier stage, and the load transistor has a W_L/L_L value of 1. Both transistors have the same channel length, so the voltage gain of the inverter stage equals 2. Assuming that C reaches 3 pF and that the input step $(E_{DD} - V_T)$ equals 15 V, one finds

$$\tau_1 = 16.7 \text{ ns} \qquad t_r = 13.3 \text{ ns} \qquad t_d = 593 \text{ ns}$$

From this example, it is obvious that IGFET technology suffers from a serious limitation relative to bipolar transistors regarding the speed of response (compare the example at end of Par. 16-17). It is not surprising, therefore, that MOS logic circuits operate generally at frequencies lower than 2 MHz, with rare exceptions as high as 10 MHz. A slight improvement can be obtained, however, if the gate of the load transistor is fed by an auxiliary source which is at a higher potential than E_{DD}, at the expense of a more complicated power supply.

20. Complementary IGFET Pulse Circuits. Complementary circuits offer interesting properties but require more sophisticated fabrication. True complementary bipolar transistors on the same chip are not readily available. Most monolithic integrated circuits use *npn* planar transistors and lateral *pnps*. The latter exhibit very poor current-gain characteristics, typically between 2 and 10. In IGFET technology it is possible to implement *p*- and *n*-channel transistors on a single chip. Among the techniques offering economical solutions are ion implantation and silicon-gate technology combined with backfilling (see Pars. **8-47** to **8-49**).

A typical complementary inverter stage is shown in Fig. 16-30. A notable feature of this circuit is that one transistor always is on while the other is off under steady-state conditions. The logical 0 and 1 are therefore respectively equal to zero and E_{DD}, and the dc power consumption of the circuit is reduced to almost zero; i.e., only the leakage currents of the reverse-biased junctions must be taken into account. The power consumption arises from the repetitive charging and discharging of the capacitor C associated with the output terminal. If f is the switching rate, the average dynamic power consumption is given by $CE^2_{DD}f$.

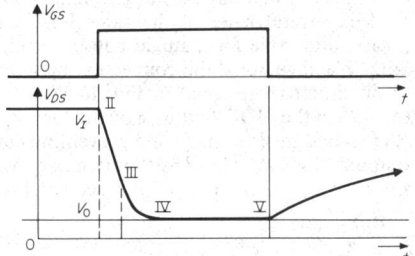

Fig. 16-29. V_{GS} and V_{DS} amplitudes vs. time in IGFET switch.

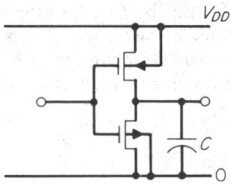

Fig. 16-30. IGFET complementary inverter stage.

The significant quantity in comparing inverters and gates, from the standpoint of dissipation, is the ratio of measured average power to the repetition rate. This ratio, which represents the energy consumed in each switching operation, is typically of the order of picojoules for complementary MOS inverters. This type of logic is therefore very attractive for micropower applications such as the quartz-controlled wrist watch.

Another reason why the energy per cycle is so low is that silicon-gate technology combined with a proper choice of substrate orientation leads to very low values of the threshold voltage V_T, typically 0.6 to 0.8 V for *p*- or *n*-channel MOS transistors. This allows source voltages to be as low as 1.3 V. The reduction of work function which results from using silicon instead of aluminum for the gate metallization explains this improvement in part.

Smaller threshold voltages lead to better switching capabilities but make relatively more important the leakage currents of the source and drain junctions, so that a trade-off exists between these effects. In practice, threshold voltages not lower than 0.6 V are appropriate. The lower threshold voltages also improve the switching speed since they increase the current-handling capability of MOS transistors without increasing at the same time the output capacitance C. A larger gate width, for instance, improves the transconductance but simultaneously increases the source and drain capacitances by the same amount, with no net change in the switching times.

21. Logic Gates. The most important application of switching transistors is logic gates, and in this field, integrated circuits have almost completely eliminated the discrete-component-circuit approach. We restrict our discussion to integrated-circuit logic gates. There are four main families of monolithic bipolar integrated-circuit logic gates, respectively called resistor transistor logic (RTL), emitter-coupled logic (ECL), diode transistor logic (DTL), and transistor transistor logic (TTL or T^2L).

RTL Gates. Resistor transistor logic is an extension of the classical inverter stage, using several parallel input transistors instead of one (Fig. 16-31). The voltage at terminal F can be high, e.g., the logical 1, only if and when all input transistors are blocked simultaneously. Any other combination, with one or more transistors saturated, leads necessarily to a low output voltage (the logical 0). The logic function this circuit performs therefore is the NOR function ($F = A + B + C$ for three-input gate considered in Fig. 16-31).

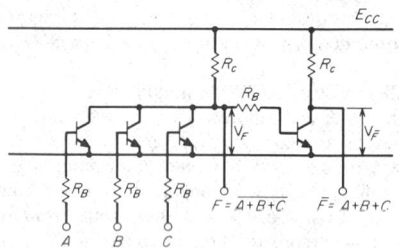

Fig. 16-31. RTL gates with NOR and OR output.

The single-input inverter stage also provides a logical function, since it negates its input signal. For instance, if we consider the second inverter in Fig. 16-31, the output signal this stage delivers negates, or complements, the previous NOR function, providing an OR output ($F = A + B + C$). Logic gates are interconnected to provide a variety of desired logic functions. Appropriate synthesis procedures are currently available (see Secs. **8** and **23**).

A fundamental requirement on all logic circuits is that any output terminal be able to feed several gates of the same logic family. The upper limit to the number of loading gates allowable for a single output terminal is known as the *fan-out* capability of the circuit. We have noted the connection between the NOR and the OR terminals of the RTL logic circuit represented in Fig. 16-31, a typical example of cascading of two identical stages. When the NOR output signal is low, V_F is of the order of magnitude of $V_{CE,sat}$, blocking the second inverter stage and preventing current from flowing through R_B. When the same output is high, V_F is fixed by the attenuator formed by R_C and R_B, and by E_{cc} and the emitter voltage drop V_{BE} of the second inverter

$$V_F = V_{BE} + \frac{R_B}{R_B + R_C}(E_{cc} - V_{BE}) \qquad (16\text{-}59)$$

If R_B were equal to zero, the output voltage V_F would be equal to V_{BE}, and the logical swing, i.e., the difference between the logical 1 and logical 0, would be very small, namely, the difference between V_{BE} and $V_{CE,sat}$ (typically 400 mV).

This situation leads to increased noise susceptibility and reduces the acceptable tolerance on the logical 1, since V_{BE} may vary by a few tens of millivolts from junction to junction. Hence, taking R_B equal to a few hundred ohms (typically 200 Ω to 1 kΩ) yields a larger logic swing, better noise immunity, and smaller tolerance on the logical 1.

The values of R_L and R_B are determined by the following considerations. When an inverter stage is saturated, the collector current is given by

$$I_{C,sat} \simeq \frac{E_{cc}}{R_L} \qquad (16\text{-}60)$$

Therefore the corresponding base current I_B must be

$$I_B \geq \frac{I_{C,sat}}{h_{FE}} = \frac{E_{cc}}{h_{FE} R_L} \qquad (16\text{-}61)$$

This inequality must be satisfied with the largest possible number of logic gates connected in parallel (Fig. 16-32). The actual value of base current will be

$$I_B = \frac{1}{N} \frac{E_{cc}}{R_L + R_B/N} \qquad (16\text{-}62)$$

Combining Eqs. (16-61) and (16-62) yields the maximum number of gates that can be connected to the same output terminal

$$N \leq h_{FE} - \frac{R_B}{R_L} \qquad (16\text{-}63)$$

This expression characterizes the fan-out capability of the circuit under consideration. Since the smaller the value of N, the more gates are overdriven, resulting in faster switching times,

16-22

it is a good practice to chose N not to exceed one-fifth of the right-hand term of the inequality Eq. (16-63).

N is determined mainly by h_{FE}. High fan-out capabilities thus require high current gain, which may be achieved by using an additional buffer output stage. A typical example is provided in Fig. 16-33. Fan-out figures of 10 to 20 are readily available in this manner, with switching times not longer than 10 ns.

RTL circuits are normally saturated when the output signal is at low level, and their switching behavior is controlled by the mechanism of minority-carrier storage in the base. Hence RTLs, like other saturated logic, are slower than nonsaturated-logic integrated-circuit gates.

ECL Gates. Emitter-coupled logic gates are typical of nonsaturated logic circuits. The basic ECL circuit, shown in Fig. 16-34, resembles the differential amplifier, and one side comprises a series of parallel transistors as in RTL circuits. Better understanding is obtained by considering the ECL circuit as a parallel combination of emitter followers comprising T_A, T_B, T_C, and T_D, with a common load R_E. The voltage across R_E always is equal to the highest of the input voltages V_A, V_B, V_C, or V_D. Thus if V_A, V_B, V_c all are simultaneously more negative than V_D, none of the three parallel transistors conducts and the output voltage V_F is high. When one of the input voltages at A, B, or C is more positive than V_D, V_F drops to a minimum voltage. The circuit thus realizes the NOR function, as in the RTL case. The second output negates the NOR output and provides an additional OR gate, also as in the RTL circuit, although the interaction between the NOR and OR sections is quite different, namely, through emitter coupling.

ECL logic circuits require input-signal swings whose range exceeds the dynamic range of the typical differential amplifier, e.g., of the order of 1 V. A typical input-output curve is shown in Fig. 16-35. Both output signals are always negative (opposite to the RTL case). Hence RTL and ECL circuits cannot be cascaded unless special interface circuits are provided. ECL circuits may be used in series without any problem, since the two output emitter-follower circuits represented in Fig. 16-34 shift the signals downward at the common collector of the triple T_A T_B T_C combination and at the collector of T_D by an amount equal to V_{BE}. Furthermore, these two output stages provide low output impedances.

An important difference between ECL logic gates and other integrated-circuit logic gates lies in the fact that all transistors in an ECL circuit are operated as emitter followers, or are blocked. Hence saturation never occurs, and switching times are among the shortest possible, typically 3 to 5 ns. The presence of the emitter-follower buffers also contributes to the rapid switching.

DTL Gates. Diode transistor logic operates in quite a different way, since the logical function is readily obtained by means of diodes, while the transistor is used only to perform negation and help reshape the output signal.

The basic DTL circuit is shown in Fig. 16-36. We first consider the situation where no current is flowing through the input diodes D_A, D_B, and D_C. Consequently, current flows through R_2, D_1, and D_2, dividing itself between R_1 and the base terminal of T. The latter current is intended to drive the transistor into saturation or to overdrive it. The voltage at point P is easy to compute, since it is equal to the sum of the voltage drops across three forward-biased junctions. Thus it amounts to approximately 2.2 V.

Next consider the case when one of the input voltages at A, B, or C is close to 0V. The current through R_2 then passes through one of the input diodes, and the voltage at point P drops to approximately V_{BE}, bringing T to cutoff because of the presence of D_1 and D_2. The logic function is thus not the NOR of the RTL and ECL cases but rather a NAND function.

The signal swing of DTL circuits is easily determined since the lowest output voltage is given by $V_{CE,sat}$ of the inverter stage and the highest level corresponds to no output current in the load, making V_F equal to E_{cc}. One of the drawbacks of this circuit is the relatively slow speed of operation. This is due partly to the fact that T saturates but also to the fact that integrated diodes are passive elements embedded with parasitic capacitances. No opportunity is available to regenerate the signal except in the output inverter stage. Typical DTL switching times are of the order of tens of nanoseconds.

T^2L gates. Transistor transistor logic uses integrated logic closely related to the DTL circuit, although a superficial look at the circuit in Fig. 16-37 reveals substantial apparent differences. The multiemitter transistor T_2 is used in the same manner as the diodes D_A, D_B, D_C and D_1, D_2 in the DTL circuit. When transistor T_1 is on, its base current flows through

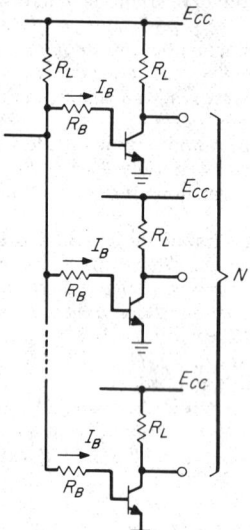

Fig. 16-32. Fan out of logic gates.

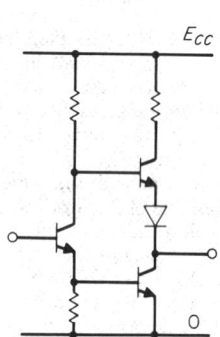

Fig. 16-33. Buffer stages used to improve fan-out capability of RTL gates.

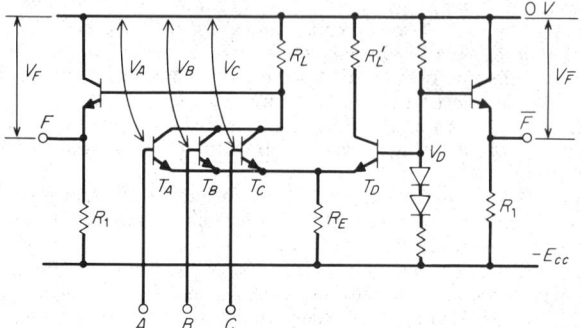

Fig. 16-34. Typical ECL logic-gate circuit. Typical values: $R_L = 270\ \Omega$, $R'_L = 300\ \Omega$, $R_E = 1.25\ \text{k}\Omega$, $R_1 = 2\ \text{k}\Omega$, $V_D = -1.15\ \text{V}$, $V_{EE} = -5.2\ \text{V}$.

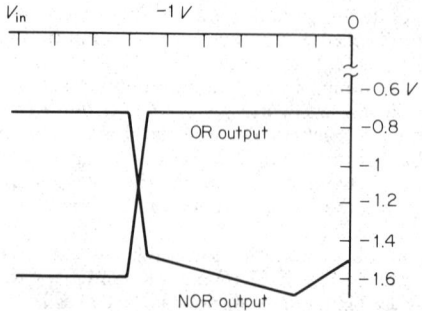

Fig. 16-35. OR and NOR outputs of ECL gates, showing voltage range between logic 1 and 0.

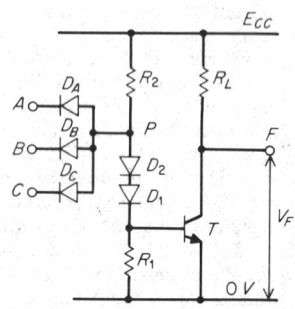

Fig. 16-36. DTL logic-gate circuit.

the forward-biased collector junction of T_2 while no current flows through any of the input emitters. Conversely, T_1 is off when one or more emitters of T_2 are grounded, blocking the collector junction in series with the base of T_1.

An important feature of this circuit is that the passage of T_1 from saturation to the blocked condition is greatly aided by the transistor T_2. To show this, suppose that all input signals are initially high; that is, T_1 is saturated. If one or more of the input signals drops suddenly to zero, T_1 must switch off, but this requires first that the base be emptied of excess minority-carrier charges, Q_S and Q_{N0} in Eq. (16-37). Consequently, reversing the base current is required to achieve fast recovery. This happens because T_2 is then in the active region and its collector current can be very high. Once the charge in the base of T_1 is emptied, no further current flows through the collector of T_2 and this transistor goes into saturation. Hence, very fast switching is possible in spite of the fact that T²L circuits fall in the saturated-logic category. Typical T²L switching times are of the order of 6 ns.

With respect to logic swing T²L logic is not different from DTL circuits, and its fan-out capabilities are comparable to those of RTL circuits unless an output buffer stage like the one shown in Fig. 16-33 is used.

Wired Logic. In the four families of logic circuits just discussed the output terminal of a logic gate feeds several other gates of the same family. Paralleling two or more output terminals is not permitted, since this would cause improper functioning of the gates. An exception, called *wired logic*, exists whereby output terminals of two or more gates are connected in parallel to a common load resistor.

RTL circuitry provides a good example of wired logic, since any number of collectors of transistors belonging to other RTL gates may be connected to the same output load R_L without trouble. Another example, with several transistor emitters connected in parallel to a common load resistor, occurs in the ECL circuit depicted in Fig. 16-34. Wired logic may result in saving components, as in the example shown in Fig. 16-38, representing two slightly modified T²L gates connected to a common load resistor R_L and to a common emitter resistor R_E. The voltage drops across R_L and R_E are used to drive the output buffer (totem-pole buffer amplifier). In this way, the NOR function of two AND gates is provided in a single circuit.

IGFET Gates. IGFETs are promising for logic gates. They are used in medium- or large-scale integrated circuits because of their small size and relative ease of fabrication. While no systematic families of logic gates exist using IGFETs, a wide variety of manufacturer-wired logic is available.

The basic NOR and NAND IGFET structures are shown in Fig. 16-39. The NOR gate

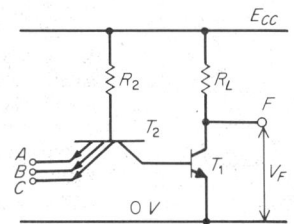

Fig. 16-37. TTL logic gate circuit.

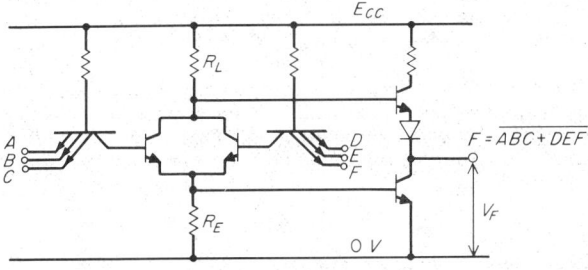

Fig. 16-38. Wired-logic gate circuit.

simply reproduces the discrete-element RTL configuration, the only difference being that the load resistor has been replaced by a saturated transistor, as in IGFET inverter stages. Less power is consumed, and higher circuit complexity is possible because of smaller size. The drawback, however, is speed of operation, i.e., switching times 10 to 20 times longer than those available with bipolar transistors. One should not overemphasize this relative sluggishness, because the increased complexity possible in medium- and large-scale integrated circuits permits a high degree of decentralization and redundancy. The multiprocessor technique used in the field of minicomputers is typical of this trend.

The basic NAND circuit shown in Fig. 16-39 is not commonly used with bipolar transistors, since the latter introduce unwanted pedestals in series with each saturated

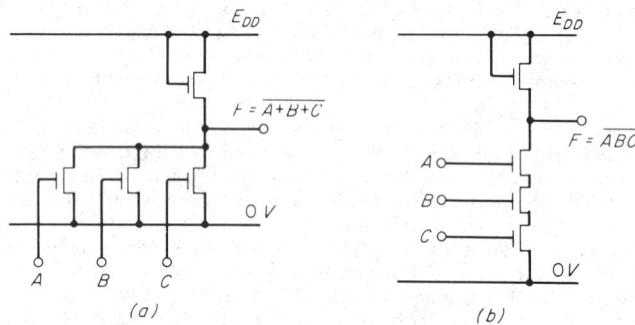

Fig. 16-39. IGFET structures: (*a*) NOR; (*b*) NAND.

transistor. IGFET transistors behave more like conventional resistors in series, and the total voltage drop across them is small compared to the normal input dynamic range.

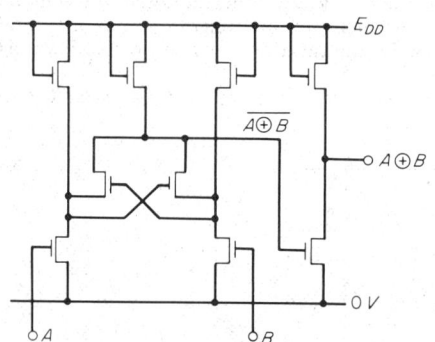

Fig. 16-40. Wired-logic IGFET circuit.

An illustration of wired-logic IGFET circuitry is shown in Fig. 16-40, which represents an EXCLUSIVE-OR gate using four switching transistors with three load transistors and one inverter stage. The same circuit using only NOR or NAND gates would be more complex.

22. Transistor Switches Other Than Logic Gates. Transistor switches are extensively used in applications other than logic gates, covering a wide variety of both digital and analog applications. A typical illustration is the circuit converting the frequency of a signal into a proportional current, the so-called *diode pump*. This circuit (Fig. 16-41) comprises a capacitor C, two diodes D_1 and D_2, and a switch formed by a transistor T and a resistor R. The transistor is assumed to be driven periodically by a square-wave source, alternatively on and off. When T is blocked, the capacitor C charges through the diode D_1, while D_2 has no effect. As soon as the voltage across C has reached its steady-state value E_{cc}, T may be turned on abruptly. The voltage with respect to ground at point A becomes negative (ideally $-E_{cc}$ if the switch is perfect), and D_1 is blocked while D_2 is forward-biased. The capacitor thus discharges itself in the load (in Fig. 16-41 an ammeter, but it could be any other circuit element that does not exhibit storage), allowing V_A to reach 0 V before T is again turned off. The charge fed to the load thus amounts to CE_{cc} coulombs. If we suppose that this process is repeated periodically, the average current in the load is given by

$$I = fCE_{cc} \qquad (16\text{-}64)$$

where f represents the switching repetition rate.

The diode-pump circuit provides a pulsed current whose average value is proportional to the frequency of the square-wave generator controlling the switching transistor T. The proportionality would of course be lost if the load exhibited storage, e.g., if the load were a parallel combination of a resistor and a capacitor in order to obtain a voltage drop proportional to the average current. Using an operational amplifier, as shown in the right side of Fig. 16-41, circumvents the problem.

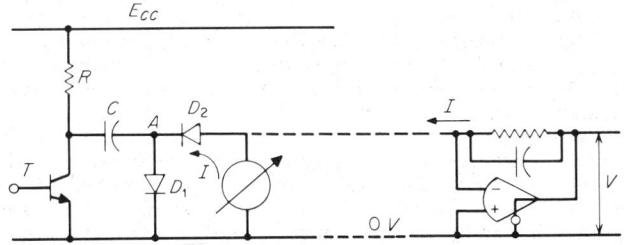

Fig. 16-41. Diode-pump circuit.

The requirements on the switching transistor in this application are different and in many respects more stringent than for logic gates.

The transistor in a logic circuit provides a way of defining two well-distinguished states, the logical 1 and 0. Nothing further is required whether these two states approach an actual short circuit or an open circuit. In the diode-pump circuit, however, the total switching characteristics are important, since the residual voltage drop across the saturated transistor of Fig. 16-41 influences the charge transfer from C to the load, thereby also introducing unwanted temperature sensitivity. The main difference lies in the fact that while T is operated as a logical element, the purpose of the circuit actually is to deliver an analog signal.

There are many other examples where the characteristics of switching transistors influence the accuracy of given circuits or instruments, as in the digital voltmeter based on the weighting principle. An even more critical problem pertains to amplitude gating, since this class of applications requires switches which correctly transfer analog signals without introducing distortion. Furthermore, positive and negative signals must be transmitted equally well, and noise introduced by the gating signals must be minimized.

Amplified Gating. One of the simplest and most effective approaches is to use balanced diode bridges. The pedestals introduced by the diodes then cancel out to the extent that symmetry is achieved.

A typical amplitude-gating circuit is shown in Fig. 16-42. We assume first that balanced gating signals are applied to the two equal resistors symmetrically connected to the bridge. With the polarities indicated in Fig. 16-42, all four diodes conduct, and the output voltage V_2 is zero, assuming that the voltage drops across the diodes are identical. The existence of a small input voltage V_1 does not alter this condition if the resulting current unbalance is kept small. Since

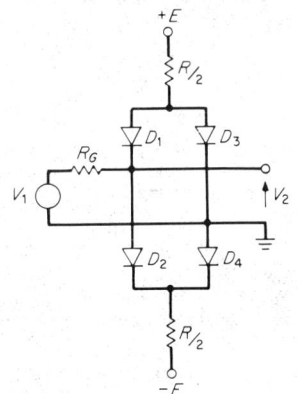

Fig. 16-42. Amplitude-gating circuit.

diodes are insensitive to current as far as forward voltage drops are concerned, the error at the output can be neglected.

In the opposite situation, when the polarities of the gating sources are reversed and the diodes are all back-biased, the output signal V_2 becomes equal to V_1 since the input generator is no longer loaded.

Many circuit alternatives are available. For example, the bridge may be put in series with the input generator and an output load, replacing the shunt approach by series transmission. The open and close control functions must be reversed with respect to the shunt case.

One of the major drawbacks of electronic gates is their sensitivity to the controlling gating signals. Fast switching requires steep control-voltage steps, which may produce unwanted transients at the output terminals, particularly those due to differences between junction capacitances of the diodes. Minority-carrier storage causes another problem because it introduces measurable delay between the appearance of the control signal and the time at which the diodes turn off.

One of the most difficult transients to eliminate is caused by the gated signal itself. This is illustrated in Fig. 16-43, showing two symmetrical control signals opening or closing the gate in the presence of a nonzero input signal V_1. The finite rise times of the gating signals cause nonsimultaneous switching of the four diodes. Therefrom, a strong unbalance exists during a very short interval of time which produces a sharp spike superimposed on the gated output signal. The amplitude and the polarity of this spurious transient depend on the gated signal, which vanishes when V_1 is equal to zero.

Choppers.   Gates are sometimes used to modulate small, slowly varying signals. Gates designed for such applications are often called choppers. To perform the switching, active elements are convenient, but the problem of switching transients and offset currents and voltages needs to be examined carefully. Interest in chopper circuits has fallen off in recent years since dc amplifiers with excellent drift characteristics are available.

A simple chopper circuit, however, based on the use of a nonsaturated field-effect transistor is shown in Fig. 16-44. It operates like the circuit of Fig. 16-42. An important feature of this circuit, besides its simplicity, lies in the fact that FETs have no offset voltage. A nonsaturated FET behaves like a resistor, not like a resistor in series with a small emf. FET choppers offer exceptional long-term stability and are capable of handling signals as low as a few microvolts. They have the disadvantage of higher switching transients, compared with bipolar choppers, but this is not a severe limitation when the repetition rate is low.

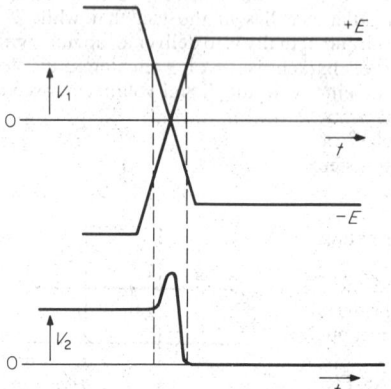

Fig. 16-43. Transient signal (bottom) caused by finite rise times of gating signals.

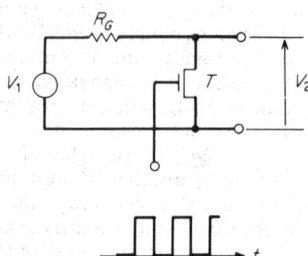

Fig. 16-44. Simple chopper circuit.

ACTIVE WAVEFORM SHAPING USING NEGATIVE FEEDBACK

23. Active Circuits.   Linear active networks used for waveshaping may employ negative feedback to improve the performance of the waveshaping circuits. Of the linear active waveshaping circuits, the operational amplifier-integrator is most widely used.

24. RC Operational Amplifier-Integrator.   In Fig. 16-45a it is assumed that the operational amplifier has infinite input impedance, zero output impedance, and a high negative gain A. The overall transfer function is

$$\frac{A}{1 + p(1 - A)T} \qquad \text{where } T = RC \qquad (16\text{-}65)$$

This function represents a first-order system with gain A and a cutoff frequency which is approximately $|A|$ times lower than the inverse of the time constant T of the RC circuit. In Fig. 16-45b, the frequency response of the active circuit is compared with that of the passive RC integrator. The widening of the spectrum useful for integration is clearly visible. For instance, an integrator using an operational amplifier with a gain of $-10,000$ and a RC network having a 0.1-s time constant has a cutoff frequency as low as 1.6×10^{-3} Hz.

In the time domain, the Taylor expansion of the amplifier-integrator response to the step function is

$$V(t) = E\frac{t}{T}\left[1 - \frac{t}{2!\,|A|T} + \frac{t^2}{3!\,(|A|T)^2} - \cdots \right] \qquad (16\text{-}66)$$

This shows that any desired degree of linearity of $V(t)$ can be achieved by providing sufficient gain in the amplifier.

25. Sweep Generators.   Sweep generators (also called time-base circuits) produce linear voltage or current ramps vs. time. They are widely used in applications such as cathode-ray displays, digital voltmeters, and television. In almost all such circuits the linearity of the ramp results from charging or discharging a capacitor through a constant-current source. The difference between circuits used in practice rests in the manner of realizing the constant-current property of the source. Sweep generators may also be looked upon as integrators with a constant-amplitude input signal. The latter point of view shows that RC operational amplifier-integrators provide the basic structure for sweep generation.

Circuits delivering a linear voltage sweep fall into two categories, the Miller time base and bootstrap time base. A simple Miller circuit is shown in Fig. 16-46. It comprises a capacitor C connected in a feedback loop around the amplifier formed by T_1 and its output-load resistor R_L. Transistor T_2 acts like a switch. When it is on, all the current flowing through the base resistor R_B is driven to ground, keeping T_1 blocked, since the voltage drop across T_2 is lower than the normal base-to-emitter voltage of T_1. The output signal V_{CE} of T_1 is thereby clamped at the level of the power-supply voltage E_{cc}, and the voltage drop across the capacitor C is approximately the same. When T_2 is suddenly turned off, it drives T_1 into the active region and causes collector current to flow through R_L. The resulting voltage drop across R_L is coupled capacitively to the base of T_1, tending to minimize the base current; i.e., the negative feedback loop is closed. The collector-to-emitter voltage V_{CE} of T_1 subsequently undergoes a linear voltage sweep downward, as illustrated in the right part of Fig. 16-46.

The circuit behaves in the same manner as the RC operational amplifier described in Par. **16-24**. Almost all the current flowing through R_B is derived through the feedback capacitor, and only a very small part is used for controlling the base of T_1. The feedback loop opens when T_1 enters into saturation, and the voltage gain of the amplifier becomes negligible.

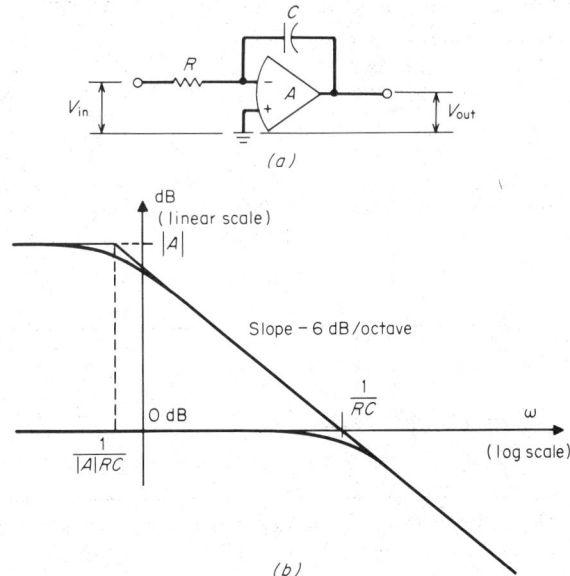

(a)

(b)

Fig. 16-45.   (*a*) Operational amplifier-integrator; (*b*) gain vs. angular frequency.

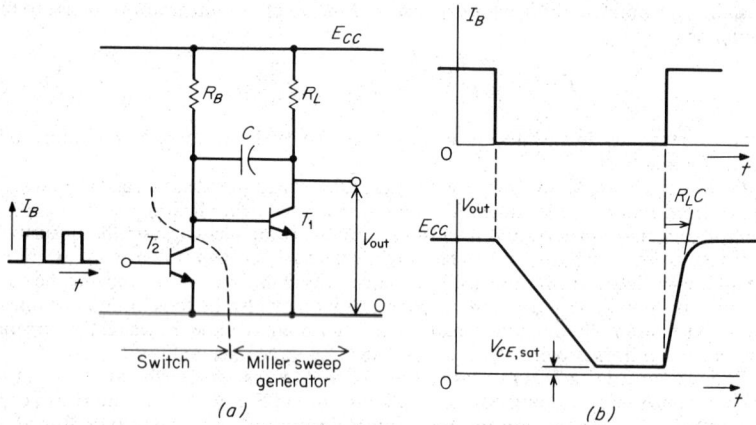

Fig. 16-46.   Miller sweep generator: (*a*) circuit; (*b*) input and output vs. time.

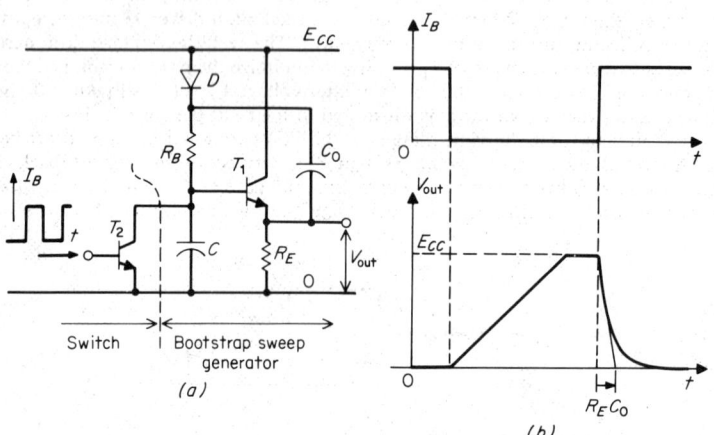

Fig. 16-47.   Bootstrap sweep generator: (*a*) circuit; (*b*) input and output vs. time.

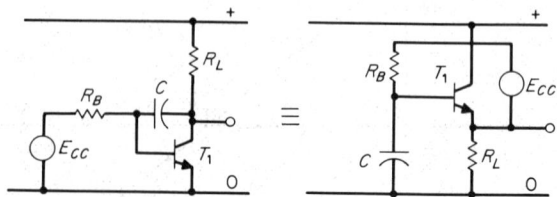

Fig. 16-48.   Equivalency of the Miller and bootstrap sweep generators.

16-30

When T_2 is subsequently turned on again, blocking T_1 and recharging C through R_L and the saturated switch, the output voltage V_{CE} rises again according to an exponential curve with time constant R_LC.

Figure 16-47 shows a typical bootstrap time-base circuit. This differs from the Miller circuit in that the capacitor C is not a part of the feedback loop. Instead the amplifier is replaced by an emitter-follower circuit delivering an output signal V_{out} which reproduces the voltage drop across the capacitor. C is charged through resistor R_B from a floating voltage source formed by the capacitor C_0 (C_0 is large compared to C).

First, we consider that the switch T_2 is on. Direct current then flows through the series combination formed by the diode D, the resistor R_B, and the saturated transistor T_2. The emitter follower T_1 is blocked since T_2 is saturated. Moreover, the capacitor C_0 can charge through the path formed by the diode D and the emitter resistor R_E, and the voltage drop across its terminals is equal to E_{CC}. When T_2 is cut off, the current through R_B flows into the capacitor C, causing the voltage drop across its terminals to rise gradually, driving T_1 into the active region. Because T_1 is a unity-gain amplifier, V_{out} is a replica of the voltage drop across C.

Since C_0 acts as a floating dc voltage source, diode D is reverse-biased almost immediately. The current flowing through R_B is supplied exclusively by C_0, and the voltage across it stays constant at the voltage drop across C_0 minus the base-to-emitter voltage of T_1.

Considering that the base current of T_1 represents only a small fraction of the total current flowing through R_B, it is evident that the charging of capacitor C occurs through a constant-current source and that therefore a linear voltage ramp is obtained as long as the output voltage of T_1 is not clamped to the level of the power-supply voltage E_{CC}.

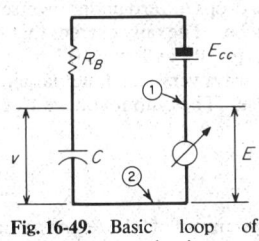

Fig. 16-49. Basic loop of sweep-generator circuits.

The corresponding output waveforms are shown in the right part of Fig. 16-47. After T_2 is switched on again, C discharges rapidly, causing V_{out} to drop, while the diode D again is forward-biased and the small charge lost by C_0 is restored. In practice, C_0 should be at least 100 times larger than C, to ensure a quasi-constant voltage source.

More detailed analysis of the Miller and bootstrap sweep generators reveals that they are in fact equivalent. We redraw the Miller circuit as shown at the left of Fig. 16-48. Remembering that the operation of the sweep generator is independent of which output terminal is grounded, we ground the collector of T_1 and redraw the corresponding circuit. As shown at the right in the figure, this is a bootstrap circuit, and the two circuits thus are equivalent.

Any sweep generator may be regarded as a simple loop (Fig. 16-49) comprising the capacitor C delivering a voltage ramp, the loading resistor R_B, and the series combination of two sources: a constant voltage source E_{CC} and a variable source whose emf E reproduces the voltage drop V across the capacitor. The voltage drop across R_B consequently remains constant at E_{CC}, and so the loop current is also constant. The voltage ramp consequently is given by

$$E = V = \frac{E_{CC}}{R_BC}t \tag{16-67}$$

Grounding terminal 1 yields the Miller network, while grounding terminal 2 leads to the bootstrap circuit.

Since the degree of linearity is one of the essential features of sweep generators, we consider the equivalent networks represented in Fig. 16-50. Starting with the Miller circuit, we determine the impedance in parallel with C

$$|A| (R_B \parallel h_{11}) \tag{16-68}$$

where $|A|$ is the absolute value of the voltage gain of the amplifier

$$|A| = \frac{h_{21}}{h_{11}}R_L$$

Next, considering the bootstrap circuit, we calculate the input impedance of the unity-gain amplifier to determine the loading impedance acting on C. This impedance is

$$R_L h_{21} \frac{R_B}{R_B + h_{11}} \qquad (16\text{-}69)$$

which turns out to be the same as given in Eq. (16-68); i.e., the two circuits are equivalent. To determine the degree of linearity it is sufficient to consider the common equivalent circuit of Fig. 16-51 and to calculate the Taylor expansion of the voltage V

$$V = \frac{E_{CC}}{R_B C} t \left[1 - \frac{t}{2! |A|(R_B \parallel h_{11})C} + \frac{t^2}{3! [|A|(R_B \parallel h_{11})C]^2} - \cdots \right] \qquad (16\text{-}70)$$

The higher the voltage gain $|A|$, the better the linearity. Thus, an integrated operational amplifier in place of T_1 leads to excellent performance in both the Miller and the bootstrap circuit. Voltage gains as high as 10,000 are readily obtained for this purpose.

26. Nonlinear Active Networks. The use of nonlinear devices with negative feedback accentuates the character of waveshaping networks. In many circumstances, this leads to an idealization of the nonlinear character of the devices considered. A good example of this is given by the ideal rectifier circuit.

27. Ideal Rectification with Nonlinear Feedback. The feedback loop in the circuit shown in Fig. 16-52, which comprises the series combination of the resistance R_2 and the diode D_1, is equivalent to an open circuit when the diode is reverse-biased and to the resistor alone under forward-bias conditions. The output voltage is taken between the resistor and the diode, which assimilates the diode as a part of the output impedance of the amplifier.

The overall gain of the amplifier, with the diode in series, drops to zero under reverse-bias condition whatever the inverse voltage across the diode may be. The only current that flows through R_2 is the leakage current of the diode, which is typically as low as 0.1 nA. The voltage drops across either resistor R_1 or R_2 are therefore always very small, not larger than 1 mV with resistors in the 100-kΩ range, as is usually the case. The output voltage V_2 under reverse-bias conditions is also small and is given by

$$V_2 = \frac{V}{A} + R_2 I \simeq \frac{V}{A}$$

Considering a gain A of 10^{-4}, R_2 equal to 100 kΩ, a leakage current of 0.1 nA, and assuming that the dynamic range of the amplifier is 15 V, this yields an output voltage V_2 equal to approximately -1.5 m V.

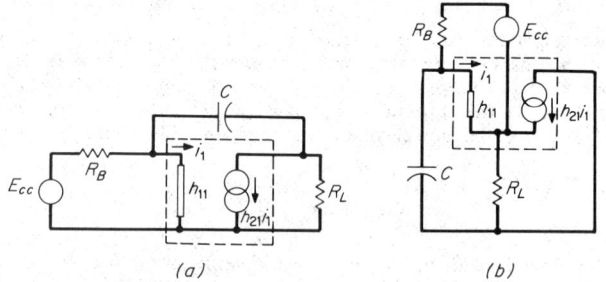

Fig. 16-50. Equivalent forms of (*a*) Miller and (*b*) bootstrap sweep circuits.

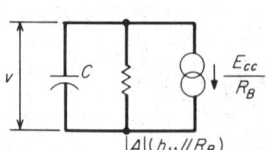

Fig. 16-51. Common equivalent circuit of sweep generators.

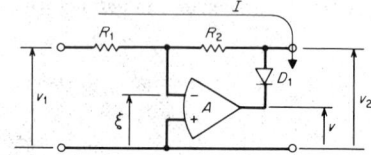

Fig. 16-52. Nonlinear-feedback rectifier circuit.

The input voltage V_1 should be of the same order of magnitude. A higher input voltage would drive the amplifier into saturation, causing the voltage ξ to rise above the usual 1-mV range and this change would be reflected in the output voltage V_2. To prevent this a bypass network may be provided for the input current, to maintain V within the dynamic range of the amplifier. This can be achieved by the introduction of a second negative-feedback loop containing another diode D_2 (Fig. 16-53). In this latter circuit, in fact, one diode is always forward-biased whether the input circuit is positive or negative, so that the feedback loop is opened only when V lies between $+0.6$ and -0.6 V; however, this does not affect the characteristics of the ideal rectifier because the corresponding input voltage of the amplifier is then extremely small.

When D_1 is forward-biased, the voltage V automatically adjusts itself to force the current flowing through D_1 and R_2 to be the same as through R_1. This means that V, in some circumstances, may be much larger than V_2, especially when V_2 (and thus also V_1) is of the order of magnitude of millivolts. In fact, V exhibits approximately the same shape as V_2 plus an additional pedestal of approximately 0.6 V. Typical waveforms obtained with a sinusoidal voltage of a few tens of millivolts are shown in Fig. 16-54.

The quasi-ideal rectification characteristic of this circuit is readily explained by replacing the operational amplifier circuit by its Norton equivalent network, a current source V_1/R_1 in parallel with a resistor $|A|R_1$ of extremely high value (Fig. 16-55). In the above example, $|A|R_1$ amounts to 1 GΩ. In other words, the nonlinear dipole D_1 in series with R_2 is fed by a current source. V_2 is given simply by $-(R_2/R_1)V_1$ independently of the forward-voltage drop across D_1. The fact that D_2 lowers the output impedance of the Norton equivalent circuit is not significant since this diode is reverse-biased when D_1 is conducting.

A measure of the wide rectification range over which this circuit performs is given by the transfer function

$$\frac{V_2}{V_1} = 1 - \frac{2}{A}\left(1 + \frac{V_{BE}}{RI}\right) \tag{16-71}$$

where V_{BE} is the voltage drop across the forward-biased diode D_1. R_1 and R_2 are assumed to

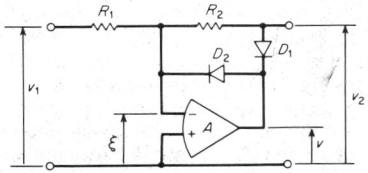

Fig. 16-53.   Rectifier circuit employing two feedback loops.

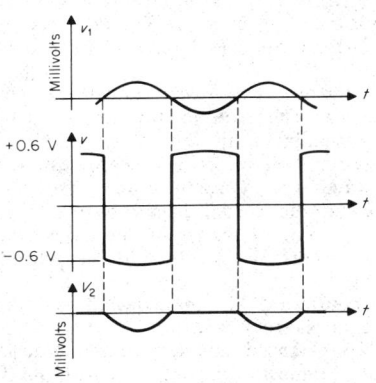

Fig. 16-54.   Waveforms of circuit in Fig. 16-53.

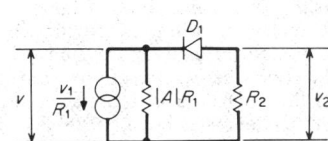

Fig. 16-55.   Norton equivalent network of operational amplifier rectifier.

be identical, and their common value is given as R, while the current I and the voltage drop V_{BE} are of course related by the diode equation (16-10).

The relative error between V_1 and V_2 is given by

$$\epsilon = -\frac{2}{A}\left(1 + \frac{V_{BE}}{RI}\right) \tag{16-72}$$

When the input voltage is small, this expression can be written

$$\epsilon \simeq -\frac{2}{A}\frac{V_{BE}}{RI}$$

This error is always very small even when RI is of the order of magnitude of 1 mV. For instance, the above example yields a relative error of only 2.4% when RI is as low as 1 mV and 0.36% when RI is equal to 10 mV.

Other negative-feedback configurations leading to very high output impedances are equally powerful in achieving ideal rectification characteristics. For instance the unity-gain amplifier used in instrumentation has wide linear ac measurement capabilities (Fig. 16-56).

28. Clipping Circuits Using Negative Feedback. The circuit of Fig. 16-57a provides clipping characteristics that are superior to those obtained with passive nonlinear elements alone. The output voltage of the circuit must be bounded by $+0.6$ to -0.6 V by the clamping action of the diodes D_1 and D_2. Only when V_2 lies between these two limits is the feedback loop around the amplifier virtually open. The input voltage ξ, however, must then be exceedingly small; so also will V_1 be, since the current flowing through the diodes is in the range of picoamperes to nanoamperes. However, when this current tends to become a few orders of magnitude larger, V_2 increases in order to drive the appropriate amount of current through the diodes.

The output characteristic of the circuit exhibits a steep rise around zero input volts, followed by flat portions approximately at $+0.6$ to -0.6 V. This provides excellent clipping characteristics (Fig. 16-57b).

ACTIVE WAVEFORM SHAPING USING POSITIVE FEEDBACK

29. Positive Feedback. Positive feedback may be employed in stable as well as unstable (free-running) circuits. Unstable networks include harmonic oscillators and free-running relaxation circuits. The latter may be self-excited or synchronized by external trigger pulses. Trigger pulses may also be used to execute transitions between stable states. A free-running relaxation circuit is called an *astable multivibrator* network. Monostable and bistable multivibrator circuits also occur; they have respectively one or two distinct stable states.

The degree to which positive feedback is used in harmonic oscillators differs substantially from that of astable, monostable, and bistable pulse circuits. In the case of the oscillator the total loop gain must be kept close to 1, and it need compensate only for the small losses in the resonating tank circuit. In the pulsed circuits, on the other hand, positive feedback permits fast switching from one state to another, e.g., from cutoff to saturation and vice versa. Before and after these transitions occur, the circuit usually is passive. Switching occurs in an extremely short time, typically less than 10 ns. After switching, the circuit operates more slowly, approaching steady-state conditions.

It is common practice to call the switching time the *regeneration time* and the time needed to reach final steady-state conditions the *resolution time*. The resolution time may range from tens of nanoseconds to several seconds or more, depending on the circuit design.

An important feature of regenerative circuits is that their switching times are essentially independent of the steepness of the trigger-signal waveshape. Once the level of instability is reached, the transition occurs at a rate fixed by the total loop gain and by the reactive parasitics of the circuit itself but independent of the rate of change of the trigger signal. Regenerative circuits thus provide means of restoring short rise times to pulses broadened by transmission through large systems.

Positive-feedback pulse circuits are necessarily nonlinear. The most conventional way to study their behavior is to use a piecewise-linear analysis technique.

30. Collector-coupled Bistable Circuits. Two cascaded common-emitter transistor stages realize an amplifier with a high positive gain. Connecting the output to the input (Fig. 16-58) produces an unstable network known as the *Eccles-Jordan bistable circuit* or *flip-flop*. Under steady-state conditions one transistor is saturated and the other is blocked. In a

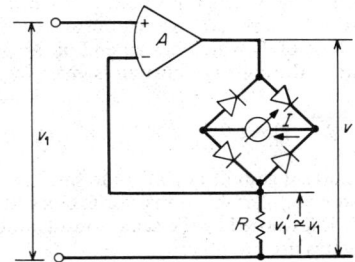

Fig. 16-56. Feedback rectification circuit used in precision measurements.

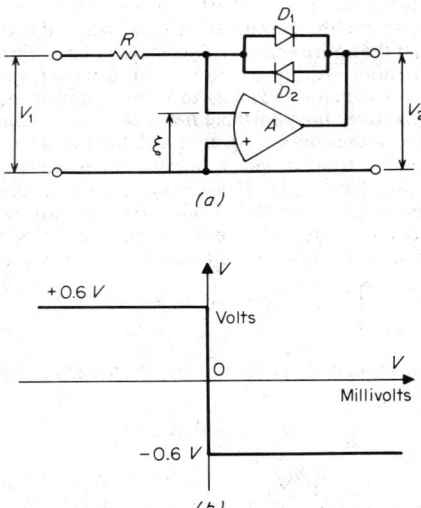

Fig. 16-57. Negative-feedback clipping circuit: (*a*) circuit; (*b*) input-output characteristic.

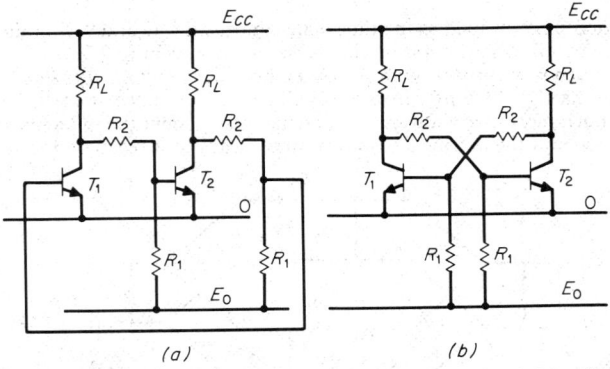

Fig. 16-58. Eccles-Jordan bistable circuit (flip-flop): (*a*) in form of cascaded amplifier with output connected to input; (*b*) circuit as customarily drawn, showing symmetry of connections.

16-35

typical case, the circuit of Fig. 16-58 has the values $R_L = 820\ \Omega$, $R_1 = 39\ k\Omega$, $R_2 = 5.6\ k\Omega$, $E_{CC} = +6\ V$, and $E_0 = -6\ V$. Suppose T_1 is at cutoff, and consider the Thevenin equivalent network connected to the base of T_2 which has an emf of $+4.3\ V$ and a series impedance of $5.5\ k\Omega$. When T_2 is saturated, the collector current is given by E_{cc}/R_L, or 7.3 mA. The base current is given by

$$I_{B2} = \frac{4.3 - 0.6}{5.5} = 0.67\ mA \tag{16-73}$$

One may conclude that a current gain of 11 suffices to saturate T_2. The base voltage of T_1 is found equal to $-0.75\ V$, showing that this transistor is blocked. The reverse situation, with T_1 saturated and T_2 cut off, is governed by the same considerations, for reasons of symmetry. Thus two distinct stable states are possible.

When one of the transistors is suddenly switched from one state to the opposite, the other transistor automatically undergoes an opposite transition. At a given time both transistors conduct simultaneously, increasing the loop gain suddenly from zero to a high positive value. This corresponds to the regenerative phase, during which the circuit becomes active.

It is difficult to compute the regeneration time since the operating points of both transistors move through the entire active region, causing large variations of the small-signal parameters of both transistors. Although determination of the regeneration time on the basis of a linear model is unrealistic and leads only to a rough approximation, we briefly examine this problem since it illustrates how unstable networks can be analyzed.

First, we introduce two capacitors in parallel with the two R_2 resistors. These capacitors provide a direct connection from collector to base under transient conditions and hence increase the high-frequency loop again. The circuit can now be described by the network of Fig. 16-59, which consists of a parallel combination of two reversed transistor hybrid equivalents without extrinsic base resistances (for calculation convenience) and with two load admittances. Starting from the admittance matrix of one of the loaded transistors,

$$\begin{bmatrix} y_{11} & y_{12} \\ y_{21} & y_{22} \end{bmatrix} = \begin{bmatrix} g_m(1 - \alpha_0) + p(C_\pi + C_{0b}) & -pC_{0b} \\ \alpha_0 g_m - pC_{0b} & Y + pC_{0b} \end{bmatrix} \tag{16-74}$$

the determinant of the parallel combination is equated to zero to find the natural frequencies of the circuit. This leads to

$$Y + g_m + pC_\pi = 0 \tag{16-75}$$

$$g_m(1 - 2\alpha_0) + Y + p(C_\pi + 4C_{0b}) = 0 \tag{16-76}$$

Only Eq. (16-76) has a zero with a positive real part causing instability in the form of an increasing exponential function with time constant

$$\tau = \frac{C_\pi + 4C_{0b}}{g_m} \tag{16-77}$$

The regeneration time (defined as the time elapsing between 10 and 90% of the total voltage excursion from cutoff to saturation or vice versa) is thus equal to 2.2τ.

A high-frequency transistor with 1-mA emitter current ($g_m = 38\ mA/V$), a 20-pF capacitor C, and a C_{0b} of 0.6 pF yields a switching time of approximately 1.5 ns. A more accurate but much more elaborate analysis, taking into account the influence of the emitter-current variation and the extrinsic base resistance, leads to a regeneration time about one

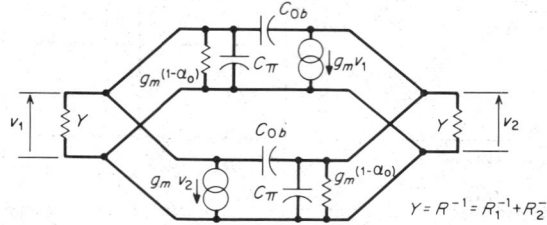

Fig. 16-59. Flip-flop circuit showing capacitances that determine time constants.

order of magnitude higher. Nevertheless, Eq. (16-77) clearly shows what factors control the regeneration time: the transconductance of the active elements and the unavoidable parasitics, in fact the merit factor of the transistor used. This conclusion is quite general and can be verified in many other positive-feedback switching circuits.

We next consider the principal time constants controlling the resolution time of this circuit (Fig. 16-59), with the same numerical constants. We suppose T_1 initially nonconducting and T_2 saturated. The sudden turning off of T_2 is simulated in Fig. 16-60 by opening switch S_2. Immediately, V_{CE2} starts increasing toward E_{cc}. The base-to-ground voltage V_{BE1} of T_1 consequently rises above its steady-state value (-0.75 V) with a time constant fixed only by the total parasitic capacitance C_0 at the collector of T_2 and base of T_1, and the parallel combination R of R_1 and R_L. Hence the time constant is

$$\tau_1 = RC_0$$

This time is normally extremely short; e.g., a parasitic capacitance of 5 pF yields a time constant of less than 5 ns. The charge accumulated across C evidently cannot change appreciably in that time and so V_{BE1} and V_{CE2} increase at the same rate. When V_{BE1} reaches 0.6 V approximately, T_1 starts conducting and a new situation arises which is illustrated in Fig. 16-60 by the passage from case b to case c. This is the moment regeneration actually takes place, forcing T_1 to go into saturation very rapidly (Fig. 16-62). Neglecting the regeneration period, case c is characterized by the time constant

$$\tau_2 = (R_L \parallel R_2)C$$

Choosing C equal to 100 pF yields $\tau_2 = 70$ ns. This time constant, although it is much longer than τ_1, is still not the longest, for we have yet to consider the evolution of V_{BE2} described in Fig. 16-61.

In this figure, the saturating transistor T_1 is replaced by the closing of S_1. The problem is the same as that of the compensated attenuator (Par. 16-5) with the difference that

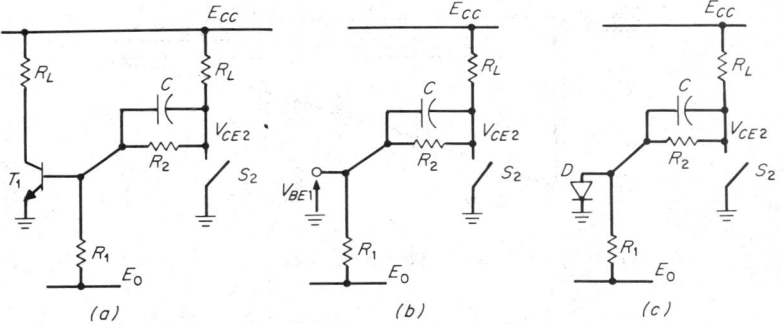

Fig. 16-60. Transfer from one side to the other of a flip-flop circuit: (*a*) circuit; (*b*) case b; (*c*) case c.

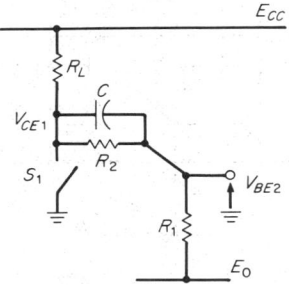

Fig. 16-61. Effect of opening the opposite side of a flip-flop circuit.

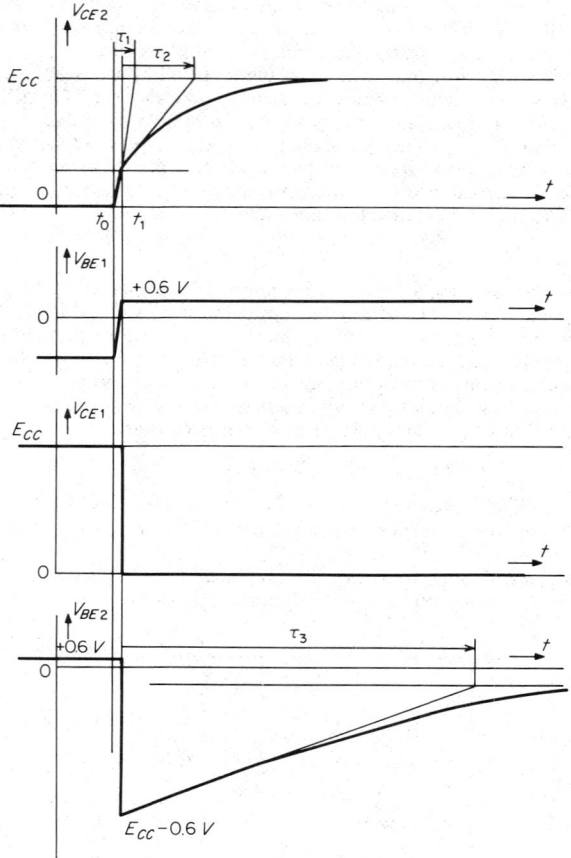

Fig. 16-62. Voltage variations vs. time of flip-flop circuit.

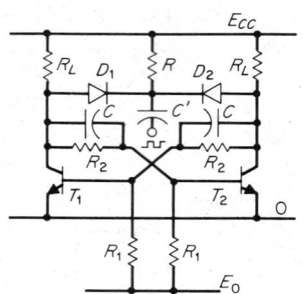

Fig. 16-63. Differentiating input to flip-flop circuit to obtain triggering from square-wave input.

overcompensation is introduced, so that V_{BE2} undergoes a large negative-voltage swing equal to E_{cc} before climbing slowly toward its final steady-state value, -0.75 V. The situation is shown in the lowest part of Fig. 16-62 (cf. Fig. 16-9 with $K < 1$).

The time constant τ_3 can be computed from Eq. (16-7) and is equal in the present case to approximately 500 ns. The capacitor C thus plays a dual role. This first is favorable since it ensures fast regeneration, but the second is unfavorable because it increases the resolution time and sets an upper limit to the maximum repetition rate at which the Eccles-Jordan circuit can be switched. The proper choice of C as well as of R_L, R_1, and R_2 must take this fact into consideration. Small values of the resistances make high repetition rates possible at the price of increased dc power consumption.

31. Triggering. Triggering usually is obtained by applying a steeply rising pulse either to the blocked transistor to make it conduct or to the saturated transistor to cut it off. The second method yields somewhat better sensitivity. The first method requires a triggering signal sufficiently large to bring the base from its steady-state negative voltage to approximately $+0.6$ V. The trigger signal may be dc- or ac-coupled.

Figure 16-63 shows a typical ac coupling network used to differentiate the input trigger when a square-wave generator is used instead of a narrow-pulse generator. The two additional diodes provide automatic channeling of the differentiated control signal. Suppose that T_1 is off and T_2 is on. Any negative control pulse is transmitted through diode D_1 to the collector of the blocked transistor and from there to the base of the saturated transistor to turn it off. Diode D_2 will not conduct in these circumstances, preventing the triggered signal from going the other way around.

If a square wave is applied to the differentiator, only the negative-going portions of the wave will provide triggering. Two such successive steps are necessary to execute a complete cycle of the bistable network. Binary frequency division can be achieved in this manner by cascading a number of identical cells, the collector of one cell providing the trigger signal controlling the next cell, etc. C' should be kept small not only to improve differentiation but also to minimize loading of the bistable circuit by the generator and thus to preserve fast regeneration times.

32. Monostable and Astable Operation—Synchronization. Figures 16-64 and 16-65 show monostable and astable collector-coupled pairs, respectively. The fundamental difference between these circuits and the bistable networks of Par. 16-30 lies in how dc biasing is achieved. In Fig. 16-64, T_2 is normally conducting except when a negative trigger pulse drives this transistor into the cutoff region. T_1 necessarily undergoes the inverse transition, suddenly producing a large negative voltage step at the base of T_2 shown in the lower part of Fig. 16-64b. V_{BE2}, however, cannot remain negative since the base is connected to the positive-voltage supply through the resistor R. The base voltage must thus rise toward E_{cc}, with a time constant RC. As soon as the emitter junction of T_2 becomes forward-biased, the monostable circuit will change its state again and the circuit remains in that state until another trigger signal is applied.

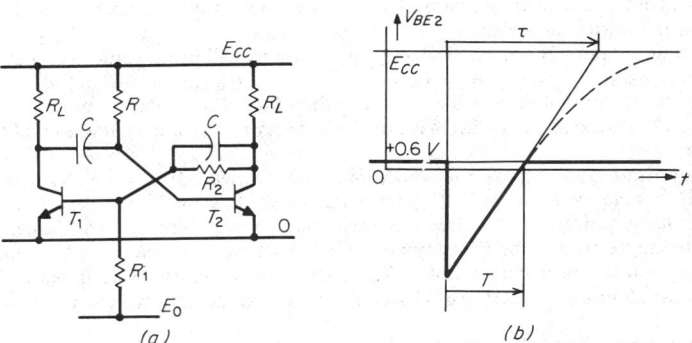

Fig. 16-64. Monostable collector-coupled pair: (*a*) circuit; (*b*) output vs. time characteristics.

The time T between the application of a trigger pulse and the instant T_2 saturates again is given approximately by

$$T = \tau \log_n 2 = 0.693\tau \tag{16-78}$$

where $\tau = RC$. The supply voltage E_{CC} must be large compared to the forward-voltage drop of the emitter junction of T_2 for this expression to apply.

The astable collector-coupled pair, or free-running multivibrator (Fig. 16-65), operates according to the same scheme except that steady-state conditions are never reached. The base-bias networks of both transistors are connected to the same positive power supply. The period of the multivibrator thus equals $2T$ if the circuit is symmetrical, and the repetition rate F_r is given by

$$F_r = \frac{1}{2\,\tau\log_n 2} = \frac{0.721}{RC} \tag{16-79}$$

An astable network need not be triggered, of course. But pulses fed to a multivibrator from an external generator may synchronize it. The mechanism can be understood on the basis of Fig. 16-66, where we assume that regularly spaced negative pulses are applied to the collectors of the multivibrator (by a circuit like that of Fig. 16-63) or to the bases through

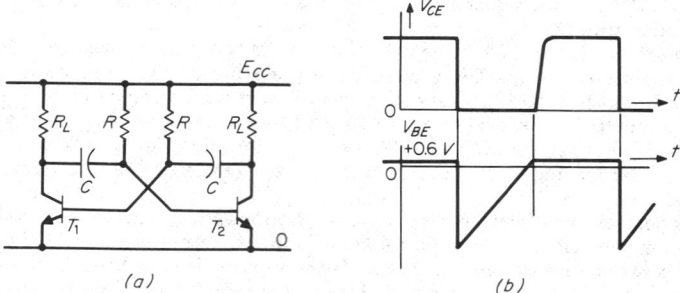

(a)   (b)

Fig. 16-65.   A stable collector-coupled pair: (*a*) circuit; (*b*) output vs. time characteristics.

small capacitances. Figure 16-66 shows typical base-voltage curves obtained when the multivibrator is synchronized by means of pulses running 6 times faster than the period of the multivibrator itself. Starting arbitrarily from a state transition, at the time t_0, we see that the pulse appearing at t_1 turns off the saturated transistor, producing a large positive spike at the collector of this transistor which is transferred to base of the blocked transistor. The resulting positive pulse is not sufficient, however, to drive the blocked transistor into the active region, because its base is still heavily reverse-biased. The second pulse does not cause switching either, but the third is successful.

If the multivibrator is symmetrical, similar situations will occur respectively at t_4, t_5, and t_6. State transitions consequently occur every third trigger pulse, and the oscillation frequency of the multivibrator is 6 times lower than that of the pulse generator. Symmetry is not absolutely necessary. In fact, a multivibrator may run at uneven fractions of the pulse-repetition rate.

Synchronization of astable networks provides a method of time division. While apparently simpler than cascading bistable networks, this method suffers from its high sensitivity to variations of the circuit elements and power supply. The larger the time division, the more critical the adjustments. Time division by a factor of 10, for instance, is easy to perform but rather unreliable. The availability of inexpensive integrated bistable networks has almost completely eliminated the synchronized-multivibrator method of time division.

33. Integrated-Circuit Bistable Collector-coupled Pairs. An integrated version of the Eccles-Jordan circuit (Fig. 16-67) shows simplifications usually introduced into the circuit of

Fig. 16-58; i.e., the negative voltage supply and the resistive dividers are omitted. Assuming that T_1 is at cutoff and T_2 is saturated,

$$V_{CE1} = V_{BE2} + \frac{R}{R + R_L}(E_{CC} - V_{BE2}) \qquad (16\text{-}80)$$

V_{CE2} is smaller than the normal forward-voltage drop across the emitter junction of T_1. Considerations of symmetry lead to the conclusion that another stable state with T_2 blocked and T_1 saturated is equally possible. The resistors R in series with the bases are not required (some versions exist with R equal to zero), but they provide increased output dynamic range according to Eq. (16-80). Each of the two inverters may be viewed as a single-input RTL NOR circuit, so that the bistable circuit shown in the left part of Fig. 16-67 can be redrawn in the form of a loop of two NOR inverters, as in the right part of Fig. 16-67.

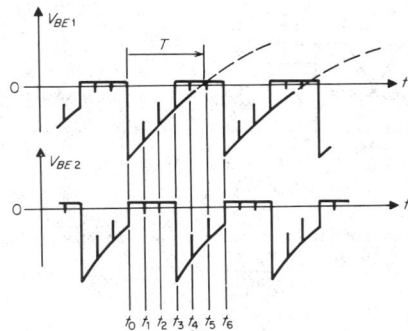

Fig. 16-66. External synchronization of an astable pair.

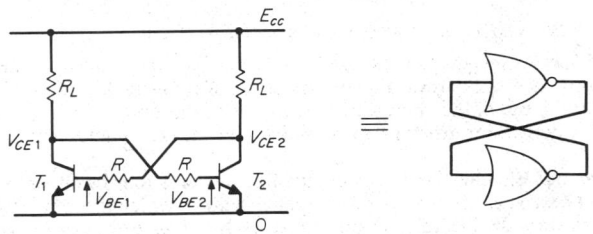

Fig. 16-67. Integrated-circuit version of Eccles-Jordan flip-flop.

Trigger circuits suitable for integrated-circuit flip-flops are almost always dc-coupled. A typical example is given in Fig. 16-68 with the corresponding symbolic representation. The triggering inputs are called *set S* and *reset R* terminals. Additional active elements are necessary to provide the actual triggering, but this is not a problem in integrated circuits because transistors are the smallest device available on the chip. Two NAND inverters in a loop also form a bistable network. Triggering occurs in the same manner by means of additional set and reset input terminals (Fig. 16-69).

The respective truth tables for the NOR and the NAND bistable circuits are

S	R	Q_1	Q_2	Line	S	R	Q_1	Q_2	Line
0	0	Q	$\bar{Q}$	1	0	0	1	1	5
0	1	1	0	2	0	1	1	0	6
1	0	0	1	3	1	0	0	1	7
1	1	0	0	4	1	1	Q	$\bar{Q}$	8
		NOR bistable					**NAND bistable**		

Lines 1 and 8 correspond to situations where the S and R inputs are both inactive, leaving the bistable circuit in one of the two possible states indicated in the tables above by the letters Q and $\bar{Q}$ (Q may be either 1 or 0). If a specified output state is required, a pair of adequate complementary dc trigger signals may be applied to the S and R inputs simultaneously.

For instance, if the output pair is to be characterized by $Q_1 = 1$ and $Q_2 = 0$, the necessary input combination, for NOR as well as for NAND bistable circuits, is $S = 0$ and $R = 1$. Changing R back from 1 to 0 does not change anything in the output state in the NOR bistable. The same is true if S is made equal to 1 in the NAND bistable. In both cases, the flip-flop exhibits infinite memory of the imposed state. The name *sequential circuit* is given to this class of network.

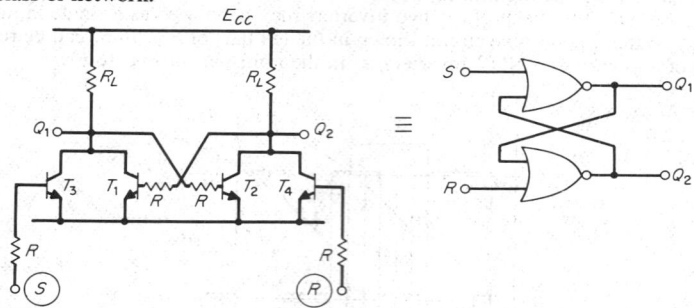

Fig. 16-68. Dc-coupled version of flip-flop, customarily used in integrated-circuit versions of this circuit.

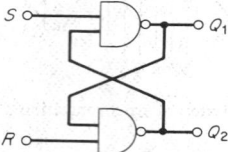

Fig. 16-69. Set S and reset R terminals of a flip-flop.

Lines 4 and 5, however, must be avoided, for the passage from line 4 to line 1 or from line 5 to line 8 leads to uncertainty regarding the final state of the bistable circuit. In fact, the final transition is entirely out of the control of the input, since in both cases it results solely from small imbalances between transistor parasitics that allow faster switching of one inverter than the other.

34. Synchronous Bistable Circuits. Sequential networks may be either asynchronous or synchronous. The asynchronous class describes circuits in which the application of an input control signal triggers the bistable circuit immediately. This is true of the circuits thus far considered. In the synchronous class, changes of state occur only at selected times, after the trigger signal has occurred.

The auxiliary signal that determines the switching times is called the *clock signal.* Synchronous circuits are insensitive to hazard conditions whereas asynchronous circuits are severely troubled by this effect, which originates in the propagation delay for which each bistable network or gate is responsible. These delays, although individually very small (typically of the order of a few nanoseconds), are responsible for introducing differential delays between signals that must travel through different numbers of logic circuits. Unwanted signal combinations may therefore appear for short periods and may be interpreted erroneously.

Synchronous circuits do not suffer from this limitation because they conform to the set and reset controls only when the clock pulse is present, usually after the transient spurious combinations are over. A simple synchronous circuit is shown in Fig. 16-70. The inhibition action provided by the absence of the clock signal is provided by a pair of input NAND circuits. Otherwise nothing is changed with respect to the classical bistable network. It should be noted, however, that the input gates reverse the roles of the set and reset controls.

Another difficulty occurs in a cascade of bistable circuits, used to realize a time-division network or a shift register. Instead of each circuit controlling its closest neighbor, when a clock signal is applied, the set and reset signals of the first bistable jump from one circuit to the next, traveling throughout the entire chain in a time which may be shorter than the duration of the clock pulse. To prevent this, a time delay must be introduced between the gating NAND circuits (Fig. 16-70) and the bistable network, so that changes of state can occur only after the clock signal has disappeared.

One approach is to take advantage of storage effects in bipolar transistors, but the so-called *master-slave* association, shown in Fig. 16-71, is preferred. In this circuit, intermediate storage is realized by an auxiliary clocked bistable network controlled by the complement of the clock signal. The additional circuit complexity is appreciable, but this approach is practical in integrated-circuit technology.

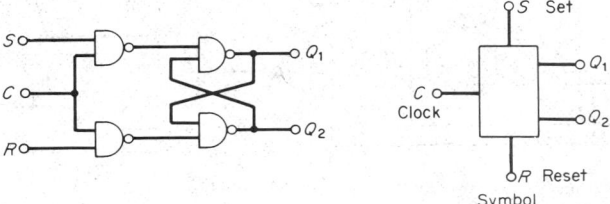

Fig. 16-70. Clock-controlled flip-flop.

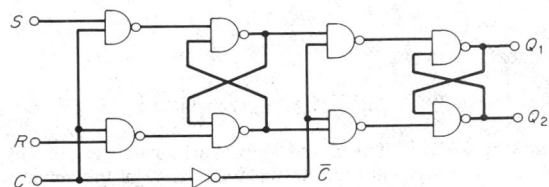

Fig. 16-71. Master-slave type of clock-controlled flip-flop.

The master-slave bistable truth table can be found from that of the synchronous circuit depicted in Fig. 16-70, which in turn can be deduced from the NAND bistable truth table in Par. 16-33. One problem remains, however, i.e., the forbidden one-one input pair which is responsible for ambiguous states of the NAND bistables each time the clock goes to zero. The solution to this problem, known as the *JK bistable* and presented in Fig. 16-72, consists in introducing additional feedback loops to link the second NAND bistable to the first clocked input gate. Analysis shows that this does not introduce any difference between the *JK* and master-slave bistable truth tables except for the critical one-one input pair which transforms the *JK* circuit in a clocked square-wave generator. The truth table then becomes

J	K	$(Q_1)_n + 1$	$(Q_2)_n + 1$	Line
0	0	Q_n	$\overline{Q}_n$	1
0	1	0	1	2
1	0	1	0	3
1	1	$\overline{Q}_n$	Q_n	4

Line 1 shows that at the end of each clock pulse, the bistable memorizes the last input introduced previously by lines 2 or 3, and line 4 indicates that state transitions occur each time a clock pulse appears. The corresponding logic equations of the *JK* bistable are

$$Q_{n+1} = J_n\overline{Q}_n + KQ_n$$
$$\overline{Q}_{n+1} = KQ_n + \overline{J}Q_n \tag{16-81}$$

Another approach, eliminating SR entries, is known as the *D bistable.* It consists of replacing the usual pair of input terminals by a single terminal (called *D*) and internally providing the complement of *D* by an additional inverter.

16-43

35. Emitter-coupled Pairs (Schmitt Circuits).   In the basic Schmitt circuit represented in Fig. 16-73, bistable operation is obtained by a positive-feedback loop formed by a common-base and common-collector transistor pair.   The circuit is similar to the Eccles-Jordan circuit except for closing the loop, which is achieved through short-circuiting the emitters of the two transistors.   The Schmitt circuit may also be considered as a differential amplifier as a part of a positive-feedback loop, which turns out to be a series-parallel association (Fig. 16-74).

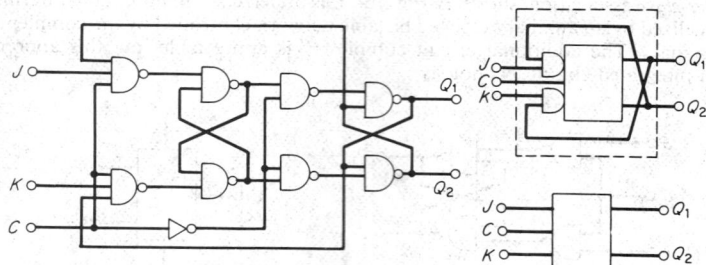

Fig. 16-72.   *JK-bistable* circuit.

For small-signal linear analysis two hybrid matrices are required, the differential amplifier (approximation):

$$\begin{bmatrix} 2\,h_{ie} & 0 \\ h_{fe} & 0 \end{bmatrix} \tag{16-82}$$

and the passive-feedback network:

$$\frac{1}{Y_1 + Y_2}\begin{bmatrix} 1 & Y_2 \\ -Y_2 & Y_1\,Y_2 \end{bmatrix} \tag{16-83}$$

where $Y_1 = R_1{}^{-1}$ and $Y_2 = R_2{}^{-1}$.   The closed-loop hybrid matrix is given by the sum of these expressions, to which one must add the internal impedance of the generator as well as the output-load admittance $Y_C = (R_C{}^{-1})$.   Hence, from

$$h_{11}^T = 2\,h_{ie} + \frac{1}{Y_1 + Y_2}$$

$$h_{12}^T = \frac{Y_1}{Y_1 + Y_2}$$

$$h_{21}^T = h_{fe} - \frac{Y_1}{Y_1 + Y_2} \simeq h_{fe}$$

and $$h_{22}^T = \frac{Y_1\,Y_2}{Y_1 + Y_2} + Y_C$$

the overall small-signal voltage gain is given by

$$A_V = \frac{v}{e} = \frac{A}{1 + A\beta} \tag{16-84}$$

with $$A \simeq \frac{-h_{fe}}{(Z_G + h_{11}^T)(h_{22}^T + Y_C)} \tag{16-85}$$

and $$\beta = \frac{Y_2}{Y_1 + Y_2} = \frac{R_1}{R_1 + R_2} \tag{16-86}$$

A represents the voltage gain of the loaded differential amplifier and β the voltage transfer function of the attenuator network.   The product $A\beta$, called the *transfer function of the open loop,* is a negative quantity which must be larger than or equal to 1, in absolute value, to ensure instability.

Switching times are very short, as in the Eccles-Jordan circuit.   Linear analysis produces the regeneration time

$$t \simeq 0.70\frac{h_{ie}}{R_C}\frac{1}{f_T} \tag{16-87}$$

16-44

where h_{ie} and f_T respectively represent the common-emitter input impedance and the transition frequency of the transistor used.

Emitter-coupled bistables are fundamentally different from Eccles-Jordan circuits, since no transistor saturates in either of their two stable states. Storage effects therefore need not be considered. This is illustrated in Fig. 16-75, which represents the two states, omitting the transistor which is blocked. In both cases, the conducting transistor operates in the common-collector configuration, provided that the voltage drop across R_C in Fig. 16-75b is not too large. Consider the numerical example $RC = 1.2\,k\Omega$, $R_1 = 3.3\,k\Omega$, $R_2 = 5.1\,k\Omega$, $I = 4.5\,mA$, and $E_{CC} = 15\,V$. The voltage V_{C1} at the collector of T_1 is found equal to 13.1 V in Fig. 16-75a and 8.4 V in Fig. 16-75b. The corresponding voltages at the base of T_2 are respectively 5.15 V in a and 3.3 V in b. Providing the input signal V_{B1} does not exceed 8 V with respect to ground, neither T_1 nor T_2 can saturate.

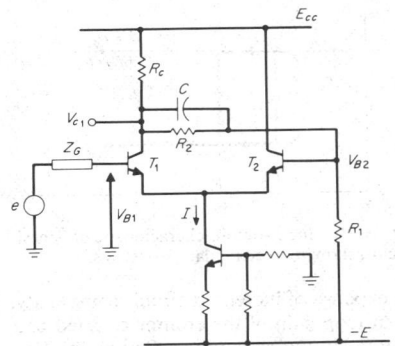

Fig. 16-73. Emitter-coupled Schmitt circuit, showing positive-feedback loop.

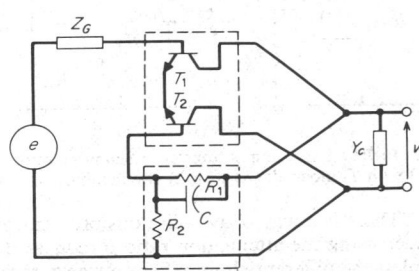

Fig. 16-74. Schmitt circuit equivalent circuit, showing operational amplifier and feedback loop.

This example also illustrates how two stable states are possible and how the Schmitt circuit executes transitions from one state to the other. In the situation depicted in Fig. 16-75a, transistor T_1 is blocked as long as the input voltage V_{B1} is kept below 5.15 V, since this is the dc voltage applied to the base of T_2 (we assume that the two transistors are identical in their forward-emitter-junction-voltage drops). In the other state, illustrated in Fig. 16-75b, T_1 is on as long as V_{B1} is above 3.3 V. A range of input voltages thus exists between 3.3 and 5.15 V where either of the two states is possible.

If we suppose that an input voltage below the smallest of these two values is applied to the Schmitt circuit, blocking T_1, and let this voltage be gradually increased, nothing happens as long as V_{B1} is smaller than 5.15 V. But when this limit is reached, the two transistors conduct simultaneously, producing an unstable situation. The current derived by T_1 increases the voltage drop across RC, lowering V_{B2}, and the emitter current of T_2 is consequently reduced, thereby increasing the current in T_1 further. The first transistor will take over all the current delivered by the current source, and the second will cut off. A new state is then reached, as shown in Fig. 16-75b, and V_{B2} now is equal to 3.3 V. Similar reasoning leads to the conclusion that switching may occur if the input voltage is decreased to as low as 3.3 V. A typical output-input voltage curve of the Schmitt bistable thus exhibits a rectangular hysteresis loop, as shown in Fig. 16-76.

Schmitt bistables, also called Schmitt triggers, are suitable for detecting the moment when an analog signal crosses a given dc level. They are widely used in oscilloscopes to provide time-base synchronization pulses. This is illustrated in Fig. 16-77, which shows a periodic signal triggering a Schmitt circuit and the corresponding output wave, which may control a sweep generator. In some Schmitt trigger circuits it is possible to modify the switching level by electrically changing the operating points of the transistors, e.g., by modifying the current delivered by the current source. This kind of control of the triggering level is frequently used in oscilloscopes.

Schmitt triggers necessarily exhibit hysteresis. In many applications, the width of the hysteresis zone, expressed in volts at the input $(V_a - V_b)$, does not play a significant role. It is easy to show that

$$V_a - V_b \simeq \frac{R_1 R_C}{R_1 + R_2 + R_C} I \tag{16-88}$$

Reducing the hysteresis is possible by decreasing either the current delivered by the current source or the attenuation ratio of the voltage divider. Hysteresis below 1 V, however, is not recommended because sensitivity to variations in circuit components or supply voltage may be experienced.

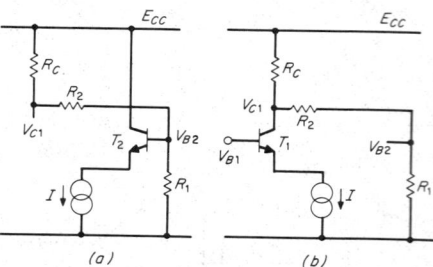

Fig. 16-75. Execution of transfer in Schmitt circuit: (a) with T_1 blocked; (b) with T_1 conducting.

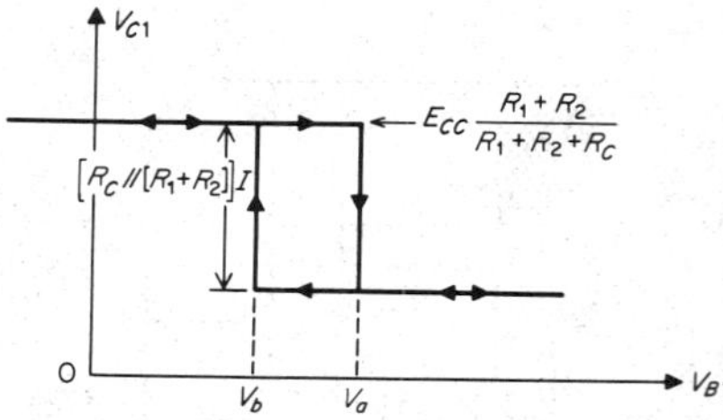

Fig. 16-76. Input-output characteristic of Schmitt circuit, showing rectangular "hysteresis."

There is a way of avoiding this problem, at the expense of increased circuit complexity. Decreasing the attenuation ratio β reduces the open-loop gain of the emitter-coupled pair unless the differential-amplifier voltage gain is increased in the same time. Combining Eqs. (16-84) and (16-85) yields the condition for instability:

$$\beta \geq \frac{(Z_G + h_{11}^T)(h_{22}^T + Y_C)}{h_{fe}} \tag{16-89}$$

In other words, more elaborate multistage amplifiers allow β to become exceedingly small without affecting the open-loop gain and without significantly lengthening the switching time. An example of such a multistage Schmitt trigger circuit is shown in Fig. 16-78. It has a β factor as small as 10^{-2}, resulting in input hysteresis of a few tens of millivolts.

Schmitt triggers often use a single resistor in place of the common-emitter current source represented in Fig. 16-73. This is a satisfactory and economical solution, but it introduces some common-mode sensitivity which changes the V_a and V_b levels slightly. The output terminals do not deliver a square wave. This effect, illustrated in Fig. 16-77 in dotted lines, can be avoided by taking the output signal at the collector of T_2 through an additional resistor, since the current flowing through T_2 is determined by V_{B2}, which is necessarily a constant when transistor T_1 is blocked. An additional advantage of the latter circuit is that the output load does not influence the feedback-loop gain.

36. **Integrated-Circuit Schmitt Triggers.** Integrated-circuit Schmitt triggers use specially designed multistage differential amplifiers called *comparators*. A typical comparator circuit is shown in Fig. 16-78. A large bandwidth, ensuring fast transient response and short recovery time, a good slewing rate and moderate gain, are among the essential differences compared to integrated-circuit operational amplifiers. Of course, the use of comparators is not restricted to integrated-circuit Schmitt triggers; they are also sensitive and rapid level detectors.

A Schmitt trigger using a comparator network is represented in Fig. 16-79. Usually, the dynamic-output range of a comparator is equal to approximately 5 V, and the input hysteresis can be approximated by

$$V_a - V_b \simeq \beta \times 5 \text{ V}$$

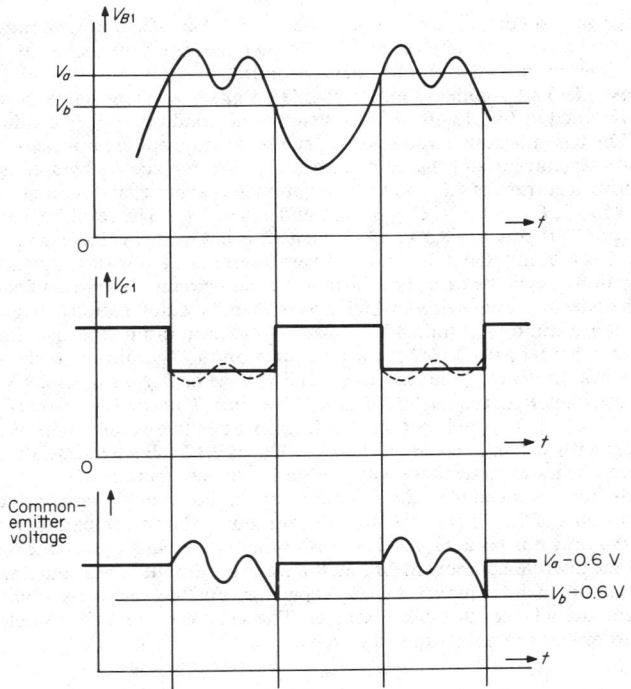

Fig. 16-77.   Trigger input and output voltage of Schmitt circuit. Solid line delivered by a current source. Broken line delivered by a resistor.

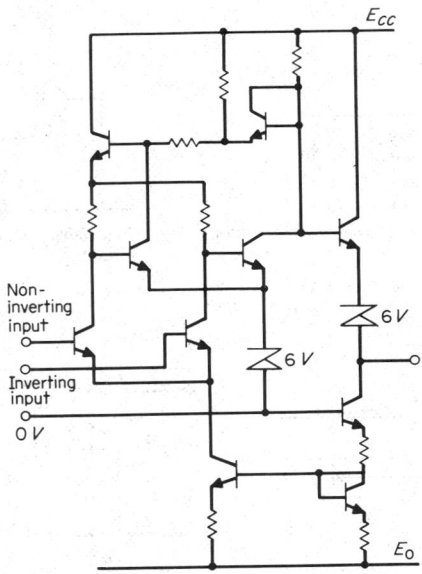

Fig. 16-78.   Comparator version of Schmitt circuit, used in integrated-circuit applications.

16-47

The voltage gain of comparators is not higher than 1,000 to 5,000. One must be careful, however, not to choose values of β below 10^{-2}, to keep the loop gain above 10. This ensures regeneration times of the order of a few nanoseconds.

37. Monostable and Astable Emitter-coupled Pairs. A monostable Schmitt trigger circuit is represented in Fig. 16-80. Under steady-state conditions, T_2 is conducting and T_1 is blocked. The base-to-ground voltage of T_2 can be found from the transistor base current or from its emitter current and the dc current gain. We replace R_E by a constant-current source delivering a current of 4.5 mA, as in the previous example, and we assume that $R_C = 1.2$ kΩ, $R_B = 150$ kΩ, $E_{CC} = 15$ V, $C = 3.3$ nF, and $h_{FE} = 100$. The resulting base-to-ground dc voltage V_{B2} of T_2 is equal to 8.3 V. As long as E is below this voltage, nothing happens. Suppose that E is 6 V and that a 4-V pulse of very short duration is suddenly added, driving T_1 into conduction. As in the case of the bistable Schmitt circuit, the current flowing through T_1 produces a collector-voltage drop, which drives T_2 in the cutoff region. The amplitude of the negative step, given by $R_c I$ (or 5.4 V), causes V_{B2} to drop instantaneously from 8.3 to 2.9 V (Fig. 16-81). E has returned to 6 V in the meantime, and so T_2 again enters the active region when V_{B2} reaches this level. The base-to-ground voltage of T_2 rises from 2.9 V toward E_{CC} with a time constant $R_B C$ (equal to 495 μs). The time T needed to make T_2 conduct is approximately 130 μs. T_1 now is cut off, resulting in a positive-voltage step of 5.4 V, which raises V_{B2} essentially instantaneously to 11.4 V. This is 3.1 V above the steady-state voltage which V_{B2} approaches exponentially with the same time constant, $R_B C$.

Integrated-circuit monostable Schmitt circuits are based upon the same principle, except that the components of the differential amplifier are replaced by a comparator similar to that described in the previous paragraph. The first of the two circuits represented at Fig. 16-82 is a replica of the discrete-component circuit, the second (with the series combination instead of parallel) has no high-frequency bypass capacitor, but the relatively high gain of the comparator counterbalances this disadvantage. The same is true for the astable integrated-circuit Schmitt circuit represented in Fig. 16-83.

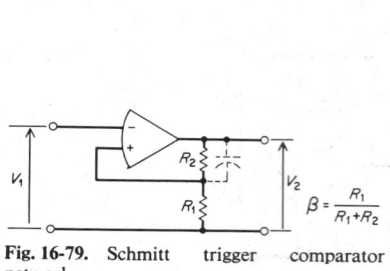

Fig. 16-79. Schmitt trigger comparator network.

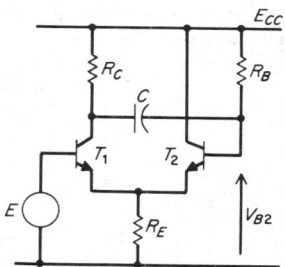

Fig. 16-80. Monostable Schmitt circuit.

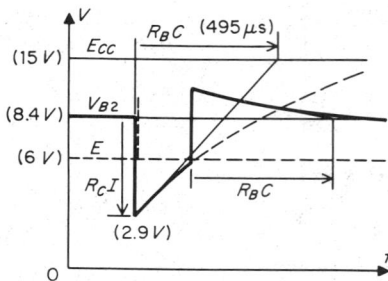

Fig. 16-81. Output waveform of circuit in Fig. 16-80.

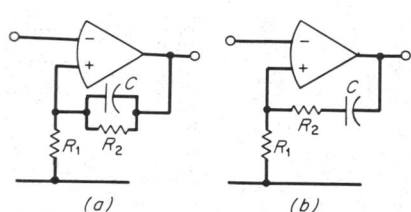

Fig. 16-82. Integrated-circuit version of monostable Schmitt circuit: (a) with shunt bypass capacitance; (b) without bypass capacitance.

Integrated-circuit monostable and astable circuits are available based upon the use of logic gates. An example of a monostable circuit using two NAND gates is shown in Fig. 16-84 with the corresponding waveforms.

An interesting discrete-component astable circuit using emitter coupling is shown in Fig. 16-85. The capacitor C provides a short circuit between the emitters of the two transistors, closing the positive-feedback loop during the regeneration time. As long as one or the other of the two transistors is cut off, C offers a current sink to the current source connected to the emitter of the blocked transistor. The capacitor thus is periodically charged and discharged by the two current sources, and the voltage across its terminal exhibits a triangular waveform.

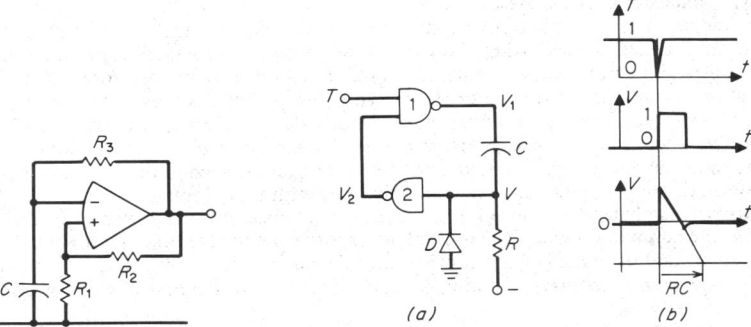

Fig. 16-83. Astable version of Schmitt circuit.

Fig. 16-84. Monostable circuit using two NAND gates: (*a*) circuit; (*b*) waveforms.

The collector current of T_1 is either zero or $I_1 + I_2$, so that the resulting voltage step across R_C is $(R_B \parallel R_C)(I_1 + I_2)$. Since the base of T_2 is directly connected to the collector of T_1, the same voltage step controls T_2 and determines the width of the input hysteresis, i.e., the maximum amplitude of the voltage sweep across C. The period of oscillation is computed from

$$T = C\left(\frac{1}{I_1} + \frac{1}{I_2}\right)(R_B \parallel R_C)(I_1 + I_2) \qquad (16\text{-}90)$$

when, as is usual, both current sources deliver equal currents, the expression for T reduces to

$$T = 4(R_B \parallel R_C)C \qquad (16\text{-}91)$$

T does not depend, in this case, on the amplitude of the current because changes in current in fact modify the amplitude and the slope of the voltage sweep across C in the same manner. A review of the waveforms obtained at various points of the circuit is given in Fig. 16-86.

38. Memory Circuits. Two-state devices and bistable circuits are ideally suited for memory (binary-storage) purposes. In the past, magnetic cores have offered the only economic solution for large-size memories, but use of solid-state memories (which rely on bistable circuits derived from the basic Eccles-Jordan configuration) has been increasing. Magnetic cores are cheap and reliable, and they offer the advantage of memory retention in the event of power-supply failure.

Magnetic cores are made either of high-permeability alloys, such as Permalloy, or of ferrites (see Sec. **6**, Pars. **6-136** to **6-145**). Ferromagnetic materials having a high permeability generally exhibit low coercive force. Ferrites have smaller permeability, but they offer higher electric resistance and thus smaller losses. The mechanism of magnetization can be understood on the basis of microscopic theory (see Sec. **23**, Pars. **23-55** to **23-63**). Within the material small domains exist where the molecular magnetic moments have the same direction. The domains are separated by walls, and their magnetic moments are distributed randomly on a macroscopic scale. An external field aligns the magnetic moments of those domains already close to the imposed direction. The domain walls then become unstable, and some of the domains grow rapidly until they encompass almost the entire material, causing a large flux-density variation followed by saturation. Further increasing the magnetization force causes only a few remaining domains, e.g., those which were oriented in the reverse direction, to rotate. After the external field is removed, the magnetic domains relax somewhat and adopt "easy" magnetization directions, which depart little from the imposed one. A reverse magnetization is required to return to the initial random situation.

When a step-function magnetizing force is applied to a magnetic core, the resulting emf developed across a test winding exhibits the shape indicated in Fig. 16-87. The initial spike is believed to be the result of rapid rotation of some domains that are already oriented close to the imposed direction. The slower and more important flux variation that follows is due to propagation of changes in the domain walls. The last negative spike corresponds to a small flux-density variation after the excitation is removed, and the magnetic moments relax toward their nearest axis of easy magnetization.

The longest time constant involved, which is associated with the wall-propagation mechanism, is usually of the order of 100 ns. When a core has been magnetized previously in some direction and a new magnetizing pulse with the same direction is imposed, propagation of walls is almost nonexistent. This means that the long pulse in Fig. 16-87 vanishes but rotation is still experienced, so that the two short spikes still remain. In applications where cores are used as memory media, some care must be taken not to interpret rotation spikes as useful bits. The use of the strobing-pulse technique or slower rise and decay times of the magnetizing force may help cope with this problem.

Spurious signals become more important as the number of interrogated magnetic cores increases. In computer memories it is usual practice to wind the reading wire of a core matrix in such a way that unwanted effects cancel out to the maximum possible extent. A practical solution used in an elementary two-dimensional matrix is shown in Fig. 16-88. Selecting the

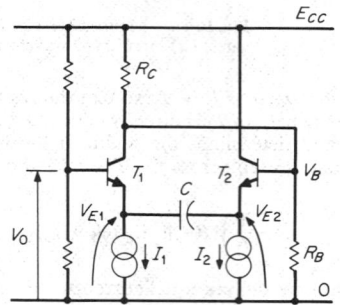

Fig. 16-85. Discrete-component emitter-coupled astable circuit.

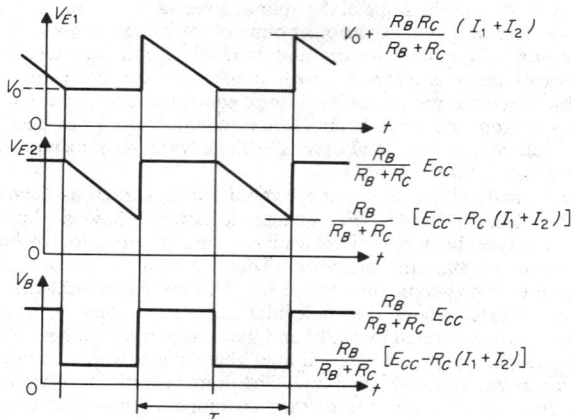

Fig. 16-86. Waveforms of circuit in Fig. 16-85.

The emitter junction of the blocking oscillator is assumed to be blocked at the time a negative pulse triggers the circuit. Under steady-state conditions C_0 thus is charged to the voltage E. It will be assumed that C_0 behaves like a floating dc source. Moreover, R and the trigger generator need not be considered further, since the common-base input impedance of the active transistor is so small.

With these simplifications, the blocking oscillator can be considered to be a common-base unity-gain current amplifier followed by a current transformer that injects in the emitter a current n times larger than the collector current. The resulting regeneration time is very short, a few nanoseconds, depending primarily on the transistor switching characteristics and the series leakage current.

The equivalent network representing the blocking oscillator during the regeneration time is depicted in Fig. 16-93. The magnetization inductance is neglected because switching occurs so fast that the magnetization current remains negligibly small. The stray capacitance of the transformer need not be considered since it is essentially short-circuited by the small input impedance of the transformer. The collector-to-ground voltage drops almost instantaneously to zero, causing saturation of the transistor before the collector current becomes appreciable. The collector junction becomes essentially a short circuit. Hence the collector loop may be represented as in Fig. 16-94, with the emitter load seen through the ideal transformer.

The current in the loop increases very rapidly until a maximum is reached:

$$\frac{1}{n} I_{E,\text{max}} = \frac{E_{CC} - nE}{n^2 h_{ib}} \tag{16-92}$$

The rise time

$$t_r = 2.2 \frac{L_l}{n^2 h_{ib}}$$

is of the order of 1 μs or less. During this period, emitter and collector current vary in the same manner, but the emitter current is n times larger than the collector current, and the transistor is thus heavily saturated.

This situation could persist indefinitely if the magnetization inductance were infinite. This not being the case, additional collector current starts flowing in the magnetization

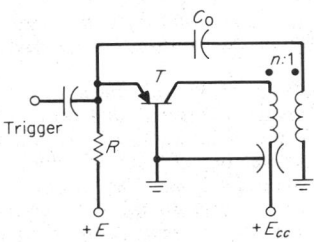

Fig. 16-91. Blocking oscillator using common-base configuration.

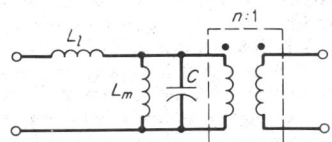

Fig. 16-92. Equivalent network of the pulse transformer used in Fig. 16-91.

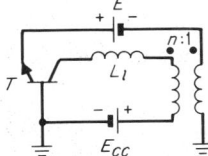

Fig. 16-93. Equivalent network of blocking oscillator during regeneration time.

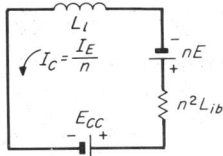

Fig. 16-94. Collector loop of circuit in Fig. 16-93.

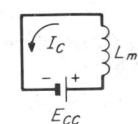

Fig. 16-95. Path of collector current through magnetization inductance.

inductance according to Fig. 16-95. The total collector current is given by the sum

$$I_c = \frac{1}{n}\left(\frac{E_{cc}}{n} - E\right)\frac{1}{h_{ib}} + \frac{E_{cc}}{L_m}t \tag{16-93}$$

This expression remains valid as long as no saturation is experienced; otherwise I_c departs suddenly from its linear increase, shown in Fig. 16-96, and increases much faster. Whether or not this happens, I_c at some point becomes equal to I_E and the transistor enters the active region. A new regeneration phase then takes place, tending to cut the transistor off as fast as occurred during the first regeneration period. The collector current vanishes almost instantaneously, and the energy accumulated in the magnetization inductance causes a rapid increase of the collector voltage, substantially above E_{CC}. The combination of L_m and C produces ringing, unless an external damping circuit is used. Avalanche breakdown is likely to occur unless steps are taken to prevent it.

NEGATIVE-RESISTANCE PULSED CIRCUITS

40. Negative Resistance. The term negative resistance applies to any two-terminal circuit element whose impedance exhibits a negative real part. Such an element delivers energy and can be used to perform amplification, oscillation, or switching functions. As in other active devices, the energy delivered cannot exceed limits imposed by the physical constraints inherent in the circuit element. This implies that the static current-vs.-voltage characteristic curve of a negative-resistance device exhibits a negative slope only locally, in one or several zones interposed between regions of positive resistance.

Most negative-resistance devices have only one active region, outside of which they behave like passive elements. In pulse circuits, it is very important to consider negative-resistance elements with respect to their positive resistance as well as their negative resistance. They are necessarily nonlinear elements and are used as such.

Negative-resistance elements belong to the two categories shown in Fig. 16-97, usually called N- or S-type devices, according to the shape of their *I-V* characteristics. Among the N-type devices are tunnel diodes, while typical S-type devices are *pnpn* diodes and avalanche transistors. The N-type characteristic can be measured only using a voltage source, while a current source is well suited for the S type. Care must be taken in either case to eliminate spurious reactive elements.

It is, in fact, meaningless to investigate the behavior of negative-resistance devices on the basis of static characteristics only. To illustrate this fact, consider the series combination of an N-type device, a resistor R, and a voltage source E, as shown in Fig. 16-98. These elements are represented by a load line in the *I-V* curve of the negative-resistance device, which intersects the latter at three points.

Two of these points, A and B, correspond to stable operation, while the point in the middle leads to instability. Unless additional reactive elements, lumped as well as distributed, are considered, there is no way of predicting the degree of instability or of computing the trajectory of the operating point between the two stable points. This difficulty is alleviated if we introduce a parallel capacitor C and a series inductor L, as shown in Fig. 16-99. The capacitor C may be as small as the unavoidable parasitic capacitance of the junction of the tunnel diode, and L may be the stray inductance of the loop. Let us define $Z_s(p)$ and $Z_p(p)$, respectively

$$Z_s(p) = R + pL \equiv \frac{N_s(p)}{D_s(p)}$$

$$Z_p(p) = \frac{R_N}{1 + pR_N C} \equiv \frac{N_p(p)}{D_p(p)} \tag{16-94}$$

where R_N represents the small-signal resistance of the tunnel diode. The total impedance is given by

$$Z(p) = \frac{V(p)}{I(p)} = \frac{N_s D_p + N_p D_s}{D_s D_p} \tag{16-95}$$

Rewriting Eq. (16-95) as

$$(N_s D_p + N_p D_s)I(p) = D_s D_p V(p)$$

and introducing the fact that the series combination of Z_S and Z_p terminates in a short circuit yields the natural frequencies of the circuit by equating $N_s D_p + N_p D_s$ to zero, as follows:

$$p^2 + p\left(\frac{R}{L} + \frac{1}{R_N C}\right) + \frac{1}{LC}\left(1 + \frac{R}{R_N}\right) = 0 \tag{16-96}$$

Stability requires that all the coefficients of this expression be positive, i.e., that the inequalities

$$|R_N| > R > \frac{L}{|R_N|C} \tag{16-97}$$

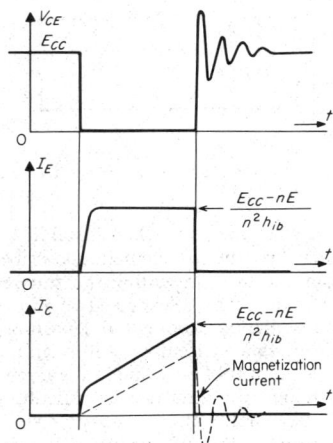

Fig. 16-96. Waveforms of blocking oscillator.

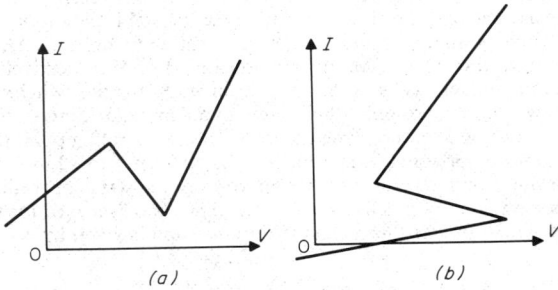

Fig. 16-97. Types of negative-resistance characteristics: (*a*) N type; (*b*) S type.

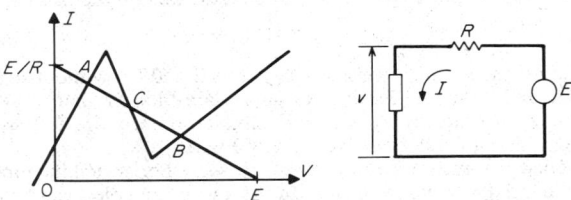

Fig. 16-98. Load-line intersection with N-type negative-resistance characteristic and resulting circuit.

16-55

be satisfied. The case of a typical tunnel diode with a R_N of -50 Ω, a junction capacitance of 5 pF, and stray inductance of 1 nH yields

$$50\ \Omega > R > 4\ \Omega \qquad (16\text{-}98)$$

It is worth noticing that if the inductance is taken slightly larger, the range of possible values of R decreases drastically. An inductance of 12.5 nH, for example, makes the circuit unstable at all values of R. A similar result occurs if the junction capacitance is smaller. Measuring the static characteristic of a tunnel diode therefore is a difficult task which must be undertaken with care regarding stray inductance.

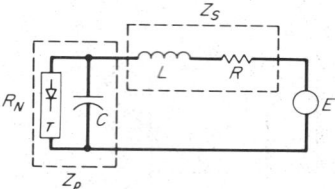

Fig. 16-99.   Use of a parallel capacitor C and series inductor L to control switching times.

This example casts light on what happens when instability occurs. From Eq. (16-96) (which is valid only for small-signal conditions and hence describes the initiation of instability only), it is obvious from the last term representing the square of the resonating angular frequency ω_0 that a sinusoidal response may be expected when R is smaller than R_N, while the opposite condition yields a nonoscillating exponential departure from the initial conditions.

41. Synthesis of Negative-Resistance Circuits.   Although the preceding example shows how an unstable situation builds up, it is of no value in the large-signal analysis of negative-resistance pulse circuits. Although this analysis is difficult, since we are dealing with nonlinear second-order networks, simplification is possible since the behavior of negative-resistance pulse circuits, in contrast to that of oscillators, is almost always dominated by a single pole. In the case of a tunnel-diode pulse circuit, for example, one can increase the series inductance substantially without changing the parasitic capacitance. This situation is illustrated in Fig. 16-100. The straight line in the I-V plane represents the load line formed by E and R. We assume that the circuit initially is in the stable conditions, e.g., at working points I or IV. Then we suddenly add or subtract a pulse of amplitude ΔE_1 or ΔE_2 to drive the operating point to the edge of the unstable region, II or V respectively. This initiates switching. Since the inductance is assumed to be large, during the switching period (which takes place in a few nanoseconds only) the current in the inductor cannot change, leaving R_N disconnected from load and source. The trajectory of the operating point thus shifts almost instantaneously along a horizontal axis from II to III or from V to VI in Fig. 16-100, ending in a non-steady-state situation, from which it approaches the stable operating point, IV or I, respectively, along a region of positive resistance. The time T needed for this last phase is appreciably longer than that of the switching duration and is given by

$$T = 2.2\frac{L}{R + R_N} \qquad (16\text{-}99)$$

where R_N represents the positive resistance between III and V or between VI and I. The actual duration of the regenerative period can be estimated on the basis of

$$t = 2.2|R_N|C \qquad (16\text{-}100)$$

where R_N represents the negative resistance between II and IV, without sign. Clearly, as C becomes smaller and L larger, T becomes appreciably longer than t, so that the curve representing the voltage across R_N vs. time exhibits a steep rise followed by a smooth exponential decay, as shown in the lower part of Fig. 16-100.

In fact, most negative-resistance circuits exhibit such short switching times that elaborate instruments are needed to measure them. In the above example t can be as small as 1 ns, while T can be made as long as several milliseconds, depending on the choice of L.

In S-type negative-resistance devices, similar reasoning leads to similar conclusions, as represented in Fig. 16-101. Here we assume that the dominating reactive element is a parallel capacitor, while the loop inductance has been minimized. Shifting the operating point from I to II, the edge of instability, causes switching of the element along a constant-voltage trajectory from II to III, because of the parallel capacitor. This is immediately followed by a smooth exponential decay toward the new stable point IV. Similarly, another trigger signal, bringing the operating point from IV to V, closes the sequence and returns the operating point to I.

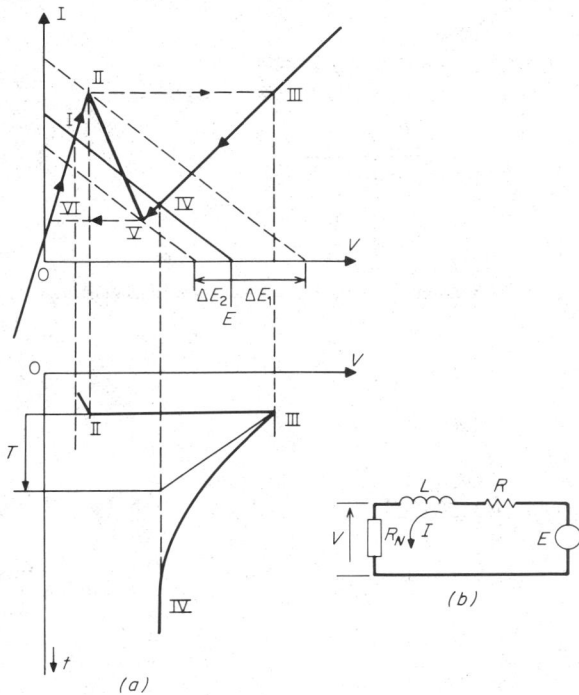

Fig. 16-100.   Sequence of operation phases in N-type negative-resistance circuit: (*a*) *IV* and *Vt* curves; (*b*) equivalent circuit.

42. Monostable and Astable Negative-Resistance Circuits.   Very little is required to change the N- and S-type circuits from bistable to monostable or astable circuits. Monostable circuits differ from bistable circuits only in the choice of the steady-state operating point, as is illustrated in Fig. 16-102. Here we note that the load line intersects the static characteristic at one point only. Provided that a sufficiently large positive-trigger signal is available to drive the operating point from I to II, fast switching occurs from II to III, followed by a slow shift from III to IV, where the operating point again reaches the edge of the unstable region, causing fast switching from IV and V. Finally the circuit returns to the stable point I at a rate which depends on L, R, and the positive resistance of R_N between V and I.

Astable operation is illustrated at Fig. 16-103 for a typical tunnel-diode circuit. It is obtained by a proper choice of the load resistor ($R < |R_N|$) and the emf E of the power supply. It is assumed that the series inductor overrules the effect of the parallel capacitor. If this condition is not met, the circuit may oscillate but the output wave will exhibit more frequent transitions, since the switching and resolution times come closer together. At the limit a quasi-sinusoidal waveform may be experienced.

16-57

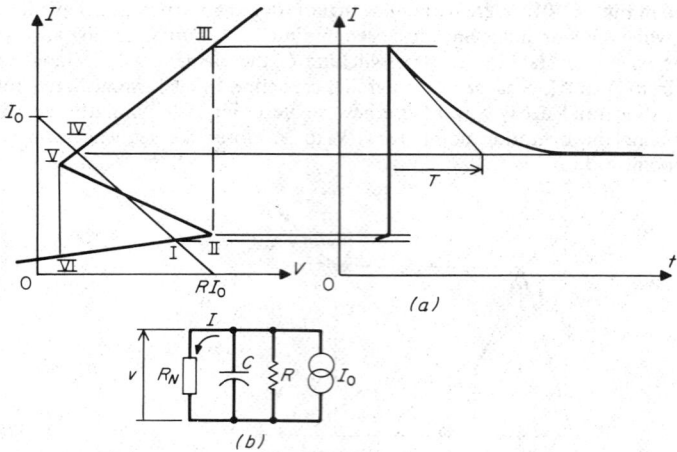

Fig. 16-101. Sequence of operating phases in S-type negative-resistance circuit: (*a*) *IV* and *Vt* curves; (*b*) equivalent circuit.

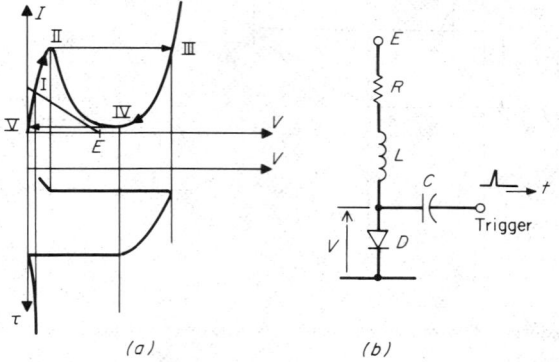

Fig. 16-102. Monostable-circuit operation with negative resistance: (*a*) *IV* and *Vt* characteristics; (*b*) equivalent circuit.

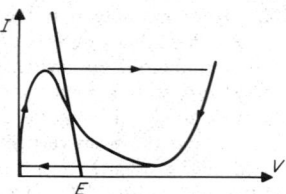

Fig. 16-103. *IV* characteristic of astable operation with negative resistance.

43. Negative-Resistance Devices. One of the most important features of negative-resistance devices is that their parasitic reactances are small and their switching times are therefore exceedingly short, typically of the order of 1 ns or less. Tunnel-diode circuits, however, offer poor dynamic range since their useful zone of operation is restricted to approximately 1/2 V. This is not the case with avalanche transistors because the zone of negative resistance occurs between two collector breakdown voltages; one with open emitter, which is usually large, typically 60 to 80 V; the other corresponding to the open-base configuration, which is smaller, typically 30 V or less. Avalanching, moreover, is a phenomenon that occurs in a fraction of a nanosecond, making this solution very attractive in many applications. A typical example is given in Fig. 16-104, showing a free-running

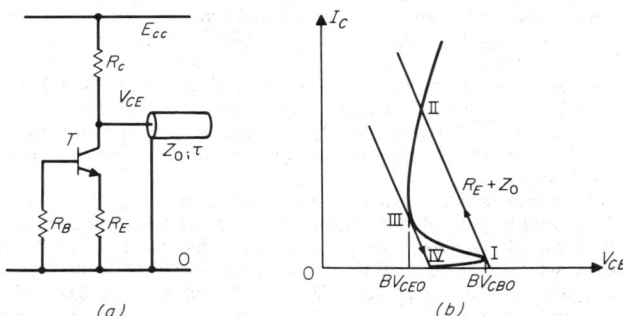

Fig. 16-104. Free-running square-wave generator using avalanche transistor action: (*a*) circuit; (*b*) *IV* characteristic.

square-wave generator based on the principles outlined in Par. **16-37.** Assuming that the transistor is initially nonconducting, the electric line charges through R_C until V_{CE} crosses the edge of the instability zone. The transistor then suddenly becomes almost a short circuit, closing the loop formed by the line in series with R_E, and causing the operating point to jump from I to II. At the same time, a steep voltage step starts traveling along the line, is reflected at the far end, and abruptly lowers the collector voltage, turning the transistor off. The line may then charge up again till another breakdown occurs. The output square wave can be obtained across R_E. The role of R_B, which at first glance does not contribute to the functioning of the circuit, is in fact critical because it creates the zone of negative resistance.

When the transistor is off, R_B clamps the base to the ground potential so that the breakdown voltage BV_{CBO} is high and determined by the properties of the collector junction only. When avalanching of the collector junction begins to occur, the current derived through R_B produces a voltage drop which tends to forward-bias the emitter junction, so that this junction also receives a portion of the injected collector current. Due to the high emitter efficiency, minority carriers are injected from the emitter into the base, causing an increased collector current to flow, which in turn contributes to more avalanche current, and so on.

From then on, the emitter junction is normally forward-biased, and full transistor action may be expected. The collector breakdown voltage drops to the usual value BV_{CEO} for a common-emitter, open-base configuration. Assuming a collector-current multiplication factor M given by

$$M = \frac{1}{1 - (V/BV_{CBO})^n} \tag{16-101}$$

it can be shown that the common-emitter breakdown voltage is related to the common-base breakdown voltage by

$$BV_{CEO} = BV_{CBO}(1 - \alpha_N)^{1/n}$$

where α_N is the common-base current-gain factor.

This explanation illustrates how a positive-feedback loop determines a zone of negative resistance. Similar situations occur in four-layer devices and unijunction transistors. The first case is shown in Fig. 16-105, a silicon four-layer device with 0 V applied to the external terminals. As this voltage is increased so that the central junction is reverse-biased while the two others are forward-biased, the entire voltage drop is concentrated at the middle junction, creating locally a high electric field which will start avalanching when the voltage becomes large enough. Current then starts flowing through the two forward-biased junctions, causing transistor effect to take place at both ends of the four-layer structure.

The emitter efficiency is largely affected by the recombination rate in the emitter junction, especially at very low currents. Hence, the common-base current gains α_{npn} and α_{pnp} of both structures (the above *npn* and the lower *pnp*) gradually approach unity, provided that a minimum avalanche current flows through the structure. Once the condition

$$\alpha_{npn} + \alpha_{pnp} = 1 \qquad (16\text{-}102)$$

is met, the overall gain of the collector-coupled transistor pair reaches a critical level and switching occurs, driving both transistors into saturation. The *pnp* device is a typical S-type negative-resistance device.

Because switching relies on Eq. (16-102) rather than on avalanching of the central junction, a more convenient way to trigger a four-layer device consists in injecting current through either one of the two forward-biased junctions, through an additional terminal. This is illustrated in Fig. 16-106, which represents the basic structure of a silicon controlled rectifier switch (SCR).

The negative-resistance concept has a much broader meaning than may appear at first glance. If we examine any of the regenerative networks considered in the previous paragraphs, it is always possible to find a connection such that when it is opened a new element is created which exhibits a negative resistance. A typical example of this is given by the Schmitt trigger circuit, represented in Fig. 16-107, which provides an S-type negative resistance between collector and base when the connection between them is opened. Consequently, the best way to utilize this element to synthesize a negative-resistance pulse circuit consists in adding a shunt capacitor across its terminals.

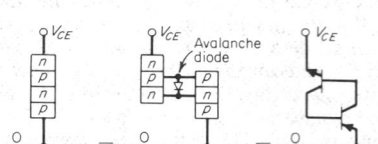

Fig. 16-105. Use of silicon four-layer device to obtain negative resistance.

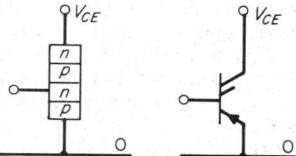

Fig. 16-106. Basic structure of silicon controlled rectifier (SCR).

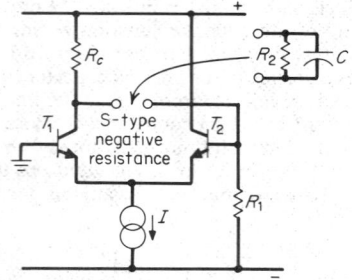

Fig. 16-107. S-type negative resistance provided by Schmitt trigger circuit.

44. Bibliography and References

Books

1. STRAUSS, L. L. "Wave Generation and Shaping," McGraw-Hill, New York, 1970.
2. MILLMAN, J., and H. TAUB "Pulse, Digital, and Switching Waveforms," McGraw-Hill, New York, 1965.
3. HARRIS, J. N., P. E. GRAY, and C. L. SEARLE "Digital Transistor Circuits," Wiley, New York, 1966.

4. PENNEY, W. M. (ED.) "MOS Integrated Circuits," Van Nostrand Reinhold, New York, 1972.

5. EIMBINDER, J. (ED.) "Semiconductor Memories," Wiley-Interscience, New York, 1971.

6. WICKES, W. E. "Logic Design with Integrated Circuits," Wiley, New York, 1968.

7. HOESCHELE, D. F., JR. "Analog-to-Digital, Digital-to-Analog Conversion Techniques," Wiley, New York, 1968.

8. MEINDL, J. D. "Micropower Circuits," Wiley, New York, 1968.

9. THORNTON, R. D., J. G. LINVILL, et al. "Handbook of Basic Transistor Circuits and Measurements," Wiley, New York, 1966.

10. KHAMBATA, A. J. "Introduction to Large-Scale Integration," Wiley-Interscience, New York, 1969.

Papers

11. ECCLES, W. H., and F. W. JORDAN A Trigger Relay Utilizing Three Electrode Thermionic Vacuum Tubes, *Radio Rev.,* 1919, Vol. 1, No. 3, pp. 143-146.

12. LINVILL, J. G. Nonsaturating Pulse Circuits Using Two Junction Transistors, *Proc. IRE,* 1955, Vol. 43, pp. 826-839.

13. SCHMITT, O. H. A Thermionic Trigger, *J. Sci. Instrum.,* 1938, Vol. 15, p. 24.

14. BEALE, I. E. A., et al. A Study of High Speed Avalanche Transistors, *Proc. IEE (Lond.),* Pt. B, July 1957, Vol. 104, pp. 394-402.

15. LINVILL, J. G., and R. H. MATTSON Junction Transistor Blocking Oscillator, *Proc. IRE,* December 1955, Vol. 43, No. 11, pp. 1632-1639.

16. CHEN, T. C., and A. PAPOULIS Domain Theory in Core Switching, *Proc. IRE,* 1958, Vol. 46, No. 5, pp. 839-849.

17. EBERS, J. J., and J. L. MOLL Large-Signal Behavior of Junction Transistors, *Proc. IRE,* December, 1954, Vol. 42, pp. 1761-1772.

18. MOLL, J. L. Large-Signal Transient Response of Junction Transistors, *Proc. IRE,* December 1954, Vol. 42, pp. 1773-1784.

19. BURNS, J. R. Switching Response of Complementary Symmetry MOS Transistor Circuits, *RCA Rev.,* December 1964, Vol. 25, No. 4, pp. 627-661.

20. MOLL, J. L., M. TANNENBAUM, J. M. GOLDEY, and N. HOLONYAK PNPN Switches, *Proc. IRE,* 1956, Vol. 44, pp. 1174-1182.

21. ESAKI, L. New Phenomenon in Narrow Ge P-N Junctions, *Phys. Rev.,* 1958, Vol. 109, p. 603.

SECTION 17

MEASUREMENT AND CONTROL CIRCUITS

BY

FRANCIS T. THOMPSON Director, Electronics and Electromagnetics Research, Research Laboratories, Westinghouse Electric Corporation; Fellow, Institute of Electrical and Electronics Engineers; Member, Instrument Society of America

CONTENTS

Numbers refer to paragraphs

SECTION 17

MEASUREMENT AND CONTROL CIRCUITS*

PRINCIPLES OF MEASUREMENT CIRCUITS

DEFINITIONS AND PRINCIPLES OF MEASUREMENT

1. Precision is a measure of the spread of repeated determinations of a particular quantity. Precision depends on the resolution of the measurement means and variations in the measured value caused by instabilities in the measurement system. A measurement system may provide precise readings, all of which are inaccurate because of an error in calibration or a defect in the system.

2. Accuracy is a statement of the limits which bound the departure of a measured value from the true value. Accuracy includes the imprecision of the measurement along with all the accumulated errors in the measurement chain extending from the basic reference standards to the measurement in question.

3. Errors may be classified into two categories, systematic and random. *Systematic errors* are those which consistently recur when a number of measurements are taken. Systematic errors may be caused by deterioration of the measurement system (weakened magnetic field, change in a reference resistance value), alteration of the measured value by the addition or extraction of energy from the element being measured, response-time effects, and attenuation or distortion of the measurement signal. *Random errors* are accidental, tend to follow the laws of chance, and do not exhibit a consistent magnitude or sign. Noise and environmental factors normally produce random errors but may also contribute to systematic errors.

The arithmetic average of a number of observations should be used to minimize the effect of random errors. The arithmetic average or mean X of a set of n readings $X_1, X_2, \ldots, X_n$ is

$$X = \frac{\Sigma X_i}{n}$$

The dispersion of these readings about the mean is generally described in terms of the standard deviation σ, which can be estimated for n observations by

$$s = \sqrt{\frac{\Sigma(X_i - X)^2}{n - 1}}$$

where s approaches σ as n becomes large.

A *confidence interval* can be determined within which a specified fraction of all observed values may be expected to lie. The *confidence level* is the probability of a randomly selected reading falling within this interval. Confidence intervals are given in Table 17-1 as a function of the number of observations and the required confidence level. Detailed information on measurement errors is given in Ref. 1, Par. **17-149.**

4. Standardization and calibration involve the comparison of a physical measurement with a reference standard. Calibration normally refers to the determination of the accuracy and linearity of a measuring system at a number of points, while standardization involves the adjustment of a parameter of the measurement system so that the reading at one specific value is in correspondence with a reference standard. The numerical value of any reference standard should be capable of being traced through a chain of measurements to a National Reference Standard maintained by the National Bureau of Standards.

* The author is indebted to I. A. Whyte, L. C. Vercellotti, T. H. Putman, T. M. Heinrich, T. I. Pattantyus, and R. A. Mathias for their suggestions and constructive criticisms. The author wishes to thank Miss Sandra Cooke and Mrs. Leslie Arthrell for their fine work in typing the manuscript.

Table 17-1.   Factors for Establishing Confidence Interval*

Number of observations	Confidence level			
	0.50	0.90	0.95	0.99
	Confidence interval			
2	$X \pm 1.00s$	$X \pm 6.31s$	$X \pm 12.71s$	$X \pm 63.66s$
3	$X \pm 0.82s$	$X \pm 2.92s$	$X \pm 4.30s$	$X \pm 9.92s$
4	$X \pm 0.77s$	$X \pm 2.35s$	$X \pm 3.18s$	$X \pm 5.84s$
5	$X \pm 0.74s$	$X \pm 2.13s$	$X \pm 2.78s$	$X \pm 4.60s$
6	$X \pm 0.73s$	$X \pm 2.02s$	$X \pm 2.57s$	$X \pm 4.03s$
7	$X \pm 0.72s$	$X \pm 1.94s$	$X \pm 2.45s$	$X \pm 3.71s$
8	$X \pm 0.71s$	$X \pm 1.90s$	$X \pm 2.37s$	$X \pm 3.50s$
9	$X \pm 0.71s$	$X \pm 1.86s$	$X \pm 2.31s$	$X \pm 3.36s$
10	$X \pm 0.70s$	$X \pm 1.83s$	$X \pm 2.26s$	$X \pm 3.25s$
11	$X \pm 0.70s$	$X \pm 1.81s$	$X \pm 2.23s$	$X \pm 3.17s$
16	$X \pm 0.69s$	$X \pm 1.75s$	$X \pm 2.13s$	$X \pm 2.95s$
∞	$X \pm 0.67s$	$X \pm 1.64s$	$X \pm 1.96s$	$X \pm 2.58s$

* Modified and abridged from Table IV of R. A. Fisher and F. Yates, "Statistical Tables for Biological, Agricultural and Medical Research," Oliver & Boyd, Edinburgh, 1963. By permission of the authors and publishers.

5. The range of a measurement system refers to the values of the input variable over which the system is designed to provide satisfactory measurements. The range of an instrument used for a measurement should be chosen so that the reading is large enough to provide the desired precision. An instrument having a linear scale which can be read within 1% at full scale can be read only within 2% at half scale.

6. The resolution of a measuring system is defined as the smallest increment of the measured quantity which can be distinguished. The resolution of an indicating instrument depends on the deflection per unit input. Instruments having a square-law scale provide twice the resolution at full scale as linear-scale instruments. Amplification and zero suppression can be used to expand the deflection in the region of interest and thereby increase the resolution. The resolution is ultimately limited by the magnitude of the signal that can be discriminated from the noise background.

7. Noise may be defined as any signal which does not convey useful information. Noise is introduced in measurement systems by mechanical coupling, electrostatic fields, and magnetic fields. The coupling of external noise can be reduced by vibration isolation, electrostatic shielding, and electromagnetic shielding. Electrical noise is often present at the power-line frequency and its harmonics, as well as at radio frequencies.

In systems containing amplification, the noise introduced in low-level stages is most detrimental because the noise components within the amplifier passband will be amplified along with the signal. The noise in the output determines the lower limit of the signal that can be observed.

Even if external noise is minimized by shielding, filtering, and isolation, noise will be introduced by random disturbances within the system caused by such mechanisms as the Brownian motion in mechanical systems, Johnson noise in electrical resistance, and the Barkhausen effect in magnetic elements. Johnson noise is generated by electron thermal agitation in the resistance of a circuit. The equivalent rms noise voltage developed across a resistor R at an absolute temperature T is equal to $\sqrt{4kTR\Delta f}$, where k is Boltzmann's constant (1.38×10^{-23} J/K) and Δf is the bandwidth in hertz over which the noise is observed.

8. The bandwidth Δf of a system is the difference between the upper and lower frequencies passed by the system (see Par. 17-41). The bandwidth determines the ability of the system to follow variations in the quantity being measured. The lower frequency is zero for dc systems, and their response time is approximately equal to $1/(3\Delta f)$. Although a wider bandwidth improves the response time, it makes the system more susceptible to interference from noise.

9. Environmental factors which influence the accuracy of a measurement system include temperature, humidity, magnetic and electrostatic influences, mechanical stability, shock, vibration, and position. Temperature changes can alter the value of resistance and capacitance, produce thermally generated emfs, cause variations in the dimensions of mechanical members, and alter the properties of matter. Humidity affects resistance values and the dimensions of some organic materials. Dc magnetic and electrostatic fields can produce an offset in instruments which are sensitive to these fields, while ac fields can introduce noise. The lack of mechanical stability can alter instrument reference values and produce spurious responses. Mechanical energy imparted to the system in the form of shock or vibration can cause measurement errors and, if severe enough, can result in permanent damage. The position of an instrument can affect the measurements because of the influence of magnetic, electrostatic, or gravitational fields.

TRANSDUCERS, INSTRUMENTS, AND INDICATORS

10. Transducers are used to respond to the state of a quantity to be measured and to convert this state to a convenient electrical or mechanical quantity. Transducers may be classified according to the variable to be measured. Variable classifications include mechanical, thermal, physical, chemical, nuclear-radiation, electromagnetic-radiation, electrical, and magnetic, as detailed in Sec. 10.

11. Instruments may be classified according to whether their output means is analog or digital. Analog instruments include the D'Arsonval (moving-coil) galvanometer, dynamometer instrument, moving-iron instrument, electrostatic voltmeter, galvanometer oscillograph, cathode-ray oscilloscope, and potentiometric recorders. Digital-indicator instruments provide a numerical readout of the quantity being measured and have the advantage of allowing unskilled people to make rapid and accurate readings.

12. Indicators are used to communicate output information from the measurement system to the observer.

MEASUREMENT CIRCUITS

13. Substitution circuits are used in the comparison of the value of an unknown electrical quantity with a reference voltage, current, resistance, inductance, or capacitance. Various potentiometer circuits are used for voltage substitution, and divider circuits are used for voltage, current, and impedance comparison. A number of these circuits and the reference components used in them are described in Pars. 17-17 to 17-24.

14. Analog circuits are used to embody mathematical relationships which permit the value of an unknown electrical quantity to be determined by measuring related electrical quantities. Analog-measurement techniques are discussed in Par. 17-36, and a number of special-purpose measurement circuits are described in Par. 17-87.

15. Bridge circuits provide a convenient and accurate method of determining the value of an unknown impedance in terms of other impedances of known value. The circuits of a number of impedance bridges and the amplifiers and detectors used for bridge measurements are described in Pars. 17-54 to 17-86.

16. Transducer amplifying and stabilizing circuits are used in conjunction with measurement transducers to provide an electric signal of adequate amplitude which is suitable for use in measurement and control systems. These circuits, which often have severe linearity, drift, and gain-stability requirements, are described in Pars. 17-44 to 17-53.

SUBSTITUTION AND ANALOG MEASUREMENTS

VOLTAGE SUBSTITUTION

17. The constant-current potentiometer, which is used for the precise measurement of unknown voltages below 1.5 V, is shown schematically in Fig. 17-1. For a constant current, the output voltage V_o is proportional to the resistance included between the sliding contacts. In this circuit all the current-carrying connections can be soldered, thereby minimizing contact-resistance errors. When the sliding contacts are adjusted to produce a null, V_o is equal to the unknown emf and no current flows in the sliding contacts. At null, no current is drawn from the unknown emf, and therefore the measured voltage is independent of the internal resistance of the source.

The circuit of a multirange commercial potentiometer is shown in Fig. 17-2. The instrument is standardized with the range switch in the highest range position as shown and switch *S* connected to the standard cell. The calibrated standard-cell dial is adjusted to correspond to the known voltage of the standard cell, and the standardizing resistance is adjusted to obtain a null on the galvanometer. This procedure establishes a constant current of 20 mA through the potentiometer. The unknown emf is connected to the emf terminals, and switch *S* is thrown to the emf position. The unknown emf can be read to at least five significant figures by adjusting the tap slider and the 11-turn 5.5-Ω potentiometer for a null

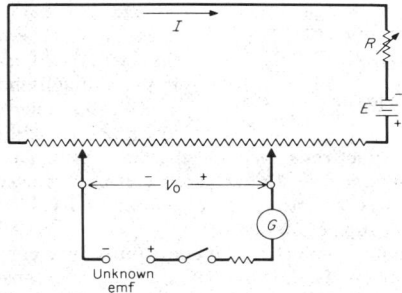

Fig. 17-1. Constant-current potentiometer.

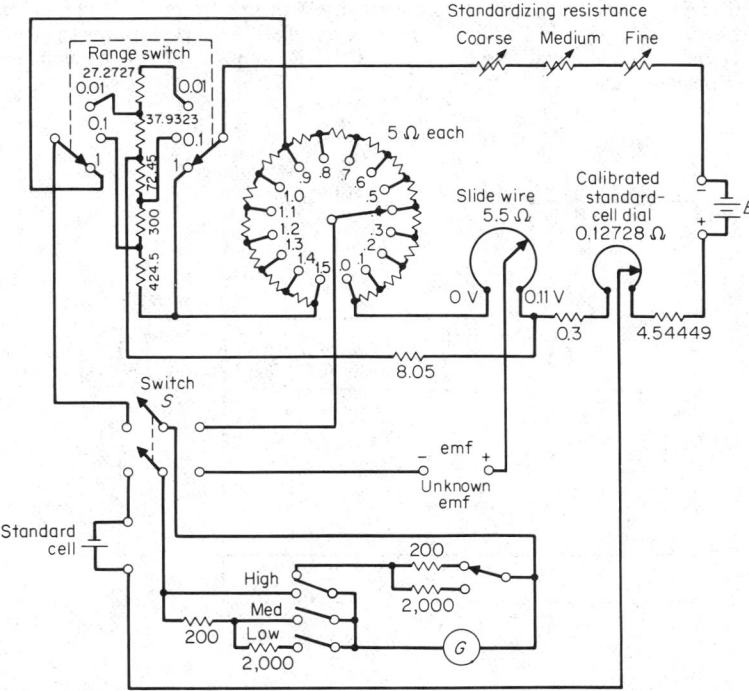

Fig. 17-2. K2 potentiometer. *(Leeds and Northrup.)*

on the galvanometer. The range switch reduces the potentiometer current to 2 or 0.2 mA for the 0.1 and the 0.01 ranges, respectively, thereby permitting lower voltages to be measured accurately. Since the range switch does not alter the battery current (22 mA), the instrument remains standardized on the lower ranges. When making measurements, the current should be checked using the standard cell to ensure that the current has not drifted from the standardized value.

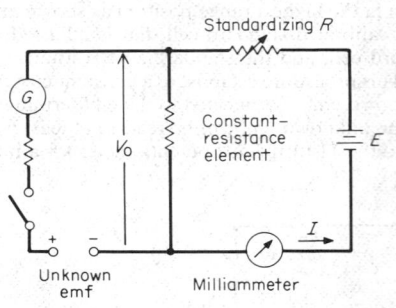

Fig. 17-3. Constant-resistance potentiometer.

18. The constant-resistance potentiometer of Fig. 17-3 uses a variable current through a fixed resistance to generate a voltage for obtaining a null with the unknown emf. The constant-resistance potentiometer is used primarily for measurements in the millivolt and microvolt range.

19. The microvolt potentiometer, or low-range potentiometer, is designed to minimize the effect of contact resistance and thermal emfs. Thermal shielding is used to minimize temperature differences. The galvanometer is connected to the circuit through a special Wenner thermo-free reversing key of copper and gold construction to eliminate thermal effects in the galvanometer circuit.

A typical microvolt potentiometer circuit consisting of two constant-current decades and a constant-resistance element is shown in Fig. 17-4. The constant-current decades use Diesselhorst rings, in which the constant current entering and leaving the ring divides between two paths. The IR drop across the resistance in the isothermal shield increases in 10 equal increments as the dial switch is rotated. The switch contacts are in the constant-current supply circuit, and therefore the effects of their IR drops and thermal emfs are minimized. A 100-division milliammeter associated with the constant-resistance element provides nearly 3 additional decades of resolution. Readings to 10 nV are possible with this type of potentiometer.

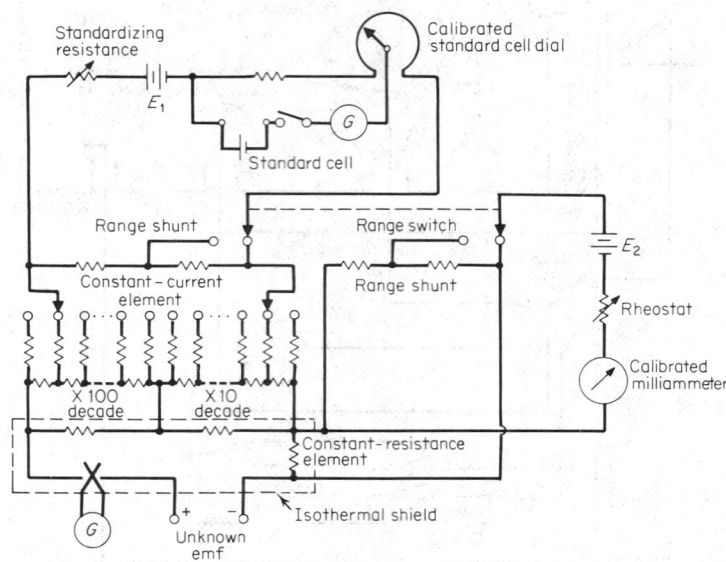

Fig. 17-4. Microvolt potentiometer.[2]

DIVIDER CIRCUITS

20. The volt box (Fig. 17-5) is used to extend the voltage range of a potentiometer. The unknown voltage is connected between 0 and an appropriate terminal, for example, X100. The potentiometer is connected between the 0 and *P* output terminals. When the potentiometer is balanced, it draws no current, and therefore the current drawn from the source flows through the resistor between terminals 0 and *P*. The unknown voltage is equal to the potentiometer reading multiplied by the selected tap multiplier. Unlike the potentiometer, the volt box does load the voltage source. Typical

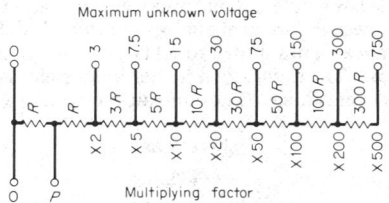

Fig. 17-5. Volt-box circuit.[2]

resistances range from about 200 to 1,000 Ω /V. The higher resistance values minimize self-heating and do not load the source as heavily. Errors due to leakage currents which could flow through the insulators supporting the resistors are minimized by using a guard circuit (see Par. **17-80**).

21. Decade voltage dividers provide a wide range of precisely defined and very accurate voltage ratios. The Kelvin-Varley vernier decade circuit is shown in Fig. **17-6**. The slide arms in the first 3 decades are arranged so that they always span two contacts. The shunting effect of the second gang resistance across the slide arms of the first decade is equal to 2*R*, thereby giving a net resistance of *R* between the slide-arm contacts. With no current drawn from the output, the resistance loading on the input is equal to 10*R* and is independent of the slide-arm settings. In each of the first 3 decades, 11 resistors are used, while only 10 resistors are used in the final decade, which has a single sliding contact. Potentiometers with 6 decades have been constructed using the Kelvin-Varley circuit.

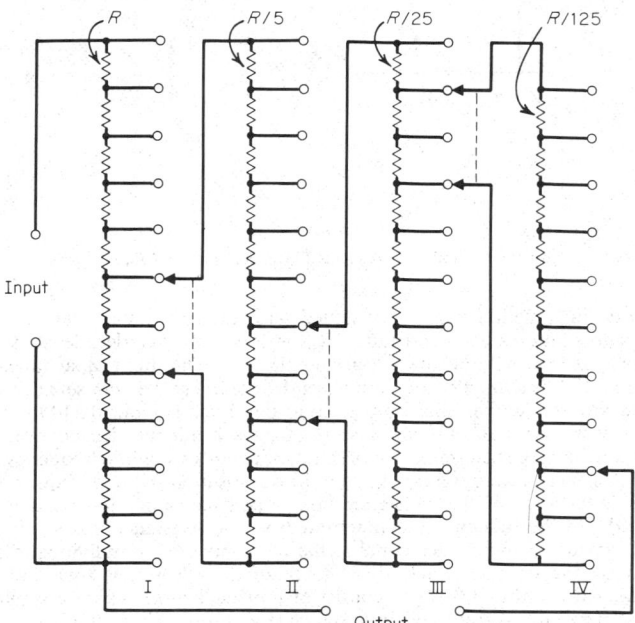

Fig. 17-6. Decade voltage divider.[2]

DECADE BOXES

22. Decade resistor boxes contain an assembly of resistances and switches, as shown in Fig. 17-7. The power rating of each resistance step is approximately constant; therefore, each decade has a different maximum current rating which should not be exceeded. Boxes having 4 to 7 decades are available with accuracies of 0.02%. Two typical 7 decade boxes provide resistance values from 0 to 1,111,111 Ω in 0.1-Ω steps and values from 0 to 11,111,110 Ω in 1-Ω steps. The accuracy at higher frequencies is affected by skin effect, series inductance, and shunt capacitance. The equivalent circuit of a resistance decade is shown in Fig. 17-8, where

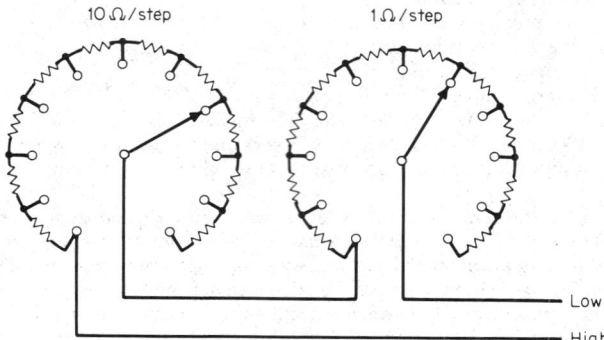

Fig. 17-7. Decade resistance box.

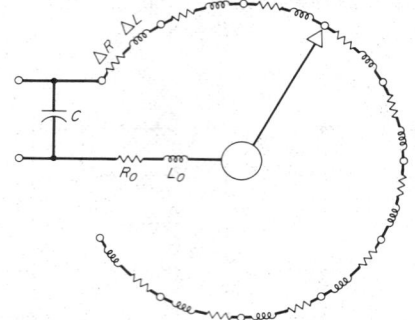

Fig. 17-8. Equivalent circuit of a resistance decade. *(General Radio Co.)*

ΔL is the undesired incremental inductance added with each resistance step ΔR. Silver contacts are used to obtain a zero resistance R_0, as low as 1 mΩ/decade at dc. Zero inductance values L_0 as low as 0.1 μH/decade are obtainable. The shunt capacitance for the configuration of Fig. 17-7 is a function of the highest decade in use, i.e., not set at zero. The shunt capacitance with the low terminal connected to the shield is typically 10 to 15 pF for the highest decade in use plus an equal value for each higher decade not in use.

Some applications, e.g., the determination of small inductances at audio frequency and the determination of resistance at radio frequency by the substitution method, require that the equivalent series inductance of the resistance box remain constant, independent of the resistance setting.[3]* In the inductively compensated decade resistance box small copper-wound coils each having an inductance equal to the inductance of an individual resistance unit are selected by the decade switch so as to maintain a constant total inductance.

23. Decade capacitor units generally consist of four capacitors which are selectively connected in parallel by a four-gang 11-position switch (Fig. 17-9). The individual capacitors and their associated switch are shielded to ensure that the selected steps add properly.[4]

---

* Superior numbers refer to References, Par. **17-149.**

Decade capacitor boxes are available with 6-decade resolution, which provides a range of 0 to 1.11111 μF in increments of 1 pF and with an accuracy of 0.05%. Air capacitors are used in the 1- and 10-pF decades, and silver-mica capacitors in the higher ranges. Polystyrene capacitors are used in some less precise decade capacitors.

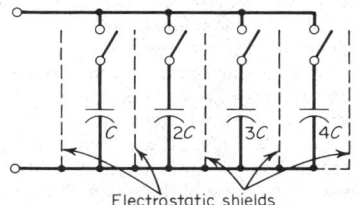

24. Decade inductance units can be constructed using four series-connected inductances of relative values 1, 2, 3, 4 or 1, 2, 2, 5. A four-gang 11-position switch is used to short-circuit the undesired inductances. Care must be taken to avoid mutual coupling between the inductances. *Decade inductance boxes* are

Electrostatic shields

Fig. 17-9. Capacitor decade.[3]

available with individual decades ranging from 1 mH to 10 H total inductance. A commercial single-decade unit consists of an assembly of four inductors wound on molybdenum-Permalloy dust cores and a switch which enables consecutive values to be selected. Typical units have an accuracy of 1% at zero frequency. The effective series inductance of a typical decade unit increases with frequency. The inductance is also a function of the ac current and any dc bias current. The Q of the coils varies with frequency.

STANDARDS

25. National Reference Standards of voltage, resistance, inductance, and capacitance are maintained at the National Bureau of Standards. Secondary standards, having a numerical value traceable to the National Reference Standards, are maintained in many laboratories.

26. The standard cell is used as a reference standard in electrical measurements and therefore must possess stability, long life, low temperature coefficient, and reproducibility. Two types of standard cells are used: the Weston saturated cell is used as an accurate reference standard in large laboratories; the Weston unsaturated cell, although less accurate (0.005%), has a lower temperature coefficient and is used as a reference in potentiometers and recording meters.

The Weston saturated cell (Fig. 17-10) has a positive electrode of metallic mercury, with mercurous sulfate as a depolarizer. The negative electrode is a cadmium-mercury amalgam (10% Cd). The electrolyte consists of a saturated solution of cadmium sulfate with an excess of cadmium sulfate crystals. This normal solution is usually acidified with sulfuric acid (0.04 to 0.08 N). The emf of acid cells is lower that that of the normal cell, $E_{20} = 1.018636 - 0.00060N - 0.00005N^2$, where E_{20} is the cell voltage at 20°C and N is the normality of the sulfuric acid.[5] The saturated cell is reproducible within a few microvolts but has a temperature coefficient of $-39.4\ \mu$V/°C at 20°C ($-52.9\ \mu$V/°C at 28°C). The formula of Vigoureux and Watts[6] is applicable to 10% amalgam cells:

$$E_t = E_{20} - 39.39 \times 10^{-6}(t - 20) - 0.903 \times 10^{-6}(t - 20)^2$$
$$+ 6.60 \times 10^{-9}(t - 20)^3 - 0.15 \times 10^{-9}(t - 20)^4$$

where t is the temperature in degrees Celsius. Large temperature gradients must be avoided since the temperature coefficient of the positive limb of the cell is about 310 μV/°C, while that of the negative limb is $-350\ \mu$V/°C at 20°C. The primary group of cells at the National Bureau of Standards is maintained at about 28°C in an oil bath in which the temperature is held constant within ±0.006°C.

The Weston unsaturated cell uses the same electrode structure, but the concentration of the electrolyte solution is chosen for saturation at 4°C and therefore is unsaturated at room temperature. Since it has a temperature coefficient of less than $-10\ \mu$V/°C, it can be used without temperature correction if the operating temperature is within a few degrees of the calibration temperature. New unsaturated cells ordinarily range between 1.0190 and 1.0194 absolute volts, and a cell comes with a calibration certificate specifying its emf at a given temperature. Because the emf of a cell may decrease by as much as 0.01%/year, a cell should be recalibrated every year and discarded when its emf has decreased to 1.0183 V.

Standard cells can be damaged by drawing excessive currents, and under no circumstances should a cell be short-circuited. It is desirable to limit the current drawn from the cell to 10

μA and to minimize the time during which this current is drawn. The current drawn from a cell should never exceed 100 μA. Cells should never be exposed to temperatures below 4°C or above 40°C. Abrupt temperature changes should be avoided, and all parts of the cell should be at the same temperature.

27. Zener diodes are used as voltage references in instruments which do not require the accuracy of a standard cell. A zener diode is a reverse-biased silicon *pn* junction having a well-controlled breakdown region, as shown in Fig. 17-11. Alloy junction zener diodes are available at nominal voltages from 2.4 to 12 V, while diffused-junction types are available from 6.8 to 200 V. The diffused-junction types exhibit a well-defined breakdown knee and have a lower dynamic impedance than alloy-junction devices under comparable operating conditions. The slope of the characteristic at the operating point in the breakdown region determines the dynamic impedance. The dynamic impedance is a function of the nominal voltage and the operating current. The temperature coefficient is a function of the zener voltage, as shown in Fig. 17-12.

28. The temperature-compensated zener diode consists of a positive-temperature-coefficient reverse-biased zener diode connected in series with one or more negative-temperature-coefficient forward-biased diodes within a single package. Specified temperature coefficients as low as 0.0002%/°C are available. The zener diode must be operated at a specified current to obtain the rated temperature coefficient. A typical circuit is shown in

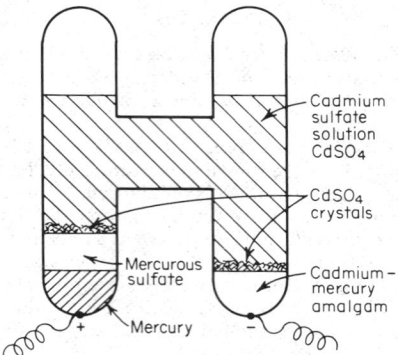

Fig. 17-10. Saturated standard cell.[19]

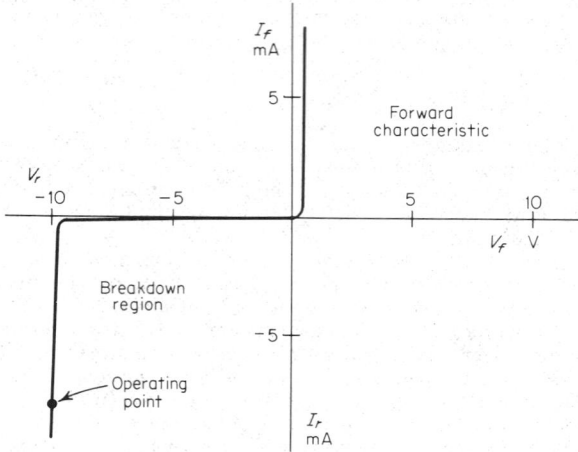

Fig. 17-11. Zener-diode characteristic.

Fig. 17-13. For small changes in the supply voltage E, the zener diode may be replaced by its dynamic resistance r_d, and the change in reference voltage is

$$\Delta E_r = \Delta E \frac{1}{1 + R_1/r_d + R_1/R_2}$$

For the values given $\Delta E_r = \Delta E/74.88$. The higher the value of R_2, the lower the reference voltage change since R_1 can be increased for a given zener-diode current. If R_2 is infinite and R_1 is selected as 1,560 Ω to maintain the specified zener-diode current, then $\Delta E_r = \Delta E/105$. Resistor R_2 can be made up of a pair of series-connected resistors, thereby permitting a portion of E_r to be obtained at their junction. If E_r is to be held within 0.001%, the voltage E of Fig. 17-13 must be regulated to better than 0.1%. Stable low-temperature-coefficient resistors must be used to achieve this degree of stability.

29. Current standards are precision four-terminal resistors having a pair of current terminals through which the current is passed. The voltage drop across the resistor is measured by a precision potentiometer connected to the voltage terminals of the resistor. A standard current is therefore a quantity that is derived from the voltage of a standard cell and a known resistance.

30. The National Reference Standard of resistance consists of a group of 10 specially constructed 1-Ω resistors maintained at the National Bureau of Standards. They are intercompared regularly and are also compared with the reference standards of other countries to ensure that their values are constant.

31. Resistance standards are made using high-resistivity metal in the form of wire or strip which is fully annealed to remove residual strains and sealed from contact with the air or other oxidizing agents. Manganin is widely used because it has a low temperature coefficient and a low thermoelectric emf at junctions with copper (see Table 17-2). Evanohm

Table 17-2. Properties of Resistance Wire[3]

Material	Typical composition	Resistivity			Thermal emf against copper, $\mu V/°C$
		$\mu\Omega\cdot cm$	$\Omega/mil\cdot ft$	Temperature coefficient, ppm/°C	
Nichrome, nichrome I to V, chromel A, tophet, etc.	Ni 80%, Cr 20%	108	650	150	22
Advance, ideal, cupron, copel, constantan, etc.	Ni 45%, Cu 55%	48	290	± 20	43
Manganin	Ni 4%, Cu 84%, Mn 12%	48	290	± 15	<3.0
Evanohm, Karma, 331 alloy	Ni 74.5%, Cr 20%, balance Al, and Fe or Cu	133	800	± 20	<2.5
Copper	Cu 99.9+%	1.724	10.37	3,930	

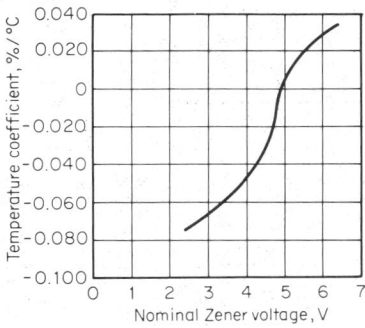 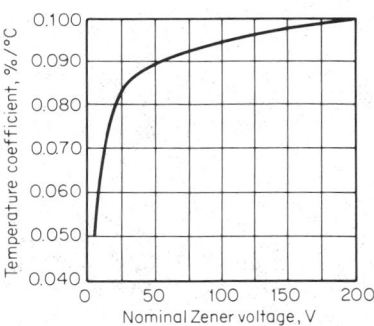

Fig. 17-12. Zener-diode temperature coefficient: (*a*) alloy-junction type; (*b*) diffused-junction type. (*Motorola, Inc.*)

is used for high-resistance standards because it exhibits similar characteristics and has a much higher resistivity. Oil baths are often used to dissipate heat and maintain a constant temperature.

The increase in resistance due to skin effect must be taken into account at high frequency. The ratio of ac to dc resistance of an isolated wire is a function of the factor $d\sqrt{f/p}$, where d is the wire diameter, f the frequency, and p the resistivity. The largest wire diameter that can be used at various frequencies and still keep the ac resistance within 1% of the dc resistance is given in Table 17-3. Wire-wound resistors have undesired series inductance and shunt capacitance associated with them. The inductance can be minimized by winding the resistance so that each turn encloses a minimum area, by selecting a diameter and resistivity which minimizes wire length, and by arranging the winding so that adjacent turns carry current in opposite directions. Low shunt capacitance is obtained by arranging the winding so that adjacent turns are widely spaced and have a minimum potential difference between them. The Ayrton-Perry winding (Fig. 17-14b) incorporates these desirable characteristics and is used for precision resistors up through 200 Ω. The more easily wound mica card resistor is used for higher resistance values.

Standard resistors of 1 Ω and below are constructed with four terminals, while some higher-value resistors have only two terminals. The standard resistor developed at the National Bureau of Standards is a four-terminal oil-cooled resistor having current lugs designed to be used in mercury cups.

32. Four-terminal resistors are used because a two-terminal resistor is subject to errors due to contact resistance. Assuming an unpredictable series resistance component of 0.001 Ω, a 1-Ω two-terminal resistor has an uncertainty of 0.1% while a 0.1-Ω two-terminal resistor has an uncertainty of 1%. This source of error is overcome by the use of four terminals, as shown in Fig. 17-15. The current enters and leaves through the outermost terminals, which are called the *current terminals*. The voltage terminals are arranged so that the voltage drop occurs between them when the current passes through the resistor. The current drawn by the potential measuring device should be negligible. Potentiometers, which draw no current at null, are used to make precise potential measurements.

33. Standard capacitors with values known to 1 part in 10[6] have been constructed at the National Bureau of Standards using four equal, closely spaced cylindrical rods in a geometry

Table 17-3. Largest Permissible Wire Diameter in Mils for Skin-Effect Ratio of 1.01[3]*

Frequency, MHz	Nichrome	Advance and Manganin	Copper
0.1	110	74	14
1	35	23	4.4
10	11	7.4	1.4
100	3.5	2.3	0.44
1,000	1.1	0.74	0.14
10,000	0.35	0.23	0.04

* For a ratio of 1.001 multiply above diameters by 0.55. For a ratio of 1.10 multiply above diameters by 1.78.

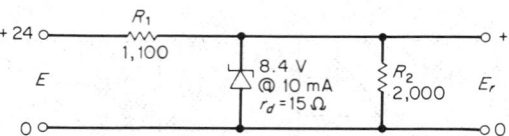

Fig. 17-13. Zener-diode reference circuit.

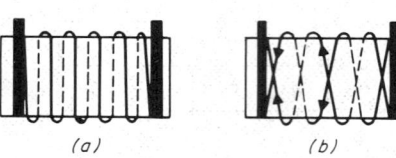

Fig. 17-14. Resistor windings: (*a*) mica cord; (*b*) Ayrton-Perry.[7]

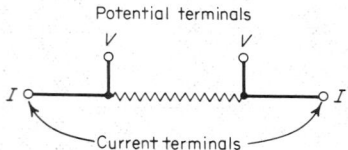

Fig. 17-15. Four-terminal resistor.

prescribed by Thompson and by Lampard.[8] The usual standard capacitors up to 1,000 pF use multiple parallel plates with a dry-air or nitrogen dielectric. Invar, a low-expansion alloy, is used to secure a low temperature coefficient, and good stability is achieved by mounting fully annealed components in a strain-free manner.

Three-terminal shielded construction is used to provide a definite value of capacitance which is independent of external objects and fields. With two-terminal construction, the capacitance from each electrode to surrounding objects forms a second capacitance in parallel with the desired capacitor. The equivalent circuit of a commercial three-terminal reference capacitor is given in Fig. 17-16. This capacitor, which has a drift of less than 20 ppm/y, is available in 10-, 100-, and 1,000-pF values calibrated against working standards whose absolute value is known to ± 5 ppm.

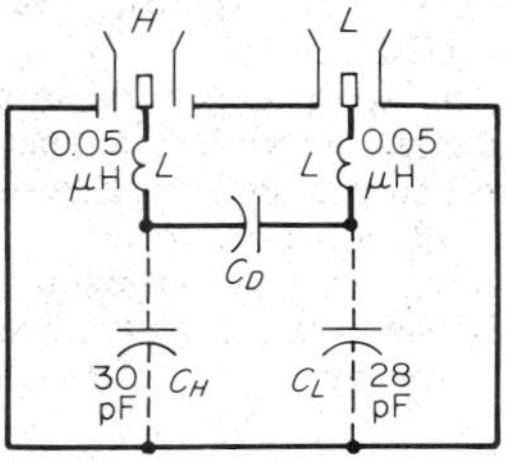

Fig. 17-16. Equivalent circuit showing direct capacitance C_D and average values of residual inductance L and terminal capacitances C_H and C_L. *(General Radio Co.)*

Precision variable capacitors, which can be set to 40 ppm, are commercially available in capacitance ranges up to 1,100 pF. A four-terminal ac-only decade capacitor is available that consists of a 1-μF polystyrene capacitor and a transformer that multiplies the effective capacitance in decade steps up to 1 F with accuracies better than 1% at 120 Hz.

34. Inductance standards used in absolute-ohm determination consist of single-layer solenoids wound on dimensionally stable forms of fused silica. The windings are laid in a lapped groove to ensure uniform winding pitch. These solenoids are large and bulky since their dimensions must be large enough to permit their values to be computed from measured dimensions.

Multilayer coils wound on ceramic, marble, or Bakelite forms are usually used as working standards. Toroidal cores are often used because they are nearly immune to external magnetic fields. The effective inductance is a function of frequency because of the capacitance associated with the winding. The frequency with which the calibrated value is associated should be given. Standard inductors, which are stable within $\pm 0.01\%$/y, are commercially available from 50 μH to 10 H. They are calibrated at 100, 200, 400, and 1,000 Hz against standards certified to $\pm 0.02\%$.

35. Mutual inductance standards, which are computable from measured dimensions, are constructed using single-layer primaries having two or three sections spaced so that there is a region in which the field gradients are very small.[5] The multilayer secondary winding is placed in this region to minimize the effect of positional errors.

ANALOG MEASUREMENTS

36. Ohmmeter circuits provide a convenient means of obtaining an approximate measurement of resistance.

The basic series-type ohmmeter circuit of Fig. 17-17a consists of an emf source, series resistor R_1, and D'Arsonval milliammeter. Resistor R_2 is used to compensate for changes in battery emf and is adjusted to provide full-scale meter deflection (0-Ω indication) with terminals X_1 and X_2 short-circuited. No deflection (infinite resistance indication) is obtained with X_1 and X_2 open-circuited. When an unknown resistor R_X is connected across the terminals, the meter deflection varies inversely with the unknown resistance. With the range switch in the position shown, half-scale deflection is obtained when the external resistance is equal to $R_1 + R_2 R_M/(R_2 + R_M)$. A multirange meter can be obtained using current-shunting resistors R_3 and R_4. A typical commercial ohmmeter circuit (Fig. 17-17b) having mid-scale readings of 12 Ω, 1,200 Ω, and 120kΩ uses an Ayrton shunt for range selection and a higher battery voltage for the highest resistance range.

In the shunt-type ohmmeter the unknown resistor R_X is connected across the D'Arsonval milliammeter, as shown in Fig. 17-18a. The variable resistance R_1 is adjusted for full-scale deflection (infinite-resistance indication) with terminals X_1 and X_2 open-circuited. The ohm scale, with 0 Ω corresponding to zero deflection, is the reverse of the series-type ohmmeter scale. The resistance range can be lowered by switching a shunt resistor across the meter. With the range switch selecting shunt resistor R_2, half-scale deflection occurs when R_X is equal to the parallel combination of R_1, R_M, and R_2. The shunt-type ohmmeter is therefore most suited to low-resistance measurements.

The use of a high-input impedance amplifier between the circuit and the D'Arsonval meter permits the shunt-type ohmmeter to be used for high-resistance as well as low-resistance measurements. A commercial ohmmeter (Fig. 17-18b) uses a field-effect-amplifier input stage which draws negligible current. The amplifier gain is adjusted to provide full-scale deflection with terminals X_1 and X_2 open-circuited. Half-scale deflection occurs when R_X is equal to the total selected tap resistance.

37. Voltage-drop (or fall-of-potential) methods for determining resistance involve measuring the current flowing through the resistor with an ammeter, measuring the voltage drop across the resistor with a voltmeter, and calculating the resistance using Ohm's law. The circuit of Fig. 17-19a should be used for low-resistance measurements since the current drawn by the voltmeter V/R_V will be small with respect to the total current I. The circuit of Fig. 17-19b should be used for high-resistance measurements since the resistance of the ammeter R_A will be small with respect to the unknown resistance R_X. An accuracy of 1% or better can

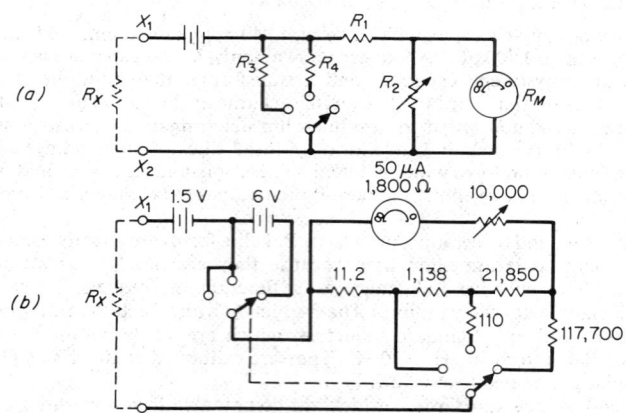

Fig. 17-17. Series-type ohmmeters. *(b. Simpson Electric Company.)*

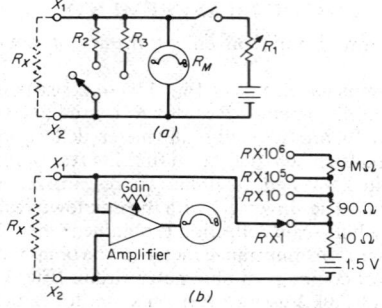

Fig. 17-18. Shunt-type ohmmeter. *(Triplett Electrical Instrument Co.)*

17-14

be obtained using 0.5% accurate instruments if the voltage source and instrument ranges are selected to provide readings near full scale.

38. The loss-of-charge method   is useful for measuring insulation resistance and other very high leakage resistances.   The unknown resistor R is connected in parallel with a known capacitor C having a leakage resistance r, as shown in Fig. 17-20. Key A is momentarily depressed to charge the capacitor. Key B is closed immediately after key A is opened, and the ballistic throw d_1 of the galvanometer recorded. The process is repeated, but now a time of t seconds is allowed to pass from the time key A is opened until key B is closed and a deflection d_2 observed. The time constant of the circuit in seconds is[2]

$$\tau = \frac{t}{2.303 \log_{10}(d_1/d_2)}$$

The time constant τ_1 is experimentally determined for the configuration of Fig. 17-20, where $\tau_1 = CRr/(R + r)$. The resistor R is disconnected, and the time constant τ_2 of the capacitor and its leakage resistance are determined by repeating the measurements, where $\tau_2 = Cr$. The resistance R is

$$R = \frac{\tau_1 \tau_2}{C(\tau_2 - \tau_1)}$$

The loss-of-charge method can also be used to determine the value of an unknown capacitance C having an unknown leakage resistance r if the value of R is known. The above measurements are performed to determine τ_1 and τ_2. The values of C and r are

$$C = \frac{\tau_1 \tau_2}{R(\tau_2 - \tau_1)} \qquad r = R\frac{\tau_2 - \tau_1}{\tau_1}$$

39. Resonance methods can be used to measure the inductance, capacitance and Q factor of components at radio frequencies. In Fig. 17-21, resistors R_1 and R_2 couple the oscillator voltage e to a series-connected known capacitance and an unknown inductance represented by effective inductance L' and effective series resistance r'. Resistor R_2 is chosen to be small with respect to resistance r', thereby minimizing the effect of source resistance of the injected voltage.

A circuit containing reactive components is in resonance when the supply current is in phase with the applied voltage. The series circuit of Fig. 17-21 is in resonance when the

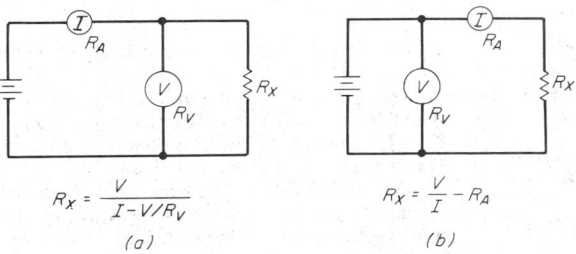

$$R_X = \frac{V}{I - V/R_V}$$

$$R_X = \frac{V}{I} - R_A$$

(a)   (b)

Fig. 17-19. Fall-of-potential method.

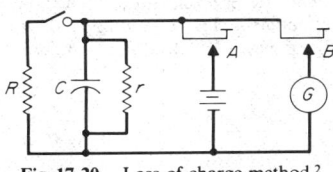

Fig. 17-20. Loss-of-charge method.[2]

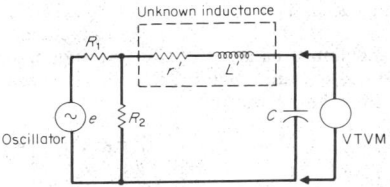

Fig. 17-21. Inductance measurement.

inductive reactance X_L is equal to the capacitive reactance X_C, which occurs when

$$\omega^2 = \omega_0^2 = \frac{1}{L'C}$$

where $X_L = \omega L'$, $X_C = 1/\omega C$, $\omega = 2\pi f$, $\omega_0 = 2\pi f_0$, and f_0 is the resonant frequency in hertz, L' is the effective inductance in henrys, and C is the capacitance in farads.

If L', r', C, and e are constant and the oscillator frequency ω is adjusted until the voltage read across C by the vacuum-tube voltmeter is maximum, the frequency ω will be slightly less than the resonant frequency ω_0:

$$\omega^2 = \frac{1}{L'C} - \frac{(r' + R_s)^2}{2L'^2} = \omega_0^2\left(1 - \frac{1}{2Q^{*2}}\right)$$

where $R_s = R_1 R_2/(R_1 + R_2)$ and $Q^* = \omega_0 L'/(r' + R_s)$. If $Q^* \geq 10$, ω and ω_0 will differ by less than 0.3%. The ratio m of the voltage across C to the voltage across R_2 can be measured while operating at ω. If $R_s \ll r'$, the value of the effective Q of the unknown inductance is related to m by

$$m = \frac{2Q'^2}{\sqrt{4Q'^2 - 1}} \qquad Q' = \sqrt{\frac{m}{2(m - \sqrt{m_2 - 1})}}$$

where $Q' = \omega_0 L'/r'$ and $m = V_C/V_{R_2}$ measured at ω. The values of m and Q' are nearly equal for high values of Q'; the difference is less than 0.4% for $Q' > 10$. If R_s is not small with respect to r', its value affects the determination of Q' only indirectly through its effect on ω. If $R_s = r$ and $Q' \geq 10$, the determination of Q' by the above equation is in error by less than 1%.

If ω, L', r', and e are constant and the capacitance C is adjusted until the voltage across it is maximum, the capacitance value C will be slightly less than the capacitance value C_R needed for resonance at the frequency ω:

$$C = \frac{1}{\omega^2 L'}\left(\frac{1}{1 + 1/Q^{*2}}\right) = C_R\frac{1}{1 + 1/Q^{*2}}$$

where $\omega = \omega_0$, $Q^* = \omega_0 L'/(r' + R_s)$, and $R_s = R_1 R_2/(R_1 + R_2)$. For $Q^* \geq 10$, C differs from C_R by less than 1%. If $R_s \ll r'$, the value of the effective Q of the unknown inductance is related to m by

$$m = \sqrt{Q'^2 + 1} \qquad Q' = \sqrt{m^2 - 1}$$

where $Q' = \omega L'/r'$, $m = V_C/V_{R_2}$, $\omega = \omega_0 = 2\pi f_0$, and f_0 is the resonant frequency in hertz for L and C_R.

The circuit of Fig. 17-21 can be used to find the value of an unknown capacitance if a stable inductance of known value is available. Similar circuits are used in Q meters (see Par. 17-87). Discussions of series resonance, parallel resonance, and Q factor are given in Refs. 3, 9, and 10.

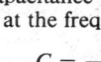

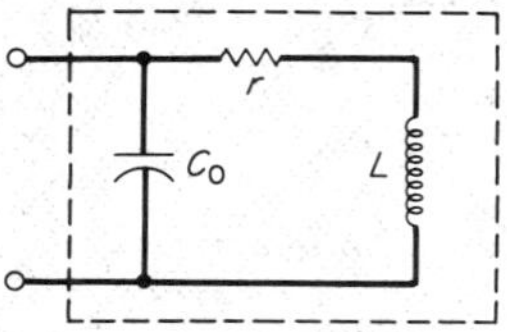

Lead lengths should be kept short when making measurements at high frequencies, and the capacitance of the vacuum-tube voltmeter must be added to capacitor C if it is not negligible.

40. The self-capacitance C_0 of an inductance is generally not negligible and can cause discrepancies in measurements. An inductance can be represented by the equivalent circuit of Fig. 17-22, where C_0 is the self-capacitance, L is the true inductance, and r is the true series resistance. The self-capacitance causes the effective inductance L' and the effective series resistance r' of the equivalent unknown inductance as measured in the circuit of Fig. 17-21 to differ somewhat from L and r, that is,

Fig. 17-22. Equivalent circuit of inductance.

$$r' = r\left[\frac{C + C_0}{C}\right]^2 \quad \text{and} \quad L' = L\frac{C + C_0}{C}$$

The value of C_0 can be measured with the aid of a Q meter (Par. **17-87**) having a calibrated variable capacitance.

41. The bandwidth of a tuned circuit can be determined by measuring the frequencies f_1 and f_2 at which the capacitor voltage in Fig. 17-23 is 0.707 times the maximum capacitor voltage. The oscillator, which maintains a constant voltage, is loosely coupled to the tuned circuit. The bandwidth Δf is equal to the difference $f_2 - f_1$ and is a function of the resonant frequency f_0 and the Q of the tuned circuit: $\Delta f = f_0/Q$.

42. The circuit Q of a tuned circuit for sinusoidal excitation is defined as 2π(maximum stored energy)/(energy dissipated per cycle). Circuit Q can be measured directly using the circuit of Fig. 17-23. In this circuit the self-capacitance C_0 appears directly in parallel with the tuning capacitor C. The bandwidth is determined as explained above, and the Q is equal to $f_0/(f_2 - f_1)$. If the circuit of Fig. 17-21 is used, the self-capacitance no longer appears in parallel with the tuning capacitor C and the effective Q is measured. The real Q can be calculated if the self-capacitance C_0 is known: $Q = Q'(C + C_0)/C$, where Q' is the effective Q.

TRANSDUCER-INPUT MEASUREMENT SYSTEMS

43. Transducers are used to convert the quantity to be measured into an electrical signal. Transducer types and their input and output quantities are discussed in Sec. 10.

TRANSDUCER SIGNAL CIRCUITS

44. Amplifiers are often required to increase the voltage and power levels of the transducer output and to prevent excessive loading of the transducer by the measurement system. The design of the amplifier is a function of the performance specifications, which include required amplification in terms of voltage gain or power gain, frequency response, distortion permissible at a given maximum signal level, dynamic range, residual noise permissible at the output, gain stability, permissible drift (for dc amplifiers), operating-temperature range, available supply voltage, permissible power consumption and dissipation, reliability, size, weight, and cost.

45. Capacitive-coupled amplifiers (ac amplifiers) are used when it is not necessary to preserve the dc component of the signal. Ac amplifiers are used with transducers that produce a modulated carrier signal. Low-level amplifiers increase the signal from millivolts to several volts. The two-stage class A capacitor-coupled transistor amplifier of Fig. 17-24

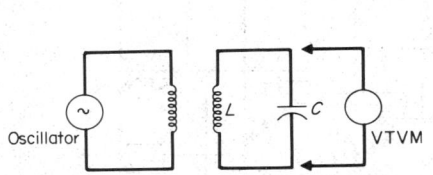

Fig. 17-23. Loosely coupled tuned circuit.

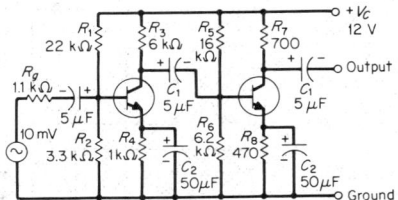

Fig. 17-24. Two-stage cascaded common-emitter capacitive-coupled audio amplifier.[11]

has a power gain of 64 dB and a voltage gain of approximately 1,000. Design information, an explanation of biasing, and equations for calculating the input impedance and various gain values are given in Ref. 11. An excellent ac amplifier can be obtained by connecting a coupling capacitor in series with resistor R_1 of the operational amplifier of Fig. 17-28. The capacitor should be selected so that $C > 1/2\pi f R_1$, where f is the lowest signal frequency to be amplified. Class B transformer-coupled amplifiers, which are often used for higher power-output stages, are also discussed in Ref. 11.

46. Direct-coupled amplifiers are used when the dc component of the signal must be preserved. These designs are more difficult than those of capacitive-coupled amplifiers because changes in transistor leakage currents, gain, and base-emitter voltage drops can cause the output voltage to change for a fixed input voltage, i.e., cause a dc-stability problem. The dc stability of an amplifier is determined primarily by the input stage since the equivalent input drift introduced by subsequent stages is equal to their drift divided by the preceding gain. Balanced input stages, such as the differential amplifier of Fig. 17-25, are widely used because drift components tend to cancel. By selecting a pair of transistors, Q_1 and Q_2, which are matched for current gain within 10% and base-to-emitter voltage within 3 mV, the temperature drift referred to the input can be held to within 10 μV/$^\circ$C. Transistor Q_3 acts as a constant-current source and thereby increases the ability of the amplifier to reject common-mode input voltages. For applications where the generator resistance r_g is greater than 50 kΩ, current offset becomes dominant, and lower overall drift can be obtained by using field-effect transistors in place of the bipolar transistors Q_1 and Q_2. Voltage drifts as low as 0.6 μV/$^\circ$C can be obtained using integrated-circuit operational amplifiers (see Fig. 17-26).

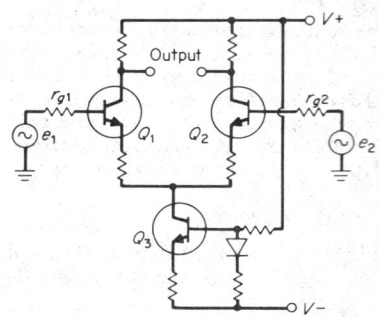

Fig. 17-25.   Differential amplifier.

47. Operational amplifiers are widely used for amplifying low-level ac and dc signals. They usually consist of a balanced input stage, a number of direct-coupled intermediate stages, and a low-impedance output stage. They provide high open-loop gain, which permits the use of a large amount of gain-stabilizing negative feedback (see Par. **17-118**). The schematic of an integrated-circuit operational amplifier which is intended for use in instrumentation applications along with some performance specifications is given in Fig. 17-26. With input voltages e_1, e_2, and e_{cm} equal to zero, the output of the amplifier of Fig. 17-27 will have an offset voltage E_{0s} defined by[12]

$$E_{os} = V_{os}\frac{R_1 + R_2}{R_1} + I_{b1}R_2 - I_{b2}\frac{R_3 R_4 (R_1 + R_2)}{R_1 (R_3 + R_4)}$$

where I_{b1} and I_{b2} are bias currents that flow into the amplifier when the output is zero and V_{os} is the input offset voltage that must be applied across the input terminals to achieve zero output. The input bias current specified for an operational amplifier is the average of I_{b1} and I_{b2}.

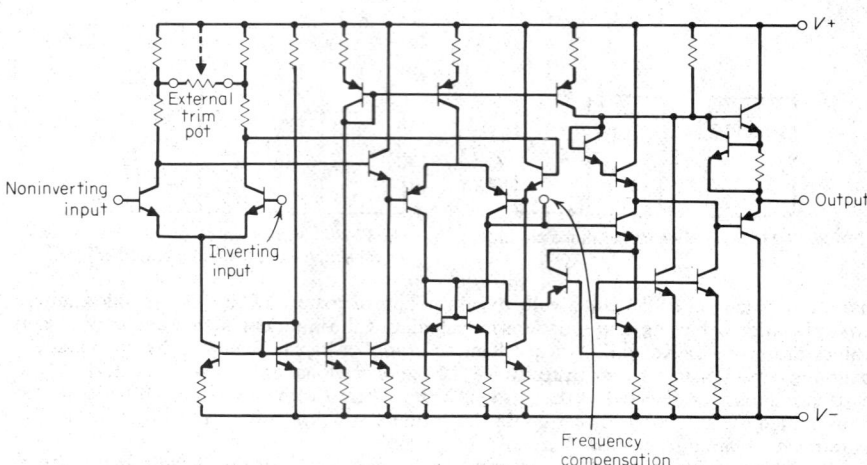

Fig. 17-26. μA 725 instrumentation operational amplifier.   *(Fairchild Semiconductor Co.)*

Typical specifications

Input offset voltage (without external trim) . 0.5 mV	Common-mode rejection ratio .... 120 dB		
Input offset current .... 2 nA	Average input offset drift,		
Input bias current .... 50 nA	Without external trim .... 2 μV/$^\circ$C		
Input Resistance .... 1.5 mΩ	With external trim .... 0.6 μV/$^\circ$C		
Open-loop gain .... 3×10^6			

Since the bias currents are approximately equal, it is desirable to choose the parallel combination of R_3 and R_4 equal to the parallel combination of R_1 and R_2. For this case, $E_{os} = V_{os}(R_1 + R_2)/R_1 + I_{os}R_2$, where offset current $I_{os} = I_{b1} - I_{b2}$.

In the ideal case, where V_{os} and I_{os} are zero, the output voltage E_0 as a function of signal voltage e_1 and e_2 and common-mode voltage e_{cm} is[12]

$$E_0 = -e_1\frac{R_2}{R_1} + e_2\frac{R_4(R_1 + R_2)}{R_1(R_3 + R_4)} + e_{cm}\frac{R_4(R_1 + R_2) - R_2(R_3 + R_4)}{R_1(R_3 + R_4)}$$

Maximum common-mode rejection can be obtained by choosing $R_4/R_3 = R_2/R_1$, which reduces the above equation to $E_0 = R_2(e_2 - e_1)/R_1$. The common-mode signal is not entirely rejected in an actual amplifier but will be reduced relative to a differential signal by the common-mode rejection ratio of the amplifier. Minimum drift and maximum common-mode rejection, which are important when terminating the wires from a remote transducer, can be obtained by selecting $R_3 = R_1$ and $R_4 = R_2$.

Where common-mode voltages are not a problem, the simple inverting amplifier (Fig. 17-28) is obtained by replacing e_{cm} and e_2 with short circuits and combining R_3 and R_4 into one resistor, which is equal to the parallel equivalent of R_1 and R_2. The input impedance of this circuit is equal to R_1.

Simarily, the simple noninverting amplifier (Fig. 17-29) is obtained by replacing e_{cm} and e_1 with short circuits. The voltage follower is a special case of the noninverting amplifier where $R_1 = \infty$ and $R_2 = 0$. The input impedance of the circuit of Fig. 17-29 is equal to the parallel combination of the common-mode input impedance of the amplifier and impedance $Z_{ia}[1 + (AR_1)/(R_1 + R_2)]$, where Z_{ia} is the differential-mode amplifier input impedance and A is the amplifier gain. Where very high input impedances are required, as in electrometer circuits, operational amplifiers having field-effect input transistors are used to provide input resistances up to $10^{12}\Omega$.

An ac-coupled amplifier can be obtained by connecting a coupling capacitor in series with the input resistor of Fig. 17-28. The capacitor value should be selected so that the capacitive reactance at the lowest frequency of interest is lower than the amplifier input impedance R_1.

Operational amplifiers are useful for realizing filter networks and integrators (see Par. 17-52). Other applications include absolute-value circuits (see Par. 17-136), logarithmic converters, nonlinear amplification, voltage-level detection, function generation, and analog multiplication and division.[12] Care should be taken not to exceed the maximum supply

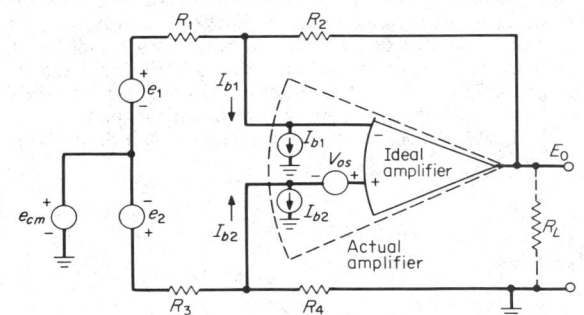

Fig. 17-27. Operational amplifier.[12]

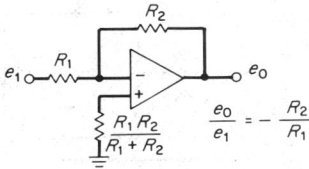

Fig. 17-28. Inverting amplifier.

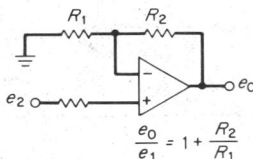

Fig. 17-29. Noninverting amplifier.

voltage and maximum common-mode voltage ratings and also to be sure that the load resistance R_L is not smaller than that permitted by the rated output.

48. The charge amplifier is used to amplify the ac signals from variable-capacitance transducers and transducers having a capacitive impedance such as piezoelectric transducers. In the simplified circuit of Fig. 17-30a, the current through C_s is equal to the current through C_1, and therefore

$$C_s \frac{\partial e_s}{\partial t} + e_s \frac{\partial C_s}{\partial t} = -C_1 \frac{de_o}{dt}$$

For the piezoelectric transducer, C_s is assumed constant, and the gain $\delta e_o/\delta e_s = -C_s/C_1$. For the variable-capacitance transducer, e_s is constant, and the gain $de_o/dC_s = -e_s/C_1$. A practical circuit requires a resistance across C_1 to limit output drift. The value of this resistance must be greater than the impedance of C_1 at the lowest frequency of interest. A typical operational amplifier having field-effect input transistors has a specified maximum input current of 2 nA which will result in an output offset of only 0.2 V if a 100-MΩ resistance is used across C_1. It is preferable to provide a high effective resistance by using a network of resistors, each of which has a value of $1 - $ MΩ or less.

The effective feedback resistance R' in the practical circuit of Fig. 17-30b is given by $R' = R_3(R_1 + R_2)/R_2$, assuming that $R_3 > 10R_1R_2/(R_1 + R_2)$. Output drift is further reduced by selecting $R_4 = R_3 + R_1R_2/(R_1 + R_2)$. Resistor R_5 is used to provide an upper frequency rolloff at $f = 1/2\pi R_5 C_s$, which improves the signal-to-noise ratio.

49. Amplifier-gain stability is enhanced by the use of feedback (see Par. **17-118**) since the gain of the amplifier with feedback is relatively insensitive to changes in the open-loop amplifier gain G provided that the loop gain GH is high. For example, if the open-loop gain G changes by 10% from 100,000 to 90,000 and the feedback divider gain H remains constant at 0.01, the closed-loop gain $G/(1 + GH) \cong 99.9$ changes only 0.011%.

If a high closed-loop gain is required, simply decreasing the value of H will reduce the value of GH and thereby reduce the accuracy. The desired accuracy can be maintained by cascading two or more amplifiers, thereby reducing the closed-loop gain required from each amplifier. Each amplifier has its own individual feedback, and no feedback is applied around the cascaded amplifiers. In this case, it is unwise to cascade more stages than needed to achieve a reasonable value of GH in each individual amplifier, since excessive loop gain will make the individual stages more prone to oscillation and the overall system will exhibit increased sensitivity to noise transients.

50. Chopper amplifiers are used for amplifying dc signals in applications requiring very low drift. The dc input signal is converted by a chopper to a switched ac signal having an amplitude proportional to the input signal and a phase of 0 or 180° with respect to the chopper reference frequency, depending on the polarity of the input signal. This ac signal is amplified by an ac amplifier, which eliminates the drift problem, and then converted back to a proportional dc output voltage by a phase-sensitive demodulator.

The chopper of Fig. 17-31 consists of a mechanical switch driven at some convenient

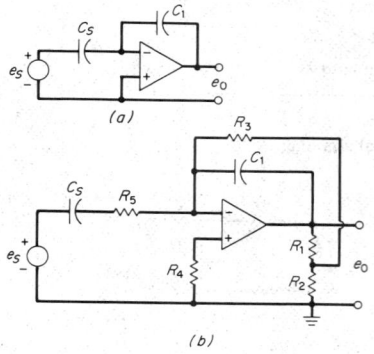

(a)

(b)

Fig. 17-30. Charge amplifier.

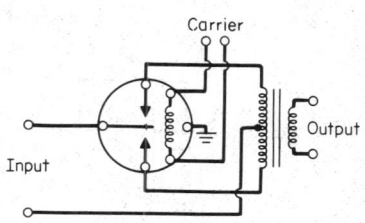

Fig. 17-31. Full-wave chopper.

reference frequency such as 60 Hz, which alternately connects the input across each half of the primary winding. The output is a square wave proportional to the input, which meets the requirement of having zero amplitude for a zero input signal. Different configurations can be used with the mechanical chopper, and other types of elements such as field-effect transistors, photoconductors, junction transistors, and diodes operated in a switching mode can be used in place of mechanical switches.

The frequency response of a chopper amplifier is theoretically limited to one-half the carrier frequency. In practice, however, the frequency response is much lower than the theoretical limit. High-frequency components in the input signal exceeding the theoretical limit are removed to avoid unwanted beat signals with the chopper frequency.

The chopper amplifier of Fig. 17-32 consists of a low-pass filter to attenuate high frequencies, an input chopper, an ac amplifier, a phase-sensitive demodulator, and a low-pass output filter to attenuate the chopper ripple component in the output signal. The frequency response of this amplifier is limited to a small fraction of the chopper frequency.

The frequency-response limitation of the chopper amplifier can be overcome by using the chopper amplifier for the dc and low-frequency signals and a separate ac amplifier for the higher-frequency signals, as shown in Fig. 17-33. Simple shunt field-effect-transistor choppers Q_1 and Q_2 are used for modulation and detection, respectively. Capacitor C_T is used to minimize spikes at the input to the ac amplifier.

51. Modulator-demodulator systems avoid the drift problems of dc amplifiers by using a modulated carrier which can be amplified by ac amplifiers (Fig. 17-34). Inputs and outputs may be either electrical or mechanical.

The varactor modulator (Fig. 17-35) takes advantage of the variation of diode-junction capacitance with voltage to modulate a sinusoidal carrier. The carrier and signal voltages applied to the diodes are small, and the diodes never reach a low-resistance condition. Input bias currents of the order of 0.01 pA are possible. For zero signal input, the capacitance values of the diodes are equal, and the carrier signals coupled by the diodes cancel. A dc-input signal will increase the capacitance of one diode while decreasing the capacitance of the other and thereby produce a carrier unbalance signal which is coupled to the ac amplifier by capacitor C_2. A phase-sensitive demodulator, such as field-effect transistor Q_2 of Fig. 17-33, may be used to recover the dc signal.

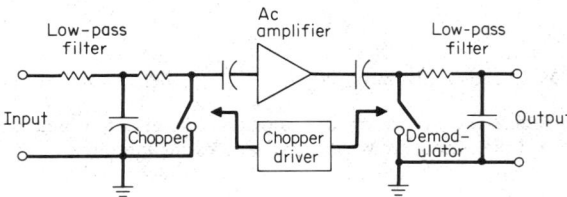

Fig. 17-32. Chopper amplifier.[12]

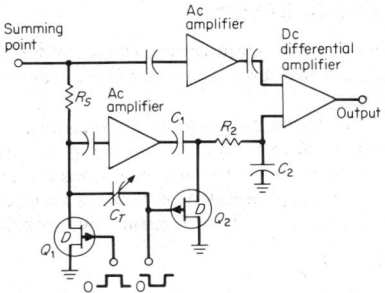

Fig. 17-33. Chopper-stabilized dc amplifier.[13]

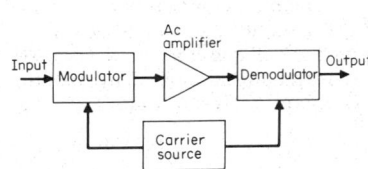

Fig. 17-34. Modulator-demodulator system.

The magnetic amplifier and second-harmonic modulator can also be used to convert dc-input signals to modulation on a carrier, which is amplified and later demodulated. Mechanical-input modulators include ac-driven potentiometers (Par. 17-104), linear variable differential transformers (Par. 17-137), and synchros (Par. 17-104). The amplified ac carrier can be converted directly to a mechanical output by a two-phase induction servomotor (Par. 17-145).

52. Integrators are often required in systems where the transducer signal is a derivative of the desired output, e.g., when an accelerometer is used to measure the velocity of a vibrating object. The output of the analog integrator of Fig. 17-36 consists of an integrated signal term plus error terms caused by the offset voltage V_{os} and the bias currents I_{b1} and I_{b2} (see Par. 17-47).

$$e_0 = -\frac{1}{R_1 C} \int e_1 \, dt + \frac{1}{R_1 C} \int (V_{os} + I_{B_1} R_1 - I_{B_2} R_2) \, dt$$

These error terms will cause the integrator to saturate unless the integrator is reset periodically or a feedback path exists which tends to drive the output toward a given level within the linear range. In the accelerometer integrator, accurate integration may not be required below a given frequency, and the desired stabilizing feedback path can be introduced by incorporating a large effective resistance across the capacitor using the technique shown in Fig. 17-30b. In this case, the integrator response is approximated by the low-pass network characteristic of Table 17-7, Par. 17-130.

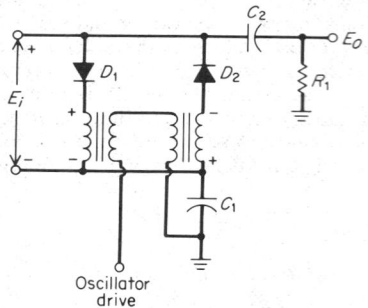

Fig. 17-35. Basic varactor modulator.[12]

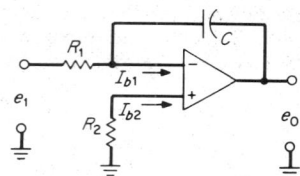

Fig. 17-36. Analog integrator.

53. Output indicators. A variety of analog and digital output indicators can be used to display and record the output from the signal-processing circuitry.

BRIDGE CIRCUITS, DETECTORS, AND AMPLIFIERS

PRINCIPLES OF BRIDGE MEASUREMENTS

54. Bridge circuits are used to determine the value of an unknown impedance in terms of other impedances of known value. Highly accurate measurements are possible because a null condition is used to compare ratios of impedances.

The most common bridge arrangement (Fig. 17-37) contains four branch impedances, a voltage source, and a null detector. Galvanometers, alone or with chopper amplifiers, are used as null detectors for dc bridges; while telephone receivers, vibration galvanometers, and tuned amplifiers with suitable detectors and indicators are used for null detection in ac bridges (see Pars. 17-82–17-86 and Ref. 19). The voltage across an infinite-impedance detector is

$$V_d = \frac{(Z_1 Z_3 - Z_2 Z_x) E}{(Z_1 + Z_2)(Z_3 + Z_x)}$$

If the detector has a finite impedance Z_5, the current in the detector is

$$I_d = \frac{(Z_1 Z_3 - Z_2 Z_x)E}{Z_5(Z_1 + Z_2)(Z_3 + Z_x) + Z_1 Z_2(Z_3 + Z_x) + Z_3 Z_x(Z_1 + Z_2)}$$

where E is the potential applied across the bridge terminals.

55. A null or balance condition exists when there is no potential across the detector. This condition is satisfied, independent of the detector impedance, when $Z_1 Z_3 = Z_2 Z_x$. Therefore, at balance, the value of the unknown impedance Z_x can be determined in terms of the known impedances Z_1, Z_2, and Z_3:

$$Z_x = \frac{Z_1 Z_3}{Z_2}$$

Since the impedances are complex quantities, balance requires that both magnitude and phase angle conditions be met: $|Z_x| = |Z_1| |Z_3| / |Z_2|$ and $\theta_x = \theta_1 + \theta_3 - \theta_2$. Two of the known impedances are usually fixed impedances, while the third impedance is adjusted in resistance and reactance until balance is attained.

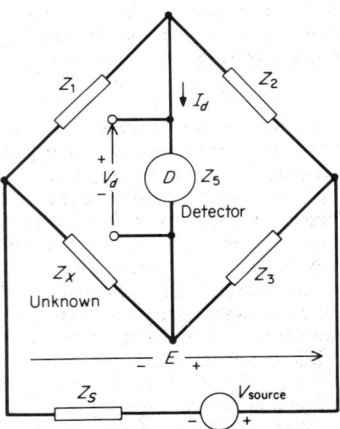

Fig. 17-37. Basic impedance bridge.

56. The sensitivity of the bridge can be expressed in terms of the incremental detector current ΔI_d for a given small per-unit deviation δ of the adjustable impedance from the balance value. If Z_1 is adjusted, $\delta = \Delta Z_1 / Z_1$ and

$$\Delta I_d = \frac{Z_3 Z_x E \delta}{(Z_3 + Z_x)^2 [Z_5 + Z_1 Z_2/(Z_1 + Z_2) + Z_3 Z_x/(Z_3 + Z_x)]}$$

where Z_5 is the detector impedance.

If a high-input-impedance amplifier is used for the detector and impedance Z_5 may be considered infinite, the sensitivity can be expressed in terms of the incremental input voltage to the detector ΔV_d for a small deviation from balance

$$\Delta V_d = \frac{Z_3 Z_x E \delta}{(Z_3 + Z_x)^2} = \frac{Z_1 Z_2 E \delta}{(Z_1 + Z_2)^2}$$

where $\delta = \Delta Z_1 / Z_1$ and ΔZ_1 is the deviation of impedance Z_1 from its balance value Z_1. Maximum sensitivity occurs when the magnitudes of Z_3 and Z_x are equal (which for balance implies that the magnitudes of Z_1 and Z_2 are equal). Under this condition, $\Delta V_d = E\delta/4$ when the phase angles θ_3 and θ_x are equal; $\Delta V_d = E\delta/2$ when the phase angles θ_3 and θ_x are in quadrature; and ΔV_d is infinite when $\theta_3 = -\theta_x$, as is the case with lossless reactive components of opposite sign. In practice, the value of the adjustable impedance must be sufficiently large to ensure that the resolution provided by the finest adjusting step permits the desired precision to be obtained. This value may not be compatible with the highest sensitivity, but adequate sensitivity can be obtained for an order-of-magnitude difference between Z_3 and Z_x or Z_1 and Z_2, especially if a tuned-amplifier detector is used.

57. Interchanging the source and detector can be shown to be equivalent to interchanging impedances Z_1 and Z_3. This interchange does not change the equation for balance but does change the sensitivity of the bridge. For a fixed applied voltage E higher sensitivity is obtained with the detector connected from the junction of the two high-impedance arms to the junction of the two low-impedance arms.

58. The source voltage must be carefully selected to ensure that the allowable power dissipation and voltage ratings of the known and unknown impedances of the bridge are not exceeded. If the bridge impedances are low with respect to the source impedance Z_s, the bridge-terminal voltage E will be lowered. This can adversely affect the sensitivity, which is proportional to E. The source for an ac bridge should provide a pure sinusoidal voltage since the harmonic voltages will usually not be nulled when balance is achieved at the fundamental frequency. A tuned detector is helpful in achieving an accurate balance.

59. Balance Convergence.[14] The process of balancing an ac bridge consists of making successive adjustments of two parameters until a null is obtained at the detector. It is desirable that these parameters not interact and that convergence be rapid.

The equation for balance can be written in terms of resistances and reactances as

$$R_x + jX_x = (R_1 + jX_1)(R_3 + jX_3)/(R_2 + jX_2)$$

Balance can be achieved by adjusting any or all of the six known parameters, but only two of them need be adjusted to achieve the required equality of both magnitude and phase (or real and imaginary components). In a ratio-type bridge, one of the arms adjacent to the unknown, either Z_1 or Z_3, is adjusted. Assuming that Z_1 is adjusted, then to make the resistance adjustment independent of the change in the corresponding reactance, the ratio $(R_3 + jX_3)/(R_2 + jX_2)$ must be either real or imaginary but not complex. If this ratio is equal to the real number k, then for balance $R_x = kR_1$ and $X_x = kX_1$. In a product-type bridge, the arm opposite the unknown, Z_2, is adjusted for balance, and the product Z_1Z_3 must be either real or imaginary to make the resistance adjustment independent of the reactance adjustment.

Near balance, the denominator of the equation giving the detector voltage (or current) changes little with the varied parameter, while the numerator changes considerably. The usual convergence loci, which consist of circular segments, can be simplified to obtain linear convergence loci by assuming that the detector voltage near balance is proportional to the numerator, $Z_1Z_3 - Z_2Z_x$. Values of this quantity can be plotted on the complex plane. When only a single adjustable parameter is varied, a straight-line locus will be produced as shown in Fig. 17-38. Varying the other adjustable parameter will produce a different straight-line locus. The rate of convergence to the origin (balance condition) will be most rapid if these two loci are perpendicular, slow if they intersect at a small angle, and zero if they are parallel. The cases of independent resistance and reactance adjustments described above correspond to perpendicular loci.

RESISTANCE BRIDGES

60. The Wheatstone bridge is used for the precise measurement of two-terminal resistances. The lower limit for accurate measurement is about 1 Ω, because contact resistance is likely to be several milliohms. For simple galvanometer detectors, the upper limit is about 1 MΩ, which can be extended to $10^{12}\Omega$ by using a high-impedance high-sensitivity detector and a guard terminal to substantially eliminate the effects of stray leakage resistance to ground.

The Wheatstone bridge (Fig. 17-39) although historically older, may be considered as a resistance version of the impedance bridge of Par. **17-54** and therefore the sensitivity equations are applicable. At balance

$$R_x = \frac{R_1 R_3}{R_2}$$

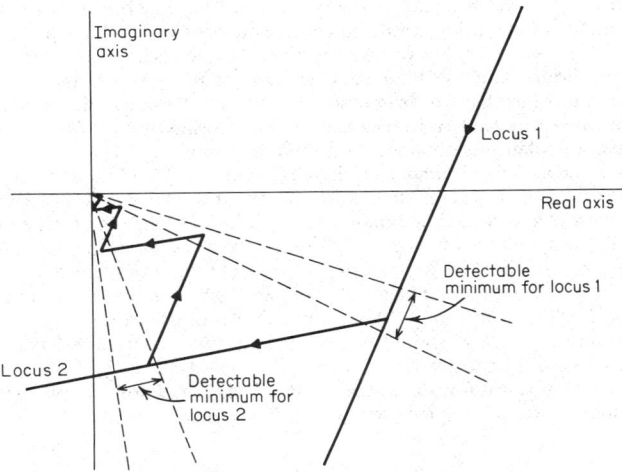

Fig. 17-38. Linearized convergence locus.[14]

Known fixed resistors, having values of 1, 10, 100, or 1,000 Ω, are generally used for two arms of the bridge, for example, R_2 and R_3. These arms provide a ratio R_3/R_2 which can be selected from 10^{-3} to 10^3. Resistor R_1, typically adjustable to 10,000 Ω in 1- or 0.1-Ω steps, is adjusted to achieve balance. The ratio R_3/R_2 should be chosen so that R_1 can be read to its full precision. The magnitudes of R_2 and R_3 should be chosen to maximize the sensitivity while taking care not to draw excessive current.

An alternate arrangement using R_1 and R_2 for the ratio resistors and adjusting R_3 for balance will generally provide a different sensitivity (see Pars. 17-56 and 17-57).

The battery key B should be depressed first to allow any reactive transients to decay before the galvanometer key is depressed. The low-galvanometer-sensitivity key L should be used until the bridge is close to balance. The high-sensitivity key H is then used to achieve final balance. Resistance R_D provides critical damping between galvanometer measurements. The battery connections to the bridge may be reversed and two separate resistance determinations made to eliminate any thermoelectric errors.

61. The Kelvin double bridge (Fig. 17-40) is used for the precise measurement of low-value four-terminal re-

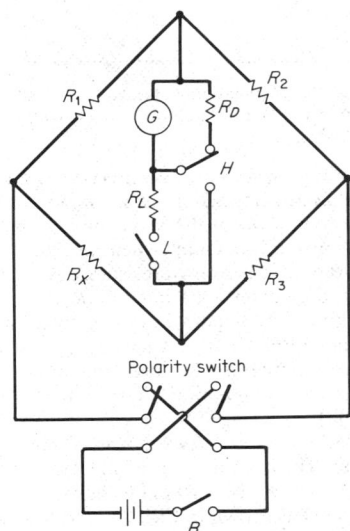

Fig. 17-39. Wheatstone bridge.

sistors in the range 1 $\mu\Omega$ to 10 Ω. The resistance to be measured X and a standard resistance S are connected by means of their current terminals in a series loop containing a battery, an ammeter, an adjustable resistor, and a low-resistance link l. Ratio-arm resistances A and B and α and β are connected to the potential terminals of resistors X and S as shown. The equation for balance is

$$X = S\frac{A}{B} + \frac{\beta l}{\alpha + \beta + l}\left(\frac{A}{B} - \frac{\alpha}{\beta}\right)$$

17-25

If the ratio α/β is made equal to the ratio A/B, the equation reduces to $X = S(A/B)$.

The equality of the ratios should be verified after the bridge is balanced by removing the link. If $\alpha/\beta = A/B$, the bridge will remain balanced. Lead resistances r_1, r_2, r_3, and r_4 between the bridge and the potential terminals of the resistors may contribute to ratio unbalance unless they have the same ratio as the arms to which they are connected. Ratio unbalance caused by lead resistance can be compensated by shunting α or β with a high resistance until balance is obtained with the link removed.

In some bridges a fixed standard resistor S having a value of the same order of magnitude as resistor X is used. Fixed resistors of 10, 100, or 1,000 Ω are used for two arms, for example, B and β, with B and β having equal values. Bridge balance is obtained by adjusting tap switches to select equal resistances for the other two arms, for example, A and α, from values adjustable up to 1,000 Ω in 0.1-Ω steps. In other bridges, only decimal ratio resistors are provided for A, B, α, and β, and balance is obtained by means of an adjustable standard having nine steps of 0.001 Ω each and a Manganin slide bar of 0.0011 Ω.

The battery connection should be reversed and two separate resistance determinations made to eliminate thermoelectric errors.

62. The dc-comparator ratio bridge (Fig. 17-41) is used for very precise measurement of four-terminal resistors. Its accuracy and stability depend mainly on the turns ratio of a

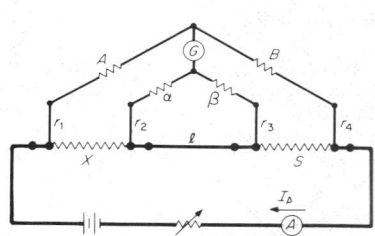

Fig. 17-40. Kelvin double bridge.[2] $A + B$ is typically 1,000 Ω, and $\alpha + \beta$ is typically 1,000 Ω.

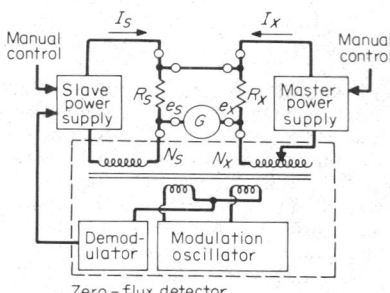

Zero – flux detector

Fig. 17-41. Comparator ratio bridge.[15]

precision transformer. The master current supply is set at a convenient fixed value I_x. The zero-flux detector maintains an ampere-turn balance, $I_x N_x = I_s N_s$, by automatically adjusting the current I_s from the slave supply as N_x is manually adjusted. A null reading on the galvanometer is obtained when $I_s R_s = I_x R_x$. Since the current ratio is precisely related to the turns ratio, the unknown resistance $R_x = N_x R_s/N_s$. Fractional turn resolution for N_x can be obtained by diverting a fraction of the current I_x as obtained from a decade current divider through an additional winding on the transformer. Turn ratios have been achieved with an accuracy of better than 1 part in 10^7. The zero-flux detector operates by superimposing a modulating mmf on the core using modulation and detector windings in a second-harmonic modulator configuration. The limit of sensitivity of the bridge is set by noise and is about 3 $\mu A \cdot$ turns.

63. Murray and Varley bridge circuits are used for locating faults in wire lines and cables. The faulted line is connected to a good line at one end by means of a jumper to form a loop. The resistance r of the loop is measured using a Wheatstone bridge. The loop is then connected as shown in Fig. 17-42 to form a bridge in which one arm contains the resistance R_x between the test set and the fault and the adjacent arm contains the remainder of the loop resistance. The galvanometer detector is connected across the open terminals of the loop, while the voltage supply is connected between the fault and the junction of fixed resistor R_2 and variable resistor R_3. When balance is attained

$$R_x = \frac{rR_3}{R_2 + R_3}$$

where r is the resistance of the loop. Resistance R_x is proportional to the distance to the fault.

In the Varley loop of Fig. 17-43, variable resistor R_1 is adjusted to achieve balance and

$$R_x = \frac{rR_3 - R_1 R_2}{R_2 + R_3}$$

where r is the resistance of the loop.

INDUCTANCE BRIDGES

64. General. Many bridge types are possible since the impedance of each arm may be a combination of resistances, inductances, and capacitances. A number of popular inductance bridges are shown in Fig. 17-44. In the balance equations L and M are given in henrys, C in farads, and R in ohms; ω is 2π times the frequency in hertz. The Q of an inductance is equal to $\omega L/R$, where R is the series resistance of the inductance.

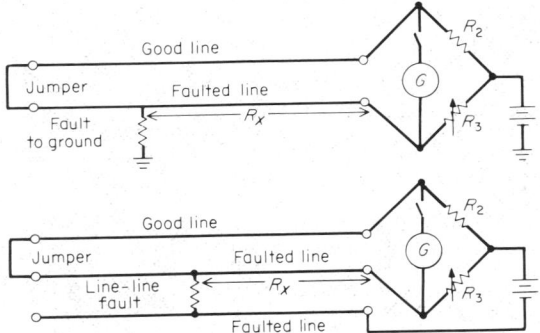

Fig. 17-42. Murray loop bridge circuits.

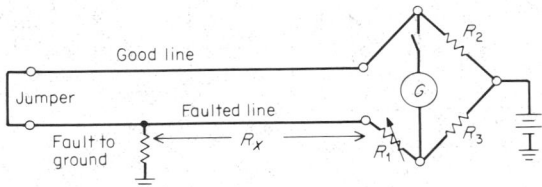

Fig. 17-43. Varley loop circuit.

65. The symmetrical inductance bridge (Fig. 17-44a) is useful for comparing the impedance of an unknown inductance with that of a known inductance. An adjustable resistance is connected in series with the inductance having the higher Q, and the inductance and resistance values of this resistance are added to those of the associated inductance to obtain the impedance of that arm. If this series resistance is adjusted along with the known inductance to obtain balance, the resistance and reactance balances are independent and balance convergence is rapid. If only a fixed inductance is available, the series resistance is adjusted along with the ratio R_3/R_2 until balance is obtained. These adjustments are interacting, and the rate of convergence will be proportional to the Q of the unknown inductance. Care must be taken to avoid inductive coupling between the known and unknown inductances since it will cause a measurement error.

66. The Maxwell-Wien bridge (Fig. 17-44b) is widely used for accurate inductance measurements. It has the advantage of using a capacitance standard which is more accurate and easier to shield and produces practically no external field. R_2 and C_2 are usually adjusted since they provide a noninteracting resistance and inductance balance. If C_2 is fixed and R_2 and R_1 or R_3 are adjusted, the balance adjustments interact and balancing may be tedious.

67. Anderson's bridge (Fig. 17-44c) is useful for measuring a wide range of inductances with reasonable values of fixed capacitance. The bridge is usually balanced by adjusting r and a resistance in series with the unknown inductance. Preferred values for good sensitivity are $R_1 = R_2 = R_3/2 = R_x/2$ and $L/C = 2R_x^2$ This bridge is also used to measure the residuals of resistors using a substitution method to eliminate the effects of residuals in the bridge elements.

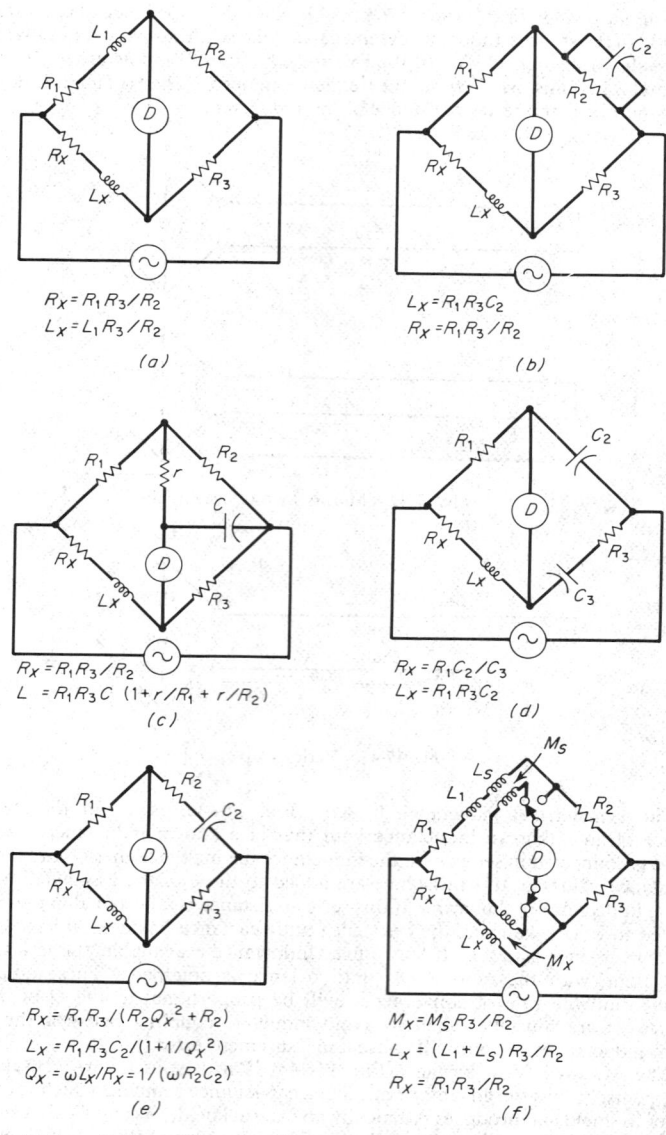

$$R_x = R_1 R_3 / R_2$$
$$L_x = L_1 R_3 / R_2$$

(a)

$$L_x = R_1 R_3 C_2$$
$$R_x = R_1 R_3 / R_2$$

(b)

$$R_x = R_1 R_3 / R_2$$
$$L = R_1 R_3 C \ (1 + r/R_1 + r/R_2)$$

(c)

$$R_x = R_1 C_2 / C_3$$
$$L_x = R_1 R_3 C_2$$

(d)

$$R_x = R_1 R_3 / (R_2 Q_x^2 + R_2)$$
$$L_x = R_1 R_3 C_2 / (1 + 1/Q_x^2)$$
$$Q_x = \omega L_x / R_x = 1 / (\omega R_2 C_2)$$

(e)

$$M_x = M_S R_3 / R_2$$
$$L_x = (L_1 + L_S) R_3 / R_2$$
$$R_x = R_1 R_3 / R_2$$

(f)

Fig. 17-44. Inductance bridges: (*a*) symmetrical inductance bridge.[3] (*b*) Maxwell-Wien bridge.[3] (*c*) Anderson's bridge.[2] (*d*) Owen's bridge.[2] (*e*) Hay's bridge.[3] (*f*) Campbell's bridge.[2]

68. Owen's bridge (Fig. 17-44d) is used to measure a wide range of inductance values in terms of resistance and capacitance. The inductance and resistance balances are independent if R_3 and C_3 are adjusted. The bridge can also be balanced by adjusting R_1 and R_3.

This bridge is useful for finding the incremental inductance of iron-cored inductors to alternating current superimposed on a direct current. The direct current may be introduced by connecting a dc-voltage source with a large series inductance across the detector branch. Low-impedance blocking capacitors are placed in series with the detector and the ac source.

69. Hay's bridge (Fig. 17-44e) is similar to the Maxwell-Wien bridge and is used for measuring inductances having large values of Q. The series R_2 C_2 arrangement permits the use of smaller resistance values than the parallel arrangement. The frequency-dependent $1/Q_x^2$ term in the inductance equation is inconvenient since the dials cannot be calibrated to indicate inductance directly unless the term is neglected, which causes a 1% error for $Q_x = 10$.

This bridge is also used for determining the incremental inductance of iron-cored reactors, as discussed for Owen's bridge.

70. Campbell's bridge (Fig. 17-44f) for measuring mutual inductance makes possible the comparison of unknown and standard mutual inductances having different values. The resistances and self-inductances of the primaries are balanced with the detector switches to the right by adjusting L_1 and R_1. The switches are thrown to the left, and the mutual-inductance balance is made by adjusting M_s. Care must be taken to avoid coupling between the standard and unknown inductances.

CAPACITANCE BRIDGES

71. Capacitance bridges are used to make precise measurements of capacitance and the associated loss resistance in terms of known capacitance and resistance values. Several different bridge circuits are shown in Fig. 17-45. In the balance equations R is given in ohms and C in farads, and ω is 2π times the frequency in hertz. The loss angle δ of a capacitor may be expressed either in terms of its series loss resistance r_s, which gives $\tan \delta = \omega C r_s$, or in terms of the parallel loss resistance r_p, in which case, $\tan \delta = 1/\omega C r_p$.

72. The series RC bridge (Fig. 17-45a) is a resistance-ratio bridge used to compare a known capacitance with an unknown capacitance. The adjustable series resistance is added to the arm containing the capacitor having the smaller loss angle δ.

73. The Wien Bridge (Fig. 17-45b) is useful for determining the equivalent capacitance C_x and parallel loss resistance R_x of an imperfect capacitor, e.g., a sample of insulation or a length of cable.

An important application of the Wien bridge network is its use as the frequency-determining network in RC oscillators (see Par. 17-95).

74. Schering's bridge (Fig. 17-45c) is widely used for measuring capacitance and dissipation factors. The unknown capacitance is directly proportional to known capacitance C_1. The dissipation factor $\omega C_x R_x$ can be measured with good accuracy using this bridge. The bridge is also used for measuring the loss angles of high-voltage power cables and insulators. In this application, the bridge is grounded at the R_2/R_3 node, thereby keeping the adjustable elements R_2, R_3, and C_2 at ground potential.

75. The transformer bridge is used for the precise comparison of capacitors, especially for three-terminal shielded capacitors.[16,17] A three-winding torroidal transformer having low leakage reactance[18] is used to provide a stable ratio, known to better than 1 part in 10^7. In Fig. 17-45d, capacitors C_1 and C_2 are being compared, and a balancing scheme using inductive-voltage dividers a and b is shown. It is assumed that $C_1 > C_2$ and loss angle $\delta_2 > \delta_1$. In-phase current to balance any inequality in magnitude between C_1 and C_2 is injected through C_5 while quadrature current is supplied by means of resistor R and current divider $C_3/(C_3 + C_4)$. The current divider permits the value of R to be kept below 1MΩ. Fine adjustments are provided by dividers a and b. N_a is the fraction of the voltage E_1 that is applied to R, while N_b is the fraction of the voltage E_2 applied to C_5. δ_1 is the loss angle of capacitor C_1 and $\tan \delta_1 = \omega C_1 r_1$, where r_1 is the series loss resistance associated with C_1. The reactance of C_3 and C_4 in parallel must be small compared with the resistance of R.

76. The substitution-bridge method is particularly valuable for determining the value of capacitance at radio frequency. *The shunt-substitution method* is shown for the series RC bridge in Fig. 17-46. Calibrated adjustable standards R_s and C_s are connected as shown, and the bridge is balanced in the usual manner with the unknown capacitance disconnected. The unknown is then connected in parallel with C_s, and C_s and R_s are readjusted to obtain balance. The unknown capacitance C_x and its equivalent series resistance R_x are determined by the rebalancing changes ΔC_s and ΔR_s in C_s and R_s, respectively: $C_x = \Delta C_s$ and $R_x = \Delta R_s (C_{s1}/C_x)^2$, where C_{s1} is the value of C_s in the initial balance.

In *series substitution* the bridge arm is first balanced with the standard elements alone, the standard elements having an impedance of Z_{s1}, and then the unknown is inserted in series with the standard elements. The standard elements are readjusted to an impedance Z_{s2} to restore balance. The unknown impedance Z_x is equal to the change in the standard impedance, that is, $Z_x = Z_{s1} - Z_{s2}$.

Measurement accuracy depends on the accuracy with which the changes in the standard values are known. The effects of residuals, stray capacitance, stray coupling, and inaccuracies in the impedances of the other three bridge arms are minimal, since these effects are the same with and without the unknown impedance. The proper handling of the leads used to connect the unknown impedance can be important.[4]

FACTORS AFFECTING ACCURACY

77. Stray Capacitance and Residuals. The bridge circuits of Figs. 17-44 and 17-45 are idealized since stray capacitances which are inevitably present and the residual inductances associated with resistances and connecting leads have been neglected. These spurious circuit elements can disturb the balance conditions and result in serious measurement errors. Detailed discussions of the residuals associated with the various bridges are given in Ref. 14.

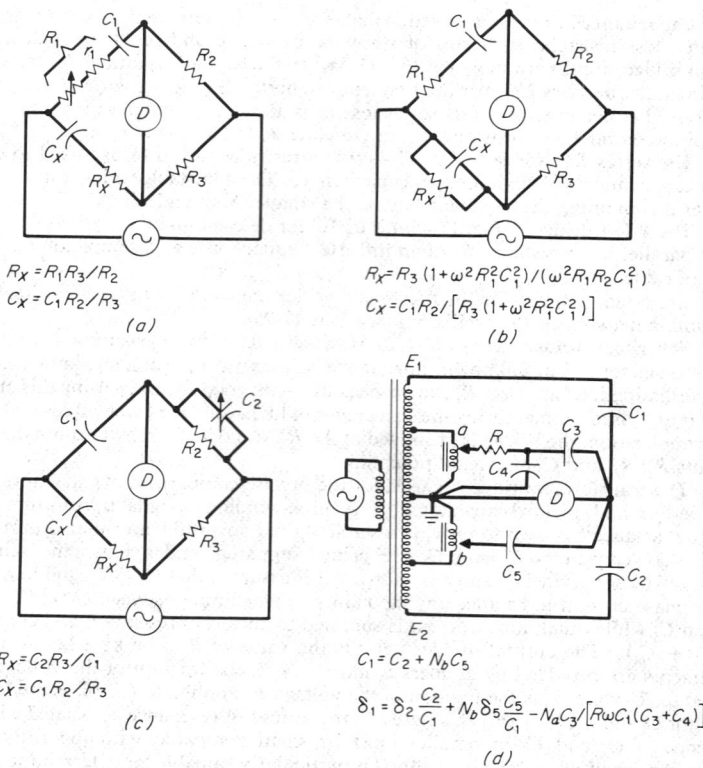

$R_x = R_1 R_3 / R_2$
$C_x = C_1 R_2 / R_3$
(a)

$R_x = R_3 (1 + \omega^2 R_1^2 C_1^2)/(\omega^2 R_1 R_2 C_1^2)$
$C_x = C_1 R_2 / [R_3 (1 + \omega^2 R_1^2 C_1^2)]$
(b)

$R_x = C_2 R_3 / C_1$
$C_x = C_1 R_2 / R_3$
(c)

$C_1 = C_2 + N_b C_5$
$\delta_1 = \delta_2 \dfrac{C_2}{C_1} + N_b \delta_5 \dfrac{C_5}{C_1} - N_a C_3 / [R \omega C_1 (C_3 + C_4)]$
(d)

Fig. 17-45. Capacitance bridges: (*a*) series-resistance-capacitance bridge,[3] (*b*) Wien bridge,[3] (*c*) Schering bridge,[3] (*d*) transformer bridge.[2]

78. Shielding and grounding can be used to control errors caused by stray capacitance. Stray capacitances in an ungrounded, unshielded series *RC* bridge are shown schematically by C_1 through C_{12} in Fig. 17-47. The elements of the bridge may be enclosed in the grounded metal shield, as shown schematically in Fig. 17-48. Shielding and grounding eliminate some capacitances and make the others definite localized capacitances which act in a known way. as illustrated in Fig. 17-49. The capacitances associated with terminal *D* shunt the oscillator and have no adverse effect. The possible adverse effects of the capacitance associated with the output diagonal *EF* are overcome by using a shielded output transformer. If the shields are adjusted so that $C_{22}/C_{21} = R_a/R_b$, the ratio of the bridge is independent of frequency. Capacitance C_{24} can be taken into account in the calibration of C_s, and capacitance C_{23} can be measured and its shunting effect across the unknown impedance can be calculated. Shielding, which is used at audio frequencies, becomes more necessary as the frequency and impedance levels are increased.

79. The Wagner ground connection (Fig. 17-50) can be used in place of shielding at lower frequencies if the utmost in precision is not required. The goal of the Wagner ground is to establish a ground connection on the oscillator in a manner that will bring the detector diagonal to ground potential. The measurement procedure is to balance the bridge as well as possible while ignoring the Wagner system. One end of the detector is then grounded as shown, and the potentiometer *P* and the balancing capacitor *C* (if present) are adjusted for a null. The detector is reconnected across the bridge, and the bridge is adjusted for a more accurate null. This process may be repeated as necessary to achieve better accuracy.

80. Guard circuits (Fig. 17-51) are often used at critical circuit points to prevent leakage currents from causing measurement errors. In an unguarded circuit surface leakage current may bypass the resistor *R* and flow through the detector *G*, thereby giving an erroneous reading. If a guard ring surrounds the positive terminal post (as in the circuit of Fig. 17-51), the surface leakage current flows through the guard ring and a noncritical return path to the

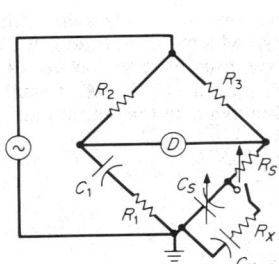

Fig. 17-46. Substitution measurement.[3]

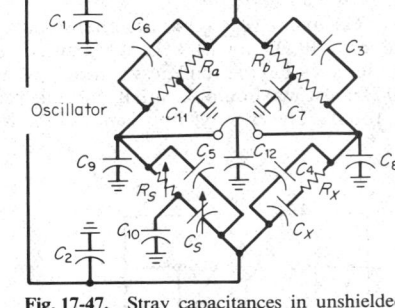

Fig. 17-47. Stray capacitances in unshielded and ungrounded bridge.[3]

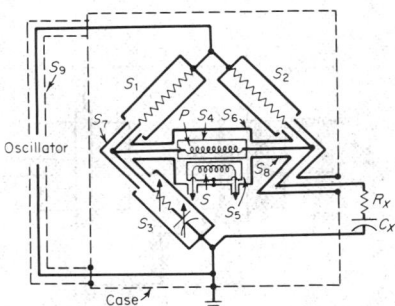

Fig. 17-48. Bridge with shields and ground.[3]

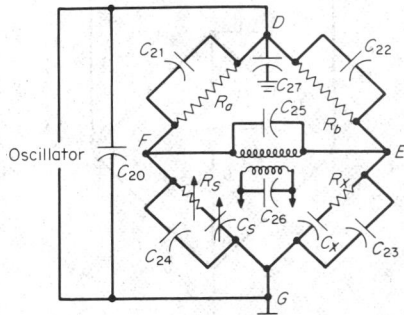

Fig. 17-49. Schematic circuit of shielded and grounded bridge.[3]

voltage source. A true reading is obtained since only the resistor current flows through the detector.

81. Coaxial leads and twisted-wire pairs may be used in connecting impedances to a bridge arm in order to minimize spurious-signal pickup from electrostatic and electromagnetic fields. It is important to keep lead lengths short, especially at high frequencies.

BRIDGE DETECTORS AND AMPLIFIERS

82. Galvanometers are used for null detection in dc bridges. The permanent-magnet moving-coil D'Arsonval galvanometer is widely used. The suspension provides a restoring torque so that the coil seeks a zero position for zero current. A mirror is used in the sensitive suspension-type galvanometer to reflect light from a fixed source to a scale. This type of galvanometer is capable of sensitivities on the order of 0.001 μA per millimeter scale division but is delicate and subject to mechanical disturbances. Galvanometers for portable instruments generally have indicating pointers and use taut suspensions which are less sensitive but more rugged and less subject to disturbances. Sensitivities are typically in the range of 0.5 μA per millimeter scale division. Galvanometers exhibit a natural mechanical frequency which depends on the suspension stiffness and the moment of inertia. Overshoot and oscillatory behavior can be avoided without an excessive increase in response time if an external resistance of the proper value to produce critical damping is connected across the galvanometer terminals.

83. Null-detector amplifiers, incorporating choppers or modulators, are used to amplify the null output signal from dc bridges to provide higher sensitivity and permit the use of rugged, less sensitive microammeter indicators. Null-detector systems are available with sensitivities of 10^{-8} V per division for a 300-Ω input impedance. Nonlinear responses can be provided so that large unbalance signals do not cause off-scale deflections. Proper design is required to avoid problems caused by zero drift, amplifier noise, thermal emfs, and stray electric and magnetic fields. Typical chopper and modulator amplifiers are described in Pars. 17-50 and 17-51.

84. Telephone receivers are often used as null detectors in ac bridges operating in the range of 200 Hz to 10 kHz. Maximum sensitivity, which depends on the receiver characteristics and the acuteness of hearing of the observer, usually occurs between 1,000 and 2,000 Hz. At maximum sensitivity, signals produced by currents of 10^{-8} A are audible in a quiet room. The effective impedance of the receiver should match the output impedance of

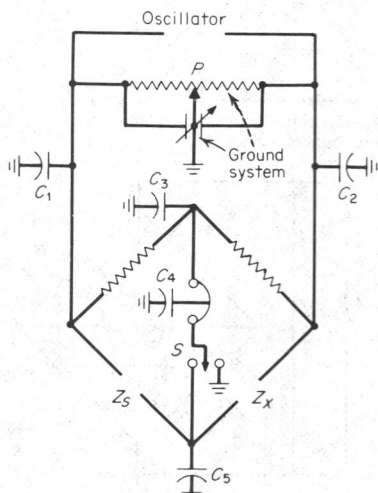

Fig. 17-50. Resistance-ratio bridge with Wagner ground connection.[3]

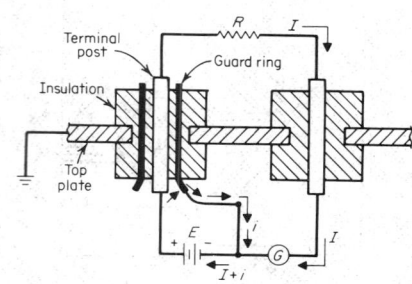

Fig. 17-51. Leakage current in guarded circuit. *(Leeds and Northrup.)*

the bridge for maximum power transfer. *Transformers* can be used for impedance matching and for overcoming the effect of stray capacitances associated with the output branch (see Par. **17-78**).

85. Vibration galvanometers provide better sensitivity than telephone receivers for frequencies below 300 Hz. The moving-coil type consists of a small coil and mirror mounted on a suspension that can be tuned by adjusting its length and tension. The coil, which is in a permanent magnet field, vibrates when alternating current flows in it. High sensitivity and frequency selectivity are obtained by tuning the lightly damped coil-suspension system to resonance at the applied frequency. The amplitude of vibration is proportional to the current.

86. Frequency-selective amplifiers are extensively used to increase the sensitivity of ac bridges. An ac amplifier with a twin-T network in the feedback loop provides full amplification at the selected frequency but falls off rapidly as the frequency is changed. Rectifiers or phase-sensitive detectors are used to convert the amplified ac signal to a direct current to drive a dc microammeter indicator. Cathode-ray-tube displays are also used to indicate the deviation from null conditions. Amplifier and detector circuits are described in Pars. **17-44** to **17-53**.

MISCELLANEOUS MEASUREMENT CIRCUITS

87. The Q meter is used to measure the quality factor Q of coils and the dissipation factor of capacitors; the dissipation factor is the reciprocal of Q. The Q meter provides a convenient method of measuring the effective values of inductors and capacitors at the frequency of interest.

The simplified circuit of a Q meter is shown in Fig. 17-52, where an unknown impedance of effective inductance L' and effective resistance r' is being measured. A sinusoidal voltage e, typically 0.01 V, is injected by L-106, in series with the circuit containing the unknown impedance and the tuning capacitor C. L-106 is selected to have a small impedance in comparison with the unknown impedance.

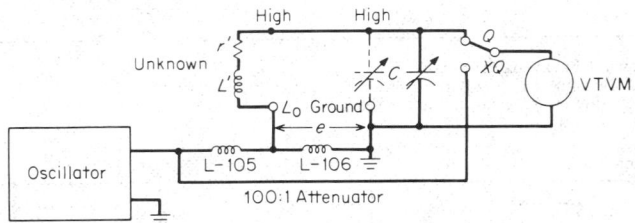

Fig. 17-52. Q meter. *(Hewlett-Packard Co.)*

Either the oscillator frequency or the tuning-capacitor value is adjusted to bring the circuit to approximate resonance, as indicated by a maximum voltage across capacitor C. At resonance $X_{L'} = X_C$, where $X_{L'} = 2\pi fL'$, $X_C = 1/(2\pi fC)$, L' is the effective inductance in henrys, C is the tuning capacitance in farads, and f is the frequency in hertz. The current at resonance is $I = e/R$, where R is the sum of the resistances of the unknown and the internal circuit. The voltage across capacitor C is $V_C = IX_C = eX_C/R$, and the indicated circuit Q is equal to V_C/e. In practice, the oscillator output voltage is adjusted to 1 V using a high-impedance voltmeter, typically a vacuum-tube voltmeter, and the indicated circuit Q is read with the voltmeter connected across C. The Q scale of the voltmeter is calibrated to read 100 for a 1-V signal. Frequency-response errors of the voltmeter tend to cancel since the ratio of V_C/e is determined. Corrections for residual resistances and reactances in the internal circuit become increasingly important at higher frequencies. It should be noted that the Q meter increases the effective Q and effective inductance of the unknown impedance (see Pars. **17-39** through **17-42**). For low values of Q, neglecting the difference between the resonance and the approximate resonance achieved by maximizing the capacitor voltage may result in an unacceptable error. Exact equations are given in Par. **17-39**.

88. Substitution-measurement methods   are used with the Q meter to cancel the effects of residual impedances (Fig. 17-53). The parallel conjunction is used for high-impedance components while the series connection is used for low impedance components. The circuit is resonated with the parallel unknown impedance removed (or series unknown impedance short-circuited), and values C_1 and Q_1 are recorded. The unknown impedance is introduced, the circuit resonated by adjusting C, and the new values C_2 and Q_2 recorded. The parameters of the unknown are given in Table 17-4.

Table 17-4.   Impedance Parameters from Parallel and Series Substitution Measurements

Parameters from Parallel Measurements	Parameters from Series Measurements
Effective Q of unknown* $$Q = \frac{Q_1 Q_2 (C_1 - C_2)}{\Delta Q C_1}$$	Effective Q of unknown* $$Q = \frac{Q_1 Q_2 (C_1 - C_2)}{C_1 Q_1 - C_2 Q_2}$$
Effective parallel resistance of unknown $$R_p = \frac{Q_1 Q_2}{\omega C_1 \Delta Q}$$	Effective series resistance of unknown $$R_s = \frac{(C_1/C_2) Q_1 - Q_2}{\omega C_1 Q_1 Q_2}$$
Effective parallel reactance of unknown† $$X_p = \frac{1}{\omega (C_2 - C_1)}$$	Effective series reactance of unknown† $$X_s = \frac{C_1 - C_2}{\omega C_1 C_2}$$
Effective parallel inductance of unknown $$L_p = \frac{1}{\omega^2 (C_2 - C_1)}$$	Effective series inductance of unknown $$L_s = \frac{C_1 - C_2}{\omega^2 C_1 C_2}$$
Effective parallel capacitance of unknown $$C_p = C_1 - C_2$$	Effective series capacitance of unknown $$C_s = \frac{C_1 C_2}{C_2 - C_1}$$

Source: Hewlett-Packard Company.
* Disregard the sign of $C_1 - C_2$ in this equation.
† A positive value indicates an inductive reactance.

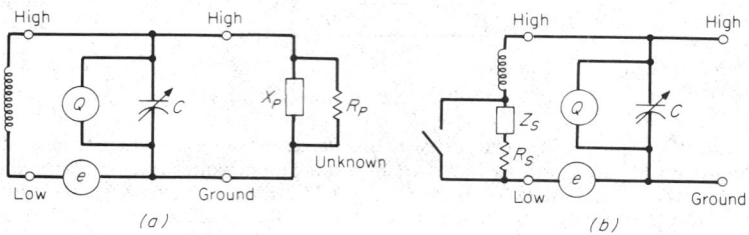

Fig. 17-53. Q-meter substitution measurements: (*a*) parallel; (*b*) series. *(Hewlett-Packard Co.)*

89. The twin-T measuring circuit   of Fig. 17-54 is used for admittance measurements at radio frequencies. This circuit operates on a null principle similar to a bridge circuit, but it has an advantage in that one side of the oscillator and detector are common and therefore can be grounded. The substitution method is used with this circuit, and therefore the effect of stray capacitances is minimized. The circuit is first balanced to a null condition with the unknown admittance $G_x + jB_x$ unconnected.

$$G_L = \omega^2 R C_1 C_2 (1 + C_a/C_3)$$
$$L = 1/[\omega^2 (C_b + C_1 + C_2 + C_1 C_2/C_3)]$$

The unknown admittance is connected to terminals a and b, and a null condition is obtained by readjusting the variable capacitors to values C'_a and C'_b. The conductance G_x and the susceptance B_x of the unknown are proportional to the changes in the capacitance settings:

$$G_x = \omega^2 R C_1 C_2 (C'_a - C_a)/C_3$$
$$B_x = \omega(C_b - C'_b)$$

90. Measurement of Coefficient of Coupling.[7] Two coils are inductively coupled when their relative positions are such that lines of flux from each coil link with turns of the other coil. The mutual inductance M in henrys can be measured in terms of the voltage e induced in one coil by a rate of change of current di/dt in the other coil; $M = -e_1/(di_2/dt) = -e_2/(di_1/dt)$. The maximum coupling between two coils of self-inductance L_1 and L_2 exists when all the flux from each of the coils links all the turns of the other coil; this condition produces the maximum value of mutual inductance, $M_{max} = \sqrt{L_1 L_2}$. The *coefficient of coupling k* is defined as the ratio of the actual mutual inductance to its maximum value; $k = M/\sqrt{L_1 L_2}$.

The value of mutual inductance can be measured using Campbell's mutual-inductance bridge (Par. **17-70**). Alternately, the mutual inductance can be measured using a self-inductance bridge. When the coils are connected in series with the mutual-inductance emf aiding the self-inductance emf (Fig. 17-55a), the total inductance $L_a = L_1 + L_2 + 2M$ is measured. With the coils connected with the mutual-inductance emf opposing the self-inductance emf (Fig. 17-55b), inductance $L_b = L_1 + L_2 - 2M$ is measured. The mutual inductance is $M = (L_a - L_b)/4$.

91. Permeameters are used to test magnetic materials. By simulating the conditions of an infinite solenoid, the magnetizing force H can be computed from the ampere-turns per unit length. When H is reversed, the change in flux linkages in a test coil induces an emf whose time integral can be measured by a ballistic galvanometer. *The Burrows permeameter* (Fig. 17-56) uses two magnetic specimen bars, S_1 and S_2, usually 1 cm in diameter and 30 cm long, joined by soft-iron yokes. High precision is obtainable for magnetizing forces up to 300 Oe. The currents in magnetizing windings M_1 and M_2 and in compensating windings A_1, A_2, A_3, and A_4 are adjusted independently to obtain uniform induction over the entire magnetic circuit. Windings A_1, A_2, A_3, and A_4 compensate for the reluctance of the joints. The reversing switches are mechanically coupled and operate simultaneously. Test coils a and c each have n turns, while each half of the test coil b has $n/2$ turns. Coils a and b are

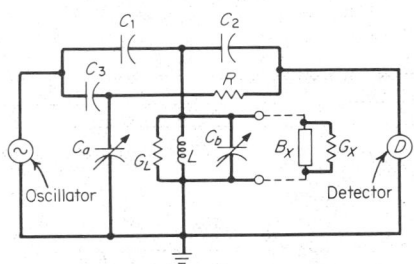

Fig. 17-54. Twin-T measuring circuit. *(General Radio Co.)*

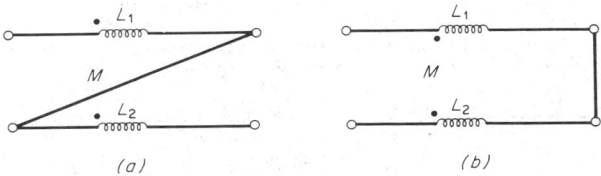

(a)   *(b)*

Fig. 17-55. Mutual inductance connected for self-inductance measurement.[2]

connected in opposing polarity to the galvanometer when the switch is in position b, while coils a and c are opposed across the galvanometer for switch position c.

Potentiometer P1 is adjusted to obtain the desired magnetizing force, and potentiometers P2 and P3 are adjusted so that no galvanometer deflection is obtained on magnetizing current reversal with the switches in either position b or position c. This establishes uniform flux density at each coil. The switch is now set at position a and the galvanometer deflection d is noted when the magnetizing current is reversed.

The values of B in gauss and H in oersteds can be calculated from

$$H = \frac{0.4\pi NI}{l} \qquad B = 10^8 \frac{dkR}{2an} - \frac{A-a}{a}H$$

where N = turns of coil M_1, I = current in coil M_1 in amperes, l = length of coil M_1 in centimeters, d = galvanometer deflection, k = galvanometer constant, R = total resistance of test coil a circuit, a = area of specimen in square centimeters, A = area of test coil in square centimeters, and n = turns in test coil a. The term $(A-a)H/a$ is a small correction term for the flux in the space between the surface of the specimen and the test coil.

Other permeameters such as the Fahy permeameter, which requires only a single specimen, the Sanford-Winter permeameter, which uses a single specimen of rectangular cross section, and Ewing's isthmus permeameter, which is useful for magnetizing forces as high as 24,000 G, are discussed in Ref. 5.

92. The frequency standard of the National Bureau of Standards is based on atomic resonance of the cesium atom and is stable to 1 part in 10^{12}. This standard, which is used to

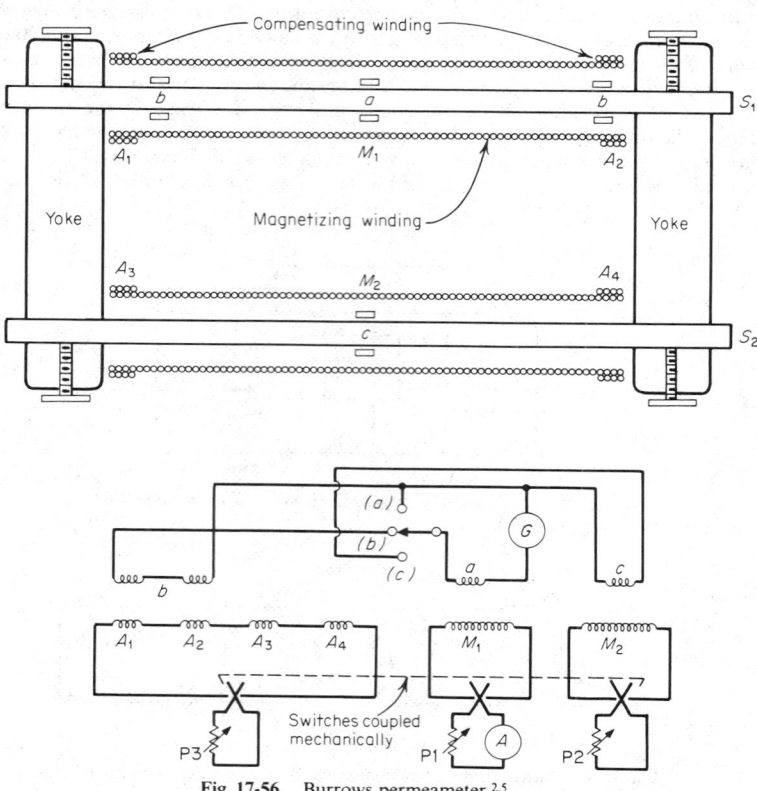

Fig. 17-56. Burrows permeameter.[2,5]

define the second, is equal to 9,192,631,770 cycles of the zero-field transition of the cesium atom. Reference frequency signals are transmitted by the National Bureau of Standards' radio station WWV at 2.5, 5, 10, 15, 20, and 25 MHz. Each frequency is modulated at 440 and 600 Hz and by pulses occurring at 1-s intervals. The carrier and modulation frequencies are accurate to better than 5 parts in 10^{11}. These frequencies are offset by a known and stable amount relative to the atomic-resonance frequency standard to reduce departures between the time scale and the astronomical scale, Universal Time 2 (see Ref. 20). Offsets, which are on the order of 3 parts in 10^8, are adjusted each year if necessary and are stated once an hour on WWV. Quartz-crystal oscillators are used as secondary standards for frequency and time-interval measurement purposes. They are periodically calibrated using the standard radio transmissions.

93. Frequency measurements can be made by comparing the unknown frequency with a known frequency, by counting cycles over a known time interval, by balancing a frequency-sensitive bridge, or by using a calibrated resonant circuit.

Frequency-comparison methods include using Lissajous patterns on an oscilloscope and heterodyne measurement methods. In Fig. 17-57, the frequency to be measured is compared with a harmonic of the 100-kHz reference oscillator. The difference frequency lying between 0 and 50 kHz is selected by the low-pass filter and compared with the output of a calibrated audio oscillator using Lissajous patterns. Alternately, the difference frequency and the audio oscillator frequency may be applied to another detector capable of providing a zero output frequency.

94. Digital frequency meters provide a convenient and accurate means for measuring frequency. The unknown frequency is counted for a known time interval, usually 1 or 10 s, and displayed in digital form. The time interval is derived by counting pulses from a quartz-crystal oscillator reference. Frequencies as high as 500 MHz can be measured by using scalers (frequency dividers). At low frequencies, for example, 60 Hz, better resolution is obtained by measuring the period $T = 1/f$. A counter with a built-in computer is available which measures the period at low frequencies and automatically calculates and displays the frequency.

95. A frequency-sensitive bridge can be used to measure frequency to an accuracy of about 0.5% if the impedance elements are known. The Wien bridge of Fig. 17-58 is commonly used, R_3 and R_4 being identical slide-wire resistors mounted on a common shaft. The equations for balance[5] are $f = 1/(2\pi\sqrt{R_3 R_4 C_3 C_4})$ and $R_1/R_2 = R_4/R_3 + C_3/C_4$.

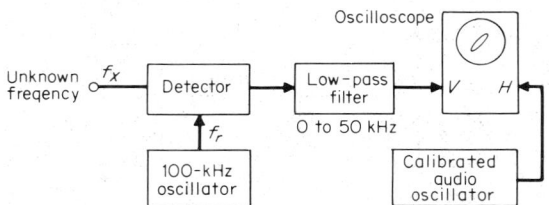

Fig. 17-57. Heterodyne frequency-comparison method.

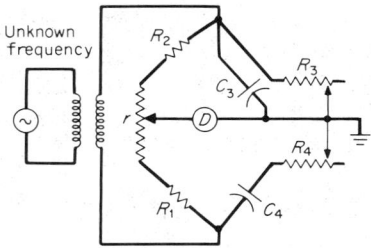

Fig. 17-58. Wien frequency bridge.[5]

In practice, the values are selected so that $R_3 = R_4$, $C_3 = C_4$, and $R_1 = 2 R_2$. Slide wire r, which has a total resistance of $R_1/100$, is used to correct any slight tracking errors in R_3 and R_4. Under these conditions $f = 1/2\pi R_4 C_4$. A filter is needed to reject harmonics if a null indicator is used since the bridge is not balanced at harmonic frequencies.

96. Time intervals can be measured accurately and conveniently by gating a reference frequency derived from a quartz-crystal oscillator standard to a counter during the time interval to be measured. Reference frequencies of 10, 1, and 0.1 MHz, derived from a 10-MHz oscillator, are commonly used.

97. Analog frequency circuits that produce an analog output proportional to frequency are used in control systems and to drive frequency-indicating meters. In Fig. 17-59, a fixed amount of charge proportional to $C_1(E - 2d)$, where d is the diode-voltage drop, is withdrawn through diode D_1 during each cycle of the input. The current through diode D_1, which is proportional to frequency, is balanced by the current through resistor R, which is proportional to e_{out}. Therefore, $e_{out} = fRC_1(E - 2d)$. Temperature compensation is achieved by adjusting the voltage E with temperature so that the quantity $E - 2d$ is constant. In the circuit of Fig. 17-60, a toroidal core having a square BH loop is alternately driven to saturation in both directions and produces a pulse during each half cycle having a fixed volt-second area equal to $2NAB_{max} \times 10^{-8}$, where N is the number of turns in each half of the secondary, A is the core area in square centimeters, and B_{max} is the saturation flux density in gauss. The pulses are rectified to provide an average voltage proportional to frequency. The rectified pulse output may be used to drive a dc meter or be filtered as shown to provide a dc output proportional to frequency.

98. Frequency analyzers are used for measuring the frequency components and analyzing the spectra of acoustic noise, mechanical vibrations, and complex electrical signals. They permit harmonic and intermodulation distortion components to be separated and measured. A simple analyzer consists of a narrow-bandwidth filter which can be adjusted in frequency or swept over the frequency range of interest. The output amplitude in decibels is generally plotted as a function of frequency using a logarithmic frequency scale. Desirable characteristics include wide dynamic range, low distortion, and high stop-band attenuation. Analog filters which operate at the frequency of interest exhibit a constant-percentage bandwidth, for example, 1%, while those using heterodyne techniques provide a constant bandwidth, for

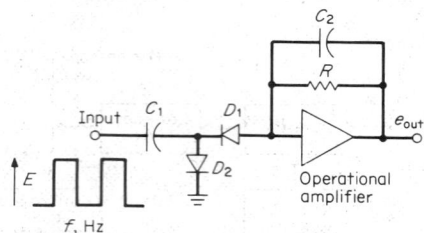

Fig. 17-59. Frequency-to-voltage converter.

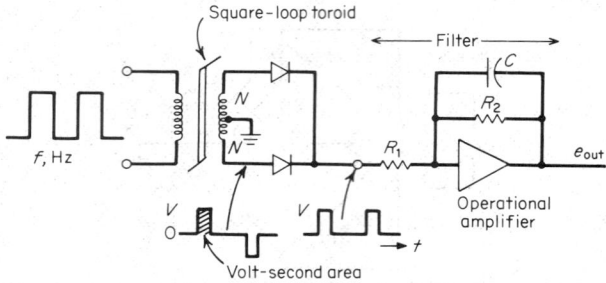

Fig. 17-60. Frequency-to-voltage converter using saturating transformer.

example, 10 Hz. The signal must be averaged over a period inversely proportional to the filter bandwidth if the reading is to be within given confidence limits of the long-time average value.[21]

99. Real-time frequency analyzers are available which perform 1/3-octave spectrum analysis on a continuous real-time basis. The analyzer of Fig. 17-61 uses 30 separate filters each having a bandwidth of 1/3 octave to achieve the required speed of response. The multiplexer sequentially samples the filter output of each channel at a high rate. These samples are converted to a binary number by the analog-to-digital converter. The true rms values for each channel are computed from these numbers during an integration period adjustable from $\frac{1}{8}$ to 32 s and stored in the memory. The rms value for each channel is computed from 1,024 samples for integration periods of 1 to 32 s.

Real-time analyzers are also available for analyzing narrow-bandwidth frequency components in real time. The required rapid response time is obtained by sampling the input waveform at three times the highest frequency of interest using an analog-to-digital converter and storing the values of a large number of samples in a digital memory. The frequency components can be calculated in real time by a dedicated digital computer using fast Fourier transforms.

100. Time-compression systems can be used to preprocess the input signal so that analog filters can be used to analyze narrow-bandwidth frequency components in real time. The time-compression system of Fig. 17-62 uses a recirculating digital memory and a digital-to-analog converter to provide an output signal having the same waveform as the input with a repetition rate which is k times faster. This multiplies the output-frequency spectrum by a factor of k and reduces the time required to analyze the signal by the same factor. The system operates as follows. A new sample is entered into the circulating memory through gate A during one of each k shifting periods. Information from the output of the memory recirculates through gate B during the remaining $k-1$ periods. Since information experiences k shifts between the addition of new samples in a memory of length $k-1$, each new sample p is entered directly behind the previous sample $p-1$, and therefore the correct order

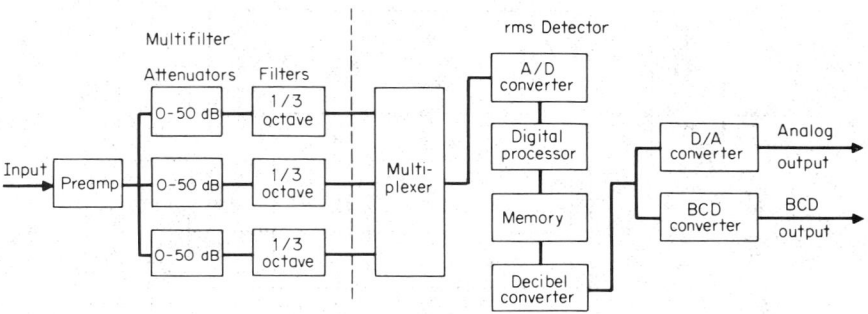

Fig. 17-61. Real-time analyzer úses 30 attenuators and filters. *(General Radio Co.)*

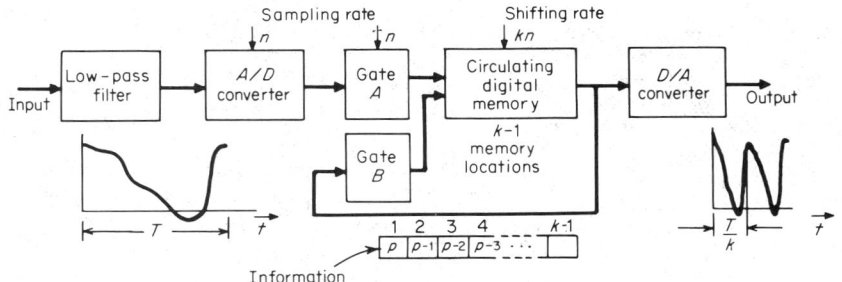

Fig. 17-62. Time-compression system.

is preserved. $(k-1)/n$ seconds is required to fill an empty memory, and thereafter the oldest sample is discarded when a new sample is entered.

101. Frequency synthesizers   provide a sinusoidal output voltage which is tunable with high resolution over a wide frequency range and yet have the stability and accuracy of a crystal oscillator reference.   They are useful for providing accurate reference frequencies and for making measurements on filter networks, tuned circuits, and communication equipment. High-precision units feature 7-decade digital frequency selection plus a continuously adjustable decade that can be manually adjusted or electrically swept over a selectable frequency range.   In the simplified block diagram of Fig. 17-63, the signal from the 5-MHz reference oscillator is sequentially processed by the digit insertion units, beginning with the least significant digit.   The frequencies in parenthesis are given for digit settings 8, 3, 5, 7, 2, 4, 6.   Phase-locked 42-MHz and 3.0-, 3.1-, . . . , 3.9-MHz signals derived from the reference oscillator are fed to all the digit-insertion units for synchronization.   All digit-insertion units are identical and process a signal near 5 MHz, as shown in Fig. 17-64.   The signal from the most significant digit-insertion unit is mixed with the 5-MHz reference to provide an output frequency between 0 and 100 kHz.   A continuously adjustable decade unit can be substituted for any digit-insertion unit if a continuously adjustable frequency is desired.   An output-frequency range of 0 to 1 MHz can be obtained by frequency-multiplying the output from the most significant decade by 10 and mixing it with a 50-MHz signal in the output mixer.

PRINCIPLES OF CONTROL SYSTEMS

TYPES OF CONTROL SYSTEMS

102. An open-loop control system   is one in which the signal controlling the output is independent of the output.   The D'Arsonval meter is an example of an open-loop system. The accuracy of an open-loop system depends on its calibration.   Changes in the characteristics of the components of the system can substantially alter the output for a given input.

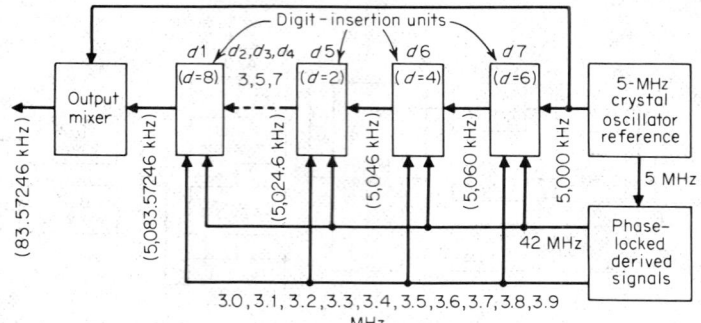

Fig. 17-63.   Coherent decade frequency synthesizer.   *(General Radio Co.)*

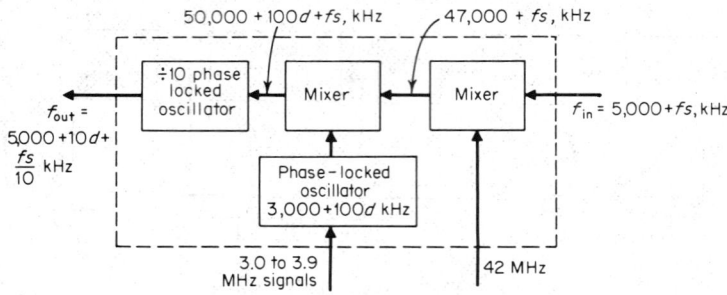

Fig. 17-64.   Digit-insertion unit.   *(General Radio Co.)*

103. A closed-loop control system is one in which the signal controlling the output depends on the output. Closed-loop systems have a number of advantages, including lower sensitivity to changes in component characteristics and partial compensation for the effects of external disturbances.

Closed-loop control systems contain a forward path and a feedback path, as shown in Fig. 17-65. The lower-case signals c (output), b (feedback signal), r (reference signal), e (error signal), m (processor output), and u (disturbance) are all functions of time. Feedback signal b, which is proportional to the output c, is compared with the reference r in the error detector to produce error signal e, where $e = r - b$. The error signal is amplified by processor g_1 to produce signal m. This signal controls the error-correcting means g_2 which produces the output c. Although the output c is relatively insensitive to disturbance u and to changes in the characteristics of the forward path elements g_1 and g_2, it is sensitive to changes in the characteristics of the output-measuring and feedback means h, the error detector, and the input-measuring means. These elements must be selected carefully if accuracy is to be achieved. Fortunately, they usually operate at low power and are usually considerably simpler and more stable than elements g_1 and g_2.

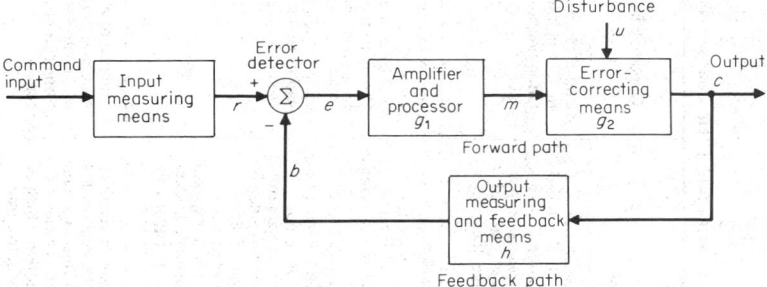

Fig. 17-65. Closed-loop control system.

The *output variable* may be any controllable quantity such as voltage, current, position, speed, torque, or temperature. When the controlled variable is a mechanical position or a time derivative of position such as velocity or acceleration, the feedback control system is called a *servomechanism*.

CLOSED-LOOP CONTROL-SYSTEM ELEMENTS

104. The error-detection subsystem consists of the input-measuring means, the output-measuring and feedback means, and the error detector. This subsystem is very important since it directly affects the accuracy of the closed-loop system. Subsystem characteristics of interest include the energy required to measure the command input, accuracy, size, reliability, linearity, noise, signal level, and resolution. The *output-measuring means* measures the output variable c or a function of it and provides an output signal which may be further processed by the feedback means before being applied to the error detector. The *input-measuring means* converts the *command input*, which may be any convenient quantity, into a suitable *reference signal* for the error detector. The *error detector* produces an error signal e proportional to the difference between the output of the input-measuring means and the feedback means. A number of error-detection subsystems are given in Table 17-5 and Fig. 17-66.

Some error-detection subsystems incorporate thermistor bridges, photoconductors, linear variable differential transformers, Hall effect devices, proximity detectors, gyroscopes, accelerometers, digital angle transducers, and heterodyne frequency-error detectors. In other error-detection subsystems, analog-to-digital converters are used to provide digital signals to error detectors which employ digital subtraction. Minicomputers can be used to perform digital subtraction and error-signal processing (see Par. 17-107).

105. The error-signal amplifier and processor increases the power level of the error signal so that it is sufficiently powerful to control the error-correcting means. The peak control-power requirement is often determined by the speed with which a mechanical device such as a throttle or valve must be operated or by the rate at which a control current must

Table 17-5. Typical Electrical Error-detecting Devices and Their Characteristics (Adapted from Ref. 22)

Number in Fig. 17-66	Type	Main application	Operation	Operating features	Accuracy limited by	Features determining energy required to vary reference quantity r
1	Dc or ac resistance bridge	Position control	Error voltage e appears when positions of moving arms of potentiometers A and B are not matched; power source E is applied across both potentiometers; A measures reference position as voltage and B regulated position as voltage, their difference being e	A and B can be remote; continuous rotation not possible	Potentiometer winding	Contact arm and bushing friction
2	Dc tachometer bridge	Speed control	Error voltage e appears when speeds of tachometers A and B vary; A measures reference speed as a voltage and B regulated speed as a voltage, their difference being e	A and B can be remote; top speed limited by commutator; A may be replaced by a potentiometer	Tachometer accuracy; commutator resistance	Brush and bearing friction
3	Ac magnetic bridge	Position control, particularly for gyro pickups, where very small forces prevail	Error voltage e appears when relative positions of rotor A and stator B do not match; rotor A measures reference position magnetically and stator B regulated position magnetically; voltage E across exciting coil L provides energy; when rotor covers unequal areas of each exposed stator pole (unbalanced magnetic bridge), pickup coils M and N have unequal voltages induced; voltage difference is e	Limited rotation; air gap usually small	Machining tolerance, magnetic fringing, and voltage-phase shift	Load taken from e; bearing friction
4	Ac synchro-system	Position control where continuous rotation is desired	Error voltage e appears whenever relative positions of rotors of synchrogenerator A and synchrocontrol transformer B are not matched; reference position is measured by A as a magnetic-flux pattern which is transmitted to the synchrocontrol transformer through interconnected stator windings; if rotor of B is not exactly 90° from transmitted flux pattern, e is produced	Unlimited rotation; synchrogenerator and control transformer can be remote	Machining tolerance, accuracy of winding distribution	Distributed or nondistributed winding of control transformer rotor; load taken from e; bearing and slipping friction

					Input impedance
5	Frequency bridge	Error voltage e appears when reference and regulated frequencies differ; transistor channel A produces a filtered sawtooth wave which gives a dc voltage inversely proportional to the reference frequency; transistor channel B produces a similar voltage as a measure of regulated frequency; difference of these dc voltages is e	A and B can be remote; transistors operate in the switching mode; wide range of frequencies can be covered	Temperature and aging effects on transistors and circuit elements	
6	Millivolt bridge	Error voltage e appears whenever regulated temperature differs from reference temperature; regulated temperature is measured as a voltage by the thermoelectric effect of two dissimilar metals B; reference temperature is represented as a voltage from battery-potentiometer source A; difference in these voltages is e	A and B can be remote; wide range of temperature can be covered	Ability to detect very low-millivolt signals	Contact arm and bushing friction; if electronic voltage source A is used, input impedance
7	Photo-transistor bridge	Error voltage e appears when movable shutter is in other than desired position; light reaching phototransistor B measures shutter position; this light is measured as a voltage by the phototransistor-current variation; A reference position of shutter is represented by battery-potentiometer voltage; difference of these voltages is e	A and B can be remote; transparent surfaces through which light travels must be kept clean	Continued accuracy of light source and phototransistor	Contact arm and bushing friction; if electronic voltage source A is used, input impedance

be changed. The frequency-response characteristic of the amplifier usually requires careful design to obtain good transient response and adequate system stability (see Par. **17-130**). Typical amplifying devices include transistors, operational amplifiers, silicon controlled rectifiers, relays, generators, and valves (see Fig. 17-67 and Pars. **17-47** and **17-141**). Amplifiers may be cascaded where high gains are required (see Table 17-6).

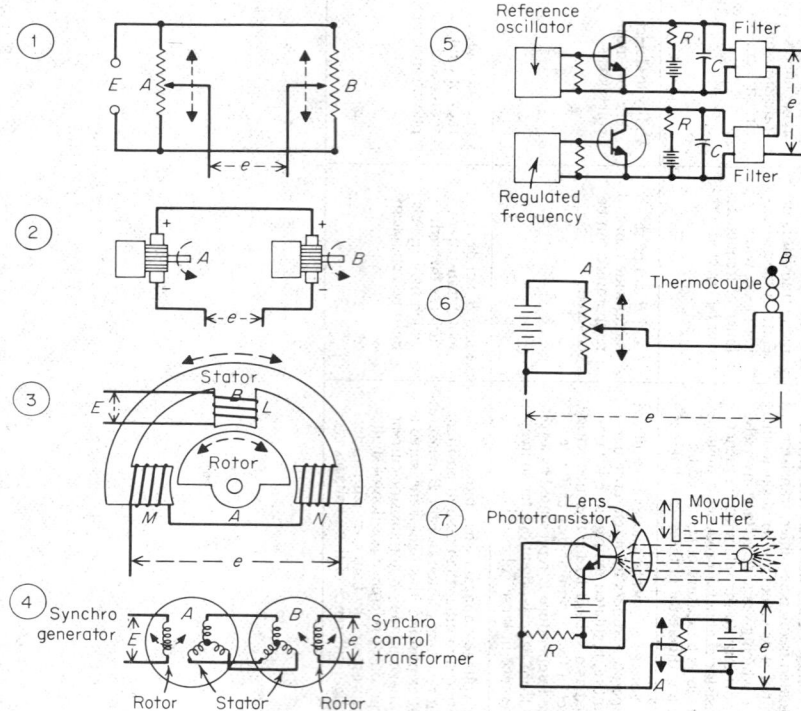

Fig. 17-66. Typical error-detection subsystems[22] (see Table 17-5).

106. The error-correcting means is a device capable of supplying the power required to change the output and thereby reduce the error. Typical devices include dc and ac motors, solenoids, stepping motors, hydraulic motors and pistons, prime movers, and fuel burners.

107. Digital signal processing may be used in implementing the error detector and processor of Fig. 17-65. An analog-to-digital converter can be used to convert the output quantity into a digital signal. The reference signal may already exist in digital form, e.g., as a number stored in a computer memory. The error is formed by digital subtraction in a minicomputer or special-purpose digital processor. Error signal processing can be performed by a digital algorithm. The processor output signal m may be provided by a digital-to-analog converter, or, alternatively, only contact-closure outputs may be provided by the processor.

TIME RESPONSE AND FREQUENCY RESPONSE

108. The time response of an element or system is often defined in terms of its response to a step input, as shown in Fig. 17-68. The *response time* is the time for the output to reach a specified value. The *rise time* is usually defined as the time required for the output to rise from 10 to 90% of the final value. *Delay time* is often specified as the time to reach 50% of the final value. The *settling time* is the time required for the output to reach and remain within

Table 17-6. **Typical Electrical Amplifiers and Their Characteristics (Adapted from Ref. 22)**

Schematic (see Fig. 17-67)	Type	Gate element	Possible input units	Possible output units	Possible power-amplification factor	Devices represented by load L	Power control
(a), (b)	Junction or field-effect transistor	Base or gate	Microwatts	Watts	1×10^5	Relay motor, generator field, impedance, solenoid	Continuous
(c)	Silicon controlled rectifier	Gate	Milliwatts	Watts or kilowatts	1×10^5	Relay motor, generator field, impedance, solenoid	Continuous
(d)	Relay	Contact	Watts	Watts or kilowatts	1×10^3	Relay motor, generator field, impedance, solenoid	On-off
(e)	Generator	Field	Watts	Watts or kilowatts	50	Motor impedance	Continuous
(f)	Saturable reactor	Dc coil	Milliwatts	Watts	3×10^2	Generator field, impedance	Continuous
(g)	Silverstat	Contacts	Grams	Watts	$1 \times 10^7 \times t$	Generator field, impedance	Stepped

a specified percentage (usually 2 or 5%) of its final value. *Overshoot* is the maximum positive value of the output minus the final output value.

A first-order system (Figs. 17-69 and 17-70) has a single pole (see Par. **17-112**), and the output response may be described in terms of the time constant τ. The *time constant* is defined as the time in seconds for the transient term to decay to $1/e$ of its initial value, where $1/e = 0.368$. A second-order system has two poles, which can be found by solving the quadratic equation. If these poles are complex, the output response may exhibit overshoot and a damped oscillatory response, as shown in Fig. 17-68. The response is a function of the location of the roots, see Par. **17-112**.

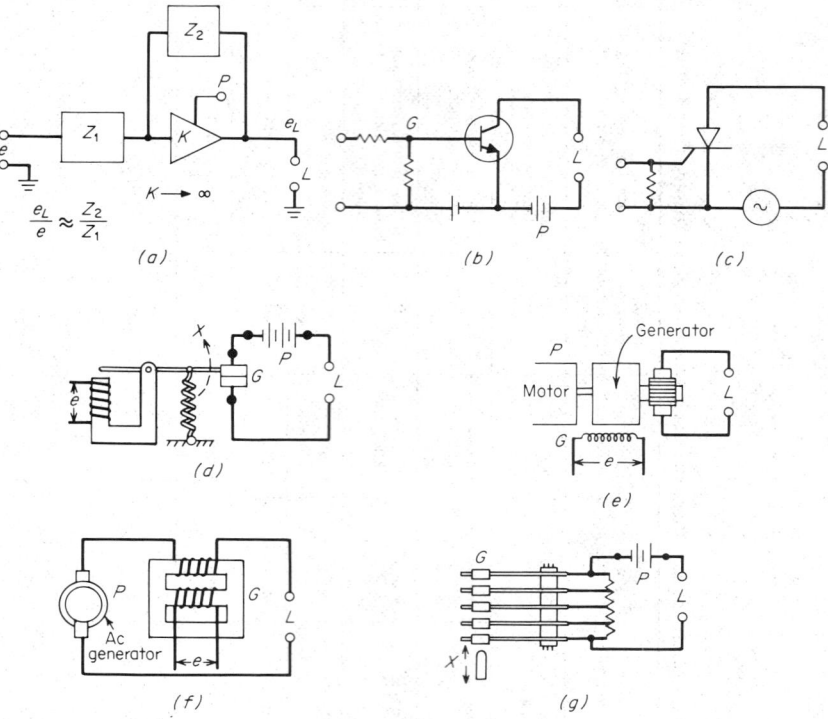

Fig. 17-67. Typical electrical amplifiers: (*a*) operational amplifier; (*b*) transistor; (*c*) silicon controlled rectifier; (*d*) relay; (*e*) generator; (*f*) saturable reactor; (*g*) silverstat[22] (see Table 17-6).

109. The frequency response of a component or system for a sinusoidal input is given by (1) the steady-state ratio of the output magnitude to the input magnitude and (2) the output-to-input phase difference for input frequencies over the range of interest. The frequency response of the *RC* circuit of Fig. 17-69 is given in Fig. 17-70. The magnitude ratio and frequency are usually plotted on logarithmic scales. The magnitude ratio M is often given in decibels M_{dB}, where $M_{dB} = 20 \log_{10} M$. The angular frequency ω in radians per second is used as the abscissa, where $\omega = 2\pi f$. The dashed lines of Fig. 17-70 show the approximate straight-line (asymptote) response. The frequency response can be found experimentally by applying a sinusoidal signal and measuring the amplitude and phase characteristics over the frequency range of interest.

110. The Laplace transform and its inverse provide a convenient means for finding the transient and steady-state response of a system. The Laplace transform $\mathscr{L}[f(t)]$ is a function $F(s)$ of the complex variable $s = \sigma + j\omega$

$$\mathscr{L}[f(t)] = F(s) = \int_{0}^{\infty} f(t)e^{-st}\, dt$$

where $f(t)$ is a function of time t defined for $t \geq 0$. The *inverse Laplace transform* $\mathscr{L}^{-1}[F(s)]$ is defined as

$$\mathscr{L}^{-1}[F(s)] = f(t) = \frac{1}{2\pi j} \int_{c-j\infty}^{c+j\infty} F(s)e^{ts}\, ds$$

where $t \geq 0$ and $c >$ real parts of all singularities of $F(s)$. [25]

It is generally not necessary to use these definitions because of the availability of *transform pairs*.[23] Table 3-4 in Sec. 3 lists a number of Laplace transform pairs. The *time response* of a system is found by (1) converting the differential equations and initial conditions describing the system into a function of the complex variable s using the Laplace transform or the transform-pair table; (2) solving for the desired output by algebraically manipulating the functions of s; (3) expressing the input signal as a function of s; (4) combining the results of steps 2 and 3 to form a final $F(s)$; (5) using the inverse Laplace transform or the transform-pair table to find the time response $f(t)$. This last step may involve using techniques (such as the partial-fraction expansion) to put $F(s)$ in a form where the terms of $F(s)$ can be matched with those of the transform-pair table (Table 3-4, Par. 3-46).

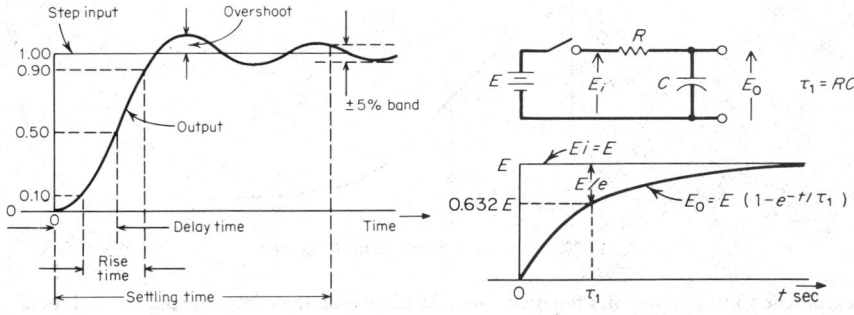

Fig. 17-68. Response to step input.   Fig. 17-69. Response of first-order system.

The step-by-step Laplace transform procedure detailed above is illustrated by finding the time response of the circuit of Fig. 17-69 using the transforms from Table 3-4, Par. 3-45.

Step 1:

$$e_0 = \frac{1}{C}\int i\, dt \rightarrow E_0 = \frac{I}{Cs} + \frac{f^{-1}(0^+)}{Cs}$$

$$e_i = iR + \frac{1}{C}\int i\, dt \rightarrow E_i = IR + \frac{I}{Cs} + \frac{f^{-1}(0^+)}{Cs}$$

where $f^{-1}(0^+)$ is the initial capacitor voltage, which for this example is assumed to be zero.

Step 2:

$$E_0 = E_i/(1 + RCs)$$

Step 3: $E_i = E/s$ for a step input of magnitude E at $t = 0$

Step 4:

$$E_0 = E/[s(1 + RCs)]$$

Step 5:   Using a partial-fraction expansion gives

$$E_0 = \frac{E}{s} - \frac{E}{s + 1/RC} \rightarrow e_0 = E - Ee^{-(t/RC)}$$

where E_i, E_0, I are functions of s, and e_i, e_0, i are functions of t.

TRANSFER FUNCTIONS

111. The transfer function $G(s)$ of a linear system is the Laplace transform of the output-to-input ratio for the condition where the initial stored energy is zero. The transfer function for the RC network of Fig. 17-70 is $E_0/E_i = 1/(1 + RCs)$, since $\tau_1 = RC$. The frequency response characteristic can be found from the transfer function by substituting $j\omega$ for s. Transfer functions of a number of networks are given in Table 17-7 of Par. 17-130.

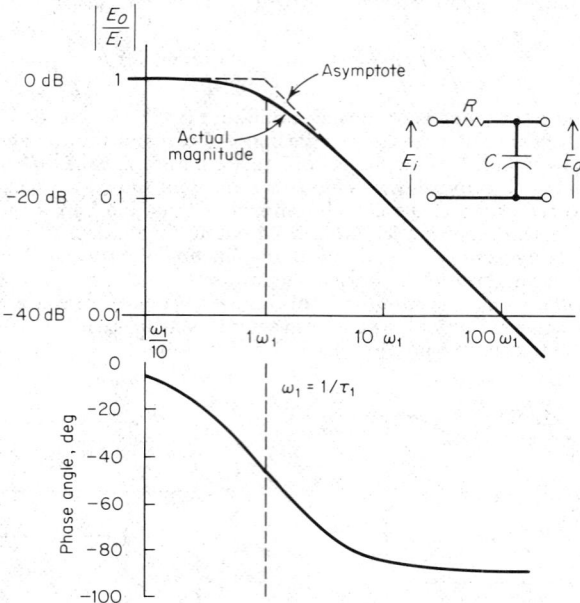

Fig. 17-70.　Frequency response of *RC* circuit.

112. The roots of a transfer function　may be plotted on the complex *s* plane. Values of *s* which make the denominator of the function equal to zero are called *poles*; values of *s* which make the numerator of the function equal to zero are called *zeros*. The first-order system of Fig. 17-69, which involves *s* to the first power, has a single pole on the real axis at $s = -1/\tau$.

A *second-order system* has a transfer function containing a denominator term which can be written as a quadratic in *s*:

$$s^2 + 2\zeta\omega_0 s + \omega_0^2$$

The poles of the transfer function are $s_1 = -\zeta\omega_0 + \omega_0\sqrt{\zeta^2 - 1}$ and $s_2 = -\zeta\omega_0 - \omega_0\sqrt{\zeta^2 - 1}$. *The undamped natural frequency* of the system is ω_0, and the *damping factor* is ζ. For $\zeta = 1$, the poles are real and equal ($s_1 = s_2 = -\omega_0$); the system is criti-

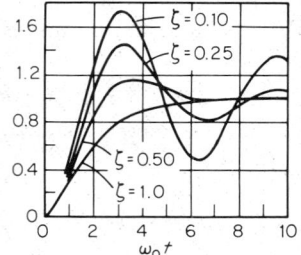

Fig. 17-71.　Unit-step-function responses.[24]

cally damped and will exhibit no overshoot. For values of $\zeta > 1$ the system is overdamped, the poles are real and unequal, and no overshoot will occur. For values of $\zeta < 1$ the poles are complex conjugates ($s = \zeta\omega_0 \pm j\omega_0\sqrt{1 - \zeta^2}$), and the system will exhibit a damped oscillatory response with overshoot. The unit-step-function responses for a second-order system having a transfer function $C/R = \omega_0^2/(s^2 + 2\zeta\omega_0 s + \omega_0^2)$ are given in Fig. 17-71. The closed-loop frequency response will exhibit a peak for low values of ζ, as shown in Fig. 17-72. For $\zeta = 0$, the undamped case, the roots lie on the imaginary axis at $\pm\omega_0$, and the response is a constant-amplitude sinusoid.

BLOCK DIAGRAMS

113. Block diagrams　containing the transfer functions of the elements of a system are useful in analyzing the characteristics of the system. The variables are given as functions of the complex operator s. The block diagram of a simple system of the form of that in Fig. 17-65 is given in Fig. 17-73. Since $C = G_1 G_2 E$ and $E = R - CH$, the closed-loop transfer function $C/R = G_1 G_2/(1 + G_1 G_2 H)$.

114. Simplification of Block Diagrams. Block diagrams can be simplified by combining blocks algebraically. Concentric feedback loops (see Fig. 17-74b) can be removed if the relationship $C/R = G/(1 + GH)$ is applied, starting with the innermost loop. If the loops

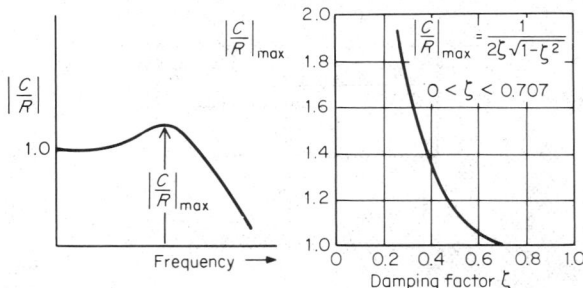

Fig. 17-72.   Frequency response corresponding to Fig. 17-71.[26]

are interconnected (see Fig. 17-74a), the diagram can usually be reduced to a concentric state by shifting s signal-pickoff point or by shifting the point at which a signal is applied to a summing junction. These rules must be followed: If a signal-pickoff branch is moved from the input to the output of a block G, the function $1/G$ must be inserted in the branch. Conversely, if the branch is moved from the output to the input of a block G, the function G must be inserted in the branch. The pickoff branch must not be moved past a summing point. A branch feeding a summing point may be moved from the input to the output of a block G if the function G is inserted in the branch. If the branch is moved from the output to the input of block G, the function $1/G$ must be inserted in the branch. The summing point must not be moved past a signal-pickoff point. An example of the reduction of a feedback system with interconnected loops is given in Fig. 17-74. Additional examples of block-diagram simplification and a systematic method for signal-flow-graph reduction are given in Refs. 25 and 26.

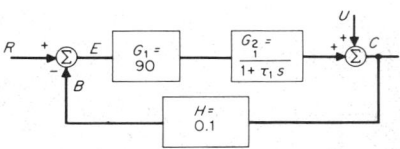

Fig. 17-73.   Closed-loop system.

SYSTEM PERFORMANCE SPECIFICATIONS

115. Performance specifications usually include the response of the system to a step input, measured in terms of response time, rise time, delay time, settling time, and overshoot, as described in Par. **17-108.** Other specifications include bandwidth, phase margin, sensitivity to gain changes, sensitivity to load disturbances, and error coefficients.

116. The bandwidth of the system is usually specified in terms of the frequency at which the magnitude of the closed-loop response is down 3 dB. Bandwidth is used as a means of specifying performance related to the speed of response. Excessive bandwidth should be avoided because noise is proportional to bandwidth.

117. Phase margin is used as a method of specifying the relative stability of the system. The phase margin is equal to 180° plus the phase angle of the open-loop transfer function GH for the frequency at which GH has unity magnitude (see Fig. 17-75).

118. The sensitivity of the system to gain changes and load disturbances depends on the loop gain at the frequency of interest. In the closed-loop system of Fig. 17-73, $C = G_1 G_2 E$ and $E = R - CH$, and therefore $C/R = G_1 G_2/(1 + G_1 G_2 H)$. If the value of the loop gain $G_1 G_2 H$ is much greater than unity, C/R is approximately equal to $1/H$, and therefore the closed-loop system is insensitive to changes in the forward-path elements G_1 and G_2. For the given example, $G_1G_2H = 9$ at low frequency, and $C/R = 9/(1 + 0.1T_1s)$. The use of feedback has reduced the time constant of the closed-loop system by a factor of $1 + G_1G_2H$. The sensitivity to changes in G_1 is reduced by the same factor; a 10% change in G_1 changes C/R by only 1%.

The sensitivity to load disturbance U in Fig. 17-73 is reduced by the factor $1 + G_1G_2H$ since the transfer function $C/U = 1/(1 + G_1G_2H)$.

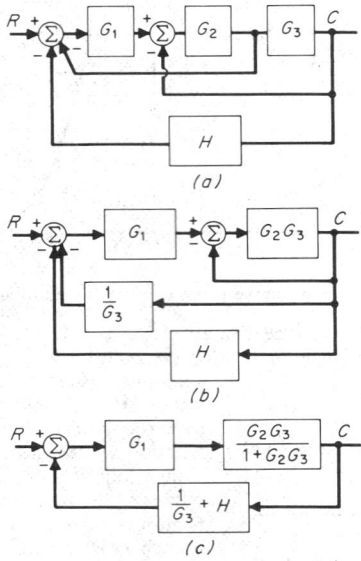

Fig. 17-74. Simplification of a block diagram.

119. Static error coefficients are used as a measure of the effectiveness of closed-loop systems for specified position, velocity, and acceleration input signals, i.e., unit-step, unit-ramp, and unit-parabola inputs, respectively. For the unity-feedback system of Fig. 17-75 these coefficients are defined as follows:

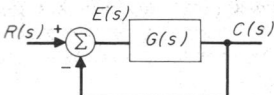

Fig. 17-75. Unity-feedback system.

$$K_p = \lim_{s \to 0} G(s) \qquad K_v = \lim_{s \to 0} sG(s) \qquad K_a = \lim_{s \to 0} s^2 G(s)$$

where K_p is the position-error coefficient, K_v is the velocity-error coefficient, and K_a is the acceleration-error coefficient.

For any given system, only one of these coefficients has a finite nonzero value. The type 0 system, which has no net poles at the origin, has a finite nonzero value of K_p, and the steady-state error for a unit-step input is equal to $1/(1 + K_p)$. The type 1 system, which has a simple pole at the origin, has a finite nonzero value of K_v, and the steady-state error for a unit-ramp input is $1/K_v$. The type 2 system, which has two poles at the origin (two integrations), has a finite nonzero value of K_a, and the steady-state error for a unit-parabola input is $1/K_a$ (see Ref. 25).

SYSTEM STABILITY

120. The stability of feedback systems is of great importance since an unstable system will not be effective in maintaining the controlled variable at approximately the desired value.

Large oscillations can have a destructive effect on the error-correcting device. Relative stability is also of importance since the performance of a system which exhibits excessive overshoot or an underdamped characteristic may be unsatisfactory in many applications.

The cause of instability can be understood by examining the closed-loop control system of Fig. 17-65. Let us open the feedback loop by disconnecting the input to block g_1 from the error-detector output. *Negative feedback* exists if in response to a positive change in the input to g_1 the error-detector output changes in a negative direction. *Positive feedback* exists if a positive change in the input to g_1 causes the error-detector output to change in the positive direction. In this latter case, if the loop is closed, any change in the input to g_1 will be reinforced by the feedback. If the loop gain $G_1 G_2 H$ with positive feedback is unity no input is required to produce an output and the system is unstable.

Phase shifts in the open-loop elements can cause a negative-feedback system to exhibit positive feedback at some frequencies. If the phase shift is $-180°$ at a given frequency, the feedback will reinforce an applied input at this frequency. The system will be unstable if the magnitude of the open loop gain $G_1 G_2 H$ is equal to unity at this frequency.

121. Transportation delays occur in distributed systems, e.g., the time delay experienced by a signal traveling along a transmission line. An ideal transportation-delay element will faithfully reproduce an input signal after a delay of T seconds. Transportation delays are detrimental to stability because they provide no attenuation and produce a phase lag that increases linearly with frequency, that is, $\theta = 2\pi f T$, where T is the transportation delay in seconds, f is in hertz, and θ is the phase lag in radians. The Laplace transform of a transportation delay of T seconds is e^{-Ts}.

122. Nonlinearities in open-loop elements will cause their characteristics to be a function of the operating point. Common forms of nonlinearity include saturation, dead band, backlash, static friction, and square-law transfer characteristics. Nonlinear systems are often analyzed by linearizations about a fixed operating point. This permits the application of linear control theory. The reader should consult Refs. 25 and 26 for techniques applicable to nonlinear systems, such as quasi linearization, describing function analysis and the phase-plane representation of system characteristics. Backlash can cause low-level oscillations to be present in systems which would otherwise be stable. Saturation can cause instability in systems which exhibit poor stability at reduced values of open-loop gain, i.e., conditionally stable systems.

123. Noise, which may be defined as any signal that does not convey useful information, must be considered in the design of closed-loop systems. Excessive noise can cause saturation in amplifying stages and thereby cause stability problems. Noise can be controlled by avoiding excessive bandwidth and by providing adequate attenuation for any frequencies at which external unwanted signals are coupled into the system.

MATHEMATICAL ANALYSIS OF LOOP STABILITY

124. The Laplace transform method of determining the stability of a linear system involves solving for the time response of the closed-loop transfer function C/R, where $C/R = G/(1 + GH)$. Simplification of the transfer functions may result in C/R being expressed in polynomial forms, i.e.,

$$\frac{C}{R} = \frac{b_0 s^m + b_1 s^{m-1} + \cdots + b_m}{a_0 s^n + a_1 s^{n-1} + \cdots + a_n}$$

The poles of C/R, that is, the values of s which make the denominator zero (which are the same values of s that make $1 + GH = 0$), determine the form of the transient terms. If all the poles lie in the left half of the s plane, i.e., have negative real parts, all exponential transient terms will decay to zero and the system will be stable. The system will be unstable if the poles lie in the right half of the s plane since the transient terms will grow exponentially. Any poles on the imaginary axis will cause a constant-amplitude sinusoidal oscillation, and the system will be unstable. The solution of the time response of the closed-loop system for a step input will provide both transient response and the steady-state response. Digital-computer routines can be used to reduce the labor involved in obtaining the solution.

125. The frequency-response method of determining stability involves plotting the magnitude and phase of the open-loop function GH for $s = j\omega$, where ω assumes all values from $-\infty$ to $+\infty$. If GH has any poles lying on the $j\omega$ axis (including the origin), these points are bypassed by making s traverse a semicircle of near-zero radius to the right of the pole. The magnitude and phase-angle information can be plotted on a polar diagram, as shown in Fig. 17-76. The plot for negative values of ω (dashed lines) can be obtained by reflecting the plot for the positive values of ω about the real axis. The mass, inertia, series inductance, and shunt capacitance of physical systems will cause GH to approach zero, as ω approaches infinity. If desired, a pole at an appropriately high frequency can be added to the GH function to represent these effects.

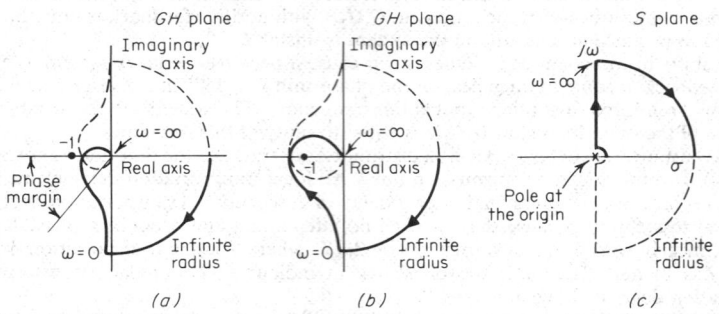

Fig. 17-76. Polar diagrams: (*a*) system stable at low gain; (*b*) system unstable at high gain; (*c*) *s*-plane contour.

126. The Nyquist criterion for a linear system which has a stable open-loop function GH can be stated as follows. The closed-loop system will be stable only if a vector from the $-1 + j0$ point to the point on the GH polar plot corresponding to a particular value of ω has a zero net counterclockwise rotation as ω varies from $-\infty$ to $+\infty$. The number of counterclockwise rotations will be equal to the number of right-half-plane poles of the closed-loop system (see Ref. 27).

If the system is open-loop unstable and GH has P poles with positive real parts, then the closed-loop system will be stable only if the vector defined above makes P counterclockwise revolutions as ω varies from $-\infty$ to $+\infty$.

127. The Bode representation, which is a direct extension of the Nyquist criterion, can be used to determine the stability of a linear closed-loop negative-feedback system which has a stable open-loop function. The magnitude and phase shift of the open-loop function GH are plotted using the same coordinates as Fig. 17-70. The phase shift should be examined at all crossover points, i.e., points at which the gain magnitude is unity (0 dB). If the magnitude passes through unity only once, the system will be stable if the phase lag at crossover is less than 180°. If the magnitude crosses unity gain at two points, the system will be stable if the phase lead is less than 180° at the low-frequency crossover point and the phase lag is less than 180° at the high-frequency crossover point. For systems having more than two crossover points, it is safer to use the Nyquist criterion.

The Bode representation provides a convenient method of analyzing simple systems and is very useful in designing stabilization networks. Detailed design information is given in Ref. 29.

128. The root-locus method is a powerful tool permitting the designer to maintain control over both the frequency response and the transient response of a feedback system. It permits the location of the closed-loop poles and zeros to be examined as a function of system gain. Various stabilization techniques can be interpreted in terms of their effect on the locations of these poles and zeros. For the negative-feedback system of Fig. 17-65, the root loci consist of all points in the *s* plane at which the phase of the open-loop transfer function GH is $180° + n360°$, where n is zero or an integer. The poles of the closed-loop system move along these loci as the open-loop gain is changed. The loci start at the open-loop poles (zero open-loop gain) and terminate at the open-loop zeros (infinite loop gain). The root loci for a simple feedback system are shown in Fig. 17-77. Rules for constructing the loci and for their use in system design are given in Ref. 25.

129. State-space methods   provide a systematic approach to the solution of large sets of differential equations describing the dynamics of a complex physical system. The matrix and vector representation is convenient for digital-computer calculations. The reader is referred to Refs. 30 to 32.

METHODS OF STABILIZATION

130. Stabilization networks   are often used to modify the open-loop frequency response to meet transient and steady-state performance requirements. A number of networks with their transfer functions and magnitude responses are given in Table 17-7.

A rule of thumb which may be used to obtain an open-loop frequency response suitable as an initial trial involves the slope of the Bode plot (gain magnitude of GH as a function of ω) near the crossover frequency ω_c. A slope of -20 dB/decade $[d(GH)/d\omega = -1]$ which extends over the frequency range from ω_c/n to $n\omega_c$ will often provide reasonable stability if $n > 3$. This rule of thumb follows from the relation between a -20 dB/decade attenuation and the resulting 90° phase lag for a minimum-phase-shift network, i.e., a network containing no right-half-plane zeros or poles. The slope at the crossover frequency ω_c has the greatest influence on the phase lag at crossover, and the slope at frequency ω becomes less important as the ratio ω/ω_c or ω_c/ω becomes large with respect to unity.

Table 17-7.   Frequency Compensation Networks

Network	Asymptotic response	Series network	Operational amplifier network
Low pass	Gain,dB ω_1 — $\omega_1 = 1/R_1C$	$\dfrac{E_{out}}{E_{in}} = \dfrac{1}{1+R_1CS}$	$\dfrac{E_{out}}{E_{in}} = -\dfrac{R_1}{R_2(1+R_1CS)}$
Lag	Gain,dB — ω_1 ω_2 — $\omega_1 = 1/(R_1+R_2)C$ — $\omega_2 = 1/R_2C$	$\dfrac{E_{out}}{E_{in}} = \dfrac{1+R_2CS}{1+(R_1+R_2)CS}$	$\dfrac{E_{out}}{E_{in}} = -\dfrac{R_1(1+R_2CS)}{R_3[1+(R_1+R_2)CS]}$
Lead	Gain,dB — ω_1 ω_2 — $\omega_1 = 1/R_1C = 1/R_aC$ — $\omega_2 = \dfrac{4R_1+R_2}{4R_1{}^2C} = \dfrac{R_a+R_b}{R_aR_bC}$	$\dfrac{E_{out}}{E_{in}} = \dfrac{R_b(1+R_aCS)}{(R_a+R_b)(1+\frac{R_aR_b}{R_a+R_b}CS)}$	$\dfrac{E_{out}}{E_{in}} = -\dfrac{4R_1R_2}{R_3(4R_1+R_2)}\left[\dfrac{1+R_1CS}{1+\frac{4R_1{}^2}{4R_1+R_2}CS}\right]$
Lag-lead	Gain,dB — ω_1 ω_2 ω_3 ω_4 — $\omega_1 = 1/(R_1+R_2)C_2 \approx 1/(R_a+R_b+R_c)C_b$ — $\omega_2 = 1/R_2C_2 = 1/R_bC_b$ — $\omega_3 = 1/(R_3+R_4)C_4 = 1/R_aC_a$ — $\omega_4 = 1/R_4C_4 \approx \dfrac{(R_a+R_b+R_c)}{R_aC_a(R_b+R_c)}$	$\dfrac{E_{out}}{E_{in}} = \dfrac{(1+R_aC_aS)(1+R_bC_bS)}{[(R_b+R_c)R_aC_aC_b]S^2 + [R_aC_a+(R_a+R_b+R_c)C_b]S+1}$	$\dfrac{E_{out}}{E_{in}} = -\dfrac{R_1}{R_3}\dfrac{(1+R_2C_2S)[1+(R_3+R_4)C_4S]}{[1+(R_1+R_2)C_2S](1+R_4C_4S)}$

The error and disturbance specifications often require a relatively high gain at very low frequency. A *lag network,* as shown in Table 17-7, is useful for providing rapid attenuation between this very low frequency and the frequency at which crossover is desired. A Bode plot illustrating the effect of the lag network on the open-loop frequency response is shown in Fig. 17-78. A *lead network,* as shown in Table 17-7, may be used to reduce the slope of the gain magnitude near crossover and thereby reduce the phase lag of the open-loop function near crossover. Lead networks must be used cautiously because they increase the gain at high frequencies and consequently increase the system susceptibility to noise. The *lead-lag network,* Table 17-7, is useful to providing a combination of the lead and lag network effects.

131. Minor-loop feedback is often used successfully to modify the frequency response of elements within the open loop. A good discussion of this technique is given in Ref. 29.

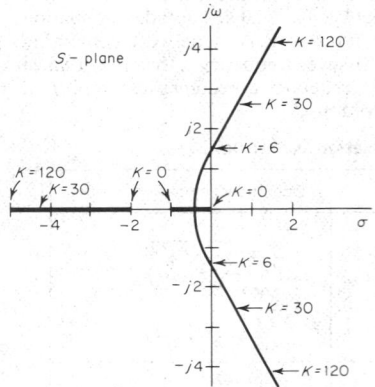

Fig. 17-77. Root-locus plot.[25]

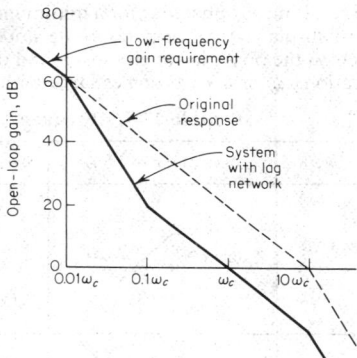

Fig. 17-78. System having lag network with breaks at $0.01\omega_c$ and $0.1\omega_c$.

AUTOMATIC CONTROL CIRCUITS

132. General. A typical feedback control system, shown in block-diagram form in Fig. 17-65, consists of an output-measuring means, an input-measuring means, an error detector, an error-signal amplifier and processor, and a final controller (error-correcting means). The components and circuits used to embody the system are discussed in this section.

133. Transducers are used to measure the input, output, and feedback quantities and to convert them into a form suitable for the error-detecting means. The resolution, accuracy, and linearity of these devices are very important because the performance of the control system can be no better than that of the measurement means. Characteristics such as input energy, size, reliability, output-signal level, and noise output are also important. The characteristics of a number of transducers are detailed in Par. **17-10**, and some typical transducers used to provide an electric error signal are shown in Fig. 17-66.

REFERENCE AND MEASURING-ELEMENT CIRCUITS

134. Reference standards are used to provide a fixed input for electrical regulator systems and transducer systems. Dc reference standards include standard cells (Par. **17-26**) and zener-diode references (Par. **17-27**). Frequency-reference standards include tuning-fork oscillators for audio frequencies and quartz-crystal oscillators for audio and radio frequencies.

135. The quartz-crystal oscillator of Fig. 17-79 is designed to operate with low-frequency duplex crystals. To ensure oscillator start-up, the transistors must be biased in the

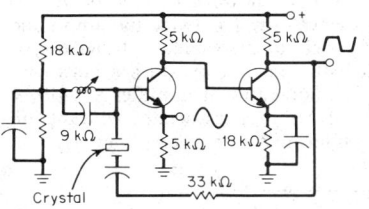

Fig. 17-79. Low-frequency crystal oscillator.

linear range under zero signal conditions and provide high loop gain. The tuned circuit provides a sinusoidal output and ensures that the crystal will oscillate only in the desired mode.

While this crystal-oscillator circuit provides an accurate frequency reference, some applications require a well-defined amplitude as well.

136. A constant-amplitude sinusoidal reference is provided by the regulator circuit of Fig. 17-80. The oscillator signal is attenuated by resistance R_1 and r_d, the dynamic impedance of diode D_1, where $r_d = kT/qI$ and k is Boltzmann's constant (1.38×10^{-23} J/°K), T is the absolute temperature in degrees Kelvin, q is the electron charge (1.6×10^{-19}C), and I is the diode current in amperes. Diode D_1 acts as a variable-gain element since its average dynamic impedance r_d is inversely proportional to the current through resistance R_2. The signal is amplified by amplifier A_1, and unwanted harmonics are attenuated by the tuned LC filter before final amplification in amplifier A_2. The output is measured by the absolute-value circuit of amplifiers A_3 and A_4 (see Ref. 33). If capacitor C_4 were removed, the waveform at point B would correspond to the instantaneous absolute value of the output. With capacitor C_4 present, the voltage at point B is equal to the average of the absolute value of the output. The voltage at point B is compared with the voltage from zener reference diode D_3 to produce an error voltage E which controls the dynamic impedance of variable-gain element D_1.

137. Measuring-element circuits in conjunction with transducers perform the function of converting outputs or inputs into electrical signals representative of measured quantities. The error-detection subsystems of Fig. 17-66 contain an input-measuring circuit and an output-measuring circuit, which are connected with their outputs in series to produce an error signal. Error signals may also be derived by parallel summing of currents, as in Fig. 17-81.

The linear variable differential transformer of Fig. 17-81 converts the mechanical-output position into sinusoidal signals e_a and e_b. The amplitude of these signals depends on the position of the magnetic core, and they are equal when the core is centered. The measuring-element circuit associated with the transducer overcomes possible summing problems, which could result from a phase difference between signals e_a and e_b, by rectifying each of these

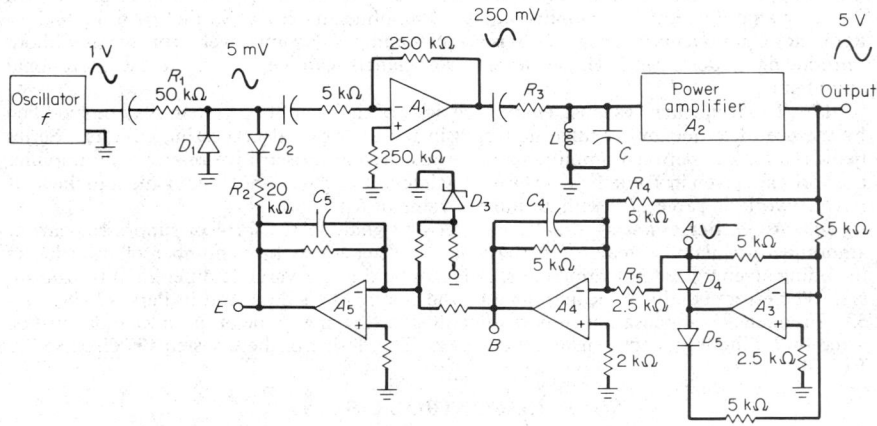

Fig. 17-80. Constant-amplitude sinusoidal reference.

signals separately before summing them. Full-wave rectification is achieved in the absolute-value circuit of operational amplifier A_1 (see Ref. 33) by combining the original sinusoidal signal with its half-wave-rectified version. The current through resistor R_2, which is proportional to the inverted half-wave-rectified waveform, has twice the amplitude as the sinusoidal current through resistor R_1. The rectifier circuit associated with signal e_a produces a positive output, while the circuit associated with signal e_b provides a negative output. These signals are applied to the summing junction of operational amplifier A_3 along with the reference-signal current from resistor R_5. Much of the ripple associated with the full-wave-rectified waveforms is removed by the low-pass filtering action of amplifier A_3.

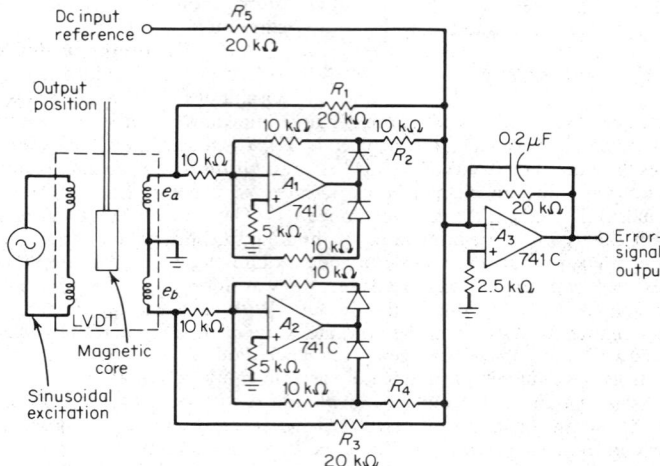

Fig. 17-81.   Linear variable differential-transformer circuit.

ERROR-SIGNAL PROCESSING CIRCUITS

138. Error-signal amplifying and processing circuits   perform the functions of increasing the voltage and power level of the error signal, providing a frequency response which is conducive to dynamic stability, and on occasion providing a specified nonlinear response characteristic. *Offset and drift* must be minimized in direct-coupled systems which operate on a dc error component, since they will cause the output to differ from the desired value. The chopper amplifiers and operational amplifiers discussed in Pars. **17-44** to **17-53** are used for error amplification. A chopper-stabilized amplifier having a parallel path for higher-frequency components (see Fig. 17-33) is useful for amplifying low-level error signals without introducing a dc offset. Higher-level error signals can be processed by operational amplifiers.

139. The frequency-response characteristics   of the open-loop system may be modified by the use of frequency-sensitive networks in the error-signal processing circuits. Series networks for use between amplifying stages and frequency-sensitive operational amplifier networks are given in Table 17-7. The analog integrator (Par. **17-52**) is equivalent to the low-pass network of Table 17-7 with an infinite value of R_1C.

140. Ac carrier systems   usually have error signals in the form of suppressed-carrier amplitude-modulated signals. These signals are generated by a number of transducers including ac-energized potentiometers, synchros, and linear variable differential transformers. The error signal may be amplified by the ac amplifiers described in Pars. **17-44** to **17-53**. Frequency compensation is more difficult since the amplitude of the sidebands must be acted on by the frequency-sensitive networks. The design of these systems is discussed in Ref. 34.

POWER-CONTROL CIRCUITS

141. Power-output circuits   are used in a wide variety of control systems (see also Sec. 15). Ac phase control is used to control the ac-power input to electric heating systems, lighting systems, arc-welding systems, solenoids, actuators, and low-power universal motors. Thyristor systems can supply regulated dc power to actuators, solenoids, battery chargers, and power supplies as well as dc motors. Inverters, which convert dc power to ac power, are used to provide controlled voltage at a controlled frequency in aircraft power systems, land-based power systems, and induction heating, as well as for induction motor drives. Details are given in Sec. **15.**

142. Thyristor firing circuits control the power output of thyristor power converters by adjusting the phase angle ϕ of the thyristor-gate firing signal with respect to the ac supply voltage. The firing angle in the circuit of Fig. 17-82 can be controlled by means of a positive input signal which determines the capacitor-charging current provided by transistor Q_2. Near the end of each half cycle the voltage V_1 falls to zero, which causes the capacitor to discharge through unijunction transistor Q_3. When voltage V_1 rises at the beginning of the next half cycle, the capacitor charges at a rate determined by transistor Q_2. When the threshold voltage of the unijunction transistor is reached, it fires and discharges the capacitor rapidly through the pulse transformer, thereby providing a pulse to the thyristor gates. The threshold voltage of the unijunction transistor V_p is equal to $\eta V_{BB} + V_D$, where $\eta = 0.75 \pm$ 10% for 2N2647, V_{BB} is the voltage between terminals $B1$ and $B2$, and V_D is a diode-voltage drop, approximately 0.5 V.[35] A simple, manually adjusted circuit can be obtained by replacing transistor Q_2 with a variable resistor.

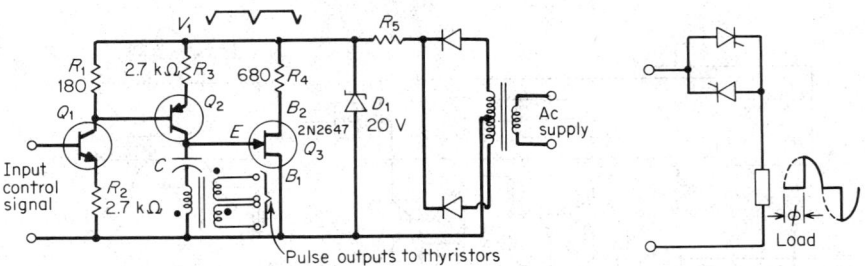

Fig. 17-82. (*a*) Unijunction firing circuit.[35] (*b*) Thyristor output circuit.

When operating with a highly inductive load or a counteremf load, the thyristor may require a longer firing pulse than available from the unijunction circuit. Furthermore, a linear transfer function from the input-control signal to the thyristor converter output can be obtained by using a cosine waveform for phase control, as shown in Fig. 17-83. When the $-\cos \omega t$ voltage to comparator A_1 becomes positive with respect to the input-control signal, differentiator C_1R_1 produces a positive pulse which triggers bistable $Q_1 - Q_2$, causing Q_1 and Q_4 to conduct, and provide a positive output to thyristor gate 1. The gate waveforms are maintained within the proper phase range by limiting the voltage excursions of the input-control signal to less than the amplitude of the end stops, which are derived from the zero crossings of the line voltage. A discussion of firing circuits for multiphase applications is given in Refs. 36 and 42.

143. Field control of a dc machine using a pair of thyristors is illustrated in the Ward-Leonard drive of Fig. 17-84. This circuit can supply only unidirectional field current but is capable of positive or negative voltage forcing to achieve rapid changes in the current. The firing pulses from Fig. 17-82 or 17-83 may be used. The *RC* network across the thyristor prevents excessive rates of change of voltage dV/dt, which could cause spurious firing. The inductance *L*, which may be supplied by transformer leakage reactance, limits the rate of rise of thyristor current to a safe value and determines the current commutation time. The dc component of the output for this two-quadrant converter, neglecting thyristor losses and commutation losses, is $(2V_{peak}\cos \phi)/\pi$, where ϕ is the firing-angle delay. The field current controls the generator output, which in turn drives the dc motor.

If a bidirectional motor drive is required, a circuit capable of supplying bidirectional field current (Fig. 17-85) should be used. The *R, L, C* transient-suppression networks associated with each thyristor have not been shown for the sake of clarity. The tapped inductance *L* is required to limit circulating current between the positive and negative current supplies. When zero field current is desired, the thyristors are fired as shown in Fig. 17-86a. For positive current the firing angles of thyristors A and B are advanced while those of thyristors C and D are retarded, as shown in Fig. 17-86b. The magnitude of the dc voltage applied to the load, neglecting commutation losses and losses in the thyristors and the current-limiting reactor, is $(2V_{peak}\cos\phi)/\pi$, where ϕ is the delay angle for firing the phase-advanced thyristor.

144. Armature control of dc motors can be accomplished by using higher-power thyristors. Motors up to 5 hp can be driven from single-phase supplies such as the center-tapped supply of Fig. 17-85 or an equivalent non-center-tapped circuit which uses a four-thyristor positive-current bridge and a four-thyristor negative-current bridge. Both these circuits and the 3-phase circuit of Fig. 17-87 are capable of driving the motor in either direction and will return power from the motor to the supply during deceleration. The circuit of Fig. 17-87 is used for driving high-power dc motors. The operation of these circuits is discussed in Refs. 36 and 37.

145. Two-phase induction motors (ac servomotors) are used as low-power actuators and are available in sizes up to 1 hp. Motors below 200 W are generally operated with a

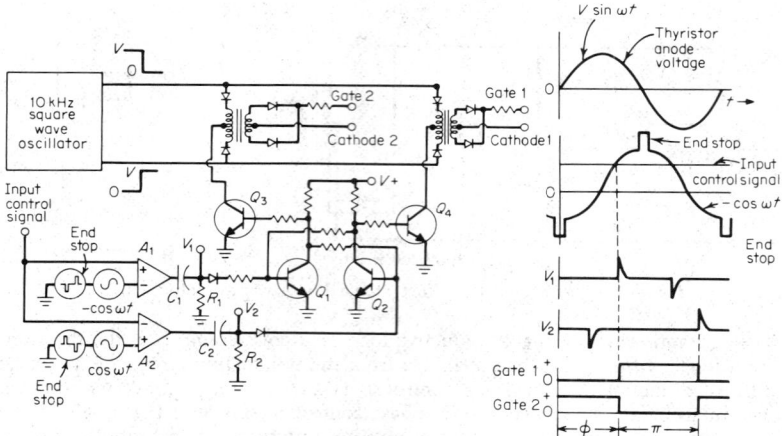

Fig. 17-83. Cosine crossing firing circuit.

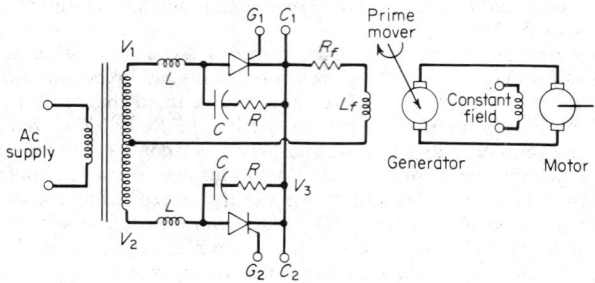

Fig. 17-84. Thyristor control of Ward-Leonard Drive.

continuously excited main winding. The control-winding voltage is applied in quadrature, with the polarity determining the direction of rotation. For a fixed main-winding voltage, the torque and speed are proportional to the control-winding voltage. The ac voltage applied to the servomotor can be adjusted by phase control of the thyristors in Fig. 17-88. Circuit *a* is suitable for low-power servomotors, while circuit *b* may be used with higher-power servomotors. Voltage control of 2-phase induction motors is inefficient because of the high slip losses.

146. **Voltage control of 3-phase induction motors** can be used in applications which have low torque requirements at low speed. The fan load, which has a load torque proportional to the square of rotor speed ω_r, is well suited to voltage control. The I^2R rotor loss is proportional to the torque multiplied by the slip S, where $S = (\omega_s - \omega_r)/\omega_s$ and ω_s is synchronous speed. Rotor loss is proportional to $S(1 - S)^2$, and maximum loss occurs for $S = \frac{1}{3}$. A detailed discussion is given in Ref. 39.

147. **Frequency control of 3-phase induction motors** provides high efficiency and full torque over a wide range of speed. The inverter of Fig. 17-89 operates from a constant dc supply, which can be obtained using a 3-phase full-wave-rectifier circuit. Dc bus commutation is provided by thyristors *A* to *D* and their associated components. Thyristors 1 to 6 and their associated diodes provide a pulse-width-modulated variable-frequency drive to the 3-

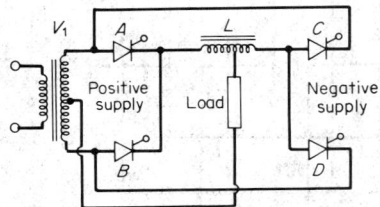

Fig. 17-85. Single-phase dual converter.

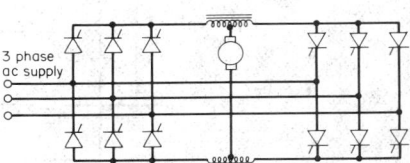

Fig. 17-87. Three-phase dual converter.

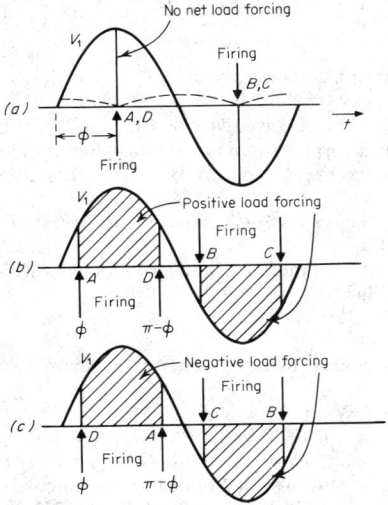

Fig. 17-86. Dual converter operation.

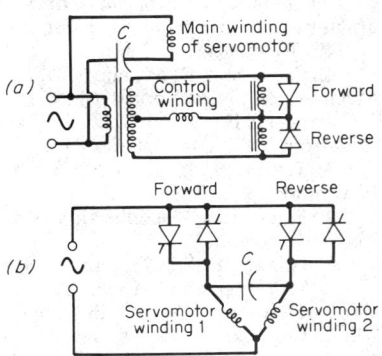

Fig. 17-88. Two-phase induction-motor servo circuits.[38]

phase induction motor. Pulse-width modulation permits voltage adjustment by time-ratio control of the conduction interval so as to maintain constant volts per hertz, which provides nearly constant motor flux. The number of pulses per cycle is increased at lower frequencies to reduce harmonics and allow higher output torque. Detailed discussions of this type of drive are given in Refs. 40 and 41.

148. Feedback signals proportional to motor speed, armature voltage, and armature current are used in torque- and speed-regulating systems. *Tachometers* can provide speed-sensing signals accurate to $\pm 0.1\%$ of full speed.

Armature voltage, from a dc motor under the condition of constant field current, can be used to derive a speed-sensing signal accurate to about $\pm 2\%$. The desired signal V_{emf}, which is proportional to speed and field current, is obtained from the armature voltage V_a by correcting for the IR drop due to armature current I_a:

$$V_{emf} = V_a - I_a R_a$$

where R_a is the resistance of the armature circuit.

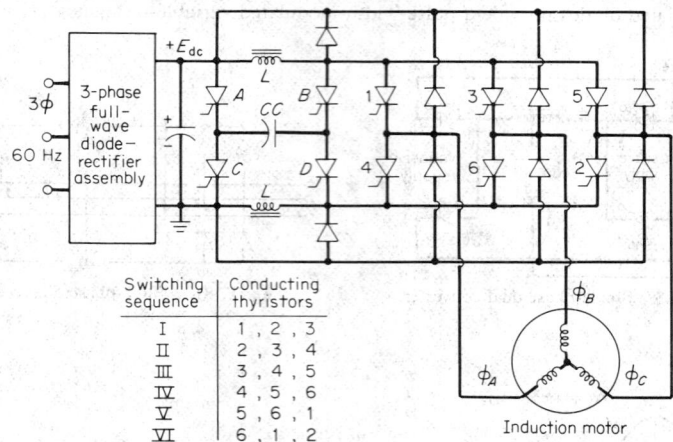

Switching sequence	Conducting thyristors
I	1 , 2 , 3
II	2 , 3 , 4
III	3 , 4 , 5
IV	4 , 5 , 6
V	5 , 6 , 1
VI	6 , 1 , 2

Induction motor

Fig. 17-89.   Forced-commutated-inverter induction-motor drive.[40]

A *current-sensing signal* for making a correction in the above equation or for use in torque control or current control can be obtained by using a current shunt or a dc transductor in the armature circuit or by using current transformers in the ac supply to the thyristors.

149. References.

1. CONSIDINE, D. M., and S. D. ROSS (eds.)   "Handbook of Applied Instrumentation," McGraw-Hill, New York, 1964.

2. FINK, D. G., and J. M. CARROLL (eds.) "Standard Handbook for Electrical Engineers," 10th ed., McGraw-Hill, New York, 1968.

3. TERMAN, F. E., and J. M. PETTIT "Electronic Measurements," 2d ed., McGraw-Hill, New York, 1952.

4. FIELD, R. F. Connection Errors in Capacitance Measurements, *Gen. Radio Exp.,* May 1947, Vol. 21.

5. HARRIS, F. K. "Electrical Measurements," Wiley, New York, 1952.

6. VIGOUREUX and WATTS, *Proc. Phys. Soc. (Lond.),* 1933, Vol. 45, p. 172.

7. TERMAN, F. E. "Radio Engineers' Handbook," McGraw-Hill, New York, 1943.

8. LAMPARD, D. G. A New Theorem in Electrostatics with Application to Calculable Standards of Capacitance, *Proc. Inst. Elec. Eng. (Lond.),* 1957, Vol. 104, Pt. C, pp. 271-280.

9. "The Royal Signals Handbook of Line Communication," H. M. Stationery Office, London, 1947.

10. Resonance Curves, *Wireless World,* January 1953, pp. 29-33.

11. WALSTON, J. A., and J. R. MILLER (eds.) "Transistor Circuit Design," McGraw-Hill, New York, 1963.

12. TOBEY, G. E., J. G. GRAEME, and L. P. HUELSMAN (eds.) "Operational Amplifiers: Design and Applications," McGraw-Hill, New York, 1971.

13. SEVIN, L. J., JR. "Field Effect Transistors," McGraw-Hill, New York, 1965.

14. HAGUE, B. "Alternating Current Bridge Methods," 5th ed., Pitman, London, 1957.

15. MACMARTIN, M. P., and N. L. KUSTERS A Direct-Current Comparator Ratio Bridge for Four-Terminal Resistance Measurements, IEEE Trans. Instrum. Meas., December 1966, Vol. IM-15, No. 4.

16. THOMPSON, A. M. The Precise Measurement of Small Capacitances, Trans. IRE, Instrumen., December 1958, Vol. I-7.

17. MCGREGOR, M. C., J. F. HERSH, R. D. CUTKOSKY, F. K. HARRIS, and F. R. KOTTER New Apparatus at the National Bureau of Standards for Absolute Capacitance Measurement, Trans. IRE, December 1958, Vol. I-7.

18. CUTKOSKY, R. D., and J. Q. SHIELDS Precise Measurement of Transformer Ratios, Trans. IRE, December 1960, Vol. I-9.

19. STOUT, M. B. "Basic Electrical Measurements," 2d ed., Prentice-Hall, Englewood Cliffs, N.J., 1960.

20. International Telephone and Telegraph Corporation "Reference Data for Radio Engineers," 5th ed., Sams, Indianapolis, 1968.

21. BLACKMAN, R. B., and J. W. TUKEY "The Measurement of Power Spectra," Dover, New York, 1958.

22. HERWALD, S. W. Forms and Principles of Servomechanisms, Westinghouse Eng., 1946, Vol. 6, pp. 149-155.

23. GARDNER, M. F., and J. L. BARNES "Transients in Linear Systems," Vol. 1, "Lumped Constant Systems," App. A, Wiley, New York, 1942.

24. JAMES, H. M., N. B. NICHOLS, and R. S. PHILLIPS "Theory of Servomechanisms," MIT Rad. Lab. Ser., Vol. 25, p. 143, McGraw-Hill, New York, 1947.

25. TRUXAL, J. G., "Automatic Feedback Control System Synthesis," McGraw-Hill, New York, 1955.

26. GRABBE, E. M., S. RAMO, and D. E. WOOLBRIDGE (eds.) "Handbook of Automation, Computation and Control," Vol. 1, "Control Fundamentals," pp. 22-02 and 22-07, Wiley, New York, 1958.

27. CHESTNUT, H., and R. W. MAYER "Servomechanisms and Regulating System Design," Vol. 1, Chap. 8, Wiley, New York, 1951.

28. THALER, G. J., and R. G. BROWN "Servomechanism Analysis," Chap. 7, McGraw-Hill, New York, 1953.

29. BOWER, J. L., and P. M. SCHULTHEISS "Introduction to the Design of Servomechanisms," Wiley, New York, 1958.

30. GANTMACHER, F. R. "The Theory of Matrices," Vols. I and II, Chelsea, New York, 1960.

31. CHEN, C. T. "Introduction to Linear System Theory," Holt, Rinehart and Winston, New York, 1970.

32. ENNS, M., J. R. GREENWOOD III, J. E. MATHESON, and F. T. THOMPSON Practical Aspects of State-Space Methods, Pt. 1, System Formulation and Reduction, IEEE Trans. Mil. Electron., April 1964, Vol. MIL-8, No, 2, pp. 81-93; Pt. 2, System Analysis and Simulation, IEEE Conf. 1964, Pap. 5, Sess. 17, pp. 501-520.

33. "Applications Manual for Operational Amplifiers," Philbrick/Nexus Research (Teledyne Philbrick), 1968.

34. IVEY, K. A. "AC Carrier Control Systems," Wiley, New York, 1964.

35. "G.E. Transistor Manual," 7th ed., 1964.

36. PELLY, B. R. "Thyristor Phase-controlled Converters and Cycloconverters," Wiley, New York, 1971.

37. BEDFORD, B. D., and R. G. Hoft "Principles of Inverter Circuits," Wiley, New York, 1964.

38. TRUXAL, J. B. "Control Engineers' Handbook," McGraw-Hill, New York, 1958.

39. PAICE, D. A. Induction Motor Speed Control by Stator Voltage Control, IEEE Trans. Power Appar. Syst. February 1968, Vol. PAS-87, No. 2.

40. DEWAN, S. B., and D. L. DUFF   Optimum Design of an Input-commutated Inverter for AC Motor Control, *IEEE Conf. Rec. 1968 IGA Group Meet., October 1968,* pp. 443-455.

41. DINGER, E. H.   Digital Applications of AC Motor Controls, *IEEE Conf. Rec. 1968 IGA Group Meet. October 1968,* pp. 271-280.

42. "G.E. SCR MANUAL," 5th ed., 1972.

43. KUSKO, A.,   "Solid-State DC Motor Drives," M.I.T. Press, Cambridge, Mass., 1969.

SECTION 18

ANTENNAS AND WAVE PROPAGATION

BY

WILLIAM F. CROSWELL   Head, Antenna Research Section, Langley Research Center, National Aeronautics and Space Administration; Senior Member, IEEE

RICHARD C. KIRBY   Director, International Radio Consultative Committee (CCIR), Geneva, Switzerland; formerly Associate Director, Office of Telecommunications, U.S. Department of Commerce, Boulder Laboratories; Fellow, IEEE

CONTENTS

Numbers refer to paragraphs

SECTION 18

ANTENNAS AND WAVE PROPAGATION

ANTENNAS

By W. F. Croswell

PROPERTIES OF ANTENNAS AND ARRAYS

1. Antenna Principles. The radiation properties of antennas can be obtained from source currents or fields distributed along a line or about an area or volume, depending upon the antenna type. The magnetic field H can be determined from the vector potential as

$$\mathbf{H} = \frac{1}{\mu} \nabla \times \mathbf{A} \qquad (18\text{-}1)$$

where μ is the permeability of the source medium. To determine the form of A first consider an infinitesimal dipole of length L and current I aligned with the z axis and placed at the center of the coordinate system given in Fig. 18-1.

The vector potential which satisfies the wave equation in this case is

$$\mathbf{A} = \mathbf{z}[\mu IL \exp(-jkr)]/4\pi r \qquad (18\text{-}2)$$

where $k = 2\pi/\lambda, j = \sqrt{-1}$, and r is the radial distance away from the origin in Fig. 18-1. Using Eqs. (18-1) and (18-2) and Maxwell's equations, the fields of a short current element are

$$H_\phi = \frac{jkIL \sin\theta}{4\pi r}\left(1 + \frac{1}{jkr}\right)\exp(-jkr)$$

$$E_\theta = \frac{jkLI\eta \sin\theta}{4\pi r}\left(1 + \frac{1}{jkr} - \frac{1}{k^2r^2}\right)\exp(-jkr) \qquad (18\text{-}3)$$

$$E_r = \frac{IL\eta}{2\pi r}\cos\theta\left(1 + \frac{1}{jkr}\right)\exp(-jkr)$$

where $\eta = \sqrt{\mu/\varepsilon}$ and ε is the permittivity of the source medium. With a transformation of coordinates, the vector potential of a short current element at an arbitrary location is

$$\mathbf{A}(x,y,z) = \frac{\mu \mathbf{I}L \exp(-jkR)}{4\pi R}$$

$$R = \sqrt{(x - x')^2 + (y - y')^2 + (z - z')^2} \qquad (18\text{-}4)$$

where I is the vector current along the dipole axis and R is the distance between the wire dipole located at (x', y', z') and the observation point (x, y, z).

By superposition, these results can be generalized to the vector of an arbitrary oriented volume-current density **J** given by

$$\mathbf{A}(x,y,z) = \frac{\mu}{4\pi}\int_{r'} \mathbf{J}(x',y',z')\frac{\exp(-jkR)}{R}dx'\,dy'\,dz' \qquad (18\text{-}5)$$

In the case of a surface current, the volume-current integral in Eq. (18-5) reduces to a surface integral of $\mathbf{J}_s[\exp(-jkR)]/R$, and for a line current reduces to a line integral of $\mathbf{I}[\exp(-jkR)]/R$. The fields of all physical antennas can be obtained from the knowledge of **J** alone. However, in the synthesis of antenna fields the concept of a magnetic volume current M is useful, even though the magnetic current is physically unrealizable.

In a homogeneous medium the electric field can be determined by

$$\mathbf{E} = -\frac{1}{\epsilon}\nabla \times \mathbf{F}, \qquad \mathbf{F} = \frac{\epsilon}{4\pi}\int_{r'}\mathbf{M}(x',y',z')\frac{\exp(-jkR)}{R}\,dx'\,dy'\,dz' \qquad (18\text{-}6)$$

The potentials for magnetic surface and line currents are determined in a manner similar to that for the electric currents. These electric and magnetic potentials are similar and are in fact duals.

Examples of antennas that have a dual property are the thin dipole in free space and the thin slot in an infinite ground plane. The fields of an electric source **J** can be determined using Eqs. (18-1) and (18-6) and Maxwell's equations. From the far-field conditions and the relationships between the unit vectors in the rectangular and spherical coordinate systems, the far fields of an electric current source **J** are

$$\eta H_\phi^J = E_\theta^J = \frac{-j\eta k\,\exp(-jkr)}{4\pi r}\int_{v'}(J_{x'}\cos\theta\cos\phi + J_{y'}\cos\theta\sin\phi -$$

$$J_{z'}\sin\theta)\exp[jk(x'\sin\theta\cos\phi + y'\sin\theta\sin\phi + \qquad (18\text{-}7)$$

$$z'\cos\theta)]\,dx'dy'\,dz'$$

$$-\eta H_\theta^J = E_\phi^J = \frac{j\eta k\,\exp(-jkr)}{4\pi r}\int_{v'}(J_{x'}\sin\phi - J_{y'}\cos\phi)\exp[jk(x'\sin\theta\cos\phi + \qquad (18\text{-}8)$$

$$y'\sin\theta\sin\phi + z'\cos\theta)]\,dx'\,dy'\,dz'$$

In a similar manner the radiated far fields from a magnetic current **M** are

$$\eta H_\phi^M = E_\theta^M = \frac{-jk\,\exp(-jkr)}{4\pi r}\int_{v'}(M_{y'}\cos\phi - M_{x'}\sin\phi)$$

$$\exp[jk(x'\sin\theta\cos\phi + y'\sin\theta\sin\phi + \qquad (18\text{-}9)$$

$$z'\cos\theta)]\,dx'\,dy'\,dz'$$

$$\eta H_\theta^M = E_\phi^M = \frac{jk\,\exp(-jkr)}{4\pi r}\int_{v'}(M_{x'}\cos\phi\cos\theta + \qquad (18\text{-}10)$$

$$M_{y'}\sin\phi\cos\theta - M_{z'}\sin\theta)\exp[jk(x'\sin\theta\cos\phi +$$

$$y'\sin\theta\sin\phi + z'\cos\theta)]\,dx'\,dy'\,dz'$$

2. Currents and Fields in an Aperture.   For aperture antennas such as horns, slots, waveguides and reflector antennas, it is sometimes more convenient or analytically simpler to calculate patterns by integrating the currents or fields over a fictitious plane parallel to the physical aperture than to integrate the source currents. Obviously, the fictitious plane can be chosen to be arbitrarily close to the aperture plane. If the integration is chosen to be an infinitesimal distance away from the aperture plane, the fields to the right of s' in Fig. 18-2 can be found using either of the equivalent currents

$$\mathbf{M}_{s'} = 2\mathbf{E}_{s'} \times \mathbf{n} \qquad (18\text{-}11a)$$

$$\mathbf{J}_{s'} = 2\mathbf{n} \times \mathbf{H}_{s'} \qquad (18\text{-}11b)$$

or

$$\mathbf{J}_{s'} = \mathbf{n} \times \mathbf{H}_{s'} \qquad \text{and} \qquad \mathbf{M}_{s'} = -\mathbf{n} \times \mathbf{E}_{s'} \qquad (18\text{-}11c)$$

If the surface s' is chosen to be away from the ground plane, the form in Eq. (18-11b) is to be used if accurate values are to be computed. The equivalent magnetic current $\mathbf{M}_{s'}$ of Eq. (18-11a) is commonly used for apertures in a very large ground plane (the effect of ground plane size is discussed later) since the tangential $\mathbf{E}_{s'}$ is zero on the ground plane and a good approximation to the aperture field can be obtained.

It should be noted that tangential $\mathbf{H}_{s'}$ is not zero on a perfectly conducting plane and therefore the equivalent $\mathbf{J}_{s'}$ form in Eq. (18-11b) is seldom used accurately in antenna problems. The combined electric and magnetic current given in Eq. (18-11c) is the general Huygens' source and is useful for aperture problems where the electric and magnetic fields are small outside the aperture; in limited cases the waveguide without a ground plane, a small horn, and a large tapered aperture can be approximated this way.

Another form of computing fields is the Stratton-Chu formulation,[1]* which in addition to currents $J_{s'}$ and $M_{s'}$ requires a knowledge of $n \cdot H_{s'}$, as shown by Tai.[2] This form therefore has no practical advantage over other forms in Eq. (18-11) although all equations give identical fields.

3. Far Fields of Particular Antennas. From the field equations stated previously or coordinate transformations of these equations, the far-field pattern of antennas can be determined when the near-field or source currents are known. Approximate forms of these fields or currents can often be estimated, giving good pattern predictions for practical purposes.

4. Electric Line Source. Consider an electric line source (current filament) of length L centered on the z' axis of Fig. 18-1 with a time harmonic-current $I(z')e^{j\omega t}$. The fields of this antenna are, from Eq. (18-7),

$$E_\theta = \frac{j\eta k \sin\theta \exp(-jkr)}{4\pi r} \int_{-L/2}^{L/2} I(z')\exp[-(jkz'\cos\theta)]\,dz'$$

$$E_\phi = 0$$

For the short dipole where $kL \ll 1$ and $I(z') = I_0$

$$E_\theta = \frac{j\eta kLI_0 \exp(-jkr)\sin\theta}{4\pi r}$$

which agrees with Eq. (18-3). Fields of other current filament antennas are given in Table 18-1.

5. Electric Current Loop. The far fields of an electric current loop of radius a, centered in the xy plane of Fig. 18-1, which has a current flowing on it can be obtained by returning to the vector potential A and deriving the expressions similar to Eqs. (18-7) and (18-8) using a potential $A_\phi = A_r$. The resulting field is

$$E_\theta = \frac{-j\eta \exp(-jkr)\cos\theta}{4\pi r} \int_0^{2\pi} I(\phi')\sin(\phi - \phi')$$

$$\exp[jka\cos(\phi - \phi')\sin\theta]d\phi'$$

$$E_\phi = \frac{-jk \exp(-jkr)\cos\theta}{4\pi r} \int_0^{2\pi} I(\phi')\cos(\phi - \phi')$$

$$\exp[jka\cos(\phi - \phi')\sin\theta]d\phi'$$

The fields for the constant current loop, $I(z') = I_0$ with a radius $a \ll \lambda$, are

$$E_\phi = \frac{k^2 a^2 \eta}{r} I_0 \exp(-jkr)\sin\theta$$

(Note that $E_\theta = 0$.)

The field of the small loop is similar to the field produced by a constant magnetic current source of length L, where $L \ll \lambda$. Due to the similarity of the pattern shape and the far-field polarization of the electric current loop and the magnetic current source, many authors have designated the loop as a magnetic source. This nomenclature is poor since it is clear that the loop is merely another wire antenna. Further characteristics of wire antennas are given in Pars. **18-22** to **18-33**.

6. Elementary Huygens' Source. Assume that constant electric and magnetic current sources $J_{x'} = J_0$ and $M_{y'} = M_0$ of equal length $L \ll \lambda$ are simultaneously placed at the origin of Fig. 18-1. If the currents are adjusted such that $\eta J_0 L = M_0 L$, then the far fields of this source are

$$E_\theta = \frac{-jk \exp(-jkr)}{4\pi r} \cos\phi\,(1 + \cos\theta)J_0 L$$

$$E_\phi = \frac{jk \exp(-jkr)}{4\pi r} \sin\phi\,(1 + \cos\theta)J_0 L$$

---

* Superior numbers correspond to numbered references in Par. **18-62.**

The unique feature of this fictitious source compared to the electric or magnetic current element alone is the obliquity factor $(1 + \cos \theta)$, which tends to cancel the far-field radiation pattern in the region $\pi/2 \leq \theta \leq \pi$ due to its cardioid shape. Aperture antennas have field distributions which can be constructed from Huygens' source elements having the patterns described above.

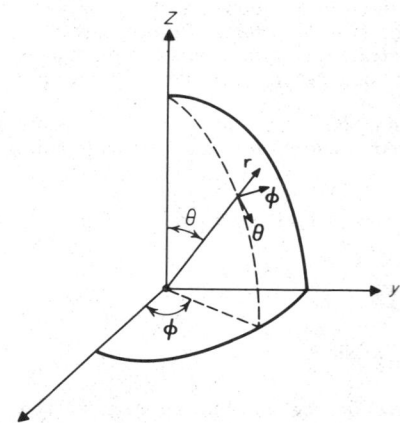

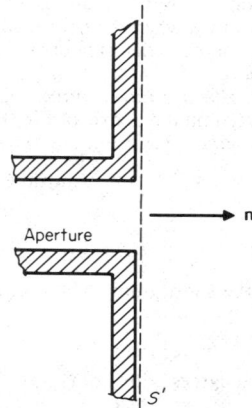

Fig. 18-1. Spherical coordinate system with unit vectors.

Fig. 18-2. Equivalent aperture plane for far-field calculations.
$$\mathbf{M} = 2\mathbf{E}'_s \times \mathbf{n}, \quad \mathbf{J} = 2\mathbf{n} \times \mathbf{H}'_s$$
$$\text{and } \mathbf{J} = \mathbf{n} \times \mathbf{H}'_s$$
$$\mathbf{M} = \mathbf{E}'_s \times \mathbf{n}$$

7. Aperture in Infinite Ground Plane. With the equivalent current $\mathbf{M}_{s'} = -2\mathbf{z} \times \mathbf{E}_{s'}$, the far field of a waveguide aperture opening onto an infinite ground plane can be obtained by integrating the aperture field since the tangential electric field is zero on the ground plane. From the magnetic current and Eqs. (18-7) to (18-10), the far-field patterns of the dominant-mode circular and rectangular waveguides can be derived as given in Table 18-1. Also given in Table 18-1 are the fields of the same antennas neglecting the ground-plane fields.

8. Simple Arrays. Consider a linear array of radiating elements which, for simplicity, are assumed to be equally spaced at a distance d apart, as illustrated in Fig. 18-3. The field at a large distance away from the mth element can be written

$$E_m = f_m(\theta, \phi) \frac{\exp(-jkr)}{r} \exp(jk_m d \cos \theta) \tag{18-12}$$

By superposition the field of an array of N elements is given by

$$E_N = \frac{-\exp(-jkr)}{r} \sum_{m=1}^{N} f_m(\theta, \phi) \exp[-j(m-1)kd \cos \theta] \tag{18-13}$$

if the element at the origin is chosen as a reference. If each element has an identical pattern, then $f_m(\theta, \phi) = a_m E(\theta, \phi)$ and

$$E_N(\theta, \phi) = E(\theta, \phi) f(\psi) \qquad f(\psi) = \sum_{m=1}^{N} a_m \exp(j\psi) \tag{18-14}$$

where a_m is a complex number representing the excitation current (voltage in the case of slots) for the mth element and $\psi = -(m-1)kd \cos \theta$. The function $f(\psi)$ is commonly called the *array factor* or *array polynomial*, and the factorization process given in Eq. (18-14) is called *pattern multiplication*.

9. Uniform Linear Array. Suppose the array in Fig. 18-3 is fed uniformly in amplitude and has a phase shift δ between adjacent elements. Here $a_m = e^{-jm\delta}$ and $\psi = kd \cos \theta - \delta$, and consequently $|f(\theta, \phi)|^2$ is

$$\text{General: } |f(\theta,\phi)|^2 = \left| \frac{\sin^2\left[\frac{N}{2}(kd \cos \theta - \delta)\right]}{N^2 \sin^2\left(\frac{kd}{2} \cos \theta - \delta\right)} \right| \tag{18-15}$$

$$\text{Broadside } \delta = 0: |f(\theta,\phi)|^2 = \left| \frac{\sin^2\left[\frac{N}{2}(kd \cos \theta)\right]}{N^2 \sin^2\left(\frac{kd}{2} \cos \theta\right)} \right| \tag{18-16}$$

$$\text{End fire } \delta = kd: |f(\theta,\phi)|^2 = \left| \frac{\sin^2\left[N\frac{kd}{2}(\cos \theta - 1)\right]}{N^2 \sin^2\left[\frac{kd}{2}(\cos \theta - 1)\right]} \right| \tag{18-17}$$

Due to the ϕ rotational symmetry in Eq. (18-15), the pattern of the line source has rotational symmetry about the z' axis. Since the cone angle of the pattern decreases with scan angle from broadside, the pattern directivity remains constant at the value N regardless of scan angle. By choosing the phase shift between elements so that $\delta = kd + 2.94/N$ the sharpest pattern in the end-fire direction ($\theta = 0$) is obtained. In this case

$$|f(\theta,\phi)|^2 = \left| \frac{\sin^2\left[\frac{Nd}{2}(k \cos \theta - k')\right]}{N^2 \sin^2\left[\frac{d}{2}(k \cos \theta - k')\right]} \right| \tag{18-18}$$

where $k' = k + 2.94/Nd$. This phase condition, which is determined graphically, is called the *Hansen-Woodward condition* for superdirectivity. This extra directivity may be several decibels in practical antennas.

10. Circular Arrays. Consider an array of N equally spaced elements about a circle of radius a in the xy plane of Fig. 18-1. If the azimuthal location of the mth element is $\phi_m = 2\pi m/N$, then for an element excitation of the form $a_m = A_{mn}e^{j\alpha m}$ the array factor is given by

$$f(\theta,\phi) = \sum_{m=1}^{N} A_m \exp\{ j[\alpha_m + ka \sin \theta \cos(\phi - \phi_m)]\} \tag{18-19}$$

For arrays having a large number of elements, $|A_m| = $ const., the array-factor pattern in the xy plane can be approximated by

$$|f(\theta, \phi = \pi/2)| \approx |J_0 [2 ka \sin \tfrac{1}{2}(\phi - \phi_0)]|$$

which is a directional beam with a maximum of $\phi = \phi_0$. If the pattern of each antenna in the circular array is of the form $F(\phi) = \sum_{m=0}^{\infty} A_m \cos^m \phi$, the pattern of a uniformly excited circular array of N such elements is approximately given by[3]

$$\Phi(\theta,\phi) \approx N \sum_{m=0}^{M} A_m(-i)^m \frac{d^m}{dz^m}[J_0(z) + 2(j)^N J_m(z) \cos N\phi] \tag{18-20}$$

where $M < N$ and $Z = ka \sin \theta$. A design curve for determining the number of sources N required in a circular array of circumference Z to produce an omnidirectional pattern with 0.5 dB ripple or less is given in Fig. 18-4. This design curve works fairly well independent of element pattern except for $F(\phi) = 1$. In this case Eq. (18-20) has infinite ripple for Z, where $J_0(Z) = 0$, independent of the number of elements in the array. This design technique has been applied successfully to several practical arrays.[4,5]

11. Planar Arrays. Now consider a planar array of equally spaced elements in the yz plane of Fig. 18-3, where the elements in the y direction are a distance d from the elements on the z axis. With the origin as a phase reference, the array factor is given by

$$f(\theta,\phi) = \sum_{n=1}^{N} \sum_{m=1}^{M} a_{mn} \exp[jk(nd\cos\theta + md\sin\theta\sin\phi)] \tag{18-21}$$

where the excitation coefficient $a_{mn} = A_{mn}\exp[jk(m\delta_y + n\delta_z)]$. If $A_{mn} = 1$ and $\delta_y = \delta_z = 0$, the array has a pattern maximum at broadside and has the array factor

$$|f(\theta,\phi)|^2 = \left| \frac{\sin^2\left(N\frac{kd}{2}\cos\theta\right)}{N^2\sin\left(\frac{kd}{2}\cos\theta\right)} \right| \left| \frac{\sin^2\left(M\frac{kd}{2}\cos\theta\sin\phi\right)}{M^2\sin^2\left(\frac{kd}{2}\cos\theta\sin\phi\right)} \right| \tag{18-22}$$

The pattern in the two principal planes of the uniformly excited array is identical in form to the linear array in the same plane. As the array is scanned from broadside, the pattern broadens and becomes asymmetrical. The directivity of the planar array (unlike that of the linear array) decreases from the broadside value by a factor $\cos\theta$.

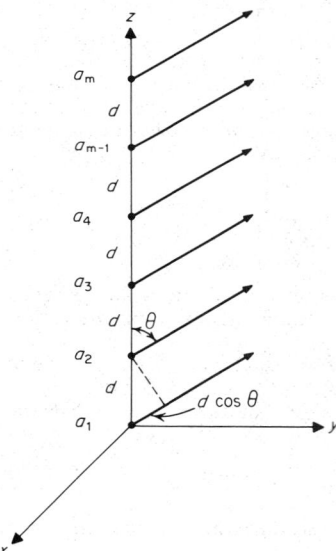

Fig. 18-3. Geometry of a linear array of equally spaced elements.

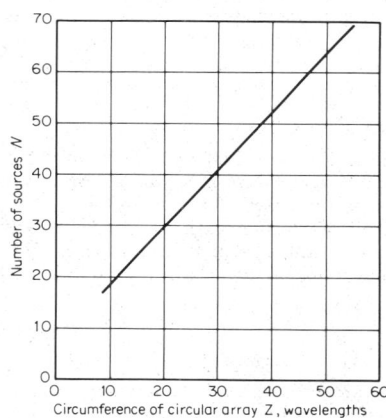

Fig. 18-4. The number of sources required in a circular array to produce an omnidirectional pattern within ±0.5 dB. For elements having circular patterns, nulls occur at $J_0(Z) = 0$, independent of the number of sources N.

12. Gain and Directivity. Antennas that are several wavelengths or larger in dimension have far-field patterns for which most of the radiated energy is restricted to narrow angular regions. Several useful measures of how the pattern is concentrated are gain, directivity, effective area, and beam efficiency. The beam efficiency is important for low-noise communication systems, microwave radiometry, and radio-astronomy antennas. The definitions of directional directivity $D(\theta,\phi)$, directivity D, gain G, and directional gain $G(\theta,\phi)$ are

$$G(\theta,\phi) = N_A(1 - |\Gamma|^2)[D(\theta,\phi)] = N_A(1 - |\Gamma|^2)\left[4\pi\frac{\text{power radiated in direction }(\theta,\phi)}{\text{total power radiated by antenna}}\right]$$

$$G = N_A(1 - |\Gamma|^2)D = N_A(1 - |\Gamma|^2)\left(4\pi\frac{\text{maximum power radiated by antenna}}{\text{total power radiated by antenna}}\right)$$

The term N_A is the antenna efficiency related to I^2R losses, and Γ is the reflection coefficient as seen at the antenna input terminals. In equation form

$$D(\theta,\phi) = \frac{4\pi|E(\theta,\phi)|^2}{\int_0^{2\pi}\int_0^\pi |E(\theta,\phi)|^2 \sin\theta\, d\theta\, d\phi} \tag{18-23}$$

where $E(\theta,\phi)$ = Electric field of the antenna

and

$$D = \frac{4\pi|E(\theta,\phi)|^2_{max}}{\int_0^{2\pi}\int_0^\pi |E(\theta,\phi)|^2 \sin\theta\, d\theta\, d\phi} \tag{18-24}$$

13. Effective Area.   The effective area of a receiving antenna is

$$A_p(\theta,\phi) = \frac{\lambda^2}{4\pi}N_A(1 - |\Gamma|^2)D(\theta,\phi)|\rho_1 \cdot \rho_2|^2 \tag{18-25}$$

where $|\rho_1, \rho_2|^2$ is the polarization loss, which accounts for the difference in polarization of the incoming wave to the polarization of the receiving antenna in direction (θ,ϕ). If the antenna polarization is matched to the incoming polarization, $|\rho_1, \rho_2|^2 = 1$; the polarization factor is described in detail in Par. **18-18**. The term Γ is the voltage-reflection coefficient observed at the antenna terminals if it is fed as a transmitting antenna. The polarization and reflection coefficient are not always included in the definition of effective area as a matter of standard definition; the standard definition is really the maximum effective area. However, in practice, these additional factors are physically present and must be included.

14. Summary: Pattern, Directivity, and Beam Width.   The radiation-pattern expressions, directivity, and beam width of typical antennas are given in Table 18-1. The factor e^{-jkr}/r is suppressed in these expressions, along with the harmonic time dependence. The half-power beam width (3 dB) is defined as the angle between half-power levels in the main lobe of a directional antenna. Another commonly used beam width is the total angle between the first nulls of the main lobe of an antenna pattern. Note that the commonly used beam width λ/D, in radians, is one-half the beam width between nulls of the main lobe of an antenna.

15. Beam Efficiency.   In addition to the gain of an antenna, the beam efficiency is a very useful parameter for judging the quality of receiving antennas intended for measurement of noise signals from an extended source. The beam efficiency is a measure of the ability of an antenna to discriminate between the received signal in the main beam and unwanted signals received through the side lobes in other directions. Assuming that the antenna aperture is in the xy plane in Fig. 18-1, the beam efficiency is defined as

$$BE = \frac{\text{power radiated in a cone angle } \theta_1}{\text{total power radiated by the antenna}}$$

$$= \frac{\int_0^{2\pi}\int_0^{\theta_1} |E(\theta,\phi)|^2 \sin\theta\, d\theta\, d\phi}{\int_0^{2\pi}\int_0^\pi |E(\theta,\phi)|^2 \sin\theta\, d\theta\, d\phi} \tag{18-26}$$

To gain some insight into how the beam efficiency varies as a function of the side-lobe level and position, calculations for a circular aperture with symmetrical aperture distributions of the form $f(\rho') = [1 - (\rho'/a)^2]^p$ are presented along with similar calculations for the rectangular aperture with $\cos^n\psi$ distribution in Fig. 18-5. It is interesting to note that the beam efficiency does not reach 90% until the angle $2\theta_1$, which is about 2 to 3 times the beam-width angle.

Table 18-1. Gains, Patterns, Side-Lobe Level, and Beam Widths of Typical Antenna

Antenna type	Description	Pattern Expressions	Directivity D	First side lobe, dB	Beam width, deg	
					3 dB	First nulls
Short dipole electric or magnetic		Electric: $E_\theta = \dfrac{j_\eta kLI_0}{4\pi r}\exp(-jkr)\sin\theta$	$3/2$	...	90	
		Magnetic: $E_\phi = \dfrac{-jk^2 L}{4\pi r}M_0\exp(-jkr)\sin\theta$	$3/2$	...	90	
Dipole, $I(z') = I_0\sin k(L-|z'|)$		$E_\theta = j_\eta \dfrac{I_0\exp(-jkr)}{2\pi r}\dfrac{\cos(kL\cos\theta)-\cos kL}{\sin\theta}$ for $L=\lambda/4$: $E_\theta = j_\eta \dfrac{I_0\exp(-jkr)}{2\pi r}\dfrac{\cos(\pi/2\cos\theta)}{\sin\theta}$	1.64	...	78	
Small loop, $I(\phi') = I_0$, $a < \lambda$		$E_\phi = \dfrac{\eta ka}{2r}I_0\exp(-jkr)J_1(ka\sin\theta)$ or $E_\phi \approx \dfrac{k^2 a^2}{r}\eta I_0\exp(-jkr)\sin\theta$	$3/2$	...	90	
Annular slot in a ground plane, $V_0 = E_0 b$, $a,b \ll \lambda$		$E_\theta = \dfrac{kV_0 a}{r}\exp(-jkr)J_1(ka\sin\theta)$ or $E_\theta \approx \dfrac{kV_0 a}{r}\exp(-jkr)\dfrac{ka}{2}\sin\theta$	$3/2$	...	90	
Thin half-wave slot in a ground plane, $V_0 = 2aE_0$, $V(x') = V_0\cos\dfrac{\pi x'}{b}$		$E_\phi = \dfrac{-jV_0}{\pi r}\exp(-jkr)\cos\left(\dfrac{\pi}{2}\sin\theta\right)$ yz plane $E_\theta = j\dfrac{V_0}{\pi r}\exp(-jkr)$ xz plane	1.64	...	78	

Pattern	Electric field	Directivity	Minor-lobe (dB)	Half-power beamwidth	Beamwidth between first nulls
Rectangular aperture, TE_{01} mode: E, H = 0 outside aperture, $2a$, $2b > \lambda$	$E_\theta = \dfrac{jk\exp(-jkr)}{r}E_0(1+\cos\theta)4ab\dfrac{\sin(ka\sin\theta)}{ka\sin\theta}$ yz plane $E_\phi = \dfrac{jk\exp(-jkr)}{2r}E_0(1+\cos\theta)2ab\dfrac{\cos(kb\sin\theta)}{\pi^2-(kb\sin\theta)^2}$ xz plane	$10.2\dfrac{ab}{\lambda^2}$	−13.2 −30.0	xz plane: $50\,\dfrac{\lambda}{b}$ yz plane: $65\,\dfrac{\lambda}{a}$	$115\ \sin^{-1}\dfrac{\lambda}{b}$ $115\ \sin^{-1}\dfrac{\lambda}{a}$
Circular waveguide, TE_{11} mode, E, H = 0 outside aperture $2a > 2\lambda$	$E_\theta = \dfrac{jk\exp(-jkr)}{2r}E_0(1+\cos\theta)kaJ_1(x'_{||})\dfrac{J_1(ka\sin\theta)}{ka\sin\theta}$ yz plane $E_\phi = \dfrac{jk\exp(-jkr)}{2r}E_0\,ka(1+\cos\theta)J_1(x'_{||})\dfrac{J_1(ka\sin\theta)}{1-\left(\dfrac{k\sin\theta}{x'_{||}}\right)^2}$ xz plane	$\dfrac{10.5(\pi a^2)}{\lambda^2}$	−17.2 −38.0	yz plane: $14.9\,\dfrac{\lambda}{a}$ xz plane: $25.1\,\dfrac{\lambda}{a}$	$115\ \sin^{-1}\dfrac{\lambda}{a}$ $115\ \sin^{-1}\dfrac{\lambda}{a}$
Uniform rectangular aperture in infinite ground plane, $E_y = E_0$	$E_\phi = \dfrac{jk\exp(-jkr)}{4\pi r}2E_0\cos\theta\,4ab\dfrac{\sin(kb\sin\theta)}{kb\sin\theta}$ xz plane $E_\theta = \dfrac{-jk\exp(-jkr)}{4\pi r}2E_0\,4ab\dfrac{\sin(ka\sin\theta)}{ka\sin\theta}$ yz plane	$\dfrac{4\pi}{\lambda^2}\times\text{area}$ $\dfrac{16\pi ab}{\lambda^2}$	−13.2	xz plane: $50.5\,\dfrac{\lambda}{b}$ yz plane: $50.5\,\dfrac{\lambda}{a}$	$115\,\dfrac{\lambda}{b}$ $115\,\dfrac{\lambda}{a}$
Circular aperture, uniform distribution	$E_\theta = \dfrac{2j\pi a^2}{\lambda r}(1+\cos\theta)\exp(-jkr)\dfrac{J_1(ka\sin\theta)}{ka\sin\theta}$	$\dfrac{4\pi}{\lambda^2}\times\text{area}$ or $\left(\dfrac{2\pi a}{\lambda}\right)^2$	−17.6	$58.5\,\dfrac{\lambda}{2a}$	$140\,\dfrac{\lambda}{2a}$
Circular aperture, $1-\left(\dfrac{\rho'}{a}\right)^2$ distribution	$E_\theta = \dfrac{2j\pi a^2}{\lambda r}(1+\cos\theta)\exp(-jkr)\dfrac{2J_2(ka\sin\theta)}{(ka\sin\theta)^2}$	$0.75\left(\dfrac{2\pi a}{\lambda}\right)^2$	−24.6	$72.7\,\dfrac{\lambda}{2a}$	$189\,\dfrac{\lambda}{2a}$
Circular aperture $\left[1-\left(\dfrac{\rho'}{a}\right)^2\right]^2$ distribution	$E_\theta = \dfrac{2j\pi a^2}{\lambda r}(1+\cos\theta)\exp(-jkr)\dfrac{8J_3(ka\sin\theta)}{(ka\sin\theta)^3}$	$0.56\left(\dfrac{2\pi a}{\lambda}\right)^2$	−30.6	$84.3\,\dfrac{\lambda}{2a}$	$232\,\dfrac{\lambda}{2a}$

Table 18-1. Gains, Patterns, Side-Lobe Level, and Beam Widths of Typical Antenna —*Concluded*

Antenna type	Description	Pattern Expressions	Directivity D	First side lobe, dB	Beam width, deg	
					3 dB	First nulls
Constant electric or magnetic line source, $J_x = J_0$ or $M_x = M_0$		Electric ($x'z'$ plane:) $$E_\theta = \frac{-j\eta \exp(-jkr)}{4\pi r} \frac{J_0 L}{2} \cos\theta \frac{\sin(\pi L/\lambda \sin\theta)}{\pi L/\lambda \sin\theta}$$ Magnetic ($x'z'$ plane:) $$E_\phi = \frac{jk \exp(-jkr)}{4\pi r} \frac{M_0 L}{2} \cos\theta \frac{\sin(\pi L/\lambda \sin\theta)}{\pi L/\lambda \sin\theta}$$	1.0 (normalized)	-13.2	$50.8\dfrac{\lambda}{L}$	$114.6\dfrac{\lambda}{L}$
Electric or magnetic line source, $\dfrac{\pi x'}{l}$ $J_x = J_0 \cos\dfrac{\pi x'}{l}$ or $M_x = M_0 \cos\dfrac{\pi x'}{L}$		Electric ($x'z'$ plane:) $$E_\theta = \frac{-j\eta \exp(-jkr)}{4\pi r} \frac{J_0 \pi L}{2} \cos\theta \frac{\cos(\pi L/\lambda \sin\theta)}{(\pi/2)^2 - (\pi L/\lambda \sin\theta)^2}$$ Magnetic ($x'z'$ plane:) $$E_\phi = \frac{jk \exp(-jkr)}{4\pi r} \frac{M_0 \pi L}{2} \cos\theta \frac{\cos(\pi L/\lambda \sin\theta)}{(\pi/2)^2 - (\pi L/\lambda \sin\theta)^2}$$	0.810	-23.2	$68.8\dfrac{\lambda}{L}$	$171.8\dfrac{\lambda}{L}$
Electric or magnetic line source, $\dfrac{\pi x'}{l}$ $J_x = J_0 \cos^2\dfrac{\pi x'}{l}$ or $M_x = M_0 \cos^2\dfrac{\pi x'}{l}$		$x'z'$ plane: $$E_\theta = \frac{-j\eta \exp(-jkr)}{4\pi r} J_0 \cos\theta$$ $$\left[\frac{L}{2}\frac{\sin(\pi L/\lambda \sin\theta)}{\pi L/\lambda \sin\theta} - \frac{\pi^2}{\pi^2 - (\pi L/\lambda \sin\theta)^2}\right]$$ $$E_\phi = \frac{jk \exp(-jkr)}{4\pi r} M_0 \cos\theta$$ $$\left[\frac{L}{2}\frac{\sin(\pi L/\lambda \sin\theta)}{\pi L/\lambda \sin\theta} - \frac{\pi^2}{\pi^2 - (\pi L/\lambda \sin\theta)^2}\right]$$	0.667	-31.5	$83.2\dfrac{\lambda}{L}$	$229.2\dfrac{\lambda}{L}$

16. Antenna Temperature. An antenna located on the earth and pointing at an angle (θ, ϕ) to the sky will receive noise from all directions. The amplitude of this noise as seen at the antenna terminals will depend upon the noise source (warm earth, cosmic noise, water vapor, radio stars, etc.), the antenna orientation, and the operating frequency and polarization.[6] The equivalent received noise power in a receiver matched to the antenna terminal impedance by a lossless transmission line is

$$P_n = kTB \qquad (18\text{-}27)$$

where k is Boltzmann's constant, T is the temperature in kelvins, and B is the band width in hertz.

Noise power for a fixed band width may be thought of as an equivalent temperature. Consequently, if it is assumed that the various noise sources that make up the antenna noise environment have an equivalent temperature $T(\theta, \phi)$, the apparent antenna temperature is given by

$$T_A = \frac{\int_0^{2\pi} \int_0^{\pi} G(\theta,\phi) T(\theta,\phi) \sin\theta\, d\theta\, d\phi}{\int_0^{2\pi} \int_0^{\pi} G(\theta,\phi) \sin\theta\, d\theta\, d\phi} \qquad (18\text{-}28)$$

The most important natural emitter of noise at microwave frequencies is the ground at 377 K compared with the sky temperature at a few degrees. Therefore, antennas which have low side and back lobes will have low apparent antenna temperatures T_A. Several low-noise antenna designs with antenna temperatures as low as 2 K have been reported.[7-13]

If the antenna has losses and is not matched to the receiver for maximum power transfer, the antenna temperature will be higher than that predicted by Eq. (18-28). The contribution due to the particular mismatch and losses of every component must be analyzed for each particular receiving system in detail. An outline of an excellent method of analysis is available.[15] For a receiving system with no mismatch loss, the apparent temperature T_a is given by

$$T_a = (1 - L)T_A + LT_0 \qquad (18\text{-}29)$$

where T_A is the antenna temperature given by (18-28) and T_0 is the physical temperature of the lossy device. For example, a 10-cm length of precision coaxial cable has a loss of about 0.013 dB ($L = 0.003$). If $T_A = 100$ K and $T_0 = 300$ K, then from (18-29) the apparent temperature $T_a = 100.6$ K. This increase, while small, is important in remote sensing with radiometric systems, where absolute temperatures are measured to $\pm$ 0.1 K.

While most reflector and horn antennas have small losses, microwave radiometer systems are being constructed which require antennas having losses known to within 0.001 dB.[14]

17. Friis Transmission Formula. Assume that there is a source antenna and an antenna under test located at a distance r apart such that

$$r \geq \frac{2(d_t)^2}{\lambda} \qquad (18\text{-}30)$$

where d_t is the maximum aperture dimension of the antenna under test. The distance specified by Eq. (18-30) is the so-called *far-field distance*. The far-field distance is commonly specified as the distance where the phase front of a spherical wave over a planar aperture will not exceed $\pi/8$ rad. For special purposes, such as the measurement of deep nulls or extremely precise side-lobe levels, the far-field distance may have to be extended further; curves using other criteria are available.[16] The power received at the terminals of one antenna located in the far field of a second antenna can be expressed as a fraction of the transmitted power as

$$P_R = P_T \left[\frac{\lambda}{4\pi r}\right]^2 N_{A_T} N_{A_R} D_T(\theta,\phi) D_R(\theta,\phi) \left(1 - \left|\Gamma_T\right|^2\right)\left(1 - \left|\Gamma_R\right|^2\right) \left|\rho_R \cdot \rho_T\right|^2 \qquad (18\text{-}31)$$

where N_{AT} and N_{AR} are the loss efficiencies of the antennas and $D_T(\theta_T, \phi_T)$ and $D_R(\theta_R, \phi_R)$ are the directivities of the antennas in the direction one antenna is pointing toward the other. $|\Gamma_T|^2$ and $|\Gamma_R|^2$ are the reflected power due to mismatch of the antenna terminals and $|\rho_R \cdot \rho_T|^2$ is the polarization loss. The term in brackets in Eq. (18-31) is the so-called *free-space loss*, which is due to spherical spreading of the energy radiated by an antenna. In the far field all antennas appear as a spherical wave emanating from a point source located at the phase center of the antenna, where the phase center may be a function of the observation angle.

18. Polarization. Consider a plane wave propagating in the z direction which has an arbitrary plane polarization with an axial ratio r_A and a tilt angle as shown in Fig 18-6. The polarization ellipse is the locus of the tip of the electric field vector as the wave propagates in space as a function of time. The axial ratio r_A is defined as the ratio of the major to minor axis of the ellipse referenced to a coordinate system. The field expression for this arbitrary plane-polarized wave is

$$E = C[\mathbf{x}(r_A \cos \phi + j \sin \phi) + \mathbf{y}(r_A \sin \phi - j \cos \phi)]\exp(-jkz)$$

When the z-phase dependence is neglected except for the sign, the normalized fields of two

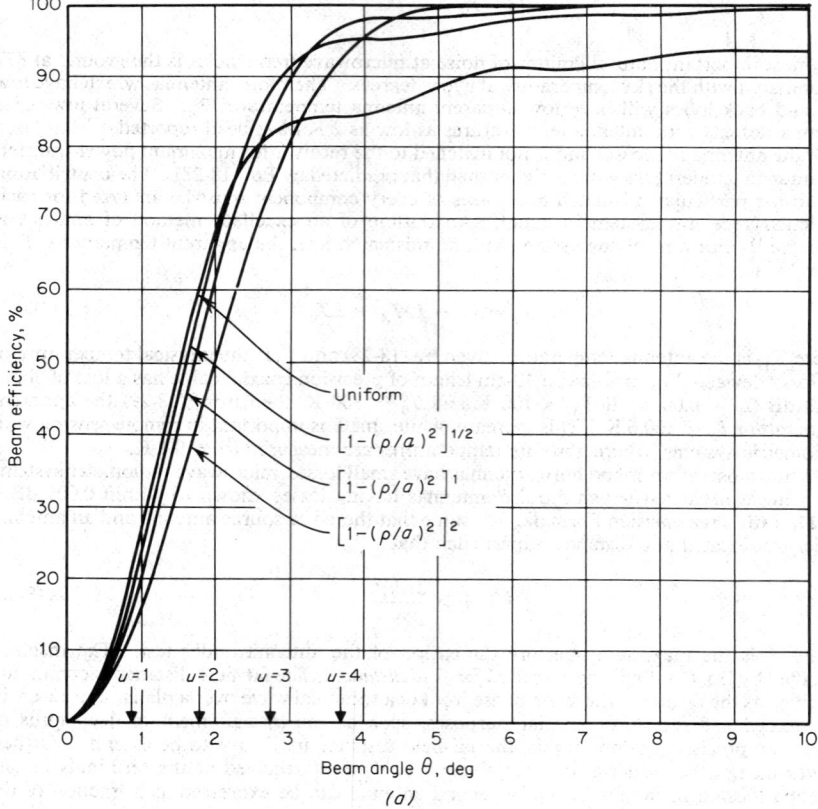

Fig. 18-5. Beam efficiency vs. angle for various distributions on a circular aperture 20λ in diameter (a) and 20λ on a side (b). Note that these data can be scaled for other apertures using the parameter $u = 2\pi a/\lambda \sin \theta$, as noted on the beam-angle scale.

different plane-polarized waves with the wave number 2 propagating in the negative z direction (two antennas pointing at one another) are given by

$$\mathbf{E}_1 = E_1\rho_1 = E_1\frac{\mathbf{x}(r_1\cos\phi_1 + j\sin\phi_1) + \mathbf{y}(r_1\sin\phi_1\cos\phi_1)}{\sqrt{r_1^2 + 1}} \tag{18-32}$$

$$\mathbf{E}_2 = E_2\rho_2 = E_2\frac{\mathbf{x}(r_2\cos\phi_2 - j\sin\phi_2) + \mathbf{y}(r_2\sin\phi_2 + j\cos\phi_2)}{\sqrt{r_2^2 + 1}} \tag{18-33}$$

The polarization loss between antennas radiating various combinations of polarizations is given in Table 18-2. The normal convention for waves traveling in the positive z direction is

$$\rho_1 = x - jy \qquad \text{right-hand circular polarization} \tag{18-34}$$

$$\rho_2 = x + jy \qquad \text{left-hand circular polarization} \tag{18-35}$$

A complete discussion of the polarization properties of antennas including measurement methods is given in Ref. 6.

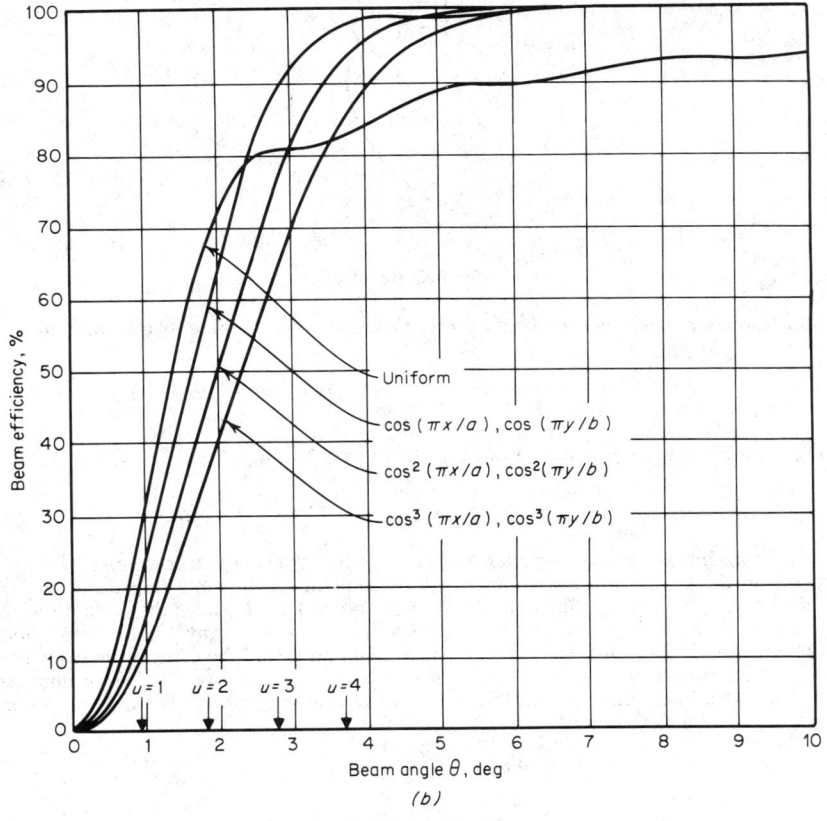

Fig. 18-5. (*Cont'd*)

19. Radiation Impedance. The complex Poynting vector $\rho = \mathbf{E} \times \mathbf{H}^*$ can be integrated over a closed surface about an antenna to give

$$\iint_{S'} \mathbf{E} \times \mathbf{H}^* \cdot ds = 4j\omega(W_m - W_e) + P_R = VI^* \tag{18-36}$$

where $W_m - W_e$ is the time-average net reactive power stored within the volume enclosed by S' and Pr is the net real power flow through S'. As a result, the real power can be related to a radiation resistance and the reactive power to a radiation reactance by equating this total complex power to an equivalent voltage V and current I at a defined set of terminals. This radiation impedance, which is a lumped-circuit-element description of an antenna, is defined at a specific set of terminals or terminal plane which may be the aperture of a waveguide antenna. The simplest term to determine in Eq. (18-36) is the radiated power P_r and the corresponding radiation resistance R_r. This is true because the surface S' may be chosen in the far field of the antenna, where fields with only $1/r$ dependence are important. The radiation reactance is determined by choosing S' close to the antenna and integrating fields where $1/r^2$ and $1/r^3$ terms are important. A convenient surface to choose for these calculations is a sphere of radius r surrounding the antenna.

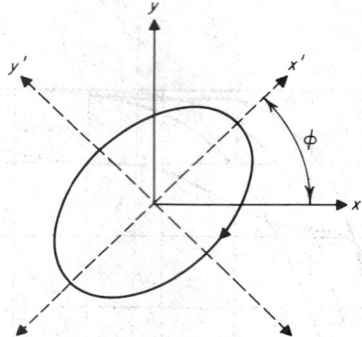

Fig. 18-6. Polarization ellipse.

20. Radiation Resistance of Short Current Filament. The far fields of a current filament from Eq. (18-3) are

$$E_\theta = \frac{jkI_0\eta L \sin\theta \exp(-jkr)}{4\pi r} \qquad H_\phi = \frac{jkI_0 L \sin\theta \exp(-jkr)}{4\pi r}$$

When these fields are used, Pr from Eq. (18-36) is given by

$$P_r = \int_0^{2\pi} \int_0^{\pi} E_\theta H_\phi^* \sin\theta \, r^2 \, d\theta \, d\phi = \frac{\eta\pi I_0^2}{3}\left(\frac{L}{\lambda}\right)^2 = I_0^2 R_r$$

21. Array Impedance. The pattern expressions for arrays neglected coupling between elements in the array. Mutual coupling will not only affect the input impedance of each element input terminal as a function of scan angle but will also change the pattern of each element in a manner dependent upon the element location in the array. The determination of the mutual and self-impedance of an element in an array and the effect of these changes upon array patterns is a specialized problem. In general, however, arrays of antennas radiating into a linear medium will have to satisfy the same equations as any linear system with n pairs of terminals, or

$$V_1 = Z_{11}I_1 + Z_{12}I_2 + \cdots + Z_{in}I_n$$

$$V_n = Z_{n1}I_1 + Z_{n2}I_2 + \cdots + Z_{nn}I_n$$

Specific examples of array pattern and impedance properties are given later in this section.

Table 18-2. Polarization Loss between Plane-polarized Waves

Type	ρ_1	ρ_2	$\lvert \rho_1 \cdot \rho_2 \rvert^2$
Linear to linear, $r_1 = r_2 \to \infty$	$\mathbf{x}\cos\phi_1 + \mathbf{y}\sin\phi_1$	$\mathbf{x}\cos\phi_2 + \mathbf{y}\sin\phi_2$	$\cos^2(\phi_1 - \phi_2)$
Linear to circular. $(r_1 = 1,\ \phi = 0 \text{ to } \pi)$	$\dfrac{\mathbf{x} \pm j\mathbf{y}}{\sqrt{2}}$	$\mathbf{x}\cos\phi_2 + \mathbf{y}\sin\phi_2$	$\dfrac{1}{2}$
Circular to circular, $r_1 = r_2 = 1$ $\phi_1 = 0 \text{ or } \pi$ $\phi_2 = 0 \text{ or } \pi$	$\dfrac{\mathbf{x} + j\mathbf{y}}{\sqrt{2}}$	$\dfrac{\mathbf{x} - j\mathbf{y}}{\sqrt{2}}$	1
	$\dfrac{\mathbf{x} + j\mathbf{y}}{\sqrt{2}}$	$\dfrac{\mathbf{x} + j\mathbf{y}}{\sqrt{2}}$	0
General case	$\dfrac{\mathbf{x}(r_1\cos\phi_1 + j\sin\phi_1)}{\sqrt{r_1^2 + 1}}$ $+\dfrac{\mathbf{y}(r_1\sin\phi_1 - j\cos\phi_1)}{\sqrt{r_1^2 + 1}}$	$\dfrac{\mathbf{x}(r_2\cos\phi_2 - j\sin\phi_2)}{\sqrt{r_2^2 + 1}}$ $+\dfrac{\mathbf{y}(r_2\cos\phi_2 + j\cos\phi_2)}{\sqrt{r_2^2 + 1}}$	$\dfrac{(1 + r_1^2 + r_2^2 + r_1^2 r_2^2 + 4r_1 r_2) + (1 + r_1^2 r_2^2 - r_1^2 - r_2^2)\cos^2(\phi_1 - \phi_2)}{2(r_1^2 + 1)(r_2^2 + 1)}$

WIRE ANTENNAS

22. Analysis of Wire Antennas. The development of wire antennas has been extensive, since such antennas are simple to analyze and construct. The classical analysis of wire antennas such as dipoles, loops, and loaded-wire antennas has been developed by Hallen and R. W. P. King and his students; a good summary of theoretical and experimental results, including specific impedance curves and design data, is given in Ref. 17. The unfortunate drawback of this analysis is that each new wire-antenna configuration presents another analytical problem which must be solved before design computations can be made. A systematic method of solving wire-antenna problems using computerized matrix methods has been developed by extending the analyses of scattering by wire objects by Richmond[18-21] and Harrington.[22-25] These matrix analysis methods have been applied to wire antennas[26-41] to determine the input impedance, current distribution, and radiation patterns by subdividing any particular wire antenna into segments and determining the mutual coupling between any one segment and all other segments. The method therefore can treat any arbitrary wire configuration, including loading and arrays, the limitation being the storage capacity of available digital computers and the patience of the programmer.

23. Numerical Method. For a single-wire antenna, the antenna is subdivided into N sections. The current on the antenna can be expressed as

$$I = \sum_1 I_n F_n \tag{18-37}$$

where F_n is a known expansion function such as a pulse, triangle, or piecewise sinusoid and the coefficients I_n are unknowns to be determined. With this current expansion for the segmented antenna a matrix can be written in the form

$$[Z][I] = [V] \tag{18-38}$$

where the column vector $[V]$ is known and the column vector $[I]$ is to be determined. The square impedance matrix $[Z]$ is completely defined by the geometry and choice of the expansion function $\mathbf{F}_n$ and a suitable testing function $\mathbf{W}_n$, where if Galerkin's method is used, $\mathbf{W}_n = \mathbf{F}_n$. By substituting Eq. (18-37) into the vector potential $\mathbf{A}$ given by Eq. (18-5), and after considerable algebraic manipulation[25] the impedance-matrix element is given by

$$Z_{mn} = \int dl \int dl' \left(j\omega\mu \mathbf{W}_m \cdot \mathbf{F}_n + \frac{\partial \mathbf{W}_n}{\partial l'} \cdot \frac{\partial \mathbf{F}_n}{\partial l} \right) \frac{\exp(-jkR)}{R} \tag{18-39}$$

Once Z_{mn} is known, $[I]$ is defined by Eq. (18-37), and hence the input impedance is known. The basic analytical methods used by Richmond and Harrington are similar except for the form of the current distribution on the fundamental wire segment.[26,37] The use of the piecewise-sinusoidal current distribution on individual wire segments by Richmond may have computational advantages. Computer programs for obtaining the radiation properties of many types of wire antennas are readily available[26-28,30-33] in the literature and at Langley Research Center.

24. Finite Gap and Feed Line. The effects of the finite gap in a linear antenna and the connecting balanced-input transmission line used as a feeder are neglected in the classical analysis. These effects have been determined by the matrix method by J. S. Chatterjee. Feeder gaps up to $\frac{1}{10}\lambda$ in a half-wave dipole with a radius $a = 0.000125\lambda$ will produce about a 2% change in the input susceptance of the antenna.

The effect of a feeder transmission line, however, is much more pronounced, as shown in Fig. 18-7. In this figure the input impedance Y_T is plotted; it is computed by transforming the dipole admittance down the transmission line d, assuming that the normal transmission-line equations hold. Y_{in} in this figure is the admittance as seen at the input terminals of the antenna, where the feeder line and dipole arms are all considered as part of the radiating system.

It appears that the best design uses feeder lines that are an odd multiple of 1/4 wavelength long. Many types of dipole antennas can be fed with an unbalanced transmission line with a balun to minimize feeder-line currents so that the effect of the feeder line is small. An excellent description of baluns is given in Ref. 42.

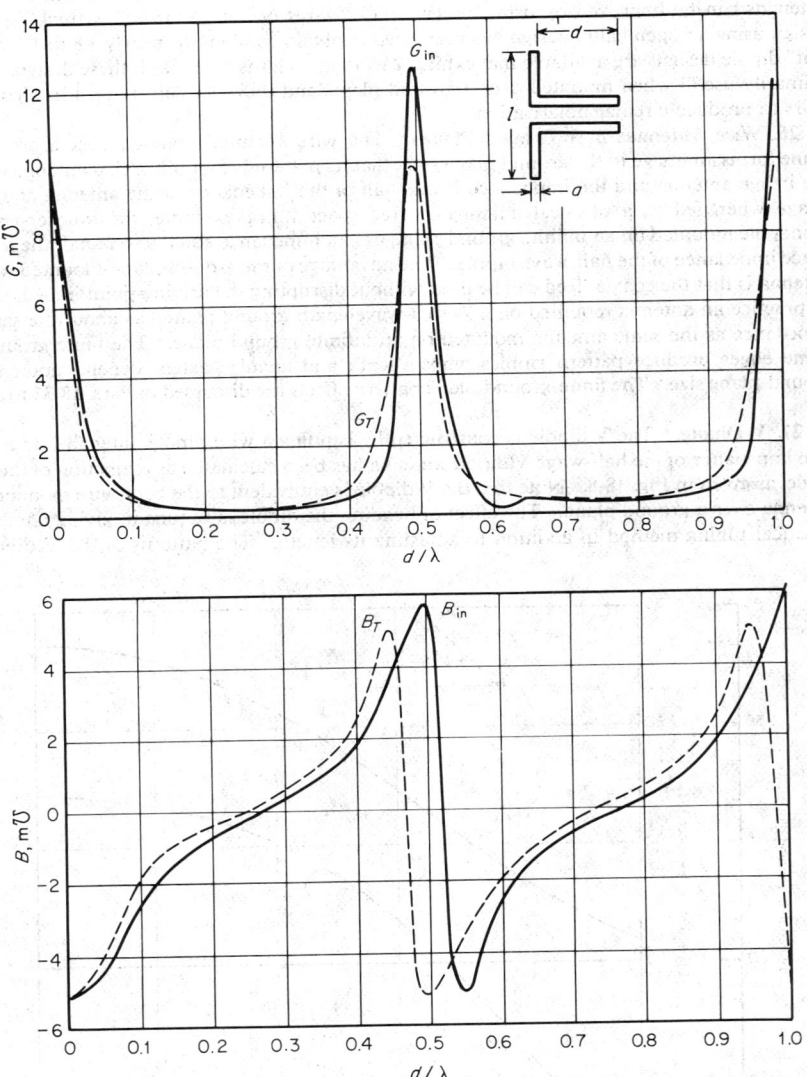

Fig. 18-7. The input admittance of a dipole with a connecting two-wire transmission line, where $Y_T = G_T + jB_T$ is the value using ordinary transmission-line theory and $Y_{in} = G_{in} + jB_{in}$ is the value obtained with the integral-equation solution. *(Courtesy of J. S. Chatterjee, Langley Research Center.)*

25. Thin and Fat Dipoles. The impedance and radiation patterns of the thin dipoles depend on the length of the wires and the location of the feed point. Detailed impedance properties and patterns of thin dipoles are available.[43] The bandwidth properties of wire antennas can be improved by using either "thick" wires or "fat" shapes. A thick-dipole design using an open balun design has been used to obtain bandwidths nearly 1.8 to 1,[44] and "fat" dipole designs are available that exhibit 2 to 1 bandwidths.[39, 45] Both these designs are primarily useful when mounted over a ground plane, and they can be arranged in crossed pairs to produce circular polarization.

26. Wire Antennas over Ground Planes. The wire antenna mounted over a ground plane forms an image in the ground plane such that its pattern is that of the real antenna and the image antenna and the impedance is one-half of the impedance of the antenna and its image when fed as a physical antenna in free space. For example, the quarter-wave monopole mounted on an infinite ground plane has an impedance equal to one-half the free-space impedance of the half-wave dipole. The advantage of the ground-plane-mounted wire antenna is that the coaxial feed can be used without disrupting the driving-point impedance. In practice an antenna mounted on a 2- to 3-wavelength ground plane has about the same impedance as the same antenna mounted on an infinite ground plane. The finite-ground-plane edges produce pattern ripples whose depth and angular extent depend upon the ground-plane size. The finite-ground-plane pattern effects are discussed in Pars. **18-31** to **18-33**.

27. V Dipole. The V dipole is constructed by bending a wire dipole antenna into a V. The impedance of the half-wave V dipole antenna has been calculated as a function of the V angle, as given in Fig. 18-8. Note that the V dipole is equivalent to the bent-wire monopole antenna over a ground plane. The effect of bending the dipole is to tune it, giving another practical tuning method in addition to adjusting its length. The patterns of the V dipole

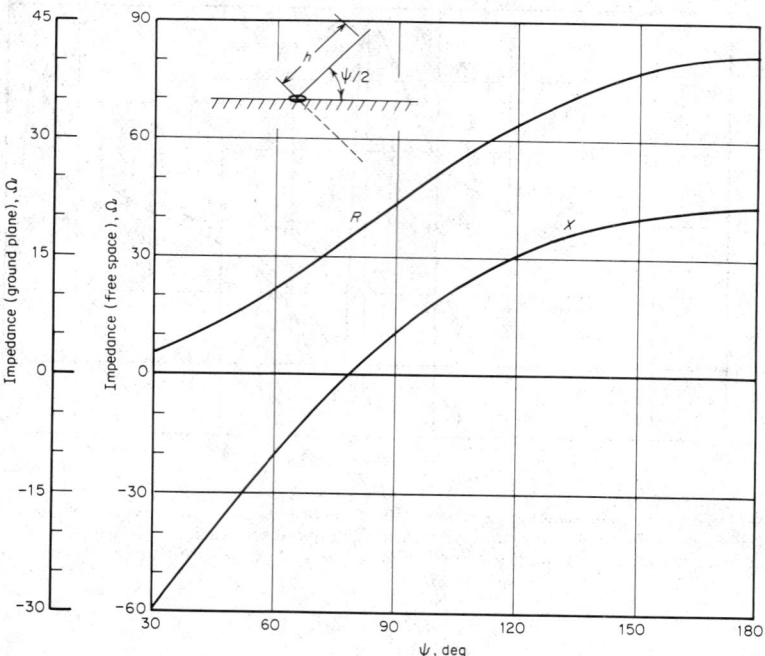

Fig. 18-8. Impedance of V-dipole antenna as a function of V angle. ($h/a = 1,000, h =$ arm length $= \lambda/4$, $a =$ wire radius.) *(Courtesy of J. E. Jones, Langley Research Center.)*

(vertical polarization) are nearly identical to those of the straight dipole for ψ angles as small as 120°. With decreasing tilt angle this antenna excites a horizontal polarized field component which tends to fill in the pattern, making the antenna a popular communications antenna for aircraft. Other calculations for V dipoles, V dipole arrays, and dipoles mounted on spacecraft are available.[26] Computer programs for the V dipole and arrays of V dipoles are available at Langley Research Center.

28. Bent Dipole. Another form of the dipole antenna that has practical application, particularly for ground-plane or airplane applications, is the bent-wire dipole formed by bending the wire 90° some distance out from the feed point. The impedance of the bent wire is given in Fig. 18-9 for both the free-space and ground-plane case. Note that this antenna can also be tuned by adjusting the lengths perpendicular and parallel to the driving point. The radiation pattern in the plane of this antenna is nearly omnidirectional for values of $H_1 \leq 0.10$, after which the pattern approaches that of the vertical half-wave dipole. Other forms of this antenna can be constructed, including loading to reduce the effective length. Indeed with these computer-analysis methods many other forms can be cheaply designed.

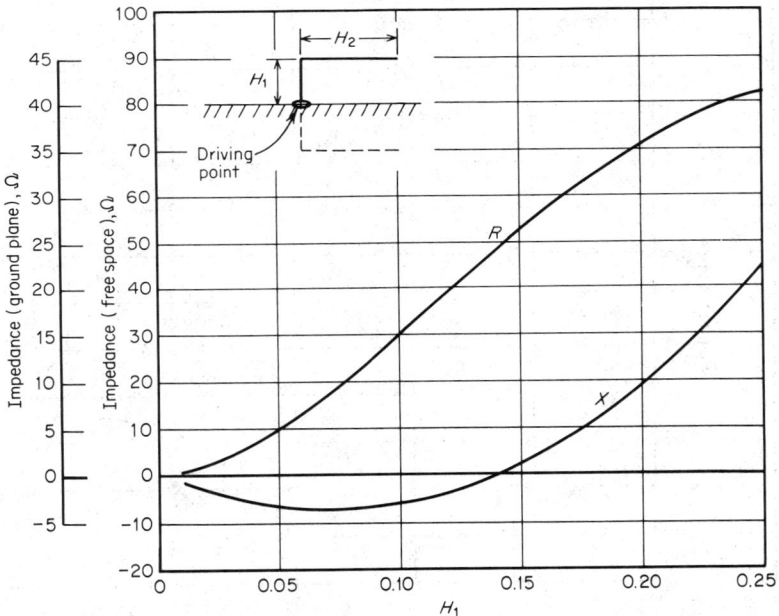

Fig. 18-9. Impedance of half-wave bent-dipole antenna. *(Courtesy of J.E. Jones, Langley Research Center.)*

29. Loop Antenna. Another useful classical antenna is the loop antenna. As stated earlier, many investigators have erroneously designated this antenna as a magnetic dipole when indeed it is just another form of the wire antenna. The admittance of the loop antenna can be computed using the matrix method by approximating the loop with a polygon having the same electrical length. The admittance of a 12-sided polygon, which is identical to the admittance of a loop of the same length, has been computed and is given in Fig. 18-10. These results have been verified in the ground-plane case experimentally.[32] Indeed the square loop or any other multisided loop with the same electrical-perimeter length has approximately the same admittance as the circular loop. Another method of improving the impedance of a loop is to add turns or load the loop with discrete lumped capacitances.[32,46-48] The patterns and impedance of loop antennas mounted on aircraft structures have been studied, and computer programs are available.[33]

30. Wire Antennas Near Ground Planes. Although the impedance of wire antennas mounted on ground planes several wavelengths in dimension for practical purposes is similar to the impedance of the same antenna mounted on an infinite ground plane, the patterns of wire antennas on finite ground planes strongly depend upon the ground-plane size. In recent years the geometrical theory of diffraction (GTD) has been successfully applied to such problems.[49-54] These published results and the method of analysis are of great interest to antenna engineers since the geometry is the practical one of interest.

31. Loop above a Finite Ground Plane. The loop above a finite ground plane is an antenna commonly used as an array element in vhf omnirange stations located near all airports as an aircraft landing aid. Since the pattern of the small loop is symmetrical in azimuth, the elevation pattern of a loop over a finite circular ground plane can be computed using a pair of closely spaced line sources fed out of phase and located over a finite-width conducting strip.[52] This simplified two-dimensional geometry illustrates one aspect of the GTD method and is briefly outlined here.

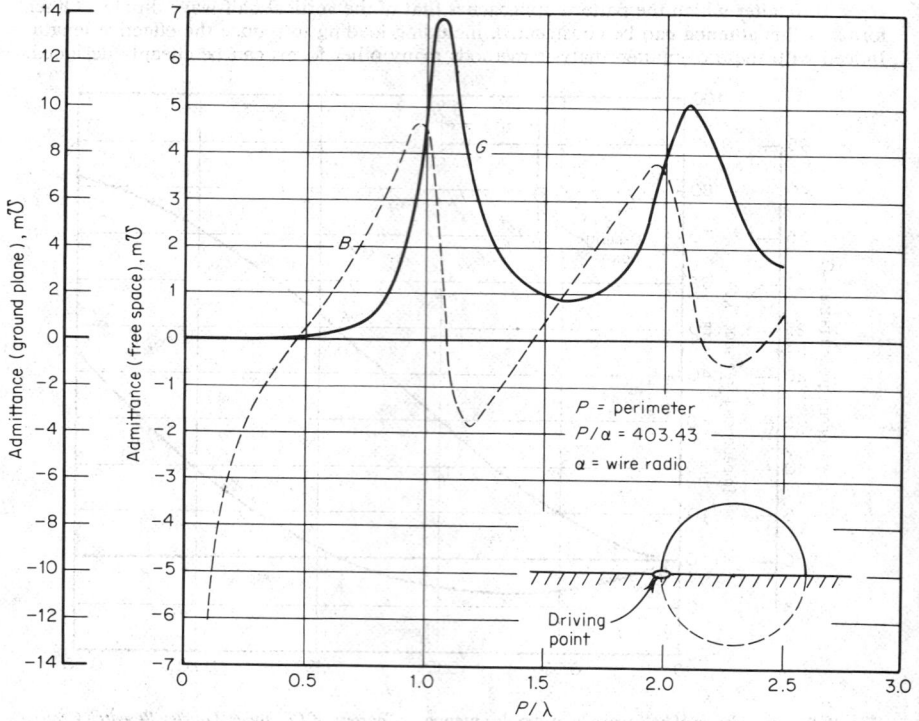

Fig. 18-10. Admittance of the loop antenna. *(Courtesy of J. E. Jones, Langley Research Center.)*

Consider the geometry of the line-source pair a distance d above a ground plane 2 x_0 in width, as shown in Fig. 18-11. For purposes of analysis the field-pattern space can be broken down into three distinct regions. Region I includes the fields of the source antenna that are reflected by the strip. It contains the incident, reflected, and diffracted fields. Region III is the *shadow* region, where a far-field observer can never see the source-region antenna. Energy reaches region III through diffraction of the line-source-pair fields by the edges of the finite ground plane. Region II is bounded by the incident shadow boundary (ISB) and the reflected shadow boundary (RSB); it contains both incident and diffracted fields, which together provide a smooth transition between fields in the lit and shadow regions. (RSB occurs at $\phi = \alpha$ and $\phi = \pi - \alpha$ from the law of reflection; ISB from geometrical optics occurs at $\phi = \alpha$ and $\phi = \pi + \alpha$.)

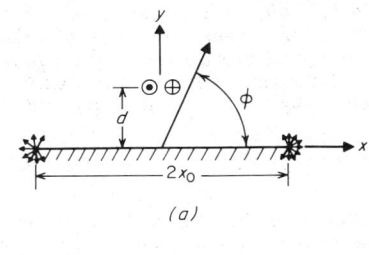

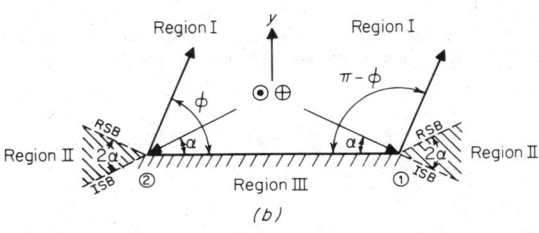

Fig. 18-11. Linear array of line sources and its diffraction mechanism: (*a*) two-element array above a ground plane; (*b*) diffraction mechanism. *(Adapted from Balanis.[52])*

To determine the fields in these three regions first assume that the ground plane is infinite. In this case, the incident radiated field E_z^i from the line-source pair will add to its image to give

$$E_z^T = j2 \cos \phi \sin(kd \sin \phi) \frac{\exp(-jk\rho)}{\rho^{1/2}} \qquad (18\text{-}40)$$

where

$$E_z^i = \cos \phi \exp(jkd \sin \phi) \frac{\exp(-jk\rho)}{\rho^{1/2}}$$

The fields diffracted from the edges of the ground plane can be determined using reported techniques.[55, 51] The fields diffracted by edges 1 and 2 are denoted E_{ZD_1} and E_{ZD_2} and may be written as

$$E_{zd_1} = \cos \alpha \,[V_B (h, \pi - \phi - \alpha, 2) - V_B(h, \pi - \phi + \alpha, 2)]$$
$$\exp(jkx_0 \cos \phi) \frac{\exp(-jk\rho)}{\rho^{1/2}} \qquad (18\text{-}41)$$

$$E_{zd_2} = -\cos \alpha \,[V_B(h, \phi + \alpha, z) - V_B(h, \phi + \alpha, z)]$$
$$\exp(-jkx_0 \cos \phi) \frac{\exp(-jk\rho)}{\rho^{1/2}} \qquad (18\text{-}42)$$

where the function $V_B(r, \beta, 2)$ is the diffraction coefficient of a two-dimensional wedge[56-59] (with the wedge angle set to zero for the thin-ground-plane case given by

$$V_B(r, \beta, z) = \pm \frac{1 + j}{2} \exp(jkr \cos \beta)\{[C(z_1) - 0.5] - j[S(z_1) - 0.5]\} \qquad (18\text{-}43)$$

where $z_1 = [(2 \, k\alpha/\pi) (1 + \cos \beta)]^{1/2}$. From Eq. (18-41) to (18-43) the fields in the three regions are

$$E_z^I = E_z^T + E_{z_{D_1}} + E_{z_{D_2}}$$
$$E_z^{II} = E_z^i + E_{z_{D_1}} + E_{z_{D_2}} \qquad (18\text{-}44)$$
$$E_z^{III} = E_{z_{D_1}} + E_{z_{D_2}}$$

The pattern of this antenna has a null at $\phi = 90°$ and a maximum value at some angle above the ground plane, which depends upon the spacing d and the ground-plane size $2 x_0$. The value of the field along the ground-plane edge ($\phi = 0$) is also of interest to the antenna designer. The variation of these field parameters is given in Fig. 18-12. For a small loop of radius r (or a large loop with the same radius r but a constant current) placed at a height d over a ground plane of dimension $2 x_0$, the diffraction field will be small if the following condition is satisfied:

$$[(2 x_0 + r)^2 + d^2]^{1/2} - [(2x_0 - r)^2 + d^2]^{1/2} = N\lambda$$

where N is an integer.

32. Horizontal Dipole over a Finite Ground Plane. The GTD method can be applied to the horizontal dipole over a finite ground plane,[53] the geometry of which is shown in Fig. 18-13. Also shown in Fig. 18-13 is the geometry of the same dipole placed over a cylinder. The purpose of this antenna design is to minimize the ripple or field variation in the pattern above the ground plane and simultaneously achieve a low back-lobe level. These field

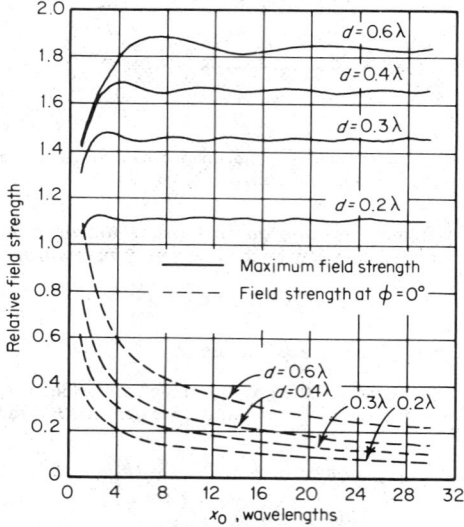

Fig. 18-12. Variations of the maximum field strength and the field strength at the angle of the ground-plane edge as a function of line-source spacing and ground-plane size. *(Adapted from Balanis.[52])*

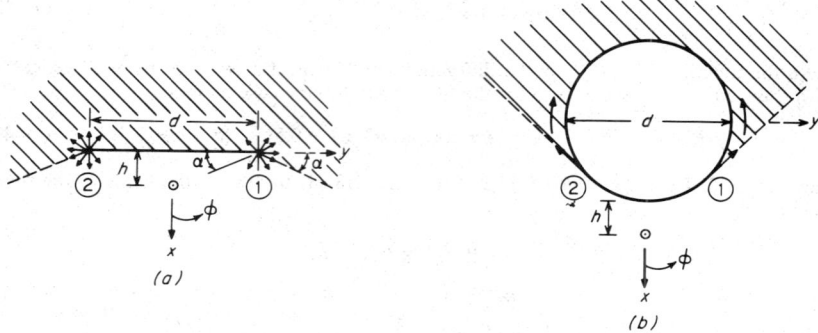

Fig. 18-13. Radiation mechanism of dipole near finite ground plane and circular conducting cylinder: (*a*) ground plane; (*b*) circular cylinder. *(From Balanis and Cockrell.[53])*

parameters are plotted as a function of ground-plane size and dipole spacing in Fig. 18-14. Also plotted in Fig. 18-14 are similar design curves for the dipole spaced the same parametric distances above a perfectly conducting cylinder having a diameter equal to the finite-ground-plane width. These calculations were made by programming available formulas.[60] The cylinder curvature allows one to obtain a better back-lobe level for a given pattern variation or ripple in the forward region. Experimentally, it has been determined that the rear part of the metal cylinder can be substantially removed with little effect.

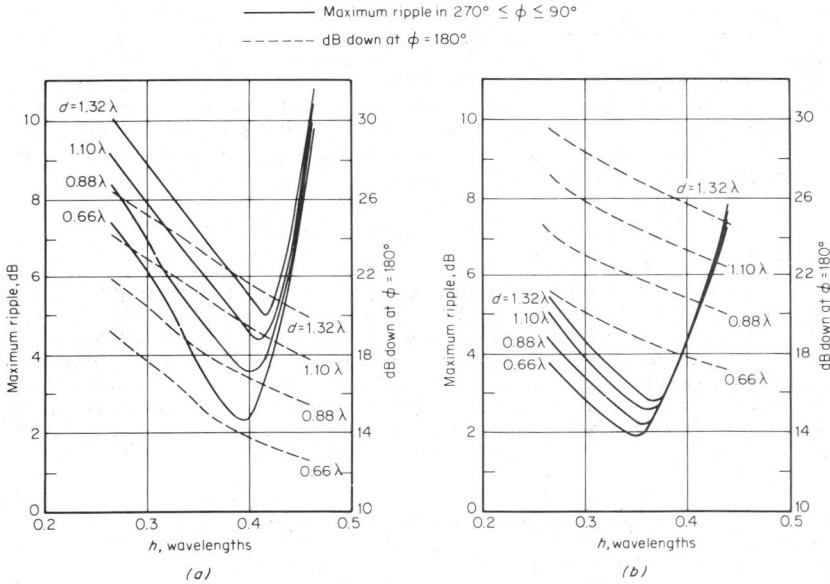

Fig. 18-14. Variations of maximum ripple in $270° \leq \phi \leq 90°$ region and radiation at $\phi = 180°$ as functions of dipole position h near the ground plane (a) and circular conducting cylinder (b). *(From Balanis and Cockrell.[53])*

33. Dipole or Monopole on a Finite Ground Plane. The radiation patterns of a half-wave dipole over a finite ground plane or the monopole mounted on a finite ground plane can be determined using GTD. These antennas have patterns which can be characterized by a null off the end of the wire, a maximum value at some angle above the ground plane, and energy diffracted about the back of the ground plane. All these field properties, except the null, depend upon the ground-plane size and (in the case of the dipole) the spacing above the ground plane. These parameters are plotted in Fig. 18-15. Note that little is to be gained by increasing the ground-plane width beyond 5λ. For circularly symmetric ground planes a caustic point is created which can significantly increase the radiation behind the ground plane.

WAVEGUIDE ANTENNAS

34. General Considerations. The waveguide antenna, which consists of a dominant-mode-fed waveguide opening onto a conducting ground plane, is very useful for many applications such as a feed for reflector antennas or a flush-mounted antenna for aircraft or spacecraft. For flush-mounting purposes it is sometimes desirable or necessary to cover the ground plane with dielectric layers to protect the aperture from the external environment or in some instances to put dielectric plugs in the feed-waveguide section. The impedance properties of waveguide antennas have been studied extensively both theoretically and experimentally, particularly for the rectangular waveguide,[61-68] the circular waveguide,[69-71] and the coaxial waveguide or so-called annular slot.[72-75] The impedance of these antennas is, for many practical purposes, relatively independent of ground-plane size so long as the ground plane is 2λ in dimension or greater.[64,70] However, the radiation pattern of the waveguide antenna mounted on finite ground planes is very dependent upon the ground-plane size. The effects of diffraction by ground-plane edges can be treated by the geometrical theory of diffraction (GTD)[76] in a manner similar to that used for wire antennas above a finite ground plane given in Pars. **18-30** to **18-33**.

35. Aperture-Admittance Theory. Using the reaction concept, Compton[62,63] has derived stationary forms of the aperture admittance of rectangular and parallel-plate wave-apertures radiating into lossy media. Combining Compton's formulation and his earlier work on plane waves propagating through inhomogeneous dielectric media,[77] Swift has determined the admittance of the rectangular waveguide radiating into an inhomogeneous dielectric slab.[65.] These two formulations have been extended by Beck[78] to include the general ground-plane-mounted aperture radiating into an arbitrary stratified layered medium. The geometry of the aperture under a layered medium is depicted in Fig. 18-16. If it is assumed that the aperture field consists of the dominant mode of the feed waveguide, the normalized input admittance of the general aperture antenna may be written

$$y = g + jb = C \int_0^\infty \left[F_{TE}(\beta) G_{TE}(\beta) + F_{TM}(\beta) G_{TM}(\beta) \right] \beta \, d\beta \qquad (18\text{-}45)$$

where the subscripts *TE* and *TM* refer to that part of the solution which can be derived from a single transverse electric or transverse magnetic vector potential in the external medium. $F_{TE}(\beta)$ and $F_{TM}(\beta)$ are the Fourier transforms of the aperture field, and the functions $G_{TE}(\beta)$ and $G_{TM}(\beta)$ are the normalized solutions to the wave equations in the outside media both evaluated at the ground-plane surface $z = 0$. A summary of the equations for the terms that go into Eq. (18-45) for various waveguides and different external media is given in Table 18-3. From Table 18-3 it can be deduced that different types of waveguides act as spatial filters to the spectrum of plane waves passing through the external media. For unloaded apertures,

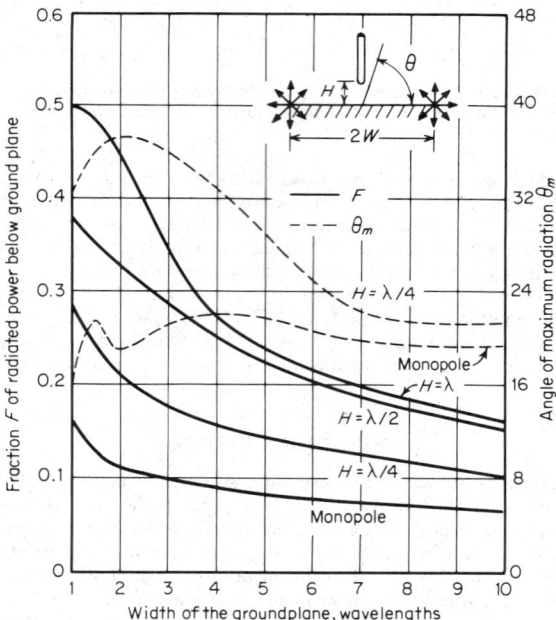

Fig. 18-15. Pattern characteristics of the half-wave dipole and the monopole over a finite ground plane. *(Courtesy of J. S. Chatterjee, Langley Research Center.)*

the assumption of the single-mode trial field in the variational solution represented by Eq. (18-45) has proved adequate for practical purposes.[64,67,70,78] However, higher-order modes can be included in these solutions if desired, by applying the Rayleigh-Ritz procedure (Ref. 60, p. 339). Computer programs for all solutions listed in Table 18-3 are available at the Langley Research Center, along with extensive calculations and measurements for specific cases.

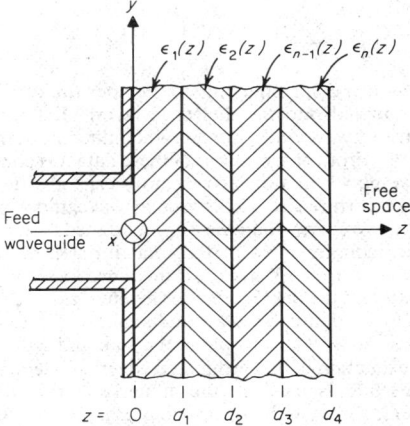

Fig. 18-16. Geometry of a waveguide of arbitrary cross section opening onto an infinite ground plane covered by a layered inhomogeneous medium.

36. Surface Waves. If we inspect the expressions given for $G_{TE}(\beta)$ and $G_{TM}(\beta)$ in Table 18-3 for the dielectric slab, poles occur in these functions for the lossless dielectric slab $(\tan \delta = 0)$ for specific values of $\beta = \beta n$. These poles occur at values of β determined by finding the roots of the transcendental equations

$$TE \text{ modes: } j\sqrt{N_2^2 - \beta n} = \sqrt{N_1^2 - \beta^2 n} \cot (k_0 d \sqrt{N_1^2 - \beta^2 n}) \qquad (18\text{-}46)$$

$$TM \text{ modes: } -jN_1^2 \sqrt{N_2^2 - \beta\, n} = \sqrt{N_1^2 - \beta^2 n} \tan (k_0 d \sqrt{N_1^2 - \beta^2 n}) \qquad (18\text{-}47)$$

and represent the cutoff conditions (eigenvalues) at the TE and TM surface waves which may propagate along a dielectric sheet, similar to that given by Collin.[79] It should be noted that only the even TM and odd TE surface-wave modes can propagate along a grounded dielectric slab.

The relationship between the regions of integration of Eq. (18-45) along the real β axis and the various terms composing the aperture admittance in a physical sense is important in understanding dielectric-covered antennas; details of the method of numerical integration of Eq. (18-45) for different waveguides are available.[64,68-70,75] The integral in Eq. (18-45) may be written symbolically as

$$y = (g_r + g_s) + ib_r$$
$$= C \int_{\beta=0}^{\beta=1} F_1(\beta)\, d\beta + C \int_{\beta=1}^{\beta=N} F_2(\beta)\, d\beta + C \int_{\beta=N}^{\beta=\infty} F_3(\beta)\, d\beta \qquad (18\text{-}48)$$

where $F_1(\beta)$, $F_2(\beta)$, and $F_3(\beta)$ are the integrands in Eq. (18-45), allowing for the root changes in different regions as discussed in Refs. 65, 70, and 75. The poles given by Eqs. (18-46) and (18-47) occur in the region between $\beta = 1$ and $\beta = N$ and are treated numerically by excluding a small region $\Delta\beta$ about each pole at β_n. The region inside $\Delta\beta$ is evaluated using the residue theorem of complex-variable theory.[80] As a result for the lossless dielectric slab, the radiation aperture conductance g_r, the aperture susceptance b_r, and the aperture surface-wave conductance g_s are obtained from the following terms in Eq. (18-48):

$$g_r = C \int_{\beta=0}^{\beta=1} F_1(\beta)\, d\beta, \qquad g_s = C\pi i \sum \text{residues}$$

$$b_r = C \int_1^N F_2(\beta)\, d\beta + C \int_N^\infty F_3(\beta)\, d\beta$$

Consequently surface waves can contribute only to the aperture conductance, and therefore g_s represents real power coupled into the dielectric layer. Extensive calculations of the aperture admittance for various waveguide antennas including the surface-wave conductance have been published.[64,68,70,75] From observations of these data it is concluded that, in general, the amount of surface-wave power in a dielectric slab is critically dependent upon the type of exciting aperture and the aperture size. The circular waveguide excited in the TE_{11} mode in turn excites less surface-wave energy than the TE_{01}-mode-fed rectangular waveguide. The annular slot in some cases couples most of the total input power into surface waves.[75]

37. Effects of Losses upon Surface Waves. If there are losses in the dielectric slab, the poles move off the real β axis and the distinct regions given in Eq. (18-48) merge. In this case the residue contribution is lumped into the total conductance. Physically this means that the surface wave will be attenuated as it travels along the slab and will be dissipated into heat. For finite-length slabs the excitation of the terminated edges will depend upon the loss. Since all real dielectric materials have some loss, the numerical problems encountered for the lossless slab are not as severe, although the integrand of Eq. (18-45) will still be rather bumpy for low-loss cases. Examples of how much dielectric loss affects the aperture admittance follow.

38. Aperture Admittance of Rectangular Waveguides. Calculations and measurements of the aperture admittance of rectangular waveguides radiating into a half space (including free space) were made in the early 1950s for a variety of width-to-height ratios.[61] Similar calculations have been made[64] using Eq. (18-45), assuming the dominant mode as a trial function. These calculations are compared with measured results in Fig. 18-17, where the waveguide flange was used as a ground plane. (Note that the X-band flange was 1.62 by 1.62 in. and the S-band flange was 6.42 by 6.42 in.) Other measurements were made with up to 10λ ground planes with similar results, as given in Fig. 18-17. Calculations and measurements of the rectangular waveguide under a dielectric slab were also performed using Eq. (18-45), and the X-band waveguide terminated in 1.62 by 1.62 in. flange.

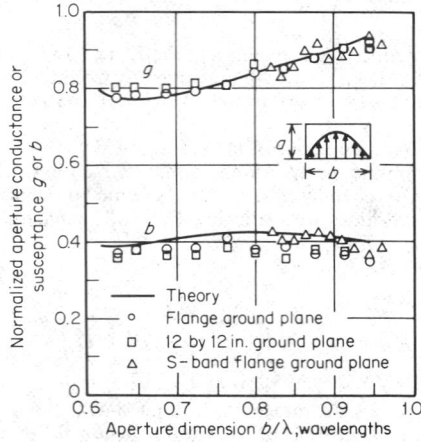

Fig. 18-17. Computed and measured aperture admittance of a rectangular waveguide, $b/a = 2.25$. *(From Croswell et al.[64])*

The results are given in Fig. 18-18, showing the effects of including the next higher mode in the aperture field along with calculations which include an air-gap tolerance in the measurements. The effects of air-gap tolerances and other slots are available,[81,82] as well as computer programs for a variety of waveguide antennas. Measurements were also performed using a 10λ ground plane with resulting data similar to those given in Fig. 18-16. The effect of losses upon the aperture admittance is given in Fig. 18-19. Further calculations for the same waveguide under various dielectric slabs are available.[83]

39. Aperture Admittance of Circular Waveguides. Calculations for the circular waveguide radiating both into free space and into dielectric slabs using Eq. (18-45) have been performed and compared with measurements.[69,70] Calculations for a 1.5-in.-diameter waveguide operating at C band are given in Fig. 18-20. Note that the circular waveguide is nearly matched when radiating into free space. Also the surface-wave conductance is a relatively small percentage of the total conductance, which means that this waveguide aperture couples less power into surface waves. Like that of the rectangular waveguide, the

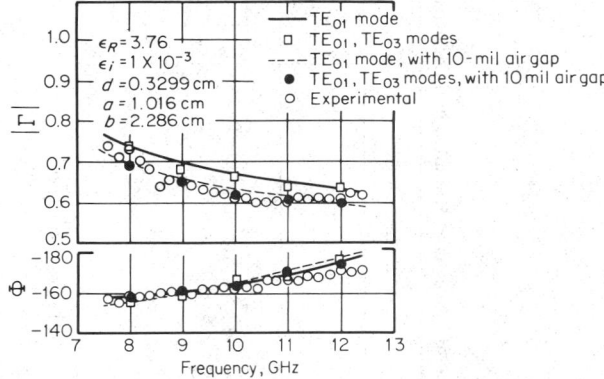

Fig. 18-18. Computed and measured reflection coefficient of a rectangular waveguide under a dielectric sheet; d = slab thickness, and b = width and a = height of the waveguide.

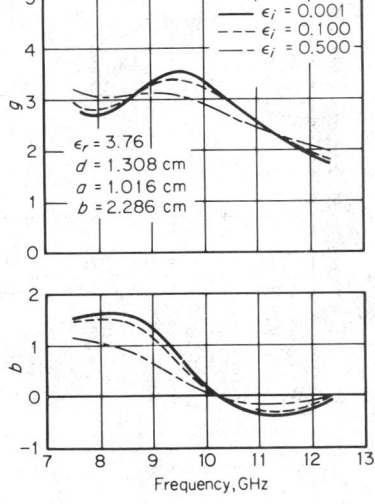

Fig. 18-19. Computed aperture admittance of a rectangular waveguide under a dielectric sheet as a function of dielectric loss; $\varepsilon_i = \tan \delta \, \varepsilon_r$, d = slab thickness, and b = width and a = height of the waveguide.

Table 18-3. Equations for the Admittance of Waveguides Radiating into Dielectric Media

Waveguide aperture (excitation mode)	C	$E_{TE}(\beta)$	$E_{TM}(\beta)$
Parallel plate (TEM) a = height	$\dfrac{k_0 a}{\pi}$	0	$\dfrac{1}{\beta}\left[\dfrac{\sin(k_0 a\beta/2)}{k_0 a\beta/2}\right]^2$
Rectangular (TE$_{01}$) a = height b = width	$\dfrac{2(k_0 a)(k_0 b)}{\sqrt{1-\left(\dfrac{\pi}{k_0 b}\right)^2}}$	$\displaystyle\int_0^{2\pi} d\alpha\,[G(\alpha,\beta)\sin^2\alpha]$	$\displaystyle\int_0^{2\pi} d\alpha\,[G(\alpha,\beta)\cos^2\alpha]$
Coaxial (TEM) a = small radius b = large radius	$\dfrac{1}{\ln\dfrac{b}{a}}$	0	$\left[\dfrac{J_0(k_0 a\beta) - J_0(k_0 b\beta)}{\beta}\right]^2$
Circular (TE$_{11}$) a = radius $x_{11} = 1.841$	$\dfrac{2}{(x_{11}^2-1)\sqrt{1-\left(\dfrac{x_{11}}{k_0 a}\right)^2}}$	$\left\{\dfrac{x_{11}}{k_0 a}\dfrac{\left[J_0(k_0 a\beta) - \dfrac{J_1(k_0 a\beta)}{k_0 a\beta}\right]}{\left(\dfrac{x_{11}}{k_0 a}\right)^2 - \beta^2}\right\}^2$	$\left[\dfrac{J_1(k_0 a\beta)}{\beta}\right]^2$

$$G(\alpha,\beta) = \left[\frac{\cos\dfrac{k_0 b\beta\sin\alpha}{2}}{\pi^2 - (k_0 b\beta\sin\alpha)^2}\right]^2 \left(\frac{\sin\dfrac{k_0 a\beta\cos\alpha}{2}}{\dfrac{k_0 a\beta\cos\alpha}{2}}\right)^2$$

External medium	$Y_{TE}(\beta)$	$Y_{TM}(\beta)$
	$\dfrac{jF'_1(\beta,0)}{k_0 F_1(\beta,0)}$	$\dfrac{k_0 N_1^2(0) A_1(\beta,0)}{jA'_1(\beta,0)}$
	$\sqrt{N_1^2-\beta^2}\;\dfrac{\dfrac{\sqrt{N_2^2-\beta^2}}{\sqrt{N_1^2-\beta^2}}+j\tan(k_0 d\sqrt{N_1^2-\beta^2})}{1+j\dfrac{\sqrt{N_2^2-\beta^2}}{\sqrt{N_1^2-\beta^2}}\tan(k_0 d\sqrt{N_1^2-\beta^2})}$	$\dfrac{N_1^2}{\sqrt{N_1^2-\beta^2}}\;\dfrac{1+j\dfrac{N_1^2\sqrt{N_2^2-\beta^2}}{N_2^2\sqrt{N_1^2-\beta^2}}\tan(k_0 d\sqrt{N_1^2-\beta^2})}{\dfrac{N_1^2\sqrt{N_2^2-\beta^2}}{N_2^2\sqrt{N_1^2-\beta^2}}+j\tan(k_0 d\sqrt{N_1^2-\beta^2})}$
	$\sqrt{N_1^2-\beta^2}\;\dfrac{1}{j\tan(k_0 d\sqrt{N_1^2-\beta^2})}$	$\dfrac{N_1^2}{\sqrt{N_1^2-\beta^2}}\,j\tan(k_0 d\sqrt{N_1^2-\beta^2})$
	$\sqrt{N_1^2-\beta^2}$	$\dfrac{N_1^2}{\sqrt{N_1^2-\beta^2}}$

N is the complex index of refraction, $N=\sqrt{\epsilon_r}$

aperture admittance at the 2λ ground-plane antenna closely approximates the infinite-ground-plane model.

40. Dielectric Plugs. The circular or rectangular waveguide aperture can be sealed with a dielectric plug and the aperture admittance computed by transforming the aperture admittance as given by Eq. (18-45). These transformations, similar to those derived by Swift,[66] are

$$Y_{in} = \frac{\tan kz_{mn} z_0 + \dfrac{jkz_{mn}\, y}{kz'_{mn}}}{y \tan kz_{mn} z_0 + \dfrac{jkz_{mn}}{kz_{mn'}}} \qquad (18\text{-}49)$$

where

$$y = \frac{kz_{mn}}{kz'_{mn}} \frac{(Yap + j \tan kz_{mn} z_0)}{(1 + jYap \tan kz_{mn} z_0)} \qquad (18\text{-}50)$$

where Yap is the admittance computed from Eq. (18-46) for the particular waveguide and $kz_{mn} = k_0 \sqrt{\varepsilon_r - (\lambda/\lambda_{cmn})^2}$, $kz'_{mn} = k_0 \sqrt{1 - (\lambda/\lambda_{cmn})^2}$, λ is the operating free-space wavelength, λ_{cmn} is the cutoff wavelength for the dominant waveguide mode, ε_r is the dielectric constant of the plug, $k_0 = 2\pi/\lambda$, and z_0 is the thickness of the plug. If the plug thickness z_0 is chosen so that $k_{zmn} z_0 = n\pi$, the input admittance equals the aperture admittance without the plug. The transformations given in Eq. (18-49) and (18-50) assume that only the dominant waveguide mode propagates in the plug-loaded section of the waveguide. For the rectangular waveguide care should be taken with dielectric plugs since resonances associated with the TE_{03} mode propagating in the plug-loaded section can occur.[66,84] These resonances can cause serious disruptions in both the admittance and radiation pattern; however, they are very narrow band and can usually be located by a swept-frequency VSWR measurement system. A similar effect has been found in plug-loaded horns used on reentry vehicles and waveguide-

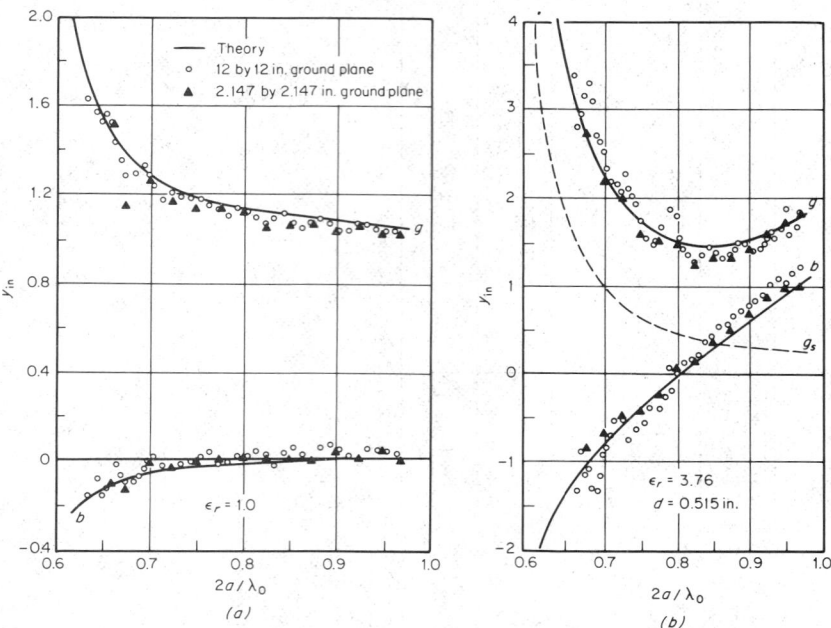

Fig. 18-20. Measured and computed aperture admittance of a circular waveguide: (a) free space, $\varepsilon_r = 1.0$; (b) dielectric-slab-covered, $\varepsilon_r = 3.76$, $d = 0.515$ in. *(From Bailey and Swift.[70])*

fed rectangular-cavity antennas. Such resonances have not been observed in dielectric-plug-loaded circular-waveguide antennas. This effect is similar to that found in dielectric-plug-loaded waveguide arrays.

41. Admittance of Waveguide Antennas and Cylinders. The aperture admittance of waveguide antennas mounted on curved surfaces is of interest for designing flush-mounted missile or aircraft antennas. Extensive calculations to determine the effect of cylinder curvature upon the aperture admittance of a TE_{01}-mode-fed axial slot have been performed.[85] These results clearly show that the axial slot mounted on a cylinder with a circumference $ka \geq 3$ has about the same aperture admittance as the same antenna mounted on an infinite ground plane. If the rectangular waveguide is rotated to be equivalent to the circumferential slot, it has been observed experimentally that this antenna is more sensitive to curvature since the H-plane dimension affects the waveguide wavelength.

42. Patterns of Waveguides on Finite Ground Planes. The problem of waveguide antennas on finite ground planes can also be treated by the geometrical theory of diffraction[86] in a manner similar to that for the wire antennas above a finite ground plane. Edge or diffraction effects will primarily occur only in the E plane (yz plane) of the circular or rectangular waveguide. In addition to the field radiated by the apertures as given in Table 18-1, the E field along the boundary of the ground plane $E_\theta(\theta = +\pi/2)$ will be diffracted by the edges of the ground plane and will add to the E_θ field radiated by the aperture, assuming an infinite ground plane. These first-order diffracted fields are given by

$$E_\theta^{(1)}(\theta) = E_\theta\left(\frac{\pi}{2}\right)\exp\left(\frac{jkl}{2}\sin\theta\right)\left[V_B\left(\frac{l}{2}, \frac{\pi}{2}+\theta, n\right)\right] \tag{18-51}$$

$$E_\theta^{(2)}(\theta) = E_\theta\left(\frac{-\pi}{2}\right)\exp\left(\frac{-jkl}{2}\sin\theta\right)\left[V_B\left(\frac{l}{2}, \frac{\pi}{2}+\theta, n\right)\right]$$

where $V_B(r,\psi,n)$ is the plane-wave diffraction defined by Eq. (3) in Ref. 51. This solution, which is the sum of the appropriate equations in Table 18-1 and Eq. (18-51), has been programmed, and the resulting patterns have been checked experimentally for both rectangular and circular waveguides. A summary of calculations for waveguides with different aperture sizes and ground-plane sizes is given in Figs. 18-21 and 18-22. It is important to note that apertures fed by waveguides larger than those where the next higher-order modes are cut off were included since small-angle horns preserve the dominant mode in the horn aperture that exists in the feed waveguide. It should also be noted that no diffractions occur in the E plane for a 1λ-wide rectangular aperture or a 1.22λ-diameter circular aperture. Indeed, for these aperture dimensions and the E- and H-plane patterns are nearly identical.

SLOT ANTENNAS

43. General Description. The slot antenna, cut in an infinite ground plane, is the complementary antenna to the strip dipole[87] and has pattern and impedance properties which can be related to the linear antenna as given in Table 18-1. Although the pattern of the thin slot on an infinite ground plane is similar for various types of slot-antenna configurations, the input impedance is highly dependent upon the type of feed network. Early work concerned the input impedance of waveguides with thin slots cut into the walls of the waveguide.[88,89] Using this work as a basis, Oliner developed a systematic design procedure for single slots in the wall of the rectangular waveguide in the form of equivalent circuits.[90,91] These equivalent circuits have been modified to account for a stratified medium outside the waveguide[92,93] and dielectric loading inside the feed waveguide.[94] Experimental data for the impedance of waveguides with slots are available,[91,95,96] including dielectric-slab covers[97-100] and plasma slabs.[101] Coupling between shunt slots in a waveguide has been considered, and extensive experimental data[102] are available. Simmons[103] has designed crossed slots in the broad wall of a rectangular waveguide to produce circular polarization. This very useful design has been extended to include dielectric covers[104] and has been constructed in arrays.[105] In addition to slotted waveguide antennas other very useful cavity-backed or waveguide-fed slot antennas have been reported, including design equations and experimental data. The dielectric-loaded waveguide-fed cavity antenna has been analyzed,[106] along with a dipole-fed cavity, to produce circular polarization,[107,108] and the T-bar-fed slot.[109] Very low profile slot antennas utilizing shallow cavities fed by coaxial cables have been designed for aircraft use[110] as well as strip-line-fed cavity-backed slots covered by dielectric slabs.[111]

The impedance properties of thin-slot antennas are relatively independent of ground-plane size if the ground plane is larger than about 1 to 2 wavelengths. Confirmation of this fact experimentally has been made.[112] Indeed it has been shown that the slot antenna mounted on a cylinder with $ka \geq 3$ has about the same impedance as a slot mounted on an infinite ground plane[85] and that the circumferential slot on a sphere has about the same impedance as the same slot mounted on an infinite cylinder, even for $ka \approx 1$ to 2.[113] Therefore there appears to be little practical justification for studying the impedance of

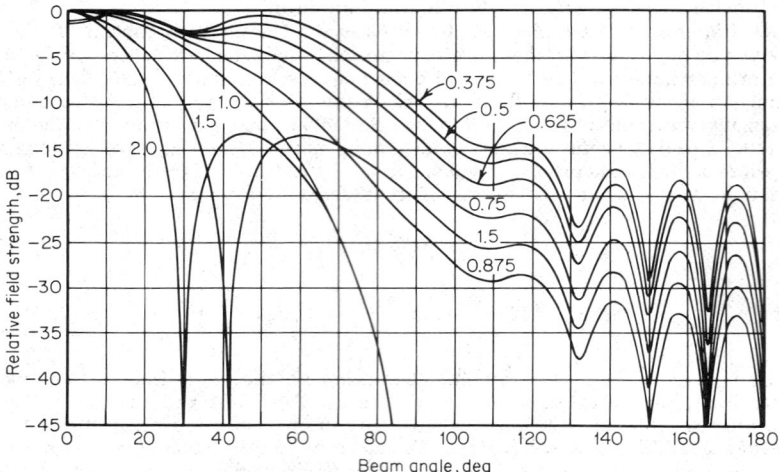

Fig. 18-21.   *E*-plane radiation patterns of a TE_{01}-mode-excited rectangular aperture opening onto a finite ground plane vs. aperture size in wavelengths. Ground-plane size = 4λ.

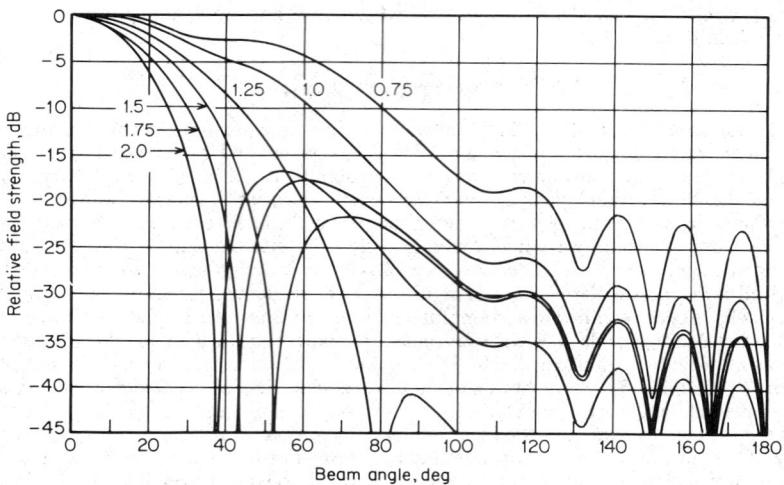

Fig. 18-22.   *E*-plane radiation patterns of a TE_{11}-mode-excited circular aperture opening onto a finite ground plane vs. aperture size in wavelengths. Ground-plane size = 4λ.

waveguide or slot antennas mounted on geometric shapes other than the plane unless the object is smaller than several wavelengths. Thus aperture admittance is sensitive only to the local region about the aperture. This is not true of the radiation patterns of slot antennas, which are very sensitive to both the size and radius of curvature of the mounting ground plane.

The patterns of slot antennas on finite ground planes have been studied both experimentally and theoretically,[55,76,112,114] including the effects of dielectric slabs.[115-117] Computer programs for many of these solutions are available at Langley Research Center. Some of the early work concerned patterns of slots on cylinders[118-122] with an excellent summary given by Wait[123] and by Compton and Collin.[124] One should be careful using these calculations[123] for precise agreement since some of these results have been found to be in error by several decibels in some angular regions. It is suggested that the formulations are generally correct and only require programming using modern computers. The patterns of slot antennas have been studied when mounted on spherical objects including spacecraft[4,5,125-127] and on cones.[128,129] From the geometrical theory of diffraction (GTD) the patterns of slots on elliptical cylinders[130,131] and on cylinders and three-dimensional objects of arbitrary convex cross section have been determined.[132]

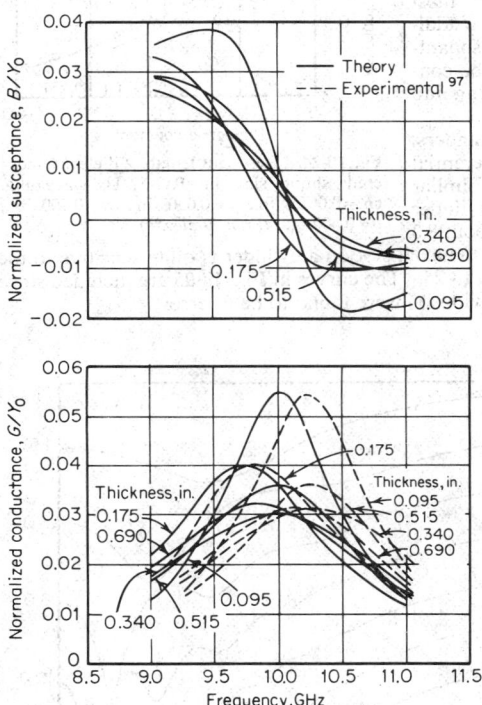

Fig. 18-23. Admittance of dielectric-covered shunt slot in RG-52/U waveguide ($\varepsilon = 2.0$, tan $\delta = 0.001$, $d = 0.093$ in., $a' = 0.475$ in., $b' = 0.0625$ in., $t = 0.050$ in., $a = 0.900$ in., $b = 0.400$ in.) *(From Bailey.[92])*

44. Narrow Slots and Covers in Rectangular Waveguides. The impedance properties of the slot in the broad wall of a waveguide can be obtained by modifying Oliner's equations[90,91] to include the addition of a stratified medium outside.[92,93] The power radiated by the slot in the broad wall of the waveguide is related to the orientation of the slot and the slot length. For maximum coupling the slot must be resonant, e.g., the input susceptance is zero. The general effect of adding a dielectric cover to a slot resonant in free space is to lower the resonant frequency, as shown for a typical case in Fig. 18-23. The free-space resonant frequency of the slot in Fig. 18-23 is 11.5 GHz. Also notice in this figure that as the thickness of the dielectric is increased, the change in resonant frequency and conductance deviates about a central value. This central value, as it turns out, is the case of a semi-infinite thickness of dielectric material. For dielectric layers thicker than about $0.2\lambda\varepsilon$ the admittance and resonant length are within 5% of the value for a semi-infinite medium. The resonant length of a shunt slot as a function of the dielectric constant of an external semi-infinite medium is given in Fig. 18-24 along with approximate values based upon empirical and quasi-static approximation.[97]

For design purposes, the slot length should be chosen using Fig. 18-24 or computed for a particular slab thickness using modifications to Oliner's formulas by Bailey.[92] If round ends are used for the slots, round ends being practical for some machining processes, the resonant length must be adjusted accordingly from experimental data. Since modern electrode-burning processes are now available, square-end slots as analyzed by Oliner can be used directly. Experimental measurements are required to determine the resonant length to an accuracy better than ±5%.

Slots cut into the narrow wall of a standard waveguide are not resonant without cutting into the top wall of the waveguide.[95] Therefore antennas employing such slots are, for the most part, designed experimentally. Since the addition of a dielectric cover reduces the resonant-slot length, resonant-slot antennas can be constructed in the narrow wall without cutting into the broad wall of the waveguide.

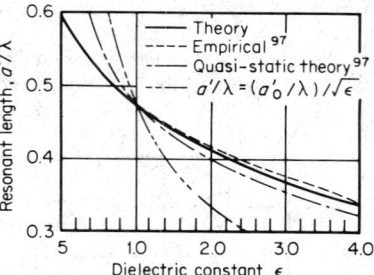

Fig. 18-24. Resonant length of dielectric-covered shunt slot in RG-52/U waveguide ($b' = 0.0625$ in., $t = 0.050$ in., $a = 0.900$ in., $b = 0.400$ in.) *(From Bailey.[92])*

45. Patterns of Narrow Slots in Cylinders. The radiation pattern of a thin circumferential slot on a cylinder is about the same as a similar slot on an infinite ground plane if the cylinder circumference $C = ka$ is greated than about 8.0 to 9.0 (Ref. 60, P. 250). The pattern of the axial slot on a cylinder is quite sensitive to the mounting cylinder size, as shown in Fig. 18-25. The curves in Fig. 18-25 are included since earlier calculations,[123] as mentioned previously, were found to be in error.

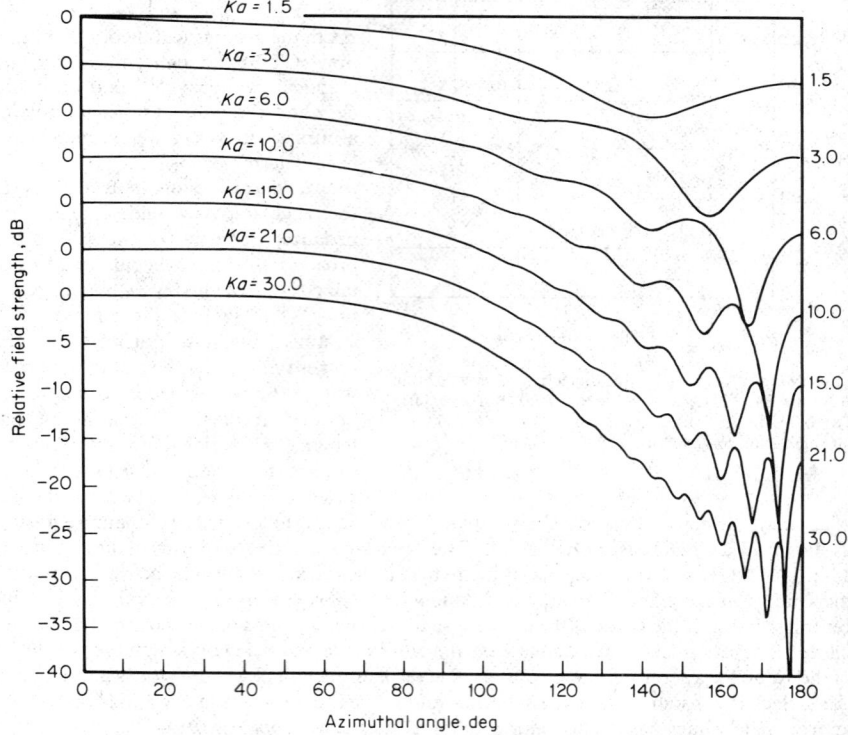

Fig. 18-25. Radiation patterns of an axial infinitesimal slot on a cylinder vs. cylinder circumference Ka in wavelengths. Note that the vertical scale for each pattern is displaced 5 dB for clarity.

46. Patterns of Slots on Finite, Coated Ground Planes.   The pattern of the thin slot on a finite ground plane is very sensitive to the ground-plane size and the thickness and dielectric properties of the coating.[115] With the coating, energy is coupled out at the slot onto a surface wave which propagates to the truncated edges and then radiates. This edge radiation interferes with the direct radiation from the slot and produces a ripple structure in the pattern. The validity of this solution has been verified experimentally for thin coatings. Typical patterns for a particular coating are given in Fig. 18-26. Note that for a large ground plane the losses in the dielectric slab damp out the surface waves.

HORN ANTENNAS

47. Horn Antennas.   The horn antenna may be thought of as a natural extension of the dominant mode waveguide feeding the horn in a manner similar to the wire antenna, which is a natural extension to the two-wire transmission line. The most common type of horns are the E-plane sectoral, H-plane sectoral, and pyramidal horn, formed by expanding the walls of the TE_{01}-mode-fed rectangular waveguide or the conical horn formed by expanding the wall of the TE_{11}-mode-fed circular waveguide. Early work concerned the determination of the forward radiation patterns, directivity, and approximate impedance of sectoral, pyramidal, and conical horns, including comprehensive experimental studies.[133-142] An excellent summary of this work is given by Compton and Collin.[143] In later work, the input impedance, wide-angle side lobes, and back lobes of certain horn antennas have been determined to a high precision using the GTD.[144-148]

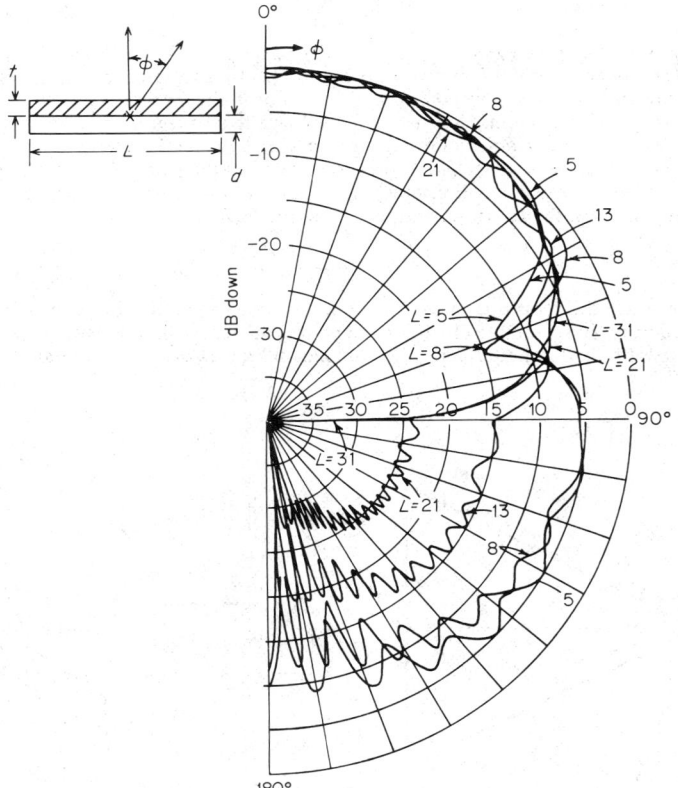

Fig. 18-26.   Radiation patterns of a dielectric-coated finite ground plane as a function of the ground-plane size L in wavelengths; $d = 1\lambda$, $t = 0.1666\lambda$, $\varepsilon_r = 2.3$, tan $\delta = 0.100$.

Special horn types, such as the multimode horn,[148-152] diagonal horn,[153] and corrugated horn,[152,154,155] have the primary feature of reducing *E*-plane side lobes to levels similar to that in the *H*-plane of the corresponding horn. This improvement in *E*-plane side-lobe level reduces the on-axis gain but results in a remarkable improvement in the beam efficiency. Horns modified with plates,[156] chokes,[157] grilles,[158] pins,[159] and wires[160] produce pattern shaping, including reduction of *E*-plane side lobes. Dielectric-loaded waveguides and horns offer improved pattern performance over unloaded horns. Ridged[167] and tapered[168] horn designs improve the bandwidth characteristics, and dual polarized configurations are also available. [169] Other useful horn designs include the horn-reflector antenna.[170] Experimental studies have determined horn-radiation properties to a great precision. The precise phase center has been determined,[171] as have the precise gain,[148,172] beam efficiency, wall losses, and input VSWR.[152]

48. Sectoral and Pyramidal Horns. Consider an *H*-plane sectoral horn, with the geometry given in Fig. 18-27, fed by a rectangular waveguide supporting the dominant TE_{01} mode. The modes that can exist in this horn are the TE_{0m} modes of the form[134]

$$E_x = A \cos \frac{m\pi\phi}{\phi_0} H_{vm}^{(1)}(k_0\rho)$$

$$H_\rho = -jA \frac{m\pi}{K_0\eta_0\phi_0\rho} \sin \frac{m\pi\phi}{\phi_0} H_{vn}^{(1)}(K_0\rho)$$

$$H_\phi = -j\frac{A}{\eta_0} \cos \frac{m\pi\phi}{\phi_0} H_{vm}'^{(1)}(K_0\rho)$$

where η_0 is the free-space wave impedance, $H_{vm}'(\psi) = d/dx\, H_{rm}(\psi)$, $vm = m\pi/\phi_0$, and *m* is the mode number. For the *H*-plane sectoral horn fed by a rectangular waveguide the throat dimension $\rho_0\phi_0 = a$ will support only the $m = 1$ mode in the throat horn section. If the angle of the horn is less than about 18°, the TE_{01}-dominant-fed waveguide mode will appear at the aperture without appreciable contribution from the higher-order modes. The curvature of the phase fronts in the horn section, however, means that the plane-horn aperture mouth will exhibit a phase error which will vary as $-k_0y^2/2\rho_1$ over the aperture. As a result, for the *H*-plane sectoral horn the aperture field will be of the form

$$E_x(x,y) = A \cos \frac{\pi y}{a} e^{-jk_0 y^2/\rho_1}$$

If the currents on the outside of the horn aperture are neglected, this aperture field can be integrated using Eqs. (18-7) to (18-10) to obtain the far-field radiation pattern. A normalized form of these radiation patterns for the *E*-plane and the *H*-plane of sectoral and pyramidal

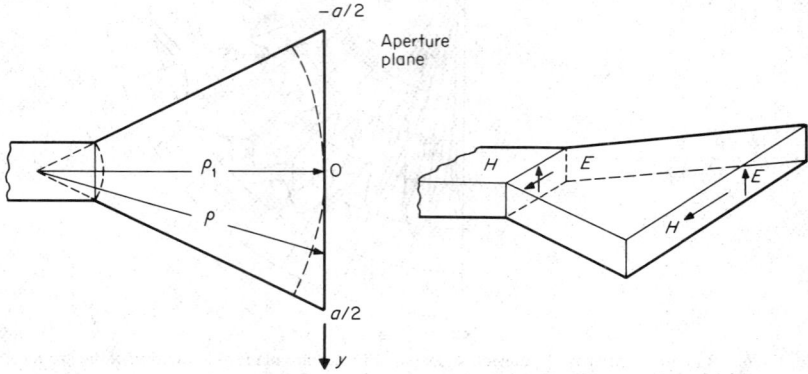

Fig. 18-27. Geometry of a *H*-plane sectoral horn. *(From Compton and Collin.[143])*

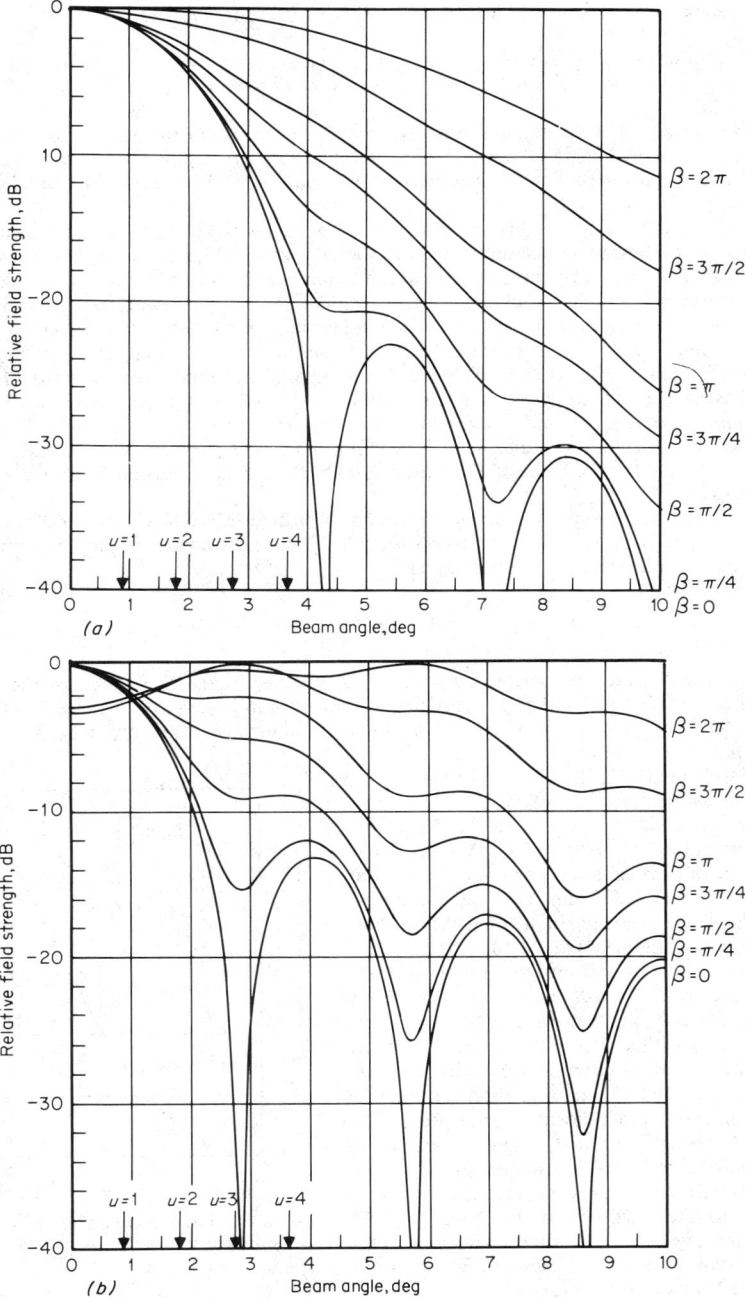

Fig. 18-28. (a) *H*-plane pattern of a 20λ horn with various amounts of phase taper of the form $\beta x^2/a^2$. These data can be scaled for other apertures using the parameter $u = 2\pi a/\lambda \sin \theta$, as noted on the beam-angle scale. (b) *E*-plane pattern of a 20λ aperture horn with various amounts of aperture phase taper of the form $\beta y^2/b^2$. These data can be scaled for other aperture sizes using the parameter $u = 2\pi b/\lambda \sin \theta$, as noted on the beam-angle scale.

horns greater than 2λ in aperture dimension is given in Fig. 18-28. Note that for small horns, with little phase error, the patterns of large-waveguide antennas given in Fig. 18-21 should be used.

For horns with wide angles the aperture may have more than one mode, and the radiation patterns given in Fig. 18-28 may be in error. The geometrical theory of diffraction can be used to compute patterns of such horn geometries as well as those with smaller plane angles.[152]

Although the E-plane sectoral horn and H-plane sectoral horn may be considered as extensions of the feed waveguide into a cylindrical region, which can then be analyzed using cylindrical modes, the pyramidal horn cannot be analyzed. For practical purposes, however, the patterns in the E and H planes of the pyramidal horn are nearly identical to those of the E- or H-plane sectoral horn with the same dimensions in corresponding planes.

The impedance of horns which have an angle less than 18° has been determined by Cockrell[173] to be the aperture impedance of a rectangular waveguide excited in the TE_{01} mode and having the same dimensions as the horn aperture. Cockrell performed calculations for H-plane sectoral horns and pyramidal horns using the equations given in Table 18-3 and verified the calculations experimentally for horns radiating into free space and dielectric slabs. The input impedance of E-plane sectoral horns can be determined either similarly or using GTD.[148]

The gain of pyramidal and sectoral horns has been determined to accuracies of about ± 0.2 dB even for frequencies as high as 38 GHz.[172] The most common gain standard is the pyramidal horn, which has a gain equal to[174]

$$\text{Gain}_{dB} = 10(1.008 + \log a_\lambda\, b_\lambda) - (L_E + L_H) \qquad (18\text{-}52)$$

where a_λ and b_λ are the aperture dimensions in wavelengths and L_E and L_H are the loss due to phase error in the E and H planes of the horn as given in Fig. 18-29. Gain curves for other horns and an excellent summary of horn-design information are given by Jakes[174] and Compton and Collin.[143]

49. Conical Horns. The conical horn formed as an extension of the circular waveguide excited in the TE_{11} mode has been thoroughly studied by King.[141] The radiation patterns of this antenna for horn angles less than about 18° can be obtained by integrating the dominant TE_{11}-mode field, with quadratic phase error, using Eqs. (18-7) to (18-10). As in rectangular-waveguide-fed horns, the fields outside the aperture are neglected, and therefore the wide-angle side lobes and back lobes will be computed incorrectly with such a procedure. The impedance of conical horns with an angle less than 18° can be obtained in a manner similar to that for pyramidal or sectoral horns. In this method the impedance is computed using the equations given in Table 18-3 and assuming that the TE_{11}-mode-excited circular waveguide has a diameter equal to the diameter of the horn aperture. Care must be used in exciting this horn since any feed asymmetry may excite the TM_{01} mode in the feed waveguide.

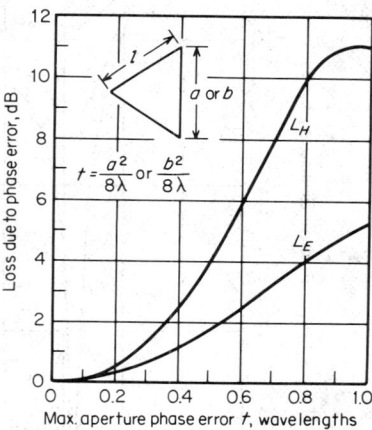

Fig. 18-29. Loss correction for phase error in sectoral and pyramidal horns. *(From Jakes.[174])*

The gain of the conical horn has been determined to be [176]

$$\text{Gain}_{dB} = 20 \log C_\lambda - L \qquad (18\text{-}53)$$

where C_λ is the circumference of the horn aperture and L is the gain loss due to phase error given by the curve in Fig. 18-30. It should be carefully noted that the gain given in Eqs. (18-52) and (18-53) neglects the losses in the conducting walls of the entire horn and the VSWR. For many applications, such as absolute radiometric calibrations, the horn-wall losses must be accounted for and measured. Available measurement techniques for determining low-wall losses include shorting the horn aperture and measuring the input VSWR using a four-probe technique.[152]

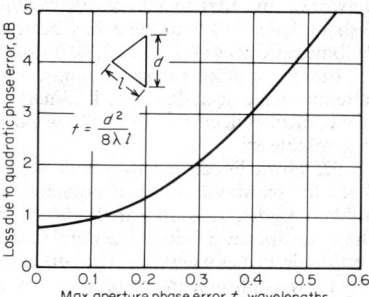

Fig. 18-30. Loss correction for phase error in conical horns. *(From Jakes.[174])*

50. **Low-Side-Lobe Design.** In addition to using a corrugated surface or loading of some kind, a simple small-aperture low-side-lobe antenna can be designed over a restricted bandwidth using a particular choice of horn-aperture dimensions. Notice in Figs. 18-21 and 18-22 that the rectangular-waveguide TE_{01} mode, with a 1λ E-plane dimension, and the circular-waveguide TE_{11} mode, with a 1.22λ diameter, have no E-plane side lobes. The patterns predicted using these dimensions have been experimentally verified using a small-angle square or conical horn to feed the particular aperture. It should be noted that the E- and H-plane patterns of these particular horns are nearly identical. The possible explanation for this performance can be obtained from GTD, in that the E-plane edge diffractions from the two edges of the horn cancel precisely for the dimensions stated. Note, however, that a similar cancellation will not occur for horn dimensions which are multiples of those stated.

REFLECTOR ANTENNAS

51. **The reflector antenna** is formed by a radiating feed antenna and some sort of reflecting ground plane. Simple forms of reflector antennas, the loop or dipole spaced over a finite ground plane, were described in detail in Pars. **18-30** to **18-33**. A further example of such an antenna is the dipole in the corner reflector, extensive experimental design data for which have been reported.[175,176] The most common type of reflector antenna is one where the reflector is parabolic or spherical. Unusual reflector antennas include the plane-surface reflector array,[177] which can be collimated and/or scanned by adjustable switches in the elements that form the reflecting surface, and the multiplate antenna[178,179] where the reflecting surface is segmented and adjusted mechanically to produce a collimated and scannable secondary-antenna pattern.

The theory commonly used for the direct-fed parabolic antenna is that of Silver,[180] using physical optics. As outlined by Silver, this theory is adequate to predict the gain, radiation pattern, and the level of the first few side lobes of the secondary pattern, neglecting blockage and scattering by feed struts. The effect of feed-strut blockage can be estimated using geometrical optics;[181] however, an analysis of the diffraction of struts using a more rigorous formulation is necessary to improve the quantitative understanding of the problem. The radiation-pattern characteristics of the offset-fed parabola have been determined approximately using an extension of Silver's formulation to include this geometry.[182,183]

The spherical reflector is a good design for a scanning reflector antenna thanks to its geometrical symmetry; however, a point source at the focal region will not produce a set of parallel rays from the secondary reflector. To correct this spherical-aberration error a line-source feed is employed.[184,185] By phasing this line source the beam can be scanned to other positions.

Another way of feeding the parabolic reflector antenna is to use a subreflector in the focal region of the parabola and illuminate the subreflector from the parabolic surface. The principal advantages of the Cassegrain system or other methods of folded optics are the increase in effective f/d ratio and the simple mechanical location of the feed so that cooled receivers used with such antennas can be serviced in a more practical manner.

The chief disadvantage is the aperture blockage of the subreflector, which restricts the application of this principal to large apertures. The Cassegrain antenna has been analyzed extensively, and computer programs are available.[186-195] This design effort resulted in the design and construction of the 210-ft dish antenna[196] used in the worldwide space-receiving network. In order to achieve all-weather operating capability a precision 120-ft Cassegrain dish under a 150-ft radome has been constructed at Millstone Hill, Massachusetts. The radome produces about 1 to 2.8 dB loss in the microwave to millimeter-wavelength region.[197]

Besides ground-based antennas, special reflector types for use as erectable spacecraft antennas have been developed. Analysis of the gored reflector[198] is available, as is the design of a conical-reflector antenna.[199] An excellent review of reflector-antenna theory and design is available.[200]

52. Horn Feeds for Parabolic-Reflector Antennas. One of the most commonly used feeds for parabolic-reflector antennas is the flared-horn antenna fed by rectangular waveguide. As discussed in Par. **18-48**, the rectangular horn with a flare angle less than 18° has the same aperture field as the dominant-mode rectangular waveguide feeding the horn. This section describes a design procedure for determining the dimensions of a TE_{01}-mode-fed horn which will optimize the resultant aperture efficiency of the parabolic-reflector antenna. The design procedure and the equations for calculating the radiation-field parameters are based upon the work of Rudge and Withers,[201] with modifications.[202] The procedure does not account for tolerance errors, aperture blockage, or scattering by support struts.

The principal components of the focal-plane electric field distribution of a parabolic reflector are related to the reflector-aperture plane-field distribution by the expression

$$E(t,\phi') = \frac{jk}{2\pi} \int_0^{2\pi} \int_0^{\hat{u}} \frac{F(u,\phi)}{(1-u^2)^{1/2}} \exp[jktu\cos(\phi-\phi')]\, u\, du\, d\phi'$$

where the coordinate system is given in Fig. 18-31 and $u = \sin\theta$, $\hat{u} = \sin\theta_{max}$, $k = 2\pi/\lambda$. An example of what the focal-plane distribution from an incoming plane wave looks like for several different parabolic reflectors is given in Fig. 18-32. By adjusting the E- and H-plane widths of the rectangular feed horn to correspond to the -10-dB lobe width of the focal-plane distribution the maximum account of energy will be collected from the parabolic reflector by the TE_{01} mode in the feed horn.

Then by definition the optimum TE_{01}-mode horn will be one whose dimensions are adjusted to fit the -10-dB levels of the focal-plane field. These optimum dimensions are

$$2\hat{x} = 0.95W_u \quad E \text{ plane}$$

$$2\hat{y} = 1.29W_u \quad H \text{ plane}$$

where W_u is the focal-plane 10-dB lobe width in wavelengths. Since the peak value of the main lobe of the focal-plane field is a function of the reflector f/d ratio due to reflector curvature, the maximum reflector-aperture efficiency, beam width, spillover, and side-lobe level are functions of the aperture diameter and the f/d ratio. The horn dimensions for the optimized horn will result in a horn pattern whose shape is symmetrical; over the illuminated part of the dish it can be approximated by $\sin[(h\sin\theta)/h]$, where $h = 2\hat{x}\pi/\lambda$. From this approximate expression, the far-field pattern of the reflector antenna fed by this horn for small angles ψ off of boresight can be expressed as

$$E(\alpha,n) = \int_0^{\theta_{max}} \frac{\sin(h\sin\theta)}{h} J_0\!\left(4\pi\alpha n \tan\frac{\theta}{2}\right) d\theta$$

where $\alpha = f/d$, $\psi = n\lambda/d$, and d is the diameter of the parabola in wavelengths. Although this field expression is certainly approximate, it has been found sufficiently accurate for design purposes. Based upon this focal-plane concept the following simple three-step design procedure can be used:

1. Given the f/d ratio, determine $\hat{u} = \sin\theta_{max}$ from Fig. 18-33.
2. Given the f/d ratio, obtain the -10-dB focal-plane lobe width W_u, from Fig. 18-33.
3. From $\hat{u}$ determine the correction T_c due to the curvature of the dish from Fig. 18-34.

Using the values of $\hat{u}$, W_u, and T_c thus determined, one can obtain the following parameters of interest:

Feed-horn dimensions: $2\hat{x} = 0.95W_u$, $2\hat{y} = 1.29W_u$.
Reradiated power (not collected by the horn) $= 2.6T_c\%$.

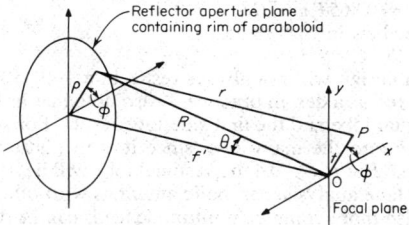

Fig. 18-31. Geometry of parabola, including focal-plane geometry. *(From Rudge and Withers.[201])*

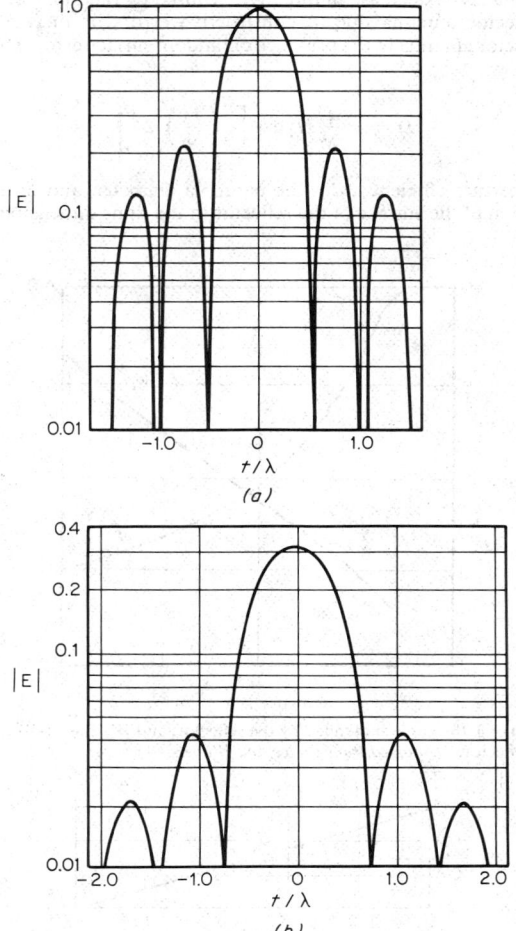

Fig. 18-32. Principal component of the focal-plane electric field distribution: (*a*) $f/d = 0.25$; (*b*) $f/d = 0.50$. *(From Rudge and Withers.[201])*

Aperture efficiency: $N_A = 86.5T_c\%$.
Gain $= N_A \pi^2 d^2/\lambda^2 = 0.865T_c\pi^2 d^2/\lambda^2$.
Side-lobe level in decibels in Fig. 18-35.
Beam width in Fig. 18-36.

Note that this optimum design will not always result in a -10-dB reflector-aperture plane taper, commonly employed as a design practice. Also note that in Fig. 18-35 a distinction is made between the vestigial lobe and the first side-lobe level. For f/d ratios less than 0.4 the vestigial lobe can merge into the main lobe, since it is in phase with the main lobe, and produce a ridged pattern. For $f/d > 0.4$ the vestigial lobe will be no higher than the first side lobe. Using the focal-plane analysis, parabolic antennas with other than "optimum horns" can be evaluated.[201] Also more complex multimode feeds can be designed which collect all the significant focal-plane energy at the expense of increased blockage; or in the case of off-axis or offset-fed parabolic reflectors the feed can be adjusted to correct for loss of gain due to coma lobes, etc.[203]

53. Random Errors. The major limitation in achieving the maximum aperture efficiency of a reflector antenna is the decrease in directivity caused by random tolerance errors in the surface of the reflector antenna due to constructive errors or thermal distortion. The directivity of a reflector antenna in the presence of random surface errors has been determined by Ruze[204] to be

$$D = \pi^2 \frac{d^2}{\lambda^2} N_A \exp\left[-\left(\frac{4\pi}{\lambda_0}\right)^2 \bar{\Delta}^2\right] \tag{18-54}$$

where N_A is the aperture efficiency, d is the parabola diameter, and $\bar{\Delta}^2$ is the mean square mechanical distortion of the surface of the reflector in the same dimensions as the free-space wavelength λ_0.

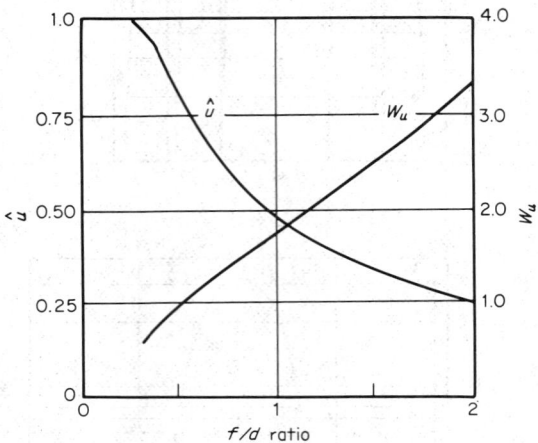

Fig. 18-33. Variation of $\hat{u}$, the angle subtended by the reflector, and W_u, the 10-dB lobe-width factor, with the f/d ratio of the reflector. *(Adapted from Rudge and Withers.[201])*

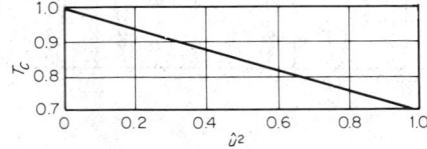

Fig. 18-34. Reflector-curvature correction factor T_c. *(From Rudge and Withers.[201])*

ARRAYS

54. Radiation Patterns. The simple equations for the radiation patterns of point sources or uncoupled antennas, using pattern multiplication, are outlined in Pars. **18-8** to **18-11**. The basic mutual-coupling impedance formulas for wire antennas are outlined in Par. **18.21**, which, with the proper definition of terms, can be applied to arrays of other antenna types. The basic design problem in antenna arrays, including the prediction of array radiation characteristics, is related to main-side-lobe and grating-lobe properties and input impedance.

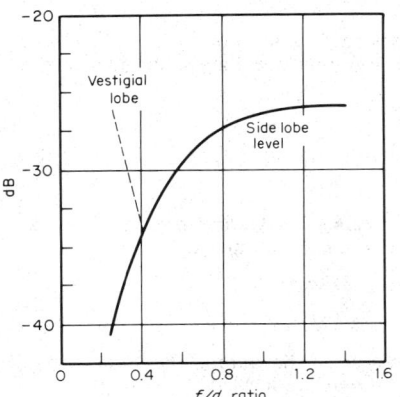

These fundamental characteristics are functions not only of the array spacing and operating wavelength but also of the array geometry and scan angle. The quantitative determination of array radiation properties can be made only by computations using an accurate analysis for a particular element type or by measurement. On a qualitative basis there are some conceptual results which are useful and give a physical insight into array design and performance. A few of these concepts will be discussed briefly here. A good summary of the work in phased arrays, including theory, methods of simulation, and design, is given in a special issue of the *Proceedings of the IEEE*.[205]

Fig. 18-35. Predicted level of first side lobes in principal planes or radiation pattern relative to peak of main beam. Circular reflector fed by optimized rectangular horn. *(From Rudge and Withers.[201])*

55. Grating Lobes in Planar Arrays. Consider a planar array located in the $x + y$ plane in Fig. 18-37 with the spacing and phasing as shown. As derived by von Aulock[206] the radiation pattern of a planar array can be written in terms of the direction cosines $\cos \alpha_x$ and $\cos \alpha_y$, where

$$\cos \alpha_x = \sin \theta \cos \phi \qquad \cos \alpha_y = \sin \theta \sin \phi$$

The direction of beam scan is measured by the angle ϕ counterclockwise from the $\cos \alpha_x$ axis as given by the direction cosines $\cos \alpha_{xs}$, $\cos \alpha_{ys}$, where

$$\phi = \tan^{-1}[(\cos \alpha_{ys})/(\cos \alpha_{xs})]$$

The angle of scan θ can be determined from $\sin \theta$, which is the distance of the point ($\cos \alpha_{xs}$, $\cos \alpha_{ys}$) from the origin. All physically observable grating lobes occur in the region

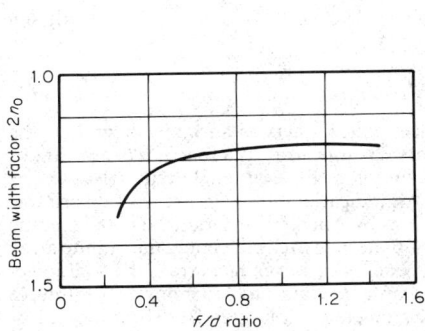

Fig. 18-36. The beam-width factor for circular reflectors. The -3-dB beam width is given by $2n_0\lambda/d$. *(Courtesy of A. W. Rudge.[201])*

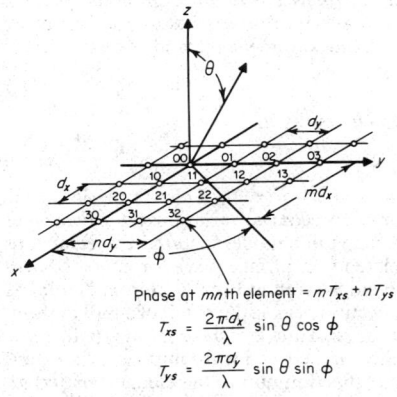

Phase at mnth element $= mT_{xs} + nT_{ys}$

$$T_{xs} = \frac{2\pi d_x}{\lambda} \sin \theta \cos \phi$$

$$T_{ys} = \frac{2\pi d_y}{\lambda} \sin \theta \sin \phi$$

Fig. 18-37. Planar-array element geometry and phasing.

$\cos^2\alpha_x + \cos^2\alpha_y \leq 1$. The array factor of a rectangular $M \times N$ array scanned to a direction given by $\cos \alpha_{xs}$, $\cos \alpha_{ys}$ can be expressed as [206]

$$E(\cos \alpha_{xs}, \cos \alpha_{ys}) = \sum_{m=0}^{M-1} \sum_{n=0}^{N-1} |A_{mn}| \exp\{ j[m(T_x - T_{xs}) + n(T_y - T_{ys})]\} \qquad (18\text{-}55)$$

where

$$T_x = (2\pi/\lambda)d_x \cos \alpha_x \qquad T_y = (2\pi/\lambda)d_y \cos \alpha_y$$
$$T_{xs} = (2\pi/\lambda)d_x \cos \alpha_{xs} \qquad T_{ys} = (2\pi/\lambda)d_y \cos \alpha_{ys}$$

and A_{mn} is the excitation amplitude of the mn^{th} element. The location of the grating lobes is highly dependent upon the array configuration. The grating lobes for a rectangular array are located at

$$\begin{aligned} \cos \alpha_{xs} - \cos \alpha_x &= \pm(\lambda/d_x)p \\ \cos \alpha_{ys} - \cos \alpha_y &= \pm(\lambda/d_y)q \end{aligned} \quad \text{for } p, q = 0, 1, 2, \ldots \qquad (18\text{-}56)$$

where the lobe for $p = q = 0$ represents the main beam. The grating lobes for a triangular lattice with elements of (md_x, nd_y), where $m + n$ is even, are located at

$$\begin{aligned} \cos \alpha_{xs} + \cos \alpha_x &= \pm (\lambda/2d_x)p \\ \cos \alpha_{ys} - \cos \alpha_y &= \pm (\lambda/2d_y)q \end{aligned} \qquad (18\text{-}57)$$

where the lobe for $p = q = 0$ represents the main beam. The grating lobes for a triangular lattice with elements of (md_x, nd_y), where $m + n$ is even, are located at

$$\lambda/d_x = \lambda/d_y \leq 1 + \sin \theta_m \qquad (18\text{-}58)$$

and for a triangular lattice array

$$\lambda/d_y = \lambda/\sqrt{3}d_x \leq 1 + \sin \theta_m \qquad (18\text{-}59)$$

Note that for linear arrays the condition given in Eq. (18-58) holds. In general for scanning arrays no grating lobes will ever exist in real space for element spacings less than or equal to $\lambda/2$. It may be observed that the triangular lattice arrays require less element population than the square array to obtain equivalent grating-lobe suppression. Other triangular grids or configurations can be obtained with even less element population[207] for the same grating-lobe performance.

56. Input Impedance as a Function of Scan Angle. As a first approximation a large array can be treated as large continuous aperture. Wheeler[208] has shown that the impedance of a large aperture approximating an array may be thought of as an impedance sheet whose normalized impedance and reflection coefficient vary as

$$\eta_{\text{aperture}} = (1 - \sin^2\theta \cos^2\phi)/\cos \theta \qquad (18\text{-}60)$$

which results in

$$|\Gamma| = \tan^2(\theta/2) \qquad (18\text{-}61)$$

for $\phi = 0°$ or $\pi/2$. This simple impedance-sheet concept acts as an upper bound on the input/impedance variation of the central element of a large array as a function of scan angle. Perhaps this model inspired the erroneous description of the scan-angle reflection peak as related to a surface wave for uncoated arrays. An indication of the accuracy of this simple bound is given in Fig. 18-38 from dipole calculations by Allen.[209] It should be noted that the reflection peak is the result of a null in the element pattern of each element in the infinite array at the scan angle corresponding to the reflection peak. For waveguide arrays the addition of fences in between waveguides[210] will extend the scan-angle range to wider angles by filling in the reflection null in the element patterns. Placing dielectric plugs and/or sheets in and/or over waveguide arrays can have the same effect as metal fences[211] for particular choices of parameters, since higher-order modes induced in the apertures fill in the element patterns in the same manner as fences. The impedance variation as a function of scan angle, for infinite arrays, can be obtained using waveguide simulators.[212,213]

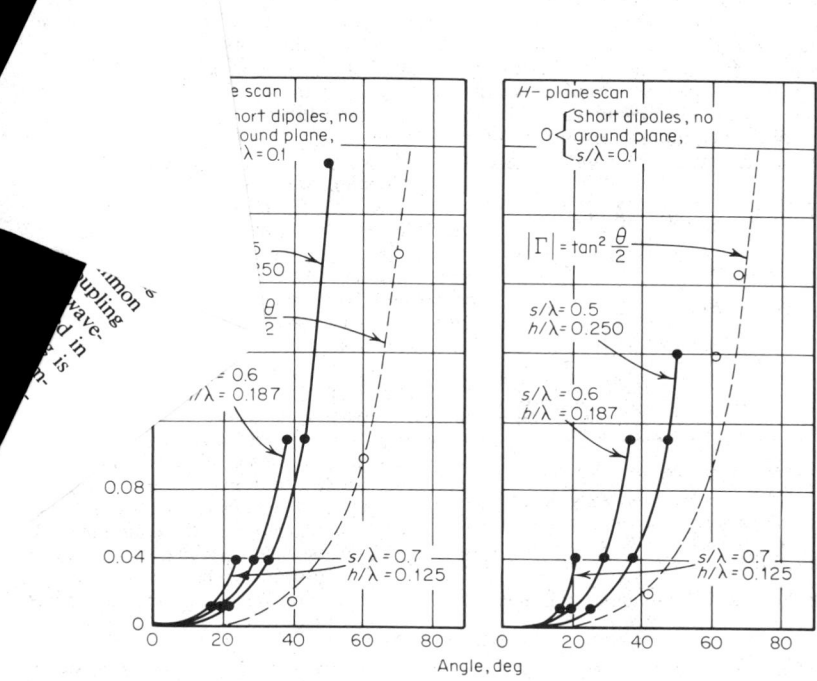

Fig. 18-38. Scanned mismatch variation for different element spacings (h/λ is the dipole spacing above a ground plane). *(Calculations by Allen.[209])*

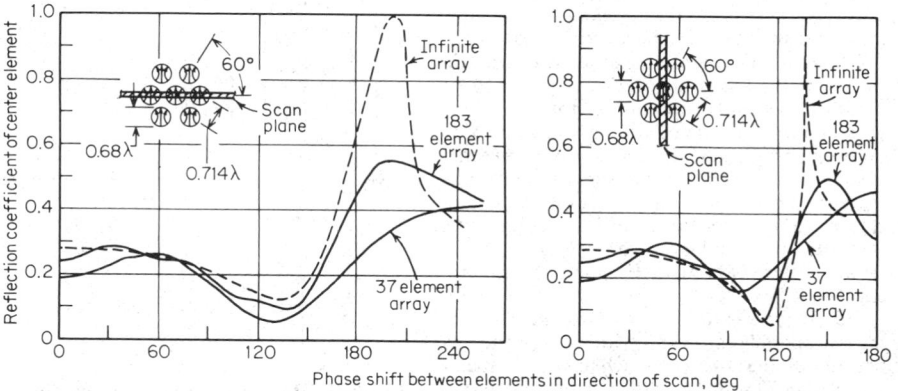

Fig. 18-39. Reflection coefficient of the central element of a triangular grating of circular waveguides excited in the TE_{11} mode, as a function of scan angle. *(Courtesy of M. C. Bailey, Langley Research Center.)*

The coupling effects of elements in finite-sized arrays of antenna ele. impedance of a single element are uniquely dependent upon the detailed tion, number of elements, and the location of the element in the array. peak observed for an element in a finite array at a particular scan angle is m for finite arrays. As an example, consider an array of dominant TE_{11}-m waveguides located in a triangular grid. The reflection coefficient of the cen finite arrays and an infinite array of circular waveguides is given in Fig. 18-39. grating calculations are by Amitay and Galindo.[214]

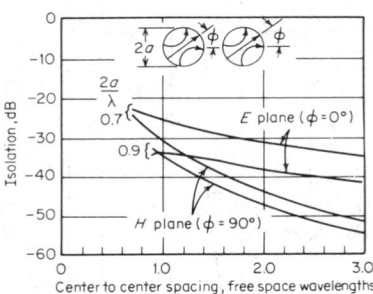

Fig. 18-40. Coupling between TE_{11} modes in two circular apertures radiating into free space. *(Courtesy of M. C. Bailey, Langley Research Center.)*

57. Coupling between Pairs of
Many times the antenna designer is with the relatively simple problem of between pairs of antennas mounted on a c ground-plane. A typical waveguide-co problem of this type is that of two circular guides excited in the TE_{11} mode, as depicte Fig. 18-40. Notice that the E-plane couplin generally higher than H-plane coupling, a co mon characteristic for many ground-plane mounted aperture antennas. A very general study of coupling (isolation) between pairs of various antenna types is available, including the-ory, nomographs, and extensive experimental data.[215] The usefulness of such flat-ground-plane data has been enhanced since the coupling (isola-tion) between aperture antennas mounted on a cylinder has been shown to be about the same as that of the identical antennas mounted at the same spacing on a flat ground plane.[216]

LOG-PERIODIC ANTENNAS

58. The frequency-independent antenna is specified only by angles. It was suggested by Rumsey[217] in 1954. The simplest form of such antennas is the equiangular spiral,[218] although early models with the frequency-independent idea included the tapered helix.[219,220] All antenna shapes which are completely specified by angles must extend to infinity; thus any physically realizable frequency-independent an-tenna has bandwidth limitations due to end effects. A simple modification of the frequency-independent antenna is the logarithmically peri-odic antenna,[221] whose properties vary periodi-cally with the logarithm of the frequency. This modification tends to minimize the end effect, although the impedance will vary as a function of frequency; such variations are sometimes small. From these early designs a number of log-periodic antennas have been developed, in-cluding conical log spirals,[222] the log-periodic V,[223] the log-periodic dipole,[224-226] and the log-periodic Yagi-Uda array,[227] among many oth-ers.

59. Log-periodic Dipole Design. One of the most popular antennas of this type is the log-periodic dipole antenna, which has the geometry depicted in Fig. 18-41. This antenna can be fed either by using alternating connections to a balanced line, as indicated in Fig. 18-41, or by a coaxial line running through one of the feeders

Direction of beam

$$\frac{R_n}{R_{n-1}} = \frac{l_n}{l_{n-1}} = \tau \qquad \frac{d_n}{2l_n} = \sigma \qquad h_n = l_n/2$$

Method of feeding

Fig. 18-41. The log-periodic dipole antenna, with definitions of parameters. *(From Carrel.[225])*

from front to back. A simple procedure determined by Carrel[225] which can be used for designing this antenna is outlined here. The number of elements is primarily determined by τ, and the antenna size is determined by the boom length, which depends primarily upon σ. The procedure is as follows.

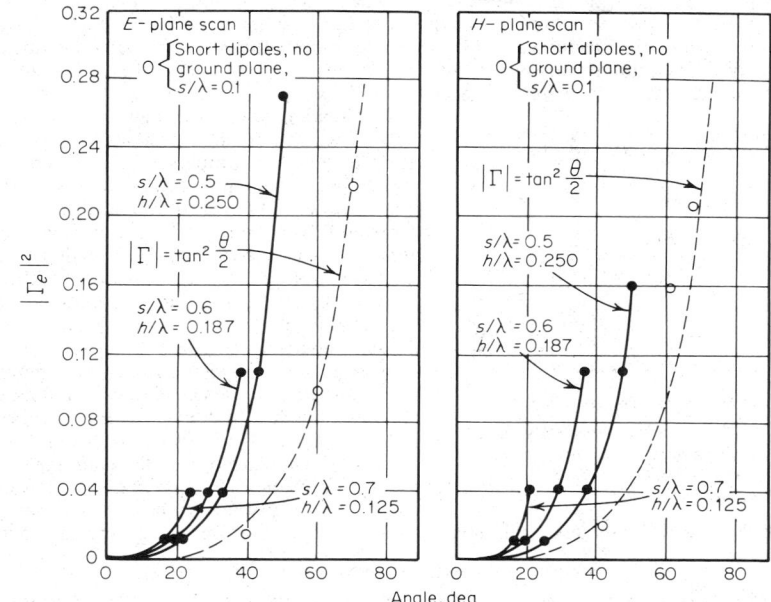

Fig. 18-38. Scanned mismatch variation for different element spacings (h/λ is the dipole spacing above a ground plane). *(Calculations by Allen.[209])*

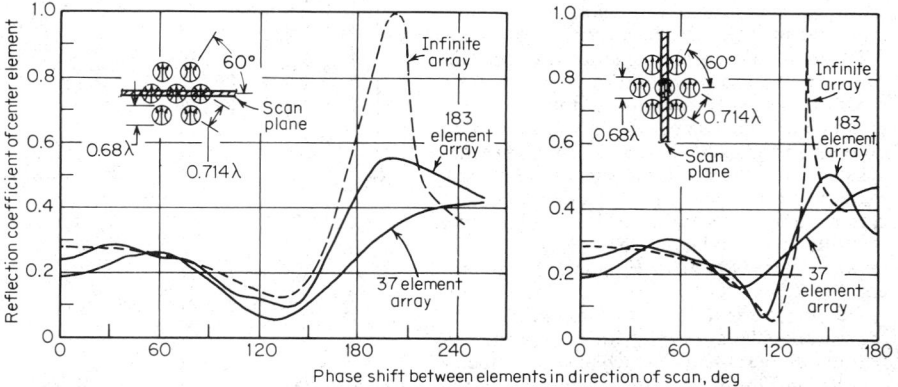

Fig. 18-39. Reflection coefficient of the central element of a triangular grating of circular waveguides excited in the TE_{11} mode, as a function of scan angle. *(Courtesy of M. C. Bailey, Langley Research Center.)*

The coupling effects of elements in finite-sized arrays of antenna elements upon the input impedance of a single element are uniquely dependent upon the detailed antenna configuration, number of elements, and the location of the element in the array. The 100% reflection peak observed for an element in a finite array at a particular scan angle is modified somewhat for finite arrays. As an example, consider an array of dominant TE_{11}-mode-fed circular waveguides located in a triangular grid. The reflection coefficient of the central element of finite arrays and an infinite array of circular waveguides is given in Fig. 18-39. The infinite-grating calculations are by Amitay and Galindo.[214]

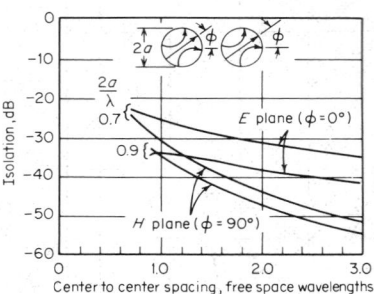

Fig. 18-40. Coupling between TE_{11} modes in two circular apertures radiating into free space. *(Courtesy of M. C. Bailey, Langley Research Center.)*

57. Coupling between Pairs of Antennas. Many times the antenna designer is concerned with the relatively simple problem of coupling between pairs of antennas mounted on a common ground-plane. A typical waveguide-coupling problem of this type is that of two circular waveguides excited in the TE_{11} mode, as depicted in Fig. 18-40. Notice that the E-plane coupling is generally higher than H-plane coupling, a common characteristic for many ground-plane-mounted aperture antennas. A very general study of coupling (isolation) between pairs of various antenna types is available, including theory, nomographs, and extensive experimental data.[215] The usefulness of such flat-ground-plane data has been enhanced since the coupling (isolation) between aperture antennas mounted on a cylinder has been shown to be about the same as that of the identical antennas mounted at the same spacing on a flat ground plane.[216]

LOG-PERIODIC ANTENNAS

58. The frequency-independent antenna is specified only by angles. It was suggested by Rumsey[217] in 1954. The simplest form of such antennas is the equiangular spiral,[218] although early models with the frequency-independent idea included the tapered helix.[219,220] All antenna shapes which are completely specified by angles must extend to infinity; thus any physically realizable frequency-independent antenna has bandwidth limitations due to end effects. A simple modification of the frequency-independent antenna is the logarithmically periodic antenna,[221] whose properties vary periodically with the logarithm of the frequency. This modification tends to minimize the end effect, although the impedance will vary as a function of frequency; such variations are sometimes small. From these early designs a number of log-periodic antennas have been developed, including conical log spirals,[222] the log-periodic V,[223] the log-periodic dipole,[224-226] and the log-periodic Yagi-Uda array,[227] among many others.

59. Log-periodic Dipole Design. One of the most popular antennas of this type is the log-periodic dipole antenna, which has the geometry depicted in Fig. 18-41. This antenna can be fed either by using alternating connections to a balanced line, as indicated in Fig. 18-41, or by a coaxial line running through one of the feeders

Direction of beam

$$\frac{R_n}{R_{n-1}} = \frac{l_n}{l_{n-1}} = \tau \qquad \frac{d_n}{2l_n} = \sigma \qquad h_n = l_n/2$$

(a) Method of feeding

Fig. 18-41. The log-periodic dipole antenna, with definitions of parameters. *(From Carrel.[225])*

from front to back. A simple procedure determined by Carrel[225] which can be used for designing this antenna is outlined here. The number of elements is primarily determined by τ, and the antenna size is determined by the boom length, which depends primarily upon σ. The procedure is as follows:

1. An estimate of τ and σ based upon the desired gain can be obtained from Fig. 18-42.
2. The bandwidth of the structure B_s is given by $B_s = BB_{AR}$, where B is the operating bandwidth and B_{AR} is determined in Fig. 18-43 using the parameter $\tan \alpha = (1 - \tau)/4\sigma$.
3. The length of the first element is always made $\lambda_{max}/2$, so that the boom length L between the largest and smallest elements can be found from $L/\lambda_{max} = \frac{1}{4}(1 - 1/B_s) \cot \alpha$.
4. The number of elements required is given by $N = 1 + [(\log B_s)/\log(1/\tau)]$.

By several iterations of this design procedure a minimum boom length can be obtained. The relative feeder impedance of the design can be found using available data.[225] The design of this antenna, including patterns and impedance, can be treated by the integral-equation methods outlined in Pars. 18-22 to 18-23. Even shorter designs of the log-periodic dipole antenna have been reported.[226]

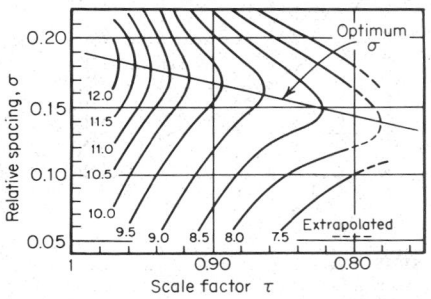

Fig. 18-42. Constant-directivity contours in decibels vs. τ and σ. Optimum σ indicates maximum directivity for a given value of τ. $Z_0 = 100\ \Omega$, $h/a = 100$, $Z_\tau = 100\ \Omega$.

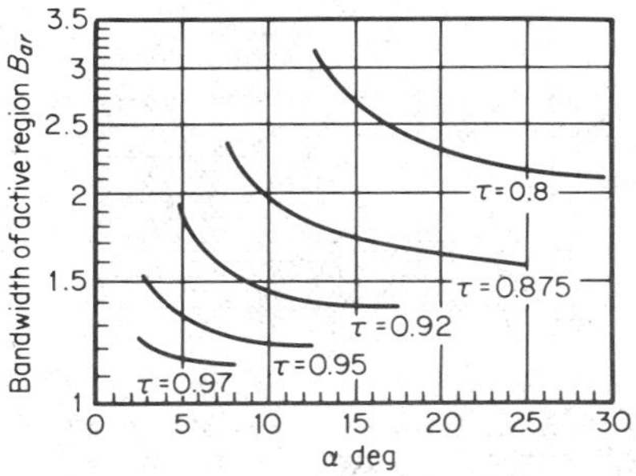

Fig. 18-43. Bandwidth of active region B_{ar} vs. α for several values of τ, for $Z_0 = 100\ \Omega$, $h/a = 125$. *(From Carrel.[225])*

SURFACE-WAVE ANTENNAS

60. General Description. A wide class of so-called *surface-wave antennas* has been devised, e.g., the Yagi, backfire, helix, cigar, and polyrod antenna. The surface-wave nomenclature is related to the idea that these antennas, if infinite in length, will support a wave which travels along the structure at a velocity slower than the velocity of light in free space. Data for the phase velocity along such antenna structures are available.[228-230] If the parameters of the antenna structure are chosen so that the resultant phase velocity causes the Hansen-Woodward condition to be met on the finite length of the antenna, an increased or supergain condition occurs, as discussed in Par. 18-9. The relative phase velocity $c/v = \lambda/\lambda_z$ to maximize the gain as a function of antenna length is given the Fig. 18-44. A typical phase-velocity variation as a function of specific antenna parameters is given in Fig. 18-45 for the

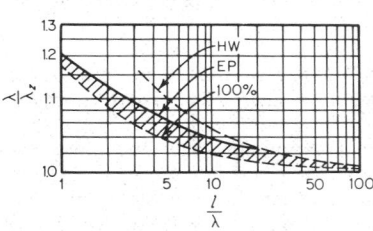

Fig. 18-44. Relative phase velocity $c/v = \lambda/\lambda_z$ for maximum-gain surface-wave antennas as a function of relative antenna length l/λ. *HW*-Hansen-Woodward condition; *EP*-Ehrenspeck and Poehler experimental values; 100%: idealized perfect excitation. *(From Zucker.[232])*

Yagi. Chosing particular antenna parameters so that the optimum phase-velocity conditions are met will result in a good first-cut design. Improved designs require extensive parametric experimental studies where the antenna elements are varied about the initial dimensions. An estimate of how much gain can be expected from surface-wave antennas is given in Fig. 18-46. An excellent summary of surface-wave antenna design and literature has been compiled by Zucker.[231,232]

The surface-wave antenna is a misnomer since surface waves do not exist on finite antennas. Consequently, aside from the use of the general concepts mentioned above, most surface-wave antennas are designed experimentally because of the importance of the feed radiation and end effects. The finite Yagi can be analyzed as an array of dipoles, where one dipole is excited and the rest are shorted, using the wire-antenna method[233] described in Pars. **18-22** and **18-23**. Indeed, all the wire versions of surface-wave antennas can be analyzed and designed in the same manner.

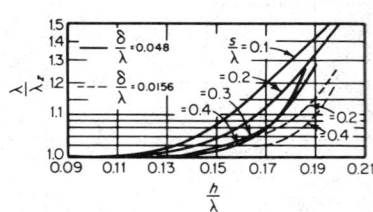

Fig. 18-45. Relative phase velocity on a Yagi antenna (data from Ehrenspeck and Poehler and Frost; see Ref. 232); $\delta =$ diameter of the wire element, $s =$ spacing between elements, and $h =$ half length of the element. *(From Zucker.[232])*

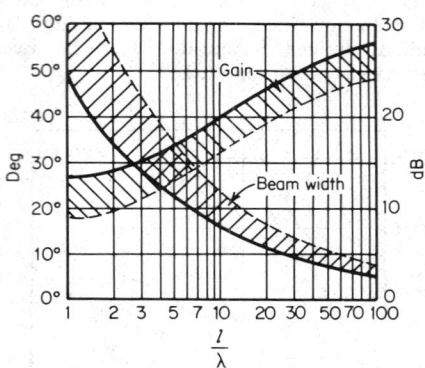

Fig. 18-46. Gain and beam width of surface-wave antenna as a function of relative antenna length l/λ. For gain (in decibels above an isotropic source), use right-hand ordinate; for beam width, left-hand ordinate. Solid lines are optimum values; dashed lines are for low-side-lobe and broad-band design. *(From Zucker.[232])*

61. Use of Feed Shields. Nearly all surface-wave antennas suffer from side-lobe and beam asymmetry if not fed from special launchers. For example, the helix mounted on a flat ground plane has pattern asymmetry and poor axial ratio (about 3 dB) on axis.[234] Short helices in particular have been found to exhibit this property. A way to improve this helix performance is to use a conical feed shield and to taper the beginning and end turns as in the helicone antenna.[235] The feed shield which improves the performance of the cigar antenna is a conical horn[236] or a square cavity or bucket.[237]

62. References on Antennas.

1. STRATTON, J. A. "Electromagnetic Theory," McGraw-Hill, New York, 1941.

2. TAI, C. T. Kirchoff Theory: Scalar, Vector, or Dyadic?, *IEEE Trans. Antennas Propag.*, January 1972, Vol. AP-20, pp. 114–115.

3. CHU, TA-SHING On the Use of Uniform Circular Arrays to Obtain Omnidirectional Patterns, *IEEE Trans. Antennas Propag.*, October 1959, Vol. AP-7, pp. 436–438.

4. CROSWELL, W. F., et al. A Dielectric-coated Circumferential Slot Array for Omnidirectional Coverage, *IEEE Trans. Antennas Propag.*, November 1967, Vol. AP-15, pp. 722–727.

5. CROSWELL, W. F., and C. R. COCKRELL An Omnidirectional Microwave Antenna for Use on Spacecraft, *IEEE Trans. Antennas Propag.*, July 1969, Vol. AP-17, pp. 459–466.

6. KO, H. C. Radio-Telescope Antenna Parameters, *IEEE Trans. Antennas Propag.*, December 1964, Vol. AP-12, pp. 891–897.

7. JELLY, J. V., and B. F. C. COOPER An Operational Ruby Maser for Observation at 21 cm with a 60 ft. Radius Telescope, *Rev. Sci. Instrum.*, February 1961, Vol. 32, pp. 166–175.

8. SCHUSTER, D., et al. The Determination of Noise Temperature of Large Paraboloid Antennas, *IEEE Trans. Antennas Propag.*, May 1962, Vol. AP-10, pp. 286–291.

9. PAULING-TOTH, I. I. K., et al. The Use of Paraboloid Reflector of Small Focal Ratio as a Low Noise Antenna System, *Proc. IRE*, December 1962, Vol. 50, p. 2483.

10. Jasik, J., and A. D. Bresler   A Low Noise Feed System for Large Parabolic Antennas in E. C. Jordan (ed.), "Electromagnetic Theory and Antennas," Pt. II, pp. 1167-1171, Pergamon, New York, 1968.

11. Degrasse, R. W., et al.   Ultra Low Noise Antenna and Receiver Combinations for Satellite and Space Communications, *Proc. Natl. Electron. Conf.*, 1959, Vol. 15, p. 370.

12. Crawford, A. B., et al.   A Horn Reflector Antenna for Space Communications, *Bell Syst. Tech. J.*, 1961, Vol. 40, pp. 1095-1116.

13. Jones, S. R., and K. S. Kelleher   A New Low Noise High Grain Antenna, *IEEE Int. Conv. Rec.*, 1963, Vol. 11, Pt. I, pp. 11-17.

14. Hidy, G. M., et al.   Development of a Satellite Microwave Radiometer to Sense the Surface Temperature of the World's Oceans, *NASA Contract. Rep.* NASA CR-1960, February 1972.

15. Otoshi, T. Y.   The Effect of Mismatched Components on Microwave Noise-Temperature Calibrations, *IEEE Trans. Microwave Theory Tech.*, September 1968, Vol. MTT-16, No. 9, pp. 675-687.

16. Hollis, J. S., et al.   "Microwave Antenna Measurements," Scientific-Atlanta, Atlanta, Ga., 1970.

17. King, R. W. P.   Cylindrical Antennas and Arrays in R. E. Collin and F. J. Zucker, "Antenna Theory," Pt. I, pp. 352-420, McGraw-Hill, New York, 1969.

18. Richmond, J. H.   Scattering by an Arbitrary Array of Parallel Wires, *Ohio State Univ. Antenna Lab. Rep.* 1522-7, Cont. N123(1953)-31663A, April 1964.

19. Richmond, J. H.   Scattering by an Arbitrary Array of Wires, *IEEE Trans. Microwave Theory Tech.*, July 1965, Vol. MTT-13, pp. 408-412.

20. Richmond, J. H.   A Wire-Grid Model for Scattering by Conducting Bodies, *IEEE Trans. Antennas Propag.*, November 1966, Vol. AP-14, pp. 782-786.

21. Richmond, J. H.   Scattering by Imperfectly Conducting Wires, *IEEE Trans. Antennas Propag.*, November, 1967, Vol. AP-15, pp. 802-806.

22. Harrington, R. F.   Theory of Loaded Scatterers, *Proc. IEE*, April 1964, Vol. 111, pp. 617-628.

23. Harrington, R. F., and J. Mautz   Matrix Methods for Solving Field Problems, *Syracuse Univ. Rep.* RADC TR-66-351, Vol. II, August 1966.

24. Harrington, R. F.   Matrix Methods for Field Problems, *Proc. IEEE*, February 1967, Vol. 55, pp. 136-149.

25. Harrington, R. F.   "Field Computation by Moment Methods," Macmillan, New York, 1968.

26. Richmond, J. H.   Theoretical Study of V Antenna Characteristics for ATS-E Radio Astronomy Experiment, *OSU Electron. Sci. Lab. Rep.* 2619-1, Contract NAS5-11543, February, 1969.

27. Richmond, J. H.   Computer Analysis of Three-dimensional Wire Antennas, *OSU Electron. Sci. Lab. Rep.* 2708-4, Contract DAAD 05-69-C-0031, December 1969.

28. Otto, D. V.   A Note on the Induced E.M.F. Method of Antenna Impedance, *IEEE Trans. Antennas Propag.*, January 1969, Vol. AP-17, pp. 101-102.

29. Richmond, J. H.   Coupled Linear Antennas with Skew Orientation, *IEEE Trans. Antennas Propag.*, September 1970, Vol. AP-18, pp. 694-696.

30. Richmond, J. H., and N. H. Geary   Mutual Impedance between Co-planar-Skew Dipoles, *IEEE Trans. Antennas Propag.*, September 1970, Vol. AP-18, pp. 414-416.

31. Thiele, G. A.   "Wire Antennas: Short Course on Computer Techniques for EM and Antennas," University of Illinois, Sept. 28-Oct. 1, 1970.

32. Richards, G. A.   Reaction Formulation and Numerical Results for Multiturn Loop Antennas and Arrays, Ph.D. dissertation, Ohio State Univ., 1970.

33. High Frequency Aircraft Antennas, *Ohio State Univ. Electrosci. Lab. Final Rep.* 2235-5, May 1968.

34. Otto, D. V., and J. H. Richmond   Rigorous Field Expressions for Piecewise-sinusoidal Line Sources, *IEEE Trans. Antennas Propag.*, January 1969, Vol. AP-17, p. 98.

35. Harrington, R. F., and J. R. Mautz   Straight Wires with Arbitrary Excitation and Loading, *IEEE Trans. Antennas Propag.*, July 1967, Vol. AP-15, pp. 502-515.

36. Harrington, R. F., and J. R. Mautz   Electromagnetic Behavior of Circular Loops with Arbitrary Excitation and Loading, *Proc. IEE*, January 1968, Vol. 175, pp. 68-77.

37. CHAO, H. H., and B. J. STRAIT   Computer Programs for Scattering and Radiation by Arbitrary Configurations of Bent Wires, *Syracuse Univ. Rep. 7*, Contract AF 19628-68-C-0180, September 1970.

38. RICHMOND, J. H., and M. H. GEARY   Mutual Impedance between Coplanar Skew Dipoles, *IEEE Trans. Antennas Propag.*, May 1970, Vol. AP-18, pp. 414–416.

39. KALAFUS, R. M.   Dipole Design Using the Method of Moments, *IEEE Trans. Antennas Propag.*, November 1971, Vol. AP-19, pp. 771–773.

40. ADAMS, A. T., and D. E. WARREN   Dipole plus Parasitic Element, *IEEE Trans. Antennas Propag.*, November 1971, Vol. AP-19, pp. 536–537.

41. MILLER, E. K., et al.   Accuracy-modeling Guidelines for Integral-Equation Evaluation of Thin-Wire Scattering Structures, *IEEE Trans. Antennas Propag.*, July 1971, Vol. AP-19, pp. 534–536.

42. BOWMAN, D. F.   Impedance Matching and Broadbanding in H. Jasik (ed.), "Antenna Engineering Handbook," Chap. 31, pp. 31-22 to 31-31, McGraw-Hill, New York, 1961.

43. TAI, C. T.   Characteristics of Linear Antenna Elements in H. Jasik (ed.), "Antenna Engineering Handbook," Chap. 3, pp. 3-1 to 3-28, McGraw-Hill, New York, 1961.

44. KING, H. E., and J. L. WONG   An Experimental Study of a Balun-Fed Open-Sleeve Dipole in Front of a Metal Reflector, *IEEE Trans. Antennas Propag.*, March 1972, Vol. AP-20, pp. 201–204.

45. ARNOLD, P. W.   A Circularly Polarized Octave-Bandwidth Unidirectional Antenna Using Conical Dipoles, *IEEE Trans. Antennas Propag.*, September 1970, Vol AP-18, pp. 696–698.

46. PUTTRE, R. E.   Study of Small Omnidirectional 250-Mc Antenna for Penetrometer, *Phase I Final Rep.*, *NASA Langley Contract* NAS1-4470, July 1965.

47. CULLEN, B. D.   Experimental Evaluation of a UHF, Dual Turnstile Omnidirectional Antenna for a Penetrometer, *Phase I Suppl. Rep. NASA Langley Contract* NAS1-4470, November 1966.

48. PUTTRE, R. E.   Development and Performance of a Small Crossed-Loop Omnidirectional UHF Prototype Antenna for a Penetrometer, *Final Rep. Wheeler Lab.*, *NASA Contract* NAS1-4470, January 1967.

49. SENGUPTA, D. L., and V. H. WESTON   Investigation of the Parasitic Loop Counterpoise Antenna, *IEEE Trans. Antennas Propag.*, March 1969, Vol. AP-17, pp. 180–191.

50. SENGUPTA, D. L., and J. E. FERRIS   On the Radiation Patterns of Parasitic Loop Counterpoise Antennas, *IEEE Trans. Antennas Propag.*, January 1970, Vol. AP-18, pp. 34–41.

51. BALANIS, C. A.   Radiation Characteristics of Current Elements near a Finite-Length Cylinder, *IEEE Trans. Antennas Propag.*, May 1970, Vol. AP-18, pp. 352–359.

52. BALANIS, C. A.   Analysis of an Array of Line Sources above a Finite Groundplane, *IEEE Trans. Antennas Propag.*, March 1971, Vol. AP-19, pp. 181–185.

53. BALANIS, C. A., and C. R. COCKRELL   Analysis and Design of Antennas for Air Traffic Collision Avoidance Systems, *IEEE Trans. Aerosp. Electron. Syst.*, September 1971, Vol. AES-7, pp. 960–967.

54. BALANIS, C. A.   Radiation from Conical Surfaces Used for High-Speed Aircraft, *Radio Sci.*, February 1972, Vol. 7, pp. 339–343.

55. BALANIS, C. A., and L. PETERS   Equatorial Plane Pattern of an Axial-TEM Slot on a Finite Size Groundplate, *IEEE Trans. Antennas Propag.*, May 1969, Vol. AP-17, pp. 351–352.

56. PAULI, W.   On Asymptotic Series for Functions in the Theory of Diffraction of Light, *Phys. Rev.*, December 1938, Vol. 54, pp. 924–931.

57. HUTCHENS, D. L., and R. G. KOUYOUMJIAN   Asymptotic Series Describing the Diffraction of a Plane Wave by a Wedge, *Ohio State Univ. Electrosci. Lab. Rep.* 2183-3, Contract AF19(628)-5929, 1966.

58. HUTCHENS, D. L.   Asymptotic Series Describing the Diffraction of a Plane Wave by a Two-Dimensional Wedge of Arbitrary Angle, Ph.D. dissertation, Ohio State Univ., 1967.

59. PATHAK, P. H., and R. G. KOUYOUMJIAN   The Dyadic Diffraction Coefficient for a Perfectly Conducting Wedge, *Ohio State Univ. Electrosci. Lab. Rep.* 2183-4, Contract AF19(628)-5929, June 1970.

60. HARRINGTON, R. F.   "Time-Harmonic Electromagnetic Fields," pp. 236–237, McGraw-Hill, New York, 1961.

61. COHEN, M. H., et al.   The Aperture Admittance of a Rectangular Waveguide Radiating into a Half-Space, *Ohio State Univ. Antenna Lab. Rep.* 339-22, Contract w33-038 ac 21114, November 1951.

62. COMPTON, R. T., JR.   The Aperture Admittance of a Rectangular Waveguide Radiating into a Lossy Half Space, *Ohio State Univ. Antenna Lab. Rep.* 1691-1, NASA Grant NsG-448, September 1963.

63. COMPTON, R. T., JR.   The Admittance of Aperture Antennas Radiating into Lossy Media, *Ohio State Univ. Antenna Lab. Rep.* 1691-5, NASA Grant NsG-448, March 1964.

64. CROSWELL, W. F., et al.   The Admittance of a Rectangular Waveguide Radiating into a Dielectric Slab, *IEEE Trans. Antennas Propag.*, September 1967, Vol. AP-15, pp. 627-633.

65. SWIFT, C. T.   The Input Admittance of a Rectangular Aperture Covered with an Inhomogeneous, Lossy Dielectric Slab, *NASA Langley Tech. Note* D-4197, September 1967.

66. SWIFT, C. T., and D. M. HATCHER   The Input Admittance of a Rectangular Aperture Antenna Loaded with a Dielectric Plug, *NASA Langley Tech. Note* D-4430, April 1968.

67. CROSWELL, W. F., et al.   The Input Admittance of a Rectangular Waveguide-fed Aperture under an Inhomogeneous Plasma: Theory and Experiment, *IEEE Trans. Antennas Propag.*, July 1968, Vol. AP-16, pp. 475-487.

68. COCKRELL, C. R.   Higher-Order Mode Effects on the Aperture Admittance of a Rectangular Waveguide Covered with Dielectric and Plasma Slabs, *NASA Langley Tech. Note* D-4774, October 1968.

69. BAILEY, M. C., et al.   Electromagnetic Properties of a Circular Aperture in a Dielectric Covered or Uncovered Groundplane, *NASA Langley Tech. Note* D-4752, October 1968.

70. BAILEY, M. C., and C. T. SWIFT   Input Admittance of a Circular Waveguide Aperture Covered by a Dielectic Slab, *IEEE Trans. Antennas Propag.*, July 1968, Vol. AP-16, pp. 386-391.

71. MISHUSTIN, B. A.   Radiation from the Aperture of a Circular Waveguide with an Infinite Flange, *Sov. Radiophys.*, November-December 1965, Vol. 8, pp. 852-858.

72. LEVINE, H., and C. H. PAPAS   Theory of the Circular Diffraction Antenna, *J. Appl. Phys.*, January 1951, Vol. 22, pp. 29-43.

73. HARTIG, E. O.   Circular Apertures and Their Effect on Half-Dipole Impedances, Ph.D. thesis, Harvard Univ., 1950.

74. CURTIS, W. L.   Calculated Values of Admittance for Annular Slot Antenna, D2-20301-1, The Boeing Co., February 1964.

75. SWIFT, C. T.   Input Admittance of a Coaxial Transmission Line Opening onto a Flat Dielectric-covered Groundplane, *NASA Langley Tech. Note* D-4158, September 1967.

76. BALANIS, C. A.   Pattern Distortion Due to Edge Diffractions, *IEEE Trans. Antennas Propag.*, July 1970, Vol. AP-18, pp. 561-563.

77. SWIFT, C. T., and J. S. EVANS   Generalized Treatment of Plane Electromagnetic Waves Passing through an Isotropic Inhomogeneous Plasma Slab at Arbitrary Angles of Incidence, *NASA Langley Tech. Rep.* 172, December 1963.

78. BECK, F. B.   Admittance Characteristics of a Circular Waveguide Radiating into a Homogeneous Dielectric Slab and Inhomogeneous Plasma, M.S. thesis, George Washington Univ., June 1971.

79. COLLIN, R. E.   "Field Theory of Guided Waves," pp. 470-476, McGraw-Hill, New York, 1960.

80. CHURCHILL, R. V.   "Introduction to Complex Variables and Applications," McGraw-Hill, New York, 1948.

81. JONES, J. EARL   The Admittance of a Parallel-Plate Aperture Illuminating a Displaced Dielectric-Dielectric Boundary, *NASA Langley Tech. Note* D-5083, February 1969.

82. JONES, J. EARL   The Influence of Air Gap Tolerances on the Admittance of a Dielectric-coated Slot Antenna, *IEEE Trans. Antennas Propag.*, January 1969, Vol. AP-17, pp. 63-68.

83. GILREATH, M. C.   Techniques for Determining the Microwave Properties of Dielectric Materials, M.S. thesis, George Washington Univ., June 1971.

84. SWIFT, C. T.   Admittance of a Waveguide Fed Aperture Loaded with a Dielectric Plug, *IEEE Trans. Antennas Propag.*, May 1969, Vol. AP-17, pp. 356-359.

85. CROSWELL, W. F., et al.   Computations of the Aperture Admittance of an Axial Slot on a Dielectric Coated Cylinder, *IEEE Trans. Antennas Propag.*, January 1972, Vol. AP-20, pp. 89-92.

86. BALANIS, C. A.   Pattern Distortion Due to Edge Diffractions, *IEEE Trans. Antennas Propag.*, July 1970, Vol. AP-18, pp. 561-563.

87. BLASS, JUDD   Slot Antennas in H. Jasik (ed.), "Antenna Engineering Handbook," Chap. 8, pp. 8-1 to 8-16, McGraw-Hill, New York, 1961.

88. STEVENSON, A. F.   Theory of Slots in Rectangular Waveguides, *J. Appl. Phys.*, 1948, Vol. 19, pp. 24-38.

89. WATSON, W. H.   Resonant Slots, *J. IEEE*, Session on Aerials and Waveguides, 1946, Vol. 93, Pt. IIIA, No. 1, pp. 747-777.

90. OLINER, A. A.   The Impedance Properties of Narrow Radiating Slots in the Broad Face of Rectangular Waveguide, Pt. I, Theory, *IRE Trans. Antennas Propag.*, January 1957, Vol. AP-5, pp. 4-11.

91. OLINER, A. A.   The Impedance Properties of Narrow Radiating Slots in the Broad Face of Rectangular Waveguide, Pt. II, Comparison with Measurement, *IRE Trans. Antennas Propag.*, January 1957, Vol. AP-5, pp. 12-20.

92. BAILEY, M. C.   Design of Dielectric-covered Resonant Slots in a Rectangular Waveguide *IEEE Trans. Antennas Propag.*, September 1967, Vol. AP-15, pp. 594-598.

93. BAILEY, M. C.   The Properties of Dielectric-covered Narrow Radiating Slots in the Broadface of a Rectangular Waveguide, *IEEE Trans. Antennas Propag.*, September 1970, Vol. AP-18, pp. 596-603.

94. LAWSON, R. W., and V. M. POWERS   Slots in Dielectrically Loaded Waveguide, *Radio Sci.*, January 1966, n.s., Vol. 1, pp. 31-35.

95. ERLICH, M. J.   Slot Antenna Arrays in H. Jasik (ed.), "Antenna Engineering Handbook," Chap. 9, pp. 9-1 to 9-18, McGraw-Hill, New York, 1961.

96. MAXUM, B. J.   Resonant Slots with Independent Control of Amplitude and Phase, *IRE Trans. Antennas Propag.*, July 1960, Vol. AP-8, pp. 384-389.

97. CROSWELL, W. F., and R. B. HIGGINS   Effects of Dielectric Covers over Shunt Slots in a Waveguide, *NASA Langley Tech. Note* D2158, 1964.

98. FRATILA, R., et al. Dielectric Covered Slot Antennas, *WADC-OSU Radome Symp.*, Wright Air Dev. Cent. Tech. Rep. 57-314, June 1957.

99. HANSON, R. L., and G. A. SHARP   Small Antenna Study, *U.S. Nav. Ordnance Lab. NAVORD Rep. 4600*, July-September 1956.

100. CROSWELL, W. F., and R. B. HIGGINS   A Study of Dielectric Covered Shunt Slots in a Waveguide, *IEEE Trans. Aerosp.*, April 1964, Vol. AS-2, pp. 278-283.

101. ADAMS, A. T.   Flush Mounted Rectangular Cavity Slot Antennas: Theory and Design, *IEEE Trans. Antennas Propag.*, May 1967, Vol. AP-15, pp. 342-351.

102. KAY, A. F., and A. J. SIMMONS   Mutual Coupling of Shunt Slots, *IRE Trans. Antennas Propag.*, July 1960, Vol. AP-8, pp. 389-400.

103. SIMMONS, A. J.   Circularly Polarized Slot Radiators, *IRE Trans. Antennas Propag.*, January 1957, Vol. AP-5, pp. 31-36.

104. BAILEY, M. C.   Effects of Dielectric Covers over Cross Slots in a Rectangular Waveguide, *NASA Langley Tech. Note* D-4194, October 1964.

105. GETSINGER, W. J.   Elliptically Polarized Leaky-Wave Array, *IRE Trans. Antennas Propag.*, March 1962, Vol. AP-10, pp. 165-172.

106. ADAMS, A. T.   Flush Mounted Rectangular Cavity Slot Antennas: Theory and Design, *IEEE Trans. Antennas Propag.*, May 1967, Vol. AP-15, pp. 342-351.

107. WILKINSON, E. J.   A Circularly Polarized Slot Antenna, *Microwave J.*, March 1961, pp. 97-100.

108. COX, R. M., and W. E. RUPP   Circularly Polarized Phased Array Antenna Element, *IEEE Trans. Antennas Propag.*, November 1970, Vol. AP-18, pp. 804-807.

109. KLEIN, C. F.   An Equivalent Circuit for the Tee-fed Slot Antenna, *IEEE Trans. Antennas Propag.*, March 1970, Vol. AP-18, pp. 280-282.

110. LINDBERG, C. A.   A Shallow Cavity UHF Crossed-Slot Antenna, *M.I.T. Lincoln Lab. Tech. Rep.* 446, March 1968.

111. CAMPBELL, T. G.   An Extremely Thin, Omnidirectional Microwave Antenna Array for Spacecraft Applications, *NASA Langley Tech. Note* D-5539, November 1969.

112. Frood, D. G., and J. R. Wait   Investigation of Slot Radiators in Rectangular Plates, *Proc. IEE,* January 1956, Vol. 103, pp. 103–110.

113. Swift, C. T., et al.   Radiation Characteristics of a Cavity Backed Cylindrical Gap Antenna, *IEEE Trans. Antennas and Propag.,* July 1969, Vol. AP-17, pp. 467–477.

114. Balanis, C. A.   Radiation from Slots on Cylindrical Bodies Using the Geometrical Theory of Diffraction and Creeping Wave Theory, *NASA Langley Tech. Rep.* R-331, February 1970.

115. Pathak, P., and R. G. Kouyoumjian   private communication, June 1971.

116. Bailey, M. C., and W. F. Croswell   Pattern Measurements of Slot Radiators in Dielectric-coated Metal Plates, *IEEE Trans. Antennas Propag.,* November 1967, Vol. AP-15, pp. 824–826.

117. Knop, C. M., and G. I. Cohn   Radiation from an Aperture in a Coated Plane, *J. Res. Natl. Bur. Stand.,* 1964, Vol. 68D, No. 4, pp. 363–378.

118. Sinclair, G.   The Patterns of Slotted-Cylinder Antennas, *Proc. IRE,* December 1948, Vol. 36, pp. 1487–1492.

119. Papas, C. H.   On the Infinitely Long Cylinder Antenna, *J. Appl. Phys.,* May 1949, Vol. 20, pp. 437–440.

120. Papas, C. H.   Radiation from a Transverse Slot in an Infinite Cylinder, *J. Math. Phys.,* January 1950, Vol. 28, pp. 227–236.

121. Silver, S., and W. K. Saunders   The External Field Produced by a Slot in an Infinite Circular Cylinder, *J. Appl. Phys.,* February 1950, Vol. 21, pp. 153–158.

122. Silver, S., and W. K. Saunders   The Radiation from a Transverse Rectangular Slot in a Circular Cylinder, *J. Appl. Phys.,* August 1950, Vol. 21, pp. 745–749.

123. Wait, J. R.   "Electromagnetic Radiation from Cylindrical Structures," Pergamon, New York, 1959.

124. Compton, R. E., Jr., and R. E. Collin   Slot Antennas in R. E. Collin and F. J. Zucker, "Antenna Theory," Pt. I, pp. 560–620, McGraw-Hill, New York, 1969.

125. Mushiake, Y., and R. E. Webster   Radiation Characteristics with Power Gain for Slots on a Sphere, *IRE Trans. Antennas Propag.,* January 1957, Vol. AP-5, pp. 47–55.

126. Bugnolo, D. E.   Quasi-isotropic Antenna in Microwave Spectrum, *IRE Trans. Antennas Propag.,* January 1957, Vol. AP-5, pp. 47–55.

127. Bangert, J. T., et al.   The Spacecraft Antennas, *Bell Syst. Tech. J.,* July 1963, Vol. 43, pp. 869–897.

128. Bailin, L., and S. Silver   Exterior Electromagnetic Boundary Value Problems for Spheres and Cones, *IRE Trans. Antennas Propag.,* January 1956, Vol. AP-5, pp. 6–16.

129. Pridmore-Brown, D. C., and G. E. Stewart   Radiation from Slot Antennas on Cones, *IEEE Trans. Antennas Propag.,* January 1972, Vol. AP-20, pp. 36–39.

130. Balanis, C. A., and L. Peters   Analysis of Aperture Radiation from an Axially Slotted Elliptical Conducting Cylinder Using Aeronautical Theory of Diffraction, *IEEE Trans. Antennas Propag.,* July 1969, Vol. AP-17, pp. 507–513.

131. Balanis, C. A.   Radiation from Slots on Cylindrical Bodies Using Geometrical Theory at Diffraction and Creeping Wave Theory, Ph.D. thesis, Ohio State Univ., 1969.

132. Yu, C. L., and W. D. Burnside   Elevation Plane Analysis of On-Aircraft Antennas, *Ohio State Univ., Electrosci. Lab. Tech. Rep.* 3188-2, January 1972.

133. Southworth, G. C., and A. P. King   Metal Horns as Directive Receivers of Ultrashort Waves, *Proc. IRE,* 1939, Vol. 27, pp. 95–102.

134. Barrow, W. L., and L. J. Chu   Theory of the Electromagnetic Horn, *Proc. IRE,* January 1939, Vol. 27, pp. 51–64.

135. Chu, L. J., and W. L. Barrow   Electromagnetic Horn Design, *Trans. AIEE,* July 1939, Vol. 58, pp. 333–338.

136. Barrow, W. L., and F. D. Lewis   The Sectoral Electromagnetic Horn, *Proc. IRE,* January 1939, Vol. 27, pp. 41–50.

137. Chu, L. J.   Calculation of the Radiation Properties of Hollow Pipes and Horns, *J. App. Phys.,* 1940, Vol. 11, pp. 603–610.

138. Schelkunoff, S. A.   "Electromagnetic Waves," Van Nostrand, New York, 1943.

139. Rhodes, D. R.   An Experimental Investigation of the Radiation Properties of Electromagnetic Horn Antennas, *Proc. IRE,* September 1948, Vol. 36, pp. 1101–1105.

140. Schelkunoff, S. A., and H. T. Friis   "Antennas: Theory and Practice," Wiley, New York, 1952.

141. KING, A. P.   The Radiation Characteristics of Conical Horn Antennas, *Proc. IRE,* March 1952, Vol. 38, pp. 249-251.

142. SCHORR, M. G., and J. J. BECK   Electromagnetic Field of the Conical Horn, *J. App. Phys.,* August 1950, Vol. 21, pp. 795-801.

143. COMPTON, R. T., JR. and R. E. COLLIN   Open Waveguides and Small Horns in R. E. Collin and F. J. Zucker, "Antenna Theory," Pt. I, pp. 621-655, McGraw-Hill, New York, 1969.

144. RUSSO, P. M., R. C. RUDDUCK, and L. PETERS   A Method for Computing *E*-Plane Patterns of Horn Antennas, *IEEE Trans. Antennas Propag.,* March 1965, Vol. AP-13, pp. 219-224.

145. YU, J. S., R. C. RUDDUCK, and L. PETERS   Comprehensive Analysis for *E*-Plane of Horn Antennas by Edge Diffraction Theory, *IEEE Trans. Antennas Propag.,* March 1966, Vol. AP-14, pp. 138-149.

146. YU, J. S., and R. C. RUDDUCK   *H*-Plane Pattern of a Pyramidal Horn, *IEEE Trans. Antennas Propag.,* Vol. AP-17, pp. 651-652.

147. THOMAS, D. T.   A Half Blinder for Reducing Certain Side-Lobes in Large Horn, Reflector Antennas, *IEEE Trans. Antennas Propag.,* November 1971, Vol. AP-19, pp. 774-776.

148. JULL, E. V.   Reflection from the Aperture of a Long *E*-Plane Sectoral Horn, *IEEE Trans. Antennas Propag.,* January 1972, Vol. AP-20, pp. 62-68.

149. POTTER, P. D.   A New Horn Antenna with Suppressed Sidelobes and Equal Beamwidths, *Microwave J.,* June 1963, pp. 71-78.

150. LUDWIG, A. C.   Radiation Pattern Synthesis for Circular Aperture Horn Antennas, *IEEE Trans. Antennas Propag.,* July 1966, Vol. AP-14, pp. 434-440.

151. TURRIN, R. E.   Dual Mode Small-Aperture Antenna, *IEEE Trans. Antennas Propag.,* March 1967, Vol. AP-15, pp. 307-308.

152. CALDECOTT, R., et al.   High Performance *S*-Band Horn Antennas for Radiometer Use, *Ohio State Univ. Electrosci. Lab. Tech. Rep.* 3033-1, NAS 1-10040, May 1972.

153. LOVE, A. W.   The Diagonal Horn Antenna, *Microwave J.,* March 1962, pp. 117-122.

154. LAWRIE, R. E., and L. PETERS   Modifications of Horn Antennas for Low Sidelobes, *IEEE Trans. Antennas Propag.,* September 1966, Vol. AP-14, pp. 605-610.

155. BOHERT, W. F., and L. PETERS   Small-Aperture Small-Flare-Angle Corrugated Horns, *IEEE Trans. Antennas Propag.,* July 1968, Vol. AP-16, pp. 494-495.

156. PEACE, G. M., and E. E. SWARTZ   Amplitude Compensated Horn Antenna, *Microwave J.,* February 1964, pp. 66-68.

157. LAGRONE, H. H., and G. F. ROBERTS   Minor Lobe Suppression in a Rectangular Horn Antenna through the Utilization of a High Impedance Choke Flange, *IEEE Trans. Antennas Propag.,* January 1966, Vol. AP-14, pp. 102-104.

158. NAIR, K. G., G. P. SRIVASTAVA, and S. HARIHARAN   Sharpening of *E*-Plane Radiation Patterns of *E*-Plane Sectoral Horns by Metallic Grills, *IEEE Trans. Antennas Propag.,* Janauary 1969, Vol. AP-17, pp. 91-93.

159. EPIS, J. J.   Compensated Electromagnetic Horns, *Microwave J.,* May 1961, pp. 84-89.

160. AJOIKA, J. S., and H. E. HARRY   Shaped Beam Antenna for Earth Coverage from a Stabilized Satellite, *IEEE Trans. Antennas Propag.,* May 1970, Vol. AP-18, pp. 322-327.

161. KING, H. E., J. L. WANG, and C. J. ZAMITES   Shaped-Beam Antennas for Satellites, *IEEE Trans. Antennas Propag.,* September 1966, Vol. AP-14, pp. 641-643.

162. TSANDOULAS, G. N., and W. D. FITZGERALD   Aperture Efficiency Enhancement in Dielectrically Loaded Horns, *IEEE Trans. Antennas Propag.,* January 1972, Vol. AP-20, pp. 69-74.

163. HAMID, M. A. K., S. J. TOWAIJ, and G. O. MARTENS   A Dielectric-loaded Circular Waveguide Antenna, *IEEE Trans. Antennas Propag.,* January 1972, Vol. AP-20, pp. 96-97.

164. QUDDUS, M. A., and J. P. GERMAN   Phase Correction by Dielectric Slabs in Sectoral Horn Antennas, *IEEE Trans. Antennas Propag.,* July 1961, Vol. AP-9, pp. 413-415.

165. SATOH, T.   Dielectric-loaded Horn Antenna, *IEEE Trans. Antennas Propag.,* March 1972, Vol. AP-20, pp. 199-201.

166. CHATTERJEE, J. S., and W. F. CROSWELL   Waveguide Excited Dielectric Spheres as Feeds, *IEEE Trans. Antennas Propag.,* March 1972, Vol. AP-20, pp. 206-208.

167. WALTON, K. L., and V. C. SUNDBERG   Broadband Ridged Horn Design, *Microwave J.,* March 1964, pp. 96-101.

168. SENGUPTA, D. L., and J. E. FERRIS Rudimentary Horn Antenna, *IEEE Trans. Antennas Propag.*, January 1971, Vol. AP-19, pp. 124–126.

169. WONG, J. Y. A Dual Polarization Feed Horn for a Parabolic Reflection, *Microwave J.*, September 1962, pp. 188–191.

170. CRAWFORD, A. B., H. C. HOGG, and L. E. HUNT A Horn Reflector Antenna for Space Communication, *Bell Syst. Tech. J.*, 1961, Vol. 40, p. 1095.

171. TEICHMAN, M. Precision Phase Center Measurements of Horn Antennas, *IEEE Trans. Antennas Propag.*, November 1970, Vol. AP-18, pp. 689–690.

172. WRIXOM, G. T., and W. J. WELCH Gain Measurements of Standard Electromagnetic Horns in the *K* and *Ka* Bands, *IEEE Trans. Antennas Propag.*, March 1972, Vol. AP-20. pp. 136–142.

173. COCKRELL, C. R. Reflection Coefficients of Pyramidal and *H*-Plane Horns Radiating into Dielectric Materials, *NASA Langley Tech. Note* D-5978, September 1970.

174. JAKES, W. C. Horn Antennas in H. Jasik (ed.), "Antenna Engineering Handbook," pp. 10-1 to 10-18, McGraw-Hill, New York, 1961.

175. COTTONY, H. V., and A. C. WILSON Gains of Finite-Size Corner-Reflector Antennas, *IEEE Trans. Antennas Propag.*, October 1958, Vol. AP-6, pp. 366–369.

176. WILSON, A. C., and H. V. COTTONY Radiation Patterns of Finite-Size Corner-Reflector Antennas, *IEEE Trans. Antennas Propag.*, March 1960, Vol. AP-8, pp. 144–157.

177. BERRY, D. G., R. G. MALECH, and W. A. KENNEDY The Reflector Array Antenna, *IEEE Trans. Antennas Propag.*, November 1963, Vol. AP-11, pp. 645–651.

178. SCHELL, A. C., et al. An Experimental Evaluation of Multiple Antenna Properties, *IEEE Trans. Antennas Propag.*, September 1966, Vol. AP-14, pp. 543–550.

179. SCHELL, A. C. The Multiplate Antenna, *IEEE Trans. Antennas Propag.*, September 1966, Vol. AP-14, pp. 550–560.

180. SILVER, S. "Microwave Antenna Theory and Design," M.I.T. Radiation Laboratory Series, Vol. 12, McGraw-Hill, New York, 1949.

181. GRAY, C. LARRY Estimating the Effect of Feed Support Member Blocking on Antenna Gain and Sidelobe Level, *Microwave J.*, March 1964, pp. 88–91.

182. RUZE, JOHN Lateral-Feed Displacement in a Paraboloid, *IEEE Trans Antennas Propag.*, September 1965, Vol. AP-16, pp. 660–665.

183. PAGONES, M. J. Gain Factor of an Offset-fed Paraboloidal Reflector, *IEEE Trans. Antennas Propag.*, September 1965, Vol. AP-16, pp. 536–541.

184. LOVE, A. W. Spherical Reflecting Antennas with Corrected Line Sources, *IEEE Trans. Antennas Propag.*, September 1962, Vol. AP-13, pp. 529–537.

185. SCHELL, A. C. The Diffraction Theory of Large-Aperture Spherical Reflector Antennas, *IEEE Trans. Antennas Propag.*, July 1963, Vol. AP-14, pp. 428–432.

186. POTTER, P. D. The Aperture Efficiency of Large Paraboloidal Antennas as a Function of Their Feed System Radiation Characteristics, *Jet Prop. Lab. Tech. Rep.* 32-149, September 25, 1961.

187. RUSCH, W. V. T. Phase Error and Associated Cross-Polarization Effects in Cassegrainian-fed Microwave Antennas, *Jet Prop. Lab. Tech. Rep.* 32-610, May 30, 1962.

188. POTTER, P. D. The Application of the Cassegrainian Principle to Ground Antennas for Space Communications, *Jet Prop. Lab. Tech. Rep.* 32-295, June 1962.

189. POTTER, P. D. A Simple Beamshaping Device for Cassegrainian Antennas, *Jet. Prop. Lab. Tech. Rep.* 32-214, Jan. 31, 1962.

190. POTTER, P. D. A Computer Program for Machine Design of Cassegrain Feed Systems, *Jet. Prop. Lab. Tech. Rep.* 32-1202, Dec. 15, 1967.

191. RUSCH, W. V. T. Edge Diffraction from Truncated Paraboloids and Hyperboloids, *Jet Prop. Lab. Tech. Rep.* 32-113, June 1, 1967.

192. LUDWIG, A., and W. V. T. RUSCH Digital Computer Analysis and Design of a Subreflector of Complex Shape, *Jet Prop. Lab. Tech. Rep.* 32-1190, Nov. 15, 1967.

193. WILLIAMS, W. F. High Efficiency Antenna Reflector, *Microwave J.*, July 1965, pp. 79–82.

194. POTTER, P. D. Application of Spherical Wave Theory to Cassegrain-Fed Paraboloids, *IEEE Trans. Antennas Propag.*, November 1967, Vol. AP-15, pp 727–736.

195. Space Program Summary, *Jet Prop. Lab. Tech. Rep.* 37-50, January 1, 1968-March 31, 1968.

196. The Deep Space Network, *Jet Prop. Lab. Space Programs Summ.* 37-52, Vol. II, July 31, 1968, pp. 78-105.

197. MEEKS, M. L., and J. RUZE   Evaluation of the Haystack Antenna and Radome, *IEEE Trans. Antennas Propag.,* November 1971, Vol. AP-19, pp 723-728.

198. INGERSON, P. G., and W. C. WONG   The Analysis of Deployable Umbrella Parabolic Reflectors, *IEEE Trans. Antennas Propag.,* July 1972, Vol. AP-20, pp. 409-415.

199. LUDWIG, A. C.   Conical-Reflector Antennas, *IEEE Trans. Antennas Propag.,* November 1972, Vol. AP-20, pp. 146-152.

200. SENGUPTA, D. L., and R. E. HIATT   Reflectors and Lenses in M. I. Skolnik (ed.), "Radar Handbook," Chap. 10, pp. 10-11 to 10-31, McGraw-Hill, New York, 1970.

201. RUDGE, A. W., and M. J. WITHERS   Design of Flared-Horn Primary Feeds for Parabolic Reflector Antennas, *Proc. IEE,* September 1970, Vol. 117, pp. 1741-1749.

202. RUDGE, A. W.   private communication, Apr. 26, 1972.

203. RUDGE, A. W., and M. J. WITHERS   New Technique for Beam Steering with Fixed Parabolic Reflectors, *Proc. IEE,* July 1971, Vol 118, pp. 857-863.

204. RUZE, J.   The Effect of Aperture Errors on the Antenna Radiation Pattern, *Nuovo Cimento Suppl.,* 1952, Vol. 9, No. 3, pp. 364-380.

205. Special Issue on Electronic Scanning, *Proc. IEEE,* November 1968, Vol. 56, pp. 1761-2048.

206. VON AULOCK, W. H.   Properties of Phased Arrays, *IRE Trans. Antennas Propag.,* October 1960, Vol. AP-9, pp. 1715-1727.

207. HSIAO, J. K.   A Broadband Wide-Angle Scan Matching Technique for Large Environmentally Restricted Phased Arrays, *IEEE Trans. Antennas Propag.,* July 1972, Vol. AP-20, pp. 415-421.

208. WHEELER, H. A.   Simple Relations Derived from a Phased-Array Antenna Made of an Infinite Current Sheet, *IEEE Trans. Antennas Propag.,* July 1965, Vol. AP-13, pp. 506-514.

209. ALLEN J. L.   On Array Element Impedance Variation with Spacing, *IEEE Trans. Antennas Propag.,* May 1964, Vol. AP-12, p. 371.

210. MALLIOUX, R. J.   Surface Waves and Anomalous Wave Radiation Nulls on Phased Arrays of TEM Waveguides with Fences, *IEEE Trans. Antennas Propag.,* March 1972, Vol. AP-20, pp 160-166.

211. AMITAY, N., and V. GALINDO   Characteristics of Dielectric Loaded and Covered Circular Waveguide Phased Array, *IEEE Trans. Antennas Propag.,* November 1969, Vol AP-17, pp. 722-729.

212. HANNAN, P. W., and M. A. BALFOUR   Simulation of a Phased Array Antenna in a Waveguide, *IEEE Trans. Antennas Propag.,* May 1965, Vol. AP-13, pp. 342-353.

213. BALFOUR, M. A.   Phased Array Simulators in Waveguide for a Triangular Array of Elements, *IEEE Trans. Antennas Propag.,* May 1965, Vol. AP-13, pp. 475-476.

214. AMITAY, N., and V. GALINDO   Characteristics of Dielectric Loaded and Covered Circular Waveguide Phased Arrays, *IEEE Trans. Antennas Propag.,* November 1966, Vol. AP-17, pp. 722-729.

215. LYON, J. A. M., et al.   Derivation of Aerospace Antenna Coupling-Factor Interference Prediction Techniques, *Univ. Mich. Radiat. Lab. Tech. Rep.* AFAL-TR-66-57, April 1966.

216. FANTE, R. L.   Calculation of the Admittance, Isolation, and Radiation Pattern of Slots on an Infinite Cylinder Covered by a Lossy Plasma, *Radio Sci.,* March 1971, Vol. 6, pp. 421-428.

217. RUMSEY, V. H.   The Equiangular Spiral, *IRE Nat. Conv. Rec.,* 1957, Pt. I, pp. 114-118.

218. DYSON, J. D.   The Equiangular Spiral, *IRE Trans. Antennas Propag.,* April 1959, Vol. AP-7, pp. 181-187.

219. SPRINGER, P. S.   End-loaded and Expanding Helices as Broadband Circularly Polarized Radiators, *Proc. Natl. Electron. Conf.,* 1949, Vol. 5, pp. 161-171.

220. CHATTERJEE, J. S.   Radiation Characteristics of a Conical Helix of Low Pitch Angles, *J. Appl. Phys.,* March 1955, Vol. 26, pp. 331-335.

221. DUHAMEL, R. H., and R. E. ISBELL   Broadband Logarithmically Periodic Antenna Structures, *IRE Natl. Conv. Rec.,* 1957, Pt. 1, pp. 119-128.

222. DYSON, J. D.   The Unidirectional Equiangular Spiral Antenna, *IRE Trans. Antennas Propag.,* October 1959, Vol. AP-7, pp. 329-334.

223. MAYES, P. E., and R. L. CARREL   Log-periodic Resonant V-Arrays, *IRE West. Conv.*, 1961.

224. ISBELL, D. E.   Log-periodic Dipole Arrays, *IRE Trans. Antennas Propag.*, May 1960, Vol. AP-8, pp. 260-267.

225. CARREL, R. L.   The Design of Logarithmically Periodic Dipole Antennas, *IRE Natl. Conv. Rec.*, 1961, Pt. 1, pp. 61-75.

226. BANTIN, C. C., and K. G. BALMAIN   Study of Compressed Log-periodic Dipole Antennas, *IEEE Trans. Antennas Propag.*, March 1970, Vol. AP-18, pp. 195-203.

227. BARBANO, N.   Log-periodic Yagi-Uda Array, *IEEE Trans. Antennas Propag.*, March 1966, Vol. AP-14, pp. 235-238.

228. MAILLOUX, R. J.   The Long Yagi-Uda Array, *IEEE Trans. Antennas Propag.*, March 1966, Vol. AP-14 p. 128.

229. COLLIN, R. E.   "Field Theory of Guided Waves," McGraw-Hill, New York, 1960.

230. BRUNSTEIN, S. A., and R. F. THOMAS   Characteristics of a Cigar Antenna, *Jet Prop. Lab. Q. Rev.*, July 1972, Vol. 1, No. 2, pp. 87-95.

231. ZUCKER, F. J.   Surface Wave Antennas in R. E. Collin and F. J. Zucker, "Antenna Theory," Pt. II, Chap 21, pp. 298-348, McGraw-Hill, New York, 1969.

232. ZUCKER, F. J.   Surface and Leaky-Wave Antennas in H. Jasik (ed.), "Antenna Engineering Handbook,"Chap. 16, pp. 16-1 to 16-57, McGraw-Hill, New York, 1961.

233. THIELE, G. A.   Analysis of Yagi-Uda Type Antennas, *IEEE Trans. Antennas Propag.*, January 1969, Vol. AP-17, pp. 24-31.

234. KRAUS, J. D.   "Antennas," McGraw-Hill, New York, 1950.

235. ANGELAKOS, D. J., and KAJFEZ DARKO   Modifications on the Axial Mode Helical Antenna, *Proc. IEEE*, April, 1967, Vol. 55, No. 4, pp. 558-559.

236. CARVER, K. R., and B. M. POTTS   Some Characteristics of the Helicone Antenna, *1970 IEEE G-AP Symp. Dig.*, pp. 142-150.

237. CROSWELL, W. F., and M. C. GILREATH   Erectable Yagi Disk Antennas for Space Vehicle Applications, *NASA Langley Tech. Note* D-1401, October 1962.

RADIO-WAVE PROPAGATION*

BY RICHARD C. KIRBY

FUNDAMENTALS OF WAVE PROPAGATION

63. Introduction: Mechanisms, Media, and Frequency Bands.   Radio waves are propagated from the point of generation to the point of reception through or along the surface of the earth, through the atmosphere, by reflection or scattering from the ionosphere or troposphere, or by artificial or natural reflectors; or they may travel through free space.   The particular propagation mechanism around which a given radio-system application is designed depends on the type of information or service to be provided, the distance to be spanned, and economic factors, including the reliability required.   Several mechanisms, depending upon frequency and distance, affect the performance of the system or mutual interference with other systems.

The conductivity and permittivity (dielectric constant) of the earth are markedly different from those of the atmosphere.   Over a line-of-sight path within the nonionized atmosphere, transmission is much as through free space, though atmospheric refraction causes bending, reflection, scattering, and fading; at extremely high frequencies there may be attenuation due to rainfall and absorption by air and water vapor.   A wave mainly diffracted along the surface of the ground encounters increasing loss with increasing frequency.   Very low frequency waves are propagated with little attenuation over thousands of kilometers.   At high frequencies losses along the ground become so great that they limit utility to short distances. At medium and high frequencies ionospheric reflections permit radio communication to great

* The author wishes to acknowledge contributions of material and review of this section by colleagues in the U.S. Department of Commerce, Office of Telecommunications, Institute for Telecommunication Sciences, Boulder, Colo., in particular A. F. Barghausen, A. P. Barsis, L. A. Berry, R. T. Disney, H. T. Dougherty, G. W. Haydon, Margo Leftin, P. L. Rice, E. K. Smith, A. D. Spaulding, J. R. Wait, Maureen Detmer, and Beverle Gibson.

distances. At frequencies much above 30 MHz, ionospheric reflections are unusual, and most communications depend upon line-of-sight propagation or tropospheric or ionospheric scattering beyond the horizon.

Because of the dependence of propagation characteristics on frequency, much of the discussion of these sections will be in terms of frequency bands, and abbreviations such as vlf for very low frequencies, vhf very high frequencies, etc., will be used.

The International Telecommunications Union has defined nine frequency bands designated by integer band numbers; for example, 10^6 Hz is the approximate midband of band 6, etc., as listed in Table 18-4.

Table 18-4

Band number*	Frequency range (lower limit exclusive, upper limit inclusive)	Corresponding metric subdivision	Abbreviation
4	3- 30 kHz	Myriametric waves	vlf
5	30- 300 kHz	Kilometric waves	lf
6	300-3,000 kHz	Hectometric waves	mf
7	3- 30 MHz	Decametric waves	nf
8	30- 300 MHz	Metric waves	vhf
9	300-3,000 MHz	Decimetric waves	uhf
10	3- 30 GHz	Centimetric waves	shf
11	30- 300 GHz	Millimetric waves	chf
12	300-3,000 GHz (3 THz)	Decimillimetric waves	

* Band number N extends from 0.3×10^N to 3×10^N Hz.

As propagation characteristics are more closely related within bands such as 10 to 150 kHz, 150 to 1,500 kHz, etc., rather than the ITU bands, the bands of frequencies are treated herein in somewhat different segments. However, ITU nomenclature is in nearly universal use and is adhered to here.

Texts on wave propagation are referred to in connection with the particular mechanisms outlined. References 1 and 2 provide introductory material; Refs. 3 to 6 are basic texts on electromagnetic propagation; Refs. 7 to 9 offer additional material on tropospheric and ionospheric propagation.* Recent results of theoretical and experimental radio science worldwide are outlined in Ref. 10.

The emphasis here is mainly descriptive, key formulas indicating the behavior of parameters and references to publications providing material for engineering calculations. Maxwell's uniform plane-wave equations are cited only to show the role of the electrical constants and the vector relationships of electric and magnetic field and power flux.

64. Wave Propagation in Homogeneous Media. Electromagnetic radiation is composed of two mutually dependent vector fields, electric and magnetic. The electric field is characterized by the vectors **E**, electric field strength in volts per meter, and **D**, dielectric displacement in coulombs per square meter. The magnetic field is characterized by **H**, the magnetic field strength in ampere-turns per meter (or amperes per meter) and **B**, flux density, in webers per square meter. The vector current density **J** is in amperes per square meter.

The relationship between the members of the various pairs of field vectors is characterized by the constitutive parameters, or *electrical constants,* of the medium:

$$\varepsilon = \text{dielectric constant, F/m}$$

$$\sigma = \text{conductivity, } \mho/\text{m}$$

$$\mu = \text{permeability, H/m}$$

In some cases these constants are functions of the coordinate. Locally, however they are always considered constant. Nearly always the time factor in these sections is exp $(+i\omega t)$, where ω is the angular frequency $2\pi f$ (f in hertz) and t the time. The *electric field*, then, is the real part of E exp $i\omega t$.

---

* For references, see Par. **18-111**.

To explain very basic notation,* a short outline of plane electromagnetic waves in a homogeneous medium[4] is given.

Ohm's law in the complex form is

$$\mathbf{J} = (\sigma + i\varepsilon\omega)\mathbf{E} \tag{18-62}$$

where $\mathbf{J}$ is the current density vector and $\mathbf{E}$ is the electric field vector.

The analogous relation for magnetic quantities is

$$\mathbf{B} = \mu\mathbf{H} \tag{18-63}$$

In source-free media the above vector quantities are related by

$$\text{curl } \mathbf{E} = -i\mu\omega\mathbf{H} \tag{18-64}$$

and

$$\text{curl } \mathbf{H} = (\sigma + i\varepsilon\omega)\mathbf{E} \tag{18-65}$$

These are Maxwell's equations.

For a homogeneous medium

$$\text{curl curl } \mathbf{E} = \text{grad div } \mathbf{E} - \text{div grad } \mathbf{E} = -i\mu\omega(\sigma + i\varepsilon\omega)\mathbf{E} \tag{18-66}$$

Since div $\mathbf{E} = 0$,

$$(\nabla^2 - \gamma^2)\mathbf{E} = 0 \tag{18-67}$$

where $\nabla^2 = $ div grad is the Laplacian operator (which operates on the rectangular components of $\mathbf{E}$) and $\gamma^2 = i\mu\omega(\sigma + i\varepsilon\omega)$. The quantity γ is called the *propagation constant*.

As a simple illustration of the role of the electrical constants of the medium, the field of a wave is assumed to vary only in the z direction in space (time factor $\exp i\omega t$ understood), and the electric field is assumed to have only an x component E_x. For this case Eq. (18-67) reduces to

$$\left(\frac{d^2}{dz^2} - \gamma^2\right)E_x = 0 \tag{18-68}$$

and the solutions are $\exp(+\gamma z)$ and $\exp(-\gamma z)$ or, in general,

$$E_x = A \exp(\gamma z) + B \exp(-\gamma z) \tag{18-69}$$

where A and B are constants. The magnetic field then has only a y component given by

$$H_y = -\frac{1}{i\mu\omega}\frac{\partial E_x}{\partial z} = -\eta^{-1}[A \exp(+\gamma z) - B \exp(-\gamma z)] \tag{18-70}$$

where

$$\eta = \left(\frac{i\mu\omega}{\sigma + i\varepsilon\omega}\right)^{1/2} \tag{18-71}$$

η is defined as the *characteristic impedance* of the medium for plane-wave propagation. Remembering that the time factor is $\exp i\omega t$, we see that the term $B \exp(-\gamma z)$ is a wave traveling in the positive z direction with diminishing amplitude and the term $A \exp \gamma z$ is a wave traveling in the negative z direction with diminishing amplitude.[†] The electric and magnetic fields are both transverse to the direction of propagation and orthogonal to each other. Such radiation is termed *plane-polarized*. It is by convention designated as *horizontal* or *vertical* according to the orientation of the plane containing the $\mathbf{E}$ vector.

The quantity η is equal to the complex ratio of the electric and magnetic field components in the x and y directions, respectively, for plane waves in an unbounded homogeneous medium, i.e.,

$$H_y = \frac{i\omega\varepsilon}{r} E_x = \frac{E_x}{\eta} \tag{18-71a}$$

$$H_x = \frac{-r}{i\omega\mu} E_y = \frac{-E_y}{\eta} \tag{18-71b}$$

---

*From Wait[4] by permission.

†The geometry of subsequent sections uses a different convention for x, y, and z directions, shown in figures.

For a perfect dielectric ($\sigma = 0$)

$$\eta = \sqrt{\frac{\mu}{\epsilon}} \quad \Omega \text{ resistance} \tag{18-72a}$$

and,

$$r = ik \tag{18-72b}$$

where $k = \omega\sqrt{\mu\epsilon} = 2\pi/\lambda$ $\quad$ λ is the wavelength

The velocity of this wave is

$$v = \frac{1}{\sqrt{\mu\epsilon}} \quad \text{m/s} \tag{18-73}$$

v is called the *phase velocity* of the wave. It presents simply the velocity of propagation of phase or a state and does not necessarily coincide with the velocity with which the energy of a wave or signal is propagated. In fact, v may exceed free-space wave velocity without violating relativity in any way.

The wavelength is defined as the distance the wave propagates in one period

$$\lambda = \frac{2\pi}{\omega\sqrt{\mu\epsilon}} = \frac{v(\text{m/s})}{f(\text{Hz})} \quad \text{m} \tag{18-74}$$

For free space

$$\epsilon = \epsilon_0 = 8.854 \times 10^{-12} \text{ F/m} \qquad \mu = \mu_0 = 4\pi \times 10^{-7} \text{ H/m} \qquad \sigma = 0$$

then

$$\gamma = ik = i\omega\sqrt{\mu_0\epsilon_0} = i\frac{2\pi}{\lambda} \tag{18-75}$$

The *velocity of the wave for free space* is

$$v_0 = \frac{1}{\sqrt{\mu_0\epsilon_0}} \approx 3 \times 10^8 \text{ m/s} \tag{18-76}$$

The *characteristic impedance of free space* is

$$\eta_0 = \sqrt{\frac{\mu_0}{\epsilon_0}} \approx 120\pi \ \Omega \quad \text{resistance} \tag{18-77}$$

$$\approx 4\pi v_0 \times 10^{-7} \ \Omega$$

Energy flow in the electromagnetic field is described by the Poynting vector

$$\mathbf{P} = \mathbf{E} \times \mathbf{H^*} \tag{18-78}$$

where the complex representation of the time-periodic quantities is implied and the asterisk denotes the complex conjugate. The *real part* of the Poynting vector represents the average power flow over a cycle of the time variation per unit area in the direction of transmission

$$\mathbf{P}_{av} = \tfrac{1}{2}\text{Re}(\mathbf{E} \times \mathbf{H^*}) = \frac{E^2}{2\eta_0} \quad \text{W/m}^2 \tag{18-79}$$

$\mathbf{P}_{av}$ is called the *power flux density or field intensity*. Note that E and H are peak values.

The permeability and dielectric constant of any medium relative to free space are called the *relative permeability* and *relative dielectric constant*. These are usually the values given in tables of physical constants; they are dimensionless and designated by μ_r and ϵ_r, respectively.

Additional material on electromagnetic radiation and propagation in waveguides, cavities, and transmission lines is given in Secs. 9 and 25.

Polarization of the Wave. *Polarization* is a term characterizing the orientation of the field vector in its travel. In radio, polarization usually refers to the electric vector. In the simplest case E_z and H_z (field components in the direction of propagation) are zero, and E and H lie in a plane transverse to the direction of propagation and orthogonal to each other. Such a plane wave is *elliptically polarized* when the electric vector E describes an ellipse in the plane perpendicular to the direction of propagation over one cycle of the wave.

When the amplitudes of the rectangular components are equal and their phases differ by some odd integral multiple of $\pi/2$, the polarization ellipse becomes a circle and the wave is *circularly polarized*. It is customary to describe as right-handed circularly polarized a clockwise rotation of E when viewed in a direction looking toward the source (opposite to direction of propagation); counterclockwise rotation is left-handed polarization.

The most important case for many radio problems is that in which the polarization is a straight line, i.e., one axis of the ellipse is zero. The wave is then *linearly polarized*. In *horizontal polarization* the electric vector lies in a plane parallel to the earth's surface. In *vertical polarization* the electric vector lies in a plane perpendicular to the earth's surface.

In order to obtain maximum transfer of power between two antennas, the polarization should match. If the transmitting antenna is horizontally polarized, the receiving antenna must likewise be horizontally polarized. If the transmitting antenna is elliptically polarized with a given degree of ellipticity and a specified direction of rotation, the receiving antenna should have the proper direction of rotation and degree of ellipticity in order to maximize the path antenna gain. Conversely, the path antenna gain for an unwanted propagation path can be minimized by introducing a maximum polarization coupling loss.

It should be noted that in the process of propagation, except through free space, the polarization may be altered.[10a] Reflections from surfaces can do this. Passage through the ionosphere in the presence of magnetic field is likely to impart elliptical polarization to a plane-polarized incident wave and rotation of the major axis. For mf or hf ionospheric propagation, the downcoming wave may be randomly polarized.

65. Reflection Most problems in wave propagation involve reflection from a boundary between media of different refractive properties, often between air and the ground, or between air and the ionosphere. In general, such a boundary may involve dissipative-media (finite-conductivity) curvature, finite dimensions, roughness, and stratification.

The *complex index of refraction* for a conducting medium is often used in radio-wave propagation[11]

$$n^2 = \frac{i\omega\mu(\sigma + i\omega\epsilon)}{i\omega\mu_0(i\omega\epsilon_0)} \qquad (18\text{-}80)$$

when $\sigma = 0$,

$$n = \sqrt{\mu_r\epsilon_r} \qquad \text{where } \mu_r = \frac{\mu}{\mu_0} \qquad \epsilon_r = \frac{\epsilon}{\epsilon_0}$$

For many applications $\mu_r = 1$, and the index of refraction is simply the square root of ϵ_r.

Figure 18-47 serves to illustrate Snell's law for refraction of plane waves at an infinite plane interface. The angle ϕ between the direction of propagation and the normal to the boundary is called the *angle of incidence*. The angle ψ between the direction of propagation and the boundary, called the *grazing angle* or *elevation angle*, often is more convenient. If the

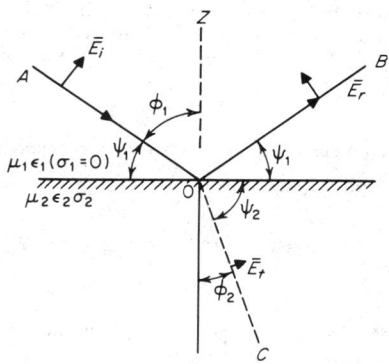

Fig. 18-47. Geometry of reflection and transmission.

medium of the incident wave is lossy, the angle of incidence is complex and can be defined in various ways.[4] At the boundary, the tangential components of **E** and **H** must be continuous; the phase of the reflected wave is in step with the phase of the incident wave to satisfy this requirement.

Snell's law of refraction for the direction of the transmitted wave toward C is

$$n_1 \cos \psi_1 = n_2 \psi_2 \qquad (18\text{-}81)$$

The *penetration depth* δ, or the depth at which the transmitted wave E_t has attenuated to $1/e$ of its incident value (for a conducting medium where $\sigma \gg \omega\varepsilon$), is

$$\delta = \frac{1}{\sqrt{\omega\mu\sigma/2}} \quad \text{m} \qquad (18\text{-}82)$$

66. Ground Reflection; Reflection Coefficients; Fresnel Zones. A wave incident upon a plane surface can be resolved into two components, one polarized normal and the other parallel to the plane of incidence. The reflection coefficients for the two components differ, and consequently the polarization of the reflected wave depends upon the angle of incidence. Consider an air-earth boundary, taking the media to be nonmagnetic; for the case where the **H** vector is parallel to the ground surface the complex reflection coefficient[11] is

$$\mathbf{R}_v = \frac{(\epsilon_r - i60\sigma\lambda)\sin\psi - (\epsilon_r - \cos^2\psi - i60\sigma\lambda)^{1/2}}{(\epsilon_r - i60\sigma\lambda)\sin\psi + (\epsilon_r - \cos^2\psi - i60\sigma\lambda)^{1/2}} \qquad (18\text{-}83)$$

where σ is the conductivity in mhos per meter, ψ is the grazing angle (Fig. 18-47), and ε_r is the relative dielectric constant of the earth to air or free space. $\epsilon_r - i60\sigma\lambda$ is referred to as the *complex dielectric constant*.

If E is parallel to the ground surface and H is in the plane of incidence,

$$\mathbf{R}_h = \frac{\sin\psi - (\epsilon_r - \cos^2\psi - i60\sigma\lambda)^{1/2}}{\sin\psi + (\epsilon_r - \cos^2\psi - i60\sigma\lambda)^{1/2}} \qquad (18\text{-}84)$$

These are reflection coefficients for vertical and horizontal polarization, respectively. Curves of values for a range of ϵ and σ are given in Refs. 11 and 12.

An important property for vertical polarization is that there exists an angle of incidence for which the reflection coefficient approaches zero (for dielectric media it equals zero). This is the *Brewster angle*, also called the *polarizing angle*, given by

$$\phi_0 = \tan^{-1}\sqrt{\frac{\epsilon_1}{\epsilon_2}} \qquad (18\text{-}85)$$

It is equal to the angle of incidence for which the reflected and refracted (transmitted) rays are at right angles. If the incidence occurs at the Brewster angle, the reflected wave is polarized entirely in the direction normal to the plane of incidence.

Wave tilt is a property frequently used to determine the electrical constants of the earth.[39] For waves traveling at nearly grazing incidence along the surface of the earth, wave tilt may be interpreted geometrically as the angle between the normal to the wave front and the tangent to the earth's surface. Wave tilt is defined to be the ratio of the horizontal to the vertical component of the electric field in the air just above the ground:

$$W = \frac{E_h}{E_v} \qquad (18\text{-}86)$$

Wave tilt is related to the electrical constants of a homogeneous earth by

$$W = \frac{\sqrt{\mu_1/\epsilon_{1c}}}{\sqrt{\mu_0/\epsilon_0}}\sqrt{1 - \frac{\mu_0\epsilon_0}{\mu_1\epsilon_{1c}}} \qquad (18\text{-}86a)$$

The subscript 1 refers to earth constants and 0 to free space; ε_{1c} is the complex dielectric constant $= \varepsilon_1 - \dfrac{i\sigma_1}{\omega}$

This procedure assumes that $\mu_0 = \mu_1$, generally a valid assumption, but if it is not, then μ_1 must be determined by some other procedure.

An important consideration in many propagation problems is the interference pattern generated by vector addition of the fields corresponding to the direct ray from an antenna to a point within line of sight plus the ground-reflected ray. In calculating such *ground reflection lobes*, the geometry of an image antenna is used,[11,13] as illustrated in Fig. 18-48. Discussion and formulas for ground reflection and Fresnel zones follow Norton and Omberg, Ref. 13, by permission.

A transmitting antenna T is at height h_1 above earth, and a receiving antenna R or a radar target is at distance d and height h_2. If d is very large, so that $r_1 \approx r_2$, the path difference $r_2 - r_1 \approx 2h_1 \sin \psi$. If e_1 is the free-space direct-path field strength at a unit distance from T in the direction of R and the free-space field in the direction of the ground reflection is also approximately equal to e_1, then the resultant field strength at R for horizontal polarization and perfectly reflecting earth is[13]

$$e_R = \frac{e_1}{r_1} \sin\left[\frac{2\pi}{\lambda}\left(\frac{r_1 + r_2}{2} - v_0 t\right)\right] 2 \sin \frac{2\pi h_1 \sin \psi}{\lambda}$$

(18-87)

$\underbrace{\qquad\qquad\qquad}_{\text{Direct field term}}$ $\underbrace{\qquad\qquad\qquad}_{\substack{\text{Interference term}\\\text{oscillating between 0 and 2}}}$

The angles at which maxima and minima occur, for horizontal polarization, are given by

$$\sin \psi = \frac{n\lambda}{4h_1} \quad \begin{array}{l}\text{maxima for } n \text{ odd}\\ \text{minima for } n \text{ even}\end{array}$$

(18-88)

For the first maximum to occur at a specified elevation ψ_1,

$$h_1 = \frac{\lambda}{4 \sin \psi_1}$$

(18-88a)

In Fig. 18-48a the ray reflections are shown as though they occurred at a point. Actually, the surface of the earth is illuminated over a wide region corresponding to the radiation patterns of the two antennas and, in accordance with Huygens' principle, reradiates elementary wavelets in all directions. In any particular direction, as toward R, these

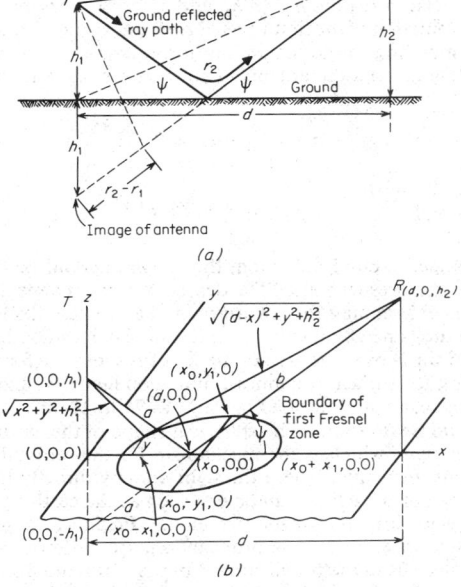

Fig. 18-48. Geometry of ground reflection, image antenna, and Fresnel zones for plane earth: (*a*) ray paths; (*b*) Fresnel zones.

elementary wavelets arrive with strength and phase such that the waves from an elliptical zone in the neighborhood of the ray reflection add nearly in phase. From successive ring areas, similarly bounded by larger ellipses, the waves alternately cancel and add.

These zones of physical reflection are called *Fresnel zones*, since they are closely related to the Fresnel zones of diffraction theory.[6] Most of the energy may be thought of as being reflected from the first Fresnel zone; it is defined, with the aid of Fig. 18-48b, for reflection paths between points such as *T* and *R*, as the area from which all the reradiated elementary wavelets arrive, according to geometric optics, within 1/2 wavelength of the phase of the direct ray. Thus the length of the geometric ray path at the edge of the *n*th Fresnel zone is *n* half wavelengths greater than the geometric ray path.

The cartesian coordinates for the extremities of the major axis of the Fresnel zone[13] are given by $x_0 - x_1$ and $x_0 + x_1$, where

$$x_0 = d_1\left[1 + \frac{h_2 - h_1}{2 h_1 [1 + (h_1 + h_2)^2/n\lambda(R + n\lambda/4)]}\right] \tag{18-89}$$

$$x_1 = \frac{(1 + n\lambda/2R)(1 + n\lambda/4R)}{\sin\theta\left[1 + \frac{n\lambda(R + n\lambda/4)}{(h_1 + h_2)^2}\right]}\left\{\frac{abn\lambda}{a + b + n\lambda/4}\left[1 + \frac{n\lambda(R + n\lambda/4)}{4h_1 h_2}\right]\right\}^{1/2} \tag{18-90}$$

$$R_n = \sqrt{x^2 + y^2 + h_1^2} + \sqrt{(d - x)^2 + y^2 + h_2^2} \tag{18-91}$$

Equation (18-89) determines the centers of Fresnel zones in terms of the location $d_1 = dh_1/(h_1 + h_2)$ of the geometric ground-reflection point.

The minor semiaxis of the ellipse is given by

$$y_1 = \left(1 + \frac{n\lambda}{4R}\right)\left[\frac{abn\lambda}{a + b + n\lambda/4}\,\frac{1 + \dfrac{n\lambda(R + n\lambda/4)}{4h_1 h_2}}{1 + \dfrac{n\lambda(R + n\lambda/n)}{(h_1 + h_2)^2}}\right]^{1/2} \tag{18-92}$$

For many applications such as radar and air-ground communication, or for radiation aimed at an ionospheric reflection region, $h_2 \gg h_1$, and $h_2 \gg \lambda$. An important use of the above formulas is the determination of the first Fresnel zone under these conditions for transmission at an angle ψ corresponding to the maximum of the first ground-reflection lobe. For this condition,[13] the distances from the antenna *T* to the near (d_n) and far (d_f) edges of the first Fresnel zone are

$$d_n = x_0 - x_1 \approx 0.688\frac{h_1^2}{\lambda} \tag{18-93}$$

and

$$d_f = x_0 + x_1 \approx 23.3\frac{h_1^2}{\lambda} \tag{18-94}$$

For a well-developed ground reflection, the ground should be flat over an area which includes at least the first Fresnel zone. The degree of flatness depends upon the wavelength and angle of incidence; assuming that phase-path changes less than $\lambda/16$ are unimportant, Rayleigh's criterion limits height deviations in terrain from a smooth surface to a magnitude less than $\Delta h = \lambda/(16 \sin\psi)$ over the area of the first Fresnel zone for waves incident at angle ψ. Methods allowing for surface roughness, finite conductivity, and divergence due to the spherical shape of the earth are outlined in Pars. **18-76** to **18-78**.

67. Diffraction and Scattering. The spherical shape of the earth and terrain irregularities give rise to *diffraction*, which is the mechanism for the redistribution of light in space according to Huygens' principle, where the light coming directly from the source may be intercepted by a screen or the edge of an object, such as the earth horizon, and is reradiated in elementary wavelets into the shadow region. Engineering methods for diffraction calculations, important in ground-wave propagation, are outlined in Par. **18-78**.

Scattering takes place from the rough surface of the earth and from small-scale irregularities in the index of refraction in the atmosphere or ionosphere. This is analogous to scattering of light in the atmosphere, although the radio problem is complicated by the wide range of relationships of radio wavelength to size of irregularity.

Tropospheric forward scattering is discussed in Par. **18-83** and ionospheric scattering in Par. **18-104**.

68. Reciprocity. Reciprocity in wave propagation means that the source and receiver may be interchanged, with the transmission loss and phase unaffected by direction of propagation. This is an application of the classical reciprocity theorem to radiation, as by J. R. Carson and others. In most radio-wave-propagation problems, with the notable exception of those involving an ionized medium with magnetic field, such reciprocity obtains. As discussed in Par. **18-92**, the refractive index of the ionosphere depends upon magnetic field effects; the direction of propagation of the wave affects attenuation, phase, and bending, and the medium is called *anisotropic*. Thus, especially at very low and medium frequencies propagated via the ionosphere, reciprocity does not obtain. Reciprocity does not in any case imply the same signal-to-noise ratio in both directions. The noise environment may be very different at the transmitting and receiving locations.

69. Transmission Loss: Free-Space Attenuation: Field Strength; Power Flux Density. The *transmission loss* of a radio circuit consisting of a transmitting antenna, a receiving antenna, and the intervening propagation medium, is defined (*CCIR Recomm.* 341) as the dimensionless ratio p_t'/p_a', where p_t' is the radio-frequency power radiated from the transmitting antenna and p_a' is the resultant radio-frequency signal power which would be available from the receiving antenna if there were no circuit losses other than those associated with its radiation resistance. The transmission loss is usually expressed in decibels:[*]

$$L = 10 \log(p_t'/p_a') = L_s - L_{tc} - L_{rc} \quad \text{dB} \tag{18-95}$$

where L_s is the system loss and L_{tc} and L_{rc} are the losses in the transmitting and receiving antenna circuits, respectively, excluding the losses associated with the antenna radiation resistances; i.e., the definitions of L_{tc} and L_{rc} are $10 \log(r'/r)$, where r' is the resistive component of the antenna circuit and r is the radiation resistance.

The *basic transmission loss* L_b of a radio circuit is the transmission loss expected between ideal, loss-free, isotropic transmitting and receiving antennas at the same locations as the actual transmitting and receiving antennas.

The *path antenna gain* G_p is equal to the realized difference in transmission loss between actual antennas used on the circuit and isotropic antennas:

$$G_p = L_b - L \tag{18-96}$$

The *propagation loss* is the system loss expected if the antenna gains and circuit resistances are the same as if the antennas were located in free space:

$$L_p = L_s - L_t - L_r \tag{18-97}$$

L_t and L_r are defined by $10 \log(r'/r_f)$, where r' is the actual antenna resistance and r_f is the resistance the antenna would have if it were in free space and there were no losses other than radiation losses. In many cases the antenna gain can be taken to be the same as in free space, so that propagation loss is practically equal to transmission loss.

The transmission loss between two isotropic antennas in free space is considered below. At a distance d very much greater than the wavelength λ, the power flux density (field intensity), expressed in watts per square meter, is simply $p_t'/4\pi d^2$ since the power is radiated uniformly in all directions. The effective absorbing area of the isotropic receiving antenna is $\lambda^2/4\pi$, and the available power at the terminals of the loss-free isotropic receiving antenna is given by

$$p_a' = \frac{\lambda^2}{4\pi} \frac{p_t'}{4\pi d^2} \tag{18-98}$$

Consequently, the *basic transmission loss in free space* can be expressed by

$$L_{bf} = 10 \log(4\pi d/\lambda)^2 \quad d \gg \lambda \tag{18-99}$$
$$= 32.45 + 20 \log f + 20 \log d$$

where L_{bf} is the basic transmission loss in free space, f is frequency in megahertz, and d is the straight-line distance between antennas in kilometers. This equation can be conveniently expressed as a nomogram (Fig. 18-49).

---

[*] From this point in the text, unless otherwise noted, capital letters are used to denote decibel quantities, and power levels P are referred to a common reference power. Unless otherwise indicated, logarithms are to base 10, and the reference power is 1 watt.

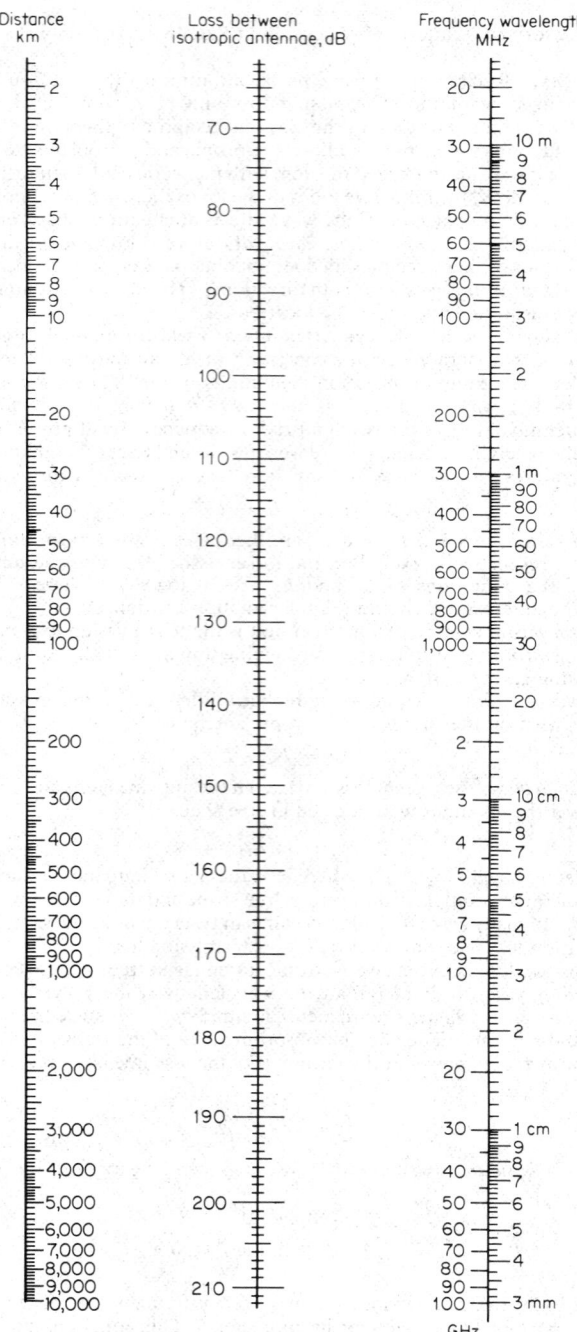

Fig. 18-49. Nomogram for determining basic transmission loss L_{bf} between isotropic antennas in free space.

18-68

We relate the available power P from the receiving antenna, the total radiated power P_r, the transmission loss L, the basic transmission loss L_b, and the path antenna gain G_p as follows:

$$P = P_r - L \qquad (18\text{-}100)$$

$$L_b = L + G_p \qquad (18\text{-}101)$$

Since the free-space gain of a short lossless electric dipole is $g_t = g_r = 1.5$, the path antenna gain for two optimally oriented short lossless electric dipoles in free space is

$$G_p = G_t + G_r = 3.52 \text{ dB} \qquad (18\text{-}102)$$

Consequently, the transmission loss between two optimally oriented short lossless electric dipoles in free space is

$$L = 10 \log(4\pi d/\lambda)^2 - 3.52 \text{ dB} \qquad (18\text{-}103)$$

The attenuation A relative to free space is defined as

$$A = L_b - L_{bf} \qquad \text{dB} \qquad (18\text{-}104)$$

For most radio links it is important to know the attenuation relative to free space A, usually a random variable with time, expressed as a function of frequency, distance, or other path parameters such as elevation angle, taking into account attenuation of the atmosphere or various propagation mechanisms; this matter is discussed further in *CCIR Rep.* 244-2.

In free space, with no absorption, the calculated transmission loss is

$$L = L_{bf} - G_t - G_r \qquad \text{dB} \qquad (18\text{-}105)$$

where G_t and G_r are free-space transmitting- and receiving-antenna gains in decibels relative to the gain of an isotropic radiator.

In many cases (especially in broadcasting) *field strength* is measured at the receiving location and conventionally used for computations or predictions of transmission. When ground losses associated with the receiving-antenna environment (usually a significant factor only at low and very low frequencies) are neglected, field strength can be related to transmission loss as follows.

First consider the relationships between field strength, power flux density (field intensity), and the available power in the receiving antenna. Let e denote the rms field strength in volts per meter. The power flux density p (field intensity) is given by

$$p = \frac{e^2}{\eta} \qquad \text{W/m}^2 \qquad (18\text{-}106)$$

where η is the characteristic impedance of the medium in which the measurement is made ($\eta_0 = 120\pi$ in free space).

The absorbing area of a receiving antenna with gain g_r relative to an isotropic antenna may be written

$$a_e = \frac{\lambda^2 g_r r_f}{4\pi r} \qquad (18\text{-}107)$$

where λ is the wavelength in the medium, r is the radiation resistance of the antenna, and r_f is the radiation resistance the antenna would have if it were in free space. Combining the above two equations, we find the following formula for the available power p'_a from a lossless receiving antenna:

$$p'_a = \frac{e^2 \lambda^2 g_r r_f}{4\pi \eta r} = \frac{v^2}{4r} \qquad (18\text{-}108)$$

The v in this equation denotes the open-circuit voltage induced in the receiving antenna. The field strength is related to the open-circuit voltage by

$$v = e\sqrt{\lambda^2 g_r r_f / \pi \eta_0} = el \qquad (18\text{-}109)$$

Field-strength meters usually are calibrated in terms of the effective length l of the antenna.

The relation between the available power p_a from the receiving antenna (**neglecting** losses) and the field strength e can be expressed in decibels as follows:

$$E = 10 \log[(4\pi\eta_0 p_a \times 10^{12})/\lambda^2 g_r]$$

$$= P_a + 20 \log f - G_r + 107.22 \text{ dB}\mu \tag{18-110}$$

where f is in MHz.

For field strength E in decibels referred to 1 μV/m, referred to 1 kW radiated from a half-wave dipole over perfectly conducting earth, propagation loss is

$$L_p = 139.4 - G_r + 20 \log f - E \quad \text{dB} \tag{18-111}$$

If the reference radiation is a short electric dipole, the constant becomes 136.0.

70. Fading; Characterization of Time-variant Multipath; Channel; Diversity. Random variations appear in the signal received via various transmission media, especially at frequencies above about 100 kHz when propagation is by the troposphere or ionosphere. Such variation is usually of two types: *one is attenuation, or power fading,* which may be quite slow (minute to minute, hour to hour, etc.) and is associated with comparatively large-scale changes in the medium, such as absorption; the other is *variable-multipath or phase-interference fading.*

Power fading is usually allowed for in the power margin designed into the system. Phase-interference fading, on the other hand, affects not only the amplitude but also the variable phase vs. frequency characteristic of the channel, limiting coherence bandwidth and introducing extraneous fluctuation in received-signal parameters. Alleviation of the effects of variable multipath is possible by diversity techniques, signal design, and signal receiving, processing, and detection operations. Fading media are discussed in Refs. 14, 16, and 19 and *CCIR Rep.* 415.

The amplitude probability distribution of the fading envelope is usually determined from samples of duration much shorter than the shortest fade duration; observation intervals over which statistical averages are taken are about 1,000 times the reciprocal of the nominal fading rate. The fit of experimental distributions of envelope fading to the Rayleigh distribution is often excellent for ionospheric and tropospheric scatter propagation, and similarly to a Nakagami-Rice[17,18] or Beckmann[19] distribution for situations where a specular component is mixed with scattered components. If long-term (power) fading is mixed with the short term, the median value of the short-term distribution changes; a log-normal distribution usually represents the long-term variation.

Most theoretical treatments of communication performance in the presence of variable multipath fading resulting from several signal components have been carried out for channels characterized by Rayleigh envelope distribution. The probability density function is given by

$$p(V) = \frac{2V}{v^2} \exp\left[-\left(\frac{V}{v}\right)^2\right] \tag{18-112}$$

where V is the fluctuating envelope and v^2 is the mean square value of V over the distribution. For the Rayleigh fading channel, the probability that the received signal envelope will fall at or below some specified value of V' is given by the cumulative distribution

$$p(V \leq V') = \int_0^{V'} \frac{2V}{v^2} \exp\left[-\left(\frac{V}{v^2}\right)^2\right] dV$$

$$= 1 - \exp\left[-\left(\frac{V'}{v}\right)^2\right] \tag{18-113}$$

The Rayleigh probability distribution function (18-113) is often used in the form

$$p(V \leq V') = 1 - \exp[-0.693(V'/V_M)^2] \tag{18-113a}$$

where V_M is the median value, about 1.6 dB below the rms value. *Fading rate* is important to certain systems. One measure of fading rate is the number of times per second (hertz) the carrier envelope crosses its median with a positive or negative slope. Another measure is the width of the received carrier-envelope spectral density.

For some time, design concern (besides its emphasis on the amplitude variation) has centered on the dispersion and multipath characteristics of the medium, in terms of linear time-variant amplitude and phase- and frequency-distortion parameters, often referred to as *multiplicative noise*, which cannot be overcome by power increase.

A more comprehensive characterization is in terms of the *system function* or *impulse function*. This approach relates the response and excitation of a channel at its input and output terminals.[14,16]

The expressions* for output are formulated in terms of operations on the input time function $x(t)$ or the spectral function (Fourier transform) $X(i\omega)$, to produce the output function $y(t)$ or $Y(i\omega)$. Each path is characterized by a system function $h(t, \tau)$ that operates on the replica of $x(t)$ traversing it; t is the time at which the observation is made, and τ is the delay or transit time for the path. The spread of delays between the input and output is determined by the system function $h(t,\tau)$, which may be called the *delay-spread system function*. For any particular elemental path x in a distribution of paths that covers some range of delays, the output for a range of delay $\Delta\tau$ centered on τ is given by $h(t,\tau)x(t - \tau)\,\Delta\tau$, and the total output of the channel is the sum of all such weighted and delayed contributing paths, namely,

$$y(t) = \int_{-\infty}^{\infty} x(t - \tau)h(t,\tau)\,d\tau \qquad (18\text{-}114)$$

For a Fourier transformable input $x(t)$, the output can be expressed in terms of the input spectral function $X(i\omega)$:

$$x(t) = \int_{-\infty}^{\infty} X(i\omega)\exp(i\omega t)\,d(\omega/2\pi) \qquad (18\text{-}115)$$

Here we characterize the channel by stating that it modifies the contribution to the structure of $x(t)$ from spectral components in the range $\Delta\omega$ centered at ω by multiplying it by the transfer function $H(i\omega, t)$. The total channel response to $x(t)$ is then

$$y(t) = \int_{-\infty}^{\infty} H(i\omega,t)X(i\omega)\exp(i\omega t)\,d(\omega/2\pi) \qquad (18\text{-}116)$$

which is the limit of the sum of the channel responses to the components of $x(t)$ from various infinitesimally wide spectral elements. The time-variant, *frequency-dependent transfer function* $H(i\omega, t)$ can be shown to be the *Fourier transform over the delay-spread variable* τ of the *delay-spread function* $h(t, \tau)$.

The randomness of the channel with time is reflected in the treatment of the system functions $h(t, \tau)$ and $H(i\omega, t)$ as sample functions of processes that are random over the space of the time variable t. The autocorrelation function of the channel response process is given by an inverse Fourier transform operation on the product of the spectral density function of the input process and the autocorrelation function of the time-variant, frequency-dependent transfer function $H(i\omega, t)$ of the channel. Further transformations produce a combined time-shift and frequency-shift correlation function and a so-called *scattering function* $S(\tau_0, f_0)$, which has the physical significance of a function that determines the weighting of the signal power as a function of the time delay τ_0 and Doppler shift f_0 incurred in transmission.

On the basis of the above functions, a set of transmission parameters for random time-variant linear filters is defined.[16]

1. The *multipath spread* or *delay spread* is determined by the relative delays of the component paths, or the "duration" of $h(t, \tau)$ over the delay variable τ.

2. The *coherence bandwidth*, the bandwidth over which correlation of amplitude fading (or coherence of phase for some applications) remains to a desired degree is usually defined in terms of specified degradation of error rate, distortion, or other parameter.

3. *Diversity bandwidth* is the frequency separation between two sinusoidal inputs which

---

* The outline of channel characterization follows E. J. Baghdady, lectures on characterization of communication channels, Boulder, Colo., November 1967, and is covered basically in Ref. 16.

results in a specified decorrelation of the fluctuating responses, usually taken to be a correlation coefficient of $1/e$.

4. The *fading rate, fading bandwidth,* * *frequency smear,* or *Doppler spread* is a measure of the bandwidth of the received signal when the input to the channel is a stable single-frequency signal.

5. The *diversity time* (or decorrelation time) is a measure of the time separation between input signals to yield correlation of less than $1/e$ between the envelopes of the responses.

These parameters are not all independent; the coherence bandwidth and delay spread are inversely proportional to each other, as are fading bandwidth and decorrelation time.

Most channel characteristics can be measured. The delay-spread response $h(t, \tau)$ can be measured directly by transmitting very short, widely spaced pulses: each received replica will correspond to one path, which can be resolved to examine the relative amplitudes and delays. The amplitude characteristic of the frequency-dependent transfer function $|H(i\omega, t)|$ can be measured for short intervals by transmitting a constant-amplitude test signal with repetitive linear sweep covering the desired frequency range. The envelope of the received signal will give a very close approximation to $|H(i\omega, t)|$.

Diversity techniques for counteracting short-term fading are used for hf ionospheric communication, forward scatter systems, and increasingly for microwave line-of-sight systems, where high reliability is required. The most common mechanism is to use *spaced antennas,* taking advantage of the fact that fading at one antenna tends to be independent of the signal fluctuation received on another antenna; provision is made to switch among signals or to combine two or more of them. Depending on the propagation mechanism, other kinds of useful diversity include *frequency, angle of arrival, polarization,* and *time* (Refs. 14 and 15 and *CCIR Rep.* 327). *Selection combining* selects the strongest signal; other linear combining methods include *maximal-ratio* and *equal-gain* combining. Diversity operation presumes the availability of n independently fading signals, referred to as diversity branches. There is no universal definition of diversity improvement, but it can be indicated for digital error probability.

For Rayleigh fading, maximal ratio combining for coherent diversity reception gives an error probability P_e for digital detection as follows (*CCIR Rep.* 195). For one receiver (no diversity)

$$P_{e_1} = \tfrac{1}{2} - \tfrac{1}{2}\sqrt{\alpha R/(\alpha R + 1)} \tag{18-117}$$

where $\alpha = 1$ for phase-reversal modulation and $\tfrac{1}{2}$ for frequency-shift or amplitude keying. R is the normalized signal-to-noise ratio, equal to the ratio of average signal energy to noise energy in each branch. For dual diversity

$$P_{e_2} = \tfrac{1}{2} - \tfrac{1}{2}\sqrt{\alpha R(\alpha R + \tfrac{3}{2})^2/(\alpha R + 1)^3} \tag{18-118}$$

The outline of fading and diversity improvement given here has assumed *flat* (nonfrequency-selective) fading and gaussian (white) noise. References 15 and 16 discuss frequency-selective fading. Paragraph **18-71** indicates nongaussian noise effects.

71. Noise: Signal-to-Noise Ratio; Service Probability.† Several types of radio noise must be considered in any design, though, in general, one type will be the dominant factor. In broad categories, the noise can be divided into two types: noise internal to the receiving system and noise external to the receiving antenna.

The *noise internal to the receiving system* is often the controlling noise in systems operating above about 300 MHz. This type of noise is due to antenna losses, transmission-line losses, and the circuit noise of the receiver itself. For receiver noise, the instantaneous noise voltage has a gaussian distribution, and the noise envelope is Rayleigh distributed.[15,20] Its effect on most communication systems can be determined mathematically with high accuracy.[15,21]

The second of the broad categories, *external radio noise,* can be subdivided further. Natural sources of radio noise are (1) atmospheric, (2) galactic, (3) solar noise from antennas

---

* *Fading bandwidth* is also often used in the same sense as diversity bandwidth, or coherence bandwidth, above.
† This material was contributed largely by R. T. Disney and A. D. Spaulding.

pointing at the sun, (4) precipitation (blowing snow or dust), (5) corona, and (6) noise reradiating from any absorbing medium through which the wanted radio signal passes. Very low noise systems used in space communications can be limited by such absorption by clouds, water vapor, and oxygen. Such *sky noise* has the gaussian characteristic of receiver noise. Examples of man-made noise sources are: (1) power lines or generating equipment, (2) automotive ignition systems, (3) fluorescent lights, (4) switching transients, and (5) electrical equipment in general.

Noise power is generally the most significant single parameter in relating the interference value of the noise to system performance. This parameter, however, is seldom sufficient in the case of impulsive noise, and a more detailed statistical description of the received-noise waveform is generally required.

Figure 18-50 shows the median value of the *available noise power spectral density* from various sources. F_a is in decibels above Kt_0 ($1\ Kt_0 = 3.97 \times 10^{-21}$ W/Hz).* While the solar noise, galactic noise, and sky noise are gaussian, the atmospheric and man-made noises are very impulsive.

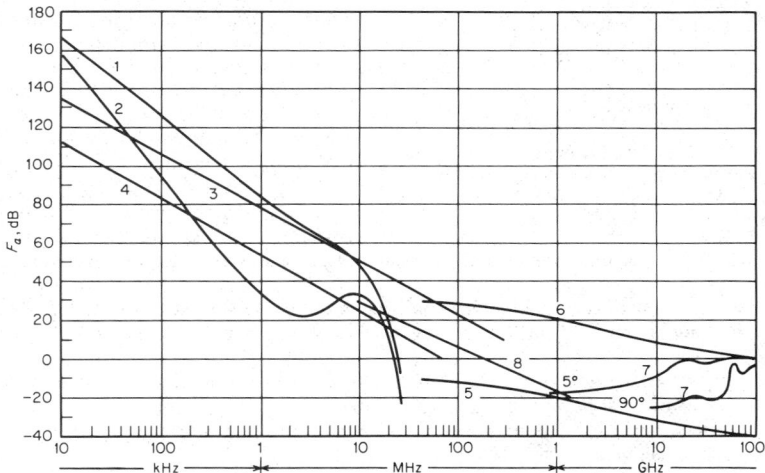

Fig. 18-50. Median radio noise-power spectral density from various sources. Curve 1, atmospheric noise, summer, 2000 to 2400 hours, Washington, D.C., omnidirectional antenna near ground; curve 2, atmospheric noise, winter, 0800 to 1200 hours, Washington, D.C., omnidirectional antenna near ground; curve 3, man-made noise, business area, omnidirectional antenna near ground; curve 4, man-made noise, quiet rural area, omnidirectional antenna near ground; curve 5, quiet sun, isotropic (0 dB gain) antenna; curve 6, disturbed sun, isotropic (0 dB gain) antenna; curve 7 sky noise, narrow-beam antenna (degrees from vertical); curve 8, galactic noise, omnidirectional antenna near ground.

CCIR Rep. 322 gives detailed definition of F_a, expected atmospheric noise level world-wide, and statistical variation as a function of geographical location, frequency, and time. Envelope probability distributions are given for atmospheric noise. Estimates of the noise-power spectral density from man-made noise expected in business, residential, rural, and quiet rural areas have been developed from measurements.[22-26] These expected values are the means of a number of location medians, with variation from location to location in each type of area and temporal variation indicated. Generally the noise below 20 MHz is associated with power lines. At 20 MHz and above, automotive electrical systems, especially ignition systems, are the dominant sources in all but rural locations.

Impulsive atmospheric or *man-made noise* disturbs communications in a way quite different

_____
* K is Boltzmann's constant; $t_0 = 300°$K.

from gaussian noise. Figure 18-51 shows the amplitude probability distribution of gaussian

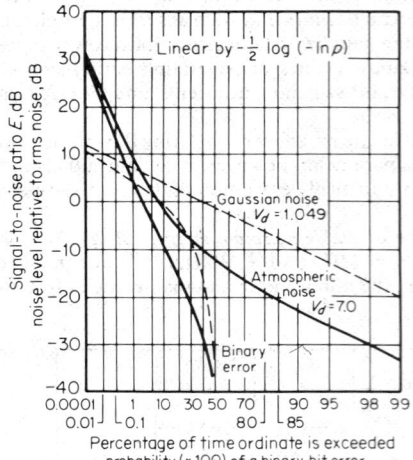

noise and a sample of atmospheric noise. The parameter V_d is the ratio in decibels of the rms voltage to the average voltage and is commonly used as an *impulsive index*. The two distributions are plotted relative to their rms level; i.e., both noises shown have the same energy or power. The probability distribution of the noise envelope determines the performance of most basic digital receivers, as indicated by the two error-rate curves for a binary coherent phase-shift-keying system.

Fig. 18-51. Comparison of noise distribution and error probabilities for gaussian and nongaussian noise (same noise power, coherent phase-shift keying).

Digital receivers are frequently designed for optimum performance in white gaussian noise. Their performance in impulsive man-made or atmospheric noise can be summarized as follows, with comparisons made on the basis of equal noise power:

1. For constant signal, at high signal-to-noise (S/N) ratio, impulsive noise causes more errors than gaussian noise; at lower S/N ratio, gaussian noise causes more errors.

2. For Rayleigh fading signals, gaussian noise causes more errors at all S/N ratios; flat-fading cases do arise, for which impulsive noise will cause more errors than gaussian noise; for diversity reception, impulsive noise is more harmful.

3. While pairing of errors in differentially coherent phase-shift keying (DCPSK) becomes more unlikely as the S/N ratio increases, pairing of errors increases as the noise becomes more impulsive.

4. For systems with time-bandwidth products on the order of unity, the standard matched filter-receiver is also optimum for impulsive noise.

5. Noise-suppression schemes such as wide-band limiting, smear-desmear, etc., are not particularly effective at high S/N ratios.

6. Receivers especially designed to reject a particular type of impulsive noise perform substantially better than receivers using the "standard" noise-suppression techniques.

The performance of analog voice systems in impulsive noise can be summarized as follows:

1. For a given articulation index, a much lower S/N ratio is required for impulsive noise than for gaussian noise.

2. Various forms of limiting in AM systems (pre-if limiting, if limiting, postdetection limiting) are quite effective in further reducing the required S/N ratio when the noise is impulsive.

Two additional measures (*CCIR Rep.* 322) are used to predict long-term performance: the percentage of time a given error rate, or better, will be achieved, termed *time availability*, and the probability that a given system will achieve a specified time availability and error rate, termed *service probability*. The service probability is designed to account for the probable errors or uncertainties in the prediction of the noise and signal distributions.

PROPAGATION OVER THE EARTH THROUGH THE NONIONIZED ATMOSPHERE

72. Introduction Mechanisms important to propagation over the earth and via the nonionized atmosphere at frequencies below about 30 MHz include free-space radiation, ground reflection, and diffraction along the surface of the earth, allowing for its finite conductivity, dielectric properties, and irregularities.

Above about 30 MHz, important mechanisms include free-space radiation, refraction, reflection from elevated atmospheric layers, reflection by the ground and various obstacles, absorption of energy by trees and buildings, diffraction over the surface of the earth and by hills, forward scattering of radio waves from atmospheric irregularities and layers, and scattering and absorption by atmospheric constituents. Terrain and meteorological conditions play important roles in determining the strength and fading properties of transmission through the troposphere.

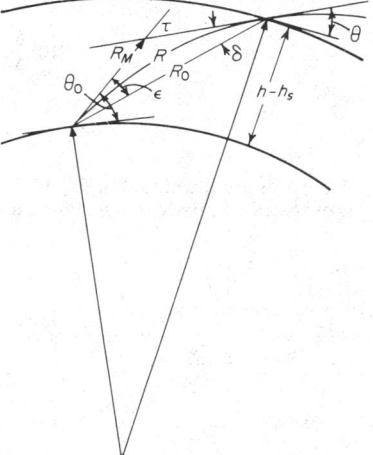

Fig. 18-52. Geometry of the refraction of radio waves. *(Bean et al.[29])*

Scattering and absorption effects by water vapor, moisture particles, and oxygen are dominant above 10 GHz.

73. Tropospheric Refraction. If a radio wave is propagated in free space (no atmosphere), the path followed by the ray is a straight line. However, in passing through the earth's atmosphere, the vertical gradient of atmospheric refractive index along the trajectory causes the ray path to become curved,[27] as shown in Fig. 18-52. The total angular refraction is τ, bending of the ray. The atmospheric radio refractive index n always has values slightly greater than unity near the earth's surface, for example, 1.0003, and approaches unity (free-space value) with increasing height. Ray paths usually (but not always) have a curvature concave downward, depending upon the gradient dn/dh.

The *radio refractive index* of the atmosphere is

$$n = 1 + N \times 10^{-6}$$

The *refractivity* is

$$N = \frac{77.6}{T}\left(P + \frac{4{,}810e}{T}\right) \quad \text{or} \quad N = \frac{77.6}{T}\left(P + \frac{4{,}810e_s RH}{T}\right) \tag{18-119}$$

where P is atmospheric pressure in millibars, e_s is saturation water-vapor pressure in millibars at temperature T, e is water-vapor pressure in millibars, T is absolute temperature in degrees Kelvin, and RH is percent relative humidity.

The elevation-angle error (ϵ in Fig. 18-52) is an important quantity in radar and other positioning or tracking systems.[27,28] It is a measure of the difference between apparent elevation angle to a terminal or target and the true elevation angle:

$$\epsilon = \tan^{-1}\frac{\cos\tau - \sin\tau\tan\theta - n/n_s}{(n/n_s)\tan\theta_0 - \sin\tau - \cos\tau\tan\theta} \tag{18-120}$$

where n_s is the refractive index at the surface

$$\tau_{1,2} = \int_{n_1}^{n_2} \cot\theta\,\frac{dn}{n} \tag{18-121}$$

To evaluate τ, n must be known as a function of height.

Many field-strength, phase, and bending calculations are made assuming a constant gradient of refractive index with height, equivalent to assuming an effective earth radius $a = Ka_0$, where a_0 is the real radius and k is determined by

$$k = \frac{1}{1 + \left(\dfrac{a_0}{n}\dfrac{dn}{dh}\right)} \tag{18-122}$$

Widespread practice uses $k = \frac{4}{3}$, corresponding to $dn/dh = -\frac{1}{4}a_0$ or a constant refractivity gradient with height of approximately $-39.3N$ units/km. The gradient is usually expressed in terms of the refractivity, $N = (n - 1) \times 10^6$, so that

$$10^6 \frac{\Delta n}{\Delta h} = \frac{\Delta N}{\Delta h} N \quad \text{unit/km} \tag{18-123}$$

and

$$k \approx \left(1 + \frac{\Delta N}{\Delta h} \Big/ 157\right)^{-1} \tag{18-124}$$

Ray paths are illustrated in Fig 18-53 for several values of k and $\Delta N/\Delta h$. The vertical scale is exaggerated relative to the horizontal scale, to make the differences in curvature apparent.

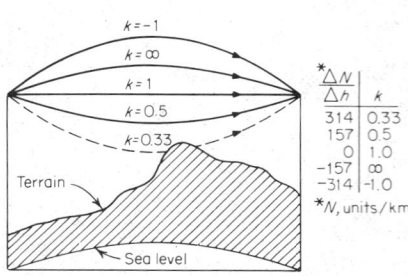

$^{*}\frac{\Delta N}{\Delta h}$	k
314	0.33
157	0.5
0	1.0
-157	∞
-314	-1.0

$^{*}N$, units/km

Fig. 18-53. Bending of radio rays for linear gradients. *(Dougherty.[57])*

For $0 < k < 1.0$, corresponding to positive gradients (subrefractive conditions), the ray curves away from the earth so that the ray joining two terminals passes close to the earth. The ray may even be interrupted by the surface so that the receiving terminal is beyond the radio line of sight. For the common situations $-157 \leq \Delta N/\Delta h \geq 0$, where $\infty \geq k \geq 1.0$, the rays are bent toward the earth's surface. At the critical value $\Delta N/\Delta h = -157N$ units/km, $|k| = \infty$, and the curvature of the ray path is equal to the curvature of the earth; the rays follow straight paths relative to the earth's surface. For $\Delta N/\Delta h < -157$, the situation is supercritical and the value of k is negative. For negative values the ray paths

are bent sufficiently toward the earth's surface for trapping of the radio rays to be possible. $\Delta N/\Delta h < -157$ is commonly referred to as a *trapping* or *ducting* condition.

The use of the constant gradient may lead to errors by overestimating refraction, especially important when one or both terminals are at high altitude, e.g., air to ground, or over long paths, and at frequencies much above vhf; at low frequencies the phase of the ground wave may be in error.

A better approximation of the mean refractive index as a function of height is given by the *exponential atmosphere* (Ref. 27, *CCIR Rep.* 231-2):

$$n(h) = 1 + N_s \exp(-bh) \times 10^{-6} \tag{18-125}$$

where N_s is the surface value of N, h is height in kilometers, and b is given by

$$\exp(-b) = 1 + \frac{\Delta N}{N_s} \tag{18-126}$$

where ΔN is the difference between the surface value of N and its value at 1 km height.

ΔN is generally correlated (for long-term means) with surface value N_s, which allows estimating ΔN in the usual case where only surface meteorological data are available:

$$-\Delta N = 7.32 \exp(0.005577\overline{N}_s) \tag{18-127}$$

(This value is for the United States: other values for other areas are given in *CCIR Rep.* 231.) An effective earth's radius a is given by

$$a = a_0[1 - 0.04665 \exp(0.005577N_s)]^{-1} \tag{18-128}$$

Worldwide seasonal charts of N_s and ΔN are published.[28,29] The maps are in terms of N_0, the sea-level value of N, and N_s is obtained from

$$N_0 = N_s \exp \frac{h}{7} \tag{18-129}$$

where h is the height above sea level in kilometers. Figure 18-54 gives long-term minimum values of N_0 useful for many propagation calculations.

The effective earth radius also depends upon frequency and polarization.[30,31] For

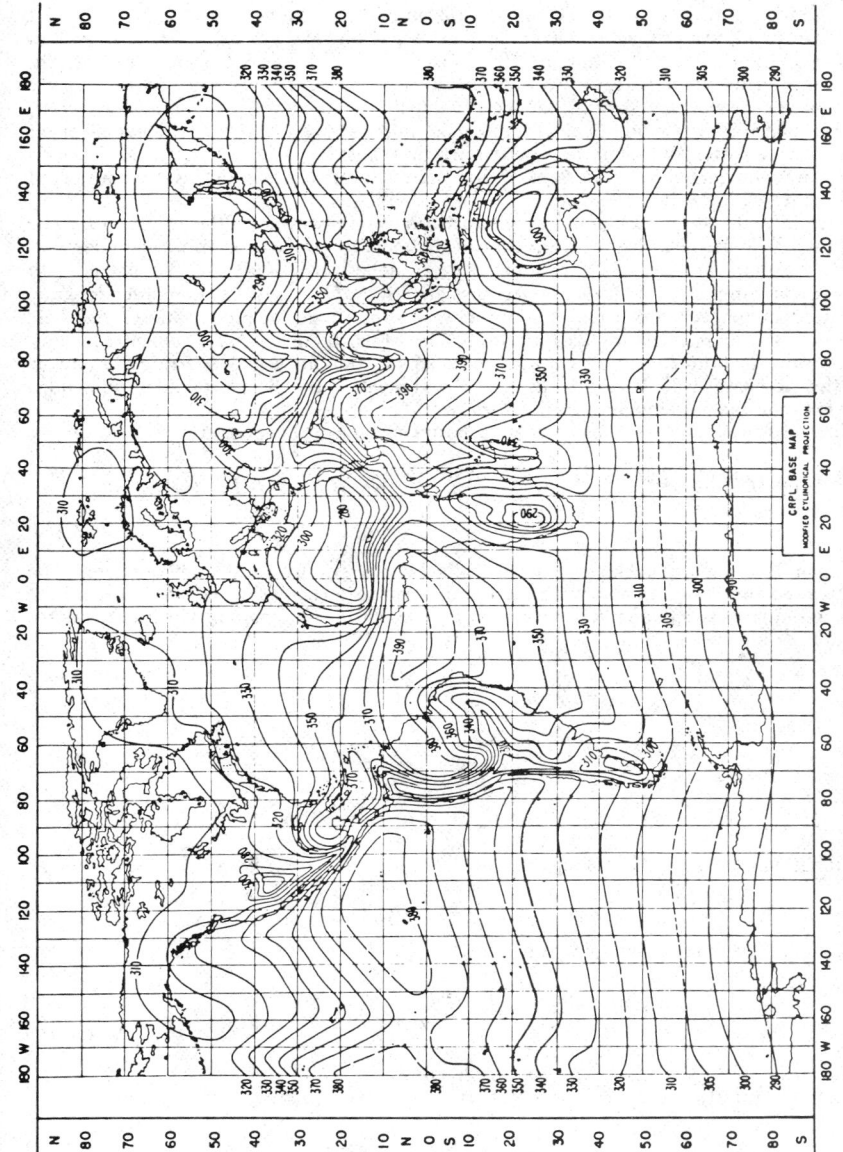

Fig. 18-54. Minimum monthly surface refractivity values N_0 referred to mean sea level. *(Bean et al.[29])*

conditions producing a ⅓ earth radius at 100 MHz, typical corresponding values of effective earth's radius at 10, 1, and 0.1 MHz are 1.3, 1.2, and 1.1, respectively.

74. Ground Wave Propagation over Homogeneous Spherical Terrain. The term ground wave refers to propagation within line of sight as well as by diffraction beyond the horizon, affected by earth conductivity, dielectric constant, and terrain, and by refraction in the lower atmosphere. Useful calculations can be obtained from the theory for a homogeneous, smooth, spherical terrain, although at frequencies much above 10 MHz additional effects of irregular terrain and variable tropospheric refraction must be considered (Pars. **18-76** to **18-84**). At frequencies below about 30 MHz, ionospheric propagation must also be considered (see Pars. **18-90** to **18-104**).

The most informative contemporary treatises on electromagnetic surface waves are those of Wait;[32,32a,32b] the historical development of the theory is traced, and practical computational equations are derived for homogeneous, spherical terrain and for mixed-path propagation. Important additional references are 3 to 5, 33, and 34. A computer program has been documented* which conveniently computes amplitude and phase of the electric field due to electric or magnetic dipoles for any values of earth constants and antenna heights.

A short radial dipole is at height h_T above a spherical, homogeneous terrain of radius a. The electric field is calculated at a distance D and height h_R. The central angle subtended by D is θ, and the arc along the earth's surface is d m.

The vertical electric field for an angular frequency $\omega = 2\pi f$ is given by

$$e_r \approx -\left[300\sqrt{P_r}\,\frac{\exp(-ikd + i\pi/4)}{d}\right] K \int_\Gamma \frac{\exp(-iv\theta t)}{W'_1(t)/W_1(t) - q_v} H_1(h_R) H_2(h_T)\, dt \qquad \text{V/m}$$

$$(18\text{-}130)$$

The quantity in square brackets is the vertical field over a perfectly conducting plane for an effective radiated power of p_r kW; the rest of the expression shows the effect of propagating along an imperfectly conducting sphere. In Eq. (18-130) Γ is a contour which encloses the poles of the integrand, and

$$K = \theta\sqrt{\frac{v}{12a\sin\theta}} \qquad k = \frac{2\pi}{\lambda}$$

$$q_v = -iv\frac{k}{k_2}\sqrt{1 - (k/k_2)^2}$$

$$v = \left(\frac{k_a}{2}\right)^{1/3} \qquad k_2 = k\sqrt{\epsilon_r - i60\sigma\lambda}$$

The functions $W_n(t)$ are Airy functions,[4,5] and

$$W'_n(t) = \frac{d}{dt}W_n(t)$$

$H_1(h_R)$ and $H_2(h_T)$ are height-gain functions:†

$$H_1(h_R) = \frac{W_1(t - y_R)}{W_1(t)}$$

and

$$H_2(h_T) = \frac{W_2(t - y_T)[W'_1(t) - q_v W_1(t)] - W_1(t - y_T)[W'_2(t) - q_v W_2(t)]}{2i}$$

where

$$y_n = \frac{kh_n}{v}$$

Note that

$$H_1(0) = H_2(0) = 1$$

so that the height-gain functions show the effect of nonzero antenna heights.

---

* L. A. Berry has contributed the computing formulas here for the ground wave.
† The form of H_1 and H_2 requires that $H_R > H_T$; since propagation is reciprocal, there is no loss of generality and the greater height can be assigned to H_R.

At distances beyond the horizon the integral in Eq. (18-130) is more easily calculated by summing the residues at the poles t_s, such that

$$W_1'(t_s) - q_v W_1(t_s) = 0$$

The classical residue series is then given by

$$e_r \approx -\left[300\sqrt{p_r}\,\frac{\exp(-ikd + i\pi/4)}{d}\right]2\pi iK \sum_{s=0}^{\infty} \frac{\exp(-iv\theta t_s)}{t_s - q_v^2} H(h_R)H(h_T) \qquad \text{V/m}$$

where $\qquad H(h_n) = \dfrac{W_1(t_s - y_n)}{W_1(t_s)}$ (18-131)

Because of the slow convergence of Eq. (18-131) when the transmitter and receiver are within line of sight, the saddle-point approximation[3,4] is used within the line-of-sight region, which has been shown to be equivalent to a geometric optical solution. An approximation useful for the diffraction region at higher frequencies is given in Par. **18-76**.

The same formulas can be used to calculate the field strength of a horizontally polarized wave by replacing q_v with

$$q_h = \frac{k_2^2}{k^2}q_v$$

Tropospheric refraction is allowed for by use of a linear profile of N, or constant height gradient, using an effective earth radius. The WKB[3,4,32] comparison-equation method, a solution for the linear profile, may be used as the basis for any profile which increases monotonically with height.

CCIR Recomm. 368 gives vertically polarized ground-wave field-strength curves from 10 kHz to 10 MHz. Figure 18-55 is an example for average soil. Curves of phase of the low-frequency ground wave are given in Ref. 35. FCC Rules[36] give field-strength curves for 500 to 1,600 kHz for use in standard AM broadcasting. Two CCIR atlases of ground-wave propagation[37,38] give curves for 30 to 10,000 MHz for vertical and horizontal polarization and elevated antennas, illustrated for 300 MHz in Fig. 18-56.

At frequencies below about 30 MHz, field strengths increase with decreasing frequency and are stronger over seawater than over land; vertical polarization gives higher fields than horizontal. Above 30 MHz there is little dependence on soil conditions, the signal strength is a more complex function of frequency, and, except over seawater, there is little difference between horizontal and vertical polarization.

75. Electrical Characteristics of the Earth. The relative permeability can usually be taken as unity. The relative importance of ε_r and σ varies with wavelength and may be judged by the ratio $\varepsilon_r/60\sigma\lambda$ (λ is in meters).

Table 18-5 shows earth constants for various types of surface.

In Table 18-5 the range of conductivity values for a specified type of surface corresponds to differences which exist in different parts of the world. In general, in fertile areas the higher values are applicable, while the conductivity of water in lakes and rivers increases with concentration of impurities. The effective values of the ground constants depend on frequency, depth of penetration, lateral spread of the wave, and the geological structure.

Table 18-5. Some Types of Surface and Their Earth Constants*

Type of surface	ε_r	σ, $\mho$/m
Seawater, at 0°C	80	4–5 (up to 1 GHz)
at 10°C	73	4–5 (up to 1 GHz)
Fresh water at 10°C	84	$1 \times 10^{-3} - 1 \times 10^{-2}$ (up to 100 MHz)
at 20°C	80	$1 \times 10^{-3} - 1 \times 10^{-2}$ (up to 100 MHz)
Very moist ground	30	$5 \times 10^{-3} - 1 \times 10^{-2}$
Average ground	15	$5 \times 10^{-4} - 5 \times 10^{-3}$
Arctic land	15	5×10^{-4}
Very dry ground and large towns (industrial areas)	3	$5 \times 10^{-5} - 1 \times 10^{-4}$
Polar ice	3	2.5–10^{-5}

* *CCIR Rep.* 229-1, 1970; this report is being revised to show frequency-dependence curves.

18-79

Moisture content is the major factor determining conductivity. Loam, which normally has a conductivity of the order of $10^{-2}\mho/m$, when dried can have a conductivity as low as $10^{-4}\mho/m$, of the same order as granite.

Measurements are made by several methods: resistance and reactance of capacitor units containing the soil samples as dielectric, resistance between probes driven into the ground, wave tilt, ground-wave attenuation, etc. (Refs. 39 and 40 and *CCIR REP.* 229-1). The ground involved in overland propagation is not usually homogeneous, so that effective ground constants are determined by several types of soil. Efforts to correlate ground constants in

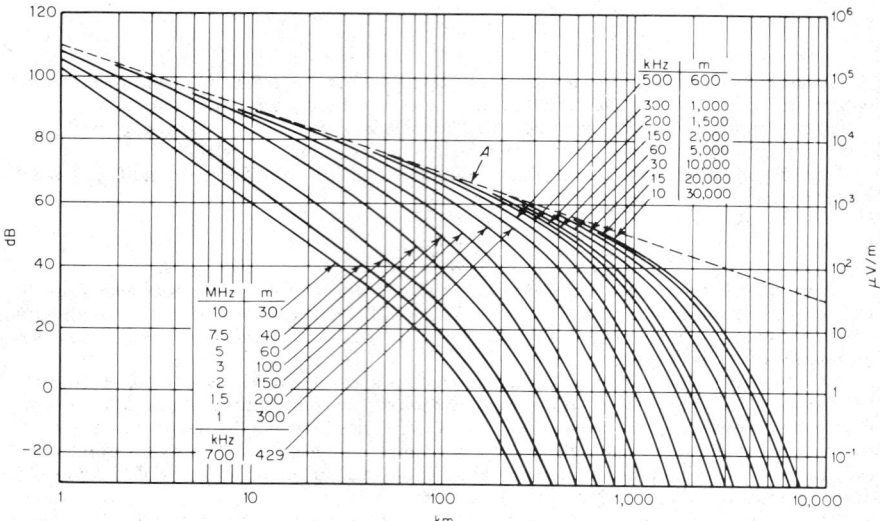

Fig. 18-55. Ground-wave propagation curves; earth, $\sigma = 3 \times 10^{-2}\mho/m$, $\varepsilon_r = 4$, vertical polarization A = inverse-distance curve. *(CCIR Rep. 368-1.)*

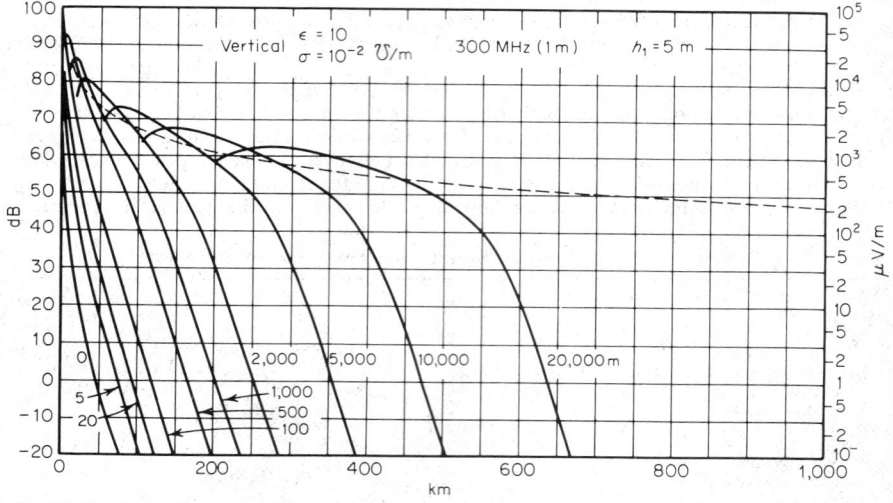

Fig. 18-56. Ground-wave field strength for 1 kW radiated; smooth earth 300 MHz; vertical polarization; average earth. *(CCIR Atlas.[38])*

detail with geological soil types have not been successful. FCC Rules[36] give estimated effective ground conductivity in the United States for the AM standard broadcast band; a similar map is available for Canada.

Inhomogeneous or Mixed Paths. Ground-wave propagation over an inhomogeneous path, e.g., a mixture of land areas of different ground constants or mixed water and land paths, requires special (usually graphical) methods of computation.

Wait[32,32a,32b,41] gives amplitude and phase factors for two-section paths for low and medium frequencies; Godzinski[42] gives curves for calculation of attenuation and phase for many-section paths. Widely accepted semiempirical methods are Millington's[43] and Kirk's.[44] Theoretical methods have been developed for high frequencies.[45] Wait [32a,32b] gives recent analytical investigations for mixed paths and applications. A general integral equation is given for the case of smooth boundaries that can be characterized by local surface impedances. A number of practical situations are considered and methods of solution are indicated. References to all important work in mixed path propagation are given.

76. Line-of Sight and Beyond-the-Horizon Propagation at 30 MHz to 10 GHz. Figure 18-57 illustrates the interrelationships between various propagation mechanisms* at frequencies above 30 MHz and the conditions under which each is dominant. The great-circle terrain profile is shown under conditions of normal refraction; the height of mountains and trees and the size of the objects are greatly exaggerated for the amount of earth curvature shown. The legend lists the propagation mechanisms likely to dominate over each of the paths illustrated.

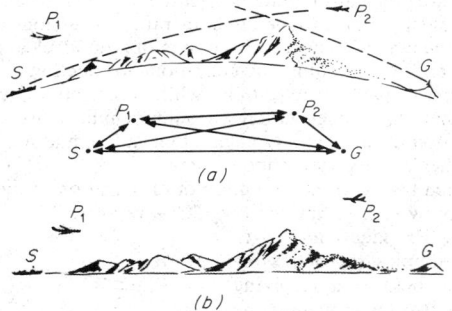

Fig. 18-57. (*a*) Propagation under conditions of normal refraction. Probably dominant mechanisms over each path are SP_1, free-space radiation and ground-reflected radio waves; P_1P_2, P_2G, vector addition of free-space and ground-reflected radio waves; SP_2, P_1G, knife-edge diffraction; SG, diffraction for distances less than 80 mi, forward scatter for greater distances. (*b*) Propagation in the presence of a ground-based duct and an elevated layer. SP_1, free-space radiation and ground-reflected radio waves; SP_2, P_1P_2, leakage through the top of the duct; weak signals further attenuated by reflection from the layer; somewhat affected by ground reflection, especially SP_2; P_2G, vector addition of free-space and ground-reflected radio waves, slightly affected by the presence of the layer; SG, P_1G, strong ducting signals. *(JTAC[1], by permission.)*

Between the ship at S and the plane at P_1, almost overhead, assuming that the ship antenna discriminates against ground reflection, transmission is virtually as in free space.

When the transmitting and receiving antenna beams also admit energy reflected from the ground, as from the plane P_2 to the ground antenna at G, or between two terminals within line of sight of each other but located near the earth, the direct and ground-reflected components must be added with the approximate relative phase. P_1G might involve smooth earth-ground reflection, and P_1P_2 might represent reflection from irregular terrain. Unless narrow antenna beams are used, reflection from hillsides or obstacles off the great-circle path often contribute a significant amount to the received signal. For transmission near or just beyond the horizon, it may be desirable to discriminate against such off-path reflections to reduce multipath fading and distortion; in other cases it may be desirable to direct antenna beams away from the great circle in order to increase the signal level by taking advantage of off-path reflection or knife-edge diffraction.

---

* This outline of mechanisms is condensed largely from Ref. 1 with permission.

As the height gradient of refractive index changes with time, the horizon rays shown in Fig. 18-57 will bend less or more than normal. With a gradient increase, for instance, ray bending will increase. If a radio ray from P_1 to P_2 more passes near the large mountain near G, the Fresnel-Kirchhoff knife-edge diffraction theory, either beyond line of sight or just within line of sight, may provide good estimates of the propagation loss between the airplanes and any elevated terminals. Knife-edge diffraction is expected to be dominant over the paths SP_2 and P_1G. Over a smoother terrain profile, the mechanism of propagation would still be diffraction, but with propagation loss increasing more rapidly with distance than for diffraction over a single knife-edge. When antennas are barely within line of sight of each other, propagation is often troubled by prolonged fading.

For the longer distances, well beyond the horizon, such as the SG path, diffraction, forward scatter, or reflection from elevated stratification (layers) are usually the dominant modes. Sometimes no distinction can be made between forward scatter from the turbulent atmosphere and the addition of incoherent reflections from patchy elevated layers. Empirical methods are used to predict statistically the characteristics observed for transhorizon paths based on observation of a great many paths over long periods of time. Tropospheric forward scatter is a useful propagation mechanism for distances between roughly 100 and 800 km, at frequencies between 30 MHz and 10 GHz.

Refraction or bending which exceeds the normal bending is called *superrefraction*, and strong superrefraction becomes ducting. Above some critical frequency and at less than some critical angle of elevation, the duct traps energy and propagates the wave efficiently over great distances. Ducts are especially common: over the sea when conditions represent a flow of warm dry air over colder water, producing temperature inversions and evaporation with lower layers, and over the land with nocturnal cooling of ground, clear skies, and calm moist air. The paths S to G and P_1 to G illustrate conditions for ducting.

Earth-space and space-space propagation, while representing essentially free-space transmission at the higher angles of elevation, encounters important refraction effects at low angles, rainfall attenuation at times at any angle of elevation, and near 10 GHz and above, attenuation by atmospheric water vapor and oxygen.

The methods outlined below for computation of propagation in the 30-MHz to 10-GHz region are given mainly by Ref. 12 and various CCIR reports.[175]

77. Line-of-Sight Propagation over Smooth Terrain. The simplest ray-optics formulas assume that the field at a receiving antenna is made up of two components, one associated with a direct ray having a path length r_0 and the other associated with a ray reflected from a point on the surface, with equal grazing angles ψ. The reflected ray has a path length $r_1 + r_2$. The field arriving at the receiver via the direct ray differs from the field arriving via the reflected ray by a phase angle which is a function of the path-length difference, $\Delta r = r_1 + r_2 - r_0$, illustrated in Fig. 18-58. The reflected-ray field is also modified by an effective reflection coefficient R_e and associated phase lag $\pi - c$, which depends on the conductivity, permittivity, roughness, and curvature of the reflecting surface. Figure 18-58a shows how the rays bend above the earth of actual radius a_0, and Fig. 18-58b shows the same rays drawn as straight lines above an effective earth radius a. The effective reflection coefficient R_e is then [46]

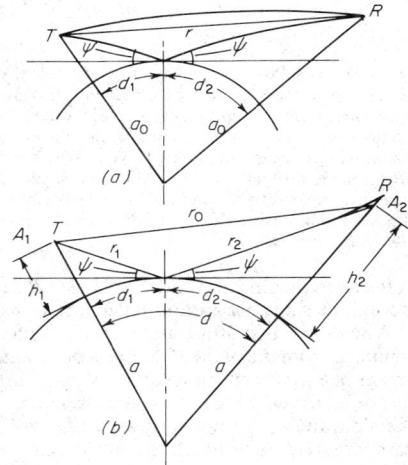

Fig. 18-58. Geometry for line-of-sight paths.

$$R_e = DR \exp\left[-\left(\frac{\pi\sigma_h}{\lambda}\sin\psi\right)^2\right] \qquad (18\text{-}132)$$

where the divergence factor D allows for the divergence of energy reflected from a curved surface, and may be approximated as

$$D = \left(1 + \frac{2d_1 d_2}{ad \tan \psi}\right)^{-1/2}$$ (18-133)

The term R represents the magnitude of the coefficient, $R \exp[-i(\pi - c)]$, for reflection of a plane wave from a smooth plane surface of a given conductivity and dielectric constant (Pars. 18-63 to 18-68 and Refs. 11 and 12). In most cases c may be set equal to zero, and R is very nearly unity. A notable exception occurs for vertical polarization over seawater. The terrain roughness factor σ_h is the rms deviation of terrain elevations.[46]

For line-of-sight transmission involving a single ground reflection, the attenuation relative to free space A can be obtained from

$$A = -10 \log\left[1 + R_e^2 - 2 R_e \cos\left(\frac{2\pi \Delta r}{\lambda} - c\right)\right]   \text{dB}$$ (18-134)

Over a smooth perfectly conducting surface, $R_e = 1$ and $c = 0$, and

$$A = -6 - 10 \log \sin^2(\pi\Delta r/\lambda)   \text{dB}$$ (18-135)

where $\Delta r = r_1 + r_2 - r_0$ and λ is the wavelength in meters.

For small grazing angles ψ, and with antennas h_1 and h_2 km above the earth,

$$\Delta r \approx 2h_1' h_2' /d$$ (18-136)

where h_1' and h_2' are the heights of the antennas above a plane tangent to the earth at the point of reflection. For equal antenna heights over a spherical earth of effective radius a,

$$\Delta r = d(\sec \psi - 1)$$

Just beyond the radio horizon of a transmitter, the dominant propagation mechanism is usually diffraction. Far beyond the horizon, the dominant mechanism is usually forward scatter.

78. Diffraction over Smooth Terrain.   Paragraphs 18-74 and 18-75 outline a method for computing diffraction over smooth terrain. For many problems in the frequency range above about 30 MHz, an approximate formula (CCIR Rep. 244-2) is useful for determining diffraction attenuation beyond the horizon, relative to free space A, over smooth terrain for horizontal polarization:

$$A = G(x_0) - F(x_1) - F(x_2) - 20.67   \text{dB}$$ (18-137)

The functions $G(x_0)$ and $F(x_2)$ and an auxiliary function $\Delta(x_{1,2})$ are plotted in Fig. 18-59.

$$x_0 = dB_0   x_1 = d_{Lt}B_0   x_2 = d_{Lr}B_0   B_0 = 670(f/a^2)^{1/3}$$

where $d_{Lt} \approx \sqrt{2ah_{te}}$ and $d_{Lr} \approx \sqrt{2ah_{re}}$ are distances from each antenna to its smooth-terrain

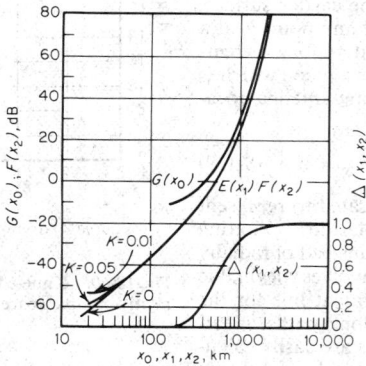

Fig. 18-59. $G(x_0)$ and $F(x_2)$ for smooth-terrain diffraction. For large values of x, $F(x) \approx G(x) - 1.356$, where $G(x) \approx 0.0575104x - 10 \log x + 2.066$. (CCIR Rep. 244-2.)

radio horizon, h_{te} and h_{re} are the effective heights of transmitting and receiving antennas, respectively, over smooth spherical earth, allowing for effective earth radius, as in Fig. 18-58*b*.* The error in A will be less than 1 dB if

$$x_0 - x_1 \Delta(x_1) - x_2 \Delta(x_2) > 320 \text{ km}$$

The CCIR atlases[37,38] illustrated in Fig. 18-56 included the effect of diffraction.

79. Knife-Edge Diffraction. A propagation path with a single isolated terrain feature which is the horizon for both terminals may often be considered as having a single diffracting wedge, as illustrated in Fig. 18-60. The diffraction attenuation (*CCIR Rep.* 244-2) relative to free space $A(\nu)$ is given as a function of the parameter ν:

$$\nu \equiv 2\sqrt{\frac{\Delta r}{\lambda}} = \sqrt{\frac{2d}{\lambda}} \tan \alpha_0 \tan \beta_0 \qquad (18\text{-}138)$$

The distance Δr is $r_1 + r_2 - r_0$. For $\nu > 3$,

$$A(\nu) = 12.953 + 20 \log \nu \qquad \text{dB}$$

For many paths the diffraction loss may be 10 to 20 dB greater than the value given by Fig. 18-60 because the obstacle is not a true knife-edge and because of other terrain effects.[47-49]

80. Propagation over Irregular Terrain, 30 MHz to 1,000 MHz. The terrain in most cases is not perfectly smooth. In the vhf and lower uhf regions, propagation for point-to-point,[†] broadcasting, and mobile services usually includes substantial terrain effects, which determine the mean attenuation below free space and affect fading characteristics.

For point-to-point services, it is useful to prepare a path profile scaled to allow for the curvature of the earth modified for refraction by effective earth radius a; this is particularly helpful in determining the distance from a transmitter or receiver to the radio horizon, which may be the bulge of the earth itself or may be a horizon obstacle, e.g., hills, buildings, or woods. The path profile is determined in the plane of the great circle between the transmitter and receiver (for paths shorter than about 70 km, a rhumb line may be used). Elevations of the terrain are read from topographic maps and tabulated vs. their distances from the transmitting antenna. The recorded elevations should include those of successive high and low points along the path. The terrain profile may be plotted on linear graph paper by modifying the terrain elevations to include the effect of the average curvature of the radio-ray path and of the earth's surface. The modified elevation y_i of any point h_i at a distance x_i from the transmitter along a great-circle path is its height above a plane which is horizontal at the transmitting-antenna location:

$$y_i = h_i - x_i^2/2a \qquad (18\text{-}139)$$

The vertical scale is exaggerated to represent the detail of terrain irregularities. Plotting terrain elevations vertically instead of radially from the earth's center involves negligible errors and allows use of straight line for the rays from antennas. For long paths, great-circle distances and bearings are easily calculated from the cosine law for oblique spherical triangles.

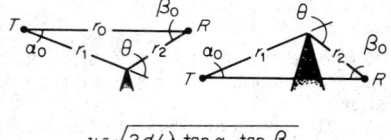

$$\nu = \sqrt{2d/\lambda} \tan \alpha_0 \tan \beta_0$$

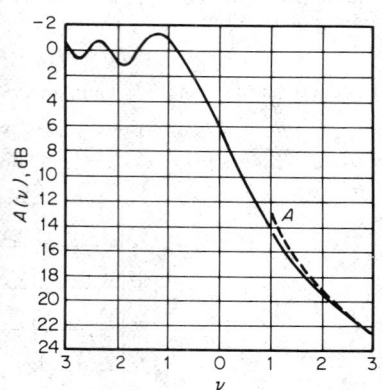

Curve *A*: asymptote, $A(\nu) = 12.953 + 20 \log \nu$

Fig. 18-60. Knife-edge diffraction, attenuation relative to free space. *(CCIR Rep.* 244-2.)

---

* If a point-to-point service uses very narrow beams, high antennas, and good terrain clearance, so that essentially free-space transmission is obtained, the outline of Par. **18-81** is applicable.

† Although this text uses metric units throughout, an approximate and convenient formula for radio horizon distance is d miles for an antenna height of h feet in $d \approx \sqrt{2h}$. This formula allows for ⅓ earth.

If two antennas are mutually visible over the effective earth, geometric optics can be used to estimate the attenuation provided it is reasonable to fit a straight line or convex curve of radius a to a reflecting portion of the terrain. However, such a procedure is of limited usefulness, being valid mainly at the lower frequencies and with low antenna heights, so that the path-length difference between direct and reflected rays is less than about 60°.

In most cases the terrain is too complex to fit a smooth curve, or only statistical descriptions of irregularities are available. Thus, empirical estimates of propagation are used for broadcasting and mobile services, and these extend well into the beyond-the-horizon region. A computer method[50] is available, and *CCIR Recomm.* 370 gives curves of field strength over irregular terrain for 1 kW radiated from a half-wave dipole, based on long series of measurement over many paths, principally in the United States and western Europe. Figure 18-61 illustrates median values for the uhf region. A parameter Δh used to define the degree of terrain irregularity is the difference in heights exceeded by 10 and 90% of the terrain in the range 10 to 50 km from the transmitter.

The height of the transmitting antenna is that above the local terrain, and the receiving antenna h_2 is taken at 10 m, typical of home-television and broadcast-receiving antennas. For mobile services the antenna will be nearer 3 m height, and the field strength will be reduced from 4 to 10 dB, depending upon frequency and terrain.

The CCIR Recommendation includes curves for additional frequency bands, corrections for various Δh, and allowances for different heights of receiving antennas. FCC Rules[51] give field strengths for the 88- to 108-MHz FM band in terms of levels exceeded 50% of the time at 50% of locations.

Additional data on radio-wave propagation for mobile communications at frequencies from 100 MHz to about 2 GHz are given in Refs. 52 to 54. An important aspect of vhf/uhf propagation in urban areas and over irregular terrain is the *time-delay spread* due to multipath reflections from buildings and terrain. These are particularly troublesome to FM broadcast and television reception and are a major factor in the design of new mobile radio systems above about 800 MHz. Typical time-delay spread exceeds $\frac{1}{2}$ μs in urban areas and $\frac{1}{4}$ μs in suburban areas.[54]

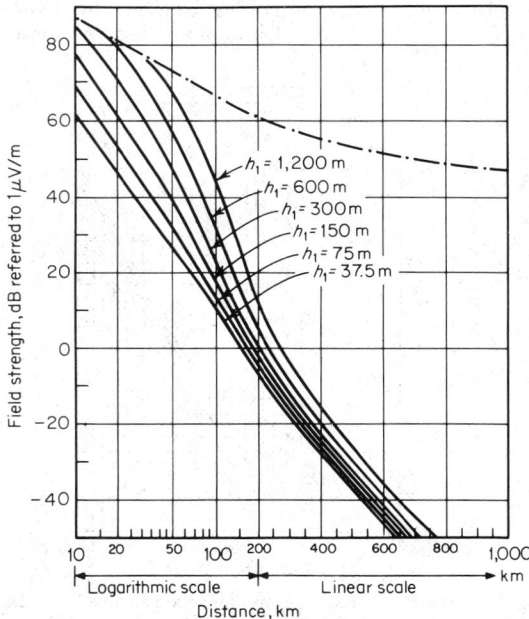

Fig. 18-61.   Field strength over irregular terrain; 1 kW radiated, frequency band 450 to 1,000 MHz, $\Delta h =$ 50 m, $h_2 = 10$ m. *(CCIR Rep. 370-1.)*

81. Line-of-Sight Propagation at Microwave Frequencies, 1 to 10 GHz. Well designed point-to-point microwave systems operating above 1 GHz use fairly narrow beam antennas so that free-space attenuation usually represents the best estimate of long-term median transmission loss. Good engineering practice for such links[55,56] assumes $\frac{2}{3}$ effective earth radius as a basis for estimating antenna heights required for terrain clearance rather than the value of $\frac{1}{3}$ associated with "standard" propagation. The $k = \frac{2}{3}$ provides against subrefractive situations, corresponding to positive refractive-index gradient near the surface and *upward* rather than downward bending of the beam (Fig. 18-53). For adequate terrain clearance *over smooth terrain*, the minimum antenna heights corresponding to the 10-dB curve in Fig. 18-62 are used. This allows for clearance of the terrain by the direct ray over an area of at least 60% of the first Fresnel zone. The clearance at path midpoint corresponding to the major axis of the full Fresnel zone is [56]

$$C = 8.66\sqrt{\frac{d}{f}} \quad \text{km} \tag{18-140}$$

where d = path length in kilometers, f is frequency in gigahertz, and the units are rationalized in the constant. Figure 18-62 gives required antenna height to protect against diffraction attenuation; Ref. 57 provides a series of such curves for different frequencies and values of gradient.

82. Fading and Diversity. Most microwave links do not experience serious fading under most meteorological conditions. However, stratification of the atmosphere and other meteorological conditions can cause severe fading. Measurements show that the range of refractivity-index gradient can be very large. At Cape Kennedy, Fla., for example, long-term measurements in the first 100 m near the surface showed that the gradient varied between 230 N/km, the value exceeded 0.5% of the time, and − 370 N/km, the value exceeded 99.9% of the time. These values correspond to k values of +0.4 and − 0.7, respectively (concave earth). A number of situations giving rise to both power fading and multipath fading are illustrated in Fig. 18-63.

Power fading includes results of beam bending, which affects terrain clearance, angle of arrival, trapping or deflection of the beam, and attenuation due to precipitation. Power fading due to loss of terrain clearance, also called *diffraction fading*, may be to depths of 20 to 30 dB. This type can be avoided except for most extreme cases by use of the design criteria of Fig. 18-62 and Ref. 57. Fading is also due to angle-of-arrival variations, up to $\pm\frac{3}{4}°$

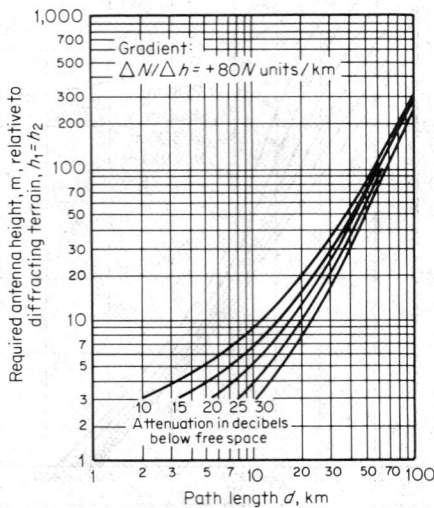

Fig. 18-62. Required antenna height to protect against diffraction attenuation at 2 GHz. *(Dougherty.*[57]*)* The 10-dB curve gives the approximate tower height required for smooth-terrain clearance over six-tenths of the first Fresnel zone for $k = 2/3$.

vertically and 0.1° horizontally. Ducts and layers cause power fades up to 20 dB or more, which may persist for hours or days. Precipitation attenuation can be important below 10 GHz but is of principal importance above 10 GHz (see Par. **18-87**). ·

Multipath fading includes phase-interference effects from ground-reflected and atmospheric paths. As the refractive index varies, interference can occur between the direct wave and the reflection from ground or water surfaces, as well as between the direct wave and partial reflections from atmospheric sheets or elevated layers. Additional direct paths can also be propagated due to surface layers of strong positive refractive gradients or horizontally distributed changes in refractive index encountered with a weather front. The frequency-selective fades extend to 20 or 30 dB below the free space, depending upon the relative amplitudes of the component waves. Specular ground reflection can produce fades persisting for minutes. Proper antenna design and siting can discriminate against terrain reflections.

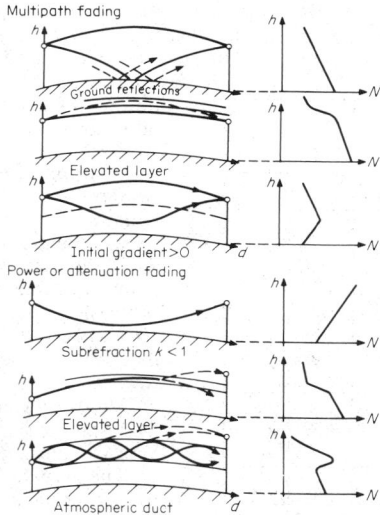

Fig. 18-63. Fading mechanisms for line-of-sight propagation. *(Dougherty.[57])*

Since one or two components usually dominate, the amplitude is not generally Rayleigh distributed but is represented better by the distribution of constant signal plus a random component. Detailed summaries of microwave-fading characteristics and mechanisms, a family of fading distributions, and a bibliography are given in Ref. 57; also important are Refs. 57a to 6la.

To design radio relay systems conforming to CCIR Recommendations, it is necessary to protect against the probability of deep fades for very small precentages of the time, e.g., about 0.0002%. *Space- and frequency-diversity techniques* are used.

A simple, generally effective space-diversity design procedure, based on Ref. 59, gives required vertical spacing, center to center:

$$\Delta h = 0.3\sqrt{\lambda d} \qquad (18\text{-}141)$$

where the path length d, wavelength λ, and spacing Δh are all in the same units. Design of diversity separations as a function of frequency, antenna height, path length, and expectation of refractive behavior is given by Ref. 57 for more difficult situations of maritime paths. See also Ref. 6l.

83. Forward Scatter. The long-term median transmission loss due to forward scatter is approximately (Ref. 12 and *CCIR Rep.* 244-2)

$$L(50) = 30 \log f - 20 \log d + F(\theta d) - G_p - V(d_e) \qquad \text{dB} \qquad (18\text{-}142)$$

where $F(\theta d)$ is shown in Fig. 18-64. The angular distance θ is the angle between radio horizon rays in the great-circle plane containing the antennas and d_e is the distance between antennas. $V(d_e)$, given by *CCIR Rep.* 244-2 for various climatic regions, may be up to $\pm$ 8 dB.

The combined gain of transmitting and receiving antennas may be less than the sum of their plane-wave gains. This apparent drop in gain, termed *gain degradation* or *antenna-to-medium coupling loss*, occurs when the beam widths of the antennas are smaller than the angle subtended by the useful scattering volume. The amount of loss depends on the antenna gain and the path length; experimentally, however, little dependence on distance is observed in the range 150 to 500 km, and an empirical estimate is

$$G_p = G_t - G_r - 0.07 \exp [0.055(G_t + G_r)] \qquad \text{dB} \qquad (18\text{-}143)$$

for values of G_t and G_r each less than 50 dB.

The siting of terminals of transhorizon links requires some care. The antenna beams must not be obstructed by nearby objects, and the basic requirement is that the antennas be directed at the horizon. If the antenna beams are tilted upward by as little as 0.5°, there may be a loss of the order of 10 dB due to decreased scattering efficiency with height.

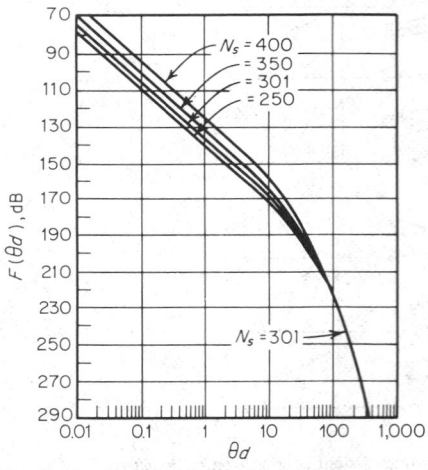

Fig. 18-64. Attenuation function $F(\theta d)$ for forward scatter; d km, θ rad. *(CCIR Rep. 244-2.)*

Theoretical and experimental information on the mechanism and characteristics of tropospheric forward scatter is given in Refs. 62-71.

To estimate *long-term variability* of forward-scatter paths, meteorological information has been used to distinguish between climatic regions (*CCIR Rep.* 244-2), and radio data from more than a thousand paths in various parts of the world provide the basis of prediction of long-term variability about the computed long-term median value in each of these regions. *CCIR Rep.* 244-2 gives methods and data.

Short-term *phase interference fading* of a tropospheric scatter signal can usually be approximated by a Rayleigh distribution. The mean fading rate is proportional to carrier frequency and is about 4 Hz at 10 GHZ. Fading rate also depends on antenna beam width, being slower for narrow beams because of restriction of the lateral phase-path extremes.

Horizontal spacing of 50 to 100 wavelengths has been found adequate for diversity throughout the frequency and distance range of interest. It is also possible to have angle diversity using multiple feeds and a common reflector.[72]

84. Ducting. When large vertical gradients are sufficient to refract a ray to the same radius of curvature as the earth (vertical decrease of N greater than $157N$ units/km, Par. 18-72), *superrefraction* is said to occur and the wave can be *trapped*. A layer of this type is called a duct, and the mode of propagation between the earth and such a layer (or between two layers) is similar to that of a waveguide. Low-loss transmission over great distances is possible via ducts, very distant radar echoes can be observed, and duct propagation constitutes a potential source of interference between satellite-earth stations and other terrestrial radio uses. Climatology of radio ducts is discussed in Ref. 27. The attenuation depends upon refractive-index gradient and thickness of the boundary layers (upper and lower in the case of an elevated atmospheric lower boundary) and how energy is coupled into the duct.[4,73-75] Adequate numerical treatment of propagation via a tropospheric duct is very difficult because of the large number of parameters involved. Reference 74 gives a method for computing an effective cutoff frequency, though the cutoff in practice is not nearly so sharp as for waveguides. Efficient coupling usually occurs only for energy (this may be side-lobe energy) grazing the duct boundary within 1° and when both transmitter and receiver are within the duct. Energy can be scattered from terrain or obstacles into the duct.

Experimental studies of ducting over ocean surfaces are given by Refs. 76 and 77. Prediction charts[29] give statistical data on refractivity gradients conducive to ducting. Figure 18-65 illustrates seasonal and geographical variation of trapping frequency. Estimates of basic transmission loss by ducting are used by CCIR (*Rep.* M/233) for estimating mutual interference in geographical areas rather crudely defined by ITU Radio Regulations.

85. Atmospheric Effects above 10 GHz, Millimeter Waves, and Earth-Space Propagation. At frequencies above about 3 GHz, the attenuation of radio waves by atmospheric gases and water and refractive scintillation and multipath effects become increasingly important. These factors become dominant considerations in the design of radio-relay or earth-satellite systems at frequencies above 10 GHz.

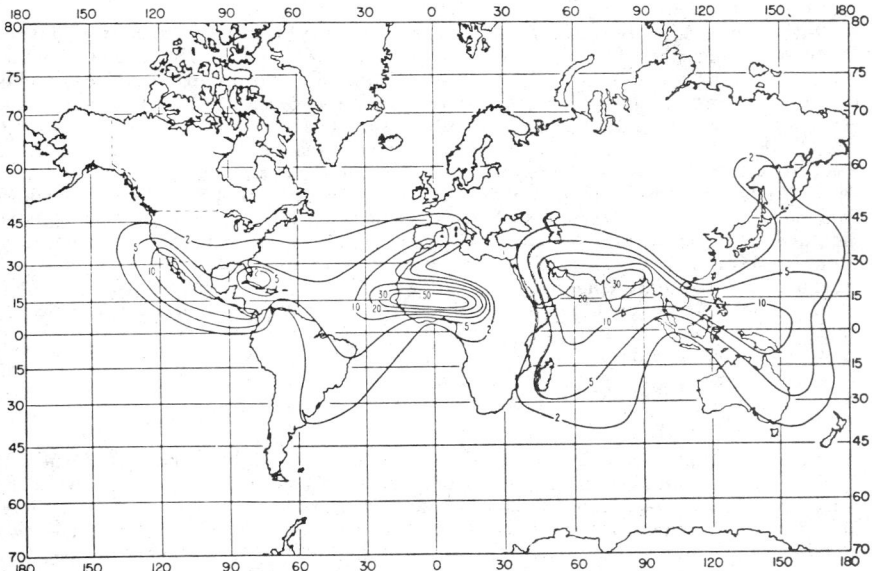

Fig. 18-65. Percent of time trapping frequency <3,000 MHz, November. *(Bean et al.[29])*

Growing use of frequencies above 10 GHz is envisaged for earth satellite systems as their information-capacity requirements grow, and the needs involve multiple steerable beams, multiple-access, and high-resolution earth-resource application. Point-to-point video and data-relay systems increasingly require the greater bandwidth capability available above 10 GHz, compared with congested lower-microwave frequencies. Other applications include high-resolution radar, precision guidance, and tracking. New aircraft-landing systems use 15 GHz and higher frequencies to obtain the required resolution. Radio-astronomy studies make use of gaseous resonance lines throughout this region of the spectrum; optical systems are being increasingly used for precise positioning and distance measurement, as well as for high-capacity information transmission.

Propagation effects of importance are absorption by the clear atmosphere due to molecular resonances; attenuation and scattering by rain or fog and phase interference and refractive scintillation by atmospheric turbulence, stratification, and terrain effects. Propagation difficulties in this frequency range should not be underestimated. Rain is a most important effect and causes noticeable attenuation at frequencies as low as 3 GHz and interruption of line-of-sight links at frequencies as low as 6 GHz; rain can restrict the reliable use of frequencies much above 30 GHz to terrestrial links of a few kilometers in length or to earth-space links near the zenith.

86. Absorption by the Clear Atmosphere. Transmission through the clear atmosphere is subject to attenuation by molecular oxygen and water vapor (Refs. 78 to 89 and *CCIR Rep.* 234-2). An attenuation peak at 22 GHz is due to the single rotational transition of the water molecule; the peak near 60 GHz results from a large number (43) of oxygen absorption lines, effects of pressure broadening in the lower atmosphere playing an important role. Each of these gases has a second absorption region below 300 GHz, oxygen at 118 GHz and water vapor at 184 GHz. Each resonance has an accompanying frequency-dependent phase-velocity, or dispersion, effect. The only serious gaseous absorption is posed by oxygen in the 55- to 75-GHz and the 118-GHz regions and by water vapor above about 125 GHz. Methods for calculating absorption and dispersion as a function of gas density, in terms of line width, strength, and frequency, are given in Refs. 81 and 82, and the parameters are available from laboratory measurements.[83,85] The line width is broadened by molecular collisions depending upon the pressure and gases involved. By modeling the pressure, temperature, and water-vapor profiles it is possible to calculate the attenuation and phase delay through the atmosphere to different heights and at different elevation angles.

Figure 18-66 shows the theoretical one-way attenuation for vertical and horizontal paths through the atmosphere for the frequency range from 1 to 160 GHz for a moderately humid climate (7.5 g/m³ at the surface). Also shown are the limits for 0 and 100% relative humidity for the horizontal path through the atmosphere. Figure 18-67 shows the fine structure of the oxygen lines and illustrates the effect of pressure broadening at low altitudes.[80]

87. Rainfall Attenuation and Scattering.* The attenuation due to rain usually exceeds the combined absorption due to oxygen and water vapor and arises from the absorption of energy in the water droplets and from the scattering of energy out of the beam of the antenna. Attenuation is usually expressed as a function of the rainfall rate R. It depends on both the liquid-water content and the velocity of fall of the drops, which in turn depends on drop size. Rain of a given rainfall rate has various distributions of drop size, and useful estimates of attenuation are obtained empirically.

The total attenuation A_r due to rainfall over a path of length r_0 can be determined by integrating the rain-absorption coefficient $\gamma_r(r)$ along the direct path between the two mutually visible antennas:

$$A_r = \int_0^{r_0} \gamma_r(r)\, dr \qquad \text{dB} \qquad (18\text{-}144)$$

Values of γ_r are given in Fig. 18-68 as a function of frequency and rainfall rate R mm/hr for a mean drop-size distribution; this figure should be used with caution in view of the tendency of the attenuation measured at some frequencies (between about 20 and 35 GHz) to exceed the maximum predicted attenuations.[90,91] Additional difficulty arises from the lack of uniformity of rainfall over actual transmission paths. Total annual rainfall data at a point, available from weather records, does not usually indicate rain rate statistics needed for attenuation estimates. Nor do surface rainfall data directly provide the high-altitude information needed for estimates of attenuation and scatter for earth-satellite links. An approximate method, based on the concept of a number of rain climatic zones, may be used.[92] This method uses detailed information for 45 United States stations from 1951 to 1960, with less detailed information for 135 worldwide stations from 1931 to 1960, to estimate long-term distributions of 1-min average rainfall rates.

Figure 18-68, in conjunction with the millimeter-per-hour rainfall rate, can be used to estimate rain attenuation for horizontal line-of-sight paths at frequencies below about 60 GHz. At frequencies much above 40 GHz, the specific attenuation depends primarily on the concentration of small drops. However, for most drop-size contributions, the attenuation for a given rainfall rate does not increase appreciably with frequency above 100 GHz. Calculation of attenuation for elevated paths requires allowance of an effective path length through the rain.[93] Path diversity on millimeter wave links is an important countermeasure for rain attenuation.[93a,93b]

Scattering by precipitation is a source of potential interference between terrestrial systems and earth-space systems because of the high elevation of the beam for earth-space systems. To calculate transmission via precipitation scatter, one must make allowance for the height and drop-size distribution and an effective path length through the precipitation (Refs. 93 to 95 and *CCIR Rep.* M/233). Figure 18-69 shows calculations of transmission loss at 4, 11, and 30 GHz for scatter from rainfall in intersecting beams on the path "beyond the earth station" as indicated in the geometry. Because the scattering volume is assumed to fill the narrow

---

* This material is taken in large part from *CCIR Rep.* 234-2.

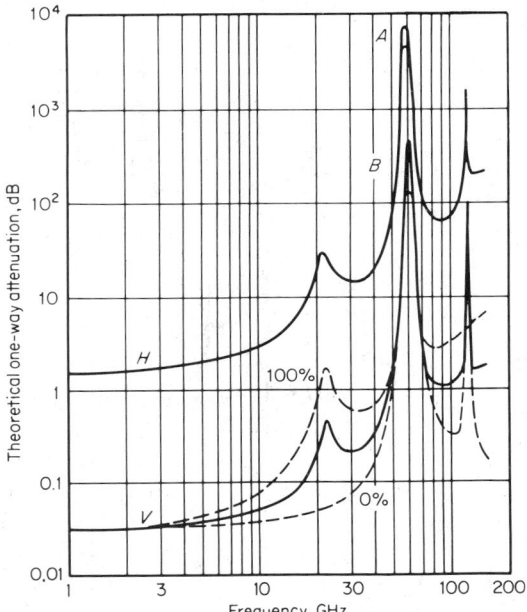

Fig. 18-66. Theoretical one-way attenuation for vertical and horizontal paths through the atmosphere (calculated using the U.S. Standard Atmosphere for July at 45°N latitude). Solid curves are for a moderately humid atmosphere; dashed curves for vertical attenuation represent the limits for 0 and 100% relative humidity. *A, B,* limits of uncertainty; *V,* vertical; *H,* horizontal. *(CCIR Rep. 234-2.)*

beams, no account need be taken of antenna gain, and transmission loss p_r/p_t is obtained directly.

Coherence bandwidth does not appear to be significantly limited by multipath scattering via rainfall; attenuation usually limits transmission before multipath effects are discernible.[96-98] Important and additional references on rainfall effects on propagation are Refs. 99 and 100.

88. Refraction Errors, Scintillation, Turbulence and Stratification; Fading, Angular, and Bandwidth Limitations.* Ray-tracing methods for refraction errors are outlined in Ref. 27. At low elevation angles (below about 5°) and for horizontal paths, amplitude-, phase-, and angular-scintillation effects can be important. These are caused in the normally turbulent atmosphere by small-scale refractive irregularities associated with random pressure, temperature, and water-vapor variations; similar effects can be produced by phase-interference effects from partial reflection from elevated layers or refractive sheets or in some cases by terrain reflections (especially on long line-of-sight paths). Theoretical methods and experimental results for engineering estimates are given in Refs. 86, 87, 89, and 100a to 103a.

For space-space or earth-space links at high angles of elevation and frequencies below about 20 GHz, small-scale variations in refractive index are generally insignificant. At higher frequencies in some meteorological conditions, their effect may be important. Assuming a source outside the troposphere and a receiving antenna of up to a few meters in diameter, the largest peak-to-peak fades expected in clear air are about ±4 and ±2 dB at 100 and 35 GHz, respectively, for elevation angles exceeding 45°. For angles of elevation of about 10°, however, the fades may occasionally reach ± 12 and ± 6 dB at 100 and 35 GHz, respectively. The standard deviation of amplitude scintillation at elevation angles below 4°

---

* Par. **18-108** discusses scintillations produced by the ionosphere.

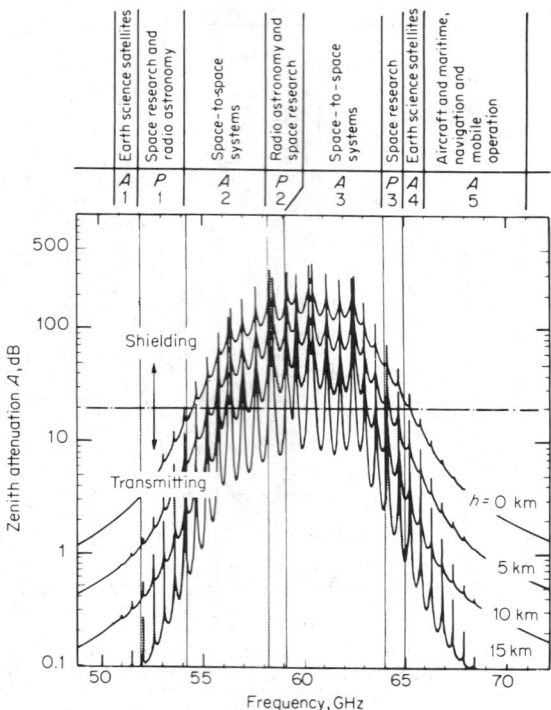

Fig. 18-67. Millimeter-wave attenuation in the 49- to 72-GHz band due to atmospheric oxygen resonances for zenith paths through the U.S. Standard Atmosphere 1962 from different initial heights h_1 to outer space. Allocated band utilizations are A(1-5), active telecommunications; P(1-3), passive systems only (WARC 1971 Conference). *(Liebe and Welch.*[80]*)*

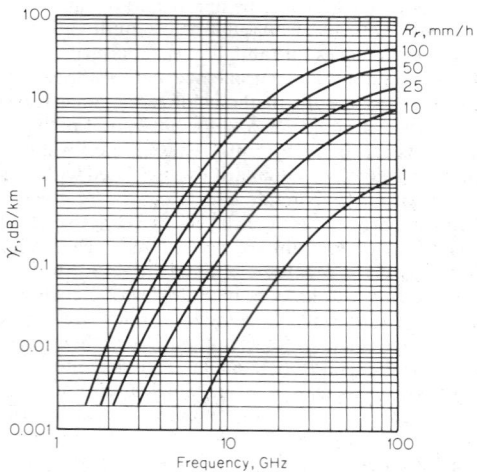

Fig. 18-68. Variation of γ_r with frequency. *(CCIR Rep. 234-2.)*

is less than 1 dB even for very large antennas (*CCIR Rep.* 234-2). At frequencies below 10 GHz this scintillation is essentially independent of frequency.

For a distance of 6 m across the wavefront (representative of the diameter of a large millimetric antenna) the rms value of phase differences caused by the troposphere on an earth-space link may sometimes reach 40 and 15° at 100 and 35 GHz, respectively, when the elevation angle exceeds 45°. At about 10° elevation, these values increase to approximately 80 and 30° at 100 and 35 GHz, respectively.

Spectra of phase fluctuations for a horizontal line-of-sight path over the sea at 10 and 35 GHz are available in Ref. 103.

Angle of arrival at 20 to 30° above the horizon shows fluctuation of the order of 0.2 to 0.3 $\times 10^{-4}$ rad/s (standard deviation for fluctuation of durations greater than $\frac{1}{10}$s). This angle may be considered as the theoretical smallest useful beam for an antenna. At lower elevation the fluctuation will be greater.[27]

Coherence bandwidth is also limited by these scintillations and multipath phenomena, especially at frequencies above 10 GHz for very wide-band systems operating over horizontal (terrestrial line-of-sight) paths. During periods of stable atmospheric stratification, multipath can limit bandwidth coherence to a few tens of megahertz.

A survey of clear-air propagation effects for optical communications is given in Ref. 104.

89. Sky-Noise Temperature Due to Atmospheric Absorption and Precipitation. At frequencies above about 10 GHz, the nonionized region of the atmosphere, as an absorbing medium, is also a source of noise radiation.[105] The effective sky-noise temperature T_s is shown in Fig. 18-70 for various angles of elevation. Rainstorms also provide significant contribution to sky-noise temperature.[106]

PROPAGATION VIA THE IONOSPHERE

90. Introduction. The ionosphere is that part of the atmosphere, at heights above about 50 km, in which free ions and electrons exist in sufficient quantities to affect the propagation of radio waves. At frequencies below about 30 MHz, regular long-distance transmission is possible by way of ionospheric reflections. At frequencies in the 30- to 100-MHz region, regular but weak propagation by ionospheric scattering is obtained, as well as strong intermittent propagation by reflection from sporadic ionospheric layers and meteoric ionization. At frequencies well above 100 MHz auroral echoes are observed, and ionospheric scintillation effects may be important for satellite-earth links.

91. Physical Description of Ionospheric Regions. The various regions of the upper atmosphere are usually described according to the nomenclature of Fig. 18-71, which also illustrates typical daytime densities and temperature.[9,107]

At levels up to about 85 km turbulent mixing keeps the relative chemical composition of the atmosphere essentially the same as at the ground, that is, N_2, O_2, Ar, and CO_2, as well as traces of water vapor, ozone, nitric oxide and hydrogen. Above about 90 km, O, O_2, and N_2 are the major constituents, and dissociation of O_2 becomes important. The important parameters of radio propagation are the concentration of free electrons (*N*, electron density) and the rate at which these electrons collide with neutral particles (*v*, collision frequency).[8,9,106a-108]

Four principal regions, or layers, affect the propagation of radio waves. The term layer identifies regions having distinct characteristic processes, heights, and densities.

The *D region* extends from 50 to 90 km height; it is a region of low electron density in collision with neutral gases, mainly causing absorption of radio waves passing through, but sufficiently reflective to provide an upper boundary for vlf and lf propagation and sufficiently irregular and turbulent to scatter waves at vhf. The cause of ionization is generally taken to be solar photoionization of NO by Lyman-α radiation (λ = 1216Å). *D*-region ionization is mainly daytime, though during disturbances and at high latitudes ionization can be caused by particle radiation. The *mesopause* at about 85 km is a region of strong turbulence and wind shear.[9]

The *E region,* at heights 90 to 130 or 140 km, is ionized mostly by solar ultraviolet and x-rays in the daytime with some small nighttime ionization by cosmic rays and meteors. Electron production in the 100- to 140-km height range is ascribed mostly to solar radiation in the 3- to 20-mm wavelength ion,[109] dependent on solar zenith angle and solar cycle. A

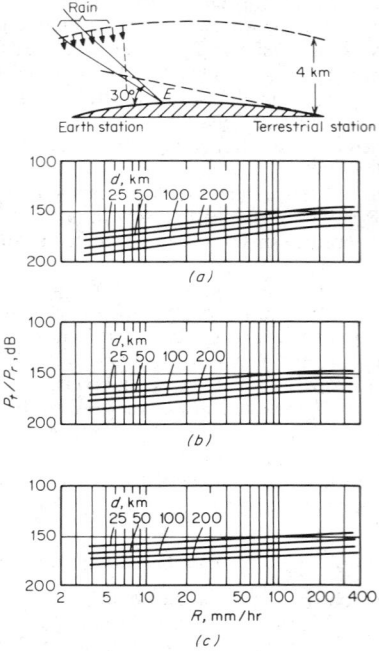

Fig. 18-69. Transmission loss vs. rainfall rate R for rain on path beyond earth station. (a) f = 4 GHz; (b) f = 11 GHz; (c) f = 30 GHz. *(CCIR SJM.[175])*

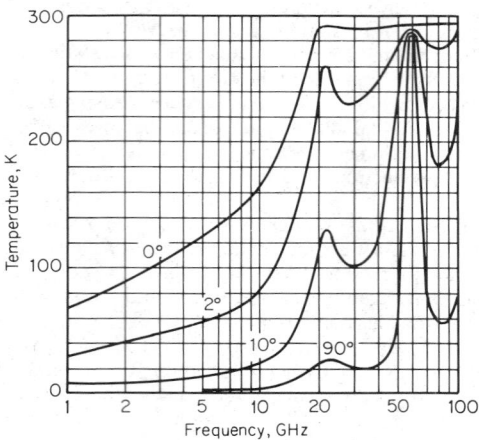

Fig. 18-70. Total sky-noise temperature; surface pressure 760 mm Hg, surface temperature 20°C, water-vapor density 10 g/m³. *(CCIR Rep. 234, rev. ed., 1972.)*

regular E layer of maximum electron density near 100 km is an important reflecting medium for daytime hf propagation, and at night for mf and lf propagation.

In addition to the regular E layer, irregular cloudlike layers of ionization, call *rundown sporadic E* (or E_s), produce partial reflections and scattering at frequencies up to 150 MHz. There are a number of types and causes of E_s.[110-112] Over the magnetic equator, E_s is a regular daytime phenomenon, attributed to the two-stream plasma instability in the presence of the equatorial electrojet. At midlatitudes E_s is commonly attributed to the concentrating effect of wind shear in the E region on positive ions, thence on electrons; there is a summer peak of occurrence. A high-latitude nighttime E_s is associated with disturbed magnetic conditions and aurora.

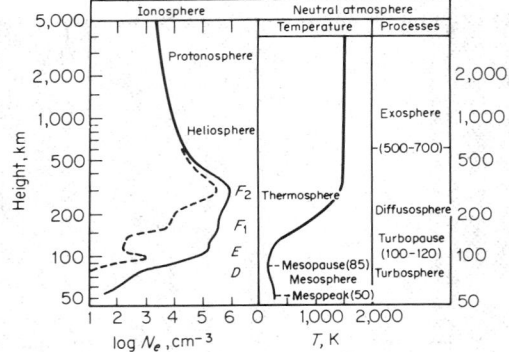

Fig. 18-71. Atmospheric nomenclature and typical daytime ionospheric electron densities and temperatures. (After VanZandt and Knecht.[107]) by permission, John Wiley & Sons.

The *F region*, above 130 to 140 km, is the most important for high-frequency propagation.

The $F1$ layer exists mainly during daylight at heights of 175 to 220 km; though fairly regular in its characteristics, it is not observable everywhere or on all days. $F1$ depends on daylight solar ionization but is also prevalent during ionospheric disturbance. At night the $F1$ layer merges into the $F2$.

The $F2$ layer, at heights from 200 to 400 km, is the principal reflecting layer for long-distance high-frequency communication day and night and has the most complex and variable characteristics. Height and electron density vary geographically, diurnally, seasonally, and with solar cycle. Unlike the E and $F1$ layers, the $F2$ does not follow solar radiation directly with zenith angle; e.g., the winter midday electron density can be 4 or more times as great as summer midday ionization. Monthly mean maximum electron density is approximately linearly proportional to 12-month running mean sunspot number R_{12}. Ionization depends on the atmospheric model, solar flux, absorption cross section, and ionization efficiency.[108,113]

Knowledge of electron and ion distribution is obtained by a number of probing techniques.[9,114]

Worldwide sounding. An international program of exchange of ionospheric data, obtained since about 1946 from ground-based radio sounding, has been coordinated by the International Radio Scientific Union (URSI) from the scientific point of view[115] and by International Radio Consultative Committee (CCIR) from the point of view of needs for radio communication (CCIR Rep. 248-2).

The *ionosonde* is a pulsed (or FM continuous-wave) radar device in which the exploring frequency varies over a wide range from about 1 to 25 MHz or higher. The equipment measures transit time (virtual height) and maximum frequency of reflection, and the results can be interpreted in terms of electron-density profiles from about 100 km to the height of maximum electron density in the F region. A typical vertical ionosonde display (Fig. 18-72) shows E-layer and $F1$ and $F2$ reflections.

Absorption measurements comparing relative amplitude of pulsed reflections, continuous-wave field-strength recordings, and measurement of galactic-noise attenuation by the ionosphere (*riometer observations*) are used to give information about absorption and electron density in the *D* region, where the ionosonde gives very little information.

Rockets and Satellites. It is now possible to explore the ionosphere by instrumentation of rockets and satellites. In particular, the region above the *F*-region maximum electron density, inaccessible to conventional ionosondes, has been studied by satellites. Rockets have provided the most satisfactory measurements of the *D* region.

"Incoherent Scatter" Radar. High-power vhf radars beamed vertically obtain echoes from electron scattering which can be related to the electron-density profile, providing a means of observing the profile continuously from below the maximum electron density to heights of 1,000 km or more.

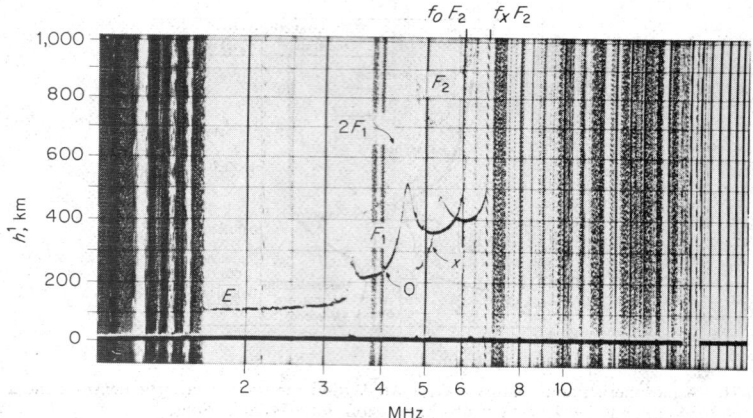

Fig. 18-72. Vertical ionosonde record (Washington, D.C., daytime).

92. Refractive Index, Polarization, Reflection, and Critical Frequency. Wave propagation in a magnetoionic medium, is discussed in Refs. 8, 9, and 116 to 118. The complex refractive index of the magnetoionic medium (the Appleton formula)[9] is given by

$$n^2 = 1 - \frac{X}{1 - iZ - \dfrac{Y_T^2}{2(1 - X - iZ)} \pm \sqrt{\dfrac{Y_T^4}{4(1 - X - iZ)^2} + Y_L^2}} \qquad (18\text{-}145)$$

In the upper regions of the ionosphere the collision frequency is sufficiently small (for frequencies greater than about 1 MHz) for us to be able to put $Z = 0$; hence the real part of the refractive index* is

$$\mu^2 = 1 - \frac{2X(1 - X)}{2(1 - X) - Y_T^2 \pm \sqrt{Y_T^4 + 4(1 - X)^2 Y_L^2}} \qquad (18\text{-}146)$$

In the absence of an imposed magnetic field ($Y_T = Y_L = 0$) and of collisions ($Z = 0$) the refractive index is given by

$$\mu^2 = 1 - X = 1\left(\frac{f_N}{f}\right)^2 = 1 - k\frac{N}{f^2} \qquad (18\text{-}147)$$

where $k = e^2 4\pi^2 \epsilon_0 m = 80.5$ and f is in hertz; f_N is the plasma frequency, where

$$X = \frac{Ne^2}{\epsilon_0 m \omega^2} \qquad Y_L = \frac{eB_L}{m\omega} \qquad Y_T = \frac{eB_T}{m\omega} \quad \text{and} \quad Z = \frac{\nu}{\omega}$$

and N is the electron density in electrons per cubic meter, e and m are the charge and mass of the electron, ω is angular frequency, Y is a parameter of the magnetic field, where B is the field in webers per square meter, and T and L subscripts refer to directions of phase propagation, transverse or longitudinal components the field, respectively; Z is a parameter associated with collisions with neutral particles, and ν is the collision frequency.

---

* Henceforth μ is used for the real part of the ionospheric refractive index and not for magnetic permeability.

An important aspect of propagation in the ionized medium is *polarization*. For the propagation of high-frequency radio waves through the E and F regions of the ionosphere, Z is usually very small and can be neglected. The *wave polarization*[9] is

$$R = \frac{i}{2Y_L}\left[\frac{Y_T^2}{1-X} \mp \sqrt{\frac{Y_T^4}{(1-X)^2} + 4Y_L^2}\right] \qquad (18\text{-}148)$$

R is the ratio of the field vectors $H_2/H_3 = E_3/E_2 = P_3/P_2$ in Fig. 18-73. The polarization R gives the amplitude ratio and the phase difference between oscillations of the displacement vector D, the electric field vector E, and the power vector P along the 2- and 3-axes. In general, the tips of these vectors describe ellipses. The H ellipse is similar to the D ellipse (and lies wholly in the 2-3 plane) but is rotated through 90° in the same direction. While the D, B, and H vectors lie in the plane of the wavefront (2-3), the P and E ellipses are tilted forward with respect to the 2-3 plane.

Equation (18-148) tells us that two, and only two, waves can propagate. Analogous to the terminology used for the propagation of light in birefringent crystals, these characteristic waves are called *ordinary* and *extraordinary* waves (o and x, respectively). In the absence of collisions the ellipse of the extraordinary wave can be obtained from the ellipse of the ordinary wave by rotating it through 90° and reversing the sense. For longitudinal propagation the two magnetoionic waves are circularly polarized. For transverse propagation the ordinary wave is polarized with the E and P vectors parallel to the magnetic field, whereas for the extraordinary wave the P and E ellipses lie entirely in the 1-3 plane of Fig. 18-73.

One of the waves (the ordinary) is reflected as in the absence of the magnetic field. The reflection of the other wave (the extraordinary) depends upon the strength (but not the direction) of the imposed magnetic field.

Reflection conditions are as follows. For a radio wave incident at angle ϕ_0 (with respect to normal) on a plane stratified ionosphere (Fig. 18-74) at the bottom of the ionosphere ($N = 0$) μ is unity. At higher levels of electron concentration, μ falls. If the electron density is

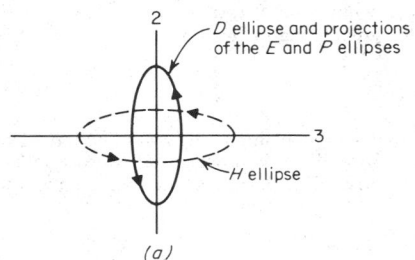

(a)

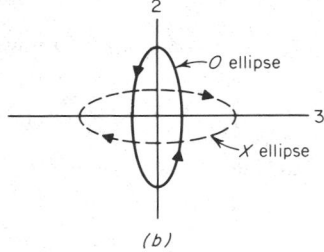

(b)

Fig. 18-73. Polarization ellipses (no collisions): (a) wave polarization, D and H ellipses; (b) relation between the corresponding ellipses of the ordinary and the extraordinary waves. *(Davies.[9])*

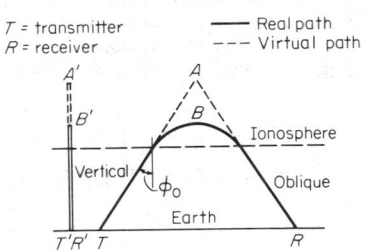

Fig. 18-74. Vertical and oblique equivalence for plane earth and plane ionosphere. *(Davies.[9])*

sufficiently high, μ will go to zero. Snell's law states that for a wave incident at an angle ϕ_0 to the normal, the angle ϕ with the normal at a level where the refractive index is μ is given by

$$\mu \sin \phi = \phi_0 \sin \phi_0 \qquad (18\text{-}149)$$

If the electron density is sufficiently large to reduce μ to zero, a wave (incident normally) will be reflected. Otherwise the wave will penetrate the layer. The highest frequency which can be reflected from the layer at vertical incidence is called the *critical frequency*. The ordinary-wave critical frequency f_o is associated with the level of maximum electron density; f_o F2 is the critical frequency for the F2 layer, etc, and is related to N by

$$N_{max} = 1.24 \times 10^{10} f_o^2 \text{ electrons/m}^3 \qquad (18\text{-}150)$$

where f_o is in megahertz.

A second critical frequency and reflection level are associated with the extraordinary wave. At frequencies above the gyrofrequency* f_H, the level of reflection is given by $X = 1 - Y$; whereas at frequencies below f_H ($Y > 1$) the condition is $X = 1 + Y$. The critical frequency for reflection of the extraordinary wave is

$$f_x = f_o + \frac{f_H}{2} \qquad (18\text{-}151)$$

A number of simplifications[8,9] of the formula for μ are important for special cases of quasi-longitudinal and quasi-transverse propagation, normal incidence, etc.

93. Absorption. Neglecting magnetic field, the *absorption* of the radio wave in passing through a medium of collision frequency v is

$$\kappa = \frac{\omega}{v_0} \frac{1}{2\mu} \frac{XZ}{1 + Z^2} = \frac{e^2}{2\epsilon_0 m v_0} \frac{1}{\mu} \frac{N_v}{\omega^2 + v^2} \qquad N_p \qquad (18\text{-}152)$$

Equation (18-152) enables us to distinguish between two types of absorption. *Nondeviative absorption* occurs in the D region of the ionosphere; it predominates when μ is near unity and when the product N_v is large. *Deviative absorption* occurs near the level of reflection and whenever there is marked bending of the ray, i.e., when μ tends to zero.

For nondeviative absorption, when ω is much greater than v, we have

$$\kappa \approx \frac{e^2}{2\epsilon_0 m v_0} \frac{Nv}{\omega^2} \qquad N_p \qquad (18\text{-}153)$$

Polarization effects on absorption due to the magnetic field are an important factor. The general case is very complex. For a variety of conditions, however, the propagation conditions are quasi-longitudinal, and the nondeviative absorption coefficient is then given, approximately, by

$$\kappa \approx \frac{e^2}{2\epsilon_0 mc} \frac{Nv}{(\omega \pm \omega_H)^2 + v^2} \qquad N_p \qquad (18\text{-}154)$$

The minus and plus signs refer to the ordinary and extraordinary waves, respectively. The ordinary wave suffers less absorption, and the extraordinary suffers more absorption, than the corresponding wave in the absence of a magnetic field.

94. Velocity of Propagation; Phase and Group Velocity. The *phase velocity* of a wave (neglecting magnetic field and collisions) is

$$v = \frac{v_0}{\mu} = v_0 \left(1 - \frac{Ne^2}{n\epsilon_0 \omega^2} \right)^{-1/2} \qquad (18\text{-}155)$$

The relationships between phase velocity, wavelength, and refractive index are

$$v\mu = v_0 \qquad \text{and} \qquad \frac{v}{\lambda} = \frac{v_0}{\lambda_0}$$

When the phase velocity of a wave in a medium is a function of the wave frequency, the medium is said to be *dispersive*.

---

* Gyromagnetic resonance frequency $f_H = 2.84 \times 10^{10} B$ MHz ≈ 1.4 MHz.

Group velocity may be regarded as the velocity with which the modulation envelope travels.

95. Propagation at Frequencies Below 150 kHz. Transmission in the frequency range below 150 kHz is in the region bounded by the earth and lower ionosphere. Elf waves ($f <$ 3 kHz) are of geophysical interest[119] because of their depth of penetration below the earth's surface and are also regarded as a vital means of communication with deeply submerged submarines.[120] Vlf is usually employed for very long distance (worldwide) communication and navigation.[121] Limited transmission to points under the sea or beneath the earth is possible at these and lower frequencies. International standard-frequency bands are allocated at 20 kHz and 60 kHz. Vlf observations of atmospheric noise give information on worldwide thunderstorm distribution; observations of phase perturbation indicate solar flares and may be used to detect nuclear explosions in the atmosphere.

A number of maritime communication and navigation services operate in the lf region; navigation services include Loran C (90 to 100 kHz), which uses pulses in a hyperbolic system. Because of high levels of atmospheric noise and physical limitations of antennas, high transmitter powers are required for transmission and applications are usually limited to relatively narrow bandwidth. At the lowest frequencies, say below 50 kHz, the dispersive characteristics of the propagation and the high Q of antennas also limit the usable bandwidth. Vertical polarization is employed. Reference 121 gives a good engineering summary.

Waveguide-Mode Theory. At vlf, propagation is usually described in terms of waveguide modes; i.e., the waves propagate between the earth and the lower boundary of the ionosphere. The waveguide-mode theory consists of a full-wave solution that includes the significant effects of diffraction and surface-wave propagation (Refs. 3, 4, 122, and 123 and *CCIR Rep.* 265-2). Such waves are considered to propagate between the earth and the ionosphere as normal waveguide modes, analogous to microwave propagation in a lossy waveguide. At frequencies above about 30 kHz the waveguide is many wavelengths high, and for short distances many propagating modes must be considered; but at vlf, especially at longer distances ($\geq$ 1,000 km) only a few modes need be considered. Vlf transmitters usually radiate a vertically polarized field.

For long paths, over smooth homogeneous terrain, the vertical field strength on the ground from a transmitter at a distance of d km on the ground (*CCIR Rep.* 265-2 and Ref. 4) is

$$e = \frac{300\sqrt{p_t}}{\sqrt{a \sin(d/a)}} \frac{\sqrt{\lambda}}{h} \exp -i(kd + \pi/4) \sum_n \Lambda_n \exp(-ikS_n d) \quad \text{mV/m} \quad (18\text{-}156)$$

where p_t is the radiated power in kilowatts (allowance must be made for antenna efficiency)[4,121], a is the radius of the earth in kilometers, λ is the free-space wavelength in kilometers, $k = 2\pi/\lambda$, Λ_n is the excitation factor for the nth mode, kS_n is a propagation constant, d is the path distance in kilometers, and h is the height of the ionosphere (70 km day, 90 km night).

In general, the terms Λ_n and S_n are complex. The excitation factor Λ_n gives the relative amplitude and phase of each mode of order n excited in the earth-ionosphere waveguide by the source. [Additional height-gain terms $G_n(y)$ and $G_n(y_0)$ for the transmitter antenna are considered when antennas are not located near the ground.] The real part of the propagation constant kS_n contains the phase information for each mode while the imaginary part determines the attenuation rate α. In order to obtain the field strength the contributions of each mode must be summed, with proper attention to the relative phases of the terms. Modification is required near the antipode when $d/a \approx \pi$.

The phase velocity $v_n = v_0/\text{Re}(S_n)$, where v_0 is the free-space velocity.

Methods for determining Λ_n and S_n are given in Ref. 4. These factors are related to the wavelength, the ionosphere height, the ground electrical properties, and the spherical reflection coefficient of the ionosphere. In turn the ionosphere reflection coefficients depend on the vertical distribution of electron density and collision frequency, the direction and magnitude of the earth's magnetic field, the frequency, and angle of incidence. The electron-density distribution is a function of latitude, season, solar cycle, time of day, and whether or not ionospheric disturbances are present. Horizontal gradients of electron density, i.e., at sunrise or sunset are also important.[4] Values of Λ, v, and α from Ref. 124 are illustrated in Figs 18-75 and 18-76. Experimental daytime field strength is illustrated in Fig. 18-77. In summary, the important effects are as follows (CCIR Rep. 265-2):

Ground Conductivity. Reducing the ground conductivity increases the attenuation rate of

Ground Conductivity.   Reducing the ground conductivity increases the attenuation rate of all modes. However, when the conductivity is very low, e.g., polar ice caps, the attenuation rate may approach a maximum and then decrease as the conductivity is lowered further. For moderate conductivities, the magnitude of the excitation factor for the first-order mode usually increases somewhat as the conductivity is reduced and the phase velocity is reduced.

Direction of Propagation with Respect to the Earth's Magnetic Field.   The earth's magnetic field causes greater attenuation for propagation to the magnetic west than for propagation to the magnetic east. Propagation in north-south directions gives intermediate attenuation rates in the daytime. At low latitudes, nighttime attenuation may be much greater for north-south paths than for other directions.[125]

Electron-Density Profiles.   The D-region electron-density profiles for vlf calculations can be approximated by [4]

$$\omega_r(z) = \omega_r(h)\exp[\beta(z - h)] \qquad (18\text{-}157)$$

where $\omega_r = \omega_0^2/\nu$ for $\nu \gg \omega$; ω, ν, and ω_0 are the wave angular frequency, the collision frequency, and plasma angular frequency, respectively; z is the height, and h is a reference height at which $\omega_r \approx 2.5 \times 10^5$. The term gives the vertical gradient of ω_r. Under daytime conditions, h is about 70 km and $\beta \approx 0.3$ km^{-1}, while at night h is around 90 km and $\beta \approx 0.5$ km^{-1}. These two parameters, h and β, provide a convenient but approximate means of describing the ionosphere for use in vlf calculations. When the magnetic field is considered, the collision-frequency profile $\nu \approx \exp\alpha_0(3 - h)$ must also be considered. Here α_0 is taken to be 0.15 km^{-1}. For a horizontal magnetic field transverse to the propagation path, a magnetic field parameter $\mho$ (see Figs. 18-75 and 18-76) assumes the values -1 and $+1$, corresponding to propagation along the magnetic equator from west to east and east to west, respectively ($\mho = 0$ corresponds to zero magnetic field).[4]

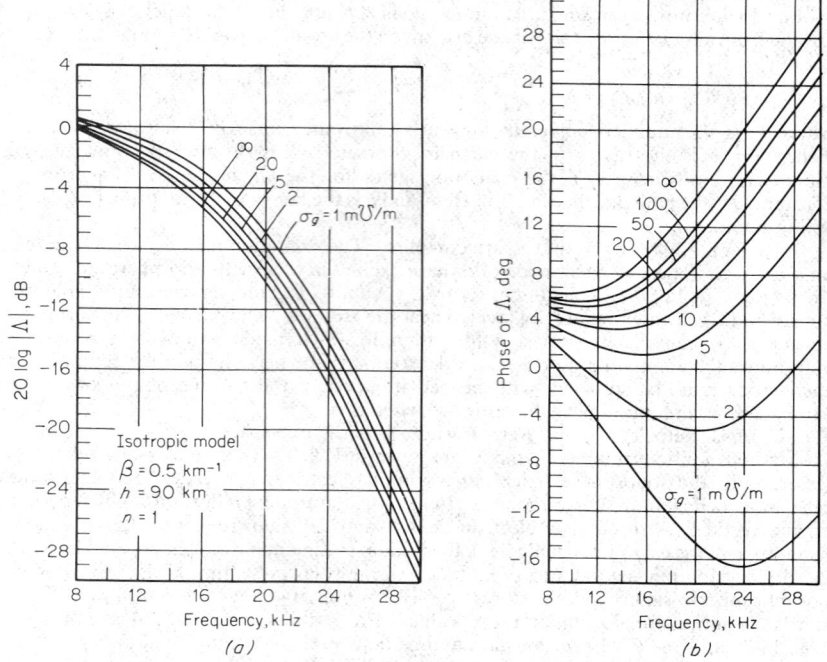

Fig. 18-75. Amplitude *(a)* and phase *(b)* of excitation factor as functions of frequency. *(Wait and Spies.*[124]*)*

Among other limitations, the calculations refer to conditions where the path properties are independent of distance. When the electron-density distribution with height along the path is constant but the ground conductivity or magnetic field angle changes, approximate calculations can be made by the methods of Ref. 4 and *CCIR Rep.* 265-2.

At sufficiently low frequencies, i.e., for *elf propagation* (< 3 kHz), the zero-order mode dominates at nearly all distances. In this case only the $n = 0$ term in Eq. (18-156) is needed, and $\Lambda_0 \approx \frac{1}{2}$. In such cases the attenuation rate is of the order of 1 dB per 1,000 km for both day and night models.

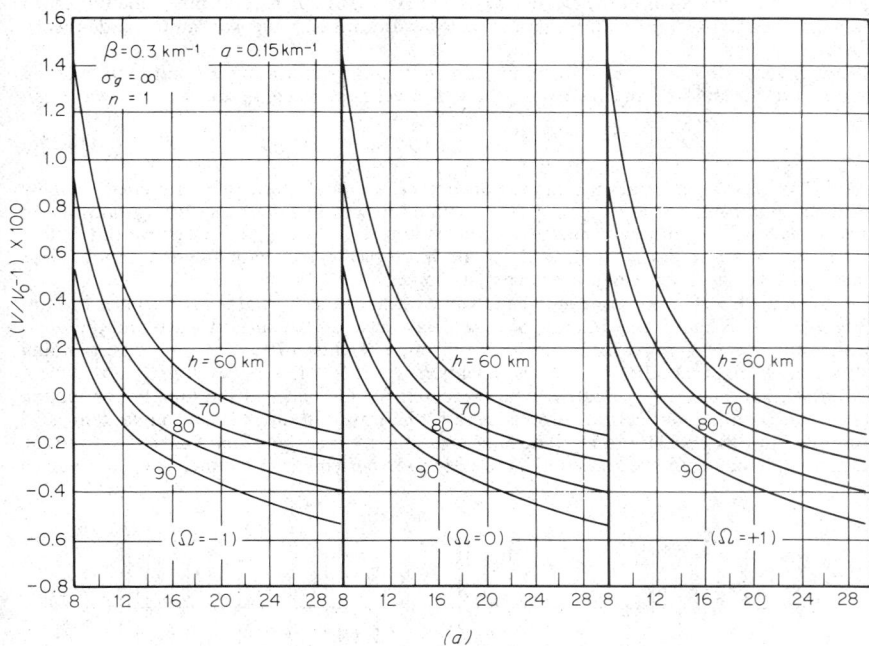

(a)

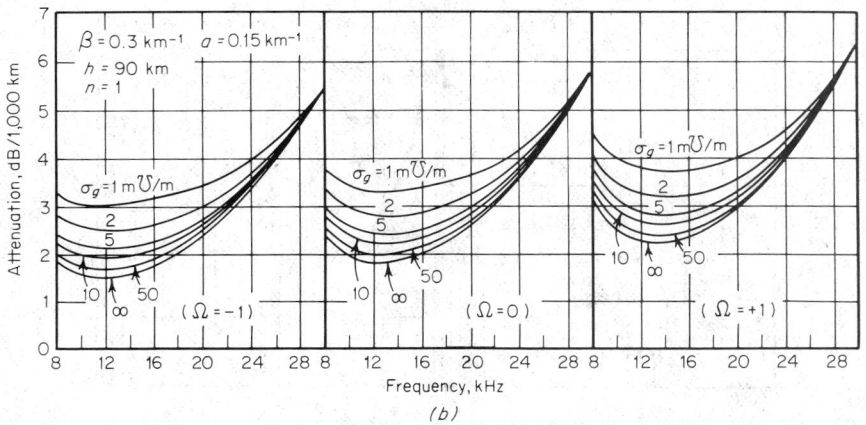

(b)

Fig. 18-76. Phase velocity *(a)* and attenuation *(b)* of $n = 1$ mode. *(Wait and Spies.[124])*

Wave-Hop Theory and Geometric-Optical Methods. Wave-hop theory interprets the resultant field strength and phase at a receiver as the sum of waves (*rays*) that have traveled via the surface wave and one or more earth-ionosphere reflections. Calculations of wave hops using full-wave theory have been shown to be equivalent to waveguide-mode theory.[126,127]

Simpler calculations using geometric-optical formulas for the ionospheric wave,[127-129] condensed here from *CCIR Rep.* 265-2, can be used to approximate the full-wave solution, providing corrections are made for diffraction. The surface-wave field is added to the ionospheric-wave field using the phase difference implied by Fig. 18-78. Application of this method is most practical in the range 50 to 150 kHz. At higher frequencies, the empirical methods of Pars. **18-96** to **18-98** are used. At lower frequencies, the waveguide-mode theory is generally used.

For a vertically polarized transmitting antenna and reception by a small loop antenna located on the surface of the earth, the effective field strength of the sky wave is

$$e_s = 2\frac{300\sqrt{p_t}}{d'}\ \cos\psi\, RDF_T F_R \quad\quad \text{mV/m} \tag{18-158}$$

where d' is the sky-wave path length in kilometers, R the ionospheric reflection coefficient for vertically polarized waves,[127-129] p_t is the radiated power in kilowatts, D the ionospheric focusing factor,[129] F_T the transmitting-antenna factor,[130] ψ the angle of departure and arrival of the sky wave at the ground, and F_R the appropriate receiving-antenna factor. For reception by a short vertical antenna $\cos\psi$ becomes $\cos^2\psi$.

To calculate d', the sky-wave path length, and estimate the diurnal phase changes, Fig. 18-78 is used. This shows the differential time delay between the surface wave and the one-, two-, or three-hop sky wave for ionospheric reflection heights of 70 and 90 km, corresponding to day and night conditions. A propagation velocity of 3×10^5 km/s is assumed.

The antenna factors F_T and F_R, which account for the effect of the finitely conducting curved earth on the vertical radiation pattern of the transmitting and receiving antennas,[130] are shown in Fig. 18-79 for land. The factors are the ratio of the actual field strength to the field strength that would be measured if the earth were perfectly conducting. Negative values

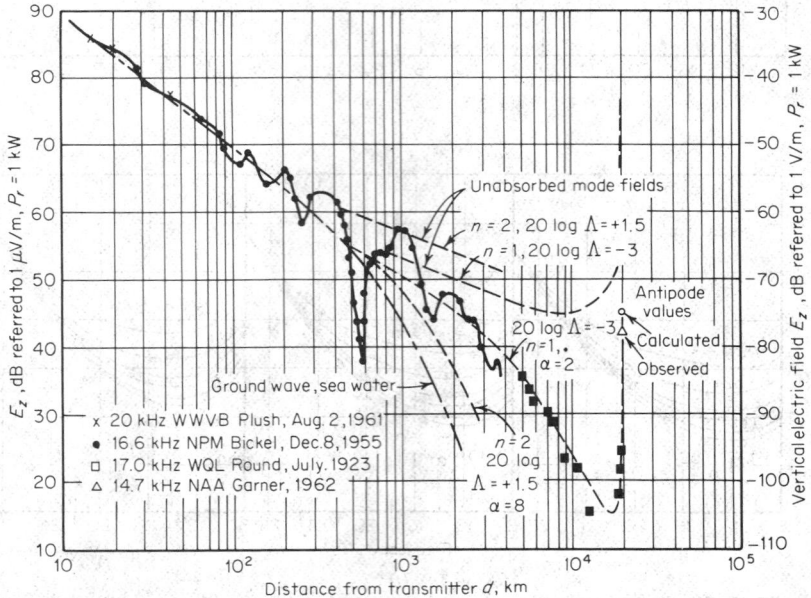

Fig. 18-77. Experimental and theoretical daytime field strength vs. distance. *(From Watt,[121] by permission of Pergamon Press, Inc.)*

of ψ refer to propagation beyond the geometric optical limiting range for a one-hop sky wave. Values of the ionospheric reflection coefficient R are illustrated in Fig. 18-80 in terms of $f \cos \phi$. The ionospheric focusing factor D for a spherical earth and ionosphere has a value between 1 and 2.5 depending on path length and frequency. D, R, and F are given in *CCIR Rep.* 265-2.

96. Propagation at 150 to 1,500 kHz. Because of the stable, only moderately attenuated ground wave and efficient nighttime ionospheric propagation, this range of frequencies is used for maritime communications and navigation and for medium-range broadcasting. Aeronautical mobile and fixed communications and radio positioning services are operated in many areas of the world.

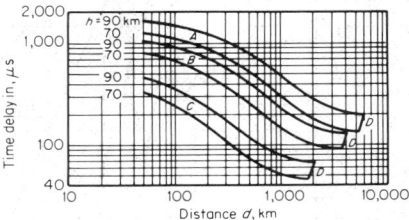

Fig. 18-78. Differential time delay between surface wave and one-, two-, and three-hop sky waves. *A*, three hop; *B*, two hop; *C*, one hop; *D*, limiting range.

Transmission is usually described by summing the *powers* of ground-wave and wave-hop reflections from the lower ionosphere. The description of propagation is similar to that above for frequencies in the 50- to 150-kHz region. However, because the phase path usually involves hundreds of wavelengths and ionospheric reflection characteristics are increasingly variable at higher frequencies, the phase of the downcoming wave is essentially random. This introduces phase-interference fading in the region where ground wave and sky wave are of comparable magnitude. Thus applications are usually designed to take advantage of the ground wave or sky wave separately, avoiding admixture. Polarization effects are important, as discussed in Par. **18-99**.

During the day, at distances less than 1,000 km, the ground wave is dominant because of intense daylight D-region absorption of the ionospheric wave. At night, ionospheric reflection coefficients can be high, and even multiple reflections are useful. Attenuation varies from one night to the next and with season. Statistical estimates of field strength are obtained empirically. The main bases are two sets of long-term measurements of mf broadcasting stations, one covering the European region by the European Broadcasting Union (EBU)[131] and the International Broadcasting and Television Organization (OIRT), and the other covering United States paths by the Federal Communications Commission (FCC). The European results and analyses are somewhat more comprehensive in frequency and temporal and geographical coverage, and are used by the CCIR (*Rep.* 264-2). However, there are some differences from the FCC results,[36,132] which represent measurements over United States paths 2 h after sunset in the 500- to 1,600-kHz band. They have been adopted for official use by North American countries for AM broadcast-frequency-sharing studies and are undoubtedly more accurate for the midcontinental United States.

97. Estimation of Sky-Wave Field Strength by the CCIR Method (Rep. 264-2). The field strength is usually represented in terms of an *annual median field strength* for local midnight at the path midpoint, with corrections for antenna, magnetic-dip latitude, local time other than midnight, seasons, and sunspot number. The available data are for the path lengths in the 300- to 3,500-km distance range and the 150- to 1,600-kHz frequency region. Shorter paths pose greater theoretical difficulty, and sufficient experimental data are lacking.

The annual median sky-wave field strength, assuming reception on a small, vertically polarized antenna, is represented by

$$E = E_0 + \Delta_A \quad \text{dB}\mu \qquad (18\text{-}159)$$

where E is the annual median value of the field strength at a reference hour for a power of 1 kW radiated by the transmitting antenna; E_0 is the annual median value of the field strength (in decibels referred to 1 μV/m) at the reference hour, for a reference transmitting antenna

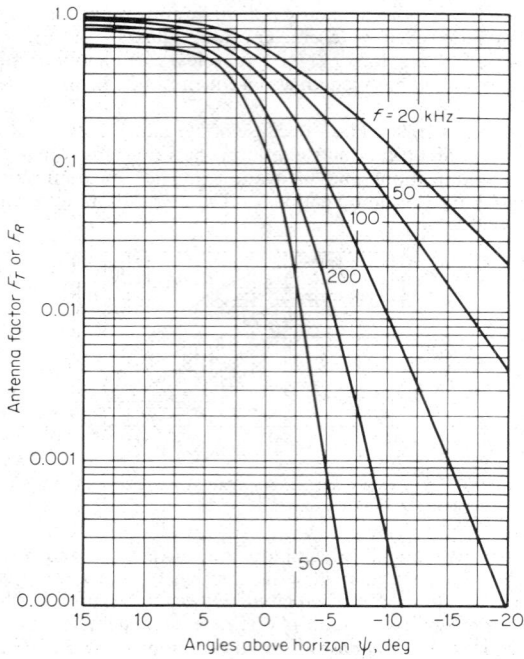

Fig. 18-79. Antenna factors (land): $\varepsilon_r = 15$; $\sigma = 2 \times 10^{-3}\ \mho/m$; $a = 4/3 \times 6{,}360$ km. *(CCIR Rep. 265-2.)*

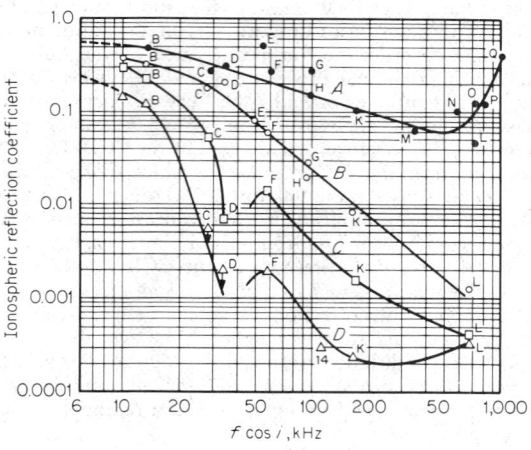

Fig. 18-80. Ionospheric reflection coefficients, solar cycle maximum. Letters designate vertical incidence measurements; numbers designate oblique incidence measurements: *A*, night (all seasons); *B*, day (winter); *C*, day (equinox); *D*, day (summer). *(CCIR Rep. 265-2.)*

giving, over perfectly conducting ground, a field strength of $3 \times 10^5 \mu V/m$, at a distance of 1 km in all directions above the horizon; Δ_A is equal to the 20 log ratio of the field strength for a given angle of departure in the vertical plane to the field strength of the reference antenna above.

Figure 18-81 gives values of Δ_A as a function of the distance between the transmitting and receiving points in the theoretical case of a lossless unloaded vertical antenna of height h/λ placed over perfectly conducting earth and, a reflection at a virtual height of 100 km above a spherical earth, for propagation via the E region of the ionosphere. The discontinuity of the curves at 2,200 km corresponds to the distance beyond which there are at least two hops. The value of E_0 is

$$E_0 = 80.2 - 10 \log d - 0.0018 f^{0.26} d \quad \text{dB}\mu \qquad (18\text{-}160)$$

where d is the distance in kilometers and f is the frequency in kilohertz.

The values of E and E_0 are for the following conditions:
The magnetic dip at the midpoint of the path is 61°.
The sunspot number is $R = 0$.
The reference hour is local midnight at the midpoint of the propagation path.

CCIR Rep. 264-2 gives methods and data for correcting the field for other values of magnetic dip, for local time, and for the percentage of nights of the year.

98. Estimation of Field Strength from FCC Data. The FCC Rules[36] give a series of propagation curves which estimate nighttime sky-wave field strengths exceeded 10 and 50% of the nights (the 50% data are shown in Fig. 18-82). Frequency dependence is not included in the FCC curves, which represent the field strength from 500 to 1,600 kHz, centered at 1,000 kHz. Frequency dependence deduced[132] from FCC data is shown in Fig. 18-83. Figure 18-81 may be used to adjust these field strengths for transmitting-antenna factor.

99. Polarization Effects. Paragraphs **18-90** to **18-92** include a description of polarization and absorption effects in the ionosphere. At medium frequencies, because of heavy absorption of the *extraordinary* wave in the E region, propagation depends almost entirely on the *ordinary* wave, which is, in general, elliptically polarized. Polarization coupling losses can occur when vertical transmitting and receiving antennas are used. Though not serious at midlatitude, they may be quite large for east-west paths near the geomagnetic equator, where the ordinary wave tends to be horizontally polarized, and for certain multiple-hop situations involving change of polarization at ground reflection.[132,133] Polarization coupling loss arises because any wave incident on the ionosphere will excite the ordinary mode to the

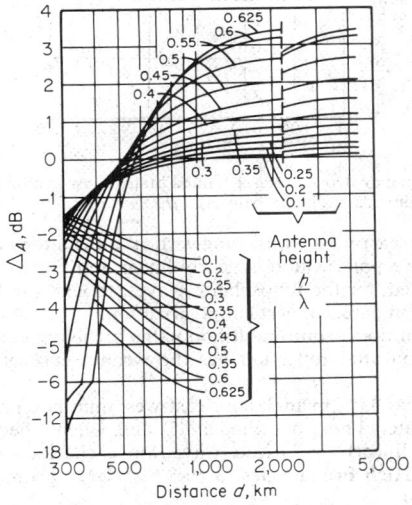

Fig. 18-81. Correction factor Δ_A of vertical transmitting antennae of various lengths as a function of the distance d from the point of reception. *(CCIR Rep. 264-2.)*

degree that the incident-wave polarization corresponds to the polarization characteristic of that mode. When the two polarizations are orthogonal, e.g., linear polarization at right angles or circular polarizations with opposite sense of rotation, there will be no coupling with the ordinary wave mode. Phillips and Knight[133] identify that serious propagation losses may be encountered in the following situations:

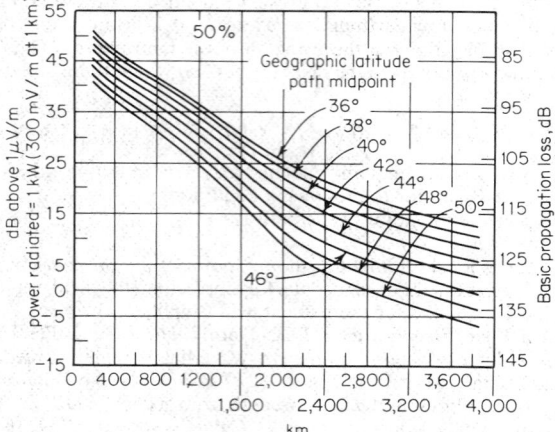

Fig. 18-82. United States sky-wave propagation from FCC data for 50% of nights at 1,000 kHz (2 h after sunset). *(From Barghausen.[132])*

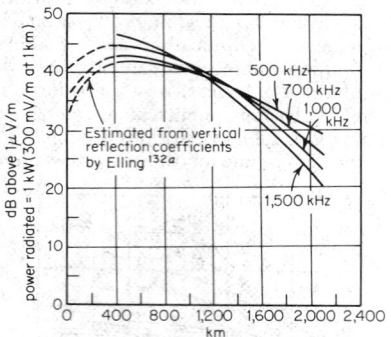

Fig. 18-83. Calculated frequency dependence of United States sky-wave propagation 500 to 1,500 kHz (vertical polarization; nighttime data 2 h after sunset). *(From Barghausen.[132])*

1. At low latitudes, on east-west paths using vertical transmitting and receiving antennas, because the ordinary wave polarization is nearly linear and horizontal. Use of horizontal antennas is to be preferred, for the ionospheric waves, though the ground wave would be appreciably attenuated compared to vertical polarization.

2. At low latitudes, on north-south multihop paths involving sea reflection, because the circular wave polarization after reflection has the wrong sense of rotation to excite the ordinary mode.

3. At midlatitudes (near 45° dip angle), for east-west multihop paths over sea. Here the polarization is approximately linear but tilted at 45° and, when reflected by the sea, emerges at right angles to the linearly polarized wave for excitation of the ordinary mode.

100. Propagation at High Frequencies, 3 to 30 MHz*. From the early 1920s to the

---

* The omission of discussion of propagation at frequencies between about 1.5 and 3 MHz reflects difficulty of adequate methods for treatment, and the relatively little use of ionospheric propagation in this frequency range because of intense daytime absorption. For nighttime propagation, one may interpolate between the results obtained by methods for the adjacent higher and lower bands. However, because of the proximity of the operating frequency to the gyromagnetic resonance frequency, attenuation is increased. Theoretical methods for this frequency range are difficult.[9]

present, high-frequency propagation has been used for economical low- to medium-power long-distance communications. While propagation at medium and low frequencies (300 kHz to 3 MHz) suffers heavy absorption during the day and frequencies above 30 MHz are not reflected from the ionosphere much of the time, the frequency range 3 to 30 MHz (hf) usually provides ionospheric reflections day and night. The range of usable frequencies is limited on the upper end by the height and maximum electron density of the controlling layer and on the low-frequency end by absorption in the *D* region.

Hf propagation is characterized by:

1. Variability of propagation conditions, requiring frequent changes in the operating frequency.

2. Interruption by ionospheric storms.

3. The large number of possible propagation paths and resulting multipath-interference effects.

4. Dispersion and frequency distortion.

5. Large and rapid phase fluctuations.

6. High interference.

101. Reflection and Absorption in Oblique Propagation. The reflection process for plane ionosphere is equivalent to mirror-type reflection at a height equal to the virtual height h' of reflection of the equivalent vertical frequency.[9,10] The equivalence theorem must be modified for presence of the earth's magnetic field or for a curved ionosphere.[8,9]

For most purposes, the highest reflection frequency for oblique incident f_{ob} is related to the vertical critical frequency by

$$f_{ob} = kf_v \sec \phi_0 \qquad (18\text{-}161)$$

where $k \approx 1.2$ and is given by transmission curves.[9] A similar equivalence applies for absorption.

The echo structure becomes complex over oblique paths, expecially over long distances. Oblique soundings have provided the most revealing insights into the characteristics of long distance hf propagation.[114,134,135]

MUF, originally *maximum usable frequency*, requires some definition. *Classical MUF* is used to designate the highest frequency propagated by ionospheric reflection alone; *operational MUF* denotes the highest frequency permitting operation between points under specified working conditions; it may be higher than the classical MUF as the result of sporadic *E*, ground scatter, or ionospheric scatter. MOF and LOF refer to highest and lowest *observed frequencies* in the oblique sounding.

Figure 18-84 gives nomenclature for reflections and illustrates the roles of the *E* and *F* regions. In the diagram *F* is used to represent both *F*1 and *F*2 layers.

*F*2 *reflections* are usually the most important, and during the day the lowest-order (lowest-angle) minimum *F* ray-path reflection is usually dominant; often even two-hop *F* reflection is more efficient than one-hop *E*. The horizon-limited distance for one-hop *F*2 propagation depends upon layer height but

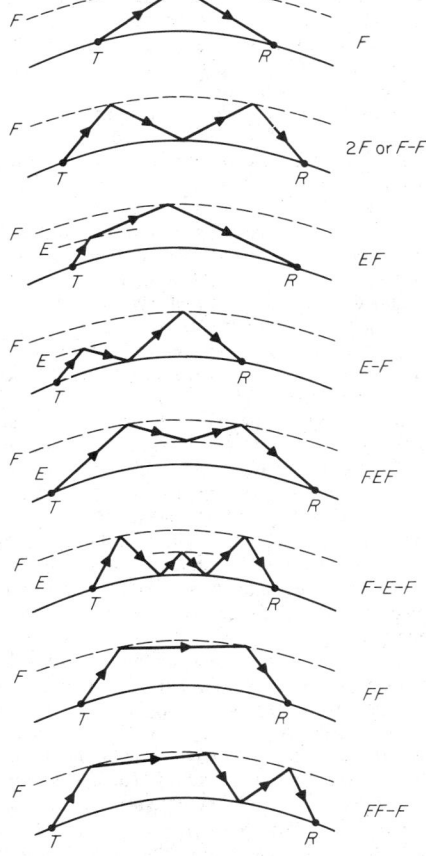

Fig. 18-84. Possible oblique reflection geometries; descriptive nomenclature for oblique ray paths. Reprinted by permission from "Ionospheric Radio Waves," by K. Davies, © 1969, Ginn and Co., published by Xerox College Publishing.[8]

is usually around 4,000 km.   On longer paths, intermediate ground reflection is usually involved, though often the high (Pederson) ray of the first hop $F2$ reflection is dominant.[134] The effects of the earth's magnetic field, involved in the double-refraction process and the separation and differential absorption of the ordinary and extraordinary rays, depend in a complicated way on path length, the magnetic latitude (dip angle), and the path direction.[8,9,118]

Rays propagated via any of the reflecting layers pass through the absorbing D *region*, encountering there the principal attenuation relative to free space.

The $F1$ *layer* is important during daylight hours and during ionospheric disturbances.   In summer at high latitudes the principal reflection for paths of 2,000 to 3,500 km is usually via $F1$.

The E *layer*, because of its low height and low nighttime density, provides useful propagation (mainly daytime) over medium distances to 2,000 km.   The ray path of a signal reflected from the $F2$ layer, however, must penetrate the E layer at one or more points, so that the frequency must be sufficiently high for penetration.   The E-layer penetration frequency for one-hop $F2$ propagation constitutes effectively the lowest useful frequency for many low- to medium-power systems in the distance range to 4,000 km.

Sporadic E (E_s) can also play a dominant role in determining maximum usable frequency. Besides providing, at times, reflection at distances up to 2,400 km, sometimes up to 150 MHz in frequency, E_s can extend the range of $F2$ propagation, as illustrated in Fig. 18-84.   One reflection illustrated provides an extended-range (second-hop) $F2$ reflection which would otherwise, via the ground, suffer D-region attenuation (*CCIR Rep.* 259-2).   Sometimes the operational MUF is higher than the calculated $F2$ MUF due to the effects of ground scatter, scatter from the ionosphere, sporadic E, or ionospheric tilts.

In the case of *ground scatter* (often called side scatter) the signals are reflected from the ionosphere obliquely to a ground region, scattered from the ground, and propagated again by ionospheric reflection to the receiver.   If the transmitter and receiver are collocated, the process is referred to as *backscatter*.   This technique has become important as a tool for studying the ionosphere, as well as in over-the-horizon radar systems.   If the receiver is located at a distance from the transmitter, the signal arrives from an off-great-circle azimuth and affords transmission under conditions where direct reflections are not possible, because of the more oblique reflections associated with the side-scatter geometry.   This phenomenon is allowed for to some extent in estimating extended hours of transmission in international hf broadcasting.   Bearing deviations from the great circle due to side scatter are important sources of error in direction-finding and radio-positioning systems (*CCIR Rep.* 429).

Ionospheric scatter can also play an important role.   In particular, under conditions represented by "spread F" on the vertical ionogram, oblique transmissions may be possible at frequencies higher than the calculated MUF.   Near the magnetic equator and in auroral regions ionospheric irregularities aligned along the magnetic field can provide returns at much higher frequencies than the calculated MUF.   Furthermore, scattering in the D region can increase the obliquity of an F reflection, making transmission at higher frequencies possible.

Ionospheric tilts can permit two ionospheric reflections without an intermediate ground reflection, especially important for a north-south transmission across the magnetic equator. The signal is propagated to long distance with only two transits of the absorbing region, thus giving high signal strength; higher frequencies are reflected than with a spherically stratified layer; there is a marked asymmetry in the path, so that the angle of elevation at departure can differ from the angle of arrival (Refs. 9, 134*a* and *CCIR Rep.* 250-2).

102. Prediction of HF Propagation.   *Atlases of Ionospheric Characteristics; Numerical Mapping.*   Because the ionosphere varies geographically as well as from hour to hour and day to day and with season and solar activity, extensive atlases of ionospheric characteristics are needed to determine the limits of useful frequencies and to perform the calculations necessary to estimate required power, angles of elevation for optimum antenna design, and so on. Comprehensive methods and data for computing high-frequency propagation are given by *CCIR Rep.* 252-2 and Refs. 136 and 137.

Computer methods are highly developed for these predictions.   Computer programs, ionospheric data in the form of coefficients for numerical mapping function,[138,139,139a] and routine computer prediction services[137] are available from the U.S. Department of Commerce, Office of Telecommunications, Institute for Telecommunication Sciences, Boulder, Colo. 80302.

The long-term variation of E, $F1$, and $F2$ parameters is tied closely to the sunspot cycle. Although no completely satisfactory measure of solar activity is available, the 12-month running average sunspot number R_{12} is most widely used. Alternatives such as the ionospheric index I_{F2} (essentially an equivalent sunspot number obtained from ionospheric observations) are useful for short-term predictions (less than 12 months in advance). Solar flux at 10 cm wavelength is especially useful for prediction of E and $F1$ region characteristics up to 6 months or so in advance. Current values of these indices are published in the *ITU Telecommunication Journal*. Predictions of sunspot number 12 months in advance are also obtainable from the above Department of Commerce address.

Basic worldwide $F2$ MUF predictions for 0 and 4,000 km distance and E MUF for 2,000 km distance are available. Reference 139 is an atlas of nearly 1,200 charts for $R_{12} = 10, 110$, and 160, representing respectively minimum solar activity, an average maximum of a solar cycle, and a very high period exceeding the peak of the average solar cycle. *CCIR Rep.* 340 contains nearly 600 worldwide charts of $F2$ MUF for all months, every 2 h, for $R_{12} = 0,100$ (Fig. 18-85a)* Charts and numerical mapping coefficients for the observed probability of occurrence of sporadic E are also available[140] (see Fig. 18-85c).

Calculation of Maximum Usable Frequency (MUF). To obtain MUF values for a given path, the great-circle path is drawn on a rectangular coordinate scale corresponding to the prediction map[139] and is used to overlay the radio circuits on the $F2$ and E MUF maps. For paths shorter than 4,000 km the path midpoint is considered to be the reflection point. Longer paths are divided into an integral number of hops and the lower MUF values are taken to be controlling. An alternative method for such long paths is more accurate in some cases and less in others: two control points for $F2$ transmission are taken, each 2,000 km from the path terminals, and the lower MUF is used.

The computer methods (Ref. 137 and *CCIR Rep.* 252-2) consider additional parameters in some detail, such as the height of the maximum electron density and the height of the bottom of the layer, reconstructing, in effect, an approximate profile of electron density along the path. A parabolic-layer assumption is made to compute the ray geometry along the path. Up to nine ray paths are evaluated in the CCIR program. The ray path must be geometrically possible for a takeoff angle equal to or greater than the minimum value given as input data.

Variability of MUF. The values of MUF referred to above are monthly medians for a given hour. Day-to-day random variation of $F2$ MUF is observed to be distributed according to a chi-squared probability distribution; the constants and computation of this distribution depend on geographical region and solar activity and are outlined in *CCIR Rep.* 252. Usually an optimum frequency for operation during a given month at a given hour is the value of MUF exceeded 90% of the time.

Transmission Loss Loss relative to free space (for the ray path) consists of ionospheric absorption, focusing and defocusing effects, and earth-reflection losses for multipath hops.

The total transmission loss is computed according to

$$L = L_{bf} + L_i + L_g + Y_p - (G_t + G_r) \qquad \text{dB} \qquad (18\text{-}162)$$

where L_{bf} is the basic free-space transmission loss expected between ideal, loss-free, isotropic transmitting and receiving antennas in free space over the distance of the ray path, L_i is the monthly median loss from ionospheric absorption; L_g is the loss caused by ground reflection (values range from 4 dB for land to 0.2 dB for seawater per reflection); Y_p is the excess transmission loss, used to account for day-to-day variability about the monthly median; and G_t and G_r are the transmitting- and receiving-antenna power gain relative to an isotropic antenna in free space, for the angles of elevation appropriate to the particular hop.

CCIR Rep. 252 gives empirical methods and data for calculating L_i and Y_p as a function of location, path length, season and hour, for each of the ionospheric reflection paths.

L_i is computed for all reflection paths which have a reasonable chance of occurring ($> 5\%$), and the smallest value, plus Y_p, is taken to represent the path loss.

These mode calculations also permit an estimate of the relative delays of the multipath signals which can cause intersymbol interference.

---

* Note that the figure, to be precise, uses the term EJF, *estimated junction frequency*, or "classical" MUF which is predicted.

The above prediction techniques are for the long term. Short-term prediction techniques generally make use of oblique sounders and provide observations of current conditions rather than predictions in the true sense (Refs. 134, 140, and 140*a*, and *CCIR Rep.* 357).

Siting considerations for hf antennas are outlined in Refs. 141 and 142. For very long paths, the lowest possible angles of radiation are advantageous.

Short-Term Variability, Fading, Multipath. Short-term-fading observations (within periods of about 5 min to exclude the long-term variability) show near Rayleigh distribution most of the time. The fading is somewhat more shallow (Rice-Nakagami distribution) under very strong signal conditions, due to the presence of a strong specular component. Duration of fades may be from 0.1 s to many seconds; the phase variation poses a limitation to phase-keying systems. In equatorial and auroral regions rapid fading is often encountered, with fading rates of 10 to 100 Hz.[143,144]

Under conditions of normal phase-interference fading, a correlation coefficient of less than 0.5 is typical for fading observed at spacings of 15λ, though these parameters depend somewhat on frequency and path length. Diversity spacings of the order of 10λ are useful in overcoming fading of narrow-band signals (bandwidth less than a few hundred hertz). The frequency-selective nature of the multipath fading shows independent fading at frequencies separated by more than a few hundred hertz (Refs. 8, 9, and 143 to 145 and *CCIR Rep.* 266-2). Polarization diversity is also useful for confined spaces, e.g., aboard ship.

Multipath propagation also leads to *intersymbol interference* in digital transmission. Differential propagation delays of several milliseconds are observed between first- and last-arriving significant signal components. Usually less than 5 ms, these relative delays can be minimized by operating at frequencies sufficiently close to the MUF to reduce multiple reflections. A *multipath reduction factor* (Fig. 18-86) has been devised to estimate the

December 00 hours R_{12} = 100; EJF (4,000) F_2, MHz

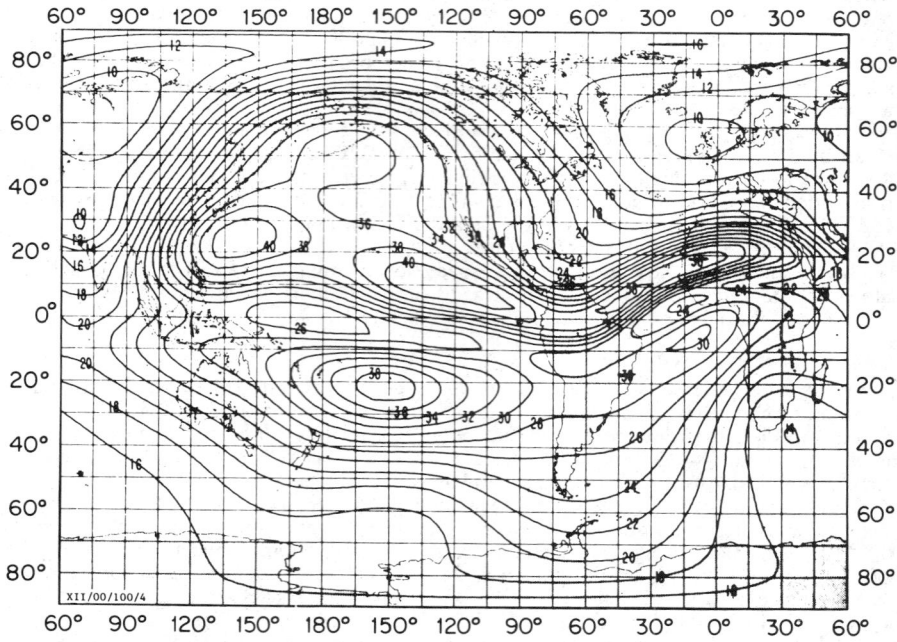

Fig. 18-85. Example of (*a*) worldwide $F2$ MUF chart *(CCIR Rep. 340)*; (*b*) worldwide E MUF chart *(Leftin et al.[139])*; (*c*) worldwide map of median value sporadic-E f_oEs occurrence *(Leftin et al.[140])*

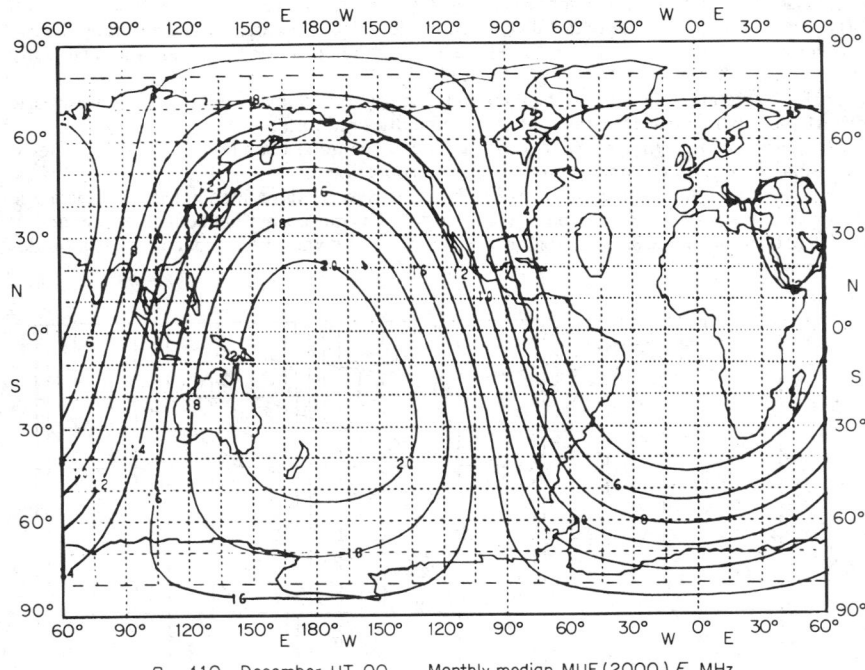

R_{12} 110   December UT 00   Monthly median MUF (2000) E, MHz

December 1958   UT = 00

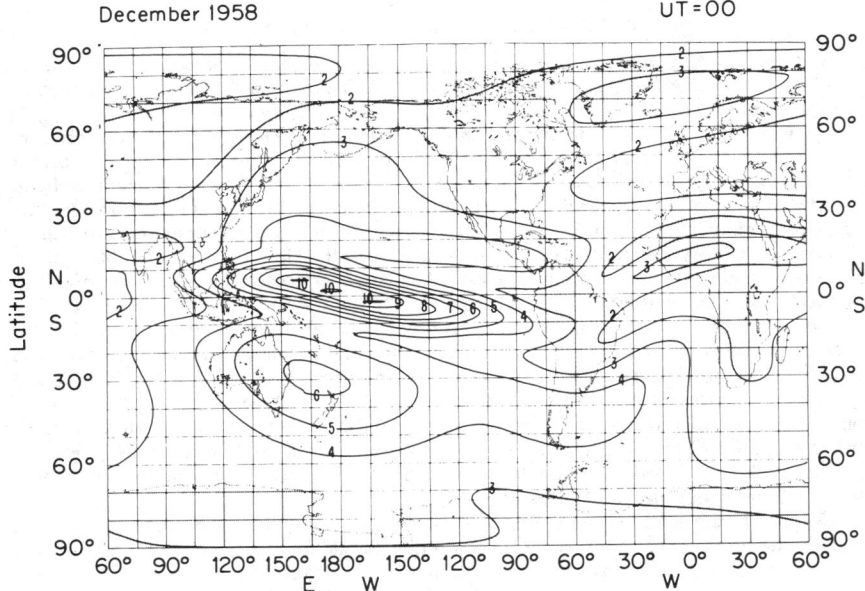

18-111

significant differential delays for given ratio of operating frequency to prevailing-path MUF.[146] For paths shorter than 1,000 km, multiple reflections can produce multipath delays exceeding 5 ms which are difficult to reduce because the reflection angles are nearly equally steep for all reflections.

103. Prediction of Disturbed Propagation Conditions. Some of the types of ionospheric disturbances which can cause propagation difficulties at hf are:

1. *Sudden ionospheric disturbances* lasting for a few minutes to an hour or more, associated with solar flares and causing intense *D*-region absorption.

2. *Ionospheric storms,* lasting for a few hours to several days, associated with magnetic disturbances, as a result of solar-particle radiation. The most prominent effects are reduction of *F*2 MUF and increase in *D*-region absorption. Aurorae often occur at high latitudes during such storms.

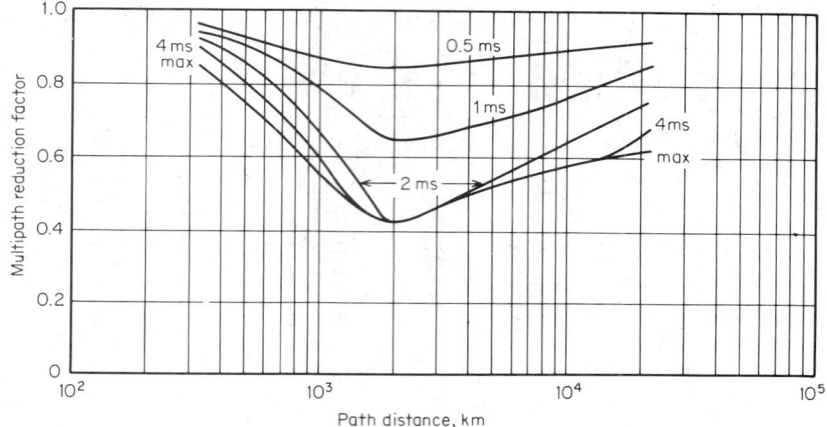

Fig. 18-86. Multipath reduction factor. f/MUF *(Salaman.*[146]*)*

3. *Polar-cap absorption*, lasting for a few hours to several days, associated with high-energy solar protons causing *D*-region ionization down to heights of the order of 50 km and intense absorption at high latitudes.

The U.S. Department of Commerce, National Oceanic and Atmospheric Administration, operates a Space Environment Forecast Center at Boulder, Colo., which issues current observations and advance forecasts of solar and geomagnetic conditions causing ionospheric disturbances. The U.S. Department of Commerce, Office of Telecommunications, Institute for Telecommunication Sciences, at Boulder, issues warnings of hf propagation disturbances by broadcast (WWV standard-frequency transmission of the National Bureau of Standards), direct-wire, and time-share computer access. The latter services include estimates of effects on MUF and transmission loss.

104. Scattering from the Ionosphere. At frequencies above about 30 to 300 MHz (vhf) electron densities of the ionosphere are rarely sufficient to reflect waves except for short periods. Transmission is mainly by scattering, although regular *F* and sporadic *E*, discussed in the previous paragraphs, do provide reflections often enough to be important sources of long-distance interference to mobile, broadcasting, and other services in this band. It is for this reason the FM broadcasting was reallocated from near 40 MHz to near 100 MHz and channel 1 TV is not used in the United States. Table 18-6 shows the main causes of long-distance interference above 30 MHz.

105. Scattering at VHF from *D*-Region Irregularities. Regular but weak scattering from irregularities in the *D*-region is useful for continuous single-frequency communication in the 30- to 60-MHz frequency range, over paths of the order of 1,000 to 2,000 km (Refs. 70 and 147 to 149 and *CCIR Rep.* 109-1 and 260-2). Approximately 1 kW per 100-band digital channel is required at 35 MHz, depending upon antenna design, modulation, and coding.

Table 18-6. Causes of Long-Distance Ionospheric Interference to Services Working in the 30- to 100-MHz Region*

Cause of interference	Latitude zone	Period of severe interference	Approximate highest frequency with severe interference, MHz	Approximate frequency above which interference is negligible, MHz	Approximate range of distances affected, km
Regular F-layer reflections	Temperate	Day, equinox winter, solar-cycle maximum	50	60	East-west paths 3,000–6,000 or north-south paths 3,000–10,000
	Low	Afternoon to late evening, solar-cycle maximum	60	70	
	Auroral	Night	70	90	
Sporadic E reflections	Temperate	Day and evening, summer	60	90	500–4,000
	Equatorial	Day	60	90	
Sporadic E scatter	Low	Evening up to midnight	60	90	Up to 2,000
Reflections from meteoric ionization	All	Particularly during showers	May be important anywhere in the range		Up to 2,000
Reflections from magnetic field aligned columns of auroral ionization	Auroral	Late afternoon and night			
Scattering in the F region	Low	Evening through midnight, equinox	60	80	1,000–4,000
Special transequatorial effects	Low	Evening through midnight	60	80	4,000–9,000

* From *CCIR Rep.* 259-2, 1970.

The *D*-region electron-density irregularities are produced by the action of turbulence, wind shears, and overlapping ionization of meteor trails, in the height range 70 to 90 km. Energy is scattered out of the incident beam, for single isotropic scattering neglecting polarization, according to[149]

$$p_a = p_t r_0^2 l^{-2} b A_r \csc(\gamma/2) S(K)$$ (18-163)

where r_0 is the classical electron radius (2.8×10^{-15}m), *l* is the distance from scattering volume to receiver, *b* is the thickness of the scattering volume, A_r is the effective area of the receiving antenna, γ is the angle through which scattering takes place, $S(K)$ is the spectrum of turbulent irregularities, p_a is the available power at the receiver, in the same units as p_t, p_t is the radiated power, and

$$K = \frac{4\pi}{\lambda} \sin \frac{1}{2}\gamma$$ (18-164)

In the case of turbulent mixing:[149]

$$S(K) = K^{-n}(dN/dh)^2$$ (18-165)

where dN/dh is the electron-density gradient.

The relationship of transmission loss to the frequency *f* and geometry of the propagation path[150] is given by

$$p_t/p_a \propto l^2 f^{n_s} (\sin \tfrac{1}{2}\gamma)^{n_s - 1}$$ (18-166)

where n_s ($n_s = n + 2$) is the frequency exponent for scaled antennas (gain constant with frequency). Observations show that n_s varies with time but is mostly between 7 and $9\frac{1}{4}$, with a median value of 8.

Figure 18-87 shows distance and frequency dependence of ionospheric scatter compared to tropospheric scattering.[70]

In the frequency range from 25 to 108 MHz, over distances of from 1,000 to 2,000 km, the transmission loss ranges from about 140 dB to about 210 dB, depending on frequency, antenna, geography, and time.

Short-term fading is Rayleigh distributed with superimposed bursts from meteor reflections. Fading rates lie between 0.2 and 3 Hz, with a mean value of 1 Hz at 50 MHz and

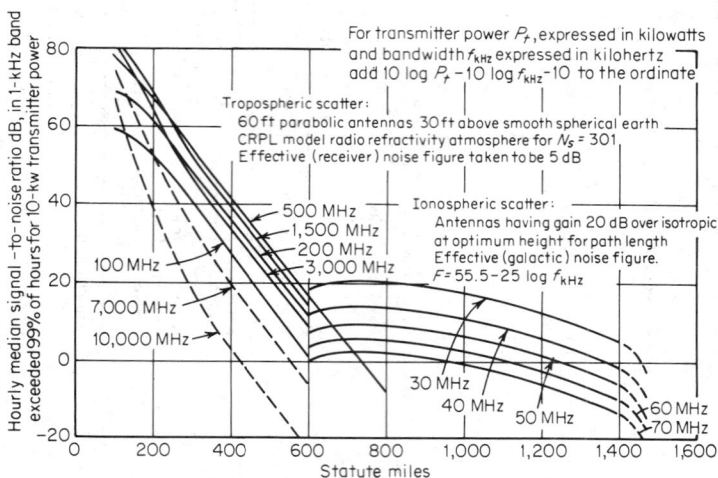

Fig. 18-87. Comparison of distance and frequency dependence of ionospheric and tropospheric scatter. (*JTAC.*[70])

varying approximately proportionately with operating frequency. Space or frequency diversity is applicable, with horizontal spacings about 5 wavelengths (or frequency spacings of about $f \times 10^{-4}$ required, where f is the operating frequency). Simple Yagi and rhombic antennas are satisfactory for some applications, while some large systems have used elaborate corner-reflector arrays. Modulation considerations include the effect of Doppler-shifted meteor reflections and long-delayed $F2$ multipath.[151]

Vhf scatter does not suffer auroral absorption, but auroral scattering may be an important source of multipath; and polar-cap absorption may attenuate the signal at frequencies below 50 MHz. During solar flares, depending upon the intensity, the signal is attenuated at frequencies below about 40 MHz and enhanced at higher frequencies. Vhf scatter is presently of little practical importance but may have a potential for reliable low-power data communications, especially in polar regions.

106. Meteor Scatter. Ionized trails are produced by meteors mostly in the height range 80 to 120 km. They diffuse rapidly and usually disappear within a few seconds. These ionized columns reflect vhf waves particularly in the frequency range 20 to 150 MHz. The lower frequency limit is set by the need to be above the regular maximum usable frequency, and the upper limit is set by weakening of the reflections and their shorter duration with increasing frequency (Refs. 152 to 153 and *CCIR Rep.* 251-1).

Meteor-burst propagation is subject to abnormal attenuation during polar-cap absorption events and sudden ionospheric disturbances.

Meteor-burst communications systems have been developed (Refs. 155 and 156 and *CCIR Rep.* 251-1) to operate duplex digital circuits in the 30- to 40-MHz range, over distances of 600 to 1,300 km with transmitter powers of 1 to 3 kW; meteor propagation is currently of little practical importance, but some applications are being suggested for low-power, intermittent data transmission from remote sensors.

107. Other Forms of Scatter. *Equatorial F Scatter.* Occurrence of F scatter is associated with equatorial spread F produced by patches of irregularities located at or below the bottom 50 km or so of the F layer.[9,157] The irregularities are elongated along the magnetic field with longitudinal dimensions of 1,000 m or more and transverse dimensions of the order of 10 m or less. The echoes are aspect-sensitive. While F scatter can permit communication at frequencies as high as 50 MHz (well above the MUF), disastrous for hf communication is the occurrence of flutter fading with rates of 10 Hz or more.[158]

Auroral Scatter. Ionization associated with auroral disturbances gives radio reflections at hf and vhf. Radio scattering is only generally correlated with visible aurora; in general the auroral reflections are aspect-sensitive. Radio amateurs communicate via auroral reflections, and radar observations have been used to study aurora for many years.[159,160] The mechanisms are:[159]

1. Weak scattering from randomly distributed gradients in electron density.
2. Strong scattering from randomly distributed clouds of ionization.
3. Weak scattering from ordered arrays of gradients in electron density produced by propagating ion-acoustic waves.

Incoherent Scatter. The existence of incoherent scatter from electrons (Thomson scattering) in the ionosphere was demonstrated in 1958.[161]

When high-power continuous-wave vhf radar is beamed vertically, weak scattering is observed from all heights in the ionosphere and the technique provides a means of determining the electron-density profile even above the $F2$ peak, which cannot be seen with conventional ionosondes. The power p_a received[161] from a distance R is

$$p_a = \frac{p_t a \sigma v_0 \tau \eta_r^2 \eta_s \eta_A}{8\pi R^2} \qquad (18\text{-}167)$$

where a is the antenna aperture area, σ is the cross section per unit volume, v_0 is the velocity of light, τ is the pulse duration in seconds, η_r, η_s, and η_A are factors to correct the aperture for the effects of resistive losses, side lobes, and tapered feed, respectively, and R is the range to the scattering volume.

p_a is proportional to R^{-2} rather than R^{-4}, as in the case of a single-scatterer. From the divergence of the beam it can be seen that the scattering volume increases proportional to the square of the range; this removes a factor of R^{-2}.

For thermal equilibrium between ions and electrons, the scattering cross section per free electron is just about one-half the classical Thomson cross section. In the more general case, when the electron and ion temperatures (T_e and T_i, respectively) differ, the relationship between the measured cross section σ_m and the classical value σ is given approximately by

$$\sigma_m = \frac{1}{1 + T_e/T_i}\sigma \tag{18-168}$$

Measurements of σ_m give values[161] less than the theoretical value of 5×10^{-29} m^2 due to D-region absorption during the day and electron-ion temperature relationship at night.

Incoherent scatter is a powerful tool for upper-atmosphere investigations and is currently the subject of much active research.[162-164]

108. Earth-Space Paths via the Ionosphere. Signals from satellites or extraterrestrial sources, upon passing through the ionosphere, undergo refractive bending and scintillation (fading), polarization rotation, absorption, frequency change and dispersion, and some path delay (Refs. 8, 9, and 165 to 168, *CCIR Rep.* 263-2, and *CCIR SJM Annex* 10-2). While the effects are of major concern at frequencies below about 300 MHz, they are of importance even at much higher frequencies to some satellite communications and to tracking and radio-astronomy applications.

Formulas for *refraction and ray-path* determination have been given.[169] At very oblique angles the rays may be reflected, but a penetration cone is defined by the semiangles $\phi_0 = \sin^{-1}\mu_m$, where $\mu_m = [\sqrt{1 - (f_c/f)^2}]$ is the refractive index at the height of maximum electron density (f_c is the critical frequency). The effect of the magnetic field is to produce two images of the source, corresponding to the ordinary and extraordinary ray, each with a different path length and apparent direction of arrival.

Scintillations occur when the waves pass through electron-density irregularities.[166-167a] Phase variations caused by these refractive-index irregularities cause an amplitude-diffraction pattern, focusing, and defocusing. Scintillations can be quite severe and may represent a practical limitation for some types of satellite communications system, especially at vhf. Scintillation characteristics depend very much on the strength and nature of the irregularities and are characteristically different for equatorial, midlatitude, and high-latitude regions. Fading of 20 dB or more is observed at frequencies below 200 MHz at equatorial and high latitudes; fading of as much as 3 to 4 dB at frequencies in the 4- to 8-GHz range has been observed in equatorial regions. Equatorial scintillation peaks near midnight and occurs mainly during equinoxes. High-latitude scintillation is often associated with auroral disturbances. Reference 167a gives a model for estimation of scintillation effects but does not account for the shf observations.

Absorption effects are small (less than about 1 dB) at frequencies above 100 MHz, except in auroral regions. A summary of important effects and their frequency dependence is given in Table 18-7.

109. Ionospheric Ducting. Long-distance hf and vhf propagation by ducting along field-aligned ionization[9] has been interpreted from observations of radar and rocket and satellite signals (Refs. 170 to 173 and *CCIR Rep.* 341-1). This type of propagation, depending on the geometry and ionization density, may lead to apparently anomalous signal transmission from satellite to earth at frequencies well above ionospheric-reflection frequencies.

Table 18-7.   Estimated Maximum Ionospheric Effects at 100 MHz for Elevation Angles of About 30°-One-Way Traversal*

Effect	Magnitude	Frequency dependence
Faraday rotation	30 rotations	$1/f^2$
Propagation delay	25 μs	$1/f^2$
Refraction	$\leq 1°$	$1/f^2$
Absorption, polar-cap	4 dB	$\sim 1/f^2$
Auroral + polar-cap	5 dB	$\sim 1/f^2$
Midlatitude	<1 dB	$1/f^2$
Dispersion	0.4 pulse/Hz	$1/f^3$
Scintillation, midlatitude	≈ 20 dB fade depth	$1/f^2$†
Equatorial	>20 dB fade depth	$1/f^2$ to $1/f$†
High-latitude	>20 dB fade depth	$\sqrt{1/f}$ to $1/f^2$†

* From *CCIR SJM Annex* 10-2, 1971.
† Applies over 100 MHz.

110. Modifying the Ionosphere with Intense Radio Waves. Induced changes in the density, temperature, and distribution of F-region ionization have been demonstrated[174] by beaming powerful ground-based radio-wave emissions upward at near the critical frequency, using 2 MW continuous-wave power, with an antenna gain of about 40. Ambient electron temperature is increased by 35% or more, anomalous reflection structure and attenuation behavior are induced, artificial spread F is generated, and air-glow emission from oxygen is increased. All the induced changes are transitory and self-reversing. Such ionospheric-modification experiments are becoming an important tool for controlled studies of the physics of naturally occurring plasmas.

111. References on Radio-Wave Propagation.

1. JTAC [Joint Technical Advisory Council, Institute of Electrical and Electronics Engineers (IEEE) and Electronic Industries Association (EIA)] "Radio Spectrum Utilization," New York, 1964.

2. RAMO, S., J. R. WHINNERY, and T. VANDUZER   "Fields and Waves in Communication Electronics," Wiley, New York, 1965.

3. BREMMER, H.   "Terrestrial Radio Waves: Theory of Propagation," Elsevier, Amsterdam, 1949.

4. WAIT, J. R.   "Electromagnetic Waves in Stratified Media," Pergamon, New York, 1962; (2d ed. 1970).

5. FOCK, V. A.   "Electromagnetic Diffraction and Propagation Problems," Pergamon, New York, 1965.

6. BORN, M., and E. WOLF: "Principles of Optics: Electromagnetic Theory of Propagation: Interference and Diffraction of Light," 3d ed., Pergamon, New York, 1965.

7. DUCASTEL, F.   "Tropospheric Wave Propagation beyond the Horizon," Pergamon, New York, 1966.

8. DAVIES, K.   "Ionospheric Radio Waves," Blaisdell, Waltham, Mass., 1969.

9. DAVIES, K.   Ionospheric Radio Propagation, *Natl. Bur. Stand. Monogr. 80,* 1965.

10. "Review of Radio Science 1969-1971," *Rep. Int. Union Radio Sci.,* Brussels, 1972.

10a. BECKMANN, P.   "The Depolarization of Electromagnetic Waves," Golem, Boulder, Colo., 1968.

11. SCHELKUNOFF, S. A., and H. T. FRIIS   "Antennas: Theory and Practice," Wiley, New York, 1952.

12. RICE, P. L., A. G. LONGLEY, K. A. NORTON, and A. P. BARSIS   Transmission Loss Predictions for Tropospheric Communication Circuits, *Natl. Bur. Stand. Tech. Note* 101 (rev.), 1967.

13. NORTON, K. A., and A. C. OMBERG   Maximum Range of a Radar Set, *Proc. IRE,* 1947, Vol. 35, No. 1, pp. 4-24.

14. BAGHDADY, E. J.   "Lectures on Communication System Theory," McGraw-Hill, New York, 1961.

15. SCHWARTZ, M., W. R. BENNETT, and S. STEIN   "Communication Systems and Techniques," McGraw-Hill, New York, 1966.

16. BAGHDADY, E. J.   Models for Signal Distorting Media, in R. E. Kalman and N. DeClaris (eds.), "Aspects of Network and System Theory," pp. 337-381, Holt, New York, 1971.

17. NAKAGAMI, M.   On the Intensity Distributions and Its Application to Signal Statistics, *Radio Sci. J. Res. Natl. Bur. Stand.,* 1964, Vol. 68D, No. 9, pp. 995-1003.

18. RICE, S. O.   Mathematical Analysis of Random Noise, *Bell Syst. Tech. J.,* 1944, Vol. 23, pp. 282-332; 1945, Vol. 24, pp. 46-156.

19. BECKMANN, P.   Rayleigh Distribution and Its Generalizations, *Radio Sci., J. Res. Natl. Bur. Stand.,* 1964, Vol. 68D, No. 9, pp. 927-932.

20. BENNETT, W. R.   "Electric Noise," McGraw-Hill, New York, 1960.

21. MIDDLETON, D.   "An Introduction to a Statistical Communication Theory," McGraw-Hill, New York, 1960.

22. DISNEY, R. T., A. D. SPAULDING, and D. H. ZACHARISEN   Electromagnetic Interference from Man-made Noise, Pt. I, Estimates of Business, Residential and Rural Area Characteristics; Pt. II, bibliography.

23. SKOMAL, E. N.   The Range and Frequency Dependence of VHF/UHF Man-made Radio Noise in and above Metropolitan Areas, *IEEE Trans. Veh. Tech.,* May 1970, Vol. VT-19, No. 2.

24. Report of Joint Technical Advisory Council, Man-Made Noise, Subcommittee 63.1.3, Unintended Radiation, Supple. 9 "Spectrum Engineering," IEEE, March 1968.

25. MIDDLETON, D.   Statistical-Physical Models of Urban Radio Noise Environments, I: Foundations, *IEEE Trans. Electromag. Compat.,* May 1972, Vol. EMC-14, No. 2.

26. THOMPSON, W. I.   III Bibliography on Ground Vehicle Communications and Control: A KWIC Index, *U.S. Dept. Transp. Urban Mass Transp. Admin., Rep.* DOT-UMTA-71-3, July 1971.

27. BEAN, B. R., and E. J. DUTTON   "Radio Meteorology," *Natl. Bur. Stand. Monogr.* 92 1966; also Dover, New York, 1968.

28. BEAN, B. R., J. D. HORN, and A. M. OZANICH, JR.   Climatic Charts and Data of the Radio Refractive Index for the United States and the World, *Natl. Bur. Stand. Monogr.* 22, November 1960.

29. BEAN, B. R., B. A. CAHOON, C. A. SAMSON, and G. D. THAYER   A World Atlas of Atmospheric Radio Refractivity, *Environ. Sci. Serv. Admin. Monogr.* 1, 1966.

30. MILLINGTON, G.   Propagation at Great Heights in the Atmosphere, *Marconi Rev.,* 1958, Vol. 21, pp. 143–160.

31. ROTHERAM, D.   Ground Wave Propagation at Medium and Low Frequencies, *Electron. Lett.,* 1970, Vol. 6, pp. 794–795.

32. WAIT, J. R.   Electromagnetic Surface Waves, in J. Saxton (ed.), "Advances in Radio Research," Academic, New York, 1964.

32*a*. WAIT, J. R.   Theory of Ground Wave Propagation, in J. R. WAIT (ed.), "Electromagnetic Probing in Geophysics," Golem, Boulder, Colo., 1971.

32*b*. WAIT, J. R.   Recent Analytical Investigation of Electromagnetic Ground Wave Propagation Over Inhomogeneous Earth, *Proc. IEEE,* Aug. 1974, Vol. 62.

33. NORTON, K. A.   The Calculation of Ground-Wave Field Intensity over a Finitely Conducting Spherical Earth, *Proc. IRE,* December 1941, Vol. 29, pp. 623–639.

34. BREMMER, H.   The Extension of Sommerfeld's Formula for the Propagation of Radio Waves over a Flat Earth to Different Conductivities of the Soil, *Physica 'sGrav.,* 1954, Vol. 20, p. 441.

35. JOHLER, J. R., W. KELLER, and L. C. WALTERS   "Phase of the Low Radio Frequency Ground Wave," *Natl. Bur. Stand. Circ.* 573, 1956.

36. FCC Rules and Regulations, Radio Broadcast Services. Secs. 73.184 and 73.190, September 1972.

37. CCIR "Atlas of Ground-Wave Propagation Curves for Frequencies between 30 and 300 MHz (Vertical and Horizontal Polarization)," International Telecommunications Union, Geneva, 1955.

38. CCIR "Atlas of Ground-Wave Propagation Curves for Frequencies between 30 and 10,000 MHz (Vertical Polarization Only)," International Telecommunications Union, Geneva, 1959.

39. MALEY, S. W.   Radio Wave Methods for Measuring the Electrical Parameters of the Earth, Chap. 2 in J. R. Wait (ed.), "Electromagnetic Probing Methods in Geophysics," Golem, Boulder, Colo., 1971.

40. KIRBY, R. S., J. C. HARMAN, F. M. CAPPS, and R. N. JONES   Effective Radio Ground-Conductivity Measurements in the United States, *Natl. Bur. Stand. Circ.* 546, February 1954.

41. WAIT, J. R., and L. C. WALTERS   Curves for Ground Wave Propagation over Mixed Land and Sea Paths, *IEEE Trans. Antenna Propag.,* 1963, Vol. AP-11, pp. 38–45.

42. GODZINSKI, Z.   A Comparison of Millington's Method and the Equivalent Numerical Distance Method with the Theory of Ground-Wave Propagation over an Inhomogeneous Earth, *Proc. IEE (Lond.), Pt. C, Monogr. 318R,* December 1958.

43. MILLINGTON, G.   Ground Wave Propagation over an Inhomogeneous Smooth Earth, *J. IEE (Lond.),* January 1949, Pt. III, Vol. 96. p. 53.

44. KIRKE, H. L.   Calculation of Ground-Wave Field Strength over a Composite Land and Sea Path, *Proc. IRE,* May, 1949, Vol. 37, pp. 489–496.

45. OTT, R. H.   A New Method for Predicting HF Ground Wave Attenuation over Inhomogeneous Irregular Terrain, U.S. Dept. Commer. Off. *Telecommun.* OT/ITS RR 7, January 1971.

46. BECKMANN, P., and A. SPIZZICHINO   "The Scattering of Electromagnetic Waves from Rough Surfaces," Pergamon, New York, 1963.

47. DOUGHERTY, H. T.   Diffraction by Irregular Apertures, *Radio Sci.,* January 1970, Vol. 5, pp. 55-60.

48. BACHYNSKI, M. P.   Propagation at Oblique Incidence over Cylindrical Obstacles, *J. Res. Natl. Bur. Stand. (Radio Propag.)*, July-August, 1960, Vol. 1, 64D, No. 4, pp. 311-315.

49. WAIT, J. R., and A. M. CONDA   Diffraction of Electromagnetic Waves by Smooth Obstacles for Grazing Angles, *J. Res. Natl. Bur. Stand. (Radio Sci.)*, 1964, Vol. 68D, No. 2, pp. 239-250.

50. LONGLEY, A. G., and P. L. RICE   Prediction of Tropospheric Radio Transmission Loss over Irregular Terrain: A Computer Method, *ESSA Tech. Rep.* ERL 79/ITS 67, 1968.

51. FCC Rules and Regulations, Radio Broadcast Services, Sec. 73.333, September 1972.

52. IEEE Communications Society "Mobile Communications," special issue of *Trans. Commun.,* 1973.

53. OKUMURA, Y., H. OMURI, T. KAWANO, and K. FUKUDA   Field Strength and Its Variability in VHF and UHF Land Mobile Radio Service, *Rev. Tokyo Elec. Commun. Lab.,* 1968, Vol. 16, pp. 825-873.

54. COX, D. C.   Doppler Spectrum Measurements at 910 MHz over a Suburban Mobile Radio Path, *Proc. IEEE,* 1971, Vol. 59, pp. 1017-1018; see also "Time and Frequency Domain Characterization of Multipath at 910 MHz in a Suburban Mobile Radio Environment," *Radio Sci.,* December 1972, Vol. 7, pp. 1069-1077.

55. "Engineering Considerations for Microwave Communications Systems," Lenkurt Electric Co., Inc., San Carlos, Calif., June 1970.

56. MIL HANDBOOK 416, "Facility Design Handbook for LOS Microwave," Superintendent of Documents, U.S. Government Printing Office, Washington, D.C. 20402.

57. DOUGHERTY, H. T.   A Survey of Microwave Fading Mechanisms, Remedies and Applications, *Environ. Sci. Serv. Admin. Tech. Rep.* ERL-69-WPL 4, 1968.

57a. DOUGHERTY, H. T., and R. E. WILKERSON   Determination of Antenna Height for Protection against Microwave Diffraction Fading, *Radio Sci.,* 1967, Vol. 2, n.s., pp. 161-165.

58. RUTHROFF, C. L.   Multipath Fading on LOS Microwave Radio Systems as a Function of Path Length and Frequency, *Bell Syst. Tech. J.,* September 1971, Vol. 50, No. 7, pp. 2375-2398.

59. VIGANTS, A.   Space-Diversity Performance as a Function of Antenna Separation, *IEEE Trans. Commun. Technol.,* 1968, Vol. COM-16, No. 6, pp. 831-836.

60. BARSIS, A. P., and M. E. JOHNSON   Prolonged Space-Wave Fadeouts in Tropospheric Propagation, *J. Res. Natl. Bur. Stand. (Radio Propag.)*, 1962, Vol. 66D, No. 2, pp. 681-694.

61. Joint Technical Advisory Council IEEE-EIA Microwave Radio Relay System Reliability, Report to the FCC, Mar. 23, 1965.

61a. BULLINGTON, K.   Phase and Amplitude Variations in Multipath Fading of Microwave Signals, *Bell Syst. Tech. J.,* July-August 1971, Vol. 50, No. 6, pp. 2039-2053.

62. BOOKER, H. G., and W. E. GORDON   Outline of a Theory of Radio Scattering in the Troposphere, *J. Geophys. Res.,* September 1950, Vol. 55, No. 3, pp. 241-246; see also *Proc. IRE,* April 1950, Vol. 38, No. 4, p. 401.

63. VOGE, J.   Radioelectricity and the Troposphere, I: Theories of Propagation to Long Distances by Means of Atmospheric Turbulence, *Onde Electr.,* 1955, Vol. 35, pp. 565-581.

64. MEGAW, E. C. S.   Fundamental Radio Scatter Propagation Theory, *Proc. IEE,* September 1957, Pt. C104, No. 6, pp. 441-455; see also *Monogr.* 236R, May 1957.

65. STARAS, H.   Tropospheric Scatter Propagation: A Summary of Recent Progress, *RCA Rev.,* March 1958, Vol. 19, pp. 3-18.

66. WHEELON, A. D.   Radio-Wave Scattering by Tropospheric Irregularities, *J. Res. Natl. Bur. Stand. (Radio Propag.).* September-October 1959, Vol. 63D, No. 2, pp. 205-234; also *J. Atmos. Terr. Phys.,* 1959, Vol. 15, No. 3 and 4, pp. 185-205.

67. FRIIS, H. T., A. B. CRAWFORD, and D. C. HOGG   A Reflection Theory for Propagation beyond the Horizon, *Bell Syst. Tech. J.,* 1957, Vol. 36, pp. 627-644, also published as *Bell Tel. Syst. Monogr.* 2823.

68. DUCASTEL, F., P. MISME, A. SPIZZICHINO, and J. VOGE   On the Role of the Process of Reflection in Radio Wave Propagation, *J. Res. Natl. Bur. Stand. (Radio Sci.)*, 1962, Vol. 66D, No. 3, pp. 273-284.

69. *Proc. IRE Special Issue on Scatter Propag.,* October 1955, Vol. 43.

70. Joint Technical Advisory Council   Radio Transmission by Ionospheric and Tropospheric Scatter, *Proc. IRE,* 1960, Vol. 48, pp. 4-44.

71. Tropospheric Wave Propagation, *IEE*, Conf. *Publ.* 48, London, 1968 (contains a number of references giving current appraisal of tropospheric propagation).

72. SURENIAN, D.   Experimental Results of Angle Diversity System Tests, *IEEE Trans. Commun. Technol.*, June 1965, Vol. COM-13, No. 2, p. 208.

73. FOCK, V. A., L. A. WAINSTEIN, and M. E. BELKINA   Radiowave Propagation in Surface Tropospheric Ducts, *Radiotechn, Elektron.*, 1958, Vol. 3, No. 12, pp. 1411-1429.

74. FREEHAFER, J. E.   in D. E. Kerr, "Propagation of Short Radio Waves," McGraw-Hill, New York, 1951; reprinted Dover, New York, 1965.

75. SODHA, M. S., A. K. GHATAK, D. P. TEWARI, and P. K. DUBEZ   Focusing of Waves in Ducts, *Radio Sci.*, November 1972, Vol. 7, No. 11, p. 1005.

76. JESKE, J., AND K. BROCKS   Comparison of Experiments on Duct Propagation above the Sea with the Mode Theory of Booker and Walkinshaw, *Radio Sci.*, 1966, Vol. 1, n.s., No. 8, pp. 891-895.

77. PIDGEON, V. W.   Frequency Dependence of Radar Ducting, *Radio Sci.*, 1970, Vol. 5, No. 3, pp. 541-550.

78. AGARD   Telecommunications Aspects on Frequencies 10 to 100 GHz, *Proc. 18th Meet. Electromagn. Wave Propag. Panel: 1972* (contains approximately 40 theoretical and experimental papers).

79. THOMPSON, M. C., JR., L. E. VOGLER, H. B. JANES, and L. E. WOOD   A Review of Propagation Factors in Telecommunication Applications of the 10 to 100 GHz Radio Spectrum, *U.S. Dept. Commer., Off. Telecommun.* OT/TRER 34, August 1972.

80. LIEBE, H. J., and W. M. WELCH   Attenuation and Phase Dispersion in the Atmosphere Due to the Microwave Spectrum of Oxygen, *AGARD Proc. 18th Meet. Wave Electromagn. Propag. Panel, 1972.*

81. VANVLECK, J. H.   The Absorption of Microwaves by Oxygen, *Phys. Rev.*, April 1947, Vol. 71, No. 7, pp. 413-424.

82. VANVLECK, J. H.   The Absorption of Microwaves by Uncondensed Water Vapor, *Phys. Rev.*, April 1947, Vol. 71, No. 7, pp. 425-433.

83. BECKER, G. E., and S. H. AUTLER   Water Vapor Absorption of Electromagnetic Radiation in the Centimeter Wavelength Range, *Phys. Rev.*, 1946, Vol. 70, No. 5 and 6, pp. 300-307.

84. BLAKE, L. V.   Radar/Radio Tropospheric Absorption and Noise Temperature, *Nav. Res. Lab. Rep.* 7461, October 1972.

85. LIEBE, H. J.   Calculated Tropospheric Dispersion and Absorption Due to the 22 GHz Water Vapor Line, *IEEE Trans. Antennas Propag.*, 1969, Vol. 17, No. 5, pp. 621-627.

86. HOGG, D. C.   Millimeter-Wave Communication through the Atmosphere, *Science 1968*, Vol. 159, No. 3810, pp. 39-46.

87. STRAITON, A. W., and C. W. TOLBERT   Factors Affecting Earth-Satellite Millimeter Wavelength Communications, *IEEE Trans.*, 1963, Vol. MTT-11, pp. 296-301.

88. CARTER, C. J., R. L. MITCHELL, and E. E. REBER   Oxygen Absorption Measurements in the Lower Atmosphere, *J. Geophys. Res.*, 1968, Vol. 73, No. 10, pp. 3113-3120.

89. LANE, J. A.   Scintillation and Absorption Fading on Line-of-Sight Links at 35 and 100 GHz, *IEE Conf. Tropospheric Wave Propag. Lond. 1968, Pub.* 48.

90. MEDHURST, R. G.   Rainfall Attenuation of Centimeter Waves: Comparison of Theory and Experiment, *Trans. IEEE Antennas Propag.* 1965, Vol. AP-13, No. 4, pp. 550-564.

91. OGUCHI, T.   Attenuation of Electromagnetic Waves Due to Rain with Distorted Raindrops, *J. Radio Res. Lab. (Tokyo)*, January 1964, Vol. 11, pp. 19-37.

92. RICE, P. L., and N. R. HOLMBERG   Cumulative Time Statistics of Surface-Point Rainfall Rates, *IEEE Trans. Commun.*, Oct. 1973, Vol. Com-21, No. 10, pp. 1131-1136.

93. DUTTON, E. J.   A Meteorological Model for Use in the Study of Rainfall Effects on Atmospheric Telecommunications, *U.S. Dept. Commer., Off. Telecommun.* OT/TRER 24, December 1971.

93a. RUTHROFF, C. L.   Rain Attenuation and Radio Path Design, *Bell Syst. Tech. J.*, January 1970, Vol. 49, No. 1, pp. 121-135.

93b. HOGG, D. C.   Path Diversity in Propagation of Millimeter Waves through Rain, *IEEE Trans. Antennas Propag.*, 1967, Vol. AP-15, No. 3, pp. 410-415.

94. GUSLER, L. T., and D. C. HOGG   Some Calculations on Coupling between Satellite-Communications and Terrestrial Radio-Relay Systems Due to Scattering by Rain, *Bell Syst. Tech. J.*, 1970, Vol. 49, p. 1491.

95. CRANE, R. K.   Propagation Phenomena Affecting Satellite Communications Systems Operating in the Centimeter and Millimeter Wavelength Bands, *Proc. IEEE*, 1971, Vol. 59, No. 2, pp. 173–188.

96. CRANE, R. K.   Coherent Pulse Transmission through Rain, *IEEE Trans. Antennas Propag.*, 1967, Vol. AP-15, No. 2, pp. 252–256.

97. GRAY, D. A.   Transit-Time Variations in Line of Sight Tropospheric Propagation Paths, *Bell Syst. Tech. J.*, July–August 1970, Vol. 49, pp. 1059–1068.

98. ROCHE, J. F., H. LAKE, D. T. WORTHINGTON, C. K. H. TSAO, and J. T. DE BETTENCOURT   Radio Propagation 27–40 GHz, *IEEE Trans. Antennas and Propag.*, July 1970, Vol. AP-18, pp. 452–462.

99. SETZER, D.   Computed Transmission through Rain at Microwave and Visible Frequencies, *Bell Syst. Tech. J.*, October 1970, Vol. 49, No. 8, pp. 1873–1892.

100. IPPOLITO, L. J.   Effects of Precipitation on 15.3- and 31.65-GHz Earth-Space Transmissions with the ATS-V Satellite, *Proc. IEEE*, 1971, Vol. 59, No. 2, pp. 189–205.

100a. TATARSKI, V. I.   "Wave Propagation in a Turbulent Medium," trans. R. A. Silverman, McGraw-Hill, New York, 1961.

101. LEE, R. W., and A. T. WATERMAN   Space Correlation of 35 GHz Transmissions over a 28 km Path, *Radio Sci.*, 1968, Vol. 3, pp. 135–140.

102. THOMPSON, M. C., JR., and H. B. JANES   Measurements of Phase-Front Distortion on an Elevated Line-of-Sight Path, *IEEE Trans. Aerosp. Electron. Syst.* 1970, Vol. AES-6, No. 5, pp. 645–656.

103. JANES, H. B., M. C. THOMPSON, JR., D. SMITH, and A. W. KIRKPATRICK   Comparison of Simultaneous Line-of-Sight Signals at 9.6 and 34.5 GHz, *IEEE Trans. Antennas Propag.*, 1970, Vol. AP-18, No. 4, pp. 447–451.

103a. MANDICS, P. A., R. W. LEE, and A. T. WATERMAN   Spectra of Short-Term Fluctuations of Line of Sight Signals, *Radio Sci.*, 1973, Vol. 8, pp. 185–201.

104. LAWRENCE, R. S., and J. W. STROHBEIN   A Survey of Clear-Air Propagation Effects Relevant to Optical Communication, *Proc. IEEE*, 1970, Vol. 58, pp. 1523–1545.

105. HOGG, D. C., and W. W. MUMFORD   The Effective Noise Temperature of the Sky, *Microwave J.*, March 1960, Vol. 3, pp. 80–84.

106. DECKER, M. T., and E. J. DUTTON   Radiometric Observations of Liquid Water in Thunderstorm Cells, *J. Atmos. Sci.*, 1970, Vol. 27, No. 5, pp. 285–290.

106a. RATCLIFFE, J. A.   "Physics of the Upper Atmosphere," Academic, New York, 1960.

107. VAN ZANDT, T. E., and R. W. KNECHT   The Structure and Physics of the Upper Atmosphere, A. Rosen and D. P. LeGallery (eds.). "Space Physics," Wiley, New York, 1964.

108. RISHBETH, H., and O. K. GARRIOTT   "Introduction to Ionospheric Physics," Academic, New York, 1969.

109. SENGUPTA, P. R.   Solar X-Ray Control of the E Layer of the Ionosphere, *J. Atmos. Terr. Phys.*, 1970, Vol. 32, pp. 1273–1282.

110. SMITH, E. K., and S. MATSUSHITA   "Ionospheric Sporadic E," Macmillan, New York, 1962.

111. WHITEHEAD, J. D.   Production and Prediction of Sporadic E, *Rev. Geophys. Space Phys.* 1970, Vol. 8, pp. 145–168.

112. Special issue on sporadic E, *Radio Sci.*, March 1972.

113. CIRA "COSPAR International Reference Atmosphere," North-Holland Amsterdam, 1965.

114. Electromagnetic Probing of the Upper Atmosphere; special issue of *J. Atmos. Terr. Phys.*, April 1970, Vol. 32.

115. PIGGOTT, W. R., and K. RAWER (eds.)   "U.R.S.I. Handbook of Ionogram Interpretation and Reduction of the World Wide Soundings Committee," Elsevier, Amsterdam, 1961.

116. RATCLIFFE, J. A.   "The Magneto-Ionic Theory," Cambridge University Press, Cambridge, 1959.

117. BUDDEN, K. G.   "Radio Waves in the Ionosphere," Cambridge University Press, Cambridge, 1961.

118. KELSO, J. M.   "Radio Ray Propagation in the Ionosphere," McGraw-Hill, New York, 1964.

119. WAIT, J. R. (ed.)   "Electromagnetic Probing in Geophysics," Golem Press, Boulder, Colo., 1971.

120. Ocean '72, *IEEE Int. Conf. Eng. Ocean Environ., IEEE Pub. 72 CHO 660-1 OCC,* 1972.

121. WATT, A. D. "VLF Radio Engineering," Pergamon, New York, 1967.

122. BUDDEN, K. G. "The Waveguide-Mode Theory of Wave Propagation," Prentice-Hall, Englewood Cliffs, N.J., 1961.

123. PAPPERT, R. A. A Numerical Study of VLF Mode Structure and Polarization below an Anisotropic Ionosphere, *Radio Sci.,* 1968, n.s., Vol. 3.

124. WAIT, J. R., and K. P. SPIES Characteristics of the Earth Ionosphere Wave Guide for VLF Radio Waves, *Natl. Bur. Stand. Tech. Note* 300, 1964 (and two supplements).

125. BICKEL, J. E., J. A. FERGUSON, and G. V. STANLEY Experimental Observation of Magnetic Field Effects on VLF Propagation at Night, *Radio Sci.* Vol. 5, No. 1, pp. 19-25.

126. BERRY, L. A. Wave Hop Radio Propagation Theory, *IEE Conf. Pub.* 36, pp. 63-69, 1967.

127. BELROSE, J. S., W. L. HATTON, C. A. MCKERROW, and R. S. THAIN The Engineering of Communication Systems for Low Radio Frequencies, *Proc. IRE,* May 1959, Vol. 47, pp. 661-680.

128. PIGGOTT, W. R., M. L. V. PITTEWAY, and E. V. THRANE The Numerical Calculation of Wave Fields, Reflexion Coefficients and Polarizations for Long Radio Waves in the Lower Ionosphere, II, *Phil Trans. Soc.,* 1965, Vol. A-257, p. 243.

129. BELROSE, J. S. The Oblique Reflection of Low-Frequency Radio Waves from the Ionosphere, in "Propagation of Radio Waves at Frequencies Below 300 kc," *AGARDO-GRAPH* No. 74, pp. 149-165, Pergamon, 1963.

130. WAIT, J. R., and A. M. CONDA Pattern of an Antenna on a Curved Lossy Surface, *Trans. IRE (N.Y.),* 1958, Vol. AP-6, pp. 348-359.

131. EBERT, W. Ionospheric Propagation on Long and Medium Waves, *EBU Rev.,* Pt. A, *Tech.* 71 to 73; also *Tech. Monogr.* 3081, 1962.

132. BARGHAUSEN, A. F. Medium Frequency Sky Wave Propagation in Middle and Low Latitudes, *IEEE Trans. Broadcast.,* June 1966, Vol. 12, pp. 1-14.

132a. ELLING, W. Scheinbare Reflexion Shöhen und Reflexionsvermögen der Ionosphäre über Tsumeb, Südwest Afrika, ermittelt mit Impulsen im Frequenzband von 350 bis 5,600 kHz, *Arch. Elekt. Ubertragung,* Vol. 15, 1961, pp. 115-124, (in German).

133. PHILLIPS, G. D., and D. KNIGHT Effects of Polarization on a Medium Frequency Sky Wave Service, Including the Case of Multihop Paths, *Proc. IEE (Lond.),* January 1965, Vol. 112, pp. 31-39.

134. HATTON, W. L. Oblique-sounding and HF Radio Communication, *IRE Trans.,* 1961, Vol. PGCS-9, p. 275.

134a. NEILSON, D. Oblique Sounding of a Transequatorial Path, in P. Newman (ed.), "Spread F and Its Effects upon Radiowave Propagation and Communication," *AGARDOGR.* No. 95, *Technivision,* 1966, pp. 467-490.

135. AGY, V., and K. DAVIES Ionospheric Investigations Using the Sweep-Frequency Pulse Technique at Oblique Incidence, *J. Res. Natl. Bur. Stand.,* September-October 1959, Vol. 63D, pp. 151-174.

136. LUCAS, D. L., and G. W. HAYDON Predicting Statistical Performance Indexes for High Frequency Ionospheric Telecommunications Systems, *Environ. Sci. Serv. Admin. Tech. Rep.* IER 1-ITSA 1, 1966.

137. BARGHAUSEN, A. F. (et al.) Predicting Long-Term Operational Parameters of High-Frequency Sky-Wave Telecommunication Systems, *Environ. Sci. Serv. Admin. Tech. Rep.* ERL 110-ITS 78, 1969.

138. JONES, W. B., and R. M. GALLET Ionospheric Mapping by Numerical Methods, *ITU Telecommun. J.,* December 1960, Vol. 27, No. 12, pp. 280-282.

139. LEFTIN, M., W. M. ROBERTS, and R. K. ROSICH "Ionospheric Predictions," 4 Vols., *U.S. Dept. Commer., Off. Telecommun.* OT/TRER 13, 1971.

139a. JONES, W. B., and R. M. GALLET Methods for Applying Numerical Maps of Ionospheric Characteristics, *J. Res. Natl. Bur. Stand. (Radio Propag.),* November-December 1962, Vol. 66D, No. 6, pp. 649-662.

140. LEFTIN, M., S. M. OSTROW, and C. PRESTON Numerical Maps of f_oE_s for Solar Cycle Minimum and Maximum, *Environ. Sci. Serv. Admin. Tech. Rep.* ERL 73-ITS 63, 1968.

140a. SANDOZ, O. A., E. E. STEVENS, and E. S. WARREN The Development of Radio Traffic Frequency Prediction Techniques for Use at High Latitudes, *Proc. IRE,* 1959, Vol. 47, p. 681.

141. UTLAUT, W. F. Effect of Antenna Radiation Angles upon HF Radio Signals Propagated over Long Distances, *Natl. Bur. Stand. J. Res.*, Pt. D., March-April 1961.

142. UTLAUT, W. F. Siting Criteria for HF Communication Centers, *Natl. Bur. Stand. Tech. Note* 139 (PB 16140), April 1962.

143. KOCH, J. W., and W. M. BEERY Observations of Radio Wave Phase Characteristics on a High-Frequency Auroral Path, *Natl. Bur. Stand. J. Res.*, 1962, Vol. 66D, pp. 291-296.

144. KOCH, J. W., and H. E. PETRIE Fading Characteristics Observed on a High-Frequency Auroral Radio Path, *Natl. Bur. Stand. J. Res.* 1962, Vol. 66D, pp. 159-166.

145. AMES, J. The Correlation between Frequency-Selective Fading and Multipath Propagation over an Ionospheric Path, *J. Geophys. Res.*, 1963, Vol. 68, pp. 759-768.

146. SALAMAN, R. K. A New Ionospheric Multipath Reduction Factor (MRF), *IRE Trans. Commun. Syst.*, June 1962, Vol. CS-10, No. 2, pp. 221-222.

147. BAILEY, D. K., R. BATEMAN, and R. C. KIRBY Radio Transmission at VHF by Scattering and Other Processes in the Lower Ionosphere, *Proc. IRE*, October 1955, Vol. 43, pp. 1181-1230.

148. KIRBY, R. C. Review of VHF Forward Scatter, *AGARD Proc. 37 Scatter Propaga. Radio Waves*, 1968.

149. WHEELON, A. D. Relation of Turbulence Theory to Ionospheric Forward Scatter Propagation Experiments, *J. Res. Natl. Bur. Stand.*, 1960, Vol. 64D, p. 301.

150. BLAIR, J. C., R. M. DAVIS, and R. C. KIRBY Frequency Dependence of D-Region Scattering at VHF, *J. Res. Natl. Bur. Stand. (Radio Propag.)*, September-October 1961, Vol. 65D, No. 3, pp. 249-263.

151. KOCH, J. W. Factors Affecting Modulation Techniques for VHF Scatter Systems, *IRE Trans. Commun. Syst.*, June 1959, Vol. CS-7, pp. 77-92.

152. MANNING, L. A., and V. R. ESHELMAN Meteors in the Ionosphere, *Proc. IRE*, February 1959, Vol. 47, pp. 186-199.

153. McKINLEY, D. W. R. "Meteor Science and Engineering," McGraw-Hill, New York, 1961.

154. SUGAR, G. R. Radio Propagation by Reflection from Meteor Trails, *Proc. IEEE*, February 1964, Vol. 52, p. 116.

155. FORSYTH, P. A., E. L. VOGAN, D. R. HANSEN, and C. O. HINES The Principles of JANET: A Meteor-Burst Communications System, *Proc. IRE*, 1957, Vol. 45, p. 1642.

156. BARTHOLOMÉ, P. J. Survey of Ionospheric and Meteor Scatter Communications, in K. Folkestad (ed.), "Ionospheric Radio Communications," *NATO Inst. Ionospheric Radio Commun. Arctic Proc.* Plenum, New York, pp. 143-154, 1967.

157. COHEN, R., and K. L. BOWLES On the Nature of Equatorial Spread F, *J. Geophys. Res.*, 1961, Vol. 66, p. 1081.

158. YEH, K. C., and O. G. VILLARD A New Type of Fading Observable on High-Frequency Radio Transmission Propagated over Paths Crossing the Magnetic Equator, *Proc. IRE*, 1958, Vol. 46, pp. 1968-1970.

159. FORSYTH, P. A. Auroral Scatter, in Scatter Propagation of Radio Waves, Pt. 2, *AGARD Conf. Proc. 37*, August 1968, pp. 34-1 to 34-10.

160. BOOKER, H. G. Radar Studies of the Aurora, in J. A. Ratcliffe (ed.), "Physics of the Upper Atmosphere," Academic, New York, 1960.

161. BOWLES, K. L. Incoherent Scattering by Free Electrons as a Technique for Studying the Ionosphere and Exosphere: Some Observations, and Theoretical Considerations, *J. Res. Natl. Bur. Stand. (Radio Propag.)*, 1961, Vol. 65D, p. 1.

162. FLOCK, W. L., and B. B. BALSLEY VHF Radar Returns from the D Region of the Equatorial Ionosphere, *J. Geophys. Res.*, 1967, Vol. 72, pp. 5537-5541.

163. ARMISTEAD, G. W., J. V. EVANS, and W. A. REID Measurement of D- and E- Region Electron Densities by the Incoherent Scatter Technique at Millstone Hill, *Radio Sci.*, 1972, Vol. 7, pp. 153-162.

164. EVANS, J. V. Theory and Practice of Ionosphere Study by Thomson Scatter Radar, *Proc. IEEE*, 1969, Vol. 57, pp. 496-530.

165. LAWRENCE, R. S., C. G. LITTLE, and H. J. A. CHIVERS A Survey of Ionospheric Effects upon Earth Space Radio Propagation, *Proc. IEEE*, 1964, Vol. 52, pp. 4-27.

166. BRIGGS, B. H. Brief Review of Scintillation Studies, *Radio Sci.*, 1966, Vol. 1, pp. 1163-1167.

167. *AGARD* Propagation Factors in Space Communications, *Conf. Proc.* 3, *Symp. Rome, Sept.* 21-25, *1965,* Mackay, London, 1967.

167*a*. FREMOUW, E. J., and C. L. RINO An Empirical Model for F-Layer Scintillation at VHF/UHF, *Radio Sci.,* 1973, Vol. 8, pp. 213-222.

167*b*. TAUR, R. R. Ionospheric Scintillation at 4 and 6 GHz, *COMSAT Tech. Rev.,* Vol. 3, 1973 (Spring), No. 1, pp. 145-163.

168. EVANS, J. V. Propagation in the Ionosphere, Chap. 2, Pt. II, in J. V. Evans and T. Hagfors (eds.), "Radar Astronomy," McGraw-Hill, New York, 1968.

169. CHVOJKOVA, E. Analytic Formulae for Radio Path in Spherically Stratified Ionosphere, *Radio Sci.,* 1965, Vol. 69D, pp. 453-457.

170. OBAYASHI, T. A Possibility of the Long Distance HF Propagation along the Exospheric Field Aligned Ionization, *Rep. Ionos. Spec. Res., Jap.* 1959, Vol. 13, p. 177.

171. GALLET, R. M., and W. F. UTLAUT Evidence of the Laminar Nature of the Exosphere Obtained by Means of Guided High Frequency Wave Propagation, *Phys. Rev. Lett.,* 1961, Vol. 6, p. 59.

172. WALKER, A. D. The Theory of Guiding of Radio Waves in the Exosphere, *J. Atmos. Terr. Phys.,* 1966, Vol. 28, p. 1039.

173. BOOKER, H. G. Guidance of Radio and Hydromagnetic Waves in the Magnetosphere, *J. Geophys. Res.,* 1962, Vol. 67, p. 4135.

174. UTLAUT, W. F., and R. COHEN Modifying the Ionosphere with Intense Radio Waves, *Science,* October, 1971, Vol. 174, pp. 245-254.

175. CCIR The following reports and recommendations are contained in Vols. I, II (Pts. 1 and 2) and Vol. III of "C.C.I.R. XII: The Plenary Assembly," 1970, International Telecommunication Union, Geneva. (Texts indicated with an asterisk are published separately.) Some of these texts have been updated in July 1974 and revised volumes are to appear in 1975. In general, the title and first three digits remain unchanged, with the -1, -2, etc. indicating a revision number.

Recommendations	Title	Volume and part
341	The Concept of Transmission Loss in Studies of Radio Systems	I
368-1	Ground-Wave Propagation Curves for Frequencies between 10 kHz and 10 MHz	II, 1
369-1	Definition of a Basic Reference Atmosphere	II, 1
370-1	VHF and UHF Propagation Curves for the Frequency Range from 30 MHz to 1000 MHz, Broadcasting and Mobile Services	II, 1
Reports:		
109-2	Radio Systems Employing Ionospheric Scatter Propagation	III
195	Signal-to-Noise Ratios in Complete System	III
229-1	Determination of the Electrical Characteristics of the Surface of the Earth	II, 1
230-1	Propagation over Inhomogeneous Earth	II, 1
231-2	Reference Atmospheres	II, 1
233-2	Influence of the Non-ionized Atmosphere on Wave Propagation; Ground-ground Propagation	II, 1
233-M and Corr. 1	From Chairman of Study Group 5: Tropospheric Propagation Data Relevant to the Planning of Space and Terrestrial Communication Services, IWP 5/1 and 5/2, Meeting at Nice, 30.11.70-4.12.70	
234-2	Influence of Tropospheric Refraction and Attenuation on Space Telecommunication Systems; Earth-Space Propagation	II, 1
238-1	Propagation Data Required for Transhorizon Radio-Relay Systems	II, 1
244-2	Estimation of Tropospheric-Wave Transmission Loss	II, 1
250-2	Long-Distance Ionospheric Propagation without Intermediate Ground Reflection	II, 2
251-1	Intermittent Communication by Meteor-Burst Propagation	II, 2
252-2	CCIR Interim Method for Estimating Sky-Wave Field Strength and Transmission Loss at Frequencies between the Approximate Limits of 2 and 30 MHz	*
259-2	VHF Propagation by Way of Sporadic E and Other Anomalous Ionization	II, 2
260-2	Ionospheric-Scatter Propagation	II, 2
263-2	Ionospheric Effects upon Earth-Space Radio Propagation	II, 2
264-2	Sky-Wave Propagation Curves between 300 km and 3500 km at Frequencies between 150 kHz and 1600 kHz in the European Broadcasting Area	II, 2
265-2	Sky-Wave Propagation at Frequencies below About 150 kHz with Particular Emphasis on Ionospheric Effects	II, 2
266-2	Fading of Radio Signals Received via the Ionosphere (see Rev. 1972)	II, 2
322	World Distribution and Characteristics of Atmospheric Radio Noise	*
327-1	Diversity Reception	I

Recommendations	Title	Volume and part
338-1	Propagation Data Required for Line-of-Sight Radio-Relay Systems	II, 1
340-1	CCIR Atlas of Ionospheric Characteristics	*
357	Operational Ionospheric Sounding at Oblique Incidence	III
415	Models of Phase-Interference Fading for Use in Connection with Studies of the Efficient Use of the Radio-Frequency Spectrum	*
424	VHF, UHF and SHF Propagation Curves for the Aeronautical Mobile Service	II, 1
425	Estimation of Tropospheric-Wave Transmission Loss; Availability of Computer Methods and Preparation of Propagation Curves for Broadcast and Mobile Services	II, 1
429	Ground and Ionospheric Side-Scatter	II, 2
SJM	Special Joint Meeting of CCIR Study Groups to Prepare Technical Bases for the World Administrative Radio Conference for Space Telecommunications, Pt I and II, 1971.	*

SECTION 19

SOUND REPRODUCTION AND RECORDING SYSTEMS

BY

DANIEL W. MARTIN Research Director, Engineering Research Department, D. H. Baldwin Company; Fellow, IEEE

CONTENTS

Numbers refer to paragraphs

SOUND REPRODUCTION AND RECORDING SYSTEMS

SECTION 19

SOUND REPRODUCTION AND RECORDING SYSTEMS
By Daniel W. Martin

STANDARD UNITS FOR SOUND SPECIFICATION[1,2*]

1. Sound Pressure. Airborne sound waves are a physical disturbance pattern in the air, an elastic medium, traveling through the air at a speed which depends somewhat upon air temperature (but not upon static air pressure). The instantaneous magnitude of the wave at a specific point in space and time can be expressed in various ways, e.g., displacement, particle velocity, pressure. However, the most widely used and measured property of sound waves is *sound pressure,* the fluctuation above and below atmospheric pressure which results from the wave.

An *atmosphere* of pressure is typically about 1 million dyn/cm^2, sometimes called a *bar.* Sound pressure is usually a very small part of atmospheric pressure. The unit of *sound* pressure is 1 dyn/cm^2, or 1 *microbar.*

2. Sound-Pressure Level. Sound pressures important to electronics engineering range from the weakest noises which can interfere with sound recording to the strongest sounds a loudspeaker diaphragm should be expected to radiate. This range is approximately 10^6. Consequently, for convenience, sound pressures are commonly plotted on a logarithmic scale called *sound-pressure level* expressed in *decibels* (dB).

The decibel, a unit widely used for other purposes in electronics engineering, originated in audio engineering (in telephony), and is named for Alexander Graham Bell. Because it is logarithmic, it requires a reference value for comparison, just as it does in other branches of electronics engineering. The reference pressure for sounds in air, corresponding to 0 dB, has been defined as a sound pressure of 0.0002 microbar. This is the reference sound pressure p_0 used throughout this section of the handbook. Thus the sound pressure level L_p in decibels corresponding to a sound pressure p is defined by

$$L_p = 20 \log (p/p_0) \qquad dB \qquad (19\text{-}1)$$

The reference pressure p_0 approximates the weakest audible sound pressure at 1,000 Hz. Consequently most decibel values for sound levels are positive in sign. Figure 19-1 relates sound pressure level in decibels to sound pressure in microbars.

Sound power and sound intensity (power flow per unit area of wavefront) are generally proportional to the square of the sound pressure. Doubling the sound pressure quadruples the intensity in the sound field, requiring 4 times the power from the sound source.

3. Audible Frequency Range. The international abbreviation Hz (hertz) is now used (instead of the former cps) for audible frequencies as well as the rest of the frequency domain. The limits of audible frequency are only approximate, because tactile sensations below 20 Hz overlap aural sensations above this lower limit. Moreover only young listeners can hear pure sounds near or above 20 kHz, the nominal upper limit.

Frequencies beyond both limits, however, have significance to audio-electronics engineers. For example, near-infrasonic (below 20 Hz) sounds are needed for classical organ music but can be noise in turntable rumble. Near-ultrasonic (above 20 kHz) intermodulation in audio circuits can produce undesirable difference-frequency components which are audible.

The audible sound-pressure level range can be combined with the audible frequency range to describe an *auditory area,* shown in Fig. 19-2. The lowest curve shows the weakest audible sound-pressure level for listening with both ears to a pure tone while facing the sound source

---

*Superior numbers refer to References, Par. **19-88**.

in a free field. The minimum level depends greatly upon the frequency of the sound. It also varies somewhat among listeners. The levels which quickly produce discomfort or pain for listeners are only approximate, as indicated by the shaded and crosshatched areas of Fig. 19-2. Extended exposure can produce temporary (or permanent) loss of auditory area at sound pressure levels as low as 95 dB.

Wavelength effects are of great importance in the design of sound systems and rooms because wavelength varies over a 3-decade range, much wider than is typical elsewhere in electronics engineering. The inverse relation of wavelength to audio frequency is shown in Fig. 19-3 for a sound speed of 344 m/s (1,127 ft/s) at 20°C (68°F). Audible sound waves vary in length from 1 cm to 15 m. The dimensions of the sound sources and receivers used in electroacoustics also vary greatly, e.g., from 1 cm to 3 m.

Sound waves follow the principles of geometrical optics and acoustics when the wavelength is very small relative to object size and pass completely around obstacles much smaller than a wavelength. This wide range of physical effects complicates the typical practical problem of sound production or reproduction.

4. Loudness Level. The simple, direct method for determining experimentally the loudness *level* of a

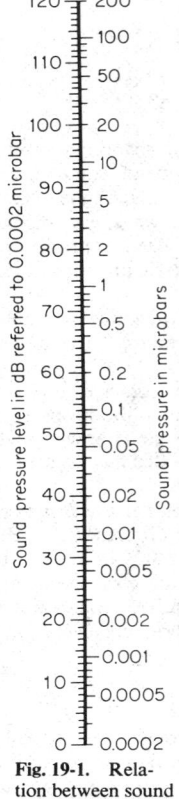

Fig. 19-1. Relation between sound pressure and sound-pressure level. (Ref. 2)

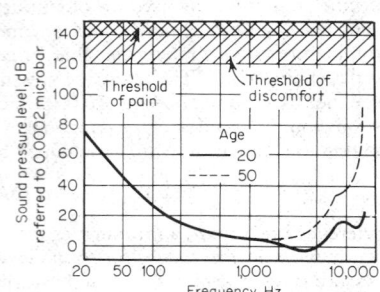

Fig. 19-2. The auditory area. (Robinson and Dadson, Ref. 3)

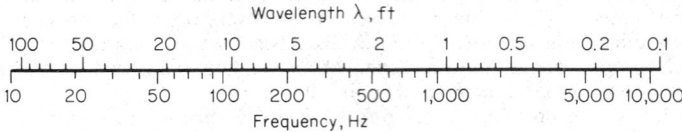

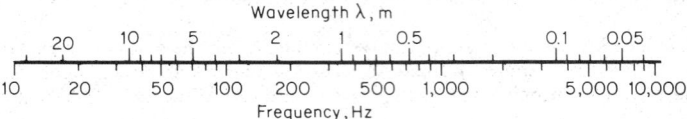

Fig. 19-3. Wavelength vs. frequency in air at 20°C. (Ref. 2)

sound is to match its observed loudness with that of a 1,000-Hz sine-wave reference tone of calibrated, variable sound-pressure level. (Usually this is a group judgment, or an average of individual judgments, in order to overcome individual observer differences.)

When the two loudnesses are matched, the *loudness level* of the sound, expressed in *phons,* is defined as numerically equal to the sound-pressure level of the reference tone in decibels. For example, a series of observers, each listening alternately to a machine noise and to a 1,000-Hz reference tone, judge them (on the average) to be equally loud when the reference tone is adjusted to 86 dB at the observer location. This makes the loudness level of the machine noise 86 phons.

At 1,000 Hz the decibel and phon levels are numerically identical, by definition. However, at other frequencies sine-wave tones may have numerically quite different sound- and loudness-levels, as seen in Fig. 19-4. The dashed contour curves show the decibel level at each frequency corresponding to the loudness level identifying the curve at 1,000 Hz.[4] For example, a tone at 80 Hz and 70 dB lies on the contour marked 60 phons. Its sound level must be 70 dB for it to be as loud as a 60-dB tone at 1,000 Hz. Such differences at low frequencies, especially at low sound levels, are a characteristic of the sense of hearing. The fluctuations above 1,000 Hz are caused by sound-wave diffraction around the head of the listener and resonances in his ear canal. This illustrates how man's physiological and psychological characteristics complicate the application of purely physical concepts.

Since loudness level is related to 1,000-Hz tones defined physically in magnitude, the loudness-level scale is not really psychologically based. Consequently, although one can say that 70 phons is louder than 60 phons, one cannot say *how much* louder.

5. Loudness. By using the phon scale to overcome the effects of frequency, psychophysicists have developed a true loudness scale based upon numerous

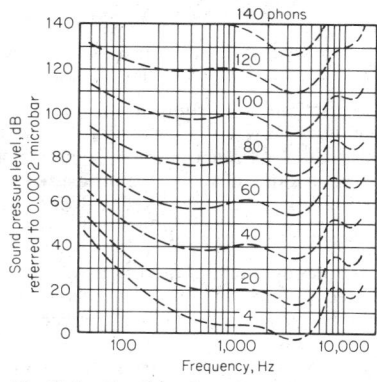

Fig. 19-4. Equal-loudness-level contours. (Robinson and Dadson, Ref. 3)

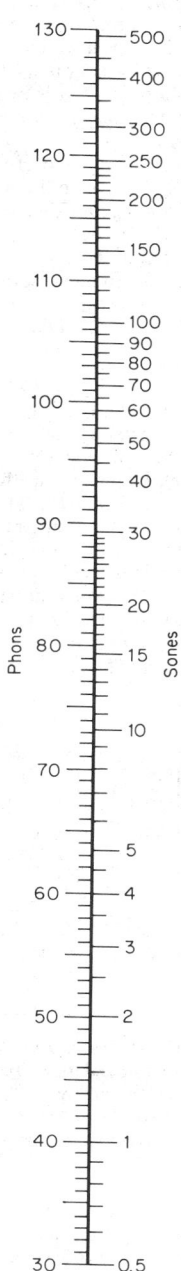

Fig. 19-5. Relation between loudness in sones and loudness level in phons. (Ref. 6)

experimental procedures involving relative-loudness judgments. *Loudness,* measured in *sones,* has a direct relation to loudness level in phons which is approximated in Fig. 19-5.[5] (Below 30 phons the relation changes slope. Since few practical problems require that range, it is omitted for simplicity.) A loudness of 1 sone has been defined equivalent to a loudness level of 40 phons. It is evident in Fig. 19-5 that a 10-phon change doubles the loudness in sones, which means *twice as loud.* Thus a 20-phon change in loudness level quadruples the loudness.

Another advantage of the sone scale is that the loudness of components of a complex sound are additive on the sone scale as long as they are well separated on the frequency scale. For example (using Fig. 19-5), two tonal components at 100 and 4,000 Hz having loudness levels of 70 and 60 phons, respectively, would have individual loudnesses of 8 and 4 sones, respectively, and a total loudness of 12 sones.

Detailed loudness computation procedures have been developed for highly complex sounds and noises, deriving the loudness in sones directly from a complete knowledge of the decibel levels for individual discrete components or noise bands.[6,7] The procedures continue to be refined.[8]

TYPICAL FORMATS FOR SOUND DATA

Sound and audio electronic data are frequently plotted as functions of frequency, time, direction, distance, or room volume. Frequency characteristics are the most common, in which the ordinate may be sound pressure, sound power, output-input ratio, percent distortion, or their logarithmic-scale (level) equivalents.

6. Sound Spectra. The frequency spectrum of a sound is a description of its resolution into components of different frequency and amplitude. Often the abscissa is a logarithmic frequency scale or a scale of octave (or fractional-octave) bands with each point plotted at the geometric mean of its band-limiting frequencies. Usually the ordinate scale is sound-pressure level. Phase differences are often ignored (except as they affect sound level) because they vary so greatly with measurement location, especially in reflective environments.

Line spectra are bar graphs for sounds dominated by discrete frequency components. Figure 19-6 is an example.

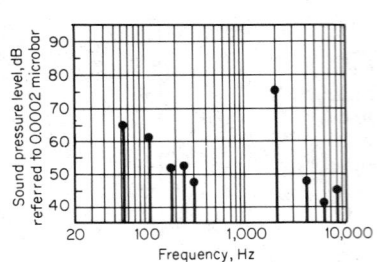

Fig. 19-6. Typical line spectrum. (Ref. 39)

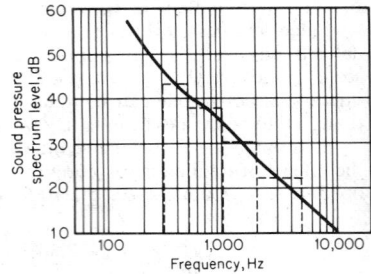

Fig. 19-7. Continuous-spectrum level curve for a motor and blower. (Ref. 2)

Continuous spectra are curves showing the distribution of sound-pressure level within a frequency range densely packed with components. Figure 19-7 is an example. Unless stated otherwise, the ordinate of a continuous-spectrum curve, called *spectrum level,* is assumed to represent sound-pressure level for a band of 1-Hz width. Usually level measurements L_{band} are made in wider bands, then converted to spectrum level L_{ps} by the bandwidth correction

$$L_{ps} = L_{band} - 10 \log_{10} (f_2 - f_1) \qquad dB \qquad (19\text{-}2)$$

in which f_1 and f_2 are the lower- and upper-frequency limits of the band.

When a continuous-spectrum curve is plotted automatically by a level recorder synchronized with a heterodyning filter or with a sequentially switched set of narrow-bandpass filters, any effect of *changing* bandwidth upon curve slope must be considered.

Combination spectra are appropriate for many sounds in which strong line components are superimposed over more diffuse continuous spectral backgrounds. Bowed or blown musical tones and motor-driven fan noises are examples.

Octave spectra, in which the ordinate is the sound-pressure level for bands one octave wide, are very convenient for measurement and for specifications but lack fine spectrum detail. Figure 19-8 is an example.

Third-octave spectra provide more detail and are widely used. Figure 19-9 shows two radically different noise spectra plotted in third-octaves. One-third of an octave and one-tenth of a decade are so nearly identical that substituting the latter for the former is a practical convenience, providing a 10-band pattern which repeats every decade. Placing third-octave

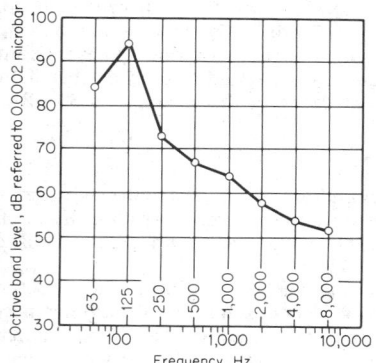

Fig. 19-8. Noise measurement by octave bands. (Ref. 2)

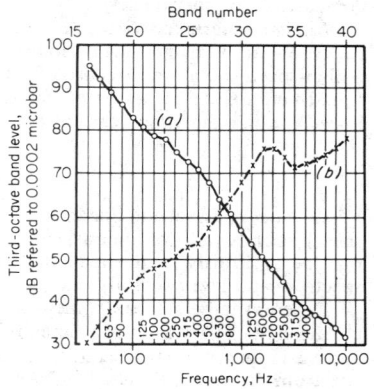

Fig. 19-9. Third-octave band spectra for two noise sources. (Ref. 2)

band zero at 1 Hz has conveniently made the band numbers equal 10 times the $\log_{10}$ of the band-center frequency; e.g., band 20 is at 100 Hz and band 30 at 1,000 Hz.

Visual proportions of spectra (and other frequency characteristics) depend upon the ratio of ordinate and abscissa scales. There is no universal or fully standard practice, but for ease of visual comparison of data and of specifications, it has become rather common practice in the United States for 30 dB of ordinate scale to equal (or slightly exceed) 1 decade of logarithmic frequency on the abscissa. Available audio and acoustical graph papers and automatic level-recorder charts have reinforced this practice. When the entire 120-dB range of auditory area is to be included in the graph, the ordinate is often compressed 2:1.

7. Response and Distortion Characteristics. Output-input ratios vs. frequency are the most common data format in audio-electronics engineering. The audio-frequency scale (20 Hz to 20 kHz) is usually logarithmic. The ordinate may be sound- or electrical-output level in decibels as the frequency changes with a constant electrical or sound input; or it may be a ratio of the output to input (expressed in decibels) as long as they are linearly related within the range of measurement.

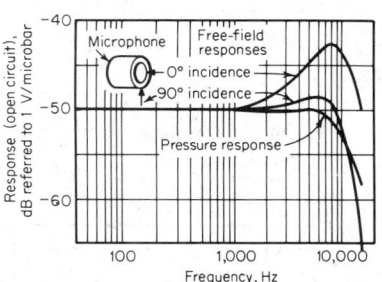

Fig. 19-10. Response vs. frequency of a Western Electric Type 640-AA condenser microphone, showing the behavior for sounds arriving along the axis and perpendicular to the axis of the microphone, as well as the pressure response. (Ref. 38)

Visual proportions (Par. 19-6) are quite important in the publication of such data. Figure 19-10 is an important (but untypical) microphone example. In this case the ordinate scale has been expanded more than usual because the response variations are so small. It is necessary for the nonuniformity to be presented accurately here, because this type of microphone has long been a laboratory and industry standard for measurement and comparison.

When the response-frequency characteristic is measured with the input frequency filtered from the output, a distortion-frequency characteristic is the result. It may be further filtered to obtain curves for each harmonic if desired.

8. Directional Characteristics. Sound sources radiate almost equally in all directions when the wavelength is large compared to source dimensions. At higher frequencies, where the wavelength is smaller than the source, the radiation becomes quite directional. Figure 19-11, an example from musical sound, shows a typical format for plotting sound level vs. angle at different frequencies. On-axis sound level is chosen as a 0-dB reference level for comparison to other angles.

9. Time Characteristics. Any sound property can vary with time. It can build up, decay, or vary in magnitude periodically or randomly. Figure 19-12 is an example showing a high-speed sound-level record of the reverberant sound decay in a room. A reverberant sound field decays rather logarithmically. Consequently the sound level in decibels falls linearly when the time scale is linear. The rate of decay in this example is 33 dB/s.

SPEECH SOUNDS

10. Speech Level and Spectrum.[10,11] Both the sound-pressure level and the spectrum of speech sounds vary continuously and rapidly during connected discourse. Although speech may be arbitrarily segmented into elements called phonemes, each with a characteristic spectrum and level, actually one phoneme blends into another.

Different talkers speak somewhat differently, and they sound different. Their speech characteristics vary from one time or mood to another. Yet in spite of all these differences and variations, statistical studies of speech have established a typical "idealized" speech spectrum shown in Fig. 19-13. The spectrum level rises about 5 dB from 100 to 600 Hz, then falls about 6, 9, 12, and 15 dB in succeeding higher octaves.

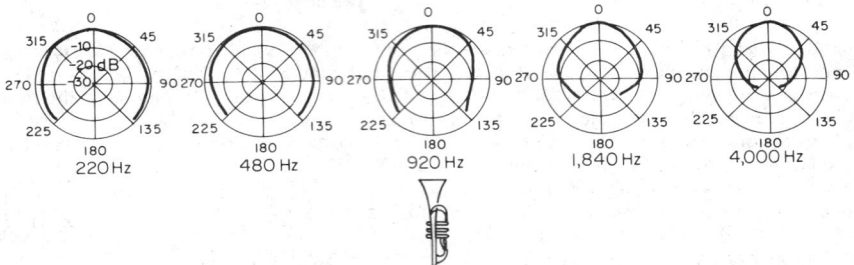

Fig. 19-11. Directional characteristics of a trumpet for five frequencies. (Martin, Ref. 9)

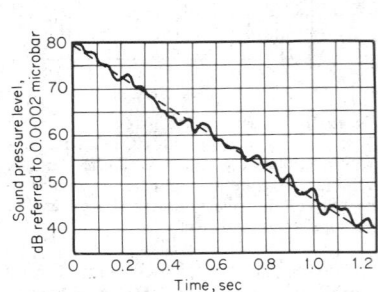

Fig. 19-12. A high-speed level record showing how the sound-pressure level in a room decays with time. Since the sound decays 40 dB in 1.2 s, the reverberation time, i.e., the time to decay 60 dB, is 1.8 s. (Ref. 36)

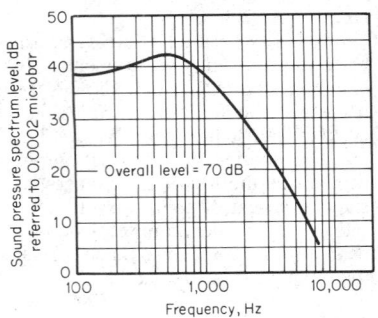

Fig. 19-13. Idealized speech spectrum for male voices at 1 m from the talker's lips. (After Dunn and White, Bell Telephone Laboratories, Ref. 10)

Overall sound-pressure levels, averaged over time and measured at a distance of 1 m from a talker on or near the speech axis, lie in the range of 65 and 75 dB when the talkers are instructed to speak in a "normal" tone of voice. Along this axis the speech sound level follows the inverse-square law closely to within about 10 cm of the lips, where the level is about 90 dB. At the lips, where communication microphones are often used, the overall speech sound level typically averages over 100 dB.

The peak levels of speech sounds greatly exceed the long-time average level. Figure 19-14 shows the difference between short peak levels and average levels at different frequencies in the speech spectrum. The difference is greater at high frequencies, where the sibilant sounds of relatively short duration have spectrum peaks.

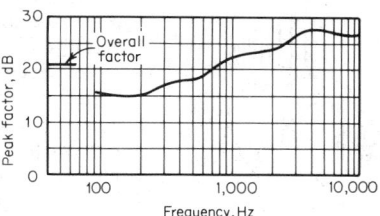

Fig. 19-14. Difference in decibels between peak pressures of speech measured in short ($\frac{1}{8}$-s) intervals and the rms pressure averaged over a long (75-s) interval. (After Dunn and White, Ref. 10)

11. Speech Directional Characteristics.[11] Speech sounds are very directional at high frequencies. Figures 19-15 and 19-16 show clearly why speech is poorly received behind a talker, especially in nonreflective environment. Above 4,000 Hz the directional loss in level is 20 dB or more, which particularly affects the sibilant sound levels so important to speech intelligibility.

12. Vowel Spectra.[12] Different vowel sounds are formed from approximately the same basic laryngeal tone spectrum by shaping the vocal tract (throat, back of mouth, mouth, and lips) to have different acoustical resonance-frequency combinations. Figure 19-17 illustrates the spectrum filtering process. The spectral peaks are called *formants,* and their frequencies are known as formant frequencies.

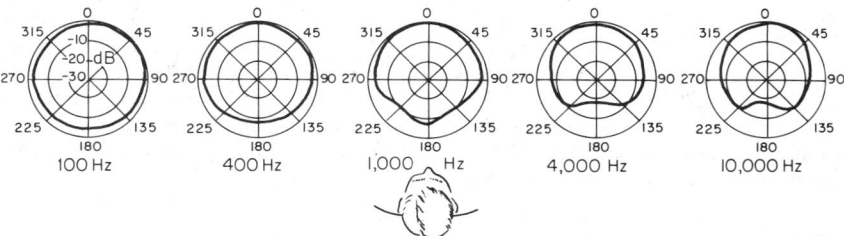

Fig. 19-15. The directional characteristics of the human voice in a horizontal plane passing through the mouth. (Ref. 22)

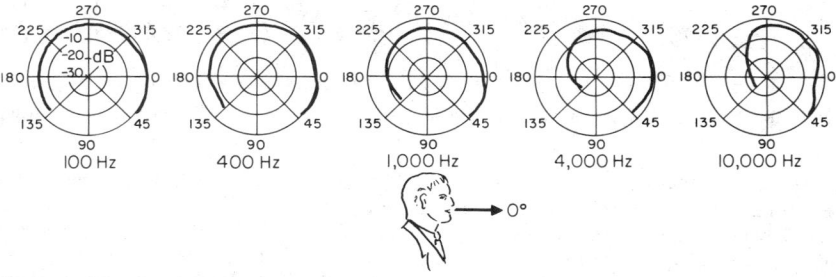

Fig. 19-16. The directional characteristics of the human voice in a bilaterally symmetrical vertical plane passing through the mouth. (Ref. 22)

Detailed vowel analyses show three or even more formants. In Fig. 19-18 the abscissa is time, the ordinate is frequency, and the darkness of shading is proportional to sound level. The dark bars are formant peaks for different vowel sounds. In diphthongs the vowels and their formants are in transition.

The shapes of the vocal tract, simplified models, and the acoustical results for three vowel sounds are shown [13] in Fig. 19-19. A convenient graphical method [14] for describing the combined formant patterns is shown in Fig. 19-20. Traveling around this vowel loop involves progressive motion of the jaws, tongue, and lips.

13. Speech Intelligibility. [15,16] More intelligibility is contained in the central part of the speech spectrum than near the ends. Figure 19-21 shows the effect upon articulation (the percent of syllables correctly heard) when low-pass and high-pass filters of various cutoff frequencies are used. From this information a special frequency scale has been developed in which each of 20 frequency bands contributes 5% to a total *articulation index* of 100%. This distorted frequency scale is used in Fig. 19-22. Also shown are the spectrum curves for speech peaks and for speech minima, lying approximately 12 and 18 dB, respectively, above and below the average-speech-spectrum curve. When all the shaded area (30-dB range between the maximum and minimum curves) lies above threshold and below overload, in the absence of noise, the articulation index is 100%.

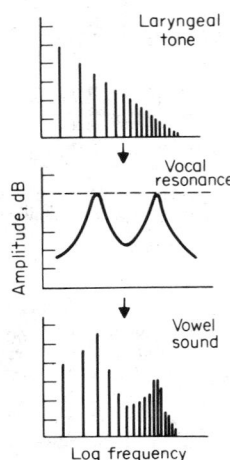

Fig. 19-17. Effects upon the spectrum of the laryngeal tone produced by the resonances of the vocal tract. (From "Language and Communication," McGraw-Hill Book Company, Inc., New York, 1951.)

If a noise-spectrum curve were added to Fig. 19-22, the figure would become an articulation-index computation chart for predicting communication capability. For example, if the ambient-noise spectrum coincided with the average-speech-spectrum curve, that is, the signal-to-noise ratio is 1, only twelve-thirtieths of the shaded area would lie above the noise. The articulation index would be reduced accordingly to 40%.

Figure 19-23 relates monosyllabic word articulation and sentence intelligibility to articulation index. In the example above, for an articulation index of 0.40 approximately 70% of monosyllabic words and 96% of sentences would be correctly received.

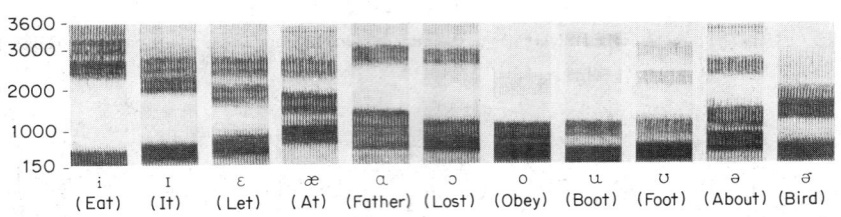

Fig. 19-18. Spectrograms of vowels from a male voice. (Courtesy of Bell Telephone Laboratories.)

However, if the signal-to-noise ratio were kept at unity and the frequency range were reduced to 1,000 to 3,000 Hz, half of the bands would be lost. Articulation index would drop to 0.20, word articulation to 0.30, and sentence intelligibility to 70%. This shows the necessity for wide frequency range in a communication system when the signal-to-noise ratio is marginal. Conversely a good signal-to-noise ratio is required when the frequency range is limited.

The articulation-index method is particularly valuable in complex intercommunication-

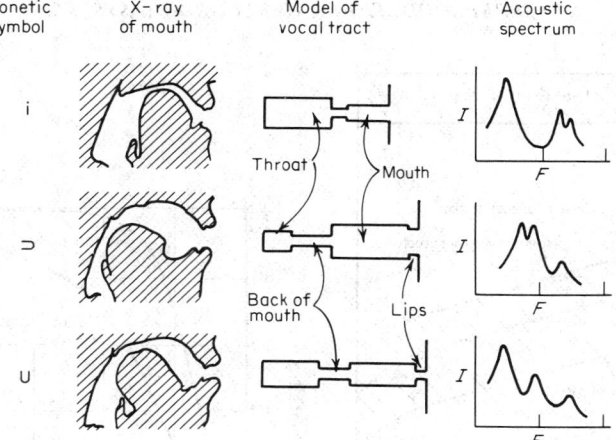

Phonetic symbol	X-ray of mouth	Model of vocal tract	Acoustic spectrum

Fig. 19-19. Phonetic symbols, shapes of vocal tract, models, and acoustic spectra for three vowels. (Ref. 13)

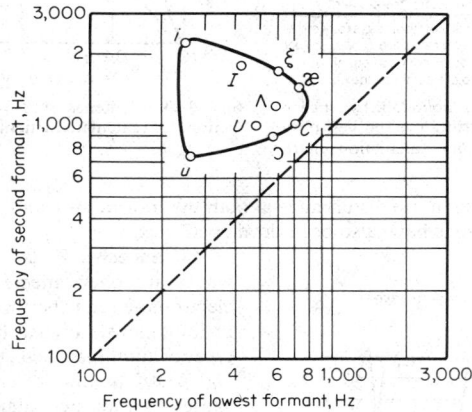

Fig. 19-20. The center frequencies of the first two formants for the sustained English vowels plotted to show the characteristic differences. (Ref. 14)

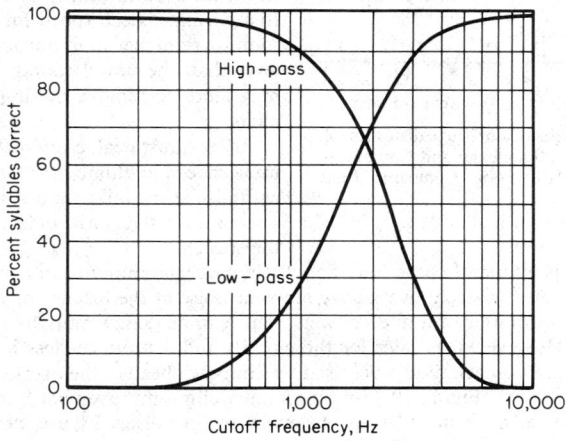

Fig. 19-21. Syllable articulation score vs. low-pass or high-pass filter cutoff frequency. (After French and Steinberg, Ref. 15)

19-11

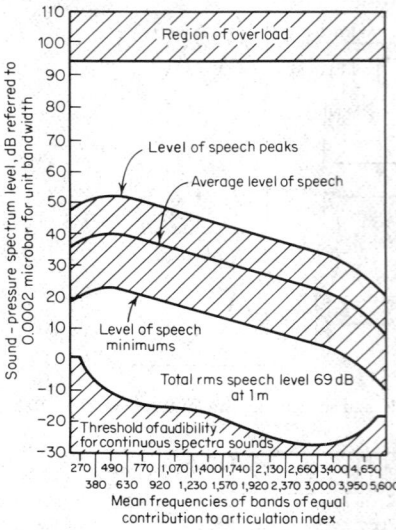

Fig. 19-22. Speech area, bounded by speech peak and minimum spectrum-level curves, plotted on an articulation-index calculation chart. (Ref. 62)

Fig. 19-23. Sentence- and word-intelligibility prediction from calculated articulation index. (Ref. 17)

system designs involving noise disturbance at both the transmitting and receiving stations.[17] Simpler effective methods have also been developed.[18]

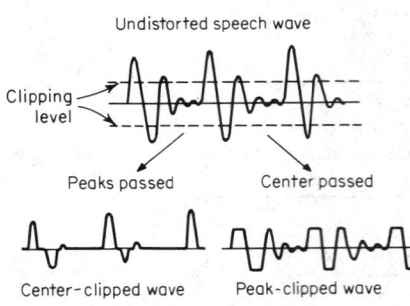

Fig. 19-24. Two types of amplitude distortion of speech waveform. ("Language and Communication," McGraw-Hill Book Company, Inc., New York, 1951.)

14. Speech Peak Clipping. Speech waves are often affected inadvertently by electronic-circuit performance deficiencies or limitations. Figure 19-24 illustrates two types of amplitude distortion, center clipping and peak clipping. Center clipping, often caused by improper balancing or biasing of a push-pull amplifier circuit, can greatly interfere with speech quality and intelligibility. In a normal speech spectrum the consonant sounds are higher in frequency and lower in level than the vowel sounds. Center clipping tends to remove the important consonants.

By contrast peak clipping has little effect upon speech intelligibility [19] as long as ambient noise at the talker and system electronic noise are relatively low in level compared to the speech.

Peak clipping is frequently used intentionally in speech-communication systems to raise the average transmitted speech level above ambient noise at the listener or to increase the range of a radio transmitter of limited power. This can be done simply by overloading an amplifier stage. However, it is safer for the circuits and it produces less intermodulation distortion when back-to-back diodes are used for clipping ahead of the overload point in the amplifier or transmitter. Figure 19-25 shows intelligibility improvement from speech peak clipping when the talker is in quiet and listeners are in noise. Figure 19-26 shows that caution is necessary when the talker is in noise, unless his microphone is shielded or is a noise-canceling type.[17]

Tilting the speech spectrum by differentiation [19] and flattening it by equalization [20] are effective preemphasis treatments before peak clipping. Both methods put the consonant and vowel sounds into a more balanced relationship before the intermodulation effects of clipping affect voiced consonants.

Caution must be used in combining different forms of speech-wave distortion, which individually have innocuous effects upon intelligibility but can be devastating when they are combined.[21]

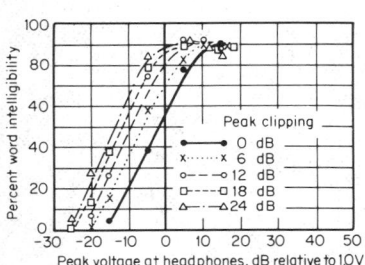

Fig. 19-25. Advantages of peak clipping of noise-free speech waves, heard by listeners in ambient aircraft noise. (After Kryter, et al., Ref. 17)

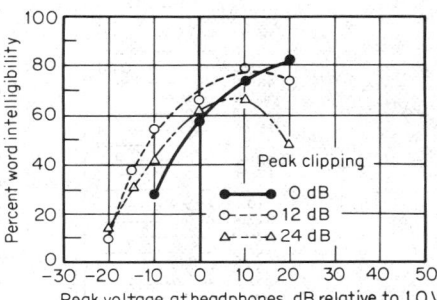

Fig. 19-26. Effects of speech peak clipping with both the talker and the listeners in simulated aircraft noise. Note that excessive peak clipping is detrimental. (After Kryter, et al., Ref. 17)

MUSICAL SOUNDS[22]

15. Musical Frequencies. The accuracy of both absolute and relative frequencies is usually much more important for musical sounds than for speech sounds and noise. The international frequency standard for music is defined at 440.00 Hz for A_4, the A above C_4 (middle C) on the musical keyboard. In sound recording and reproduction the disk-rotation and tape-transport speeds must be held correct within 0.2 or 0.3% error (including both recording and playback mechanisms) to be fully satisfactory to musicians.

The mathematical musical scale is based upon an exact octave ratio of 2:1. The subjective octave slightly exceeds this,[23] and piano tuning sounds better when the scale is stretched very slightly.[24]

The equally tempered scale of 12 equal ratios within each octave is an excellent compromise between the different historical scales based upon harmonic ratios.[25] It has become the standard of reference, even for individual musical performances which may deviate from it for artistic or other reasons. Figure 19-27 shows the fundamental frequencies corresponding to ANSI standard[1] musical notation and locations on the musical staff, omitting sharps and flats. The bottom octave is chiefly for the lowest (32-ft) organ pedal tones, and the upper two octaves for the fundamental frequencies of the highest organ stops and mixtures.

Different musical instruments play over different ranges of *fundamental* frequency, shown in Fig. 19-28. However, most musical sounds have many harmonics which are audibly significant to their tone spectra. Consequently high-fidelity recording and reproduction need a much wider frequency range, given in Fig. 19-29. When the musical frequency range must be restricted in recording and reproduction, the result can be estimated from Fig. 19-30.

16. Sound Levels of Musical Instruments. The sound level from a musical instrument varies with the type of instrument, the distance from it, which note in the scale is being played, the dynamic marking in the printed music, the player's ability, and (on polyphonic instruments) the number of notes (and stops) played at the same time.

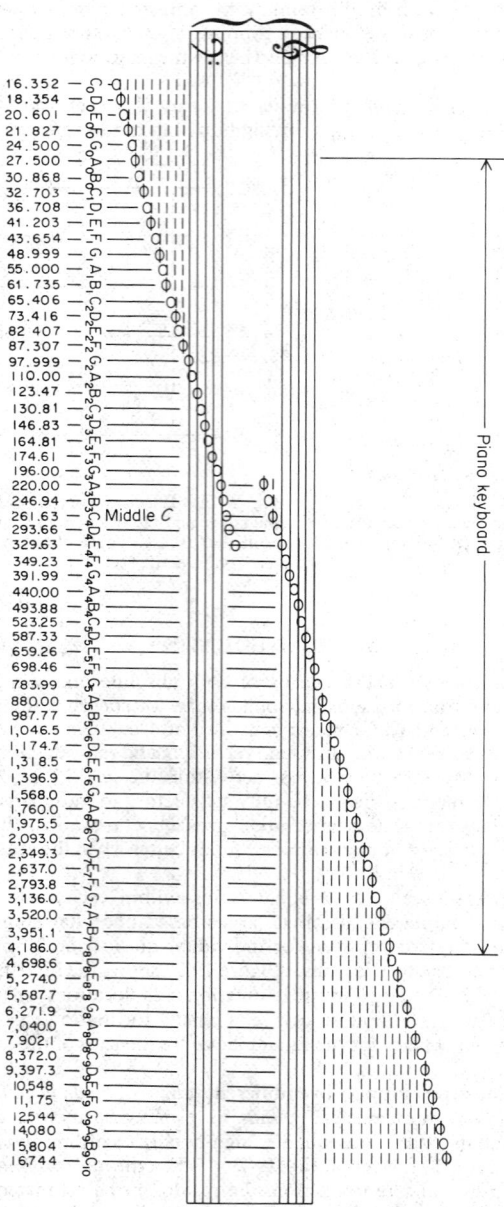

Fig. 19-27.   Equally tempered scale frequencies and musical notation.   (Ref. 22)

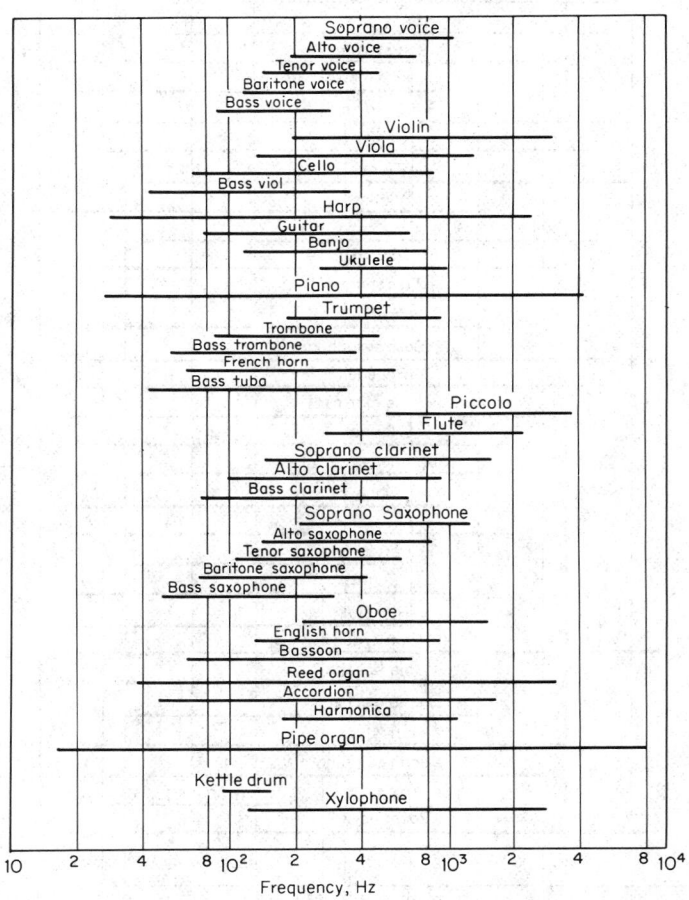

Fig. 19-28. Range of the fundamental frequencies of voices and various musical instruments. (After Olson, Ref. 31)

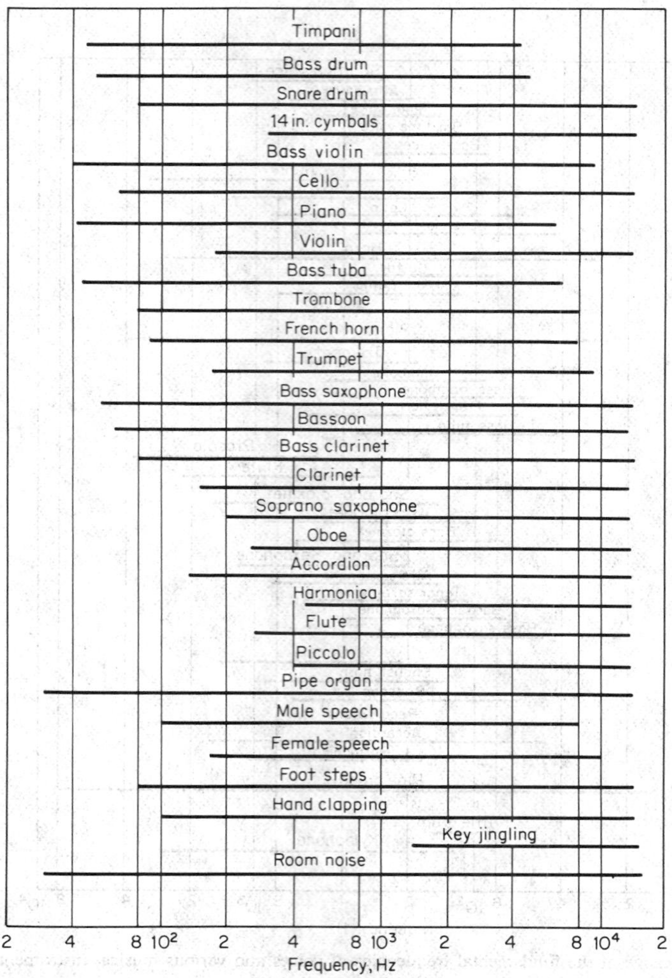

Fig. 19-29. Frequency ranges required for speech, musical instruments, and noises so that no frequency discrimination will be apparent. (After Snow, Ref. 26)

Orchestral Instruments. The following sound levels are typical at a distance of 10 ft in a nonreverberant room. Soft (pianissimo) playing of a weaker orchestral instrument, e.g., violin, flute, bassoon, produces a typical sound level of 55 to 60 dB. Fortissimo playing on the same instrument raises the level to about 70 to 75 dB. Louder instruments, e.g., trumpet or tuba, range from 75 dB at pianissimo to about 90 dB at fortissimo.[27]

Certain instruments have exceptional differences in sound level of low and high notes. A flute may change from 42 dB on a soft low note to 77 dB on a loud high note, a range of 35 dB. The French horn ranges from 43 dB (soft and low) to 93 dB (loud and high).[27]

Sound levels are about 10 dB higher at 3 ft (inverse-square law) and 20 dB higher at 1 ft. The louder instruments, e.g., brass, at closer distances may overload some microphones and preamplifiers.

Percussive Instruments. The sound levels of shock-excited tones are more difficult to specify because they vary so much during decay and can be excited over a very wide range. A bass drum may average over 100 dB during a loud passage with peaks (at 10 ft) approaching 120 dB. By contrast a triangle will average only 70 dB with 80-dB peaks. A single tone of a grand piano played forte will initially exceed 90 dB near the piano rim, 80 dB at the pianist, and 70 dB at the conductor 10 to 15 ft away. Large chords and rapid arpeggios will raise the level about 10 dB.

Instrumental Groups. Orchestras, bands, and polyphonic instruments produce higher sound levels since many notes and instruments (or stops) are played together. Their sound levels are specified at larger distances than 10 ft because the sound sources occupy a large area; 20 ft from the front of a 75-piece orchestra the sound level will average about 85 to 90 dB with peaks of 105 to 110 dB. A full concert band will go higher. At a similar distance from the sound sources of an organ (pipe or electronic) the full-organ (or crescendo-pedal) condition will produce a level of 95 to 100 dB. By contrast the softest stop with expression shutters closed may be 45 dB or less.

17. Growth and Decay of Musical Sounds. These characteristics are quite different for different instruments. Examples are given in Fig. 19-31. Piano or guitar tones quickly rise to an initial maximum, then gradually diminish until the strings are damped mechanically. Piano tones have a more rapid decay initially than later in the sustained tone.[28] Orchestral instruments can start suddenly or smoothly, depending upon the musician's technique, and they damp rather quickly when playing ceases. Room reverberation affects both growth and decay rates when the time constants of the room are greater than those of the instrument vibrators. This is an important factor in organ music, which is typically played in reverberant environment.

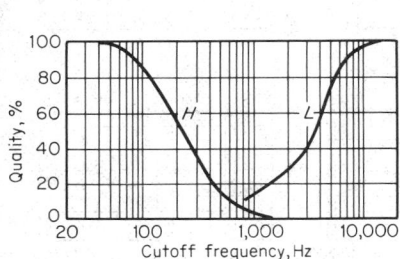

Fig. 19-30. Orchestral music-quality judgment vs. high-pass (*H*) and low-pass (*L*) filter cutoff frequencies. (Ref. 26)

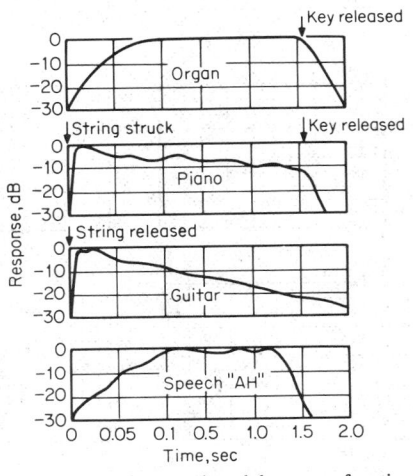

Fig. 19-31. The growth and decay as a function of the time of an organ, piano, and guitar tone and the speech sound "ah." (Ref. 22)

Many types of musical tone have characteristic transients which influence timbre greatly. In the "chiff" of organ tone the transients are of different fundamental frequency. They appear and decay before steady state is reached. In percussive tones the initial transient is the cause of the tone (often a percussive noise), and the final transient is the result.

These transient effects should be considered in the design of audio electronics such as "squelch," automatic gain control, compressor, and background-noise reduction circuits.

18. Spectra of Musical Instrument Tones.[29]   Figure 19-32 displays time-averaged spectra for a 75-piece orchestra, a theater pipe organ, a piano, and a variety of orchestral instruments, including members of the brass, string, woodwind, and percussion families.

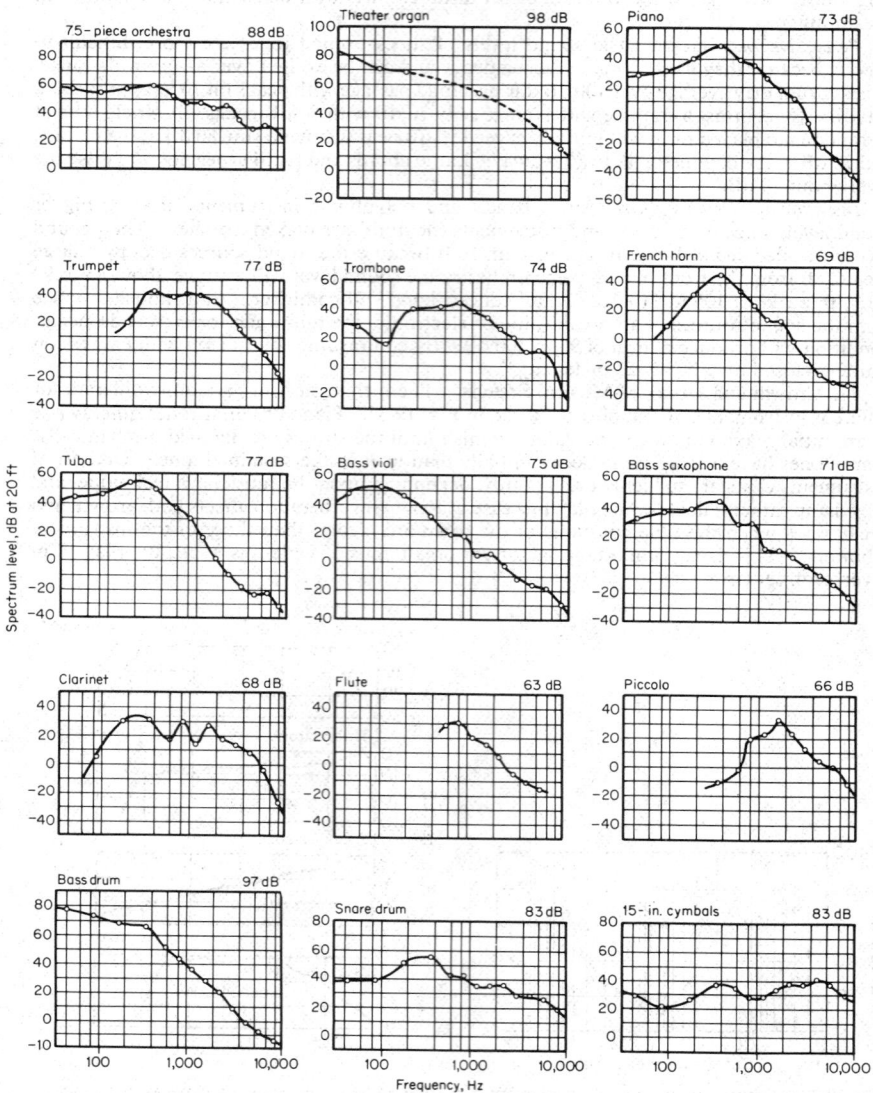

Fig. 19-32.   Time-averaged spectra of musical instruments.   (After Sivian, et al., Ref. 29)

19-18

These vary from one note to another in the scale, from one instant to another within a single tone or chord, and from one instrument or performer to another. For example, a concert organ voiced in a baroque style would have lower spectrum levels at low frequencies and higher at high frequencies than the theater organ shown.

The organ and bass drum have the most prominent low-frequency output. The cymbal and snare drum are strongest at very high frequencies. The orchestra and most of the instruments have spectra which diminish gradually with increasing frequency, especially above 1,000 Hz. This is what has made it practical to preemphasize the high-frequency components, relative to those at low frequencies, in both disk and tape recording. However, instruments which differ from this spectral tendency, e.g., coloratura sopranos, piccolos, cymbals, create problems of intermodulation distortion and overload.

Spectral peaks occurring only occasionally, for example, 1% of the time, are often more important to sound recording and reproduction than the peaks in the average spectra of Fig. 19-32. The frequency ranges shown in Table 19-1 have been found[29] to have relatively large instantaneous peaks for the instruments listed.

Table 19-1. Frequency Band Containing Instantaneous Spectral Peaks

Band Limits Hz	Instruments
20–60	Theater organ
60–125	Bass drum, bass viol
125–250	Small bass drum
250–500	Snare drum, tuba, bass saxophone, French horn, clarinet, piano
500–1,000	Trumpet, flute
2,000–3,000	Trombone, piccolo
5,000–8,000	Triangle
8,000–12,000	Cymbal

19. Directional Characteristics of Musical Instruments. Most musical instruments are somewhat directional. Some are highly so, with well-defined symmetry, e.g., around the axis of a horn bell[9] (see Fig. 19-11). Other instruments are less directional because the sound source is smaller than the wavelength, e.g., clarinet, flute. The mechanical vibrating system of bowed string instruments is complex, operating differently in different frequency ranges, and resulting in extremely variable directivity (Fig. 19-33). This is significant for orchestral seating arrangements both in concert halls and recording studios.[30] The large size and complex shape of piano soundboards and bridges and the unsymmetrical shape of grand piano reflecting lids create a complex set of directional characteristics (Fig. 19-34) at high frequencies. Audiences and recording microphones are usually located opposite the inclined reflecting lid.

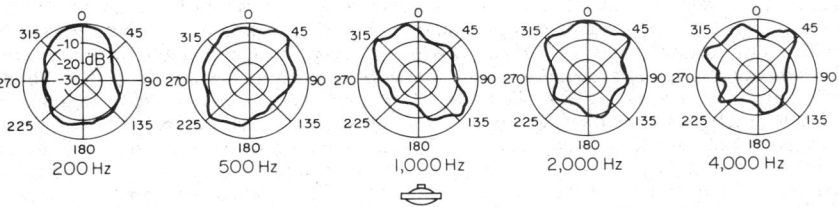

Fig. 19-33. The directional characteristics of a violin for five different frequencies. (Ref. 22)

20. Audible Distortions of Musical Sounds. The quality of musical sounds is more sensitive to distortion than the intelligibility of speech. A chief cause is that typical music contains several simultaneous tones of different fundamental frequency in contrast to typical speech sound of one voice at a time. Musical chords subjected to nonlinear amplification or transduction generate intermodulation components which appear elsewhere in the frequency spectrum.

Difference tones are more easily heard than summation tones because the summation tones are often hidden by harmonics which were already present in the undistorted spectrum and because auditory masking of a high-frequency pure tone by a lower-frequency pure tone is much greater than vice versa.

Factors affecting the threshold of audible distortion in music include the nature of the musical sounds, the frequency range of the reproducing system, the degree to which the listener can control the sounds, and the listener's familiarity with the sounds.

Figure 19-35 shows perceptible, tolerable, and objectionable amounts of controlled nonlinear distortion for five different system high-frequency cutoffs (shown in the figure) for both speech and music and for two types of distortion. (The pentode distortion type has a more extended harmonic series.) Both speech and music were live (recording and playback were not involved).[31]

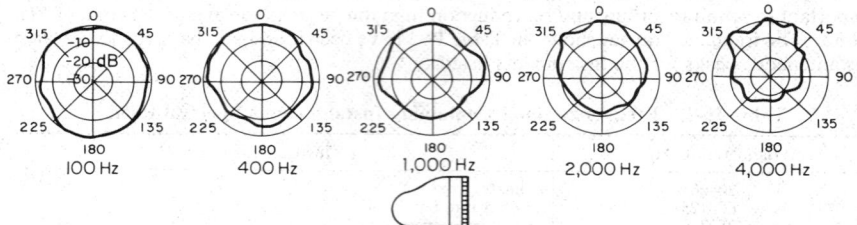

Fig. 19-34. The directional characteristics of a grand piano for five different frequencies. (Ref. 22)

When a critical listener controls the sounds heard (an organist playing an electronic organ on a high-quality amplification system) and has unlimited opportunity and time to listen, even lower distortion (0.2%, for example) can be perceived.[32]

Frequency-range division into multiple channels[33] is the successful method for preventing intermodulation distortion between widely separated frequencies (such as 100 and 4,000 Hz).[34] The method has been extended to frequencies which lie within the same octave,[35] using a larger number of dividing networks and as many as four loudspeaker channels.

AMBIENT NOISE AND ITS CONTROL[36,37]

21. Nature of Noise. In audio-electronics engineering, noise is unwanted sound or audio disturbance. Speech or music can be noise if they are cross talk between audio lines or are received from an adjacent recording studio. The term *noise* is also used sometimes when referring to the very weak spectrum components or to the low-level transients or to the second-order modulations which seem superficially to be in the background of the main signal or the natural sound but which in many cases turn out to be vital although subtle. The treatment here concerns audible-noise measurement, noise criteria for rooms and for hearing conservation, and the reduction and isolation of airborne and structure-borne noise.

22. Noise Measurement.[38,39] Figure 19-36 is a general system for noise measurement. The simplest system is a sound-level meter (with microphone) which indicates sound-pressure level as previously defined [Eq. (19-1)]. Its amplifier output can also be supplied to an oscillograph, a graphic level recorder, or a sound-spectrum analyzer, or all three. If the noises are to be stored for reference, a magnetic-tape recorder can be used. For later detailed study the tape can be played into the oscillograph, graphic-level recorder, or spectrum analyzer. This assumes that the recorder acts linearly, without automatic gain control, with wide frequency range and low internal-noise level, and is calibrated or has a calibrated reference sound recorded on the tape.

The actual shape of the microphone output wave can be seen on the oscillograph, assuming zero phase-shift in the sound-level-meter amplifier. Magnetic-tape recorder-playback phase shift must be considered when tape is used if phase is significant. Special instrumentation tape recorders are available.

An example involving an oscillograph and a graphic level recorder is shown in Fig. 19-37. The oscillograms on the left have corresponding level graphs on the right. In (a) the single-frequency logarithmic decrement gives a smooth, straight graph of sound-level decay. In (b) the two frequencies beat as shown in both the wavy envelope and the oscillating level

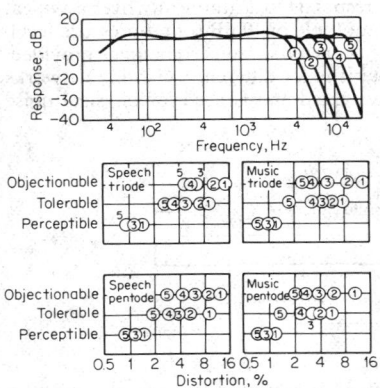

Fig. 19-35. Perceptible, tolerable, and objectionable amounts of controlled nonlinear harmonic distortion of speech and music for various high-frequency cutoffs. (Ref. 31)

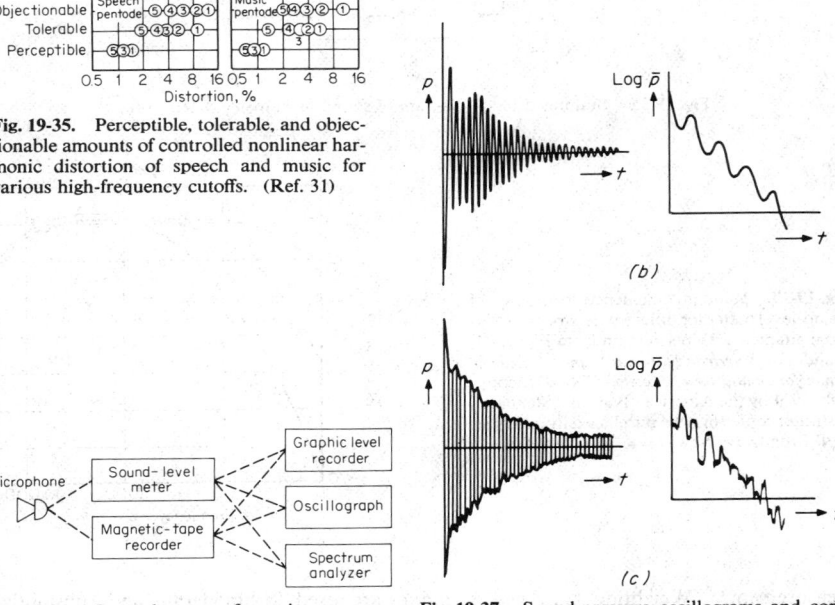

(a)

(b)

(c)

Fig. 19-36. General system for noise measurement. (Ref. 38)

Fig. 19-37. Sound-pressure oscillograms and corresponding graphs of sound-pressure level vs. time. (a) single frequency decay; (b) two adjacent frequencies with the same decay constant; (c) many closely spaced frequencies with the same decay constant. (Ref. 37)

decay. In *(c)*, which is more typical of room reverberation, numerous modes contribute a rather random aperiodic modulation to the wave envelope and graph.

23. Sound-Level Meter.   A simplified block diagram is shown in Fig. 19-38. The first attenuator prevents the microphone output from overloading the high-gain amplifier when the noise level is very high.   The weighting networks are inserted when the response-frequency characteristic is switched from uniform to predetermined standard curves (see Fig. 19-39).   The amplified output is available for oscillograph, etc., and is also rectified (full-wave) for the indicating meter (usually rms), which responds logarithmically over a typical range of -4 to 10 dB relative to the switch-selected multiple of 10 dB sound-pressure level. The meter response time is typically 0.5 s.   A choice of meter speed is sometimes provided.

The standard overall response-frequency characteristic for different weighting networks is shown in Fig. 19-39.   Curve *C* is essentially unweighted and is used for physical noise

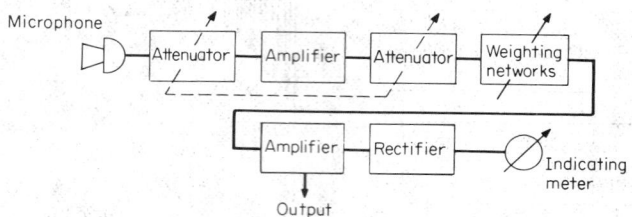

Fig. 19-38.   Simplified block diagram of sound-level meter.   (Ref. 38)

Fig. 19-39.   Random incidence response of sound-level meter for different networks. (Material adapted with permission from Figures 1, 2, and 3 of *American National Standard Specification for Sound Level Meters, S1.4-1971*, copyright 1971 by the American National Standards Institute; copies may be purchased from ANSI, 1430 Broadway, New York, N.Y. 10018.)

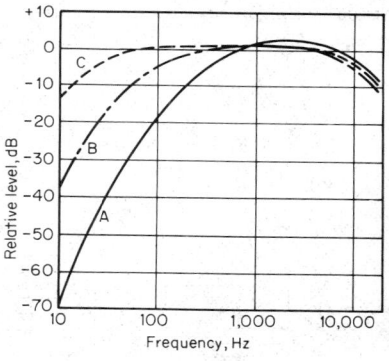

measurements.   Weighting networks *B* and *C* are used for moderate and substantial attenuation of the low-frequency noise components.   Weighting is used when the noise level is low, and the *C* scale meter reading would correlate poorly with loudness level (see the shape of the 20- and 40-phon equal-loudness contours in Fig. 19-4).   The *A* weighting has also been found useful in connection with hearing-conservation criteria described later.

A number of precautions should be taken in making noise-level measurements.

1. If the noise is loud, e.g., louder than your own loud speech sounds, wear ear protection during sustained exposure.

2. If spectrum analysis is not contemplated, make measurements on all three scales (*C* and *A* at least) to get some rough knowledge of the noise spectrum.

3. Calibrate the meter electrically (according to the manufacturer's procedure) before each measuring session and calibrate the microphone periodically.

4. Listen to the noise in various locations and measure where the most significant noises are heard.   If in doubt, measure in several locations, noting each for later data analysis.

5. Listen on a headset to the amplified output of the sound-level meter to be certain you are measuring what you intended to and that you are not measuring meter noise.

6. Turn off the noise to be measured, if possible, and measure the ambient background noise. Or if the noise fluctuates greatly, watch the lowest values and listen for correlation of these values with ambient-noise dominance. Knowing the level of the ambient noise, you can correct for its contribution to the total reading by using Fig. 19-40.

7. When the noise is dominated by steady components of fixed frequency, especially at low frequencies and within small rooms, the data will probably be position-dependent. If so take enough data for averaging.

8. Note Fig. 19-41. The size and shape of the sound-level meter and the proximity of the observer can influence the reading somewhat. If these differences are important to your data, use a microphone stand and extension cable (with cable correction) or a smaller meter or binoculars to read the meter. In any case hold the meter away from you and stand to one side of the line from the source to the meter.

9. If the noises are caused by sharp or sudden impact, or if they sound percussive or clicky, an impact noise meter[40] may be needed for significant data.

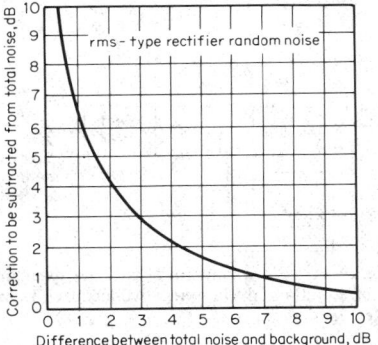

Fig. 19-40. Correction for ambient background level in terms of the difference between the total noise (including the one being measured) and the ambient noise. The solid line indicates the correction where both noises are essentially random in character and the rectifier characteristic is of the rms type. (Ref. 39)

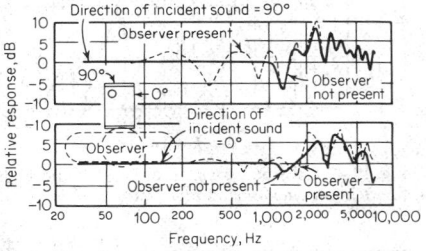

Fig. 19-41. The effect on the frequency response of using the microphone mounted directly on a rectangular sound-level meter with and without an observer present in a free field. (Ref. 39)

24. Noise-Spectrum Analyzers. The least expensive and most available noise analyzer is an experienced analytical listener. He is not a substitute for a good analyzing instrument, but he can save many hours and dollars of data analysis by helping choose between a narrow- or a broad-band analyzer and by focusing quickly on the pertinent frequency range.

The bandpass filter set is the most rapid tool for simple spectrum analysis. Response-frequency characteristics for a good octave bandpass filter set are shown in Fig. 19-42. (Recent standards have centered the passbands at 62.5, 125, 250, 500, 1,000, etc.) Note the uniformity of response within the passband, the 40-dB rejection one octave from the edge of the band, and the greater rejection beyond.

Automatic switching of a level indicator (meter, graphic level recorder, or oscillograph) to each of the filter outputs in a frequency sequence gives an octave spectrum plot. When the sequence is rapidly cycled, one can monitor time variations of the spectrum. Separate indicators on each filter output allow continuous monitoring rather than intermittent. Half- and third-octave bands are also used for greater spectrum detail.

Searching for very fine detail on the frequency axis of the noise spectrum is accomplished by a heterodyne analyzer. Figure 19-43 gives a block diagram. The selectivity is determined by an ultrasonic fixed-frequency filter (usually quartz crystal or inductor-capacitor with electronic feedback). The noise signal is mixed (in a balanced modulator) with a variable ultrasonic oscillator signal in order to translate the audio-frequency noise spectrum upward to the filter frequency. Varying the oscillator frequency causes different spectral components to appear at the filter output indicator. Since the filter is fixed, the bandwidth in hertz is constant and spectrum-level data in decibels are automatically provided (after a constant correction for filter bandwidth in hertz). This type of analyzer is very useful in the search for spectral-line components within a continuous-spectrum background.

A good compromise between fractional-octave filters and the heterodyne analyzer is the narrow-band proportional-bandwidth analyzer shown in Fig. 19-44. It uses an electric null circuit of variable frequency in a negative-feedback loop to convert an amplifier stage into a

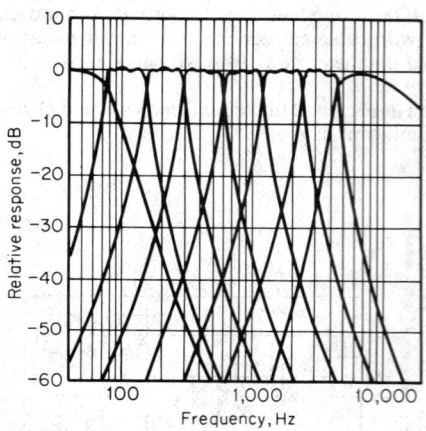

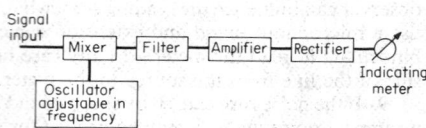

Fig. 19-43. Basic elements of a heterodyne analyzer. (Ref. 38)

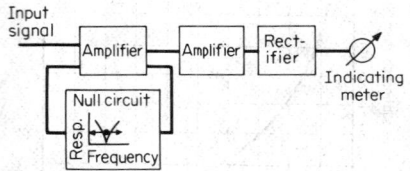

Fig. 19-42. Response vs. frequency of the eight filters in an octave-band filter set. (Ref. 38)

Fig. 19-44. Block diagram of analyzer using an electric null circuit in a negative-feedback loop. (Ref. 38)

narrow-bandpass filter. Since its bandwidth is proportional to frequency, for example, 2%, the output-level data are of the same type as octave spectra, requiring -3 dB/octave tilt for conversion to spectrum level. This type of analyzer is very useful when the search is for discrete-frequency components at lower frequencies and when such narrow selectivity (in hertz) is neither needed nor desired at high audio frequencies.

Figure 19-45 compares the response-frequency characteristics of all three types of noise-spectrum analyzers in the three principal decades of the audio-frequency range.

A special analyzer[41] for magnetically recorded speech samples has also been found useful for noise. The sample recorded on a loop is played back repeatedly at high speed for heterodyne analysis. The final spectrogram is a plot of frequency vs. time, with spectrum level indicated by shading on a white-black-gray scale (see Fig. 19-18).

Another special automatic analyzer of the monitored filter-set type has been developed for loudness analysis of noises.[42] Filter bandwidth in fractional octaves is greater at low frequencies to assist in proper loudness summation. Chopped rectified output levels of the set of filters are sequentially sampled every 25 ms and presented on a cathode-ray-tube screen. The vertical loudness scale (see Fig. 19-46) is calibrated in sones.

25. Hearing-Conservation Criteria. Government regulations have established permissible noise exposures for occupational safety and health on government contracts[43] and for all businesses engaged in interstate commerce.[44] Table 19-2 duplicates Table G-16 of the 1970 Act. Note that the meter is read with the A weighting network (see Fig. 19-39) and slow meter response.

The effect of the use of the A scale is to permit somewhat higher sound-pressure levels at low audio frequencies than at medium and high frequencies.

There is a separate restriction concerning impact noise peaks, which are not permitted to exceed 140 dB sound-pressure level. Impact-noise peak levels are to be measured only with an impact meter or an oscilloscope.

Table 19-2. Permissible Noise Exposures* [44]

Duration per Day, h	Sound Level dB(A), Slow Response
8	90
6	92
4	95
3	97
2	100
1½	102
1	105
½	110
¼ or less	115

* When the daily noise exposure is composed of two or more periods of noise exposure at different levels, their combined effect should be considered, rather than the individual effect of each. If the sum of the fractions $C_1/T_1 + C_2/T_2 + \cdots + C_n/T_n$ exceeds unity, the mixed exposure should be considered to exceed the limit value. C_n indicates the total time of exposure at a specified noise level, and T_n indicates the total time of exposure permitted at that level.

These regulations are based upon many years of psychoacoustic research, clinical observation, and industrial hygiene records, through the cooperation of scientific, medical, industrial, and government laboratories. The intent is to protect the hearing acuity of the population through protective legislation, industrial cooperation, and government inspection and enforcement.

26. Room-Noise Criteria. In typical background-noise spectra for rooms the octave-band sound-pressure level usually decreases with increasing frequency, for the following reasons:

1. The noise of most machinery in good operating condition is dominated physically by the lower-frequency components.

2. Intervening partitions attenuate the high-frequency noises more than the low-frequency noises.

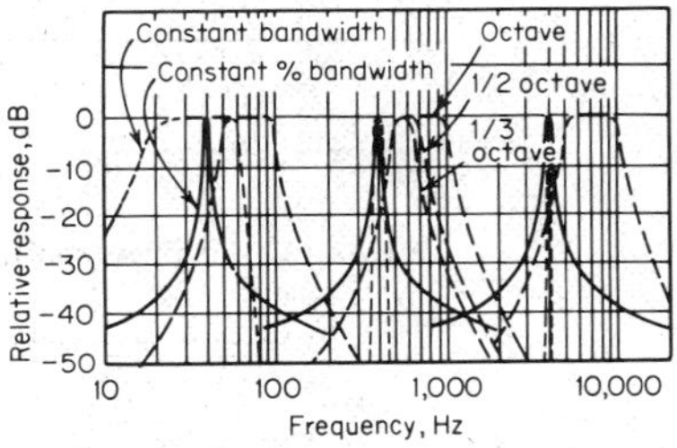

Fig. 19-45. Comparison of response curves of typical sound analyzers having constant bandwidth, constant-percentage bandwidth, octave, and fractional-octave bandwidth. (After Snow, Ref. 39)

3. The sound-absorption coefficient of conventional acoustical materials is greater at medium and high frequencies than at low frequencies. Thus room absorption of sound tends to accentuate the downward spectrum trends of machinery noise.

Fortunately the equal-loudness contours curve upward at low frequencies, especially at low sound levels (see Fig. 19-4). Thus the characteristics of typical building noise and of hearing-response curves tend to complement each other. This convenient (or evolutionary) relationship has led to the widely used set of noise-criterion (NC) curves in Fig. 19-47.

The NC level of the noise in a room is determined by measuring sound-pressure levels in each octave band and then comparing them to the grid of NC curves. The band having the highest interpolated NC value establishes the NC level for the entire spectrum. Consider, for example, a noise having a uniform octave-band level of 40 dB. It would have a rating of NC 43 because at 4,800 to 10,000 Hz its highest octave band (relative to the NC curves) lies three-fifths of the way between NC 40 and NC 45.

When NC 30 is specified for a room in the planning stage, this means that none of the octave-band levels in the completed room should lie above the NC 30 curve, with all normally operating equipment in operation and with the room unoccupied.

Typical recommended noise criteria are given in Table 19-3. These are not invariant, being subject to economic necessity and varying local conditions. In some instances, e.g., offices with poorly attenuating partitions, a higher level of broadband, midfrequency noise is tolerated, or even welcomed for its masking effect upon the intelligibility of distant conversations. However, dependence upon masking to compensate for inadequate construction requires considerable experience, judgment, and care in planning and execution. The

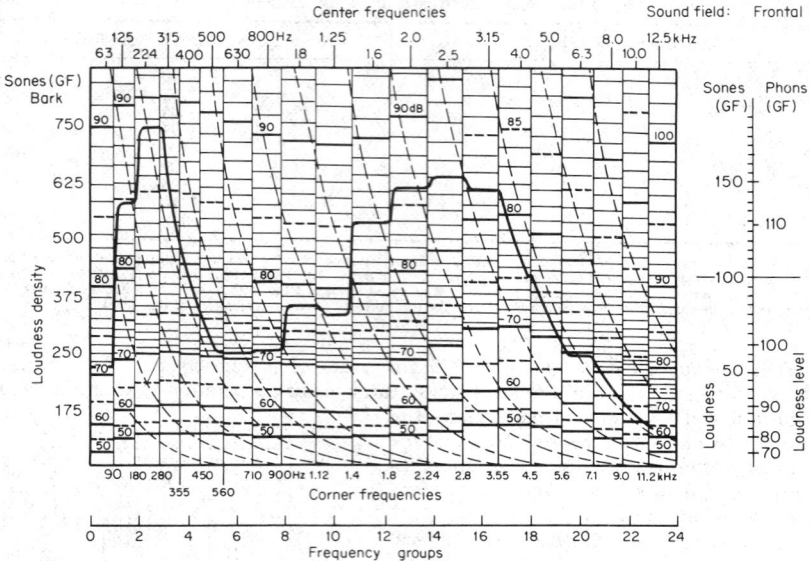

Fig. 19-46.   Loudness analysis of a single typewriter stroke.   (Ref. 42)

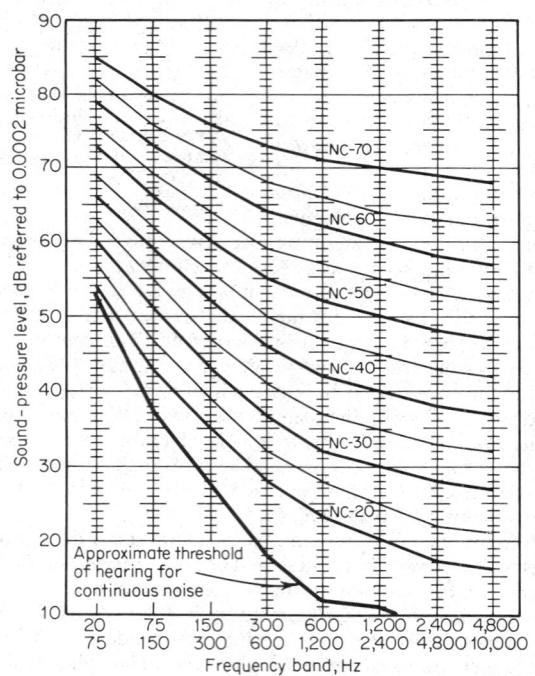

Fig. 19-47.   Noise-criterion curves for rooms.   Octave-band-level measurements should not exceed the specified criterion curve in any band when the unoccupied room is in normal operation, e.g., ventilation system, business machinery.   (Beranek, Ref. 37)

recommendations of Table 19-3 are backed by a large amount of acoustical experience and should not be compromised without expert consultation.

Measurement of octave-band spectra is strongly recommended in checking compliance with NC curves. However, if no spectrum-analysis equipment can be obtained, make a rough check on the *A* scale of a standard sound-level meter. Many room noises (there are exceptions) have a decibel *A* level *approximately* 10 dB greater than the corresponding NC value.

Table 19-3. Typical Recommended Noise Criteria for Rooms

Type of Space	Recommended NC Curve*
Broadcast studios	15–20
Concert halls	20
Legitimate theaters (no amplification)	20–25
Music rooms	25
Schoolrooms	25
Large conference rooms	25
Apartments and hotels	25–30
Assembly halls (amplification)	25–30
Homes (sleeping areas)	25–35
Small conference rooms	30
Motion picture theaters	30
Hospitals	30
Churches	25
Courtrooms	30
Libraries	30
General offices	30–40
Restaurants	45
Coliseums for sports only (amplification)	50
Stenographic offices (business machines)	40–50

* Each NC curve is a code for specifying permissible sound-pressure levels in eight octave bands. The curve should not be exceeded in any band.

27. Noise Reduction. In audio-electronics engineering the first step in reducing noise is to determine with certainty where the problem noise originates. There are many possibilities including (1) any recorded media being used, (2) long lines for incoming or outgoing audio signals, (3) the electronic system (either the wiring or the components), (4) input transducers, (5) airborne noise generated within the room, (6) airborne noise generated elsewhere, (7) structure-borne vibration.

In general listening (by high-fidelity headset) to the noise output of an audio system while checking various source possibilities is an important analytical supplement to any visual indicators used such as meters and oscillographs. (This obvious advice is too often ignored by electronics engineers.)

System Noise. Recorded media are easily checked by stopping them and long lines by disconnecting them temporarily (while replacing them with dummy sources of equivalent nominal impedance).

The *electronic system* probably contains the noise problem if the symptoms persist when incoming and outgoing lines are replaced by dummy-load connectors. However, oscillations from input-output coupling could be caused by wiring proximity at the connectors or lack of shielding near the connectors. Too many ground points, poorly soldered and intermittent ground connections, or lack of a good system ground are common causes of hum and television interference. After volume controls, switches, and disconnects have been tried, some judicious ac shorting of sequential signal points to ground will usually pinpoint which stage of the circuit is the noise source. Then trial component replacement is an obvious remedy.

If the noise problem disappears when the electronic system is isolated and it cannot be found in the incoming and outgoing lines by themselves, the lines may be serving as antennas for the audio electronics at higher than audio frequencies. A good high-frequency oscillograph is a necessity here because listening clues will only be by-products of the main action. Judicious use of decoupling components or networks is suggested if the problem is really at radio frequencies, with full consideration for response and phase conservation at high audio frequencies.

If a noise problem occurs only when the sounds to be recorded, transmitted, or measured are being produced in the room and the problem persists when substitute input transducers are used, try replacing the input transducers with dummy loads and producing the sounds to check for microphonics in the electronic components. These may be caused either by sound or by vibration.

Lifting the equipment or giving it a resilient mechanical support will show whether vibration is the problem. Moving the equipment outside temporarily or enclosing it in a box will check whether the excitation of the electronic components is acoustical.

Input transducers can also be overloaded by excessive sound or recorded levels. Microphones can be affected aerodynamically from air conditioning and close breathing, as well as the wind, and by building vibration transmitted through microphone stands. Microphone lines very high or very low in impedance are also susceptible to electrostatic or magnetic field pickup. Microphone-line impedance of 50 to 600 Ω is a good compromise. Balanced lines with the shield grounded only at the amplifier end are recommended for low noise.

Room Noise. When the noise problem is definitely in the room, not the audio electronics system, listening is again an obvious analytical aid. If the noise source is within the room and cannot be moved out, e.g., essential operating equipment, it can be dynamically balanced or isolated mechanically to keep the floor or walls from radiating its vibration. Possibly it can be enclosed acoustically, totally or partially. If totally, the possible degree of attenuation is greatly improved. This will be discussed below under noise isolation.

Absorption of noise within a room by the installation of absorptive acoustical treatment is so well known that it is often assumed to be more effective than it is in practice. Absorptive treatment of room boundaries can be quite effective for *reflected* sound but cannot reduce *direct* sound. When the noise source is distant in an acoustically absorptive environment, the noise level decreases approximately 6 dB when the distance is doubled. This occurs in anechoic chambers and is approximated in large-area, low-ceilinged rooms with absorptive acoustical ceilings and carpeted floors. However, when the noise source is nearby, where the direct sound dominates the generally reflected sound, absorptive treatment is not very effective in noise-level reduction. It does reduce reverberation, however, and thereby reduces the spatial impression of noisiness.

The reduction of generally reflected sound in a room is given by the equation

$$\text{Reduction} = 10 \log_{10} \frac{a_2}{a_1} \quad \text{dB} \tag{19-3}$$

when the total sound absorption in the room is increased from a_1 to a_2. The total absorption is obtained by adding together each surface (or object) area in the room multiplied by its respective absorption coefficient (percentage). For example, doubling the total absorption reduces the built-up noise level 3 dB. Figure 19-48 provides the reduction for other values of the absorption ratio. Bear in mind that this *does not* affect the level of the noise received *directly* without reflection.

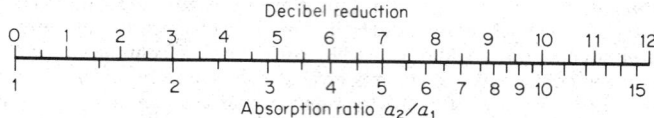

Fig. 19-48. Reduction of generally reflected sound level caused by increasing total sound absorption in the room from a_1 to a_2. (Ref. 36)

Figure 19-49 shows the effect of absorptive treatment upon the loudness of noise buildup (after a noise source is turned on) in a previously reflective room. Reverberation after the noise is turned off is also shown.

In Fig. 19-50 both the direct noise level (dashed line) and the total (direct plus reflected) noise level (solid curves) are shown at different distances from an omnidirectional sound source. The source was arbitrarily selected to produce a direct noise level of 100 dB at a distance of 1 ft. The parameter is total sound absorption in the room in sabins. For example, at 10 ft the total noise is 20 dB higher than the direct noise when the total absorption is only 50 sabins but only 2 dB higher for 10,000 sabins.

28. Noise Isolation. When a noise source in an adjacent space is the problem (even with connecting doors and windows closed) look and listen first for sound-leakage paths. Measure their noise contribution by placing the microphone end of a small sound-level meter as close as possible to the suspected leakage openings. (Use the *A* weighting scale on the meter in this test.)

Small clearance cracks, e.g., under doors, transmit considerable amounts of noise. Their contribution can easily be checked. Close the cracks by any temporary expedient. If this makes an audible or measurable difference, a practical mechanical solution becomes worthwhile.

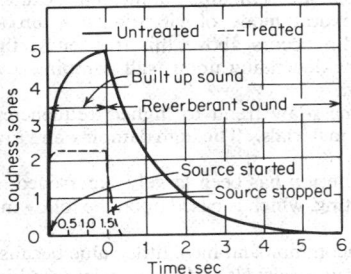

Fig. 19-49. Loudness of sound buildup and reverberant sound in a highly reflective room before and after absorptive acoustical treatment. (Ref. 36)

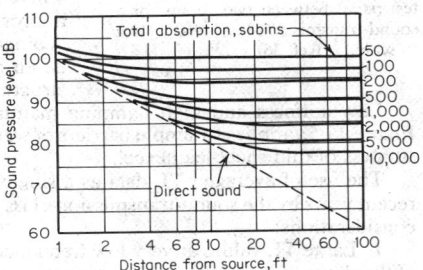

Fig. 19-50. Comparison of total noise level with the direct noise level at different distances from the noise source with different amounts of sound absorption. (Ref. 36)

Partitions. If noise transmission by the wall itself is suspected, an easy way to check it is to rest one ear on the wall while sealing the other ear with a finger. What appears to be a solid wall, e.g., concrete block, may actually be acoustically transparent if it is highly porous.

Surface porosity can be checked by blowing into the wall with lips sealed against the wall. If air flows freely into the wall surface, it can probably be improved as an acoustical barrier by plastering or by a sealing coat of grout or thick paint. However, if the air escapes between the lips and the wall, or if a back pressure builds up, wall porosity is not the problem.

The *transmission loss* (TL) of a wall is defined by

$$TL = 10 \log_{10} \frac{1}{\tau} \quad dB \tag{19-4}$$

where τ is the percentage of the incident-sound power which is transmitted. For example, if 1% of the sound reaching a boundary passes through, the TL is 20 dB.

Figure 19-51 shows the two-room test method for measuring the TL of a partition test panel. When the receiving room is anechoic (totally nonreflective acoustically) the measured difference between sound-pressure levels L_1 and L_2 on the source and receiving sides of the panel provides the TL rating. However, when the receiving room is somewhat reverberant, the generally reflected sound will contribute (as in Fig. 19-50) to the sound level at the microphone. Then the equation for determining TL experimentally, including the room-correction term, is

$$TL = L_1 - L_2 + 10 \log_{10} \frac{S}{a} \quad dB \tag{19-5}$$

where S is the area of the test panel in square feet and a is the total receiving room absorption in sabins.

The TL of a partition varies with the frequency of the sound, sometimes in a very complex manner. An average of the TL at a number of frequencies was long used as a single-number rating for a partition. Figure 19-52 compares this rating for a wide variety of wall, ceiling, floor, door, and window types. Also shown is a reference line for an empirical mass law, having a slope of 4.4 dB per doubling of the area density. Doors and porous wall materials generally lie below the mass-law line. Double-layer boundaries, with mechanically unbridged air space between, provide higher TL than the mass law.

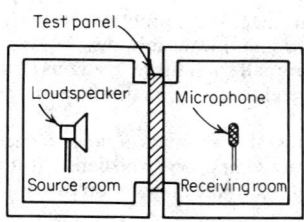

Fig. 19-51. Mounting of a partition test panel between two rooms for a sound-transmission-loss test. (Not to scale.) (Ref. 36)

A mass law also applies to the TL vs. frequency of the noise. Theoretically the TL should increase 6 dB/octave. It usually does so (see Fig. 19-53) up to where the sound at a certain angle of incidence excites a prominent flexural mode of vibration. A broad "coincidence" dip occurs above this frequency, the depth and extent depending upon wall damping and structural design.

Low stiffness and high damping maintain the mass-law trend to higher frequencies. Figure 19-53 applies to simple barriers of well-known materials. The high damping and low stiffness of thin sheet lead excel.

The use of averaged TL data as a single-value criterion has been largely superseded in recent years by the sound-transmission-class (STC) rating, which is based upon the following considerations:

1. Large TL values at very low frequencies are uncommon and have little value because of the shape of the equal-loudness contours of hearing at low levels and frequencies (see Fig. 19-4).

2. Many boundary materials and constructions provide increasingly large TL values at high audio frequencies. There is little additional audible advantage beyond the midfrequency range, since typical transmitted noise spectra diminish beyond inaudibility at high frequencies.

3. The most important TL values are at the central frequencies, where walls typically have TL coincidence dips in frequency bands where noise and audibility are both appreciable.

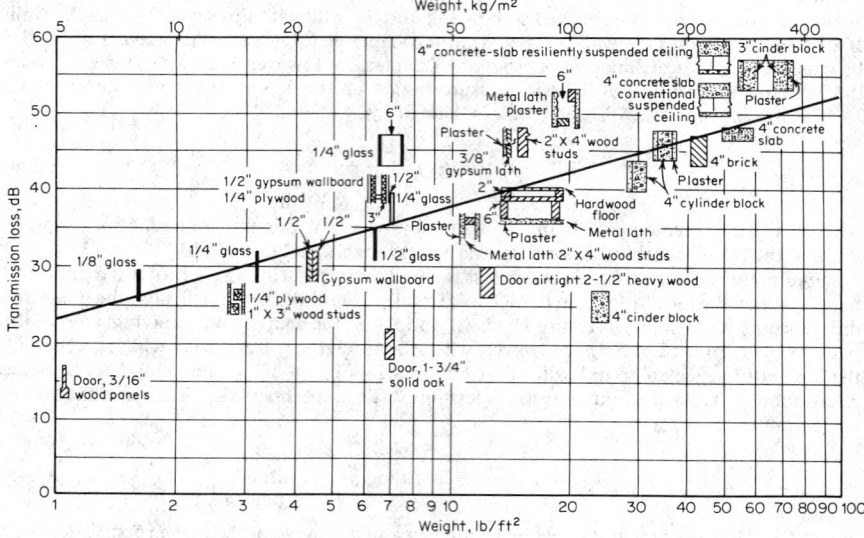

Fig. 19-52. The mass-law relation between average sound-transmission loss and mass per unit area of partition. (Ref. 36)

Based upon these factors, the standard STC contour[45] shown in Fig. 19-54 has been developed. To rate a partition follow this procedure:

1. Place an overlay bearing the STC contour upon the 16-frequency TL test data for the partition.

2. Slide the overlay upward until the sum of the deficiencies (deviations below the contour) equals 32 dB, with no deficiency exceeding 8 dB.

3. Read the STC rating in decibels where the STC contour crosses the 500-Hz ordinate line.

The choice of materials and construction is based upon TL data (or STC values), depending upon the NC criteria selected for a room and the noise or sound spectra expected in adjacent spaces (including those above and below).

Leakage Effect. It is good practice in partition design to match the STC ratings of doors and windows to the associated walls. Tight, careful construction is assumed but frequently is not obtained, especially around doors and windows. Figure 19-55 emphasizes the importance of leakage. The greater the STC value of the partition, the greater the loss of effectiveness produced by the same amount of leakage.

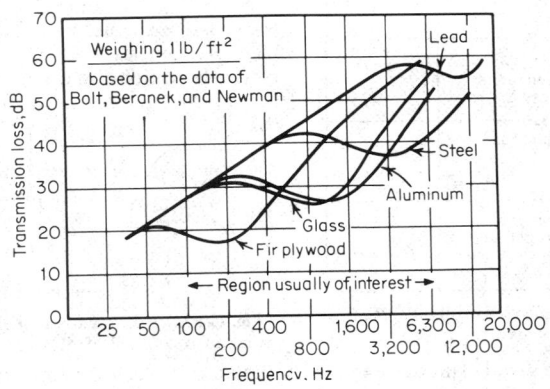

Fig. 19-53. Transmission loss of ideal single-layer barriers having equal area density. *(SAE Reprint 833C, Lead Industries Association, Inc., New York, 1964.)*

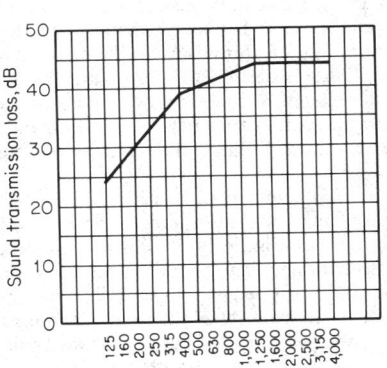

Fig. 19-54. Typical STC contour (STC-40). (ASTM E413-70T.)

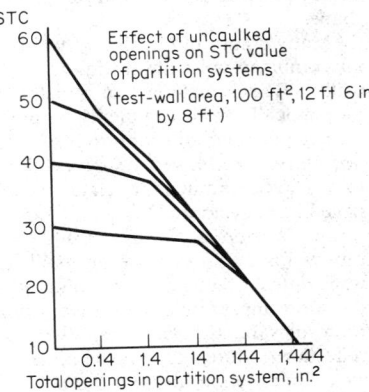

Fig. 19-55. The effect of leaks. ("Sound, Noise and Vibration Control," L. F. Yerges, Van Nostrand Reinhold Company, New York, 1969. Reprinted by permission.)

Leaks can best be avoided by care during construction because they may be inaccessible after construction. Figure 19-56 shows the cumulative effect of sequential caulking lines in a multiple-layer partition.

Ventilation Noise.[46,47]   In modern buildings the air-distribution systems often contribute significantly to ambient room noise and sometimes transmit sounds from one space to another through otherwise well-constructed partitions and floors. Figure 19-57 summarizes the acoustical essentials of a ventilation system. Return air ducts are omitted only because their quieting principles are the same as those for supply ducts. Sound travels against airflow, as well as with it.

In Fig. 19-57 the motor and blower are supported on a separate platform mechanically isolated on vibration mounts from the building slab. Canvas sleeves attenuate mechanical vibration transmitted to the plenum and the duct system. The plenum reduces acoustical noise transmission to the duct system, especially at high frequencies and if treated internally. Noise absorption also occurs at bends, especially if the duct is acoustically treated at the bends.

Edge detail	Lab STC performance
Unsealed	29
Single bead of sealant at base	49
Both base layers sealed	53
Sealant applied between runner and base layers	53
Base and face layers sealed	53
Six rows of sealant, two at each side and under runner	53

Fig. 19-56. The effect of caulking. *(United States Gypsum Company.)*

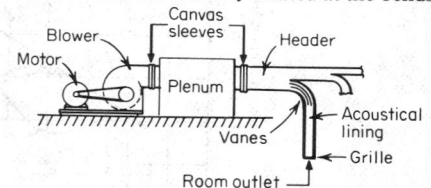

Fig. 19-57. Acoustical essentials of a typical ventilation system. (Ref. 46)

Vanes can cause aerodynamic noise to be generated at bends if the air velocity in the duct is too high.

Splitting the duct cross section into parts can increase noise attenuation in the duct if the splitters are also acoustically lined. The duct cross section is often increased where splitting occurs to avoid constriction of the air flow. Special duct attenuators are of this expanded shape.

Many brands of duct liner (having absorptivity in the range of 0.70 to 0.90) are available for combined noise absorption along the duct length and heat insulation through the duct walls. These materials also help damp the vibration of thin duct walls responding mechanically to air turbulence within. Absorptive lining is particularly important in ducts connecting adjacent rooms, but sufficient treated-duct length and absorptive bends are also necessary. Grilles and diffusers at the openings into the rooms can generate disturbing noise locally if the terminal air velocity is too high. Standards and diffuser ratings to prevent this have been developed.[48,49]

Structure-borne Noise.[50]   Buildings can have partitions and floors with excellent STC ratings for airborne noise and still have noise problems because of shock and vibration transmitted directly through solid materials and structure. Figure 19-58 illustrates some of the ways impact noise can travel from one room to another. The sound of impact in one room may be airborne to another room through walls or open windows or by multiple reflection. More often impact noise goes through the solid material of a floor and is radiated by ceiling motion in the room below. Even when the lower room has a separate ceiling resiliently supported from the floor above, impact traveling through the structure can shake the walls of the room.

There are structural solutions, but the reduction of impact at the source is most effective. One common solution to the problem is to carpet the floor. Figure 19-59 compares the impact sound insulation of ⅛-in. linoleum and ⅜-in. carpet on a concrete floor slab. Floating-floor constructions are very effective where a padded surface is impractical.

Machinery Quieting.[51] If a machine is well balanced dynamically and relatively rigid in construction, the principles used in further noise reduction are much the same as for reduction of airborne and structure-borne noise.

To reduce the acoustical noise from purely mechanical or electromechanical machinery, reduce the motion and the area of the exposed surface or enclose it within mechanically isolated and internally acoustically absorptive walls. If the machinery is aerodynamic in operation, reduce the speed of airflow, if possible, and expand the flow cross section through tortuous paths lined with sound-absorbing materials.

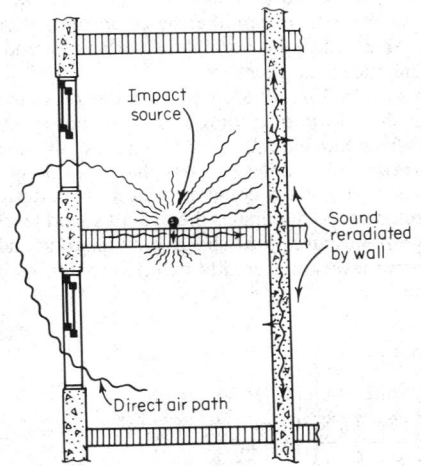

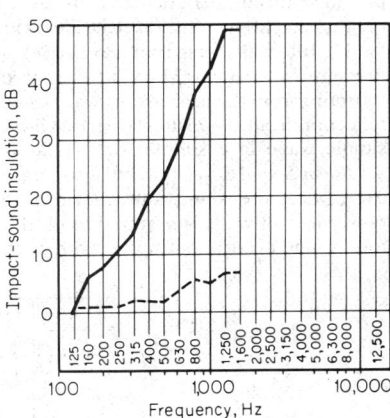

Fig. 19-58. How impact sound travels from one room to another. (Ref. 50)

Fig. 19-59. Impact-sound insulation provided by two different types of floor finish on a concrete floor, (*a*) ⅛ in. linoleum (dashed line), and (*b*) ⅜ in. Wilton carpet (solid line). (Ref. 50)

Machinery vibration can be isolated from the supporting floor or walls by vibration mounts chosen to resonate (with the machinery mass) at a frequency well below the fundamental frequency of machinery rotation or reciprocation. Figure 19-60 shows the reduction of vibration transmission above the natural resonance frequency of the mass-spring supported (or suspended) system with damping as the parameter. Several machines are often mounted on a common slab (resiliently supported on vibration isolators) in order to increase the total mass and reduce the number of isolation devices required.

ACOUSTICAL ENVIRONMENT CONTROL

29. Introduction.   Acoustical needs within a room vary with its functions. Acoustical conditions in a room depend upon:
1. Internal factors.
    a. Room size and shape.
    b. Surface types and their locations.
    c. Furnishings and their locations.
    d. Sound and noise sources within the room.
    e. Number and location of auditors.
2. External factors.
    a. Noise and vibration sources outside the room.
    b. Their mechanical and acoustical isolation from the building structure and room walls.
    c. The transmission characteristics of the walls and structure.
    d. Service-connected equipment, e.g., lighting, heating, and ventilating.

The external factors have been described previously. Internal acoustical factors should also be considered because they strongly influence the input program material to audio electronic systems and the output sound reproduction from them.

30. Acoustical Functions of Rooms. The specialized acoustical purposes of some rooms are clearly indicated by their names, such as lecture rooms, conference rooms, symphony halls, opera houses, recording studios, motion picture theatres, anechoic chambers, reverberation chambers, and audiometric rooms. For such rooms the acoustical conditions which are optimum for the special purpose can usually be specified and obtained by various fixed solutions to the problem.

Other rooms have more varied acoustical purposes, such as convention halls, stadiums, churches, multipurpose auditoriums, and even general-purpose rooms. The optimum internal acoustics for some of the functions of such rooms must either be compromised or provided through sound controls such as variable reflectors, variable absorbers, sound-reenforcement systems, or electroacoustically enhanced reverberation.

31. Internal Acoustical Properties of Rooms. In Fig. 19-61 a pulse wave of sound pressure in a room (24 by 40 ft floor plan) is shown *(a)* expanding from the source, *(b)* reflecting from the near (front) wall, *(c)* reflecting from the side walls, and *(d)* just after the first reflection from the distant (rear) wall has returned to the source. In less than $\frac{1}{20}$ s a very simple wave in a simple rectangular room has developed a complex reflection pattern.

Room Sound Level. In most rooms the reflected sound dominates the overall sound level except near the sound source (refer to Fig. 19-50). Figure 19-62 shows how much sound power in milliwatts is required to produce different levels of generally reflected sound, with total sound absorption in the room as the parameter.

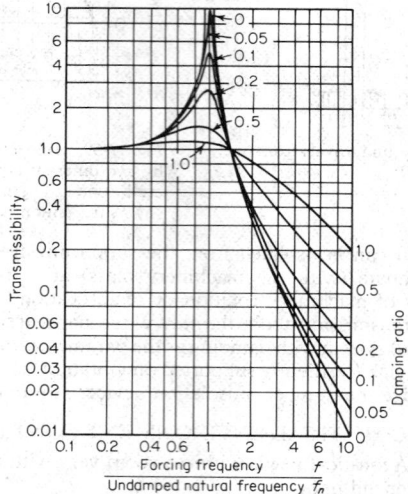

Fig. 19-60. Transmissibility of a vibrating system.

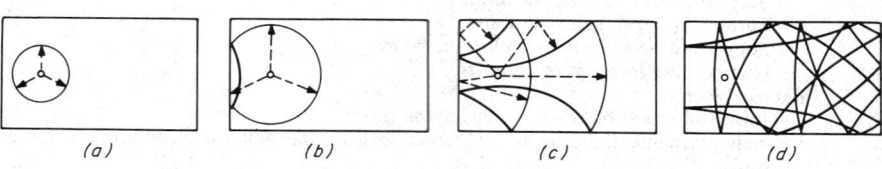

| *(a)* | *(b)* | *(c)* | *(d)* |

Fig. 19-61. Progress of single sound pulse in a closed room: (*a*) $\frac{1}{200}$ s, (*b*) $\frac{1}{100}$ s, (*c*) $\frac{1}{50}$ s, (*d*) $\frac{1}{17}$ s. (Ref. 36)

The sound powers are quite small. However, in sound-reproduction systems the loudspeaker efficiency ranges from the order of only 0.1% (for some small bookshelf enclosures) to 20% for highly efficient theater-type sound systems. This factor and the 12- to 15-dB peak factors needed for unclipped music reproduction lead to audio-electronic power requirements 80 to 30,000 times the acoustic powers shown.

Room Reverberation. Reverberation in the room is caused by the room's multiple reflective acoustical properties and is the resulting tendency for sound level in the room to persist after direct sound ceases. The measure of reverberation is the time t_{60} for the sound level to decay 60 dB from its steady-state level after it is turned off. The commonly used Sabine reverberation equation is

$$t_{60} = 0.049 \frac{V}{a} \qquad (19\text{-}6)$$

where t_{60} is the reverberation time in seconds, V is the room volume in cubic feet, and a is the total sound absorption in sabins (square-foot units). In metric units

$$t_{60} = 0.161 \frac{V}{a} \qquad (19\text{-}7)$$

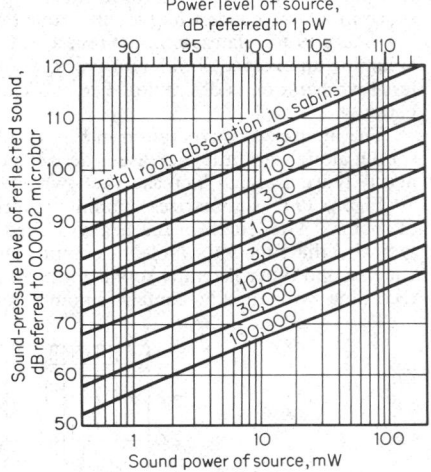

Fig. 19-62. Sound levels for different sound powers in rooms having different total absorption. (After Young, Ref. 36)

where V is in cubic meters and a is the total absorption in square meters. The value of the constant in the equations depends to a small degree upon the shape of the room, but this is infrequently considered.

The total sound absorption is calculated from

$$a = \Sigma \alpha_i S_i = \alpha S \qquad (19\text{-}8)$$

where α_i is the absorption coefficient (percentage of incident sound absorbed) for each surface area S_i, S is the total interior surface area in the room, and α is the average absorption coefficient of the surfaces.

As an example consider (at 500 Hz) a living room 20 ft long, 13 ft wide, and 8 ft high, with a plaster ceiling ($\alpha_1 = 0.02$), a carpeted floor ($\alpha_2 = 0.30$), a wood-paneled side wall ($\alpha_3 = 0.12$), an opposite glass wall ($\alpha_4 = 0.03$), an end wall of medium drapery ($\alpha_5 = 0.40$), and a brick fireplace ($\alpha_6 = 0.02$) for the other end wall. With no additional furnishings or occupants the total sound absorption would be

$$a = (0.02 + 0.30)(260) + (0.12 + 0.03)(160) + (0.40 + 0.02)(104)$$
$$= 151 \text{ sabins} = 0.144(1,048)$$

The average absorption coefficient is 0.14.

The reverberation time at 500 Hz would be approximately

$$t_{60} = 0.049 \frac{2080}{151} = 0.68 \text{ s}$$

The optimum reverberation time for a room depends upon room volume, sound frequency, and the type of sound which is most important to the room function, e.g., conversation, recorded music, or instrumental music. Larger rooms need greater reverberation in order to reinforce the loudness of sound at typically greater distances from the sound source. Low-frequency sounds need a longer 60-dB reverberation time than medium- or high-frequency sounds in order to have equivalent audible duration of reverberation, because of the higher threshold of audibility at low frequencies (refer to Fig. 19-2). Speech intelligibility can be degraded somewhat by the same amount of reverberation needed for maximum appreciation of some types of music such as classical organ. Conductors and musicians prefer a crowded rehearsal studio to be less reverberant than an equivalent stage space in a large concert hall for the same type of music.

Figure 19-63 relates optimum reverberation time to room volume at different frequencies, assuming an average of different types of sound. For special speech and recording studios less reverberation is desirable. For churches and concert halls more reverberation sounds better.

The increased reverberation in Fig. 19-63 at low frequencies is also a compromise. For speech alone very little increase is desired at low frequencies, and music performance needs nearly twice as much increase as shown.

Figure 19-64 summarizes optimum reverberation (at 500 to 1,000 Hz) for different types of rooms. The length of the heavy dashed line allows somewhat for different room sizes typical of that function. Note the compromise for combined speech and music requirements. Cinema and other listening rooms for recorded sounds require less reverberation, since the recording will already contain suitable reverberation from the recording environment.

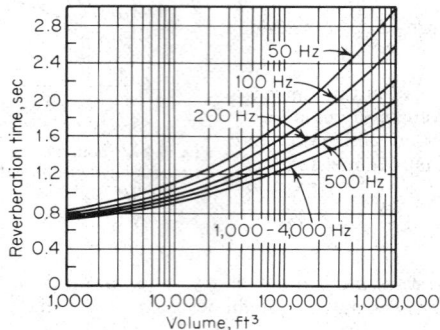

Fig. 19-63. Optimum reverberation for different room sizes at different frequencies (average for speech and music). (Ref. 22)

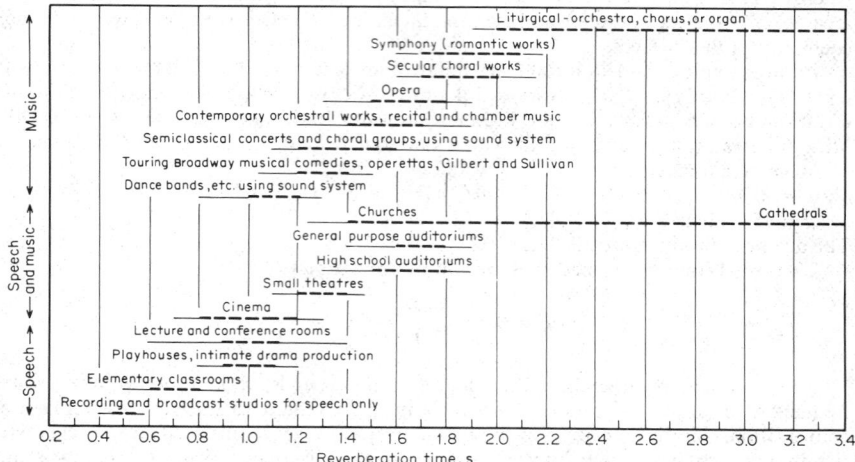

Fig. 19-64. Optimum reverberation (at 500 to 1,000 Hz) for auditoriums and similar facilities for speech and music. (After Russell Johnson, Ref. 54)

Although optimum reverberation times and time-frequency characteristics have been derived largely from subjective observations and empirical evidence, they are well founded and documented[52] and very useful for specification purposes.

Sound Distribution within the Room. Acoustical requirements for sound distribution also depend upon the room function. For example a lecture room needs outstanding one-way speech distribution from a rostrum (usually near one end of the room) to an audience seated toward the other end of the room. Figure 19-61 illustrates this case with reflective surfaces near the lecturer and a need for absorption at the opposite end to minimize reverberation.

By contrast, a conference room has many interchangeable source and receiver locations. A concert hall (Fig. 19-65) is the musical equivalent of combining the conference (a musical

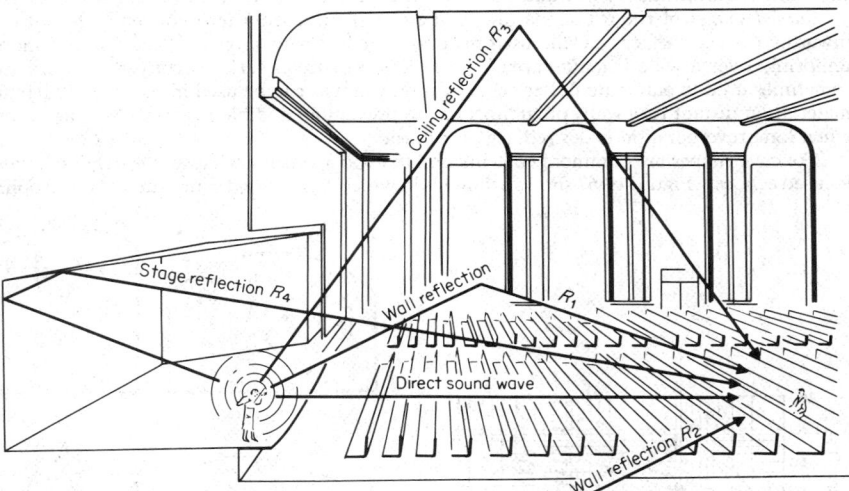

Fig. 19-65. Paths of direct and early reflected sound from an orchestral performer to a listener in a concert hall. (After Beranek, Ref. 52)

ensemble) with chiefly one-way communication from the performing group to the audience. A courtroom is an example having several scattered but well-defined source locations (judge, witness, attorneys) and seated groups of listeners (jury, public).

All these situations require a combination of beneficially shaped sound-reflecting surfaces near the sound sources, acoustically absorptive audience areas, and acoustically absorptive or diffusing surfaces beyond the audience areas.

Room echoes are discrete, separately heard sound reflections occurring too late to provide beneficial reenforcement to the direct sound. Beneficial early reflections arrive within about 20 ms of direct-sound arrival. A concentrated echo arriving more than 50 ms late is a serious acoustical defect.

A *flutter echo* is a rapid (usually regular) succession of reflected pulses resulting from a single initial pulse.

General reverberation, resulting from many superimposed reflections, tends to mask echo effects. Consequently the *reduction* of reverberation, without correcting the cause of an echo, can make the echo more *audible*.

32. Acoustical Shape Factors and Effects.[55] Room boundary shapes can be either beneficial or detrimental to room acoustics. Shape determines the frequency distribution of room resonances and can also produce undesirable sound-focusing effects. Detailed calculation of room resonances and of geometric focusing effects is extremely complicated, but some general understanding of them is very important to room-acoustics design and understanding.

Simple geometric shapes, e.g., spheres, cylinders, and cubes, give the least complex room patterns and consequently the most obvious and undesirable acoustical resonances and poorest sound distribution. The scale of dimensions determines where in the audible frequency range the resonance patterns occur. The smaller the room the higher the frequency of the lowest-pitched resonance modes, which are the most noticeable because they are well separated on the frequency scale.

Dimension Ratios. There is no unique set of "perfect" acoustical dimension ratios for a room. However, to help prevent a poor choice and as a guide toward good practice, Fig. 19-66 is an example applicable to rectangular rooms.[56] The three room dimensions are separated by third-octave steps in order to spread the resonance frequencies uniformly on a logarithmic frequency scale. Shifts of one dimension by an octave provide for the needed variety of room types (small, average, low, long). The living room in the reverberation-calculation example above had dimension ratios chosen from the "average" set of Fig. 19-66.

Curved shapes of room boundaries can be classified broadly into convex and concave inward. *Convex* surfaces can be used near sound sources to spread reflected sound more uniformly over a wide listening area. On reflective parallel surfaces convex diffusers are sometimes used to eliminate flutter echo. Convex shapes can be used in place of absorbing material on distant rear walls of auditoriums to prevent an audible rear-wall-to-stage echo when long reverberation is desired, e.g., for music.

Concave shapes are commonly detrimental to good acoustics because they tend to create focused echoes. Figure 19-67 shows examples in which S is a sound source and S' is the point

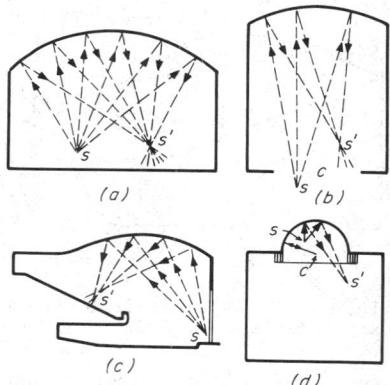

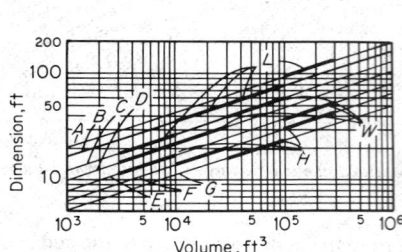

Fig. 19-66.   Preferred studio dimensions. In the graph, H = height, W = width, L = length. Small rooms, $H:W:L$ = 1:1.25:1.6 = $E:D:C$. Average shape rooms, $H:W:L$ = 1:1.6:2.5 = $F:D:B$. Low-ceiling rooms, $H:W:L$ = 1:2.5:3.2 = $G:C:B$. Long rooms, $H:W:L$ =1:1.25:3.2 = $F:E:A$. (After Volkmann, Ref. 56)

Fig. 19-67.   Examples of acoustical defects resulting from concave internal surfaces. (*a*) Barreled ceiling; (*b*) concave rear wall; (*c*) domed ceiling; (*d*) cylindrical stage plan. (Ref. 53)

where a listener or a microphone would receive the maximum echo effect. (They are interchangeable.) These are the most difficult and expensive acoustical defects to cure because they usually require a change of surface shape. Absorptive treatment of the concave surface can reduce the echo level as much as 10 dB, but this seldom makes it inaudible enough.

In some special circumstances concave curvature is not detrimental. When a dome has a radius of curvature small in comparison to the height of its base above the listening space, its focal points will be outside the listening range. A radius of curvature large compared to room dimensions will not focus sounds from remote points, but it can produce whispering-gallery effects at grazing incidence when the source and receiver are both near the curved surface. Close scrutiny of concave curvature is always necessary in acoustical planning.

33. Absorptive Control.   The principal design factor for internal acoustics of a room, in addition to shape and size, is the control of sound absorption and reflection. The total absorption required to give optimum reverberation can be calculated from Eq. (19-6) or (19-

7). Decisions on the amounts and types of absorptive treatment are based upon Eq. (19-8), i.e., which surfaces are preferably absorptive or reflective, the area of these surfaces, and the selection of acoustical materials which will meet economic, visual, and operational, e.g., fireproof, criteria.

Absorption and Reflection Materials. Table 19-4 shows typical approximate absorption coefficients for a variety of general building materials at three frequencies. Materials having coefficients less than 10% are useful for reflecting sound. Most commercial absorptive materials have coefficients greater than 50%. Moderate absorbers are in the 10 to 50% range.

Table 19-4. Sound-Absorption Coefficients for General Building Materials

Material	125 Hz	500 Hz	2,000 Hz
Brick, unglazed	0.03	0.03	0.05
Painted	0.01	0.02	0.02
Carpet, heavy, on concrete	0.02	0.15	0.60
On heavy padding	0.08	0.50	0.70
Concrete block, porous, unpainted	0.35	0.30	0.40
Pores painted full	0.10	0.10	0.10
Concrete, poured	0.01	0.02	0.02
Covered by floor tile	0.02	0.03	0.03
Floor, wood	0.15	0.10	0.07
Glass, heavy plate	0.18	0.04	0.02
Window	0.35	0.18	0.07
Gypsum board, ½-in. on 2 by 4 in., 16 in. o.c.	0.30	0.05	0.07
Plaster, gypsum or lime, smooth	0.02	0.03	0.04
Plywood paneling, ⅜-in. thick	0.25	0.15	0.10

For more detail see bulletins of the Acoustical Materials Association and data of individual manufacturers and building materials associations.

A widely used single-number coefficient for commercial acoustical materials is the *noise reduction coefficient* (NRC), which averages the coefficients at 250, 500, 1,000, and 2,000 Hz.

Sound absorption is largely a result of viscous air damping in porous materials, mechanical damping in acoustically driven diaphragms or panels, acoustical damping within acoustical resonators, or combinations of these.

Absorptivity by a porous material is very dependent at low frequencies upon its spacing from rigid boundaries, where particle velocity is too low for viscous absorption. Figure 19-68 shows the importance of air space behind the material.

Porous absorbing materials may be covered, for mechanical protection or appearance, by thin highly perforated facings without much loss in absorptivity. However, the holes must be closely spaced, as shown in Fig. 19-69.

Absorptive Structures. A widely used mechanical absorber is plywood paneling with air space (Fig. 19-70). This low-frequency absorption can be used to balance or supplement the greater absorption of porous materials at the higher frequencies. Rooms paneled throughout may be lacking in low-frequency reverberation (see Fig. 19-63).

Absorptive structures can also be suspended away from room boundaries for greater absorption efficiency when it is acceptable visually. Figures 19-71 and 19-72 give the absorption in sabins per structural unit for two different types of porous absorbers. Figure 19-71 refers to a circularly symmetrical unit formed of two cones face to face. Figure 19-72 is for flat 2 by 4 ft absorptive baffles. Note that scattering the units increases absorption per unit and that exposing two sides improves total absorption.

Acoustical resonators are another type of absorptive structure. They can be built into room walls, e.g., hollow tile with circular or slot openings into the room. They can be room furnishings (vases or bottles) but this will seldom be of practical significance. They are best used for selective absorption at their acoustical resonance frequencies.

Seating and Audience Absorption. The audience is one of the largest absorbers of sound in an auditorium. Audience size variation can be the most variable part of the total room absorption. In order to stabilize room acoustics while the audience varies from none to capacity, acousticians often recommend auditorium chairs with upholstered back rests and seat cushions. A person occupying this type of chair covers (subtracts) most of the chair absorption and adds his own absorption. The net increase depends upon his size and attire and the acoustical rating of the chair.

Published acoustical data on upholstered chairs are useful for chair comparisons, but the values are often larger (in sabins per chair) than the effective absorption in an auditorium installation,[57] for the same reason that spacing affects absorption in Figs. 19-71 and 72.

Studies of audience and seating absorption in large halls have led to a recent practice of calculating seating-area absorption, with or without audience, on the basis of sabins per square foot of seating floor space. Evidently the same number of chairs and auditors absorb more sound when spread out than when compactly arranged. Table 19-5 compares seating-area coefficients with audience and without, the latter for two typical chair coverings.[58]

Distribution of Materials and Structures. The location and distribution of functional materials and furnishings having acoustical absorptive properties are often determined on a functional rather than acoustical basis. (Examples are seating, carpeting, window draperies, and stage curtains.)

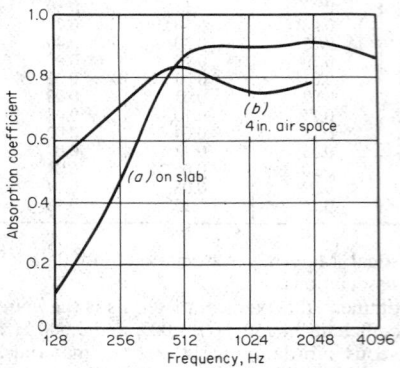

Fig. 19-68. Absorption of 1-in. rock-wool blanket (*a*) mounted against reflective ceiling, and (*b*) suspended 4 in. below. (Ref. 53)

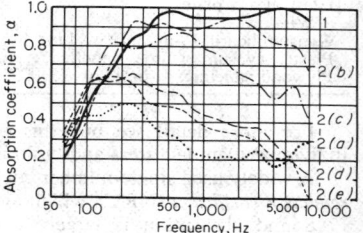

Fig. 19-69. Variation of absorption with spacing of holes in a sheet-metal cover: (1) 3-in. rock wool, (2) 3-in. rock wool covered with 22 BWG steel sheet (*a*) unperforated; (*b*) ⅛-in. holes at 5⁄16-in. centers; (*c*) ⅝-in. centers; (*d*) 1¼-in. centers; (*e*) 1⅞-in. centers. *(From National Physical Laboratory, England.)*

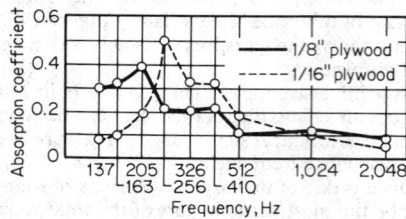

Fig. 19-70. Absorption characteristics of ply-wood panels, 2 by 9 ft with transverse braces 3 ft apart, with 2¼-in. air space. *(From P. Sabine and L. Ramer,* Ref. 53)

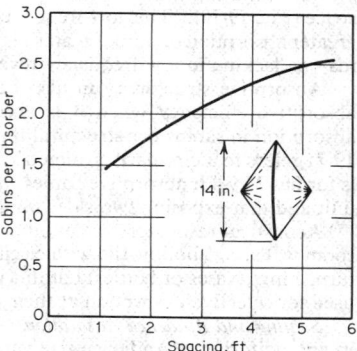

Fig. 19-71. Absorption of illustrated suspended unit absorber as a function of spacing. Absorption is averaged from 250 to 2,000 Hz. *(Johns-Manville Research Center,* Ref. 36)

Table 19-5.   Sound-Absorption Coefficients* for Audience and Seating Areas

Condition	125 Hz	500 Hz	2,000 Hz
Audience	0.60	0.88	0.93
Unoccupied "average" cloth-upholstered chairs	0.49	0.80	0.82
Unoccupied leather-upholstered chairs	0.44	0.60	0.58

* In sabins per square foot of seating floor space.

Before deciding where to put supplementary absorption (if any), at least the following factors should be considered:

1. Acoustical functions of the room.
2. Optimum reverberation time for the functions.
3. Probable location of sound sources (talkers, musical instruments, loudspeakers, noisy equipment) and receivers (listeners, microphones).
4. The room shape planned.
5. Whether electronic sound reenforcement or reproduction is planned.

If the goal is acoustical privacy for individuals or small groups within a large area (library, open-plan school, public dining area), absorptive treatment of the ceiling, the floor, or both is important, and wall treatment is relatively unimportant.

If the goal is airborne communication, either one-way or two-way, an acoustically reflective ceiling is necessary when the sources and receivers are separated very much. Near source locations the walls should usually be reflective. Near receiver locations wall absorption is desirable. When the communication is definitely one-way, as in a typical lecture room or theater, efficient rear-wall absorption comes first, then rear side walls, and rear ceiling, in that order, if necessary for reverberation control. When opposite surfaces are parallel, the absorptive material can advantageously be distributed or installed in a staggered manner to prevent flutter echo. In a

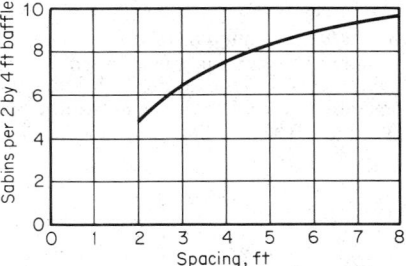

Fig. 19-72. Absorption of continuous rows of 2 by 4 ft baffle-type suspended absorber as a function of spacing. Absorption is averaged from 250 to 2,000 Hz. *(Owens-Corning Fiberglas Corp, Ref. 36)*

classroom or large conference room some ceiling treatment near the edges provides reverberation control, allowing a central reflective ceiling to aid two-way communication.

Because rooms have some acoustical modes of vibration that lie in one direction or plane, it is unwise to concentrate all the sound absorption in one place or along one axis of the room. This allows some isolated modes to reverberate while all the rest are properly damped. A good check is to compare the absorption coefficient averaged over each pair of opposite surfaces (ceiling and floor, front and rear, side walls).

All these suggestions are general and not necessarily the answer to a specific acoustical problem.

Variable Absorbers.   In multipurpose rooms variable sound absorbers can provide optimum acoustical conditions for each purpose. However, ease of control is a necessity, and the operation must be understood. Otherwise a good fixed compromise is preferable to random control.

Mechanisms for variable absorption include absorptive areas with reflective hinged covers or rotating shutters, absorptive draperies which can be extended from slot openings, absorptive blankets which roll up into reflective enclosures, and rotating panels with reflective and absorptive sides.

Electronic Reverberation Systems.   Recordings made in studios lacking sufficient reverberation are often enhanced by the addition of reverberation. Figure 19-73 shows two methods. In Fig. 19-73*a* the audio signal to be recorded or rerecorded is reproduced in a reverberant chamber, picked up, and mixed with direct signal. In Fig. 19-73*b* a parallel group of delay lines substitutes for the room. Figure 19-74 shows magnetic-tape loop and acoustic-tube systems which have been used for delay with multiple-pickup means. Metallic plates and spring systems are also used with appropriate input and output transducers.

The same electronic systems can be used with an extensive loudspeaker system to enhance the natural reverberation of a room. The loudspeakers must be well distributed over the normally reflective surfaces of the room and oriented for diffuse rather than directional radiation, for the additional reverberation to blend with the natural reverberation.

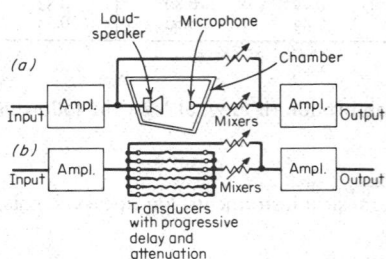

Acoustical resonators, each containing a microphone and loudspeaker in a stable, controlled acoustical-feedback relation, can be tuned to different frequency ranges and positioned in various parts of an auditorium to extend the natural reverberation time.

Fig. 19-73. Reverberation-enhancement systems: (*a*) reverberation chamber; (*b*) delay lines. (*"Music, Physics and Engineering," Dover Publications, Inc., New York, 1967.*)

34. Room-Acoustics Specifications and Predictions. The complex theories of physical acoustics, the uncertainties in predicting variations in the properties of both natural and manufactured building materials, and the difficulty in controlling installation procedures such as painting, sealing, bracing, furring, and draping all keep the specification and prediction of room acoustics from being an exact science. The frequent need to evaluate the final result by listening rather than by purely quantitative methods gives acoustical planning a somewhat subjective aspect. This is not unique. The solutions to many engineering problems involve human experience and statistical factors.

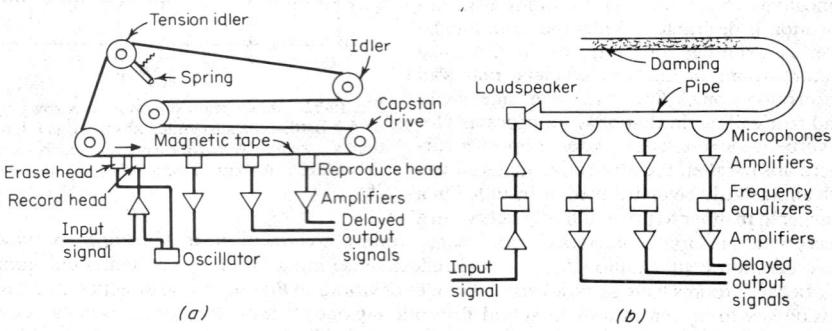

Fig. 19-74. Audio delay systems: (*a*) magnetic tape; (*b*) electroacoustic line. (*"Music, Physics and Engineering," Dover Publications, Inc., New York, 1967.*)

Examples of *building acoustical properties* often specified include:
1. Noise criteria.
2. STC ratings of partitions, doors, and windows.
3. Precautions concerning leakage, bridging of double walls, and machinery isolation.
4. Ductwork ratings.
5. Structural isolation requirements.

Examples of *room acoustical properties* often specified include:
1. Reverberation-time goals at various frequencies, room unoccupied and occupied.
2. Room size and shape needed for the acoustical functions.
3. Room volume needed to obtain the maximum reverberation specified.
4. Shape requirements for sound diffusion, for good acoustical dimension ratios (in small rooms) and good sound-level distribution (in large rooms).
5. Shaping precautions to avoid sound focusing and flutter echo.
6. Types, amounts, and locations of seating, absorptive and reflective surface materials, and structures expected to fulfill the acoustical goals.

7. Means proposed for any final adjustment after construction where the requirements are critical.

The adjustment means can be variable absorption or reflection devices, or if variation is not desired, a reserve amount of material, e.g., draperies, which can be omitted or added, depending upon the final result.

A comprehensive and creative study of the dimensional and acoustical properties of concert halls and opera houses[52] includes rating scales intended for correlation of measurable physical properties with expert subjective judgment. Both positive factors, e.g., liveness, and negative factors, e.g., echo, are included. Experience with this approach and evolution of the scales will assist future specification and prediction.

Electroacoustic system specifications are often included with acoustical specifications. They are discussed later under microphones and loudspeakers.

35. Acoustical Engineering Tests. Tests suggested *before* acoustical specifications are prepared for a room or building include noise levels and spectra at the location of construction or renovation (with noise sources operating), noise data on equipment of the general types to be installed, and reverberation-time frequency characteristics and ambient-noise spectra in similar rooms considered satisfactory by the owner or client. Test equipment for these purposes has been described under Noise Measurement.

Inspection must be carried out *during construction* for inadvertent bridging (across double walls and vibration isolators), acoustical leakage, improper installation of sound-rated doors and windows, insecure mounting of panels (causing subsequent rattle), omission of the acoustical lining of ducts, and any other items hidden during final inspection. The porosity of acoustical plasters and concrete block should be checked as they are first installed so that materials and installation techniques can be corrected early. Interim acoustical tests of sound-transmission loss and reverberation time (while construction work is shut down) can advantageously be made on completed sample rooms or on incompleted rooms if allowances are made.

Final acoustical tests include compliance with specifications on noise criteria, STC ratings, reverberation characteristics, sound-level distribution, and the effect of any variable absorption. Echoes are usually detected by listening while traversing the room with a mechanical-impulse generator. Oscillographic methods are helpful in their identification. Maintenance instructions are useful, e.g., on the painting of porous materials, the adjustment and repair of sound-rated door mechanisms, and the caulking of any separations or cracks which may occur during settling or use. Operation instructions or schedules are needed for optimum use of variable absorbers.

Articulation Tests.[59,60] Both rooms and sound systems can be given a final test by articulation methods. These tests use talkers, listeners, standardized word lists, and simple procedures. However, the results depend somewhat upon talker and listener selection, ability, and training. The articulation score is the average percentage of words correctly heard by all listeners for all talkers. Phonetic spelling is more important than dictionary spelling.

Each word is spoken in a carrier sentence such as "You will write ——— now." An example of a phonetically balanced list of 50 words of one syllable follows: ask, bid, bind, bolt, bored, calf, catch, chant, chew, clod, cod, crack, day, deuce, dumb, each, ease, fad, flip, food, forth, freak, frock, front, guess, hum, jell, kill, left, lick, look, night, pint, queen, rest, rhyme, rod, roll, rope, rot, shack, slide, spice, this, thread, till, us, wheeze, wig, yeast. Additional lists and rearrangement of words are necessary to prevent memorization.

Accuracy in articulation testing is achieved only with a trained test crew and statistical analysis of the data. A cursory test with a novice crew can give an indication, but not a definitive answer.

MICROPHONES AND ACCESSORIES[61,62,31]

36. Sound-responsive Elements. The sound-responsive element in a microphone may have many forms (Fig. 19-75). It may be a stretched membrane *(a)*, a clamped diaphragm, *(b)*, or a magnetic diaphragm held in place by magnetic attraction, *(c)*. In these the moving element is either an electric or magnetic conductor, and the motion of the element creates the electric or magnetic equivalent of the sound directly.

Other sound-responsive elements are straight *(d)* or curved *(e)* conical diaphragms with various shapes of annular compliance rings, as shown. The motion of these diaphragms is transmitted by a drive rod from the conical tip to a mechanical transducer below.

Other widely used elements are a circular piston *(f)* bearing a circular voice coil of smaller diameter and a corrugated-ribbon conductor *(g)* of extremely low mass and stiffness suspended in a magnetic field.

37. Transduction Methods. Microphones have a greater variety of transduction methods currently in use than other types of electroacoustic and electromechanical transducers. The variety is shown in Fig. 19-76.

The loose-contact transducer (Fig. 19-76a) was first achieved by Bell in magnetic form and later made practical by Edison's use of carbonized hard-coal particles. It is widely used in telephones. Its chief advantage is its self-amplifying function, in which diaphragm amplitude variations directly produce electric resistance and current variations. Disadvantages include noise, distortion, and instability.

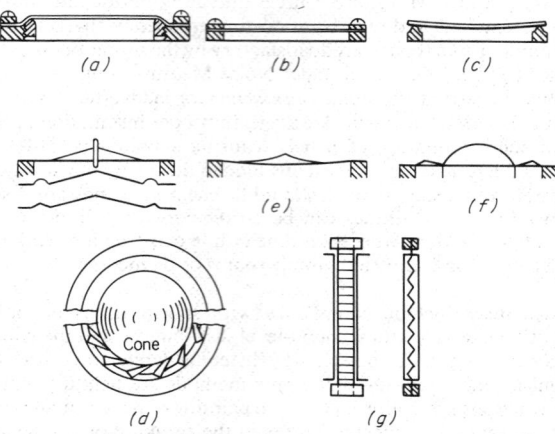

Fig. 19-75. Sound-responsive elements in microphones. *(B. B. Bauer, Proc. IRE, 1962, Vol. 50, p. 719.)*

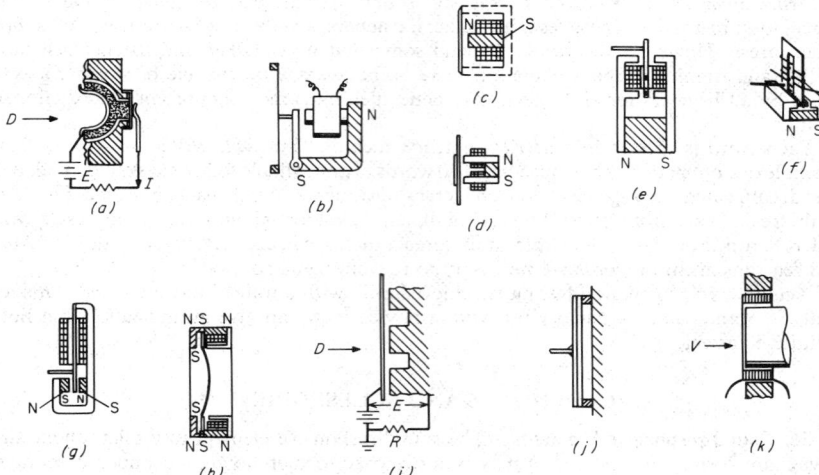

Fig. 19-76. Microphone transduction methods. *(B. B. Bauer, Proc. IRE, 1962, Vol. 50, p. 719.)*

Moving-iron transducers have great variety, ranging from the historic pivoted armature (Fig. 19-76*b*) to the modern ring armature driven by a nonmagnetic diaphragm (Fig. 19-76*h*). In all these types a coil surrounds some portion of the magnetic circuit. The reluctance of the magnetic circuit is varied by motion of the sound-responsive element, which is either moving iron itself (Fig. 19-76*c* and *d*) or is coupled mechanically to the moving iron (Fig. 19-76*b, e,* to *h*). In some of the magnetic circuits (Fig. 19-76*e* to *h*) that portion of the armature surrounded by the coil carries very little steady flux, operating on differential magnetic flux only. Output voltage is proportional to moving-iron velocity.

Electrostatic transducers (Fig. 19-76*i*) use a polarizing potential and depend upon capacitance variation between the moving diaphragm and a fixed electrode for generation of a corresponding potential difference. An old, rarely used type of electrostatic microphone, the electret microphone, has again become practical. The new electret microphone holds its polarization almost indefinitely without continued application of a polarizing potential.[63]

Piezoelectric transducers (Fig. 19-76*j*) create an alternating potential through the flexing of crystalline elements which, when deformed, generate a charge difference proportional to the deformation on opposite surfaces. Rochelle salt was the crystalline material commonly used in low-cost microphones for many years. Because of climatic effects and very high electric impedance it has been largely superseded by polycrystalline ceramic elements.

Moving-coil transducers (Fig. 19-76*k*) generate potential by oscillation of the coil within a uniform magnetic field. The output potential is proportional to coil velocity.

38. Equivalent Circuits.[64] Electronics engineers understand electroacoustic and electromechanical design better with the help of equivalent or analogous electric circuits. Microphone design provides an ideal base for introduction of equivalent circuits because microphone dimensions are small compared to acoustical wavelengths over most of the audio-frequency range. This allows the assumption of lumped circuit constants.

Figure 19-77 shows equivalent symbols for the three basic elements of electrical, acoustical, and mechanical systems. In acoustical circuits the resistance is air friction or viscosity, which occurs in porous materials or narrow slots. Radiation resistance is another form of acoustical damping. Mechanical resistance is friction. Mass in the mechanical system is analogous to electric inductance. The acoustical equivalent is the mass of air in an opening or constriction divided by the square of its cross-sectional area. The acoustical analog of electric capacitance and mechanical-spring compliance is acoustical capacitance. It is the inverse of the stiffness of an enclosed volume of air under pistonlike action. Acoustical capacitance is proportional to the volume enclosed.

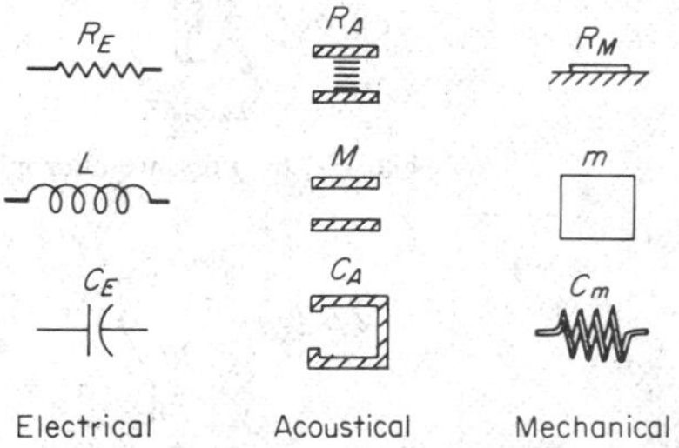

Figure 19-78 is an equivalent electric circuit for a Helmholtz resonator. Sound-pressure and air-volume current are analogous to electric potential and current, respectively. Other analog systems have been proposed. One frequently used has advantages for mechanical systems.[65]

Fig. 19-77. Equivalent basic elements in electrical, acoustical, and mechanical systems. (Ref. 22)

39. Microphone Types and Equivalent Circuits. Different types of microphone respond to different properties of the acoustical *input* wave. Moreover, the electrical *output* can be proportional to different internal mechanical variables.

Pressure Type, Displacement Response. Figure 19-79 shows a microphone responsive to the sound-pressure wave acting through a resonant acoustical circuit upon a resonant

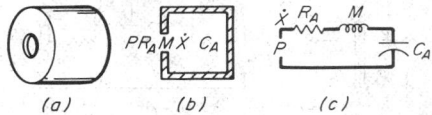

(a)   (b)   (c)

Fig. 19-78. Helmholtz resonator in (*a*) perspective and (*b*) in section and (*c*) equivalent electric circuit. (Ref. 22)

diaphragm coupled to a piezoelectric element responsive to displacement. (The absence of sound ports in the case or in the diaphragm keeps the microphone pressure responsive.) In the equivalent circuit the sound pressure is the generator. L_a and R_a represent the radiation impedance, L_s and R_s are the inertance and acoustical resistance of the holes; C_s is the capacitance of the volume in front of the diaphragm; L_m, C_m, and R_m are the mass, compliance, and resistance of the piezoelectric element and diaphragm lumped together; and C_b is the capacitance of the entrapped back volume of air. The electrical output is the potential differential across the piezoelectric element. It is shown across the capacitance in the equivalent circuit because microphones of this type are designed to be stiffness-controlled throughout most of their operating range.

Pressure Type, Velocity Response. Figure 19-80 shows a moving-coil pressure microphone which is a velocity-responsive transducer. In this microphone three acoustical circuits lie behind the diaphragm. One is behind the dome and another behind the annular rings. The third acoustical circuit lies beyond the acoustical resistance at the back of the voice-coil gap and includes a leak from the back chamber to the outside. This microphone is resistance-controlled throughout most of the range, but at low frequencies its response is extended by the resonance of the third acoustical circuit. Output potential is proportional to the velocity of voice-coil motion.

Pressure-Gradient Type, Velocity Response. When both sides of the sound-responsive element are open to the sound wave, the response is proportional to the *gradient* of the pressure wave. Figure 19-81 shows a ribbon conductor in a magnetic field with both sides of the ribbon open to the air. In the equivalent circuit there are two generators, one for sound

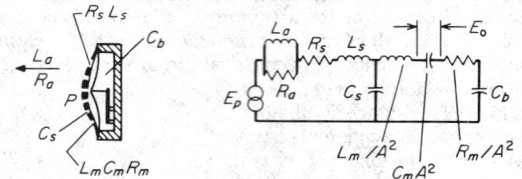

Fig. 19-79. Pressure microphone, displacement response. (Ref. 61)

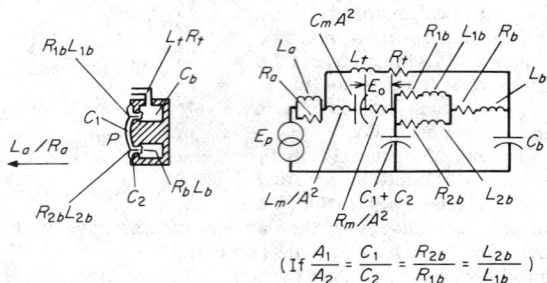

$$\left(\text{If } \frac{A_1}{A_2} = \frac{C_1}{C_2} = \frac{R_{2b}}{R_{1b}} = \frac{L_{2b}}{L_{1b}}\right)$$

Fig. 19-80. Pressure microphone, velocity response. (Ref. 61)

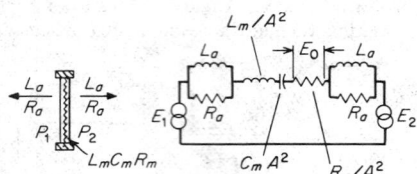

Fig. 19-81. Gradient microphone, velocity response. (Ref. 61)

pressure on each side. Radiation resistance and reactance are in series with each generator and the circuit constants of the ribbon. Usually the ribbon resonates at a very low frequency, making its mechanical response mass-controlled throughout the audio-frequency range. The electrical output is proportional to the conductor velocity in the magnetic field. Gradient microphones respond differently to distant and close sound sources (see Par. **19-42**).

40. Directional Patterns and Combination Microphones. Because of diffraction, a pressure microphone is equally responsive to sound from all directions as long as the wavelength is larger than microphone dimensions (see Fig. 19-82*a*). (At high frequencies it is somewhat directional along the forward axis of diaphragm or ribbon motion.)

By contrast a pressure-gradient microphone has a figure-eight directional pattern (Fig. 19-82*b*), which rotates about the axis of ribbon or diaphragm motion. A sound wave approaching a gradient microphone at 90° from the axis produces balanced pressure on the two sides of the ribbon and consequently no response. This defines the *null plane* of a gradient microphone. Outside this plane the microphone response follows a cosine law.

If the pressure and gradient microphones are combined in close proximity (see Fig. 19-83) and are connected electrically to add in equal (half-and-half) proportions, a heart-shaped cardioid pattern (Fig. 19-82*c*) is obtained. (The back of the ribbon in the pressure microphone is loaded by an acoustical resistance line.) By combining the two outputs in other proportions other limaçon directional patterns can be obtained.

41. Phase-Shift Directional Microphones. Directional characteristics similar to those of the combination microphones can also be obtained with a single moving element by means of equivalent circuit analysis using acoustical phase-shift networks. Figure 19-84 shows a moving-coil, phase-shift microphone and its simplified equivalent circuit. The phase-shift network is composed of the rear-port resistance R_2 and inertance L_2, the capacitance of the volume under the diaphragm and within the magnet, and the impedance of the interconnecting screen. The microphone has a cardioid directional pattern.

42. Special-Purpose microphones include two types which are superdirectional, two which overcome noise, and one without cables.

Line microphones (Fig. 19-85) use an approximate line of equally spaced pickup points connected through acoustically damped tubes to a common microphone diaphragm. The

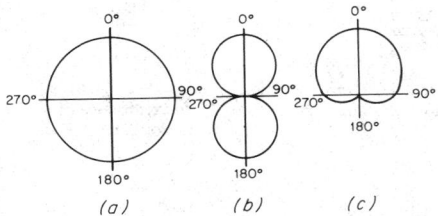

(*a*)   (*b*)   (*c*)

Fig. 19-82. Directional patterns of microphones: (*a*) nondirectional, (*b*) bidirectional, (*c*) unidirectional. (Ref. 22)

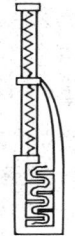

Fig. 19-83. Combination unidirectional microphone. (Ref. 61)

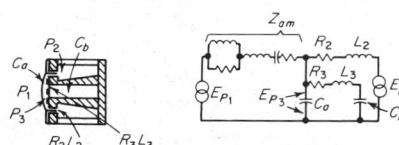

Fig. 19-84. Phase-shift unidirectional microphone. (Ref. 61)

phase relationships at these points for an incident plane wave combine to give a sharply directional pattern along the axis if the line segment is at least one wavelength.

Parabolic microphones face a pressure microphone unit toward a parabolic reflector at its focal point, where sounds from distant sources along the axis of the parabola converge. They are effective for all wavelengths smaller than the diameter of the reflector.

Noise-canceling microphones[66] are gradient microphones in which the mechanical system is designed to be stiffness-controlled rather than mass-controlled. For distant sound sources the resulting response is greatly attenuated at low frequencies (see Fig. 19-86). However, for a very close sound source the response-frequency characteristic is uniform because the *gradient* of the pressure wave near a point source decreases with increasing frequency. Such a microphone provides considerable advantage for nearby speech over distant noise on the axis of the microphone. In addition, there is an advantage from the effect of the figure-eight directional pattern. Figure 19-87 shows the combined advantage of on-axis close speech over distant randomly incident noise.

Contact microphones are used on string and percussion musical instruments, on seismic-vibration detectors, and for pickup of body vibrations including speech. The throat microphone was noted for its convenience and its rejection of airborne noise. Most types of

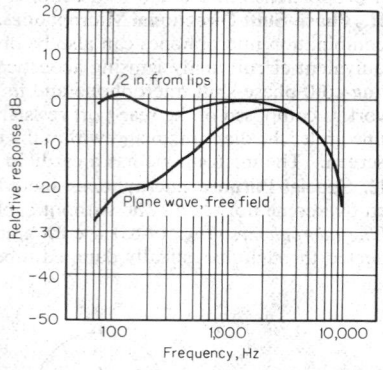

Fig. 19-85. Line microphone. ("Music, Physics and Engineering," Dover Publications, Inc., New York, 1967.)

Fig. 19-86. On-axis response-frequency characteristics of an RCA pressure-gradient microphone. (Ref. 17)

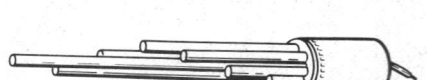

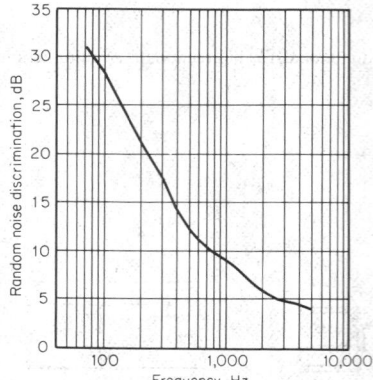

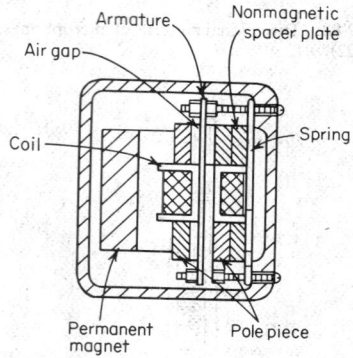

Fig. 19-87. Random-noise discrimination of the microphone in Fig. 19-86. (Ref. 17)

Fig. 19-88. Balanced-armature magnetic throat microphone (inertia operated). (Ref. 67)

throat microphone are inertia-operated (Fig. 19-88), the case receiving vibration from the throat walls actuated by speech sound pressure within the throat. The disadvantage is a deficiency of speech sibilant sounds received back in the throat from the mouth.[67]

Wireless microphones have obvious operational advantages over those with microphone cords. A wireless microphone contains a small, low-power radio transmitter with a nearby receiver connected to an audio communication system. Any of the microphone types can be so equipped. The potential disadvantage is in rf interference and field effects upon transmitted level.

43. Microphone Use in Recordings. The choice of microphone type and placement greatly affects the sound of a recording. For speech and dialogue recordings pressure microphones are usually placed near the speakers in order to minimize ambient-noise pickup and room reverberation. Remote pressure microphones are also used when a maximum room effect is desired. The farther from the sound source the less direct the pickup and the greater the microphone pickup of generally reflected sound.

In the playback of monophonic recordings room effects are more noticeable than they would have been to a listener standing at the recording microphone position because single-microphone pickup is similar to single-ear (monaural) listening, in which the directional clues of localization are lost. Therefore microphones generally need to be closer in a monophonic recording than in a stereophonic recording.

In television pickup of speech, where the microphone should be outside the camera angle, unidirectional microphones are often used because of their greater ratio of direct to generally reflected sound response.

Both velocity (gradient) microphones and unidirectional microphones can be used to advantage in broadcasting and recording. Figure 19-89a shows how instruments may be placed around a figure-eight directivity pattern to balance weaker instruments 2 and 5 against stronger instruments 1 and 3 with a potential noise source at point 4. In Fig. 19-89b source 2 is favored, with sources 1 and 3 somewhat reduced and source 4 highly discriminated against by the cardioid directional pattern. In Fig. 19-89c an elevated unidirectional microphone aimed downward responds uniformly to sources on a circle around the axis while discriminating against mechanical noises at ceiling level. Fig-

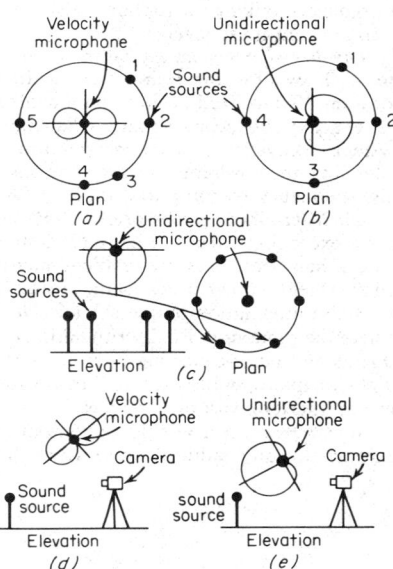

Fig. 19-89. Use of directional microphones. (Ref. 22)

ure 19-89d places the camera noise in the null plane of a figure-eight pattern, and Fig. 19-89e shows a similar use for the unidirectional microphone. Camera position is less critical for the cardioid microphone than for the gradient microphone.

Early classical stereo recordings used variations of two basic microphone arrangements. In one scheme two unidirectional microphones were mounted close together with their axes angled toward opposite ends of the sound field to be recorded. This retained approximately the same arrival time and phase at both microphones, depending chiefly upon the directivity patterns to create the sound difference in the two channels.

In the second scheme the two microphones (not necessarily directional) were separated by distances of 5 to 25 ft, depending upon the size of the sound field to be recorded. Microphone axes (if directional) were again directed toward the ends of the sound field or group of sound sources. In this arrangement the time of arrival and phase differences were more important, and the effect of directivity was lessened. Each approach had its advantages and disadvantages.

With the arrival of tape recorders having many channels a trend has developed toward the use of more microphones and closer microphone placement. This offers much greater flexibility in mixing and rerecording, and it largely removes the effect of room reverberation from the recording. This may be either an advantage or a disadvantage depending upon the viewpoint. Reverberation can be added later (see Fig. 19-73).

In sound-reenforcement systems for dramatic productions and orchestras the use of many microphones again offers operating flexibility. However, it also increases the probability of operating error, increased system noise, and acoustical feedback, making expert monitoring and mixing of the microphone outputs virtually mandatory.

44. Microphone Mounting. On podiums and lecterns microphones are typically mounted on fixed stands with adjustable arms. On stages they are mounted on adjustable floor stands. In mobile communication and in other situations where microphone use is occasional, hand-held microphones are used during communication and are stowed on hangers at other times. For television and film recording, where the microphone must be out of camera sight, the microphones are usually mounted on booms overhead and are moved about during the action to obtain the best speech-to-noise ratio possible at the time. In two-way communication situations which require the talker to move about or to turn his head frequently, the microphone can be mounted on a boom fastened to his headset. This provides a fixed close-talking microphone position relative to the mouth, a considerable advantage in high-ambient-noise levels.

Lavalier microphones are worn on the chest, suspended by a cord around the speaker's neck. They allow the wearer great flexibility of movement without concern for microphone location. Their fixed position on the talker and their proximity to the speech source give a better acoustical feedback margin than most stand mountings provide in sound reenforcement systems. Lavalier microphones require a special response-frequency characteristic which attenuates the low frequencies to compensate for the chest-baffle effect and which accentuates high-frequency response to compensate for the microphone position off the axis of speech.

45. Microphone Accessories. *Noise shields* are needed for microphones in ambient noise levels exceeding 110 dB. Figure 19-90 shows a removable noise shield installed on a military type of hand-held pressure-gradient microphone with a talk switch. The thin rubber flange on the shield is pressed against the face around the mouth during speech. Figure 19-91 shows the noise attenuation by the shield. Noise shields are quite effective at high frequencies, where the random-noise discrimination of noise-canceling microphones diminishes. Noise shields and noise-canceling microphones complement each other. Noise shields are also useful on microphones used for dictation or commentary in situations where the talker's voice would interfere with proceedings.

Windscreens are available for microphone use in airstreams or turbulence. Without them aerodynamically induced noise is produced by turbulence at the microphone grille or

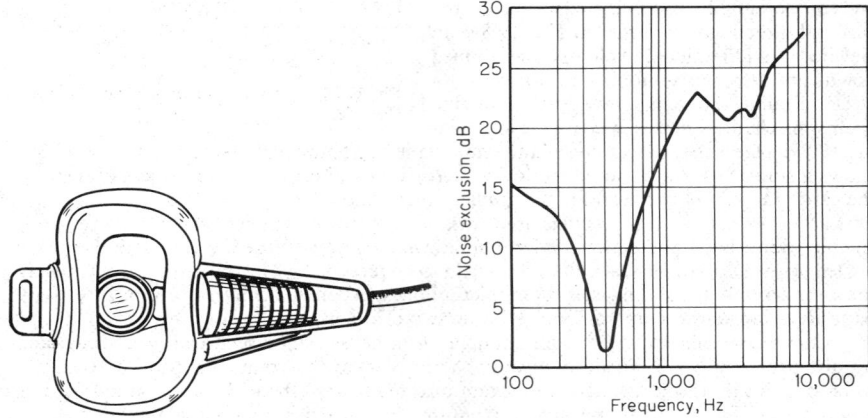

Fig. 19-90. Noise shield on a military pressure-gradient moving-coil RCA microphone (M-34/AIC). (Ref. 17)

Fig. 19-91. Exclusion of external noise when the shield shown in Fig. 19-90 is worn. (Ref. 17)

openings. Figure 19-92 shows a cylindrical windscreen which lowers wind noise by over 20 dB. Large windscreens are more effective than small ones because they move the turbulence region farther from the microphone.

Figure 19-92 also shows special sponge-rubber mountings for the microphone and cable to reduce extraneous vibration of the microphone. Many microphone stands and booms have optional suspension mounting accessories to reduce shock and vibration transmitted through the stand or boom to the microphone.

46. Special Properties of Microphones. The source impedance of a microphone is important not only to the associated preamplifier but also to the allowable length of microphone cable and the type and amount of noise picked up by the cable. High-impedance microphones (10,000 Ω or more) cannot be used more than a few feet from the preamplifier without pickup from stray fields. Microphones having an impedance of a few ohms or less are usually equipped with step-up transformers to provide a line impedance in the range of 30 to 600 Ω, which extensive investigation has established as the most noise-free line-impedance range.

The microphone unit itself can be responsive to hum fields at power-line frequencies unless special design precautions are taken. Most microphones have a hum-level rating based upon measurement in a standard alternating magnetic field.[68]

For minimum electrical noise balanced and shielded microphone lines are used, with the shield grounded only at the amplifier end of the line.

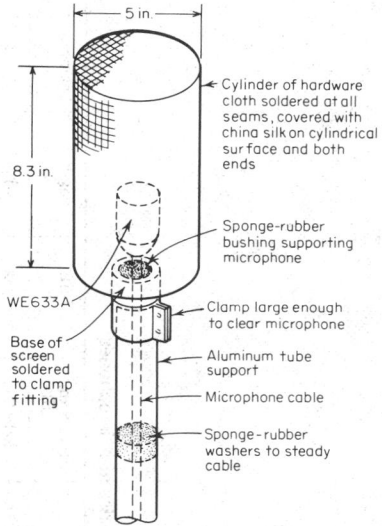

Fig. 19-92. Cylindrical experimental microphone windscreen. (Ref. 46)

Microphone linearity should be considered when the sound level exceeds 100 dB, a frequent occurrence for loud musical instruments and even for close speech. Close-talking microphones, especially of the gradient type, are particularly susceptible to noise from breath and plosive consonants.

47. Specifications. Microphone specifications typically include many of the following items: type or mode of operation, directivity pattern, frequency range, uniformity of response within the range, output level at one or more impedances for a standard sound-pressure input, (for example, 10 dynes/cm^2), recommended load impedance, hum output level for a standard magnetic field (for example, 10^{-3} G), dimensions, weight, finish, mounting, power supply (if necessary), and accessories. Table 19-6 compares several microphone types.

48. Microphone Tests. Acoustical tests of microphones are generally run under anechoic conditions, either in an anechoic chamber or on an outdoor tower. This is necessary for the measurement of directional characteristics. Directional characteristics of microphones are plotted on polar-coordinate level recorders while the microphone rotates upon its axis in front of the standard sound source.

Microphone response curves are usually smoother than loudspeaker response curves. For this reason the sound pressure at a microphone under test is sometimes maintained constant by feedback from a nearby standard laboratory microphone to an automatic gain control in the loudspeaker amplifier circuit. This minimizes the effect of loudspeaker variations upon the appearance of the microphone test curve.

LOUDSPEAKERS, EARPHONES, AND ACCESSORIES[62,69,70]

49. Introduction. A loudspeaker is an electroacoustic transducer intended to radiate acoustic power into the air, with the acoustic waveform equivalent to the electrical input waveform. An earphone is an electroacoustic transducer intended to be closely coupled acoustically to the ear. Both the loudspeaker and earphone are receivers of audio electronic

Table 19-6. Characteristics of Microphones

Company and model	Transducer	Output, dB re 1 V/microbar	Impedance	Pattern	Freq. Range ±1 dB	±3 dB	Length, in.	Diameter, in.	Weight, lb
Altec 633A	Moving-coil	−90	30 Ω	Circular (no baffle)	200-600	50-5,000	3½	2	⅝
B & K 4134	Condenser	−58	20 pF	Circular	10-25,000	3-30,000	½	½	On preamp
E-V RE 50	Moving-coil	−83	150 Ω	Circular	120-4,000	80-13,000	7¾	2	⅝
RCA 77-D	Ribbon	−79	250 Ω	Circular cosine, or cardioid	120-3,000	50-10,000	11½	2½	3
Shure 300	Ribbon	−87	150 Ω	Cosine	50-5,000	40-7,000	9¼	1½	1
W.E. 640AA	Condenser	−50	50 pF	Circular	10-10,000	1-12,000	1	1	On preamp

signals. The principal distinction between them is the acoustical loading. An earphone delivers sound to air in the ear. A loudspeaker delivers sound indirectly to the ear through the air.

The transduction methods of loudspeakers and earphones are historically similar and are treated together. However, since loudspeakers operate primarily into radiation resistance and earphones into acoustical capacitance, the design, measurement, and use of the two types of electroacoustic transducers will be discussed separately.

50. Transduction Methods. Early transducers for sound reproduction were of the mechanoacoustic type. Vibrations received by a stylus in the undulating groove of a record were transmitted to a diaphragm, placed at the throat of a horn for better acoustical impedance matching to the air, all without the aid of electronics. Electroacoustics and electronics introduced many advantages and a variety of transduction methods including moving-coil, moving-iron, electrostatic, magnetostrictive, and piezoelectric (Fig. 19-93).

Most loudspeakers are moving-coil type today, although moving-iron transducers were once widely used. Electrostatic loudspeakers are used chiefly in the upper range of audio frequencies, where amplitudes are small. Magnetostrictive

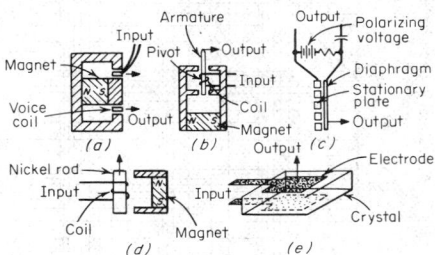

Fig. 19-93. Loudspeaker (and earphone) transduction methods: (*a*) moving-coil; (*b*) moving-iron; (*c*) electrostatic; (*d*) magnetostrictive; (*e*) piezoelectric. (Ref. 69)

and piezoelectric loudspeakers are used for underwater sound. All the transducer types are used in earphones except magnetostrictive.

Moving-Coil. The mechanical force on the moving coil of Fig. 19-93*a* is developed by the interaction of the current in the coil and the transverse magnetic field disposed radially across the gap between the magnet cap and the iron housing which completes the magnetic circuit. The output force along the axis of the circular coil is applied to a sound radiator.

Moving-iron transducers reverse the mechanical roles of the coil and the iron. The iron armature surrounded by the stationary coil is moved by mechanical forces developed within the magnetic circuit. Moving-iron magnetic circuits have many forms (see Fig. 19-93). In the balanced armature system (Fig. 19-93*b*) the direct magnetic flux passes only transversely through the ends of the armature centered within the two magnetic gaps. Coil current polarizes the armature ends oppositely, creating a force moment about the pivot point. The output force is applied from the tip of the armature to an attached sound radiator. In a balanced-diaphragm loudspeaker the armature is the radiator.[71]

Electrostatic. In the electrostatic transducer (Fig. 19-93*c*) there is a dc potential difference between the conductive diaphragm and the stationary perforated plate nearby. Audio signals applied through a blocking capacitor superimpose an alternating potential, resulting in a force upon the diaphragm, which radiates sound directly.

Magnetostrictive transducers (Fig. 19-93*d*) depend upon length fluctuations of a nickel rod caused by variations in the magnetic field. The output motion may be radiated directly from the end of the rod or transmitted into the attached mechanical structure.

Piezoelectric transducers are of many forms using crystals or polycrystalline ceramic materials. In simple form (Fig. 19-93*e*) an expansion-contraction force develops along the axis joining the electrodes through alternation of the potential difference between them.

51. Sound Radiators. The purpose of a sound radiator is to create small, audible air-pressure variations. Whether they are produced within a closed space by an earphone or in open air by a loudspeaker, the pressure variations require air motion or current.

Pistons, Cones, Ports. Expansion and contraction of a sphere is the classical configuration, but most practical examples involve rectilinear motion of a piston, cone, or diaphragm. In addition to the primary direct radiation from moving surfaces there is also indirect or secondary radiation from enclosure ports or horns to which the direct radiators are acoustically coupled.

Attempts have been made to develop other forms of sound radiation such as oscillating airstreams and other aerodynamic configurations with incidental use, if any, of moving mechanical members (see Par. **19-55**).

Air Loading.   Figure 19-94 shows the resistive and reactive components of the air load on one side of a circular piston of radius *r* mounted in a very large flat baffle. The abscissa is proportional to frequency, *k* being $2\pi/\lambda$. (For example, when the diameter equals the wavelength, $kr = \pi$.) The ordinate is mechanical impedance per unit area of piston divided by the characteristic acoustical impedance of air ρc. Note that below $kr = 1.0$ the reactance is directly proportional to frequency and the resistance is proportional to the square of frequency. Beyond $kr = 2$ the resistance is approximately unity, and the reactance falls off inversely with frequency. At low frequencies reactance dominates, and at high frequencies resistance does. These curves are a general basis for understanding the air loading of direct radiator loudspeakers. Not all loudspeaker mountings correspond to a piston in a large flat baffle, but the slopes and critical points of the curves in Fig. 19-94 vary only slightly when the baffle surrounding the piston changes all the way from flat to a long tube extending directly behind the piston.[72]

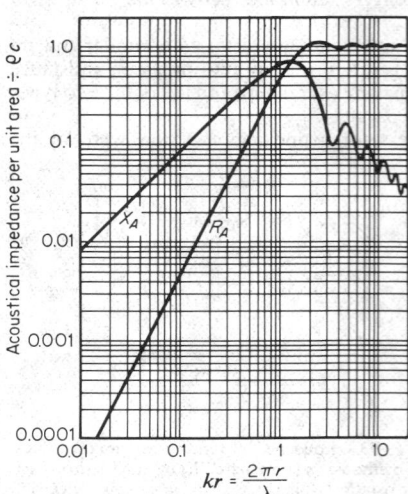

Fig. 19-94. Acoustical resistance R_A and reactance X_A per unit area of air load on a piston of radius *r* in an infinite baffle. (Ref. 22)

Directivity.   Figure 19-95 shows the directional characteristics of a rigid circular piston for different ratios of piston diameter and wavelength of sound. (In three dimensions these curves are symmetrical around the axis of piston motion.) For a diameter of one-quarter wavelength the amplitude decreases 10% (approximately 1 dB sound level) at 90° off axis. For a four-wavelength diameter the same drop occurs in only 5°. (The beam of an actual loudspeaker cone is less sharp than this at high frequencies, where the cone is not rigid.) Note that all the polar curves are smooth when the single-source piston vibrates as a whole.

Radiator Arrays.   When two separate, identical small-sound sources vibrate in phase, the directional pattern becomes narrower than for one source. Figure 19-96 shows that for a separation of one-quarter wavelength the two-source beam is only one-half as wide as for a

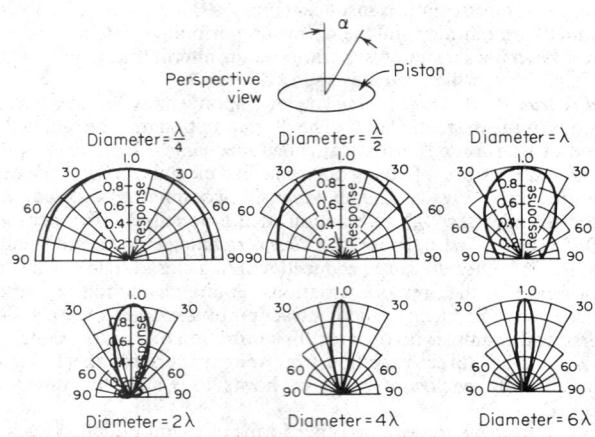

Fig. 19-95. Directional characteristics of rigid circular pistons of different diameters or at different sound wavelengths. (Ref. 22)

single piston. At high frequencies the directional pattern becomes very complex. (In three dimensions these curves become surfaces of revolution about the axis joining the two sources.)

Arrays of larger numbers of sound radiators in close proximity are increasingly directional. Circular-area arrays have narrow beams which are symmetrical about an axis through the center of the circle. Line arrays, e.g., column loudspeakers, are narrowly directional in planes containing the line and broadly directional in planes perpendicular to the line.[73]

52. Direct-Radiator Loudspeakers.[74] Most direct-radiator loudspeakers are of the moving-coil type because of simplicity, compactness, and inherently uniform response-frequency trend. The uniformity results from the combination of two simple physical

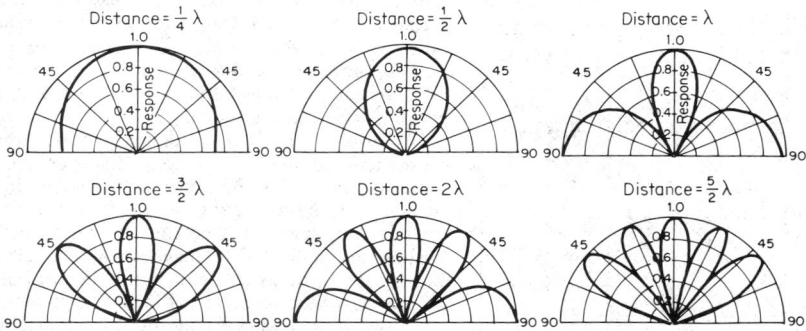

Fig. 19-96. Directional characteristics of two equal, small in-phase sound sources separated by different distances or different sound wavelengths. (Ref. 22)

principles: (1) the radiation resistance increases with the square of the frequency (Fig. 19-94), and hence the radiated sound power increases similarly for constant velocity amplitude of the piston or cone; (2) for a constant applied force (voice-coil current) the mass-controlled (above resonance) piston has a velocity amplitude which decreases with the square of the frequency. Consequently a loudspeaker designed to resonate at a low frequency combines decreasing velocity with increasing radiation resistance to yield a uniform response within the frequency range where the assumptions hold.

Equivalent Electric Circuits.[64] Figure 19-97 shows a cross-sectional view of a direct-radiator loudspeaker mounted in a baffle, the electric voice-coil circuit, and the equivalent electric circuit of the mechanoacoustic system. In the voice-coil circuit e is the emf and R_{EG} the resistance of the generator, e.g., power-amplifier output. L and R_{EC} are the inductance and resistance of the voice coil. Z_{EM} is the motional electric impedance from the mechanoacoustic system.

F_M is the driving force resulting from interaction of the voice-coil current field with the gap magnetic field. M_C is the combined mass of the cone and voice coil. C_{MS} is the compliance of the cone-suspension system. R_{MS} is the mechanical resistance. The mass M_A and radiation resistance R_{MA} of the air load complete the circuit.

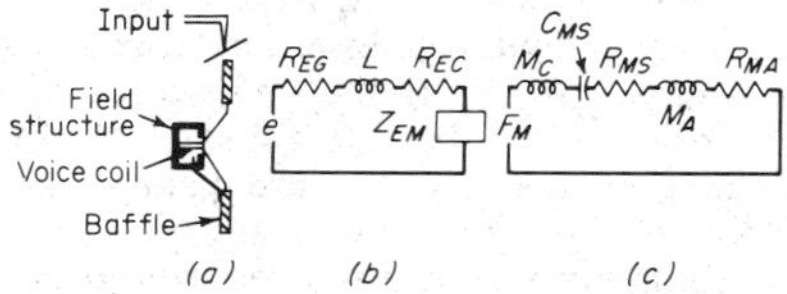

Fig. 19-97. (*a*) Structure, (*b*) electric circuit, and (*c*) equivalent mechanical circuit for a direct-radiator moving-coil loudspeaker on a baffle. (Ref. 69)

Figure 19-98 summarizes these mechanical impedance factors for a 4-in. direct-radiator loudspeaker of conventional design. Above resonance (where the reactance of the suspension system equals the reactance of the cone-coil combination) the impedance-frequency characteristic is dominated by M_C. From the resonance frequency of about 150 Hz to about 1,500 Hz the conditions for uniform response hold.

Efficiency. Since R_{MA} is small compared to the magnitudes of the reactive components, the efficiency of the loudspeaker in this frequency range can be expressed as

$$\text{Efficiency} = \frac{100(Bl)^2 R_{MA}}{R_{EC}(X_{MA} + X_{MC})^2} \% \qquad (19\text{-}9)$$

in which B is gap flux density in gauss, l is the voice-coil conductor length in centimeters, and R_{EC} is the voice-coil electric resistance in abohms. Since R_{MA} is proportional to the square of the frequency and both X_{MA} and X_{MC} increase with frequency, the efficiency is theoretically uniform.

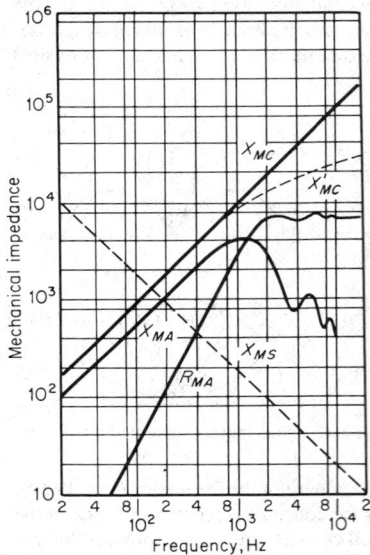

Fig. 19-98. Components of mechanical impedance of a typical 4-in. loudspeaker. (Ref. 69)

All this has assumed that the cone moves as a whole. Actually wave motion occurs in the cone. Consequently at high frequencies the mass reactance is somewhat reduced (as shown in the dashed curve of Fig. 19-98), tending to improve efficiency beyond the frequency where radiation resistance becomes uniform.

Magnetic Circuit. The magnet may be either permanent or field-coil type. However, most magnets now are a high-flux, high-coercive permanent type, either an alloy of aluminum, cobalt, nickel, and iron, or a ferrite of iron, cobalt, barium, and nickel. The magnet may be located in the core of the structure or in the ring, or both. However, magnetization is difficult when magnets are oppositely polarized in the core and ring.

Air-gap flux density varies widely in commercial designs from approximately 3,000 to 20,000 G. Since most of the reluctance in the magnetic circuit resides in the air gap, the minimum practical voice-coil clearance in the gap compromises the maximum flux density. Pole pieces of heat-treated soft nickel-iron alloys, dimensionally tapered near the gap, are used for maximum flux density.

Voice Coils. The voice coil is a cylindrical multilayer coil of aluminum or copper wire or ribbon. Aluminum is used in high-frequency loudspeakers for minimum mass and maximum efficiency. Voice-coil impedance varies from 1 to 100 Ω with 4, 8, and 16 Ω standard. For maximum efficiency the voice-coil and cone masses are equal. However, in large loudspeakers the cone mass usually exceeds the voice-coil mass. Typically the voice-coil mass ranges from tenths of a gram to 5 g or more.

Cones. Cone diameters range from 1 to 18 in. Cone mass varies from tenths of a gram to 100 g or more. Cones are made of a variety of materials. The most common is paper deposited from pulp upon a wire-screen form in a felting process. For high-humidity environment cones are molded from plastic materials, sometimes with a cloth or fiber-glass base. Some low-frequency loudspeaker cones are molded from low-density plastic foam to achieve greater rigidity with low density.

So far piston action has been assumed in which the cone moves as a whole. Actually at high frequencies the cone no longer vibrates as a single unit. Typically there is a major dip in response resulting from quarter-wave reflection from the circular rim of the cone back to the voice coil. For loudspeaker cones in the range of 8 to 15 in. diameter this dip usually occurs in the range of 1,000 to 2,000 Hz. Figure 19-99 combines typical impedance and response-frequency characteristics for a baffle-mounted 8-in. loudspeaker without edge damping. The quarter-wave dip is at 1,050 Hz. Other dips are at integral multiples of a wavelength along the cone radius. Peaks occur at odd multiples of a half wavelength.[75]

Figure 19-100 shows corresponding nodal patterns of cone vibration at numbered frequencies of interest in Fig. 19-99. It is this complex pattern of cone breakup that sustains the efficiency of a cone loudspeaker at high frequencies. However, the resulting nonuniformity of response and extreme complexity of directional characteristics make two- and three-way loudspeaker systems with frequency-range dividing networks necessary for high-quality sound reproduction.

Typical Commercial Design Values. Figure 19-101 shows typical values for several cone and voice-coil design parameters for a range of loudspeaker diameters.[62] These do not apply to extreme cases, such as high-compliance loudspeakers or high-efficiency horn drivers. The effective piston diameter (Fig. 19-101a) is less than the loudspeaker cone diameter because the amplitude falls off toward the edges. A range of resonance frequencies is available for any cone diameter, but Fig. 19-101b shows typical values. In Fig. 19-101c typical cone mass is M including the voice coil and M' excluding the voice coil. Figure 19-101d shows typical cone-suspension compliance.

Impedance. The impedance-frequency characteristic of a typical baffle-mounted loudspeaker is also shown in Fig. 19-99. A major peak results from motional impedance at primary mechanical resonance. Impedance is usually uniform above this peak until voice-coil inductance becomes dominant over resistance.

Power Ratings. Different types of power rating are needed to express the performance

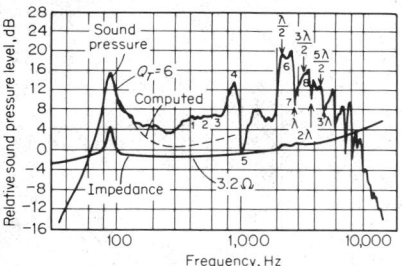

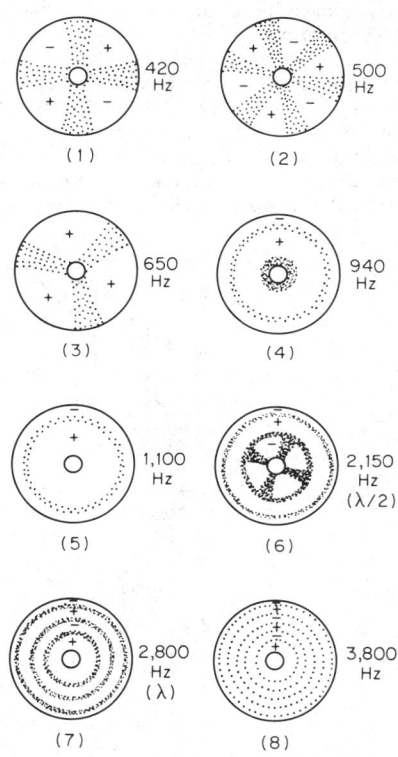

Fig. 19-99. Impedance and response-frequency characteristics for an 8-in.-diameter loudspeaker mounted in a very large baffle. Note the response nonuniformity caused by cone breakup. (Ref. 75)

Fig. 19-100. Nodal patterns of the loudspeaker cone of Fig. 19-99 at selected significant frequencies (numbered points along the curve). (Ref. 75)

capabilities of loudspeakers. The large range of typical loudspeaker efficiency makes the acoustical power-delivering capacity quite important. The electrical power-receiving capacity (without overload or damage) determines the choice of power amplifier.

Loudspeaker efficiencies are seldom measured but are often compared by measuring the sound-pressure level at 4 ft on the loudspeaker axis for 1-W audio input. High-efficiency direct radiators provide 95 to 100 dB. Horn loudspeakers are typically higher by 10 dB or more, being both more efficient and more directional.

Loudspeakers are also rated by the maximum rms power output of amplifiers which will not damage the loudspeaker or drive it into serious distortion on peaks. Such ratings usually assume that the amplifier will seldom be driven to full power. For example a 30-W amplifier will seldom be required to deliver more than 10 W rms of music program material. Otherwise music peaks would be clipped and sound distorted.

However, in speech systems for high-ambient-noise levels the speech peaks may be clipped intentionally, causing the loudspeaker to receive the full 30 W much of the transmission time. Then the loudspeaker must handle large excursions without mechanical damage to the cone suspension and without destroying the cemented coil or charring the form.

Distortion. Nonlinear distortion in a loudspeaker is inherently low in the mass-controlled range of frequencies. However, distortion is produced by nonlinear cone suspension at low frequencies, voice-coil motion beyond the limits of uniform air-gap flux, Doppler shift modulation of high-frequency sound by large cone velocity at low frequencies, and nonlinear distortion of the air near the cone at high powers (particularly in horn drivers). Methods for controlling these distortions follow.

1. When a back enclosure is added to a loudspeaker, the acoustical capacitance of the enclosed volume is represented by an additional series capacitor in the mechanical circuit of Fig. 19-97. Insufficient volume stiffens the cone acoustically, raising the resonance frequency and limiting the low-frequency range of the loudspeaker. It is convenient to reduce nonlinear distortion at low frequencies by increasing the cone-suspension compliance and depending upon the back enclosure to provide the system stiffness. Since an enclosed volume is more linear than most mechanical springs, this lowers low-frequency distortion.

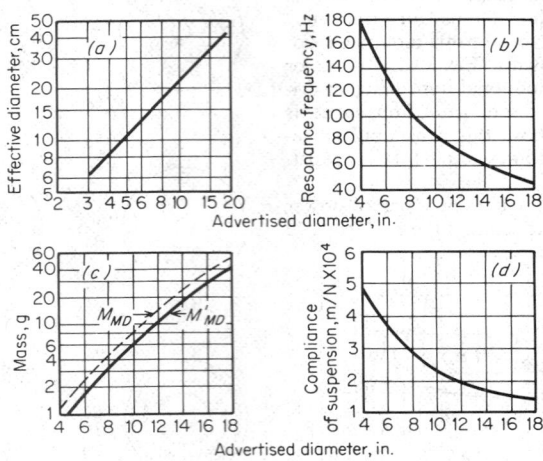

Fig. 19-101.   Typical cone and coil design values. (Ref. 62)

2. Distortion from inhomogeneity of the air-gap flux can be reduced by making the voice-coil length either considerably smaller or larger than the gap width. This stabilizes the total number of lines passing through the coil, but it also reduces loudspeaker efficiency.

3. Doppler distortion can be eliminated only by separating the high and low frequencies in a multiple loudspeaker system.

4. Air-overload distortion can be avoided by increasing the radiating area.

Transient distortion is a factor near cone resonance frequencies. Figure 19-102 shows the opposite effects at a cone peak and a dip caused by quarter-wave reflection from the cone rim. Edge damping or angularly unsymmetrical cone properties, e.g., elliptical rim, will modify this condition.

53. Loudspeaker Mountings and Enclosures.   Figure 19-103 shows a variety of mountings and enclosures. An unbaffled loudspeaker is an acoustic doublet for wavelengths greater than the rim diameter. In this frequency range the acoustical power output for constant cone velocity is proportional to the fourth power of the frequency.

Baffles.   In order to improve efficiency at low frequencies it is necessary to separate the front and back waves. Figure 19-103a is the simplest form of baffle. The effect of different baffle sizes is given in Fig. 19-104. Response dips occurring when the acoustic path from front to back is a wavelength are eliminated by irregular baffle shape or off-center mounting.

Enclosures.   The widely used open-back cabinet (Fig. 19-103b) is noted for a large response peak produced by open-pipe acoustical resonance. A closed cabinet (Fig. 19-103c) adds acoustical stiffness at low frequencies where the wavelength is larger than the enclosure.

At higher frequencies the internal acoustical resonances create response irregularities requiring internal acoustical absorption.

Ported Enclosures (Fig. 19-103d). Enclosure volume can be minimized without sacrificing low-frequency range by providing an appropriate port in the enclosure wall. Acoustical inertance of the port should resonate with the enclosure capacitance at a frequency about an octave below cone resonance frequency. B, Fig. 19-105, shows that this extends the low-frequency range. This is most effective when the port area equals the cone-piston area. Port inertance can be increased by using a duct. An extreme example of ducting is the acoustical labyrinth (Fig. 19-103e). When duct work is shaped to increase cross section gradually, the labyrinth becomes a low-frequency horn (Fig. 19-103f).

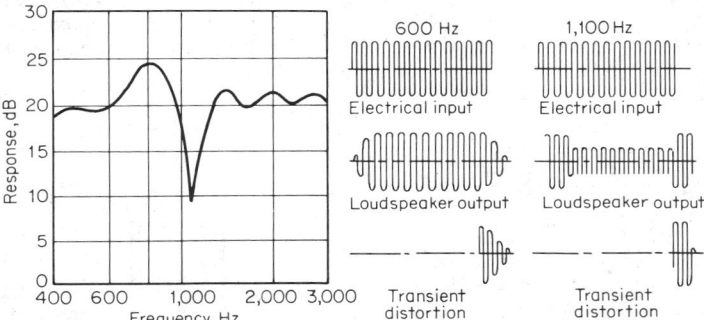

Fig. 19-102. Response-frequency fluctuation in a direct-radiator loudspeaker and transient response to tone bursts at the peak and dip frequencies. (Ref. 69)

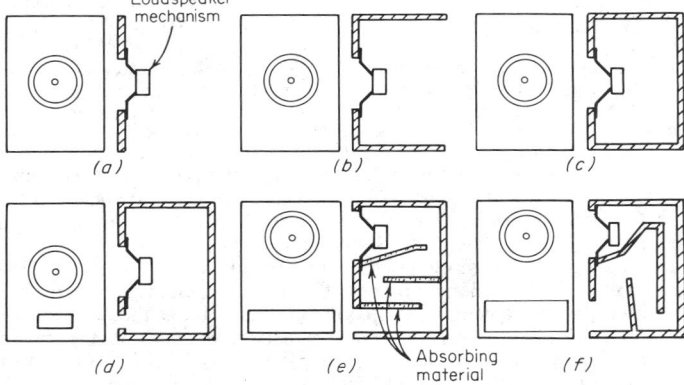

Fig. 19-103. Mountings and enclosures for direct-radiator loudspeakers: (a) flat baffle; (b) open-back cabinet; (c) closed cabinet; (d) ported closed cabinet; (e) labyrinth; (f) folded horn. (Ref. 69)

Direct-radiator loudspeaker efficiency is typically 1 to 5%. Small, highly damped types with miniature enclosures may be only 0.1%. Transistor amplifiers easily provide the audio power for domestic loudspeakers. However, in auditorium, outdoor, industrial, and military applications much higher efficiency is required.

54. Horn Loudspeakers.[76] Higher efficiency is obtained with an acoustic horn, which is a tube of varying cross section having different terminal areas to provide a change of acoustic impedance. Horns match the high impedance of dense diaphragm material to the low air impedance. Horn shape or taper affects the acoustical transformer response. Conical, exponential, and hyperbolic tapers have been widely used. The potential low-frequency cutoff of a horn depends upon its taper rate. Impedance transforming action is controlled by the ratio of mouth to throat diameter.

Horn Drivers. Figure 19-106 shows horn-driving mechanisms and straight and folded horns of large- and small-throat types. A large-throat driver (Fig. 19-106a) resembles a direct-radiator loudspeaker with a voice-coil diameter of 2 to 3 in. and a flux density around 15,000 G. A small-throat driver (Fig. 19-106b) resembles a moving-coil microphone structure. Radiation is taken from the back of the diaphragm into the horn throat through passages which deliver in-phase sound from all diaphragm areas. Diaphragm diameters are 1 to 4 in. with throat diameters of $\frac{1}{4}$ to 1 in. Flux density is approximately 20,000 G.

Large-Throat Horns. These are used for low-frequency loudspeaker systems. A folded horn (Fig. 19-106c) is preferred over a straight horn (Fig. 19-106d) for compactness.

Small-Throat Horns. A folded horn (Fig. 19-106e) with sufficient length and gradual taper can operate efficiently over a wide frequency range. This horn is useful for outdoor music reproduction in a range of 100 to 5,000 Hz. Response smoothness is often compromised by segment resonances. Extended high-frequency range requires a straight-axis horn (Fig. 19-106f).

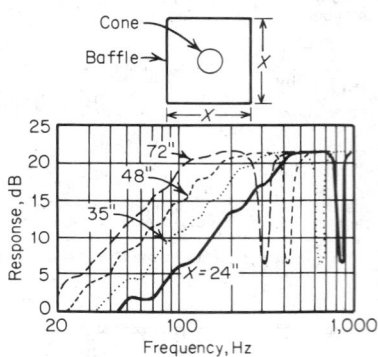

Fig. 19-104. Response-frequency for loudspeaker in 2-, 3-, 4-, and 6-ft square baffles. (Ref. 31, after Olson, "Acoustical Engineering.")

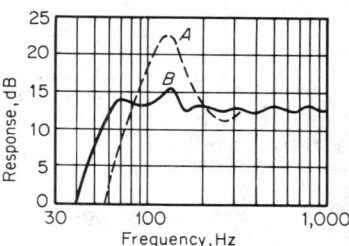

Fig. 19-105. Response-frequency for loudspeaker in closed (*A*) and ported (*B*) cabinets. (Ref. 31, after Olson, "Acoustical Engineering.")

Horn Directivity. When horns have large mouths and well-defined wavefronts, horn radiation is highly directional. This is an advantage in outdoor applications and in large auditoriums. A problem with directivity of simple horns is the narrow beam at high frequencies.

Horn Arrays. For controlled directivity over a broad angle a horn array shown in Fig. 19-107a can be used. It has a large number of small horn mouths spread over a spherical surface with horn throats converging to a common point. It has the directional characteristics shown in Fig. 19-107b. In mid-range the beam narrows somewhat then broadens again at the high frequencies. Horns with radial symmetry and without cellular construction provide cylindrical wavefronts with smoother directional characteristics.

55. Special Loudspeakers. Special types of loudspeakers for limited applications include the following.

Electrostatic high-frequency units have an effective spacing of about 0.001 in. between a thin metalized coating on plastic and a perforated metal backplate.[77] This spacing is necessary for sensitivity comparable to moving-coil loudspeakers, but it limits the amplitude and the frequency range. Extension of useful response to the lower frequencies can be obtained with larger spacing, for example, $\frac{1}{16}$ in., with a polarizing potential of several thousand volts.[78] This type of unit employs push-pull operation.

Modulated-airflow loudspeakers have an electromechanical mechanism for modulating the airstream from a high-pressure pneumatic source into a horn. Low audio power controls large acoustical power in this system. A compressor is also needed. Nonlinear distortion in the air and reduced speech intelligibility have been limitations of this high-power system.[79]

Thermoelectronic loudspeakers produce an ionized "cloud" of air by corona discharge in the throat of a horn. Audio modulation of the electric field in the throat causes the air to

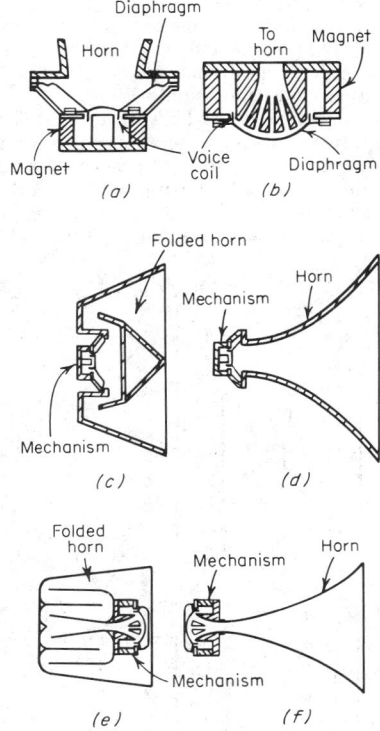

Fig. 19-106. Horns and horn drivers: (*a*) large-throat driver; (*b*) small-throat driver; (*c*) folded large-throat horn; (*d*) straight large-throat horn; (*e*) folded small-throat horn; (*f*) straight small-throat horn. (Ref. 69)

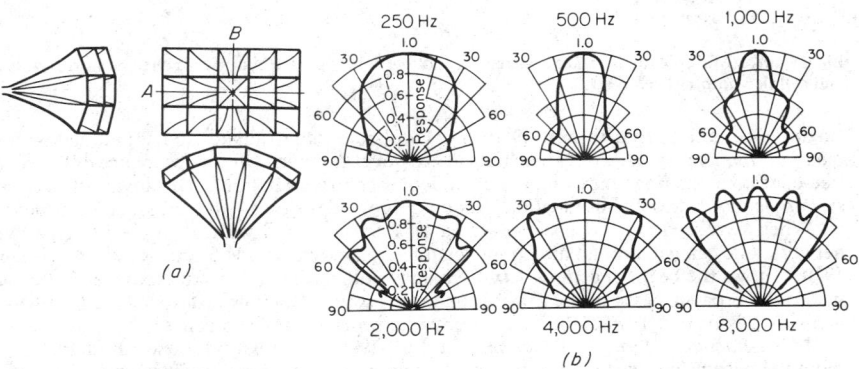

Fig. 19-107. Horn array (cellular) and directional characteristics: (*a*) array; (*b*) horizontal directional curves. (Ref. 31, after Olson, "Acoustical Engineering.")

expand and contract accordingly.[80]   Large amplitudes of vibration dissipate the concentrated charge, making the device useful chiefly at high frequencies.

56. Loudspeaker Systems.   The audio range covers so many octaves that more than one sound radiator is required for the best combination of efficiency, response smoothness, and broad directivity.   Dividing the frequency range into parts and assigning each part a suitable sound radiator gives a superior acoustical result.

Frequency-range division is possible either between the power amplifier and the several loudspeaker or driver units (using a low-impedance, high-power dividing network), or ahead of individual power amplifiers for each of the loudspeaker units.   The choice depends upon the number of power amplifiers planned.   High-power networks in the output are most common, but low-level bandpass filters ahead of the power amplifiers are also used.

Multiple direct-radiator systems are shown in Fig. 19-108a and b.   A double cone with a

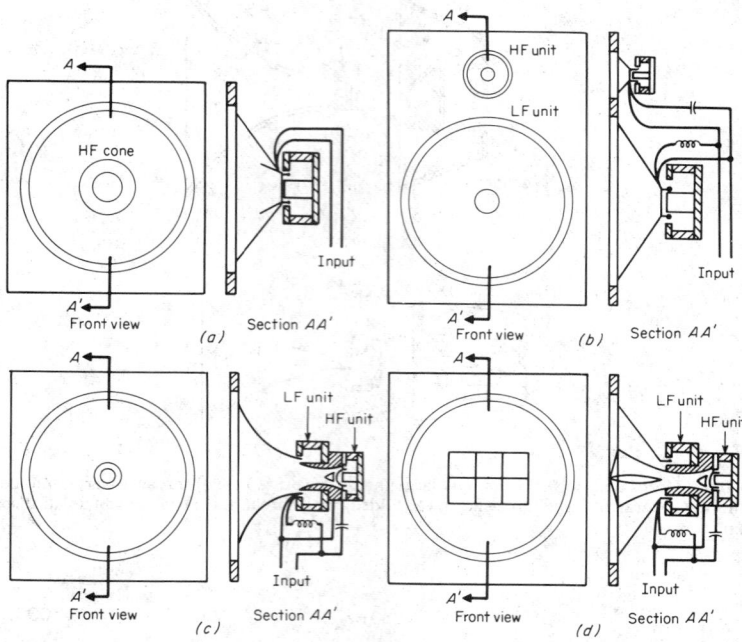

Fig. 19-108.   Multiple loudspeaker systems: (a) double cone; (b) two cones; (c) cone and horn; (d) cone and cellular horn.   (Ref. 69)

single voice coil is used in Fig. 19-108a.   The larger cone radiates the sound at low and medium frequencies.   Above a mechanical crossover frequency at the resonance of the small free-edge cone the frequency range is extended by central radiation.   This low-cost two-way system is used widely although a response dip occurs near the crossover frequency.

When two separate loudspeakers (Fig. 19-108b) are used with a simple *LC* crossover network, one can get a broad directional characteristic over a wide frequency range.   Figure 19-109 shows the beam of a 15-in. low-frequency unit getting narrow at crossover frequency just as the broader beam of the 2½-in. cone takes over.   Doppler distortion in Fig. 19-108a is missing from Fig. 19-108b because of physical separation of the cones.

Direct-Radiator Horn Combination.   In Fig. 19-108c a high-frequency horn driver is mounted within the magnet structure of the low-frequency direct-radiator unit.   The horn forms part of the magnet structure.   The cone of the low-frequency unit is the high-frequency horn.   In Fig. 19-108d a small cellular horn radiates the high-frequency sound.   These coaxial combinations provide smooth transition in the crossover range with a minimum of distortion from phase difference.

Multiple Horn.   When space permits, e.g., behind a movie screen, a two-horn system takes full advantage of horn efficiency.   Figure 19-110 shows such a system using a cellular high-frequency horn and a folded low-frequency horn.   The frequency-range dividing network shown has a slope of 12 dB/octave outside the passband.

Stereophonic loudspeaker systems use a matched pair of any loudspeaker type.   For maximum benefit the stereophonic loudspeaker systems should be separated.   When the stereophonic pickup is three-channel, the center channel should seem to come from a phantom loudspeaker between the real loudspeakers.

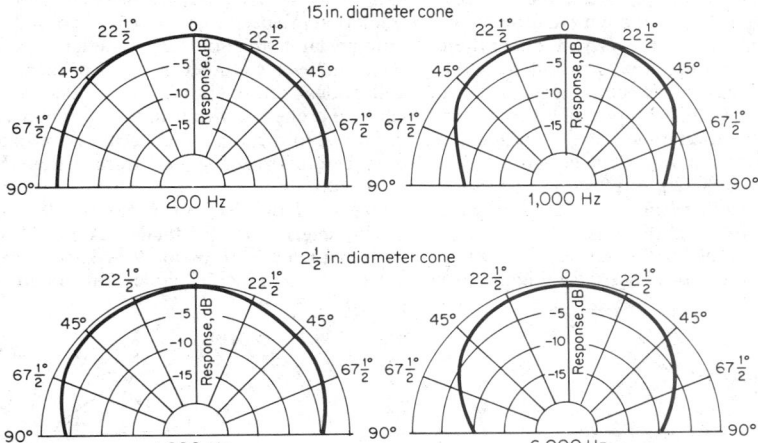

Fig. 19-109.   Directional characteristics of two-cone system.   (Ref. 69)

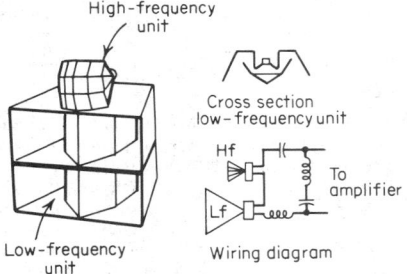

Fig. 19-110.   Two-horn theater loudspeaker system.   (Ref. 31, after Olson, "Acoustical Engineering.")

It is beneficial to aim stereo loudspeaker axes to cross at the rear of the listening area. This allows listeners at one side of the listening area to hear the output from the opposite side better.

Four-channel recording and playback systems have not yet been sufficiently standardized to be described dependably.

57. Loudspeaker Use in Rooms.   Although loudspeakers are designed and measured on an idealized basis, they are used in rooms.   There are a number of practical considerations.

Location Effects.   Loudspeaker response at low frequencies depends upon source location within the room.   At specific frequencies response depends upon room dimensions and the associated low-frequency vibration modes.   Smoothed loudspeaker response curves of Fig. 19-111 show response trends for the same loudspeaker in four different locations.   Maximum low-frequency response is obtained with the loudspeaker in the corner. The next best location is the center of a wall at floor level. A mid-wall location is next, and suspension in the center of the room provides the least low-frequency sound.

If a high-frequency loudspeaker or a wide-range loudspeaker must be mounted in a location where most listeners will be off the axis, acoustic-lens accessories are available which broaden the directional beam at high frequencies.[81]

Localization and Orientation. Different sound systems have different purposes. If the intent is to gain attention (announcing systems) or provide realistic sound reproduction (sound motion pictures), the sources should easily be localized by listeners. For such purposes high directivity and point-source radiation are desirable. For background music, the simulation of reverberation, or the simulation of large-area sources it is better to use broad directional patterns, distributed sound sources, and even to direct the loudspeakers toward scattering and reflecting surfaces. Some applications require both types of sound radiation. For example, the reproduction of a studio-recorded orchestra within an auditorium of limited reverberation might use a multichannel group of directional loudspeakers on stage in combination with delayed and artificially reverberated mixed signals from scattered loudspeakers aimed at reflecting surfaces.

Level Distribution. When sound is only reproduced, the distribution of sound level can be predicted from the sensitivity and directional characteristics of the loudspeakers. However, when some of the sound is direct and the remainder is from loudspeakers, a realistic reenforcement is desired. Figure 19-112 illustrates the basic principles for an auditorium

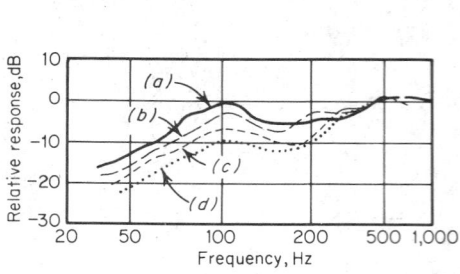

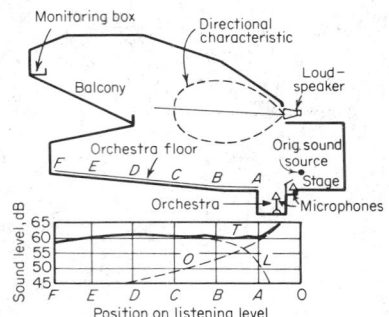

Fig. 19-111. Loudspeaker location effects on low-frequency response: (*a*) corner; (*b*) wall center at floor level; (*c*) center of wall; (*d*) center of room. (Ref. 62)

Fig. 19-112. Sound reenforcing system in a theater. The graph shows the direct sound at the points indicated on the orchestra floor: curve *O*, the original sound; curve *L* the level due to the loudspeaker; curve *T* the total sound. (Ref. 31, after Olson, "Acoustical Engineering.")

with balcony. The loudspeaker is situated directly above the original sound source so that direct and amplified sound arrive at distant listeners simultaneously. The loudspeaker height and its directivity are so chosen that nearby listeners receive chiefly direct sound. The loudspeaker beam is directed toward the front part of the balcony so that the curvature of the loudspeaker distribution pattern minimizes amplified sound in the front rows and increases the amplified sound level gradually toward the rear of the main floor. This gives a combination which is nearly uniform across the entire audience. The geometry of the beam and ceiling reflection of the upper side of the beam assist the sound level at the balcony rear.

58. Loudspeaker Specifications and Measurements. Typical loudspeaker specifications are shown in Table 19-7 for a variety of loudspeaker types.

Loudspeaker impedance is proportional to the voltage across the voice coil when driven by a high-impedance constant-current source. Continuous power ratings are obtained from sustained life tests with typical audio-program material restricted to the frequency range appropriate for the loudspeaker type. Sensitivity, response-frequency characteristics, frequency range, and directivity are most effectively measured under anechoic conditions using

Table 19-7.   Characteristics of a Variety of Loudspeaker Types

Company	Altec	Altec	Bozak	RCA
Model no.	755C	1505B horn 290D driver	CM-109-23	LC1B
Type	Direct radiator	Cellular horn (3 × 5)	Three-way column	Duo-cone
Sensitivity (at 4 ft for 1 W), dB	95	110	106	95
Frequency range, Hz	40–15,000	300–8,000	65–13,000	25–16,000 (±4 dB)
Impedance, Ω	8	4	8	15
Power rating, W	15	100	200	20
Distribution angle, deg	90	105 horizontal 60 vertical	90 horizontal 30 vertical	120
Voice-coil diameter, in.	2	2.8	(3 sizes)	(2 cones)
Cone resonance, Hz	52	...	(3 sizes)	22
Crossover frequency, Hz	...	500	800, 2,500	1,600
Diameter, in.	8⅜	18½ high 30½ wide	57 in. high 22¾ wide	17
Depth, in.	2¼	30	15¾	7½
Weight, lb	3¾	43	250	21

calibrated laboratory microphones and high-speed level recorders. (See Pars. 19-1 to 19-9 on standard units, typical formats for sound data, and noise measurement.)

Distortion measurements in audio electronic systems are generally of three types shown in Fig. 19-113. For harmonic distortion a single sinusoidal signal A is supplied to the loudspeaker and wave analysis at the harmonic frequencies determines the percent distortion.

Both intermodulation methods supply two sinusoidal signals of different frequency to the loudspeaker. In the older Society of Motion Picture and Television Engineers (SMPTE) method the frequencies are widely separated, and the distortion is expressed in terms of sum and difference frequencies around the higher test frequency. This method is meaningful for wide-range loudspeaker systems.

The CCIF (International Telephone Consultative Committee) method is more applicable to narrow-range systems and loudspeakers receiving input at high frequencies. It supplies two high frequencies to the loudspeaker and checks the low difference frequency.

59. Earphones.[82] The transduction methods are the same as for loudspeakers (Fig. 19-93). Telephone and hearing-aid receivers are usually moving iron. Most military communication headsets are now moving coil. Piezoelectric, moving-coil, and electrostatic types are used for professional and private listening to recorded music.

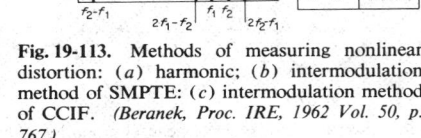

Fig. 19-113. Methods of measuring nonlinear distortion: (*a*) harmonic; (*b*) intermodulation method of SMPTE; (*c*) intermodulation method of CCIF. (*Beranek, Proc. IRE, 1962 Vol. 50, p. 767.*)

Equivalent Electric Circuits. Figure 19-114 shows a cross section of a moving-coil earphone and the equivalent electric circuit.[83] The voice-coil force drives the voice coil and diaphragm. (Mechanical resonance of earphone diaphragms occurs at a high audio frequency in contrast to loudspeakers.) Diaphragm motion creates sound pressure in several spaces behind the diaphragm and the voice coil and between the diaphragm and the earcap. Inertance and resistance of the connecting holes and clearances combine with the capacitance of the spaces to add acoustical resonances. *Z* is the acoustical impedance of the ear.

Idealized Ear Loading. The ear is approximately an acoustical capacitance. However, acoustical leakage adds a parallel resistance-inertance path affecting low-frequency response. At high frequencies the ear canal-length resonance is a factor.

Since the ear is a capacitance, the goal of earphone design is a constant diaphragm amplitude throughout the frequency range. This requires a stiffness-controlled system or a high resonance frequency. The potential across the ear is analogous to sound pressure within the ear cavity. This sound pressure is proportional to diaphragm area and inversely proportional to enclosed volume. Earphone loading conditions are extremely varied for different types of earphone mountings.

Earphone Mountings. The most widely used earphone is the single receiver unit on a telephone handset. It is intended to be held against the ear but is often tilted away, leaving considerable leakage. An artificial ear developed for telephones reflects this condition.[84]

Headsets provide better communication than handsets because they supply sound to both ears and shield them. Four types of headset mounting for earphones are shown in Fig. 19-115. A fifth type for the deaf is the insert earphone worn within the outer ear and supported by it.

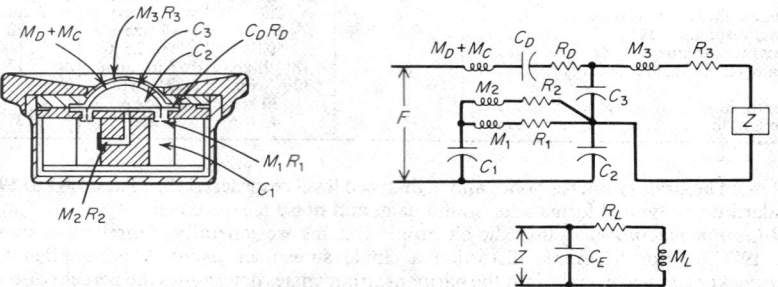

Fig. 19-114. Moving-coil earphone cross section and equivalent electric circuit. (Ref. 83)

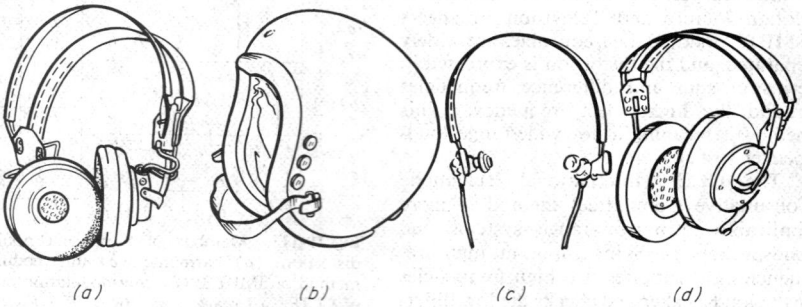

Fig. 19-115. Four types of headset mounting of earphones: (*a*) headband with earcushions; (*b*) helmet; (*c*) light headband with semi-insert tips; (*d*) headband with circumaural earmuffs. (After J. Zwislocki, Ref. 6)

Ear cushions on a headband (Fig. 19-115*a*) press against the ear and enclose approximately 6 cm³. Semi-insert phones (Fig. 19-115*c*) provide a smaller contact area, require less headband force, and enclose approximately 2 cm³. Earmuffs (Fig. 19-115*d*) have larger contact area, require stronger headbands, and enclose the entire ear (100 to 300 cm³). Military helmets (Fig. 19-115*b*) may be adapted to any of the ear coupling devices. ANSI Standard S3.7 covers the use of four coupler types.

Insert earphones are attached to individually fitted inserts. There is no applied force. The enclosed volume is 2 cm³ or less.

A remote earphone can drive the ear canal through a small acoustic tube. The length may be an inch or two for hearing aids and several feet for music listening on aircraft.

Efficiency, Impedance, and Driving Circuits. Moving-iron earphones and microphones can be made efficient enough to operate as sound-powered (battery-less) telephones. Efficient magnet structures, minimum mechanical and acoustical damping, and minimum volume of acoustical coupling are required for this purpose. In some earphone applications overall efficiency is less critical, and wearer comfort is important.

Insert earphones need less efficiency than external earphones because the enclosed volume is much smaller; however, they require moderate efficiency to save the amplifier batteries.

Circumaural earphones are frequently driven by amplifiers otherwise used for loudspeakers. Here efficiency is less important than power-delivering capacity.

Typically 1 mw of audio power to an earphone will produce 100 to 110 dB in a standard 6-cm³ coupler. The same earphone will produce less sound level in an earmuff than in an ear cushion and more when coupled to an ear insert.

The shape of the enclosed volume also affects response. The farther the driver is from the eardrum the lower the frequency of standing-wave resonance. Small tube diameters produce high-frequency attenuation.

The response-frequency characteristic of moving-iron or piezoelectric earphones is quite dependent upon source impedance. A moving-iron earphone having uniform response when driven at constant power will have a rising response (with increasing frequency) at constant current and a falling response at constant voltage[85] (Fig. 19-116).

Real-Ear Response. The variety of earphone-coupling methods and the variability of outer-ear geometry (among different listeners) make response data from artificial ears only indicative, not definitive. Out of necessity a real-ear response-measuring technique was developed. A listener adjusts headset input to match headset loudness to an external calibrated sound wave in an anechoic chamber. From matching data at numerous frequencies an equivalent free-field sound-pressure level can be plotted for constant input to the earphone. This curve usually differs from a sound-level curve on a simple earphone coupler for two reasons: (1) probe measurements of sound at the eardrum and outside the ear in a free field differ because of ear amplification and diffraction about the head (Fig. 19-117);[86] and (2) for equal loudness an earphone must produce a higher sound pressure at the eardrum than if the sound originated in the outside air (Fig. 19-118).[87] This paradox is only partially understood.

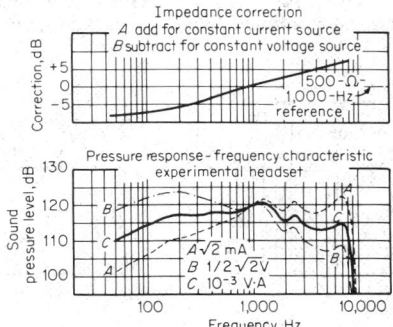

Fig. 19-116. Effect of source impedance upon earphone response curve: (*a*) constant current; (*b*) constant voltage; (*c*) constant power. (Ref. 85)

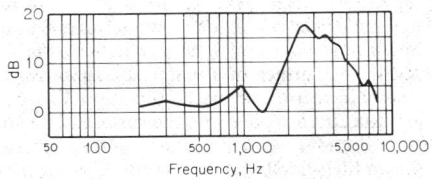

Fig. 19-117. Relative level of sound pressures at the listener's eardrum and in the free sound field. (After Wiener and Ross, Ref. 86)

Acoustic attenuation by earphones can be measured either by threshold shift or by matching the loudness of tones heard from an external loudspeaker, with and without the headset on. The sound-level difference is plotted as attenuation in decibels (Fig. 19-119). The solid curve is for the headset (Fig. 19-115a), and the dashed curve is Fig. 19-115d. The third curve is for a special earphone socket using wax inside the sealing ring. Even greater attenuation can be obtained below 1,000 Hz by combining larger mass, larger contact area, a stronger headband, and larger enclosed volume.[88] Attempts to improve attenuation above 1,000 Hz are defeated by bone conduction to the inner ear unless rigid helmets are worn.

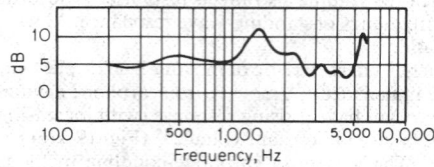

Fig. 19-118. Relative level of sound pressures at the eardrum of closed and open ears for equal loudness. (After Wiener and Filler, Ref. 87)

Monaural, Diotic, and Binaural Listening. A handset earphone provides monaural listening. Diotic listening with the same audio signal in both earphones localizes sound within the head. This is not unpleasant and may actually be an aid to concentration. In natural binaural listening the ears receive sound differently from the same source unless it is directly on the listening axis. Usually there are differences in phase, arrival time, and spectrum (because of diffraction about the head).

Recordings provide true binaural effects only if the two recording microphones are on an artificial head. Stereophonic microphones are usually separated much farther, so that headset listening gives an exaggerated effect. For some listeners this is an enhancement, but for listeners who prefer natural binaural listening a circuit has been developed to convert stereo information into binaural (Fig. 19-120). This circuit produces cross talk between stereo channels with phase and amplitude differences at each ear to simulate sound diffraction around the head.[89]

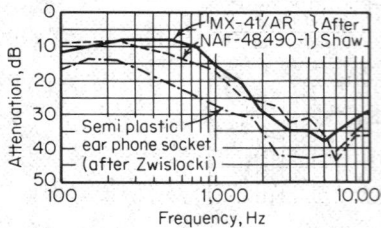

Fig. 19-119. Attenuation of external noise by different headsets. (After J. Zwislocki, Ref. 6)

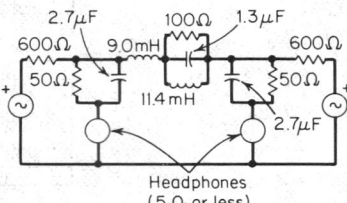

Fig. 19-120. Network for transforming stereophonic playback for binaural headset listening. (Ref. 89)

DISK RECORDING SYSTEMS

60. Principles and Characteristics of Disk Recording.[90,91] "A sound recording system is a combination of transducing devices and associated equipment suitable for storing sound in a form capable of subsequent reproduction."[1] In mechanical sound recording the vibrations are engraved or embossed in the surface of a material. Embossing deforms the surface by rubbing it and is used chiefly in recording speech dictation for direct reproduction without further processing. Engraving cuts a groove into the surface of the material. It has now been developed to provide the full audio-frequency range and the freedom from distortion and noise which are needed for high-fidelity disk recording and for mass reproduction after processing.

An ideal record is one which when reproduced generates from the output of the pickup an electric wave identical to that originally supplied to the recorder, with a minimum of noise and of harmonic, frequency, and phase distortion. This technical goal is independent of the record producer's goal, which may be either to reproduce or recreate an original sound or to create a new or modified sound or sound effect.

Figure 19-121 shows a stereophonic disk recording system with a choice of source material from either microphones in a studio or a two-channel stereophonic magnetic tape. Other source possibilities include electronic musical instruments and synthesizers. Very often the original recording is made on tape because of the ease of starting, stopping, editing, and playback without damage to the original. Then the finally edited and perhaps enhanced tape original is rerecorded on the disk original. Previous subsections have treated audio program material, studio acoustical environment, microphones, and monitoring loudspeakers (see Pars. **19-72–19-86** for magnetic-tape sound recording and reproducing systems). The part of the overall system concerned specifically with disk recording includes the disk material; the turntable and drive, which move the material during recording; the stylus, which cuts the groove; the recording head or cutter, which moves the stylus in a plane perpendicular to the grooves, the drive for the cutter head, and the electronic preemphasis of the electrical input wave by the equalizer of Fig. 19-121.

61. Disk Materials. For many years original recordings were made on a wax surface. The actual material was a blend of waxes with metallic soaps, first in the form of a thick disk of wax, and later a thin layer of wax melted and flowed onto a smooth, flat metal base. Before recording began, the solid wax disks were shaved to a highly polished surface while rotating at high speed. For flowed wax the smoothness and flatness of the recording surface depended largely upon the metal base.

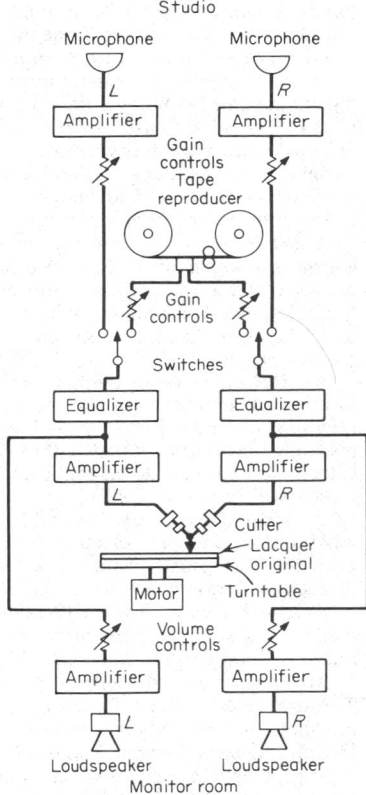

Fig. 19-121. Stereophonic-disk recording system. (Ref. 22a)

Now original recordings are commonly made on a lacquer-coated disk of metal or glass. The basic lacquer compound has been combined with new and superior coating techniques to meet the requirements of microgroove stereophonic recording.[92]

62. Drive Mechanisms and Speeds. The National Association of Broadcasters (NAB) standard recording speeds are 33-1/3 and 45 r/min (the older 78.26 r/min speed is no longer considered standard). Speeds of 16-2/3 and 8-1/3 r/min are useful for speech, especially talking books for the blind. The 33-1/3 r/min speed can be obtained by 54:1 reduction and 45 r/min by a 40:1 reduction from 1,800 r/min. The NAB standard requires that recording speed be held to 0.1% tolerance. On the time scale this amounts to less than 2 s in ½ h of program, and on the frequency scale less than one-fiftieth of a musical semitone.

Recording Turntables. The NAB accuracy specification necessitates the use of synchronous-driven turntables. Whether the connection between the motor and the turntable is by gears, friction drive, or a belt, mechanical filtering is usually required to prevent transmission of motor vibration to the turntable and to avoid audible flutter. Among the synchronous drives which have been used are single-phase selsyn motors, 3-phase power selsyn motors, and synchronous motors of the variable-reluctance type.

Flutter, Wow, and Rumble. "Flutter is any deviation in frequency of reproduced sound from the original frequency."[1] Wow is a colloquial term commonly applied to flutter frequency of the order of 1 Hz. Rumble is low-frequency noise transmitted from the motor (or the coupling) through the turntable to the disk and stylus, whether the system has constant speed or not. Since the job of the turntable is to rotate the disk quietly at accurate, constant speed, all these faults must be minimized.

Research has shown[93] that flutter in the wow range is barely audible at 0.04% frequency modulation, the NAB upper limit during recording, but that high-frequency tones with flutter rates of 3 to 10 Hz have a perceptibility threshold an order of magnitude lower. This is reflected in the weighting curve of the IEEE standard 193-1971, Method of Measurement for Weighted Peak Flutter of Sound Recording and Reproducing Equipment, shown in Fig. 19-122 and based on a test frequency of 3,150 Hz. Only the highest-quality recording equipment is free of flutter by these standards.

Cutter Drive Speed. For microgroove recording the lathe screw bearing the cutter head advances it nominally 1/260 in. per disk revolution. However, between a 4.75-in. minimum diameter and a 11.5-in. maximum diameter this provides only 22½ min recording time. Variable-pitch recording[94] decreases the cutter-drive speed during low-level passages, extending recording time to 30 min. Special speech recordings (talking books) have up to 650 grooves per inch.

63. Recording-Head Design. In disk recording the cutter must have a high enough mechanical impedance at the top of the stylus to control the recording medium. This requirement leads directly to high stiffness and density for the stylus and indirectly to a resonance centrally located within the frequency range to be recorded. For the stylus velocity to be proportional to the applied force, the resonance peak must be eliminated by mechanical resistance control.

Recording Transduction Methods. Recording heads have operated chiefly on moving-coil, moving-iron, and piezoelectric principles of transduction. In the past the moving-coil principle was usually adopted for vertical (hill-and-dale) recording. Moving-iron and piezoelectric transducers were well suited to lateral groove modulation.

Lateral Recorder. Figure 19-123 is a sectional view of a magnetic disk recorder which modulates the groove in the plane of the disk. Alternating flux in the armature, corresponding to the coil current, interacts with the balanced permanent-magnetic field, applying an alternating force to the upper end of the armature which is pivoted in the lower gap. Through lever action a force of opposite direction moves the cutting stylus laterally to modulate the groove being cut in the record surface. An equivalent electric circuit of this recording head is shown in Fig. 19-124. For constant-velocity recording the circuit constants must provide a current I_2 which is independent of frequency. This requires the damping resistance to be relatively large.

Stereophonic Recorder.[95] The recorder of Fig. 19-125 contains two moving-coil drive units, one for each stereophonic channel. Each unit also contains a feedback coil located in

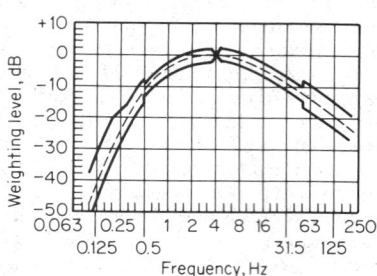

Fig. 19-122. Weighting curve for flutter measurement at 3,150 Hz. *(IEEE Standard 193-1971.)*

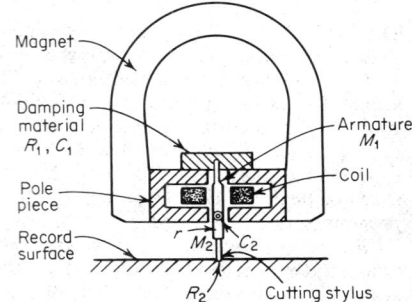

Fig. 19-123. Lateral disk-recording head (magnetic). (Ref. 91)

a separate annular branch of the magnetic circuit. Copper shields between the two coils reduce the inductive cross talk. (The purpose of the feedback coil is given below.) V-shaped beryllium-copper coil-support springs hold the coil assemblies, allowing them to move only along their axes. Motion is transmitted from the coils to a tubular stylus-support member through wire links braced with magnesium sleeves. A single magnet supplies flux to the magnetic circuit through series-parallel gaps, ensuring equal flux densities in the corresponding gaps. The orthogonal drive system controls stylus motion within a plane approximately perpendicular to the axis of the groove. For practical reasons the plane is actually tilted from the vertical at an angle standardized at 15°.

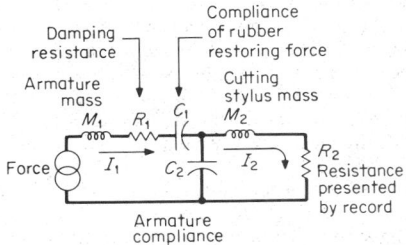

Fig. 19-124. Equivalent electric circuit for Fig. 19-123. (Ref. 91)

Feedback Recording.[96] The output of each feedback coil is used for negative-feedback control (in its channel amplifier) of resonance in the vibratory system. An advantage of the negative-feedback method is the damping provided to the electromechanical system without resistive losses.

Recording-Stylus Design.[97] Disk recording styli are ground from sapphire. The lack of grain and cleavage planes permits grinding to very acute angles without breakage. Figure 19-126 shows the shape of a cutting stylus, with a magnified view of the stylus point and the principal angles affecting the properties of the recorded groove. The dubbing facet was found necessary for satisfactory cutting of lacquer disks.[98] An inherent loss in high-frequency range, caused by the finite size of the dubbing facet relative to the wavelength, has been overcome by the hot-stylus recording technique,[99] which allows extremely small facets on the cutting edge in combination with a quiet groove and good high-frequency response. The heating coil is wrapped directly around the stylus just above the cutting portion, as shown in Fig. 19-125. The effect of heating upon noise can be optimized by current control. Excessive heating produces groove-burning noise. A typical heating coil consists of six turns of No. 40 AWG nichrome wire carrying a current of approximately ½A.

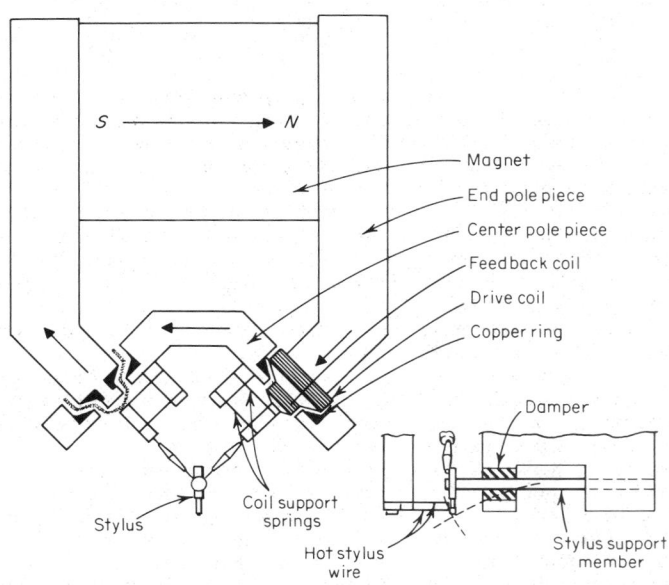

Fig. 19-125. Stereophonic disk-recording head. (Ref. 95)

The clamping of the stylus in its holder is critical. Minute misfitting or looseness can seriously affect recording quality. The grinding of styli is very scientific, but stylus use has both technical and interpretive aspects.[100]

64. Groove Modulation. The groove velocity varies with the diameter of the spiral and the rotation rate of the record, as shown in Fig. 19-127. A 6-mil groove width was standard for the old 78 r/min records, with a radius of 1.5 mils at the bottom of the groove. The playback stylus had a radius in the range of 2.5 to 3.0 mils. When 33-1/3 and 45 r/min records became standard, the groove width was reduced to 2.6 mils with a bottom radius of 0.2 mil. The playback-stylus radius was reduced to the range of 0.8 to 1.1 mils. Stereophonic recording decreased the stylus radius to 0.7 mil to reduce the tracing distortion associated with the vertical component of stylus motion in the groove.

Stereophonic Modulation. Figure 19-128 compares the theoretical groove shape and dimensions for lateral recording and for stereophonic recording. The solid lines define the unmodulated groove, and the dashed lines show the maximum lateral excursions for monophonic recording and the maximum vertical and lateral excursions for stereophonic recording. The dimensions are very similar except for reduced amplitude of motion in 45-45, amounting to a reduction of 3 dB in output level for each stereophonic channel. Thus the total power remains the same.

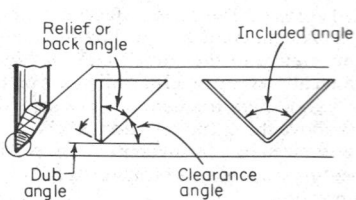

Fig. 19-126. Cutting stylus point. (Ref. 97)

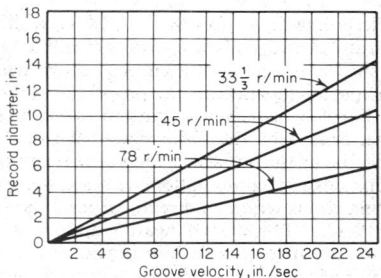

Fig. 19-127. Groove velocity vs. diameter for different record speeds. *(M. S. Corrington and T. Murakami, 1958 IRE Conv. Rec., Pt. 7.)*

The choice of 45-45 operation for stereo (over vertical-lateral channels) was based upon symmetry and the compatibility of the stereophonic mode with the previous standard lateral monophonic mode. Monophonic playback of stereo combines the horizontal components of both channels. The relative phase of the stereophonic channels was standardized for equal in-phase signals in the two channels to give lateral motion.

Quadriphonic Recording.[101-104] Just as stereophonic recording and reproduction provide a line of virtual sound sources between two loudspeakers separated in front of a listener, quadriphonic sound systems, which use four loudspeakers spaced horizontally around a listener, can surround him with a plane of virtual sound sources. The direct logical extension of two-channel recording in a single groove would be four-channel recording in two parallel grooves. However, this uneconomical method would require twice the recording space and a pair of coordinated styli in both recording and reproducing.

Numerous and diverse approaches to quadriphony, all intended to be compatible with stereophony, can be classified either 4-4-4, which uses carrier techniques for two additional channels just above the audio range recorded in the same groove with two audio channels, or 4-2-4, which uses matrix encoding of four audio channels into two before recording and decoding back to four during playback. Currently both are in commercial use, and neither standardization of one nor compatibility between the two has been arranged.

The CD-4 system of Japan Victor and RCA is 4-4-4. It superposes all left side information (both front and back) in one of the 45-45 components of recording stylus motion and all right side information in the other. In this sense it is "discrete" or separated. However, for stereo playback compatibility each audio (20 Hz-15 kHz) channel contains the sum of front and back information; the superposed carrier (30 kHz) is frequency modulated by the difference between front and back. (Disregarding the carrier, stereo playback recovers the left and right sum signals only.) Quad playback demodulates the left and right difference signals, for addition or subtraction with their respective sum signals, to yield left-front, left-back, right-front and right-back signals. The channels are theoretically discrete, but left and right channel separations have the usual compromising factors found in stereo, and front-back separation depends upon perfect modulation, demodulation, and equalization before addition and subtraction. The extended frequency range (to 40 kHz) has required improved record material, playback styli and transducers; and (currently) recording the master at half speed.

The SQ method of CBS and the QS method of Sansui are the currently active 4-2-4 systems, which combine left-front, left-back, right-front and right-back signals for later separation. Unlike early matrix methods of simple addition and subtraction of inputs in various proportions, both SQ and QS use all-pass, phase-shift networks to get quadrature components. This minimizes undesired cancellation effects. The two audio signals recorded for SQ are

$$L_T = L_F - j \cdot 0.707 L_B + 0.707 R_B$$
$$R_T = R_F + j \cdot 0.707 R_B - 0.707 L_B$$

This allows the two front signals to remain isolated from each other to the degree that they are in stereo systems. During SQ quadriphonic playback decoding the L_T and R_T signals are reproduced separately for front channels. Each is also mixed with minus 90° phase-shifted versions of the other to obtain the back channels. All four outputs contain predominantly the appropriate input signal plus two side-effect signals which are symmetrical (same combination in front and back on a given side). The back-channel separation is compromised and the back images are drawn toward each other by the SQ system. Psychoacoustic factors, such as normally reduced discrimination and localization in the back hemisphere of listening, are used to justify this compromise which assumes that listeners face the front. In the stereo mode, SQ presents the back sources as virtual images between the front loudspeakers and seemingly farther away.

An advanced version of SQ uses a logic-directed matrix decoder to sense relative levels in signal channels, and to adjust relative gains for minimum side effects in the reproduction of highly localized original sources.

The QS system of 4-2-4 has a symmetrical amplitude and phase matrix defined by

$$L = (L_F + jL_B) \cos \theta + (R_F + jR_B) \sin \theta$$
$$R = (R_F - jR_B) \cos \theta + (L_F - jL_B) \sin \theta$$

Decoding equations are

$$L_F = L \cos \theta + R \sin \theta$$
$$R_F = R \cos \theta + L \sin \theta$$
$$R_B = R \cos \theta - L \sin \theta$$
$$L_B = L \cos \theta - R \sin \theta$$

Because of its symmetry, QS inherently provides identical front–back and left–right separation. The listener has no preselected listening axis. For any playback channel the crosstalk is down only 3 dB in adjacent channels, but is theoretically eliminated in the diagonally opposite channel.

Competitive quadriphonic system results and costs are still being evaluated. Currently it appears that CD-4 provides the greatest four-channel separation, with slightly reduced frequency range, at higher equipment cost for both recording and playback. SQ provides good front-channel (stereo) separation with full frequency range, but with some back-channel compromises. Both matrix 4-2-4 systems have considerably lower equipment cost for both recording and playback. QS is symmetrical and provides good diagonal-channel separation with full frequency range, but it compromises adjacent-channel separation.

65. Preemphasis and Deemphasis. Disk recorders are designed for constant-velocity recording, but typical input spectra (Figs. 19-13 and 19-32) have steep negative slopes. To prevent low-frequency overload and improve high-frequency signal-to-noise ratio the input spectrum is tilted upward before recording, then tilted back down after playback, as shown in Fig. 19-129. The numbers along the curve are network time constants for high-frequency rolloff, low-frequency boost, and very low frequency rolloff. The standard level for stereophonic disks is 5 cm/s peak at 1 kHz.

66. Electronic Control Circuits.[105] Response-frequency characteristics can be varied automatically with recording level if desired. The dynamic spectrum equalizer boosts low frequencies at low level and the 2- to 8-kHz range at high level to make typical home listening more like a concert hall. Another recording control circuit reduces tracing distortion (see below) by introducing complementary distortion through the use of a tapped delay line in conjunction with sampling gates.[106] Recording overload indicator or control circuits are used with a differentiator to avoid curvature overloading at high frequencies and with an integrator to anticipate groove overload at low frequencies.

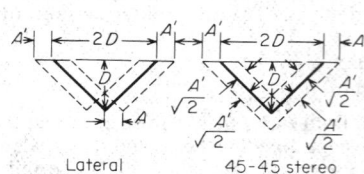

Fig. **19-128.** Comparison of 45-45 stereophonic and lateral monophonic grooves. (Ref. 95)

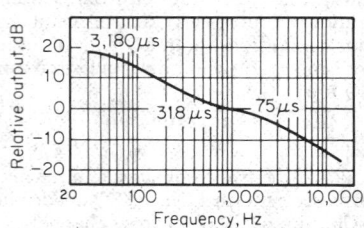

Fig. **19-129.** NAB standard disk-reproducing characteristic: relative output level vs. frequency for constant velocity input. Tolerance: +2 dB, 50 Hz to 10 kHz; +2 dB, −3 dB below 50 Hz, above 10 kHz. *(NAB Disc Recording and Reproducing Standard, 1964.)*

67. Practical Recording Techniques.[100,107] Examples beyond the scope of this book and described in the references include recording stylus mounting and heating, panoramic potentiometers, microphone techniques for smooth response, and monitoring practices.

68. Disk Processing and Specifications. Figure 19-130 shows the steps in the mass production of records. The lacquer original is usually cellulose nitrate with a variety of resins, oils, lacquers, and volatile solvents. The master is produced by spray-silvering the original surface, then plating with nickel and then with copper to a thickness of 0.030 to .040 in. The master has ridges where the original had grooves. After a film coating to ensure separability the master is copper-plated to produce the mother, which has grooves like the original. In like manner stampers are made from the mother.

The final record is thermoplastic material pressed between two stampers heated by steam within the press. After the grooves are formed, the stampers are rapidly cooled to set the plastic material and the record is removed.

The NAB standard specifies a center-hole diameter of 0.286 in. +0.001 −0.002 in. Hole concentricity should be ±0.005 in. For disk flatness the total indicator reading should not exceed $\frac{1}{16}$ in. over the disk surface with a $\frac{1}{32}$-in. limit within any 45° sector.

DISK REPRODUCTION SYSTEMS

69. Reproducer Heads, Styli, and Arms. A stereophonic system consists of a record on a motor-driven turntable, a stylus on a reproducer pivoted on an arm, a two-channel amplifier with ganged volume controls, each channel including a deemphasis equalizer, a power amplifier, and loudspeaker.

Successful transduction of the groove modulation into an electric output wave depends upon the interactive combination of stylus, pickup, and arm mounting.

Stylus Design. The playback stylus is a sapphire or diamond cone with a spherical tip of 0.0007-in. radius. Although the walls of the record groove wear well under typical tracking forces, there is some stylus penetration of the groove walls because of elastic and plastic deformation.[108] For a disk of Vinylite (a copolymer of vinyl chloride acetate) the pressure from a stylus with a typical tracking force, for example, 1 g, exceeds elastic flow limits and causes some plastic flow. A short duration of stress is important.

Stereophonic Pickups. Figure 19-131 shows a moving-coil stereophonic pickup. The two coils on Mylar hinges have their axes at 45° from the horizontal. Each coil is linked to a beam bearing the stylus. The beam rear is secured in a flat spring to prevent rotation and to provide uniform stylus compliance for all motions in the plane of groove modulation.

Figure 19-132 shows stereophonic pickups of magnetic and ceramic types. The magnetic pickup has low mechanical impedance. Stylus motion is transmitted through the arm to the magnet. Pole pieces on four sides of the magnet are surrounded by coils, which when connected in proper phase, generate potentials corresponding to the magnet velocity along the $+45°$ and $-45°$ axes.

In the ceramic pickup the stylus displacement bends the ceramic bars along the $+45°$ and $-45°$ axes, generating output potentials. Because the ceramic pickup is displacement-responsive, the response curve approximates the standard reproducing characteristic directly.

Tracing distortion [109] occurs in playback because the spherical stylus does not trace an exact replica of the modulated groove. In Fig. 19-133 the groove modulation is sinusoidal,

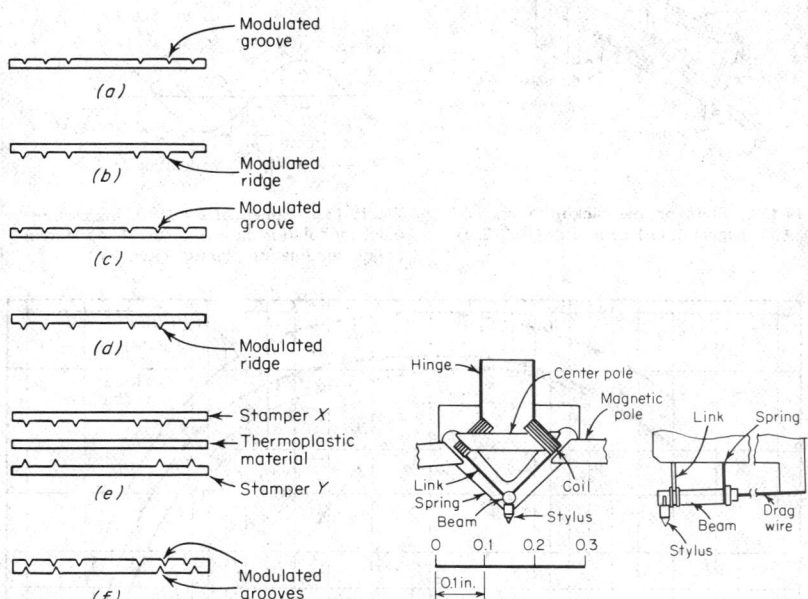

Fig. 19-130. Steps in the mass-production process for disk records: (*a*) original, (*b*) master, (*c*) mother, (*d*) stamper, (*e*) stamping, (*f*) record. (Ref. 22)

Fig. 19-131. Section of moving-coil stereophonic pickup. (Ref. 95)

but the path of the sphere center is not. This type of distortion can be minimized by complementary distortion of the recording stylus path.[106]

Pickup Trackability. [110] The stylus must not wear the groove unduly but must remain in contact with the record at maximum groove-modulation velocity. The dashed line in Fig. 19-134 is the theoretical maximum limit for recorded velocity (sometimes exceeded). The smooth curves show the velocity to which a specially designed pickup tracks for forces of 1 and ¾ g.

Arm Design. [111] The arm holds the pickup, applies the stylus tracking force, and provides damping for the resonance of the arm mass with the stylus-support compliance. The dimensions and angle of the arm-pickup-stylus combination determine any tracking error (see Fig. 19-135), which is the angle α between the planes in which the recording and playback styli move. To a first approximation (see Fig. 19-136) this angle can be adjusted for a given arm length l and groove radius R by adjusting (1) the angle β between the pickup axis and the

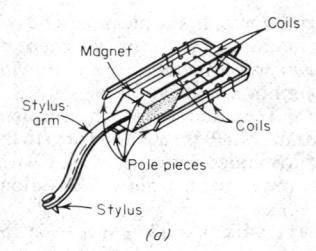

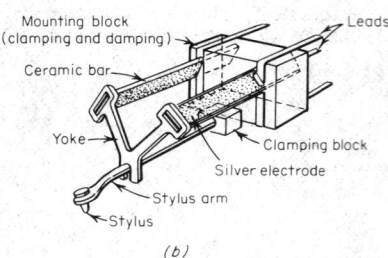

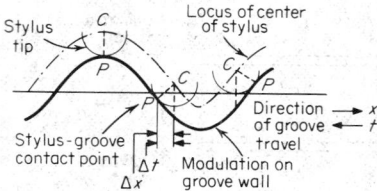

Fig. 19-132. Stereophonic pickups (perspective): (*a*) magnetic; (*b*) ceramic. (Ref. 22a)

Fig. 19-133. Spherical stylus tracing a sinusoidal modulation in one of the 45-45 stereophonic modulation planes. (Ref. 106)

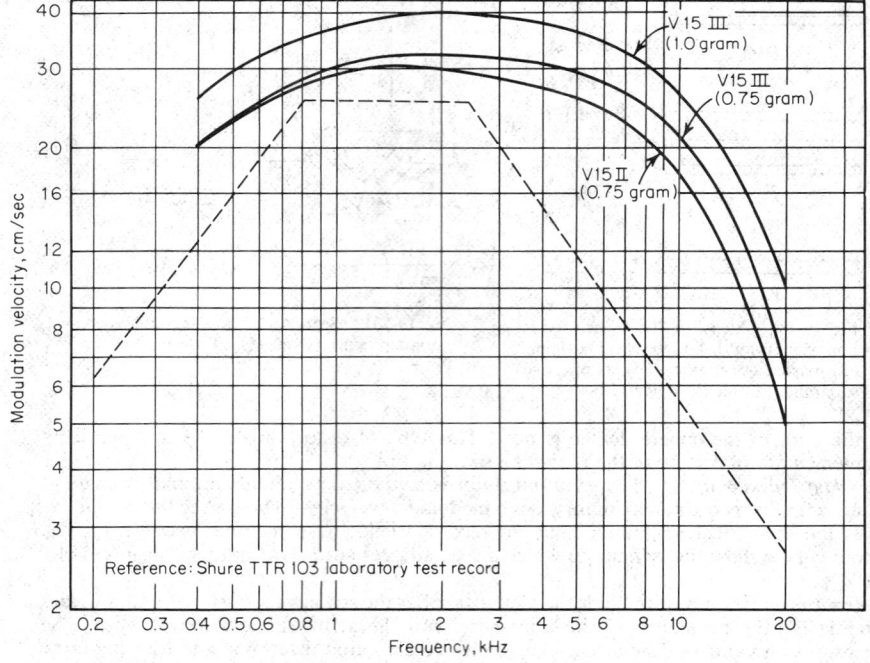

Fig. 19-134. Pickup trackability characteristics. *(Shure Bros., Inc.)*

line between the pivot and stylus point and (2) the amount of overhang *D* by which the stylus passes the record center. The head offset angle creates friction between the stylus and groove, resulting in a tendency to skate toward the record center. Some pivoted arms have an adjustable antiskating compensation weight or spring.

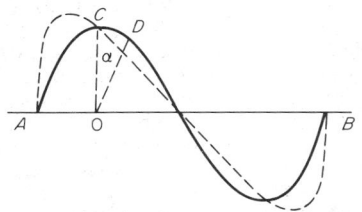

Tracking error can be eliminated by sliding the pickup on a radial arm and using an automatically articulated arm to maintain tangent tracking.

Figure 19-137 shows calculated distortion vs. tracking angle at different radii at 33⅓ r/min.

70. Reproducer Turntables, Drives, and Changers. Record-playing units play either a single disk at a time or a stack of disks in sequence. Either the arm or the changer can be automatic. Turntables have the same basic principles whether recording or reproducing, but design practices differ somewhat.

Fig. 19-135. Distortion of laterally recorded sine wave by the error α in tracking angle. (Ref. 91)

Turntables. The average speed of a reproducing turntable is typically held only to ±0.3%, but some turntable drives contain provision for fine speed adjustment. This is necessary for musical applications. Wow and flutter are held within 0.1 to 0.2% in playback equipment. Standard center-pin diameter is 0.285 in. ±0.0005 in.

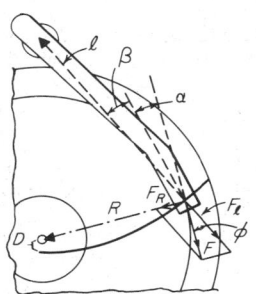

Fig. 19-136. Arm design for minimum tracking error. (Ref. 111)

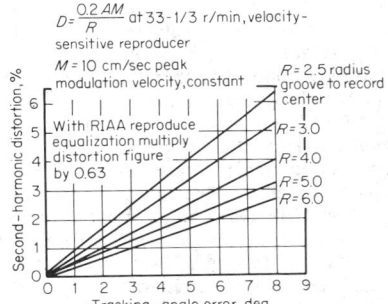

Fig. 19-137. Calculated second-harmonic distortion vs. tracking-angle error at different radii on the disk. (Ref. 95)

The NAB standard for low-frequency noise, or rumble, covers a frequency range of 10 to 500 Hz. For maximum rumble in a stereo turntable it is −35 dB relative to 1 cm/s peak velocity at 100 Hz. Turntables with good drive systems meet this specification easily. However, the bass boost in the disk reproducing characteristic accentuates the inherent rumble.

Drive Mechanisms. In a majority of record changers and single-play turntables an idler wheel presses against the motor shaft and the inside of the turntable rim, transmitting motion from the former to the latter. Idler material and bearings and the design of the disengagement mechanism control the amount of flutter, wow, and rumble transmitted through or contributed by the idler system. Multiple-speed drives may use idlers with stepped diameters or an idler with a conical rather than a cylindrical shape. A conical idler may be moved up and down continuously for fine speed adjustment.

A flexible belt drive can also be used between the motor shaft and the turntable. Belt flexibility reduces motor flutter. If the belt is continuous and smooth, it should not contribute noise itself.

Synchronous motors provide good speed accuracy and independence of line-voltage variations. Induction motors have greater torque and inherently less flutter. Hybrid motor designs used now on many record players combine both sets of advantages. Dc drive is also being used, both in dc motors coupled directly to the turntable shaft and in servo electronic circuits.

Record Changers.   Changer mechanisms generally require more power than turntables do and use induction motors.   Some systems have two motors, synchronous for the belt drive and induction for the changer mechanism.

Record changers have a special problem in the vertical tracking angle.   On a single-play turntable the vertical angle can be corrected by one arm adjustment.   On a changer the vertical tracking angle depends upon the number of records in the stack.   Generally the angle is compromised and preset for three records in the stack.   Some changers with a single-play mode provide one adjustment for the single record and another for a record stack.

A widely used system for changing records has a long center spindle with a retracting member and a small supporting platform at the outer edge of the disk.   Other changers use central support only with small retractable members on each side of the spindle, using the disk edge only for diameter sensing in arm control.

Shock and Vibration Isolation.   Record players are isolated from external shock and vibration by spring supports which resonate with the changer mass at a very low frequency (see Fig. 19-60).   Care must be taken to lock the turntable to its base during transportation; otherwise large excursions at the low-resonance frequency can cause serious damage.

71. Specifications and Measurements.   Specifications were previously given for turntable speed, accuracy, wow and flutter, and center-pin dimension; for disk flatness; center-hole diameter and concentricity; groove width, shape, and separation; for stylus tip shape and vertical tracking angle; response-frequency characteristics for the electrical playback system; standard velocity reference levels; minimum rumble level and channel phasing.   Another factor is level separation between recorded and unrecorded channels.   It should be more than 20 dB (preferably 25 dB) from 100 to 7,500 Hz.   Above and below these frequencies the separation should not diminish faster than 6 dB/octave.   At 1 kHz the channel balance should match within 0.25 dB.

Record Measurements.   Test records can be calibrated by optical-interference patterns[112] in addition to microscopic measurement.   Distortion measurements by test recordings[113] are similar to those used for other purposes (see Fig. 19-113).   Figure 19-138 is an example of harmonic-distortion measurement for different stereo-pickup modes.   Distortion is twice as great for the vertical mode as for the others.

Figure 19-139 is an example of record-noise measurement for a single channel of a Vinylite pressing at 5 and 10 in. diameters.   The third-octave band spectrum includes the effect of standard playback deemphasis.   Also shown (dashed lines) are the single-channel maximum velocity limits set by the physical groove excursion, the maximum slope of a cut, and the minimum radius of curvature of groove modulation.   The upper

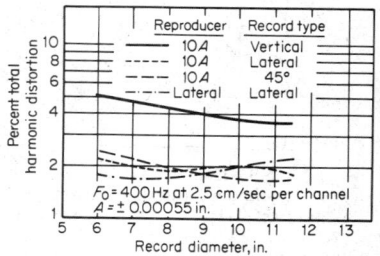

Fig. 19-138.   Measured harmonic distortion vs. record diameter for different stereo pick-up modes.   (Ref. 95)

solid curves correspond to the dashed lines after deemphasis.   The area between the upper curves and the lower curves defines the available dynamic range for disk recording.

MAGNETIC-TAPE SOUND SYSTEMS

72. Principles and Characteristics of Tape Systems.[115-120]   In Fig. 19-121 a stereophonic recording system is shown in which the master recording is made on magnetic tape. Facilities are provided either for sound reproduction from the tape or for rerecording upon a disk master.   Figure 19-140 shows a typical magnetic-tape sound recorder-reproducer for use in such a system.

In magnetic recording a magnetizable medium (usually tape) from a supply reel is moved at constant speed across a recording head, which induces in the medium a magnetization proportional to the current in the coil of the recording head.   Thus the variation of recording current with time is recorded on the tape as a variation of magnetization with distance.

When the tape later passes by the reproducing head, flux from the tape passes through the pickup coil of the reproducing head in proportion to the tape magnetization. Thus during playback (at the same speed) flux variations along the tape induce an electromotive force in the coil of the reproducing head.

In some tape recorders a separate reproducing head is used for continuous monitoring of the recorded result. In other recorders a single head serves as either recorder or reproducer at different times. The erasing head obliterates anything previously recorded and prepares the tape for new magnetization by the recording head.

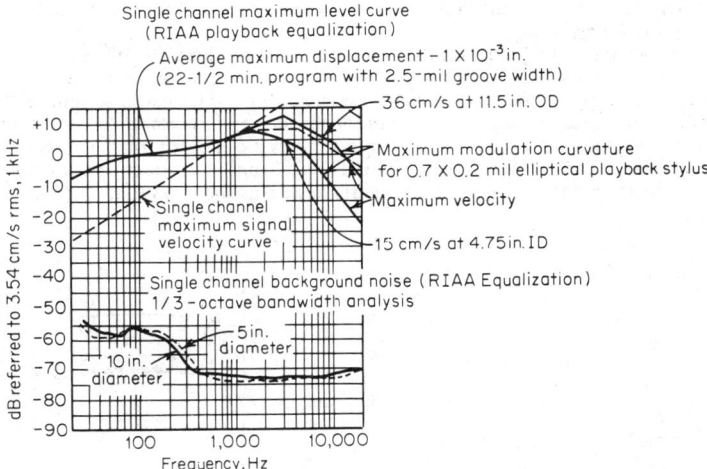

Fig. 19-139. Dynamic range of 33⅓ r/min disks (RIAA equalized) in terms of rms velocity vs. frequency. (Ref. 114)

Constant tape speed, an important requirement, is controlled chiefly by the drive capstan against which the tape is pressed by a roller. Guideposts at each end of the tape segment being driven and recorded isolate the segment from vibrations and variations occurring at the supply and takeup reels, which are generally driven and braked separately to maintain moderate tension in the tape at all times. The reels may be either open or within cartridges or cassettes. When the reels are packaged together, windows are provided (in the cartridge or cassette) where the heads and drive-roller contact the tape segment being recorded or played.

73. Playback Process. The playback process is simpler and easier to understand than the recording process. In Fig. 19-141 the tape record is a multipolar permanent magnet moving past a transducer head. The coil of the head provides an output voltage corresponding to the magnetic field of that part of the magnetic layer in closest proximity to the air gap. Many types of head have been used, but this type with confronting pole heads is the most common.

If the coil has N turns of wire and the recorded flux is ϕ_r, the induced voltage is

$$E = mN \frac{d\phi_r}{dt} \tag{19-10}$$

where m is the fractional part of the recorded flux which threads the coil; m varies with frequency, wavelength λ, gap size g, geometry of the pole pieces, permeability of the head material, permeability of the tape, thickness δ of the magnetic layer on the tape, and the clearance α between the head and the tape. Of the various portions of recorded flux shown (Fig. 19-141) d is far from the gap, and c is somewhat removed. Part a, located just outside the gap, threads the pole pieces and coil. When the recorded wavelength is large compared to the width W of the head, the rate of flux change is small and the output is low. As the wavelength approaches the gap size g, the rate and output increase. However cancellation occurs when the wavelength coincides with the gap dimension, as shown in Fig. 19-142. The solid curve is more exact but has a complex function. The approximate dashed curve corresponds to

$$m = \frac{\sin (\lambda/g)}{\lambda/g} \qquad (19\text{-}11)$$

In practice the output at high frequencies rolls off sooner than predicted by the finite gap width. Other factors include the depth of the magnetic layer and the clearance between the tape and the recording head.

Figure 19-143 shows the effect on the response curve produced by different factors. An ideal tape (curve 1) would have a rising response (6 dB/octave) for constant velocity in which the output is inversely proportional to the wavelength. Finite gap size (curve 4) has its characteristic null dip as the wavelength approaches the gap dimension. Curve 2 represents loss caused by the thickness of the layer. Curve 3 shows the effect of small clearance between the tape and head. The combined effects of curves 2, 3, and 4 subtract from the ideal curve 1 to give a typical response-frequency characteristic such as shown in Fig. 19-144 for 7½ in./s tape speed.

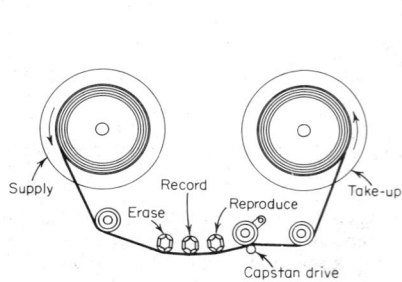

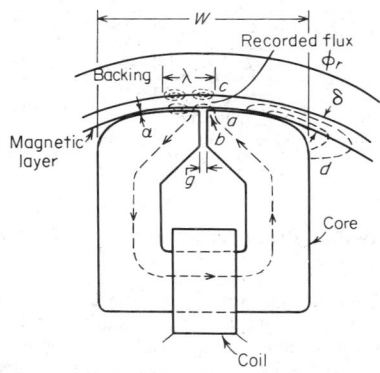

Fig. 19-140. Typical magnetic-tape sound recording and reproducing system. (Ref. 91)

Fig. 19-141. Head-tape configuration. (Ref. 120)

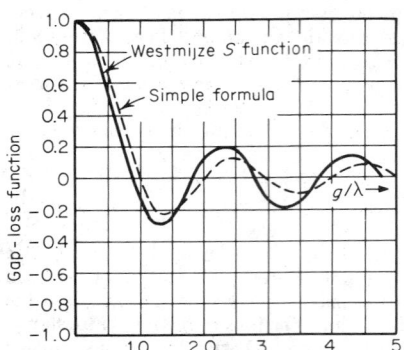

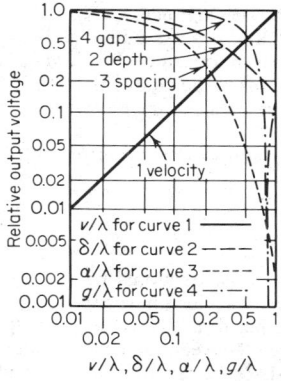

Fig. 19-142. Theoretical output loss at short wavelengths caused by finite gap size. (Ref. 120)

Fig. 19-143. Effect of tape velocity, thickness, spacing, and gap size on output. (Ref. 120)

74. Tape Materials.[121-124] A review of tape materials and their properties precedes discussion of the recording process. The main function of magnetic tape is to produce at the surface of the reproducing head a magnetic field having the same time variation as that which occurred during the recording process. The tape surface should be smooth and relatively soft to give maximum mechanical contact between the magnetic material and the head. The magnetic layer should be thin to concentrate the magnetic forces near the gap. However, it should be thick enough to prevent amplitude modulation and dropout of the reproduced signal. Magnetic layers on successive wrappings of the tape upon a reel should be separated enough to prevent magnetic interaction known as *print-through*. This layered construction, consisting of a magnetic coating on a nonmagnetic carrier, allows each layer to be optimized separately to fulfill its function.

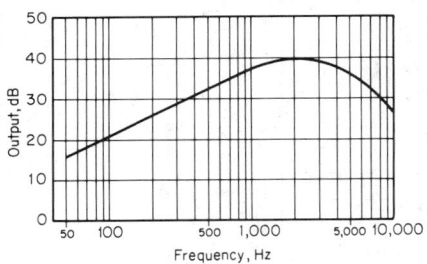

Fig. 19-144. Typical response-frequency characteristic for playback head output from 7½ in./s tape.

Cellulose acetate and polyester materials are widely used as magnetic-tape backings. Acetate tape is slightly smoother. Both have a small temperature coefficient. Polyester is much less sensitive to changes in relative humidity. For equal tape thickness the polyester is a little harder to stretch by a small percentage. However, after permanent yield begins, the acetate breaks at approximately 25% elongation while polyester will elongate 100%. This can be an advantage or a disadvantage. In some audio applications breakage is preferable to continued elongation before subsequent splicing. Acetate tapes contain a chemical plasticizer for flexibility, which ages gradually over a period of years. Polyester does not contain a plasticizer.

Gamma ferric oxide in the form of needle-shaped particles has been the principal magnetic material used in magnetic tape coatings for a number of years. Recently chromium dioxide has also found a firm place in audio recording. The particles of both material types are similar in size, but the chromium dioxide particles are single crystals and have smoother edges. They are freer from porosity and do not form branches. The coercivity of chromium dioxide is approximately one-third higher than for gamma ferric oxide. Chromium dioxide offers a superior high-frequency response, an advantage in low tape-speed applications.

75. Tape Dimensions and Properties. The standard width for nominal ¼-in. tape is 0.246 in. ± 0.002 in. Thickness should not exceed 0.0022 in. Standard base thicknesses are 1.5 and 1.0 mil. A full-track monophonic recording has a width of 0.238 in. (Fig. 19-145). For half-track monophonic recordings each track is 0.082 in. wide, and the track separation is 0.074 in. These dimensions also hold for two-track stereo, in which the upper track is the left channel. For four-track stereo and four-channel recordings each track is 0.043 in. wide, and the spacing between the centerlines of adjacent tracks is nominally 0.067 in. Thus the outside tracks reach the tape edges.

For cassette tape the standard width is 0.150 in. (Fig. 19-146). The maximum tape thickness is 0.0008 in. The tape is wound with the magnetic layer facing outward to contact the external heads. The four recorded tracks are 0.025 in. The central blank stripe is wider than the other two separation stripes. When the tape moves from left to right with the magnetic layer facing the observer, the bottom track is numbered 1 and the top 4. Track 1 is the left stereo channel and track 2 the right. (Track 4 left and track 3 right become tracks 1 and 2 when the cassette is turned over.)

The tape for an eight-track cartridge is standard ¼-in. width, and the eight individual tracks are of the same width as in a cassette.

Figure 19-147 summarizes the tracks and playback head locations for open-reel, eight-track cartridge and cassette tapes for both stereo and quadriphonic operation. In the quadriphonic mode L is left, R is right, F is front, and B is back. The two quadriphonic cassette proposals represent different approaches. One uses present standard cassette track widths, and the other halves them.

Other mechanical properties[125] affecting tape interaction with mechanical drive systems include static and dynamic coefficients of friction, tape-head abrasion, and adhesion of lubricant where it is used. These factors are especially important to cartridge performance.

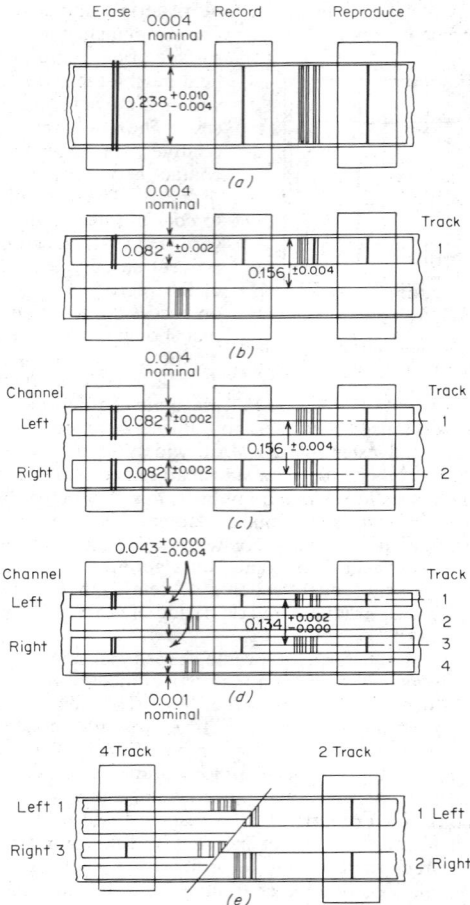

Fig. 19-145.   Comparison of track dimensions and locations for different recording modes on ¼-in. tape: (*a*) full-track monaural, (*b*) half-track monaural, (*c*) two-track stereo, (*d*) four-track stereo and four-channel, (*e*) comparison of tracks.

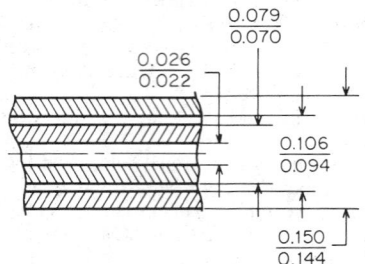

Fig. 19-146.   Track dimensions for cassette tape.

76. Recording Process.[126-130] The requirements for magnetic recording are a strong enough field to leave a permanently magnetized record on the tape, concentration of the field into a narrow region (for high definition), and magnetization proportional to the input signal (for low distortion). Proper selection of the medium and reverse use of the head-tape configuration of Fig. 19-141 satisfy the first two requirements. However, meeting the third requirement requires a special technique, because magnetic recording depends upon ferromagnetic materials, which have nonlinear and hysteresis properties.

In Fig. 19-148 curve *abocd* relates the remanent induction in the tape to the maximum magnetizing force in the gap when pure audio-frequency current is supplied to the recording head. The nonlinearity in this transformation would give a highly distorted playback waveform. The ingenious method of high-frequency bias recording overcomes this inherent limitation of magnetic recording.

In this method the unmagnetized tape receives at the recording head a magnetic field which is compounded of the audio signal and an ultrasonic (50- to 150-kHz) component called the *bias*. These two signals are simply added in the recording coil without modulation of one frequency by the other. Electrical addition can be accomplished in various ways. The ultrasonic oscillator output can be connected either in parallel or in series with the audio input to the head, or it can be supplied to a separate winding in the record head. A separate head for the ultrasonic field may also be placed opposite the record head in close proximity to the tape.

A simple (incomplete) explanation of bias recording is based on Fig. 19-148. The algebraic sum of the audio and bias signals is applied along the *H* axis, the bias signal being larger than the audio. This keeps the envelope away from the origin region, in which

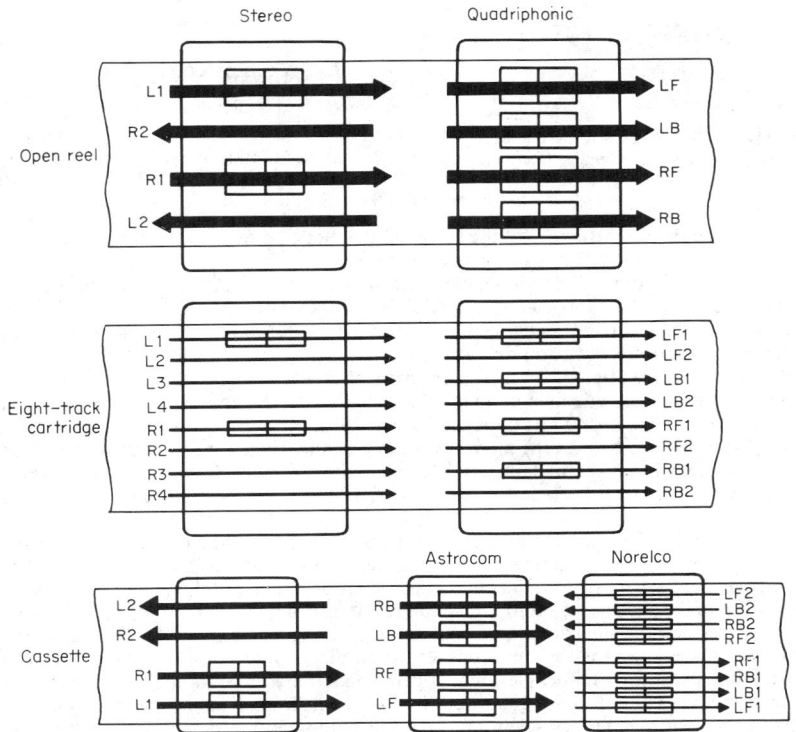

Fig. 19-147. Comparison of track configurations for reel, cartridge, and cassette tapes in stereo and quadriphonic modes.

nonlinearity occurs. The output variations along the B axis have a similar envelope shape. As the tape leaves the gap, demagnetization takes place at the bias frequency. The remaining net induction is the difference between the positive and negative half cycles of the wave, curve *ef*, which closely resembles the original audio signal. In the absence of sound the tape is almost completely demagnetized for low noise level.

Proper adjustment of the bias amplitude and suitable limitation of the audio amplitude transfer the audio to the tape in the linear portions *b* and *c* of the characteristic curve, minimizing distortion. The shortcomings of this explanation are detailed in the references, which also give alternative explanations in terms of Preisach diagrams and statistical theories and

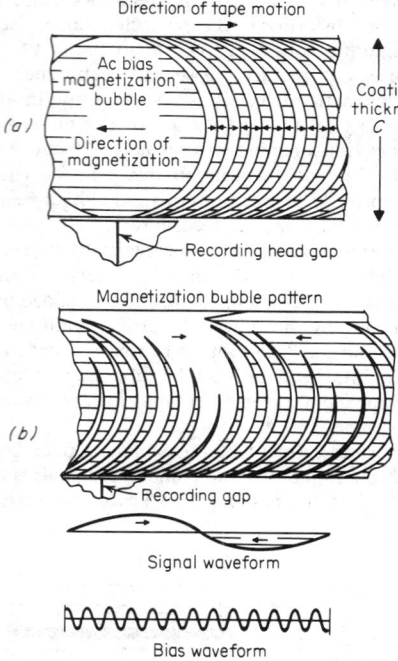

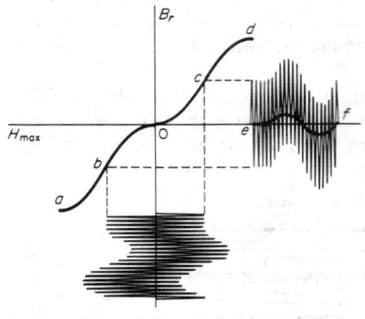

Fig. 19-148.   Linear recording by addition of a high-frequency bias signal.   (Ref. 91)

Fig. 19-149.   Model for 100 kHz ac bias recording of 0.3-mil tape coating at 3¾ in./s: (*a*) zero audio signal; (*b*) audio signal. (After Bauer and Mee, Ref. 126)

experiments on the fine structure of the recorded magnetic pattern on the tape (Fig. 19-149). Understanding of the process is not yet complete.

The optimum amount of ultrasonic bias current depends upon the magnetic properties of the tape used. Figure 19-150 is an example of the variation of tape output level and percent distortion as the bias current is changed. The current is adjusted to give a good compromise between signal-to-noise ratio and distortion.

77. Erasure and Noise.[131,132]   Complete reels of tape can be erased by gradual removal or reduction of the magnetic field of a bulk demagnetizer. Also just before recording occurs the erase head of Fig. 19-140 saturates the tape at an ultrasonic frequency (usually the bias frequency to avoid beats). As the tape leaves the gap, the fringes of flux around the edges of the wide erasing gap subject the tape to successive magnetizing cycles of decreasing intensity. Even when bulk-erased, the tape has some background noise because the medium is not continuous.

In the example of Fig. 19-151 the lowest noise spectrum is that generated in the playback equipment when no tape is running. The erased-tape noise is slightly higher in well-designed equipment except at very high frequencies. Introduction of the bias signal increases the noise further, even in the absence of audio. This can result from asymmetry in the bias field or from amplitude-modulation sidebands of the bias frequency occurring within the audio band. Modulation noise is a random amplitude variation in the audio output for constant input, occurring chiefly at low frequencies, which can be attributed to nonuniform distribution of coating particles. Other low-frequency or dc noises come from a magnetized recording head, nearby permanent magnets or solenoids, or small bumps on the tape or head which cause temporary separation.

Another common noise is print-through, the printing of recorded signals onto adjacent layers of a tape reel during storage. Print-through is sensitive to time and temperature of storage. It depends upon both proximity to another magnetized layer and magnetic instability in the recorded layer. Print-through tendency can be reduced through control of particle size and crystalline smoothness within the medium.

An important factor in selecting track width is signal-to-noise ratio. Playback signal is proportional to track width, but noise increases with the square root of the width. Thus halving track width reduces the signal-to-noise ratio 3 dB.

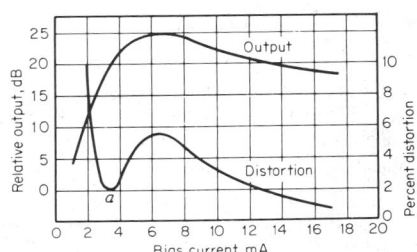

Fig. 19-150. Dependence of output and distortion on high-frequency bias current. (Ref. 91)

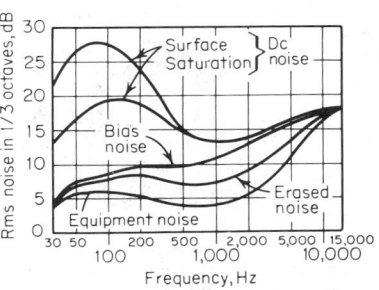

Fig. 19-151. Comparison of noise spectra from audio cassette recorder at 1-⅞ in./s with standard playback equalization. (Daniel, Ref. 132)

78. Tape Standardization and Packaging. The wide variety of tape-recording formats and equipment is based upon a family of standards (from various standards organizations) too numerous to list here.[133] Several standards give detailed support to open-reel, cartridge, and cassette packaging of tapes and associated record-reproduce equipment.[134-136] Tape and track dimensions quoted previously were from these standards. In addition the NAB standards cover hub and reel diameters, winding instructions, recorded level and uniformity, reel dimensions, tape speeds, and the dimensions and shapes of cartridges and their mechanical features. Signal-to-noise ratio (unweighted and weighted), distortion, cross talk, and stereophonic channel separation are also included. The specified response-frequency characteristics for standard reproducing systems and for standard recording are reviewed later under Equalization. (Par. **19-84**).

Special-purpose magnetic-tape cartridges have been developed for continuous-loop operation [137] in which the tape is drawn from the center of the reel. After it passes the heads and pinch roller, it returns to the outside of the reel. The necessity for continuous slippage between adjacent turns of tape requires coating the base material with a lubricant which wears well and will not disintegrate.

79. Tape Speeds and Related Properties. Standard preferred tape speed is 7½ in./s ± 0.2%. Supplementary tape speeds are 15 and 3¾ in./s with the same tolerance. Recent improvements have expanded the use of 3¾ in./s. Cassette tape speed is 1⅞ in./s.

Flutter. Perceptibility of flutter varies with flutter rate, leading to the weighting curve shown in Fig. 19-122. With a 3-kHz test tone the standard limits for weighted rms flutter are 0.05% at 15 in./s, 0.07% at 7½ in./s, and 0.10% at 3¾ in./s, measured over a flutter frequency range of 0.5 to 200 Hz.

Frequency Limits. NAB standard tapes contain test tones from 30 Hz (for all tape speeds) to 15 kHz for 7½ and 15 in./s; to 10 kHz for 3¾ in./s; and to 5 kHz for 1⅞ in./s. However, recent tape material and technique improvements extend measurements to 10 kHz and even higher at 1⅞ in./s.

80. Tape Transport Mechanisms.[138,139] The single-capstan system shown in Fig. 19-140 is typical. Figure 19-152 identifies its dynamic elements for theoretical analysis. Semiprofessional machines often use a single motor for the capstan drive. It also operates both the supply and take-up reels through friction idler wheels. Better recorders generally use three motors, a multispeed hysteresis-synchronous motor for the capstan and bidirectional torque motors for the supply and take-up reels.

In the record and playback modes the motor on the supply reel (although rotating counterclockwise) exerts a light clockwise torque to provide even tape tension as it unwinds. The take-up motor rotates counterclockwise with just enough torque to supply the correct tension for smooth tape wind. During rewind the supply motor has a high torque in a clockwise direction and the take-up motor a low torque in the opposite direction. For fast forward the situation reverses. Electromechanical brakes on the torque motors stop and hold the reels to prevent spilling the tape.

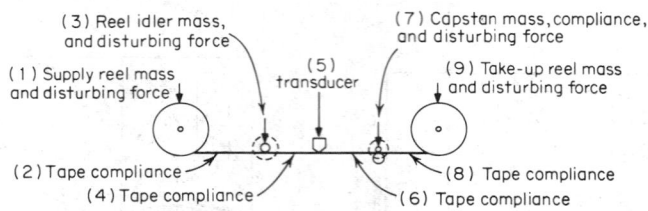

Fig. 19-152.   Physical system for a simplified tape transport. (Ref. 139)

Another transport system uses a dual-capstan drive, illustrated in Fig. 19-153. Here the capstans tend to turn at slightly different speed, and the tape tension is controlled by the differential action. This system provides better isolation of the active tape segment from the supply and take-up reels. A third type of transport eliminates the pinch-roller completely by using a capstan with high surface friction and a large tape-wrap angle.

In all these systems the goal is constant tape tension, but in some the result is constant torque, so that the tension varies with the radius of the tape winding. In fully professional equipment servo controls have been used to sense tape tension and correct reel torque accordingly.

Mechanical Damping.   Figure 19-152 shows only inertial and compliance elements, omitting resistance elements such as tape friction at heads and guides and tape damping by the rubber puck at the capstan. Without damping, the dynamic system would be unduly excited by mechanical transients caused by tape splices, tape-reel contacts, mechanical imperfections or power fluctuations. Figure 19-154 demonstrates a transient-response improvement produced by addition of a viscous damper to a reel idler.

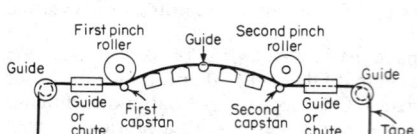

Fig. 19-153.   Dual-capstan tape-transport system.

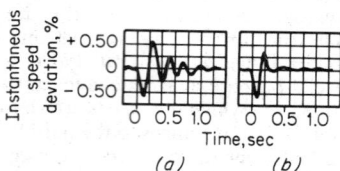

Fig. 19-154.   Measured speed deviation vs. time for an impulse at the supply reel, MR-70 tape transport with ½-in.-wide tape: (*a*) undamped flywheel on reel idler, $Q = 4.5$; (*b*) viscous damper on reel idler, $Q = 1.7$. (Ref. 139)

81. Magnetic-Tape Heads.[140]   Figure 19-141 shows the general shape and design factors. Typically the magnetic circuit consists of two high-permeability laminated cores which are butted together with a thin gap-defining shim between them to form a symmetrical ring-type structure. Balanced coil windings cancel external magnetic fields. After terminals are connected, the assembly is encapsulated. The tape-contacting surface is ground and lapped for curvature and smoothness.

One method of multitrack head assembly precisely stacks individually encapsulated wafer head elements. Selective assembly of tested wafers gives well-matched groups. Shields are selected for thickness to compensate for wafer-thickness variations.

In another assembly method cast case halves are precisely milled to receive shields and laminated core halves which with their coils are epoxied in place. The face of the entire half-head assembly is ground and lapped to achieve a high degree of flatness. A quartz spacer of less than 0.0001 in. is evaporated on the pole-tip area before the halves are joined. The laminations are grounded electrically. The entire assembly is shielded for minimum pickup of extraneous fields.

Figure 19-144 shows a typical response-frequency characteristic for a playback head. Typical cross talk between adjacent heads is down 50 to 60 dB, degraded slightly by the presence of bias. This is compromised in recording formats using very narrow tracks and very close spacing.

82. Head Wear and Tape Contact.[141,142] Abrasive action by the tape oxide surface wears heads and tape guides. If heads extend beyond tape edges, grooves are worn which damage the edges and cause intermittent tape separation from the head. (The effect of separation was shown in Fig. 19-143.) High-frequency losses caused by head wear can be partially regained by relapping the head.

When tape under tension wraps around a curved surface, a transverse tape curvature is induced. The smaller the wrap curvature and the higher the tension, the greater the tendency for tape cupping, edge separation, and formation of concave grooves.

83. Special Heads.[143-145] Recording resolution can be improved by sharpening the field at the recording gap. When the semicircular field of a conventional head is modified by adding a field normal to the tape thickness, a sharper gradient is produced at one edge of the gap. An X-field head using this principle, shown in Fig. 19-155, uses the erase gap for the X

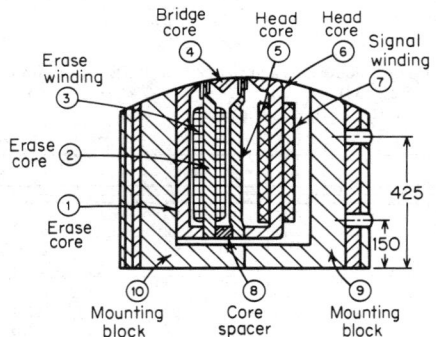

Fig. 19-155.   Stereo X-field head.   (After Camras, Ref. 143)

field. Thus only two coils are required per channel. Erase and record gap dimensions are 4 and 0.005 mil, respectively. Elimination of the separate erase head compensates for the additional X-field structure.

Another special magnetic-gap technique is to deposit the magnetic material near the gap by electroplating and to form the gap by engraving. This is appropriate for slow tape speeds or noncontacting heads, where wear is not a problem.

Flux-sensitive heads using thin semiconductors sensitive to the Hall effect have been developed. Their inherent frequency response is flat into the ultrasonic range, but there are still limitations in the associated head structure.

84. Equalization. The basis for tape equalization is the NAB standard reproducing characteristic shown in Fig. 19-156. This assumes that the recorder circuits will preemphasize low-frequency response and that typical high-frequency losses must be equalized in playback by different amounts, depending upon tape speed. Although curve B is identified with $1\frac{7}{8}$ in./s, the standard reproducing characteristic for cassette tapes assumes greater bass preemphasis (1,590-μs time constant) and greater treble loss (120-μs time constant). This corresponds to approximately $+10$ dB at 30 Hz and -18 dB at 10 kHz. Improved cassette tape material is expected to reduce 120 to 70 μs.

Figure 19-156 assumes constant flux in the reproducing head core, theoretically removing the recorded tape from the measurement. Calibrated standard NAB tapes made under laboratory conditions are available to manufacturers and users. Their test-tone levels are preemphasized so that the output level of a tape reproducer meeting the requirements of Fig. 19-156 should be uniform within the limits of Fig. 19-157.

After the recorder-reproducer designer has met the reproducer limits of Fig. 19-157, the

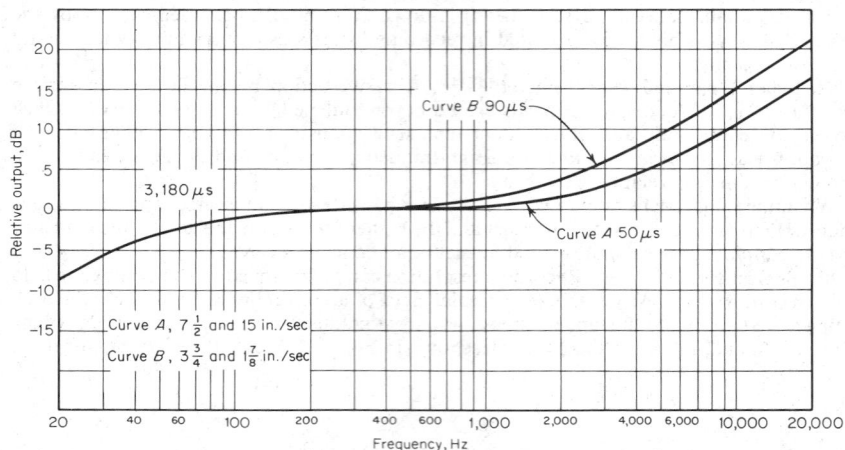

Fig. 19-156. NAB standard reproducing characteristic. Amplifier output for constant flux in the core of an ideal reproducing head. (Ref. 134)

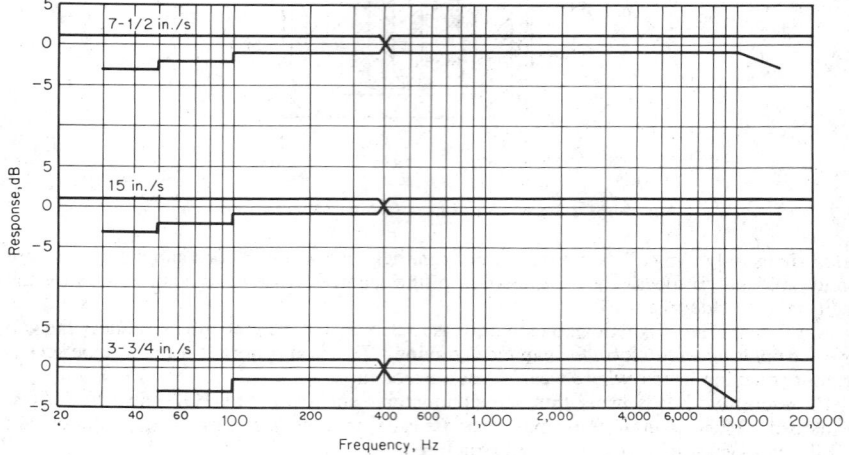

Fig. 19-157. NAB standard reproducing-system response limits. (Ref. 134)

preemphasis circuits of the recorder are designed to give an overall record-reproduce response which is similarly uniform. This equipment specification must then identify the tape type used.

Although the standard reproducing characteristic assumes specific resistance-capacitance networks, preemphasis and postequalization circuits can be designed which are complementary to other system parameters in order to obtain the desired result for a standard tape.

85. Noise-Reduction Circuits.[146-149] At low recording levels the need to suppress background noise from the recording medium has led to various signal-processing systems, some acting only on signal played back and others involving both pre- and postprocessing. Two systems are finding increasing use.

Figure 19-158 shows a system differentially operating oppositely on the signal before recording and after playback (Dolby system). Signal multipliers G_1 and G_2 are controlled by the amplitudes and dynamic properties of the signals fed into them. Network G_2 during reproduction sends low-level components back to the subtractor, partially canceling these components in the reproduced signal. To compensate, identical network G_1 adds an identical component ahead of the recording. Originally G_1 and G_2 comprised identical sets of four filters and low-level compressors. The four frequency divisions are 80 Hz low-pass, 80 to 3,000 Hz bandpass, 3 kHz high-pass, and 9 kHz high-pass. At high sound levels all bands are reproduced normally. At low levels the hum, rumble, and hiss noises (and spectrally associated signals) are deemphasized. A by-product in tape recording is the reduction of high-frequency modulation noise and of high-frequency sidebands of low-frequency signals resulting from scrape flutter. A recent version (type B) uses only two channels with a 1-kHz crossover frequency.

The block diagram in Fig. 19-159 shows a noise limiter operating in the playback amplifier after playback equalization. High sound levels are reproduced only through the all-pass amplifier. The phase splitter supplies opposite-phase signals to a 5-kHz high-pass amplifier. When its output falls more than 40 dB below reference level, the level-triggered amplifier output cancels the high-frequency output from the all-pass amplifier. The level-triggered amplifier is normally open during silent passages, effectively canceling interim hiss.

86. Measurements. Knowing how much flux is actually recorded on a tape aids in determining and standardizing reference levels, specifying the properties of magnetic media, and measuring the sensitivity of magnetic heads. Special high-efficiency symmetrical heads have been constructed, calibrated, and verified by magnetometer flux measurements for these purposes.[150]

Overall tape-system measurements are highly complex involving many factors. Their data reflect the state of the art and indicate directions for needed improvements. They also provide guidance for the users of recording equipment. Figure 19-160 is an example of a comprehensive study of the dynamic range of tape cartridges and cassettes.

87. Specifications. The data in the comparison of two tape record-reproducers shown in Table 19-8 were published by independent testing laboratories. Recorder A is for semiprofessional use. Recorder B is an advanced amateur recorder. The price ratio is approximately 3:1.

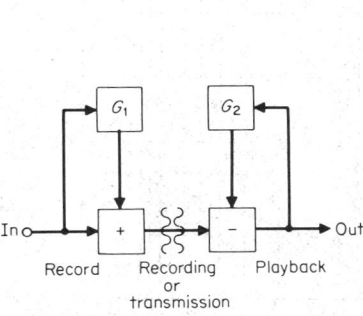

Fig. 19-158. Basic block diagram of Dolby noise-reduction system. Operators G_1 and G_2 are identical set of frequency-range dividers and low-level compressors. (Ref. 146)

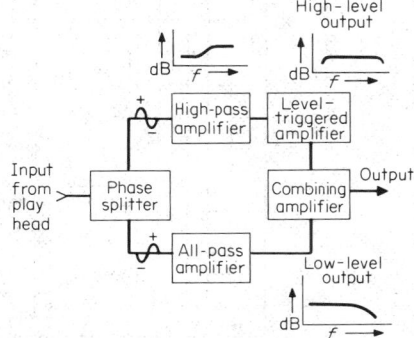

Fig. 19-159. Basic block diagram of hiss-noise limiter system.

19-89

Table 19-8.   Comparison of Tape Recorders

	Recorder A (semiprofessional)	Recorder B (advanced amateur)
Type	Reel to reel	Cassette
Number of tracks	2	2
Tape width, in.	0.246	0.150
Tape speeds, in. /s	3¾, 7½, 15	1⅞
Speed accuracy	±0.2%	+0.1 to 0.6% fast
Wow and flutter	0.10, 0.07, 0.05% (NAB unweighted)	0.13% (unweighted)
Reel size..............................	5 to 10½ in. NAB	Cassette
Start time, s	0.1	
Rewind or fast forward time, s	45 (1,200 ft, 7 in. reel)	43 (C-60 cassette)
Signal-to-noise ratio (re 0 vu), dB	60	52
Erasure (400 Hz), dB	60	50
Cross talk (400 Hz), dB	60	46
Percent third harmonic	<1	<1.4 to 5 kHz
Record/playback response	±2 dB; 20 Hz to 30 kHz	+1, −5 dB; 30 Hz to 14 kHz
Capstan drive motor	Hysteresis-synchronous	
Reel motors	Torque	
Drive belt	Ground neoprene over flywheel	
Braking system	Electrodynamic	
Head assembly	Erase, record, reproduce (independently adjustable)	

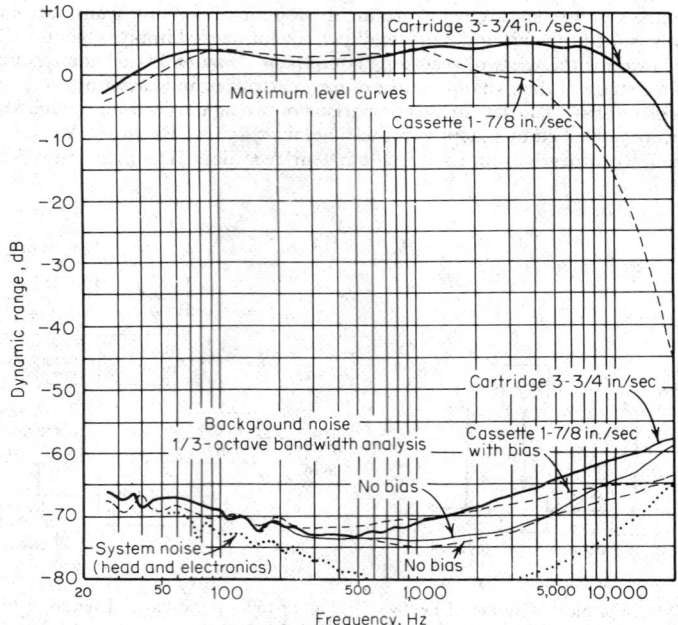

Fig. 19-160.   Dynamic range of tape cartridges and cassettes.   (After Gravereaux, Gust, and Bauer, Ref. 114)

88. References
1. Acoustical Terminology (Including Mechanical Shock and Vibration), S1.1-1960, American National Standards Institute, Inc., New York.
2. YOUNG, R. W.   Physical Properties of Noise and Their Specification, Chap. 2 in C. M. Harris (ed.), "Handbook of Noise Control," McGraw-Hill, New York, 1957.
3. ROBINSON, D. W., and R. S. DADSON   *Br. J. Appl. Phys.*, 1956, Vol. 7, p. 166.
4. ISO Recommendation 226, International Organization for Standardization, 1961.
5. Expression of the Physical and Subjective Magnitudes of Sound or Noise, ISO Recommendation 131, International Organization for Standardization, 1959 (supplemented by R357-1963).
6. MUNSON, W. A.   The Loudness of Sounds, Chap. 5 in C. M. Harris (ed.), "Handbook of Noise Control," McGraw-Hill, New York, 1957.
7. Method for Calculating Loudness Level, ISO Recommendation 532, International Organization for Standardization, 1966.
8. STEVENS, S. S.   *J. Acoust. Soc. Am.*, 1972, Vol. 51, p. 575.
9. MARTIN, D. W.   *J. Acoust. Soc. Am.*, 1942, Vol. 13, p. 309.
10. DUNN, H. K., and S. D. WHITE   *J. Acoust. Soc. Am.*, 1940, Vol. 11, p. 278.
11. DUNN, H. K., and D. W. FARNSWORTH   *J. Acoust. Soc. Am.*, 1939, Vol. 10, p. 184.
12. POTTER, RALPH K., G. A. KOPP, and H. C. GREEN   "Visible Speech," Van Nostrand, New York, 1947.
13. DUNN, H. K.   *J. Acoust. Soc. Am.*, 1950, Vol. 22, p. 740.
14. POTTER, RALPH K., and G. E. PETERSON   *J. Acoust. Soc. Am.*, 1948, Vol. 20, p. 528.
15. FRENCH, N. R., and J. C. STEINBERG   *J. Acoust. Soc. Am.*, 1947, Vol. 19, p. 90.
16. BERANEK, L. L.   *Proc. IRE*, 1947, Vol. 35, p. 880.
17. HAWLEY, M. E., and K. D. KRYTER   Effects of Noise on Speech, Chap. 9 in C. M. Harris (ed.), "Handbook of Noise Control," McGraw-Hill, New York, 1957.
18. WEBSTER, J. C.   *J. Audio Eng. Soc.*, 1970, Vol. 18, p. 114.
19. LICKLIDER, J. C. R.   *J. Acoust. Soc. Am.*, 1946, Vol. 18, p. 429.
20. MARTIN, D. W.   *J. Acoust. Soc. Am.*, 1950, Vol. 22, p. 614.
21. MARTIN, D. W., R. L. MURPHY, and ALBERT MEYER   *J. Acoust. Soc. Am.*, 1956, Vol. 28, p. 597.
22. OLSON, H. F.   "Musical Engineering," McGraw-Hill, New York, 1952.
22a. OLSON, H. F.   "Music, Physics and Engineering," 2d ed., Dover, New York, 1967.
23. WARD, W. D.   *J. Acoust. Soc. Am.*, 1954, Vol. 26, p. 369.
24. MARTIN, D. W., and W. D. WARD   *J. Acoust. Soc. Am.*, 1961, Vol. 33, p. 582.
25. MARTIN, D. W.   *Sound*, 1962, Vol. 1, p. 22.
26. SNOW, W. B.   *J. Acoust. Soc. Am.*, 1931, Vol. 3, p. 155.
27. CLARK, MELVILLE, and DAVID LUCE   *J. Audio. Eng. Soc.*, 1965, Vol. 13, p. 151.
28. MARTIN, D. W.   *J. Acoust. Soc. Am.*, 1947, Vol. 19, p. 535.
29. SIVIAN, L. J., H. K. DUNN, and S. D. WHITE   *IRE Trans. Audio*, 1959, Vol. AU-7, p. 47; revision of paper in *J. Acoust. Soc. Am.*, 1931, Vol. 2, p. 33.
30. MEYER, J.   *J. Acoust. Soc. Am.*, 1972, Vol. 51, p. 1994.
31. OLSON, H. F.   "Elements of Acoustical Engineering," Van Nostrand, New York, 1947.
32. WAYNE, W. C., A. B. BERESKIN, and D. W. MARTIN, unpublished work.
33. "Motion Picture Sound Engineering," Chaps. 29 and 30, Van Nostrand, New York, 1938.
34. BEERS, G. L., and H. BELAR: *Soc. Motion Pict. Eng.*, 1943, Vol. 40, p. 207.
35. MARTIN, D. W.:   U.S. Patent No. 3,467,758 (1969).
36. HARRIS, C. M. (ed.)   "Handbook of Noise Control," McGraw-Hill, New York, 1957.
37. BERANEK, L. L.   "Noise Reduction," McGraw-Hill, New York, 1960.
38. PETERSON, A., and P. V. BRUEL   Instruments for Noise Measurements, Chap. 16 in C. M. Harris (ed.), "Handbook of Noise Control," McGraw-Hill, New York, 1957.
39. SCOTT, H. H.   Noise Measuring Techniques, Chap. 17 in C. M. Harris (ed.), "Handbook of Noise Control," McGraw-Hill, New York, 1957.
40. PETERSON, A. P. G.   *Gen. Radio Exp.*, 1956, Vol. 30, p. 1.
41. KOENIG, W., H. K. DUNN, and L. Y. LACY   *J. Acoust. Soc. Am.*, 1946, Vol. 18, p. 19.
42. BLASSER, H., and H. FINCKH   *Hewlett-Packard J.*, 1967, Vol. 19, p. 3.
43. Paragraph 50-204.10, Walsh-Healey Public Contracts Acts, *Fed. Reg.*, May 20, 1969.

44. Pt. 1910, Tit. 29, Occupational Safety and Health Act of 1970, *Fed. Reg.*, Vol. 36, No. 105, May 29, 1971.
45. Determination of Sound Transmission Class, ASTM Designation E413-70T, 1970.
46. LEONARD, R. W.   Heating and Ventilating System Noise, Chap. 27, in C. M. Harris (ed.), "Handbook of Noise Control," McGraw-Hill, New York, 1957.
47. Sound Control chapter, "ASHRAE Guide and Data Book," biennial publication.
48. ASHRAE Standard 36B-63.
49. Air Diffusion Council Test Code 1062R1.
50. INGERSLEV, F., and C. M. HARRIS   Control of Solid-Borne Noise, Chap. 19, in C. M. Harris (ed.), "Handbook of Noise Control," McGraw-Hill, New York, 1957.
51. HARRIS, C. M., and C. CREDE   "Shock and Vibration Handbook," McGraw-Hill, New York, 1961.
52. BERANEK, L. L.   "Music, Acoustics and Architecture," Wiley, New York, 1962.
53. KNUDSEN, V. O., and C. M. HARRIS   "Acoustical Designing in Architecture," Wiley, New York, 1950.
54. "Music Buildings, Rooms and Equipment," Music Educators National Conference, Washington, D.C., 1966.
55. MORSE, P. M.   Chap. 8, in "Vibration and Sound," McGraw-Hill, New York, 1948.
56. VOLKMANN, J. E.   *J. Acoust. Soc. Am.,* 1942, Vol. 13, p. 324.
57. LANE, R. N.   *J. Acoust. Soc. Am.,* 1956, Vol. 28, p. 101.
58. BERANEK, L. L.   *J. Acoust. Soc. Am.,* 1960, Vol. 32, p. 661.
59. S3.2-1960, American National Standards Institute, 1960.
60. BERANEK, L. L.   Chap. 17, in "Acoustic Measurements," Wiley, New York, 1949.
61. BAUER, B. B.   *Proc. IRE,* 1962, Vol. 50, 50th Anniversary Issue, p. 719.
62. BERANEK, L. L.   "Acoustics," McGraw-Hill, New York, 1954.
63. SESSLER, G. M., and J. E. WEST.   *J. Audio Eng. Soc.,* 1964, Vol. 12, p. 129.
64. OLSON, H. F.   "Dynamical Analogies," Van Nostrand, New York, 1943.
65. FIRESTONE, F. A.   *J. Acoust. Soc. Am.,* 1956, Vol. 28, p. 1117.
66. OLNEY, B., F. H. SLAYMAKER, and W. F. MEEKER   *J. Acoust. Soc. Am.,* 1945, Vol. 16, p. 172.
67. MARTIN, D. W.   *J. Acoust. Soc. Am.,* 1947, Vol. 19, p. 43.
68. ANDERSON, L. J.   *IRE Trans. Audio.,* 1953, Vol. 1, p. 1.
69. OLSON, H. F.   *Proc. IRE,* 1962, Vol. 50, 50th Anniversary Issue, p. 730.
70. KNOWLES, H. S.   Sec. 13 in "Electrical Engineers' Handbook," Wiley, New York, 1950.
71. HANNA, C. R.   *Proc. IRE,* 1925, Vol. 13, p. 437.
72. LEVINE, H., and J. SCHWINGER   *Phys. Rev.,* 1948, Vol. 73, p. 383.
73. WOLFF, I., and L. MALTER   *J. Acoust. Soc. Am.,* 1930, Vol. 2, p. 201.
74. RICE, C. W., and E. W. KELLOGG   *Trans. AIEE,* 1925, Vol. 44, p. 461.
75. CORRINGTON, M. S.   *Proc. IRE,* 1951, Vol. 39, p. 1021.
76. WEBSTER, A. G.   *Proc. Natl. Acad. Sci.,* 1919, Vol. 5, p. 275.
77. JANSZEN, A. A.   *J. Audio Eng. Soc.,* 1955, Vol. 3, p. 87.
78. MALME, C. I.   *J. Audio Eng. Soc.,* 1959, Vol. 7, p. 47.
79. FIALA, W. T., J. K. HILLIARD, J. A. RENKUS, and J. J. VAN HOUTEN   *J. Acoust. Soc. Am.,* 1965, Vol. 38, p. 956.
80. KLEIN, S.   *Acustica,* 1954, Vol. 4, p. 77.
81. KOCK, W. E., and F. K. HARVEY   *J. Acoust. Soc. Am.,* 1949, Vol. 21, p. 471.
82. WENTE, E. C., and A. L. THURAS   *J. Acoust. Soc. Am.,* 1931, Vol. 3, p. 44.
83. ANDERSON, L. J.   *J. Soc. Motion Pict. Eng.,* 1941, Vol. 37, p. 319.
84. INGLIS, A. H., G. H. G. GRAY, and R. T. JENKINS   *Bell Syst. Tech. J.,* 1932, Vol. 2, p. 293.
85. MARTIN, D. W., and L. J. ANDERSON   *J. Acoust. Soc. Am.,* 1947, Vol. 19, p. 63.
86. WIENER, F. M., and D. A. ROSS   *J. Acoust. Soc. Am.,* 1946, Vol. 18, p. 401.
87. WIENER, F. M., and A. S. FILLER   *Harv. Rep.* PNR-2, 1945.
88. SHAW, E. A. G., and G. J. THIESSEN   *J. Acoust. Soc. Am.,* 1958, Vol. 30, p. 24.
89. BAUER, B. B.   *J. Audio Eng. Soc.,* 1961, Vol. 9, p. 148.
90. BACHMAN, W. S., B. B. BAUER, and P. C. GOLDMARK   *Proc. IRE,* 1962, Vol. 50, 50th Anniversary Issue, p. 738.

91. FRAYNE, G., and H. WOLFE   "Elements of Sound Recording," Wiley, New York, 1949.
92. JACKSON, J. E.   *J. Audio Eng. Soc.,* 1965, Vol. 13, p. 134.
93. ALBERSHEIM, W. J., and D. MACKENZIE   *J. Soc. Motion Pict. Eng.,* 1941, Vol. 37, p. 452.
94. BACHMAN, W. S.   U.S. Patent No. 2,738,385, March 13, 1956.
95. DAVIS, C. C., and J. G. FRAYNE   *1958 IRE Natl. Conv. Rec.,* Pt. 7 Audio, p. 62.
96. VIETH, L., and C. F. WIEBUSCH   *J. Soc. Motion Pict. Eng.,* 1938, Vol. 30, p. 96.
97. MARCUCCI, R.   *J. Audio Eng. Soc.,* 1965, Vol. 13, p. 130.
98. LEBEL, C. J.   *J. Acoust. Soc. Am.,* 1942, Vol. 13, p. 265.
99. BACHMAN, W. S.   *Audio Eng.,* 1950, Vol. 34, p. 11.
100. MOURA, C. E.   *J. Audio Eng. Soc.,* 1960, Vol. 8, p. 228; 1961, Vol. 9, p. 60.
101a. SCHEIBER, P.   *J. Audio, Eng. Soc.,* 1971, Vol. 19, p. 267; also 1971, Vol. 19, p. 647.
101b. EARGLE, J. M.   *J. Audio Eng. Soc.,* 1971, Vol. 19, p. 552.
101c. JURGEN, R. K.   *IEEE Spectrum,* July, 1972, p. 55.
102a. INOUE, T., N. TAKAHASHI, and I. OWAKI   *J. Audio Eng. Soc.,* 1971, Vol. 19, p. 576.
102b. OWAKI, I., T. MURAOKA, and T. INOUE   *J. Audio Eng. Soc.,* 1972, Vol. 20, p. 361.
102c. INOUE, T., N. SHIBATA, and K. GOH   *J. Audio Eng. Soc.,* 1973, Vol. 21, p. 166.
102d. INOUE, T., I. OWAKI, Y. ISHIGAKI, and K. GOH   *J. Audio Eng. Soc.,* 1973, Vol. 21, p. 625.
103a. BAUER, B. B., D. W. GRAVEREAUX, and A. J. GUST   *J. Audio Eng. Soc.,* 1971, Vol. 19, p. 638.
103b. BAUER, B. B., G. A. BUDELMAN, and D. W. GRAVEREAUX   *J. Audio Eng. Soc.,* 1973, Vol. 21, p. 19.
103c. BAUER, B. B., R. G. ALLEN, G. A. BUDELMAN, and D. W. GRAVEREAUX   *J. Audio Eng. Soc.,* 1973, Vol. 21, p. 342.
104a. ITOH, R.   *J. Audio Eng. Soc.,* 1972, Vol. 20, p. 167.
104b. COOPER, D. H., and T. SHIGA   *J. Audio Eng. Soc.,* 1972, Vol. 20, p. 346.
105. OLSON, H. F.   *J. Audio Eng. Soc.,* 1964, Vol. 12, p. 98.
106. FOX, E. C., and J. G. WOODWARD   *J. Audio Eng. Soc.,* 1963, Vol. 11, p. 294.
107. EARGLE, J. M.   *J. Audio Eng. Soc.,* 1969, Vol. 17, p. 276.
108. BASTIAANS, C. R.   *J. Audio Eng. Soc.,* 1967, Vol. 15, p. 389.
109. PIERCE, J. A., and F. V. HUNT   *J. Soc. Motion Pict. Eng.,* 1937, Vol. 29, p. 493.
110. ANDERSON, C. R., and P. W. JENRICK   *J. Audio Eng. Soc.,* 1972, Vol. 20, p. 162.
111. BAUER, B. B.   *Electronics,* 1945, Vol. 18, p. 110; IEEE Trans., 1963, Vol. AU-11, p. 47.
112. BAUER, B. B.   *J. Acoust. Soc. Am.,* 1946, Vol. 18, p. 387; 1955, Vol. 27, p. 586.
113. ROYS, H. E.   *J. Audio Eng. Soc.,* 1953, Vol. 1, p. 78.
114. GRAVEREAUX, D. W., A. J. GUST, and B. B. BAUER   *J. Audio Eng. Soc.,* 1970, Vol. 18, p. 530.
115. POULSEN, V.   *Ann. Phys. (Leipz.),* 1900, Vol. 3, p. 754.
116. BEGUN, S. J.   "Magnetic Recording," Murray Hill Books, New York, 1949.
117. WILSON, C. F.   *IRE Trans. Audio.,* 1956, Vol. AU-4, p. 53.
118. WESTMIJZE, W. K.   Studies on Magnetic Recording, *Phillips Res. Rep.,* 1953, Vol. 8, pp. 161-183, 245-269.
119. FUJII, M., G. REHKLAU, J. G. MCKNIGHT, and W. MILTENBURG   *J. Audio Eng. Soc.,* 1960, Vol. 8, p. 245.
120. CAMRAS, M.   *Proc. IRE,* Vol. 50, 1962, 50th Anniversary Issue, p. 751.
121. EILERS, D. W.   *J. Audio Eng. Soc.,* 1969, Vol. 17, p. 303; 1970, Vol. 18, p. 540.
122. NESH, F., and R. F. BROWN   *IRE Trans. Audio,* 1962, Vol. AU-10, p. 70; 1964, Vol. AU-12, p. 55.
123. MEE, C. D.   *IRE Trans. Audio,* 1964, Vol. AU-12, p. 72.
124. NAUMANN, K. E., and E. D. DANIEL   *J. Audio Eng. Soc.,* 1971, Vol. 19, p. 822.
125. FINGER, R. A., P. MURPHY, and E. J. FOSTER   *J. Audio Eng. Soc.,* 1972, Vol. 20, p. 549.
126. BAUER, B. B., and C. D. MEE   *IRE Trans. Audio,* 1961, Vol. AU-9, p. 139.
127. WOODWARD, J. G., and E. DELLA TORRE   *J. Appl. Phys.,* 1960, Vol. 31, p. 56.
128. PREISACH, F.   *Z. Phys.,* 1935, Vol. 94, p. 277.
129. ELDRIDGE, D. F.   *IRE Trans. Audio,* 1961, Vol. AU-9, p. 155.

130. RADOCY, F., and A. KRAMER   *J. Audio Eng. Soc.*, 1957, Vol. 5, p. 76.

131. ELDRIDGE, D. F.   *IEEE Trans. Audio*, 1964, Vol. AU-12, p. 100.

132. DANIEL, E. D.   *J. Audio Eng. Soc.*, 1972, Vol. 20, p. 92.

133. Audio Standards Listings: Magnetic Tape Sound Recording: *J. Audio Eng. Soc.*, 1970, Vol. 18, p. 319.

134. NAB Standard, Magnetic Tape Recording and Reproducing (Reel-to-Reel), April 1965.

135. NAB Standard, Cartridge Tape Recording and Reproducing, October 1964.

136. HANSON, E. R.   *J. Audio Eng. Soc.*, 1968, Vol. 16, p. 430; 1971, Vol. 19, p. 24.

137. KNOX, A.   *J. Audio Eng. Soc.*, 1964, Vol. 12, p. 32.

138. BIXLER, O. C.   *IRE Trans. Audio*, 1954, Vol. AU-2, p. 15.

139. MCKNIGHT, J. G.   *J. Audio Eng. Soc.*, 1964, Vol. 12, p. 140.

140. SARITI, A. A.   *J. Audio Eng. Soc.*, 1960, Vol. 8, p. 243.

141. ARNOLD, R. R., L. J. ANANKA, and S. V. MARSOV   *J. Audio Eng. Soc.*, 1972, Vol. 20, p. 470.

142. TABER, W. D.   *J. Audio Eng. Soc.*, 1968, Vol. 16, p. 61.

143. CAMRAS, M.   *IEEE Trans. Audio*, 1964, Vol. AU-12, p. 41.

144. PETERS, C. J.   *IRE Trans. Audio*, 1962, Vol. AU-10, p. 79.

145. CAMRAS, M.   *IRE Trans. Audio*, 1962, Vol. AU-10, p. 84.

146. DOLBY, R. M.   *J. Audio Eng. Soc.*, 1967, Vol. 15, p. 383.

147. MULLIN, J. T.   *IEEE Trans. Audio*, 1965, Vol. AU-13, p. 31.

148. OLSON, H. F.   *Electronics*, 1947, Vol. 20, p. 118.

149. SCOTT, H. H.   *Electronics*, 1947, Vol. 20, p. 96.

150. MCKNIGHT, J. G.   *J. Audio Eng. Soc.*, 1970, Vol. 18, p. 250.

SECTION 20

TELEVISION AND FACSIMILE SYSTEMS

BY

WILLIAM L. HUGHES Professor and Head of the School of Electrical Engineering, Oklahoma State University; Fellow, IEEE

HENRY N. KOZANOWSKI Manager Advanced Development, Commercial Electronic Systems Division, RCA Corporation; Fellow IEEE

CHARLES W. RHODES Manager, Television Products Engineering, Communications Division, Tektronix, Incorporated; Senior Member, IEEE; Fellow, SMPTE; member, Royal Television Society

NORMAN W. PARKER Staff Scientific Advisor, Motorola Incorporated; Senior Member, IEEE

HOWARD W. TOWN Manager Product Planning, Video Systems Division, Ampex Corporation

KEITH Y. REYNOLDS VTR Product Manager, International Video Corporation

FREDERICK M. REMLEY, JR. Technical Director, Television Center, University of Michigan; Member, IEEE

C. H. EVANS Research Associate, Research Laboratories, Eastman Kodak Company; Fellow, Society of Motion Picture and Television Engineers

ROBERT A. CASTRIGNANO Department Manager, EVR Technology Department, CBS Laboratories; Senior Member, IEEE

E. G. RAMBERG Fellow of the Technical Staff, RCA Laboratories, David Sarnoff Research Center, RCA Corporation; Fellow, IEEE

EARLE D. JONES Manager Electronics and Optics Group, Stanford Research Institute; Member, IEEE

DAVID G. FALCONER Research Physicist, Engineering Sciences Laboratory, Stanford Research Institute

HENRY BALL Corporate Staff Engineer, Research and Engineering, RCA Corporation, David Sarnoff Research Center

CONTENTS

Numbers refer to paragraphs

TELEVISION AND FACSIMILE SYSTEMS

SECTION 20

TELEVISION AND FACSIMILE SYSTEMS

TELEVISION FUNDAMENTALS AND STANDARDS

By William L. Hughes

1. Criteria for Picture Reproduction. In the design of a picture-reproduction system, the basic criteria to be met are that the reproduced images shall be acceptable to the human eye and that the technical details of the system shall not be obtrusively evident to the viewer.

The first quality of a picture judged by the eye is its *sharpness,* or *pictorial clarity.* If the image is out of focus or the details or edges of objects are not clear and sharp, the eye will attempt without success to focus the picture and eyestrain will result. The second quality of importance to the eye is the *contrast* between light areas, dark areas, and the related *background illumination.* Background lighting affects the contrast observed by the eye and can introduce contrast changes not present in the original scene.

The third quality of importance is *continuity of motion.* Reproduced pictures in motion are created by a succession of still frames, and the illusion of motion is created, in part, by the fact that the human eye briefly retains any image impinging on it. Early work in motion pictures revealed that a rate of 16 pictures per second is sufficient to preserve the sense of continuity of motion. The first movies used 16 pictures per second as a standard, and most amateur movie cameras still do. The standard for motion pictures later became 24 frames per second. European television standards call for 25 frames per second, while the American, Canadian, and Japanese television standards call for 30 frames per second.

The fourth picture quality of importance is *flicker.* Even when continuity of motion is preserved, the picture may flicker. To eliminate flicker at useful image brightnesses, the pictures must be presented at a rate considerably greater than that required for continuity of motion. Flicker is a function of picture brightness (see Par. 20-4). Flicker can cause eyestrain, and it makes the reproduced picture unpleasant to view. Some tolerance for flicker can be built up over an extended period of viewing, as Americans exposed to European television can testify.

The fifth characteristic is that *color values,* if present, must be accepted as *realistic.* The eye is particularly critical of flesh-tone colors. Color reproduction need not be *accurate* when compared with the original scene, because the colors observed are greatly influenced by the surroundings and illumination and the eye compensates for such variations.

In establishing the standards of a picture-reproduction system, the problem is to satisfy these requirements of the human eye, adequately and economically.

2. Luminance and Brightness. The eye does not respond equally to radiated energy of all visible wavelengths. There is wide variation between observers, and the response is also a function of light intensity. Based on thousands of measurements on human observers, the average eye is considered to respond according to the *luminosity function of the standard observer* (Fig. 20-1).

The *luminance* of a surface is the effect on the average eye of the light emitted by a unit area of the surface. It is the integrated effect of the eye response $y(\lambda)$ (Fig. 20-1) and the visible light power radiated by the surface $E(\lambda)$, both of which are functions of the wavelength λ. The integration is expressed by

$$\text{Luminance} = 680 \int E(\lambda)\, y(\lambda)\, d\lambda \quad \text{lm/unit area}$$

where lm is the abbreviation for lumen and the radiated power $E(\lambda)$ is in watts per unit area. The constant 680 lm/W is the *luminosity* of radiant power at the peak of the luminosity curve, at 546 nm. The *luminous efficacy* is defined as the lumens emitted per watt radiated.

The *brightness* of a surface is defined in terms of a surface which reflects the light in perfectly diffuse fashion. Such a surface has a brightness of 1 foot-lambert (fL) for each lumen incident upon it if it does not absorb any energy, i.e., has reflectivity of unity.

A perfectly *diffuse reflector* reflects light such that the luminance flux density falls off as the cosine of the angle θ measured from the vertical to the surface. Such a surface displays no apparent change in brightness to the eye as the viewing angle changes (Fig. 20-2).

A *specular reflector* is one that favors particular directions of reflection. Perfect specular reflectors do not exist, but optically flat mirrors approach this condition. Diffuse reflection is used in movie screens so that the audience can view the images adequately over wide areas.

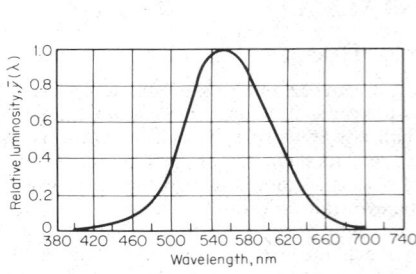

Fig. 20-1. Standard luminosity function.

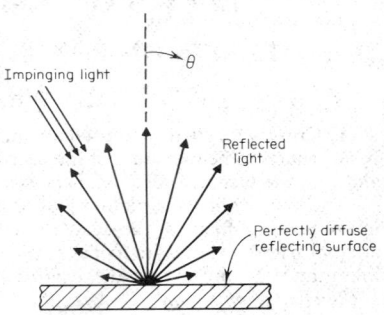

Fig. 20-2. Cosine-law reflector.

3. Contrast (Tonal Range). A characteristic to which the eye is particularly sensitive is *contrast ratio*. This parameter is defined in terms of a diffuse, flat, spectrally neutral reflector, i.e., one that scatters the light falling on it according to the cosine law and reflects all wavelengths of the visible spectrum equally. If two areas in such a reflector have different reflection coefficients, the contrast ratio displayed between the two areas is equal to the ratio of the reflection coefficients. For example, if the reflection coefficients are 80 and 4%, and if both areas are uniformly illuminated, the contrast ratio is

$$\frac{0.80}{0.04} = 20 \qquad \text{contrast ratio} = 20{:}1$$

When the illumination falling on the two areas is different, the contrast ratio is modified proportionately by the ratio of the illumination. The contrast ratio of that picture is further affected by ambient illumination and by light scattered from one area to another.

In practice in a well-designed, darkened movie theater, contrast ratios of 100:1 or more can be achieved, but in television the situation is quite different. The reflectance of a television picture tube to light falling on it from the room may be 25% or more. Such light is usually present since observation of television in a totally darkened room is not usual and in fact is not recommended. Light from one part of the television image to another is scattered, inside the picture tube and between phosphor grains.

The result is that television receivers do not produce contrast ratios (under normal viewing conditions) of much more than 20:1. This range of contrast is adequate, but it demands close control of gamma (particularly when film is used) and of brightness levels and black (clamping) levels in the video signal handling if the image is to offer a reasonable range of grays from white to black.

In color television pictures, the definition of contrast ratio, as in monochrome, is the luminance of the brightest area divided by the luminance of the darkest area, independently of whether they are of the same color. High-luminance areas tend to not be highly saturated colors (see Par. **20-15**). Dark areas may have a high color saturation, but they tend to display a dark or blackish appearance.

4. Flicker, Fields, and Frames. The flicker effect has the following characteristics:
1. It is independent of motion in the picture.
2. For a given brightness, it becomes less pronounced as the number of flashes per second is increased.
3. It is a function of brightness. If the large-area highlight brightness is 10 fL (equivalent

to a luminance of 10 lm/ft^2 scattered according to the cosine law), the flicker effect disappears at approximately 40 flashes per second. This is a typical condition encountered in a motion picture theater. If the large-area brightness is 100 fL (as it may be in a very bright television picture), flicker disappears at 50 or more flashes per second.

4. Small areas have a lower critical flicker frequency than large areas of the same brightness.

For motion pictures, it would be uneconomical to provide enough film for 40 pictures per second merely to overcome flicker, since continuity of motion is preserved at a much lower rate. Consequently 24 still pictures (frames) are projected per second, but each frame is flashed twice, producing a flash rate of 48 fields per second. Thus both the continuity of motion and flicker requirements are satisfied with one-half the film consumption.

At the higher brightnesses occurring in television, the 24-frame-48-field rate is not entirely adequate. In the United States, a 30-frame-60-field rate has been chosen, originally to minimize hum effects in receivers operated on 60-Hz power. At this high rate, flicker is not evident at any brightness produced by home television receivers.

In Europe, where the predominant power frequency is 50 Hz, a 25-frame-50-field rate has been standardized. The permissible (flicker-free) highlight brightness is not as high as in the United States, but satisfactory flicker-free performance is obtained.

In recent years, the hum effects in receivers have been minimized to the point where the power frequency is not a major factor. In color television in the United States a field rate of 59.94 Hz has been standardized. The reason for this slight variation is related to the requirements for coding of the color signals, as discussed in Par. 20-20.

5. **Aspect Ratio.** In extensive subjective tests, observers have been found to prefer a rectangular picture with slightly greater width than height. In motion picture standards, the picture *aspect ratio* (width to height) was adopted as 4:3, and this ratio prevailed until the advent of the wide screen.

When the standards for black-and-white television in the United States were set up by the National Television Systems Committee in 1940, it was decided that the motion picture standard was valid and that little would be gained by changing it. It was adopted in the United States in 1941 and subsequently in all television systems throughout the world. Wide-screen movies have been adapted to television transmission by cropping their edges to the 4:3 aspect ratio, in most cases with little subjective loss.

6. **Viewing Distance.** In subjective testing to establish moving picture and television standards, considerable attention has been paid to the distance at which viewers choose to view the picture. Most observers prefer to sit at a distance ranging from 4 to 8 times the picture height and close to the centerline if they sit close to the picture.

This range of preferred viewing distances has a primary effect on the number of scanning lines (and indirectly on the video bandwidth) required in the reproduced pictures. The eye cannot resolve the fine structure of an image viewed from too great a distance. In television images having 525 or 625 lines (the present standards) this limit is reached at a distance of about 10 times the picture height under ideal conditions. No serious loss of visible detail is evident under typical home viewing conditions when the image is viewed at approximately 4 times the picture height (see Par. 5-8).

7. **Scanning Patterns and Apertures.** It has been established as standard in broadcast television that scanning starts in the upper left corner of the picture and proceeds across to the right and slightly downward. When the right-hand side is reached, the scanning spot retraces rapidly to a position below its starting position and again proceeds to the right and slightly downward, and so on, ultimately reaching the bottom of the picture (see Fig. 20-3). At that point, the spot returns to the top and repeats the process, except that the lines of the second scanning field fall between the lines of the first field. Thus, successive fields are *interlaced* (Fig. 20-4). This arrangement permits two picture flashes (fields) for each frame and thus greatly reduces the tendency to flicker (see Par. 20-4).

The electron beams that create the scanning spots are approximately circular, but their intensity is not uniform, their energy falling off in an error-function distribution, as shown in Fig. 20-5. The *effective width* of the spot is the diameter of an equivalent spot of uniform intensity.

When the spot is moving, the brightness distribution at right angles to the line produced is not the same as that observed in a stationary spot, because the energy radiated is a function not only of intensity at the point in question but also of time of exposure. The result is that the effective width of a scanning line generated by a moving spot is about 88% of the equivalent stationary spot.

8. Number of Scanning Lines. The choice of the number of scanning lines in the image hinges on the resolving capability of the human eye and the viewing distance (Par. 20-6). From physiological testing, it has been determined that if a pair of parallel lines is viewed at such a distance that the angle subtended by them at the eye is less than 2 minutes of arc, the eye sees them as one line. This fact is used to select the number of lines for a television

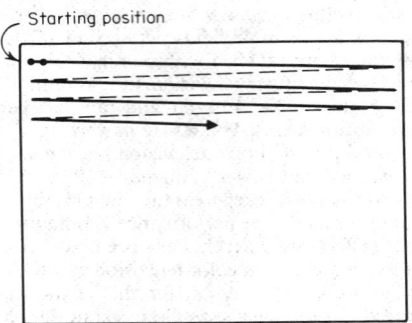

Fig. 20-3. Scanning directions and sequence.

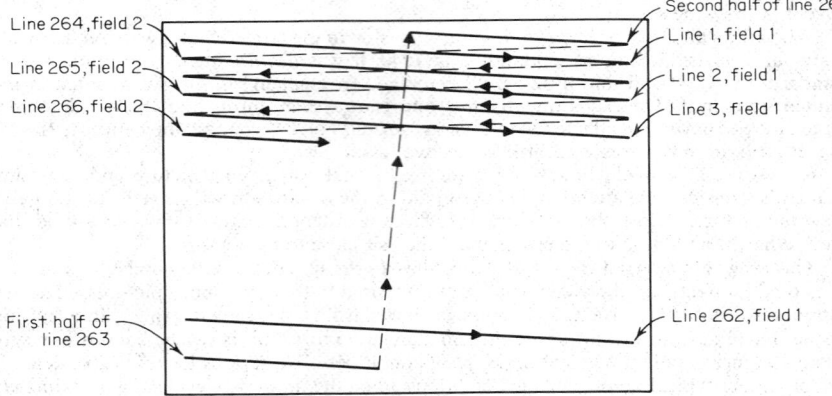

Fig. 20-4. Interlaced scanning pattern (raster).

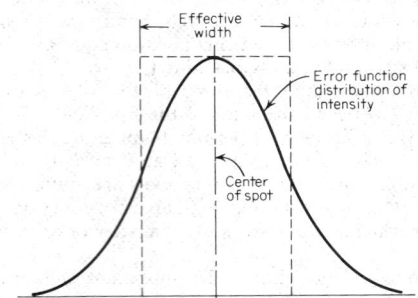

Fig. 20-5. Typical scanning-spot distribution.

system. If the closest preferred viewing distance is 4 times the picture height, as shown in Fig. 20-6, two parallel lines closer than $d = 0.00232h$ cannot be separately resolved by the observer. The number of lines contained in the picture height at this limit is $1/0.00232 = 431$ lines. Thus, approximately 430 lines is the minimum figure for television scanning. A sharp-eyed observer can resolve the line structure at a distance of 4 times the picture height, but the average observer cannot. While there is no "correct" number of lines, the choice should be in excess of 400 for reasons of resolution but not too much higher for reasons of economics.

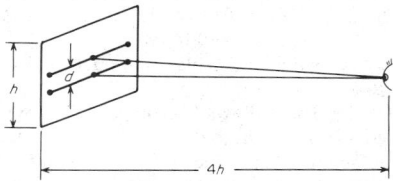

Fig. 20-6. Resolution at 4 times picture height.

The United States has standardized on 525 lines at 30 frames and 60 fields, whereas the European standard is 625 lines at 25 frames and 50 fields. The rest of the world has settled on one or the other of these standards.

9. Interlaced Scanning. As stated in Par. **20-4**, large areas of high brightness have a higher critical flicker frequency than small areas of the same brightness. Thus an individual scanning line flickers much less in the critical range than does the larger area of the image as a whole. This fact allows the use of interlaced scanning in television.

Interlaced scanning is achieved by making the horizontal (line-scanning) rate an odd multiple of one-half the vertical (field-scanning) rate. In the United States standards, the horizontal rate is $15,750 = 525(^{60}/_2)$ lines per second. In other words at 30 frames per second, the scanning pattern has 525 lines per frame and 262.5 lines per field (Fig. 20-4).

An equivalent statement is that interlacing is achieved when the number of lines per frame is an odd number, thus requiring each field to have an even number of lines plus one-half line. The half line, left over at the end of a field scan, displaces the next field downward by a full line, and interlacing is achieved. In the United States standard there are 30 frames of 525 lines, that is, 60 fields of $262\frac{1}{2}$ half-lines each. By this scheme, large area flicker is avoided, the number of lines scanned per second is reduced by 2:1, and the resolution is essentially unaffected. The European system achieves interlace in the same way, although the scanning standards are slightly different.

The diagram in Fig. 20-4 is idealized to show the mechanism of interlace. In practice, the vertical-retrace and synchronizing periods occupy the time of several line scans so that less than 500 lines are actually available for picture information in each frame.

10. Blanking. To prevent the retrace lines from being observed on television receivers and monitors, a blanking pulse is applied to the video signal, the leading edge of which precedes the leading edge of the horizontal synchronizing pulse. The time relationship of the blanking and synchronizing pulses is illustrated in Fig. 20-7. The region between the leading

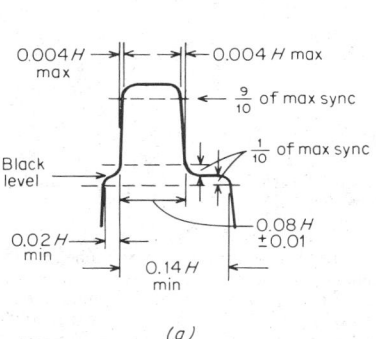

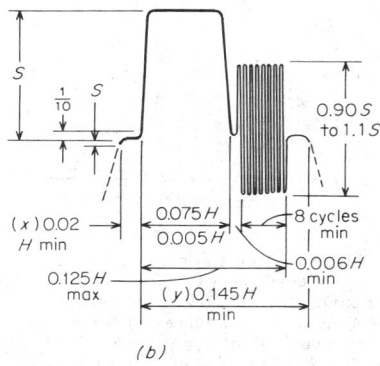

Fig. 20-7. Horizontal blanking and synchronizing pulses: (*a*) black and white; (*b*) National Television System Committee color.

edge of the blanking signal and the leading edge of the synchronizing pulse is termed the *front porch*, while the region between the synchronizing-pulse trailing edge and the blanking-pulse trailing edge is termed the *back porch*.

Color television standards use the back porch to position the *color burst*, an eight-cycle burst of color subcarrier (Fig. 20-7b) that synchronizes the color-subcarrier oscillator at the end of each scanned line.

A vertical video blanking pulse is also used, of length to 21 horizontal lines, shown in Fig. 20-8.

11. Black-Level Clamping. The average level of the television signal varies as the picture goes from bright to dark. These dc variations are lost when the signal is passed through any *RC*-coupled amplifier. To provide a reference for reinserting the dc level, it is necessary to employ a clamping circuit, which sets the synchronizing pulses at a constant dc level, independently of picture-brightness variations. The clamping circuit may be keyed by the synchronizing pulses themselves or by specially generated driving pulses. The circuits are variously called clampers, keyed clampers, restorers, etc.

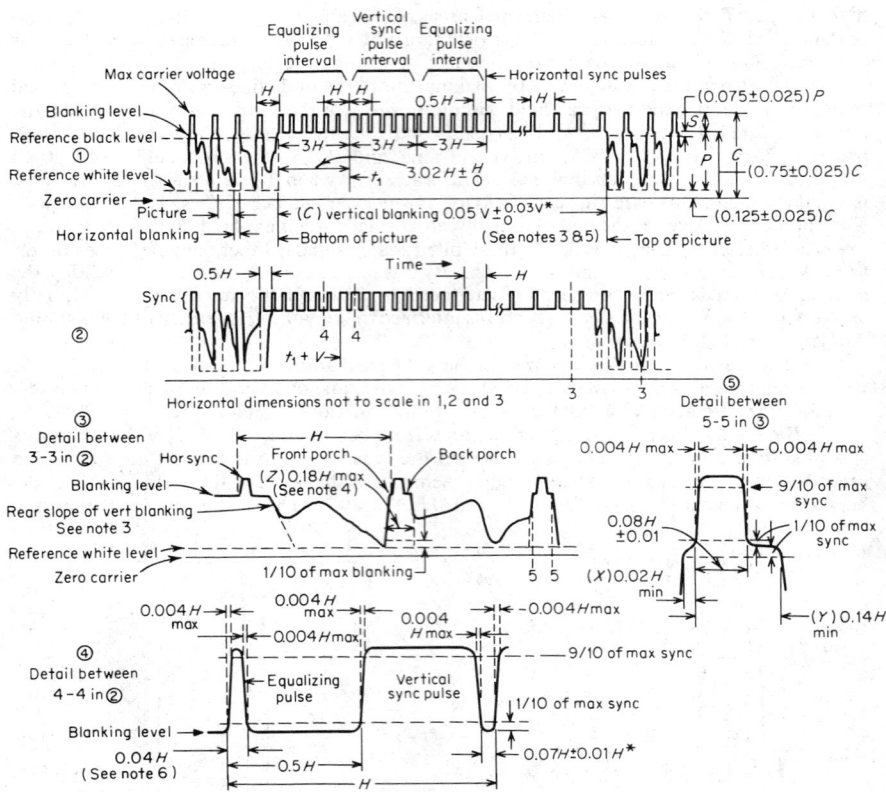

Fig. 20-8. Monochrome synchronizing-signal waveform.
Notes: (1) H = time from start of one line to start of next line. (2) V = time from start of one field to start of next field. (3) Leading and trailing edges of vertical blanking should be complete in less than $0.1H$. (4) Leading and trailing slopes of horizontal blanking must be steep enough to preserve minimum and maximum values of $x + y$ and z under all conditions of picture content. (5) Dimensions marked with asterisk indicate that tolerances given are permitted only for long time variations and not for successive cycles. (6) Equalizing pulse area shall be between 0.45 and 0.5 of area of a horizontal synchronizing pulse. (7) Refer to text for further explanations and tolerances.

The keyed clamper, operating at the horizontal-line rate, can remove power-line hum interference without disturbing the picture information if the hum is additive to the video signal but not if the signal has become modulated by a nonlinear process.

12. Synchronizing and Blanking Signals.   The standard television synchronizing and blanking signals in the United States are as illustrated in Fig. 20-8 for monochrome television and in Fig. 20-9 for color television.

13. Colorimetry.   Color television standards are based on the manner in which the eye perceives colored light.   It is a fortunate fact that a wide range of colors can be reproduced, to the satisfaction of the eye, by the addition of only three monochromatic light sources, e.g.,

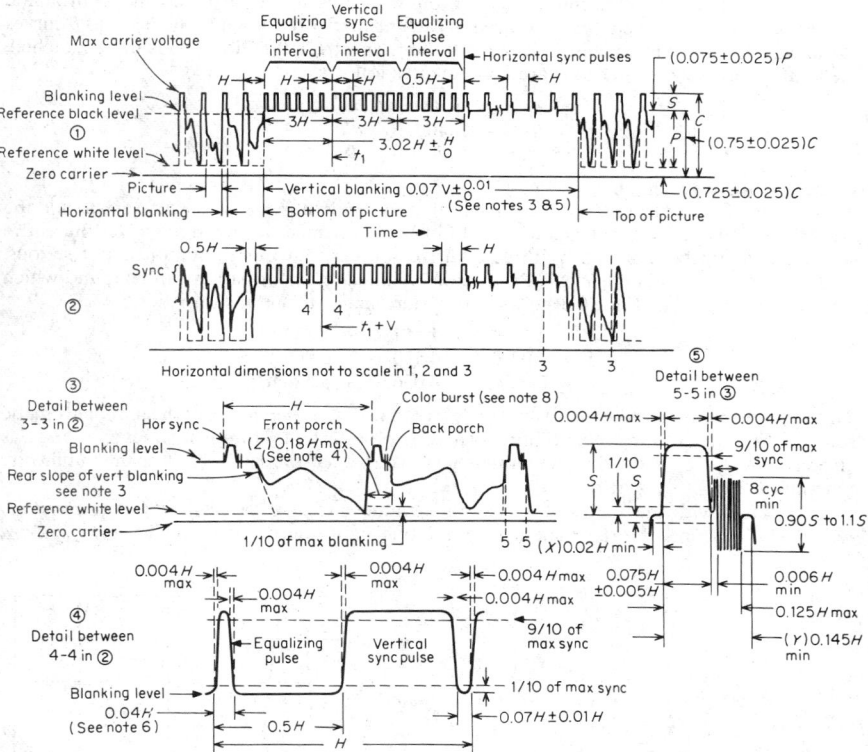

Fig. 20-9.   National Television System Committee color synchronizing-signal waveform.

Notes: (1) H = time from start of one line to start of next line. (2) V = time from start of one field to start of next field. (3) Leading and trailing edges of vertical blanking should be complete in less than $0.1H$. (4) Leading and trailing slopes of horizontal blanking must be steep enough to preserve minimum and maximum values of $x + y$ and z under all conditions of picture content. (5) Dimensions marked with asterisk indicate that tolerances given are permitted only for long time variations and not for successive cycles. (6) Equalizing pulse area shall be between 0.45 and 0.5 of area of a horizontal synchronizing pulse. (7) Color burst follows each horizontal pulse but is omitted following the equalizing pulses and during the broad vertical pulses. (8) Color bursts to be omitted during monochrome transmission. (9) The burst frequency shall be 3.579545 MHz. The tolerance on the frequency shall be ± 10 Hz with a maximum rate of change of frequency not to exceed $\frac{1}{10}$ Hz/s. (10) The horizontal scanning frequency shall be 2/455 times the burst frequency. (11) The dimensions specified for the burst determine the times of starting and stopping the burst, but not its phase. The color burst consists of amplitude modulation of a continuous sine wave. (12) Dimension P represents the peak excursion of the luminance signal from blanking level but does not include the chrominance signal. Dimension S is the synchronizing amplitude above blanking level. Dimension C is the peak carrier amplitude.

red, green, and blue. The study of colorimetry is based on this property of the eye. Only the elements essential to color television are discussed here.

The three CIE* standard primaries are monochromatic light of wavelength 700 nm (red), 546.1 nm (green), and 435.8 nm (blue). A *colorimeter* is an optical system that mixes these lights *additively*, i.e., by superimposing each primary on the other two. Adjacent to the additive area, another area displays monochromatic light at wavelengths selected individually throughout the visible spectrum. The added primaries and monochromatic sources are compared, and a match in color and brightness is sought by adjusting the intensity (power) of each of the primaries.

When such matching is performed by thousands of observers, the results are as shown in Fig. 20-10. Figure 20-10 is interpreted as follows: to match a particular monochromatic wavelength, the relative energies required of each primary are shown in the $\bar{r}$, $\bar{g}$, and $\bar{b}$ curves at the wavelength to be matched. If a continuous-spectrum function $I(\lambda)$ is to be matched, the relative primary proportions are computed as follows:

$$R = \int \bar{r}(\lambda) I(\lambda)\, d\lambda$$
$$G = \int \bar{g}(\lambda)\, I(\lambda)\, d\lambda$$
$$B = \int \bar{b}(\lambda)\, I(\lambda)\, d\lambda$$

where λ is the wavelength.

The negative values in the $\bar{r}$ curve mean that to obtain a match it is necessary to add the red primary to the monochromatic sample in the colorimeter. For this reason the entire spectrum cannot be matched by three additive sources, but this limitation is not serious.

14. X, Y, and Z (Nonphysical) Primaries. A linear transformation can be found which transfers the curves of Fig. 20-10 to positive coordinates; that transformation is

$$X = 2.7690R + 1.7518G + 1.1300B$$
$$Y = 1.0000R + 4.5907G + 0.0601B$$
$$Z = 0.0000R + 0.0565G + 5.5943B$$

This transformation has the interesting property that Y is the *luminance* of the monochromatic sources while X and Z have zero luminosity. A plot of the tristimulus values $\bar{x}$, $\bar{y}$, and $\bar{z}$ is given in Fig. 20-11. Note that the luminance ($\bar{y}$) curve has the same shape as the luminosity function in Fig. 20-1, as it must.

Associated with the $\bar{x}$, $\bar{y}$, and $\bar{z}$ tristimulus values are a set of *artificial* (or *nonphysical*) *primaries* X, Y, and Z. To match a continuous light source of wavelength vs. energy,

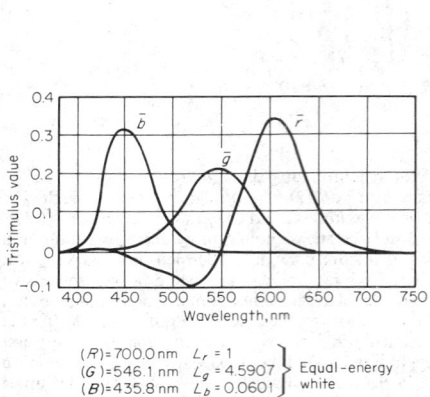

$$(R) = 700.0\,nm \quad L_r = 1$$
$$(G) = 546.1\,nm \quad L_g = 4.5907$$
$$(B) = 435.8\,nm \quad L_b = 0.0601$$
Equal-energy white

Fig. 20-10. Tristimulus values for equal-energy spectrum.

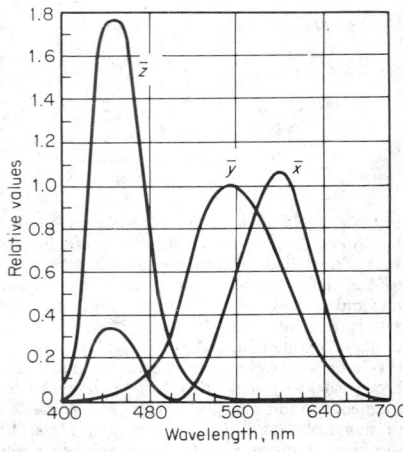

Fig. 20-11. Tristimulus values of CIE nonphysical XYZ primary colors.

---

* Comité International d'Éclairage.

represented by a function $I(\lambda)$, the amounts of the primaries X, Y, and Z required are computed from the following integrals:

$$X = \int I(\lambda)\; \bar{x}(\lambda)\; d\lambda \qquad Y = \int I(\lambda)\; \bar{y}(\lambda)\; d\lambda \qquad Z = \int I(\lambda)\; \bar{z}(\lambda)\; d\lambda$$

where λ is the wavelength.

The nonphysical primaries are mathematical transformations to avoid negative numbers. While they cannot exist as light sources, *they can exist as electric signals.* Using the nonphysical primaries, we can define any light source by two quantities, x and y, such that

$$x = \frac{X}{X + Y + Z} \quad and \quad y = \frac{Y}{X + Y + Z}$$

The third quantity $z = Z/(X + Y + Z)$ is redundant.

15. CIE Chromaticity Diagram. The plot of x and y for all light sources, including monochromatic sources, is the CIE chromaticity diagram (Fig. 20-12). All colors lie within the spectrum line of Fig. 20-12. The CIE chromaticity chart is the worldwide standard method of representing color.

The chromaticity diagram displays the hue and saturation of colors. The *hue* describes the intrinsic nature of the color, i.e., red, green, cyan, purple, etc. *Saturation* is a measure of color intensity, i.e., its pastel vs. vivid quality. Desaturated colors are washed out or whitish. The hue varies on the chromaticity diagram with the angle measured with the white point (illuminant *C*) as the vertex. Saturation is measured by the radial distance from the white point at the center of the chart.

16. Television Standard Primaries. Since the primaries used in color television reproduction are phosphors that are not monochromatic, additional standards must be specified. In setting up the United States color television standards the FCC has specified the x, y coordinates of the standard red, green, and blue primaries (based on practical phosphors) as:

	x	y
Red	0.67	0.33
Green	0.21	0.71
Blue	0.14	0.08

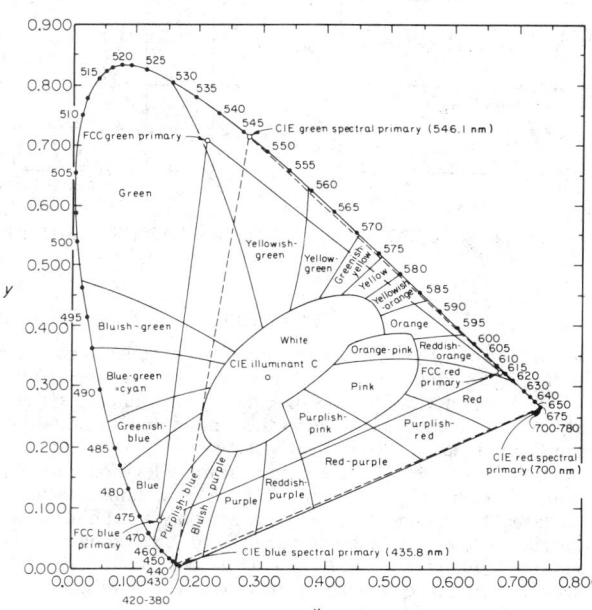

Fig. 20-12.   CIE chromaticity diagram.

These primaries, shown in Fig. 20-12, form a triangle that bounds the color gamut covered by the color system. The transformations to obtain the X, Y, and Z primaries for color television, based on these FCC primaries, are

$$X = 0.608R + 0.174G + 0.200B$$
$$Y = 0.299R + 0.587G + 0.114B$$
$$Z = 0.000R + 0.0662G + 1.112B$$

These X, Y, and Z primaries, like those previously defined, are nonphysical and do not represent real colors. They can represent real electric signals but must be electrically transformed (using an electrical analog of the transformation equations) to R, G, and B signals before being displayed.

The tristimulus values for the standardized FCC primaries are given in Fig. 20-13. Since negative values occur, an exact match would require a color television camera with six camera tubes, but as a practical matter, such cameras are not used. Instead, a compromise set of all-positive *taking primaries* is used. A typical set of such primaries is illustrated in Fig. 20-14. Cameras are also built with a luminance channel and two color channels. Typical taking sensitivities for such cameras are illustrated in Fig. 20-15.

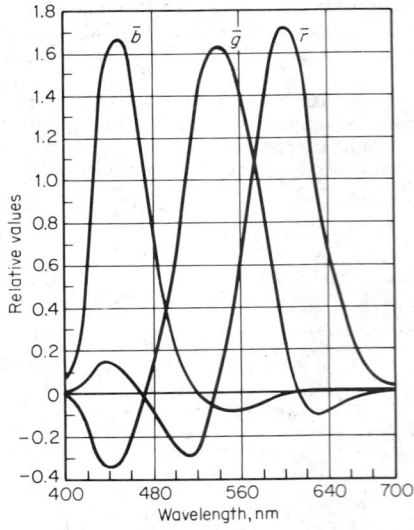

Fig. 20-13.   Tristimulus values of FCC *RGB* primaries.

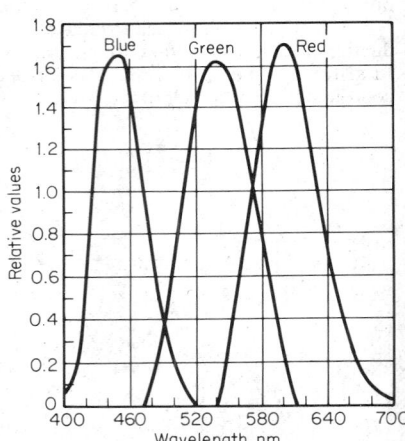

Fig. 20-14.   Practical *RGB* taking sensitivities.

COLOR SYSTEMS

17. Field Sequential Color Systems.   The simplest color television system was the first one devised, the field sequential system. This system was developed to a high level of performance by CBS Laboratories and was designated as a standard system in the United States in the early 1950s. It was supplanted as the United States standard by the National Television System Committee (NTSC) system in 1954.

The field sequential system employs a monochrome television camera, with a color-scanning disk mounting near the focal plane. The disk is rotated synchronously with the vertical sweep, such that during one field, light through only a red filter falls on the camera-tube photocathode, during the next field light through only a green filter, and during the next field light through only a blue filter, and so on. The video signal derived from the camera tube thus consists of sequential color fields in the order that the primary light filters appear in front of the camera tube.

The display device was, originally, a black and white cathode-ray tube in front of which another synchronous color filter disk was placed, the camera and receiver color disks being synchronized. Satisfactory color pictures generally resulted.

In many industrial and medical applications, the field sequential color system is still used because of its trouble-free nature. The size of the picture tube is limited by the size of the rotating wheel required. This limit has been overcome by modern three-gun picture tubes. A major advantage is the fact that there is no camera-tube-registry problem because only one camera tube is used.

The most serious disadvantage of the field sequential system for broadcast use is the fact that it is not compatible with the scanning standards of black-and-white sets; i.e., a black-and-white receiver cannot use a sequential color transmission. Another disadvantage is that fast-moving, high-brightness objects, e.g., white gloves and Ping-Pong balls, have a tendency to display *color breakup,* i.e., they appear as red, green, and blue flashes instead of superimposed colors giving neutral white. Finally, there are minor problems with color desaturation due to residual images on the camera tube. That is, during the blue scan, for example, the blue stored image is not completely dissipated before the next (say green) scan appears, and so on. After efforts to develop a compatible color television system were successful, these difficulties precluded further development of the sequential system for broadcast purposes.

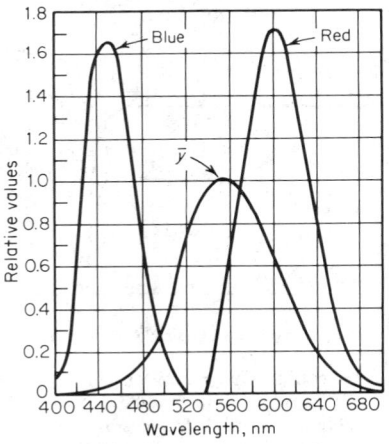

Fig. 20-15.   Practical *RYB* taking sensitivities.

18. Simultaneous (Compatible) Color System. To be feasible for broadcast use in the United States in a market where black-and-white receivers exceeded 50 million, a compatible color system was required. This requirement was met in the development work of the second NTSC which brought together the talents of the electronic industry in the United States from 1950 to 1954.

The problem of transmitting the three primary-color channels simultaneously, in the same bandwidth, and with essentially similar scanning standards as black-and-white transmissions was difficult indeed. To require further that the images observed on black-and-white receivers during color broadcasts be subjectively undisturbed seemed to pose an insurmountable task. By 1954, however, these objectives were met. For a detailed account of this work the reader is referred to the references in Par. 20-26.

19. Constant-Luminance Principle. All the standard color television systems of the world are now based on the principle that if signals in the color-carrier channel do not appreciably affect the luminance of the reproduced picture, the signals are of very low visibility on a television receiver. This requires that the luminance of a given picture area shall be essentially unaffected by the presence of the signals carrying the color information for that area. The design of all composite color television signals is directed toward this end because it takes advantage of characteristics exhibited by the human eye.

20. Composite Color Signals, *XYZ* System. One form of composite color television signal possessing the constant-luminance characteristic (not now used, but illustrative of the principles involved) is formed from three wideband color signals R, G, and B. A resistive matrix network changes these to the signals X, Y, and Z according to the transformations previously listed (Par. **20-16**):

$$X = 0.608R + 0.174G + 0.200B$$
$$Y = 0.299R + 0.587G + 0.114B$$
$$Z = 0.000R + 0.0662G + 1.112B$$

The Y signal is the luminance signal. It is representative of the black-and-white television signal that would be derived from the same subject matter by a high-performance monochrome camera. The X and Y signals carry the color information, i.e., the *nonluminance content*. The X and Z signals are imposed on a subcarrier in the upper part of the video passband (in the vicinity of 3.5 MHz, using a synchronous, suppressed-carrier quadrature modulation system, as shown in Fig. 20-16).

In this arrangement there appear to be several engineering defects.

1. The X and Z color information signals have much less bandwidth than the Y luminance signal.

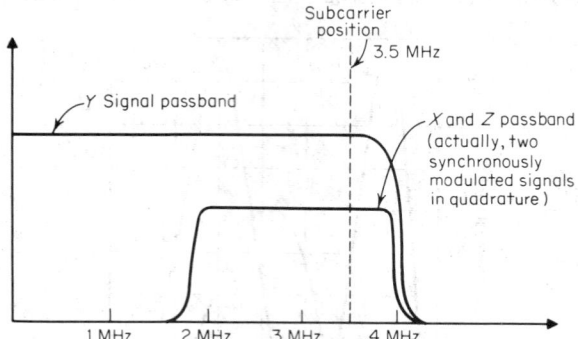

Fig. 20-16. Possible method of coding a color signal (not in use).

2. The X and Z signals have high energy content in the upper-frequency part of the luminance channel, which will produce interference with the fine detail of the picture.

3. The upper and lower sidebands of the X and Z signals are not equal. Therefore, when they are detected, quadrature cross talk between them will cause color distortions on sharp edges.

4. To demodulate the X and Z signals, the reference phase of the suppressed carrier frequency must be preserved and reinserted during demodulation.

These problems were in fact solved in the NTSC system by methods described in Par. **20-24**. To recover the red, green, and blue signals, an inverse *decoding transformation* is used, as follows:

$$R = 1.191X - 0.532Y - 0.288Z$$
$$G = -0.982X + 2.00Y - 0.0283Z$$
$$B = 0.0585X - 0.119Y + 0.900Z$$

When the decoded R, G, and B signals are displayed on a color receiver tube, even with the limited color information present, acceptable reproduction is obtained. This occurs if the luminance information is kept at full bandwidth because the eye is insensitive to lack of detail in color. Quadrature cross talk does occur, and there are colored halos or fringes at the edges of sharp color transitions.

When the decoded signals are observed on a black-and-white receiver, slanted interference lines occur. The black-and-white picture is acceptable, although the interference effects are objectionable.

21. Reduction of Dot Interference. By selecting the color subcarrier frequency so that it and its sidebands fall at odd multiples of half the line-scanning frequency, successive dots (picture elements) on one line interleave with dots on the next scanning of that line. This reduces the visibility of the slanted interference lines, and the interference, previously noted on the black-and-white receiver, disappears at normal viewing distances. The interference is still actually present, but it appears as a fine dot structure below the threshold of the eye's limiting resolution capability at normal viewing distances. This technique, known as *frequency interleaving,* is used in the design of most of the compatible color television systems in the world (the exception is the French SECAM system, which employs a frequency-modulated color signal, thus precluding the possibility of frequency interleaving).

22. Quadrature Cross Talk. The problem of the colored edges caused by quadrature cross talk has been alleviated by intensive subjective testing with various coded signals. Instead of using the X and Z signals as originally suggested, it was found possible to code the color information in a *luminance signal* Y and two *color-difference signals* $R - Y$ and $B - Y$. The third color-difference signal $G - Y$ is derived from the Y, $R - Y$, and $B - Y$ signals at the receiver and hence need not be transmitted. Early in the NTSC development, in fact, experimentation was done with $R - Y$ and $B - Y$ signals rather than with the X and Z signals because the former are easier to derive.

For a neutral color (black, gray, or white), $R = G = B = Y$, and the color-difference signals disappear. Since neutral colors at high brightness are precisely those for which the eye is most critical of objectionable interference, it is good practice for the color signals to go to zero as the saturation of the colors decreases.

It has been found that the cross-talk fringes are reduced to insignificance by the following steps:

1. The color subcarrier is precisely synchronized with the scanning rates.

2. Two color signals are used, derived from the color-difference signals as follows:

$$I = 0.74(R - Y) - 0.27(B - Y)$$
$$Q = 0.48(R - Y) + 0.41(B - Y)$$

3. The I and Q signals are produced in such a way that the Q signal has a bandwidth of around 0.6 MHz and is transmitted by double sideband suppressed carrier. The I signal has a lower sideband of 1.2 MHz and an upper sideband of 0.6 MHz and is transmitted by suppressed carrier in quadrature with the Q signal. The spectral pattern is shown in Fig. 20-17. This is the channel arrangement of the U.S. standard (NTSC) system.

This video-signal-coding configuration eliminates half the quadrature cross-talk problem (Q into I) and was selected such that the other half (I into Q) is of low visibility.

23. Subcarrier Phase Recovery. To preserve the subcarrier phase information for synchronous demodulation, a few cycles of the subcarrier frequency (called the *color burst*) are added immediately, trailing each horizontal synchronizing pulse as follows. The burst positioning is shown in Fig. 20-7b.

At the receiver the burst is passed through a gating circuit and thereafter is used to synchronize a crystal oscillator periodically at the end of each line scan. The oscillator signal is used in a synchronous demodulator to decode the color information.

The frequency for the color subcarrier in the United States is 3.579545 MHz ± 10 Hz. This frequency meets two frequency-interleaving requirements. When multiplied by the submultiple $\frac{2}{455}$, the horizontal scanning frequency becomes 15,734.3 Hz and the vertical scanning frequency is 59.94 Hz. These are values so close to the black-and-white standards that receivers have no difficulty in synchronization. In addition, the beat frequency between

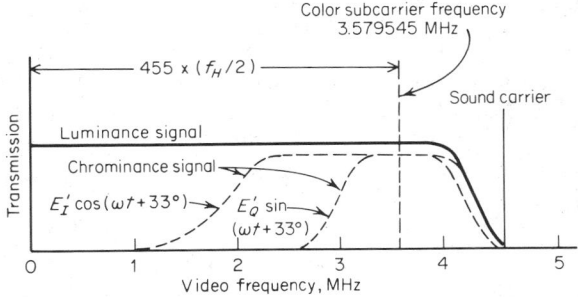

Fig. 20-17. Spectrum pattern of NTSC color signal.

3.579545 MHz and the sound carrier at 4.25 MHz is also frequency-interleaved, thus making the visibility of the beat frequency very low.

24. American Standard (NTSC) Color System. Table 20-1 lists the characteristics of the American (NTSC) Color System.

25. PAL and SECAM Systems of Color Television. In Europe, two color-television systems have evolved, PAL and SECAM. Both operate on the constant-luminance principle. The PAL system is very similar to the NTSC system except that the color-subcarrier phase is reversed every other line and simple color-difference signals are used in place of the I and Q signals. These modifications make the system less sensitive to subcarrier phase errors and minimize quadrature cross talk. The receiver is somewhat more complicated, however, and thus more expensive.

The SECAM system transmits one color signal on one line and the other color signal on the subsequent line. The receiver has a delay line, with delay equal to one horizontal-line period, and an electronic switching system. These elements allow the three types of color

Table 20-1. Characteristics of the Color-Television System in Use in the United States

Scanning and Video Characteristics

Number of lines per picture (frame)	525
Field frequency, fields/second	59.94
Interlace	2:1
Picture (frame) frequency, pictures/second	29.97
Line frequency, lines/second	15,734.264
Tolerance, lines/second	± 0.044
Aspect ratio, width/height	4:3
Scanning sequence, line	Left to right
Field	Top to bottom
System capable of operating independently of power-supply frequency	Yes
Approximate gamma of picture signal	0.45
Nominal video bandwidth, MHz	4.2
Chrominance subcarrier frequency, MHz	3.579545
Tolerance, Hz	± 10

A burst of at least 8 cycles at the frequency of the chrominance subcarrier occurs during each horizontal blanking period after the line-synchronizing pulse and at least $0.006H$ from the trailing edge of that pulse and lasts until not more than $0.125H$ from the leading level, and its peak-to-peak amplitude about the blanking level is from 0.90 to 1.1 times the difference between the levels of the synchronizing pulses and the blanking level. The color burst is omitted during the field-blanking period.

Composition of the Color-Picture Signal

$$E_M = E'_Y + [E'_Q \sin(\omega t + 33°) + E'_I \cos(\omega t + 33°)]$$

where
$$E'_Q = 0.41(E'_B - E'_Y) + 0.48(E'_R - E'_Y)$$
$$E'_I = -0.27(E'_B - E'_Y) + 0.74(E'_R - E'_Y)$$
$$E'_Y = 0.30E'_R + 0.59E'_G + 0.11E'_B$$

For color-difference frequencies below 500 kHz the signal is

$$E_M = E'_Y + \{ 1/1.14 [1/1.78 (E'_B - E'_Y) \sin \omega t + (E'_R - E'_Y) \cos \omega t]\}$$

where E_M is the total video voltage, corresponding to the scanning of a particular picture element, applied to the modulator of the picture transmitter, E'_Y is the gamma-corrected voltage of the monochrome portion of the color picture signal, corresponding to the given picture element, E'_Q and E'_I are the amplitudes of two orthogonal components of the chrominance signal corresponding respectively to narrow-band and wide-band axes, E'_R, E'_G, and E'_B are the gamma-corrected voltages corresponding to red, green, and blue signals during the scanning of the given picture element, and ω is the angular frequency, 2π times the frequency of the chrominance subcarrier.

The portion of each expression between brackets represents the chrominance subcarrier signal which carries the chrominance information.

The phase reference in the E_M equation is the phase of the burst $+180°$. The burst corresponds to amplitude modulation of a continuous sine wave.

Bandwidths

The equivalent bandwidths assigned prior to modulation to the color difference signals E'_Q and E'_I are as follows:

Q-channel bandwidth:

At 400 kHz	<2 dB down
At 500 kHz	<6 dB down
At 600 kHz	At least 6 dB down

I-channel bandwidth:

At 1.3 MHz	<2 dB down
At 3.6 MHz	At least 20 dB down

video information to be displayed simultaneously. The color signals are frequency-modulated on a subcarrier high in the video passband. This arrangement precludes the use of frequency interlace but makes it possible to ignore subcarrier phase accuracies in recording and in microwave transmission.

These systems operate on 625 lines at 50 fields. Detailed specifications are given in Table 20-2.

Table 20-1 (Continued)

Primary Colors

The gamma-corrected voltages E'_R, E'_G, and E'_B are suitable for a color picture tube having primary colors with the following chromaticities in the CIE system of specification:

	x	y
Red (R)	0.67	0.33
Green (G)	0.21	0.71
Blue (B)	0.14	0.08

and having a transfer gradient (gamma exponent) of 2.2 associated with each primary color. The voltages E'_R, E'_G, and E'_B may be respectively of the form $E_R^{1/\gamma}$, $E_G^{1/\gamma}$, and $E_B^{1/\gamma}$, although other forms may be used with advances in the state of the technique.

Subcarrier Characteristics

The radiated chrominance subcarrier vanishes on the reference white of the scene. The numerical values of the signal specification assume that this condition will be reproduced as standard illuminant C ($x = 0.310$, $y = 0.316$) of the CIE. E'_Y, E'_Q, E'_I and the components of these signals match each other in time to 0.05 μs. The angles of the subcarrier measured with respect to the burst phase, when reproducing saturated primaries and their complements at 75% of full amplitude, are within $\pm 10°$, and their amplitudes are within $\pm 20\%$ of the values specified above. The ratios of the measured amplitudes of the subcarrier to the luminance signal for the same saturated primaries and their complements fall between the limits of 0.8 and 1.2 of the values specified for their ratios.

Radio-Frequency and Modulation Characteristics

Nominal radio-frequency bandwidth	6 MHz
Sound carrier relative to vision carrier	+4.5 MHz
Sound carrier relative to nearest edge of channel	−0.25 MHz
Nominal width of main sideband	4.2 MHz
Nominal width of vestigial sideband	0.75 MHz
Type of polarity of vision modulation A5C	Negative
Synchronizing level as a percentage of peak carrier	100
Blanking level as a percentage of peak carrier	72.5-77.5
Difference between black level and blanking level as a percentage of peak carrier	2.875-6.75
Peak-white level as a percentage of peak carrier	10-15
Type of sound modulation	F3, ± 25 kHz, 75 μs preemphasis
Ratio of effective radiated powers of vision and sound	10:1-5:1

Synchronizing Signals

Line synchronization:	Percent of Line Period (H)	μs
Line period H	100	63.556
Line-blanking interval	16.5018 . . .	10.5-11.4
Interval between time datum H_0 and back edge of line-blanking signal	12.7-16 . . .	8.06-10.3
Front porch	>2	>1.27
Synchronizing pulse	6.6-8	4.2-5.1
Buildup time (10-90%) of the edges of the line-blanking signal	<0.75	<0.48
Buildup time (10-90%) of line-synchronizing pulses	<0.4	<0.25

Field synchronization:		Duration
Field period V		16.683 ms
Line period H		63.556 μs
Field-blanking period		1168-1335 ms = (0.07-0.08)V $\approx$ (18-21) H
Buildup times (10-90%) of the edges of field-blanking pulses		<6.36 μs
Duration of first equalizing pulse sequence		3H
Duration of synchronizing pulse sequence		3H
Duration of second sequence of equalizing pulses		3H
Duration of equalizing pulse		2.29 μs
Duration of field-synchronizing pulse		26.4-28.0 μs
Interval between field-synchronizing pulses		3.8-5 μs
Buildup times (10-90%) of edges of synchronizing signals		<0.25 μs

Table 20-2. Specifications of the PAL and SECAM III Systems

Characteristic	Standard		
	5.5 MHz (G)	6.0 MHz (I)	6.5 MHz (L)
1. General specifications			
Luminance component (both)	Amplitude modulation of the picture carrier		
Chrominance component:			
PAL	Simultaneous pair of components transmitted as amplitude-modulated sidebands of a pair of suppressed subcarriers in quadrature having a common frequency		
SECAM	A pair of components transmitted alternately on successive lines as the frequency modulation of a subcarrier		
2. Color subcarrier f_{sc}			
PAL	4.3361875 MHz ± 10 Hz	4.3361875 MHz	4.3361875 MHz
SECAM	4.4375 MHz ± 2 kHz	4.4375 MHz ± 2 kHz	4.4375 MHz ± 2 kHz
3. Frequency spectrum of composite color picture and sound signals, MHz			
Vision-to-sound spacing:			
PAL	5.5	6	6.5
SECAM	5.5	6	6.5
Main sideband (luminance):			
PAL	5.0	5.5	6
SECAM	5.0	5.5	6
Vestigial sideband:			
PAL	0.75	1.25	1.25
SECAM	0.75	1.25	1.25
Chrominance sidebands f_{sc}:			
PAL: E'_U, F'_V signal	+0.57 max −1.3 max	+1.07 max −1.6 max	+1.57 max ...
SECAM: D'_R, D'_B	+0.57 −1.40	+1.07 −1.40	... ±1.40

20-18

4. Transmitted color-picture-signal waveform

Color synchronization:

PAL:

Subcarrier burst, duration	13 ± 1 cycle
Start	5.5 ± 2 μs after the leading edge of the line-synchronizing pulses
Amplitude	0.5 ± 0.1 of line-synchronizing amplitude
Omission	Omitted during field-blanking periods for 9 lines
Phase sequence (see Phase reference in item 9)	First field (even) starts on line 7, + 135° on odd lines Second field (odd) starts on line 319, − 135° on even lines Third field (even) starts on line 6, + 135° on even lines Fourth field (odd) starts on line 320, − 135° on odd lines
SECAM	Subcarrier signal modulated in frequency and amplitude to correspond to a sawtooth color-difference signal D'_B or D'_R during six lines of each field-blanking period
Deviation and amplitude	Correspond to maximum D'_B or D'_R
Duration	Active line period
Sequence	First field (even), $-D'_B$ on lines 11, 13, 15; $-D'_R$ on lines 10, 12, 14 Second field (odd), $-D'_B$ on lines 323, 325, 327; $-D'_R$ on lines 324, 326, 328 Third field (even), $-D'_B$ on lines 10, 12, 14; $-D'_R$ on lines 11, 13, 15 Fourth field (odd), $-D'_B$ on lines 324, 326, 328; $-D'_R$ on lines 323, 325, 327
Duration of color subcarrier protection interval	Starts 5.7 ± 0.3 μs after leading edge of line-synchronizing pulse

5. Delay characteristic of the transmitted signal

Both	0.08 μs at f_{sc} 0.27 μs at 5.5 MHz

6. Luminance component

Attenuation-frequency characteristics, both ..	Uniform 0-5.0 MHz	Uniform 0-5.5 MHz (notch filter at f_{sc} permissible)	Uniform 0-6.0 MHz (notch filter at f_{sc} permissible)

7. Scanning

Line-scanning frequency f_{line}:

PAL	15.625 Hz = $(4 f_{sc} - 2 f_{field})/1135$
SECAM	15.625 Hz = $f_{sc}/284$
Gamma, both	Corresponds to display gamma of 2.2

8. Synchronizing and blanking waveforms

Both	Similar to characteristics of monochrome television systems

Table 20-2. Specifications of the PAL and SECAM III Systems (*Concluded*)

Characteristic	Standard		
	5.5 MHz (G)	6.0 MHz (I)	6.5 MHz (L)
PAL†	**9. Equation of complete color signal** $$E_M = E'_Y + E'_U \sin \omega t + E'_V \cos (\omega t \pm \pi/2)$$ where $E'_U = 0.493(E'_B - E'_Y)$ and $E'_V = 0.877(E'_R - E'_Y)$ The argument of the cosine is $\omega t + \pi/2$ during the odd lines of the first and second fields and during the even lines of the third and fourth fields. It is $\omega t - \pi/2$ during the even lines of the first and second fields and during the odd lines of the third and fourth fields, as under item 4 (color synchronization). The color-difference signals E'_U and E'_V can be formed by matrixing E'_Y, E'_I, and E'_Q video signals having the bandwidths specified for the NTSC system and the appropriate standard I, G, or L. The bandwidth of either of the color-difference signals E'_U and E'_V, respectively, will therefore not exceed the bandwidth of the appropriate E'_I signal.		
Phase reference	E'_U axis $E'_U = -0.545E'_I + 0.839E'_Q$ $\qquad$ $E'_V = 0.839E'_I + 0.545E'_Q$		
SECAM†	In large area of color $$E_M = E'_Y + A \cos (\omega_{sc} + E'_c \tfrac{\Delta\omega_{sc}}{t})$$ where E'_c is a color-difference signal D'_R or D'_B and $\Delta\omega_{sc}/2\pi$ is the frequency deviation corresponding to unit amplitude of the preemphasized color-difference signal; $D'_R = -1.9(E'_R - E'_Y)$ and $D'_B = 1.5(E'_B - E'_Y)$; A is a function of $E'_c \Delta\omega_{sc}$ and determines the amplitude of the chrominance signal. In the absence of color, for E'_Ymax $= 1.0$, this function has the following values:		
Equivalent bandwidths of color-difference signals D'_B and D'_R before preemphasis and modulation	$A = 0.115$ 1.4 MHz	$A = 0.1$ At 1.0 MHz < 2 dB down At 1.5 MHz > 5 dB down At 2.0 MHz < 20 dB down	$A = 0.1$ 1.4 MHz
Transmission-time difference of E'_Y, D'_R, and D'_B	...	± 40 ns	± 50 ns
	10. SECAM color-subcarrier modulating signal		
Preemphasis of signal D'_R and D'_B before modulation	Time constant $= 1.12$ μs ($+ 14$ dB at 1 MHz)		
Frequency modulation of color carrier	$f_{sc} = 230$ kHz per unit amplitude of D'_R and D'_B after preemphasis		
Maximum subcarrier deviation	$f_{sc} \pm 500 \pm 50$ kHz		

11. SECAM supplementary amplitude modulations of chrominance subcarrier

Radio-frequency preemphasis of chrominance subcarrier	0 dB at f_{sc} 5–6.2 dB at $f_{sc} \pm 230$ kHz 10.8–11.6 dB at $f_{sc} \pm 500$ kHz		
Cross-color corrective modulation of subcarrier for luminance-signal components near subcarrier frequency‡	3.5–6 dB for $E'_Y = 1.0$ 0 dB for $E'_Y < 0.5$	6 dB for $E'_Y = 1.0$ 0 dB for $E'_Y < 0.2$	3.5–6 dB for $E'_Y = 1.0$ 0 dB for $E'_Y < 0.5$
Equalization of chrominance subcarrier amplitude on successive lines	Permissible		

12. SECAM phase § of chrominance subcarrier during protection interval

| Polarity, of chrominance subcarrier .... | When phase, as defined above, is 0°, the polarity is positive; when 180°, the polarity is negative |
| Of composite signal .... | Reversed during every third line and additionally during every alternate field |

Source: C.C.I.R., "Green Book," *Doc. 11th Plenary Assem. Int. Radio Consul. Comm.*, Vol. 5, Sec. E4, 1971.
† E'_M = total video voltage applied to modulator of transmitter; E'_Y = voltage of luminance component of composite signal; E'_R, E'_G, and E'_B = gamma-corrected voltages corresponding to the red, green, and blue signals.
‡ The value of E'_Y corresponding to white = 1.0.
§ Either 0 or 180° with respect to a reference oscillator f_R.

26. Bibliography

General Reference Texts

FINK, DONALD G.   "Television Standards and Practice," McGraw-Hill, New York, 1943.

——— "Television Engineering," McGraw-Hill, New York, 1952.

——— "Color Television Standards," McGraw-Hill, New York, 1955.

McILWAIN, K., and CHARLES E. DEAN   "Principles of Color Television," Wiley, New York, 1956.

Scanning

MERTZ, P.   Television: The Scanning Process, *Proc. IRE,* October 1941, Vol. 29, pp. 529-537.

——— and F. GRAY   A Theory of Scanning and Its Relation to the Characteristics of the Transmitted Signal in Telephotography and Television, *Bell Syst. Tech. J.,* July 1934, Vol. 13, pp. 464-515.

SCHADE, O. H.   Electro-Optical Characteristics of Television Systems, Pt. I-IV, *RCA Rev.,* March, June, September, and December, 1948.

Colorimetry

HARDY, A. C.   "Handbook of Colorimetry," Technology Press, Cambridge, Mass., 1936.

JUDD, D. B.   The 1931 I.C.I. (C.I.E.) Standard Observer and Coordinate System for Colorimetry, *J. Opt. Soc. Am.,* 1933, Vol. 23, pp. 359-374.

WINTRINGHAM, W. T.   Color Television and Colorimetry, *Proc. IRE,* October 1951, Vol. 39, pp. 1135-1172.

Sequential Color System

GOLDMARK, P. C., J. N. DYER, E. R. PIORE, and J. M. HOLLYWOOD   Color Television, Pt. I, *Proc. IRE,* April 1942, Vol. 30, pp. 162-182.

———, E. R. PIORE, J. M. HOLLYWOOD, T. H. CHAMBERS, and J. J. REEVES   Color Television, Pt. II, *Proc. IRE,* September 1943, Vol. 31, pp. 465-478.

Compatible Color System

ABRAHAMS, I. C.   The Frequency Interleaving Principle in the NTSC Standards, *Proc. IRE,* January 1954, Vol. 42, pp. 81-83.

HIRSCH, C. J., W. F. BAILEY, and B. D. LOUGHLIN   Principles of NTSC Compatible Color Television, *Electronics,* February 1952, Vol. 25, No. 2, pp. 88-95.

——— and W. F. BAILEY   Quadrature Cross Talk in NTSC Color Television, *Proc. IRE,* January 1954, Vol. 42, pp. 84-90.

LOUGHREN, A. V.   Recommendations of the National Television System Committee for a Color Television Signal, *J. Soc. Motion Pict. Telev. Eng.,* April 1953, Vol. 60, pp. 321-336; May 1953, p. 596.

European Color Systems

C.C.I.R.   "Green Book," *Doc. 11th Plenary Assem. Int. Radio Consult. Comm.,* Vol. 5, Sec. E4, 1971.

TELEVISION CAMERAS

BY HENRY N. KOZANOWSKI

27. Monochrome Cameras.   The history of monochrome television cameras in broadcasting and for closed-circuit or industrial duty covers four distinct phases, each based on the availability of advances in pickup-tube technology.   The tubes involved are photoemissive, (iconoscope and image orthicon) or photoconductive (vidicon and the plumbicon).

The first commercially applied cameras were based on the iconoscope.   This was the first camera tube to employ photoelectric storage, i.e., to integrate the effects of incident light on a stationary scene in a raster charge image.   By present-day standards the scene illumination required with iconoscopes was very high, but the gray-scale rendition was pleasing.   The iconoscope was particularly useful for reproduction of motion-picture film.

The principal limitation of iconoscopes was their inability to produce video waveforms undistorted by spurious variable shading signals, strongly influenced by scene content.   This required constant monitoring and addition of cancellation waveforms to produce satisfactory film reproduction.   Nevertheless, the iconoscope was universally used by broadcasters for over 15 years in monochrome film reproduction.

28. Monochrome Image-Orthicon Cameras. The image-orthicon tube, which was invented and developed about 1940, furnished the impetus for a new generation of cameras which practically monopolized the broadcast field for more than 20 years (see Fig. 20-18). The extreme highlight sensitivity and a self-adjusting knee in the exposure-transfer characteristic enabled the tube to accommodate to practically all lighting conditions, including outdoor scenes of high contrast.

In operation, e.g., with the type 5820 image orthicon, lighting and lens openings are sufficient to place the highlight exposure well into the knee portion of the transfer characteristic. This produces a self-stabilized image-electron redistribution around the brightest objects in the scene, outlining them in black and giving an effect of sharpness not actually present. In addition, this effect provides a stretching of the low-light information. The net result is a video signal with limiting in the highlights and excellent visibility of low-light information. This mode of operation was standard in monochrome broadcasting for many years.

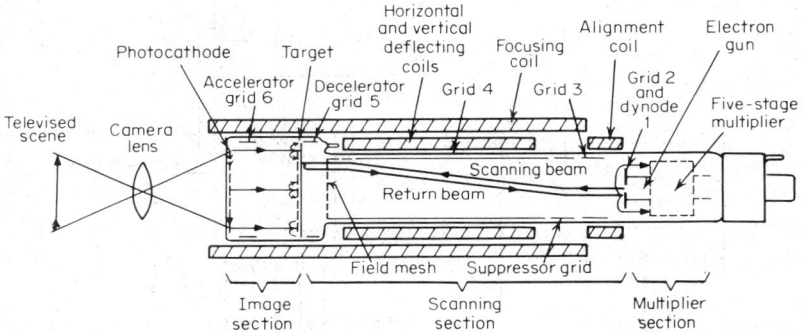

Fig. 20-18. Image-orthicon camera tube with focusing and deflection system.

29. Image-Orthicon Camera Head. The image-orthicon tube and its focus and deflection yoke assembly are mounted on ways, or guides, so that it can easily be moved longitudinally for focus of the optical image on the photocathode of the image orthicon. A four-position lens turret-mounted on a central-shaft structure accessible to the camera operator provides a choice of lens focal length from 25 to 500 mm (or more) for wide-angle and telephoto coverage of the televised scene, respectively. Calibrated irises are provided in these lenses but are not normally varied during operation.

Horizontal and vertical drive signals are provided to the camera to generate sweep and output waveforms for horizontal and vertical deflection of the image orthicon tube and the cathode-ray viewfinder unit. All circuits use vacuum tubes.

A vacuum-tube preamplifier amplifies the image-orthicon signal of approximately 5 μA across 22,000 Ω to approximately 0.5 V into a 50-Ω load represented by the coaxial line which is part of the multiconductor camera cable connecting the camera to the camera-control unit. The video signal is flat-equalized within a 7-MHz passband to compensate for the shunt capacity of image-orthicon output-signal electrode.

30. Image-Orthicon Camera-Control Unit. All major electrical setup and operating controls for the image-orthicon tube are provided at the camera-control unit and sent to the camera-head circuits by assigned conductors in the multiconductor camera cable. Infrequently used camera and tube setup controls are located at the rear of the camera head behind protective hinged covers.

Using the picture monitor and oscilloscope at the camera control position, the control operator can align and normalize the video output of the camera chain. A processing amplifier in the camera control accepts the video signal output from the coaxial camera cable and carries out the functions of gain control, blanking, black-level setting, clamping, transfer-characteristic modification, and synchronizing wave addition and delivers the composite wave signal in standardized form and amplitude for program assembly or broadcast use.

A block diagram of a typical chain is given in Fig. 20-19.

31. Image Orthicon 4½-in. Studio Cameras. In 1958, as a result of investigations by the British Broadcasting Corporation, the technical quality of monochrome television pictures was greatly improved. Detailed comparisons of 3- and 4½-in. image-orthicon signal-to-noise ratio, gray-scale characteristic, and resolution and of lighting and staging techniques, showed that a large improvement in picture quality could be realized by using 4½-in. image-orthicon tubes, operated with primary emphasis on correct light exposure. Careful control of scene lighting and using exposure control or variable iris as the only operating variable made it possible consistently to produce pictures with excellent gray-scale rendition, signal-to-noise ratio, and sharpness, eliminating the distortions generated by the uncontrolled operating modes used in earlier cameras.

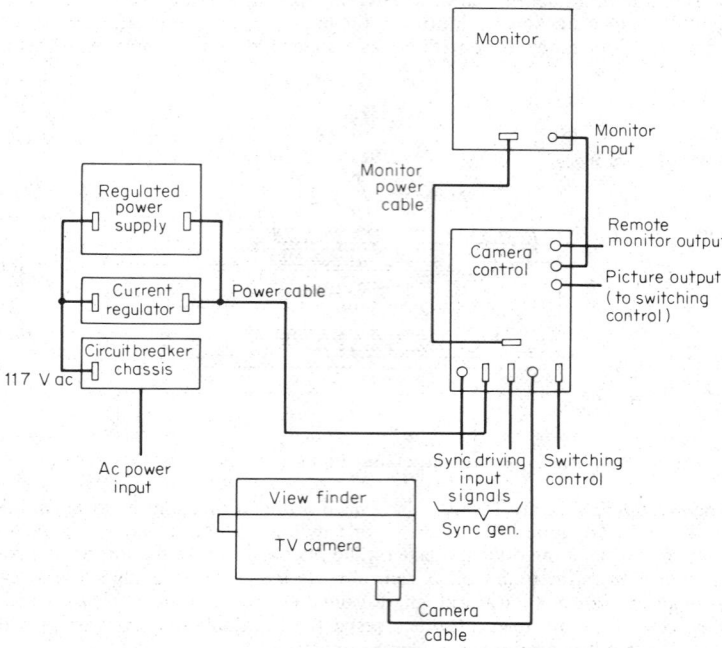

Fig. 20-19. Camera, control, monitor, and power-supply connections.

In the new form of operation, the camera is set up so that normal lens exposure on a gray-scale chart brings the highest highlights just to the knee of the transfer characteristic for close-spaced image orthicons and into the knee for the wide-spaced higher-sensitivity tubes. By careful attention to lighting and scene reflectance one can operate such cameras with practically no changes in iris setting, since the setup conditions are closely duplicated in the scenes themselves.

A further requirement in the new operating techniques is the *electronic gamma correction* of the signal since the new mode of operation gives substantially a linear gray-scale transfer characteristic. The addition of external gamma correction of 0.7 to 0.5 slope produces pictures that are smooth and pleasing, with excellent face-tone rendition, and because of the greater target area, the signal-to-noise ratio is entirely adequate, even after gamma correction.

Cameras designed around the 4½-in. image-orthicon tube and intended for operation under controlled exposure conditions were made available by major equipment manufacturers by 1960. The cameras are oriented to vacuum-tube circuits but contain numerous solid-state elements and transistor circuits.

32. Monochrome Vidicon Cameras. Vidicon camera tubes, first introduced in 1951, employ photoconductive pickup devices. In comparison with the image orthicon, the vidicon stands out in its directness and simplicity of operation. A composite layer of photoconductive material, such as antimony sulfide, is evaporated onto a transparent conductive substrate sealed to the end of a cylindrical glass tube. An electron gun, with the usual limiting aperture, focus electrodes, and field mesh, completes the essential components of the assembly. An optical image focused on the semitransparent photoconductive layer causes the conductivity to vary from point to point in accordance with the light-intensity variations. The charge pattern so developed is scanned by the electron gun. The electrons replaced by this scanning process flow through a load resistor connecting the photoconductive layer to a positive potential source of about 30 V, producing the video signal. For normal operation the peak video-signal current is approximately 300 nA, and the load resistor is approximately 50 kΩ.

The vidicon-tube operation is quite insensitive to misadjustment, whereas with the image-orthicon tube it is a much more involved procedure to obtain optimum results. Because of the ease of setup, reasonable sensitivity, the moderate cost of the vidicon, and the simplicity of the circuits, vidicon cameras are widely used for industrial and closed-circuit operation.

33. Operating Features of Vidicon Camera. Vidicon cameras are capable of good operation, with adequate signal-to-noise ratio, at illumination levels of 200 lm/ft^2 or more. Since the gamma (slope of the log-log signal output to light input) is approximately 0.65, the camera is noncritical of input lighting; i.e., an increase in highlight illumination intensity by a factor 2 will raise the signal by only 50%. This transfer characteristic also improves the visibility of low-light detail, compared with a linear gamma function.

With low values of target voltage, the dark current or "no-light" scanning signal is negligibly small, producing excellent black-level performance. At larger values of target voltage, the dark current rises rapidly and may become a significant part of the total signal. Since the photoconductive layer is a semiconductor, the dark current doubles in value for every 10°C rise in operating temperature.

A very important feature of vidicons having antimony sulfide targets is the increase in effective sensitivity with increase of target potential. The sensitivity varies approximately as the 2.5 power of the change of target voltage. Thus, automatic gain or sensitivity controls can easily be provided by feedback circuits which increase the target voltage as the incident lighting decreases. A very simple device for obtaining such sensitivity control is a by-passed high resistance in series with the target voltage source.

Lag and Retentivity. Under threshold operating conditions, with high-target voltages and low scene-light levels, vidicons may show a lag effect with motion in the scene and high retentivity (afterimage burn) on the raster. These effects can be minimized by operating the vidicon at the highest available light level on the photo surface and at the lowest target voltage consistent with generating the required video signal.

While the vidicon camera has been almost universally applied to industrial or closed-circuit problems, it has had only limited application in live broadcast TV work. Even though stationary pictures having excellent signal-to-noise ratio, resolution, and gray scale can be produced at moderate lighting levels, the signal lag, producing smear with motion, is sufficiently troublesome to limit the use of such cameras to programs where the amount of motion is moderate.

34. Vidicon Motion-Picture Reproduction. In United States practice, vidicon tubes find their widest application in the reproduction of monochrome and color motion-picture film. They are operated at high light levels and low target voltages, using intermittent-motion 16- or 35-mm film transports. Exposure of the raster is carried out during the active scanning sequence and occupies 35% or more of the total scanning time.

Conversion of the standard motion picture 24 frames per second showing rate to the TV 60 fields per second display rate is carried out by using a 3:2 intermittent mechanism in the motion-picture projector. One motion-picture frame is held stationary for three television-field raster scans, and the next is held for two TV field scans. Thus during two film frames ($\frac{1}{12}$ s at the 24-frame rate) five raster fields are produced in the same time, each at intervals of $\frac{1}{60}$ s (60 fields per second).

The storage characteristics of vidicon film chains are excellent and allow for nonsynchronous operation of projector and TV system. This is especially important in color TV since the raster is then not locked to 60 Hz. When, as is usual, the motion-picture projector is operated from the local power supply, as long as the light-application time exceeds 30% and the difference between power-line frequency and raster rate does not exceed 0.5 Hz, there are no problems with transition bars in the reproduced picture.

The low dark current, steady black level, inherent low gamma or transfer characteristic, good resolution, and excellent signal-to-noise ratio have made the vidicon very attractive to the broadcaster; under the operating conditions described, vidicons have low lag and low retentivity and burn characteristics. Because of their long life and moderate initial cost they are the most economical means of reproducing color film.

Since there are variations in film highlight density which will change the peak video output, the system is operated to give "constant performance" by increasing the light through dense film. A neutral-density optical disk in the projector light path, positioned by a video-level-sensitive servoamplifier feedback loop, utilizes the light reserve of the motion-picture incandescent lamp to reestablish a reference output video level without any deterioration of signal-to-noise ratio. A close electronic approximation to this constant-performance characteristic is obtained by feedback control of vidicon target potential, to increase the effective sensitivity with dense film.

For high utilization of film chains, multiplying optical systems are used. Three (sometimes four) picture sources are arranged on a film island. By using movable front-source mirrors any one of the pictures from these sources can be projected as a real image at a single field-lens position. The field-lens image is viewed by the vidicon film chain. Thus, 2 by 2 slide projectors and 16- or 35-mm film projectors can be used. A functional diagram of the multiplexing system is shown in Fig. 20-20.

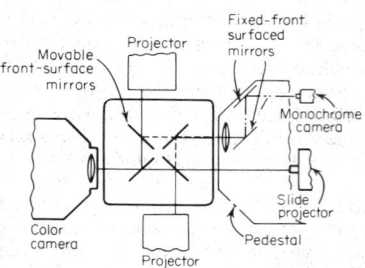

Fig. 20-20. Multiplex use of several projectors with monochrome and color cameras (type TP-15 multiplexer). *(RCA Corporation.)*

35. Processing Camera Signals. The fundamental processes involved in generating picture signals involve the use of synchronizing signals, which control the deflection of the scanning raster on the photosensitive pickup tube, and the generation of the basic video output wave from the scanning of the charge image. The video signal must be compensated for attenuation of high frequencies due to shunt capacitance in the pickup tube. The signal is *amplified* to standard level, *blanked* to clean up the waveforms during horizontal- and vertical-sweep return time, *clipped* at a reference black level, *clamped* to preserve the dc reference, and delivered to a 70-Ω coaxial-cable distribution point at 0.7 V peak amplitude. At this point the synchronizing signal is added to the video for transmission or general distribution. Frequently aperture correction, in both the horizontal and vertical directions, is carried out electronically to enhance the sharpness of the transmitted picture. Modification of the transfer characteristic *(gamma correction)* is generally introduced in monochrome broadcast equipment, is always used in color cameras, and is rarely used in closed-circuit or industrial equipment.

36. Synchronizing Generators (see also Pars. **20-53** to **20-58**). Broadcast camera chains require horizontal and vertical *drive pulses,* a *composite blanking waveform,* and synchronizing signals for proper operation. These are usually produced in a separate unit to conform to FCC specifications. The composite video signal thus produced is distributed through the broadcast facility on coaxial lines, through distribution amplifiers. Two additional signals, color subcarrier and burst-flag pulse, are required for color systems.

While the distribution of the six timing signals within the broadcast plant is straightforward, it requires a large investment in coaxial cable, amplifiers, and connectors. The most recent installations use simplified single-cable systems in which the timing information is coded in a single complex waveform. At a distribution point this waveform is decoded and reconstituted into the required drive pulses by solid-state circuits. This procedure simplifies distribution and offers greater flexibility in system interconnections and time-delay correction.

Closed-circuit systems are generally provided with free-running oscillators or multivibrators to produce the required waveforms, and the vertical-horizontal frequency relationship provides random interlace in the transmitted signal. The free-running characteristics make the camera and pickup tube relatively immune to damage from lack of drive pulses and to deflection-circuit malfunction.   The camera is generally arranged to accept external "standard" pulses, in which case the output conforms to broadcast practice.

Advanced designs of portable color broadcast cameras contain very compact solid-state synchronizing generators which produce completely acceptable timing and subcarrier waveforms.

In an elaborate installation involving several studios and remote cameras, all synchronizing generators are tied together (gen-locked) to achieve system stability in switching and assembling programs.

37. Focus of Camera Tubes.   To obtain a video signal from a pickup tube the charge image is removed by scanning (Par. **20-9**).

In magnetically focused tubes, such as the image orthicon and vidicon, the motion of an electron from the scanning gun is *cycloidal.* The axial magnetic-focus field strength and the wall voltage are chosen so that an integral number of cycloidal loops is produced between the gun aperture and the surface being scanned.   This produces a well-defined scanning spot for removing the video information.   Normally, the magnetic field of image orthicons is approximately 70 gauss and the wall-focus voltage is about 300 V, with respect to the electron-gun cathode (see Fig. 20-18).

The photoelectrons in the image section also travel in cycloidal motion from the photocathode to the secondary-emission target.   Electron focus is obtained by adjustment of photocathode voltage, which is approximately -500 V with respect to the target and electron gun.   Special precautions are taken to minimize the horizontal- and vertical-deflection fields at the target since these will blur the magnetostatic charge image of the target.

38. Scanning of Camera Tubes.   Linear raster scan requires the generation of a sawtooth of current in the horizontal and vertical deflection windings of the pickup tube yoke.   Such waveforms are generated from the horizontal and vertical drive pulses which periodically discharge suitable capacitor-resistance networks.   The vertical output stage is required to furnish a current of 200 mA peak to peak into the vertical deflection winding, which typically has an inductance of about 30 mH and a resistance of 35 Ω.   The horizontal deflection winding requires a peak-to-peak sawtooth current of 500 mA.   For vacuum-tube circuits, the horizontal yoke inductance is about 1.0 mH and the dc resistance is 5 Ω.   For transistor operation, the yoke inductance is 100 μH with a dc resistance of a fraction of an ohm.

COLOR CAMERAS

39. Three-Tube Image-Orthicon Cameras.   The first generation of color studio cameras used three image-orthicon tubes.   The camera is essentially three identical monochrome camera channels with provisions for superposing the three output-signal rasters mechanically and electrically.   The optical system consists of a taking lens which is part of a four-lens turret assembly.   The scene is imaged in the plane of a field lens using a 1.6-in. diagonal-image format.   An alternative arrangement uses a 10:1 range zoom lens.   The real image in the field lens is viewed by a back-to-back relay-lens assembly of approximately 9 in. focal length.   At the rear conjugate distance of the optical relay is placed a dichroic-prism beam splitter with color-trim filters.

In this manner, the red, blue, and green components of the scene lens are imaged on the photocathodes of the three image-orthicon tubes.   A remotely controlled iris located between the two relay-lens elements is used to adjust the exposure of the image orthicons.   This iris is the only control required in actual studio operation.   A schematic diagram of the optical system and pickup tubes is shown in Fig. 20-21.

Successive developments in the three-tube image-orthicon cameras have provided refinements in precision of registry, electrical stability, and simplified control procedures. Many of these cameras are still in use.

40. Four-Tube Color Cameras.   Four-tube (luminance-channel) cameras were introduced when color receivers served a small fraction of the audience; the viewer of color programs in monochrome became aware of lack of sharpness.   Using a high-resolution luminance channel to provide the brightness component in conjunction with three chrominance channels for the *R, G,* and *B* components produces images that are sharp and independent of registry errors.

Such cameras use zoom lenses instead of turret assemblies, are all solid-state, and produce images having fully compatible colorimetry.   Typical broadcast cameras use either a 4½-in.

image orthicon in the luminance channel with three vidicons for chrominance, or four identical 30-mm plumbicons. A large number of these cameras are in use in studio and outdoor color broadcasting. The four-tube luminance-channel principle is also used in the reproduction of color film. Nearly all film-reproduction cameras in the United States are of the four-tube vidicon type.

41. Three-Tube Plumbicon Cameras. The three-tube plumbicon camera, developed and introduced by N. V. Philips of Eindhoven, Holland, marks a major advance in color-camera technology. It represents the first successful design to combine high technical performance, small size, low motional lag, and a signal-to-noise ratio limited only by the performance of the external video amplifier.

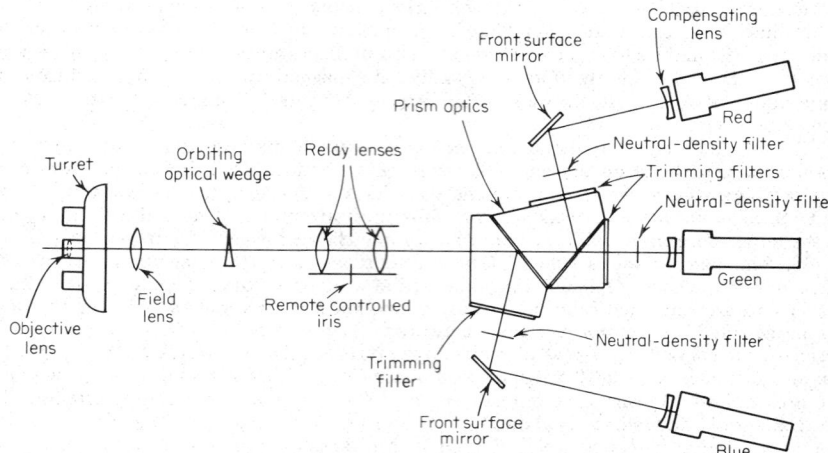

Fig. 20-21. Optical system of three-tube image-orthicon camera.

The camera is based on the 30-mm plumbicon, which uses a 0.8-in. (20-mm) image raster diagonal, has a linear transfer characteristic, and a normal signal current of approximately 300 nA. Under these conditions the motional lag or smear is small.

An important factor in the camera design is the simple optical system. A unique beam splitter, which utilizes high-efficiency dichroic color-separation surfaces and air-gap optical reflection for low polarization and compactness, is located in the back-focal distance of a specially designed zoom lens. By placing the plumbicons in locations dictated by the optical-design criteria, it is possible to eliminate all reimaging optics and to focus the three color-separation images directly on the three plumbicons. A schematic diagram of the optical system is shown in Fig. 20-22.

Gamma Correction. Since the plumbicon has a linear transfer characteristic, it is necessary to correct the video signal to a log-log slope of 0.5 to compensate for the reproducing picture tube (effective gamma of 2.4). Under these conditions, with careful design of the external video amplifiers it is possible to produce color pictures which have signal-to-noise ratios of 48 dB. This achievement is extremely important because nearly all color broadcasts are tape-recorded. The requirement of preparing multiple-generation recordings in program assembly and editing makes excellent video signal-to-noise mandatory.

A useful feature of plumbicon cameras is the possibility of trading signal-to-noise ratio for an increase in sensitivity. Thus a doubling of a sensitivity by increasing the external video gain will introduce a decrease of 6 dB in signal-to-noise ratio. The process can be extended to the point where motional lag becomes a limiting factor in camera performance. Techniques are now available for improving the lag characteristics of the plumbicon under low light conditions. These use uniform *bias lighting* from an external source, falling on all three rasters to shorten the buildup and decay time of the photoconductive surfaces.

Precision techniques in fabricating the plumbicon tubes and the focus and deflecting assemblies make it possible to provide excellent registry with a three-tube camera. In addition, electrical comb-filter or delay-line techniques are used to enhance plumbicon aperture response in both the horizontal and vertical direction.

These developments have all resulted in simpler, more compact cameras and have set a pattern for practically all present-day color camera designs. Examples of such three-tube cameras are the RCA TK-44A, the Marconi Mark VIII, the EMI (2001), and the IVC.

42. Special-Purpose Color Cameras. For field and studio pickup special self-contained color cameras have been developed. These include all synchronizing generators and

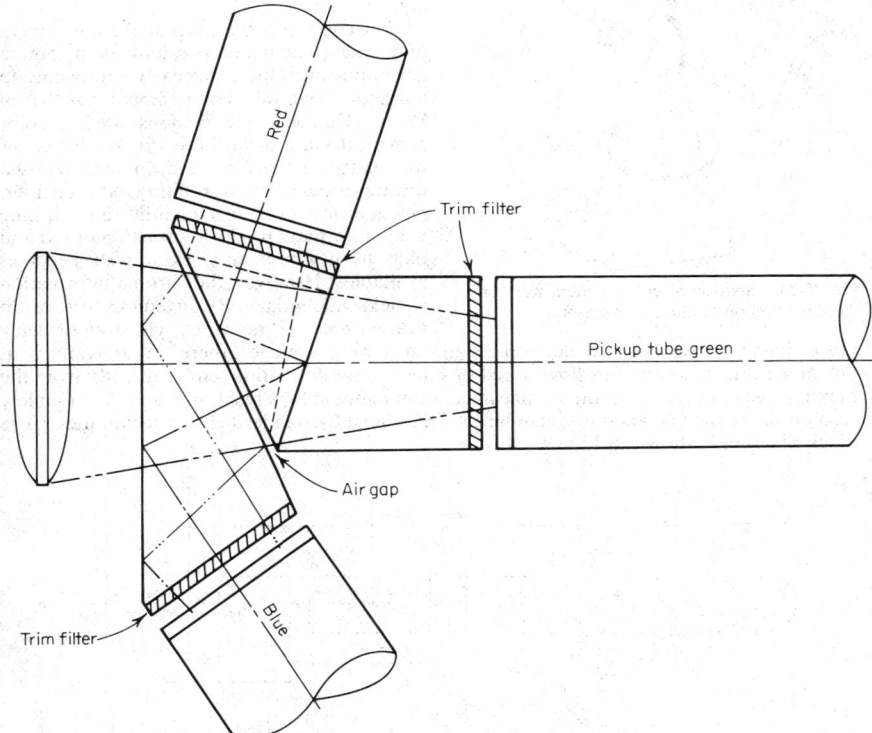

Fig. 20-22. Optical system of three-tube plumbicon camera.

processing and encoding circuits; they are connected to a control point by means of inexpensive, small, and light triax cable. Digital-command circuits transmitted along this cable permit setup, adjustment, and operation of such cameras from a remote location. The fundamental developments preparing for such cameras were carried out by the CBS Laboratories.

43. Rotating-Filter Color Cameras. Color systems have been devised to generate sequentially the video signals corresponding to the red, blue, and green components of a scene being televised (see Par. **30-17**). This is done by exposing a single pickup tube to the focused optical information of the scene through a sequence of color-separation filter sectors arranged to rotate synchronously at the vertical scanning rate. The charge information stored on the pickup tube is then removed by the raster scanning sequence. For good color performance without undue dilution or desaturation of colors, the charge information must be removed in a single scanning field, so that residual information does not dilute or contaminate information deposited on successive fields. It is difficult to arrive at good compromises between sensitivity, signal-to-noise ratio, color saturation, and lag performance.

With present pickup tubes and techniques it is possible to produce sequential color-television camera systems with nonstandard line and frame scan rates which can provide adequate colorimetry and acceptably low color breakup or fringing on motion.

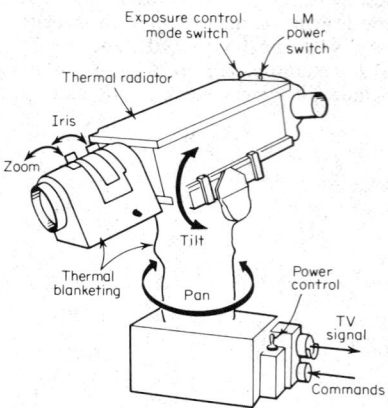

Fig. 20-23. Sequential color camera used on Apollo 14 moon-landing broadcasts.

Special sequential color cameras have been introduced for industrial and medical use and for the Apollo space program. They represent the smallest and simplest color cameras that can be produced with available pickup tubes. Figure 20-23 shows the *Apollo 14* color camera.

The concept of the sequential color camera offers many intriguing possibilities in camera development. One of the early approaches to a practical two-tube color camera was that of W. L. Hughes. He demonstrated a color camera using a pickup tube for brightness or luminance information and another exposed through a synchronous rotating red-blue filter, giving a flow of field-sequential red and blue video information. By subtraction of red and blue from luminance, green also becomes available. However, the chroma information is field-sequential. By ingenious use of an iron-wire delay line which will store or delay video signals for $\frac{1}{60}$ s and an electronic-signal switching sequence, there is thus available at any given time a continuous flow of red or blue video information, either directly from the pickup-tube output or from the storage line. Luminance at high resolution and R, B, G at low resolution are then coded into a standard NTSC signal by standard circuit techniques. The block diagram is shown in Fig. 20-24.

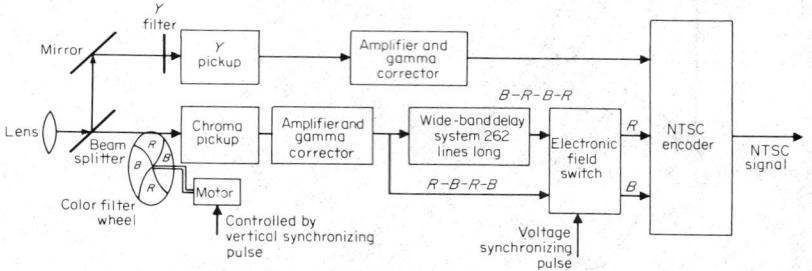

Fig. 20-24. Hughes two-tube sequential color camera, used in broadcast of 1968 Olympic Games.

The concept was further refined by American Broadcasting Corporation (ABC) engineers working with Ampex, to produce a color camera which is simple, small, and portable and uses a small-diameter camera cable. This camera was used extensively by ABC in broadcasts of the Olympic games of 1968. Video delay or storage in this latest development was obtained by rotating video disks or special delay lines.

44. Flying-Spot Scanners. Flying-spot scanners have been used for many years in reproduction of transparencies. The system uses a high-brightness kinescope-scanning raster optically imaged on a monochrome or color transparency. The light transmitted by the transparency is gathered by photocells and amplified to form the video signal. Since the raster is generated by a single focused electron beam or spot under the influence of horizontal and vertical deflection fields, the flying-spot designation is very descriptive of the system.

The high-brightness, high-resolution raster is generated at second-anode voltages of 25 to 30 kV. Usually 5- or 7-in.-diameter kinescopes are employed. Special phosphors having short buildup and decay characteristics are used. For color use the phosphors must have high light outputs in the red, green, and blue portions of the visible spectrum. The light transmitted through the televised transparency is separated into red, green, and blue components by a dichroic mirror system and is sufficient to produce simultaneous R, G, B signals with adequate signal-to-noise ratio.

Since there is at any instant only a single point of light in the scanning raster which analyzes the transparency, there is practically no optical flare in the system; this is highly desirable for accurate reproduction of wide-range gray scales. Because of the inherent nature of the flying-spot scanning process, the system has no memory or storage capacity. This imposes special restrictions on its use with motion-picture-film reproduction.

The flying-spot transfer characteristic, relating video-signal output to light input or film transmission, is strictly linear. This requires the use of extensive electronic gamma correction of the output video signal to accommodate the transfer characteristic of the color-display kinescope, which has an effective log-log slope of approximately 2.4.

A block schematic of a flying-spot scanner is shown in Fig. 20-25.

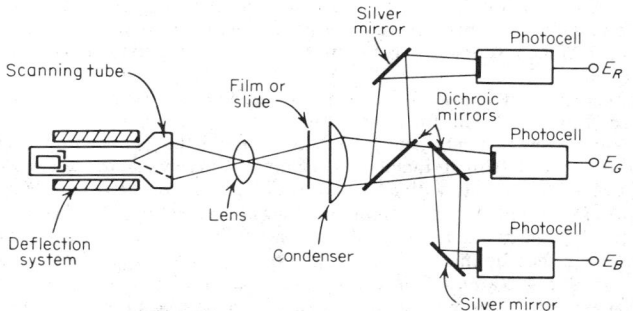

Fig. 20-25. Flying-spot scanner for film or slides.

Flying-spot scanners are used in broadcast stations in Europe for both color and monochrome film reproduction. In the United States they have had very little application for broadcast service but are used for scientific scanning instruments. This difference in use is due primarily to the absence of any storage property in flying-spot scanners and the difference in operational standards. In Europe, the raster has a 50-Hz field rate. Standard motion pictures are shown nominally at 24 pictures per second, but for television they are shown at 25 pictures per second. It is relatively easy to design and construct motion-picture projectors to immobilize the picture in the optical system, so that the flying spot can develop the video signal at the 50 field per second rate, as is essential because the scanner has no electrical storage.

Experiments in the United States on such systems showed that with the added requirement of converting 24 picture frames per second to 60 television fields per second, the inherent simplicity of the flying-spot scanner and its freedom from the raster registration problem are outweighed by the complexity and need for very high precision tolerances in projector construction and maintenance.

Since slides present no special problem for flying-spot scanners they can be used in American TV systems with complete satisfaction. However, because of the poor equipment-utilization factor, it is not economical to use flying spots for slides alone.

45. Camera Performance. Objective evaluation of camera performance for broadcast, closed-circuit, or industrial applications is a useful tool for comparison of various designs of equipment and for maintaining picture quality standards. The factors which enter into such an evaluation are sensitivity, resolution, signal-to-noise ratio, transfer characteristic (gamma), flare, raster-field uniformity, and scanning stability and uniformity.

Color cameras require additional performance specifications such as color-sensitivity curves in the red, green, and blue channels, together with required colorimetric tolerances. In addition, since the final picture depends on the accurate registry of three of four rasters, the factors which influence raster accuracy and geometry are especially important both in specification and evaluation. Since some of these factors are interrelated, special charts, tests, and procedures have evolved which provide meaningful data for evaluation.

46. Sensitivity. Camera sensitivity describes the video-signal output in terms of incident illumination required on the scene being viewed. A lens opening of $f/8$ on an image-orthicon raster diagonal of 1.6 in. (40 mm) is generally used as a reference base for adequate depth of field. With plumbicons, the same depth of field corresponds to $f/4$ since the plumbicon raster diagonal is 0.8 in. (20 mm). Incident illumination, usually specified in lumens per square foot (foot-candles), is based on the use of a 60% reflectance neutral target as the brightest object in the scene.

The light intensity on the photosensitive raster of the pickup tube required to produce the necessary signal-current level compatible with the desired signal-to-noise ratio is

$$I_s = \frac{4f^2 \, I_{pc}}{TR}$$

where I_s is the incident-scene illumination in lumens per square foot, f the f-number setting of the camera-lens aperture, I_{pc} the photocathode illumination of the pickup tube in lumens per square foot, T the transmission factor of lens, generally 80% with modern lenses, and R is the reflectance of the test object in the scene (in color pickup this is a matte white card with reflectance of 60%).

For color cameras, an additional factor is required in the denominator giving the dichroic optical transmission coefficient for the red, blue, or green channel.

From this expression it can be determined, for example, that at $f/4$ a three-tube plumbicon camera requires about 150 lm/ft^2 incident on a 60% neutral reflectance card or chip to develop a 300-nA green video signal with a signal-to-noise ratio of 49 dB. Measurements are conventionally made with a unity gamma transfer characteristic. Studio lighting for color TV is furnished by quartz-halogen incandescent-lamp fixtures operating at 3200 K. Because of colorimetry shifts, dimming of lighting fixtures is rarely used in studio practice.

47. Signal-to-Noise Ratio. This factor, which is the ratio of peak video-signal amplitude to the rms value of the random noise in the raster, is generally stated in decibels. While the peak-video-signal measurement is easy, the determination of noise is more difficult and requires special techniques. An oscilloscope display of peak noise components is observed for a period of a few seconds, and conversion to rms values is obtained by dividing by a factor of 6. Thus, if the white signal has 100 units amplitude and the peak noise has a 6-unit amplitude, the signal-to-noise ratio is $100/(6/6) = 100$, or 40 dB. Thermal integrating meters are sometimes used to obtain noise rms values.

In the image-orthicon tube, the noise is evenly distributed through the spectrum (white noise). It is determined inherently by the effective percentage modulation of the scanning beam. To attain the best signal-to-noise ratio possible, the beam current should be adjusted to just discharge the highest scene highlights. Any excess beam current decreases the effective beam percentage of modulation and hence the signal-to-noise ratio. In an image orthicon the maximum signal-to-noise ratio is obtained with full exposure into the knee of the 400 nA.

Effect of First Video Amplifier. The signal-to-noise ratio depends almost entirely on the figure of merit of the first video-amplifier stage, not directly on the noise generated by the picture tube itself. Because video amplifiers for vidicons and plumbicons must compensate for the shunt-input capacity of the pickup tube at the rate of 6 dB increased gain per octave, they have a *triangular noise characteristic* in which higher noise amplitudes are produced at higher frequencies in the video band. These are correspondingly less subjectively visible to the viewer. This produces a net gain in *effective signal-to-noise* ratio in the video display. The gain is usually stated as the *weighting* factor between flat and triangular noise distribution.

Present video-amplifier circuits using the best available field-effect transistors in the head end can produce signal-to-noise ratios of about 50 dB for a 5-MHz bandwidth and signal currents of 300 nA. Video displays having such signal-to-noise performance show no visible noise in the picture structure and are practically noiseless in the black areas of the raster. "Velvet blacks" are characteristic of such pictures.

Since the signal-to-noise capability of the photoconductive tube and external video amplifier is so high, it is practical to trade improved sensitivity for lower signal-to-noise ratio. Thus by decreasing the scene lighting by a factor of 2 and increasing the video gain by 2, the signal-to-noise ratio is decreased from 50 to 44 dB, which is still highly acceptable. Reducing scene lighting by a further factor of 2 reduces the signal-to-noise ratio to 38 dB, which can be tolerated in many situations. Hence the scene lighting can be reduced from 150 to 37 lm/ft^2 at full video-signal output. At this point, while the signal-to-noise ratio is marginally acceptable, the lag or smear of lead oxide photoconductive surfaces for motion in the televised scene becomes a dominant factor in picture quality.

48. Gamma Considerations. Shadow-mask color picture tubes have a nonlinear transfer characteristic; i.e., the light is produced as a function of the control-grid video drive. In a log-log display of the screen brightness plotted against video excitation the slope of the transfer function is about 2.4 over a brightness range of 100 to 1. To produce a pleasing picture in color with consistent colorimetric qualities, it is necessary that the transfer characteristic of the pickup camera be the reciprocal of the picture-tube characteristic.

Since the plumbicon is a linear device (the video signal is directly proportional to the light in the scene) gamma correction in the form of nonlinear signal amplification, to a power of 1/2.4, or approximately 0.40, must be introduced into R, G, B channels of the camera before the signal is encoded. The low signal levels are stretched and the high signal levels compressed in accordance with the power curve selected. Stretching the black signal by a factor of 2 decreases the signal-to-noise ratio by 6 dB, whereas compressing the white signal improves the signal-to-noise ratio. Since significant stretching of blacks must be used, the black signal-to-noise may be degraded by 12 dB or more. Various band-limiting and signal-coding techniques to minimize noise effects and stretch the low-light signal have been successfully used in plumbicon cameras.

49. Raster Linearity, Registration, and Stability. The starting point for generating color pictures is the optical and electronic superposition of the red-, green-, and blue-tube scanning rasters. In high-quality broadcast service, precision focus and scanning components make possible substantially identical conditions for horizontal and vertical deflection and precise geometric orientation of the three rasters with the optical picture input.

Solid-state circuits designed for the highest possible linearity and stability are provided to drive the deflection circuits and to maintain magnetic-focus fields constant to within 0.25%.

By adjustments made with special test charts it is practical to obtain scanning linearity to better than 0.5% within a circle of diameter equal to picture height (zone I, as shown in Fig. 20-26) and within 1% anywhere else in the raster (zones II and III). The registration accuracy of the three rasters can be set to closer than 0.1% in zone I and closer than 0.25% in zone II.

Standard test charts for testing linearity and registration test charts are shown in Figs. 20-27 and 20-28. These are obtainable as opaque charts or as 2 by 2 in. glass-mounted transparencies from Electronic Industries Association, 2001 Eye St., N.W., Washington, D.C. 20006, and Society of Motion Picture and Television Engineers, 9 E. 41st St., New York 10017.

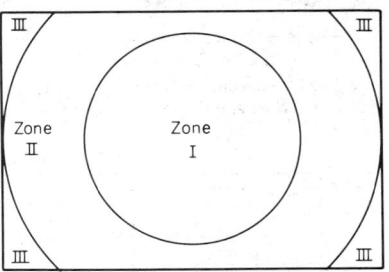

Fig. 20-26. Zone chart for checking scanning linearity of cameras.

The procedure generally adopted is to use the green channel as a reference, adjust it for correct size and linearity, using the EIA Ball chart (Fig. 10-28), and then to match red and green rasters to it, using both the Ball chart and the registration chart (Fig. 10-27). It is important that differential errors in size and linearity be avoided, since they produce color fringing and low resolution in the portions of the raster which do not coincide or register. Small errors in absolute linearity or size go relatively unnoticed.

A technique frequently used to help linearity, size, and registry adjustments is that of reversing the polarity of the red and blue video signals with respect to the green video reference. The narrow vertical and horizontal lines which constitute the registration pattern produce pulse waveforms and raster display signals which cancel completely when there is coincidence between red and green or blue and green rasters. The operator's eye is thus drawn to areas where the positive and negative pulses do not cancel and where adjustments in linearity, size, or centering are needed.

The same technique is useful in pickup-tube rotation to obtain parallel scanning of rasters. Such adjustments are necessary on initial setup and require little if any change with change of pickup tubes. Skew control or electrical compensation is provided for residual magnetic and electrical cross talk in the deflection components. Such cross talk causes errors in the

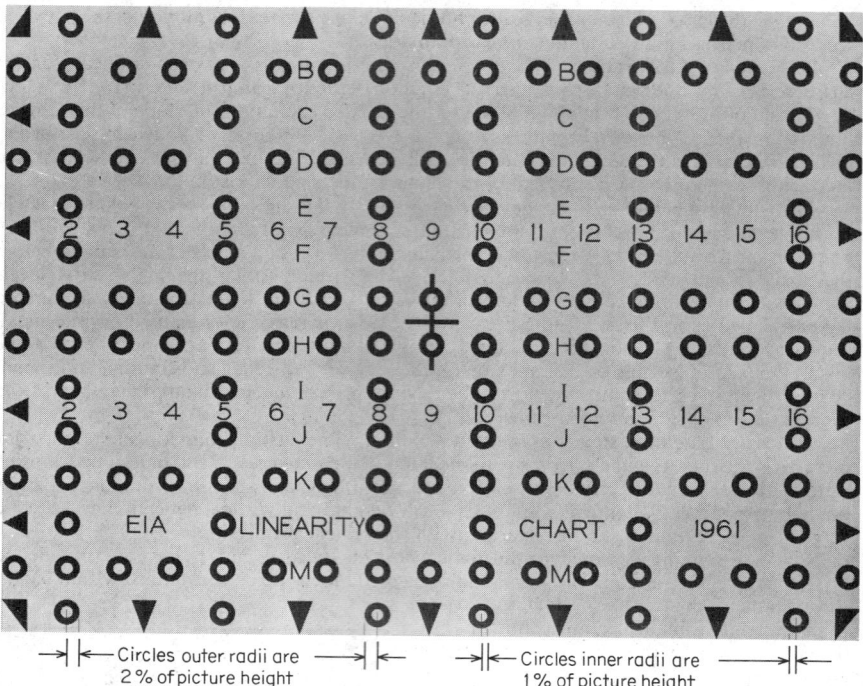

Fig. 20-27. EIA linearity test chart. Aspect ratio is 4:3. Horizontal blanking 17.5%. Vertical blanking 7.5%. Electrical grating pattern generator frequencies: 315 kHz horizontal, 900 Hz vertical.

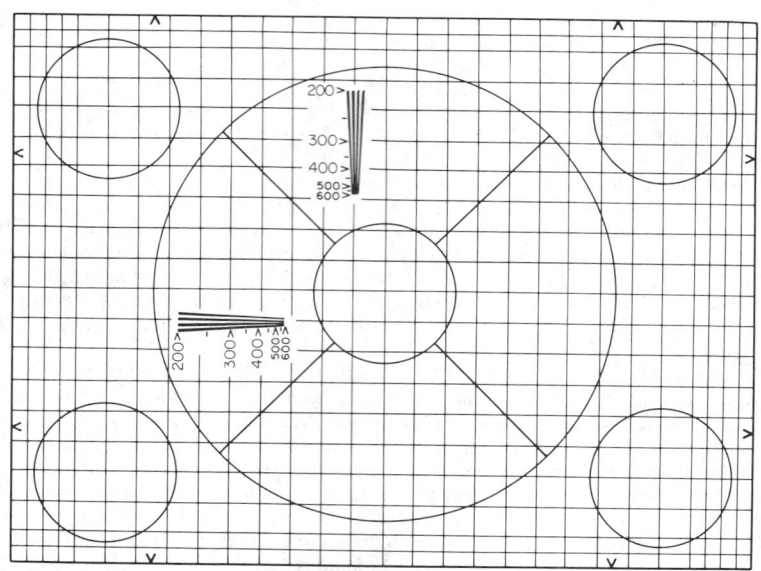

Fig. 20-28. EIA registration test chart.

orthogonality of horizontal and vertical scan. In well-designed deflection components these errors are small and need little attention after the initial compensation setting. The circuits are designed with long-term stability characteristics.

A camera initially registered and then turned off will return to the registry condition within 15 min after being turned on again. Minor adjustments in horizontal and vertical centering may be required to obtain precise registration since the electron-gun structures in the pickup tubes show long-term *thermal-wandering* effects which produce small centering-registry errors even though all electric circuits are completely stable.

50. Resolution. Resolution, or picture sharpness, depends on the optical performance of the camera lens and on the electromagnetic focus of the scanning beam on the pickup-tube raster. The EIA (RETMA) or SMPTE resolution charts are generally used to evaluate resolution. A typical chart is shown in Fig. 20-29.

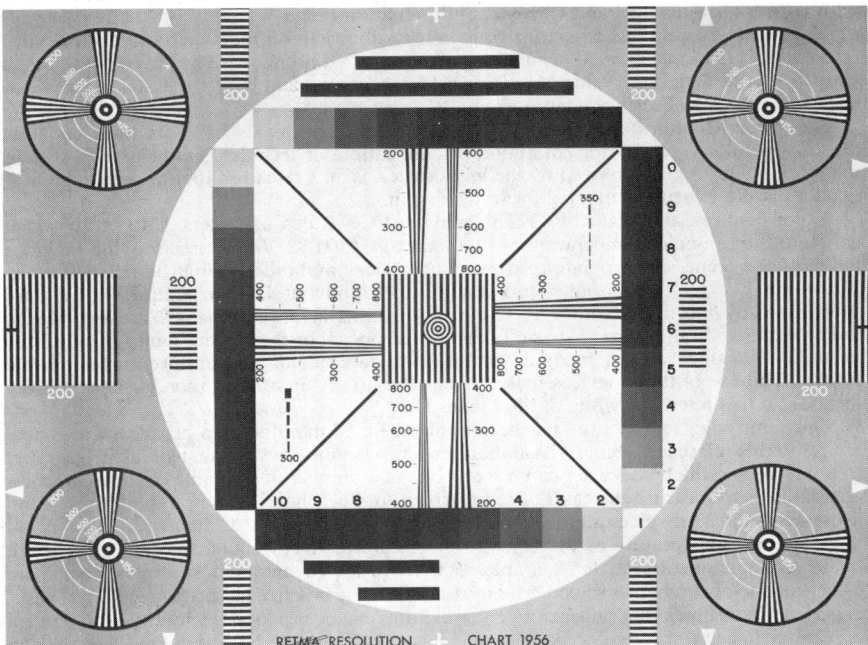

Fig. 20-29. EIA resolution and gray-scale chart.

The charts indicate resolution in television lines. They consist of square-wave black-to-white transitions arranged in wedge form or as discrete bursts corresponding to specific frequencies or television line numbers. To make observations as objectively as possible, resolution is given as an aperture-response percentage of low-line number performance. Thus, a 40% aperture response at 400 lines (5 MHz) is the measured peak-to-peak video signal at 400 lines, referred to 100% response at approximately 50 lines.

Line-selector oscilloscope displays make such measurements relatively easy and accurate. The optical and electric focus are optimized before response values are measured. In pickup tubes using electromagnetic focus, the raster will rotate very slightly with electric-focus adjustment. For monochrome TV the amount can be completely ignored. In three-tube color cameras, electric focus must be set before the registry procedure is carried out if consistent results are expected.

51. Gray-Scale (Gamma) Charts. Neutral gray-scale charts available from EIA as opaques 24 in. wide and 18 in. high are used to test the tonal setup of monochrome and color cameras. Two horizontal nine-step wedges having a maximum reflectance of 60% and a range of 20 to 1 are provided in either logarithmic or linear step increments. The logarithmic chart is most useful in camera alignment since the oscilloscope display after appropriate gamma correction is an equal-increment series of steps.

The gray-scale chart is essential in alignment of the gamma-correction circuits in the individual *R, B, G* color channels. The technique used calls for accurate setting of black level, gain, and gamma correction or differential gain in each channel. The goal is to generate a gray-scale video display in which the colorplexed subcarrier vanishes for every gray step level. The color monitor becomes a very sensitive device for checking adjustments since any differential variation in the three transfer characteristics will become evident as color in the gray-scale wedge display.

Small variations in absolute value of the overall gamma have little effect on the color picture, whereas small differential variations in gamma from channel to channel can be prominent and annoying. If the adjustment of the combined pickup tube and electronic transfer characteristics is carried out meticulously and effectively, the camera will track well with iris exposure control and with scene-illumination range.

The most stringent test of a color camera is its ability to reproduce accurately a neutral or colorless gray scale on a color monitor or receiver. If proper care has been taken in the colorimetric design, the production of consistent, accurate, and highly pleasing color video signals becomes a necessary consequence.

52. Color Monitors and Receivers. Color television signals have, as their end use, display on color receivers for entertainment, education, or technical application. A color television monitor, as employed by the broadcaster, is an evaluation tool to determine how good a picture is being generated for transmission.

Color pictures, as specified by NTSC and the FCC Rules, are intended to be viewed at a "white" screen-color temperature of illuminant D, 6500 K. More precisely the intended white of the scene, corresponding to zero subcarrier amplitude, should be reproduced as illuminant D. A close visual approximation to illuminant D is a neutral (white) card illuminated by "north sky" daylight. Reference standards for illuminant D are available to the broadcaster in the form of specially controlled phosphor fluorescent lamps or regulated incandescent-lamp sources modified by suitable stable color-temperature-raising optical filters. The use of these devices makes it possible to set up a color monitor at any given location to this reference white of illuminant D.

Once this color temperature has been established, all monitors in a broadcasting system are set to this reference. Since visual monitor setup is time-consuming, tedious, and subject to human error, the broadcaster can use any one of a number of available commercial color comparators to measure the ratio of red to green to blue phosphor output and rapidly and accurately adjust any given number of monitors to the illuminant D standard.

The color temperature of the color receiver "white" has frequently been set at much higher color temperatures, for example 9500 K, to match monochrome phosphors more closely and to obtain higher screen brightness. While the practice has no firm technical basis, experience has shown that satisfactory color pictures can be produced by intuitive adjustment of receiver hue and chroma controls. As long as there is no side-by-side comparison with an illuminant D monitor, the viewer is well satisfied with the color picture.

SYNCHRONIZING SIGNAL GENERATION

By Charles W. Rhodes

53. Performance Requirements. Television sync-pulse generators (SPGs) provide the timing pulses required for the system. In monochrome signals, these pulses are:

1. Composite sync (horizontal sync, vertical serrated sync, and equalizing pulses).
2. Composite blanking (horizontal and vertical sync combined).
3. Camera horizontal drive pulses.
4. Camera vertical drive pulses.

The horizontal and vertical blanking pulses are combined within the SPG, and they appear at its output as composite blanking (see Figs. 20-30 and 20-31, upper traces). The horizontal sync, serrated vertical sync pulses, and equalizing pulses are also combined within the SPG and appear at the composite sync output, as in Figs. 20-30 and 20-31, lower traces. The time relationships between sync and blanking are shown.

In early camera designs, the practice was to control the scanning circuits by means of horizontal and vertical drive pulses. These are now combined within the SPG; that is, individual coaxial cables carry these pulses to the camera. The leading edges of the drive pulses coincide with the leading edges of the blanking pulses, as shown in Figs. 20-32 and 20-33. Many cameras do not utilize drive pulses, thus reducing the complexity and weight of the cable from the camera to its camera control unit (CCU).

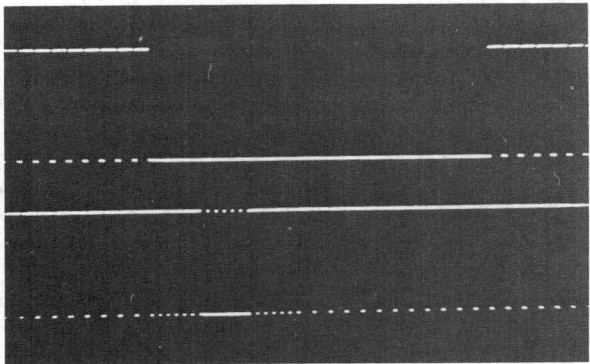

Fig. 20-30. Detail or oscilloscope presentation of composite vertical blanking *(top)* and vertical synchronization signal *(bottom)*.

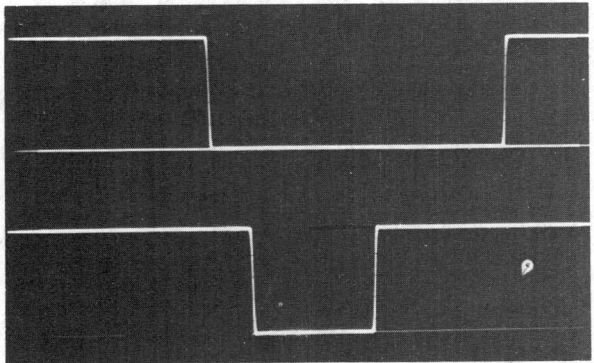

Fig. 20-31. Oscilloscope presentation of composite horizontal blanking *(top)* and composite horizontal synchronizing signals *(bottom)*.

Monochrome sync generators usually have provision to be synchronized to the power line, an internal crystal oscillator, or an external source of 31.5 kHz; or they can be *gen-locked* to a composite video signal. In the NTSC color system, the power-line and internal crystal modes of operation are not feasible. Monochrome SPGs are converted to color use by driving them with a 31.5-kHz signal derived by counting down from the color subcarrier. The subcarrier is generated by a highly stable crystal oscillator held at constant temperature. Such subcarrier sources, including the countdown circuits to produce the locked 31.5-kHz signal, are called *color standards.*

The FCC Rules require the absolute frequency of the color subcarrier to remain within ± 10 Hz of 3.579545 MHz. The short-time frequency stability (drift rate) is also specified. Better than 3 ppm stability is readily achieved by temperature-stabilized quartz-crystal oscillators. Network operations frequency require a much more stable subcarrier.[1]

---

[1]Superior numbers refer to References, Par. **20-59**.

Rubidium frequency standards have been developed to provide the color subcarrier and 31.5-kHz signals with a stability of $1\text{-}5 \times 10^{-11}$.

54. NTSC Synchronizing Signal Waveform. Details of the NTSC sync waveform as radiated are shown in Fig. 20-9. This waveform differs from that of the SPG, due to subsequent distortions introduced principally by the transmitter. The group-envelope delay characteristic specified for the transmitter introduces a 170-ns delay of the luminance and sync signals, relative to the subcarrier color burst (see Fig. 20-41). The *breezeway* is specified at 379 ns minimum in the transmitter's output but must be greater by at least 170 ns at the SPG output. The maximum rise time of the sync pulses is specified. Good engineering practice requires that the rise time of the SPG outputs be somewhat faster than specified in the FCC Rules.

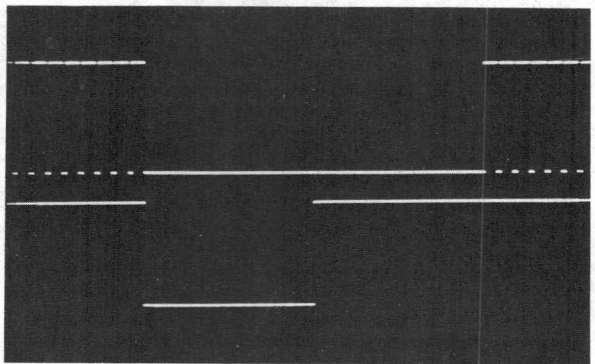

Fig. 20-32. Composite vertical blanking *(top)* and vertical drive *(bottom)*.

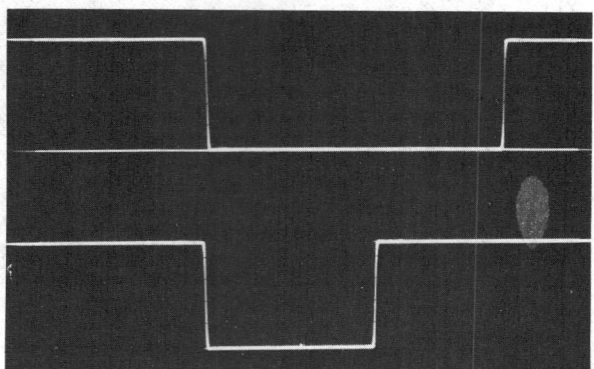

Fig. 20-33. Composite horizontal blanking *(top)* and horizontal drive *(bottom)*.

It is desirable to control the shape of the sync pulse *transitions* produced by the SPG, using *sine-squared filters.*[2] These are phase-equalized, low-pass filters which limit the frequency spectrum of these pulses. A sync pulse having such a controlled transition is shown in Fig. 20-34. One particular advantage in band-limiting the outputs of an SPG is that ringing and cross talk are reduced, as the signals are distributed within the broadcast plant. This is especially important in current SPG units, in which rise times may be as short as 10 ns. The sine-squared filters have a cutoff frequency approximately the inverse of the half-amplitude duration (or step rise time) of the pulse. Rise times are always measured from the 10 to 90% amplitude point on the transition, as shown in Fig. 20-34. While the usual practice in measuring the pulse durations is to measure between 50% points, the FCC Rules require measurement between the 10% points on the rising and trailing edge of a pulse.

The color burst is not transmitted during the first nine lines in each vertical blanking interval. A nine-line *keyout* of the burst is produced within the SPG by inhibiting generation of the burst-flag pulses. Figure 20-35 compares the burst-flag pulses with the composite video, in the vicinity of the vertical blanking pulse. The absence of the nine color bursts is apparent. The timing of the burst flag relative to both horizontal sync and the burst envelope is shown in Fig. 20-36. The flag precedes the burst, to accommodate the delay in the bandpass filters through which the burst flag and burst envelope must pass in the encoder.

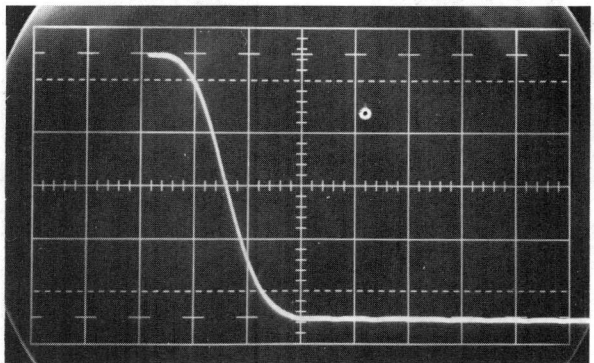

Fig. 20-34. Trailing edge of sine-squared-shaped sync pulse (125 ns from 10 to 90% levels).

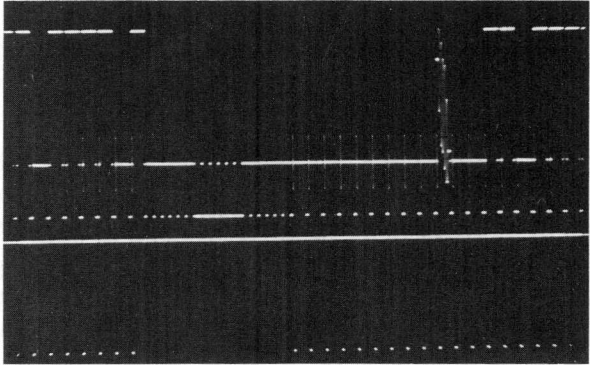

Fig. 20-35. Composite video sync *(top)*, with burst flag *(bottom)* during vertical blanking. The nine-line keyout interval is shown.

Black-Burst Technique. The cost of distributing the individual outputs of an SPG throughout the broadcast plant on multiple coaxial cables has led to the development of the *black-burst signal,* shown in Fig. 20-37. This signal carries all the required timing information from one SPG to another on a single coaxial cable. Many SPGs can thus be slaved to the master SPG via its black-burst output. The slaved SPG units can thus be timed relative to a common point within the plant so that encoded video signals from any source are in precise time synchronism with respect to any other. This condition, accurate to a few degrees at subcarrier frequency, is essential during switches, fades, wipes, and dissolves. The *differential pulse cross* display may be used to measure the relative timing of two video signals. Waveform monitors are also used to measure relative timing errors, a differential input being required.

A color picture monitor, equipped with both pulse cross display and differential input amplifiers producing signals of opposite polarity, can perform the entire timing operation. The two opposite color bursts will cancel each other when the two bursts are precisely in phase with each other. Oscilloscope techniques can readily measure both the rise time and pulse width. Figure 20-34 shows a rise-time graticule used for this purpose.

55. Pulse Jitter. Pulse jitter must be measured with respect to the color subcarrier. Jitter between successive horizontal-sync-pulse leading edges will completely destroy the frequency interlace of the NTSC signal. Sync pulse jitter also establishes performance limits in video-tape recording (VTR), especially in multiple-generation rerecording. Time-base correction circuits in the VTR correct jitter in the picture signal by measuring the jitter of the sync pulses (see Par. 20-88). This consideration places severe restrictions upon permissible jitter in the SPG output.

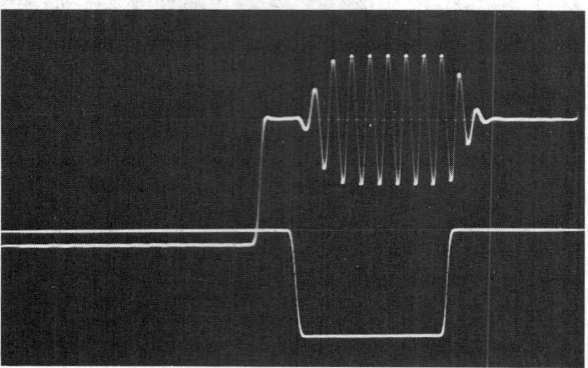

Fig. 20-36. Composite video sync *(top)* and burst flag *(bottom)* during horizontal blanking.

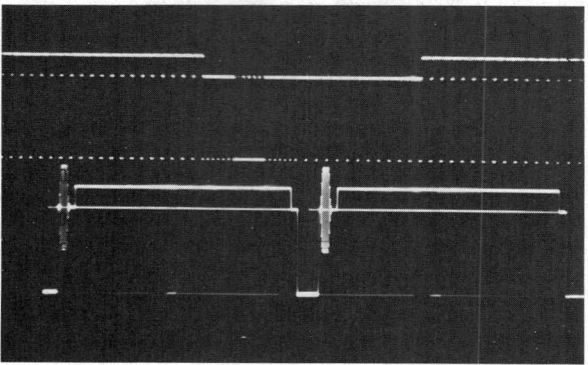

Fig. 20-37. Black burst *(bottom, center)*. Vertical blanking detail at top, horizontal blanking detail at bottom.

Jitter can be measured to 3 to 5 ns by comparing the leading edges of all horizontal sync pulses against the color subcarrier; this can best be done with a *vectorscope*. The time resolution of an NTSC vectorscope is

$$\frac{10^{-6}}{3.58 \times 360°} = \frac{\mu s}{1.288 \times 10^3} = 0.77 \text{ ns/deg}$$

The leading edge of horizontal sync pulses generates phase transients which may be observed on the vectorscope, as in Fig. 20-38. The 2:1 frequency interlace of NTSC causes the double transient display. This figure shows severe jitter.

Another technique for verifying the stability of an SPG is the *pulse cross display* as seen on a picture monitor. Figure 20-39 shows a typical display. The vertical blanking may be expanded for improved resolution in this critical area. Jitter will affect the vertical alignment of the leading edges of horizontal sync pulses and equalizing and serrated pulses. Measurement of jitter is facilitated by using the delayed sweep trigger from a laboratory oscilloscope, the delayed trigger being added to the video signal to modulate the display in intensity. In this way, quantitative measurements can be made with a high degree of accuracy.

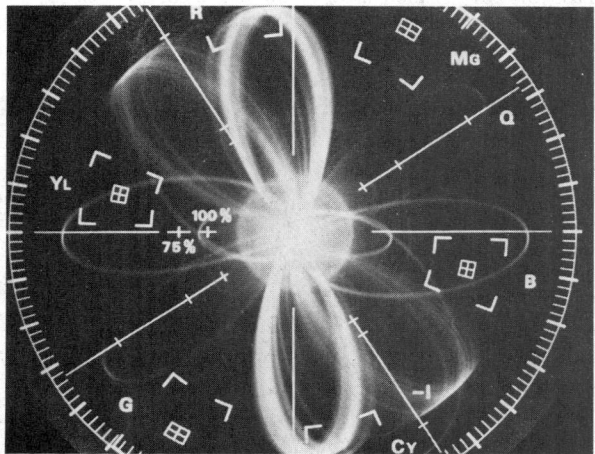

Fig. 20-38. Vectorscope display, showing phase jitter.

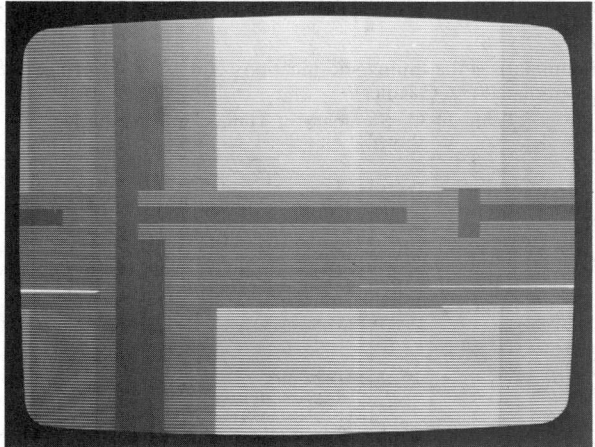

Fig. 20-39. Pulse cross display on a picture monitor.

56. Burst Timing. One of the most difficult measurements of the SPG is that of burst timing, at which the SPG provides the burst flag. The significant parameters are the breezeway and the burst width. The burst is usually generated in the color encoder to ensure phase stability with respect to the chrominance signal. The critical timing is the time between the sync trailing edge and the burst envelope (breezeway). Figure 20-40 shows the breezeway. The phase of the burst, relative to leading edge of sync, is unspecified in the FCC-NTSC Rules, at present. Breezeway varies with the phase, as shown in Fig. 20-40. One technique to resolve the ambiguity is to temporarily void the frequency coherence between sync and subcarrier. The FCC Rules do not specify the rise time of the burst envelope or the point on the envelope of burst to be measured; a reasonable compromise, otherwise within the rules, is to make the burst-envelope half-amplitude duration nominally 8 cycles (2.31 μs).

In NTSC color, a four-field sequence exists that poses problems in VTR operation*. The four different fields exist because of the phase reversal (dot interlace) of the subcarrier in each line, with respect to the leading edge of sync. Also a given line in two successive frames has a subcarrier phase reversal; i.e., there are four different successive fields in the NTSC system. Sync generators should be designed so that the phase of subcarrier output is stable and unambiguous with respect to the leading edge of horizontal sync. This benefits VTR operations by reducing the lockup time.

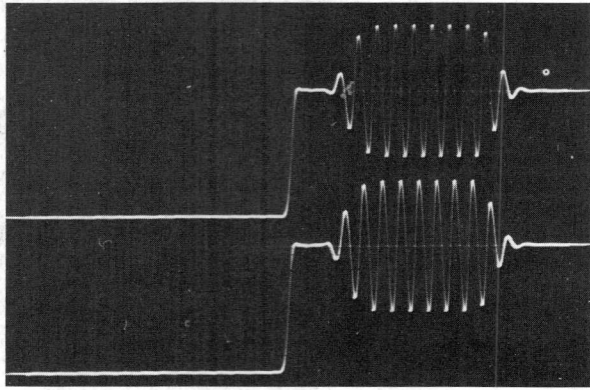

Fig. 20-40. Breezeway display with lower display shifted 180°, showing the change in breezeway with subcarrier phsase.

Table 20-3 summarizes the timing specifications of the NTSC signal as generated (SPG nominal) and as radiated (FCC Rules).

Table 20-3. Summary of Timing Specifications
(See also Figs. 20-41 and 20-42)

	FCC Rules[a]	SPG nominal[b]
Color subcarrier, 3.579545 MHz	± 10 Hz, 0.1 Hz/s	± 5 Hz
Pulse widths:		
Horizontal sync..................	4.49–5.09 μs	4.71 μs ± 50 ns
Equalizing pulses	2.54 μs nominal	2.33 μs ± 50 ns
	0.45–0.50 area horizontal sync	
Serrations in vertical sync	3.81–5.09 μs	4.5 μs ± 200 ns
Rise time	254 ns max	115 ns[c] ± 10%
Amplitude	...	4 V ± 5% into 75 Ω
Pulse jitter[d]	...	4 ns
Breezeway	379 ns min	750 ns ± 50 ns[e]
Burst duration	8 cycles min	2.31 μs + 70 ns[f]
	11 cycles max	...
Horizontal blanking	10.48 μs min	11.1 μs

   [a] Pulse widths measured at 10% point to 10% point.
   [b] Pulse widths measured at half-amplitude duration.
   [c] Transient response determined by sine-squared filter, overshoot <2%.
   [d] Pulse jitter refers to maximum time error between zero crossing of subcarrier cycle and leading edge of any sync pulse in a field.
   [e] Measured from 10% point, trailing edge horizontal sync to 50% point, leading edge, burst envelope; horizontal sync not coherent with subcarrier.
   [f] Measured half-amplitude duration of burst (2.31 μs = 8 cycles), horizontal sync not coherent with subcarrier.

57. Organization of SPG Circuits. Interlaced scanning in television systems is achieved by the design of the sync signal. The essential element is that each frame consists of an odd integral number of lines so that each field contains precisely half that number (see Fig. 20-4). Any jitter in the vertical sync with respect to horizontal sync tends to destroy interlace;

   *The PAL system has an eight-field sequence, and a twelve-field sequence is employed in SECAM.

e.g., a time jitter of 30 μs completely destroys interlace. For this reason, sync generators rely upon the 31.5-kHz pulses to form this leading edge of horizontal sync, vertical serrated, and the equalizing pulses. The functions of the counters (Fig. 20-43) are to select which of the 31.5-kHz pulse edges are to form the sync-pulse components. Within the divide-by-525 counter, it is possible, using digital logic, to develop gating pulses at any 31.5-μs interval throughout the television field.

There are three techniques available to time the durations of the sync-pulse components: delay lines, analog-delay pickoffs, and digital counters. The delay-line method is of historical interest only, due to the physical bulk of a suitable delay line. The analog-delay method provides simple and independent adjustments of pulse durations without affecting leading-edge timing. It has excellent long-time stability and freedom from jitter or temperature effects.

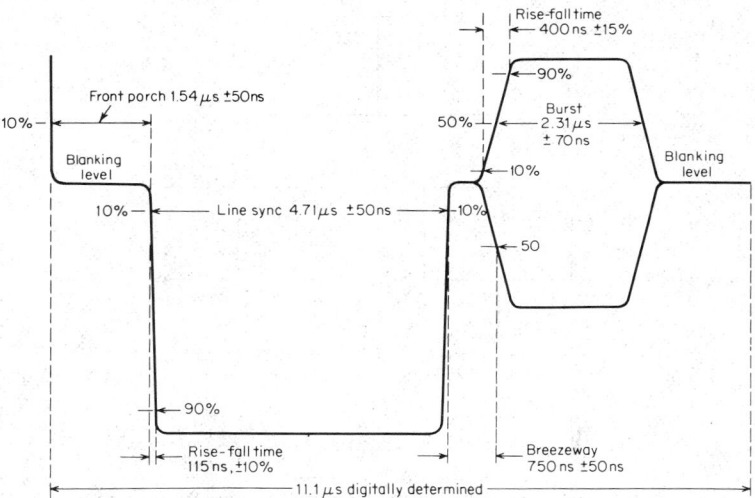

Fig. 20-41. Specification of horizontal blanking detail (see Table 20-3).

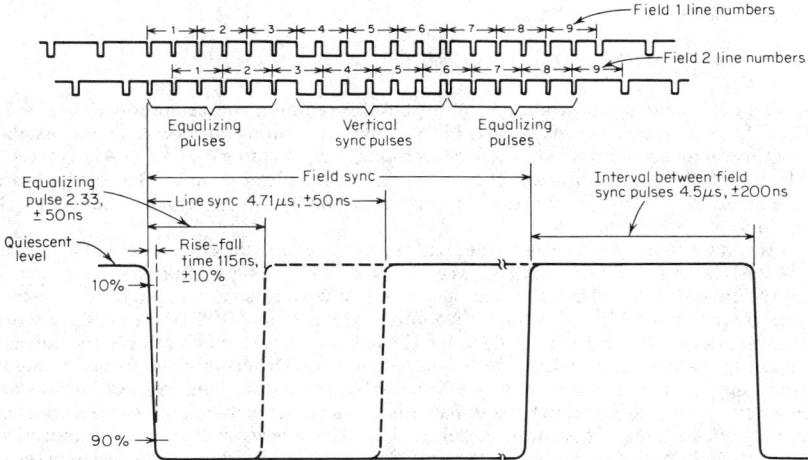

Fig. 20-42. Specification of composite sync blanking detail (see Table 20-3). All measurements are with respect to 10% points.

20-43

58. Digital Techniques. The digital-counting method requires a subcarrier oscillator operating at a harmonic of 3.579545 MHz to provide reasonable time resolution. When the fourth harmonic is chosen, the time resolution of the counter is 70 ns. In practice, this is an appropriate compromise between oscillator stability and time resolution. Higher-frequency quartz oscillators offer less long-term stability. While analog delay is continuously variable over a small range, digital delay provides pulse-width increments of 70, 140 ns, etc. Both techniques have found commercial acceptance, using integrated circuits to overcome the problems of complexity.

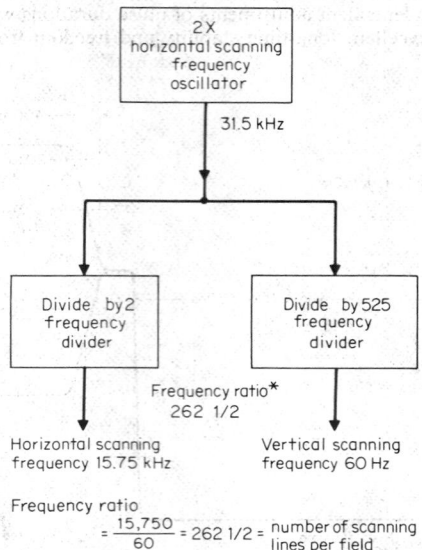

Fig. 20-43.   Analog counting technique for generating monochrome sync signals.

In the NTSC system, the color-subcarrier frequency is related to the horizontal-scanning frequency by the ratio

$$f_{sc} = \frac{455 \times 2F_h}{2 + 2}$$

Division by 455 and multiplication by 4 satisfies this requirement, i.e., produces $2F_h = 31.5$-kHz. This technique was used in early NTSC sync-generation equipment, but the instability of the times 4 multiplier has led to better techniques. By dividing 4×3.579545 MHz ($= 14.4$ MHz) by 455, the desired $2F_h$ frequency of 31.5 kHz is readily obtained. The color subcarrier is obtained with a divide-by-4 divider driven by the 14.4-MHz crystal oscillator. This technique is used commercially.

A second technique avoids both the times 4 multiplier and the crystal oscillator operating at 14.4 MHz in favor of a 3.579545-MHz crystal, which offers improved long-term and temperature stability. This technique, shown in Fig. 20-44, also eliminates the divide-by-455 digital frequency divider, to improve stability. Here, the 1.00693-MHz crystal oscillator (operating at $64F_h$) is digitally divided by 128 to form pulses which sample the subcarrier frequency, a sample being taken once every 455 cycles. The error signal from the sampling gate is the frequency error of the 1.00963-MHz oscillator, with respect to subcarrier frequency. This passes through a low-pass filter to a varactor that controls the frequency of the oscillator with respect to the color subcarrier. The subcarrier frequency is controlled by a proportional-control oven housing the crystal and oscillator circuit. The design of the low-pass filter and the inherent frequency stability of the crystal prevent the 1.00693-MHz crystal from being locked to the wrong multiple of the subcarrier frequency.

Once frequency lock has been established, the system becomes a *phase-locked loop*. It operates to hold the leading edge of the sampling pulses at $F_h/2$ coincident with the positive-going zero crossing of the subcarrier cycles. This defined relationship between the F_h pulses and the subcarrier phase is fundamental to this precise synchronization required in the NTSC system. The timing of leading edges of the horizontal sync pulses, equalizing pulses, and vertical serrated pulses is made absolutely precise by forming the leading edges in the same integrated circuit, in each case triggered by a 31.5-kHz pulse selected by digital logic circuits. Pulse widths are controlled by the reset timing. Pulse widths may be timed using either analog or digital circuits.

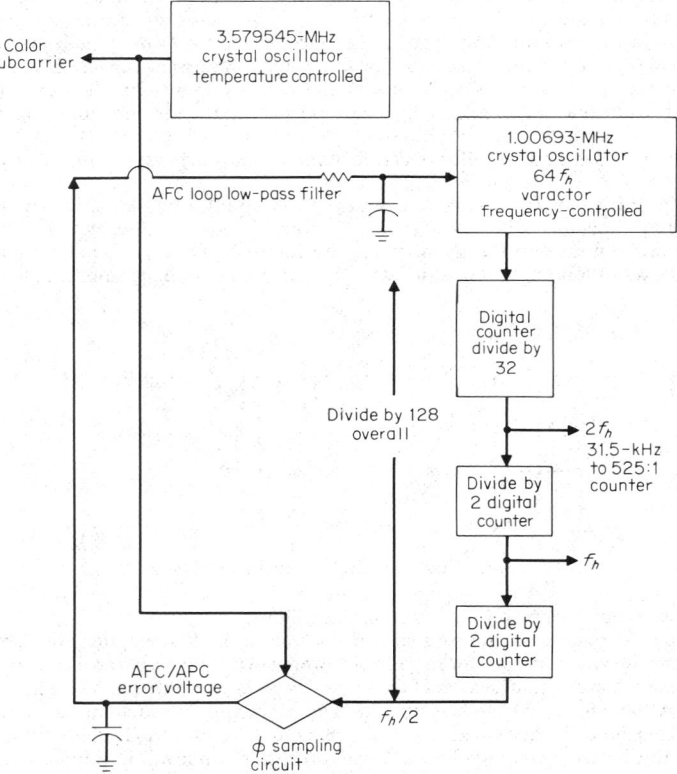

Fig. 20-44. Digital technique for generating NTSC color sync signals.

59. References on Synchronizing Pulse Generators

1. DAVIDOFF, FRANK   The CBS Automatic Color Wire-Lock System, *J. Soc. Motion Pict. Telev. Eng.*, August 1969, Vol. 78, pp. 621–625.

2. KASTELEIN, A.   A New Sine-squared Pulse and Bar Shaping Network, *IEEE Trans. Broadcast.*, December 1970, Vol. BC-16, No. 4, pp. 84–89.

TELEVISION IMAGE-REPRODUCING EQUIPMENT

BY NORMAN W. PARKER

60. Introduction.   In the following paragraphs, the components and circuits employed in monitors and receivers to reproduce monochrome and color images are described. The discussion is limited, in general, to those portions of receivers and monitors which follow the second detector, i.e., sync separation, deflection, subcarrier regeneration, dc restoration, color-signal decoding and matrixing, video amplification, picture tubes, and their auxiliary components. The rf and if aspects of broadcast receivers are treated in Sec. 21.

61. Synchronizing Signal Separation. The synchronizing signals, transmitted at a signal voltage higher than the active picture content, are separated from the picture content by *amplitude separation* or *clipping*. Since the frequency distribution of the vertical and horizontal signals differs widely, it is possible to separate the vertical, horizontal, and color components by filtering.

The method of sync separation most generally used employs an overloaded amplifier to provide selective transmission of the sync tips, in two steps. The first step provides a signal to the sync separator of sufficient magnitude for the input-signal *operating window* to be less than the magnitude of the sync signals. The second step tracks the sync level and keeps the sync signal in the operating window. The latter step is necessary because the operating level at which the sync signals occur varies at the receiver even with the most sophisticated automatic gain control. In most sync separators now in use the grid current or base-emitter current provides diode action in a peak-detector circuit which, combined with the overloaded amplifier, provides plate or collector conduction only during the sync pulses. A typical basic circuit of this type is shown in Fig. 20-45.

62. Typical Clipping Circuits (Amplitude Separation). In Fig. 20-45, the base-emitter diode circuit, together with R_1, R_2, and C_1, forms a peak rectifier circuit in which base-emitter current flows only during the sync-pulse interval. During this interval sufficient base current flows to force the transistor into collector-emitter voltage saturation.

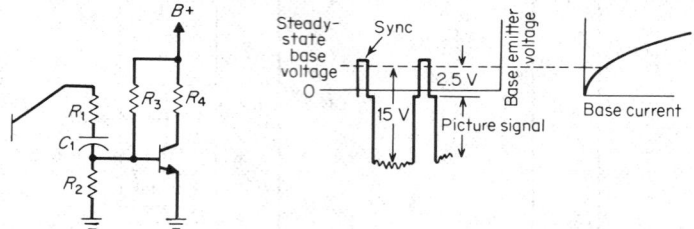

Fig. 20-45.   Transistorized sync separator. *(Motorola, Inc.)*

The essential design considerations are as follows.

1. The video signal at the base should not exceed the base-emitter breakdown voltage. This limits the video voltage, for most low-power general-purpose transistors, to less than 10 to 15 V, since the clamping action of the input circuit applies the total video as a reverse-bias signal. Although the reverse-bias limitation does not apply to vacuum-tube circuits (which usually have much higher video voltages applied to the grid-cathode diode circuit), vacuum-tube circuits have a larger cutoff-to-zero-bias operating window. This requires much higher video drive to obtain satisfactory separation.

2. The charging time constant of R_1C_1 and the saturation impedance of the base emitter diode should be much longer than the charging interval of the horizontal sync pulse, and the discharging time constant $R_1R_2C_1$ should be long compared to a horizontal interval. Maintaining appropriately long time constants establishes constant base current and provides symmetrical horizontal pulses, giving a minimum of tilt on the vertical pulses.

3. The resistor R_1 must be chosen large enough to prevent rapid charging of C_1 on short noise pulses coincident with the sync pulses. However, as R_1 is increased, more of the sync pulse appears across R_1 instead of the base-emitter circuit. Making R_1 too large eventually causes the blanking signals to replace the sync pulses.

4. The value of R_3 is chosen to provide some forward bias so that the separator continues to operate when the video goes to black, and the ac signal applied to the base-emitter circuit may be of the order of 1 V. The slight forward bias also helps prevent a reduction in amplitude of the vertical sync signals due to their increased duty cycles reducing the required charging current. The circuit can be operated with the emitter grounded and without temperature-compensating bias since the base current during the conduction period is many

times the steady collector-base current and the transistor is switched into saturation during sync pulses.

To prevent noise signals from overcharging C_1 during sync intervals, a second RC network is usually included in the charge path for C_1. This uses a small capacitor which charges up rapidly on large pulses. Since it is shunted by a resistor which discharges the capacitor in approximately one line interval, a single charging pulse is prevented from causing large changes in the voltage on C_1.

63. Vertical Sync Separation. The vertical signal is a series of pulses approximately three lines in duration and occurring at the field-repetition rate. This signal has most of its energy in a frequency band below the spectrum of the horizontal sync pulses. The method in most common use provides simple low-pass filtering, typically two sections of an RC low-pass filter, each designed to cut off at 4 kHz. A separator network of this type is usually referred to as a *vertical integrator* since it places the horizontal pulses on an attenuation characteristic which increases at a rate of 6 dB/octave and can be viewed as integrating the individual pulses to derive the vertical sync pulses.

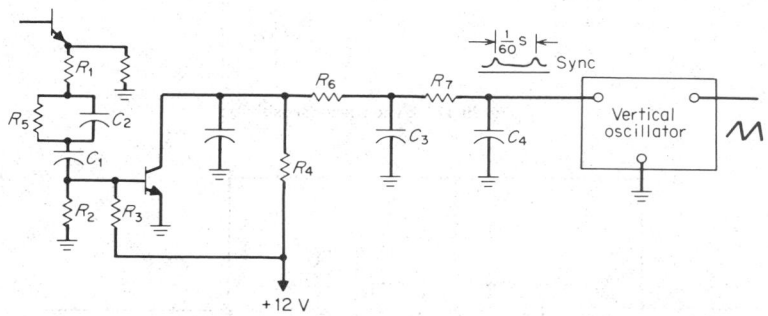

Fig. 20-46.   Vertical sync separator. *(Motorola, Inc.)*

Since two sections of 6 dB/octave filtering are used, the circuit provides double integration of the sync signals. A filter of this type passes the first 10 harmonics of the field-frequency signal without attenuation or significant phase shift. This recovered signal is not a replica of the transmitted vertical-pulse envelope, since the three-line vertical-pulse train represents a little over 1% of the frame interval and the energy distribution of the pulse train extends beyond the first terms of the horizontal-envelope signals, having more than one-half its pulse energy above 6 kHz.

64. Vertical Deflection Oscillators. The directly triggered (multivibrator type) relaxation oscillator provides a simple and convenient means of generating vertical sawtooth voltages which give excellent phase accuracy when impulsively triggered by the integrated sync signal. The output of the multivibrator circuit can be used directly to drive the vertical-sweep power amplifier.

The relaxation type of sweep oscillator has the disadvantage that synchronism is accomplished by shortening the relaxation cycle by the action of the sync pulse. The effect is to require that the free-running frequency of the oscillator be lower than the vertical repetition rate. In the absence of sync signals, the video signal precesses with respect to the oscillator and sweep, which produces a continuously rolling picture on the picture tube screen. As shown in Fig. 20-47, the multivibrator becomes increasingly sensitive to pulse synchronism as the pulse time approaches the point where the oscillator spontaneously changes mode. As a result, the circuit possesses some inherent noise-gating effects.

The time required for the system to reach synchronism when a sync signal is applied at random is determined by the rate of precession of the point where the circuit can be triggered. During this time, the picture rolls slowly until the blanking bar reaches the top of the picture, at which time synchronism occurs. The closer the sync rate is to the free-running frequency, the more slowly the picture rolls when sync has been lost. On the other hand, the more rapid the out-of-sync precession is made, the higher the sync-pulse amplitude required to trigger in the proper phase, which generally makes the vertical jitter more violent in the presence of noise pulses accompanying the sync pulse. The sync-pulse level is often adjusted to satisfy the taste of the designer. A typical vertical oscillator and driver circuit are shown in Fig. 20-48.

65. Horizontal Sync Separation. Direct triggering of horizontal scanning, although used in the earliest television receivers, was abandoned because of the high signal-to-noise ratio required to ensure that the horizontal oscillator was sufficiently free of phase

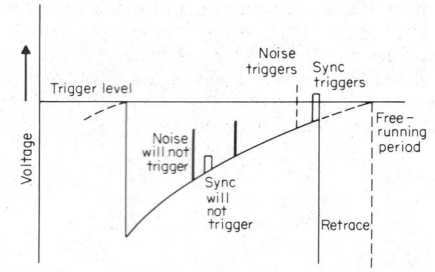

Fig. 20-47.   Sync trigger sensitivity.

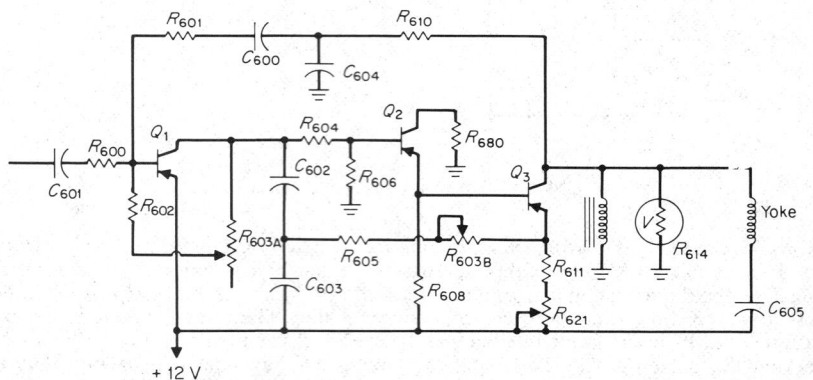

Fig. 20-48.   Typical vertical oscillator and driver. *(Motorola, Inc.)*

modulation. Since the limits of permissible phase modulation are between 3 and 5°, an oscillator with automatic phase control is used universally. The automatic-phase-control loop is characterized by a phase-comparison detector which provides a control voltage proportional to the phase difference between the horizontal oscillator and the horizontal sync signal followed by a low-pass filter that couples the output of the phase comparator to the oscillator. This oscillator is generally of the relaxation type, which can be shifted in frequency by applying a control voltage directly to the oscillator.

66. Typical Horizontal Sync Circuits. A common phase detector is the single-ended balanced diode circuit. It has the advantage of using single-ended sync and reference sawtooth voltages, while at the same time deriving a control voltage free from offset voltages proportional to either input signal.

A circuit of this type is shown in Fig. 20-49. The flyback pulses are integrated to form a 2.5-V peak-to-peak sawtooth having a voltage ramp during retrace. The sawtooth voltage is applied in series with the synchronizing signals, which are much larger than the sawtooth wave (11 V peak to peak). The voltage developed by diode D1 appears across capacitors C501 and C500. When the automatic-phase-control system is in synchronism, the sync pulses are stationary on the retrace ramp of the sawtooth wave; they produce a voltage on

C500 and C501 which is proportional to the phase between the sync signal and the sawtooth wave derived from the sweep. The positive voltage of the junction of D1 and D2 tends to bias D2 into a cutoff condition, and the positive voltage is transferred to C516, C503, and C504 by R500. The negative sync pulses rapidly discharge C516, C503, and C504, keeping the voltage at the automatic-phase-control output at reference ground. The control-voltage variation is limited to the height of the ramp, or ± 1.25 V.

67. Impulse-Noise Protection. The reaction of the sync separator to impulse noise is extremely important in the performance of the deflection oscillators. Impulse noise has two important effects on synchronization: the presence of a noise impulse in the separated sync may cause modulation of the sweep oscillators, and noise impulses higher than sync level may produce temporary errors in the separation levels, causing blocks of sync pulses to be missing. The most critical of these effects is the latter, the loss of sync pulses due to separator charge-up. To prevent this effect it is necessary to ensure that any noise pulses which accompany the sync signal are not appreciably higher in amplitude than the sync signals.

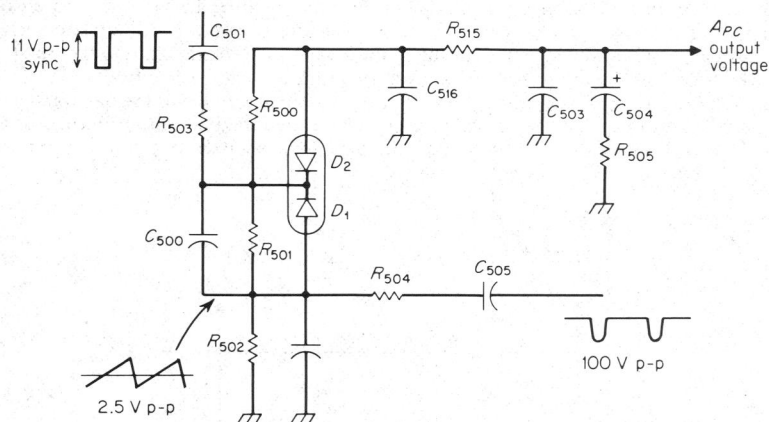

Fig. 20-49. Horizontal automatic-phase-control circuit. *(Motorola, Inc.)*

Noise limiting for this purpose can be accomplished by taking the sync signals from a video stage which is designed to limit at a level above sync tips. In addition, a series resistor and a double-time constant prevent sharp pulses of high amplitude but relatively low energy from shifting the bias level on the separator.

In some receivers the sync separator is gated by the sweep-retrace pulses to exclude noise pulses which occur during the active scanning time. This system, however, can have severe difficulties when the receiver is not synchronized. The sync pulses then appear as sampled fractions of the total sync, causing the system to remain locked out of synchronism. As a consequence, gated sync separators are now rarely used.

A more sophisticated system, used in expensive receivers, employs the principle of noise inversion. This principle uses a distinct separator which is biased to operate above the sync level and which separates the noise pulses and delivers them to a noise-inverting circuit. The inversion circuit amplifies the noise pulses and subtracts them from the sync signal. The resulting sync signal has sync pulses which are serrated by noise but do not have spikes higher than the inversion level. A circuit of a typical noise inverter is shown in Fig. 20-50.

68. Color-Burst Separation. Separation of the color-synchronizing burst requires time gating. The gate requirements are largely determined by the horizontal sync and burst specifications illustrated in Fig. 20-7b. It is essential that all video information be excluded, and it is desirable that both the leading and trailing edges of the burst be passed so that the complementary phase errors, introduced at these points by quadrature distortion, average to zero.

Widening the gate pulse to minimize the required gate-timing accuracy has negligible effect on the noise performance of the reference system and may be beneficial in the presence of echoes. The 1.2-μs spacing between trailing edges of burst and horizontal blanking (Fig. 20-7b) determines the total permissible timing variation. Noise modulation of the gate timing should not be permitted; i.e., noise excursions must not be allowed to encroach upon the burst, since the resulting cross modulation has the effect of increasing the noise power delivered to the chrominance-frequency reference system.

The gate-pulse generator must provide steady-state phase accuracy and reasonable noise immunity. When a high level of chrominance is available, a single-diode disabling gate may be employed, but in some applications it is necessary to remove the burst from the subcarrier before demodulation (burst suppression). The traditional monochrome horizontal-scanning oscillator system meets the noise-immunity requirements and, with some redesign, can approximate the steady-state requirements. Accordingly, the horizontal-flyback pulse is widely used for burst gating.

Although the horizontal-flyback pulse is relatively noise-free, its phase may vary with the adjustment of the horizontal-hold control. The effect of gate-phase variation may be to cause the burst to be clipped, or the gate may slide into the picture; i.e., the picture chroma information tends to serve as the reference phase. For this reason it is desirable to derive a gating signal from a delayed sync pulse which has a fixed time relationship to the burst. A circuit which derives the appropriately delayed sync pulse is shown in Fig. 20-51.

The clipped sync signal is applied to an amplifier which is coupled to the burst-gate stage through a resonant circuit. The resonant frequency of the tuned circuit is adjusted so that the rise time of the translated pulse delays the gate long enough to gate the burst only.

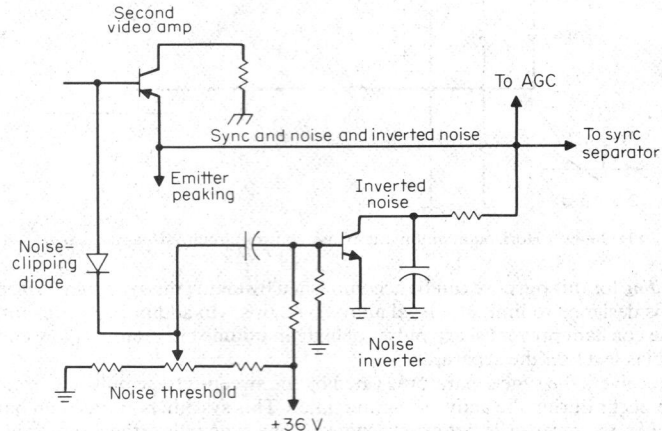

Fig. 20-50. Noise inverter. *(Motorola, Inc.)*

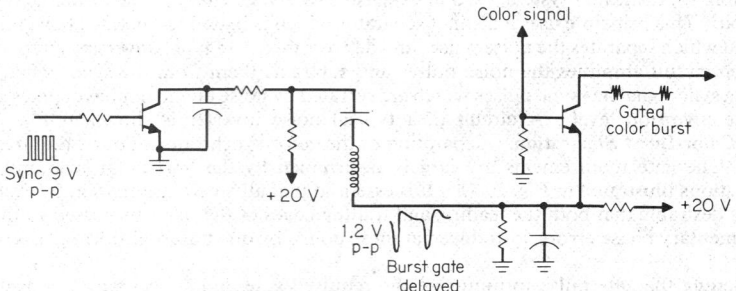

Fig. 20-51. Typical burst gate. *(Motorola, Inc.)*

69. Color Subcarrier Regeneration. The color burst provides a reference phase for correct operation of the synchronous detection process, which in turn restores the color-difference signals to baseband from their quadrature-modulation components in the color subcarrier. The method of recovering an oscillator signal with the correct phase for synchronous detection involves filtering the 3.58 MHz chrominance-subcarrier frequency from the carrier and sidebands that make up the color burst. Two techniques have been used to recover the reference phase oscillator signal. The first uses a narrow-band filter (quartz crystal) which passes the carrier component while substantially attenuating the sidebands of the color burst. The derived carrier component is amplified and limited to give a constant-amplitude reference signal. The second method uses an oscillator in a phase-locked-loop relationship with the burst signal. The phase-locked loop forms an effective filter-limiter combination. The filter width is determined by a low-pass filter in baseband combination with the phase detector and oscillator control element.

While the phase-locked loop was used almost exclusively in early receiver designs, the less complicated *ringing circuit* (crystal filter) is now a popular form of subcarrier regeneration. The ringing circuit may also be used to lock the phase of a free-running oscillator which acts as a regenerative limiter, removing any amplitude modulation which would otherwise result from incomplete attenuation of the sidebands of the burst signal.

In a ringing circuit the burst energy is spread over the entire line period, giving a ringer output voltage of about the average value of the burst level. To maintain a constant amplitude of reference signal it is desirable that the crystal continue to ring over the entire line period, so that a simple limiting amplifier can provide a constant signal. The decay that can be expected from a simple resonant circuit of a given Q, after n cycles of free oscillation, is

$$E_0/E_n = e^{-n\pi/Q}$$

where E_0 is the level of voltage at the beginning of free oscillation and E_n is the level of voltage after n cycles of free oscillation.

At the 3.58-MHz color subcarrier and 63-μs scanning period, the circuit rings freely for about 215 cycles, so that the Q required to maintain 90% of the initial level is about 7,000. To maintain passive circuit Q's of this level requires a crystal filter.

A typical ringing-type subcarrier-regeneration circuit is shown in Fig. 20-52. The burst is applied to a phase-splitter circuit that provides in-phase and reverse-phase burst signals. The reverse-phase burst signal is coupled from the collector and used to cancel the burst signals, through the parallel crystal capacitance. The remaining signal at the junction of the crystal and capacitor C1 is a result of the burst passing through the series resonant circuit made up of the mechanical vibration and piezoelectric effects of the crystal. The low-amplitude continuously ringing signal is amplified in the following transistor stage and coupled to a free-running Colpitts oscillator, which is locked by the amplified ringing signal. The output of the oscillator is a phase-locked reference signal of constant amplitude.

70. Subcarrier Amplitude and Phase Control. Since the color information in the subcarrier contains two separate signals amplitude-modulated on the in-phase and quadra-

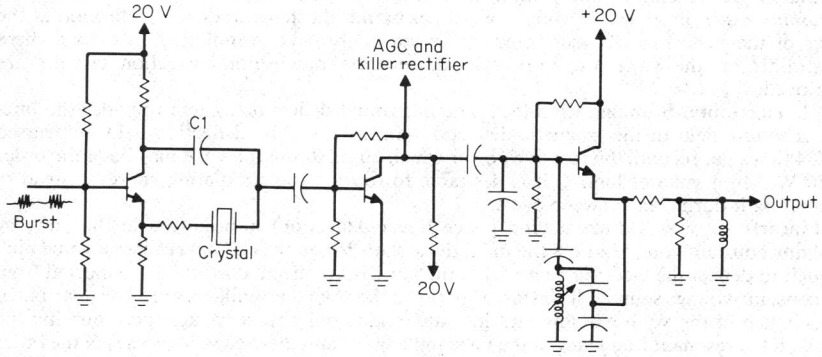

Fig. 20-52. Crystal ringing and subcarrier-oscillator circuit. *(Motorola, Inc.)*

20-51

ture components of the subcarrier, it is necessary to preserve precisely the phase and amplitude relationships of both the lower sideband and upper sideband of the color subcarrier. Otherwise phase and amplitude distortion of the in-phase and quadrature components will produce cross talk between the components.

The subcarrier amplifier has two principal functions: to restore the subcarrier to a form appropriate for decoding by the synchronous subcarrier detectors and to control the amplitude of the subcarrier, to provide control of the luminance-to-chrominance ratio. In the process of detecting the composite luminance and chrominance signal, the color subcarrier appears on the high-frequency slope at the edge of the luminance response curve, resulting in severe distortion of the upper and lower sidebands of the color subcarrier. The response of the color-subcarrier amplifier must be complementary to this distortion to restore the subcarrier to decodable form. Such a complementary characteristic requires critical tuning of the receiver to maintain the match. However, this technique provides the best compromise since the so-called flat-if response, which does not distort the color subcarrier in the detection process, is also sensitive to tuning adjustments. Variations in the picture carrier level in the latter method cause the received signal to be overmodulated with a resulting rectification of the envelope of the color subcarrier in the video detector and a loss of saturation in the reproduced color picture.

The response of the chrominance bandpass amplifier places the color subcarrier on the rising portion of a sharply peaked response. When combined with the response of the video detector, this produces a flat response around the color subcarrier while operating the video detector with a reduced level of chrominance subcarrier. The bandpass amplifier is also designed with adjustable gain so that the level of the chrominance signal can be adjusted to provide the appropriate color saturation in the color picture. The amplifier gain may also be made to vary in response to the nominally constant level of the burst signal, to provide automatic control of color saturation.

The phase of the demodulation axis controls the hue of the reproduced image. Manual control of the hue can be accomplished by shifting the phase of either the color subcarrier or the subcarrier reference signal. Since it is more difficult to shift the phase of the subcarrier without distorting the sideband components, hue shift is usually accomplished by shifting the phase of the subcarrier reference signal. RC, RL, or RLC phase shifting may be used. Where the design allows for large shifts in hue, the phase-shift network may introduce amplitude-level shifts, which should be removed by limiting the subcarrier signal.

71. Monochrome Tube Focus and Deflection. In the past, magnetic and electrostatic methods of focusing the electron beam have been used in monochrome picture tubes, but because of its lower cost and inherent simplicity, the electrostatic method is now used exclusively. See Par. 11-77 for details. In deflection of the beam to form the picture raster, both electrostatic and magnetic methods were used, but magnetic means are now used exclusively, a change made necessary to meet the need for wider viewing area with shorter tube length.

If the center of deflection were identical with the center of curvature of the inside glass surface of the picture tube, a rectilinear raster would be obtained. However, the shape of the faceplate is determined by many other factors, and so some compromise is necessary. The *pincushion raster,* produced in wide-angle tubes by the shortened radius of deflection at the edges of the raster, can be corrected by design of the yoke windings. In color, where constraints on the yoke are more critical, separate pincushion-correction circuits are employed.

72. Horizontal-Scanning Circuits. The horizontal deflection system provides the line-rate scanning field in the magnetic-deflection yoke. Since the deflection field is reversed 15,734 times per second, the reactive power which circulates in the yoke may be of the order of 50 W. In a yoke of high Q it is desirable to recover the circulating energy instead of dissipating it during each sweep cycle.

Linearly increasing current (and hence linear deflection) is produced in the yoke by applying constant voltage across the inductive yoke. When the current reaches a value high enough to deflect the electron beam fully, the yoke is, in effect, abruptly disconnected from the constant-voltage source. The stored energy in the yoke then collapses into the distributed capacitance of the yoke winding and into any additional circuit capacitance shunting the yoke. If the resonant frequency of the yoke inductance and these capacitances is of the order of 60 kHz, the energy in the capacitance has sufficient time during the retrace interval to flow back into the yoke. If the voltage source is then, in effect, reconnected, the energy is returned to the yoke during the initial part of the scanning period, while in the last part of the cycle energy is supplied to the yoke. Such a circuit can thus supply large circulating energy for scanning, with little real power dissipated in the process. This type of deflection system is used in all modern receiver designs.

A transistor horizontal-deflection circuit of this type is shown in Fig. 20-53. Q_1 saturates to place the battery voltage across the autotransformer input terminals, placing a slightly higher, but constant, voltage across the yoke-size coil combination. The current increases linearly in the yoke until the beam reaches the end of the horizontal period, at which time the transistor Q_1 is switched off. The energy stored in the yoke is transferred to the two capacitors and the distributed capacitance by resonance, resulting in a large voltage across

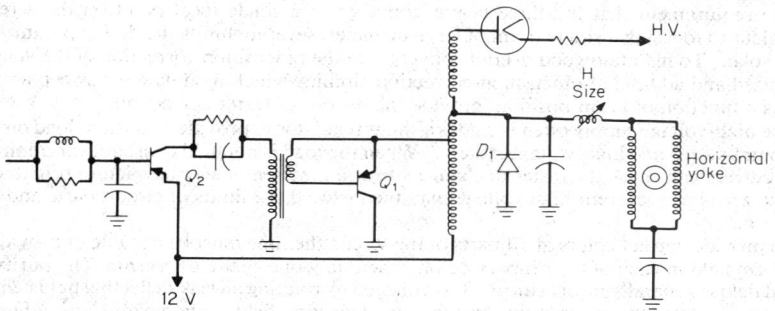

Fig. 20-53. Typical horizontal-deflection system. *(Motorola, Inc.)*

Q_1. As the voltage swings back to zero, the current in the yoke reverses and the diode D_1 conducts, returning the energy taken on the last part of the scanning cycle to the 12-V source. The inductance of the deflection coil is chosen to provide a retrace voltage across Q_1 which does not exceed the breakdown voltage of the transistor while at the same time providing sufficient circulating energy for beam deflection.

The transistor Q_1 is chosen to provide as rapid a switch-off characteristic, from its saturated condition, as possible. Any lag in switch-off causes current in the transistor during the rapidly rising retrace portion of the collector voltage wave, which causes large power losses at the collector of Q_1. The power-supply voltage and yoke impedance are chosen to match the available transistor-switch characteristics.

The output transformer provides a convenient source of obtaining high-voltage power for the picture tube. A power supply of this type has three main advantages: (1) the short retrace-energy exchange results in high pulse voltages which, with a minimum of step-up, can be directly rectified to produce second-anode voltage; (2) since the pulse-repetition rate is high (15,734 Hz line-scanning frequency), high capacitance values of high-voltage filter capacitors, which are both expensive and dangerous when charged, are not required; and (3) the pulse supply is synchronous with the scanning, which avoids interference due to nonsynchronous high-voltage fields.

73. Vertical-Deflection Circuits. Two factors minimize the required scanning power for vertical deflection: (1) the television picture has a 4:3 aspect ratio with the smaller dimension in the vertical direction, which requires a proportionally smaller scanning angle for vertical deflection, and (2) the rate at which reactive energy must be exchanged between the yoke and the source is 1/262.5 of that required for horizontal scanning, with the power reduced accordingly. The result is that vertical-scanning reactive power is insignificant. Instead the sweep system is governed by the resistive characteristics of the yoke, which is driven by a waveform derived from a class A power amplifier. A circuit of this type is shown at the right in Fig. 20-48, where the transistors Q_1, Q_2, and Q_3 form an asymmetrical multivibrator with the capacitors in the collector of Q_1 forming a sawtooth signal which drives the base of the power-amplifier output stage Q_3 with the emitter follower Q_2. The yoke is isolated by C605 to prevent the dc component caused by the operating point of class A operation from shifting the vertical position of the picture. The rate of voltage rise in the yoke is slow enough to provide essentially resistive loading during the vertical-trace period. However, during retrace an inductive pulse distorts the sawtooth wave at the collector of the power amplifier, and the pulse is resistively damped to limit the peak voltage level that the power transistor must withstand, consistent with short retrace time.

74. Color-Tube Focus and Deflection. Essentially all color tubes in current use have three beams for producing the color picture (see Sec. 11, Par. **11-94**). Such tubes use electrostatic focus (no external focus elements are required) and magnetic deflection. The operation of the yoke and deflection circuits is similar to that used in black-and-white television, but larger yoke fields are used to accommodate the three beams and to provide a highly uniform field for each. The second-anode voltage supply is higher for color tubes to provide additional beam power to compensate for the beam energy intercepted by the shadow mask or grille.

The requirement that the three beams converge to a single focal point on the screen, equivalent to focusing a single beam of large diameter, severely limits the design parameters of the yoke. To maintain good overall convergence the pincushion correction of the yoke is minimized and additional pincushion-correction circuits, which modulate the sweep amplitude as a function of beam position, provide the necessary raster correction.

The high-voltage supply often includes a shunt regulator to provide a constant load on the horizontal-sweep and high-voltage system. When the load is not held constant, variations in beam current can cause the raster to change size as a function of average picture brightness, and the zero-beam-current high voltage may then exceed the limits of circuit parts and the color tube.

To provide correct colors at all parts of the screen the tube must be capable of providing a uniform field in each of the primary colors when only one beam is present. The purity of the red field is generally most critical. It is adjusted by rotating a weak deflecting field behind the yoke to center the three beams in the yoke-deflecting field. The yoke is also adjusted axially so that the beams arrive at the shadow mask from the proper center of deflection and only the intended phosphor is excited.

To assure that the color picture has a minimum of misregistration (color fringing) the three beams must be *converged* so that they impinge together at all parts of the screen. Static convergence is accomplished by adjusting separate magnets, located so as to affect each beam separately and thus to provide convergence at the center of the picture. In addition, dynamic electromagnetic fields are applied to each beam to ensure convergence at the edges of the picture. The dynamic-field waveforms are derived from the deflection voltages. They control the strength of the convergence action appropriately at all parts of the screen. The components on the color-tube neck are shown in Fig. 20-54.

75. Trinitron Color Tube. The Trinitron is an in-line three-gun tube using electrostatic convergence. The Trinitron differs from the three-gun shadow-mask color tubes in both the screen and the gun. The three in-line closely spaced electron beams are derived from a single-barrel electron lens. A grid structure having three holes, one for each beam, is used, and the grids are connected together. The three separate cathodes are separately driven by the *R, G,* and *B* signals. The Trinitron uses low-voltage electrostatic focus (Einzel lens). The focus is adjusted in a manner commonly used in monochrome tubes using Einzel lens focusing; i.e., the focus electrode is returned to any of several low-voltage points. By observing the overall focus, the proper fixed voltage is chosen.

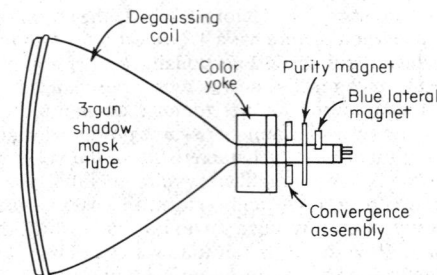

Fig. 20-54. Shadow-mask color-picture-tube assembly.

The in-line horizontal-beam arrangement must be maintained as the beam passes through the yoke to the screen if purity and convergence are to be maintained. The fringe fields of the deflection yoke tend to twist the beam since they contain components parallel to the tube axis. The twisting effect is counteracted by a coil on the neck carrying yoke current which produces a countertwist and maintains the horizontal beam alignment as the beam passes to the screen.

The three beams emerging from the spaced grid apertures tend to produce three horizontally spaced spots on the screen. However, the outer beams pass through a pair of deflection plates similar to those used in electrostatically deflected tubes. When the proper voltage is applied, these plates deflect the outer beams to converge with the center beam. To maintain convergence over the tube face, the voltage on the convergence deflection plates is modulated by the scan voltages as the beam is deflected. Figure 20-55 shows gun elements and circuits necessary to operate the Trinitron tube.

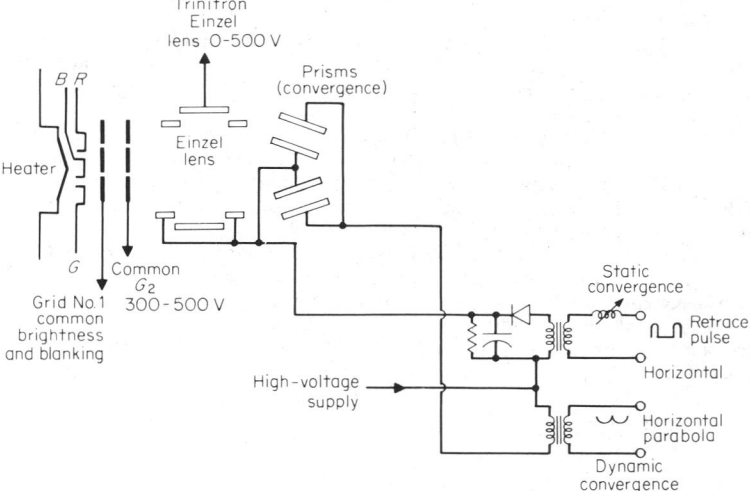

Fig. 20-55.   Trinitron gun and convergence circuit.

Although the Trinitron gun principle can be used with a dot screen or a line screen, current Trinitron tubes use a line screen. An aperture grille, a curved plate of formed slots, provides the shadow apertures for the line screen.

76. Monochrome Video Amplifiers. A typical television receiver has its signal amplification distributed in three frequency ranges. The tuner, with modest gain, raises the low-level rf signal by 20 to 50 times; the if amplifier, operating at a fixed frequency, raises the signal to several volts; an envelope detector reduces the signal to baseband; finally the video amplifier raises the signal to from 50 to 100 V to drive the picture tube. To minimize the danger of regeneration it is desirable to distribute the gain in these sections as equally as possible, but it is difficult and expensive to increase the gain in the tuner. Since the envelope detector uses a diode which requires signals of at least 1 V to provide reasonable linearity, the largest portion of the gain falls to the if amplifier. Tuners and if amplifiers are treated in Sec. 21.

The video amplifier may be used to drive the picture tube in either a grounded-grid or grounded-cathode configuration. When the tube is driven in the grounded-cathode circuit, the effects of beam current do not load the video amplifier and the loading is caused only by the capacitive input impedance of the picture tube. With grounded-cathode drive, using a vacuum tube or *npn*-transistor video-output stage, the operating point of the video stage is near cutoff when the picture tube approaches zero bias (maximum white). Since the drive requirements for the picture tube represent an appreciable portion of the power-supply voltage, the video stage is not linear over the entire range of the output-voltage change. The signal is compressed in the region of cutoff or (for grounded-cathode drive) in the highlights of the picture. At the same time, the picture-tube beam current shunts the video amplifier, tending to minimize white stretching. In this circuit the drive voltage is also effectively applied to the screen grid of the picture tube which reduces the curvature and gamma of the picture tube, resulting in higher average brightnesses for the same contrast ratio.

Comparisons of the two drive systems show that the cathode-drive circuit provides a brighter picture with less compression of the whites than the grid drive does. Most monochrome receivers use cathode drive.

The video amplifier has adjustable gain to provide customer control of contrast ratio. This is usually done by using a variable degeneration control in the emitter circuit (or cathode of a vacuum-tube circuit). A typical monochrome video amplifier is shown in Fig. 20-56.

77. DC Restoration. Restoration of the dc component of the video signal can be accomplished in several ways, and it may be omitted entirely for the sake of simplicity of design and low cost. In the latter case, the manual brightness control is used to set the average signal level at the preferred setting.

The simple diode dc restorer shown in Fig. 20-57 is most frequently used. It maintains black level by reinserting the dc component on the sync tips, using the relative values required by signal standards. This type of restorer is very susceptible to error because of noise impulses and is not accurate.

Clamping circuits, although more accurate, faster-acting, and less susceptible to noise, are infrequently used in receivers because of their greater complexity and cost. In the usual circuit, the signal is sampled during the back porch to obtain an output which will bring the level of the sync tips to the correct value. This type of circuit is immune to noise except that present during the sampling interval.

78. Color Decoding and Matrixing Techniques. The color signals corresponding to red, blue, and green in the camera, that is, E_R, E_B, and E_G, are coded before being transmitted. The received signal is in three parts, E_Y, the luminance information, and E_I and E_Q, which must be transformed back into the primary signals. Usually this is done in two steps, first, deriving color-difference signals, $E_Y - E_R$, $E_Y - E_B$, and $E_Y - E_G$, then regaining E_R, E_B, and E_G.

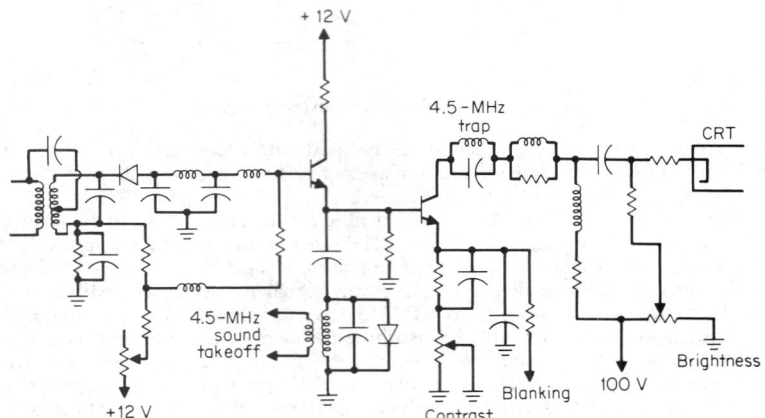

Fig. 20-56. Monochrome video amplifier. *(Motorola, Inc.)*

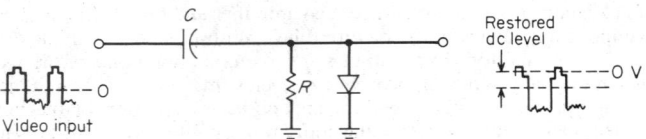

Fig. 20-57. Basic diode dc restoration circuit.

The E_I and E_Q components of the chrominance information are demodulated by synchronous detectors referenced to the phase of the original subcarrier upon which they are modulated in transmission. The luminance and chrominance signals may be recombined in matrix networks to obtain the primary signals, or the composite signal may be directly demodulated.

Four arrangements are available, depending on the combination of the following choices: (1) *IQ* vs. equiband demodulation; (2) picture-tube vs. pre-picture-tube chrominance and luminance matrixing.

The *IQ* receiver design utilizes two signals in quadrature. The *I* signal lags 57° behind the color-burst phase; the *Q* signal lags by 147°. The required bandwidth for the *I* signal is 1.25 MHz and for *Q*, 0.5 MHz. To compensate for the different delays associated with the different bandwidths, a time-delay circuit is necessary in the *I* channel. Following synchronous demodulation (Par. **20-79**) along the two axes to obtain E_I and E_Q, matrixing is necessary to obtain the color-difference signals. Both polarities of the *I* and *Q* signals are necessary for this matrix operation.

Almost all current color receivers employ *equal-bandwidth* designs, using color demodulators which decode the color information along the *RY* and *BY* axes, with the demodulation angles shifted sufficiently to provide acceptable color reproduction.

Vacuum-tube equiband receivers, using matrix recombination to derive the primary signals, employ *picture-tube recombination* of the luminance and chrominance signals, while receivers using solid-state elements generally employ *pre-picture-tube recombination* to provide R, B, and G signals for the picture tube. Picture-tube recombination is generally less complicated, but the color-difference voltages must often exceed twice the luminance voltage, and video-output stages for color must be capable of more than 200 V of video drive. To avoid the use of high-voltage-rating transistors, at the expense of increased complexity, transistor circuits generally use video-output stages that form the R, G, and B drive voltages before application to the picture tube.

79. Synchronous Detection. The synchronous detector is a means of obtaining a vector product of a reference signal (which defines the axis along which the color signal is detected) and the color subcarrier.

$$E_r \cos \omega t \, E_{ct} \cos (\omega t + \phi_c) = E_r E_{ct} \cos \phi_c = (E_B' - E_Y')/2.03$$
$$E_r \sin \omega t \, E_{ct} \cos (\omega t + \phi_c) = E_r E_{ct} \sin \phi_c = (E_R' - E_Y')/1.14$$

The color signal $E_{ct} \cos (\omega t - \phi)$ can be derived from the color-difference signals $E_R - E_Y$ and $E_B - E_Y$, where E_{ct} represents the instantaneous value of color saturation and ϕ_c represents the instantaneous phase of the subcarrier (proportional to the hue of the reproduced image). $E_R - E_Y$ and $E_B - E_Y$ can be derived from the subcarrier by multiplication with unit-amplitude sine and cosine terms.

Large-Signal Demodulation. There are two ways of performing the electrical equivalent of the multiplication process. In the first method a large signal of reference phase is added to the subcarrier color signal. When the reference signal greatly exceeds the value of the color signal, the components which are in phase with the reference signal add directly to its magnitude, while the quadrature components add as the square root of the sum of the squares, which remains constant when one signal dominates. When the combined signal is applied to an envelope detector, the output is proportional to the modulation in the color subcarrier along the reference axis. A balanced-diode synchronous detector using this technique is shown in Fig. 20-58. By using diodes oppositely poled and with the reference signal of the same phase applied to both while opposite-phase color signals are applied to each, the signals from the subcarrier are recovered without an offset voltage from the rectified reference signal.

Product Demodulation. A second method consists of producing a current proportional to one of the signals, i.e., either the color subcarrier or the reference signal, and modulating the developed current with the second signal. When a vacuum-tube circuit is used, it is customary to use a tube designed for mixer service. The cathode current is modulated by grid 1 with the color-subcarrier signal. Cathode current flows under the influence of the screen electrode, which is held at constant voltage to a point preceding the second control grid, where a virtual cathode is formed. Grid 3, to which the reference voltage is applied, modulates the current which flows past grid 3 to the plate circuit. The plate current is then the product of the voltages applied to grids 1 and 3. A circuit of a vacuum-tube synchronous-detector circuit is shown in Fig. 20-59.

In solid-state circuits, since suitable multicontrol-element devices are not readily available, their function is simulated by interconnecting transistors with single control elements. A widely used circuit is shown in Fig. 20-60, where a current which is a function of the chroma voltage is generated by TR_1. The current from TR_1 provides a constant-current supply for transistors TR_2 and TR_3, which are connected as a differential amplifier. The reference oscillator alternately switches the current from TR_2 to TR_3. The switched output of TR_2 contains the demodulated color signal. To avoid the effects of the switching voltage, the circuits are usually formed in full-wave pairs. An example of a complete solid-state color synchronous detector is shown in Fig. 20-61.

80. Matrix Circuits. Figure 20-62 shows a block diagram of a decoder using *picture-tube matrixing* to obtain the *R, G,* and *B* signals. The synchronous detector may be any of the types previously discussed. The oscillator is similar to the subcarrier-regeneration system described in Par. 20-69. The phase-shift network is used to obtain the desired axis of demodulation. The matrix provides a resistive mixing of $R-Y$ and $B-Y$ to obtain $G-Y$.

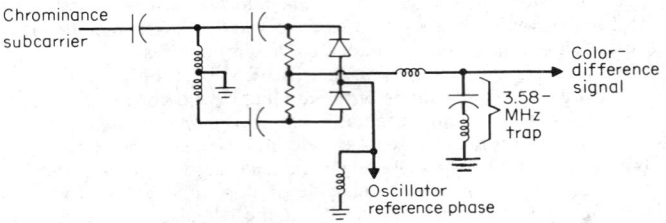

Fig. 20-58. Diode synchronous detector.

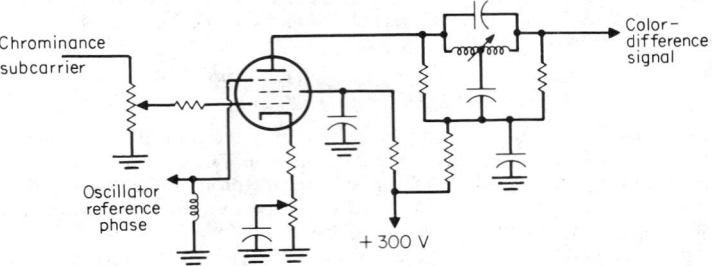

Fig. 20-59. Vacuum-tube form of synchronous detector.

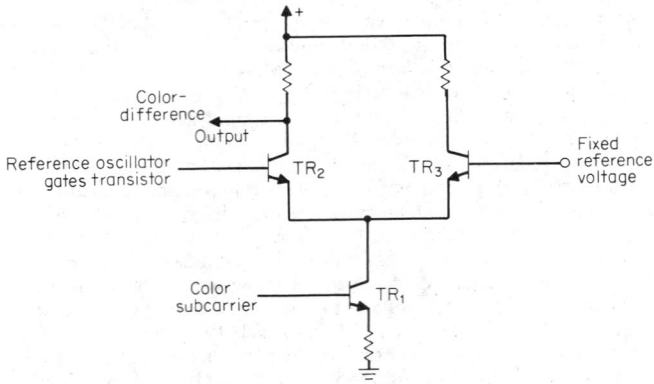

Fig. 20-60. Transistor synchronous detector.

20-58

Since $G-Y$ is made up of reverse-polarity components of $R-Y$ and $B-Y$, the $G-Y$ video amplifier provides one less phase reversal than the others. The color-difference signals are applied to the grids of the picture tube while the luminance signal is applied to the cathodes. The composite voltage applied between grid and cathode in each case is R, G, and B.

Figure 20-63 shows a block diagram of a decoder using *pre-picture-tube* matrixing. The

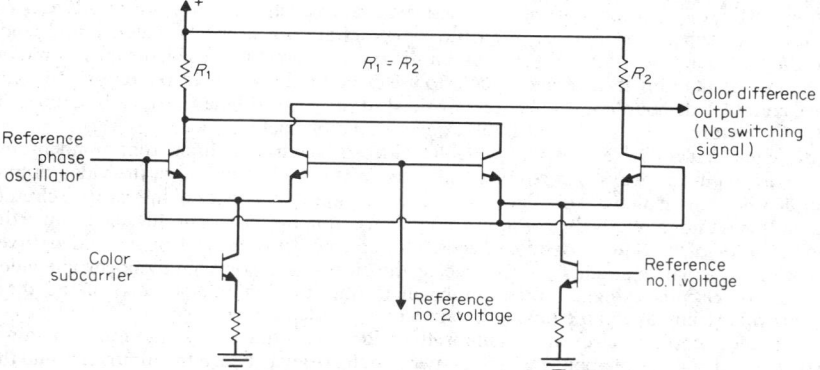

Fig. 20-61. Full-wave transistor synchronous detector.

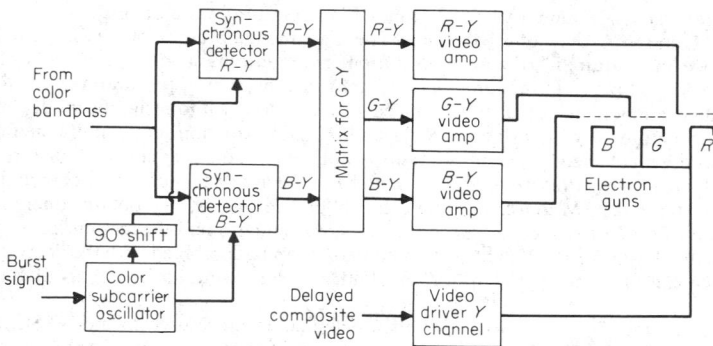

Fig. 20-62. Color-decoding circuit using picture-tube matrixing.

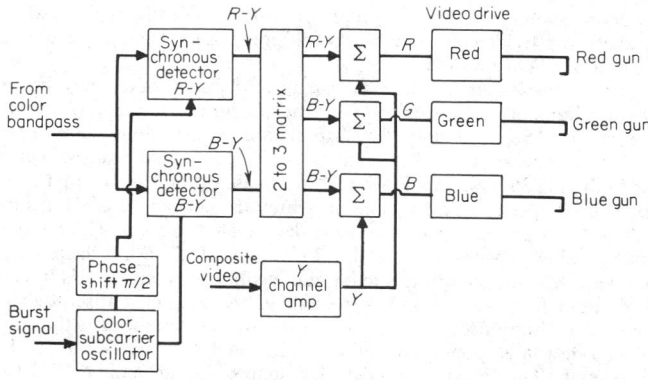

Fig. 20-63. Color decoding using pre-picture-tube matrixing.

functions are essentially the same except that a separate $G - Y$ detector is often used to replace the matrix method of $G - Y$ derivation and separate adders are used to combine the Y signal with the color-difference signals before application to the video-output driver stages.

81. DC Restoration for Color. Dc restoration has special significance for color reproduction since the control of the average brightness applies to each of the primary-color components of the picture. Since color scenes may produce differing video signals for each primary color, the average value of brightness may vary widely on each primary. If each dc value is not properly maintained, the primary colors are not properly matched and color errors are introduced. Although dc clamping circuits are sometimes used, the most common method of maintaining the average value is to use dc coupling in the video circuits. In some cases, complete dc coupling is purposely avoided to produce a higher average brightness in the color picture. This is done at some expense to color fidelity.

82. Color Video-Drive Amplifiers. Color video-drive circuits differ from monochrome drivers in significant ways. In circuits which use picture-tube matrixing, the video-output stage drives all three picture-tube cathodes. The low-input impedance of the cathode circuits in parallel requires a relatively low source impedance from the video-output stage. Since the cutoff in color-picture tubes usually exceeds 100 V, the power required from the video driver in picture-tube matrix circuits greatly exceeds the video-drive power in monochrome video drivers. In circuits using pre-picture-tube matrixing, the loading on each of the three videodrivers is similar to an equivalent monochrome video driver.

The video amplifier circuits containing the wideband luminance signal must contain a delay line with a delay of about 1 μs, to compensate the delay between the luminance and the chrominance signals introduced by the bandpass filter. The bandwidth of the luminance channel is a compromise between good resolution and the visibility of the subcarrier. If the bandwidth is extended to provide full resolution, the visible presence of the subcarrier will introduce erroneous luminance information, which desaturates the colors.

83. PAL and SECAM Color Systems. The fundamental premise of the SECAM system is that since the required horizontal resolution in the chrominance components of color pictures is not as high as in the luminance component, the *vertical* resolution devoted to the chrominance components can also be reduced, e.g., by 2 to 1, without significant degradation of the color picture. By imposing the $R - Y$ and $B - Y$ information sequentially on alternate lines, the problem of encoding is substantially simplified; i.e., the carrier is no longer required to be modulated in quadrature by the $B - Y$ and $R - Y$ components of the color signal. The SECAM system uses FM to transmit the color-difference signals. To obtain coherent color signals in a SECAM receiver it is necessary to store the chroma information for a full line period, so that when the $B - Y$ signal is transmitted, it can be combined with the delayed $R - Y$ signal from the previously transmitted line and thus form simultaneously available color signals.

The video amplifier and matrix circuits are similar to those used in the NTSC system. When a shadow-mask tube is used, the high-voltage and convergence circuits are also similar. The chroma-detector circuits differ in that the synchronous detectors are replaced with a single discriminator circuit which converts the FM subcarrier into sequential $R - Y$ and $B - Y$ signals. A reversing switch activated by color-sync signals selectively alternates the direct and delayed signals to provide continuous color-difference signals. Figure 20-64 shows the block diagram of a SECAM decoder.

The PAL system uses alternate line averaging and line-period delay but avoids the problems of the FM transmission system by using the quadrature-modulation method, similar to the NTSC system. The $R - Y$ and $B - Y$ signals are simultaneous pairs of components transmitted as amplitude-modulated sidebands of a pair of suppressed subcarriers in quadrature, as in the NTSC system. However, the phase of the $R - Y$ signal is reversed on alternate lines. Unlike the SECAM system, in which the subcarrier is always present, the subcarrier of the PAL system disappears on fully desaturated signals. The phase reversal of the $R - Y$ signal on alternate lines causes the $R - Y$ signal to lose interlace with the $B - Y$ signal. Hence the $R - Y$ dot pattern has maximum visibility when $B - Y$ is interlaced at an odd multiple of one-half the line-scanning frequency. As a compromise, the subcarrier is chosen at one-quarter line offset, i.e., at an odd multiple of one-fourth the line-scanning frequency. This compromise does not provide as accurate color interlace as can be obtained with the NTSC signal. The decoder for the PAL system is similar to the NTSC decoder with the addition of a one-line delay and a reversing switch activated by the color-burst signal. A block diagram of a PAL type decoder is shown in Fig. 20-65. The output of the delay line is added to the direct signal to obtain the subcarrier components of $B-Y$ only. The output

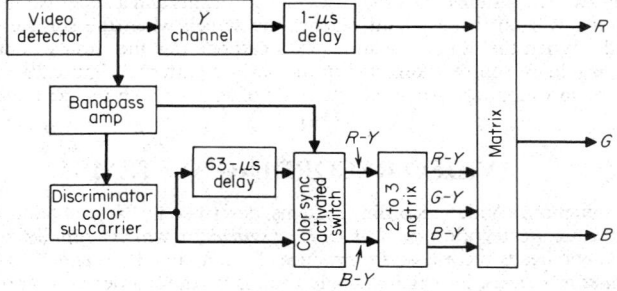

Fig. 20-64. Decoder for SECAM system.

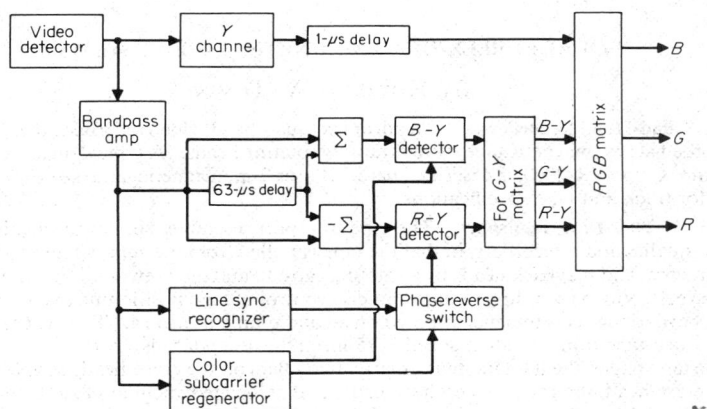

Fig. 20-65. Decoder for PAL system.

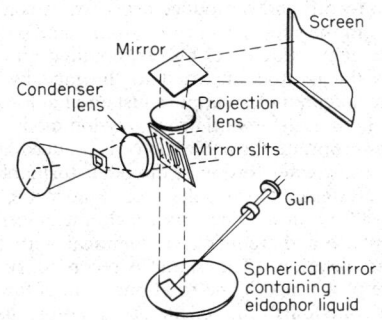

Fig. 20-66. Raster formation in Eidophor system. *(After Baumann, J. Soc. Motion Pict. Telev. Eng.)*

of the delay line is subtracted from the direct signal to obtain $R-Y$ free of $B-Y$ components. The reinserted color subcarrier is reversed on alternate lines to provide continuous, properly phased $R-Y$ signals. The $R-Y$, $B-Y$, and Y signals are matrixed to provide equivalent R, G, and B signals.

84. Eidophor Systems. A number of systems have been devised for large-screen projection of television pictures, among them the Eidophor system. This system relies on a liquid, controlled in such a way as to produce a raster on a screen of the desired size. One method uses two slit systems; the other uses one set of slits and a spherical mirror. In either case, the setup is such that no light reaches the screen when the surface of the liquid is undisturbed. When the liquid is scanned by a cathode ray, the surface is deformed so that light from an intense source is directed in the desired pattern. Figure 20-66 is a schematic illustration of an Eidophor system using one set of slits for both incident and reflected rays.

VIDEO RECORDING SYSTEMS

85. Introduction. Video recording systems, described in Pars. **20-86** to **20-110**, fall into the following major categories: *quadruplex-head systems,* in which a rapidly rotating assembly of four magnetic heads records and reproduces from a laterally scanned track on magnetic tape; *helical scan systems,* in which a single head is used; recorder and reproducer systems, using *magnetic cartridges or cassettes; electron-beam (EVR)* recording systems; *holographic* recording and reproducing systems and *video disk systems.*

QUADRUPLEX-HEAD MAGNETIC-TAPE SYSTEMS

By Howard W. Town

86. Quad-Head Recorders. In quadruplex (quad-head) video recorders, the video signal is recorded across magnetic tape by four rapidly rotating heads. A typical quad-head system, the Ampex type VR-2000, is described here. It uses 2-in. magnetic tape (see Figs. 20-69 and 20-70 for track and tape specifications).

87. VR-2000 Tape Transport. The tape transport resembles similar mechanisms found in high-quality audio recorders, in that the tape is pulled from the supply reel and across the erase, record, and playback heads by a rotating capstan and then rewound by a motor-driven takeup reel. Video recording requires precise control of tape positioning and of head drum phase, beyond the requirements imposed on an audio-tape transport. The primary nominal linear video-tape transportation speed is 15 in./s; the secondary is 7.5 in./s.

The tape passes the left tape-tension arm, two idlers on the erase-head assembly, the full-width erase head, the rotary video-head drum and associated vacuum block tape guide, the control track record-reproduce head, and the audio and cue erase-record-reproduce heads before passing the capstan and its associated pressure idler.

Tape motion in the record and reproduce processes is controlled solely by the capstan, which pulls the tape from the supply reel through the tape path and drives the tape toward the takeup reel on the right. Each reel hub is mounted on a torque motor. The supply torque motor opposes the motion imparted to the tape by the capstan, thus providing holdback tension. The takeup motor supplies just enough torque to take up the tape as it is supplied by the capstan. In fast-forward or fast-rewind modes, the capstan idler is removed and the torque of the appropriate turntable motor is reduced, allowing the tape to be pulled from that turntable by the greater torque of the other turntable motor.

88. Servo Control System. During the recording mode, the video head-drum motor rotates at 240 Hz (14,400 r/min). The drum tachometer generates a square-wave signal whose instantaneous phase and frequency is identical with the instantaneous phase and frequency of head-drum rotation. This signal is recorded longitudinally along the control track at the lower edge of the tape. The rotational rate of the capstan motor is determined by the drive frequency, one-fourth that of the drum motor, that is, 60-Hz rate.

During the playing mode, the video head drum again rotates at the 240-Hz rate and the drum tachometer generates a square-wave signal indicative of the instantaneous head-drum phase and frequency. The latter signal is phase-compared with the 240-Hz-reproduced control track. The resulting error signal is used to establish and maintain the exact relationship between the head-drum phase position and the longitudinal position of the tape that existed during recording. Table 20-4 shows the relationship of recorder-system factors, depending on the television standard in use.

The head-drum servo employs a phase servo loop and a velocity servo loop. The phase

Table 20-4

Factor	Television standard		
	525/30	625/25	405/25
Power-line frequency, Hz	60	50	50
Frames/second	30	25	25
Vertical rate, fields/second	60	50	50
Horizontal rate, pulses/second	15,750	15,625	10,125
Capstan-motor power frequency, Hz	60	62.5	62.5
Drum-motor power frequency, Hz	240	250	250
Drum rate, motor r/min	14,400	15,000	15,000
Nominal tape speed, in./s	15	15.625	15.625

servo loop locks the head-drum synchronous motor to the video signal being recorded, the station reference sync, or the power line, depending on the mode of operation. The velocity servo loop suppresses the tendency of the synchronous motor to hunt. In the recording mode, the equipment is normally locked to the video signal; in the reproducing (playing) mode, it may be locked to the power-line frequency or to station reference sync.

For the corrections, the phase servo loop uses the error signal resulting from phase comparison of the drum tachometer signal with the power-line frequency or with vertical sync, depending upon whether the operating mode is in record or play. If the phase of the incoming video signal suddenly shifts, which can occur during signal discontinuity or when the program source is switched, the phase comparator generates an error voltage which acts to gradually shift the frequency generated by the voltage-controlled drum oscillator until the drum phase matches the phase of the incoming sync.

89. Frame and Line Locks. Switching to and from the recorder can produce vertical-image disturbances and video-signal discontinuities at the studio output. To avoid these effects, two types of servo locking are employed. In *frame locking*, the recorder is made sufficiently synchronous to ensure that vertical transfer switching will produce no visible jump in the image. This can be accomplished by a drum servo lock, whose error should not exceed approximately half the line scan, that is, ± 30 μs.

The more complex problem is to achieve full *line lock*. For this purpose the signals recorded must be within $1/10$ μs accuracy, relative to the studio reference signals. This is accomplished by the use of phase modulation of the drum motor, allowing its instantaneous position to be corrected on a line-by-line basis, as shown in Fig. 20-67.

An early technique used multiphase sine-wave signals to drive the capstan and drum motors, but there were inherent defects in this approach. The phase splitting, accomplished through Scott wound transformers, did not always produce precise 120° displacement of the three phases needed for the drum motor. The individual amplitudes of these three phases were difficult to maintain in perfect balance, and the time required for phase correction was limited by the low-pass filtering action of the power transformers.

The current approach to this problem is to use transistor power-switching circuits which gate on at the designated times so that phase-to-phase amplitude balance is very precise. The 120° phase displacement is accurately maintained by logic circuitry and binary countdowns, and the holding response to phase modulation is virtually instantaneous (see Fig. 20-68).

Phase shifts can be accomplished by altering the switching time of the binaries. The drum-motor windings are driven by square waves which add algebraically between any two terminals and result in a staircase waveform with almost complete attenuation of the odd harmonics. This is a particularly desirable condition, since the smoothness of angular velocity depends upon precise equality of the angularity of pulses applied to the rotor, on whose shaft the video head drum is mounted.

The use of these methods in driving the rotary head assembly has resulted in a long time-base stability in the order of 200 ns or better when the signals from video-tape playbacks are referenced to a stable studio source.

90. Intersync Servo Control System. Magnetic-tape recording and reproduction of television signals involves the scanning of a moving tape by four video heads mounted in precise quadrature on the periphery of a rapidly rotating head drum. The longitudinal movement of the tape is synchronized with the rotation of the head drum by the action of two

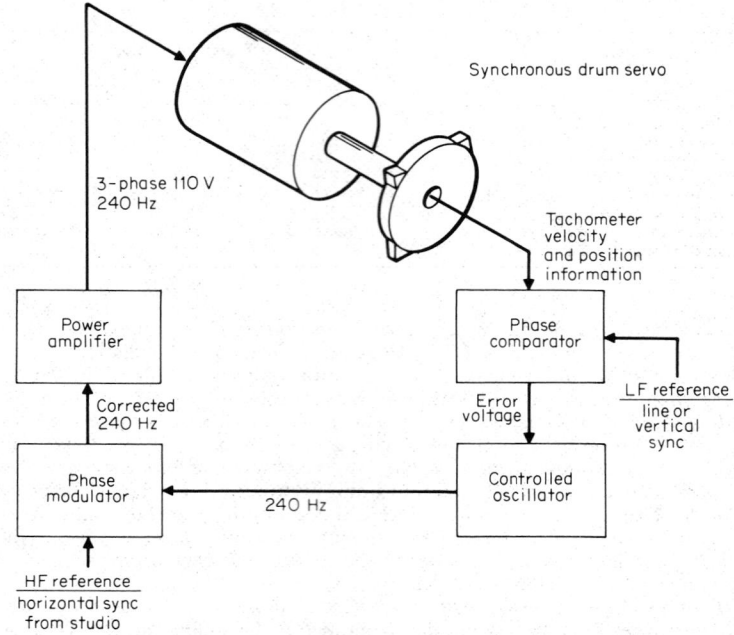

Fig. 20-67. Synchronous drum servo for quad-head magnetic recorder.

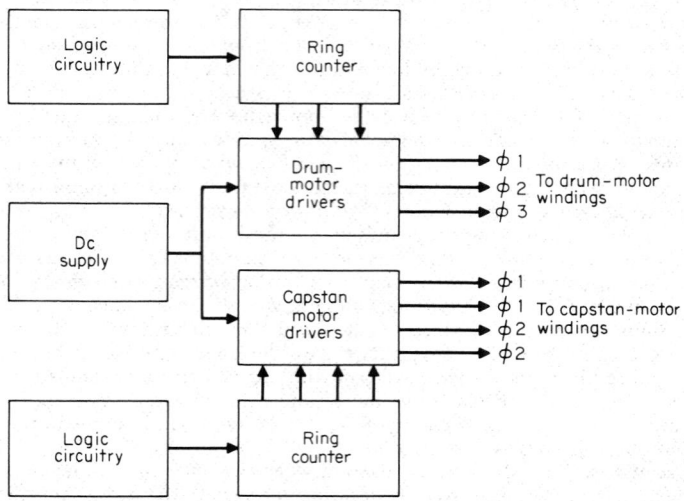

Fig. 20-68. Diagram of 2- and 3-phase power-switching circuits.

mutually synchronized electronic servos. In the basic system, the tape speed is controlled by the action of the capstan servo system; the head-drum rotation is governed by the head-drum servo. By precisely controlling both servos, the intersync servo system provides improved stability of the reproduced video signal.

The vertical synchronizing pulses are recorded midway across the area occupied by the video information and recur on every sixteenth path of video information recorded across the width of the tape. The vertical pulses that initiate the first field scan of each frame coincide in time with a 30-pulse-per-second frame pulse that is recorded simultaneously on the longitudinal control track at the lower edge of the tape. The frame pulse recorded on the control track always appears on the tape in the same relative position with respect to the corresponding vertical sync information, which is recorded as a part of the composite video signal.

The vertical synchronizing information that initiates a new frame is preceded by a half line of video information; the similar information that appears on the sixteenth path following completes the frame and is preceded by a full line of video information. Thus, complete frames are presented 30 times per second. Figure 20-69 shows the vertical synchronizing

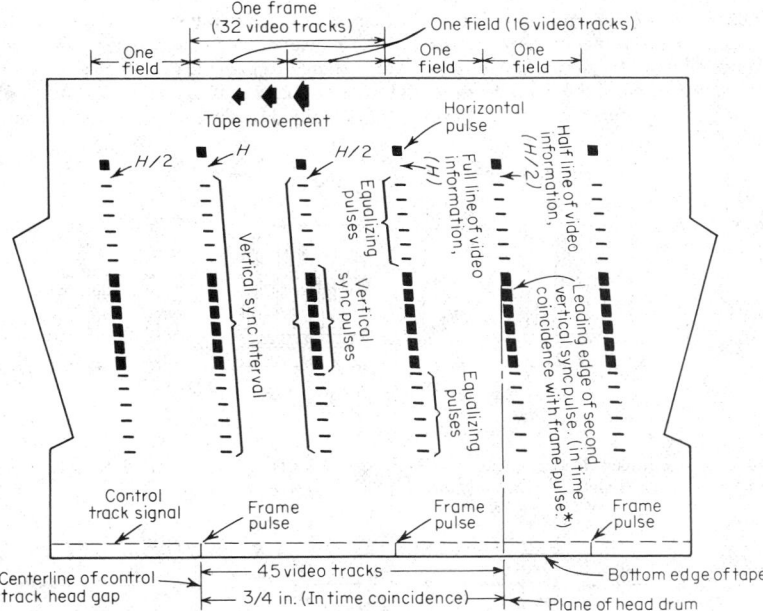

Fig. 20-69. Position of synchronizing information on 2-in. video tape. (Not to scale.)

information which appears at the middle of every sixteenth transverse path and includes (in sequence) six equalizing pulses, six vertical sync pulses, and six equalizing pulses. The recorded 240-Hz control-track signal and the superimposed 30-pulse-per-second control-track frame pulses appear at the bottom edge of the tape. Figure 20-69 also shows that the stationary control-track record-reproduce head is located approximately 3/4 in. "downstream" from the plane of the video heads and is mounted on the video-head assembly. It records and reproduces the 30-pulse-per-second frame pulses in addition to the 240-Hz control-track signal. Because of its downstream location, the vertical sync information associated with each frame pulse is 45 video tracks upstream from it. SMPTE Recommended Practice RP-16, Specifications of Tracking Control Record for 2-in. Video Magnetic Recordings, gives details*. Figure 20-70 shows the recording pattern standardized for quad-head recorders.

---

* Obtainable from the Society of Motion Picture and Television Engineers, 9 East 41st St., New York, N.Y. 10017.

91. Limitation of Timing Errors. Timing errors from the following causes are the most common.

1. Momentary changes of head-drum velocity caused by variations of head-drum loading.
2. Momentary phase transients in the reference signal or the video signal.
3. Momentary loss or degradation of horizontal or vertical sync.
4. Instability of the reference signal caused by a malfunction of the plant sync-generator system.

The principal function of the intersync servo system is the maintenance of synchronization between the reproduced video signal and other video sources to a degree permitting program switching between these sources without interruption of picture continuity. The intersync servo unit accomplishes this by locking the reproduced signal to the plant sync reference, to which the other video signals are locked, and by holding the timing errors in the reproduced signal within very narrow limits. Figure 20-71 shows the intersync servo system in block-diagram form.

92. Servo Record-Mode Functions. During the record mode, the intersync servo unit provides two closed servo loops that control the head-drum rotation and maintain maximum timing stability of the recorded signal.

The *positional loop*, shown in Fig. 20-72, determines the average rotational rate of the head drum and the instantaneous angular position (or phase) of the video heads with respect to the reference signal. Its gain and bandwidth are designed to provide a "soft" servo control that resists the reaction of the head drum to high rate-of-change disturbances that may appear in the video or the reference signal. Control is maintained by the phase, i.e., time, comparison

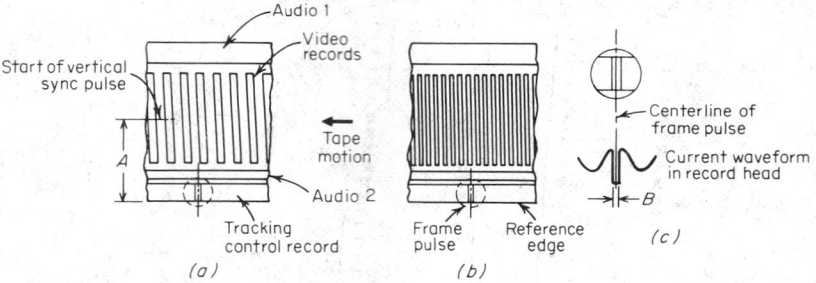

Fig. 20-70. Quadruplex video magnetic recording specification: (*a*) 15 in./s, (*b*) 7.5 in./s, (*c*) enlargement of frame-pulse area. *(From SMPTE Standard RP 16-1970 by permission of SMPTE.)*

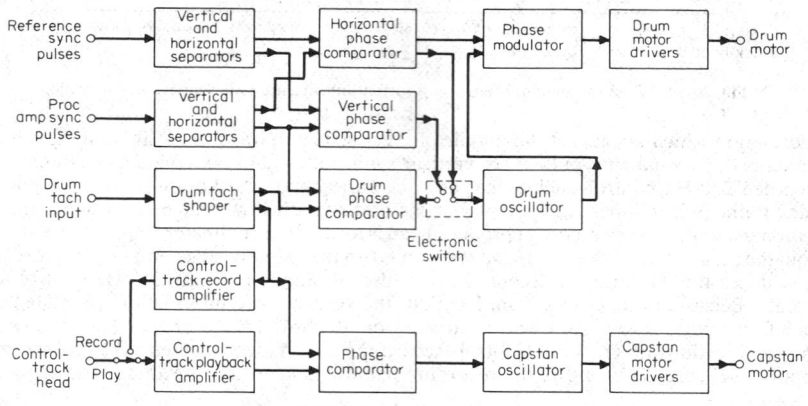

Fig. 20-71. Intersync servo system.

of the tachometer signal (representative of head-drum phase) with the sync components of the incoming video signal. The resulting measure of timing error is represented by a proportional voltage that controls the frequency of the drum-oscillator output. This action places video head 4 in position to record the vertical sync information at the precise center of the tape width and in accordance with SMPTE Recommended Practice, RP-16.

If a program transition to a nonsynchronous video-signal source is required during the recording, the positional servo must correct the angular position of the head drum at a controlled rate in order to minimize a corresponding instability that will occur during the subsequent reproduction. In the event of severe interruptions of the video sync component, the intersync servo unit may be switched to the power line or other external reference.

The second loop, shown in Fig. 20-73, provides damping and is used during all modes of operation. Its damping action minimizes high rate-of-change disturbances that affect the instantaneous frequency of head-drum rotation, including those resulting from the natural tendency of the drum motor to hunt at a 7- to 10-Hz rate, or momentary changes of head-to-tape pressure (drum loading) caused by the passing of a tape splice. Minimizing these disturbances establishes a flat frequency characteristic within the bandpass of the servo system.

During the record mode, the capstan servo system maintains constant velocity of the tape in its movement past the rotating video-head drum. The frequency of the capstan drive signal is 60 Hz derived from an oscillator and locked to the drum tachometer signal. The capstan synchronous motor is thus caused to drive the tape at the primary nominal linear rate of 15 in./s but is electronically locked to the rotation of the video-head drum.

93. Servo Play-Mode Functions. Reproduction of the recorded composite video signal requires accurate servocontrols that will position the tape longitudinally and phase the video heads transversely to reestablish the positional relationship existing during the recording process.

The tape is positioned longitudinally by the capstan, which is controlled by comparison of the reproduced control-track signal with a reference signal. Simultaneously the phase of the video-head drum is precisely indicated by the phase of the tachometer signal which is

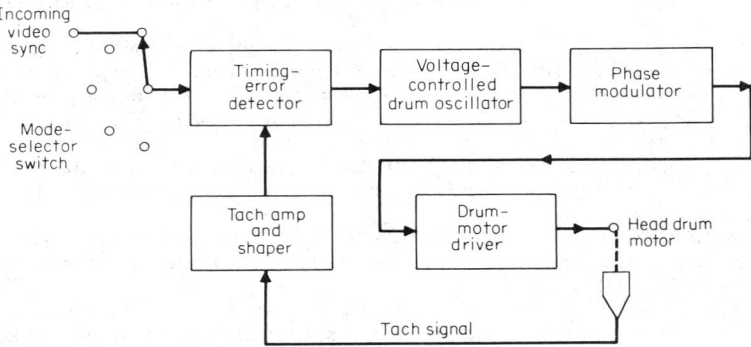

Fig. 20-72.   Head-position servo system (positional loop).

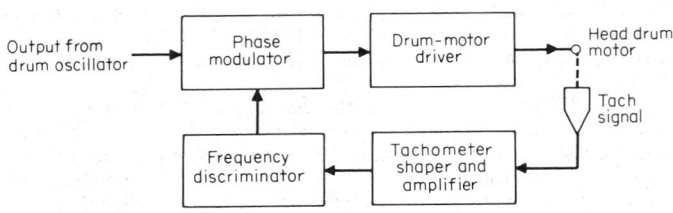

Fig. 20-73.   Damping servo loop.

compared with that of the reference signal. An error in the longitudinal position of the tape or in the phase of the head drum results in an error-correcting voltage that acts on the respective servo system to cancel the error.

94. Intersync Servocontrol modes. The intersync servo unit allows manual selection of five operating modes providing great flexibility in every operational situation, color or monochrome.

Automatic Mode. This mode locks the recorder to the incoming video vertical-sync signals in record and permits full intersync operation which tightly locks the recorder to reference horizontal and vertical sync in the playback mode (see Fig. 20-74).

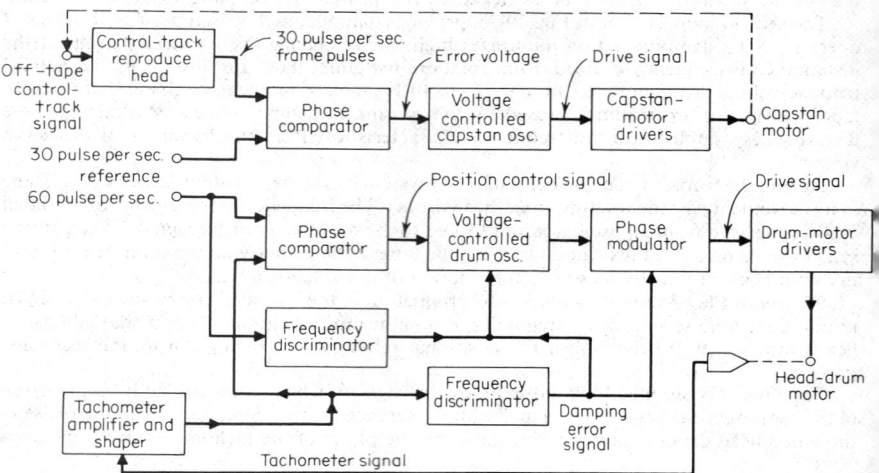

Fig. 20-74. Diagram of complete servo system, shown in the initial condition of the automatic mode.

Horizontal Mode. This mode locks the recorder to the incoming video vertical-sync signal in record and tightly locks to external reference horizontal sync in playback. Use of the horizontal mode minimizes effects from signal discontinuity and permits improved quality when playing back physically spliced color and monochrome tapes. The reproduced video will not necessarily be vertically framed with respect to reference sync.

In this mode, capstan rotation is controlled by phase comparison of the reproduced control-track signal with the tachometer signal. There is no other correction of the longitudinal position of the tape.

Vertical Mode. This mode locks the recorder to the incoming video vertical-sync signal in record and to vertical reference sync in playback and maintains accurate frame lock when no-roll switching is desired.

Frame synchronism within ± 10 μs of the reference is achieved rapidly in two sequential operations. Initially, the capstan positions the tape to place the beginning of a frame, in the recorded signal, in line with the head drum coincidentally with the beginning of a frame of information in the reference signal. This is accomplished by phase comparison of a reproduced frame pulse with a reference frame pulse and is normally completed within 3 to 4 s following the appearance of the reproduced control-track information. During this period head-drum phase is corrected to place video head 4 at the center of the tape width coincidentally with the occurrence of reference vertical sync. This is accomplished by phase comparison at the tachometer signal.

Normal Mode. For monochrome use, this mode locks the recorder to the incoming video-signal vertical sync in record and to the power line in playback. The normal mode is recommended for monochrome reproduction when the recording is known to contain sync-timing discontinuities such as when a tape is improperly spliced. Reproduced video is soft-locked to the power-line frequency. However, reproduced sync timing is totally ambiguous with the station sync-generator reference. Tape reproduction results without regard to framing and is based on phase comparison of the 240-Hz control-track signal with the drum tachometer signal, which in turn is phase-locked to the power line.

Preset Mode. In the preset mode the controls of the intersync servo unit may be set to any desired record and/or reproduce reference. This mode is used mainly for checking various areas of the circuitry and to meet unusual requirements of the user. Preset mode is similar to normal mode except that its timing reference source is chosen by means of two selector switches.

95. Video-Head Assembly. The video-head assembly includes four video heads that are mounted 90° apart on the periphery of the video-head drum. The drum is approximately 2 in. (50.8 mm) in diameter and rotates during record, play, and ready modes at 240 r/sec, or 14,400 r/min. In recent years major design refinements in precision rotating heads have been introduced. New approaches in the construction of unitized transducers and innovations in head-drum design have eliminated to a great extent the time-consuming and complicated mechanical and electrical quadrature maintenance adjustments. Improved head-tip materials have provided high-efficiency performance, markedly better signal-to-noise ratio, and longer head-tip life.

Air Bearings. The air-bearing system for rotary heads was developed mainly for color-television signal recording. Air bearings provide improved rotational stability with a reduction of mechanical friction, and the time-base stability is superior, requiring less electrical correction and minimizing the accumulative effect of time-base error, particularly in multiple-generation recordings. Geometric distortion inherent in the earlier ball-bearing counterpart head assemblies has for all practical purposes been eliminated by air bearings.

Rotary Transformers. Modern rotary-transformer developments provide for low-noise coupling of the video head-tip transducer to the preamplifier input, with improved signal-to-noise ratio and lower cross talk from channel to channel of the rf signal. A substantial reduction in operational maintenance has come with the elimination of slip rings and mechanical commutators.

Preamplification. Development of tube (nuvistor) and transistor preamplifiers has led to the use of the cascode low-noise circuit. The preamplifiers are located in close proximity to the video-head rotary transformer, further improving signal-to-noise ratio. Individual piston capacitors for accurate matching of electrical characteristics are also used to equalize the performance of each head-channel system.

Video-Head Testing. To test head alignment, sweep-test modules are available to plug into line-standard module sockets. The sweep modules generate a sawtooth test signal that modulates the recorder's built-in modulator. A head test-loop assembly (rf coupling device) is mounted on the vacuum guide of the video-head assembly. This induces a sweep-test signal through the individual head tips under test while the head-drum motor is off and the drum is stationary, allowing separate head checkout. The rf coupling device is connected to a detector, and the output is viewed on an oscilloscope for adjustment of the Q-compensation and frequency-compensation controls. This permits accurate compensation of video-head resonance.

Video-Head Tips. In wideband color-video recording, considerable engineering has been required to provide more efficient video-head transducers and to achieve a higher-frequency recording. To provide higher output level and improved signal-to-noise ratio, new transducer designs have been developed. The length of the video-tip gap has been reduced from roughly 100 μin. to a nominal 50 μin., thus accommodating the higher frequencies associated with high-band recording.

SMPTE provides a recommended practice, RP-11, Tape Vacuum Guide Radius and Position for Recording Standard Video Records on 2-in. Magnetic Tape. All video tapes should be recorded to a standard *tip engagement* as specified in that document.

HELICAL-SCAN (SLANT-TRACK) VIDEO RECORDERS

By Keith Y. Reynolds

96. Introduction. An alternative to quad-head recording is known as *helical scan* or *slant-track* recording. The signal-to-noise ratio, bandwidth, and time-base stability approach quadruplex performance at significantly lower costs. Helical-scan recorders are designed to fulfill three design goals: ability to record a full NTSC color signal in its original form, portability, and minimum use of tape.

In International Video Corporation (IVC) 800 series of helical-scan recorders, the tape speed is 6.91 in./s, and the head writing speed is 723 in./s. The video signal is recorded on the tape at an angle of 4°45' and produces a long slanting track across the tape. The track is sufficiently long to record one complete television field. The track width is 6 mils, and the guard band between tracks is 3.6 mils. The first audio signal (number 1) is near the top edge of the tape, is recorded at an angle of 25°, and is 39 mils wide. This angle and the angle of the video track make a total of almost 30° angular difference between the two tracks. The center of this audio 1 track is 100 mils in from the edge of the tape. Figure 20-75 shows the geometry of the helical-scan recording.

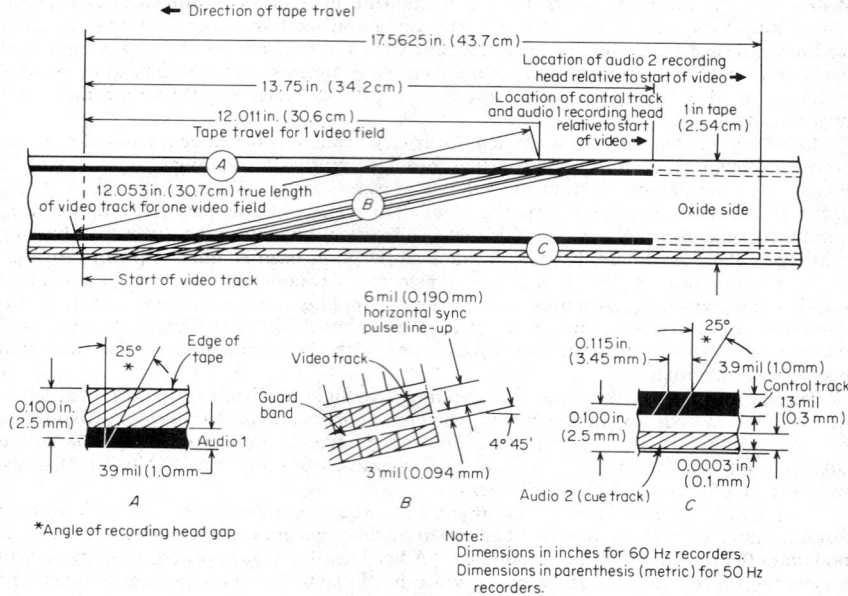

Fig. 20-75. Geometry and specifications of helical-scan (slant-track) video-tape recording.

The control track is near the bottom edge of the tape and like the audio track 1, is recorded at a 25° angle, 39 mils wide, with the center 100 mils in from the edge.

Audio track 2 is also at the bottom edge of the tape but is not recorded at an angle.

Using the low writing speed of 723 in./s, this format allows the recording of 1 h of video information on an 8-in. reel containing 2,150 ft of 1-in. tape. The combination of small reels and small scanning drum sizes allows such recorders to be mounted in a rack or in a carrying case.

97. Time-Base Stability. To permit the helical recording of NTSC color (without resorting to pilot tone systems or converting the NTSC signal to another color system) requires sufficient time-base stability in the reproduced signal to allow recovery of a color signal using the information contained in the color bursts. Since there is only one color burst per line scan, any velocity change during one line produces an uncorrectable hue shift. To achieve such time-base stability, with a 360° tape wrap, an air bearing is used to make the tape flow smoothly around the drum. The air bearing is generated by the rotating member of the scanning assembly which carries the video head. The air is applied to both the top and bottom parts of tape as it passes around the scanning assembly. Without this air cushion, the tape would tend to stick to the drum surface, causing motion instabilities.

The capstan is placed ahead of the scanning assembly and serves to feed the tape at a precise speed into the scanning assembly. The take-up reel drive provides sufficient tension for smooth, undisturbed tape movement across the surface of the drum and also provides speed control, with low torque from the capstan drive motor.

VIDEO TAPE CARTRIDGES AND CASSETTES

By Frederick M. Remley, Jr.

98. Cartridge and Cassette Recording. In cartridge video recording, the video tape is wound on a single spool for storing the tape outside the recorder. In use, the unattached end of the tape is extracted from the cartridge by the recorder and threaded into the mechanism automatically. It is wound onto a take-up spool which is integral to the recorder. After use, the tape is rewound into the cartridge and removed from the machine. This type of container cannot be removed from the recorder without rewinding the entire tape into the cartridge.

The video-tape cassette is a container with supply and take-up spools. The ends of the tape are permanently affixed to the hubs of these spools. When the cassette is inserted in the player, a loop of tape is extracted by the machine and guided into position. Some machines are designed to permit the cassette to be removed without rewinding the tape. The spools may be placed side by side (coplanar cassette) or one above the other (coaxial cassette).

Some cartridge and cassette systems are intended primarily for domestic use, i.e., for attachment to a television receiver in the home. These systems are designed for the simplest possible operation and have such features as stereophonic sound. Other systems are more specifically intended for institutional use, e.g., in education and training. These units have more sophisticated control features, such as still-frame and slow-motion operation, automatic cueing of tape sections, and remote control.

Cartridge and cassette tape containers are used both with quadruplex-head and with helical-scan recording systems. Quadruplex cartridge or cassette systems are used primarily for broadcasting purposes, usually for short recordings of less than 8 min duration. Helical-scan cartridge and cassette systems are primarily intended for educational, industrial, and home use. Helical-format tape containers usually hold enough tape for 1 h of recording or playing time.

99. Quadruplex Systems. The RCA TCR-100 system makes use of a two-spooled cassette (termed a cartridge by RCA) in a two-station recorder-player. The system is capable of presenting 22 consecutive recordings, each from a separate cassette. In each case, the machine cues the recording to the proper point and begins to reproduce the signal on a command from a television system external to the machine or from programmable internal logic. Each cassette holds enough tape for 3 min of recorded program. Cueing and some machine functions are controlled by audio-frequency tones recorded on the audio record track 2 (cue track) of the standardized quadruplex recording. The transport mechanisms extract and guide the tape through mechanical means. Tape speed is 15 in./s, and 2-in. tape is used. The SMPTE standard spool can be used.

The Ampex ACR-25 system uses a two-spooled cassette, with sufficient video tape for 6 min of recorded material. The use of two tape-transport stations permits continuous random sequencing of 25 cassettes. Control is by external command or internal preset logic. The tape transport uses vacuum chambers to extract the tape from the cassette and to place it in the proper position for recording or reproduction. The SMPTE standard spool can be used. Tape speed is 15 or $7\frac{1}{2}$ in./s. Cueing and machine functions are controlled by audio-frequency tones recorded on the audio record track 2 of the standardized quadruplex recording.

The RCA and Ampex quadruplex cassette systems produce and use standard quadruplex recordings of suitable duration to suit their capacity. As a result, short recordings prepared on more conventional reel-to-reel quadruplex recorders can be reproduced on either cassette system when mounted on the SMPTE spool. In addition, since both machines are recorders as well as reproducers, SMPTE standard tape spools may be used for recording on either machine and reproduced on the other.

100. Helical Systems. A system originated by the Sony Corporation of Japan has been adopted by a number of Japanese manufacturers. It is intended for use on 525-line-60-field television systems. The cassette is of coplanar design and holds up to 100 min of tape, operating at a tape speed of $3\frac{1}{4}$ in./s. The player automatically loads the tape into playing-recording position after insertion of the cassette. The cassette can be removed from the player without rewinding the tape. The system is designed to provide excellent color reproduction of NTSC color-television signals. The heterodyne system of color recovery is used. Equipment currently produced for this system is intended to be attached to a television receiver for institutional use.

The *Instavideo system,* originated by the Ampex Corporation, makes use of tape ½ in. wide recorded in conformity with the EIAJ type I format specifications and contained in a single-spool cartridge. The free tape end is automatically guided into the player-recorder. Two sound tracks are available, as are stop motion and still frame. The tape can be provided with electronic indexes which find specific portions of the tape automatically. The cartridge holds sufficient tape for 30 min of material or 60 min with optional half-speed operation. Color recording and playback are provided for, and color recovery is heterodyne.

The *Cartrivision system* has been designed especially for domestic use as part of a television-receiver–tape-recorder system. It uses a coaxial cassette loaded with ½-in. tape, sufficient for 2 h of recording at a tape speed of 3.8 in./s. The video signal is recorded using the skip-field system, where every third field is recorded and the recorded fields are scanned 3 times on replay to achieve a playback rate of 60 fields per second.

The *VCR system,* designed by N.V. Philips of Holland, makes use of a coaxial cassette loaded with ½-in.-wide chromium dioxide tape. It is especially designed to operate with 625-line–50-field television systems. Color recording and reproduction of the PAL television system is provided for. The equipment is intended for connection to domestic television receivers. The European version of the system can record or reproduce 60 min of program at a tape speed of 5.6 in./s.

TELEVISION FILM RECORDING

By Charles H. Evans

101. Television Film Recording. The transcription of pictorial information from its electrical form to motion-picture film is called *television film recording.* It may be used in black and white or in color in broadcasting or special-purpose systems. For broadcasting, the final form of the recording must be suitable for projection at the worldwide standard rate for professional motion pictures, 24 frames per second. Such recordings can be replayed on the film chains of any broadcast television system, despite differing national standards, and can be viewed by optical projection.

Television film is recorded by photographing a visual display or by writing directly on the film without intermediate imaging. In either case the elements of each picture frame are presented to the film sequentially, point by point, rather than simultaneously, as in ordinary motion-picture photography.

The interlaced nature of television scanning means that to preserve the full vertical detail the recording beam must scan each film frame twice. The intermittent nature of the television display and the discrepancy between television and film framing rates have made it necessary to develop special recording cameras. For television systems operating at 50 fields per second it is customary to operate the recording camera at 25 frames per second, accepting the small errors of motion and sound that occur when the recordings are optically projected using standard 24 frame per second projectors. In the same systems, all motion pictures are televised at 25 frames per second. Conventional motion-picture cameras advance the film intermittently from frame to frame, and nearly half of each frame period is required for the film advance. During this pulldown time the film is shielded from light by a rotating shutter. With such a mechanism only alternate television fields can be recorded (Fig. 20-76*a*). The spaces normally occupied by the lines of the missing field can be filled by

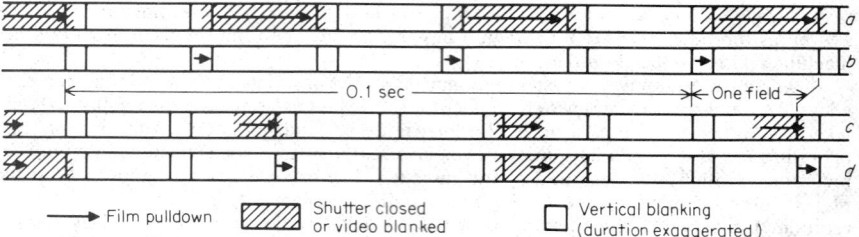

Fig. 20-76. Television film-recording cycles. Frame rates for (*a*) and (*b*) television and film, 25 frames per second; for (*c*) and (*d*) television 30 frames per second, film 24 frames per second.

artificially broadening the lines of the recorded field, but the loss of detail is objectionable. Intermittent mechanisms with ultrafast pulldown during television vertical blanking have been developed for 16-mm film, allowing fully interlaced recordings to be made in 25 frame per second systems (Fig. 20-76b). This can also be done on continuously moving film by employing a jump-raster display or a special compensating optical system.

For 60 field per second systems the discrepancy between television and film frame rates is much too great to be ignored. Television film recordings are made at an average rate of 24 frames per second by discarding one-fifth of the information gathered by the television system. The method of Fig. 20-76c discards half fields, allowing up to $\frac{1}{120}$ s for the intermittent film pulldown but introducing a picture splice which is visible if not perfectly made. The method of Fig. 20-76d discards full fields, eliminating the picture splice (if the camera is synchronized and properly phased to the television signal) but requiring ultrafast pulldown. Several other ways of converting the frame rate are possible, but all involve losing at least one-fifth of the available information.

102. Exposing Means. Various devices utilize the video signals of the television system to control a flow of energy in a form suitable for recording on photographic materials. The most familiar of these is a specialized form of cathode-ray tube. A focused beam of high-energy electrons, intensity-modulated by the video signal and positionally directed by the scanning circuits, writes the picture on a phosphor screen. The visible picture thus created is photographed. Cathode-ray tubes especially made for recording are generally small (screen diameter 5 to 10 in.) and employ accelerating voltages in the neighborhood of 27 kV. For recording on black-and-white films, the blue-emitting phosphor P11 (peak emission at wavelength 460 nm) is widely used.

For recording in full color on color film, three separate cathode-ray tubes are used, each tube controlled by the corresponding color video signal. As seen from the camera, the three primary-color pictures are superposed by two color-selective mirrors, one of which reflects red and transmits green and blue and the other reflects blue and transmits green and red. This triniscope arrangement allows free choice of the color phosphors, and to help match the spectral sensitivity of the recording film, colored filters may be used in the individual optical channels.

A less expensive alternative is a color picture tube of the type used in television sets. Modern tubes can be driven to a highlight luminance of 40 to 50 fL (roughly one-tenth the luminance of a triniscope). With careful adjustment and control, good-quality recordings can be made.

Another exposure device is the laser. In the television recording applications announced so far no use is made of the coherent properties of the beam; it serves merely as a high-intensity light source. The beam is directed to the film through an intensity modulator controlled by the video signal and through beam deflectors which produce a raster scan on the film. At present the most practical deflectors for the television application are moving mirrors, galvanometer-mounted for the relatively slow vertical scan, and the faces of a multisided polygon rotating at high speed for the horizontal scan. To record on color film, the light from three selected laser lines may be independently modulated and caused to pass along a common optical axis before being deflected and converged to a small spot on the film. Once aligned, the three beams remain in register over the entire picture.

Color television pictures for visual use can be photographed as if they were daylight-illuminated scenes. However, a corrective filter may be required to achieve a proper color balance on the film. In particular, the rare-earth P22R phosphor requires a red filter such as the Kodak color-compensating filter CC40R, in addition to any filter specified for daylight use of the film. If the display is set up strictly for film recording, all required color balancing should be made by adjusting the individual red, green, and blue picture luminances instead of using filters.

Color pictures can also be obtained through black-and-white film recording. Each color signal from the television camera may be recorded as a black-and-white picture, either on different pieces of film or on different areas of the same film. Full-color pictures are then made from these color-separation records by printing them in register on color film, using lights of the appropriate color in the printer.

103. Gamma and Resolution. Because both television and photography involve elements with nonlinear-amplitude transfer characteristics, it is important to adjust the overall transfer characteristic of the recording system. Linearity is generally achieved by inserting a controlled compensating nonlinear amplifier. It is likewise necessary to consider the modulation transfer function (MTF) of each element in the recording system because this is related to the sharpness and resolution of the picture.

The resolution of a television picture is limited by the number of scanning lines, the picture-repetition rate, and the bandwidth of the video signal. No matter what its size, a television picture contains at most the number of picture elements set by these parameters. The resolution of a film, however, is measured in terms of cycles per millimeter or, equivalently, line pairs per millimeter, and the limiting number of elements in a photographic picture is directly related to picture size.

When reduced to 16-mm film, the 490+ scanning lines per picture height specified by United States television standards correspond to about 34 cycles/mm, and the upper frequency limit of the video signal corresponds to about 22 cycles/mm across the width of the picture. However, a film with a limiting resolving power even as high as 100 cycles/mm still has losses at these lower frequencies, and the loss in the overall MTF of the display device, recording lens, and film may be quite significant. It is possible to compensate for this loss to some extent by boosting the high frequencies in the electronic portion of the system. Noise and other effects limit this process, and it is important to reduce optical losses as much as possible. There is a loose reciprocal relationship between film speed and grain on the one hand and film resolving power on the other. In recording from relatively weak sources, such as the shadow-mask tube, some compromise may be necessary.

104. Sound Recording. A complete television film recording carries a sound track as well as the picture. In single-system recording the picture and sound track are recorded simultaneously on a single camera film. The sound may be recorded optically, as a variable-area or variable-density photographic record; it may be recorded magnetically if the film is prestriped. In double-system recording, picture and sound are handled separately until combined on one film at a later printing stage. Again, optical or magnetic methods and combinations thereof may be used.

105. Standards. Obviously some standardization is required if motion pictures (including television film recordings) are to be readily interchangeable among different users. American National Standards in this field are generated through the activities of the Society of Motion Picture and Television Engineers (SMPTE) and its sponsorship of Standards Committee PH22, Motion Pictures. The SMPTE also issues various recommended practices. The standards and practices are published at the time of issue in the *Journal* of the SMPTE. Copies of the documents may be purchased from the American National Standards Institute, 1430 Broadway, New York 10018. These publications deal with film dimensions and usage, size and location of picture and sound track, test films, and other related matters.

ELECTRON-BEAM RECORDING

By Robert A. Castrignano

106. Electronic video recording (EVR) is a mechanism of storing video information on black-and-white film by scanning with a modulated electron beam. After the film is developed, it can be copied in conventional optical printing systems. The advantage of direct electron-beam recording is that much higher resolution can be achieved than when a cathode-ray-tube face is photographed directly, as in television film recording. EVR requires that the original film pass through a vacuum chamber, and the vacuum must be sufficient to sustain a sharply focused electron beam. The EVR system has been developed by the CBS Laboratories. A diagram of the EVR system is shown in Fig. 20-77.

The three basic components of the EVR system are the electron-beam recorder, which transcribes the original TV signal to a master film; the printing process, which produces multiple copies of the master film; and the player, which produces the original TV signal on a TV set.

The EVR film format is a double-track system; each track can carry a separate black-and-white program, or both tracks can be combined to provide complete information for a color picture. For color TV, luminance information is carried primarily on one track and chroma information, specially encoded, is carried on the other. The film is scanned optically in the playback device through the use of a flying-spot scanner with double optics, as shown in Fig. 20-78. Special scanning and preemphasis techniques are used to give maximum resolution with minimum noise.

HOLOGRAPHIC RECORDING

By E. G. Ramberg

107. Introduction.　Holographic recording is a method of permanent video recording which offers significant advantages in economy of recording material, ease of duplication, scratch resistance, and simplicity of playback.　The hologram is a recorded interference pattern between light from an object and a *reference beam,* both derived from the same

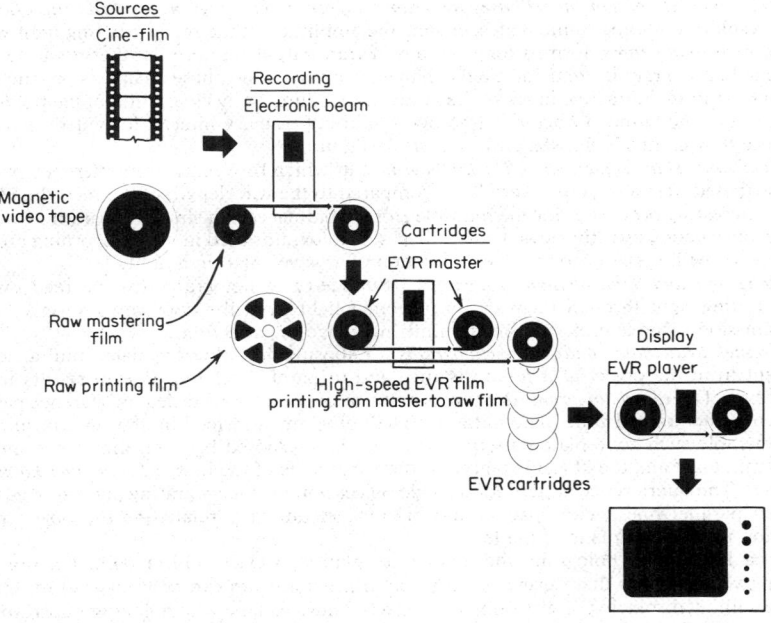

Fig. 20-77.　Electronic video recording (EVR) system.

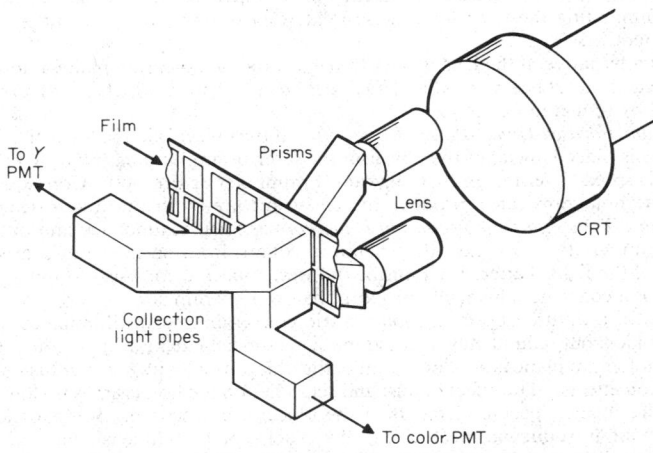

Fig. 20-78.　Optical system of EVR player.

coherent-light source (laser). If a *reconstructing beam* similar to the reference beam falls on the hologram, it is diffracted to form an image beam which appears to diverge from a replica of the object located in the object position. A viewer intercepting this image beam thus sees the original object framed by the periphery of the hologram. If the reconstructing beam is reversed relative to the reference beam, the reconstructed image (at the location with respect to the hologram of the original object) is a real image and can be recorded on a photographic plate; or if it is incident on a camera-tube target it can be transmitted over a television system.

108. Types of Holograms. Depending on the recording medium and the relative convergence of the object beam and reference beam, holograms are classified as follows.

Absorption Holograms and Phase Holograms. *Absorption holograms,* usually recorded on high-resolution photographic plates, change the amplitude of the reconstructing light wave, while *phase holograms,* formed for example in transparent photoresist, thermoplastics, or bleached photographic emulsions, only change its phase. The phase change is produced by a variation in the refractive index of the hologram medium or by a variation of the hologram thickness. The latter, surface-relief holograms, are of primary interest for video recording because they can be embossed on low-cost plastic film.

Thick and Thin Holograms. *Thick holograms,* in which the recorded interference pattern is distributed over a thickness very large compared to the wavelength of the recording light, have interesting properties for the multiple storage of images in a single hologram as well as for reconstructing directly viewed color images. Holograms used in video recording are *thin holograms,* with surface modulations a fraction of a wavelength in amplitude.

Reflection and Transmission Holograms. Surface-relief holograms can be read out by transmitting light through them or by reflecting light from the corrugated surface. The transmission mode is preferable for minimizing image imperfections.

Fresnel, Fraunhofer, and Image Holograms. Exposing the hologram plate simultaneously to light from the object at a finite distance and to parallel reference beam results in the recording of a *Fresnel hologram.* For any single object point the recorded interference pattern is then a fine-grained zone plate pattern. The hologram is formed by the superposition of innumerable such zone plate patterns. If a lens is interposed between the object and the hologram plate and the object is placed in the focal plane of the lens, a *Fraunhofer hologram* results. The interference pattern for a single object point is then a grating pattern of equally spaced parallel lines, with spacing and orientation uniquely related to the object point position in the focal plane of the lens.

In a Fraunhofer hologram, the unique correlation between object point position and grating vector means that during reconstruction the hologram can be translated in its own plane without displacing or distorting the image. Thus, if a tape with holograms corresponding to successive film frames is moved through the reconstructing beam, a stationary image is obtained, with successive frames blending into each other. This permits a great simplification of the tape-transport mechanism used in playback. An added advantage of the Fraunhofer hologram is that translation of the hologram normal to its plane has no effect on the image, eliminating the need for precise relative positioning of the components of the playback system.

As a final alternative, if the lens images the object on the hologram plane, interference of the reference and object beams produces an *image hologram,* from which a normal image can be recovered by schlieren optics.

Diffused- and Focused-Object-Beam Holograms. If the object is in the form of a film frame and a diffuser is placed ahead of the object in the laser beam illuminating it, light from any object point is spread over the entire hologram (Fraunhofer or Fresnel). Consequently, any portion of the hologram can reproduce the entire picture, with resolution depending on hologram size. Thus *diffused-object-beam hologram* has ideal redundancy and optimal dust and scratch immunity. On the other hand, it suffers from speckle noise arising from interference of the light scattered within the diffuser. Speckle consists of random intensity fluctuations of a coarseness inversely proportional to hologram size.

If, instead of a diffused beam, a monocentric converging beam illuminates the object, speckle is avoided, but redundancy is lost at the same time; the light distribution of the object beam at the hologram plane is similar to that of the object itself, with some diffusion resulting from diffraction effects. The effect of dust and scratches on the hologram is no longer spread over the entire image; specific elements of the object are obscured. Redundancy can be restored without introducing speckle, e.g., by placing a two-dimensional phase grating generating nine diffracted beams of equal intensity just ahead of the object, forming nine overlapping, mutually displaced holograms. The grating pattern appears superposed on the image but can be rendered innocuous by making the grating-line spacing too small to be resolved by the playback system.

109. Holographic Color Recording. The holographic recording process here described reconstructs a monochrome image. Color transparencies are translated into black-and-white transparencies on which the color is encoded. In effect, the color transparency may be printed through minus-red and minus-blue dye gratings with different space frequencies, resulting in a black-and-white image with a fine superposed stripe structure, the amplitude of which is related to the color content. During reconstruction with a television camera the high-frequency content of the picture signal is separated and envelope-detected to yield the red and blue picture signals, while the low-frequency content delivers the luminance signal. In practice the encoding is carried out with an electron-beam recorder.

110. Holographic Recording System. The complete holographic recording and play-back process may comprise the following steps:

1. Preparation of a black-and-white film with encoded color from an original color film, color TV camera output, or video-tape recording.

2. Recording of a sequence of Fraunhofer holograms of successive film frames on photoresist-coated tape (see Fig. 20-79).

3. Development of the exposed hologram tape, followed by a plating process to form a metal master.

4. Replication of the metal master on inexpensive, thin plastic tape by passing them together through a pair of rollers.

5. Playback by reconstruction with a low-power laser of the encoded image on the target of a vidicon camera with output circuits for delivering a standard color-television signal (see Fig. 20-80).

VIDEO DISK SYSTEMS

By Henry Ball

110a. Video Disk Systems. In the mid-70s a number of video disk recording systems were under development, but none had been introduced commercially. The following paragraphs review developments as of 1974.

Storage Medium. The replicated disk is usually made of plastic: a homogeneous sheet or pressing of vinyl, or a sandwich consisting of a substrate coated with layers to permit signal recovery or to provide handling protection. Disks currently range in diameter from 21 to 30 cm (8.25 to 12 in.) and in thickness from sheet stock to several mm. To achieve the required information density, spiral track or groove pitches up to 500 per mm (approx. 13,000 per in.) have been used. This corresponds to groove or track widths as small as 2 μm. The information contained is most often in the shape of a vertical structure (hill and dale modulation) of sub-micrometer amplitudes.

Electrical Conversion. Electrical conversion of the video information currently falls into two classes:

(1) Mechanical (Grooved Disk). Detection is by means of a stylus tracking a groove, much in the manner of a phonograph record. The primary difference lies in the inability of the stylus to physically follow the undulations of the information-carrying surface.

In one of the grooved systems, the stylus is affixed to piezoelectric material which senses and converts the changes in instantaneous pressure resulting from the washboard-like surface in the groove into an electrical signal. The stylus mass itself is immobile, but transmits the pressure waves. Due to the large number of grooves per millimeter, insufficient tracking force is available to move the pickup arm. External tracking forces are supplied by a worm gear, which moves the pickup assembly to maintain tracking continuity between the stylus and the disk.

In another grooved system, the undulations contained in a groove are covered by two layers, a conductive base coating and an insulating coating. The base coating is grounded and a conductive coated stylus riding on the insulating layer is used to detect the changes in capacitance between the undulating conductive coating and the stylus coating.

In both cases the stylus has a low-incident-angle bearing surface which rides in the groove, simultaneously covering several information elements. This bearing area is needed to allow the distribution of tracking forces. The signal-detecting part of the stylus is in each case the back edge, which is nearly perpendicular to the disk surface. In the first above-described mechanical playback system, the back edge produces the pressure waves resulting from the relatively sudden compression release of the individual information elements. In the second system, the conductive surface coating acts as a knife-edge-like electrode of small aperture, which detects the undulations of the grounded coating on the disk.

(2) Optical (Ungrooved Disk). A light source, typically a HeNe laser, is focused on the information-carrying surface. The undulations (or other surface deformations) result in changes of reflection, transmission, refraction, or other optically detectable change of state. In these systems, differential sensing is usually used to serve as a tracking servo input reference, since there is no physical groove on the surface.

Storage Method. Video disks, unlike magnetic video tapes where audio is carried on separate tracks, are restricted to a single information-carrying channel. Composite color video as well as one or more sound tracks must be accommodated. Encoding the video information on a frequency-modulated carrier, similar to the video magnetic tape, is a widely followed practice. Where NTSC- or PAL-encoded color video signals are used, the color subcarrier is often reduced in frequency and recombined with the FM-encoded luminance information to accommodate the limited bandwidth capability. In one system, the color information (R,G,B) is sequenced as a low-frequency luminance signal (less than 0.5 MHz) at the TV line rate. The high-frequency information (above 0.5 MHz) is transmitted at the normal TV line rate. On playback, the sequenced information is restored, by means of two delay lines, to be available simultaneously. The color information is then encoded in the conventional manner. Design targets frequently call for luminance bandwidths of less than 3 MHz and chrominance bandwidths of less than 0.5 MHz.

The audio signal is combined with the video signal. In one system, a low-frequency (0.25 MHz) audio carrier is added to the high-frequency (0.5 MHz) FM video carrier. The resulting sinusoidal excursions of the FM carrier are then removed by clipping. The audio information remains detectable as a zero-crossing symmetry shift. This is distinguished from other systems where the audio carrier amplitude modulates the FM carrier. The AM is retained in the storage medium and recovered by AM detection for playback. In another method, the repetition of the information-carrying elements for luminance information is varied in width for chrominance information and in element-to-space symmetry for audio information.

The TV frame repetition rates have thus far determined the rotational rates of video disks. At present, all European disk systems operate at 1500 rpm (25 rps), which is equal to the TV frame repetition rate. In the U.S., developments are following a similar path and the corresponding rate is 1800 rpm (30 rps). One development uses a disk operating at 450 rpm (7.5 rps), which represents four TV frames per rotation.

These rotational rates have been chosen to produce minimum discontinuity of the visual presentation on the receiver in case of accidental groove or track skipping. In addition, it provides for the implementation of special effects obtained by other than steady-state tracking. Since the information on adjacent tracks has the same timebase (i.e., vertical blanking occurs in the same radial on the disk) the pickup method can include provisions such as still frame, slow or accelerated motion, and random access of individual frames.

In all cases the storage method requires a recoverable resolution capability which is a function of the FM carrier frequency and the groove or track velocity at the innermost diameter, while maintaining the desired signal-to-noise ratio. For example, the linear groove velocity at 10 cm diameter is 9.4 m/s for 1800 rpm, 7.8 m/s for 1500 rpm, and 2.4 m/s for 450 rpm. Given a top carrier frequency of 5 MHz, it is necessary to store and recover 1.9, 1.5, or 0.5 micrometer elements respectively.

These video disk processes are compared with other recording systems in their information storage areas in the following table:

Storage Medium	Storage Means	Storage Area Usage
Quadruplex 2″ VTR	Magnetic	19,350 mm²/s
EIA-J 1/2″ VTR	Magnetic	2,420 mm²/s
U-Matic 3/4″ VTR	Magnetic	1,810 mm²/s
Super 8 Film	Photographic	475 mm²/s
Video Disk	Plastic	22 to 45 mm²/s

Recording and Replication Process. The video information, in the form of sub-micrometer surface disturbances, is recorded by one of the following means: (*a*) vertically moving, heated styli on lacquer surfaces, (*b*) exposure of photoresist and other coatings by coherent light or electron beams in a vacuum chamber, and (*c*) exposure of conventional photographic emulsions. Depending upon limitations of stylus mass or available power density from the light or electron beam source, recordings have been made at reduced rates, (e.g., 1/20th of real time) as well as in real time. It has been found that, although the primary information is recorded as an essentially binary signal, some analog capability is also available. This makes it possible to recover small amounts of amplitude modulation placed on the FM carrier, as previously described.

Master disks are typically made of highly stable substrates such as glass or other prepared material. Where photoresist or electron-beam resist is used, it is developed after exposure to produce a three-dimensional pattern which may be plated or processed to produce one or more stampers for use in pressing or embossing copies in large quantities.

Reference. Video Disc Systems (special issue), *J. Soc. Motion Picture Television Engs.* (SMPTE), vol. 83, No. 7, July 1974.

FACSIMILE SYSTEMS

SCANNERS AND RECORDERS

By David G. Falconer

111. Introduction. Hard-copy facsimile systems generate a signal by systematically scanning the subject copy and producing a current corresponding to its light-intensity variations by means of a photocell. The field of view of the photocell is restricted to an area about $1/100$ in. in diameter. During or after each scan across the document the subject copy is advanced (or the photocell translated) a distance equal to the width of one scan line.

The recording equipment converts the transmitted signal into a facsimile copy by moving a printing mechanism across the recording medium synchronously with the photocell scan. Synchronism is achieved by the power line or by sending timing signals. As the printing mechanism moves, the signal produces small marks on the recording medium in the same relative positions as in the original document. The information on the subject copy is thus built up in the facsimile reproduction from narrow strips which blend together so closely that they are nearly invisible to the naked eye.

112. Scanning Techniques. Scanning the subject copy is accomplished mechanically, electronically, or acoustically.* In the past, mechanical scanners dominated the market since they were cheaper to construct and telephone lines were much too slow for high-speed electronic devices. On the other hand, electronic devices, e.g., the cathode-ray-tube scanner and television camera, offer high-speed, low-maintenance service and are attractive whenever wide-band communication links are available. Acoustical scanners, though promising, remain at the laboratory stage.

Two methods exist for analyzing the subject copy (Fig. 20-81). In the *spot-projection (flying-spot)* technique a small light spot is created on the subject copy and a photocell measures the light reflected from this elemental area. In *flood projection (image projection)* a relatively large area of the subject copy is illuminated and imaged onto an *aperture plate.* The photocell measures the amount of light passing through the aperture.

113. Mechanical Scanners. Mechanical scanners scan the subject copy with rotating, oscillating, and/or translating mechanical parts, using drums, mirrors, slits, or fiber-optic arrays.

Rotating-Drum Technique. In rotating-drum equipment the optical system is mounted on the base of the scanner and the subject copy attached to a rotating drum. In one configuration, shown in Fig. 20-82, rotation of a lathe-type lead screw causes the drum to

---

* Scanners are classified according to the technique used to effect the transverse (fast) rather than the longitudinal (slow) portion of the scan.

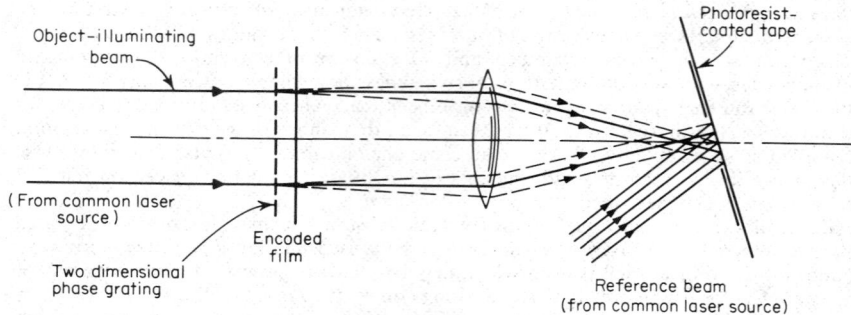

Fig. 20-79. Holographic recording system.

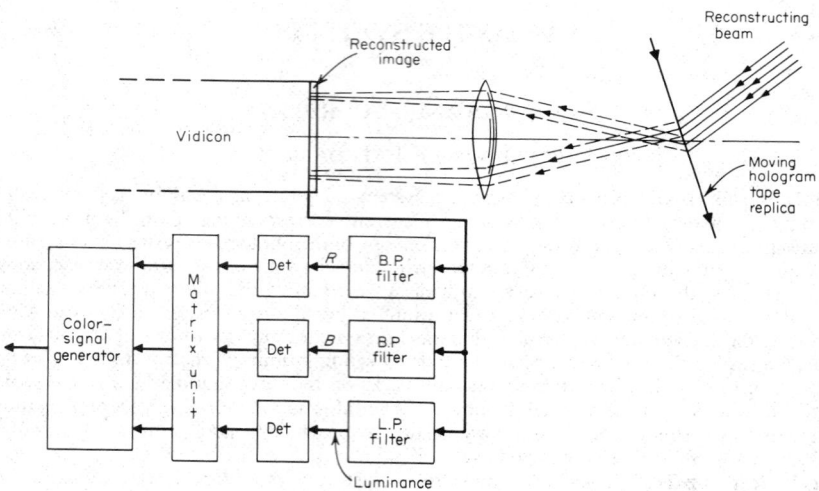

Fig. 20-80. Holographic playback system.

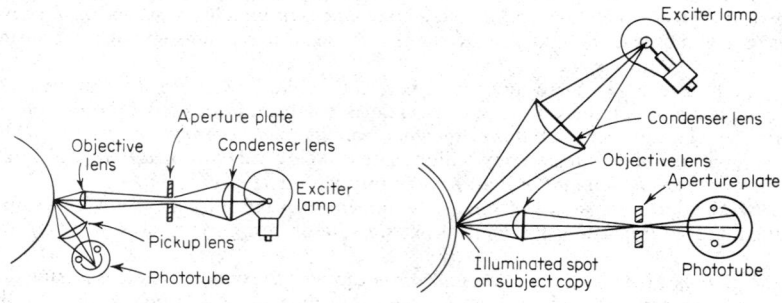

Fig. 20-81. Spot projection *(left)* and flood projection.

rotate and to translate along its cylindrical axis. Alternatively, the drum can be rotated at a fixed position on its shaft and the optical system translated parallel to the drum axis.

The facsimile signal is generated by illuminating the subject copy with an exciter-lamp and condenser-lens assembly. An objective lens, placed at an angle to avoid gathering specularly reflected light, collects light scattered away by the document and focuses it onto an aperture plate. A photocell converts the light passing through the aperture to an equivalent signal, and the signal is transmitted to a facsimile recorder for printing.

Stationary-Drum Technique. In this technique the subject copy is mounted facing inward on a stationary Lucite cylinder. A front surface mirror, mounted inside the cylinder, reflects the beam from a spot-projecting lens system onto the outer surface of the transparent cylinder. As the mirror rotates, the spot sweeps across the subject copy. Line feed is obtained by moving the cylinder along its axis. The mirror collects light reflected from the copy and passes it through a condensing lens to the phototube. The mirror angle is such that specular reflections from the cylinder surfaces and subject copy are directed back into the cylinder so that they cannot reach the phototube.

Flying-Spot Oscillating-Mirror Scanner. In this scanner the copy lies flat, and the light spot moves back and forth across the document. As shown in Fig. 20-83, a front-surface spherical mirror, driven by a special cam, sweeps the spot from a concentrated arc lamp at uniform rate across the subject copy. At the end of each sweep, the cam contour changes abruptly, and the beam moves back across the sheet at a very rapid rate. During the mirror-return interval, the output of the phototube is blanked. The return time can be limited to about 5% in scanners operating over voice-band channels.

Image-dissecting Oscillating-Mirror Scanner. In a similar mechanism, using an image-dissecting scan, the subject copy is floodlighted and fed at the line-feed rate by rollers. An image of the illuminated area of the copy is focused onto an aperture plate and the light

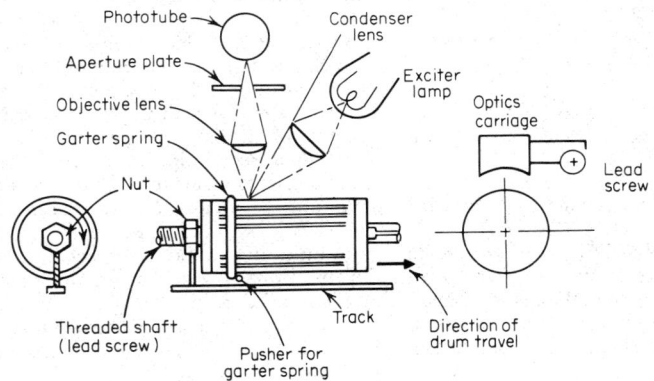

Fig. 20-82. Rotating-drum scanner.

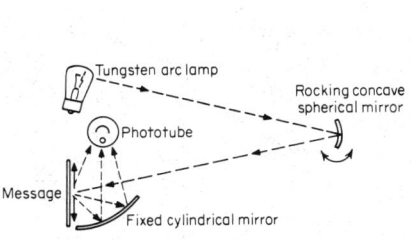

Fig. 20-83. Flat-bed flying-spot scanner using oscillating mirror.

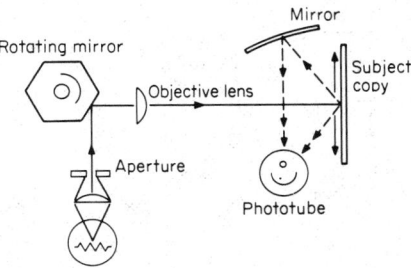

Fig. 20-84. Flying-spot scanner with rotating multifaced mirror.

passing through the aperture detected by a phototube. The mirror serves to sweep the image across a masking aperture, rather than a spot across the subject copy.

Flying-Spot Rotating-Mirror Scanner. In this arrangement, shown in Fig. 20-84, a rotating multifaced mirror sweeps the light spot across the subject copy. Since consecutive scanning lines are formed by different mirror faces, extreme precision is required in fabricating the mirror and its drive mechanism, particularly the gearing. Minute deviations from absolute flatness of the mirror faces and in the angle each makes with adjoining faces or any eccentricities in the drive will cause the successive scan lines to be displaced horizontally from their correct position. These displacements produce a repetitive pattern of distortion called *jitter.*

Image-dissecting Rotating-Mirror Scanner. In this scanner, shown in Fig. 20-85, a strip of the subject copy is illuminated by fluorescent lamps as it is advanced at the line-feed rate by rollers. The objective lens collimates the light before it strikes the mirror face. The face of the mirror, which makes an angle of approximately 45° with the optical axis of the lens, reflects the beam emerging from the objective lens toward a scan-limiting slit. A collimating lens then focuses light from the slit on an image-dissecting aperture. The scan-limiting slit prevents the phototube from seeing beyond the edges of the subject copy while a new mirror face is rotated into position.

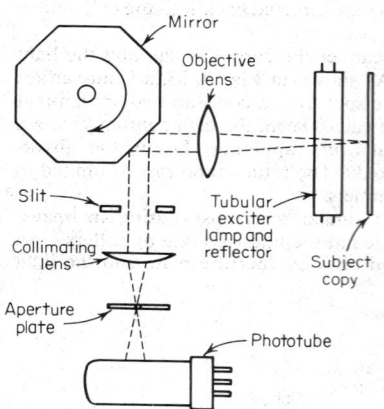

Fig. 20-85. Image-dissecting scanner with rotating mirror.

Spiral-and-Slit Technique. The spiral-and-slit arrangement, shown in Fig. 20-86, uses a flat bed to hold the subject copy. An exciter lamp illuminates the subject copy via a curved mirror, and an objective lens images the reflected light to an aperture plate in front of a scanning disk. The scanning disk is opaque except for a transparent spiral, which curves outward from the center of the disk. As the disk turns, the spiral and slit dissect the image formed by the objective lens. Light passing through the spiral and slit is then imaged via a spherical mirror to a phototube, where an electronic facsimile signal is generated.

Helix-and-Slit Technique. The helix-and-slit configuration is especially suitable for scanning large documents, such as weather maps. A narrow strip across the subject copy is illuminated by long fluorescent lamps, and an image of the floodlighted strip is focused by means of the objective lens upon an aperture plate having a long narrow slit. An opaque cylinder, containing a transparent helix of approximately the same width as the slit in the

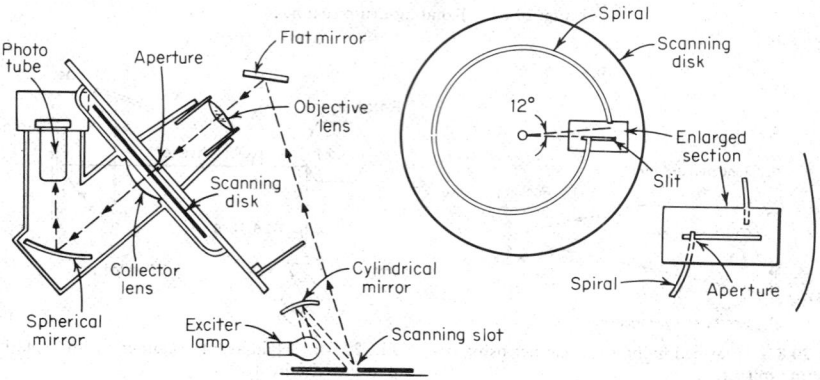

Fig. 20-86. Image-dissecting scanner using spiral and slit.

aperture plate, lies immediately behind the aperture plate. As the cylinder rotates, the intersection of the helix and slit becomes a moving aperture which dissects the image of the subject copy. The subject copy is transported at the line-feed rate so that the image is completely scanned line by line. A curved mirror directs light from the aperture out one end of the cylinder to a phototube. A square aperture would be preferable to the parallelogram formed in this manner, but the difference can be made relatively insignificant by reducing the size of the image optically.

Slit-and-Slit Technique. A slit-and-slit scanning technique is illustrated in Fig. 20-87. Light from an exciter lamp illuminates the subject-copy transparency via a flat mirror. An objective lens then images the transmitted light to a scanning disk containing a set of radial slits. As the disk rotates, the radial slits turn past a stationary slit and thereby dissect the image. Light passing through the slit system is imaged to a photomultiplier, where a facsimile signal is generated. A sync light source and photomultiplier provide the necessary timing signals for the recorder. The viewing screen allows the operator to select the desired frame for transmission.

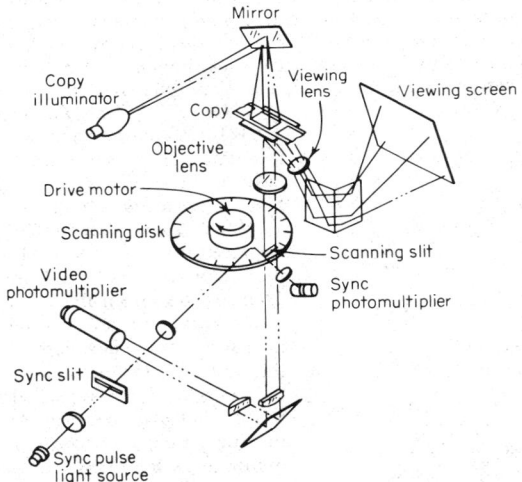

Fig. 20-87. Slit-and-slit optical system.

Fiber-Optic Circle-to-Line Arrays. In recent years, fiber-optic technology has advanced to a point where its techniques are applicable to facsimile imaging. A possible scanner configuration uses a circle-to-line array. The subject copy is mounted on a flat bed and illuminated by a pair of flood lamps. Light reflected by the subject copy enters the linear end of the fiber-optic array and is transmitted to the circular end by a series of total internal reflections at the side walls of the glass fibers. A fiber on a rotating crank then sweeps past the circular end of the array, bringing the light from each to a photomultiplier in turn.

The advantages of the fiber-optic arrangement include higher light levels (since fiber optics have high numerical apertures) and compact mechanical layout (since optical projection of the subject copy is obviated). In the present state of the art, broken fibers or the accumulation of dirt or paper on the ends of the glass fibers can cause a variation in the light level along the scanning line. Consequently this technique has not achieved wide use.

114. Electronic Scanners. With the possible exception of the multifaced mirror, the mechanical scanners described above are limited to speeds between 180 and 3,600 transverse scans per minute. For higher speeds, electronic scanners become more practical and economical and offer quieter operation and lower maintenance costs. Such devices take advantage of the low-inertia scans possible with cathode-ray tubes, television cameras, and silicon photodiode arrays. Because voice-grade telephone lines are several orders of magnitude slower than electronic scanning rates, electronic scanners serve mainly in special-purpose applications where wide-band in-house or common-carrier lines are available.

High-Resolution Cathode-Ray-Tube Systems. High-resolution cathode-ray tubes (CRTs) can scan an 8½ by 11 in. document at a resolution of 100 lines per inch in a fraction of a second. Light emitted by the CRT scanning spot is imaged to the subject copy by an objective lens. Light reflected by the subject copy is collected by a collimating lens and projected on a photomultiplier tube, where the facsimile signal is generated. Although often used in the laboratory, such arrangements are too bulky and expensive for commercial use.

High-Resolution Vidicon System. By using a high-resolution vidicon and a high-quality photographic objective, an ordinary television camera becomes a high-speed facsimile scanner. In this arrangement a photographic objective focuses an image of the subject copy upon the mosaic surface of the photocathode, and the electron beam of the camera tube sweeps the photocathode in raster fashion. The electron beam picks up the charge from each area of the mosaic of the photocathode. These charges are proportional to the intensity of the light striking each area.

Flat Fiber-Optic Faceplate CRT Scanner. Fiber-optic faceplate CRT scanners eliminate the imaging lens and improve the light efficiency of the conventional CRT scanner. In this technique the CRT is fabricated with a fiber-optic (rather than clear-glass) faceplate. Light generated at the phosphor is conducted by the adjacent fiber to the surface of the subject-copy transparency. Light passing through the transparency is collected by the collimating lens and projected onto a photomultiplier tube. The document is converted into a facsimile signal as the electron beam executes its raster scan. Although it is an improvement on the conventional CRT system, the technique cannot scan reflective subject copy.

Prism Fiber-Optic Faceplate CRT Scanner. A CRT outfitted with a prism-shaped fiber-optic faceplate can scan reflective subject copy at electronic speeds. As shown in Fig. 20-88, the fiber-optic faceplate has two surfaces inclined at an obtuse angle to each other. The manuscript is placed on one surface of the faceplate, and illumination is supplied through the other. Light emitted from the phosphor surface proceeds through the adjacent fiber and appears on the surface of the faceplate, as shown. Because this surface is inclined relative to the fiber axis, the emerging light is refracted toward the manuscript as it leaves the fiber. Light reflected from the subject-copy surface falls onto the photomultiplier, where it is converted into an electronic facsimile signal. With this technique, resolutions of 6 lines per millimeter have been obtained.

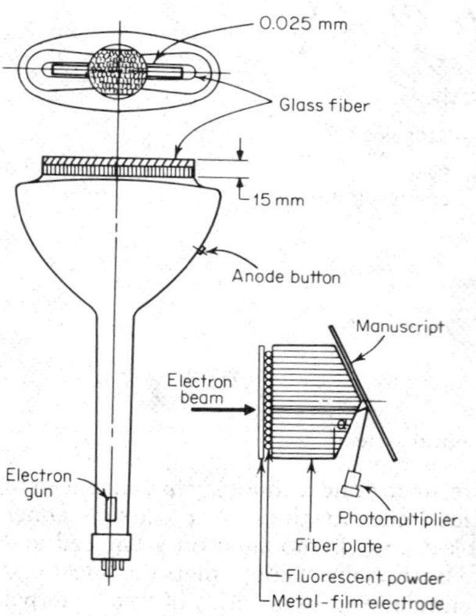

Fig. 20-88. Prism-fiber reading tube.

115. Solid-State Scanners. Solid-state scanners using linear or matrix arrays of silicon photodiodes offer important advantages over conventional scanning mechanisms. In particular, such devices feature fewer moving parts, rapid scanning, and random access to the subject copy.

Linear-Array Solid-State Scanners. Linear silicon-photodiode arrays scan the subject copy electronically in one dimension and mechanically in the other. As indicated in Fig. 20-89, the subject copy is flood-illuminated with exciter lamps and then imaged to the photodiode array with an objective lens. The photodiodes become conductive when illuminated by light from the subject copy. An electronic system generates the facsimile signal by measuring the conductivities of the photodiodes in turn and outputting a corresponding electronic signal. To effect the longitudinal scan, the document is translated at right angles to the photodiode array with feed rollers and the conductivities of the photodiodes again outputted to the facsimile recorder.

Matrix-Array Solid-State Scanning. Although more difficult to construct, the two-dimensional photodiode array offers the possibility of all-electronic, random-access scanning. The subject copy is illuminated by exciter lamps and then imaged to a two-dimensional photodiode array. An electronic-system raster scans the conductivity of each array element and outputs a corresponding electronic facsimile signal. Although attractive, large two-dimensional photodiode arrays remain in the future, while staggered linear arrays with thousands of elements can be fabricated with current technology.

Acoustical Scanning Techniques. Acoustical scanners are becoming attractive as a result of technological advances brought about by the development of optical memories for digital computers. In acoustical scanning, a high-frequency acoustic signal is injected into a liquid or solid medium (Fig. 20-90). The longitudinal acoustic wave produces alternating regions of rarefaction and condensation in the medium, thus creating a diffraction grating. The angle of deflection θ (see Fig. 20-90) depends linearly on the acoustic frequency f. The sound velocity v and the optical wavelength λ are constants. The time for deflection to occur is identical to the time τ required for the sound wave to cross the light beam; the number of well-resolved elements equals the bandwidth of the transducer (about 25 MHz for water) multiplied by transit time τ (about 20 μs/in.).

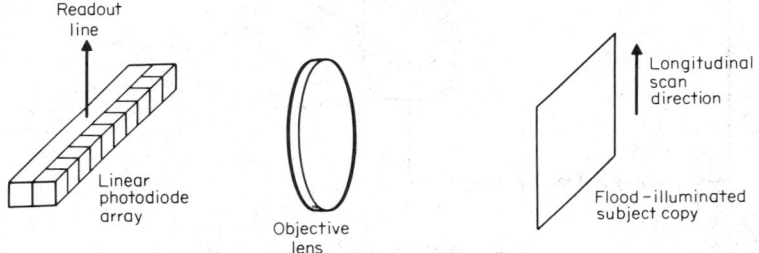

Fig. 20-89. Linear-array solid-state scanner.

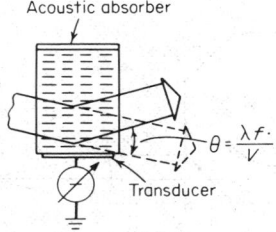

Fig. 20-90. Acoustic deflector.

116. Mechanical Imaging (Recording) Techniques. Facsimile equipment, particularly the recorder, must be constructed with utmost precision so that the scanning lines are exactly contiguous at all times. For example, even the slightest overlap of the scanning lines will cause dark strips to appear in gray areas due to the double exposure of the recording film. Likewise failure of the scanning lines to touch each other throughout their entire length will cause light strips (where the film remains unexposed) to appear in dark areas.

Rotating-Drum Recorder. Figure 20-91 shows the basic drum recorder. Photographic film is attached to one surface of the drum with its emulsion side facing outward. Facsimile signals received on the telephone line are amplified and converted to a form appropriate for controlling the brightness of the light source, usually a crater lamp (glow-modulator tube). Light from the source illuminates an aperture plate, and an objective lens forms an image of the aperture on the photographic film. A drive-motor system rotates and translates the drum so that a sequential pattern of elemental exposures is traced on the photographic film. The received facsimile signal controls the brightness of the light source in step with the brightness of the scanned copy. At the end of the recording, the entire surface of the film has been exposed, thereby forming a latent image of the subject copy.

Multifaced-Mirror Recorder. Figure 20-92 shows a rotating multifaced-mirror recorder. As with the drum recorder, electronics converts the facsimile signal to a form appropriate for driving a crater lamp. A condenser lens images light from the lamp onto an aperture plate.

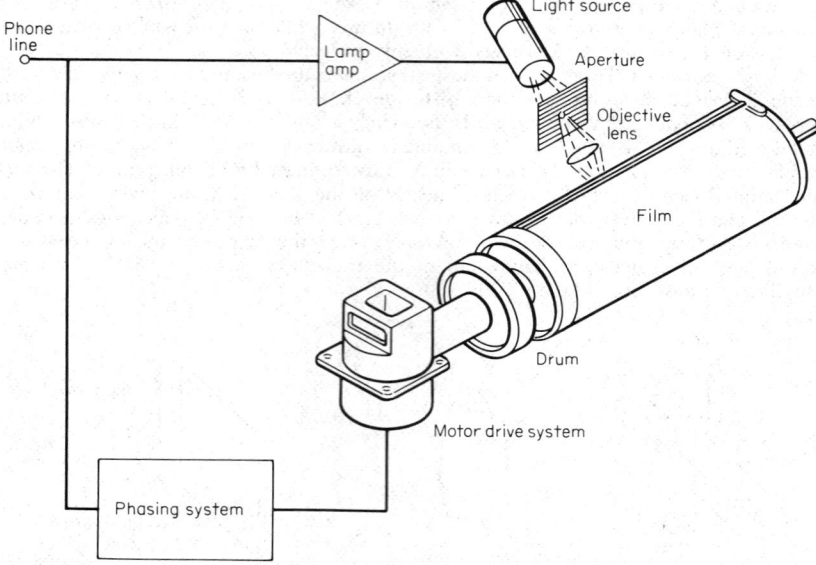

Fig. 20-91.   Drum recorder.

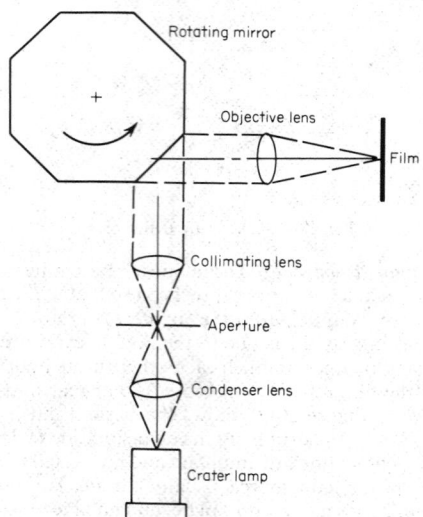

Fig. 20-92.   Recorder using multifaced mirror.

A collimating lens then renders the light emerging from the aperture as a parallel beam. An objective lens takes light reflected from the mirror surface and focuses it on the surface of photographic film. As the mirror rotates, the angle of the illuminated face changes, causing the spot to move across the film. When the spot reaches one edge, the next mirror face becomes illuminated and the spot again sweeps across one photographic film, thereby forming a facsimile image of the subject copy.

Laser Rotating-Mirror Recorder. Figure 20-93 shows a laser recorder which employs a rotating pyramidal mirror. Light from a helium-neon gas laser passes through a light modulator and an optical system. The beam expander enlarges the beam diameter, and the imaging lens forms the recording spot. The rotating mirror causes the recording spot to sweep across the photographic film. The photographic film is held in a cylindrical configuration to keep the spot in focus throughout its sweep. In addition, the film is uniformly translated at right angles to the spot motion to accomplish the longitudinal portion of the recording operation.

Oscillating-Mirror Technique. An oscillating-mirror system using a mechanical-cam configuration is shown in Fig. 20-94. Light from the recording lamp illuminates an aperture

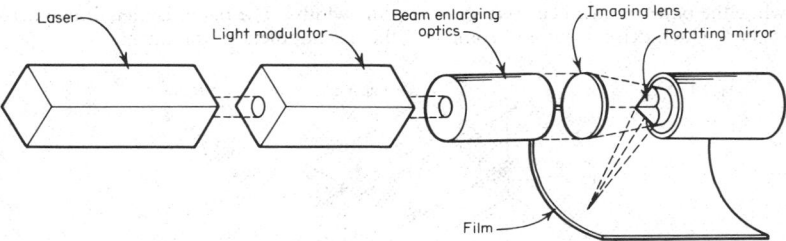

Fig. 20-93. Rotating-mirror laser recorder.

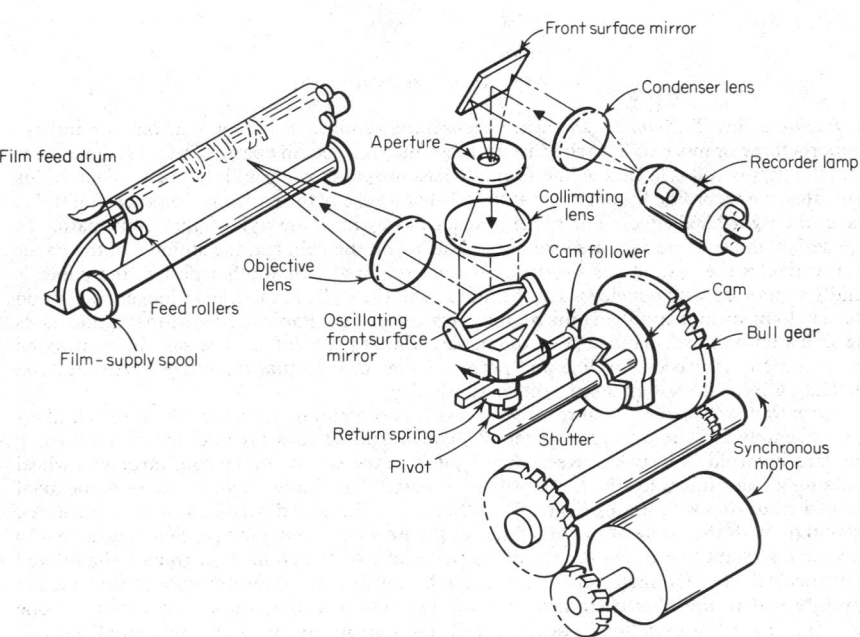

Fig. 20-94. Oscillating-mirror optical system.

20-87

which is imaged on the recording material through a light path that includes an oscillating mirror. The spot location on the recording material depends upon the mirror angle, controlled by a cam-and-follower arrangement. The spot starts at one edge and sweeps across the recording material at a constant rate. The accuracy of the drive system for the oscillating mirror may be considerably lower than that for the rotating-drum mirror. Any positional error in the low-gear cam surface repeats for every recording line and does not cause a jitter pattern in the recorded copy. On the other hand, the oscillating-mirror system is speed-limited; i.e., the cam follower will not stay on the cam surface during the return trace if the speed is too high.

Helix-and-Bar Technique. In this system electrolytic paper is drawn by feed rollers from a container and over a metallic spiral mounted on an insulating cylinder. As shown in Fig. 20-95, the spiral sweeps across the back of the paper as the cylinder rotates. A wire metal bail presses lightly against the front of the paper. A current whose amplitude is controlled by the facsimile signals passes from the bail through the damp paper to the metallic helix at the point of intersection, depositing (by electrolytic action) minute amounts of the material of which the bail is composed. The helix consists of a single turn. Hence as its intersection with the bail passes beyond the right edge of the paper, it starts again at the left edge. Meanwhile the paper is advanced one scanning-line width. The paper is then drawn across a heater bar, which dries it out and completes the development of the image.

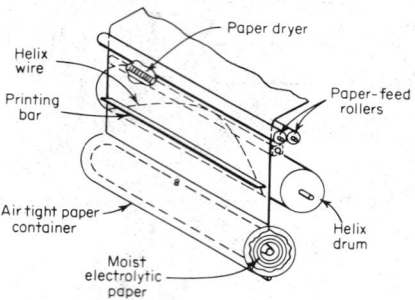

Fig. 20-95. Helix-and-bar recorder.

Bug-on-a-Belt Technique. In these recorders an endless belt mounted on two pulleys supports three or more styli, each spaced a recording-line length apart. The recording paper from the supply roll is fed at the line-feed rate past the paper guide and styli belt. Each stylus generates one recording line, starting at the left-hand edge of the paper as the preceding stylus leaves the right-hand edge. The paper is spaced away from the styli on the return path. In one design, the styli are mounted solidly to holders on the belt, and the stiffness of the paper itself provides the light stylus-to-paper pressure required. In another design, the paper is solidly supported with a metal platen behind it, and the styli are mounted plunger fashion on the belt, light springs supplying the required pressure. A suitable hardened-metal skid eases the stylus onto the left-hand edge of the paper. Great precision is necessary in positioning the styli along the recording line and perpendicular to it so that repetitive horizontal and vertical patterns do not appear in the facsimile copy.

Fiber-Optics Recorder. Figure 20-96 shows a photographic recorder which uses a fiber-optic assembly as a circle-to-line converter. The illustrated device records in full color, using standard Polaroid Colorpaks. Recording light from the xenon lamp passes through a wheel containing color filters for the three primary colors. The filtered light illuminates the axial end of a crank-shaped rotating fiber. The other end of the fiber describes a circle as it rotates, coinciding with the circle at the other end of the fiber-optic assembly, which is arranged to move in a straight line across the recording material. As the crank fiber rotates, the filtered light successively illuminates fibers in this circle, causing the recording spot to traverse the opposite end of the fiber-optic array. Each revolution of the crank corresponds to one recording line. For each line recorded, a lead-screw arrangement moves the entire Colorpak up one line width. A line-sequential recording technique is used to record at 120 lines per inch for each of the primary colors, the first line red, the second (displaced one-third of a recording line width) blue, and the third (displaced two-thirds of a recording line width) green. The recorder contains circuits which automatically adjust the recording level to the color sensitivity of the film.

117. Electronic Recording Techniques. Recorders for high-speed applications commonly utilize a flying-spot cathode-ray tube (CRT) with a beam current modulated in accordance with received facsimile signals. Other newer techniques for high-speed electronic recording include multistylus electrode arrays, light-emitting diode arrays, and ink-jet printers.

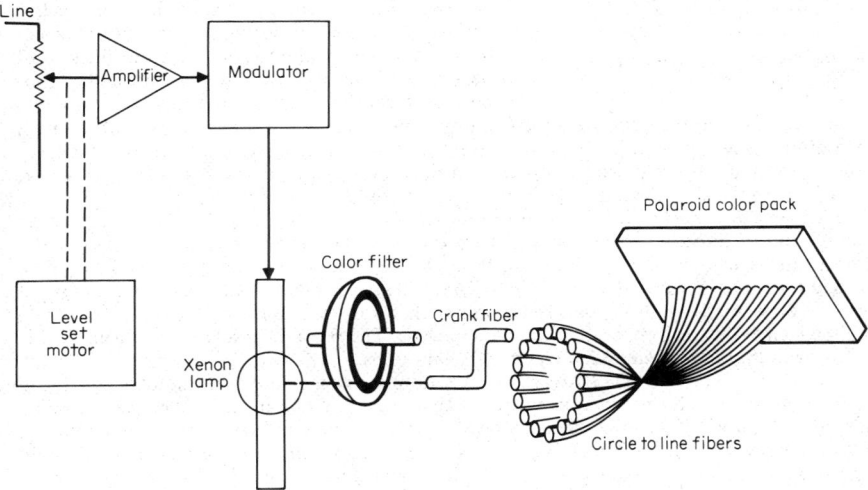

Fig. 20-96.   Fiber-optics color recorder.

Flying-Spot CRT Recorder. Cathode-ray tube recorders may generate either a full raster or single line scan. The raster scan is simpler to understand, since it amounts to taking a picture of a television screen. However, instead of using two interlaced half-resolution images, a single sweep of the tube face generates the complete facsimile copy. An objective lens forms an image of the CRT pattern on the recording material and thereby effects the exposure. After completion of the frame, the paper moves to an automatic processor, and a new sheet of paper is brought into the recording area.

Cathode-ray tubes are capable of very high speed recording so that the loading time of the photographic paper may be an appreciable percentage of the total recording time. The objective lens must have a focal length long enough to produce an acceptably sharp image in the corners of the frame. For good-quality recording, dynamic focus must be used to keep the electron beam from diverging as it leaves the center of the tube. Dynamic brightness compensation may also be required.

Fiber-Optic Faceplate CRT Recorder. This system uses a line, rather than a raster scan, and requires no lens between the CRT and photosensitive paper. A fiber-optics faceplate consisting of many fine glass fibers sits in direct contact with the photosensitive surface of the recording paper (see Fig. 20-97). When the electron beam stimulates the fluorescent layer, light enters one or more of the glass fibers and appears at the surface of the faceplate. Light-sensitive printing paper (silver halide printing paper, electrofax paper, or the like) is held in close contact with the tube to record a latent image.

Thin-Window CRT Recorder. Another form of contact-writing CRT is the thin-window type. A very narrow strip of mica, only a few thousandths of an inch thick, replaces the fiber-optics faceplate of Fig. 20-97. Figure 20-98 shows a thin glass-window printing tube developed in Japan employing a 180-μm glass window as a face plate. Fluorescent light, stimulated by the impinging electron beam, appears on the surface of the faceplate. Owing to the thinness of this window, the generated light has little chance to diverge or scatter while passing through the faceplate. On emerging from the window, the light strikes a light-sensitive material, such as photographic film, and thus generates the desired latent image. The efficiency in exposing the photosensitive material is reported to be even higher than the fiber-optics faceplate tubes.

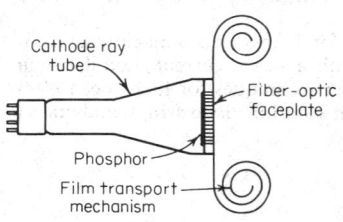

Fig. 20-97. Fiber-optics faceplate recorder.

Thin-Film Penetration CRT Recorder. Thin-film penetration tubes use an electron beam to expose the recording medium without converting it into any other form of energy, such as light or electric charge. The penetration window prevents the electron beam from scattering in the atmosphere and hence degrading the resolution. The nickel support protects the penetration window from being worn away by direct friction with the printing paper. The main feature of the tube is high recording speeds. In addition, the tube has no afterglow, as in fiber-optics printing tubes, or time lag from the charge-discharge cycle, as in electrostatic-charge printing tubes. However, recording media with adequate electron-beam sensitivities remain to be developed.

Electron-Beam Recorder. In the electron-beam recorder (see Fig. 20-77), the recording film is exposed directly by the electron beam in a vacuum. The unexposed film passes into the vacuum, past the recording line, and through a separate vacuum seal to the take-up reel. A vacuum pump is required to make up for the air leakage at these seals. The recording spot may be very small—with proper design less than 1 μm. Writing rates of the order of 200 million elemental areas per second can be achieved. The recorder is relatively expensive due to the vacuum-pump system and electron-beam optics.

Wire-Matrix Faceplate CRT Recorder. A wire-matrix faceplate printing tube, developed by Stanford Research Institute, uses electric charge instead of light to expose the recording medium. Figure 20-99 shows one such printing tube. Small metallic pins are embedded in a matrix array in the faceplate of the tube. The pins are 35 μm in height and about 1 mm in length, with a center-to-center spacing of 70 μm. During recording, the electron beam strikes and charges the pins. The charge subsequently emerges on the surface of the tube. Electrostatic printing paper moves past the faceplate, and the paper is charged by gaseous

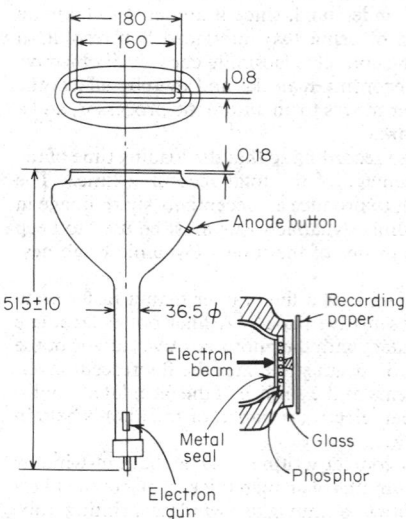

Fig. 20-98. Thin-window cathode-ray-tube recorder. Dimensions are in millimeters.

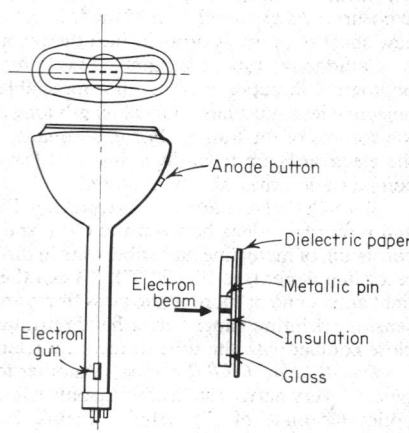

Fig. 20-99. Wire-matrix faceplate recorder.

ionic conduction between the pins and the paper. The charged paper is toner-developed to convert the electrostatic image into a visual one. The maximum resolution is about 7 lines per millimeter and is influenced by capacity between the pins. Printing density vs. beam current is generally contrasty. Tube sensitivity is about one-third that of a fiber-optics printing tube if the performances of the two electron guns are comparable.

Multielectrode-Array Recorders. This technique uses a high-voltage electrode array to write on dielectric or electrosensitive paper. Facsimile signals from the transmitter are amplified and rectified in the receiver and then applied sequentially to the elements of the electrode array. (The sequencing is accomplished by timing and coincidence circuits.) The electrode array itself consists of about 1,000 thin conductors insulated from one another but in electrical contact with the dielectric paper. As the paper moves past the printing station, an electrostatic latent image is formed on the paper. Following imaging, the paper is rubbed with toner particles and fixed by heating. The technique features medium speed and easy digital-computer interface.

LED Array Recorders. Light-emitting diodes (LED) are attractive for photofacsimile recording. The recording power required is quite low, typical LEDs requiring 5 to 50 mA at 1.5 V, and the response time quite fast, of the order of 1 μs. These diodes are already used for sound recording on film by mounting the diode in near contact with recording material. For facsimile recorders, it has been proposed that a series of light-emitting diodes be interfaced with a fiber-optics array. Another possibility is to use an imaging lens. This system, however, has a lower effective f number and consequently a lower light efficiency. The latter problem can be serious in either configuration, since the LED features moderate brightness at the longer optical wavelengths (red and green).

Acoustical Techniques. These devices use an acoustically driven deflector to effect the horizontal portion of the recording raster. The laser beam is first modulated by a Kerr cell, an electrooptical shutter, or an electromechanical-galvanometer arrangement using the incoming facsimile signal. The beam is then expanded and collimated by a beam-expanding lens system. The acoustical deflector angles the expanded beam to the right or left, after which the beam condenser focuses it onto photosensitive paper. Longitudinal scanning is accomplished by moving the paper perpendicular to the deflection direction. The technique remains at the laboratory stage for facsimile but has found commercial application in digital-computer memories.

118. Recording Materials and Processes. Facsimile systems generate card-copy imagery by applying light, heat, electricity, or pressure to a sensitized recording medium. The medium can be a specially prepared paper or an auxiliary screen or drum. In the past, electrolytic techniques and burn-off papers dominated the commercial market, with carbon-transfer processes also having some importance. More recently, electrostatic and electro-photographic methods have gained popularity, mainly because of improved copy quality. In the future all-electronic, dry-stabilized organic photographic systems may be expected.

Table 20-5 lists the principal materials and processes available for facsimile recording.

FACSIMILE TRANSMISSION

BY EARLE D. JONES

119. Analog and Binary Facsimile Signals. Of the two fundamental classes of facsimile systems in use, the more common, the *analog type,* accepts as input copy of pictorial information displaying a continuous scale of grays, as well as black-and-white line drawings and documents. Such systems must preserve the gray scale throughout the facsimile process to the output copy.

The second class of systems, the *binary type,* restricts the input copy to drawings and documents where the scanned material is either white or black; no in-between gray-scale values are allowed. This type of system is restricted to typewritten and printed documents, line drawings, and the like.

The binary type of facsimile signal can be *clocked,* i.e., sampled at regular intervals, or *unclocked.* When clocked, the facsimile signal is affected by the sampling process since black-white transitions are forced to occur at the time of the sampling clock. This may introduce a horizontal displacement (jitter) of edges in the facsimile copy. If the clock frequency is sufficiently high, the effects of this horizontal jitter are minimal. On the other hand, if the clocking frequency is low, the signal transitions are shifted laterally by an appreciable amount

Table 20-5

A. Inorganic photographic techniques
1. Conventional silver-halide photography, with separate development
2. Stabilized silver halide photography, developer included in emulsion
3. Reversible silver process (Itek RS) using photoconductor layer for recording
4. Dry-processing technique (Kodak) heat stabilized, developed by exposure to ultraviolet radiation
5. Dry-silver process (3M Corp.), a combination of silver halide photography and thermography
6. Diffusion transfer reversal (Polaroid DTR process)

B. Organic photographic techniques
1. Photoimaging materials (DYLUX); organic dyes developed by ultraviolet radiation
2. Photothermographic materials (Warren Type 1264); colored dyes with ultraviolet exposure and heat development (American Can nitrone process); similar procedure based on aryl nitrone
3. Photopolymerization imaging; laboratory stage only
4. Free-radical materials (Horizon); dyes combined with polyhalide and accelerator; controllable gamma

C. Electrophotographic techniques
1. Xerox process (see Fig. 20-100)
2. Electrolytic process (Nashua) employs photoconductive and dye layers
3. Electrofax process (RCA); photoconductive paper
4. Electroprint process (ElectroPrint); electrostatically modulated aperture for transport of toner to print media
5. Persistent internal polarization; photoconducting electret to store latent image, with toner powder development
6. Electrostatic reproduction (SCM); reverse photoelectric effect; electrostatic latent image with toner power development
7. Photoelectrophoretic technique (Xerox); photoconducting toner particles, with preferential absorption of light; applicable to color reproduction
8. Reflex photoconductor (Friden) corona charge of zinc oxide–coated sheet in direct contact with document

D. Electrical techniques
1. Electrolytic recording
2. Electrosensitive recording
3. Conductive papers
4. Capacitive papers
5. Dielectric recording
6. Electrochemical recording

E. Pressure techniques
1. Carbon transfer papers
2. Blush papers
3. Chemical papers

F. Heat techniques
1. White wax on black base; heated stylus removes wax

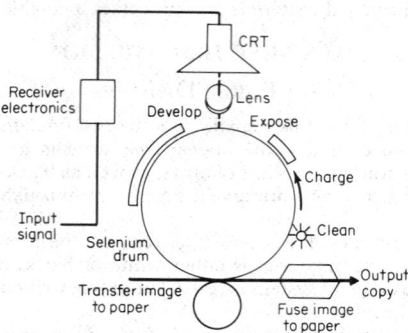

Fig. 20-100. Xerox electrophotographic system.

and appear as unacceptable distortion in the copy produced. The use of a higher sampling frequency represents more information bits along the scanning line, requiring either a larger communication bandwidth or more document-transmission time. In clocked binary facsimile systems, therefore, a trade-off must be made between the appearance of the output copy and the time of transmission.

In an unclocked system, black-white transitions may occur anywhere along the scanning lines; i.e., they are not forced to coincide with a periodic sampling pulse. Such a system is capable, at least in theory, of making an exact reproduction of the input copy without horizontal displacement or jitter. On the other hand, clocking the binary facsimile permits the economics of digital signal processing to be employed.

Where the input copy must reproduce the gray levels, as in Wirephoto or weather-satellite photographic-facsimile systems, it is desirable to duplicate as nearly as possible all the levels of gray present in the input and multilevel digital processing must be used. For example, if 64 distinct levels of gray are desired, each sample can be encoded with 6 binary bits ($2^6 = 64$). This results in a very high bit rate, since each picture element of the input copy corresponds to 6 bits of information in the output data stream.

120. Encoding. Raw data from a scanner may undergo an encoding process in two steps. The first, *source encoding*, takes advantage of the statistics of the source (input copy) to achieve an increase in transmission efficiency. The second, *channel encoding*, depends on the statistics of the communication channel to effect more efficient transmission.

Most facsimile systems in use today do not employ source encoding. Some use run-length coding, by which considerable savings can be made for certain types of input copy.

Figure 20-101 shows a typical clocked output from a black-and-white input document. The scanned output is sampled, and a binary stream of data is generated, each white sample denoted by a 0, each black sample by a 1. The 61 samples in the segment are encoded as 61 bits. In run-length encoding the number of bits in each uninterrupted run of black or white is sent as a binary number. For example, the first four samples are white. This is encoded as 0100 (binary 4). This is followed by eight black samples. This is encoded as 1000 (binary 8). The process is continued, encoding the length of each run until the end of the data segment is reached. In the example shown, 61 data bits are run-length-encoded as 32 data bits. At the receiver, the encoded data are separated into groups of four bits each, then decoded into specific run lengths for black-and-white runs. Thus a significant reduction in the number of data bits is achieved.

Channel encoding, normally incorporated in the modem, tailors the facsimile signal to the exact characteristics of the communication channel, seeking out the maximum possible transmission-efficiency limit of the channel itself, its frequency response, noise, distortion, etc. The most common type of channel encoder accepts a binary (two-level) signal and converts it into a multilevel signal. The resulting signal can be sent over a narrower-bandwidth communication facility or alternatively at a higher rate of speed over the same bandwidth communication link. The simplest type of digital channel encoder converts groups of input binary samples into a single multilevel value.

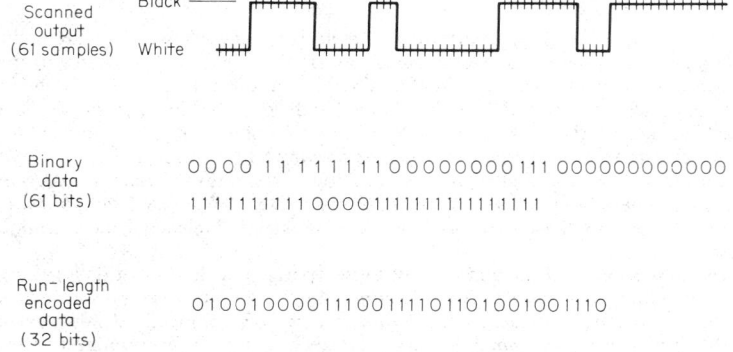

Fig. 20-101.   Run-length encoding.

121. Facsimile Modems. Before delivery to the communication link, the output signal from the facsimile scanner must undergo a modulation process. The system component that performs this modulation function (and its associated demodulation at the receiving terminal) is usually referred to as a *modem* (from *mo*dulation-*dem*odulation). For very short transmission distances the scanner and printer may be connected by a pair of wires, in which it is necessary only to provide the appropriate voltage levels for proper interconnection. On the other hand, when long-distance facilities are employed, where circuit losses and distortion are substantial or the characteristics of the communication link vary with time, the performance requirements placed on the modem become severe.

A facsimile modem is shown in simplified form in Fig. 20-102. The processed video signal from the scanner is fed to a modulator, whose function it is to shift the signal spectrum of the scanner output to the appropriate band determined by the communication link. The filter removes any undesired signal components produced by the modulation process. A switch delivers the signal to the transmission facility. In the receiving mode the incoming data are switched to the demodulator, whose function is to restore the facsimile signal to its original form before modulation. An equalizer follows; its purpose is to compensate for any distortion in the communication line. The equalized signal is then fed to the facsimile printer.

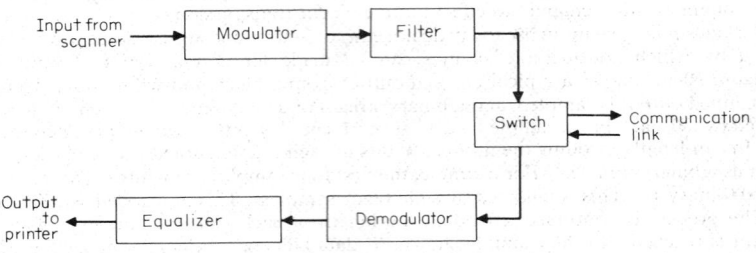

Fig. 20-102. Elements of a facsimile modem.

122. Modulation Schemes. The choice of modulation method strongly affects the performance of the facsimile system since it influences document transmission time, error performance, and the cost and complexity of the system. Three basically different modulation methods are in current use: amplitude modulation (AM), frequency modulation (FM), and phase modulation (PM). Variations on these basic schemes include *frequency-shift keying* (FSK), in which a binary signal is represented by two signaling frequencies denoting 0s and 1s; *vestigial-sideband* amplitude modulation (VSB), in which only a single sideband of an AM signal is transmitted, with a vestige of the modulated carrier; and *four-phase modulation*, in which each pair of input bits is represented by a single phase of the modulated carrier. Another approach employs four different carrier phases and four amplitude levels to achieve—in effect—16 different signaling states, each of which represents 4 information bits.

Experiments have been conducted with *duo-binary encoding* followed by VSB modulation. In duo-binary processing, a two-level signal is converted into a three-level signal by inverting alternate pulses. The advantages of the duo-binary technique are that a two-to-one increase in speed can be achieved for a given transmission bandwidth, the binary facsimile signal need not be clocked, and the technique is very easy to implement with simple hardware.

When the data rate in bits per second is approximately equal to the available channel bandwidth measured hertz and high accuracy (low error rate) is desired, *double-sideband techniques* are generally employed. For higher data rates (3,600, 4,800, and 9,600 bits per second) (b/s) over the dial-telephone network, vestigial or single-sideband techniques are preferred.

123. Equalization. All modems except those operated at the lowest data rates require some degree of line equalization. For a facsimile system working over a dedicated line (one that is not periodically switched), a fixed equalizer can be employed. At data rates up to about 1,200 b/s *frequency-domain equalizers* are in general use. These equalizers are passive networks that flatten the amplitude and phase response of the overall system.

Above 1,200 b/s *time-domain equalizers* are generally used. The most common configuration of a time-domain equalizer employs a long delay line tapped at equal delay increments along its entire length. At each tap the input signal is made available in both positive and negative form. A weighted sum is taken of all of these signals at the output of the equalizer. The length of the delay line and the required number of taps are determined by the severity of the distortion one is attempting to compensate. Adjustment of both the frequency-domain and time-domain equalizers is a slow process, requiring up to 10 or 15 min. For this reason equalizers of this type are generally used with dedicated facilities.

For use in a switched network, where the exact nature of the distortion encountered varies with each dialed connection, *adaptive equalization* is used; i.e., the equalizer performs its own adjustment. For example, a test pulse of known shape and duration may be transmitted; at the receiving end, any distortions are identified and used to adjust the time-domain equalizer until the test pulse is restored to its original known form. Other modems use the transmitted data to adjust the equalizer continuously.

124. Control of Errors. Unless they are encoded by such schemes as the run-length encoding described previously, facsimile data generally do not require any special error-control system. If 1 bit on the transmission facility corresponds to one picture element on the facsimile copy, a very high error rate can be tolerated since 1-bit errors appear as small specks against the background. Error rates of 1 in 10^3 to 10^4 are tolerable for unencoded facsimile data.

When the data are heavily encoded, a rate better than 1 error in 10^5 is usually required. In such cases either *forward error correction* or *feedback error correction* is used. In forward error correction, data efficiency is sacrificed by adding extra bits to the data stream, to detect and correct errors caused by noise on the transmission line. It is possible to detect and correct all errors if a sufficient number of control bits are added to the data stream. Usually some compromise is made between data efficiency and the ability to check and correct errors. Typically the error-correcting bits constitute 10 to 40% of the bits transmitted.

Feedback error-control systems detect but do not attempt to correct errors at the receiver. When an error is detected in a block of data, the transmitter is notified by the use of a narrowband reverse channel. The block of data in error is then retransmitted to the receiver. This system also causes reduction in data rate since time is consumed in retransmission of data. However, the system is 100% efficient when no errors are present.

125. Coupling to the Channel. Another important consideration in the performance of the facsimile-system modem is the means used to couple it to the transmission line. For dedicated facilities, a wire connection can be used directly to the transmission line. For the switched telephone network, three types of coupling equipment are currently in use. The Bell System 602 Dataphone, in use since 1959, includes a modem and dial-telephone handset. The scanner output is fed directly to the data set, and the user dials any other terminal so equipped. A more convenient access to the dial-telephone network is the data access arrangement (DAA). The DAA allows the facsimile user to connect his modem through the standard telephone handset after establishing the link by voice. Models of the DAA are available that provide an automatic answering function for unattended reception.

The other coupling modes are in use that connect the facsimile signal directly through a telephone handset by *acoustic coupling* and by *inductive coupling.* The acoustic coupler uses the output of the facsimile modem to modulate a small source of sonic energy. This sonic energy is coupled into the mouthpiece microphone of the handset. At the receiving end the sound emerging from the earpiece is detected by a small microphone that provides the signal for the receiving modem. In the inductive coupler the same function is performed by delivering the modem output to a coil placed in close proximity to the handset. Acoustic and inductive couplers are used chiefly at the low data rates associated with teletype terminals.

126. Facsimile Transmission Facilities. The most readily accessible means for telecommunications is the public dialed telephone network. Via this network, practically any country in the world with some telephone service can be reached. The International Telecommunication Union (ITU) sets standards for telecommunications to provide compatibility between telephone sets throughout the world. Since the telephone is the most accessible telecommunication instrument, most facsimile systems are designed to take advantage of existing telephone lines.

The line characteristic of a voice communications channel are determined by the following parameters: impulse noise, nonimpulse noise, frequency response (attenuation distortion), envelope delay distortion, echo and echo suppression, harmonic distortion, frequency offset, and incidental-angle modulation (phase jitter). Any or all of these parameters may have a significant effect on the performance of facsimile equipment.

Impulse Noise. Impulse noise is characterized by relative short bursts of high amplitude. It is basically a transient phenomenon appearing on the telephone channel as a result of switching in central-office equipment, electrical storms, or other causes. Impulse noise is measured by short-time-constant peak-reading instruments. Instruments presently used register impulse noise on an accumulating counter each time a pulse at the receiver terminal exceeds a threshold setting.

Impulse noise usually occurs in bursts. Therefore, the measured impulse noise is usually based on 15-min measured intervals and on the signal-to-impulse-noise threshold of 5 dB below the received signal power. This level was established because it provides a good evaluation of the impairment of data transmission caused by impulse noise.

For voice and unencoded facsimile transmission, the effects of impulse noise are usually less critical because of the inherent redundancy. Table 20-6 shows some typical impulse-noise measurements.

Nonimpulse Noise. In a standard voice circuit, the steady background noise is measured on a weighted basis using a filter having a characteristic as shown in Fig. 20-103. This filter limits the noise-power contributions below 600 and above 3,400 Hz considerably.

Table 20-6. Impulse Noise Performance*
(Number of Impulses in a 15-min Interval)

Connection links	Observed count distribution		
	Mean	Median	Standard deviation
Short (0–180 mi)	32±15	3	105
Medium (180–725 mi)	48±44	7	150
Long (725–2900 mi)	74±48	15	219
All (0–2900 mi)	39±21	4	128

* From Duffy and Thatcher, *Bell Syst. Tech. J.*, April 1971.

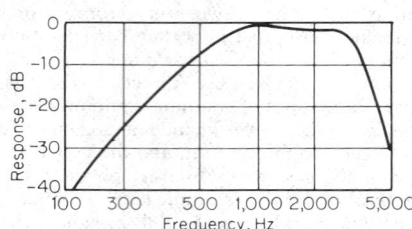

Fig. 20-103. Characteristic of noise-weighting filter (C-message weighting).

This characteristic is deliberately introduced in order to suppress any 60-Hz harmonics of the power-line network, which may be significant up to 300 Hz.

Noise measurements can be made on idle channels. However, for the facsimile user, the signal-to-noise ratio is of greater interest. Unfortunately, an idle-noise measurement does not allow the computation of the signal-to-noise ratio because the line noise is usually signal-dependent. This is due to compandors on analog channels and quantizing noise on digitized voice circuits. Special equipment is used for direct measurement of the signal-to-noise ratio. Table 20-7 gives some typical figures for the ratios experienced on a switched line.

Frequency Response (Attenuation Distortion). The frequency response or attenuation distortion of a channel is usually referenced to the 1,000-Hz test-tone level. The mean values and standard deviation as measured in 1969–1970 are published by the American Telephone and Telegraph Company. Figure 20-104 shows a plot of attenuation distortion.

Envelope Delay. The envelope-delay distortion is measured as a function of frequency with reference to 1,700 Hz. It should be noted that no absolute value for delay is specified; however, the relative distortion is a large contributor to intersymbol interference and limits the maximum data rate which can be transmitted over the voice line. Figure 20-105 shows measured data on the mean values.

Echoes. Impedance irregularities along the transmission line or at its end may cause a portion of the signal energy to be reflected back toward the sending end, causing it to appear as an echo to the sender. The echo will not affect facsimile or data transmission if the transmission is performed in a half-duplex mode; i.e., transmission takes place only in one direction at a time.

Table 20-7. Voice-Channel Signal-to-Noise Ratio in Decibels*

Connection length	Mean	Standard deviation
Short	41.7	11.8
Medium	40.1	5.5
Long	36.7	4.1

* From Duffy and Thatcher, *Bell Syst. Tech. J.,* April 1971.

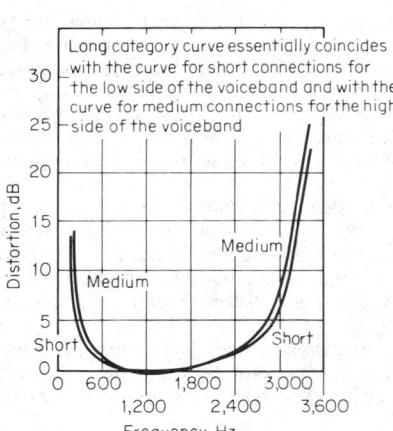

Fig. 20-104. Mean values of attenuation distortion relative to 1,000 Hz.

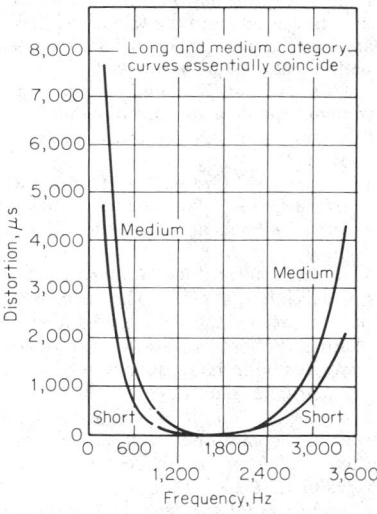

Fig. 20-105. Mean values of envelope-delay distortion relative to 1,700 Hz.

More severe is the problem that occurs when an impedance irregularity reflects the signal back with a certain time delay to the receiver. For data transmission, the receiver echo can produce direct interference with the received data if the time delay is significant and the echo is of sufficient amplitude. Usually long-haul DDD circuits are equipped with echo suppressors that attenuate both the listener and talker echo at the same time. The effects of echo suppression on facsimile transmission are inconsequential except for the time delay required in reversing the direction of signal flow.

Harmonic Distortion. Measurements made for harmonic distortion are referenced to a fundamental frequency of 525 Hz. Table 20-8 gives the test results of measurements made by American Telephone & Telegraph.

Frequency Offset. Medium- and long-haul telephone systems utilize frequency-division carrier multiplex equipment which may introduce minor frequency deviations to the signal. In the last few years improvements in the carrier system have reduced the frequency offset to less than 1 Hz per facility section.

Incidental-Angle Modulation (Phase Jitter). Both analog and pulse code modulation (PCM) carrier systems may introduce a repetitive shift in phase or frequency of the signal. The impairment caused by phase jitter to data transmission depends on its magnitude and

Table 20-8. Results for the Ratio of Received 525-Hz Fundamental to the Second and Third Harmonics on Toll Connections*

Connection length airline miles	Second		Third	
	Mean dB	Standard deviation, dB	Mean dB	Standard deviation, dB
All	41.8 ± 2.5	11.6	45.7 ± 2.4	12.3
0–180	42.6 ± 3.3	12.4	47.3 ± 2.6	13.3
180–725	38.2 ± 1.9	7.8	39.7 ± 2.4	6.3
725–2900	41.6 ± 1.4	7.6	42.0 ± 1.6	5.3

* From Duffy and Thatcher, *Bell Syst. Tech. J.*, April 1971.

frequency. Phase-jitter frequencies are typically below 1,000 Hz and cause sidebands with magnitudes no greater than 18 dB below the level of the carrier.

127. Leased-Line Facilities. A facsimile network for very high volume traffic would include a transmission facility dedicated to the network. The modems for a dedicated line may be leased from the telephone company or owned by the user. If user-owned modems are used, an interface device has to be leased unless an acoustic coupler is used in conjunction with a standard telephone set.

Narrowband Facilities. In general, narrowband facilities are defined by the capability of transmitting data speeds no higher than 300 b/s. At present, the majority of computer-data terminals make use of these facilities, but they are of limited value for transmission of facsimile.

Series 1000 Channels. Most of the series 1000 services are offered by both AT&T and Western Union. The channels are unconditioned and capable of transmitting dc mark-space or binary signals at rates up to 150 bauds. They are used for remote-metering supervisory control.

Series 2000 and 3000 Channels. These services are defined as channels with a bandwidth not exceeding 4 kHz. They are used for voice, supervisory control, remote operation of mobile radiotelephone systems, civil defense, data transmission, and facsimile. The 3002 channel is offered with several grades of line conditioning. This conditioning or equalization improves both the frequency response and the envelope-delay distortion of the channel and allows higher data rates.

Series 4000 Channel. The series 4000 is an expansion of the 3000 series, providing specially conditioned high-grade circuits for data and telephoto service. These channels show useful bandwidth from 300 to 3000 Hz and exhibit error rates less than 1 in 10^5 at data rates of 1,300 to 1,600 b/s.

Wideband Facilities. In the communication industry, wideband channels are usually referred to as those with channel bandwidths of more than 4 kHz. AT&T offers five different series of bandwidth channels. Some of these are specifically for broadcast radio and television and therefore of little interest to the general industrial user.

Series 5000 Channel. The 5000 series is a broadband channel commonly known as the TELPAK series. The channels have the following characteristics:

Type	Maximum Channel Bandwidth, kHz	Equivalent Voice Channels
5700	240	60
5800	1,000	240

TELPAK channels are used with high-speed facsimile equipment. Transmission speed of 6 s per 8½ by 11 in. page is achieved using type 5700 channel.

Series 6000 and 7000 Channels. This series offers medium and wideband channels for the transmission of audio and video on a simplex or half-duplex basis. The service was created to satisfy the needs of the broadcast industry.

Series 8000 Channels. The 8000 series offers a 48-kHz equivalent to 12 voice channels for the transmission of high-speed data, facsimile, or individual voice-grade channels.

128. Privately Owned Communications Systems. Under FCC Rules, industry users can own and operate their own communication system. Typically the railroads are the largest operators of privately owned systems followed by pipeline and oil companies. These users have the advantage of having the right-of-way for their facilities which enables them to use wireline or cables as well as microwave systems. Non-right-of-way companies are restricted to microwave systems.

There are several types of microwave systems available, low-density and high-density systems. Low-density systems usually have a capability of 12 or 24 voice channels with expansion capabilities up to 120 channels. High-density systems can have as many as 960 or 1,800 voice channels.

In a privately owned communication system the channel assignment can be made to suit the needs of the owner. The system can accommodate facsimile, high-speed data, voice or teletype, or similar circuits.

129. Government Owned Communication Systems. Since the middle 1960s state and local governments have made large investments in their own communication facilities. Many states and foreign countries have backbone microwave facilities serving public safety and education. Again the wideband capability of these systems varies according to the need of the users. Applications include teletype, voice, facsimile, digital-data, and television channels.

SECTION 21

BROADCASTING SYSTEMS

BY

JOSEPH L. STERN   Vice President and Director of Engineering, Goldmark Communications Corporation; formerly Vice President - Engineering, CBS Television Services, Columbia Broadcasting System; Senior Member, IEEE

NORMAN W. PARKER   Staff Scientific Advisor, Motorola Incorporated; Senior Member, IEEE

CONTENTS

Numbers refer to paragraphs

* Tuners and IF Amplifiers; for Detectors and Video Systems, see Sec. 20.

SECTION 21

BROADCASTING SYSTEMS

BROADCAST TRANSMISSION PRACTICE

By Joseph L. Stern

STANDARD-BROADCAST (AM) PRACTICE

1. Standard-Broadcast Allocations. The band 535 to 1,605 kHz is used for standard amplitude-modulation (AM) sound broadcasting. This band is divided into 107 channels each 10 kHz wide; carrier frequencies are assigned at 10-kHz intervals from 540 to 1,600 kHz.

Table 21-1 lists the standard broadcast channels allocated in the United States by the Federal Communications Commission (FCC) with their service classes.

Table 21-1. U.S. Standard-Broadcast Carrier Frequencies and Service Classes*

Channel, kHz	Classification	FCC Class
540	Clear	II
550-630	Regional	IIIA, IIIB
640-680	Clear	I, II
690	Clear	II
700-720	Clear	I, II
730-740	Clear	II
750-780	Clear	I, II
790	Regional	IIIA, IIIB
800	Clear	II
810-850	Clear	I, II
860	Clear	II
870-890	Clear	I, II
900	Clear	II
910-930	Regional	IIIA, IIIB
940	Clear	I, II
950-980	Regional	IIIA, IIIB
990	Clear	II
1,000	Clear	I, II
1,010	Clear	II
1,020-1,040	Clear	I, II
1,050	Clear	II
1,060-1,140	Clear	I, II
1,150	Regional	IIIA, IIIB
1,160-1,210	Clear	I, II
1,220	Clear	II
1,230-1,240	Local	IV
1,250-1,330	Regional	IIIA, IIIB
1,340	Local	IV
1,350-1,390	Regional	IIIA, IIIB
1,400	Local	IV
1,410-1,440	Regional	IIIA, IIIB
1,450	Local	IV
1,460-1,480	Regional	IIIA, IIIB
1,490	Local	IV
1,500-1,530	Clear	I, II
1,540	Clear	II
1,550-1,560	Clear	I, II
1,570-1,580	Clear	II
1,590-1,600	Regional	IIIA, IIIB

* For details on special operating procedures and assignment criteria, see FCC Rules, beginning at Sec. 73.25.

2. Allocation Standards.   Section 73.21 of the FCC Rules establishes in the United States three classes of channels in the standard broadcast band: *clear channels*, for high-powered stations; *regional channels*, for medium-powered stations; and *local channels*, for low-powered stations (see Table 21-1).   These stations have three service areas*, primary, secondary, and intermittent (see Pars. **21-8** to **21-10**).   Class I stations (Par. **21-3**) render service to all three areas.   Class II (Par. **21-4**) stations render service to a primary area, but the secondary and intermittent service areas may be materially limited or destroyed by interference from other stations.   Class III and IV (Pars. **21-5** and **21-6**) stations serve primary-service areas, as interference from other stations usually prevents secondary service and may limit intermittent service.

3. Class I stations are dominant stations operating on clear channels with powers of not less than 10 nor more than 50 kW.   These stations are designed to render primary and secondary service over an extended area and at relatively long distances.   They are so allocated (Table 21-1) that their primary-service areas are free from objectionable interference from other stations on the same and adjacent channels.   Their secondary-service areas are free from objectionable interference from stations on the same channels but not from those on adjacent channels.   If it is desired to determine the area in which adjacent-channel ground-wave interference (10 kHz removed) to sky-wave service exists, it may be considered as the area within which the ratio of the desired 50% sky wave of the class I station to the undesired ground wave of a station 10 kHz removed is 1:4.

From an engineering point of view, class I stations may be divided into two groups.   In group IA are stations on whose channels (except to the extent provided by that section and by Sec. 73.22, of the FCC Rules) duplicate nighttime operation is not permitted.   The power of these stations is required to be 50 kW.

Stations in group IA are afforded protection as follows:

Daytime: To the 0.1 mV/m ground-wave contour from stations on the same channel and to the 0.5 mV/m ground-wave contour from stations on adjacent channels.

Nighttime: To the 0.5 mV/m 50% sky-wave contour from stations on the same channel and to the 0.5 mV/m ground-wave contour from stations on adjacent channels.

Stations in group IB are those assigned to channels on which duplicate operation is permitted; i.e., other class I or class II stations operating on unlimited time may be assigned to such channels.   During nighttime hours, stations in group IB are protected to the 500 μV/m 50% sky-wave contour and during daytime hours of operation to the 100 μV/m ground-wave contour from stations on the same channel.   Protection is given to the 500 μV/m ground-wave contour from stations on adjacent channels for both day and night operation.   The operating powers of stations in group IB must not be less than 10 kW or more than 50 kW.

4. Class II stations are secondary stations which operate on clear channels with powers not less than 0.25 kW nor more than 50 kW, except that class IIA stations may not operate during nighttime hours with less than 10 kW.   Class II stations are required to use a directional antenna or other means to avoid causing interference within the protected service areas of class I stations or other class II stations.   For special rules and standards concerning class IIA stations, see the FCC Rules, Sec. 73.22.

These stations normally render primary service only, the area of which depends on the geographical location, power, and frequency.   The service area may be relatively large but is limited by, and subject to, interference from class I stations.   It is recommended that class II stations be so located that the interference received from other stations will not limit the service area to greater than the 2.5 mV/m ground-wave contour at night and the 0.5 mV/m ground-wave contour during the day (the values for the mutual protection of this class of stations with other stations of the same class).   Class IIA stations are normally protected to their 0.5 mV/m ground-wave contour (daytime) and at night to the limit imposed by the cochannel class IA station.

5. Class III stations operate on regional channels and normally render primary service to the larger cities and to contiguous rural areas.   These stations are subdivided into two classes: class IIIA stations, which operate with powers not less than 1 kW nor more than 5 kW, are normally protected to the 2,500 μV/m ground-wave contour at night and to the 500 μV/m ground-wave contour during the day.   Class IIIB stations, which operate with powers

---

* Definitions of these service areas are given in Sec. 73.11 of the FCC Rules.

not less than 0.5 kW nor more than 1 kW at night and 5 kW daytime, are normally protected to the 4,000 μV/m ground-wave contour at night and 500 μV/m ground-wave contour during the day.

6. **Class IV stations** operate on local channels, normally rendering primary service only to a city or town and the contiguous suburban or rural areas, with power not less than 250 W. The upper limit is 250 W at night and 1 kW daytime. Such stations are normally protected to the 0.5 mV/m contour daytime.

On local channels the separation required for the daytime protection also determines the nighttime separation. Where directional antennas are employed by class IV stations operating with more than 250 W power, the separations must not be less than those necessary to afford protection, assuming nondirectional operation with 250 W.

The actual nighttime limitation must be calculated. An approximate method is based on the assumption of a quarter-wavelength antenna height and 88 mV/m at 1 mi effective field for 250 W power, using the 10% sky-wave field-intensity curve*. Zones defined by circles of various radii specified in the accompanying table are drawn about the desired station, and the interfering 10% sky-wave signal from each station in a given zone is considered to be the value tabulated. The effective interfering 10% sky-wave signal is taken to be the square root of the sum of the squares (rss value) of all signals originating within these zones. Stations beyond 500 mi are not considered.

Zone	Inner radius	Outer radius	10% skywave signal, mV/m
A	...	60	0.10
B	60	80	0.12
C	80	100	0.14
D	100	250	0.16
E	250	350	0.14
F	350	450	0.12
G	450	500	0.10

Where the power of the interfering station is other than 250 W, the 10% sky-wave signal should be adjusted by the square root of the ratio of the power to 250 W.

7. **Service Reclassification.** The class of any station is determined by the channel assignment, the power, and the field-intensity contour to which it renders service free of interference from other stations as determined by the FCC Rules. No station is permitted to change to a class protected to a contour of less intensity than the contour to which the station actually renders interference-free service. Any station of a class normally protected to a contour of less intensity than that to which the station actually renders interference-free service is automatically reclassified by the FCC. Likewise, any station to which the interference is reduced, so that service is rendered to a contour normally protected for a higher class, is automatically reclassified.

8. **Signal Levels for Different Service Areas—Primary Service.** The signals necessary to render primary service to different types of service areas are as follows:

Area	Field intensity ground wave[1], mV/m
City business or factory areas	10-50
City residential areas	2-10
Rural, all areas during winter or northern areas during summer	0.1-0.5
Rural, southern areas during summer	0.25-1.0

[1] Section 73.184 of the FCC Rules gives curves showing distance to various ground-wave field intensity contours for different frequency and ground conductivities.

These values are based on the usual noise levels in the respective areas, assuming no objectionable finding or limiting interference from other broadcast stations. The values apply both day and night, but fading or interference from other stations usually limits the primary service at night in rural areas to higher values of field intensity than the values given.

The FCC will authorize a directive antenna for a class IV station for daytime operation only with power in excess of 250 W. In computing the degrees of protection which such an

*FCC Rules, Par. 73.190, Fig. 2.

antenna will afford, the radiation produced by this antenna must be assumed to be no less, in any direction, than that which would result from nondirectional operation utilizing a single element of the directional array with 250 W.

Standards are not stated for interference from atmospherics or man-made electric noise, as no uniform method of measuring these effects has been established. In an individual case objectionable interference from any source, except other broadcast signals, can be determined by comparing the actual noise interference reproduced during reception of a desired broadcast signal to the degree of interference that would be caused by another broadcast signal within 20 kHz of the desired signal and having a carrier power ratio of 20:1 with both signals modulated 100% on peaks of usual programs.

9. Secondary Service.   Secondary service is delivered in the areas where the sky wave for 50% or more of the time has a field intensity of 500 μV/m or greater. It is not considered that satisfactory secondary service can be rendered to cities unless the sky wave approaches the ground-wave value required for primary service. The secondary service is necessarily subject to interference and fading, whereas the primary-service area of a station is not. Only class I stations are assigned on the basis of rendering secondary service.

Standards have not been established for objectionable fading as such standards necessarily depend on receiver characteristics, which have changed considerably during the years. Only selective fading causing audio distortion and signal fading below the noise level are objectionable in modern receivers, since the automatic-volume-control circuits usually maintain the audio output sufficiently constant to be satisfactory during conditions of fading.

10. Intermittent Service.   Intermittent service is rendered by the ground wave. It begins at the outer boundary of the primary-service area and extends to the value of signal that has no service value. This limit may extend down to a few microvolts in certain areas and up to several millivolts in areas of high noise level, interference from other stations, or objectionable fading at night. The intermittent-service area may vary widely from day to night and from time to time, as the name implies. Only class I stations are assigned protection from interference from other stations in the intermittent-service area.

11. Objectionable Interference.   Objectionable nighttime interference from another broadcast station is defined as the degree of interference produced when at a specified field-intensity contour with respect to the desired station the field intensity of an undesired station (or the rss value of field intensities of two or more stations on the same frequency) exceeds, for 10% or more of the time, the values set forth in the FCC standards.

With respect to the rss values of interfering field intensities (except in the case of class IV stations on local channels) calculation is accomplished by considering the signals in order of decreasing magnitude, adding the squares of the values, and extracting the square root of the sum, excluding those signals which are less than 50% of the rss value of the higher signals already included.

The rss value is not considered to be increased when a new interfering signal is added which is less than 50% of the rss value of the interference from existing stations and which at the same time is not greater than the smallest signal included in the rss value of interference from existing stations. The application of this "50% exclusion" method of calculation may result in anomalies. For example, the addition of a new interfering signal or the increase in value of an existing interfering signal may cause the exclusion of a previously included signal and may cause a decrease in the calculated rss value of interference.

In such cases, an alternate method of calculating the rss values of interference should be employed.

As an example of rss interference calculations, assume that the existing interferences are:

Station	Interference, mV/m
1	1.0
2	0.60
3	0.59
4	0.58

The rss value from stations 1 to 3 is 1.31 mV/m; therefore interference from station 4 is excluded, for it is less than 50% of 1.31 mV/m.

Station A receives the following interferences:

Station	Interference, mV/m
1	1.0
2	0.60
3	0.59

It is proposed to add a new limitation, 0.68 mV/m. This is more than 50% of 1.31 mV/m, the rss value of stations 1, 2, and 3. The rss value of station 1 and of the proposed station would be 1.21 mV/m, which is more than twice as large as the limitations from station 2 or 3. Under the above provision of a new signal, the three existing interferences are calculated for purposes of comparative studies, resulting in an rss value of 1.47 mV/m. However, if the proposed station is ultimately authorized, only station 1 and the new signal are included in all subsequent calculations for the reason that stations 2 and 3 are less than 50% of 1.21 mV/m, the rss value of the new signal and station 1.

12. Coverage Estimates. For the purpose of estimating the coverage and the interfering effects of stations in the absence of field-intensity measurements, use must be made of Fig. 8 of FCC Rules, Par. 73.190, which describes the estimated effective field for 1 kW power input of simple vertical omnidirectional antennas of various heights with ground systems of at least 120 quarter-wave-length radials. Certain approximations, based on the curve or other appropriate theory, may be made when other than such antennas and ground systems are employed, but in any event the effective field to be employed shall not be less than that tabulated:

Class of station	Effective field, mV/m
I	225
II, III	175
IV	150

If a directional antenna is employed by the interfering station, the interference varies in different directions, being greater than the tabulated limiting values in certain directions and less in others. To determine the interference in any direction the measured or calculated radiated field (unabsorbed field intensity at 1 mi from the array) must be used in conjunction with the appropriate propagation curves. The existence or absence of objectionable interference due to sky-wave propagation must be determined by reference to the appropriate propagation curves (Fig. 1a or Fig. 2 of Sec. 73.190 of the FCC Rules).

A summary of FCC regulations applicable to standard broadcast stations appears in Table 21-2.

According to the FCC Rules, Table 21-3 is to be used for determining the minimum ratio of the field intensity of a desired to an undesired signal for interference-free service. For a desired ground-wave signal interfered with by two or more sky-wave signals on the same frequency, the rss value of the latter is used. From Table 21-3 it is apparent that in many cases stations operating on channels 10 and 20 kHz apart may be operated with antenna systems side by side or otherwise in proximity without any indications of interference if the interference is defined only in terms of the permissible ratios listed. As a practical matter, serious interference problems may arise when two or more stations with the same general service area are operated on channels 10, 20, and 30 kHz apart.

13. Ground-Wave Field Intensity. The computed values of ground-wave field intensity as a function of the distance from the transmitting antenna are given in the FCC Rules, Sec. 73.184, Graph 12. The ground-wave field intensity is considered to be that part of the vertical component of the electric field received on the ground which has not been reflected from the ionosphere or the troposphere. These charts were computed for a dielectric constant of 15 for land and 80 for seawater (referred to air as unity) and for the ground conductivities (expressed in millimhos per meter) given on the curves.

The curves show the variation of the ground-wave field intensity with distance to be expected for transmission from a short vertical antenna at the surface of a uniformly conducted spherical earth with the ground constants shown on the curves; the curves are for

Table 21-2. Summary of FCC Regulations for Standard Broadcast Stations

Class of station	Class of channel used	Permissible power, kW	Signal-intensity contour of area protected from objectionable interference,[a] μV/m		Permissible interfering signal on same channel,[b] μV/m	
			Day[c]	Night	Day[c]	Night[d]
IA	Clear	50	100 same channel 500 adjacent channel	500 same channel (50% sky wave)[e] 500 adjacent channel[c]	5	25[e]
IB	Clear	10–50	100 same channel 500 adjacent channel	500 same channel (50% sky wave) 500 adjacent channel[c]	5	25
IIA	Clear	0.25–50 day 10–50 night	500	500	25	25
IIIB, IIID	Clear	0.25–50	500	2,500[f]	25	125
IIIA	Regional	1–5	500	2,500[c]	25	125
IIIB	Regional	0.5–1 night 5 day	500	4,000[c]	25	200
IV	Local	0.25 night 0.25–1 day	500	[g]	25	[g]

[a] When a station is already limited by interference from other stations to a contour of higher value than that normally protected for its class, this contour shall be the established standard for such station with respect to interference from all other stations.

[b] For adjacent channel, see paragraph (w) of FCC Rules.   [c] Ground wave.

[d] Sky-wave field intensity for 10% or more of the time.

[e] Class IA stations on channels reserved for the exclusive use of one station during nighttime hours are protected from cochannel interference on that basis. On the frequency 770 kHz, two class I stations may be assigned.

[f] These values are with respect to interference from all stations except class IB, which stations may cause interference to a field-intensity contour of higher value. However, it is recommended that class II stations be so located that the interference received from class IB stations will not exceed these values. If the class II stations are limited by class IB stations to higher values, then such values shall be the established standard with respect to protection from all other stations.

[g] Not prescribed; see paragraph (a)(4) of FCC Rules.

an antenna power and efficiency such that the inverse-distance field is 100 mV/m at 1 mi. The curves are valid at distances large compared to the dimensions of the antenna for other than short vertical antennas. A typical FCC set of ground-wave field-intensity curves, computed for 1,000 kHz, is shown in Fig. 21-1.

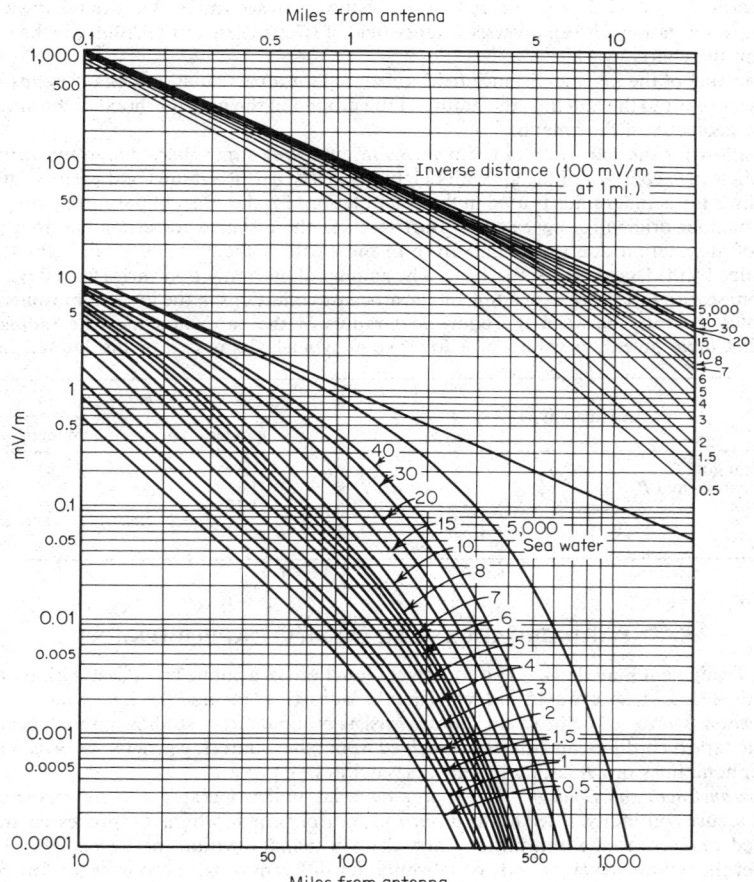

Fig. 21-1. Curves for ground-wave field intensity vs. distance computed at 1,000 kHz for various values of ground conductivity. The upper curves apply to the upper scale of distance, lower curves to the lower scale. *(From FCC Rules, Sec. 73.184.)*

Table 21-3. Desired-to-Undesired Signal Ratios

Frequency separation of desired to undesired signals, kHz	Desired ground wave to:		Desired 50% sky wave to undesired 10% sky wave
	Undesired ground wave	Undesired 10% sky wave	
0	20:1	20:1	20:1
10	1:1	1:5	*

* The secondary service area of a class I station is not protected from adjacent channel interference. However, if it is desired to make a determination of the area in which adjacent channel ground-wave interference (10 kHz removed) to sky-wave service exists, it may be considered as the area where the ratio of that desired 50% sky wave of the class I station to the undesired ground wave of a station 10 kHz removed is 1:4.

The *inverse-distance field* (100 mV/m divided by the distance in miles) corresponds to the ground-wave field intensity to be expected from an antenna with the same radiation efficiency when it is located over a perfectly conducting earth. To determine the value of the ground-wave field intensity corresponding to a value of inverse-distance field other than 100 mV/m at 1 mi, the field intensity as given on the charts is multiplied by the desired value of inverse-distance field at 1 mi divided by 100. For example, to determine the ground-wave field intensity for a station with an inverse-distance field of 1,700 mV/m at 1 mi, multiply the values given on the charts by 17.

The value of the inverse-distance field to be used for a particular antenna depends upon the power input to the antenna, the nature of the ground in the neighborhood of the antenna, and the geometry of the antenna.

To allow for the refraction of radio waves in the lower atmosphere due to the variation of the dielectric constant of the air with height above the earth, a radius of the earth equal to four-thirds the actual radius is used in the computations for the effect of the earth's curvature in the manner originally suggested by Burrows; i.e., the distance corresponding to a given value of attenuation due to the curvature of the earth in the absence of air refraction is multiplied by the factor $(4/3)^{2/3} = 1.21$. The amount of this refraction varies from day to day and from season to season, depending on the air-mass conditions in the lower atmosphere. If k denotes the ratio between the equivalent radius of the earth and the true radius, the following table gives the values of k for several typical air masses in the United States.

Air-mass type	k	
	Summer	Winter
Tropical gulf T_c	1.53	1.43
Polar continental P_c............................	1.31	1.25
Superior S	1.25	1.25
Average	1.33	

STANDARD-BROADCAST TRANSMITTING EQUIPMENT

14. Equipment Functions. The equipment functions of a public broadcast station (which apply to AM, FM, shortwave, and television broadcasting) include the following:

Program Source. The program source consists of inputs from studios, remote locations, disk and tape recordings, and material received from other sources, e.g. network originations, by telephone lines or cables, radio relay, or satellites.

Program Input and Control. The source material, under operator or computer supervision to assure continuity, is fed to the input and control system, where it is processed, mixed, switched, and sent to the transmitter plant. The principal functions performed are to raise or lower the volume level, provide equalization for different program sources (or for special effects), to mix multiple sources, to provide switching and continuity, and to monitor program activity and quality.

Studio-to-Transmitter Link. For audio sources, the link usually consists of an audio cable connecting the program input and control facilities to the transmitter. If the distance to be covered is large, a high-quality equalized telephone line or radio relay link may be used. In FM broadcasting, off-the-air pickup may be used under favorable reception conditions.

Transmitter Input and Control. These facilities and controls bring the level of the signal from the link up to the point that the transmitter can accept. Included are limiting and clipping amplifiers to maintain a high average signal level, thus keeping the modulation percentage and output power as high as is consistent with good audio quality and the FCC Rules. At the transmitter, facilities are also provided for the measurement of the input level and output monitors to check the overall quality of the signal, modulation percentage, and carrier-frequency tolerance. The high degree of stability exhibited by modern AM transmitters has led to widespread use of remote control. The transmitter is unattended, and all control and monitoring functions are performed at the program-source location.

15. Transmitter Performance Standards.   The Electronic Industries Association (EIA, formerly RETMA) has formulated standards of performance for AM broadcast transmitters, published as Standard TR-101A, Electrical Performance Standards for Standard Broadcast Transmitters.   This document, which should be consulted for detailed numerical values and measurement criteria, lists definitions, standards, and methods of measurement of the following items:

Carrier output rating.
Carrier-power output capability.
Carrier-frequency range.
Carrier-frequency stability.
Carrier shift.
Carrier noise level.
Magnitude of rf harmonics.
Normal load.
Transmitter-output-circuit adjustment facilities.
Output voltage and impedance for rf and audio monitor connections.
Modulation capability.
Audio-input level for 100% modulation.
Audio-frequency response.
Audio-frequency harmonic distortion.
Rated power supply.
Power-supply variation.
Power input.

16. Typical Transmitter.   The simplified block diagram of a typical 5- to 10-kW AM transmitter* is shown in Fig. 21-2.   The manufacturer's electrical specifications are given in Table 21-4.   The transmitter, which is air-cooled, utilizes a high-efficiency, class C plate-modulated power amplifier permitting one or two type 5762 tubes to deliver the nominal 5 or 10 kW, respectively.   The plate efficiency, according to the manufacturer, appreciably exceeds that of a conventional class C amplifier.   The circuit arrangement is similar to that of a conventional class C amplifier except for the presence of two high-efficiency resonators.   The transmitter is designed for a minimum of tuning adjustment, and after initial adjustment, all tuning can be performed from the front panel using only two controls.   The transmitter is designed for either manual or remote-control operation.

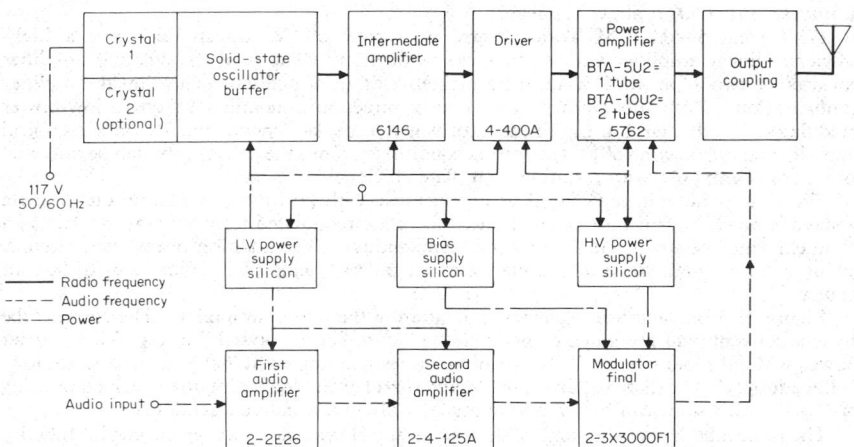

Fig. 21-2.   Block diagram of a typical 5- to 10-kW standard-broadcast transmitter employing a high-efficiency class C plate-modulated final amplifier.   *(RCA Corporation.)*

*RCA type BTA-5U2/10U2.

Table 21-4.   Specifications of a Typical 5- to 10-kW AM Transmitter

Af input level (100% modulation)	+ 10 ± 2 dBm
Af response:	
50-7500 Hz	± 1 dB
30-10,000 Hz	± 1.5 dB
Af distortion (95% modulation):	
50-10,000 Hz	2.5%
Noise (below 100% modulation)	60 dB
Frequency range	535-1620 kHz
Frequency stability	± 5 Hz
Carrier shift (0-100% modulation), 400 Hz	3% at constant line voltage 5% at normal line-voltage regulation
Rf voltage, for frequency monitoring	10 V rms 75 Ω
For modulation monitoring	10 V 75 Ω
Power output (nominal):	
BTA-5U2	5,000 W
BTA-10U2	10,000 W
Power-output capability:	
BTA-5U2	5,500 W
BTA-10U2	10,600 W
Tube complement:	
1 6146 intermediate-power amplifier	
1 4-400A driver	
2 2E26 first audio-frequency amplifier	
2 6155/4-125A second audio-frequency amplifier	
2 3X3000F1 modulator	
1 5762 power amplifier for BTA-5U2	
2 5762 power amplifier for BTA-10U2	

The modulator consists of a pair of 2E26 tubes located in the exciter portion, resistance-coupled to drive two 4-125A second audio-frequency amplifiers, which, in turn, are resistance-coupled to drive a pair of 3X3000 F1 modulators. The modulator tubes are low-mu triodes, drawing no grid current.

Broadband neutralization is achieved through the use of a slug-tuned coil which is motor-driven for local or remote control. The second-harmonic trap also utilizes a slug-tuned coil, eliminating the possibility of contact pitting from high rf currents. Neutralization of the power amplifier is achieved by a broadband transformer and a variable vacuum capacitor.

The power supply for the transmitter incorporates silicon rectifiers in the high-voltage circuits. The rectifiers are hermetically sealed to minimize effects of varying weather conditions. The units can operate at ambient temperatures ranging from -20 to +45°C, at altitudes up to 7,500 ft above sea level.

17. Typical 50-kW AM Transmitters. A typical 50-kW transmitter* uses a high-efficiency linear amplifier with screen-grid modulation. The final (modulated) amplifier consists of two type 4CX35000C ceramic tetrodes in a high-efficiency linear amplifier configuration. Both tubes are simultaneously screen-grid-modulated from a low-power modulator. In this circuit, the linearity of negative peaks depends mainly on screen-grid linearity and not on control-grid operating conditions. Thus the carrier tube can be operated as a class C amplifier with resulting high plate efficiency.

Since the plate-voltage swing does not increase with positive modulation, the dc plate voltage is much higher than normally used for plate-modulated transmitters. At 16 kV, a plate efficiency of 80% is claimed by the manufacturer. The addition of resonant circuits, giving a trapezoidal shape to the plate-current pulses, yields plate efficiencies of 90% or higher.

Figure 21-3 is a simplified schematic diagram of the output amplifier. The carrier tube V_2 is a conventional grounded-cathode class C amplifier supplying the full 50-kW carrier power without modulation. The screen voltage is maintained at 700 V by a separate low-voltage supply. The plate voltage is 16 kV, and rf grid excitation maintains a peak plate swing of 15 kV. This stage can be screen-modulated only in a negative direction.

The peak tube V_1 has the same plate voltage and rf grid excitation as the carrier tube V_2. As a positive-going voltage is applied to the peak-tube screen, the tube delivers power into the load until it reaches crest condition, i.e., until the instantaneous screen voltage of V_2 equals the carrier-tube-V_2 screen voltage.

*Continental Electronics Manufacturing Company, Type 317C.

At the crest condition, both tubes are operating with equal split plate swing, load impedance, screen voltage, and grid drive. Since the carrier-tube load impedance is half what it is at carrier level and the plate swing is the same, the carrier-tube output is doubled from 50 to 100 kW. Since the peak tube is operating under identical conditions, it also provides 100 kW. The combined output is therefore 200 kW on positive peak, as required for 100% amplitude modulation.

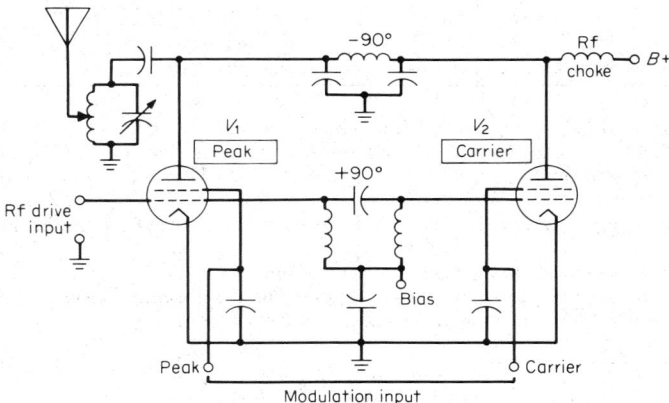

Fig. 21-3. Schematic of a screen-grid-modulated 50-kW final amplifier, for standard-broadcast service, with 90° phase shift between carrier and peak amplifiers. *(Continental Electronics Manufacturing Company.)*

Since the voltage contributed by the carrier tube undergoes a 90° phase lag by the time it appears across the load, it is necessary to introduce a 90° phase advance in the carrier-tube grid driving voltage for the power output of both tubes to combine in proper phase. This is accomplished by the leading 90° grid network shown in the figure.

Another 50-kW transmitter* utilizes a phase-to-amplitude form of modulation which the manufacturer identifies as Ampliphase. In this type of modulation, phase-modulated rf is applied at low level to two separate amplifier chains, so that when the phase is advancing in one chain, it is retarding in the other. The resultant outputs, in a common load circuit, combine so that at the crest condition the sum of the two output voltages is equal to twice their carrier value. At the trough condition, these voltages are 180° out of phase, and the resultant output is zero. The impedance presented to each of the two amplifiers varies over the modulation cycle from a low value at crest condition to an infinite value at trough condition. The rf driver stages are grid-modulated to maintain proper grid drive and assure grid linearity during the wide excursions of loading.

Figure 21-4 is a simplified block diagram of the BTA-50J transmitter, showing the two parallel stages. Each stage has its own tank circuit, and each network can be adjusted to provide proper loading to the power-amplifier stage. The manufacturer's electrical specifications (Table 21-5) indicate that the transmitter is capable of being modulated over the frequency range of 10 to 30,000 Hz. The transmitter can be continuously modulated at 100% at any frequency between 30 and 15,000 Hz without detrimental effects. The transmitter has been designed for remote control. All power supplies utilize solid-state rectifiers, and the transmitter is capable of operating at temperatures as low as -20°C. Each output tube is capable of delivering in excess of 25 kW power to the common load. The tube is rated at 35 kW dissipation, but under average modulation is required to dissipate only 14 kW. The solid-state exciter-modulator provides mirror-image symmetrical feeds from a single audio input, to feed the two branches of the output circuits. Services are provided for a second exciter-modulator which can be energized from a cutover switch or operated by remote control.

---

* RCA 50-kW Ampliphase AM transmitter, type BTA-50J.

Table 21-5. Specifications of the BTA-50J AM Transmitter

Power output (at transmitter terminals) .... 56 kW (max)
Frequency .... Any specified between 535 and 1,620 kHz
Frequency stability .... Assigned frequency ±5 Hz
Modulation .... Phase to amplitude
Audio-input level .... +10 ±2 dBm
Audio response .... ±1.5 dB 30–10,000 Hz
Af distortion .... <3% rms 50–7,500 Hz
Noise level .... 60 dB below 100% modulation
Carrier shift .... <5% negative 100% modulation
Spurious emission (second harmonic and above) .... 83 dB down

A third type of 50-kW transmitter* is outlined in the block diagram of Fig. 21-5. This is
a high-level plate-modulated air-cooled transmitter using a novel pulse-duration modulation
scheme. The transmitter proper uses only five tubes. The pulse-duration modulator
operates as follows (see Fig. 21-6):

1. The audio input (*a*) is added to a 70-kHz sawtooth wave (*b*) to form the threshold input
(*c*).

2. The threshold level (power-controlled) determines the point on the sawtooth wave at
which the pulse amplifier conducts. After clipping and amplification, squared pulses (Fig.
21-6*d*), which vary in duration with the audio, are formed at (*d*).

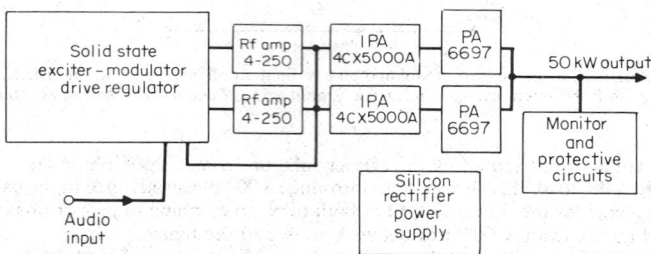

Fig. 21-4. Block diagram of a 50-kW standard-broadcast transmitter using phase-to-amplitude (Ampliphase) low-level modulation. *(RCA Corporation.)*

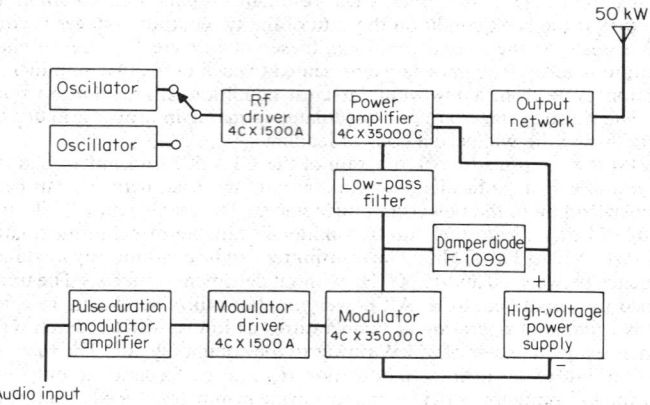

Fig. 21-5. Block diagram of a 50-kW standard-broadcast transmitter using audio-controlled pulse-
duration circuits for modulation. *(Gates Division, Harris Intertype Corp.)*

* Gates Model MW-50.

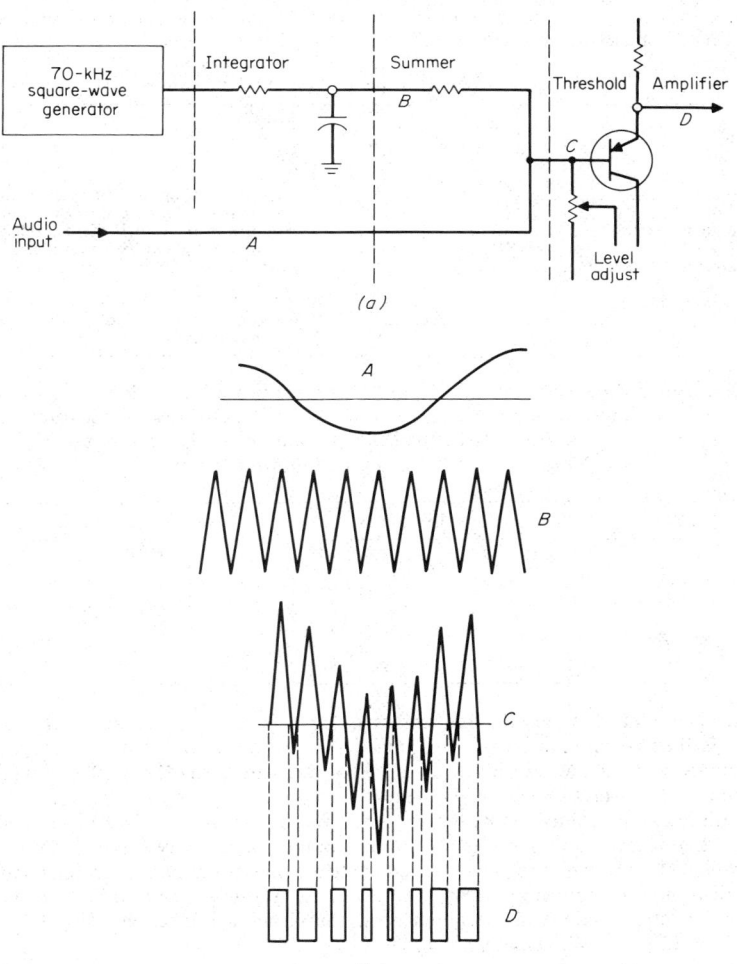

Fig. 21-6. Circuit and waveforms of the pulse-duration modulation employed in the transmitter of Fig. 21-5. The waveforms (*b*) present in the circuit (*a*) are marked with corresponding letters, as described in the accompanying text.

21-15

3. A low-pass filter removes the 70-kHz pulse frequency. Over 25,000 W of audio power is thus produced to modulate the power-amplifier stage. No modulation transformer or reactor is used.

The duty cycle of the pulse determines the voltage at the plate of the power amplifier, e.g., a 50% duty cycle produces 12 kV, whereas a 100% duty cycle places the full supply voltage of approximately 25 kV on the power amplifier, conforming to the 100% positive-modulation peak. At 0% duty cycle, zero voltage appears at the power amplifier, forming the 100% negative-modulation tip. The variation of the pulse width is determined by the audio signal. Table 21-6 lists the manufacturer's specifications for this transmitter.

Table 21-6. Specifications of MW-50 Transmitter

Power outputs, rated	50,000 W
Capable	60,000 W
Rf frequency range	535–1,620 kHz
Rf frequency stability	±5 Hz
Rf harmonics	Exceeds FCC and CCIR specifications
Carrier shift	Less than −2% at 100% modulation
Audio-frequency response	±1.5 dB from 20 to 10,000 Hz referenced to 1,000 Hz at 95% modulation
Audio-frequency distortion	Less than 3%, 20 to 10,000 Hz at 95% modulation
Noise (unweighted)	−60 dB or better below 100% modulation
Audio input	600/150 Ω at +10 dBm ±2 dB for 100% modulation

18. Shortwave Broadcast Service. An international broadcast station utilizes frequencies allocated to the broadcasting service between 5,950 and 26,100 kHz whose transmissions are intended to be received directly by the general public in foreign countries.

The frequencies assigned by the FCC lie in the following bands:

Band	Frequency, kHz	Meter band, m
A	5,950–6,200	49
B	9,500–9,775	32
C	11,700–11,975	25
D	15,100–15,450	19
E	17,700–17,900	16
F	21,450–21,750	14
G	25,600–26,100	11

Assignments are made for specific frequencies and specific hours of operation for transmission to specified geographic areas. Frequencies are assigned only if they will provide a delivered median field intensity, either measured or calculated, exceeding 150 μV/m for 50% of the time at the distant foreign target area.

The minimum transmitter output power is 50 kW. Transmitters must be equipped with automatic frequency control capable of maintaining the operating frequency within 0.003% of the assigned frequency. Frequency assignments provide a minimum cochannel delivered median field-intensity protection ratio of 40 dB to the transmissions of other broadcasting stations, at reference points in the target area. Similarly, a protection ratio of 11 dB is provided for adjacent channel assignments.

FREQUENCY-MODULATION BROADCAST PRACTICE

By Joseph L. Stern

19. FM Allocation Standards. The FM broadcast band comprises the radio-frequency spectrum from 88 to 108 MHz, divided into 100 channels of 200 kHz each. The channels available (including those assigned to noncommercial educational broadcasting) are given numerical designations by the FCC, from channel 201 (carrier frequency 88.1 MHz) to channel 300 (107.9 MHz).   The channel number N is related to the carrier frequency f in megahertz by

$$N = 5(f - 47.9)$$
$$f = N/5 + 47.9 \text{ MHz}$$

Field Strengths in FM Broadcast Service. Figure 21-7 shows estimated field strengths at distances from 1 to 100 mi, with transmitting antenna heights from 100 to 5,000 ft, for 1 kW of effective radiated power (50% of the receiving locations, 50% of the time).

20. Class A Service.   Except as provided in Sec. 73.204 of the FCC Rules, the following frequencies are designated as class A channels and are assigned to class A stations only:

Frequency, MHz	Channel no.	Frequency, MHz	Channel no.	Frequency, MHz	Channel no.
92.1	221	97.7	249	103.1	276
92.7	224	98.3	252	103.9	280
93.5	228	99.3	257	104.9	285
94.3	232	100.1	261	105.5	288
95.3	237	100.9	265	106.3	292
95.9	240	101.7	269	107.1	296
96.7	244	102.3	272		

A class A station is designed to render service to a relatively small community, city, or town and the surrounding rural area.   Class A stations are not authorized to operate with effective

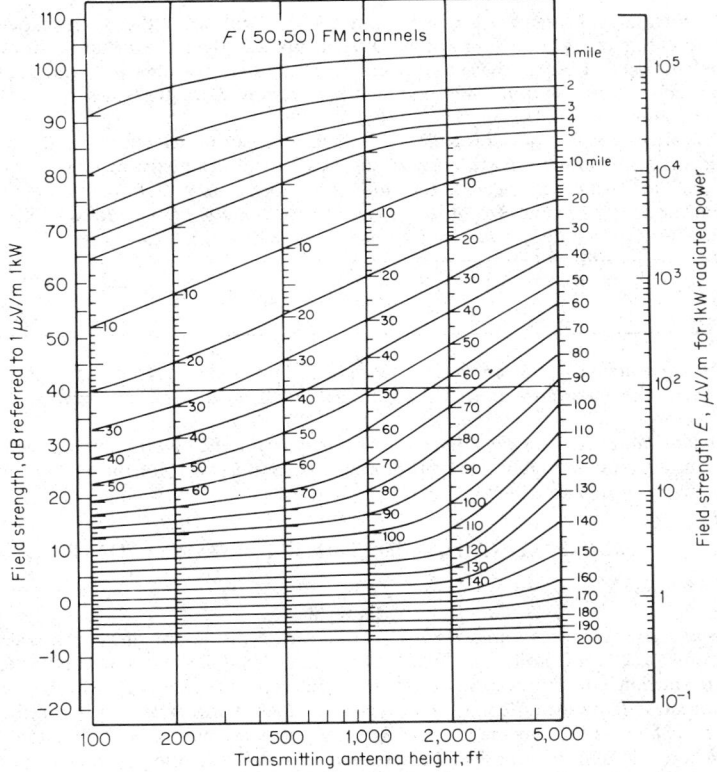

Fig. 21-7.   Field-intensity curves vs. antenna height for various distances in FM broadcast service.   The F(50,50) designation gives the estimated field strength exceeded at 50% of the potential receiver locations for at least 50% of the time (receiving antenna height 30 ft). *(From FCC Rules, Sec. 73.333.)*

radiated power greater than 3 kW (4.8 dBk), and the coverage of a class A station must not exceed that obtained from 3 kW effective radiated power and antenna height above average terrain of 300 ft. For provisions concerning minimum facilities and reduction in power where antenna height above average terrain exceeds 300 ft, see Sec. 73.211 of the FCC Rules.

21. Class B-C Service. Except for the class A channels listed above, all channels from 222 through 300 (92.3 through 107.9 MHz) are classified as class B-C channels. Subject to the restrictions set forth in Sec. 73.204 of the FCC Rules, they are assigned for use in zones I and IA by class B stations only and for use in zone II by class C stations only.

Class B and class C stations are designed to render service to a sizable community, city, or town (or to the principal city or cities of an urbanized area) and to the surrounding area. *Class B stations* authorized to operate after Sept. 10, 1962, may not operate with effective radiated power greater than 50 kW (17 dBk), and the coverage of a class B station authorized after that date may not exceed that obtained from 50 kW effective radiated power and 500 ft antenna height above average terrain. For provisions concerning minimum power and reduction in power where antenna height above average terrain exceeds 500 ft, see Sec. 73.211 of the FCC Rules.

Class C stations authorized after Sept. 10, 1962, are not permitted to operate with effective radiated power greater than 100 kW (20 dBk). The coverage of a class C station authorized after that date may not exceed that obtained from 100 kW effective radiated power and antenna height above average terrain of 2,000 ft. For provisions concerning minimum power and reduction in power where antenna height above average terrain exceeds 2,000 ft, see Par. **21-23.**

22. Cochannel and Adjacent-Channel Separations. Stations of the classes shown in the left-hand column of Table 21-7 must be located no less than the distance shown from cochannel stations and first adjacent-channel stations (200 kHz removed) and second and third adjacent-channel stations (400 and 600 kHz removed) of the classes shown in the remaining columns of the table.

The distances shown apply regardless of which is the proposed station under consideration, e.g., distances shown between a new class A station and an existing class C station are also the distances between a new class C and an existing class A station.

23. Power and Antenna-Height Requirements. *Minimum Power Requirements.* The minimum effective radiated power for FM broadcast stations is:

Class A .... 100 W (-10 dBk)
Class B .... 5 kW (7 dBk)
Class C .... 25 kW (14 dBk)

No minimum antenna height above average terrain is specified. However, if the antenna height exceeds the maximum for each class, the permitted effective radiated power is reduced as noted below.

Maximum Power and Antenna Height. The maximum effective radiated power in any direction and maximum antenna height for equivalence purposes, for the various classes of stations, are given in Table 21-8.

TECHNICAL STANDARDS FOR FM BROADCASTING

24. Definitions*

General

Antenna Height Above Average Terrain. The average of the antenna heights above the terrain from 2 to 10 mi from the antenna for the eight directions spaced evenly for each 45° of azimuth starting with true North. In general, a different antenna height will be determined in each direction from the antenna. The average of these various heights is considered the antenna height above the average terrain. In some cases fewer than eight directions may be used. Where circular or elliptical polarization is employed, the antenna height above average terrain must be based upon the height of the radiation center of the antenna which transmits the horizontal component of radiation.

* Adapted from Sec. 73.310 of the FCC Rules.

Table 21-7. Required Mileage Separations by Class of Station and Frequency Separation (kHz)

Class of station	Class A				Class B				Class C				10-W educational			
	Cochannel	200	400	600	Cochannel	200	400	600	Cochannel	200	400	600	Cochannel	200	400	600
A...........	65	40	15	15	...	65	40	40	...	105	65	65	...	30	15	15
B...........	...	...	...	...	150	105	40	40	170	135	65	65	...	...	40	40
C...........	...	...	...	...	...	...	...	...	180	150	65	65	...	...	65	65
10-W educational																

NOTE: Stations or assignments separated in frequency by 10.6 or 10.8 MHz (53 or 54 channels) will not be authorized unless they conform to the following separation table:

Class of stations	Required spacing, mi
A to A	5
B to A	10
B to B	15

Class of stations	Required spacing, mi
C to A	20
C to B	25
C to C	30

Table 21-8.   Maximum Power and Antenna Height

Class	Maximum power		Maximum antenna height, ft above average terrain
	kW	dBk	
A	3	4.8	300
B	50	17.0	500
C	100	20.0	2,000

NOTE: Antenna heights exceeding those specified may be used provided the effective radiated power is reduced. The amount of reduction is specified in Sec. 73.211 and Fig. 3 of Sec. 73.333 of the FCC Rules.

Antenna power gain. The square of the ratio of the rms free-space field strength produced at 1 mi in the horizontal plane, in millivolts per meter for 1 kW antenna input power, to 137.6 mV/m. This ratio should be expressed in decibels.

Center frequency. (1) The average frequency of the emitted wave when modulated by a sinusoidal signal, or (2) the frequency of the emitted wave without modulation.

Effective radiated power. The product of the antenna power (transmitter-output power less transmission-line loss) times (1) the antenna power gain or (2) the antenna field gain squared. When circular or elliptical polarization is employed, the term effective radiated power is applied separately to the horizontal and vertical components of radiation. For allocation purposes, the effective radiated power authorized is the horizonally polarized component of radiation only.

Free-space field strength. The field strength that would exist at a point in the absence of reflected waves.

Frequency swing. The instantaneous departure of the frequency of the emitted wave from the center frequency resulting from modulation.

Multiplex transmission. The simultaneous transmission of two or more signals within a single channel. Multiplex transmission as applied to monophonic FM broadcast stations means the transmission of facsimile or other signals in addition to the regular broadcast signals.

Percentage modulation. The ratio of the actual frequency swing to the frequency swing defined as 100% modulation, expressed in percentage. For FM broadcast stations, a frequency swing of ±75 kHz is defined as 100% modulation.

Definitions Applying to FM Stereophonic Broadcasting

Cross talk. An undesired signal occurring in one channel caused by an electric signal in another channel.

FM stereophonic broadcast. The transmission of a stereophonic program by a single FM broadcast station utilizing the main channel and a stereophonic subchannel.

Left (or right) signal. The electrical output of a microphone or combination of microphones placed so as to convey the intensity, time, and location of sounds originating predominantly to the listener's left (or right) of the center of the performing area.

Main channel. The band of frequencies from 50 to 15,000 Hz which frequency-modulate the main carrier.

Pilot subcarrier. A subcarrier serving as a control signal for use in the reception of FM stereophonic broadcasts.

Stereophonic separation. The ratio of the electric signal caused in the right (or left) stereophonic channel to the electric signal caused in the left (or right) stereophonic channel by the transmission of only a right (or left) signal.

Stereophonic subcarrier. A subcarrier having a frequency which is the second harmonic of the pilot subcarrier frequency, employed in FM stereophonic broadcasting.

Stereophonic subchannel. The band of frequencies from 23 to 53 kHz, containing the stereophonic subcarrier and its associated sidebands.

25. FM Broadcast Equipment Standards. [*] Under FCC regulations, the design of FM broadcast transmitting systems, from input terminals of microphone preamplifier, through the audio facilities at the studio, lines or other circuits between studio and transmitter, audio facilities at the transmitter, and the transmitter must be in accordance with the following principles and specifications.

Modulation Percentage and Bandwidth. The transmitter shall operate satisfactorily in the

---

[*] Section 73.317 of the FCC Rules.

operating power range with a frequency swing of 75 kHz defined as 100% modulation. The transmitting system shall be capable of transmitting a band of frequencies from 50 to 15,000 Hz.

Preemphasis shall be employed in accordance with the impedance-frequency characteristic of a series inductance-resistance network having a time constant of 75 µs. The deviation of the system response from the standard preemphasis curve shall lie between two limits (shown in Fig. 2 of Sec. 73.333 of the FCC Rules). The upper of these limits shall be uniform (no deviation) from 50 to 15,000 Hz. The lower limit shall be uniform from 100 to 7,500 Hz, and 3 dB below the upper limit; from 100 to 50 Hz the lower limit shall fall from the 3-dB limit at a uniform rate of 1 dB/octave (4 dB at 50 Hz); from 7,500 to 15,000 Hz the lower limit shall fall from the 3-dB limit at a uniform rate of 2 dB/octave (5 dB at 15,000 Hz).

Harmonic Distortion. At any modulation frequency between 50 and 15,000 Hz and at modulation percentages of 25, 50, and 100%, the combined audio-frequency harmonics measured in the output of the system shall not exceed the following rms values:

Modulating frequency, Hz	Distortion, %
50-100	3.5
100-7,500	2.5
7,500-15,000	3.0

Measurements shall be made employing 75-µs deemphasis in the measuring equipment and 75-µs preemphasis in the transmitting equipment (without compression if a compression amplifier is employed). Harmonics shall be included to 30 kHz. It is recommended that none of the three main divisions of the system (transmitter, studio-to-transmitter circuit, or audio facilities) contribute over one-half of these percentages since at some frequencies the total distortion may become the arithmetic sum of the distortions of the divisions.

Noise. The transmitting-system output noise level (frequency modulation) in the band of 50 to 15,000 Hz shall be at least 60 dB below 100% modulation (frequency swing ±75 kHz). The measurement shall be made using 400-Hz modulation as a reference. The noise-measuring equipment shall be provided with standard 75-µs deemphasis; the ballistic characteristics of the instrument shall be similar to those of the standard volume-unit meter.

The transmitting-system output noise level (amplitude modulation) in the band of 50 to 15,000 Hz shall be at least 50 dB below the level representing 100% amplitude modulation. The noise-measuring equipment shall be provided with standard 75-µs deemphasis; the ballistic characteristics of the instrument shall be similar to those of the standard volume-unit meter.

Carrier-Frequency Control. Automatic means shall be provided in the transmitter to maintain the assigned center frequency within the allowable tolerance of 2,000 Hz.

Out-of-Band Emissions. Any emission appearing on a frequency removed from the carrier by between 120 and 240 kHz, inclusive, shall be attenuated at least 25 dB below the level of the unmodulated carrier. Compliance with this specification will be deemed to show the occupied bandwith to be 240 kHz or less.

Any emission appearing on a frequency removed from the carrier by more than 240 kHz and up to and including 600 kHz shall be attenuated at least 35 dB below the level of the unmodulated carrier.

Any emission appearing on a frequency removed from the carrier by more than 600 kHz shall be attenuated at least 43 + 10 $\log_{10}$ (power, in watts) decibels below the level of the unmodulated carrier, or 80 dB, whichever is less.

26. Subsidiary FM Communications Authorizations (SCA). An FM broadcast station may be issued a Subsidiary Communications Authorization (SCA) to provide limited types of subsidiary services on a multiplex basis. Permissible uses fall within the following categories:

1. Transmission of programs which are of a broadcast nature but which are of interest primarily to limited segments of the public wishing to subscribe thereto. Examples include background music, storecasting, detailed weather forecasting, special time signals, and other material of a broadcast nature expressly designed and intended for business, professional, educational, religious, trade, labor, agricultural, or other groups.

2. Transmission of signals which are directly related to the operation of FM broadcast stations, e.g., relaying broadcast material to other FM and standard broadcast stations, remote cueing and order circuits, remote-control telemetering functions, and similar uses. SCA operations may be conducted without time restriction so long as the main channel is programmed simultaneously.

Subsidiary communications multiplex operations are governed by the following engineering standards: Frequency modulation of SCA subcarriers must be used. The instantaneous frequency of SCA subcarriers must at all times be within the range 20 to 75 kHz, provided, however, that when the station is engaged in stereophonic broadcasting (Par. 21-27) the instantaneous frequency of SCA subcarriers must be within the range 53 to 75 kHz.

The arithmetic sum of the modulation of the main carrier by SCA subcarriers must not exceed 30%, provided, however, that when the station is engaged in stereophonic broadcasting the arithmetic sum of the modulation of the main carrier by the SCA subcarriers must not exceed 10%. The total modulation of the main carrier, including SCA subcarriers, must meet the requirements of Sec. 73.268 of the FCC Rules. Frequency modulation of the main carrier caused by the SCA subcarrier operation, in the frequency range 50 to 15,000 Hz, must be at least 60 dB below 100% modulation, provided, however, that when the station is engaged in stereophonic broadcasting frequency modulation of the main carrier by the SCA subcarrier operation, in the frequency range 50 to 53,000 Hz, must be at least 60 dB below 100% modulation.

27. Stereo Transmission Standards. (See definitions in Par. **21-24**.) The modulating signal for the main channel consists of the sum of the left and right signals.

The *pilot subcarrier* at 19,000 Hz ± 2 Hz is transmitted and frequency modulates the main carrier between the limits of 8 and 10%.

The *stereophonic subcarrier* is the second harmonic of the pilot subcarrier and must cross the time axis with a positive slope simultaneously with each crossing of the time axis by the pilot subcarrier. Amplitude modulation of the stereophonic subcarrier must be used, and the stereophonic subcarrier must be suppressed to a level less than 1% modulation of the main carrier.

The stereophonic subcarrier must be capable of accepting audio frequencies from 50 to 15,000 Hz, and the modulating signal for the stereophonic subcarrier must be equal to the difference of the left and right signals.

The *preemphasis characteristic* of the stereophonic subchannel must be identical with those of the main channel with respect to phase and amplitude at all frequencies.

The sum of the sidebands resulting from amplitude modulation of the stereophonic subcarrier must not cause a peak deviation of the main carrier in excess of 45% of total modulation (excluding SCA subcarriers) when only a left (or right) signal exists; simultaneously in the main channel, the deviation when only a left (or right) signal exists must not exceed 45% of total modulation (excluding SCA subcarriers).

At the instant when only a positive left signal is applied, the main-channel modulation must cause an upward deviation of the main-carrier frequency; and the stereophonic subcarrier and its sidebands signal must cross the time axis simultaneously and in the same direction.

The ratio of peak main-channel deviation to peak stereophonic subchannel deviation when only a steady-state left (or right) signal exists must be within $\pm 3.5\%$ of unity for all levels of this signal and all frequencies from 50 to 15,000 Hz.

The phase difference between the zero points of the main-channel signal and the stereophonic subcarrier sidebands envelope, when only a steady-state left (or right) signal exists, must not exceed $\pm 3°$ for AM frequencies from 50 to 15,000 Hz.

If the stereophonic separation between left and right stereophonic channels is better than 29.7 dB at AM frequencies between 50 and 15,000 Hz, it will be assumed that the requirements of the preceding two paragraphs have been complied with.

Crosstalk into the main channel caused by a signal in the stereophonic subchannel must be attenuated at least 40 dB below 90% modulation, and crosstalk into the stereophonic subchannel caused by a signal in the main channel must be attenuated a like amount.

28. FM Transmitter Performance Standards. The Electronic Industries Association (EIA, formerly RETMA) has issued standards of performance for FM transmitters, in Standard TR-107, Electrical Performance Standards for FM Broadcast Transmitters (88–108

MHz). This standard lists definitions, standards, and methods of measurements of the following items:

Carrier power-output rating.
Carrier power-output capability.
Normal load.
Rf output coupling-impedance range.
Carrier-frequency range.
Carrier-frequency stability.
Spurious emissions.
Modulation capability.
Audio-input level for 100% modulation.
Audio-frequency response.
Audio-frequency harmonic distortion.
FM noise level on carrier.
AM noise level on carrier.
Output voltage and impedance for audio and rf monitoring.
Rated power supply.
Power supply variation.

This standard should be consulted for detailed requirements.

FM BROADCAST EQUIPMENT

29. Typical 1- and 2.5-kW FM Transmitters. Figure 21-8 shows a simplified block diagram of a typical 1-kW FM transmitter[*], and Table 21-9 outlines the manufacturer's specifications. It is designed to meet the requirements of monaural, multiplex, and stereo transmission.

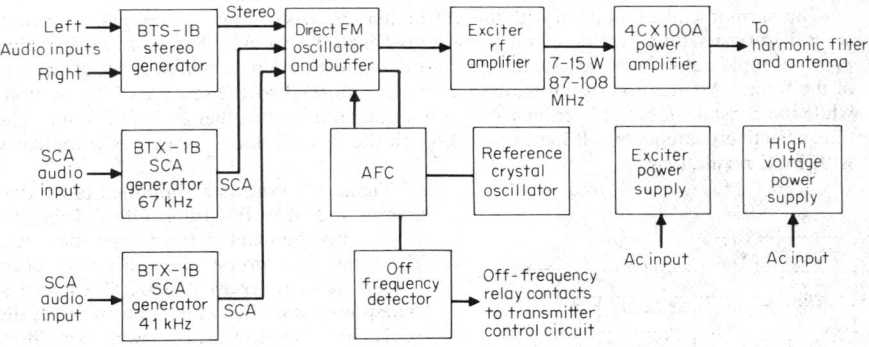

Fig. 21-8.   Block diagram of a 1-kW FM broadcast transmitter. *(RCA Corporation.)*

Modulation of the temperature-compensated on-frequency oscillator is achieved by applying the composite stereo or SCA signals to a pair of push-pull varicap diodes which are coupled to the basic frequency-determining resonant circuit. A 10-dB resistive attenuator couples this circuit to the buffer amplifier, and the 500-mW output of the buffer drives a 15-W three-stage rf amplifier, as well as a binary divider chain in the automatic-frequency-control (AFC) circuit.

AFC is achieved by sampling the buffer output frequency. A reference crystal operates at 1/1,024 of the desired output frequency. The outputs from the reference and the basic oscillator binary dividers are phase-compared in a comparator. The output of this circuit, the AFC error voltage, is filtered and applied to another pair of varicap diodes coupled to the basic oscillator circuit. Thus the basic oscillator is phase-locked to the 1,024th harmonic of the oven-controlled reference crystal.

The output of the exciter is fed to the input of the ceramic 4CX-1000A amplifier tube. The amplifier input circuit is a simple parallel resonant circuit, tuned by a variable inductance

---

[*] RCA type BTF-1E2.

Table 21-9.   Specifications of Type BTF-1E2 FM Transmitter

Frequency range .... 88 to 108 MHz
Power output .... 250 to 1,000 W
Frequency deviation, 100% modulation .... ± 75 kHz
Modulation capability .... ± 100 kHz
Carrier-frequency stability .... ± 1,000 Hz max
Audio-input impedance .... 600/150 Ω
Audio-input level—(100% modulation) .... + 10 ± 2 dBm
Audio-frequency response—(50 Hz to 15 kHz) .... ± 1 dB max
Preemphasis-network time constant .... 75 or 50 μs or flat as desired
Harmonic distortion (50 Hz to 15 kHz) .... 0.5% or less
FM noise level (referred to 100% FM modulation) .... − 65 dB max
AM noise level (referred to 100% AM modulation) .... − 50 dB max
Subcarrier-input level (100% modulation) .... − 15 to + 10 dBm adjustable
Subcarrier-input impedance .... 600/150 Ω balanced
Subcarrier frequency .... 20 to 67 kHz
Main-to-subchannel cross talk .... − 50 dB referred to ± 7.5 kHz deviation
                                                       of the subcarrier by a 400-Hz tone;
                                                       main-channel modulation 70% by 50-
                                                       to 15,000-Hz tones
Subchannel-to-main-channel cross talk .... − 60 dB referred to ± 75 kHz deviation of
                                                       the main carrier by a 400-Hz tone;
                                                       subchannel modulated 100% (± 7.5
                                                       kHz/s) by 50- to 6,000-Hz tones;
                                                       subcarrier modulated 30% on main
                                                       carrier

with resistance swamping for stability. This stage is neutralized by varying inductance in series with the screen. The output circuit is a modified π network, having a variable inductance across the tube capacity, which is used to adjust the loading.

The harmonic filter supplied with the transmitter consists of an *m*-derived half-T section, several low-pass filter sections, and a constant-*K* half-T section. The *m*-derived section provides rapid cutoff in the second harmonic region and a termination impedance at one end of the filter. Attenuation of the harmonics is accomplished by the low-pass filter sections while the constant-*K* half-T section serves as a termination at the other end of the unit. The filter effectively attenuates all harmonics through the seventh and thus assures compliance with FCC regulations.

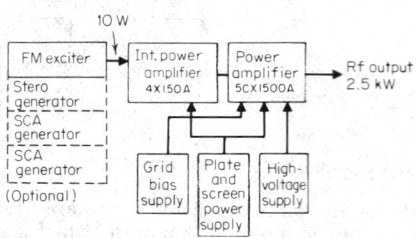

Fig. 21-9. Block diagram of a 2.5-kW FM transmitter using only two tubes following the exciter. *(Gates Division, Harris Intertype Corp.)*

Figure 21-9 shows a simplified block diagram of a 2.5-kW FM transmitter.* Table 21-10 presents the manufacturer's specifications. The transmitter proper uses only two tubes. A solid-state exciter drives a 4X150A intermediate power amplifier which in turn drives the 5CX1500A single-ended power amplifier. The plate circuit of the final amplifier is a shorted quarter-wavelength configuration, with the plate line operated at dc ground potential. Coarse plate tuning is preset for the operating frequency, on the quarter-wave plate circuit. Fine adjustment is made with the plate tuning knob on the front panel. Amplifier loading is changed by a variable-output loading control. A built-in motor-operated rheostat connected to the screen supply adjusts the power output. A built-in reflectometer with a voltage standing-wave ratio power meter is used in making power adjustments.

The transmitter is normally provided for standard monaural operation. Plug-in units are available for the exciter to convert to stereo and/or SCA. An internally mounted harmonic filter is provided to meet FCC requirements for spurious radiations.

----

* Gates type FM-2.5H3.

30. Typical 40-kW FM Transmitter. Shown in the simplified block diagram of Fig. 21-10 is a 40-kW transmitter made up of two 20-kW transmitters fed by a solid-state exciter, both feeding into a combining network to give the desired 40-kW output. The manufacturer's specifications are shown in Table 21-10. The use of two transmitters offers circuit simplicity, lower parts inventory, and transmission continuity in case one of the units should fail. The output of each 20-kW unit is fed through its own harmonic filter to the combining network. This hybrid network adds the two 20-kW signals to produce a 40-kW output to the transmission line, while isolating one transmitter from the other.

Should one transmitter fail, the other continues feeding the combining network, and the combiner continues to operate as a power divider, 10 kW being fed to the output transmission line and 10 kW to the dummy load connected to the combiner. This division is necessary to provide isolation between transmitters and to allow the nonoperating unit to be serviced without rf coupling.

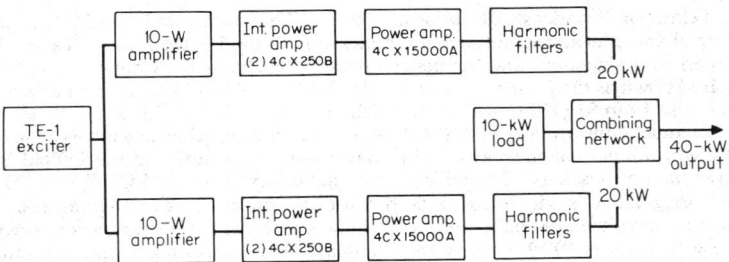

Fig. 21-10. Block diagram of a 40-kW FM transmitter composed of two identical 20-kW transmitters feeding a combining network. If one unit fails, it can be repaired while the other is fully operative. *(Gates Division, Harris-Intertype Corp.)*

Table 21-10. Specifications for Type FM-2.5H3 and FM40H Transmitters

Power output:
  FM2.5H3 .... 2.5 kW
  FM-40H .... 40 kW
Frequency range .... 87.5-108 MHz
Rf output impedance .... 50 Ω
Frequency stability .... 0.001% or better
Type of modulation .... Direct-carrier frequency modulation
Modulation capability .... ±100 kHz
Rf harmonics .... Suppression meets all FCC requirements
Monaural mode:
  Audio-input impedance .... 600 Ω balanced
  Audio-input level .... +10 dBm ±2 dB for 100% modulation at
                                                      400 Hz
  Audio-frequency response .... Standard 75-μs, FCC preemphasis curve
                                                      ±1 dB, 30-15,000 Hz
  Distortion .... 0.5% or less, 30-15,000 Hz
  FM noise .... 65 dB below 100% modulation referred to
                                                      400 Hz
  AM noise .... 55 dB below reference carrier AM
                                                      modulation 100%
Stereophonic mode:
  Pilot oscillator .... Crystal-controlled
  Pilot stability .... 19 kHz ±1 Hz
  Audio-input impedance .... Left and right 600 Ω balanced
  Audio-input level .... Left and right +10 dBm ±1 dB for 100%
                                                      modulation at 400 Hz
  Audio-frequency response .... Left and right standard 75-μs FCC
                                                      preemphasis curve ±1 dB, 50-15,000
                                                      Hz
  Distortion .... Left or right 1% or less, 50-15,000 Hz
  FM noise .... Left or right 60 dB minimum below 100%
                                                      modulation, referred to 400 Hz
  Stereo separation .... 35 dB minimum, 50-15,000 Hz
  Subcarrier suppression .... 42 dB minimum, 50-15,000 Hz
  Cross talk (main to subchannel or vice versa) ..... 42 dB below 90% modulation

The solid-state exciter employs direct carrier-frequency modulation. The modulated oscillator operates at carrier frequency and has an output of 10 W. Separate 10-W isolation amplifiers are used to feed the two 20-kW units. The isolation amplifiers drive a pair of 4CX250B tubes, which produce 400W of drive for a single 4CX15000 power-amplifier stage. The output tuning of the power-amplifier stage is accomplished with an inductively tuned plate-tank circuit, eliminating the need for vacuum capacitors. The transmitter is equipped for remote operation, including a motor-operated screen voltage-supply control for power-output adjustment.

TELEVISION BROADCASTING PRACTICE

By Joseph L. Stern

31. Television Broadcast Allocations. The FCC has authorized 82 6-MHz channels for commercial and educational television broadcasting in the United States. Table 21-11 lists the numerical designations and frequency limits of the channels. Channels 2 to 6 (54 to 88 MHz) are known as the *low-band vhf* channels, 7 to 13 (174 to 216 MHz) as the *high-band vhf* channels and 14 to 83 (470 to 890 MHz) as the *uhf channels.*

32. Channel Utilization. The FCC has prepared a comprehensive table of the commercial and educational channels assigned to particular communities in the United States, its territories, and possessions. The table appears in Sec. 73.606 of the FCC Rules. Each local channel assignment is identified as to the use of "offset" carrier frequencies. Certain assignments carry the nominal picture and sound carrier frequencies, others are required to operate with carriers 10 kHz above the nominal values (plus offset), and still others with carriers 10 kHz below the nominal values (minus offset). The offset system of carrier assignments substantially reduces the visible effects of cochannel interference.

33. Cochannel and Adjacent-Channel Separations. Table 21-12 states the minimum cochannel mileage separations set up by the FCC.

The minimum adjacent-channel separations applicable to all zones are 60 mi for channels 2 to 13 and 55 mi for channels 14 to 83. Due to the greater than normal frequency spacing

Table 21-11. Designations and Frequency Limits of Television Channels in the United States

Channel designation	Frequency band, MHz	Channel designation	Frequency band, MHz	Channel designation	Frequency band, MHz
2	54–60	30	566–572	57	728–734
3	60–66	31	572–578	58	734–740
4	66–72	32	578–584	59	740–746
5	76–82	33	584–590	60	746–752
6	82–88	34	590–596	61	752–758
7	174–180	35	596–602	62	758–764
8	180–186	36	602–608	63	764–770
9	186–192	37	608–614	64	770–776
10	192–198	38	614–620	65	776–782
11	198–204	39	620–626	66	782–788
12	204–210	40	626–632	67	788–794
13	210–216	41	632–638	68	794–800
14	470–476	42	638–644	69	800–806
15	476–482	43	644–650	70	806–812
16	482–488	44	650–656	71	812–818
17	488–494	45	656–662	72	818–824
18	494–500	46	662–668	73	824–830
19	500–506	47	668–674	74	830–836
20	506–512	48	674–680	75	836–842
21	512–518	49	680–686	76	842–848
22	518–524	50	686–692	77	848–854
23	524–530	51	692–698	78	854–860
24	530–536	52	698–704	79	860–866
25	536–542	53	704–710	80	866–872
26	542–548	54	710–716	81	872–878
27	548–554	55	716–722	82	878–884
28	554–560	56	722–728	83	884–890
29	560–566				

between channels 4 and 5, 6 and 7, and 13 and 14, the minimum adjacent-channel separations specified above are not applicable to these pairs of channels.

The minimum stations separations between stations on the uhf channels 14 to 83, inclusive, must meet conditions set forth in Table IV of Sec. 73.698 of the FCC Rules. This table sets up separations based on considerations of intermediate-frequency beat interference, intermodulation, adjacent-channel interference, oscillator interference, sound-image interference, and picture-image interference, with separations from 20 to 75 mi, depending on the type of interference. Each uhf channel (14 to 83 inclusive) is paired with another channel or channels, in each category, and the minimum separation for each type of interference must be met or exceeded in each category. Section 73.698 of the FCC Rules should be consulted for details.

Table 21-12. Minimum Cochannel Separations in Miles

Zone (see Fig. 21-11)	Channels 2 to 13	Channels 14 to 83
I	170	155
II	190	175
III	220	205

* The minimum cochannel mileage separations between a station in one zone and a station in another zone shall be that of the zone requiring the lower separation.

34. Definitions Applicable to Television Broadcasting.* See also Sec. 20, Pars. 20-1 to 20-24.

Antenna electrical beam tilt. The shaping of the radiation pattern in the vertical plane of a transmitting antenna by electrical means so that maximum radiation occurs at an angle below the horizontal plane.

Antenna height above average terrain. The average of the antenna height above the terrain from 2 to 10 mi from the antenna for the eight directions spaced evenly for each 45° of azimuth starting with true north. (In general, a different antenna height will be determined in each direction from the antenna. The average of these various heights is considered the antenna height above the average terrain.) In some cases fewer than eight directions may be used.

Antenna mechanical beam tilt. The installation of a transmitting antenna so that its axis is intentionally off vertical in order to change the normal angle of maximum radiation in the vertical plane.

Antenna power gain. The square of the ratio of the rms free-space field intensity produced at 1 mi in the horizontal plane, in millivolts per meter for 1-kW antenna input power, to 137.6 mV/m. This ratio should be expressed in decibels. (If specified for a particular direction, antenna power gain is based on the field strength in that direction only.)

Aural center frequency. The average frequency of the emitted wave when modulated by a sinusoidal signal; the frequency of the emitted wave without modulation.

Blanking level. The level of the signal during the blanking interval, except the interval during the scanning synchronizing pulse and the chrominance-subcarrier synchronizing burst.

Chrominance. The colorimetric difference between any color and a reference color of equal luminance, the reference color having a specific chromaticity.

Chrominance subcarrier. The carrier which is modulated by the chrominance information.

Effective radiated power. The product of the antenna input power and the antenna power gain. This product should be expressed in kilowatts and the decibels above 1 kW (dBk).

Free-space field intensity. The field intensity that would exist at a point in the absence of waves reflected from the earth or other reflecting objects.

Negative transmission. Transmission in which a decrease in initial light intensity causes an increase in the transmitted power.

Peak power. The power over a radio-frequency cycle corresponding in amplitude to synchronizing peaks.

Percentage modulation. As applied to frequency modulation, the ratio of the actual frequency swing to the frequency swing defined as 100% modulation, expressed in percentage.

---

* Adapted from Sec. 73.681 of the FCC Rules.

For the aural transmitter of television broadcast stations, a frequency swing of ± 25 Hz is defined as 100% modulation.

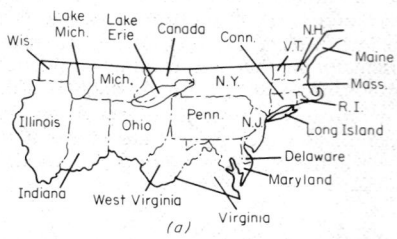

(a)

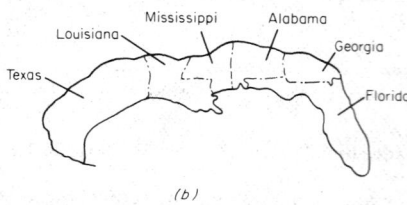

(b)

Fig. 21-11.   Geographical zones set up by the FCC to take into account different requirements for television cochannel mileage separations: (*a*) zone I, (*b*) zone II. The remainder of the United States is designated as zone III.

Polarization. The direction of the electric field as radiated from the transmitting antenna.

Reference black level. The level corresponding to the specified maximum excursion of the luminance signal in the black direction.

Reference white level of the luminance signal. The level corresponding to the specified maximum excursion of the luminance signal in the white direction.

Vestigial-sideband transmission. A system of transmission wherein one of the generated sidebands is partially attenuated at the transmitter and radiated only in part.

Visual-carrier frequency. The frequency of the carrier which is modulated by the picture information.

Visual-transmitter power. The peak power output when transmitting a standard television signal.

35. Television Transmission Standards in the United States. See also Sec. 20. The following standards have been adopted by the FCC.

Channel Standards. The width of the television broadcast channel is 6 MHz, with the visual-carrier frequency nominally 1.25 MHz above the lower boundary of the channel and the aural center frequency 4.5 MHz higher than the visual-carrier frequency.

The visual-transmission amplitude characteristic is in accordance with Fig. 21-12 (for stations operating on channels 15 to 83 and employing a transmitter with maximum peak visual power output of 1 kW or less, the visual-transmission amplitude characteristic given in Fig. 5a of FCC Sec. 73.699 may be used).

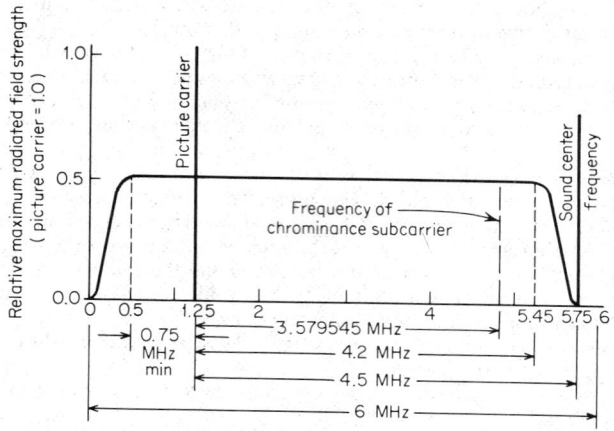

Fig. 21-12.   The standard FCC channel for monochrome and color television picture transmissions in the United States (not to scale), shown as an idealized amplitude characteristic of the picture transmission. *(From FCC Rules, Sec. 73.699.)*

The chrominance subcarrier frequency is 3.579545 MHz ± 10 Hz with a maximum rate of change not to exceed $\frac{1}{10}$ Hz/s.

For monochrome and color transmissions the number of scanning lines per frame is 525, interlaced two to one in successive fields. The horizontal scanning frequency is 2/455 times the chrominance subcarrier frequency; this corresponds nominally to 15,750 Hz (the actual value is 15,734.264 ± 0.044 Hz). The vertical scanning frequency is 2/525 times the horizontal scanning frequency; this corresponds nominally to 60 Hz (the actual value is 59.94 Hz). For monochrome transmissions only, the nominal values of line and field frequencies may be used. Other aspects of the scanning and modulation standards are treated in Sec. **20**, Pars. **20-7** to **20-12**.

Frequency Tolerance. The carrier frequency of the visual transmitter must be maintained within ± 1,000 Hz of the assigned frequency, while the center frequency of the aural transmitter must be maintained at 4.5 MHz ± 1,000 Hz above the assigned visual-carrier frequency.

The signals radiated are horizontally polarized. The effective radiated power of the aural transmitter must not be less than 10% nor more than 20% of the peak radiated power of the visual transmitter.

The peak-to-peak variation of transmitter output within one frame of video signal due to all causes, including hum, noise, and low-frequency response, measured at both scanning synchronizing peak and blanking level, must not exceed 5% of the average scanning synchronizing peak-signal amplitude (this provision is subject to change but is considered the best practice under the present state of the art).

The waveforms and transfer characteristics set up by the FCC for monochrome and color transmissions are stated in Sec. **20**, Pars. **20-9** to **20-12**.

36. Test and Identification Signals. The interval beginning with the last 12 μs of line 17 and continuing through line 20 of the vertical blanking interval of each field may be used for the transmission of test signals subject to the conditions set forth by the FCC. Test signals include signals used to supply reference modulation levels so that variation in light intensity of the scene viewed by the camera will be faithfully transmitted, signals designed to check the performance of the overall transmission system or its individual components, and cue and control signals related to the operation of the television broadcast station.

The use of test signals must not result in significant degradation of the program transmissions of the television broadcast station or create emission components in excess of those permitted for normal program transmissions.

They may not be transmitted during horizontal blanking.

The intervals within the first and last 10 μs of lines 21 through 23 and 260 through 262 (on a field basis) may contain coded patterns for the purpose of electronic identification of television broadcast programs and spot announcements, provided the coded patterns do not exceed 1 s in duration. The transmission of these patterns must not result in significant degradation of broadcast transmission.

37. Coverage Determinations and Standards. In the authorization of television broadcast stations, two field-intensity contours, specified as grade A and grade B, indicate the approximate extent and area of the coverage over average terrain in the absence of interference from other television stations. Under actual conditions, the true coverage may vary greatly from these estimates when the terrain is different from the average terrain on which the field-strength charts were based. The required field intensities $F(50,50)$ (see Fig. 21-13) in decibels above 1 μV/m (dBμ) for the grade A and grade B contours are given in Table 21-13.

Table 21-13. Grades of Television Service

Channels	Grade A, dBμ	Grade B, dBμ
2-6	68	47
7-13	71	56
14-83	74	64

In predicting the distance to the field-intensity contours, the $F(50,50)$ field-intensity charts (Figs. 9 and 10 of FCC Sec. 73.699) are used. Figure 21-13 applies to channels 2 to 6 and 14 to 83. The 50% field intensity is defined as that value exceeded for 50% of the time. The $F(50,50)$ charts give the estimated 50% field intensities exceeded at 50% of the locations, in decibels above $1~\mu V/m$. The charts are based on an effective power of 1 kW radiated from a half-wave dipole in free space, which produces an unattenuated field strength at 1 mi of about 103 dB above $1~\mu V/m$ (137.6 mV/m).

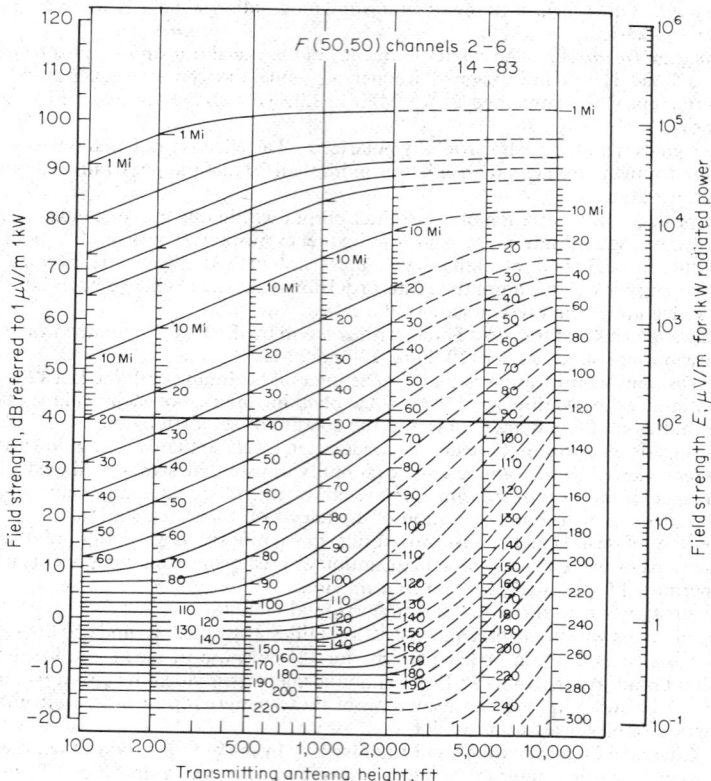

Fig. 21-13. Field-intensity curves vs. antenna height for television transmissions at various distances. For the significance of the designation $F(50,50)$, see Fig. 21-7. Receiver antenna height is 30 ft. *(From FCC Rules, Sec. 73.699.)*

Depression Angle. In predicting the distance to the grade A and grade B field-intensity contours, the effective radiated power used is that radiated at the vertical angle corresponding to the depression angle between the transmitting-antenna center of radiation and the radio horizon as determined individually for each azimuthal direction concerned. The depression angle is based on radiation above the average terrain and the radio horizon, assuming a smooth spherical earth with a radius of 5,280 mi, determined by.

$$A_h = 0.0153\sqrt{H}$$

where A_h is the depression angle in degrees and H is the height in feet of the transmitting-antenna radiation center above average terrain of the 2- to 10-mi sector of the pertinent radial. The antenna height used with the $F(50,50)$ charts is the height of the radiation center of the antenna above the average terrain along the radial in question. In determining the average elevation of the terrain, the elevations between 2 and 10 mi from the antenna site are employed. Profile graphs are drawn for eight radials beginning at the antenna site and extending 10 mi, for each 45° of azimuth starting with true north. At least one radial must include the principal community to be served even though it may be more than 10 mi from the antenna site. For further details of coverage prediction, see FCC Rules, Sec. 73.684.

38. Transmitter Location and Antenna System. The transmitter location chosen must provide, on the basis of the effective radiated power and antenna height above average terrain employed, the following minimum field intensity in decibels above 1 μV/m (dBμ) over the entire principal community to be served: channels 2 to 6, 74 dBμ; channels 7 to 13, 77 dBμ; channels 14 to 83, 80 dBμ.

Location of the antenna at a point of high elevation is necessary to reduce to a minimum the shadow effect on propagation due to hills and buildings which may materially reduce the intensity of the station's signals. In general, the transmitting antenna of a station should be located at the most central point at the highest elevation available. To provide the best degree of service to an area, it is usually preferable to use a high antenna rather than a low antenna with increased transmitter power. The location should be so chosen that line of sight can be obtained from the antenna over the principal community to be served; in no event should there be a major obstruction in this path. The antenna must be constructed so that it is as clear as possible of surrounding buildings or objects that would cause shadow problems. When the shape of the desired service area and population distribution make the choice of a transmitter location difficult, consideration may be given to the use of a directional antenna system, although it is generally preferable to choose a site where a nondirectional antenna can be used.

An antenna designed or altered to produce a noncircular radiation pattern in the horizontal plane is considered to be a directional antenna. Antennas purposely installed to give mechanical beam tilting of the major vertical radiation lobe are included in this category. Stations operating on channels 2 to 13 are not permitted to employ a directional antenna having a ratio of maximum to minimum radiation in the horizontal plane in excess of 10 dB. Stations operating on channels 14 to 83 and employing transmitters delivering a peak visual power output of 1 kW or less are not limited as to the ratio of directional characteristics.

TELEVISION BROADCASTING EQUIPMENT

By Joseph L. Stern

39. TV Broadcast-Equipment Performance Standards. The Electronic Industries Association (EIA, formerly RETMA) has issued a 21-page set of television-equipment standards, Standard RS-240: Electrical Performance Standards for Television Broadcast Transmitters, Channels 2-6, 7-13, and 14-83. This comprehensive standard should be consulted for the detailed definitions, standards, and methods of measurement it contains. The topics covered fall into four parts: the television transmitter, the visual transmitter, the aural transmitter, and safety standards, as follows:

Television transmitter:
 Television transmitter.
 Power rating.
 Rated power supply.
 Power-supply variation.
 Carrier-frequency range.
 Transmitter harmonic and subharmonic output.
 Cabinet radiation.
Visual Transmitter:
 Visual transmitter.
 Power-output rating.
 Peak power-output adjustment.
 Variation of output.
 Regulation of output.
 Carrier-frequency stability.
 Amplitude vs. frequency response.
 Upper and lower sideband attenuation.

Linearity.
Differential gain.
Differential phase.
Envelope delay.
Transmitter input polarity.
Transmitter input impedance.
Transmission polarity.
Transmitter input level for rated modulation.
Carrier pedestal level.
Carrier-reference white level.
Output polarity and voltage for composite picture-signal monitor connections.
Output voltage for rf monitor connections.
Hum and noise.
Incidental frequency modulation.
Aural Transmitter:
Aural transmitter.
Carrier power-output rating.
Center-frequency stability.
FM noise level on carrier.
AM noise level on carrier.
Output voltage and impedance for audio and rf monitor connections.
Modulation capabilities.
Audio-input impedance and input level for 100% modulation.
Audio-frequency response.
Audio-frequency harmonic distortion.
Intermodulation distortion.
Rf output-coupling impedance range.
Safety:
Transmitter enclosure.
Grounding.
Grounding switches.
Grounding sticks.
Interlocks.
Access to voltages.
High-voltage metering devices.
Disconnect device.
Radiation.
Electromagnetic exposure.
Mechanical safeguards.

40. Signal Processing. The generation of the composite television signal, before its arrival at the transmitter, is treated in Sec. 20, Pars. 20-27 to 20-59. The transmitter receives the composite signal via cables, rebroadcast receivers, or microwave relay links. The distortion that can occur, especially to the synchronizing pulses, must be taken into account. To convert the incoming signal into a standard television waveform, a synchronizing regenerator may be introduced at the input of the visual transmitter, to regenerate the synchronizing pulses and stabilize the video level. For color transmission, it is also necessary to comply with two additional requirements: (1) the color burst must be transmitted without distortion, even when using clamping circuits operating during the back-porch interval, and (2) the amplitude of the color subcarrier must be regulated to the nominal value. Either the color-synchronizing burst or a corresponding element in the insertion signal can be used as reference. A typical arrangement for regeneration and stabilization is shown in Fig. 21-14.

41. TV-Transmitter Design Principles. *High- vs. Low-modulation levels.* In high-level modulated transmitters, modulation is carried out in the final rf power stage. Linearity and high video drive-power requirements are of concern in high-level modulation. In addition, the vestigial-sideband filter must handle the total transmitter-power output. With low-level modulation, one or more linear power amplifiers follows the modulated amplifier stage. These linear amplifiers must be designed for wide-band operation. In this case the vestigial-sideband filter can be located immediately after the modulated amplifier and thus operate at relatively low power.

The visual and aural transmitters are normally connected to the antenna through a combining-unit diplexer. The use of separate antennas for visual and aural transmitters is now rare.

The block diagram of the various stages of a typical vhf transmitter modulated in the power-amplifier stage is shown in Fig. 21-15. The vestigial-sideband filter and diplexer are integrated in a single unit, known as a filterplexer.

Triodes or tetrodes are used principally in vhf transmitter-output stages, tetrodes or klystrons for the uhf band. The tetrode is more efficient than the klystron, but three or four tetrode stages are required to equal the gain of a klystron, and the risks of tube failure are greater. The life of a klystron is greater than that of currently available tetrodes. The greater amplification available from the klystron also permits low-power, semiconductor drive stages to be used. The cooling system required for a klystron transmitter must handle a heavy cooling load, and the power-supply voltages are considerably higher.

Intermediate-Frequency Modulation. In if-modulated transmitters, however, the video signal is modulated at an intermediate carrier at low power, which is then heterodyned to the output frequency. The advantages of the if-modulated transmitter include the facts that the distortion is reduced, thanks to the low power at which modulation takes place (although the linearity of the subsequent stages must be of a high order), and that group-delay equalization can be applied at the intermediate frequency under optimum conditions.

Direct Modulation. The alternative to if modulation is direct modulation. In the earliest transmitters modulation was applied to the grid of the final rf tube. For large output powers,

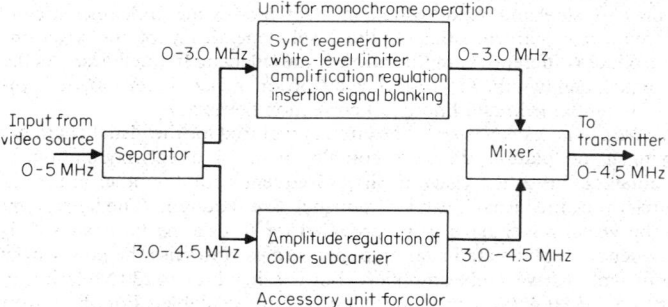

Fig. 21-14. Block diagram of signal-processing circuits for monochrome and color operation.

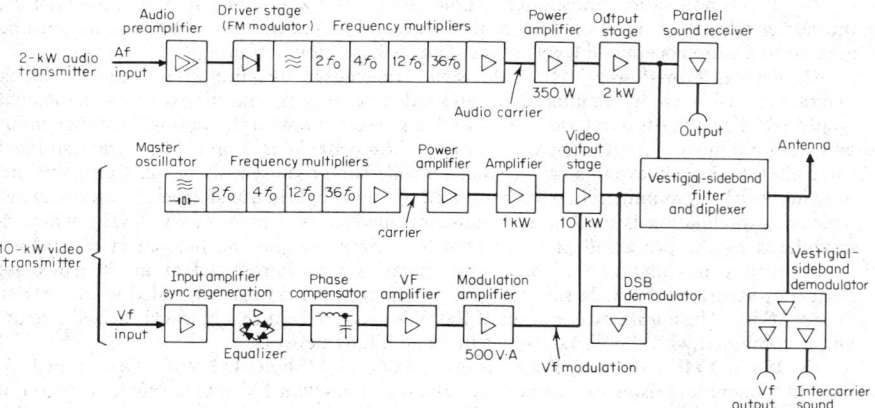

Fig. 21-15. Block diagram of a typical high-level modulated transmitter of 10-kW visual and 2-kW aural output.

the video modulation-amplifier requirements were quite onerous. Video voltages in the 400- to 500-V range, with reactive currents approaching 2 A at 5 MHz, were not uncommon.

Improvements in vacuum tubes and the introduction of sweep-frequency techniques for adjustment permitted wide-band rf amplifiers to be designed and high-power transmitters to be made with one or more rf linear amplifiers following the modulated stage.

Improvements in vacuum tubes have continued, and long-life tubes for high-power rf linear amplifier service are available with power gains exceeding 16 dB. This makes possible a single-tube or two-tube linear amplifier to be driven from a solid-state modulated rf drive.

The solid-state drive equipment takes several basic forms, e.g., a low-level modulator followed by transistor linear amplifiers or a higher-level modulator at the power level necessary to drive a vacuum tube. Investigations have also been made into the possible use of PIN diodes and rf transistors for high-power modulators.

Uhf transmitters use linear output amplifiers employing multicavity klystrons or traveling-wave tubes. With the high gains provided by these devices, the modulated-drive-level requirements are modest, usually a few watts, and are readily obtained from a solid-state modulator. Because of the low output-power requirements, the continuous-wave (cw) drive level is acceptably small. However, the incidental phase modulation introduced by the modulator is of sufficient magnitude to require prephase modulation of the cw drive signal to cancel incidental phase modulation.

The vestigial-sideband shaping filter may be placed between the modulator and the output linear amplifier. If it is placed at the input to a klystron amplifier, two undesirable effects are introduced: (1) intermodulation products generated in the klystron produce signals in the lower (suppressed) sideband region which, in effect, modify the slope response of the overall amplitude-frequency characteristic; (2) the level dependency of the klystron frequency response introduces differential amplification of the sideband frequencies. As the sidebands are not of equal bandwidth, the resulting distortion makes independent high- and low-frequency (differential gain and line-time) correction necessary.

Comparative Circuit Complexity. The circuits required for the visual chain of a transmitter employing if modulation are more complex than for direct modulation. Two high-stability oscillators, one at the visual if carrier frequency and the other at the sum of visual radiated carrier plus the visual if carrier frequency, are required. The heterodyne oscillator chain and the visual mixer have to be provided for translating the visual if signal to the radiated frequency. Solid-state linear amplifiers are used to raise the power from the mixer to a level sufficient to drive a tube amplifier. For vhf signals up to 230 MHz, linear solid-state amplifiers providing 60 W (sync level) output power are available. For uhf transmitters, only 2 or 3 W of output power is required to drive a klystron or traveling-wave amplifier, a level readily obtained from uhf transistor circuits.

For direct-modulation transmitters employing high-power solid-state visual modulators, the carrier-wave drive level is, of necessity, large, and pre- or postphase modulation must be applied to ensure acceptable levels of incidental phase modulation.

42. Typical Low-Power VHF Television Transmitter. Figure 21-16 shows a block diagram of a 1-kW vhf transmitter* using all solid-state circuits, with the exception of the final amplifier. The latter is a tetrode in a grounded-grid circuit, with single-tuned cathode input and a double-tuned output. The drive power of 70 to 100 W is supplied by a transistorized drive circuit. To achieve this power over the wide band required, six power transistors are used in the driving amplifier, in pairs paralleled by ferrite hybrid transformers. Intermediate-frequency modulation is used at a crystal-controlled vision carrier of 38.9 MHz, which is modulated by the processed picture signal in a circuit employing field-effect transistors.

Frequency modulation for the sound signal is also introduced at an intermediate frequency separated (for U.S. standards) 4.5 MHz from the crystal-controlled vision carrier of 38.9 MHz. The aural power output of 100 W is produced entirely in transistorized circuits and is combined with the visual carrier in a combining network.

43. 25-kW VHF Television Transmitter. Figure 21-17 shows the block diagram of a 25-kW vhf transmitter† using low-level if modulation. Less than 1 V of video signal is required to modulate the visual carrier. A highly linear broadband diode-ring modulator is used, with

---

* Designed and manufactured by the Marconi Broadcasting Division of the English Electric Company.
†Model BT-25L/BT-25H of the Gates Division of Harris Intertype Corporation.

active delay compensation and low-level vestigial sideband filtering. The transmitter proper
is fed by transistorized exciters for the aural and visual inputs. The master oscillator is in the
visual exciter.

Ceramic tetrode tubes are used in the final amplifier stages, operating in grounded-grid
and grounded-screen configurations. Neutralization is not required. All control functions
and metering are arranged for remote-control operation. Table 21-14 lists the specifications
of this transmitter.

44. 50-kW Television Transmitter for High-Band VHF. Figure 21-18 shows the block
diagram of one 25-kW unit of a 50-kW transmitter intended for use in the high-band vhf

Table 21-14. Specifications of the Type BT-25L/H Television Transmitter

Visual:
Power output .... 25 kW peak (FCC); 20 kW peak (CCIR "B")
Output impedance .... 50 Ω
Frequency range:
  BT-25L .... 48 to 88 MHz (channels 2 to 6)
  BT-25H .... 174 to 230 MHz (channels 7 to 13)
Carrier stability .... ±250 Hz (maximum variation over 30 days)
Regulation of rf output power (black to white picture) ... <3%
Variation of output over one frame .... <2%
Visual sideband response:
  +475 MHz and higher .... −20 dB or better
  Carrier to +4.18 MHz .... +0.5, −1 dB
  Carrier .... 0 dB reference
  Carrier to −0.5 MHz .... +0.5, −1 dB
  −1.25 MHz and lower .... −20 dB or better
  −3.58 MHz .... −42 dB or better
Frequency response vs. brightness .... ±0.75 dB (measured at 65 and 15% of modulation; reference 100% peak of sync)
Visual modulation capacity .... 3% or better
Differential gain .... 0.5 dB or better (maximum variation of subcarrier amplitude from 75 to 10% of modulation; subcarrier modulation percentage 10% peak to peak)
Linearity (low-frequency) .... 0.5 dB or better
Differential phase .... ±3° or better (maximum variation of subcarrier phase with respect to burst for modulation percentage from 75 to 10%; subcarrier modulation percentage 10% peak to peak)
Signal-to-noise ratio .... −50 dB or better (rms) below sync level
Envelope delay:
  0.05 to 2.1 MHz .... +70 ns*
  At 3.58 MHz .... ±35 ns*
  At 4.18 MHz .... ±70 ns*
Video input .... Bridging, loop through input with −30 dB or better return loss up to 5.5 MHz, 75-Ω system
Harmonic radiation .... −80 dB
Aural:
Power output .... 5 kW at diplexer output
Audio input .... +10 dBm, ±2 dB into 600 Ω
Input impedance .... 600/150 Ω
Preemphasis .... 75 μs
Frequency response .... ±0.5 dB relative to preemphasis (30 to 15,000 Hz)
Distortion:
  After 75 μs deemphasis with ±25 kHz deviation .... 0.5% or less
  After 50 μs deemphasis with ±50 kHz deviation 0.7%
FM noise .... −60 dB relative to ±25 kHz deviation
AM noise .... −52 dB relative to 100% modulation measured after deemphasis
Output impedance .... 50 Ω output connector 3⅛-in. EIA standard
Frequency stability .... ±250 Hz (maximum variation over 30 days)

* Reference to standard FCC curve.

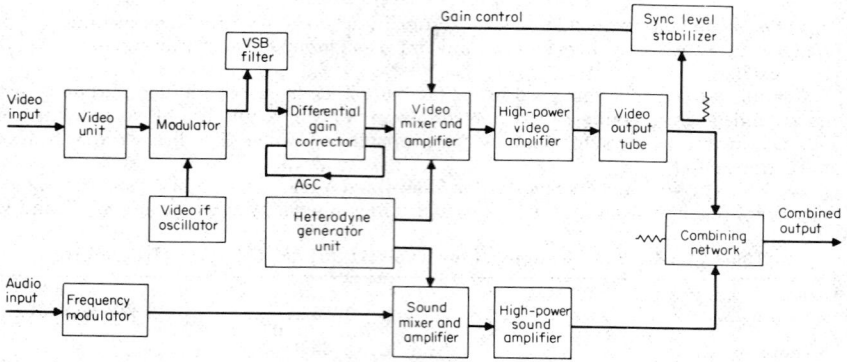

Fig. 21-16. Block diagram of 1-kW vhf television transmitter employing low-level modulation. *(Marconi Broadcasting Division.)*

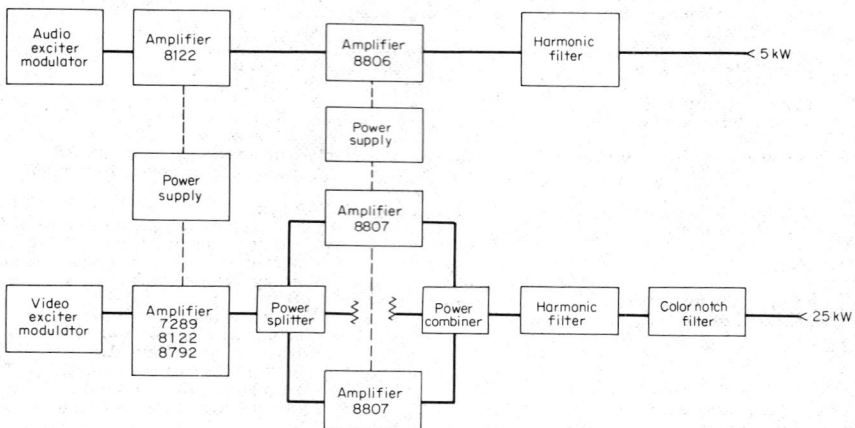

Fig. 21-17. Block diagram of 25-kW vhf television transmitter. *(Gates Division, Harris Intertype Corp.)*

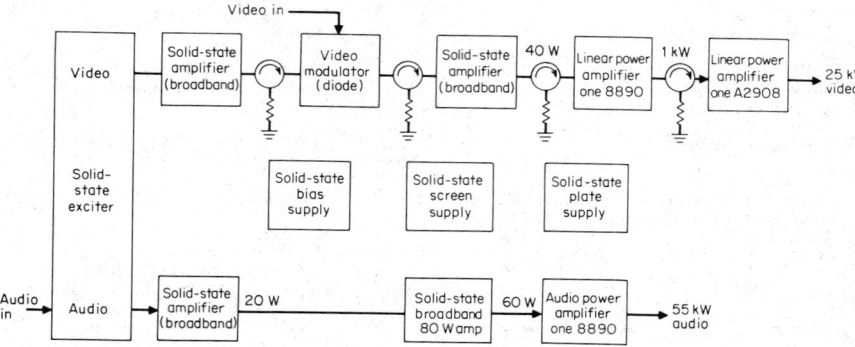

Fig. 21-18. Block diagram of one 25-kW unit of a 50-kW high-band vhf television transmitter. *(RCA Corporation.)*

channels.* The two identical 25-kW units are used so that one can continue operation in the event the other fails. The two units provide 11 kW of aural carrier output. The circuits employ solid-state devices, except the intermediate power amplifier and the final power amplifier, which are type 8890 tetrodes. The solid-state exciter and modulators provide power levels of 40 W (visual) and 60 W (aural) to the intermediate power amplifiers. Two exciters are provided, one of which drives both transmitters, the other serving as a standby unit which is automatically switched into operation should the first exciter fail. The exciters provide 10-W aural and 15-W visual carriers at one-sixth the respective carrier output frequencies; they are followed by six-times frequency multipliers. The aural center frequency is locked to the crystal-controlled visual-carrier reference source by an automatic frequency-control system. Noteworthy is the use of ferrite isolators between successive stages of the visual-signal chain, permitting each stage to be tuned with negligible reaction on the preceding or following stages.

This transmitter uses only two tuned linear-amplifier stages following the visual modulator. Sideband attenuation is performed by a filter at the transmitter output, which employs temperature-compensated elements to assure precise tuning at rated power levels over the temperature range from -20 to +50°C. Table 21-15 lists the specifications of this transmitter.

45. 55-kW UHF Television Transmitter. Figure 21-19 shows a block diagram of a transmitter† designed for high-power operation in the uhf television band. It employs two identical five-cavity vapor-cooled klystrons as the visual and aural final power amplifiers.

---

* RCA type TT-50H.
† Model BT-55U, Gates Division of Harris Intertype Corporation.

Table 21-15.   Specifications of the Type TT-50H Television Transmitter

Frequency range	Channels 7–13
Rated power output:	
Visual	50 kW
Aural	11 kW
Input level:	
Visual ....,....................	0.5 to 2.0 V peak to peak
Aural (for ±25 kHz deviation)	+10 ±2 dBm
Amplitude vs. frequency response:	
Aural	±1 dB, 30 Hz–15 kHz, of 50- or 75-μs preemphasis response curve
Visual sideband response:	
At carrier +0.5 to 3.58 MHz	±0.75 dB
At carrier −0.5 MHz	+0.5, −0.75 dB
At carrier −0.75 MHz	+0, −1.5 dB
Relative to response at +3.58 MHz between +2 and +4.18 MHz	+0.5, −0.75 dB
Variation in frequency response with brightness	±0.75 dB
Carrier frequency stability:	
Visual	±250 Hz
Aural	±500 Hz
Modulation capability:	
Visual	3%
Aural	±50 kHz
Audio-frequency distortion	0.5% max (30 to 15,000 Hz)
FM noise (below ±25 kHz deviation)	−60 dB
AM noise rms:	
Visual	−50 dB below sync level
Aural	−50 dB
Amplitude variation over one picture frame	<2.0%
Regulation of output	3%
Burst vs. subcarrier phase	±3°
Subcarrier amplitude	0.7 dB
Subcarrier phase vs. brightness (differential phase)	±2°
Linearity (differential gain)	0.5 dB
Linearity (low-frequency)	1.0 dB
Envelope delay vs. frequency:	
0.2 to 2.0 MHz	±60 ns
At 3.58 MHz	±30 ns
At 4.18 MHz	±60 ns
Harmonic and spurious radiation	−80 dB
Spurious radiation	−100 dB

These klystrons develop full power output with less than 1 W of drive power. The exciters deliver nominally 2 W of visual drive and 5 W of aural drive. Low-level if modulation is used. All circuits except the final amplifiers employ solid-state devices exclusively. The klystrons are controlled by digital logic circuits and operate completely independently, with separate magnet supplies and overload sensors. To remove heat from the klystrons, a unitized heat exchanger is used, containing cooling cores, blower and motor, circulating pump, storage tank, and control devices. The transmitter is fully equipped for remote-control operation. Table 21-16 lists the specifications of the BT-55U transmitter.

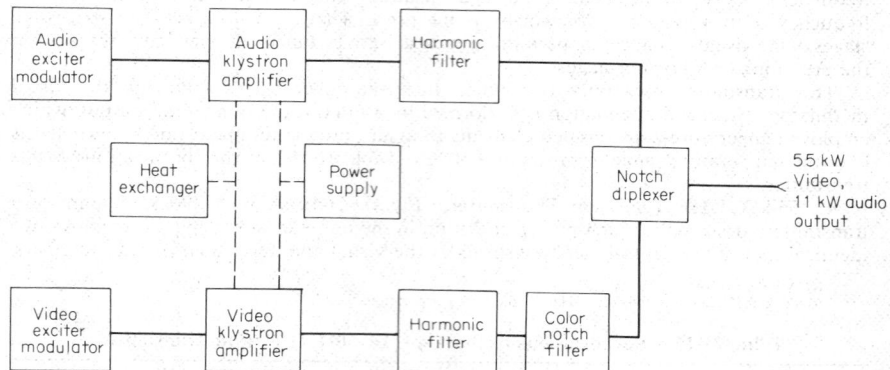

Fig. 21-19.   Block diagram of 55-kW uhf television transmitter.   *(Gates Division, Harris Intertype Corp.)*

AM BROADCAST RECEIVERS

By Norman W. Parker

46. Introduction.   AM broadcast receivers are designed to receive amplitude-modulated signals between 540 and 1,600 kHz (555 to 185 m wavelength), with channel assignments spaced 10 kHz. To enhance ground-wave propagation the radiated signals are transmitted with the electric field vertically polarized.

AM broadcast transmitters are classified, according to the input power supplied to the power amplifier, from a few hundred watts up to 50 kW. The operating range of the ground-wave signal, in areas where the ground conductivity is high, is up to 200 mi for 50 kW transmitters. During the day the operating range is limited to the ground-wave coverage. At night, refraction of the radiated waves by the ionosphere causes the waves to be channeled between the ionosphere and the earth, resulting in sporadic coverage over many thousands of miles. The nighttime interference levels thus produced impose a restriction on the number of operating channels that can be used at night.

The signal-selection system is required to have a constant bandwidth (approximately 10 kHz), continuously adjustable over a 3:1 range of carrier frequencies. The difficulty of designing cascaded tuned rf amplifiers of this type has resulted in the universal use of the superheterodyne principle in broadcast receivers.

A block diagram of a typical design is shown in Fig. 21-20. In this figure the signal is supplied by a vertical monopole (whip) antenna in automobile radio receivers or by a ferrite-rod loop antenna in portable and console receivers. An rf amplifier is used in most automobile designs but not in small portable models.

In some receivers the local oscillator is combined with the mixer, which simplifies the rf portion of the receiver. An intermediate frequency of 455 kHz is used in portable and console receivers, while 262.5 kHz is common in automobile radio designs. Diode detectors are almost universally used for detection of the if signal. Push-pull class B audio-power amplifiers are used to minimize current drain. A moving coil or dynamic speaker is used as an output transducer.

Table 21-16.   Specifications of the Model BT-55U UHF Television Transmitter

Visual:
Output power .... 55 kW (peak of sync)
Frequency range .... 470 to 890 MHz (channels 14 to 83)
Carrier stability .... ± 250 Hz (maximum variation over 30 days)
Regulation of rf output power (black to white picture) ... Less than 3%
Variation of output over one frame .... <2%
Visual sideband response
+4.75 MHz and higher .... −20 dB or better
Carrier to +4.18 MHz .... +0.5, −1 dB
Carrier .... 0 dB reference
Carrier to −0.5 MHz .... +0.5, −1 dB
−1.25 MHz and lower .... −20 dB or better
−3.58 MHz .... −42 dB or better
    Corner frequencies scaled to meet CCIR standards.
Frequency response vs. brightness .... ±0.75 dB (measured at 65 and 15% of modulation; reference 100% = peak of sync)
Visual modulation capability .... 3% or better
Differential gain .... 0.5 dB or better
Linearity (low frequency) .... 0.5 dB or better
Differential phase .... ±4° or better
Signal-to-noise ratio .... −50 dB or better (rms) below sync level
Envelope delay:
.05 to 2.1 MHz .... ±70 ns*
At 3.58 MHz .... ±35 ns*
At 4.18 MHz .... ±70 ns*
Video input .... Bridging, loop through input with −30 dB or better return loss up to 5.5 MHz, 75-Ω system.
Video input level .... 1.0 V peak to peak ±3 dB, sync negative
Harmonic radiation .... −80 dB
Aural:
Audio input .... +10 dBm, ±2 dB into 600 Ω
Preemphasis .... 75 μs
Frequency response .... ±0.5 dB relative to preemphasis (30 to 15,000 Hz)
Distortion .... 0.5% or less after 75 μs deemphasis with ±25 kHz deviation
FM noise .... −59 dB relative to ±25 kHz deviation
AM noise .... −55 dB relative to 100% modulation (measured after deemphasis)
Frequency stability .... ±250 Hz (maximum variation over 30 days)
Output power .... 5.5 to 11 kW

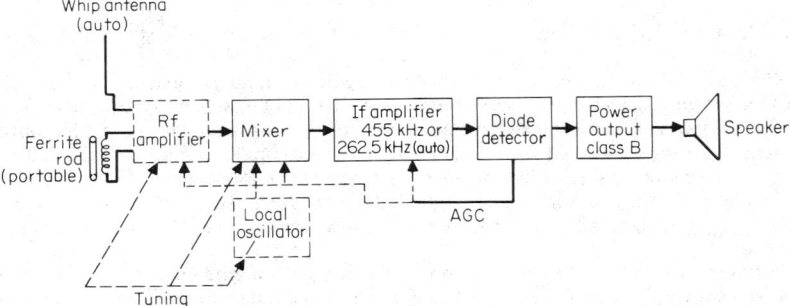

Fig. 21-20.   Block diagram typical of AM receivers.

47. Design Categories. AM receiver designs currently fall into three categories.

Portable Battery-powered Receivers without External Power Supply. These units vary in size from small pocket radios operating on penlite cells to larger hand-carried units using D cells for power. The battery life of the pocket radio depends on the type of cells used and the duty cycle of receiver operation. In a typical pocket radio using carbon-zinc penlite cells with operation of 2 h/day the life expectancy is about 70 h, to an end point at 80% of the initial battery voltage. Failure usually occurs with loss of power output from the audio-power amplifier or from loss of local oscillation. The audio-power output is usually about 75 mW for the pocket receiver. The larger portable units using D cells operate for longer periods of time without shortening the battery life. When the average operation is less than 2 h/day, the battery life is limited by shelf-life degeneration. The power output in the larger portable units is about 250 mW.

Console and Component Type AM Receivers Powered by the Power Line. These units are usually a part of an AM-FM receiver, with high audio-power output capability. The power output ranges from several watts to more than 100 W. Most audio systems use push-pull class B operation. However, since quiescent power drain is not a limitation, some units use single-ended class A power amplifiers to save transistor costs. Most such systems are equipped with two amplifier systems for FM stereo operation.

Automotive Receivers Operated on the 12-V Battery-Generator System of the Automobile (Usually Positive Ground). The primary current used in transistorized receivers usually does not exceed 1 A. Because of operation in the high ambient noise of the vehicle, the power output required is relatively high (2 to 3 W).

48. Sensitivity and Service Areas. The required sensitivity is governed by the expected operating field strengths. Typical field strengths for primary (ground-wave) and secondary (sky-wave) service are as follows:

	Field strength, mV/m
Primary service:	
Central urban areas (factory areas)	10-50
Residential urban areas	2-10
Rural areas	0.1-1.0
Secondary service:	
Areas where sky-wave signals exceed 0.5 mV/m for at least 50% of the time	

Cochannel protection is provided for signals exceeding 0.5 mV/m. The receiver sensitivity and antenna system should be adjusted to provide usable outputs with signals of the order of 0.1 mV if the receiver is to be used over the maximum coverage area of the transmitter.

The required circuit sensitivity is controlled by the efficiency of the antenna system. An auto receiver vertical antenna is adjustable to about 1 m in length. Since the shortest wavelengths are of the order of 200 m, the antenna can be treated as a nonresonant short-capacitive antenna. The open-circuit voltage of such a short monopole antenna is

$$E_a = 0.5\ l_{eff}E_f\ \text{mV}$$

where l_{eff} is the effective length of antenna in meters and E_f is the field strength in millivolts per meter.

The radiation resistance of the short monopole is small compared with the circuit resistance of the receiver input circuit, but the antenna is not matched to the input impedance since matching is not critical, adequate antenna voltage being available at the minimum field strength (0.1 mV/m) needed to override noise. The car-radio antenna is coupled to the receiver by shielded cable. This, with the receiver input capacitance, forms a capacitive divider, reducing the antenna voltage applied to the receiver. To ensure adequate operation the receiver should offer 20 dB signal-to-noise ratio when 10 to 15 μV is applied to the input terminals.

Portable and console receivers use much shorter built-in antennas, usually horizontally polarized coils wound on ferrite rods. The magnetic antenna can be shielded from electric field interference. Although the effective length of a ferrite rod is shorter than that of a whip antenna, the higher Q of the ferrite rod and coil provide approximately the same voltage to the receiver. The unloaded Q of a typical ferrite-rod antenna coil is of the order of 200. The voltage at the terminals of the antenna coil is QE_a.

49. Selectivity. Channels are assigned in the broadcast band at intervals of 10 kHz, but adjacent channels are not assigned in the same service area. Class 1 stations are protected from interference on adjacent channels to a ground-wave signal-level contour of 0.5 mV. In superheterodyne receivers, selectivity is required not only against interference from adjacent channels but also to protect against image and direct if signal interference.

The primary adjacent channel selectivity is provided by the if stages, whereas image and direct if selectivity must be provided by the rf circuits. In receivers using a ferrite-rod antenna and no rf stage, the rf selectivity is provided entirely by the antenna. High Q in the antenna coil thus not only provides adequate signal to override the mixer noise but also protects against image and if interference. With a Q of 200 the image rejection at 1,400 kHz is about 40 dB while the direct if rejection at 600 kHz is about 24 dB. With an rf stage added, the image rejection is about 50 dB and the if rejection about 46 dB.

Since auto-radio receivers are subjected to an extreme dynamic range of signal levels, the selectivity must be slightly greater to accommodate strong signal conditions. The image rejection at 1,400 kHz is typically 58 dB in spite of the lower if frequency, the if rejection is typically 50 dB and the adjacent-channel selectivity is about 20 dB. Figure 21-21 shows the overall response of a typical portable receiver using a ferrite rod without an rf amplifier.

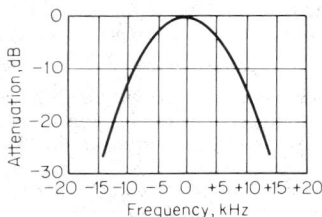

Fig. 21-21. Typical selectivity curve of an AM receiver using a ferrite-rod antenna without rf amplifier.

50. High-Signal Interference. When strong signals are present, the distortion in the rf and if amplification stages can generate additional interfering signals. The transfer characteristic of an amplifier system can be expressed in a power series as follows:

$$E_{out} = G_1 E_{in} + G_2 (E_{in})^2 + G_3 (E_{in})^3 + \cdots + G_n (E_{in})^n$$

where E_{out} is the output voltage, E_{in} the input voltage in the same units, and $G_1, G_2, \cdots G_n$ are the voltage gains of the successive amplifier stages.

When the input signal consists of two or more modulated rf signals, E_{in} becomes

$$E_{in} = e_1 \cos \omega_1 t + e_2 \cos \omega_2 t$$

where

$$e_1 = E_1 \left[1 + \frac{Ea_1(t)}{E_1} \right]$$

$$e_2 = E_2 \left[1 + \frac{Ea_2(t)}{E_2} \right]$$

$$E_1 = \text{signal 1 carrier}$$

$$E_2 = \text{signal 2 carrier}$$

$$Ea_1(t) = \text{audio modulation first signal}$$

$$Ea_2(t) = \text{audio modulation second signal}$$

IF Beat. When two strong signals are applied to the amplifier with carrier frequencies separated by a difference equal to the intermediate frequency, a difference-frequency signal appears which is independent of the local oscillator frequency. Because of the wide frequency spacing, interference of this kind can take place only in the mixer or rf amplifier and only with strong signals (signal strengths of several volts per meter). These signals are derived from the G_2 term:

$$E_{\text{if,beat}} = G_2 e_1 e_2 \cos{(\omega_1 - \omega_2)t}$$

where $e_1 e_2$ and $\omega_1 \omega_2$ are the respective amplitudes and angular frequencies.

Cross-Modulation. When a strong interfering signal is present, the modulation on the interfering signal can be transferred to the desired signal by third-order distortion components. This type of interference does not occur at a critical frequency of the interfering signal, provided that it is close enough to the desired signal frequency not to be attenuated by the selectivity of the receiver. This type of distortion can take place in the rf, mixer, or if stages of the receiver. These signals are derived from the G_3 terms:

$$E_{\text{crossmod}} = \frac{G_3 (e_2)^2 e_1}{4} \cos{\omega_1 t}$$

The cross modulation is proportional to the square of the strength e_2 of the interfering signal.

Intermodulation. Another type of interference due to third-order distortion is caused by two interfering carriers. When these signals are so spaced from the desired signal that the first signal is arbitrarily spaced by Δf and the second signal is $2\Delta f$, third-order distortion can create a signal on the desired carrier frequency. These signals are derived from the G_3 terms:

$$E_{\text{intermod}} = G_3 e_1{}^2 e_2 / 4 \cos{(2\omega_1 - \omega_2)t}$$

The interference is proportional to the square of the amplitude of the closer carrier times the amplitude of the farther carrier. Intermodulation is sometimes masked by cross modulation; it occurs only when the e_2 signal is stronger than the desired carrier after attenuation by the rf selectivity of the receiver.

Harmonic Distortion. Harmonic-distortion interference usually arises from harmonics generated by the detector following the if amplifier, which radiate back to the tuner input.

51. Choice of Intermediate Frequency. Two intermediate frequencies are used in broadcast receivers: 455 kHz and 262.5 kHz. The 455 kHz intermediate frequency has an image at 1,450 kHz when the receiver is tuned to 540 kHz, thus allowing good image rejection with simple selective circuits in the rf stage. At 540 kHz the selectivity must be sufficient to prevent if feedthrough since the receiver is particularly sensitive at if frequencies because converter circuits typically have higher gain at if than at the converted carrier frequency. The choice of the higher if frequency also makes if beat interference less likely. The second harmonic falls at 910 kHz and the third harmonic at 1,365 kHz.

The 262.5 if has a lower limit of image frequency at 1,065 kHz, which requires somewhat more rf selectivity than is needed when the higher if is used. The second harmonic of the 262.5-kHz frequency falls below the broadcast band (at 525 kHz) and hence does not interfere. On the other hand, there are more higher-order responses in the passband (787.5, 1050, 1,312.5, and 1,575 kHz). Sensitivity to if feedthrough when the receiver is tuned to 540 kHz is greatly reduced by the use of the lower if frequency.

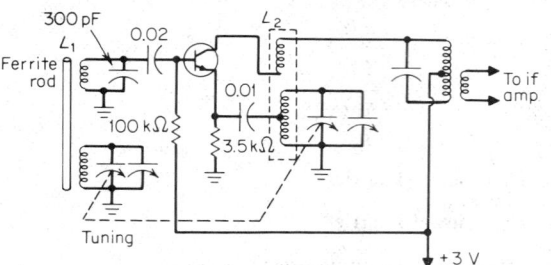

52. Tuners. There are three basic circuit configurations used in AM receiver tuners. The least complicated uses a self-oscillating mixer stage. Figure 21-22 shows a typical circuit of a self-oscillating mixer. The most critical design factor is tracking the oscillator with the antenna tuning to maintain a fixed frequency difference of 455 kHz below the oscillator. The oscillator covers the range from 995 kHz to 2075 kHz as the antenna circuit covers the band from 540 to 1,600 kHz. Since the frequency ratio covered by the oscillator is less than the antenna circuit, the total capacitance change and the rate of capacitance change are less in the oscillator than in the antenna circuit. Since the oscillator and antenna are ganged together on the same control,

Fig. 21-22. Typical self-oscillating mixer for AM service.

the plates of the oscillator must be shaped to provide tracking. The shortcoming of this circuit lies in its lack of rf selectivity and the inability to employ automatic gain control (AGC) to control overload. In some circuits a separate mixer is used to allow the application of AGC to the mixer stage.

Figure 21-23 shows a circuit with a separate mixer and an rf amplifier stage, using a three-gang variable capacitor for tuning. The capacitors in the base and collector circuit of the rf amplifier are identical. When the trimmer capacitors are adjusted to provide the same effective minimum capacitance in both circuits and the collector coil is adjusted to match the base-circuit resonance, the rf tuning adjustments track. The oscillator relies on capacitor-plate shaping to track the rf stages to produce a constant 455 kHz difference frequency.

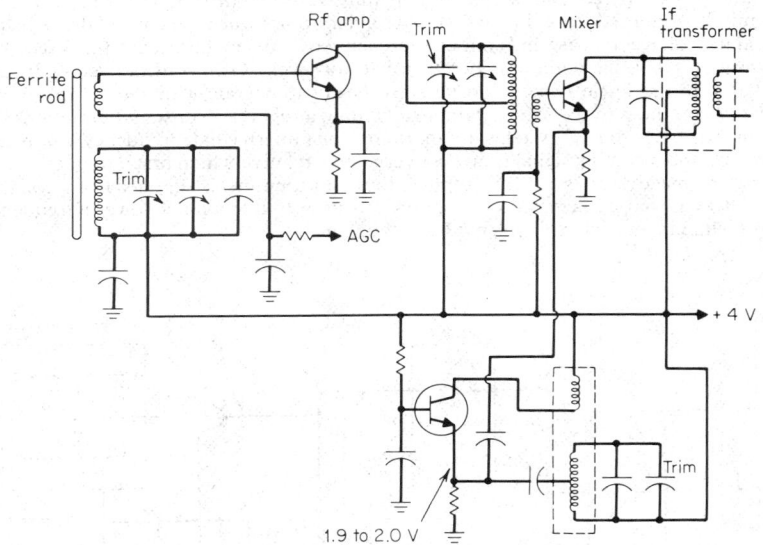

Fig. 21-23. Tuner with rf stage and separate oscillator. *(Motorola, Inc.)*

The forward bias on the rf transistor is reduced by the negative voltage at the diode detector. Reducing the forward bias reduces transistor beta and reduces the gain to the rf amplifier. Controlling the gain of the rf stage tends to reduce spurious signals which would otherwise occur in the mixer stage when strong signals are received. Since the tuning of the rf circuits relies on the antenna circuit, the Q of this circuit must be as high as possible. Unloaded Q's of more than 200 are desirable. When the loaded Q is adjusted to one-half the unloaded value, the antenna coil is matched for optimum power transfer from the antenna circuit. Since the radiation resistance is a small fraction of the resistance that determines the Q, the antenna source impedance is never matched to the input. Using the highest possible Q with the greatest effective rod length provides the best possible compromise antenna match.

Transistor noise is low enough for satisfactory signal-to-noise ratios to be obtained with 0.1mV/m when an rf stage is not used.

Automotive radio equipment relies on push-button tuners, which use slide mechanisms that adjust core positions in coils rather than resetting the exact angle on the rotating shaft of a variable capacitor. Figure 21-24 shows a circuit of a typical automotive radio tuner. Since the tuner uses the 262.5 kHz if, the tracking problem is greatly simplified. In the slug-tuned receiver the tracking is accomplished with coil geometry so that a relatively constant frequency difference is maintained throughout the slug travel.

53. IF Amplifiers. The if amplifier is primarily responsible for the gain and selectivity of the receiver. Since the if amplifier operates at a fixed frequency, the bandwidth of the receiver is independent of the tuning adjustment over the full tuning range from 540 to 1,600 kHz. Although it is tempting to design essentially all of the gain into the if amplifier, this practice may cause instability from regeneration unless expensive shielding is used. Hence total gain is distributed among the rf, if, and audio stages.

The if designs usually employ two stages of single-tuned amplification, with about 60 dB gain overall. A typical gain distribution is as follows: 10 to 30 dB in the rf section, 60 dB in the if amplifier, and 30 to 60 dB in the audio section, the total power gain being 100 to 150 dB. Figure 21-25 shows the circuit of a typical two-stage if amplifier with AGC applied to the first stage. The circuit derives its selectivity from three single-tuned transformers with tightly coupled secondary windings to match the output impedance of the collector and the input impedance of the base-emitter circuit of the following amplifier. Ideally, the Q of the circuit is determined by the loading of the amplifiers, but in practical coil designs the attainable Q is limited and some mismatching (2 to 3 dB) of the amplifier occurs.

Significant gain reduction in the first if amplifier occurs when the current drops below $\frac{1}{4}$ mA and the voltages on base and emitter approach zero. When this occurs, the detector level is slightly over 1 V, the optimum operating detector level. Gain control over 40 dB can be achieved with the single stage. Typical AGC begins to operate with about 1 mV from the mixer and continues to reduce the gain, maintaining a relatively constant output at 200 mV.

Some superheterodyne systems use more than one intermediate frequency. In receivers of this type, the incoming signal is first converted to a relatively high first if frequency chosen to minimize image responses. The output of the first if amplifier is supplied to a second mixer stage, where a fixed-frequency oscillator converts its output to a much lower if frequency, at which high gain and selectivity can be easily obtained.

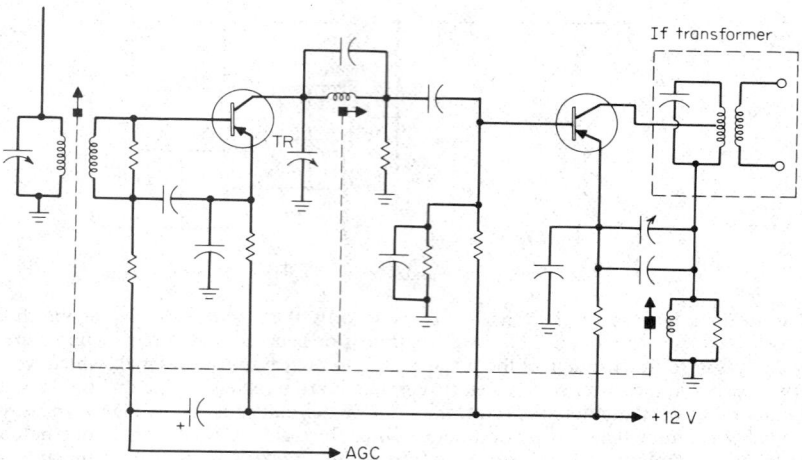

Fig. 21-24. Tuner for automobile AM radio. *(Motorola, Inc.)*

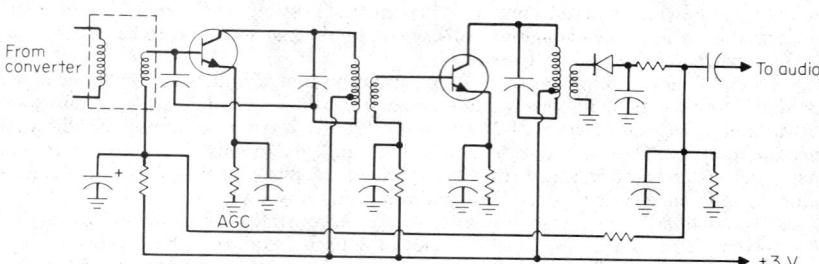

Fig. 21-25. Typical two-stage AM receiver if amplifier. *(Motorola, Inc.)*

Mechanically vibrating elements can be designed with Q's much higher than electric circuits, and it is feasible to use mechanical vibrators in if filter circuits to provide excellent selectivity in low-frequency if circuits. There are two basic types of electromechanical filter circuits. (1) Piezoelectric vibrator plates made of quartz or (for low and medium frequencies) barium titanate form the equivalent of a high-Q coil-and-capacitor combination which can be used singly or in combination to form if filters with steep band-edge selectivity characteristics. The piezoelectric vibrator converts energy from electrical to mechanical and back to electrical with relatively little loss. (2) Passive mechanical filters are made of mechanically vibrating elements, coupled to form a composite filter of extremely high selectivity. Low-loss filters of this type are often used in communication receivers, where good selectivity is more important than cost. A typical filter of this type uses several lumped masses coupled together by torsion rods. The low-loss rod-and-mass resonant combinations are used as elements in Butterworth or Chebyshev type filters. The signal is converted from electrical form to mechanical form to drive the filter, and the filter output converted back to electrical form by piezoelectrical or magnetostrictive converters. The conversions may introduce losses of 20 dB or more.

Barium titanate resonators have been used in some broadcast receiver designs. While they have the advantage of small size (at 455 kHz), the disadvantage is the numerous spurious responses caused by multiple modes of vibration. Hence those resonators must be used in combination with coils and capacitors to suppress the spurious responses. The need for supplementary filtering has resulted in limited use of resonators of this type.

54. IF Detectors. The series-connected diode detector is almost universally used in detecting the if signal. The diode detector is simple, efficient, and relatively free of distortion when properly used. The ac diode output constitutes the audio signal, while the dc component provides the control bias for AGC on the if (and the rf stage if one is present). To operate the diode at low signal levels it is necessary to provide forward bias to overcome the diode-contact potential. In Fig. 21-25 the AGC circuit is connected to the power supply to forward-bias the first if stage, while a 3,300-Ω resistor connects the detector to the first if stage and thus provides forward bias for the detector.

The effective loading of the diode on the last if stage is determined by the energy dissipated in the diode-load resistor. If the efficiency of the detector is high, the voltage across the diode load is equal to the peak voltage driving the detector. The input resistance R_i of the diode detector at carrier frequency is given by

$$R_i = R_d/2N_0$$

where R_d is the diode-load resistance and N_0 the efficiency of detector (maximum 1).

The diode-load capacitor is charged by the if output coil through the diode and discharged by the resistive diode load. The discharge time constant must be short enough to follow downward changes in the carrier level without causing the diode to drop out. When diode dropout occurs, the output wave follows the RC discharge curve and a type of distortion known as *diagonal clipping* occurs.

The capacitive reactance should be chosen to accommodate the relationship

$$Z_m/R_d > m$$

where Z_m is the capacitor reactance at the highest modulating frequency, R_d is the diode-load resistance, and m is the modulation factor at the highest frequency (modulation percentage divided by 100).

55. Audio Amplifiers. The audio amplifier raises the detected signal from a level of about 100μW to the required audio-power level. In portable receivers, where the power output is 100 mW, the gain required is about 30 dB with additional reserve gain for low signal levels. In console or component-type receivers, where the audio output may be 100 W, gains in excess of 60 dB are required.

Figure 21-26 shows a typical battery-operated portable-radio audio and power-output circuit. The combined current drain of the first two stages is about 3 mA, with class A biasing. The first two stages provide sufficient power gain to drive the power stages, which supply about 75 mW of audio-power output. Since peak currents of 50 mA would be required to supply a class A audio stage, operation of the output stage would require a large battery or short battery life. To minimize current drain (particularly zero signal-current drain) the audio is operated in push-pull, class B.

In battery-operated portable sets the small speaker enclosures limit the low-frequency response. The tone control consists of a simple *RC* filter which reduces the response to high-frequency audio signals. In console receivers with wide-band acoustical response, the audio response is varied by more complex tone controls. Figure 21-27 shows a circuit of a tone control used in such receivers. This circuit provides independent attenuation or boost of the bass and treble responses. In addition to the tone control it is customary to provide a tone-compensated volume control on the larger console or component systems. The tone-compensated volume control provides a bass boost when the volume is reduced, to compensate for the normal drop in the sensitivity of the ear to low levels of the low frequencies. The presence of the 0.009-μF capacitor and series resistor to ground from the tap on the volume control provides progressively increasing bass boost until the control is retarded beyond the tap point, below which the boost remains constant.

56. Loudspeakers. The important element in an audio system affecting the quality of the reproduced sound is the loudspeaker and its enclosure (see also Sec. **19**). Loudspeakers fall into two general categories: direct-radiator piston types, using a moving coil in a permanent magnet field, and horn-type speakers, in which a wide-band acoustical-impedance transformation network (exponential horn) is used to couple a small piston to the free-space radiation impedance.

Although the horn loudspeaker is much more efficient than the direct-radiator type (efficiencies up to about 50%), such high efficiency is available only in horns of very large size if low-frequency signals are to be reproduced. Therefore horn units are used principally in large theater-type installations, where both high power and high efficiency are required.

In radio receivers the direct-radiator speaker is universally used for middle- and low-frequency reproduction. Horn-type tweeter units, designed to reproduce frequencies above 1 kHz only, and hence of moderate size, are used in some combination speaker systems.

In direct-radiator loudspeakers a coil of wire suspended in a magnetic field (voice coil) provides the driving force for a paper-cone piston supported at the edges by a springlike suspension. The force which moves the piston against the spring loading of the suspension is proportional to the voice-coil current. Since the coil is resistive over most of the audio frequencies, the force applied to the piston is constant with frequency for a given voltage applied to the speaker driving element.

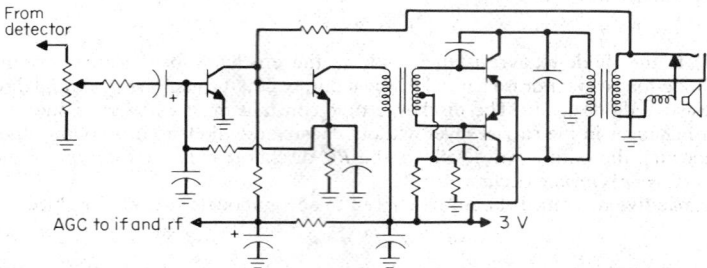

Fig. 21-26. Typical driver and class B audio output stages of a battery-operated portable receiver.

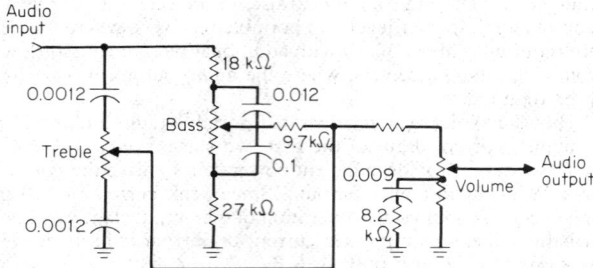

Fig. 21-27. Wide-range tone control and compensated volume control.

A piston enclosed on one side, so that its vibration changes the instantaneous air pressure near the piston, couples acoustical power to the air. For the radiated power to be independent of frequency, the velocity of the piston must be inversely proportional to frequency at low and medium frequencies, where the piston diameter is smaller than the wavelength of the radiated sound. The mechanical system represented by the piston and the elastic suspension is arranged to operate above resonance, in the audio range covered by the speaker. In the region above resonance, the load on the driving force is the mass of the piston, the velocity of which decreases directly with frequency with constant force applied. Below the mechanical resonance the velocity decreases with frequency such that the power output drops 12 dB/octave.

The radiated power is constant for constant drive power from the mechanical resonant frequency up to a frequency at which the distance around the periphery of the piston is equal to the wavelength of the sound in air. Above this point the radiation resistance of the piston is independent of frequency. Since the velocity of the piston is inversely proportional to frequency and the radiated power is proportional to the square of the velocity times the radiation resistance, the output above this point drops 12 dB/octave.

Two effects counter the tendency of the speaker to radiate less power at frequencies where the periphery of the piston exceeds a wavelength. The first effect is the directivity pattern of the radiating piston, above the frequency of constant radiation resistance, which increases the power radiated along the axis perpendicular to the piston. This increase counteracts the loss in total radiated power by concentrating the power radiated directly in front of the speaker.

The second effect which alters the radiated-power pattern at high frequencies is the loss of integrity of the moving piston. The piston is a cone made of light, stiff paper. At high frequencies the cone ceases to move as a single unit and breaks into separate cone resonance patterns. The cone breakup greatly alters the radiation pattern, resulting in substantial changes from the theoretically predicted radiation levels.

Since these effects are undesirable in a wide-range loudspeaker, concentric elastic rings are often built into the cone, which progressively alter the cone motion so that as the frequency is increased, the outer rings of the cone become decoupled and no longer move with the driving force applied at the center of the cone. The result is that the speaker operates over a wide frequency range without reaching a point where the radiation resistance becomes independent of frequency, with the resulting erratic behavior in the sound-power radiation pattern.

In high-quality audio systems it is customary to use several speakers, each covering a separate frequency range, instead of relying on decoupling in the cone. These systems usually employ a bass speaker system using a 12- or 15-in. speaker (woofer) to cover the range from 30 to 500 Hz, an 8- to 10-in. mid-range speaker to cover from 500 to about 2,000 Hz, and a tweeter to cover from 2,000 to 15,000 Hz.

Small speakers may be designed to act as bass sources when space is at a premium and large speakers cannot be used. When the piston is small and enclosed in a small sealed enclosure, the motion required of the piston increases inversely as the square of the piston diameter for a given power output. To radiate significant power at low frequencies the force driving the cone must be increased, resulting in a lower acoustical efficiency. A significant effect, in using small speakers in small enclosures to reproduce low frequencies, is the elastic effect of the trapped air in the enclosure behind the piston. The trapped air provides most of the cone-restoring force, determining the resonant frequency of the speaker system. Speakers in which the restoring force is principally the trapped air in the enclosure are referred to as *air-suspension speaker* systems.

When a speaker is operated in a small enclosure, the enclosure compliance tends to stiffen the suspension. As a result, the lowest usable frequency which can be reproduced is raised. In speakers designed for air suspension, the cone mass is increased to reduce the resonant frequency in the enclosure. The increased mass in the cone requires increased driving force for the same power output and a corresponding loss in efficiency. In battery-operated receivers, little attempt is made to reproduce at full level the frequencies below 150 Hz. In such cases the cone resonance is relied upon to enhance the response at the mechanical resonance of the speaker, which is the lowest frequency where reliable response can be obtained. In this case, the magnetic circuit is purposely weakened to minimize damping of the speaker resonance by the electromagnetic coupling between the mechanical system of the speaker and the source impedance of the power amplifier. This also provides a weight reduction in the speaker since a smaller magnet can be used.

TELEVISION BROADCAST RECEIVERS*

BY NORMAN W. PARKER

57. General Considerations. Broadcast television receivers are designed to receive signals in two vhf bands and one uhf band according to the United States and Canadian standards (for other standards, see Sec. **20**). The lower vhf band (channels 1 to 6) extends from 54 to 88 MHz in 6-MHz channels, with the exception of a gap between 72 and 76 MHz. The higher vhf band (channels 7 to 13) extends from 174 to 216 MHz in 6-MHz channels. The uhf channels are spaced 254 MHz above the highest vhf channel, comprising 70 6-MHz channels extending from 470 to 890 MHz.

The television tuner is thus required to cover a frequency range of more than 16:1. TV tuners use separate units to cover the uhf and vhf bands.

The signal coverage of TV transmitters is generally limited to line-of-sight propagation, with coverage extending from 30 to 100 mi, depending on antenna height and radiated power. The coverage area is divided into two classes of service, depending on the signal level. The service area labeled class A is intended to provide essentially noise-free service and specifies the signal levels shown.

Channels	Peak signal level, μV/m	Peak open-circuit (half-wavelength) antenna voltage, μV
Class A Service		
2–6	2,500	3,500
7–13	3,500	11,800
14–83	5,000	700
Class B Service		
2–6	225	300
7–13	600	300
14–83	1,600	225

For the limiting area of fringe service the signal levels are defined as shown for class B service. The typical level of the sound signal is from 3 to 20 dB below the picture level, due to radiated sound power and antenna gain.

58. Tuners. The TV tuner comprises two sections, one to cover the vhf bands from 54 to 88 MHz and from 174 to 216 MHz, the second to cover the uhf band from 470 to 890 MHz. The vhf tuner supplies gain at the selected transmitter carrier frequency, matches the antenna and impedance, and isolates the receiver local oscillator from the antenna. The uhf tuner consists of input circuits for matching and selectivity together with a crystal mixing circuit and local oscillator to provide an output at the if frequency, which is supplied to the vhf tuner. Because the uhf tuner has a conversion loss instead of a gain, the vhf tuner is designed to provide additional gain at the intermediate frequency. When so operated, the local oscillator in the vhf tuner is turned off and the rf amplifier and mixer are tuned to the if frequency when the receiver is switched to the uhf position.

Most receiver designs use a 300-Ω balanced input to the tuner terminals. The recent trend toward optional operation on closed-circuit television (e.g., CATV) systems, which generally use shielded 72-Ω coaxial cable, has resulted in tuners designed both for 300-Ω balanced and 72-Ω coaxial inputs.

Although balanced 300-Ω input is universally used, the rf amplifier stages are single-ended and so the transmission line must be coupled to the receiver through a balun transformer which converts the balanced input to a matched unbalanced load. The balun transformer usually consists of a two-window ferrite core with two windings of four turns each, coupled to provide a balance-to-unbalance match and an impedance transformation from 300 Ω balanced to 72 Ω unbalanced.

---

* The following paragraphs on television receivers deal only with the tuner and if-amplifier sections of monochrome and color broadcast receivers. The portions of receivers that handle second detection, synchronization, video amplification, color-signal processing, picture tubes, and their auxiliaries (which are common to other forms of television-reproduction equipment) are treated in Sec. **20**, Pars. **20-60** to **20-82**.-Ed.

The rf amplifier must have the lowest noise figure possible, consistent with cost, bandwidth, and dynamic range. The typical vhf tuner employs a bipolar transistor in a grounded-base amplifier circuit. To prevent signal overload in the collector circuit and mixer stage, forward-bias AGC is applied to the base of the rf amplifier stage. Channel tuning is accomplished by switching coils in series, on a wafer switch. A typical tuner of this type has four wafers ganged together. The first wafer tunes the input of the rf amplifier, with single tuned circuits. The rf-mixer coupling circuit, with a double-tuned circuit, uses two wafers, and the local oscillator is tuned with the fourth wafer.

The circuit of a complete vhf and uhf tuner of this type is shown in Figs. 21-28 and 21-29. This type of tuner has a typical noise figure of 3 dB and an overall gain of about 30 dB. The intermediate frequency is taken from the collector circuit by a low-impedance coupling network, using a capacitance divider to reduce the collector load impedance by a factor of 170 and feeding the signal through a coaxial coupling cable to the if amplifier. This method of coupling has become widely used since it allows the tuner to be located at a distance from the if amplifier, thus providing flexibility in cabinet and chassis locations; e.g., the channel selector can be located at any convenient place in the cabinet without regard to the location of the if amplifier stages.

59. IF Amplifiers. The choice of intermediate frequency is an important factor in determining the ability of the receiver to operate without interference from spurious responses. The local oscillator frequency is positioned above the desired signal frequency such that the frequencies are 45.75 MHz for the picture carrier, 41.25 MHz for the sound carrier, and 42.17 MHz for the color subcarrier. The images on channels 2 to 6 then fall in the gap between channels 6 and 7, while the images for channels 7 to 13 are above channel 13. In the uhf band (channels 14 to 83) each image falls 15 channels higher than the desired channel, so that the FCC uhf channel-assignment plan avoids assigning transmitters with 15-channel separation in the same area.

Numerous other combinations of signals can provide significant interference. The most important is the if beat, which occurs when the carriers of two strong channels are spaced seven channels apart, the mixture of the two causing a spurious if signal proportional to the

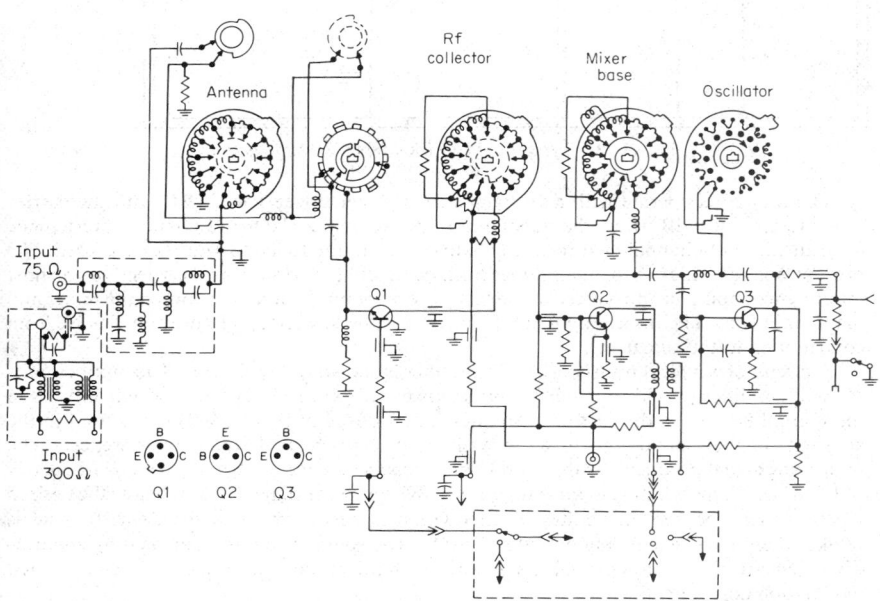

Fig. 21-28. Typical vhf television tuner. *(Motorola, Inc.)*

21-49

product of the two signals. In the uhf band, where seven-channel spacing is possible, the assignment to any one area of channels with seven-channel spacing is avoided. In the vhf band, the beat does not occur, since the channel assignments cannot be spaced seven channels. In the vhf band intermodulation and cross-modulation products are avoided by spacing the channels so that two strong signals do not fall on adjacent channels.

In spite of the care with which the intermediate frequency was chosen, there is at least one significant source of interference. This occurs on channel 6, where harmonics of the carrier generated in the rf amplifier can beat with the local oscillator to produce an interference with no additional signal present. On that channel the second harmonic of the sound channel beats with the local oscillator to produce a 0.75-MHz interfering signal, and the color subcarrier provides additional interference at 1.15 MHz.

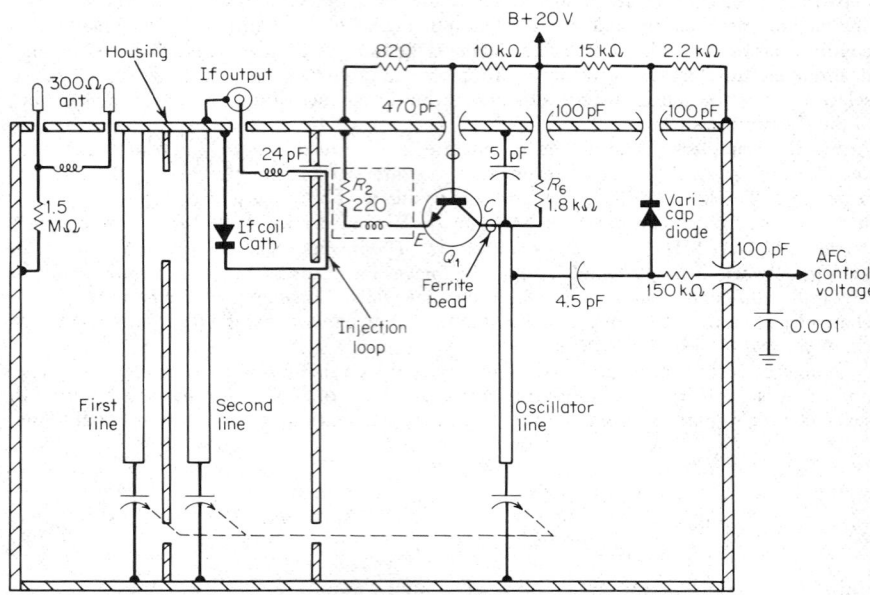

Fig. 21-29. Typical uhf television tuner. *(Motorola, Inc.)*

The if circuits operate with a vestigial-sideband signal (see Par. 20-24) with the carrier tuned to a point 6 dB below the maximum response point. To avoid possible interference from an adjacent-channel picture carrier, a fixed-tuned trap tuned to 47.25 MHz is used. To prevent nonlinearities in the if amplifier from generating spurious signals, selectivity against interference should be introduced at the lowest-level point, preferably between the mixer and the if amplifier. The insertion loss of the filter must be low enough to prevent if noise from contributing to the signal.

A circuit of a typical transistorized if amplifier is shown in Fig. 21-30. The input circuits of the if amplifier provide attenuation of the lower adjacent sound carrier, which appears in the if amplifier above the carrier. An attenuation of the order of 60 dB is required in the adjacent-channel sound trap to avoid visible adjacent-channel sound interference. The interfering signal produced by the sound signal produces a beat with a mean video frequency of 1.5 MHz. Signals falling in this range are visible when the signal is 50 dB below the desired video carrier. Moreover, additional attenuation is necessary when the desired signal is weaker than the adjacent-channel interference. The sound-signal is attenuated by about 26 dB to ensure that nonlinear effects in the if amplifier do not produce spurious cross-modulation components.

60. Intercarrier Sound Circuits. Prior to the video detector circuit the intercarrier FM sound signal, with a carrier frequency of 4.5 MHz, is removed from the video signal and is fed to a separate detector for sound only. The separate sound detector is required in color receivers, since the presence of the color subcarrier at 3.58 MHz would produce a 920-kHz beat if nonlinearities existed in the detector transfer characteristic. Since beats as low as 50 dB below the carrier level are perceptible, the required linearity would be impractical if the sound signal were to be recovered from the video detector.

The separate diode detector is connected to the last if amplifier stage, ahead of the video detector sound trap, for the purpose of developing the 4.5-MHz intercarrier sound beat. The sound-detector load circuit is tuned to 4.5 MHz. Although the intercarrier design makes tuning less critical (since the sound signal is present whenever the composite signal is received), the advent of color has made the tuning adjustment of the picture carrier frequency in the if amplifier more critical, since mistuning causes the color subcarrier to change amplitude rapidly.

In receivers with intercarrier sound, the output of the sound detector is not a function of the tuner oscillator frequency shift, and this method of automatic tuning oscillator control (AFC) cannot be used. To maintain correct tuning in color receivers of intercarrier design, a separate discriminator tuned to the picture carrier provides the control voltage for the tuner AFC. A typical AFC circuit is shown in Fig. 21-31. The sound channel supplies additional gain at 4.5 MHz, bringing the intercarrier sound signal to a level suitable for operation of the FM detector. One of the most popular sound-detector circuits is the ratio detector, since its inherent AM rejection characteristics minimize the need for limiting stages.

In most recent designs, composite amplification and FM detection circuits have been combined on integrated-circuit chips with a minimum number of external connections. In circuits of this type the amplifiers are generally of the wide-band type with no interstage selectivity. Instead, the selectivity is placed ahead of the integrated circuit to minimize the number of external connections. The discriminator is also designed to include a minimum number of coils and external connections.

Since in integrated circuits the number of external connections required is more critical than the number of stages used, it is possible to provide additional gain and limiting at 4.5 MHz and to provide an FM detector less complex than the ratio detector, thus reducing the external connections. The circuit usually employed is the quadrature detector, which uses a single coil to provide a quadrature voltage at the carrier frequency. Since the phase of the voltage derived from the tuned circuit varies with frequency, by comparing the tuned-circuit voltage with the voltage obtained directly from the amplifier-limiter, an audio-output voltage proportional to input frequency can be obtained. For additional discussion of circuits used in TV broadcast receivers, see Pars. **20-60** to **20-83**.

FM BROADCAST RECEIVERS

By Norman W. Parker

61. General Considerations. Broadcast FM receivers are designed to receive signals between 88 and 108 MHz (3.5 to 2.8 meters wavelength). The broadcast carrier is frequency-modulated with audio signals up to 15 kHz, and the channel assignments are spaced 200 kHz. The FM band is primarily intended to provide a relatively noise-free radio service with wide-range audio capability for the transmission of high-quality music and speech. The service range in the FM band is generally less than that obtainable in the AM band, especially when sky-wave signals are relied on for extending the AM coverage area.

Vhf signals are limited to usable service ranges of less than 70 mi. Since sky-wave signals do not materially affect the transmission of FM signals, there is no equivalent night effect and licenses are not limited to daylight hours only, as with many AM operations. In the past, all FM signals in the United States were horizontally polarized. However, the rules have been changed to allow maximum power to be radiated in both the horizontal and vertical planes.

Unlike AM broadcasting, where the station power is measured by the power supplied to the highest-power rf stage, FM transmitters are rated in *effective radiated power* (ERP), i.e., the power radiated in the direction of maximum power gain. This method of power measure-

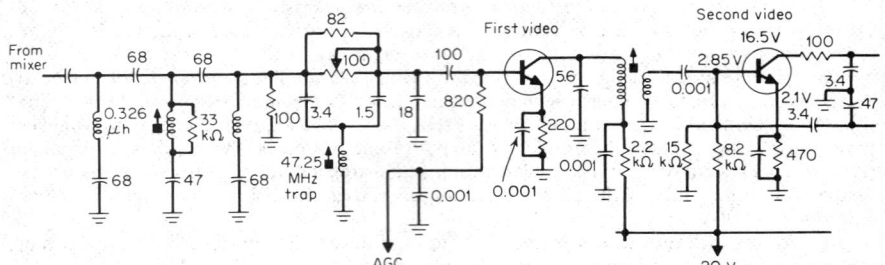

Fig. 21-30. Typical three-stage

ment is used since the transmitting antenna has significant power gain, resulting in an ERP many times the input power supplied to the rf output stage of the transmitter.

Although FM receivers have been principally used in high-fidelity audio installations, more recently small FM receivers have been designed with limited audio-output capabilities. Also, an increasing number of FM broadcast sets have been included in automobile installations.

In 1960 the FCC amended the broadcast rules to allow the transmission of stereophonic signals on FM stations (see Par. **21-68**). This transmission is equivalent to the transmission of two audio signals, each having 15 kHz bandwidth, transmitted on the same carrier as is used for monophonic FM signals. Since the FM signal was initially designed to have sufficient additional bandwidth to achieve improved signal-to-noise ratio at the receiver, there is room to multiplex the second component of the stereophonic signal with no increase in the radiated bandwidth. However, the signal-to-noise ratio is reduced when the multiplexing technique is employed. Most console and component-type FM receivers are currently designed for the reception of multiplex stereo.

62. Sensitivity. The field strength for satisfactory FM reception in urban and factory areas is about 1 mV/m. For rural areas, 50 μV/m is adequate signal strength. These signal levels are considerably lower than the equivalent levels for AM reception in the standard broadcast bands. Three effects make satisfactory reception with these lower signal levels possible: (1) the effects of lightning and atmospheric interference (static) are negligible at 100 MHz, compared with the interference levels typical of the standard broadcast band; (2) the antenna system at 100 MHz can be matched to the radio-receiver input impedance, providing more efficient coupling between the signal power incident on the antenna and receiver, and (3) the use of the wide-band FM method of modulation reduces the effects of noise and interference on the audio output of the receiver.

The open-circuit voltage of a dipole for any length up to one-half wavelength is given by

$$E_{oc} = E_f(5.59\sqrt{R_r})/)F_s \text{ mV}$$

where E_{oc} is the open-circuit antenna voltage, E_f the field strength at antenna in millivolts per meter, F_s the received signal frequency in megahertz, and R_r the antenna radiation resistance in ohms. For a half-wave dipole $R_r = 72 \Omega$; for a folded dipole $R_r = 300 \Omega$. For antennas substantially shorter than one-half wavelength, $R_r = 8.75l^2F_s^2 \times 10^{-3}$, where l is the total length of the dipole in meters. For a folded dipole one-half wavelength long, operating at 100 MHz, the open-circuit voltage is $E_{oc} = 0.97E_f$. The voltage delivered to a matched transmission and receiver input is one-half of this value, $0.48E_f$.

The noise in the receiver output is caused by the antenna noise plus the equivalent thermal noise at the receiver input. The input impedance generates an excess of noise when compared to the noise generated by an equivalent resistor at room temperature. The noise generated (Fig. 21-32) is given by

$$E_{nr} = E_n\sqrt{2NF - 1} \text{ volts}$$

where E_{nr} is the equivalent noise generated in receiver input, E_n the equivalent thermal noise

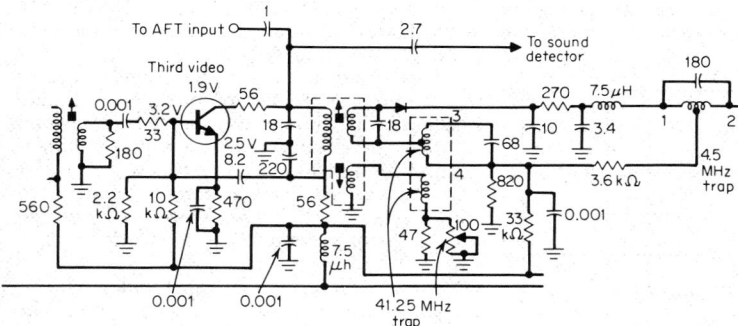

television if amplifier. *(Motorola, Inc.)*

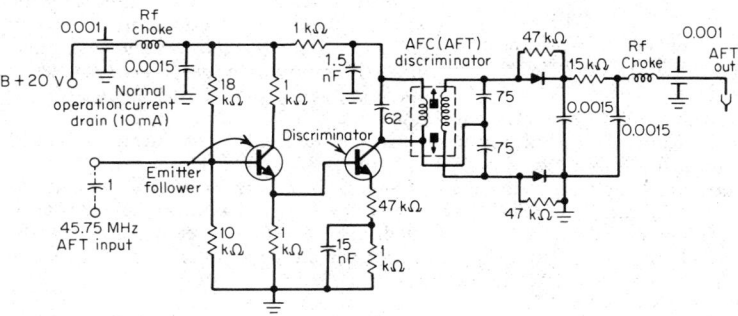

Fig. 21-31.   Discriminator for automatic frequency control.

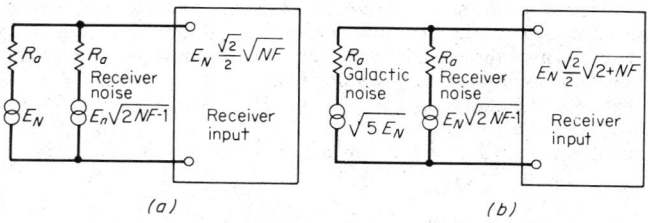

(a)   *(b)*

Fig. 21-32.   Equivalent sources of receiver noise: (*a*) with resistive generator; (*b*) with galactic noise source included.

in volts ($\sqrt{4R_{in}\,kT\,\Delta f}$), R_{in} the receiver input resistance, k is Boltzmann's constant (1.38×10^{-23} J/K), T the absolute temperature in degrees Kelvin (290 K), Δf the half-power bandwidth of the receiver response taken at the discriminator, and NF is the receiver noise figure. (If the noise figure is given in decibels, $NF_{dB} = 10\,\log_{10} NF$.)

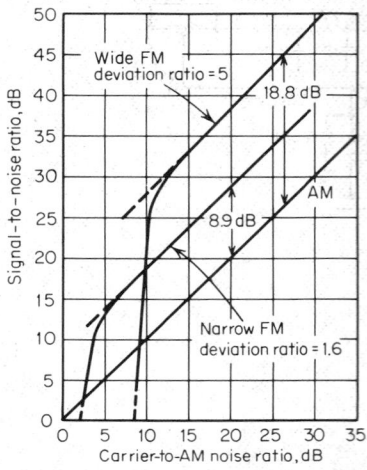

Fig. 21-33. Typical signal-to-noise ratios for FM compared with AM for deviation ratios of 1.6 and 5.

The generator noise and receiver noise add as the square root of the sum of the squares. Figure 21-32 shows that the equivalent receiver input noise is $0.707\,E_n\,\sqrt{NF}$. For a receiver with 300-Ω input resistance and 200-kHz noise bandwidth, $E_n = 0.984\ \mu V$. A typical noise factor for a well-designed receiver is 3 dB, or 2 times power ($\sqrt{2}$ voltage) increase, giving an equivalent noise input of 1.39 μV.

In an AM receiver the signal-to-noise (S/N) ratio at the receiver input is a direct measure of the S/N ratio to be expected in the audio output. In an FM receiver using frequency deviation greater than the audio frequencies transmitted, the S/N ratio in the output may greatly exceed that at the rf input. Figure 21-33 shows typical output S/N ratios obtained with receiver bandwidths adjusted to accommodate transmitted signals with modulation indexes of 1.6 and 5.0, compared with the audio S/N ratio when AM modulation is used and the bandwidth of the receiver is adjusted to accommodate the AM sidebands only. As shown in this figure, the S/N ratio for a properly designed receiver operating with a modulation index of 5 is 18.8 dB higher than that of an AM receiver with the same rf S/N ratio.

For rf S/N ratios in FM higher than 12 dB, the audio S/N ratio increases 1 dB for each 1-dB increase in rf S/N ratio. For FM S/N ratios lower than 12 dB, the S/N ratio in the audio drops rapidly and falls below the AM S/N ratio at about 9 dB. The point at which the ratio begins to fall rapidly is called the *threshold signal level*. It occurs where the carrier level at the discriminator is equal to the noise level. The threshold level increases directly as the square root of the receiver bandwidth, i.e., approximately the square root of the modulation index. The equation for S/N improvement using FM is

$$\frac{(S/N)_{FM}}{(S/N)_{AM}} = \sqrt{3}\,\Delta$$

where Δ is the deviation ratio. Since broadcast standards in the United States for FM call for a modulation index of 5 for the highest audio frequency, the direct S/N improvement factor is 18.8 dB for rf S/N ratios exceeding 12 dB.

In the FM system a second source of noise improvement is provided by preemphasis of the high frequencies at the transmitter and corresponding deemphasis at the receiver. The preemphasis network raises the audio level at a rate of 6 dB/octave above a critical frequency, and a complementary circuit at the receiver decreases the audio output at 6 dB/octave, thus producing a flat overall audio response. Figure 21-34 shows simple *RC* networks for preemphasis and deemphasis.

The additional S/N improvement using deemphasis in an FM receiver is

$$\frac{(S/N)_{out}}{(S/N)_{in}} = \frac{f_a{}^3}{3[\,f_a f_0{}^2 - f_0{}^3 \tan^{-1}(f_a/f_0)\,]}$$

where $(S/N)_{out}$ is the signal-to-noise ratio at deemphasis output, $(S/N)_{in}$ the signal-to-noise ratio at deemphasis input, f_a the maximum audio frequency, and f_0 the frequency at which the deemphasis network response is down 3 dB. For $f_a = 15$ kHz and a 75-μs time constant ($f_0 = 2.125$ kHz), the S/N improvement is 13.2 dB. The total S/N improvement over AM is 32 dB when the carrier is high enough to override noise plus 12 dB for the 75-kHz deviation used in United States broadcast stations. The minimum coherent S/N ratio is therefore 44 dB.

When a dipole receiving antenna is used, an additional noise component is produced by *galactic noise*, because the dipole pattern does not discriminate against sky signals. The ratio of signal to galactic noise can be improved by using an antenna array with gain in the horizontal direction, i.e., one that discriminates against sky-wave signals. The additional noise is shown in Fig. 21-32. Using the calculated value of E_n and assuming a 3-dB noise factor in the receiver gives an equivalent noise input to the receiver of 1.39 μV. The required field strength to produce a 12-dB S/N ratio at the receiver and a 44-dB S/N ratio at the audio output is 11.5 μV/m with a half-wave dipole.

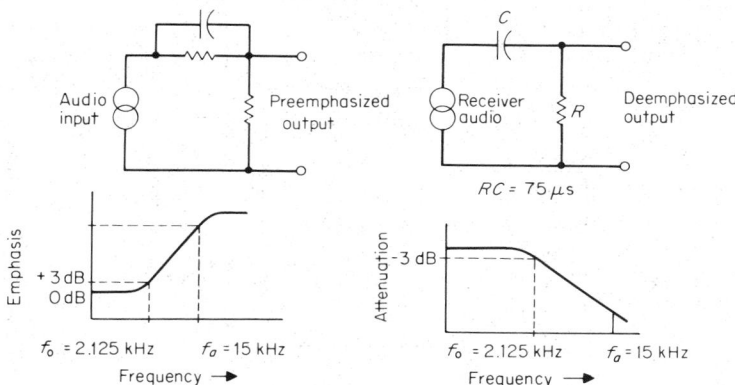

Fig. 21-34.   Preemphasis and deemphasis circuits and characteristics.

63. Selectivity. When the FM system uses a high modulation index, the system is not only capable of improving the S/N ratio but will reject an interfering cochannel signal. The FM signal modulation involves very wide phase excursions, and since the phase excursion which can be imparted to the carrier by an interfering signal is less than 1 rad, the effect of the interference is markedly reduced. The cochannel-interference-suppression effect requires that the interfering signal be smaller than the desired signal, since the larger signal acts as the desired carrier, suppressing the modulation of the smaller signal. This phenomenon is called the *capture* effect since the larger signal takes over the audio output of the receiver.

The capture effect produces well-defined service areas, since signal-level differences of less than 20 dB provide adequate signal separation. Although it is useful in suppressing undesired signals of a level less than the desired signal, the capture effect can also produce an annoying tendency for the receiver to jump between cochannel signals when fading, e.g., that caused by airplanes, causes the desired signal to drop below the interfering signal by only a few decibels. This effect also occurs in FM radios used in automobiles when motion of the antenna causes the relative signal levels to change.

64. Tuners. The FM tuner is matched to the antenna input. An rf amplifier is used to override the mixer noise. The mixer provides a 10.7-MHz if signal. Most FM tuners contain a single stage of rf amplification. The mixer may be self-oscillating or employ a separate oscillator. Since the tuning range is from 88 to 108 MHz a range of only 1.23:1, the tuner may be tuned by any convenient means. In auto receivers tuning usually is accomplished inductively, using variable-permeability slugs, suitable for push-button operation. The rf stage must have a low noise figure to reach the minimum threshold-signal level, but its most important requirement is to provide the mixer and if amplifier with signals free of distortion.

When the rf amplifier is overloaded, the signal supplied to the if amplifier may be distorted or suppressed. For single interfering signals there are three significant sources of difficulty: (1) image signals may capture the receiver, suppressing the desired signal; (2) strong signals at one-half the if frequency (5.35 MHz) above the desired signal may capture the receiver; or (3) a strong signal outside the range of the if beat but strong enough to cause limiting in the rf stage may drastically reduce the output of the rf amplifier at the carrier frequency.

When two strong signals are present, three conditions may produce unsatisfactory operation: (1) cross modulation of two adjacent upper- or lower-channel signals may produce an on-channel carrier, (2) two strong signals spaced 10.7 MHz in the rf passband may produce an if beat, or (3) submultiple if beats may be produced by strong signals spaced at if submultiple spacings in the rf band. To minimize the effects of distortion and provide a low noise figure, most FM tuners employ an FET type transistor rf stage. A circuit of a typical FM tuner employing an FET rf amplifier and a separate-oscillator mixer is shown in Fig. 21-35.

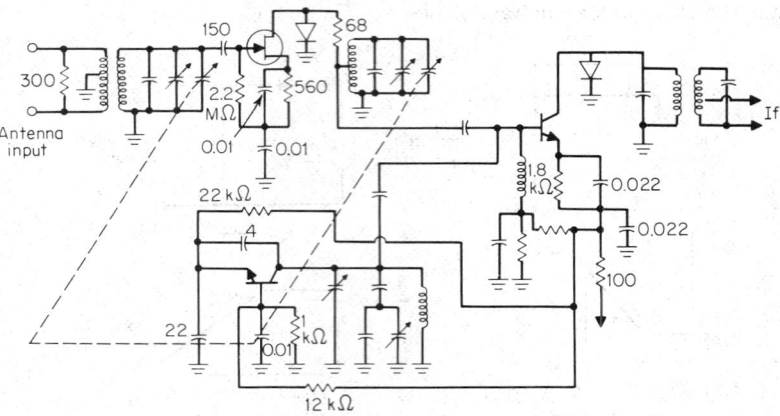

Fig. 21-35. Typical FM tuner using capacitive tuning.

65. IF Amplifiers. To provide sufficient image separation, a higher intermediate frequency (10.7 MHz) is used in FM than in standard broadcast AM. Since the if frequency is higher and the bandwidth greater (200 kHz), the amplifier design is more demanding in FM receivers. The if amplifier must provide sufficient gain for the noise generated by the rf amplifier to saturate the limiting stages fully if the benefits of wide-band FM are to be obtained at low signal levels. The high gain should be supplied with a low noise figure in the first if amplifier, so that the noise introduced by the if is small compared to the noise from the rf amplifier.

One of the most important characteristics of the if amplifier is phase linearity, since envelope-delay distortion in the passband is a principal cause of distortion in FM receivers. Care must also be taken to avoid regeneration since this would cause phase distortion and hence audio distortion in the detected signal.

Although AGC is theoretically unnecessary, it is sometimes used in the rf stage to avoid overload. Such overload, coming before sufficient selectivity is present, could produce cross modulation, causing capture by an out-of-band signal. The requirements of high gain and good phase linearity are generally met by using amplifiers with double-tuned circuits adjusted to operate at critical coupling. A circuit of a typical if amplifier is shown in Fig. 21-36. The gain is provided by integrated-circuit differential amplifiers which provide symmetrical limiting.

66. Limiters. The design of the limiter is critical in determining the operating characteristics of an FM receiver. The limiter should provide complete limiting with constant-amplitude signal output on the lowest signal levels which override the noise. In addition, the limiting should be symmetrical, so that the carrier is never lost at low signal levels. This is essential if the receiver is to capture on the strongest signal when there is little difference in signal strengths between the weaker and stronger signals. Finally, the bandwidth in the output must be wide enough to pass all the significant sideband terms associated with the carrier, to prevent spurious amplitude modulation due to the lack of sufficient bandwidth to provide the constant-amplitude FM signals. The differential amplifier with dc coupling can be made to provide highly symmetrical limiting.

67. FM Detectors. The FM detector should provide an output voltage which changes linearly with frequency. Most balanced detectors provide zero voltage output at the center frequency (when the carrier is unmodulated), and the voltage varies plus and minus as the carrier is modulated above and below the carrier frequency. This provides a means of converting frequency modulation into an audio voltage. The bandwidth of the linear portion of the detector response should be wider than the expected frequency swing to provide protection from demodulation of rapid excursions in frequency generated by interfering signals.

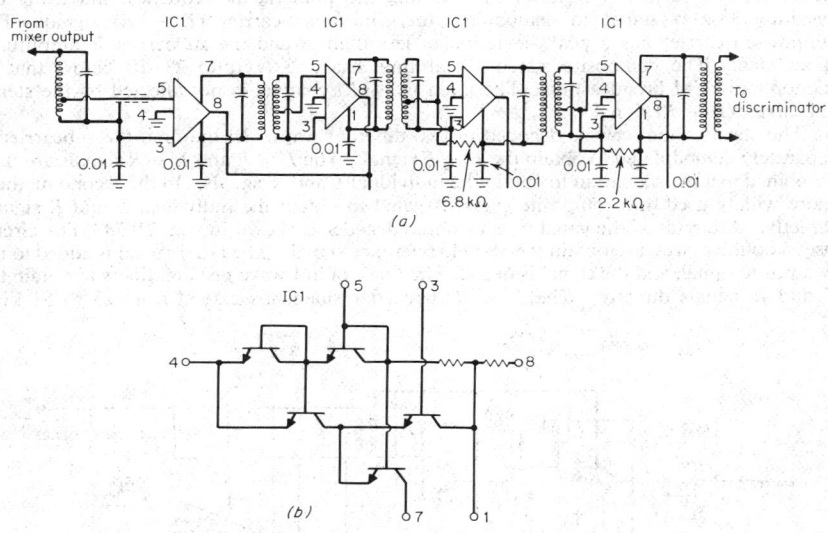

Fig. 21-36. If amplifier for FM receiver using integrated circuits: (*a*) complete circuit; (*b*) detail of integrated circuit.

There are five well-known types of FM detectors: (1) the balanced-diode discriminator (Foster-Seeley circuit); (2) the ratio detector using balanced diodes; (3) the slope-detector-pulse-counter circuit using a single diode with an *RC* network to convert FM to AM; (4) the locked-oscillator circuit, which uses a variation in current as the frequency is varied to convert the output of an oscillator (locked to the carrier frequency) to a voltage varying with the modulation; and (5) the quadrature detector circuit that produces the electrical product of the two voltages, the first derived from the limiter, the second from a tuned circuit which converts frequency variations into phase variations. The output of the product device is a voltage which varies directly with modulation. A typical ratio detector is shown in Fig. 21-37.

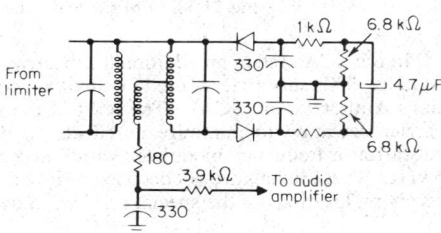

Fig. 21-37. Ratio detector circuit for FM demodulation.

68. FM Stereo and SCA Systems. Since FM broadcasting uses a bandwidth of 200 kHz and a modulation index of 5 for the highest audio frequency transmitted, it is possible by using lower modulation indexes to transmit information in the frequency range above the audio. This method of using a supersonic carrier to carry additional information is used in FM broadcasting for a number of purposes, most notably in FM stereo broadcasting.

In broadcasting stereo the main-channel signal must be compatible with monophonic broadcasting. This is accomplished by placing the sum of the left- and right-hand signals (*L*

+ R) on the main channel and their difference (L - R) on the subcarrier. The subcarrier is a suppressed-carrier AM signal carrying the L - R signal.

The suppressed-carrier method causes the carrier to disappear when L and R vary simultaneously in identical fashion. This occurs when the L - R signal goes to zero and allows the peak deviation in the monophonic channel to be unaffected by the presence of the stereo subcarrier.

In the receiver it is necessary to restore the subcarrier. In the U.S. standards, the technique used provides a pilot signal at 19 kHz, one-half the suppressed-carrier. The frequency sub-carrier is restored by doubling the pilot-signal frequency and using the resulting 38-kHz signal to demodulate the suppressed-carrier (L − R) signal. The suppressed carrier has a peak deviation of less than 2, and the subcarrier is amplitude-modulated. The composite stereo signal thus has a S/N ratio 23 dB below that of monophonic FM broadcasting. The main (L + R) channel is not affected by the stereo subcarrier (L − R) signal.

The stereo signal can be decoded in two different ways. In the first, the subcarrier is separately demodulated to obtain the L − R signal. The L + R and L − R signals are then combined in a matrix circuit to obtain the individual L and R signals. In the second method, more widely used, the composite signal is gated to obtain the individual L and R signals directly. A circuit of the gated type of stereo decoder is shown in Fig. 21-38. The circuit uses a doubler circuit to obtain the 38-kHz reference signal. The latter signal is added to the composite signal, and the signal is decoded in a pair of full-wave peak rectifiers to obtain the L and R signals directly. The L − R subcarrier sidebands extend from 23 to 53 kHz.

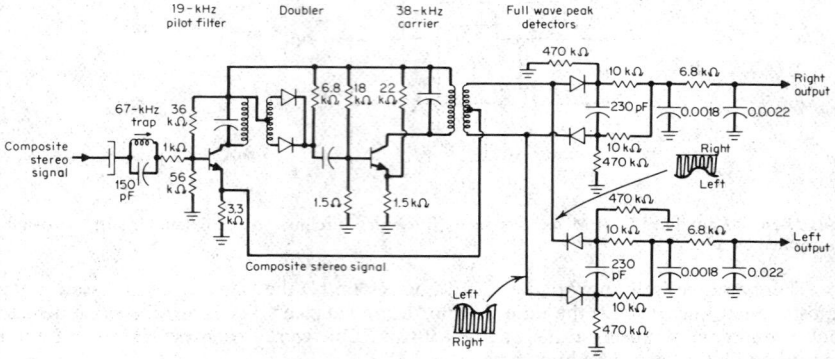

Fig. 21-38. FM stereo decoder using the gating principle.

In the SCA system an additional subcarrier is placed well above the stereo sidebands at 67 kHz. This subcarrier is used for auxiliary communication services (Special Communications Authorization, SCA). The level of the subcarrier is kept to about 10% of the peak carrier deviation to minimize its effects on the broadcast service signal. The auxiliary subcarrier is frequency-modulated with 8-kHz peak deviation; the audio is usually limited to 5 kHz. These signals are not decoded by broadcast receivers; special receivers are used at the receiving locations of the special services to decode the subcarrier signal.

CABLE TELEVISION (CATV) SYSTEMS

By Joseph L. Stern

69. General Considerations. Cable television systems, also known as community-antenna television (CATV) systems, use coaxial cable to distribute standard TV signals to homes or establishments subscribing to the service. The program material may originate at a distant vhf or uhf broadcast station, or it may be locally generated within the facilities of the CATV system. Signals received off the air are picked up by a specially designed antenna system (community antenna) which provides freedom from noise, interference, and multipath distortion not obtainable directly at the subscribers' homes. In the early 1970s nearly 3,000 CATV systems were in use in the United States, serving more than 8 million subscribers.

70. Elements of a CATV System. The typical CATV system comprises four main elements: a *head end*, in which the signals are received and processed; a *trunk system*, the main artery carrying the processed signals; a *distribution system*, which is bridged from the trunk systems and carries signals to subscriber areas; and *subscriber drops*, fed from taps on the distribution system to feed into the subscriber's TV receiver. Figure 21-39 shows a diagram of a typical CATV system.

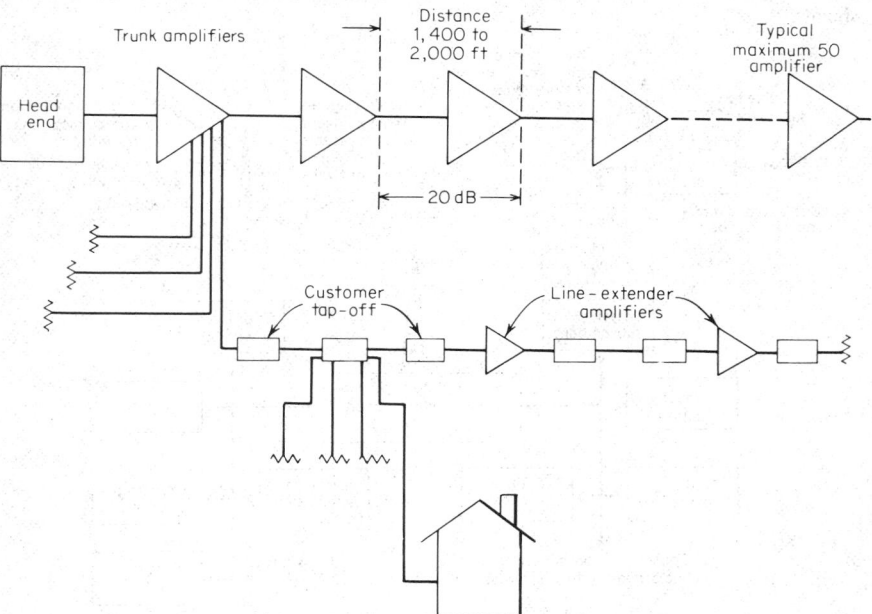

Fig. 21-39. Block diagram of typical CATV system.

While cable television systems still have the primary goal of bringing high-quality signals to home viewers, another basic function, *interactive communications*, has been developed to allow the subscriber to interact with the program source to provide or request various types of information capable of being transmitted over the wide-band system. Cable television systems may carry many more channels of television programming than could be received off the air, as well as new communication services. Figure 21-40 illustrates a two-way interactive system and some of its potential uses.

Since the primary subscriber device connected to the CATV system is a standard home TV receiver, the distribution of signals normally occurs in the TV broadcast band. On occasion, frequencies are converted from the original channels to other channels to permit carriage of more than 12 channels, to overcome direct pickup interference from a local station, or to provide for selective viewing by subscribers equipped with special conversion devices.

The cable is usually installed on leased telephone poles. While aerial construction is the most common, some cable systems are installed underground in conduit or by direct burial. Undergrounding is chosen to meet local regulations or, in some cases, to minimize damage from local environmental conditions.

71. Head End. The head end is the originating point of signals for the cable television system, where signals are received off the air, by microwave relay or where locally generated signals are introduced into the system. The units located at the head end are antenna systems, antenna preamplifiers, heterodyne repeaters, video modulators, FM heterodyne repeaters, pilot-carrier generators, and the associated mixing networks required to combine the individual outputs of various program sources to feed the coaxial cable. Figure 21-41 is a block diagram of a typical head-end configuration.

The head-end equipment is usually located at a high elevation in an area of low ambient electrical noise where it is possible to receive the desired channels with a minimum of interference and at sufficiently high level to obtain a high-quality signal. High-gain directional receiving antennas are utilized to provide sufficient gain to pick up a distant signal and provide discrimination against unwanted adjacent-channel, cochannel, and reflected signals.

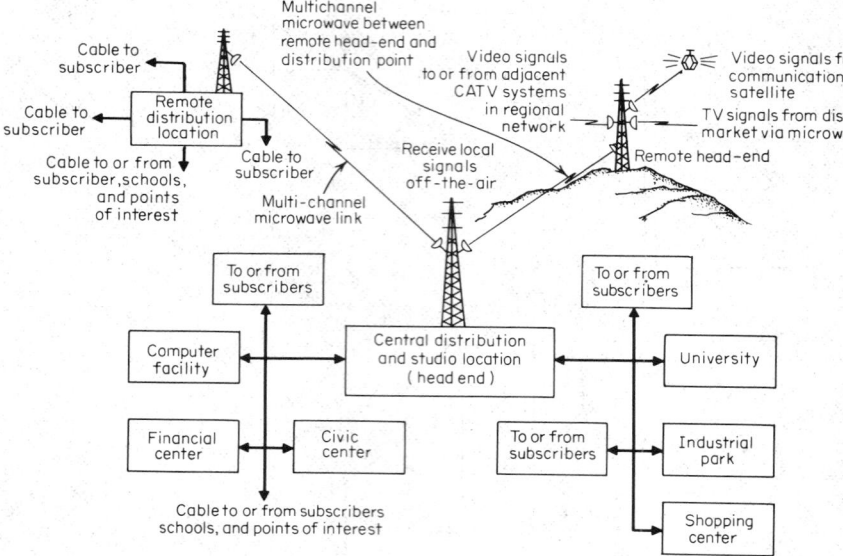

Fig. 21-40. Layout and functions of a two-way (interactive) wide-band cable system.

Signal processing is performed at the head end to fix the signal-to-noise ratio at the highest practical value, to control the output level of the signal to a close tolerance automatically, to reduce the aural-carrier level relative to visual-carrier level to avoid interference with adjacent cable channels, and to suppress undesired out-of-band signals.

Processing is also used to convert the received signal to a different channel, to convert signals to vhf, to introduce signals received by microwave into the system, and to introduce locally originated video services into the cable. FM broadcast signals are converted to respace the FM signals so that the cable system can accommodate as many FM stations as possible.

In some systems the head end feeds a *hub center*, where locally generated signals are inserted in the system. In such cases, depending on the configuration of the trunk and distribution system, control and processing for signals other than distant off-the-air signals may be located at the hub center. Signals from the head end are usually carried to this point by *supertrunks*, large-diameter cables carrying signals in the 5- to 95-MHz portion of the spectrum. This frequency range is called *sub-low*.

72. Trunk System. The trunk system carries a multiplicity of channels through coaxial cable with minimum distortion. Amplifiers and (where required) equalizers overcome the losses in the coaxial cables. From the output of a given repeater amplifier, through the span of coaxial cable and the equalizer, to the output of the next repeater amplifier, unity gain is required so that the same signal level is maintained on all channels at the output of each trunk unit.

As indicated in Fig. 21-39, repeater amplifiers are spaced from 1,400 to 2,000 ft, depending on the diameter of the coaxial cable. This represents an electrical loss of about 20 dB. Systems with trunk-amplifier cascades up to 50 amplifiers are possible, depending on the number of channels the system carries and the specifications adopted for the system.

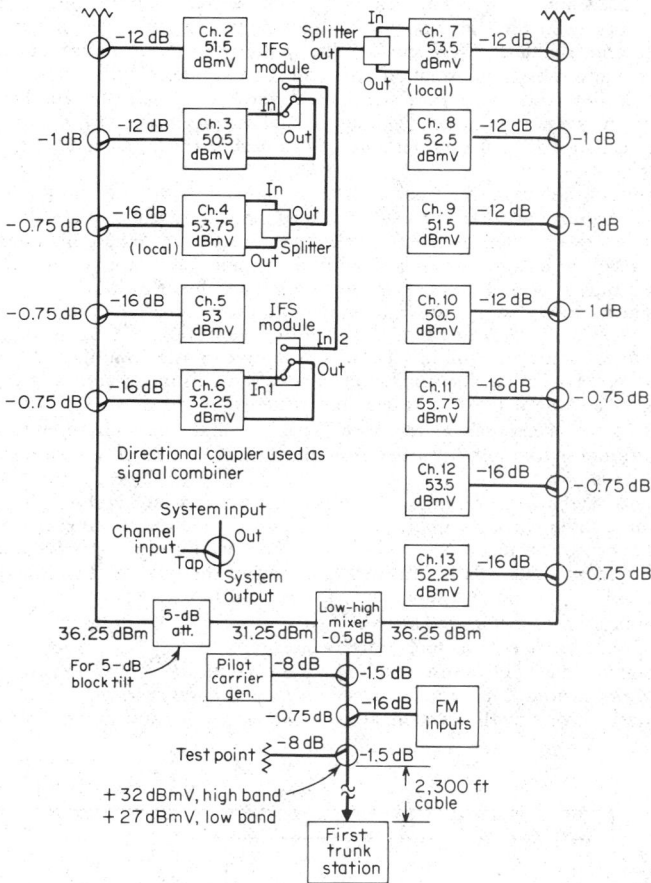

Fig. 21-41. Typical CATV head-end configuration and signal-processing functions. Equipment is interconnected for channel mixing to obtain a single broadband output feed to the trunk amplifiers. *(Jerrold Manufacturing Co.)*

Bridging amplifiers are used to feed signals to the distribution system, en route to the subscriber drops. Using a wide-band directional coupler, a portion of the signal on the trunk is tapped off and fed to the distribution amplifiers. This bridging amplifier acts as a buffer, isolating the distribution system from the trunk system while providing the level required to drive the distribution lines.

73. Distribution Lines. As shown in Fig. 21-39 up to four distribution lines are fed from a bridging station. Distribution lines are routed through the subscriber area. In addition to the coaxial cables, amplifiers and customer tap-off devices are provided to meet the needs of the subscriber density of the particular area.

A commonly used tap-off device is the *directional coupler multitap*, which allows up to four subscribers to be attached to one unit. Individual taps, called pressure taps, fastened directly to the distribution cable were once common, but the undesirable signal reflections produced have limited them to small systems. They are no longer in general use.

The multiple-output tap device in common use samples the appropriate amount of energy from the distribution cable through a directional coupler and splits this energy into multiple paths, each proceeding from its tap location via a subscriber-drop line into the subscriber's home. The tap unit introduces some attenuation, but its output is adequate for good performance on a standard TV receiver. The splitter portion of the tap unit is of hybrid design to introduce substantial isolation from reflection or interference coming from a home-subscriber location and to prevent such interference from affecting another subscriber connected to the same tap device. The directional-coupler tap is also designed to have a much higher isolation from the output port feeding back to the tap than it has from the input port feeding into the tap.

The output of the tap device feeds a 75-Ω coaxial-cable drop line into the home. Since many TV receivers are restricted to 300-Ω balanced input, a transformer is utilized to convert the 75-Ω single-ended system to a 300-Ω balanced impedance. Where high ambient-signal levels exist, excellent balance is required on the 300-Ω side. Transformers are available with good balance plus a form of Faraday shield to minimize direct pickup.

As the signal proceeds along the distribution line, the attenuation of the coaxial cable and the insertion loss of the customer tap-off devices reduce its level to a point where small *line-extender amplifiers* may be required. These inexpensive booster amplifiers, having a gain of 25 to 30 dB, are convenient for small extensions of an existing system to provide for new subscribers added after a system has been completed.

74. Head-End Processors—Heterodyne Type. Several types of head-end signal processor are in current use: the *heterodyne processor*, the *demodulator-modulator pair*, and the *single-channel strip amplifier*.

The heterodyne processor, a simplified block diagram of which is shown in Fig. 21-42, is the principal head-end processor currently in use. A typical specification is shown in Table 21-17. The processor heterodynes the incoming signal to an intermediate frequency, amplifies, filters, and controls levels, then heterodynes the processed signal to the desired output channel. The following functions are performed:

1. Amplification of the desired signal.
2. Rejection of adjacent channels (filtering).
3. Automatic level control of visual and aural carriers.
4. Setting of desired level ratios between visual and aural carriers.
5. Channel conversion (if the channel to be applied to the cable differs from the received channel).

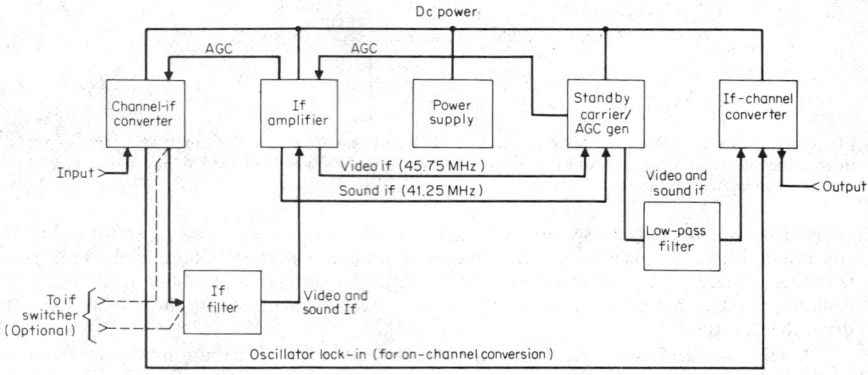

Fig. 21-42. Block diagram of typical heterodyne-type head-end processor. *(Jerrold Manufacturing Co., model CHC-*.)*

The third and fourth functions are carried out to permit operation of subscriber TV receivers with adjacent channels with minimum interference, since most receivers are not designed for adjacent-channel operation. Aural-carrier levels are reduced at the head end to reduce adjacent-channel sound interference in the subscribers' sets. A visual-to-aural carrier ratio of 15 to 17 dB is typical in cable systems as a compromise between intolerable adjacent-sound-carrier interference and poor sound quality.

Channel conversion (function 5, above) provides a change in the received signal to a transmission channel optimized for application to the cable system. Processors are designed so that by using appropriate input and output modules any input channel can be accepted, processed, and translated to any other location in the 5- to 300-MHz spectrum. Conversion may be necessary to eliminate problems with direct pickup at the TV set or when two signals received on the same channel from different directions are to be applied to the same cable system. Other applications of this function are to provide for the midband and superband channels of cable systems (see Table 21-18) and to supply the output for subband systems.

The visual-signal if passband of a typical heterodyne converter is shown in Fig. 21-43. Note that this curve is not like that of a TV set, where the visual carrier is set at a point 6 dB

Table 21-17. Specifications of a Typical Heterodyne Head-End Processor*

Input-frequency range	Single, standard vhf TV channel
Output-frequency range	Single, standard vhf TV channel
Intermediate frequency	45.75 MHz, video carrier, 41.25 MHz, sound carrier
Frequency response:	
Video	0.75 MHz below video carrier to 4.2 MHz above video carrier
Sound	Sound carrier ±0.1 MHz
Response flatness	±0.5 dB nominal, ±1.0 dB maximum
Sensitivity	−10 dBm input level to yield at least +60 dBmV output level
Adjacent-carrier rejection	50 dB or better
Image rejection	60 dB or better
Input level dynamic range	At least 40 dB (−10 dBm to +30 dBm)
Noise figure:	
Low-band	6 dB maximum at full gain
High-band	7.5 dB maximum at full gain
Input impedance	75 Ω, unbalanced
Output impedance	75 Ω, unbalanced
Output-level range	+50 to +60 dBm
Maximum operational output level:	
With external filter	+60 dBm, video carrier, +45 dBm, sound carrier; spurious signals down at least 60 dB in any band (standard, sub-, mid-, or mid-high)
Without external filter	+54 dBm, video carrier, +39 dBm, sound carrier; spurious signals down at least 60 dB in standard vhf TV band
AGC regulation	±0.5 dBm maximum output level variation for input level dynamic range of −10 to +30 dBm
AGC type:	
Channel if converter and if amplifier modules	Keyed, sync referenced
if channel converter	sync tip referenced
Sound limiting	10 dB for rated output level with −10 dBm minimum input level
Standby carrier:	
Delay on	~20 s
Delay off	~4 s
Range	operates in absence of air signal; incorporates facility for manual override
Standy-carrier mode	Unmodulated carrier; carrier modulated with internal 15 kHz, 0 to 37.5%; carrier modulated with video, 0 to 87.5%; carrier modulated with 4.5 MHz

* Jerrold model CHC-*.

down on the response curve. A linear phase characteristic is easier to achieve in the flat portion of the passband. This is one of the reasons the heterodyne processor has better differential phase characteristics than the demodulator processor.

75. Demodulator-Modulator Pair. In a demodulation-modulator pair, the demodulator is basically a high-quality television receiver to the point of video and audio output, whereas the modulator is essentially a low-power television transmitter. The demodulator-modulator pair provides increased selectivity and better AGC control and interconnection flexibility compared with that attainable with a strip amplifier. The demodulator (Fig. 21-44) consists of a tuner, an if amplifier section, a video detector, a 4.5-MHz amplifier, and a discriminator. The tuner utilizes either a high-quality conventional TV tuner or one with a crystal-controlled local oscillator. It has high gain and a good noise figure and a wide range of AGC control.

The if amplifier is very similar to a high-quality TV receiver. A typical if response curve is shown in Fig. 21-45. Note the point at which the visual carrier is located on the passband. The 4.5-MHz intercarrier sound is taken off prior to video detection, amplified, and limited to remove video components.

The demodulator must be carefully designed to minimize phase and amplitude distortion. This can be done by linearizing the detector, but quadrature distortion is inherent in a system using an envelope detector on a vestigial-sideband signal and can be corrected for only by video processing. Typical demodulator specifications are shown in Table 21-19. In the modulator, shown in Fig. 21-46, the composite input is applied to the separation section, where the video and sound subcarriers are separated and the video fed to a video amplifier. From this point the video signal is processed, mixed with a carrier oscillator to obtain the desired output frequency, filtered, and amplified to obtain the necessary power level and remove any undesired products. Following the rf amplifier is the vestigial-sideband filter required to remove most of one sideband and allow adjacent-channel operation. The

Table 21-18.   CATV Channel Designations and Carrier Frequencies

Channel designation	Visual carrier, MHz	Aural carrier, MHz
Low band:		
2	55.25	59.75
3	61.25	65.75
4	67.25	71.75
5	77.25	81.75
6	83.25	87.75
High band:		
7	175.25	179.75
8	181.25	185.75
9	187.25	191.75
10	193.25	197.75
11	199.25	203.75
12	205.25	209.75
13	211.25	215.75
Midband:		
A	121.25	125.75
B	127.25	131.75
C	133.25	137.75
D	139.25	143.75
E	145.25	149.75
F	151.25	155.75
G	157.25	161.75
H	163.25	167.75
I	169.25	173.75
Superband:		
J	217.25	223.75
K	223.25	227.75
L	229.25	233.75
M	235.25	239.75
N	241.25	245.75
O	247.25	251.75
P	253.25	257.75
Q	259.25	263.75
R	265.25	269.75
S	271.25	275.75
T	277.25	281.75

Table 21-19.   Specifications of a Typical Demodulator-Modulator Pair

Demodulator

Input level .... −20 to +30 dBm
Input frequency .... Any vhf or uhf channel
Input impedance .... 75 Ω
Noise figure .... Vhf, 7 dB maximum; uhf, 12 dB
                         maximum
Image rejection .... 60 dB minimum
If response flatness .... ±0.5 dB 41.6 to 46.5 MHz
Adjacent video-carrier rejection .... >40 dB
Adjacent sound-carrier rejection .... >40 dB
AGC sensitivity .... ±0.5 dB
Video output level .... 1.5 V peak to peak max
Audio output level .... +6 vu
4.5-MHz output level .... 0.2 V peak to peak max
Audio-frequency response .... ±1.5 dB, 50 to 15,000 Hz

Modulator

	Visual	Aural
Frequency range	Channels 2 to 13	Channels 2 to 13
Input level	0.5 V peak to peak	−10 vu for full deviation
Input impedance	75 Ω	600 Ω
Output impedance	75 Ω	75 Ω
Output level	+50 dBm	+40 dBm
Output control range	20 dB	20 dB
Frequency response	30 Hz to 4.0 MHz, ±1 dB	50 Hz to 15 KHz ±1 dB (75 *μs* preemphasis)
Hum and noise	60 dB down at 100% modulation	60 dB down with ±50 kHz deviation
Color response	±2 dB max differential gain ±4° max differential phase	
Vestigial-sideband response:		
Picture carrier, −1.25 MHz ..	−20 dB	
Picture carrier, −3.58 MHz ..	−42 dB	
Adjacent-channel	−22 dB	

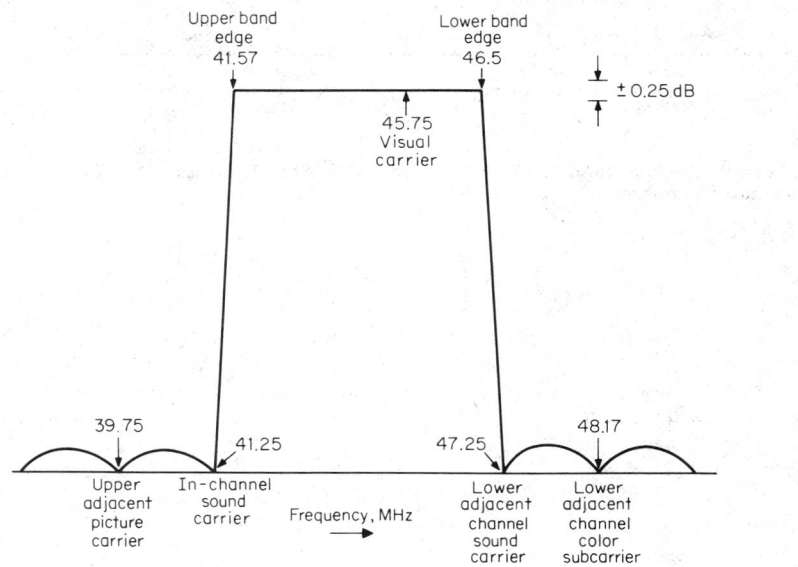

Fig. 21-43. Typical idealized video if response curve of heterodyne processor. Note that the visual carrier is located on the flattop portion of the curve to permit improved phase response.

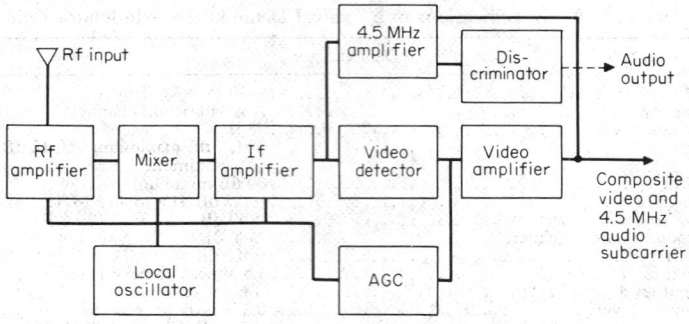

Fig. 21-44. Block diagram of demodulator portion of a demodulator-modulator pair.

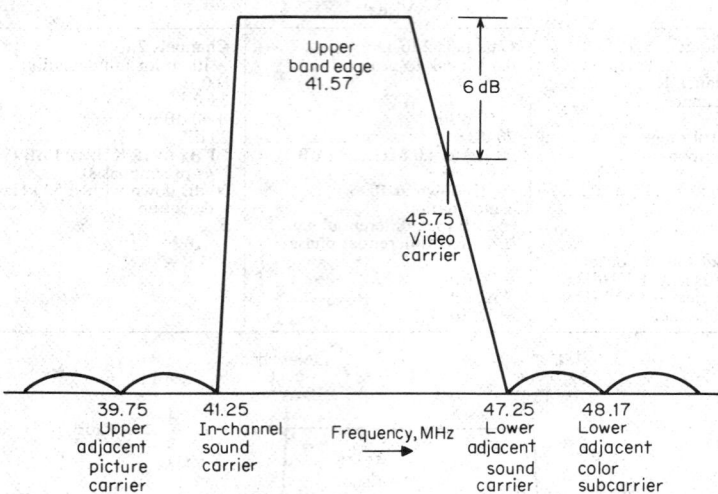

Fig. 21-45. Idealized video if response curve of a demodulator. The visual carrier is located 6 dB below the maximum response.

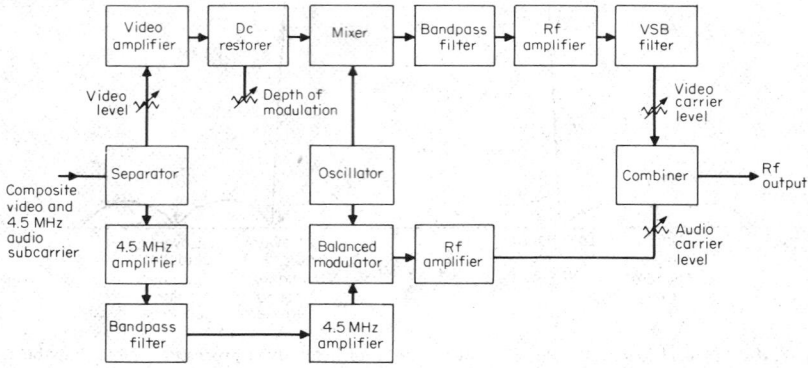

Fig. 21-46. Block diagram of the modulator portion of a demodulator-modulator pair.

21-66

characteristics of this filter are designed to meet the same requirements as that of a TV transmitter. Following the filter the carrier-level adjustment occurs, and then the visual carrier is combined with the aural carrier.

In the modulation process, a high percentage of modulation is desirable for high signal-to-noise ratio, but this produces differential gain and phase on the color subcarrier. The vestigial-sideband filter must be optimized to minimize phase distortion near the picture carrier. The compensation required for the demodulator-modulator pair is usually provided in a video equalizer in the modulator. Typical modulator specifications are shown in Table 21-19.

76. Single-Channel Amplifier. Single-channel amplifiers, or *strip amplifiers*, are the simplest head-end processors. They amplify one channel and reject the others. In simplest form they consist of a filter, amplifier, and power supply. More elaborate types provide the above functions plus AGC in some form and usually better input and output filtering. A representative type is shown in block-diagram form in Fig. 21-47. Typical specifications are shown in Table 21-20.

Strip amplifiers are used where the desired signal levels are fairly high and the undesired levels low or absent. They do not offer the selectivity of the more complicated heterodyne and demodulator-modulator processors. They also lack such features as independent control of sound- and picture-carrier levels and the ratio between them, separate AGC, and limiting for picture and sound channels, respectively. Their use is restricted to the 12 channels because they cannot be translated to other channels. They are also subject to leading ghost problems if the signal on a given channel can be picked up by sets in the distribution system. The off-the-air signal arrives far enough ahead of the signal through the cable to create a ghost ahead of the cable-transmitted signal.

77. Trunk and Distribution Amplifiers. Since the trunkline is the main artery of the CATV system, the trunk amplifiers must provide minimum degradation to the system. Specifications of a typical trunk amplifier are contained in Table 21-21. Figure 21-48 shows the block diagram of a typical trunk/AGC amplifier and the distribution of gain. The first stage is designed for low-noise figure and the remainder for low cross modulation.

78. Cross Modulation. The maximum output level allowable in CATV system amplifiers is almost always determined by cross modulation in the picture signal. Where cross modulation exists, it results in a variation in the peak voltage of an otherwise unmodulated signal substituted for the wanted carrier. The contribution by the third stage to the total

Table 21-20. Typical Single-Channel Amplifier Specifications

Minimum gain	51 dB (channels 2 to 13 and FM)
Maximum output for 0.5-dB gain compression	+66 dBm (2 V)
AGC range	40 dB
AGC capability	±0.5 dB for 40-dB input change
Minimum input for TASO* "excellent" picture	0 dBm
Bandpass	6 MHz, ±0.5 dB for TV
Skirt selectivity	25 dB down ±9 MHz from midchannel

* Television Allocations Study Organization.

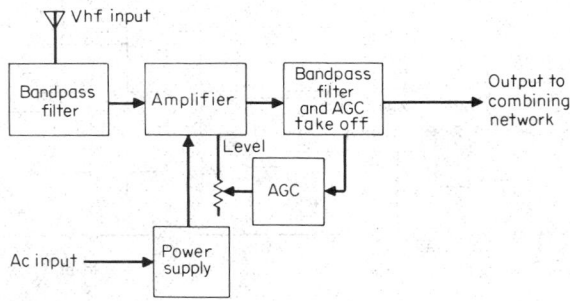

Fig. 21-47. Block diagram of single-channel amplifier (strip amplifier) with AGC.

cross-modulation distortion is only 1.27 dB; that is, the output stage principally determines the amplifier cross-modulation distortion.

Cross modulation is most likely to occur in the output stage, where the signal levels are high. The cross-modulation distortion products at the output of a typical amplifier (Fig. 21-48) are 93 dB below the desired signal at an output level of +32 dBm. Note from Fig. 21-48 that the gain of the third stage is 8 dB. The output level of the third stage would therefore be

$$P_3 = P_4 - G_4 = +32 - 8 = +24 \text{ dBm}$$

Figure 21-49 illustrates the relationship between cross-modulation distortion and amplifier output level. As shown, a 1-dB decrease in output level produces a corresponding 2-dB decrease in the cross-modulation distortion. Using this relationship, we find the output cross-modulation distortion of the third stage to be −109 dB.

79. Frequency Response and Beat Distortion. The most important design consideration in CATV amplifiers is the frequency response. An amplifier frequency response flat within ±0.25 dB over 40 to 300 MHz is required of an amplifier carrying 20 or more 6-MHz channels to permit a cascade of 32 or more amplifiers. To meet this requirement special attention must be paid not only to the high-frequency parameters of the transistors and associated components but to good high-frequency layout and packaging techniques as well.

Table 21-21.   Specifications of a Typical Trunk Amplifier

Operating levels:	
Input channel 13	+10 dBm
Channel 2	+16 dBm
Output:	
Channel 13	+32 dBm
Channel 2	+27 dBm
Control range:	
Recommended gain:	
Channel 13	22 dB
Channel 2	11 dB
Spacing	22-dB cable to 14-dB cable plus 8 dB flat
AGC temperature compensation:	
Input change	±3 dB
Output change	±0.5 dB
Cross modulation	−95 dB
Second-order beat	−85 dB
Hum modulation	−60 dB
Return loss:	
Trunk in	18 dB
Trunk out	18 dB
Noise figure:	
Channel 2	16 dB
Channel 13	10.5 dB
Frequency range	50 to 270 MHz
Ripple:	
Trunk	±0.25 dB
Bridger	±0.5

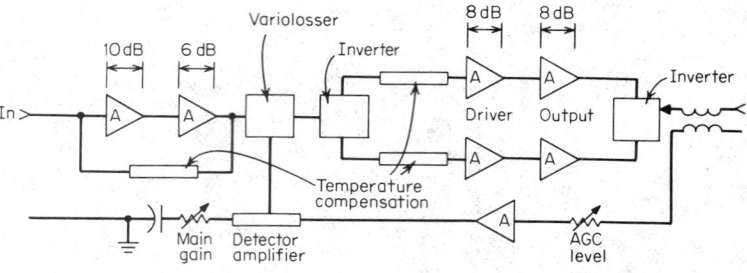

Fig. 21-48.   Functional block diagram of a trunk-AGC amplifier, and its gain distribution. A = amplifier.

The circuit shown in Fig. 21-50 is representative of amplifiers designed to achieve flat response. By properly designing the feedback network comprising C_1, R_1, and L_1, sufficient negative feedback can be utilized to maintain a nearly constant output over a wide frequency range. The collector transformer T_2 and the splitting transformers T_1 and T_3 play an important part in the amplifier's performance. In a representative amplifier of this type transformer T_2 is bifilar wound on a ferrite core and presents an essentially constant 75-Ω impedance over the entire frequency range. Transformers T_1 and T_3 are similar in construction but have the additional function of providing the required 180° phase shift for the push-pull pair Q_1, Q_2 while maintaining a 75-Ω input-output impedance.

The sweep-response patterns shown in Fig. 21-51 illustrate the wide bandwidth and amplitude linearity necessary for cascade signal processing. The response irregularity indicated in this figure is, by itself, of little concern, but if this small irregularity (perhaps 0.2 dB) occurs at the same frequency in each amplifier, it becomes a response "signature" and accumulates to a magnitude of 6.4 dB at the end of 32 amplifiers. The degree of signature in a high-quality CATV trunk amplifier is typically no more than 0.1 dB.

Second-Order Beat. As the number of CATV television channels has been increased from 12 nonharmonically related to 20 or more harmonically related, the second-order-beat distortion characteristics of an amplifier have become important to CATV equipment manufacturers.

A number of approaches has been used to reduce amplifier second-order distortion, but the most successful is the push-pull configuration. The circuit illustrated in Fig. 21-50 is representative of the 75-Ω push-pull building-block approach. Each push-pull stage is of this

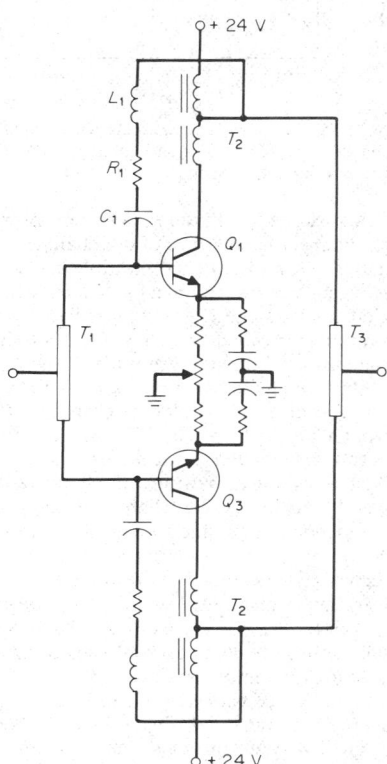

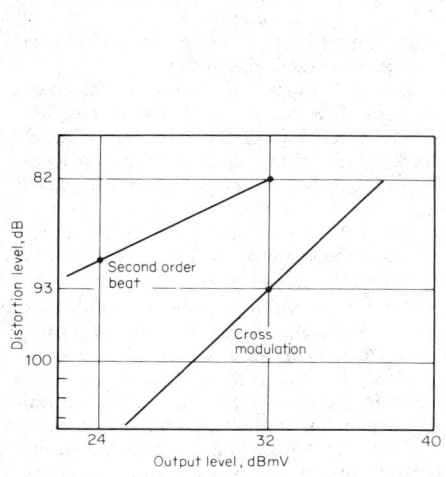

Fig. 21-49. Cross-modulation and second-order beat distortion vs. output level of a trunk amplifier.

Fig. 21-50. High-performance push-pull wide-band amplifier stage for a trunk amplifier.

basic configuration and differs essentially in component values only. The splitting transformers T_1 and T_3 are the key to the operation of the push-pull amplifier, since for maximum second-order cancellation to occur the proper 180° phase relationship must be maintained over the full amplifier bandwidth. Additionally, it is necessary that the gain be equal in both push-pull halves and that the individual transistors be optimized for maximum second-order linearity.

80. Channel Capacity. CATV system design has advanced to the point where it is feasible to provide 20 or more channels with two-way capability. The FCC Rules require that CATV systems located in whole or in part within a major television market have at least 120 MHz of bandwidth (the equivalent of 20 television broadcast channels) available for immediate or potential use. In addition, the Rules stipulate that each system maintain a plant having technical capacity for return communications.

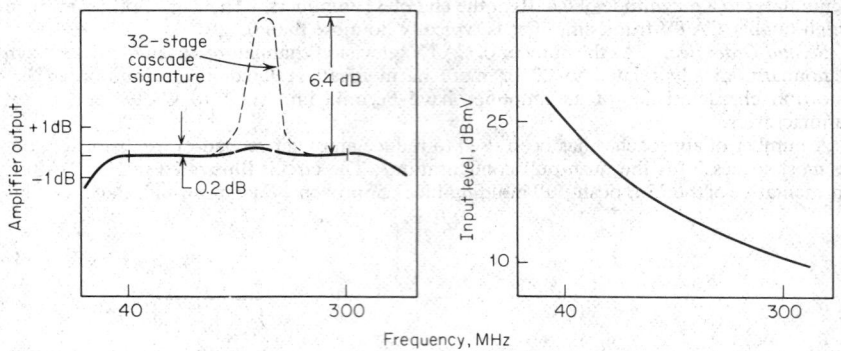

Fig. 21-51. Input and output levels of the amplifier shown in Fig. 21-50 as functions of frequency over the band 40 to 300 MHz. The dashed curve shows the cumulative effect of the 0.2-dB hump when the signal is passed through 32 such stages in cascade.

A variety of techniques have been proposed for the provision of more than the usual 12 vhf channels, such as use of uhf channels, frequency-division multiplexing of a single cable, dual cables, and switched multiple-cable networks with or without multiplexing. Each of these schemes has technical problems which are much more troublesome than those of the 12-channel systems. The standard 12 TV channels were allocated by the FCC in a manner to minimize second-order distortion products and interference from nonbroadcast signals. As additional channels are utilized, these problems become most significant.

One scheme for distributing more than 12 channels is to utilize dual cables. In this case, two subscriber drop cables are brought to the subscriber's TV receiver and fed to the receiver through an A-B switch. The subscriber is able to receive 12 channels on each cable, operating the switch to select the desired channel group. In practice, direct pickup from local TV stations prevents the use of the full 12 channels per cable but generally, in major markets, 14 to 18 channels are available using this scheme.

A second technique to provide more than 12 channels involves the use of a converter. The output of the converter feeds the receiver at some unused vhf channel position. The receiver remains tuned to this unused local channel, and the converter is used to tune in the desired vhf channels. Converters are also utilized to avoid direct-pickup problems. Converters are commercially available providing for 31 channels. In conjunction with a dual-cable system and an A-B switch, they can provide as many as 62 channels.

Uhf distribution on cable has been considered but has not proved satisfactory. The combination of very high cable losses at these high frequencies and the limitations of uhf tuners in commercial TV receivers have made such a distribution scheme impractical.

81. Two-Way Systems. Since most existing CATV systems utilize a single cable, the majority of conversions to provide additional channel capacity and two-way operation have involved retrofitting the single-cable systems. To do this, the earlier limited-bandwidth single-ended amplifiers have been replaced with broadband push-pull amplifiers and all other components provided with extended frequency ranges. To provide two-way service, filters have been installed to provide frequency separation and feed return signal amplifiers.

A dual-cable trunk and feeder system providing for up to 44 channels (using a switched converter), 12 private return channels, and 4 return channels from subscribers is shown in block form in Fig. 21-52. This system contains no frequency-splitting filters in cable A, minimizing group-delay problems for the main TV channels. It also has the advantage of utility for private channel distribution in the reverse direction, meeting a number of the requirements outlined in Fig. 21-40.

82. Technical Standards. The Federal Communications Commission has issued definitions and technical standards governing CATV systems. These appear in the FCC Rules, Sec. 76.601-76.617 (subpart K, Technical Standards) and Sec. 76.5 (subpart A, General, Definitions).

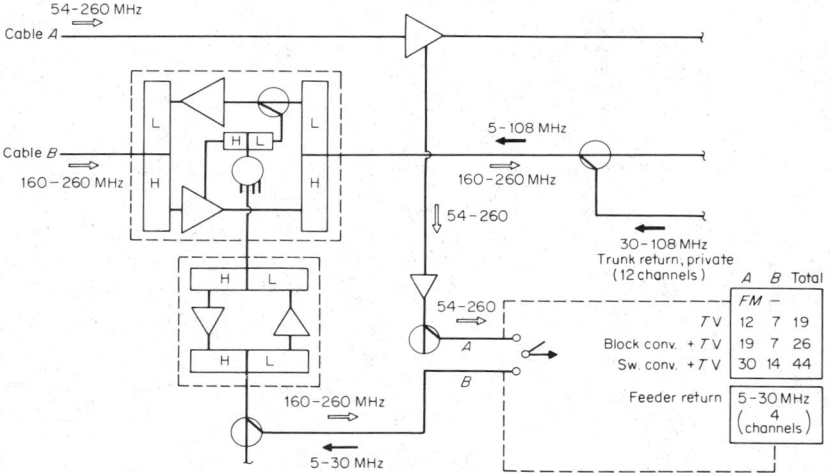

Fig. 21-52. Block diagram of a two-way trunk and feeder return for a dual-cable system.

SECTION 22

POINT-TO-POINT AND MOBILE COMMUNICATIONS SYSTEMS

BY

BYRON S. ANDERSON Chief, Information Acquisition Technical Area (Ret.), Communications/ADP Laboratory, U.S. Army Electronics Command; Senior Member, IEEE

CONTENTS

Numbers refer to paragraphs

SECTION 22

POINT-TO-POINT AND MOBILE COMMUNICATIONS SYSTEMS

BYRON S. ANDERSON

COMMUNICATIONS SYSTEMS

1. Introduction. This section deals with communications by electrical and electronic means in point-to-point and mobile systems. The first organized communications network was an optical telegraph system developed by Claude Chappe in France in the latter part of the eighteenth century. Electrical methods displaced this slow system when the first operating telegraph lines were developed by Wheatstone in England in 1839 and by Samuel F. B. Morse in the United States in 1845. Morse also invented a practical telegraph code still in use. In 1858 the first telegraph message was sent across the Atlantic by cable, and in 1875 the telephone was invented by Alexander Graham Bell.

In 1888 Heinrich Hertz demonstrated the generation of radio waves, and in 1896 Marconi received his first radio patent. The first trans-Atlantic transmission and reception of radio signals occurred in 1901. In 1907 Lee De Forest invented the three-element electron tube, and the subsequent development of electronics gave great impetus to communications development and expansion.

In 1915 the first transcontinental telephone conversation took place, in the United States over land lines. In the 1920s radio amateurs demonstrated the usefulness of the high-frequency radio spectrum for long-distance radio communication, and commercial exploitation quickly followed. Stimulated in part by the military development and experience in World War II, a great expansion developed in the 1940s and 1950s in the use of the vhf and uhf regions of the radio spectrum for vehicular communications and of the uhf and microwave regions for broadband radio relay communications. In 1956 the first trans-Atlantic telephone cable system was placed in service, and in 1963 the first telephone conversation was held via a communication satellite in synchronous orbit above the earth.

2. Analog and Digital Communication Signals. Communication between human beings takes place primarily by speech or written letters and numbers. Within limits, speech can be considered as a continuously variable quantity; when a transducer such as the telephone microphone translates the speech energy into electric energy, an electrical analog of the basic speech signal is produced, as shown in Fig. 22-1a. The telephone receiver reproduces the original speech energy. Facsimile is another example of communication usually transmitted by an analog signal.

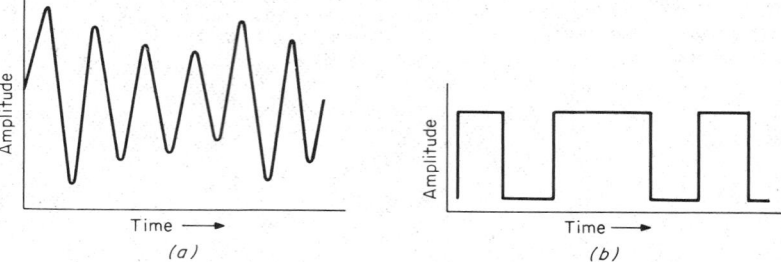

Fig. 22-1. Representation of (*a*) analog and (*b*) digital signals.

The symbols of written and printed communications are discrete items of information, best transmitted as correspondingly discrete digital signals. The digital signal is usually achieved by quantization in amplitude and time. Figure 22-1b illustrates the two-amplitude level (binary) signal, the most common form. The present rapidly expanding demand for computer data communications has made digital systems the subject of intensive study and development.

3. **The Switched Telephone Network.** In 1970 there were over 105 million telephones in the United States, most of which could be interconnected with any other on demand, as well as to a large percentage of the telephones in the rest of the world. This achievement is made possible by a network of switching and transmission facilities developed for efficient low-cost communication of speech as an analog signal. A discussion of the principles of the network switching plan is given in Par. 22-12. The switching network also handles the interconnection of audio and television broadcast program facilities and facsimile and telegraph transmission, among other services.

The increasing demand for computer data communications has generated modifications and additions to the circuits and equipment of the network to make it more efficient in handling high-speed digital signals. The network is by far the most readily available source of communication circuits.

4. **Digital Data and Telegraph Systems.** The terminal instruments of digital and telegraph systems include teletypewriters, data-card transceivers, data input-output equipment, and data sets. The switching methods and transmission facilities and their characteristics are suitable for forming complete digital communication systems. Digital systems are treated in Pars. 22-44 to 22-58.

5. **Radio Communications, the Radio Spectrum, and Regulatory Authorities.** The radio spectrum was first used for point-to-point and mobile communications; it is now also used by such services as audio and television broadcasting, meteorological aids, radio navigation, radio location, radio astronomy, standard-frequency broadcasts, and amateur radio.

Since the rf spectrum is a resource in the public domain, it has long been recognized that its use must be regulated. Internationally, the nations of the world have set up the International Telecommunications Union (ITU) and have agreed on the assignment of blocks of frequencies for each of the major services using the rf spectrum throughout its range, in three major regions of the world. The International Frequency Registration Board (IFRB) coordinates national frequency assignments to control interference. In the United States the rf assignment and regulatory authority for nongovernment users is the Federal Communications Commission (FCC). Table 22-1 shows the terminology of the rf spectrum bands.

TELEPHONE SETS

6. **The Telephone Set.** Figure 22-2 shows a simple one-way telephone communication system illustrating the principles of the telephone-set transmitter and receiver. Sound waves impinge on the diaphragm of the carbon microphone transmitter, the movement of which causes variation in its resistance. An alternating current, an analog of speech, is thus

Table 22-1. Bands of the Radio-Frequency Spectrum

Band number	Frequency range	Band name	Main uses
1	10-30 kHz	Very low frequency (vlf)	Maritime and military
2	30-300 kHz	Low frequency (lf)	
3	300-3,000 kHz	Medium frequency (mf)	Maritime and radio broadcasting
4	3-30 MHz	High frequency (hf)	All services
5	30-300 MHz	Very high frequency (vhf)	Radio and TV broadcasting, mobile communications, point-to-point links
6	300-3,000 MHz	Ultrahigh frequency (uhf)	TV broadcasting, mobile communications, radio-relay, satellite
7	3-30 GHz	Superhigh frequency (microwave) (shf)	Broadband microwave radio relay
8	Above 30 GHz	Extremely high frequency (ehf)	Waveguide systems

generated by changes in the direct current furnished by the battery. This current activates the receiver coil, and the resulting varying magnetic field causes movement of the receiver diaphragm and generation of sound waves which are an approximate reproduction of the sound wave impinging on the transmitter.

Since a telephone set must use a single circuit to transmit and receive, an induction-coil hybrid circuit is used. Means for alerting the user to an incoming call and for signaling the central office for an outgoing call (and in automatic systems a means for transmitting the desired number code for switching) must also be provided.

There are two types of telephone sets, the magnetic, or local-battery, type and the common-battery type (the latter being by far the most prevalent). In the *local battery set* the transmitter current is furnished by a battery at the local telephone set. Alerting the operator or distant telephone is accomplished by a hand-cranked magneto, furnishing about 80 V at 16 to 24 Hz. This type of set is used in rural areas and in military field systems.

In the *common-battery set* a dc voltage is provided over the wire loop from the switchboard or central-office battery which provides the transmitter current. A circuit closure, activated by the switch hook, serves to alert the central office. Dialing is accomplished subsequently through controlled interruption of the loop current. The user is alerted by superimposing 85 V at 20 Hz on the circuit at the central office to operate a ringer bridged across the loop. Both types of telephone sets employ an induction coil antisidetone hybrid circuit for connecting the transmit and receive circuits.

A typical common-battery telephone set is the Bell System type 500. The set must operate with various types of central offices, manual and automatic, and over different lengths of wire loop, of the order of 1,000 to 15,000 ft of 26-gauge copper-cable pair.

The requirements for a typical telephone instrument to meet these conditions are shown in Table 22-2. The circuits and functional elements of this telephone set are discussed in succeeding paragraphs.

7. Antisidetone Circuit. An induction-coil hybrid circuit of the antisidetone type is required to permit talking and listening simultaneously. This circuit also prevents too much sidetone in the receiver, which would cause the talker to speak too softly for effective transmission.

Figure 22-3 shows a generalized form of such a circuit. Turn ratio and impedances are selected such that when voltage is generated in the transmitter, adequate current flows to the line and network but very little in the receiver. Conversely, when speech current is received from the line, current flows to the receiver with very little to the remainder of the circuit. Figure 22-4 shows a schematic circuit diagram of the 500 set arranged in the fashion of Fig. 22-3. The ringer is bridged across the line to receive the 20-Hz ringing current. The switchhook contacts are open except when the set is in use. The dial interrupts the line current to produce the number code. It is bridged by a capacitor to suppress radio interference.

The antisidetone network contains varistor V_2, which compensates for changes in line impedance to keep the network within reasonable matching limits. The receiver circuit also contains a varistor which limits the receiver output to levels below those which would be objectionable to the listener, regardless of the length of wire loop from the central office.

8. Transmitter, Receiver, and Handset. The *transmitter* of the telephone set is of the carbon-granule type more fully described in Par. 19-37. The carbon is specially selected and processed and is enclosed in a housing composed of concentric electrodes and a specially shaped diaphragm. It operates in any position and covers the frequency range of 200 to 3,000 Hz. Since it generates the ac speech current by modulating a dc voltage, it is actually an electromechanical amplifier, with a gain of 20 to 30 dB.

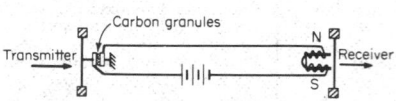

Fig. 22-2. Elementary telephone set.

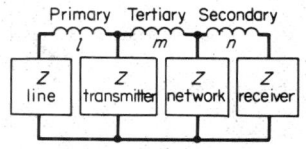

Fig. 22-3. Common antisidetone circuit.

Table 22-2.   Requirements for Typical Telephone Instrument

	Short loop	Long loop
Line current, mA	200	20
Impedance, Ω	600	900
Transmit level, dBV	−10	−20
Receive level dB (Reference Acoustic Pressure = 0.0002 mb)	95	86
Side tone, dB(RAP)	95	85
Rotary dial	10 + 0.5 pulses per second	
Make-break ratio	61 ± 3%	

For special purposes transmitters using piezoelectric rochelle salt crystals, condensers, or an electromagnetic mechanism resembling the receiver described below have been used in telephone sets to give certain desired response characteristics, but none provides the output level of the carbon type without the addition of amplifiers and consequent higher cost and complexity.

The receiver of the 500 set is of the electromagnetic type, with coil, permanent-magnet core, and diaphragm, and is also more fully described in Par. 19-59.   The coil impedance is of the order of 100 to 200 Ω.

The handset housing the transmitter and receiver is diagrammed in Fig. 22-5.   Its function is to place the transmitter and receiver in proper relationship to mouth and ear and to activate the switch hook.   Critical dimensions are the center-to-center distance R to T, and the R and T angles, which must compensate for the efficiencies of the transmitter and receiver used.

The overall transmission characteristics of a typical type 500 set transmitter-receiver combination over the frequency range are shown in Fig. 22-6 for wire loops of two different lengths.   The response is reasonably flat from about 250 to 3,000 Hz.

Available telephone sets with special audio equipment are the amplifier telephone, which contains an adjustable transistor amplifier in the receiver circuit for the hard of hearing, and

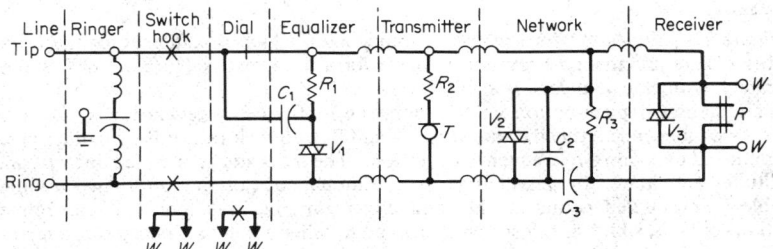

Fig. 22-4.   Typical common-battery telephone instrument.

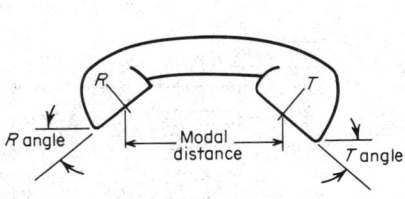

Fig. 22-5.   Telephone handset.

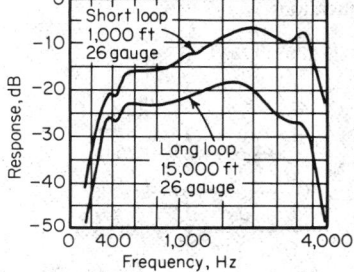

Fig. 22-6.   Transmission characteristics of type 500D telephone set.

the hands-free speakerphone set. In using the speakerphone precautions must be taken in placing the speaker and adjusting amplifier gain to avoid setting up a feedback path.

9. Supervision, Ringing, and Dialing. *The ringer,* which alerts the user to an incoming call, is usually bridged across the line, as shown in Fig. 22-4. For multiparty lines other arrangements are required. Placing the ringer on either side of the line to ground return provides two-party selective ringing. In polarized ringing the ringer is placed in a bridge circuit with a cold-cathode electron tube arranged for either positive or negative superimposed ringing signal on ground return, from either side of line; it will give up to four-party selective ringing per line.

The ringing signal is typically 85 V at 20 Hz, superimposed on the 48 V dc wire loop. The ringer, a bell-like device, has its major acoustic output in the 1,000- to 2,500-Hz region. In rural areas, coded ringing on multiparty lines is sometimes used. Buzzers and horns may replace the bell; amplified tone ringing through a loudspeaker is also available.

The switch hook connects the telephone-set circuit to the two sides of the line when the handset is off the hook and opens the connection when the handset is on the hook. The weight of the handset actuates the switch. The off-hook position places a dc closure on the loop circuit and prepares the central-office switching equipment to receive a call by lighting a lamp at the operator's position in manual switching centers and by obtaining dial tone and a circuit in preparation for dialing in an automatic switching center. In the local-battery telephone set, a switch actuated by the handcrank alerts the operator by operating a drop at the switchboard position.

The rotary dial, generally of the single-lobe cam-and-pawl type, is driven by a spring motor which is wound by the user as he dials each digit. The return motion is controlled by a governor to maintain the 10 ± 0.5 pulse per second output required.

Push-button dialing (Touch-Tone as it is termed in the Bell System) is a new method of transmitting the number code inspired by the higher switching speeds of electronic switching systems and the consequent need for more accurate and rapid dialing. Each number is represented by a pair of audio frequencies generated in the telephone set and transmitted simultaneously. The number 5, for instance, is signaled by the transmission of tones of 770 and 1,336 Hz.

Figure 22-7 shows the 4×3 matrix of frequencies, which is sufficient for the usual 10-number dial requirement. Provision is made for a 4×4 matrix (16 pairs of frequencies) ultimately, with extended code possibilities.

A circuit used in push-button dial telephone sets for frequency generation is shown in Fig. 22-8. Two independent tuned transformers A and B, with their respective windings in series, are employed to obtain two-frequency operation. The transistor is operated linearly, and the amplitudes are limited by varistors to make simultaneous oscillation at two frequencies possible. The circled points on the transformers in Fig. 22-8 represent the frequency-selection contacts, which are closed on each winding when a button is depressed, interrupting the direct current and shock-exciting the oscillator. Since the tones generated by the push-button dial are in the voice-frequency range, they can be transmitted throughout the telephone communication system and used for other purposes besides dialing the telephone number code. Much use is made of this device by subscribers in data communications applications.

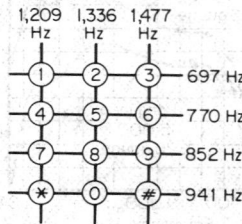

Fig. 22-7. Basic arrangement of push buttons used for Touch-Tone dialing.

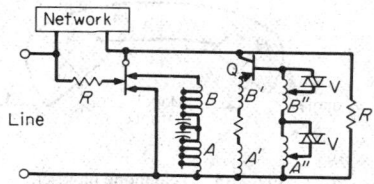

Fig. 22-8. Touch-Tone dial circuit for two-frequency operation.

Dialers, in which the manual process of number dialing is replaced by machine processing using cards, tapes, or drums on which the number code is recorded, are available for rotary or push-button dial sets. The processing is usually initiated by pressing a start button; the record medium is then read and the proper pulses or tones placed on the line.

10. Key Telephone Sets. When a given location has more than one line from the central office or private branch exchange (PBX) and a number of telephone stations, it is often desirable to equip the latter with key sets to provide local switching functions and other features such as (1) ability to pick up one or more of several central-office lines, (2) ability to hold one or more lines and use another, (3) ability to signal an associate, (4) ability to have local intercommunication. A wide variety of arrangements and instruments are available, ranging from single-line desk set with line pickup through 2-, 4-, or 6-line pickup multibutton key sets (Fig. 22-9) to large 10-, 18-, 20-, 30-, and upward lever or push-button, key boxes, turrets, and consoles. One type is illustrated in Fig. 22-10. Audible and visual signals are required if there is more than one line and multiaccess thereto. The audible signals can be similar to central-office ringing supply; the visual-signal device in newer installations has small lamps under plastic key caps.

Either of two basic line circuit-relay configurations is generally used to provide station control of line pickup and hold functions in key telephone-set arrangements. In one

Fig. 22-9.   Typical station apparatus.

Fig. 22-10.   A 30-button call-director telephone with self-designating keys. (American Telephone and Telegraph Co.)

arrangement, illustrated in Fig. 22-11, the central-office or PBX line current is used to operate and release the holding bridge. A balance lead to compensate for the unbalanced transmission characteristics of the relay circuit and a hold lead in addition to the two line leads are required for each line so equpped.

Where auxiliary functions are furnished, such as local visual and audible signals, additional leads to provide these functions are required. The second configuration illustrated in Fig. 22-12 requires an auxiliary conductor per line designated the *A* lead in addition to tip and ring lead of the line circuit. Local power supply is required for relay operation, lamps, and audible signals.

Visual-signal indications in key-set arrangements have been generally standardized so that a rhythmical (60 per minute) flashing signal (sometimes augmented by an interrupted audible signal) indicates an incoming call and a steady lamp signal indicates a line in use or a held line. The latter is sometimes indicated by a fluttering signal.

TELEPHONE SWITCHING SYSTEMS

11. Switched-Telephone-Network Concepts. A switched network is the only practical way to make universal telephone calling possible. A community of 10 telephones each

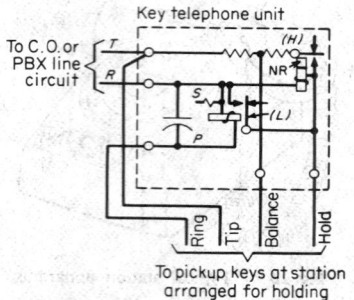

Fig. 22-11. Line pickup and hold circuit (operates and releases on line current).

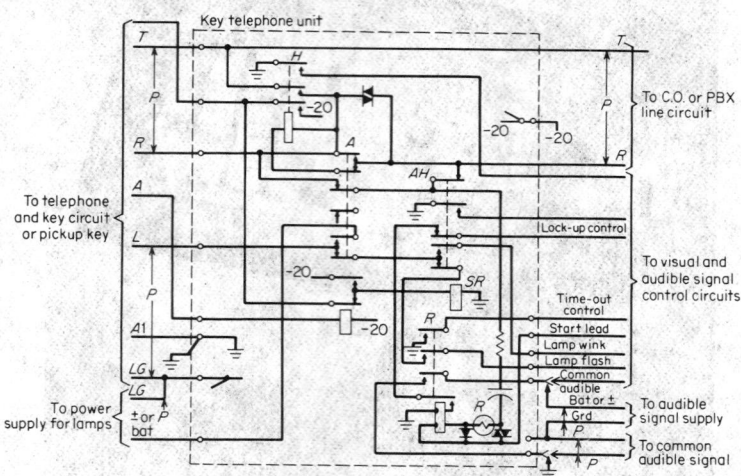

Fig. 22-12. "*A*" lead-type central-office or PBX line circuit.

directly connected to the other 9 at its own location would require a total of 90 wire connections, each of which would have limited use. In contrast, a simple switchboard with 10 wire connections, one to each telephone, allows any party to call any other.

The local switching point is usually termed the *local switching center* or *central office*. The wire connection between the local switching center and the user's telephone is generally termed the *loop* or *line*.

The reasoning used for justifying switched interconnection of telephone sets also applies to interconnection of switching centers, by direct interconnection of local switching centers or through switching centers which are used exclusively for such interconnections. The latter are called *toll switching centers* in the national long-distance network and *tandem switching centers* in those urban networks where they are required. The single message circuit capable of conveying one telephone conversation between switching centers is called a *trunk*. A collection of such trunks between two switching centers is called a *trunk group*.

12. Trunk Switching Plan. The organization of a telephone-system network is exemplified by that employed in the United States for interconnecting the various local and trunk switching centers. The principle is illustrated by Fig. 22-13. The structure is hierarchical with five classes of switching centers, the local switching center identified as class 5 and the regional centers as class 1.

The number of centers in each class increases as the class number increases. There are 10 class 1 regional offices in the United States network as contrasted to about 22,000 local switching (class 5) offices. Each switching center connects to a center of the next highest rank by a trunk connection which is termed the *final* or *via trunk* route, shown by solid lines in the diagram. These trunk routes are also provided from each class 1 center to every other.

Direct trunk interconnections are provided between centers of the same or different classes wherever traffic and economy dictate, as shown by the dotted line in the diagram. This type of interconnection is called a *high-usage trunk group*. The most direct high-usage trunk route is the first choice in routing a telephone call at all stages of its progress, the final trunk route being the last choice except where it is the most direct route. As indicated by the name, the high-usage trunk groups are large enough and have a great enough traffic-carrying potential to be operated at high efficiency in the busy hours but are unable to accommodate all calls in the busiest hours.

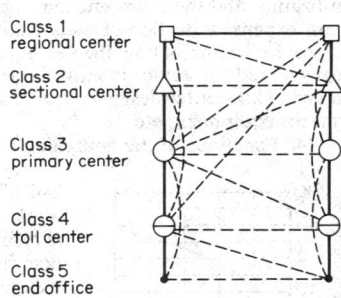

Class 1
regional center

Class 2
sectional center

Class 3
primary center

Class 4
toll center

Class 5
end office

Fig. 22-13. Hierarchical structure and routing in toll switching plan.

The final trunk routes are designed to handle practically all calls offered, including overflow calls, during a normal busy hour. The maximum number of trunks required to complete a call is nine in this plan, and only a very small percentage of calls requires this number of trunks for completion.

Transmission considerations are discussed in Pars. 22-22 to 22-27. The ultimate goal is to assure that over the longest transmission route the speech-to-noise ratio at the end telephones is satisfactory under all but catastrophic conditions. As discussed in Par. 22-22, the two-wire termination of a 4-wire trunk circuit limits the transmission level at which it can operate, due to transhybrid return loss. Therefore, 4-wire switching is used at Bell System class 1 and class 2 switching centers to improve the overall transmission performance. It is also used in some government and military switching systems using commercial facilities.

13. Rudiments of Telephone Traffic Engineering. Telephone traffic engineering is a complex subject rooted in the mathematics of statistics and probability and network theory. It plays a key part in the determination of the routings and size of trunk groups in a trunk-switching plan and in determining the number of cross-connecting circuits and the size of the control equipment in switching-equipment design.

A few of the important terms used and their meaning are introduced here. The fundamental data required for any traffic study are a determination or estimate of the number of telephone *calls per hour* offered, particularly in the busy hour, and the *average holding time* per call. The fundamental traffic unit is the *Erlang*, defined as

$$E = \frac{nh}{3,600}$$

where n is the number of calls originated during a specific hourly period, h is the average holding time per call in seconds, and E (Erlang) is the average number of simultaneous calls during a specific hourly period. This formula gives the traffic density for a given trunk group.

The number of trunks required in a given trunk group is found by first assigning the *grade of service* we wish to provide. Grade of service is defined as the probability of finding all trunks busy in the given trunk group. A usual figure for grade of service is $P = 0.01$; that is, the probability of finding all trunks busy in the given trunk group during the busy hour will be no greater than 1 in 100 on the average. Tables have been constructed (such as those in Ref. 9) which give, for the Erlang number and grade of service desired, the number of trunks required in the trunk group.

For a given grade of service, the larger the trunk group the greater the percentage utilization and the more efficient the system, since transmission facilities are generally the most expensive portion of the system.

An indication of how the percent occupancy of the trunk group increases with its size for a given grade of service is shown in Fig. 22-14. Similar traffic-engineering methods are used in switching-center design to determine the number of cross-connecting circuits, size of the control equipment, etc.

14. Functions of the Switching Center. The ubiquitous local switching center (class 5 office) and the toll and tandem trunk switching centers (class 1–4 offices) are the nodes of the commercial telephone network. Another type of switching center is the private branch exchange (PBX), which includes those numerous switchboards and switching centers installed in commercial establishments, institutions, government agencies, etc., providing local intercommunication as well as access to the commercial telephone network via connection through local switching centers.

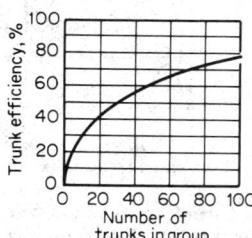

Fig. 22-14. Trunk efficiency relationships. *(From David Talley, "Basic Telephone Switching Systems," Hayden, New York, 1969.)*

In carrying out the mission of connecting a calling party to a called party at a local or distant destination, the following functions must be performed in the switched telephone network:

1. Identify the calling line when the user raises his handset from the switch hook.
2. Make a connection to the calling line.
3. Pick out an available transmission path to the called party, who may be connected to the same switching center or to some distant switching center reached over one or more trunk circuits.
4. Determine whether the called line is already in use.
5. Signal the called station.
6. Record the calling party, called party, and duration of call where required.
7. Restore all circuits to a state of readiness for other calls when the parties hang up.

In the commercial telephone network all these functions are performed by the local switching center with the toll and tandem trunk switching centers providing assistance in function 3. The PBX in its intercommunication role performs all the functions except possibly function 6.

15. Classification of Switching Equipment. The first major classification of switching equipment relates to manual vs. machine methods of making cross connections.

Manual switching systems are operator-controlled, and connections are established with cords (or with keys in cordless switchboards). Before the widespread installation of automatic switching equipment, a large part of the commercial switched network employed cord-type manual switchboards in which the operator supervised and performed the control functions listed in Par. 22-14. Operator assistance is still required in many situations, e.g., person-to-person calling, emergency services, coin telephones, reverse-charge, and credit-card calls. Many of these services are performed at consoles using semiautomatic methods.

Automatic switching systems can be classified, by the method of control in the switching, as progressive or centralized control.

Progressive-control automatic switching implies that the path through the switching matrix is set up one step at a time in accordance with the information in each successive dialed digit. A further subdivision of progressive control relates to whether the control of the switch is direct or by register.

In *direct control* the movement of the switch contacts is controlled directly by the user's dial and advances as the dial produces its pulses. In *register control* the dial pulses are registered in an electric counting mechanism, stored briefly, and translated if necessary; the information is then released to the switch-operating mechanism, usually after each dialed digit. Information released to the switches may not be in decimal form but must have a one-to-one correspondence.

The original Strowger step-by-step switching mechanism was a direct progressive automatic switching system, with switches under the direct control of the user's dial. It is still used in this manner in many installations, particularly in PBX systems.

The advent of direct distance dialing (DDD) in the commercial network required that a register be interposed in local switching centers where step-by-step switching equipment was installed or retained. A step-by-step direct progressive automatic switching system and modification of such a system to a register progressive system are discussed in Par. **22-17**.

Other types of direct progressive-control automatic switching systems are the *all-relay type*, used in smaller offices, the *X-Y switch*, and the *register progressive control* systems (the panel system in the United States and the rotary system in Europe), which use motor-driven switches and revertive pulsing to the register for control of switch position.

The term *centralized common-control automatic switching systems* is used here to designate systems in which essentially all control of the switching function is placed in a central mechanism shared by many switches. From the dialed information stored in a register, the central control mechanism finds the called party's line or a trunk, finds an idle path for the connection, and sets up the necessary testing and ringing functions and then drops out of the circuits to attend to another call. A further subdivision of centralized common-control systems relates to the type of control equipment—whether predominantly electromechanical or electronic.

An electromechanical centralized common-control system, the Bell System no. 5 crossbar is described in Par. **22-18**. The Pentaconta system is a similar crossbar automatic switching system used extensively in Europe.

An electronic centralized common-control system, the Bell System no. 1 electronic switching system (ESS) is described in Par. **22-19**. The electronic system has the advantage of higher speed, many operations being performed at speeds 10,000 times faster than in electromechanical systems. In effect, in ESS the central control is a special-purpose digital computer.

A classification of switching systems more fundamental than those discussed above involves the method of establishing the interconnecting switching path. The methods previously described use *physical circuits* to establish the switched path. Historically this method has been the most practical, and until quite recently it has been the only method considered. It is termed *space-division switching* in contrast to two other methods, *time-division switching* and *frequency-division switching*.

In *time-division switching* the information offered is sampled at a high enough rate so that no information is lost and the samples are interleaved (multiplexed) in time with other simultaneously transmitted samples on a common bus or highway. The interleaving is controlled by electronic gates, and for a given call, the gates of the calling and called party are keyed to the same time slot in the multiplexed signal for a given individual call, with a different time slot for each other call being serviced during the call period.

The assignment of time slots and control of electronic gates is under a common control mechanism. Time-division multiplexing, much used in telephone and data transmission, is more fully treated in Pars. **22-39** to **22-41**. Some experimental switching systems employing this method have been developed and tested, but there has been no extensive use. The bandwidth of the system is limited by the sampling rate established for individual channels in system design.

In *frequency-division switching* the information for an individual call is separated from that of other calls by occupying a different frequency band on a common bus. In this system the entrance and exit circuitry of each calling and called party must have the same carrier frequency established by a common control mechanism. It resembles time-division switching in that it is also band-limited, dependent on the carrier-frequency spacing. While it is a technique long used in telephone transmission and is discussed more fully in Pars. **22-33** to **22-37**, the method has had only minor experimental use in switching.

16. Manual Switching Equipment. Figure 22-15 illustrates in cross section a typical cord-type manual-switchboard operator position. The subscriber lines and trunks to the other offices are terminated on jacks in front of the operator. By inserting cords in the proper jacks on receipt of a proper signal and receipt of information from the calling party, the call connection is established. In a large office with several hundred or thousand lines and trunks the jacks terminations appear in multiple at several operators' positions.

Figure 22-16 is a schematic of the line and cord circuit of a typical manual switchboard. When the calling party goes off hook with his telephone set, the L relay is operated, lighting lamp L at the switchboard position. When the operator plugs in the answering cord the CO relay operates from the cord battery. The S_1 relay is operated through the local loop to keep the cord lamp extinguished. When calling information is obtained, the operator plugs the calling cord into the jack of the called party, first testing for busy by touching the cord tip to the jack sleeve. When plugged up, ringing current is applied by operation of ringing key. When the called party goes off hook, lamp S_2 is extinguished; and when he hangs up, it is lighted until the cord is removed from the jack. When a calling party goes on hook, the S_1 relay releases and the S_1 lamp lights until the cord is removed from the jack.

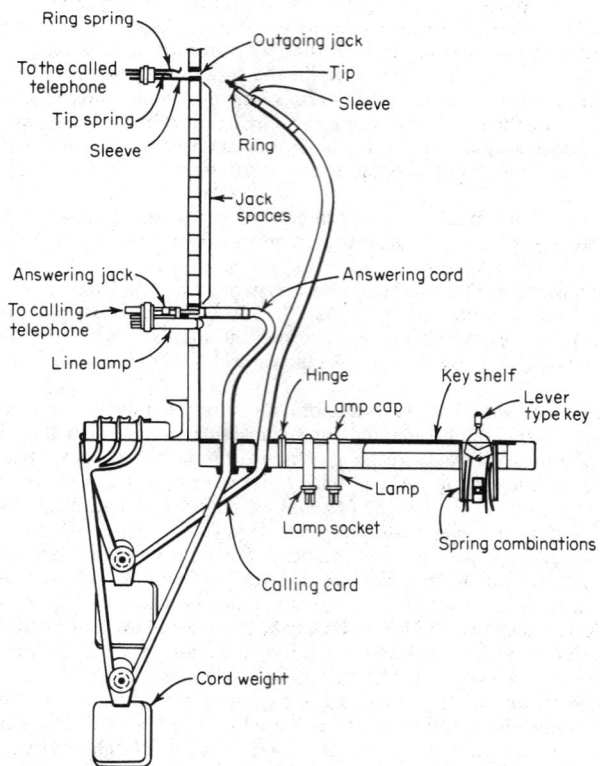

Fig. 22-15. Cross section of typical manual switchboard.

Manual cord-type switchboards have been used in all types of central offices and extensively in PBX applications. Automatic switching equipment has supplanted manual equipment in most installations.

For small PBXs a cordless-type manual switchboard (Fig. 22-17) is often used. The lines and trunks are arranged in a matrix and connections established by operating keys to establish the cross connections.

17. Step-by-Step Automatic Switching Systems. The basic *Strowger step-by-step switch* is shown in Fig. 22-18. The basic switch has 100 outputs. Other variations are available, providing up to 200 outputs.

A train of pulses operates the vertical magnet labeled 392 to move the contact fingers vertically to the proper level. The next train of pulses operates the rotary magnet labeled 408 to the desired output in that level. The rotary pulses are often obtained from control relays mounted on the switch.

A double dog provides a single mechanical latch for both vertical and rotary motions. Operation of the release magnet labeled 396 disengages the double dog and permits a spring to restore the rotary motion, followed by gravity restoration of the vertical motion. Control relays are mounted on the switch to perform various functions according to whether the switch functions as a line finder, selector, or connector.

A *direct progressive-control step-by-step automatic switching system* is shown in outline form in Fig. 22-19. When the handset is removed from the cradle of the user's telephone set, the line circuit causes the line-finder switch to sweep through the different line appearances until it reaches the calling-line output. Immediately a first selector is seized and dial tone applied to the line. The first dialed digit causes the first selector to reach the pulsed level, and the control relay circuit causes rotary hunting at that level for an idle second selector. This process is repeated for the second digit dialed, and the second selector finds an idle connector. The third digit dialed raises the connector to the desired level, and the fourth digit dialed should rotate the connector switch to the called party output. The connection then tests the line and if it is not busy applies a ringing tone which is tripped when the called party goes

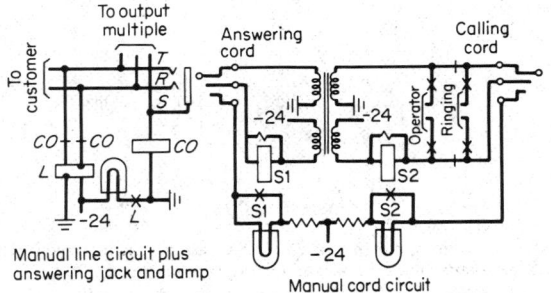

Fig. 22-16. Manual-switchboard cord and line circuits.

Fig. 22-17. Nonmultiple manual cordless PBX.

off hook. The scheme shown is for a four-digit number and is more applicable to a large PBX of 1,000 to 10,000 lines. Smaller PBXs have fewer selectors.

With the three- to four-digit numbering scheme used in the commercial telephone system today, more selectors are required; most step-by-step local switching centers now employ register progressive control. However, if the three-digit office code is that of a distant office, the third selector in this case will hunt for an idle trunk to that office, and the remaining digits enter the switching train of the distant office at the proper selector stage to complete the call.

A *register progressive-control step-by-step automatic switching system* is illustrated in Fig. 22-20. In this case, when the calling party's handset is removed from the cradle of the telephone set, the line finder provides a link to the register and the register seizes a first selector and applies dial tone to the line circuit. The register (or, in its more sophisticated

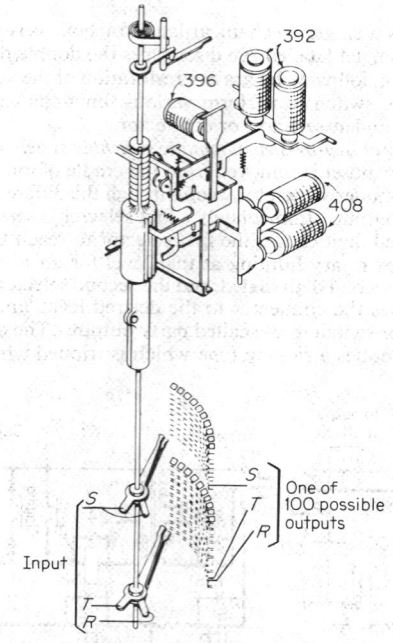

Fig. 22-18. Early patent drawing of a step-by-step (SXS), switch showing mechanical operation. The vertical magnet is 392, the rotary magnet is 408, and the release magnet is 396.

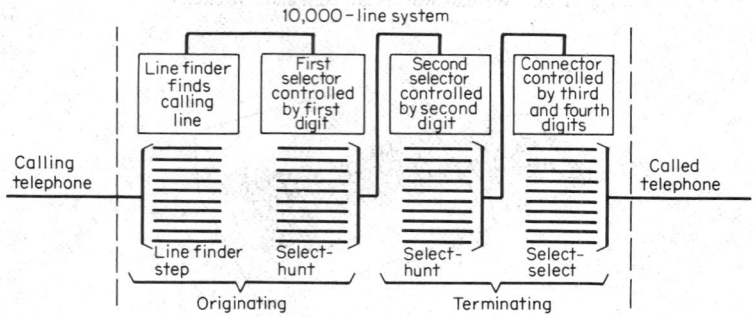

Fig. 22-19. Direct-control step-by-step switching system. *(From David Talley, "Basic Telephone Switching Systems," Hayden, New York, 1969.)*

versions, the director) receives the dialed information, stores it, and acts on it in accordance with the wired program of instructions before releasing the information to the switch train. For instance, the program may call for absorption of digits on the local office code and release the register for direct dialing of the last four digits, the called-party code, to the selectors and connector in turn.

On calls to other central offices it may cause alternate trunk routes to be selected where first-choice routes are found busy, and it may call in multifrequency pulse-signaling equipment to repeat dial information to distant offices over long-distance trunks. The control circuitry is so arranged that when the register has completed its functions in establishing a call, the link circuit operates to remove the register from the circuit so that it will be available for processing other calls.

The more sophisticated versions using the *director* are designed so that they can be placed in a step-by-step direct-control installation, without requiring modification of the switches. They are constructed using both relay and electronic components.

18. Crossbar Automatic Switching System. The crossbar switch (Fig. 22-21) operates in a rectangular coordinate system in which contacts are made by contact springs operated by bars in both vertical and horizontal planes to close one of 200 possible contacts. The sequence of operation is that operating magnet A for the horizontal bars operates and tilts all selecting fingers C at that level into one of two possible operate positions. Following this, the holding-bar magnet B is operated and rotates the vertical holding bar E into position. At the point where the select finger operated by the horizontal bar and the vertical holding bar intersect the sets of contacts associated therewith are held closed by the pressure of the holding bar on the select fingers.

The horizontal operating magnet A may then be released and the switch made available for handling other connections not involving the same vertical coordinate. The contact springs are connected by multiple wiring which runs horizontally across the rear of the switch.

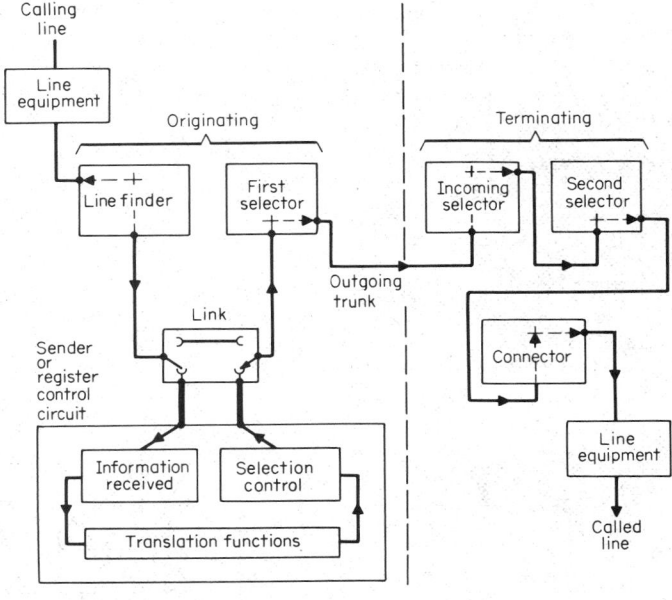

Fig. 22-20. Common-control switching as applied to the step-by-step system. *(From David Talley, "Basic Telephone Switching Systems," Hayden, New York, 1969.)*

The basic interconnecting arrangement for crossbar switch frames (Fig. 22-22) shows eight crossbar switches in frames 1 and 2, respectively, as might be mounted in a central office. Only two lines have been shown from each side of a switch to simplify the drawing; in an actual central office there would be 10 lines. The geometric pattern of interconnection is shown by the lines in the diagram. For instance to get from point *A* to point *B* in the diagram two possible paths exist, as shown, but in an actual system with 20 vertical unit switches 10 possible paths would exist between *A* and *B*, illustrating the flexibility achieved by use of the crossbar switch with centralized control.

The *centralized-control no. 5 crossbar system* of the Bell System is illustrated in simplified form in Fig. 22-23. As the user removes the handset from the cradle, he is connected to an outlet on a crossbar switch in the line link frame, and in a fraction of a second the common-control marker is alerted, sets up a connection from the calling line to an originating register, and then disconnects itself to handle other calls. Dial tone is then transmitted to the calling line and the dialed information stored in the register.

At the completion of dialing the originating register calls in an idle completing marker and transmits the dialed information to it. The marker then queries the number-group frame, which is a sort of mechanical telephone directory listing the number of every telephone line in the central-office area, the line-link switch, and position on the switch where that line is connected. When the marker obtains this information, the register and number-group frame are released.

The marker then sets up a connection between the calling-party switch position and called-party switch position in the line-link frames by way of an intraoffice trunk in the trunk-link frame. The connection is then complete, and the marker releases itself from this call and stands ready to handle other calls. Ringing is then applied by the intraoffice trunk circuit.

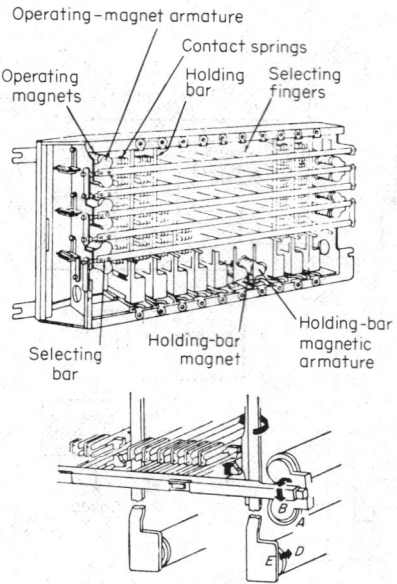

Fig. 22-21. Partial perspective of crossbar switch with 10 vertical units.

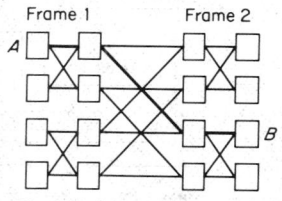

Fig. 22-22. Basic switching-frame arrangements in crossbar system.

If the call is to a party located in another central office, the procedure differs in that when the dialed information is transferred to the marker, it disconnects the register from the calling line, connects an outgoing sender to the proper idle interoffice trunk in the trunklink frame, and interconnects the calling line and outgoing trunk through the line-link and trunk-link frames. The sender than transmits the number of the called party in suitable code to an incoming register in the distant central office. The incoming register then calls in a completing marker, and call connection is established as previously discussed for a local call. Outgoing senders are available in several varieties to transmit the proper signals to different types of central offices to which connection might be required such as step-by-step, crossbar, electronic, and occasionally manual central offices.

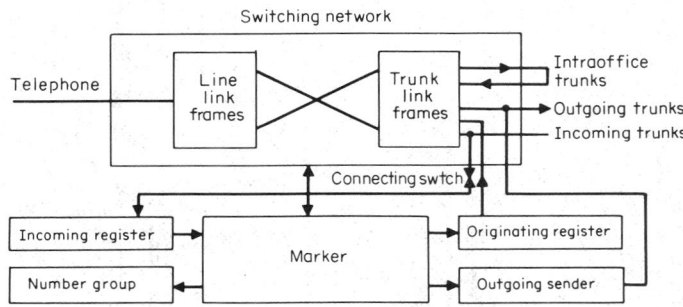

Fig. 22-23. Path of call in crossbar system.

While the operations described and the equipment involved are complex, they operate at high speed, and the switching time is many times shorter than in step-by-step systems.

There are about 1,800 crossbar switches and some 70,000 relays of all types in a typical no. 5 crossbar central office. One marker in itself has some 2,000 relays, but is in use only a very short time on each call.

19. Electronic Automatic Switching Systems. The Bell System no. 1 ESS uses a stored-program synchronized digital-computer centralized system for controlling essentially all switching-system functions. The switching network is set up through physical connections in space division, using a new type of switch which is lighter and smaller than those previously discussed. The speed of handling calls is increased many times over that achieved in electromechanical systems, and the stored-program central control allows much greater flexibility in accommodating system changes and offering additional services. The switch and line sensor of the switching network and the semipermanent and temporary memories of the central control are to some extent peculiar to this system. The central control uses extensive transistor logic and other circuitry basic to digital computers, as discussed in Secs. 8 and 23.

The ferreed switch, shown in Fig. 22-24, is an adaption of the reed relay. The reeds, made of a flexible magnetic material, form part of the magnetic circuit or flux path and also provide the electrical contacts. The switch is made of two magnets of a material exhibiting a square hysteresis loop. When the excitation is such that the magnetomotive forces (mmfs) in the two halves add, the flux is continuous through magnets and reeds and closure results. When the mmfs of the two halves are opposite the net magnetomotive force is low and the reeds release.

An array of four such switches, as used in the no. 1 ESS, is shown in Fig. 22-25. Differential excitation is used to operate the desired switch while leaving the others in nonoperate status. In this case the battery is applied to the top row and ground to the left-hand column. Only in switch 1 do the mmfs of both sets of winding add to produce contact closure. In switches 2 and 3 the current flows through only one set of windings, and in switch 4 through none. Very short pulses suffice to change the magnetic state and operate the switch but the switch contacts require a time of the order of milliseconds to close. Such switches are typically mounted in an 8×8 array, forming an assembly only slightly larger than one crossbar switch in the vertical dimension. These switches are the basic equipment in the line-link and trunk-link networks of the system.

Ferrod sensors are associated with each line appearance in the line-link network. They are scanned periodically by the line scanner, under the direction of central control, to determine whether the line is in an on-hook or off-hook condition. This type of sensor can be thought

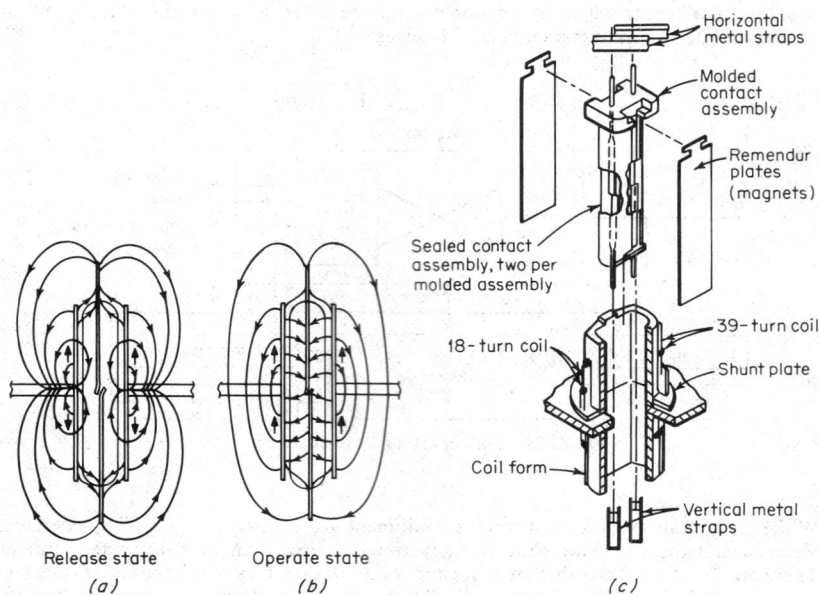

Fig. 22-24. The ferreed switch: (*a*) and (*b*) show the magnetic flux patterns produced by the magnets to permit the reed contacts to open and to cause them to close, respectively. The exploded view (*c*) shows the construction of the switch and the method of mounting in a shunt plate of magnetic material. *(Bell Telephone Laboratories.)*

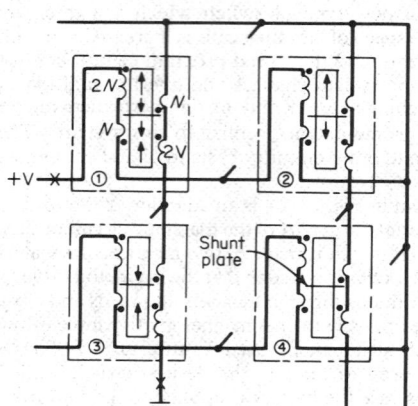

Fig. 22-25. Differential excitation of ferreed switch magnets.

of as a three-winding transformer, with a saturable core. Two windings perform the interrogate and answer functions associated with the scanner, while the third is in series with the line. When the interrogate winding is pulsed and there is no line current, the answer winding generates a similar response pulse, indicating an on-hook condition. When line current flows through the third winding, if the telephone set has gone off hook, the current saturates the core rod so that on interrogation no corresponding pulse is generated in the answer winding and this condition is identified by central control.

A *ferrite sheet memory* is used as the temporary memory in the central-control call store. The basic ferrite sheet is 1 in. square, with 256 holes punched in it. It provides the equivalent of a 256-core memory, the material surrounding the hole constituting the basic memory cell as in a core. A *memory stack* consists of 64 such sheets; three adjacent stacks comprise a memory module.

Inhibit and sense wires are threaded through the corresponding holes in all 64 sheets to form the *X* direction. The *Y* conductors are formed by interconnecting plated conductors on three ferrite sheets on each level of the module. The *X* and *Y* wires carry the half-strength pulses which switch the cell at their intersection. The inhibit wire is used to prevent switching during the writing of a bit.

A *twister memory* is used as the semipermanent memory constituting a major portion of the program store in central control. This memory can be read repeatedly without changing its state; moreover loss of power supply does not affect the information in storage. The basic element is an aluminum card containing 2,815 tiny dots of a magnetic alloy. A magnetized dot represents a binary 0, and an unmagnetized dot represents a binary 1.

The twister wire, a copper wire wrapped spirally with very thin Permalloy tape, is run under the alloy dots and is used as the sensor or read wire. The taped wire forms a miniature transformer, giving a voltage step-up at each magnetized dot.

To change the information stored in the twister memory, the card is removed and placed in a writer which magnetizes or demagnetizes the dots as desired. In this system, a module consists of 128 cards, which use the common sensing equipment. A complete memory consists of 16 modules, i.e., a total of 130,000 words of 44 bits each.

The *central control* with its associated *program store* and *call store* are the heart of the system and control the operation to a far greater extent than the central control in the crossbar system. The central control uses the logic circuitry, which carries out the operations called for from the information in the semipermanent memory of the program store aided by the transitory information in the temporary memory of the call store—all under the scheduled supervision of an executive program and timed by a master clock at 2 MHz.

The programs carried in the *program store* and the associated information receptacles in the cell store are as follows:

1. *Input programs* gather information such as the state of all lines and trunks. This involves line-scanning information placed in the line-service request hopper of call store.

2. *Operational programs* examine the information received and decide what, if any, output actions are required. This includes digit-scanning, dial-connection, digit-analysis, and ringing-connection operation programs. It also involves information deposited in and read from the digit hopper, originating register, and ringing register of call storage.

3. *Subroutines* contain data on translation of dialed digits and trunk and network path requirements for use of output programs. The information in the path memory and the idle trunk lists in the call store are used by these programs.

4. *Output programs* make and release connections. These include ringing-trip-scanning and talking-connection programs. Required associated information is deposited in the output buffer and ringing trip hopper of call store.

Each ESS central office has two central controls which process all operations simultaneously and compare results at key points as an error check, so that one can carry on if the other is disabled or under maintenance. Associated with central control in an installation are from two to six program stores and up to 14 call-store units.

A representation of the no. 1 ESS system, featuring the switching network, is shown in Fig. 22-26. The call-processing function is as follows: The ferrod sensors associated with each line appearance are scanned at 200-ms intervals to determine the status of the line. When a party, party B in this instance, goes off hook, this information is determined by the line scanner; the line location is noted, and the information is placed in call store as a *service request*. Information from program store shows the type of station equipment in use on this line, and the line-link and trunk-link networks are directed to connect the proper service trunk to the line and dial tone is sent by this trunk. The digit-scanning program at 10-ms intervals is then initiated and the data recorded in the register in the call store.

After all digits are dialed, the program store determines the type of call being made. For an intraoffice call, central control refers to the network map in the call store to find idle links and paths to set up the connection to called line C. Having received the data, central control directs the line-link network to make these connections through a junctor circuit. If the called line is busy, the ringing-connection program will direct a busy tone to be returned to calling party B. If the line is idle, the junctor circuit will connect ringing current to the called line and an audible ringing tone is returned to calling party B.

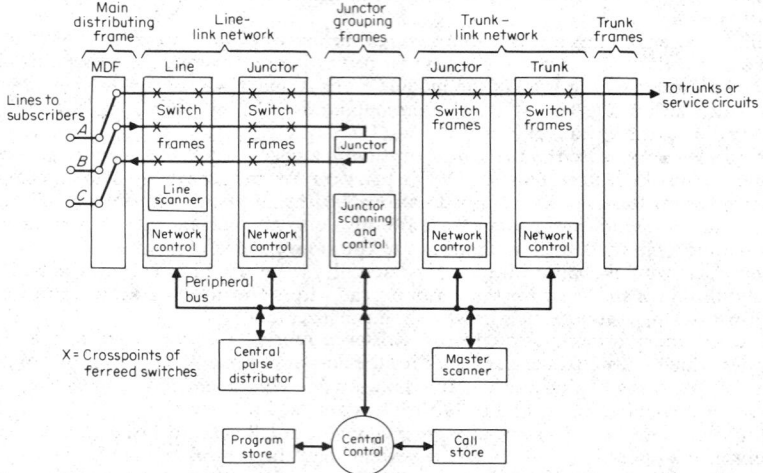

Fig. 22-26. No. 1 ESS switching network. *(From David Talley, "Basic Telephone Switching Systems," Hayden, New York, 1969.)*

When the called party answers, the change in state is noted in call store. The calling line is then scanned at regular 200-ms intervals by the line scanner until a change in state to on hook is detected, at which time central control directs the line-link network to release all connections.

Interoffice or trunk calls are handled in a similar manner, as indicated by the path from calling party A in Fig. 22-26. In this case, however, both the line-link and trunk-link networks are involved in the connection to obtain an idle trunk in the proper trunk group and associated supervisory equipment. The central control handles in a given time interval the detailed operations for several calls but performs only one operation on one call at a given time; i.e., it interleaves the individual operations of several calls in time. This is possible because of the very high speed at which it performs operations and gives instructions.

Electronic PBX equipment is available in which the central control and the associated program store and call store are located in the telephone company central office. The switching network, scanners, etc., are assembled in *switch units* located on the customer's premises. Interconnection is via data links for transmission of control information and instructions. Installations of up to 3,000 lines are made. The Bell System type 101 ESS is a system of this type.

20. Trunk Signaling and Dialing. Many forms of trunk supervision signaling have been invented over the years. *Ringdown signaling,* the simplest method, was used between manual central offices. It relied on the application of 20-Hz ringing current to the trunk leads by the operator to signal the distant operator that a trunk call was waiting. It was also necessary for the operator to ringdown the trunk to signal for disconnect. The scheme was also adaptable to carrier-derived trunks with suitable voice-frequency ringers for transmission of the ringdown signal over the carrier trunks.

With the advent of common-battery trunk switching and automatic trunk-switching systems other more automatic means were devised, including the high-low and reverse battery schemes. The current universal method for two-way trunk supervision is known as E and M lead supervision.

E and M lead trunk supervision derives its name from designated leads in Bell System toll trunk circuit. A schematic of this type of circuit is shown in Fig. 22-27. The M lead sends information to the distant office. Battery current through a resistance lamp indicates off hook and ground on hook. The E lead receives information from the distant office; and ground means off hook and open means on hook. The battery on M operates the distant SS relay but not the relay in the local office.

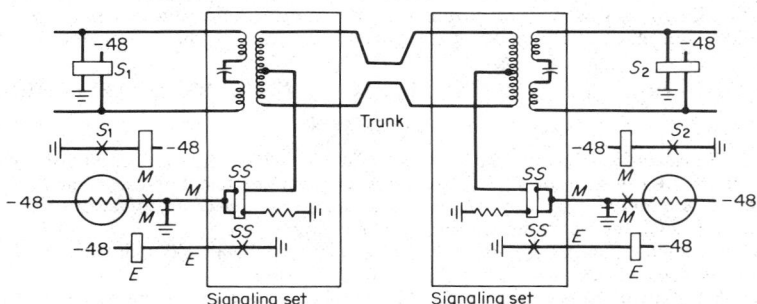

Fig. 22-27. E and M lead supervision. Battery on M operates the distant SS (signaling-set) relay but not the nearby SS relay.

The E and M leads connect to special signaling sets. Where physical trunk circuits are used, these sets reapply the signals to the tip and ring or talking path of the trunk on a dc basis. Where carrier-derived trunks are involved, the leads are connected to voice-frequency signaling sets for transmission over the trunk circuit. These signaling sets are discussed in Par. 22-41.

For *trunk dialing*, as for trunk signaling, many methods have been devised over the years, including dial pulsing, revertive pulsing, panel-call indication, and dc key pulsing. The E and M lead supervisory circuit can be used for 10 pulse per second dialing between step-by-step offices, but this method is far too slow for modern trunk systems.

Multifrequency key pulsing is now the most widely used trunk dialing method. It is applicable both to operator dialing using a key set and to outpulsing by senders in centralized-control crossbar and electronic automatic switching centers. Each digit, with a start and stop signal, is represented by a unique pair of frequencies between 700 and 1,700 Hz as follows:

Digit	Frequencies, Hz	Digit	Frequencies, Hz
1	700 + 900	7	700 + 1,500
2	700 + 1,100	8	900 + 1,500
3	900 + 1,100	9	1,100 + 1,500
4	700 + 1,300	0	1,300 + 1,500
5	900 + 1,300	KP	1,100 + 1,700
6	1,100 + 1,300	ST	1,500 + 1,700

The KP signal is the start signal and serves to prepare the receiver by increasing sensitivity. The ST signal denotes the end of dialing and serves to desensitize the receiver. Since signals are in the voice-frequency range and travel over the talking path, the scheme can be used on all types of trunks including repeatered and carrier-derived trunks. The speed of sending is at least 10 times faster than direct-dial pulsing.

21. Central-Office Power. The central office, with its switching, transmission, and test-equipment assemblies, requires substantial amounts of power at several voltages and frequencies: 48 V dc for the main switching and talking functions, 24 V dc for some carrier and microwave systems, 130 V dc for plate voltage for vacuum-tube and cable-repeater power supplies, 86 V at 20 Hz for ringing current, and various ac tones for supervisory signals.

The main power source is usually the commercial power supply, which operates rotary converters or solid-state rectifiers to convert to the 48 to 50 V dc main power. A 48-V battery of sufficient capacity to carry essential operations during a reasonable period in event of commercial power failure is floated across the main dc power bus. It also acts as a voltage regulator and filter for the rectified supply.

The ringing generator and tone generators are driven by the main dc supply. Ringing generators of various types, such as rotary generators, electronic oscillators, vibrating-reed converters, or subharmonic generators, driven by 60-Hz commercial power, are used. All equipment power supplies and main buses are generally fused and provided with failure alarms. Emergency diesel-, gas-turbine- or gasoline-engine-driven generators are usually installed in the office to take up the essential load automatically when commercial power fails.

WIRE TRANSMISSION AND MULTIPLEXED (CARRIER) TELEPHONE

SYSTEMS

22. Transmission Factors in Switched Telephone Systems. The equipment which permits the transmission of a telephone message from one telephone set to any other on a switched network with satisfactory results is designed by reference to various transmission factors, of which the following are the most important:

1. The attenuation of the path in the voice-frequency band, 200 to at least 3,000 Hz, and the related factor of circuit linearity.

2. The relative phase change of the different frequencies in the speech band.

3. The amount of unwanted current or noise produced in the circuit, including cross talk from other telephone units.

4. The degree of impedance matching throughout the circuit. Serious impedance mismatches can cause amplitude-frequency distortion through reflection, and echoes.

Since in a switched telephone network it is not possible to adjust each circuit for each call, certain standards in design, initial adjustment, and circuit lineup must be followed to attain satisfactory performance. In telephone transmission large ratios of power occur. The *decibel* expresses this type of relationship in communications transmission. If two powers P_o and P_i are expressed in the same units, their ratio in decibels is defined by:

$$dB = 10 \log (P_o/P_i)$$

In the discussion of transmission parameters, the decibel is used frequently in several contexts.

A telephone transmission circuit can be considered as a complex tandem buildup of many two-port networks. The total gain or loss of a circuit made up of a tandem connection of such networks is the algebraic sum of the gains and losses of each of the component networks expressed in decibels.

The *volume unit* (vu) is another unit of measurement used in telephone engineering. Since speech is a complex, nonperiodic function, the average rms and peak values are irregular functions of time. Hence one number cannot characterize them. An arbitrary unit, the volume unit, has been devised to measure speech magnitude by the indications of a specific meter with specified damping. This device, known as a volume indicator, has a base-10 logarithmic scale, and its readings bear the same relationship to each other as decibel readings do. Relationships have been developed between the volume of a talker and his long-term average power and peak power.

23. Telephone-System Attenuation Factors. Many factors affect the efficiency of an interconnection, including the talker and listener's telephone habits, efficiency of the telephone set, the talker and listener's noise environment, noise and cross talk in the various interconnected circuits, etc. Statistical studies over many years have enabled certain assumptions to be made so that a uniform transmission plan approaching the ideal is possible.

The type 500 telephone set previously discussed can be assumed to have an average output of -15 vu. Required receiving level depends on many factors but should not be lower than -40 vu; excellent service requires -27 vu. The overall attenuation of the connecting circuit therefore need not be less than 12 dB and should not exceed 25 dB. Because of stability limits in two-wire voice connections, the circuit attenuation should not be much less than 12 dB.

The *via net loss (VNL)* (a concept more fully developed in the discussion of echo) is the lowest loss at which a circuit can be operated and remain stable. The overall loss objective is VNL + 4 dB, the additional 4 dB arising from 2 dB loss in each station or subscriber line. Trunks are designed to VNL, as are high-velocity facilities such as radio or cable multiplex.

Typical values for attenuation of components of the system designed under these criteria are:

```
Station lines ......................................2.5-8 dB
Trunks with two-wire terminations .................2.0-2.5 dB
Trunks with four-wire terminations ....................0 dB
Interoffice trunks (between class 5 offices) ..........5-7 dB
```

The attenuation (loss) values cited above are determined with a 1,000-Hz test tone.

The *attenuation-frequency characteristics* of a telephone circuit are usually specified with reference to the 1,000-Hz attenuation. From 500 to 2,500 Hz the attenuation should not vary more than -1 to 4 dB from the 1,000-Hz loss; at 200 and 3,300 Hz the loss should not be more than + 10 dB greater than the 1,000-Hz loss. The first figure defines the slope, while the latter defines the limiting frequencies in the passband.

Control of slope characteristics is important for good-quality voice transmission and even more important for low-error-rate data transmission. Overall slope objectives are usually assigned for a system. The contributions of the individual component circuits (mainly the trunk circuits) are related to the overall slope as the square root of the sum of the squares of the individual slopes. Station lines are usually assigned a 3-dB slope value. Slope equalization is usually provided on all trunks. Where severe conditions must be faced, additional equalization is provided, at the receive leg of the station.

24. Transmission Levels. Transmission level is defined as the ratio in decibels of the power of a signal at a given point to the power of the same signal at a reference point. The reference point is assigned a 0-dB transmission-level point; such a point was at one time accessible for measurements. In current practice, it is customary to consider the outgoing side of the toll transmitting switch as the -2-dB transmission-level point (-2 TLP). The test tone used for measurement is 1,000 Hz (unless otherwise stated) or the corresponding frequency obtained by modulation with 1,000 Hz in carrier systems.

The test-tone power at 0 TLP is defined as 1 mW and is abbreviated as 0-dBm (0 dB with reference to 1 mW). At the -2 TLP, -2 dBm in test-tone power is present.

Besides the -2 TLP, two other important transmission-level points are the input and output TLP of four-wire transmission systems. In the United States these are designed and operated at -16 and +7 dB, respectively. Adjustments to obtain these levels are made by inserting attenuator pads.

Standardization of levels is necessary for proper administration and operation of the telephone network, particularly with reference to interchange of multiplex systems in emergencies. Figure 22-28 illustrates the TLPs in a typical four-wire trunk circuit with four-wire switching operated at 0 dB loss. Figure 22-29 shows the TLPs in a typical four-wire trunk with two-wire terminations and switching operating at 2 dB loss. In this case the hybrid terminating loss is taken as 4 dB and is accounted for in the figures −6, −2 or 0, −4. The 2-dB loss is assumed to be the approximate VNL for this circuit.

In private-wire or direct circuits, where no public switching network is involved, as in many data installations, the reference at the transmitter (the user's station) is considered the 0 TLP. Levels must be adjusted to obtain −16 TLP at the input to the four-wire transmission facility in this case.

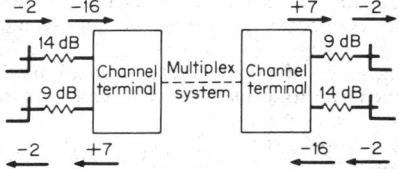

Fig. 22-28. A typical 4-wire zero-loss trunk.

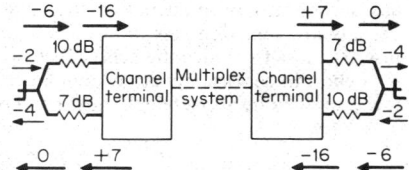

Fig. 22-29. A typical 2-wire trunk with loss.

25. Noise. Noise, in the most general sense, is any signal present in the communication channel other than the desired signal. Two specific types of noise are significant in telephone practice. The first is *circuit noise* (sometimes called *white noise*), which consists of thermal noise in the circuit, intermodulation noise, and cross talk. This type of noise is relatively continuous. The second type is *impulse noise*. It occurs at random intervals, is of high amplitude, and usually of short duration.

Circuit noise is measured with a weighted (noise) meter such as the Western Electric Co. type 3A. The meter readings are indicated in decibels above reference noise (dBrn). For telephone circuit measurement, a C-message weighting network typical of the acoustic properties of the type 500 telephone set is included in the meter circuit to measure the disturbing effect as it would appear to the telephone user. When this C-message weighting is used, the readings are specified as dBrnc, and 0 dBrn = -90 dBm.

In addition to the C-weighting network, a 3-kHz flat-weighting network is available which can be used to measure the actual power density of the circuit noise. Noise can be measured at any TLP and referred to 0 TLP by subtracting the level from the meter reading. A noise-meter reading taken at 0 TLP is abbreviated dBrn0. Similarly readings in dBrnc, referred to 0 TLP, are abbreviated as dBrnc0.

Circuit-noise requirements in switched systems are stated by reference to the 0 TLP. Consider the trunk circuit of Fig. 22-29. The noise measured at -6 TLP (35 dBrnc) would be translated to 41 dBrnc0 by adding TLP equivalent of 6 dB to the reading in order to translate to the 0 TLP.

Circuit noise is higher on longer circuits and adds on a power basis in built-up connections. That is, if a circuit is rated at 41 dBrnc0, two such circuits connected in tandem would have an overall rating 3 dB greater, or 44 dBrnc0. Practical noise objectives range from 31 dBrnc for shorter circuits to 48 dBrnc0 for circuits up to 6,000 mi long. Noise on circuits tends to increase with age of equipment, and requirements are usually stated for busy-hour traffic conditions when cross talk, supervisory-tone noise, and switching noise are highest.

Impulse noise, consisting of random pulses of noise at high amplitude, usually does not affect voice communications seriously, but it can cause serious trouble with other forms of communications, particularly data systems (where it can obliterate or insert bits causing serious errors in the information stream). Impulse noise is usually measured by a meter circuit arranged to operate a counter whenever a signal exceeds a predetermined level. Impulse-noise requirements are imposed primarily for data transmission (see Pars. **22-44** to **22-56**).

26. Echo and Losses. Echo occurs when a signal meets an impedance mismatch and is reflected back toward the source, the reflected signal being an echo of the desired signal. The difference in decibels between the transmitted-signal level and the level of the echo at any point in the circuit is the *echo return loss* at that point. This return loss includes all gains and losses between the echo-reflection point and the point of measurement.

Sources of echo are circuit terminations, line irregularities, and junctions between dissimilar facilities, particularly the hybrid termination of four-wire transmission to two-wire switching facilities. Echo paths in a typical switched connection with two station lines and two trunks with four-wire facilities terminating on two-wire switches are shown in Fig. 22-30. The talker echoes labeled *T* are those reflected toward the talker; listener echoes labeled *R* are those reflected toward the listener. It will be noted that both types occur in such a connection, and if the sum of the gains exceeds the sum of the losses including return loss in any section at any frequency in the passband, singing can occur. High echo return loss permits low insertion loss in circuits, and conversely poor echo return loss requires high insertion loss to reduce the echo effect.

Echo can seriously affect telephone communication. If the echo delay is small, it will increase sidetone at the telephone set and cause the talker to speak at lower volume. As the

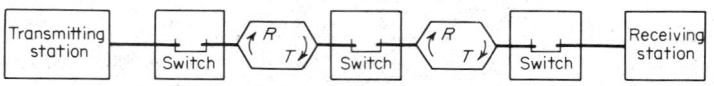

Fig. 22-30. Some echo paths in a typical connection.

delay increases, it appears to the talker as an acoustic echo, disconcerting to the speaker. In four-wire systems echo is not a problem, except at the terminating telephone sets in voice communications, where the sidetone circuit provides a return path.

The disturbing effect of echo is related to the delay in its receipt and to its magnitude. The relationship of the round-trip delay in milliseconds to the minimum echo path loss in decibels which will be tolerated by the average user is shown in Table 22-3. In a switched communication system it is uneconomical to design two-to-four-wire terminations to achieve the high return losses required to avoid echo on the longer circuits and avoid impedance mismatches. Accordingly losses are intentionally inserted in the various circuits to increase the echo path loss without causing the signal to be received at unacceptably low levels.

Table 22-3. User Tolerance to Talker Echo
(Loss Required in the Round-Trip Echo Path to Satisfy the Average User with Respect to Round-Trip Delay)

Round-Trip Delay, ms	Average Tolerance to Loss in Echo Path, dB
0	1.4
20	11.1
40	17.7
60	22.7
80	27.2
100	30.9

The assignment of losses has been discussed in Par. 22-23. In trunk circuits the via net loss is related to the length of the circuit and the velocity of propagation of the trunk facility, which together produce the echo delay time on the circuit in question. The relationship is usually given in terms of length of circuit and the via net loss factor (VNLF) as follows:

$$\text{VNL} = \text{VNLF} \times \text{one-way distance} + 0.4 \text{ dB}$$

where

$$\text{VNLF} = 2 \times 0.102/\text{velocity of propagation}$$

For most multiplex facilities VNLF = 0.0015 dB/mi.

Where long delay paths are involved, as in satellite communications, and the calculated VNL is excessive (2.6 dB max is the limit used in Bell System), *echo suppressors* may be installed in the four-wire portion of a circuit. They serve to operate switchable pads to increase the loss in the return path upon detecting a signal in a given direction. They must react fast on operate and release, to prevent loss of any portion of speech in either direction, and they must not chatter on impulse noise.

27. Transmission-Line Parameters. The transmission lines of primary interest in communications are twisted-pair cables, coaxial cable, and (to a small extent today) open-wire lines. The transmission characteristics of these lines are determined by such properties as the conductivity, diameter, and spacing of conductors and the dielectric constant of the insulation. These properties in turn determine the constants $R, L, G,$ and C, which are the uniformly distributed series resistance, series inductance, shunt conductance, and shunt capacitance of the line, in ohms, henrys, mhos, and farads per mile, respectively. The transmission characteristics or secondary constants can be calculated from these constants using the following relations:

$$\text{Characteristic impedance} = Z_0 = \sqrt{\frac{R + j\omega L}{G + j\omega C}} \quad \Omega$$

where $\omega = 2\pi f$ and f is the frequency in hertz. Characteristic impedance is a complex quantity, independent of length, and expressed in ohms.

$$\text{Propagation constant} = \gamma = \alpha + j\beta = \sqrt{(R + j\omega L)(G + j\omega C)}$$

$$\text{Attenuation} = \alpha = \sqrt{1/2} \sqrt{(R^2 + \omega^2 L^2)(G^2 + \omega^2 C^2)} + RG + \omega^2 LC$$

Attenuation is in nepers (Np) per mile, converted to decibels by multiplying by 8.686.

$$\text{Phase constant} = \beta = \sqrt{\tfrac{1}{2}}\sqrt{(R^2 + \omega^2 L^2)(G^2 + \omega^2 C^2)} - RG + \omega^2 LC \qquad \text{rad/mi}$$

$$\text{Velocity of propagation} = V_P = \frac{\omega}{\beta} \qquad \text{in mi/s}$$

28. Open-Wire Transmission Lines. Open wire refers to that form of line construction in which bare conductors are supported on insulator-equipped crossarms mounted on poles at uniform intervals, typically 130 ft apart. The conductors are usually hard-drawn copper or copper-clad steel of 104, 128, or 165 mils diameter. Two typical arrangements for mounting the pairs of conductors on crossarms are shown in Fig. 22-31.

Open-wire lines have the advantage of substantially lower transmission loss than cable but the disadvantage of considerable variation in loss due to temperature changes, wet weather, snow and ice, and susceptibility to all manner of electrical interference from other communication lines, power lines, lightning, and many other man-made and natural causes. Today, open wire is rarely installed except on light routes (10 circuits or less), but a large number of such facilities are still in use. To reduce cross talk and other interference, transpositions of the wire pairs have been worked out in standard patterns depending on the type of transmission facility used.

29. Twisted-Pair Cable. Two insulated conductors of high-purity copper twisted together constitute a *cable pair*. The insulation may be either wood pulp formed on the conductors in a process similar to papermaking or plastic (usually polyethylene) formed by extrusion. Conductor sizes are generally 19-, 22-, 24-, and 26-gauge, and in rare instances 16-gauge. The pairs are stranded in *units*, neighboring pairs being twisted with different pitch to cut down cross talk. Unit sizes are 6 to 50 pairs for polyethylene-insulated cable (PIC) and 25 to 100 pairs for pulp-insulated cables. *Cores* consist of one or more units and range in size from 6 to 900 pairs for PIC and 300 to 2,700 pairs for pulp-insulated cables. The core is covered by a moistureproof sheath whose composition depends upon conditions of use.

The inductance L of such cable is of the order of 1 mH/mi for low frequencies, decreases to about 70% of this value as the frequency increases from 50 kHz to 1 MHz, and remains level at higher frequencies. The capacitance C generally has either of two standard values, 0.066 and 0.083 μF/mi, each relatively independent of frequency. The conductance G is very small for PIC and roughly proportional to frequency for pulp insulation. The resistance R is approximately constant over the voice band and proportional to the square root of frequency at higher frequencies.

Simplified equations for attenuation constant and phase constant, at voice frequencies, where ω is large in relation to G/C but is small compared to R/L, are to a close approximation

$$\alpha \cong \frac{R}{2}\sqrt{\frac{C}{L}} \qquad \text{Np/mi}$$

$$\beta \cong \omega \sqrt{LC} \qquad \text{rad/mi}$$

At higher frequencies (multiplex and video frequencies), where ω is large compared to R/L and G/C and R/L is much larger than G/C, the following approximate expressions, which also apply to coaxial cable, may be used:

$$\alpha \cong \sqrt{\frac{\omega RC}{2}} \qquad \text{Np/mi}$$

$$\beta \cong \sqrt{\frac{\omega RC}{2}} \qquad \text{rad/mi}$$

Typical attenuation-frequency characteristics for 19-gauge cable pairs up to 150 kHz at three temperatures are shown in Fig. 22-32. It should be noted that the slope is relatively constant from 20 to 150 kHz.

Shielded pairs consist of two balanced 16-gauge conductors, insulated with expanded polyethylene, twisted together, and surrounded by a longitudinal-seam copper shield. They are sometimes included in multipair cables for use in transmitting signals at frequencies of the order of 1 MHz or higher, particularly television signals. The balance and the shielding are needed for interference reduction at the lower frequencies of the transmitted band. The

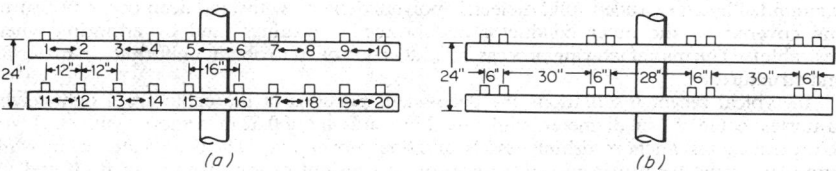

Fig. 22-31. Open-wire pole-line arrangements: (*a*) for voice frequency; (*b*) for multiplexed circuits.

heavy-gauge conductor and the low capacitance result in relatively low loss (-18 dB/mi at 4.5 MHz, the maximum frequency of the television signal).

30. Loading. Study of the attenuation of cable pairs based on typical values of the primary constants (Par. 22-27) shows that a substantial reduction of attenuation can be obtained by increasing the value of L. Attaining the desired value on a uniformly distributed basis has been found to be impractical, but a substantial improvement is obtained by adding series inductances, called *loading coils*, at regular intervals in the circuit pair. The loading arrangement is described by a code letter designating the distance between loading/coil insertions and a number giving the value of the loading in millihenrys. Loading arrangements in use are H-44, H-88, D-88, and B-135, where H designates 6,000-ft, D 4,500-ft, and B 3,500-ft spacings. Terminations of loading systems are normally either at midcoil or midsection.

A loaded circuit approximates a low-pass filter. The impedance, attenuation, and cutoff frequency of a loaded circuit can be approximated from the following formulas:

$$\text{Impedance} = L/C \qquad \Omega/\text{mi}$$

$$\text{Attenuation} = \frac{R}{2}\sqrt{\frac{C}{L}} + \frac{G}{2}\sqrt{\frac{L}{C}} \qquad \text{Np/mi}$$

$$\text{Cutoff frequency} = f_c = \frac{1}{2\pi\sqrt{LC}} \qquad \text{Hz}$$

The improvement in voice-frequency transmission achieved by loading is indicated by the fact that the 1,000-Hz attenuation of 19-gauge unloaded cable is 1.27 dB/mi while with H-88 loading the attenuation is reduced to 0.42 dB/mi.

31. Coaxial Cable. A coaxial cable consists of a center conductor surrounded by a concentric outer conductor and separated from it by a dielectric, preferably dry air. In commercial-telephone long-haul practice the center conductor is 10-gauge copper supported by insulating polyethylene disks at 1-in. intervals, and the outer conductor has a diameter of 0.375 in. The shielding afforded by the outer conductor is particularly effective at the frequencies above the voice range (the usual operating range) but is poor in the range of voice frequencies, since the depth of current paths (skin depth) becomes comparable to conductor thickness at low frequencies.

Cables comprise as many as 20 coaxial conductors combined with groups of twisted pairs which are used to pass control and alarm signals to and from remote repeaters in the circuits. The makeup of a coaxial conductor and a cable with eight coaxial conductors is sketched in Fig. 22-33. The consistent mechanical structure and the shielding qualities of the coaxial cable give better control and less frequency dependence in the primary constants than are available in other types of cable in the normal operating ranges.

The capacitance C and inductance L are practically independent of frequency in this operating range; conductance G for the air dielectric is negligible; while resistance R due to skin effect increases as the square root of frequency in the usual operating range. The expressions in the equations on page 22-25 give very close approximations for attenuation and phase constant for this type of cable in the usual operating range. As a checkpoint the attenuation of the 0.375-in. coaxial conductor at 1 MHz is approximately 4 dB/mi.

Submarine cable. Transoceanic communications by coaxial cable place severe requirements for high reliability in a demanding service. In this case, the central conductor is surrounded by an extruded solid dielectric polyethylene, to withstand deep-ocean pressures, and covered by the outer conductor and protective coatings (and by armoring where desirable). The manufacturing process is rigidly controlled to obtain uniformity in materials and structure.

In typical recent installations the deep-water portions of the cable are a lightweight, armorless cable of 1 in. diameter, while the shore ends are of 0.62 in. armored with steel wire to protect against anchors, fishing vessels, and tidal movement. The deep-water cable, while armorless, must be strong enough to support the weight of long lengths of itself and the submarine repeaters which are inserted in the circuit at regular intervals (see Par. 22-38).

32. Voice-Frequency Transmission Devices. Several devices are available to improve voice-frequency transmission over wire and cable. These include negative-impedance repeaters, repeater amplifiers, and 2- to 4-wire hybrid terminating circuits, used with amplifier circuits.

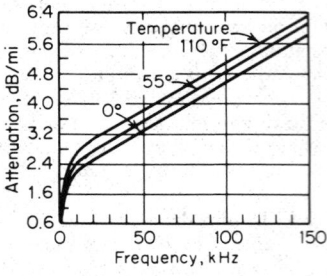

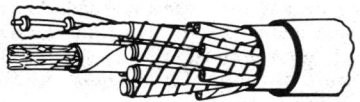

Fig. 22-33. An assembly of eight coaxial cables.

Fig. 22-32. Typical attenuation of 19-gauge cable pairs.

The *negative-impedance repeater* is the gain device most often used in 2-wire circuits at voice frequencies. Such a device is shown in Fig. 22-34. The arrangement shown is a combination of two such devices: a series converter, which inserts series negative impedance in the line, and a shunt converter, which inserts negative impedance across the line. The former is open-circuit-stable while the latter is short-circuit-stable.

Either can be used by itself to provide bilateral gain, but the combination gives better performance and stability with proper network selection. In each case the converter has a controlled-feedback transistor-amplifier circuit which in combination with the associated network provides the necessary frequency phase control to give the desired degree of negative impedance and resultant gain. Gains of up to 10 dB are usual. The negative-impedance repeater offers a through path for dc supervisory and signaling currents.

The *hybrid-coil 2- to 4-wire terminating set* is used wherever a 4-wire amplified circuit must be transformed into a 2-wire circuit, as at a 2-wire switching center. Such a device is shown in Fig. 22-35. This hybrid is essentially a bridge circuit employing two transformers having turns ratios such that the impedance of the network (NET) and the 2-wire circuit match over the frequency range as closely as possible. When power enters the circuit from the receive amplifier, it divides between the network and the 2-wire circuit equally (if they are in balance) and no power enters the transmitting amplifier. If the balance is disturbed, some of the power enters the transmit amplifier and is reflected back to the source as echo (Par. 22-26). The ratio in decibels between received and reflected power is known as the return loss of the hybrid circuit. It is a function of the impedance match between the 2-wire circuit and the network at all frequencies of interest and expressed as

$$\text{Return loss} = 20 \log_{10} \frac{Z_n + Z_L}{Z_n - Z_L}$$

where Z_n is the network impedance and Z_L is the 2-wire circuit impedance.

Power entering the hybrid from the 2-wire circuit is divided equally between the transmit and receive legs. The portion in the receive leg is dissipated, while the portion in the transmit leg is transmitted. The transmission loss in either direction can never be less than 3 dB plus the coil losses. The network never is an exact reproduction of the 2-wire circuit and often is required to match a variety of such circuits, as in a switching center. Accordingly it is often a compromise network. In 2-wire central offices a 900-Ω network is often shunted by appropriate values of capacitors to provide a good match with the office wiring.

The *type 22 repeater* shown in Fig. 22-36 is an arrangement for utilizing amplifier-type repeaters to achieve gains in 2-wire voice-frequency circuits. It consists of two amplifiers connected to the 2-wire circuit by two hybrid coils. The gain is limited by the return loss of the hybrid network at all frequencies within the voice-frequency range transmitted. A precision network is used to match the line, but compromise networks are used at switching-center terminal ends.

When the gain exceeds the return loss around the circuit, oscillation (singing) results. Echo considerations limit the number of such repeaters used in a circuit. The maximum gain is limited to about 12 dB by cable-characteristic changes (due to temperature variations) termination variations, and switching and cross-talk considerations. Since the repeater amplifier passes only signals in the range 200 to 3,500 Hz, there is the added disadvantage that some other means must be found for transmitting dc and low-frequency ringing and supervisory signals if required.

The *type 24 and 44 repeaters* illustrated in Fig. 22-37 show the application of the voice-frequency amplifier repeater to *4-wire circuits*. The type 24 repeater is the terminal repeater; it includes a hybrid coil to connect with the 2-wire circuit and equalization in the receive-amplifier circuit to reshape the attenuation-frequency characteristics of the signal.

The type 44 repeater is the intermediate repeater; it includes equalization in both amplifier circuits.

In 4-wire operation, the singing path is extended around the full circuit, and the number of echo reflection points is reduced. Hence such repeaters can be operated at higher gains,

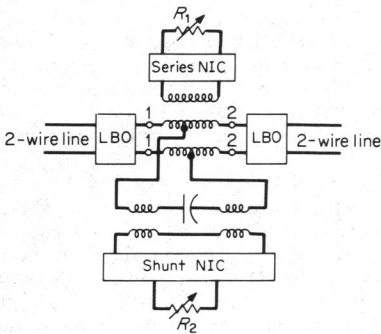

Fig. 22-34. Negative-impedance repeater.

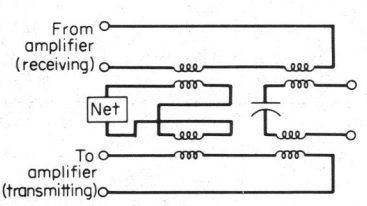

Fig. 22-35. The typical hybrid coil circuit.

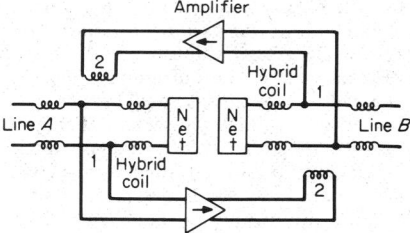

Fig. 22-36. Type 22 repeater.

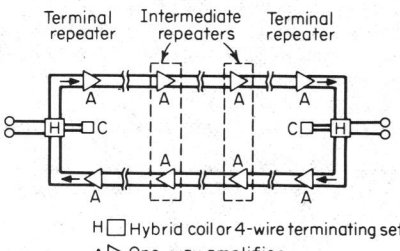

H ☐ Hybrid coil or 4-wire terminating set
A ▷ One-way amplifier
C ☐ Compromise balancing network

Fig. 22-37. Typical 4-wire repeatered line.

particularly at intermediate points, and more repeaters and longer circuits may be used. The 4-wire circuit or its equivalent is used exclusively in multiplexed systems.

The design of the *station line* or *loop* is generally done on a resistance basis and depends on the central-office equipment and types of telephone sets. With the type 500 telephone set, a limit of 1,200 Ω for the loop resistance without use of additional facilities has been established. This corresponds to about 15,000 ft of 16-gauge cable. Where greater line length is required, loading can be added to give improved transmission and a higher voltage battery provided for supervisory current.

The operating telephone companies have evolved a set of rules governing the loop limit and facilities to be used for extended ranges, including heavier-gauge cable, loading, negative-impedance repeaters, etc. In the Bell System this is known as the unigauge concept. In Par. 22-23 it is indicated that station lines are usually engineered to have a 1,000-Hz attenuation of 2.5 to 8 dB, but the dc resistance must also be low enough to give satisfactory supervisory operation. Table 22-4 gives the dc resistance and attenuation at 1,000 Hz for common nonloaded and loaded cable facilities. H-88 is the usual loading for circuits used for telephone communication.

Interoffice or exchange area trunks between class 5 offices are usually engineered for 5 to 7 dB attenuation; use of nonloaded, loaded, or repeatered cable facilities as the situation demands is usual.

33. Frequency-Division Multiplexing (FDM). In frequency-division multiplexing each channel of a system is assigned a discrete portion of the transmitted-frequency spectrum so that many channels can be transmitted over a single transmission medium. The translation of the individual voice circuits to their assigned portion of the spectrum is accomplished by modulating a carrier frequency and is illustrated in principle in Fig. 22-38. The process was originally termed *carrier transmission,* but the term *frequency-division multiplexing* is a more accurate description, distinguishing it from the process of time-division multiplexing (TDM).

Amplitude modulation of the carrier frequency is the process commonly employed in telephone communication systems. A discussion of this subject appears in Sec. 14. In the modulating process two sidebands, one on either side of the carrier frequency, are formed. When both sidebands and the carrier are transmitted, it is called *double-sideband transmitted-carrier* (DSBTC) multiplexing. This multiplexing process results in simpler transmitting and receiving equipment but requires twice as much frequency space for the same number of channels. Double-sideband suppressed-carrier (DSBSC) modulation is sometimes employed but requires reinsertion of the carrier at precisely the original carrier frequency if distortion is to be avoided in demodulation.

In the most widely used form of FDM, one sideband and the carrier frequency are removed at each modulation step. This is termed *single-sideband suppressed-carrier* (SSBSC) multiplexing and is illustrated by Fig. 22-39. The initial carrier frequency is 12/kHz, and the lower sideband is selected for transmission. In most SSBSC systems the carrier frequencies are spaced 4kHz, for the initial voice-channel formation. Due to the filtering process required the voice channel itself usually has a passband width of 3,100 Hz, with attenuation-frequency characteristics as shown in Fig. 22-40. This form of FDM multiplexing uses the frequency spectrum more economically and requires less power per channel since the carrier is not transmitted. It has the disadvantage that the carrier must be reinserted at the receiver at a precise value (within a few hertz of the original carrier) for reasonably accurate reproduction of voice and data.

Table 22-4. Attenuation and DC Resistance for Nonloaded (NL) and Loaded Cable Pairs

Cable gauge and loading	Attenuation at 1,000 Hz, dB/mi	Dc resistance, Ω/mi
26 NL	2.8	440
26 H88	1.8	448
24 NL	2.3	274
24 H88	1.2	282
22 NL	1.8	172
22 H88	0.8	180
19 NL	1.3	85
19 H88	0.42	93

Another modulation process sometimes employed in FDM systems is called *twin-channel,* in which the same carrier frequency is employed to form two voice channels; the lower sideband is transmitted for one channel, and the upper sideband is transmitted for the second channel. This in effect yields similar characteristics as SSBSC for each channel.

The SSBSC form of FDM is widely used throughout the world, particularly in high-traffic-density routes. A 12-channel group occupying the frequency range of 60 to 108 kHz has been accepted as standard. The composite signal in this basic group is further modulated so that five groups are combined into a supergroup of 60-channel capacity, and so on, as shown in Fig. 22-45.

34. Factors in FDM System Design. An FDM system is a 4-wire system with transmitter-receiver terminals and 4-wire intermediate repeaters much like the repeater system shown in Fig. 22-37 except that the modulation-demodulation function of the transmitter-receiver terminal must be added and the amplification requires broadband repeaters.

An alternative is an equivalent 4-wire system in which the *A-B* direction is at one (low) frequency range and the *B-A* direction at another (high) frequency range. Directional filters are used to separate the bands at terminal and repeater points. In most cable systems, power for the repeaters is transmitted over the signal wires to the repeater power supply and also transmitted around the repeater to the next repeater.

Repeater spacing is governed by line attenuation, particularly that at the high-frequency end of the transmitted band, and by noise and cross talk. Equalization is provided at terminal and repeater points so that the overall attenuation over the frequency band is relatively flat. Where the facility is subject to wide changes in attenuation due to weather conditions, such as open wire and aerial cable, automatic regulation of amplifier gain is provided by pilot signals within the transmitted frequency band or at either end.

Frequency frogging is a technique used when the two directions of transmission use two different frequency allocations, and the allocation is changed at each repeater point by modulation at that point, as shown in Fig. 22-41. Thus the low group in one repeater section

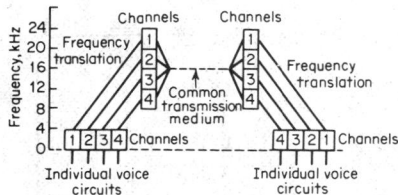

Fig. 22-38. In frequency-division multiplexing, each circuit is translated to a separate position in the frequency spectrum before being applied to a common transmission medium.

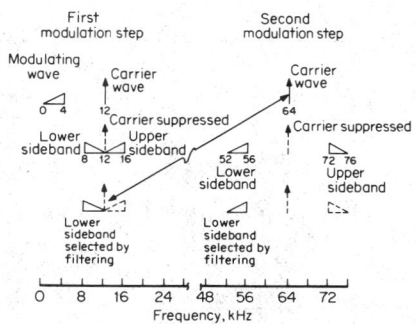

Fig. 22-39. Single-sideband suppressed-carrier multiplexing.

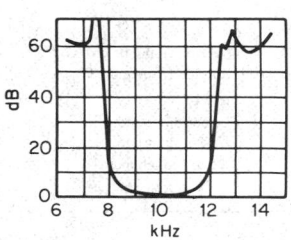

Fig. 22-40. Typical attenuation-frequency characteristic of a voice channel in a multiplex system.

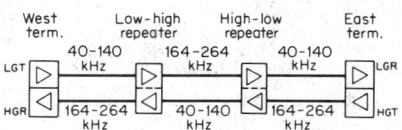

Fig. 22-41. Carrier-frequency bands interchanged by frequency frogging at repeater stations.

becomes the high group in the next repeater section and vice versa. This results in nearly constant attenuation in the two directions of transmission and minimizes singing and cross-talk problems; i.e., the high-level output of a repeater is at a different frequency from the low-level input to other repeaters.

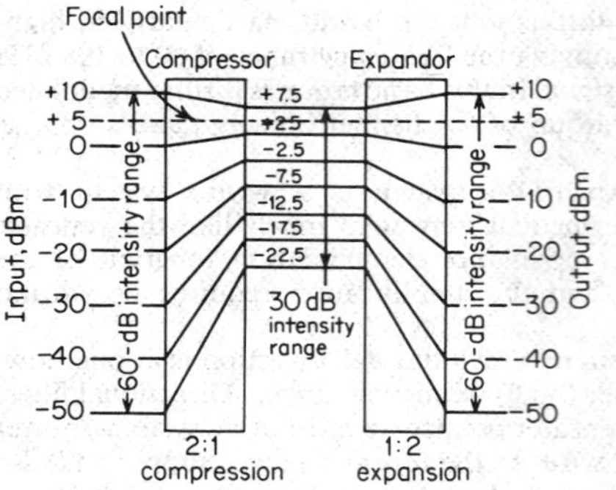

The *compandor* is a device for obtaining an effective improvement in signal-to-noise ratio over a voice circuit permitting operation over noisier circuits than would otherwise be possible. It consists of a volume compressor at the transmitter end of a multiplex channel and volume expander at the receiver end. By imparting more gain to low-intensity signals than to high-intensity signals, the compressor reduces the dynamic range of the signals, as shown in Fig. 22-42. It permits greater repeater spacing in multiplex systems and increases system loading. It is not effective on data circuits.

Fig. 22-42. The effect of a typical compandor on various power levels. *(Lenkurt Electric Co., Inc.)*

System design has been influenced greatly by the proposed field of use. In general, systems designed for long-haul use (over 250 mi in the Bell System) use elaborate terminal arrangements to place as many channels as possible on the line facility. In short-haul systems, the cost of the terminal is kept as low as possible even at the expense of less efficient use of the line facility. DSBTC modulation, compandors, and frequency frogging are often used for short-haul systems.

35. FDM Systems for Open-Wire Lines. Representative FDM systems for open-wire facilities are shown in Fig. 22-43. The Western Electric J system is the original long-haul

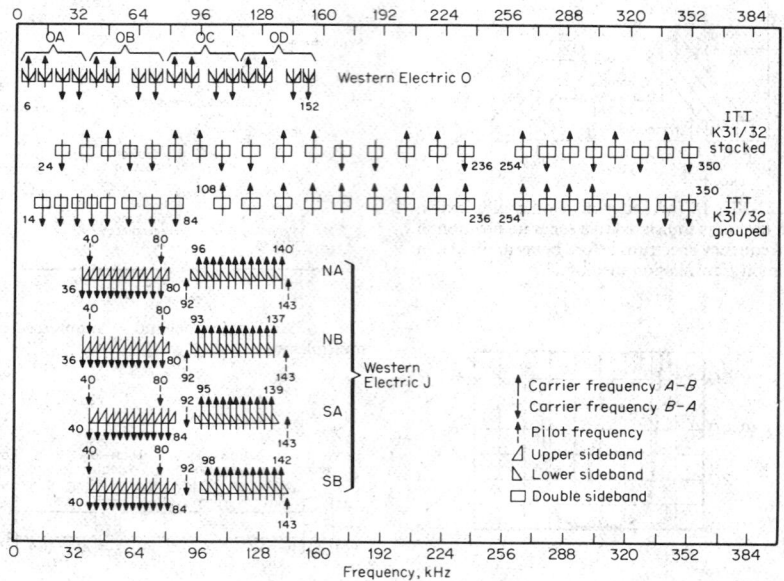

Fig. 22-43. Open-wire carrier systems (4-wire); the numbers are in kilohertz. *(From "Reference Data for Radio Engineers," 5th ed., ITT and Sams, Indianapolis, Ind., 1968.)*

carrier facility. It utilizes SSBSC modulation and has 12-channel capacity on one open-wire lead with different frequency bands in each direction. The other two systems are primarily for short-haul, slowly growing routes with channel add-on capacity as required. The Western Electric O system consists of three add-on units of four-channel capacity each, utilizing twin-channel modulation on an open-wire lead. The ITT K31/32 system has several arrangements for grouping and stacking channels up to a capacity of 12 channels, using DSBTC modulation on an open-wire lead.

36. FDM Systems for Multipair Cable Facilities. Representative FDM systems for use on multipair cable facilities are shown in the frequency-allocation diagram of Fig. 22-44. These systems are intended primarily for short-haul traffic, up to 250 mi. The Western Electric N system provides 12 two-way voice channels on two cable pairs using different frequency bands for each direction of transmission. Frequency frogging is used at repeater points. The repeater spacing is about 6 mi on 19-gauge cable. The Lenkurt LN system is similar. The Western Electric ON system differs in that twin-sideband modulation is used to obtain 24 two-way voice channels; it can be used on somewhat longer systems. The Lenkurt 45 BN system has similar capacity and use except that straight SSBSC modulation is used. The Lenkurt 81A and ITT-Kellogg K24A are two examples of short-haul 24-channel systems using DSBTC and DSBSC modulation, respectively, with transmission in the two directions in approximately the same frequency range except that the channel carriers are displaced with respect to each other, thus reducing intelligible cross talk between systems. The Lenkurt X and Lynch B630 use FM techniques to form the channels. Some advantage in signal-to-noise ratio is obtained from FM at the expense of greater channel bandwidth, but the technique is little used for FDM systems.

37. FDM Systems for Coaxial Cable and Microwave Radio Facilities. The modulation plan for deriving systems suitable for high-density, long-haul circuits is illustrated in Fig. 22-45. SSBSC modulation is used throughout. The first step is the formation of the basic 12-voice-channel group in the frequency range 60 to 108 kHz. This group is used in the Bell

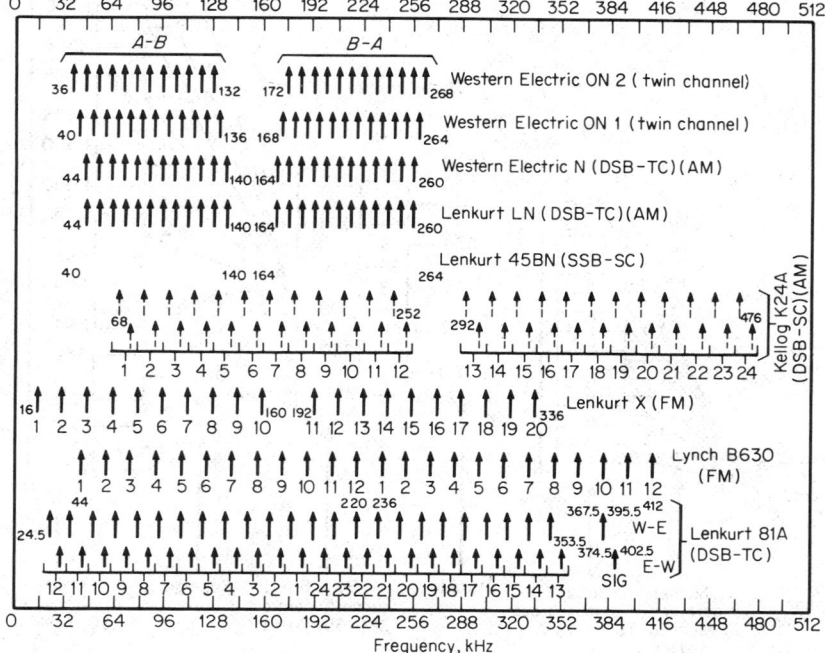

Fig. 22-44. Frequency allocations of several typical cable multiplex systems. *(Lenkurt Electric Co., Inc.)*

System type J open-wire system previously discussed and in the Bell System type K system transmitting in the band 12 to 60 kHz on two cable pairs (the original cable system).

Five of the basic groups can then be modulated to form a 60-channel supergroup occupying the band 312 to 552 kHz. In the next step, up to 10 supergroups are modulated to form a system bank of 600 channels in the band 60 to 2,540 kHz (in the Lenkurt 46A system) suitable for radio relay or coaxial-cable transmission. In the Bell system this grouping of 10 supergroups is called a master group. There are two types. The L-600 is similar to that shown in the figure except that supergroups SGR-9 and SGR-10 are 120 and 240 kHz higher in frequency than those shown in the figure, giving a band of 60 to 2,788 kHz.

This mastergroup forms the signal used on the L1 coaxial and TD-2, TJ, TL, and TM radio relay systems. In the other mastergroup arrangement, U-600 occupies the band 564 to 3,084 kHz and is used to form higher-capacity systems. Three such master groups and a supergroup form the signal for L-3 coaxial and TH radio relay systems of 1,860-channel capacity and six such master groups with capacity of 3,600 channels form the L-4 coaxial system.

Any number of systems with channel capacities from 12 to 3,600 channels can be formed from these basic groups and supergroups for use on coaxial cable and radio relay facilities.

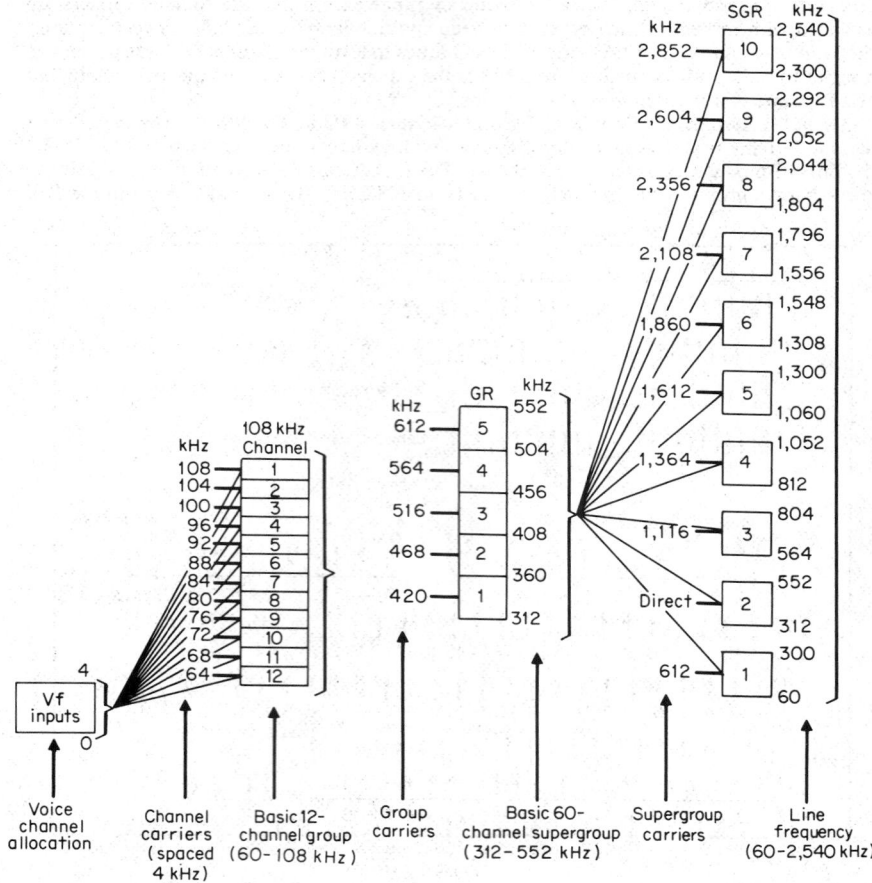

Fig. 22-45. Three-stage modulation plan for Lenkurt 46A multiplex system. *(Lenkurt Electric Co., Inc.)*

The Lenkurt 46A system, for instance, is available in 12-, 60-, 120-, and 600-channel capacity. The basic group and supergroup arrangements of Fig. 22-45 are also CCIT standards, used by European administrations.

Wideband facilities can be introduced at the group, supergroup, and master-group levels to replace the voice channels with data or television signals.

38. Submarine Cable Systems. Submarine cable systems for transmitting speech are FDM systems with shore-based terminals having special power feeding and supervisory facilities and submerged repeaters inserted in specially designed submarine coaxial cable (Par. 22-31). Such systems employ specially designed FD multiplexers with 3-kHz carrier spacing, which gives performance almost equivalent to the usual 4-kHz system. Equivalent 4-wire transmission is employed, i.e. the low-frequency band for *A-B* direction and high-frequency band for *B-A* direction. The capacity and repeater spacing of two representative long-haul systems are (1) 80-channels with 29.5 nautical mile repeater spacing in bands 60 to 300 and 360 to 609 kHz, (2) 360 channels with 9.5 nautical mile repeater spacing transmitting in bands 312 to 1,428 and 1,848 to 2,964 kHz. Both employ 1.00-in. lightweight cable.

The *submerged repeater* is shown in Fig. 22-46. Directions of transmission are separated on a frequency basis by directional filters; i.e., one amplifier is employed for both directions

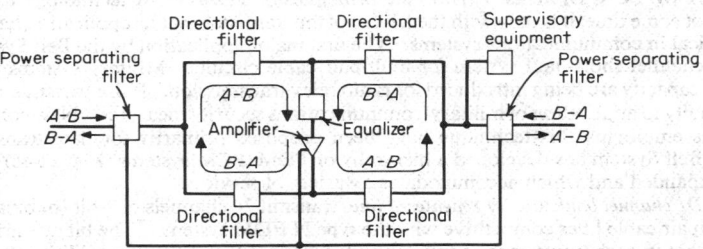

Fig. 22-46. Block schematic of a two-way submarine repeater. *(From "Reference Data for Radio Engineers," 5th ed., ITT and Sams, Indianapolis, Ind., 1968.)*

of transmission. Construction is such that maximum electrical and mechanical reliability is assured. All components are enclosed in high-tensile-steel housing with pressure-resistant seals and suitable glands for cable entry. On long systems a shipboard-adjustable equalizer is inserted at intervals of every 10 or 12 repeaters to adjust for differences between cable characteristics and repeater-equalization characteristics.

Power feeding is generally done from both ends on a constant-current basis. On the busier cables *time-assignment speech interpolation* (TASI) is used, whereby talkers seize a circuit when they begin to speak and relinquish it when silent. TASI approximately doubles circuit capacity for telephone use.

39. Time-Division Multiplexing (TDM). In time-division multiplexing message information from many channels is sampled briefly in time sequence, the sample from each channel being placed in its assigned time slot and transmitted over the medium. At the receiver the samples are separated and the original message information reconstructed. This implies (1) that each channel is sampled often enough to ensure that no significant information is lost in the construction and (2) that the frequency of sampling is fast enough for all channels to be sampled in turn in each sampling cycle. The message sample is a pulse or train of pulses modulated in amplitude, width, or position by the message information. These modulated pulses may be further processed by being pulse-code-modulated. Pulse-modulation techniques are discussed in Sec. 14.

The modulation technique most generally used for TDM systems in telephone communications networks is pulse-code modulation (PCM). The fully descriptive title for these systems would be TDM/PCM systems, but they are generally referred to as PCM systems.

In voice-frequency applications the sampling rate is generally 8,000 per second ($2 \times 4,000$, where 4,000 Hz is the highest frequency in the channel band). A seven-digit code is used in Bell System PCM systems. This gives 128 discrete amplitude levels for describing the sample amplitude. The difference (less than 0.8%) between the reconstructed signal and original signal produces quantizing noise; it is not very significant with a seven-digit code. The process is illustrated in Fig. 22-47 for four message channels showing pulse-amplitude sampling, interleaving of PAM samples, and conversion to PCM in a coding process with a seven-digit code.

The interval between samples for each channel is $T = 1/8,000$. With N channels to be multiplexed, the sampling time at each channel must be something less than T/N. As the number of channels increases, the number of pulses in a given time interval and the bandwidth for transmission also increase.

From sampling theory we can deduce that if F is the highest frequency per message channel to be transmitted and n the number of pulses per code group describing the message sample, then the minimum bandwidth required for the system is nF Hz. This is 7 to 9 times the bandwidth for a SSBSC/FDM system, but the increased bandwidth is offset by the noise advantage resulting from regenerated and essentially noise-free pulses at each repeater.

Instantaneous companding is a process used to improve the signal-to-distortion ratio in speech circuits reducing quantizing noise. It takes advantage of the fact that in typical speech waveforms, smaller amplitudes are much more prevalent than large ones. Each PAM sample is subject to instantaneous compressor action, and the coder codes the resulting compressed amplitudes in uniform steps. The reverse occurs at the receiver by decoding expander action.

40. TDM/PCM Systems. While the principles of TDM/PCM technology have been known for some time, it is only with the advent of the transistor that its application has proved economical in communications systems. The first major application by the Bell System was in a 24-channel short-haul system for multipair cable circuits. Military systems up to 96 channel capacity are being introduced for radio relay transmission. Pulse transmission lends itself readily to application of military-communications security measures. New commercial networks employing this technique have been proposed primarily for data transmission.

The Bell System has developed a hierarchy of TDM/PCM systems (Fig. 22-48) which is being expanded and which accommodates a variety of services.

The *D1 channel bank* and *T1 repeatered lines* transmit 24 channels of 7-bit (b) binary PCM on multipair cable lines competitive with the type N FDM systems. The bit rate arises from the fact that in each frame each channel is coded into a 7-b word and an additional signaling bit multiplexed with it. At the end of each 24-channel sampling cycle a framing bit is multiplexed into the stream, resulting in $[(7+1) \times 24] + 1 = 193$ b per sampling cycle (frame). With 8,000 sampling cycles per second the resulting digital capacity required for the system is 1.544 megabits per second (Mb/s).

The *D2 channel bank*, which encodes 96 voice channels into 8-b binary PCM on four each T1 lines, is superseding and will replace four D1 banks. The *M-12 digital multiplex* combines four T1 pulse streams into one 6.3 Mb/s pulse stream to be transmitted over the *T2 repeatered system*. The T2 system is also designed to be used on multipair cable at closer repeater spacings than the T1. Even larger multiplex groupings are provided in the hierarchy, and provisions for handling wideband data, television, and Picturephone signals are included in the scheme.

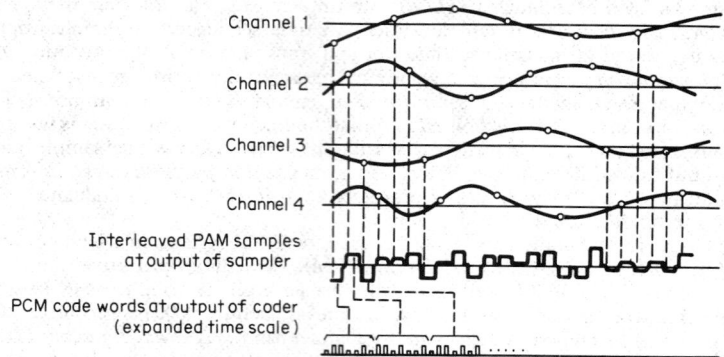

Fig. 22-47. Signal waveforms in a channel bank. *(Copyright, 1970, Bell Telephone Laboratories, Inc.; reprinted by permission.)*

22-36

The *regenerative repeater* is an essential element in the successful application of TDM/PCM to communications systems. The three essential elements are the amplifier-equalizer, timing circuit, and regenerator, shown in Fig. 22-49. The diagram also shows the noisy incoming signal, the signal after reshaping by the amplifier-equalizer, the timing pulses recovered from the signal, the retimed signal, and the regenerated outgoing signal.

Precision timing and synchronization are essential in the operation of TDM/PCM systems. Four methods of synchronization are available: (1) by a system *master clock*, (2) *phase averaging* of all signals, (3) *local stable clocks* in each office, and (4) *pulse stuffing*. In the latter, the transmission rate is somewhat faster than the information rate, thus allowing the addition of bits to complete a frame in the allowed time and stripping these additional bits at the receiver. This method, in conjunction with local stable clocks, is used in the Bell System T1 and T2 systems. Information on the stuffed pulses is transmitted on a separate multiplexed channel.

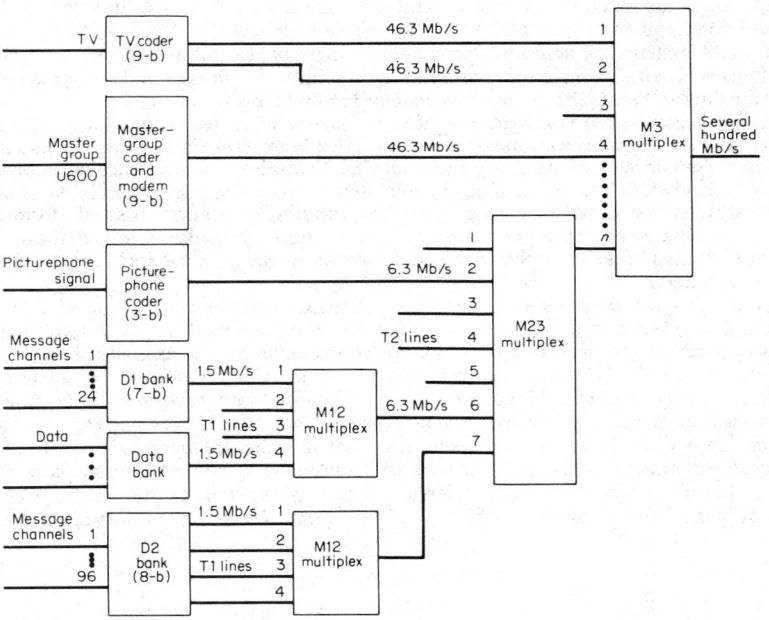

Fig. 22-48. PCM hierarchy. *(Copyright, 1970, Bell Telephone Laboratories, Inc.; reprinted by permission.)*

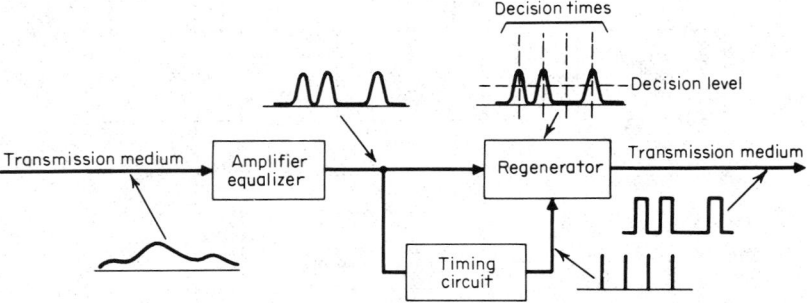

Fig. 22-49. Regenerative repeater block diagram. *(Copyright, 1970, Bell Telephone Laboratories, Inc.; reprinted by permission.)*

Timing jitter at regenerative repeaters, which may arise from several causes, including intersymbol interference and threshold effects, is often systematic and cumulative. The resulting jitter is a limiting factor on the number of repeaters which can be included in a system. Ultimately it can cause cross talk and distortion in the reconstructed analog signal.

41. Signaling in Multiplexed Systems. The most common form of signaling on FDM systems is the extension of the E and M lead signaling system described in Par. 22-20 by means of a single-frequency (SF) in-band signaling system. The SF system uses the M lead status to key a 2,600-Hz sinusoid on the line. The presence of the 2,600-Hz signal denotes an on-hook circuit condition, and its absence indicates an off-hook condition. There is, therefore, a signal on the line when it is idle at a nominal level of -20 dBm0 which must be included as part of the system loading.

Guard circuits operate on the principle that when other frequencies are present, a 2,600-Hz tone is not a system supervisory signal; they are used to prevent talkdown of the signaling system. Separate channel and out-of-band signaling systems are also used but lack the flexibility of the in-band system and have other disadvantages.

In TDM systems, as noted in Par. 22-40, a single bit is multiplexed in each channel sampling cycle to indicate the supervisory signal condition and thus extend the M lead status. This signaling increases the system bit-capacity requirements by some 14%.

42. Television-Signal Transmission. The production of the television signal is discussed in Sec. 20. In television transmission, the objective is to transmit the signal without any significant impairment of its picture-reproducing qualities. Picture impairment can be caused by bandwidth limitations affecting resolution; transmission deviations in flat gain and delay characteristics producing smear, overshoot, ringing, or echoes; cross talk from other systems causing erratic unwanted images; random noise, causing jitter and flicker; and nonlinear distortion (as in system amplifiers) causing intensity or monochrome changes in color tone values.

Standards have been adopted for these characteristics in the design of television transmission systems to preserve picture quality. The bandwidth of the NTSC standard signal requires the transmission of important components as low as 30 Hz to an upper-frequency limit of 4.2 MHz, with gentle roll-off. A signal with insignificant degradation well above and beyond the 3.58-MHz color subcarrier is required for color transmission. The Bell System has standards for other impairments, based on weighted curves since low-frequency interference is generally more detrimental than that at higher frequencies.

Local television transmission in urban areas between the broadcaster's master control, the studios, the radio transmitter, and the telephone company's operating center occurs over A2A facilities in the Bell System, as shown in Fig. 22-50. An A2A facility consists of transmitter,

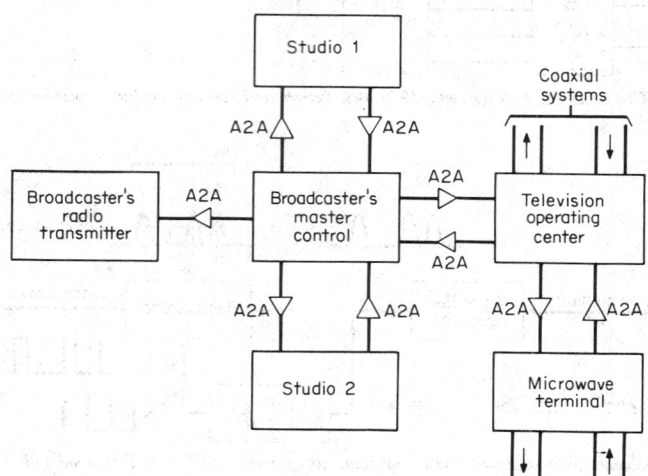

Fig. 22-50. Typical television transmission interconnection with A2A facilities.

receiver, and line amplifiers transmitting over the shielded-cable pair described in Par. **22-29.** Long-haul transmission occurs over microwave or coaxial-cable facilities capable of handling at least a master group on analog facilities. In PCM systems it is contemplated that 9-b encoding will be required resulting in two 46.3 Mb/s streams to an M-23 multiplex, as shown in Fig. 22-48.

43. Visual-Telephone Transmission. Trials are now being held of a video adjunct to voice service for telephone subscribers, under the registered trade name Picturephone in the Bell System. While primarily for face-to-face communication, it also permits limited-resolution graphic transmission. The switching network associated with this service also permits computer access and wide-band data services. The picture is 5½ by 5 in. and is formed in a 2:1 interlaced raster, at a rate of 30 frames per second at 250 lines per frame. The bandwidth is essentially flat from 0 to 1 MHz, permitting an adequate head and shoulders view.

The plan for the picturephone transmission network is shown in Fig. 22-51. The baseband video signal is transmitted on a cable pair in analog form in the local area. Two video pairs and one voice pair are required for each subscriber. The video pairs are unloaded 19-, 22-, 24-, or 26-gauge cable pairs equipped with special equalizers at about 1-mi intervals. In the long-haul plant PCM trunks will be used, and the audio will be multiplexed with the video at 6.3 Mb/s, as shown in Fig. 22-48. Dialing, signaling, and supervision will be over the voice pair in the local area. Eventually the system is expected to result in a new transmission plan providing fixed loss from local switching center to local switching center on all connections.

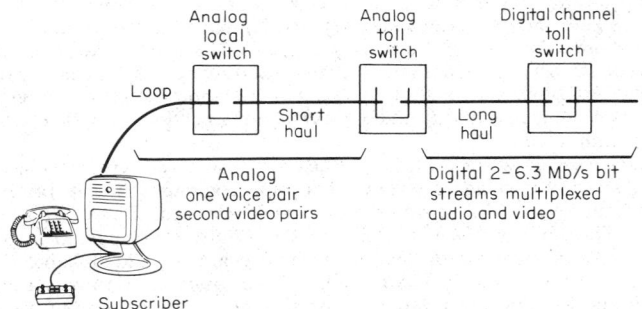

Fig. 22-51. Picturephone transmission network. *(Copyright, 1970, Bell Telephone Laboratories, Inc.; reprinted by permission.)*

DATA AND TELEGRAPH COMMUNICATION SYSTEMS

44. Digital Communications. In this subsection, communications systems using digital signals as a representation of letters, numbers, and other symbols used in business or technical language are discussed. Digital signals were first used in telegraph communications and now form the basic language for computer data communications. The transmission of digital signals occurs, for the most part, over the telephone communications systems originally designed for analog signal transmission, since it is the available source of communication circuits and facilities.

45. Baseband Digital Signals. A digital signal is quantized in both amplitude and time. When quantized in two amplitude levels, it is called a *binary signal*, three levels a *ternary signal*, four levels a *quaternary* signal, etc. Binary signals are preferred since detection between two states, particularly in a noisy transmission system, is much simpler than discriminating among many levels. Binary devices are correspondingly simpler, more reliable, and less expensive.

The signal must also be quantized in time. The shortest (unit) signal element that represents a discrete piece of information is designated as a digit. In a binary signal it is called a *bit*, a contraction of *binary digit*, and abbreviated b. In data systems the speed is usually designated in bits per second (b/s). The *baud* (Bd), a unit of the number of signal elements generated per second, originated in telegraph parlance. The signaling rate in bits per second is equal to the *speed* in bauds.

Three basic methods of generating baseband binary signals are shown in Fig. 22-52. The amplitude levels are designated 0 and 1 in data parlance and *space* and *mark* in telegraph parlance. In the *neutral* or *unipolar* scheme no current indicates 0 and positive battery a 1. In the *polar* scheme negative battery indicates a 0 and positive battery a 1. In the *bipolar* scheme no current indicates a 0 and each 1 bit has opposite polarity from its predecessor 1 bit. The latter scheme reduces the dc component to be transmitted. These signals may be generated electromechanically by mechanical contacts on a switch or electronically by transistor flip-flop circuits.

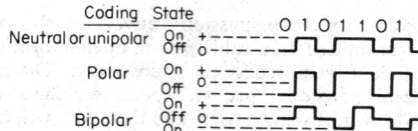

Fig. 22-52. Forms of baseband binary signals.

46. Codes. To give universal meaning to groupings of pulses that can be recognized wherever received, it is necessary to devise codes assigning to each letter, number, symbol, or function a specific and discrete grouping of pulses called a *character*.

The *Morse codes* shown in Table 22-5 were used in the earliest telegraph systems and are still in limited use today. Long and short pulses, called *dots* and *dashes*, make up the codes. The space is equal to a dot. A dash has the length of three dots. The characters in this code are of unequal transmission length. These codes are employed primarily on circuits with hand-operated sending keys and an audible sounder for reception. The International version is most frequently used.

The *Baudot code*, shown in Table 22-6, is designed so that each character has a 5-b length. The code has been the standard code for telegraph communication in printing-telegraph systems for many years. The version shown has been designated by CCIT as International Alphabet 2. The 5-b code has a capability of $2^5 = 32$ discrete characters or combinations. Since this is not sufficient for all "plain language" characters (26 alphabetic, 10 numeric, punctuation marks, plus special characters), it is necessary to use two special characters (figures shift and letters shift) to indicate that all that follows has one of the two possible meanings for each code combination until a new shift symbol appears. Allowing for the two

Table 22-5. International and American Morse Codes

	International	American		International	American
A	.–	.–	R	.–.	. ..
B	–...	–...	S	...	...
C	–.–.	.. .	T	–	–
D	–..	–..	U	..–	..–
E	.	.	V	...–	...–
F	..–.	.–.	W	.––	.––
G	––.	––.	X	–..–	.–..
H	....	....	Y	–.––	.. ..
I	..	..	Z	––..	... .
J	.–––	–.–.			
K	–.–	–.–	1	.––––	.––.
L	.–..	———	2	..–––	..–..
M	––	––	3	...––	...–.
N	–.	–.	4	....–	....–
O	–––	. .	5	.....	–––
P	.––.	.....	6	–....	......
Q	––.–	..–.	7	––...	––..
Period	......	..–..	8	–––..	–....
Comma	.–.–.–	.–.–	9	––––.	–..–
Interrogation	..––..	–..–.	0	–––––	———

shift characters, the all 1s (erase of null) and the all 0s (blank), we obtain $2 \times (32 - 4) = 56$ characters total. The differences between the communication, weather, and military versions occur only in the nonalphanumeric symbols in figure-shift operation.

The *ARQ 7-Unit (Moore) Code* shown at the right in Table 22-6 is related to the Baudot code in that each has the same figures- and letters-shift functions and related character arrangements. It is, however, a 7-b code such that each character has four spaces or 0s to three marks or 1s. This constant ratio is particularly useful in radio telegraphy and noisy circuits for detecting invalid combinations so that corrections can be obtained on mutilated transmissions. This code has three additional characters, as indicated in the table, which are assigned to functions peculiar to radio-telegraph systems, where the code is primarily used.

The American Standard Code for Information Interchange (ACSII), the proposed revised edition of which is shown in Table 22-7, is rapidly becoming the standard code for communications and data processing. It is a 7-b code with 128 discrete characters. An

5-unit Baudot code 1	2	3	4	5	Letters	International	Western Union	Military	TWX	Weather	ARQ 7-unit Moore code 1	2	3	4	5	6	7
O	O	–	–	–	A	–	–	–	–	↑	–	–	O	O	–	O	–
O	–	–	O	O	B	?	?	?	5/8	⊕	–	–	O	O	–	–	O
–	O	O	O	–	C	:	:	:	1/8	O	O	–	–	O	O	–	–
O	–	–	O	–	D	Idf.	$	$	$	↗	–	–	O	O	O	–	–
O	–	–	–	–	E	3	3	3	3	3	–	O	O	O	–	–	–
O	–	O	O	–	F	Opt.		!	1/4	→	–	–	O	–	–	O	O
–	O	–	O	O	G	Opt.	&	&	&	⬎	O	O	–	–	–	–	O
–	–	O	–	O	H	Opt.	#	Stop	Stop	↓	O	–	O	–	–	O	–
–	O	O	–	–	I	8	8	8	8	8	O	O	O	–	–	–	–
O	O	–	O	–	J	Bell	Bell	'		⬈	–	O	–	–	O	O	–
O	O	O	O	–	K	(	(	(	1/2	←	–	–	–	O	–	O	O
–	O	–	–	O	L	)	)	)	3/4	⬊	O	O	–	–	–	O	–
–	–	O	O	O	M	.	.	.	.	.	O	–	O	–	–	–	O
–	–	O	O	–	N	,	,	,	7/8	ⅅ	O	–	O	–	O	–	–
–	–	–	O	O	O	9	9	9	9	9	O	–	–	–	O	O	–
–	O	O	–	O	P	0	0	Ø	0	Ø	O	–	–	O	–	O	–
O	O	O	–	O	Q	1	1	1	1	1	–	–	O	O	–	–	O
–	O	–	O	–	R	4	4	4	4	4	O	O	–	–	O	–	–
O	–	O	–	–	S	'	'	Bell	Bell	Bell	–	O	–	O	–	O	–
–	–	–	–	O	T	5	5	5	5	5	O	–	–	–	O	O	–
O	O	O	–	–	U	7	7	7	7	7	–	O	O	–	–	O	–
–	O	O	O	O	V	=	;	;	3/8	Φ	O	–	–	O	–	–	O
O	O	–	–	O	W	2	2	2	2	2	–	O	–	–	O	–	O
O	–	O	O	O	X	/	/	/	/	/	–	–	O	–	O	O	–
O	–	O	–	O	Y	6	6	6	6	6	–	–	O	–	O	–	O
O	–	–	–	O	Z	+	"	"	"	+	O	O	–	–	–	O	O
–	–	–	O	–			Carriage return				O	–	–	–	–	O	O
–	O	–	–	–			Line feed				O	–	O	O	–	–	–
O	O	O	O	O			Letters				–	–	–	O	O	O	–
O	O	–	O	O			Figures				O	O	–	–	O	O	–
–	–	O	–	–			Word space				O	O	–	O	–	–	–
–	–	–	–	–			Blank				–	–	–	–	O	O	O
										RQ signal	–	O	O	–	O	–	–
										Idle alpha	–	O	–	O	–	–	O
										Idle beta	–	O	–	O	O	–	–
Opt. = Optional										Idf. = { Identification, Answer back, Who are you							

Table 22-6. Teleprinter Codes and Typical Character Assignments

eighth parity bit is often added for error-checking purposes. The International Organization for Standardization (ISO) is preparing to recommend a 7-b code very similar to ACSII for worldwide use.

47. Synchronization and Character Transmission. When digital signals are sent synchronously, the pulses or bits occur at an unvarying rate and in step with a clock. The receiver is also aware of and recognizes the clock rate. Thus, the grouping of pulses of bits within a given time period determines the character sent and received. Digital signals can also be sent *asynchronously*, no fixed time pattern being used. In this case either (1) a particular digit is used to indicate separation between message digits or (2) coding ensures that no two successive digits are alike. Compared with synchronous transmission, asynchronous transmission is inefficient because extra digits are needed to represent a given message.

b_4	b_3	b_2	b_1	Row	0	1	2	3	4	5	6	7
0	0	0	0	0	NUL	DLE	SP	0	'	P	@	p
0	0	0	1	1	SOH	DC1	!	1	A	Q	a	q
0	0	1	0	2	STX	DC2	"	2	B	R	b	r
0	0	1	1	3	ETX	DC3	#	3	C	S	c	s
0	1	0	0	4	EOT	DC4	$	4	D	T	d	t
0	1	0	1	5	ENG	NAK	%	5	E	U	e	u
0	1	1	0	6	ACK	SYN	&	6	F	V	f	v
0	1	1	1	7	BEL	ETB	'	7	G	W	g	w
1	0	0	0	8	BS	CAN	(	8	H	X	h	x
1	0	0	1	9	HT	EM	)	9	I	Y	i	y
1	0	1	0	10	LF	SS	*	:	J	Z	j	z
1	0	1	1	11	VT	ESC	+	;	K	[	k	(
1	1	0	0	12	FF	FS	,	<	L	~	l	─
1	1	0	1	13	CR	GS	—	=	M	]	m	)
1	1	1	0	14	SO	RS	.	>	N	⌃	n	∣
1	1	1	1	15	SI	US	/	?	O	_	o	DEL

Table 22-7. Proposed Revised American Standard Code for Information Interchange

Transmission of digital signals can be either *serial* (serial by bit or serial by character, where each character is sent bit by bit) or *parallel* (parallel by bit and serial by character). In parallel transmission, all the bits of a given character are sent at the same time on parallel channels, but the characters are sent one after another.

START-STOP is a form of serial binary digital transmission in which a START pulse precedes each group of bits forming a character and a STOP pulse ends it. The character bits are transmitted in synchronous time from the initiation of the START bit. The STOP and START pulses are of opposite values so that there is always a transition at end of the STOP and beginning of the START pulse. Thus the STOP pulse may be of indefinite length in time above a certain minimum; i.e. the characters are transmitted asynchronously. START-STOP transmission is used widely in teleprinter and teletypewriter communications systems. Figure 22-53 illustrates the application of START-STOP coding to the 5-b Baudot and 7-b ACSII codes. In the former, the STOP pulse is usually 1.42 or 1.5 times the length of the character-forming bits, while the START bit is of same length and there is a mark to space transition between the two.

48. Printing-Telegraph Systems. Printing systems have replaced essentially all Morse telegraph manual equipment because of their greater speed, the written record they provide, their relative ease of operation, and adaptability to automation. They are also used extensively for low-speed data transmission. Messages are usually initiated by an operator using a keyboard. The keyboard in one system prepares a paper tape by punching holes corresponding to the code. The other method places the signals directly on the line by the teleprinter-keyboard sending mechanism. When signals are received, they either perforate a paper tape or operate a printer through a selecting mechanism.

Most commercial systems operate on the START-STOP Baudot or ASCII codes, described in Par. 22-47. The principle of the system operation on this basis, shown in Fig. 22-54, involves segmented distributors and revolving brushes operated by motors through friction clutches. When a character is to be sent (by depressing a key on a keyboard or starting a paper tape through a tape reader), the START magnet at the transmitter is activated, releasing the brush arm, which rotates from the STOP segment and across the START segment, thus opening the line and releasing the START magnet at the receiver. The transmitter and receiver brushes then rotate together to transmit and detect the pulses forming the transmitted character, and the cycle is then repeated for the length of the message. START-STOP operation was an economical choice for electromechanical teleprinters since timing starts afresh with the start pulse of each character and speed regulation of motors and precision of oscillators is not critical to proper operation.

The common types of tape and page user equipment in the teleprinter family, with their abbreviated initial designations are as follows:

Transmitter distributor (TD) reads and transmits punched paper tapes.

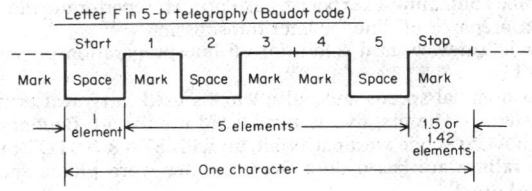

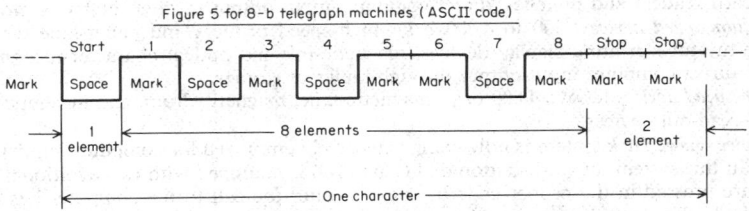

Fig. 22-53. START-STOP characters in Baudot and ASCII codes.

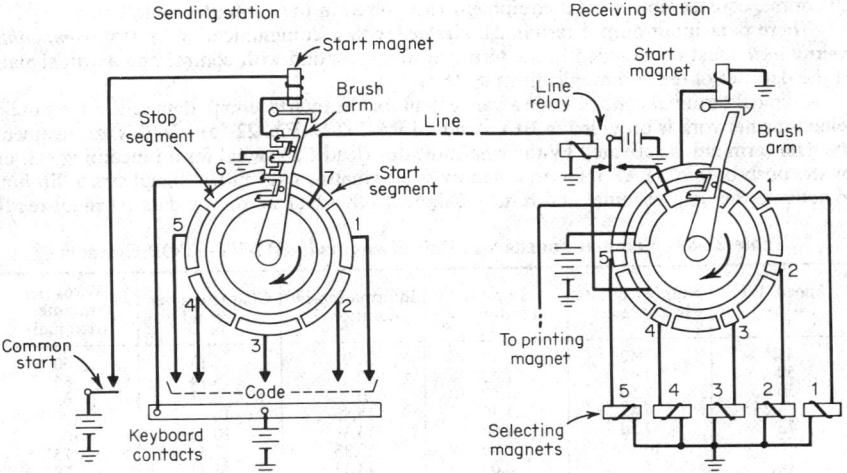

Fig. 22-54. Simplified diagram of START-STOP systems.

Receive-only typing reperforator (ROTR) receives the signal and punches paper tape, printing characters on the edge of the tape.

Reperforator-transmitter (RT) is a combination of a transmitter distributor and typing reperforator.

Receive-only page printer (RO) receives the signal and prints on page-width paper from paper roll.

Keyboard Sending and Receiving Unit (KSR) receives signal and prints (as with an RO) and transmits data by operation of keyboard, usually providing home page copy.

Keyboard typing reperforator (KTR) receives signal and punches paper tape as in ROTR. Operation of the keyboard transmits the message while providing punched paper-tape copy.

Automatic sending and receiving unit (ASR) combines the functions of all the devices listed above in one machine containing a keyboard, page printer, reperforator, and tape transmitter. Paper tape is often prepared off-line for later transmission.

Tape perforator is a device used solely for off-line preparation of paper tape for later transmission by a TD.

Table 22-8 lists nominal speeds and pulse lengths used in typical printing-telegraph or low-speed data systems. The first five systems listed use the 5-b Baudot code; the last two systems use the 7-b ACSII code with parity bit, all with START-STOP pulses added. The words-per-minute ratings are based on a five-character word plus a space character (six characters per word, total).

49. Data Input-Output Devices. Devices providing input to and output from a data communications system are numerous (see Sec. 23). The common devices, according to a frequently employed but arbitrary speed of transmission classification are as follows:

Low-speed devices (0 to 300 b/s): all those teleprinters and paper-tape devices listed in Par. 22-48, card readers and punches, direct computer input-output through buffer or storage.

Medium-speed devices (300 to 4,800 b/s): high-speed printers and paper-tape devices, cathode-ray-tube printing-display devices, magnetic-ink and optical character-recognition devices, direct computer input-output through buffer or storage.

High-speed devices (above 4,800 b/s): magnetic-tape, magnetic-drum, and magnetic-card read-in-read-out devices.

A *voice answerback* system is now being extensively employed for computer inquiry and reply. In this system the push-button dial (Par. 22-9) is equipped with two additional keys which are pressed in the proper code by the user after the call to the computer has been established. The reply is composed in speech generated by the computer and transmitted as such. Since all the signals in this system are compatible with the telephone system, no special devices are required; i.e., the transmission requirements are the same as for telephone circuits. Of course, considerable special equipment is required in the computer installation.

Where data input-output terminals *interface* with a communication system, *compatible control logic* must be included in the terminal to correspond with control and alarm signals of the data set of the communications system.

A typical terminal control-logic arrangement for the interface with data sets of the public telephone network is provided in EIA standard RS-232-B (Fig. 22-55). In this arrangement the data terminal is activated by the ring-indicator (lead *CE*) signal for an incoming call or by the push-button START for an outgoing or originating call. Either signal sets a flip-flop to activate the data terminal and return a signal to the data set on the data-terminal-ready

Table 22-8. Nominal Speeds and Pulse Lengths in START-STOP Systems

Speed, Bd (= b/s)	Number of bits in character	Stop-bit duration, b	Information-bit duration, b	Characters per second	Words per minute (nominal)
45.5	7.42	1.42	21.97	6.13	60
50	7.42	1.42	20	6.74	66
50	7.50	1.50	20	6.67	66
74.2	7.42	1.42	13.48	10	100
75	7.50	1.50	13.33	10	100
75	10	1.00	13.33	7.5	75
75	11	2.00	13.33	6.82	68
150	10	1.00	6.67	15	150

(*CD*) lead. On sensing this signal the data set turns on the data-set-ready (*CC*) lead and goes into either the sending or receiving condition, depending on the status of the request-to-send (*CA*) lead from the data terminal. If this lead (*CA* lead) is not turned on, a signal returned from the data set on the data-carrier-detector (*CF*) lead tells the data terminal to prepare to receive data.

At the end of transmission the data terminal must tell the data set that transmission is complete by turning off the data-terminal-ready (*CD*) lead. This requires some logical arrangement in the data terminal for recognizing an end-of-transmission character in the receive mode or a similar signal internally generated when it is in transmit mode.

50. Data and Telegraph Transmission at Baseband. Low-speed data and telegraph circuits operating at baseband on private-line facilities are served by communication systems although they are diminishing in number. Systems are in use with operating speeds of 75 Bd or less using neutral signals (mark = full line current, space = no line current) with line currents of 20 or 62.5 mA obtained from keying dc voltages of + 130 and − 130, + 130 and − 48 or + 130 and ground. When cable pairs are used, the length of line for satisfactory transmission without repeaters is limited to a few miles. On open-wire pairs, line lengths up to 20 mi can be used without undue distortion.

Systems having operating speeds of 75 to 180 Bd cannot use 60- or 20-mA line currents on pairs in the same cable with voice- or program-grade channels. Systems using 6-mA balanced neutral or 3-mA polar signals can be economically used up to about 2,000 Ω loop resistance.

For in-house data transmission at speeds up to 300 Bd baseband transmission on wire pairs or even coaxial cable to reduce interference is often economical and satisfactory for distances up to a few thousand feet.

51. Data Transmission on Analog Circuits. The devices discussed in Pars. 22-48 and 22-49 generate the baseband signals discussed in Par. 22-45. These signals have both dc and high-frequency components and are not compatible with the voice circuit of the public

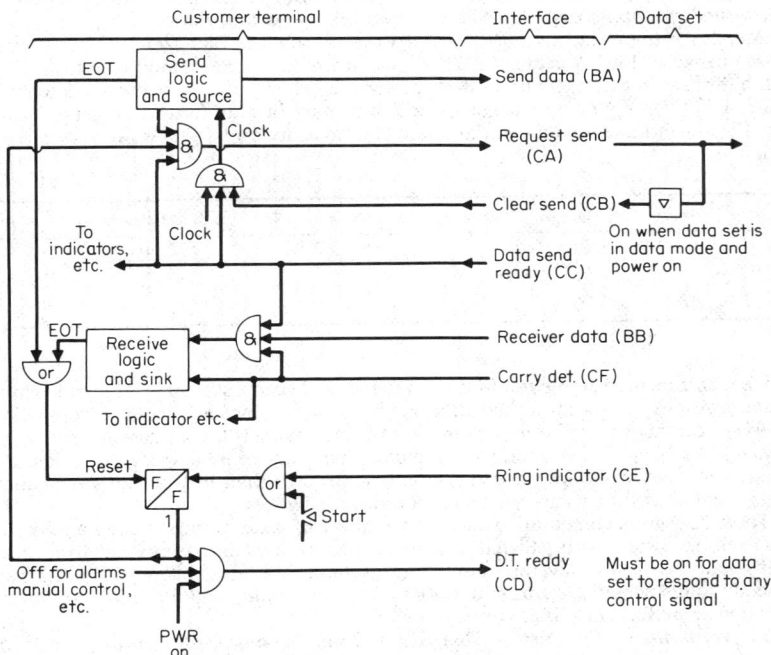

Fig. 22-55. Typical control and logic requirements in a customer terminal.

telephone network, which nominally has a bandwidth of about 300 to 3,500 Hz. Compatible transmission is obtained by modulating a carrier frequency within the channel passband by the baseband signal in a modem (modulator-demodulator) usually incorporated in a data set.

In a band-limited system, Nyquist and Shannon have shown that the maximum numbers of bits that can be transmitted is

$$C = W \log_2 n \quad b/s$$

where W is the usable bandwidth and n the number of recognizable states of the signal. The expression can also be written

$$C = W \log_2 (1 + S/N)$$

where S is the average signal power and N the average noise power in the band.

Present techniques do not permit attaining anything like the speeds theoretically possible from this formula. It will be noted from these expressions that the maximum speed possible for binary transmission, with $n = 2$, is W. The usable channel bandwidth is somewhat less than the nominal bandwidth W due to filter-shaping characteristics and resulting distortions at the channel edges.

Data transmission employs all three modulation methods (amplitude, frequency, and phase). A description of these methods is given in Sec. **14**.

An on-off *amplitude-modulated* (AM) signal is shown in Fig. 22-56a. This modulation scheme is not much used for voice-band modems but is used in wide-band modems with raised cosine pulse shaping. Envelope-detection schemes are not common in data transmission, since the envelope is quite often severely distorted by the transmission medium. A variation of the method in which one sideband and a portion of the other are transmitted, called vestigial-sideband (VSB) AM, is sometimes employed to give increased transmission speed over double-sideband AM while permitting carrier recovery for detection purposes.

An example of a binary *frequency-modulated* (FM) carrier wave, sometimes called frequency-shift keying (FSK), is shown in Fig. 22-56b. While FM or FSK requires somewhat greater bandwidth than AM for the same transmission speed, it gives much better performance in the presence of impulse noise and voltage-level change. It is used extensively in low- and medium-speed (voice-band) telegraph and data systems.

A *binary phase-modulated* (PM) carrier wave is shown in Fig. 22-56c. In this example a phase change of 180° is depicted. However, this modulation method is usually employed with a 4-phase and sometimes an 8-phase change, which must be accommodated within a range of 180°. In a 4-phase-change system, the binary bits are formed in pairs called *dibits*. The dibits determine the differential phase change as follows:

Bits	Phase change
0,0	−135°
0,1	− 45°
1,1	+ 45°
1,0	+135°

In effect, this method employs a four-state signal, and such a system is inherently capable of greater transmission speeds for the same bandwidth, as is obvious from the Nyquist-Shannon criteria stated above. With improvement of voice channels the 4-phase (*quaternary-phase-modulated*) scheme is being employed increasingly for medium-speed (voice-band) data-transmission systems to give higher transmission speeds than FM for the same bandwidth. The system is useful only in synchronous transmission.

Table 22-9 gives comparative data for a number of systems (both binary and multilevel) on a fixed-bandwidth basis. Signal-to-noise ratio is given on the basis of both fixed maximum steady-state power and average power. This table indicates theoretical performance of perfectly implemented systems with optimum signal shaping and filtering. Actual systems offer poorer performance by 1 dB or more.

52. Transmission Parameters That Affect Data. The transmission parameters of telephone communication systems are discussed in Pars. 22-22 to 22-35 with emphasis on the provision of economical voice communications. Certain of these parameters can display

Table 22-9. Comparisons of Various Data Systems under Limitations of Average and Maximum Steady-State Signal Power[23]

System	Number of states	Speed, b/s per hertz of bandwidth	Signal-to-noise ratio for 10^{-4} b error rate	
			Average signal power	Maximum steady-state signal power
Unipolar baseband	2	2	14.4	17.4
	4		22.8	26.9
Bipolar baseband	2	2	14.4	17.4
Polar baseband	2	2	11.4	11.4
	4	4	18.3	20.8
	8	6	24.3	28.0
	16	8	30.2	34.4
Full-carrier AM:				
Envelope detection	2	1	11.9	14.9
Coherent detection	2	1	11.4	14.4
	4	2	19.8	23.8
	8	3	26.5	31.0
Suppressed-carrier AM, coherent detection	2	1	8.4	8.4
	4	2	15.3	17.8
	8	3	21.3	25.0
	16	4	27.2	31.4
PM, coherent detection	2	1	8.4	8.4
	4	2	11.4	11.4
	8	3	16.5	16.5
	16	4	22.1	22.1
	32	5	28.1	28.1
	64	6	34.1	34.1
Differential detection	2	1	9.3	9.3
	4	2	13.7	13.7
	8	3	19.5	19.5
FM	2	1	11.7	11.7
	4	2	21.1	21.1
	8	3	28.3	28.3
VSB, 50% modulation	2	2	16.2	17.9

VSB, suppressed-carrier, coherent detection (see polar baseband)
Quadrature AM, suppressed-carrier, coherent detection (see polar baseband)

wide tolerances without seriously affecting voice communications but must meet more severe limits where data transmission is concerned.

Delay distortion is probably one of the most inhibiting characteristics in successful use of voice channels for data transmission. Differences in velocity of propagation of various frequencies in the signal passband cause relative phase shifts that spread out the pulse in time. This effect is related to the derivative of the phase-shift-vs.-frequency curve at frequencies of interest. The derivative, called *envelope delay*, is usually expressed in microseconds. Figure 22-57 shows the spread of envelope delay on telephone circuits in the public telephone network in 1958.

It is considered that the safe operating speed for data transmission, for a random choice of circuits in the public telephone network, is about 1,200 b/s. Envelope delay is improved by the use of equalizers having characteristics inverse to the delay-distortion curve. The Bell System and Western Union provide three classes of such conditioned circuits for private-line data communications with limits as shown in Table 22-10. Data-transmission speeds of 1,400, 2,400, and 4,800 b/s are possible on these conditioned circuits.

Attenuation or amplitude-frequency distortion occurs when the signal is attenuated differently as a function of frequency over the band of interest. Figure 22-58 shows a generalized attenuation-frequency characteristic of a typical switched telephone-network circuit. At the low-end cutoff, below 300 Hz, a slope of 15 to 25 dB/octave occurs. The range 300 to 1,100 Hz is relatively flat. From 1,100 to 2,900 Hz, a linearly rising loss occurs such that, on the average, the loss at 2,600 Hz is 8 dB above that at 1,000 Hz. Above 3,000 Hz the loss rapidly increases, at a rate of 80 or 90 dB/octave.

Figure 22-59 shows the variation in loss between 1,000 and 2,600 Hz for a number of circuit measurements in the Bell System switched network in 1958. At that time 10% of the circuits could have an attenuation slope in this region of as much as 13 to 14 dB on all types of connections. Within certain limits equalization can be applied to correct this attenuation-frequency distortion. The conditioned voice-band circuits available from the Bell System and Western Union are equalized for this characteristic to specifications shown in Table 22-10.

Noise. As noted in Par. 22-25, there are two types of noise: circuit (gaussian) and impulse noise. The circuit-noise limits established for voice circuits are satisfactory for data service, but *impulse noise* is another matter. Tolerable levels in voice communications can seriously degrade service on data circuits. Impulse noise is generated by switches and other man-made electrical disturbances and by high-frequency natural phenomena such as lightning or fading in radio circuits. Its effect in data systems is to cause clusters of errors. Error-detection and correction schemes can overcome the occasional difficulty from impulse noise. When impulse noise is serious, circuit rerouting, shielding, change in modulation method, and tightening of all connections may help.

Echo, discussed in Par. 22-26, is not usually a serious matter for data transmission if the circuit is satisfactory for voice. Listener echo, which could adversely affect data transmis-

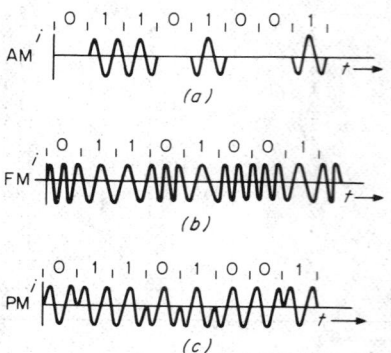

Fig. 22-56. Binary amplitude; frequency; and phase-modulated carrier waves.

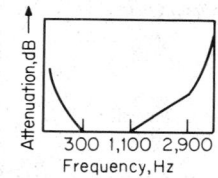

Fig. 22-58. Relative attenuation of typical telephone circuit.

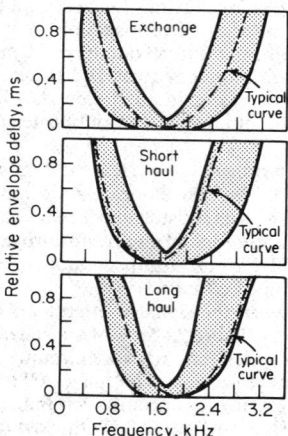

Fig. 22-57. Typical envelope delay in the public telephone network.

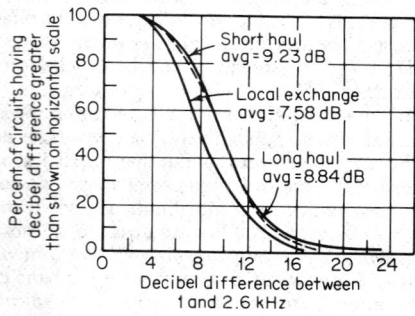

Fig. 22-59. Attenuation distortion in the public telephone network in 1958.

22-48

Table 22-10.   Specifications for Conditioned Voice Lines

C1 conditioning* (class E lines)†		C2 conditioning* (class F lines)†		C4 conditioning* (class H lines)†	
Amplitude Variation					
Amplitude variation, dB	Frequency range, Hz	Amplitude variation, dB	Frequency range, Hz	Amplitude variation, dB	Frequency range, Hz
−2 to +6	300 to 999	−2 to +6	300 to 499	−2 to +6	300 to 499
+1 to +3	1,000 to 2400	−1 to +3	500 to 2,800	−2 to +3	500 to 3,000
−2 to +6	2,401 to 2,700	−2 to +6	2,801 to 3,000	−2 to +6	3,001 to 3,200
Envelope Delay					
Envelope delay, μs	Frequency range, Hz	Envelope delay, μs	Frequency range, Hz	Envelope delay, μs	Frequency range, Hz
<1,000	1,000 to 2,400	<500	1,000 to 2,600	< 300	1,000 to 2,600
		<1,500	600 to 2,600	<1,500	800 to 2,800
		<3,000	500 to 2,800	<1,500	600 to 3,000
				<3,000	500 to 3,000

* Telephone-company terminology.
† Western Union terminology.

sion, is usually sufficiently attenuated not to be troublesome. The use of automatic-gain-control circuits, which reject signals 12 dB or less than the incident signal in data sets, usually takes care of any problem of this nature.

Frequency offset occurs when a particular frequency is transmitted in a suppressed-carrier circuit but after carrier insertion at the receiver is translated to a different frequency, e.g., in a FDMSSBSC system, the predominant system for long-distance transmission. The translation (offset) in frequency, if large enough, may cause errors, particularly in narrow-band telegraph systems. The CCITT has recommended that frequency offset be limited to ±2 Hz per link or ±10 Hz for a five-link connection. Modern systems are designed to fall within this limit. Some older systems may display as much as 25 to 50 Hz offset. This is unsatisfactory for narrow-band data or telegraph transmission.

Sudden amplitude changes occur due to faulty amplifier components, substitution of different facilities, and operation and maintenance errors. While these changes are seldom more than 6 dB, they can seriously disturb data transmission, in AM systems in particular.

Sudden phase changes also occur due to switching of carrier supplies or change in transmission facility.

Hits (momentary loss of signal that occurs due to open or short circuit resulting from accident, storm, construction, or maintenance activity) can cause errors and possible loss of synchronization.

Nonlinear distortions can occur, causing harmonic distortion and generation of spurious frequencies, but this is seldom of serious concern. However, nonlinear amplitude distortion introduced by compandors may affect data signals, particularly high-speed AM signals.

53. Data Sets. The data set (incorporating a modulator-demodulator, or modem) is the equipment which accepts the telegraph or data baseband signal, modulates it, transmits it over the communication system, and performs the reverse when receiving the data- or telegraph-modulated signal. It also incorporates some type of interface for control of the telegraph or data user equipment, conforming to that shown in Fig. 22-55.

Choice of Modulation System. Factors influencing the choice of the modulation scheme are shown in Table 22-11. In all columns except the second and third the systems are given numbers indicating the order of preference; the same number is given to more than one where there is insufficient basis for a choice. AM is the simplest method and gives satisfactory operation when channel amplitude characteristics are stable and there is assurance of little noise or interference.

FM or FSK is reasonably simple, performs well over a wide variety of channels, and permits asynchronous operation. It is the most widely used method at present.

Binary PM is somewhat more complex, has greater tolerance for noise, but is limited to synchronous operation with differential detection (with coherent detection asynchronous

Table 22-11. Characteristics to Be Considered in Choosing a Data System[23]

System	Complexity of instrumentation	Bits per cycle of nominal band	Variable-speed asynchronous operation	Signal-to-noise performance		Tolerance to parabolic delay distortion	Tolerance to linear delay distortion	Tolerance to amplitude change	Tolerance to frequency offset	Tolerance to phase jumps
				Equal bit rate	Equal bandwidth					
Binary polar baseband	1	2	Yes	1	3	2	—	1	—	—
Binary on-off AM	2	1	Yes	5	5	2	1	3	1	1
Binary FM	3	1	Yes	4	4	1	2	1	3	2
Binary PM differentially coherent	4	1	No	2	2	3	3	2	3	2
Binary PM coherent	5	1	Yes	1	1	2	1	1	1	4
VSB AM suppressed carrier	6	2	Yes	1	3	2	5	1	2	4
Quaternary PM differentially coherent	7	2	No	3	5	4	4	2	4	3

operation is possible, but carrier recovery and polarity ambiguity lead to increased complexity).

Suppressed-carrier VSB combines double the speed capability with high tolerance to noise, but carrier recovery involves complexity and there is less tolerance to unsymmetrical delay distortion, with carrier located near the edge of band.

Differential quaternary PM has double the speed capability, without the problems of carrier recovery but is more complex and less tolerant of delay distortion and noise. It is evident that these systems have their strong and weak points. The choice depends on the most economical solution for the specific application.

Low- and Medium-Speed Data Sets and Modems. A partial but representative list of available data sets is shown in Table 22-12.

Data sets having speeds up to 300 b/s are used for teletypewriter and low-speed data communications, with suitable interfaces. Sets having speed up to 1,800 b/s can be used on a dial-up basis on the public telephone network. At higher speeds, conditioned circuits, such as those listed in Table 22-10, usually are required, although some sets with manual or automatic equalization networks can provide good service at higher speed on typical voice circuits. Sets requiring conditioned circuits are intended primarily for private-wire facilities and assigned circuits. Most data sets are intended for direct wired connection to the communication network.

Some low-speed modems can be acoustically coupled through the handset of the telephone set to the system. This type of connection may produce variable results, due to ambient-noise conditions, efficiency of the handset, and acoustic-coupling efficiency. Use of this method at speeds above 600 b/s is not recommended.

High-Speed Data Modems. In Pars. 22-34 to 22-37 are shown the modulation processes for forming wideband groups and supergroups. The wideband channels thus formed are used for high-speed bulk data transmission. Table 22-13 lists Bell System data sets for such high-speed data transmission. The *restored polar* modulation method, used in the higher-speed sets, is one in which a high-pass filter, with cutoff at about 2,000 Hz, is used to eliminate the dc components of the baseband signal. At the receiving end, the signal is restored by a quantized feedback and slicer circuit.

These data sets are associated with appropriate line facilities, including the wide-band portion of those described in Pars. 22-35 to 22-38, replacing 12 and 60 voice channels. These facilities are designated as types 8803 (19.2 kb/s), 8801 (40.8 kb/s), and 5800 (230.4 kb/s) in

Table 22-12. Partial List of Low- and Medium-Speed Data Sets and Modems

Available from:	Model	Speed, b/s	Modulation method	Channel requirements	Synchronous or asynchronous	Equalization
Bell System	103A	300	FSK	Voice	Async	N/R
	103E	150	FSK	Voice	Async	N/R
	201A	2,000	PM	Cond	Sync	
	202C	1,800	FSK	Cond	Async	
	403D*	10 char		Voice	Async	
Collins Radio	TMX-202 Series	1,800	FSK	Voice	Either	
	TE-216 series	4,800	PM-AM	Cond	Sync	N/R
General Dynamics	EDX-1403	4,800	AM/VSB	Cond	Sync	Auto
	EDX-1402	1,200	FSK	Voice	Async	
Lenkurt Electric	25B	600	FSK	Voice	Async	N/R
	26C	2,400	FSK	Cond	Either	Man
	26D	4,800	FSK	Cond	Sync	Man
Rixon Electronics	SEB it series	4,800 7,200 9,600	AM/VSB	Cond	Sync	Man
	FM-18	1,800	FSK	Voice	Async	N/R
	PM-24	2,400	PM	Voice	Sync	N/R
Western Union	2121-B	1,200	FSK	Voice	Async	N/R
	2247-A	2,400	PM	Voice	Sync	Fix
	2481-A	4,800	PM	Voice	Sync	Man

* Bell System 403D is intended for use with low-speed punched-card or remote metering system.

Bell System tariffs. Similar wide-band data sets are offered by other manufacturers, and other public communications systems can supply similar line facilities.

54. Subchannel FDM Systems for Telegraph and Data. The voice channel can accommodate more than one low-speed telegraph or data signal by applying FDM methods. Table 22-14 shows the three most common systems; they employ FSK modulation. The 120- and 170-Hz channel-spacing systems have been used for telegraph transmission for many years, originally with amplitude modulation. They are extensively used today. The 340-Hz

Table 22-13. Western Electric High-Speed Channel Terminals

Type	Data-set capability			Channel required nominal bandwidth, kHz	Modulation
	Bit rate, kb/s	Minimum (set is asynchronous)	Synchronous		
301B	40.8	...	X	48	Phase
303B*	52-μs element length	X	...	24	Vestigial-sideband amplitude
303B	19.2	...	X	24	Vestigial-sideband amplitude
303C*	20-μs element length	X	...	48	Restored polar-line signal
303C	50	...	X	48	Restored polar-line signal
303D*	4.3-μs element length	X	...	250	Restored polar-line signal
303D	230.4	...	X	250	Restored polar-line signal

* For binary facsimile signals.

Table 22-14. Typical Voice-Frequency Telegraph-Carrier Multiplex Channel Assignments and Signaling Rates

120-Hz spacing, ±30-Hz frequency shift (signaling rates to 60 Bd)		170-Hz spacing, ±35-Hz frequency shift (signaling rates to 75 Bd)		340-Hz spacing, ±70-Hz frequency shift (signaling rates to 150 Bd)	
Channel number	Midband frequency	Channel number	Midband frequency	Channel number	Midband frequency
1	420	1	425	1	680
2	340	2	597	2	1,020
3	660	3	765	3	1,360
4	780	4	935	4	1,700
5	900	5	1,105	5	2,040
6	1,020	6	1,275	6	2,380
7	1,140	7	1,445	7	2,720
8	1,260	8	1,615	8	3,060
9	1,380	9	1,785		
10	1,500	10	1,955		
11	1,620	11	2,125		
12	1,740	12	2,295		
13	1,860	13	2,465		
14	1,980	14	2,635		
15	2,100	15	2,805		
16	2,220	16	2,975		
17	2,340	17	3,145		
18	2,460				
19	2,580				
20	2,700				
21	2,820				
22	2,940				
23	3,060				

channel-spacing system is suitable for medium-speed data using parallel transmission (serial by character, parallel by bit) with an 8-b code (such as ASCII with parity bit). This system has the advantage of greater tolerance for many channel-transmission impairments than serial transmission for the same character-transmission rate. However, means must be established to preserve synchronization on each channel with respect to the other.

55. Data Transmission in TDM Systems. The TDM (time-division-multiplexed) systems, discussed in Pars. **22-39** and **22-40**, are digital in nature and therefore compatible with data and telegraph systems. One of the major system design problems is the maintenance of system synchronization and bit integrity in all channels throughout the system.

Stable clocks in each office and the *bit stuffing* process are described in Par. **22-40**. This method is generally used for synchronizing and multiplexing synchronous data streams into the TDM system.

Transition encoding is a method whereby only the transitions are encoded, by a 3-b character. When there is no transition, all 1s are transmitted. The first bit of the transition character is a 0 and is called the *address bit*. The second bit is a code bit indicating in which half of the time interval the transition occurred, while the third bit is a sign bit which indicates whether the transition is from 0 to 1 or vice versa. The efficiency is about 3 b for each transition bit. Thus a 50 kb/s data signal displaces three 64 kb/s channels when transition encoding is used. This form of encoding is necessary when it is to multiplex asynchronous data signals.

The Bell System plans to have available in most metropolitan areas a digital data service (Dataphone Digital Service®) that will be integrated with the Bell System network (Fig. 22-60) and will use the equipment in the digital hierarchy (Fig. 22-48). This will be a private-line offering. The ultimate aim is to provide service to 96 cities. The block labeled data bank is tentatively identified as the T1 data multiplexer (T1DM). It will accept twenty-three 56 kb/s signals. A data submultiplexer can be appended to any of the 56 kb/s ports of the T1DM and will accept twenty 2.4 kb/s, ten 4.8 kb/s, or five 9.6 kb/s synchronous data signals for multiplexing into the 56 kb/s data stream (see Fig. 22-60a). The circuits are extended to the

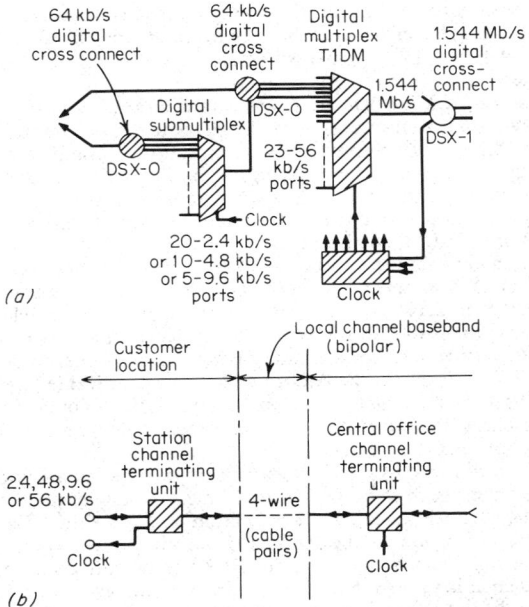

Fig. 22-60. Bell System digital data service multiplexing and circuit arrangements: (*a*) central-office multiplexing arrangements; (*b*) circuit extension, central office to customer's premises.

customer's premises over 4-wire cable circuits using modified bipolar baseband transmission, as shown in Fig. 22-60b.

Suitable data sets with provision for the EIA RS-232-C or CCITT V-35 interface will be available for circuit termination. Stable clock signals will be used at the central office and made available at the customer's premises by derivation from the bit stream.

Western Union plans to offer a tariffed digital data service by 1975 both via terrestrial microwave and via satellite. Transmission of digital signals at 1.544 Mb/s will be the basis of the system with a T1 data multiplexer to produce synchronous data ports at 56 kb/s. The 56 kb/s ports will be equipped with submultiplexers to produce lower-speed synchronous channels at various combinations of 2.4, 4.8, and 9.6 kb/s. A master clock system will maintain system timing to one part in 100 billion. Customer channels will be compatible with Bell System Dataphone Digital Service® interfaces.

56. Error Rates and Error Control. When any of the transmission impairments discussed in Par. 22-52 (particularly hits, impulse noise, or phase or amplitude distortion) exceeds the tolerance of the data-transmission system, the resulting distortion of the signal can produce errors and usually does. The error rate varies widely with time on any given transmission medium. The CCITT Study Group on Data Transmission, in tests on leased circuits and the international telex circuits in Europe operating at a speed of 50 Bd, has found error rates as follows:

Point-to-point service: one to two bit errors in 100,000 transmitted; one to eight characters per 100,000 transmitted.

Switched telex circuits: one to two bit errors per 100,000 transmitted and four to five character errors per 100,000 transmitted.

A considerable number of the errors occurred in bursts, 50 to 60% being isolated single-bit errors, 10 to 20% two-error bursts, 3 to 10% three-error bursts, and 2 to 6% four-error bursts. This experience was primarily on subchannel FDM multiplex circuits and tended to show greater error rates as the number of channels in tandem increased.

Experience on voice-channel data circuits in the Bell System, at 1,200 and 600 b/s transmission speeds in 1958, is shown in Fig. 22-61. These data tend to show that in the long run data-error rates on a voice channel with 600 b/s transmission should be better than 1 in 10^5, whereas at 1,200 b/s, the error rate should be in vicinity of 1 in 10^5. Wide-band systems exhibit similar error rates. Improvements in system parameters and data sets are resulting in lower error rates.

Error control is of vital importance in data-transmission systems handling data of a quantitative nature. On systems transmitting language text messages, it is not so important, since the recipient can usually supply the error corrections. The first step in error control is detection of errors. All systems for error detection require that redundant information be added to the basic code and therefore reduce the data throughput rate. A detailed study of error detection and correction methods is beyond the scope of this section; only the more common methods are cited. (See also Sec. 4.)

Parity Codes. In this simplest form of error detection, a bit is added to the ACSII character code to make the number of 1s or 0s in every character an odd number (odd parity) or an even number (even parity). The error correction breaks down when bit errors occur in pairs. A further refinement is to add horizontal parity in a block of characters in a stored matrix so that both vertical and horizontal parity are checked in each block.

Constant-ratio codes, such as the Moore code, shown in Table 22-6, have a constant number of 1s and 0s in each character. So long as there is not a double switch of 1 for 0 and 0 for 1 in an error character, the error will be detected. This type of code is much used in high-frequency radio transmission.

More powerful codes employ much higher degrees of redundancy and are capable of detecting multiple errors, such as occur in error bursts. Examples are the Hamming, Bose-Chaudhuri and Hagelbarger codes. Implementation employs shift registers and modulo-2 adders. Such error correction, at the receiver, is called forward error correction (FEK).

In other error-detection schemes (without FEK) error correction is accomplished by repeating the transmission. On half-duplex circuits (one-way transmission-reversible) transmission is stopped at the end of each code block for receipt of (1) an ACK (positive acknowledgement) signal if no error is detected at the receiver or (2) NAK (negative acknowledgement, i.e., repeat request) signal if an error is detected at the receiver. On a full duplex circuit, two-way transmission, an ARQ (automatic repeat request) signal is transmitted even while data are being transmitted.

Figure 22-62 shows the improvement in error rate achieved on Bell System circuits by FEK with simple single-error detection and correction and with a scheme for burst-error correction.

57. Low-Speed Switched Telegraph and Data Communications Systems. *TWX and TELEX Systems.* There are two circuit-switched public teleprinter networks available in the United States. TWX, teletypewriter exchange service, was originated by the Bell System and other telephone companies but is now operated by Western Union. It provides a primary 75-Bd (10 characters per second Baudot or 6 characters per second ACSII) teletypewriter circuit, with automatic dial switching, using modified equipment of types examined in Pars. **22-17** and **22-18** and provides its own directory.

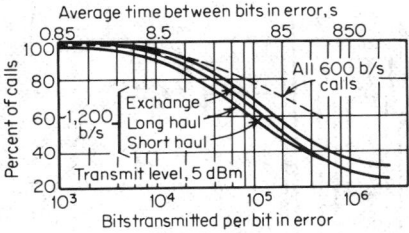

Fig. 22-61. Error rates typical in a telephone network.

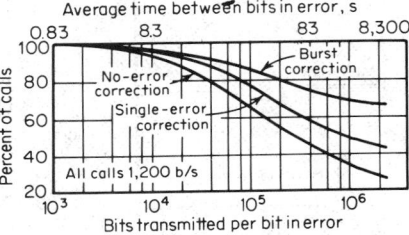

Fig. 22-62. Error-rate improvement with error correction.

The user equipment is rented from the utility or provided by the customer if it meets electrical specifications. Alternately the customer may provide equipment operating up to 150-Bd speed. In this case it can communicate only with similar equipment. The circuits usually include the 75- or 150-Bd subchannel FDM channels described in Par. **22-54**.

Telex is a similar system operating at 50 Bd, 66 words per minute Baudot code, on a worldwide basis. Western Union provides this service in the United States on a dial-up automatic switched basis. Automatic station-answer-back identification is a feature of this service. The circuits provided are the 50-Bd subchannel FDM channels described in Par. **22-54**.

Message Switched Telegraph Systems. Telegraph companies, government agencies, and many business enterprises handle large-volume bulk-message traffic with punched-tape systems. The basic equipment at the switching centers is the *reperforator-transmitter* (RT), which terminates each end of each circuit, and manual or automatic switching to effect *cross-office connection* of one circuit to another. Cross-office transmission is usually faster than line transmission speeds. The tape is used for storage as well as for transmission.

Destination and routing indicators must be affixed to the message when automatic cross-office switching is used. Intercept positions are required for lost or mutilated messages and emergency routing. Many refinements can be included in these systems, including priority controls, multiple-address handling, provision for automatic alternate routing, etc. The message is printed on a *receive only* (RO) *page printer* at the destination. Trunk circuits are usually provided by the subchannel FDM channels described in Par. **22-54**.

58. Data-Communications Systems. A typical arrangement of a data-communications system employed by many medium and large business firms with many locations is shown in Fig. 22-63. The system may include a comprehensive data processing center in the main plant, a large scientific computer in the laboratories, and small computers in subsidiary plants. At the heart of the system is the on-line real-time computer center in the main plant, which must possess a large communications-control function (to control access between the communication lines), the data processing functions, and the data base store.

Among the communication facilities incorporated in such a system are (1) a leased high-speed data link, (2) many leased voice-grade medium-speed lines, with data-set terminations as described in Par. **22-53**, and (3) a message-switching system with leased low-speed data

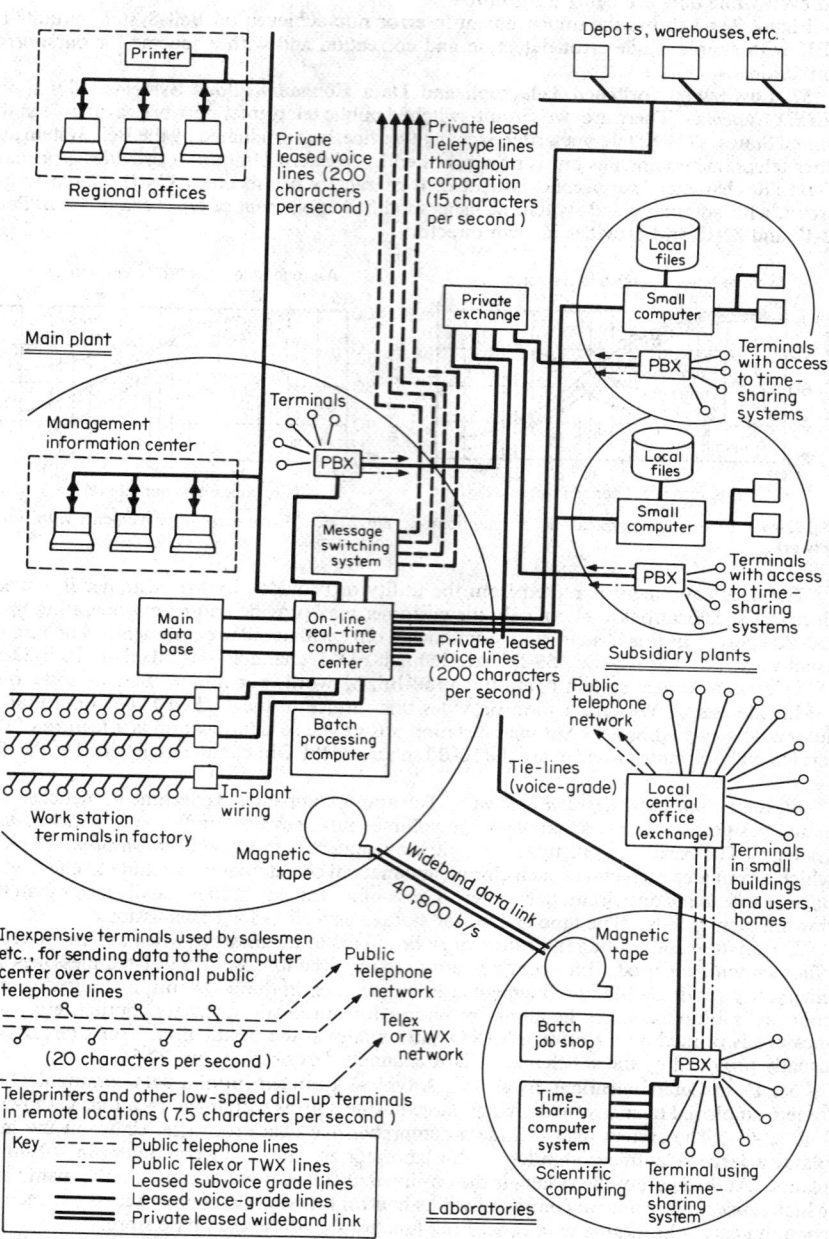

Fig. 22-63. Large computer center in main plant with links to computers in subsidiaries and laboratory; an in-plant data-collection system and on-line terminals in sales offices and depots. *(From J. Martin, "Telecommunications and the Computer," Prentice-Hall, Englewood Cliffs, N.J., 1969.)*

circuits, as described in Par. 22-57. Access may also be provided over the TWX or telex public networks described in Par. 22-57. Voice-answerback service can be provided over the public telephone network (Par. 22-49).

In systems having far-flung data communications networks, e.g., those of the airlines, multiplexers and data concentrators are used at various junction points in the system to combine circuits and thus reduce costs.

Data concentrators are devices which store data from several low-speed sources, perform certain manipulations on the data, retransmit at a higher speed, and vice versa. Logical functions which data concentrators usually perform are:

1. Buffer storage of messages.
2. Allocate storage and message queues.
3. Sample low-speed lines at a high enough rate to ensure bit integrity.
4. Translate codes (when different codes are used on low- and high-speed lines) including conversion from asynchronous to synchronous modes and vice versa.
5. Affix indications giving the sending-terminal address, whether the message segment is a complete message, and end of message.

If a part of a multipoint line, the concentrator logic must provide for response when polled.

More complex data concentrators are available, which are in effect small stored-program computers. These provide additional application-oriented functions, such as producing formatted phrases or sentences from simple character codes (or the reverse) to reduce the bits per message transmitted on the line facilities.

RADIO RELAY, TROPOSPHERIC SCATTER, AND SATELLITE COMMUNICATIONS SYSTEMS

59. Radio Relay (Microwave) Communication. Wide-band multichannel communications in the vhf, uhf, and shf radio-frequency bands was first experimentally introduced in the 1930s. It received considerable impetus through military use in World War II and has progressed steadily in frequency, capacity, and usage ever since. Today this type of circuit constitutes about half of the Bell System long-distance message-circuit mileage, and there are numerous other users of this type of system.

Most systems today operate at frequencies above 1,000 MHz (in the microwave region). The public telephone networks use the frequency bands 3,700 to 4,200, 5,925 to 6,425, and 10,750 to 11,750 MHz. Operation at these frequencies is characterized by (1) line-of-sight propagation, (2) absence of atmospheric noise, (3) ability to use a wide modulating bandwidth to accommodate many voice channels and TV traffic, (4) the use of highly directional antennas with high power gain, and (5) limitation in range to 20 to 40 mi per hop due to earth curvature and limits of tropospheric diffraction.

The low power requirements, moderate antenna size, and reliable propagation at these frequencies have justified the establishment of multihop circuits, with relay points at 20- to 40-mi intervals, in systems up to 4,000 mi long. The capacities of such circuits range from 12 to 22,000 voice circuits.

60. Components of Microwave Radio Systems. The components of microwave radio systems include the entrance links, baseband and intermediate-frequency repeaters (including a radio transmitter and receiver), the antenna with associated cable or waveguide feed, and the power supply. Frequency modulation is universally employed for microwave radio systems.

The *entrance links* connect the multiplex equipment to the radio equipment at distances from 50 ft to several miles. They consist of suitable cable (usually coaxial), baseband amplifiers with system drive-gain adjustment, cable equalizers, and transmission-level-shaping networks (pre- and de-emphasis networks). The multiplex terminals are usually FDM terminals, described in Par. 22-37.

The *baseband radio repeater,* shown schematically in Fig. 22-64, is the type usually employed for short-haul systems (up to 25 mi). The signal information is transferred within the repeater between the receiver and the transmitter at baseband frequencies and transmitted on a radio-frequency channel. Alternatively, at the entrance links, a connection can be established to multiplex terminals for terminating channels and introducing new channels. More than one radio system is usually established on a given antenna system, so that channel-separation and combining networks are required. A given system has two transmitters and two receivers at each repeater point and one transmitter and receiver at each terminal point.

The *intermediate-frequency repeater,* commonly employed on long-haul systems (up to 4,000 mi long), is shown in Fig. 22-65. In this type of repeater, the receiver output is heterodyned to an intermediate frequency, amplified, and coupled to the transmitter, where it is translated upward to the radio frequency for radio transmission to the next repeater. Elimination of the baseband-to-FM modulation step avoids several sources of noise and misalignment problems, since an if repeater does not change the deviation of the signal passing through it. Alternatively, at the points marked "to FM terminals" in Fig. 22-65 the required equipment for conversion to a baseband repeater can be added for terminating or introducing baseband channels. Typical transmitter output powers for this type of system range from 0.5 to 10 W. The received signals range from -39 to -20 dBm (an attenuation of 1 million to 1).

Antennas. There are two principal types of *microwave antennas,* the parabolic and horn reflector (see Sec. **18**). The *parabolic,* or *dish antenna,* 5 to 10 ft in diameter, is widely used

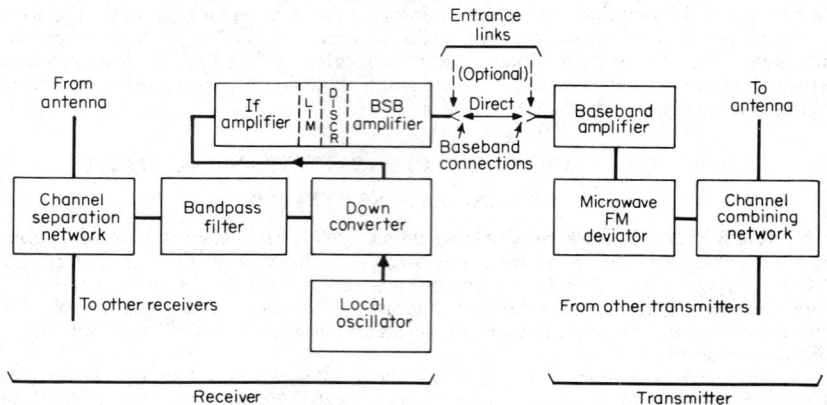

Fig. 22-64. Baseband radio repeater. *(Copyright, 1970, Bell Telephone Laboratories, Inc.; reprinted by permission.)*

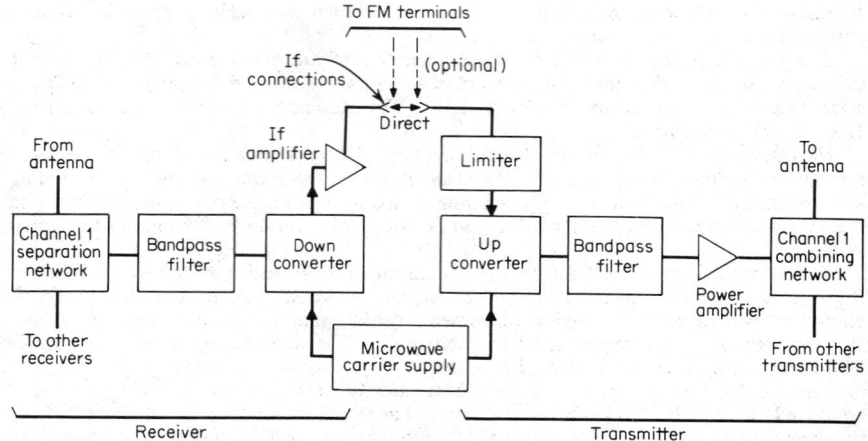

Fig. 22-65. Intermediate-frequency radio repeater. *(Copyright, 1970, Bell Telephone Laboratories, Inc.; reprinted by permission.)*

in short-haul systems. The *horn-reflector antenna* has excellent impedance match between antenna and waveguide and wide-band characteristics. It is generally used in long-haul systems. A suitable form of waveguide is usually used between transmitter, receiver, and antenna. Good impedance match at waveguide terminations is essential to keep losses low.

Power-supply reliability is a very important requirement for radio relay systems. Diesel-engine-driven power supplies, designed to function when commercial power fails, are required at relay installations. Where even momentary interruptions are intolerable, a floating battery system may be used to span the interval between commercial power failure and starting of the auxiliary power source.

Supervisory and alarm systems are required to monitor unattended relay or repeater installations. These systems vary in complexity from simple on-off tone signals to indicate trouble to highly sophisticated systems to report faults automatically. A relatively simple system using a 2,600-Hz monitor tone in the baseband, below 4,000 Hz, is often used, supplemented by interrogation tones to pinpoint the cause of the fault more closely. Major faults determined in this way are failure of commercial power, failure of transmission, low battery, lightning-arrestor failure, and failure of tower air-navigation warning lights. The clearance of failure by maintenance personnel is also signaled.

61. Engineering Factors in Microwave Radio Systems. In the public telephone network, microwave radio relay systems have performance criteria which are the equivalent of conventional wire-telephone circuits. In other commercial applications, where interconnection is not a problem, different standards have been worked out. These are incorporated in a number of EIA Standards on this subject.

The most important factor is the maintenance of a suitable signal-to-noise ratio over the system, on the noisiest channel, for 99.9% of time. In radio relay systems there is considerably more tolerance in determining repeater or relay-point spacing than in cable systems since the path loss increases only 6 dB as path length doubles. Other factors, such as tower cost, path clearance, geography, fading, rain attenuation, interference, and system requirements, play a significant part in the spacing determination and location of repeater points.

An expression for the signal-to-noise ratio at the baseband output, which sums up the losses and gains in decibels from transmitter output to the receiver baseband output for a single link, is

$$\text{Baseband S/N} = P_t + G_t - L_t - L_{fs} - L_{fad} + G_r$$
$$- L_r - 10 \log_{10} F_s K T_r B + 10 \log_{10} \text{MNI} \qquad (22\text{-}1)$$

where baseband S/N is the signal-to-noise ratio at receiver baseband output in decibels, P_t the transmitter output power in decibels with reference to 1 W, G_t the transmitter antenna gain in decibels, L_t the transmitter antenna line loss in decibels, L_{fs} the free-space propagation loss in decibels, L_{fad} the fading loss in decibels, G_r the receiving-antenna gain in decibels, L_r the receiving-antenna line loss in decibels, B the receiver noise bandwidth, $T_r = 290$ K, F_s is the system noise figure, K is Boltzmann's constant = 1.38×10^{-23} W·s/K, and MNI is the modulation noise improvement. For data regarding the terms of this expression see Secs. **9, 14,** and **18.** The free-space path loss L_{fs} can be ascertained from the nomograph, Fig. 22-66.

The *modulation-noise-improvement (MNI) factor* is defined as the signal-to-noise power ratio at the output of the receiver divided by the carrier-to-noise ratio at the input of the

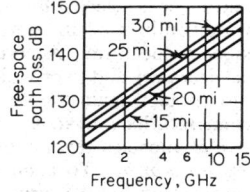

Fig. 22-66. Path loss as a function of frequency.

receiver. For FM systems, the most common type, the MNI factor relative to full baseband can be shown as a first approximation to be

$$\frac{3m^2B}{2b}$$

where B is the receiver bandwidth before demodulation, b is the bandwidth at receiver after demodulation (baseband), $m = \Delta F/b$ is the modulation index or deviation ratio, and ΔF is the peak frequency deviation for peak modulation by the full baseband signal.

In FM/FDM multichannel systems, where the number of channels is greater than 100, it can be shown that the MNI factor relative to the signal-to-noise ratio in the noisiest (highest) channel is

$$\frac{m^2B}{2b}$$

These relationships are valid only when the threshold power of the carrier at receiver input is 10 times the total noise power at the receiver input. Under other conditions, more complex relationships exist, which are treated in the references.

Thermal noise for the most part originates in the first stages of the radio receiver and sets the minimum signal amplitude acceptable for a given signal-to-noise ratio.

Intermodulation noise arises from transmission imperfections such as echoes in waveguide terminations, noise added by repeater local oscillators, and tones produced by beats between desired radio carrier frequency and interfering frequencies. Noise is generally considered to add on a power basis.

The received signal level can be determined from

$$P_r = P_t - L_t + G_t - L_{fs} - L_{fad} + G_r - L_r \qquad (22\text{-}2)$$

where P_r is the received signal level in decibels with reference to 1 W and P_t, L_t, G_t, L_{fs}, L_{fad}, G_r, and L_r are as defined for Eq. (22-1). Gains of from 25 to 50 dB are obtainable with the paraboloid and horn-reflector antennas generally used at these frequencies.

The theoretical free-space path loss L_{fs} is substantially that actually experienced for a large percentage of the time if the path provides an optical line-of-sight path with adequate clearance with respect to surrounding objects. For a small percentage of the time, however, abnormal atmospheric conditions cause deviations in propagation. Such fading may be (1) relatively fast and frequency-selective, caused by interference between two or more rays, or (2) multipath, caused by sizable atmospheric irregularities collecting under windless conditions in the upper atmosphere.

Another deviation from free-space path loss is the additional attenuation caused by atmospheric absorption (rain and oxygen). In the 4- to 6-GHz region this attenuation is small relative to other losses, but in the 10-GHz region and higher it is sufficiently great to be a factor in path engineering.

Space and frequency diversity are the techniques used to assure reliability of radio relay circuits. They can overcome severe fading effects not compensated by the automatic-gain-control circuits in the radio receivers. In *space diversity* two receiving antennas are separated vertically by 100 to 200 wavelengths. The signal output of the associated receivers is fed to the following transmitter. Taller and stronger (and hence more expensive) towers are required in this scheme.

Frequency diversity employs two radio channels of sufficient frequency separation to ensure that fading is essentially uncorrelated between them. The receiver outputs of the two channels are monitored, and the stronger is selected for onward transmission. In the Bell System this type of diversity is frequently used in short-haul systems. It is, of course, expensive in use of the radio-frequency spectrum. Accordingly, on long-haul systems it is customary to use space diversity as necessary and to provide only one standby radio channel for several operating channels to protect against both fading and equipment failures.

62. Tropospheric-Scatter Communications Systems. It has been found possible to communicate at uhf and microwave frequencies beyond the horizon, for distances up to 600 mi, by means of *tropospheric forward scatter*. To achieve this form of communication, appreciably higher transmitter powers (over 10 kW), larger antennas (up to 120 ft diameter), and narrower modulation bandwidths are required than in standard radio-relay communications. For modulation bandwidths of about 100 kHz, it is possible to cover distances up to

600 mi; for bandwidths of about 2 MHz, 100 mi is possible. Experience has shown that line-of-sight circuits are more economical than beyond-the-horizon tropo-scatter circuits, where sites are accessible for relay points. In many situations, e.g., island hopping, tropo-scatter operation is the only feasible mode.

The expression for received signal level, Eq. (22-2), must be revised for this type of transmission to include several additional attenuation loss factors. One expression is

$$P_r = P_t + G_t - L_t - L_{fs} - L_s - L_{cpl} - L_{ref} - L_{fad} + G_r - L_r$$

where P_t, G_t, L_t, L_{fs}, G_r, and L_r are as defined in Eq. (22-1) and P_r is the received signal level in decibels referred to 1 W, P_{fs} the free-space path loss, L_{cpl} the aperture-to-medium coupling loss at antenna, L_s the median scatter loss, L_{ref} the correction factor due to variation in the refractive index of air near the surface, and L_{fad} the fading margin. The significance of these terms is more fully discussed in Sec. 18 under tropospheric forward scatter propagation, and in the references. The great variability in tropospheric scatter propagation requires the use of dual or quadruple diversity, achieved by combinations of frequency and space diversity, to obtain reliable communications.

63. Satellite Communications Systems. The orbiting communications satellite permits line-of-sight communications between points on the earth's surface separated by several thousands of miles. Many believe it to be the ultimate radio-relay communications system.

Passive satellites, such as Echo I and II, were among the first types tried with success. This type of satellite is a reflector body illuminated by the two earth stations communicating with each other. Since the orbit altitude is relatively low, the time available for common viewing by the stations is relatively short and the number of satellites required for continuous viewing is unrealistically large. Therefore the passive satellite type is not considered to be a feasible communications system.

Active, real-time, communications satellites have proved very successful, particularly when placed in 24-h ("stationary") orbit, with the orbital plane including the equator of the earth. A satellite placed in such an orbit remains nearly stationary over a point on the earth's equator; i.e., its transorbit time about the earth corresponds to the 24-h period of the earth's rotation. To achieve such an orbit the satellite must be placed at an altitude of 22,300 mi above the earth.

Three such satellites, equally spaced in the orbital plane, provide continuous coverage for essentially the entire inhabited earth, with some overlap. The geometry of this arrangement, shown in Fig. 22-67, is the basis for international communications satellite operations and planning.

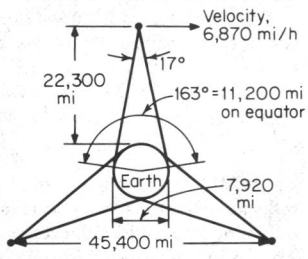

Fig. 22-67. Geometry of 24-h satellites.

Satellite systems of this type require sophisticated launching techniques, precise orbit control of the satellite, and spin stabilization, to maintain a constant attitude toward the earth and thus to permit the use of high-gain directional antennas and oriented solar-power decks. The communications satellite is a radio relay, consisting of a receiver and transmitter, plus a command receiver and transmitter, to control the satellite and to provide information on the status of its component systems.

The *ground stations* in a communications satellite system consist of a sensitive communications receiver, transmitter, and high-gain antenna. At least one ground station must also contain the command transmitter and receiver for control of the satellite.

The Intelsat (international communications satellite) satellites have employed frequencies in the range 3,700 to 4,200 MHz for the down direction (satellite to ground) and in the band 5,925 to 6,425 MHz for the up direction (ground to satellite). The capacities and bandwidths of present systems are such that over 1,000 telephone channels or several television channels can be transmitted over the system simultaneously.

In the engineering of ground-satellite communications links, the expressions for baseband signal-to-noise ratio, Eq. (22-1), and received power level, by Eq. (22-2), apply. The free-space path loss L_{fs} is the significant term, only atmospheric absorption loss has a modifying effect. The down-link, satellite-to-ground, calculation is the significant one in system engineering, since the necessarily relatively small size of the satellite antennas and limited power sources restrict antenna gain and transmitter output power. The ground stations, in contrast, have large, high-gain, precision-guided antennas. Examples are the 68-ft horn type at Andover, Maine, and the 60-ft parabolic type at Goonhilly Downs, U.K. Very sensitive, low-noise-temperature receivers are used, often of the *maser* type. Transmitters use traveling-wave tubes or klystrons at 50 to 20 kW output.

The distance traveled in a synchronous satellite circuit is some 45,000 mi, which represents a time delay approaching 0.3 s. Since echoes arising in telephone circuits of this length are annoying, echo suppressors are usually required (see Par. 22-26).

HIGH-FREQUENCY RADIO COMMUNICATIONS SYSTEMS

64. High-Frequency Radio Communications. In the 1920s the high-frequency (hf) band (3 to 30 MHz) was found to be a very useful means for long-distance radio communications. Since then the band has been extensively exploited. It is now extremely crowded but closely regulated to maintain some order in allocations. It is primarily used for long-distance international radio telegraph and telephone services, for communications to and between ships at sea, for communications to airplanes flying long-distance routes, for naval-command communications with the fleet at sea, and for military communications as initial contacts and primary backup in command communications.

The development in recent years of submarine-cable telephone communications systems, microwave radio relay, tropospheric forward-scatter communications systems, and satellite communications systems has considerably reduced dependence on hf radio as a long-distance communications medium.

65. Propagation at HF. The most important single factor in engineering hf radio circuits is knowledge and understanding of the propagation phenomenon in this band (see Sec. 18). Propagation over distances greater than approximately 100 mi is obtained by sky waves reflected from the ionosphere. During the day, the ionosphere is considered to consist of D, E, F_1, and F_2 layers at heights of 50, 90, 100, 200, and 300 km, respectively. At night, only the F_2 layer is present. The F_2 layer is the most important, both for day and night transmission, although the E layer is dependable for daylight transmission up to 2,000 km. The state of the ionosphere varies with the geographic zone, time of day, season of year, and the period in the 11-year sunspot cycle.

Waves reflected from the ionosphere return to earth and are there reflected again. Hence waves may arrive at the receiver in more than one hop, and several waves may arrive at the receiver by more than one path. The multipath effect with consequent fading must be taken into account in system design.

The working frequencies for a given path at a given time can be determined from

$$\text{MUF} = K f_c \sec \phi$$

where MUF is the maximum usable frequency, f_c the critical frequency (maximum frequency reflected from the ionospheric layer by vertical incident wave), ϕ the angle of incidence at the reflecting layer, and K the correction factor for earth curvature and vertical ionization density. The optimum working frequency (OWF) is usually taken as 0.85 times the maximum usable frequency. Predictions of worldwide ionospheric conditions are available on a regular basis from the Central Radio Propagation Laboratory of the National Bureau of Standards, Boulder, Colo. They are used to draw schedules for working hf circuits. Propagation is also affected by unpredictable factors, e.g., auroral activity about the geomagnetic poles, solar flares, and noise from thunderstorm activity.

66. HF Radio Transmitters and Receivers. The radio components of the hf radio communications systems are the transmitter, receiver, transmitting and receiving antennas, and transmission lines connecting antennas to transmitters and receivers. Normal transmitted bandwidths are confined to 3, 6, or 12 kHz for this portion of the spectrum.

Hf radio transmitters in greatest use are of two general types. The simplest type, the *frequency-shift-keyed* (FSK) *transmitter*, is shown in block-diagram form in Fig. 22-68. It is primarily used for telegraph transmission. The carrier is shifted about 400 Hz when the signal changes from a mark to a space. The bandwidth can be calculated from the approximate formula

$$F_w = 2(F_d + 3F_h)$$

where F_w is the required bandwidth, F_d is one-half the total shift, and F_h is the fundamental keying speed in hertz (one-half the speed in bauds). This type of transmitter is also used for facsimile transmission.

The *single-sideband transmitter* is the prevalent installation today because of its efficient utilization of the frequency band. It is shown in Fig. 22-69. The communication input can be one 3-kHz channel occupying one sideband of the radio spectrum or two 3-kHz channels separately occupying the sidebands on either side of the radio carrier. The 6-kHz channel is generally formed by FDM conversion of two 3-kHz channels. The sidebands are usually formed at intermediate frequency in the single-sideband generator.

The 3-kHz inputs may be voice, multiplexed telegraph or data, or facsimile signals (see Par. 22-67). Single-sideband generators, balanced modulators, and mixer exciters are discussed in Sec. **14.** This type of transmitter is generally rated on peak envelope power. Typical values range from 1 to 30 kW.

The *superheterodyne receiver* used to receive FSK signals is shown in Fig. 22-70. Dual and triple space diversity, employing two or three antennas suitably spaced and each with its own receiver, is frequently used in this system. High-frequency radio receivers of this type

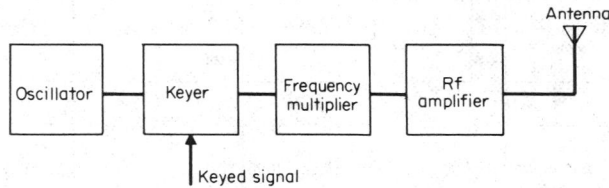

Fig. 22-68. Frequency-shift-keyed (FSK) transmitter.

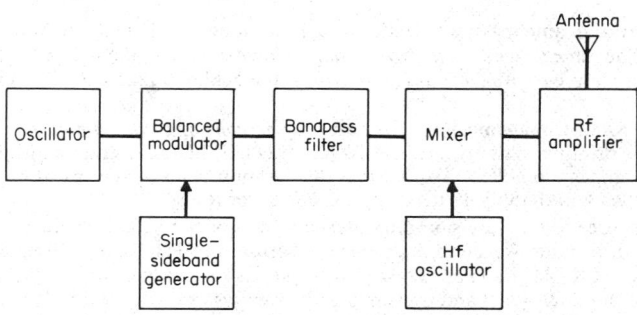

Fig. 22-69. Single-sideband transmitter.

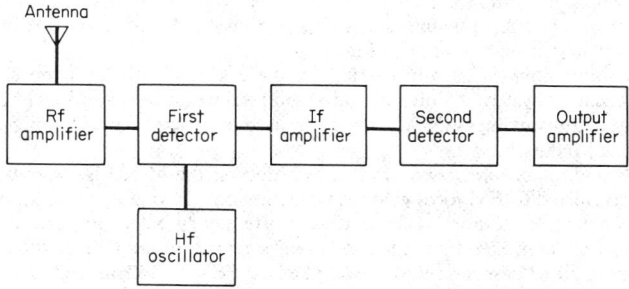

Fig. 22-70. Superheterodyne radio receiver.

typically have noise figures of 4 to 10 dB. Similar receivers, with appropriate intermediate- and audio-frequency bandwidths, are used for facsimile and audio reception.

The *single-sideband* (SSB) *receiver* is shown in Fig. 22-71, in dual-diversity form with two antennas, amplifiers, converters, filters, and demodulators. The two outputs are fed to combining or selecting circuits for obtaining the stronger signal. A typical noise figure for this type of receiver is 6 dB.

An SSB receiver has a 9-dB advantage over an equivalent double-sideband (DSB) receiver. The audio-frequency amplitude is twice that for DSB operation and uses only half the bandwidth.

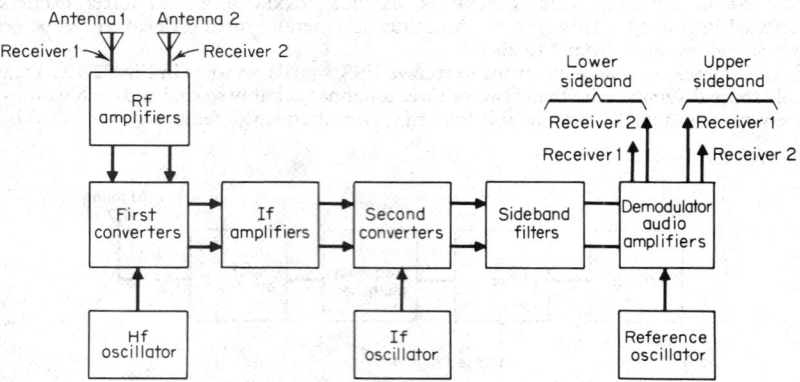

Fig. 22-71. Space-diversity single-sideband radio receiver.

Antennas and antenna-site selection are important to efficient operation of hf radio systems. The *rhombic antenna* is in wide use. Other types include half-wave dipole arrays, the logarithmic type with folded-dipole arrays, the fishbone receiving antenna, and the V antenna.

67. HF Radio Communications Services. *Telegraph and data communications,* particularly machine printing systems, are the largest users of hf radio communications systems. Multipath propagation effects require that the duration of a signal element be 5 ms or more (200 Bd or less), preferably 10 ms or more (100 Bd or less).

Single-source or time-division-multiplexed signals of 50 to 200 Bd spread can be applied to the FSK transmitter-receiver radio system discussed in Par. 22-66. Frequency-division-multiplexed FSK signals (Par. 22-54 and Table 22-14) are generally used in the 3-kHz channels of the single-sideband transmitter-receiver system described in Par. 22-66. Codes and error control (Pars. 22-54 and 22-64) are used. The constant-ratio Moore code is quite frequently used for telegraph transmission. For data transmission the ASCII code (with simple parity or more complex error-control schemes) is used. In data transmission parallel-by-bit, serial-by-character transmission is most common. Morse code is still used to some extent, particularly in marine applications.

Voice communication is usually carried on the 3-kHz channels of the single-sideband transmitter-receiver system. Voice-operated gain-adjusting devices (VOGAD) and voice-operated device antisinging (VOSAD) are often used in this service to compensate for fading and echo impairments.

Facsimile and radiophoto services are transmitted on the hf bands. For best results the signals are converted to FM form prior to transmission. In one method, facsimile at audio frequency is converted to an FM form at audio frequency before application to the input channels of an hf radio SSB transmitter-receiver system. In another method, the facsimile signal is converted to baseband and used to key an FSK radio transmitter-receiver system (Par. 22-66). In both methods limiter-discriminator FM detection is used at the receiver.

MOBILE-RADIO COMMUNICATIONS SYSTEMS

68. Mobile-Radio Communications Services. Mobile services occupy about 41 MHz of the frequency spectrum in the 25- to 50-, 150- to 162-, and 450- to 470-MHz ranges, commonly referred to as low band, high band, and 450-MHz band respectively, in mobile-radio parlance. From 1946 to 1959 these services had a growth rate of better than 20% per year, but the rate of growth has declined somewhat since. In 1963 the total number of vehicular mobile transmitters installed exceeded 1 million. In 1970 the FCC authorized the use of 115 MHz of bandwidth in the 900-MHz region for the mobile land radio services particularly in urban areas. It is expected that this allocation will expand the radio-telephone service and lead to more utilization of the spectrum for mobile services where spectrum crowding and interference are particularly bothersome.

The mobile services are divided into five basic groups by FCC Rules and Regulations, each comprising services having similar usage or from related industries, as follows:

Maritime mobile services include coastal, inland-waterway, and Great Lakes ship-to-shore communications.

Public-safety radio services include forestry conservation; highway maintenance; special emergency for hospitals, doctors, disaster relief, and ambulances; fire; state guard; and local government.

Industrial radio services include forest products, motion picture, relay press, special industrial, manufacturers, and telephone maintenance.

Land-transportation radio services include motor carrier, railroad, taxicab, and automobile emergency.

Domestic public radio services include domestic fixed and mobile, for use by general public to provide an extension to the telephone system of the office and home, and rural, to provide telephone facilities in remote areas.

Frequency allocations to each service are made by assigning several blocks of closely spaced frequencies in the mobile radio-frequency bands, so that each service has available for its own use only a small fraction of the total frequencies. The closely spaced channels within a block, coupled with the limited number of available frequency assignments, create intersystem and intrasystem radio-frequency interference problems.

69. Mobile-Radio Propagation. Propagation at the vhf and uhf frequencies used in mobile-radio services is treated in Sec. 18. Propagation in the mobile-radio case is very complex, due to the disparity in antenna heights between base station and the mobile station and the changing transmission paths. The propagation is (1) free-space for short distances, where Fresnel zone clearance is maintained, and (2) plane-earth for longer distances, modified by various losses as follows:

Diffraction loss occurs due to the transmission path following the earth's curvature. The magnitude of this loss depends on frequency and distance when antenna heights are less than a limiting height, as shown in Fig. 22-72.

Shadow loss covers the transmission loss through or around hills, buildings, trees, etc., in the transmission path. It increases with frequency and is to some extent offset by reflections from objects not located in the direct path. Egli has treated this loss as a terrain factor, arriving at median terrain losses of 0, 12, and 22 dB for the low, high, and 450-MHz bands respectively. A statistical treatment, applicable when average height of hills is known, is shown in Fig. 22-73. This gives the estimated shadow loss exceeded at 10 and 50% of the locations for each of the three frequency bands.

Ionospheric scatter is not used for communicating in mobile services but can be a potent source of interference during periods of high sunspot activity.

Tropospheric scatter is seldom a cause of interference in these services. It is sometimes used for communications between fixed stations.

The expected *base-station propagation losses* are shown in Fig. 22-74 for the three bands. For ranges of 1 to 5 mi, free-space loss controls, increasing at a rate of 6 dB as the distance doubles. For 5 to 10 mi, plane-earth loss is controlling, increasing at a rate of 12 dB per double distance. From 10 to 60 mi, plane-earth loss is controlling, with diffraction loss gradually increasing from 12 to 30 dB per double distance. From 60 to 100 mi scatter loss reduces the slope to about 10 dB per double distance. Beyond 100 mi the attenuation slope gradually increases to 40 dB per double distance.

The expected *base-to-mobile-station losses* are shown in Fig. 22-75. Plane-earth plus diffraction losses are controlling up to 170 dB attenuation. Above this point the losses are controlled by scatter. Where obstructions occur in the path, shadow loss must be added.

Expected *mobile-to-mobile-station losses* are shown in Fig. 22-74. Again plane-earth plus diffraction losses are controlling up to 170 dB, but the ranges are substantially shorter since both antenna heights are low.

70. Transmitter and Receiver for Mobile-Radio Systems. The basic equipment design for mobile service, particularly that of the transmitter and receiver, is substantially the same in all services. However, options mounting into the hundreds are available to meet specific service demands. Practically all equipment manufactured for these services is built to conform to EIA Standards. A comparison of EIA Standards with a typical manufacturing specification for receivers and transmitters is shown in Table 22-15.

Transmitters used in the mobile-radio services shown schematically in Fig. 22-76 consist of a crystal oscillator, an audio section, a phase modulator, frequency multipliers, and power amplifiers.

The *crystal oscillator* is designed to hold maximum frequency change to better than 0.0005%. Close frequency control minimizes impulse-noise degradation. Multifrequency operation on up to four channels is usually provided by a crystal-changing switch or by switching to other oscillators. In the *audio section* the audio must often be preamplified 20 to 30 dB before passing to the limiter. A preemphasis peak limiter and deemphasis receiving limiter using simple *RC* circuits, with time constants of 75 μs, are often used. Limiting is

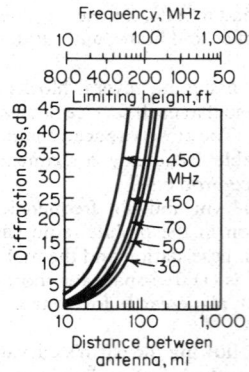

Fig. 22-72. Diffraction loss caused by bending of radio waves around the earth.

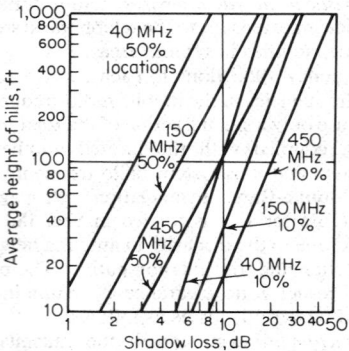

Fig. 22-73. Estimated shadow loss exceeded at 10 and 50% of locations in areas where the average height of hills is known.

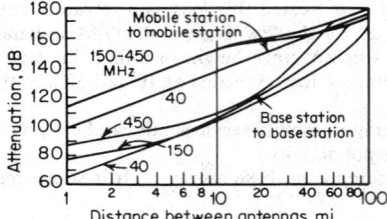

Fig. 22-74. Attenuation between base-station antennas with antenna heights 100 ft above average terrain. Attenuation between base-station and mobile-station antennas for base-station antenna height of 100 ft.

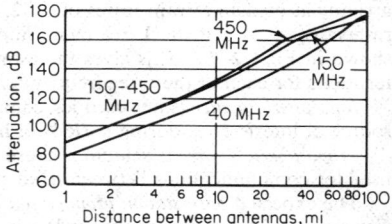

Fig. 22-75. Attenuation between base station and mobile station with antenna height of base station at 100 ft. Effective antenna height of 40 MHz mobile assumed to be 20 ft.

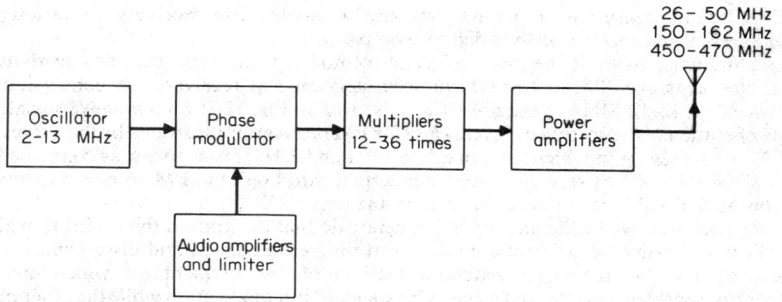

Fig. 22-76. Typical block diagram for a phase-modulated radio transmitter used in the mobile services.

Table 22-15.   EIA   Standards and Manufacturers' Specifications

Quantity measured or specified	EIA Standards			Typical manufacturers' specifications		
	25–54 MHz	144–174 MHz	400–470 MHz	25–54 MHz	144–174 MHz	400–470 MHz
Mobile Receiver						
Sensitivity, μV	1.0	1.5	2.5	0.3	0.5	1.0
Selectivity, dB	70	70	70	80	80	80
Spurious attenuation, dB	85	85	80	100	100	90
IM spurious attenuation, dB ..	50	50	50	60	60	60
Audio power output, W	1	1	1	2	2	2
Mobile Transmitter						
Power output	*N/A*	*N/A*	*N/A*	80 W	80 W	60 W
Frequency stability, %	±0.002	±0.0005	±0.0005	±0.0005	±0.0005	±0.0005
Spurious radiation, dB	43 + 10 log (power output)			53 + 10 log (power output)		

applied to reduce the maximum deviation in the modulator to 5 or 15 kHz, as required by the FCC. A *phase modulator* is used since the carrier can be generated by a stable crystal oscillator. The resulting modulated signal is substantially FM when the audio response falls off at a rate of 6 dB/octave. The *frequency multipliers* serve the dual purpose of amplifying the voltage from the oscillator and multiplying the phase deviation of the modulator. Multiplication factors normally used provide peak deviations of 12, 18, 24, or 36 rad at the last stage, with power output nominally in the 1- to 3-W range. The *power amplifiers,* normally class C, receive the multiplier output and amplify it in one or more stages producing from 30 to 330 W output at base stations and from 10 to 100 W output at mobile stations. This output is coupled to the antenna through a low-pass or bandpass filter to reduce the amplitude of spurious responses generated (or amplified in the power-amplifier stage) and also to reduce transmitter noise.

The *radio receivers* commonly used in the mobile-radio services are solid-state, double- or triple-conversion superheterodyne types, designed for the reception of FM signals. The design of these receivers is well covered in the *IEEE-PGVC Transactions.* Receiver performance is critical to the system. The more important performance requirements and methods of measurement are covered by EIA and IEEE Standards.

Receiver sensitivity is generally limited by radio-frequency noise. The most widely accepted method of specifying sensitivity is the 12-dB SINAD or *usable-sensitivity* method. SINAD refers to the ratio of signal + noise + distortion to noise + distortion expressed in decibels. The 12-dB ratio is used as a convenient reference. Referring to the values listed on the sensitivity line in Table 22-15, it would be required (when the signal at the antenna is as shown) that the SINAD ratio be 12 dB or better. The curves of Fig. 22-77 illustrate

SINAD ratio measurements for narrow- and wide-band FM receivers for various signal inputs. An AM receiver is included for comparison.

While noise figure is the most universal method of measuring receiver sensitivity, such measurements are difficult to perform accurately on FM receivers. A conversion graph between the 12-dB SINAD and noise figure is given in Fig. 22-78 for a range of signal inputs. The operation of the so-called *capture effect* for FM receivers is illustrated by the curves of Fig. 22-77, in the region in which the slopes are 1.9/1 and 2.4/1, respectively, as contrasted to the 1/1 slope for the AM receiver. Above a certain threshold, the FM receivers follow a 1/1 slope, as the AM receiver does throughout the range.

Receiver selectivity is defined by a characteristic that determines the extent to which the receiver is capable of differentiating between the desired signal and disturbances at other frequencies. The two-signal-generator method of measurement (in which one signal generator represents the desired signal with standard test modulation while the other provides the disturbance at other frequencies) has been standardized. Figure 22-79 illustrates a typical selectivity curve for a narrow-band fm receiver. The position from 0 to about 16 kHz off-channel frequency is produced by the low i-f filter. Above the 16-kHz point other circuits closer to the antenna play the major role.

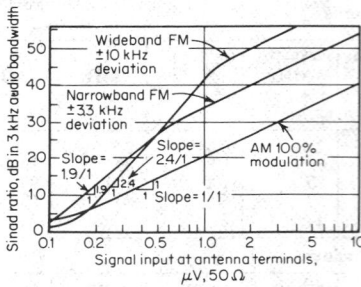

Fig. 22-77. Comparison of typical sensitivity performance of wide-band FM and narrow-band FM and AM receivers. All three receivers are compared on the basis of a 10-dB noise figure.

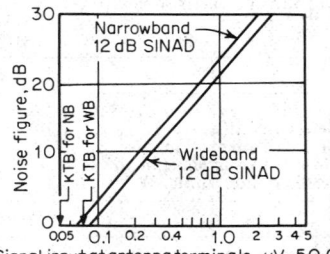

Fig. 22-78. Conversion of sensitivity to noise figure for typical wide-band and narrow-band receivers.

Multifrequency operation can be achieved by use of switched crystals or separate front-end switching. If frequency separations are greater than 0.4%, dual-front-end receivers are recommended since the loss of sensitivity increases from 3 dB for 0.4% separation to 14 dB for 0.7% separation.

A *squelch circuit* is normally provided on all fixed tuned receivers. Its function is to render the audio amplifier and speaker inoperative when no signal or carrier is being received. This cuts out background noise and resulting annoyance. The squelch circuit is usually controlled by the noise or noise and signal from the limiter. When noise is high, the squelch circuit acts to disable the audio amplifier; when signal is present (and the noise low), the squelch opens and the audio amplifier becomes operative. The squelch circuit is critical and can cause trouble when circuit values change for any reason.

Tone- or coded-tone-operated squelch systems are frequently used. The tones used are in the band 67 to 250 Hz, outside the audio passband of the communications system. Means for applying the tone to the transmitter and for opening the local receiver squelch must be provided. Tone-coded squelch allows a variety of functions in setting up communication nets, in addition to their normal use.

71. Radio-Frequency Interference. Interference has become the most important factor to consider in the design, operation, and maintenance of mobile communications systems. Interference problems are arising at a faster rate than the ability of the equipment designer to cope with them. The major types of interference are classified as *cochannel* and *off-frequency* interference. Sources of cochannel interference include local cochannel transmitters, skip cochannel transmitters, incidental radiation devices, galactic and atmospheric noise, transmitter noise, transmitter spurious emissions and harmonics, and transmitter intermodulation. Sources of off-frequency interference include receiver desensitization, receiver selectivity, receiver spurious responses and images, and receiver intermodulation.

Transmitter and receiver intermodulation spurious responses have become a serious interference problem. Intermodulation interference is generated in any nonlinear circuit by the mixing of two or more signals whose products fall on or near the desired frequency. *Receiver intermodulation* is controlling when the transmitter antennas are separated by more than 500 ft; it is usually produced in the first converter. Expressions for typical third- and fifth-order equivalent intermodulation interference levels for high-band receivers are

$$IM_3 = 7 + 2A + B - 60 \log_{10} (f_A - f_B) \quad \text{dBW}$$

$$IM_5 = -60 + 3A + 2B - 135 \log_{10} (f_A - f_B) \quad \text{dBW}$$

where IM_3 and IM_5 are third- and fifth-order equivalent intermodulation interference level in decibels referred to 1 W, f_A and f_B are frequency of interfering transmitters in megahertz, A the signal level at receiver input of nearest frequency transmitter in decibels referred to 1 W, and B the signal level at receiver input of farthest frequency transmitter in decibels referred to 1 W.

Transmitter intermodulation is usually controlling when the transmitter antennas are less than 500 ft apart; it is usually produced in the power amplifier. Transmitter-intermodulation products are usually expressed in decibels below the level of the injected signal, as shown in Fig. 22-80. For example, if two high-band transmitters have 50 dB attenuation between their antennas and are spaced 1 MHz apart, the third-order product from each transmitter would be 63 dB below their respective carrier outputs since the injected signal is 50 dB below the carrier output. *Cavity filters* can be used to reduce or eliminate transmitter and receiver intermodulation products. In some cases loose connections in the transmitter antenna or poor bonding of metal parts near the antenna may be the cause.

Receiver desensitization occurs when an off-frequency signal which enters on the upper or if portion of the selectivity curve (Fig. 22-79) causes limiting in certain amplifier or mixer stages of the receiver. Improvements in receiver performance, particularly the use of helical tuned circuits in the rf portion, have virtually eliminated this problem.

Modulation splatter occurs when an off-frequency transmitter has sidebands which lap over into the acceptance band of the selectivity curve in adjacent channels. It adds to the noise and reduces the channel interference ratio.

Receiver and transmitter spurious responses and radiation have gradually diminished as important interference problems. The use of low-pass filters in transmitters and improvements in receiver selectivity have been major factors in eliminating this problem.

Incidental radiation from devices operating in the range from 30 to 100 MHz are polluting the rf spectrum and are a factor to be considered in system planning, although no reliable data

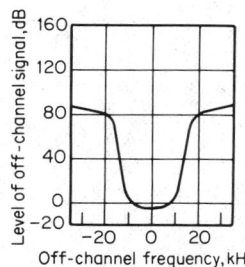

Fig. 22-79. Typical two-signal selectivity curve of narrow-band receiver.

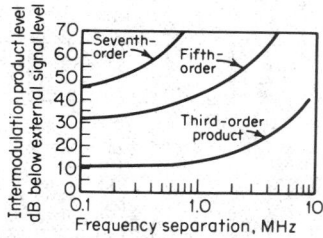

Fig. 22-80. Typical intermodulation curves for high-band transmitter. Each curve represents the level of intermodulation product radiated from a transmitter in decibels below the level of external signal injected into the transmitter.

are available on average and peak noise level. Measurements have shown levels (from white noise to peaks) of 80 dB above average noise level.

72. Base-Station Equipment. A base-station installation comprises the transmitter and receiver combination, antenna and transmission line, control equipment, and cavity filters.

The *transmitter and receiver* are of the types discussed in Par. 22-70. Cabinets may be mounted on floor, wall, pole, or desk and should be arranged for easy access to equipment.

The *antennas* are normally mounted on a tower of suitable height to give coverage desired. In the low-band service, the usual types of antennas are ground-plane and coaxial. For high-band and 450-MHz operation, stacked arrays of folded dipoles are frequently used.

Where transmitters and receivers are operated simultaneously from the same locations, it is necessary to provide the maximum possible attenuation between transmitting and receiving antennas by vertical or horizontal spacing. Ten times greater horizontal spacing is required for the same attenuation relative to vertical spacing. For vertical spacing of 25 ft, attenuation between antennas is about 32 dB at 40 MHz, 52 dB at 150 MHz, and 65 dB at 450 MHz.

The *transmission lines* are of the coaxial or balanced-pair types suitable for the frequency range and installation environment and are discussed in Sec. 9. Losses should be kept to the minimum to obtain maximum transmitter-power output.

Cavity filters are often used where two or more transmitters or a transmitter and receiver are operated simultaneously from the same location to provide additional selectivity for both the transmitters and receivers and thereby reduce interference to receivers. One effective means is to insert the cavity filter in the transmission line between the antenna and the transmitters and receivers.

A *control unit* is required when operation is not conducted by manipulating the controls at the transmitter and receiver itself but at some distance. A local control unit can extend control up to several hundred feet. It usually consists of an audio amplifier, indicator lights, microphone, speaker, push-to-talk switch, squelch control, and level controls. The level controls maintain optimum modulation level at the transmitter and adjust the receiver audio to meet operating needs.

Remote-control units are provided where operation is at a considerable distance from the transmitter-receiver location and is conducted through one or two telephone pairs between locations. Three methods for providing telephone-line control circuits are shown in Fig. 22-81. Method *(a)* uses two telephone pairs, one pair for the audio and the other (a metallic pair)

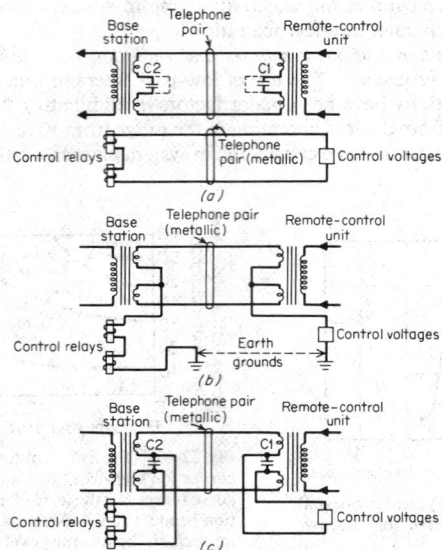

Fig. 22-81. Telephone-line control circuits.

for the controls. Method *(b)* uses one pair with the audio applied directly to the pair through a transformer or repeating coil and the controls applied at the center of the transformer secondary and ground. In method *(c)*, the audio is applied to the pair through the transformer primary, but the control voltages are applied to the pair through split-transformer secondary windings, thus avoiding using ground where earth currents are a problem.

The policy of the telephone companies recently has been to furnish audio pairs only, so that methods *b* and *c* must be used in new installations. Telephone companies have set limits on transmission levels to reduce mutual interference. The following are typical: maximum dc voltage on line of 270 V, maximum dc voltage line to ground of 135 V, maximum line current ac or dc of 0.35 A, and maximum speech level, 8 vu.

73. Mobile-Station Equipment. A mobile installation usually consists of receiver and transmitter, antenna, power supply, control head, microphone, speaker, and in mobile telephone service a handset and dial. In passenger vehicles, the main unit, consisting of transmitter, receiver, and power supply, is usually mounted in the trunk while the control head and speaker are mounted on the dash. In trucks and other large vehicles, the entire installation may be mounted under and on the dash.

The *power supply for the mobile station* is almost universally of solid-state design, operating from the vehicle power system. Most vehicles and small boats have a nominal 6 or 12 V dc battery, charged by an engine-driven alternator with a voltage regulator. This power system produces wide fluctuations in output voltage. For 6- or 12-V dc primary power systems it is usually designed to accommodate a 20% variation in primary power-source voltage in the station-radio power supply. Some form of voltage regulation is generally required in the supply for oscillator and audio stages.

Higher-voltage mobile power systems are also in use, particularly nominal 72 V diesel-locomotive battery-power systems. The higher-voltage primary sources usually have greater than $\pm 20\%$ variation in voltage, so that a voltage regulator in the power supply is a necessity in all applications.

Vehicular noise, primarily produced in the vehicle's ignition system, is often a major source of radio interference. Standards of SAE are designed primarily to reduce interference to home television reception. Levels can be as much as 25 to 50 dB above front-end receiver noise. Vehicular noise is primarily of the impulse type, and its effect on received signal depends on the particular receiver design and frequency stability of transmitter and receiver.

Impulse-noise blankers may be necessary where other measures do not reduce impulse-noise interference to a satisfactory level. These devices may be an integral part of the receiver or an outboard device. Three major types, each with its own advantages and disadvantages, are shown in Fig. 22-82. Blanker A is applied to the audio portion of the receiver and has the advantage that it can be adjusted to trigger only on those impulses which would eventually reach the audio. It has the disadvantage that since the noise impulse becomes considerably stretched in the low if stage, the required repetitive blanking pulses will remove an appreciable part of the audio. An upper blanking limit of about 200 pulses per second has been established.

Blanker B is applied to the high if stage and is triggered by voltage derived after rf selectivity. It has the advantage of having considerably shorter blanking pulses operating at rates of greater than 2,000 pulses per second. It has the disadvantages that blanking may occur on strong desired signals, may generate intermodulation signals and desensitize the receiver, and may cause interference by not accommodating to a change in the operating frequency.

Blanker C has a completely separate rf circuit for deriving the trigger voltage, with the result that blanking pulses may be less than 4 μs in length and the rate greater than 100,000 pulses per second. It has the disadvantage that while removing all undesired impulses it continues blank unless turned off and can thus degrade the receiver's performance because of intermodulation in the blanking channel or through blanking a very strong adjacent channel signal.

Mobile antennas are available in two basic types: quarter-wavelength ground-plane and half-wavelength bumper-mount coaxial antennas. High-band and 450-MHz ground-plane antennas can be mounted on the rooftop because of their limited height. Low-band antennas are 5 to 8 ft long and must be mounted on the bumper or lower part of the fender. Loading coils are used to reduce the physical length of low-band antennas, allowing trunk-lid mount. Pattern distortion can result from any of the mountings except the rooftop type. Coaxial antennas are well suited to high-band and 450-MHz use when mounted on rear bumper. Little pattern distortion results provided a considerable percentage of the antenna radiator is above the rooftop.

74. Mobile-Radio Systems. In mobile-system design, the first consideration is communication range. Factors influencing this parameter are the frequency band used, transmitter power, base-station antenna height, man-made noise levels, terminal variations, and receiver sensitivity. A typical well-engineered system in an average environment with antenna heights of 100 and 20 ft and transmitter powers of 250 and 20-W each in base and mobile stations, respectively, has a maximum transmission range of 40 mi for the low- and 450-MHz bands, and 35 mi for the high band. The lower inherent noise in the 450-MHz range makes it competitive in noisy environments.

In the quieter environments, the low and high bands have much greater range, up to 50 to 70 mi. Reliable communication ranges under average conditions are usually quoted as 25 to 50 mi in the low band, 15 to 25 mi in the high band, and 10 to 15 mi in the 450-MHz band.

Single-frequency simplex systems are illustrated in Fig. 22-83. Base and mobile stations transmit on a common frequency on a push-to-talk basis. When two or more systems operate on the same frequency in the same general area, considerable interference results. This effect reduces the effective mobile-to-base range in proportion to the distance between base stations.

Two-frequency simplex systems are illustrated in Fig. 22-84. Base and mobile stations operate on a push-to-talk basis but use different transmitting frequencies. Large systems can operate on a zone basis, with a base station near the center of each zone and adoption of usage rules to take care of overlap areas.

Two-frequency base duplex systems are illustrated in Fig. 22-85. The base station transmits on one frequency while simultaneously receiving on a second frequency. The mobile station operates in two-frequency duplex on a push-to-talk basis. Attenuation must be provided between base-station transmitting and receiving antennas by cavity filters, antenna separation, or both. The base duplex station can also operate as a relay for retransmitting mobile-mobile conversations and thus extend mobile-mobile communication ranges considerably. For this type of operation, the base station must be equipped with a carrier-operated or coded-signal-operated relay (or both) to bring the relay station on line.

Two-frequency duplex systems (Fig. 22-86) are similar to the base duplex, but the mobile station also operates duplex. To achieve this the mobile station must be equipped with a second antenna (or necessary filters between transmitter and receiver) to prevent interference.

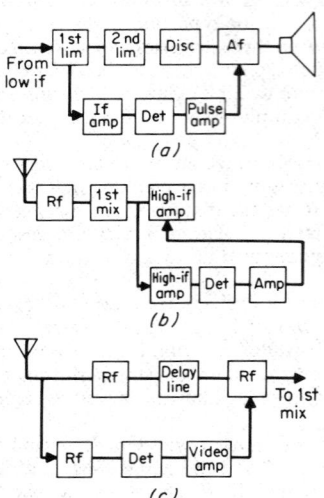

Fig. 22-82. Impulse blankers used in mobile equipment.

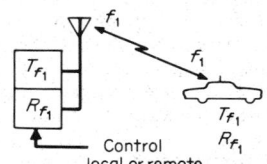

Fig. 22-83. Single-frequency simplex system. Both base and mobile stations operate push-to-talk on a single frequency.

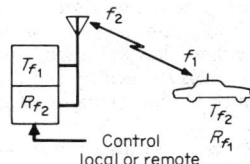

Fig. 22-84. Two-frequency simplex system. Base and mobile stations operate push-to-talk on different frequencies.

Since duplex systems operate on a heavier-duty cycle, the mobile equipment is of larger size and requires heavier components. This system has found only limited application, except in mobile telephone systems (Par. 22-75).

75. Mobile Telephone Systems. These are two-frequency duplex systems. A simplified schematic of such a system is shown in Fig. 22-87. Considerable flexibility is required in the equipment, since a mobile-telephone subscriber is not identified with any one base station but may use his mobile station in any area where the service is given. The base-station receiver and transmitter are tied to the land telephone system through an automatic level-control device to assure proper signal levels.

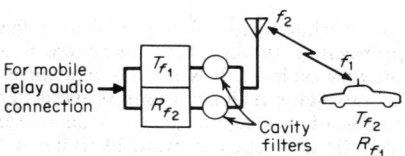

Fig. 22-85. Two-frequency base duplex systems. Base operates duplex, while mobile operates push-to-talk two-frequency simplex.

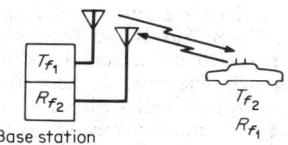

Fig. 22-86. Two-frequency duplex system. Both base and mobile stations operate duplex. Both base and mobile stations must have either antenna filters or two antennas each.

In the simplest form of system, the mobile-telephone operator alerts the mobile-telephone station by dialing a coded signal over the appropriate base transmitter, usually employing 600- and 1,500-Hz frequency shift tones which actuate a decoder in all mobile units on the channel. When the code combination assigned a particular mobile unit is received, the decoder actuates circuits which cause a buzzer to operate and a lamp to light indicating a waiting call. The decoder was originally based on the Gill selector, a polar relay actuating a stepping mechanism. Mechanical selectors have been largely replaced by solid-state logic devices. In placing or acknowledging a call the mobile subscriber takes his handset from the cradle and presses the push-to-talk switch which activates the transmitter alerting the operator.

A recently implemented system, the improved mobile-telephone system (IMTS), features full automatic call handling in both directions and provides access to several channels by each mobile station.

When the mobile station is activated, the receiver hunts over its assigned range for a *marked-idle-channel*, identified by a steady 2,000-Hz tone. It remains tuned to this channel until the channel is seized. It then hunts and finds another such channel. All other mobile stations not engaged in a call continue this hunting routine.

When a mobile station seizes an idle channel by activating the transmitter, coded tones automatically transmit its identifying code, which is recorded for billing purposes. The mobile subscriber then receives dial tone and uses his dial to place the call as from home or office.

These mobile systems employ several geographically separated base-receiver locations for each base transmitter. Some form of voting logic is included in the base-receiver audio control to select the receiver location providing the best signal-to-noise ratio and locking out all others during the transmission.

76. Backbone Mobile-Radio Systems. This name is given to systems used in the operation and maintenance of limited-access highways, railroads, and pipelines. They are most often operated as two-frequency base-duplex systems and are characterized by the fact that the system is required to cover only a narrow strip along a right of way. This configuration needs more than one base station along the route to provide coverage. Interference can thus occur at the mobile receiver in the overlap areas between adjacent base stations. The uncertainty in selecting the base receiver providing the best signal-to-noise ratio must also be overcome. Techniques developed to overcome these problems are:

1. Installation of manually or automatically switched directional antennas in the mobile unit. This allows the mobile unit the possibility of selecting the stronger signal while attenuating the weaker signal.

2. Maintaining accurate frequency control at the base stations with the view of reducing the carrier difference to less than 300 Hz so that the beat note of the two carriers will be below the audio passband of the mobile receiver.

3. Phase-equalizing the audio to maintain synchronized deviation on all base stations. Interference is a minimum when the instantaneous deviation between each base station as received at the mobile receiver is a minimum.

4. Equipping the base stations with directional antennas having the greatest possible front-to-back ratio, thus reducing the overlap area considerably.

5. Installing voting-logic devices in the audio portion of the control system to automatically select the base receiver providing the best signal-to-noise ratio, locking out all other receivers for the duration of transmission.

77. Open-Wire Radio-Transmission Systems. Such systems are employed in tunnels, buildings, and underground facilities where propagation through the atmosphere is too limited to be effective. In these environments, attenuation in decibels at radio frequencies is directly proportional to distance. Attenuation in a two-lane vehicular tunnel is about 10 dB per 100 ft, limiting the range of mobile-unit communications to about 1,000 ft. Concrete and steel buildings have attenuations of 10 dB per 100 ft (in large open areas) to 20 dB per 100 ft (for partially partitioned areas). Attenuation through 8-in. concrete walls is 8 to 10 dB and through solid metallic walls from 10 to 40 dB.

To provide adequate received signal level in these environments it is necessary to install distributed lumped-radiator feed from low-loss transmission lines. Lumped radiators in the form of dipole antennas are satisfactory if the area can be covered with no more than 10 antennas. Where greater coverage is required, unshielded, twin-conductor transmission lines such as RG-86/U are a better choice. The attenuation of RG-86/U varies with the spacing from the dielectric surface of the wall and with frequency, as shown in Fig. 22-88.

A design calculation can be made for such a tunnel communication system by assuming 55 dB attenuation between the transmission-line radiator and the mobile equipment. Assume a 50-W 160-MHz transmitter at the middle of the line, with 25 W (14 dBW) output in each line section and a minimum received signal level of − 130 dBW. The system gain is 144 dB. The line-to-mobile unit attenuation of 55 dB allows a line loss of 89 dB before the radiated signal becomes unsatisfactory. From Fig. 22-88 this gives a transmission-line length of about 8,900 ft per section, or 17,800 ft (3.25 mi) total length.

78. Personal Portable Radio Systems. These systems have been fabricated in smaller and smaller packages through advances in solid-state circuitry, batteries, and other components, to the point that a transmitter-receiver can fit conveniently in a man's clothing. Each reduction in size is accompanied by a reduction in range for talk-back, due to reduced antenna efficiency and battery power. More base-station receiver locations are required to compensate for this.

Personal portable receivers capable of being carried or worn by a person have been the result of advances in solid-state circuitry. The EIA performance specifications for these receivers are shown in Table 22-16 (compare those for mobile receivers in Table 22-15). The most significant performance reduction is in sensitivity, which in this case is the average

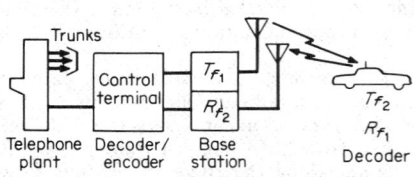

Fig. 22-87. Mobile-telephone system. The base station operates duplex, while most mobile stations operate push-to-talk.

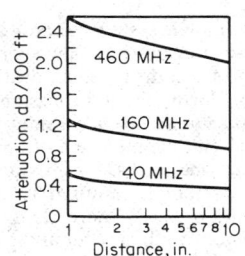

Fig. 22-88. Attenuation of RG-86/U at various distances from a concrete surface.

radiating sensitivity including the antenna and the influence of man on its efficiency. The reduction in audio power is not significant since the receiver is usually close to the ear or is an insert in the ear. The receiver may be operated as a separate unit in a one-way signaling system or combined with a transmitter for two-way communications. Ranges are limited to a few miles in the open and much less inside buildings.

Personal portable transmitters also employ solid-state circuit design. The power output ranges from 0.5 to 20 W in the 25- to 54-MHz band and from 0.2 to 8 W in the 144- to 174-MHz band. (The 450-MHz band has not found much use in this service.) The effective radiated power of the transmitters is considerably reduced from these ratings since the transmitter is often operated with the telescoping antenna in the down position. With the antenna in the extended position at head height the effective radiated power is less than 3 dB below rated power output.

Portable transceiver equipment with transmitter and receiver in one compact package (often called a walkie-talkie) has found widespread use in outdoor work, where ranges of less than 1 mi are satisfactory. Typical applications are in all types of construction, harbor pilots to bridges, and outside maintenance on power and communications lines and rights-of-way. In these applications the package is often carried by a belt clip.

Antennas for portable equipment are necessarily small, with the largest dimension less than one-half wavelength. The radiation impedance of small antennas is usually less than their self-reactance, so that they operate at less than unity power factor and store an appreciable amount of energy. This reduces the radiated power, compared to a half-wave dipole. High-Q antennas of this type have restricted bandwidth compared with a half-wave dipole. The most common forms of these small antennas and their average loss, compared with a half-wave dipole, are:

1. High-Q ferrite-core resonant coils: 20 dB loss.
2. Earpiece cord and short wires: 25 dB loss.
3. Combination of 1 and 2: 15 dB loss.
4. Quarter-wave whips without adequate ground plane: 10 dB loss.

The ferrite core is the compact antenna with least loss.

Batteries for personal portable equipment include the zinc-carbon, alkaline, and mercury nonrechargeable and the nickel-cadmium rechargeable types. The alkaline and mercury types give the longer life before discharge; the nickel-cadmium type, with repeated recharges, has the longer overall life. By EIA standards, the minimum power-supply life is 1 day consisting of 8 h of operation with duty cycle of 6 s receive at rated audio power output, 6 s transmit at rated rf power output, and 48 s standby. The rechargeable batteries normally provide several days of life between charges while the nonrechargeable batteries provide up to 3 weeks of life.

Personal paging systems are one-way signaling systems which include a voice message to the personal receiver, following an alerting signal. The alerting signal consists of one or more tones, which operate reed selectors at the receiver. Combinations of two or more tones and two or more reeds in the receiver selector are required when more than 30 codes are required in the system.

One type of system provides wide-area general-use subscription service to the public, operated by the telephone company in conjunction with a telephone-answering service. Messages are transmitted over a 250-W base station. Coverage is available over a 3- to 5-mi

Table 22-16. EIA Standards and Manufacturers' Specifications for Personal Receivers

Quantity measured or specified	EIA Standards (revision proposed)		Typical manufacturers' specifications	
	25–54 MHz	144–174 MHz	25–54 MHz	144–174 MHz
Average sensitivity, μV	*	30	12	15
Selectivity, dB	30	30	45	45
Spurious attenuation, dB	35	35	45	45
IM spurious attenuation, dB	30	30	45	45
Audio power output, mW	250	250	100	100

* No standard given.

radius, depending upon whether the receiver is indoors or out. A similar system is the *in-plant system,* with the base station located in the center of the area of coverage and contact established through the plant's automatic telephone exchange.

79. References

General

1. HAMSHER, D. H. (ed.) "Communication System Engineering Handbook," McGraw-Hill, New York, 1967.
2. Bell Telephone Laboratories, Inc., "Transmission Systems for Communication," 4th ed., Western Electric Co., Winston-Salem, N.C., 1970.
3. "Reference Data for Radio Engineers," 5th ed., ITT and Sams, Indianapolis, Ind., 1968.

Telephone Switching

4. CALDWELL, SAMUEL H. "Switching Circuits and Logical Design," Wiley, New York, 1958.
5. KEISTER, W., A. E. RITCHIE, and S. H. WASHBURN "The Design of Switching Circuits," Van Nostrand, Princeton, N.J., 1969.
6. SCUDDER, F. J., and J. N. REYNOLDS Crossbar Dial Telephone Switching System, *Bell Syst. Tech. J.,* January 1939, Vol. 18, p. 76.
7. BELL LABORATORIES STAFF ESS-19 Pap. 1, *Bell Syst. Tech. J.,* September 1964, Vol. 43.
8. PEARCE, J. G. An Electronic Register Sender for Step-by-Step Automatic Telephone Systems, *IEEE Trans. Commun. Technol.,* June 1965, Vol. COM-13, p. 149.
9. WILKINSON, R. I. The Interconnection of Telephone Systems, *Bell Syst. Tech. J.,* October 1931, Vol. 10, pp. 531–564.
10. YOUNG, J. S. Common Control for Step-by-Step Offices (Series 100 Director), *Autom. Electron. Tech. J.,* October 1963, Vol. 8, p. 258.
11. KORN, F. A., and J. G. FERGUSON The No. 5 Crossbar Dial Telephone Switching System, *Trans. AIEE,* 1950, Vol. 69, Pt. 1, p. 244.
12. BELL LABORATORIES STAFF TOUCH-TONE Calling, Pt. 1, Application, Pt. 2, Central Office Arrangements, Pt. 3, Signal System and Receiver, Pt. 4, Caller for Station Sets, *IEEE Trans. Commun. Electron.,* March 1963, Vol. 82, pp. 1, 5, 9, and 17.
13. TALLEY, DAVID "Basic Telephone Switching Systems," Hayden, New York, 1969.

Transmission

14. ANDREWS, F. T., JR., and R. W. HATCH National Telephone Network Transmission Planning in the American Telephone and Telegraph Co., *IEEE Trans. Commun. Technol.,* June 1971, Vol. COM-19, No. 3, p. 302.
15. HALLENBECK, F. J., and J. J. MAHONEY, JR. The New L Multiplex-System Description and Design Objectives, *Bell Syst. Tech. J.,* March 1963, Vol. 43, No. 2, pp. 207–221.
16. HOLBROOK, D. B., and J. T. DIXON Load Rating Theory for Multi-channel Amplifiers, *Bell Syst. Tech. J.,* October 1939, Vol. 18.
17. EVERETT, W. L. "Communication Engineering," 2d ed., McGraw-Hill, New York, 1937.
18. JOHNSON, WALTER C. "Transmission Lines and Networks," McGraw-Hill, New York, 1950.
19. HENNENBERGER, T. C., and M. D. FAGEN Comparative Transmission Characteristics of Polyethylene Insulated and Paper Insulated Communication Cables, *Trans. AIEE Commun. Electron.,* March 1962, Vol. 59, pp. 27–33.
20. MERRILL, J. L., A. F. ROSE, and J. O. SMETHURST Negative Impedance Telephone Repeaters, *Bell Syst. Tech. J.,* September 1954, Vol. 33, No. 5, pp. 1055–1092.
21. BLEISCH, G. W. The N3 Carrier System Plan, *Bell Lab. Rec.,* March 1965, Vol. 43, pp. 77–83.
22. FULTZ, K. E. T1 Carrier System, *Bell Syst. Tech. J.,* September 1965, Vol. 44, pp. 1405–1451.

Data and Telegraph

23. BENNETT, W. R., and J. R. DAVEY, "Data Transmission," McGraw-Hill, New York, 1965.
24. LUCKY, R. W., J. SALZ, and E. J. WELDON "Principles of Data Communication," McGraw-Hill, New York, 1965.

25. MARTIN, J. L. "Telecommunications and the Computer," Prentice-Hall, Englewood Cliffs, N.J., 1969.

26. MARTIN, J. L. "Teleprocessing Network Organization," Prentice-Hall, Englewood Cliffs, N.J., 1970.

27. ALEXANDER, A. A., R. M. GRYB, and D. W. NAST Capability of the Telephone Network for Data Transmission, *Bell Syst. Tech. J.,* May 1960, Vol. 39, pp. 431-476.

28. FRANCO, A. G., and M. E. WALL Coding for Error Control: An Examination of Techniques, *Electronics,* Dec. 27, 1965, Vol. 38, No. 26.

29. FREEBODY, J. W. "Telegraphy," Pitman, London, 1959.

Microwave Radio

30. GUENTHER, R. Radio Relay Design Data 60 to 600 Mc, *Proc. IRE,* September 1951, Vol. 39, pp. 1027-1034.

31. ROETKIN, A. A., K. D. SMITH, and R. W. FRIIS The TD-2 Microwave Radio Relay System, *Bell Syst. Tech. J.,* October 1951, Vol. 30, pp. 1041-1077.

32. FRIIS, R. W., J. J. JANSEN, R. M. JENSEN, and H. T. KING The TM-1/TL-2 Short Haul Microwave Systems, *Bell Syst. Tech. J.,* January 1966, Vol. 45, pp. 1-95.

33. RICE, S. O. Properties of a Sine Wave plus Random Noise, *Bell Syst. Tech. J.,* January 1948, Vol. 27, pp. 109-157.

34. CROSS, T. G. Intermodulation Noise in FM Systems Due to Transmission Deviations and AM/PM Conversion, *Bell Syst. Tech. J.,* December 1966, Vol. 45, pp. 1749-1773.

35. PARRY, C. A. A Formalized Procedure for the Prediction and Analysis of Multichannel Tropospheric Scatter Circuits, *IRE Trans. Commun. Syst.,* September 1959, pp. 211-221.

36. DE GRASSE, R. W., D. C. HAGG, E. A. OHM, and H. E. B. SCOVIL Ultra Low Noise Antenna and Receiver Combination for Satellite or Space Communications, *Proc. Natl. Electron. Conf., 1959,* Vol. 15, pp. 370-379.

HF Radio

37. GOLDSTINE, H. E., G. E. HANSELL, and R. E. SCHOCK SSB Receiving and Transmitting Equipment for Point-to-Point Service on HF Radio Circuits, *Proc. IRE,* December 1956, Vol. 44, pp. 1789-1794.

38. LYONS, W. Design Considerations for Frequency Shift Keyed Circuits, *RCA Rev.,* June 1954, Vol. 15, No. 2.

39. SALAMAN, R. K., W. B. HARDING, and G. E. WASSON Fading, Multipath and Direction of Arrival Studies for High Frequency Communications, *Natl. Bur. Stand. Rep.* 7206, 1961.

Mobile Radio

40. BAILEY, A. Developments in Vehicular Communications, *Proc. IRE,* May 1963, Vol. 50, pp. 1415-1420.

41. TALLEY, D. A Prognosis of Mobile Telephone Communications, *IRE Trans.,* August 1962, Vol. VC-11, No. 2, pp. 27-39.

42. MYERS, R. T. A Transistorized Receiver for 150 Mc Mobile Service, *IRE Trans.,* August 1960, Vol. VC-9, No. 2, pp. 70-79.

43. EGLI, J. J. Vehicular Transmission, *IRE Trans.,* July 1958, Vol. VC-11, pp. 86-90.

44. FARMER, R. A. SINAD System Design, *IRE Trans.,* April 1961, Vol. VC-10, No. 1, pp. 103-108.

45. GISSELMAN, A. C. The Meat of the Backbone, *IRE Trans.,* July 1958, Vol. VC-11, pp. 64-70.

46. MITCHELL, D., and K. G. VAN WYNEN 150 Mc Personal Radio Signaling System, *IRE Trans.,* August 1961, Vol. VC-10, No. 2, pp. 57-70.

47. DOUGLAS, V. A. The MJ Mobile Radio Telephone System, *Bell Lab. Rec.,* December 1964, pp. 382-388.

SECTION 23

ELECTRONIC DATA PROCESSING

BY

RODGER L. GAMBLIN   Area Manager, Chemical Products Development, Office Products Division, IBM Corporation

WILLIAM P. HEISING   Senior Programmer, Advanced Systems Architecture, Systems Development Division, IBM Corporation

GERHARD E. HOERNES   IBM Scholar, Advanced Education, Systems Products Division, IBM Corporation; Member, IEEE

CONTENTS

Numbers refer to paragraphs

SECTION 23

ELECTRONIC DATA PROCESSING*

By Rodger L. Gamblin, William P. Heising, and Gerhard E. Hoernes

PRINCIPLES OF DATA PROCESSING

1. Definition of a Computer. A computer is a device that determines the solution of some problem by calculation, i.e., by the application of a set of *logical* or *mathematical* *operations* to *data* or *information* to achieve a result called a *solution.* An example of a computer is the slide rule. Numerical values are laid out on the slide rule so as to be proportional to the logarithm of the number shown. When the operator adds the lengths on the two scales, the sum of the logarithms of the numbers is obtained. Since the sum of the logarithms of two numbers is the logarithm of the product, the sum of the lengths indicates the product on the rule. The accuracy is limited by the length of the rule and by the ability of the operator to set and read the scales.

Another type of computer is the abacus. Each of the beads on the abacus is assigned a numerical value. A multiplace decimal number is entered by pulling down (or pushing up) appropriate beads in successive columns. Other numbers are added to the first entry by displacing appropriate beads successively in each column and by following rules for the *carry* *operation* whenever the sum of the digits of the first and second numbers exceeds the number of bead values in a column. The accuracy of the abacus is determined by the number of columns used to represent a quantity. For example, a 20-column abacus can represent a decimal digit to an accuracy of 20 places. If appropriate procedures are followed, the abacus can be used to perform a wide variety of arithmetic operations.

The slide rule and abacus are examples of the two basic classes of computational machines, *analog* and *digital,* respectively. In an analog device, a substitution of a physical quantity for a number occurs, the substituted quantity (length, current, etc.) being used to represent the number. In the digital computer, computations are carried out upon objects that represent a *code* for numerical values. The coded quantity can represent any quantifiable entity.

2. Desk Calculator. A desk calculator is another type of digital computer. Conceptually, it is only slightly more complex than an abacus. A keyboard, with buttons for each of the 10 decimal digits, is used for entry of numerical information into mechanical or electrical *registers* arranged for *display.* Addition is accomplished by successively entering numbers into the keyboard and depressing an add key, which initiates a sequence of electrical or mechanical actions corresponding directly with the motions of an abacus.

Desk calculators capable of multiplication and division have a *shifting function,* such that an upper carriage is shifted to the left or right relative to the lower carriage. Multiplication is performed by adding the multiplicand a number of times equal to a digit of the multiplier. Then, after the carriage is shifted left, the multiplicand is added the number of times specified by the next multiplier digit, and so on.

For multiplication, the machine follows two prescribed procedures, shifting and addition, to realize the product. Division takes place by a similar routine of subtractions and shifts.

3. Stored-Program Methods. A punched paper tape or similar device can be used to control the sequence of buttons pressed on a desk calculator, thus allowing more complicated computations, e.g., extracting a square root or other such functions, to take place automati-

* The authors express their appreciation to the many individuals in the IBM Corporation whose support and advice contributed to this section. H. T. Marcy's aid, encouragement, and criticisms were especially helpful. Members of the *IBM Systems* and *R & D Journal* staffs, in particular Homer Sarasohn, Charles Sconce, and Robert Neudecker, aided by reading and criticizing the manuscript. Others such as Charles Gold and William Aul shared their insights in the computer field. Nancy Schordine contributed the major portion of the typing, and Donald Mierisch prepared most of the illustrations.

cally. Such a tape is said to contain a *program* for the solution of the given problem. Modern desk calculators incorporate magnetic or electronic storage that operates in this way.

The next step in automating a calculator is to provide a second paper tape for the automatic *entry of information* in the keyboard and a third tape to collect and record the *computed result*. In digital-computer language, the desk calculator has been provided with *input* and *output* equipment.

A calculator equipped with storage tapes can realize any algebraic equation whose solution depends on a straightforward sequence of the fundamental operations of arithmetic. However, certain functions cannot be so formed. For example, if the absolute value of a number is desired, different machine actions are implied depending upon whether the number is positive or negative. Such an operation means that the machine must operate as the result of a *decision*, depending upon a specific result at an intermediate stage of computation. A single paper tape, operated in strictly sequential fashion, cannot determine such an operation, since it specifies only one specific sequence. But if *two* control tapes are available so that depending upon the outcome of a specific operation specified on the first tape, the second can be activated to contribute an alternative program, the absolute value of a number can be realized and its sign identified. In the terminology of digital computers, the two program tapes provide a *branch* operation.

The control and input-output functions associated with such tapes can be initially read into a unit, called a *store* (or *memory*), prior to the initiation of procedures for a solution. Furthermore, if the store can be readily altered during the course of operation, intermediate results of complex operations can be stored temporarily for use in later sequences of a program.

4. Internal Modification of Stored Programs. If a suitable alterable store is used in conjunction with a calculator, and if provision is made for programming variable sequences of operations and for automatic input and output of data, a powerful device is available for the solution of a wide class of computational problems. In addition, a fundamental factor in the power of the modern digital computer is its ability to modify its stored program *within the computer itself.*

To accomplish this, a numerical code stored in the computer memory is used to specify the particular operation of the calculator at any one time. For example, the code number 1 may be used to specify addition, the code number 2 to specify subtraction, and these codes may be followed by a numerical code to determine what data are to be operated upon. This latter numerical code, called an *address,* specifies the location of data in the physical store. The operation at different parts of the program can be changed from addition to subtraction by adding the number 1 to the code that specifies addition.

Similarly, if the data on which the operation is performed are to be varied during the computation, the numerical field that specifies the location of the information associated with the instruction need only be changed from one value to another by adding a suitable constant. Such operations imply that the machine is capable of treating its own program as *data subject to manipulation.*

Suppose, for example, that a problem calls for the execution of a series of arithmetic steps on each of a list of a thousand numbers. These numbers can be stored sequentially in storage. The problem is to perform the sequential operations 1,000 times, without having to provide 1,000 separate program steps, each differing only in the location of data upon which it operates. An option is to provide a means, at the termination of each sequence of operation, for the data portion of each instruction in the program to be modified to call upon data at successive locations. The ability of the computer to operate numerically upon its program permits such a procedure.

5. Memory, Processing, and Control Units. A desk calculator performs arithmetic operations, but it can be arranged to receive *alphabetic* as well as *numeric* information by assigning internal codes to represent the alphabetic characters (letters). The letters, in turn, can be manipulated by applying the rules of logic or arithmetic to the codes that represent them. For example, a calculator could compare letters and determine which precedes the other and thus could place in alphabetical order a set of letters from A to Z. It is characteristic of modern computers that the arithmetic section can manipulate more general codes than purely arithmetic data. This section is therefore commonly called the *arithmetic logic unit* (ALU). Another name applied to the ALU is the *processing unit.*

Implicit in the above discussion of the operation of the desk calculator is a mechanism by which the program is sequenced. In computer language the section that translates the stored program codes into an appropriate sequence of operations in the ALU, memory, or input-output equipment is called the *control unit*.

In summary, the basic subsystems in a computer are the input and output sections, the store, the ALU, and the control section. Each of these units is described in detail in this section. Generally, a computer operates in the following way. An external device such as a magnetic tape, card reader, or disk file delivers a program and data to specific locations in the computer store. Control is then transferred to the stored program, which manipulates the data (and the program itself) to generate the output. These output data are delivered to a device, such as a tape, printer, or display, where the information is used in accordance with the purpose of the computation.

6. Historical Background. In the early 1800s, almost 150 years before their widespread use, stored-program digital computers were conceived and partially implemented by a remarkable Englishman, Charles Babbage.* Babbage's conception included a unit (corresponding to the ALU) that he called the *mill*, an internal storage device, control units, and input-output devices. Babbage understood the requirements for transfer of control depending upon the results of intermediate computations and the need for program storage. Furthermore, he realized the potential and power of the computer's treating its own program as data to modify the procedure as the calculation proceeded. In his work Babbage was aided and supported, especially in the area of programming, by Byron's daughter, Lady Lovelace. Babbage's ideas were never fully reduced to practice because of the primitive technology that existed at that time.† There were no electronics, and the mechanical devices used were neither precise enough nor reliable enough to realize a practical machine.

In the 1890s Hollerith developed card-tabulating equipment to cope with the growing demands of the United States census. His work laid the basis for the subsequent development of electromechanical methods that were used in commerce and industry during the early part of the twentieth century.

Starting in the 1930s, many contributions were made that set the stage for modern computer systems. The high-speed electromechanical relay was developed for application to card-tabulating equipment and to telephone and telegraphic switch gear. The development of radio and other electronic equipment contributed to the widespread availability of electronic components and methods. In particular, cosmic-ray and radar research led to pulsed and digital circuits.

During the early 1940s at the Moore School of Electrical Engineering at the University of Pennsylvania, at Harvard, and at IBM, a number of projects were initiated. At Harvard the Mark I machine was developed in conjunction with IBM.‡ This was an electromechanical sequence calculator programmed by paper tape and mechanical switches. The Eniac§ at the Moore School and the IBM card-programmed calculator were built with electronic components to achieve high-speed operation. At the Moore School of Engineering, a team fortunate enough to have the services of John von Neumann, of the Princeton Institute of Advanced Study, produced a fully programmable, electronic digital computer, the Edvac. This machine incorporated most of the elements of modern systems. The description of the machine written at the time discussed in detail many problems of digital computers that are still pertinent.¶ Besides von Neumann others from Princeton contributed to the Moore School program under the sponsorship of the U.S. Army Ordnance Department.

7. Digital Computation Functions. As implied above, the input to a digital computer is restricted only by the basic character set accepted by the machine. The operations performed upon the information delivered from the input are limited only by the capabilities of logical decision making represented in the ALU. Finally, the program that uses these operations is limited only by the capabilities of the programmers and their ability to devise procedures for the solution of specific problems.

More specifically, a digital computer can be used for the conversion of one form of data to another, for logical operations upon information, for sorting, alphabetical ordering, payroll preparation, information storage, process control, retrieval of documents, accounting opera-

---

* Charles Babbage "Passages from the Life of a Philosopher," Longmans, London, 1864.
† H. P. Babbage "Babbage's Calculating Engines," E. and F. N. Spon, London, 1889.
‡ A. H. Aiken and G. M. Hopper The Automatic Sequence Controlled Calculation, *Elec. Eng.*, 1946, Vol. 65, p. 384.
§ J. G. Brainerd and T. K. Sharpless The ENIAC, *Elec. Eng.*, February 1948, pp. 163–172.
¶ A. W. Burks, H. H. Goldstine, and J. von Neumann "Preliminary Discussion of the Logical Design of an Electronic Computing Instrument," Institute for Advanced Study, Princeton, N.J., 1946 (reprinted in *Datamation*, September 1962).

tions, mathematical and scientific procedures, language translation, engineering design, and playing checkers. The general name for this type of computation is *data processing.*

8. Data Processing Speed and Integrity. Modern digital computers are required to operate at extremely high speed, in part because the economic justification of the computer lies in its replacement of repetitive and tedious human thought processes. Thus the computer is in competition with man and must have a lower cost per computation. Since large expense is associated with the development of a machine, a large volume of work is required to justify the cost.

Because a data processing system operates by means of a sequence of programmed steps, each step depends on the successful and accurate completion of prior steps. Since the program sequence is usually long, any errors that occur early in the computational process will usually propagate throughout the program. Moreover, the computer is insensitive to the data that it manipulates, so that a relatively small error at some point in a program may cause the generation of complete nonsense at the output. Thus the *integrity* of data in a system is paramount.

These requirements for high-speed operation and extreme accuracy limit the technologies that can be used in computer systems and tend to direct the course of their application. In the early 1940s electromechanical devices, i.e., relays and stepping switches, were developed for computer functions that operated in 1 to 10 ms. After World War II, electron-tube devices were developed that could realize logical functions in times of the order of a microsecond. Toward the end of the 1950s transistor circuits became available. At present these operate in the nanosecond range.

9. Binary and Decimal Numbers. Most transistors display random variations of their operating parameters over relatively wide limits. Similarly, passive circuit elements experience a considerable degree of variation, and noise and power-supply variations, etc., limit the accuracy with which quantities can be represented. As a result, the preferred method is to use each circuit in the manner of an on-off switch, and representation of quantities in a computer is thus almost always on a *binary* basis. Figure 23-1 shows the binary numbers equivalent to the decimal numbers between 1 and 10. Figure 23-2 shows the addition of binary 6 to binary 3 to obtain binary 9.

The process of addition can be dissected into digital, logical, or Boolean operations upon the binary digits, or bits (b). For example, a first step in the procedure for addition is to form the so-called EXCLUSIVE-OR addition between bits in each column. This function of two binary numbers is expressed in Fig. 23-3*a* in tabular form. This table is called a *truth table.* In Fig. 23-3*b* is the table used to generate the *carries* of a binary bit from one column to another. This latter function of two binary numbers is variously called the AND function, *intersection,* or *product.* The entries at each intersection in each table are the result of the combination of two binary numbers in the respective row and column. Figure 23-3 also shows the decimal addition tables. They illustrate the relative simplicity of the binary number system.

The names truth table and logical function arise from the fact that such manipulations were first developed in the *sentential calculus,* a subsection of the calculus of logic, dealing with the truth or falsity of combinations of true or false sentences (see Par. **23-36**).

Binary Encoding. Information in a digital processing machine need not be restricted to

Decimal	Binary
0	0
1	1
2	10
3	11
4	100
5	101
6	110
7	111
8	1000
9	1001
10	1010

Fig. 23-1. Binary and decimal numbers between 1 and 10.

Decimal	Binary
6	110
+3	+ 11
9	1001

Fig. 23-2. Addition of 6 and 3 in binary: $1 + 0 = 0 + 1 = 1$; $0 + 0 = 0; 1 + 1 = 0$ plus carry 1. The binary is added to the next column to the left, as in decimal addition.

numerical information since a different specific numeric code may be assigned to each letter of the alphabet. For example, A in the EBCDIC code is given by the binary sequence 11000010. When alphanumeric information is specified, such a code sequence represents the symbol A, but in the numeric context the same entry is the binary number equal to decimal 194.

10. Internal and External Information. The fact that a data processing system operates internally in binary form raises the problem of communication between the system and the operators who provide input-output and programming. It is difficult for most users to write or recognize numbers or codes presented in binary, but the machine can use no other type of material.

A number of approaches to the problem of internal-external communication with machines have been followed. For example, a 4-bit (4-b) binary code can be used inside a machine to represent the 10 decimal numbers. By assigning suitable weights to each bit, it is a relatively straightforward task for an operator to translate such a code into decimal equivalents. Such computers have been called binary-coded decimal (BCD) machines. The use of BCD in a machine implies underutilization (see Par. 23-28) of the circuits and storage elements in the system.

A second option is for the machine to work internally in binary fashion but at each juncture of input or output to translate the codes, either by programming or by hardware, so as to accept and offer decimal numeric information. Such systems are complicated, and failures in the coding and decoding system can prevent interaction with the program.

The third approach is a compromise between the human requirement for a decimal system and the machine requirement for binary. An example is the use of the base-16 (hexidecimal) system, which is relatively amenable to human recognition and manipulation (see Par. 23-26).

11. Error-Correction Codes. Though the circuits in modern data processing systems have reached degrees of reliability undreamed of in the relatively recent past, errors can still arise. Hence it is desirable to detect and, if possible, correct such errors. As discussed in Par. 23-33, it is possible by appropriate selection of binary codes to detect errors. For example, if a 6-b code is used, a seventh bit is added to maintain the number of 1 bits in the group of 7 as an odd number. When any group of 7 with an even number of 1s is found by appropriate circuits in the machine, an error is detected. Such a procedure is known as *parity checking*. More complex codes have been developed to detect and correct errors to almost any desired degree (see Par. 23-34).

12. Boolean Functions. Figure 23-4 illustrates truth tables (Par. 23-9) for functions of

	0123456789
0	0123456789
1	1234567890
2	2345678901
3	3456789012
4	4567890123
5	5678901234
6	6789012345
7	7890123456
8	8901234567
9	9012345678

Decimal addition table
excluding the carry

	01
0	01
1	10

Binary addition table
excluding the carry
(a)

	0123456789
0	0000000000
1	0000000001
2	0000000011
3	0000000111
4	0000001111
5	0000011111
6	0000111111
7	0001111111
8	0011111111
9	0111111111

Decimal addition
carry generation

	01
0	00
1	01

Binary addition
carry generation
(b)

Fig. 23-3. Addition tables for decimal and binary numbers. The binary addition table (a) is called the EXCLUSIVE-OR or modulo-2 truth table; the carry table (b) performs the AND or *intersection* operation 2.

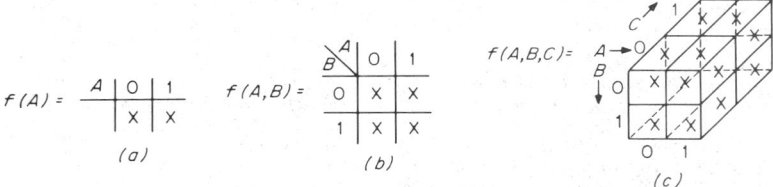

Fig. 23-4. Binary functions of (a) one, (b) two, and (c) three binary variables.

one, two, and three binary variables. Each x entry in each table can be either 0 or 1. Hence for one variable x four functions $f(x)$ can be formed; for two variables x_1 and x_2, 16 functions $f(x_1,x_2)$ exist; for three variables, 256 functions, etc. In general, if $f(x_1,\cdots,x_n)$ is a function of n binary variables, 2^{2^n} such functions exist.

For functions of one variable, the most important is the *inverse function*, defined in Fig. 23-5. This is called NOT A or $\bar{A}$, where A is the binary variable. Also illustrated in Fig. 23-5 are the two most important functions of two binary variables, the AND (*product* or *intersection*) and the OR (*sum* or *union*). If A and B are the two variables, the AND function is usually represented as AB and the OR as $A + B$.

Figure 23-6 shows how the products AB, $\bar{A}B$, $A\bar{B}$, and $\bar{A}\bar{B}$ are summed to yield any function of two binary variables. Each of these products has only one 1 in the four positions in their respective truth tables, so that appropriate sums can generate any function of two binary variables. This concept can be expanded to functions of more than two variables; i.e., any function of n binary variables can be expanded into a sum of products of the variables and their negatives. This is the general theorem of Boolean algebra. Such a sum is called the *standard sum* or the *disjunctive normal form*.

The fact that any binary function can be so realized implies that mechanical or electrical simulations of the AND, OR, and NOT functions of binary variables can be used to represent any binary function whatever.

13. Logical-Function Realization. Relays or toggle switches are logical devices. Figure 23-7 shows how the contacts on a relay or switch are interconnected to form a logical function. The inverse Boolean function of one variable (NOT) is realized by using a *normally closed* contact of a switch or relay in a given circuit. The AND function of two variables is realized by a series connection of contacts and the OR function by a parallel connection.

In the example of the implementation of the EXCLUSIVE-OR function given in Fig. 23-7e, no current flows in the left branch of the circuit unless relay A is off and relay B is on. Similarly, no current flows in the right branch unless relay A is on and relay B off. Since either branch can carry current, the condition under which current is transmitted is $A\bar{B} + \bar{A}B$, where the symbols refer to the state of current in the relay coils. For mechanical switches, the same considerations apply except that the position of the switch is determined by a mechanical force. With relays or switches, the logical-function operations are expressed by interconnecting wires, the variables being associated with the physical contacts.

14. Electronic Realization of Logical Functions. Logical functions may be realized using electronic circuits. Figure 23-8 illustrates the realization of the OR function and Fig. 23-9 the realization of the AND circuit using diodes. Each input lead is associated with a Boolean variable, and the upper level of voltage represents the logical 1 for that variable; in the OR circuit, any input gives rise to an output. Thus for a three-variable input, the output

$$f(A) = \bar{A} = \begin{array}{c|c} A & 01 \\ \hline & 10 \end{array}$$

(a)

$$f(A,B) = A+B = A\cup B = A\vee B$$

$$\begin{array}{c|c} B\backslash & 01 \\ \hline 0 & 01 \\ 1 & 11 \end{array}$$

(b)

$$f(A,B) = AB = A\cap B = A\wedge B$$

$$\begin{array}{c|c} B\backslash & 01 \\ \hline 0 & 00 \\ 1 & 01 \end{array}$$

(c)

$$AB = \begin{array}{c|c} B\backslash & 01 \\ \hline 0 & 00 \\ 1 & 01 \end{array} \qquad A\bar{B} = \begin{array}{c|c} B\backslash & 01 \\ \hline 0 & 00 \\ 1 & 10 \end{array}$$

$$\bar{A}B = \begin{array}{c|c} B\backslash & 01 \\ \hline 0 & 01 \\ 1 & 01 \end{array} \qquad \bar{A}\bar{B} = \begin{array}{c|c} B\backslash & 01 \\ \hline 0 & 10 \\ 1 & 00 \end{array}$$

$$A\bar{B} + \bar{A}B = A \oplus B = \begin{array}{c|c} B\backslash & 01 \\ \hline 0 & 01 \\ 1 & 10 \end{array}$$

Fig. 23-5. Significant functions of one and two binary variables. (*a*) The negation (NOT) function of one binary variable; (*b*) the OR function of two binary variables; (*c*) the AND function of two binary variables.

Fig. 23-6. The four products of two binary variables (*top*). The realization of the EXCLUSIVE-OR function is shown below.

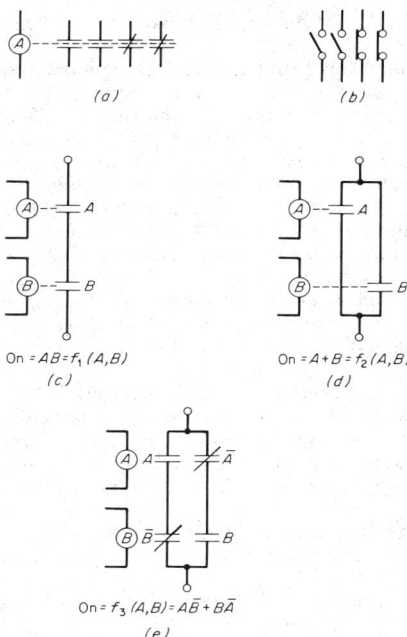

On $= AB = f_1(A,B)$
(c)

On $= A+B = f_2(A,B)$
(d)

On $= f_3(A,B) = A\bar{B} + B\bar{A}$
(e)

Fig. 23-7. Relay or switch contacts for realization of logical functions. (*a*) Relay and (*b*) switch with normally open and closed contacts. A normally off contact corresponds to an inverse function of one variable. A series connection (*c*) realizes the AND function and a parallel connection (*d*) the OR. (*e*) is the EXCLUSIVE-OR circuit. The contacts correspond to the binary variables, while the interconnections express the functions realized.

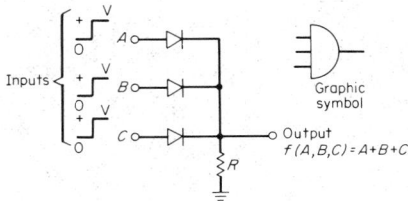

Fig. 23-8. Diode realization of an OR circuit. A positive input on any line produces an output.

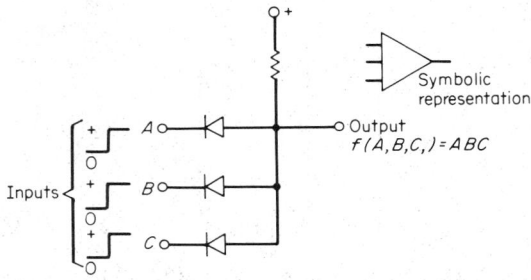

Fig. 23-9. Diode realization of an AND circuit. All inputs must be positive to produce an output.

is $A + B + C$. With the AND function no output is realized unless all inputs are positive; the output function generated is ABC.

The inverse function (NOT) of a Boolean variable cannot be readily realized with diodes. The circuit shown in Fig. 23-10 uses the inverting property of a grounded-emitter transistor amplifier to perform the inverse function. Also shown in Fig. 23-10 is an example of how the OR function and the NOT function are combined to form the NOT-OR (NOR) function. In this case, since the transistor circuit provides both voltage and current gain, the signal-amplitude loss associated with transmission through the diode may be compensated, so that successive levels of logic circuits can be interconnected to form complex switching nets.

Figure 23-11 illustrates the realization of the EXCLUSIVE-OR function. Note that the variables are represented by the wiring of interconnected circuit blocks while the function is realized by the circuit blocks themselves.

15. Levels of Operation in Data Processing. A detailed sequence of operations is generally required in a data processing system to realize even simple operations. For example, in carrying out addition, a machine typically performs the following sequence of operations:

1. Fetch a number from a specific location in storage.
   a. Decode the address of the program instruction to activate suitable memory lines. Such decoding is accomplished by activating appropriate AND and OR gates to apply voltage to the lines in storage specified by the instruction address.
   b. Sequence storage to withdraw the information and place it in a storage output register.
   c. Transmit information from the storage output register into the appropriate ALU.
2. Withdraw a number from storage and add it to the number in the ALU. These operations break down into:
   a. Decode the instruction address, activate storage lines, and transmit the information to the ALU input for addition.
   b. Form the EXCLUSIVE-OR of the second number with the number in the ALU to form the sum less the carry. Form the AND of the two numbers to develop the first level carry.
   c. Form the second-level EXCLUSIVE-OR sum.

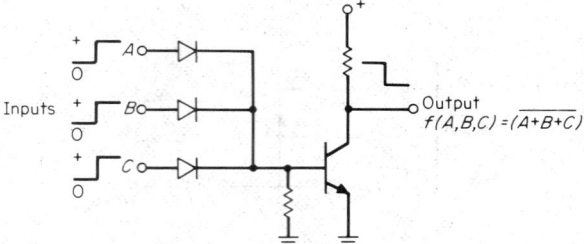

Fig. 23-10. Use of a transistor circuit for inverting a function. The circuit shown forms the NOT-OR (NOR) of the inputs.

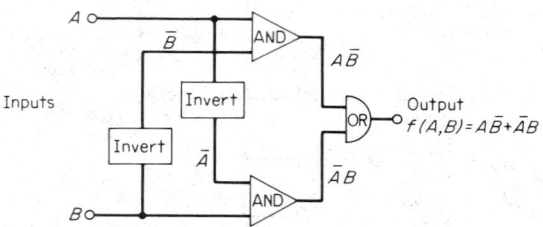

Fig. 23-11. Circuit realization of the EXCLUSIVE-OR function.

   d. AND the first-level carry with the first-level EXCLUSIVE-OR sum to form the second level carry.
   e. Generate the third-level EXCLUSIVE-OR by forming the EXCLUSIVE-OR of the second-level carry with the second-level EXCLUSIVE-OR sum. AND the second-level carries with the second-level EXCLUSIVE-OR for the third-level carry and so forth until no more carries are generated.
   3. Store the result of the addition into a specified location in storage.

This sequence illustrates two basic types of operation in a data processing machine. Operations denoted above by numbers are of specific interest to the programmer, since they are concerned with the data stored and the operations performed thereupon. The second level, denoted above by letters, are operations at the logical-circuit level within the machine. These operations depend upon the particular configurations of circuits and other hardware in the machine at hand.

If only the higher-level (numbered) instructions are used, some flexibility in machine operation is lost. For example, only an add operation is possible at the higher level. At the lower-level (lettered operations) the AND or EXCLUSIVE-OR of the data words can be formed and placed in storage.

The organization of current digital computers follows the lines of these two divisions (numbered and lettered, above). The *macroinstruction set* associated with each machine can be manipulated by the programmer. These instructions are usually implemented in a numerical code. For example, the instruction "load ALU" might be 01 in binary. "Add ALU" might be given by 10 and "store ALU" by 11. Similarly, each instruction has an associated storage address to provide source data. Other instruction formats are discussed in Pars. 23-101 to 23-109. The microinstruction set comprises a series of suboperations that is combined in various sequences to realize macroinstruction.

Two methods of realizing the sequence of suboperations specified by the operations portion of the instruction have been used in machine design. In one such method a direct decoding of the information from the instruction occurs when it is placed in an instruction address register. Specific clock sequences turn on the successively required lines that have been wired in place to realize the action sought.

An alternative for actuating a subprogram is to store a number of information bits, called *microinstructions,* that are successively directed to the appropriate control circuits to activate selectively and sequentially individual wires to gate sequential actions for the realization of the requisite instruction.

The first method of computer design is called *hard-wired,* and the second is the *microprogrammed.* The microprogram essentially specifies a sequence of operations at the individual circuit level to specify the operations performed by the macroinstruction. Microprogramming is discussed in more detail in Par. **23-105.** It is preferred in modern computer design.

   16. Programming with Instruction Sets. Fundamental to the operation of a digital computer system is the generation of a sequence of *machine instructions* that manipulate data to achieve the desired solution. Thus a program might consist of the following sequence:
   1. Initiate action in the input device to provide for loading of memory with program and data.
   2. Perform the required sequence of instructions to manipulate and find the solution to the given problem. For example, if the sine of A is wanted, each value of A is subjected to the arithmetic operations that are specified in the series of expansion of the sine function, to some predetermined accuracy.
   3. Provide for an output of the data to a suitable receiving machine, e.g., a printer or display, of each value of the result for each A.

The large number of steps in such a program and the fact that each operation must use the machine instruction format make the writing of even a relatively simple program difficult, onerous, and error-prone.

   Direct Translation of Machine Code. A step in alleviating this programming problem is to provide direct translation of the machine-code instruction (in one-for-one correspondence) to a format more readable by operators and programmers. For example, the internal machine code that specifies "load ALU" is given a mnemonic such as "L ALU." Similarly, the data-storage field, instead of involving a specific storage location in binary, is given a code name such as "A1."

Another program is developed to translate the man-readable input into machine code. The translating program may operate, for example, by making a comparison of the operation-code field with a stored table. Similarly, data words can be recognized and storage locations automatically selected from a table of unused locations. This process is widely used in computer systems. The one-for-one language is called an *assembly language,* and the program that translates the assembly language into machine language is called an *assembler program.*

Subroutines. Another way of reducing the tedium of programming is to write and reuse frequently used routines. For example, a program can be written for the solution of the sine function; then this program is repeatedly incorporated as a *subroutine* within any program that requires the generation of this function.

Higher-Level Languages. Other approaches to the alleviation of programming problems are widely used. For example, if the algebraic equation $a = b + c$ is specified, this equation may be entered directly into the computer for translation. As in the assembler program, the individual symbols are recognized by suitable tables. Furthermore, the program can generate a "load ALU" instruction for b and "add c," with the result $b + c$ returned to a selected location in memory.

The specific statements that are acceptable to the computer form a sublanguage, with which human communication to the computer is made easier. Ideally, such a language should be completely independent of the machine used and be similar to human thought processes. A number of such *higher-level languages* have been developed including FORTRAN, BASIC, ALGOL, and COBOL, discussed in Pars. 23-122 to 23-126.

17. Numerical Methods. A data processing system uses a finite word size, performs a finite number of operations in problem solutions, and contains storage for a limited number of words. In its operations, therefore, a continuous function must be represented by a sequence of discrete values that, of necessity, are approximations to the actual function. Approximate numerical procedures must be adopted in dealing with operations upon continuous functions, e.g., performing an integration.

A branch of mathematics called *numerical analysis* deals with such questions as error generation, error stability, error propagation, algorithmic methods, and so forth. This subject is discussed in Par. 23-35.

18. Information Flow and the Job Stream. In a computer system a high rate of information flow occurs between the ALU and memory, and, at times, large flows can come from the input or be delivered to the output. A basic problem in design is to balance these flows to prevent idleness of any specific unit with relation to others. For example, if a very fast ALU is provided, with slow memory-access circuits, the ALU will usually be waiting for information from the memory or for access to it. Similarly, bottlenecks may occur on the input or output when high volumes of information are delivered across these interfaces.

This problem is complicated somewhat by the different types of application to which a computer may be put. In business applications a considerable amount of input-output information is delivered to and from the machine, with little internal manipulation. In scientific applications there is considerably less input-output information flow but very heavy use of the memory-ALU system.

19. Types of Computer Systems. There is a wide variety of computer-system arrangements, depending upon the type of application. One type of installation is that associated with *batch processing.* A computer in a central job location receives programs from many different sources and runs the programs sequentially at high speed. An overall supervisory program, called an *operating system,* controls the sequence of programs, rejecting any that are improperly coded and completing those which are correct.

Another type of system, the *time-shared system,* provides access to the computer from a number of remote input-output stations. The computer scans each remote station at high speed and accepts or delivers information to that location as required by the operator or by its internal program. Thus a small terminal can gain access to a large high-speed system.

Still another type of installation, the *minicomputer,* involves an individual small computer that, though limited in power, is dedicated to the service of a single operator. Such applications vary widely from those associated with a small business, with limited computational requirements, to an individual engaged in scientific operations.

These three types of systems are representative of the major classifications. Other computers are used for dedicated control of complex industrial processes. These are individual, once-programmed units that perform a *real-time* operation in *systems control,* with sensing elements that provide the inputs. Highly complex interrelated systems have been developed in which individual computers communicate with and control each other in an overall major *systems network.* Among the first of such systems was the SAGE network, developed in the 1950s for defense against missile or aircraft attack.

Computers that are interconnected to share workload or problems are said to form a *multiprocessing system.* A computer system arranged so that more than one program can be executed simultaneously is said to be *multiprogrammed.*

20. Internal Organization of Digital Computers. The internal organization of a data processing system is called the *system architecture.* Such matters as the minimum addressable field in memory, the interrelations between data and instruction word size, the instruction format and length or lengths, parallel or serial (by bit or set of bits) ALU organization, decimal or binary internal organization, etc., are typical questions for the system architect. The answers depend heavily upon the application for which the computer is intended.

Two broad classes of computer systems are *general-purpose* and *special-purpose* types. Most systems are in the general-purpose class. They are used for business and scientific purposes. General-purpose computers of varying computer power and memory size may be grouped, sharing a common architecture. These are said to constitute a *computer family.*

A computer scientifically designed for, and dedicated to the control of, say, a complex refinery process is an example of a special-purpose system.

A number of design methods have been adopted to increase the speed and functional range for a small increase in cost. For example, in the instruction sequence, the next cell in storage is likely to be the location of the next instruction. Since an instruction can usually be executed in a time that is short compared with storage access, the store is divided into subsections. Instructions are called from each subsection independently at high speed and put into a queue for execution. This type of operation is called *look-ahead.* If the instructions are not sequential, the queue is destroyed and a new queue put in its place.

Since instructions and data tend to be clustered together in storage, it is advantageous to provide a small, high-speed store (local store) to work with a larger, slower-speed, lower-cost unit. If the programs in the local store need information from the larger store, a least-used piece of the local store reverts to the larger store and a batch of data surrounding the information sought is automatically brought into the high-speed unit. This arrangement is called a *hierarchical memory.*

Other systems concepts relate to the optimization of information flow between external equipment and storage. Such considerations are discussed in Pars. **23-101** to **23-109.**

21. Analog Computation. Digital computers can be programmed to perform nearly all the tasks encompassed by analog devices. In many applications, however, the analog device is less expensive, simpler, and easier to use. The analog computer is of use, for example, in the simulation and design of servo systems, electric circuits, acoustical devices, mechanical systems, and other systems governed by linear (or essentially linear) differential equations.

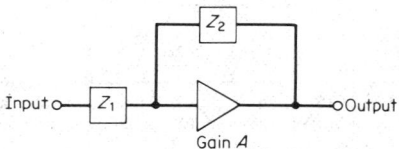

Fig. 23-12. Operational amplifier used in analog-computer circuits.

A basic circuit that is used in analog-computer realization is shown in Fig. 23-12. An input impedance Z_1 is connected to the input of an amplifier whose output is fed back to the input through an impedance Z_2. If the gain of the amplifier is A, the ratio of the output voltage to input voltage is given by

$$e_i/e_o = -(Z_1/Z_2)(1 - 1/A) + 1/A$$

If the gain of the amplifier is very large compared to 1, and if Z_1 and Z_2 are comparable in magnitude, the ratio of the input to output is given by $-Z_1/Z_2$. If Z_1 is a resistor and Z_2 a capacitor, and if the Laplace transform variable s is used, then

$$e_i/e_o = -(RC)s$$

This equation states that the output is the negative integral of the input. Such a circuit has been termed a *Miller integrator* or merely an *integrator.*

If multiple inputs are used with the operational amplifier, as in Fig. 23-13, the device can be used to form a sum of various voltages. Similarly, in Fig. 23-12 if Z_1 and Z_2 are both resistances, the unit will multiply the input by a negative constant equal to the ratio of Z_1 to Z_2.

Figure 23-14 shows an analog device that multiplies two voltages. A servomechanism whose rotation is proportional to an input voltage e_1 drives a potentiometer with voltage e_2 impressed across its resistance. The output from the potentiometer arm is proportional to the product of e_1 and e_2.

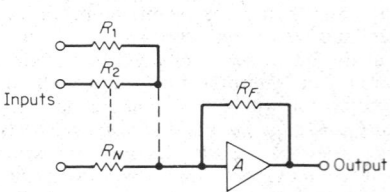

Fig. 23-13. Operational amplifier used to sum a series of voltages. The output is proportional to the negative of the weighted sum of the inputs.

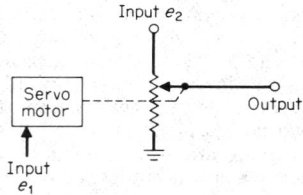

Fig. 23-14. Servomultiplier using a servomotor (whose displacement is proportional to an input voltage) in conjunction with a linear potentiometer.

Figure 23-15 illustrates how a set of integrators is interconnected to yield the solution of the general second-order differential equation of one variable. The voltage on the line x is used to drive a chart recorder or other device that displays the solution.

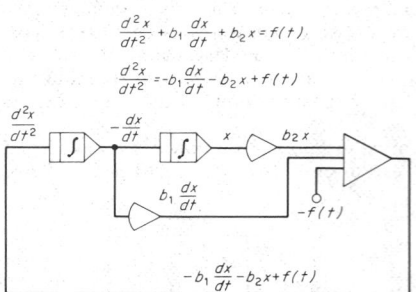

Fig. 23-15. Analog circuit elements to solve a second-order differential equation in one variable.

In an *analog computer system,* several operational amplifiers are provided along with suitable capacitors and resistors to form integrators and summing networks. In addition, provision is made for putting an initial charge on the integrators to represent initial conditions and for certain types of input functions. The response of the analog system gives both a transient and steady-state response.

22. Amplitude and Time-Scale Compensation. Most commercial analog computers are limited in bandwidth. At the high-frequency end, servomultipliers and output recorders are limited by the mechanical motions involved. Integrators, on the other hand, tend to have a low-frequency limit due to systematic input errors that are magnified over long time periods and to the approximation regarding amplifier gain, which is poor at very low frequencies. It is usually necessary to adopt an initial estimate of the time scale and later to change the time scale of the problem.

In second-order differential equations, an estimate of the natural scale of the independent variable can usually be obtained by taking the square root of the ratio of the coefficient of the second-order term to that of the zero-order term. For third-order systems the same principle usually applies. To change the time scale, a substitution such as $y = at$ can be made. Since $d^n/d(at)^n = 1/a^n\, d^n/dt^n$, suitable coefficient adjustments can be made.

Analog computers operate properly over a limited amplitude range. An amplitude scale change is implied between the physical problem and the analog simulation in any case, and the amplitude of any particular time can be estimated from the relation

$$\text{Amplitude} \approx a_n \omega^n U$$

where a_n is the coefficient of the nth derivative, ω is the inverse of the independent-variable scale, and U is the dependent-variable amplitude.

23. Analog-Computer Components. The *operational amplifiers* used in analog computers present a difficult design problem. They must possess a high degree of stability with high gain at low frequency. The most successful approach to this problem has been to chop the input signal with a high-frequency chopper, amplify the chopped signals, and recombine them at the output. Low-frequency drift can thereby be largely eliminated.

A number of substitutes for servomultipliers have been developed to perform similar functions. For example, in a cathode-ray-tube system, the input signals e_1 and e_2 are applied simultaneously to the magnetic deflection yoke and to the electrostatic deflection plates. An optical detection system provides a feedback signal to keep the beam in a fixed position, and the feedback signal is proportional to the product of e_1 and e_2.

Other devices use the square-law characteristics of such devices as rf ammeters to square the sum and difference of two signals. The difference between these squares is proportional to the product.

24. Function Generation. A number of methods have been developed to introduce input functions *(forcing functions)* into analog systems. If the functions are solutions of linear differential equations in one variable, the analog computer itself may be used for the generation of the input function. For example, sinusoids, damped sinusoids, exponentials, parabolas, and so forth can be generated by putting together appropriate analog elements. Repetitive impulse functions, square waves, and sawtooths are easily generated using a pulse generator in conjunction with operational amplifiers.

For input functions that do not have simple analytic form, methods have been developed to generate arbitrary functions of one variable. One such method utilizes a chart recorder. The function to be generated is drawn on the chart with a conducting ink. The stylus associated with the chart is driven forward linearly in time, and feedback is used to maintain the stylus on the conducting line. The feedback signal is the function desired.

A similar scheme can be developed using a cathode-ray tube with a mask representing the function to be generated on its face. Optical feedback in conjunction with a linear sweep can generate a voltage proportional to the displacement of the mask opening from a reference position.

Another means of function generation involves the use of selectively sampled voltages in conjunction with an integrator circuit. During any time segment of the function to be generated, a specific voltage is fed to the integrator. This generates a straight-line segment whose slope, as determined by the input voltage, approximates the function desired.

Analog methods are not restricted to operational amplifier systems. For example, if a thin rubber membrane is stretched over pegs placed at heights proportional to voltages at points in space, a small ball rolling over the surface of the membrane represents the path of an electron in motion between pins whose potential is proportional to heights of the supports for the sheet.

Sheets of conductive paper, in conjunction with appropriate voltage inputs on a sensitive potentiometer, can be used to find solutions to Laplace equations in two dimensions. Another option is to use an electrolytic bath to generate sheets whose resistance varies with the coordinates in a prescribed way.

As the realization of analog systems becomes more complex and difficult, digital solutions using standard analytic techniques become increasingly attractive. In the case of servo or other linear systems, analog methods may be preferred due to ease of implementation, speed, and economy of solution. But the use of complex tanks, paper sheets, etc., and the cost and inconvenience of setup and output generation may well tip the balance to digital methods.

25. Hybrid Systems. In some applications it is desirable to combine analog and digital methods. Such systems are called *hybrids,* to indicate that they operate in a mixed mode. See Refs. 15 to 23, Par. 23-131.

NUMBER SYSTEMS, ARITHMETIC, CODES, AND NUMERICAL METHODS

26. Representation of Numbers. A distinction must be drawn between the concept of a *number,* an abstraction concerning the quantity or order of elements in a given set, and the *numerals* that are used to *name* these quantities. Considerable latitude exists in the selection of names, so that a choice can be made according to a definite purpose. Though the word "numeral" should properly be used whenever the names of numbers are being discussed, in common practice "number" is acceptable unless a distinction is necessary.

In primitive societies, where there is little need for computation, unique symbols are applied only to the first few natural integers. Such a number system is not extendable and fails to meet the requirements of a sophisticated society. The Roman number system was not infinitely extendable, though economy of symbol utilization was achieved by assigning symbols to represent aggregates of lesser units.

A set of codes and names for numbers that meets the requirements of extendability and convenience of operation was introduced into western civilization from the Arabian culture. This system is based upon the properties of a particular set of members of the class of the power series

$$N = A_n X^n + A_{n-1} X^{n-1} + \cdots + A_1 X + A_0 + A_{-1} X^{-1}$$
$$+ \cdots + A_{-m} X^{-m} \tag{23-1}$$

Here the number is represented by the sum of the powers of an integer X, each having a coefficient A_i. A_i may be any integer equal to or greater than zero and less than X. In the Arabic system, X equals 10 and the coefficients A_i range from 0 to 9.

A property of the power series that makes it useful in the representation of a number system is that if the series is cut at a particular point, the set of all values of the lower-order sum is everywhere dense to a unit value of the term that precedes the point of the cut; i.e., if the cut is taken subsequent to the term $A_m X^m$, the sum of all terms represented by $A_{m-1} X^{m-1} + A_{m-2} X^{m-2} + \cdots$, may range from zero to X^m, the degree of approach to X^m being as close as desired.

Another useful property of the power series is the fact that its multiplication by X^k may be viewed as a shift of the coefficients of any given term by the number of positions specified by the value of k. These results are independent of the choice of X in the series representation.

There is little reason to write the value of the number in the form shown in Eq. (23-1) since complete information on the value can be readily deduced from the coefficients A_i. Thus, a number can be represented merely by a sequence of the values of the coefficients. To determine the value of the implied exponents on X, it is customary to mark the position of the X_0 term by a period immediately to the right of its coefficient. Equation (23-2) shows the power series for a number represented in the system $X = 10$ and its normal decimal notation.

$$3 \times 10^3 + 0 \times 10^2 + 2 \times 10^1 + 4 \times 10^0 + 6$$
$$\times 10^{-1} + 2 \times 10^{-2} = 3{,}024.62 \tag{23-2}$$

The value of X is called the *radix* or *base* of the number system. Where ambiguity might arise, a subscript to indicate the radix is attached to the low-order digit, as in $1000_2 = 8_{10} = 10_8$ (1000 binary equals 8 decimal equals 10 octal). The power series for a number in base 2 and its representation in binary notation is

$$1 \times 2^4 + 1 \times 2^3 + 0 \times 2^2 + 1 \times 2^1 + 1 \times 2^0$$
$$+ 0 \times 2^{-1} + 1 \times 2^{-2} + 1 \times 2^{-3} = 11011.011 \tag{23-3}$$

27. Number-System Conversions. Since computer systems, in general, use number systems other than base 10, conversion from one system to another must be carried out frequently. Equation (23-4) shows the integer N represented by a power series in base 10 and base 2.

$$\sum_{i=0}^{n} A_i 10^i = \sum_{j=0}^{m} B_j 2^j \tag{23-4}$$

The problem is to find the correlation between the coefficients A_i and B_j. In the binary series, if N is divisible by 2, then B_0 must be 0. Similarly, if N is divisible by 4, B_1 must be 0, and so forth. Thus if the decimal coefficients A_i are given, successive divisions of the decimal number by 2 will yield the binary number, the binary digits depending on the value of the remainder of each successive division. This process is shown in Fig. 23-16.

The conversion of a binary integer to a decimal integer is

$$100011011 = 1 \times 2^8 + 0 \times 2^7 + 0 \times 2^6 + 0 \times 2^5$$
$$+ 1 \times 2^4 + 1 \times 2^3 + 0 \times 2^2 + 1 \times 2$$
$$+ 1 \times 2^0$$
$$= 283$$

In the case of conversion of an integer in binary to an integer in decimal, the powers of 2 are written in decimal notation and a decimal sum is formed from the contribution of each term of the binary representation. For conversion from a binary fraction to a decimal fraction, a similar procedure is used since the value of terms as multiplied by the A_i can be added together in decimal form to form the decimal equivalent.

The conversion of a decimal fraction to a binary fraction is defined by

$$0.5764_{10} = A_{-1}2^{-1} + A_{-2}2^{-2} + \cdots + A_n 2^{-n} \tag{23-5}$$

To determine the values of the A_i, first multiply both sides of Eq. (23-5) by 2 to give

$$1.1528_{10} = A_{-1} + A_{-2}2^{-1} + \cdots + A_{-n}2^{n-1} \tag{23-6}$$

Since the position of the decimal point (more accurately called the *radix point*) is invariant, and since in a binary series each successive term is at most half of the maximum value of the preceding term, the leading 1 in the decimal number in Eq. (23-6) indicates that A_{-1} must have been 1 prior to the multiplication. A second multiplication by 2 can similarly determine the coefficient A_{-2}. This process of conversion of a base-10 fraction to a base-2 fraction is illustrated in Fig. 23-17.

Conversion from binary integers to octal (base 8) and the reverse can be handled simply since the octal base is a power of 2. Figure 23-18 shows such a conversion. As the table in that figure shows, binary to octal conversion consists of grouping the terms of a binary number in threes and replacing the value of each group with its octal representation. The process works on either side of a decimal point. The octal-to-binary conversion is handled by converting each octal digit, in order, to binary and retaining the ordering of the resulting groups of three bits.

Since there are not enough symbols in decimal notation to represent the 16 symbols required for the hexadecimal system, it is customary in the data processing field to use the first

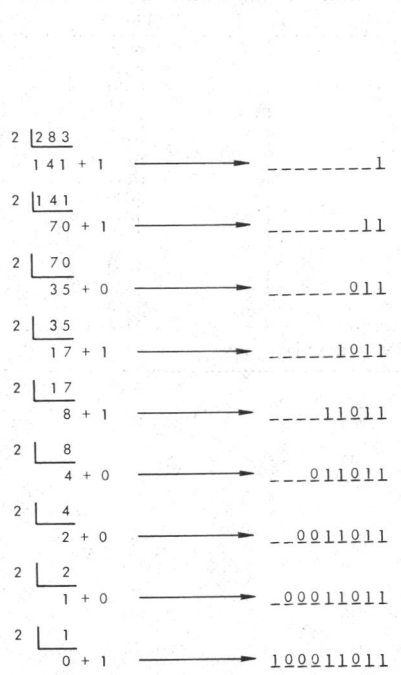

Fig. 23-16. Conversion from a decimal to binary by repeated division of the decimal integer. At each division the remainder becomes the next higher-order binary digit.

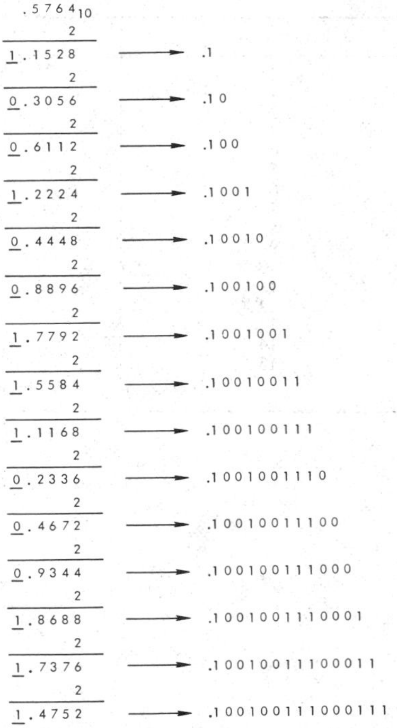

Fig. 23-17. Conversion of a decimal fraction to a binary fraction. At each stage the number to the right of the decimal is multiplied by 2. The resulting number to the left of the decimal point is entered as the next available lower-order position of the binary fraction to the right of the binary radix point.

six letters of the alphabet to complete the set. The hexadecimal symbols and their decimal and binary equivalents are shown in Table 23-1.

This table, in a manner similar to that of an octal system, is used to convert from hexadecimal to binary. Since 16 is a power of 2, the conversion consists of direct substitution of groups of four binary symbols for the hexadecimal equivalents, and vice versa, on either side of the decimal point.

Conversions from decimal to octal or hexadecimal can proceed indirectly by first converting decimal to binary and then binary to octal or hexadecimal. Similarly, a reverse path from octal or hexadecimal to binary to decimal can be used. Direct conversions, however, between hexadecimal and octal and decimal exist and are widely used. In going from hexadecimal or octal to decimal, each term in the implied power series is expressed directly in decimal, and the result is summed. In converting from a decimal integer to either hexadecimal or octal, the decimal is divided by either 16 or 8, respectively, and the remainder becomes the next higher-order digit in the converted number.

28. Number-System Efficiency. Some number systems have advantages over decimal other than simplicity of mechanical implementation, e.g., in such operations as multiplication, division, and subtraction. Another advantage of some number systems is their *efficiency of representation.*

In the design of keyboards and of storage-cell structures, the cost of a system is approximately proportional to the number of keys in each column, the number of states that must be associated with each cell, and the number of columns on the calculator or cells in the

Table 23-1.

Decimal	Hexadecimal	Binary
0	0	0000
1	1	0001
2	2	0010
3	3	0011
4	4	0100
5	5	0101
6	6	0110
7	7	0111
8	8	1000
9	9	1001
10	A	1010
11	B	1011
12	C	1100
13	D	1101
14	E	1110
15	F	1111

$$155.90625_{10} = 9B.E8_{16} = |\,0001\,|\,1011\,|\,1110\,|\,1000\,|_2$$

Octal	Binary
0	000
1	001
2	010
3	011
4	100
5	101
6	110
7	111

$$|\,7\,|\,5\,|\,4\,.\,6\,|\,2\,|\,1\,\xrightarrow{\;\;}\,|\,111\,|\,101\,|\,110\,.\,110\,|\,010\,|\,001\,|_2$$

Fig. 23-18. Conversion from binary to octal is carried out by grouping the binary digits in groups of three, on either side of the radix point, and substituting the appropriate octal digit. Conversion from octal to binary is performed by a reverse substitution.

store. Thus in a system with radix r and number of positions n the equation for cost C might be

$$C = nr \tag{23-7}$$

The total number of states that can be expressed is given by

$$N = n^r \tag{23-8}$$

Since utility is proportional to N, the efficiency of a number system may be defined in terms of a particular choice for a radix. For a given N, Eq. (23-7) may be minimized with respect to r. Utilizing the log of Eq. (23-8) to eliminate n from Eq. (23-7), it can be seen that the least costly number system has a radix equal to 3. Radixes 2 and 4 are equal in efficiency to each other and are near the optimum. Radix 2 is used in digital computers for the reasons given in Par. 23-9.

29. Binary-Arithmetic Operations. Figure 23-19 shows an example of the addition of two binary numbers, 1001 and 1011 (9 and 11 in decimal). The rules for manipulation are similar to those in decimal arithmetic except that only the two symbols, 1 and 0, are used and the addition and carry tables are greatly simplified.

Figure 23-20 shows an example of binary multiplication with a multiplication table. This process is also simple compared with that used in the decimal system. The rule for multiplication in binary is as follows: if a particular digit in the multiplier is 1, place the multiplicand in the product register; if 0, do nothing; in either case shift the product register to the right by one position; repeat the operations for the next digit of the multiplier.

Figure 23-21 shows an example of binary subtraction and the subtraction and borrow tables. The subtraction table is the same as the addition table, a feature unique to the binary system. The borrow operation is handled in a fashion analogous to that in decimal. If a 1 is found in the preceding column of the subtrahend, it is borrowed, leaving a 0. If a 0 is found, an attempt is made to borrow from the next higher-order position, and so forth.

Binary	Decimal
1011	11
× 1001	× 9
1011	99
101100	
1100011	

Binary		Decimal
1011	=	11
+ 1001	=	9
10100	=	20

Fig. 23-19. Binary addition and corresponding decimal addition.

	0	1
0	0	0
1	0	1

Fig. 23-20. Binary multiplication. The binary multiplication table is the AND function of two binary variables. The process of multiplication consists of merely replicating and adding the multiplicand, as shown, if a 1 is found in the multiplier. If a 0 is found, a single 0 is entered and the next position to the left in the multiplier is taken up.

Binary	Decimal
100110	38
1001	9
11101	29

$A - B$, less borrow

	01
0	01
1	10

$A - B$, borrow

	01
0	00
1	10

Fig. 23-21. Binary subtraction and corresponding decimal subtraction. The subtraction table is the same as the addition table. The borrow operation is handled in analogy to decimal subtraction.

An example of binary division is

$$
\begin{array}{r}
110 \\
101 \overline{\big)\ 11110} \\
\underline{101} \\
101 \\
\underline{101} \\
0
\end{array}
\qquad
\begin{array}{r}
6 \\
5 \overline{\big)\ 30}
\end{array}
$$

The procedure is as follows:

1. Compare the divisor with the leftmost bits of the dividend.
2. If the divisor is greater, enter a 0 in the quotient and shift the dividend and quotient to the left.
3. Try subtraction again.
4. When the subtraction yields a positive result, i.e., the divisor is less than the bits in the dividend, enter a 1 in the quotient and shift the dividend and the quotient left one position.
5. Return to step 1 and repeat.

Binary division, like binary multiplication, is considerably simpler than the decimal operation.

30. Subtraction by Complement Addition. If subtraction and addition were handled in a data processing system in the manner described above, the borrow and carry functions would require separate circuits. Since it is desirable that the circuits for addition be used without modification for subtraction, most computers perform subtraction by *complement addition*.

The complement of a number is the difference between the number and some larger number called the *reference*. For computational purposes the reference number is chosen to be a power of the radix. In the decimal system, for example, one might choose the reference number $10^4 = 10,000_{10}$. The complement of a number, say 432, with respect to 10,000 is then 9,568. The number 9,568 is called the *true complement* of 432 to distinguish it from the *nines complement* of the same number. The nines complement is easier to form since it never requires a borrow operation. Thus the nines complement of 432 with respect to 10,000 is 9,567; that is, $10,000 - 1 = 9,999$, and $9,999 - 432 = 9,567$. Arithmetic can be done with either type of complement. For example, subtract 6,437 from 8,594 by true-complement addition. The true complement of 6,437 is 3,563. Then $8,594 + 3,563 = 12,157$, and, ignoring the leading 1, we have the correct answer to the problem, 2,157.

Binary complements are formed in a manner analogous to that in the decimal system. Let the reference number be chosen to be 2^4, or 10000_2. The true complement of the binary number 0101 with respect to 10000 is 1011. The binary analog of the decimal nines complement is called the *ones complement*. The ones complement of 0101 with respect to 10000 is 1010; that is, $10000 - 1 = 1111$, and $1111 - 0101 = 1010$. Figure 23-22 shows the process of forming a ones complement for a reference $2^{12} = 1000000000000_2$.

The binary ones complement is extremely simple to form since it requires only that the 1s and 0s of the number be interchanged (inverted). The mathematical basis for subtraction by complement addition is

$$
\begin{aligned}
N_A - N_B &= N_A + [(N_{ref} - 1) - N_B] + 1 - N_{ref} \\
&= N_A + \bar{N}_B + 1 - N_{ref}
\end{aligned}
\tag{23-9}
$$

where N_{ref} is the reference number, N_A is the minuend, N_B is the subtrahend, and $\bar{N}_B$ is the ones complement of N_B.

When arithmetic is performed by finding true complements, the terms of Eq. (23-9) are grouped as follows:

$$
\begin{aligned}
N_A - N_B &= N_A + \{[(N_{ref} - 1) - N_B] + 1\} - N_{ref} \\
&= N_A + \bar{N}_{B,true} - N_{ref}
\end{aligned}
\tag{23-10}
$$

The *maximum word size* in a data processing system is defined by the number of positions available in the addition register. The size is ordinarily 1 b greater than the numerical-field portion of a word size in memory and may be taken to be N_{ref} in Eq. (23-9).

In Eq. (23-9) if after adding N_A and N_B, so that the difference is positive, the 1 can be accounted for by adding 1 back into the sum $N_A + N_B$. In this case the quantity represented by $-N_{ref}$ appears as a 1 bit in the complement position after the complement addition. If this 1 is suppressed by a logical operation, $-N_{ref}$ is, in effect, subtracted away (Fig. 23-22). If N_A

$\geq N_B$, then $N_A + \bar{N}_B$ is negative or zero, but since $1 - N_{\text{ref}}$ represents a field in the register with all 1s, the result is automatically given in complement form (Fig. 23-23).

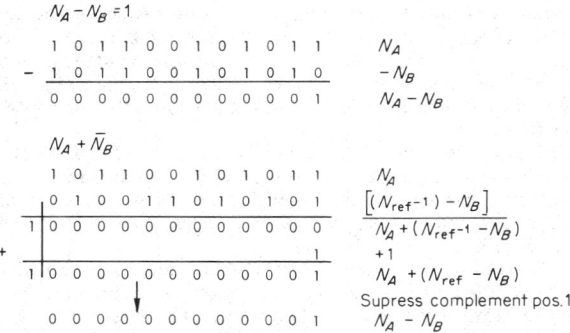

Numerical field

$$\boxed{1\,|\,0\,|\,0\,|\,0\,|\,0\,|\,0\,|\,0\,|\,0\,|\,0\,|\,0\,|\,0\,|\,0\,|\,0\,|\,0\,} \quad = N_{\text{ref}}$$

$$\boxed{0\,|\,1\,|\,1\,|\,1\,|\,1\,|\,1\,|\,1\,|\,1\,|\,1\,|\,1\,|\,1\,|\,1\,|\,1\,|\,1\,} \quad = N_{\text{ref}}-1$$

$$\boxed{1\,|\,0\,|\,1\,|\,1\,|\,0\,|\,0\,|\,1\,|\,0\,|\,1\,|\,0\,|\,1\,|\,1\,} \quad = N_A$$

$$\boxed{0\,|\,1\,|\,0\,|\,0\,|\,1\,|\,1\,|\,0\,|\,1\,|\,0\,|\,1\,|\,0\,|\,0\,} \quad = \left[(N_{\text{ref}}-1)-N_A\right] = \bar{N}_A$$

$N_A - N_B = 1$

```
  1 0 1 1 0 0 1 0 1 0 1 1     N_A
- 1 0 1 1 0 0 1 0 1 0 1 0    - N_B
  0 0 0 0 0 0 0 0 0 0 0 1     N_A - N_B
```

$N_A + \bar{N}_B$

```
    1 0 1 1 0 0 1 0 1 0 1 1      N_A
    0 1 0 0 1 1 0 1 0 1 0 1    [(N_ref-1) - N_B]
  1 0 0 0 0 0 0 0 0 0 0 0 0     N_A + (N_ref-1 - N_B)
+                       1     +1
  1 0 0 0 0 0 0 0 0 0 0 0 1     N_A + (N_ref - N_B)
```
                                              Supress complement pos.1
```
    0 0 0 0 0 0 0 0 0 0 0 1     N_A - N_B
```

Fig. 23-22. Subtraction of N_B from N_A by complement addition ($N_A + \bar{N}_B$ positive). $N_{\text{ref}} - 1$ is effectively added to the difference $N_A - N_B$ in forming the complement. This requires that 1 be added and N_{ref} suppressed in the result.

The fact that the two cases, i.e., whether the answer is positive or negative, demand different procedures might appear to preclude complement addition as a reasonable alternative in computer design. But if all negative numbers are stored and used in complement form, with each complement suitably tagged with a 1 bit in the complement position, a consistent procedure exists for summing either complement or noncomplement numbers as long as the absolute value of the result does not exceed the word-size limit of the addition register.

31. Floating-Point Numbers. In a computer having a fixed number of bits that define a word, the bits represent the maximum size of a numerical value. For example, if 40 bit positions are provided for a word, the maximum decimal number that can be represented is of the order of 1.099×10^{12}. Though this number is large, it does not suffice for many applications, especially in science, where a greater range of magnitudes may be routinely encountered. To extend the range of values that can be handled, numbers are represented in floating-point notation. In floating point the most significant digits of the number are written with an assumed radix point immediately to the left of the highest-order digit. This number is called the *fraction*. The intended position of the radix point is identified by a second number, called the *characteristic*,

$N_A - N_B = 0 = N_A - N_A$

$$\begin{array}{l} 101100101011 \\ 010011010100 \\ \hline 0\ \boxed{111111111111} \end{array}$$

Fig. 23-23. Complement addition when $N_A - N_B \leq 0$. When a complement addition would result in a negative or zero value, the result is given automatically in complement form, and a 1 bit need not be added in the low-order position.

which is appended to the fraction. The characteristic denotes the number of positions that the assumed radix point must be shifted to achieve the intended number. For example, the number 146.754 in floating point might be 146754.03 where 146754 would be equivalent to 0.146754 and the .03 would denote a shift of the decimal point three places to the right. In binary notation the number 11011.011 (27.375 in decimal) might be represented in floating point as 11011011.101 with the fraction again to the left of the decimal and the characteristic to the right.

With floating-point addition and subtraction, a shift register is required to align the radix points of the numbers. To perform multiplication or division, the fraction fields are appropriately multiplied or divided and the exponents summed or subtracted, respectively. As with fixed-point addition or subtraction, provision is usually made to detect an overflow condition in the characteristic fields. In some systems provision is made to note when an addition or subtraction occurs with such widely differing characteristics that justification destroys one of the two numbers (by shifting it out the end of a shift register).

32. Numeric and Alphanumeric Codes. The numeric codes used to represent numerical values previously discussed include the hexadecimal, octal, binary, and decimal codes. In many applications the need arises for the coding of nonnumeric as well as numeric information, and such coding must use the binary scheme. A code embracing numbers, alphabetic characters, and special symbols is known as an *alphanumeric code.*

A widely used code with its roots in the past is the telegraph code (the Baudot code). Figure 23-24 illustrates this code, which is still used in some major communication networks. Other alphanumeric codes have been devised for special purposes. One of the most significant of these, because of its present use and its contribution to the design of other codes, is the Hollerith code, developed in the 1890s. Hollerith's equipment (Par. **23-6**) contributed to the development of electromechanical accounting machines that provided the foundation for electronic computers.

In the Hollerith code bits (holes) located in rows 0, 11, or 12 of a punched-card column are used in conjunction with one or more holes in rows 1 to 9 of the same column to represent an alphabetic or special character. The Hollerith code is generally used for character sets consisting of 48 alphabetic, numeric, and special characters. Figure 23-25 shows a tabulating card encoded in Hollerith.

Figure 23-26 shows the binary-coded decimal (BCD) code, an outgrowth of the Hollerith code. In this 6-b code, the four lower-order bits are a binary representation of the numeric portion of a Hollerith coded character. The two higher-order bits correspond to the presence or absence of holes in the 0, 11, or 12 rows of the Hollerith code. For example, the letter Q is coded in BCD as 101000, whereas in Hollerith it is represented by holes in rows 8 and 11. With the development of more powerful computers an extension of the BCD code was required, so that more symbols could be represented. Such a code is shown in Fig. 23-27. Another code of importance in the United States is the American Standard Code for Information Interchange (ASCII) (see Fig. 23-28).

This code, developed by a committee of the American National Standards Institute (ANSI), has the advantage over most other codes of being *contiguous,* in the sense that the binary combination used to represent alphanumeric information is sequential. Hence alphabetic sorting can be easily accomplished by arithmetic manipulation of the code values.

Codes used for data transmission generally have both data characters and *control characters.* The latter perform control functions on the machine receiving information. For example, in the Baudot code (Fig. 23-24) the characters for space, carriage return, and line feed do not generate a character but operate mechanisms associated with the receiving printer. In more sophisticated codes, such as ASCII, these control functions are greatly extended and hence applicable to machines of different design.

By sending special characters called *escape characters* the mode of operation of the receiving machine can be changed to generate a different character set. Such characters or groups of such characters can extend the scope of any coding system. For example, with the Baudot code shown in Fig. 23-24, transmission of the up-shift code ↑ causes the machine to print in uppercase until a down-shift ↓ code is received. The addition of these two escape characters almost doubles the character set of the device.

Other Numeric Codes. Not all numeric information is represented by binary numbers. Other codes are also used for numeric information in special applications. Figure 23-29 shows a widely used code called the *reflected* or *Gray* code. It has the property that only 1

● Denotes positive current

Code signals: Start	1	2	3	4	5	Stop	Lowercase	Uppercase: CCITT standard international telegraph alphabet 2	United States teletype commercial keyboard	A T & T fractions keyboard	Weather keyboard
	●	●				●	A	-	-	-	↑
	●			●	●	●	B	?	?	5/8	⊕
		●	●	●		●	C	:	:	1/8	○
	●			●		●	D	Who are you?	$	$	↗
	●					●	E	3	3	3	3
	●		●	●		●	F	Note 1	!	1/4	→
		●		●	●	●	G	Note 1	&	&	↘
			●		●	●	H	Note 1	#		↓
		●	●			●	I	8	8	8	8
	●	●		●		●	J	Bell	Bell	'	↙
	●	●	●	●		●	K	(	(	1/2	←
		●			●	●	L	)	)	3/4	↖
			●	●	●	●	M	.	.	.	.
			●	●		●	N	,	,	7/8	⦷
				●	●	●	O	9	9	9	9
		●	●		●	●	P	0	0	0	Ø
	●	●	●			●	Q	1	1	1	1
		●		●		●	R	4	4	4	4
	●		●			●	S	'	'	Bell	Bell
					●	●	T	5	5	5	5
	●	●	●			●	U	7	7	7	7
		●	●	●	●	●	V	=	;	3/8	⦶
	●	●			●	●	W	2	2	2	2
	●		●	●	●	●	X	/	/	/	/
	●		●		●	●	Y	6	6	6	6
	●				●	●	Z	+	"	"	+
						●	Blank				-
	●	●	●	●	●	●	Letters shift				↓
	●	●		●	●	●	Figures shift				↑
			●			●	Space				■
				●		●	Carriage return				<
		●				●	Line feed				≡

Fig. 23-24. The Baudot telegraphers' code, a 5-b code. The code can be extended by using a shift character.

b is changed between any two successive values, irrespective of number size. This code is used in digital-to-analog systems since there is no need for propagation of carry integers in sequential counting as in a binary code.

Decimal Codes. Though most modern computers use the binary number system, other machines have a decimal code. Since four bit positions can accommodate 16 different code words, 4 b are more than adequate to represent the 10 symbols of the decimal system. In fact, since there are 16 code words and any 6 can be discarded, there are $16!/10!6!$, or about 3×10^{10} ways of assigning 4-b code words to the set of 10 decimal numerals. One subset of such codes comprises the *weighted codes,* in which the bit positions are assigned numerical weights such that when they are combined with their respective coefficients and summed, they yield the decimal number sought. Figure 23-30 shows three such codes. The first is the BCD

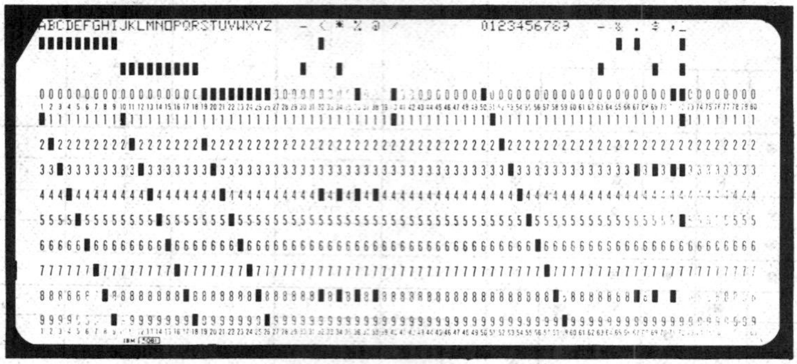

Fig. 23-25. A Hollerith code punched card, first developed in the 1890s and still widely used. The 0, 11, and 12 punches are called zone punches; punches 1 to 9 are the numeric field.

	00	01	10	11
0000	blank	ƀ	—	&+
0001	1	/	J	A
0010	2	S	K	B
0011	3	T	L	C
0100	4	U	M	D
0101	5	V	N	E
0110	6	W	O	F
0111	7	X	P	G
1000	8	Y	Q	H
1001	9	Z	R	I
1010	0	‡	!	?
1011	# ≡	,	$	.
1100	(1)'	% (	*	□)
1101	:	ɣ	]	[
1110	>	\	;	<
1111	√	+++	△	≢

Identification of
control codes and special meanings

≢ record mark	ɣ word separator
√ tape mark	ƀ substitute blank
△ mode change	+++ segment mark
≢ group mark	

Fig. 23-26. The IBM BCD interchange code, a 6-b code based upon the Hollerith code. The last four bits represent a numeric portion on a Hollerith card in BCD, while the first two bits represent the zone punches.

code, which has weights 8, 4, 2, and 1. This code cannot be used for complement subtraction since the first six digits, when complemented, form combinations that are not allowed by the code. The 2*421-weighted code shown in Fig. 23-30b is useful in complement arithmetic. The 5421-weighted code is also useful because of the appeal of the 5 weight in forming numbers from the bit code symbols. Many other codes have been developed for representation of decimal numbers. A summary is found in Ware.[7]

33. Error Detection and Correction Codes. The integrity of data in a computer system is of paramount importance because the serial nature of computation tends to propagate errors. Internal data transmission between computer-system units takes place repeatedly and at high speed. Data may also be sent over wires to remote terminals, printers, and other such equipment. Because imperfections of transmission channels inevitably produce some erroneous data, means must be provided to detect and correct errors whenever they occur.

A basic procedure for error detection and correction is to design a code in which each word contains more bits than are needed to represent all the symbols used in a data set. If a bit sequence is found that is not among those assigned to the data symbols, an error is known to have occurred.

One such commonly used error-detection code is called the *parity check*. Suppose that 8 b are used to represent data and that an additional bit is reserved as a check bit. A simple electronic circuit can determine whether an even or odd number of 1 bits is included in the eight bit positions. If an even number exists, a 1 bit can be inserted in the check position. If an odd number of 1s exists, the check position contains a 0. As a result all code words must contain an odd number of 1 bits. If a 9-b sequence is found to contain an even number of 1s, an error can be presumed.

There are limitations in the use of the simple parity check as a mechanism for error detection, since in many storage systems, there is a tendency for a failure to produce simultaneous errors in two adjacent positions. Such an error would not be detected since the parity of the code word would remain unchanged.

To increase the power of the parity check, a list of code words in a two-dimensional array may be used, as shown in Fig. 23-31. The code words in the horizontal dimension have a parity bit added, and the list in the vertical dimension also has an added parity bit, in each column. If one bit is in error, errors appear in both the row and the column. If simultaneous errors occur in two adjacent positions of a code word, no parity error will show up in that row, but the column checks will detect two errors.

It is possible to design codes that can detect directly whether errors have occurred in two bit positions in a single code word. Figure 23-32 shows such a code, an example of a *Hamming* code. The code positions in columns 1, 2, and 4 are used to check the parity of the respective bit combinations. Since the codes overlap, if an error occurs in any two bits, both errors will be detected. The pattern formed by the particular parity bits that show errors indicates which bit is in error in the case of a single bit failure.

If a code has a 2-b *detection* property, it is implied that it has a 1-b *correction* property. For example, if one error has occurred in a word, it will be detected. If bits are then changed sequentially throughout the word, a double error occurs if the bit changed is not the bit in error. Thus the particular bit that has been modified can be changed back to its original state and a new bit tried until the faulty bit is changed, at which point no error will be detected. This reasoning can be applied to the case of 4-b detection with 2-b correction. In general, if one can detect errors in $2n$ b, n-b correction is implied.

34. Polynomial Codes. There is a general procedure for deriving error-detecting codes by means of binary polynomials. For example, the Hamming, Bose-Chaudhuri, and Fire codes the parity check and the codes of Melas can all be viewed as polynomial codes.*

Polynomial codes can be described in terms of the *modulo-2 division* of binary polynomials. Modulo-2 division is defined in terms of modulo-2 addition. In modulo-2 addition binary numbers are added without regard to the carry operation; i.e., each column is treated independently. In a modulo-2 subtraction the borrow operation is suppressed, and since the

---

* W. W. Peterson and D. T. Brown   Cyclic Codes for Error Detection, *Proc. IRE*, January 1961; R. W. Hamming   Error Detecting and Error Correcting Codes, *Bell Syst. Tech. J.*, April 1950; R. Bose and D. Ray Chaudhuri   A Cease of Error-correcting Binary Group Coded, *Inform. Control*, March 1960, Vol. 3; P. Fire   A Cease of Multiple-Error-Correcting Binary Codes for Non Independent Errors, *Stamford Electron. Lab. Tech. Rep.* 55, April 29, 1959; M. Melas   A New Group of Codes for Correction of Dependent Error in Data Transmission, *IBM J. Res. Dev.*, 1960, Vol. 4, p. 58.

Data characters

	Normal shift									Upper shift							
Character	S	B	A	8	4	2	1	Parity	Character	S	B	A	8	4	2	1	Parity
1	0	0	0	0	0	0	1	0	=	1	0	0	0	0	0	1	
2	0	0	0	0	0	1	0	0	¢	1	0	0	0	0	1	0	1
3	0	0	0	0	0	1	1	1	;	1	0	0	0	0	1	1	1
4	0	0	0	0	1	0	0	0	:	1	0	0	0	1	0	0	0
5	0	0	0	0	1	0	1	1	°C	1	0	0	0	1	0	1	1
6	0	0	0	0	1	1	0	1	'	1	0	0	0	1	1	0	0
7	0	0	0	0	1	1	1	0	-	1	0	0	0	1	1	1	0
8	0	0	0	1	0	0	0	0	+	1	0	0	1	0	0	0	1
9	0	0	0	1	0	0	1	1	(	1	0	0	1	0	0	1	0
0	0	0	0	1	0	1	0	1	)	1	0	0	1	0	1	0	0
a	0	1	1	0	0	0	1	0	A	1	1	1	0	0	0	1	1
b	0	1	1	0	0	1	0	0	B	1	1	1	0	0	1	0	1
c	0	1	1	0	0	1	1	1	C	1	1	1	0	0	1	1	0
d	0	1	1	0	1	0	0	0	D	1	1	1	0	1	0	0	1
e	0	1	1	0	1	0	1	1	E	1	1	1	0	1	0	1	0
f	0	1	1	0	1	1	0	1	F	1	1	1	0	1	1	0	0
g	0	1	1	0	1	1	1	0	G	1	1	1	0	1	1	1	1
h	0	1	1	1	0	0	0	0	H	1	1	1	1	0	0	0	1
i	0	1	1	1	0	0	1	1	I	1	1	1	1	0	0	1	0
j	0	1	0	0	0	0	1	1	J	1	1	0	0	0	0	1	0
k	0	1	0	0	0	1	0	1	K	1	1	0	0	0	1	0	0
l	0	1	0	0	0	1	1	0	L	1	1	0	0	0	1	1	1
m	0	1	0	0	1	0	0	1	M	1	1	0	0	1	0	0	0
n	0	1	0	0	1	0	1	0	N	1	1	0	0	1	0	1	1
o	0	1	0	0	1	1	0	0	O	1	1	0	0	1	1	0	1
p	0	1	0	0	1	1	1	1	P	1	1	0	0	1	1	1	0
q	0	1	0	1	0	0	0	1	Q	1	1	0	1	0	0	0	0
r	0	1	0	1	0	0	1	0	R	1	1	0	1	0	0	1	1
s	0	0	1	0	0	1	0	1	S	1	0	1	0	0	1	0	0
t	0	0	1	0	0	1	1	0	T	1	0	1	0	0	1	1	1
u	0	0	1	0	1	0	0	1	U	1	0	1	0	1	0	0	0
v	0	0	1	0	1	0	1	0	V	1	0	1	0	1	0	1	1
w	0	0	1	0	1	1	0	0	W	1	0	1	0	1	1	0	1
x	0	0	1	0	1	1	1	1	X	1	0	1	0	1	1	1	0
y	0	0	1	1	0	0	0	1	Y	1	0	1	1	0	0	0	0
z	0	0	1	1	0	0	1	0	Z	1	0	1	1	0	0	1	1
.	0	1	1	1	0	1	1	0	.	1	1	1	1	0	1	1	1
$	0	1	0	1	0	1	1	1	!	1	1	0	1	0	1	1	0
,	0	0	1	1	0	1	1	1	,	1	0	1	1	0	1	1	0
/	0	0	1	0	0	0	1	1	?	1	0	1	0	0	0	1	0
:	0	0	0	1	0	1	1	0	±	1	0	0	1	0	1	1	1
&	0	1	1	0	0	0	0	1	+	1	1	1	0	0	0	0	0
-	0	1	0	0	0	0	0	0	-	1	1	0	0	0	0	0	1
@	0	0	1	0	0	0	0	0	*	1	0	1	0	0	0	0	1

(a)

Fig. 23-27. The extended IBM BCD 8-b code, used for data transmission. (a) Data characters, (b) control characters.

Control characters		(Either shift) (Either setting of S bit)					
Backspace		1	0	1	1	1	0
End of transfer		0	0	1	1	1	1
Delete		1	1	1	1	1	1
Down-shift		1	1	1	1	1	0
Carriage return		1	0	1	1	0	1
Prefix		0	1	1	1	1	1
Idle		1	0	1	1	1	1
Reader stop		0	0	1	1	0	1
Space		0	0	0	0	0	0
End of block		0	1	1	1	1	0
Up-shift		0	0	1	1	1	0
Line feed		0	1	1	1	0	1
Tab		1	1	1	1	0	1
Restore		1	0	1	1	0	0
Bypass		0	1	1	1	0	0
End of heading		0	0	1	0	1	1
Punch on		0	0	1	1	0	0
Punch off		1	1	1	1	0	0

(b)

modulo-2 subtraction and addition tables are identical, modulo-2 addition and subtraction coincide. Both operations are denoted by the single symbol $\oplus$.

Modulo-2 addition of two numbers, a and b, is as follows:

$$
\begin{array}{r}
110010110101101 \\
101101101011101 \\
\hline
11111011110000
\end{array}
$$

The process of modulo-2 division is

$$
\begin{array}{r}
11001100011 \quad \text{quotient} \\
110101 \overline{) 1011010110100000} \quad \text{dividend} \\
110101 \\
\hline
110000 \\
110101 \\
\hline
101110 \\
110101 \\
\hline
110111 \\
110101 \\
\hline
100000 \\
110101 \\
\hline
101010 \\
110101 \\
\hline
11111 \quad \text{remainder}
\end{array}
$$

The divisor is initially subtracted modulo 2 from the leftmost bits of the dividend and a 1 entered in the quotient. The dividend and quotient are then shifted one position to the left. If the leading number in the dividend is a 1, the process is repeated. If the leading number is 0, a 0 is placed in the quotient and the dividend and quotient are again shifted to the left by one position. The process is repeated until the numbers in the dividend are exhausted.

A polynomial code is generated in the following manner. Let C be the binary number 10110101101 to be encoded. Let $C(x)$ be a binary polynomial in x, derived in the following way:

$$C(x) = 1 \times x^{10} + 0 \times x^9 + 1 \times x^8 + 1 \times x^7 + 0 \times x^6 + 1 \times x^5 + 0 \times x^4 + 1 \times x^3 + 1 \times x^2 + 0 \times x^1 + 1$$
$$= x^{10} + x^8 + x^7 + x^5 + x^3 + x^2 + 1 \tag{23-11}$$

Let $G(x)$ be another binary polynomial called the *generating polynomial*, given by

$$G(x) = x^5 + x^4 + x^2 + 1$$

The binary code word corresponding to this polynomial $G(x)$ is 110101.

Multiply $C(x)$ by x^r, where r is the degree of $G(x)$, in this case 5, and divide $x^r C(x)$ by $G(x)$ to yield

$$\frac{x^r C(x)}{G(x)} = Q(x) + \frac{R(x)}{G(x)} \tag{23-12}$$

	000	001	010	011	100	101	110	111
0000	NULL	① DC_0	ᵇ	0	⁽ᵃ⁾	P		
0001	SOM	DC_1	!	1	A	Q		
0010	EOA	DC_2	"	2	B	R		
0011	EOM	DC_3	#	3	C	S		
0100	EOT	DC_4 (STOP)	$	4	D	T		
0101	WRU	ERR	%	5	E	U		
0110	RU	SYNC	&	6	F	V		
0111	BELL	LEM	'	7	G	W		Unassigned
1000	FE_0	S_0	(	8	H	X		
1001	HT SK	S_1	)	9	I	Y		
1010	LF	S_2	*	:	J	Z		
1011	V_{TAB}	S_3	+	;	K	[		
1100	FF	S_4	,	<	L	\		ACK
1101	CR	S_5	-	=	M	]		②
1110	SO	S_0	.	>	N	↑		ESC
1111	SI	S_7	/	?	O	←		DEL

Identification of control symbols and some graphics

NULL	Null/idle	V_{TAB}	Vertical tabulation	S_0-S_7	Separator (information)
SOM	Start of message	FF	Form feed	ᵇ	Word separator (space, normally nonprinting)
EOA	End of address	CR	Carriage return		
EOM	End of message	SO	Shift out	<	Less than
EOT	End of transmission	SI	Shift in	>	Greater than
WRU	"Who are you?"	DC_0	Device control 1 Reserved for data link escape	↑	Up arrow (exponentiation)
RU	"Are you...?"			←	Left arrow (implies/ replaced by)
BELL	Audible signal	DC_1-DC_3	Device control	\	Reverse slant
FE_0	Format effector	DC_4 (STOP)	Device control (stop)	ACK	Acknowledge
HT	Horizontal tabulation	ERR	Error	②	Unassigned control
SK	Skip (punched card)	SYNC	Synchronous idle	ESC	Escape
LF	Line feed	LEM	Logical end of media	DEL	Delete/idle

Fig. 23-28. The ASCII code has a contiguous alphabet, so that numeric ordering permits alphabetic sorting.

where $Q(x)$ and $R(x)$ are, respectively, the quotient and remainder from the division. Equation (23-12) can be multiplied through by $G(x)$ to give

$$x^r C(x) = Q(x)G(x) \oplus R(x) \qquad (23\text{-}13)$$

Since modulo-2 addition and subtraction are equivalent,

$$x^r C(x) \oplus R(x) = Q(x)G(x) \qquad (23\text{-}14)$$

The above operations define the following procedure: add r 0s to the right of C. Divide C by G, modulo 2, generating a remainder. Place the remainder in the positions occupied by the 0s that were added to the right side of C.

Decimal	Grey code
1	0000
2	0001
3	0011
4	0010
5	0110
6	0111
7	0101
8	0100
9	1100
10	1101
11	1111
12	1110
13	1010
14	1011
15	1001
16	1000

Fig. 23-29. The Gray code, used in analog-to-digital encoding systems. There is only a 1-b change between any two successive integers.

The result, the left-hand side of Eq. (23-14), is the encoded form of the code word C, which we call $M(x)$. From Eq. (23-14) it can be seen that $M(x)$ is divisible by $G(x)$. After transmission, if $M(x)$ is divided by $G(x)$, the division should take place without a remainder.

In summary, polynomial encoding consists of utilizing a generating polynomial $G(x)$ of degree r to form a message. The message can be checked by division by the polynomial $G(x)$. Decoding occurs by removing the final r bits from M.

To show how this scheme operates, consider that in M an error pattern occurs that is n b long. The error pattern can be defined as that polynomial $E(x)$ which when added modulo 2 to $M(x)$ duplicates the errors in $M(x)$. The detailed modulo-2 division shown (page 23-21) above was selected to be $x^5 C(x)/G(x)$, where $C(x)$ and $G(x)$ are as given above. Thus $M(x)$ is 1011010110111111. If errors occur in bit positions 2, 3, 6, and 9 counting from the left, $E(x)$ is given by $x^{14} + x^{13} + x^{10} + x^7$ or $x^7(x^7 + x^6 + x^3 + 1)$. The error pattern is 7 b long in this case.

The message that is transmitted is $Q(x)G(x) + x^n E(x)$. For the error to go undetected $E(x)$ must be divisible by $G(x)$. If the degree of $E(x)$, which for generality may be taken as n, is less than r [the degree of $G(x)$], the error must be detected. If $r = n$, the error can go undetected only if $G(x) = E(x)$. If $n > r$, the error can go undetected only if $E(x)$ is divisible by $G(x)$, an unlikely occurrence if E is of reasonable degree.

Polynomial Coding Implementation. The implementation of a polynomial checking scheme is relatively simple, as shown in Fig. 23-33. A set of shift registers is used to represent the generating code. A number of modulo-2 adders (equal to one more than the 1s in the generating polynomial) are interspersed with the shift registers in the relative positions of the 1s in the generating polynomial. The polynomial to be encoded is entered into the shift register from left to right until a 1 is found in the leftmost column. A 1 is then added at each modulo-2 adder point as a shift left occurs. If the leftmost bit is still a 1, the process is repeated. Otherwise only a shift occurs. When all members of the input code point have been entered, the result in the shift register is the remainder that can then be substituted for the 0s in $C(x)x^r$.

35. Numerical Methods of Computation. In the internal operations of a digital computer, a finite number of processing steps occurs on a finite member of data words stored with a finite precision, due to limited word length. These factors imply that, in general, the operations of a computer are *digital approximations* of any continuous functions applied as inputs or supplied as outputs. Thus a set of discrete points must represent the solution to a differential equation. A continuous function such as sin x must of necessity be stored as a sequence of values Y_i, where i is an integer.

Since many of the operations of calculus and algebra are continuous, digital procedures must be found to deal with them. For example, for integration of a continuous function represented by a finite set of data points, a polynomial may be used to approximate a few data points and the integral of the polynomial taken over successive wider intervals until all data points are included. Such a procedure is an example of problems dealt within *numerical analysis.*

A basic question in the operation of a digital computer is the finite precision of numerical representation. For example, most numerical values suffer from *truncation errors* when restricted to a finite field. Similarly, *translation errors* occur in going from binary or hexadecimal to decimal representation, and vice versa. Since data may be subject to manipulation several times throughout a program, the propagation of errors that have arisen earlier must be dealt with.

In some cases, e.g., methods developed for the solution of certain types of differential equations, *feedback* is used in finding a solution. In such cases, earlier errors may be magnified as the program progresses, and the solution procedure is then said to be *unstable.*

To illustrate such a situation, consider the differential equation $dy/dx = x$, with $y = 0$ at $x = 0$. The analytic solution of this equation is $y = x^2/2$. To obtain the solution by means of a digital computer, a stepwise procedure can be used. Since $y = 0$ at $x = 0$, at $x = \delta$, $y = 0$ but $dy/dx = \delta$. At $x = \delta$ the change in the interval $\delta < x < 2\delta$ can be computed with $dy/dx = \delta$ but $y = 0$. Such a solution will always fall below the analytic solution, due to cumulative error in the result. Here a more appropriate strategy would be to predict the slope at $x = \delta$ and correct the slope over the interval before computing y at $x = \delta$. The procedure for defining the predicted slope is called a *predictor* and the procedure for adjusting it the *corrector.*

Decimal digit	8421	Decimal digit	2*421	Decimal digit	5421
0	0000	0	0000	0	0000
1	0001	1	0001	1	0001
2	0010	2	0010	2	0010
3	0011	3	0011	3	0011
4	0100	4	0100	4	0100
5	0101		0101		0101
6	0110		0110	Unused	0110
7	0111	Unused	0111		0111
8	1000		1000	5	1000
9	1001		1001	6	1001
	1010		1010	7	1010
Unused	1011	5	1011	8	1011
	1100	6	1100	9	1100
	1101	7	1101		1101
	1110	8	1110	Unused	1110
	1111	9	1111		1111
(a)		*(b)*		*(c)*	

Fig. 23-30. Weighted codes for expressing decimal digits in binary. The first, 8241 (*a*), is the most direct, the binary numbers corresponding to decimal numbers, but it cannot be used in a complement-add scheme. The second, 2*421 (*b*), is self-complementing. The third, 5421 (*c*), is easy to remember and is used when converting from binary representation to decimal.

In more general situations, e.g., with second- or higher-order equations, the Taylor series expansion $f(x + \delta) = f(x) + f'(x)\delta + f''(x)\delta^2/2 + \cdots$ can be used to develop intermediate predicted values that, in turn, increase the precision of each step. However, as the feedback introduced by such methods increases, random errors, subject to growth at each step, may destroy the solution. The references in Par. **23-131** deal with the stability of various types of differential equations, notably Ref. 24-28.

Since in a digital computer functional representation must be in terms of discrete values, an algorithmic representation of the function usually involves a power series of some kind. The series must have properties of convergence that permit finding a solution point in a reasonable number of computations. Thus, methods of increasing the convergence properties of a series, the expansion of functions in various types of orthogonal functions with good convergence properties, and other methods of summing series to minimize computation are important topics in numerical analysis as applied to digital computers.

Other important topics in numerical analysis include the *difference calculus* and the *summation calculus*. These are discrete-point methods of forming analytic solutions of finite-difference and summation problems, analogous to differentiation and integration with continuous functions.

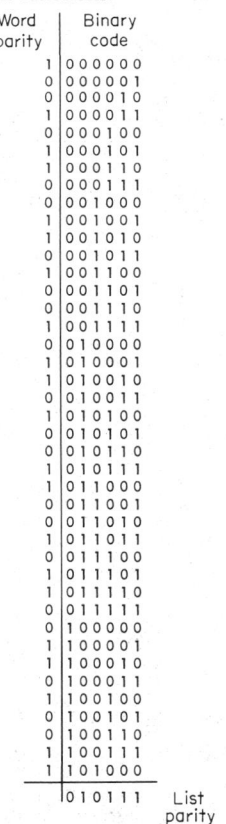

Word parity	Binary code
1	0 0 0 0 0 0
0	0 0 0 0 0 1
0	0 0 0 0 1 0
1	0 0 0 0 1 1
0	0 0 0 1 0 0
1	0 0 0 1 0 1
1	0 0 0 1 1 0
0	0 0 0 1 1 1
0	0 0 1 0 0 0
1	0 0 1 0 0 1
1	0 0 1 0 1 0
0	0 0 1 0 1 1
1	0 0 1 1 0 0
0	0 0 1 1 0 1
0	0 0 1 1 1 0
1	0 0 1 1 1 1
0	0 1 0 0 0 0
1	0 1 0 0 0 1
1	0 1 0 0 1 0
0	0 1 0 0 1 1
1	0 1 0 1 0 0
0	0 1 0 1 0 1
0	0 1 0 1 1 0
1	0 1 0 1 1 1
1	0 1 1 0 0 0
0	0 1 1 0 0 1
0	0 1 1 0 1 0
1	0 1 1 0 1 1
0	0 1 1 1 0 0
1	0 1 1 1 0 1
1	0 1 1 1 1 0
0	0 1 1 1 1 1
0	1 0 0 0 0 0
1	1 0 0 0 0 1
1	1 0 0 0 1 0
0	1 0 0 0 1 1
1	1 0 0 1 0 0
0	1 0 0 1 0 1
0	1 0 0 1 1 0
1	1 0 0 1 1 1
1	1 0 1 0 0 0
0	0 1 0 1 1 1 List parity

Fig. 23-31. Two-dimensional parity checking, in which a single error can be corrected and a double error detected.

Decimal digit	Position
	1 2 3 4 5 6 7
0	0 0 0 0 0 0 0
1	1 1 0 1 0 0 1
2	0 1 0 1 0 1 0
3	0 0 0 0 0 1 1
4	1 0 0 1 1 0 0
5	0 1 0 0 1 0 1
6	1 1 0 0 1 1 0
7	0 0 0 1 1 1 1
8	1 1 1 0 0 0 0
9	0 0 1 1 0 0 1

Parity checks   8, 4, 2, 1 code

Fig. 23-32. A Hamming code. The parity bit in column 1 checks parity in columns 1, 3, 5, and 7; the bit in column 2 checks 2, 3, 6, and 7; and the bit in column 4 checks 4, 5, 6, and 7. The overlapping structure of the code permits the correction of a single error and the detection of single or double errors in any code word.

Another class of problem involves searches for optimum methods among a set whose numbers are so large that they prevent an exact analysis. For example, in the computation of the diffusion of thermal neutrons the amount of computation for an exact solution is prohibitive, but repeated solutions of the motion of one particle subject to random collisions can be obtained. As the number of trials increases, a picture of average behavior emerges. Such approaches are called *Monte Carlo methods*, due to the random element used to approach the solution. Such methods are called *heuristic*. The specific strategy used in forming a solution is called an *algorithm*.

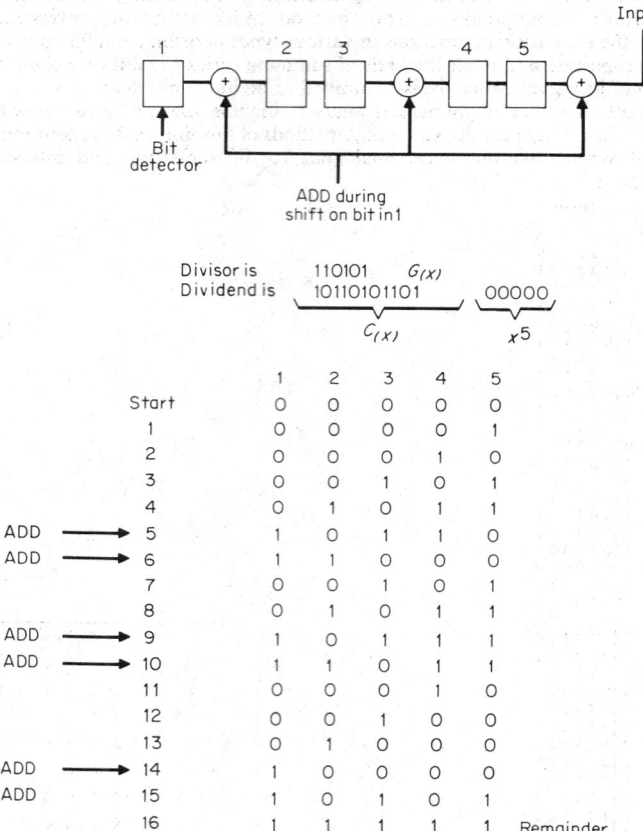

		1	2	3	4	5	
	Start	0	0	0	0	0	
	1	0	0	0	0	1	
	2	0	0	0	1	0	
	3	0	0	1	0	1	
	4	0	1	0	1	1	
ADD →	5	1	0	1	1	0	
ADD →	6	1	1	0	0	0	
	7	0	0	1	0	1	
	8	0	1	0	1	1	
ADD →	9	1	0	1	1	1	
ADD →	10	1	1	0	1	1	
	11	0	0	0	1	0	
	12	0	0	1	0	0	
	13	0	1	0	0	0	
ADD →	14	1	0	0	0	0	
ADD	15	1	0	1	0	1	
	16	1	1	1	1	1	Remainder

Fig. 23-33. Generation of a polynomial code.

BOOLEAN OPERATIONS, CIRCUIT REALIZATIONS, INTERCONNECTIONS, AND COMPUTER ORGANIZATION

36. Symbolic Logic and the Sentential Calculus. Before the advent of the electronic computer, a branch of philosophy called *symbolic logic* was developed to explore the logic of mathematical thought. A subset of this subject, called the *sentential calculus*, deals with the truth or falsehood of statements or sentences combined by various logical connectives, e.g., "and," "or," or "if . . . then." The sentential calculus is also commonly termed the *calculus of propositions*. One of the major architects of the symbolic manipulations that underlie the calculus of propositions was George Boole.* The manipulation of binary variables is commonly called *Boolean algebra*.

* G. Boole "The Mathematical Analysis of Logic," 1847, reprinted by Blackwell, Oxford, 1951.

The Sentential Calculus. The subject matter treated by the sentential calculus can be illustrated by an example. Suppose that two sentences are given that are either true or false. Consider that one sentence, represented symbolically by *A,* is "It rains" and the other, represented by *B,* is "It freezes." The question is to explore the truth value of these sentences when combined into more complex statements. For example, suppose a combined statement *C* "It rains or it rains and it freezes" is given. If "It rains" is true, then statement *C* is always true, since a choice is given between "It rains" and "It rains and freezes." Thus, even if statement *B* were false, statement *C* would be true. On the other hand, if the statement *A* "It rains" is false, the combined statement *C* is false since statement *B* "It rains and freezes" must also be false (since the connective "and" implies that both *A* and *B* must be true). It thus can be seen that the combined sentence *C* is equivalent only to the statement "It rains."

Symbolically, if we let the connective "or" be represented by the symbol v and the "and" by the centered dot, the combined statement becomes $A \ v \ A \cdot B$. The logical analysis given above of the combined statement shows that it is equivalent only to *A.* Thus we can write $A \ v \ A \cdot B = A$, a statement deducible from considerations of Boolean algebra alone.

37. Boolean Algebra. The importance of symbolic logic to computer design arises because the behavior of current flow in complex switching networks is governed by the mathematical framework of Boolean algebra. This association was first discussed by Shannon in 1938,[*] in connection with the analysis of relay switching networks. As stated in Par. **23-13,** sentences and their negations may be associated with normally open and normally closed relay contacts, respectively, while the AND and OR connectives of logical analysis may be associated with series and parallel connections of these contacts.

A relationship also exists between Boolean algebra and the properties of *sets of objects,* under the operations of *union, intersection,* and *inversion.* This relationship is also useful in switching-system design, since a graphical representation of the interrelationship between sets can be used to simplify the corresponding switching problem.

To illustrate, let *A* and *B* be sets with members *a* and *b,* respectively. Let a set *C* be formed such that *c* is in *C* if *c* is a member of *A* and a member of *B.* *C* is called the *intersection* of *A* and *B,* and the operation is represented as $A \cap B = C$. Another set *D* can be defined such that *d* is a member of *D* if *d* is contained in *A* or *B* or both. *D* is called the *union* of *A* and *B,* denoted by $D = A \cup B$.

As a third case, let a set *A* be defined that has as its members *a* all members of all sets. Then a set *Ā,* called the *inverse* of *A,* can be defined that has as its members *ā,* that is all members that are not in *A.* The operations of intersection, union, and inverse are analogous to the AND, OR, and NOT in sentential calculus and to series, parallel, and normally closed contacts in relay-circuit design, respectively. Figure 23-34 shows the concept of union, intersection, and inverse represented graphically on *Venn diagrams.*

38. Boolean Functions; Minterms and Maxterms. If *A, B, C,* . . . are binary variables, any function $f(A,B,C, \dots)$ can be formed from the variables using the AND, OR, and NOT connectives. In particular, any function can be formed in a unique way (except for the ordering of terms) from the set of ORs of products of the variables and their negations. Such a sum of products is called the *disjunctive normal form* (cf. Par. **23-12**).

---

[*] C. E. Shannon Symbolic Analysis of Relay and Switching Circuits, *Trans. AIEE,* 1938, Vol. 57, p. 713.

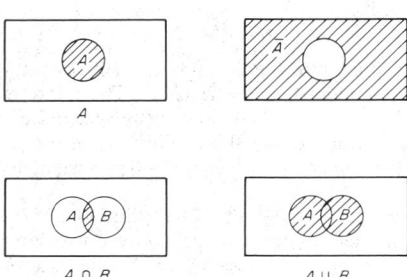

Fig. 23-34. Venn diagrams. Graphical representations of relationships between binary variables that also illustrate the relationships between set theory and Boolean algebra.

Suppose a function of n binary variables $f(A_1, A_2, \ldots, A_n)$ is given. A *minterm* (following Ware[7]) is defined as a *product* of each variable or its negation, so that there are 2^n possible minterms. A *maxterm* is similarly defined as the *sum* of each variable or its negation. As with the minterms, there are 2^n maxterms. Table 23-2 illustrates minterms and maxterms.

Table 23-2. $f(A_1, A_2, A_3, A_4)$

Minterms of f	Maxterms of f
$\bar{A}_1 \bar{A}_2 \bar{A}_3 \bar{A}_4$	$A_1 + A_2 + A_3 + A_4$
$\bar{A}_1 \bar{A}_2 \bar{A}_3 A_4$	$A_1 + A_2 + A_3 + \bar{A}_4$
$\bar{A}_1 \bar{A}_2 A_3 \bar{A}_4$	$A_1 + A_2 + \bar{A}_3 + A_4$
$\bar{A}_1 \bar{A}_2 A_3 A_4$	$A_1 + A_2 + \bar{A}_3 + \bar{A}_4$
$\bar{A}_1 A_2 \bar{A}_3 \bar{A}_4$	$A_1 + \bar{A}_2 + A_3 + A_4$
$\bar{A}_1 A_2 \bar{A}_3 A_4$	$A_1 + \bar{A}_2 + A_3 + \bar{A}_4$
$\bar{A}_1 A_2 A_3 \bar{A}_4$	$A_1 + \bar{A}_2 + \bar{A}_3 + A_4$
$\bar{A}_1 A_2 A_3 A_4$	$A_1 + \bar{A}_2 + \bar{A}_3 + \bar{A}_4$
$A_1 \bar{A}_2 \bar{A}_3 \bar{A}_4$	$\bar{A}_1 + A_2 + A_3 + A_4$
$A_1 \bar{A}_2 \bar{A}_3 A_4$	$\bar{A}_1 + A_2 + A_3 + \bar{A}_4$
$A_1 \bar{A}_2 A_3 \bar{A}_4$	$\bar{A}_1 + A_2 + \bar{A}_3 + A_4$
$A_1 \bar{A}_2 A_3 A_4$	$\bar{A}_1 + A_2 + \bar{A}_3 + \bar{A}_4$
$A_1 A_2 \bar{A}_3 \bar{A}_4$	$\bar{A}_1 + \bar{A}_2 + A_3 + A_4$
$A_1 A_2 \bar{A}_3 A_4$	$\bar{A}_1 + \bar{A}_2 + A_3 + \bar{A}_4$
$A_1 A_2 A_3 \bar{A}_4$	$\bar{A}_1 + \bar{A}_2 + \bar{A}_3 + A_4$
$A_1 A_2 A_3 A_4$	$\bar{A}_1 + \bar{A}_2 + \bar{A}_3 + \bar{A}_4$

A binary function can be expanded as either a sum of minterms or a product of maxterms, as follows:

$$
\begin{aligned}
f(A_1, A_2, A_3, \cdots, A_n) = &\ f(A_1 A_2 A_3 \cdots A_n)(A_1 A_2 A_3 \cdots A_n) \\
&+ f(\bar{A}_1 A_2 A_3 \cdots A_n)(\bar{A}_1 A_2 A_3 \cdots A_n) \\
&+ f(A_1 \bar{A}_2 A_3 \cdots A_n)(A_1 \bar{A}_2 A_3 \cdots A_n) \\
&+ f(\bar{A}_1 \bar{A}_2 A_3 \cdots A_n)(\bar{A}_1 \bar{A}_2 A_3 \cdots A_n) \\
&+ f(A_1 A_2 \bar{A}_3 \cdots A_n)(A_1 A_2 \bar{A}_3 \cdots A_n) \\
&\ \cdot \\
&\ \cdot \\
&\ \cdot \\
&+ f(\bar{A}_1 \bar{A}_2 \bar{A}_3 \cdots \bar{A}_n)(\bar{A}_1 \bar{A}_2 \bar{A}_3 \cdots \bar{A}_n)
\end{aligned}
$$

$$
\begin{aligned}
f(A_1, A_2, A_3, \cdots, A_n) = &\ [A_1 + A_2 + A_3 + \cdots A_n + \bar{f}(A_1, A_2 A_3 \cdots A_n)] \\
&[A_1 + \bar{A}_2 + \bar{A}_3 + \cdots \bar{A}_n + \bar{f}(\bar{A}_1, A_2 A_3 \cdots A_n)] \\
&[\bar{A}_1 + A_2 + \bar{A}_3 + \cdots \bar{A}_n + \bar{f}(A_1, \bar{A}_2 A_3 \cdots A_n)] \\
&[\bar{A}_1 + \bar{A}_2 + \bar{A}_3 + \cdots \bar{A}_n + \bar{f}(\bar{A}_1, \bar{A}_2 A_3 \cdots A_n)] \\
&[\bar{A}_1 + \bar{A}_2 + A_3 + \cdots \bar{A}_n + \bar{f}(A_1, A_2, \bar{A}_3 \cdots A_n)] \\
&\ \cdot \\
&\ \cdot \\
&\ \cdot \\
&[A_1 + A_2 + A_3 \cdots A_n + \bar{f}(\bar{A}_1, \bar{A}_2, \bar{A}_3 \cdots \bar{A}_n)]
\end{aligned}
$$

Since for any individual function, the 2^n min- or maxterms may either be present or absent, there are 2^{2^n} binary functions of n variables. The minterm sum expansion given above is the disjunctive normal form. The result of the maxterm product given above is called the *conjunctive normal form.* Either the conjunctive or disjunctive normal form is a unique representation of any binary function (provided a standard ordering of terms is accepted).

Graphical Representations. In Fig. 23-4, Par. 23-12, functions of one, two, and three binary variables are shown in truth tables of one, two, and three dimensions, respectively. Such a representation breaks down for four variables. Figure 23-7 gives a representation of a binary function using relay contacts, and in Fig. 23-34 Venn diagrams are presented. A fourth way of showing a binary function is the *Karnaugh map.* One or two variables are represented in each dimension by a Gray code, so that a planar map can represent up to four variables. The five- and six-dimensional cases can be treated by using respectively two or four planar 4 × 4 frames. Figure 23-35 illustrates Karnaugh maps of different dimensions.

A fifth representation of a binary function (Fig. 23-36) is called a *vertex diagram.* A set of circular dots represents each code word possible in the space of the Boolean variables represented. The dots are then connected to all other code words that differ only in one bit position. The function is represented by blackening each dot that corresponds to the particular code word represented by the appropriate minterm in the disjunctive normal form. Figure 23-37 is an example of a particular function of four binary variables represented by a Venn diagram and a Karnaugh map.

39. Algebraic Properties of Binary Variables. A set of variables and operations with the properties listed in Table 23-3 is called a Boolean algebra (see Birkhoff and MacLane[40]). These postulates can be visualized by Venn diagrams or by circuits. Several cases are shown in Fig. 23-38.

From the properties listed in Table 23-3 a number of theorems can be established, among the most useful being

$$x + xy = x \tag{23-15}$$

$$x + \bar{x}y = x + y \tag{23-16}$$

De Morgan's Theorem. In the examples given in Par. **23-14** of the AND function implemented with electronic circuits, it can be seen that if the logical assignment of input

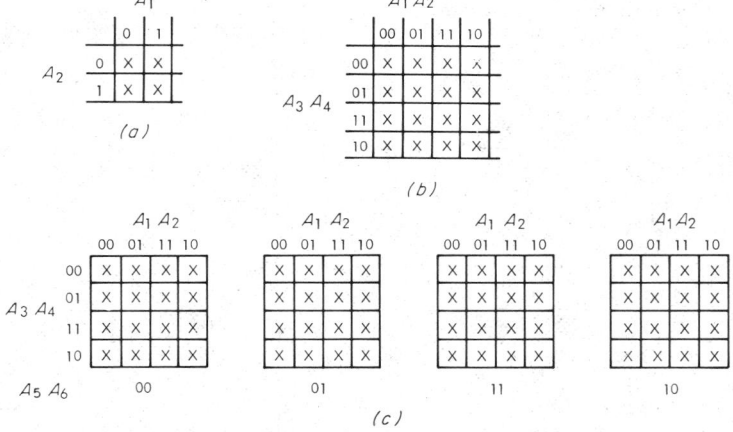

Fig. 23-35. Representative Karnaugh maps. Each X may be either a 0 or a 1, depending on the particular terms in the disjunctive normal form of the binary function. (a) Two-dimensional, (b) four-dimensional, (c) six-dimensional map.

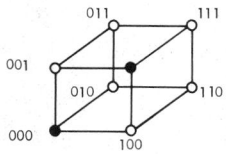

Fig. 23-36. Vertex diagram, in which all points are interconnected to adjacent points, i.e., points that differ in only one binary position, in the binary space. Points included in the disjunctive normal forms are filled in in solid black to represent a particular function.

$$f(A,B,C,D) = A\bar{B}\bar{C}D + ABCD + A\bar{B}C\bar{D} + ABC\bar{D} +$$
$$\bar{A}\bar{B}\bar{C}D + \bar{A}\bar{B}CD + \bar{A}B\bar{C}\bar{D} + \bar{A}\bar{B}C\bar{D} +$$
$$A\bar{B}CD + \bar{A}\bar{B}\bar{C}D$$

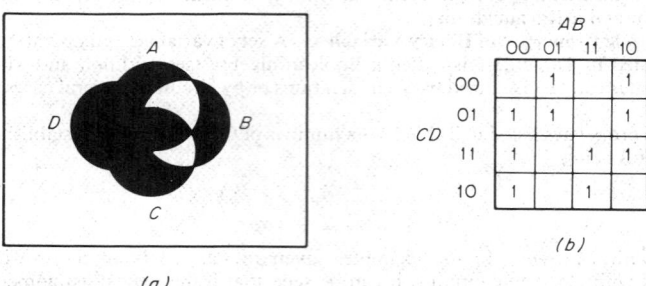

(a)   (b)

Fig. 23-37. (*a*) Venn diagram and (*b*) Karnaugh map of a binary function of four variables. The clustering of entries on the Karnaugh map indicates that a simpler expression can be found to express the function. The regularity of the spaces included in the Venn diagram gives a similar indication.

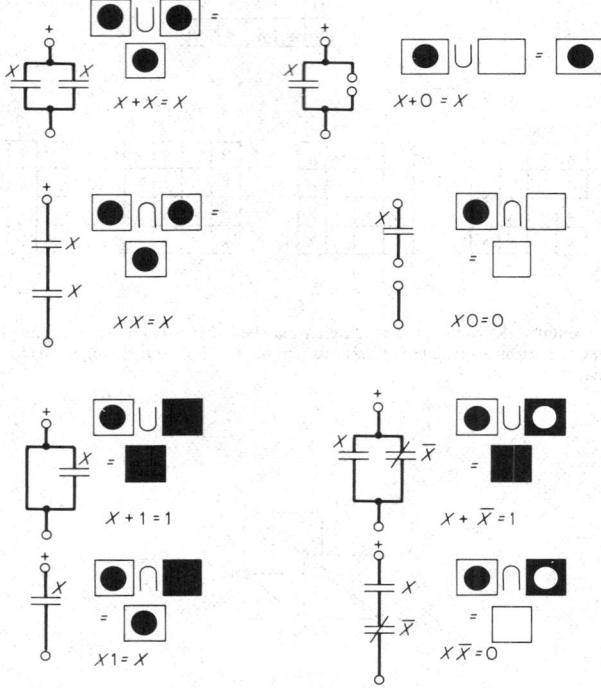

Fig. 23-38. Relay and Venn diagram analogs to postulates of Boolean algebra.

levels is reversed, i.e., a 0 level is associated with the 1 and a positive level with the 0, the AND function becomes an OR. The converse is also true. In Boolean algebra, the question of the interpretation of the complement of a function is addressed by De Morgan's theorem, which is illustrated by the following relationships:

$$\overline{xy} = x + y \tag{23-17}$$

$$\overline{x+y} = xy \tag{23-18}$$

$$\overline{\overline{x}} = x \tag{23-19}$$

where the superscript bars indicate negation or "the negative of." These relationships can be used to construct more complex relationships so that each subsum can be treated as an independent variable and reduced under the bar symbol in an inductive process. In general, all variables are complemented under the bar; and each product is changed to a sum and each sum to a product.

Table 23-3. Properties of Boolean Variables

1. $xx = x$	5. $x\bar{x} = 0$	9. $x + y = y + x$
2. $x0 = 0$	6. $x + x = x$	10. $x + \bar{x} = 1$
3. $x1 = x$	7. $x + 0 = x$	11. $x(y + z) = xy + xz$
4. $xy = yx$	8. $x + 1 = 1$	

40. Minimization of Boolean Functions. In computer systems Boolean expressions are realized by means of wiring and relay contacts or logic circuits. The theorems in Par. **23-39**, relating to equivalent Boolean expressions, can be used to reduce the number of variables and terms that are associated with a given Boolean function. The process of the reduction of a Boolean expression is called *minimization*. For example, the following expressions are equivalent:

$$a\bar{b}c + \bar{a}bc + \bar{a}\bar{b}\bar{c} + ab\bar{c} + a\bar{b}\bar{c} + \bar{a}b\bar{c} = a\bar{b}(c + \bar{c}) + \bar{a}b(c + \bar{c})$$
$$+ (\bar{a}\bar{b} + ab)\bar{c} = a\bar{b} + \bar{a}b + \bar{c} \tag{23-20}$$

Figure **23-39** shows the corresponding circuit reduction in relay contacts. The effect of minimization on the cost of computer equipment is evident.

One method of minimization is to use the theorems of Boolean algebra directly. For complex problems, a number of other methods have been developed. For example, the values represented by the minterms in the disjunctive normal form can be listed as a set of 1s or 0s; prescribed rules based upon the theorems are then used to eliminate redundant terms. Figure **23-40** illustrates a reduction of Eq. (23-20) in graphical form. In this case the basic rule used (based on Table 23-3, postulate 10) is that if two terms exist such that all positions

$$f(A,B,C) = A\bar{B}\bar{C} + \bar{A}B\bar{C} + \bar{A}\bar{B}\bar{C} + AB\bar{C} + A\bar{B}\bar{C} + \bar{A}B\bar{C}$$
$$= A\bar{B} + \bar{A}B + \bar{C}$$

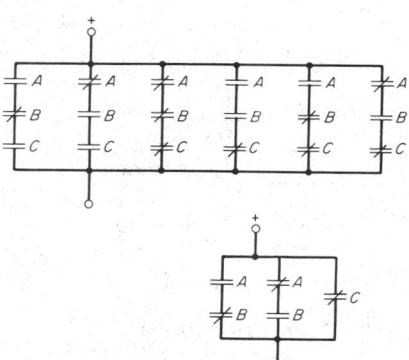

Fig. 23-39. Two circuit realizations of equivalent Boolean expressions. The disjunctive normal form (*top*) of a binary expression can often be reduced to a simpler form (*bottom*) by using the theorems of Boolean algebra.

are the same but one, the dissimilar variable can be eliminated. For example in Fig. 23-40, lines 3 and 5 can be combined to delete the A variable, as can lines 4 and 6. The resultant can then be combined to eliminate the B variable leaving only $\bar{C}$. Another rule is based on Eq. (23-16). A variable plus its negation times a second variable can be used to eliminate the negation. Thus, in the last step in Fig. 23-40, C can be eliminated from lines 1 and 2.

A number of other graphical methods have been used for circuit reduction. For example, the Venn diagram shown in Fig. 23-41 is used to achieve the above reduction. Individual boxes representing each minterm are shaded in. Subsequently, either the total area covered or uncovered can be expressed more simply by visual inspection. In the latter case De Morgan's theorem is used to invert the function to the result.

Another graphical procedure is defined by the use of the Karnaugh map, as in Fig. 23-42. A characteristic of a Karnaugh map is that variables that can be eliminated are associated with map entries that cluster in subspaces of 2, 4, or 8 dimensions. Such clustering indicates independence of the variables within the subspace.

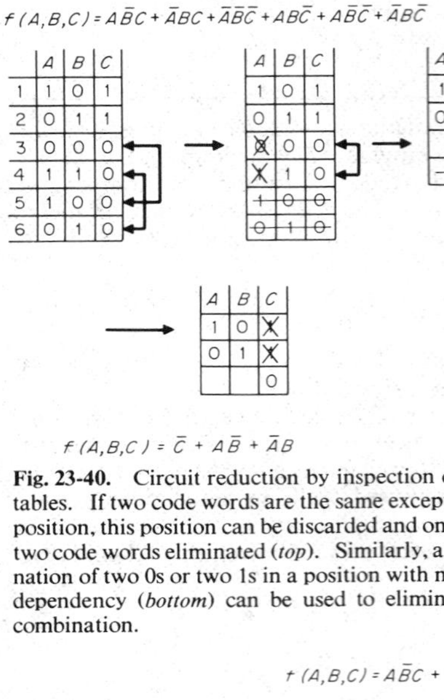

$$f(A,B,C) = A\bar{B}C + \bar{A}BC + \bar{A}\bar{B}C + AB\bar{C} + A\bar{B}\bar{C} + \bar{A}B\bar{C}$$

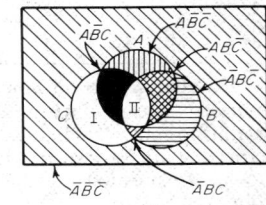

$$f(A,B,C) = A\bar{B}C + \bar{A}BC + \bar{A}\bar{B}\bar{C} + AB\bar{C} + A\bar{B}\bar{C} + \bar{A}B\bar{C}$$

Area $\text{I} = C\bar{A}\bar{B}$
Area $\text{II} = CAB$

Thus

$$f(A,B,C) = C(\bar{A}\bar{B} + AB)$$
$$= \bar{C} + A\bar{B} + \bar{A}B$$

$$f(A,B,C) = \bar{C} + A\bar{B} + \bar{A}B$$

Fig. 23-40. Circuit reduction by inspection of truth tables. If two code words are the same except in one position, this position can be discarded and one of the two code words eliminated (*top*). Similarly, a combination of two 0s or two 1s in a position with no other dependency (*bottom*) can be used to eliminate the combination.

Fig. 23-41. Use of a Venn diagram for circuit minimization. The areas on the diagram, represented by the terms in the disjunctive normal form, are shaded. In this case only two unshaded areas are left, and the sum of these, $f(A,B,C)$, is shown in black.

$$f(A,B,C) = A\bar{B}C + \bar{A}BC + \bar{A}\bar{B}\bar{C} + AB\bar{C} + A\bar{B}\bar{C} + \bar{A}B\bar{C}$$

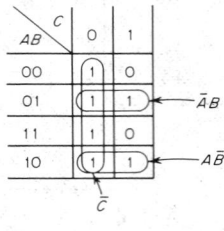

Thus
$$f(A,B,C) = \bar{C} + \bar{A}B + A\bar{B}$$

Fig. 23-42. Use of Karnaugh map for circuit minimization. Entries are made from the disjunctive normal form. Linear or rectangular clusters of 2, 4, 8, etc. variables indicate that a function can be reduced. In the example shown, the vertical cluster of four 1s indicates that the cluster depends only upon C, while the two horizontal clusters indicate independence from C.

41. Minimization in Code Translation. A common problem in computer systems is the translation from one code to another, i.e., from n input wires carrying a set of signals representing a code to m output wires that carry signals representing a different code. Figure 23-43 illustrates the procedure for translation from one code to another, for a three-position binary coded decimal (BCD) code to a three-position Gray code. The input variables are labeled a, b, and c. All possible combinations of these variables are associated with eight

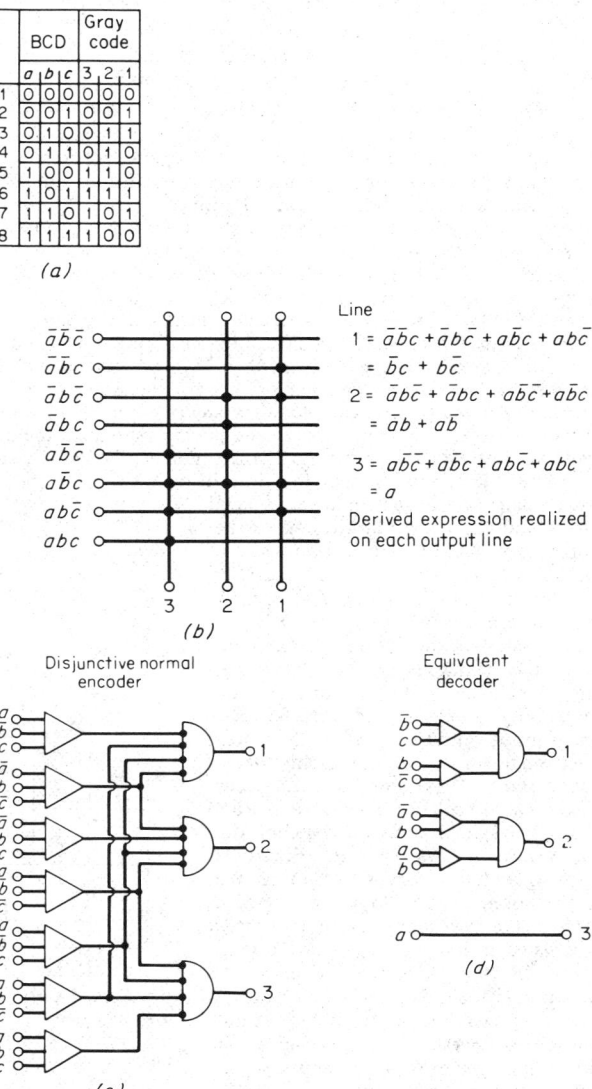

	BCD			Gray code		
	a	b	c	3	2	1
1	0	0	0	0	0	0
2	0	0	1	0	0	1
3	0	1	0	0	1	1
4	0	1	1	0	1	0
5	1	0	0	1	1	0
6	1	0	1	1	1	1
7	1	1	0	1	0	1
8	1	1	1	1	0	0

(a)

Line

$$1 = \bar{a}\bar{b}c + \bar{a}b\bar{c} + ab\bar{c} + ab\bar{c}$$
$$= \bar{b}c + b\bar{c}$$
$$2 = \bar{a}b\bar{c} + \bar{a}bc + a\bar{b}\bar{c} + a\bar{b}c$$
$$= \bar{a}b + a\bar{b}$$
$$3 = a\bar{b}\bar{c} + a\bar{b}c + ab\bar{c} + abc$$
$$= a$$

Derived expression realized on each output line

3 2 1

(b)

Disjunctive normal encoder

Equivalent decoder

(d)

(c)

Fig. 23-43. Translation from BCD to Gray code. The BCD inputs from the table (a) appear as inputs to the left of the code translator (b). Inputs that give rise to outputs on the Gray code lines are marked with a dot. By reading upward on each output line (1, 2, or 3) the disjunctive normal form of that line as a function of the input variables can be written down. The realization of the decoder is shown for the disjunctive normal case (c) and the minimized case (d).

horizontal lines on a graph. Three lines, 1, 2, and 3, representing the low- to high-order positions of the Gray code, are drawn perpendicular to the eight input lines. When one of the input lines is activated, a connection from this line to the output lines is symbolically represented by a dot provided that the particular input line gives rise to an output on a specific output line. The signals at the output lines arise from an OR from a number of the inputs, each of which is a specific minterm of the input variables. Thus each output variable can be expressed as a binary function of the three inputs. In the case at hand

$$\bar{a}\bar{b}c + \bar{a}b\bar{c} + ab c + ab\bar{c} = \text{line 1}$$
$$\bar{a}b\bar{c} + \bar{a}bc + a b\bar{c} + abc = \text{line 2}$$
$$a\bar{b}\bar{c} + a\bar{b}c + ab\bar{c} + abc = \text{line 3}$$

These reduce respectively to $b\bar{c} + c\bar{b}$, $\bar{a}b + \bar{b}c$, and a.

The Decoder as a Memory Unit. The decoder shown in Fig. 23-43 may be viewed in another way. Three separate inputs are expanded through a decoding tree into all possible code lines, which in turn can be selectively interconnected with three output lines by means of electronic elements such as diodes, in accord with the decoding scheme desired. This configuration may be viewed in terms of a store. The inputs correspond to address lines, and the output corresponds to the word selected in the storage unit. Thus a decoder in disjunctive normal form, as in Fig. 23-43, may be viewed as a read-only storage (ROS) unit. More generally, a store of any kind may be viewed as a type of variable decoding structure.

42. Sequential Circuits. A decoder is an example of a *combinatorial circuit*, in which a set of static inputs is used to create a specific function of the input variables at an output, irrespective of time. But time is inevitably a factor. Most circuits in a computer operate in sequence; e.g., serial-by-number adders, memory-access systems, counting circuits, and shift registers all act in sequential stages. Even for a decoder or other combinatorial circuits, some finite delay occurs between the input voltages and the appearance of the output. In a relay system, outputs are not available for times of the order of 10^{-3} s. In electronic circuits, outputs also follow the inputs, though the time delay may be of the order of 10^{-9} s.

When a multistage logical system is so designed that the logical functions proceed through successive stages in an uncontrolled fashion, the action is said to be *asynchronous.* But if each stage of logic is *gated* by timing pulses derived from an external source, the action is said to be *synchronous.*

In multilevel decoding the action is combinatorial, in the sense that the outputs are fully determined by the inputs, even if delay through one or more circuits occurs. But if an output is used as a feedback signal to a prior input within the system, the system output may depend not only upon the combination of inputs but upon the time sequence of their occurrence. Such systems are called *sequential circuits* and are classed as either synchronous or asynchronous.

Analysis of Sequential Logic Systems. Figure 23-44 shows a simple latch structure with the table that describes its operation. The table is constructed as follows. Each of the two inputs (with its current stable state) is listed in the eight rows of the table. The last row is the new state, realized under the combination of inputs and the current stable state. If both inputs are energized, the output is 0 for as long as both inputs are energized; but the new stable state is indeterminate upon removal of the inputs, so that question marks are used as table entries.

In more complex systems, a similar approach is used in characterizing a sequential circuit. By means of a timing chart or other graphical aids, the detailed time sequence of the desired system states is developed. This chart in turn is translated into an input and state chart, with the desired transitions at each juncture derived from the current system

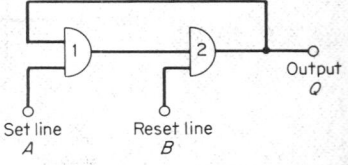

Set line   Reset line
  A   B

A	B	Q	Q'
0	0	0	0
0	0	1	1
1	0	0	1
1	0	1	1
0	1	0	0
0	1	1	0
1	1	0	?
1	1	1	?

Fig. 23-44. Analysis of a latch circuit. The condition of the latch is characterized by the state of the output Q and of the inputs A and B. Certain combinations of the inputs lead to a changed output Q', but some combinations of inputs do not definitely determine the new output, depending upon which input is removed first. This latter condition is indicated in the table by a question mark.

inputs and stable states. Since these transitions depend only upon the combination of internal states and inputs at a specific time, the combinatorial problem is subject to the methods of analysis illustrated in Fig. 23-44.

A number of mechanical methods have been developed to analyze problems in sequential logic.[43] In some cases problems exist in sequential circuits when a circuit action depends critically upon which relay or logic element completes its operations first. When such an ambiguity in circuit action exists, a *critical race* is said to be present.

In asynchronous circuits and when feedback paths exist, false circuit actions may occur. Such conditions are called *hazards*. The problem arises not only with relays but also in transistor-logic circuits when both an a and an $\bar{a}$ line may be zero during a transition, thus creating an unwanted state. The problem of *hazards* is discussed in Refs. 43 to 48.

43. Circuit Realization with Relays. The basic wiring patterns used with relays are discussed in Par. **23-13**, Fig. 23-7. The current input to a relay determines the state of the Boolean variable expressed by that relay. The normally open and closed relay contacts express the binary variable and its negation. A combination of such contacts creates current paths that express some Boolean function of the relay-coil current inputs. Relays used in computer systems must have high reliability.

Figure 23-45 shows the structure of a widely used type of relay. A magnetic yoke with a movable armature is attracted from the closed condition when the relay coil is energized.

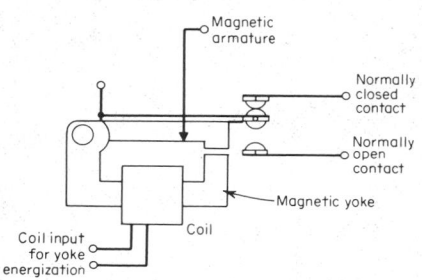

Fig. 23-45. Typical electromechanical relay of the moving-armature type.

Figure 23-46 shows the *reed relay*, combining small size and high reliability. A magnetic field induced by an external coil follows the path of the encapsulated contact arm, causing a force to pull the two arms together. The device works at high speed, due to the low inertia of the relay contacts. The controlled atmosphere surrounding the contact points makes it a highly reliable device (see also Sec. 7, Par. **7-159**).

Some relays employ *pick and hold coils* on each relay yoke, with alternate voltages to the coils to maintain continuous power on the relay and provide for controlled synchronous opening and closing of the contacts. Figure 23-47 shows a system using a three-relay ring. If a relay is selected on an A cycle, its contacts are used only to initiate action in a B cycle, and vice versa. The relay contacts do not make or break current. The action of the circuit is sequenced by the relative spacing of the pulses on the A and B lines.

44. Circuit Realization with Electron Tubes. Electron tubes were once used almost exclusively for logical-circuit realization, but at present almost all computers use transistor

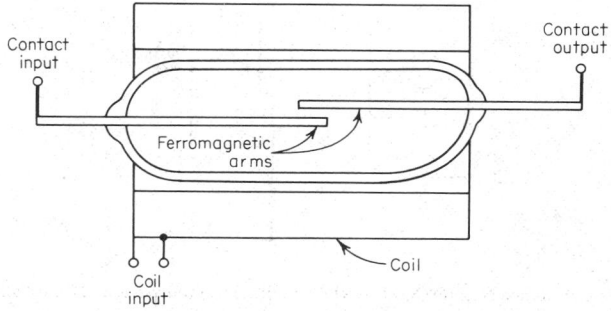

Fig. 23-46. Typical reed relay. An externally supplied magnetic field causes the contact arms to pull together, thus closing the normally open circuit.

or transistor-diode combinations for logic circuits. Occasionally a need still arises for tube devices in input-output gear.

Figure 23-48 shows a basic circuit used in tube switching, realizing the logical NAND, i.e., the NOT of a and b and c. By De Morgan's theorem this function is equal to $\bar{a} + \bar{b} + \bar{c}$. Figure 23-49 shows three other circuits used in electron-tube logic. The first realizes the NOR, which is the NOT of the OR of the inputs; the second the OR function. The third circuit is a *flip-flop*, which has two stable states of operation.

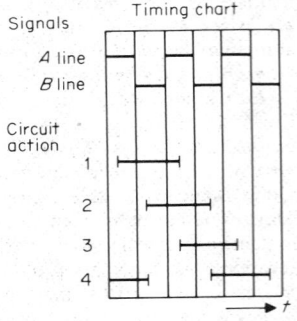

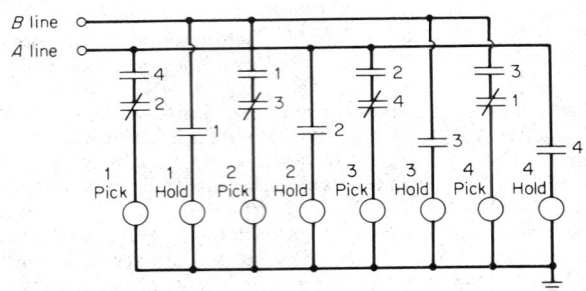

Fig. 23-47. Synchronous sequential relay ring. By using a pick coil and a hold coil on each relay yoke, with alternately energized A and B lines, problems associated with current interrupt at the relay contacts are largely eliminated. When a relay is opened or closed on an A cycle, its contacts are used only in a B cycle.

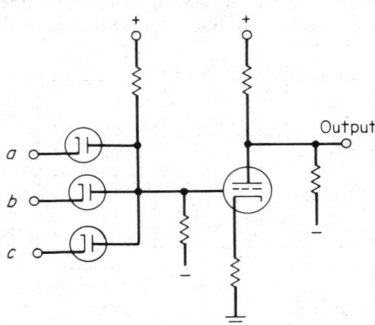

Fig. 23-48. An electron tube NAND (NOT-AND) circuit. The inputs a, b, and c must all be positive for the tube to become conducting.

The basic problem of electron-tube devices in computers is reliability. In computer applications it is customary to derate filament temperature to improve tube life. Other limitations of tube circuits in computers are physical size, power consumption, speed of operation, and cost.

45. Circuit Realization with Semiconductor Devices. The majority of switching circuits used in modern computers employ semiconductor devices (diodes and transistors) because of their small size, low cost, reliability, low power consumption, and high speed. The two basic types of transistors used for the majority of circuits in data processing are the *bipolar* and *field-effect* devices (FETs), discussed in Secs. 7 and 8.

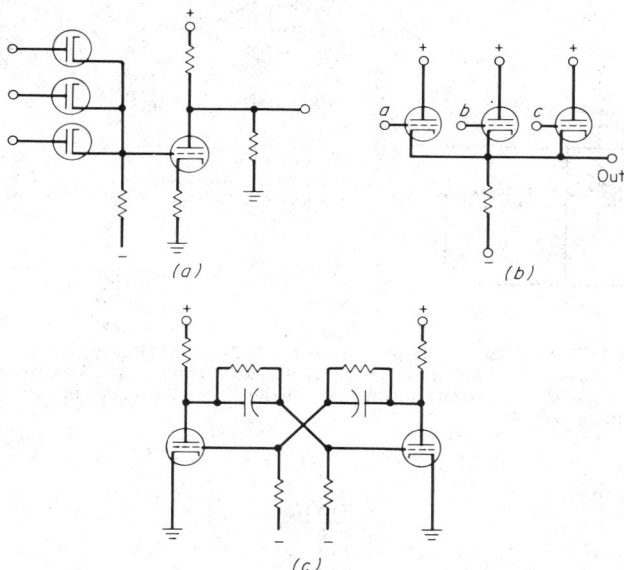

Fig. 23-49. Electron-tube computer circuits: (*a*) NOR, (*b*) OR. (*c*) Storage element, or flip-flop.

If a bipolar transistor has forward bias across its collector-base junction, saturation occurs and recovery of the device is delayed due to charge storage in the collector. Hence, computer circuits attempt to avoid saturation. In addition, high-speed transistors for computer application are often made with gold doping in the collector. This provides short carrier lifetime and minimizes the effects of charge storage.

Field-effect transistors are typically slower than comparable bipolar devices, but may have the advantage of requiring fewer processing steps in production. In data processing applications, bipolar devices are used where speed of operation is essential and cost is not a prime factor. FETs, on the other hand, are used where price is of primary importance.

46. Transistor Logic Circuits. Figure 23-50 shows a basic NOR circuit using FETs. The load resistor for the switching elements is an FET instead of a resistor, since this is convenient in FET transistor fabrication. Figure 23-51 shows an FET AND circuit using series components. Also shown is the basic FET latch (storage cell) used in memory applications.

A number of logic families have been developed for bipolar transistors. The *current-logic* family uses low voltage swings with relatively fixed current levels that represent logical 1s and 0s. The other circuit, *voltage logic*, is characterized by fixed voltage swings. Either current or voltage logic may be used in designs of different families of computer circuits. The various families of transistor logic circuits (DCTL, LLL, DTL, TTL, CML, etc.) are described in Sec. **8**, Pars. **8-78** to **8-86**.

47. Representative Computer Circuits. *Adders.* The logic circuits discussed here are building blocks used to realize computer subsystems. Figure 23-51 defines the various graphical symbols that represent AND, OR, NOR, NAND, and inverter logic blocks.

Basic to any computer system is the addition of two binary variables. The truth table for an augend A, addend B, sum, and carry of a 7-b adder is shown in Fig. 23-52. For the sum function, the logical EXCLUSIVE-OR of A and B is formed. For the carry, the function is the logical AND. Figure 23-53 shows a number of circuit realizations using AND and OR or NAND and NOR blocks to realize these functions. Since the carry from the next lower-order position is not included, the units shown in Fig. 23-53 are known as *half adders.*

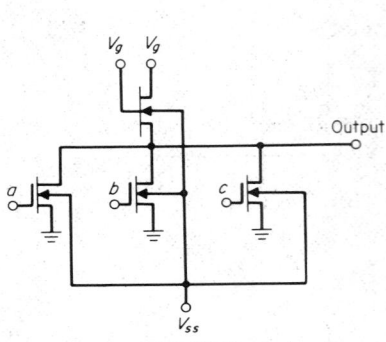

Fig. 23-50. An FET NOR circuit. A positive-going input at any of the three inputs a, b, or c causes a negative-going voltage at the output, thus forming NOT-OR.

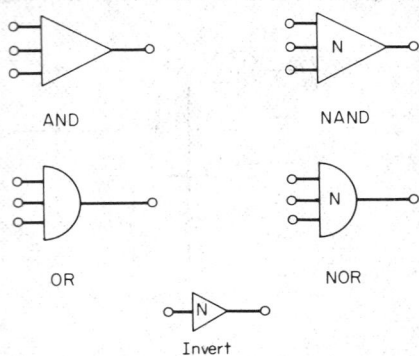

Fig. 23-51. Graphical symbols for representation of logical-circuit blocks. The symbols shown represent but one convention of many in use.

A	B	S	C
0	0	0	0
0	1	1	0
1	0	1	0
1	1	0	1

Fig. 23-52. Truth table for a half adder. A = augend, B = addend, S = sum and C = carry.

A *full-adder* truth table is shown in Fig. 23-54. It accepts the carry C_i from the next low-order position in a binary adder. The disjunctive normal forms for the sum S and the output carry C_o are

$$S = \bar{A}\bar{B}C_i + \bar{A}B\bar{C}_i + A\bar{B}\bar{C}_i + ABC_i \qquad (23\text{-}21)$$

$$C_o = \bar{A}BC_i + A\bar{B}C_i + AB\bar{C}_i + ABC_i \qquad (23\text{-}22)$$

These functions can be reduced by the application of appropriate minimization theorems to

$$S = C_i(AB + \bar{A}\bar{B}) + \bar{C}_i(\bar{A}B + A\bar{B}) \qquad (23\text{-}23)$$

$$C_o = (A + B)C_i + AB \qquad (23\text{-}24)$$

Figure 23-55 shows how these functions can be realized with NOR and NAND blocks. These adder circuits, in conjunction with suitable shift registers, are central to computer operation.

Serial Adders. In some computer designs, addition occurs in a serial-by-bit fashion, illustrated in Fig. 23-56. Three shift registers are used in the example shown. At the initiation of operation, the addend and augend are placed in their respective shift registers. Upon command the registers are gated into the full adder by an in-phase clock pulse. The results are stored in an output shift register. Any carry is carried back into a carry latch, to be available for entry into the next cycle. The out-of-phase clock pulse is used to shift the sum and set the carry latch at the appropriate time. The addend shift register may be used to accumulate the sum, in which case the output from the full adder is fed back into the addend shift register so that the sum is accumulated as the operation progresses.

Parallel Adders. A parallel adder can be constructed using a set of full adders, as shown in Fig. 23-57. When a parallel add is performed, a carry might be propagated from the low-order to high-order positions. If n is the word length of the adder and τ the time delay of a single-bit add, $n\tau$ delay might be experienced. Logic circuitry can be developed that generates carry bits from four or more positions and transmits them to higher-order positions. Such circuits can reduce carry-propagation times by a factor as great as 4.

Machine organization is not restricted to serial (bit-by-bit) or full parallel add but may use a serial-parallel adder; e.g., addition may occur in groups of 4 (or more) bits in parallel.

Multiplication Circuits. Given adders and shift registers, a multiplier can readily be

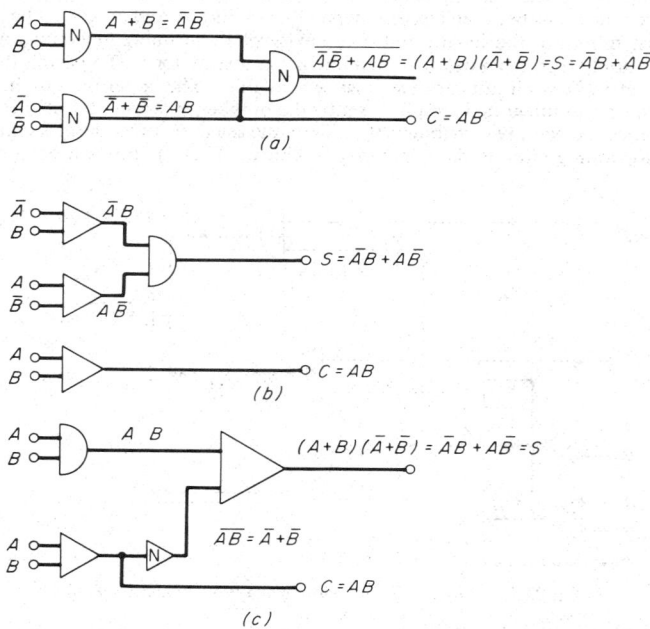

Fig. 23-53. Circuits for realization of the half adder. Circuit (a) uses NOR blocks exclusively; (b) is a direct expression of the Boolean functions; (c) does not require the input of the complement of the addend or the augend.

A	B	C_i	S	C_o
0	0	0	0	0
0	0	1	1	0
0	1	0	1	0
0	1	1	0	1
1	0	0	1	0
1	0	1	0	1
1	1	0	0	1
1	1	1	1	1

Fig. 23-54. Truth table for a one-position full adder. A = augend, B = addend, C_i = input carry, S = sum, and C_o = output carry.

designed. Figure 23-58 shows the basic scheme for a multiplier. If the low-order bit of the multiplier is a 1, the multiplicand is added to the accumulator and the multiplier is shifted to the left. If the low-order bit in the multiplier is a 0, only a shift occurs. In both cases, upon the termination of the shift, the multiplier is shifted right one position and a new multiplier bit becomes active.

Microelectronic Elements and Interconnections. The basic techniques, construction, and processing of integrated circuits and microelectronic components are treated in Sec. **8.** Computers, perhaps more than other devices, depend upon microcircuitry. Even a small computer may have as many as 10,000 such circuits, whereas a large system may have as many as a million.

Nanosecond circuit speeds are common in high-performance computer systems. Since light travels approximately $\frac{1}{3}$ m in 1 ns, system configurations must be kept small to take advantage of the speed potential of available circuits. Thus the emphasis is on the use of microcircuit fabrication and packaging techniques.

48. Design and Packaging of Computer Systems. A computer system contains from 10^5 to 10^7 interconnections between circuits, depending on its size and power. The layout of the system must minimize the length and complexity of these interconnections and must be realized, without error, from detailed manufacturing instructions. To permit these requirements to be met a *basic* circuit package must be available. The upper limit to the size of such a basic unit, e.g., an integrated circuit, is set by the number of crystal defects per unit area of silicon. If these defects are distributed at random, the selection of too large a chip size results in some inoperative circuits on a majority of chips. There is thus an economic balance

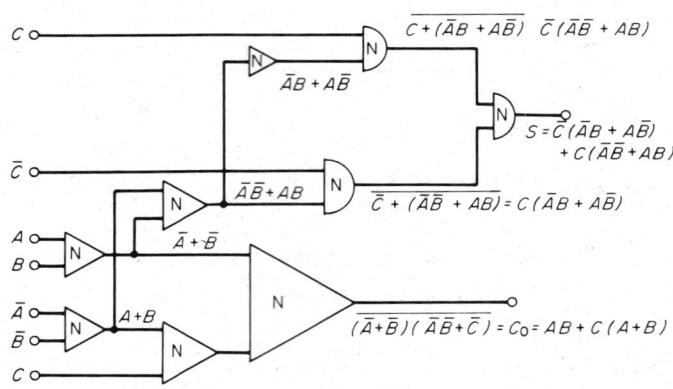

Fig. 23-55. Circuit of a full adder using NAND and NOR blocks.

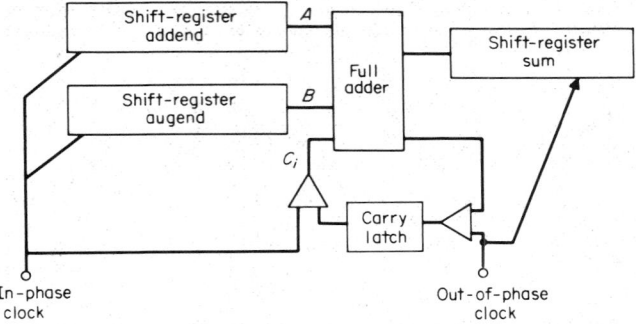

Fig. 23-56. A serial-by-bit adder. The in-phase and out-of-phase clock pulses time the storage of the carry pulse from a previous add.

between the number of circuits that can be fabricated in an integrated circuit and the yield of the manufacturing process.

Another limit on the size of the basic package is set by the number of interconnections between the integrated circuits. An empirical relation *(Rent's rule)* exists between the number m of circuits in a package and the number n of external connections required. The relation is $n = km^a$, where a is a constant in the range of 1.4 to 1.6 and k is a constant depending on the fan-in and fan-out characteristics of the particular circuit family used.

49. Design Automation. To accommodate the complexity of computer-system design, methods have been developed that automate the design procedure. One element in such a system is devoted to checking the designer's logical layout. A program simulates the logic operation to determine if the design represents the logical functions desired.

A second level of design automation involves the automatic arrangement of logical functions into standard modules and the assignment of these modules into groups, to minimize external connections. A third level of automation involves the placement of modules on cards, to minimize the amount of interconnected wiring within the card. Other aspects of design automation concern the preparation of documentation and field manuals for the repair of a system subsequent to manufacture.

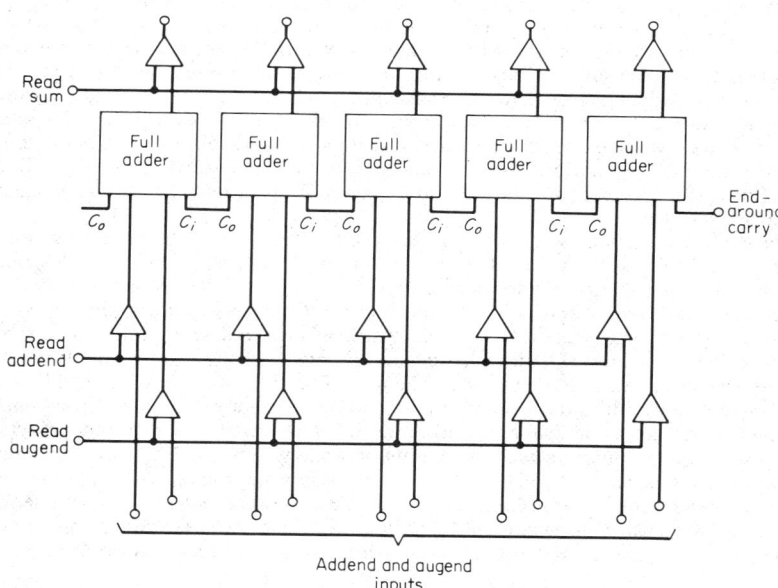

Fig. 23-57. A parallel adder.

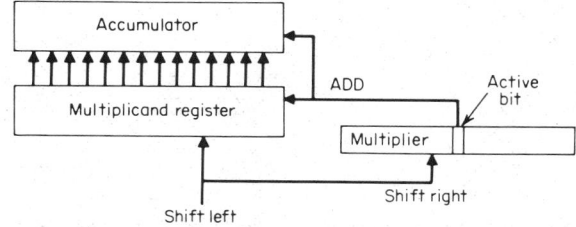

Fig. 23-58. A multiplier using two shift registers and an accumulator.

Since many operational steps in the preparation of modules, cards, etc., are automated, design automation may also generate information to drive the manufacturing machines that realize the structure. The design automation program can also generate test patterns, to determine whether the output of the manufacturing process is operational.

50. Other Logic Elements. A number of phenomena have been proposed for the realization of logical functions other than transistors, electron tubes, and relays. Most have received only limited use.

Magnetic-Core Logic. Single-aperture magnetic-core elements may be used for logic. The device is basically a magnetic amplifier circuit with a drive signal transmitted if the core is in a set position. Logical functions can also be realized with multiaperture devices. Figure 23-59 shows a two-aperture element. Winding A can be used to set both legs 2 and 3 in agreement so that little coupling is achieved between windings B and C. Alternatively, A can set leg 2 in opposition to 3 to induce coupling. The AND, OR, and NOT functions can thus be achieved. More complex functions can be realized with multiple windings in each aperture.

Cryogenic Devices. A number of cryogenic elements have been proposed and partially developed for use in computer systems. Figure 23-60 shows a loop of material such as lead or tin that becomes superconducting at low temperatures. In the particular circuit shown the loop current, in combination with the input current, creates a superimposed system such that current either appears to flow in the right-hand or left-hand leg of the superconducting element. In the presence of a magnetic field in proximity to a second element, the device current paths can be switched from the superconducting to the conducting states, and a rapid shift in current direction through the loop structure is induced. Since the transition from the superconducting to the conducting state is from zero to a rather low resistance, the inductance in the superconducting loop can maintain a current for a considerable time. Thus cryogenic elements are not fast devices by the standard set by semiconductor devices. Recent developments in cryogenics have introduced the possibility of providing a Josephson junction to increase speed.

The Paramatron. Most logical circuits involve two fixed voltage levels to represent the logical 1 and 0 states. The paramatron is novel in that the logical information is carried by the *phase* of a continuous-wave (ac) signal.

The parametric amplifier devices discussed in Secs. 7 and 9, e.g., varactor diodes, produce an interaction between two or more frequency sources through a nonlinear load, to achieve energy input at the sum or difference frequency. Figure 23-61 shows a parametric amplifier proposed for use in a computer system. The input frequency 2ω drives matched magnetic cores in such a way that the voltages from each core cancel in the common secondary. However, the nonlinearities in the core provide drive at frequency ω that, in conjunction with the tuned circuit, produces an oscillation in the secondary circuit. The phase of the output can exist in either of two states (0 or 180°). These phase/states can represent logical 1 and 0.

51. Threshold Logic. Conceptually, the stored-program computer operates numerically upon a fixed sequence of data points. In certain applications, such as pattern recognition, the problem is to compare an input pattern, usually in binary, to a sequence of stored references.

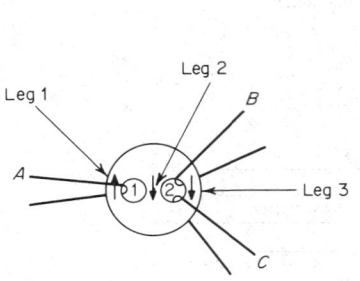

Fig. 23-59. Multiaperture core logical element.

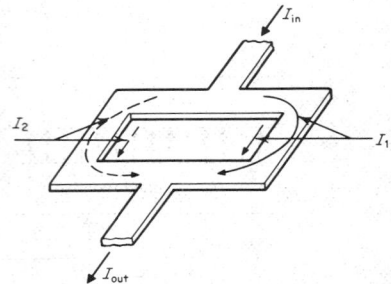

Fig. 23-60. Cryogenic logical element.

These comparisons involve computing the cross correlation between an input and each reference. In large, complex reference patterns the processing requirements are time-consuming and relatively inefficient.

In contrast, the human mind is capable of recognizing subtle pattern differences rapidly and without difficulty. The neural construction of the brain is organized differently from a conventional computer. Figure 23-62 shows a simplified model of a neuron. A series of

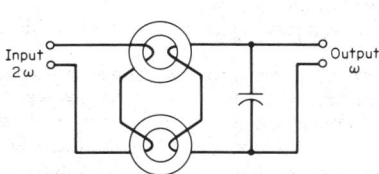

Fig. 23-61. Degenerate paramagnetic amplifier that produces a binary change in the phase of its output. (*After Gschwind.*[2])

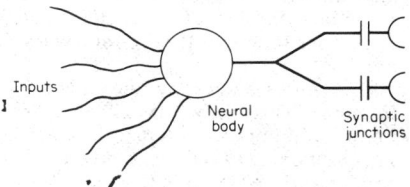

Fig. 23-62. Schematic diagram of a neuron.

inputs, summed on the neural body, give rise to outputs that cross synaptic junctions on subsequent neural bodies that either transmit or terminate the signal. A threshold effect appears to be associated with the inputs to the neural body, in that an output is delivered when at least some (usually a majority) of the inputs are activated.

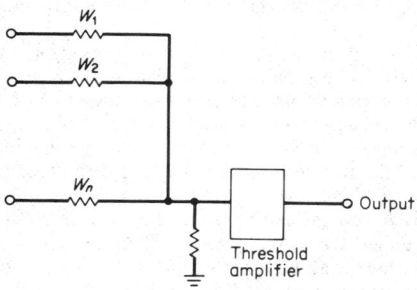

Fig. 23-63. Threshold logic device. The input voltages form a weighted sum that activates the output if the sum is above a certain threshold.

Threshold memory elements were developed to imitate the action of a neuron. Figure 23-63 shows such a device. The input pattern of binary variables are *weighted*, with a series of weights whose value ranges from 0 to 1. These weights in turn are summed in the logical body, and a response is transmitted if the input sum reaches a critical value. The mathematical form of this operation is $\sum_i \omega_i x_i = c$ where the x_i are the input variables, ω_i are the weights, and c is a constant. This is the equation of a plane in the hyperspace defined by the x_i. A series of such planes can be used to divide the hyperspace into subspaces, each of which may encompass a class of patterns. Such classes are defined by the set of outputs from the threshold elements so that the pattern of outputs can be used to recognize new members of a class. Threshold devices can also be used in the *learning-machine* configuration. If the outputs are correct for a particular pattern, no action occurs; but incorrect weights are adjusted to give a more correct response. By this means a series of input patterns can be used to "teach" the system to recognize correctly.

STORAGE METHODS, DEVICES, AND CIRCUITS

52. Functions of Storage. There are two basic kinds of instruction in a computer: *control instructions,* which act to shift a program sequence or change the conditions of operation of a program, and *data instructions,* which manipulate data, e.g., add a number in the ALU, move data from one storage location to another, or compare two numbers. Since each instruction is obtained from storage and may specify operations on data in storage, each elemental machine operation provides for at least one storage access (and usually many more). Storage is therefore a machine component whose cost, speed, and size are basic to a computer system's power and economy.

In early computer design, it was customary to use one ALU and one store. As systems have evolved, however, it has been found advantageous to use multiple processers and storage subsystems. For example, a subsystem may control a set of input-output gear and within itself have most of the elements of a central CPU; another subsystem may handle a group of communications gear; and yet another be dedicated to higher-level programming-language translation.

There are two basic types of retrieval of information; *direct addressing* and *content addressing*. In direct addressing, each storage location in a group of storage cells is associated with an address. When the specific address of the information sought is transmitted to the store, the contents of the addressed storage cells are delivered for use by the machine. There is complete freedom of choice in the nature and type of information at each storage location in question so long as the code and word size match the physical characteristics of the storage unit.

In content addressing, the address is keyed to the information content at the storage location. This method is similar to the indexing schemes that are used, for example, in libraries. Suppose a storage location contains information about an employee of some organization. His name is stored in a particular field of the data word in storage, and the word is retrieved by searching through successive words until the desired name is found. The information associated with the name is presented to the system for utilization. This concept can be expanded by considering any of the fields associated with the word such as name, employee serial number, social security number, wage rate, location, etc., as components of a multidimensional vector, any component of which may be used in a search mode.

In a given group of employees, each can be given an arbitrary serial number to be used as an address for information about the employee. If the serial numbers have been appropriately selected, the set of employee names can be stored in sequential locations in the computer store. The storage address is then the specific serial number. In this case, the direct-addressing scheme has been implemented by the assignment of content-addressable fields.

Grouping Storage Cells. The information content in a storage system is proportional to the number of bits stored. If each bit is separately addressable in an n-b memory, m b, where m is an integer greater than or equal to $\log_2 n$, are required in a machine instruction to specify the storage cell. Since increasing the lengths of instruction words and the numbers of storage addresses involves additional cost, it is advantageous to group storage cells together into individual addressable groups. The selection of the size of this group is a basic architectural feature in the design of computer subsystems.

The minimum grouping of bits at each storage address can be determined by the binary-character length of the basic symbols manipulated by the computer. For example, in a numeric system in BCD 4 b are required, whereas for alphanumeric and special characters 6 to 8 b (depending upon the size of the character set) are required for a complete specification of the symbols used. In modern systems, the demand for a large character set has led to a minimum grouping of bits in storage of 8 b. This unit is called a *byte*. In an 8-b addressable store, with a total of n b, the number m of required instruction address bits is $m = \log_2 n - 8$.

Systems have been designed with groupings of up to 40 b, called a *word,* as the addressable unit in storage. In such machines, in applications involving manipulation and storage of individual symbols, most of the capacity of the store is wasted. Since underutilization of storage is inordinately expensive, most modern systems use variable word lengths with instructions from 16 to 48 b long and data words from 8 to 256. Word-size variability is provided for by permitting storage addressing to occur at the 8-b (byte) level.

53. Random Access vs. Serial Storage. In a *serial store,* a specific address is found by sequencing through successive locations. The address for each location may be generated by a counter that counts from the datum address in synchronism with the passage of each word location. In the *random-access store,* stored information is found at different physical locations that are addressed directly by electrical means, i.e., by *address lines.* Information from any location in the store is accessible at equal speed, irrespective of its location. The distinction between random access and serial storage is not significant from the viewpoint of the user, if the time of access is longer than the maximum recycle time of the serial store.

In serial devices, addressing is simplified. A single synchronous counter or clock serves all addresses, and sensing elements determine the information stored at each location. Since these act on only one set of data at a time, the interconnection cost is lower than that of a comparable random-access store. In random access, addressing is achieved through electrical selection, i.e., by direct interconnection of each storage cell. In addition, sensing is complicated since the output from each cell must be connected to a common sensing device. These factors make serial storage cheaper than random access.

Computer subsystems are often implemented by using a high-speed central random-access device, called a *local store*, in conjunction with a larger-capacity, lower-cost, slower device called a *backup* or *backing* store. For example, the backup store may contain a block of data equal to some submultiple of the size of the central random-access unit. To give the appearance of random access to all data in the total storage structure, a bit address specifies which block and what location within the block contains the information sought; provision is made to transfer the block from the backup to the local storage automatically. In this method bits are required (within the address portion of each instruction) to address the total store.

In another scheme, the external program is arranged to deliver blocks from the backup store in time sequence. The program operations are restricted to that block, within a limited time. The latter scheme is more conservative of bit addressing, since each program address need only contain bits to specify the storage location within each block. The first scheme, i.e., where the data address specifies the total store, is called *virtual memory* since, with appropriate automatic equipment, the overall memory structure appears as if it were much larger than the size of the high-speed direct-access device. In the second scheme, the *program* is said to have been *phased*.

54. Nomenclature of Storage Devices. A store is called *nondestructive read out* (NDRO) if the act of reading information from the store does not destroy the stored information. In *destructive readout* (DRO), the physical process of reading the memory destroys the information content of the cells addressed. In some DRO systems the storage system provides for an automatic rewrite, and the system appears to the user as if it were NDRO.

A *read-only store* (ROS) is one in which the information content is not intended to be changed by the user. A *read-write storage* system (RWS) provides for both the withdrawal and writing back of information with approximately equal convenience.

A storage system is classified as *volatile* if information retention depends upon continuous system operation. For example, if power is removed from a volatile store, the information content is destroyed. In a *nonvolatile* system, information is destroyed only by an external command.

55. Magnetic-Core Storage. A widely used storage element is the *ferrite toroid*, shown schematically in Fig. 23-64a. The toroid's azimuthal magnetic characteristics are such that the response of the core is reversible by the application of a low coercive field H. The portion of the hysteresis loop (Fig. 23-64b) from points A to E is a linear region over which the core can be driven by a suitable H field in a reversible way. If sufficient coercive force is applied to drive the core beyond the knee of its magnetization loop, beyond point C, the core switches from point C through point D and a permanent change in the direction of magnetization in the core takes place. Upon removal of the driving field, the state of the core will be represented by point F rather than A.

Since a voltage is induced in a wire that threads a core in response to a change in an encircling B field of a core, the flux change associated with switching can be detected by suitable equipment attached to the threading wire. Figure 23-65 shows a core that is threaded by an external wire, with typical signals on the wire (a) when the core does not switch and (b) when it does. The signal associated with switching has larger amplitude and longer duration than when no switching occurs.

Figure 23-66 shows a typical layout of a core memory plane. The individual cores are threaded by three lines, two to select a particular core and the third to sense an output. If both selection lines are energized, and if the core is in the appropriate magnetization state, the core undergoes a change of state and induces a voltage on the sense line. This system is called *half-select* since each wire carries one-half of the current required for switching.

Figure 23-67 shows an alternative means of selection. Only ⅓ unit of select current threads any unselected core, thus reducing the possibility of partial switching of a nonselected core.

Writing a 1 in a core is accomplished by reversing the direction of the select currents to reset the core to its initial state. Writing a 0 is accomplished by leaving the core in the as-read condition, entering a reverse current on the sense line to inhibit the writing of a 1, or by selection with the halt-select lines.

Characteristics of Core Storage. Even when a core is not switched, there is a finite transfer of energy from the select lines to the sense line. Some of this transfer arises because there is a finite excursion of the B field in a core in response to even a nonswitching H field. Another part arises from magnetic coupling through leakage paths of the select to the sense line, and the third arises from capacitive coupling. These signals on the sense line result in noise that tends to limit the size of an individual plane of cores.

56. Random-Access Selection Methods. When an n-b binary input (the address) is presented to the store and a k-b binary output results, the input may be considered as an n-b code translated into a k-b output code. The output on each of the k lines can be expressed as a Boolean function of the input code words in disjunctive normal form. Each of the 2^n possible terms in the function is either present or absent, depending upon the information stored. Thus logically a memory storage cell can be modeled as in Fig. 23-68. An n-b AND circuit performs the selection of the information stored, and the storage cell provides an output if the cell contains a 1.

Figure 23-69 shows several arrangements for selection of 1 b out of m, where $m = 2^n$. In Fig. 23-69a the input to the store is decoded externally, and each storage cell is energized separately. In this arrangement, there is no necessity for a selection function to be associated with a cell.

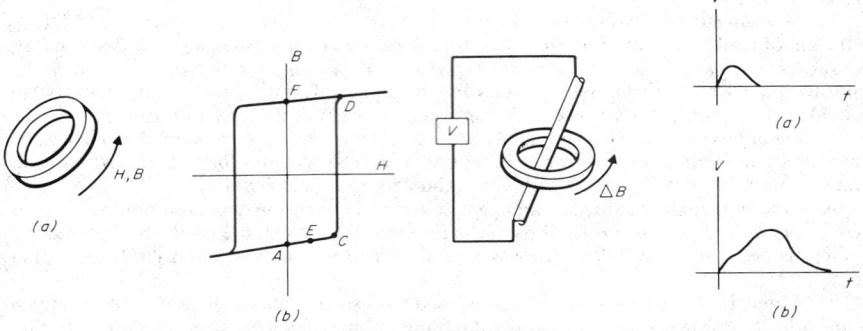

Fig. 23-64. Magnetic-toroid core and its azimuthal magnetic characteristics.

Fig. 23-65. Voltages developed on a sense line threading a magnetic core. When the core does not switch, some magnetic coupling still exists (a). When switching does occur, the output signal (b) is longer and of greater amplitude.

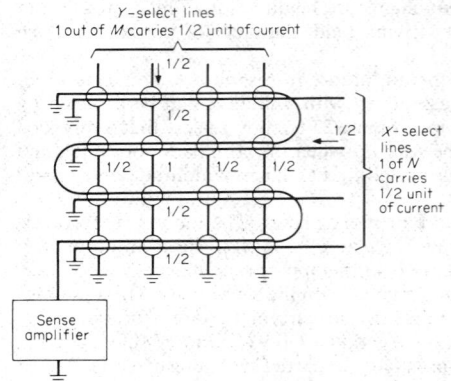

Fig. 23-66. Schematic layout of a typical magnetic-core plane. The current input on a selected line is one-half that required to switch the core. Only the core at the intersection of two selected lines receives enough current to switch the magnetic state.

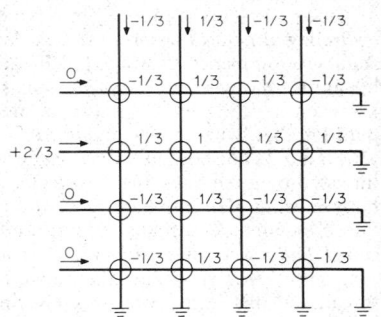

Fig. 23-67. One-third select-core plane. No core, except the one selected, receives a net current of more than one-third that required for switching.

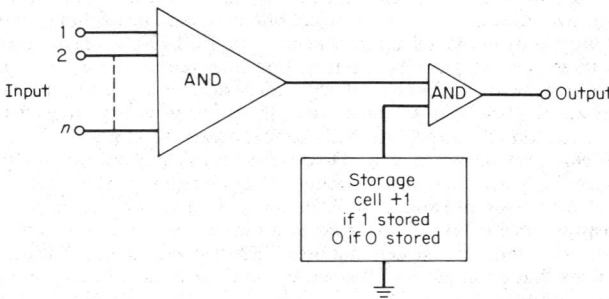

Fig. 23-68. Functional model of a storage cell. The logical operation of random-access storage may be divided into a selection and a storage function. The selection lines are decoded by the AND of the inputs after the cell is selected; the AND is formed of the content of the cell and the select line.

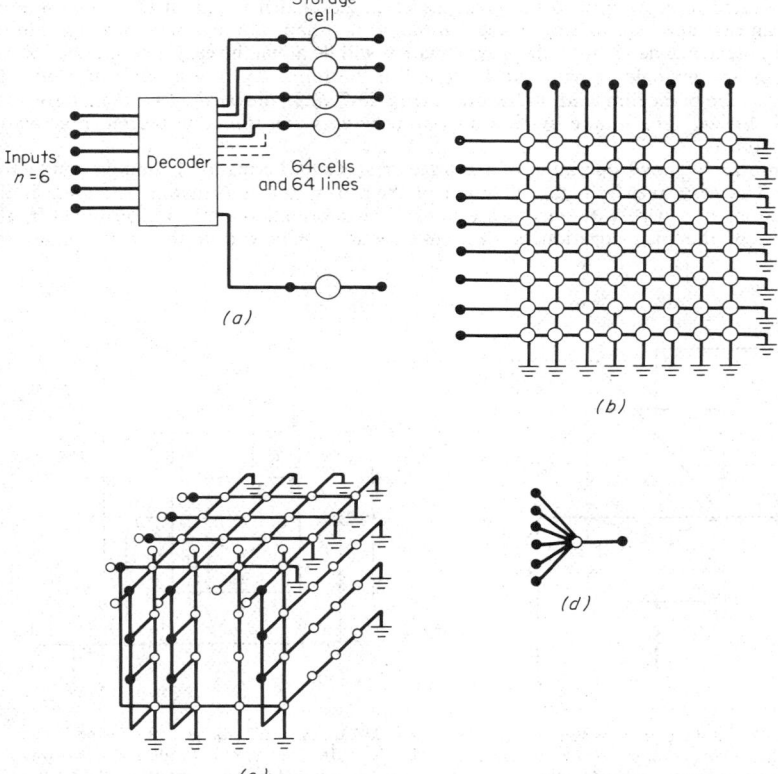

Fig. 23-69. Various arrangements of a random-access memory. In (a) no AND function is associated with the storage cells, and selection to all 64 lines occurs in a decoding tree. In (b) each storage cell is associated with a two-way AND so that two sets of eight lines can be used for selection. In (c) the cells have a three-way AND so that external decoding must be performed to three sets of three lines. In (d) each cell both decodes and performs the storage function; a six-way AND in each cell with no external decoding. Two sets (each input variable and its negation) of six lines, only one cell shown.

In Fig. 23-69*b* each storage cell contains a two-way AND, e.g., as in magnetic cores. In this system each of two lines is energized to select one core, and instead of *m* interrogation lines, $2m^{1/2}$ lines are used in the selection scheme. In Fig. 23-69*c*, each storage cell is associated with a three-way AND. The interrogation lines required are $3m^{1/3}$. In the final case (Fig. 23-69*d*), a six-way AND is used with each storage cell and only $6m^{1/6}$, or 12, interrogation lines are required. In the latter case, the number of drive lines is minimized, and no decoding function is required externally to the storage system.

57. Semiconductor Memories. Figure 23-70 shows a basic circuit for realization of a storage function using a bipolar transistor circuit. The select line performs the function of selection, and if selected, the sense and write lines can be used for the read or write function. This bipolar transistor store is NDRO, volatile, and random access.

Figure 23-71 shows a static storage cell that uses FET devices. Selection is accomplished through *x* and *y* lines that open gates to the sense and reset lines. The system is NDRO, volatile, and random access.

Bipolar circuits in a storage system are generally faster, more wasteful of power, and more expensive than FET devices. The difference in cost arises primarily because bipolar devices require more complex processing steps in manufacture.

58. Thin-Film Magnetic Storage. Figure 23-72 shows a thin magnetic film, such as Permalloy, with two labeled hysteresis loops. An isotropy is present in the film if it is plated or evaporated on a support in the presence of an external magnetic field. This isotropy defines an easy and hard axis of magnetization. If the shape of the film is long and thin (to prevent pole demagnetization), the magnetization will lie along the easy axis in either of two directions and a single domain will be found in the film. In the easy-axis direction the hysteresis loop of the film tends to be square (Fig. 23-72*b*); in the hard-axis direction the loop is nearly linear. The square hysteresis loop in the easy-axis direction permits nonvolatile binary storage.

Figure 23-73 shows operation of a storage array using Permalloy plated on conductive wires. The easy axis is in the direction of the wires, and information is stored at the intersections of a wire with each sense line. The information state is determined by the direction of the magnetization at each intersection. Current in the wire rotates the

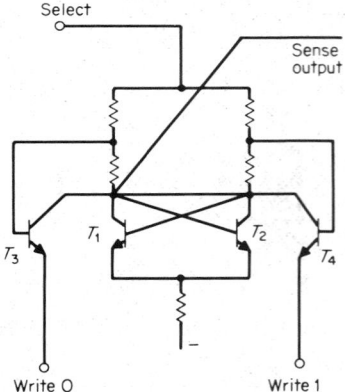

Fig. 23-70. A bipolar transistor storage cell. Storage is accomplished by means of a cross-coupled transistor flip-flop. The normal state of the select line is such that the output is below detection levels in either state of the flip-flop. When the cell is selected, the sense output becomes sufficiently positive for detection if T1 is off. Similarly the select line needs to be positive while writing; otherwise T3 and T4 remain off.

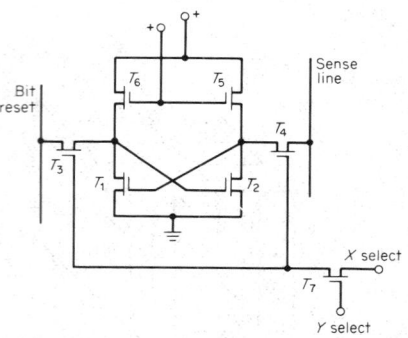

Fig. 23-71. An FET storage cell. T_5 and T_6 are FETs used as load resistors while T_7 is a two-way AND circuit that connects the cell to the bit and sense lines when selection occurs. T_1 and T_2 form a cross-coupled flip-flop. The state when T_2 is on represents a 1. The sense line is used for set and the bit reset line for reset.

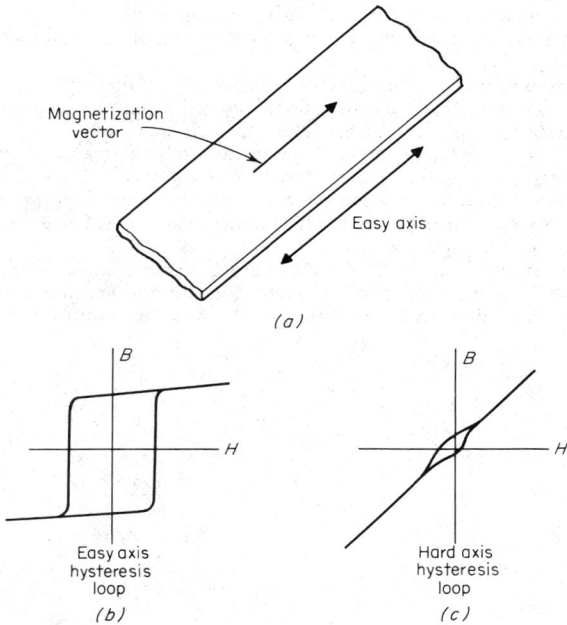

Fig. 23-72. Magnetic storage properties of a ferromagnetic film. When a magnetic film (*a*), such as Permalloy, is deposited in the presence of a magnetic field, the film develops a preferred (easy) axis of magnetization. If the film is thin and long in the easy-axis direction, the magnetization of the material along the easy axis can occur in either of two directions, with a square hysteresis loop (*b*). The loop (*c*) in the hard direction is not square.

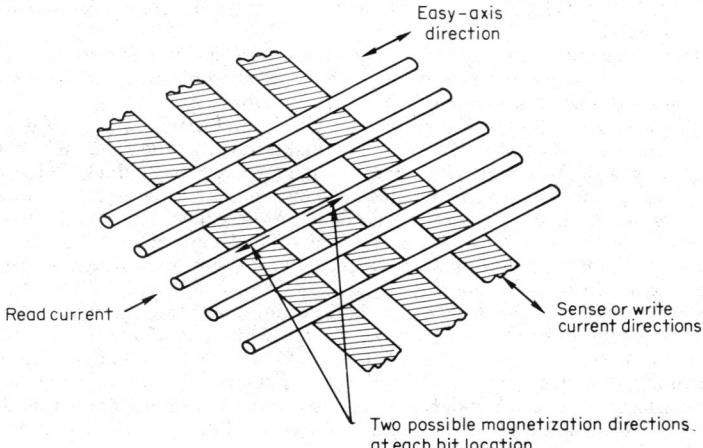

Fig. 23-73. Storage system using wires plated with magnetic material. The easy axis is in the direction of the wires, so that the film magnetization at each bit location is either parallel or antiparallel to the read-current direction. When a read current is applied, the magnetization is rotated into the hard-axis direction, causing coupling to the sense lines. The signal on the sense line is either positive- or negative-going, depending upon the prior direction of the magnetization vector.

magnetization of the film in the hard-axis direction. This magnetization change develops a component that couples to the perpendicularly running sense wires. The polarity of the output signal on a sense wire is determined by the direction of magnetization along the easy axis.

Writing is accomplished by a current down the plated wire that produces a rotation of the magnetization vector in the hard-axis direction, followed by an easy-axis current of suitable polarity to drive the system into the particular easy-axis state desired.

Thin-film stores are generally of high speed due to the rotational method used for switching the magnetization vector from one state to another. The cost of such units is comparable to core. Thin-film stores have experienced difficulties in process control and manufacture and suffer from relatively low signal levels due to the small amount of material that is perturbed upon interrogation.

59. Magnetic Drums. Figure 23-74 shows a drum storage system. The cylindrical drum is coated with a film of magnetic material such as ferrite powder embedded in an organic binder. A magnetic field created by a current in the winding of an external magnetic head

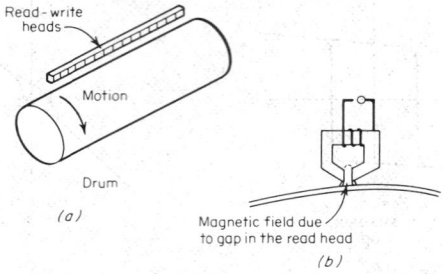

Fig. 23-74. (*a*) Magnetic-drum storage system. A set of magnetic heads writes and reads information to and from a magnetic coating of the drum. (*b*) Read-head structure.

element, called a *write head,* is used to align the magnetization of the film. A transition in the direction of magnetization within the film as written by the write head creates an effective magnetic pole along the line of transition. This pole in turn acts as a source of external magnetic field. Such fields may be used to induce a voltage in a second *read head* when the drum is in motion.

The drum rotates at high speed with respect to fixed read and write heads. A timing track is permanently recorded on the drum to provide a synchronizing signal for reading and writing. Drum storage is sequential, NDRO and nonvolatile.

There are two methods of recording called *saturation* and *nonsaturation.* With saturation recording, material under the head is fully saturated throughout the material thickness; this implies a coating depth comparable to the spacing of bits on a track. Nonsaturation recording uses a thick coating, and only the upper surface layer is fully saturated. In nonsaturation recording cross talk may arise between subsequent recordings unless there is an erasure field that removes any magnetization associated with writing. For erasure, an ac field is used to create a state of zero magnetization. A strong dc field may be used for saturating the medium in one direction.

60. Storage on Thin Magnetic Media. In the drum system the read signal is very nearly the derivative of the write signal. Thus if the write signal is a square wave, it saturates a portion of the drum surface. The read head picks up the transition from one state of magnetization to the other and thus delivers positive-going and negative-going outputs that mark the start and termination of the recorded square wave. The most widely used methods of recording information on a drum take account of the difference in the writing and read signals.

Figure 23-75a illustrates *return-to-zero* (RZ) recording. In each time period a positive-going or negative-going current is used to record information. Since the readout signal is the time derivative of the input, the pickup signal in any time slot is a minus-plus or plus-minus spike depending upon whether a 0 or a 1 is recorded.

Another type of recording is *nonreturn to zero* or NRZ (Fig. 23-75b). Information is represented by flux levels. That is, where 0 is to be stored, the substrate direction is maintained in its initial condition; otherwise the flux is reversed. As can be seen in Fig. 23-75*b*, the read head produces positive-going output pulse whenever a 0-to-1 transition is encountered and a negative-going pulse at a 1-to-0 transition. In 1-1 or 0-0 transitions, no signal is produced by the read head.

The *nonreturn-to-zero invert* (NRZI) system records 1s only. Each 1 is represented by a transition, in either direction, as shown in Fig. 23-75*c*. On readback the pulses in the read head, when rectified, represent the 1s recorded.

61. Magnetic-Disk Recording. Figure 23-76 shows a magnetic-disk system. The recording medium is spread over a disk-shaped substrate, rotated about its center axis. Information is written into and read from the magnetic medium by means of read and write heads, as in drum recording. The volumetric efficiency of a disk system tends to be higher than that of a drum since several disks can be stacked vertically on the same spindle. Otherwise drum and disk systems are similar. Disk systems are usually accessed with movable arms, as shown in the figure. The time delay associated with the arm movement is usually called a *seek time* (sometimes *access time*); the rotational delay is called the *latency time*. Multiple fixed heads may also be used, in which case the disk system is called a *fixed-head file*.

62. Characteristics of Drums and Disk Systems. Typical drum and disk systems have rotation speeds from 1,800 to 20,000 r/min, and the average latency times are of the order of 2 to 17 ms. Seek times are generally of the order of a few tens of milliseconds. The information density on a track in a system ranges from 1,000 to 30,000 flux reversals per inch. Track densities approaching 1,000 tracks per inch have been reported. The capacity of a typical disk system is 10^9 bytes.

To achieve high performance in disk and drum systems, the heads must be within a few microinches of the surface, and close design tolerance is required of the air bearing that spaces the head with respect to the surface.

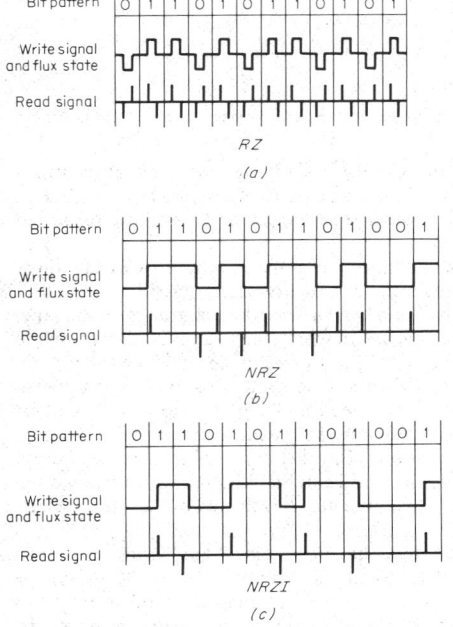

Fig. 23-75. Methods of magnetic recording: (*a*) Return to zero; (*b*) nonreturn to zero; (*c*) nonreturn to zero invert method.

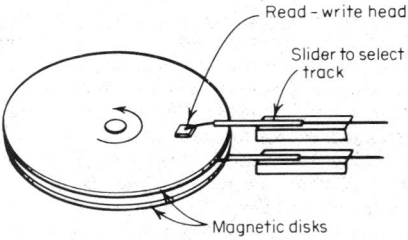

Fig. 23-76. Magnetic-disk recording. Information is read from, and written onto a magnetic film on the surface of a rotating disk. Multiple disks (two are shown) may be attached to the same spindle, and each disk may be accessed by a movable magnetic head.

63. Magnetic-Tape Systems.   Figure 23-77 shows the design of a magnetic-tape transport.   The thin plastic (usually polyester) tape, coated with magnetic material, is moved past a read head by capstans at constant speed.   The tape is accumulated in a vacuum column prior to transfer to or from reels.   Most such devices are made with a removable reel so that the unit can be used as an input-output device as well as for storage.   The drives are usually self-threading.

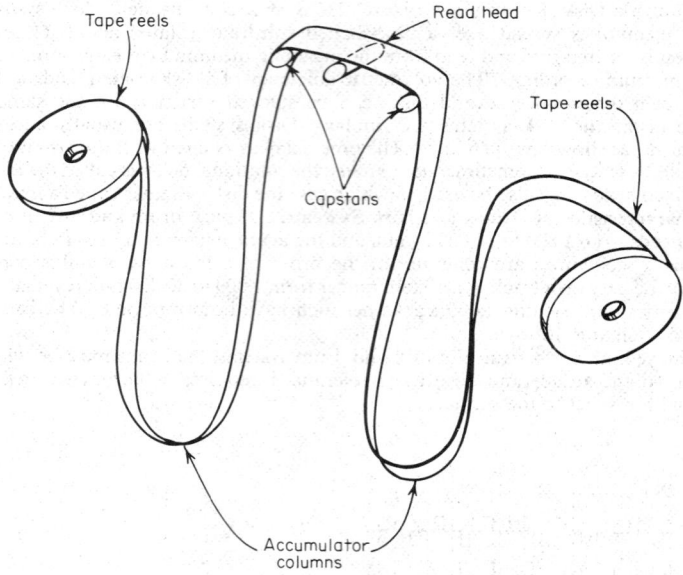

Fig. 23-77.   Magnetic-tape drive system.   The tape is driven by capstans between accumulator columns.

Nonsaturation recording is generally used in tape systems.   Parity checking is customarily used in a two-dimensional mode (Par. 23-33).   Storage capacity of a reel of tape may reach 100 million bytes, and usually tape recording offers the lowest cost per unit of stored information of all the common media.

Tape drives generally use ½-in. tape, but wider tapes are sometimes used.   Cassette units have been developed for low-cost systems, with ¼-in. tape in a two-reel cartridge.   Another type of magnetic-tape unit uses a transversely rotating head to form helical tracks on the tape.   Such a system has two modes of operation: a *skip mode,* in which the head is stationary while the tape moves at high speed, as on a conventional drive, and a *read mode,* in which the rapidly rotating head draws information at high speed while the tape itself is moving at low speed.

64. Read-Only Storage Systems.   Some systems need massive storage of information that seldom changes.   For example, archival records are infrequently accessed and rarely changed, except for the addition of new information.   In other types of storage, e.g., those associated with the control store of a computer, it is undesirable for the user to change the store contents, since the basic architecture of the machine may be affected in ways unforeseen by the person initiating the change.

As a result of such requirements, storage systems have been developed whose contents, once written, can be changed only with difficulty.   Such systems are called *read-only storage* (ROS).   Figure 23-78 shows such a system.   Linear passive electrical elements are used to couple a word to a sense line.   Power is fed onto a word line and, depending upon the presence or absence of the passive coupling device, information is developed on the sense lines for interpretation.   Resistors, linear cores, capacitors, and inductors have been used as coupling elements.   The ratio of the energy received at a sense amplifier to that at the input at a word line is, at best, 1 over the number of bits stored in a linear passive system.   Such considerations limit the size of the array in this class of device.

Diode or Transistor ROS.   To avoid the limitations of linear passive systems, diodes or transistors have been used to connect the word and sense lines in a ROS.   ROS structures provide for initial writing and perhaps for a limited degree of changeability.   With diodes, fuses may be provided in the input or output leads.   Suitable currents can thus burn in 0 bits in an array initially fully populated with 1s.   In capacitor memories, one plate of an individual cell may be removed mechanically or by chemical processing.

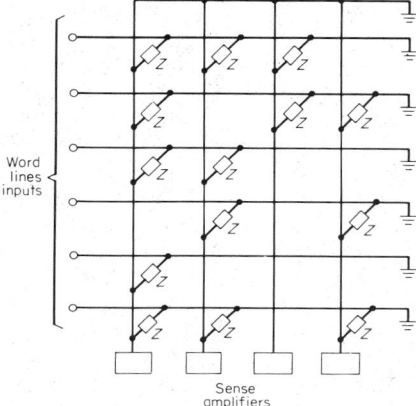

Fig. 23-78.   Schematic of a read-only store (ROS).   A decoder circuit, not shown, delivers energy to the word lines.   These feed energy onto the sense lines if a 1 is stored at a particular bit location.   The storage elements Z may be linear passive or active or nonlinear circuit devices.

Optical Stores.   A number of ROS systems have been proposed or developed that use light beams for storage and readout of information.   In one type, a laser beam evaporates a small spot on a metallized tape to register a 1.   Readout is accomplished by focusing the image of previously written spots on a photodetection array.

In the photographic disk system, information is stored by a photographic emulsion on circular tracks on a transparent substrate.   Light beams, servoed to the track, are focused on a photodetection diode to determine the presence or absence of bits in particular tracks.

Holographic Memories.   Another type of optical memory uses holography.   The holographic process is shown in Fig. 23-79.   The information to be recorded is placed on a bit mask that is illuminated by a portion of a laser beam.   Another section of the beam is used as a reference, and the interference pattern between the two is recorded by a photographic plate.   On development of the plate, the reference beam alone can be used to reconstruct the image.   As opposed to conventional image storage, holography offers a substantial advantage in the media-location tolerance, since the tolerance is determined by the large projected image rather than by the actual storage density on the hologram itself.   Small imperfections in the hologram film do not destroy individual bits in the output image.

65. Read-Write Optical Storage.   A number of systems have been proposed for optical stores that use an alterable storage medium.   Manganese bismuthide, europium oxide, and some other materials are optically active if magnetized.   Depending upon the state of magnetization, the surfaces of such materials rotate the polarization vector of an illuminating light beam.   If a uniformly magnetized disk coated with such materials is selectively heated by a laser beam to a value above its Curie temperature, to destroy the magnetization in selected areas, the readout beam can be suitably filtered to make an image appear.   Since the disk surface can be remagnetized, a read-write system is feasible.   Europium oxide operates at liquid-nitrogen temperature.   Manganese bismuthide is useful at room temperature, but considerable heating power is required to record information since its Curie temperature is high.

66. Electrostatic Storage.   A number of other methods have been proposed for the implementation of storage systems.   One of these uses charge storage on the surface of a dielectric film.   An electron beam in a cathode-ray tube, after passing through a signal screen, is used to address spots on a dielectric surface on the face of the tube.   During writing a high-voltage beam causes the emission of secondary electrons and leaves a positive charge on the dielectric surface. Reading is accomplished by readdressing the spot and observing the secondaries collected upon the signal screen.

Problems are encountered with such electrostatic storage schemes due to the spread of charge from cell to cell (via the finite paths of the dielectric) or by redistribution of secondary

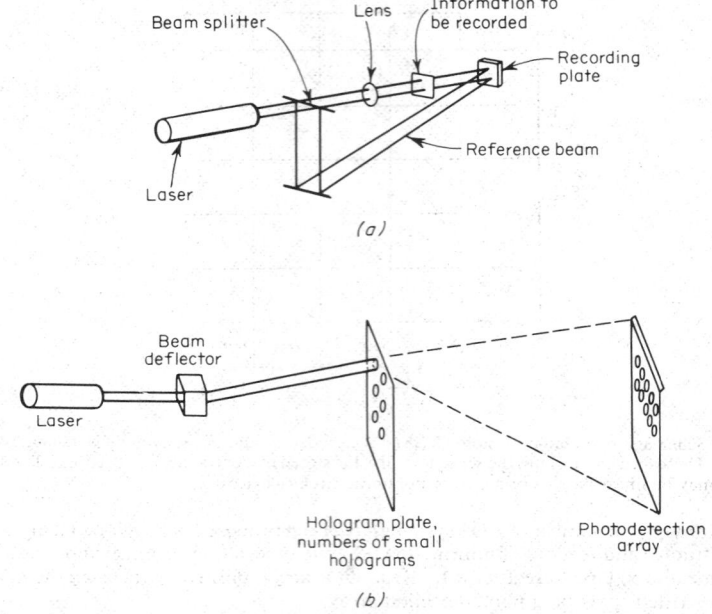

Fig. 23-79. Hologram storage. A hologram is formed from the interference pattern between two laser beams as shown at (*a*). The laser beam is split. One portion passes through or illuminates the image to be recorded. The other (the reference beam) forms an interference pattern with the first beam. The interference pattern (stationary in time and space because the two beams are coherent with each other) is recorded on photographic film. The object beam can be recreated by illuminating the developed hologram with the reference beam. (*b*) Holographic information storage system.

electrons. Accurately directing an electron beam when the information densities are high is also difficult.

67. Cryogenic Devices. A superconducting loop can carry persistent currents in either direction and thus can exist in either of two stable states that can be used for binary information storage. Figure 23-80 shows a basic cell of a cryogenic memory. If current is flowing clockwise and the word line is activated, a signal is transmitted along the sense lines, but not if the current flows counterclockwise. Writing is accomplished by activating the write or sense lines, after selection of the word line. The unit is NDRO and is nonvolatile so long as the temperature of the cells remains below the critical temperature of the superconductor.

Two basic types of cryogenic storage cells are under development. In the first, a continuous superconducting loop element is used; control is achieved by current lines that create a magnetic field at some point in the loop. This field depresses the critical temperature of the loop at that point so that a return to the normal

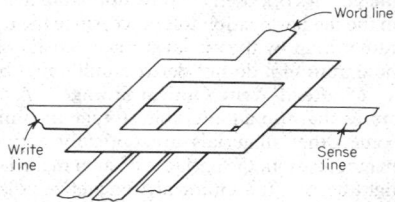

Fig. 23-80. Cryogenic memory element. When current flows in the word line, current can be selected to flow solely in either the left or right leg of the loop. If the current is in the right-hand loop, the sense line is made normally conducting and this state is sensed at the output.

conducting state occurs. In the second type of cell, a Josephson junction is used in conjunction with an external magnetic field source that can lower the junction threshold. Switching is associated with an active voltage source in the loop due to the junction. The Josephson junction device is more complex in fabrication, but has much greater speed potential than the continuous loop.

68. Magnetic-Bubble Devices. In certain thin-film magnetic materials, e.g., the ortho-ferrites, the magnetization vector points outward from the plane of the film. The minimum-energy condition is achieved by the film when it breaks up into a set of magnetic stripe domains. Under the stimulus of an external magnetic field, the stripe domains shrink to circular stable regions of magnetization called *magnetic bubbles*. These bubbles, by external field guidance, can be moved, created, or destroyed, and logical functions can be performed by interactions between bubbles.

69. Amorphous Semiconductor Devices. The *chalcogenide glasses* exhibits properties useful for information storage. A small block of such material can be used to interconnect two conducting lines and to switch from a high to a low resistance state by the application of voltage and current pulses of suitable shape and duration. Configurations similar to ROS structures but with provision for switching can thus provide a read-write store.

Chalcogenides can also be used with light beams. Upon exposure to suitable focused external light beams, a chalcogenide glass can be changed from a state of low to high reflectance and vice versa. An image can be read out by a low-intensity beam and suitable photodetectors.

70. Storage Organization. In accordance with the two basic methods of memory organization discussed in Par. 23-52, address bits may be provided for the store as a whole and blocks of data may be transferred automatically from the backing store to the local store. In such automatic systems a table can be kept of the use of subsections of the local store and the least used subsection transferred to the backing store when a block of data is called for from the backing store. This system can be made totally automatic, so that to the user it appears to have the capacity of the total storage system, with a speed approaching that of the local store.

By means of the operating system, in conjunction with compilers of higher-level language, it is possible to provide for transfers in and out of local store without having an automatic feature. Such transfers under the control of the operating system are called *phasing*. The function of a hierarchical storage is thus achieved with less requirement for wide data-address fields within the machine instruction.

71. Look-ahead and Partition Methods. Even with high-speed local stores, instruction-execution times may be considerably less than the access time to storage. It then is advantageous to retrieve future instructions from storage before executing the current instruction. The machine is then said to employ a *look-ahead* procedure. In such a case, the storage organization must provide for sequential instructions to be stored in separate modules, and each module must be subject to separate addressing, to maintain high speed in the instruction stream. Provision must be made in the event of a branch instruction to break off the queue of current instructions and establish a new queue with respect to the branch condition.

Storage Partitions. Many computer systems provide for multiprogramming; i.e., the system works upon a number of programs at the same time. Provision must be made to prevent the operation of one program from destroying data in an area reserved for another. By providing that one program contain addresses only of a portion of total memory, e.g., only odd or even addresses or other such means, *partitions* can be set up to prevent interference.

72. List Stores. In some classes of problem a list of required program steps is generated sequentially, such that the first step generated is the last performed. The evaluation of complex arithmetic sequences is an example of such a sequence. In such an evaluation, levels of parentheses are generally analyzed from the outermost to the innermost, but computation proceeds in the innermost to the outermost. In such cases, it is convenient to use the concept of a *push-down* list, which operates as follows. As each entry is made in the list, all other items move down one step until the full sequence is loaded. On playback the last item in is the first out, and all other items move up one, so that the second to the last entered becomes the first, and so forth.

Devices constructed to implement such a function are called *push-down* stores. An alternative for generating such a store is to add an address field to each word in a list that specifies the location of the next word. This arrangement, called *linking*, effectively implements a push-down store, since an access to the first item in the store permits addressing the second.

INPUT-OUTPUT EQUIPMENT AND SYSTEMS

73. Input-Output (I/O) Equipment. Input-output equipment includes printers, console typewriters, card readers and punches, paper tapes, keyboards, character-recognition units, process-control sensors, and cathode-ray tubes or other display devices. Such equipment presents a wide range of characteristics that must be taken into account at their interface with the CPU and the ALU. A magnetic-tape unit operates serially by byte, writing or reading information of varying record length. A printer ordinarily uses information one line at a time. Different types of I/O gear operate over widely different speed ranges, from a few bytes per second to millions of bytes per second. In addition to these variables, magnetic drums, printers, card equipment, etc., present differences in access and start-up times, and the operational speed and capabilities of CPUs vary over several orders of magnitude.

74. I/O Configurations. Figure 23-81 shows a configuration used in past systems that presently is found with a small CPU. The program gains access to the I/O gear through the logic circuitry of the ALU. Thus to transfer information to the I/O equipment, the program must first extract the information from storage, edit it, and interact with the receiving equipment. Time must be spent by the CPU waiting for delivery of information to the I/O gear. If several types of I/O equipment are used, the program must take into account variations in formatting and control of each type.

Larger systems use more varied types of I/O, and their speed of operation is higher, so that waiting time is correspondingly expensive. The power of a large system ALU is not fully used in specialized control functions associated with specific I/O equipment. As a result, large systems employ more complex methods of handling I/O than that shown in Fig. 23-81.

A method of alleviating problems at an I/O interface is *I/O buffering*, shown in Fig. 23-82. The buffer accumulates information at machine speeds, so that subsequent information transfers to the I/O gear can take place in an orderly fashion without holding up the CPU. A buffer system is most appropriate when the average I/O equipment information rate is less than the program rate of delivery. Such a system lacks flexibility in the types of I/O equipment accommodated and must use some of the capability of the ALU for control and formatting.

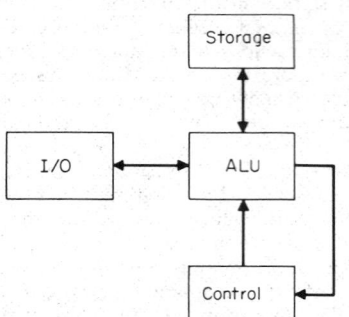

Fig. 23-81. Organization of a small machine. Transfers into and out of external equipment take place through the ALU. A program must first transfer information from the store to the ALU and from there to the I/O gear. The organization provides for data manipulation and editing in the ALU and permits the control unit to handle special I/O requirements.

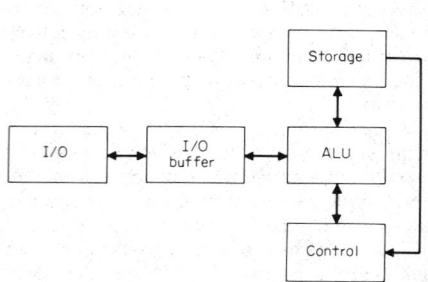

Fig. 23-82. Machine organization with an I/O buffer. Data can be accumulated in the buffer at machine speeds and transferred at a different rate to the I/O gear, and vice versa.

75. I/O Memory-Channel Methods. The arrangements shown in Figs. 23-81 and 23-82 involve information transfer from the I/O gear into the ALU and thence to main store. With a modularized main store, I/O data can be directly entered in, and extracted from, main storage. Direct access to storage implies control interrelationships at the interface between the I/O and the CPU to assure coordination of accesses between the two units. The basic configuration used for direct access of I/O into storage is shown in Fig. 23-83. The connecting unit between the main store and the I/O gear is called a *channel.* A channel is not merely a data path but a logical device that incorporates control circuits to fulfill the relatively complex functions of timing, editing, data preparation, I/O control, etc.

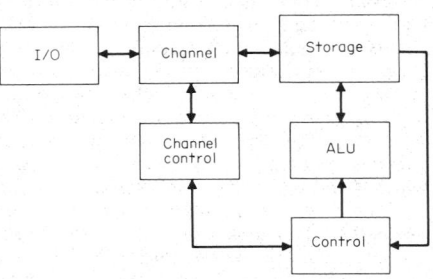

Fig. 23-83. Machine organization with a channel, a separate logical device that can be used to aid in solving problems associated with maintaining CPU speed when working with I/O gear.

76. Channel Functions. When a program calls for writing into a specific piece of equipment, the program instruction first initiates a channel address. The channel in turn brings forth a sequence of commands by means of its own control circuits so that it can specify the location of the information to be transferred from main storage, specify the commands for the initiation of attachment of the device sending or receiving information, and edit, modify, or otherwise correlate the data into a format suitable for transmission. In medium-sized systems a channel may have limited computational facilities, but in a larger system the channel may have a complexity approaching that of a medium-sized data processing machine.

Figure 23-84 shows an arrangement in which several channels are incorporated in a system. The principal I/O signal paths are multiplexed into a sequence of channels, which

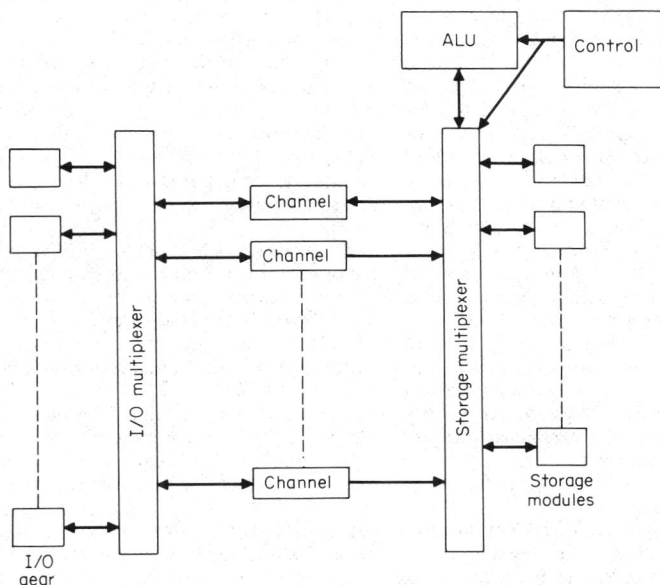

Fig. 23-84. Multiple multiplexed channels. In large systems a set of channels may be used in conjunction with multiplexers for storage and I/O gear. By such a means the CPU can initiate multiple I/O operations and pursue other internal programs while waiting for an I/O response.

in turn may be multiplexed into the CPU. Once the CPU initiates action upon a particular channel, it is free to pursue other programs that may initiate other channel actions. In particular, a CPU might operate upon a program until the program specifies an I/O access. Command would then be transferred to a specific channel and a new program initiated in the CPU until the information transfer has occurred with the prior program. If more than one channel is available, the second program could initiate action on a new channel and a third program be undertaken until the results were available in a prior program.

Multiprogramming is the ability of a system to operate on more than one program at a time. Multiprogramming permits maintenance of the internal speed of operation, the CPU and ALU, while accommodating the slower speed of the I/O equipment. The flexibility of channel architecture also permits many different types of I/O gear to be incorporated, without unduly complicating the programming system of the CPU.

Channels may operate in either the *burst mode* or the *unit-message mode*. In the burst mode, the total message is sent with the information emitted continuously until the message is completed. In the unit message mode, information may be sent separately to different devices on the same channel, one byte at a time.

Some channels (*selector* channels) may attach only to one I/O device at a time. Other channels can attach simultaneously to a number of subchannels on a time-shared multiplexed basis. Such channels are called *multiplexer* channels. Selector channels operate in the burst mode to be efficient.

77. Channel Attachment. Many pieces of I/O gear, e.g., printers, must have special electric signals to control specific operations. For example, in one type of printer (Par. 23-85), an image of a unit called a *chain* must be maintained and periodically matched in time with the contents of the information to be printed. Such functions, specific to the device addressed, usually are handled by a *controller* that interfaces between the channel and the I/O gear itself.

In the operation of a direct-access disk device, the channel may call for a specific track on a particular disk as the repository of the information from the CPU. The disk controller generates the electric signals that position the read head at the appropriate track and receives feedback signals when the appropriate record mark had been found.

A specialized but essential function of I/O gear is to provide enough information for the CPU to begin operation. In volatile storage, the store is blank when power is first applied. Since program storage is required for even the simplest operations, an input of information is needed to permit the central processor to initiate operations. The problem of initial program load is solved by a disk or tape subsystem that subsequent to the power-on condition initiates an *initial program load* (IPL) into portions of the channel and the main store.

78. Card Unit Record Equipment. The Hollerith card has historically formed a basic unit input record to computer systems. A keyboard input causes the card to be punched in accord with the Hollerith code. Provision is made for automatic card feed and duplication and for the insertion of a program card to skip or duplicate certain fields and thus relieve the operator of some routine operations.

To ensure accuracy of information in a keypunch operation, *verification* is commonly performed. An operator on a second piece of equipment, called a *verifier,* repeats the input operation. If the information keyed in the second operation disagrees with the original keyed information, a mark is made on the card and the error thus indicated.

The cards are fed into the computer through a reading machine. Up to 2,000 cards per minute can be read. The holes in the card are sensed by wire contacts or by optical scanning. A CPU may deliver output to cards using a *card punch.*

Equipment of the card variety permits flexibility of insertion and deletion of records, ease of correction, and permanence of information. Subsidiary equipment used with such unit record installations includes *interpreters* that can print information on a card or *sorting equipment* that can sort or resort a deck in accordance with information on the cards themselves.

79. Key-to-Tape and Key-to-Disk Systems. In applications involving a number of key entry operators, it is convenient to use tape or disk magnetic recording directly as the system input. In some systems a clustered key-entry magnetic-tape cassette system is used. Each individual keyboard records on its own cassette unit. These in turn may be used as an input to an intermediate tape or disk unit. *Magnetic cards* have been developed that can store more information than the punched cards while maintaining their flexibility.

80. Other Key-Entry Devices. A number of *terminal* systems have been developed that use a key-entry device with a printer for direct access to a computer on a time-shared basis. The programming system in the CPU is arranged to scan a number of such systems, through a set of subchannels. Each character from each terminal is entered into a storage area in the computer until an appropriate action signal is received. Then by means of an internal program the computer generates such responses as error signals, requests for data, or program outputs. By terminal systems, the power of a large computer can be made available to many users as if each were operating the system independently.

81. Process-Control Entry Devices. In process control, inputs to the computer generally come from sensors that measure such physical quantities as temperature, pressure, rate of flow, or density (see Secs. 10 and 24). These units operate with a suitable analog-to-digital converter that forms direct inputs to the computer. The CPU may in turn communicate with such devices as valves, heaters, refrigerators, pumps, etc., on a time-shared basis, to feed the requisite control information back to the process-control system.

82. Magnetic-Ink Character-Recognition Equipment. A number of systems have been developed to read documents by machine. *Magnetic-ink character recognition* is one. Magnetic characters are deposited upon paper, or other carrier materials, in patterns designed to be recognized by machines and by operators. A change in reluctance, associated with the presence of magnetic ink, is sensed by a magnetic read head. The form of the characters is selected so that each yields a characteristic signature.

83. Optical Character Recognition. Information to be subjected to data processing comes from a variety of sources, in much of which it is not feasible to employ magnetic-ink characters. To avoid retranscription of such information by key-entry methods, devices to read documents optically have been developed.

Character recognition by optical means occurs in a number of sequential steps. First the characters to be recognized must be located and initial starting points on the appropriate characters found. Second, the characters must be scanned to generate a sequence of bits that represents the character. The resulting bit pattern must then be compared to prestored reference patterns to identify the pattern in question.

The scanning function may be performed in a number of ways. In one method, an image of the character is projected on the face of an image orthicon tube. An electron beam scanning across the tube derives a bit pattern in accord with the image on the tube. In another method, the character image is translated with respect to a linear photodiode array and the output of the array sampled at periodic intervals to determine the bit pattern. In a third method, a scanning mirror is used in conjunction with a light-emitting diode (LED) array. A laser may also be used in conjunction with a photodiode.

Finding the initial character in the group to be recognized and sequencing from that character involves control of the character locations on the document. In another technique a *dark-space search* locates an upper left dark area printed on the document.

Pattern recognition can be accomplished by using threshold circuits, as described in Par. 23-51. Each threshold circuit is used to partition a space of binary patterns into two parts, and a set of such circuits can be used to partition the total space of patterns into subspaces. The input pattern is classified according to the subspace partitions and thus recognized. Implementation of such character-recognition routines often occurs by simulation of the operation of threshold circuits, a digital system performing the simulation. Since character sets have evolved to meet needs that bear little relation to machine requirements, most recognition schemes require additional algorithmic search routines to differentiate fine features. For example, if an O and a Q are recognized as the same letter, a sequential search routine may be used to distinguish between them.

Another character-recognition technique involves computation of a *correlation function* between the input and the stored patterns. Optical spatial filtering may be used although digital techniques are more common.

84. Batch-Processing Entry. One computer can be used to enter information into another. For example, it is often advantageous for a small computer system to communicate with a larger one. The smaller system may receive a higher-level language problem, edit and format it, and then transmit it to a larger system for translation. The larger system, in turn, may generate machine language that is executed at the remote location when it is transmitted back.

Remote systems may receive data, operate upon them, and generate local output, with portions transmitted to a second unit for filing or incorporation into summary journals. In other cases, two CPUs may operate generally upon data in an independent fashion but be so interconnected that in the event of failure, one system can assume the function of the other. In other cases, systems may be interconnected to share the work load. Computer systems that operate to share CPU functions are called *multiprocessing* systems.

85. Printers. Much of the output of computers takes the form of printed documents, and a number of *printers* have been developed to produce them. The two basic types of printer are *impact printers,* which use mechanical motion of type slugs driven against a carbon ribbon or paper, and *nonimpact* printers, which use various physical materials to produce the characters.

Impact printers are divided into two classes, *line printers* and *serial printers.* Line printers print a full line at one time and operate at from 50 to 2,000 lines per minute. Serial printers include typewriters and other such devices that print a character at a time. Some serial printers use engraved type for formation of solid character images. Others use a matrix of wires to print dots that form a character. The latter are called *matrix printers.*

Impact printers generally use a high-speed electromechanical hammer to drive a print slug into contact with a ribbon and paper. In the *front printer* the hammer drives the print slug against the ribbon and paper. In the *back printer* the hammer drives the paper and ribbon against the type slug.

Figure 23-85 shows the basic arrangement of a *drum printer,* in which a drum is engraved with a set of characters for each column to be printed. As the drum rotates, the full set of characters is successively presented at each print position. The unit shown in Fig. 23-85 is a back printer, with the drive hammers mounted behind the paper. The paper is driven against the drum at an appropriate time to print the characters desired. A full rotation of the drum is required in order to print a line. The paper stays stationary while printing takes place.

Figure 23-86 shows the basic configuration of a chain printer. The device shown is a back printer. The chain moves characters horizontally across the paper. A hammer at each print

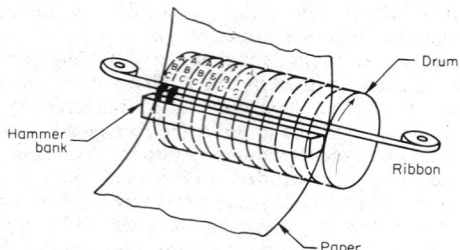

Fig. 23-85. A drum printer. The rotating drum carries a sets of characters, one set for each column. As the drum rotates, successive characters move under the print hammers. As the required character in each column is displayed to the hammer, the hammer is fired, driving the paper and ribbon against that character.

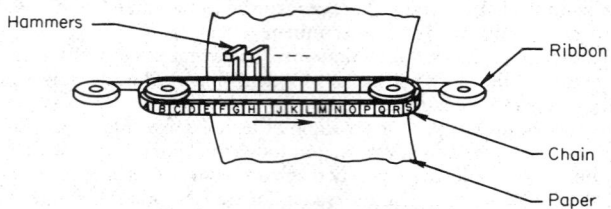

Fig. 23-86. A chain printer. The characters, on an endless belt, are moved transversely with respect to the paper. A hammer at each print position is fired as the character to be printed at that position appears.

position behind the paper is fired at the appropriate time to print the required character as the belt moves by. Chain printers have an advantage over drum devices in that any mistiming of the hammer results in a horizontal displacement of the character printed. Such a displacement is less discordant to the eye than the vertical displacement produced by a drum printer. The *train printer* is similar to the chain device, except that the type slugs are pushed through a guide instead of being pulled on a belt.

Two basic types of print hammers are used in high-speed printers. One operates by means of an electromagnetic circuit with a movable iron armature. The armature itself forms the hammer-drive mechanism. The other uses the current in a coil in a magnetic field as a motive mechanism and operates in a manner similar to the voice coil of a loudspeaker.

Carriage movement in printer systems is provided by hydraulic actuators, stepping motors, dc servomotors, or by mechanical escapements under relay control. Motion of the carriage from one line to the next may occur in times as short as 4 ms, and the speed of paper movement may range up to 100 in/s. Line printers are equipped to skip at high speed past lines not to be printed so as to increase the *throughput,* i.e., the total length of paper printed per unit time.

Electronic circuits, such as power drivers for the carriage and hammers, are usually contained in a separate control unit. In chain or drum devices, an electronic image of the physical position of the chain or drum is stored and compared with the characters desired at each location. When a match is found, the hammer is fired and the appropriate character is printed. The logical devices required for such operations are often contained in a separate control unit but may be located in the CPU box.

Serial-Impact Printers. A number of mechanisms have been developed for serial-impact printing. One such device consists of a ball covered with characters that is rotated and tilted before moving forward to imprint a character. This device is widely used in typewriters. Ordinary electric typewriters, using typebars and horizontal carriage motion, are also used as serial printers.

Another type of serial printer is shown in Fig. 23-87a. A helical wheel is rotated at high speed, with a hammer unit mounted in the back-printing configuration. When the appropriate character is lined up, the hammer is fired. The hammer, which moves continuously across the page, is two positions wide and therefore always in front of the rotating helical wheel. The fact that the wheel is helical permits a continuous carriage motion while maintaining the character set at a particular position. Many other similar serial devices have been proposed involving disks (Fig. 23-87b) or drum sections (Fig. 23-87c). The limited speed of the motion of the character set past the print head usually restricts such serial devices to speeds below 30 characters per second.

Nonimpact Printers. Nonimpact printers use various mechanisms to imprint characters on a page. There are two major classes of such devices, operating with *special paper* and with *ordinary paper.* In one type of special-paper device, the paper is coated with an opaque waxy substance on a contrasting background. When the paper is heated, an image appears. Characters are formed by contact with selectively heated wires or resistors that form a matrix character. Another type of special paper is coated with a thin aluminum or other metal film

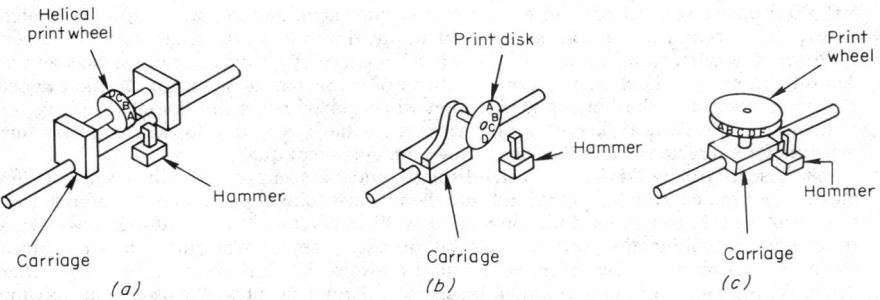

Fig. 23-87. Types of serial character printers: (*a*) helical-wheel mechanism; (*b*) disk mechanism; (*c*) wheel device.

that can be evaporated with laser light. The laser beam is switched on and off under electronic control while being scanned across the paper by means of a mirror. The metal film may also be eroded away by means of a spark discharge from a suitably actuated array of wires.

In other systems, electrophotographic methods are used, with papers coated or filled with a photoconductive substance such as zinc oxide. The sheet is charged by corona wires and is selectively discharged by the optical image of the character. Small electrostatically charged particles, called *toners,* can be made to adhere to the paper at the positions of discharge and fixed by heat or pressure or both. Another method of nonimpact printing on special paper is called *electrography.* It uses a dielectric-coated paper that is directly charged by ion generation. The charged paper can be toned and fixed as in electrophotography. Nonimpact printing using chemically treated optically sensitive papers is also available. The characters are generated by exposure to shaped light beams. In diazo systems, for example, the paper is optically illuminated by the character images and developed in a bath of gaseous ammonia.

Nonimpact printing can also be accomplished using plain paper, and is an advantage in a large computer installation, where the cost of paper is a major factor in the overall cost of printing. In *transfer electrophotography,* used with plain paper, a photoconductive surface (such as selenium or certain organic materials) is charged and selectively discharged as with electrophotography. A toner image is developed directly from the photoconductor, but this image is subsequently electrostatically transferred to plain paper, which is then fused (see Sec. 20, Par. 20-118).

86. Image Formation. Electrophotography, transfer electrophotography, and thermal methods require that optical images of characters be generated. In some types of image generation, the images of the characters to be printed are stored. Other devices use a linear sweep arrangement with an on-off switch. The character image is generated from a digital storage device that supplies bits to piece together images. In the latter system any type of material can be printed, not just the character sets stored; the unit is said to have a *noncoded-information* (NCI) capability.

One method of optical character generation employs a cathode-ray tube. A character mask within the tube selectively intercepts the electron stream. The resulting image on the tube face can be projected upon a suitable nonimpact printing receptacle. In another method, the electron beam undergoes linear scanning and is switched on and off by an electronic character generator. Each line of the sweep produces a section of a line of characters. Repeated sweeps are used to generate a full line of information. Characters may be also generated using a laser and a rotating mirror either by switching the laser off and on or by projecting through a standard or holographic character mask. Sets of light-emitting diodes may also be used for character generation.

87. Ink Jets. Another method of direct character formation on plain paper uses ink droplets. When a stream of ink emerges from a nozzle vibrated at a suitable rate, the droplets tend to break up in a uniform, serial manner. Figure 23-88 shows droplets emerging from a nozzle and being electrostatically charged by induction as they break off from the ink stream. In subsequent flight through an electrostatic field, the droplets are displaced according to the charge they receive from the charging electrode. Droplets generated at high rates (up to 500,000 droplets per second), are guided in one dimension and deposited upon untreated paper. The second dimension is furnished by moving the nozzle relative to the paper.

A set of nozzles, as shown in Fig. 23-89, can be used with vertical paper displacement to deposit characters. Each nozzle prints sections of characters in sequential fashion under electronic control in the horizontal direction while paper movement completes character formation in the vertical direction. Ink-jet systems have been developed using very fine matrices for character generation, producing high document quality.

88. Visual-Display Devices. Visual-display devices associated with a computer system may range from console lights that indicate the internal state of the system to cathode-ray-tube displays that can be used for interactive problem solving. In the cathode-ray tube, a raster scan, in conjunction with suitable bit storage, generates the output image. Fixed character sets may also be generated by masks within the cathode-ray tube. The latter devices require less stored information, since a coded input signal can be used to display the appropriate character. Such displays are often used with a keyboard for information entry, so that the operator and computer can operate in an interactive mode.

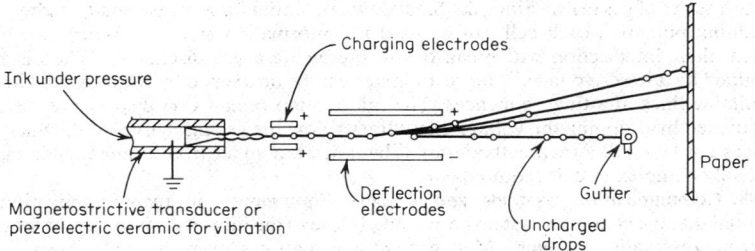

Fig. 23-88. Ink-jet printing. Ink under pressure is emitted from a vibrating nozzle, producing droplets which are charged by a signal applied to the charging electrodes. After charging, each drop is deflected by a fixed field, the amount of deflection depending on the charge previously induced by the charging electrodes.

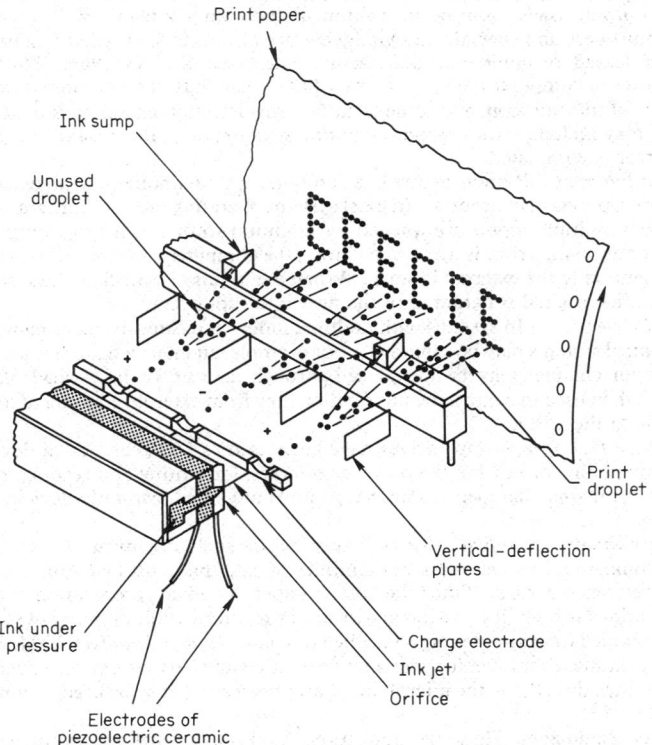

Fig. 23-89. A printing system using ink jets. Individual nozzles are used to print characters in sets of two. Horizontal electrostatic deflection forms line sections of each character, while the paper movement supplies the vertical dimension.

Another type of display device uses two plain parallel glass plates in conjunction with an intersecting transparent matrix of embedded conductors. The intersections of the conductors define a series of gas cells. Since the breakdown potential for a gas is usually higher than the sustaining potential, each cell can be used for information storage. When two lines are pulsed, their intersection will break down, producing a gas discharge. The discharge is sustained by a lesser voltage. Hence an image can be developed by selection and activation of suitable lines in a time sequence. The information remains on display until the unit is deactivated by dropping the voltage on all lines below the gas-discharge level. Such display devices are less bulky than cathode-ray-tube units, but suffer from limited color range and relatively complex circuit requirements.

89. Communication Systems and Remote Equipment. In many applications, it is essential that the computer system be extended close to the information source, as for example in airline reservation systems. Management information systems have also been developed that permit the collection of information from the diverse activities of a large corporation. Within such retail establishments as supermarkets and department stores, computer systems are used in conjunction with terminals and cash registers.

These extensions of computer services require a communications net. A number of specialized pieces of equipment have been developed for use in such nets. One such piece of equipment, the *modem,* interrelates with multiple low-bandwidth communication-line inputs and by time or frequency multiplexing concentrates the multiple inputs into a single wide-band output. *Switch gear* permits automatic hookup to a telephone system. Communication controllers and specialized switch gear are also used for special function interconnection on leased or interconnected facility lines (see Sec. 22, Pars. **22-44** to **22-57**).

The nature of computer data puts extraordinary demands upon communication systems for accuracy of transmission, and some form of error-detection encoding is used. Recovery techniques may include error correction; but more commonly, the message is retransmitted until the error is eliminated.

Program Interrupt. Remote access to a computer system implies that an external inquiry can interrupt a program in process. In most systems operating in a multiprogramming mode, accesses from multiple inputs are queued for attention so that as a program in progress is completed or as an interrupt is made mandatory, the computer can store its current program and give attention to the external inquiry. When the inquiry is satisfied, the system resumes operation on its original program until the next interrupt occurs.

Channel Extension. In systems with multiple inputs and extensive communication links, channel control systems may be extended. For example, an intermediate storage mechanism for each input channel may be introduced through a drum or fixed-head file. A CPU subsystem is then used to control the information flow from external equipment to the file and from the file to the CPU.

Paper Tape I/O. Paper-tape drives have been used as I/O gear. Such devices usually operate by punching or sensing the presence of holes, for writing and reading, respectively. A sprocket maintains alignment. Paper-tape units are slow, read-only devices but usually inexpensive.

Computer-Output Microfilm. Record keeping in such organizations as finance and insurance companies involves massive quantities of data that must be retained, even though individual reference is rare. Since the cost of paper document preparation and storage in such applications is high, it is customary to store the information on microfilm. Microfilm printers developed for computer output, called *computer output microfilm* (COM) printers, use nonimpact printing techniques for character generation and lens systems to reduce the optical image for storage directly on the microfilm. Laser beams may also be used, scanning directly on microfilm.

90. Audio Equipment. In some applications, audio devices deliver a recognizable speech signal in response to an inquiry to the computer. The simplest approach is to use recorded words or phrases, sequenced to form messages. More complex devices can synthesize speech sounds into words and phrases. Equipment has also been developed that can discriminate between human voice patterns. Words are broken down into frequency bands, and the intensity of the bands as a function of time forms a recognizable pattern. Recognition of such patterns is accomplished as in character recognition and the word information thus extracted.

CENTRAL PROCESSING METHODS AND EQUIPMENT

91. Functions of the Central Processing Unit. The central processing unit (CPU) executes instructions in the computer program and thus controls the operation of the system. The CPU loads an appropriate set of programs, executes them, and initiates the output. Its operation sequence includes interruptions, detection of abnormal conditions, recovery procedures, and termination of operations. The basic classes of operation of the CPU are fetching and decoding of instructions from storage and executing the content of these instructions on data. Fetching and decoding operations are said to occur during an instruction cycle (I cycle) and execution during an execution cycle (E cycle).

The CPU comprises two major elements, the *arithmetic logical unit* (ALU), which performs arithmetic operations, such as add, subtract, multiply, and divide, and the *control section*, which moves data within the CPU and to and from the main store and interfaces with the remaining sections of the computer.

92. Arithmetic Logical Unit (ALU). The ALU performs arithmetic and logical operations between two operands, such as OR, AND, EXCLUSIVE-OR, ADD, MULTIPLY, SUBTRACT, or DIVIDE. The unit may also perform operations such as INVERT on only one operand, and it tests for minus or zero and forms a complement.

Adders and multipliers are at the heart of the ALU. In Fig. 23-90*a*, one bit position of an ALU is shown as part of a wider data path. One latch (part of a register A) feeds an AND circuit that is conditioned by a CONTROL A. The output feeds INPUT A of the adder circuit. One latch of register B is also ANDed with CONTROL B and feeds the other input into the adder. A true-complement circuit is shown on the B line. This latter circuit has to do with subtraction and can be assumed to be a direct connection when adding. Each adder stage is a combinatorial circuit that accepts a carry from the stage representing the next lower digit in the binary number (assumed to be on the right). The collection of outputs from all adder stages is the sum. This sum is ANDed into register D by CONTROL D.

All bit positions of each of the registers are gated by a single control line. If the gate is closed (control equal to 0), all outputs are 0s. If the gate is open (control equal to 1), the bit pattern appearing in the register is transmitted through the series of AND circuits to the input of the adders. Thus, a gate is a two-way AND circuit for each bit position. The diagram of Fig. 23-90 illustrates all positions of an *n*-position adder since all positions are identical. In such a case heavy lines, as shown, indicate that this one line represents a line in each bit position.

93. Binary Addition. At the outset of an addition, it is assumed that registers A and B (Fig. 23-90*a*) contain the addends. An addition is performed by pulsing the control lines with signals originating in the control section of the CPU. Time is assumed to be metered into fixed intervals by an oscillator (clock) in the control section. These time slots are numbered for easier identification (Fig. 23-90*b*). At time 1 the inputs are gated to the adder, and the adders begin to compute the sum. At the same time, register D is *reset* (all latches set to 0). At time 2 the outputs of the adders have reached steady state, and control line D is raised, permitting those bit positions for which the sum was 1 to set the corresponding latches in register D. Between times 2 and 3, the result is latched up in register D, and at time 3, control D is lowered. Only after the result is locked into D and cannot change, may control A and B be lowered. If they were lowered earlier, the change might propagate through the adder to produce an incorrect answer.

The length of the pulses depends on the circuits used. The times from 2 to 3 and 3 to 4 are usually equal to a few logic delays (time to propagate through an AND, OR, or INVERTER). The time from 1 to 2 depends on the length of the adder and is proportional to the number of positions in a parallel adder due to potential carry-propagation times. This delay can be reduced by *carry look-ahead* (sometimes called *carry by-pass*, or *carry anticipation*).

94. Binary Subtraction. Subtraction can be accomplished using the operation of addition, by forming the complement of a number. Negative numbers are represented throughout the system in complement form. To subtract two numbers such as *B* from *A*, a set of logic elements may be put into the line shown in Fig. 23-90*a* as input *B*. Following the process discussed in Par. 23-30, the sign of a number is changed by complementing each bit and adding 1 to the result. The inversion of the bit is performed by the logic element interposed on the input *B* line in Fig. 23-90*a* known as a *true-complement* (T/C) *gate*. This unit gates the unmodified bit if the control is 0 and inverts each bit if the control is 1. The Boolean equation for the output of the T/C gate is

$$\text{Output} = \overline{T/C} \cdot B + T/C \cdot \bar{B}$$

The T/C gate is a series of EXCLUSIVE-ORs with one leg, common to all bit positions, connected to the T/C control line. The other leg of each EXCLUSIVE-OR is connected to one bit of the circuit containing the number to be complemented.

The T/C gate produces the 1s complement; a 1 must be added in the low-bit position to produce the true complement. The low stage of an adder may be designed to have an input for a carry-in, designed to accommodate the 1 bit automatically produced from the high-order position of the T/C gate. Such a logical interconnection accomplishes the required 1 input for a true-complement system when a positive number B is subtracted from a positive number and is called an *end-around carry*. Consistency of operation is obtained by entering the appropriate high-order T/C gate into the low-order carry position.

95. Decimal Addition. In some systems the internal organization of a computer is such that decimal representations are used in arithmetic operations. In binary coded decimal (BCD), a conventional binary adder can be used to add decimal digits with a small amount of additional hardware. Adding two 4-b binary numbers produces a 4- or 5-b binary result.

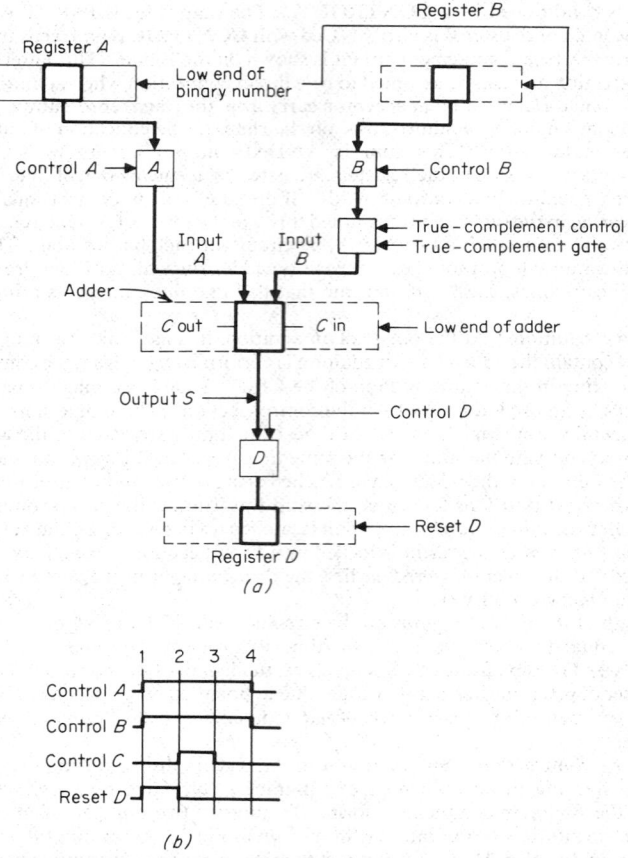

Fig. 23-90. Basic addition logic. The heart of the ALU is the adder circuit shown in functional form in (*a*). The control section applies the appropriate time sequence of pulses (*b*) on the control lines to perform addition. Heavy lines indicate a repeat of circuitry in each bit position to form a machine adder.

When two BCD numbers are added, the result is correct if it lies in the range 0 to 9. If the result is greater than 9, that is the resulting bit pattern is 1010 to 10010, the answer must be adjusted by adding 6 to that group of 4 b, this number being the difference between the desired base (10) and the actual base ($2^4 = 16$). The binary carry from a block of 4 b must also be adjusted to generate the appropriate decimal carry.

The circuits to accomplish decimal addition are shown in Fig. 23-91. A test circuit generates an output if the binary sum is 1010 or greater. This output causes a 6 to be added into the sum and also is ORed with the original binary carry to produce a decimal carry. The added circuits needed to perform decimal additions with a binary adder represent one-half to two-thirds of the circuits of the original binary adder.

96. Decimal Subtraction. In most computers that provide for decimal operation, decimal numbers are stored with a sign and magnitude. To perform subtraction the true complement is formed by subtracting each decimal digit in the number from 9 and adding back 1. Once the complement is formed, addition produces the desired difference.

In a machine that provides for decimal arithmetic a *decimal true-complement* switch may be incorporated in each group of four BCD bits to form the complement. As with binary, provision must be made for the addition of an appropriate low-order digit and for the occurrence of overflows.

97. Shifting. All computers have shift instructions in their instruction repertoire. Shifting is required, for example, for multiply and divide operations, as discussed in Par. 23-29. The minimum is a shift of one position left or right, but most computers have circuits permitting a shift of one or more bits at a time, that is, 1, 4, or 8.

In shifting, a problem arises with the bit(s) shifted out at the end of the register and the new (open) bit position(s). Those shifted out at one end can be inserted in the other end (referred to as *end-around shift* or *circulate*), or they can be discarded or placed in a special register. The newly created vacancies can be filled with all 0s, all 1s, or from another special register.

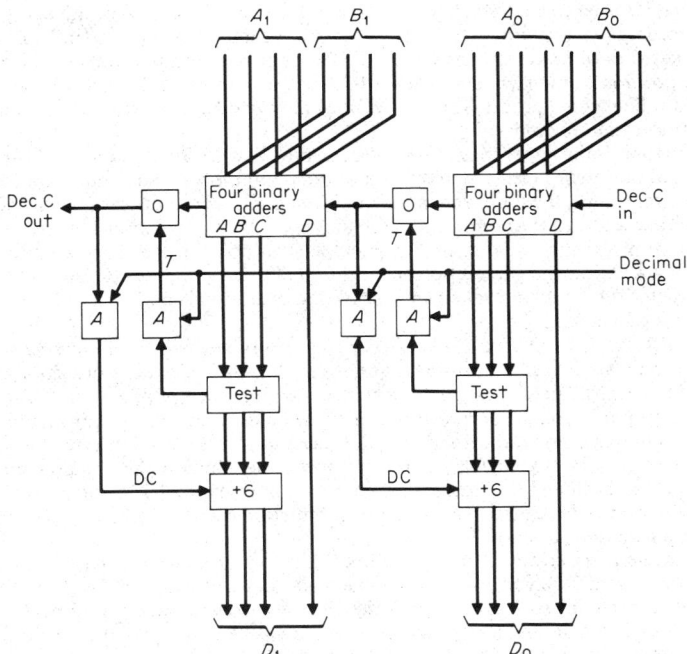

Fig. 23-91. Binary adder with decimal-add feature. If the input of the binary add is 1010 or greater, the addition of a binary 6 produces the correct result by modulo-10 addition.

In a typical computer, shifting is not performed in a shift register but with a set of gates that follow the adders. The outputs of the shift gates are connected to the output register, with an offset by providing a separate gate for each distance and direction of shift. One bit position of the output register can therefore be fed from several adder outputs, but only one gate is opened at a time, as in Fig. 23-92. The pattern shown is repeated for every bit position.

98. Multiplication. Figure 23-93 shows a possible data-flow system for multiplication, i.e., the data flow of the adder shown in Fig. 23-90 with the addition of register C to hold the multiplier and the shift registers as shown in Fig. 23-92. An extra register E holds any extra bits that might be generated in the process of multiplication. Register E in the particular system shown also receives the contents of C after transmission through C's shift register.

The process of binary multiplication involves decoding successive bits in the multiplier, starting with its lowest-order position. If the bit is a 1, the multiplier is added into an accumulating sum; if 0, no addition takes place. In either case the sum is moved one position to the right and the next higher-order bit in the multiplier considered.

In Fig. 23-93 the multiplier is stored in register C, the multiplicand in A, and all others are reset to zero. If the low-order position of C is a 1, the contents of B and C are added and shifted one position to the right into register D. After the addition, register C is shifted one position to the right and stored in register E. If the low-order bit in C had been a zero, only register B would have been gated into the adders; i.e., the addition of the contents of A would have been suppressed, but subsequent operations would have remained the same.

Each add and shift operation subsequent to an add may generate low-order bits that may be shifted into register E, since as the contents of C are shifted to the right, unused positions become successively available. After the add and shift cycles, registers D and E are transferred into B and C, respectively, and the process is repeated until all positions of the multiplier are used. The content of D and E is the product.

99. Division. To provide for the division of two numbers the functions of the registers in Fig. 23-93 must be rearranged as shown in Fig. 23-94. A gating loop is provided from the shift register to register A. Initially the divisor is placed in register B and the dividend in A. The T/C gate is used to subtract B from A, and if the result is zero or positive, a 1 is placed in the low-order position of register E. If the result is negative, a 0 is placed in E and the input from B ungated to reset to the contents of A. The shift register then shifts the output of the adder one position to the right and gates it back to A. E is gated through C and shifted on to the right. The whole process is repeated until the dividend is exhausted. E contains the quotient and A the remainder.

100. Floating-Point Operations. As discussed in Par. 23-31, in some applications, it is convenient to represent numerical values in floating-point form. Such numbers consist of a *fraction* that represents the number's most significant digits, in a portion of the field of a word in the machine, and a *characteristic* in the remaining portion. The characteristic denotes the decimal-point position relative to the assumed decimal point in the characteristic field. In floating-point addition and subtraction, justification of the fractions according to the contents of the characteristic field must take place before the operation is performed; i.e., the decimal points must be lined up.

In an ALU such as in Fig. 23-93, the operation proceeds in the following way: Two numbers A and B are placed in registers A and B, respectively. The control section then gates only the characteristic fields into the adder, in a subtract mode of operation. The absolute difference is stored in an appropriate position of the control section. Controlled by the sign of the subtraction operation, the fraction of the least number is placed through the adder into the shift register. The control section then shifts the fraction the required number of positions, i.e., according to the stored difference of characteristic fields, and places the result back in the appropriate register. Addition or subtraction can then proceed according to the generating machine instruction.

This procedure is costly of machine time. Instead of using the ALU adder for the characteristic-field difference, provision can be made for subtraction of the characteristics in the control section. In such a case, only the characteristics A and B need be entered into registers A and B and the lesser number can be placed in B so that shifting can be accomplished by the circuits normally used in multiplication.

101. Control Section. The control section is the part of a CPU that initiates, executes, and monitors the system's operations. It contains clocking circuits and timing rings for

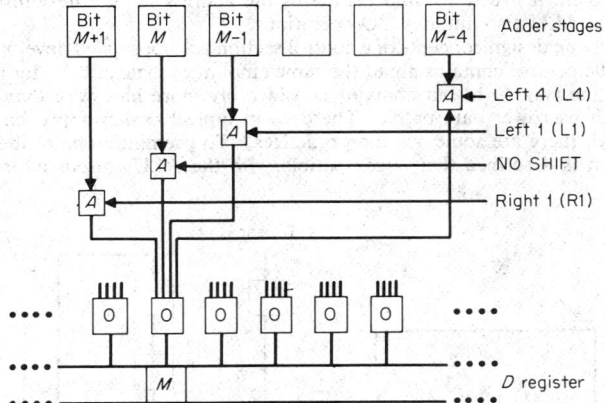

Fig. 23-92. Shift gates. In many systems shifting is accomplished in conjunction with the output of the adder; i.e., position shifts can be accomplished with only one circuit delay.

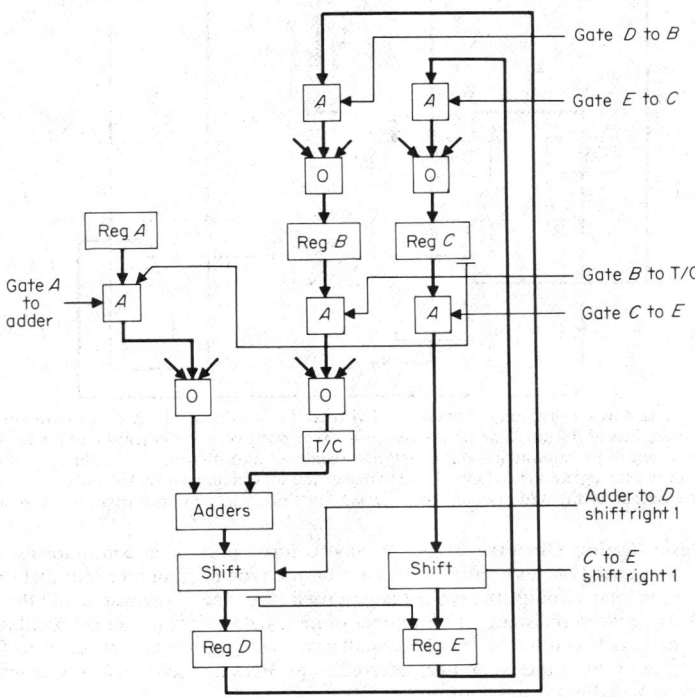

Fig. 23-93. Data flow for multiplication.

opening and closing gates in the ALU. It fetches instructions from main storage, controls execution of these instructions, and maintains the address for the next instruction to be executed. The ALU also initiates I/O operations.

The underlying design concepts of a control section are not so well developed as those of the ALU. The control contains about the same amount of logic circuits for a medium-size parallel machine (say 30 b) but contains considerably more hardware than the ALU for machines with narrower data paths. There is no typical design approach for a control section, though there are some common practices. To exemplify one of these practices a control section is described that is compatible with the ALUs discussed in the previous paragraphs.

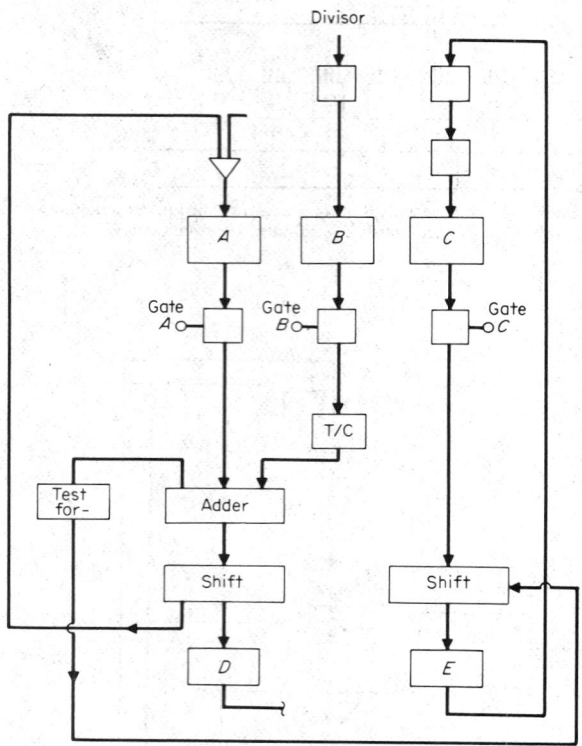

Fig. 23-94. Data flow for division. *B* holds the divisor and *A* the dividend. A trial subtraction is made of the high-order bits of *B* from *A*, and if the results are 0 or positive, a 1 is entered into the low-order bit of *E* and the result of the subtraction shifted left and reentered into *A* (with gate *A* closed). If the result had been negative, the *B* gate would have closed to negate the subtraction, and a 0 would have been entered into *E*. The output of the adder would then have been shifted left one position and reentered in A.

102. Basic Timing Circuits. Basic to any control unit is a continuously running oscillator. The speed of this oscillator depends on the type of computer (parallel or serial), the speed of the logic circuits, the type of gating used (the type of registers), and the number of logic levels between registers. The number of logic delays for an average oscillator cycle time is typically between 6 and 15. The oscillator pulses are usually grouped to form the basic operating cycle of the computer, referred to as *machine cycles*. In this example four pulses are combined into such a group.

In Fig. 23-95 an oscillator is shown that derives a four-stage ring. At any one time, only one stage is on. Suppose an addition is to be performed between registers *A* and *B* and the result is to be placed back in *B*. The addition circuitry described in Fig. 23-93 uses three registers, two for the addends (registers *A* and *B*), and one for temporary storage (register *D*). The operation to be performed is to add the content of register *A* to the content of register *B*, and store the result in register *B*. Register *D* is required for temporary storage since if *B* were connected back on itself, an unstable situation would exist; i.e., its output (modified by *A*) would feed its input.

The operation of the four-stage clock ring in controlling the addition and transfer is shown in Fig. 23-96. Action is initiated by the coincidence of an ADD signal and clock pulse *A*, which starts the *add ring*. This latter ring has two stages and advances to the next stage upon occurrence of the next *A* clock pulse. The timing chart in Fig. 23-96 describes one sequence of actions required and the gates needed for the addition. The circuit diagram shows a realization of these gates. Each register is reset before it is reused, and all pulses are derived from the four basic clock pulses shown in Fig. 23-95. The add ring initiates the add by opening the gates between registers *A* and *B* and the adder. The add latch is then reset, *D* is reset, the ring stage transferred, and so forth. An ADD-FINISHED signal is furnished, and may be used elsewhere in the system.

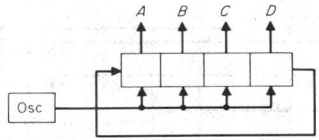

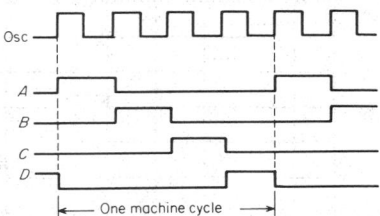

Fig. 23-95. Oscillator and ring circuit. Many control functions can be performed by using an oscillator in conjunction with a timing ring that sequentially sends signals on separate lines, in synchronism with the oscillator.

In the timing diagram, it is assumed that the time required for transmission through the adder is about one machine cycle and that one clock cycle is sufficient for the signal to propagate through the necessary gating and set the information in the target register. These times must include the delay in the circuits and any signal-propagation delay.

103. Control of Instruction Execution. The following approach for the design of a control section is straightforward but extravagant of hardware. For each instruction in the computer a timing diagram is developed similar to the one shown for the add operation in Par. **23-102**. These timings are implemented in rings that vary in length, according to the complexity of the instruction. The concept is simple and has been widely used, but it is costly. To reduce cost, rings are used repeatedly within one instruction and/or by several instructions. Subtraction, for example, might use the ADD ring, except for an extra latch that might be set at ring time 1, clock time *A* (denoted 1.*A*) and reset at 2.*C*. This new latch feeds the T/C gates. Another latch that might be added to denote decimal arithmetic would also be set at 1.*A* time and reset at 2.*C* time. The addition of two latches and a few additional ANDs and ORs permits the elimination of three two-stage rings and associated logic. Further reductions in the number of required circuits can be achieved by considering the iterative nature of some instructions, such as multiplication, division, or multiple shifting.

104. Controls Using Counters (Multiplication). To exemplify this approach multiplication is considered. Multiplication can be implemented as many add-shift cycles, one per digit in the multiplier (say *N*). The control for such an instruction can be implemented using one

$2N + 1$ position ring, the first position initializing and the next $2N$ positions for N add-shift cycles (two positions are needed per add cycle). Such an approach requires not only an unnecessarily long ring but is relatively inflexible for other than N-b multipliers.

The alternate approach, requiring considerably less hardware, uses the basic operation in multiplication, an add and a shift. Therefore a multiply can be implemented by using the controls for the add-shift instruction, plus a binary counter, plus some assorted logic gates to control the gates unique to the multiply, as in Fig. 23-97. In the figure some of the less important controls needed for multiply have not been included to simplify the presentation. For example, the NO SHIFT signal for add must be conditioned with a signal that multiply is not in progress, and during multiplication an ADD-FINISHED signal should be ignored (nor start an I cycle), the reset B also resets C, the reset D also resets E, gate D to B also gates E to C, gating of A to the adder is conditioned on the last bit of C, etc.

In Fig. 23-97 the action is started by raising the MPY line that sets a *multiply* latch. This in turn sets the binary counter to the desired cycle count. Also set is the ADD-LATCH, and

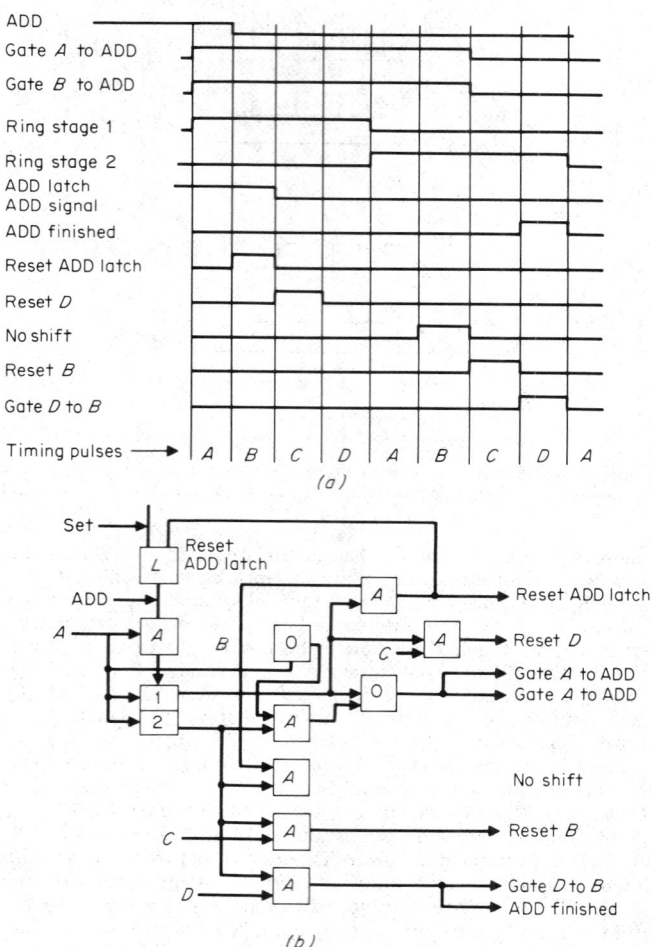

Fig. 23-96. Control circuit (b) and timing chart (a) for addition. The timing ring shown in Fig. 23-95 is used to control the adder circuit shown in Fig. 23-93 with the aid of additional switching circuits.

the addition cycles starts. When the counter goes to 0, the ADD-LATCH is no longer set and the MULTIPLY-FINISHED signal is raised.

105. Microprogramming. The control of the E cycle, in the preceding descriptions, is performed by a set of sequential circuits designed into the machine. Once an execution is initiated, the clocking circuits complete the operation, through wired paths in a specific manner. An alternative method of design for a control unit, *microprogramming,* is described briefly in Par. 23-15. The concept is not sharply defined but has as its objective implementation of the control section at low cost and with high flexibility.

In many cases it is desirable to design compatible computers, i.e., units with the same instruction set, with widely varying performance and cost. To provide a slower version at a lower cost, the width of the data path is reduced to lower circuit counts in the ALU. On the other hand, to operate with the same instruction set, the reduced data-path width usually implies more iterative operations in the control section, at added cost.

Considerable investment is required for programming development, and normally such systems run only on computers with identical instruction sets. If appropriate flexibility is provided, one computer can mimic the instruction set of another. Microprogramming provides for such operations. The process by which one system mimics another is called *emulation.*

In the control-section operation for an add given above in the example of a wired system (Figs. 23-90 and 23-91), the ring circuit delivers a series of electric signals in time sequence

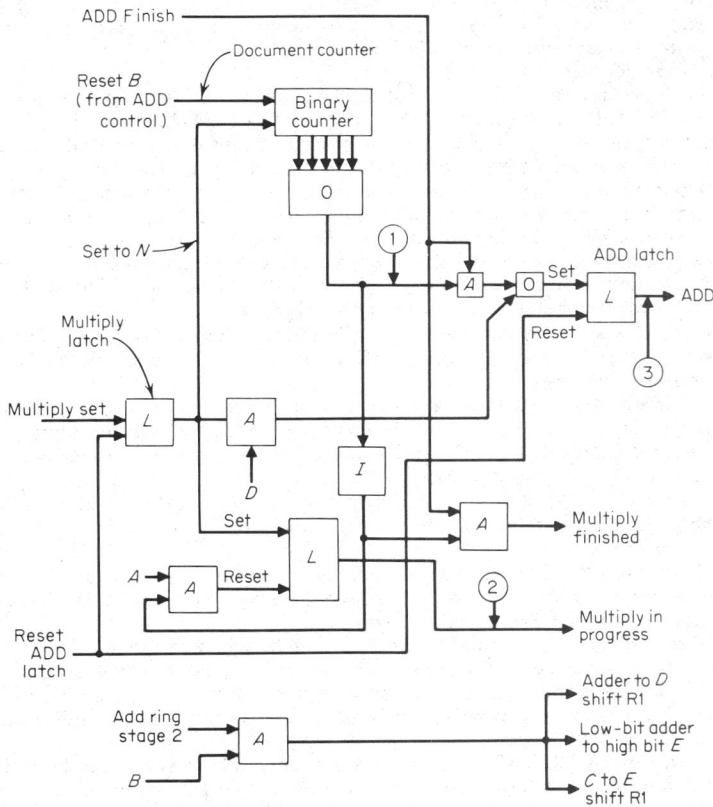

Fig. 23-97. Control circuit for multiplication. Repetitive operations such as add and shift are combined in cyclical fashion. The timing diagram of Fig. 23-96*a* is assumed.

23-79

on specific logical control lines. An alternative approach follows. Upon decoding the instruction code, a sequence of words is called forth from a special storage unit and put into a device that converts the bit positions of each word directly into signals that selectively activate the control wire. The sequence of words in storage then controls the sequence of operations in the ALU at the gate level.

This procedure is the essence of microprogramming. Control lines are activated not by logic gates in conjunction with counters but by words in a storage system that are translated directly into electric signals. The words selected are under the control of the instruction decoder, but the sequence of words may be controlled by the words themselves by provision of a field that does not directly control gates but specifies the location of the next word. The name given to these control words is *microinstruction* as opposed to machine instruction or instruction. *Macroinstruction* is a term given to machine instructions when it is essential to distinguish the two.

When two computers are designed for the same instruction set but with different data-path widths, the microinstruction sets of the two computers are radically different. For the small computer, the program for a given macroinstruction is considerably longer than that for the large computer. The same macroinstructions can be used in both computers. The difference in control-system cost between the two is not large. Although the microprogram is longer with the smaller computer, the difference is in the number of storage places provided, not in the control-section hardware.

The design of the sequence of microprogramming words is conceptually little different from other programming. The microprogram implementation, however, requires a thorough knowledge of a machine at the gate level. A *microinstruction counter* is used to remember the location of the next microinstruction, or a field can be provided to allow branching or specification of a next instruction.

A microprogram generally does not reside in main store but in a special unit called a *control store.* This may be a part of main store, not addressable by the user, or it may be a separate storage unit. In many cases, the microprogram is stored in an ROS storage medium; i.e., it is written at the factory. ROS units are faster and may be cheaper per bit of storage and do not require reloading when power is applied to the computer. There is also an advantage in preventing an unsophisticated user from manipulating the bits in a microprogram. Alternatively the microprogram may be stored in medium that can be written into, called a *writable control store* (WCS). By reloading the WCS, entirely different macroinstructions can be implemented, using the same microinstruction set in a different microprogram. By such means emulation is achieved at minimal expense.

106. CPU Microprogramming. Figure 23-98 shows an ALU and Fig. 23-99 a control section with microprogram organization. The microinstructions embodied in these two units are shown in Table 23-4.

To simplify the program several provisions have been made:

1. Each microinstruction contains the control-store address of the next microinstruction. If omitted, the address in one microinstruction is the current address of the one written just below it. It may *not* be next in numeric sequence.

2. An asterisk at the beginning of a line in the program indicates a comment. This means that the entire line contains information about the program and does not translate into a microinstruction.

3. To simplify the drawing, a gate in a path is indicated by placing an X in the line and omitting from the drawing any control lines controlling these gates. It is also assumed that where two lines join, OR circuits are implied.

4. Rather than listing a numeric value for each field, a shorthand description for the desired action is invented. All actions not so described are assumed to be zero. For example, A to ADD implies that register A is gated to the adder, or T/C means raise T/C gate. These changes do not in any way modify the concept of microprogramming but make the result more readable.

107. Instruction FETCH. In the preceding paragraphs the execution of instructions (E cycles) is discussed. In these cases, operations are initiated by setting an appropriate latch for the function to be performed. The signals that set the latch are in turn generated by circuits that interpret the information of the operation-code part of an instruction cycle (I cycle).

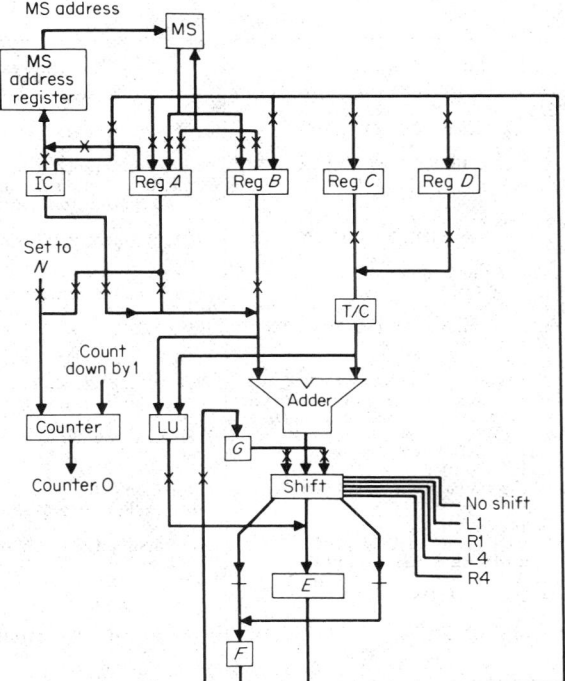

Fig. 23-98. Microprogrammed ALU.

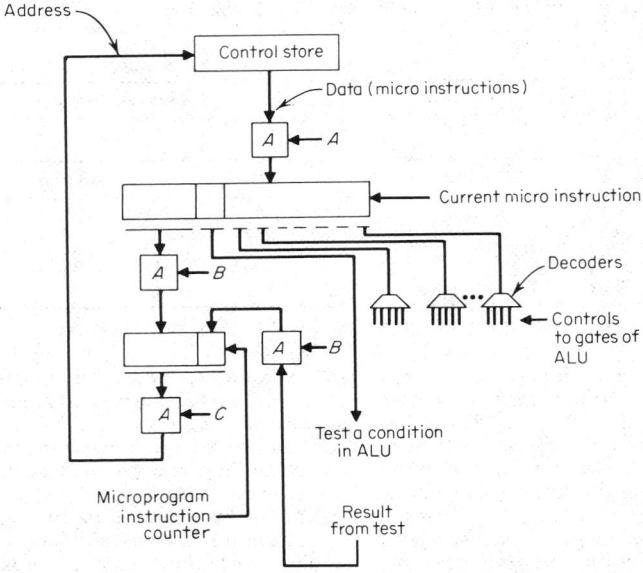

Fig. 23-99. Microprogrammed control unit.

Table 23-4.

Current address	Microinstruction	Comment
	*ADD	
51	B to ADD, C to ADD, NO-SHIFT	Add two operands
8	E to B	
9	A(2) to MS-ADR-REG, B to MS, GO TO 1	Store result branch to 1, next microinstruction executed to be taken from control store location 1
	*SUBTRACT	
52	B to ADD, C to ADD, NO-SHIFT T/C, GO TO 8	Go to 9, where result is stored
	*BRANCH (unconditional)	
53	A(3) to IC, GO TO 1	This is the macroinstruction branch
	*MULTIPLY	
54	Set an N into counter, C to ADD, NO-SHIFT	Initialize
17	E to D, set C to 0s	
18	If last bit D = 1, then (B to ADD), C to ADD shift-R1, 0 to input of high end of shifter, output of low end of shifter to F	Perform one add shift if last bit D = 1, only shift if last bit D = 0.
19	E to C, F to G, COUNT down by 1	Increment counter
20	D to ADD, SHIFT-R1, G to input of high end of shifter	Shift D
21	E to D, if counter is not 0 then GO TO 17	Close loop
22	C to ADD	Store result in two MS locations
23	E to B, force 1 into C	
24	A(2) to ADD, C to ADD NO-SHIFT; A(2) to MS-ADR-REG B to MS	Store first half of result, increment result address
25	E to A(2)	
26	D to ADD, NO-SHIFT	
27	E to B	
28	A(2) to MS-ADR-REG, B to MS GO TO 1	Store second half of result
		Branch to 1-fetch
	*SHIFT LEFT	
55	A(2) to COUNTER, C to ADD, NO-SHIFT	Operand 1 is number of bits operand 2 is to be shifted
10	If COUNTER = 0 then GO TO 9 E to B	Test if shift count was 0
11	B to ADD, SHIFT L1, COUNT down by 1	Shift loop
12	If COUNT not 0, then GO TO 11 E to B	
13	GO TO 9	Completed shift

NOTE 1: (Location 7) This special test places the op-code bit pattern into the low part of the address for the next instruction, causing a branch to the appropriate microroutine for each op code. Branches are as follows:

Op code	Instruction	Address
1	ADD	51
2	SUBTRACT	52
3	BRANCH	53
4	MULTIPLY	54
5	SHIFT LEFT	55

NOTE 2: (Location 1) It is assumed that START button sets microinstruction counter to 1.

Whenever an instruction has completed execution (or when the computer operator presses the start button on the console), a *start I-cycle* signal is generated at the next *A* time of the master-clock timing ring. This signal starts the I-cycle ring. The first action of this ring is to fetch the instruction to be executed from the main store by gating the instruction counter (IC) (sometimes called *program counter*) to the address lines of main storage and initiating a main-store cycle by pulsing a line called *start MS*.

These operations are illustrated in Fig. 23-100. At ring time 2, the instruction arrives back and is placed in the instruction register (IR). The instruction typically contains three main fields: the *operation code* (op code), that determines what instruction is to be executed (ADD, SUBTRACT, MULTIPLY, DIVIDE, BRANCH), and the two addresses of the two operands participating in the operation. For certain classes of instruction, the operands must be delivered to appropriate locations before the E cycle begins.

During ring time 3 and 4, the first operand is fetched from main store and stored in A. During ring time 5, this operand must be transferred from A to an alternate location, depending upon the nature of the instruction being executed. During ring time 5 and 6, the second operand is fetched and stored in A. Ring time 6 is also used to gate the op code to the *instruction decoder*, which is a combinatorial logic circuit accepting the *operation code*, or *instruction code*, from the P b in the op code. The decoder has 2^P input wires, one for each unique input combination. Thus for each bit combination entering the decoder, one output line is activated. These signals represent the start of an E cycle with each wire initiating the execution of one instruction by setting some latch, e.g., an add or multiply latch, or by initiating an appropriate microprogram.

Some op-code bit combinations may not be valid instruction codes. The outputs of the decoder are ORed together and fed into a circuit that ultimately interrupts the normal processing of the computer and indicates the invalid condition. At the beginning of the I cycle, the content of the instruction counter (IC) points to the instruction to be executed next. An *increment-counter* signal is generated during the I cycle in order to increment the counter, so that the address stored in IC points to the next sequential instruction in the program.

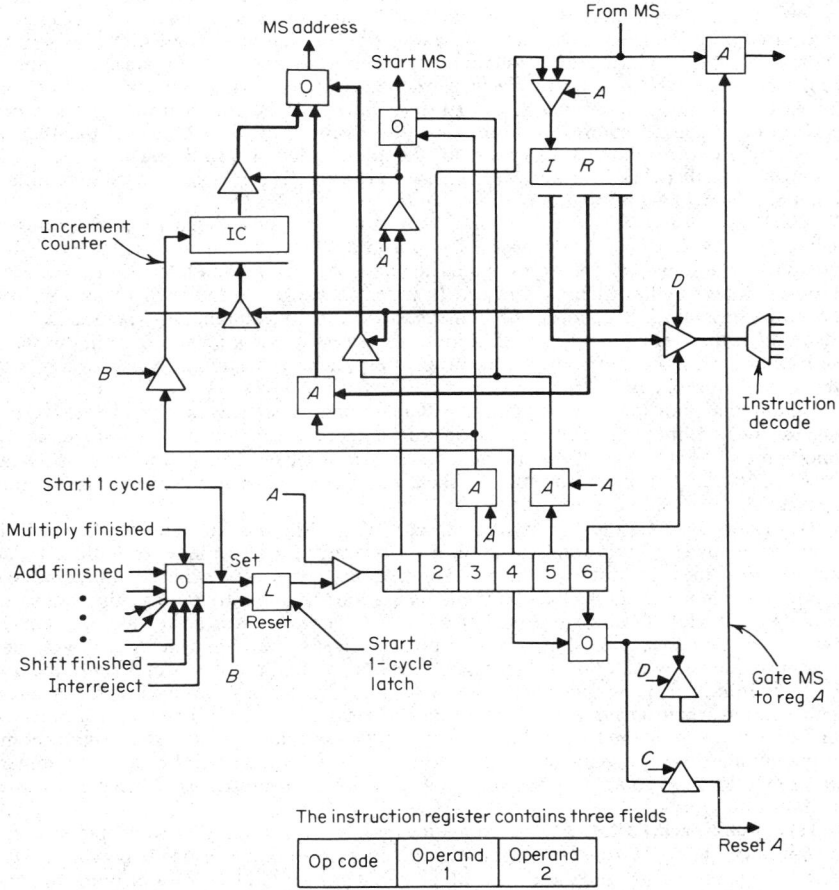

Fig. 23-100. Implementation of equipment for an I cycle in a two-address machine. The instruction is first brought from main store and deposited in the instruction register. The operation proceeds by successively gating the information associated with the two address fields into register A. The first is moved out of A while the second is being sought from main store.

108. Instruction and Execution Cycle Control. In normal program operation, I and E cycles alternate. The I cycle brings forth an instruction from storage, sets up the ALU for execution, resets the instruction counter for the next I cycle, and initiates an E cycle. A number of conditions serve to interrupt this orderly flow, as follows:

1. When no more work is to be done, the end of a program is reached and the computer goes to a WAIT state. Such a state is reached, for example, by a specific instruction that terminates operations at the end of the I cycle in question, by a signal from the instruction counter when a predetermined limit is reached.

2. A STOP button may prevent the next setting of the *Start I cycle* latch. A START button resets this latch.

3. When starting up after a shutdown, e.g., in the morning, activity is usually initiated by depressing an INITIAL PROGRAM LOAD button on the operator's console (the name varies from system, e.g., IPL, LOAD, START). The button usually performs three functions: a reset of all rings, latches, and registers to some predefined initial condition; a read-in operation from an I/O device into main store (usually the first few locations) of a short program; and an initiation of an I cycle. Program execution generally starts at a fixed location so that the IC initially is set to this value.

4. In multiprogramming, i.e., concurrent operations upon more than one program, only one program at a time is in operation in the CPU, but transfers occur from one to another as required by I/O accesses, etc. The program to be transferred is handled by an *interrupt*. Under interrupt, an address is forced into the instruction counter so that upon completion of the E cycle of the current program, a new instruction is referenced that starts the interruption. This instruction initiates program steps that store the data of the old program, e.g., in special registers or in special main-store locations. The contents of the IC, part of the IR, the contents of any registers, and the reason for the interruption are stored. The collection of these fields together is called a *program status word* (PSW). It can be referenced to reinitiate action at a later time on the program interrupted.

109. Branch Instructions. Two kinds of instructions permit change in program sequence: *conditional* and *unconditional.* The purpose of such instructions is to permit the system to make some decision, so as to alter the flow of program, and to continue execution at some point not in the original sequence. In systems that use two address fields as described above, the program instruction to be branched to is usually contained in operand 2. The execution of the branch instruction thus involves merely moving operand 2 from the IR to the IC. In nonconditional branches the original program instruction provides for a branch whenever the particular instruction occurs.

Conditional branches take the extra step of determining if some condition to be tested has been satisfied. Either the op code or the operand 1 field normally defines the test and/or the condition to be tested for. If the specified test has been satisfied, the branch is executed as described above. Otherwise no action is taken, and the next normally sequenced instruction is executed.

110. Low-Speed CPUs. To reduce the cost of a CPU, the width of the data path is frequently reduced. For example, a simple addition translates into a series of additions. On each iteration, the carry out of the high end of the adder on the $(N-1)$st iteration becomes the carry into the low end of the adder on the Nth iteration. In the extreme case, the data paths reduce to a single bit being processed at a time. When one bit is so treated, the storage register may be replaced by shift registers and all registers *revolve* in synchronism with one another. The gating is performed at, say, the low bit, and all registers shift toward the low end, end-around. Placing data in a register is performed by feeding the data into the high end as it rotates in synchronism with other registers. Delaying each bit by one bit cycle shifts the number left. Feeding the high end from the second least significant digit shifts right. Some older computers were based on this principle, and today small and special-purpose computers still use it. In these computers registers are ordinarily implemented as acoustic delay lines or MOS shift registers.

111. High-Speed CPUs. A fast computer may be built in several ways, but speed is usually gained by *parallel operation.* Two functions that can be performed in parallel are the I fetch of one instruction and the E cycle of the previous instruction. Most medium-size and large computers use this method of speedup. Also common is the speedup of the adder circuit by *carry look-ahead* (*carry by-pass* or *carry anticipation*), discussed previously in Par. **23-47.**

A common instruction, especially in scientific work, is multiply. This operation can be speeded up in two ways. In the first method, groups of 2 or 3 b in the multiplier are decoded simultaneously, and appropriate switching is developed between the multiplicand and the accumulator. If, for example, 2 b are decoded simultaneously and they are respectively 00, 01, 10, and 11 the circuit can (1) perform a double shift; (2) add the multiplier and shift twice; (3) shift, add the multiplier, and shift; or (4) perform the logical function on the multiplicand that corresponds to adding it to itself shifted one position, add the result, and shift left.

The second method takes advantage of the fact that in binary a group of three or more binary 1s corresponds to a number that is a power of 2, less 1. For example, binary 111 (decimal 7) can be replaced by $1000 - 1$ (decimal $8 - 1$), and similarly 1111 becomes $10000 - 1$, 11111 becomes $1000000 - 1$, etc. In multiplication any grouping of 3 b or more can be replaced by one subtraction of the multiplier from the accumulator and an addition into the accumulator after a shift of the appropriate number of spaces.

An effective but costly approach to increased speed is to design special processors for frequently used instructions. The instruction fetch cycle passes the operands to the appropriate special processor that operates independently to perform the instruction. At the same time, I fetch starts for the next instruction, and if this instruction does not use a result of the preceding instruction, it passes the second instruction together with its operands to some other processor. To gain maximum speed, some computers contain several processors that can handle the same instructions.

This approach can be carried to the point where 20 or 30 instructions are at various stages of execution at one time. This type of processing is called *pipeline processing*. No difficulties arise during uninterrupted processing. When an interrupt does occur, however, it is difficult to determine which instruction has caused the interrupt since the interrupt may arise in a subsystem sometime after the IC has initiated action. In the meantime, the IC may have started a number of subtasks elsewhere by stepping through subsequent cycles. Operands within subunits may not be saved in an arbitrary intermediate state, since information is in the process of being generated for return to the main program. Because of the requirement that no further I cycles be started, interrupt is signaled when the pipeline is empty. At the time the interrupt is signaled, the IC does not point at the instruction causing the interrupt but somewhat past it. This type of interrupt is called *imprecise*.

Multiprocessing (Par. **23-76**) is a further step in using parallelism. It permits multiple CPUs to operate out of one main store. In such a configuration, the programmer issues instructions that start other processors by providing them with the start of other instruction streams. One problem that plagues multiprocessors is that one processor must not modify what some other processor has in its registers or in storage.

112. Polish Notation. In the CPUs discussed thus far, the instructions store all results, so that the next time an operand is used it has to be fetched. For example, a program to add *A*, *B*, and *C* and put result into *E* appears as

MOVE A to E   1 fetch, 1 store
ADD E to B store in E   2 fetch, 1 store
ADD E to C store in E   2 fetch, 1 store

In languages such as PL/I or FORTRAN (Par. **23-123**) this program might be written as the single statement

$$E = A + B + C$$

This equation describes a sequence of actions, as in the case of the program, but the specific sequence is not described. Since addition is commutative, a correct result is achieved by $E = ((A + B) + C)$, $E = (A + (B + C))$, or $E = ((A + C) + B)$, each step occurring in any order. The computer, however, uses a specific program in achieving a result so that the method of writing the equation must generate a specific sequence of actions.

A method of writing an equation that specifies the order of operation is called *Polish notation*. For the above example of addition, one possible Polish string would be

$$AB + C + E =$$

In this string, the system would find A and B and, as determined by the plus sign *following* the two operands, add them. The result is then combined with *C* under addition called for by the second plus sign. The E = symbols indicate that the result is to be stored in E. The plus sign appears *after* the A and B, and the specific string shown is called *postfixed*. An equivalent convention could place the operator first and would be called *prefixed*.

Any complex expression can be translated into a Polish string. For example in PL/I language, the statement

$$M = (A + B) * (C + D * E) - F;$$

means evaluate the right-hand side of the equation and store the result in the main-store location corresponding to variable M (asterisks indicate multiplication). The Polish string translation for this statement is

$$AB + DE * C + * F - M =$$

In translation from the types of expression permitted by higher-level languages (Pars. 23-121 to 23-123) a machine can be programmed to analyze successively an arithmetic expression of the types shown above. In so doing, first, the outermost expressed or implied parentheses are aggregated and successively broken down until no more quantities remain. The first such quantities analyzed are generally the last computed, so that in the development of a Polish string from an algebraic expression, a first-in, last-out situation prevails.

113. Stacks. Evaluation of a Polish string in a machine is best performed using a *stack* (push-down list). A stack has the property that it holds numbers in order. A PUSH command places a value on the stack; i.e., it stores a number and an operation at the top of the stack and, in the process, lowers all previous items by one position. Numbers are retrieved from the stack by issuing a POP command. The number returned by the stack on a POP command is the most recently PUSHED one. The following example illustrates the behavior of a stack. The value in parentheses is the value *placed* on the stack for PUSH and returned by the stack for POP (assume the stack is initially empty):

PUSH (A)	stack contains	A
PUSH (B)	stack contains	B A
POP (B)	stack contains	A
PUSH (C)	stack contains	C A
POP (C)	stack contains	A
POP (A)	stack contains	nothing

Such a stack lends itself very well to the evaluation of Polish strings. The rules for evaluation are:

1. Scan the string from left to right.

2. If a variable (or constant) is encountered, fetch it from main store and place its value on the stack.

3. If an operator is encountered, POP the operands and PUSH the result.

4. Stop at the end of the string. If executed correctly, the stack is in the same state at the end of execution as it was at the start.

The advantage of using a stack is that intermediate results never need storing and therefore no intermediate variables are needed. In sequences of instructions where there are no branches, the operations can be stored in a stack. A program becomes a series of such stacks put together between branches.

This approach is called *stack processing*. In stack processing, a program consists of many Polish strings that are executed one after another. In some cases the entire program may be considered to be one long string.

Stacks are implemented by using a series of parallel shift registers, one per bit of the character code. The input is placed into the leftmost set of register positions. PUSH moves an entry to the right, and POP moves it to the left. The length of the shift registers is finite and fixed. The stack, however, usually must appear to the user as though it were infinitely deep. The stack is thus implemented so that the most active locations are in the shift register, and if the shift register overflows, the number shifted out at the right on a PUSH is placed in main storage. There the order is maintained by hardware, microprogramming, or a normal system program.

114. Parallel Processing. In the discussion of channel operations (Par. 23-76) it was pointed out that the problems of the interface between a CPU and I/O equipment may be solved by providing one or more channel processors that operate simultaneously. This concept of parallel processing leads to a machine organization different from those discussed in Pars. 23-111 to 23-113. Figure 23-101 shows a machine organization that is highly parallel.

A number of cellular elements are shown, each capable of storing a word of data or an instruction and also of performing simple arithmetic instructions on data within itself or in conjunction with data from other cells.

The cells can communicate with each other for program branches or data manipulation. For example, an instruction in one cell might specify that the program transfer to another cell for a branch. The second cell might specify that the contents of a third cell be added into the contents of a fourth. Shift instructions could involve data transfer between groups of cells. Such an organization can perform all the basic instructions provided by a more conventionally organized system.

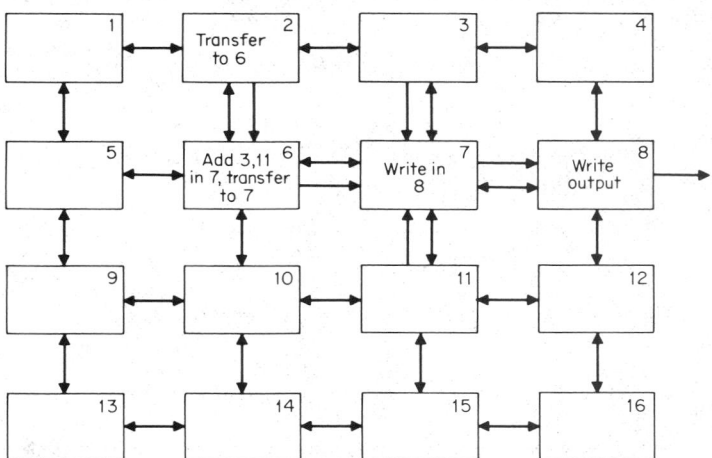

Fig. 23-101. Cellular computer organization, intended to use a large number of small low-speed elements that communicate with one another. In the case shown a single simple program is initiated and stepped through elements of the structure.

Another organization is shown in Fig. 23-102. A set of computer, memory, I/O, and operator modules are interconnected through a switching network. Except for the extra computer modules, such an organization is similar to the ones described in Par. 23-76 for channel organization.

The addition of separate computer modules adds considerable flexibility to program execution. For example, in the description of Polish strings and stack processing (Par. 23-113) it was stated that the purpose of the stack is to arrange the computation into a specific sequential order that produces a correct result on a single processor. Suppose that a Polish string is set up for the addition of 512 separate numbers. The string would be a program for the sequential addition of these numbers. But if 256 processors are available, the addition might proceed by first adding 256 pairs, then 128, and so forth, so that nine sequential steps could produce the sum. Of course, in a mixed expression of successive multiplications and additions, problems exist in which only one order produces a correct result. In such cases no possibility exists for parallel processing.

115. Lists, Stacks, and Associative Processing. When a program operates with an instruction counter, successive instructions must be placed contiguously in storage. Similarly for a program to operate serially upon a set of data, successive storage locations are required. Such sequential sets of data are called *lists*. Suppose, however, items on such a list require change; i.e., some items are to be inserted and others deleted. All entries below the point of insertion must be read and rewritten into the next lower storage location. To permit deletions, gaps of unused storage would have to be left or all items below the point of depletion moved up one position.

One alternative procedure is to assign ordered, but not necessarily sequential, address fields to each item on the list and insert them in an *associative store* in any order. An item can be inserted in the list by rearrangement of the address fields until a suitable gap is found, but the word itself can be physically located in any available location covered by the content addressing. By this means all physical storage locations are used with great freedom.

Another alternative organization is to provide each information storage location with a *tag*, to specify the next item in the list. By such a means neither program nor data lists need be contiguous in physical store. An item can be deleted or inserted in a list merely by modifying the tags of two items. The provision for such a tag is called *linking*. It can be used at the end of contiguous chains to specify the location of another chain of contiguous elements. The concept of linking can be expanded to include two dimensions under conditional control for branching or the specification of links in two or more directions in more complex programming structures. Using list structures in computation is sometimes called *list processing*.

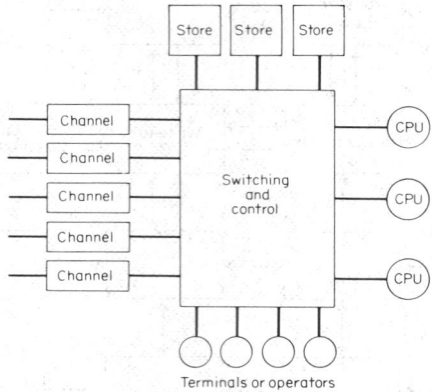

Fig. 23-102. The extended computer concept, a multiprocessing arrangement in which all elements can communicate with each other through the switching system. This type of organization has become popular because with suitable programming systems it has a high degree of inherent flexibility.

In some associative storage systems, the bits that tag words may not only be compared but manipulated until an appropriate match is found. By such a means, words of the same characteristic fields may be subjected to arithmetic operations that are useful in certain types of problem solutions. Units that have an associative feature and are capable of bit manipulation are often called *functional memories*.

116. Instruction Formats. The instruction format thus far described includes an op-code field and fields for two data addresses. Other formats have been used in different machine configurations, as follows.

One-Address System. The instruction includes an op-code field and only one address field. Only one access to storage is permitted with each operation so that, for example, to add two numbers an instruction must call for a number and place it in a suitable register. A second instruction is used to add a second number from storage, and yet a third to store the result. For a conditional branch, the address must specify the location of the alternate instruction; otherwise the original sequence would be maintained.

Two-Address Systems. In such machines, the two addresses most commonly specify the locations of data involved with the op code. For example, in an ADD instruction, the addresses specify the locations of the two numbers to be added. In another embodiment of the two-address format one field is used as in the one-address system and the other specifies the location of the next instruction in storage. In this case the program is arranged in a list.

Three-Address Machines. In a three-address machine, fields exist for three locations that specify two locations in which the operands are found and a location for storing a result. The first two addresses might also be used as operands, as in the two-address format, and the third for the next instruction specification in a list-processing type of operation.

COMPUTER PROGRAMMING

117. Alternative and Loop Programming Techniques. When a stored-program digital computer operates, its store contains two types of information: the data being processed and program instructions controlling operation. The ALU makes references to storage to gain access to data and modify it and to gain access to instructions and possibly to modify them. The conditional branch instruction provides the basis for two basic programming techniques: selection from *alternative processes* and the *program loop.* These techniques are illustrated here using a computer whose storage consists of 10,000 words, each word containing eight decimal digits (bytes) numbered 0 to 7 and sign. The instruction format is

S	1	2	3	4	5	6	7
Op code		0	0		Address		

Op code	Name	Description
+01	LOAD	Loads value from addressed word into data register
+02	COMP	Compares value of addressed word with data-register value
+03	ADD	Adds value of addressed word to data register
+08	STORE	Copies value of data register into addressed storage word
+20	BRLO	Branches if data-register value from last previously executed COMP was less than comparand

The computer used in this example contains a separate nonaddressable data register that contains one word of data. Further, each instruction is accessed at an address 1 more than that of the previously executed instruction unless that instruction was a BRLO instruction with a low COMP condition, in which case the address part of the BRLO instruction is the address at which the next instruction is to be accessed.

Consider the following program instructions (beginning at address 0100) to select the higher value of two items (in words 0950 and 0951), and place the selected value in a specific place (word 0800).

Address	Instruction	Effect
0100	+01000950	Place first item value in data register
0101	+02000951	Compare second-item value with data-register value
0102	+20000104	Branch to next instruction at address 0104 if data-register value was lower
0103	+01000951	Place second item value in data register
0104	+08000800	Store higher value in result (word 0800)

118. Flow Charts. A *flow chart* exhibits graphically the logical structure of a program. The program of the preceding example is depicted by the flow chart shown in Fig. 23-103.

The flow chart contains boxes representing processes (rectangular boxes) and decisions (diamond-shaped boxes). The arrows connecting the boxes represent the paths and

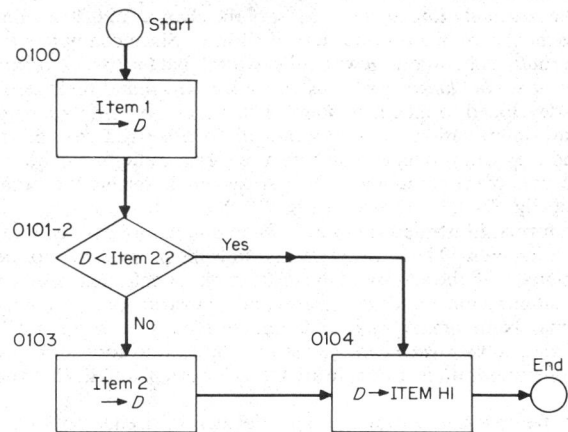

Fig. 23-103. Flow chart of a simple program. The boxes represent processes, the diamonds decisions.

sequences of instruction execution. A decision box represents an instruction (or a sequence of instructions) with more than one possible successor depending on the result of some processing test (this is commonly a conditional branch). In the example, instruction 0103 is or is not executed depending on the values of the two items.

If the example is extended to require finding the largest of four item values, the flow chart is that shown in Fig. 23-104. If the example is further extended to find the largest value of 1,000 items (in words 0950 through 1949 inclusive), the flow chart and the corresponding program become very large if analogous extensions of the flow charts are used.

The alternative is to use the technique known as the *program loop*. A program loop for this latter example is:

Address	Instruction	Effect
0100	+01000950	Move first item as initial value of result (ITEMHI)
0101	+08000800	
0102	+01000700	Initialize loop to begin with item 2
0103	+08000104	
0104	(+00000000)	Loop: Nth item to data register
0105	+02000800	Compare with prior ITEMHI value
0106	+20000108	Branch to 108 if Nth item value low
0107	+08000800	Store Nth item value as ITEMHI
0108	+01000104	Increment value, of N by 1
0109	+02000701	
0110	+08000104	
0111	+02000702	Compare against $N = 1,001$
0112	+20000104	Branch for looping if $N < 1,001$
0113	end	
0700	+01000951	Load item 2; initial instruction
0701	+00000001	Address increment of 1
0702	+01001050	Limit test; load 1,001st item

The corresponding flow chart appears in Fig. 23-105.

The corresponding flow chart appears in Fig. 23-105. The loop proper (instructions 0104 to 0112) is executed 999 times. The instruction at 0104 accesses the Nth item and is altered each time the program flows through the loop so that on successive executions successive words of the item table are obtained. After each loop execution, a test is made to determine if processing is complete or a branch should be made back to the beginning of the loop to repeat the loop program.

The loop proper is preceded by several instructions that *initialize* the loop, presetting ITEMHI and the instruction 0104 value for the first time through. A loop customarily has a *process* part, an *induction* part to make changes for the next loop iteration, and an *exit test* to determine if an additional iteration is required.

119. Symbolic Assembly Languages. Most of the changes in instructions resulting from storage rearrangements are of a routine clerical nature. Since computers are well suited to performing such routine operations, it was quite natural that the first automatic programming aids, the *symbolic assembly languages* (or *assembly languages*) and their associated assembly programs, were developed to take advantage of that fact. Assembly languages permit the critical addressing interrelations in a program to be described regardless of the storage arrangement, and they can produce therefrom a set of machine instructions suitable for the specific storage layout of the computer in use. An assembly-language program for the 1,000-value program of Fig. 23-105 is shown in Fig. 23-106.

The program format illustrated is typical. Each line has four parts: location, operation, operand(s), and comments. The location part permits the programmer to specify a symbolic name to be associated with the address of the instruction (or datum) defined on that line. The operation part contains a mnemonic designation of the instruction operation code. Alternatively, that line may be designated to be a datum constant, a reservation of data space, or a designation of an assembly *pseudo operation* (a specification to control the assembly process itself). Pseudo operations in the example are ORG for origin and END to designate the end of the program.

The operand field(s) give the additional information needed to specify the machine instruction, e.g., the name of a constant, the size of the data reservation, or a name associated with a pseudo operation. The Comment part serves for documentation only; it does not affect the assembly-program operation.

After a program is written in assembly language, it is entered as input data to its *associated assembly program,* either by direct keyboard entry or indirectly via punched cards, paper tape, or magnetic tape. The assembly program reads the symbolic assembly-language input and produces (1) a machine instruction program with constants, usually in a form convenient for subsequent program loading, and (2) an assembly listing that shows in typed or printed form each line of the symbolic assembly-language input, together with any associated machine instructions or constants produced therefrom.

The assembly pseudo operation ORG specifies that the instructions and/or constant entries for succeeding lines are to be prepared for loading at successive addresses, beginning at the specified load origin (value of operand field of ORG entry). Thus the 13 symbolic instructions following the initial ORG line in Fig. 23-106 are prepared for loading at addresses 0100 through 0112 inclusive, with the following symbolic associations established:

Location Symbol	(Local) Address
START	100
LOOPST	104
LOOPINC	108

Four instructions of this group of 13 contain the symbol LOOPST in the operand field, and the corresponding machine instructions will contain 0104 in their address parts.

The operation of a typical assembly program therefore consists of (1) collecting all location symbols and determining their values (addresses), called *building the symbol table,* and (2) building the machine instructions and/or constants by substituting op codes for the OP mnemonics and location symbol values for their positions in the operand field. The symbol

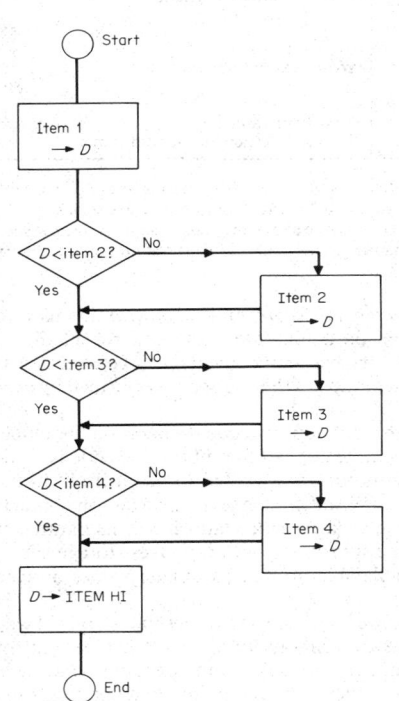

Fig. 23-104. Flow chart for a repetitive task.

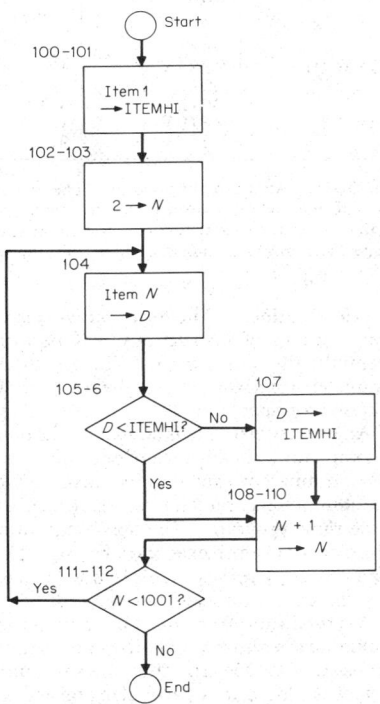

Fig. 23-105. Flow chart showing a program loop.

table must be formed first since, as the first instruction in the example shows, a machine instruction may refer to a location symbol that appears in the location field near the program end. Thus most assembly programs process the program twice; the *first pass* builds the symbol table, and the *second pass* builds the machine-language program. Note in the example the use of the operation RESRV to reserve space (skipping in the load-address sequence) for variable data.

Assembly language is specific to a particular computer instruction repertoire. Hence, the basic unit of assembly language describes a single machine instruction (so called one-for-one assembly process).

LOC	OP	Operand	Comment
	ORG	100	Start at ADDR 100
START	LOAD	ITEM 1	Move first item value to ITEMHI
	STORE	ITEM HI	
	LOAD	CONST 1	Set LOOP to start with second item
	STORE	LOOP ST	
LOOP ST	CONST		* Load *N*TH item
	COMP	ITEM HI	Compare with previous HI
	BRLO	LOOP INC	Skip if lower
	STORE	ITEM HI	Store on equal or high
LOOP INC	LOAD	LOOP ST	Increment *N* by 1
	ADD	ONE	
	STORE	LOOP ST	Modify storage
	COMP	CONST 2	Exit test
	BRLO	LOOP ST	Repeat if *N* < 1001
	. . .		(End of example)
	ORG	700	
CONST 1	LOAD	ITEM 1 + 1	Initial LOOP ST instruction
ONE	CONST	+ 1	Incrementation constant
CONST 2	LOAD	ITEM 1 + 1000	LOOP end test constant
	. . .		
	ORG	800	
ITEM HI	RESRV	1	Word where high value left
	. . .		
	ORG	950	
ITEM 1	RESRV	1	First item value
	RESRV	999	Second through thousandth items

Fig. 23-106. An assembly program. The program statements are in one-for-one correspondence with machine instructions. Hence the procedure is fully supplied by the programmer, according to the particular macrostruction set of the system. The assembly language alleviates housekeeping routines, such as specific storage assignments, and permits the use of man-oriented symbols instead of numeric or binary code.

120. Routines. The term *routine* (also *subroutine, procedure,* and *subprogram*) is used to refer a group of instructions that perform some particular function used repeatedly in essentially the same *context.* The quantities that vary between contexts may be regarded as parameters (or arguments) of the routine. The adaptation of the routine is performed by *open* or *closed* routines.

An open subroutine is adapted to its parameter values during code preparation (assembly or compilation) in advance of execution, and a separate copy of the subroutine code is made for each different execution context. An open subroutine is written to adapt itself during execution to its parameter values; hence, a single copy suffices for several execution contexts in the same program. The open subroutine executes faster since tailoring to its parameter values occurs before execution begins. The closed subroutine not only saves storage space, since one copy serves multiple uses, but is more flexible, in that parameter values derived from the execution itself can be used.

A closed subroutine must be written to determine its parameter values in a standard way (including the return point after completion). The conventions for finding the values and/or addresses of values are called the subroutine linage *conventions.* Quite commonly, a single address is placed in a particular register, and this address in turn points to a consecutively addressed list of addresses and/or values to be used. Subroutines commonly use (or *call*) other open subroutines, so that there are usually a number of levels of subroutine control available at any point during execution. That is, one routine is currently executing, and others are waiting at various points in partially executed condition.

121. High-Level Programming Languages. On general-purpose digital computers, *high-level programming languages* have largely superseded assembly languages as the predominant method of describing application programs. Such programming languages are said to be *high-level* and *machine-independent.* High-level means that each program function is such that several or many machine instructions must be executed to perform that function. Machine-independent means that the functions are intended to be applied to a wide range of machine-instruction repertoires and to produce for each a specific machine representation of data.

The high-level program is known as a *compiler,* i.e., a program that converts an input program written in a particular high-level language *(source program)* to the machine language of a particular machine type *(object program).*

The advantages of high-level languages, as opposed to assembly languages, are that it is possible to use the same program for execution on several different machine types with few if any changes; fewer errors are to be expected, and checkout is simpler because the higher-level functions require less detailed coding; programming is easier, faster, and more economical; programming skills in a particular language are transferable to another machine; and better programs result since compilers use specialized techniques that it is not economical for an assembly-language programmer to employ on an extensive basis.

122. High-Level Procedural Languages. Most of the high-level programming languages are said to be *procedural.* The programmer writing in a high-level procedural language thinks in terms of the precise sequence of operations, and the program description is in terms of sequentially executed *procedural statements.* Most high-level procedural languages have statements for *documentation, procedural execution, data declaration,* and various compiler and execution *control specifications.*

The program in Fig. 23-107, written in the FORTRAN high-level language (Par. **23-123**), describes the program function given in Fig. 23-106 in assembly language. The first two lines are for documentation only, as indicated by C in the first column. The DIMENSION statement defines ITEM to consist of 1,000 values. The assignment statement ITEMHI = ITEM (1) is read as "set the value of ITEMHI to the value of the first ITEM." The next statement is a loop-control statement meaning: "do the following statements through the statement labeled 1 for the variable *N* assuming every value from 2 through 1,000." The statement labeled 1 causes a test to be made to "see if the *N*th ITEM is greater than .GT. the value of ITEMHI, and if so, set the ITEMHI value equal to the value of the *N*th item."

123. FORTRAN. The high-level programming languages most commonly used in engineering and scientific computation are FORTRAN, ALGOL, and BASIC. FORTRAN, the first to appear, was developed during 1954 to 1957 by a group headed by Bachus*

* J. W. Bachus et al. The FORTRAN Automatic Coding System, *Proc. West. Joint Compu. Conf.,* 1957, Vol. II, p. 188.

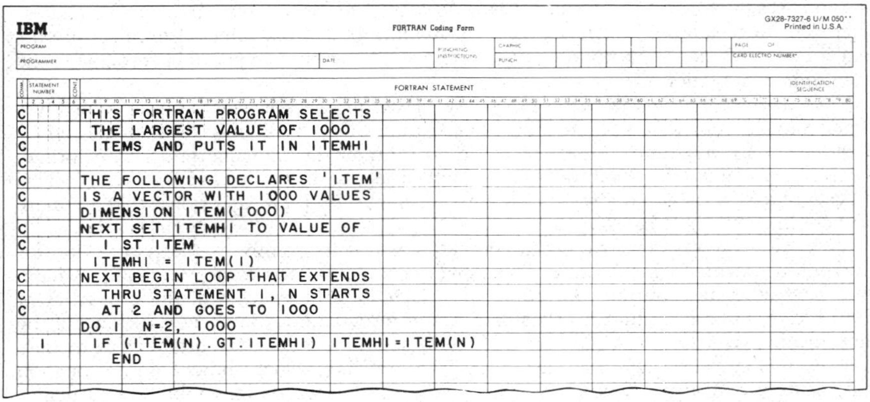
IBM FORTRAN Coding Form GX28-7327-6 U/M 050** Printed in U.S.A.

```
C    THIS FORTRAN PROGRAM SELECTS
C      THE LARGEST VALUE OF 1000
C      ITEMS AND PUTS IT IN ITEMHI
C
C    THE FOLLOWING DECLARES 'ITEM'
C    IS A VECTOR WITH 1000 VALUES
      DIMENSION ITEM(1000)
C    NEXT SET ITEMHI TO VALUE OF
C       1ST ITEM
      ITEMHI = ITEM(1)
C    NEXT BEGIN LOOP THAT EXTENDS
C       THRU STATEMENT 1, N STARTS
C       AT 2 AND GOES TO 1000
      DO 1  N=2, 1000
    1   IF (ITEM(N).GT.ITEMHI) ITEMHI=ITEM(N)
          END
```

Fig. 23-107. An example of a FORTRAN program, corresponding to the flow chart of Fig. 23-105 and the assembly program of Fig. 23-106.

of IBM. Based on algebraic notation, it allows two types of numbers: integers (positive and negative) and floating point. Variables are given character names of up to six positions. All variables beginning with the letters I, J, K, L, M, or N are integers; otherwise they are floating point. Integer constants are written in normal fashion, 1, 0, -4, etc. Floating-point constants must contain a decimal point, 3.1, -0.1, 2.0, 0.0, etc. For example, 6.02×10^{24} is written 6.02E24. This standard notation was adopted to accommodate the limited capability of computer input-output equipment.

READ and WRITE statements permit values of variables to be read into or written from the ALU, from or to input, output, or intermediate storage devices. The latter may operate merely by transcribing values or may be accompanied by conversion or editing specified in a separate FORMAT statement. Some idea of the range of operations provided in FORTRAN is shown by the following value-assignment statement:

$$\text{ROOT} = (-(B/2.0) + \text{SQRT}((B/2.0) **2 - A*C))/A$$

This is the formula for the root of a quadratic equation with coefficients A, B, and C. The asterisk indicates multiplication, / stands for division, and ** exponentiation.

The notation: *name (expression)* and *name (Expression, expression)* etc., is used in FORTRAN with two distinct meanings depending on whether or not the specific name appears in a DIMENSION statement. If so, the *expression*(s) are subscript values; otherwise the *name* is considered to be a function name, and the expressions are the values of the arguments of the function. SQRT$((B/2.0)**2-A*C)$ in the preceding assignment statement requires the expression $(B/2.0)**2-A*C$ to be evaluated, and then the function (square root here) of that value is determined. Square root and various other common trigonometric and logarithmic functions and their respective inverses are standardized in FORTRAN, typically as closed subroutines.

The same notation may be employed for a function defined by a FORTRAN programmer in the FORTRAN language. This operation is performed by writing a separate FORTRAN program headed by the statement:

FUNCTION name (arg 1, arg 2, etc.)

where arg represents the name that stands for the actual argument value at each evaluation of that function. Similarly, any action or set of actions described by a closed FORTRAN subroutine is called for by "CALL subroutines (args)" together with a defining FORTRAN subroutine headed by "SUBROUTINE subroutine name (args)."

A FORTRAN program, with all necessary subprograms (SUBROUTINE and FUNCTION programs), defines an execution process. However, it has not been customary to compile the entire executable program in one step. It is usual instead to compile each program or subprogram as a separate process into a *relocatable* form and to use a relocating loader (also called *linking loader, link editor,* etc.) to combine the relocatable forms to the absolute machine-language form. The relocatable form permits all symbolic-address references to be adjusted relative to a small number of address values determined by the loader in combining the appropriate cross-references between the relocatable programs, one from each compilation.

Two advantages follow from this procedure: (1) during checkout, each correction involves changes to only one (or a small number of) programs that must then be recompiled, and (2) if the assembly program is designed to produce output in the same relocatable format, particular functions and/or subroutines need not be necessarily written in FORTRAN. This procedure allows the use of processes not expressible in FORTRAN or ones that are more efficient in assembly language.

124. ALGOL was developed by a committee appointed jointly by the Association for Computing Machinery in the United States and its counterparts in France, Germany, Britain, and the Netherlands during 1957 to 1960. ALGOL includes most of the properties of FORTRAN but has a number of features new to programming languages. Probably the most distinctive feature of ALGOL is its *block* structure. An ALGOL program, as written, consists of blocks of contiguous statements. Two contiguous blocks either have no common statements or one block is wholly contained in the other. Declarations apply to a particular block (and to any blocks nested within it). Names of variables, statements, and procedures are declared (defined) and have that declared meaning only for statements within that block.

The Revised Report on the Algorithmic Language (ALGOL 60*) presents a new method of specifying the syntax of ALGOL, introducing a formal notation to eliminate ambiguities, inconsistencies, or misunderstandings. Although not part of ALGOL proper, the use of such a description of ALGOL syntax serves both to demonstrate the underlying unity of the ALGOL language and to show the usefulness of such descriptive methods in dealing with the syntax of programming languages. ALGOL is more widely used in Europe than the United States, probably because FORTRAN had become firmly established in the United States at an early date.

125. BASIC is a high-level programming language based on algebraic notation that was developed for solving problems at a terminal; it is particularly suitable for short programs and instructional purposes. The user normally remains at the terminal after entering his program in BASIC, while it compiles, executes, and types the output, a process that typically requires only a few seconds.

BASIC is similar to FORTRAN. Many of the important differences relate to the mode of usage, i.e., immediate compilation and execution, with the terminal serving as the sole input-output device. Hence subroutines are treated differently: BASIC compiles a single program to produce a self-contained, complete object program. Every statement in BASIC is numbered, and the statements are arranged in ascending order, before compilation. This feature facilitates program correction during checkout. For example, if more than one statement is entered with the same number, only the last is retained. The other statement is deleted by entering a statement number only (a null statement replaces the statement being deleted). Statements are inserted by using a statement number between the two statements at which insertion is to be made. Since statement numbers need not be consecutive (in ascending order only) it is good programming practice in BASIC to leave gaps in the initial statement numbering to allow for later insertion.

126. High-Level Commercial Languages. High-level programming languages used for business data processing applications emphasize description and handling of files for business record keeping. Two widely used programming languages for business applications are COBOL (*common business-oriented language*) and RPG (*report program generation*). Compilers for these languages, with generalized sorting programs, form the fundamental automatic programming aids of many computer installations primarily used for business data processing. COBOL and RPG have comparable file, record, and field-within-record descriptive capabilities, but the specification of processing and sequence control derive from basically different concepts.

COBOL is a procedure-oriented language like the scientific programming languages FORTRAN and BASIC. A COBOL program has a *procedure divisor* part, in which executable statements control not only computation and summarization but also input, output, loop, and branching (GO TO) statements. Thus the COBOL programmer specifies the precise execution sequence, and the COBOL language in the *procedure division* reflects the notion of what might be called a procedure statement instruction counter.

RPG reflects the concepts and uses the terminology of punched-card accounting-machine methods. Four different forms are used to describe different aspects of the application: the files involved, the fields of each input record by file, the fields of working storage and how their contents are established, and the output record (and its editing) in terms of which input and/or working storage fields are to be copied.

After an RPG programmer has completed the forms describing the application, their content is entered on punched cards to an RPG compiler. This compiles an object program to perform the programmed task.

COBOL was developed by the Conference on Data Systems Languages, beginning in 1959. Besides a procedure division with GO TO, IF, and loop control statements (analogous to FORTRAN or ALGOL), COBOL has an *environment division* and a *data division* which specifies fields in input records, output records, and working storage.

127. Sorting Programs. Programs that accept parameters describing an input sequential file to be sorted are important in commercial applications. Parameters for fixed-length record files include not only file-descriptive information, e.g., record length and blocking, but also the positions within each record of the sorting keys. For example, to sort by man

---

* Peter Naur (ed.)   Revised Report on the Algorithmic Language ALGOL 60, *Commun. Assoc. Comput. Mach.,* 1963, Vol. 6, p. 1.

number within department number, two keys are used in each record. The major key is the department number, and the second key is the man number.

While there is some variation in the details of sorting methods, the following steps are typical of sort programs:

1. Rearrange the records so that keys are in standard position, ordered major to minor (usually at the beginning of each record).

2. Read in sufficient records to fill main storage. Sort them in sequence. Write them out in order.

3. Repeat step 2 until the original input file has been entirely processed. Merge the passes, using $2N$ intermediate files where $N \geq 2$. (We assume eight tape drives are available for intermediate files.)

4. Arrange the output of step 3 with one-quarter of sequences on each of four files.

5. Read the first record of each of four sequences from four separate input tapes.

6. Select the lowest of the first records from four current input blocks for output.

7. Repeat step 6, selecting the lowest unprocessed record from four input blocks.

8. Repeat. If any input block is exhausted, read another from the same input drive. These steps will form a single merged sequence of records with four times as many records as existed in each input sequence. The entire output sequence is written on one output drive.

9. Repeat steps 6 to 8, building additional output sequences except that each succeeding sequence goes on another output drive, on a round-robin basis.

10. Repeat steps 6 to 9 using the output drives as input and vice versa. Each repetition creates an output sequence 4 times longer than the preceding merge pass.

11. When only one sequence exists, rewrite file with keys rearranged in their original position.

In practice, the steps are combined as follows: *phase I* (executed once) combines steps 1 to 4 in a single pass over the file. *Phase II* (executed a number of times until the number of sequences n is such that $2 \leq n \leq N$) performs merge passes comprising steps 6 to 10, and *phase III* is a final merge pass, executed once, combining steps 6 to 11.

128. Checkpoint-Restart. In commercial business records, it is important to be able to reconstruct data that have been destroyed accidentally (tape breakage, fire, etc.). One common technique used for periodically updated sequential files is *checkpoint-restart.* Copies of the files for succeeding periods are called successive *generations* of the file. Several generations of a file and their associated input *transaction files* are customarily retained.

For lengthy files, it may be too time-consuming to rerun a file generation and its associated transactions entirely to regenerate the entire output file. In such cases, during processing a *checkpoint* is taken periodically (say once per output reel or every fifty-thousandth record). A checkpoint consists of writing the entire contents of main storage on a medium known as a *checkpoint record.* The main storage must contain sufficient information to indicate the position of every input file, and this information must be in some standard location. A checkpoint represents a point in processing where processing could be restarted using a *restart program.* The restart program reads the checkpoint information, repositions the original input files which must have been made accessible, restores the contents of main storage, and branches back into storage to resume processing at the restart point. This operation may be performed days, weeks, or months after the original processing.

129. Control Programs. To minimize idle computer time between machine runs, programs variously called *control programs, monitors,* or *executive programs* have come into wide use. In its simplest form, the basic input for a succession of jobs is placed on a single sequential input file and each basic input consists of control information for the control program and optional programs and/or data input. A control program remains continuously in overall control. It calls each job and regains control at successful completion or after error conditions. In the latter case, selected storage information is produced for off-line analysis. Any unprocessed input information is skipped over, and processing of the next job begins. If files must be mounted and/or removed between jobs executed, this requirement must be communicated via the control program to the machine operator, who acknowledges completion of such acts to the control program in some standardized way (machine restart, input message, etc.). Only one job is handled at a time under such a control program, and each job goes to successful and/or unsuccessful conclusion before another job is started.

Electromechanical devices, such as card readers, punches, and line printers, are sufficiently slow compared to large computer internal speeds to constitute a potential bottleneck under the type of control program just described.   Instead such peripheral equipment can be attached to small computers *(off-line systems)* which perform transcription to or from tapes or disk packs.

130. Multiprogramming.   Control programs can direct a central computer system to divide its time among several jobs, the control going back and forth intermittently from one job to another.   This technique is called *multiprogramming.*   Successful multiprogramming depends on hardware that prevents one job from destroying another, by modifying the respective programs or data.   Multiprogramming permits the use of distinct peripheral and processing programs, and it may also allow a computer to handle intermittently used telecommunication programs and on-line terminal programs, e.g., those using BASIC.

Control programs for multiprogramming are probably the most complex programs in use. They depend heavily on hardware interrupts, absolute control by the control program of input-output handling, and hardware-software combination techniques involving storage probing and lockout and/or address relocation to prevent mutual interference.   Some control programs reallocate storage and/or channels and I/O equipment dynamically.   Such resource management can greatly improve performance in certain operating environments.

A key aspect of *resource allocation* is the manner in which the ALU time is allotted first to one job then another.   Priorities may be fixed or varied slowly.   Alternatively, particularly where on-line terminal systems are used with operators waiting directly for computer response, control may be shifted regularly on a short-interval basis (in fractions of a second) from one job to another.

131. Bibliography

Digital Computer Fundamentals

1. BARTEE, T. C.   "Digital Computer Fundamentals," 2d ed., McGraw-Hill, New York, 1966.

2. GSCHWIND, H. W.   "Design of Digital Computers," Springer-Verlag, New York, 1967.

3. CHU, Y.   "Introduction to Computer Organization," Prentice-Hall, Englewood Cliffs, N.J., 1970.

4. MARSEL, H.   "Introduction to Electronic Digital Computers," McGraw-Hill, New York, 1969.

5. FLORES, I.   "Computer Organization," Prentice-Hall, Englewood Cliffs, N.J., 1969.

6. HELLERMAN, H.   "Digital Computer System Principles," McGraw-Hill, New York, 1967.

7. WARE, W. H.   "Digital Computer Technology and Design," Vols. I and II, Wiley, New York, 1963.

8. RICHARDS, R. K.   "Electronic Digital Systems," Wiley, New York, 1966.

9. MALEY, G. A., and E. J. SKIKO   "Modern Digital Computers," Prentice-Hall, Englewood Cliffs, N.J., 1964.

10. RICE, J. K., and J. R. RICE   "Introduction to Computer Science: Algorithms, Languages, and Information," prelim. ed., Rinehart and Winston, New York, 1967.

11. ARDEN, B. W.   "An Introduction to Digital Computing," Addison-Wesley, Reading, Mass., 1963.

12. HULL, T. E.   "Introduction to Computing," Prentice-Hall, Englewood Cliffs, N.J., 1966.

13. BROOKS, F. P., JR., and K. E. IVERSON   "Automatic Data Processing," Wiley, New York, 1963.

14. DAVIS, G. B.   "An Introduction to Electronic Computers," McGraw-Hill, New York, 1965.

Analog Computers

15. TRUITT, T. D., and A. E. ROGERS   "Basics of Analog Computers," Rider, New York, 1960.

16. JOHNSON, C. L.   "Analog Computer Techniques," McGraw-Hill, New York, 1956.

17. KORN, G. A., and T. M. KORN   "Electronic Analog and Hybrid Computers," 2d ed., McGraw-Hill, New York, 1972.

18. SOROKA, W. W.   "Analog Methods in Computation and Simulation," McGraw-Hill, New York, 1954.

19. KARPLUS, W. J., and W. W. SOROKA   "Analog Methods: Computation and Simulation," 2d ed., McGraw-Hill, New York, 1959.

20. ASHLEY, J. T. "Introduction to Analog Computation," Wiley, New York, 1963.
21. FIFER, S. "Analog Computation," Vols. I to IV, McGraw-Hill, New York, 1961.
22. JACKSON, A. S. "Analog Computation," McGraw-Hill, New York, 1960.
23. LEVINE, L. "Methods for Solving Engineering Problems Using Analog Computers," McGraw-Hill, New York, 1964.

Numerical Analysis
24. MACON, N. "Numerical Analysis," Wiley, New York, 1963.
25. MILNE, W. E. "Numerical Calculus," Princeton University Press, Princeton, N.J., 1949.
26. CONTE, S. "Elementary Numerical Analysis: An Algorithmic Approach," McGraw-Hill, New York, 1965.
27. JENNINGS, W. "First Course in Numerical Methods," Macmillan, New York, 1964.
28. STIEFEL, E. L. "An Introduction to Numerical Mathematics," trans. by W. C. Rheinbaldt and C. J. Rheinbaldt, Academic, New York, 1963.
29. McCORMICK, J. M., and M. G. SALVADORIE "Numerical Methods in FORTRAN," Prentice-Hall, Englewood Cliffs, N.J., 1964.
30. PENNINGTON, R. H. "Introductory Methods and Numerical Analysis," Macmillan, New York, 1965.
31. HILDEBRAND, F. B. "Introduction to Numerical Analysis," McGraw-Hill, New York, 1956.
32. WILKINSON, J. H. "Rounding Errors in Algebraic Processes," Prentice-Hall, Englewood Cliffs, N.J., 1964.
33. RALSTON, A. A. "A First Course in Numerical Analysis," McGraw-Hill, New York, 1965.
34. FORSYTHE, G. E., and W. R. WASOW "Finite Difference Methods for Partial Differential Equations," Wiley, New York, 1960.
35. HAMMING, R. W. "Numerical Methods for Scientists and Engineers," McGraw-Hill, New York, 1962.
36. HOUSEHOLDER, A. S. "Principles of Numerical Analysis," McGraw-Hill, New York, 1964.
37. FRÖBERG, C. E. "Introduction to Numerical Analysis," Addison-Wesley, Reading, Mass., 1965.
38. HORN, F. "Applied Boolean Algebra," 2d ed., Macmillan, New York, 1966.
39. KORFHAGE, R. "Logic and Algorithms with Applications to the Computer and Informational Sciences," Wiley, New York, 1966.
40. BIRKHOFF, G., and S. MacLANE "A Survey of Modern Algebra," Macmillan, New York, 1953.
41. MacLANE, S., and G. BIRKHOFF "Algebra," Macmillan, New York, 1967.
42. WHITESITT, J. E. "Boolean Algebra and Its Applications," Addison-Wesley, Reading, Mass., 1961.
43. MARCUS, M. P. "Switching Circuits for Engineers," Prentice-Hall, Englewood Cliffs, N.J., 1962.
44. CALDWELL, S. H. "Switching Circuits and Logical Design," Wiley, New York, 1958.
45. PHISTER, M., JR. "Logical Design of Digital Computers," Wiley, New York, 1958.
46. MALEY, G. A., and J. EARLE "The Logic Design of Transistor Digital Computers," Prentice-Hall, Englewood Cliffs, N.J., 1963.
47. McCLUSKEY, E. J., JR. "Introduction to the Theory of Switching Circuits," McGraw-Hill, New York, 1965.
48. HUMPHREY, W. S., JR. "Switching Circuits with Computer Applications," McGraw-Hill, New York, 1958.
49. HARRISON, M. A. "Introduction to Switching and Automata Theory," McGraw-Hill, New York, 1965.
50. KREIGER, M. "Basic Switching Circuit Theory," Macmillan, New York, 1967.
51. MILLER, R. E. "Switching Theory," Vol. 1, "Combinatorial Circuits," Wiley, New York, 1965.
52. PRATHER, R. E. "Introduction to Switching Theory: A Mathematical Approach," Allyn & Bacon, Boston, Mass., 1967.
53. WARFIELD, J. N. "Principles of Logic Design," Ginn, Boston, 1965.

54. DERTOUZOS, M. L. "Threshold Logic: A Synthesis Approach," M.I.T. Press, Cambridge, Mass., 1965.

55. HU, S. T. "Threshold Logic," University of California Press, Berkeley, Calif., 1965.

56. LEWIS, P. M., and C. L. COATES "Threshold Logic," Wiley, New York, 1967.

57. PHISTER, M., JR. "Logical Design of Digital Computers," Wiley, New York, 1958.

58. RICHARDS, R. K. "Arithmetic Operations in Digital Computers," Van Nostrand, Princeton, N.J., 1965.

Circuit Implementation

59. HURLEY, R. B. "Transistor Logic Circuits," Wiley, New York, 1961.

60. SHEA, R. F. "Transistor Circuit Engineering," Wiley, New York, 1957.

61. MOTOROLA, INC. "Integrated Circuits: Design Principles and Fabrication," McGraw-Hill, New York, 1965.

62. GRAY, H. J. "Digital Computer Engineering," Prentice-Hall, Englewood Cliffs, N.J., 1963.

63. LYNN, D. K., C. S. MEYER, and D. J. HAMILTON "Analysis and Design of Integrated Circuits," McGraw-Hill, New York, 1967.

64. XLANDER, M. "Fundamentals of Reliable Circuit Design," Vols. I and II, ILIFFE Books, London, 1966.

65. MOTOROLA, INC. "Analysis and Design of Integrated Circuits," McGraw-Hill, New York, 1967.

66. KEONJIAN, E. "Micro-electronics: Theory, Design, and Fabrication," McGraw-Hill, New York, 1963.

Optical Computing

67. TIPPETT, J. T., D. A. BERKOWITZ, L. C. CLAPP, C. J. KOESTER, and A. VANDERBURGH, JR. "Optical and Electro-Optical Information Processing," M.I.T. Press, Cambridge, Mass., 1965.

Information Retrieval

68. BECKER, J., and R. M. HAYES "Introduction to Information Storage and Retrieval: Tools, Element, Tolerances," Wiley, New York, 1963.

69. SALTON, G. "Automatic Information Organization and Retrieval," McGraw-Hill, New York, 1968.

70. CHADRA, C. (ed.) *Annu. Rev. Inf. Sci. Technol.*, Interscience, New York, 1966, Vol. 1, and 1967, Vol. 2.

71. WEGNER, P. "Programming Languages, Information Structures and Machine Organization," McGraw-Hill, New York, 1968.

Computer Graphics

72. FETTER, W. A. "Computer Graphics in Communicating," McGraw-Hill, New York, 1965.

73. GRUENBERGER, F. (ed.) "Computer Graphics: Utility/Production/Art," Thompson, Washington, D.C., 1967.

Computer Organization

74. BUCHHOLZ, W. (ed.) "Planning a Computer System," McGraw-Hill, New York, 1962.

75. GRAY, H. J. "Digital Computer Engineering," McGraw-Hill, New York, 1963.

76. RICHARDS, R. K. "Electronic Digital Systems," Wiley, New York, 1966.

77. DESMONDE, W. H. "Computers and Their Uses," Prentice-Hall, Englewood Cliffs, N.J., 1964.

78. LADEN, H. N., and T. R. GILDERSLEEVE "System Design for Computer Application," Wiley, New York, 1963.

79. McCARTHY, E. J., J. McCARTHY, and D. HUMES "Integrated Data Processing Systems," Wiley, New York, 1959.

80. ELLIOTT, C. O., and R. S. WASLEY "Business Information Processing Systems," Erwin, Homewood, Ill., 1965.

81. FEIGENBAUM, E. A., and J. FELDMAN (eds.) "Computers and Thought," McGraw-Hill, New York, 1966.

82. UHR, L. (ed.) "Pattern Recognition," Wiley, New York, 1966.

83. ARBIB, M. "Brains, Machines, and Mathematics," McGraw-Hill, New York, 1964.

84. NILSSON, N. J. "Learning Machines," McGraw-Hill, New York, 1965.

85. FOGEL, L. J., A. J. OWENS, and M. J. WALSH "Artificial Intelligence through Simulated Evolution," Wiley, New York, 1966.

86. YOUNG, J. Z. "A Model of the Brain," Clarendon Press, Oxford, 1964.

Computer Theory

87. GILL, A. "Introduction to the Theory of Finite-State Machines," McGraw-Hill, New York, 1962.

88. GINSBURG, S. "An Introduction to Mathematical Machine Theory," Addison-Wesley, Reading, Mass., 1962.

89. CAIANELLO, E. R. (ed.) "Automatic Theory," Academic, New York, 1966.

90. GLUSHKOV, V. M. "Introduction to Cybernetics," Academic, New York, 1966.

91. SHANNON, C. E., and J. McCARTHY (eds.) "Automatic Studies," Princeton University Press, Princeton, N.J., 1956.

92. MOORE, E. F. (ed.) "Sequential Machines: Selected Papers," Addison-Wesley, Reading, Mass., 1964.

93. MINSKY, M. "Computation: Finite and Infinite Machines," Prentice-Hall, Englewood Cliffs, N.J., 1967.

94. HENNIE, F. C., III "Iterative Arrays of Logic Circuits," M.I.T. Press, Cambridge, Mass., and Wiley, New York, 1961.

95. HARTMANIS, J., and R. E. STEARNS "Algebraic Structure of Sequential Machines," Prentice-Hall, Englewood Cliffs, N.J., 1966.

Computer Programming

97. IVERSON, K. E. "A Programming Language," Wiley, New York, 1962.

98. FISCHER, F. P., and G. F. SWINDLE "Computer Programming Systems," Holt, Rinehart and Winston, New York, 1964.

99. FLORES, I. "Computer Programming," Prentice-Hall, Englewood Cliffs, N.J., 1966.

100. HASSITT, A. "Computer Programming and Computer Systems," Academic, New York, 1967.

101. STARK, P. A. "Digital Computer Programming," Macmillan, New York, 1967.

102. STEIN, M. L., and W. D. MUNRO "Computer Programming: A Mixed Language Approach," Academic, New York, 1964.

103. ROSEN, S. "Programming Systems and Languages," McGraw-Hill, New York, 1967.

Microprogramming

104. HUSSON, S. S. "Microprogramming Principles and Practices," Prentice-Hall, Englewood Cliffs, N.J., 1970.

Programming Languages

105. GOODMAN, R. (ed.) *Annu. Rev. Autom. Program.,* 1960–1964, Vols. 1 to 4.

106. HALSTEAD, M. H. "Machine-independent Computer Programming," Spartan, New York, 1962.

107. AMERICAN NATIONAL STANDARDS INSTITUTE "Standards X39—1966 FORTRAN, and X3.10—1966, Basic FORTRAN," American National Standards Institute, Inc., 1430 Broadway, New York, 1966.

108. INTERNATIONAL COMPUTATION CENTRE "Symbolic Languages in Data Processing," *Proc. Symp. Rome, Mar. 26–31, 1962,* Gordon and Breach, New York, 1962.

109. WEGNER, P. (ed.) "Introduction to System Programming," Academic, New York, 1965.

110. FLORES, I. "Computer Software," Prentice-Hall, Englewood Cliffs, N.J., 1965.

Advanced Programming

111. DESMONDE, W. H. "Real-Time Data Processing Systems: Introductory Concepts," Prentice-Hall, Englewood Cliffs, N.J., 1964.

112. CHOROFAS, D. N. "Programming Systems for Electronic Computers," Butterworth, London, 1962.

113. MARTIN, J. "Design of Real-Time Computer Systems," Prentice-Hall, Englewood Cliffs, N.J., 1967.

114. ERDWINN, J. D. (chairman) "Executive Control Programs, Sess. 8," *Proc. AFIPS 1967 Fall Joint Comp. Conf.,* Vol. 31, Thompson, Washington, D.C., 1967.

115. M.I.T. COMPUTATION CENTER "Compatible Time-sharing System: A Programmer's Guide," 2d ed., M.I.T. Press, Cambridge, Mass., 1967.

116. GINSBURG, S.   "The Mathematical Theory of Contest-free Languages," McGraw-Hill, New York, 1966.

117. FEIGENBAUM, E. A., and J. FELDMAN (eds.)   "Computers and Thought," McGraw-Hill, New York, 1966.

118. BAR-HILLEL, Y.   "Language and Information: Selected Essays on Their Theory and Application," Addison-Wesley, Reading, Mass., 1964.

119. CHOMSKY, N.   "Aspects of the Theory of Syntax," M.I.T. Press, Cambridge, Mass., 1965.

120. POLYA, G.   "Mathematics and Plausible Reasoning," Vol. I, "Induction and Analogy in Mathematics," Vol. II, "Patterns of Plausible Inference," Princeton University Press, Princeton, N.J., 1954.

Applications

121. BEMER, R. W.   "Computers and Crisis," ACM 70, Association for Computing Machinery, New York, 1971.

122. UTSUMI, TAKESHI (chairman)   *Summer Comput. Simulation Conf., Boston, 1971,* Board of Simulation Conferences, 5975 Broadway, Denver, Colo., 1971.

123. MEACHAM, A. D., and V. B. THOMPSON   "Computer Applications Service," Vols. 1-8, American Data Processing, Inc., 2200 Book Tower, Detroit, Mich., 1962-1971.

124. KLERER, M., and G. A. KORN   "Digital Computer User's Handbook," McGraw-Hill, New York, 1967.

125. GILBERT, C. P.   "The Design and Use of Electronic Analogue Computers," Chapman and Hall, London, 1964.

126. TAU, J. T.   "Computer and Information Sciences," Vol. II, Academic, New York, 1967.

127. LEVIN, H. S.   "Office Work and Automation," Wiley, New York, 1956.

128. BRANDON, D. H.   "Management Standards for Data Processing," Van Nostrand, Princeton, N.J., 1963.

129. SCHMIDT, R. N., and W. E. MEYERS   "Electronic Business Data Processing," Holt, Rinehart and Winston, New York, 1963.

130. GREGORY, R. H., and R. L. VAN HORN   "Business Data Processing and Programming," Woodsworth, Belmont, Calif., 1963.

131. DESMONDE, W. H.   "Real-Time Data Processing Systems," Prentice-Hall, Englewood Cliffs, N.J., 1964.

132. GRUENBERGER, F. (ed.)   "Computers and Communications: Toward a Computer Utility," Prentice-Hall, Englewood Cliffs, N.J., 1968.

133. SPENCER, D. G.   "Game Playing with Computers," Spartan, New York, 1968.

134. WYMAN, F. P.   "Simulation Modeling: A Guide Using Simscript," Wiley, New York, 1970.

135. MOON, B. A. M.   "Computer Programming for Science and Engineering," Van Nostrand, Princeton, N.J., 1966.

136. MARTIN, F. F.   "Computer Modeling and Simulation," Wiley, New York, 1968.

137. SUSSKIND, A. K. (ed.)   "Notes on Analog-Digital Conversion Techniques," M.I.T. Press, Cambridge, Mass., and Wiley, New York, 1960.

138. BROTTON, D. M.   "The Application of Digital Computers to Structural Engineering Problems," Spon, London, 1962.

139. SATTINGER, I. J.   "Applying Computers," Allen, Cleveland, Ohio, 1963.

140. SACKMAN, H.   "Man-Computer Problem Solving," Auerbach, Princeton, N.J., 1970.

141. TAVISS, I. (ed.)   "The Computer Impact," Prentice-Hall, Englewood, Cliffs, N.J., 1970.

142. FARINA, M. V.   "Flowcharting," Prentice-Hall, Englewood Cliffs, N.J., 1970.

143. FREIBERGER, W. F., and W. PRAGER   "Applications of Digital Computers," Ginn, Boston, 1963.

SECTION 24

ELECTRONICS IN PROCESSING INDUSTRIES

BY

C. A. McKAY Corporate Director of Research Development and Engineering, The Foxboro Company; Member, Institute of Electrical and Electronics Engineers

P. D. HANSEN Senior Technical Consultant, The Foxboro Company

D. A. RICHARDSON Development Project Engineer, The Foxboro Company

R. A. WILLIAMSON, JR. Senior Research Engineer, The Foxboro Company

G. A. ROSICA Manager of Electronic Development & Engineering, The Foxboro Company; Member, Institute of Electrical and Electronics Engineers

W. CALDER Chairman, Corporate Products Safety Committee, The Foxboro Company

J. L. LEE Manager, Program Management Services, The Foxboro Company

CONTENTS

Numbers refer to paragraphs

SECTION 24

ELECTRONICS IN PROCESSING INDUSTRIES

OVERVIEW

By C. A. McKay

1. General. The purpose of this section is to discuss requirements of electronic devices as applied in process industries, in particular those factors that impose design constraints different from other areas of application.

The process industries include chemical, petroleum, power, food, textile, paper, metallurgical, among others. In general, these are industries that continuously or semicontinuously process gases, liquids, or solids.

Electronic devices are used for measuring, indicating, recording, and controlling the flow, pressure, temperature, level, and composition of these process materials.

Applications for measurement and control of process variables range from the indication and/or regulation of a single process variable to the optimization of hundreds of variables in an entire plant.

The trend is increasingly one of process *management*, in which the objective is to provide the maximum profit per unit time of operation. This emphasis demands larger aggregates of equipment assembled in *systems* configurations in order to:

a. Monitor more data simultaneously.

b. Provide more efficient control of interactive variables.

c. Present more information "by exception" (see Par. 24-22).

d. Ensure a high level of availability of the process.

e. Allow for lower-cost expansion both "vertically" (compatibility with higher-level computing equipment) and "horizontally" (system-size expansion).

f. Facilitate the entry and change of more exogenous data.

In response to these demands, process control equipment has evolved continually over the last thirty years. The rapid advances in electronic component technology and design techniques have facilitated innovative development of signal processing equipment. Control system accuracy is now paced by the basic measurement devices. Some present trends that will probably continue for process control systems include:

1. Control room consolidation. Centralizing the control room functions provides the opportunity for coordinated control strategies (it also increases the necessity for remote communications capability).

2. Increased use of process control computers.

3. Continued standardization of field signal levels.

4. Increasing emphasis on reliability of system performance.

5. Increasing requirements for safety by design. These include safety of operating personnel as well as safe operation in hazardous areas.

6. The demand for more flexibility of system configurations. This enables the user to experiment and reconfigure his control system as he learns more about his process.

Electronic process control systems tend to be designed as one-of-a-kind applications. This occurs because:

a. The process-dynamics determining factors vary widely between industries and among processes within industries (e.g., vessel capacities, transportation lags, etc.).

b. The relative importance of system features depends on the individual user (e.g., which is more important: control enforcement or continuous operator information, equipment reliability or system cost, etc.).

c. The environmental requirements vary dramatically.

Although different control strategies emphasize different performance characteristics of the system, the functions that are normally required to be performed by the control system include:

   a. Process data acquisition.

   b. Alarm for abnormal conditions.

   c. Indicating and recording process measurement, set point, and output values.

   d. Single-variable control using standard feedback algorithms.

   e. Multivariable control including cascade, ratio, feedforward, and interacting configurations.

   f. Communications to supervisory, control, and/or information system computers.

The three major interfaces that must be considered in the equipment design or application are as follows (see Fig. 24-1):

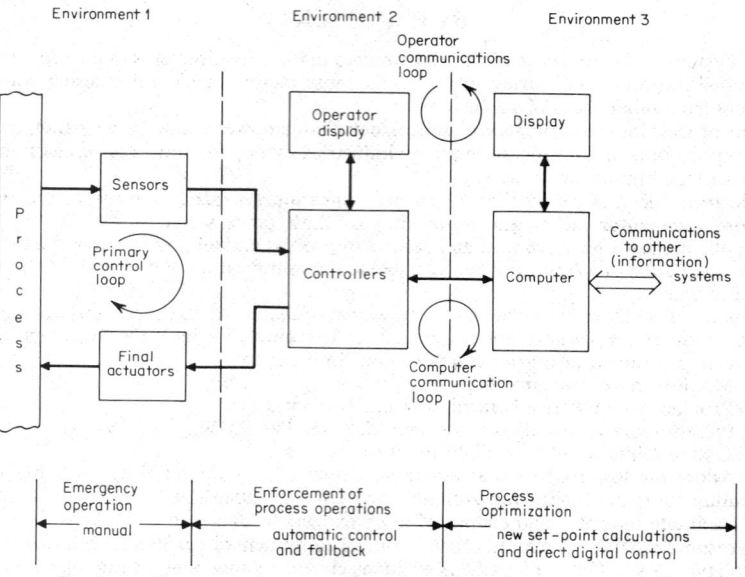

Fig. 24-1.   Three major interface loops important in design or application.

2. Process Interface.   Paragraph **24-20** et seq. discuss the major issues involved with the basic measurement and transmission of process variables. The signal levels involved are becoming standard on a worldwide basis. Table 24-1 indicates some of the more common types of process measurement transmitters and signal levels.

3. Operator Interface.   The window through which the operator views his process is critical and variable. Some of the presently available options are delineated in Par. **24-20** et seq.

4. Digital Computer Interface.   The increasing use of digital computers in process control demands a less costly and more efficient interface than has been available previously. Computers are used as supervisory devices (providing optimizing calculations and outputting set-point information to controllers), direct digital control computers (comparing process measurements with set points, calculating the control action required, and outputting to final actuators), and as communications processors. Paragraph **24-26** et seq. discuss present trends in providing a more compatible interface to digital computers.

Other specifics of designing and applying electronic devices to process measurement and control include the use of intrinsic safety designs for hazardous-area applications and the service conditions encountered in various plants. These factors are under study for international standardization. Paragraphs **24-47** and **24-58** discuss these important process control parameters.

Table 24-1.   Typical Transmitters and Signal Levels

Measurement	Transmitter type	Signal type
Flow	Orifice plate/differential pressure transmitter	Current (proportional to flow squared)
	Target meters	Current (proportional to flow squared)
	Magnetic flowmeters	Current (linear)
	Positive-displacement meters	Frequency (linear)
	Turbine meters	Frequency (linear)
Pressure and differential pressure	Force balance	Current (linear)
	Strain gage	Millivolts (linear)
Position and level	Differential pressure transmitter	Current (linear)
	Displacers	Current (linear)
	Shaft encoders	Digital code (linear)
Temperature	Thermocouple	Millivolts (nonlinear)
	Resistance temperature detector	Variable resistance (linear)
	Thermistor	Variable resistor (linear and nonlinear)
Speed	Magnetic pickups	Frequency (linear)
	Photodiode pickups	Frequency (linear)
	Tachometers	Ac and dc voltage (linear)
Force	Strain gage	Millivolts (linear)
Analytical	Amplifiers	Volts (linear)
Power, current, and voltage ..	Hall effect transducers	Current (linear)
	Thermal converters	Millivolts (linear)
	Transformers	Ac volts (linear)

Note: All current values can be 10 to 50, 4 to 20, or 1 to 5 mA. All millivolt values can be ± 50 to ± 500 mV. Voltage signals can be 0 to 1 to 0 to 10 V. Frequency signals can be 0 to 100 Hz to 0 to 4 kHz.

An extensive bibliography at the end of the section furnishes assistance in obtaining related information in detail.

PROCESS SIGNAL TRANSMISSION

By P. D. Hansen

5. Introduction.   Centralized control of chemical plants, petroleum refineries, and other process industries may require transmission of hundreds of measurement and control signals over several thousands of feet. The cost of the installed transmission system is usually significant, often exceeding that of the measuring instruments. Furthermore, the signal transmission system can significantly affect plant safety, control reliability, and product quality. Paragraphs **24-6** to **24-11** outline many factors that should be considered in designing and specifying instruments and cable for electrical-process signal transmission.

6. Interference.   Most process variables change very slowly, their measurements containing no useful information at frequencies beyond 1 Hz. In an analog system, pickup from power-line and higher-frequency sources is usually filtered adequately by signal converters, controllers, recorders, and indicators, as well as by the process. As a result, special precaution is required only for low-level signals or when nonlinear operations are involved.

When signals are periodically sampled, Fourier components of the original signal with frequencies greater than half the sampling frequency cannot be distinguished from lower-frequency components. In particular, any signal component that is an integer harmonic of the sampling frequency appears after sampling as if it were direct current. Consequently, for digital signal processing it is much more important than for analog that high-frequency signal components be very small. This distinction can be accomplished in several ways: through use of shielding, twisted leads, and other measures, to minimize pickup; by filtering, to remove high-frequency components; and by counting the zero crossings of the frequency signal, to average high-frequency noise (digital conversion).

Low-frequency (dc) information is commonly transmitted in several forms, e.g., by dc voltage, direct current, or ac frequency. In each case, the transmission signal is proportional to the quantity to be transmitted. The following paragraphs discuss considerations affecting the design of transmitters, receivers, and cable for these signals.

7. Voltage Signal. Dc voltage signals are conveyed with a pair of leads (Fig. 24-2). Each lead has a potential with respect to local earth ground. The average of these potentials is called the *common-mode voltage.* The difference in potentials is the *signal,* or *normal-mode, voltage.* Twisting the leads causes nearly identical electromagnetic (inductively coupled, series) voltage pickup on each lead. Also, electrostatic (capacitively coupled) current pickup from an external high-voltage source tends to be the same into both leads. If the impedances are matched from each lead to earth ground, the lead voltage resulting from these currents will be equal. Consequently, twisted leads minimize inductive and possible capacitive normal-

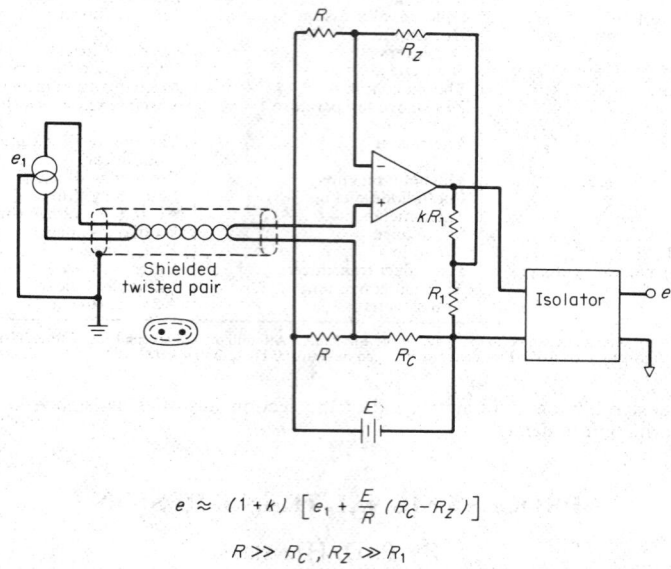

$$e \approx (1+k)\left[e_1 + \frac{E}{R}(R_C - R_Z)\right]$$

$$R \gg R_C,\ R_Z \gg R_1$$

Fig. 24-2.   Low-level voltage-signal transmission.

mode pickup, but do not affect common-mode pickup. Therefore a receiver with high common-mode rejection (10^6 or better) is essential for low-level-voltage measurement. Capacitively coupled pickup can be further reduced by shielding. Resistively coupled pickup is the result of a common conduction path. If two circuits have a common node, there should be minimum length of shared conductor. Also, ground loops (parallel-shared conductors) should be avoided because induced currents can cause variations in potential.

Many voltage sources (e.g., thermocouples and ion-selective electrodes) are grounded in the field. Their transmission lines should be isolated with high impedance from the control-room ground to prevent current flow through the signal leads, owing to possible differences between local ground potentials.

Because of cable series resistance, the normal-mode current flow is usually small. The receiver should have a high normal-mode input impedance and low normal-mode input current. Consequently, power to operate a voltage signal field transmitter should be conveyed over neither signal wire. Often a very small normal-mode current (perhaps 10 nA) is intentionally introduced at the amplifier input to cause the output to saturate if the measurement circuit opens. In a thermocouple application, this feature is called *burnout indication.* An open thermocouple can cause the amplifier output to go full high or low, depending on the direction of the normal-mode bleed current.

Resistance measurement (Figs. 24-3 and 24-4) using three-wire transmission requires current flow through the wires. However, if the wire resistances match, the voltage drops cancel when the bridges are balanced. With additional receiver circuitry, it is possible to achieve first-order cancellations of the effect of cable resistance on span as well. When transmission distances are longer than several hundred feet, a field-mounted transmitter should be employed. Cable should be twisted to eliminate normal-mode inductive pickup. A shield or grounded conduit will minimize capacitive pickup.

8. Current Signal. A current signal is conveyed through a single conductor (Fig. 24-5). The return path could be shared with other circuits without significant error. However, independent return paths are used to avoid excessive and unpredictable line drop that would

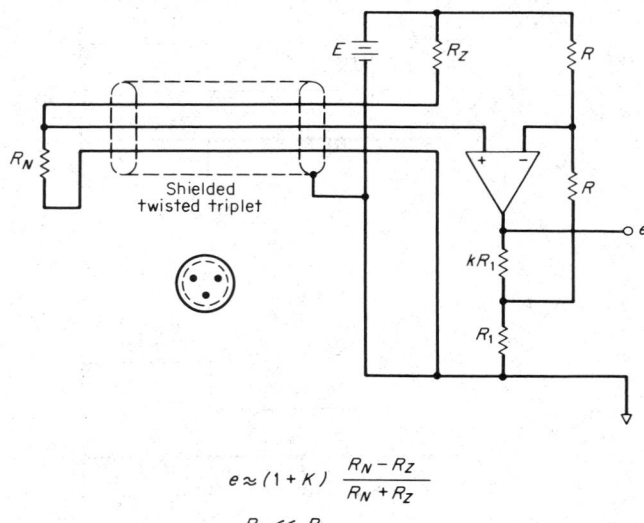

$$e \approx (1 + K) \frac{R_N - R_Z}{R_N + R_Z}$$

$$R_1 \ll R$$

Fig. 24-3. Series-excitation transmission from resistance bulb.

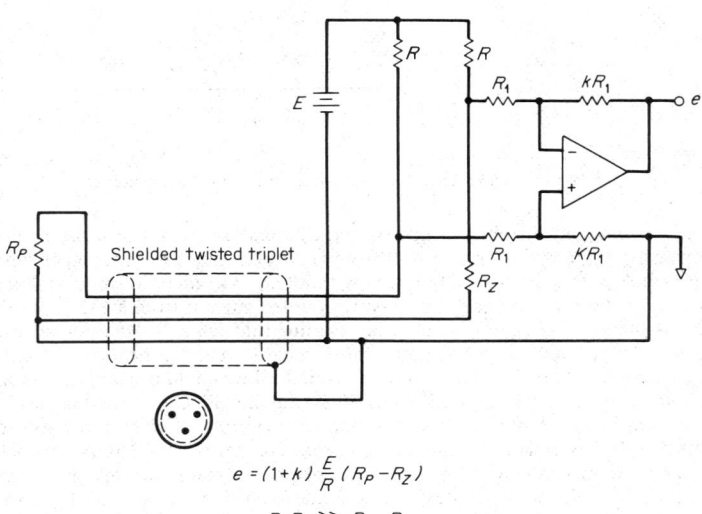

$$e = (1+k) \frac{E}{R} (R_P - R_Z)$$

$$R, R_1 \gg R_P, R_Z$$

Fig. 24-4. Parallel-excitation transmission from resistance bulb.

occur in a shared common. Also, a pair of wires form a loop of small area, tending to intercept a small number of magnetic flux lines. A twisted pair may further help to eliminate inductive coupling. The inductive voltage pickup divides between the transmitter and receiver according to the ratio of their impedances. (Consider half the cable capacitance in shunt with each.) Thus, if the transmitter output impedance at power-line frequency is 25 kΩ and the receiver impedance is 250 Ω, 1% of the inductive pickup appears at the receiver. As a result, a twisted pair is seldom necessary.

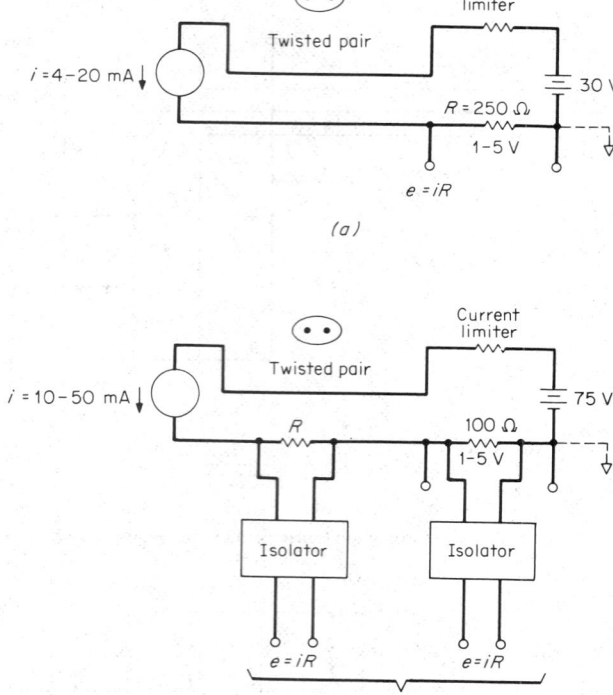

(a)

(b)

Fig. 24-5. High-level current transmission from field transmitter.

Capacitively coupled current pickup is generally small compared with the signal current span. Unusually large pickup may result from running signal cables together with power lines. Nevertheless, this practice is followed without adverse effect in some analog systems. For example, for the peak-to-peak capacitively coupled current from a 115-V 60-Hz power line to be less than ¼% of span (16 mA) after passing through a 10-Hz first-order low-pass filter, the unbalanced mutual capacitance must be less than 1.7 nF. If the unbalanced coupling capacitance were 10 pF/ft, power and signal cables could be run together for 170 ft.

Cable shunt conductance (approximately 50 μs/mile for PVC insulated cable) limits the length for accurate dc current transmission in unguarded cable. For example, a 4- to 20-mA system with a 30-V dc supply has a maximum leakage current of 1.5 μA/mi, 0.01% of span per mile. A 10- to 50-mA system with a 75-V supply has a maximum leakage current of 4 μA/mile; also 0.01% of span per mile. By using a driven shield or guard, this error can be reduced by one to four orders of magnitude, depending on the current range (Figs. 24-6 and 24-7).

If the current signal is to be transmitted to several locations, the receivers may be connected in series between the transmitter and power supply (Fig. 24-5). In order that the presence or absence of one receiver shall not affect the signal at another, each receiver input should be isolated with high impedance from its output, other inputs, the circuit power supply, and ground. Also, the transmitter signal should be isolated with respect to ground. Shunt paths are thus eliminated that could otherwise cause signal current to bypass some of the receivers. If the circuit power supply is also isolated from ground and other circuits, one fault, a short circuit between two circuits or to ground, may be tolerated without error.

An important property of an isolated receiver is the allowable (common-mode) voltage range between its input and output, between its inputs, and between its input and ground. Some applications require rejection of 500 V or more of common mode. To achieve this, the signal must be transmitted across a nonconducting (galvanically isolated) barrier separating the receiver's input and output. This is commonly done with an isolation transformer or magnetic amplifier requiring that the primary-side dc input signal modulate an ac carrier and that the secondary ac signal be demodulated or rectified to achieve the dc output.

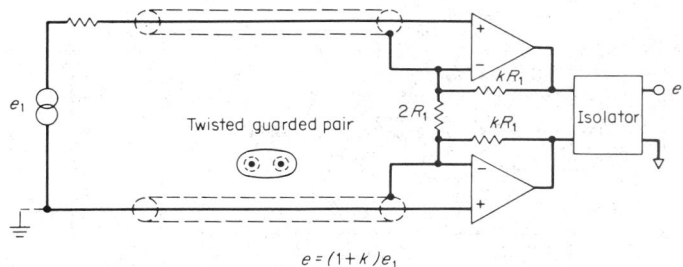

$$e = (1 + k)e_1$$

Fig. 24-6. High-source impedance voltage transmission.

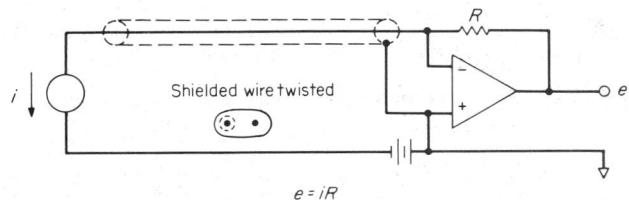

$$e = iR$$

Fig. 24-7. Current signal from photomultiplier.

A biased positive unidirectional dc current signal (typical ranges being 4 to 20 or 10 to 50 mA) permits a small amount of quiescent power (0.1 W) to be conveyed from the receiver to the transmitter over the same pair of wires. It can be used to power a transducer that senses force, deflection, strain, temperature, or other process variable. An open transmission circuit causes the receiver to overrange low, and a normal-mode short circuit causes the receiver to overrange high.

A series diode blocks current flow in the reverse direction to prevent damage if the power-supply leads are reversed. Two common transmitter concepts are shown in Figs. 24-8 and 24-9. In Fig. 24-8, the output current is sensed resistively and fed back as a voltage to rebalance a bridge network. Accuracy depends upon the stability of the reference voltage, the sensing resistor, and the bridge resistances.

In Fig. 24-9, the output is passed through a *voice coil,* a moving coil in a permanent magnet field. This action generates a force that rebalances a mechanical lever system. Balance is sensed with a differential transformer that controls the amplitude of a sinewave oscillator, which in turn controls the output current. Accuracy depends on the current-force relationship of the voice coil and the mechanical stability of the lever system. Both these circuits can have a very high output impedance and can operate over a wide range of power-supply voltage and external series resistance. Typical operating limits are shown in Fig. 24-10. Limited dc power and high internal frequency (requiring small internal-energy storage) permit application of these transmitters to intrinsically safe circuits, which are incapable of igniting an explosive mixture under fault conditions.

Three- or four-wire transmitters are used when more than 0.1-W quiescent power is required. Examples include the magnetic flowmeter, which requires a high current field excitation, and the pH sensor, which may employ a dedicated ultrasonic cleaner to prevent electrode fouling. Both use four wires, two to transmit the signal to the receiver and two to connect directly to the ac mains. This arrangement enables the signal to be galvanically isolated from ground.

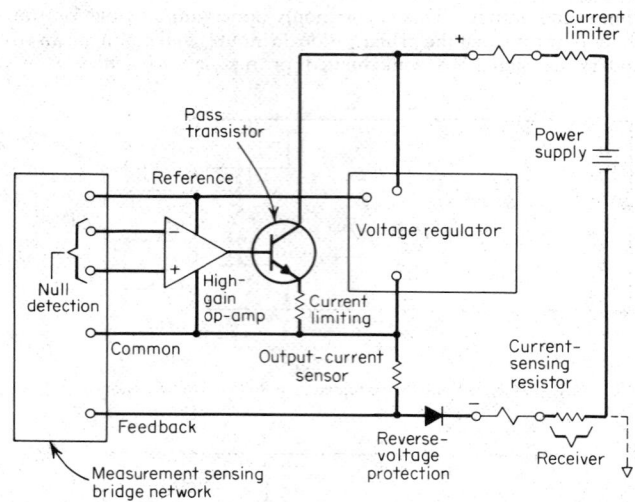

Fig. 24-8. Two-wire current-output, voltage-feedback configuration.

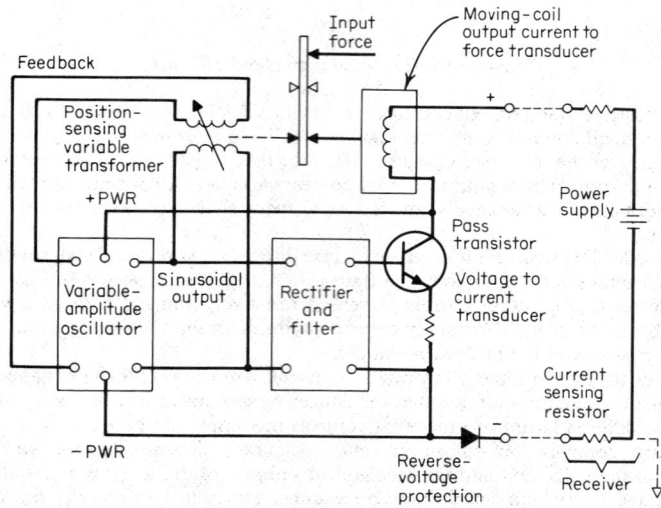

Fig. 24-9. Two-wire current-feedback or force-balance transmitter.

A three-wire transmitter may be used to achieve a bidirectional (or "live zero") signal current. In this case, two wires connect to positive and negative dc supplies and the third carries the bidirectional signal current to a receiver or series connection of receivers referenced to the power supply common (Fig. 24-11).

Low-level-current measurement (Fig. 24-7) requires shielding to reduce capacitive pickup and leakage current owing to cable conductance. The receiver input resistance is made virtually zero to reduce inductive pickup as well as cable leakage. Very high source-impedance voltage measurement (Fig. 24-6) requires independent guarding of each input lead as well as a receiver with much higher input impedance and high common-mode rejection.

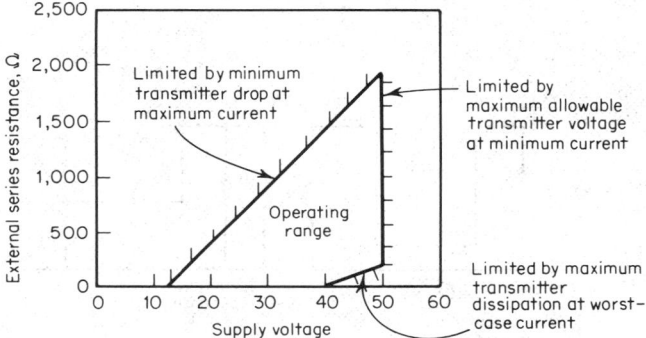

Fig. 24-10. Output operating limits of force-balance transmitter.

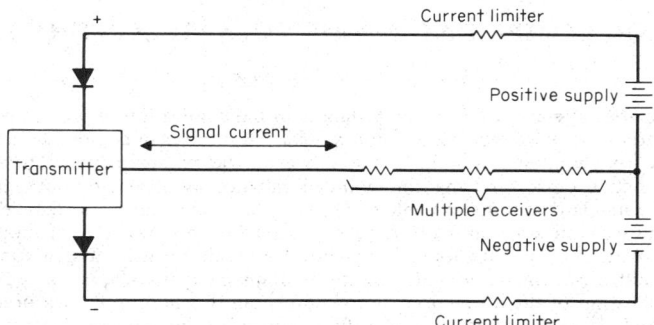

Fig. 24-11. Three-wire bidirectional current transmitter.

9. Frequency Signal. The only property of an ac wave that is not altered by transmission is its frequency. Amplitude and duty cycle are affected by attenuation and distortion. An audio-frequency signal can be readily transmitted through a galvanic isolator. Also, it can be converted to a digital value by counting zero crossings for a sufficient time interval to achieve the desired resolution. Dc power can be conveyed to a transmitter over the same pair of wires provided the transmitter produces a variable-frequency current signal added to a dc quiescent current (Fig. 24-12). For minimum distortion at the receiver, the terminal resistance should equal the line's characteristic impedance $\sqrt{L/C}$ minus half the series resistance. For example, a mile of typical two-conductor 19-gage cable is represented by L = 1 mH, C = 0.1 μF, and R = 100 Ω. The characteristic impedance is 100 Ω, and the optimum terminal impedance is 50 Ω. The voltage amplitude at the receiver will be down 50% from that at the transmitter, but the current amplitude will be virtually unchanged for frequencies up to 10 kHz. A substantially higher terminal resistance can be used if maximum bandwidth and minimum distortion are not of concern. The low-frequency dynamics of the transmission system are controlled by terminal resistance (plus half the cable series resistance) and the line capacitance. For a terminal resistance of 250 Ω, a 5-kHz current signal will be attenuated 3 dB at the receiver 1 mile away.

A twisted pair is used to prevent this current signal from causing a coherent electromagnetic field that could result in inductive pickup in a low-level signal pair. A coaxial cable also avoids pickup and in addition provides a higher characteristic impedance because of lower shunt capacitance. A sinusoidal or triangular waveform is preferred because it induces less pickup than a square wave or pulse train of the same amplitude.

By using a high-level current signal, a large amount of pickup can be rejected at the receiver. The receiver is a hysteretic zero-crossing detector (Schmitt trigger) with its gap set to exceed the maximum pickup voltage.

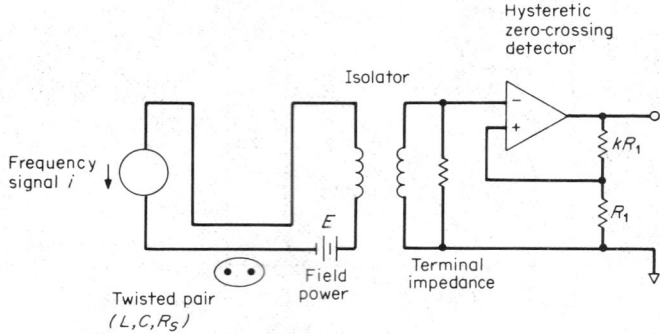

Fig. 24-12. High-level frequency signal from field transmitter.

ANALOG COMPUTING ELEMENTS AND CONTROLLERS

By D. A. Richardson

10. Computing Elements. In many instances in the application of process control, the measured variable is not directly usable for control. A number of computational elements are available from equipment suppliers to modify available measurements. These elements are single-function devices having common-denominator input and output signal levels. They provide standard continuous analog computing functions that the system designer can use to configure complex measuring techniques. Typically, they have the capability of being scaled to provide proper scans for different process requirements. Similar functions are provided in either pneumatic or analog electronic implementations.

One application of these functions is to compensate for nonlinear measurements. A typical example is in the measurement of flow, where the most common technique is to measure the difference in pressure across an orifice plate in a flow line. This differential pressure is equal to the square of the flow rate. In some cases, as when the desired measurement is the sum of several flows, the square root of the differential pressures must be taken before the signals can be added. This requires a square-root extracting function that solves the equation $E_{out} = \sqrt{E_{in}} \times 10$. This function requires no scaling because the measurement is always zero-based, and the proper span is provided by the transmitter scaling.

A second application of computing elements is the case where the variable to be controlled is not easily measurable.

A measurement of mass flow, for instance, can be calculated from the expression

$$M = K \sqrt{\frac{hp}{T}}$$

where M = mass flow,
      h = absolute pressure,
      p = differential pressure across an orifice plate,
      T = temperature, and
      K = constant.

Here a multiplier-divider is required, with each input scalable to suit the process equation. The selection of transmitter spans can greatly influence the accuracy of the calculated result in this case, inasmuch as low denominator levels and a divider greatly magnify small system errors.

A further application of computing elements is the case in which the controlled variable is dependent on another measurement. A typical example is the control of composition by the introduction of additives. Here the amount of additive (controlled flow) should be in a specific relationship to the main (wild) flow rate.

A ratio unit is used to set the proper proportioning in the form $E_{out} = RE_{in}$, where R is the ratio of controlled to wild flow.

Typical functions provided by most equipment manufacturers include:

Square-root extractors.

Summing units having two to eight inputs, for addition and subtraction. Scaling is provided for input spans and biasing.

Multiplier-dividers, often in the form

$$D = \frac{A \times B}{C}$$

with all variables scaled to suit the process requirements.

Selector units, to select the highest or lowest of several variables.

Signal characterizers, primarily static nonlinear function generators that are typically characterized to the process by factory adjustment.

Dynamic compensators, primarily lead and lag units with adjustable time functions.

Time-function generators, such as ramp generators or signal programmers, for batch operations.

The circuits used in these blocks are fundamentally similar to that used and described under analog computers. In practice, however, the industrial environment imposes the requirement for operation over a wide range of temperatures, and supply-voltage fluctuations not encountered in typical analog computing applications. Additional constraints include the requirement for long-term static accuracy and a design life of 5 to 10 years of continuous operation. Careful attention must be paid to component selection, derating, and packaging, and circuit design must be conservative.

11. Controllers. In any industrial process in which a variable is to be held near a fixed value *(set point),* a controller compares the measurement signal with the desired value and operates on the process to minimize the difference. Although controllers range from the simple ON-OFF type to those employing complex mathematical functions, all have certain elements in common. The control function can be implemented in many forms, such as pneumatic, mechanical, hydraulic, electronic analog, and digital. This section is restricted to analog devices; digital implementation is described in Par. 24-42 et seq. Schematic representations are in electrical terms, although pneumatic equivalents of the described functions are readily available.

Figure 24-13 shows a basic controller having two inputs, one representing the measured variable and the other representing the set point, a desired value of the measurement. These are compared, and the difference operated on by some control function. Typical functions or sections of a process controller are shown in Fig. 24-14. Means for generating the set point is often a simple manual control, but may include switching between several sources, such as a computer or another analog signal source.

Manual control of the output signal is usually provided to assist in start-up and troubleshooting procedures. Various means are provided for the transfer between the automatic and manual modes of operation. The signal levels from each mode must be equal

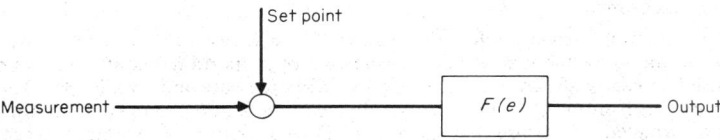

Fig. 24-13. Basic controller with two inputs.

at the time of transfer to avoid process upsets. Some controllers require a manual balancing procedure, but newer designs have automatic balancing circuits permitting this transfer to be effected by either the operator or an external binary signal.

Display of the measurement, set point, and output levels is normally provided. The measurement and set-point displays are usually so arranged that their difference can be easily seen, for this deviation is the primary measure of the control effectiveness.

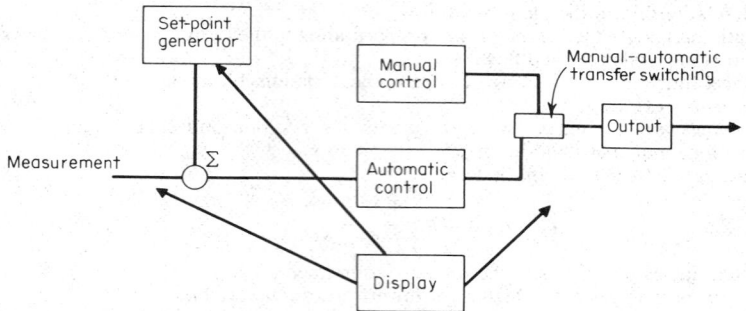

Fig. 24-14. Typical functions of a process controller.

CONTROL MODES

The control mode is the method whereby the controller counteracts the deviations between measurement and set point. This section discusses a number of the commonly used functions.

12. Two-Position (ON-OFF) Control. This is the simplest control mode, in which the output is on when the measurement is below the set point, and off when the measurement is above the set point. It is used primarily for large single-capacity processes such as tank level or room heating. ON-OFF controllers often include some hysteresis, called *differential gap*, to reduce the cycle rate inherent in this type of control.

13. Proportional Control. Proportional control is the oldest and simplest form of continuous analog control. The output is proportional to the deviation in the form $E_o = (100/PB)\,e + B$, where e is the deviation between measurement and set point, B is a bias, and PB is the *proportional band*, a frequently used term defined as:

$$PB = \frac{1}{\text{gain}} \times 100\%$$

The low gain often necessary to maintain process stability results in large deviations from load changes. For this reason, it is most commonly used for applications such as surge tanks where tight control is not necessary or desirable, or on essentially single-capacity processes which have a maximum phase shift of 90° and thus will be stable even at very high gains.

14. Proportional Plus Reset Control (P + I). The most commonly used control function is a combination of proportional and integral action in the form

$$E_o = \frac{100}{PB}\left(e + \frac{1}{R}\int edt\right)$$

where e = deviation,
   PB = proportional band, and
   R = reset time.

This combination provides a low dynamic gain to achieve process stability and a high static gain to minimize the load error characteristic of proportional control. This control mode is used on nearly all flow control loops and many pressure and level loops. The range of dynamic gains is from 0.2 to 50, and reset time constants from 1 s to 60 min. Figure 24-15 shows the frequency response characteristic of a P + I controller. A simplified schematic of one configuration of P + I control is shown in Fig. 24-16.

The very long time constants required to match the dynamics of physically large processes dictates the use of extremely high gain, high impedance amplifiers with very low offset deviations. The typical design life of 5 to 10 years in industrial atmospheres further requires that extreme care be taken in the selection and derating of components, packaging techniques, and circuit reliability.

For these reasons, chopper-stabilized amplifiers are often used, and high-input-impedance modulators using varactors or FETs are common.

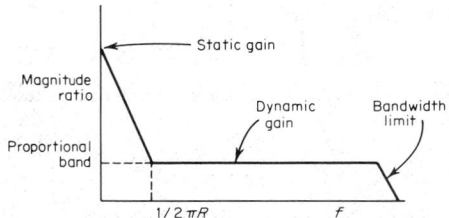

Fig. 24-15. Frequency response of a P + I controller.

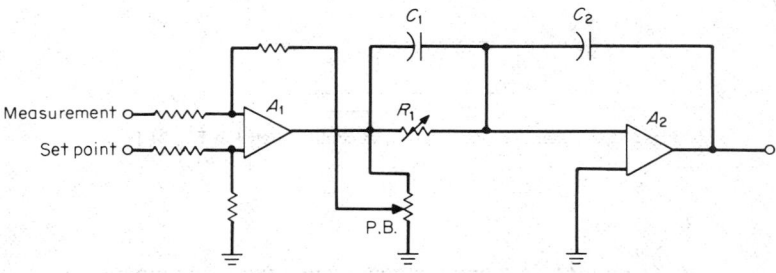

Fig. 24-16. Simplified schematic of a representative P + I control.

15. Reset Windup. Any control function containing an integrating mode is subject to reset windup, or saturation of the integrating stage, if a deviation is sustained for a long time period. This effect is not normally encountered on continuous processes, but is a problem on batch processes that are started up under automatic control. In a controller configuration, shown in Fig. 24-17 for instance, on start-up, the output is saturated and no control action occurs until after the measurement has reached the set-point value. This causes the measurement to overshoot the set point by an appreciable amount, as shown in Fig. 24-18.

In Fig. 24-19 an amplifier provides a bias to the feedback circuit when the integrator output reaches a limit value. Thus, on start-up, control action begins immediately provided the process rate of change is faster than the time-constant RC.

16. Proportional + Reset + Derivative (PID). Some process having a number of capacities in series or having dead time can be more successfully controlled by the addition of a derivative or phase-lead mode of control to the $P + I$ controller. The ideal form of such an addition would be

$$E_o = \frac{100}{PB} \left(e + \frac{1}{R} \int e\,dt + D\,\frac{de}{dt} \right)$$

where PB = proportional band,
   R = reset time,
   D = derivative time, and
   e = deviation.

As a practical matter, however, the rate gain is limited to the range of 10 to 20 by noise and the phase lead to about 60°. Frequency response of a PID controller is shown in Fig. 24-20.

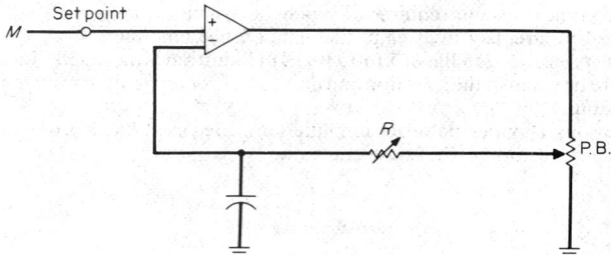

Fig. 24-17.   No control action occurs until set point is reached.

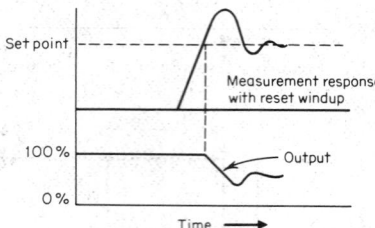

Fig. 24-18.   Characteristic response of control shown in Fig. 24-17.

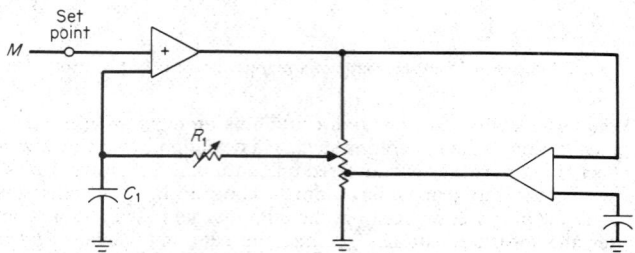

Fig. 24-19.   Control action starts when process change rate is faster than *RC*.

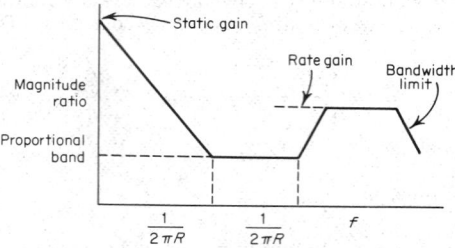

Fig. 24-20.   Frequency response of PID controller.

The derivative mode is typically incorporated in the feedback circuit, as shown in Fig. 24-21. A better method is to apply derivative action to the measurement only, and to isolate its action from the reset mode. This implementation is found in few controllers, but is a significant feature, particularly when the set point is manipulated by a computer.

17. Cascade Control. Many processes have an intermediate variable that responds both to the manipulated variable and to process disturbances. In Fig. 24-22 a temperature is controlled by the manipulation of fuel flow.

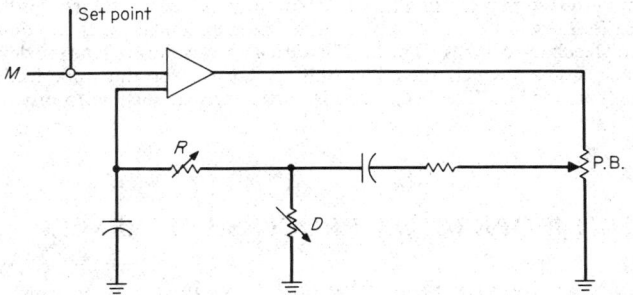

Fig. 24-21. Derivative mode used in feedback circuit.

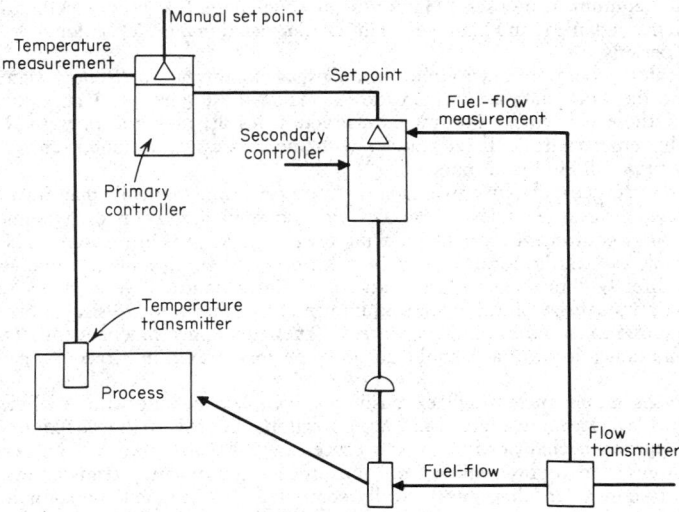

Fig. 24-22. Temperature controlled by manipulation of fuel flow.

Pressure changes in the fuel line (intermediate variable) may be much faster than the time constant of the temperature change. Thus appreciable fuel-rate changes could occur before corrective action is taken by the temperature controller. The addition of a secondary loop that directly controls fuel flow will minimize fuel-pressure effects and provide significant improvement in loop stability.

18. Adaptive Control. In many processes there are nonlinearities of process gain or time constant. Thus the fixed proportional band and reset time setting of the controller will be correct for optimum loop dynamics only over a limited range of operation. Adaptive control is the automatic adjustment of a control mode as a function of some measured variable.

Relatively few control loops have adaptive control either because the nonlinearities encountered in the normal operating range are not severe or because proper mode settings cannot be obtained from measurable parameters.

19. Feedforward Control. In a feedback control system, a deviation between measurement and set point must exist before control action will occur. If the major sources of load variation can be measured, it is possible to act directly on the manipulated variable so as to compensate for the changes before a deviation occurs between the set point and the controlled variable. This action is known as feedforward control.

The feedforward system is essentially a model of the process. Although it is theoretically possible to use feedforward control only, it is far more practical to use a combination of feedforward and feedback control. The feedforward elements thus reduce the deviation seen by the feedback system, which then has only to correct for the imperfections in the feedforward-process model. This technique is particularly applicable to processes having significant dead time.

THE OPERATOR/PROCESS INTERFACE

By R. A. Williamson, Jr.

20. Evolution. The two ingredients of the operator/process interface are the human being and the equipment he uses. Human beings contribute two unique skills to process control systems: judgment and highly efficient tactile/visual manipulation for short concentrated time periods.

The operator/process interface equipment consists of information displays, manipulative controls, and the work station at which they are located. Figure 24-23 briefly traces the evolution of these elements. The major characteristics are that the operator has been progressively removed from direct contact with the process and, concurrently, that his supervisory responsibility has expanded.

In early days, process dynamics were manually observed and manually manipulated. The tools were sight glasses, gages, levers, handles, etc., mounted directly on and reading directly from the process equipment (boilers, cracking towers, tanks, pipelines, etc.). The control equipment was awkwardly located (atop high towers) and cumbersome to operate (large, heavy, and directly connected to large valves). Related controls were widely separated owing to the physical size of the process equipment. Considerable physical effort of many men was required to tend the complete process. They functioned independently for lack of effective communications. Each could do no more than watch and react to part of the process.

As processes became increasingly expensive and complex, higher operative efficiency was required to achieve acceptable return and safe operation. Coordination was required among remote but interconnected portions of the process. It was achieved by centralizing the control equipment in rooms remote from the process equipment. Transmitters relayed information to controllers that automatically computed and executed control action, with only occasional manual attention. Operators could now function in real time with the process, and begin to anticipate its behavior.

Modern control strategy emphasizes prevention of upsets and optimization of normal running conditions. This is implemented by using a combination of unique human skills for *critical* tasks, automatic control for *routine* tasks, and computerized communications. Present technology emphasizes:

    a. Anticipation, using predictive control.

    b. Greater extent and degree of supervisory responsibility for one man.

    c. High information density.

    d. Supervision by exception, a tactic wherein only *nonnormal* conditions are brought to the operator's attention.

    e. Flexible, adaptive display and control equipment.

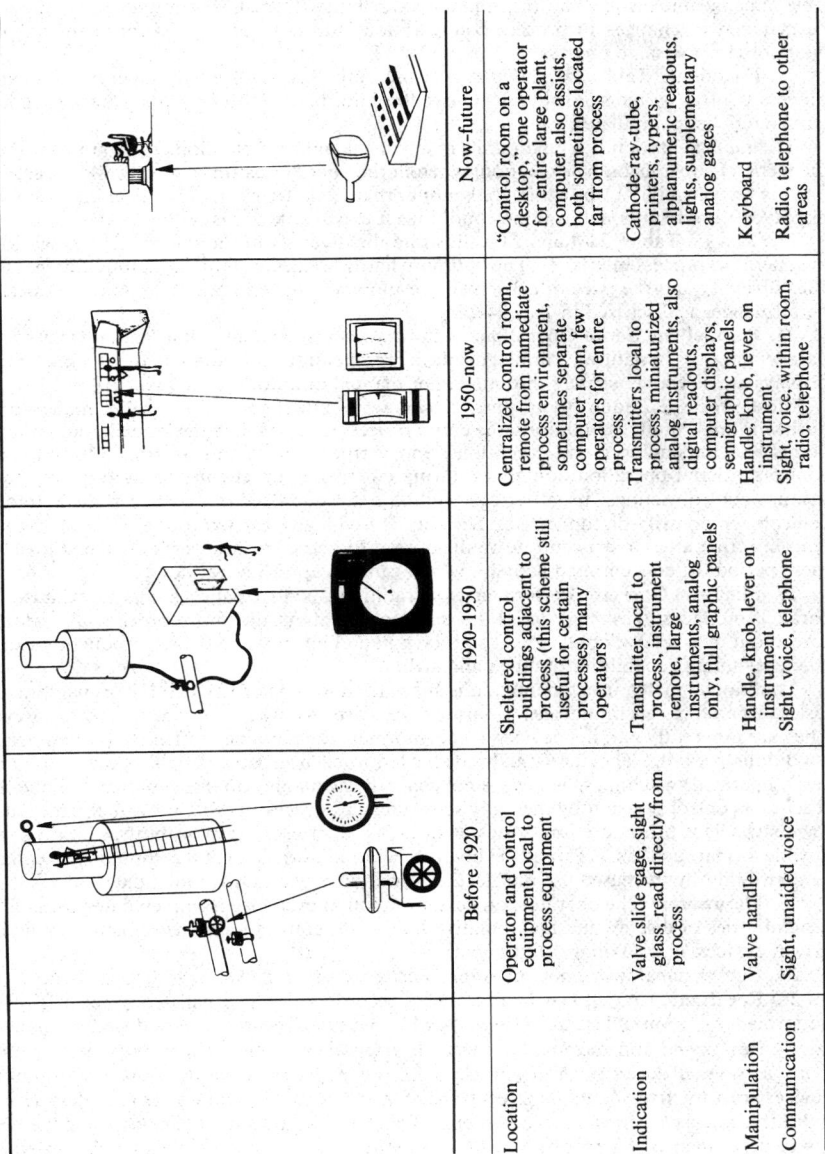

	Before 1920	1920–1950	1950–now	Now–future
Location	Operator and control equipment local to process equipment	Sheltered control buildings adjacent to process (this scheme still useful for certain processes) many operators	Centralized control room, remote from immediate process environment, sometimes separate computer room, few operators for entire process	"Control room on a desktop," one operator for entire large plant, computer also assists, both sometimes located far from process
Indication	Valve slide gage, sight glass, read directly from process	Transmitter local to process, instrument remote, large instruments, analog only, full graphic panels	Transmitters local to process, miniaturized analog instruments, also digital readouts, computer displays, semigraphic panels	Cathode-ray-tube, printers, typers, alphanumeric readouts, lights, supplementary analog gages
Manipulation	Valve handle	Handle, knob, lever on instrument	Handle, knob, lever on instrument	Keyboard
Communication	Sight, unaided voice	Sight, voice, telephone	Sight, voice, within room, radio, telephone	Radio, telephone to other areas

Fig. 24-23. Evolution of operator and process interface.

24-19

f. Expansion of raw measured data for interpretation, diagnosis, and prediction.

g. Changeable display and manipulation capability (format, resolution, etc.), which can be matched to changes in process configuration, and tailored to preferences of different operators for the same process.

21. Function. Table 24-2 summarizes the activities involved in supervising modern process control systems. Table 24-3 lists the information the operator receives and the parameters he manipulates.

22. Implementation. Human skills in judgment and manipulation are exploited through the tactic of supervision by exception, wherein the operator assumes more on-line decision-making responsibility by attending only nonnormal conditions. Only critical information is displayed to him, although routine conditions are available at his request.

23. Tasks. Tables 24-4 and 24-5 illustrate the diversity of the information exchange at the operator/process interface. The optimum hardware device is different for different tasks. This diversity and the pace of equipment development have imposed limits upon operator performance, as discussed in Par. 24-25.

24. Parallel and Serial Operation. Information display and manipulation tasks can be parallel or series in nature. Parallel displays present all the information all the time. Serial displays are selective, usually through either manual or automatic sequencing.

Most processes require both parallel and serial attention. The parallel mode occurs during normal running conditions; the entire process is scanned for deviations from set target points and impending alarm conditions, and vernier adjustments are made to automatic controls. Serial operation dominates during emergency upsets, in the form of sequential manual control actions, usually according to a fixed procedure, with certain automatic controls temporarily disconnected. Seriality is frequently carried to the point of virtually ignoring other areas to concentrate on the upset. Increased manpower is often employed for these periods of concentrated activity, which can last up to several hours.

25. Limits on Operator Performance. Traditional control panels are large, exhaustively parallel interfaces; they provide all the information, about the entire process, all the time, spread out over a large area. Panel size, layout, and information density frequently exceed efficient human capabilities of reach and sight.

Traditional analog instrumentation is specialized for one or two display or manipulative tasks. Therefore several devices of various types are required to service a process variable. They are inherently parallel devices, since information cannot be "turned off" when normal conditions prevail. They are usually dedicated to each process variable because reliable, easily operated switching schemes are not generally available. In this situation, the operator reaches a control decision by mentally screening and processing the individual information segments. This procedure limits the extent of his supervisory responsibility. Devices with flexible format such as CRTs, and keyboards operating through digital computers, extend this responsibility by increased use of selective, serial display and manipulation.

26. Hardware. The operator/process information exchange is implemented using three general types of display and manipulative hardware: conventional instruments, computer-driven devices, and auxiliary equipment.

27. Conventional instrumentation includes recorders, indicators, and controllers.

28. Recorders. As shown in Fig. 24-24, recorders produce hard-copy records of the performance of a variable against time. A wide variety of recorders is produced in small and large cases, round and rectangular charts, horizontal or vertical chart movement, with as many as several dozen recording tracks. Ink on paper is generally used, but noninking devices are attractive, including direct transfer from CRTs. Operators use recorders as *trend* indicators and as *historical* record keepers. Trend indication is used to extrapolate the near-future performance of a process. History is useful for evaluating efficiency. Recorders are usually dedicated to particular variables. Sometimes, however, one recorder is shared among several variables by means of manually switchable connection devices.

29. Indicators. As shown in Fig. 24-25, indicators provide momentary, real-time notice of measurement value.

30. Controllers. The indicating controller (Fig. 24-26) concentrates the operator's major supervisory functions into one instrument. Some provide optional recording in the same case. Display includes value of measurement, set point and output; status of set point and

Table 24-2. Operator Supervision Activities

Phase	Characteristics	Tasks	Observation	Manipulation
Maintenance of normal running order	Parallel; comparative (deviation, trend); qualitative (analog, gross scale)	Monitor deviation from set target point; observe alarm; establish whether process is in or out of control; determine and execute corrective action	Relative position of measurement at set point Measurement position in control range Relative position of measurement at alarm range	Vernier adjustment of automatic controls (set point) Acknowledge alarms
Emergency upset	Serial information display Quantitative (precise analog scale, alphanumeric readout) Concentration on a few critical loops Normal override of automatic controls Voice communication with personnel at the process trouble point	Determine exact condition for individual loops Determine if dangerous or not If dangerous, determine and undertake corrective action	Exact position of measurement in control range Rate and direction of change of measurement Amount of control capacity remaining	Vernier adjustment of control output signals Acknowledge alarm Initiate supplementary displays

Table 24-3.   Process and Control System Information

	Item	Indication
Process information	Set point*	Value (engineering units or percent of scale)
		Mode (local, remote)
	Measurement	Value (engineering units or percent of scale)
		Trend of valve (curve of up to 1 h of variable performance)
	Output*	Value (percent in 0 to 100% range)
(* Manipulated)		Mode (automatic, normal, computer)
	Alarms	
	Absolute (high, low)	ON or OFF
	Deviation (high, low)	ON or OFF
	Loop identification	Type of loop (flow, pressure, level temperature, motor)
		Number of loop
		Name of loop
Auxiliary	Flow diagrams	Show elements of loops, connections to other loops
Process information	Procedures	
Control system information	Alarm range	Values (engineering units or percent of scale)
	Absolute	Different high and low values
	Deviation	Symmetrical percentage about set point
(All manipulated)	Control range	Values
	Tuning constants	Proportional (percent scale)
		Integral (counts per unit time)
		Deviation (time)
	Ratio	Linear multiplying factor
	Bias offset	Value (engineering units or percent of scale)

Table 24-4.   Indication

Display contents	Display classification	Optimum hardware
A. Plant-area identifiers	A. Alphanumeric (A/N) readout	A. CRT
B. Plant-area alarms	B. Binary (ON/OFF) indicator	B. Flashing indicator light
Control group names	A/N readout	CRT
Loop identifiers	A/N readout	CRT
Loop variables (measurement, set point, output)		
A. Gross value	A. Moving pointer against fixed scale (analog)	A. Analog scale meter
B. Exact value	B. Digital readout	B. CRT
C. Status (stable; increasing; decreasing, and if so, how fast?)	C. Graphic symbology (arrows, etc.)	C. CRT
A. Loop alarm limits	A1. Moving pointer against fixed scale (analog)	A1. Analog scale meter
1. Gross value		
2. Exact value	A2. Digital readout	A2. CRT
B. Loop alarm conditions (absolute, deviation; high, low)	B. Four-state readout	B. CRT
Loop operating mode (local, remote, automatic, manual, computer)	Binary (A/N) indicator	CRT
Trend recorder assignment, recorder and pen identifiers, pen/variable matching	A/N readout	CRT

Table 24-5.   Manipulation

Operator tasks	Task classification	Optimum hardware
Address and acknowledge plant area alarms	Selection from a variable group	Switch in fixed location dedicated to each item in group
Address control groups	Selection from a fixed group (several groups, composition of each being different)	Switch in fixed location dedicated to each item in group
Address loops	Selection from a fixed group (several groups, composition of each being different)	Switch in fixed location dedicated to each item in group
Change loop set points, outputs	Magnitude adjustments of values within fixed ranges *A.* Gross adjustment—relative positioning on a fixed scale	*A.* Potentiometer with analog scale meter feedback
Change loop alarm limits	*B.* Exact adjustment—precise numerical-value selection	*B.* Ten-key numeric keyboard with digital readout feedback
Change loop operating mode	Selection between alternatives	Labeled indicator switch (position or lighted feedback)
Assign trend pens to loops	*A.* Recorder choice—selection from a fixed numerical list *B.* Pen choice—three-state selection *C.* Pen/variable matching	*A,B.* Switch in fixed location dedicated to each item in group *C.* Jack/plugboard (patchpanel)

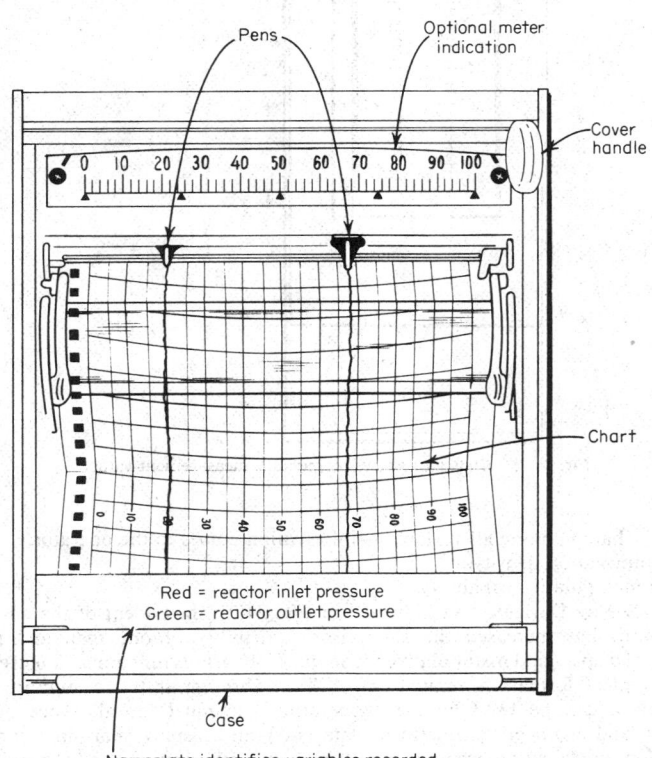

Fig. 24-24.   Recorder produces hard-copy history of a variable.

output; and alarms.   Manipulation includes changing value and status of set point and output.

31. Computer-driven Devices.   These include cathode-ray tubes, other alphanumeric displays, printers, typers, and input devices such as keyboards, light pens, etc.   Major characteristics are:

    a. Quantitative (alphanumeric) or qualitative (active graphics) information.

    b. Flexible format of display and control.

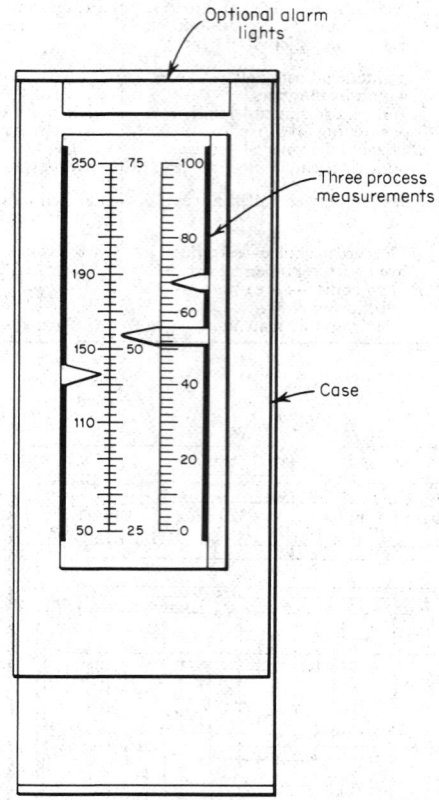

Fig. 24-25.   Indicator shows momentary measurement value.

    c. Facility shared among all variables—information comes to the operator.

    d. High information density.

    e. Shared manipulative capability.

32. Cathode-Ray Devices.   As indicated in Fig. 24-27, the advent of shared hardware (CRT/keyboard) has increased the supervisory capability of one man and prompted commonality of display and manipulative techniques for various functions.   Figures 24-28 to 24-31 show typical formats generated on CRTs.   Through software support, this one hardware device can be used by plant operating personnel for all types of control, manipulation, and engineering functions: data entry and display; alarming; control-loop implementation, modification, trending, and tuning; and supervisory control functions such as execution of programs, data file changes, and debugging.

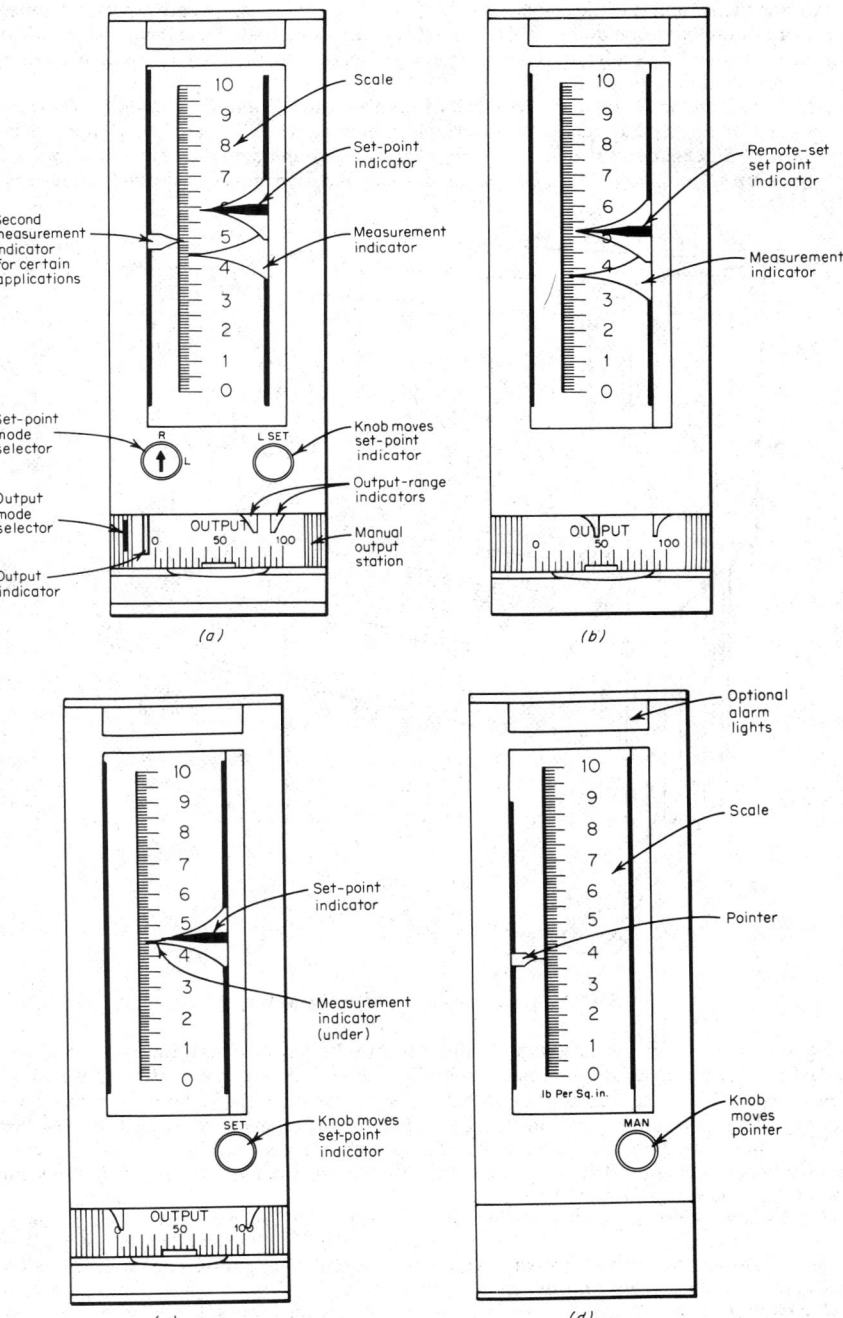

Fig. 24-26. Indicating controller can be remote- or local-set; remote-set; local-set; and manual.

Alphanumeric and graphic display capabilities, format changes, etc., are especially useful in elaborating information about process variables above that obtainable from conventional loop instruments. This advantage aids the predictive, diagnostic mode of modern process control supervision.

33. Alphanumeric Readouts. Displays of numbers and letters (Fig. 24-32) are extensively used when precise values of a variable are desired. They are most effective as indicators of process performance when complemented by analog indicators and recorders. Examples are gas discharge tubes, projection devices, and segmented- or dot-matrix devices.

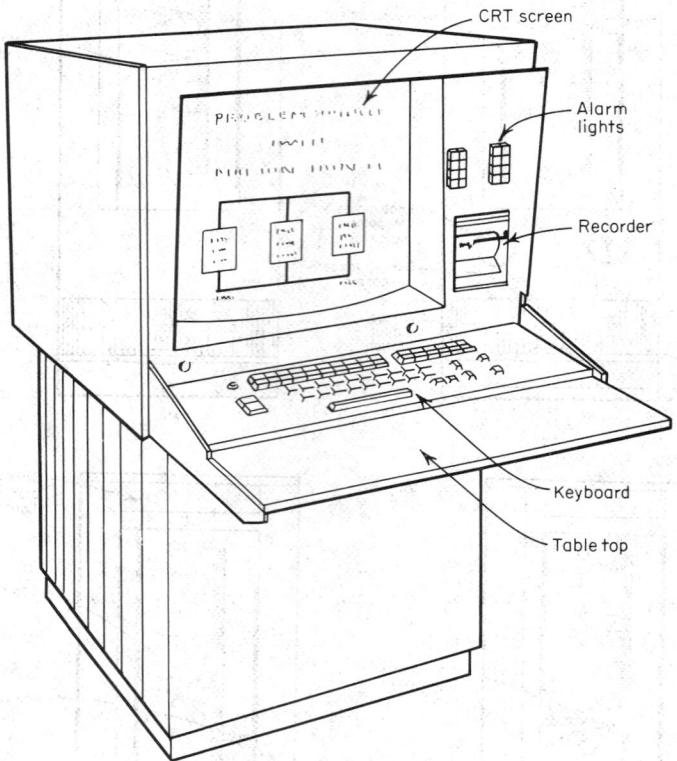

Fig. 24-27.   Typical cathode-ray-tube display console.

34. Keyboards utilizing lighted and unlighted pushbuttons are conventional means for manually entering data into computer-controlled process control systems. Figure 24-33 shows a special keyboard arrangement that gives the operator the same capability as the knobs and levers on conventional instruments. USASCII formats are also used (Fig. 24-34).

35. Printers and Loggers. Hard-copy devices such as typers and line printers are often used with computer control systems to record alarm history (loop on which alarm occurred, time of day, and type of alarm).

36. Auxiliary equipment supplements the basic display and manipulative devices, sometimes providing otherwise unobtainable specialized capability. Typical devices are graphic panels, alarm-monitoring subsystems, temperature-monitoring subsystems, and miscellaneous gages, meters, lights, and switches.

37. Graphic panels are a type of mimic display utilizing the process flow diagram. A *full-scale* graphic panel (Fig. 24-35) is essentially an enlargement of the process flow sheet with instruments installed at the point of use therein. To retain simplicity, only equipment

PROCESS DISPLAY 2 FEED SPLITTER

LOOP	BLOCK	INPUT	MEAS	UNITS	SETPT	OUTPUT	SCAN	CNT	MODE	ALARMS ABS	ALARMS DEV
FSM1ØØ	FSC1ØØ	FSP1ØØ	34Ø.5	PSI	3ØØ.Ø		ON	ON	AUTO	HI	
	FSC1Ø1	FSF1Ø1	5.8	TCFT/H	5.6		ON	ON	COMP		
FSM3ØØ ▲	FSC3ØØ	FSL3ØØ	12.8	FT	13.Ø		ON	ON	AUTO		
	FSC3Ø1	FSF3Ø1	12Ø.Ø	TGPH	124.Ø		ON	ON	BKUP		
FSM4ØØ	FSC4ØØ	FST4ØØ	55Ø.7	DEGF	555.Ø		ON	ON	AUTO		
	FSC4Ø1	FSF4Ø1	128.5	TP/H	125.Ø		ON	ON	COMP		
FSM45Ø	FSC45Ø	FSF45Ø	132.7	TGPH	134.Ø		ON	ON	COMP		

Fig. 24-28. Alphanumeric status report displayed on CRT.

PROCESS DISPLAY 1 FEED ACCUMULATOR

BLOCK	MEAS	UNITS	ALARM LIMITS LOW	ALARM LIMITS HIGH	BLOCK STATUS	ALARM STATUS
ACP2ØØ	32Ø.6	PSI	25Ø.Ø	3ØØ.Ø	ON	HI
ACT2ØØ ▲	487.8	DEGF	N/A	5ØØ.Ø	ON	

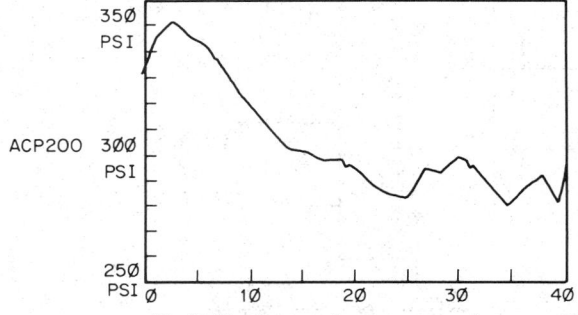

Fig. 24-29. Loop-performance curve shown on CRT.

PROCESS DISPLAY 6 RCTR TEMP SUP CONT LOOP

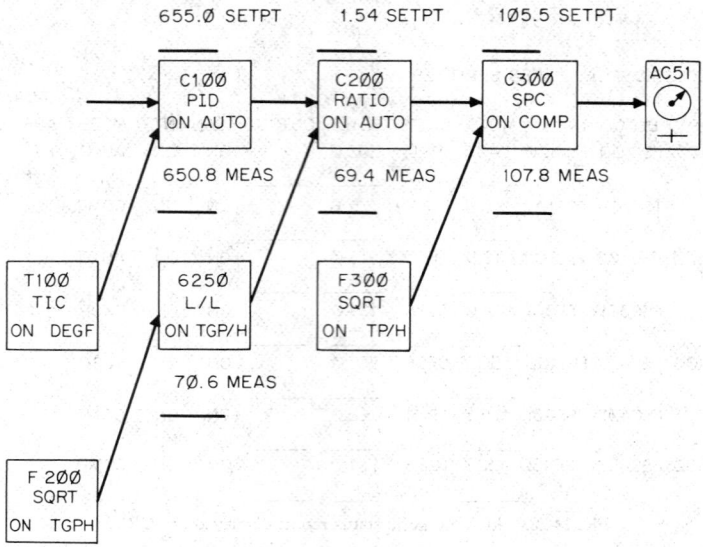

Fig. 24-30. Loop configuration diagram displayed on CRT.

PROCESS DISPLAY 7   BATCH REACTOR 5
UNIT: POLY 5 REC: 14 BATCH: 337 PHASE: RUN   CONTROL STATE: HOLD
SELECT FUNCT: EMSHTDN   HOLD MAINT   RESTART   JUMP   TRACE   TRACE OFF

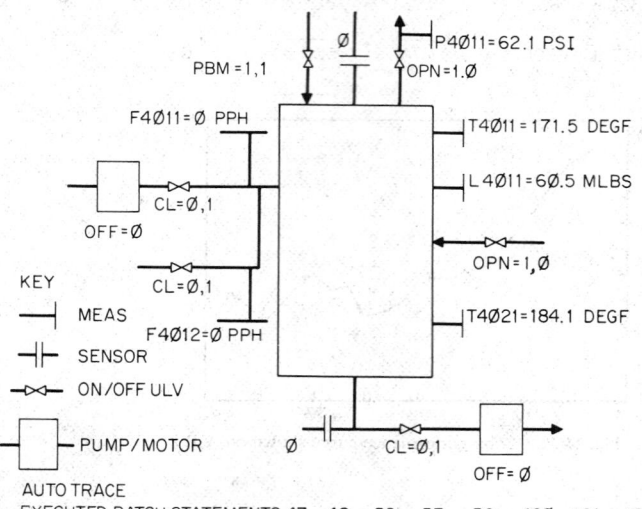

Fig. 24-31. Process equipment diagram shown on CRT.

directly involved in the process control system is usually shown; auxiliary equipment such as relief valves, extra pumps, blowdown lines, etc., is omitted.

38. Semigraphic panels employ separate areas for flow diagram and instruments (Fig. 24-36). Instruments are uniformly grouped in a grid below the flow diagram, located as closely as possible beneath their points of application. Semigraphics are smaller than full-scale graphics.

Graphic panels were widely used during the initial transition to centralized control rooms. Modern alphanumeric/graphic display hardware (CRT, projector, etc.) precludes the need for separate graphic panels. The information provided on graphic panels is most useful during initial operation of a new process, and is virtually ignored after the operators are familiar with the process.

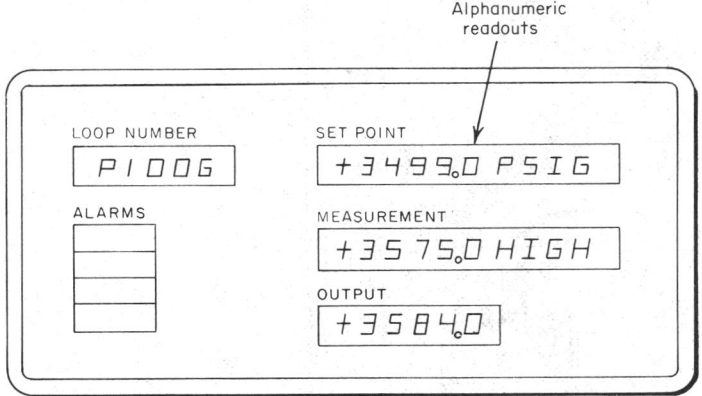

Fig. 24-32. Panel displays alphanumeric readout.

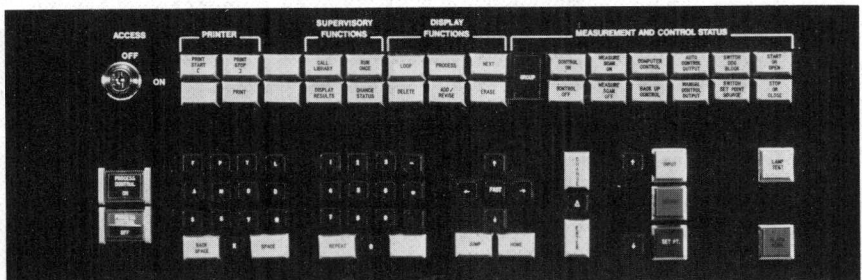

Fig. 24-33. Typical process-manipulation keyboard.

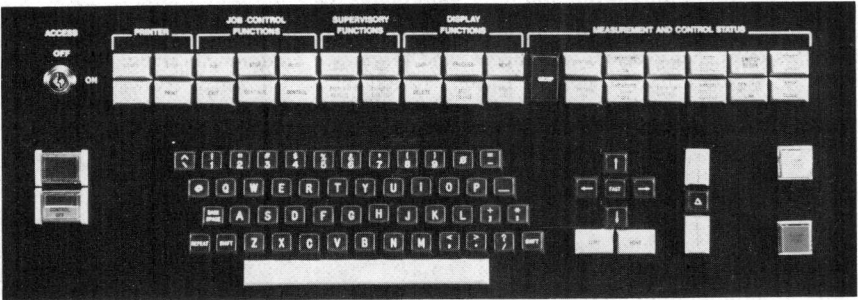

Fig. 24-34. Keyboard for entering computer data.

39. Alarm Annunciator Subsystems. Annunciators are self-contained devices used to indicate, but not record, alarm conditions (Fig. 24-37). By collecting all alarm indications in a single location, they supplement and elaborate upon the alarm indication provided on individual loop indicators and controllers. They provide visual and audible notice of alarm occurrences, and means by which the operator "acknowledges" the alarm. They can interpret and indicate alarm hierarchy (absolute, deviation, criticality), and frequently are physically arranged to represent the grouping of related variables.

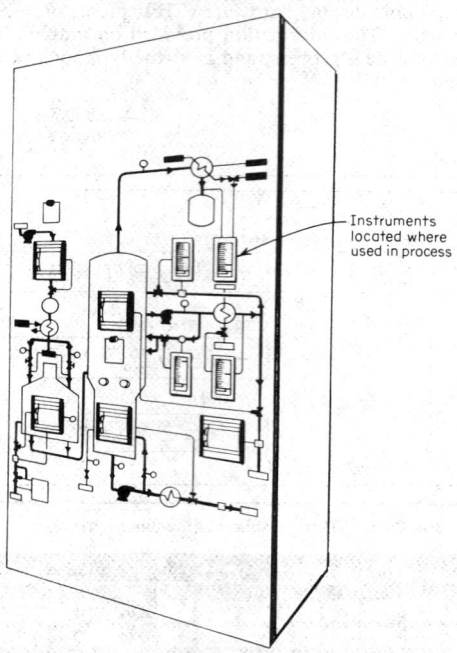

Instruments located where used in process

Fig. 24-35. Full-scale graphic panel.

40. Miscellaneous Equipment. Figure 24-38 shows the variety of devices used to augment the basic process instrumentation. This category includes meters, gages, lighted indicators, and all kinds of manual switches. Use of such equipment is prompted by cost (a lighted, alternate-action pushbutton is sufficient to turn a motor on and off), the nature of the display or control function (optimum hardware is usually different for each different function), hardware limitations (most conventional instrumentation has fixed display and manipulative capability), and customer preference (personalized ways of supervising the process).

41. Projectors, Workbooks, Other Reference Devices. Procedural and instructional information peculiar to an individual process, company preferences, etc., is available to operators through projectors, workbooks, and placards. Traditionally, these devices are physically separate from the control instrumentation and control panel. However, modern alphanumeric/graphic devices (CRTs, automatically controlled slide or microfilm projectors) provide the capability to integrate all operator aids.

THE DIGITAL COMPUTER INTERFACE

By G. A. Rosica

42. Control Using Computers. Computers have been used in process control in many different ways. A primary use, and one of the earliest, is data logging. In this application data are collected from the process, analyzed, tabulated, and outputted, either on request or as periodic reports. The resulting data are then utilized by individuals, who in turn control the process directly. The computer performs functions such as converting raw measurements to engineering units, compensating flow measurements for temperature and/or pressure, and alarm-limit checking.

A second major function of computers is supervisory control. In this case, process data are used for performance and optimization calculations. These calculations provide the appropriate set points for analog controllers performing the first line of control on the process. For example, an analog controller might control a flow by operating a valve. The value of the flow is read from a flow transmitter (inferred from the differential pressure) and compared by the analog controller with a set point provided by the supervisory computer. The valve is manipulated by the analog controller to maintain the flow at the specified value. The process operator observes the process and the changes taking place. He has the option of overriding the computer and adjusting the set points himself if he deems it necessary.

The third major use of computers is for direct digital control (DDC). Here the computer gathers data from the process and uses it to solve equations that are equivalent to the analog control functions. The computer then adjusts the final actuator (valve, etc.) to effect the appropriate regulatory control. The set points of these control loops may be entered by the operator or supplied by the computer performing supervisory calculations. The advantage of DDC over analog control lies in the area of flexibility for implementing more sophisticated control concepts. The computer allows a wider selection of more complicated control algorithms which can be interconnected easily for advanced control approaches. Adaptive tuning can improve the dynamic response of the plant to upsets under changing conditions.

Unlike analog control, however, where the regulatory control function for each loop is performed by a separate instrument, a failure of the computer in DDC results in all loops going off control. For this reason DDC has not been widely accepted without analog control

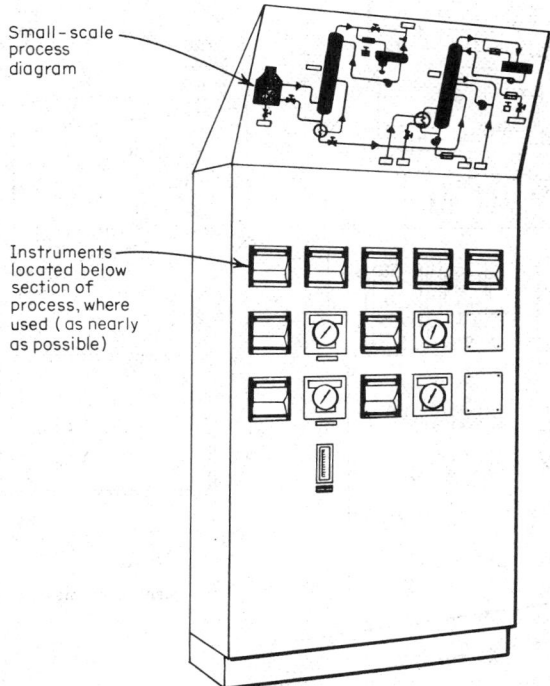

Fig. 24-36. Semigraphic panel with instruments grouped by function.

backup.   In each of these cases it is necessary to provide an interface between the process and the computer.   This typically consists of *input* and *output multiplexers,* as shown in Fig. 24-39.

43. Input Multiplexer.   The primary job of the input multiplexer is to scan many analog inputs.   These inputs are conditioned by filtering and amplification to make them compatible with an analog-to-digital converter, which provides data in the digital form that can be used by the computer (Fig. 24-40).   Input multiplexers are constructed as three basic types: single-ended, double-ended, and flying-capacitor.   They are typically designed to receive signals from flow or pressure transmitters or measurement devices such as thermocouples.

Single-ended multiplexers switch one wire only (Fig. 24-41).   This means that one wire of each input is connected to a signal common.   This approach is low-cost because only one switch per input is used.   This switch is usually a solid-state type, and therefore this type of multiplexer is capable of scanning thousands of points per second.   The solid-state switch, however, has the disadvantage of a much higher voltage drop (millivolts) than relay switches, and therefore solid-state single-ended multiplexers are usually limited to use with high-level signals (1 to 10 V full scale).   The main source of problems with single-ended multiplexers comes from having more than one common conductor in the system.   In most cases this second common is unintentional or results accidentally, as with the grounding of thermocouples.   A second problem occurs when any of the switches fails in the closed position,

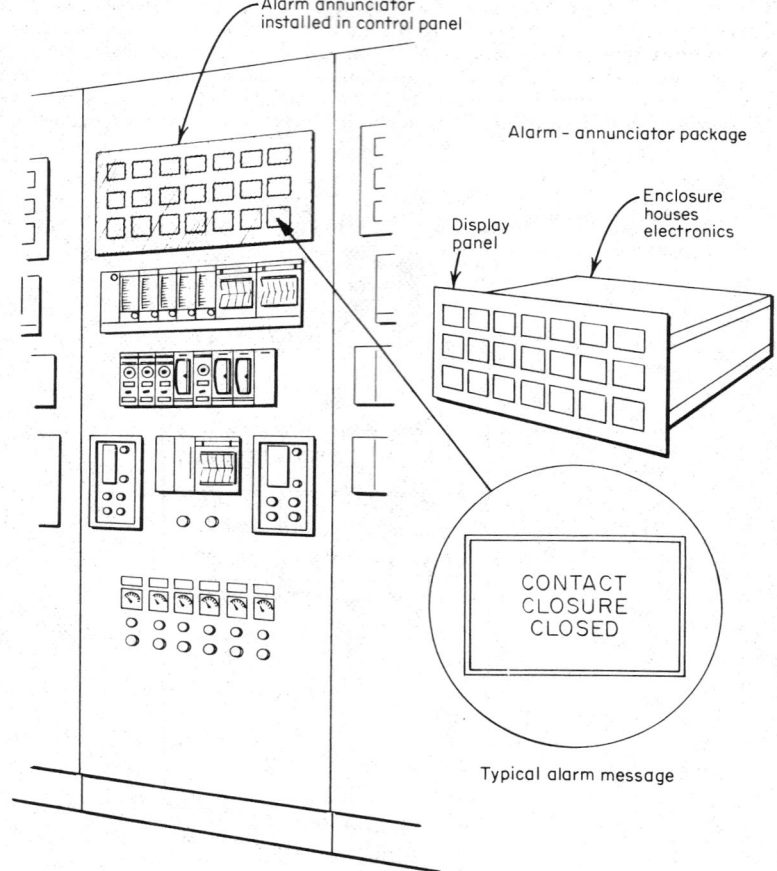

Fig. 24-37.   Typical alarm annunciator displays.

24-32

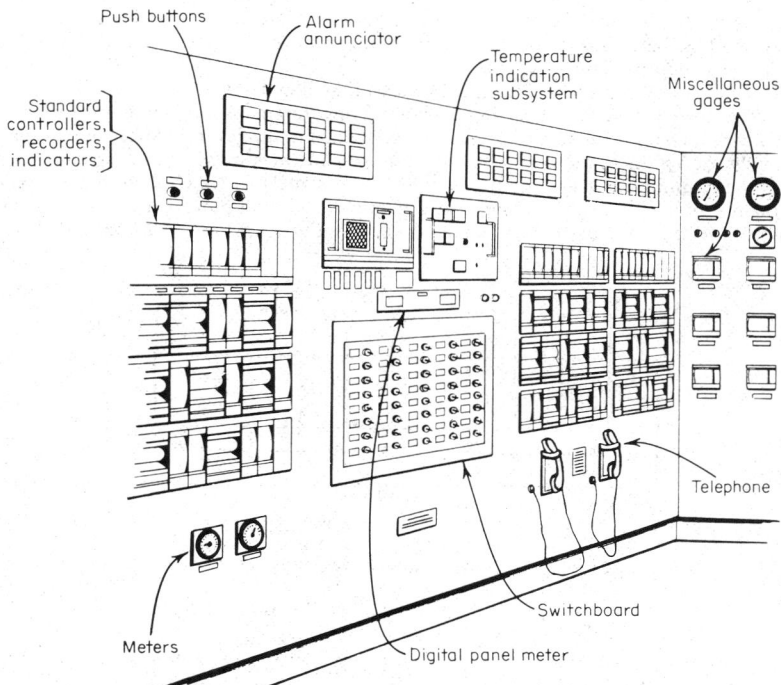

Fig. 24-38. Auxiliary equipment in control center.

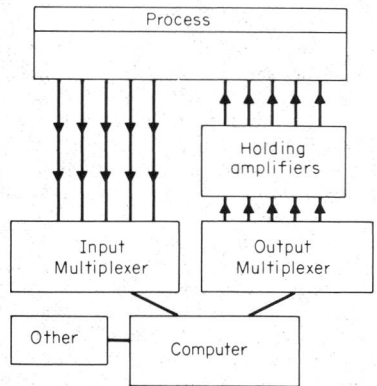

Fig. 24-39. Interface using input and output multiplexers.

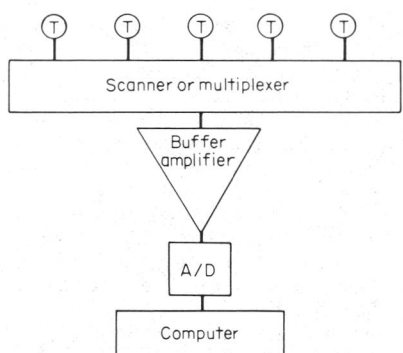

Fig. 24-40. Analog-to-digital converter feeds the computer.

disabling the entire multiplexer. This occurs because there will always be two input points connected together. Ironically, the only point that will result in a correct reading is that on which the switch has failed.

This problem is somewhat alleviated by use of the double-ended multiplexer (Fig. 24-42). This version switches both leads, thereby eliminating the signal common found in the single-ended case. To take full advantage of a double-ended multiplexer, a differential amplifier is generally used prior to the A/D conversion to provide maximum common-mode rejection. This type of multiplexer is often used for low-level signals where common-mode voltage is the main problem. For this reason, most double-ended multiplexers use relay switches for minimum voltage drop across the switch. Relay units are capable of scanning only hundreds

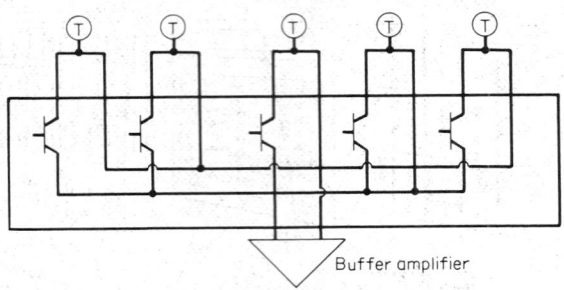

Fig. 24-41.   Single-ended multiplexer.

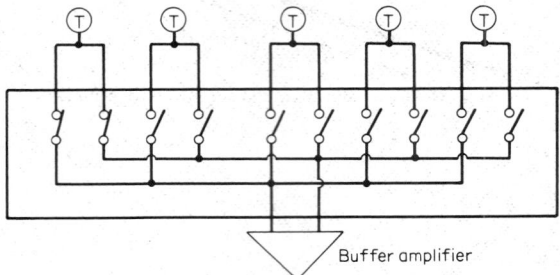

Fig. 24-42.   Double-ended multiplexer.

of points per second as compared with thousands of points for solid-state units. Double-ended multiplexers are becoming more available in solid-state versions as the cost of semiconductor switches decreases. The solid-state units are, however, usually limited to high-level signals only. A switch failure on a double-ended multiplexer can result in an undesired common between the point associated with the failed switch and each other point as selected. This is usually not as serious as a failure in a single-ended multiplexer where two points are being measured at once, but it can cause a reduction in accuracy that is difficult to detect.

The flying-capacitor multiplexer was developed to deal with the problems of the double-ended multiplexer (Fig. 24-43). Both leads are still switched, but there is no direct connection (even when the switches are on) between the field and the A/D converter. The field signal is connected to a capacitor when a point is not selected. In this way signal voltage charges the capacitor. When a point is then selected, the charged capacitor, not the field lines, is switched to a high-input-impedance A/D converter. This type of multiplexer provides the best common-mode rejection and isolation and limits the effect of a single failure to one point only, but is the most expensive because of the switching complexity.

44. Output Multiplexer.   An output multiplexer takes the result of a computer calculation and directs it toward the proper control device. This may be the set point of a controller for supervisory control or the output of a controller for direct digital control with analog backup. A given output multiplexer usually handles one of two major types of outputs. These are positional and velocity. The positional output tells the device what position to assume, for example, 23% of full scale, 45% of full scale, 92% of full scale, etc. A positional output requires a sample-and-hold device to maintain the output at the required position between command changes from the computer. This is usually accomplished by an integrating amplifier, which stores the signal on a capacitor. A velocity output tells the device how much to move and in what direction, for example, up 12%, down 14%, etc. Thus the output is incremented by the specified amount.

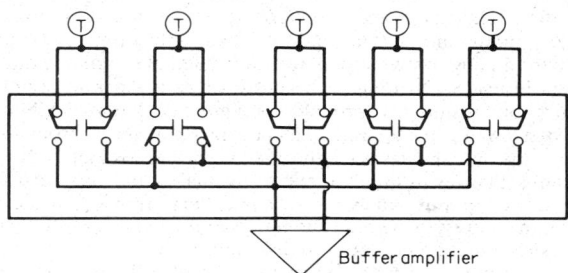

Fig. 24-43. Flying-capacitor multiplexer.

Positional and velocity outputs have typically been handled by the electronic integration of a pulse signal from the output-multiplexer addressing circuits. The integrator calculates the area under the pulse. In the positional case, the previous values stored are entirely replaced by the integrated area of the new pulse. In the velocity case, the previous values stored are incremented by the value of the output pulse. Different computers have different output multiplexers and use different parameters of the pulse signals to vary the outputs. Some examples of the types of pulse signals used are constant-width, variable-height voltage pulses; constant-width, variable-height current pulses; variable-width, constant-height voltage pulses; variable-duration contact closures; pulse trains and contact closure trains. Each type of output multiplexer typically requires its own unique circuitry, so that it is difficult, if not impossible, to change the method used, once it has been initially selected.

45. Controller Interface. Electronic analog controllers are interfaced to a computer for two major reasons. The first is supervisory set-point control; the second is analog backup for direct digital control. Although this interface represents a combination of input- and output-multiplexer connections, it is treated as a separate function because the trend is to optimize this interface. The evolution of plant control computers used in conjunction with analog instrumentation has been such that neither the computers nor their associated instrumentation have been ideally suited to the tasks required of them. In the field of analog/digital interfacing, standardization has been almost nonexistent. Also, although most computer control applications have been on a large scale, the interfacing techniques used require that each analog controller be treated as a separate entity when connecting it to the digital computer. Experience has shown that the problems associated with interfacing analog instrumentation to plant control computers increase almost linearly with the size of the job. The trend is to provide the control equipment and necessary interfacing hardware on a subsystem basis.

Considering the controller interface as a single functional interface problem allows three major improvements over previous approaches. The first is to minimize controller-to-computer interface wiring. The second is to provide an interface that does not require changes in the controllers themselves, when they are being interfaced to different computers. The third is to reduce the cost of the system interface by eliminating duplication of functions in the controller system and the computer system.

Previous interfacing methods treated each analog controller as an independent entity and tied it to the computer as a separate unit. This was done by using appropriate input points on the input multiplexer and output points from the output multiplexer. Thus a multiplicity of wires between the computer and the analog controller was necessary. By integrating the input and output multiplexing functions into the controller equipment, it is possible to reduce drastically the number of wires involved. Typically, 8 to 10 wires were required between the computer and each controller. Thus, for 200 controllers, there would be between 1,600 and 2,000 discrete wires. Incorporating the input and output multiplexing into the analog control system can reduce the wiring to six time-shared conductors. This not only reduces the system cost, but drastically simplifies the task of installation and decreases the chance of error.

The standardization of the controller interface among different computers can be greatly enhanced by approaching the problem in a manner similar to the way the problem of interfacing peripherals to computer systems has been approached. Peripherals interface to a computer in digital language and therefore are more universal than pseudo-analog interfaces such as input and output multiplexing. Integrating the input and output multiplexing into the analog control system provides a digital communication medium for all analog controllers instead of treating each controller as a separate pseudo-analog entity. Such a communication medium also provides for a great deal more flexibility than has been provided in the past. With this approach, a controller can accept from a computer either positional or incremental updates to either set points or outputs. It can also accept commands for status changes from automatic to manual or manual to automatic. The set point, measurement, or output values for all controllers are easily made available to the computer. This is in contrast to the rigid and limited functional capability typically provided by the use of classical input and output multiplexing.

The elimination of duplicate functions in the control and computer systems is accomplished by taking advantage of field signal conditioning that has already taken place in the analog control system prior to introducing signals into the computer. In the past, computer input multiplexers have provided all the necessary buffering of field signals prior to the conversion of these signals into digital form. This was typically done in parallel with similar conditioning provided by the control equipment. Integrating the input multiplexer into the analog control equipment helps eliminate duplication of equipment.

This approach also allows the multiplexing equipment to be segmented. Since the multiplexing equipment can take advantage of the signal conditioning provided for by the control element, it is inherently less expensive than previous computer multiplexers. This reduced cost makes possible the segmentation of this function, that is, the use of multiple multiplexers and A/D converters. Typically, a single multiplexer and A/D converter might serve only 16 controllers. This segmentation with multiple A/D converters then reduces the required speed of these units to allow a further cost reduction.

Controllers are placed in a nest-type arrangement with the communications hardware, as shown in Fig. 24-44. A typical nest handles 16 of these controllers. The controllers are connected to the digital communications logic, which is contained in the same nest assembly by means of factory-installed backplane wiring, eliminating the need to run this wiring at the job site. Multiples of these modular nest assemblies can be connected to the same serial communications cable, and thus to a single data communication channel of a digital computer (Fig. 24-45).

This segmented, or modular, configuration allows placing the control equipment close to the process rather than near the computer center. In those instances where the computer center is remote from the process or where there are several process unit areas and only one computer center, a significant saving in field wiring can be achieved. The many field wires otherwise required between the field and the computer center are replaced by a few wire serial data communications links. In such an approach it is possible to provide photooptic or other forms of isolation in the serial communication link so as to separate completely the grounds of the process-located control equipment and the remote computer installation. Such isolation eliminates many noise problems that have plagued process computer installations in the past.

46. System Security. With the conventional computer interface to the process, the security problem is one associated with the computer system and its input and output multiplexer. In the approach that integrates the multiplexing with the control equipment, security must be provided in both areas. In either case the security considerations fall into two broad categories. The first is to ensure that the particular control or measurement point desired was correctly addressed. The second is to ensure that the value transmitted to or from the control equipment was transmitted accurately.

If the measurement read into the system is one from a different transmitter than the one intended, the computer system will think the state of process is quite different from that actually existing. This error may cause it to take unreasonable control action to restore the desired operating level. A similar problem can occur when the computer system accurately addresses the measurements desired, computes the appropriate changes to be made, but then outputs them to the wrong controller or actuator. In either case it is not sufficient merely to know that an error has occurred. It is necessary to check for errors on a real-time basis and prevent any unwanted changes to the process from taking place when an error occurs. This error check can be accomplished by the use of parity, check sums, or any of the more sophisticated redundant data codes that are in use, to check the functioning of the hardware equipment prior to the movement of data in the desired direction. Security can also be achieved by having the program check the result of a command to change a controller set point or output value. One of the most effective ways of ensuring appropriate changes to process outputs from the computer or the operator is to provide a *live* joy stick. This approach uses a gradual change to the parameter selected, coupled with a live reading of the parameter under change, to ensure that the change is taking place to the appropriate parameter and in the appropriate magnitude.

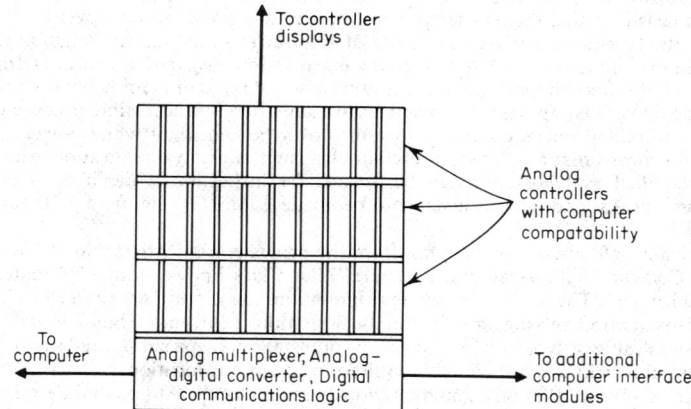

Fig. 24-44. Typical computer interface module with integral analog controllers.

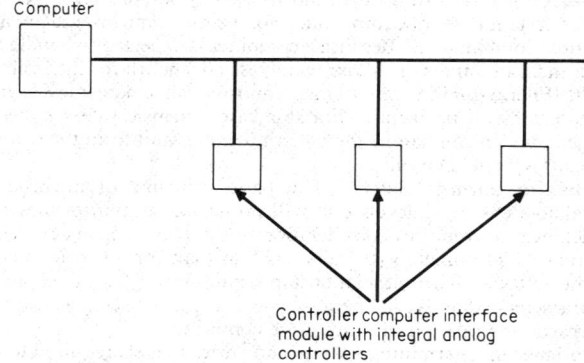

Fig. 24-45. Computer control system interfacing with analog controllers.

INTRINSIC SAFETY AND PROCESS CONTROL
INSTRUMENTS

By W. Calder

47. Instrument Location. In the application of process control instruments it is not unusual for them to be installed in hazardous locations; this is true particularly of the transmitting or final control devices. A hazardous location is one in which the presence of combustible gases or vapors presents an explosion hazard. These locations are fully defined in Article 500 of the National Electrical Code, and summarized in Table 24-6. Intrinsic safety is one of several techniques for making electrical equipment safe for use in such locations. It deserves special attention because it is the accepted method for protecting process controls.

48. Intrinsic Safety Defined. A circuit is considered to be intrinsically safe if it limits energy, even under abnormal equipment conditions, to levels that cannot possibly initiate an explosion. This does not mean that the circuit cannot produce a spark; but any spark that does occur will have insufficient energy to cause a flame to propagate. Intrinsically safe systems are used because they are reliable, convenient for the user, and economical to install. The techniques employed render the system safe, almost in spite of the user. The system is rendered unsafe not by neglect, but only by a deliberate act.

Other approved systems require constant vigilance if they are to be maintained safe. Neglect can easily render them unsafe. For instance, an explosionproof cover is safe only so long as all the bolts are properly tightened. If an electrician tightens only half of the bolts, the housing may no longer be able to contain an explosion. Also, if the cover is dropped or corrodes or if the smooth mating surface is otherwise damaged, it is probably no longer safe.

Intrinsically safe equipment is convenient to maintain. It is accessible because there are no heavy covers and no requirements that the cover be kept tight while power is on. In practice, this means that a differential pressure transmitter or a valve positioner installed in an area classified as Division 1 can be opened for inspection while it is in operation. Calibration checks and adjustments can be made safely in the field without special precautions.

Intrinsically safe circuits are exempt from the provisions of Article 500 of the National Electrical Code and fall instead under Article 725 as Class 2 power-limited, remote-control, or signal circuits. The code imposes few limitations on circuit wire size, insulation, or method of installation, relying for safety on the limitation of current. Thus a designer can use No. 22 wire in cable or No. 18 wire in any type of installation at a substantial saving over the No. 14 wire and rigid conduit usually required in hazardous locations.

In practice, this installation economy is not always realized. Although the Code does not call for it, conduit or thin wall tubing may be necessary to minimize pickup or to provide mechanical protection.

49. Suitable Applications for Intrinsic Safety. In general, any system that uses a low power signal level is potentially suitable for intrinsic safety design. However, this technique is not practical for certain types of equipment. For example, intrinsic safety and computers are not a compatible combination. Because a computer is expected to handle hundreds and, sometimes thousands, of inputs in a reasonably sized enclosure, intrinsic safety design becomes difficult. Energy-limiting requirements and allowable power levels and mechanical design details pose difficult problems. For this case, intrinsic safety is best provided in external equipment, which can handle the worst potential fault from the computer and still provide intrinsically safe field circuits.

50. Approaches to Intrinsic Safety. The basic principle of intrinsic safety is the limitation of available energy to levels that will not cause ignition of the vapors or gases involved. To achieve this requires consideration of the two elements of electrical energy: voltage and current. Unfortunately, there is no Ohm's law for intrinsic safety because, for a given circuit, the voltage-to-current relationship is nonlinear. Also, the nonlinearity differs with circuit parameters; that is, if energy-storage elements such as capacitors and/or inductors are present, the relationships are more complex.

To aid the designer in determining voltage and current levels required to cause ignition of circuits with particular characteristics, curves have been developed based on careful testing. The most current curves have been prepared by the Safety in Mines Research

Table 24-6.   Hazardous Classifications, Groupings, and Divisions*

Class I.   Locations in which flammable gases or vapors are, or may be, present in the air in quantitites sufficient to produce explosive or ignitable mixtures.

    Division 1 locations

       a. Hazardous concentrations exist continuously, intermittently, or periodically under normal operating conditions.

       b. Hazardous concentration may exist frequently because of repair or maintenance operations or because of leakage.

       c. Where breakage or faulty operation of equipment or processes which might release hazardous concentrations of flammable gases or vapors might also cause simultaneous failure of electrical equipment.

    Division 2 locations

       a. Where hazardous volatile liquids, vapors, or gases are normally confined within closed containers or closed systems from which they can escape only in case of accidental rupture or breakdown of such containers or systems, or in case of abnormal operation of equipment.

       b. Hazardous concentrations are normally prevented by positive mechanical ventilation but might become hazardous through failure or abnormal operation of the ventilating system.

       c. Areas adjacent to Division 1 locations to which hazardous concentrations of gases or vapors might occasionally be communicated.

    Group A   Atmospheres containing acetylene.

    Group B.   Atmospheres containing hydrogen or gases, or vapors of equivalent hazard such as manufactured gas.

    Group C.   Atmospheres containing ethyl ether vapors, ethylene, or cyclopropane.

    Group D.   Atmospheres containing gasoline, hexane, naphtha, benzine, butane, propane, alcohol, acetone, benzol, lacquer solvent vapors, or natural gas.

Class II.   Locations which are hazardous because of the presence of combustible dust.

    Division 1 locations

       a. Areas in which combustible dust is or may be in suspension in the air continuously, intermittently, or periodically under normal operating conditions, in quantities sufficient to produce explosive or ignitable mixtures.

       b. Where mechanical failure or abnormal operation of machinery or equipment might cause a combustible dust mixture to be produced, and might also provide a source of ignition through simultaneous failure of electrical equipment, operation of protective devices, or from other causes.

       c. Areas where dusts of an electrically conducting nature may be present.

    Division 2 locations

       a. Areas where dangerous concentrations of suspended dust are not likely, but where dust accumulations might form on, or in the vicinity of, electrical equipment in sufficient quantities to produce explosive or ignitable mixtures except under conditions:

          1. Where deposits or accumulation of dust may interfere with the safe dissipation of heat from electrical equipment or apparatus.

          2. Where deposits or accumulation of dust on, in, or in the vicinity of electrical equipment might be ignited by arcs, sparks, or burning material from such equipment.

    Group E   Atmospheres containing metal dust (magnesium, aluminum, bronze powder, etc.).

    Group F   Atmospheres containing carbon black, coal, and coke dust.

    Group G   Atmospheres containing grain dust (flour, starch, pulverized sugar and cocoa, dairy powders, dried fat, etc.).

Class III.   Locations where easily ignitable fibers or flyings are present but not likely to be in suspension in quantities sufficient to produce ignitable mixtures.

    Division 1 locations

       a. Areas where easily ignitable fibers or materials producing combustible flyings are handled, manufactured, or used.

          1. Such locations include some parts of rayon, cotton, combustible-fiber manufacturing and processing plants; cotton gins and cottonseed mills; flax processing plants; clothing manufacturing plants; woodworking shops; and establishments and industries involving similar hazardous processes or conditions.

    Division 2 locations

       a. Areas where easily ignitable fibers are stored or handled (except in the process of manufacture).

          1. Such fibers include rayon, cotton (cotton linters and waste), hemp, kapok, excelsior, sisal or henequen, istle, jute, tow, cocoa fiber, oakum, baled waste, Spanish moss, and other materials of similar nature.

---

* Essential information defining limitations as given in NEC Art. 500.

NOTE: Some plant areas in the manufacture, handling, and storage of explosives or ammunition and nitrocellulose products such as celluloid photographic films, etc., involve conditions that are not covered by NEC classifications. This is particularly true where black powders, smokeless powder, dust from TNT, and other explosives are present. These areas require special equipment and installation methods (refer to Par. 3.6 and appropriate Engineering Standards).

Establishment in the United Kingdom. These are included with the British Standards Institute draft for the requirements for intrinsic safety (see References, Par. 24-71). Another set of curves was developed by The Foxboro Company; these are included in Fig. 24-46 and agree almost exactly with the SMRE curves. However, they are for resistive circuits only.

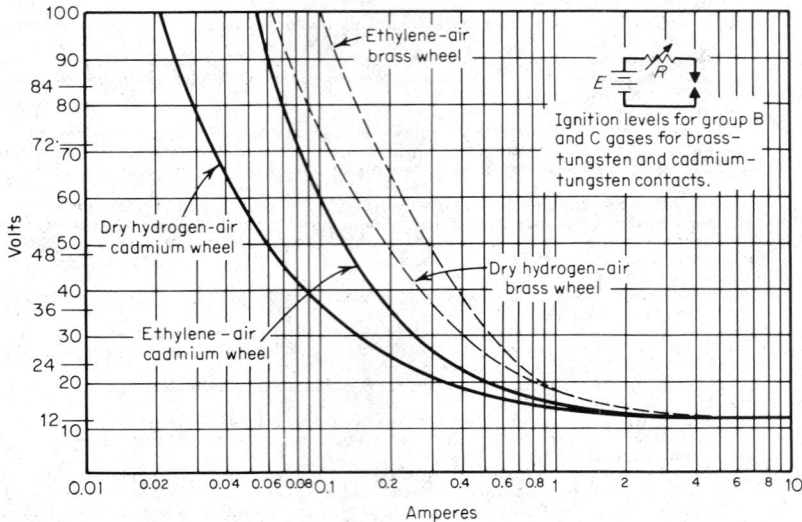

Fig. 24-46. Ignition curves for several mixtures under different tests.
Tests were performed for group B (hydrogen) and group C (ethylene) gases. The voltage source consisted of 12-V wet-cell automobile storage batteries connected in series to reach the desired voltage levels (12, 24, 36, 48, 60, and 84 V dc). Circuit inductance was minimized in the wiring methods, and energy-limiting resistance was used. Test apparatus was a Cocasset Corporation Incendivity Tester using the German designed wheel-and-whisker contact mechanism. The whiskers were of tungsten in all cases. The wheel was of brass in one case and cadmium in the second. The gas-air mixture was adjusted to produce ignition in a circuit consisting of a low source impedance (less than 1 Ω) 24 V dc supply through a 0.095 Hy air core inductor to the test contacts. The currents used in the reference circuit for each condition were as tabulated below.

Current, mA

Contact Material	Gas—air mixture	
	Group B: hydrogen—air	Group C: ethylene—air
Brass-tungsten	65	95
Cadmium-tungsten.............	36	65

The reference circuit proved the gas-air mixture before and after each test run when ignition was not achieved in the circuit under test.

Several approaches to intrinsic-safety design are in use today. Each has the basic elements: voltage and current limiting. They are as follows:

51. The loop approach provides a power package that positively prevents line voltage from reaching secondary circuits, such as by using a properly constructed power transformer. This is followed by whatever additional voltage and current limiting is necessary to limit energy, such as current-limiting resistors and voltage-clamping zener diodes. This approach is explained in detail in several papers by Hickes[1-6*] and in Magison's book.[7] Figure 24-47 shows a schematic representation.

52. The passive-barrier approach makes use of a single package of components that

---

*Superior numbers correspond to numbered references, Par. **24-71.**

provide the required energy limiting. It is inserted in the control loop at the interface between the safe and hazardous areas. The passive barrier uses a combination of zener diodes, resistors, and fuses and a low (less than 1 Ω) ground connection to achieve the required energy limiting.[8] Figure 24-48 gives a schematic representation.

53. The active-barrier approach also serves as a barrier between the control room and the field at the safe-unsafe area interface. The barrier in this case is high resistance in each path leading to the field, preventing dangerous energies from reaching the hazardous area. The measurement signal is fed into a high-input-impedance amplifier and simply repeated. A similar arrangement is used for the control output signal. This device also supplies power for the field circuits from an internal supply, using a properly constructed transformer and current-limiting resistors (Fig. 24-49).

54. A hybrid approach uses both the active and passive barriers in combination. Voltage limiting is provided using SCRs in a crowbar circuit in place of the zener diodes in the passive-barrier portion. The active barrier uses high-input-impedance amplifiers with high-value resistors in each path to the hazardous area. Current-limiting resistors are provided for the supply voltage to the field. See Fig. 24-50 for a schematic representation. Any of the

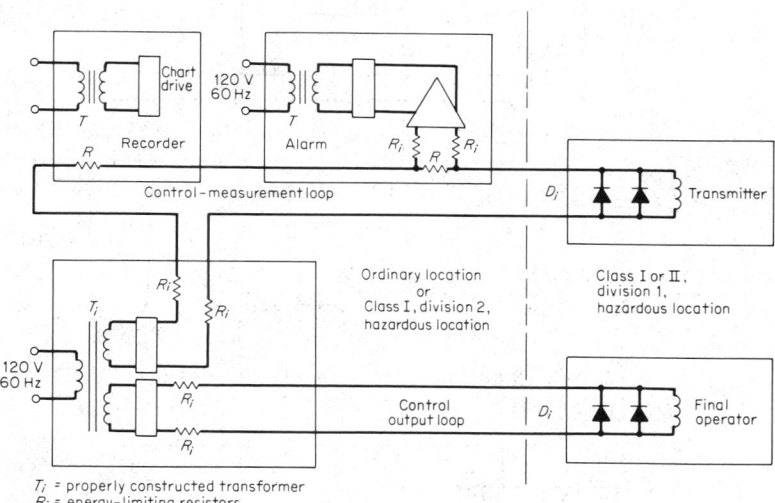

T_i = properly constructed transformer
R_i = energy-limiting resistors
D_i = Arc suppression diodes

Fig. 24-47. Loop approach designed for intrinsic safety.

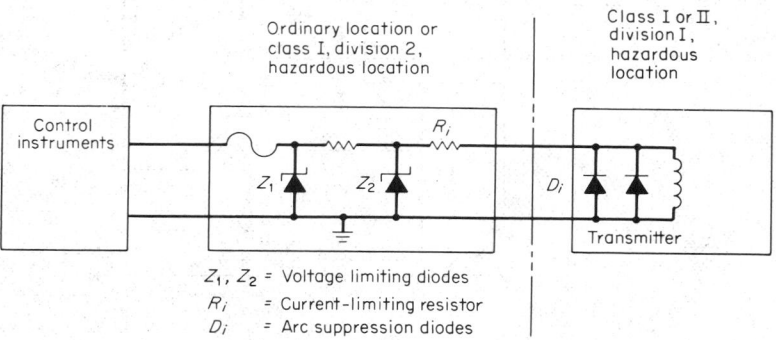

Z_1, Z_2 = Voltage limiting diodes
R_i = Current-limiting resistor
D_i = Arc suppression diodes

Fig. 24-48. Passive-barrier approach for intrinsic safety.

approaches listed provides equivalent safety, but one may be preferred over another, depending on the application.

55. Requirements for Intrinsic Safety. Several standards have been prepared covering the requirements of intrinsic safety. The designer must determine which requirements to meet, based on the market area of the equipment. A representative sampling of existing standards is listed in Refs. 9 to 16, Par. **24-71.**

56. Design Considerations. In designing process controls to function intrinsically safely in hazardous areas, there are several basic considerations. The intrinsic safety is determined

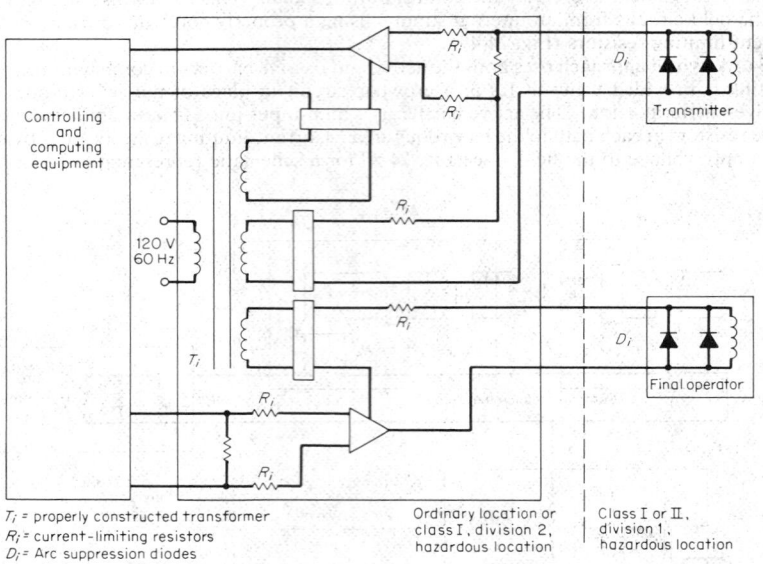

T_i = properly constructed transformer
R_i = current-limiting resistors
D_i = Arc suppression diodes

Ordinary location or class I, division 2, hazardous location

Class I or II, division 1, hazardous location

Fig. 24-49. Active-barrier configuration for intrinsic safety.

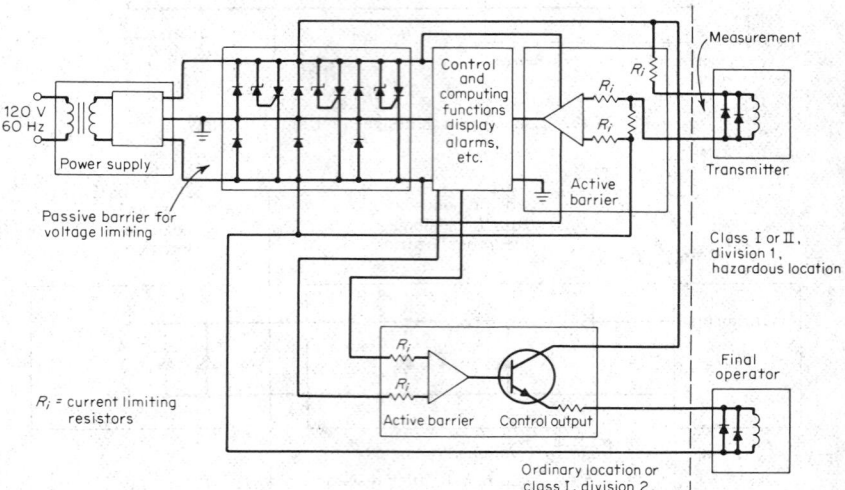

R_i = current limiting resistors

Fig. 24-50. Approach to intrinsic safety using hybrid barrier.

by a combination of the power source, field equipment, any other devices connected in that circuit, and the interconnecting wiring. The fewer devices needed for intrinsic safety, the less complicated the design becomes and the more flexible the process control system can be without affecting intrinsic safety. An important design consideration is to try to place all intrinsic-safety dependency as near the hazard as possible.

Flexibility is important. If some kind of hazardous-area interface equipment is provided, it should be capable of safely limiting the maximum expected fault voltage that could occur. Such voltage is normally considered some nominal line-voltage level, such as 250 V. By placing full protection at the hazardous-area interface, or in that proximity from the circuit point of view, applications engineers will have complete freedom in setting up the controlling and computing functions, provided the maximum stated voltage allowed is not exceeded.

Another basic consideration is the mechanical layout from printed-wiring boards out to the field. The most important consideration is separation. Intrinsically safe circuits must not be placed too close to nonintrinsically safe circuits. This concept must be applied out to and including the final field termination.

Other considerations that must be carefully analyzed involve design tradeoffs. In general, the ideal approach for intrinsic safety is diametrically opposed to the usual functional design requirements. As a result, the designer must deal in areas of uncertainty and ambiguity where interpretations of the requirements govern. For this reason, there is a certain amount of guesswork involved even after much experience has been gained in dealing with approval agencies. Some of the tradeoff considerations are discussed in Par. 24-57.

57. Design Tradeoffs. The layout of the device in terms of providing proper spacing and separation can influence package size.

Because one of the advantages of intrinsic safety is to be able to use smaller wire sizes, this needs to be considered in determining lead-length allowances and the additional load smaller wire can present.

Maximum open-circuit voltage determines allowable lead length owing to the distribution capacity of cables. If the open-circuit voltage to the field is too high, it can severely limit allowable lead length. This is particularly true of the West German requirements. It is not as critical in other areas (e.g., the United Kingdom and North America).

Safety-dependent components must be conservatively rated to handle fault-power situations (for metal-film resistors, five times fault power). This means using what may appear to be extremely highly rated devices in normally low power situations.

Temperature rise must be considered such as that developing in transistors under thermal runaway conditions. This applies only to transistors in the hazardous area, but it can alter the requirements considerably by making necessary a heat sink that would not be required under normal working conditions.

Energy-storage components must be treated properly in designing field-mounted devices. Arc-suppression diodes must be placed across inductors, and either the value of capacitors must be limited to low levels (1 μF) or a discharge path provided through a reliable resistor.

The maximum specified voltage to the field must not be exceeded. For field connections where both intrinsically safe and nonintrinsically safe circuits are involved, the requirements for separation are conservative and may mean providing separate wiring components or separation of up to 2 in.

This list represents the kind of things that deserve careful study. When starting a design project, it is advisable to contact a test agency early. For a minimal investment, circuits can be tested and evaluated and a comprehensive test report provided. The test agency will not dictate how to design, but it will point out in the design the problems that will adversely affect safety. When striving for certification, disagreements in the safety area can result in much time-consuming discussion.

SERVICE CONDITIONS AND INSTRUMENT RELIABILITY

By J. L. Lee

58. General Requirements. This section explains the need for control system reliability and for careful application of process control instrumentation. The manufacturer must select the appropriate electronic components and circuit packaging techniques and use sophisticated reliability assurance procedures. The user must be able to understand and evaluate these in order to assess the quality of new instrument lines and systems, and most importantly, he must apply these instruments correctly.

The need for process control system reliability, together with the special service conditions encountered in the process industries, has affected the design and manufacturing practices for high-quality process control instrumentation. Process control instruments are usually required to operate continually for periods of months or years; they are expected to have a life of over ten years. In addition, they are required to provide reliable operation under environmental conditions that can be extremely harsh. In general, service conditions in the process industries are more severe than those in most other industrial environments.

NEED FOR PROCESS CONTROL SYSTEM RELIABILITY

59. Trends. The pressure for improved control reliability has grown steadily for a number of reasons:

a. The trend in the industry is toward fewer and larger plants, and so plant shutdown has a more immediate effect on operating company profit.

b. To reduce the capital cost of new plants, they have less buffer storage capacity, so that the effect of a plant section failure is transmitted upstream and downstream within a short time.

c. The increasing integration of the chemical industry, the growth of huge chemical complexes containing many interdependent plants with interrelated processes, means that the failure of one plant will affect many others.

d. The growing awareness of pollution calls for measurement and control of emissions, and often for tighter and more reliable control of the process, than would be justified on a purely economic basis.

e. Legislative action, such as the Occupational Hazards and Safety Act, is continually demanding higher standards of safety.

f. Increasing sophistication among buyers and ultimate consumers is putting pressure on the process industries for better control of product quality and purity.

60. Impact of Reliability. The costs of unreliable instrumentation appear in several ways. There is the problem of delayed start-up of new plants, and higher rates of scheduled and unscheduled maintenance. There are the costs associated with off-specification product and lost production. There is the cost of plant and process equipment damage. And there is the problem of human accidents and the resulting financial liabilities.

The two main applications of reliability engineering techniques to plant instrumentation are in plant safety and plant availability. The emphasis is different. In a plant having a high risk of explosion, the emphasis, particularly in the design of a shutdown system, is toward low probability of failure. Where safety is not a factor, the criterion is the fraction of time for which the plant is capable of producing the product, that is, its availability. The difference between reliability and availability can be explained as follows: The reliability of a system is the probability that it will operate without failure for a defined period of time. The availability of a system is the probability of finding the system in an operating state. It is a function of mean time to repair as well as mean time between failures.

It is also necessary to recognize that there are two kinds of failure. There is the normal failure wherein an instrument ceases to perform its function, and there is also the possibility of degradation of performance of the instrument. Degradation is much more difficult to recognize and correct. Zero shifts, span changes, or gradual loss of calibration in critical instruments can cause serious loss of plant efficiency and can even put the plant into an unsafe condition.

61. Plant Safety. Plant safety relates ultimately to human safety, but it usually is effected by preventing fires and explosions or the release of toxic liquids or gases which could injure plant operators or persons in the surrounding community.

The need for the highest reliability is in automatic shutdown systems in high-risk chemical plants. The reliability of alarm systems is also important in a subtle way. If an alarm system is continually signaling false alarms, the operator will find some way of disabling it so that, when a real danger occurs, the system gives no warning.

One of the difficulties in designing for plant safety is deciding how much safety is enough. This problem can be approached by considering the average risk to a person working in the chemical industry. On the basis of 25 killed per year out of 360,000 employees subject to the U.K. Factories Act (1965 figures), there is a fatality risk per hour of exposure to the risk of 0.035×10^{-6}. Assuming that the risk from a new high-risk plant could be allowed to be as high as the additional average plant and that two persons were continually present in the plant, the risk that one person would be killed would be 6×10^{-4} per year.

The risk that a potentially lethal event occurs must, of course, be minimized. On the basis of the calculation above the probability of occurrence of an event, which would be expected to kill one person on average if it does occur (an explosion, a fire, etc.), should, by virtue of the safety system design, be much less than 6×10^{-4} per year. The sum of the probabilities of all such events should be so much less than 6×10^{-4} that when the probabilities of all other unspecified lethal events are taken into account, the risk of killing one of those two persons per year does not exceed 6×10^{-4}. The fatality rates for several comparable types of risk are given in Table 24-7.

Table 24-7. Fatality Rates for Various Risks

Source of Risk*	Estimated Fatality Rate (Deaths per Person per Hour of Exposure)
Industry:	
Manufacturing (total)	0.021×10^{-6}
Chemical and allied	0.035×10^{-6}
Construction	0.091×10^{-6}
Quarries	0.36×10^{-6}
Transport (drivers and passengers):	
Rail...	0.07×10^{-6}†
Road..	0.43×10^{-6}
Air	1.7×10^{-6}
General:	
Home accidents	0.036×10^{-6}
All accidents	0.047×10^{-6}
Suicide	0.012×10^{-6}
Lightning	0.000022×10^{-6}

*The information given applies to Great Britain and has been extracted from Annual Abstract of Statistics, No. 103, 1966 (H.M.S.O.); Annual Report of H.M. Chief Inspector of Factories, 1965 (H.M.S.O.); Report of H.M. Chief Inspector of Mines and Quarries for 1965 (H.M.S.O.).
†Based on a 10-year average.

62. Data on Instrument Reliability. Little failure-rate data are available in the open literature. Table 24-8 is a compilation of available data based on field failure data collected by instrument users. Instrument failure rates can also be predicted based on the failure rates of the components used in the design. This is particularly true of electronic instruments: a great deal of data are available on the failure rates of electronic components.

For new instrument types for which there has been no opportunity to accumulate field failure data, calculated reliability figures are the only figures available. Most manufacturers of high-quality instruments will supply these figures on request.

Availability is a useful parameter for evaluating process control systems having redundant (backup) elements. Starting from the fully operating state, the availability of such a system decreases with time, until it reaches a value known as the steady-state availability. It is the steady-state value of availability that is of interest to the user. Availability is calculated by a computer manipulation of matrixes known as *transition rate analysis.* Most suppliers of instrumentation systems will provide predicted availability figures. However, no industry standards exist that make predicted reliability or availability figures comparable between system suppliers.

63. Service conditions are those to which equipment is likely to be exposed during storage, transportation, and operations:

a. Environmental service conditions, related to physical environment such as temperature, vibration, etc.

b. Service conditions related to system power supply, including frequency, voltage, transients, etc.

64. Environmental parameters must be considered. In some cases, it is necessary to consider not only the range over which that parameter is expected to vary but also its rate of change. The environmental parameters are:

    a. Temperature (and rate of change).
    b. Humidity and air pollution.
    c. Barometric pressure (and rate of change).
    d. Vibration and repetitive shock.
    e. Electromagnetic fields.
    f. Radioactivity (in nuclear power applications).
  65. Effects of environmental parameters are described as follows:
    a. Temperature (and rate of change). Temperature affects the parameters of nearly all semiconductor devices. Too high a temperature will cause out-of-specification performance; beyond that, thermal runaway will cause permanent damage. Low temperatures can cause out-of-specification performance, but usually do not cause permanent damage except when ice forms where water is allowed to penetrate. Temperature changes can cause cracking of conductors or protective seals when materials having different thermal expansion coefficients are bonded together—even plastic-to-metal seals which normally do not crack because the plastic will flow, but the plastic will crack if the temperature changes too fast.

Table 24-8. Data on Instrument Reliability

Instrument	Failure rate, faults/year	
	Actual	Assumed
Control valve (*p*)	0.25	
Solenoid valve	...	0.26
	...	0.1 (design)
Controller (*p*)	0.38	
Differential pressure transmitter (*p*)	0.76	
Variable area flowmeter transmitter (*p*)	0.68	
Thermocouple	...	0.088
Temperature trip amplifier:		
Type A	2.6	
Type B	1.7	
Type C	0.1	
Pressure switch	0.14	
O₂ analyzer	2.5	
Tachometer	...	0.044
Stepper motor	...	0.044
Pressure gage	...	0.088
Relay (*p*)	0.17	
Indicator (moving-coil meter)	...	0.026
Recorder	...	0.22

p = pneumatic.
SOURCE: Anyakora, Engel, and Lees, "Some Data on the Reliability of Instruments in the Chemical Plant Environment."

    b. Humidity and atmospheric contamination. High humidity can cause *tracking*—short-circuiting between conductors on printed-wiring boards and/or chemical attack on conductors, leading to catastrophic failure. These effects are most severe when condensation occurs. Where low-level analog signals are being handled, the reduction of the area resistivity of printed-wiring boards can cause zero drift, span, or calibration errors.
    High humidity causes problems only when there are contaminants present. These come from the manufacturing process or from atmospheric contamination. On top of "normal" air pollution, chemical plants and paper mills, for example, can have exceptionally high levels of specific contaminants. Adverse effects are generally accelerated by high temperature.
    Tables 24-9 and 24-10 relate to atmospheric contamination. They list the type of contaminants to be considered and give an indication of the concentration likely to be encountered.
    Low humidity permits electrostatic-charge buildup; the resultant random discharges cause spurious pulses that can interfere with the operation of digital equipment. This effect can interfere so completely with digital computers used for process control that they will fail to control the plant at all.
    c. Barometric pressure and rate of change. Instruments designed to withstand normal changes in barometric pressure in service could be destroyed if transported in the unpressurized cargo hold of an airplane.

Table 24-9. Levels of Atmospheric Contamination
(Ppm unless otherwise specified)

	H_2S	SO_2	NH_3	O_3	CO_2	NO_2	CO	Cl_2	Lead	Dust
Air conditioned (max. values)	≤1.0	≤0.5	≤0.5	NDA	NDA	≤0.05	≤1.0	≤0.004	NDA	NDA
Heated enclosure (max. values)	≤5.0	≤5.0	≤1.0	NDA	NDA	≤0.12	≤8.0	≤0.005	NDA	NDA
Sheltered or outside*	≤10.0	≤5.0	≤50.0	≤0.1	≤5000	≤5.0	≤50.0	≤1.0	≤0.2mg/m³	†
Extreme (in excess*)	>10.0	>5.0	>50.0	>0.1	>5000	>5.0	>50.0	>1.0	>0.2mg/m³	†

*Threshold limit values; "represents conditions under which nearly all workers may be repeatedly exposed day after day without adverse effect." (N.I. Sax, "Dangerous Properties of Industrial Materials," 3d ed. Reinhold, New York, 1968.)

†Greater than 50×10^6 particles/ft³ of air.

NDA = no data available at date of publication.

SOURCE: The Foxboro Company.

Table 24-10. Air-Pollution Alert Levels

Air-pollution Alert Levels (in Order of Increasing Severity)	Recommended Action (Each Level Including Action from Preceding Level)
1. Forecast: Indicates weather condition, such as an inversion, which could lead to an air-pollution episode.	None.
2. Alert: Declared for any one of the following pollutant conditions: SO_2—0.3 ppm, 24-h avg. Particulate—3.0 Coh units, 24-h avg. Combined—product of above equals 0.2. CO_2—15 ppm, 8-h avg. O_3—0.1 ppm, 1-h avg. NO_2—0.15 ppm, 24-h avg., or 0.6 ppm, 8-h avg. Meteorological conditions such that these levels are expected for 12 or more hours.	Oil or coal-fired electric power and steam generation required to reduce emission substantially via low-ash and -sulfur fuel and by using generation outside the episode area. No open burning; incineration, boiler lancing, and soot blowing only in early afternoon; and curtailing, postponing, or deferring manufacturing operations which produce these pollutants.
3. Warning: If air quality continues to deteriorate, it is declared for any one of the following conditions: SO_2—0.6 ppm, 24-h avg. Particulate—6.0 Coh units, 24-h avg. Combined—product of above equal to 1.0. CO_2—30 ppm, 8-h avg. O_3—0.4 ppm, 1-h avg. NO_2—12 ppm, 1-h avg., or 0.3 ppm, 24-h avg.	Maximum reduction of pollution from steam and power generation required. Incineration eliminated. Manufacturing which is easy to curtail is asked to cut back as much as possible. All other industry asked for maximum possible reduction in activity, even at the expense of economic hardship.
4. Emergency: Declared for substantial health danger when any one of the following conditions is imminent: SO_2—1.0 ppm, 24-h avg. Particulate—10 Coh units, 24-h avg. Combined—product of above, 2.4. CO_2—50 ppm, 8-h avg.; 75 ppm, 4-h avg.; 125 ppm, 1-h avg. O_3—0.4 ppm, 4-h avg.; 0.6 ppm, 2-h avg.; 0.7 ppm, 1-h avg. NO_2—2.0 ppm, 1-h avg., or 0.5 ppm, 24-h avg.	All industry required to eliminate contaminants to the greatest extent possible without causing personal injury or property damage. Motor-vehicle use prohibited, except for emergencies. Offices, schools, etc., required to close immediately.

SOURCE: Environmental Protection Agency.

d. Vibration and repetitive shock. Process control instruments are usually mounted on steel structures near rotating machinery or on process pipework that is subject to mechanical shock. A sharp blow from a 2-ft-long, $3/4$-in.[2] steel bar known as *plant iron* is occasionally used to clear blocked pipes.

e. Electromagnetic fields. Surges of current to heavy machinery can induce voltages in instrument wires and in structural steelwork. The use of structural steelwork as ground is to be avoided wherever possible.

f. Radioactivity, for example, in nuclear power applications, is treated as another environmental condition. Gamma rays and neutrons can displace atoms in the crystal structures of semiconductor devices, causing changes in characteristics that can be temporary (there is a self-healing effect) or permanent. Sufficient change can cause catastrophic failure of the component.

Irradiation of certain insulating materials causes them to soften and flow or become brittle. Certain metals, even if present only as impurities, can acquire sufficient radioactivity of their own so that after removal from the radioactive area, the instrument will continue to emit gamma rays, making it dangerous to handle.

66. Service Conditions in Selected Industries.   Some process control industries and their differences in service conditions are as follows:

Chemical: Plant-mounted instruments usually exposed to the weather and to all kinds of combinations of corrosive gas and vapors. Process lines may be very hot or very cold.

Oil and gas: Pipelines in inaccessible areas calling for high-reliability instruments. Plant-mounted instruments usually exposed to the weather.

Paper: Permanent high-humidity and pervasive airborne particulate matter. Instruments may be subjected to hosing down with water.

Power: Few extremes of ambient temperature. High concentration of sulfur compounds.
Iron and steel: High temperature, heavy concentrations of dust.
Food: Steam used for cleaning—frequent high humidity.

Within each industry, and indeed within each plant, there is a wide range of environmental conditions. There are no "standardized" service conditions nor any extensive published data. Data that certain large manufacturers have compiled for their own use are considered proprietary. Each manufacturer must choose the set of environmental conditions within which his equipment is to operate, and design accordingly. The user of that equipment must assure himself that the service conditions in his plant are no more severe than those within which the equipment will operate according to the manufacturer's specifications.

An international draft standard on service conditions classifies areas as follows:

a. Air-conditioned areas.

b. Heated and/or cooled areas.

c. Sheltered areas.

d. Outdoor areas.

e. Extreme-condition areas.

This standard then defines several ranges of temperature for storage and operating ambient for each type of area. Certain other things should be considered, such as the rate of change of temperature possible when equipment that has become quite hot in the sun is suddenly drenched by a thundershower.

67. Power-Supply Characteristics. The characteristics of power supplies to be considered when designing process control instrumentation are nominal voltage and expected variation, nominal frequency and expected variation, and the characteristics of outages and overvoltages likely to be experienced. Alternating-current power is often available as a three-phase supply, but instrumentation systems rarely use this. Power-supply continuity is especially important in digital-system installations.

MANUFACTURING RELIABLE INSTRUMENTS

68. Reliability in Design. The reliability of an instrument depends on the intrinsic reliability of its components, the process to which those components were subjected in instrument assembly, and the environment within which it operates. Reliability must be designed in, by judicious selection of parts and processes and by circuit design that includes proper component application and derating.

The types of components used in process control instrumentation are those that will operate reliably under adverse environmental conditions. While components that get hot in service are generally not desirable in electronic equipment, a hot instrument is less likely than a cold one to suffer condensation. The severe environmental conditions make hermetic sealing of semiconductor-device packages almost essential. Great care must be taken in selecting parts whose metallic surfaces have a finish (generally a plating of specific materials and minimum thicknesses) that is corrosion-resistant. Insulating surfaces must be resistant to tracking, which can be caused by wetting or moisture absorption in conditions of high humidity or by chemical reaction of the insulating material itself with atmospheric contaminants. Metallic surfaces are particularly important when selecting connectors, switches, and relays.

Electronic-component packaging provides physical support, interconnection, and protection of the electronic components. The need for physical support and interconnection is self-evident and is not unique to process control equipment. However, process control equipment is normally expected to operate over a wide temperature range, and so proper thermal design is essential. Also, the packaging must provide protection against the ingress of moisture and atmospheric contaminants. Sealed containers, potting, and conformal coating are methods used. Methods that keep out moisture in face of varying ambient temperature and atmospheric pressure are emphasized.

69. Reliability in Manufacturing. Quality can easily be degraded in assembly. Because of the high humidity prevalent in process control industries, the removal of contaminants by effective cleaning methods is essential.

It is possible to increase the reliability of instruments by "burning them in" in a high ambient temperature under power-on conditions and shipping the survivors. However, it is much more cost-effective to burn-in the critical electronic components that are found to contribute most to the failure rate of the final instrument. Linear and digital integrated circuits contribute more to the failure rate of an electronic assembly than any other part. Screening of these parts is essential for high quality. See Table 24-11. *Screening* in this context is 100%, 125°C, 168-h power-on testing, a process that assures the failure rate of a batch of electronic components is less than a predefined value.

Table 24-11. Electronic-Component Quality Levels for High-Quality Instruments

Electronic-component type	Target failure rate, failure/10^6 h	Survivors OK, fallout rate, %*
Linear integrated circuits	0.48	1.0
MSI digital circuits	0.48	2.0
SSI flip-flops	0.35	1.5
SSI gates	0.16	0.5

* If the fallout rate after 125°C, 168 h burn-in of 100% tested components is less than or equal to this figure, the survivors have a failure rate less than or equal to the target failure rate.

70. Designing for Serviceability. Serviceability is an important element in achieving high availability. Since process control instruments are usually mounted in areas where they cannot be tested and repaired, the plug-in replacement approach is favored and design must be modular to permit this. Almost all process control instrumentation is purchased, not leased, and the user traditionally performs his own repairs. The trend is toward the manufacturer offering a new or factory-repaired instrument with cash adjustment for a failed instrument; therefore the importance of designing instruments so that they can be easily repaired in the field is declining.

In large systems, because they are required to operate continuously, it is important to put critical and noncritical elements in separate modules. The purpose of this is that when a noncritical element fails and its module must be removed and replaced, the serviceman is not obliged to pull the critical element and shut down the plant.

71. References

1. HICKES, W. F. Intrinsic Safety and the CPI, *Chem. Eng. Prog.,* Vol. 64, No. 4, pp. 67-74, 1968.
2. HICKES, W. F. "Instrumentation in the Chemical and Petroleum Industries," Vol. 6, "International Standards of Intrinsic Safety," Instrument Society of America, Plenum, New York, 1970.
3. HICKES, W. F. Fundamentals of Intrinsic Safety, *Chem. Eng.,* June 30, 1969, pp. 139-140.
4. HICKES, W. F. New Developments in Intrinsic Safety, *Proc. Texas A & M Symp.,* 1969, pp. 53-57.
5. HICKES, W. F. "Instrumentation in the Chemical and Petroleum Industries," Vol. 3, "Evaluation of Intrinsic Safety," pp. 121-136, Instrument Society of America, Plenum, New York, 1967.
6. HICKES, W. F. Intrinsic Safety and Instrument Design, *18th Annu. Symp. Proc.,* Instrument Society of America, Plenum, New York, 1966.
7. MAGISON, E. C. "Electrical Instruments in Hazardous Locations," 2d ed., Instrument Society of America, Plenum, New York, 1972.
8. REDDING, R. J. "Intrinsic Safety," McGraw-Hill, New York, 1971.
9. INSTRUMENT SOCIETY OF AMERICA Electrical Safety Abstracts, 4th ed., 1972.
10. NATIONAL FIRE PROTECTION ASSOCIATES Intrinsically Safe Process Control Equipment for Use in Hazardous Locations, NFPA 493-1969.
11. UNDERWRITERS' LABORATORIES, INC. Standard for Intrinsically Safe Electrical Circuits and Equipment for Use in Hazardous Locations, Subject 913, 1971.
12. VERBANDE DEUTSCHE ELECTROTECHNIQUE Specifications for the Construction and Testing of Electrical Apparatus for Use in Explosive Gas Atmospheres, VDE 0170/0171/2.65, 1965.
13. FACTORY MUTUAL ENGINEERING CORPORATION Approval Standard for Electrical Equipment as Intrinsically Safe for Class I, Division 1 Areas on Nonincendive for Class I, Division 2 Areas, 1967.

14. CANADIAN STANDARDS ASSOCIATION Intrinsically Safe and Nonincendive Equipment for Use in Class I, Groups A, B, C, D Hazardous Locations, Draft Standard C22.2, No. 157, 1971 (based on 1971 IEC draft).

15. DEPARTMENT OF TRADE AND INDUSTRIES BRITISH APPROVAL SERVICE FOR ELECTRICAL EQUIPMENT IN FLAMEABLE ATMOSPHERES Intrinsic Safety, Certification Std. SFA 3012, 1972.

16. INTERNATIONAL ELECTROTECHNICAL COMMISSION Draft of Recommendation for the Construction and Test of Intrinsically Safe Apparatus and Associated Apparatus, IEC TC 31/SC31G, 1972.

72. Bibliography

CONSIDINE, DOUGLAS M. (ed.) "Process Instruments and Controls Handbook," pp. 8-57, McGraw-Hill, New York, 1957.

LIPTAK, GELA G. (ed.) "Instrument Engineers Handbook," Vol. 1, "Process Measurements," Chilton Book Co., New York, 1969.

DOEFELIN, ERNEST O. "Measurement Systems," McGraw-Hill, New York, 1969.

BECKWITH, THOMAS G., and N. LEWIS BUCK "Mechanical Measurements," Addison Wesley, Reading, Mass., 1969.

FINK, DONALD G. (ed.) "Standard Handbook for Electrical Engineers," 10th ed. McGraw-Hill, New York, 1968.

HAMSHER, DONALD H. (ed.) "Communication System Engineering Handbook," McGraw-Hill, New York, 1967.

THE FOXBORO COMPANY Technical Information 39-5b, Interference in Electronic Consotrol Circuits (1963); 39-5c, Selection of Electronic Consotrol Transmission Cable (1965).

SECTION 25

RADAR, NAVIGATION, AND UNDERWATER SOUND SYSTEMS

BY

DAVID K. BARTON Consulting Scientist, Missile Systems Division, Raytheon Company; Fellow, Institute of Electrical and Electronics Engineers

HAROLD R. WARD Principal Engineer, Missile Systems Division, Raytheon Company; Senior Member, Institute of Electrical and Electronics Engineers

SVEN H. DODINGTON Assistant Technical Director, International Telephone and Telegraph Company; Fellow, Institute of Electrical and Electronics Engineers

STANLEY L. EHRLICH Consulting Engineer, Submarine Signal Division, Raytheon Company; Senior Member, Institute of Electrical and Electronics Engineers

DONALD A. FREDENBERG Principal Engineer, Submarine Signal Division, Raytheon Company

JACK H. HEIMANN Principal Engineer, Submarine Signal Division, Raytheon Company; Senior Member, Institute of Electrical and Electronics Engineers

JOSEPH A. KUZNESKI Senior Engineer, Submarine Signal Division, Raytheon Company; Senior Member, Institute of Electrical and Electronics Engineers

PAUL SKITZKI Technical Consultant, Submarine Signal Division, Raytheon Company; Member, Institute of Electrical and Electronics Engineers

CONTENTS

Numbers refer to paragraphs

This section of the Handbook is a contribution of the IEEE Aerospace and Electronic Systems Society. The editor is indebted to David B. Dobson, editor of *IEEE Transactions on Aerospace and Electronic Systems,* for his assistance in planning its contents and arrangements with the contributors.—D.G.F.

SECTION 25

RADAR, NAVIGATION, AND UNDERWATER SOUND SYSTEMS

RADAR PRINCIPLES

By David K. Barton

1. Basic Functions. The basic functions of radar are inherent in the word, whose letters stand for *r*adio *d*etection *a*nd *r*anging. Measurement of target angles has been included as a basic function of most radars, and doppler velocity is often measured directly as a fourth basic quantity. Resolution of the desired target from background noise and clutter is a prerequisite to detection and measurement, and resolution of surface features is essential to mapping or imaging radar. The radar resolution cell is a four-dimensional volume, bounded by antenna beamwidths, width of the processed pulse, and bandwidth of the receiving filter. Within each such resolution cell, a decision may be made as to presence or absence of a target, and if a target is present, its position may be interpolated to some fraction of the cell dimensions.

The block diagram of a typical pulsed radar is shown in Fig. 25-1. The equipment has been divided arbitrarily into seven subsystems, corresponding to the organization of the paragraphs on radar technology (Pars. 25-59 to 25-83). Radar operation is initiated by a synchronizer, which controls the time sequence of transmissions, receiver gates and gain settings, signal processing, and display. When called for by the synchronizer, the modulator

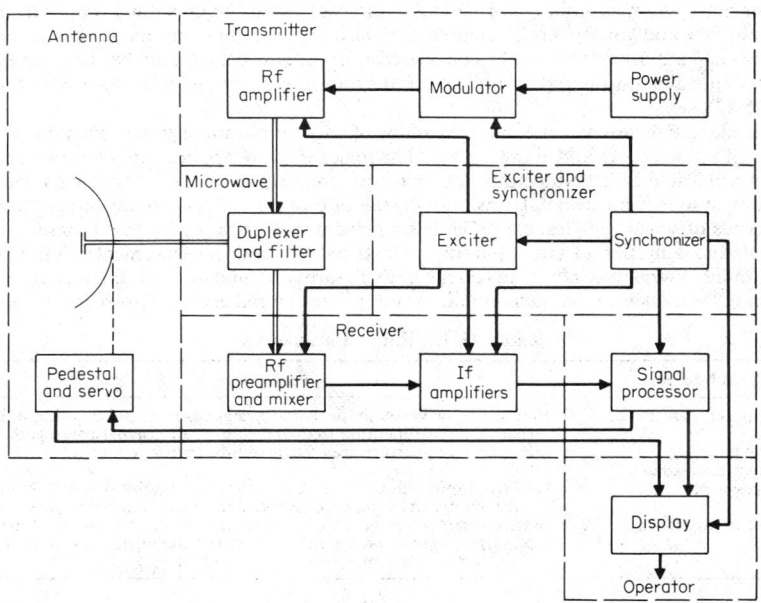

Fig. 25-1. Block diagram of typical pulsed radar.

applies a pulse of high voltage to the rf amplifier, simultaneously with an rf drive signal from the exciter. The resulting high-power rf pulse is passed through a transmission line or waveguide to the duplexer, which connects it to the antenna for radiation into space. The antenna shown is of the reflector type, steered mechanically by a servo-driven pedestal. A stationary array may also be used, with electrical steering of the radiated beam.

After reflection from a target, the echo signal reenters the antenna, which has been connected to the receiver preamplifier or mixer by a duplexer. A local-oscillator signal furnished by the exciter translates the echo frequency to if, which can be amplified and filtered in the receiver prior to more refined signal processing. The processed if signal is passed through an envelope detector and displayed, with or without video processing. Data to control the antenna steering and to provide outputs to an associated computer are extracted from the time delay and modulation on the video signal.

2. Radar Parameters. A radar system can be described in terms of several basic parameters listed in Table 25-1, which determine its performance as a detection and measurement device. Other parameters and specifications are needed to describe subsystem design characteristics (antenna type, signal stability, input power source, etc.), but those shown in the table establish fundamental constraints on performance. From these parameters, the maximum operating range R_m on typical targets can be found, resolution properties can be described, and the signal-to-noise and signal-to-clutter ratios which constrain detection and measurement performance can be calculated for different environments.

3. Target Characteristics. Given a set of radar parameters, the detection and location performance will depend also on target characteristics: (a) target cross section (average or median) and fluctuation with time and frequency; (b) target size, shape, distribution of scatterers, aspect angle, and rate of turn; (c) target position in range, azimuth, and elevation and its velocity and acceleration in all three coordinates.

4. Noise and Background. The input noise level, determined by the receiving-system noise factor under standard conditions, is not an accurate measure of radar performance. When connected to the antenna, the receiver sees a low-noise background of empty space, modified by surrounding terrain or sea surfaces and atmosphere, at about 290 K, the sun at several thousand degrees Kelvin, galactic noise (at low frequencies), and various man-made sources of interference. Noise received through antenna side lobes must be added to internal receiver noise and main-lobe-noise. Finally, the background of unwanted echoes from the earth's surface and atmosphere (including rain, birds, insects, and man-made objects such as buildings and automobiles) must be considered. Since these echoes may be either targets or clutter, depending on the radar application, they are described under Radar Cross Section (Par. 25-12).

5. Radar Applications. A complete catalog of radar applications would extend for many pages, with new entries added each year. The major fields of application, however, remain as shown in Table 25-2. Numerous miscellaneous applications, not readily categorized, can also be mentioned: intrusion alarms, monitoring of bird migration, ground vehicle control, rendezvous of space vehicles, etc. The basic principles of radar apply to all these systems with suitable definition of target parameters and resolution or measurement requirements.

6. Radar Frequencies.[9]* In essence, there are no fundamental bounds on radar frequency. Any device that detects and locates targets by radiating electromagnetic energy

Table 25-1.   Radar Parameters

Subsystem	Parameters
Transmitter	Frequency, wavelength, peak and average power, pulse width, pulse energy, pulse repetition frequency, duty ratio, signal bandwidth
Antenna	Size, effective aperture area, beamwidths, scan rate and coverage
Receiver and signal processor	Effective input temperature, bandwidth, input noise power, output pulse width, MTI improvement factor, video integration gain
System performance	System losses, received power and energy on standard target, time on target, signal-to-noise ratios in power and energy, measurement slope factors, data rate, instrumental errors

---

*Courtesy of Dr. Merrill I. Skolnik. Superior numbers correspond to numbered references, Par. **25-161.**

and utilizes the echo scattered from the target can be classed as a radar, no matter what its frequency. Radars have been operated at wavelengths of 100 m (short waves) or longer to wavelengths of 10^{-7} m (ultraviolet) or shorter. The basic principles are the same at any frequency, but the implementation is widely different. In practice, most radars operate within the microwave-frequency range, but there are many notable exceptions.

A set of letter designations exists for the frequency bands commonly used for radar (see Table 25-3). The original code letters, (P, L, S, X, and K) were introduced during World War II. After the need for secrecy no longer existed, these designations remained. Others were added later (C, K_u, and K_a) as new bands were opened, and some were seldom used (P and K). There have been attempts to subdivide the entire microwave-frequency spectrum by letter codes and to extend the letter nomenclature to the millimeter-wave region, but this has not gained wide acceptance.

7. Radar Propagation.

Atmospheric Attenuation. The frequency bands used for radar were selected to minimize the effects of the atmosphere while achieving adequate bandwidth, antenna gain, and angular resolution. Attenuation is introduced by the air and water vapor, by rain and snow, by clouds and fog, and (at some frequencies) by electrons in the ionosphere.

Attenuation in the clear atmosphere is seldom a serious problem at frequencies below 16 GHz (Fig. 25-2). The initial slope of the curves shows the sea-level attenuation coefficient k in decibels per kilometer, falling off as the path reaches higher altitude. Above 16 GHz, atmospheric attenuation is a major factor in system design (Fig. 25-3). The absorption lines of water vapor (at 22 GHz) and oxygen (near 60 GHz) are broad enough to restrict radar operations above 16 GHz in the lower troposphere to relatively short range, even under clear-sky conditions. Attenuation vs. frequency for two-way paths through the entire atmosphere is shown in Fig. 25-4.

Table 25-2. Radar Applications

Air surveillance	Long-range early warning; ground-controlled intercept; acquisition for weapon system; height finding and 3D radar; airport and air-route surveillance
Space and missile surveillance	Ballistic missile warning; missile acquisition; satellite surveillance
Surface-search and battlefield surveillance ..	Sea search and navigation; ground mapping; mortar and artillery location; airport taxiway control
Weather radar	Observation and prediction; weather avoidance (aircraft); cloud-visibility indicators
Tracking and guidance ...	Antiaircraft fire control; surface fire control; missile guidance; range instrumentation; satellite instrumentation; precision approach and landing
Astronomy and geodesy ..	Planetary observation; earth survey; ionospheric sounding

Table 25-3. Radar Frequency Bands

Nomenclature	Frequency range	Radiolocation bands based on ITU assignments in region II
Vhf	30–300 MHz	137–144 MHz 216–225 MHz
Uhf	300–1,000 MHz	420–450 MHz 890–940* MHz
P-band†	230–1,000 MHz	
L-band	1,000–2,000 MHz	1,215–1,400 MHz
S-band	2,000–4,000 MHz	2,300–2,550 MHz 2,700–3,700 MHz
C-band	4,000–8,000 MHz	5,255–5,925 MHz
X-band	8,000–12,500 MHz	8,500–10,700 MHz
K_u-band	12.5–18 GHz	13.4–14.4 GHz 15.7–17.7 GHz
K-band	18–26.5 GHz	23–24.25 GHz
K_a-band	26.5–40 GHz	33.4–36 GHz
Millimeter	>40 GHz	

* Sometimes included in L-band.
† Seldom used nomenclature.

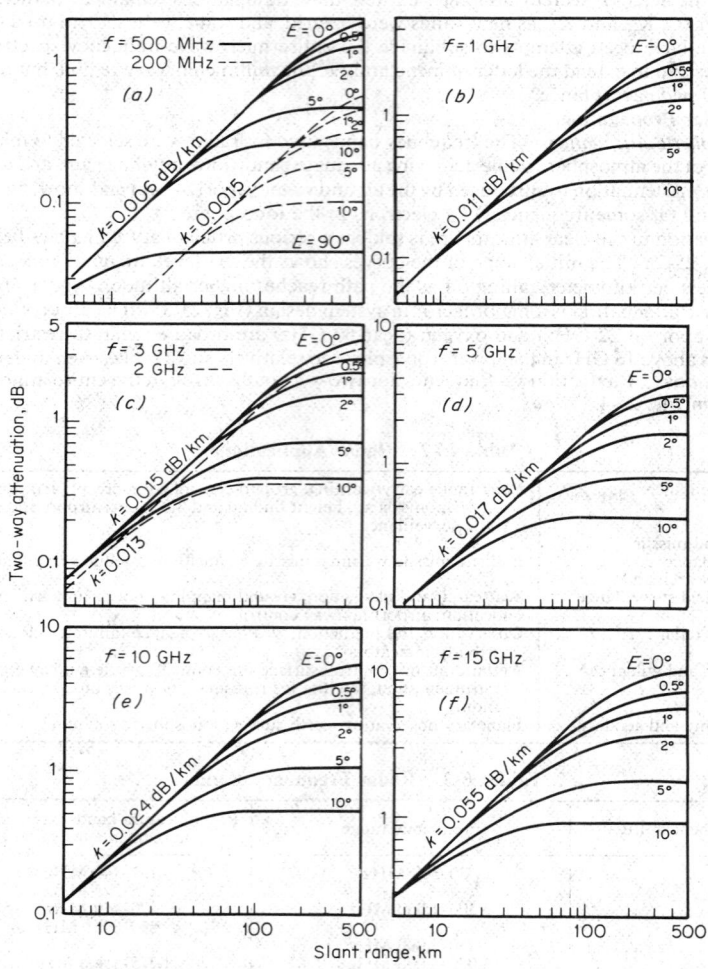

Fig. 25-2. Atmospheric attenuation (0.2 to 15 GHz) vs. range and elevation angle *E*. *(Based on data from Blake, Ref. 10.)*

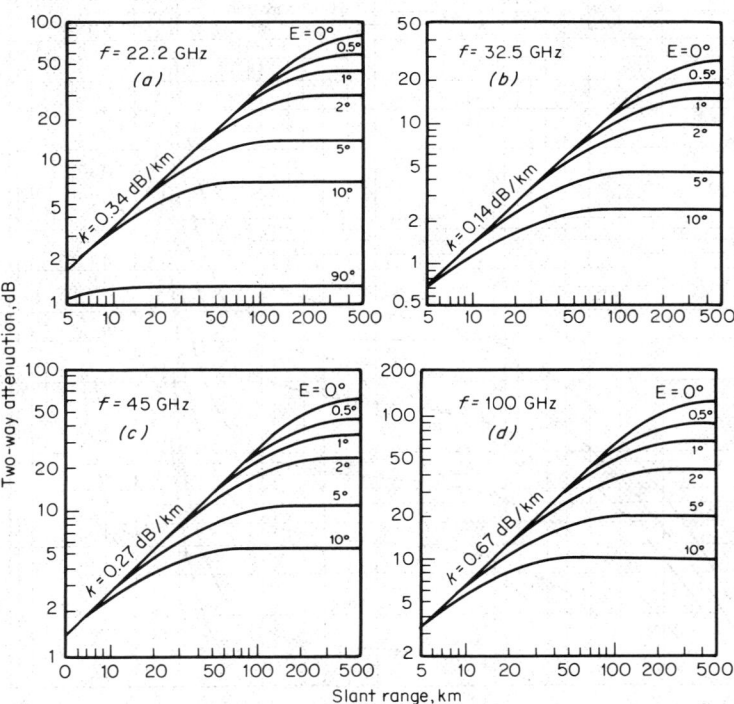

Fig. 25-3. Atmospheric attenuation (20 to 100 GHz) vs. range and elevation angle *E.* *(Based on data from Blake, Ref. 10.)*

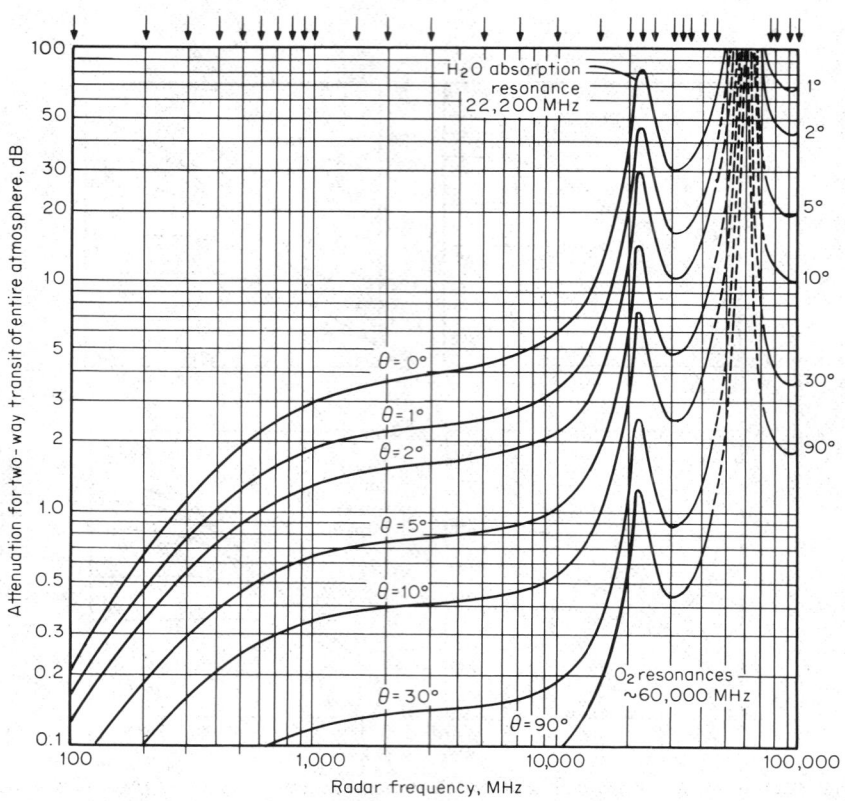

Fig. 25-4. Absorption loss for two-way transit of the entire troposphere, at various elevation angles (Ref. 10).

Precipitation Effects.   Above 2 GHz, rain causes significant attenuation, with k roughly proportional to rainfall rate r and to the 2.5 power frequency.   The classical data[1] on rain attenuation[11,12] were based on drop-size distributions given by Ryde and Ryde,[13] which gave generally accurate results, except for a 40% underestimate of the loss between 8 and 16 GHz, at low rainfall rates.   Later data were derived by Wexler and Atlas[14] from a modified Marshall-Palmer distribution.[15]   At high rates (100 mm/h) the loss coefficient k/r is doubled between 8 and 16 GHz, giving better agreement with measurements and matching the estimates of Medhurst[16] above 16 GHz.   The Wexler and Atlas data provide the most satisfactory estimates for general use, and these were used in preparing Fig. 25-5.

Very small water droplets, suspended as clouds or fog, can also cause serious attenuation, especially since the affected portion of the transmission path can be tens or hundreds of kilometers.   Attenuation is greatest at 0°C (Fig. 25-6).   Transmissions below 2 GHz are affected more seriously by heavy clouds and fog than by rain of downpour intensity.

Water films that form on antenna components and radomes are also sources of loss. However, such surfaces can be specially treated to prevent the formation of continuous films[17,18].

Apparent Sky Temperature.   Associated with the atmospheric loss is a temperature term, which must be added to the radar receiver input-temperature (Par. **25-73**).   Figure 25-7 shows this loss temperature T'_a as a function of frequency.   Also included in T'_a is the galactic background noise for $f \leq 1$ GHz.

Ionospheric Attenuation.   In the lowest radar bands, the daytime ionosphere may introduce noticeable attenuation.[19]   However, above 100 MHz, this attenuation seldom exceeds 1 dB.

8. Surface Reflections.   The radar antenna illuminates the target with direct rays and indirectly, with energy reflected from the surface (Fig. 25-8a).   The direct and reflected rays combine to produce a pattern of lobes and nulls (Fig. 25-8b).   The effect of specular reflection from a plane surface is described most conveniently by assuming an image antenna below the

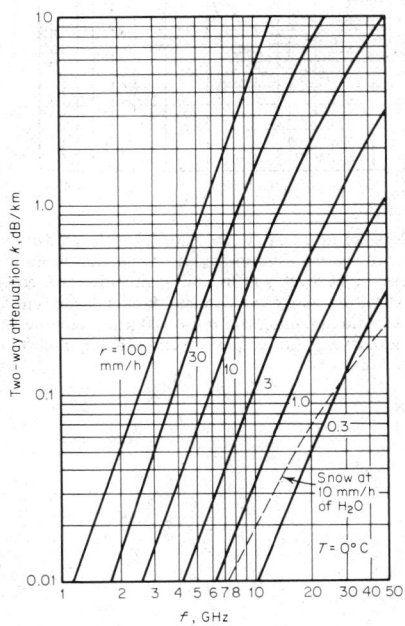

Fig. 25-5. Attenuation in rain.

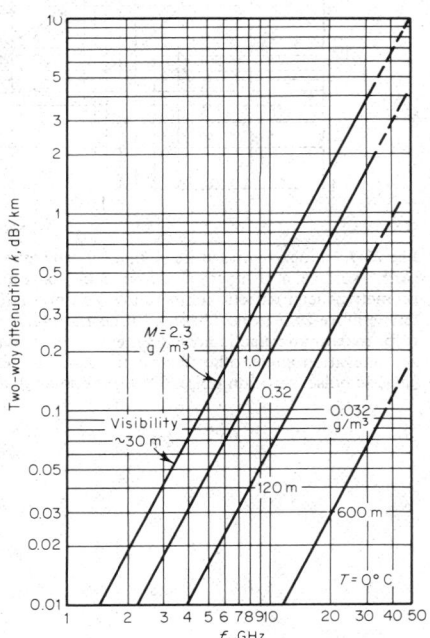

Fig. 25-6. Attenuation in clouds or fog (values of k are halved at $T = 18$°C).

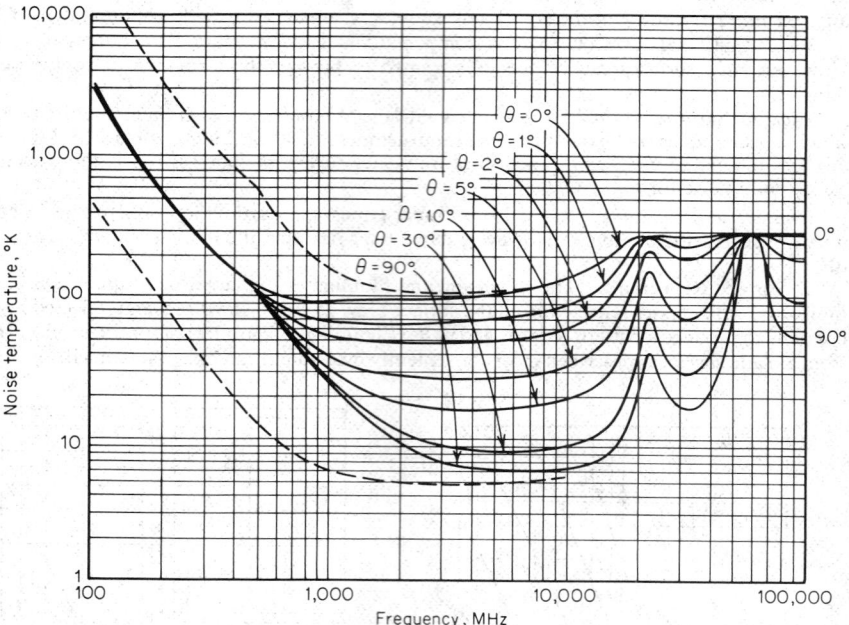

Fig. 25-7. Noise temperature of an idealized antenna (lossless, no earth-directed side lobes) located at the earth's surface, as a function of frequency, for a number of beam elevation angles. Solid curves are for geometric-mean galactic temperature, sun noise 10 times quiet level, sun in unity-gain side lobe, cool-temperature-zone troposphere, 2.7 K cosmic blackbody radiation, zero ground noise. Upper dashed curve is for maximum galactic noise (center of galaxy, narrow-beam antenna), sun noise 100 times quiet level, zero elevation angle, other factors the same as the solid curves. Lower dashed curve is for minimum galactic noise, zero sun noise, 90° elevation angle (Ref. 10).

reflecting plane, tilted downward by the same angle as the real antenna tilts upward. If the free-space voltage pattern of the antenna, as a function of elevation angle E, is $f(E)$, the pattern when operated over the reflecting plane is

$$f'(E) = f(E) + \rho(E)f(-E)\exp[-j(\frac{4\pi h_r}{\lambda}\sin E + \phi)] \qquad (25\text{-}1)$$

$$= Ff(E)$$

where ρ = magnitude of surface reflection coefficient, h_r = antenna height, λ = radar wavelength, ϕ = phase angle of reflection coefficient, and F = pattern-propagation factor.[10] Equation (25-1) is valid for targets at long range ($R \gg h_r^2/\lambda$), as is normally the case in land-based or shipborne radar.

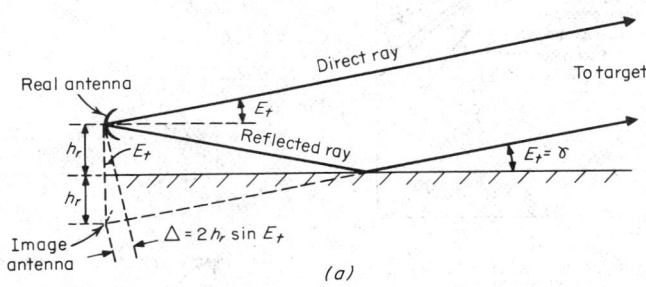

(a)

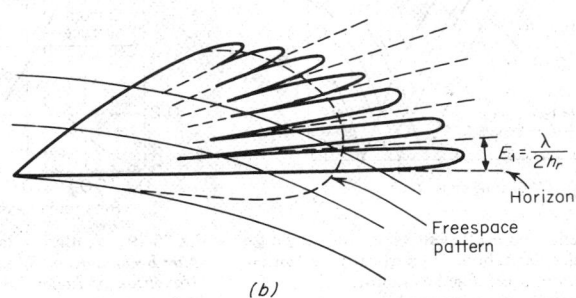

(b)

Fig. 25-8. Effect of surface reflections. *(From "Radar System Analysis," Copyright 1964. Reprinted by permission of Prentice-Hall, Inc., Englewood Cliffs, N.J.)*

At zero elevation, where $f(0) = f(-0)$, then $\rho \approx 1.0$ and $\phi \approx 180°$ for either polarization, leading to a total null in the pattern. Partial nulls with gains $f(E) - \rho f(-E)$ will occur when $\sin E$ is an even multiple of $\lambda/4h_r$, over elevations where $\phi \approx 180°$. Figure 25-9 shows the magnitude of ρ for vertical and horizontal polarizations and for different surfaces. Reflection from seawater is a function of frequency below 2 Ghz, but freshwater and ground reflections are essentially the same for all radar frequencies. As shown in Fig. 25-8, the grazing angle γ is equal to target elevation angle E_t for a long-range target over a level surface. For all practical purposes, $\phi \approx 180°$ for horizontal polarization at all grazing angles and for vertical polarization below the angle of minimum ρ (the Brewster angle). More precise data for all frequencies down to 30 MHz are given in Ref. 20.

Rough-Surface Reflection. In the preceding discussion, an idealized smooth reflecting surface was assumed. Actual land and water surfaces are irregular, multiplying the smooth-surface specular reflection coefficient (Fig. 25-9) by a factor

$$\rho_s = \exp\left[-2\left(\frac{2\pi\sigma_h \sin \gamma}{\lambda}\right)^2\right] \tag{25-2}$$

as shown in Fig. 25-10. Here, σ_h is the rms deviation in height of the surface from its average plane. The energy lost from the specular reflection is scattered at other angles, including backscatter to the source.

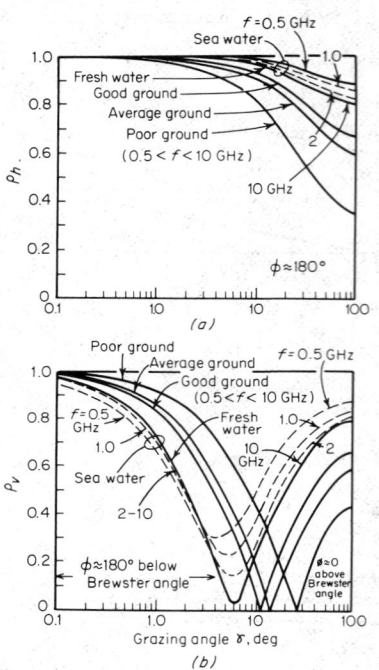

Fig. 25-9. Reflection coefficient vs. grazing angle for (*a*) horizontal polarization, (*b*) vertical polarization. Surface conditions are defined as follows:

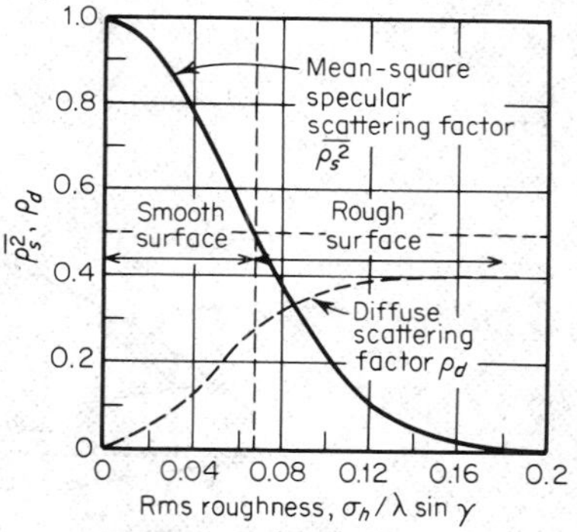

Fig. 25-10. Scattering factors vs. roughness. *(After Beckmann and Spizzichino, Ref. 21. From "Handbook of Radar Measurement," copyright 1969. Reprinted by permission of Prentice-Hall, Inc., Englewood Cliffs, N.J.)*

	Dielectric Constant	Conductivity, $\mho$/m
Poor ground............	4	0.001
Average ground........	15	0.005
Good ground..........	25	0.02
Seawater (10 GHz).....	65	15
Seawater (0.5 GHz)	81	5
Freshwater (10 GHz)...	65	10
Freshwater (0.5 GHz) ..	81	0.01

9. Tropospheric Refraction. The refractive index of the troposphere, for all radar frequencies, can be expressed as

$$N \equiv (n - 1) \times 10^6 = \frac{77.6}{T}\left(P + \frac{4,810p}{T}\right) \tag{25-3}$$

where T = temperature in degrees Kelvin, P = total pressure in millibars, p = partial pressure of the water-vapor component, n = refractive index, and N = a scaled-up deviation in n, known as the refractivity. Dry air at sea level can have a value of N as low as 270, but normal values lie between 300 and 320.

The Central Central Radio Propagation Laboratory (CRPL) of the National Bureau of Standards established a family of exponential approximations to the normal atmosphere in which the average U.S. conditions are represented by[22]

$$N(h) = 313.0 \exp(-0.14386h) \tag{25-4}$$

where h = altitude in kilometers above sea level. The velocity of propagation is $1/n$ times

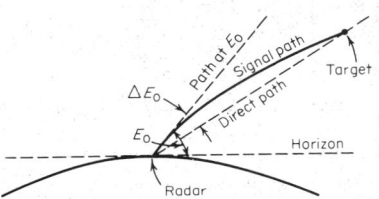

Fig. 25-11. Geometry of tropospheric refraction.

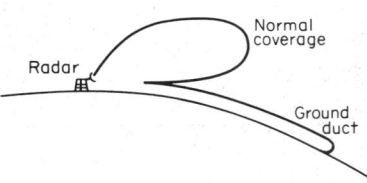

Fig. 25-12. Low-angle ducting effect (Ref. 2).

the vacuum velocity, introducing an extra time delay in radar ranging and causing radar rays to bend downward relative to the angle at which they are transmitted. Figure 25-11 shows, on an exaggerated scale, the geometry of tropospheric refraction.

For air-surveillance radar, the effects of tropospheric refraction are adequately expressed by plotting ray paths as straight lines above a curved earth whose radius is 4/3 times the true radius.

A special problem can arise when the ray is transmitted at elevations below 0.5° into an atmosphere whose refractivity has a rapid drop, several times greater than the standard 45 N-units per kilometer. Under these conditions, the ray can be trapped in a surface duct (Fig. 25-12) or in an elevated duct bounded by layers of rapidly decreasing N. The result is a great increase in radar detection range for targets within the duct (surface targets and clutter, in most cases) at the expense of coverage just above the ducting layer. Although there is some leakage of energy from the top of the duct, increasing at lower frequencies, a duct will usually trap all radar frequencies, leaving a coverage gap just above the horizon.

10. Tropospheric Fluctuations. Deviations from average atmospheric conditions at a given site will fluctuate over a continuous spectrum from several hertz down to one cycle per year, with increasing spectral density at lower frequencies.[23] The average spectral density $W(f)$ can be expressed[8] in parts per million (ppm) of the path length as

$$W(f) = \begin{cases} 32/f \, \text{ppm}^2/\text{Hz}, & 10^{-8} \leq f \leq 10^{-5} \, \text{Hz}, \\ 10^{-6}/f^{2.5} \, \text{ppm}^2/\text{Hz} & f \geq 10^5 \, \text{Hz} \end{cases} \tag{25-5}$$

Integration of this spectrum over frequencies which can be observed during a measurement interval t_0 (i.e., down to about $1/t_0$) gives the variance of the path length in ppm^2 occurring during that interval.

11. Ionospheric Refraction. The refractivity of the ionosphere at radar frequencies is given by

$$N_i = (n-1) \times 10^6 = -\frac{40 N_e}{f^2} \times 10^6 = \frac{-1}{2}\left(\frac{f_c}{f}\right)^2 \times 10^6 \tag{25-6}$$

where N_e = electron density per cubic meter, and f_c = critical frequency in hertz ($f_c \approx 9\sqrt{N_e}$). Since f_c seldom exceeds 14 MHz, the refractivity at 100 Mhz is less than 10^4 N-units, and above 1 GHz it does not exceed 100 N-units. Figures 25-13 and 25-14 give the range and elevation-angle errors for normal ionospheric conditions for targets at different altitudes. Ionospheric errors are not significant for radars operating in the gigahertz region, but can dominate the error analysis in the 200- to 400-MHz band.

12. Targets and Clutter. Radar Cross Section.   The primary parameter describing a radar target is its radar cross section, or backscattering coefficient, defined as 4π times the ratio of reflected power per unit solid angle in the direction of the source to the power per unit area in the incident wave.   If a target were to scatter power uniformly over all angles, its radar cross section would be equal to the area from which power was captured from the incident wave.   A large sphere (whose radius $A \gg \lambda$) captures power from an area πa^2, scattering it uniformly in solid angle, and hence has a radar cross section $\sigma = \pi a^2$ equal to its projected area.

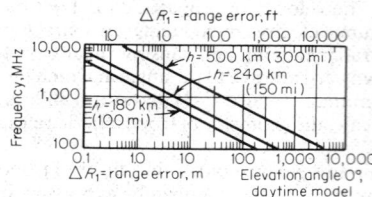

Fig. 25-13.   Ionospheric range error vs. frequency. *(After Pfister and Keneshea, Ref. 24.)*

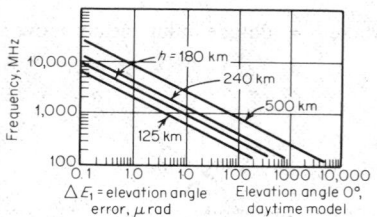

Fig. 25-14.   Ionospheric angle error vs. frequency. *(After Pfister and Keneshea, Ref. 24. From "Radar System Analysis," copyright 1964.   Reprinted by permission of Prentice-Hall, Inc., Englewood Cliffs, N.J.)*

The variation of sphere cross section with wavelength (Fig. 25-15) illustrates the division into three regions of the spectrum, for any smooth target:

a. The optical region ($a \gg \lambda$), where cross section is essentially constant with wavelength.

b. The resonant region ($a \approx \lambda/2\pi$), where cross section oscillates about its optical value, due to interference of the direct reflection with a creeping wave, propagated around the circumference of the object.

c. The Rayleigh region ($a \ll \lambda/2\pi$), where the cross section drops rapidly below its optical value, varying as a/λ^4.

Although the sphere cross section varies with wavelength, it is constant over all aspect angles, and hence can serve as a reference for radar testing and evaluation.   The cross sections of all other objects vary with aspect angle, requiring more complex descriptions: amplitude probability distributions, fluctu-

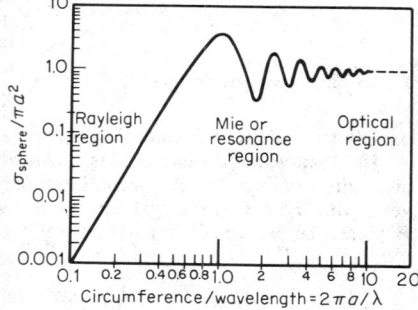

Fig. 25-15.   Radar cross section of a sphere. a = radius, λ = wavelength (Ref. 2).

ation frequency spectra (or time correlation functions), and radar frequency correlation functions.   For relatively simple objects, it is possible to write analytical expressions for cross section as a function of wavelength and geometry (see Table 25-4).   Reflectivity patterns can be plotted for such objects as cylinders and flat plates, and lobe widths can be found as for antenna patterns.   For cylinders and plates of length L (measured normal to the radar beam and to the axis of rotation), there will be a main (specular) reflection lobe whose half-power width is

$$\Delta = 0.44\lambda/L \quad (\text{rad})$$

whose side lobes are of width

$$\Delta_0 = \lambda/2L$$

and whose null-to-null main-lobe width is $2\Delta_0 = \lambda/L$.

13. Amplitude Distributions. When the target aspect angle is unknown, or the target is too irregular to permit cross section to be expressed as a function of geometry, the amplitude distribution can be used to describe cross section statistically. Figure 25-16 plots the probability density and cumulative probability functions for several types of target, on decibel scales. The first two plots are the Swerling fluctuation models:[25]
Cases 1,2:

$$dP = \frac{1}{\bar{\sigma}} \exp\left(\frac{-\sigma}{\bar{\sigma}}\right) d\sigma \qquad \sigma \geq 0 \tag{25-7}$$

Cases 3,4,:

$$dP = \frac{4\sigma}{\bar{\sigma}^2} \exp\left(\frac{-2\sigma}{\bar{\sigma}}\right) d\sigma \qquad \sigma \geq 0 \tag{25-8}$$

where $\bar{\sigma}$ = arithmetic mean of the distribution. The median value σ_{50} is used as the center of each plot. Cases 1 and 3 describe slowly fluctuating targets such that all pulses integrated within a single scan are correlated but successive scans give uncorrelated values. Cases 2 and 4 describe fast (pulse-to-pulse) fluctuation.

Table 25-4. Radar Reflectivity Characteristics of Simple Bodies

Object	σ_{max}	σ_{min}	Number of lobes	Major lobe width
Sphere	πa^2	πa^2	1	2π
Ellipsoid $(k = a/b)$	πa^2	$\dfrac{\pi b^2}{k^2}$	2	$\sim \dfrac{b}{a}$
Cylinder	$\dfrac{2\pi a L^2}{\lambda}$	Null	$\dfrac{8L}{\lambda}$	$\dfrac{\lambda}{L}$
Flat plate	$\dfrac{4\pi A^2}{\lambda^2}$	Null	$\dfrac{8L}{\lambda}$	$\dfrac{\lambda}{L}$
Dipole	$0.88\lambda^2$	Null	2	$\dfrac{\pi}{2}$
Infinite cone (half-angle α)	$\dfrac{\lambda^2 \tan^4 \alpha}{16\pi}$	Null		
Convex surface	$\pi a_1 a_2$			
Square corner reflector	$\dfrac{12\pi a^4}{\lambda^2}$	. . .	4	$\dfrac{\pi}{4}$
Triangular corner reflector	$\dfrac{4\pi a^4}{3\lambda^2}$	. . .	4	$\dfrac{\pi}{4}$

SOURCE: D. K. Barton, "Radar System Analysis," copyright 1964. Reprinted by permission of Prentice-Hall, Inc., Englewood Cliffs, N.J.

The lower plots in Fig. 25-16 show two analytical models used to describe clutter or targets with a broad spread of amplitudes, such as the distribution of a cylinder. Physically, the Swerling case 1 or 2 model corresponds to any target composed of more than three scatterers of comparable size (including aircraft at most aspect angles and at frequencies above a few hundred megahertz, rain clutter, and surface clutter viewed with grazing angles larger than about 5°). The case 3 or 4 model corresponds to a target in which one constant-amplitude scatterer dominates the echo but is modulated by two or more smaller scatterers. Weibull and log-normal distributions describe the statistics of surface clutter viewed at low grazing angles by medium- or high-resolution radar.[6,26,27,28]

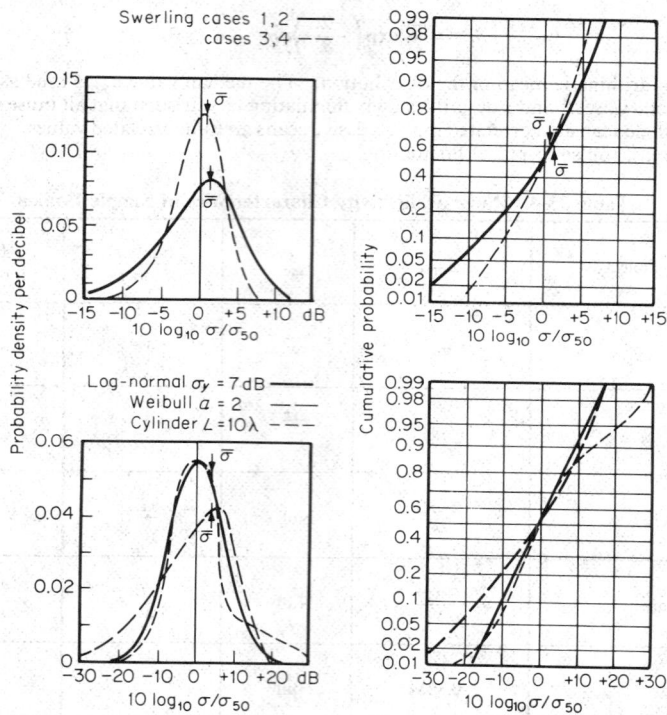

Fig. 25-16. Cross-section amplitude distributions.

14. Spectra and Correlation Intervals. Radar performance is sensitive to the time scale of cross-section fluctuation, which is related to the rate of rotation in aspect angle and to internal vibration or rotation of target components. Considering a rigid target whose aspect angle changes at a rate ω rad/s, the width of the fluctuation power spectrum will be proportional to the product ωL_x, where L_x is the spread of scatterers normal to the radar line of sight and the axis of rotation. If the scatterers are distributed uniformly over L_x, the spectrum will be uniform over a band of width $f_{max} = 2\omega L_x/\lambda$. Nonuniform scatterer distributions will produce corresponding-shaped spectra,[8] but any object can be approximated closely by an equivalent spread L_x and bandwidth B, analogous to the equivalent noise bandwidth of a filter. The corresponding correlation interval in time is $t_c = 1/f_{max} = \lambda/2\omega L_x$. The bandwidth may be increased (and t_c decreased) by internal vibrational and rotational components such as aircraft propellers or turbines. Typical values of f_{max} for aircraft-body echoes are on the order of a few hertz at X-band.

Spread of target scatterers over a length L_r along the radar line of sight leads to frequency sensitivity of cross section, with a correlation interval $f_c = c/2L_r$. Radars using frequency diversity can take advantage of this to obtain independent samples of target echo at different frequencies, all transmitted and received within one target correlation time interval t_c. Thus the Swerling case 1 (or 3) target model (Fig. 25-16), applicable when $t_c \geq t_0$, may be changed to case 2 (or 4) by rapid change in radar frequency, reducing the echo correlation time to a new value $t_c' \ll t_0$. Typical aircraft targets, for which $L_r \approx 10$ m, give correlation frequency intervals on the order of 15 MHz, so that many independent samples are available in the tuning band of a microwave radar.

15. Target Glint and Scintillation. Targets composed of multiple scattering elements whose varying phase relationships cause fluctuations in signal amplitude are subject to errors in radar position measurement.[29] The apparent source of the composite echo signal wanders back and forth across the target, and at times the signal appears to originate from points well beyond the physical spread of the target itself. In principle, the variance in position measurement is infinite, for a measuring system with unlimited dynamic range and bandwidth. However, for practical systems this "glint" error is closely approximated by a gaussian distribution with standard deviation $\sigma_r = 0.35 L_r$ (in range), or $\sigma_\theta = 0.35 L_x/R$ rad (in angle).[8] On typical aircraft targets, the distribution of scatterers is equivalent to a uniformly scattering object with L from one-third to two-thirds of the physical span, leading to rms errors from 0.1 to 0.25 times the tip-to-tip aircraft dimensions. The apparent doppler spread of target echoes (frequency glint) is a scaled replica of the cross-range glint, with $\sigma_f = 0.35 L_x(2\omega/\lambda)$ for a target aspect angle rate ω rad/s (or 0.1 to 0.25 times the tip-to-tip difference in doppler frequency). The fact that the peak-to-peak doppler error exceeds the band of frequencies actually present in the signal is the result of nonlinearities inherent in the measurement process.

16. Clutter Echoes. Unwanted radar echoes (clutter) may originate from land or sea surfaces (characterized by a dimensionless surface reflectivity σ^0), from weather, from chaff occupying a volume of space (with a volume reflectivity η_v in m^2/m^3), or from discrete objects described by cross sections in square meters. The cross section of surface clutter is found by multiplying $\sigma°$ by the surface area A_c within the radar resolution cell, while for volume clutter, η_v is multiplied by the resolution volume (see Par. 25-47). These parameters have statistical distributions and spectra describing variation in space and time. Sea and land clutter as functions of grazing angle are shown in Figs. 25-17 and 25-18.

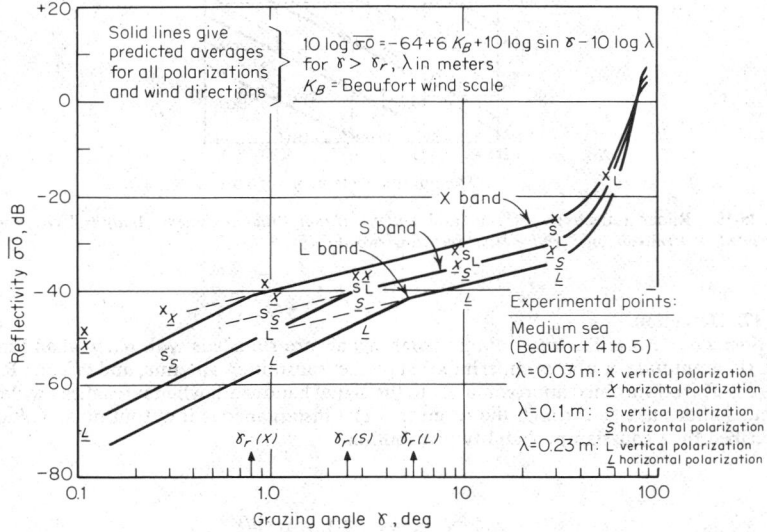

Fig. 25-17. Sea-clutter reflectivity vs. grazing angle.

Rain and snow reflectivity, based on the classical (and generally accurate) Gunn and East data,[12] is plotted in Fig. 25-19. Although rainfall statistics averaged over periods of an hour show negligible probability of encountering rain rates greater than 50 mm/h, instantaneous (1-min) values at the surface and sustained densities aloft may reach 100 to 300 mm/h.

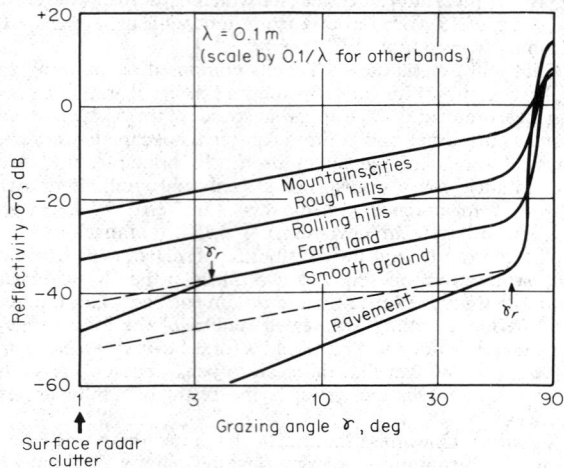

Fig. 25-18. Land-clutter reflectivity vs. grazing angle.

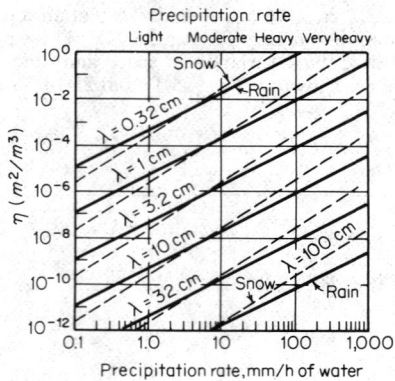

Fig. 25-19. Radar reflectivity of rain and snow. *(From "Radar System Analysis," copyright 1964. Reprinted by permission of Prentice-Hall, Inc., Englewood Cliffs, N.J.)*

17. Detection.

Signal and Noise Statistics. Most radar signals are sinusoids with narrowband modulation (in amplitude or phase) superimposed by the transmitter, antenna, and target. Random noise is also constrained approximately to the signal bandwidth when it reaches the detection circuits after passage through the receiver. The instantaneous if output noise voltage V is described by a gaussian probability distribution

$$dP_v = \frac{1}{\sqrt{2\pi N}} \exp\left(\frac{-V^2}{2N}\right) dV \tag{25-9}$$

where dP_v = probability that voltage lies between V and $V + dV$, and N = mean-square noise voltage (average noise power). The video noise voltage E_n out of a linear envelope detector has a Rayleigh distribution

$$dP_e = \frac{E_n}{N} \exp\left(\frac{-E_n^2}{2N}\right) dE_n \qquad E_n \geq 0 \tag{25-10}$$

The output voltage of a square-law detector follows the exponential distribution of if noise power,

$$dP_\psi = \frac{1}{N} \exp\left(\frac{-\psi}{N}\right) d\psi \qquad \psi \geq 0 \tag{25-11}$$

These noise distributions are shown in Fig. 25-20. The probability that the noise envelope will exceed a given threshold voltage level E_t is the shaded area to the right of E_t in Fig. 25-20c.

$$P_n = \int_{E_t}^{\infty} \frac{E_n}{N} \exp\left(\frac{-E_n^2}{2N}\right) dE_n = \exp\left(\frac{-E_t^2}{2N}\right) \tag{25-12}$$

If a sinusoidal signal of peak amplitude E_s is present with the noise, the distribution of envelope voltage E_r is Rician,

$$dP_s = \frac{E_n}{N} \exp\left(-\frac{E_n^2 + E_s^2}{2N}\right) I_0\left(\frac{E_n E_s}{N}\right) dE_n \tag{25-13}$$

where I_0 = Bessel function with imaginary argument. The probability of detection P_d for a sample taken at the peak of the signal is the area under this curve and above the threshold E_t, as shown in Fig. 25-21 for different signal-to-noise power ratios $S/N = E_s^2/2N$.

  Single-Sample Detection Probability. When a single sample is available for detection, the threshold is set to give the desired false-alarm probability according to Eq. (25-12), and the detection probability follows from the integral of Eq. (25-13). The results are plotted in Fig. 25-22, where each curve gives P_d versus S/N for the theshold setting corresponding to the P_n value shown. The value of S/N required to achieve a given P_d with fixed P_n is denoted by

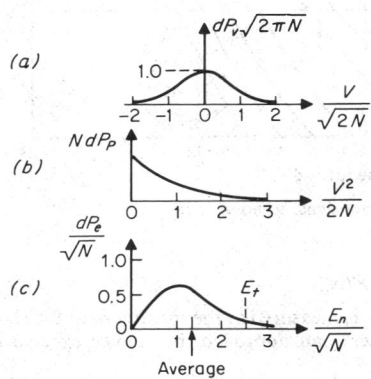

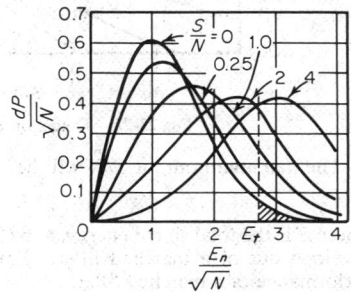

Fig. 25-20. Probability distributions of noise. (a) Gaussian distribution of noise voltage; (b) exponential distribution of noise power; (c) Rayleigh distribution of detected envelope. *(From "Radar System Analysis," copyright 1964. Reprinted by permission of Prentice-Hall, Inc., Englewood Cliffs, N.J.)*

Fig. 25-21. Probability distributions of envelope of signal plus noise. *(From "Radar System Analysis," copyright 1964. Reprinted by permission of Prentice-Hall, Inc., Englewood Cliffs, N.J.)*

$D_0(1)$, the single-pulse detectability factor for the steady-signal case. Data from this figure are used as the basis for establishing detectability factors in multiple-pulse systems with different target types.

18. Filters and Signal Spectra. Detection performance for a single sample has been shown to depend on the S/N ratio at the if output. Noise power at this point is a function of the receiver gain, the bandwidth, and the noise spectral density in the early stages of the receiver, before bandwidth is established. It is customary in detection analysis to assume an ideal (noise-free) receiver of unity gain, and to replace all actual noise sources with an equivalent broadband source of spectral density N_0, added to the input signal. The output noise power is then $N = N_0 B_n$, where B_n is the equivalent noise bandwidth of the receiver. The signal output power will also depend upon the receiver bandpass characteristics, and on the input energy.

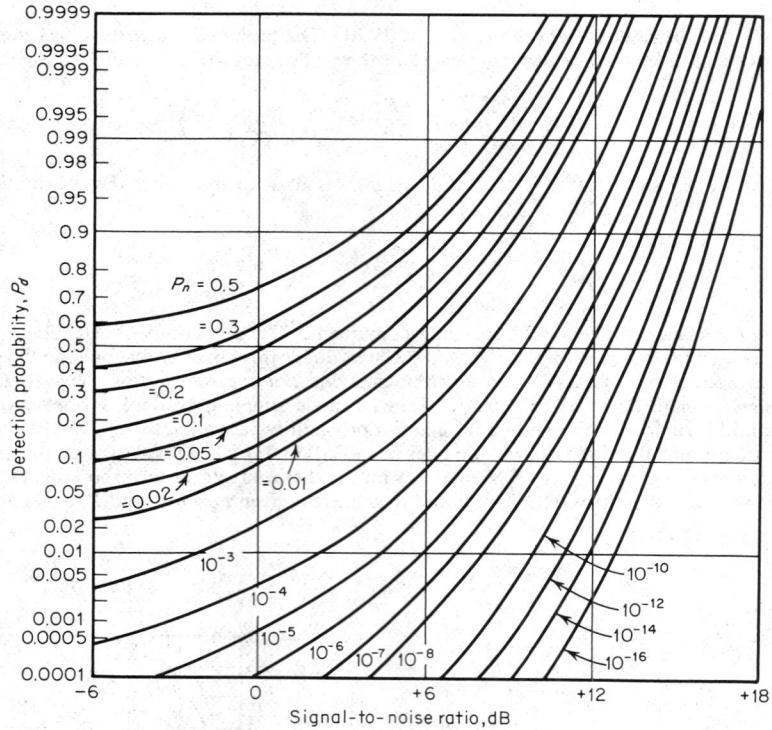

Fig. 25-22. Detection probability vs. signal-to-noise ratio.

The maximum output S/N will be

$$(S/N)_{\max} = E/N_0$$

where E is the total signal energy, and $(S/N)_{\max} =$ is measured at the maximum of the signal envelope out of a matched filter. Practical filters can approach, but never exceed, the performance of a matched filter.

The matched filter for a single, uncoded pulse of duration τ can be approximated by a conventional bandpass filter whose bandwidth is $B \approx 1/\tau$. A matching loss is defined for this case as

$$L_m = \frac{E/N_0}{S/N}$$

25-20

(see Fig. 25-23). Pulses with internal coding or modulation require more complex filters (see Par. 25-74).

A train of pulses produces a spectrum consisting of many separate lines (Fig. 25-24), and requires a *comb filter* with a matched series of response bands, properly phased to add all signal components into one output. Although such a filter is readily synthesized for a selected point in the time-frequency domain, using a range gate and narrow-band filter, it is difficult to match to signals of unknown time delay and doppler shift. Hence the integration of pulse trains is usually carried out at video (after envelope detection). The if filter is matched approximately to a single pulse, giving $S/N = (E_1/N_0)/L_m$ at the envelope detector, where $E_1 = (E/N_0)/n$ is the average energy in one pulse of the n-pulse train. Successive video pulses are then added on the radar display or in some other range-ordered storage device, prior to threshold detection.

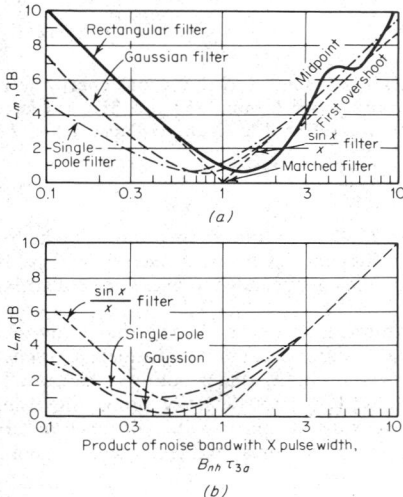

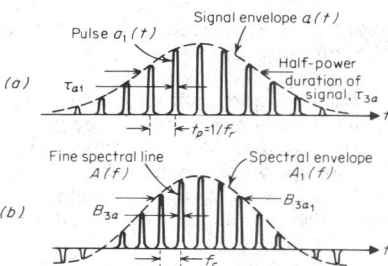

Fig. 25-23. Intermediate-frequency filter loss. (*a*) Rectangular pulse with different filters; (*b*) gaussian pulse with different filters. *(From "Handbook of Radar Measurement," copyright 1969. Reprinted by permission of Prentice-Hall, Inc., Englewood Cliffs, N.J.)*

Fig. 25-24. Waveform and spectrum of coherent pulse train. (*a*) Waveform of pulse train; (*b*) spectrum of pulse train. *(From "Handbook of Radar Measurement," copyright 1969. Reprinted by permission of Prentice-Hall, Inc., Englewood Cliffs, N.J.)*

19. Video Integration. Pulse train integration at video reduces the required energy of each pulse but introduces a loss in performance relative to predetection (matched-filter) integration. This integration loss is defined as the increase in total signal energy required for a given P_d and P_n, relative to that needed with predetection integration. The detectability factor for n-pulse video integration as defined by Blake[10] and the IEEE[30] is

$$D_0(n) = \frac{L_i D_0(1)}{n} = \frac{E_1/N_0}{L_m} \tag{25-14}$$

Curves of integration loss L_i versus n are shown in Fig. 25-25, with $D_0(1)$ as a parameter. For example, if $P_d = 0.90$ and $P_n = 10^{-6}$ are needed, $D_0(1)$ is 13.1 dB from Fig. 25-22, and L_i is 5.7 dB for $n = 100$ pulses. From these two figures, the steady-target data may be generated with an accuracy of about 0.1 dB.

20. False-Alarm Time. The D_0 curves of Fig. 25-22 correspond to a given probability P_n of a false alarm from each independent sample applied to the threshold. If the detection circuits are operative for $t_g \leq t_p$ seconds in each repetition interval t_p, there will be $\eta = t_g B_n$ independent samples applied to the threshold during the integration interval nt_p, producing an average of ηP_n false alarms. The ratio of total time to average number of alarms is the false-alarm time. In some discussions, following Marcum's usage,[31] the false-alarm time t_{fa} is defined as the interval in which the probability of at least one false alarm is 0.50, and a false-alarm number n' is defined as the number of independent samples applied to the threshold during t_{fa}. Since $t_{fa} = 0.69 - t_{fa}$, Marcum's fale-alarm number n' is

$$n' = \frac{0.69 B_n t_g f_r \overline{t_{fa}}}{n} = \frac{0.69}{P_n} \tag{25-15}$$

21. Collapsing Loss. Practical radars seldom preserve their full rf signal resolution through the integration and thresholding processes, where insufficient video bandwidth or broadened range gates may prove economical. The n video samples of signal plus noise are then integrated, along with m extra samples of noise alone, giving a collapsing ratio ρ:

$$\rho = 1 + \frac{m}{n} \tag{25-16}$$

The effect, when using a square-law detector, is the same as if the signal energy were redistributed over $\rho n = m + n$ pulses, leading to the larger L_i value (Fig. 25-25) associated with integration of ρn pulses. The additional loss is referred to as the collapsing loss L_c:

$$L_c(\rho, n) = \frac{L_i(\rho n)}{L_i(n)} \tag{25-17}$$

Formulas for ρ in different cases are given in Par. **25-42**, with rules for determining whether the number of independent threshold samples remains constant or varies inversely with ρ. In the latter case, P_n may be allowed to increase, and the energy added to overcome collapsing effects need not be as great.

22. CRT Integration. Signal detection by human operators, using cathode-ray-tube displays, cannot be characterized by a definite false-alarm probability. Instead, a *visibility factor* V_0 is used to give approximate requirements for E_1/N_0 under optimum bandwidth, display, and viewing conditions. Figure 25-26 shows V_0 for 0.50 detection probability using PPI and A-scope displays, assuming rectangular pulses and approximately matched rectangular filters ($B\tau \approx 1.2$). The visibility factor V_0 cannot be compared directly with D_0, but it

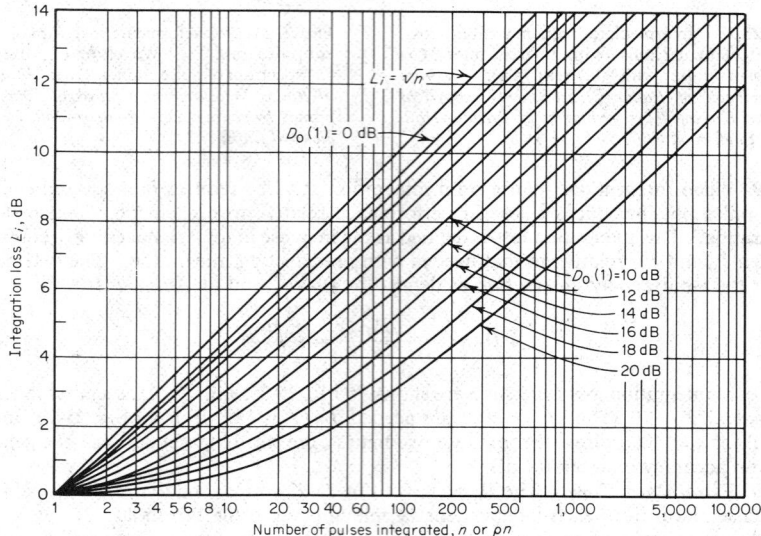

Fig. 25-25. Integration loss vs. number of pulses (Ref. 33).

is approximately equal to $L_m L_c D_0$ for $P_n = 10^{-4}$, assuming a minimum product $L_m L_c = 1.7$, or 2.3 dB, and an averaging time of 3 s for the combination of A-scope and operator. An operator can be expected to achieve the performance shown in Fig. 25-26 only under ideal conditions involving short attention spans and at least some a priori knowledge of where the signal should be expected on the display. Signal energy required for initial detection of an unexpected target during an extended watch may be higher by several decibels (operator loss).

23. Detection of Fluctuating Signals. The single-pulse detectability factor $D_1(1)$ for Swerling case 1, or case 2 (Fig. 25-16), is plotted in Fig. 25-27, showing a considerable increase in the required S/N ratio for high detection probabilities.

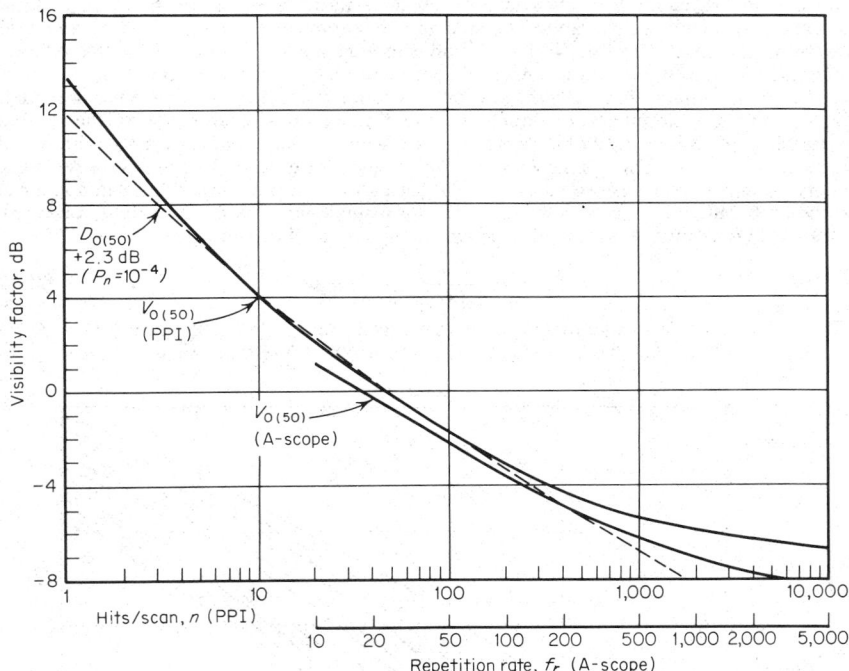

Fig. 25-26. CRT display visibility factors. *(After Blake, Ref. 10.)*

The difference in decibels between $D_0(1)$ and $D_1(1)$ for a given P_d and P_n is known as *fluctuation loss*, L_f, for the case 1 target, and L_f remains essentially constant for all values of n on this target. However, when n pulses from a case 2 target are integrated, the loss in decibels is $1/n$ times as great as for case 1. In general, when n pulses containing n_e independent Rayleigh target samples are integrated, the detectability factor $D_e(n, n_e)$ can be found by adding to $D_0(n)$ the loss L_f / n_e decibels. A system with n_e-fold diversity will then have a *diversity gain* of $L_f(1 - 1/n_e)$. The case 3 target has an equivalent $n_e = 2$ for all values of n, while case 4 has $n_e = 2n$. Values of n_e for intermediate cases are computed[33] using ratios of the observation time t_0 to target correlation time t_c and of tuning bandwidth Δf to correlation frequency f_c.

24. Detection with Log-Normal Statistics. In response to a growing awareness of actual signal and clutter distributions which are broader than the gaussian and Rayleigh models, detection curves have been derived for log-normal signals in gaussian noise[34] and for steady and Rayleigh signals in log-normal clutter.[35]

The defining equation for log-normal distribution of signal or noise power, ψ, is

$$dP_\psi = \frac{1}{\sqrt{2\pi}\,\sigma\psi} \exp\left[-\frac{(\ln\psi/\psi_m)^2}{2\sigma^2}\right] d\psi \tag{25-18}$$

where ψ_m = median power, and σ = standard deviation of $\ln \psi$. The width of the distribution is also described by the standard deviation in decibels, $\sigma_y = 4.34\sigma$. The ratio of average to median power in the log-normal distribution increases rapidly with σ or σ_y.

Because a large portion of the average power is contributed by the low-probability tail of the distribution, detection of log-normal signals is characterized by a large fluctuation loss. This loss can be reduced substantially by integration of independent signal samples, but the diversity gain is not as large as given by Ref. 34 for the Swerling models.

The large ratios of average and peak power to median power also affect detection of steady and fluctuating signals in log-normal clutter.[28,35] In this case, the threshold must be set well above the average clutter level to achieve low false-alarm rates. Since the average clutter power can be many decibels above the median, the required ratio of signal to median clutter can be several tens of decibels. The solution is to use a *median detector,* of which the binary integrator is an example, with as many independent clutter samples as possible.

25. Binary Integration.   An integration procedure which has found wide use in modern radar systems is the binary, or double-threshold, integrator. In this system, the receiver output is applied to a threshold, generating a sequence of binary ones and zeros within each repetition interval. These binary signals are summed for each range cell over n repetition intervals, and an alarm is generated when at least k out of n ones are accumulated in a cell. If the probability of a threshold crossing is p for a single interval, the probability of exactly j threshold crossings in n intervals is given by the binomial distribution

$$P(j) = \frac{n!}{j!(n-j)!}p^j(1-p)^{n-j} \tag{25-19}$$

The probability that j will equal or exceed the second threshold k is the sum $P(k) + P(k + 1) + \cdots + P(n)$. The median detector is instrumented by setting $k/n \approx 1/2$.

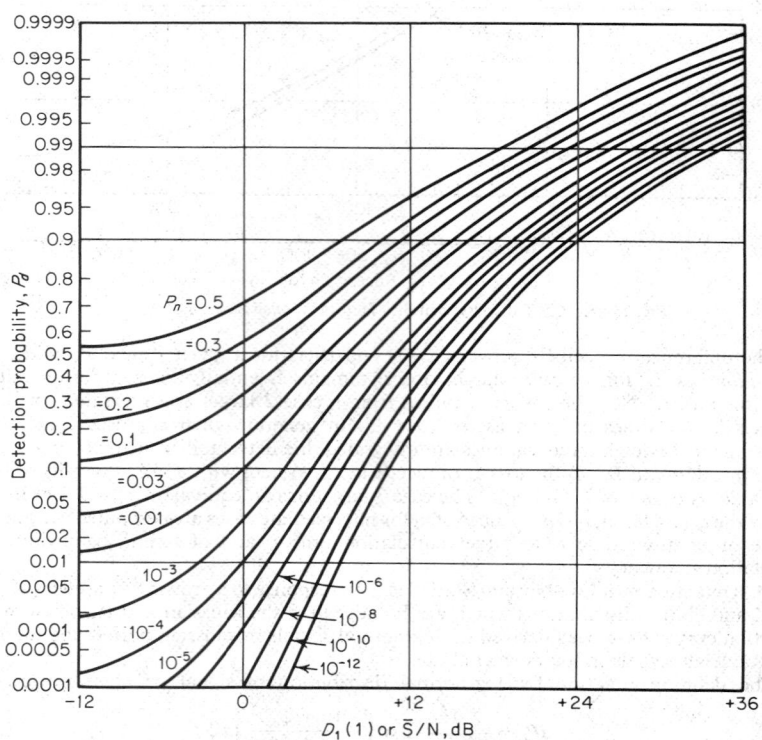

Fig. 25-27.   Detectability factor for Rayleigh target (Swerling case 1).

The performance of the optimum video integrator is approximated (with about 1 dB extra loss) for $0.25 \le k/n \le 0.75$, depending upon the type of signal fluctuation. For a given value of k/n, Eq. (25-19) is used to calculate the required single-pulse probability on noise or clutter to give the required low false-alarm probability, and the probability on signal plus noise or clutter to give detection probability. Figure 25-22 can then be used (after application of an appropriate fluctuation loss) to find single-pulse S/N required for detection.

In Eq. (25-19) it is assumed that the single-pulse probability p is the same for each pulse, but that the occurrence of threshold crossings is independent from pulse to pulse. When the S/N ratio varies over a train of n pulses (as with a scanning antenna), the average p can be used. However, if a fluctuating signal is correlated over the n pulses, the probabilities are not independent. The proper procedure is then to compute S/N required for a steady target, and to add the fluctuation loss.

26. Resolution. Definitions and Measures. A target is said to be resolved if its signal is separated by the radar from those of other targets, in at least one of the coordinates used to describe it. For example, a tracking radar may describe a target by two angles, time delay, and frequency. A second target signal from the same angle and at the same frequency, but with different time delay, may be resolved if the separation is greater than the delay resolution of the radar.

Resolution, then, is determined by the relative response of the radar to targets separated from the target to which the radar is matched. The antenna and receiver are configured to match a target signal at a particular angle, delay, and frequency. The radar will respond with reduced gain to targets at other angles, delays, and frequencies. This *response function* can be expressed as a surface in a four-dimensional coordinate system. Because four-dimensional surfaces are impossible to plot, and because angle response is almost always independent of the delay-frequency response, these pairs of coordinates are usually separated.

In angle, the response function $\chi(\theta, \phi)$ is simply the antenna pattern. It is found by measuring the system response as a function of the angle from the beam center. It has a main lobe in the direction to which it is matched, and side lobes extending over all visible space. Angular resolution, i.e., the main-lobe width in the θ and ϕ coordinates, is generally taken to be the distance between the 3-dB points of the pattern. The width, amplitude, and location of the lobes are determined by the aperture illumination (weighting) functions in the two coordinates across the aperture.

Because the matched antenna is uniformly illuminated, its response has relatively high side lobes, which are objectionable in most radar applications. To avoid these, the antenna illumination may be mismatched slightly, with resulting loss in gain and broadening of the main lobe.

Time delay and frequency can also be viewed as if they were two angular coordinates, i.e., as a two-dimensional surface which describes the filter response to a given signal as a function of the time delay t_d and the frequency shift f_d of the signal relative to some reference point. Points on the surface are found by recording the receiver output voltage, while varying these two target coordinates. The response function $\chi(t_d, f_d)$ is given, for any filter and signal, by

$$\chi(t_d, f_d) = \int_{-\infty}^{\infty} H(f)A(f - f_d)\exp(j2\pi ft_d)df \tag{25-20}$$

or

$$\chi(t_d, f_d) = \int_{-\infty}^{\infty} h(t_d - t)a(t)\exp(j2\pi f_d t)dt \tag{25-21}$$

where the functions $A(f)$ and $a(t)$, $H(f)$ and $h(t)$, are Fourier transform pairs describing the signal and filter, respectively.

The transform relationships, Eqs. (25-20) and (25-21), governing the time-frequency response function are similar to those which relate far-field antenna patterns to aperture illumination. Hence data derived for waveforms can be applied to antennas, and vice versa, by interchanging analogous quantities between the two cases. There is a significant difference between waveform and antenna response functions and constraints, however, because the two waveform functions (in time delay and frequency) are dependent upon each other through the Fourier transform. The two antenna patterns (in θ and ϕ coordinates) are essentially independent of each other, depending on aperture illuminations in the two spatial coordinates x and y. Further differences arise from the two-way pattern and gain functions applicable to the antenna case.

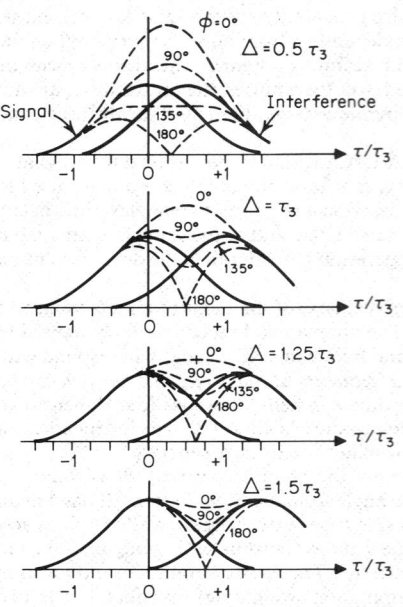

Fig. 25-28. Resolution between adjacent targets with separation Δ and rf phase difference ϕ (Ref. 41).

The ability of the radar to form distinguishable response peaks on closely spaced targets can be illustrated by plotting the composite waveform of two equal targets with varying rf phase difference (Fig. 25-28). With 0.5-pulse-width separation, separate peaks appear only when the phase difference is near 180°. Separations between 1.0 and 1.25 give distinguishable peaks over most phase angles, and at 1.5 pulse-widths the two targets are always distinguishable (separately detectable and measurable with small interaction). For the smoothly shaped pulse with no side-lobes, shown in the figure, the targets become resolvable at separations between 1.25 and 1.5 times the half-power pulse-width. In angle, with a scanning two-way antenna pattern, the corresponding resolution is between 0.9 and 1.1 times the half-power (one-way) beamwidth.

When a large target is present, greater separation is needed to resolve a smaller target. However, because of the steep skirts attainable in most radar response functions, the separation need seldom exceed twice the pulse-width, or 1.5 times the beamwidth, unless the side-lobe response from the larger target obscures the smaller one. Thus the half-power width of the response function can provide a measure of resolution, subject to modification by a factor near unity to account for targets of different amplitudes. Other measures of resolution, such as Woodward's time and frequency resolution constants[36] (extended, by analogy, to angle) will also be shown to be closely related to the half-power widths.

27. Antenna and Waveform Functions. The resolution properties of response functions generated by continuous apertures and waveforms can be summarized by a few simple parameters. In a coordinate z, which may represent time delay, frequency, or angle, the response function $\chi(z)$ can be described by its width at significant levels below the peak response χ_m:

z_3, the half-power (3-dB) width;

z_6, z_{10}, z_{20}, at the 6-, 10-, or 20-dB points.

Figure 25-29 shows plots of three typical waveform functions, from which these widths can be measured.

Other measures of resolution are

Effective width for noise or random clutter:

$$z_n \equiv \frac{1}{\chi_m^2}\int_{-\infty}^{\infty} \chi^2(z)\,dz \tag{25-22}$$

Rms response width:

$$z_{rms} \equiv 2\pi\left[\frac{\int_{-\infty}^{\infty} z^2\chi^2(z)\,dz}{\int_{-\infty}^{\infty} \chi^2(z)\,dz}\right]^{1/2} \tag{25-23}$$

Woodward resolution constant:

$$z_w \equiv \frac{\int_{-\infty}^{\infty} |R(z)|^2 dz}{\chi_m^2} \qquad (25\text{-}24)$$

Equation (25-24) is a measure of the total amount of response (or ambiguity) of the autocorrelation of χ, $R(z)$, in the z coordinate, where

$$R(z) = \int_{-\infty}^{\infty} \chi(s)\chi^*(s + z)ds$$

Table 25-5 lists the resolution widths for several functions in terms of waveform coordinates t and f. These coordinates may be interchanged, or antenna coordinates θ and x/λ may be substituted.

If the first three functions of Table 25-5 are excluded, a remarkable similarity between the several tapered functions can be seen. These have almost identical main lobes (at least out to the -10-dB points), with $\tau_n \approx 1.05\tau_3$ and $\tau_w \approx 1.3\tau_3$. The significant differences are side-lobe levels and presence of extended lobes in frequency or time domains (or equivalent antenna coordinates).

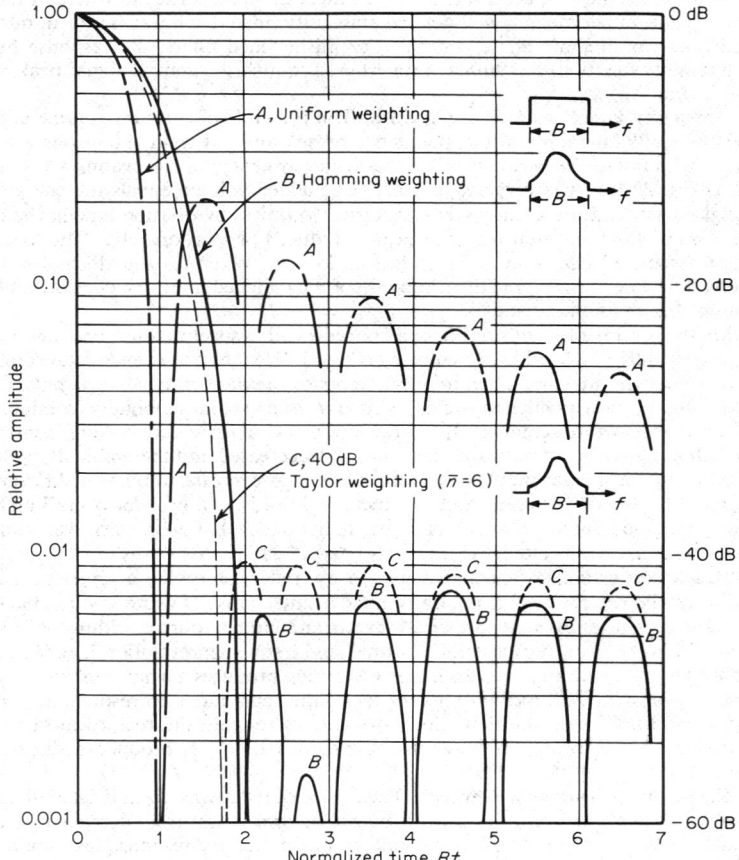

Fig. 25-29. Comparison of waveform functions (Ref. 38).

28. Ambiguity Functions of a Single Pulse.   The time-frequency response function for a system whose filter is matched to the waveform is the ambiguity function for that waveform. This function describes the response to objects at different time delays and frequencies (doppler shifts). The ambiguity function of a simple pulse of width τ (without phase modulation or coding) consists of a single main lobe whose base width is 2τ along the time axis and whose frequency response has a lobe structure dependent on the discontinuities in the time function or its derivatives (Fig. 25-30a).

Introduction of phase modulation during the pulse broadens the frequency spread of the function and narrows the response along the time axis. This is the principle of pulse compression, of which the most common form is linear fm (or chirp), shown in Fig. 25-30b. With the linear FM function, very low side lobes can be obtained in regions both sides of the main, diagonal response ridge. Along this ridge, however, the response falls off slowly from its central value, and targets separated by almost one transmitted pulse width will be detected if they are offset in frequency by the correct amount. Pseudo-random phase coding can generate a single narrow spike in the center of the ambiguity surface (Fig. 25-30c), at the expense of large-amplitude side lobes elsewhere on the ambiguity surface.

The plots of Fig. 25-30 illustrate an important property of the ambiguity function: the total volume under the surface describing the square of the function, $|\chi(t_d,f_d)|^2$, is equal to the signal energy (or to unity, when normalized) for all waveforms. Compression of the response along the time axis must be accompanied by an increase in response elsewhere in the time-frequency plane, either along a well-defined ambiguity ridge (for linear fm) or in numerous random lobes (for random-phase codes). For mismatched filters, the response function (cross-ambiguity function) has similar properties, although the central signal peak may be reduced in amplitude.

29. Ambiguity Functions of Pulse Trains.   The principle of constant volume under the squared ambiguity function is also applicable to pulse trains. A train of noncoherent pulses of width τ, with interpulse period $t_p = 1/f_r \gg \tau$, merely generates a repeating set of surfaces similar to Fig. 25-30a at intervals t_p in time. The added volume equals the energy of the additional pulses, and if the energy is normalized to unity (by division among the several pulses), the amplitude of each response peak is reduced proportionately. The location of peaks and associated side lobes is shown in Fig. 25-31a. When the signal is coherent over an observation time $t_0 = nt_p$, the time response of the matched filter stretches the ambiguity function to $\pm t_0$ along the time axis.

Within the central lobes, this response is concentrated in spectral lines separated by f_r and approximately $1/t_0 = f_r/n$ wide in frequency (Fig. 25-31b). Near the ends of the ambiguity function, where the matched-filter impulse response overlaps only $n' < n$ pulses of the received train, the lines broaden to width f_r/n', and at the end of the ambiguity function, where $n' = 1$, no line structure remains. If the repetition rate is increased, holding constant the pulse width and number of pulses in the train, t_0 is decreased and the ambiguity volume is redistributed into a smaller number of broader lines. A decrease in pulse width, such that t_0 is restored to its original value and n is increased, leads to a broader overall ambiguity function, with the original number of lines in frequency but with narrower and more numerous response bands along the time axis (Fig. 25-31c).

30. Resolution with CW Transmissions.   A cw radar, observing a target over a time interval t_0, can be regarded as having transmitted a single pulse of width $\tau = t_0$. Generally, this interval is sufficiently long so that all targets and clutter sources within the beam are included within the main response lobe in time, and frequency resolution is relied upon to distinguish targets. Introduction of a linear FM sweep produces a single ambiguity ridge of the type shown in Fig. 25-30c, with a very long time span and a correspondingly narrow frequency width. Time resolution, along the axis, is roughly the reciprocal of the total frequency sweep. Repeated FM sweeps at intervals $t_p \ll t_0$ produce a diagonal-line structure.

31. Resolution of Targets in Clutter.   The choice of radar waveform is often dictated by the need to resolve small targets (aircraft, buoys, or projectiles) from surrounding clutter. The clutter power at the filter output is found by integrating the response function over the clutter region, with appropriate factors for variable clutter density, antenna response, and the

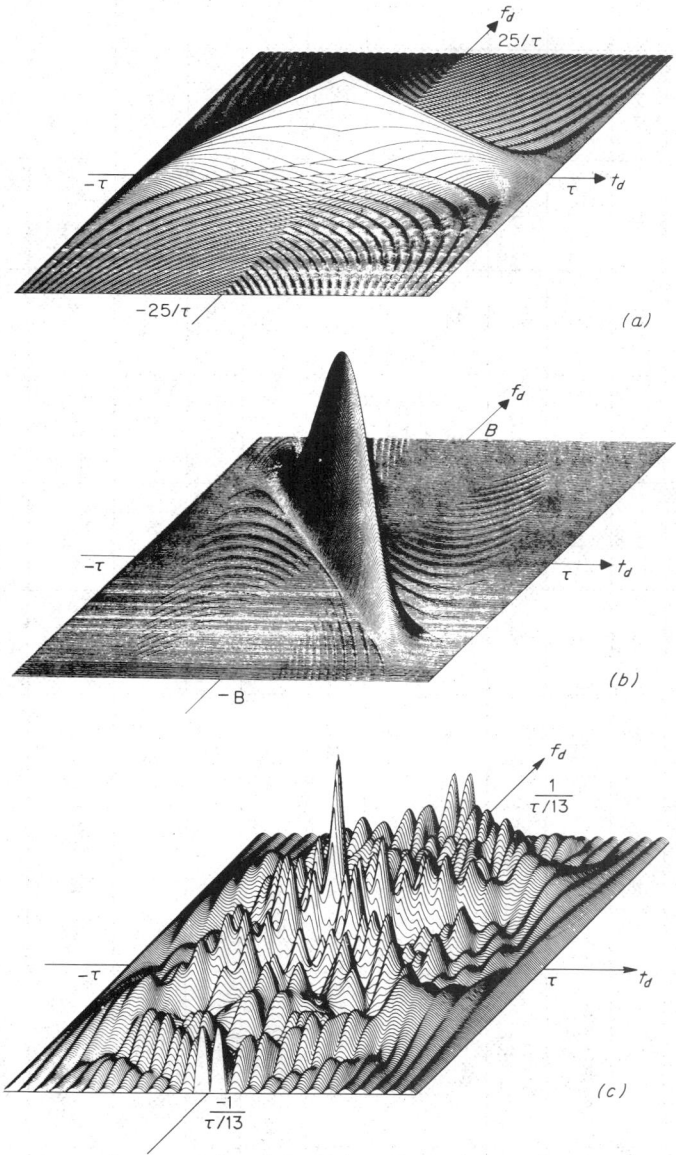

Fig. 25-30. Ambiguity functions of single pulses. (*a*) Response to constant carrier pulse with a rectangular envelope; (*b*) chirp response for bilateral Hamming weighting; (*c*) response for the 13-element Barker code (Ref. 37).

Table 25-5. Resolution Properties of Different Waveforms

Waveform	Spectrum	Side-lobe level, dB	$\tau_3 B_3$	$\tau_3 B$	τ_6/τ_3	τ_{10}/τ_3	τ_{20}/τ_3	τ_n/τ_3	$\alpha/\tau_3 = \tau_{rms}/\tau_3$	τ_w/τ_3
Rectangular (width τ)	$\dfrac{\sin \pi f\tau}{\pi f\tau}$	∞	0.89	∞	1.00	1.00	1.00	1.00	1.81	0.67
$\dfrac{\sin \pi tB}{\pi tB}$	Rectangular (width B)	14	0.89	0.89	1.36	1.66	*	1.13	∞	1.13
$\dfrac{\sin^2(\pi tB/2)}{(\pi tB/2)^2}$	Triangular (width B)	26	0.37	1.28	1.39	1.72	2.3	1.05	2.70	1.41
$\dfrac{\cos(\pi t/\tau)}{1 - 4t^2/\tau^2}$ ($t_3 = 0.5\tau$)	$\dfrac{\cos \pi f\tau}{1 - 4f^2\tau^2}$	∞	0.59	∞	1.33	1.60	1.87	1.00	2.27	1.07
$\cos^2(\pi t/\tau)$ ($\tau_3 = 0.37\tau$)	$\dfrac{\sin \pi f\tau}{(1 - f^2\tau^2)\pi f\tau}$	∞	0.53	∞	1.37	1.70	2.18	1.02	2.42	1.30
$\dfrac{\cos \pi tB}{1 - 4t^2 B^2}$	$\cos(\pi f/B)$	23	0.59	1.19	1.40	1.70	2.18	1.05	2.66	1.27
$\dfrac{\sin \pi tB}{(1 - t^2 B^2)\pi tB}$	$\cos^2(\pi f/B)$	32	0.53	1.44	1.39	1.73	2.28	1.08	2.52	1.36
Taylor	Taylor weighted (max. width B)	30	0.52	1.12	1.40	1.74	2.26	1.05	∞	1.34
Taylor	Taylor weighted (max. width B)	40	0.50	1.24	1.40	1.76	2.35	1.05	∞	1.39
Hamming	$0.54 + 0.46 \cos\left(\dfrac{2\pi f}{B}\right)$	43	0.51	1.12	1.43	1.77	2.35	1.06	∞	1.39
Gaussian $\exp(-t^2/2\sigma_t^2)$	Gaussian $\exp(-f^2/2\sigma_f^2)$	∞	0.44	∞	1.41	1.82	2.58	1.06	2.67	1.50

* Side-lobe level less than 20 dB down.

inverse-fourth-power range dependence included in the integrand. Signal-to-clutter ratio (S/C) for a target on the peak of the radar response is then given by

$$\frac{S}{C} = \frac{\sigma G_t(0) G_r(0) |\chi(0,0)|^2}{\int_v \eta_v(\theta,\phi,f_d,t_d) G_t(\theta,\phi) G_r(\theta,\phi) |\chi(f_d,t_d)|^2 (R/R_c)^4 \, dv} \tag{25-25}$$

where σ = target cross section, η_ω = clutter reflectivity, R/R_c = target-to-clutter range ratio, and v = four-dimensional volume containing clutter. The usual equations for S/C ratio (Par. 25-41) are simplifications of Eq. (25-25) for various special cases (e.g., surface clutter, homogeneous clutter filling the beam, etc.). Clearly, the S/C ratio is improved by choosing a waveform and filter such that $\chi(f_d, t_d)$ is minimized in clutter regions while maintaining a high value $\chi(0, 0)$ for all potential target positions. In a search radar, a two-dimensional bank of filters would be constructed to cover the intervals in delay and doppler occupied by targets, and the clutter power for each of these filters would then be evaluated using Eq. (25-25).

   32. **Basic Process of Measurement.** A radar measures target position in a given coordinate by identifying one or more resolution cells containing detectable target signals and then interpolating to refine the estimate of location. A very simple estimator consists of a bank of contiguous resolution cells (range gates, doppler filters, or beam positions) in which the center of the cell containing the strongest signal above a detection threshold is identified

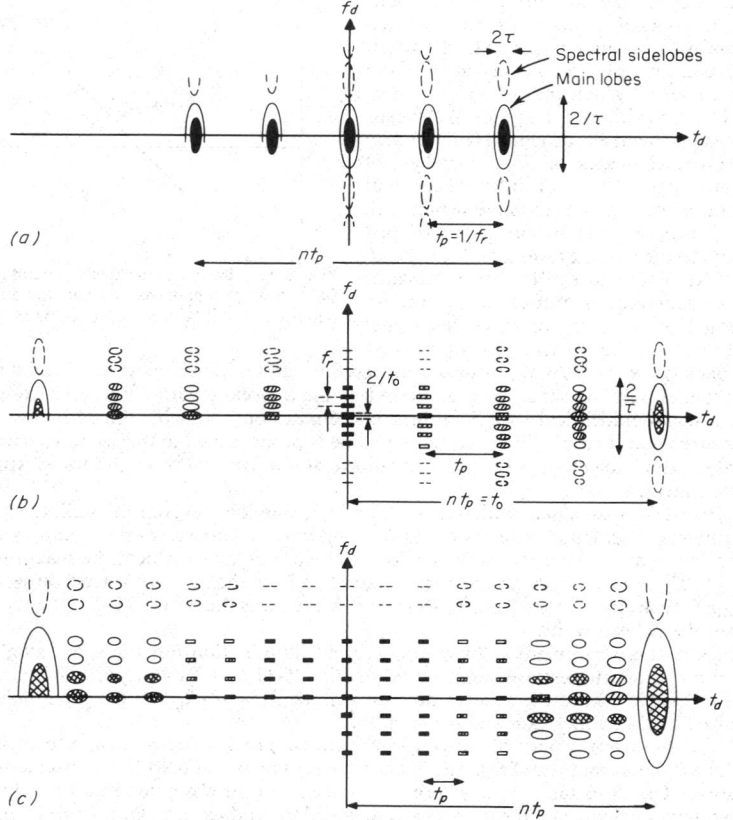

Fig. 25-31. Location of response lobes for uniform pulse trains. (*a*) Noncoherent pulse train; (*b*) coherent pulse train; (*c*) coherent, reduced τ, higher f_r.

as the target location. Although noise may lead to identification of the wrong cell in cases where a target is in a region of overlapping response, the error seldom exceeds ± one-half cell width, and an rms error of 0.2 to 0.3 times the 3-dB cell width can be expected on targets strong enough to be detected reliably.

More accurate estimates are made by interpolating between overlapping cells, or by placing the peak of the response function at the target location Ideally, this is done by forming a function which is the derivative of the response function in the measured coordinate, and adjusting its position until the target gives a null output. If this difference-channel null occurs within the region of peak response in the main response, or Σ, channel, the target must lie exactly at the peak of the Σ channel, where the derivative is zero. Since the derivative is an S-shaped (odd) function with a linear region each side of the null, small deviations in target positions may be estimated directly from the difference-channel output. In an idealized monopulse angle estimator, two offset beam patterns (Fig. 25-32a) are formed by feed networks. Outputs of these beams are combined in a hybrid to produce an on-axis Σ pattern and a Δ pattern which is the derivative of the Σ (Fig. 25-32b). When the Δ voltage is normalized to Σ, a calibrated curve for Δ/Σ versus target displacement θ is produced (Fig. 25-32c) which gives estimates of θ independent of signal strength. Noise or interference will introduce errors in the estimate which are inversely proportional to the slope of this curve and to the voltage ratio $\sqrt{S/N}$ or $\sqrt{S/I}$.

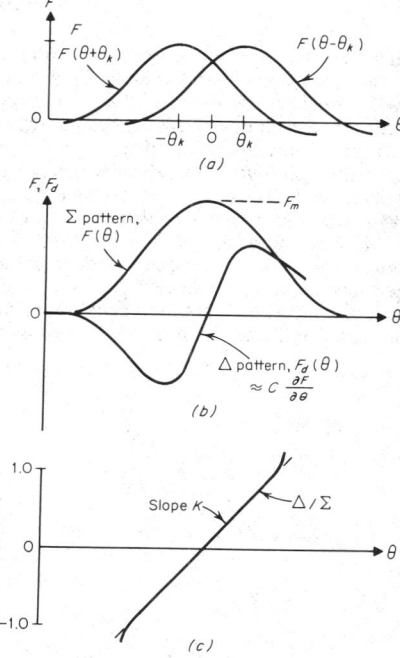

Fig. 25-32. Basic angular measurement process. (*a*) Offset beam patterns; (*b*) sum and difference patterns; (*c*) calibration curve for Δ/Σ.

33. Ideal Estimators. The ideal estimator for noise achieves an output signal-to-noise ratio equal to the ratio of twice the signal energy reaching the aperture to input noise power density, $\mathscr{R}_0 = 2 E/N_0$, using a matched filter and uniform aperture illumination for the Σ channel. At the same time, it maximizes the Δ slope separately in each coordinate, using a Δ channel identical with Σ in three coordinates but following the derivative of Σ in the measured coordinate. The maximum Δ slope is proportional to the second derivative at the peak of the Σ channel, which is determined by the rms width of the transform of the response function.

34. Practical Monopulse Estimators. The high side-lobe level of the uniformly illuminated antenna is seldom satisfactory in radar, and horn-fed apertures must employ tapered illumination to avoid excessive spillover loss. The use of taper reduces the maximum gain from F_0^2 to $F_m^2 = \eta_a F_0^2$, reduces the energy extracted from the incident wave from $\mathscr{R}_0$ to $\mathscr{R}_m = \eta_a \mathscr{R}_0$, reduces the Δ slope from K_0 to $K = K_r K_0$, reduces the Σ and Δ side-lobe levels, and increases the Σ beamwidth.

Tables of these parameters, for many different illumination functions, are available.[8,39] Cases of common interest are summarized in Fig. 25-33 and Table 25-6. The parameters plotted in Fig. 25-33 are the relative difference slope $K_r = K/K_0$ and the monopulse slope normalized to Σ-channel gain and beamwidth.

Table 25-6 shows the performance characteristics of practical horn-fed apertures, in terms of *H*-plane (*x*-coordinate) and *E*-plane (*y*-coordinate) slopes and both Σ- and Δ-channel side-lobe ratios. Use of rectangular apertures is assumed, but the absolute levels of performance are essentially unchanged if the corners are removed to produce circular apertures (η_a and K_r values will be increased because these are referred to the reduced potentialities of the smaller circular aperture). Additional noise errors are produced if the Σ beam is not kept exactly on the target.

Table 25-6. Monopulse Feed Horn Performance

Type of horn	η_a	H-plane		E-plane		G_{sr}, dB	G_{se}, dB	Feed shape
		$K_r\sqrt{\eta_y}$	k_m	$K_r\sqrt{\eta_x}$	k_m			
Simple four-horn	0.58	0.52	1.2	0.48	1.2	19	10	
Two-horn dual-mode	0.75	0.68	1.6	0.55	1.2	19	10	
Two-horn triple-mode	0.75	0.81	1.6	0.55	1.2	19	10	
Twelve-horn	0.56	0.71	1.7	0.67	1.6	19	19	
Four-horn triple-mode	0.75	0.81	1.6	0.75	1.6	19	19	

SOURCE: D. K. Barton and H. R. Ward, "Handbook of Radar Measurement," copyright 1969. Reprinted by permission of Prentice-Hall, Inc., Englewood Cliffs, N.J.

35. Scanning Systems. Angle estimates can also be made by scanning a single beam around (conical scan) or across (linear scan) the target, obtaining sequentially the offset samples which are gathered simultaneously in monopulse radar. Two-dimensional measurement by sequential scanning is much less efficient than monopulse, because the target is not fully illuminated by the transmitting and receiving Σ patterns, the difference slope is lower, and the energy must be shared between two orthogonal measurements. Figure 25-34 shows values of conical-scan errors for one- and two-way conical-scan systems, using gaussian and $(\sin x)/x$ beam patterns, as functions of offset angle.

The one-dimensional linear-scan case, encountered in search and height-finding radars, gives the following accuracy, expressed as a function of actual signal energy ratio, number of hits per beamwidth, and on-axis signal-to-noise ratio $(S/N)_m$:

$$\sigma_\theta = \frac{\theta_3}{k_p \sqrt{\mathcal{R}}} \approx \frac{\theta_3}{2\sqrt{n(S/N)_m}} \tag{25-26}$$

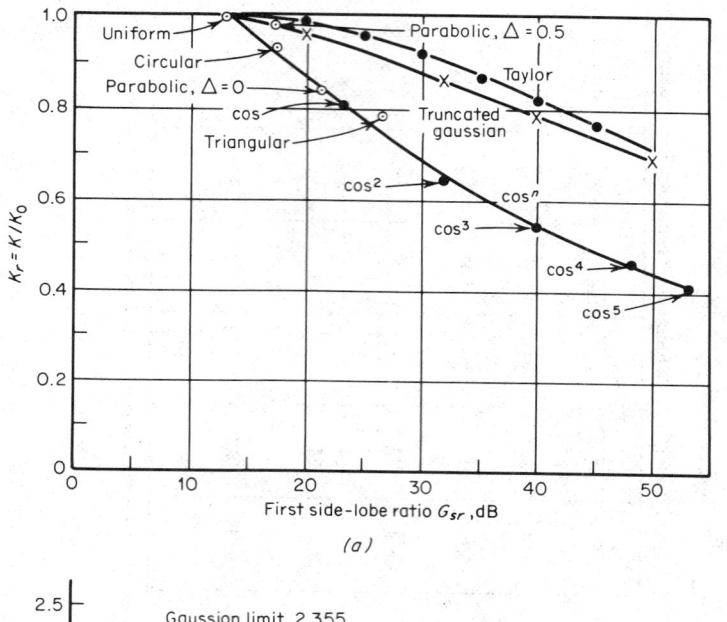

(a)

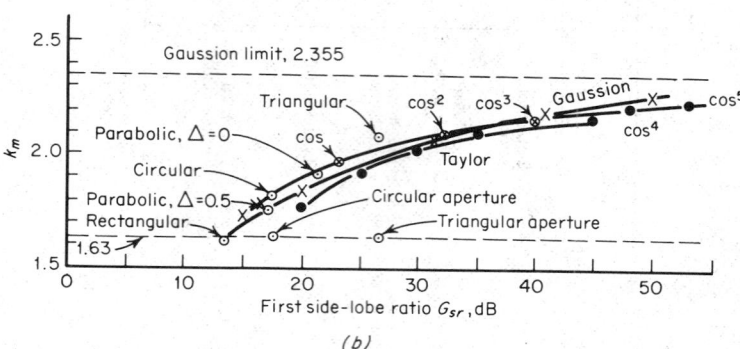

(b)

Fig. 25-33. Difference slopes vs. Σ-channel side-lobe ratio. *(a)* Relative-difference slope; *(b)* normalized monopulse slope. *(From "Handbook of Radar Measurement," copyright 1969. Reprinted by permission of Prentice-Hall, Inc., Englewood Cliffs, N.J.)*

The constant k_p varies from 1.18 for one-way operation with a gaussian pattern to 1.76 for two-way operation with a $(\sin x)/x$ pattern, but the associated beam-shape loss varies so that the approximation shown is accurate to within 15% in σ_θ for all cases. It is assumed that an optimum estimation process is used on the received pulse train, and that n is large (≥ 10).

36. Range Estimators. Radar-range (R) measurements are made by estimating the round-trip time delay t_d between transmission and reception of a signal, and converting to range, using the velocity of light in vacuum, c:

$$R = \frac{t_d c}{2}$$

$$c = 2.997925 \times 10^8 \text{ m/s}$$

$$= 0.9835692 \times 10^9 \text{ ft/s}$$

$$= 1.618750 \times 10^5 \text{ naut mi/s}$$

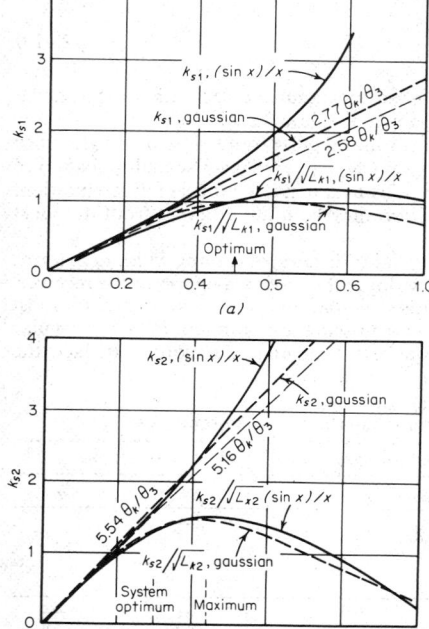

Thus the delay per unit range is

$$\frac{t_d}{R} = \frac{2}{c} = 6.671281 \times 10^{-9} \text{ s/m}$$

$$= 2.033410 \times 10^{-9} \text{ s/ft}$$

$$= 12.35521 \times 10^{-6} \text{ s/naut mi}$$

Fig. 25-34. Conical-scan error slopes. (*a*) one-way case; (*b*) two-way case. (*From "Handbook of Radar Measurement," copyright 1969. Reprinted by permission of Prentice-Hall, Inc., Englewood Cliffs, N.J.*)

According to the ideal process described earlier, the arrival time of a signal should be estimated by passing the signal through a matched filter, differentiating it, and measuring the point at which the derivative passes through zero. The accuracy will then be determined by the energy ratio and the rms bandwidth. Matched filters for many waveforms and spectra can be closely approximated by practical receivers, with resulting bandwidths, 3-dB pulse widths, and spectral widths, as shown in Table 25-7.

In pulse-compression radar, the transmission is often a rectangular pulse of width τ and bandwidth $B \gg 1/\tau$, with an approximately rectangular spectrum. Weighting is used in the receiver to reduce side lobes, giving a mismatched filter with reduced S/N and Δ slope. The range error with mismatched filter is

$$\sigma_t = \frac{1}{K\sqrt{\mathcal{R}}} = \frac{\tau'}{k_t \sqrt{2S/N}} = \frac{(\sigma_t) \min}{K_r} \tag{25-27}$$

where τ' is the 3-dB width of the output waveform, and the slope factors are the waveform analogs of the corresponding angular slopes shown in Fig. 25-33. Thus, for the special case of uniform-spectrum pulse compression, k_t values between 1.6 and 2.3 may be used to describe the difference slope normalized to the 3-dB width of the compressed pulse.

Another case of special interest is that of the band-limited (but uncoded) rectangular pulse

(Ref. 2, p. 468) for which the rms bandwidth shown in Table 25-7 is $\beta = \sqrt{2B/\tau}$. Then the range error is

$$\sigma_t = \sqrt{\frac{\tau}{2\,B\mathcal{R}}} = \frac{\tau}{\sqrt{2\,B\tau\mathcal{R}}} \tag{25-28}$$

Other studies[40] have shown a theoretical limit for unrestricted bandwidth, given by

$$\sigma_t = \frac{\sqrt{2}\,\tau}{\mathcal{R}} \tag{25-29}$$

To realize the optimum accuracy of a band-limited rectangular pulse with $B\tau \gg 1$, the receiving system must adapt its bandwidth to the actual energy ratio.[8]

Practical range estimators, especially those operating on pulse trains with low single-pulse energy and S/N, take the form of a split-gate tracker (see Par. 25-56). The multiplication of a waveform $a(t)$ by the split-gate function, followed by averaging, is a form of differentiation, although the composite filter-differentiator function may be quite different from the ideal matched-filter estimator.

Figure 25-35 shows the performance of rectangular split gates, of varying width τ_g, in terms of the product $K\tau_{3a}$, where τ_{3a} refers to the 3-dB width of the input waveform to the receiver, and τ_g is the total width of the split-gate pair. It is seen that optimum performance for most pulse shapes requires $\tau_g \approx 1.4\tau_{3a}$, but that very narrow gates are optimum for a rectangular pulse which has been optimally filtered to produce a triangular waveform. In fact, the

Table 25-7. Waveform and Spectrum Measurement Parameters

Input-signal description		Half-power widths		Rms widths	
Spectrum	Waveform	B_3	τ_3	β	α
$\dfrac{\sin \pi f\tau}{\pi f\tau}$ ($\lvert f\rvert < B/2$)	Band-limited rectangular (width τ)	$\dfrac{0.89}{\tau}$	τ	$\sqrt{\dfrac{2B}{\tau}}$	$1.81\,\tau$
Time-limited rectangular (width B)	$\dfrac{\sin \pi tB}{\pi tB}$ ($\lvert t\rvert < \tau/2$)	B	$\dfrac{0.89}{B}$	$1.81\,B$	$\sqrt{\dfrac{2\tau}{B}}$
$\dfrac{\sin^2(\pi f\tau/2)}{(\pi f\tau/2)^2}$	Triangular (width τ)	$\dfrac{1.27}{\tau}$	$0.29\,\tau$	$\dfrac{3.45}{\tau}$	$1.28\,\tau$
Triangular (width B)	$\dfrac{\sin^2(\pi tB/2)}{(\pi tB/2)^2}$	$0.29\,B$	$\dfrac{1.27}{B}$	$1.28\,B$	$\dfrac{3.45}{B}$
$\dfrac{\cos \pi f\tau}{1 - 4f^2\tau^2}$	$\cos\dfrac{\pi t}{\tau}$	$\dfrac{1.18}{\tau}$	$0.50\,\tau$	$\dfrac{3.14}{\tau}$	$1.14\,\tau$
$\cos\dfrac{\pi f}{B}$	$\dfrac{\cos \pi tB}{1 - 4t^2B^2}$	$0.50\,B$	$\dfrac{1.18}{B}$	$1.14\,B$	$\dfrac{3.14}{B}$
$\dfrac{\sin \pi f\tau}{(1 - f^2\tau^2)\pi f\tau}$	$\cos^2\dfrac{\pi t}{\tau}$	$\dfrac{1.43}{\tau}$	$0.37\,\tau$	$\dfrac{3.65}{\tau}$	$0.89\,\tau$
$\cos^2\dfrac{\pi f}{B}$	$\dfrac{\sin \pi tB}{(1 - t^2B^2)\pi tB}$	$0.37\,B$	$\dfrac{1.43}{B}$	$0.89\,B$	$\dfrac{3.65}{B}$
Gaussian $\exp(f^2/2\sigma_f^2)$ $= \exp(-2\pi^2 f^2\sigma_t^2)$	Gaussian $\exp(-t^2/2\sigma_t^2)$ $= \exp(-2\pi^2 t^2\sigma_f^2)$	$1.66\,\sigma_f$	$1.66\,\sigma_t$	$4.45\,\sigma_f$ $= 2.67\,B_3$	$4.45\,\sigma_t$ $= 2.67\,\tau_3$

narrow-gate pair approaches an ideal differentiator of unlimited bandwidth as its width shrinks to zero, reproducing the ideal estimator for any signal which has already passed through a matched filter.

37. Doppler Estimators. A transmission at frequency f_0, reflected from a target moving with radial velocity v_r, will be received at $f_0 + f_d$. The change in frequency f_d is known as the doppler shift:

$$f_d = f_0\left(\frac{c - v_r}{c + v_r} - 1\right) = \frac{-2 f_0 v_r}{c}\left(1 - \frac{v_r}{c} + \frac{v_r^2}{c^2} - \cdots\right)$$

$$\approx -\frac{2 f_0 v_r}{c} = -\frac{2 v_r}{\lambda} \tag{25-30}$$

A measurement of f_d can be translated to radial velocity:

$$v_r = -\frac{f_d c}{2 f_0}\left(1 - \frac{f_d}{2 f_0} + \frac{f_d^2}{4 f_0^2} - \cdots\right) \approx -\frac{f_d c}{2 f_0} = -\frac{f_d \lambda}{2} \tag{25-31}$$

In most cases, the signal bandwidth is small enough relative to f_0 so that the doppler shift can be regarded as a simple displacement of the spectrum relative to that transmitted.

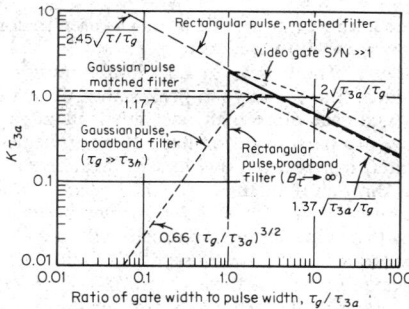

Fig. 25-35. Normalized slope for split-gate discriminator. *(From "Handbook of Radar Measurement," copyright 1969. Reprinted by permission of Prentice-Hall, Inc., Englewood Cliffs, N.J.)*

The spectrum of a typical coherent pulse train is shown in Fig. 25-24. The spectral envelope is determined by the waveform of an individual pulse, and can provide a coarse frequency estimate if applied to an if discriminator. This discriminator would, ideally, be matched to the derivative of the spectral envelope $A_1(f)$, in which case the error in frequency estimate would be

$$\sigma_f = \frac{1}{\alpha\sqrt{\mathscr{R}}}$$

with α representing the rms width of an individual pulse. Frequency error on rectangular pulses with mismatched filters can be estimated using the analogy to antennas, as described in connection with Eq. (25-27) for time measurements, but with time and frequency interchanged.

Of greater practical significance is the measurement of doppler shift on the fine-line spectrum, where the line width is a function of a pulse-train duration t_0 and its envelope shape, or of phase-stability factors in the radar equipment. For stable radar, an rms observation time α may be calculated using the pulse-train envelope (as determined by antenna pattern in a scanning radar, for instance). Table 25-7 shows values of α for typical functions. Values of slope factor K_f, analogous to K in angular or range measurement with mismatched filters, are near $1/B_3$ for most signals and discriminators.

38. Signal-Processing Losses. The preceding discussion of thermal-noise errors in measurement has been based on the optimum use of signal energy, so that the signal-to-noise ratio is well above unity before envelope detection or similar nonlinear processing, and extraneous noise samples are excluded from the averaging process. Increased error will result if nonoptimum conditions apply, as in the following cases:

Detector Loss. In most radars, it is impractical to integrate all the signal energy coherently in a matched filter prior to the detectors. The S/N ratio at the detector is then

$$\frac{S}{N} = \frac{\mathcal{R}_1}{2 L_m} = \frac{\mathcal{R}}{2 n L_m} \tag{25-32}$$

where L_m = loss in matching the if filter to a single pulse (Fig. 25-23), and n = number of pulses which are averaged after detection in performing the measurement. The reduction in effective S/N ratio caused by signal suppression in the detector for monopulse, time, or doppler measurement can be described as L_x:

$$L_x = \frac{(S/N) + 1}{S/N} = \frac{S + N}{S} \tag{23-33}$$

For conical scan,

$$L_x = \frac{2(S/N) + 1}{(S/N)} = \frac{2S + N}{2S} \tag{25-34}$$

Thus, even when a postdetection integrator or data filter is used to combine all the received samples, the effective energy ratio will be reduced by L_x.

Matching or Collapsing Loss. The loss L_m in Eq. (25-32) implies that the if filter passes more noise than would an ideal filter, relative to peak signal. Usually, the if bandwidth is wider than optimum, in which case more than one independent sample of noise error is available on each pulse. Use of a narrow video filter following the detector, or of a matched range gate and low-pass filter, can recover this loss in data, except to the extent L_m has contributed to detector loss through reducing S/N. The effective number of measurement samples becomes $nB\tau$ for these cases. However, if B is too narrow or if the range gate is wider than the pulse, noise samples from the adjacent range cells will be included in the output data, and error will increase proportionately. The average S/N ratio during each output sample should be used in the error expressions for these cases: $S/N = \mathcal{R}_1 / L_m$.

Time-Sharing of Signals. It was noted that the energy available in conical-scan measurements of each angular coordinate was only half the received energy. A similar situation exists in some monopulse configurations, where a single Δ channel is shared between the two angular-error signals, and $\mathcal{R}$ should be reduced accordingly.

39. Other Sources of Error. Thermal noise is only one component of error in radar measurement, although it has received much attention because it lends itself to mathematical analysis. Other errors are sometimes random and noiselike, but may also appear as fixed bias, slow drifts from a calibrated setting, sinusoids, or other functions of time or target motion.

Errors from random clutter and noiselike interference can often be analyzed as though the interfering signals were thermal noise. Care must be taken to ensure that the interference is approximately normally distributed (Rayleigh amplitude distribution) and homogeneous over the resolution cells surrounding the target. Also, since clutter may be correlated over time intervals longer than one repetition period, the number of independent samples may be less than the number of pulses averaged, n. When a target is observed in land clutter or certain types of interference, signal-to-interference (S/I) may follow a broad, log-normal distribution, and large peak errors are possible.[41] Special editing procedures may be necessary to maintain tracking even when the median S/I is large.

Multipath errors, caused by reflection of target signals from ground or sea surfaces, become a serious problem for low-angle targets. Primarily affecting elevation angle, these reflections may also cause significant errors in other coordinates of precision tracking systems.[8]

For targets within a beamwidth of the horizon, and for any interfering target near enough in range and angle to be unresolvable, the angular error is

$$\sigma_\theta = \frac{\theta_i}{\sqrt{2(S/I)n_e}} \tag{25-35}$$

where θ_i = angular separation of interference (image or reflection region) from the target, and

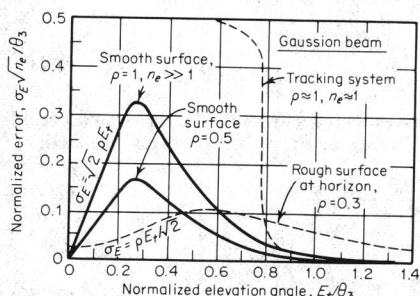

Fig. 25-36. Elevation multipath error vs. target elevation. *(From "Handbook of Radar Measurement," copyright 1969. Reprinted by permission of Prentice-Hall, Inc., Englewood Cliffs, N.J.)*

S/I = signal-to-interference power ratio. Figure 25-36 shows the low-angle elevation error over smooth and rough reflecting surfaces, for a gaussian beam and derivative Δ pattern. Below $0.8\theta_3$ over a smooth surface, the tracking systems become unreliable because they may, at times, lock on the image, and will generally oscillate randomly between target and image. However, over rough or absorbing surfaces, tracking remains possible, with large elevation errors, to the horizon.

Errors from target glint (Par. **25-15**) are described in detail in the literature.[8,29] While glint is a factor in all measurements of extended targets, scintillation error depends upon the sequential sampling procedure and its relationship to target correlation time.

The scintillation error on a Rayleigh target is

$$\sigma_\theta \approx \frac{\theta_3}{k_s}\sqrt{\frac{\beta_n}{2\pi^2 t_c f_s^2}} = \frac{0.225\theta_3}{f_s k_s}\sqrt{\frac{\beta_n}{t_c}} \tag{25-36}$$

For typical values ($f_s = 30$ Hz, $k_s = 1.5$, $\beta_n = 4$ Hz, $t_c = 0.1$s) the error is about $0.03\theta_3$. In search radar, the error depends on the ratio of time on target to correlation time, as shown in Fig. 25-37. The same curves apply to frequency-scanned antennas, where the ratio of frequency shift Δf, required for one-beamwidth scan, to correlation frequency f_c is substituted for t_0/t_c.

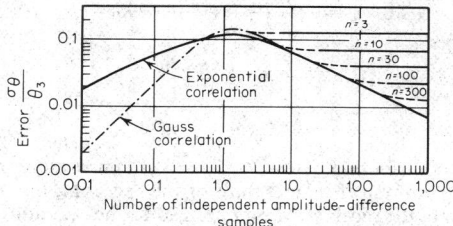

Fig. 25-37. Scintillation error in a scanning radar. Horizontal lines on the right indicate limits derived by Swerling for the case 2 target with different values of n. *(From "Handbook of Radar Measurement," copyright 1969. Reprinted by permission of Prentice-Hall, Inc., Englewood Cliffs, N.J.)*

40. Combination of Errors. An estimate of overall radar accuracy in any coordinate can be made by calculating separately the several error components discussed above, along with instrumental errors in the equipment and lag errors caused by target motion, and combining them in root-sum-square fashion:

$$\sigma^2 = \sigma_1^2 + \sigma_2^2 + \cdots$$

The process is repeated for each measured coordinate, and the major axes of an ellipsoid of error described in terms of a radial error, $\sigma_r = \sigma_t c/2$, and two orthogonal components $\sigma_n = R\sigma_\theta$ for elevation and traverse errors. Where smoothing or differentiation of radar outputs is performed, it may be necessary to form a composite error spectrum for each coordinate, so that the effect of the data filters on bias and noise at different frequencies can be evaluated.

41. Radar-Range Equations. Signal-to-Noise Ratio. The *signal power* S received by the radar antenna is calculated from transmitted power P_t and a series of factors that describe, basically, the geometry of the radar beam and target:

$$S = \frac{P_t G_t A_r \sigma}{(4\pi)^2 R^4} \tag{25-37}$$

where G_t = transmitting antenna gain, A_r = effective receiving aperture area, σ = target cross section, and R = target range. Since the receiving aperture and gain are related by

$$A_r = \frac{G_r \lambda^2}{4\pi} \tag{25-38}$$

the signal power can also be given as

$$S = \frac{P_t G_t G_r \lambda^2 \sigma}{(4\pi)^3 R^4} \tag{25-39}$$

In pulsed radar, P_t and S are the so-called peak power levels (actually the mean power over an rf cycle at the peak of the pulse envelope).

Radar noise is a combination of receiver-generated and environmental noise, extending over a broadband with a power density N_0:

$$N_0 = kT_s = kT_0 F_{n0} \tag{25-40}$$

where k = Boltzmann's constant, 1.38×10^{-23} W/(Hz · K), T_s = system noise temperature referred to the antenna terminals, T_0 = standard temperature (290 K) used in measuring noise factor, and F_{n0} = operating noise factor ($F_{n0} \equiv T_s/T_0$). The system noise temperature[10] is calculated from receiver noise factor F_n, line losses L_r, and antenna temperature T_a:

$$T_s = T_a + T_r + L_r T_e \tag{25-41}$$

$$T_a = \frac{0.876 T_a' - 254}{L_a} + 290 \tag{25-42}$$

$$T_r = T_{tr}(L_r - 1) \tag{25-43}$$

$$T_e = T_0(F_n - 1) \tag{25-44}$$

where T_r = temperature contribution of loss L_r, T_a' = sky temperature from Fig. 25-7, L_a = antenna ohmic loss, T_{tr} = physical temperature of the receiving line (290 K), and T_e = temperature contribution of the receiver. For $T_a \approx 290$ K, the operating noise factor $F_{n0} \approx F_n$, and $T_s \approx 290 F_n$. For antennas directed into space, F_{n0} can be much less than F_n, and T_s much less than 290 K.

The noise power at the if output of the receiver will depend upon receiver bandwidth and gain, but this noise power is equivalent to an input power at the antenna terminals of

$$N = N_0 B = kT_s B_n$$

where B_n = noise bandwidth of the if filter. For a wideband filter ($B_n \tau \gg 1$) the signal peak is not affected by the filter and $S/N = S/kT_s B_n$. In general, however, the S/N ratio at the receiver if output is calculated from the ratio of received pulse energy $S\tau$ to noise density:

Ideal energy ratio for single pulse:

$$\frac{E_1}{N_0} = \frac{P_t \tau G_t G_r \lambda^2 \sigma}{(4\pi)^3 R^4 kT_s} \tag{25-45}$$

Intermediate-frequency power ratio for single pulse:

$$\frac{S}{N} = \frac{E_1}{N_0 L_m} = \frac{P_t \tau G_t G_r \lambda^2 \sigma}{(4\pi)^3 R^4 kT_s L_m} \tag{25-46}$$

where L_m = if filter matching loss shown in Fig. 25-23 and τ = pulse width. For a cw or coherent pulse radar which integrates over an observation interval t_0 in a predetection filter, the if output S/N ratio is

Intermediate-frequency power ratio over interval t_0:

$$\frac{S}{N} = \frac{E}{N_0 L_m} = \frac{P_{av} t_0 G_t G_r \lambda^2 \sigma}{(4\pi)^3 R^4 k T_s L_m} \tag{25-47}$$

where P_{av} = average transmitter power ($P_t \tau f_r$ for pulsed radar), and L_m = matching loss of the filter to the entire waveform over t_0 seconds.

42. Loss Factors. Equations (25-41) to (25-47) consider free-space transmission conditions and ideal radar operation. In practice, a number of other factors must be included.

(a) *Signal Attenuation prior to Receiver:* Transmission line loss L_t; antenna losses (included in G_t, G_r, T_s); receiving line and circuit losses at rf (included in T_s); atmospheric attenuation L_a (from Figs. 25-2 to 25-6); atmospheric noise (included in T_s for clear air; $T'_a \rightarrow 290$ for larger values of L_a).

(b) *Surface Reflection/Diffraction Effects:* Pattern-propagation factor F, calculated from Eqs. (25-1) to (25-3) with data from Figs. 25-9 and 25-10, appears as F^4 in the numerator ($F > 1$ implies extra gain); details for the diffraction case appear in Ref. 10.

(c) *Antenna Pattern and Scanning:* For tracking and searchlighting case, G_t and G_r are defined for beam axis; for one-coordinate scan at ω rad/s, a reference energy is calculated using $t_0 = \theta_3/\omega$, where θ_3 is one-way half-power beamwidth in radians, and gains G_t and G_r at the point in the scan nearest the target; the effective energy is this reference level reduced by a beam-shape loss $L_{p1} \approx 1.45$ (or 1.6 dB); for two-coordinate scan, the reference energy is based on maximum gains and $t_0 = t_s \theta_a \theta_e/\psi_s$, where θ_a and θ_e are azimuth and elevation beamwidths, and ψ_s is the solid angle searched in time t_s; effective energy is this reference reduced by $L_{p2} \approx L_p^2 \approx 2.1$ (or 3.2 dB).

(d) *Signal-processing Losses:* For noncoherent integration of $n = t_0 f_r$ pulses, effective energy is reduced by $L_i(n)$ (Fig. 25-25) relative to the matched-filter value in Eq. (25-47); for loss of resolution in signal processing, a collapsing loss L_c defined by Eq. (25-17) is included; collapsing ratio ρ is found from Table 25-8; losses from nonoptimum threshold settings or

Table 25-8 Equations for Collapsing Ratio $\rho = \dfrac{m+n}{n}$

Cases for which P_n/ρ remains constant:

(a)	Restricted CRT sweep speed s, where d = spot diameter and τ = pulse width	$\rho = \dfrac{d + s\tau}{s\tau}$
(b)	Restricted video bandwidth B_v, where $B = 1/\tau$ is intermediate-frequency signal bandwidth	$\rho = \dfrac{2B_v + 1/\tau}{2B_v} = \dfrac{2B_v + B}{2B_v}$
(c)	Collapsing of coordinates onto the display, where $2\Delta_r/c$ = time-delay interval displayed per display cell, $\omega_e t_v$ and $\omega_a t_v$ = elevation and azimuth scans during integration time t_v, and θ_e and θ_a = beamwidths	$\rho = \dfrac{\frac{2\Delta_r}{c}}{c\tau}$ or $\rho = \dfrac{\omega_e t_v}{\theta_e}$ or $\rho = \dfrac{\omega_a t_v}{\theta_a}$

Cases for which P_n remains constant:

(d)	Excessive intermediate-frequency bandwidth $B_n > 1\tau$ followed by matched video	$\rho = \dfrac{B + 1/\tau}{B}$ (Use L_c in place of L_m.)
(e)	Receiver ouputs mixed at video, where M = number of receivers	$\rho = M$
(f)	Intermediate-frequency filter followed by gate of width τ_g and by video integration	$\rho = \dfrac{1}{B\tau} + \dfrac{\tau_g}{\tau}$

operator factors are also included as a factor L_x. These several factors are incorporated into the radar equation for S/N ratio as a factor F^4/L, where L is the product of loss factors for a given case, e.g.,

Intermediate-frequency power ratio for single pulse, in tracking radar:

$$\frac{S}{N} = \frac{P_t \tau G_t G_r \lambda^2 \sigma F^4}{(4\pi)^3 R^4 k T_s L_m L_t L_\alpha} \tag{25-48}$$

Effective energy ratio for noncoherent search radar:

$$\frac{E}{N_0} = \frac{P_{av} t_0 G_t G_r \lambda^2 \sigma F^4}{(4\pi)^3 R^4 k T_s L_m L_t L_\alpha L_p L_i L_c L_x} \tag{25-49}$$

43. Calculation of Detection Range. When the requirement for signal-to-noise ratio in power or energy is known, Eqs. (25-45) to (25-49) can be solved for maximum radar range R_m. In Par. **25-19** the S/N ratio required for each of n equal signal pulses in a train was defined by Eq. (25-14) as the detectability factor $D_0(n)$. For fluctuating targets the required average S/N, computed using $\bar{\sigma}$ in Eq. (25-46) is denoted by D_1, D_2, D_3, D_4, or D_e, depending on the target case and number of diversity samples n_e. For a given case 1, 2, 3, 4, or e, relationships for $D_e(n)$, $L_i(n)$, and fluctuation loss L_{fe} are shown in Figs. 25-25 to 25-27. The maximum range equation can be written

$$R_m^4 = \frac{P_t \tau G_t G_r \lambda^2 \bar{\sigma} F^4}{(4\pi)^3 k T_s D_j(n, n_e) L_m L_t L_\alpha L_p L_c L_x} \tag{25-50}$$

$$R_m^4 = \frac{P_{av} t_0 G_t G_r \lambda^2 \bar{\sigma} F^4}{(4\pi)^3 k T_s D_0(1) L_s} \tag{25-51}$$

where $L_s = L_m L_t L_\alpha L_p L_c L_x L_i L_{fe}$, and fluctuation loss $L_{fe} = $ a function of target case, n_e, and P_d. The definition of fluctuation loss is

$$L_{fe} = \frac{D_e(n, n_e)}{D_0(n)} \tag{25-52}$$

Equation (25-50) is essentially the form given by Hall in his classic 1956 paper,[42] while Eq. (25-51) is the more universal form used in a later work.[43]

If detection is performed visually on a CRT display, the product $D_0 L_m$ for a steady target in Eq. (25-50) can be replaced by $V_0 C_B$, where $V_0(n)$ is the visibility factor from Fig. 25-26, and C_B is a bandwidth correction factor.[10]

$$C_B = \frac{B\tau}{4.8}\left(1 + \frac{1.2}{B\tau}\right)^2 \tag{25-53}$$

Since $C_B = 1$ for the optimum $B\tau = 1.2$, an allowance for minimal matching and collapsing loss $L_m L_c \approx 1.7$ is included within V_0. Curves for V_0 are available only for $P_d = 0.50$ at an effective $P_n \approx 10^{-4}$ on steady targets. Adjustments to other cases can be made by assuming that $V_e \approx 1.7 D_e$ for optimum viewing conditions.

A work sheet for maximum-range calculation, devised by L. V. Blake of the Naval Research Laboratory (Blake Chart) has been widely accepted as a means of standardizing such calculations. The procedure is based on Eq. (25-50), with λ replaced by c/f and $D_0 L_m$ replaced by $V_0 C_B$. It can be applied to any target, with or without diversity, if $D_e(n, n_e)$ is entered in place of V_0, $\bar{\sigma}$ in place of σ, and L_m in place of C_B. Radar and target parameters are converted to decibel form relative to common engineering units, and the conversion constants are combined into a single constant. The Blake Chart appears in NRL Report 6930, published by the Naval Research Laboratory and in Ref. 10.

For a tracking radar, the Blake Chart or Eq. (25-50) may be used to calculate maximum range for a given performance level by finding the required single-pulse S/N ratio and entering it in place of D_e or V_0. Alternatively, if the requirement for doubled energy ratio $\mathcal{R}$ is known, Eq. (25-51) may be used with $D_0(1) L_i L_{fe} = \mathcal{R}/2$.

44. Search-Radar Equation. The potential performance of a search radar can be determined from its average power, receiving aperture, and system temperature, without regard to its frequency or waveform. The steps in deriving optimum search performance from Eq. (25-51) are:

(a) *Assume uniform search, without overlap,* of an assigned solid angle χ_s in a time t_s using a rectangular beam whose solid angle is

$$\psi_b = \theta_a \theta_e = \frac{4\pi}{G_t L_n} \ll \psi_s = A_m (\sin E_m - \sin E_0) \tag{25-54}$$

where θ_a and θ_e = 3-dB beamwidths, A_m = azimuth sector searched, and E_m and E_0 = upper- and lower-elevation search limits.

(b) *Express the observation time t_0 for a target as*

$$t_0 = \frac{t_s \psi_b}{\psi_s} = \frac{4\pi t_s}{G_t \psi_s L_n} \tag{25-55}$$

and assume that all signal energy reaching A_r during t_0 is integrated for one detection decision. Note that the definition of two-coordinate beam-shape loss, which is included in L_s, is consistent with Eq. (25-55).

(c) *Substitute* Eqs. (25-38), (25-54), and (25-55) into Eq. (25-49) and assume $F = 1$, to obtain the search radar equation:

$$R_m{}^4 = \frac{P_{av} A_r t_s \overline{\sigma}}{4\pi \psi_s k T_s D_0(1) L_s L_n} \tag{25-56}$$

Neither frequency nor waveform appears directly in Eq. (25-56) although frequency and aperture must permit Eq. (25-54) to be satisfied to concentrate energy within ψ_s, and the loss terms will vary with frequency, waveform, and scan procedure. The new loss term L_n appearing in Eq. (25-54) is an antenna beam loss which accounts for energy outside the main lobe and not available for integration. Spillover and side-lobe energy, for example, reduce G_t without adding to the beam angle ψ_b or time t_0. The conventional gain expression $G = 25,000/\theta_a \theta_e$, with θ in degrees, corresponds to a 40% loss of useful energy, or $L_n = 2.0$ dB. Reduced directivity caused by illumination taper does not appear in L_n but enters the search equation through A_r. Practical minimum values for $L_s L_n$ are about 10 dB on steady targets (for $P_d \approx 0.90$) and 11 to 13 dB on fluctuating targets with optimum diversity and scan procedure.

45. Cumulative Probability of Detection. Search and acquisition radars normally scan their assigned volumes more than once in order to achieve high detection probability. If a single-scan probability P_1 is obtained on each of k scans, the cumulative detection probability P_c is

$$P_c = 1 - (1 - P_1)^k \tag{25-57}$$

This procedure gives earlier detection of some penetrating targets, and minimizes effects of fluctuation, pattern lobing, and patches of large clutter. However, the distribution of energy into k scans without scan-to-scan integration is less efficient than matched-filter integration (or even postdetection integration). The loss in effective energy relative to the matched filter can be expressed[3] as a scan distribution loss L_d.

$$L_d \equiv \frac{k D_0(1, \text{ for } P_d = P_1)}{D_0(1, \text{ for } P_d = P_c)} \tag{25-58}$$

For example, to obtain $P_c = 0.90$ in $k = 4$ scans, $P_1 = 0.44$ and $D_0(1) = 12.7$ dB (at $P_n = 2.5 \times 10^{-9}$). The same result is obtained in one scan with $D_0(1) = 14.2$ dB (at $P_n = 10^{-8}$), giving $L_d = 4.5$ dB. All or a portion of L_d is recovered if the target fluctuates, because L_f is lower at $P_1 =$ than at $P_d = 0.90$; also, using postdetection integration, $L_i(n/k)$ is less than $L_i(n)$. The optimum k usually approximates 4.

46. Beacon-Range Equations. One-way transmission from *a radar to a beacon* gives the interrogation power level S_b at the beacon receiver:

$$S_b = \frac{P_t G_t G_b \lambda^2}{(4\pi)^2 R^2 L_t L_{\alpha 1} L_b} \tag{25-59}$$

where G_b = beacon antenna gain in the radar direction, L_{a1} = one-way atmospheric loss, and L_b = loss between the beacon antenna and receiver. On the return link, a beacon peak power P_b determines the signal power at the radar antenna terminal.

The beacon response power is

$$S = \frac{P_b G_b G_r \lambda^2}{(4\pi)^2 R^2 L_b L_{a1}} \tag{25-60}$$

from which the single-pulse signal-to-noise ratio is

$$\frac{S}{N} = \frac{P_b \tau G_b G_r \lambda^2 F^2}{(4\pi)^2 R^2 k T_s L_m L_b L_{a1}} \tag{25-61}$$

47. Radar Range in Clutter. The echo power returned by clutter is found from Eq. (25-37) or (25-39) when the clutter range R_c and cross section σ_c are used in place of target σ. The cross section of homogeneous clutter is found by multiplying reflectivity by the area or volume of the resolution cell.

Surface clutter:

$$\sigma_c = A_c \sigma^0 = \frac{R_c \theta_a}{L_p} \times \frac{\tau_n c}{2} \times \sigma^0 \tag{25-62}$$

Volume clutter:

$$\sigma_c = V_c \eta_v = \frac{R_c^2 \theta_a \theta_e}{L_p^2} \times \frac{\tau_n c}{2} \times \eta_v \tag{25-63}$$

where θ_a and θ_e = 3-dB beamwidths in azimuth and elevation, τ_n = effective (noise) width of the processed pulse (Table 25-5), and $L_p \approx 1.45$ = beam-shape loss. In writing expressions for signal-to-clutter ratios, the approximation $\tau_n \approx 1/B$ will be used, and an improvement factor I will be included to describe the increase in signal-to-clutter (S/C) output, due to MTI or doppler processing:

$$I \equiv \frac{(S/C)_{\text{out}}}{(S/C)_{\text{in}}} = \frac{\sigma_c}{\sigma} \left(\frac{R}{R_c}\right)^4 (S/C)_{\text{out}} \tag{25-64}$$

In systems without range ambiguity, the clutter competing with a target is at the same range; so $R/R_c = 1$. The processed clutter or clutter residue will be assumed noiselike, so that required S/C ratios are given approximately by V_0, D_0, or D_e. If the output clutter power competing with the signal is correlated from pulse to pulse, an increased integration loss term will be needed to describe the reduced number of independent samples: $L_{ic} \geq L_i$. Thus the subclutter visibility (SCV) is defined and related to D_e and D_0 by

$$SCV = (C/S)_{\text{in}} \qquad \text{when } (S/C)_{\text{out}} = D_e(n)$$

$$= \frac{I}{D_e(n)} = \frac{In}{D_0(1) L_f L_{ic}} \tag{25-65}$$

Substituting Eq. (25-62) into (25-65), with $\tau_n = 1/B$, $n = t_0 f$, and and unambiguous range $R_u = c/2f_r = t_0 c/2n$, the maximum range becomes

$$R_m = \frac{B I \sigma t_0 (R_c/R_m)^3}{R_u \theta_a \sigma^0 D_0(1) L_f L_{ic} L_c L_x} \tag{25-66}$$

(losses L_p, L_f, and L_x have been applied to the target).
The equivalent form for volume clutter

$$R_m{}^2 = \frac{B I \sigma t_0 (R_c/R)^2}{4\pi R_u \psi_b \eta_v D_0(1) L_f L_{ic} L_c L_x} \tag{25-67}$$

48. Search-Radar Techniques. Control of Coverage Patterns. The most basic problem in search radar is to establish reliable detection coverage over the assigned volume (e.g., for air surveillance, a volume extending from the horizon to a given altitude within a maximum range). Above 100 MHz, propagation paths are reliable only above the horizon, restricting coverage on targets in the troposphere.

The horizon range limit is

$$R_h \approx \begin{cases} 4.15(\sqrt{h_r} + \sqrt{h_t}) & [\text{km } (h \text{ in meters})] \\ 1.23(\sqrt{h_r} + \sqrt{h_t}) & [\text{naut mi } (h \text{ in feet})] \end{cases} \qquad (25\text{-}68)$$

where h_r = radar antenna height, and h_t = target height. These are the conventional values based on the "4/3 earth's curvature" approximation. More accurate coverage estimates for targets at all altitudes can be made using range-height-angle charts such as Figs. 25-38 and 25-39, based on ray tracing through the exponential reference atmosphere.[2,22,45]

Search-radar vertical coverage is conventionally plotted on linear scales as in Fig. 25-38, with contours showing R_m versus E_t for different values of P_d. A value of R_m for arbitrary E_{t0} is computed from Eq. (25-50) or (25-51), and this scaled at other elevations:

$$\frac{R_m(E_t)}{R_m(E_{t0})} = \frac{F(E_t)}{F(E_{t0})} \left[\frac{(G_t G_r / L_a) \text{ at } E_t}{(G_t G_r / L_a) \text{ at } E_{t0}} \right]^{1/4} \qquad (25\text{-}69)$$

For example, as shown in Fig. 25-40a, an air-surveillance radar provides coverage to 30,000 ft (10 km) at 80 mi (150 km), with a cosecant-squared antenna pattern to maintain approximately constant-altitude coverage at shorter range.

$$G_t = G_r = G(3°)\csc^2 E_t$$
$$\frac{R_m(E_t)}{R_m(3°)} = \frac{F(E_t)}{F(3°)} \left[\frac{L_a(3°)}{L_a(E_t)} \frac{\csc^4 E_t}{\csc^4 3°} \right]^{1/4} \qquad (25\text{-}70)$$
$$\approx \frac{\csc E_t}{\csc 3°}$$

since F and L_a are approximately constant for $E_t > 3°$. Below 3°, the antenna gain peaks and falls off to its half-power point approximately at the horizon. A small increase in atmospheric attenuation is included in the average ($F = 1$) pattern at low angle, and the effect of lobing is shown as a series of lobes whose spacing depends on antenna height, as in Eq. (25-1). The average range near the horizon can be increased by directing the beam axis nearer the horizon, but the depth of nulls will be greater and they will extend further into the high-angle coverage.

Apart from lobing problems, heavy illumination of the ground can introduce intolerable clutter at short range, both from fixed objects and from moving clutter such as birds and insects. To maintain visibility of aircraft at high elevation angles, it may be necessary to direct substantial energy into these regions while using a pattern which cuts off rapidly at and

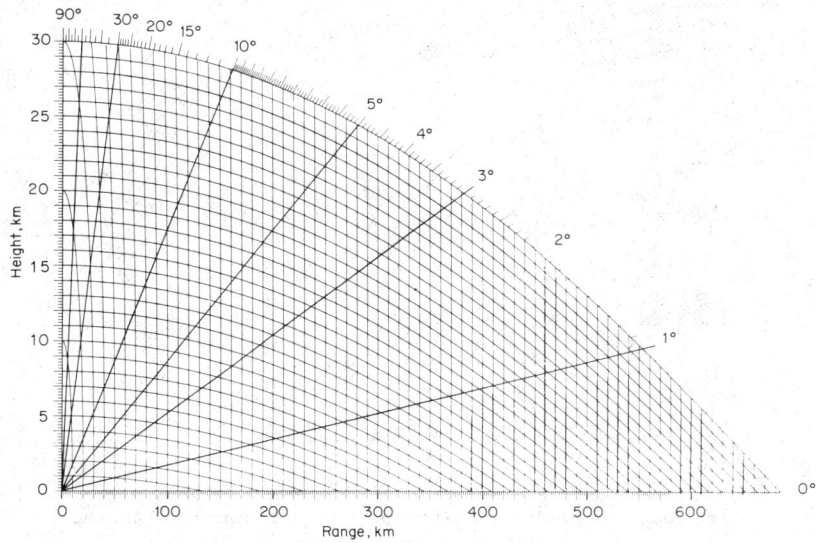

Fig. 25-38. Range-height-angle chart for aircraft (Ref. 45).

below the horizon (Fig. 25-40*b*).   Horn-fed reflector systems have been designed that vary the receiving pattern as a function of range delay, in order to obtain the benefits of good high-angle visibility and low response to short-range clutter without sacrificing long-range coverage near the horizon (Fig. 25-40*c*).

Search coverage can be controlled even more completely when a narrow pencil beam is scanned in a raster, or when several such beams are stacked one above the other to cover an elevation sector with multiple receiver channels.   With the agile beam, both transmitted energy and receiver gates can be varied to select the desired coverage in each beam position, but the number of beam positions may not permit long enough dwells for MTI or doppler processing.

49. Search-Radar Detection.   Early radars depended entirely on CRT displays with human operators for target detection, and this procedure remains one of the most efficient and adaptable.   The curves for visibility factor (Fig. 25-26) show signal integration performance near the optimum limit set by information theory.   Such performance cannot be expected under field conditions, where the operator may be fatigued or distracted by his surroundings. However, it remains true that a trained operator, using his experience to recognize targets and reject interference on the basis of complex visual patterns and scan-to-scan memory, may outperform the most sophisticated automatic detector in a difficult environment.   To overcome the factors of fatigue and inattention in early-warning applications, as well as minimizing effects of random interference and collapsing on a display, the CRT-operator

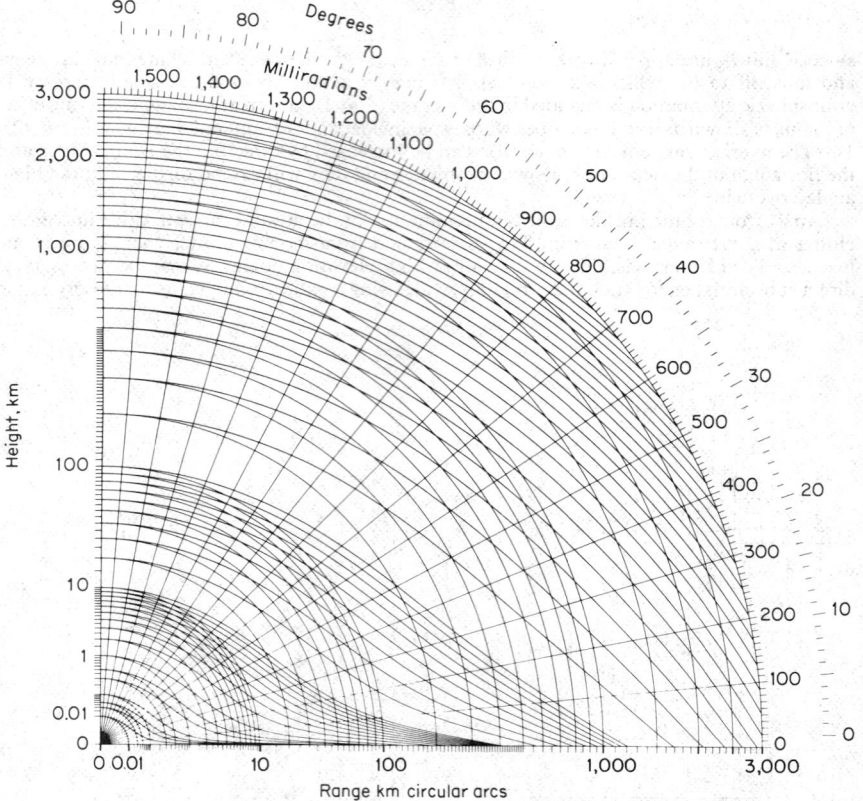

Fig. 25-39.   Range-height-angle chart for missiles and satellites (Ref. 45).

integration may be replaced by a video sweep integrator (Par. **25-80**). The output of the integrator provides a bright blip on an otherwise uncluttered display, and may also operate an audible alarm to attract attention in early-warning use. Performance is restricted by the resolution and stability of the integrator memory device and by ability to set and hold the threshold precisely relative to random noise (see CFAR discussion, Par. **25-53**).

50. Moving-Target Indication. A moving-target indicator (MTI) is a device which limits the display of radar information primarily to moving targets. The sensitivity to moving targets is provided primarily through their doppler shifts, although *area MTI* systems have been built which cancel targets on the basis of overlap of their signal envelopes in both range and angle.

In the usual pulsed-amplifier coherent MTI system (Fig. 25-41*a*), two cw oscillators in the radar are used to produce a phase and frequency reference for both transmitting and receiving, so that echoes received over the train of pulses from fixed targets have a constant phase at the detector. These echoes will be canceled, leaving in the output only those signals whose phase varies from pulse to pulse. Coherent MTI can also be implemented with a pulsed oscillator transmitter (Fig. 25-41*b*), in which the coherent oscillator is locked in phase to remember the random phase with which each successive pulse is transmitted. Although the transmitted signal is noncoherent, the if signal is coherent and has the same line structure as in the pulsed amplifier system. Both systems attenuate targets in a band centered at zero radial velocity, the depth and width of the rejection notch depending on design of the canceler and stability of the received signals.

Two variations on the coherent MTI are available for rejection of clutter with nonzero radial velocity. In the clutter-locked MTI, the average doppler shift of a given volume of clutter is measured and used to control an offset frequency oscillator in the receiver, shifting the clutter into the rejection notch. Short- or long-term averages may be used to obtain rapid adaptation to varying clutter velocity (as in weather clutter) or better rejection of selected parts of a complex clutter background. The alternative is *noncoherent MTI*, in which the clutter surrounding a target provides the phase reference with which the target signal is mixed to produce a doppler signal. Although simpler to implement, noncoherent MTI does not cancel as completely and may lose target signals when the clutter is too small (as well as when it is too large).

Cancelers for MTI radar are designed to pass as much of the target spectrum as possible while rejecting clutter. Since search-radar MTI must cover many range cells without loss of resolution, canceling filters are implemented with delay lines, with multiple-range gates feeding bandpass filters, or with range-sampled digital filters which perform both these functions. The response of several typical cancelers is shown in Fig. 25-42. A wide variety of response shapes is available through use of feedback and multiple, staggered repetition rates.[46] In particular, through proper use of stagger (Fig. 25-42*d*), it is possible to maintain detection of most targets with nonzero radial velocities, even those which would fall in one of the blind speeds v_{bj} (ambiguous rejection notches) of an MTI with a single repetition rate:

$$v_{bj} = j\frac{\lambda f_r}{2} \qquad j = \pm 0, 1, 2, \ldots \qquad (25\text{-}71)$$

51. Performance of MTI. The basic measure of MTI performance is the MTI improvement factor *I*, defined in Eq. (25-64). This is equal to the clutter attenuation when the canceler gain is set to unity on noise or targets uniformly distributed in velocity. The basic relationships between *I* and radar/clutter parameters can be expressed in terms of the ratio of rms clutter spread to repetition rate or blind speed:

$$z = \frac{2\pi\sigma_f}{f_r} = \frac{2\pi\sigma_v}{v_{b1}} = \frac{4\pi\sigma_v}{\lambda f_r} \qquad (25\text{-}72)$$

where σ_f = standard deviation of clutter power spectrum in hertz, f_r = repetition rate, σ_v = standard deviation in meters per second, and v_{b1} = first blind speed from Eq. (25-71). For a scanning gaussian beam,

$$z = 2\sqrt{\ln 2}\,\frac{\omega}{f_r\theta_3} = \frac{1.665}{n} \qquad (25\text{-}73)$$

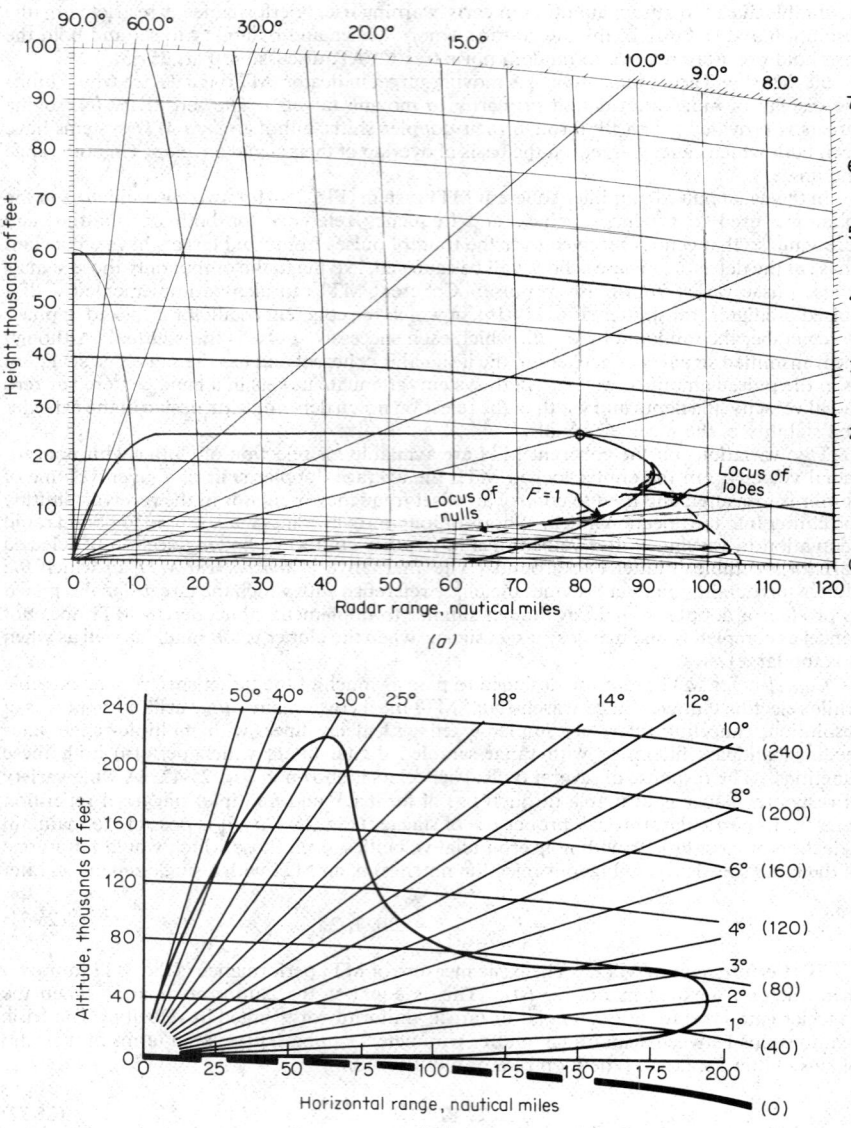

Fig. 25-40. Search-coverage patterns for aircraft detection. (*a*) Cosecant-squared air-search coverage; (*b*) free-space-coverage diagram for the ARSR-2 antenna; (*c*) two-way-coverage diagram for adjustable receiving beam (Ref. 47).

For a $(\sin x)/x$ beam, the constant 1.665 should be changed to 1.760, giving slightly poorer MTI performance.

In general, the rms spread should be calculated using the sum of components σ_s due to scanning, σ_i due to internal motion of the clutter, σ_m due to relative motion between the radar and the clutter, and σ_x due to low-frequency instabilities in the radar.

52. CW Radar. A cw transmission has no velocity ambiguity, and so cw radar equipment can easily be designed to provide 80 to 100 dB rejection of fixed or moving clutter. Coherent integration of target signals in selected doppler bands is also provided by narrow-band filters of relatively simple construction. Three problems, however, restrict the performance and usefulness of cw radar for search:

(a) *Isolation of receiver from transmitter.* Direct feed-through of transmitter power to the receiver must be minimized, requiring separate antennas in high-power systems and careful design in all systems to avoid receiver saturation.

(b) *Magnitude of short-range clutter echo.* The echo power received from clutter in a cw radar is the integrated product of the reflectivity, (range)$^{-4}$, and antenna gain factors of the radar equation (25-39) over the common volume of the transmitting and receiving beams. In both volume and surface clutter, the echo power is controlled by the clutter at the shortest range in the common volume, and the effective clutter cross section is a function of beamwidth and range R_c to the point where the beams substantially overlap. The required clutter improvement for a cw radar may therefore be very high because of the $(R/R_c)^4$ term.

(c) *Transmitter noise.* Both the direct feed-through from transmitter to receiver and the echoes from short-range clutter will contain random-noise components from the transmitter. Special circuits may be designed to cancel the direct feed-through and low-frequency components of reflected noise, but the higher-frequency components will appear with phase shift from the range delay and cannot be canceled completely. Subclutter visibility in cw systems is generally controlled by these noise components.

53. Control of False Alarms. The signal-detection statistics previously described are based on normally distributed random noise (Rayleigh envelope distribution) with fixed (and known) rms level. The false-alarm probability for that case, given in Eq. (25-12), is a function

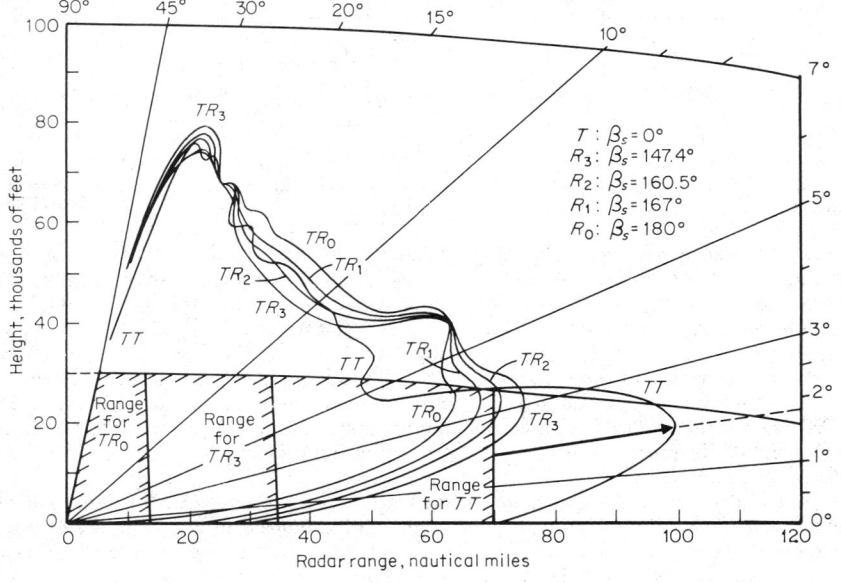

(c)

Fig. 25-40. *(Continued)*

of threshold setting. In the real environment, however, the background against which targets must be detected has a varying rms level and may also depart from the Rayleigh assumption. This is especially true of clutter, and even the residue after MTI filtering may be non-Rayleigh and partially correlated from pulse to pulse. Many techniques are used to minimize false alarms under these conditions, and these are generally (if inaccurately) described as *constant false-alarm rate* (CFAR) techniques.

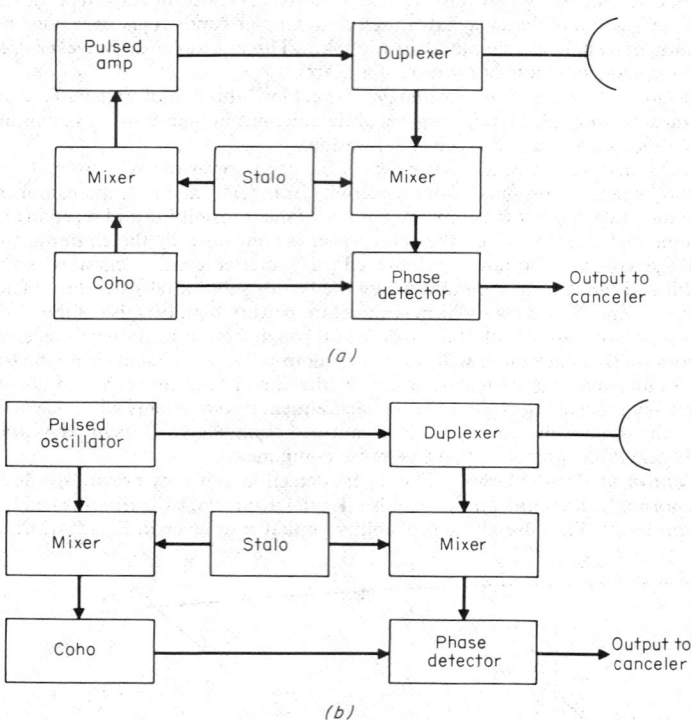

Fig. 25-41. Block diagrams of MTI radars. (*a*) Simplified block diagram for power-amplifier; (*b*) pulsed oscillator.

Early radars used a simple programmed variation in receiver gain, as a function of range, to reduce alarms from small short-range objects. This procedure, known as *sensitivity time control* (STC) remains an effective means of rejecting echoes from birds and insects and of minimizing saturation on larger clutter sources. If gain is programmed to maintain a constant signal level for targets near the center of the beam, the noise level is sharply reduced at short range, and false alarms in that region should result only from clutter. Targets not in the main beam (e.g., aircraft flying at constant altitude into the cosecant-squared portion of search coverage) will also be lost. Some compromise in gain program and antenna pattern will normally be used to give increased target-detection probability at short range without excessive clutter alarms.

Another well-established technique is the log-FTC receiver, in which a logarithmic if or video response is followed by a video differentiator of fast time constant. This is effective in compressing the range of output fluctuation from random clutter whose rms level is varying slowly (as in rain-cloud echoes or clutter from hilly terrain). Target echoes in the clutter region are also suppressed, but targets which exceed the local noise or clutter fluctuation by the visibility factor V_0 remain detectable. This technique is especially useful in non-MTI systems where clutter would otherwise saturate the display or video processor elements. It is compatible with frequency diversity and video integration techniques, and can also be used with MTI outputs. A moderate loss (1 or 2 dB) in detectability is usually incurred.

Examples of CFAR techniques are:

(a) *Guard-band receivers.* Filter channels adjacent to the signal spectrum are used to estimate the broadband noise level (Fig. 25-43a).

(b) *Dicke fix or wideband limiter.* As with example (a) the level over a band broader than the signal spectrum is used to control if gain, using in this case a limiter to hold total output constant (Fig. 25-43b). The effective number of samples is the ratio of total bandwidth to signal bandwidth. In coherent systems, this procedure can be applied within the ambiguous doppler interval, f_r, which is wider than the spectral-line width of an individual target.

(c) *Range-averaged AGC.* Automatic gain control based on a local average of range cells near the detection cell can provide a measure of noise plus clutter within a single repetition interval (Fig. 25-43c). It may be combined with time averages over several repetition intervals to increase the accuracy of the estimate or increase the rate at which the estimate will follow local changes in clutter as a function of range.

(d) *Side-lobe blanking.* Averages in the angular coordinates may be used to recognize side-lobe response to strong clutter or active sources. These averages are generally taken with an auxiliary, broad-beam antenna and receiving channel (Fig. 25-43d).

54. Search-Radar Measurements. The theoretical limits on angular measurement, applicable to scanning-search radar, are summarized in Fig. 25-32. The optimum beam-splitting process can be approached closely by differentiating the range-gated output of a video integrator to find the peak of the beam-pattern envelope. This process is the angular analog of the optimum time-delay estimator, which consists of a matched filter and differentiator. As with time-delay measurement, approximations are available which locate the centroid of the signal envelope using split angular gates or interpolation between a selected point on the leading and trailing edges. The presence of scintillation error (Fig. 25-32) usually limits accuracy to a level near 0.05 beamwidth, regardless of S/N ratio or processor efficiency.

An important function in some surveillance systems is target-height measurement on multiple targets detected by the search radar. Since range is known to high accuracy, the elevation angle error becomes the dominant problem. One of the following procedures may be used to measure elevation.

A separate height finder, designated to target azimuth by the 2D search radar, scans a narrow beam in elevation (Fig. 25-44a) with accuracy expressed by Eq. (25-26) and Fig. 25-37 relative to the elevation beamwidth. One or two such height finders are sequenced among many targets to achieve a low data rate (e.g., one reading per 30 s).

Separate Receivers. The search-radar elevation coverage is divided among several stacked receiving beams feeding separate receivers (Fig. 25-44b). With sufficient overlap to give smooth coverage, the receiver outputs may be compared (at if or video) to produce a monopulse estimate. The number of receivers is minimized at some cost in accuracy if integrated video outputs are compared. Scintillation does not affect accuracy.

The V-beam approach may be used,[8] wherein two broad beams scan in azimuth, one of them tilted to produce an output whose delay (in the azimuth scan cycle) is proportional to elevation angle (Fig. 25-44c). The elevation error is proportional to the difference between the two azimuth readings multiplied by the cotangent of the tilt angle. The azimuth readings are subject to independent scintillation errors, leading to rather large errors in the elevation estimate.

A scanning pencil beam (Fig. 25-44d) can be used to measure both angular coordinates, with monopulse instrumentation if scintillation error is to be avoided. If frequency scan is used in elevation, the scintillation error is found from Fig. 25-37, using the correlation frequency f_c of the target and the frequency shift Δf per beamwidth to give $n_e = \Delta f / f_c$.

For low-elevation targets ($E_t < 1.5\theta_e$), the multipath error from Fig. 25-36 will be a significant contributor to height-finder error. In a V-beam system, the effective elevation beamwidth for evaluation of multipath depends on the vertical aperture used to generate the tilted beam.

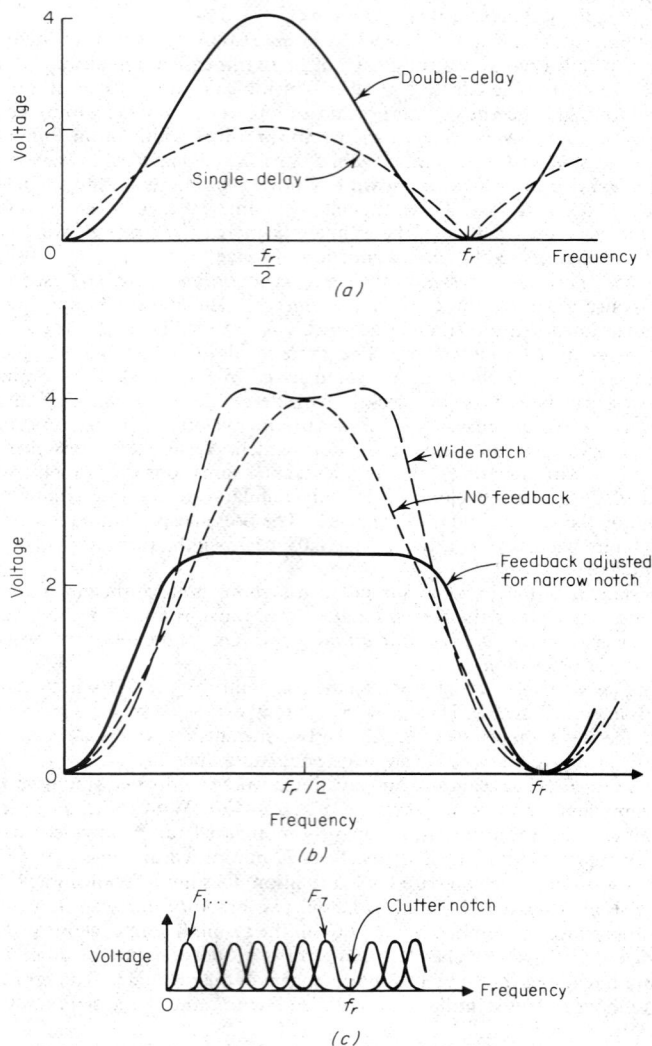

Fig. 25-42. Frequency response of MTI filters. (*a*) Single and double delay without feedback; (*b*) double delay with feedback; (*c*) range-gated filter bank; (*d*) double and triple delay with staggered prf and feedback (Ref. 46).

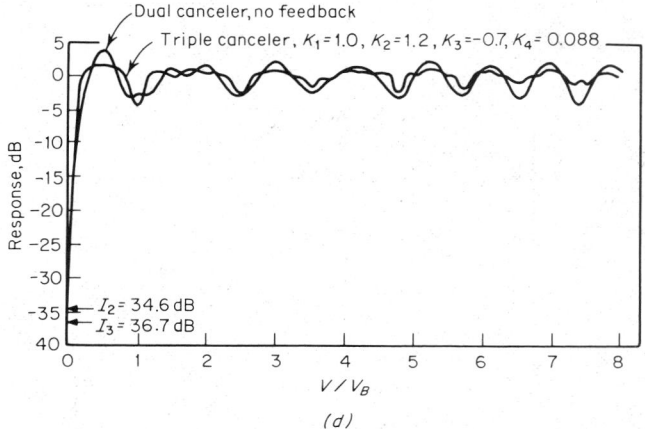

Fig. 25-42. *(Continued)*

TRACKING-RADAR TECHNIQUES

55. Angle Tracking. In tracking radar, a narrow (and generally circular) beam is directed at a selected target, either continuously (with a mechanically steered reflector or lens) or with a time-shared array beam. The electromechanical or electrical servo loop is controlled to minimize the angle errors, as measured by an error-sensing antenna and receiver system. The error-sensing sensitivity of monopulse antennas is described in Par. **25-34** in terms of the normalized slope of the Δ/Σ pattern ratio, while conical-scan sensitivity depends on the fractional modulation of the pattern per beamwidth of target displacement.

In a typical *monopulse radar* (Fig. 25-45), Σ and Δ patterns are formed with a four-horn feed and hybrid network before conversion of signals to intermediate frequency. Normalization (formation of the Δ/Σ ratio) is performed using a common automatic gain control voltage, derived from the range-gated Σ-channel output and applied to Σ and Δ receivers in parallel. Phase-sensitive error detectors, using the Σ output as a reference, produce bipolar video outputs proportional to the Δ/Σ ratios, and hence to the off-axis error angles in each coordinate, After range gating to select the target at the range of interest, these video signals can be stretched and smoothed to produce dc error signals for the servo loop. Apart from the choice of monopulse feeds and receiver designs, the major variations between monopulse radar techniques appear in the procedure for normalization, error detection, and calibration.

The *common-AGC* normalization technique works well for a single-target tracker, where pulse-to-pulse signal variations are relatively small and longer-term fluctuations can be followed by a fast AGC loop.[50] Multiple-target trackers require single-pulse normalization, which can be provided by IAGC, logarithmic receivers, or limiter systems. Details on these different error sources and effects will be found in the literature.[8,9]

The *conical-scan tracker* (Fig. 25-46) samples the beam positions around the tracking axis sequentially, obtaining a train of amplitude-modulated pulses (Fig. 25-47) from which elevation and azimuth (traverse) errors are extracted. Only a single receiver channel is used, with AGC to normalize the error signals. In place of the mechanically rotating or nutating feed, some systems use electronic scan to produce higher scan rate or to scan on receive only. This latter approach is sometimes referred to as "pseudo-monopulse" because it employs a fixed, four-horn feed with Σ and Δ outputs, but its sensitivity to noise and scintillation is the same as given for conical scan with one-way patterns (Fig. 25-34a). The primary perform-

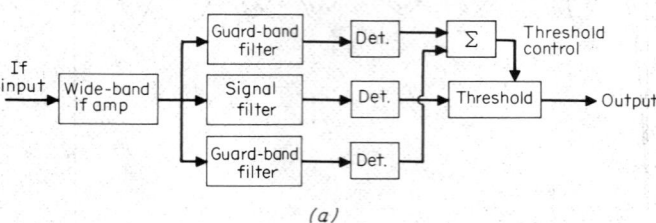

(a)

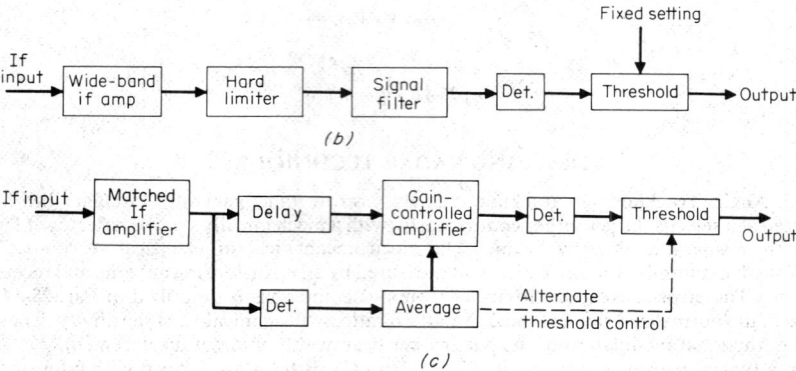

(b)

(c)

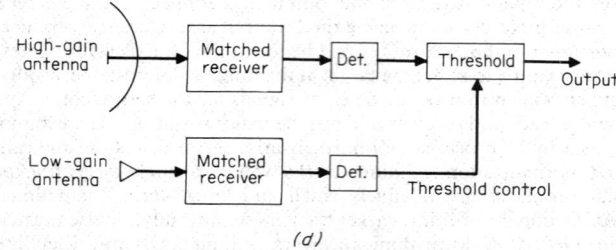

(d)

Fig. 25-43. CFAR techniques. (*a*) Guard-band system; (*b*) Dicke fix system; (*c*) range-averaged AGC; (*d*) sidelobe blanker.

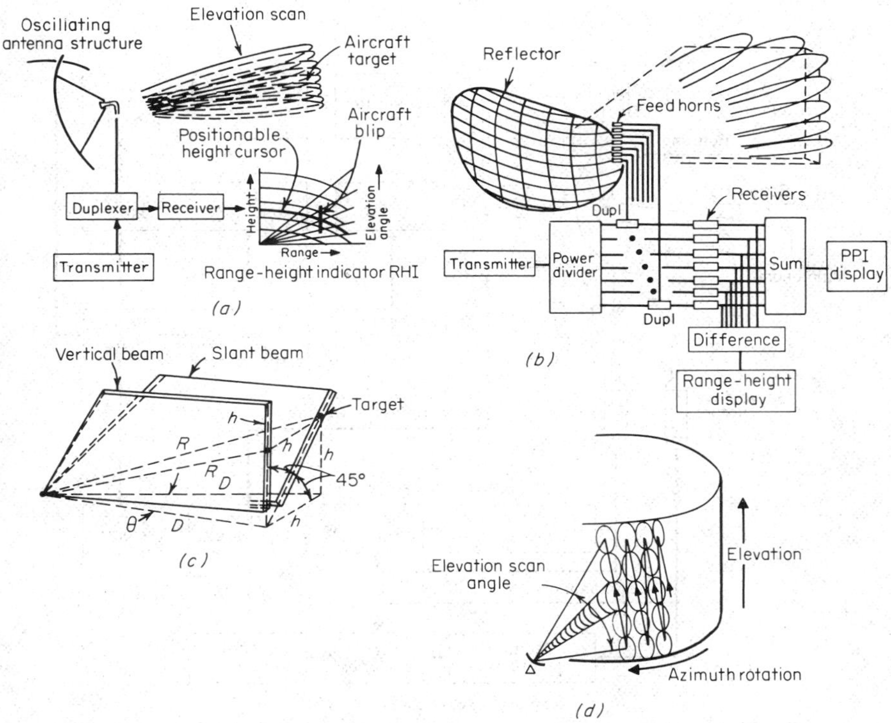

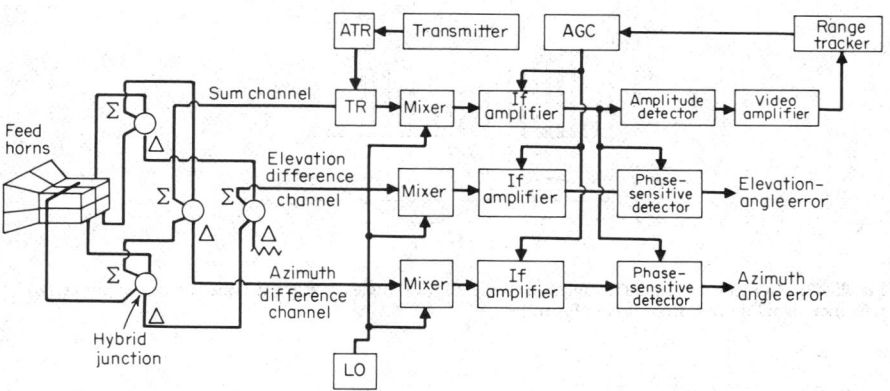

Fig. 25-44. Height finding in surveillance radar system. (*a*) Nodding-height-finder representation; (*b*) stacked-beam-radar representation; (*c*) V-beam geometry; (*d*) azimuth-elevation raster produced by a three-dimensional scanning radar (Ref. 49).

Fig. 25-45. Block diagram of a conventional monopulse tracking radar (Ref. 9).

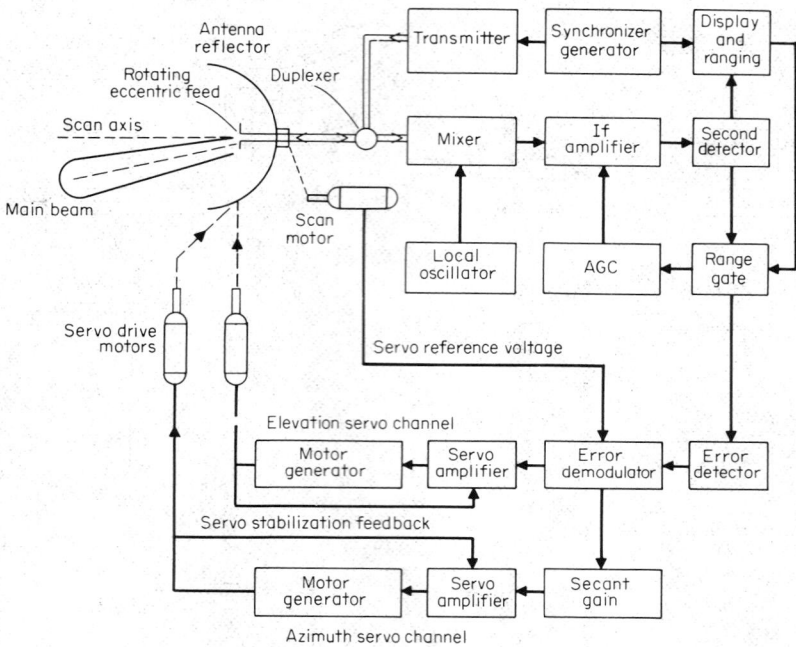

Fig. 25-46.   Block diagram of a conical-scan radar (Ref. 29).

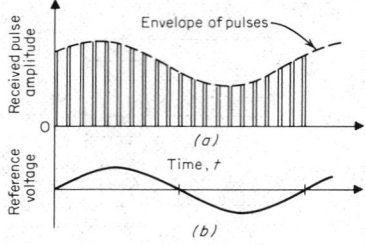

Fig. 25-47.   Conical-scan signal modulation (Ref. 29).   (*a*) Angle error in envelope of received pulses; (*b*) reference signal derived from drive of scan feed.

25-56

ance differences between conical-scan and monopulse trackers are the increased tracking noise from target scintillation (amplitude noise) in conical scan and the greater thermal noise error at long range.

56. Range Tracking. A simple video-signal range tracker is shown in Fig. 25-48. Video pulses are applied to a time discriminator which measures the position of the centroid, peak, or leading edge of the signal relative to the delay stored in the tracker. Any difference between these two positions is used as the servo error signal to reposition the tracker output, after suitable smoothing and filtering to control servo bandwidth and lag. The range delay output is used to gate the angle tracker and AGC circuit, so that ideally all three coordinates are controlled by the single selected target. The techniques used in the time discriminator and the delay generator determine accuracy and sensitivity of the range tracker. The split-gate discriminator process is similar to the use of two offset beams in monopulse in that the difference in voltage between the offset gates is found on each pulse, and normalized to the total signal in the main range gate (Σ channel).

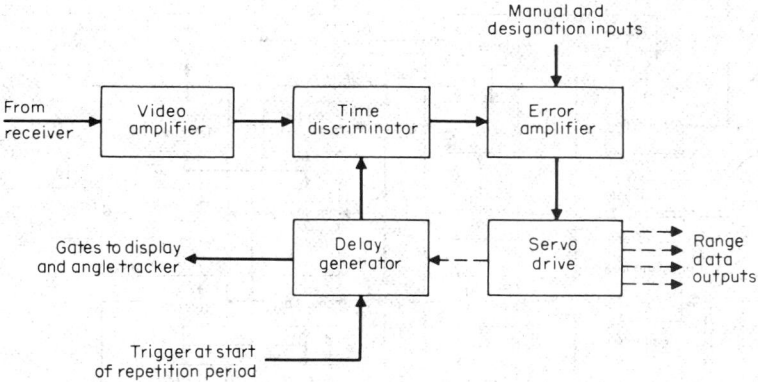

Fig. 25-48. Basic elements of automatic range tracker. *(From "Radar System Analysis" copyright 1964. Reprinted by permission of Prentice-Hall, Inc., Englewood Cliffs, N.J.)*

An alternative process forms a signal which approximates the derivative of the video pulse, and locates its zero crossing by placing an error detector gate to obtain zero average output. The leading-edge discriminator consists of two differentiators in cascade, the first producing a short pulse width $\tau' \approx 1/B$ on the leading edge of an extended echo, and the second locating the peak of this pulse (hence the point of greatest slope on the original input signal). Any of these techniques will work well on strong signals, and the first two procedures can be used on signals near or below noise level if the error-detecting gate outputs are smoothed over many pulses.

57. Doppler Tracking. In cw or high-prf (high-pulse-rate-frequency) pulsed radar, doppler tracking is used instead of range tracking as a primary means of resolving targets. It is also used in low- and medium-prf pulsed radar to supplement range resolution in clutter environments and to measure radial velocity with high accuracy. Used only for measurement, the doppler tracker can operate in parallel with (and independently of) the range- and angle-tracking loops, which do not require coherent data. If the doppler resolution is needed to obtain adequate S/C or S/N ratio, however, all four loops must be interdependent, with coherent signals being filtered and processed in a common Σ channel and four Δ channels.

Figure 25-49 shows such a system, in which the monopulse Σ channel feeds a discriminator for doppler error sensing as well as providing reference inputs for AGC and error detectors in range, azimuth, and elevation. The discriminator, operating on the selected fine line of the signal spectrum, controls the voltage-controlled oscillator (VCO), which heterodynes the first if signal into the center of the doppler filter at the second if. To obtain doppler resolution in the range Δ channel, the split-gate function is applied as an if phase reversal (commutator), and the narrow-band second if amplifier averages the two reversed halves of the signal to produce zero output with a centered gate and opposite-polarity error outputs on either side of center.

58. Tracker-Acquisition Procedure. Targets for high-resolution trackers must be designated with reasonable accuracy if they are to be acquired when near the threshold of sensitivity. In the absence of four-dimensional designation, one or more coordinates must be scanned to acquire the target, and sensitivity will be degraded. Since it is possible to observe all range and doppler cells simultaneously, a two-dimensional bank of acquisition cells is usually established (as in some search radars) and the region of angular uncertainty is scanned sequentially until the target is detected. Targets strong enough to be acquired can then be tracked with high accuracy, since the search losses associated with overlap and straddling of multiple gates and filters, and with beam scanning, are eliminated in track.

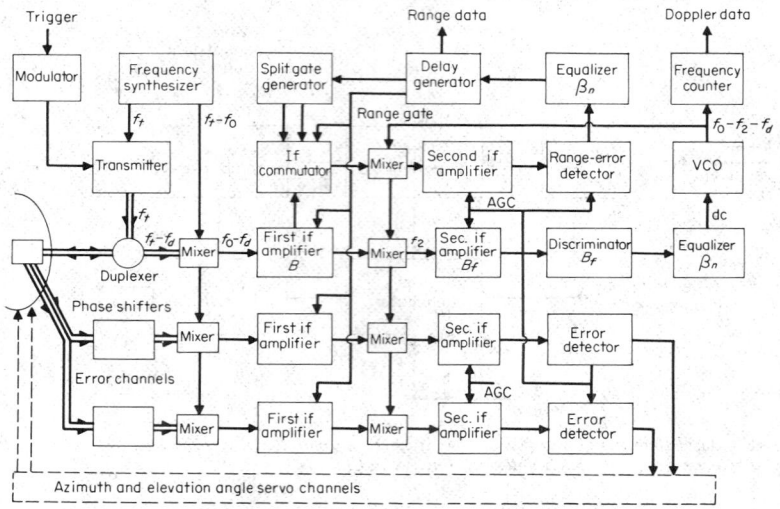

Fig. 25-49. Four-dimensional tracker, monopulse type. *(From "Radar System Analysis," copyright 1964. Reprinted by permission of Prentice-Hall, Inc., Englewood Cliffs, N.J.)*

RADAR TECHNOLOGY

By Harold R. Ward

Radar Technology Radar development, since its beginning during World War II, has been paced by component technology. As better rf tubes and solid state devices have been developed, radar technology has advanced in all its applications. This subsection presents a brief overview of radar technology. It emphasizes the components used in radar systems that have been developed specifically for the radar application. Since there are far too many devices for us to mention all of them, we have selected only the most fundamental to illustrate our discussion.

In this subsection, the sequence in which each subsystem is discussed parallels the block diagram of a radar system. Pictures of various radar components give an appreciation for their physical size, while block diagrams and tabular data describe their characteristics. The material for this subsection was taken largely from Skolnik's *Radar Handbook*, Ref. 51. We suggest the reader see Ref. 51 for more detail and its references for still greater depth.

59. Radar Transmitters. The requirements of radar transmitters have led to the development of a technology quite different from that of communication systems. Pulse radar transmitters must generate very high power, pulsed with a relatively low duty ratio.

Peak powers ranging from a few kilowatts to a few megawatts are needed to produce detectable echoes from the desired targets. The duty ratio of the pulse transmission is typically between 0.1 and 1%. These requirements most often arise in the frequency range from 1 to 10 GHz.

Radar transmitters are of two types, power oscillator and power amplifier. *Power oscillator transmitters* use a single high-powered oscillator to generate pulses of rf energy. This is the simpler, conventional radar transmitter in which the power oscillator is usually a magnetron. In a *power amplifier transmitter*, a pulse generated at low power level is amplified to the desired output level by a pulsed amplifier.

Power oscillators and power amplifier stages consist of three basic components: a power supply, a modulator, and a tube. The power supply converts the line voltage to dc voltage of from a few hundred to a few thousand volts. The modulator supplies power to the tube during the time the rf pulse is being generated. Although the modulation function can be applied in many different ways, it must be designed to avoid wasting power in the time between pulses. The third component, the rf tube, converts the dc voltage and current to rf power. The devices and techniques used in the three transmitter components are discussed in the following paragraphs.

60. RF Tubes. The tubes used in radar transmitters are classified as *crossed-field, linear-beam,* or *gridded* (see Sec. 9). The crossed-field and linear-beam tubes are of primary interest because they are capable of higher peak powers at microwave frequencies. Gridded tubes such as triodes and tetrodes are sometimes used at uhf and below. Since these applications are relatively few, gridded tubes will not be described here (see Pars. 9-9 to 9-19).

61. Modulators. If a pulsed radar transmitter is to obtain high efficiency, the current in the output tube must be turned off during the time between pulses. The modulator performs this function by acting as a switch, usually in series with the anode current path. Some rf tubes have control electrodes that can also be used to provide the modulation function. There are three kinds of modulators in common use today: the line-type modulator, magnetic modulator, and active-switch modulator. Their characteristics are compared in Table 25-9.

The line-type modulator is the most common and is often used to pulse a magnetron transmitter. A typical circuit including the high-voltage power supply and magnetron is shown in Fig. 25-50. During the time between pulses, the pulse-forming network, (PFN) is charged. A trigger fires the thyratron $V1$, shorting the input to the PFN, which causes a voltage pulse to appear at the transformer $T1$. The PFN is designed to produce a rectangular pulse at the magnetron cathode, with the proper voltage and current to cause the magnetron to oscillate. The line-type modulator is relatively simple, but has an inflexible pulse width.

Active-switch modulators are capable of varying their pulse width within the limitation of the energy stored in the high-voltage power supply. A variety of active-switch cathode pulse

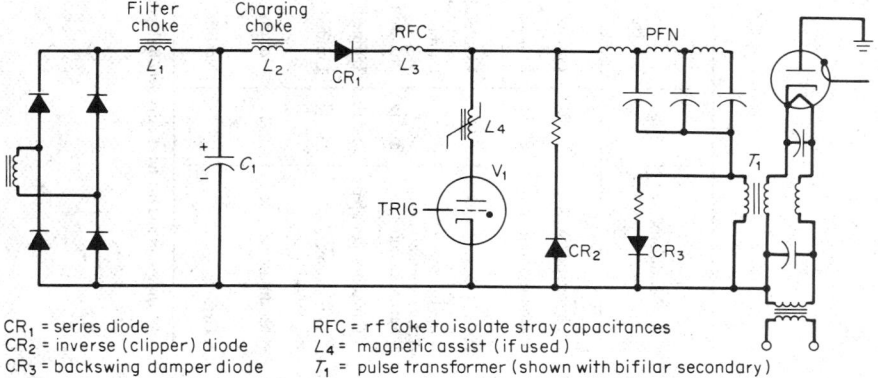

CR$_1$ = series diode   RFC = r f coke to isolate stray capacitances
CR$_2$ = inverse (clipper) diode L_4 = magnetic assist (if used)
CR$_3$ = backswing damper diode T_1 = pulse transformer (shown with bifilar secondary)

Fig. 25-50. Line-type modulator (Ref. 52).

Table 25-9. Comparison of Modulators

| Modulator | Fig. | Flexibility | | Pulse-length capability | | Pulse flatness | Crowbar required | | Modulator voltage level |
		Duty cycle	Mixed pulse lengths	Long	Short		Load arc	Switch arc	
Line-type: Thyratron/SCR	25-50	Limited by charging circuit	No	Large PFN	Good	Ripples		No	Medium/Low
Magnetic modulator	25-52	Limited by reset and charging time	No	Large C's and PFN	Good	Ripples		No	Low
Hybrid SCR–magnetic modulator	...	Limited by reset and charging time	No	Large C's and PFN	Good	Ripples		No	Low
Active switch: Series switch	25-51a	No limit	Yes	Excellent; large capacitor bank	Good	Good	Maybe	Yes	High
Capacitor-coupled	25-51b	Limited	Yes	Large coupling capacitor	Good	Good	Maybe	Yes	High
Transformer-coupled	25-51c	Limited	Yes	Difficult; XF gets big; large capacitor bank	Good	Fair	Maybe	Yes	Medium–high
Modulator anode	...	No limit	Yes	Excellent; large capacitor bank	OK, but efficiency low*	Excellent	Yes	Yes	High
Grid	...	No limit	Yes	Excellent; large capacitor bank	Excellent	Excellent	Yes	...	Low

* Unless ON and OFF tubes carry very very high peak current or unless modulator anode has high mu. After Weil. Ref. 52. Par. 25-161.

modulators is shown in Fig. 25-51. Active-switch modulators may use a vacuum tube free of gas but capable of passing high current and holding off high voltage; these are called *hard-tube modulators.*

The magnetic modulator, a third type of cathode pulse modulator, is shown in Fig. 25-52. It has the advantage that no thyratron or switching device is required. Its operation is based on the saturation characteristics of inductors L2, L3, and L4. A long, low-amplitude pulse is applied to L1 to charge C1. When C1 is nearly charged, L2 saturates, and the energy in C1 is transferred resonantly to C2. The process is continued to the next stage, where the transfer time is about one-tenth that of the stage before. The energy in the pulse is nearly maintained so that at the end of the chain a short-duration high-amplitude pulse is generated.

62. Power Supplies and Regulators. The power supply converts prime power from the ac line to dc power, usually at a high voltage. The dc power must be regulated to remove the effects of line-voltage and load variation. A comparison of various regulators is given in Table 25-10.

Protective circuitry is usually included with the high-voltage power supply to prevent the rf tube from being damaged in the event of a fault. Improper triggers and tube arcs are detected and used to trigger a crowbar circuit that discharges the energy stored in the high-voltage power supply. The crowbar is a triggered spark gap capable of dissipating the full energy of the power supply. Thyratrons, ignitrons, ball gaps, and triggered vacuum gaps are used.

Stability. Radar systems with moving-target-indicator (MTI) place unusually tight stability requirements on their transmitters. Small changes in the amplitude, phase, or frequency from one pulse to the next can degrade MTI performance. In the transmitter, the MTI requirements appear as constraints on voltage, current, and timing variations from pulse

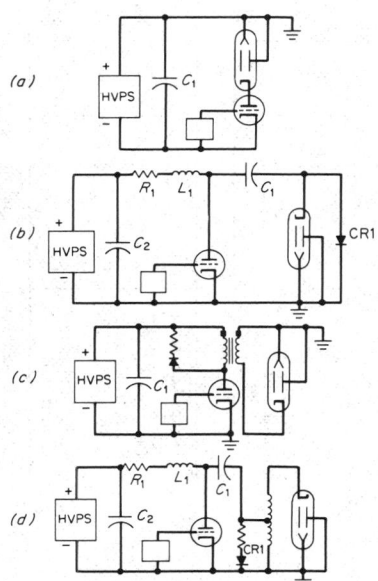

Fig. 25-51. Active-switch cathode pulsers. Circuit *A*, direct-coupled; circuit *B*, capacitor-coupled; circuit *C*, transformer-coupled; circuit *D*, capacitor- and transformer-coupled (Ref. 52).

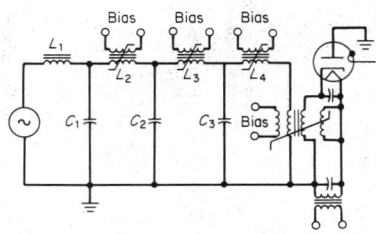

Fig. 25-52. Magnetic modulator (Ref. 52).

Table 25-10. Regulators for Modulators

Type of regulator	Input or output control range	Efficiency	Speed	Accuracy, %	Ripple reduction	Notes
Primary control						
Motor-driven Variac*	Full	Very good	Very slow	0.3	No	Moving parts, heavy.
Ferroresonant regulator	Fixed output	Good	Fast	1.0	No	Heavy, fixed frequency, single phase only. May be square-wave or sinewave type.
SCR or ignitron ..	Full	Very good (lowers power factor)	Medium	0.1	No	Small, raises ripple, some RFI.
Dc series regulator	Full	Poor	Very fast	0.01	Yes	Uses tubes. Tube must handle peak charging current (in line type) or peak load current (in active-switch modulator).
Constant-current hard-tube modulator	Full	Poor	Fast	1.0	Yes	
Line type only						
Damped charging choke	30%	Poor	Medium	0.5	No	Tubes.
De-Q-ing	30%	Poor	Fast	1.0	Yes	Tubes or SCRs.
Series triode and return diode	Full	Good	Fast	1.0	Yes	Tubes. Clipper circuit can be omitted.
Series diode and return SCR	30%	Good	Fast	1.0	Yes	Tubes or SCRs.

* Or equivalent Powerstat, Inductrol, etc.
SOURCE: Weil, Ref. 52, Par. **25-161**.

Table 25-11. Stability Factors

	Frequency- or phase-modulation sensitivity	Impedance, ratio	Current or voltage change for 1% change in HVPS voltage	
		Dynamic static	Line-type modulator	Low-impedance* hard-tube modulator, or dc operation
Magnetron	$\dfrac{\Delta f}{f} = \left(\begin{array}{c}0.001 \\ \text{to} \\ 0.003\end{array}\right)\dfrac{\Delta I}{I}$	0.05–0.1	$\Delta I = 2\%$	$\Delta I = 10\text{-}20\%$
Stabilotron or stabilized magnetron	$\dfrac{\Delta f}{f} = \left(\begin{array}{c}0.0002 \\ \text{to} \\ 0.0005\end{array}\right)\dfrac{\Delta I}{I}$	0.05–0.1	$\Delta I = 2\%$	$\Delta I = 10\text{-}20\%$
Backward-wave CFA	$\Delta\phi = 0.4$ to $1°$ for 1% $\Delta I/I$	0.05–0.1	$\Delta I = 2\%$	$\Delta I = 10\text{-}20\%$
Forward-wave CFA	$\Delta\phi = 1.0$ to $3.0°$ for 1% $\Delta I/I$	0.1–0.2	$\Delta I = 2\%$	$\Delta I = 5\text{-}10\%$
Klystron	$\dfrac{\Delta\phi}{\phi} = \dfrac{1}{2}\dfrac{\Delta E}{E} \quad \phi \approx 5\lambda$ $\Delta\phi \approx 10°$ for 1%$\Delta E/E$	0.67	$\Delta E = 0.8\%$	$\Delta E = 1\%$
TWT	$\dfrac{\Delta\phi}{\phi} \approx \dfrac{1}{3}\dfrac{\Delta E}{E} \quad \phi \approx 15\lambda$ $\Delta\phi \approx 20\%$ for 1%$\Delta E/E$	0.67	$\Delta E = 0.8\%$	$\Delta E = 1\%$
Triode or tetrode	$\Delta\phi = 0$ to $0.5°$ for 1%$\Delta I/I$	1.0	$\Delta I = 1\%$	$\Delta I = 1\%$

* A high-impedance modulator is not listed because its output would (ideally) be independent of HVPS voltage.
SOURCE: Weil, Ref. 52, Par. **25-161.**

to pulse. The relation between voltage variations and variation in amplitude and phase shift differs with the tube type used. Table 25-11 lists stability factors for the various tube types used in a high-voltage power supply (HVPS).

63. Radar Antennas. The great variety of radar applications has produced an equally great variety of radar antennas. These vary in size from less than a foot to hundreds of feet in diameter. It is not feasible even to mention each of the types here, but rather we will discuss the three basic antenna categories, search antennas, track antennas, and multifunction array antennas, after first reviewing some basic antenna principles.

A radar antenna directs the radiated power and receiver sensitivity to the azimuth and elevation coordinates of the target. The ability of an antenna to direct the radiated power is described by its antenna pattern. A typical antenna pattern is shown in Fig. 25-53. It is a plot of radiated field intensity measured in the far field (a distance greater than twice the diameter squared divided by the wavelength from the antenna) and is plotted as a function of azimuth and elevation angle. Single cuts through the two-dimensional pattern, as shown in Fig. 25-54, are more often used to describe the pattern. The principle of reciprocity assures that the antenna pattern describes its gain as a receiver as well as transmitter. The gain is defined relative to an isotropic radiator.

The gain used as a defining parameter is the gain at the peak of the beam or main lobe (see Fig. 25-54). This is the one-way power gain of the antenna:

$$G_p = \frac{4\pi A \eta_a}{\lambda^2}$$

where A = area of antenna aperture (reflector area for a horn-fed reflector antenna), λ = radar wavelength in the units of A, and η_a = aperture efficiency, which accounts for all losses inherent in the process of illuminating the aperture. Tapered-aperture-illumination functions designed to produce low side lobes also result in lower aperture efficiency η_a and larger beamwidth θ_3, as shown in Fig. 25-55. A second gain definition sometimes used is directive gain. This is defined as a maximum radiation intensity, in watts per square meter, divided by the average radiation intensity, where the average is taken over all azimuth and elevation angles.

Directive gain can be inferred from the product of the main-lobe widths in azimuth and elevation, over a wide range of tapers (including uniform). For example, an array antenna, with no spillover of illumination power, gives

$$G_d = \frac{36,000}{\theta_{3a}\theta_{3e}} \tag{25-73a}$$

where θ_{3a} = 3-dB width of the main lobe in the azimuth coordinate in degrees, and θ_{3e} = 3-dB main-lobe width in the elevation coordinate in degrees. For horn-fed antennas the constant in Eq. (25-73a) is about 25,000.

64. Search Antennas. Two examples of search-radar antennas illustrating the variety of shapes and sizes are shown in Fig. 25-56 (an airborne weather radar antenna used to locate storms in the path of the aircraft) and Fig. 25-57 (a conventional air-search radar antenna).

Conventional surface and airborne search radars generally use mechanically scanned horn-fed reflectors for their antennas. The horn radiates a spherical wavefront that illuminates the reflector. The shape of the reflector is designed to cause the reflected wave to be in phase at any point on a plane in front of the reflector. This focuses the radiated energy at infinity. Mechanically scanning search radars generally have fan-shaped beams that are narrow in azimuth and wide in the elevation coordinate. In a typical surface-based air-search radar the upper edge of the beam is shaped to follow a cosecant-squared function.

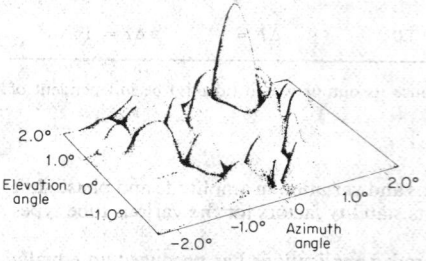

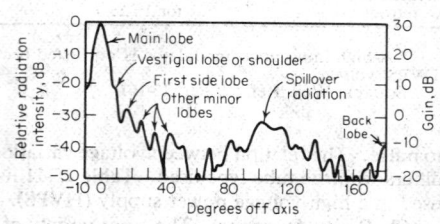

Fig. 25-53. Three-dimensional pencil-beam pattern of the AN/FPQ-6 radar antenna. *(Courtesy D. D. Howard, Naval Research Laboratory, Ref. 54.)*

Fig. 25-54. Radiation pattern for a particular paraboloid reflector antenna illustrating the main-lobe and the side-lobe radiation. *(Skolnik, Ref. 53.)*

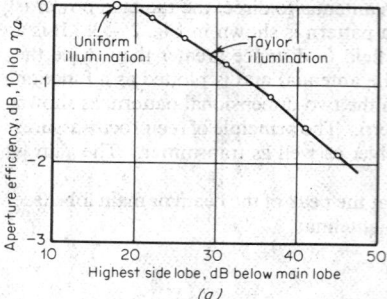

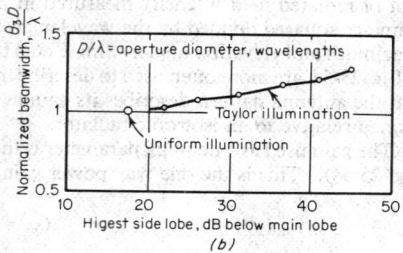

Fig. 25-55. Aperture efficiency and beamwidth as a function of highest side-lobe level for a circular aperture. *(a)* Aperture efficiency; *(b)* normalized beamwidth (Ref. 55).

Fig. 25-56. RDR-1200 airborne weather radar system. *(Courtesy Bendix Corporation. Skolnik, Ref. 53.)*

Fig. 25-57. AN/TPN-19 airport surveillance radar. *(Courtesy Raytheon Co.)*

This provides coverage up to a fixed altitude. Figure 25-58 illustrates the effect of cosecant-squared beam shaping on the coverage diagram as well as on the antenna pattern. In the horn-fed reflector the shaping can be achieved by either the reflector or the feed, and the gain constant in Eq. (25-73a) is reduced to about 20,000.

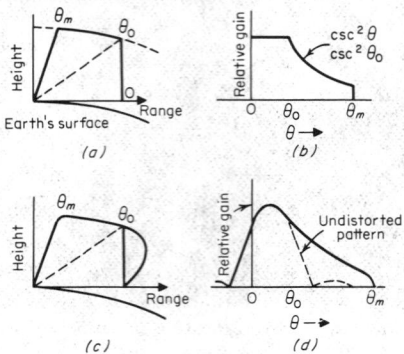

Fig. 25-58. Elevation coverage of a cosecant-squared antenna. (*a*) Desired elevation coverage; (*b*) corresponding antenna pattern desired; (*c*) realizable elevation coverage with pattern shown in *d*; (*d*) actual cosecant-squared antenna pattern. (*Freedman, Ref. 58.*)

65. Tracking-Radar Antennas. The primary function of a tracking radar is to make accurate range and angle measurements of a selected target's position. Generally, only a single target position is measured at a time, as the antenna is directed to follow the target by tracking servos. These servos smooth the errors measured from beam center to make pointing corrections. The measured errors, along with the measured position of the antenna, provide the target-angle information.

Tracking antennas like that of the AN/ FPS-16 use circular apertures to form a pencil beam about 1° wide in each coordinate. The higher radar frequencies (S-, C-, and X-band) are preferred because they allow a smaller aperture for the same beamwidth. The physically smaller antenna can be more accurately pointed. In this section we discuss aperture configurations, feeds, and pedestals.

One of the simplest methods of producing an equiphase wavefront in front of a circular aperture uses a parabolic reflector. A feed located at the focus directs its energy to illuminate the reflector. The reflected energy is then directed into space focused at infinity. The antenna is inherently broadband because the electrical path length from the feed to the reflector to the plane wavefront is the same for all points on the wavefront.

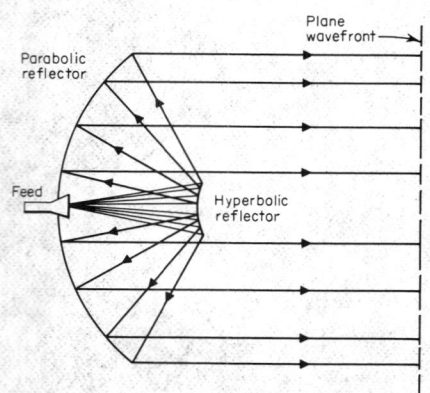

Fig. 25-59. Schematic diagram of a Cassegrain reflector antenna. (*Sengupta and Hiatt, Ref. 59.*)

Locating the feed in front of the aperture is sometimes inconvenient mechanically. It also produces spillover lobes where the feed pattern misses the reflector. The Cassegrain antenna shown in Fig. 25-59 avoids these difficulties by placing a hyperbolic subreflector between the parabolic reflector and its focus. The feed now illuminates the subreflector, which in turn illuminates the parabola and produces a plane wavefront in front of the aperture.

Lenses can also convert the spherical wavefront emanating from the feed to a plane wavefront over a larger aperture. As the electromagnetic energy passes through the lens, it is focused at infinity (see Fig. 25-60). Depending on the index of refraction n_g of the lens, a concave or convex lens may be required. Lenses are typically heavier than reflectors, but they avoid the blockage caused by the feed or subreflector.

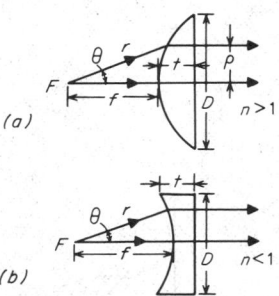

(a)

(b)

Fig. 25-60. Geometry of simple converging lenses. (a) $n > 1$; (b) $n < 1$. *(Sengupta and Hiatt, Ref. 59.)*

A single feed providing a single beam is unable to supply the angle-error information necessary for tracking. To obtain azimuth and elevation-error information, feeds have been developed that scan the beam in a small circle about the target (conical scan) or that form multiple beams about the target (monopulse). Conical scanning may be caused by rotating a reflector behind a dipole feed or rotating the feed itself. It has the advantage compared with monopulse that less hardware is required in the receiver, but at the expense of somewhat less accuracy.

Modern trackers more often use a monopulse feed with multiple receivers. Early monopulse feeds used four separate horns to produce four contiguous beams that were combined to form a reference beam and azimuth and elevation-difference beams. More recently, multimode feeds have been developed to perform this function more efficiently with fewer components. See Fig. 25-45.

Shaft angle encoders quantize radar pointing angles through mechanical connections to azimuth and elevation axes. The output indicates the angular position of the mechanical bore-site axis relative to a fixed angular coordinate system. Because these encoders make an absolute measurement, their outputs contain 10 to 20 bits of information. A variety of techniques is used, the complexity increasing with the accuracy required. Atmospheric errors ultimately limit the number of useful bits to about 20, or 0.006 mrad. In less precise tracking applications, synchros attached to the azimuth and elevation axes indicate angular position within a fraction of a degree.

66. Multifunction Arrays. *Array antennas* form a plane wavefront in front of the antenna aperture. These points are individual radiating elements which, when driven together, constitute the array. The elements are usually spaced about 0.6 wavelength apart. Most applications use planar arrays, although arrays conformal to cylinders and other surfaces have been built.

Phased arrays are steered by tilting the phase front independently in two orthogonal directions called the *array coordinates*. Scanning in either array coordinate causes the beam to move along a cone whose center is at the center of the array. The paths the beam follows when steered in the array coordinates are illustrated in Fig. 25-61, where the z axis is normal to the array. As the beam is steered away from the array normal, the projected aperture in the beam's direction varies, causing the beamwidth to vary proportionally.

Arrays can be classified as either active or passive. *Active arrays* contain duplexers and amplifiers behind every element or group of elements; *passive arrays* are driven from a single feed point. Only the active arrays are capable of higher power than conventional antennas.

Both passive and active arrays must divide the signal from a single transmission line among all the elements of the array. This can be done by an optical feed, a corporate feed, or a multiple-beam-forming network. The *optical feed* is illustrated in Fig. 25-62. A single feed, usually a monopulse horn, illuminates the array with a spherical phase front. Power collected by the rear elements of the array is transmitted through the phase shifters that produce a planar front and steer the array. The energy may then be radiated from the other side of the array, as in the lens, or be reflected and reradiated through the collecting elements, where the array acts as a steerable reflector.

Corporate feeds can take many forms, as illustrated by the series-feed networks shown in Fig. 25-63 and the parallel-feed networks shown in Fig. 25-64. All use transmission-line

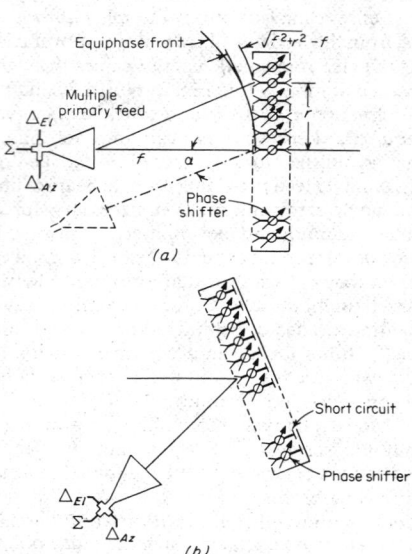

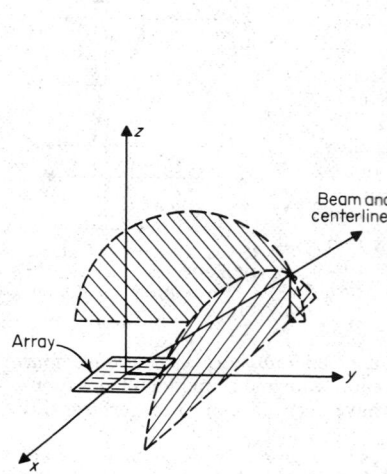

Fig. 25-61. Beam-steering contours for a planar array.

Fig. 25-62. Optical-feed systems. (*a*) Lens; (*b*) reflector. *(Cheston and Frank, Ref. 61.)*

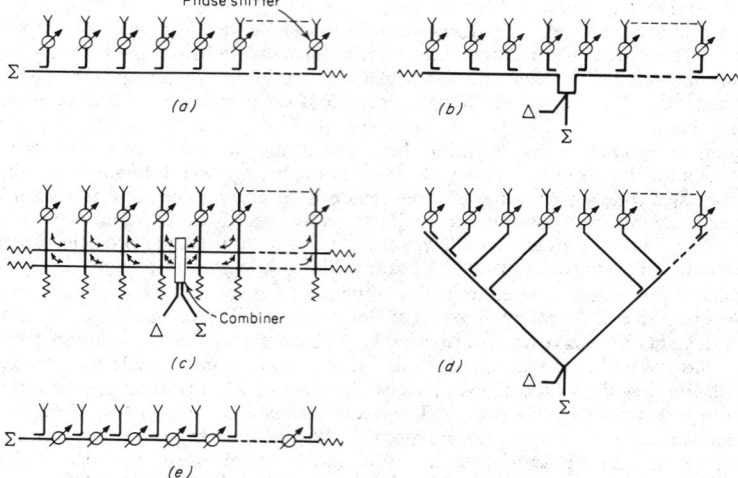

Fig. 25-63. Series-feed networks. (*a*) End feed; (*b*) center feed; (*c*) separate optimization; (*d*) equal path length; (*e*) series phase shifters. *(Cheston and Frank, Ref. 61.)*

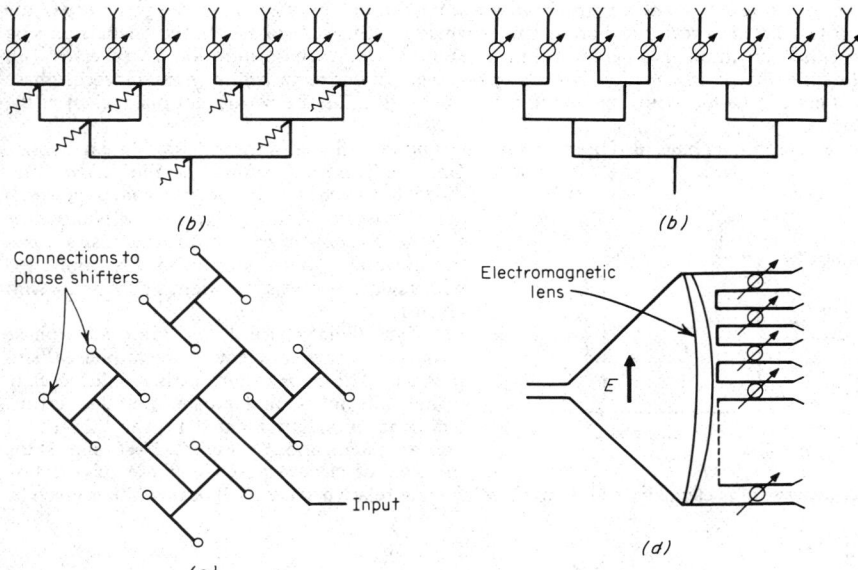

Fig. 25-64. Parallel-feed networks. (*a*) Matched corporate feed; (*b*) reactive corporate feed; (*c*) reactive stripline; (*d*) multiple reactive divider. *(Cheston and Frank, Ref. 61.)*

components to divide the signal among the elements. Phase shifters can be located at the elements or within the dividing network.

Multiple-beam networks are capable of forming simultaneous beams with the array. The Butler matrix shown in Fig. 25-65 is one such technique. It connects the N elements of a linear array to N feed points corresponding to N beam outputs. It can be applied to two-dimensional arrays by dividing the array into rows and columns.

The phase shifter is one of the most critical components of the array. It produces controllable phase shift over the operating band of the array. Digital and analog phase shifters have been developed using both ferrites and PIN diodes. Phase shifter designs always strive for a low-cost, low-loss, and high-power-handling capability.

The Reggia-Spencer phase shifter consists of a ferrite inside a waveguide, as illustrated in Fig. 25-66. It delays the rf signal passing through the waveguide. The amount of phase shift

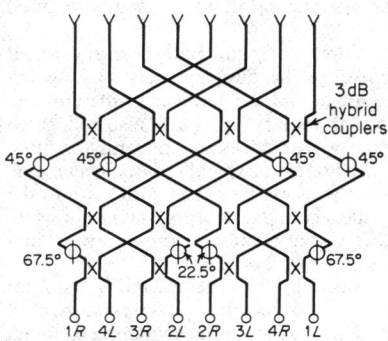

Fig. 25-65. Butler beam-forming network. *(Cheston and Frank, Ref. 61.)*

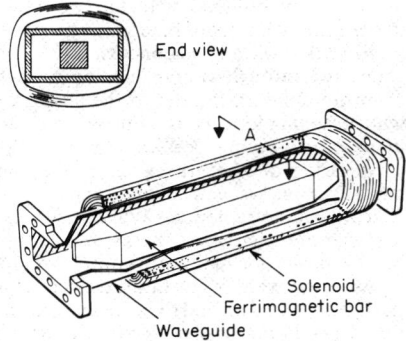

Fig. 25-66. Typical Reggia-Spencer phase shifter. *(Stark et al., Ref. 62.)*

can be controlled by the current in the solenoid, through its effect on the permeability of the ferrite. This is a reciprocal phase shifter which has the same phase shift for signals passing in either direction. Nonreciprocal phase shifters (where phase-shift polarity reverses with the direction of propagation) are also available. Either reciprocal or nonreciprocal phase shifters can be locked or latched in many states by using the permanent magnetism of the ferrite.

Phase shifters have also been developed using PIN diodes in transmission-line networks. One configuration, shown in Fig. 25-67, uses diodes as switches to change the signal path length of the network. A second type uses PIN diodes as switches to connect reactive loads across a transmission line. When equal loads are connected with a quarter-wave separation, a pure phase shift results.

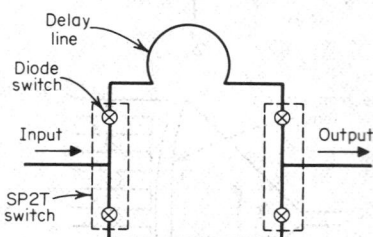

Fig. 25-67. Switched-line phase bit. *(Stark et al., Ref. 62.)*

When digital phase shifters are used, a phase error occurs at every element due to phase quantization. The error in turn causes reduced gain, higher side lobes, and greater pointing errors. Gain reduction is tabulated in Table 25-12 for typical quantizations. Figure 25-68 shows the rms side-lobe levels caused by phase quantization in an array of N elements. The rms pointing error relative to the 3-dB beamwidth is given by

$$\frac{\sigma_\theta}{\theta_3} \approx \frac{1.12}{2^m \sqrt{N}}$$

where m = number of bits of phase quantization, and N = number of elements in the array.[55]

Frequency scan is a simple array-scanning technique that does not require phase shifters, drivers, or beam-steering computers. Element signals are coupled from points along a

Table 25-12.   Gain Loss in a Phased Array with m-Bit Digital Phase Shifters

Number of Bits, m	Gain Loss, dB
3	0.228
4	0.057
5	0.0142
6	0.00356
7	0.0089
8	0.0022

transmission line as shown in Fig. 25-69. The electrical path length between elements is much longer than the physical separation, so that a small frequency change will cause a phase change between elements large enough to steer the beam. The technique can be applied only to one array coordinate, so that in two-dimensional arrays, phase shifters are usually required to scan the other coordinate.

67. Microwave Components. The radar transmitter, antenna, and receiver are all connected through rf transmission lines to a duplexer. The duplexer acts as a switch connecting the transmitter to the antenna while radiating, and the receiver to the antenna while listening for echoes. Filters, receiver protectors, and rotary joints may also be located in the various paths. See Sec. 9 for a description of microwave devices and transmission lines.

A variety of other transmission-line components are used in a typical radar. Waveguide bends, flexible waveguide, and rotary joints are generally necessary to route the path to the feed of a rotating antenna. Waveguide windows provide a boundary for pressurization while allowing the microwave energy to pass through. Directional couplers sample forward and reverse power for monitoring, test, and alignment of the radar system.

68. Duplexers. The duplexer acts as a switch connecting the antenna and transmitter during transmission and the antenna and receiver during reception. Various circuits are used that depend on gas tubes, ferrite circulators, or PIN diodes as the basic switching element. The duplexers using gas tubes are most common. A typical gas-filled TR tube is shown in Fig. 25-70. Low-power rf signals pass through the tube with very little attenuation. Higher power causes the gas to ionize and present a short circuit to the rf energy.

Figure 25-71 shows a balanced duplexer using hybrid junctions and TR tubes. When the transmitter is on, the TR tubes fire and reflect the rf power to the antenna port of the input hybrid. On reception, signals received by the antenna are passed through the TR tubes and to the receiver port of the output hybrid.

69. Circulators and Diode Duplexers. Newer radars often use a ferrite circulator as the duplexer. A TR tube is required in the receiver line to protect the receiver from the

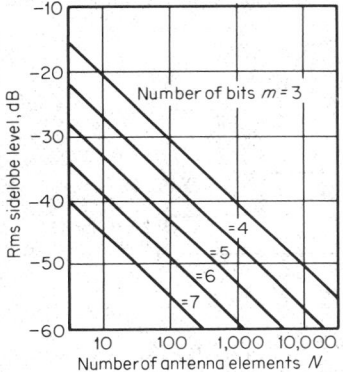

Fig. 25-68. Rms side lobes due to phase quantization. *(Cheston and Frank, Ref. 61.)*

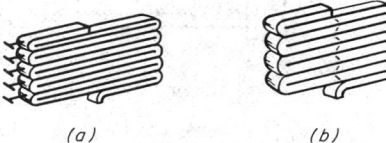

(a)   (b)

Fig. 25-69. Simple types of frequency-scanned antennas. (a) Broad-wall coupling to dipole radiators; (b) narrow-wall coupling with slot radiators. *(Hammer, Ref. 64.)*

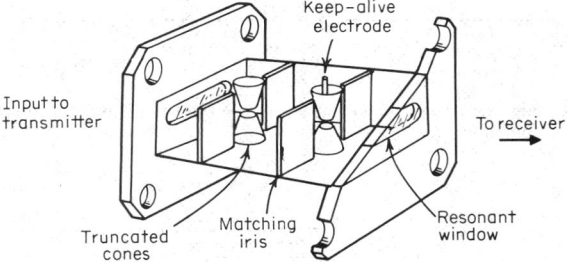

Fig. 25-70. Typical TR tube. *(Skolnick, Ref. 53.)*

transmitter power reflected by the antenna due to an imperfect match. A four-port circulator is generally used with a load between the transmitter and receiver ports so that the power reflected by the TR tube is properly terminated.

PIN diode switches have been used in duplexers in place of the TR tubes. These are more easily applied in coaxial circuitry and at lower microwave frequencies. Multiple diodes are used when a single diode cannot withstand the required voltage or current.

70. Receiver Protectors. TR tubes with a lower power rating are usually required in the receive line to prevent transmitter leakage from damaging mixer diodes or rf amplifiers in the receiver. A keep-alive ensures rapid ionization, minimizing spike leakage. The keep-alive may be either a probe in the TR tube maintained at a high dc potential or a piece of radioactive material. Diode limiters are also used after TR tubes to further reduce the leakage.

71. Filters. Microwave filters are sometimes used in the transmit path to suppress spurious radiation or in the receive signal path to suppress spurious interference. Because the transmit filters must handle high power, they are larger and more difficult to design.

Narrow-band filters in the receive path, often called *preselectors,* are built using mechanically tuned cavity resonators or electrically tuned YIG resonators. Preselectors can provide up to 80 dB suppression of signals from other radar transmitters in the same rf band but at a different operating frequency.

Harmonic filters are the most common transmitting filter. These absorb the harmonic energy to prevent it from being radiated or reflected. Since the transmission path may provide a high standing-wave ratio at the harmonic frequencies, the presence of harmonics can increase the voltage gradient in the transmission line and cause breakdown. Figure 25-72 shows a harmonic filter where the harmonic energy is coupled out through holes in the walls of the waveguide to matched loads.

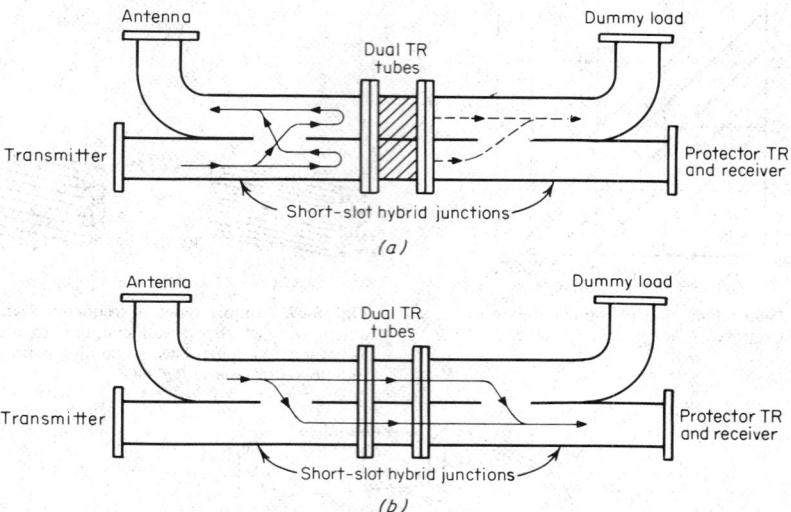

Fig. 25-71. Balanced duplexer using dual TR tubes and two short-slot hybrid junctions. (*a*) Transmit condition; (*b*) receive condition. *(Skolnik, Ref. 53.)*

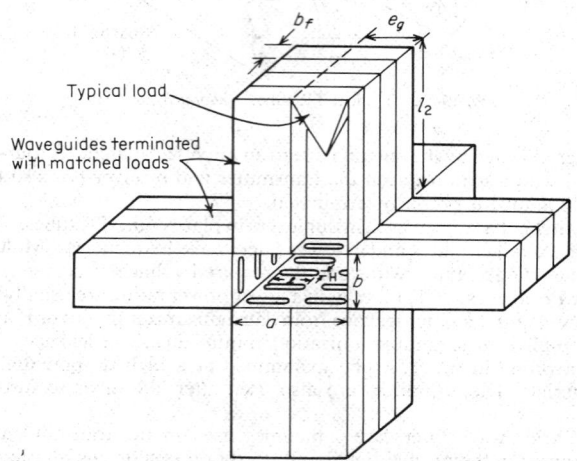

Fig. 25-72. Typical construction of a dissipative waveguide filter. *(Matthaei et al., Ref. 65.)*

72. Radar Receivers. The radar receiver amplifies weak target returns so that they can be detected and displayed. The input amplifier must add little noise to the received signal, for this noise competes with the smallest target return that can be detected. A mixer in the receiver converts the received signal to an intermediate frequency where filtering and signal decoding can be accomplished. Finally, the signals are detected for processing and display.

73. Low-Noise Amplifiers. Because long-range radars require large transmitters and antennas, these radars can also afford the expense of a low-noise receiver. Considerable effort has been expended to develop more sensitive receivers. Some of the devices in use will be described here after a brief review of noise-figure and noise-temperature definitions.

Noise figure and noise temperature measure the quality of a sensitive receiver. Noise figure is the older of the two conventions and is defined as

$$F_n = \frac{(S/N) \text{ at the output}}{(S/N) \text{ at the input}}$$

where S is signal power, N is noise power, and the receiver input termination is at room temperature. Before low-noise amplifiers were available, a radar's noise figure was determined by the first mixer, which would be typically 5 to 10 dB. For these values of F_n, it was approximately correct to add the loss of the waveguide to the noise figure when calculating signal-to-noise ratio. As better receivers were developed with lower noise figures, these approximations were no longer accurate, and the noise-temperature convention was developed.

Noise temperature is proportional to noise-power spectral density through the relation

$$T = \frac{N}{kB}$$

where k = Boltzmann's constant, and B = the bandwidth in which the noise power is measured. The noise temperature of an rf amplifier is defined as the noise temperature added at the input of the amplifier required to account for the increase in noise due to the amplifier. It is related to noise figure through the equation

$$T = T_0(F_n - 1)$$

where T_0 = standard room temperature, 290 K.

The receiver is only one of the noise sources in the radar system. Figure 25-73 shows the receiver in its relation to the other important noise sources. Losses, whether in the rf transmission line, antenna, or the atmosphere, reduce the signal level and also generate thermal noise. The load presented to the rf transmission line by the antenna is its radiation resistance. The temperature of this resistance, T_a, depends on where the antenna beam is pointed. When the beam is pointed into space, this temperature may be as low as 50 K. However, when the beam is pointed toward the sun or a radio star, the temperature can be much higher. All these sources can be combined to find the system noise temperature T_s, according to the equation

$$T_s = T_a + (L_r - 1)T_{tr} + T_e L_r$$

where T_a = temperature of the antenna (see **Par. 25-7**), L_r = transmission-line loss defined as the ratio of power in to power out, T_{tr} = temperature of the transmission line, and T_e = receiver noise temperature.

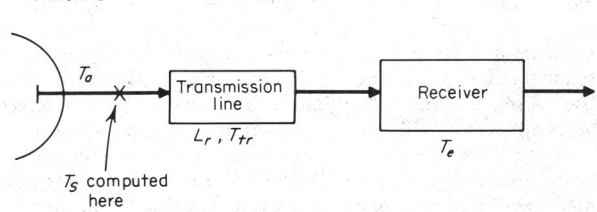

Antenna

Fig. 25-73. Contributions to system noise temperature.

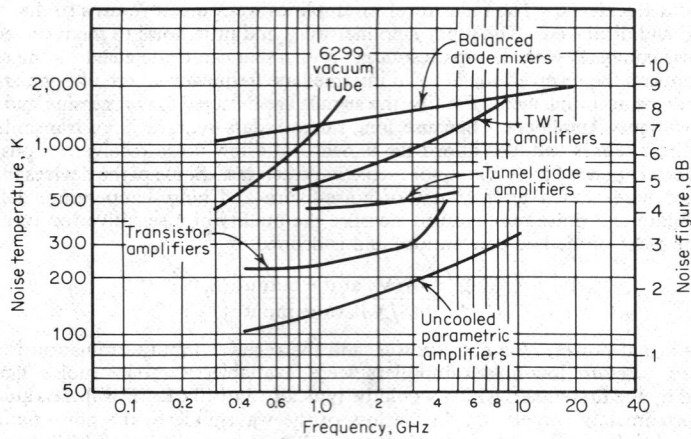

Fig. 25-74.   Noise characteristics of radar front ends.   *(After Taylor and Mattern, Ref. 66.)*

Figure 25-74 shows the noise temperature as a function of frequency for radar-receiver front ends. All are noisier at higher frequencies. Transistor amplifiers and uncooled parametric amplifiers are finding increased use in radar receivers. *Transistor amplifiers* have been improved steadily, with emphasis on increased operating frequency. Although the transistor amplifier is a much simpler circuit than the parametric amplifier, it does not achieve the parametric amplifier's low-noise temperature.

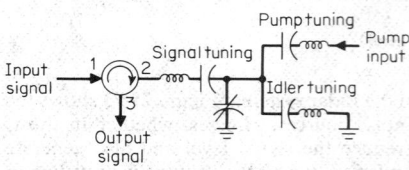

Fig. 25-75.   Single-port   parametric   amplifier. *(Taylor and Mattern, Ref. 66.)*

The *parametric amplifier* is unusual in that it requires a high-frequency pump. Figure 25-75 shows a parametric amplifier circuit. A varactor diode acts as a capacitance, varying at the pump frequency, typically K-band for a microwave amplifier. In this condition, the varactor presents a negative resistance to the circulator port 2. This in turn produces an amplified reflected signal from that port which leaves the circulator from port 3. Gains of 10 to 20 dB per stage and bandwidths up to 10% are typical.

A balanced mixer is often used to convert from rf to if. Balanced operation affords about 20 dB immunity to amplitude noise on the local-oscillator signal. Intermediate frequencies of 30 and 60 MHz are typical, as are 1.5 to 2 dB intermediate-frequency noise figures for the if preamplifier. Double conversion is sometimes used with a first if at a few hundred megahertz. This gives better image and spurious suppression.

The *matched filter* (Par. **25-18**) is usually instrumented at the second if frequency. This filter is an approximation to the matched filter and therefore does not achieve the highest possible signal-to-noise ratio. This deficiency is expressed as mismatch loss. Table 25-13 lists the mismatch loss for various signal-filter combinations when the optimum bandwidth is used.

74. Pulse Compression. Pulse compression is a technique in which a rectangular pulse containing phase modulation is transmitted. When the echo is received, the matched-filter output is a pulse of much shorter duration. This duration approximately equals the reciprocal of the bandwidth of the phase modulation. The compression ratio (ratio of transmitted to compressed pulse lengths) equals the product of the time duration and bandwidth of the transmitted pulse. The technique is used when greater pulse energy and range resolution are required than can be achieved with a simple uncoded pulse.

A pulse-compression radar can be instrumented in three basic ways, as illustrated in Fig. 25-76. In Fig. 25-76a two conjugate networks are used, one on transmit and the other on receive to generate and compress the coded pulse. The same network can be used for generation and compression if time inversion can be realized, as shown in Fig. 25-76b. The

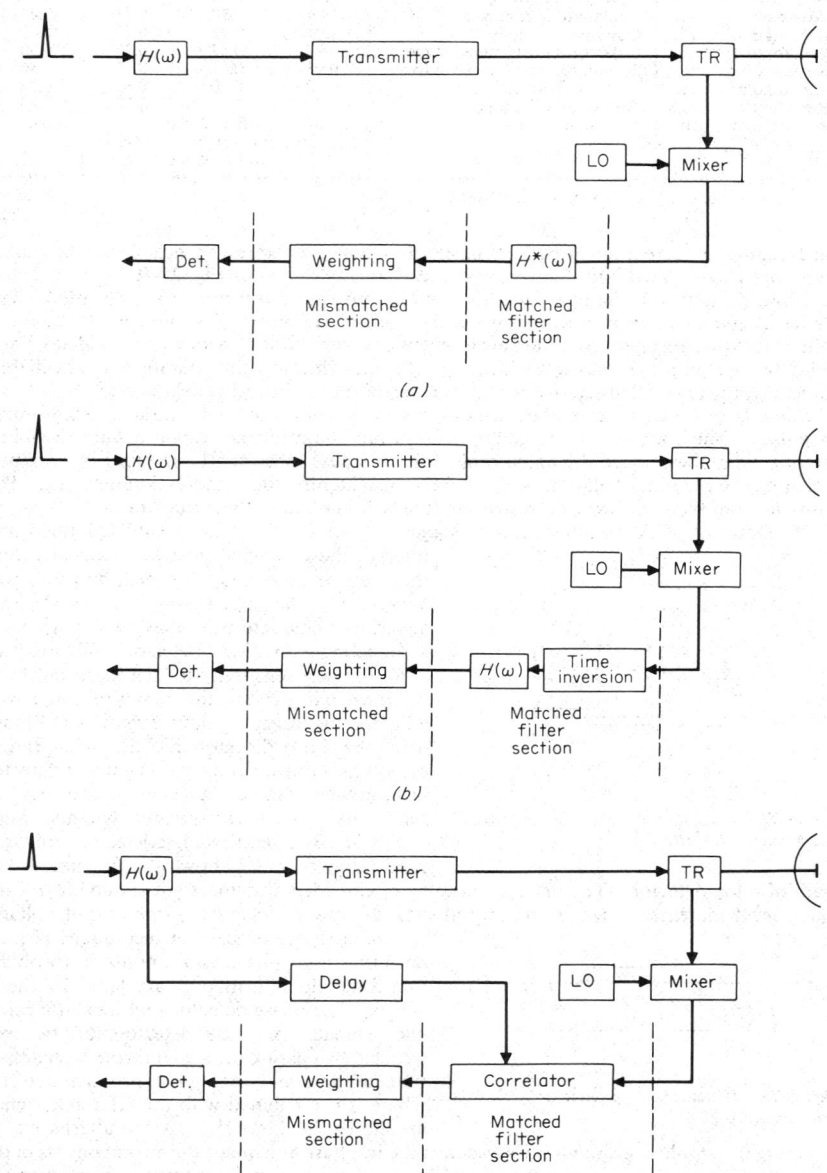

Fig. 25-76. Pulse-compression radar using (a) conjugate filters; (b) time inversion; (c) correlation. *(Farnett et al., Ref. 67.)*

Table 25-13.   Approximations to Matched Filters

Pulse shape	Filter	Optimum bandwidth-time product			Mismatch loss, dB
		6 dB	3 dB	Energy	
Gaussian	Gaussian bandpass	0.92	0.44	0.50	0
Gaussian	Rectangular bandpass	1.04	0.72	0.77	0.49
Rectangular	Gaussian bandpass	1.04	0.72	0.77	0.49
Rectangular	5 synchronously tuned stages	0.97	0.67	0.76	0.50
Rectangular	2 synchronously tuned stages	0.95	0.61	0.75	0.56
Rectangular	Single-pole filter	0.70	0.40	0.63	0.88
Rectangular	Rectangular bandpass	1.37	1.37	1.37	0.85
Rectangular chirp	Gaussian	1.04 × 6 dB width of equivalent (sin x)/x pulse, (0.86 × width of spectrum)			0.6

SOURCE: Taylor and Mattern, Ref. 66, Par. 25-161.

third technique, correlation, uses the transmit network to generate a replica of the transmitted pulse that is correlated with the received waveform, as shown in Fig. 25-76c.

Linear fm (chirp) is the phase modulation that has received the widest application. The carrier frequency is swept linearly during the transmitted pulse. The wide application has both caused and resulted from the development of a variety of dispersive analog delay lines. Delay-line techniques covering a wide range of bandwidths and time durations are available. Table 25-14 lists the characteristics of a number of these dispersive delay lines.

Range lobes are a property of pulse-compression systems not found in radar using simple cw pulses. These are responses leading and trailing the principal response and resembling antenna side lobes; hence the name range lobes. These lobes can be reduced by carefully designing the phase modulation or by slightly mismatching the compression network. The mismatch can be described as a weighting function applied to the spectrum.

75. Detectors.   Although bandpass signals on an if carrier are easily amplified and filtered, these signals must be detected before they can be displayed, recorded, or processed. When only the signal amplitude is desired, square-law characteristics may be obtained with a semiconductor diode detector, and this provides the best sensitivity for detecting pulses in noise when integrating the signals returned from a fluctuating target. Larger if signal amplitudes drive the diode detector into the linear range, providing a linear detector. The linear detector has a greater dynamic range with somewhat less sensitivity. When still greater dynamic range (up to 80 dB) is required, log detectors are often used. Figure 25-77 shows the functional diagram of a log detector. The detected outputs of cascaded amplifiers are summed. As the signal level increases, stages saturate, reducing the rate of increase of the output voltage.

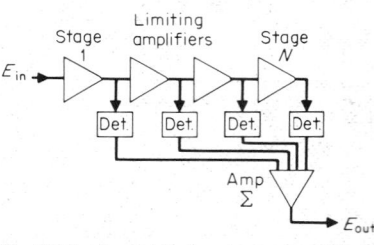

Fig. 25-77.   Logarithmic detector.   *(Taylor and Mattern, Ref. 66.)*

Some signal processing techniques require detecting both phase and amplitude to obtain the complete information available in the if signal. The phase detector requires an if reference signal. A phase detector can be constructed by passing the signal through an amplitude limiter, and then to a product detector where it is combined with the reference signal, as shown in Fig. 25-78. An alternative to detecting the amplitude and phase is to detect the in-phase and quadrature components of the if signal. The product detector shown in Fig. 25-78 can also provide this function when the input signal is not amplitude-limited. Quadrature detector circuits differ only in that the reference signal is shifted by 90° in one detector relative to the other.

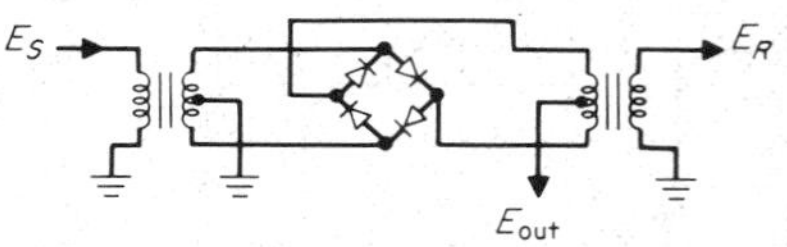

Fig. 25-78.   Balanced-diode detector.   *(Taylor and Mattern, Ref. 66.)*

76. Analog-to-Digital Converters. Digital signal processors require that the detected if signals be encoded by an analog-to-digital converter. A typical converter may sample the detected signal at a 1-MHz rate and encode the sampled value into an 8-bit binary word. Encoders operating at higher rates have been built, but with fewer bits in their output. Encoders typically have errors that about equal the least significant bit.

77. Exciters. Two necessary parts of any radar system are an *exciter* to generate rf and local-oscillator frequencies and a *synchronizer* to generate the necessary triggers and timing pulses.

The components used in exciters are oscillators, frequency multipliers, and mixers. These can be arranged in various ways to provide the cw signals needed in the radar. The signals required depend on whether the transmitter is an oscillator or a power amplifier.

Transmitters using power oscillators such as magnetrons determine the rf frequency by the magnetron tuning. In a noncoherent radar, the only other frequency required is that of the local oscillator. It differs from the magnetron frequency by the if frequency, and this difference is usually maintained with an automatic frequency control (AFC) loop. Figure 25-79 shows the circuit of a simple magnetron radar, illustrating the two alternative methods of tuning the magnetron to follow the *stable local oscillator* (*stalo*) or tuning the stalo to follow the magnetron.

If the radar must use coherent detection (as in MTI or pulse doppler applications), a second oscillator, called a *coherent oscillator* (*coho*), is required. This operates at the if frequency and provides the reference for the product detector. Because an oscillator transmitter starts with random phase on every pulse, it is necessary to quench the coho and lock its phase with that of the transmitter on each pulse. This is accomplished by the circuit shown in Fig. 25-80.

When an amplifier transmitter is used, coho locking is not required. The transmit frequency may be obtained by mixing the stalo and coho frequencies, as shown in Fig. 25-81. The stalo and coho are not always oscillators operating at their output frequency. Figure 25-82 shows an exciter using crystal oscillators and multipliers to produce the rf and local-oscillator frequencies. Crystals may be changed to select the rf frequency without changing the if frequencies.

The stability required of the frequencies produced by the exciter depends on the radar application. In a simple noncoherent radar a stalo frequency error shifts the signal spectrum in the if passband, and an error which is a fraction of the if bandwidth can be allowed. In MTI or pulse doppler radars, phase changes from pulse to pulse must be less than a few degrees. This requirement can be met with crystal oscillators driving frequency multipliers or fundamental oscillators with high-Q cavities when sufficiently isolated from electrical and mechanical perturbation. Instability is often expressed in terms of the phase spectrum about the center frequency.

Crystal oscillators driving frequency multipliers are finding increased use as stalos. A typical multiplier might multiply a 90-MHz crystal oscillator frequency by 32 to obtain an S-band signal. This source has the long-term stability of the crystal oscillators, but with

Table 25-14. Characteristics of Passive Linear-fm Devices

	B, MHz	T, μs	BT	f_0, MHz	Typical loss, dB	Typical spurious, dB
Aluminum strip delay line ...	1	500	200	5	15	−60
Steel strip delay line	20	350	500	45	70	−55
All-pass network	40	1,000	300	25	25	−40
Perpendicular diffraction delay line	40	75	1,000	100	30	−45
Surface-wave delay line	40	50	1,000	100	20	−50
Wedge-type delay line	250	65	1,000	500	50	−50
Folded-tape meander line	1,000	1.5	1,000	2,000	25	−40
Waveguide operated near cutoff	1,000	3	1,000	5,000	60	−25
YIG crystal	1,000	10	2,000	2,000	70	−20

Source: Farnett et al., Ref. 67, Par. **25-161.**

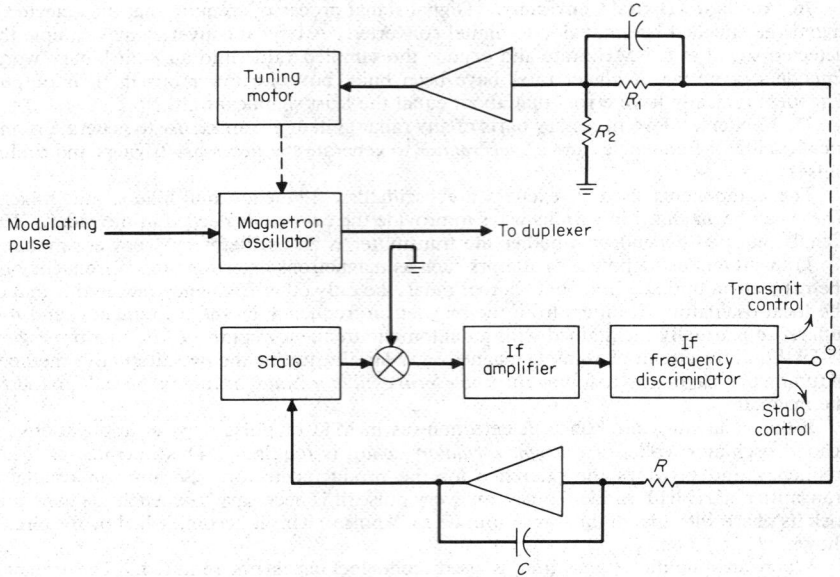

Fig. 25-79. Alternative methods for AFC control. *(Taylor and Mattern, Ref. 66.)*

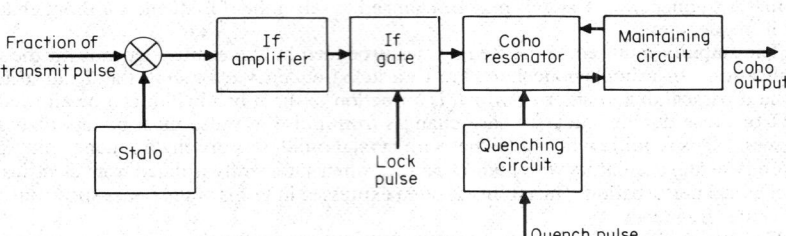

Fig. 25-80. Keyed coho. *(Taylor and Mattern, Ref. 66.)*

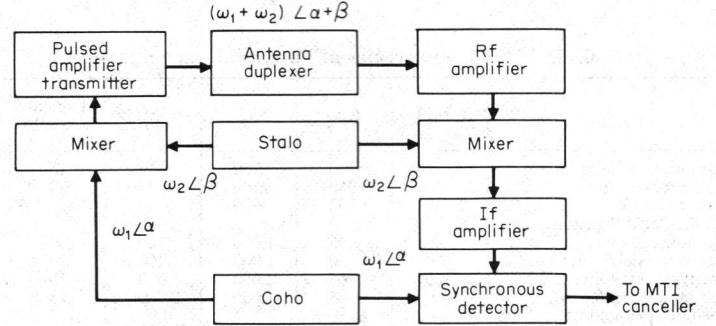

Fig. 25-81. Coherent radar. *(Taylor and Mattern, Ref. 66.)*

degraded short-term stability. This is because the multiplier increases the phase modulation on the oscillator signal in proportion to the multiplication factor; i.e., each doubler stage raises the oscillator sidebands 6 dB. Frequency may be varied by tuning the crystal oscillator (about 0.25%) or by changing crystals.

78. Synchronizers. The synchronizer delivers timing pulses to the various radar subsystems. In a simple marine radar this may consist of a single multivibrator that triggers the transmitter, while in a larger radar 20 to 30 timing pulses may be needed. These may turn on and off the beam current in various transmitter stages; start and stop the rf pulse time attenuators; start display sweeps; etc.

Timing pulses or triggers are often generated by delaying a pretrigger with delays that may be either analog or digital. New radars are tending toward digital techniques, with the synchronizer incorporated into a digital signal processor. A diagram of the delay structure in a digital synchronizer is shown in Fig. 25-83. A 10-MHz clock moves the initial timing pulse through shift registers. The number of stages in each register is determined by the delay required. Additional analog delays provide a fine delay adjustment to any point in the 100-ns interval between clock pulses.

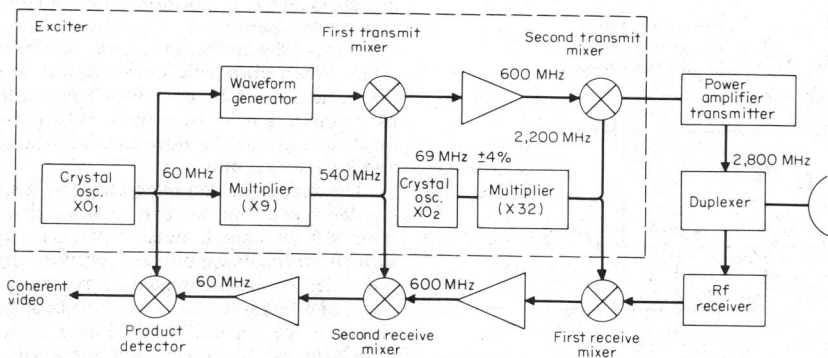

Fig. 25-82. Coherent-radar exciter.

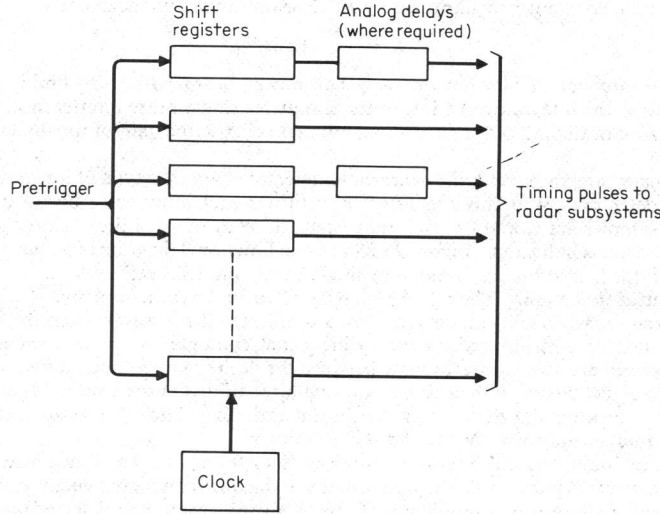

Fig. 25-83. Digital synchronizer.

The synchronizer will also contain a range encoder, in radars where accurate range tracking is required or where range data will be processed or transmitted in digital form. Range is usually quantized by counting cycles of a clock, starting with a transmitted pulse and stopping with the received echo. Where high precision is required, the fine range bits are obtained by interpolating between cycles of the clock, using a tapped delay line with coincidence detectors on the taps.

79. Signal Processing. The term signal processing describes those circuits in the signal path between the receiver and the display. The extent to which processing is done in this portion of the signal path depends on the radar application. In search radars, postdetection integration, clutter rejection, and sometimes pulse compression are instrumented in the signal processor. The trend in modern radar has been to use digital techniques to perform these functions, although many analog devices are still in use. The following paragraphs outline the current technology trends in postdetection integration, clutter rejection, and digital pulse compression.

80. Postdetection Integration. Scanning-search radars transmit a number of pulses toward a target as the beam scans past. For best detectability, these returns must be combined before the detection decision is made. In many search radars the returns are displayed on a plan-position indicator (PPI), where the operator, by recognizing target patterns, performs the postdetection integration. When automatic detectors are used, the returns must be combined electrically. Many circuits have been used, but the two most common are the video sweep integrator and binary integrator.

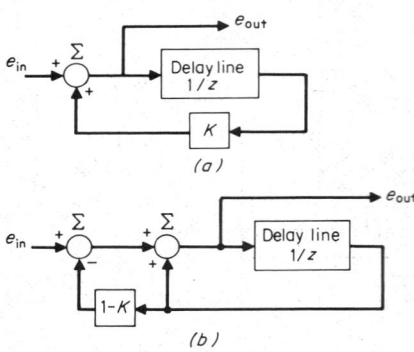

Fig. 25-84. Two forms of sweep integrators. *(Shrader, Ref. 68.)*

The simplest video integrator uses a single delay line long enough to store all the returns from a single pulse. When the returns from the next pulse are received, they are added to the attenuated delay-line output. Figure 25-84 shows two forms of this circuit. The second (Fig. 25-84b) is preferred because the gain factor K is less critical to adjust. The circuit weights past returns with an exponentially decreasing amplitude where the time constant is determined by K. For optimum enhancement

$$K = 1 - 1.56/N$$

where N = number of hits per one-way half-power beamwidth. By limiting the video amplitude into the integrator, the integrator eliminates single-pulse interference. The delay may be analog or digital, but the trend is to digital because the gain of the digital loop does not drift.

The binary integrator or double-threshold detector is another type of integration used in automatic detectors. With this integrator, the return in each range cell is encoded to 1 bit and the last N samples are stored for each range cell. If M or more of the N stored samples are ones, a detection is indicated. Figure 25-85 shows a functional diagram of a binary integrator. This integrator is also highly immune to single-pulse interference.

81. Clutter Rejection. The returns from land, sea, and weather are regarded as clutter in an air-search radar. They can be suppressed in the signal processor when the spectrum is narrow compared with the radar's prf. Filters that combine two or more returns from a single-range cell are able to discriminate between the desired targets and clutter. This allows the radar to detect targets with cross section smaller than that of the clutter. It also provides a means of preventing the clutter from causing false alarms. The two classes of clutter filters are moving target indicator (MTI) and pulse doppler.

MTI combines a few pulse returns, usually two or three, in a way that causes the clutter returns to cancel. Figure 25-86 shows a functional diagram of a digital vector canceler. The in-phase and quadrature components of the if signal vector are detected and encoded. Stationary returns in each signal component are canceled before the components are rectified and combined. The digital canceler may consist of a shift register memory and a subtractor to take the difference of the succeeding returns. Often only one component of the vector canceler is instrumented, thereby saving about half the hardware, but at the expense of signal detectability in noise.

A pulse doppler processor is another class of clutter filter where the returns in each range resolution cell are gated and put into a bank of doppler filters. The number of filters in the bank approximately equals the number of pulse returns combined. Each filter is tuned to a different frequency, and the passbands contiguously positioned between zero frequency and prf. Figure 25-87 shows a functional diagram of a pulse doppler processor. The pulse doppler technique is most often used in either airborne or land-based target-tracking radars, where a high ambiguous prf can be used, thus providing an unambiguous range of doppler frequencies. The filter bank may be instrumented digitally by a special-purpose computer wired according to the fast Fourier transform algorithm.

82. Digital Pulse Compression. Digital pulse compression performs the same function as the analog pulse compression described in Par. 25-74, except that it is instrumented in the signal processor. Digital pulse compression is easiest to instrument with binary-phase-coded waveforms. The transmitted pulse is divided into equal-length subpulses, where the phase of each subpulse is set at 0 or 180° according to the phase code, and the number of subpulses equals the compression ratio. Figure 25-88 shows a functional diagram of a digital pulse compression processor. The quadrature components of the if vector are encoded to 1 bit, and shift registers, having one stage per subpulse, store a length of data equal to the transmitted pulse. Each of the subpulse bits is summed after reversing the bits corresponding to the positions of 180° phase shift on transmission. A maximum output is obtained when the expanded pulse just fills the shift register. Shifting the pulse in the register by one subpulse destroys the match between the signal and the tap coding, causing the output to drop to a low value. This form of digital pulse compression has the advantage that the code can be more easily changed than in analog pulse compression, where the coding is determined by a dispersive network.

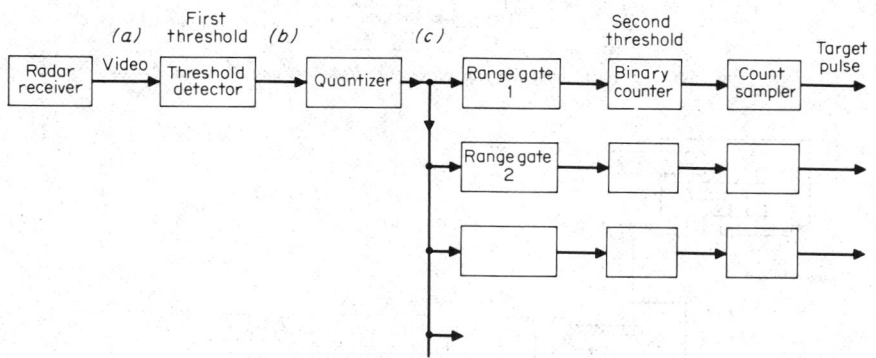

Fig. 25-85. Binary integrator. *(Skolnik, Ref. 53.)*

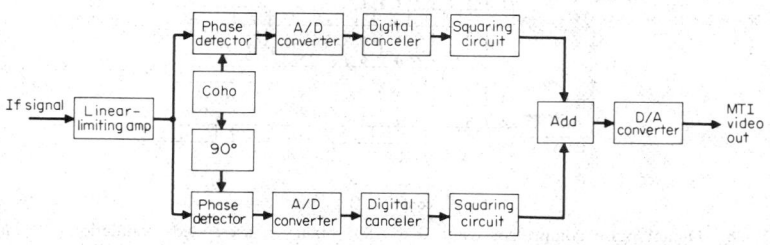

Fig. 25-86. Digital vector canceler. *(Shrader, Ref. 68.)*

25-81

83. Displays. Radar indicators are the coupling link between radar information and a human operator. Radar information is generally a time-dependent function of range or distance. Thus the display format generally uses a displacement of signal proportional to time or range. The signal may be displayed as an orthogonal displacement (such as an oscilloscope) or as an intensity modulation or brightening of the display. Most radar signal presentations have the intensity-modulated type of display where additional information such as azimuth, elevation angle, or height can be presented.

A common format is a polar-coordinate, or plan-position, indicator (PPI), which results in a maplike display. Radar azimuth is presented on the PPI as an angle, and range as radial distance. In a cartesian coordinate display, one coordinate may represent range and the other may represent azimuth elevation, or height. Variations may use one coordinate for azimuth and the other coordinate for elevation, and gate a selected time period or range to the display. The increasing use of processed radar data can provide further variations. In each case, the display technique is optimized to improve the information transfer to the operator. Fifteen display formats with their scanning patterns are shown in Fig. 25-89, with the types designated below the displays described in the legend, e.g., (a) represents A-scope, etc.

The types J, K, L, M, and N formats are modifications of the type A display and use deflection modulation for signal presentation. Type R is an expanded-scale version of type A for precision range measurement. The types C, D, E, H, and I formats are modifications

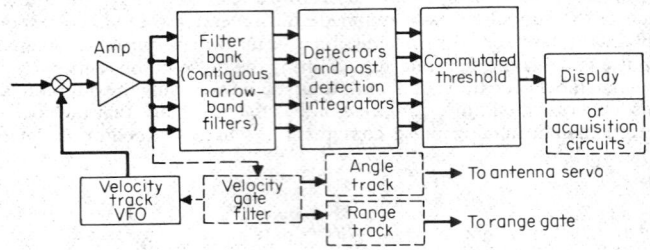

Fig. 25-87. Typical pulse doppler processor. *(Mooney and Skillman, Ref. 69.)*

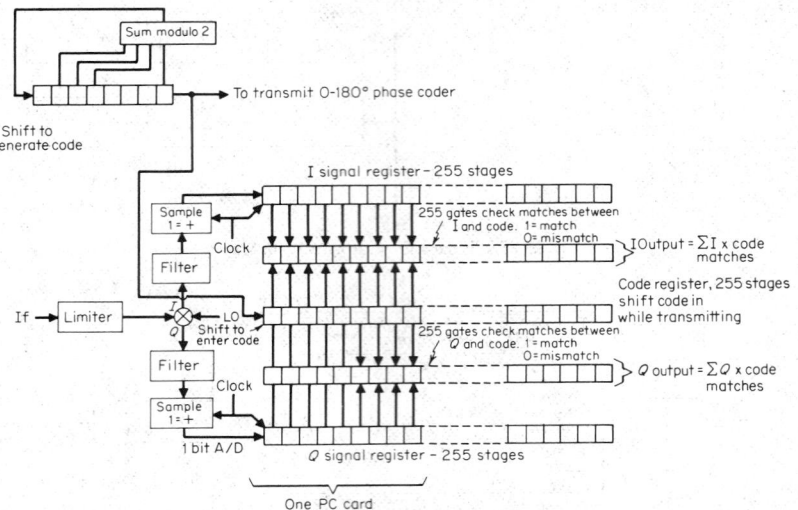

Fig. 25-88. Digital pulse compressor using 0 to 180° binary phase-coded modulation. *(Nathanson, Ref. 70.)*

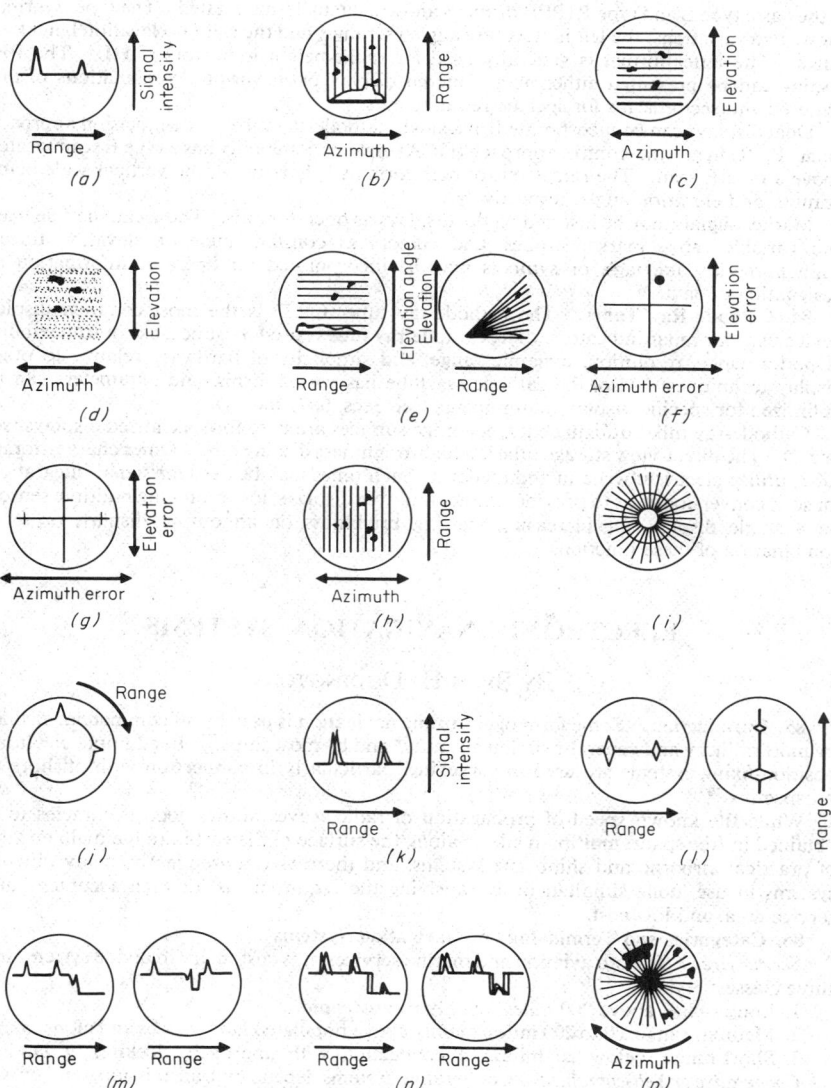

Fig. 25-89. Types of radar data presentation. *(Berg, Ref. 71.)* (*a*) A-presentation, (*b*) B-presentation, (*c*) C-presentation, (*d*) Coarse range information provided by position of signal in broad azimuth trace. (*e*, right) Range-height indication. (*f*) Single signal only; in the absence of a signal, the spot may be made to expand into a circle. (*g*) Single signal only; signal appears as wing spot, the position of which gives azimuth error and elevation error; length of wings is inversely proportional to range. (*h*) Signal appears as two dots; left dot gives range and azimuth of target; relative position of right dot gives rough indication of elevation. (*i*) Antenna is conical; signal appears as circle, the radius of which is proportional to the range; the brightest part of the circle indicates the direction from the axis of the cone to the target. (*j*) Same as type (*a*) except that the time base is circular and signals appear as radial deflections. (*k*) Type (*a*) with antenna switching; spread voltage displaces signals from two antennas; when pulses are of equal size, antenna assembly is directed toward target. (*l*) Same as type (*k*) but signals from two antennas are placed back to back. (*m*) Type (*a*) with range step or range notch; when pulse is aligned with step or notch, range can be read from dial or counter. (*n*) A combination of types (*k*) and (*m*). (*p*) Plan position: range is measured rapidly from the center.

of the basic type B and type P (PPI) displays and are intensity-modulated. The type E format shows two variations: the left is elevation angle vs. range, and the right is elevation height vs. range. The latter format is generally used for range-height indicators (RHI). The PPI display can be presented either as a centered display or in various combinations of off-centered and sector scans for specific requirements.

Dual displays can be incorporated on a single cathode-ray tube. The precision approach radar (PAR) in ground-control approach (GCA) systems commonly has a type B scan located above a type E scan. The range axis of both formats is horizontal, the vertical scale being azimuth and elevation angle, respectively.

Marker signals may be inserted on the displays as operator aids. These can include fixed and variable range marks, strobes, and cursors as constant-angle or elevation traces. Alphanumeric data, tags, or symbols may be incorporated for operator information or designation of data.

84. Cathode-Ray Tubes. The cathode-ray tube (CRT) is the most common display device used for radar indicators. The cathode-ray tube is used most because of its flexibility of performance, resolution, dynamic range, and simplicity of hardware relative to other display techniques. Also, the cathode-ray tube has varied forms, and parameters can be optimized for specific display requirements (see Secs. 6, 7, and 11).

Cathode-ray tubes utilizing charge-storage surfaces are used for specialized displays (see Sec. 7). The direct-view storage tube is a high-brightness display tube. Other charge-storage tubes utilize electrical write-in and readout. Such tubes may be used for signal integration, for scan conversion so as to provide simpler multiple displays, for combining multiple sensors on a single display, for increasing viewing brightness on an output display, or for a combination of these functions.

ELECTRONIC NAVIGATION SYSTEMS

By Sven H. Dodington

85. Introduction. Some form of electronic navigation is used by all commercial airlines, by most military and general-aviation aircraft,[76] and by most ships.[77] In addition, electronic position-fixing systems are used in surveying, particularly in connection with offshore oil prospecting.[78]

While the known speed of propagation of radio waves allows good accuracies to be obtained in free space, multipath effects along the surface of the earth are the main enemies of practical airborne and shipborne systems, and there are, consequently, many different systems in use, none simultaneously satisfying the requirements for high accuracy, large service area, and low cost.

86. Categories and Terminology of Navigation Systems

Service Area. In both aviation and marine services, it is customary to divide systems into three classes:

1. Long-range, above 200 miles, mainly transoceanic.
2. Medium-range, 20 to 200 miles, mainly above populated land masses and along coasts.
3. Short-range, below 20 miles, in connection with approach, docking or landing.

Cooperative or Self-contained. Cooperative systems depend on transmission, one- or two-way, between one or more ground stations and the vehicle. These are capable of providing the vehicle with a fix, independent of its previous position.

Self-contained systems are entirely contained in the vehicle and may be radiating or nonradiating. In general, they measure the distance traveled and have errors that increase with time or distance.

Data Rate. For map making, it may be acceptable to take a multiplicity of readings at one spot hours apart on different days, to achieve the highest possible redundancy and accuracy. At the other extreme, a fast-moving aircraft, during the approach and landing phase, may need positional updating at a rate of ten times a second.

Cost. Systems designed for small, privately owned boats or aircraft have vehicular equipment costs similar to those of household electrical appliances, while those designed for high-accuracy military-weapons delivery often have vehicular equipment costs in the hundreds of thousands of dollars. In cooperative systems, it is generally the object to maximize the ground equipment cost and minimize the vehicular equipment cost, for a given total cost.

Accuracy. For transoceanic navigation, errors of a few miles may be acceptable, while for docking and landing, the required accuracies are measured in feet.

Safety. It has long been standard design philosophy that when a navigation aid fails, it shall give *no* information rather than false information. Designs are therefore favored which, as far as possible, automatically fail safely. Where this cannot be achieved, elaborate monitoring and alarm circuits become mandatory. These latter are sometimes more costly than the device they monitor.[70]

Vehicle-derived or Ground-derived. While the majority of navigation aids produce a direct readout on the vehicle, an important exception is a class of aids in which the position of the vehicle is read out on the ground and the resulting information transmitted to the vehicle by ordinary communication (voice, telegraphy, or data link). While this results in low vehicular cost (since the communication link is already available), such systems have not found favor, in part due to complexity, time delay, division of responsibility, and allocation of cost. However, in earth-controlled space navigation it has, so far, been standard practice to track the vehicle from the ground and then to send various corrective instructions to the vehicle.

87. Standardizing Agencies. Since aircraft and ships may travel to any part of the world and, obviously, do not wish to carry a different equipment for each country, a large degree of international standardization is in effect for those navigation aids which depend on cooperation between a shore and a vehicular station. These standards, typically, take a decade to become established, and then remain in effect for several decades. Major agencies are:

International Civil Aviation Organization (ICAO, Montreal, Canada), an agency of the United Nations. Defines the signal characteristics, but not the hardware, for standard civil aviation systems, worldwide.

Federal Aviation Agency (FAA), Washington, D.C. Operates ground-based navigation aids and traffic control systems in the United States, for both civil and military aircraft.[80]

U. S. Coast Guard (USCG), Washington, D. C., operates shore-based navigation aids in the United States for shipping.

In each country, the military services operate additional navigation aids. Some of these are compatible with the international civil aids. For surveying, the systems are largely of a proprietary nature, each firm supplying the complete system.

88. Glossary of Navigation Terms.

Accelerometer. A device that senses the force per unit mass along a given axis, due to acceleration of vehicle.

Angle of cut. The angle at which two lines of position intersect, preferably a right angle.

Approach path. The portion of the flight path between start of descent and touchdown.

Azimuth. The angle in the horizontal plane with respect to a fixed reference, usually true North, measured clockwise (refer to definition of bearing below).

Back course. In ILS, the course located on the opposite end of the runway behind the localizer.

Base line. The line joining two points between which electrical time is compared. Large base lines produce high instrument accuracy but may also introduce instrument ambiguities.

Bearing. An angle in the horizontal plane with respect to a reference, usually expressed in degrees measured clockwise from the reference. *Relative* bearing is to some arbitrary reference; *absolute* bearing is to North—usually magnetic North in navigation systems.

Bend. A departure of a course line from a straight line. It is usually oscillatory and generally caused by interference between direct and multipath signals (refer to definition of scalloping below).

Bore sighting. The process of aligning a directional antenna system, often by optical means.

CEP. Circular error probability. In a two-dimensional error distribution, the radius of a circle encompassing half the errors.

Chain. A network of stations operating as a group.

Clearance sector. In ILS, the sector from the course to the back course, in which sector it is desirable to maintain the left-right needle off scale.

Coherent pulses. Pulses used in navigation systems in which the phase of the radio-frequency cycles within the pulse is retained for measurement purposes (as in Loran C, as contrasted with Loran A, which uses only the envelope).

Cone of ambiguity. In VOR and Tacan, the conical volume of airspace above the beacon in which bearing information is unreliable.

Cone of silence. The conical volume above an antenna where field strength is relatively low.

Course. The intended direction of travel. Also, the direction defined by a navigation aid.

Course-line computer. A vehicle-carried device which converts navigation signals into courses not generated directly by the signals themselves (for example, hyperbolic to straight-line, rho-theta to straight-line).

Course softening. Intentional decrease in course sensitivity as the navigation aid is approached.

CPE. Circular probable errors. Same as CEP.

Crab angle. Correction angle to compensate for wind drift. The angular difference between course and heading.

Dead reckoning. Determination of position at one time with respect to known position at a previous time, by the application of course and distance information derived without reference to external aids.

Decision gate. In ILS, the point at which the pilot must decide to land or to execute a missed-approach procedure.

Drift angle. Angle between heading and track, due to effect of wind or water currents.

Error, attitude. Varies with attitude of vehicle. Often related to polarization error.

Error, instrument. Error caused by the equipment itself.

Error, polarization. Varies with polarization of antenna at one or both ends of signal path. Often related to attitude error.

Error, propagation. Error caused by variations in the propagation medium.

Error, readout. Error caused by failure of navigator to read his instrument properly.

Error, site. Error caused by reflections from obstructions close to the site of the navigation aid. Of major concern in directional systems.

Fix. A position determined without reference to a former position.

Flag alarm. An indicator on a navigation instrument to show when a reading is unreliable.

Flare-out. That part of the approach path which rapidly decreases the glide angle by nosing up the aircraft at touchdown.

Geoid. The shape of the earth as defined by the hypothetical extension of mean sea level through all land masses.

Heading. The horizontal direction in which a vehicle is pointed with respect to a reference, often magnetic North, usually expressed in degrees, clockwise from the reference.

Homing. The process of approaching a desired point by directing the vehicle toward that point.

Instrument approach. An approach using navigation instruments rather than direct visual reference to the terrain.

LOP. Line of position. A line plotted on the earth's surface representing the locus of constant indication of navigational information (in VOR, straight radial lines; in DME, circles; in Decca, Loran, and Omega, hyperbolas).

Most probable position. A computed position based on several lines of position, all adjusted to a common time and weighted in accordance with their estimated probable errors.

Night effect. An error occurring mainly at night, when ionospheric reflection is at a maximum. A major limitation to the useful range of continuous-wave systems in the lf/mf bands.

Octantal error. A bearing error, usually due to departure of an antenna pattern from an ideal shape. It varies sinusoidally throughout the 360° and has four positive and four negative maxima.

Pitch. The angular displacement between the longitudinal axis of the vehicle and the horizontal.

Quadrantal error. An error in measured bearing, frequently due to antenna or goniometer characteristics, which varies sinusoidally throughout the 360° and has two positive and two negative maxima.

Reciprocal bearings. The opposite direction to a bearing (bearing ± 180°).

Rho-rho. A generic term for navigation systems that derive position by measurement of distance to two stations (DME/DME).

Rho-theta. A generic term for navigation systems that derive position by measurement of distance and bearing from a single station (VOR/DME, Tacan).

Roll. The angular displacement between the transverse axis of the vehicle and the horizontal.

RVR. Runway visual range. The forward distance visible along the runway during a landing approach.

Scalloping. Oscillatory-course bends occurring at a rate higher than can be followed by the vehicle.

Slant distance. Distance between two points not at the same elevation. Also called *slant range.*

Theta-theta. Generic term for navigation systems that derive position by measurement of bearing from two stations (VOR/VOR).

Track. Actual path traveled.

Yaw. Angular displacement between the normal axis of the vehicle and the course line.

89. Navigation Techniques.

Direction Finding. Direction finding (DF) is the oldest and most widely used form of navigation aid. The direction of a transmitter may be determined by comparing the arrival time of its transmission at two or more known points. In the simplest practical system, these two points are the vertical arms of a loop antenna connected to a receiver. As the loop is rotated, the received signals cancel each other when the plane of the loop is at right angles to the direction of the station. Two such nulls, 180° apart, exist if the loop is less than half a wavelength wide. At greater widths, additional nulls appear. These ambiguities may be resolved by use of additional antennas.[81]

The weakness of direction finding systems is their susceptibility to site errors (see Fig. 25-90). As shown, an obstruction is located close to the DF site in such a way as to reflect a signal having 90° error. When direct and reflected signals are equal and in phase, the resulting error is 45°. If, on the other hand, the receiving and transmitting stations were reversed, the error would be 5°. This error is therefore associated with the *site* of the DF antenna, and is particularly serious on ships and aircraft, where large reflecting objects (funnels, masts, vertical stabilizers, wings) are present.

The chief weapon against site error is the use of large DF antenna aperture. In most cases, a multiplicity of antennas, suitably combined, can be made to favor the direct path and discriminate against indirect paths. However, most vehicles cannot afford the space for such large antenna arrays. A reverse form of DF is therefore employed whereby the directional antenna system is at the transmitter and a simple omnidirectional antenna is at the receiver. The site error then occurs at the transmitter, but there is usually sufficient space there to allow

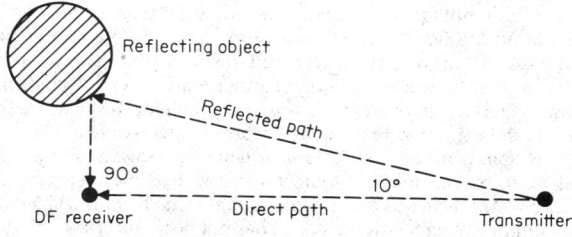

Fig. 25-90. Site error.

reduction by the use of large antenna aperture. This principle is used in instrument landing systems and in omnidirectional ranges, such as VOR and Tacan.

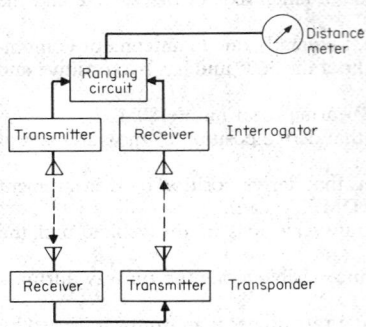

Fig. 25-91. Two-way distance ranging.

Two-Way Distance Ranging. For automatic distance measuring, it is customary to use a *transponder* on the desired target (Fig. 25-91). This device receives the interrogator pulse and replies to it with a much stronger pulse, usually on a different frequency. The distance in meters between interrogator and transponder is then equal to 150 times the elapsed round-trip time in microseconds (less any built-in fixed equipment delays). Various codes can be employed to limit responses to a single target or class of target. While pulse-type distance measuring is the most common, there are also cw systems in which received phase is compared with transmitted phase, ambiguities being solved by the use of different subcarriers and subsubcarriers. These latter can produce high instrumental accuracies, but usually with only one target at a time.

Differential Distance Ranging. Two-way ranging requires a transmitter at both ends of the link. To avoid carrying a transmitter on the vehicle, two transmitters are placed on the ground. One is a master, and the other a slave repeating the master, as shown in Fig. 25-92. The receiver measures the *difference* in the arrival of the two signals. For each time difference, there is a hyperbolic line of position, and such systems are therefore called *hyperbolic systems.* They may be either pulsed, using a common-carrier frequency, or continuous-wave (cw), using different (but related) carrier frequencies. At least two pairs of stations are needed to produce a fix.

One-Way Distance Ranging. If both stations are provided with stable, synchronized clocks, the distance between them may be established by a one-way transmission whose elapsed time is measured with reference to the two clocks (see Fig. 25-93). The

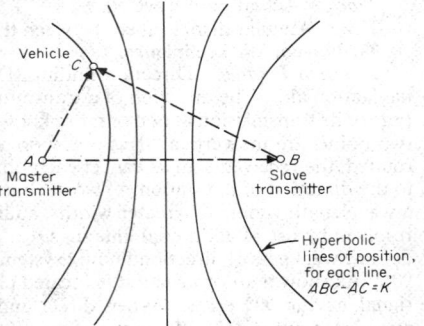

Fig. 25-92. Differential distance ranging (hyperbolic).

distance in meters is equal to 300 times the elapsed time in microseconds (less any built-in equipment delays). An error of approximately one mile per elapsed hour occurs for a frequency difference between the clocks of one part in 10^9, and so this system is little used, except where there are external means for periodically synchronizing the clocks.

90. Radar Navigation Systems.

Ground-based. Ground-based radar navigation stations are positioned along the airways and at the entrance to harbors. In the case of the airways, they are primarily for the purpose of air traffic control[82] and not for navigation, although aircraft are occasionally "vectored" by command from the air traffic controller, the latter, in effect, temporarily navigating the aircraft. In the case of harbors, the radar PPI display is sometimes televised to ships, allowing them to see themselves in relation to other traffic, even though they do not have radar of their own. This has also frequently been proposed for the airways, but has not found favor due in part to the extensive interpretation problems involved.

Ship-based.[77] Radar is a very valuable tool when it allows mapping of coastlines and detection of other ships; in those circumstances it becomes a major navigation aid.

Airborne Radar.[83] While it was at one time thought that airborne ground-mapping radar would become an important navigation aid, this has not been the case in civil aviation, largely due to problems of interpretation. Instead, civil airborne radar has been limited to three major areas: (*a*) Nose-mounted *weather radar,* which detects the presence of thunderheads and is mandatory on airliners. (*b*) *altimeters*, which read distance to the terrain below the aircraft. These are mainly used to improve the flare-out characteristic during landing. They operate in 4,200-MHz band, either with frequency-modulated cw signals at less than 1 W or with pulses of about 100 W peak power. (*c*) *Doppler radar*,[84] which uses relatively low-power transmissions that are reflected from the ground and return to the aircraft with a frequency difference proportional to the speed of the aircraft. Typical frequencies are in the 10- to 20-Ghz region, with power levels below 1 W, available from solid-state (transistorized) equipment. Integration of speed gives distance traveled, to an accuracy of better than 1%. However, the direction of travel must be determined by other means. The system is much used by transoceanic airlines and is substantially less costly than inertial systems.

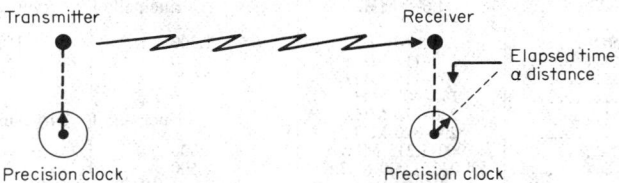

Fig. 25-93.   One-way distance ranging.

91. Sonar.[85,86]   Acoustic waves travel in water at about 1,500 m/s and are attenuated much less than in air. They have consequently long been regarded as offering the possibility of radarlike navigation aids. However, development has lagged behind corresponding radar activity. The most common civil application is for depth sounding, equivalent to the radar altimeter. This is of navigational value when approaching harbors whose depths have been previously charted. Doppler sonar has also been developed to give ships the same benefits as are provided to aircraft by doppler radar, at least in relatively shallow water. In all sonar work, accuracy is degraded by changes in speed of propagation (about $\pm 3\%$), caused by change in water temperature and salinity. To a limited extent, bottom-mounted acoustic transponders have also been employed, to give similar results to radio transponders used in two-way distance ranging. Sonar frequencies are generally in the acoustic to kilohertz range.

92. Other Systems.

Inertial Navigation.[87]   This is the most widely used "self-contained" technology. Accelerometers constantly sense the vehicle's movements and convert these, by double integration, into distance traveled. To reduce errors caused by vehicle attitude, the accelerometers are mounted on a gyroscopically controlled stable platform. Accuracy is limited by gyroscopic drift and is typically a few miles per elapsed hour. Since this is the most expensive vehicular navigation technology, it is used only by military aircraft and transoceanic airlines.

Celestial Observation.[88]   Position fixing with reference to the stars has been standard marine practice for centuries and has been aided in recent years by electronic star trackers and by vehicular computers. Ships, however, still have to cope with a weather problem when using this technique, and while high-flying aircraft are less troubled by this problem, the considerable human judgment needed has limited the use of this technology.

Earth Satellites.[89]   The limitation of celestial observation by the weather has led to consideration of artificial satellites using radio, and one such system has seen several years of operational use. The chief stumbling blocks have not been technical, but political, concerning such questions as what agency provides and maintains the satellite system.

93. Radio Frequencies Used in Electronic Navigation.   Virtually all radio frequencies have been used in navigation systems. At the low-frequency end, systems are limited by the massive antenna systems needed and by low data rates. Above 10 GHz there are limits on the amount of rainfall that can be penetrated.[15]   With few exceptions, frequencies and technologies have been chosen to avoid dependence on ionospheric reflection; while such reflection has been of much value in communication, it is deliberately used only in navigation systems where its effects are well understood and predictable. Table 25-15 lists the principal frequency bands used for navigation.

94. Internationally Standardized Systems.   The systems described in Pars. 25-95 to 25-107 are used by hundreds of thousands of vehicles, throughout the world. Standardization refers principally to the radiated signal characteristics. Each country, manufacturer, or user

Table 25-15. Radio Frequencies Used in Electronic Navigation

System	Frequency-band	No. of stations	No. of vehicles
Omega	10–13 kHz	8	1,000
Decca	70–130 kHz	150	20,000
Loran–C/D	100 kHz	30	1,000
Lf range	200–400 kHz	70	1,000
ADF/NDB*	200–1,600 kHz	2,000	106,000
Coastal DF*	285–325 kHz	1,000	100,000
Consol	250–350 kHz	15	5,000
Loran A	2 MHz	100	10,000
Hi-Fix			
Raydist	1.6–5 MHz	Principally for surveying	
Toran			
Marker beacon*	75 MHz	1,500	150,000
ILS localizer*	108–112 MHz	600	150,000
VOR*	108–118 MHz	2,000	150,000
Transit	150, 400 MHz	6	100
Trident	230–270 MHz	Principally for surveying	
Shoran	290–320 MHz		
DME, Tacan*	960–1,215 MHz	2,000	70,000
ATCRBS*	1,030, 1,090 MHz	500	80,000
CAS	1,600 MHz	In development	
Altimeter	4,200 MHz	...	5,000
Talking beacons	9 GHz	3	1,000
MLS	5, 15 GHz	In development	
Weather radar	5, 9 GHz	...	10,000
Doppler radar	10–20 GHz	...	5,000
Hydrodist	36 GHz	Principally for surveying	

* Internationally standardized systems,

decides individually on the detailed equipment design which best fits its needs, market, and resources. These systems have been in use for at least two decades and can be expected to remain in use for at least two more decades.

95. Direction Finding.

Shipboard DF and Coastal Beacons.[91] While the more than 10,000 broadcasting stations in the world can be used as "targets" for a shipboard direction finder, special coastal beacons operate in the 285- to 325-kHz band, specifically for the benefit of ships. This frequency band provides ground-wave coverage over seawater to about 1,000 miles, and the assigned channels are such that interference between beacons is minimized out to that distance. The beacons also transmit notices of interest to mariners. Power varies from 100 W up to 10 kW.

The simplest shipboard antenna consists of a loop about 2 ft in diameter which can be rotated around its vertical axis, by hand (Fig. 25-94a). It is connected to a receiver whose output shows a sharp null when the plane of the loop is at right angles to the direction of the

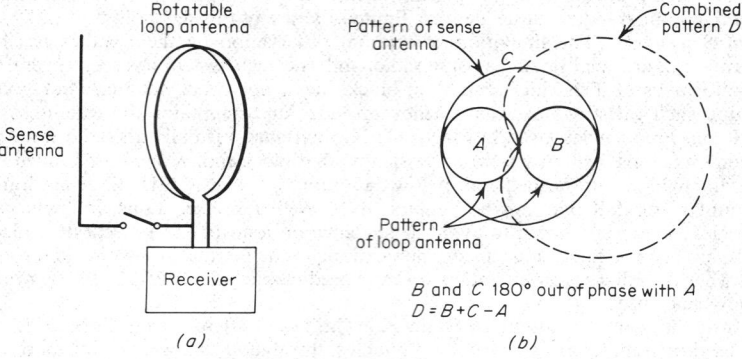

Fig. 25-94. Sense antenna applied to direction finding. (*a*) Antennas; (*b*) patterns.

transmitter. If the resulting 180° ambiguity causes doubt (it rarely does), a "sense" antenna is switched in; in conjunction with the loop, this generates a cardioid pattern (Fig. 25-94b) having a single null, but less sharp.

The chief disadvantages of this system are site errors caused by the ship's superstructure, ionospheric reflections (sky waves) which may not always travel along the vertical plane joining the transmitter and the receiver, atmospheric noise, and polarization errors. The latter occur when the polarization of the arriving signal is not precisely vertical. The relatively slow speed of ships, and the relatively long intergration time that can be employed, allow a well-designed DF system to show an accuracy of about ±2° under typical conditions.[92]

While the single rotating loop is the simplest system, most large ships employ two fixed loops, at right angles to each other, with a rotatable goniometer below decks feeding their outputs to the receiver. More recent designs use digitally controlled electronic goniometers and digital displays. While these ease the task of the operator, they do little to alleviate the propagation effects previously listed.

Airborne ADF and Ground NDB.[93] Early aircraft used the same system as described for ships. Since, on most aircraft, the navigator is also the pilot, a greater degree of automation became necessary, culminating around 1935 in the *automatic direction finder,* abbreviated ADF. Typically two crossed ferrite-core loops are used, projecting about 1 in. from the skin of the aircraft, with an electronic goniometer generating a direct-reading display of bearing to the selected transmitter.

The tuning range is usually 200 to 1,600 kHz, taking in not only the broadcast stations and maritime coastwise beacons, but also specially designed inland beacons operating from 200 to 1,600 kHz. These are known as *nondirectional beacons,* abbreviated NDB. They radiate from 10 W to 2 kW, into a vertical radiator usually at least 100 ft high. An important feature of these beacons is the sharp reduction in signal strength obtained directly over the beacon due to vertical polarization of both transmitting and receiving antennas. This effect is frequently used to obtain a position fix.

The low ground-wave attenuation occurring over the seawater is not present over land. Depending on soil conditions, the effective ground-wave range may be only a few hundred miles, while site error, sky-wave interference, and atmospheric and polarization errors are as serious as in the shipboard case.[94] Consequently, accuracy is usually not better than ±5°, and the useful range not much above 200 miles at 200 kHz and 50 miles at 1,600 kHz. This range may be reduced to zero during heavy atmospherics. Nevertheless, the ADF navigation aid is one of the most widely used of all airborne electronic aids, and NDBs are to be found in practically every country of the world.

96. Instrument Landing System (ILS).[95] Early work on instrument landing in Germany and England prior to World War II was consolidated into a standard U.S. system in 1942 and became an ICAO standard in 1947. The ground equipment is made up of three separate elements: the *localizer,* giving left-right guidance; the *glide slope,* giving up-down guidance; and the *marker beacons,* which define progress along the approach course. Using line-of-sight frequencies, ILS is free of atmospherics and sky-wave effects, but, it is still much subject to site effects.

The localizer operates on 40 channels spaced 50 kHz apart, 108 to 112 mHz, radiating two antenna patterns which give an equisignal course on the centerline of the runway, the transmitter being located at the far end of the runway. The left-hand pattern is amplitude-modulated by 90 Hz, the right-hand pattern by 150 Hz. The airborne receiver detects these tones, rectifies them, and presents a left-right display on a zero-center dc meter in the cockpit. The accuracy is better than ± 0.1°.

Minimum ICAO performance calls for the airborne meter to remain hard left or hard right to a minimum of ± 35° from the centerline; i.e., there must be no ambiguous or "false" courses within this region (Fig. 25-95a). More sophisticated systems exist in which a usable "back course" (with reverse sense) is obtained, and a separate transmitter, offset by about 10 kHz, provides "clearance" (Fig. 25-95b), so that no ambiguities exist throughout ± 180°. Total rf power is of the order of 25 W. The localizer may be voice-modulated, but this feature is seldom used.

The glide-slope transmitter, of about 7 W power, is located at the approach end of the runway and up to about 500 ft to the side. It operates in the 330- to 335-mHz band, each channel being paired with a localizer channel. In the airborne receiver, both channels are selected by the same control. Two antenna patterns are radiated, giving an equisignal course about 3° above the horizontal. The lower pattern is modulated with 150 Hz, and the upper pattern by 90 Hz (Fig. 25-95c). The airborne receiver filters these tones, rectifies them, and presents the output on a horizontal zero-centered meter mounted in the same instrument case as the localizer display, the two together being called a *cross-pointer display*. When one needle is horizontal and the other vertical, the aircraft is exactly on course. The accuracy is better than ±0.1°.

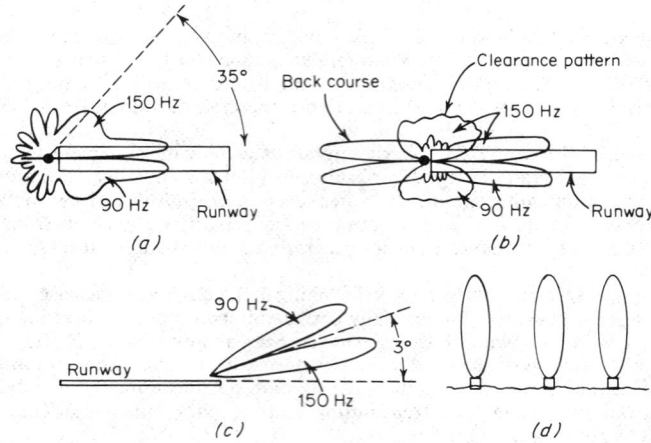

Fig. 25-95. Instrument landing system. (*a*) Minimum ICAO localizer pattern; (*b*) localizer pattern with back course and clearance; (*c*) glide-slope pattern; (*d*) marker beacons.

In this frequency band it has not been found possible to generate the required antenna patterns without use of either an excessively tall array (100 ft or more, which would be dangerous to aircraft) or of deliberate ground reflection. The glide slope therfore suffers from course bends due to the terrain in front of the array, and is generally not depended on below 50 ft of altitude.

Marker beacons operate at a fixed frequency of 75 MHz, and radiate about 2 W upward toward the sky with a fan-beam antenna pattern whose major axis is across the direction of flight (Fig. 25-95d). At each ILS installation there is an"outer" marker about 5 miles from touchdown, and a "middle" marker about 3,500 ft from touchdown. A few runways also have "inner" markers just before touchdown. Each type is modulated by audio tones which are easily recognized as the aircraft passes through their antenna pattern. Alternatively, differently colored lamps are set to light in the cockpit as each marker is passed.

At one time, marker beacons were also used on the U.S. airways for enroute flying, but this use is gradually being phased out. Eventually, marker beacons may be replaced by DME, particularly for overwater approaches.

ICAO has established categories of *ILS performance*, dependent on the quality of the installation and the qualifications of the air crew. These place minimum limits, as follows, on how close an aircraft may approach the touchdown point:

Category I. 200-ft ceiling and 1/2-mile visibility. About 600 runways in the world.

Category II. 100-ft ceiling and 1/4-mile visibility. Less than 100 runways in the world.

Category III. Zero ceiling and 700-ft visibility. As of 1974, only a half dozen runways in the world.

ILS installations are usually dual (except antenna), with elaborate monitoring and immediate changeover to the standby unit when the course is found to deviate excessively. If the standby unit also fails, the facility is shut down.

97. VOR System. The very high frequency omnidirectional range, often called Omni-Range or just Omni, was standardized in the United States in 1946 and became an ICAO standard in 1949 for en route flying. Like ILS, it uses line-of-sight frequencies and and is thus free of atmospherics and sky-wave distortions.   Also, like ILS, it places the directional burden on the ground, rather than in the aircraft, where more extensive means can be employed to alleviate site errors.   Line-of-sight limits its service area to about 200 miles for high-flying aircraft, and some stations are intended for only 25-mile service to low-flying aircraft.   There are more than 1,000 stations in the United States and about an equal number in the rest of the Western world.   There are three variations: conventional, doppler (which provides increased site-error reduction), and "precision" (which requires modified airborne equipment).

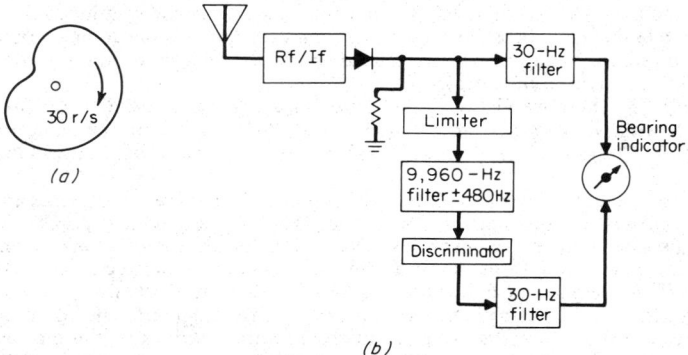

Fig. 25-96.   VOR system.   (*a*) Conventional VOR antenna pattern; (*b*) receiver for conventional and doppler VOR.

Conventional VOR[96]   operates on 40 channels, 50 kHz apart, between 108 and 112 MHz (interleaved between ILS localizer channels), and on 120 channels, spaced 50 kHz, between 112 and 118MHz.   The airborne receiver is frequently common with the airborne localizer receiver, and may use the same airborne antenna.   Power output from the ground transmitter varies from 25 to 200 W, depending on antenna design and on the desired service area.

The ground-antenna pattern forms a cardioid (Fig. 25-96*a*) in the horizontal plane which is rotated 30 times per second.   The cw transmission is amplitude-modulated by a 9,960-Hz tone which is frequency-modulated ±480 Hz at a rate of 30 Hz.   This latter 30-Hz "reference" tone, when extracted in the airborne receiver, is compared with the 30-Hz amplitude modulation provided by the rotating antenna.   The phase angle between these two 30-Hz tones is the bearing of the aircraft, with respect to North (Fig. 25-96*b*).

Instrument accuracy of the complete system is of the order of ±1°.   However, site errors may often degrade this to ±3° or worse.   Generally speaking, a sector is declared "unflyable" if errors exceed ±4.5°.

Doppler VOR[97]   reduces site error about tenfold by using a large-diameter antenna array at the ground station.   This array consitutes a 44-ft-diameter circle of antennas (52 in the U.S. version, 39 in the German version).   Each antenna is sequentially connected to the transmitter in a manner to simulate the rotation of a single antenna around a 44-ft-diameter circle at 30 r/s.   The receiver sees an apparant doppler shift in the received rf of ±480 Hz at a 30 r/s rate, and at a phase angle proportional to the receiver's bearing with respect to North.   This signal is therefore identical with the conventional VOR reference tone.   It remains merely to transmit at 30-Hz AM tone, separated 9960 Hz, as the doppler reference, to radiate an identical signal to the conventional one, receivable in an identical receiver but benefiting from a tenfold increase in ground-antenna aperture.

Since the doppler VOR costs substantially more than the conventional VOR, it is used only at sites that cannot be properly served by the conventional VOR.   In the United States this currently amounts to about 3% of the sites, with plans to reach about 10%.   In some other countries, the doppler system is being installed at all new sites, its extra cost being somewhat compensated by lower site cost.   (The conventional VOR needs a flat, cleared area of at least 1,500-ft radius and even then can be seriously disturbed by buildings a mile or more away.)

Precision VOR[98]   is aimed at improving the performance of airborne receivers using the doppler VOR and requires a relatively minor receiver modification. In the doppler VOR, the reference consists of an AM 30-Hz modulation of a carrier separated 9,960 Hz from the doppler carrier. This is to allow complete compatibility between conventional and doppler receivers. This 30-Hz modulation is sometimes cross-modulated in the receiver by the doppler signal, leading to cyclical bearing errors. In the "precision" system, therefore, the 30-Hz reference is also transmitted as a frequency modulation of a 6.5kHz subcarrier which can be limited, prior to detection, to remove cross-modulation. The overall accuracy, with this feature, is on the order of ±0.25°. The system was still in evaluation in 1974.

All three kinds of VOR can be voice-modulated, though little use of this is made except for identification purposes. The VOR presentation in the cockpit is usually on a radiomagnetic indicator (RMI), on which the angle between needle and rotating card is the VOR bearing, while the angle between rotating card and the frame is the compass output. The needle thus represents the angle between heading and station if the magnetic deviation is the same at the station as at the aircraft.

98. DME[98], the abbreviation for **distance measuring equipment**, is an interrogator-transponder two-way distance-ranging system (Fig. 25-91). It became an ICAO standard in 1959. Some 2,000 ground stations and 70,000 airborne equipments are in use in the Western World.

The airborne interrogator transmits 1-KW pulses of 3.5 μs duration, 30 times a second, on one of 126 channels 1 MHz apart, 1,025 to 1,150 MHz. The ground transponder replies with similar pulses on another channel 63 MHz above or below the interrogating channel. (This allows both equipments to use the transmitter frequency as the receiver local-oscillator frequency if the intermediate frequency is 63 MHz.) All channels are crystal-controlled, and a single antenna is used at both ends of the link. To reduce interference from other pulse systems (e.g., ATCRBS) and to allow addition of future functions by pulse coding, paired pulses are used in both directions, their spacing being 12, 30, or 36 μs. The fixed delay in the ground transponder is 50 μs.

In the airborne set, the received signal is compared with the transmitted signal, their time difference derived, and a direct digital reading of miles is displayed. The typical accuracy is ±0.2 mile. Ground transponders are arranged to handle interrogation from up to 100 aircraft simultaneously, each aircraft recognizing the replies to its own interrogation by virtue of the pulse repetition frequency being identical with the interrogation. Early analog models required some 20 s for this identity to be initially established (after which a continuous display was provided), but newer digital models perform the search function in less than a second.

In ICAO practice, the DME is nearly always associated with a VOR, the two systems forming the basis for a rho-theta area navigation system. Eighty DME frequencies are paired with 160 VOR channels, the same selector in the cockpit operating both sets.

DME is particularly immune to site and propagation errors, and better accuracy is readily obtainable if required.

99. TACAN[99,100] is a NATO military system which adds a bearing function to DME, on the same frequencies, allowing greater portability of the ground station than is the case with ICAO's VOR/DME, particularly on aircraft carriers.

A standard ICAO DME beacon is arranged to operate at "constant duty cycle"; i.e., when 100 sets of interrogations are lacking, the gain of its receiver is increased until an equivalent set of "squitter" pulses (from receiver noise) is generated. The DME antenna is then replaced by rotating directional antenna generating two superimposed patterns (Fig. 25-97). One of these is a cardioid, as in VOR, but rotating at 15 Hz. The other is a nine-lobe pattern, also rotating at 15 Hz. The squitter pulses and replies are amplitude-modulated as the antenna rotates. Reference pulses are transmitted at 15 and 135 Hz.

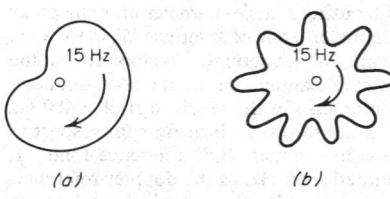

(a)   (b)

Fig. 25-97. Tacan antenna pattern.

In the aircraft, a "coarse" phase comparison can then be made at 15 Hz, supplemented by a "fine" comparison at 135 Hz, the overall instrumental accuracy being on the order of ±0.2°. Since the DME frequency is about ten times greater than the VOR frequency, the antenna system is about ten times smaller than that of the VOR, and thus more portable.

A major problem in Tacan has been the design of sufficiently good vertical directivity in the ground antenna to preclude the generation of vertical nulls by ground reflections.

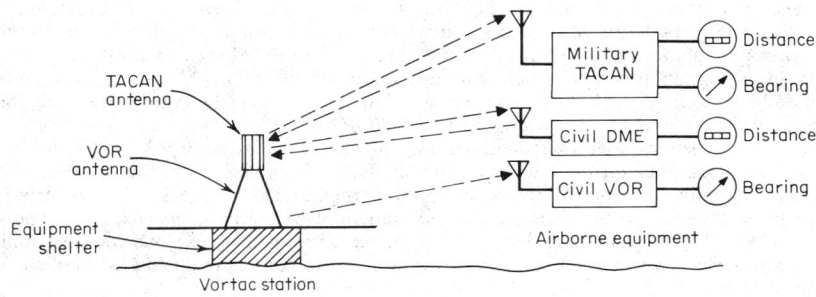

Fig. 25-98. Vortac system.

100. Vortac. In NATO countries having a common air traffic control system for the civil and the military (e.g., the United States, Germany), the ICAO rho-theta system is implemented by the use of Tacan rather than DME. Tacan transponders are colocated with VOR stations, and civil aircraft get their CME service from the Tacan station, as shown in Fig. 25-98. In the United States, about 700 VORs have colocated Tacan transponders. About 40,000 airborne Tacan sets have been built.

101. IFF (Identification of Friend from Foe) Systems. To distinguish friend from foe,

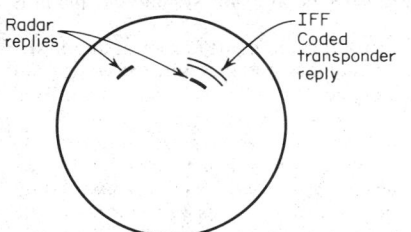

Fig. 25-99. PPI display with IFF.

early radars employed an interrogator-transponder system operating at a different set of frequencies, the "friend" being transponder-equipped and the "foe" not. The interrogator was pulsed at about the same time as the radar, and the airborne transponder produced coded replies shortly after the direct radar reply from the aircraft skin, as shown in Fig. 25-99. In theory, even if the foe used the same transponder equipment, he would not know the code of the day. This identification of friend from foe system became known as IFF, and after World War II, became stabilized among NATO countries as a system in which all interrogation takes place at 1,030 MHz and all replies at 1,090 MHz. Typical pulse powers are 500 W, with 1-μs length for interrogation and 0.5-μs length for reply. The ground antennas are typically 20 ft long and rotate at 15 r/min.

102. ATCRB System. In 1958 IFF became an ICAO standard, known as *SSR* (secondary surveillance radar) or *ATCRBS* (air traffic control radar beacon system). It is used by air traffic control authorities to track aircraft and is not primarily intended for navigation. In the United States, there are over 300 ATCRBS interrogators and 70,000 airborne equipments.

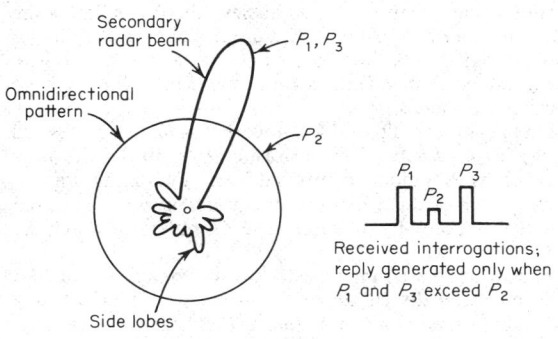

Fig. 25-100. Side-lobe suppression.

Paired pulses are used for interrogation, and a third pulse between the two is radiated omnidirectionally to reduce triggering by side lobes (Fig. 25-100). The airborne transponder replies only when the directional pulses are stronger than the omnidirectional pulse. The reply pulses comprise a train of up to 14 pulses, lasting 21 μs, which can be combined into 4,094 codes. These can be used to identify the aircraft or to communicate its altitude to the ground controller (height-finding radar has been found impractical for this purpose).

The chief problem besetting the ATCRBS system is the interference ("garbling") which occurs when two or more aircraft are at about the same azimuth and distance from the interrogator. To alleviate this effect, the FAA has plans in the United States to institute a system of interrogation coding which will allow each aircraft to be addressed by a discrete code, and thus only "speak when spoken to." This system will be compatible with the present ATCRBS, to allow an orderly transition.[101]

103. Mechanical vs. Electronic modulation. Most of the systems just described depend in one way or another on the amplitude modulation of a carrier or on the rotation of a directional antenna pattern. While these objectives might have been reached by electronic means, by far the majority of existing ground equipments use *mechanical* means; and it is only in the 1970-1971 time frame that electronic means have received any serious consideration. Since these systems will probably be in use for decades to come, the distinction is of some importance.

104. ILS Modulation. For both localizer and glide slope, it is necessary to modulate the carrier with 90- and 150-Hz sine waves. More than 90% of the world's installations use the mechanical system. Here, the carrier is generated in a single source and then split into two equal parts by an Alford bridge. The main feature of this bridge, which has a 180° phase shift in one arm, is that energy entering any one "corner" does not appear at the opposite "corner." Thus the two carrier portions may now be separately modulated without effect on each other. This modulation is performed by two "paddlewheels," one with three paddles and one with five, both rotated at 30 r/s by a constant-speed motor. The two signals are then combined in another Alford bridge to give the signals needed by the antenna system, carrier plus sidebands and sidebands only.

The claimed advantages for this system are that variations in the rf generator and in the motor speed affect the 90- and the 150-Hz components equally, and therefore the course can be shifted only by variations in the modulators, which, being mechanical, can remain fixed for years at a time.

105. VOR Modulation. This calls for a rotating cardioid pattern, and one way to generate this has been by a mechanically rotating dipole—which generates a figure-eight pattern—superimposed on an omnidirectional pattern. On the same shaft as the rotating dipole is also mounted the 30-Hz reference generator. A variation uses a fixed array of the four loop antennas, but feeds them from a mechanically rotating goniometer whose shaft also drives the reference generator.

In both cases, phase lock between the two 30-Hz signals is maintained because they both are generated by the same motor shaft.

These systems, used worldwide for some twenty years, were challenged in 1969 by a 100% solid-state alternative[96] in which all 30-Hz modulation is by electronic means. This is done at low level, to preserve linearity of modulation; the sidebands are then separated from the carrier and from each other, amplified in class C power stages, and recombined and fed to a pair of fixed crossed dipoles, to generate a rotating figure eight. As in the case of ILS, the result is lower first cost, less power consumption, and easier maintenance.

106. Tacan Modulation. This calls for a cardioid superimposed on a nine-lobe pattern, both rotating at 15 Hz. All operational stations in the world achieve this by radiating the pulsed carrier omnidirectionally and then mechanically rotating a series of parasitic dipoles around the central radiator. The cardioid is generated by a single parasite at about 3 in. radius, and the nine-lobe pattern is generated by nine parasites at about 15 in. radius. On the same shaft are generators for the reference signals.

No operational electronic version had appeared by 1974, but work was underway toward several solutions. One of these uses a large number of fixed parasites (over 100 for the nine-lobe array) and switches them on and off by PIN diodes controlled from a multiphase square-wave generator. By suitable programming of this generator, the granularity of the on-off modulation is held to a level below that imposed by the discrete DME pulses on which the Tacan bearing signal is modulated.

107. ATCRBS Modulation. The ground-based interrogator for this system uses a 20-ft directional antenna which rotates mechanically at about 15 r/min. This speed is a compromise; if it were higher, to give a higher data rate, the antenna would have to be smaller, have less aperture, and provide less azimuth resolution.

It has therefore been proposed that a large fixed array be subsituted. This would comprise a vertical cylinder, at least 30 ft in diameter, on whose surface would be mounted a large number of electronically switched antennas, to give a narrower beam which yet could move almost instantaneously. Such a beam could move quickly between targets, yet stop when it found a target.

108. Nonstandard Cooperative Systems. The following systems have existed, at least in developmental form, for a decade or more, and some have seen extensive use. They have failed to win international standardization for a variety of reasons.

109. Ground-Based Direction Finding.[92]   The site-error limitations of vehicular direction finding (DF) led to the concept of placing the direction-finding station at a good site on the ground, taking bearings on the vehicle's transmitter, and then communicating the result to the vehicle. As recently as 1961, this concept was being seriously proposed for a national air-navigation system in one country, aircraft position to be obtained by triangulation from a multiplicity of ground stations connected by telephone lines. The system obviously has great economy as far as the vehicle is concerned. The chief disadvantages appear to be low data rate and relatively expensive ground operations, including much manpower. The principle is still in use, but mainly as an aid in the location and rescue of vehicles whose other navigational functions have been lost.

Major technologies used are as follows:

LF-MF. The "crossed-loop" principle, but using vertical elements only, spaced several hundreds yards apart, in a so-called *Adcock array.*[102]

HF. The *Wullenweber system,* introduced in Germany during World War II. Here, a 1,000-ft-diameter circle of antennas feeds a rotating commutator which connects each antenna to an appropriate delay line so that all signals coming from a given direction reach the receiver with the same phase, while those from other directions arrive with random phase. A British version, known a CADF (for *c*ommutated *a*ntenna *DF*), operates in a similar manner.[103]   Instrumental accuracy is better than 1°, but is adversely affected by the errors caused by ionospheric reflection. Systems of this type are in use by military and intelligence agencies for nonnavigational purposes.

VHF-UHF. Here the doppler principle, used in the doppler VOR, is most effective. A circle of antennas, about 10 ft in diameter, is commutated in such a manner that the receiver sees, in effect, a single antenna rotating around that circle. As the receiver moves away from the signal, the frequency appears to be lowered, and as it moves toward the signal, the frequency appears to be raised. An FM discriminator at the output of the receiver delivers a sinewave whose phase depends on the direction of the transmitter. The display is usually on a cathode-ray tube with circular scan, radial lines being produced in the direction of received signals. The FAA has about 200 such installations.

The Wullenweber, CADF, and doppler systems are basically schemes for using a wide antenna aperture, without ambiguity, and thereby reducing site errors.

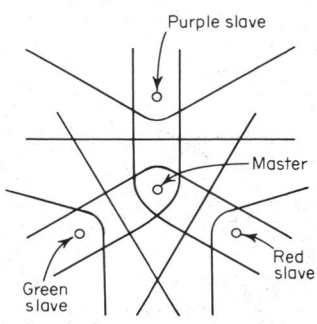

Purple slave

Master

Red slave

Green slave

Fig. 25-101. Decca chain.

110. LF Four-course Range.[93]   This system was introduced in the United States in 1927, reached its zenith during World War II, and was overtaken by the VOR before it could achieve international standardization. Since about 1950, it has gradually been phased out of existence, one of the very few radio navigation aids to have suffered this fate. Four 125-ft towers are arranged in a 400-ft square, and each diagonal pair fed with about 200 W in the 200- to 400-kHz band. The rf is modulated with a 1,020-Hz tone, which is keyed on a diagonally located pair of antennas with the Morse code letter A and, on the other pair, with the letter N, the phases of these Morse codes being such that, when they are received at equal signal strength, a constant 1,020-Hz tone results. Four equisignal courses are thus generated, each course being characterized by the reception of A on one side, N on the other side, and a pure tone on course. The system is frequently referred to as the *A-N range.*

Aside from the disadvantage of only four courses, the LF range suffers from sky-wave reflections and atmospherics, limiting its service range to no more than is obtained with line-of-sight systems. At its peak, there were about 300 stations in the United States, Canada, and Mexico. As of 1975, fewer than 50 remained in use.

111. Decca.[103] This system has been developed in the United Kingdom starting in 1939. It is a continuous-wave hyperbolic system operating in the 70- to 130-kHz band. A typical chain comprises four stations, one master and three slaves, separated about 70 mi, arranged as in Fig. 25-101. Each station is fed with a signal which is an accurately phase-controlled multiple of a base frequency f, in the 14-kHz region, the master at $6f$, the "red" slave at $8f$, the "green" slave at $9f$, and the "purple" slave at $5f$. At the receiver, these four signals are received, multiplied, and phase-compared, as shown in Fig. 25-102. Each of the three phase meters, called *decometers*, thus provides a position along a hyperbolic line, and by plotting these positions on a map, a unique fix is obtained. Considerable ambiguity exists with this system, since equiphase hyperbolic lines are obtained as close as 1 mi apart. Additional complexity is added to resolve the ambiguity.

There are about 25 Decca chains in Europe and about 10 elsewhere in the world. The system is used by about 20,000 ships, half of which are fishing vessels.[104] Accuracy varies from under 100 yd on a summer day, close to the chain, to several miles on a winter night 200 mi from the chain. The chief advantage of Decca is that it is non-line-of-sight and that the simplest receiver is relatively inexpensive. Disadvantages are those common to other cw lf aids: sky-wave propagation and atmospherics. From time to time, attempts have been made to justify Decca for airborne applications. These have not been successful, a major difficulty being the high cost of converting the hyperbolic readouts to information directly usable by an aircraft pilot.

112. Loran A.[93] The original Loran system (Loran A) was developed in the United States during World War II and has since been maintained by the U.S. Coast Guard as an aid to ships and aircraft. The name is an abbreviation of *long-range navigation*. It is a hyperbolic system operating around 2 MHz and using pulses, to allow discrimination against sky waves. Receivers use a cathode-ray-tube display and the earlier arrival of the ground wave can readily be observed.

A chain normally comprises a master and two slaves, spaced 200 mi on each side of the master along a coastline, for the benefit of ships and aircraft crossing the ocean. Discrimination between chains is by use of four radio frequencies, 50 kHz apart, and 24 sets of pulse repetition frequencies in the 20 to 35 pulses/s region. The pulse rise time is 21 μs.

The peak power is about 100 kW into an antenna about 100 ft high. At the receiver, the pulse positions are matched on a cathode-ray oscilloscope to an accuracy of about 1 μs, and navigational accuracy (which depends on the angle of cut of the hyperbolic lines), is typically 1,000 ft between the stations and about 1/2 mi at the extreme ground-wave range.

There are 25 chains in the North Atlantic and North Pacific, and no further expansion is

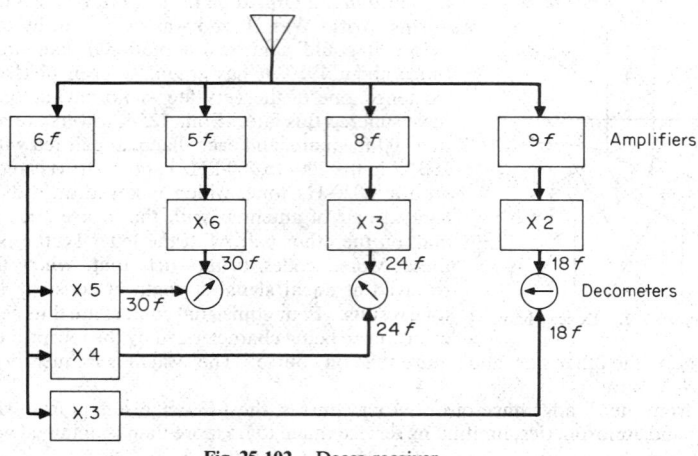

Fig. 25-102. Decca receiver.

planned. Users include merchant ships and transoceanic airliners which carry Loran A receivers as backups to their doppler radar or inertial navigators. Coverage extends to about 600 mi, leaving some gaps in transoceanic coverage.[105]

113. Loran C.[106] Loran C is a U.S. military hyperbolic system, with most of the fixed chains maintained by the Coast Guard. It uses pulses at a fixed frequency of 100 kHz and, unlike Loran A, matches the individual rf cycles within each pulse, thereby gaining added resolution and accuracy. The present signal characteristics were established around 1960. Since all stations operate on the same frequency, discrimination between chains is by the pulse repetition frequency. A typical chain comprises a master and two slaves, about 600 miles from the master, along a coastline. Each antenna is 1,300 ft high and is fed 5-MW pulses which build up to peak amplitude in about 50 μs and then decay to zero in about 100 μs. The slow rise and decay times are necessary to keep the radiated spectrum within the assigned band limits of 90 to 110 kHz. At the receiver, the first three rf cycles are used to measure the time of arrival. At this point, the pulse is at about half amplitude. The rest of the pulse is ignored, since it may be contaminated by sky-wave interference.

To obtain greater average power at the receiver without resorting to higher peak power, the master transmits groups of 9 pulses, 1,000 μs apart, and the slaves transmit 8 pulses, also 1,000 μs apart. These groups are repeated at rates ranging from 10 to 25 per second. Within each pulse, the rf phase may be varied for communication purposes.

At the receiver, phase-locked loops track the master and slave signals and present their time differences on a digital display, from where they can be transferred to a map on which hyperbolic lines of position have been printed. Alternatively, at the cost of doubling the size and price of the equipment, a digital computer can provide direct readouts in latitude and longitude or in left-right steering information and distance to go. To reduce interference from the numerous cw stations in the band, narrow notch filters are employed at the front end of the receiver. These automatically track and eliminate the strongest interference; another filter than tackles the second strongest; and so on.

The advantages of Loran C are that it extends beyond the line of sight, and has long range (up to 1,000 miles) and high accuracy (better than 0.1 mi). The disadvantages are high cost, typically $30,000 to $100,000 for the receiver, slow acquisition of the signal (up to 10 min), inconvenient readout, and lack of worldwide coverage.

Coverage presently extends to the North Pacific, North Atlantic, and Mediterranean, using about 30 transmitters. About 90 would be needed to cover all oceans, and 200 to cover the world.

114. Loran D is a tactical version of Loran C, with shorter base lines, lower power, and smaller masts. To compensate for the necessarily reduced performance, 16 pulses are employed per group, and measurements are made at the peak of the pulse, rather than at the 50% point. This is justified by the reduced sky wave interference at short ranges. Receivers are the same as for Loran C. Little civil use has yet been made of Loran C, although it has been proposed as a backup for transoceanic airliners' doppler and inertial navigators.

115. Omega.[107] This is a U. S. Navy system aimed at providing worldwide coverage from only eight stations. It is a hyperbolic system, using the VLF band at 10- to 13 kHz. At this low frequency, sky-wave propagation is relatively stable. There is, however, a marked difference in propagation between day and night, as shown in Fig. 25-103, but this is predictable and can be compensated for if the observer at the receiver has a rough idea of his location. Overall accuracy is consequently of the order of 1 mile, even at ranges of 5,000 mi.

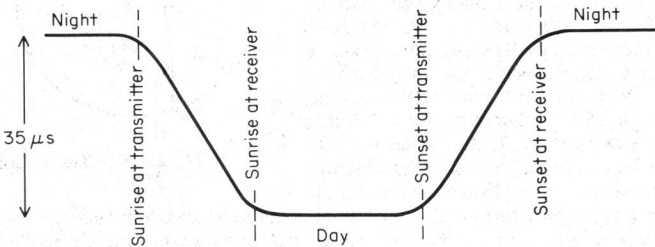

Fig. 25-103. Typical Omega diurnal propagation change.

There are no masters or slaves, each station transmitting according to its own standard. The signal format is shown in Fig. 25-104. Each station transmits on one frequency at a time, for a minimum of about one second, the cycle being repeated every 10 s. These slow rates are necessitated by the high Q's of the necessarily inefficient transmitting antennas.

The simplest type of receiver receives only the 10.2-kHz signals, comparing those from one station against those of another by use of a medium-stability internal oscillator. The phase differences are transferred to a map with hyperbolic coordinates. Typical receivers, for two frequencies, sell for below $10,000.

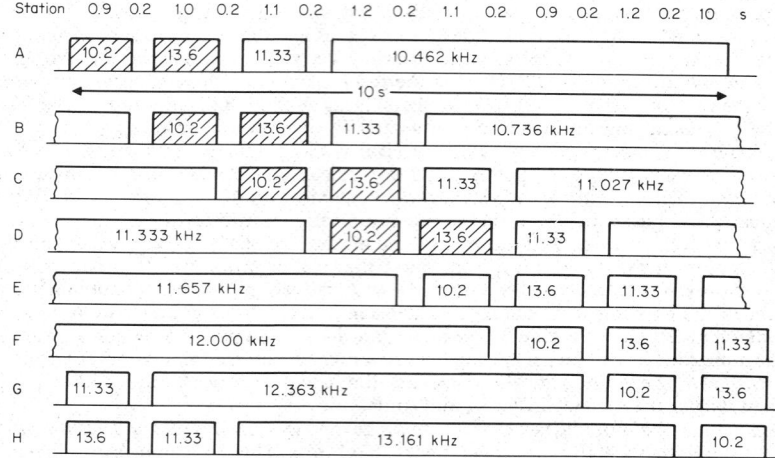

Fig. 25-104.   Omega signal format.

At the transmitted VLF frequencies, lane ambiguity occurs every 8 miles or so (half a wavelength). However, by using the beats between the VLF frequencies, these ambiguities can be extended to 24 miles for a two-frequency receiver and 72 mi using a three-frequency receiver.

For several years, Omega has operated with only the signals shown shaded in Fig. 25-104. Full coverage is now under way and is expected to lead to further use by international merchant shipping. The low data rate and inconvenient readout are handicaps in airborne use.[108]

116. Transit.[109]   This is a U.S. Navy satellite-based system and is the only satellite system in operational use. Its first test was in 1959, and in 1967 it was released for civil use; however, if the total cost of the system is divided by the number of civil users to date, this is one of the most costly civil systems yet devised.

In its simplest form, shown in Fig. 25-105, one satellite in polar orbit at 600 miles altitude circles the earth every 1 3/4 h, radiating two cw frequencies near 150 and 400 MHz. As it passes an observer on the surface of the earth, these frequencies undergo a doppler shift, the magnitude of which is dependent on the distance to the satellite, as shown in Fig. 25-106. From the rate of change (slope) of this doppler shift, the distance to the satellite may be computed, and since the satellite position is predicted in published tables, the ob-

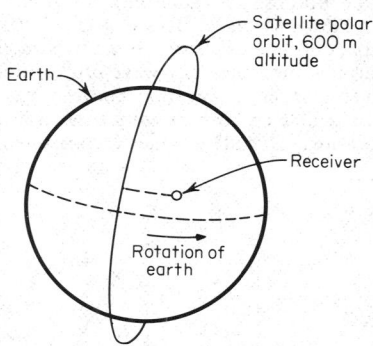

Fig. 25-105.   Transit orbit.

server's position may be determined. For a single satellite at least four such fixes are obtained per day for all positions on the earth. The system was therefore the first to provide worldwide coverage, albeit at a rather low data rate.

While a single radiated frequency would allow an accuracy of about one mile, two frequencies allow errors due to ionospheric refraction to be reduced. Other errors are caused by air friction and variations in the earth's gravitational field. These errors are observed by four ground-based tracking stations, and correctional notes are sent to the satellite, which rebroadcasts them. Total system error on board a ship traveling at known speed is below 0.25 mi. Even with several satellites in orbit (typically five), the data rate is too low to be of interest in most airborne applications.

117. Consol.[110]   This system, originally known as Sonne, was developed in Germany for guidance of submarines during World War II. It operates in the 300-kHz region and consequently is subject to atmospherics and to skywave propagation effects. Three vertical antennas are stationed in a line about 2,700 m apart, the central one being fed cw at about 3 kW and the outer ones cw at 750 W each. The phases of the signals from the outer ones are varied in time in steps of 1 s for 30 s with alternate transmission of the Morse letters E and T, producing a multilobe rotating pattern.

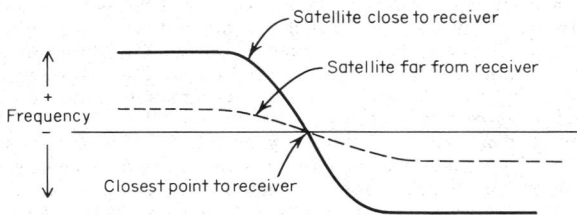

Fig. 25-106.   Doppler shift or transit signal.

The observer, using a simple receiver with audio output, may estimate his direction by counting the number of E's and T's he has heard since the start of the transmission, which repeats every minute. The range over seawater is of the order of 1,000 mi, and the accuracy better than 1° under favorable conditions. Although the system has frequently been declared obsolete, the simplicity of the receiving equipment is attractive, and as recently as 1971, the Norwegian government installed four new sets of stations for the benefit of fishing vessels. Its low data rate and indirect readout have precluded its use in airborne applications.

118. Talking Beacons.   Since the advent of highly directional microwave beams, it has frequently been proposed that a simple directional system could be devised in which such a directional beam rotated slowly while constantly announcing, by recorded voice, the direction in which it was pointing. The U.S. Coast Guard started several such developments in the early 1960s, but they were unsuccessful, due mainly to mistakes in hardware design. However, since 1966 Japan has had a successful three-station system in operation in the Straits of Tsushima.[111] Stations are at least 100 m above sea level, transmit 7-kW pulses at 9,300 MHz, and are pulse-duration-modulated by voice. The pulse repetition frequency is 10 kHz, and rotational rate is 1/3 r/min, with three beams per station 120° apart. The voice modulation is recorded on magnetic drums which are geared to the antenna rotation. Antenna beamwidth is 2°, and bearing announcements are made every 2°. The receiver cost is below $200. The accuracy is about 1°, and the range about 70 mi.

119. Cooperative Systems Used for Surveying.[78]   *Shoran* is a two-way pulse system developed in the United States during World War II. It transmits 0.25 μs interrogation pulses, 10 times per second in the 210-to-260 MHz region, which are replied to by transponders in the 290-to-320-MHz region. Observation is on an oscilloscope. *Trident* is a French variation of Shoran with interrogation at 230 MHz and reply at 270 MHz. *Raydist* is a U.S. proprietary system using a cw phase comparison in two-way and hyperbolic modes in the 1.6-to-5-MHz band. *Hi-Fix* is a British proprietary system (owned by the Decca Co.) using bursts of cw in the 1.6-to-2.6-MHz band. Sea-Fix is a higher-power version with shorter transmissions. *Toran* is a French cw hyperbolic system in the 1.4-to-2-MHz band. Major French ports have permanent installations. *Hydrodist* is a South African proprietary two-way system using cw phase comparison, superimposed on a 36-GHz carrier, with parabolic dishes to reduce multipath effects. Display is on an oscilloscope.

MAJOR SYSTEMS IN DEVELOPMENT

120. Satellite Systems.   Many believe that only a system using earth satellites can provide worldwide navigational coverage with the accuracy associated with line-of-sight frequencies.   Aside from Transit, whose data rate is considered too low, many other schemes have been proposed, many relying on the use of synchronous satellites, successfully used for communication.[112]   The problems are primarily political, stemming chiefly from the question of the country or agency responsible for the high costs of satellite manufacture, launch, and maintenance.

Because of the relatively inaccurate navigation aids and the absence of direct surveillance in midocean, transatlantic aircraft have long been forced to fly on parallel paths that are 120 miles apart at the same altitude, compared with similar paths 5 miles apart in areas having VOR/DME service.   This forces some aircraft either to take circuitous routings or to fly at uneconomic altitudes.   However, the advent of larger aircraft has tended to keep the number of aircraft fixed, despite traffic growth.   Thus the pressure to solve this problem has not been great.   The airlines would like better communication with transoceanic aircraft but are not particularly concerned with navigation or surveillance.   The FAA is primarily interested in surveillance.   Only the military, so far, have shown much interest in satellite navigation. One such system in development uses four synchronous satellites in a three-dimensional hyperbolic system.   Planned for the 1,600-MHz band for satellite-to-aircraft transmission, potential accuracies of 100 ft are claimed over about one-third of the earth's surface per chain of satellites, with an airborne equipment cost below $100,000.[113]

121. Microwave Landing System.   While the present ICAO ILS has served well for over 30 years, requirements for the future are believed to necessitate more channels, more flexible approach paths, and greater freedom from site effects.   These can readily be obtained at microwave frequencies where greater antenna directivity and a wider frequency spectrum are available.   Since a range of only 20 miles or so is needed, line-of-sight limitations pose no problem.   Two systems are under active consideration: 5 GHz for a general-purpose system and 15 GHz for a short-range, highly portable precision system (rain attenuation limits the range at 15 GHz).   At both frequencies, the choice appears to lie between two systems: scanning-beam and doppler.[114]

In the scanning beam system (Fig. 25-107*a*), the directional signal along the runway is formed into a narrow beam—of the order of 1/2°—and scanned through the required

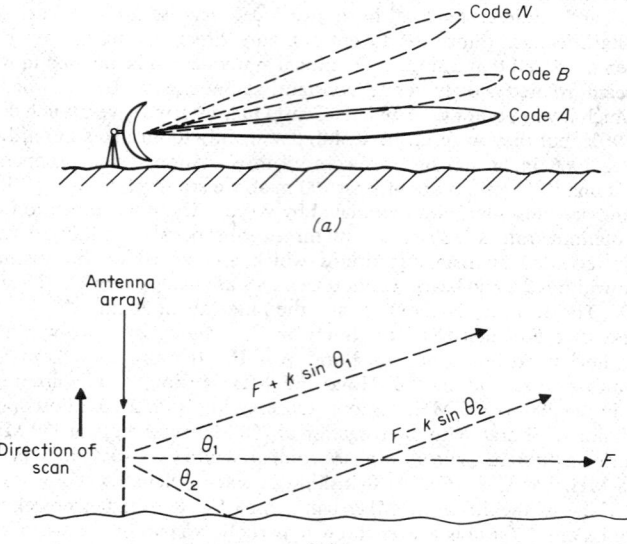

Fig. 25-107.   Microwave landing system.   (*a*) Scanning-beam; (*b*) doppler.

approach path, 0 to 20° for elevation and ±60° for azimuth. As it scans, the beam is coded with its position. On the aircraft it is only necessary to decode this information at the moment of maximum signal strength to determine the aircraft's angle with respect to the transmitter. A data rate of at least 10 scans/s seems necessary, making it doubtful that mechanical scanning will be adequate. Electronic scanning, on the other hand, raises questions concerning reliability of monitoring.[115,116]

122. Doppler System. (Fig. 25-107b). This system, invented in the United Kingdom,[117] is claimed to provide the lowest-cost electronic scan, with straightforward monitoring. A cw signal is sequentially commutated along an array of antennas, so as to simulate the movement of a single antenna. The frequency seen at the aircraft then appears shifted, by the doppler effect, by an amount proportional to the sine of the angle between the aircraft and the perpendicular bisector of the antenna array. Thus measurement of this frequency gives the aircraft angle directly. Since the shift is small, the transmitter also radiates a reference frequency from a stationary antenna, and the beat between the two is measured in the aircraft. This method also cancels out doppler shift due to aircraft motion. In general, signals traveling along reflective paths have different doppler shifts than those traveling by the direct path.

It is believed that scanning-beam and doppler systems should give about the same performance for a given frequency and antenna size. The choice between the two will therefore be based on other factors. It is not expected to result in an operational system until 1980.[118]

123. Collision-Avoidance System. Since about 1955 the scheduled airlines have been actively looking for a system that would protect against midair collisions. Most of the schemes proposed, while recognizing the threat of another approaching aircraft, have suffered from excessive "false alarms"; this is not surprising when one considers that two aircraft on parallel tracks may pass each other very closely, yet with complete safety. A major difficulty has been that no practical airborne antenna has been available with sufficient directional discrimination to distinguish between a collision course (no change of angle) and a passing course (small change of angle). A further difficulty has been that all practical systems have required equipment on all aircraft, not just those willing and able to pay for the protection. Thus low cost has been an essential requirement.[119]

One system which has been extensively flight-tested by the Air Transport Association calls for each aircraft in the system to radiate cw at 1,600 MHz during a unique time slot every 3 s. During this period, all other aircraft listen and compare the received frequency with their own internal frequency standard. Precision of frequency control is sufficiently high to allow one-way distance ranging (Fig. 25-93), and the doppler shift of frequency gives a direct reading of the rate of closure. Aircraft altitude is also taken into consideration. When the range rate divided by the range exceeds a certain number, standardized evasive maneuvers are indicated to the pilots. The system costs about $50,000 per aircraft.[120]

Not the least of the problems is what happens to other traffic in a dense environment when two aircraft take such evasive action, unknown to the air traffic controller on the ground. While alleviating one problem, it may generate others. A growing school of thought consequently believes that the only long-term solution is an accurate and complete traffic control system in which all aircraft are under surveillance and control by a single authority on the ground.

UNDERWATER SOUND SYSTEMS

By S. T. Ehrlich, D. A. Fredenberg, J. H. Heimann, J. A. Kuzneski,
and Paul Skitzki

124. Principles and Functions. Sound energy travels in water as a result of particle motion initiated by the application of physical forces to the particles from a vibrating diaphragm, collapsing air bubbles, or other energy sources with sufficient mechanoacoustical coupling for the transfer of the energy. It can be controlled, directed, and transmitted for many useful purposes.[121-125]

Water is an excellent medium in which to transmit compressional sound waves. Liquids have higher specific acoustic impedances by several orders of magnitude than gases. The high acoustic impedance of water (1.5×10^6 N·s/m³ for seawater) makes it possible to design transducers whose internal mechanical impedance approaches the radiation load impedance, with conversion efficiency on the order of 50% over a band of an octave, or over 80% over a narrow band.[121]

The transmission and reception of underwater sound can be controlled and directed to perform the functions of communications, navigation, detection, tracking, classification, etc., which in aerospace are accomplished with electromagnetic energy.[122] The wavelengths of underwater sound systems and radar systems are of the same order of magnitude, since the frequencies employed differ by the ratio of the speed of sound to the speed of electromagnetic waves. The term *sonar*, derived from sound navigation and ranging, is used synonymously with *underwater sound* and *underwater acoustics*.

125. Applications of Underwater Sound. The applications of underwater sound for defense purposes, both pro- and antisubmarine, have advanced with development of the nuclear submarine and other platforms. In military applications, underwater sound is used for depth sounding; navigation; ship and submarine detection, ranging, and tracking (passively and actively); underwater communications; mine detection; and for guidance and control of torpedoes and other weapons. Most systems are monostatic, but bistatic systems are also employed.

Civilian applications of underwater sound are numerous and are continuing to increase as attention is focused on the hydrosphere, the ocean bottom, and the subbottom. These applications include depth sounding; bottom topographic mapping; object location; underwater beacons (pingers); wave-height measurement; doppler navigation; fish finding; subbottom profiling; underwater imaging for inspection purposes; buried-pipeline location; underwater telemetry and control; diver communications; ship handling and docking aid; antistranding alert for ships; current flow measurement; and vessel velocity measurement.

PROPAGATION

126. Propagation of sound in water may be represented by the sound pressure, the sound particle velocity, and/or the sound intensity as a function of position, time, and frequency. Because the sound pressure can usually be measured more nearly directly, it is the preferred parameter for most experimental data.[135] The *sound pressure amplitude p* in water is expressed in pascals.* The logarithmic unit of a *sound pressure level, L_p,* is expressed in decibels with respect to the reference sound pressure amplitude, i.e., $1 \mu Pa = p_0$, where $L_p = 20 \log(p/p_0)$. The phase of the sound pressure is expressed in degrees or in radians with respect to a specified reference.

The difference between the sound pressure level at the reference position and the sound pressure level at a point in the sound field is called the *propagation loss N_W* for that point. For a small sound source, the reference position may be at a standard distance of 1 m in the direction of the maximum response. For a larger source, far-field data may be extrapolated back to the reference distance r_0.

The propagation loss may be considered[129-132] to consist of two basic components, one due to the spreading of sound energy with increasing radial distance r from the sound source (N_{spr}), and the other to attenuation of sound as it propagates through the medium (N_{att}):

$$N_W = N_{spr} + N_{att}$$

The *spreading loss* is given by $10n \log(r/r_0)$, where n is dependent on the spreading law and is equal to 2 for the theoretical case of *spherical spreading*. The *attenuation loss* is given by $10^{-3} \alpha r$, where α, the *attenuation coefficient*, is as discussed in Par. **25-128**.

Common propagation paths are illustrated[129] in the ray diagrams of Fig. 25-108. The paths of the direct ray and a ray with a single surface reflection in water with constant sound speed, are shown in Fig. 25-108*a*. In *b* a surface-layer sound channel confines the ray near the surface, with leakage rays due to diffraction and reflected waves from a rough surface. A ray which experiences a single bottom reflection is shown in *c*, while two bottom reflections with an intermediate surface reflection are shown in *e*. A pair of rays that diverge and return to a crossover in the convergence zone is shown in *d*. The three diverging rays, trapped in a deep sound channel and crossing each other several times before converging at the receiver, are shown in *f*. The reliable acoustic path, RAP (not shown), exists between a source at moderate depth and a surface receiver.

---

* 1 pascal = 1 newton/meter² = 10 dynes/centimeter² ≈ 10 microbars.

The theoretical treatment of sound propagation in water depends on assumptions used to simplify the mathematical formulation. Typical assumptions used by various authors include combinations of one or more of the following:[129,132,139]

One sound source with constant frequency and spherically isotropic radiation.

Propagation medium with linear transmission characteristics and sound speed dependent only on depth.

An ideal horizontal plane sea surface with zero acoustic impedance.

An ideal horizontal plane sea bottom with infinite acoustic impedance.

A sound receiver with zero rise time, flat frequency response, and spherically isotropic reception.

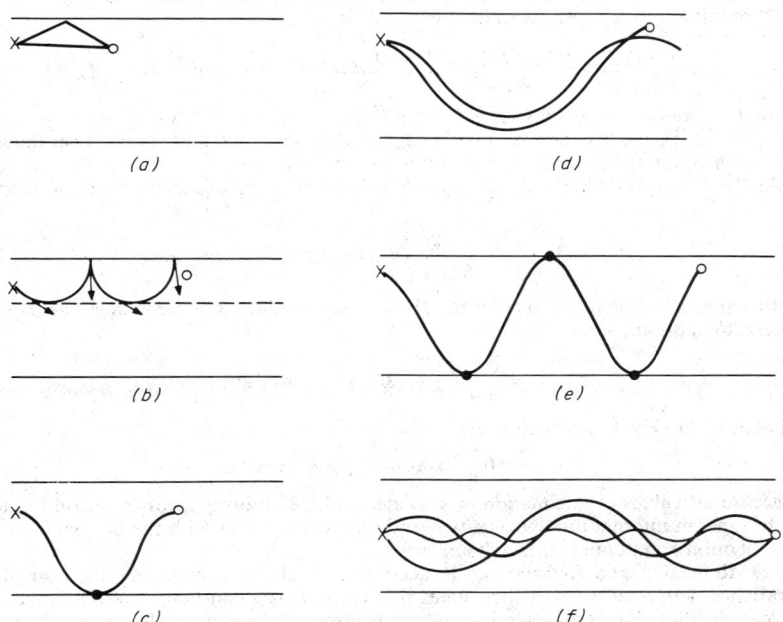

Fig. 25-108. Propagation paths.

127. Speed of Sound. Essentially, all seawater may be represented by the following conditions:

Temperature T:	$-3°$ to $+30°C$
Depth d:	0 to 10,000m
Salinity S:	33 to 37 parts per thousand

with atmospheric pressure (absolute pressure) 0.1013 MPa at zero depth at sea level. For these conditions, the speed of sound c in seawater is given by an empirical formula due to Wilson,[126-128] here simplified to

$$c = 1449.3 + 4.572T - 0.0445T^2 + 0.0165d + 1.398(S - 35) \quad \text{m/s}$$

At zero depth the accuracy of c is about 3 m/s, or 0.2%; for extreme conditions with comparable accuracy a more complete equation, including higher-order and cross-product terms, becomes necessary. At $T = 0°C$, $d = 0$ m, and $S = 35$ parts per thousand, $c = 1,449.3$ m/s. The nominal value $c = 1,500$ m/s, corresponding to about $T = 13°C$, is convenient for engineering calculations.

128. Attenuation of Sound. The attenuation of sound in seawater has been studied by many investigators to determine its variation in different frequency bands as well as its dependence on temperature, salinity, and depth. At frequencies greater than 1 MHz, the attenuation mechanism is generally attributed to shear and dilatational viscous losses.[132] In the frequency band between 10 and 40 kHz, the increased attenuation is almost solely due to a relaxation-type mechanism in the $MgSO_4$ salts dissolved in the seawater.[138] Recent work indicates that between 0.1 and 1.0 kHz another relaxation-type mechanism, not yet identified, dominates the attenuation.[133] Below 50 Hz other attenuation mechanisms are of greater importance.

An expression for the attenuation coefficient of seawater from 0.1 kHz to 100 MHz includes three components resulting from the three principal attenuation mechanisms,[129,132,133,136] multiplied by a depth-dependent term:[137]

$$\alpha = \left[\frac{0.11 f^2}{1 + f^2} + \frac{0.70 f_T f^2 (S/35)}{f_T^2 + f^2} + \frac{0.03 f^2}{f_T} \right] (1 - 65 \times 10^{-6} d) \quad \text{dB/km}$$

where f = frequency in kilohertz, f_T = relaxation frequency[136] in kilohertz = $21.9 \times 10^{[6 - 1,520/(T + 273)]}$, and T = temperature in degrees Celsius, S = salinity in parts per thousand, d = depth below the air-to-water boundary surface in meters.

At $T = 4°C (f_T \approx 71 \text{ kHz})$, $d = 0$ m, and $S = 35$ parts per thousand, the equation simplifies to

$$\alpha = \frac{0.11 f^2}{1 + f^2} + \frac{50 f^2}{5,000 + f^2} + 0.0004 f^2 \quad \text{dB/km}$$

Further simplification is possible for the above listed conditions, at frequencies below about 20 kHz, to approximately

$$\alpha = \frac{0.11 f^2}{1 + f^2} + 0.010 f^2 \quad \text{dB/km}$$

and above 200 kHz, to approximately

$$\alpha = 50 + 0.0004 f^2 \quad \text{dB/km}$$

Measured values of attenuation in seawater include absorption losses, scattering losses due to random internal inhomogeneities, and interaction losses with the bottom boundary, the subbottom, and upper surface boundary.

129. Reflection and Refraction. Reflection and refraction of sound are normally in accordance with Snell's law, which states that $\cos \theta_i / c_i$ is a constant, where θ_i is the angle between the direction of propagation and the horizontal plane and c_i is the sound velocity, at the point i of the ray. At a boundary where the sound velocity is discontinuous, the angle θ_i is the angle of incidence with respect to the plane tangent to the boundary, which is not necessarily horizontal.

A special case of interest occurs in a region with a constant *sound speed gradient* ∇c. This results in a circular sound ray path with radius equal to $c_i / \nabla c \cos \theta_i$. The center of the circle corresponds to a position where c_i becomes zero. In a surface layer, the resultant upward refraction leads to shadow zones which theoretically contain no propagated sound energy.

A second special case occurs at a boundary where the incident sound speed is lower than the refracted sound speed c_r, such as a plane interface between water and air. The well-known phenomenon of total internal reflection results when θ_i is less than a critical angle θ_{crit}, at which $c_i / \cos \theta_{crit}$ is equal to c_r.

Another important case occurs when the sound in water is propagated by a direct path and also by a single reflection from a boundary of slower sound speed, such as air. Because of a 180° phase shift in the sound pressure at such a reflecting boundary, the resultant sound pressure exhibits maxima and minima as a function of the position of the receiver. This is called the *Lloyd mirror*, an image interference effect, since the condition may be represented by an additional sound source of equal amplitude and opposite phase as a virtual image that provides the constructive and destructive interference with the real sound source.

Other important cases include a multiple-layered medium where the sound field at each discontinuity must be accounted for; irregular and nonstationary boundaries which tend to randomize the reflection and refraction properties; moving boundaries which tend to modify the frequency of the sound energy because of the doppler effect; and intentional discontinuities introduced in transducer and array designs.

130. Reverberation. Reverberation of sound in water produces energy usually unwanted at the receiver. It is caused by scattering, i.e., reflection and refraction of sound from discontinuities other than those of primary interest. When the sound source and sound receiver are at the same location, reverberation is produced primarily by backscattering. When the scatterers are boundaries of the medium, the effect is called *surface reverberation,* which may be subdivided into *sea-surface* and *sea-bottom reverberation.* When the scatterers are contained within the medium, the effect is called *volume reverberation,* which may be due to fine particles, fish, or other inhomogeneities, including the structure of the sea.

NOISE

131. Background Noise. Underwater sound systems operate in a medium which has a very low acoustic-noise level under quiet conditions. Stimulation from natural and manmade causes, however, can generate and propagate acoustic noise at various levels and frequencies. The acoustic-noise background consists of an *ambient-acoustic-noise level* and a *self-noise level* caused by the sonar-platform presence and its movement. The factors contributing to the acoustic-noise background are shown in Fig. 25-109. Passive sonar systems detect target-generated noise while active sonar systems detect target echoes in the presence of the noise background.

132. Ambient Noise. The main contributions to ambient noise are shown in Fig. 25-110. Tides, waves, and seismic disturbances predominate at the very low frequencies, in the region of 1 Hz. At frequencies used for sonar systems, the main sources are sea surface agitation due to meteorological effects, noise from marine life, and man-made noises from shipping and other activities.[140,141] Of these, surface agitation due to wind and wave actions is most significant. The acoustic spectrum levels for various sea states are shown in Fig. 25-111.

At the high frequencies (above 100 kHz), thermal molecular motion of the water is the principal noise contributor, but does not limit sonar performance at lower frequencies. It has

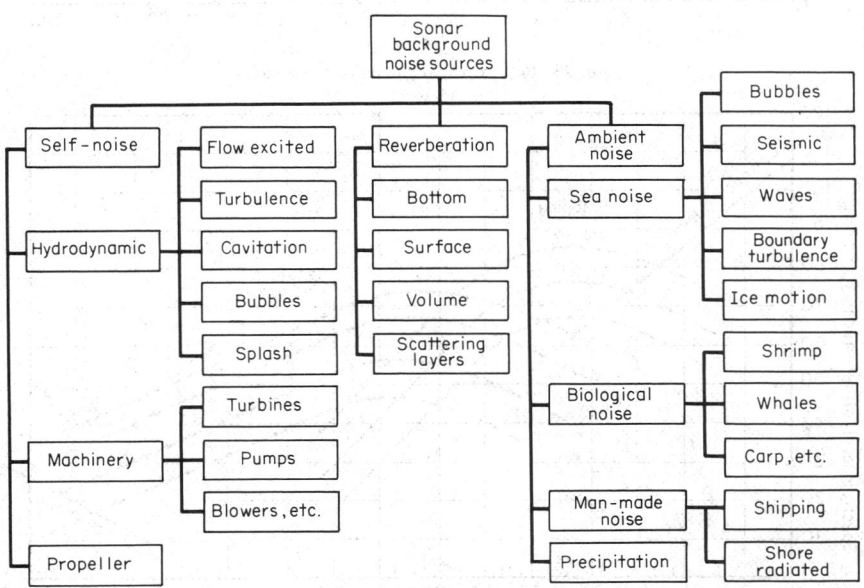

Fig. 25-109. Sonar background-noise sources.

a spectrum level that rises with frequency at a rate of 6 dB/octave. The mean-square sound pressure in a 1-Hz band is

$$p_T^2 = 4\pi kT_{\rho c}/\lambda^2$$

where k = Boltzmann's constant, T = absolute temperature of the water, ρc = specific acoustic impedance, and λ = wavelength. Acoustic thermal noise is consistent with its more familiar electrical counterpart. It is far below sea state levels at frequencies below 10 kHz. At higher frequencies than 100 kHz, it sets a threshold on the minimum detectable pressure levels in the medium.[142] The ambient-noise level, as a sonar parameter, is the intensity of the

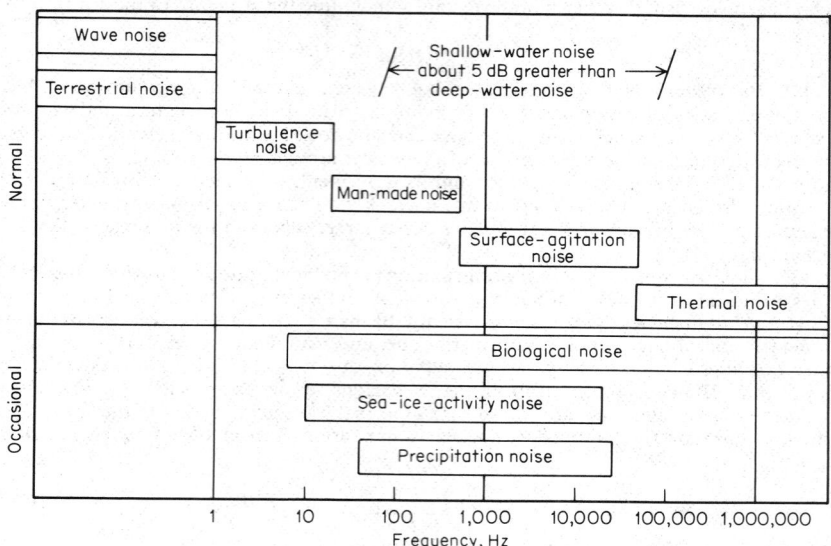

Fig. 25-110. Principal ambient-noise sources.

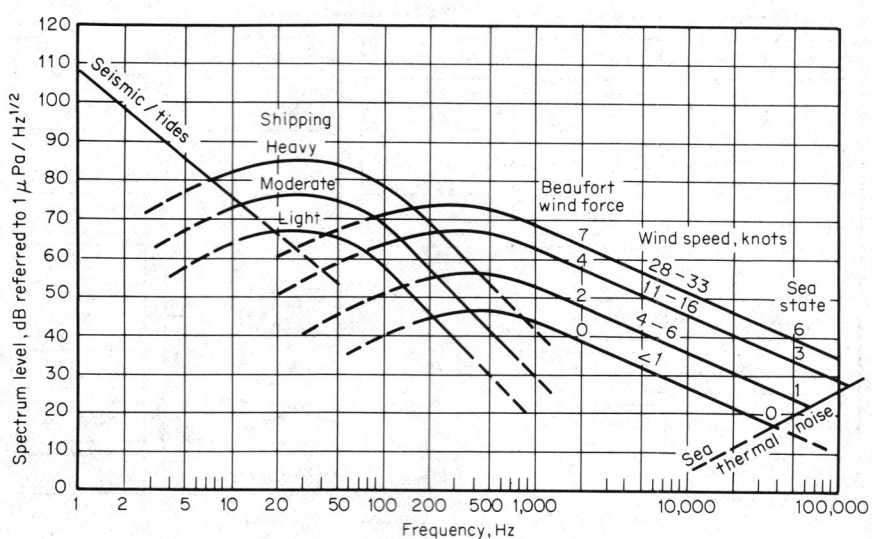

Fig. 25-111. Average deepwater noise.

25-108

noise background as measured with a nondirectional hydrophone, referred to the intensity of a plane wave having an rms pressure of 1 μPa.

133. Platform Noise. Platform noise is a degrading factor in the peformance of underwater sound systems, particularly in the case of mobile platforms such as ships, submarine, aircraft, torpedoes, and other sonar-carrying vehicles. Fixed-position platforms, including sonobuoys, moored structures, bottom-mounted structures, mines, etc., are also plagued with platform-noise problems, primarily induced by hydrodynamic flow or motion. Good sonar performance with mobile platforms at any speed above approximately 10 knots is achievable only by most careful attention to platform-noise reduction and by design of the sonar system for minimum susceptibility to local noise. Figure 25-112 shows the dominant sources and relative levels at various ship speeds.

Platform noise may enter the sonar system via radiation in the medium or by conduction through the platform structure. Generally, conducted noise can be reduced below the level arriving at the array via the medium. The techniques involve sensor design for minimum response to acceleration forces, isolation of the sensor elements from the mounting structure, and isolation and location of the array away from hull-borne and structural vibrations. The noise radiated into the medium may reach the array directly or via reflected paths, as shown in Fig. 25-113. The ship's propellers and machinery are the dominant noise sources. Since

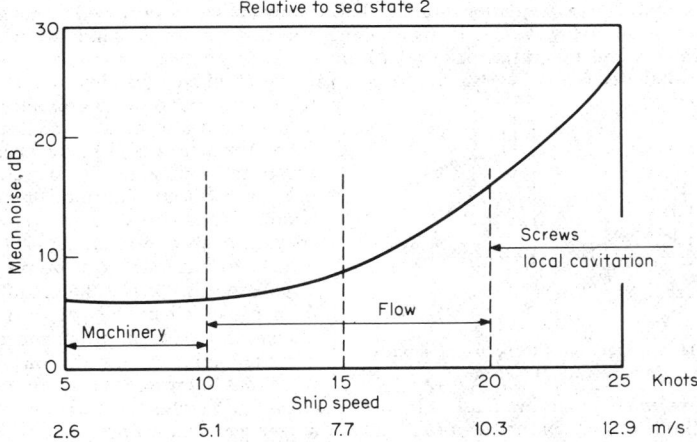

Fig. 25-112. Self-noise behavior.

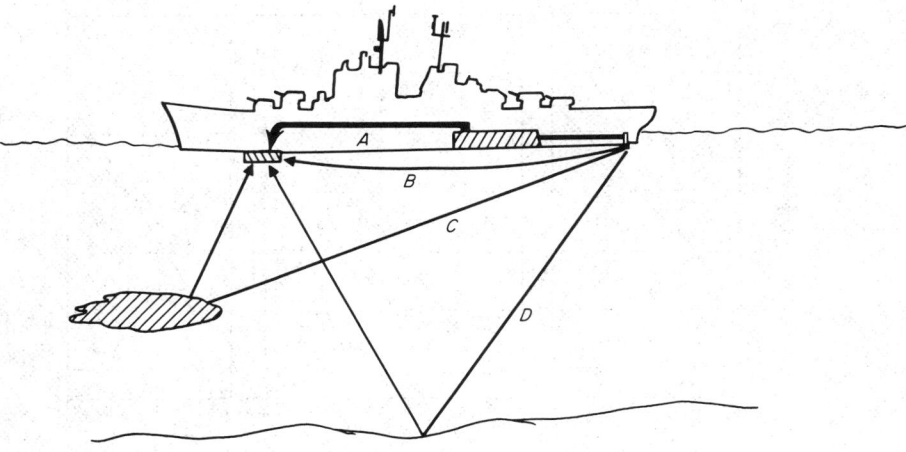

Fig. 25-113. Paths of self-noise.

the noise level is highest on stern bearings, baffles are generally employed behind the sonar array to minimize stern noise, even though this results in loss of sonar performance over a portion of the azimuth.

134. Radiated Noise. Platform-generated noise, radiated into the medium, produces an acoustic signature that can be detected by passive sonar systems. The principal radiated noise sources on ships, submarines, and torpedoes are listed in Table 25-19. Machinery noise is defined as noise caused by propulsion and auxiliary machinery on the vessel. The noise produced by the various machines, generators, pumps, actuators, etc., travels by diverse paths to the hull structure, where it is introduced into the medium. Propeller noise originates outside the hull and is mainly due to cavitation at the propeller blades. Cavitation-produced bubbles generate acoustic noise, the acoustic spectrum of which differs from machinery noise and varies with speed and depth. In addition to the cavitation noise, propellers produce amplitude-modulated noise modulated at a frequency equal to the shaft rotation speed times the number of blades. Such *propeller beats* are most pronounced just beyond the onset of cavitation and are swamped by cavitation noise at higher speeds. Propellers may also produce a "singing" noise due to vibrational resonance of the blades.[141]

Hydrodynamic noise results from the flow of fluid past the moving platform. It increases with hydrodynamic structural irregularities of the platform and the fluid flow rate. Breaking bow and stern waves can excite the hull or structural members. Hydrodynamic noise is a minor contributor to the platform-radiated noise, and is usually swamped by machinery and propeller noise. However, it is an important element in consideration of self-noise for underwater sound systems associated with the platform.

135. Radiated Noise Levels. Radiated noise consists of broadband noise and tonal noise (line components). Measurements of radiated noise are made at some distance (e.g., 200 yd) from the vessel and reduced to source spectrum level values by correction for the test distances and the measurement bandwidths. Tonals are determined by fine-grain spectral analysis. Figure 25-114 illustrates broadband and tonal noise from a submarine at two speeds. In (a) the broadband noise from cavitation at the propeller begins to appear, and the tonals from machinery noise are predominant. In (b) the broadband noise has increased as a result of higher speed, and many tonals are masked by the broadband noise while other tonals are changed in amplitude or frequency.[141] Figure 25-115 shows average radiated-noise spectra for several classes of ships, and Fig. 25-116 illustrates noise from running torpedoes.

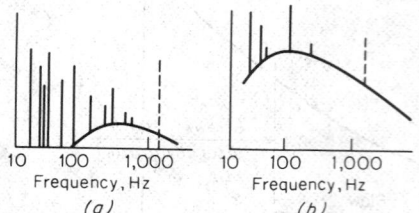

Fig. 25-114. Diagrammatic spectra of submarine noise. (a) Low speed; (b) high speed.

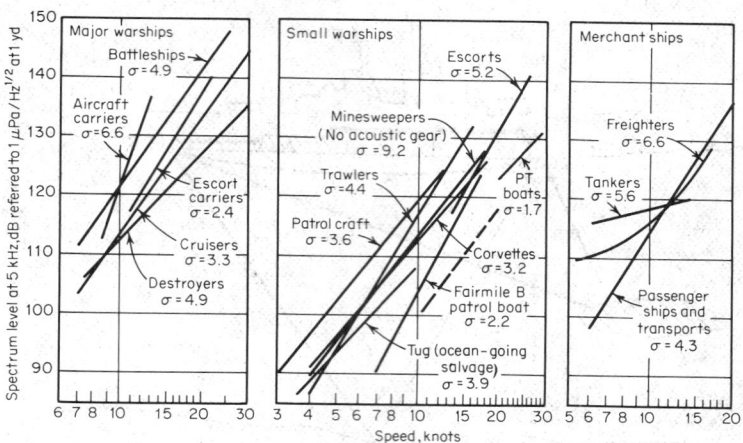

Fig. 25-115. Average radiated spectrum levels of ships.

TRANSDUCERS AND ARRAYS

136. Transducers for underwater sound applications perform the functions of generating a sound wave in the medium or detecting the existence of a sound wave and its properties (e.g., amplitude and phase) in the medium. In the generating case, the transducer is commonly referred to as a *source,* or *projector,* and in the detecting case, as a *hydrophone.* Often the transducer is required to perform both functions. Single transducers or arrays of transducers may be designed to control the directional properties (i.e., directivity or beam pattern) of the generated acoustic enery, and to discriminate against the noise in the receiving case.

The function of a projecting transducer is to convert the input energy (usually electrical) to acoustic energy in a manner that is efficient and compatible with the other components of the transmitting system (e.g., amplifiers). In the hydrophone, linear conversion of the acoustic signal to an electrical signal is the basic function, and compatibility must be maintained with the other components of the receiving subsystem.

The conversion of energy is accomplished by any of a variety of physical phenomena, e.g., piezoelectricity and electrostriction; piezomagnetism and magnetostriction; electrodynamics and magnetodynamics; and chemical transformations and hydrodynamics.

The selection of the transduction mechanism and the design of the transducer are based on the following considerations: operating frequency; bandwidth; power (acoustical and electrical); directional properties; the characteristics of available energy-converting materials; the characteristics of, and materials used for, packaging (such factors as stability with static pressure, temperature, and time and resistance to corrosive effects of the medium); cavitation and other nonlinear effects of the medium; and the effect of static pressures encountered at great depths on the overall transducer design and its operating characteristics.

Significant developments have been achieved in the calibration and performance testing of transducers and arrays both in the laboratory and in the ocean environment, and standard techniques have been established.[146,147]

GENERAL PROPERTIES OF TRANSDUCERS

137. Types of Projectors. There are two main transducer types used as underwater sound sources: those that operate with a continuous-wave or modulated (amplitude, frequency) input and those that operate as impulse sources. The former are used for most military and many commercial applications, while the latter are used mainly for oceanographic and geophysical applications.

138. CW and Modulated Sources. Transducers designed for cw or modulated input underwater applications utilize piezoelectric, electrostrictive, or magnetostrictive energy-conversion materials. Piezoelectric crystals (e.g., quartz) possess the property of a linear relationship between strain and electric field. However, their application is limited by low dielectric constant, low electromechanical coupling coefficient (the ratio of the converted energy to the total input energy in a transducer), narrow bandwidth, low power-handling capability, and limited availability of geometrical shapes. They can yield very high conversion efficiencies (quartz transducers having efficiencies in excess of 90% have been built).[148]

Ferroelectric crystals, in either single-crystal or polycrystalline ceramic form, are often electrostrictive and have higher-order nonlinear properties in their natural state. With the application of a polarizing electric field, the electromechanical processes in these materials can be linearized over a wide range of operating conditions. These materials have the advantage of high dielectric constant (which results in low impedance), high electromechanical coupling coefficients, broad bandwidth, and high power-handling capability when properly used, and are available in a wide variety of shapes (e.g., plates, cylinders, rings, spherical zone sections). Efficiencies in excess of 70% can be achieved. Operating frequencies from less than 1 Hz to more than 10 MHz can be achieved.

The most widely used materials in the ferroelectric class are modified lead zirconate titanate and modified barium titanate. Idealized transmitting current and voltage responses of a transducer that utilizes these materials is shown in Fig. 25-117(a) and 25-117(b). In this figure f_0 is the frequency of mechanical resonance of the transducer. Figure 25-118 shows a typical impedance characteristic.

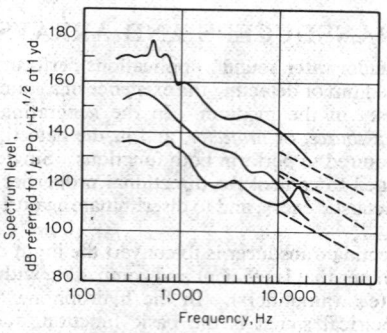

Fig. 25-116. Noise spectra of torpedoes.

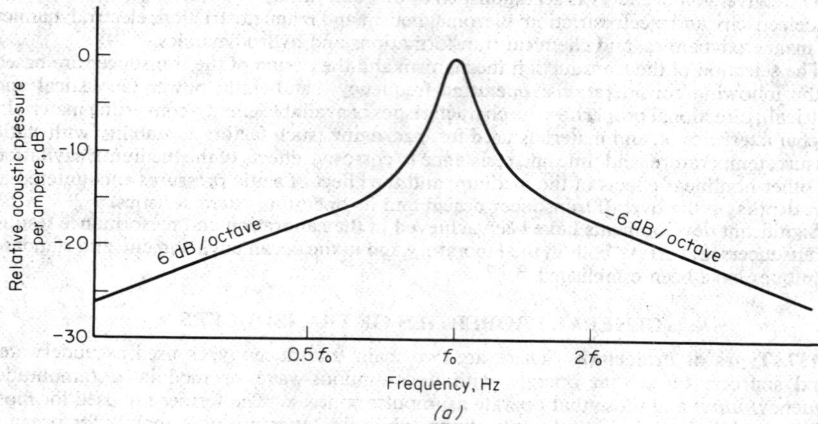

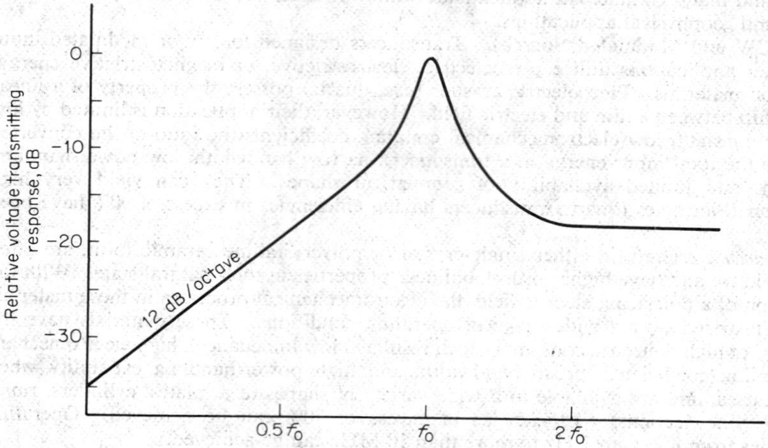

Fig. 25-117. Idealized transmitting current and voltage responses for a piezoelectric transducer: (*a*) current response; (*b*) voltage response.

Magnetostrictive transducers depend upon the interchange of energy between magnetic and mechanical forms. In an unpolarized state, such transducers are nonlinear, frequency-doubling devices. However, in a polarized state (achieved by the use of permanent magnets, direct current, or operation at remanence) they are linear (i.e., piezomagnetic). Commonly used materials include various nickel alloys and ferrites.

Properly designed magnetostrictive transducers can achieve radiated acoustic powers up to several kilowatts with efficiencies in excess of 50%. Operating frequency is usually limited to frequencies below 100 kHz. Figure 25-119 shows a typical idealized transmitting voltage characteristic.

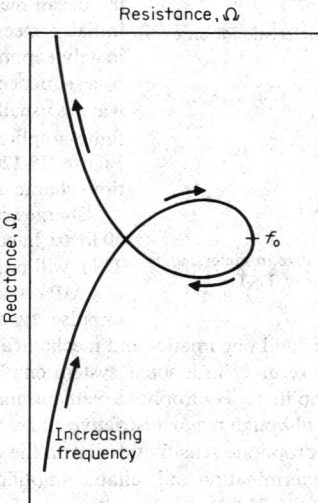

Fig. 25-118. Idealized impedance locus for a piezoelectric transducer.

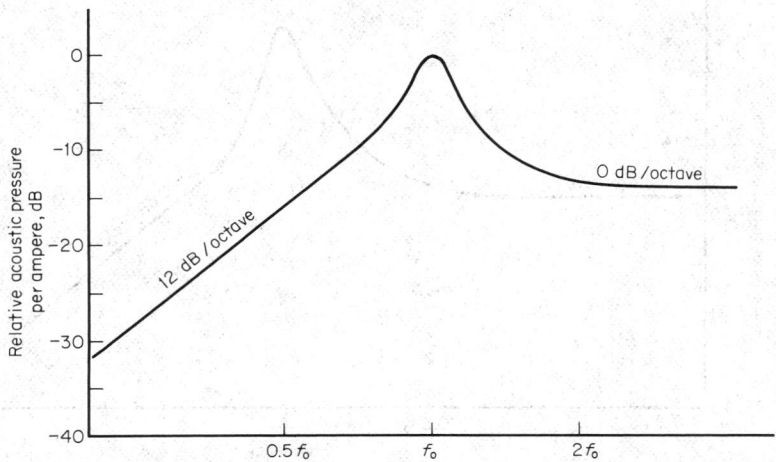

Fig. 25-119. Idealized transmitting current response for a magnetostrictive transducer.

Electrodynamic sources have been used in underwater acoustics for low-frequency applications.   One noteworthy application is the low-frequency *standard projector*.[149]

139. Impulse Sources.   Impulse sound sources produce a short-time duration, high-amplitude transition of more or less regular waveform.   For example, explosives (e.g., TNT or other high-burning-rate chemical) with provision for hydrostatic, electrical, or fused detonation produce in the ocean medium a pressure wave that is initially steep-fronted, displays approximately exponential decay, and is followed by a sequence of bubble pulses.   The shock wave is usually so intense that the resulting finite-amplitude effects are appreciable.   Figure 25-120 shows a typical pressure-time characteristic.

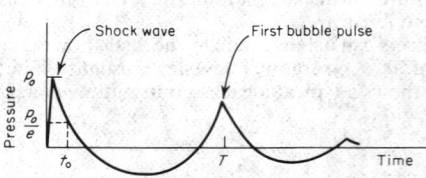

Fig. 25-120.   Pressure-time characteristic for an explosive source.[150]

$$p_0 = 4.22 \times 10^8 \left(\frac{w^{193}}{r}\right)^{1.13} \quad \text{(Pa)}$$

$$t_0 = 36.4 \times 10^{-6} w^{1/3} \left(\frac{r}{w^{1/3}}\right)^{0.22} \quad \text{(s)}$$

$$T = \frac{3.35 w^{1/3}}{(0.305d + 10.1)^{5/6}} \quad \text{(s)}$$

where d is depth in meters, r is range in meters, and w is equivalent yield in kilograms of TNT.

Charges ranging in weight from 1 oz to 50 lb are in common use.   A 4-lb charge of TNT will produce a pressure level at 1km of 4 MPa (that is, 252 dB re 1 μPa).   Other impulse type sources include implosive devices, spark-gap generators, and pneumatic- and mechanical-impact mechanisms.

140. Hydrophones.   The receiver in a sonar system employs a hydrophone or hydrophone array coupled to an amplifier.   Hydrophone elements most often employ piezoelectric energy-conversion materials, although magnetostrictive and electrodynamic mechanisms are sometimes used.   Typical hydrophone sensitivities are on the order of -180 to -200 dB re 1 V/μPa.   Proper impedance termination and suitable amplification are necessary to obtain useful electrical levels.   Figure 25-121 shows a typical idealized open-circuit receiving response for a piezolectric hydrophone.

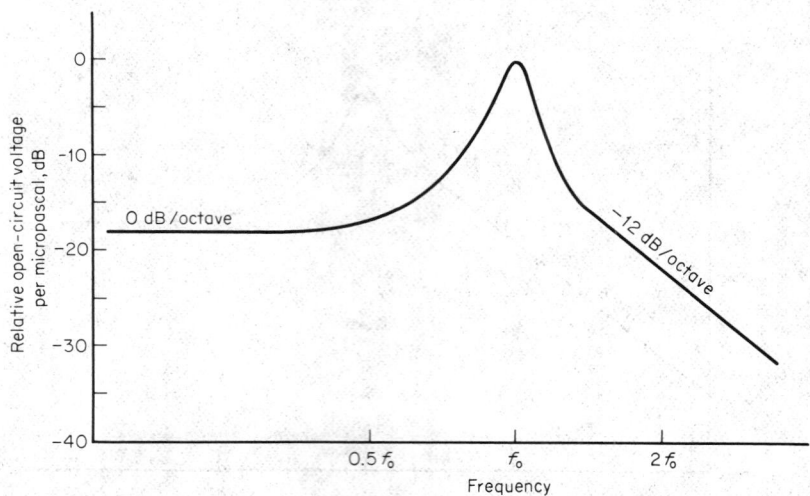

Fig. 25-121.   Idealized open-circuit receiving response for a piezoelectric transducer.

PRODUCTION OF SOUND FIELDS

141. Acoustical Principles. The production of a sound field in the medium involves an electrical imput that is converted by the transducing mechanism to the motion of a surface in contact with the medium. The motion initiated by the moving surface of the transducer is communicated to the adjacent water particles, and a sound wave is propagated from its surface.

The *far-field sound pressure* produced by the source may be described in terms of the radiated acoustic power P_a by the following equation:

$$p^2(r,\phi,\theta) = \frac{\rho c P_a D_i R(\phi,\theta)}{4\pi r^2} \tag{25-74}$$

where $p^2(r,\phi,\theta)$ = mean-square acoustic pressure in pascals; r,ϕ,θ = spherical coordinates, r in meters; ρc = product of density and speed of sound of the medium, that is, the specific acoustic impedance of the medium in $N \cdot s/m^3$; $R(\phi,\theta)$ = normalized pattern function; and D_i = directivity factor of the source.

The directivity factor is defined as

$$\frac{1}{D_i} = \frac{1}{4\pi r_0^2} \iint \frac{p^2(r,\phi,\theta)}{p_0^2} dS \tag{25-75}$$

where p_0 = pressure at distance r_0 in the direction of maximum response, and dS = element of surface area on a sphere having radius r_0.

An example of a normalized beam pattern is shown in Fig. 25-122. This figure shows a plot of $10 \log p^2(\theta,\phi_k)$ versus θ for a particular value of $\phi = \phi_k$. Dividing Eq. (25-74) by the square of the input current to the transducer, I^2, and rearranging,

$$\frac{p^2(r,\phi,\theta)}{I^2} = \frac{\rho c D_i R(\phi,\theta) R_e}{4\pi r^2} \frac{P_a}{P_e} \tag{25-76}$$

where P_e = electrical input power to transducer in watts, and R_e = electrical input resistance of transducer in ohms.

The current transmitting response ($20 \log S_0$) is given by

$$20 \log S_0 = 10 \log R_e + 10 \log D_i + 10 \log \eta_{ea} + 170.8$$
$$\text{(dB } re \text{ 1 } \mu\text{Pa/A at 1 m)} \tag{25-77}$$

The term $10 \log D_i$ is referred to as the *directivity index* N_{DI}, or gain, of the transducer; $\eta_{ea} = P_a/P_e$.

The free-field open-circuit receiving response M_o in volts per micropascal is related to the current transmitting response for a reciprocal transducer by

$$20 \log M_o = 20 \log S_0 - 20 \log f - 294 \quad \text{(dB } re \text{ 1 V/}\eta\text{Pa)} \tag{25-78}$$

where f is the frequency in hertz, and nominal conditions in the water are assumed.

In determining the reaction of the medium on the moving surface of the transducer, it is assumed that the vibrating surface of the source has a velocity u, and that the surface exerts a force F_r on the water, and the force exerted by the water on the moving surface of the source is $-F_r$. The radiation impedance Z_r is expressed as[151]

$$Z_r = \frac{-F_r}{u} = R_r + jX_r \tag{25-79}$$

where R_r = radiation resistance, and X_r = radiation reactance.

In a linear system consisting of a continuous source, the value of Z_r is frequency-dependent, but is a constant at constant frequency. If the radiation impedance is known, calculating the acoustic power P_a of the source is greatly simplified, since

$$P_a = \tfrac{1}{2}u_{\text{peak}}^2 R_r = u_{\text{rms}}^2 R_r \tag{25-80}$$

Table 25-16 lists the radiation impedances for various radiating surfaces, and Fig. 25-123 shows plots of radiation impedance for typical surfaces.

Table 25-16.   Radiation Impedance for Simple Geometries

Type of Radiator	Radiation Impedance
Rigid circular piston in infinite baffle	$Z = \pi a^2 \rho c \left[1 - \dfrac{J_1(2ka)}{ka} + j\dfrac{S_1(2ka)}{ka} \right]$ where J = Bessel function S = Struve function
Vibrating Strip of infinite length in an infinite baffle	Per unit length: $Z = 2\rho ca \left[2\Lambda(2ka) - H_1^{(2)}(2ka) + \dfrac{j}{\pi ka} \right]$ where $\Lambda(x) = \dfrac{1}{2}\displaystyle\int_0^x H_0^{(2)}(x)\,dx$ H = Hankel function
Sphere Pulsating Oscillating	$Z = \dfrac{4}{3}\pi a^2 \rho c\,\dfrac{(ka)^2 + jka}{1 + (ka)^2}$ $Z = \dfrac{4}{3}\pi a^2 \rho c\,\dfrac{(ka)^4 + jka(1 + k^2 a^2)}{4 + (ka)^4}$
Pulsating cylinder of infinite length	Per unit length: $Z = 2\pi a\rho cj\,\dfrac{J_0(ka) - jN_0(ka)}{J_1(ka) - jN_1(ka)}$ where J = Bessel function N = Neumann function

TRANSDUCER MATERIALS

142. Piezoelectric Materials.   Since 1950, man-made piezoelectric ceramics as transducing materials have reached maturity by achieving reproducibility for a given composition and by diversification.[156]

The electromechanical nature of these materials in a polarized state is described by linear equations.   With stress and electric field as the independent variables,

$$S = s^E T + dE$$
$$D = dT + \varepsilon^T E \tag{25-81a}$$

With stress and charge density (electric displacement) as the independent variables,

$$S = s^D T + gD$$
$$E = -gT + \beta^T D \tag{25-81b}$$

where S = strain, T = stress, E = electric field, D = electric displacement, s^E elastic compliance at constant electric field, S^D = elastic compliance at constant electric displacement, ε^T = dielectric constant at constant stress, β^T = dielectric impermeability at constant stress, and d and g are piezoelectric constants that are defined as:

$$d = \left(\frac{\partial S}{\partial E}\right)_T = \left(\frac{\partial D}{\partial T}\right)_E$$
$$g = \left(\frac{-\partial E}{\partial T}\right)_D = \left(\frac{\partial S}{\partial D}\right)_T \tag{25-82}$$

The electromechanical coupling coefficient is an important measure of the effectiveness of the energy conversion mechanism and is defined as:

$$k = \frac{U_m}{U_e U_d} = \frac{d}{\epsilon^T S^E} \tag{25-83}$$

where U_m = mutual elastic and dielectric energy density, U_e = elastic self-energy density, and U_d = dielectric self-energy density. This is a quasistatic parameter that may be related to the fundamental material constants for one-dimensional transducers. The definition is not necessarily applicable to all geometrics.

A more complete treatment of the piezoelectric equations and the measurement of the various constants may be found in References 154–160.

Although piezoelectric ceramics can be operated in various modes, two types are of major importance to underwater sound transducers. In the parallel, or *33-mode* type, the stress, strain, and electric field are in the same direction. In the transverse, or *31-mode* type, the stress and strain are the same direction but are orthogonal to the electric field. Each of these mode types is characterized by its associated constants, for example d_{31} or d_{33}, g_{31} or g_{33}, k_{31} or k_{33} respectively. The dielectric constant of interest is in the direction of the electric field for both cases and is denoted ϵ_{33}. For the above constants, the direction of the electric vectors are denoted by the 3 direction, and the direction of the mechanical variable, if different, is the other subscript. An additional electromechanical coupling coefficient k_p is useful for some applications. It is related to k by

$$k_p = \sqrt{\frac{2}{1 - \sigma^E}}\, k_{31} \tag{25-84}$$

where σ^E = Poisson's ratio at constant electric field.

The most important piezoelectric ceramic materials for underwater sound transducers are the modified lead zirconate titanate (PZT) compositions, and to a lesser extent modified barium titanate compositions. Table 25-17 lists some of the more important properties of these materials. A more comprehensive table of properties is available in Berlincourt et al., Ref. 156, Chap. 3. These materials are characterized as being "very hard" lead zirconate

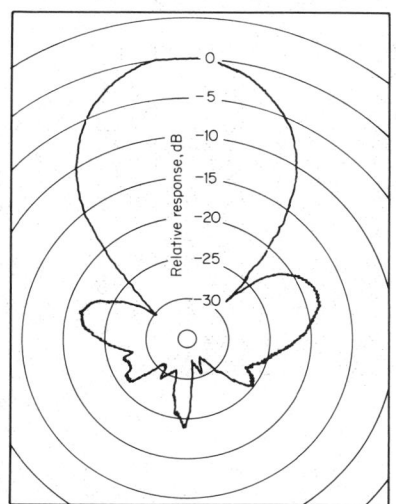

Fig. 25-122. Typical beam pattern of an underwater sound transducer.

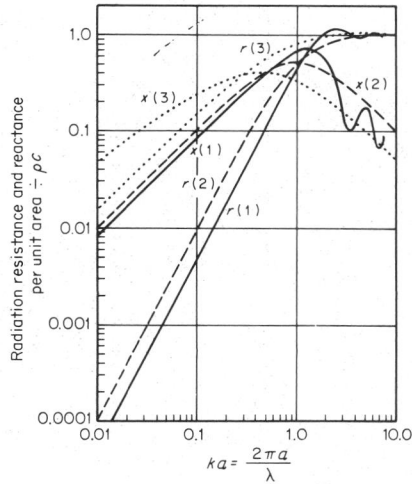

Fig. 25-123. Radiation resistance and reactance per unit area divided by π as a function of ka (a = radius) for (1) a circular piston in a rigid baffle; (2) a pulsating sphere; (3) a pulsating cylinder of infinite length.

Table 25-17. Characteristics of Commonly Used Piezoelectric Ceramics—Low-Signal Properties at 25°C

Quantity	PZT-4*	PZT-5*	PZT-8*	95% wt BaTiO₃, 5% wt CaTiO₃
k_p	0.58	0.60	0.51	0.36
k_{31}	0.334	0.344	0.30	0.212
k_{33} (T = at constant stress)	0.70	0.705	0.64	0.50
$\varepsilon_{33}^T/\varepsilon_0$ (T = at constant stress)	1,300	1,700	1,290	1,700
$\varepsilon_{33}^S/\varepsilon_0$ (s = at constant strain)	635	830	580	1,260
tan δ	0.004	0.02	0.004	0.006
d_{33}, 10⁻¹² C/N	289	374	225	149
d_{31}	-123	-171	-97	-58
g_{33}, 10⁻³ V·m/N	26.1	24.8	25.4	14.1
g_{31}	-11.1	-11.4	-10.9	-5.5
s_{11}^E, 10⁻¹² m²/N (E = at constant electric field)	12.3	16.4	13.5	8.6
s_{33}^E	15.5	18.8	11.5	9.1
s_{11}^D (D = at constant displacement)	10.9	14.4	10.4	8.3
S_{33}^D	7.90	9.46	8.0	7.0
Q_M	500	75	1,000	400
ρ, 10³ kg/m³	7.5	7.75	7.6	5.55
N_1, Hz·m†	1,650	1,400	1,700	2,290
N_3‡	2,000	1,770	2,070	2,740
Curie point, °C	328	365	300	115
Heat capacity, J/kg·°C	420	420	420	500
Thermal conductivity, W/m·°C	2.1	2.1	2.1	3.5
Static tensile strength, lb/in.²	13,000	13,000	12,000	12,000
Rated dynamic tensile strength, lb/in.²	6,000	4,000	7,000	7,500

* Trademark, Vernitron Piezoelectric Division.
† N_1 = frequency constant of a thin bar with electric field perpendicular to length, $f_r \cdot l$.
‡ N_3 = frequency constant of a thin plate with electric field parallel to thickness, $f_r \cdot t$.

titanate, "hard" lead zirconate titanate, and "soft" lead zirconate titanate. Progress has recently been made in specifying and classifying the various ceramic compositions.[161] The properties of the piezoelectric ceramics vary as functions of time, static stress, stress cycling, and electric field strength.[156]

143. Magnetostrictive Materials. Magnetostrictive materials offer certain advantages for some underwater sound transducer applications. One example is that of a large low-frequency source that is submerged to a great depth and must operate unattended for long periods of time. Two forms of magnetostrictive materials are used: metal alloys and ceramic compositions.

The physical quantities regarding the magnetic and mechanical state of a polarized material are the magnetic field strength (H), the magnetic flux density (B), the mechanical stress (T), and the mechanical strain (S). For a sinusoidal variation, these are related by

$$S = s^H T + dH$$
$$B = dT + \mu^T H$$

(25-85a)

or

$$T = c^B S - hB$$
$$H = -hS + \nu^S B$$

(25-85b)

where s^H = elastic compliance at constant magnetic field strength, c^B = elastic stiffness at constant magnetic flux density, μ^T = permeability at constant stress, ν^S = reluctivity at constant strain, and d and h are the piezomagnetic constants that are defined as

$$d = \left(\frac{\partial S}{\partial H}\right)_T = \left(\frac{\partial B}{\partial T}\right)_H$$
$$h = -\left(\frac{\partial T}{\partial B}\right)_S = -\left(\frac{\partial H}{\partial S}\right)_B$$

(25-86)

More detailed information regarding the magnetostrictive equations and constants may be found in References 153, 156, 162, and 163.

The electromechanical coupling factor k of magnetostrictive materials has the same physical meaning as for piezoelectric materials, with the same limitations.

Eddy currents can play an important role in the efficiency of magnetostrictive materials. For this reason, magnetostrictive assemblies are often constructed from thin laminations cemented together, usually in an annealed state. Eddy current losses can be taken into account by multiplying the permeability by a complex eddy current factor χ. This factor depends on the geometry. Modifying the analysis in Ref. 163, a skin-effect parameter m^2 can be defined as

$$m^2 = -\omega\mu(0)/p_e$$

(25-87)

where $\mu(0)$ = permeability with no eddy current effect, ω = angular frequency, and p_e = resistivity.

The apparent permeability is given by

$$\mu = \mu(0)\frac{\tanh mt/2}{mt/2}$$

(25-88)

where t = thickness of the sheet of material.

Figure 25-124 shows a plot of the magnitude and phase angle of the complex correction factor χ versus $mt/2$.

Table 25-18 contains values of important properties of a number of magnetostrictive materials.[156] Reference 156 contains a table that includes more materials but is limited in the number of characteristics shown.

Table 25-18. Properties of Magnetostrictive Materials

	Nickel	Alfenol	Ferroxcube 7A1	Ferroxcube 7A2
k_{33} (opt)	0.15 to 0.31	0.25 to 0.31	0.25 to 0.30	0.21 to 0.25
d_{33} (opt), 10^{-9} Wb/N	~3.1	~7.1	−2.8 to −4.4	−1.6 to −2.9
μ_{33}^S/μ_0 (opt)	22	58	15 to 25	8 to 15
$1/s_{33}^H$, 10^{10} N/m²	~20	~14	15.1	16.1
$Q_M H$	50 to 250		2,500 to 5,000	2,500 to 5,000
tan δ			0.001 to 0.002	0.001 to 0.002
H_{opt}, 10 A/m*	7 to 10	7 to 10	15 to 24	11 to 19
B_{bias}, Wb/m²†	~0.4	~0.6	0.22 to 0.24	0.22 to 0.24
B_R, Wb/m²	~20		0.11 to 0.16	0.15 to 0.17
μ_{33}^S/μ_0,rem	~0.14		30 to 45	30 to 50
k_{33}, rem	~1.5		0.15 to 0.20	0.15 to 0.19
d_{33}, rem			−2.3 to −3.8	−2.2 to −3.7
Resistivity, Ω·m	7×10^{-8}	9×10^{-7}	>10	>10
Curie point, °C	358	~500	530	530
H_c, 10^2 A/m	~0.3	~0.1	2.5 to 5.0	2.0 to 4.0
ρ, 10^3 kg/m³	8.8	6.5	5.35	5.35
v^B, 10^3 m/s	c5.0	~4.8	~5.65	~5.75
v^H, 10^3 m/s	c4.85	~4.55	~5.45	~5.6

* Field required for highest k_{33}.

† $B_{opt} = 0.7 B_{sat}$.

SOURCE: Berlincourt et al., Ref. 156, Par. **25-161**.

TRANSDUCER ARRAYS

144. Beam Formation. Transducers and hydrophones can be arranged individually or in arrays to possess omnidirectional or directional characteristics, depending upon effective aperture dimensions, geometrical shape, and vibrational modes utilized. At high frequencies, since the wavelengths are short, highly directional individual units can be designed. At lower frequencies, multiple transducers or hydrophones are used in arrays of planar, cylindrical, spherical, or volumetric configuration.

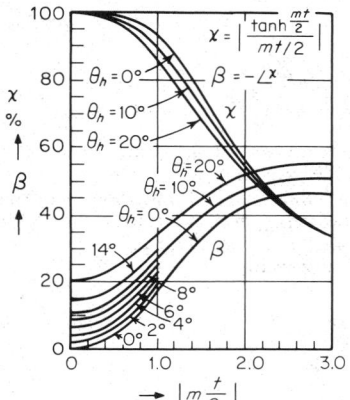

Directionality is highly desirable in underwater sound detection systems because it enables directional transmission as well as the determination of the direction of arrival of a signal. As in directional radar or communications antennas, this reduces the noise relative to the signal from other directions. Arrays can be steered mechanically by physical rotation or electrically by phasing or time delay networks. The direction of maximum sensitivity of a plane array of elements can be rotated into a direction lying at angle θ_0 to a reference direction by delaying differentially the output of each element. In this way, an irregular array can be effectively converted into a line array.

In their simplest form, arrays are arranged with elements along a line or distributed along a plane. The acoustic axis of such line or plane arrays, when unsteered, lies at right angles to the line or plane. The beam pattern of a line array may be visualized as a doughnut-shaped figure having supernumerary attached doughnuts formed by the side lobes of the pattern. The three-dimensional pattern of a plane array is a searchlight type of figure with rotational symmetry about the perpendicular to the plane plus side lobes.

Fig. 25-124. Eddy current loss factor, magnitude, and phase angle of magnetic materials (Ref. 163). $\phi_h < \mu(0)$

145. Lines of Equally Spaced Elements.[150] The beam pattern of a line of equally spaced, equally phased (i.e., unsteered) elements is derived as follows. Let a plane sinusoidal sound wave of unit pressure be incident at an angle θ to a line of n such elements, each spaced from the next a distance d. The output of the mth element relative to that of the zeroth element is delayed by the time necessary for sound to travel the distance $l_m = md \sin \theta$. The corresponding phase delay for sound of wavelength λ, at frequency $\omega = 2\pi f$, is

$$u_m = mu$$

where the phase delay between adjacent elements, in radians, is

$$u = \frac{2\pi d}{\lambda} \sin \theta$$

The output voltage of the mth element of voltage response R_m is

$$V_m = R_m \cos(\omega t + mu)$$

and the array voltage is the sum of such terms:

$$V = R_0 \cos \omega t + R_1 \cos(\omega t + u) + \cdots$$
$$+ R_m \cos(\omega t + mu) + \cdots + R \cos(\omega t + nu)$$

In the complex notation, the array voltage will be

$$V = (R_0 + R_1 e^{iu} + R_2 e^{2iu} + \cdots + R_n e^{(n-1)iu})e^{i\omega t}$$

If the array elements all have unit response ($R = 1$),

$$V = (1 + e^{iu} + e^{2iu} + \cdots + e^{(n-1)iu})e^{i\omega t}$$

Multiplying by e^{iu} and subtracting,

$$V = \frac{e^{inu} - 1}{e^{iu} - 1} e^{i\omega t}$$

Neglecting the time dependence, this becomes

$$V = \frac{\sin(nu/2)}{\sin(u/2)}$$

Finally, expressing u in terms of θ, the beam pattern, the square of the function-normalized to unity at $\theta = 0$ is

$$b(\theta) = \left(\frac{V}{n}\right)^2 = \left[\frac{\sin(n\pi d \sin\theta/\lambda)}{n\sin(\pi d \sin\theta/\lambda)}\right]^2 \tag{25-89}$$

146. Continuous-Line and Plane Circular Arrays.[150] When the array elements are so close together that they may be regarded as adjacent, the array becomes a continuous-line transducer and the beam pattern is found by integration rather than by summation. For this case, let the line transducer be of length L and have a response per unit length of R/L. The contribution to the total voltage output produced by a small element of line length dx located a distance x from the center is (neglecting the time dependence)

$$dv = \frac{R}{L} e^{(i2\pi/\lambda)\sin\theta} dx$$

The beam pattern, the square of V normalized so that $b(\theta) = 1$, is

$$b(\theta) = \left(\frac{V}{R}\right)^2 = \left\{\frac{\sin[(\pi L/\lambda)\sin\theta]}{(\pi L/\lambda)\sin\theta}\right\}^2 \tag{25-90}$$

In a similar manner, the beam pattern of a circular-plane array of diameter D of closely spaced elements can be shown to be

$$b(\theta) = \left\{\frac{2J_1[(\pi D/\lambda)\sin\theta]}{(\pi D/\lambda)\sin\theta}\right\} \tag{25-91}$$

where $J_1[\pi D/\lambda)\sin\theta]$ is the first-order Bessel function of argument $(\pi D/\lambda)\sin\theta$.

Generalized beam patterns for continuous-line and circular-plane arrays are drawn in Fig. 25-125 in terms of the quantities $(L/\lambda)\sin\theta$ and $(D/\lambda)\sin\theta$.

Figure 25-126 is a nomogram for finding the angular width between the axis and the -3-dB and -10-dB points of the beam pattern of continuous-line and circular-plane arrays. The dashed lines indicate how the nomogram is to be used. Thus a circular-plane array of 500 mm diameter at a wavelength of 100 mm (corresponding to a frequency of 15 kHz at a sound speed of 1,500 m/s) has a beam pattern 6° wide between the axis of the pattern and the -3 dB points.

PASSIVE SONAR SYSTEMS

147. Introduction. Passive sonar systems, also referred to as *listening sonar systems*, are designed to respond to acoustic energy radiated by sources in the band of the sonar receiver. These systems are designed to accentuate the response to wanted signals while suppressing unwanted background noise. Passive systems are designed to maximize the signal-to-noise ratio to the degree that the characteristics of the signal and noise are known or can be predicted.

148. Passive Sonar Equations. The fundamental relation of passive sonar can be written in terms of the signal-to-interfering-background ratio. Since in sonar the usual prectice is to write equations in decibel notation, the ratio is defined as the *signal differential* $\Delta L_{S/N}$, in decibels, between the equivalent plane-wave levels of signal and interfering noise at the passive sonar receiving array:

$$\Delta L_{S/N} = (L_{Sf} - N_W + N'_{BW}) - [(L_{Nf} - N_{DI} + N_{BW}) + N_\delta] \tag{25-92}$$

where

L_{Sf} = target-radiated-noise spectrum level vs. frequency at 1 m from the effective center of the radiating source (dB re 1 μPa/Hz$^{1/2}$).

N_W = one-way acoustic propagation loss between the radiating source and the passive sonar array (dB).

N'_{BW} = 10 log (signal bandwidth) = signal-bandwidth-level correction.

L_{Nf} = equivalent-plane-wave interfering-noise spectrum level vs frequency at the passive sonar array resulting from the summation of noise from all sources (dB re 1 μPa/Hz$^{1/2}$).

N_{DI} = 10 log D, where D = effective directivity of the passive sonar array and beam former against isotropic noise (dB). If the noise background has directional components, the effectiveness of the array in decreasing background noise is modified.

N_{BW} = 10 log (noise bandwidth) = noise-bandwidth-level correction.

N_δ = receiving deviation loss (dB).

The value of the signal differential required at the array output varies with the application, e.g., detection, classification, or localization.

The signal differential can be considered to be the sum of two terms: the detection threshold N_{DT}, defined as that value of the signal differential required to just detect the signal, and the signal excess N_{SE}, which is the amount in decibels by which $\Delta L_{S/N}$ exceeds N_{DT}. A detection threshold N_{DT} adjustment in the system is usually set to a value consistent with a sufficiently low false-alarm rate on the display in use. Substituting this sum into Eq. (25-92).

$$N_{SE} = (L_{Sf} - N_W + N'_{BW}) - [(L_{Nf} - N_{DI} + N_{BW}) + N_\delta + N_{DT}] \qquad (25\text{-}93)$$

Equation (25-93) is arranged to show that signal excess in decibels is the differential between a set of signal terms and a set of effective-noise terms, i.e.,

$$N_{SE} = L_I - L_{MD} \qquad (25\text{-}94)$$

where L_I = incident signal level, and L_{MD} = minimum detectable signal level, given by

$$L_{MD} = L_{Nf} - N_{DI} + N_{BW} + N_\delta + N_{DT} \qquad (25\text{-}95)$$

Another useful measure is the figure of merit N_{FM}, defined as the maximum allowable one-way propagation loss under the condition of zero signal excess. From Eq. (25-93),

$$N_{FM} = (L_{Sf} + N'_{BW}) - [(L_{Nf} - N_{DI} + N_{BW}) + N_\delta + N_{DT}] \qquad (25\text{-}96)$$

and from Eq. (25-95), the figure of merit can also be written

$$N_{FM} = L_{Sf} + N'_{BW} - L_{MD} \qquad (25\text{-}97)$$

An example of the use of the passive sonar detection equation is detailed in Fig. 25-127.

149. Sonar Parameters. The radiated-noise level of a target is usually composed of broadband and narrow-band noise from the propeller, machinery, and possibly echo-ranging pings from the target, as well as from other ships. Since this radiation can vary considerably in frequency and transmitted intensity as a function of time, the signal excess, and consequently the maximum detection range, can also vary over a wide range.

Acoustic propagation loss, also called *transmission loss,* varies according to an applicable spreading law modified by absorption, refraction, and reflection. In deepwater, spherical spreading applies, and the propagation loss N_W in decibels is given by

$$N_W = 20 \log R + aR + 60 \qquad (25\text{-}98)$$

where R = range in kilometers; $\alpha(R)$ = absorption in decibels per kilometer, which varies with frequency f; and the number 60 represents the conversion from meters to kilometers. For shallow sources in deep oceans when a surface layer is present, spherical spreading applies for the first kilometer and cylindrical spreading thereafter, and the propagation loss then becomes

$$N_W = 10 \log R + aR + 60 \qquad (25\text{-}99)$$

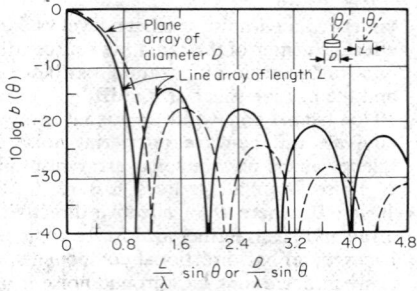

Fig. 25-125. Beam patterns of a line array of length *L* and a circular plane of diameter *D*.

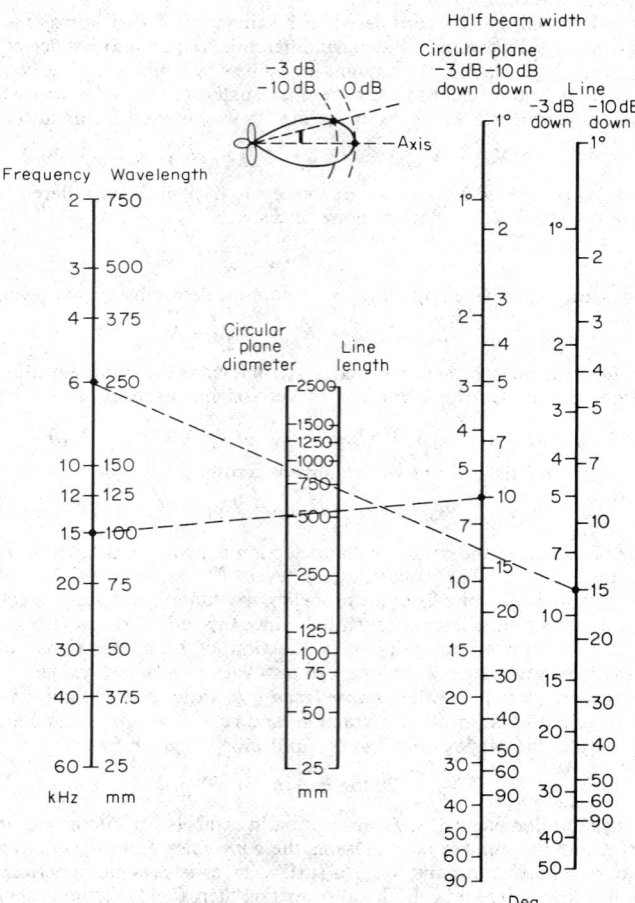

Fig. 25-126. Nomogram for finding width of beam pattern.

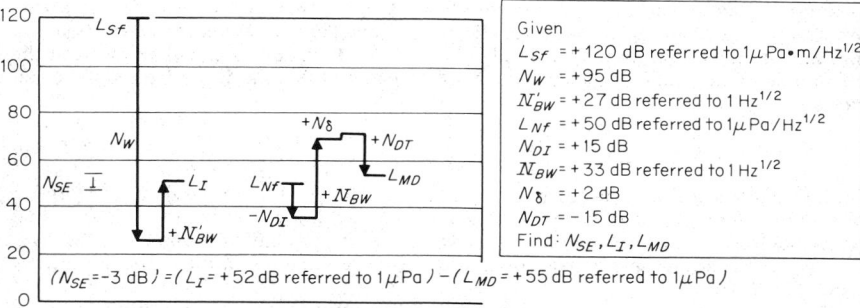

Fig. 25-127. Passive sonar system analysis.

The interfering background noise places a limit on detectability. However, depending on the character of the noise, on the array interelement correlation, and on the signal processing used, the threshold of detection can be lowered considerably below the rms noise band level. Background noise is composed of self-noise, ambient noise, and, when present, reverberation. The noise components in these categories are listed in Table 25-19.

Self-noise is generated by all those sources associated with the listening vessel and its interaction with the surrounding water. *Ambient noise* is caused by both natural and man-made sources, while *reverberation* results from reflections of sonar pings by the ocean surface, bottom, volume, and scattering layers.

Self-noise has many directional characteristics (near-field effects); ambient noise generally has an omnidirectional distribution, with exceptions because of man-made and precipitation components; and reverberation is largely directional in nature.

The effective directivity index is an indicator of the degree to which the receiving array and beam former discriminates against background noise, assuming that this noise is isotropic in character and that the array geometry results in low correlation from element to element in the operating frequency band.

Examination of Eq. (25-96) shows that it is possible to distinguish three groups of parameters that determine figure of merit. First, associated with own ship, its sonar, background noise, and the operator are N_{DI}, N_{BW}, N_δ, N_{DT}, and those components of L_N associated with own ship and the surrounding sea. The second group of parameters is associated with the acoustic properties of the sea (or freshwater) medium and its boundaries, and it consists of N_W and the ambient components of L_N, including surface noise and reflections of noise from the sea surface. The third group of parameters, summarized in the terms L_S and N'_{BW}, is concerned with the radiating acoustic source or target, i.e., the description of the characteristics of the target in terms of radiated-noise spectrum level as a function of frequency, including both broadband and narrow-band spectra, and the band-width of the radiated noise.

150. System Configuration and Parameters. A generalized passive sonar detection model is shown in Fig. 25-128, which depicts all the parameters associated with a passive

Table 25-19. Sources of Background Noise

1. Self-noise	2. Ambient noise	3. Reverberation
a. Hydrodynamic	a. Sea	a. Bottom
(1) Flow-excited	(1) Bubbles	b. Surface
(2) Turbulence	(2) Seismic	c. Volume
(3) Cavitation	(3) Waves	d. Scattering layers
(4) Bubbles	(4) Boundary turbulence	
(5) Splash	(5) Ice motion	
b. Machinery	b. Biological	
(1) Turbines	(1) Shrimp	
(2) Pumps	(2) Whales	
(3) Blowers etc.	(3) Carp etc.	
c. Propeller	c. Man-made	
	(1) Shipping	
	(2) Shore-radiated	
	d. Precipitation	

sonar system and its operator. To account for system hardware/operator gains and losses, terms must be added to Eqs. (25-92), (25-93), (25-95), and (25-96). While these equations are applicable at the output of the idealized beam former, the modified equations apply at the operator decision point indicated in Fig. 25-128. Specifically, at the decision point the signal excess is

$$N'_{SE} = (L_{Sf} + N_S - N_W + N'_{BW} + N_P)$$
$$- [(L_{Nf} - N_{DI} + N_{BW}) + N_M + N_\delta + N_T + N_{SD})] \tag{25-100}$$

The added terms are N_S to account for more than one ray path from the acoustic source to the sonar N_p, for signal processing gain; N_M, for hardware design margin (loss); N_p, for signal processing loss, as in a clipping processor; N_T for the effect of the threshold level, and N_{SD}, to allow for the signal differential required for signal recognition on the particular display used.

The sonar detection system shown in Fig. 25-129 is placed between the hydrophone array and the decision point. It consists of a receiver, a visual and/or aural display, and an operator. The detection threshold N_{DT} is defined[164] as the ratio in decibels of the signal power S in the receiver band to the noise power N_0 in a 1-Hz band, measured at the receiver terminals; i.e.

$$N_{DT} \equiv 10 \log \frac{S}{N_0} \tag{25-101}$$

The signal and signal-plus-noise values are taken in the receiver band, and N_{DT} computed

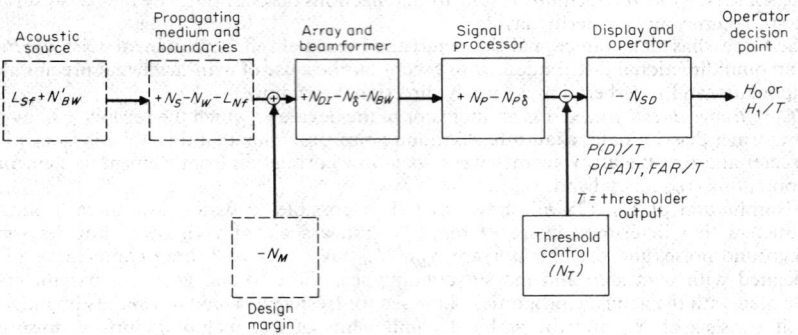

At operator decision point, N_{SE} is given in Eq. (25-100)

Fig. 25-128. Passive sonar system detection model.

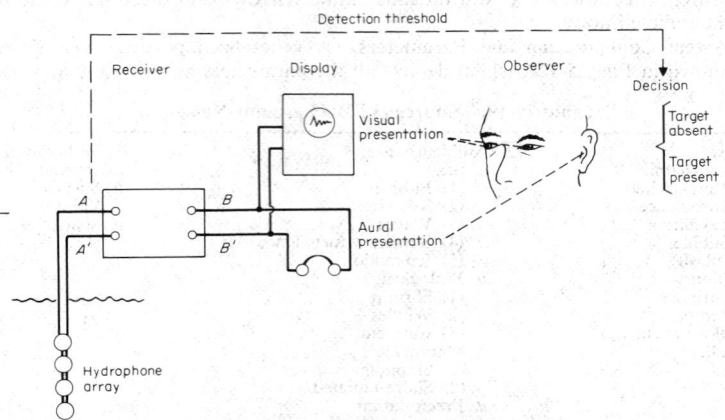

Fig. 25-129. Elements of a sonar receiving system.

for use in the sonar equation. Noise backgrounds are expressed as power spectrum levels, i.e., as powers in 1-Hz bands.

Detection decisions are binary in nature; i.e., signal is present or signal is absent. But since a signal can actually be present or actually absent at the receiver input, there are four possible situations, summarized in Table 25-20. The correct decisions are shown on the diagonal of the matrix, viz., the detection and null decisions, with probabilities $P(D)$ and $1 - P(FA)$, respectively, where $P(D)$ is the probability of detection and $P(FA)$ is the probability of false alarm. False-alarm and miss decisions can occur with respective probabilities $P(FA)$ and $1 - P(D)$.

To implement the detection threshold criterion, it is necessary to apply a threshold voltage to a comparison circuit, at the receiver output. The threshold voltage is either fixed at a level corresponding to the desired N_{DT} or it can be controlled and calibrated in terms of N_{DT}. Whenever the level of the waveform at the comparison point exceeds the threshold voltage, the decision of signal present is made, unless a rule is adopted or circuit implemented, which, for example, counts the number of times the threshold is exceeded during a fixed time interval and bases the detection decision on this count.

The effects of setting the threshold voltage at various levels are shown in Fig. 25-130, at three target signals in a noise background, with two possible threshold voltages. High settings like T_1 allow only strong targets to be detected and extremely few, if any, false alarms; consequently, both $P(D)$ and $P(FA)$ are low. Low settings like T_2 allow many more possible signals to be detected and also many false alarms to occur; consequently both $P(D)$ and $P(FA)$ are high. Both of these threshold settings are unsatisfactory. A threshold setting should be

Table 25-20. Binary Decision Matrix

When at the input:	And the decision is:	
	Signal present	Signal absent
Signal present	Correct decision: detection with probability $P(D)$	Incorrect decision: miss with probability $1-P(D)$
Signal absent	Incorrect decision: false alarm with probability $P(FA)$	Correct decision: null with probability $1-P(FA)$

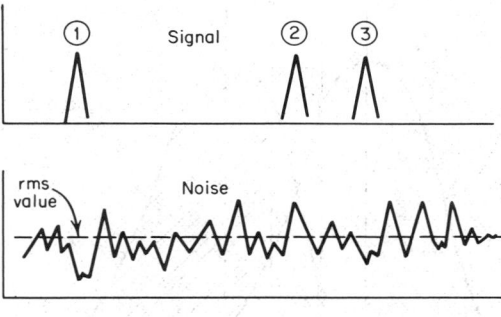

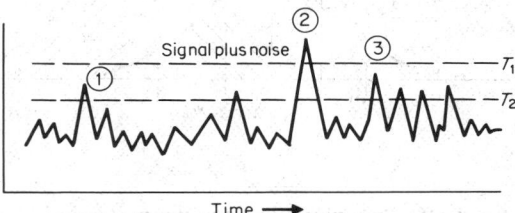

Fig. 25-130. Signal and noise at two threshold settings.

found that produces a display with a uniformly distributed (gray) background in the presence of noise only, and which prevents the buildup of the number of noise markings as a function of time and the decay rate of the display storage system in use; i.e., the false-alarm rate must be bounded.

Detection decisions depend upon the independent probabilities $P(D)$ and $P(FA)$, and consequently on the distributions of noise and signal-plus-noise at the receiver output. Figure 25-131 shows the probability density of noise alone and signal-plus-noise plotted as a function of amplitude[164] a. The curves are assumed to have a gaussian distribution with variance σ^2, the first with mean noise amplitude $M(N)$ and the second with mean signal-plus-noise amplitude $M(S+N)$. Then the parameter detection index d is defined as

$$d \equiv \frac{(M_{S+N} - M_N)^2}{\sigma^2} \qquad (25\text{-}102)$$

which is equivalent to the signal-to-noise ratio of the envelope of the receiver output effectively at the terminals where the threshold voltage T is established. The area (integral) under the probability density curve of signal-plus-noise to the right of T in Fig. 25-131 is the probability that an amplitude in excess of T is due to signal-plus-noise and is equal to the probability of detection $P(D)$. Similarly, the area under the probability curve of noise alone to the right of T is equal to the probability of false alarm $P(FA)$. Since $P(D)$ and $P(FA)$ vary as the threshold T is changed, their values depend upon the parameter d.

The probabilities associated with various threshold voltage levels can be plotted, as shown in Fig. 25-132, with the detection index d as a parameter. Thus, for high-T settings, for example, the corresponding low $P(D)$, $P(FA)$ points are plotted in the lower left portion of the figure above the diagonal. These are known as receiver operating characteristic (ROC) curves.

If the curves of Fig. 25-131 are imagined to refer to the receiver input, the likelihood ratio for an input sample of amplitude a is

$$L = \frac{B}{A} \qquad (25\text{-}103)$$

Thus, the ROC curves are related to the signal-to-noise ratio at the receiver input required for detection.

Following the detector, a postdetection averaging filter is used to smooth out fluctuations

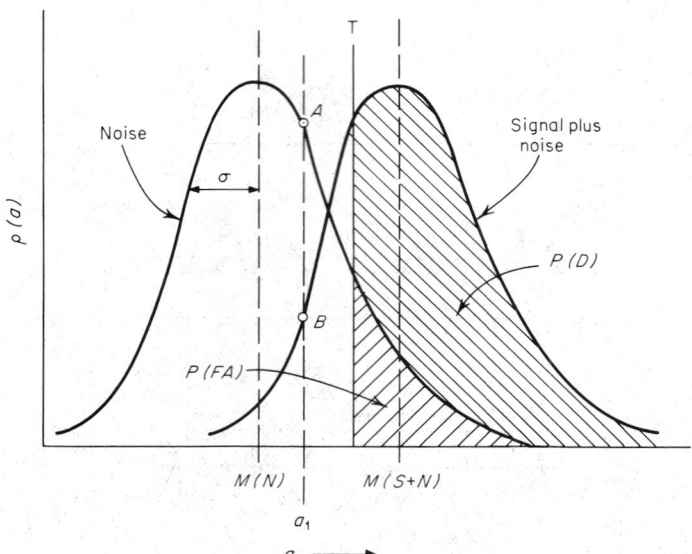

Fig. 25-131.   Probability density distributions of noise.

25-128

due to detector output noise. The optimum integration time T of this filter is the time equal to the signal duration t. In passive sonar systems t can be as long as it is operationally useful to integrate the incoming signal in the noise and/or reverberation background. Here, factors such as stabilization of own ship's receiving beams (for rotating or preformed beams), rotating receiving beam RPM (which determines the time on target during each rotation), receiving beamwidth, possible range of the target, and possible speed of the target must be considered.

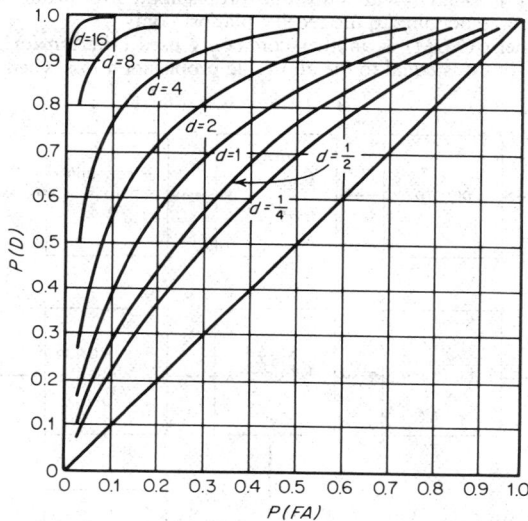

Fig. 25-132. Receiver operating characteristic (ROC) curves.

ACTIVE SONAR SYSTEMS

151. Introduction. Active sonar systems, like radar, make use of *reflected energy reception,* referred to as *echo ranging.* Active sonar systems are used primarily to determine target range and bearing. In addition, active systems may be used for determining target depth, aspect angle, course, and speed. Target motion may be computed from successive echo returns or measurement of target-generated doppler on each ping.

A wide variety of signal waveforms, pulsed or continuous, may be used in echo-ranging operations. Pulsed continuous waves of 1-ms to several seconds duration are commonly used, short pulses providing maximum range resolution and long pulses providing maximum energy return from the target, hence greater detection range. Pulsed frequency-modulated waves with up or down sweeps and pseudo-random noise waves are used to obtain better target discrimination in a reverberant background. Combinations of the foregoing waveforms may be used to reduce mutual interference.

Continuous transmission systems (cw or FM) are employed in short-range work, such as navigation sonars. Continuous transmission is not used in long-range detection systems because of the difficulty of acoustic isolation between the receiving and transmitting arrays.

152. Active Sonar Equation. The active sonar equation, corresponding to the model shown in Fig. 25-133, is used for performance prediction. The following equation is commonly used for applications with two-way propagation loss:

$$N_{SE} = L_S - (N_{\delta T} + N_{\delta R}) + N_{TS} - 2N_W - L_B - N_{DT} \qquad (25\text{-}104)$$

where
N_{SE} = signal excess (dB)
L_S = source level (dB re 1μ Pa at 1 m) on transmit beam axis
$N_{\delta T}, N_{\delta R}$ = transmit and receive deviation losses due to target ray being off axis of transmit and receive beams, respectively (dB)
N_{TS} = target strength (dB)

N_W = one-way propagation loss (dB)

L_B = noise level in receiver band, including ambient noise, self-noise, and reverberations (dB re 1 μPa)

N_{DT} = detection threshold in receiver band and at the receiver input to produce desired display statistic (dB)

Equation (25-104) also applies to applications with one-way propagation losses, such as communications or telemetry, with only one propagation loss taken into account and background noise corresponding to the receive platform only.

The figure of merit (FOM) is another commonly used criterion for evaluating sonar performance. The FOM is equal to the allowable propagation loss when the signal excess is zero.

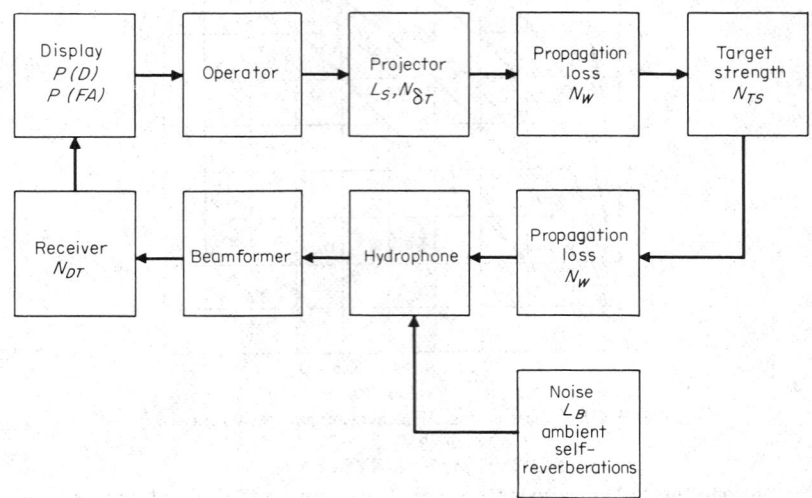

Fig. 25-133. Echo-ranging block diagram.

153. Source Level. The radiated acoustic powers of shipboard sonars[164] range from a few hundred watts to tens of kilowatts, with transmitting directivity indexes between 10 and 30 dB. The equation for the value of the source level L_S is

$$L_S = 170.8 + 10 \log P_A + N_{DI} - N_D \qquad (25\text{-}105)$$

where P_A = acoustic radiated power in watts, N_{DI} = directivity index in decibels, and N_D = dome loss in decibels where applicable.

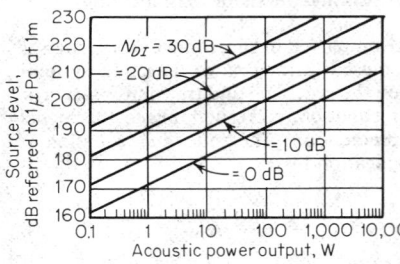

Fig. 25-134. Plot of Eq. (25-105).

A plot of L_S versus P_A is shown in Fig. 25-134 with N_{DI} as a parameter. The efficiency of converting electric to acoustic power varies with the transducer design, the match between transmitter source and transducer load impedance, and array-mutual-interaction effects.

There are two major acoustic limitations that limit achievable source level; cavitation and array mutual interaction. Cavitation limits occur[164] when power applied to a transducer is such that bubbles begin to form on the face and just in front of the transducer. These bubbles are a manifestation of cavitation and are caused by the rupture of water through negative pressure of the generated sound field. Cavitation causes a deterioration of transducer performance and life by erosion of the transducer surface; loss of radiated acoustic power in absorption and scattering by cavitation bubbles; deterioration in beam pattern; and reduction of the acoustic impedance into which the transducer must radiate. Cavitation threshold occurs at the point of departure from linearity of the input/output power curve and is a function of frequency, pulse length, and operating depth.

The effect of increased depth of operation is to increase the cavitation threshold:

$$P_C(h) = P_c(0) (1 + 0.1h)^2 \qquad (25\text{-}106)$$

where $P_C(h)$ = cavitation threshold in watts per square meter at the operating depth h in meters. The cavitation threshold increases with decreasing pulse lengths below 5 ms, as shown in Fig. 25-135.

Unequal loading effects occur in arrays of transducer elements because of mutual radiation impedances[173,174] between independent sound sources. These effects cause hot spots in the array (i.e., one element receives many times the average power per element), so that it may be driven to destruction. Some elements can also absorb the acoustic output of others even when all elements are driven in like manner. Other effects are a gradual deterioration of beam pattern (or directivity index) and reduction of electrical power into the array, with corresponding loss of L_S. Mutual-

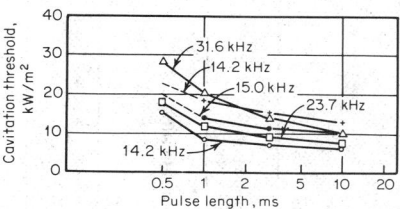

Fig. 25-135. Cavitation threshold vs. pulse duration.

interaction effects can be controlled by proper spacing between array elements, tuning the transducer element to reduce mutual-radiation impedance, increasing the self-radiation impedance and controlling individual transducer velocity and phase.

Sonar domes may have a significant transmission loss. Expressions for the transmission loss and specular reflections have been obtained theoretically and verified experimentally. Both increase with frequency and thickness, as well as density of the dome wall.

154. Target Strength. Target strength relates the echo intensity returning from a target to the incident intensity. The target strength, a concept equally useful for submarine, mine detection, and fish location, is defined as 10 times the logarithm to the base 10 of the ratio of the intensity of the sound returned by the target, at a distance of 1 m from its acoustic center to the incident intensity from a distant source. The theoretical target strength of a number of geometric shapes and forms is shown in Table 25-21. Nominal values of measured target strength for typical underwater targets are tabulated in Table 25-22. Target strength is a function of aspect angle, frequency, and pulse length.

155. Deviation and Propagation Losses.

Deviation loss results when the target is not on the maximum response axis of either the transmitting or receiving beam or both. Deviation loss varies from 0 dB on the maximum response axes to several decibels off the axis, depending upon the beam pattern responses at the specific angles involved.

The propagation loss is a highly variable function, and only average values can be predicted. Average propagation-loss data have been published as a function of ocean model, propagation path, wind state, and season.[164,168] In the absence of empirical data, several theoretical models may be used: spherical or cylindrical spreading loss as a function of range, sound absorption loss as a function of range and frequency, and sound wave or ray propagation.

The propagation from the target to the transducer is the same as that from the transducer to the target but may include several propagation paths. Propagation paths of interest for active sonar include in-layer direct path, across-layer direct path, bottom bounce, and convergence zone. Separate elevation angles may be required for the different propagation paths, depending on the ocean model.

156. Background Noise. Background noise in an active sonar system[164,168] is the random addition of ambient and self-noise as well as reverberations. Ambient and self-noise are assumed to be stationary during the ping cycle, while reverberations vary with detection range. Ambient and self-noise are usually assumed to be isotropic in azimuth and elevation

Table 25-21 Target Strength of Simple Forms

Form	Target Strength t $= 10 \log t$	Symbols	Direction of incidence	Conditions
Any convex surface	$\dfrac{a_1 a_2}{4}$	$a_1\, a_2$ = principal radii of curvature r = range $k = 2\pi/$wavelength	Normal to surface	$ka_1,\ ka_2 \gg 1$ $r > a$
Sphere Large	$\dfrac{a^2}{4}$	a = radius of sphere	Any	$ka \gg 1$ $r > a$
Small	$61.7 \dfrac{V^2}{\lambda^4}$	V = volume of sphere λ = wavelength	Any	$ka \ll 1$ $kr \gg 1$
Cylinder Infinitely long Thick	$\dfrac{ar}{2}$	a = radius of cylinder	Normal to axis of cylinder	$ka \gg 1$ $r > a$
Thin	$\dfrac{9\pi^4 a^4}{\lambda^2}\, r$	a = radius of cylinder	Normal to axis of cylinder	$ka \ll 1$
Finite	$\dfrac{aL^2}{2\lambda}$	L = length of cylinder a = radius of cylinder	Normal to axis of cylinder	$ka \gg 1$ $r > L^2/\lambda$
	$\dfrac{aL^2}{2\lambda}\left(\dfrac{\sin\beta}{\beta}\right)^2 \cos^2\theta$	a = radius of cylinder $\beta = kL \sin\theta$	At angle θ with normal	
Plate Infinite (plane surface)	$\dfrac{r^2}{4}$		Normal to plane	

Finite Any shape	$\left(\dfrac{A}{\lambda}\right)^2$	A = area of plate L = greatest linear dimension of plate l = smallest linear dimension of plate	Normal to plate	$r > \dfrac{L^2}{\lambda}$ $kl \gg 1$
Rectangular	$\left(\dfrac{ab}{\lambda}\right)^2 \left(\dfrac{\sin\beta}{\beta}\right)^2 \cos^2\theta$	a, b = side of rectangle $\beta = ka\sin\theta$	At angle θ to normal in plane containing side a	$r > \dfrac{a^2}{\lambda}$ $kb \gg 1$ $a > b$
Circular	$\left(\dfrac{\pi a^2}{\lambda}\right)^2 \left(\dfrac{2J_1(\beta)}{\beta}\right)^2 \cos^2\theta$	a = radius of plate $\beta = 2ka\sin\theta$	At angle θ to normal	$r > \dfrac{a^2}{\lambda}$ $ka \gg 1$
Ellipsoid	$\left(\dfrac{bc}{2a}\right)^2$	a, b, c = semimajor axes of ellipsoid	Parallel to axis of a	$ka, kb, kc \gg 1$ $r \gg a, b, c$
Conical tip	$\left(\dfrac{\lambda}{8\pi}\right)\tan^4\psi\left(1 - \dfrac{\sin^2\theta}{\cos^2\psi}\right)^{-3}$	ψ = half angle of cone	At angle θ with axis of cone	$\theta < \psi$
Average overall aspects Circular disk	$\dfrac{a^2}{8}$	a = radius of disk	Average overall directions	$ka \gg 1$ $r > \dfrac{(2a)^2}{\lambda}$
Any smooth convex object	$\dfrac{S}{16\pi}$	S = total surface area of object	Average overall directions	All dimensions and radii of curvature large compared with λ
Triangular corner reflector	$\dfrac{L^4}{3\lambda^2}(1 - 0.00076\theta^2)$	L = length of edge of reflector	At angle θ to axis of symmetry	Dimensions large compared with λ

SOURCE: Urick, Ref. **164**, Par. **26-161**.

for performance predictions (i.e., in the active sonar equation), so that the ambient-noise band level L_B at the receiver input is

$$L_B = L_{Nf} + N_{BW} - N_{DI} \qquad \text{(dB)} \qquad (25\text{-}107)$$

where L_{Nf} = equivalent isotropic spectrum level of ambient or self-noise (dB re 1 μPa) in a 1-Hz band, N_{BW} = receiver input band level (dB re 1 Hz), and N_{DI} = directivity index (dB).

Neither the ambient nor the self-noise is in fact isotropic, so that performance predictions based on equivalent isotropic noise levels predict average conditions. Ambient noise varies with elevation angle, as shown in Fig. 25-136, and varies in azimuth as well as with the maximum noise arriving from the direction of the wave swells. In addition to being anisotropic, ambient noise varies with the ocean configuration, wind, current, diurnal heating and cooling cycles, and widely varying velocity structures.

Self-noise is also highly directional, and as might be suspected, the highest self-noise on ships and submarines originates at the propeller. Self-noise can be separated into platform and equipment noise. Sonar systems are usually designed so that equipment noise is small compared with ambient or platform noise.

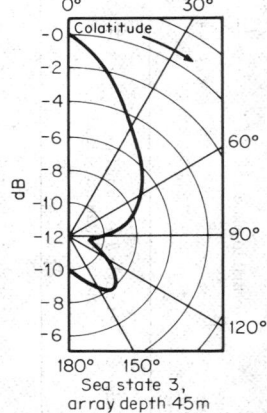

Fig. 25-136. Vertical directivity of ambient sea noise.

The receiving transducer may be the same as the transmitting transducer or can be completely independent. When the same transducer is used, a transmit/receive switch connects the transducer to either the transmitter or beam former as required. The receive transducer (or set of transducers) accepts the echo plus noise and furnishes these to a beam former. Beam formers reject noise and have either an adaptive or fixed configuration. Adaptive beam formers are designed to reject variable directional noise sources such as self-noise and reverberations by monitoring the noise field and adjusting internal parameters to minimize the background noise. Fixed beam formers are designed to reject isotropic noise and take advantage of the directional property of the echo.

157. Reverberation.[164] Reverberation often limits performance on modern high-power sonars. The boundaries of the ocean at the surface and bottom and inhomogeneities in the ocean, such as schools of fish, layers of air bubbles near the surface, the deep scattering layer, as well as pinnacles and seamounts on the sea bed, all reradiate a portion of acoustic energy incident on them. This reradiation of sound is called *scattering,* and the sum-total acoustic energy at the receiving transducer from all the scatterers is called *reverberations.* There are many different types of reverberations, and it is important to visualize the kinds of reverberations occurring during the ping cycle and what azimuth and elevation angle these will be coming from.

Certain characteristics of reverberations can be used to good advantage by the sonar designer. Knowledge of the directions of reverberations may be used to reduce their effect on sonar operations. For instance, with a direct propagation path, a narrow vertical

Table 25-22. Nominal Values of Target Strength

Target	Aspect	Target strength, dB
Submarines	Beam	+25
	Bow-stern	+10
	Intermediate	+15
Surface ships	Beam	+25 (highly uncertain)
	Off-beam	+15 (highly uncertain)
Mines	Beam	+10
	Off-beam	+10 to −25
Torpedoes	Bow	−20
Fish of length L, ft	Dorsal view	−31 + 30 log L

SOURCE: Urick, Ref. 164, Par. 25-161.

transmit-and-receive beam pattern will first decrease the incident acoustic energy on the bottom, and second, reject reverberations coming from the bottom outside the main lobe of the receive beam pattern.

A knowledge of the reverberation spectrum can also be used to minimize the effect. The sonar system then becomes self-noise or ambient noise limited with a significant drop in the background-noise level and corresponding improvement in system performance. Long cw pulses have reverberation spectrums which are small compared with the receiver input bandwidth and lend themselves readily to this scheme. Own-ship doppler nullification (ODN) circuits compensate for platform velocity-induced shifts in center frequency of the reverberation spectrums. The notch filter needed for rejection of the reverberation has to be wide enough for the transmitted pulse spectrum convolved with the scatterer motion. Homogeneous scatterer motion causes center-frequency shifts, and random scatterer motion causes freqency spreading. For echo-ranging applications, it is usually possible to reject the reverberations with a notch filter with a bandwidth corresponding to a $\pm$ 0.25 m/s random scatterer motion.

The doppler shift of the center frequency of the reverberations from that of the transmitted pulse spectrum can be used to measure ship speed with respect to the immediate surrounding water.

158. Detection Threshold. The detection threshold N_{DT} is the signal-to-noise ratio, S/N, required in the receiver input band at the receiver input to produce the desired display statistics.[164] The term N_{DT} is usually applied to surveillance displays and refers to the desired single-ping probability of detection and false alarm. For other sonar modes of operation, such as fine-grained measurement of target range, doppler range rate, bearing, and elevation angle, the target has already been detected, and the required S/N at the receiver input is related to the desired standard deviation of the estimator.

The desired display statistics for a surveillance display are usually a 50% probability of detection and an acceptably low number of false alarms on the display. A typical surveillance display is a B-scan with linear presentation of range along the vertical axis and linear presentation of bearing along the horizontal axis. The number of independent range cells is usually taken as the echo-ranging time required for the display range gate, in seconds, times the receiver output bandwidth in hertz. The number of independent bearing cells is usually the number of beams required to give continuous bearing coverage for the desired azimuthal sector. The false-alarm probability of any one of the independent range-bearing cells is just the number of allowed false alarms on the display divided by the total number of such range-bearing cells.

Once the desired probability of detection $P(D)$ and probability of false alarm $P(FA)$ have been established, the required S/N at the display input can be derived through the use of receiver operating curves (ROC) as described in Par. 25-150.

The N_{DT} at the receiver input can be established through the use of S/N transfer functions for a given signal processor.[164,175-177] It should be noted that the desired $P(FA)$ of independent range-bearing cells is usually also a function of the signal processor because of different receiver output bandwidths for different signal processors.

A sonar receiver for surveillance has to work over a wide range of conditions, which, because of the nature of the targets, are generally unknown. A *robust receiver* (i.e., one that works well over all the expected variations of input parameters) or a combination of several receivers to cover such variations is usually desired. The "ideal" N_{DT} for any particular receiver degrades with target doppler, multipath arrivals, and nonstationary noise background.

The selection of the receiver for a particular application cannot be based on the lowest N_{DT} alone, but has to take many other receiver characteristics into account, such as cost, reliability, maintainability, logistics, power consumption, cooling requirement, operator training, weight, space, plus additional items peculiar to the application.

The optimum receiver for a signal known exactly, with white, stationary, gaussian noise, is a matched filter. The signal processing gain for this type of receiver is 10 times the logarithm to the base 10 of the time-bandwidth product of the signal, and the resulting N_{DT} is the lower limit. Cross-correlators and comb-filter receivers are matched-filter-type receivers.

Incoherent receivers are optimum for unknown signals, with white, stationary, gaussian noise. The signal processing gain for this type of receiver is 5 times the logarithm to the base

10 of the time-bandwidth product. Energy and envelope detectors, as well as autocorrelators and spatial correlators, are examples of this type of receiver.

Since the echo for an active sonar is neither known exactly nor completely unknown, semicoherent receivers are sometimes used which have matched-filter features for those echo characteristics which are relatively well known and incoherent receiver characteristics for those echo characteristics which are likely to change significantly. The post detection pulse-compression receiver is an example of this kind of receiver. The signal processing gain of this type of receiver is between that of the matched-filter type and the incoherent receiver.

159. Surveillance Receivers. A typical active surveillance display is shown in Fig. 25-137. This is a rectangular B-scan format with linear presentation of range along the vertical axis and linear presentation of bearing along the horizontal axis. The center bearing of the B-scan display can usually be selected to be true North or ship's bearing by the operator. It is driven by a thresholded output of a multichannel surveillance receiver. The signal excess over the threshold controls the display intensity in each bearing-range cell. The ability to threshold relies on noise-power normalization, so that changes in background-noise levels caused by variations in ambient noise and/or self-noise, as well as reverberations, do not cause an excessive number of false alarms.

A number of different surveillance receivers are available to drive the display: linear or clipped correlators, autocorrelator, spatial correlator, comb-filter receivers (with or without reverberation suppression), energy or envelope detectors, and pulse compression receivers. Performance prediction for the various types of receivers is covered in the literature on signal processing techniques. It is important to remember that these receivers should be compared on the basis of equal display statistics and nonstationary reverberations, self-noise, and ambient background noise, as well as "ideal" conditions.

Noise-power variations may be reduced by clipping, nonlinear amplification such as logarithmic amplifiers, or variable linear amplification such as time variable gain (TVG), automatic gain control (AGC), or step gain control (SGC) ahead of the receiver. Clipping is inexpensive but has the disadvantage of possible capture of the clipper by strong interfering noises. Logarithmic or other nonlinear amplifications can hold the noise power constant but cause frequency distortion and spreading. TVG is generally used early in the ping cycle to prevent system saturation right after transmission and then to increase the gain vs. time in a predetermined manner as a function of the transmitted source level and pulse length. AGC is closed-loop gain control where the noise power is estimated and the gain inversely controlled to maintain constant noise power. SGC is like AGC except that the gain is

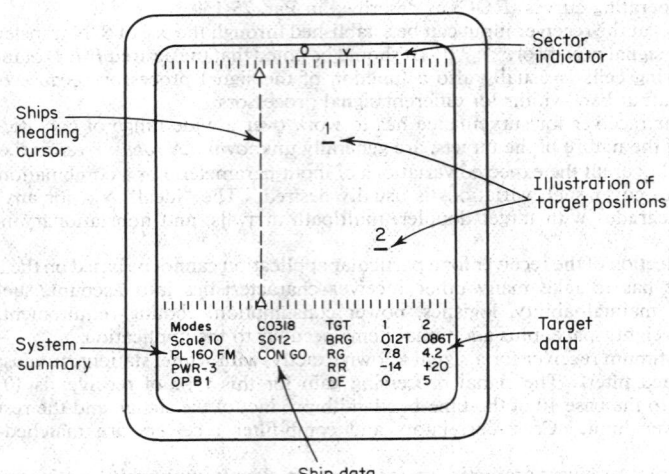

Fig. 25-137. B-scan surveillance display.

controlled in steps, so that inverse proportional gain control is not needed. Both AGC and SGC are sometimes referred to as reverberation gain control (RGC).

Another type of active surveillance display is the plan position indicator (PPI) display. The PPI format has the ship at the display center, with bearing linearly presented as display angle and the ship's heading presented vertically upward. Range is linearly presented as the radius from the display center.

Track and localization displays are usually presented in A-scan format, with a range gate linearly presented along the horizontal axis and either azimuth or elevation angle gate or doppler range rate linearly presented along the vertical axis. Localization and track displays and receivers are generally rated according to the resolution in range, bearing, elevation, and doppler range rate available to the operator as a function of S/N.

160. Sonobuoy Systems. Sonobuoy systems are miniature passive and/or active sonar systems designed for deployment from aircraft, making use of vhf or uhf radio channels for transmission of information from the buoy to the aircraft. In its simplest configuration, a passive sonobuoy consists of a cable-connected omnidirectional hydrophone which is released from the buoy after water entry and which sinks to a predetermined operating depth; an amplifier to raise the level of the hydrophone output; a vhf or uhf transmitter; antenna system; and batteries for power. These items are often contained within a cannister less than 4 ft in length and 5 in. in diameter, fitted with a rotochute or other device for control of descent after release from the aircraft. The buoy is designed to float so as to maintain its antenna system above the water surface. On the aircraft the sonobuoy signal is extracted from the radio channel and applied to a processor/recorder and operator's headphones. Multichannel receiving and processing equipment is used to monitor the outputs of several sonobuoys simultaneously (Fig. 25-138).

More complex sonobuoy systems result when directional arrays are used in lieu of the simple omnidirectional hydrophone, necessitating the transmission of array orientation and

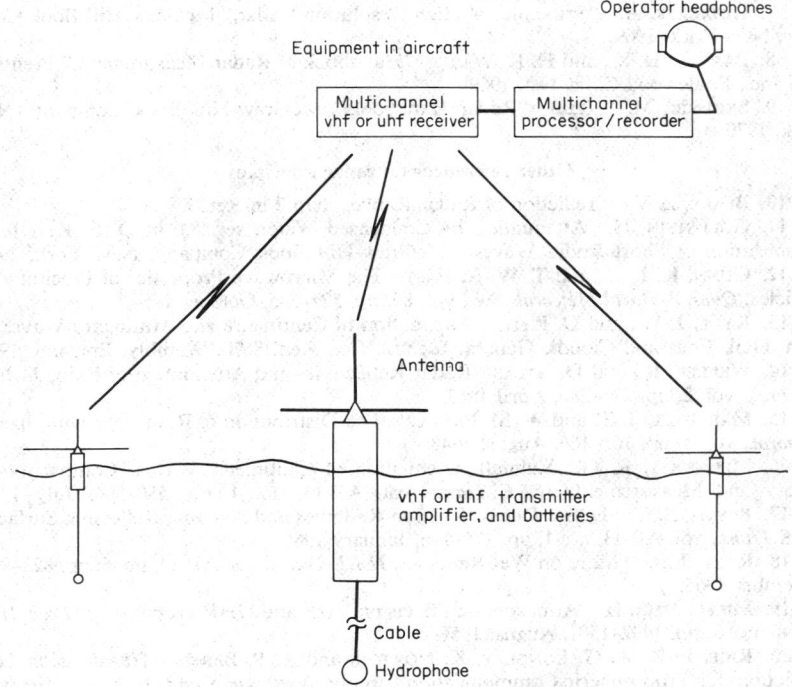

Fig. 25-138. Passive sonobuoy system.

target-bearing information via the radio link. Greater complexity, posing a severe challenge to the sonobuoy designer, results from combination of passive/active systems and control of the sonobuoy functions from the aircraft via a command radio link. Because of the advances made in microminiaturization, integrated circuits, multiplexing, miniature acoustic arrays, digital signal processing, and high-density packaging, it is now possible to design sophisticated multifunction sonobuoy systems.

The design of such systems includes heavy emphasis on sonobuoy performance, cost, size, weight, life, and reliability factors—in view of the end usage of the buoy as an expendable item. System partitioning, in which the functions to be performed in the buoy and in the aircraft equipment are defined, is an important element of the performance/cost consideration, particularly for the more sophisticated sonobuoy systems. Sonobuoy system design includes the application of passive and active sonar system design techniques described in this section of the Handbook, plus aerodynamic and hydrodynamic technologies.

161. Bibliography

Reference books on radar

1. RIDENOUR, L. N. (ED.)   "Radar System Engineering," M.I.T. Radiation Laboratory Series, vol. 1, McGraw-Hill Book Company, New York, 1947.
2. SKOLNIK, M. I.   "Introduction to Radar Systems," McGraw-Hill Book Company, New York, 1962.
3. BARTON, D. K.   "Radar System Analysis," Prentice-Hall, Inc., Englewood Cliffs, N.J., 1964.
4. BERKOWITZ, R. S. (ED.)   "Modern Radar," John Wiley & Sons, Inc., New York, 1965.
5. COOK, C. E., and M. BERNFELD   "Radar Signals," Academic Press, Inc., New York, 1967.
6. NATHANSON, F. E.   "Radar Design Principles: Signal Processing and the Environment," McGraw-Hill Book Company, New York, 1969.
7. RIHACZEK, A.   "Principles of High Resolution Radar," McGraw-Hill Book Company, New York, 1969.
8. BARTON, D. K., and H. R. WARD   "Handbook of Radar Measurement," Prentice-Hall Inc., Englewood Cliffs, N.J. 1969.
9. SKOLNIK, M. I. (ED.)   "Radar Handbook," McGraw-Hill Book Company, New York, 1970.

Other references on radar principles

10. BLAKE, L. V.   Prediction of Radar Range, chap 2 in Ref. 9.
11. GOLDSTEIN, H.   Attenuation by Condensed Water, sec. 8.6 in D. E. Kerr (ed.), "Propagation of Short Radio Waves," McGraw-Hill Book Company, New York, 1951.
12. GUNN, K. L. S., and T. W. R. EAST   The Microwave Properties of Precipitation Particles, *Quar. J. Royal Meteorol. Soc.,* vol. 80, pp. 522–545, October 1954.
13. RYDE, J. W., and D. RYDE   Attenuation of Centimetre and Millimetre Waves by Rain, Hail, Fogs and Clouds, General Electric Co., Rep. 8670, Wembly, England, 1945.
14. WEXLER, R., and D. ATLAS   Radar Reflectivity and Attenuation of Rain, *J. Appl. Meteorol.,* vol. 2, pp. 276–280, April 1963.
15. MARSHALL, J. S., and W. M. PALMER   The Distribution of Raindrops with Size, *J. Meteorol.,* vol. 5, pp. 165–166, August 1948.
16. MEDHURST, R. G.   Rainfall Attenuation of Centimeter Waves: Comparison of Theory and Measurement, *IEEE Trans.,* vol. AP-13, no. 4, pp. 550–564, July 1965.
17. BLEVIS, B. C.   Losses Due to Rain on Radomes and Antenna Reflecting Surfaces, *IEEE Trans.,* vol. AP-13, no. 1, pp. 175–176, January 1965.
18. RUZE, JOHN   More on Wet Radomes, *IEEE Trans.,* vol. AP-13, no. 5, pp. 823–824, September 1965.
19. MILLMAN, G. H.   Atmospheric Effects on VHF and UHF Propagation, *Proc. IRE,* vol. 46, no. 8, pp. 1492–1501, August 1958.
20. RICE, P. L., A. G. LONGLEY, K. NORTON, and A. P. BARSIS   Transmission Loss Predictions for Tropospheric Communication Circuits, *Natl. Bur. Stand. Tech. Note* 101 (vol. 2), 1967, rev. U.S. Govt. Printing Office.

21. BECKMAN, P., and A. SPIZZICHINO   "The Scattering of Electromagnetic Waves from Rough Surfaces," The Macmillan Company, New York, 1963.

22. BEAN, B. R., and G. D. THAYER   CRPL Exponential Reference Atmosphere, *Natl. Bur. Stand. Monog.* 4, Oct. 29, 1959.

23. THOMPSON, M. C., H. B. JONES, and R. W. KIRKPATRICK   An Analysis of Time Variations in Tropospheric Refractive Index and Apparent Radio Path Length, *J. Geophys. Res,* vol. 65, no. 1, pp. 193–201, January 1960.

24. PFISTER, W., and T. J. KENESHEA   "Ionospheric Effects on Positioning of Vehicles at High Altitudes," Air Force Survey in Geophysics No. 83, Cambridge Research Center, Cambridge, Mass., March 1956 (DDC document AD 98 777).

25. SWERLING, P.   Probability of Detection for Fluctuating Targets, *IRE Trans.,* vol. IT-6, no. 2, pp. 269–308, April 1960.

26. BOOTHE, R. R.   "The Weibull Distribution Applied to the Ground Clutter Backscatter Coefficient," U.S. Army Missile Command, Redstone Arsenal, Ala., Rep. RE-TR-69-15, June 12, 1969 (DDC document AD 691 109).

27. BARTON, D. K.   Radar Equations for Jamming and Clutter, *IEEE Trans.,* vol. AES-3, no. 6, pp. 340–355, November 1967 (EASTCON Suppl.).

28. TRUNK, G. V., and S. F. GEORGE   Detection of Targets in Non-Gaussian Sea Clutter, *IEEE Trans.,* vol. AES-6, no. 5, pp. 620–628, September 1970.

29. DUNN, J., and D. D. HOWARD   "Target Noise," chap. 21 in Ref. 9.

30. IEEE Standard Definitions, No. 172: Navigation Aid Terms, March 1971.

31. MARCUM, J. I.   A Statistical Theory of Target Detection by Pulsed Radar, *IRE Trans.,* vol. IT-6, no. 2, pp. 59–267, April 1960.

32. DiFRANCO, J. V., and W. L. RUBIN   "Radar Detection," Prentice-Hall, Inc., Englewood Cliffs, N.J., 1968.

33. BARTON, D. K.   Simple Procedures for Radar Detection Calculations, *IEEE Trans.,* vol. AES-5, no. 5, pp. 837–846, September 1969.

34. HEIDBREDER, G. R., and R. L. MITCHELL   Detection Probabilities for Log Normally Distributed Signals, *IEEE Trans.,* vol. AES-3, no. 1, pp. 5–13, January 1967.

35. TRUNK, G. V.   Detection of Targets in Non-Rayleigh Sea Clutter, *IEEE EASCON 1971 Rec.,* pp. 239–245.

36. WOODWARD, P. M.   "Probability and Information Theory with Applications to Radar," Pergamon Press, New York 1953.

37. RIHACZEK, A. W.   "Principles of High Resolution Radar," McGraw-Hill Book Company, New York, 1969.

38. FARNETT, E. C., T. B. HOWARD, and G. H. STEVENS   Pulse-compression Radar, chap. 20, in Ref. 9.

39. HANNAN, P. W.   Optimum Feeds for All Three Modes of a Monopulse Antenna, *IRE Trans.,* vol. AP-9, no. 5, pp. 444–461, September 1961.

40. MANASSE, R.   Range and Velocity Accuracy from Radar Measurements, *M.I.T. Lincoln Lab Rep.* 312–326, Feb. 3, 1955 (DDC document AD 236 236).

41. BARTON, D. K.   Radar Measurement Accuracy in Log-Normal Clutter, *IEEE EASCON 1971 Rec.,* pp. 246–251.

42. HALL, W. M.   Prediction of Pulse Radar Performance, *Proc. IRE,* vol. 44, no. 2, pp. 224–231, February 1956.

43. HALL, W. M.   General Radar Equation in "Space/Aeronautics R and D Handbook," 1962–1963.

44. BARTON, D. K., and W. W. SHRADER   Interclutter Visibility in MTI Systems *IEEE EASCON 1969 Rec.,*pp. 294–297.

45. BLAKE, L. V.   Radio Ray (Radar) Range-Height-Angle Charts, *Microwave J.,* vol. 4. no. 10, pp. 49–53, October 1968.

46. SHRADER, W. W.   "MTI Radar" chap. 17 in Ref. 9.

47. WINTER, C. F.   Dual Vertical Beam Properties of Doubly Curved Reflectors, *IEEE Trans.,* vol. AP-19, no. 2, pp. 174–180, March 1971.

48. WARD, H. R., and W. W. SHRADER   MTI Performance Degradation Caused by Limiting, *IEEE 1968 EASCON Rec.,* pp. 168–174.

49. BROWN, B. P.   Radar Height Finding, chap. 22 in Ref. 9.

50. DUNN, J. H., and D. D. HOWARD The Effects of Automatic Gain Control of Performance on the Tracking Accuracy of Monopulse Radar Systems, *Proc. IRE*, vol. 47, no. 3, pp. 430–435, March 1959.

References on radar technology

51. SKOLNIK, M. I. (ED.) "Radar Handbook," McGraw-Hill Book Company, New York, 1970.

52. WEIL, T. A. "Transmitters," chap. 7 in Ref. 51.

53. SKOLNIK, M. I. "Introduction to Radar Systems," McGraw-Hill Book Company, New York, 1962.

54. SHERMAN, J. W. "Aperture-Antenna Analysis," chap. 9 in Ref. 51.

55. BARTON, D. K., and H. R. WARD "Handbook of Radar Measurement," Prentice-Hall, Inc., Englewood Cliffs, N.J., 1969.

56. ASHLEY, A., and J. S. PERRY "Beacons," chap. 38 in Ref. 51.

57. CRONEY, J. "Civil Marine Radar," chap. 31 in Ref. 51.

58. FREEDMAN, J. "Radar," chap. 14 in "System Engineering Handbook," McGraw-Hill Book Company, New York, 1965.

59. SENGUPTA, D. L. and R. E. HIATT "Reflectors and Lenses," chap. 10 in Ref. 51.

60. DUNN, J. H., D. D. HOWARD, and K. B. PENDLETON "Tracking Radar," chap. 21 in Ref. 51.

61. CHESTON, T. C., and J. FRANK "Array Antennas," chap. 11 in Ref. 51.

62. STARK, L., R. W. BURNS, and W. P. CLARK "Phase Shifters for Arrays," chap. 12 in Ref. 51.

63. KEFALAS, G. P., and J. C. WILTSE "Transmission Lines, Components, and Devices," chap. 8, in Ref. 51.

64. HAMMER, I. W. "Frequency-scanned Arrays," chap. 13 in Ref. 51.

65. MATTHAEI, G. L., L. YOUNG, and E. M. T. JONES "Microwave Filters, Impedance Matching Networks, and Coupling Structures," McGraw-Hill Book Company, New York, 1964.

66. TAYLOR, J. W., and J. MATTERN "Receivers," chap. 5 in Ref. 51.

67. FARNETT, E. C., T. B. HOWARD, and G. H. STEVENS "Pulse-Compression Radar," chap. 20 in Ref. 51.

68. SHRADER, W. W. "MTI Radar," chap. 17 in Ref. 51.

69. MOONEY, D. H., and W. A. SKILLMAN, "Pulse-Doppler Radar," chap. 19, in Ref. 51.

70. NATHANSON, F. "Radar Design Principles: Signal Processing and the Environment," McGraw-Hill Book Company, New York, 1969.

71. BERG, A. A. "Radar Indicators and Displays," chap. 6 in Ref. 51.

References on electronic navigation

72. KAYTON, M., and W. FRIED "Avionics Navigation Systems," John Wiley & Sons, Inc., New York, 1969. (Includes chapters on radio, doppler, inertial, radar, celestial, and satellite navigation systems, together with other chapters on computers, ILS, air traffic control, and displays.)

73. SANDRETTO, P. C. "Electronic Avigation Engineering," International Telephone and Telegraph Corp., New York, 1958.

Journals on electronic navigation

74. *Navigation,* published quarterly by the Institute of Navigation, Washington, D. C.

75. *IEEE Transactions on Aerospace and Electronic Systems,* published six times a year by the IEEE, New York.

References on electronic naviation

76. ANDERSON, E. W. Air Navigation Techniques, *Navigation,* vol. 18, no. 1, Spring 1971.

77. DUNLAP, G. D. Major Developments in Marine Navigation During the Last 25 Years, *Navigation,* vol. 18, no. 1, Spring 1971.

78. FRENCH INSTITUTE OF NAVIGATION "Etude Comparative des Systemes Hyperboliques et Circularies pour la Navigation et la Localisation," Paris, April 1970.

79. BRAVERMAN, N.   Aviation System Design for Safety and Efficiency, *Navigation*, vol. 18, no. 3, Fall. 1971.

80. JACKSON, W.   The Federal Airways System, *IEEE*, 1970.

81. KEEN, R.   "Wireless Direction Finding," Iliffe and Sons, London, 1938.

82. ASTHOLZ, P.   Air Traffic Control, in M. KAYTON and W. FRIED (EDS.) "Avionics Navigation Systems," John Wiley & Sons, Inc., New York, 1969.

83. WILEY, C.   Radar Navigation, in M. KAYTON and W. FRIED (EDS.) "Avionics Navigation Systems," John Wiley & Sons, Inc., New York, 1969.

84. FRIED, W.   Doppler Navigation in M. KAYTON and W. FRIED (EDS.) "Avionics Navigation Systems," John Wiley & Sons, Inc., New York, 1969.

85. GOULET, T. A.   The Use of Pulsed Doppler Sonar for Navigation of Manned Deep Submergence Vehicles, *Navigation*, vol. 17, no. 2, Summer 1970.

86. TURNER, E. E., ET AL.   The Raytheon Acoustic Doppler Navigator, *Navigation*, vol. 13, no. 3, Autumn 1966.

87. KAYTON, M.   Inertial Navigation, in M. KAYTON and W. FRIED (EDS.) "Avionics Navigation Systems," John Wiley & Sons, Inc., New York, 1969.

88. QUASIUS, G.   Celestial Navigation in M. KAYTON and W. FRIED (EDS.)   "Avionics Navigation Systems," John Wiley & Sons, Inc., New York, 1969.

89. DUNCAN, R. C.   Satellite Navigation in M. KAYTON and W. FRIED (EDS.)   "Avionics Navigation Systems," John Wiley & Sons, Inc., New York, 1969.

90. HAWKINS, H., ET AL.   Radar Performance Degradation in Fog and Rain, *IRE Trans. ANE*, vol. ANE-6, no. 1, March 1959.

91. JANSKY, C. M.   The Current State of the Science of Marine Navigation, *Navigation*, vol. 12, no. 1, Spring 1965.

92. HOPKINS, H. G., ET AL.   Current D/F Practice, *Proc. IEE*, vol. 105, part B, no. 9, March 1959.

93. SANDRETTO, P.   "Electronic Avigation Engineering," International Telephone & Telegraph Corp., New York, 1958.

94. BUSIGNIES, H.   Evaluation of Night Errors in Aircraft Direction Finding, *Electr. Commun.*, vol. 23, no. 2, June 1946.

95. KAYTON, M.   Landing Guidance, in M. KAYTON and W. FRIED (EDS.) "Avionics Navigation Systems," John Wiley & Sons, Inc., New York, 1969.

96. POPP, H.   Solid-State VOR, *Elect. Commun.*, vol. 44, no. 4, December 1969.

97. ANDERSON, S., ET AL.   The CAA Doppler Omnirange, *Proc. IRE*, vol. 47, no. 5, May 1959.

98. PORITSKY, S. (ED.)   Special issue on VOR/DME, *IEEE Trans. ANE*, vol. ANE-12, no. 1, March 1965.

99. COLIN, R., ET AL.   Principles of Tacan, *Electr. Commun.*, vol. 33, no. 1, March 1956.

100. DODINGTON, S.   Recent Developments of the Tacan Navigation System, *Electr. Commun.*, vol. 44, no. 4, December 1969.

101. RENICK, R. C.   A Next Generation Beacon System, *IEEE EASCON 1969 Rec.*

102. ADCOCK, F.   British Patent No. 130490, 1919.

103. POWELL, C.   The Decca Navigator System for Ship and Aircraft Use, *Proc. IEE*, vol. 105, part B, no. 9, March 1958.

104. RÖRHOLT, B. A.   Electronic Aids to Navigation for Fishing Vessels and Other Open Sea Users, *Navigation*, vol. 16, no. 3, Fall 1969.

105. KUEBLER, W.   Marine Electronic Navigation Systems: A Review, *Navigation*, vol. 15, no. 3, Fall 1968.

106. VAN ETTEN, J. P.   Loran C System and Product Development, *Electr. Commun.*, vol. 45, no. 2, June 1970.

107. SWANSON, E. R.   Omega, *Navigation*, vol. 18, no. 2, Summer 1971.

108. KLASS, P. J.   Omega Navaid Aviation Use Studied, *Aviation Week*, Nov. 29, 1971.

109. STANSELL, T.   Transit, the Navy Navigational Satellite System, *Navigation*, vol. 18, no. 1, Spring 1971.

110. BROWN, A. H.   Consol Navigation System, *J. IEE*, part 3A, vol. 94, no. 15, 1947.

111. TADARO, T., ET AL.   Talking Beacon System in Japan, *Navigation*, vol. 16, no. 4, Winter 1969.

112. KEANE, L. M.   *Prog. Nav. Satellites, Navigation,* vol. 15, no. 4, Winter 1968–1969.
113. WOODFORD, J. B., ET AL.   Satellite Systems for Navigation Using 24-hour orbits, *IEEE EASCON 1969 Rec.*
114. RADIO TECHNICAL COMMISSION FOR AERONAUTICS   "A New Guidance System for Approach and Landing," DO-148, Washington, D.C., Dec. 18, 1970.
115. LITCHFORD, G.   Low Visibility Landing, *Astronaut. Aeronaut.,* vol. 6, nos. 11 and 12, November-December 1968.
116. POGUST, F.   The Status of Microwave Scanning Beams Landing System Developments, *IEEE EASCON 1969 Rec.*
117. EARP, C. W., ET AL.   Doppler Scanning Guidance System, *Electr. Commun.,* vol. 46, no. 4, December 1971.
118. EDWARDS, J. A.   Microwave Landing System, *IEEE EASCON 1971 Rec.*
119. Special issue on CAS, *Trans. IEEE Group AES,* vol. AES-4, no. 2, March 1968.
120. BORROK, M. J., ET AL.   Results of ATA CAS Flight Test Program, *Navigation,* vol. 17, no. 3, Fall 1970.

References on underwater sound detection systems: Principles and Functions

121. BATCHELDER, L. B.   Sonics in the Sea, *IEEE Proc.,* vol. 53, no. 10, October 1965.
122. BECKEN, B. A.   Sonar, *Adv. Hydrosci.,* 1964.
123. HORTON, J. W.   "Fundamentals of Sonar," U.S. Naval Institute, Annapolis, Md., 1959.
124. GRAY, D. E.   "American Institute of Physics Handbook," 2d ed., McGraw-Hill Book Company, New York, 1963.
125. URICK, R. J.   "Principles of Underwater Sound for Engineers," McGraw-Hill Book Company, New York, 1967.

Propagation

126. WILSON, W. D.   *J. Acoust. Soc. Amer.,* vol. 32, p. 1357, 1960.
127. *Ibid.,* p. 641, 1960.
128. *Ibid.,* p. 866, 1962.
129. URICK, R. J.   "Principles of Underwater Sound for Engineers," McGraw-Hill Book Company, New York, 1967.
130. TOLSTOY, I., and C. S. CLAY   "Ocean Acoustics," McGraw-Hill Book Company, New York, 1966.
131. ALBERS, V. M.   "Underwater Acoustics Handbook," Penn State University Press, University Park, Pa., 1960.
132. HORTON, J. W.   "Fundamentals of Sonar," U.S. Naval Institute, Annapolis, Md., 1957.
133. THORP, W. H.   *J. Acoust. Soc. Amer.,* vol. 42, p. 270, 1967.
134. BATCHELDER, L.   *Proc. IEEE,* vol. 53, p. 10, 1965.
135. BOBBER, R. J.   "Underwater Electroacoustic Measurements," Naval Research Laboratory, Washington, D.C., 1970.
136. SCHULKIN, M., and H. W. MARSH   *J. Acoust. Soc. Amer.,* vol. 34, p. 864, 1962.
137. FISHER, F. H.   *J. Acoust. Soc. Amer.,* vol. 30, p. 442, 1958.
138. LIEBERMANN, L. N.   *Phys. Rev.,* vol. 76, p. 1520, 1949.
139. STEPHENS, R. W. B. (ED.),   "Underwater Acoustics," John Wiley & Sons, Inc.-Interscience Publishers, London, 1970.

Noise

140. BECKEN, B. A.   Sonar, *Adv. Hydrosci.,* 1964.
141. URICK, R. J.   "Principles of Underwater Sound for Engineers," McGraw-Hill Book Company, New York, 1967.
142. BATCHELDER, L.   Sonics in the Sea, *IEEE Proc.,* vol. 53, no. 10, October 1965.
143. BARTBERGER, C. L.   "Lecture Notes on Underwater Acoustics," U.S. Naval Air Development Center, Johnsville, Md., May 17, 1965.
144. BOBBER, R. J.   "Underwater Electroacoustic Measurements," Naval Research Laboratory, Washington, D.C., 1970.
145. "Introduction to Sonar Technology," NavShips Publ. 0967-129-3010, Navy Dept., Bureau of Ships, Washington, 1965.

Transducers and arrays

146. AMERICAN NATIONAL STANDARDS INSTITUTE, ANSI S1.20-1972, Procedures for Calibration of Underwater Electroacoustic Transducers, New York.

147. BOBBER, R. J. "Underwater Electroacoustic Measurements," Naval Research Laboratory, Washington, D.C., 1970.

148. HUETER, T. F., and R. H. BOLT "Sonics," John Wiley & Sons, Inc., New York, 1955.

149. SIMS, C. C. High-Fidelity Underwater Sound Transducers, *Proc. IRE*, vol. 47, p. 866, 1959.

150. URICK, R. J. "Principles of Underwater Sound for Engineers," McGraw-Hill Book Company, New York, 1967.

151. KINSLER, L. E., and A. R. FREY "Fundamentals of Acoustics," John Wiley & Sons, Inc., New York, 1962.

152. OLSON, H. F. "Acoustical Engineering," D. Van Nostrand Company, Inc., Princeton, N.J., 1957.

153. KIKUCHI, Y. "Ultrasonic Transducers," Corona Publishing Company, Ltd., Tokyo, 1969.

154. CADY, W. G. "Piezoelectricity," vols. 1 and 2, Dover Publications, Inc. New York, 1964.

155. MASON, W. P. "Piezoelectric Crystals and Their Application to Ultrasonics," D. Van Nostrand Company, Inc., Princeton, N.J., 1950.

156. BERLINCOURT, D. A., D. R. CURRAN, and H. JAFFEE "Piezoelectric and Piezomagnetic Materials and Their Function in Transducers," in W. P. Mason (ed.), "Physical Acoustics," vol. I, part A, Academic Press, Inc., New York, 1964.

157. Piezoelectric Crystals, IEEE Standard 176, 1949.

158. Definitions and Methods of Measurements of Piezoelectric Vibrators, IEEE Standard 177, 1966.

159. Piezoelectric Crystals: Determination of the Elastic, Piezoelectric, and Dielectric Constants: The Electromechanical Coupling Factor, IEEE Standard 178, 1958.

160. Measurement of Piezoelectric Ceramics, IEEE Standard 179, 1961.

161. Piezoelectric Ceramic for Sonar Transducers, MIL-STD-1376 (SHIPS).

162. Magnetostrictive Materials: Piezomagnetic Nomenclature, IEEE Standard 319, 1971.

163. KIKUCHI, Y. Magnetostrictive Metals and Piezomagnetic Ceramics as Transducer Materials, in O. E. Mattiat (ed.), "Ultrasonic Transducer Materials," Plenum Press, New York, 1971.

Passive Sonar Systems

164. URICK, R. J. "Principles of Underwater Sound for Engineers," McGraw-Hill Book Company, New York, 1967.

165. PETERSON, W. W., and T. G. BIRDSALL The Theory of Signal Detectability, *Univ. Mich. Eng. Res. Inst. Rep.* 13, 1953.

166. PETERSON, W. W., T. G. BIRDSALL, and W. C. FOX The Theory of Signal Detectability, *Trans. IRE*, vol. PGIT-4, p. 171, 1954.

167. LAWSON, J. L., and UHLENBECK Threshold Signals, M.I.T. Radiation Laboratory Series, vol. 24, McGraw-Hill Book Company, New York, 1950.

Active Sonar Systems

168. *Adv. Hydrosci.*, vol. 1, 1964.

169. Special Issue on Detection Theory and Its Applications, *Proc. IEEE*, May 1970.

170. SCHWARTZ, MISHA "Information Transmission, Modulation and Noise," 2d ed., McGraw-Hill Book Company, 1970.

171. BROCH, J. T. Effects of Spectrum Non-Linearities upon Peak Distributions of Random Signals, *Brüel and Kjaer Tech. Rev.*, no. 3, 1963.

172. CRONEY, J. Clutter on Radar Displays: Reduction by Use of Logarithmic Receivers, *Wireless Eng.*, vol. 33, 1956.

173. SHERMAN, CHARLES H., and DONALD F. KASS, Radiation Impedances, Radiation Patterns, and Efficiency for Large Array on a Sphere, U.S. Navy Underwater Sound Laboratory, Fort Trumbell, New London, Conn., Research Rep. 429, July 17, 1959.

174. CARSON, DAVID L.   Diagnosis and Cure of Erratic Velocity Distributions in Sonar Projector Arrays, *J. Acoust. Soc. Amer.,* vol. 34, no. 9, September 1962.

175. SKOLNICK, MERRILL J.   "Introduction to Radar System," McGraw-Hill Book Company, New York, 1962.

176. DAVENPORT, W. B., and W. L. ROOT,   "Introduction to Random Signals and Noise," McGraw-Hill Book Company, New York, 1958.

177. GERLACH, ALBERT A.   "Theory and Application of Statistical Wave-Period Processing," Report, Cook Electric Co., Chicago, Ill.

SECTION 26

ELECTRONICS IN MEDICINE AND BIOLOGY

BY

HOWARD P. APPLE, Ph.D. Assistant Professor of Biomedical Engineering, Case Western Reserve University

CHARLES H. GIBBS, Ph.D. Assistant Clinical Professor and Senior Research Associate, School of Dentistry and Engineering Design Center, Case Western Reserve University

PETER G. KATONA, Sc.D. Associate Professor of Biomedical and Electrical Engineering, Case Western Reserve University

RAYMOND J. KIRALY, M.S. Director of Engineering, Department of Artificial Organs, Research Division, Cleveland Clinic Foundation

WEN H. KO, Ph.D. Director, Engineering Design Center, Professor of Electrical Engineering, Professor of Biomedical Engineering, Case Western Reserve University

FLORO MIRALDI, Sc.D., M.D. Professor of Engineering, Case Western Reserve University; Chief, Nuclear Medicine, Cleveland Metropolitan General Hospital

J. THOMAS MORTIMER, Ph.D. Associate Professor of Biomedical Engineering, Case Western Reserve University

MICHAEL R. NEUMAN, Ph.D. Associate Professor of Biomedical Engineering, Case Western Reserve University

YUKIHIKO NOSÉ, M.D., Ph.D. Department Head, Department of Artificial Organs, Research Division, Cleveland Clinic Foundation; Professor of Biomedical Engineering, Case Western Reserve University

ROBERT PLONSEY, Ph.D. Professor of Biomedical Engineering, Case Western Reserve University

CONTENTS

Numbers refer to paragraphs

SECTION 26

MEDICAL ELECTRONICS AND BIOMEDICAL ENGINEERING

INTRODUCTION

BY WEN H. KO, PH.D.

1. Biomedical Engineering. Biomedical engineering is a new discipline that combines physical and life sciences to seek engineering solutions and understanding of biomedical problems. There are no clearly defined boundaries between similar activities in biophysics, biomaterials, and certain sections of medical branches such as applied physiology, biometric radiology, etc. Biomedical engineering generally refers to the use of technology, knowledge, and skills of physical sciences for biological and medical problems. However, the interplay of the two major disciplines has also generated many fruitful results in using life-science knowledge to solve engineering problems. An example is in the field of bionics; the design of airplanes, submarines, and radar can benefit from the study of birds, fish, and bats. The major textbooks, journals, and papers devoted to biomedical engineering are listed in the references, Par. **26-66**.

Biomedical electronics relates to electronic theory, technology, and circuit application to biomedical problems. The field is so diversified that no comprehensive presentation can be contained in a single volume. This section of the Handbook presents a summary of selected topics in biomedical electronics with survey references.

BIOMEDICAL-INSTRUMENTATION ELECTRICAL SAFETY

BY HOWARD P. APPLE, PH.D.

2. Safety Considerations. The increasing use of electrical instruments in hospitals makes electric shock, burns, and fire potential hazards for patients and personnel. Figure 26-1 and Table 26-1 illustrate the severity of the problem. Currents as low as 20 μA at 60 Hz directly through the heart can cause ventricular fibrillation and death.[11,12]* Faults and/or errors in individual parts or interaction between parts may cause problems. Figure 26-2 shows the schematic diagram of a biomedical instrumentation system.

3. The Patient. Patients are susceptible to electrical shock, burns, fire, and/or explosions. Burns occur from high current density and power dissipation and are possible at measuring, stimulating, or ground plate electrodes if small contacts and/or high current occur. Fire or explosions can occur in anesthetic gas or high-oxygen-content environments. There are extensive standards that apply in this area.[13]

Electric shock has two categories: *macroshock* and *microshock*. Macroshock refers to body shock where electrical stimulation of skeletal muscle is predominant, and microshock refers to the existence of direct low-resistance paths to the heart and the possibility of ventricular fibrillation occurring at low current levels (20 μA).[14]

Ventricular fibrillation refers to malfunctioning of the ventricular musculature, which will interfere with the normal blood-pumping action and cause death. The use of conductive catheters into the heart, for example, provides a low-resistance path in which 20-μA level of current through the myocardium may trigger ventricular fibrillation. It is for this reason (microshock hazard) that special precautions are needed with surgical, intensive care, and coronary care patients.

---

*Superior numbers correspond to numbered references, Par. **26-66**.

Table 26-1. Bodily Effects of 1-Second Contact with 60-Hz Alternating Current

Path	Amperes	Results
Current Through Body Trunk	1 mA, 0.001	Perception threshold, tingling sensation
	16 mA, 0.016	"Let go" current
	50 mA, 0.050	Pain, possible fainting, exhaustion, mechanical injury
	100 mA, 0.100 to 3 full A, 3.000	Ventricular fibrillation
	4 full A, 4.000 to 10 full A, 10.000	Severe burns and physiological shock
Current Directly Through Heart	20 μa, 0.000020	Possible ventricular fibrillation
	Leakage current in excess of 20 μa can be found in electrical and electronic equipment.	

SOURCE: "Electrical Safety Test Procedures for Hospitals," Instrutek, Inc., Annapolis, Md., 1972.

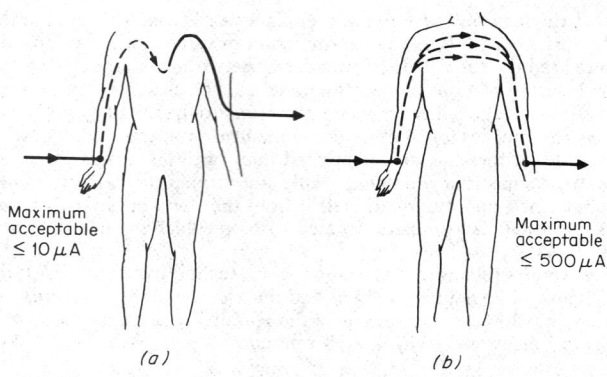

Fig. 26-1. Schematic of different lead configurations. (*a*) Body surface to heart; (*b*) body surface to body surface.

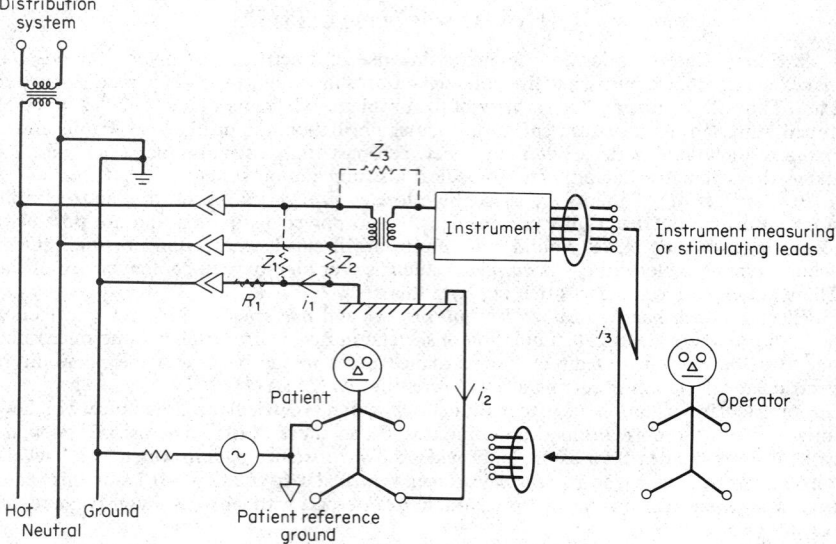

Fig. 26-2. Line-powered biomedical instrumentation system.

4. The Operator. The operator is also susceptible to hazards, except that he is not exposed to the microshock hazard and is in better health. Education is most important in this area. Electrical hazards can be markedly reduced when the operator understands the technical operation of the instruments, and how these instruments may be dangerous to himself and the patient.

5. Distribution System and Equipotential Environment. If one assumes that a patient with a pacing catheter presents a 500-Ω resistance to the flow of electric current and, to be safe, no more than 10 μA of direct current to 60-Hz current will be allowed to pass through the patient, then no voltage difference greater than 5 mV across the patient can be permitted. The best test methods measure the current that would flow through a 500-Ω resistance or another assumed model put in place of the patient. If no source yielding more than 10 μA collectively is available within reach of the patient or of anyone else touching the patient, the patient is relatively safe from electrical shock. In general patient areas, a reduced standard of 500 μA is generally accepted.

Achieving this condition results in the equipotential environment. It requires that possible sources of hazardous currents be connected by low-resistance conductors to one reference ground point within the patient's area and as close to him as possible. The sensitive area should be considered as including any conductors that may be contacted by the patient or those attending him.

An extremely dangerous situation can arise if multiple instruments are driven off different transformers with separate grounds. Checks should be made to always ensure a common power source per patient and room. Specific approaches, procedures, and test instruments are covered in the references, Par. **26-66.**

6. Checking Electrical Safety of Line-operated Equipment. All line-operated equipment has a power side and a patient side with leads or metal surfaces that may touch the patient. Referring to Fig. 26-2, the paths of the leakage currents i_1, i_2, and i_3 are through impedances Z_1, Z_2, and Z_3, which are inherent in the cord and instrument. Suppose a patient's heart is connected to instrument ground through a grounded blood pressure transducer or a grounded pacemaker lead, and that the third line is broken due to a faulty plug and/or wiring. The current i_1 now becomes zero, and the leakage current path is through the heart of the patient. Fibrillation could be induced if $Z_1 \leq 5 \times 10^6 \ \Omega$ (a capacitor of 480 pF, equivalent to the capacitance of 12 ft of normal line cord).

A patient may contact two or more "grounded" instrument cases where the so-called "grounds" may not be at the same potential. Leakage currents flowing in long ground lines can develop sizable potential differences along these lines. An example is illustrated in Fig. 26-3 in which a patient is connected to a grounded blood pressure transducer and a grounded

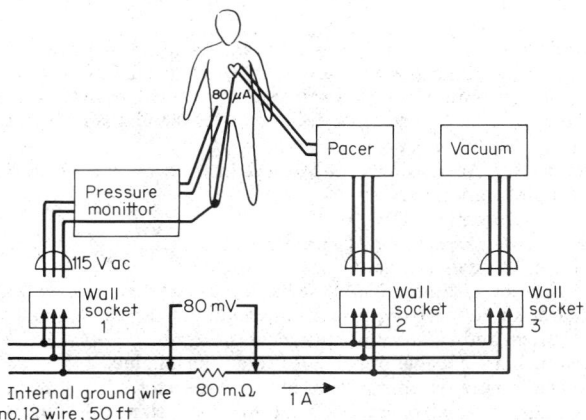

Fig. 26-3. Illustration of possible fatality due to a change in ground potentials as the result of an unrelated use of power.

pacemaker, together forming a low-impedance path through the heart. The pacemaker is referred to ground through 50 ft of No. 12 wire with a wire resistance of 80 mΩ. If a vacuum cleaner, for example, is plugged into another outlet, the pacer and vacuum cleaner share the same ground wire. Vacuum cleaners are notoriously "leaky" machines (a typical machine has a leakage current of 1 A). In such a case, 80 mV may readily be developed between two internal points on the patient. Death may occur due to 160 μA ac-induced fibrillation.

Following is a brief outline of electrical safety checking of instruments. Detailed procedures and test instruments are given in the references, Par. 26-66.

Classification as to Use. (a) For general patients or intensive care/cardiac care patients; (b) for portable or nonportable use.

Visual Inspection. Cracked damaged plugs, cords, case, controls, patient leads, etc.

Grounding Resistance; Checking Grounding Integrity. Verifies low-resistance connection between the instrument case and the line plug's ground lug, or of other surfaces and connections on the instruments which are to be grounded, i.e., reference electrode on some heart monitors.

Electrical Isolation-to-Ground Test. Verifies whether the power side, the patient side, and some of the ungrounded metal surfaces are sufficiently isolated from the patient under all conditions.

Leakage Current Tests. Any unwanted current flow (i_1, i_2, i_3) is usually referred to as leakage current. Controlled leakage current (i_1) is carried away to ground through the green ground wire of the line cord. The current should be as low as possible, since it could pass through the patient if the ground wire opened. This leakage current also causes buildup of voltage gradient along the ground bus (see Table 26-2).

Table 26-2.   Comparison of Proposed Maximum Leakage Currents in Microamperes*

Measurement	NFPA	CHA	AAMI	VA	UL
From instrument case to ground (third line)................................	100 ac	500 ac†	10 rms	...	10 ac
From instrument case to ground (ungrounded)	100 ac 100 dc	10 ac 10 dc	10 ac 10 dc	...	10 ac 14 dc
From any or all patient electrodes to ground	10 ac 50 dc	10 ac 10 dc	10 ac 10 dc	10 ac	5 ac 7 dc
Between patient leads	10 ac 50 dc	10 ac 10 dc	10 ac 10 dc	...	5 ac 7 dc
120 V ac applied to all electrodes or transducers	20 ac	10 ac	10 ac		

* See Par. 26-7 for explanation of abbreviations heading the columns.
† With ground-loss monitor.

7. Safety Standards, Testing and Procedures. As of 1974 there were no uniform national standards. The standards generally referred to are the National Electrical Code (1971) and that of Underwriters' Laboratory. Recommended leakage current maxima are listed in Table 26-2. A list of some other recommended standards includes:

1. National Fire Protection Association (NFPA).[15,16]
2. Association for the Advancement of Medical Instrumentation (AAMI).[17]
3. Veterans Administration (VA).[18,19]
4. Underwriters' Laboratory (UL).[20]
5. California Hospital Association (CHA).[21]

The central issues of safety standards are:

1. Absolute safety levels (10 to 20 μA), including dc–100-kHz frequencies.
2. One standard for all instruments or differentiation as to use and environments.
3. Human electrical and/or susceptibility models.
4. Standards which are capable of being tested and maintained.
5. Use of isolated power systems and ground fault detectors.
6. Standards under normal operating conditions and standards under possible fault conditions.

ELECTROCARDIOGRAPHY AND BIOPOTENTIALS

By Robert Plonsey, Ph.D.

8. Electrophysiology of the Heart. In electrocardiography (ECG), as in electroencephalography (EEG) and other bioelectric phenomena, electrical signals of interest are generated by the body itself. For this reason the ECG has diagnostic value, since it reflects the biological (hence, clinical) behavior of the heart.

The human heart, illustrated in Fig. 26-4, is composed primarily of a specialized muscle (cardiac muscle). Individual muscle cells are around 15 μm in diameter and 100 μm in length and are stacked together somewhat like bricks. Collectively, their long dimensions lie parallel to each other, and in the aggregate they constitute bands of muscle fibers which make a broad spiral in forming the heart walls.

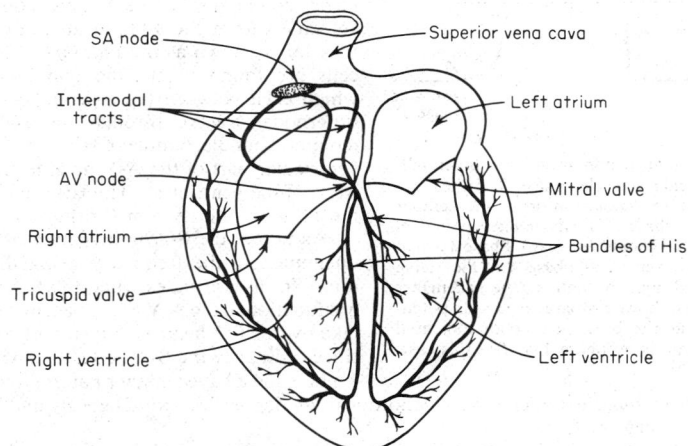

Fig. 26-4.   Structure of the heart.

Individual heart cells are surrounded by a plasma membrane which separates the intracellular from the extracellular space. For a typical heart cell, metabolic energy is used to create an internal environment rich in potassium but low in sodium compared with an extracellular composition of high sodium and low potassium. Because of this existing imbalance, a resting potential exists across the cell membrane, with the inside some 90 mV negative relative to the outside. When the cell is stimulated (by passing an electric current which momentarily increases the transmembrane potential), the membrane properties go through a cyclic change, during which time the membrane becomes highly permeable to sodium. A large inward sodium current results from the diffusion and electrical gradients. This inflow constitutes precisely a generation of electric current, and during the transient the cell behaves, essentially, like a current dipole source. This transient sodium current is responsible for (and is part of) a local circuit current which links with adjoining cells and serves as a stimulus to initiate their activity. In this way activity (once started) spreads continuously to adjacent cells. When the membrane recovers (returns to resting properties), the *action potential* of the cell is ended, and it is once again at rest and capable of being restimulated.

Figure 26-5 shows a typical ventricular action potential. The *upstroke* corresponds to activation *(depolarization)*, and during this period the cell constitutes precisely a source of emf representable as a dipole. During the plateau (phase 2 in Fig. 26-5) the cell is electrically quiescent. Recovery occupies a relatively longer interval of time, during which the cell behaves as a weak emf source; this is phase 3 of Fig. 26-5.

In addition to muscle cells of the type described, specialized *pacemaker cells* (P cells) exist in the S-A nodal region of the heart and in the A-V node (see Figs. 26-4 and 26-5). While the resting potential can be maintained indefinitely in ordinary muscle cells, for the pacemaker cells the transmembrane potential spontaneously increases until the threshold for excitation is reached, giving rise to an action potential such as previously described. Thus pacemaker cells are self-excitatory. Their oscillations communicate to neighboring cells by local circuit currents, and in this way cyclic excitation is established in all portions of the heart.

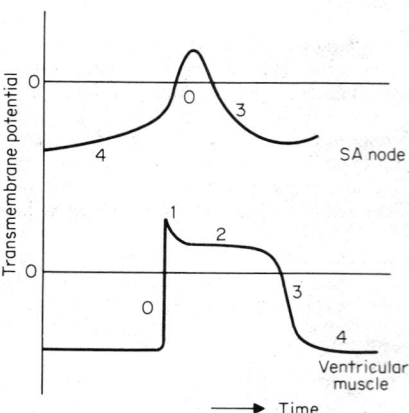

Fig. 26-5. Transmembrane action potentials. For ventricular muscle, phase 0 is the "upstroke" due to depolarization of the cellular membrane. Phase 2 is a prolonged period of little potential change (plateau). Phase 3 is that of rapid recovery, and phase 4 is the stable resting condition. In contrast, the pacemaker cell's phase 4 shows a progressive rise in potential until, at the break, an action potential ensues. There is no phase 1 or 2 in this case.

An additional specialized cell type which forms conduction tissue (where conduction velocity is from 2 to 10 times that of ordinary muscle) is known as the *Purkinje cell.* These cells are found in specific bundles (tracts) which exist between the S-A and A-V node (internodal tracts), through the A-V node (common bundle, bundle of His), and from the ventricular part of the A-V node to the inner walls of the ventricles. The latter constitutes the His-Purkinje system (Purkinje network).

As noted, the S-A node causes a cyclical phenomenon, the frequency being the heart rate. In disease states when the S-A node is nonfunctional, the A-V nodal pacemakers may take over. The heart rate then will be lower, since ordinarily the S-A node, to drive the A-V node, must have a higher natural frequency. Ordinary heart muscle, under certain conditions, can depolarize spontaneously and become a pacemaker (ectopic focus).

The activation of all cells in the heart, initiated by the S-A node, forms the basis for ECG. As noted, cells undergoing activation (depolarization) behave as dipole current sources. The resulting local currents not only contribute to the spread of activity, but also result in the presence of currents everywhere in the torso (which behaves as a passive resistive medium). During recovery, each cell again behaves as an emf source, though the magnitude is much less and the duration much longer than during activation.

9. The Electrocardiogram. If electrical potential measured between the right and left arm is amplified, the ECG signal will have the appearance shown in Fig. 26-6. Each of the three wave complexes, designated as P, QRS, and T, corresponds to a particular electrical event. Initiation of the electrical cycle takes place by the pacemaker cells of the S-A node. Their spontaneous "firing" is communicated by conduction throughout the atrial tissue. Thus the first major event is depolarization of both right and left atria, and this collective electrical activity gives rise to the P wave.

Activity is prevented from spreading directly from the atria to the ventricles by the presence of a ring of nonconducting fibrous tissue separating these two regions, except in the vicinity of the A-V node. At the A-V node, a pathway for conduction exists; however, a very low velocity is characteristic of this region. Thus initiation of activity in the ventricles is delayed by the time required for activity to propagate first from the S-A to A-V node (mainly through the speedier internodal tracts) and then through the common bundle of the A-V node. The interval gives rise to the designated P-R segment, which is a measure of the composite delay.

Once the ventricular conduction (Purkinje) system is activated, rapid spread occurs and general ventricular activity results. This gives rise to electrocardiographic QRS complex. QRS waveform reflects the composite depolarization of ventricles, and its duration measures the total time required. As noted in Fig. 26-6, the Q wave is the first downward deflection before any upward deflection, the R wave is the first upward deflection, and the S wave is the first downward deflection after an upward deflection. One or more parts can be missing.

Recovery is also a process that generates a potential field, generally of lower magnitude. Ventricular recovery produces the T wave. The S-T segment is ordinarily at the base line as determined by successive TP segments; however, in the case of cell injury or death, this may be elevated or depressed. Recovery in the atrial tissue also results in an electrocardiographic wave (T_p); however, it is of low magnitude and ordinarily masked by the QRS with which it overlaps.

Normal values of electrocardiographic intervals, segments, and waveform durations are given in Table 26-3.

In recording ECG, a standard-time and potential-amplitude base is utilized. Corresponding to a paper speed of 25 mm/s, major 5-mm divisions represent intervals of 0.2 s.

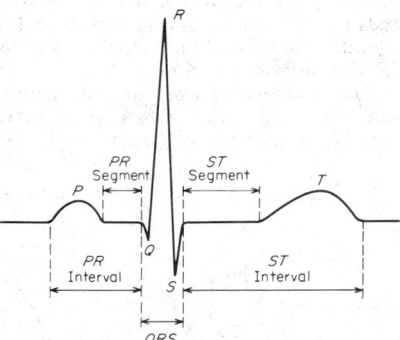

Fig. 26-6. Standard scalar electrocardiogram.

Amplification is adjusted ordinarily to 0.1 mV/mm. Peak ECG signals are generally under 5 mV.

10. Mechanical Activity of the Heart. The function of the heart is to pump blood. Electrical activity is important only because it initiates muscular contraction. Abnormal conduction pathways or heart rates that are too high or low can cause reduced cardiac output and may produce clinical symptoms. Conversely, diseases affecting the strength of muscular contraction may be reflected in abnormalities in the ECG. Examples of the former are conduction defects and atrial flutter or tachycardia, while the latter may arise from loss of blood flow to portions of the heart, resulting in areas of tissue necrosis and subsequent replacement with fibrous (noncontractile) tissue (myocardial infarction).

Table 26-3. Normal Electrocardiographic Values for Adults*

P-R interval	0.12–0.20 s
QRS interval	0.07–0.10 s
P amplitude . . . . Lead II	$\leq$0.25 mV
P duration . . . . Lead II	$\leq$0.11 s
R amplitude . . . . V_1-V_6	$\leq$2.7 mV
T amplitude . . . . Lead II	$\leq$0.8 mV

Heart rate	Q-T interval, s	S-T segment, s
60	0.33–0.43	0.14–0.16
70	0.31–0.41	0.13–0.15
80	0.29–0.38	0.12–0.14
90	0.28–0.36	0.11–0.13
100	0.27–0.35	0.10–0.11

* See Fig. 26-8 for definition of leads.

Under normal conditions, the correlation between electrical and mechanical events is as depicted in Fig. 26-7. Mechanical and circulatory values connected with the heart's behavior as a pump are given in Table 26-4.

11. Electrocardiographic Lead Systems. Figure 26-8 shows the placement of electrodes on the body for standard clinical ECG *leads*. Voltages are recorded between selected pairs of electrodes, for example:

V_I = potential at left arm–potential at right arm
V_II = potential at left leg–potential at left arm
V_III = potential at left leg–potential at right arm

These leads (electrode connections) are referred to as the extremity, or limb, leads. Note that because of Kirchhoff's voltage law, $V_\text{I} + V_\text{III} = V_\text{II}$, so that only two are independent.

A synthetic lead used as a reference is formed by connecting the limbs through 5,000-Ω resistors to a common point called the *Wilson central terminal* (CT). Precordial leads (V_1 to V_6) are voltages between designated chest electrodes (see Fig. 26-8) and the CT. Augmented leads result from measuring each limb lead against the central terminal, except that the respective limb contribution to the CT is disconnected. (This results in augmentation of the signal by a factor of 1.5.)

The resultant 12 leads are conventionally taken in clinical electrocardiography. A summary of the connections is given in Table 26-5. Figure 26-9 shows a typical normal record from each of the 12 leads.

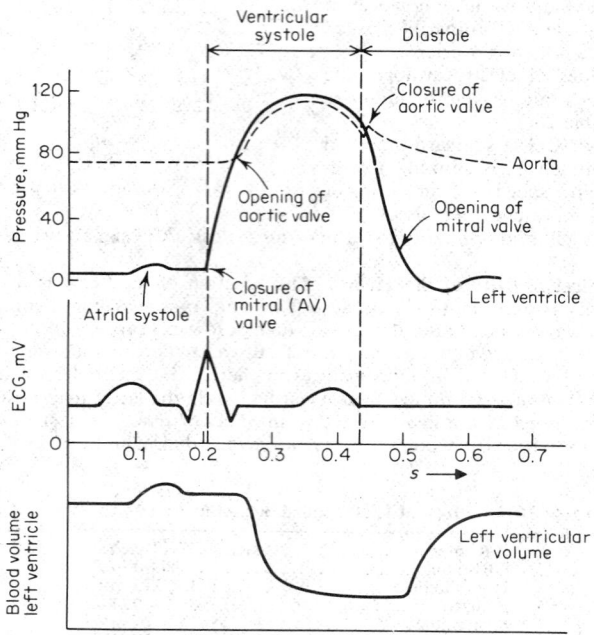

Fig. 26-7. Diagram of hemodynamic and ECG events in the cardiac cycle.

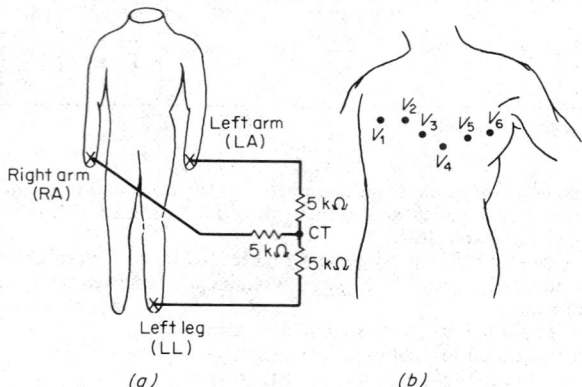

Fig. 26-8. Location of leads for standard 12-lead electrocardiogram. (Further details in Table 26-5.) CT = Wilson central terminal. (*a*) Limb leads; (*b*) precordial leads.

Table 26-4. Normal Heart Parameters

Stroke volume (resting) .... 75 ml	Heart volume .... 700–750 ml
Left ventricular end diastolic volume . 135 ml	Heart weight .... 250–390 g
Heart rate (resting) .... 69 min⁻¹	Left ventricular wall thickness .... 1.2 cm
Blood flow .... 5.5 l/min	Mean heart height .... 9.7 cm
Left atrial volume (max.) .... 60 ml	Mean heart width .... 10.7 cm
Atrial systole .... 0.11 s	Ventricular systole .... 0.27 s
Atrial diastole .... 0.72 s	Ventricular diastole .... 0.56 s
Isometric contraction .... 0.06 s	Rapid inflow .... 0.16 s
Isometric relaxation .... 0.05 s	Diastasis .... 0.23 s

Peak aortic pressure .... 120 mm Hg
Peak pulmonary pressure .... 25 mm Hg
End diastolic pressure (aorta) .... 80 mm Hg
End diastolic pressure (pulmonary aorta) .... 10 mm Hg

Table 26-5. Standard Electrocardiographic Leads

Standard or Limb Leads

$V_I = \Phi(LA) - \Phi(RA)$ $\Phi(LA)$ = potential of left arm
$V_{II} = \Phi(LL) - \Phi(LA)$ $\Phi(RA)$ = potential of right arm
$V_{III} = \Phi(LL) - \Phi(RA)$ $\Phi(LL)$ = potential of left leg

Precordial Leads

Wilson central terminal (CT) is the junction of three 5,000 Ω resistances, each connected to a limb lead. Precordial leads utilize CT as reference. Precordial leads are located as follows:

V_1 Fourth right intercostal space at sternal edge
V_2 Fourth left intercostal space at sternal edge
V_4 Fifth left intercostal space at the mid-clavicular line
V_3 Midway between V_2 and V_4
V_5 Same level as V_4 at anterior axillary line
V_6 Same level as V_4 at mid-axillary line

Augmented Limb Leads

aV_L Left arm with respect to junction of 5-kΩ resistors, one to *RA*, the other to *LL*
aV_R Right arm with respect to junction of 5-kΩ resistors, one to *LA*, the other to *LL*
aV_F Left leg with respect to junction of 5-kΩ resistors, one to *RA*, the other to *LA*

To the extent that the heart behaves as a dipole, all 12 leads can be synthesized. The dipole nature of the heart, as an electrical generator, appears to be a good approximation. Current research has shown that significant nondipolar behavior is exhibited during portions of the QRS. Practical measurement and clinical utilization of higher-order multipole terms are still under investigation.

12. The ECG Signal. Typical signals seen in electrocardiography are shown in Fig. 26-9 for the 12-lead ECG. Under normal conditions the waveform is periodic at the heart rate (nominally around 70 beats/min). A single cycle is on the order of 0.85 s, the QRS duration is around 0.1 s, while the R-wave spike might cover an interval of 0.03 s. One might suppose a high-frequency content up to perhaps 60 Hz, mainly due to the QRS complex. The low-frequency content might be estimated from the periodicity itself, which is approximately 0.8 Hz.

For diagnostic purposes the time intervals and segments as well as amplitudes are utilized. In addition, the waveform of the P, QRS, and T waves can be important. These waveforms may be abnormal in the event that an ectopic pacemaker should initiate a cardiac cycle. This

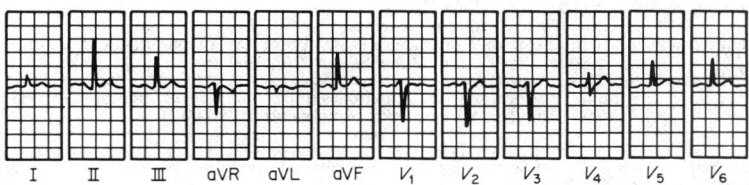

I   II   III   aVR   aVL   aVF V_1 V_2 V_3 V_4 V_5 V_6

Fig. 26-9. Twelve-lead electrocardiogram from normal heart.

comes about because the pathway of activation is abnormal; hence the temporal development of cardiac sources and their potential fields are quite different.

13. ECG Instrumentation. The essential part of a typical ECG machine is a high-input-impedance differential amplifier. Amplifier input may be ac-coupled to the electrode with 0.05-Hz low-frequency cutoff, or may be dc-coupled with a dc voltage balance control to balance off the difference in dc base-line voltages of each electrode.

One of the most persistent problems in ECG measurement is the reduction or elimination of ac interference. If the body is considered as a conducting sphere of 0.5-m radius, its capacitance to infinity is given by

$$C = 4\pi\varepsilon_0(0.5) = 55 \text{ pF}$$

The capacitance to surrounding power-line sources varies due to proximity, but is approximated as 100 pF. At the power-line frequency of 60 Hz, this constitutes an impedance of 26 mΩ. If the right leg is connected to the common ground of the amplifier through an electrode with contact resistance of, say, 2000 Ω, then a power-line potential of 110 V will be reduced by the ratio of $2,000/(26 \times 10^6)$ to a value of around 10 mV as a noise input. This is a substantial signal, in excess of the ECG itself. However, by using differential amplifiers, it will be mostly eliminated, because normally each lead develops essentially the same ac signal.

A second major problem in ECG recording is that of *base-line drift,* which may arise from physiological tension or poorly connected electrodes. The patient should be lying down when recording ECGs, and there should be freedom from noise or other distractions. Such conditions also reduce interference from muscle noise. In spite of this care, an *RC* network is still necessary to block residual direct current. A time constant of 3 s corresponds to a low-frequency half-power point of 1/20 Hz, which appears low enough to retain normal waveshape. Another solution to this problem is the use of *base-line clamper.*

A third important problem is electrical safety during continuous monitoring in the hospital (see Pars. **26-42** to **26-45**).

Inadequate high-frequency response can result in waveform distortion. In particular, the amplitudes and durations of a pertinent wave can be in error. It has been noted that a high-frequency response between 60 and 500 Hz is the desirable range. Studies on characteristic ECGs have shown that amplitude errors in Q, R, and S of less than 0.1 mV require cutoff frequencies above 100 Hz. (Reduction of amplitude errors below 0.05 mV would require an even higher cutoff frequency of perhaps 200 Hz.) Timing errors are less affected and appear satisfactory at cutoffs of 60 to 80 Hz.

The current Recommendations for Standardization of Leads and of Specifications for Instruments in Electrocardiography and Vectorcardiography by the American Heart Association are available. Summarized in Table 26-6 are some salient features.

14. Other Electrophysiological Systems. In addition to cardiac muscle, striated (skeletal) muscle, smooth muscle, and nerve are also capable of undergoing electrical activity. In the same way as described for the heart, an action potential can be elicited from each such

Table 26-6. Direct-writing Electrocardiographs

Input impedance (between any lead and ground) . . . 500k Ω
Leakage current to patient . . . 1 μA
Frequency response: 0.14–50 Hz . . . Flat to $\pm$0.5 dB
        0.05–0.14 Hz . . . <3 dB loss in response at 0.05 Hz (from 0.14 Hz); requires time constant > 3.2 s
        50–100 Hz . . . For 5 mm peak-to-peak output at 50 Hz, <3 dB decrease in strength at 100 Hz
Common-mode rejection of 1,000:1 . . . 45–65 Hz
        100:1 . . . other frequencies
Electrode impedance . . . Skin-to-electrode impedance (with appropriate preparation and use of paste or jelly), <5,000 Ω at 60 Hz for current <100 μA
System performance . . . <5% deviation from linearity for peak-to-peak amplitudes between 5 and 50 mm. For peak-to-peak amplitudes under 5 mm deviation, not to exceed 0.25 mm. (Input signal frequency components, 0.05–100 Hz.)

Source: Committee on Electrocardiography of the American Heart Association, Selected Summary of Recommendations for Standardization of Leads and Specifications for Instruments in Electrocardiography and Vectorcardiography, *Circulation,* Vol. 35, pp. 583–602, 1967.

excitable cell.   During the time-varying phase the cell behaves precisely as a dipole source of electricity, and surface or implanted electrodes record the collective contribution to the electric potential field from all such sources (gross activity).   Since these measurements reflect the electrophysiological character of the responsible cells, diagnostic information can be deduced concerning them.

Electroencephalography (EEG) is concerned with the measurement of electrical signals on the scalp which arise from the underlying neural activity in the brain.   Such signals are therefore a reflection of the state of the central nervous system and prove useful in the diagnosis of epilepsy, brain tumors, and other disease states.   Table 26-7 lists a number of physiological systems in which muscle or nerve generates a waveform that has been identified and characterized.   The problem of noise is more critical in electroencephalography since the desired signal is of the order of 50 μV (Fig. 26-10).   Bandwidth requirements for electromyography range from 20 to 500 Hz, while EEG requires direct current to 150 Hz.   Not all signals are available at the body surface, and some type of deep electrode becomes necessary. Electrodes specially adapted to each system under study have been developed which provide good contact, correct geometrical location, and low impedance (see Table 26-7).

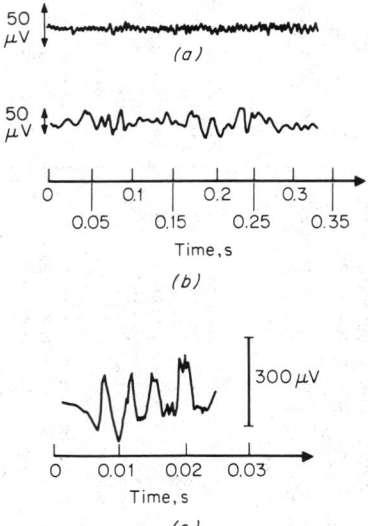

Fig. 26-10.   Biopotential waveforms. (*a*) EEG under excited conditions; (*b*) EEG under sleep conditions; (*c*) EMG wave.

THERAPEUTIC AND DIAGNOSTIC RADIOLOGY

By Floro Miraldi, Sc.D., M.D.

15. Production of X-Rays.   Radiology consists of three major divisions: diagnostic radiology, therapeutic radiology, and nuclear medicine.   The main function of the diagnostic radiologist is to produce and interpret shadow images of internal organs of the body using x-radiation.   Therapeutic radiology concerns itself primarily with the treatment of disease by the destruction of tissue with high-energy radiation.   Nuclear medicine has both diagnostic and therapeutic aspects but differs from the others in that the radiation is usually internally distributed in the body and arises from the decay of radioactive materials which have been ingested or injected.

In *diagnostic radiology* and for superficial therapy purposes, the energy spectrum of radiation varies from about 10 to 100 keV.   In *therapeutic radiology* and in *nuclear medicine,*

Table 26-7. Electrode Characteristics

Type of bioelectric phenomena	Type and placement of electrodes	Biological source	Physical characteristics of signal
Electro-encephalography	Small metal disk, approx. 7 mm (preferably silver-silver chloride) placed on scalp. Needle electrodes may be used—inserted into the subcutaneous tissue. [Array of 16 electrodes over top of skull roughly at intersection of four longitudinal and four transverse columns. (Bipolar electrodes.)]	Gross electrical activity of brain nerve cells	Amplitude under 100 μV, average 10–50 μV; frequency range dc to 150 Hz (see text)
Electro-cardiography	Metal plate-silver electrodes 3.5 × 5 cm placed on extremities, concave "suction-cup" type for chest (4.75 cm dia.).	Gross electrical activity of the heart	Amplitude under 10 mV, average 1–5 mV; frequency range 0.14–100 Hz (see text)
Electro-myography	Surface electrodes superficial to muscle: silver-silver chloride type. Needle electrodes: concentric steel outersheath 0.65 mm, platinum-wire core. 0.04 mm diameter, beveled. (Bipolar or unipolar, use wire core only.) Inserted into muscle bundle.	Extracellular recording of one or more motor unit (groups of muscle fibers) due to their action potentials arising from muscle contraction	Amplitude under 3 mV, usually 0.1–1.0 mV; frequency range 40–3,000 Hz (see text)
Electro-retinography	One electrode consists of chlorided silver wire contacting the physiological saline solution held by contact lens placed on cornea. Indifferent electrode is a plate-type (Ag-AgCl) on forehead (use skin preparation and electrode paste).	Summation of receptor potentials generated by light-sensitive cells (rods and cones) in the retina due to light stimulation	Amplitude under 1 mV; frequency range dc to 20 Hz
Electro-occulography	Electrode pairs equidistant from pupil above and below and/or laterally and medially (at outer and inner canthi). Electrodes are small (10-mm diameter), plate silver (preferably chlorided silver) for surface contact (use skin preparation and electrode paste).	Eyebulb in orbit establishes a dc current dipole of a galvanic type	Amplitude 0.05–3.5 mV (0.2–0.4 mV/20° rotation); frequency range dc to 125 Hz

the energies of interest range from about 100 to 10,000 keV. Corresponding wavelengths range from 1 to 0.001 Å and constitute the x-ray region of the electromagnetic spectrum.[35]*

Production of x-rays is accomplished by bombarding material with high-speed electrons, using the principle that accelerated charged particles radiate.[36,37] Figure 26-11 shows spectral distributions of x-rays produced at different energy electron bombardment. The continuous distribution which results is called *bremsstrahlung,* or *white radiation.* The superimposed peaks on the white radiation are called *characteristic radiation* because they are characteristic of the atoms in the target. Also shown in Fig. 26-11 are characteristic radiations of tungsten, which is the most commonly used target material. Even in the most favorable cases the useful radiation output is only a few percent of the total electron energy; that is, most of the energy is dissipated in the target as collisional energy or heat.

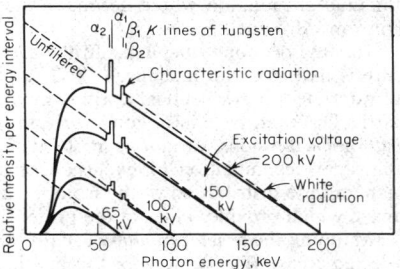

Fig. 26-11. Spectral distributions of x-rays produced at different energy electron bombardment. *(Courtesy H.E. Johns, "The Physics of Radiology," 3d ed., Fig. II-14, p. 43, Charles Pummis, Springfield, Ill., 1969.)*

Characteristic radiation occurs when electrons from higher-atomic-energy orbits go into lower-energy orbits. This occurs when the lower-energy electrons in an atom are ejected by a collisional process.[38] Hence *k* radiation is produced when the *k* electron is ejected from the atom; *l* radiation when the *l* electron is ejected from the atom; etc. The principal emission lines for tungsten are shown in Table 26-8.

Table 26-8. Principal Emission Lines for Tungsten

Transition	Symbol	Energy, keV
$N_{II}N_{III}$-K	$K\beta_2$	69.089
M_{III}-K	$K\beta_1$	67.236
L_{III}-K	$K\alpha_1$	59.310
L_{II}-K	$K\alpha_2$	57.972

SOURCE: H. E. Johns, "The Physics of Radiology," 3d ed., Table II-2, p. 51, Charles Pummis, Springfield, Ill., 1969.

The production of x-rays is straightforward, but the practical requirements of constructing instruments to deliver x-rays of specific quality and quantity are quite severe. A conventional x-ray tube consists of an anode-and-cathode assembly placed in an evacuated glass envelope.[38] The anode is usually a massive piece of copper in which is placed a small tungsten target. The cathode assembly usually consists of a filament of tungsten wire placed in a shallow focusing cup. The hot tungsten filament provides the source of electrons, which are accelerated toward the anode by applying a high voltage between the anode and cathode. Although such tubes are usually excited by an alternating voltage, the electron currents through the tube can flow only when the anode is positive with respect to the cathode. Such tubes are *self-rectifying,* and the current through the tube consists of a half-wave.

Under severe electron bombardment, the tungsten target of the cathode may reach a temperature at which electron emission from its surface begins. On the half cycle in which the filament is positive, electrons can then travel from anode to filament. If conduction occurs during the inverse cycle, the filament will be destroyed by the electron bombardment. To avoid this, reverse electron flow is prevented by the use of rectifiers.

Because of the low efficiency of conversion of the electron energy into radiation energy, the heat load of the anode becomes severe in high-current machines. Since most diagnostic machines operate with tube currents of the order of hundreds of milliamperes, at 10^4 to 10^6 V, the power load can easily reach the order of 10^4 to 10^5 W. Thus cooling of the target is required, which is accomplished in different ways, such as rotating the anode to expose different surfaces to the beam, immersion of the tube in liquid coolant, requiring intermittent

---

*Superior numbers correspond to numbered references, Par. **26-66**.

use of the machine, etc.[39] Shielding and direction of a beam with the use of diaphragms and ports are required to obtain a usable film exposure or a satisfactory treatment.[40] Problems of radiation tissue damage must be considered, as well as the necessary deliverance of proper amounts of radiation where needed. The problem of radiation dosimetry is discussed by Hine and Brownell.[41]

16. Radioisotope Imaging. In the area of nuclear medicine the most important procedure is that of organ imaging.[42-46] The patient ingests a gamma-emitting radioisotope in appropriate pharmaceutical form which allows the concentration of the isotope in a particular organ system (or conversely, the exclusion of an isotope from a particular organ). Pathological lesions, variations in blood perfusion, or anatomical variations will produce an abnormal accumulation or exclusion of the isotope in particular organs. Accordingly, mapping the distribution of the isotope via detection of the emitted gamma rays yields an image which provides clues of the presence of pathological lesions.

Mapping the *time distribution* usually requires following the activity over one or several selected areas. The devices for this are the simplest types of radiation detector systems and usually employ scintillation crystals, because they are the most efficient for gamma detection. The general configuration of such a probe consists of a collimator to limit the field of view, followed by a scintillator, which in turn is followed by a photomultiplier with associated electronics for recording. Output curves are handled in many ways, such as oscilloscope display, use of scalers, count rate meters, or strip chart recorders.

Spatial mapping, or *imaging,* of an organ is a somewhat more complex problem. At the present time, two basic approaches are commonly used, rectilinear scanning and the scintillation-camera scanning.

17. Rectilinear Scanning. The rectilinear scanner is primarily a scintillation probe which is moved over the region of interest in a linear motion. The system is a point-by-point detection method which places a dot on film or paper for each incident recorded at a particular point. Most rectilinear scanners use a focusing collimator in order to limit the field of view without undue sacrifice of sensitivity. The most significant disadvantage of the rectilinear system is that, for high resolution, the sensitivity is small and scanning time required is long. The low sensitivity is partly the result of observing only a small region at a time. Depending on the organs scanned, times required may range from 15 min to several hours.

The sensitivity of these instruments, measured as the number of data points displayed per gamma ray emitted in the body during the scan, is typically of the order of 10^{-4} to 10^{-6}. Increasing the amount of radioisotopes to increase the speed with which a scan may be performed has its limitations because of radiation dose delivered. Consequently, for better sensitivity and better scans, other approaches are needed.

18. Scintillation Cameras. The scintillation camera is an attempt to bypass the problem of low sensitivity. In this concept, a large field is viewed simultaneously. Obviously, if 10 detectors were placed in parallel, the speed would increase by an order of magnitude. A multiplicity of detectors of the rectilinear scanner type ganged to run in parallel paths has been proposed, and such an instrument is available commercially.

A second approach employs a matrix of scintillation crystals covering the entire field of interest, each with its own collimator system, has been suggested and developed. When systems such as this do not involve motion over the body, the system is called a *camera,* e.g., the autofluoroscope of Bender and Blau. It suffers the disadvantage that, to cover a reasonably sized field, the number of crystals and photomultipliers required is large, yielding a rather complex system.

The most promising camera system to date, and also the most widely used, is the scintillation-camera system of Anger. This instrument consists of a large scintillation crystal behind a straight-bore collimator and backed by an array of 16 photomultipliers. Using a passive network to triangulate on the position of a scintillation as viewed by several photomultipliers, this system is capable of resolving scintillations well enough for most clinical applications. It is many times faster and more sensitive than the rectilinear scanners because it exposes much more crystal to the field of view than the other systems. Its sensitivity appears to be approximately four to seven times faster than the rectilinear scanner for comparable resolution. It suffers the disadvantage that its resolution becomes poor for gamma rays below about 100 keV and at high energies the sensitivity drops rapidly. This system is of great interest at the present time because many pharmaceuticals can be tagged with Tc-99m, which emits a gamma ray of 140 kV.

19. Other Devices for Radiation Mapping. Recently, interests have developed in the multiple-wire proportional counter. In this instrument, the anode of a conventional gas-proportional counter is replaced by a grid. By blanking the wires of the grid in timed sequence, the position of an interaction in the photomultiplier tube can be determined. In this manner, a position-dependent proportional counter system is obtained. The resolution is extremely high and can be of the order of millimeters. Sensitivity is also high for very-low-energy photons, but becomes completely inadequate at energies above about 30 kV.

Another system which has been developed uses an image amplifier tube situated behind a large scintillation crystal or an array of scintillation crystals. The potential resolution of these systems is very high, but they also suffer sensitivity problems which are inherent in the image-amplifier portion of the system.

A recent innovation is a hybrid scanner which combines both rectilinear and camera concepts to produce a line detector. Using a bar of scintillation crystal, it has been shown that the light output from the ends of the bar falls off exponentially with distance from the end. A photomultiplier situated at the end of the bar, followed by logarithmic amplifiers, then has an output which is directly proportional to position measured from the end of the bar. The bar detector is moved across the field of view, and position of a scintillation is determined by noting the position of the bar and position along the bar where the scintillation occurred.

ARTIFICIAL ORGANS

By RAYMOND J. KIRALY, M.S., and YUKIHIKO NOSÉ, M.D., PH.D.

20. Artificial Kidney (Hemodialysis). Since 1960, chronic or long-term hemodialysis has become a routine procedure. About 10,000 people in the United States and an equal number in Europe now use chronic hemodialysis. Sixty percent are being treated at medical centers, while the remainder use their own or rented equipment in their homes. Dialysis is usually performed 2 or 3 times per week for a duration of 6 to 14 h.

21. Function of the Kidney. The function of the kidney is to maintain a chemical and water balance in the body by removing excess or waste material from the blood. Dialysis based upon the semipermeable properties of thin membrane is used as the artificial kidney. A semipermeable membrane such as cellulose membrane has multiple pores averaging 40 Å and is permeable for the waste products of urea, creatinine, and uric acid ions, but not other large molecules of proteins and blood cells. Waste products pass through the membrane due to the concentration gradient between the blood and the dialysate fluid on the other side of the membrane.

Chemical balance in the human body is maintained by adding specific chemicals to the dialysate fluid to make the concentration the same as that in the blood.

22. Hemodialysis Membrane and Dialysate. The membrane most commonly used is a *cellulose membrane* on the order of 1 mil, or 25 μm, thick. This material is used in sheet form to assemble flat dialyzers. It is also available in tubular form, which is flattened and formed into a coil, resulting in a compact dialyzer. An artificial kidney utilizing multiple hollow cellulose fibers is also in use.

Dialysate is the solution that exchanges the solute through the cellulose membrane with the blood. It is mainly water. To control what is removed from the blood, it is necessary to control the concentration of the dialysate solution. In a normal dialysate fluid, electrolyte concentration is equal to corresponding concentrations in the blood of a normal human being.

23. Description of Hemodialyzers. Dialyzers can be divided into three general types: the coil, the plate, and the hollow fiber. Coil-type dialyzers are the two-coil and disposable-cartridge type. The twin-coil artificial kidney, shown in Fig. 26-12, is formed by winding two cellulose tubes together with a scrim to form two blood flow paths in parallel while maintaining a passage to control the blood film thickness. Cellophane tubing 45 mm wide is used for this purpose. The surface areas for coils range from 0.9 to 1.9 m². The coil is put inside a canister, which is then placed inside the dialysate tank. The dialysate is pumped through the coil axially. A blood pump is necessary for the arterial side because the twin-coil has a high resistance to blood flow. For safety purposes, a pressure monitoring system is connected to an automatic shutdown of the system. The twin-coil dialyzer system is shown in Fig. 26-13.

The original flat-plate type, the Kiil dialyzer, has relatively large dimensions and uses three basic boards, as shown in Fig. 26-14. Two membrane sheets are inserted between these three boards and establish two separate blood compartments with a surface area of $\frac{1}{2}$ m^2 each, resulting in approximately 1 m^2 of surface area for the dialyzer. Blood-flow resistance is relatively small, so that a blood pump is not required. The hollow fiber dialyzer is the most compact type employing thousands of capillary tubes in a shell-and-tube configuration. The blood passes through the tubes while the dialysate passes over the tubes in the shell of the unit. The pressure drop is low and a blood pump is generally not required.

24. Access to Blood. To connect the dialyzer to the blood supply, it is necessary to have repeated cannulations to the artery and vein. The Scribner-Quinton shunt gives easy access to the blood without any surgical procedure. Usually, this silastic shunt is implanted between the radial artery and an adjacent subcutaneous vein and can be maintained for more than one year.

25. Dialysate Supply and Monitoring Systems. The functions of a dialysate delivery system are to prepare and deliver dialysate of the required chemical makeup for use in the hemodialyzer. Monitoring and control equipment are sometimes included as part of the system to ensure that the dialysate is of the proper makeup and ready for use by the hemodialyzer.

Methods of preparing and delivering are generally categorized as either batch or continuous. In a continuous dialysate supply system, concentrate and tap water are continuously mixed during the course of the dialysis and delivered to the dialyzer. This type of system eliminates the space required for mixing an entire batch of dialysate. However, to be effective, the system must be closely controlled, since any malfunction can result in an improperly mixed dialysate.

Dialysate can be recirculated, although fresh dialysate is most effective because of its high-concentration gradient with the blood. In a single-pass system, the flow rate of dialysate is

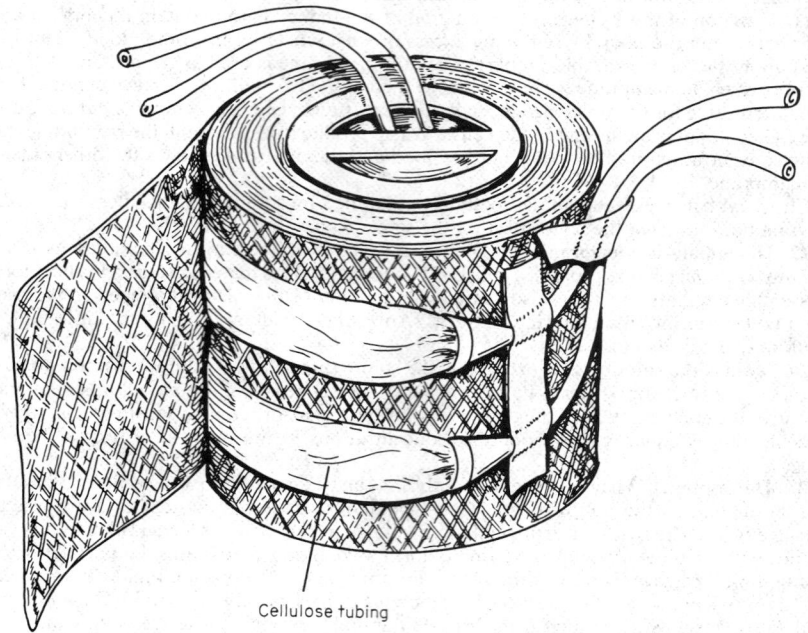

Cellulose tubing

Fig. 26-12. Ultra-Flo coil. Two loops of cellophane tubing are wound around the core, together with crisscross polypropylene screen.

usually kept as low as possible to limit the amount of dialysate required. A recirculation system where the dialysate is pumped from a tank through the coil dialyzer is shown in Fig. 26-13.

Dialysate delivery systems also include monitoring, controlling, and safety equipment. These devices range from simple apparatus to automated systems capable of operating without an attendant. Monitoring equipment includes flowmeters, temperature gages, dialysate conductivity probes, and display meters. Control equipment includes thermostat-controlled heaters or mixing valves for regulating dialysate temperatures and composition, valves for regulating flow rates, and adjustable high/low limits on various safety-monitoring devices. Safety equipment includes devices to correct any factor in dialysis which exceeds the desired operating limits for safe operation. This includes audio/visual alarms and fail-safe shutdown procedures that would be accomplished during the course of dialysis. Figure 26-15 and Table 26-9 indicate the most commonly monitored items during hemodialysis.

26. Artificial-Heart System. The complete artificial-heart system is composed of four essential elements: (1) energy source, (2) power-conversion device, (3) system controls, and (4) the pump. Figure 26-16 shows the physical interrelationship of these four elements and the six possible arrangements for locating elements implanted in the body and external to the body. With the skin line shown in position 1, the entire system is implanted in the body, which would be the ultimate goal. At present, with most experiments in animals, the skin line is represented by position 6, with only the pumping element implanted and the energy source, power-conversion equipment, and system controls external to the body. When the pumping element is integrated into the human system, the other devices can be miniaturized and added to the implanted package. The progression would then be from position 6 to position 1, as shown in Fig. 26-16.

The specific circulatory parameters to be measured in the human body and fed back to the control system are not all known. Figure 26-17 shows the normal circulation system. The natural heart has two pumping chambers. The right heart is fed from the venous system at

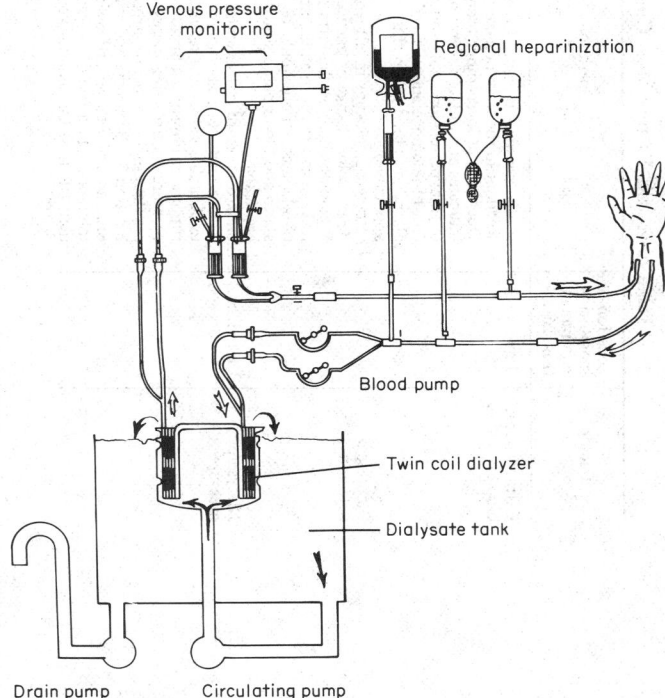

Fig. 26-13. Schematic of original twin-coil dialyzer setup.

Table 26-9. Six Factors Most Commonly Monitored During Hemodialysis

Factor	Location of monitoring equipment (Fig. 26-15)	Operation	Remarks*
Extracorporal blood pressure	1 and/or 2	Usually measures pressure in drip chamber. Readout in the form of gage or manometer. Operation at levels other than normal indicates any one of several malfunctions (e.g., hypertension, clots and blood leaks). Out-of-tolerance action should be audiovisual alarm and blood pump shut off (if applicable).	Installation in location 1 is generally regarded as providing the most meaningful information with respect to changes in flow (clots) or blood-line leaks. Installation in location 2 is mandatory if a blood pump is used.
Blood-leak detector	3	Photoelectric pickups in effluent dialysate line detect color change or particulate matter. Readout usually in the form of alarm. Blood leakage can result in shock and death. Out-of-tolerance action should be audio-visual alarm and blood pump shut off (if applicable).	Some devices currently in use claim ability to detect 0.1 ml blood/l fluid. These units are affected by power loss.
Dialysate pressure	3,4 and/or 4	Standard-pressure instrumentation (usually gage, sometimes manometer). Operation at undesirable values can result in membrane rupture or improper ultrafiltration. Negative pressure control often incorporated to regulate pressure, below atmospheric, at location 3. Both positive- and negative-pressure relief valves are desirable. Out-of-tolerance action should be relief and stop dialysate flow through dialyzer.	Various types of negative-pressure controls are used. Their operation and potential malfunctions should be understood. The operation of relief and check valves, desirable features in any pressure-regulated system, should also be understood. Pumps, used to pull dialysate through the dialyzer, that are electrically driven will be affected by power loss.

Dialysate temperature	5 or 6	Thermostatic measurement generally used to control electric heaters or hot-cold water-mixing valves. Dial thermometer or thermocouple gage readout. Operation at undesirable temperature can result in blood cell damage or chilling. Out-of-tolerance action should be to stop dialysate flow through dialyzer.	Heater controls are not always used with batch-type delivery systems. Depending on the size of batch (originally mixed at proper temperature), heat loss can be held within reasonable limits during the course of dialysis. Electrical heating apparatus is affected by power loss.
Dialysate flow rate	5	Generally measures and displays flow in a variable-area flow tube (called rotameter). Unless extremely out-of-tolerance, this variable cannot result in undue harm to the patient. Action should be to stop dialysate flow through dialyzer.	In through-flow systems, dialysate is normally used at the rate of 500 ml/min. Some systems incorporate valvular adjustments; others maintain essentially constant flow through the use of positive-displacement pumps. Electrically operated pumps will be affected by power loss.
Dialysate concentration	5 or 6	Measures electrical conductivity of dialysate. Output meter is required to display value. Improper concentration can result in blood cell and central nervous system damage. Out-of-tolerance action should be to stop dialysate flow through dialyzer.	Conductivity of an electrolyte, such as dialysate, is affected by temperature; therefore it is recommended that any conductivity probe be temperature-compensated. This measurement will also be affected by power failures. Continuous concentration measurement of some sort is a necessity in a delivery system utilizing continuous proportioning of dialysate. However, this equipment is not required in batch-type delivery systems where predialysis concentration measurements can be made.

* Due consideration should be given to equipment that is affected by power loss, to shutdown sequence, and to safety aspects.

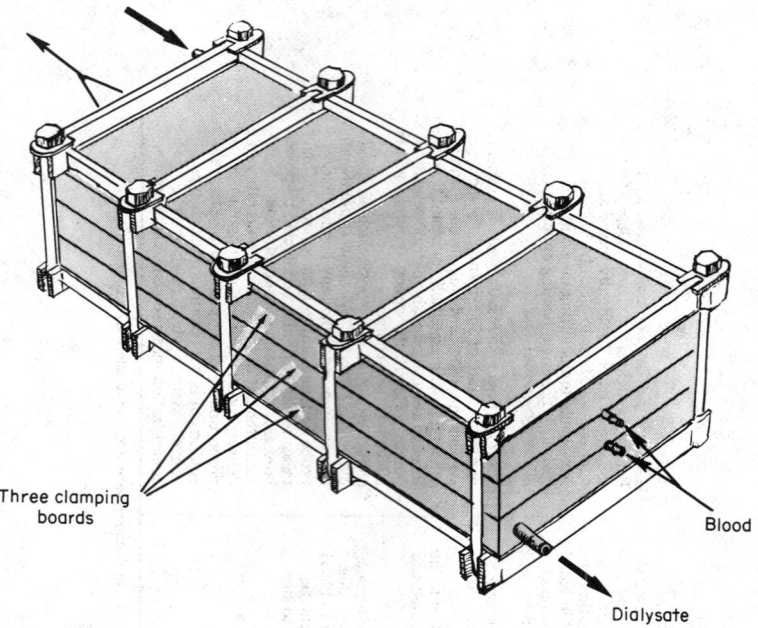

Fig. 26-14. Kiil dialyzer.

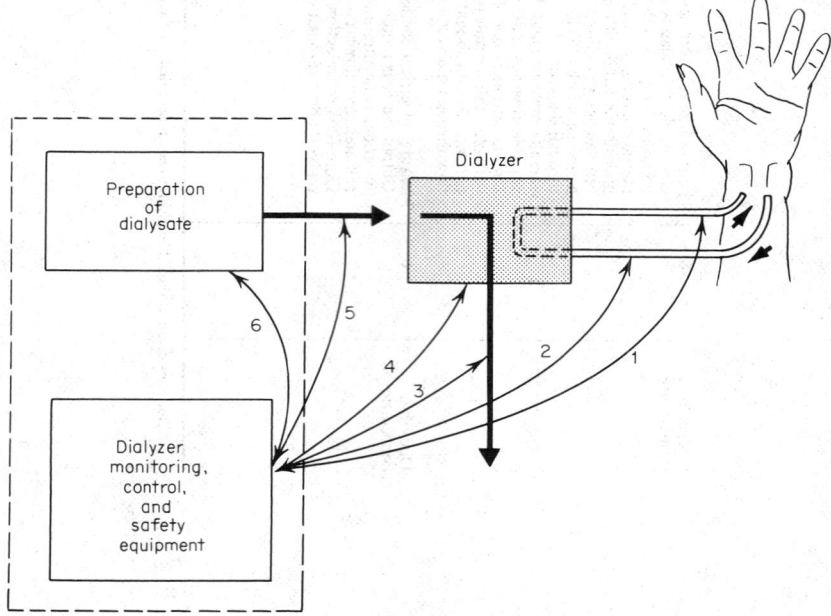

Fig. 26-15. Dialysate delivery system. The numbers refer to the factors most commonly monitored during hemodialysis (see corresponding numbers in Table 26-9).

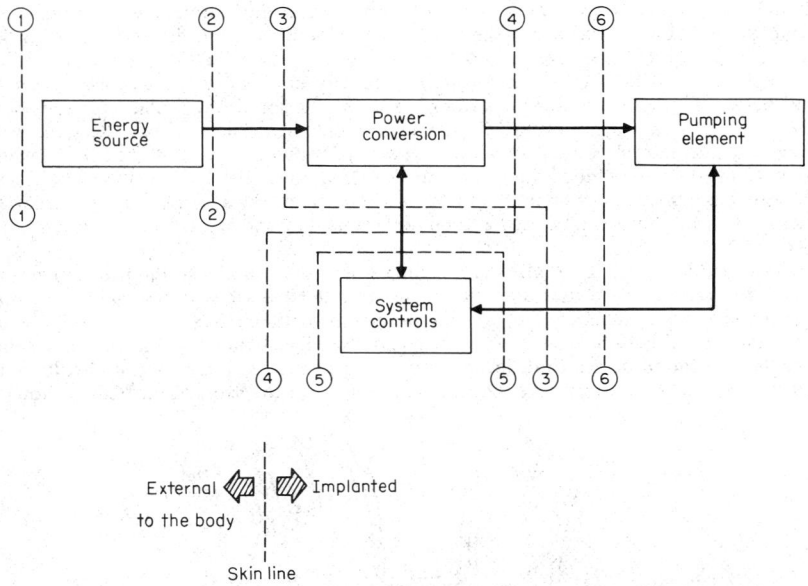

External to the body | Implanted

Skin line

Fig. 26-16.   Artificial-heart-system concepts.

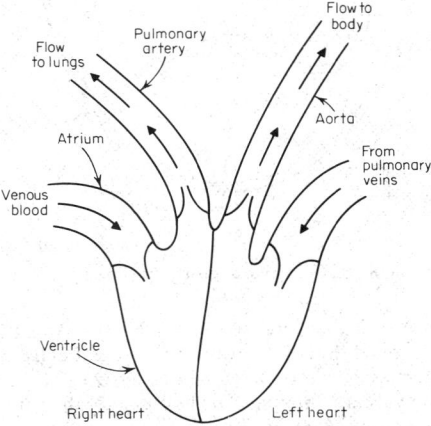

Fig. 26-17.   Natural-heart circulation.

near ambient pressure. Contraction of the heart pumps blood from the right ventricle through the pulmonary artery into the lungs. The blood, having been oxygenated in the lungs while rejecting CO_2, returns to the left ventricle through the pulmonary veins. Pressure in the pulmonary arteries is normally 20 mm Hg; pulmonary venous pressure is near atmospheric. The left ventricle then pumps blood into the aorta for circulation in the body. Aortic pressure ranges from 60 to 150 mm Hg. If the artificial heart has the same beat frequency as the natural heart, it should be capable of between 60 and 140 beats/min while pumping 3 to 14 l/min of blood flow. The average fluid power delivered to the blood is 3 to 5 W. These parameters can be used in designing an artificial heart which has two separate pumps. Artificial hearts can be used as total replacement for the natural heart or as an assist device.

27. Description of Artificial Hearts. Artificial hearts are similar to the natural heart in size and geometry and are generally constructed of a flexible sac within a rigid container. Either active or passive check valves are incorporated on both the inlet and outlet sides of the heart. The space between the flexible sac and the rigid container may contain some mechanical actuation device to deflect the sac, or this space is pressurized cyclically with either a gas or a liquid to achieve the pumping motion of the sac. Figure 26-18 shows a

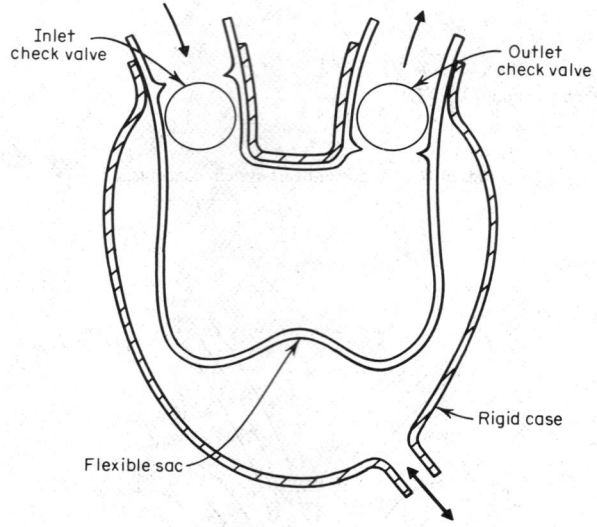

Fig. 26-18. Typical fluid-driven artificial heart, showing one pumping chamber.

typical heart using a fluid to actuate the sac in the heart. Actuating fluid flow and/or pressures would be controlled by the power-conversion device and system controls. Differences in various heart designs involve variations in size and shape of the sac, and to a larger extent design of the valves. Internal blood volume of the ventricles for the artificial hearts generally lies in the range of 100 to 200 ml.

Figure 26-19 shows the construction of an artificial heart with biological components. The use of aldehyde-preserved valves results in good hemodynamic performance. The aldehyde-preserved natural tissue lining of the blood contact surface provides resistance to clotting. Figure 26-20 shows a completed pneumatically actuated heart used in research.

28. Artificial-Heart Driving Systems. Figure 26-21 shows a simple, yet effective pneumatic driving unit developed in 1964 for driving and testing artificial hearts on a laboratory basis. This mechanism provides a pneumatic output that alternates between pressure and vacuum to operate the heart. Pulse rates and pressure-vacuum durations can be adjusted electrically. The electrical controls provide on/off signals to the solenoid valves. A more

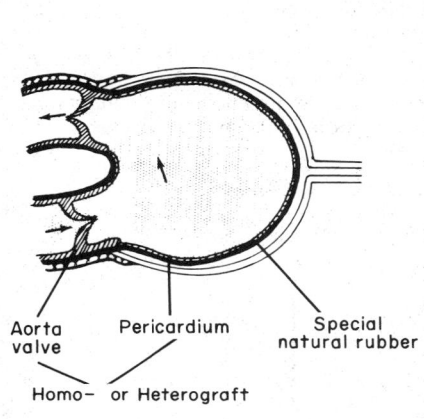

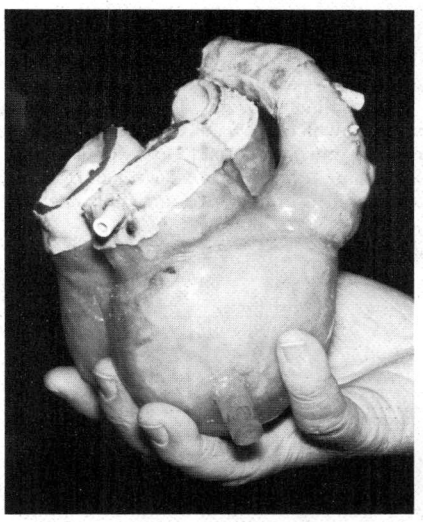

Fig. 26-19. Pneumatically actuated artificial heart in which all blood contact surfaces are natural tissue. Natural aortic valves are used at both the inlet and outlet. Natural rubber reinforces the pericardium forming the pumping sac. The outside case is polyurethane. (Arrows indicate blood flow.)

Fig. 26-20. Total cardiac prosthesis fabricated for in vivo evaluation in a 90 kg calf.

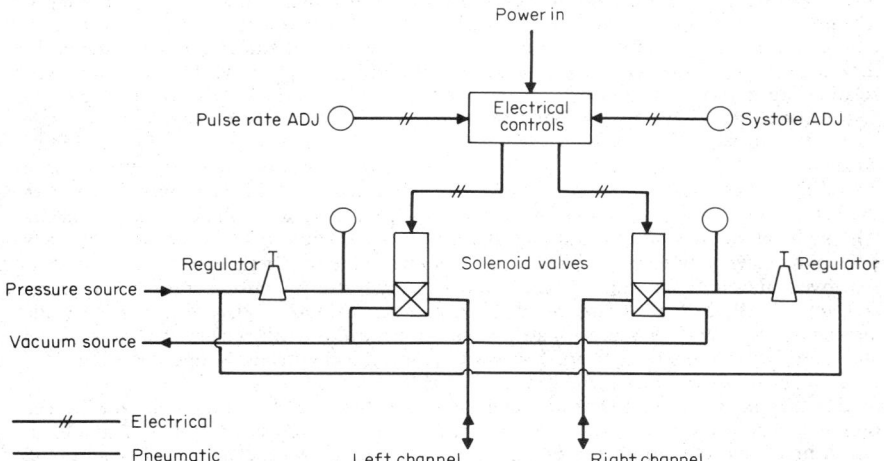

Fig. 26-21. Schematic of Detroit coil timer used for pneumatically driven hearts.

sophisticated and versatile unit developed by the Cleveland Clinic and NASA Lewis Research Center is shown in Fig. 26-22. It has great versatility as a research tool in that it can be synchronized with the normal heart, has closed-loop control on selected parameters, and employs multiple feedbacks from the heart and circulatory system.

Implantable energy-conversion devices include the Rankine cycle and the Stirling cycle systems, which allow conversion from heat energy to mechanical energy. Both employ heat engines utilizing thermal energy sources, the most practical at present being a radioisotope for a long-term implant. Mechanical energy can, in turn, be controlled to operate the artificial heart directly or through some intermediate conversion device.

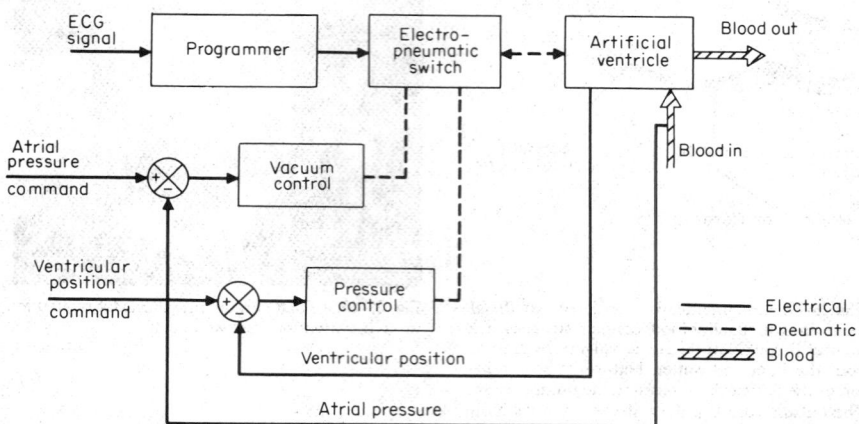

Fig. 26-22. Block diagram of multipurpose artificial-heart actuation system.

29. Artificial-Heart Control. The heart circulation system in Fig. 26-17 shows that there is one flow path through the two halves of the heart which may operate slightly differently. The first obvious control criterion is that both right and left hearts pump equal flow rates on a time-average basis; otherwise the lungs will be overfilled or drained of blood. One of the most important parameters affecting blood flow rate through the heart is atrial pressure. The relationship between cardiac output and atrial pressure has been termed *Starling's law.* Experimental data have shown that small changes in atrial pressure result in fairly large changes in cardiac output, as illustrated in Fig. 26-23. This type of control alone would be adequate for the artificial heart. If the artificial heart could be made sensitive to atrial pressure, then as the atrial pressure would rise, indicating a large blood supply, flow rate of the heart would increase to pump this away to satisfy continuity in the circulatory system. Although automatic control based on atrial pressure measurements has been used in many pneumatic driving systems, a simpler technique has been used whereby the heart itself is sensitive to the atrial or filling pressure. One of the early hearts to incorporate such a control is shown in Fig. 26-24. The heart-filling process maintains atrial pressure near the atmospheric reference pressure. If atrial pressure rises, filling will be faster and output flow will increase. Although the heart shown was designed and tested to be operated on air, the pumping chamber can be actuated by mechanical means as well.

Another approach to a pneumatically operated heart is shown in Fig. 26-25. The pumping chamber incorporates a control-pressure port for feedback to the fluid amplifier, which functions so as to alternate from suction to pressure whenever the sac reaches full travel, and alternately closes and opens the feedback control port in the heart case.

30. Artificial Lung. The artificial lung is a gas-exchange device for removing CO_2 and adding oxygen to the blood. Advances in surgery and organ transplantation have increased the need for short-term heart-lung bypass equipment. In addition, there is a growing application for oxygenators for pulmonary respiration support in case of natural-lung failure. An artificial lung should be able to oxygenate a blood flow of 5 l/min and remove CO_2 while minimizing blood damage and have a relatively small blood volume.

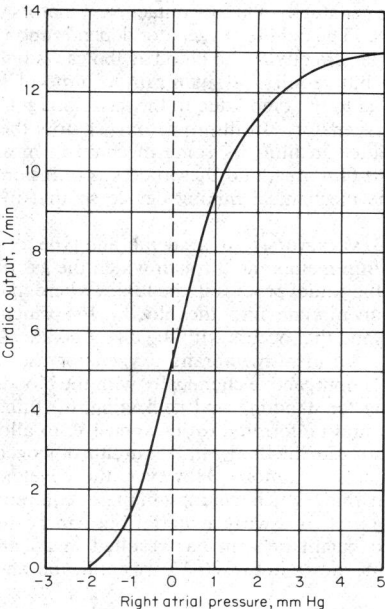

Fig. 26-23. Relationship between flow rate and atrial pressure for the natural heart.

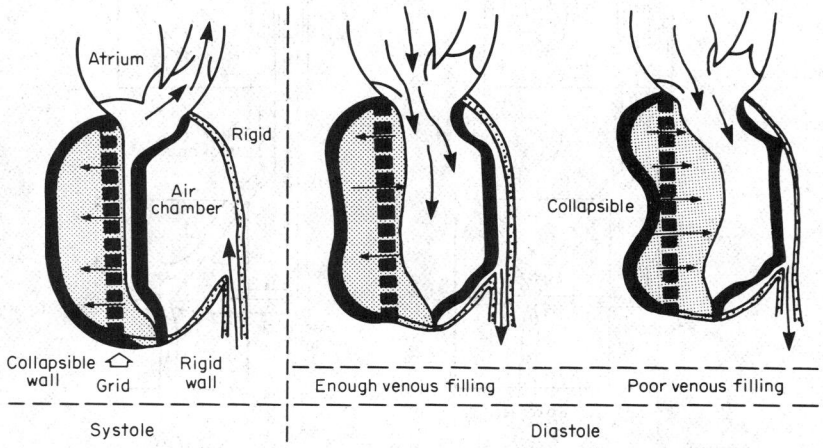

Fig. 26-24. Artificial heart with ambient pressure reference. The filling pressure is maintained near atmospheric pressure by the collapsible outside wall. Flow rate is sensitive to filling pressure even though there is full travel of the pumping chamber on each beat.

31. Direct-Contact Oxygenators. Direct-contact oxygenators can be classified as either bubble- or film-type devices. The bubble oxygenator depends upon the transport of gas from the surface of oxygen bubbles directly to the blood as the gas is bubbled through the blood. This type is quite simple but requires a significant volume of blood. It also causes a significant amount of trauma to the blood due to the gas-blood mixing process. Figure 26-26 shows a bubble-type oxygenator. In film-type oxygenators the gas contacts the blood, which is spread or distributed in films by some mechanical means. The gas transfer is proportional to the exposed film area, and the process is diffusion-limited. These devices generally incorporate some mechanical mixing device so that fresh blood is continually exposed to the gas.

32. Membrane-Type Oxygenators. In a membrane-type oxygenator, the gas-blood interface is separated by a thin membrane through which the gas diffuses. The membrane-type oxygenator simulates the actual process in the lungs, where the respiratory gases diffuse through the natural-lung membrane into the blood. Its principal advantage is that it eliminates damage to the blood that occurs with the direct-contact type and allows practical usage for periods of days. An ideal membrane oxygenator uses a membrane having the following characteristics: (1) nonreactive chemically with the blood; (2) high permeability to respiration gases; (3) strong for handling and packaging; (4) thin, to reduce the diffusion resistance; (5) high quality and uniformity; (6) inexpensive, to allow disposal after use; (7) readily accepts current sterilization methods; and (8) nonthrombogenic. One of the materials that meets these requirements is Teflon. However, the development of silicone-rubber membranes 0.002 to 0.005 in. thick which met most of these requirements resulted in practical membrane-type oxygenators. It is a weak material but can be supported and reinforced. Treatment with the anticoagulant heparin has resulted in an antithrombogenic surface. Membrane-type oxygenators, shown in Fig. 26-27, are applicable and used as a long-term life-support device.

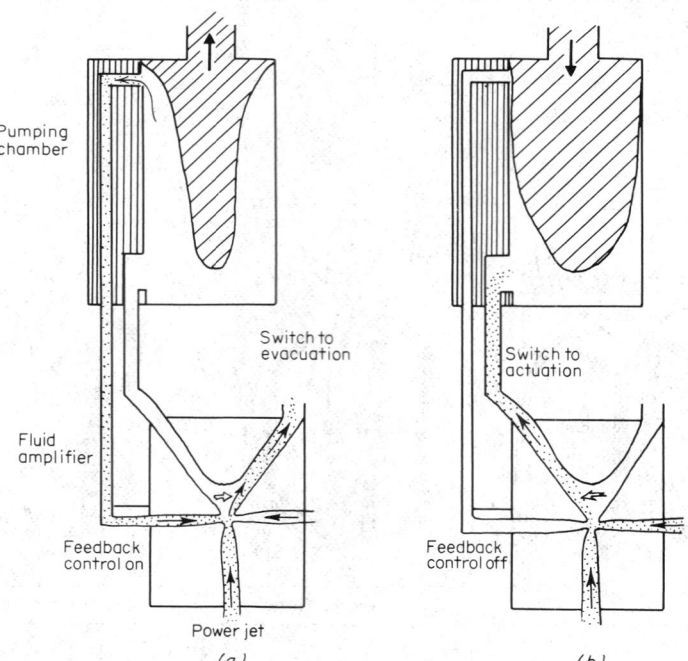

Fig. 26-25.   Schematic diagram of the function of the fluid-amplifier-controlled pump. Pumping rate is a function of filling rate, since the feedback signal changes as the control port is alternately covered and exposed by the flexible sac. (*a*) End of systole; (*b*) end of diastole.

33. Liquid-Liquid Oxygenators. An oxygenator undergoing development is the liquid-liquid type shown in Fig. 26-28. It uses a liquid in which oxygen and CO_2 are highly soluble. Liquid is mixed with blood, and the oxygen diffuses from the liquid to the blood, eliminating problems found with the direct liquid-gas contact and membrane types of oxygenators. The liquid must be chemically and biologically inert and have a high carrying capacity for oxygen and CO_2. Silicone and fluorocarbon fluids have been used in liquid-liquid oxygenators. Testing shows they are efficient and quite competitive in performance with other types.

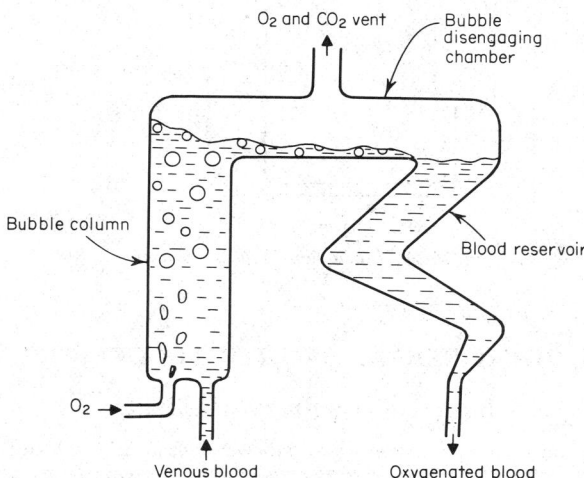

Fig. 26-26.   Typical bubble-type oxygenator.

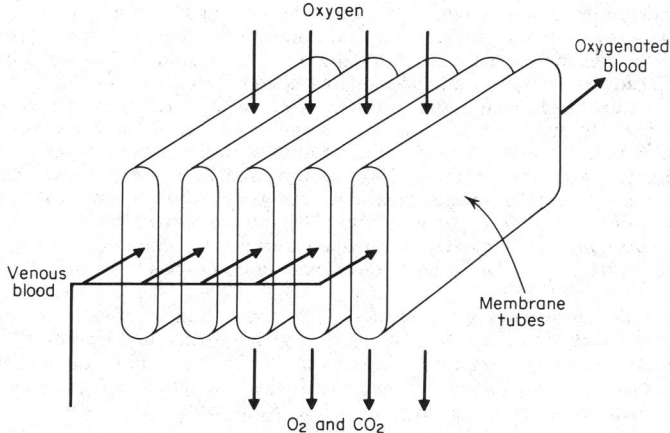

Fig. 26-27.   Membrane-type oxygenator.

26-29

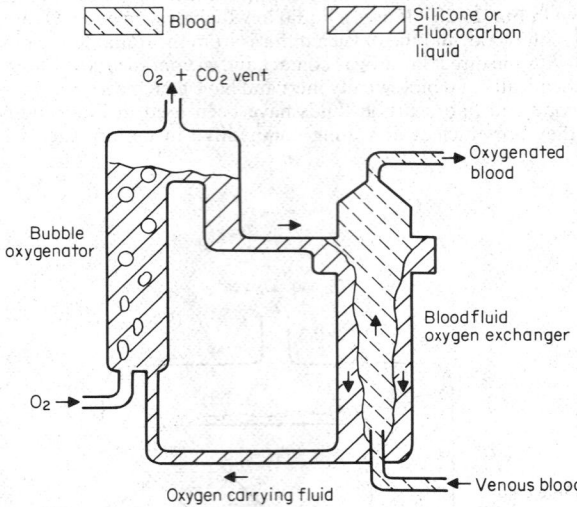

Fig. 26-28. Liquid-liquid-type oxygenator.

PROSTHETIC AND ORTHOTIC DEVICES

By J. Thomas Mortimer, Ph.D.

34. Orthotic and prosthetic devices are used to overcome functional losses resulting from limb paralysis or amputation. Prosthetic systems refer to assistive devices replacing absent body parts. Originally, an orthotic device[57]* was used to straighten a distorted part of the body; more recently, it has taken on the additional meaning of bracing or splinting for functional purposes. In both cases, an increase in function is the important objective; the ideal device is both cosmetically pleasing and functional. In fitting a person with an orthosis or prosthesis, the objectives of the device should be clearly established.

Upper-extremity orthotics stabilize or guide arm and hand segments into functional positions and can be roughly grouped into body-powered and externally powered devices.

Lower-extremity orthotics stabilize and guide the leg and foot segments to provide the subject with a more functional gait. Virtually all lower-extremity bracing utilizes body power which can be voluntarily controlled or controlled by electrical stimulation. A clinical example of a stimulated brace is the *functional electronic peroneal brace* first described by Liberson et al.[58] In this system, an electrical stimulus applied to the nerve innervates the muscles which raise the foot. Applying the stimulus at the right time during the gait cycle eliminates toe drag when the foot is being brought forward. More complicated systems under study involve stimulation of several muscle groups at appropriate times to effect a more functional gait in patients with extensive loss of leg muscle control.[59]

Spinal orthotics are used for stabilization and deformity correction of the head, neck, and trunk regions of the body. The application of external forces has been virtually limited to springs.

The two basic categories of prosthetic devices are upper- and lower-extremity prostheses. There are also prosthetic devices for sensory aids, such as vision and hearing.[60]

For general information on limb prosthesis, refer to Wilson.[61] For more information on orthotic and prosthetic systems, address inquiries to the Committee on Prosthetics Research and Development, National Academy of Sciences, Washington, D.C. 20418.

---

*Superior numbers correspond to numbered references, Par. 26-66.

UPPER-EXTREMITY ORTHOTIC AND PROSTHETIC DEVICES

35. The Hand. The primary task of the human upper extremity is to place the hand where it can perform its primary function of grasp. The function of the normal hand is very complex. Theoretically, each muscle controlling hand movements could be replaced by gears or cables and a motor; however, more than twenty motors would be required.

Restoration of hand function is providing hand prehension (grasp) patterns to the patient. The basic patterns are (1) Palmar prehension: *(a)* opposition of the thumb with the second and third digits, *(b)* opposition between thumb and digits 2 and 3; (2) tip prehension: opposition between tips of thumb and second or third digits; (3) lateral prehension: ball of thumbs opposes lateral surface of the second digit; (4) other grips used for holding larger objects.

There is no single grasp pattern suitable for all tasks; however, Palmar prehension has been the one most widely adapted. Figure 26-29 shows a powered hand splint capable of producing finger prehension and flexion-extension of the wrist.

The method of operating the prehensile device varies; the splint may be powered in both opening and closing or may utilize power in one direction and rely on a spring for return. When electric power is employed in an orthotic hand device, the motors must be located away from the splint because of their size and weight. This requires cables. For powered prehension in both directions, two flexible cables may be employed to pull in either direction (see Fig. 26-29), or a single stiff cable which both pushes and pulls is required. Examples of electric-powered hand splints can be found in Karchak et al.[62] and Grahn.[63] Pneumatic power with the McKibben muscle has been used.[64]

36. The Wrist. Basic motions of the wrist are flexion-extension and adduction-abduction. Although not strictly correct, forearm supination-pronation is often lumped in with wrist movements and termed wrist rotation. Wrist movements are generally considered secondary to prehension. Consequently, orthotic and prosthetic devices with wrist movement generally have been research devices. A hand splint with powered wrist flexion and extension is shown in Fig. 26-30. This device was used in conjunction with a powered-arm assist device having a total of seven degrees of freedom.[65,66] The prosthetic device with wrist movement is close to clinical application. The motor and mechanical apparatus may be easily concealed within the forearm housing. An example of a powered wrist rotation system is that developed by Grahn.[67]

37. The Elbow and the Arm. There have been relatively few attempts at powered elbow orthotic devices,[65,66] because of the complexity of such a system compared with mechanically coupled devices, such as the *balanced forearm orthosis* (BFO). Under controlled conditions (as in a hospital) patients fitted with BFOs are able to use them for a variety of tasks, including self-feeding.

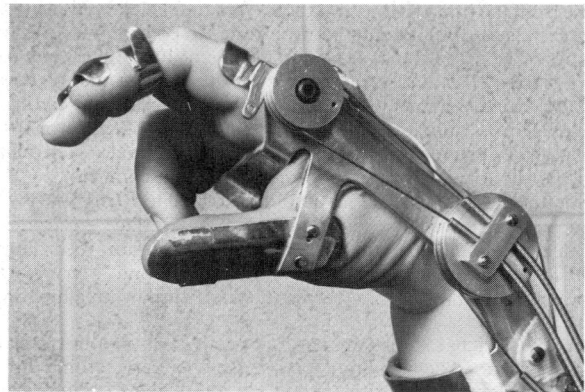

Fig. 26-29. Powered hand splint capable of producing finger prehension and flexion-extension of the wrist.

The basic movements performed by the arm are difficult to enumerate. There are over fifty muscles involved in arm movements; duplication of such a system is virtually impossible. There have been two attempts at an orthotic arm; both provided seven degrees of freedom. One was developed at Case Western Reserve University, Cleveland, Ohio; the other at Rancho Los Amigos Hospital, Downey, California. The Case arm was pneumatically powered, while the Rancho arm was electrically powered. Electric- or pneumatic-powered[71,72] prosthetic arms have met with success.

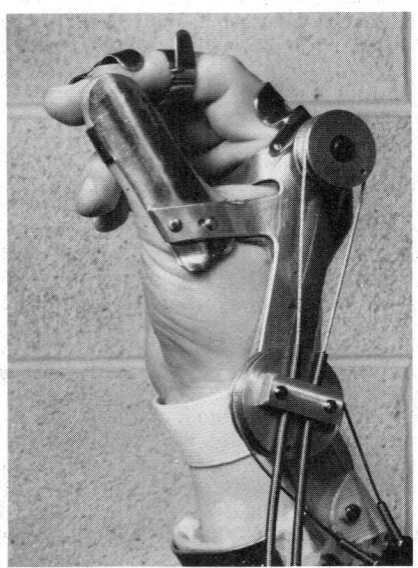

Fig. 26-30. Hand splint with power wrist flexion and extension.

38. Control Techniques. Control techniques include the communication channel the individual uses to transmit his desires to the device, as well as hardware for processing the signal for control. The communication channels used are either the bioelectric signal or the physical displacement of the body. Common bioelectric signals used for assistive-device control are those from the brain (EEG) and those from the skeletal muscle (EMG). The skin impedance charge was also reported to be promising for control.[73]

Characteristics of the EEG signal are *(a)* amplitude of 10 to 300 μV; *(b)* frequency: alpha wave is 8 to 13 Hz normal awake, resting, and mentally inactive, and beta wave is 14 to 100 Hz alert, mentally active. Surface electrodes on the subject's scalp can be used to record EEG. The relatively few attempts on EEG control are restricted to on-off systems, using switching from alpha to beta or beta to alpha as the control mode. The main disadvantage is that the device control and concentration may come into conflict.

Characteristics of EMG signals are *(a)* amplitude of 100 to 10 mV and *(b)* frequency of 20 Hz to 1 kHz, spectrum peak between 50 and 100 Hz. Surface electrodes and needle or implant wire electrodes may be used for recording. Surface recordings have smaller amplitudes and lose some energy in the high-frequency band. Dry electrodes made of stainless steel or gold-plated metal have been successfully used with a high-input-impedance amplifier.

The recorded EMG signal is the summation of activity from a number of muscle fibers firing nonsynchronously. (The firing rate for a single fiber varies from a low steady state of 2-Hz frequency to a high of approximately 50 Hz, the highest occurring during maximal contraction.) Duration of the action potential is approximately 5 ms.

At low contractions one may record from such a small number of active fibers that the recording appears as a series of regularly spaced pulses. As the contractile force is increased, the firing rate increases until other fibers are recruited. Eventually, the individual action potentials are lost, and the signal appears as random pulses.

To a fair approximation, the rms value is linearly related to the muscle force. Therefore two possible control modes exist: first, the low contraction range where single-fiber activity can be identified and coded by frequency; second, the somewhat higher contraction mode where rms value is used as the control signal.

In the low contraction range, the mode of control is called *single motor unit* control. The name implies that the recording is made from a single group of fibers innervated by a common motor neuron. The usable frequency range lies between 1 and 15 Hz. Frequencies below 1 Hz are difficult to maintain. Craig[74] reports that continuous control over the discharge rate may not be possible.

The higher contraction mode, referred to as *EMG control,* appears more promising. This has been utilized in a number of clinical devices. The control signal is the rms value of electrical activity; signal processing consists of rectification and filtering. The lag introduced by the filter must be within a reasonable range, and the signal must be quantized into the maximum number of independent levels. For an entire system, 0.5 s is generally considered the maximum tolerable delay. Approximately half this time may be for the decision and generation of command signal; therefore a time delay of 0.25 s may be considered a reasonable limit for the filters. The minimum number of independent signal levels (quanta) needed depends on the system. For an on-off system, only one active level is required. If proportional control is desired, three independent levels, excluding the inactive state, would be the minimum.

39. Control of Orthotic Devices. In the case of orthotic applications, there are no available nonfunctioning limb muscles under voluntary control. Thus EMG signals must be derived from other functional muscles. Interference with normal function when providing a control signal generally will not be tolerated by the patient. Additionally, location of the pickup electrodes over muscle to provide reliable signals is difficult and requires assistance from other individuals. The most common usage of EMG control to date has been for on-off control of devices with a single degree of freedom. These devices require two signals to indicate the desired direction of motion (forward or backward). Two muscles have been used to generate the signals, one muscle signal for each direction. One-muscle systems have also been adapted to control the on-off systems by using signal levels to indicate control direction; this is called *three-state control.* The off state corresponds to the inactive state of the muscle. A low-level contraction signifies motion in one direction, and a high-level contraction signifies motion in the other direction. A small deadband is required between the two states.

There are also ways to code or combine electrical signals from several sources to perform many functions. To gain additional control, one muscle generally sacrifices rapid decision time. A unique version of a two-muscle system controlling two degrees of freedom was reported by Childress.[75]

Development of pattern-recognition schemes to control multiaxis systems has been shown to be a potentially useful scheme.[76] This scheme avoids the problems of isolated muscle contraction by utilizing patterns from several muscles. At present, pattern-recognition schemes employ a computer for movement identification. However, if the present rate of success continues, miniaturization of the necessary electronics could make such a scheme clinically applicable.

40. Physical Movement as the Control Signal. Past experience has shown that systems employing body movement as a control signal are more easily adopted by the patient than systems employing EMG control. Furthermore, these systems generally exhibit finer control than do the EMG systems. Major disadvantages are the difficulty in transduction and unstable reference levels. Transducers tend to be bulky and difficult to mount in consistent positions. Reference levels are usually unstable because the subject's posture changes. In patients with high-level spinal cord injury, this is a major problem because these individuals retain very little trunk control, causing them to slide or slouch.

One of the simplest methods of transducing body movement is to use a potentiometer mounted in a stable position with a lever attached to the body where the movement is to be transduced. However, potentiometers are subject to wear, resulting in signal noise due to intermittent contact. Seamone, Hoshall, and Schmeisser[77] utilized a movable magnet and a stationary semiconductor element which responds to changes in magnetic field strength to detect the relative movement of a surgical scar. Semiconductor strain gages have been used to detect muscle bulge. Kadefors[73] suggested using changes in limb impedance during muscle contraction for control.

One of the most precise position-control sites is the shoulder, where two possible axes of motion are available: vertical and horizontal. Attempts were made to utilize these two axes for control. The first system[72] transduces shoulder motion relative to the center of the chest through measurement of the flexion of a fine rod. One end of the rod is mounted on the shoulder, and the other at the center of the chest. Two semiconductor strain elements are mounted on the rod to measure vertical and horizontal flexions.

41. Sensory Feedback. In a normal human being, information about the physical state of the limb and its immediate environment is continually transmitted to the central nervous system (CNS), which regulates movement based on desires and sensory feedback. Sensory feedback takes many forms, such as limb position through joints and skin sensors, tactile information through skin and hair receptors, temperature through thermal receptors, and pain through receptors which respond to noxious stimuli. When a portion of the limb is amputated or no longer operational, sensory feedback is no longer available to the CNS. Importance of the loss varies with the type of patient. For example, pain and temperature information from the artificial extremity may not be as important to the prosthesis user as to the user of an orthosis. Without this information, a person may incur severe cuts and burns which do not heal readily and are subject to infection.

PATIENT MONITORING

By Michael R. Neuman, Ph.D.

42. Patient monitoring is concerned with the continuous observation of the seriously ill patient, including appearance, physical examination, records of physiologic parameters, and intervention and administration of therapy when necessary. Electronic devices can monitor, display, record, and make elementary decisions concerning a patient's hospital course. A generalized scheme for an electronic monitoring system is shown in Fig. 26-31. The transducers convert physiologic parameters into electrical signals manipulated by the signal processor to allow qualitative or quantitative display and recording. Computing algorithms can be applied to data so that abnormalities are recognized and the clinical staff is alerted.

43. Coronary Care Monitoring. Most deaths occur in coronary disease as a result of cardiac arhythmias. For example, a myocardial infarct resulting from blockage of blood flow to a portion of the heart muscle can cause heart muscle fibers to contract randomly. This is known as *fibrillation,* and if not rapidly detected and corrected, it results in the patient's death. Electronic monitoring devices can be applied to detect fibrillation and some of its precursors. A typical cardiac monitoring system is shown in Fig. 26-32. The electrocardiogram (ECG), as sensed by electrodes on the anterior chest wall, is the main parameter monitored (refer to Pars. **26-8** to **26-14**).

The signal is amplified through a high-input impedance amplifier (5 to 10 mΩ). Frequency response of the amplifier should range from 0.1 to 100 Hz within 3 dB for an undistorted signal; however, poorer-fidelity amplifiers often are used to reduce artifact. The amplified signal is displayed on a cardioscope with a long-persistence phosphor or with a continuously updated memory. A recording oscillograph may also be used to provide a permanent record. The amplifier output is also fed to a short-term memory loop which records the ECG of the most recent 15 s. If the operator or the alarm circuitry detects an arhythmia, the chart recorder can be switched to the memory loop to record the ECG for 15 s, leading up to the observed arhythmia. The memory loop can be a magnetic tape loop or a shift-register type of solid-state memory.

Heart rate is calculated by cardiotachometer circuits, either of the averaging or the beat-to-beat type. The former determines average heart rate over a period of several seconds, while the latter determines heart rate as the reciprocal of the interval between individual beats. Output from the cardiotachometer is fed to analog or digital display devices, and a heart-rate alarm circuit sets off an alarm when the heart rate exceeds a maximum or falls below a minimum as determined by the operator. In many coronary care monitors, the alarm also connects the recording oscillograph to the memory loop to automatically record the event which set off the alarm.

Most coronary care units have displays at the patient's bedside as well as at a central monitoring station. Many monitoring units also can continuously monitor other physiologic parameters, such as blood pressure, peripheral pulse pressure, and respiration.

Radio telemetry and computers are used in coronary monitoring. A radio transmitter, worn by the patient and having a range of several hundred feet, can broadcast ECG and other parameters to a receiver at the central station, allowing greater patient mobility. In this system, the patients are completely isolated from the monitoring apparatus connected to the power mains, thereby eliminating the shock hazard from the monitor and reducing the power mains' artifact.

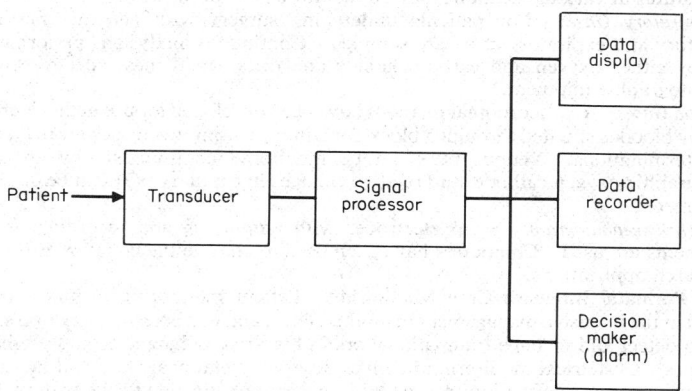

Fig. 26-31. Generalized electronic monitoring system.

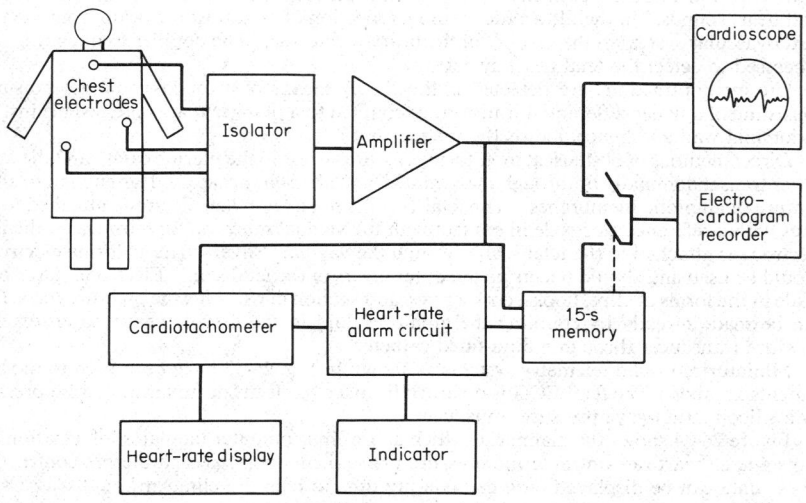

Fig. 26-32. Typical cardiac monitoring system.

44. Intensive Care, Operating Room, Recovery Room Monitoring. Besides the coronary care unit, patients with other acute illnesses, in shock, or suffering from severe burns, as well as those undergoing and recovering from surgery, require intensive monitoring. In these patients several vital signs must be monitored continuously.

Heart Rate and ECG.   See Coronary Care Monitoring, Par. 26-43.

Arterial and Venous Blood Pressure.   The direct method involves the cannulation of an artery or vein with either a miniature pressure transducer or a fluid-filled catheter for hydrostatically coupling the vessel to an external pressure transducer. The indirect method uses either an automatic sphygmomanometer cuff which determines the systolic and diastolic pressures in the same way as is done manually, or uses the cuff to obtain sphygmograms by inflating the cuff to a pressure intermediate between systolic and diastolic levels and continuously recording the pressure within the cuff.

Peripheral Pulse Pressure.   Changes in peripheral pulse pressure can be determined by plethysmographic methods. A finger plethysmograph consisting of a small cuff fits over the finger and determines changes in finger volume due to the blood pressure pulse.

Peripheral and Core Temperatures.   Temperatures in the rectum, the esophagus, and skin temperatures at various locations may be monitored by thermistors.

Respiratory Gases.   For patients undergoing surgery with general anesthesia, both inspiratory and expiratory gases are sampled. Continuous analysis is performed on these gases by either oxygen and carbon dioxide analyzers, small mass spectrometers, or gas chromatograph equipment.

Blood Gases.   Extracorporeal methods have been developed for use in the operating room whereby blood is shunted through a block containing sensing electrodes for pO_2, pCO_2, and pH determinations. Venous pO_2 can be monitored continuously by an intravenous polarographic oxygen cathode, and relative changes in capillary pO_2 can be detected by the ear oximeter.

Electroencephalogram.   Scalp electrodes with amplifying and recording or displaying instruments are used. Computers have been used to analyze the complex EEG waveforms in research applications.

45. Perinatal Intensive Care Monitoring. Patient monitoring techniques have been applied in the intensive management of mother, fetus, and newborn in the period surrounding birth to detect and manage traumatic labors in high-risk pregnancies and premature or ill newborns. Obstetrical monitoring involves observing fetal heart rate and uterine contractions during labor. In the indirect method, sensors are attached to the maternal abdomen. Fetal heart activity can be detected either through phonocardiographic or ultrasonic methods. In the former, a sensitive microphone is placed on the maternal abdomen to detect fetal heart sounds. In the ultrasonic method, reflection of an ultrasonic beam from the fetal cardiovascular system in the vicinity of the heart is detected. The doppler frequency shift is processed to detect the fetal heart cycle.

Uterine contractions are detected indirectly by means of a tokodynamometer usually consisting of a linear differential transformer coupled to a plunger that is pressed against the abdominal wall and hence, indirectly, the uterus.

Direct methods of obstetrical monitoring require access to the uterine cavity and the fetus, either transabdominally or through the vagina when labor has progressed far enough to allow rupture of amniotic membranes. The fetal ECG is obtained by an electrode attached to the fetus with a reference electrode in contact with the vaginal mucosa. In most cases, the fetal electrode is attached to the fetal scalp through the vagina. Silver–silver chloride electrodes should be used and should penetrate the outer layers of the fetal skin. Electrodes have been made in the forms of clips, hooks, corkscrews, and suction cups. Uterine pressure recordings can be made directly by coupling the amniotic fluid in the uterine cavity to an external pressure transducer through a fluid-filled catheter.

Miniaturized radio telemetry systems as shown in Fig. 26-33 have been used to monitor patients in labor. The fetal ECG is obtained from a clip electrode and intrauterine pressure by a silicon-strain-gage pressure transducer.

Figure 26-34 shows the manner in which an on-line computer tabulates information and plots a fetal heart rate–intrauterine pressure phase plane loop for each uterine contraction. These data can be displayed on a cathode-ray tube to help the clinician rapidly assess the patient's condition.

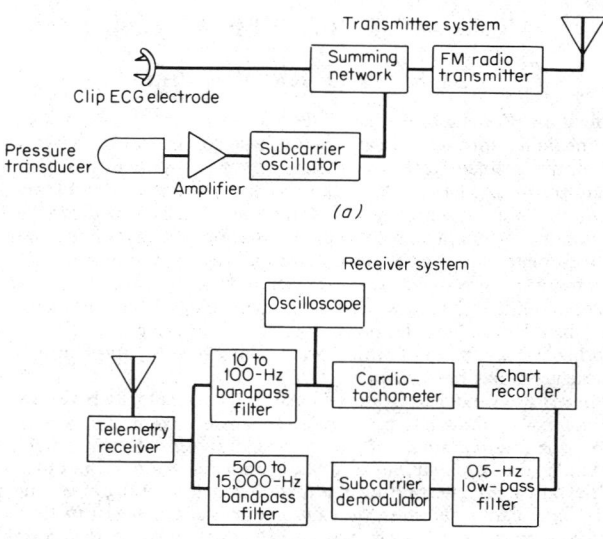

Fig. 26-33. Block diagram of a radio telemetry system. (*a*) Transmitter system; (*b*) receiver system.

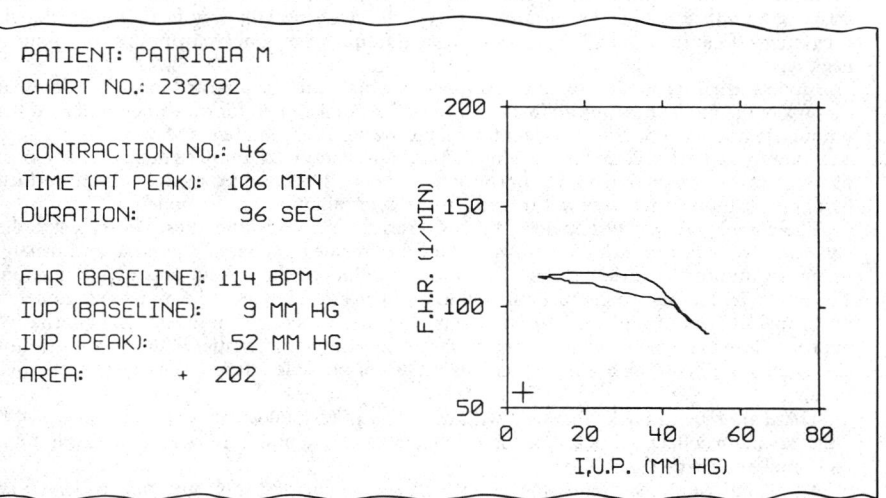

Fig. 26-34. Example of computer display of monitored obstetrical data with a phase plane loop of fetal heart rate vs. uterine pressure for a single contraction.

COMPUTER APPLICATIONS

BY PETER G. KATONA, PH.D.

46. Classification of Computer Uses. Computer technology was introduced into life sciences and medicine through research. Analysis of neural activity and modeling of physiological control systems were typical early computer applications. Instead of bringing about a revolution as predicted years ago, computers have brought about a slow and sometimes painful evolution which has only gradually affected medicine.[81,82,83*]

First successes have been achieved in areas where problems are well defined and easily quantitated. Automated analysis of electrocardiograms, computerized clinical laboratories, and automated monitoring systems are commercially available and have received acceptance in the medical community. Diagnosis, medical record keeping, hospital information systems, and community health-care facilities pose problems of pattern recognition, complex systems, economics, and sociology, none of which are easily solved by automation. Consequently, progress in these areas has been slower.

47. Modeling of Physiological Systems. Although a similarity between engineering and physiological control systems has been long recognized, computers provided the greatest impetus to the quantitative study of physiological systems by modeling (or simulation) techniques. The goal of such studies is a better understanding of the physiological system. It may be achieved by *(a)* interpreting the model structure and/or parameters in terms of physiologically significant quantities; *(b)* using the model to suggest further experiments on the physiological system; or *(c)* constructing the model to serve as a component within a quantitative description of a more encompassing physiological system.

Modeling starts with the assumption of a structure for the model, including inputs and outputs. If possible, this structure, usually consisting of an interconnection of compartments or network elements, should correspond to known or hypothesized physiological mechanisms operating within the system. Linear components are characterized by transfer functions (or equivalently, linear differential equations or impulse responses). The most common nonlinearities are multiplicative interaction of variables and input/output relationships showing both threshold and saturation. Thresholds are especially significant in physiological systems because chemical concentrations and frequencies of nerve firings cannot assume negative values.

Models usually contain unknown parameters which must be adjusted in such a way that outputs of the model best approximate outputs of the real system for the same inputs. If the outputs do not match within a certain tolerance, the model structure may have to be changed and new parameters computed. This iteration is continued until structure and model parameters are obtained that give appropriate outputs. The obtained model must be verified by applying inputs that were not utilized in the determination of the model.

The complexity of physiological systems poses a variety of difficulties. Foremost is the variability of data caused by inability of the experimenter to control the large and usually unknown number of variables that continuously influence the system under investigation. The model builder may construct a model to fit the averaged data, or he may use a model in which parameters are allowed to vary from one experiment to the next. An alternative approach is to use a model which accounts for only gross characteristics of the response. For steady-state analysis, such a model assigns ranges of possible outputs to ranges of possible inputs.

There are several books that deal with modeling of physiological systems in general.[85,86] Others contain collections of articles describing models of neural, cardiovascular, respiratory, and renal-endocrine systems.[84,87]

48. Signal Analysis. Neurophysiology. Development of LINC, the first widely used laboratory computer, was largely motivated by the need for on-line computation in the neurophysiology laboratory. Since then, the small computer has become a common tool of the neurophysiologist.

As an illustration of how computers can be utilized to analyze neurophysiological data, consider the spontaneous activity of two single neurons, schematically shown in Fig. 26-35*a* and *b*. Interval histograms *c* and *d* and joint interval histograms *e* and *f*, generated and

---

* Superior numbers correspond to numbered references, Par. **26-66.**

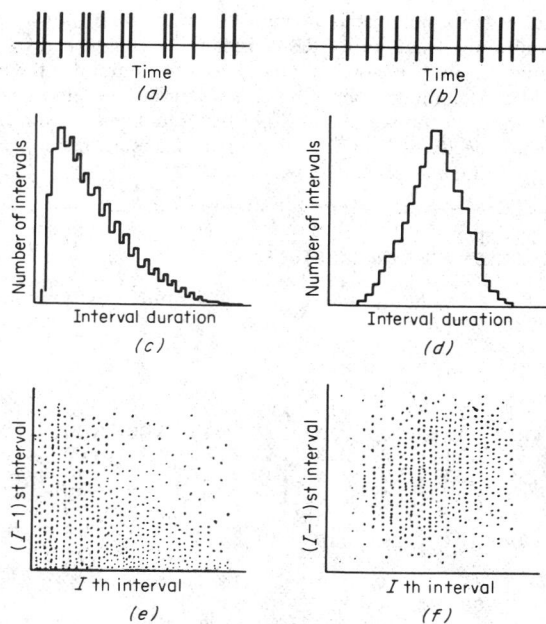

Fig. 26-35. Interval- and joint-interval histograms for two neural impulse trains. *(Based on Figs. 3, 4, and 7 of Ref. 92, by permission.)*

displayed by the computer, are shown schematically for both impulse trains. The interval histogram is equivalent to the probability density function of the time interval between neighboring nerve impulses and is obtained by plotting, for a long stretch of data, the number of intervals (as the vertical axis) having the time duration shown on the horizontal axis. The joint interval histogram provides information about the dependence of one interval on the previous one. It is obtained by plotting, for each pair of consecutive interspike time intervals, a point whose horizontal coordinate is the duration of the first interval and whose vertical coordinate is the duration of the second interval in the pair.

49. Patient Monitoring. In addition to automating routine monitoring functions (see Pars. 26-42 to 26-45), computers also provide a display of trends in selected variables (see Fig. 26-36). A desirable feature for such a display is an indication of the reliability of collected data as a function of time. Variability in readings and the percentage of lost data are such possible indicators. Computers can also derive clinically significant information by considering the relationship between recorded variables.

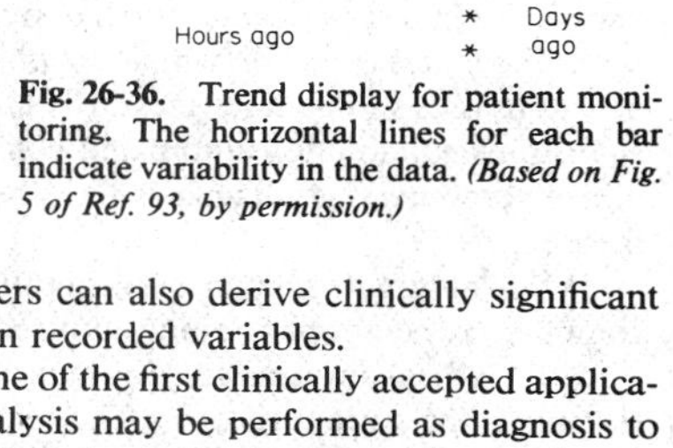

Bed 4
Cardiac output
Maximum: 52
Minimum: 40

1/4 1/2 1   3   6   12 * 1   2   4
Hours ago   * Days
                             * ago

Fig. 26-36. Trend display for patient monitoring. The horizontal lines for each bar indicate variability in the data. *(Based on Fig. 5 of Ref. 93, by permission.)*

50. ECG Analysis. ECG analysis[88] provided one of the first clinically accepted applications of computerized signal processing. ECG analysis may be performed as diagnosis to determine if any cardiac abnormalities are present in a short stretch of recording, or it may be performed as part of continuous monitoring of a patient requiring intensive care. Both types of analysis require sampling of the ECG signal recorded from the patient as one, three, six, or twelve waveforms (see Par. 26-11). Since standard electrocardiographic recorders have a bandwidth of 100 Hz, a sampling rate of 200 to 500 is recommended.

Diagnosis requires the simultaneous examination of the shape of waveforms on several leads. It starts with identification of QRS complexes, accomplished by detection of the largest rates of change in the waveform, and followed by a search for the location of the P, Q, R, S, and T waves. Magnitudes, slopes, and time intervals are determined to characterize morphologic and timing features of the ECG. After obtaining required parameters, a sequence of branching decisions is used to arrive at a suggested diagnosis. An example of computer-generated ECG analysis is shown in Fig. 26-37.

```
TELEMED CORPORATION
COMPUTER PROCESSED ELECTROCARDIOGRAM
MSDL APPROVED VERSION D 41-42-25-11

PATIENT 007986205
DATE    111770
TIME    0905
CODE    1216
        I    II   III  AVR  AVL  AVF  V1   V2   V3   V4   V5   V6
PR     .29  .25  .13  .00  .00  .14  .13  .00  .00  .00  .00  .00
QRS    .11  .10  .11  .08  .12  .11  .08  .09  .10  .10  .14  .10
QT     .33  .32  .35  .27  .34  .40  .47  .39  .38  .44  .48  .33
RATE    66   66   65  102  102  104  81   77   81   64   64   64

CODE   4A   4A   4A   2A   4A   2A   2    3    2    2    2    2

AXIS IN    P   QRS   T    Q    R    S   STO                 ST-T QRS-T
DEGREES       -66  222       17  -63  243                    21   72

2631 QRS VARIABLE, P ABSENT      . ATRIAL FIBRILLATION
8311 QRS AXIS RANGE -30 TO -90   . LEFT AXIS DEVIATION
6413 ST DEPRESSION, -.10MV OR    . EXCLUDE DIGITALIS EFFECT
     MORE NEGATIVE               .
6212 NEGATIVE T WAVES            . CONSISTENT WITH ISCHEMIA
                                 . ABNORMAL ECG

                                 . ------------ M.D.
```

Fig. 26-37. Computer-generated analysis of a 12-lead electrocardiogram. *(TELEMED Corporation.)*

While diagnostic ECG analysis may be performed off-line, analysis of ECG in monitoring applications must be performed in real time. In coronary care units, for example, the ECG is continuously scanned to detect any occurrence of premature ventricular contractions.

51. Medical Diagnosis. The essence of the diagnostic process is to sift a large amount of data (symptoms) presented by the patient and to decide, based on these data and accumulated knowledge and experience, which of many possible diseases (if any) the patient has. In most cases, the logical steps by which the physician arrives at a diagnosis are obscure. Computer programs thus should not necessarily attempt to mimic human thought processes but use new algorithms suitable for automated computation. With a few exceptions, most attempts to utilize computers for diagnosis have been based on statistical approaches involving Bayes' theorem.[89]

While automated diagnosis generally has not received wide acceptance, ECG analysis by computer is being increasingly utilized in clinical applications (see ECG analysis, above).

52. Medical Information Systems. Medical Record. The core of medical information systems is the patient record. The record may be that of an essentially healthy person or it may belong to the acutely ill patient. Traditionally, it has been a chronologic series of hard-to-read notes, impressions, and test results without logical connection. Yet patient care often depends on physicians who exchange information primarily through such a record. The major problem in developing computer-based medical records is that while computers are effective in dealing with well-structured information, the medical record in its most general form does not have a well-defined structure.

Diseases, symptoms, test results, and treatments constantly change and do not provide a solid basis for organization. It has been suggested that all entries in the record should be linked to the chief complaint of the patient, but implementation in general cases is difficult. Further problems arise from narrative information that is often used. Storing and retrieving such information in the proper context are a formidable task. To obviate this problem, information must be captured in a structured format at the time of entry. This is facilitated by a question-and-answer dialogue between user and computer, utilizing interactive terminals.

Interactive terminals contain a display tube and some means of accepting information from the user. Information is entered by responding to questions displayed by the computer. Response is made by pushing a function button, typing an answer on a keyboard, touching a portion of the screen, or pointing to a spot on the display by a light pen (see Fig. 26-38). The computer acknowledges the answer and provides immediate feedback if the entered information is conflicting or is in the wrong format.

53. Multiphasic Screening. The automated multiphasic screening facility[90] screens individuals by subjecting them to a variety of tests and examinations. Most of the data, including medical history, blood tests, urinalysis, and physical measurements, are collected by computer. Nonautomated measurements are manually entered. The system analyzes the data and produces a report. If abnormalities are detected, additional tests are recommended or the patient is sent to a physician.

54. Clinical Laboratory. The work load of clinical laboratories has been rapidly increasing, not only because of the increased number of patients, but also because of new test procedures. Commercial, high-volume clinical laboratories not affiliated with hospitals routinely rely on computer processing of determinations. Automation and electronic

Fig. 26-38. Use of an interactive display terminal. *(Courtesy Dr. G. Octo Barnett.)*

data processing have received acceptance in hospital laboratories as well. Some hospitals use punch cards in conjunction with their business computer; others prefer small, dedicated computer systems. These systems are usually equipped with analog/digital converters, enabling them to automatically collect data from analytic instruments. The most promising concept involves utilizing a central computer facility for storing patient information, including test results, and using a small peripheral computer for collecting readings from the instruments.

55. Hospital Information Systems. It has been estimated that more than 25% of a hospital budget is spent on information processing, indicating a large potential for automation. Initially, computers were used off-line for accounting purposes only; more recently, emphasis has shifted toward on-line systems with terminals located throughout the hospital. In these newer systems the computer acts not only as an information processor, but also as the center of a communications network.

Assume, for example, that the physician orders blood tests to be performed on a particular patient. A clerk enters this information into the computer at the nursing station. When the sample is drawn, the computer generates a machine-readable label that is attached to the test tube before it is sent to the clinical laboratory where tests are performed by automated analyzers. Results are stored in the patient's file. If the label indicates an emergency, results are immediately reported to the nursing station; otherwise they are reported at the end of the day as part of a summary. Simultaneously, proper charges for the tests are entered into the patient's account.

Progress in developing such systems has been slower than expected. Three interface problems have been identified as sources of difficulties.[91]

The first interface problem stems from a lack of understanding of the complexities involved in developing a viable system serving the entire hospital. The hospital system is so complex, and there is so much inertia to change, that practicality often limits development of an all-inclusive information system in existing hospitals. In new institutions, automated information processing can be designed into the physical plant and administrative organization. The most promising solution for existing hospitals is a stepwise, modular development of computer systems.

The second interface problem arises from the necessity of capturing information in a reliable yet convenient manner. Interactive display terminals are providing at least a partial solution. Consider, for example, a case when the computer requires the user to specify the dosage for a particular drug. Instead of expecting a typed answer, it is desirable to display the most commonly used dosages for that drug and allow the user to choose by pointing at or touching the desired portion of the display.

The third interface problem is caused by the often conflicting expectations placed upon the programmer. He must develop a feeling and competence for medical problems, yet he usually spends most of his time writing code and debugging programs. An interdisciplinary team of physicians, nurses, systems analysts, and programmers is indispensable for developing a viable hospital information system.

IMPLANT INSTRUMENTATION

By Wen H. Ko, Ph.D.E.E.

56. Applications of Implant Instrumentation. With the development of microelectronics and solid-state electronic devices and technology, electronic instruments can be made small and light enough to be implanted in a living subject to perform measurements, stimulation, and control functions.[94-101*] The implant instrument is a valuable experimental technique for animal study and medical research and is becoming an acceptable device for diagnostic and therapeutic purposes, e.g., the implant heart pacemaker. Advantages of implant instruments include minimum restriction or interference with the subject and long-term operation without cross-skin connection.

Animal Study for Physiology, Pharmacology, and Disease. Implant instruments can provide nearly normal environment while the subject assumes regular activity. Continuous telemetering of body information can be obtained over a period of weeks, months, or even years. Desired body response can be controlled by electrical stimulation over a long period at a remote location.

Clinical Monitoring. After surgery, physiological parameters deep in the body of many patients need to be monitored over a long period. Implant instruments, when properly developed, can fulfill the need. For example, an implant device will eliminate the need for periodic skull surgery now practiced for a periodic check on intracranial fluid of hydrocephalic children with an implanted shunt.

Implant stimulators may be used as a clinical device to regulate body function. The implant heart pacemaker to regulate heart rate and a blood pressure regulator device to suppress high blood pressure are now commercially available. Many clinically oriented stimulators are being developed, including stimulators for bladder, brain, muscle, implantable visual aid for the blind, electroanesthesia machines, etc.

An implant closed-loop control system has many applications for biological investigation, such as the study of body regulations or substitution of a damaged control loop in the body to provide simple body functions or control of paralyzed limbs.[101]

Implant manipulators can be used to release drugs or provide mechanical or electrical stimuli to perform therapeutic functions. For a certain type of cancer, a nearly continuous

---

* Superior numbers correspond to numbered references, Par. 26-66.

release of a small amount of drugs into the bloodstream over a period of months is desirable; the implant instrument will be a preferred method.

57. Implant Telemetry. Implant telemetry is used to obtain information from within the body of an unrestrained living organism at a remote location through electronic transmission linkage. The block diagram of a typical telemetry system is shown in Fig. 26-39. Location of the implant transmitter may be intracavitary, such as within the intestines, mouth, bladder, etc., or may be inside the body between the internal as well as external surfaces, such as subdermally or deep in the tissues. Surely, the same transmitter can also be mounted on the surface of the body when desired. The radio receiver and signal processor are remotely located to receive, record, and display the information.

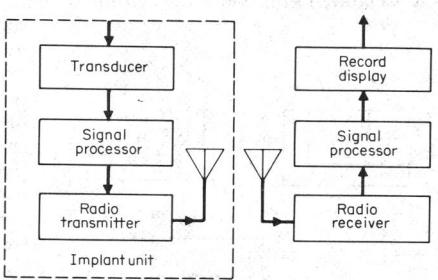

Fig. 26-39. Block diagram of a telemetry system.

Major considerations in selecting a telemetry system are (1) characteristic of signal and performance required; (2) size, weight, and package as determined by the required life span of the implant device; (3) implant location and techniques; (4) environmental interference and system requirements; (5) reliability and cost.

Surveys and bibliographies on biotelemetry are also available.[95-97] Journals on biomedical engineering, medical instrumentation, and telemetry usually carry recent developments in biotelemetry. Many physiological signals have been telemetered. The characteristics of these signals are summarized in Table 26-10.

Active power supply may be chemical primary, secondary, or nuclear batteries, or energy converters to convert body energy into electrical energy. The passive implant telemeter may contain a resonant circuit in which the resonant frequency, made to vary with body signal,

Table 26-10. Signal Characteristics of Telemetered Parameters

Physiological parameters	Transducer	Amplitude range	Frequency range
Electrocardiogram (EKG)	Electrodes	0.05–4 mV pulse	0.1–100 Hz
Electroencephalogram (EEG)	Electrodes	10–75 μV	0.5–200 Hz
Electrogastrograph	Surface Electrodes	10–350 μV	0.05–0.2 Hz
Electromyogram (EMG)	Electrodes	0.1–4 mV pulse	2–10⁵ Hz (10–500 Hz clinical)
Eye potential, EOG or ERG	Electrodes	500 μV	0–250 Hz
Nerve potentials	Electrodes	3 mV peak	Up to 1,000 pulses/s rise time 0.3 s
Bladder pressure	Strain gages	0–100 cm H_2O	0–10 Hz
Blood flow	Electromagnetic flowmeter; ultrasonic flowmeter	1–300 cm/s	1–20 Hz
Blood pressure	Strain gage on artery; hydraulic coupling to transducer; cuff gages	0–400 mm Hg	0.5–100 Hz
Gastrointestinal pressure	Variable inductance	20–100 cm H_2O	0–10 Hz
Intestinal forces	Strain gages	1–40 g	0–1 Hz
Respiration rate	Electrode impedance; piezoelectric devices; pneumograph	...	0.15–6 Hz
Stomach pH	Glass electrode; antimony electrode	3–13	0–0.1 Hz
Temperature	Thermistor; thermal expansion	90°–100°F	0–0.1 Hz
Tidal volume	Impedance; pneumograph	50–100 ml/breath	0.15–6 Hz

can be detected by a grid dip meter [103] or may contain transmitters powered by pulses of radio energy stored on a capacitor, or the transmitter may be powered by continuous radio energy supplied at other frequencies. [104]

Many single-channel telemetry units have been used. Compelled by limitations on size and weight, most present transmitting units use one or two transistors to obtain transducer, conditioner, and transmitter functions. With new micropower techniques these limitations may be removed. Figure 26-40 shows five simple transmitter circuits from the literature. [95,98] The circuit in *(a)* is crystal-controlled and is used for animal tracking and narrow-band signal transmission. Both cw and pulsed operation can be obtained by varying R_1.

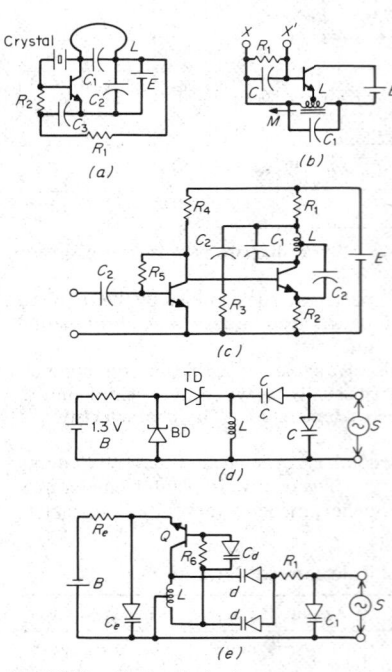

The circuit in Fig. 26-40*b* is used as a pulsed oscillator; the rate of rf pulses is modulated by signals. For temperature measurements R is a thermistor; for pressure, the core M is moved by pressure change; for pH, electrodes may be connected across input X. Circuit *c* uses the voltage-sensitive emitter-to-base capacitance of the transistor to modulate the frequency of the carrier. Circuits *d* and *e* use tunnel diodes or varactors and transistors for continuous-wave FM transmitters. Table 26-11 lists the performance of transmitters *d* and *e*. Transmitter *e* has been integrated into a single silicon chip.

For long-term monitoring, to improve the base-line stability, a subcarrier is needed even for single-channel transmission. Both AM/FM and FM/FM multiplex systems with four to eight channels [106] have been reported.

In the time-division multiplex system, signal channels are sampled and transmitted at different time slots. PAM/FM and PDM/PAM systems were designed, [107] and the modular blocks were integrated into chips or flat packs. Depending on the signal frequency band, these systems will cover 2 to 30 channels.

Fig. 26-40. Five typical telemetry transmitter circuits.

58. Receiving Systems. Due to restrictions on size and weight, implant transmitters are designed with minimum components and operate at low power level. The transmitting frequency is generally not stabilized and drifts with time and motion of the subject due to change of loading. Therefore noise and interference become serious problems at the receiving end. To maintain system reliability and performance, the receiver and demodulator should incorporate various noise-discriminating schemes, such as *(a)* automatic gain

Table 26-11. Performance of Transmitters (Fig. 26-40 *d* and *e*)

Item	Figure 26-40*d*	Figure 26-40*e*
Size, less battery	0.8 cm diam. × 0.2 cm	0.9 × 0.6 × 0.2 cm
Weight, less battery	0.44 g	0.6 g
Power consumption	0.2 V, 1.2 mA	1.3 V, 0.5 mA
Rf frequency	100–250 MHz	100–250 MHz
System noise (10 kΩ, 1 kHz)	3 μV	3 ~ 5 μV
Dynamic range (limited by receiver)	10^3	10^3
Frequency response	0.1–20 kHz	0.1–20 kHz
Input impedance	300 kΩ to several MΩ	300 kΩ to several MΩ
Transmission range	5 μV at 1 to 2 m	5 μV at 1 to 2 m
Carrier frequency temperature stability	Better than 0.05%°C	Better than 0.05%°C

control (AGC) for fading; (b) automatic frequency control (AFC) and phase-locked loop for frequency drift; (c) noise limiter or amplitude discrimination for noise and interference; and (d) pulse amplitude discrimination, pulse width discrimination, and pulse frequency discrimination for pulsed carrier systems.

59. Implant Stimulation. Electrical stimulation of the nervous system for remote control of animals was tried in 1934, but its clinical application is relatively recent. The heart pacemaker is the best-known clinical device in use. Additionally, stimulation has been applied to the bladder for voiding; the phrenic nerve for diaphragmatic contraction; baroreceptors for reduction of blood pressure; the spinal cord for pain suppression; and paralyzed muscle for regaining motor functions. With recent developments in brain stimulation[99,100] it is possible that it can be used to bypass sensory organs, as well as to temporarily alter personality or induce sleep.

The stimulating circuit is either a simple radio-powered passive resonant circuit with rectified output or an active pulse generator powered by batteries, rechargeable batteries, radio power, and energy converters. The stimulating parameters for heart, bladder, nerves, and muscles are summarized in Table 26-12.

Table 26-12.   Stimulating Parameters

	Heart	Bladder	Nerve	Indirect muscle (via nerve)	
				Intramuscular	Surface
Pulse width, ms	0.1-10	0.1-10	0.05-1	0.1-10	0.1-10
Pulse rate*	30-200 ppm	15-35 pps	5-100 pps	1-50 pps	1-50 pps
Pulse voltage, V	0-9	0-10	0-10	0-10	10-60
Pulse current, mA	4-45	50	0.1-10	0.5-10	5-50

pps = pulses per second; ppm = pulses per minute.

Major problems in stimulators are power supply, packaging, electrodes, and tissue damage. The first two problems will be discussed later. Requirements for electrodes include (a) minimal corrosion in body, (b) good fatigue resistance, and (c) minimum body reaction to limit the increase of threshold.

Single- or multistrand wires of gold, platinum, stainless steel, and Elgiloy are used. Coils or helical-wound multistrand wires provide better flexibility and less fatigue. Elgiloy is reported to have the best fatigue resistance but corrodes when used as an anode. Platinum electrodes can be successfully used either as an anode or a cathode.

Tissue damage from electrical stimulation using "nontoxic electrodes" can be caused by (a) gas generation due to large current density; (b) heat generated at the electrode site; (c) toxic material generated by electrode-chemical reaction at the electrode site; (d) mechanical stress caused by the electrode assembly.

60. Heart Pacemaker. The heart pacemaker is the best-known implant stimulator.[108] Referring to Par. 26-8, the heart rate is normally generated at the S-A node. In a diseased state the S-A node is nonfunctional or the conduction passage is blocked. The heart either will not contract or will be controlled by A-V node at a much lower rate; an artificial pacemaker then is needed. Implant pacemakers can be categorized as follows:

A synchronous pacemaker, where the rate, set by electronic circuits, is either fixed or adjustable by some electronic means and is not synchronized with the natural-heart pacemaker.

Demand pacemaker, which stimulates with a preset interval after the preceding spontaneous or pacemaker-induced ventricular depolarization. When the R-R interval is shorter than the escape interval of the artificial pacemaker, the artificial pacemaker remains dormant. When the interval is longer than this interval, the artificial pacemaker will be activated, with resulting stimulation. This stimulation continues with the same escape interval until a spontaneous beat occurs.

Atrial synchronous pacemaker senses the signal from the S-A node to deliver a synchronous pacing pulse to the ventricle. This type of pacemaker is shown schematically in Fig. 26-41.

The electronic pacemaker can be attached endocardially or myocardially. It can be mounted on the surface with a catheter across the body boundary or can be totally implanted as shown in Fig. 26-42. Much research on improved pacemakers is being carried out. The main effort is in the power supply, where different types of primary batteries, such as the solid

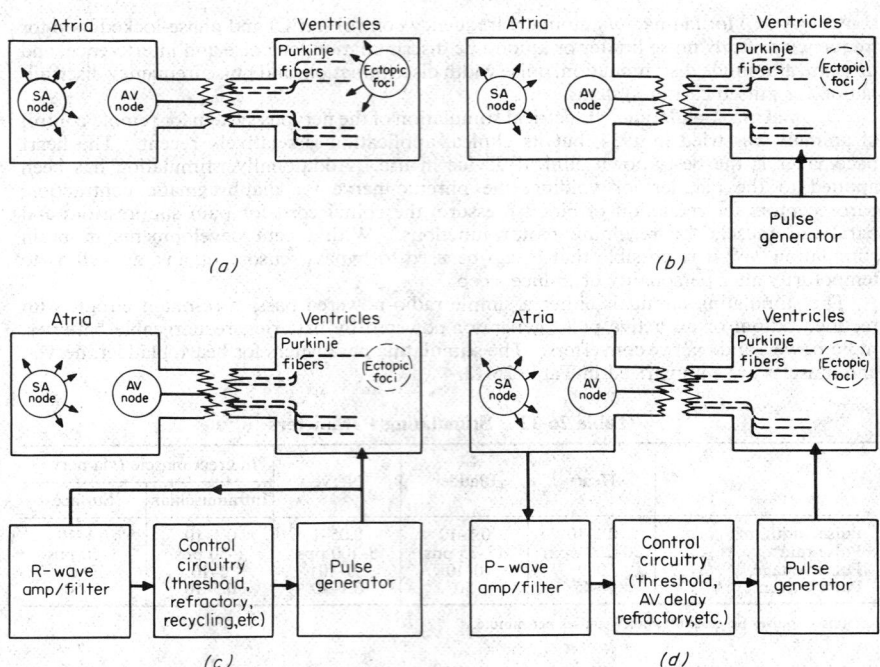

Fig. 26-41. Block diagrams of four types of heart pacemakers.

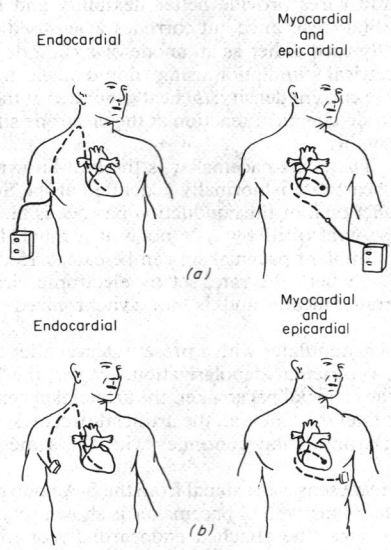

Fig. 26-42. Placement of heart pacemaker. (*a*) External asynchronous pacer connected to either endocardial or myocardial electrodes; (*b*) internal asynchronous pacer connected to either endocardial or myocardial electrodes in a unipolar configuration.

electrolyte battery, nuclear battery, and fast-recharging battery, have been investigated. Radio-frequency induction-powered pacemaker has been developed and is available commercially. The conventional nonsynchronous pacemaker is basically a free-running multivibrator with an output stage for constant-current stimulation. Demand pacemakers will have a more complex circuitry. The parameters are given in Table 26-12.

61. Closed-Loop Control. Although implant closed-loop control systems were proposed many years ago, no clinical systems presently exist. Research projects have demonstrated the use of implant telemetry to transmit shoulder EMG to control arm or hand operation.[101] Problems such as internal feedback through body conduction, regenerative feedback through the nervous system, and the matching of external circuitry to characteristics of the signal source need to be studied. It is believed that implantable control systems will become clinically useful in the near future.

ELECTRONIC EQUIPMENT USED IN DENTISTRY

By Charles H. Gibbs, Ph.D.

62. Electronic Applications in Dentistry. Dental equipment utilizing electric or electronic devices is listed in Table 26-13. The electrosurgical unit, tooth-vitality tester, and ultrasonic prophylaxis unit are selected as examples of electronic instruments in dentistry. The literature [109]* is to be consulted for a review of developments in dental x-ray equipment, motorized dental chairs, saliva-ejector units, and rotary hand pieces. Fiber optic lighting systems are being used in dentistry for oral examination, diagnosis, and intraoral illumination during operating procedures.

Table 26-13. Dental Equipment Utilizing Electric or Electronic Devices

Electric welder	Fiber optic lighting systems
Hydrobath	Tooth-vitality tester
Laboratory engine	Electrosurgical instrument
Dental lathe (grinder)	Ultrasonic prophylaxis unit
Model trimmer	Ultrasonic cleaner
Vacuum investing machine	Electronic muscle stimulator
Vibrator	Sterilizer, dry heat
Electric ovens	Adjustable chair
Autoclave	X-ray view box
Temperature-regulated water bath	X-ray machine (single head)
Electric amalgamator	X-ray machine (panoramic head)
Incubator	X-ray film processors
Operating light	Dental unit with air-driven handpieces
Pneumatic condenser	and saliva evacuator

Electrosurgical Unit. The basic diagram of electrosurgical equipment is shown in Fig. 26-43. The output of an rf oscillator is applied between a thin-wire electrode and a ground plate. Frequencies of 2 to 5 MHz have been found to be optimum for surgical work. Depending on the shape of the handpiece electrode, the unit may be used for surgery (cutting), coagulation (desiccation), or fulguration.

In surgery, the electrode consists of a straight wire or wire loop; wire diameter varies from 7 to 20 mils and is usually made of tungsten. Magnitude of the rf voltage generally is in the range of 250 to 350 V ac rms. Tissue impedance during surgery [110] varies from 50 to 200 Ω.

Electrosurgery in dentistry was made practical when Oringer [110] found that electrosurgery on the gums depended on two considerations: (1) a fine wire must be used and moved rapidly through the tissue as it is parted, and (2) rf amplitude should be kept as high as possible without causing sparking. Shielding should be employed to minimize capacitive coupling to the handpiece.

In *coagulation,* a small ball-tip electrode is used. When the ball tip is momentarily touched to the tissue with rf voltages similar to those of surgery, the tissue turns white. This is used to stop surface bleeding from small capillaries.

Fulguration is sparking to the tissue using a spikelike electrode. High temperature of the arc plays a major role in this mode; rf voltage is approximately twice the value used for the

---

* Superior numbers correspond to numbered references, Par. **26-66.**

other two modes. Fulguration is generally used to stop extensive bleeding when larger capillaries or small veins have been cut.

63. Tooth-Vitality Tester.   To determine tooth vitality, voltage pulses are applied to the tooth. If the tooth is vital, the patient will feel a shocking sensation, but none is felt otherwise. Pulse amplitude required to elicit sensation varies considerably for different teeth. The severity of the shock felt by the patient depends upon amplitude and duration of the pulse.

One commercially available tooth-vitality tester produces 400 pulses/s of approximately 70 μs duration. This device consists of a battery-powered oscillator, amplifier, and transformer. Maximum amplitude of the pulses is + 500 V at no load. The positive electrode is placed in contact with the tooth being tested. Current flows from tooth to gums, to the dentist's hand touching the patient, through the dentist to the dentist's other hand, which is in contact with the case of the tester, where the circuit is completed.

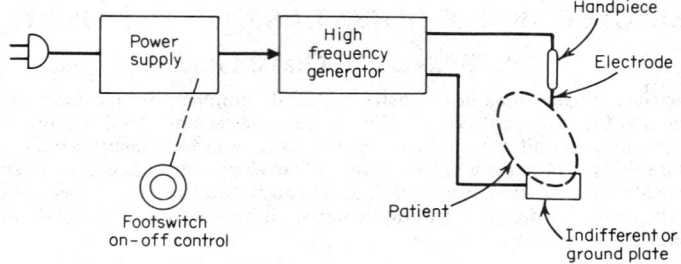

Fig. 26-43.   General block diagram of electrosurgical equipment.

64. Ultrasonic Prophylaxis Unit.   One of the major problems faced by dentistry in prophylaxis and periodontal treatment is the removal of plaque from teeth. Plaque accumulation facilitates growth of bacteria, which leads to caries, infection of gums, and ultimately contributes to tooth loss. Ultrasonic prophylaxis units are used in dental practice. They require little force, are faster and easier to use than hand scalers, and produce less discomfort to the patient.

Frequencies above the audible range are preferable; the higher the frequency, the shorter the stroke for the same total energy. At 25,000 Hz, the 0.025-mm stroke produces an equivalent scraping rate of 1,250 mm/s. The heat from friction at this scraping rate can be considerable, and water cooling of the working surface is employed. The Cavitron, an ultrasonic prophylaxis unit, operates at 25 kHz, powered by a 110 V ac line. The ultrasonic current is converted, via a handpiece with insert, into 25-kHz mechanical strokes/s. These strokes are transmitted to a scaling tip which has a lateral movement of 0.025 mm. The tip is supplied with a water jet for cooling and the removal of debris.

A magnetostrictive stack is used to convert the 25-kHz current into mechanical strokes. The 25-kHz current supplied to the coil and the stack, housed in the handpiece, constricts the stack. This movement is transmitted to the prophylaxis tip.

65. Research Equipment.   Exciting and interesting changes are taking place in the field of dental equipment. Electronic instrumentation gives insight into needs for future dental equipment. There is a particular research interest in evaluating factors relating to loss of teeth from periodontal disease. Forces of tooth contact generated during jaw movements are believed to be a contributing factor to periodontal disease and joint dysfunction. Forces of tooth contact result in looseness of teeth; therefore the possibilities of an electronic tooth-mobility gage for diagnosing trauma to teeth and periodontal involvement are entertained.

Instrumentation to Measure Jaw Movements.   Primary considerations in designing instrumentation to measure jaw movement are the physiological restrictions of the jaw. Vertical jaw movements are usually limited to 5 cm, and a maximum jaw velocity rarely exceeds 25 cm/s. Velocity obtained during normal chewing is usually less than 8 cm/s. A sampling of instruments being used in research to measure jaw motion includes digital encode system [111] (see Fig. 26-44); analog potentiometer method [112]; resolver method [113]; capacitor method [114]; photographic method [115-117]; and telemetry method. [118]

Measurement of Tooth Mobility. Tooth mobility has attracted the interest of clinical and scientific investigators. For clinicians, mobility has become an important diagnostic parameter of the function state, integrity, and disease of the periodontium. Also, the mechanism by which masticatory forces are transmitted to the supporting structure offers an attractive scientific challenge. Mobility is related to the hydrodynamic and biophysical properties of a complex connective tissue system, in part unmineralized, in part mineralized.

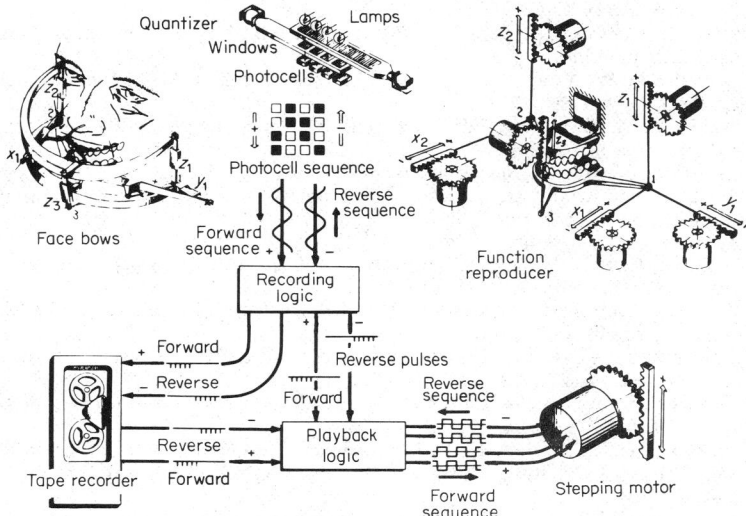

Fig. 26-44. Case gnathic replicator.

Mechanical systems such as Muhlemann's microperiodontometer and macroperiodontometer[119] and an improved instrument by O'Leary and Rudd,[120] consisting of a rod connected to a mechanical gage for applying force and a dial indicator attached to an impression tray for measuring movement, are now in use. Electrical static-type systems such as Parfitt's[121] differ only by substituting for the dial indicator an electrical transducer such as a differential transformer or strain gage.

The drawback of static systems is the rigid attachment to teeth required in order to establish a static reference from which tooth movement can be measured. The cementation of a clutch to the teeth is not practical when reference teeth are very mobile or missing. To overcome this drawback, dynamic vibratory-type systems appear promising, though at present they are not developed beyond the experimental stage.

The need for expanded dental research is emphasized by the National Institute of Dental Research[122] in five fields: pain control, dental caries (tooth decay), periodontal (gum) disease, herpetic virus infections, and oral facial development.

66. Bibliography

General references

1. SCHWAN, H. P. (ed.) "Biological Engineering," McGraw-Hill Book Company, New York, 1969.

2. CLYNES, M., and MILSUM, J. (eds.) "Biomedical Engineering," McGraw-Hill Book Company, New York, 1970.

3. BROWN, J., JACOB, J., and STARK, L. "Biomedical Engineering," F. A. Davis, Philadelphia, Pa., 1971.

4. BUGLIARELLO, G. (ed.) "Bioengineering: An Engineer's View," San Francisco Press, San Francisco, Calif., 1968.

5. (a) *IEEE Transactions on Biomedical Engineering.*

    (b) *Biomedical Engineering Journal,* Culver City, Calif.

    (c) *Journal of American Association of Medical Instrumentation.*

(d) *BioMedical Engineering,* United Trade Press, London.

(e) *Biomedical Engineering (Transactions of Meditsinskaya Tekhnike).*

(f) *Proceedings of Annual Conference of Engineering in Medicine and Biology* (since 1959).

6. (a) Biophysics, Bioengineering and Medical Instrumentation, International Medical Abstracting Service, sect. 27, *Excerpta Medica,* Amsterdam.

(b) Medical Electronics and Communications Abstract, MultiScience Publishing Company, Essex, England.

(c) *Index Medicus,* U.S. Dept. of H.E.W., Public Health Service, National Library of Medicine (monthly).

7. DUMMER, G. W. A., and ROBERTSON, J. M. (eds.)   "Medical Electronic Equipment, Pergamon Press, New York, 1970.

8. WEYER, E. M. (ed.)   Materials in Biomedical Engineering, *Ann. N.Y. Acad. Sci.,* Vol. 146, Part 1, 1968.

9. LEVINE, S. N. (ed.)   *J. Biomed. Material Res.,* March 1967.

10. STARK, L., and AGARWAL, G. (eds.)   "Biomaterials," Plenum Press, Plenum Publishing Corporation, New York, 1969.

Safety

11. WHALEN, R., et al.   Electrical Hazards Associated with Cardiac Pacemaking, *Ann. N.Y. Acad. Sci.,* Vol. 111, pp. 922-931, 1964.

12. Electrical Safety Test Procedures for Hospitals, Instrutek, Inc., Publ. TP-871, Lincoln Park Center, Annapolis, Md., 1972.

13. Standard for the Use of Inhalation Anesthetics, National Fire Protection Association, NFPA 56A, Boston, Mass., 1970.

14. Electrical Hazards in Hospitals, National Academy of Sciences, Card No. 74-606017, Washington, D.C., 1971.

15. 1971 National Electrical Code, National Fire Protection Assn., Boston, Mass.

16. Safe Use of Electricity in Hospitals, 1971, NFPA 76BM National Fire Protection Assn., Boston, Mass.

17. Recommended AAMI Safety Standard for Electromedical Apparatus, Part 1, Safe Current Limits, Mar. 17, 1971, Association for Advancement of Medical Instrumentation.

18. Biomedical Monitoring Systems, X-1414 (with amendment) Veteran's Administration, Sept. 1, 1970.

19. Test Instrument, Medical Electrical Safety X1432, Veteran's Administration, Oct. 15, 1970.

20. 4th Draft of Standard for Medical and Dental Equipment Intended for Use in Non-hazardous Areas, Underwriters' Laboratories, Inc., May 26, 1971.

21. Manual for Safe Use of Electrical Equipment in Hospitals, California Hospital Association, May 6, 1971.

Electrocardiography, etc.

22. HOFFMAN, I. (ed.)   "Vectorcardiography, 2," North-Holland Publishing Company, Amsterdam, 1971.

23. POZZI, L.   "Basic Principles in Vector Electrocardiography," Charles C Thomas, Springfield, Ill., 1961.

24. PLONSEY, R.   "Bioelectric Phenomena," McGraw-Hill Book Company, New York, 1969.

25. GESELOWITZ, D. B., and SCHMITT, O. H.   Electrocardiography, in H. P. Schwan (ed.), "Biological Engineering," McGraw-Hill Book Company, New York, 1969.

26. McFEE, R., and BAULE, G. M.   Electrocardiography and Magnetocardiography, *IEEE Trans. BME,* 1972.

27. KOSSMANN, C. E. (chairman)   Recommendations for Standardization of Leads and of Specifications for Instruments in Electrocardiography and Vectorcardiography, *Circulation,* Vol. 35, pp. 583-602, 1967.

28. HURST, J. W., and LOGUE, R. B. (ed.)   "The Heart," McGraw-Hill Book Company, New York, 1970.

29. WALTER, C. W. (ed.)   "Electric Hazards in Hospitals," National Academy of Sciences, Washington, 1970.

30. GEDDES, L. A., and BAKER, L. E.   "Principles of Applied Biomedical Instrumentation," John Wiley & Sons, Inc., New York, 1968.

31. BRAZIER, M. "The Electrical Activity of the Nervous System," The Macmillan Company, New York, 1958.

32. DONCHIN, E., and LINDSLEY, D. B. (ed.) "Average Evoked Potentials," NASA, Washington, 1969.

33. PINELLI, P. (ed.) "Progress in Electromyography," Elsevier Publishing Company, Amsterdam, 1962.

34. WULFSOHN, N. L., and SANCES, A., JR. "The Nervous System and Electric Currents," Plenum Press, New York, 1970.

Radiology

35. GLASSER, O., et al. "Physical Foundations of Radiology," 2d ed., Paul B. Hoeber, Inc., New York, 1952.

36. EVANS, R. D. "The Atomic Nucleus," McGraw-Hill Book Company, New York, 1956.

37. JOHNS, H. E., and LAUGHLIN, J. S. Interaction of Radiation with Matter, Chap. 2 in Hine and Brownell (eds.), "Radiation Dose Symmetry," Academic Press, Inc., New York, 1956.

38. HEITLER, W. "Quantum Theory of Radiation," 3d ed., Oxford University Press, Fair Lawn, N.J., 1954.

39. JOHNS, H. E. "The Physics of Radiology," 3d ed., Charles Pummis, Springfield, Ill., 1969.

40. GLASSER, O., QUIMBY, E. H., TAYLOR, L. S., and WEATHERWAX, J. L. "Physical Foundations of Radiology," Paul B. Hoeber, Inc., New York, 1954.

41. HINE, G., and BROWNELL, G. "Radiation Dosimetry," Academic Press, Inc., New York, 1956.

42. WAGNER, H. B. "Principles of Nuclear Medicine," W. B. Saunders Co., Philadelphia, Pa., 1968.

43. BLAHD, W. H. "Nuclear Medicine," 2d ed., McGraw-Hill Book Company, New York, 1971.

44. MAYNARD, D. C. (ed.) "Clinical Nuclear Medicine," Lea & Febiger, Philadelphia, Pa., 1969.

45. GROSS, W., FEITELBERG, S., and QUIMBY, E. H. "Radioactive Nuclides in Medicine and Biology: Basic Physics and Instrumentation," 3d ed., Lea & Febiger, Philadelphia, Pa., 1970.

46. EARLY, P. J., RAZZAK, M. A., and SODEE, D. B. "Textbook of Nuclear Medicine Technology," C. V. Mosby Company, St. Louis, Mo., 1969.

Artificial organs

47. *Transactions of the American Society of Artificial Internal Organs.*

48. NOSÉ, Y. "Manual on Artificial Organs," Vol. 1, "The Artificial Kidney," C. V. Mosby Company, St. Louis, Mo., 1969.

49. HAMPERS, C. L., and SCHUPAK, E. "Long Term Hemodialysis," Grune & Stratton, Inc., New York, 1967.

50. GUYTON, A. C. "Circulatory Physiology: Cardiac Output and Its Regulation," W. B. Saunders Company, Philadelphia, Pa., 1963.

51. BREST, A. N. "Heart Substitutes, Mechanical and Transplant," Charles C Thomas, Springfield, Ill., 1966.

52. LEE, H., and NEVILLE, K. "Handbook of Biomedical Plastics," Pasadena Technology Press, Pasadena, Calif., 1971.

53. NOSÉ, Y. "Cardiac Engineering," Interscience Publishers, New York, 1970.

54. GALLETTI, P. M., and BRECHER, G. A. "Heart-Lung Bypass," Grune & Stratton, Inc., New York, 1962.

55. BARTLETT, R. H., DRINKER, P. A., and GALLETTI, P. M. "Advances in Cardiology," Vol. 6, "Mechanical Devices for Cardiopulmonary Assistance," S. Karger, New York, 1971.

56. LEFAUX, R. "Practical Toxicology of Plastics," Chemical Rubber Press, Cleveland, Ohio, 1968.

Prosthetic devices

57. LICHT, S. "Orthotics Etcetera," Elizabeth Licht, New Haven, Conn., 1966.

58. LIBERSON, W. T., HOLMQUIST, H. J., SCOT, D., et al. Functional Electrotherapy: Stimulation of the Peroneal Nerve Synchronized with Swing Phase of the Gait of Hemiplegic Patients, *Arch. Phys. Med. Rehabil.,* Vol. 42, pp. 101–105, 1961.

59. VODOVNIK, L.   Functional Stimulation of Extremities, *Adv. Electron. Electron Phys.,* Vol. 30, pp. 282–297, 1971.

60. FRIEBERGER, H.   Research in Sensory Aids for the Blind—State of the Art, Reprinted from *Blindness,* 1966. *AAWL Annual.*
    BUSHOR, W. E.   Medical Electronics, Part IV, Prosthetics-Hearing Aids and Blind Guidance Devices, *Electron. Mag.,* June 23, 1961.

61. WILSON, A. B.   Limb Prosthetics—1970, *Artificial Limbs,* Vol. 14, pp. 1–52, Spring 1970.

62. KARCHAK, A., ALLEN, J. R., et al.   The Electric Hand Splint, *Orthop. Prosthet. Appliance J.,* June, 1965, pp. 135–136.

63. GRAHN, E. C.   A Power Unit for Functional Hand Splints, *Bull. Prosthet. Res.,* Spring 1970, pp. 52–56.

64. ENGEN, T. J., and OTTNAT, L. F.   Upper Extremity Orthotics: A Project Report, *Orthoped. Appliance J.,* June, 1967, pp. 112–127.

65. APPLE, H. P.   Engineering Design Studies of Cybernetic Orthotic/Prosthetic Systems, Engineering Design Center Rep. EDC 4-69-26, Case Western Reserve University, Cleveland, Ohio, 1969.

66. KARCHAK, A., ALLEN, J. R., et al.   Applications of External Power to Orthotic and Prosthetic Devices, *Arch. Phys. Med. Rehabil.,* Vol. 48, pp. 341–344, 1967.

67. GRAHN, E. C.   A Powered Wrist Rotator, *Bull. Prosthet. Res.,* Spring 1970, pp. 46–51.

68. LeBLANC, M. A.   Externally Powered Prosthetic Elbows, Committee on Prosthetics Research and Development Rep. E-4, National Academy of Sciences, 1970.

69. NICKEL, V. L., KARCHAK, A., and ALLEN, J. R.   Electrically Powered Orthotic Systems, *J. Bone Joint Surg.,* Vol. 51-A, pp. 343–351, 1969.

70. McLAURIN, C. A.   Prosthetic Research and Training Unit, *Inter-Clinic Inf. Bull.,* Vol. 6, pp. 13–28, November 1966.

71. MONGEAU, M.   Our Experience with the Thalidomide Children: An Interim Report, *Inter-Clinic Inf. Bull.,* Vol. 6, No. 4, pp. 3–12, January 1967.

72. SIMPSON, D. C.   The Choice of Control System for the Multimovement Prosthesis, International Symposium on Control of Upper Extremity Orthoses and Prostheses, Göteborg, Sweden, 1971.

73. KADEFORS, R.   Electrical Impedance: A New Control Signal in Orthotics and Prosthetics, International Symposium on Control of Upper Extremity Prostheses and Orthoses, Göteborg, Sweden, 1971.

74. CRAIG, P. M.   Isolation of Training of Motor Units, in L. Vodovnik (ed.), Some Topics on Myo-electric Control of Orthotic/Prosthetic Systems, EDC Rep. 4-67-17, Case Western Reserve University, Cleveland, Ohio, 1967.

75. CHILDRESS, D. S.   A Control Principle for Above-Elbow Artificial Limbs, *Abstr. 55th Ann. Meet. FASEB,* 709, Chicago, Ill., 1971.

76. WIRTA, R. W., and TAYLOR, D. R.   Multiple-Axis Myoelectrically Controlled Prosthetic Arm, Final Rep. SRS, Moss Rehabilitation Hospital, Philadelphia, Pa., September 1970.

77. SEAMONE, W., HOSHALL, C. H., and SCHMEISSER, G.   Modular Externally-powered System for Limb Prostheses, *APL Tech. Digest,* Vol. 10, pp. 14–23, 1971.

78. MORTIMER, J. T., BAYER, D. M., LORD, R. H., and SWANKER, J. W.   Shoulder Position Transduction for Proportional Two Axis Control of Orthotic/Prosthetic Systems, International Symposium on Control of Upper Extremity Orthoses and Prostheses, Göteborg, Sweden, 1971.

79. SALISBURG, L. L., and COLMAN, A. B.   A Mechanical Hand with Automatic Proportional Control of Prehension, U.S. AMBRL, Walter Reed Army Medical Center, Technical Rep. 6611, Washington, D.C.

80. MANN, R. W., and REIMERS, S. D.   Kinesthetic Sensing for the EMG Controlled "Boston Arm," *IEEE Trans. Man-Machine Systems,* Vol. MMS-11, pp. 110–115, 1970.

Computer applications

81. BARNETT, G. O.   Computers in Patient Care, *New England J. Med.,* Vol. 279, pp. 1321–1327, 1968.

82. Special Issue on Technology and Health Services, *Proc. IEEE,* No. 57, 1969.

83. STACY, R. W., and WAXMAN, B. D. (eds.)   "Computers in Biomedical Research," Vol. 3, Academic Press, Inc., New York, 1969.

84. BROWN, J. H. U., JACOBS, J. E., and STARK, L. (eds.) "Biomedical Engineering," F. A. Davis Co., Philadelphia, Pa., 1971.

85. GRODINS, F. S. "Control Theory and Biological Systems," Columbia University Press, New York, 1963.

86. MILSUM, J. H. "Biological Control Systems Analysis," McGraw-Hill Book Company, New York, 1966.

87. BROWN, J. H. U., and GANN, D. S. (eds.) "Applications of Engineering Principles in Physiology," Vol. II, Academic Press, Inc., New York, 1973.

88. CACARES, C. A., and DREIFUS, L. S. (eds.) "Clinical Electrocardiography and Computers," Academic Press, Inc., New York, 1970.

89. LEDLEY, R. S. Practical Problems in the Use of Computers in Medical Diagnosis, *Proc. IEEE*, Vol. 57, pp. 1900-1918, 1969.

90. COLLEN, M. F. Periodic Health Examinations Using an Automated Multitest Laboratory, *J. Am. Med. Assoc.*, Vol. 195, pp. 142-145, Mar. 10, 1966.

91. BARNETT, G. O., and GREENES, R. A. Interface Aspects of a Hospital Information System, *Ann. N.Y. Acad. Sci.*, Vol. 161, Art. 2, pp. 756-768, 1969.

92. RODIECK, R. W., KIANG, N.Y.-S, and GERSTEIN, G. L. Some Quantitative Methods for the Study of Spontaneous Activity of Single Neurons, *Biophys. J.*, Vol. 2, pp. 351-368, 1962.

93. WARNER, H. R., and MORGAN, J. D. High-Density Medical Data Management by Computer, *Computers Biomed. Res.*, Vol. 3, pp. 464-476, 1970.

Implants

94. References (1) to (4).

95. MACKAY, R. S. "Bio-medical Telemetry," John Wiley & Sons, Inc., New York, 1970. With comprehensive reference list.

96. CACERES, C. A. (ed.) "Biomedical Telemetry," Academic Press, Inc., New York, 1965.

97. FRYER, T. B. Implantable BioTelemetry Systems, NASA Rep. SP-5094, 1970.

98. KO, W. H., and NEUMAN, M. R. Implant Biotelemetry and Microelectronics, *Science*, Vol. 156, No. 3773, pp. 351-360, April 1967.

99. DELGADO, J. M. R. Aggressive Behavior Evoked by Radio-Stimulation in Monkey Colonies, *Am. Zool.*, Vol. 6, pp. 669-683, 1966.

100. BRINDLEY, G. S., and LEWIN, W. S. The Sensation Produced by Electrical Stimulation of the Visual Cortex, *J. Physiol. Cond.*, Vol. 196, pp. 479-493, 1968.

101. VODOVNIK, L., et al. Myo-electric Control of Paralyzed Muscles, *IEEE Trans. BME-12*, Nos. 3 and 4, pp. 169-172, 1965.

102. FULLER, J. L., and GORDON, T. M. The Radio-Inductograph: A Device for Recording Physiological Activity in Unrestrained Animals, *Sci.*, Vol. 108, p. 287, 1948.

103. COLLINS, C. C. Miniature Passive Pressure Transensor for Implanting in the Eye, *IEEE Trans. BME-14*, pp. 74-83, April 1967.

104. KO, W. H., and YON, E. T. RF Induction Power Supply for Implant Circuits, *Proc. 6th Int. Conf. Med. Electr. Biol. Eng.*, Tokyo, Japan, August 1965.

105. KO, W. H., and HYNECEK, J. Micropower Pulse Modulated Multiplex Telemetry Systems, *Proc. 9th Int. Conf. Med. Biol. Eng.*, Melbourne, Australia, August 1971.

106. ZWEIZIG, J. R., et al. Design and Use of an FM/AM Radio-Telemetry System for Multichannel Recording of Biological Data, *IEEE Trans. BME-14*, No. 4, pp. 230-238, 1967.

107. RAMSETH, D. J. An Optimal Monolithic Integrated Circuit Multiple-Channel Biomedical Telemetry System, Ph.D. thesis, Case Western Reserve University, Cleveland, Ohio, September 1970.

108. FURMAN, S. (ed.) Advances in Cardiac Pacemaker, *Ann. N.Y. Acad. Sci.*, Vol. 167, Art. 2, pp. 515-1075, October 1969.

Dentistry

109. PETTIT, G. G. Panoramic Radiography.

KILPATRICK, H. C. Auxiliary Equipment for Dental Practitioners.

WEINERT, A. M. An Evaluation of the Modern Dental Lounge Chair.

SNEDAKER, R. F. High Volume Evacuation.

SOCKUELL, C. L. Dental Handpieces and Rotary Cutting Instruments.

BOMBA, J. L. Fiber Optic-lighting Systems: Their Role in Dentistry.

(All from *Dental Clinics of North America*, Vol. 15, No. 1, January 1971.)

110. ORINGER, M. F. "Electrosurgery in Dentistry," W. B. Saunders Company, Philadelphia, Pa., 1968.

111. MESSERMAN, T., RESWICK, J. B., and GIBBS, C. H. Investigation of Functional Mandibular Movements, *Dent. Clin. North America,* Vol. 13, No. 3, pp. 629-642, July 1969.

112. KNAP, F. J., RICHARDSON, B. L., and BOGSTAD, J. Study of Mandibular Motion in 6 Degrees of Freedom, *J. Dent. Res.,* March-April, 1970.

113. BEWERSDORFF, H. J. Electrognathographic-Electronische dreidimensionale Messung und Registrierung von Kieferbewegungen, *Odontol. Tidskr.,* Vol. 77, 1969.

114. OISHI, S., ISHIWARA, T., and AI, M. A Study of the Mandibular Movement by Means of the Arm-Type Motion Analyzer and the Condenser Method, *Bull. Tokyo Med. Den. Univ.,* Vol. 15, No. 4, pp. 359-360, December 1968.

115. KOVUMAA, K. Cinefluorographic Analysis of the Masticatory Movements of the Mandible, *Suom. Hannaslaak. Toim.,* Vol. 57, No. 4, pp. 306-368, 1961.

116. AHLGREN, J. Mechanism of Mastication, *Odontol. Scand.,* Vol. 24, Suppl. 44, 1966.

117. GLICKMAN, I., PAMEIJER, J. H. N., and ROBER, F. Intraoral Occlusal Telemetry, Part I, A Multifrequency Transmitter for Registering Tooth Contacts in Occlusion, *J. Prosthet. Dent.,* Vol. 19, p. 60, January 1968.

118. MILLER, S. C. "Textbook of Periodontia," 3d ed., McGraw-Hill Book Company, New York, 1950.

119. MUHLEMANN, H. R. Periodontometry, a Method for Measuring Tooth Mobility, *Oral. Surg., Oral Med., Oral Pathol.,* Vol. 4, p. 1220, 1951.

120. O'LEARY, T. J., and RUDD, K. O. An Instrument for Measuring Horizontal Tooth Mobility, *Periodontics,* Vol. 1, p. 249, 1963.

121. PARFITT, G. J. Development of an Instrument to Measure Tooth Mobility, *J. Den. Res.,* Vol. 37, p. 64, 1958.

122. Oral Disease: Target for the 70's: Five Year Plan of the National Institute of Dental Research, National Institutes of Health, Washington, D.C., 1970.

SECTION 27

ELECTRONIC ENERGY CONVERSION METHODS

BY

P. D. DUNN Professor of Engineering Science, Department of Applied Physical Sciences, The University of Reading, England

CONTENTS

Numbers refer to paragraphs

SECTION 27

ELECTRONIC ENERGY CONVERSION METHODS

THERMOELECTRIC CONVERSION

1. Introduction. The phenomenon of thermoelectricity has been used for many years for the measurement of temperature by means of the thermocouple. More recently, advances in solid-state materials have resulted in greatly improved thermal efficiency, and thermoelectric heat-to-electricity conversion has become of interest. The reversed generator or refrigerator has also been developed, and such devices are now available commercially.

2. Seebeck Effect. A temperature gradient in an electrical conductor causes charge carriers to diffuse to the colder end. The electric field set up owing to these charge carriers tends to limit further diffusion, and an equilibrium point is reached at which a steady open-circuit voltage is produced. This is the Seebeck effect; and the magnitude of the voltage will depend on the material of the conductor and on the temperature difference.

The *Seebeck coefficient* α is defined as the open-circuit volts observed for unit temperature difference. α is sometimes referred to as *thermoelectric power*.

$$\alpha = \frac{dV}{dT} \tag{27-1}$$

Typical values for α range from 10×10^{-6} for metals, 100×10^{-6} V/K for highly doped semiconductors, to $1{,}000 \times 10^{-6}$ V/K for lightly doped semiconductors.

The sign convention adopted is such that the Seebeck coefficient is positive if the cold end is positive and negative if the cold end is negative. In doped semiconductors the difference in sign is related to the type of doping since, if the semiconductor is n-type, conduction is by free electrons and a negative charge accumulates at the cold end; in p-type material, conduction is by positively charged holes and there is an unneutralized positive charge at the cold end. Both electrons and holes contribute significantly in an intrinsic semiconductor, and the net effect depends on the mobility of each type of carrier.

Metals are generally thought of as electronic conductors, but in some cases a positive Seebeck coefficient is observed. This arises where conduction takes place in overlapping energy bands. In this way a hole contribution is possible and it can be sufficient to produce a net positive coefficient.

3. Peltier Effect. When electric current flows through a junction between two different conductors, heat is absorbed or released. The amount of heat Q_π evolved is found to be proportional to the current I and is given by

$$Q_\pi = \pi_{12} I \tag{27-2}$$

where π_{12} is called the *Peltier coefficient* of the couple. The current I is assumed to flow from material 1 to material 2.

The energy change arises from the difference in the heat-carrying capacity of the electrons in the two materials, i.e., a change in the electronic specific heat. Since electrons arrive and depart at the same rate, a net transport of heat to or from the junction occurs.

The Peltier coefficient for a particular junction depends on the junction temperature. Peltier coefficients, like Seebeck coefficients, vary widely and typical values at room temperature range from 3×10^{-3} V for metals, 3×10^{-2} V for highly doped semiconductors, to 3×10^{-1} V for lightly doped semiconductors. The sign of the Peltier coefficient for a particular substance is the same as for the Seebeck coefficient for the same substance.

4. Thompson Effect. It is observed that heat is released or absorbed along the length of a homogeneous conductor carrying a current if a temperature gradient exists along the conductor. This is the Thompson effect. It can be thought of as arising due to the specific heat of the current carriers as they move in the temperature gradient dT/dx.

The rate at which heat is absorbed or released is given by

$$Q_T = \tau I \frac{dT}{dx} \tag{27-3}$$

where I = electric current, dT/dx = temperature gradient along the conductor, and τ = Thompson coefficient. The sign convention is such that the Thompson coefficient is positive if heat is absorbed when conventional current flows from the cold to the hot region.

Thompson coefficients vary from 10×10^{-6} V/K in metals to 100×10^{-6} V/K in semiconductors. At large values of dT/dx the Thompson effect may be comparable with the Seebeck effect and must be taken into account in the design of generators and refrigerators.

5. Thermodynamic Relationships. Kelvin showed that the Seebeck, Peltier, and Thompson coefficients are related by the following expressions:

$$\pi_{12} = \alpha_{12} T \tag{27-4}$$

$$\frac{d\alpha_{12}}{dT} = \frac{\tau_1 - \tau_2}{T} \tag{27-5}$$

where T = absolute temperature. In his derivation Kelvin neglected irreversible losses due to Joule heating and heat conduction. It is possible to give a full treatment, including these irreversibilities, using the method of irreversible thermodynamics. This treatment leads to the same expressions.

These relationships lead to the concept of an absolute Seebeck coefficient for a material. We can define

$$\frac{d\alpha}{dT} = \frac{\tau}{T} \quad \text{or} \quad \alpha = \int_0^T \frac{\tau}{T} dT \tag{27-6}$$

Also, for material 1,

$$\frac{d\alpha_1}{dT} = \frac{\tau_1}{T} \tag{27-7}$$

and for material 2,

$$\frac{d\alpha_2}{dT} = \frac{\tau_2}{T} \tag{27-8}$$

Hence

$$\alpha_1 - \alpha_2 = \int_0^T \frac{\tau_1 - \tau_2}{T} dT = \alpha_{12} \tag{27-9}$$

Thus the Seebeck coefficient for the junction between two materials is equal to the difference between the absolute coefficients.

6. Effects of Transverse Magnetic Field. The thermoelectric effects discussed above form part of a class of phenomena known as *coupled diffusion effects.* These effects arise from the coupling between heat flow and electric current. The Lorentz, or $J \times B$, force on moving conductors due to a transverse magnetic field produces additional effects, notably those of Nernst and Ettinghausen.

7. Nernst-Ettinghausen Effect. It is observed that an electric field E_y exists in a direction mutally perpendicular to a longitudinal heat flow Q_x and a transverse magnetic field B_z. The sign is such that the field appears in the y direction when the heat flow is in the x direction and the magnetic field is in the z direction.

Hence

$$E_y = +A_{EN} B_z Q_{ix} \tag{27-10}$$

This equation is sometimes written in terms of temperature gradient.

$$E_y = -A'_{EN} B_z \frac{dT}{dx} \tag{27-11}$$

Clearly,

$$A_{EN} = -\frac{1}{Q_x}\frac{dT}{dx}A'_{EN} \qquad (27\text{-}12)$$

With the restraints that $J_y = J_x = dT/dy = 0$, these equations serve as definitions of the coefficients A_{EN}, A'_{EN}.

8. Ettinghausen Effect. This effect describes the appearance of a thermal gradient dT/dy mutually perpendicular to the direction of a longitudinal electric current J_x and a transverse magnetic field B_z.

The Ettinghausen coefficient A_E is defined

$$\frac{dT}{dy} = -A_E B_z J_x \qquad (27\text{-}13)$$

with the conditions

$$J_y = Q_y = \frac{dT}{dx} = 0$$

The Ettinghausen coefficient A_E is related to the Nernst-Ettinghausen coefficient A_{EN} by the relationship

$$A_E = TA_{EN} \qquad (27\text{-}14)$$

THERMOELECTRIC GENERATORS

9. General. The basic arrangement of a thermoelectric generator is shown in Fig. 27-1. The device is a heat engine operating on a Rankine cycle and employing an electron gas as the working fluid. The upper limit to conversion efficiency therefore is the Carnot efficiency η_c.

$$\eta_c = \frac{T_h - T_c}{T_h} = \frac{\Delta T}{T_h} \qquad (27\text{-}15)$$

where the hot junction is maintained at a temperature T_h, and the cold junction at a temperature of T_c.

10. Design of Thermoelectric Generators. In the derivation of design formulas we neglect any heat losses from the sides of the elements and thermal and electrical losses in the connecting link joining the two hot ends of the couple. In addition we assume that the properties of the thermoelectric materials (the Seebeck coefficient α_{12}, the electrical conductivity $\sigma = 1/\rho$, and the thermal conductivity k) are independent of temperature. From Eq. (27-5) we see that the assumption of a temperature-independent Seebeck coefficient implies a zero Thompson coefficient τ. L and A refer to the length and area of a limb, and subscripts 1 and 2 specify the limb.

This simple theory provides a useful basis for designing thermoelectric generators. It may be extended for use with materials having temperature variation by substituting averaged values (see Par. 27-11).

First consider the time rate of heat input at the hot junction Q.

$$Q = Q_\pi + Q_k \qquad (27\text{-}16)$$

$$Q_\pi = \text{Peltier heat component} = \pi_{12}I = \alpha_{12}T_hI \qquad (27\text{-}17)$$

$$Q_k = \text{thermal conduction component} = -K\frac{dT}{dx} + \frac{I^2R}{2} \qquad (27\text{-}18)$$

where

$$K = \frac{k_1 A_1}{L_1} + \frac{k_2 A_2}{L_2} \qquad (27\text{-}19)$$

$$R = \frac{\rho_1 L_1}{A_1} + \frac{\rho_2 L_2}{A_2} \qquad (27\text{-}20)$$

$$\text{Output voltage } V = \alpha_{12}\Delta T - IR \tag{27-21}$$

$$\text{Output power } P = VI = I\alpha_{12}\Delta T - I^2 R \tag{27-22}$$

$$\text{Efficiency } \eta = \frac{P}{Q} = \frac{\alpha_{12}\Delta T - IR}{\alpha_{12}T_h + K\Delta T/I - \frac{1}{2}IR}$$

$$= \eta_c \frac{m}{1 + m - \dfrac{\Delta T}{2T_m} + \dfrac{(1+m)^2 RK}{\alpha_{12}^2 T_h}} \tag{27-23}$$

where

$$m = \frac{R_L}{R} = \frac{\text{load resistance}}{\text{generator resistance}} \tag{27-24}$$

For maximum efficiency the limb dimensions should be chosen to satisfy the relation

$$\sqrt{\frac{k_1 \rho_1}{k_2 \rho_2}} = \frac{L_1/A_1}{L_2/A_2} \tag{27-25}$$

in which case

$$\eta = \eta_c \frac{m}{1 + m - \dfrac{\Delta T}{2T_h} + \dfrac{(1+m)^2}{ZT_m}} \tag{27-26}$$

where

$$Z = \frac{(\alpha_1 - \alpha_2)}{\left|(\rho_1 k_1)^{1/2} + (\rho_2 k_2)^{1/2}\right|^2} \tag{27-27}$$

Z, the figure of merit, depends only on the material properties and should be as high as possible. If $\alpha_1 = -\alpha_2 = \alpha$, $\rho_1 = \rho_2 = \rho$, and $k_1 = k_2 = k$, Z becomes $\alpha^2/\rho k$, or $\alpha^2/\sigma/k$. The electrical load may be chosen to maximize efficiency.

$$m_{\text{opt}} = \sqrt{1 + ZT_m} = M$$

In this case the efficiency expression reduces to

$$\eta = \eta_c \frac{M - 1}{M + T_c/T_h} \tag{27-28}$$

η/η_c as a function of ZT_m is given in Fig. 27-2 for the cases T_c/T_h equal to 0.33 and 0.5 when $T_m = (T_h + T_c)/2$.

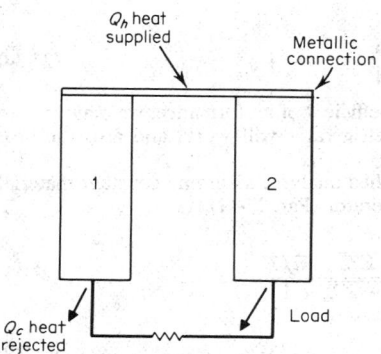

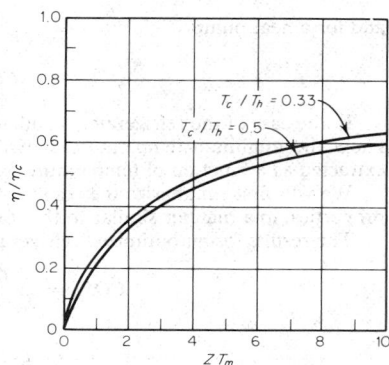

Fig. 27-1. Basic elements of a thermoelectric generator.

Fig. 27-2. Relative efficiency of thermoelectric generators as a function of the figure of merit ZT_m, for two values of the hot-to-cold temperature ratio.

$$T_m = \frac{T_h + T_c}{2}$$

Maximum Power Output. The maximum power output from a given generator is obtained when $M = 1$, that is, when $R_L = R$. This condition does not, in practice, differ very greatly from $m = M$.

11. Materials with Temperature-dependent Properties. The materials used in thermo-electric generators have properties that are strongly temperature-dependent (Par. **27-14**). It is necessary to determine the temperature variation along each limb by solving

$$\frac{d}{dr}\left(k\frac{dT}{dr}\right) - \tau I \frac{dT}{dx} + \rho I^2 = 0 \tag{27-29}$$

The value of the Thompson coefficient τ is obtained from the Kelvin relation, Eq. (27-6). The heat input Q_h and electrical resistance R can then be evaluated.

The open-circuit voltage V_{oc}, which does not require a knowledge of the temperature distribution, is given by

$$V_{oc} = \int_{T_c}^{T_h} (\alpha_1 - \alpha_2)dT \tag{27-30}$$

Hence we have efficiency

$$\eta = I\frac{V_{oc} - IR}{Q_h} \tag{27-31}$$

Such calculations are best carried out by using a digital computer. However, reasonably accurate design may be attained by using the average material properties to obtain a value for

$$Z_{av} = \left[\frac{\alpha_1 - \alpha_2}{(\rho_1 k_1)^{1/2} + (\rho_2 k_2)^{1/2}}\right]^2 \tag{27-32}$$

This value may then be substituted in the equations in Par. **27-10**.

Reference 3 (Par. **27-18**) shows that this procedure predicts efficiencies which are within about 6% of the values obtained by exact calculations.

THERMOELECTRIC REFRIGERATORS AND HEAT PUMPS

12. General. As is the case with other heat engines, the thermoelectric generator may be reversed and operated either as a refrigerator or a heat pump. To compare performance of these devices we define a coefficient of performance (COP).

For a refrigerator

$$COP_R = \frac{Q_c}{W} \tag{27-33}$$

and for a heat pump

$$COP_{hp} = \frac{Q_h}{W} \tag{27-34}$$

In the case of the refrigerator, in addition to coefficient of performance, we may wish to know the minimum temperature at which the refrigerator will work, and also the heat extracted as a function of temperature.

We will first summarize the results of a simplified analysis, assuming constant material properties, in a manner similar to that for the generator (Par. **27-11**).

The results for an optimized refrigerator are

$$COP_R = \frac{T_c}{T_h - T_c}\frac{\sqrt{1 + ZT_m} - T_h/T_c}{\sqrt{1 + ZT_m} + 1} \tag{27-35}$$

$$Q_c = \alpha_{12} I T_c \tag{27-36}$$

$$T_{max} = \tfrac{1}{2}T_c^2 Z \tag{27-37}$$

For optimum performance the requirements of Eqs. (27-25) and (27-27) must be satisfied.

An analysis taking account of the temperature dependence of material properties may be carried out in a manner similar to that for the generator.

13. Nernst-Ettinghausen Devices. The Nernst-Ettinghausen effect may be employed for generation, and the Ettinghausen effect for refrigeration. Both possibilities have been investigated by a number of authors. The Nernst-Ettinghausen generator does not appear to offer any promise, except perhaps at low temperatures, where applications are unlikely. The Ettinghausen refrigerator may prove to be of more interest where large temperature differences are required. In both cases the cost and bulk of the necessary magnetic field arrangement must be taken into account.

THERMOELECTRIC MATERIALS

14. Practical Generator Materials (See also Sec. 6.) We may summarize the current position as follows:

Low-temperature Range, Up to 250°C (Type I). The bismuth telluride family of materials has the highest conversion efficiency in this temperature range. They cannot be used much above 250°C owing to the increase in intrinsic conduction resulting from the small energy gap. The practical result of this effect is an increased thermal conductivity and reduced voltage output leading to a pronounced fall in conversion efficiency. Conversion efficiencies of 5% are achievable.

Intermediate Temperature Range, Up to 500° to 600°C (Type II). Materials based on lead telluride, germanium telluride, silver antimony telluride, lead tin telluride, and various alloys of these are used in this temperature range. The upper temperature limit is fixed by the chemical stability of these materials, particularly in air. In an inert atmosphere and with suitable encapsulation, temperatures of up to 500°C are possible. A conversion efficiency of 9% is theoretically possible.

High-temperature Range, Up to 1000°C (Type III). In this range germanium-silicon alloy and other silicides are used. From Table 27-1 we see that germanium and silicon have good values of $\alpha^2\sigma$ but are unsuitable for thermoelectric generation due to their high phonon conduction (K_L) values. By alloying germanium and silicon, K_L may be reduced considerably, resulting in a useful material. Indium arsenide, with the addition of phosphorus, has possible interest as an *n*-type material. An advantage of these materials is that they are strong mechanically. A conversion efficiency of 9% is possible for a generator operating between 15° and 1000°C.

The efficiencies quoted are conversion efficiencies. Overall generator efficiencies are lower than this because of heat losses and voltage inversion losses. Typical values are 3 to 5% for power levels greater than a few watts, using either Type I or II materials. Cascading of these materials is possible, and overall generator efficiencies of up to 10% can be achieved.

Figure 27-3 summarizes the performance of the best thermoelectric materials. The curves of $ZT_m = 1$ and $ZT_m = 2$ are also plotted on Fig. 27-3. Together with Fig. 27-2 they indicate the upper limit on performance we may expect to achieve. These data represent the results of a considerable program of development over a number of years, and there does not appear to be any immediate prospect of any significant improvement on these figures.

15. Selection of Limb Length. In the design of thermoelectric devices a number of practical considerations arise. We have shown that, for optimum performance, it is necessary to arrange that Eq. (27-25) is satisfied. This requirement merely determines the length-to-area ratio, but does not fix the length of the element. To fix the length of the thermoelectric elements, we must consider thermal insulation thickness and contact resistance.

The electrical resistance of a limb $\rho L/A$ should be large compared with the contact resistance ρ_c/A; that is,

$$\frac{\rho_c}{A} \ll \frac{\rho L}{A} \tag{27-38}$$

or

$$L \gg \frac{\rho_c}{\rho}$$

Typical values are $\rho_c \approx 10 \times 10^{-6}\,\Omega/\text{cm}^2$.

$$\rho \approx 10^{-3}\,\Omega \cdot \text{cm}$$

Table 27-1. Thermoelectric Properties of Semiconductors

Semiconductor	E_g, eV	Melting point, °C	Type of semi-conductor	$\alpha^2\sigma$	K_e, W /cm-°C	K_L, W /cm-°C	Z_{max}	Temp., K	Max. operating temp., K
	(1)	(2)	(3)	(4)	(5)	(6)	(7)	(8)	(9)
Bi₂Te₃	0.15	575	n or p	4×10^{-5}	0.004	0.016	2×10^{-3}	300	450
BiSb₄Te₇.₅	…	…	p	4.6×10^{-5}	0.004	0.010	3.3×10^{-3}	300	450
Bi₂Te₂Se	0.3	…	n	3.6×10^{-5}	0.003	0.013	2.3×10^{-3}	300	600
PbTe	0.3	904	n or p	2.6×10^{-5}	0.003	0.02	1.2×10^{-3}	300	900
GeTe(+Bi)	…	725	p	3.2×10^{-5}	0.006	0.016	1.6×10^{-3}	800	900
ZnSb	0.6	546	p	2×10^{-5}	…	0.17	1.2×10^{-3}	500	500
AgSbTe₂	0.6	576	p	1.4×10^{-5}	0.003	0.006	1.8×10^{-3}	700	600
CuTe₁S	…	930	…	18×10^{-6}	…	0.012	1.5×10^{-3}	1100	900
GeSi	0.8	…	n	3×10^{-5}	0.003	0.033	9×10^{-4}	900	1200
			p	2×10^{-5}			6×10^{-4}		
Ge	0.65	958	n or p	4×10^{-5}	0.006	0.63	6.3×10^{-5}	300	
Si	1.15	1420	n or p	2×10^{-5}	0.003	1.13	1.5×10^{-5}	300	
InSb	0.17	527	n	8×10^{-5}	0.01	0.16	4.7×10^{-4}	300	

NOTE: The values of $\alpha^2\sigma$ in column 4 are the maximum, obtainable by optimum doping, to give $\alpha = 200$ to $250\ \mu\text{V/K}$. The correct value of α in this range is higher, the higher is $K_e K_L$, and is near 250 for AgSbTe₂, 200 for InAs. The values for K_e in column 5 apply also to the case of optimum doping. The figures quoted apply to the temperatures listed in column 8.
SOURCE: Wright. Ref. 4. Par. 27-18.

27-8

Hence L should not be less than about 1 cm.   Thermal insulation thickness of 1 or 2 cm is found to be suitable in practical generator designs.   Hence L is normally chosen at about 1 or 2 cm.

The theory given in the preceding paragraphs enables the generator efficiency to be calculated.   To predict the overall device efficiency it is necessary to take account of bypass losses due to heat conduction from the source through the thermal insulation.   For flame-heated generators, not all the energy in the fuel can be used, and burner efficiency must be included in the overall efficiency calculation.

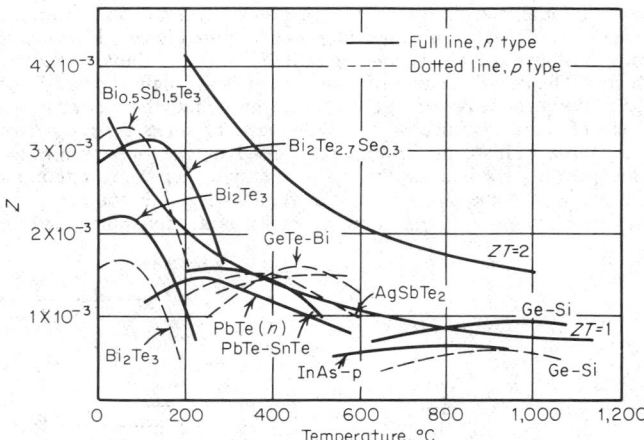

Fig. 27-3.   Figure of merit Z vs. temperature for various thermoelectric materials.   *(Wright, Ref. 4.)*

Multi-junction Devices.   Since the voltage generated by a single junction is low, typically 0.05 V, it is usual to arrange for a number of junctions to be connected electrically in series (Fig. 27-4a).   Where the generator output is less than 10 W, this results in limbs of low cross section, and hence of low mechanical strength.   Much effort has gone into the production of modules like those shown in Fig. 27-4b, in which the limbs are bonded together while maintaining electrical insulation between them.   Similar considerations apply to refrigerator design.   If the materials of the n and ρ limbs have markedly different thermal expansion, the differential thermal expansion may require separately sprung contact arrangements.

16. Segmented and Cascaded Devices.   It is found that practical thermoelectric materials operate most efficiently over a restricted temperature range.   It is possible to build up a limb from different segments of material so as to use a given material only over its most effective temperature range.   Such an approach, known as *segmenting,* is shown in Fig. 27-4c.

A number of practical problems arise in segmenting, e.g., the contact resistance between the dissimilar materials may be large, and there are also difficulties in electrical, thermal, and mechanical matching.

An alternative approach is to make a complete device that operates over part of the available temperature range and to construct further devices for other parts of the range.   This arrangement, known as *cascading,* is shown in Fig. 27-4d.

The overall efficiency η_N of N cascaded generator stages is given by the expression

$$\eta_N = 1 - \sum_{J=1}^{J=N} (1 - \eta_j) \tag{27-39}$$

where η_j is the efficiency of the jth stage.

The coefficient of performance of an N-stage refrigerator is given by the expression

$$COP_R = \sum_{J=1}^{J=N} \left[(1 + \frac{1}{COP_j}) - 1 \right]^{-1} \tag{27-40}$$

17. Applications.   Thermoelectric generators are used where the silence, long life, absence of moving parts, long periods between maintenance, and low weight and size are important. They can be used with any heat source, for example, solar radiation, flames, or radioactive isotopes. The low conversion efficiency of practical generators, about half the theoretical thermoelectric conversion efficiency, restricts their use to low power outputs, normally less than 100 W.

Thermoelectric generators are particularly suitable for use in remote or inaccessible locations, and for such applications, a radioactive-isotope heat source is often chosen. Typical applications include power supplies for communication satellites, underwater repeater stations, microwave repeater stations, navigational buoys, and weather stations. Power levels fall within the 5- to 100-W range. The initial cost is high, but such generators compare well with other electrical power sources on a watthour basis. Cardiac pacemakers have also been constructed; in this case the power required is very low, 1 mW.

Radioactive isotopes are now becoming available in considerable quantities from fission waste products. The power from such sources can be as high as several watts per cubic centimeter, and although the power level decreases throughout the life of the isotope, the half-life can be tens of years. Isotopes from fission waste products are β emitters and require substantial shielding. This is not an important disadvantage for such uses as navigational buoys, but for space and medical use the more expensive α emitters are required. Table 27-2 lists the properties of some of the more important radioactive isotopes.

Thermoelectric refrigerators are used where long life, reliability, and small size are

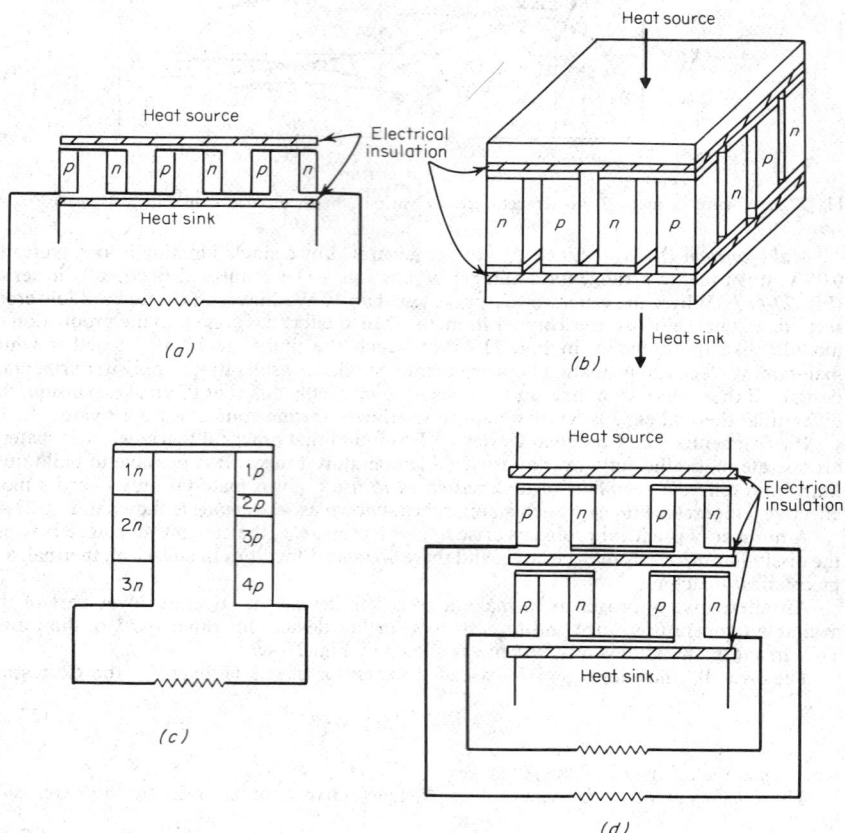

Fig. 27-4. Thermoelectric generator arrangements. (*a*) multiple junction series connection; (*b*) multiple junction series connection, module construction; (*c*) segmented type; (*d*) cascade type.

Table 27-2. Properties of Some Radioisotopes Suitable for Power Generators

Radio-isotope	Half-life, years	Ci/W(t)	Specific power, W(t)/g	Com-pound	Power density (theor.), W/cm^3	Power density (pract.), W/cm^3	Radia-tion	Source	Comments
Sr^{90}/Y^{90}	28	150	0.95	SrTiO$_3$	2.35	1.05	β	Fission product; fission yield 5.77%	Available in large quantities at low cost from fission waste. Currently used in SrTiO$_3$ form, which has a very low solubility in water and brine (an extremely valuable property for safety considerations). Later Sr$_2$TiO$_3$ and SrO may be used. Heavy shielding is required.
Cs^{137}	30	207	0.42	Cs glass	...	1.24	β, γ	Fission product; fission yield 6.15%	Available in large quantities at a cost similar to Sr^{90}. There are some difficulties in fabrication as a glass. Power density is only one-quarter that for SrTiO$_3$. Heavy shielding is required. It is unlikely to compete with Sr^{90}.
Pm^{147}	2.67	2,270	0.37	Pm$_2$O$_3$	2.03	1.9	β	Fission product; fission yield 2.6%	Short half-life but a better power density than Sr^{90}, and requires only one-sixth as much shielding. However, it is more expensive and unlikely to compete with Sr^{90} except for special applications.
U^{232}	74	26	4.8	UO$_2$	46	33	α, γ	Irradiation of Th^{230} in reactor	Hard gamma radiation. Expensive. Good power density. Can operate at high temperature. Suitable for thermionic converter. Unlikely to compete with Sr^{90}.
Th^{228}	1.91	24	170	ThO$_2$	1,530	1,270	α, γ	Irradiation of Ac^{227} or, more likely, combined with U^{232} as decay product of latter	Short half-life. Requires shielding. Expensive. Unsuitable for medical applications. Very high power density. Unlikely to compete with Sr^{90}.
Cm^{244}	18	30	2.8	Cm$_2$O$_3$	27	26.4	α, n	Reactor irradiation of Pu^{242}	Some shielding required; so unsuitable for medical applications. Expensive. Good power density. Unlikely to compete with Sr^{90}.

Source: Report of the IAEA Study Group on Isotope Powered Electrical Generators, July 1966.

required. With present materials they are not used below around 250 K for single stage and 200 K for cascaded stages. A good account of small commercially available refrigerators for use with infrared detectors is given in Ref. 7, Par. **27-18**.

18. References on Thermoelectric Conversion

General references

1. SPRING, K. H. "Direct Energy Conversion," Academic Press, Inc., New York, 1965.
2. SUTTON, G. W. "Direct Energy Conversion," McGraw-Hill Book Company, New York, 1966.
3. HEIKES, R. R., and URE, R. W., JR. "Thermoelectricity: Science and Engineering," Interscience Publishers, Inc., New York, 1961.

Text references

4. WRIGHT, D. A. *Metall. Rev.,* No. 148, Institute of Metals, London, 1970.
5. CORLISS, W. R., and HARVEY, D. G. "Radioisotopic Power Generation," Prentice-Hall, Inc., Englewood Cliffs, N.J., 1964.
6. International Symposium on Industrial Application of Isotopic Power Generators, Harwell, U.K.A.E.A., September 1966, European Nuclear Energy Agency (ENEA) and Organisation for Economic Co-operation and Development (OESD).
7. WOLFE, W. L. (ed.) "Handbook of Military Infra-red Technology," Department of the Navy, Office of Naval Research, Washington, 1965.

THERMIONIC GENERATION

19. Introduction. In a thermionic generator (Fig. 27-5) electrons are emitted from the heated cathode (emitter) and collected by a cooler anode (collector). The electrons return to the cathode by means of an external load. Like the thermocouple, the thermionic generator is a heat engine using an electron gas as the working fluid. Two useful features of the thermionic generator are its high power density (up to several tens of watts per square centimeter) and its high-temperature heat rejection (700 to 1000°C). The latter is particularly valuable in space applications. Single units may be constructed from a few watts output to several hundred watts. Thermionic generators may be operated in series-parallel connection to give higher power levels.

Laboratory generators have been operated at efficiencies of 10 to 18% and with development may be expected to achieve 25%. Lifetimes of greater than 10,000 h have been reported.

20. Ideal Thermionic Generators. Space-charge effects are neglected in the following treatment. In Fig. 27-6a, the current density that may be drawn in a space-charge neutralized generator is plotted against the retarding potential V_c, measured from the Fermi level in the emitter. If $V_c < \phi_e$, the current density will be independent of V_c and the saturation current density j_s is as given by Richardson's equation.

$$j_s = AT_h^2 \exp\left(-e\phi_e/kT_h\right) \tag{27-41}$$

where T_h and ϕ_e = emitter temperature and work function, respectively, A = a constant, e = charge on the electron, and k = Boltzmann's constant. For $V_c > \phi_e$

$$j = j_s \exp\left[-e\left(V_c + \phi_e\right)/kT_h\right] \tag{27-42}$$

Any operating point P may be chosen, provided that $V_c > \phi_e$, where ϕ_c is the collector work function.

Figure 27-6b shows the potential-energy diagram for the case $V_c = \phi_c$, and Fig. 27-6c that at $V_c > \phi_e$. In the latter case a negative sheath is formed at the emitter surface, although there is no potential gradient across the greater part of the interelectrode spacing.

The output voltage V_0 is given by

$$V_0 = V_c - \phi_c \tag{27-43}$$

The output power is

$$P_0 = j(V_c - \phi_c) \qquad \text{(per unit area)} \tag{27-44}$$

In practical generators, the collector temperature and work function are such that back emission from the collector may be neglected. The input power is

$$P = j(V_c + 2kT_h/e) + P_L \qquad \text{(per unit area)} \tag{27-45}$$

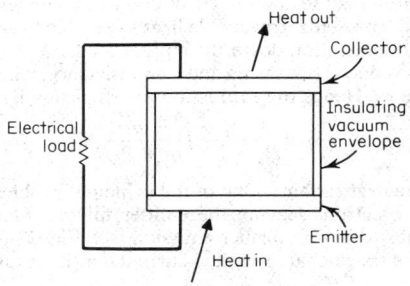

Fig. 27-5. Basic thermionic generator.

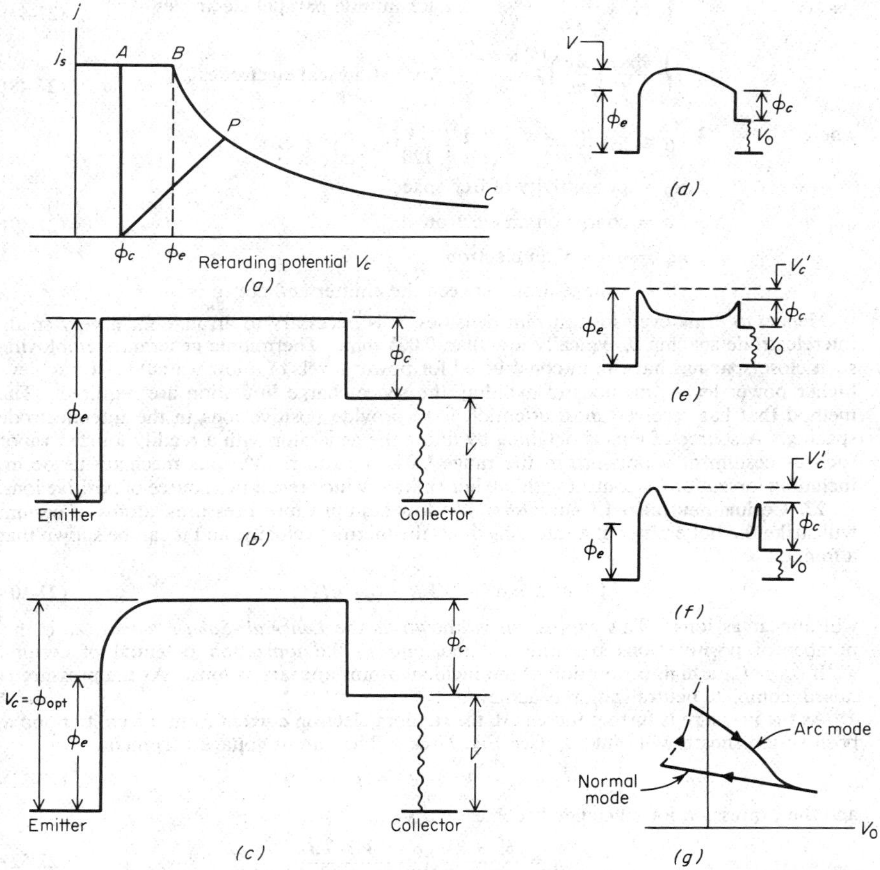

Fig. 27-6. Retarding potential graph and potential energy diagrams (*a*) retarding potential graph; (*b*) potential energy diagram of fully centralized generator $V_c = \phi_e$; (*c*) of fully neutralized thermionic generator $V_c > \phi_e$; (*d*) space-charge-limited; (*e*) unignited mode; (*f*) ignited mode; (*g*) current-voltage characteristic.

The term $2kT_h/e$ represents the cooling effect due to the higher-energy electrons in the velocity distribution. P_L represents irreversible heat losses due to radiation from the emitter to the collector and heat conduction down the emitter leads.

The output voltage is reduced by plasma and lead resistance, which may be accounted for by an internal resistance r. Hence the expression for efficiency is

$$\eta = \frac{V_c - \phi_e - jr}{V_c + 2kT_h/e + P_L/j} \tag{27-46}$$

21. Space-Charge Neutralization. One of the principal problems in thermionic generators is space charge. Electrons leaving the emitter fill the interelectrode space with a negative charge that tends to inhibit further emission (see Fig. 27-6d).

In a vacuum diode the space-charge-limited current density is given by Child's equation:

$$j = \begin{cases} \dfrac{4\epsilon_0}{9}\left(\dfrac{2e}{m_e}\right)^{1/2}\dfrac{V^{3/2}}{d^2} & \text{for infinite parallel electrodes} & (27\text{-}47) \\[3mm] \dfrac{8\pi\epsilon_0}{9}\left(\dfrac{2e}{m_e}\right)^{1/2}\dfrac{V^{3/2}}{2\beta} & \text{for cylindrical electrodes} & (27\text{-}48) \end{cases}$$

where

$$\beta = \log\frac{b}{a} - \frac{2}{5}(\log\frac{b}{a})^2 + \frac{11}{120}(\log\frac{b}{a})^3 + \cdots$$

$$\begin{aligned}
\epsilon_0 &= \text{permittivity of free space} \\
e &= \text{charge on an electron} \\
m_e &= \text{mass of an electron} \\
V &= \text{voltage drop between the emitter and collector}
\end{aligned} \tag{27-49}$$

Hence, to draw even low current densities, it is necessary to arrange for a very small interelectrode spacing d, typically less than 0.005 mm. Thermionic generators employing such close spacings have been constructed for power levels of a few watts.[3,4]* To achieve higher power levels, methods of avoiding the space-charge limitation are required. The method that has received most attention is to provide positive ions in the interelectrode spacing. A source of ions is obtained by filling the generator with a readily ionized vapor such as cesium at a pressure in the range 0.01 to 10 torr. Various mechanisms occur, including ionization by contact with the hot emitter, which result in a source of positive ions.

22. Cesium-neutralized Generators. At low cesium-vapor pressures, atoms of cesium will strike the hot emitter at a rate $\frac{1}{4}n_0$ times the thermal velocity, and it can be shown that a fraction

$$\left\{1 + 2\exp\left[-e\left(V_i - \phi_e\right)/kT_h\right]\right\}^{-1} \tag{27-50}$$

will appear as ions. This expression is known as the *Langmuir-Saha relation*. n_0 is the number of positive ions per unit volume, and V_i the ionization potential of cesium.

If $\phi_e > V_i$, a high proportion of the incident atoms appears as ions. As the pressure is raised, complete neutralization is achieved.

As the pressure is further increased, the random electron current from the emitter and a positive-ion sheath will build up (see Fig. 27-6e). The output voltage V is then

$$V = \phi_e - V_e - \phi_c + V'_c - Jr \tag{27-51}$$

and the expression for efficiency becomes

$$\eta = \frac{\phi_e - V_e - \phi_c + V'_c - Jr}{\phi_e + 2kT_h/e + P_L/J} \tag{27-52}$$

For high-work-function materials, e.g., tungsten, Langmuir showed that the presence of cesium vapor can modify considerably the emission properties of the surface by reducing the effective work function. In practical generators the optimum work function of 2.5 to 3 eV can be achieved using refractory emitters, e.g., tungsten, rhenium, and a cesium pressure in the range 1 to 10 torr.

---

*Superior numbers correspond to numbered references, Par. 27-24.

The advantage of this arrangement is that refractory cathode materials having a very low rate of evaporation, and correspondingly long life at the working temperature, can be used. The disadvantage is that the high-pressure plasma has appreciable resistivity and necessitates small interelectrode spacings (of the order of 0.25 to 0.5 mm). A further advantage of the cesium vapor is the modification of the collector work function to the value corresponding to that of bulk cesium, $\phi_c = 1.8V$.

It is found that the high-pressure cesium generator can operate in two distinct modes, the *normal mode* and the *arc mode* (see Figs. 27-6e–g). The two modes merge at higher emitter temperatures. Typical performance characteristics of a cesium-neutralized generator operating in the arc mode are shown in Figs. 27-7 to 27-10.

Figure 27-7 shows how the voltage-current characteristic depends on cesium pressure, expressed in terms of saturation temperature T_R, in a generator operated with emitter and collector temperature held constant. The envelope to these curves gives the performance when the cesium pressure is optimized for each output voltage. Figure 27-8 shows how such envelopes depend on emitter temperature. Figure 27-9 shows how the envelopes depend on electrode spacing, for a fixed emitter temperature. Figure 27-10 plots the power density as a function of emitter temperature for several values of current density, using different emitter materials.

23. Applications.[1,2]

Space Power Supplies. The relatively high reject temperatures of the collector (approximately 1000 K) is an attractive feature for space applications because it simplifies reduction cooling.

Thermionic generators may be used in association with nuclear heat sources, e.g., fission reactors and isotope heat sources. Solar heating is also feasible. In nuclear heating, the thermionic generators may be placed inside the reactor (in the core) or the heat may be removed by the reactor by heat pipes (Pars. 27-33 to 27-38) and the generator situated outside the reactor. A considerable amount of development work has gone into this application, and a number of design studies have been published (see Ref. 2, Par. 27-24).

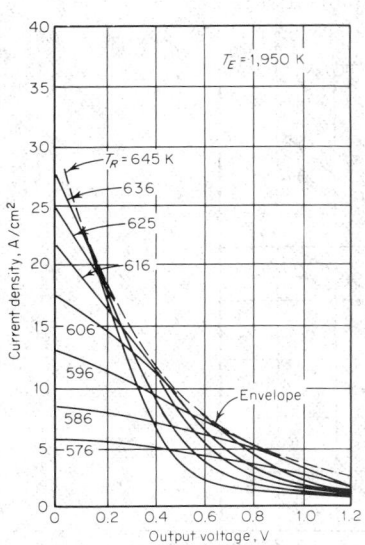

Fig. 27-7. Typical current-voltage curves of a cesium-neutralized thermionic generator. *(Rujeh, Lieb and Van Someren. Ref. 5.)*

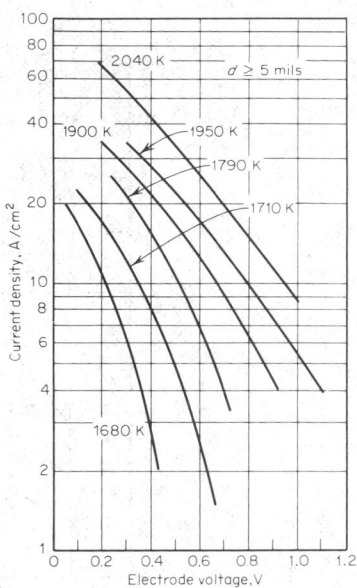

Fig. 27-8. Optimized envelopes at various emitter temperatures. *(Rujeh, Lieb and Van Someren. Ref. 5.)*

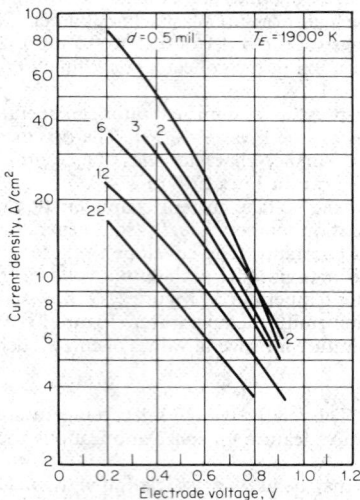

Fig. 27-9. Envelopes for various spacings. *(Rujeh et al., Ref. 5.)*

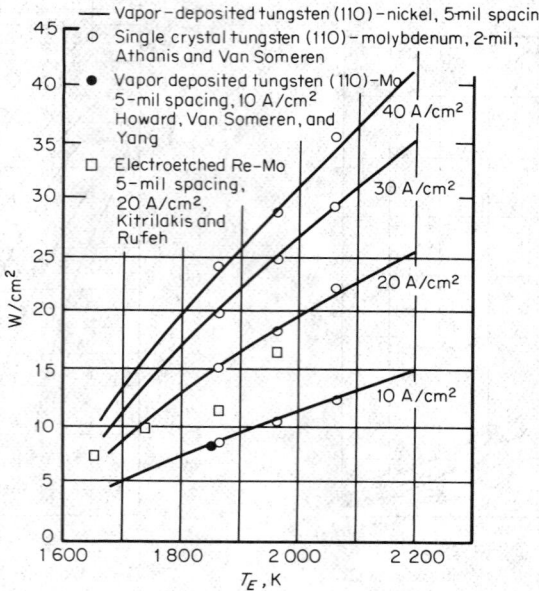

Fig. 27-10. Comparison of four thermionic converters. *(Wilson, Ref. 6.)*

Central Power Stations. Thermionic generators may be used as toppers for both nuclear- and fossil-fired-powered stations, but the economic justification is difficult. Problems include lifetime, reliability, and cost.

Small Generators. Application of fossil-fuel-fired thermionic generators in the 100-W range has been suggested. The integrity of the vacuum envelope to prevent hydrogen diffusion is a problem with flame heating.

24. References on Thermionic Generation
General references
SPRING, K. H. "Direct Energy Conversion," Academic Press, Inc., New York, 1965.
SUTTON, G. W. "Direct Energy Conversion," McGraw-Hill Book Company, New York, 1966.

Specialist references
Advances in the subject are reported in the proceedings of various specialist conferences. A comprehensive account of developments over the international field is to be found in the following publications:

1. *Proceedings of International Conference on Thermionic Electrical Power Generation,* IEE and ENCA, London, September 1965, published by Institution of Electrical Engineers, London.

2. *Proceedings of International Conference on Thermionic Electrical Power Generation,* ENEA and Euratom, Stresa, May 27-31, 1968, published by Euratom, Stresa, Italy.

Text references
3. BEGGS, J. E. *Advanced Energy Conversion,* Vol. 3, p. 447, 1963.
4. RASOR, N. S., GASPER, K. A., and DE STEESE, J. G. *Proceedings of the International Conference on Thermionic Electrical Power,* (B-15), Stresa, May 27-31, 1968, published by Euratom, Stresa, Italy.
5. RUFEH, F., LIEB, D., and VAN SOMEREN, L. *Ibid.* (A-1).
6. WILSON, V. *Ibid.* (A-2).

MAGNETOHYDRODYNAMIC POWER GENERATION

25. Basic Principle. A high-velocity, electrically conducting fluid stream crossing a magnetic field may be regarded as taking the place of the moving conductors of a conventional dynamo. This arrangement of basic elements is used in magnetohydrodynamic (MHD) generation.

The operation of an MHD generator is illustrated in Fig. 27-11.

If a jet of conducting fluid, velocity u, moves through a magnetic field of flux $\bar{B}$ at right angles to $\bar{u}$, an electric field $\bar{E}$ exists.

$$\bar{E} = \bar{u} \times \bar{B} \qquad (27\text{-}53)$$

If electrodes are suitably placed in contact with the jet, energy can be extracted and delivered to an external load. Thermodynamically, the system is analogous to that of a turbine with electromagnetic braking taking the place of the mechanical braking of the turbine blades.

Assuming that the working fluid behaves as a normal electrical conductor of conductivity σ, we see that the current density j is given by

$$j = \sigma(E - uB) \qquad (27\text{-}54)$$

and the electrical power output per unit volume of duct is

$$-jE = \sigma E(uB - E) \qquad (27\text{-}55)$$

If the ratio of load resistance to total resistance is $K = E/uB$, the electrical power generated per unit volume of duct is

$$W = -jE = K(1 - K)\sigma u^2 B^2 \qquad (27\text{-}56)$$

This power is obtained by the work done by the moving stream against the body force.

$$jB = -(1 - K)\sigma uB^2 \qquad \text{per unit volume of duct} \qquad (27\text{-}57)$$

Then work done by the stream,

$$-jBu = (1 - K)\sigma u^2 B^2 \qquad \text{per unit volume of duct} \qquad (27\text{-}58)$$

and the difference between Eqs. (27-58) and (27-56) represents ohmic heating in the fluid and is given by

$$(1 - K)^2 \sigma u^2 B^2 \tag{27-59}$$

We see that with this method the thermal energy of the source must first be converted to kinetic energy, and the energy-conversion chain thus involves a mechanical-energy link.

From Eq. (27-56) it is seen that power output is proportional to the electrical conductivity σ. It is necessary for σ to lie in the range 1 to 100 S. Such values may be achieved either by thermal ionization in combustion gases seeded by a small amount of readily ionizable material such as sodium or potassium or by nonequilibrium discharge in an inert gas seeded in a similar manner.

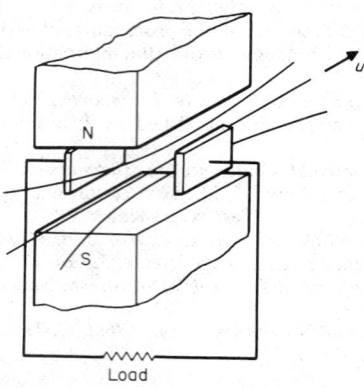

Fig. 27-11. Basic principle of the magnetohydrodynamic generator.

26. MHD Generation. Unlike thermoelectric and thermionic generation, MHD generation is not suitable for low power levels, and development work has been directed to levels in the multi-megawatt range.

MHD generators are characterized by high temperature, high power density, and absence of moving parts. The principal advantage over the conventional power plant is the higher conversion efficiency of the combined MHD-steam plant. To be of practical interest, low unit cost, long life, and reliability must be achieved.

All the conventional fuels (nuclear, coal, oil, natural gas) may in principle be used; the particular choice depends on the conditions applying in a given country. The required temperature levels are readily obtained from fossil fuels, but further development is required if nuclear reactors are to be used.

27. Open-Cycle Plant. The open-cycle plant uses a fossil fuel and atmospheric oxygen. A schematic of a central power station is shown in Fig. 27-12. This typical design operates on a regenerative Brayton cycle with a conventional steam plant. The flow is linear (being either sonic or subsonic). Seeding is required. The magnetic field is provided by a superconducting magnet.

Practical problems include seed recovery, removal of oxides of nitrogen, and development of long-life components, including duct, electrodes, and air preheater. Extensive pilot studies are being carried out in various countries. Descriptions of these studies, together with accounts of current developments, are given in the *Proceedings* of the MHD international conferences held biannually (Par. **27-29**).

28. Closed-Cycle Plant. Closed-cycle MHD systems are normally associated with nuclear heating (see Fig. 27-13). Either a seeded noble gas or a liquid metal may be used as the working fluid, the former operating on a Brayton and the latter on a Rankine cycle. Work has been carried out on the noble-gas-closed-cycle systems, but is less advanced than for the fossil-fired open-cycle system. Suitable high-temperature reactors must be developed, and the problems of plasma stability (associated with the use of nonequilibrium conductivity) must be solved. Developments in closed-cycle gas turbines may permit the same conversion

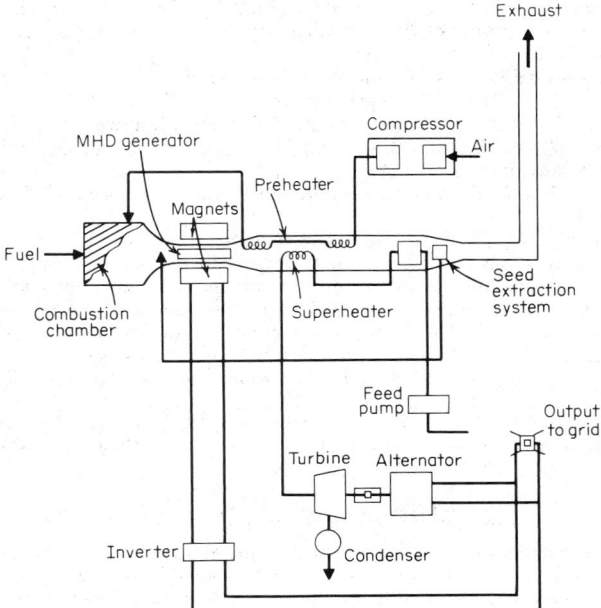

Fig. 27-12.   Open-cycle fossil-fuel **MHD** system.

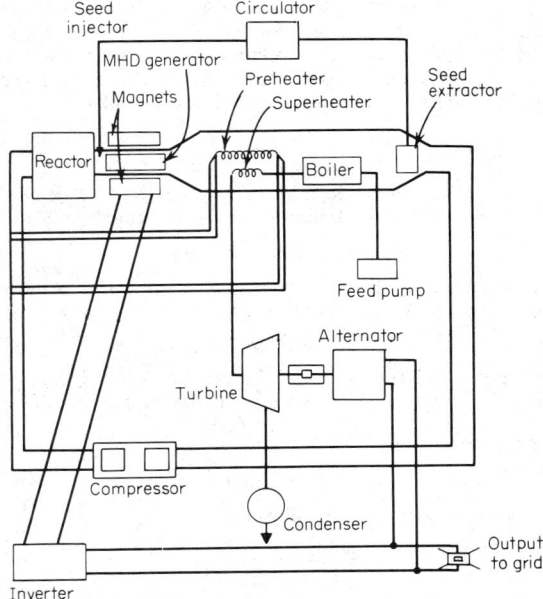

Fig. 27-13.   Closed-cycle nuclear **MHD** system.   *(After Dunn and Wright, Proc. IEE No 10 Oct 63, Vol. 110.)*

efficiency to be achieved at lower gas outlet temperatures. The relative advantages of the two systems have yet to be quantitatively established. Current developments are reported in Refs. 1-4, Par. 27-29.

29. References on Magnetohydrodynamic Generation

General references

SPRING, K. H. "Direct Energy Conversion," Academic Press, Inc., New York, 1965.

SUTTON, G. W., and SHERMAN, A. "Engineering Magnetohydrodynamics," McGraw-Hill Book Company, New York, 1965.

WOMACK, G. J. "M.H.D. Power Generation Engineering Aspects," Chapman & Hall, London, 1969.

Specialist references

Advances in the subject are reported in the proceedings of various specialist conferences. A comprehensive account of developments over the international field is to be found in the *Proceedings of the International Conferences on Magnetohydrodynamic Power Generation* organized jointly by the ENEA (European Nuclear Energy Agency) and from 1966 jointly with IAEA (International Atomic Energy Authority). Recent conference locations and dates include Paris, July 6-11, 1964 (Ref. 1); Salzburg, July 4-8, 1966 (ref. 2); Warsaw, July, 1968 (Ref. 3); Munich, Apr. 19-23, 1971 (Ref. 4).

A full account of the British MHD Collaborative Committee investigation of open-cycle systems is given in J. B. HEYWOOD and G. J. WOMACK, "Open Cycle M.H.D. Power Generation," Pergamon Press, New York, 1969.

ELECTROHYDRODYNAMIC GENERATION

30. Basic Principle. In an electrohydrodynamic generator, charged particles are transported by a fluid from a region of low electrical potential to a region of high electrical potential. The situation is very similar to that in a Van de Graaff generator, except that the fluid takes the place of the moving belt as a means of transporting the charge against electrostatic forces. Either a gas or a liquid may be used. Most work has gone into gas generators (electrogasdynamics, or EGD).

When a gas is used, the generator may be incorporated into a thermodynamic cycle, normally the Brayton cycle, and the system used to convert heat directly into electricity. The EGD generator differs from the MHD generator in that it is a high-impedance device, typically producing in the voltage range 5 to 500 kV and current of a few microamperes whereas the MHD generator produces a relatively low voltage and very high current.

Consider a long cylinder, radius r, containing a charge density ρ. The axial electric field at the ends, E_a, and the radial field in the center, E_r are given by

$$E_a = E_r = \frac{\rho r}{2\epsilon} \tag{27-60}$$

where ϵ = permittivity of the dielectric, approximately 8.85×10^{-12} F/m.

The current I transported by the stream moving with velocity u is given by

$$I = \rho u \pi r^2 = 2\pi \epsilon E_1 r u \tag{27-61}$$

where E_1 = limiting field set by breakdown considerations. If V is the voltage developed across the generator section, the power output P is

$$P = 2\pi \epsilon E_1 r u V \tag{27-62}$$

The power output is proportional to velocity, whereas the frictional loss is proportional to the cube of the velocity. For this reason it is not usual to exceed a velocity of 50 m/s in practical generators.

Equation (27-62) may be written

$$P = \Delta p u \pi r^2 \tag{27-63}$$

where Δp is the pressure drop per stage. Hence

$$\Delta p = \frac{P}{u \pi r^2} = \frac{2\epsilon E_1 V}{r} \tag{27-64}$$

Substituting typical values shows that $\Delta p/p$ is of the order of 1 or 2%. Hence a multistage generator is required.

31. Production of Charged Particles. Equation (27-61) assumes that there is no slip between the particles and the gas. This requires that the particle mobility should not be too high. With an axial field of the order 1 MV/m, the particle size should not be less than 10^{-8} m. Other factors, such as deposition on the walls and erosion from the wall, influence optimum particle size. Musgrove and Wilson (Ref. 2, Par. 27-32) conclude that particles should lie in the size range 10^{-6} to 10^{-5} m. Such particles may be charged triboelectrically and by corona charging. The latter appears to be most favorable.

32. References on Electrohydrodynamic Generation.

1. GOURDINE, M. C. *Proceedings of the International Symposium on Electrohydrodynamics,* Massachusetts Institute of Technology, Cambridge, Mass., 1969, pp. 164–168.

2. MUSGROVE, P. J., and WILSON, A. D. 1st Static Electrification Conference, Vienna, May 1970.

THE HEAT PIPE

33. Introduction. The heat pipe is a development of the two-phase *thermosyphon,* illustrated in Fig. 27-14a. A fluid heated at the bottom of a tube evaporates, taking away the heat and condensing at the cooler upper end. The condensate is returned to the reservoir at the base by gravitational forces. A typical thermosyphon contains 5% liquid by volume, the remaining volume being occupied by vapor.

The thermosyphon is characterized by a very high rate of heat transfer, but it suffers from the gravity feed return, which requires that the evaporating section should be lower than the condensing section. The *heat pipe,* shown in Fig. 27-14b, is like the thermosyphon; i.e., it operates under saturation conditions and is a two-phase device having a boiler and condenser. It differs from the thermosyphon in that the condensate is returned by capillary forces, which removes the restriction on the relative position of the boiler and condenser.

The heat pipe principle was patented by Gaugler of General Motors in 1944 (Ref. 3, Par. 27-39). The recent interest in the heat pipe followed its independent invention by Grover (Ref. 4, Par. 27-39) in 1962, as a solution to heat-transfer problems in nuclear-powered thermionic generators, intended for space power supplies.

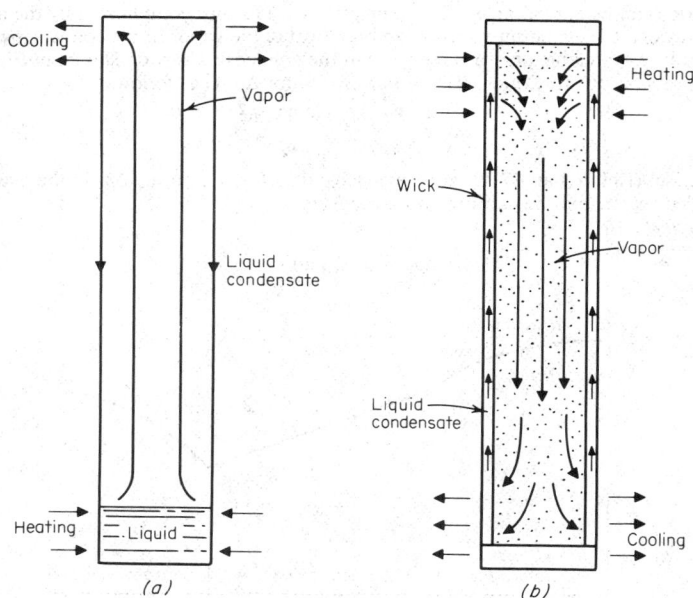

Fig. 27-14. Heat-transfer devices. (*a*) Thermosyphon; (*b*) heat pipe.

Heat pipes can be designed to have one or more of the following characteristics:
a. Very high thermal conductance, which may be a thousand times that of copper.
b. Ability to achieve an isothermal surface of low thermal impedance.
c. Ability to act as a thermal flux transformer.
d. Ability to act as a variable thermal impedance (gas-buffered heat pipe, Par. 27-33).

By suitable selection of the working fluid and the containment material, heat pipes can be constructed to operate over the range from cryogenic temperatures to temperatures at least as high as 2000°C. Power levels of 20 kW/cm² of cross section have been achieved with liquid-metal heat pipes, and the theoretical limits are considerably higher than this.

34. Basic Design. (For symbols, see Par. 27-38.) For a heat pipe to operate, the available capillary pumping head Δp_c must be greater than the total pressure drop in the heat pipe. This pressure drop is made up of the viscous drop in the liquid, Δp_l, the viscous and inertial drop in the vapor, Δp_v, and the gravitational head, Δp_g. Thus

$$\Delta p_c \geq \Delta p_l + \Delta p_v + \Delta p_g \tag{27-65}$$

Capillary Head

$$\Delta p_c = \frac{2\sigma \cos \theta}{r} \tag{27-66}$$

where r is chosen as the average pore radius. For rectangular channels

$$r = \frac{\text{cross-sectional flow area}}{\text{wetted perimeter}}$$

Liquid/wick combinations are chosen such that the liquid wets the wick, in which case cos $\theta = 1$.

Liquid Pressure Drop. The flow in the wick is laminar, and Δp_l may be calculated from Darcy's law.

$$\Delta p_l = \frac{V_l l_{\text{eff}} \dot{m}}{KA} \tag{27-67}$$

Vapor Pressure Drop. Analytical expressions for Δp_v are complex because viscous and inertial terms can be comparable. The expressions are further complicated by the addition of vapor through the evaporator section and removal of the vapor in the condensing region. There may be appreciable pressure recovery in the condensing section shown in Fig. 27-15. An expression due to Cotter (Ref. 5, Par. 27-39) for Δp_v is as follows:

$$\Delta p_v = \frac{8\mu_v \dot{m} l_s}{\pi \rho_v r_v^4} - \frac{0.074 \dot{m}^2}{\rho_v r_v^4} \tag{27-68}$$

Van Andel (Ref. 6, Par. 27-39) gives extensive data for computing Δp_v in the evaporator and shielded regions but neglects pressure recovery.

Gravitational Head

$$\Delta p_g = g \rho_l l_{\text{eff}} \sin \alpha \tag{27-69}$$

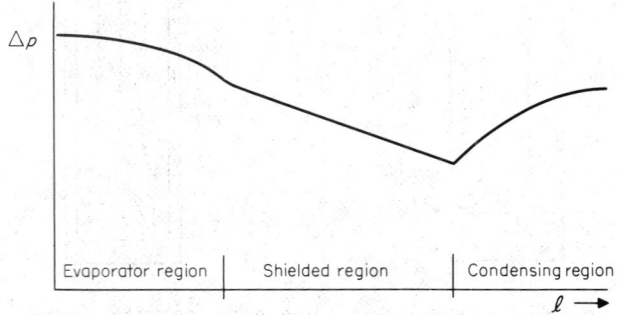

Fig. 27-15. Vapor-pressure variation along a heat pipe.

In a practical heat pipe other factors to be considered include the selection of combinations of chemically compatible structure and working fluid.

Upper limits to the power-handling capability of a heat pipe may be set by one or more of the following factors:

a. *Sonic limit.* At low pressures the velocity of a sound u_v may limit the total power capability of the pipe. It may also set a limit when the pipe has a constriction at some point along its length.

$$Q = A_v p_v u_v l \qquad (27\text{-}70)$$

b. *Entrainment of liquid by vapor.* The onset on entrainment may be estimated from the Weber number We. When We > 1, entrainment occurs.

$$W_e{}^n = \frac{\rho u^2 d}{2\pi\sigma} \qquad (27\text{-}71)$$

Hence,

$$Q = A\left(\frac{2\pi\sigma\rho_v^2}{d}\right)^{0.5} \qquad (27\text{-}72)$$

c. *Capillary Head.* As the heat input is increased, a point will be reached at which the capillary forces are unable to pump sufficient fluid around the pipe and dry-out will occur in the evaporator section.

d. *Burnout.* Burnout will occur in the evaporator at high radial heat flux. A similar limitation on peak flux occurs in the condenser.

35. Fluids for Heat Pipes. Desirable properties of working fluids are high surface tension σ, low kinematic viscosity μ_l/ρ_l, and high enthalpy of vaporization L. These properties, together with the wetting requirement, may be combined in a figure of merit $\rho_l L\sigma \cos\theta/\mu_l$, which is plotted against temperature for a number of liquids in Fig. 27-16.

In operation and at start-up, the temperature of the fluid must not approach either the freezing point, the pourpoint, or the critical point.

36. Gas-buffered Heat Pipe. It is sometimes required to maintain a surface at constant temperature over a range of thermal input. This may be achieved by maintaining a constant pressure in the heat pipe, while at the same time varying the condensing area in accordance with the change in thermal input. *Gas buffering* is a convenient method of achieving this

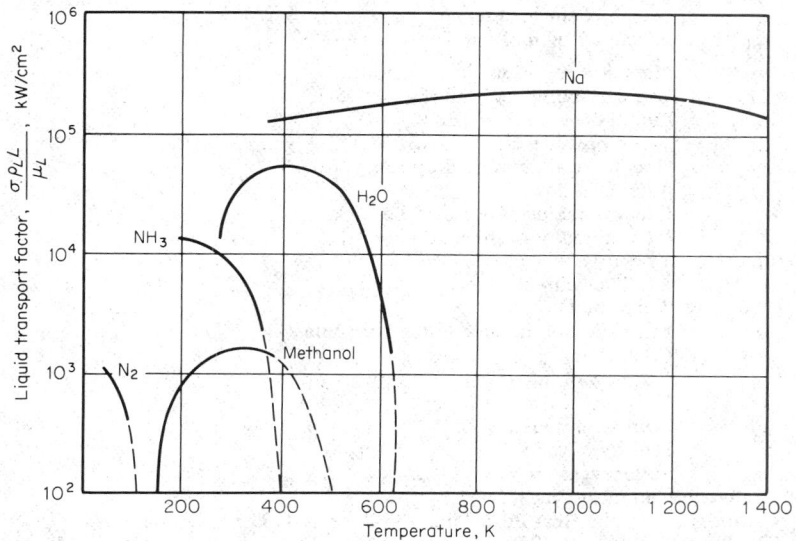

Fig. 27-16. Liquid transport factor for various working fluids. *(Ref. 1, Par. 27-29.)*

variation of the condensing area. The heat pipe is connected to a reservoir having a volume much larger than that of the heat pipe. The reservoir is filled with an inert gas, which is arranged to have a pressure corresponding to the saturation vapor pressure of the fluid in the heat pipe. In normal operation a heat-pipe vapor will tend to pump the inert gas back into the reservoir, and the gas-vapor interface will be situated at some point along the condenser surface.

The operation of the gas buffer is as follows: Assume that the heat pipe is initially operating under steady-state conditions. Now let the heat input increase by a small increment. The saturation vapor temperature will increase and with it the vapor pressure. Vapor pressure increases rapidly for very small increases in temperature; for example, the vapor pressure of sodium at 800°C varies as the tenth power of the temperature. The small increase in vapor pressure will cause the inner gas interface to recede, thus exposing more condenser surface. Since the reservoir volume has been arranged to be large compared with the heat pipe volume, a small change in pressure will give a significant movement of the gas interface.

Gas buffering is not limited to small changes in heat flux but can accommodate considerable heat-flux changes. The temperature that is controlled in the gas-buffered heat pipe, as in other heat pipes, is that of the vapor in the pipe. Normal thermal drops will occur when heat passes through the wall of the evaporating surface, and also through the wall of the condensing surface.

37. Applications of Heat Pipes. Heat pipes find their application in situations in which one or more of the characteristics listed in Par. 27-33a to d may be required, e.g., transporting heat from electronic equipment, furnaces, transport, and flux transformation in the heating and cooling of nuclear-heated thermionic generators and temperature control of test rigs in nuclear reactors.

38. Symbols in Heat Pipe Equations

A	Cross-sectional area of the wick.
A_v	Cross section of the pipe for vapor flow.
d	Characteristic dimension of vapor-liquid interface.
g	Acceleration due to gravity.
K	Wick permeability defined by Eq. (27-67).
L	Enthalpy of evaporation of the working fluid.
l_e	Length of evaporator section.
l_c	Length of condenser section.
l_s	Length of shielded section.
l_{eff}	Effective length of the pipe $= l_s + \dfrac{l_e + l_s}{2}$
m	Mass flow in the shielded section.
Δp_c	Capillary pumping head [Eq. (27-66)].
Δp_l	Liquid pressure drop [Eq. (27-67)].
Δp_v	Vapor pressure drop [Eq. (27-68)].
Δp_g	Gravitational pressure drop [Eq. (27-69)].
Q	Axial heat flow.
r	Effective radius of pores in the wick.
r_i	Internal radius of the pipe.
u_s	Velocity of sound.
u	Axial velocity of vapor.
We	Weber number.
α	Inclination of the pipe to the horizontal.
θ	Contact angle.
μ_l	Viscosity of liquid.
μ_v	Viscosity of vapor.
ρ_l	Density of liquid.
ρ_v	Density of vapor.
σ	Surface tension.

39. References on Heat Pipes

1. CHISHOLM, D. The Heat Pipe, Mills and Boon Ltd., London, 1971.
2. The Heat Pipe: A List of Pertinent References, Department of Trade and Industry, National Engineering Laboratory, London, November, 1970.

Text references
3. GAUGLER, R. S.   U.S. Patent No. 2,350,348, application date Dec. 21, 1942, accepted June 6, 1944.
4. GROVER, G. M., COTTER, T. P., and ERIKSON, G. F.   *J. App. Phys.,* Vol. 35, No. 6, 1990–1991, 1964.
5. COTTER, J. P.   USAEC Rep. LA - 3246 - MS, University of California, Los Alamos Scientific Laboratory, 1965.
6. VAN ANDEL, D.   International Conference on Thermionic Electrical Power Generation, Stresa, May 27–31, 1968, published by Euratom, Stresa, Italy.

INDEX

1